FOURTEENTH EDITION

Mechanical and Electrical Equipment for Buildings

FOURTEENTH EDITION

Mechanical and Electrical Equipment for Buildings

Walter Grondzik
Architectural Engineer
Freelance Consultant

Daniel Overbey
Associate Professor of Architecture
Ball State University

Alison Kwok
Professor of Architecture
University of Oregon

WILEY

Published by John Wiley & Sons, Inc., Hoboken, New Jersey.

Library of Congress Cataloging-in-Publication Data Applied for

Hardback ISBN: 9781394257119
ePDF ISBN: 9781394257133
epub ISBN: 9781394257126

Cover Design: Wiley
Cover Image: Courtesy of Douglas Adams Photography

Set in 9.5/12pt PhotinaMTStd by Straive, Chennai, India

Part opener images are from the drawing set for the Lillis Business Complex at the University of Oregon, designed by SRG Partnership, Portland, OR.

DISCLAIMER
The information in this book has been derived and extracted from a multitude of sources including building codes, fire codes, industry guidelines and standards, manufacturers' literature, engineering reference works, and personal professional experience. It is presented in good faith. Although the authors and the publisher have made every reasonable effort to make the information presented accurate and authoritative, they do not warrant, and assume no liability for, its accuracy, completeness, or fitness for any specific purpose. The information is intended primarily as a learning and teaching aid, and not as a final source of information for the design of building systems by design professionals. It is the responsibility of users to apply their professional knowledge in the application of the information presented in this book, and to consult original sources for current and detailed information as needed, for actual design situations.

SKY10150013_032026

Contents

PART III BUILDING ENCLOSURE 233

PART IV PASSIVE ENVIRONMENTAL SYSTEMS 313

PART V ACTIVE ENVIRONMENTAL SYSTEMS 503

Preface

NINE DECADES AND A FEW GENERATIONS have passed since the first edition of *Mechanical and Electrical Equipment for Buildings* was published in 1935. At its birth, this book was 429 pages long. Now, in its 14th edition, the book includes around 1800 pages, an increase of 400%. This expansion gives pause to our publisher and strengthens the arms of students—but, more seriously, it reflects the growing complexity of building design and the burgeoning knowledge base that confronts today's designers. Many new topics have been added over the years, and a few have disappeared. The book has increasingly added discussions of "why" to its historic focus upon "how-to." Computer simulations are now routinely used in system design; new codes, standards, and guidelines offer advice and challenges to designers; equipment and distribution systems have undergone major changes; mechanical cooling is commonplace across building types; fuel choices have shifted. Energy efficiency, greenness, and decarbonization are key design concerns.

Most of the systems presented in this book consume energy and embody materials. Some systems consume water. As global society has moved from its early reliance on renewable energy sources (wind, water, and horsepower) to today's seemingly unbreakable addiction to nonrenewable fossil fuels, it has also added vastly to its population and increased its global per capita energy use. The resulting environmental degradation (primarily evident in air and water quality) has spurred efforts to reverse this decline. Governmental regulations are a part of such efforts, but this book emphasizes the investigation of alternative fuels and design approaches that go beyond those solutions currently deemed "minimally acceptable" to society. Designers are encouraged to take an aggressive leadership role in mitigating environmental degradation.

It is becoming increasingly clear that climate change is well under way. The distressing value of 425 ppm of atmospheric CO_2 has been surpassed. We may not know the precise extent to which our insatiable carbon-based energy consumption is responsible for pushing the world to this new reality. It seems professionally irresponsible, however, to believe that human actions have had no effect. It is very clear that the world's supply of fossil fuel is diminishing, that the consumption of these fuels dumps massive quantities of CO_2 (and other greenhouse gases) into the air, and that there will be future consequences for all buildings (and their occupants) that today rely so thoroughly on nonrenewable energy sources.

The buildings of today contribute to negative global consequences that will impact future generations, and our approach to mechanical and electrical systems must consider how best to minimize and mitigate—if not negate—such detrimental environmental impacts. Thus, on-site resources—daylighting, passive solar heating, passive cooling, solar water heating, rainwater capture, wastewater treatment, photovoltaic electricity—share the spotlight with traditional off-site resources (natural gas, oil, the electrical grid, water and sewer lines). On-site processes can be spatially intensive and labor-intensive and can involve increased first costs that require years to recover through savings in energy, water, and/or material consumption. Off-site processes are usually subsidized by society, often with substantial environmental costs that accrue to the commons. On-site energy use requires us to look beyond the building, to pay as much attention to a building's context as to the mechanical and electrical spaces, equipment, and systems within.

Throughout the many editions of this book, another trend has emerged. Society has slowly moved from infrastructure that centralizes the provision of heating, cooling, water, and electricity toward approaches that encourage more localized production and control. The increased sophistication of digital control systems has supported this trend. Further encouragement comes from multipurpose buildings whose schedules of occupancy are fragmented and from clients with varying work schedules that result in partial occupancy on weekends. Another factor in the move to decentralization is worker satisfaction; there is solid evidence that productivity increases with a sense of individual control of one's work environment. Residences are commonly being used for office work. Expanding communication networks have made this possible. As residential designs thus become more complex (with office-quality lighting, zoning for heating/cooling, sophisticated communications, noise control), our nonresidential work environments must become more attractive and individualized.

Air and water pollution problems stemming from buildings (and their systems and occupants) are widely recognized and generally condemned. An interest in environmentally responsible design on the part of clients and designers may help to mitigate such problems. Today's green design is hopefully just an intermediate step in the journey to truly sustainable solutions. And, no, we are not designing sustainable solutions today—despite the claims about sustainable this and sustainable that filling the internet, conference presentations, and professional journals.

Another pervasive pollutant affecting our quality of life is noise. Noise will impact building siting, space planning, exterior and interior material selections—even the choice of cooling systems (as with natural ventilation). Air and water pollution can result in physical illness, but so can noise pollution, along with its burden of mental stress.

This book is written with a focus on the North American building design community and has always emphasized examples from this region. Yet other areas of the world, some with similar traditions and fuel sources, have worthy examples of new strategies for building design utilizing on-site energy and energy conservation. Thus, buildings from Europe and Asia are featured herein, along with many North American examples. The names of these buildings (and associated researchers and designers) have been included in the index. Design standards presented in this book are generally reflective of practice in the United States; this is a result of our experiences, but also reflects a desire not to burden readers by listing the many variations in design standards that exist across the globe.

Building system design is now widely undertaken using computers, often through proprietary software that includes hundreds of built-in assumptions. This book encourages the designer to take a rational and pragmatic approach to system design: to verify intuitive design moves and assumptions and to use computers as tools to facilitate such verification, but to use patterns and approximations to point early design efforts in the right direction (and catch the occasional garbage-in/garbage-out simulation result). Hand calculations have the benefit of exposing all pertinent variables and assumptions to the designer. This in itself is a valuable rationale for conducting some portion of an analysis manually. Rough hand-calculated results should point in the same direction as results obtained with a computer; the greater the disparity, the greater the need to check both approaches. This is not to disparage the use of simulations, which are indispensable in maximizing the benefits of complex and sometimes counterintuitive systems.

This book is written with the student, the architect- or engineer-in-training, and the practicing professional in mind. Basic theory, preliminary design guidelines, and detailed design procedures allow the book to serve both as an introductory text for the student and as a more advanced reference for both professional and student. This work is intended to be used as a textbook for a range of courses in architecture, architectural engineering, and building/construction management.

A "MEEB 14" website provides supporting materials to enhance learning about and understanding of the concepts, equipment, and systems dealt with in this book. As with previous editions, a 14th edition instructor's manual has been developed to provide additional support for those teaching with the book. That manual, updated by Prof. Troy Peters, outlines the contents and terminology in each chapter; highlights concepts of special interest or difficulty; and provides sample discussion, quizzes, and exam questions. The manual is available to instructors who have adopted this book for their courses.

Mechanical and Electrical Equipment for Buildings continues to serve as a reference for architectural registration examinees in the United States and Canada. In this age of information abundance (not all of which is dependable), we hope that this 14th edition carries on the longstanding tradition of MEEB as a trustworthy reference for students as well as the offices of architects, engineers, and building managers.

Walter Grondzik
Daniel Overbey
Alison Kwok

Acknowledgments

Many people and organizations have contributed knowledge, materials, and insights to several editions of this book (affectionately called MEEB). We begin this acknowledgment with thanks to those from whose work we have borrowed at length: Baruch Givoni, John Tillman Lyle, Murray Milne, William McGuinness, and Victor Olgyay; Architecture 2030, ASHRAE (the American Society of Heating, Refrigerating and Air-Conditioning Engineers), the American Solar Energy Society (ASES), the Illuminating Engineering Society (IES), the International Code Council (ICC), the National Fire Protection Association (NFPA), the International Living Future Institute (ILFI), the International WELL Building Institute (IWBI), Phius, the US Green Building Council (USGBC), collective wisdom from the Society of Building Science Educators (SBSE), the creative talents and expertise of numerous architects, engineers, consultants, and other design professionals, and the many equipment manufacturers whose product information and photographs illustrate the book.

We owe a great debt to the insight and technical knowledge shared through MEEB over many years by Benjamin Stein and John Reynolds. Many of their contributions continue in this edition, serving as a core around which we have revised, updated, and developed new material.

Alison Kwok, an active co-author of the 13th, 12th, 11th, and 10th editions of MEEB (wow), has assumed the role of inactive author for this edition. Thus, Kwok, Reynolds, and Stein are responsible for much of what you find informative and inspiring in this work—while Grondzik and Overbey are the final arbiters of content.

Numerous professionals provided valuable assistance in assembling materials and clarifying ideas and details. These include Michael Bonn (Ankrom Moisan); Mike Steffen (Walsh Construction); Jeff Becksfort (PAE); Walker Macy (Landscape Architects, Orchards at Orenco); Charles Salter (Charles M. Salter Associates, Inc., Skywalker Ranch Technical Building); Carl Sterner (Sterner Design, LLC, Iowa Nest); James Lowe (Ball State University Facilities Planning and Management, BSU ground source heat pump system); Pete Munoz and Jean Wisenbaugh (Biohabitats); Kyle Andersen (GBD Architects); Mitch Dec, Brightworks (formerly at Glumac); Chris Lowen and Rob Schnare (Glumac); Charles Brucker (PLACE, Hassalo on 8th); Michael Utzinger and Joel Krueger (Aldo Leopold Legacy Center), Vikram Sami and Jim Nicolow (Blue Ridge Parkway Destination Center), Craig Christiansen (NREL research reports), William Lowry (climate of cities), Daniel Panetta (AIWPS and recycling at California Polytechnic, San Luis Obispo), Dr. Jonathan Stein (computer applications), John A. Van Deusen (vertical transportation), and Martin Yoklic (cooltower performance analysis). In addition, Peter Clegg (Feilden Clegg Bradley Studios); Guy Nevill and Max Fordham (The Hive); Robert Schnare and THA (Mercy Corps); Ron Rochon and Jim Hanford of Miller Hull Partnership; Denis Hayes, Bullitt Foundation (Bullitt Center); Julia Lau and David Lung (International Commerce Centre). Donald Corner, Lee Eckert, Tisha Egashira, Bruce Haglund, Toshi Woudenberg, Erica Ling, and Thomas Collins kindly contributed images that enhance the book.

We are very pleased to continue to include illustrations by talented students. For this edition: Evelyn Dinh. For the 13th edition: Sharon Alitema, Kelli Kimura, Alisa Kwok, Nathan Leigh, Martin Lee, Achyuthan Ramaswamy, and Alexa Thornton. For the 12th edition: Zeta Fernando, Tyler Mavichien, Wesley Thompson, Karen Tse, and Ayush Vaidya. For the 11th edition: Lisa Leal, Nathan Majeski, and Jonathan Meendering. For the 10th edition: Dain Carlson, Amanda Jo Clegg, Eric Drew, Erik Winter, and (again) Jonathan Meendering. For the 8th edition: Michael Cockram. These are students (many now professionals) who embraced the principles and concepts of environmental technology in their design work and therefore clearly understood what they were drawing. We also acknowledge the many architects and engineers who provided illustrations of their buildings and design artifacts that appear in many of the chapters—citations to these firms and individuals are found throughout the book.

We are indebted to colleagues for updates, revisions, and complete overhauls to various parts of previous editions of this book: University of Oregon Professor Siobhan Rockcastle (major update of electric light sources); University of Oregon Professor Alex Rempel (total rethinking of passive heating); Ball State University Professor Thomas Collins (chapter updates to elevators, escalators, and development of case studies for the Rocky Mountain Institute and Ball State University's ground source heat pump system); Professor Omar Al Hassawi (updates to the downdraft cooltowers strategy); Fred Tepfer (Hassalo on 8th case study about waste water treatment); and Rick Diamond (design process and paths to a low carbon future).

Special appreciation goes to Professor Thomas Collins (Ball State University) for a thorough revision and update in this 14th edition of Chapter 32: Signal and Communication Systems. Likewise, we greatly appreciate the efforts of Professor Keith Simon (University of Texas at Austin) in rethinking and reframing Chapter 9: Building Enclosure and Systems.

Thanks also to Wentworth Institute of Technology Professor Troy Peters for updating the companion materials for instructors and students.

Over the years, we have received valuable suggestions and feedback from readers and users of MEEB. Student supporters include Troy Anderson, Rachel Auerbach, David Bartley, Mike Beamer, David Bisers, Christina Bollo, Gabriel Brown, Thomas Collins, Christopher Deel, Charlie Deese, Ryan Dirks, Mathieu Deraspe, Sophia (Duluk) Beavis, Chelsey Effinger, Alfredo Fernández-González, Alyssa (Franco) Dirks, Erin Fox, Sara Goenner, Jeff Guggenheim, Susie Harriman, Diana Hogard, Daniel Hundley, Thu Jenni Hyunh, Daniel Irurah, Eric Issertes-Carbonnier, Reza Javandel, Jeff Joslin, Jake Keeler, Anna Liu, Angela Matt, Julia May, Nicklaus McGuire, Jonathan Meendering, James Max Moore, Heather (Nelson) Jauregui, Niyati Naik, Christopher Nielson, Tobin Newburgh, Roger Ota, Therese Peffer, Troy Peters, David Posada, Nick Rajkovich, Barbara Reed, Amanda Rhodes, Anupam Satumane, Eric Schmidt, Devin Smith, Ted Shriro, Jonathan Thwaites, Nick Venezia, Michael Walsh, Paul Wolfe, Emily Wright, and Hannah Zalusky.

Recent faculty commenters include Michael Ermann (Virginia Tech), Patricia Andrasik (Catholic University), and Jeff Haberl (Texas A&M University). A special acknowledgment goes to Michael Ober, who provided unrestrained encouragement and lasting pleasure with his video extolling the virtues of "MEEB" (check it out on YouTube, "MEEB Like This").

Much of the work involved in producing a manuscript is accomplished by dedicated supporting personnel. We are grateful to the ongoing benefits of having engaged an experienced and trusted production team for the 13th edition of MEEB: Ayush Vaidya (file coordination, chapter updates, project and database management) and Karen Tse (chapter updates, on-call assistance, file management).

Finally, we are seriously indebted to the staff at John Wiley & Sons for their diligent and highly professional work, especially Amanda Miller, group vice president and general manager; Kalli Shultea, senior editor; Katherine Wong, senior managing editor; Isabella Proietti, editorial assistant; Kavin Shanmughasundaram, content refinement specialist, Susan Dunsmore, copyeditor extraordinaire; and Straive, compositors.

About the Companion Website

This book is accompanied by a student and instructor website.

The URL for the site is **www.wiley.com/go/meeb14e**

This website provides resources for instructors and students, detailed as follows,

- For instructors,
    - Instructor's guide
    - PowerPoint slides for lectures
- For both students and instructors,
    - Student's guide
    - Guide to the fourteenth edition

Prof. Troy Peters, Wentworth Institute of Technology, is acknowledged as curator of these companion materials.

PART I

THE BUILDING DESIGN CONTEXT

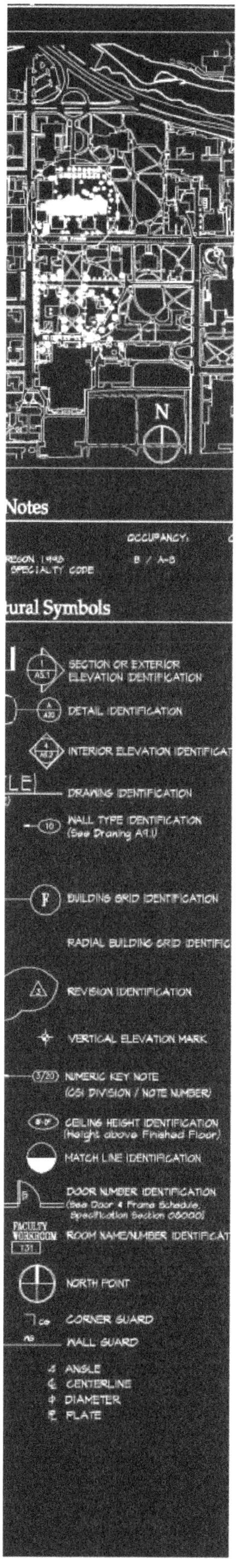

The design of mechanical and electrical equipment for buildings is often not considered until many other important design decisions have already been made and locked into place. This is the result of a long-established relay-race type of design process, whereby decisions are handed off sequentially from architect to consulting engineer. This process is changing—with the evolution of integrated design and delivery—but is still exceptionally common. For some design teams, active systems are seen as corrective interventions, permitting a building design to work on a site and in a climate that were essentially ignored during early design phases. Such is the power of fossil-fueled systems. However, such power comes with a price—both economic and environmental.

The chapters that comprise Part I encourage designers to use the wonderfully flexible building design process to full advantage, and to include site and climate—as well as the owner's project requirements that will impact mechanical and electrical systems—in their earliest design thinking. Chapter 1 discusses the building design process and the important roles played by codes, costs, and project objectives in shaping a final building design. The critical importance of clear design intents and criteria is emphasized. Principles to guide environmentally responsible design are given. Chapter 2 discusses the relationship of energy, water, and material resources to buildings, from design through demolition. The concept of the environmental footprint is introduced as an ultimate outcome of design decision making. Chapter 3 encourages designers to view a building site as a collection of renewable resources, to be appropriately used in the lighting, heating, ventilation, and cooling of buildings.

CHAPTER 1

Design Process

IN MARCH 1971, VISIONARY ARCHITECT MALCOLM WELLS published a watershed article in *Progressive Architecture*. It was intriguingly and challengingly titled "The Absolutely Constant Incontestably Stable Architectural Value Scale." In essence, Wells argued that buildings should be benchmarked (to use a current term) against the environmentally regenerative capabilities of wilderness (Figure 1.1). This seemed a radical idea then—and remains so even now, more than 50 years later. Such a value scale, however, may be just what is called for as the design and construction professions slowly but inevitably move environmental focus from energy-efficient to green, to decarbonized, to sustainable. The main problem with Wells' "Incontestably Stable" benchmark is that most buildings fare poorly (if not dismally) against the environment-enhancing characteristics of wilderness. But perhaps this is more a wake-up call than a problem.

As we enter the second quarter of the twenty-first century, *Progressive Architecture* is no longer in business, Malcolm Wells has sadly passed away, mechanical and electrical equipment has improved, simulation techniques have radically advanced, and information exchange has been revolutionized. In broad terms, however, the outcomes of the design process have evolved only incrementally since the 1970s.

The immediacy of climate change on the design, construction, and operation of buildings mandates a change in the underlying values and philosophy of the design process. The design professions must intend to build a low-carbon, more resilient future. This change will start in the earliest days of a project—when the owner's project requirements (markers for a successful building) are established.

To meet the challenges of the coming decades, it is critical that design teams and clients consider and adopt values appropriate to the nature of the problems being confronted—both at the individual project scale and globally. Nothing less makes sense.

1.1 INTRODUCTION

The design process is an integral part of the larger and more complex building procurement process through which an owner defines facility needs, considers architectural possibilities, contracts for design and construction services, and ultimately uses the resulting facility. Numerous decisions (literally thousands) made during the design process will determine the need for specific mechanical and electrical systems and equipment, and very often will determine owner and occupant satisfaction. Discussing selected aspects of the design process seems a good way to start this book.

A building project typically begins with predesign activities that establish the need for, feasibility of, and proposed scope for a facility. If a project is deemed feasible and can be funded, a multiphase design process follows. The design phases are typically described as conceptual design, schematic design, and design development. If a project remains feasible as it progresses, the design process is followed by the construction phase and then occupancy. In fast-track approaches (such as design-build), design efforts and construction activities may substantially overlap.

Mechanical and Electrical Equipment for Buildings, Fourteenth Edition. Walter Grondzik, Daniel Overbey, and Alison Kwok.

Companion website: www.wiley.com/go/meeb14e

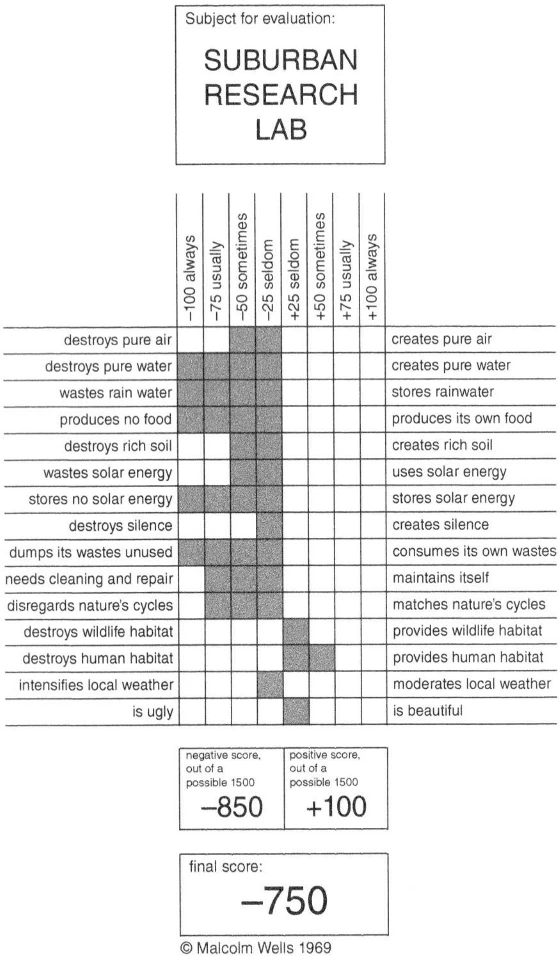

Figure 1.1 *Evaluation of a typical project using Malcolm Wells's "absolutely constant incontestably stable architectural value scale." The value focus was wilderness; today it might well be sustainability. (Used with permission from Malcolm Wells, 1981.)*

Predesign activities may be conducted by the design team (often under a separate services contract), by the owner, or by a specialized consultant. The product of predesign activities should be a clearly defined scope of work for the design team to act upon. This product is variously called a program, a project brief, or the owner's project requirements (OPR). The design process converts the owner's requirements into drawings and specifications that permit a contractor to then convert the owner's (and designer's) wishes into a physical reality.

The various design phases are the primary areas of focus in this book. The design process may span weeks (for a simple building or system) or years (for a large, complex project). The design team may consist of a sole practitioner for a residential project, or hundreds of people located in different offices, cities, or even countries for a large project. Decisions made during the design process, especially during the early stages, will affect the project owner and occupants for many years—influencing operating costs, maintenance needs, comfort, enjoyment, and productivity. Decisions made for a single project can ripple outward and affect those external to the project.

The scope of work accomplished during each of the various design phases varies from firm to firm and project to project. In many cases, explicit expectations for the phases are described in professional service contracts between the design team and the owner. Often there is close collaboration between the client and design firm. A series of images illustrating the development of the Net Zero Iowa Nest Residence (Figure 1.2), a high-performance residence in southeast Iowa, is used to illustrate the steps of a building project that will offer a durable, ultra-low-energy, and comfortable residence.

The story of this remarkable project and its design process is jointly chronicled by the designer and owners in the blog (iowanest.com). The team began design with five primary goals: (1) Net-Zero Energy: the building will produce as much energy as it uses on an annual basis; (2) Conventional Cost: the house will cost the same (per unit floor area) as conventional new residential construction in Iowa; (3) Passive Cooling/No Air Conditioning: achieving thermal comfort during the hot and humid summer season using natural resources; (4) Integrated with the Land: increasing biohabitat and aging gracefully in the landscape; (5) 200-Year House: self-explanatory (but not the norm).

Conceptual design (Figure 1.3) attempts to outline a solution to the owner's program that can meet the budget and capture the owner's imagination so that design can continue on the right track. For the Iowa Nest, conceptual design incorporated key passive strategies that were identified using early "simple box" building energy modeling (Figure 1.4).

Fundamental decisions about the proposed project should be made during the conceptual design

Figure 1.2 *The Iowa Nest. (Photo © Sterner Design, LLC; used with permission.)*

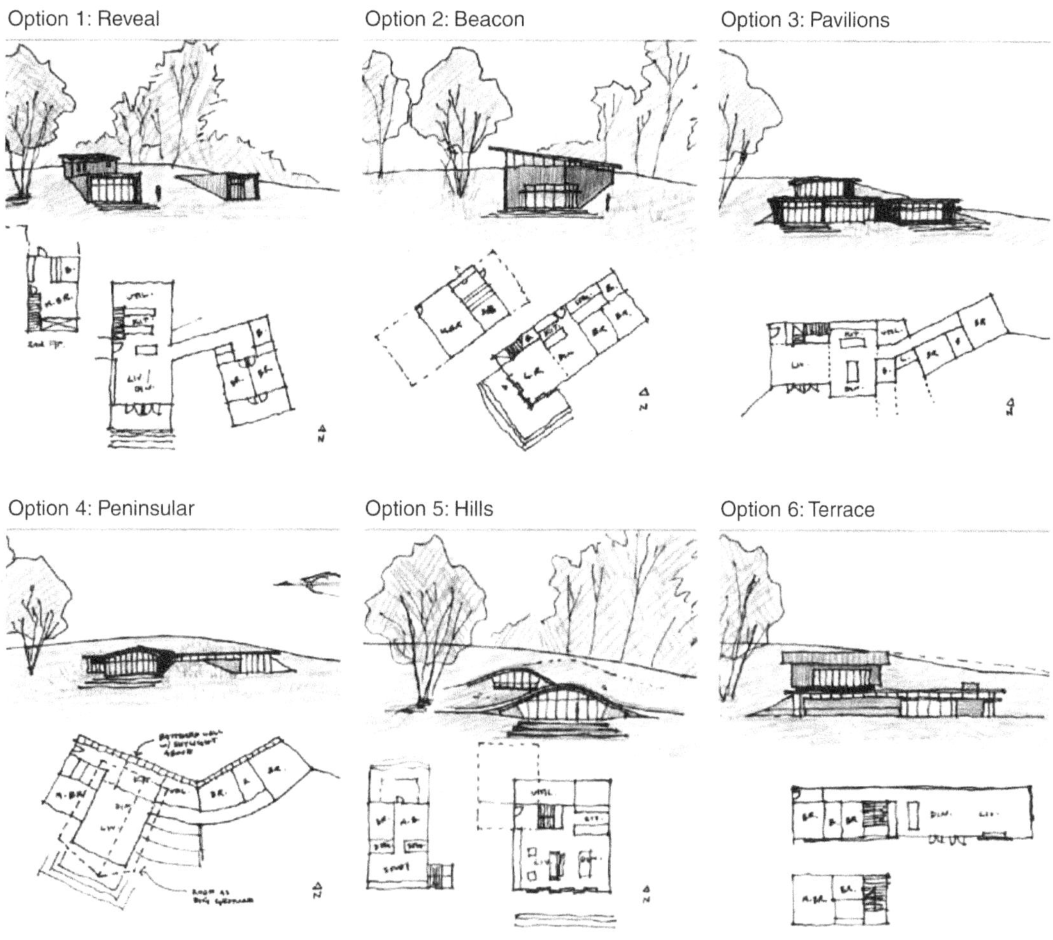

Figure 1.3 *Design iterations began with hand sketches showing attention to earth berming and daylighting zoning. (© Sterner Design, LLC; used with permission.)*

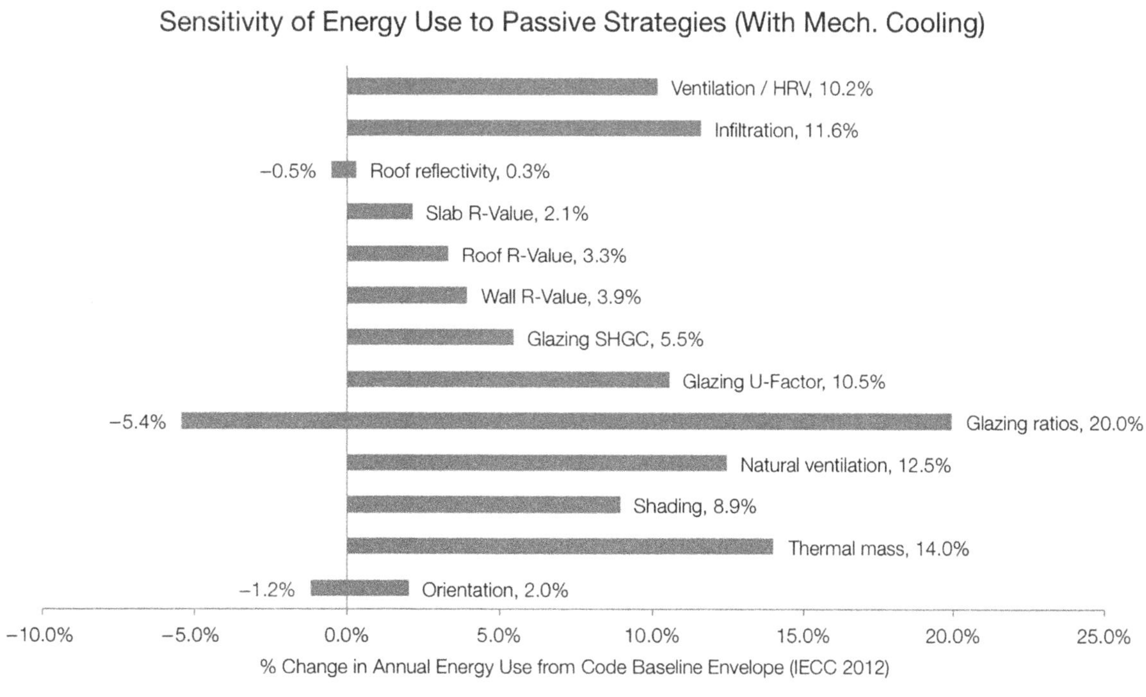

Figure 1.4 *Several analyses compared the impact of passive strategies against a code-compliant baseline enclosure including mechanical cooling for percentage change in annual energy use and without mechanical cooling for percentage change in annual hours where the air temperature is below 82.4 °F [28 °C]. (© Sterner Design, LLC; used with permission.)*

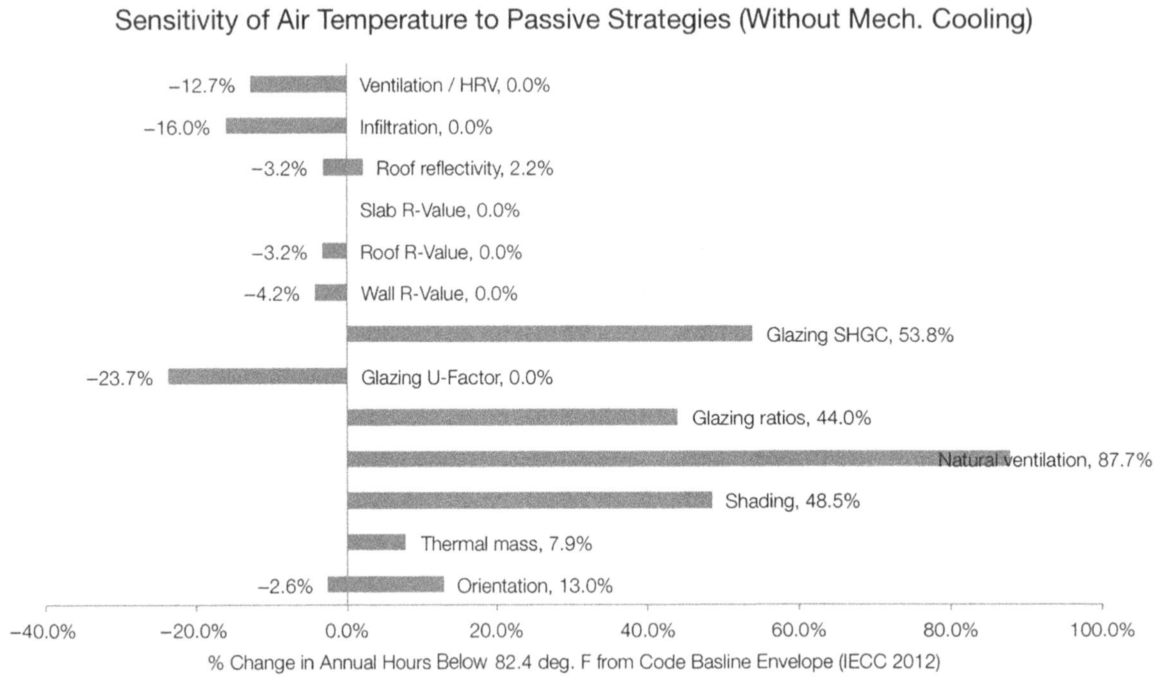

Figure 1.4 *(continued)*

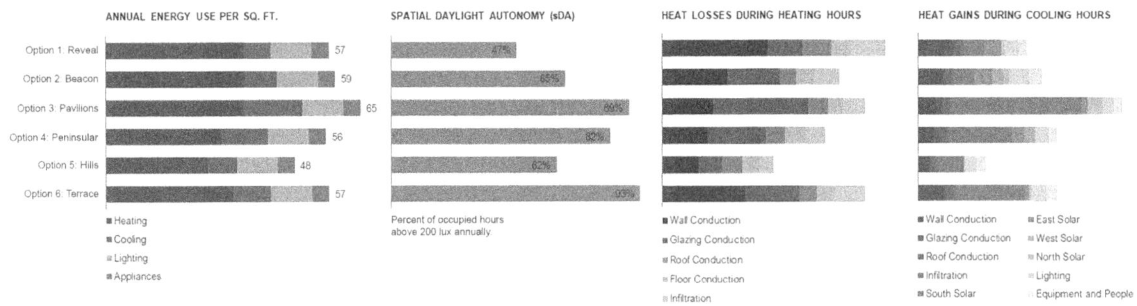

Figure 1.5 *The Iowa Nest design concept was further refined by comparing the design iterations for daylight and energy performance using Sefaira software. Early analysis rarely dictates the selection of a concept but can nevertheless inform design moving forward. This conceptual study showed a strong correlation between earth berming and low energy use. (© Sterner Design, LLC; used with permission.)*

(Figures 1.5 and 1.6). During schematic design, the conceptual solution is further developed and refined. While the “waterfall” diagram (Figure 1.7) shows a ground-source heat pump, the team ultimately replaced this with a more cost-effective electric radiant system. This worked because of good enclosure and passive design. The heating loads were too small for a ground-source heat pump to be practical. The best path to net-zero use turned out to be a less efficient system and a slightly larger photovoltaic array—demonstrating the impact of integrated design. During design development and construction documentation (Figure 1.8), all decisions regarding a design solution are finalized, and construction drawings and specifications detailing those innumerable decisions are prepared.

The construction phase (Figure 1.9) is primarily in the hands of the contractor (and, in this case, the owner and friends). Although design decisions have determined what will be built, these decisions may dramatically affect constructability. One specific example of this: airtightness is often treated as a “detail” to be figured out during construction documentation. The experience of designing the Iowa Nest confirmed that basic design moves can have a significant impact in making airtightness goals more or less easy to achieve. In this case, the air barrier was defined early in design, and the design ensured that critical areas would be accessible for air sealing during construction.

The building owners/occupants are the key players during the occupancy phase. Their experiences with the building (utility bills, comfort, and beauty) will clearly be influenced by design decisions

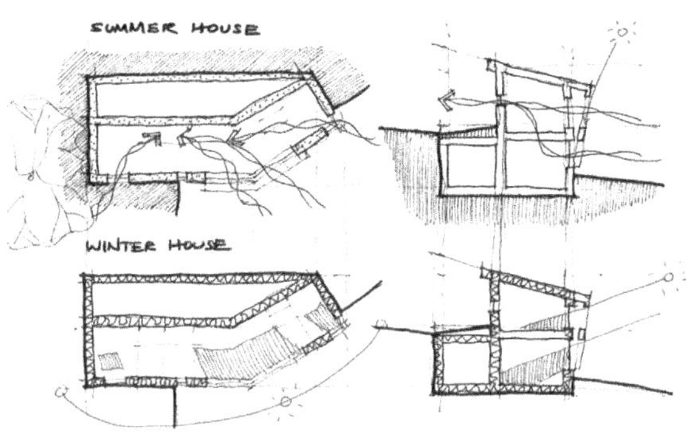

Figure 1.6 Conceptual sketches for shading and natural ventilation strategies at the Iowa Nest suggest in fairly strong terms the "first, best moves" for design direction, yet details are left to be developed in later design phases. There is a clear focus on shading, ventilation, earth berming, insulation, and passive solar design even at this stage—a focus that was carried throughout the project. (© Sterner Design, LLC; used with permission.)

and construction quality, as well as by maintenance and operation practices. A feedback loop that allows construction and occupancy experiences (lessons learned—both good and bad) to be used by the design team on future projects is essential to good design practice. With this residence, the designer and owner are actively tracking temperature, relative humidity, and energy use to understand whether the building is performing as intended (Figure 1.10). The results of this ongoing monitoring are documented on the blog.

1.2 DESIGN ISSUES

Design issues are those things that need to be considered and addressed on a project. For a typical building, there are literally thousands of issues to be resolved as part of the design process. Many issues (such as building code requirements) will be dealt with through process memory. Other issues (such as budget and schedule) will be engaged through dialog with the client. Emerging issues (such as decarbonization, resilience, social justice) that are not yet part of the conventional design process can be forgotten until it becomes difficult to incorporate a resolution into the design proposal. Brainstorming issues very early in the design process—ideally during predesign—can help ensure that issues of importance to a project are on the radar for the design team.

The following is a very brief list of possible design issues for a building project. This list pairs with the examples of design intent seen in Section 1.3:

- comfort
- IT (information technology)
- carbon emissions

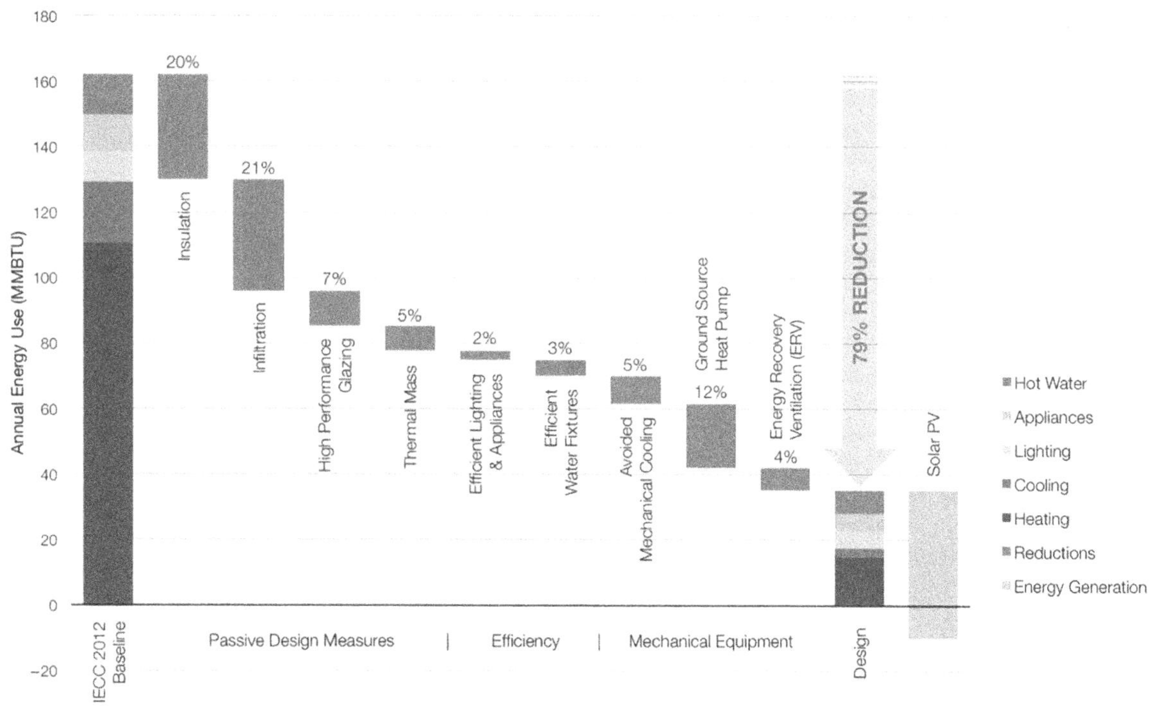

Figure 1.7 Further analysis examined the impact of specific passive strategies (insulation, reducing infiltration, use of high-performance glass, and thermal mass), efficiency strategies (electric lighting, appliances, and plumbing fixtures), and high-performance mechanical equipment (no air-conditioning, ground-source heat pump, and ERV), which drove the predicted annual energy use to almost 80% less than a baseline building. (© Sterner Design, LLC; used with permission.)

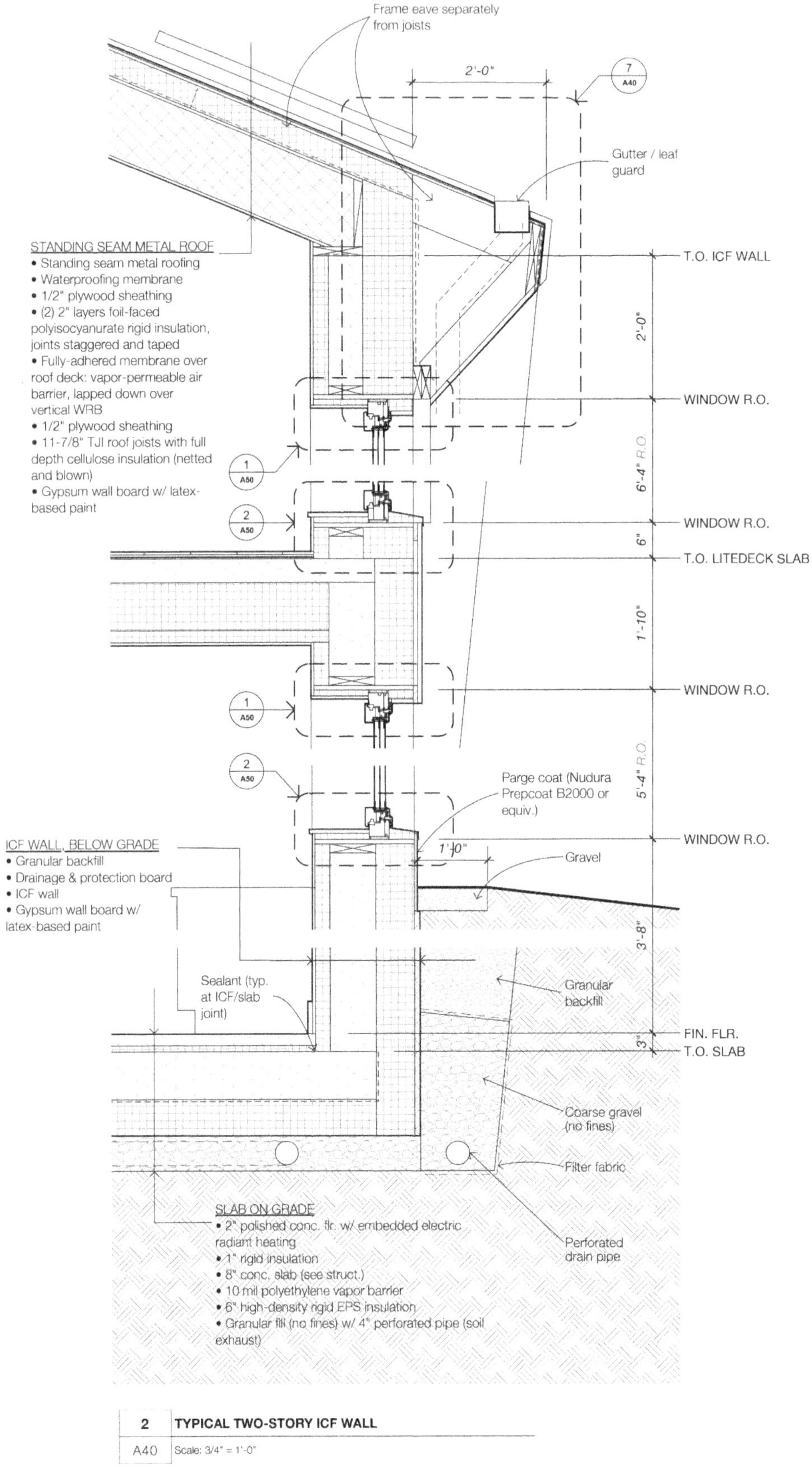

Figure 1.8 *During design development, construction details that convert an idea into a building evolve. This drawing illustrates the development of working details for the Iowa Nest enclosure, resulting in highly insulated assemblies: R-28 slab, R-45 walls, and an R-70 roof. [R-SI: 4.9, 7.9, and 12.3 respectively] (© Sterner Design, LLC; used with permission.)*

(a) (b)

Figure 1.9 (a) Construction-phase photo of the west wall of the Iowa Nest, showing shading from a deciduous tree on the west window on a summer day and (b) shading of the high-performance, triple-pane sliding doors on the south-facing façade. Design intent becomes reality during construction. (Photo © Carl Sterner; used with permission.)

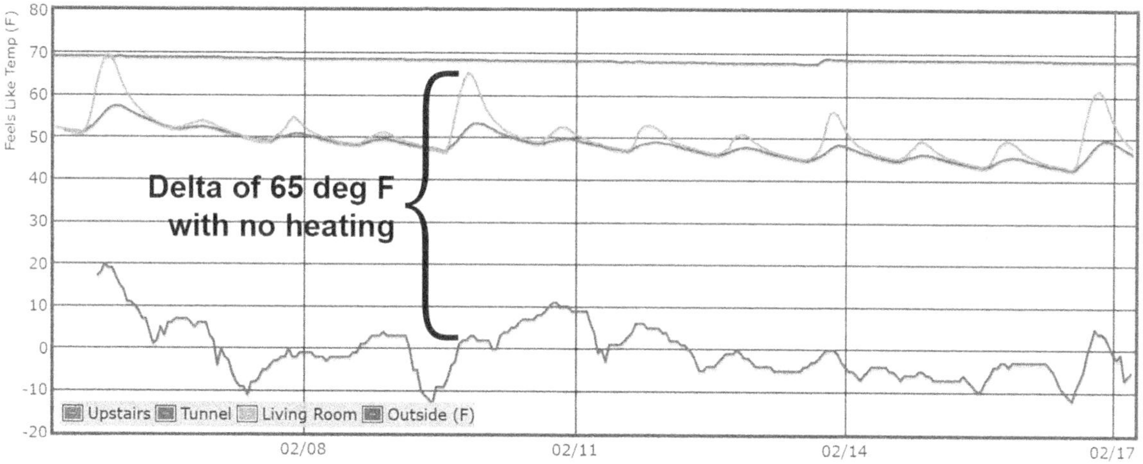

Figure 1.10 Monitoring data from the Iowa Nest during a significant winter event in February 2021. The difference between the outdoor and indoor temperatures demonstrates the capability of this residence to maintain safe temperatures without mechanical systems running. (Diagram © Carl Sterner; used with permission.)

- functional flexibility
- health and wellness
- adaptability
- resilience

1.3 DESIGN INTENT

Design efforts should focus on achieving a solution that will meet the expectations of a well-thought-out and explicitly defined design intent. A design intent is simply a statement that outlines an expected high-level outcome of the design process. Making such a fundamental statement is critical to the success of a design, as it points to the general direction(s) that the design process must take to achieve success. Design intent should not try to capture the totality of a building's character; this will come only with the completion of the design. It should, however, adequately express the defining characteristics of a proposed building solution. Example design intents (from among thousands of possibilities) might include the following:

- The building will provide outstanding comfort for its occupants.
- The building will feature the latest in information technology.
- The building will be carbon-neutral.
- The building will provide a high degree of flexibility for its occupants.
- The building will support health and wellness for the occupants.
- The building will be adaptable.
- The building will be resilient.

Clear design intents are important because they set the tone for design efforts, allow all members of the design team to understand what is considered truly critical to success, provide a general direction for early design efforts, and put key and/or unusual design concerns on the table. Professor Larry Peterson, former director of the Florida Sustainable Communities Center, has described the earliest decisions in the design process as an attempt to make the "first, best moves." Strong design intent will inform such moves. Weak intent will typically result in a so-so building. Great moves too late will be futile. The specificity of the design intent will evolve throughout the design process. Outstanding comfort during conceptual design may become outstanding thermal, visual, and acoustic comfort during schematic design.

1.4 DESIGN CRITERIA

Design criteria are the targets against which success or failure in meeting design intent is measured. For example, Bill Bordass (Usable Buildings Trust, United Kingdom) offers a clear "multiplier effect" to achieving low-carbon buildings. The approach began in the early 1990s when he was reporting on office energy case studies, which then led to the first edition of the *Energy Consumption Guide 19* for offices. The idea further developed for the PROBE studies (1995–2002) and was formalized in 1997 (Field et al., 1997), in *CIBSE TM22—Energy Assessment and Reporting Methodology: The Office Assessment Method* (1999), and in the report *Flying Blind* (Bordass, 2001).

Bordass discovered that most decisionmakers were receptive to a simple three-step hierarchy. We embrace this finding by introducing chapters in this book (as Bill Bordass has summarized) in three steps:

1. Reduce energy demand by 50% by designing a passive building that is well oriented for thermal and lighting (MEEB, Chapters 2–13);
2. Reduce energy supply another 50% by specifying energy-efficient active systems (MEEB, Chapters 14–17); and, finally
3. Reduce the carbon content by 50% for the remaining energy supplied to the building (MEEB, Chapter 29).

In addition to providing a basis against which to evaluate success, design criteria will ensure that all involved parties seriously address the technical and philosophical issues underlying a project's design intent. Setting design criteria demands the clarification and definition of many intentionally broad terms used when crafting design intent statements. For example, what is really meant on a project by green, by flexibility, by comfort? If such terms cannot be benchmarked, then there is no way for the success of a design to be evaluated—essentially anything goes, and all solutions are potentially equally valid. Setting design criteria for qualitative issues (such as exciting, relaxing, or spacious) can be challenging but equally important (and possible). Design criteria should be established as early in the design process as possible—but certainly no later than the schematic design phase. Because design criteria will define success or failure in a specific area of the building design process, they should be realistic and not subject to whimsical change. In many cases, design criteria will be used both to evaluate the success of a design approach or strategy and to evaluate the performance of a system or component in a completed building. Examples of design criteria might include the following:

- Thermal conditions will meet the requirements of the *ASHRAE Standard 55, Thermal Environmental Conditions for Human Occupancy*.
- The power density of the lighting system will be no greater than 0.7 W/ft^2 [7.5 W/m^2].
- The building will achieve a LEED v.5 Platinum certification.
- Fifty percent of building water consumption will be provided by rainwater capture.
- Background sound-pressure levels in classrooms will not exceed RC 35.

Figure 1.11 suggests the importance of establishing design issues, intent, and criteria for a project before embarking on a quest for design solutions.

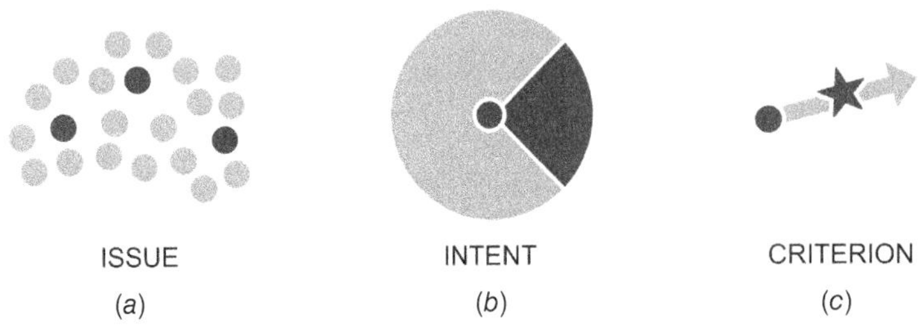

Figure 1.11 *Schematic representation of (a) issues—identifying those of importance to the project from a huge range of topics; (b) intent—setting direction for each project issue to clearly focus design efforts; and (c) criterion—determining the design target for each project intent.*

1.5 METHODS AND TOOLS

Methods and tools are the means through which design intent is accomplished and design criteria met. They include design methods and tools, such as a heat loss calculation procedure or a sun angle calculator. They also include the components, equipment, and systems that collectively define a building. It is important that an appropriate method or tool be used for a particular purpose. It is also critical that methods and tools (as means to an end) never be confused with either design intent (a desired end) or design criteria (benchmarks for success).

For any design situation, there are typically many valid and viable solutions available to the design team. It is important that no reasonable solution be overlooked or ruled out because of design process short circuits. Although this may seem unlikely, methods (such as fire sprinklers, electric lighting, and sound absorption) are often included as part of a design intent statement. Should this occur, all other possible (and perhaps more desirable) solutions are ruled out by exclusion—if electric lighting is seen as an intent, then there is no place for daylighting. This does not serve the client or occupants well and is also a disservice to the design team.

This book is a catalog of design recommendations and methods with equipment and systems information to support implementation efforts. Sorting through this extensive information will be easier with specific design intent and criteria in mind. Owner expectations and designer experiences will typically inform design intent. Sections of the book that address fundamental principles will assist with the establishment of appropriate design criteria. Table 1.1 provides examples of the relationships between design issues, intent, criteria, design tools, and implementation methods.

1.6 VALIDATION AND EVALUATION

To function as a knowledge-based profession, design (architecture and engineering) must reflect upon previous efforts and learn from existing buildings. Except in surprisingly rare situations, most building designs are unique, comprising a collection of elements not previously assembled in precisely the same way. Most buildings are a design team hypothesis: "We believe that this solution will work for the given situation." Unfortunately, most buildings exist as untested hypotheses. Little in the way of performance evaluation or structured feedback from the owner and occupants is typically sought. This is not to suggest that designers do not learn from their projects, but rather that little research quality, publicly shared information is captured for use on other projects. This is not an ideal model for professional practice from the perspective of society at large.

Bill Bordass (mentioned above), with the Usable Buildings Trust in the United Kingdom, has occasionally presented the Society of Building Science Educators (SBSE) with summaries of lessons learned through extensive post-occupancy evaluation (POE) studies of buildings. This chapter is an appropriate place to digest some of the design recommendations that flow from these findings. Bordass notes that

TABLE 1.1 Example Design Issues, Intent, Criteria, and Tools/Methods

Issue	Intent	Possible Criterion	Potential Design Tools	Potential Implementation Methods
Thermal comfort	Acceptable thermal comfort	Compliance with ASHRAE Standard 55	ASHRAE Standard 55 graphs/tables or comfort software	Passive climate control and/or active climate control systems
Lighting level (illuminance)	Acceptable illuminance	Compliance with recommendations in the *IESNA Lighting Handbook*	Hand calculations or computer simulations	Daylighting and/or electric lighting
Energy efficiency	Outstanding energy efficiency	Meet the requirements of the ASHRAE 50% *Advanced Energy Design Guide* for the building type	Handbooks, simulation software, manufacturer's data, experience	Envelope strategies and/or system and equipment strategies
Carbon emissions	Reduce greenhouse gas emissions	Reduce overall greenhouse gas emissions by 50% from current average for building type	Guidelines, databases, case studies, software	Materials specifications, ultra-low energy systems, renewable energy
Green project	Obtain green building certification	Meet the requirements for LEED v.5 Gold	LEED supporting materials, case studies, submission forms	Any acceptable combination of strategies to obtain sufficient points

building design features tend to have four attributes, sometimes possessing these attributes simultaneously:

- *Physical:* Fit and forget—if the designer and contractor have done a good job, the feature does its job and users can take it for granted.
- *Administrative:* Fit and manage—the feature needs looking after, and the question arises: Are the vigilance demands clear to the client and the operator? Often design features turn out to be more demanding on the operator than is realized at the time of design.
- *Behavioral:* Implement and internalize—the users must understand the feature to make effective use of it. Often, however, the design intent is not clear, the feature has not been properly delivered, how it should be used has not been explained to the occupants, and use does not make sense or go with the flow of occupancy, even if explained.
- *Perverse:* Risk and freedom—often, design features have both good and bad effects; it is easy for designers to get excited by the good ones and forget about the bad ones.

An intriguing recommendation, based upon the results of the Usable Buildings Trust POE studies, is: "Keep it simple and do it well, and only after that begin to be clever." This guidance can be illustrated in the following sets of words to guide the wise designer (Bordass et al., 2001):

- process before product—then product and back to process
- passive before active
- simple before complicated
- better before more
- 80 before 20 (use design time wisely)
- robust before fragile
- self-managing before managed
- efficient before elaborate
- trickle before boost
- intelligible before intelligent
- usable before alienating
- forgiving before demanding
- assets before nuisances
- response before provision
- off before on
- cellular before open
- experience before hope
- thought before action
- horses before carts

(a) Conventional Validation/Evaluation Approaches

Design validation is very common, although perhaps more so when dealing with quantitative concerns than with qualitative issues. Many design validation approaches are employed, including hand calculations, computer simulations and modeling, physical models (of various scales and complexity), and opinion surveys. Numerous design validation methods are presented in this book. Simple design validation methods (such as broad approximations, lookup tables, or nomographs), requiring few decisions and little input data, are typically used early in the design process. The latter stages of design see the introduction of more complex methods (such as computer simulations) requiring substantial and detailed input.

Building validation is much less common than design validation. Structured evaluations of occupied buildings are rarely carried out. Historically, the most common means of validating building performance is a post-occupancy evaluation (POE). Published POEs have typically focused upon some specific (and often nontechnical) aspect of building performance, such as wayfinding or productivity. The building commissioning process and evaluative case studies of projects are seeing more use as building validation tools. Third-party validations, such as ENERGY STAR certified buildings and the Leadership in Energy and Environmental Design (LEED) rating system, are popular approaches to performance verification.

(b) Commissioning

Building commissioning is a proven approach to quality assurance. An independent commissioning provider (an individual or, more commonly, a team) verifies that the design decisions and related building assemblies, equipment, and systems can meet the OPR (design intent and criteria). Verification is accomplished through review of design documents, observation of component installation, and detailed testing of equipment and systems under conditions expected to be encountered by the building in use. Historically focused upon mechanical and electrical systems, commissioning is currently being applied to numerous building systems—including envelope, security, fire protection, and information systems. Active involvement of the design team is critical to the success of the commissioning process (ASHRAE, 2019; Grondzik, 2009).

(c) Case Studies

Case studies represent another approach to design/construction validation and evaluation. The underlying philosophy of a case study is to capture information from a particular situation and convey the information in a way that makes it useful to a broader range of situations. A building case study attempts to present the lessons learned from one case in a manner that can benefit other cases (future designs). In North America, the Vital Signs and Agents of Change projects focused upon disseminating a building performance case study methodology for design professionals and students—with an intentional focus upon occupied buildings (*à la* POEs). The American Institute of Architects (AIA) and the U.S. Green Building Council (USGBC) have developed case studies dealing with design process/practice. In the United Kingdom, numerous case studies have been conducted under the auspices of the PROBE (Post-Occupancy Review of Buildings and Their Engineering) project.

1.7 HISTORIC INFLUENCES ON THE DESIGN PROCESS

The design process may appear to revolve primarily around the needs of a client and the capabilities of the design team—as exemplified by the establishment of design intent and criteria. There are several other notable influences, however, that affect the conduct and outcome of the building design process. Some of these influences are historic and affect virtually every building project; others represent emerging trends and affect only selected projects. Key design-influencing factors are discussed below.

(a) Codes, Standards, Ratings, and Guidelines

The design of virtually every building in North America (and most of the world) will be influenced by codes and standards. Codes are government-mandated and enforceable documents that stipulate minimum acceptable building practices. Designers usually interface with codes through an entity known as the authority having jurisdiction. There may be several such authorities for any given locale or project (fire protection requirements, for example, may be enforced separately from general building construction requirements or energy performance requirements). Codes essentially define the minimum performance that society deems acceptable for dealing with a particular building design issue. In no way is code compliance—by itself—likely to be adequate to meet the needs of a client. On the other hand, code compliance is indisputably necessary as codes represent and protect the health, safety, and welfare of the general public.

Codes are written in normative (mandatory) language and usually offer a prescriptive and/or performance-based pathway for compliance. A prescriptive compliance path mandates that something be done in a certain way. Examples of prescriptive code requirements include maximum U-factors for glazing assemblies, minimum pipe sizes for a roof drainage system, or a minimum number of hurricane clips per length of roof. The majority of codes adopted throughout the United States maintain a prescriptive compliance path.

By contrast, a performance-compliance path defines outcomes—an objective that must be met. Examples of performance code language include a maximum permissible design heat flow through a building envelope, a minimum design rainfall that can be safely drained from a building roof, or a defined wind speed that a roof construction must resist without damage. Some primarily prescriptive codes offer performance "options" for compliance. This is especially true of energy codes and for smoke control requirements in fire protection codes.

Codes are continually in transition. Each jurisdiction (city, county, and/or state, depending upon legislative provisions) is generally free to adopt whichever model code it deems most appropriate. Historically, four model codes (the *Uniform Building Code*, the *Standard Building Code*, the *Basic Building Code*, and the *National Building Code*) were used in various regions of the country. This regional code pattern has changed, with the development and widespread use of a single model, the *International Building Code*, to provide a more uniform and standardized set of code requirements. Some jurisdictions (typically large cities) use homegrown codes instead of a model code. Canada has its own *National Building Code*. Knowledge of current code requirements for a project is a critical element of the design process.

Standards are model documents that present a set of minimum requirements for some aspect of building design. Such requirements have been developed by a recognized authority (such as Underwriters Laboratories, the National Fire Protection Association, or the American Society of Heating, Refrigerating and Air-Conditioning Engineers). Standards do not carry the weight of government enforcement that codes do, but they are often incorporated into codes via reference. Standards play an important role in building design and are often used by legal authorities to define the level of care expected of design professionals. Standards are typically developed under a consensus process with substantial opportunity for external review and input.

Rating systems are documented frameworks used to evaluate and certify the environmental and/or social performance of buildings, interiors, and communities. Rating systems are usually voluntary systems used to gauge a project's achievement beyond minimum regulatory requirements. The ultimate "rating" of a project will be determined by an independent third-party certification body (such as Green Business

Certification Inc.). Certification may be awarded based upon design requirements, construction achievements, or the measurement and/or verification of actual project performance data. Certification programs are generally not as robustly consensus in nature as standards.

Guidelines and handbooks are less formal than codes, standards, and rating systems. General practice, the least formalized basis for design, captures the norm for a given locale or discipline. Table 1.2 provides examples of codes, standards, and related design guidance documents.

(b) Costs

Costs are a historic influence on the design process and are just as pervasive as codes. Typically, one of the earliest and strictest limits on design flexibility is the maximum construction budget imposed by the client. First cost (the cost for an owner to acquire the keys to a completed building) is the most used cost factor. First cost is usually expressed as a maximum allowable construction cost or as a cost per unit area. Life-cycle cost (the cost for an owner to acquire and use a building for some defined period of time) is generally

TABLE 1.2 Codes, Standards, and Other Design Guidance Documents

Document Type		Characteristics	Examples
Codes		Government-mandated and government-enforced (typically via the building and occupancy permit process); may be a legislatively adopted standard	*Florida Building Code*; California Title 24; *Chicago Building Code*; *International Building Code (IBC)*; *International Energy Conservation Code (IECC)*
Standards		Usually a consensus document developed by a professional organization under established procedures with opportunities for public review and input. Standards are developed to serve either as a code or as a technical reference for codes or rating systems.	*ASHRAE Standard 90.1, Energy Standard for Sites and Buildings Except Low-Rise Residential Buildings; ASTM E413–87, Classification for Rating Sound Insulation; ASME A17.1, Safety Code for Elevators and Escalators*
Rating systems		Usually a consensus document developed by a professional organization under established procedures with opportunities for public review and input. Intended for voluntary adoption for projects that wish to surpass regulated minimum standards.	Fitwel, LEED (Leadership in Energy and Environmental Design), Just Communities, Living Building Challenge, Phius Core, SITES (Sustainable Sites Initiative), the WELL Building Standard
Guidelines		Usually a consensus document developed by a professional organization, but within a looser structure and with less stringent public review. Written in informative (permissive)—versus normative (mandatory)—language.	ASHRAE Guideline 0, *The Commissioning Process*; NEMA Lighting Systems Division (LSD) documents
Handbooks		Development can vary widely—involving formal committees and peer review or single/multiple authors with no formal external review	*ASHRAE Handbook—Fundamentals*; *IES Lighting Library*; *NFPA Fire Protection Handbook*
General practice		The prevailing norm for design within a given community or discipline; least formal of all modes of guidance	System sizing approximations; generally accepted flashing details (i.e., *Architectural Graphic Standards*)

Image sources: Codes—with permission of the International Code Council; standards, guidelines—with permission of the American Society of Heating, Refrigerating and Air-Conditioning Engineers; rating systems—used with permission of the U.S. Green Building Council; handbooks—with permission of the National Fire Protection Association; general practice—with permission of John Wiley & Sons.

Acronyms: ASHRAE = American Society of Heating, Refrigerating and Air-Conditioning Engineers; ASME = American Society of Mechanical Engineers; ASTM = ASTM International (previously American Society for Testing and Materials); ICC = International Code Council; IES = Illuminating Engineering Society; NEMA = National Electrical Manufacturers Association; NFPA = National Fire Protection Association.

as important as, or more important than, first cost, but is often ignored by owners and not well understood by designers.

Over the life of a building, operating and maintenance costs can far exceed the cost to construct or acquire a building. Thus, whenever feasible, design decisions should be based upon life-cycle cost analyses and not simply first cost. The math of life-cycle costing is not difficult. The primary challenge in implementing a life-cycle cost analysis is estimating future expenses and the uncertainty naturally associated with projecting future conditions. These issues are not as difficult, however, as they might seem. Well-regarded life-cycle cost methodologies have been developed and are readily available. Appendix J provides basic information on life-cycle cost factors and procedures. The design team may find life-cycle costing a persuasive ally in the quest to convince an owner to make important, but apparently expensive, decisions.

(c) Passive and Active Approaches

"Passive" and "active" building systems refer to different approaches used to regulate indoor environmental conditions, such as temperature, air quality, and light. Passive systems manage environmental conditions without mechanical or electrical devices. Passive systems rely on the natural behavior of materials and the environment to regulate conditions. By contrast, active systems use mechanical or electrical devices to regulate conditions. The distinction between passive and active systems may mean little to the average building owner, but it can be critical to the building designer and occupant. Development of passive systems must begin early in the design process and requires early and continuous attention from the architectural designer. Passive system operation will often require the earnest cooperation and involvement of building occupants and users. Table 1.3 summarizes the defining characteristics of passive and active systems. These approaches are conceptually opposite in nature. Individual systems that embody both active and passive characteristics are often called hybrid systems. Hybrid systems are commonly employed as a means of tapping into the best aspects of both approaches.

Passive and active systems represent opposing philosophical concepts. Design is seldom so straightforward as to permit the exclusive use of one philosophy. Thus, the hybrid system. Hybrid systems are a composite of active and passive approaches, typically leaning more toward the passive. For example, single-purpose, electricity-consuming (active) ceiling fans might be added to a natural ventilation (passive) cooling system to extend the performance of the system and thus reduce energy usage that would otherwise occur if a fully active air-conditioning system were turned on instead of the fans.

The typical building will usually include both passive and active systems. Passive systems may be used for climate control, fire protection, lighting, acoustics, circulation, and/or sanitation. Active systems may also be used for the same purposes and for electrical distribution and signaling. A key characteristic of passive systems is that they are inherently zero-carbon emitting.

(d) Energy Efficiency

Efficiency is the ratio of system output to system input. The greater the output for any given input, the higher the efficiency. The concept of energy efficiency is linked to the design process through model codes and standards that specify minimum efficiencies for many items of mechanical and electrical and enclosure equipment for buildings. Code requirements for energy-efficient building solutions arose during the energy crises of the 1970s and have been updated on a periodic basis since then. As with all code requirements, mandated energy efficiency levels represent the minimum performance level that is considered acceptable—not an optimal performance level. What is considered acceptable minimum performance has evolved over time in response to changes in energy costs and availability, and in response to changes in the costs and availability of building technology.

In the United States, *ANSI/ASHRAE/IES Standard 90.1* (published by the American Society of Heating,

TABLE 1.3 Differentiating Passive and Active Systems

Characteristic	Passive System	Active System
Energy source	Uses no purchased energy (no electricity, natural gas, fuel oil, etc.); example: daylighting system	Uses primarily purchased (and nonrenewable) energy; example: electric lighting system
System components	Components play multiple roles in system and in the building as a whole; example: concrete floor slab that is structure, walking surface, and solar collector/storage	Components are commonly single-purpose elements; example: gas furnace
System integration	System is usually tightly integrated (often inseparably) with the overall building design; example: natural ventilation system using windows	System is usually not well integrated with the overall building design, often seeming an add-on; example: window air-conditioning unit

Refrigerating and Air-Conditioning Engineers, co-sponsored by the Illuminating Engineering Society, and approved by the American National Standards Institute) is one of the most commonly encountered energy efficiency benchmarks for commercial/institutional buildings. Some states (such as California and Florida) utilize state-specific energy codes. Residential energy efficiency requirements are addressed by several model codes and standards (including the *International Energy Conservation Code*, the *International Green Construction Code*, and *ANSI/ASHRAE Standard 90.2*).

Energy efficiency requirements for residential buildings tend to focus upon envelope (walls, floors, roofs, doors, windows) and mechanical equipment (heating, cooling, domestic hot water) performance. Energy efficiency requirements for commercial/institutional buildings address virtually every building system (including lighting and electrical distribution). Most energy codes present a set of prescriptive minimum requirements for individual building elements, with an option for an alternative means of compliance to permit innovation and/or a systems-based design approach.

The concept of energy conservation implies saving energy by using less. This is conceptually different from efficiency but is an integral part of everyday usage of the term. Energy efficiency codes and standards include elements of conservation embodied in equipment control requirements or thermal insulation levels. Because of negative connotations that some associate with "conservation" (doing without), the term energy efficiency is generally used to describe both conservation and efficiency efforts in buildings.

Most energy efficiency standards deal solely with on-site energy usage. The reason for this approach lies in the controversy surrounding assigning site-source energy adjustment factors that do not commercially disadvantage one fuel over another (there is no such controversy regarding renewable energy sources). Off-site energy consumption (for example, that required to transport fuel oil or natural gas, or the substantial process losses from electrical generation plants) is not typically addressed in energy efficiency regulations. A site-source focus can seriously skew thinking about energy efficiency design strategies, and this should be recognized by the design team. Off-site energy consumption that is directly tied to on-site consumption is real, can be substantial, and contributes to carbon emissions and fossil fuel depletion.

Passive design solutions usually employ renewable energy resources. Several active design solutions, however, also utilize renewable energy. Energy conservation and efficiency concerns are typically focused upon minimizing depletion of nonrenewable energy resources—even when not explicitly stated. The use of renewable energy sources (such as solar radiation and wind) changes the passive versus active discussion, should change the perspective of the design team, and may affect the way compliance with energy efficiency codes/standards is evaluated.

Energy codes define legally acceptable energy efficiency/conservation performance. Design teams and clients may wish to exceed such code-minimum performance. If so, they have a wide range of possible energy targets from which to select. A team/client may seek to do slightly better than the energy code in terms of efficiency; it may seek to do much better than the energy code; or it may seek to design a net-zero energy building. Between code and net-zero energy targets lies a range of performance that can be called low-energy and another (more stringent) realm of performance called ultra-low-energy. Architecture 2030 visualizes this range and provides guidance in selecting an appropriate energy target on its Zero Tool website. Figure 1.12 illustrates Zero Tool results for a proposed office building in Tallahassee, Florida. The baseline is representative of a typical modern building of the same space type, size, and occupancy. To summarize, meeting the baseline is driven by the regulatory

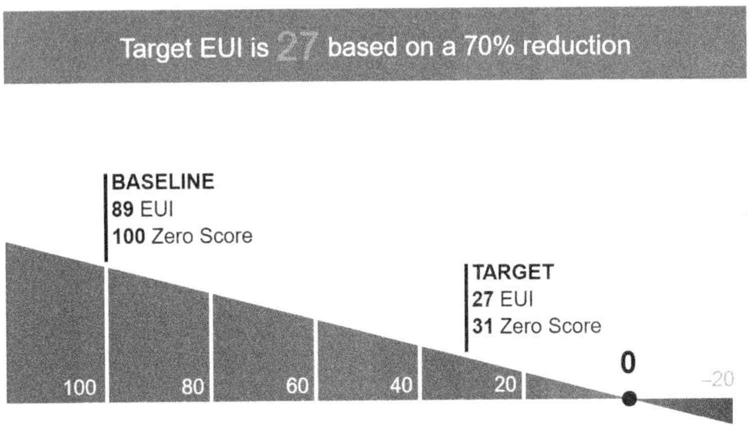

Figure 1.12 *Sample result from the Architecture 2030 Zero Tool showing the range of possible energy performance outcomes between code-compliant (100 Zero Score) and net-zero (the right-hand dot). Beyond the dot lies energy positive performance. EUI stands for energy use intensity. (Architecture 2030, n.d.c.)*

minimum standard; all performance points to the right of the baseline represent ethical decisions.

Project teams will find several well-regarded guidelines and certifications to assist them in designing low-energy buildings. The *Advanced Energy Design Guides*, published by ASHRAE, provide design recommendations for several building typologies with 30% better than code and 50% better than code energy outcomes. These guides are somewhat dated relative to current code requirements, but nonetheless provide useful recommendations. USGBC's LEED green building rating systems give a sense of the value USGBC affords to energy efficiency and renewable energy through the energy points available toward project certification.

Resources are also available to teams looking beyond low-energy to ultra-low-energy. Three *Advanced Energy Design Guides* focus on net-zero energy building design. *Passive Building* certification will result in ultra-low-energy performance. Such approaches are discussed below. ASHRAE is currently developing *Standard 227P: Passive Building Design Standard* to provide design requirements for buildings that have exceptionally low energy usage and that are durable, resilient, comfortable, and healthy. The roll-out of *Standard 227* will provide a code-enforceable path to very energy efficient buildings.

Multiple meanings of the term "passive" demand further explanation. A passive system is an approach to design that fits the discussion of passive versus active given above—a system that does not consume purchased energy is deemed passive. A passive building will consume purchased energy and will include active systems—but has been certified as an ultra-low-energy building by an independent third party. A passive building may also include passive systems, but these are not mandated.

Drawing from North American experiences with super-insulated houses, passive building certification originated in Germany with the Passivhaus Institute and was enhanced by Phius (formerly Passive House Institute US). Passive buildings provide a rigorous, voluntary standard for exceptional energy efficiency. Passive building projects are so efficient that they require little energy for space heating or cooling. Passive building certification places stringent limits on both site and source energy.

The fundamental precepts of a passive building include:

- high-thermal resistance envelope components (floor, walls, roof, and windows)
- low air leakage (infiltration) with indoor air quality control provided by an air-to-air heat exchanger
- highly efficient heat and cooling systems, water heating equipment, and appliances
- concern for occupant thermal comfort and indoor air quality
- confirmation of design and construction efforts by an independent verifier

This carefully orchestrated set of design and construction requirements leads to buildings with measured performance outcomes that approach net-zero energy. Often the addition of a renewable energy system (such as photovoltaics) can deliver a net-zero energy project.

HERS. The energy performance of a passive building single-family home is around 30–40 on the HERS Index (Figure 1.13), depending on the size of the home, the climate, and whether a solar thermal

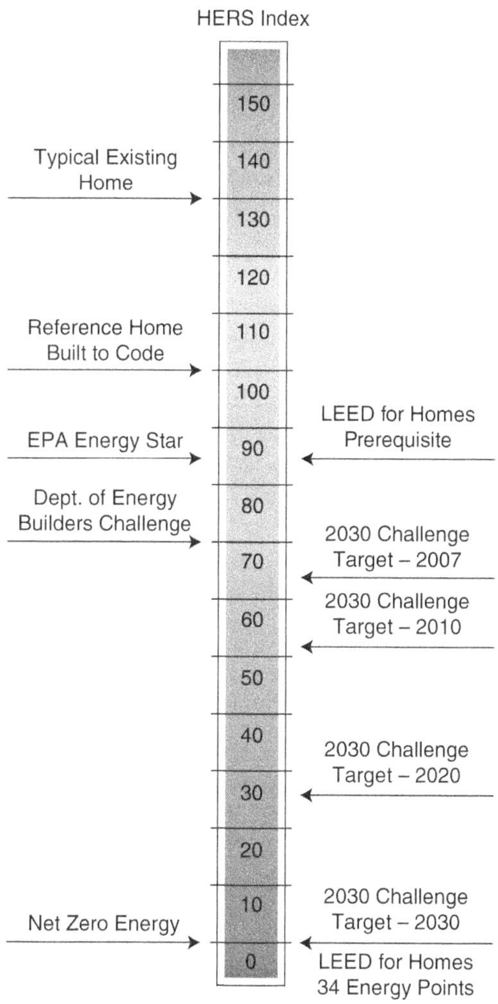

Figure 1.13 *HERS (the Home Energy Rating System) is a relative comparison scale for residential energy performance. It sets baseline performance as 100 (which is linked to compliance with the 2006 International Energy Conservation Code) and sets exemplary performance at 0, which is a net-zero energy residence. (Courtesy BuildingGreen, Inc.; used with permission.)*

system is installed (without added photovoltaics). The HERS Index is an energy-scoring system developed by the Residential Energy Services Network (RESNET). The HERS Index rates the energy efficiency of a home compared to a standardized reference home. It is a relative scale where the reference home (i.e., a standard new home) has a score of 100. A lower HERS Index value indicates greater energy efficiency. A HERS Index score of 0 means a home produces as much energy as it consumes (i.e., it's a net-zero site energy home).

The HERS Index is an energy rating index (ERI). An ERI typically compares the energy efficiency of a specific home to a reference home. The reference home performance is defined according to a set of standards. For instance, the 2021 *International Energy Conservation Code (IECC)* allows for an Energy Rating Index alternative compliance path based on a project's climate zone. The 2021 IECC specifies that the ERI shall be determined in accordance with *RESNET/ICC 301–2019 Standard for the Calculation and Labeling of the Energy Performance of Dwelling and Sleeping Units using an Energy Rating Index*. The HERS Index uses the methodologies and specifications outlined in *RESNET/ICC 301*. In this sense, the ERI alternative compliance path utilizes the HERS Index.

EUI. The energy performance of any type of building can be expressed using an energy utilization index value—also referred to as energy use intensity. EUI is calculated as the annual overall site energy consumption of a building divided by the conditioned floor area of the building. Units for EUI are Btu/ft^2 yr [MJ/m^2 yr]. It is common to see building energy performance expressed as EUI—several instances of this have been discussed above (for example, in Figure 1.12).

(e) Green Buildings

Green building considerations—whether part of a formal building rating or just a matter of better project outcomes—go beyond energy-efficient design in order to address both local and global imperatives related to building energy, water, and materials usage. Energy efficiency is a key, but not the sole, element of green design. The concept that is broadly called "green building" arose from concerns about the wide-ranging environmental impacts of design decisions. Although there is no generally accepted concise definition of green, the term is typically understood to incorporate concern for the health and well-being of building occupants/users and respect for the larger global environment. A green building should maximize beneficial impacts on its direct beneficiaries while minimizing negative impacts on the site, local, regional, national, and global environments.

Several rating systems have found wide acceptance as benchmarks for "greenness." These include the U.S. Green Building Council's (USGBC) Leadership in Energy and Environmental Design (LEED) rating system, the Green Building Initiative's Green Globes Environmental Assessment system, and an international evaluation methodology entitled GBTool. Green building rating systems are in active use throughout the world. LEED, for example, has been utilized in 186 countries and territories. Most green building ratings systems are voluntary, would be correctly termed guidelines, and are generally ill-suited for adoption as code. A set of green building requirements written in normative language appropriate for code adoption, however, was developed by a coalition of professional organizations under the auspices of *ASHRAE Standard 189.1, Standard for the Design of High-Performance Green Buildings Except Low-Rise Residential Buildings*. *ASHRAE Standard 189.1* has since been incorporated as the technical basis for the ICC's International Green Construction Code (IgCC).

Typical of green building guidelines, every variation of the LEED rating system (there are adaptations of LEED for a variety of project types) presents a palette of design and construction options from which the project team can select strategies appropriate for a particular building and its site development. LEED certification (Figure 1.14) is achieved through a formalized third-party verification process by which points are awarded to projects for achieving certain green building qualities or outcomes. Based on the number of points earned (in addition to several prerequisites for the design and construction of the project), either basic, Silver, Gold, or Platinum certification is awarded.

Recently, USGBC released a suite of performance-based LEED certification models, dubbed LEED Zero, based on achievement toward net-zero energy, carbon, water, and/or waste measured over a 12-month performance period. Certification based on verified net-zero goals appears poised as the future of green building rating systems. Other performance-based rating systems that require measured data for certification include BREEAM (Building Research Establishment Environmental Assessment Method) by the Building Research Establishment (BRE), ENERGY STAR by the U.S. Environmental Protection Agency (EPA), the Living Building Challenge (LBC) and Zero Energy Certification by the International Living Future Institute (ILFI), and the RESET Standard by GIGA.

The emergence of green building rating systems has greatly rationalized design intent and design criteria in this particular realm of architecture. Prior

(a) (b)

Figure 1.14 *(a) The Jean Vollum Natural Capital Center, Portland, Oregon. A warehouse from the industrial era was rehabilitated by Ecotrust to serve as a center for conservation. (b) LEED plaque on the front façade of the Vollum Center. The plaque announces the success of the design team (and the owner) in achieving a key element of their design intent. (© Alison Kwok; all rights reserved.)*

to the advent of these aforementioned rating systems, anyone could claim greenness for his/her designs. Although green building is entered into voluntarily (very few jurisdictions have adopted the IgCC, although a number of municipalities require new public buildings to be LEED certified), there are now several generally accepted standards against which green building may be assessed—and validated.

1.8 EMERGING INFLUENCES ON THE DESIGN PROCESS

(a) Net-Zero Energy

A net-zero energy building (NZEB) can be defined in several ways, depending upon the project goals. The National Renewable Energy Laboratory (Torcellini et al., 2006) maintains four basic definitions of NZEB:

- *Net-Zero Site Energy:* A site NZEB produces at least as much energy as it uses in a year, when energy is accounted for at the site.
- *Net-Zero Source Energy:* A source NZEB produces at least as much energy as it uses in a year, when energy is accounted for at the source.
- *Net-Zero Energy Costs:* In a cost NZEB, the amount of money a utility pays the building owner for the energy the building exports to the grid is at least equal to the amount the owner pays the utility for the energy services and energy used over the year.
- *Net-Zero Energy Emissions:* A net-zero emissions building produces at least as much emissions-free renewable energy as it uses from emissions-producing energy sources.

The concept of net-zero allows for a convenient target that design teams and owners can attempt to meet. Net-zero is the point where consumption of fossil fuel-based, carbon-emitting energy is balanced with non-carbon-emitting renewable energy. A calendar year is a convenient accounting period, with accounting being done either at the site or at the source (the source is more scientifically honest from a greenhouse gas perspective) and using either Btu [kWh], cost, or emissions as the basis for calculations. The term zero energy (promoted by the U.S. Department of Energy) should be avoided—a zero energy building is likely abandoned.

Site energy refers to the amount of energy used by a building and everything within its site boundary, typically measured through utility meters. By contrast, source energy refers to the total energy expenditure required to generate and deliver energy to the site, including losses that occur during production, transmission, and distribution. According to the U.S. Energy Information Administration (EIA), the energy lost from source to site varies depending on the energy resources and the characteristics of the generation, transmission, and distribution systems. For electricity, typical losses in the United States can be over 60%. This means only about a third of the energy generated at a power plant reaches the end-user as usable electricity. To approximate a building's net source energy, both imported and exported energy can be multiplied by the appropriate site-to-source conversion multipliers (Table 1.4).

Recently, ASHRAE published its first net-zero—or to use their preferred term, "zero net"—energy and carbon standard. *ANSI/ASHRAE Standard 228, Standard Method of Evaluating Zero Net Energy and Zero Carbon Performance Buildings* sets detailed requirements

TABLE 1.4 EPA Source-Site Ratios for All Energy Types

Energy Type	U.S. Ratio	Canadian Ratio
Electricity (Grid Purchase)	2.80	1.83
Electricity (Onsite Solar or Wind)	1.00	1.00
Natural Gas	1.05	1.06
Fuel Oil (No. 1, 2, 4, 5, 6, Diesel, Kerosene)	1.01	1.00
Propane and Liquid Propane	1.01	1.00
Steam	1.20	1.33
Hot Water	1.20	1.33
Chilled Water	0.91	0.57
Wood	1.00	1.00
Coal/Coke	1.00	1.00
Other	1.00	1.00

Source: U.S. Environmental Protection Agency (EPA), 2023.

for evaluating whether a building project (or group of buildings) meets a specific zero net definition during operation.

Pushing energy efficiency toward its limits will lead to the realm of building performance associated with net-zero energy buildings. High efficiency alone is not sufficient to produce a net-zero building, but it is a practical prerequisite. Such a building will, despite aggressive energy-efficiency efforts, still use energy (for things such as domestic water heating, electric lighting, space heating/cooling, and appliances). Any such residual energy requirements will, however, be provided by renewable energy resources that match the magnitude of fossil-fuel-based energy consumption.

Looking at a net-zero energy building from another perspective—such a building may use energy derived from fossil fuels (such as electricity from a coal-fired power plant) to meet its programmatic and occupancy needs. But every Btu [kWh] of energy from a nonrenewable resource will be matched by a Btu [kWh] of energy from a renewable resource. A net-zero energy building is not a no-energy building, and it is not a no-nonrenewable-energy building. It is, however, a low-energy building that annually employs at least 50% renewable energy. This is a big step on the road to sustainability. Sustainability (on the energy front) may lie in what some designers are describing as plus-energy buildings. Also known as net-positive energy buildings or energy-plus buildings, these are structures that produce substantially more renewable energy than they consume over the course of a year.

Designers have some flexibility in defining a net-zero energy building within the limits of energy balance described above. This flexibility lies in the setting of system boundaries (i.e., the conceptual limits that define what is included within a system and subject to analysis). The system boundaries may be spatial, temporal, and/or organizational. Some examples follow:

- Today, the most common perception of a net-zero energy building is one that is net-zero considering operational energy measured at the site boundary.
- The system boundaries may be expanded back to the proximate source of the building's energy, such that source (versus site) energy is balanced; this is roughly three times more challenging for an all-electric building (due to generation and transmission losses that are not included in a site-based analysis).
- Rather than considering only operational energy, the net-zero analysis boundary might be extended backward in time to include construction process energy (and perhaps design process energy).
- An owner might want to consider not just the building as the system, but also the organizational efforts supported by (or perhaps required by) the building; employee commuting energy might be considered, and/or the energy required to clean and maintain the building.

The source of renewable energy inputs to a net-zero energy building may also be addressed as a function of the site boundary. For example, the renewable energy component might come from a green power purchase agreement (with the energy production occurring remotely), or the energy might be produced from systems located on or adjacent to a building. The authors' philosophical preference is for site-based renewables—such that the design team is directly responsible for necessary energy production. In this case, the design process (relative to energy, and perhaps also water) will be seen as a job of balancing demand with supply.

(b) Decarbonization

Climate change and global warming are growing concerns in the design community, as evidenced by the positive response of many professional organizations to the 2030 Challenge since its introduction in 2006 by Architecture 2030 (Architecture 2030 n.d.a). Design to reduce carbon emissions is becoming a concern on many building projects. The term carbon-neutral design is generally used to express this focus, and it accurately represents a primary design intent in a number of innovative projects. The Aldo Leopold Legacy Center in Baraboo, Wisconsin, is an exciting example of such a project.

Greenhouse gases (GHG) trap heat below the Earth's atmosphere in more or less the same way that glass traps heat from solar radiation in a greenhouse (or in a passive solar heating system). This trapping of heat increases temperatures and leads to climate change. Carbon dioxide (CO_2) is a major greenhouse gas; others include methane (CH_4), nitrous oxide (N_2O), hydrofluorocarbons (HFCs), perfluorocarbons (PFCs), and sulfur hexafluoride (SF_6). Carbon dioxide equivalent (CO_2e) serves as a metric by which to compare the emissions of various GHGs by converting their quantities to an equivalent amount of carbon dioxide with the same global warming potential. The term "carbon" is often informally used as shorthand for "carbon dioxide"—which often serves as a proxy to discuss total GHG emissions. Buildings are important contributors to carbon dioxide emissions and are therefore logical targets for mitigation in an attempt to reduce climate change potential. For example, see Figure 1.15 for an estimate of the role buildings play in producing CO_2 emissions in the United States.

At a conceptual scale, carbon dioxide (and other greenhouse gas) emissions may be classified into three broad categories (EPA), termed scopes. These scopes apply at the scale of a single project:

Scope 1: All direct GHG (greenhouse gas) emissions (such as from a gas-fired boiler or a wood-burning stove).

Scope 2: Indirect GHG emissions from consumption of purchased electricity, heat, or steam.

Scope 3: Other indirect emissions, such as from the extraction and production of purchased materials and fuels, transport-related activities in vehicles not owned or controlled by the reporting entity, electricity-related activities (e.g., transmission and distribution) not covered in Scope 2, outsourced activities, waste disposal, etc.

In parallel with net-zero energy analyses, it might be useful to consider that buildings produce (or are

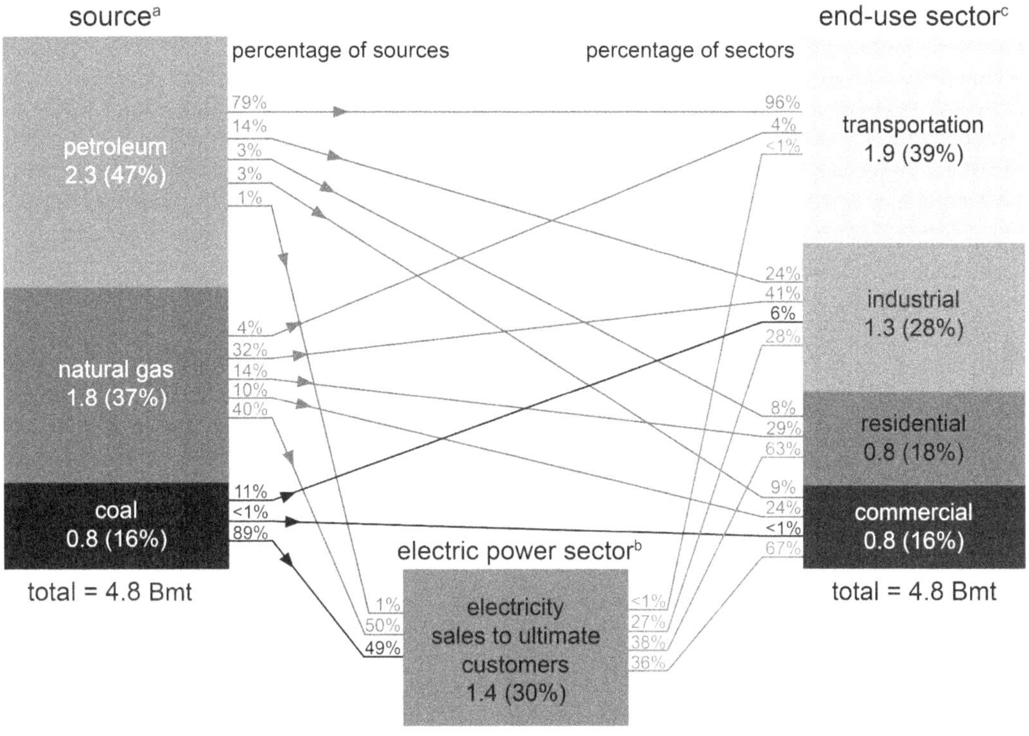

Note: Sum of components may not equal total due to independent rounding. Includes the relatively small amount of carbon dioxide (CO_2) emissions from geothermal and nonbiomass waste for electric power sector use not shown elsewhere, See EIA's Monthly Energy Review, Section 11.

[a]CO_2 emissions from primary energy consumption. Each energy source is measured in different physical units and converted to metric tons of CO_2.

[b]The electric power sector includes electricity-only and combined-heat-and-power (CHP) plants whose primary business is to sell electricity, or electricity and heat, to the public. CO_2 emissions from electricity sales to ultimate customers in each end-use sector are equal to the weighted average of fuels used to generate electricity and allocated proportionally to the amount of sales to each end-use sector.

[c]Industrial and commercial sectors include primary energy consumption by CHP and electricity-only plants in the sector. Includes the CO_2 emissions from the electricity sales allocated to each end-use sector.

Figure 1.15 *U.S. CO_2* emissions from energy consumption by source and sector, 2023. *(U.S. Energy Information Administration (EIA), 2024.)*

linked to) carbon dioxide emissions in several distinct ways that may be of concern to an owner:

- as a result of fossil fuel energy consumed during the design process (computer use, printing, site visits, etc.)
- as a result of fossil fuel energy consumed during the construction process (by equipment, worker commutes, site preparation, etc.)
- through the disposal of organic construction waste that decomposes
- as a result of ongoing fossil fuel energy consumption for heating, cooling, lighting, and building support operations
- as a result of vehicle use associated with building functions and siting (including fossil fuels used for employee commuting, product deliveries, etc.)
- as a result of waste produced by a building in operation

Of these various carbon release mechanisms, energy consumption for building operation is normally the largest contributor and the most readily available target for reductions. A reminder: Energy use itself is not the carbon culprit, but rather the use of fossil fuels as an energy source.

Options for reducing carbon emissions from the operation of building systems include improving the efficiency of building envelopes and systems (the ultimate, but unrealistic, goal being a zero-energy project); using localized renewable energy resources to meet the energy needs that remain after aggressive efficiency moves (the goal being a net-zero-energy building); and purchasing or obtaining "green power" to mitigate the effects of residual carbon emissions not stemmed by efficiency and local renewables.

Two common forms of green power credit are renewable energy credits and carbon offsets. Renewable energy credits (RECs) represent defined percentages of an off-site renewable energy generation project, such as a wind or solar installation. Typically, one REC is issued for each megawatt-hour of electricity generated from renewable sources. RECs allow a building owner to offset a carbon footprint by purchasing these credits—effectively claiming the environmental benefits of the green power generation associated with the REC. This enables a building owner to reduce the facility's net environmental impact without directly changing the building energy infrastructure. The notion of claiming RECs is controversial because doing so allows building projects to claim "net-zero energy emissions" without actually reducing on-site emissions. The critique is that RECs enable building owners to pay for renewable energy generated elsewhere, without necessitating any direct change to their own energy infrastructure or consumption patterns.

Carbon offsets are a financial instrument used by building owners to mitigate their facility's net contribution to greenhouse gas emissions. Carbon offsets work by funding off-site projects that reduce, avoid, or sequester emissions through reforestation, renewable energy developments, and/or energy efficiency improvements. When a ton of CO_2e is mitigated or sequestered by such an off-site project, the "credit" can be purchased by building owners to "offset" the emissions contributed by their facility. This concept can be used to achieve net "carbon neutrality" (at both the building scale and at the enterprise level). Carbon credits are also controversial, being akin to buying one's way out of trouble—but all green power purchases help to promote the proliferation of clean energy resources. Both RECs and carbon offsets should be certified (e.g., Green-e certified or equivalent). Industry best practice is that RECs only account for the effects of Scope 2 electricity use, while carbon offsets may be used to mitigate Scope 1 or Scope 2 emissions on a metric ton of carbon dioxide-equivalent basis.

As jurisdictions begin requiring environmental performance benchmarking for buildings, it is important for designers and building owners who need to create energy and carbon reduction goals and quantify savings to understand energy use and the associated emissions metrics. Energy use intensity (EUI), introduced above, is one such benchmark metric. Since EUI is expressed per unit of floor area, it lends itself as a normalized metric for comparison across similar building types of varying sizes. As such, in 2007, Architecture 2030, AIA, ASHRAE, IES, and USGBC agreed to use EUIs reported to the Energy Information Administration's 2003 Commercial Buildings Energy Consumption Survey (CBECS) as the baseline for their common target goals. The CBECS data represent the national average energy consumption of existing U.S. commercial buildings (Architecture 2030 2007). A building's metered energy data can be input into the ENERGY STAR® PortfolioManager®, which is made publicly available by the U.S. Environmental Protection Agency (EPA), to determine the building's EUI, based on site energy. The EPA's Target Finder calculator allows design teams and owners to assess the energy performance of commercial building designs and existing buildings against a range of "what-if" scenarios. Using building energy modeling (BEM), design teams can determine a project's projected EUI (pEUI) and compare it against comparable CBECS data to determine the percent improvement toward the 2030 Challenge goals. Refer to Appendix L for more information about BEM tools.

The metric carbon dioxide equivalent (CO_2e) emissions (introduced above) can be used to assess a building's operational GHG contributions

(Bryan and Trusty, 2008). Consider the following example: we are designing a large commercial office building for Indianapolis, Indiana, USA, and we need to determine the projected operational GHG emissions for the building. The latest energy modeling iteration indicates that the 281,078 ft^2 [26,113 m^2] building's pEUI is 62 kBtu/ft^2 yr [704 MJ/m^2 yr] with an estimated 5744 million British thermal units (MBtu) [6,060,242 MJ] of natural gas use per year and 11,677 MBtu [12,319,889 MJ] of electricity use per year.

The operational carbon attributed to the natural gas would be determined by multiplying the MBtu/yr [MJ/yr] of site natural gas use by the EPA's direct GHG emissions factor of 53.11 kg/MBtu [0.050 kg/MJ] (Table 1.5), which equals 305,071 kg CO_2e/year.

Regarding the operational carbon attributed to the electricity, since the building is located in Emissions & Generation Resource Integrated Database (eGRID) subregion RFCW (Figure 1.16), multiply

TABLE 1.5 Direct Greenhouse Gas Emission Factors for the United States and Canada

	CO_2e Emissions			
	United States	Canada		
Fuel Type	kg/MBtu [kg/MJ]	kg/MBtu [kg/MJ]	g/L	kg/tonne
Natural Gas	53.11 [0.050]	By Province		
Propane	61.95 [0.059]	64.37 [0.061]	1544	–
Fuel Oil (No. 1)	73.49 [0.070]	75.10 [0.071]	2762	–
Fuel Oil (No. 2)	74.20 [0.070]	75.10 [0.071]	2762	–
Fuel Oil (No. 4)	75.28 [0.071]	75.10 [0.071]	2762	–
Fuel Oil (No. 5, 6)	75.26 [0.071]	78.81 [0.075]	3175	–
Diesel Oil	74.16 [0.070]	73.98 [0.070]	2689	–
Kerosene	75.44 [0.072]	71.93 [0.068]	2569	–
Coal (anthracite)	104.42 [0.099]	122.43 [0.116]	–	3214
Coal (bituminous)	94.01 [0.089]	100.50 [0.095]	–	2381
Coke	114.40 [0.108]	116.34 [0.110]	–	3179

Source: U.S. Environmental Protection Agency (EPA), 2024.
Note: SI unit conversions added by the authors.

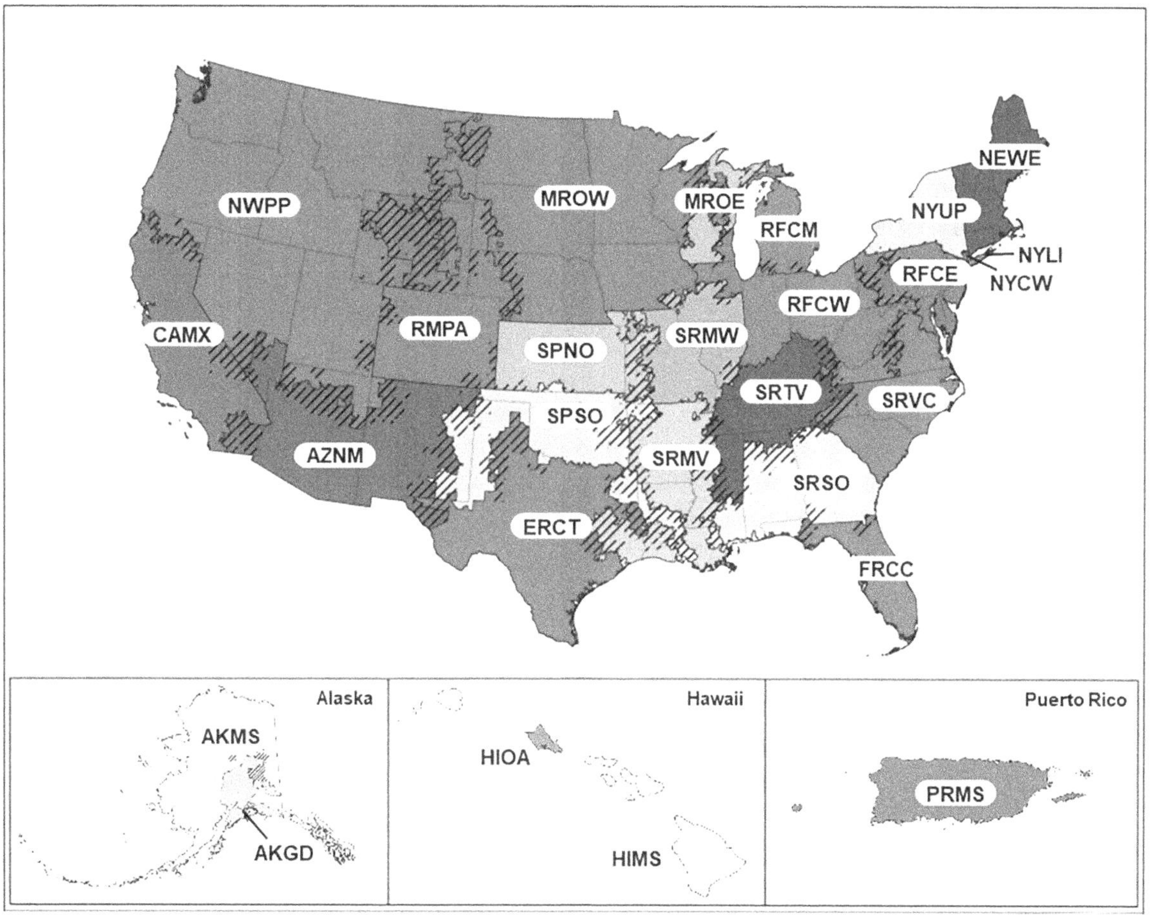

Note: Crosshatching indicates that an area falls within overlapping eGRID subregions due to the presence of multiple electric service providers. Visit the EPA's Power Profiler to definitively determine the eGRID subregion associated with your location and electric service provider: www.epa.gov/egrid/power-profiler

Figure 1.16 *Map of the eGRID Subregions. (U.S. Environmental Protection Agency (EPA), 2024.)*

TABLE 1.6 Indirect Greenhouse Gas Emission Factors for Electricity in the United States

eGRID Regional Description	eGRID Acronym	CO_2e Emissions (kg/MBtu [kg/MJ])
South/Central Alaska	AKGD	140.61 [0.133]
Most of Alaska	AKMS	66.14 [0.063]
Southwest US	AZMN	103.60 [0.098]
Southwest Coast	CAMX	66.36 [0.063]
Most of TX	ERCT	102.92 [0.098]
Most of Florida	FRCC	108.58 [0.103]
HI excluding Oahu	HIMS	154.58 [0.147]
Oahu Island	HIOA	210.91 [0.200]
Eastern WI	MROE	197.87 [0.188]
Upper Midwest	MROW	125.38 [0.119]
New England	NEWE	71.84 [0.068]
Northwest US	NWPP	80.53 [0.076]
New York City	NYCW	117.86 [0.112]
Long Island, NY	NYLI	160.73 [0.152]
Upstate NY	NYUP	36.61 [0.035]
Puerto Rico	PRMS	212.66 [0.202]
Mid Atlantic	RFCE	87.76 [0.083]
Most of Michigan	RFCM	162.71 [0.154]
Ohio Valley	RFCW	133.71 [0.127]
CO-Eastern WY	RMPA	150.43 [0.143]
KS-Western MO	SPNO	127.51 [0.121]
TX Panhandle-OK	SPSO	129.62 [0.123]
Lower Mississippi	SRMV	106.83 [0.101]
Middle Mississippi	SRMW	183.44 [0.174]
SE US, Gulf Coast	SRSO	119.31 [0.113]
Tennessee Valley	SRTV	124.76 [0.118]
Virginia/Carolina	SRVC	83.19 [0.079]
National Average		**109.99 [0.104]**

Source: U.S. Environmental Protection Agency (EPA), 2024.
Note: SI unit conversions added by the authors of this book.

the MBtu/yr [MJ/yr] of site electricity by the EPA's indirect GHG emissions factor of 133.71 kg/MBtu [0.133 kg/MJ] (Table 1.6) to obtain 1,561,332 kg CO_2e/year.

Note that as the RFCW electricity grid decarbonizes, these emissions could be reduced precipitously over the life of the building, whereas the natural gas GHG emissions are "locked in" for the life of its utility. This hope for the future grid is the basis for the design community's advocacy of building electrification.

Whereas the EPA's GHG emissions factors are based on recent data, project teams may use NREL's Cambium dataset to derive projected emissions factors that incorporate the potential impacts of different energy policies, technologies, and market conditions on the electric grid.

(c) Embodied Energy/Embodied Carbon

According to data aggregated by the non profit research organization Architecture 2030, the built environment is responsible for about 40% of annual global CO_2 emissions. Of those total emissions, building operations are responsible for approximately 27% of annual global emissions. As such, operational energy and the associated operational carbon emissions are the typical focus, respectively, of net-zero energy and carbon-neutral design.

However, beyond the energy necessary to power our buildings, energy resources are required for the production, transport, construction, maintenance, and disposal of building materials and equipment. This energy is termed embodied energy, and it comes with direct and indirect associated carbon emissions—which are termed embodied carbon. Embodied carbon constitutes the remaining 13% of the global 40% contribution from the building sector.

In typical, code-compliant buildings with a reasonable life span, embodied carbon (and energy) may be 10–15% of the total carbon emissions attributable to a building during its life. In highly energy-efficient buildings, operational energy and carbon magnitudes decrease, while embodied energy and carbon—if not addressed through design—remain the same as in a non-efficient project. The proportion of operational to embodied energy/carbon thus decreases as buildings become more efficient. The case has been made that embodied carbon mitigation—since it occurs now, rather than in the future—is one of the most effective ways to quickly stem climate change.

A word about terminology. "Embodied" carbon is commonly used to refer to the carbon emissions related to building materials and equipment. This follows the concept of embodied energy—wherein energy is "baked into" materials during their manufacturing, transport, and installation. The problem with carbon is actually the opposite—greenhouses gases are disembodied (emitted) as materials are made and placed in a building. Some materials (such as wood), however, do in fact hold carbon. This is commonly called sequestration. It is too late to change the usage of embodied carbon, but it is important to remember that this actually means net emitted carbon equivalent.

A client or design team truly interested in environmentally responsive projects will look at the embodied energy and carbon impacts of design decisions. Though rapidly evolving, resources and design tools to assist with reducing embodied carbon and embodied energy are not as well developed as resources for reducing operational energy. Most resources dealing with embodied carbon/energy are found in the life cycle assessment field. This is a maturing and complex area of building design—requiring, for example, clear definitions of system analysis boundaries and tracking of environmental inputs across time and space. Several resources, however, can open the door to engagement with this issue. *Life Cycle Assessment* by Simonen (2014) provides an excellent overview of embodied carbon. The Carbon Leadership Forum (2024) continues to develop an online repository of information and resources including its whole-building life-cycle assessment (WBLCA) benchmark study. Appendix L in this book lists several life cycle assessment tools.

There are at least three rationales for considering the carbon and energy required to place building materials into a project. First, even at, say, 10% of total energy/carbon, the magnitude of environmental effects is worthy of consideration—particularly if no- or low-cost changes to materials specifications can be enacted. Second, at some point in the journey to low-energy buildings, more economical effects can be had by mitigating embodied carbon than by further reducing operational energy—it will be cheaper to save the next unit of carbon emissions by reducing embodied rather than operational carbon. Third, an intriguing argument for better addressing and mitigating embodied carbon lies in efforts to stem atmospheric carbon dioxide concentrations sooner rather than later to constrain global temperature rise. This design context suggests that the operational life of a building of concern to society is much shorter than the life as seen by an owner. If 10 or 20 years are seen as the window of opportunity for constraining average global temperature, then emitted carbon associated with construction materials becomes critical in designing mitigating buildings. The ratio of embodied to operational carbon increases substantially as the building life assumed for analysis decreases.

Increasingly, building design and construction teams are assessing both operational carbon and embodied carbon throughout the delivery of a project. Tracking both on a normalized basis offers the opportunity to develop baselines and targets. As such, the following metrics are notable:

- *Operational Carbon Intensity (OCI):* The global warming potential (GWP) associated with the operational emissions of a building, expressed as kg CO_2e/m^2.
- *Embodied Carbon Intensity (ECI):* The GWP intensity associated with the embodied emissions of a building, expressed as kg CO_2e/m^2.
- *Whole Life Carbon Intensity (WLCI):* The GWP intensity associated with all emissions of a building, expressed as kg CO_2e/m^2.
- *Whole Life Carbon (WLC):* These emissions are defined as the entire amount of carbon produced by any particular built asset. WLCI is sometimes referred to as total carbon intensity (TCI).

(d) Sustainability

Unlike green building, the meaning of "sustainability" in architecture has not yet been rationalized. The term "sustainable" is used freely—and often mistakenly—to describe a broad range of intents and performances. This is unfortunate, as it tends to make sustainability a meaningless term—and sustainability is far too important to be rendered meaningless by baseless claims. For the purposes of this book, sustainability is defined as follows (paraphrasing the Brundtland Commission): Sustainability involves meeting the needs of today's generation without detracting from the ability of future generations to meet their needs.

Sustainability is essentially long-term survival under an assumed standard of living. In architectural terms, sustainability involves ensuring the survival of an existing quality of life for future generations. From the standpoint of energy, water, and materials, it can be argued that sustainability requires zero use of nonrenewable resources. Any long-term removal of nonrenewable resources from the environment will surely impair the ability of future generations to meet their needs (with fewer resources being available, due to our actions). Because sustainability is so important a concept and objective, the term should not be used lightly. It is highly unlikely that any single building built in today's economic environment can be sustainable (yielding no net resource depletion). Sustainability at the community scale is more probable; examples, however, are rare.

(e) Regenerative Design

Energy efficiency is an attempt to use less energy to accomplish a given design objective (such as thermal comfort or adequate lighting). Green design is an attempt to maximize the positive effects of design while minimizing the negative ones—with respect to energy, water, and material resources. Sustainable design is an attempt to solve today's problems while reserving adequate resources to permit future generations to solve their problems. Energy efficiency is a necessary constituent of green design. Green design is a necessary constituent of sustainable design. Regenerative design goes beyond sustainability.

Regenerative design is an approach that does more than just reducing the negative impacts of building construction and operation. Rather than being "less bad" by reducing the negative impacts of a building, regenerative design aims to be "good" by creating structures that positively contribute to the environment and surrounding community. The goal of regenerative design is to produce a net-positive environmental impact—to leave the world better off with respect to energy, water, and materials. Regenerative design has served as the philosophical underpinning of the IFLI's Living Building Challenge since its inception.

If sustainable design is difficult, then regenerative design is even more difficult. Nevertheless, there are some interesting examples of regenerative design projects, including the Eden Project in the United Kingdom and the Center for Regenerative Studies (Figure 1.17) in the United States. Both projects involve substantial site remediation and innovative design solutions.

(f) Extreme Events

A series of notable extreme events, such as Hurricane Harvey's impact on Houston or Tropical Storm Sandy's destruction in the New York area, have raised questions about the role of buildings and building design in buffering owners and occupants from the effects of such events. There are several distinct concepts associated with design for extreme-event mitigation.

Resilience. Relative to infrastructure, resilience has been described as "the ability to reduce the magnitude and/or duration of disruptive events" (NIAC, 2009). There is no formal definition of resilient design or consensus list of characteristics that would identify such a building. Herein, we define a resilient building (or system or assembly) as one that can take a beating (during a flood, hurricane, ice storm, earthquake) and keep on working. The project-specific limits of "beating" and "working" will be established under the OPR.

The basic design principles of resilient design are common sense, hardening, and redundancy. There are a range of issues and potential solutions—such as air-conditioning condensers that are not washed away by storm surge; windows that are not blown out in a hurricane; key electrical equipment that is not under water in a flood. In some cases, the answer lies in better placement of equipment (emergency generators above flood stage); in some cases, hardening systems (storm shutters) is a solution; in other cases, it is redundancy (municipal and PV power systems).

BuildingGreen, Inc. provides resources to assist designers engaging in design for extreme events (buildinggreen.com). Also, the New York State Energy Research and Development Authority (NYSERDA) and the University at Buffalo developed *Climate Resilience Strategies for Buildings in New York State*, which provides general descriptions of climate resilience strategies that are accessible to all audiences. The document gives readers an overview of climate-resilience strategies, with clickable links within the text that allow readers to dig deeper and access more specific information.

Passive Survivability. Coined by Alex Wilson, the fundamental precept of passive survivability is that habitable conditions be maintained in a building (or portion of a building) during an extreme event. This would allow sheltering in place—assuming a resilient project—rather than abandonment or attempting to endure under severe interior conditions. The basic design question for passive survivability is: what can be done to allow occupants to stay in place, without risking their health, in the event of failure of normal utilities (electricity, natural gas, water, sewer)?

Tools to assist a designer in addressing passive survivability exist. They are generally the same tools used for conventional building systems analysis—but use the assumption that active systems are shut off. Passive survivability has recently been called "climate-safe" design.

Adaptability. Adaptability refers to the ability of a building to perform reasonably well under changing (but not extreme) conditions. One example is the design of a building that can perform well without radical renovations under changing climate conditions. Design for adaptability might involve overdesigning for today's conditions to better align with expected future conditions. This can be expensive but might make sense for an element that would be very difficult to change (such as wall insulation). Another approach is to allow space for future installation of systems or system modules—such as preparing for a rooftop photovoltaic (PV) system in anticipation of lower-cost equipment in the future.

1.9 DESIGN TO REDUCE ENVIRONMENTAL IMPACT

From a design process perspective, the philosophy of this book is that development of appropriate design intent and criteria is critical to the successful design of buildings and their mechanical and electrical systems. Passive systems should generally be used before active systems (this in no way denigrates active systems, which will be necessary features of almost any large-scale building); life-cycle costs should be considered instead of simply first cost; and green building is a

(*a*)

(*b*)

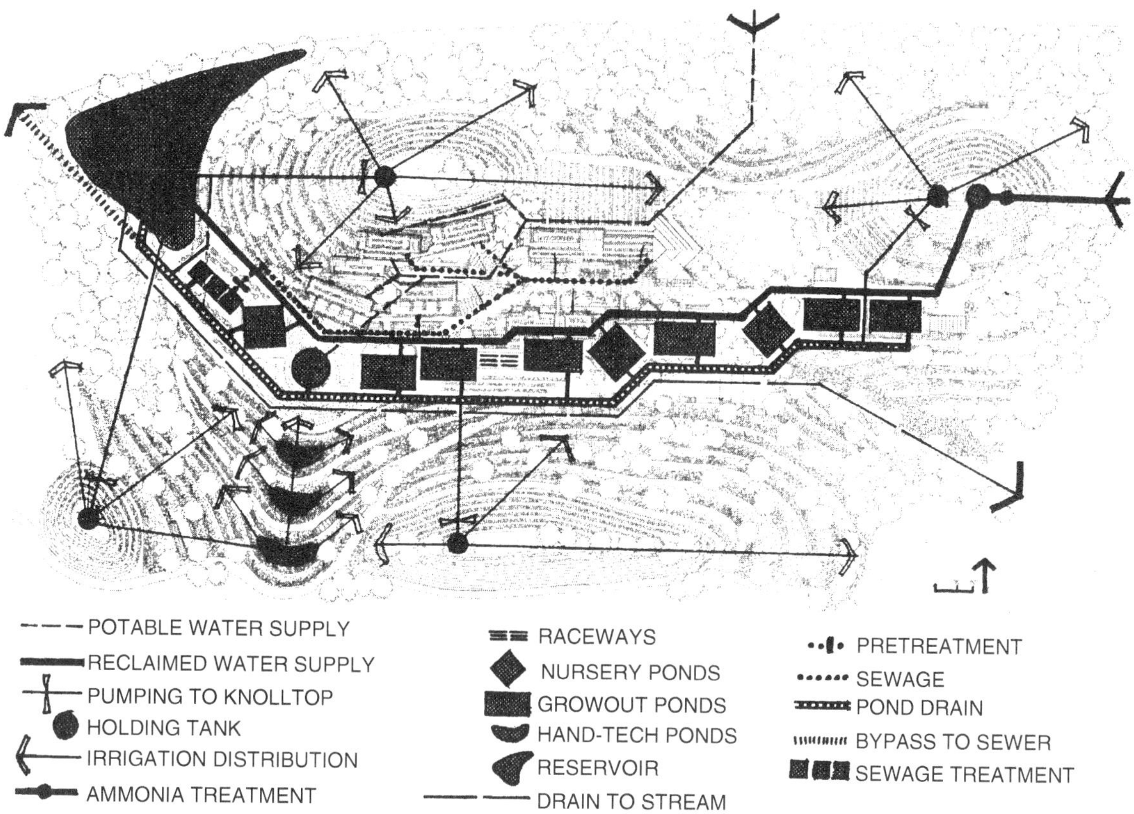

(*c*)

Figure 1.17 *(a) The Center for Regenerative Studies (CRS), California Polytechnic State University–Pomona. (b) Plants provide water treatment and generate biomass in an aquacultural pond at the Center for Regenerative Studies, Cal Poly–Pomona. (c) Site plan for the CRS. It's not easy being regenerative—the highlighted elements relate only to the water reclamation aspects of the project. (Photos © Terri Meyer Boake; used with permission; drawing from Lyle, 1996.)*

desirable intent that will ensure energy efficiency and provide a pathway toward sustainability. Design validation, commissioning, and post-occupancy evaluation should be aggressively pursued.

John Lyle presented an interesting approach to design (that elaborates upon this general philosophy) in his foundational book, *Regenerative Design for Sustainable Development*. The following section of this chapter presents an overview of Lyle's approach. The strategies provide design teams with varied opportunities to integrate site and building design with components and processes. Those strategies most applicable to the design of mechanical and electrical systems are presented here. This approach guided the

Figure 1.18 Letting nature do the work—via daylighting. Mt. Angel Abbey Library, St. Benedict (Mt. Angel), Oregon, designed by Alvar Aalto. (© Tyler Mavichien; used with permission.)

design of the Center for Regenerative Studies at the California Polytechnic State University at Pomona, California (see Figure 1.17).

(a) Let Nature Do the Work

This principle expresses a preference for natural/passive processes over mechanical/active processes. Designers can usually find ways to use natural processes on site (Figure 1.18), where they occur, in place of dependence upon services from remote/nonrenewable sources. Smaller buildings on larger sites are particularly good candidates for this strategy.

(b) Consider Nature as Both Model and Context

A look at this book reveals a strong reliance upon physical laws as a basis for design. Heat flow, water flow, electricity, light, and sound follow rules described by physics. This design principle, however, suggests looking at nature (see Figure 1.17) for biological, in addition to the classical physical, models for design. The use of a Living Machine to process building wastes, as opposed to a conventional sewage treatment plant, is an example of where this strategy might lead.

(c) Aggregate Rather Than Isolate

This strategy recommends that designs focus upon systems holistically, and not just upon the parts that make up a system—in essence, seeing the forest through the trees. The components of a system should be highly integrated to ensure workable linkages among the parts and the success of the whole. An example would be optimizing the solar heating performance of a direct-gain system involving glazing, floor slab, insulation, and shading components, even though such optimization might reduce the performance of one or more constituent parts of the system (Figure 1.19).

(d) Match Technology to the Need

This strategy seeks to avoid using high-grade resources for low-grade tasks (Figure 1.20). For example, it is

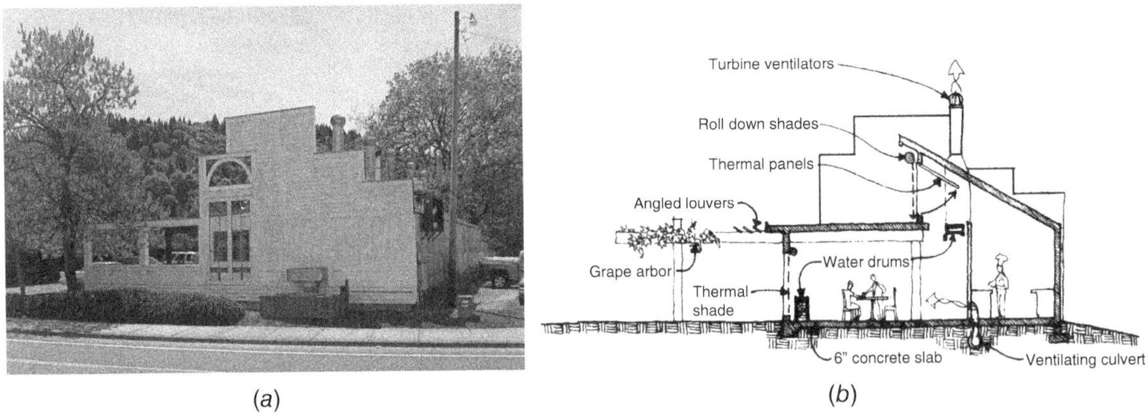

Figure 1.19 Aggregating, not isolating. (a) The former Cottage Restaurant, Cottage Grove, Oregon, operated successfully with passive strategies for 30 years. (b) This section through the restaurant illustrates the substantial integration and coordination (aggregation) of elements typical of passive design solutions. (Photo by Lisa Leal; drawing by Michael Cockram; © by John S. Reynolds; used with permission.)

Figure 1.20 Match technology to the need. Sometimes it's the simple things that count. Keeping cool with a solar-powered fan cap.

wasteful to flush toilets with purified water, but perhaps less clearly wasteful (but equally a mismatch) to use electricity (a very high-grade energy form) to heat water for bathing. The concept of exergy (discussed in a subsequent chapter) relates to this design strategy.

An online tool offered by Architecture 2030 (n.d.b), called the 2030 Palette, offers an interactive platform that gives the designer guiding principles, information, and resources to select appropriate technologies for a variety of scales: building, site, district, city, and region (2030palette.org). The 2030 Palette offers a great first step for design teams looking for passive strategies that might be useful.

Figure 1.21 Seek common solutions. The "atrium" of the Hood River County Library, Hood River, Oregon, provides a central hub for the library, daylighting, views (spectacular), and stack ventilation. (© Alison Kwok; all rights reserved.)

(e) Seek Common Solutions to Disparate Problems

This approach requires breaking out of the box of categories and classifications. An understanding of systems should lead to an increased awareness of system capabilities—which will often prove to be multidisciplinary and multifunctional. Making a design feature (Figure 1.21) serve multiple tasks (perhaps mechanical, electrical, and architectural in nature) is one way to counteract the potential problem of a higher first cost for green design features. Solutions can be as simple and low-tech as using heat from garden composting to help warm a greenhouse.

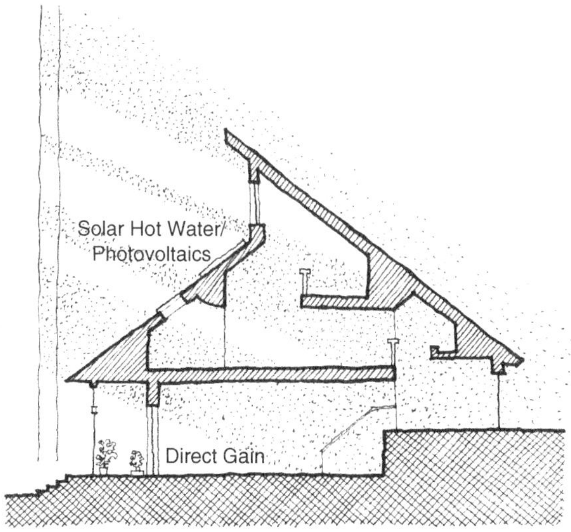

Figure 1.22 Shaping the form to the flow. Using a "band of sun" analysis as a solar form giver (see Chapter 3 for further details). (Redrawn by Jonathan Meendering.)

(f) Shape the Form to Guide the Flow

The best examples of this strategy are solar-heated buildings that are shaped (Figure 1.22) to gather winter sun, or naturally ventilated buildings shaped to collect and channel prevailing winds. Daylighting is another place to apply the "form follows flow" strategy, which can have a dramatic impact upon building design efforts and outcomes.

(g) Shape the Form to Manifest the Process

This is more than a variation on the adage "If you have got it, flaunt it." This strategy asks that a building inform its users and visitors about how it works both inside and out (Figure 1.23). In passive solar-heated and passively cooled buildings, much of the thermal performance is evident in the form of the exterior envelope and the interior space, rather than hidden in a closet or mechanical penthouse. Professor David Orr of Oberlin College addresses this issue succinctly by asking, "What can a building teach?"

(h) Use Information to Replace Power

This strategy addresses both the design process and building operations. Knowledge is suggested as a substitute for brute force (and associated energy waste). Designs informed by an understanding of resources, needs, and system capabilities will tend to be more effective (successfully meeting intent) and efficient (meeting intent using fewer resources) than uninformed designs. Building operations informed by feedback and learning (Figure 1.24) will tend to be more effective and efficient than static, unchangeable operating modes. Users of buildings can play a leading role in this approach by being allowed to make decisions about when to do what it takes to maintain the desired conditions. Reliance on a building's users is not so much a direct energy saver—most controls use very little power—as it is an education. A user who understands how a building receives and conserves heat in cold weather is likely to respond by lowering the indoor temperature and reducing heat leaks. Furthermore, some studies of worker comfort indicate that with more personal control (such as operable windows), workers express feelings of comfort across a wider range of temperatures than with centrally controlled air conditioning.

(i) Provide Multiple Pathways

This strategy celebrates functional redundancy as a virtue—for example, providing multiple and separate fire stairs for emergency egress. There are many other examples, from backup heating and cooling systems, to multiple water reservoirs and piping pathways for fire sprinklers, to emergency electrical and lighting systems. This strategy also applies to climate-site-building interactions in which one site-based resource may temporarily weaken but can be replaced by another (Figure 1.25).

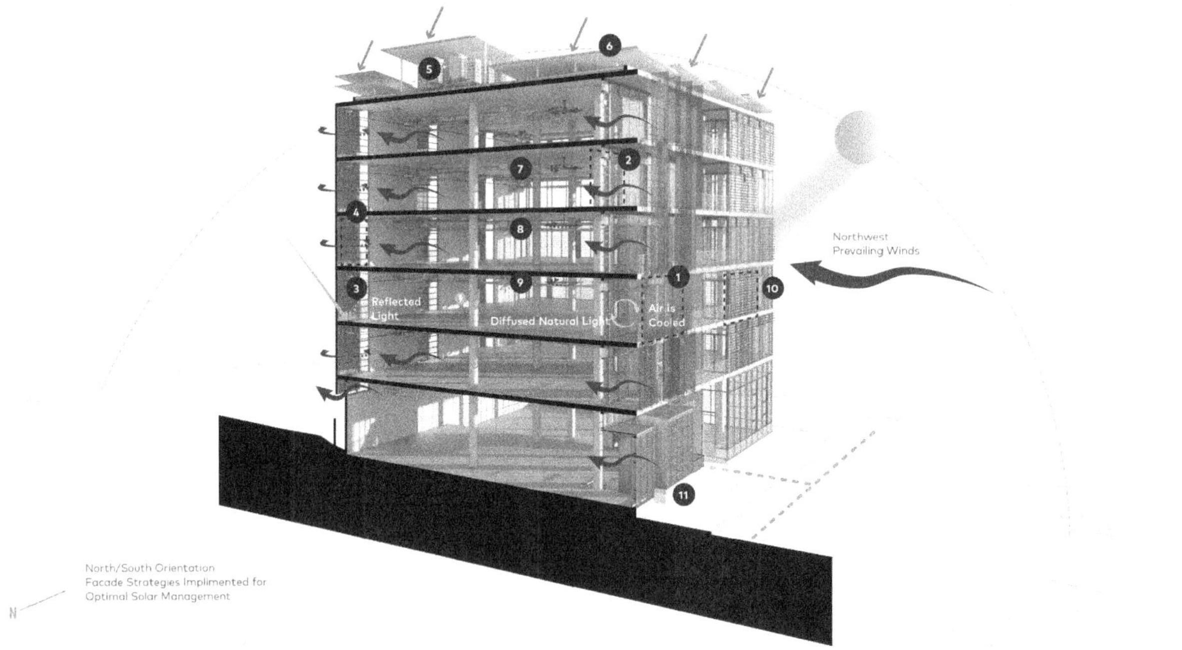

Figure 1.23 *Building system and form work together at the LEED Platinum net zero energy Makers Quarter Block D, which utilizes (1) operable sun shades along the west façade, (2) overhead sectional garage doors for access to outdoors and cross-ventilation, (3) daylight-harvesting light wells that control views while allowing light deep into the building, (4) operable windows and doors on each façade, (5) high-performance modular mechanical system, (6) rooftop photovoltaic panels, (7) modular daylight, air quality, and acoustic sensors that connect to a learning Building Automation System (BAS), (8) fans and fresh air systems to circulate air, (9) low-carbon concrete optimized to store and purge heat, (10) automated exterior sunshades to control daylight and glare, and (11) a modular wetland to store and treat stormwater. (© BNIM; used with permission.)*

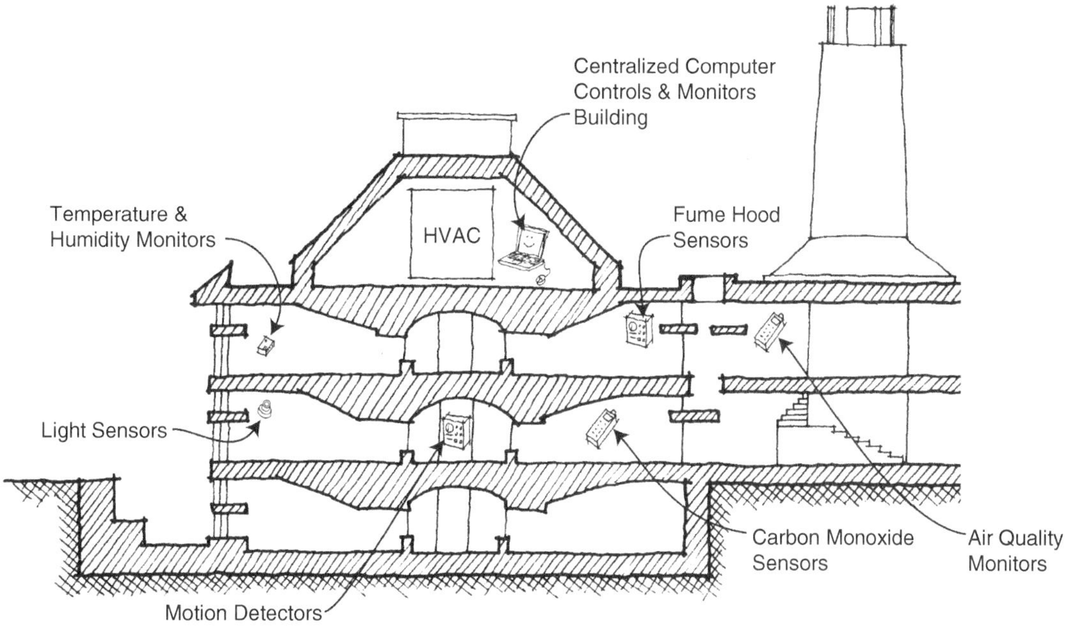

Figure 1.24 *Use information to replace power. Section showing intelligent control system components for the proposed EPICenter project, Bozeman, Montana. (Courtesy of Place Architecture LLC, Bozeman, Montana, and Berkebile Nelson Immenschuh McDowell Architects, Kansas City, Missouri. Redrawn by Jonathan Meendering.)*

Figure 1.25 *Providing multiple pathways. Three distinct sources of electricity are projected in this conceptual diagram for the proposed EPICenter project, Bozeman, Montana. (Courtesy of Place Architecture LLC, Bozeman, Montana, and Berkebile Nelson Immenschuh McDowell Architects, Kansas City, Missouri. Redrawn by Jonathan Meendering.)*

(j) Manage Storage

Storage is used to help balance needs and resources across time. Storage appears as a topic throughout this book. The greater the variations in the resource supply cycle, the more critical storage management becomes. Rainwater can be stored in cisterns, balancing normal daily demands for water against variable rainfall patterns. The high variability of wind-generated electricity output can be managed with hydrogen storage, providing a combustible fuel that can be drawn on at a rate and time independent of wind speed.

On sunny winter days, excess solar energy reaching a room can be stored in thermally massive materials (Figure 1.26), to be released at night. On cool summer nights, coolth (the conceptual opposite of heat) can be stored in these same materials and used to condition the room by day. Most storage solutions will impact building architecture.

Figure 1.26 *Manage storage. The 2007 MIT Solar Decathlon house features a Trombe wall made of translucent tiles to capture and store heat. (© Alison Kwok; all rights reserved.)*

1.10 CASE STUDY—DESIGN PROCESS

Rocky Mountain Institute Innovation Center, Basalt, Colorado

PROJECT BASICS

- Location: Basalt, Colorado, United States
- Latitude: 39.40°N; longitude: 107.0°W; elevation: 6611 ft [2015 m]
- Heating degree days: 8549 base 65 °F [4801 base 18.5 °C]; cooling degree days: 1716 base 50 °F [953 base 10 °C] at Aspen Airport; annual precipitation 16 in. [427 mm]
- Building area: 15,610 ft^2 [1450 m^2] conditioned space
- Completed December 2015
- Client: Rocky Mountain Institute (RMI)
- Core team members: ZGF Architects (architect of record), JE Dunn (general contractor), Architectural Applications (high-performance design consultant), PAE Consulting Engineers (mechanical, electrical, plumbing, IT), David Nelson & Associates (lighting); True North Management (owner's representative), Graybeal Architects, TG Malloy (land planner), Sopris Engineering (civil engineer), KPFF Consulting Engineers (structural engineers), DHM Design (landscape architect), Resource Engineering Group (commissioning), Collective Sun (solar power purchase agreement), SunPower (solar panels)

Figure 1.27 South façade showing solar shading devices and daylight apertures. (Courtesy of ZGF Architects; © Tim Griffith.)

Background. The RMI Innovation Center, located near Aspen, Colorado, was designed to demonstrate how an "aggressively green" office building should be designed, built, and operated. The project incorporates a wide variety of passive and active design strategies and systems, and it has earned an impressive number of high-performance building certifications including LEED Platinum, Living Building Challenge Energy Petal, and Passive House. The client sees these positive outcomes as resulting from a deliberate and robust integrated design process that sought exceptional levels of building performance in one of the coldest climates in the United States.

Context. The Rocky Mountain Institute (RMI) was founded in 1982 by Amory and Hunter Lovins to promote sustainability and energy efficiency through research, education, and consultancy. The Innovation Center (Figures 1.27 and 1.31) is a direct extension of RMI's organizational mission that showcases high-performance green building design and provides space for collaboration with communities, thought leaders, and industry. From the onset, RMI had a clear vision for the building that included defined performance targets as well as ways in which the design team would work together to achieve them (Figures 1.28 and 1.29). Perhaps the most important lesson from this project is the value that a client brings to the design process by clearly outlining requirements, expectations, and goals to guide the design team in their efforts to achieve challenging building performance outcomes that will be verified upon completion.

Design Intent. The RMI Innovation Center sought to demonstrate that moderately sized commercial office buildings, a common and energy-intensive building type in the United States, can be designed to achieve very high energy efficiency and outstanding comfort conditions for its users. The design process and strategies employed aimed to provide a viable precedent and guidance to assist the building industry in achieving similar results for office buildings going forward.

Design Criteria and Validation. Although RMI is a staunch advocate of energy-efficient building and renewable energy, the design criteria outlined in the initial request for proposals (RFP) emphasized integrated design and the Integrated Project Delivery (IPD) acquisition model as ways to ensure that the design team focused on collaboration and finding design synergies rather than on discrete systems, technologies, or components. The Owners' Project Requirements (OPR) document provided specificity on

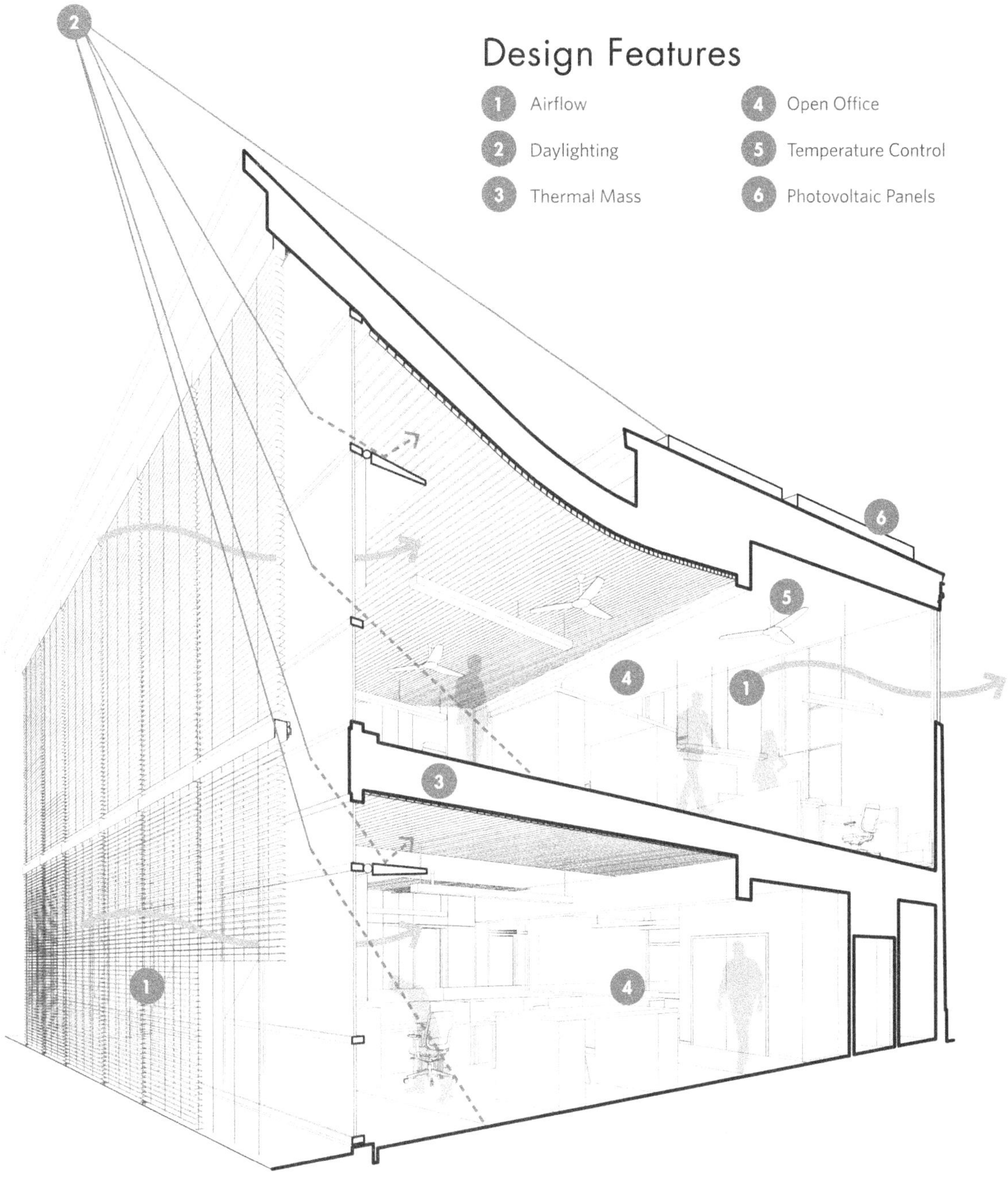

Figure 1.28 *Key design features of RMI Innovation Center. (Courtesy of ZGF Architects.)*

which systems would be tested and validated using the commissioning process and gave the performance expectations for those systems. RMI then went a step further and tied fees for the design team to achievement of performance goals.

Key Design Features

- Long façades face south and north minimizing solar exposure on the east and west for logical passive heating, solar shading, and glare control.
- Heating, ventilating, electrical, and fire protection systems are integrated into cross-laminated timber (CLT) floor panels.
- 83 kW peak photovoltaic system producing over 114,000 kWh annually, 50% more than the facility uses (see Figure 1.30).
- 30 kW, 45-kWh battery storage system that reduces the building's peak demand.
- Dual-pipe plumbing system configured for gray-water reuse (sink and shower water for toilet

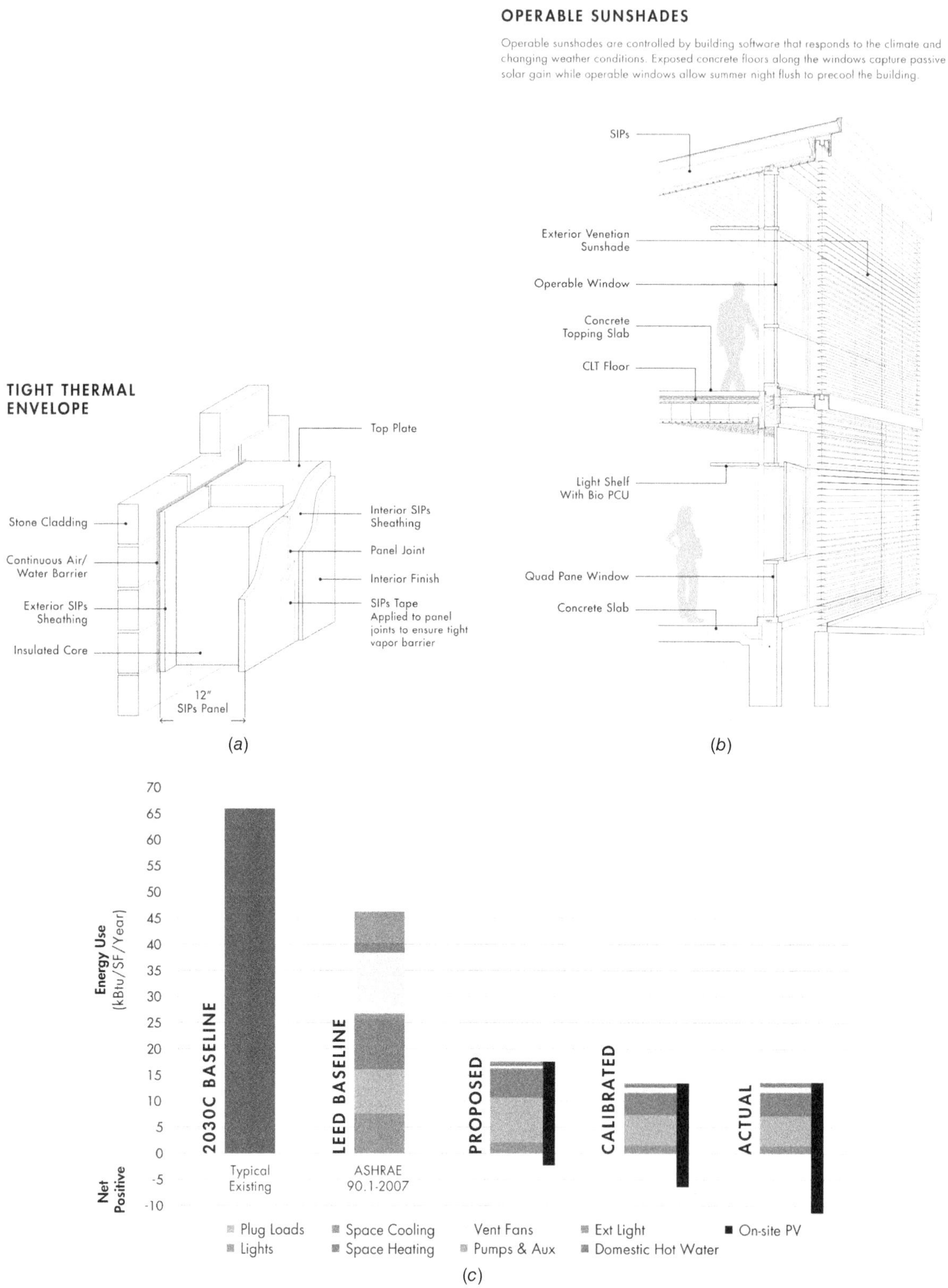

Figure 1.29 *The design team anticipated the need for a range of high-performance systems (a, b) to achieve ambitious project goals –including air-to-air heat pumps with ground-source loops and a hydronic radiant heating system. After careful analysis (c) it was discovered that a high-performance enclosure and efficient electric resistant system were enough to achieve project goals. (Courtesy of ZGF Architects.)*

Figure 1.30 *PV array being installed on the roof. (Photo by RMI.)*

Figure 1.32 *Interior view of the daylit office workspace. (Courtesy of ZGF Architects; © Tim Griffith.)*

flushing) and put in place for when applicable regulations are established in Colorado.

- Resiliency considerations allowing for future expansion, placing the building above the 500-year flood plain, and aiming for a 100-year lifespan.
- An energy-monitoring system tracks building performance and makes the information available to building users.
- Daylight as the primary light source (Figure 1.32).

The Integrated Project Delivery (IPD) model helped to find synergies and savings, as shown in Figure 1.33.

Step 1: At the conclusion of the validation phase, base target cost (BTC) and incentive compensation layer (ICL) are set.

Step 2: The projected costs are lowered through innovation, additional scope is added, and a portion of the savings is added to the ICL. The BTC is adjusted (– savings and + scope). Some 25% to the ICL and 75% toward the Add List item.

Step 3: The final target cost (FTC) is set at the conclusion of the Design period.

Step 4: If the actual cost exceeds the FTC, the ICL is reduced dollar for dollar but capped at the ICL amount. If the actual cost is less than the FTC, the ICL is increased by a percentage of the savings, with 50% going to the ICL, 50% going to the owner.

Post-Occupancy Validation Methods. The Innovation Center has now been operational for nearly a

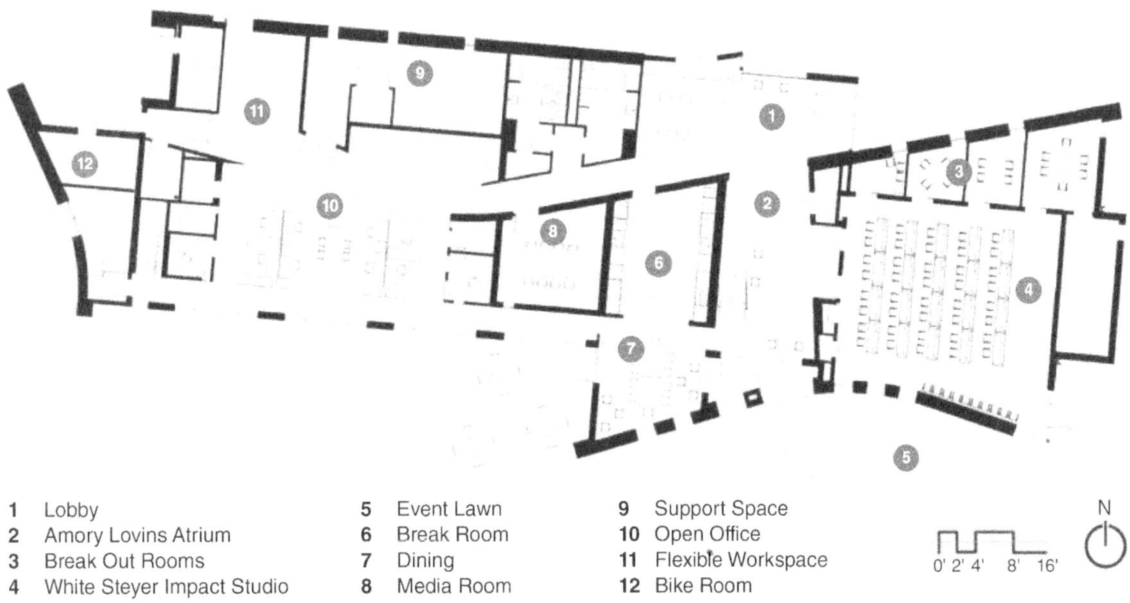

Figure 1.31 *Floor plan of the RMI Innovation Center. (Courtesy of ZGF Architects.)*

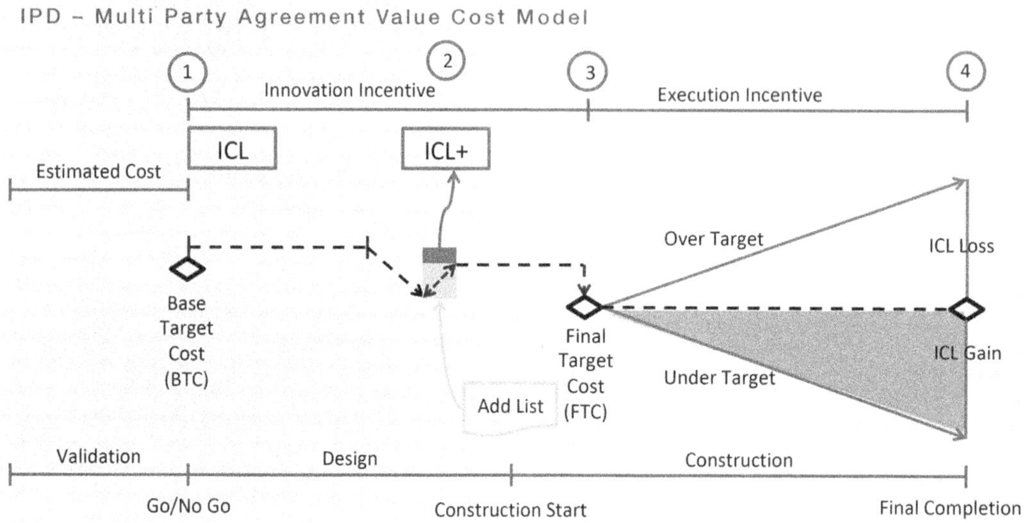

Figure 1.33 *The design team emphasized integrated design and the Integrated Project Delivery (IPD) model. (© Rocky Mountain Institute.)*

decade—long enough for RMI to evaluate how the building is working (Figure 1.34). Staff and occupants are more satisfied than expected, with 74% saying that the building increases their productivity, 78% saying they spend more time at the office because they like the work environment, and 83% saying that the building helps them to articulate the RMI brand through their work. The design team took a significant risk during the design process by expanding thermal comfort set points and shifting focus to thermal comfort parameters such as mean radiant temperature (MRT) and air velocity. Hyperchairs (climate-controlled chairs), sit/stand desks, and personally controlled fans offer occupants opportunities to adjust their comfort at a local level. These risks appear to have paid off because occupants report excellent comfort in the building, which is in one of the coldest climates in North America.

The building is extensively submetered, which has been a challenge in terms of maintenance and a benefit in terms of operations. RMI now has an additional consultant to maintain the submeters and to ensure that the meters are working effectively with the other building systems. The submetering was a great help in validating that the systems were performing optimally during the commissioning verification process. The submetering also allowed RMI to troubleshoot an occupant comfort problem by isolating an area of the building where wall heaters were incorrectly wired and not producing adequate heating.

The Innovation Center has been awarded several third-party certifications and awards (see the list in the Performance Data). RMI reports that the certifications did not require much additional coordination due to the fact that the integrated design team members were involved early in the design process. Also, the project had ambitious goals from the start, and the certifications were more about seeing which credits or points fit these goals rather than the certifications driving the goal-making process. RMI also received assistance from Phius on the energy modeling for the building in an effort to expand the Phius commercial passive building portfolio.

Performance Data. In the first year of occupancy, the measured EUI was lower than predicted by the energy modeling—16.8 kBtu/ft^2 yr [191 MJ/m^2 yr] measured compared with 17.2 kBtu/ft^2 yr [195 MJ/m^2 yr]. Annual electricity usage was 76,840 kWh with a peak of approximately 10,320 kWh in December and 4400 kWh in July. Measured airtightness was 0.175 ACH50, which is 71% lower than Phius benchmarks. The building has received the following certifications and awards since completion:

- LEED Platinum Certified (NC v2009)
- Phius+2015 Source Net Zero
- Architecture 2030 Challenge target exceeded
- Energy Star (targeting a score of 100)
- International Living Futures Institute Zero Energy Certified
- Phius Design Award: Winner of Best Overall Project and best in Office/Institutional category, 2016
- Engineering News Record Mountain States Award: Winner, Best Green Project and Safety Project of the Year, 2016
- AIA Portland 2030 Challenge Award: Winner, 2016
- CBE Livable Building Award: 2018

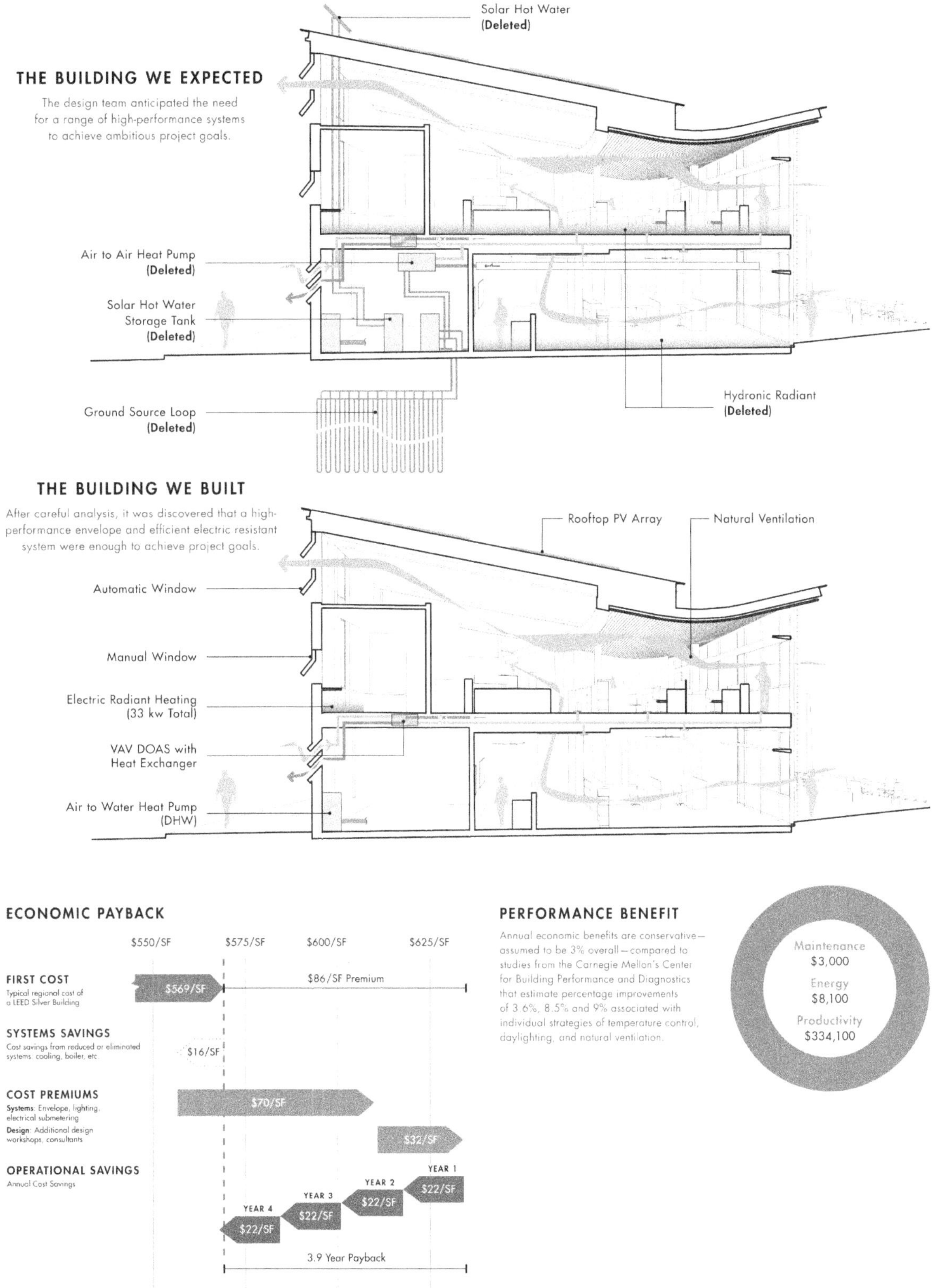

Figure 1.34 *Summary of the economic payback and performance benefits of the RMI Innovation Center's high-performance building solutions. (Courtesy of ZGF Architects.)*

As predicted, the PV production is approximately three times the electrical usage in the summer months, but insufficient to meet energy demands in the winter months (with electric heating). However, balancing summer and winter yields has allowed the building to meet its net-positive energy goals and it produced 37.7 MWh of surplus power in the first year. Extensive submetering has allowed RMI to monitor energy end uses at different times of the year. For example, plug loads drive energy consumption in the summer months (57% of total electricity), but heating drives energy use in the winter (66% of total electricity use). Battery storage for the PV system was designed to manage electricity demand charges (based upon peak consumption). However, since RMI carefully managed electricity use and demand to well below 50 kW, the battery storage may be more effective in buffering electrical-load increases in the future.

FOR FURTHER INFORMATION

Green Building News. 2014. "Rocky Mountain Institute Breaks Ground on Innovation Center." greenbuildingnews.com/2014/10/22/rocky-mountain-institute-breaks-ground-on-innovation-center

Rocky Mountain Institute. n.d. "The RMI Innovation Center." rmi.org/our-work/buildings/scaling-zero-net-carbon/rmi-innovation-center

References and Resources

Aldo Leopold Foundation. 2023. "The Leopold Center." aldoleopold.org

Architecture 2030. n.d.a. architecture2030.org

Architecture 2030. n.d.b. "2030 Palette." 2030palette.org

Architecture 2020. n.d.c. "Zero Tool." zerotool.org

Architecture 2030. 2007. "The 2030 Challenge Benchmark Set." architecture2030.org/wp-content/uploads/2018/09/Press_Release_5_4_07.pdf

ASHRAE. 2004–2012. *30% Advanced Energy Design Guides* (multiple published components). Atlanta, GA: American Society of Heating, Refrigerating and Air-Conditioning Engineers.

ASHRAE. 2011–2012. *50% Advanced Energy Design Guides* (multiple published components). Atlanta, GA: American Society of Heating, Refrigerating and Air-Conditioning Engineers.

ASHRAE. 2018–2019. *Zero Energy Advanced Energy Design Guides* (multiple published components). Atlanta, GA: American Society of Heating, Refrigerating and Air-Conditioning Engineers.

ASHRAE. 2019. *Guideline 0–2019: The Commissioning Process*. Atlanta, GA: American Society of Heating, Refrigerating and Air-Conditioning Engineers.

ASHRAE. 2022. *ANSI/ASHRAE/IES Standard 90.1–2022: Energy Standard for Buildings Except Low-Rise Residential Buildings*. Atlanta, GA: American Society of Heating, Refrigerating and Air-Conditioning Engineers.

ASHRAE. 2023a. *ANSI/ASHRAE Standard 228–2023: Standard Method of Evaluating Zero Net Energy and Zero Net Carbon Building Performance*. Atlanta, GA: American Society of Heating, Refrigerating and Air-Conditioning Engineers.

ASHRAE. 2023b. *ANSI/ASHRAE Standard 55–2023: Thermal Environmental Conditions for Human Occupancy*. Atlanta, GA: American Society of Heating, Refrigerating and Air-Conditioning Engineers.

Bordass, W. 2001. *Flying Blind: Things You Wanted to Know about Energy in Commercial Buildings but Were Afraid to Ask*. London: Association for the Conservation of Energy and the Oxfordshire Energy Advice Centre.

Bordass, W., A. Leaman, and P. Ruyssevelt. 2001. "Assessing Building Performance in Use 5: Conclusions and Implications." *Building Research & Information* 29(2): 144–157.

Bryan, H. and W. Trusty. 2008. "Developing an Operational and Material CO_2 Calculation Protocol for Buildings." Presented at Sustainable Building 2008 (SB08) Conference (September 21–25), Melbourne, Australia," in *Proceedings of the 2008 World Sustainable Building Conference*, G. Foliente (ed.). Balnarring, Australia: ASN Events Pty.

Center for Environmental Design Research. *Vital Signs*. Berkeley, CA: University of California, Berkeley.

CIBSE. 1999. *CIBSE TM22: Energy Assessment and Reporting Methodology: The Office Assessment Method*. London: Chartered Institution of Building Services Engineers. usablebuildings.co.uk/fp/OutputFiles/FR2MainText.html

CLF. 2024. *Carbon Leadership Forum*. Oakland, CA: Carbon Leadership Forum. carbonleadershipforum.org

DOE. 2018. *2017 Buildings Energy Data Book*. Washington, DC: U.S. Department of Energy.

EPA. 2024. *Technical Reference: Greenhouse Gas Emissions*. Washington, DC: U.S. Environmental Protection Agency. portfoliomanager.energystar.gov/pdf/reference/Emissions.pdf

EPA. 2024a. *Benchmark Your Building with PortfolioManager*. Washington, DC: U.S. Envimental Protection Agency. portfoliomanager.energystar.gov

EPA. n.d.b. www.epa.gov

EPA. n.d.c. *Use EPA's Target Finder Calculator*. Washington, DC: U.S. Environmental Protection Agency. www.energystar.gov/buildings/resources-audience/service-product-providers/commercial-new-construction/target-finder

Field, J., J. Soper, P. Jones, W. Bordass, and P. Grigg. 1997. "Energy Performance of Occupied Non-Domestic Buildings: Assessment by Analysing End-Use Energy Consumptions." *Building Services Engineering Research and Technology* 18(1): 39–46.

Fuller, S. K. and S. R. Peterson. 1995. *Life-Cycle Costing Manual for the Federal Energy Management Program*. Gaithersburg, MD: National Institute of Standards and Technology. www.nist.gov

GBI. 2024. *Green Globes Certification*. Portland, OR: Green Building Initiative. thegbi.org/greenglobes

Grondzik, W. 2009. *Principles of Building Commissioning*. Hoboken, NJ: John Wiley & Sons.

ICC. 2021a. *2021 International Building Code*. Washington, DC: International Code Council.

ICC. 2021b. *2021 International Energy Conservation Code*. Washington, DC: International Code Council.

IWBI. 2020. *WELL Building Standard Version 2*. New York: International WELL Building Institute. v2. wellcertified.com

Kutscher, C. (ed.). 2007. *Tackling Climate Change in the U.S.* Boulder, CO: American Solar Energy Society.

Lyle, J. T. 1996. *Regenerative Design for Sustainable Development*. New York: John Wiley & Sons.

McLennan, J. R. 2004. *The Philosophy of Sustainable Design*. Kansas City, MO: ECOTone Publishing Company, LLC.

NIAC. 2009. *NIAC Critical Infrastructure Resilience: Final Report and Recommendations*. Washington, DC: National Infrastructure Advisory Council.

NRC. 1995. *National Building Code of Canada*. Ottawa, ON: National Research Council Canada.

NREL. National Renewable Energy Laboratory. www.nrel.gov

NREL. *Cambium*. Golden, CO: National Renewable Energy Laboratory. www.nrel.gov/analysis/cambium.html

NYSERDA. 2018. "Climate Resilience Strategies for Buildings in New York State." Report Number 18–11, June. Buffalo, NY: New York State Energy Research and Development Authority and the University at Buffalo.

Passivhaus Institute. www.passivehouse.com

Phius. www.phius.org

Preiser, W. F. E. (ed.). 1989. *Building Evaluation*. New York: Plenum Press.

Schaeffer, J. et al. 1997. *A Place in the Sun: The Evolution of the Real Goods Solar Living Center*. White River Junction, VT: Chelsea Green Publishing Company.

Simonen, K. 2014. *Life Cycle Assessment*. New York: Routledge.

Torcellini, P., S. Pless, M. Deru, and D. Crawley. 2006. *Zero Energy Buildings: A Critical Look at the Definition*. Golden, CO: National Renewable Energy Laboratory.

University of Oregon. 2005. *Agents of Change*. Eugene, OR: University of Oregon. aoc.uoregon.edu

Usable Buildings Trust. usablebuildings.co.uk

U.S. Energy Information Administration (EIA). 2024. *Monthly Energy Review, April. U.S. Department of Energy, Office of Energy Efficiency and Renewable Energy*. Washington, DC: EIA

USGBC. *LEED Rating System*. Washington, DC: U.S. Green Building Council. www.usgbc.org/leed

Wells, M. 1981. *Gentle Architecture*. New York: McGraw-Hill Education.

Wells, M. n.d. "Wilderness-Based Checklist for Design and Construction." www.sbse.org/resources/docs/wells_checklist_explanation.pdf

World Commission on Environment and Development. 1987. *Our Common Future* (the Brundtland Report). Oxford: Oxford University Press.

CHAPTER 2

Environmental Resources

TODAY'S BUILDING DESIGNER WILL ADD OR reshape environments on a planet that has, for eons, evolved and thrived by using renewable energy arriving from the sun at a generally fixed and limited rate. Suddenly (in geologic time) our planet is experiencing population growth, nonrenewable resource depletion, and measurable global climate change. Today's designers have available a declining reservoir of materials, fossil fuel energy, and water resources. Do we continue to support a rapidly increasing population, where each person consumes more resources than did her/his grandparents? Eventually, our planet must live within a fixed budget of renewable energy, water, and materials. The question is: How can our building design philosophies best move toward a necessary accommodation with sustainability? How can building designers be environmentally proactive instead of simply reactive?

Population growth is both the source of much of our work as building professionals and the underlying source of our greatest problems. Our planet did not support a human population of 1 billion until about 1830, at which time the United States depended almost entirely upon renewable energy provided by fuel wood and work animals; interior lighting was provided by candles or small oil or gas devices. In less than 200 years, 7 billion more people were added to our planet, and the United States shifted to a predominant dependence upon nonrenewable fuels: coal, oil, and natural gas (Figure 2.1). The building design process plays an active role in deciding where these people will live and work, and how much of what kinds of resources they will use. The mechanical and electrical systems that support our new buildings can be part of a growing problem or an important start to a necessary solution.

2.1 INTRODUCTION

Buildings depend upon energy and matter for their very existence and must pay heed to several fundamental rules of science. The first law of thermodynamics establishes the conservation of energy and matter (energy/matter can neither be created nor destroyed) and essentially states that you cannot get something for nothing. The second law of thermodynamics expresses the tendency toward disorder that is part of the normal nature of things. Entropy is a measure of such disorder; as disorder increases, so does entropy. The second law is a declaration against perpetual motion, and basically states that not only can't you get something for nothing, you can't even break even, due to unavoidable losses (disorder) that contribute to increased entropy.

The construction and operation of buildings are ordering processes. Materials are mined or harvested, refined or shaped, placed in manufactured products, transported to a building site, and assembled. All of these processes consume energy and materials as various building systems are established. The operation and maintenance of a building consume further resources, as conditions (temperatures, light levels, flows of water) are provided that would not occur under less-ordered natural conditions. Maintaining building materials and systems over time against the forces of nature requires additional inputs of energy

Mechanical and Electrical Equipment for Buildings, Fourteenth Edition. Walter Grondzik, Daniel Overbey, and Alison Kwok.

Companion website: www.wiley.com/go/meeb14e

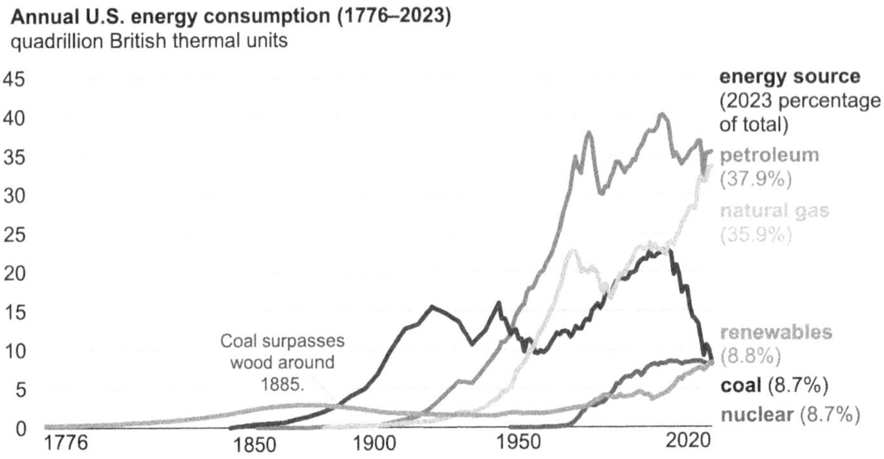

Figure 2.1 *Annual U.S. energy consumption (1776–2023). (Data source: U.S. Energy Information Administration, Monthly Energy Review. Pre-1949 data based on Energy in the American Economy, 1850–1975: Its History and Prospects and U.S. Department of Agriculture Circular No. 641, Fuel Wood Used in the United States 1630–1930.) Note: Data use captured energy approach to account for wind, hydro, solar, and geothermal. (Figure adapted for clarity by the authors.)*

and materials. Buildings are inherently antientropic. This is not necessarily bad (learning and evolving are also antientropic), but it should be considered during the design process. As will be seen later in this chapter, buildings have a substantial collective impact on our patterns of energy, water, and materials consumption.

The design process should consider various scales of concern relative to the impacts of buildings upon the environment. One such scale is geographic. Geographic scales of concern include the microscale, the site scale, and the macroscale, with the building design focus historically being at the site scale. The terms micro (small) and macro (large) are not absolutely defined but are often referenced to a particular site. Thus, the area of influence of a microclimate is smaller than that of a macroclimate—but is usually also smaller than that of the site scale, often applying to one part of a site (perhaps with a steeper slope, lower elevation, or greater shading than other parts). The site scale is normally self-explanatory, running from property line to property line. Energy-efficiency issues are typically and historically addressed at the site scale and often ignore (unfortunately) energy consumption off-site (such as electric power production losses or natural gas transportation losses). Renewable/passive energy systems must consider microscale effects (such as orientation) to be successful. Nonrenewable/active systems often are oblivious to any scale of concern.

Time is another, and very interesting, scale of concern to building design. The time frames typically addressed include now and the future—although the past is of concern with adaptive reuse and historic preservation projects. The concept of the future is usually left quite nebulous unless life-cycle costing is undertaken for a project, in which case the expected lifetimes of systems and equipment are explicitly estimated. Most buildings have a useful life of 25, 50, perhaps 100 years (or more). Stuart Brand provides an interesting look at buildings over time in *How Buildings Learn: What Happens after They're Built* (Brand, 1994). The problem with design for the future is that we don't know precisely what the future holds. Nevertheless, design for sustainability requires that the design team and the design process consider the needs of future generations. This makes sustainability a very challenging concept and highly subjective—but no less important than design for today.

2.2 ENERGY

Energy resources are broadly classified as renewable or nonrenewable. Renewable resources are those that are available on a recurring cycle without substantial depletion. They are, however, typically diffuse and can be consumed only at a rate generally controlled by nature. For example, the influx of solar energy varies from day to day, but on average it should be available forever and at a reasonably predictable rate. Likewise, a woodlot produces a limited amount of wood per year but can do so for centuries if properly managed. An analogy for using renewable energy sources is living on a fixed annual salary—with no hope for spectacular annual raises or unexpected bonuses, but with long-term stability if wisely managed.

Nonrenewable energy resources are those that, once exhausted, cannot be replaced in a time frame that is meaningful to humans. Coal, oil, and natural gas are examples of nonrenewable energy resources. Using nonrenewable fuel sources is analogous to living off a one-time lottery win that can be spent in 1 year or over 50 years, depending upon needs and planning, but that is gone for good when it is all spent.

The United States, like other industrialized countries, has experienced an energy transition since the mid-1800s (Figure 2.1) beginning with renewable energy sources, obtained locally, that provided relatively low-grade work. Animals pulled or pushed, and wood was burned to provide heated air, water, or steam. Buildings and people were directly affected by energy sources. Work animals needed to be fed, tended, and housed; wood for fuel was cut nearby and stacked in large sheds (Figure 2.2); fireplaces were social centers of buildings; and smoke from chimneys indicated activity inside. The side effects of energy use were also directly sensed—animal wastes, deforestation, and polluted air. Fuel use depended upon human labor. Architects of the era responded to the visual and spatial organization potentials of fireplaces and chimneys (Figure 2.3).

North America is now almost entirely dependent upon nonrenewable energy resources, a proportion of which consists of imported oil and natural gas and electricity transported across substantial distances (Figure 2.4). This situation is partly due to rapid growth in both population and per capita energy consumption. It is also due to the allure of highly concentrated energy available from fossil fuels, which encourages the use of high-quality energy sources—such as electricity and natural gas—for buildings and gasoline for transportation. People are now largely oblivious to their sources of energy. Electricity is generated in far-off power plants; natural gas arrives through buried pipelines, and fuel oil via supertankers. Buildings account for a large percentage of the world's total energy consumption. The impact of occupant experiences on energy use (and on building design and operation) tends to be diluted. Energy consumption is regulated by automatic controls, and heating and cooling equipment is hidden from sight. Building occupants/users—who often do not personally control a thermostat, see climate control equipment, or pay a utility bill—have no direct contact with or concern about energy resources. Clients hire architects to provide function and comfort. Architects then pay engineers to design (and usually to successfully hide) mechanical and electrical equipment.

Figure 2.3 The fireplace and the more efficient wood stove can inspire architectural form. This chimney symbolizes permanence as well as protection against the cold. The major social space of the house is marked both by the arched window and by the fireplace chimney. (Photo by William Johnston.)

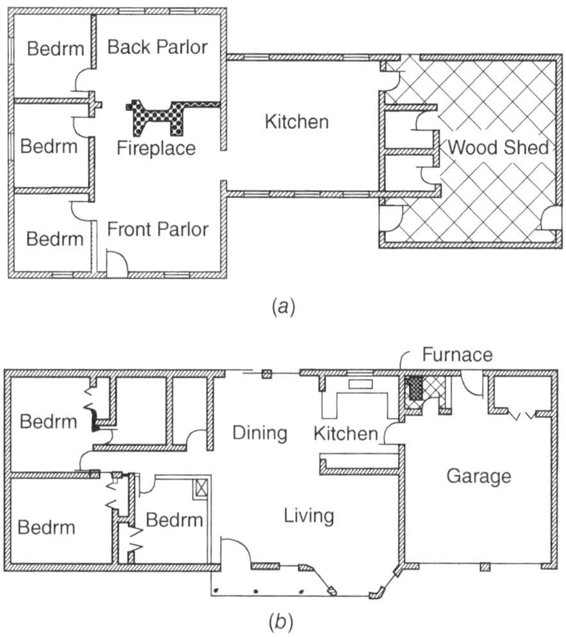

Figure 2.2 Residential heating: past and present. (a) A house dependent on fireplaces or wood stoves also depends on someone to tend the fire. The warmer area near the fire in this early Oregon farmhouse was used for social purposes; the colder extremities served as sleeping areas and for storage of food and fuel. (Based upon a plan drawn by Philip Dole.) (b) The contemporary suburban home has either a small area for heating/cooling equipment or electric heat built into each room. Climate control equipment is no longer a major influence on building form.

For several reasons, buildings designed today are likely to rely heavily upon electricity (Figure 2.5), a situation that carries serious implications for resource depletion and environmental quality:

1. Consumption of electricity is expected to rise about twice as fast as overall energy demand, and we are often using electricity in place of other energy

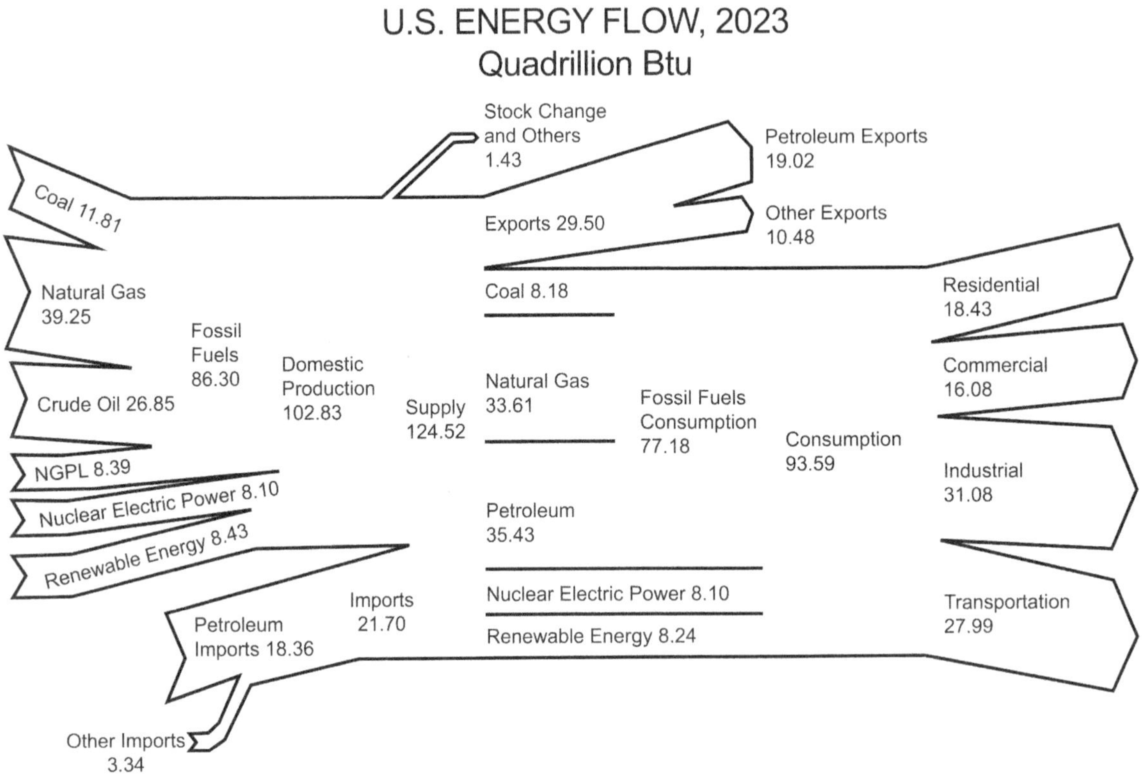

Figure 2.4 U.S. energy flow, 2023: sources and end uses. These data are updated on a regular basis, but the general patterns shown in this figure change slowly. Fuel types and sources are shown to the left and end use sectors to the right. Note the importance of residential and commercial consumption to total U.S. consumption—and the currently minuscule contribution of renewable energy sources to the whole. (Adapted from EIA, 2024.)

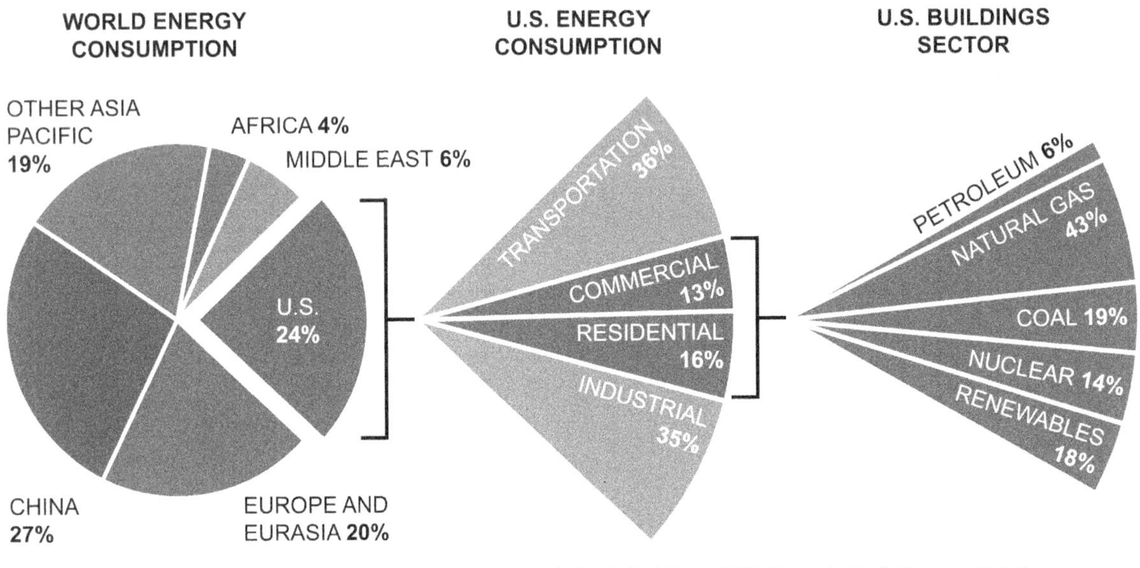

Figure 2.5 Energy resources as consumed by various end-use sectors in the United States, 2022. (Drawn by Daniel Overbey. Data from EIA, 2023.)

forms. Part of the reason for this is that, for some primary energy sources (such as coal, heavy fuel oil, or a nuclear reactor), generation of electricity for subsequent (secondary) distribution to buildings is the only convenient usage option. Efforts to electrify buildings (by replacing on-site combustion with electric equipment in an attempt to reduce greenhouse gas emissions) will only accelerate this trend.

2. Other than daylighting (and quality daylighting is unfortunately rare in today's buildings), electricity is the only source for building illumination.

Heat produced by electric lighting may reduce a building's need for space heating, but it increases its need for space cooling—and mechanical cooling is almost universally provided by electric air-conditioning equipment.

3. Plug loads in all types of buildings are increasing, while heating and cooling loads tend to be decreasing. Plug loads (appliances and equipment served by convenience receptacles) are totally electric.
4. Electricity is a convenient and versatile energy form; it not only serves such high-quality and highly concentrated (or high-grade) tasks as lighting and providing drive power via electric motors, it can also serve low-quality, low-temperature (or low-grade) tasks such as cooking, water heating, and space heating. All-electric buildings are commonplace, even though they are subject to paralysis in blackouts—as any building dependent upon a single energy source is vulnerable to disruptions. Table 2.1 offers a breakdown of U.S. energy consumption for the commercial and residential sectors.
5. Electricity generated by thermal processes (except for cogeneration) delivers to the end user less than one-third of the total energy that goes into its production; more than two-thirds is usually lost as waste heat at the generating plant (Figure 2.6).

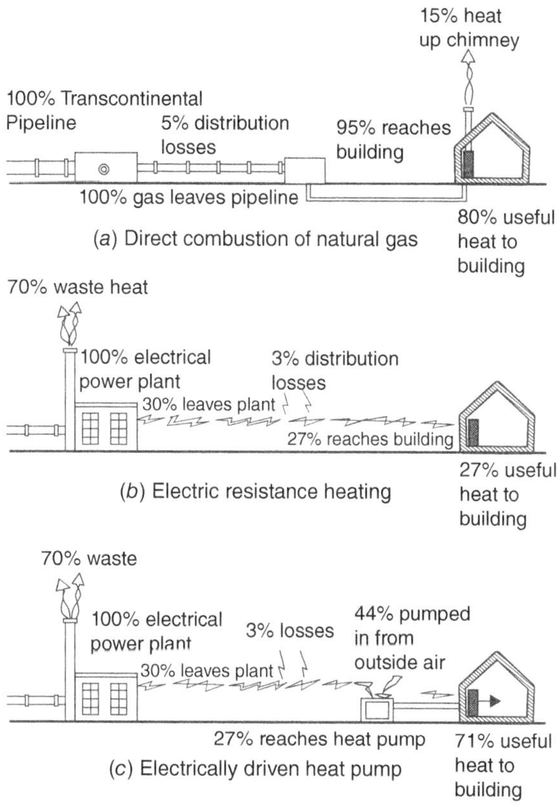

Figure 2.6 *Variations on higher-grade energy and lower-grade tasks. (a) Natural gas (a fossil fuel) is often burned in furnaces to provide low-grade space heating. With today's high-efficiency furnaces, well over 90% of the energy in the gas is delivered to the building as space heat. (b) However, when that natural gas is used instead to generate electricity, and electric resistance is used for space heating, the inefficiencies at the electric power plant cut deeply into the available useful energy: Only about 27% is delivered to the space as heat. (c) On the other hand, when the electricity generated by natural gas is used to drive a heat pump, and the outdoor air is above freezing, about 71% of the energy in the gas is delivered as space heat. (Drawing by Michael Cockram; © by John S. Reynolds; all rights reserved.)*

As we consume our planet's resources, including fossil fuels, designers might want to return to reliance on renewable and sustainable energy. Such a future would be partially implemented by solar energy converted directly to electricity through photovoltaic (PV) modules placed on or near the building consuming the electricity. Solar collectors (some for heating water, others for producing electricity) and daylighting apertures would shape the roofs and façades of buildings. Building design professionals will decide whether to embrace or ignore such opportunities.

Hydrogen may be stored and distributed as a high-grade fuel, produced from water using electricity generated from renewable resources, such as solar energy and wind. The lifetimes of mechanical and electrical equipment specified today will probably overlap such a future. Lower-tech processes such as biomass conversion (combustion of wood and waste products) may develop faster than higher-tech processes. PV is growing very rapidly; its growth curve is as steep as that of IT during the early years of computer technology. Some energy scenarios anticipate approximately half of global energy demand being met by alternatives to fossil fuels by the year 2050. Other scenarios see no end to our consumption of fossil fuels and the related production of greenhouse gas emissions.

Today's fossil-fuel-based economy seems so entrenched as to defy a transition to renewable energy. Some doubt the capacity of renewables and ask: Is solar energy adequate for our energy needs? Table 2.2 compares the amount of solar energy received at the Earth's surface in a single day with other energy phenomena. There is more than adequate global resource availability—given a will to move toward renewable and site-based resources. Energy efficiency and conservation efforts will play a critical role in making such a transition feasible.

To date, societal focus has been primarily upon using less energy—involving energy efficiency (using energy more effectively) and energy conservation (simply using less energy). The oil embargo of the 1970s spurred the development of building energy-efficiency standards, which have remained a fixture of building design ever since. In general, such standards seek to reduce building energy

TABLE 2.1 U.S. Commercial and Residential Energy Consumption, 2022 (Trillion Btu)[a]

	End-use Energy Consumption															
	Primary Consumption[b]															
	Fossil Fuels				Renewable Energy											
	Coal	Natural Gas[c]	Petroleum[d]	Total	Hydro-electric Power	Geothermal	Solar[e]	Wind	Biomass	Total	Total Primary	Electricity[f]	Total End Use	Electrical System Losses[g]	Total
Commercial sector	14	3633	947	4595	1	20	63	1	190	274	4868	4746	9614	6961	16,574
Residential sector	NA	5140	992	6132	NA	40	200	NA	422	662	6793	5150	11,943	7553	19,496
Total	14	8773	1939	10,727	1	60	263	1	612	936	11,661	9896	21,557	14,514	36,070

Source: EIA, 2024.

[a]SI unit conversion: 1 Btu = 1055 J.

[b]Energy consumed in the form that it is first accounted for, before any transformation to secondary or tertiary forms of energy.

[c]Natural gas only; excludes the estimated portion of supplemental gaseous fuels.

[d]Residential sector: does not include biofuels that have been blended with petroleum—biofuels are included in "Biomass."

[e]Includes small-scale solar photovoltaic (PV) electricity and solar thermal energy in the sector.

[f]Electricity sales to ultimate customers reported by electric utilities and, beginning in 1996, other energy service providers.

[g]Total losses are calculated as the primary energy consumed by the electric power sector minus the energy content of electricity sales to ultimate customers. Total losses are allocated to the end-use sectors in proportion to each sector's share of total electricity sales to ultimate customers.

TABLE 2.2 Daily Arrival of Solar Energy on Earth Compared to Other Energy Quantities

Solar energy received each day	1
Melting of an average winter's snow during the spring	1/10
A monsoon's circulation between ocean and continent	1/100
Use of energy by all mankind in a year	1/100
A mid-latitude cyclone	1/1000
A tropical cyclone	1/10,000
Kinetic energy of motion in Earth's general circulation	1/100,000
The first hydrogen bomb	1/100,000
A squall line containing severe thunderstorms	1/1,000,000
A thunderstorm	1/100,000,000
The first atomic bomb	1/100,000,000
The daily output of Boulder Dam	1/100,000,000
A typical local rain shower	1/10,000,000,000
A tornado	1/100,000,000,000
Lighting New York City for one night	1/100,000,000,000

Source: From Lowry, 1988.

consumption—not to shift energy resources from nonrenewables to renewables. The green building movement has provided momentum for a serious look at both reduced energy use and the use of energy from renewable resources.

2.3 WATER

The efforts of architectural and engineering professionals toward more resource-efficient buildings have, for the past several decades, focused primarily upon energy. This focus has been warranted by the economic, environmental, and societal effects of consuming nonrenewable energy resources, amplified by the physical limits imposed by entropy. There is no option for the reuse or recycling of fossil fuels (as may be done with water and materials) and little opportunity for downstream use of the waste energy produced by such fuels.

Water, however, must be reused. A quote generally attributed to *National Geographic* (October 1993) made this point succinctly: "All the water that will ever be is right now." Although water is a recyclable resource, it is not a renewable resource (no new daily supplies are being delivered to Earth). Water concerns have reached crisis level in many parts of the world, and water may well be the emerging limit to growth and development—especially locally and regionally—rather than energy. Periodic water rationing is an unpleasant fact in many areas.

Legal entitlements may regulate access and use of water resources that flow through a site. Where water is found is not necessarily where it is needed or where it may be used. Concerns about a viable supply of potable water have dominated politics and civil engineering in the arid western United States for a century. Longstanding "water rights" in the western United States are critical due to the region's arid climate and the importance of water for agriculture, industry, and urban development. Table 2.3 compares regional water resources with sustainable water usage capacity; it is clear that some areas of the United States now have serious water shortages.

As with energy, per capita use of water involves more than consumption within a building. In the case of energy, transportation and industrial uses influence per capita consumption; with water, energy production and agricultural uses play a role. About 40% of all U.S. fresh and saline water withdrawals in 2015 were used in conjunction with thermoelectric power generation. Most of this was surface water used for once-through cooling at power plants. Withdrawals for this use have been relatively stable since 1985. (In a quirky energy–water connection, the California Energy Commission estimates that 19% of California's total electricity use is related to pumping and treating water.)

Irrigation is the second largest use of freshwater in the United States, accounting for slightly less than 40% of consumption. Historically, more surface water than groundwater has been used for irrigation. The percentage of total irrigation withdrawals from groundwater has fluctuated, from 23% in 1950 to 42% in 2000 to 48% in 2015. Irrigated acreage more than doubled between 1950 (25 million acres) and 1980 (58 million), then increased to over 63 million acres by 2015 (USGS, 2017).

Public water supply withdrawals in 1950 were 14 Bgal/d [53 GL/d]; in 2015, this use had reached approximately 39 Bgal/d [163 GL/d]. In 2015, about 87% of the U.S. population obtained drinking water from public suppliers, compared to 62% in 1950. Surface water provided the majority of this resource. Private water sources, such as a well, are less common

TABLE 2.3 Comparison of Regional Water Use Versus Resources for the Continental United States

Water Resources Region[a]	Consumptive Use	Renewable Water Supply[b]	Ratio (Use as a % of Supply)
New England	0.6	78.4	0.8
Mid-Atlantic	1.3	80.7	1.5
South Atlantic/Gulf	6.1	233.5	2.6
Great Lakes	1.9	74.3	2.6
Ohio	2.3	139.6	1.7
Tennessee	0.3	41.2	0.7
Souris-Red-Rainy	0.5	6.5	7.7
Upper Mississippi	2.3	77.2	3.0
Lower Mississippi	40.3[c]	484.8	8.3
Missouri	17.5	52.9	33.1
Arkansas-White-Red	9.6	68.7	14.0
Texas Gulf	9.1	33.1	27.5
Rio Grande	3.5	5.4	64.8
Upper Colorado	4.2	13.9	30.2
Lower Colorado	10.6[c]	10.3	103.0
Great Basin	3.5	10.0	35.0
Pacific Northwest	11.2	276.2	4.1
California	25.8	74.6	34.6

Source: Adapted from United States Geological Survey.

[a]These are water resource areas generally independent of state boundaries.

[b]Renewable water supply represents a long-term sustainable resource: precipitation plus imports less evaporation, exports, and water needed to maintain minimal stream flows.

[c]These values are for the entire river system.

than in the past, used by about 12% of the U.S. population in 2015. As with energy, population growth is increasing overall water consumption, while conservation efforts provide some counterbalancing effect.

Water-efficiency and conservation standards are not nearly as extensive or ubiquitous as energy-efficiency standards, although most building users are aware of low-flow toilet requirements, flow restrictors for showers, and self-closing bathroom faucets in public facilities. Surprisingly, given the few design restrictions that exist, per capita water use in the United States has declined 16% from 2005 to 2015. At 82 gal [310 L] per capita per day (GPCD), domestic water use (from both public and private sources) in the U.S. is about 16% lower than the 98 GPCD [371 L] rate from 2005. This is partly because overall per capita use involves important sectors other than buildings (agriculture and power generation, for example) and partly due to increasing awareness of the value of water.

As with renewable energy sources, green building efforts have also increased awareness of and design for water supply savings and alternatives. Several chapters in this book address alternative design approaches for water and waste that would likely be considered for adoption in a green building.

2.4 MATERIALS

A global or even countrywide perspective on materials resources is more difficult to obtain than is the case for energy and water. In the United States, there appears to be a general upward trend in per capita consumption of materials. In the early 1900s, per capita consumption was around 3 metric tons per person, about 50% of which was construction materials; in 1950, it was about 6 tons, with about 60% in construction materials; and in 1995, about 10 tons, with about 65% in construction materials. By 2020, nonfuel material consumption in the United States was about 9 tons, with about 75% in construction materials.

Given the scarcity of quantitative data, a qualitative comparison with water must suffice. Many materials used for building construction and repairs come from a generally fixed resource base. Like water, materials (at least many) may be recycled, but there is a fixed quantity of resources available on Earth. For many materials, what we have now is what we will have in the future, a major exception being those organic materials (such as fiber products) that are renewable.

Finding adequate materials resources for a building directly from the building site is rare: wood, straw, and earthen construction systems, for example, require large land areas to supply the materials for even a small building. The most common construction systems involve materials brought to a site from some distance. A designer can generally select between imported renewable or nonrenewable materials and between imported virgin or recycled materials. The common thread is *imported*—not necessarily from overseas but from a distance. Reducing the transportation distance and finding local materials when available are emerging design strategies.

Wood is the only renewable construction material currently in wide use in North America. It is easily worked, supports a wide variety of finishes, has moderate structural strength, requires regular maintenance for long life when exposed, catches fire easily, and has only moderate value either as thermal mass or as insulation. This common building material illustrates the impact a rapidly increasing population can have on a fixed, even if renewable, resource base. As huge old trees are harvested to the point of disappearance, growing demand for wood outstrips the supply available from younger, smaller trees. New production methods are devised (such as laminates, particle board, and engineered lumber) to allow large wood members to be constructed from smaller timber. The value of salvaged older wood members increases.

Nonrenewable materials are, by far, the most common constituents of mechanical, electrical, and plumbing systems; metals and plastics predominate. Their advantages include strength, durability, fire resistance, and conductivity or resistivity as required. Most such materials are obtained, however, at a significant energy and carbon cost for mining, manufacturing, transportation, and installation.

With rapidly increasing demand for both renewable and nonrenewable materials, how should a designer on a planet with fixed resources respond? What are the key materials issues that the design team ought to consider?

(a) Embodied Energy and Emitted Carbon

One issue is embodied energy, a complex (and therefore elusive) indicator of how much energy has been invested to mine/harvest/produce, fabricate, and transport a unit of building material. Embodied energy data never truly developed as a mainstream building design tool. It is now seemingly a back-burner consideration overtaken by interest in and development of emitted carbon tools. The UN Environment Programme (UNEP) provides a good explanation for this shift.

The original ICE (Inventory of Carbon and Energy) database contained embodied energy and embodied carbon factors (UNEP, 2019). However, since 2019, embodied energy factors are no longer included. The data in the wider literature, which the ICE database relies upon, generally no longer reports the embodied energy of construction products. Instead, embodied carbon has become the main metric, with the term "carbon" standing as shorthand for carbon dioxide (CO_2). Carbon emissions give a better indicator of the contribution of that energy to global warming and climate change.

Embodied energy and emitted carbon are not identical issues. There is a strong cause and effect connection between the two, however, that supports the UN ICE statement. The embodied energy resources (whatever their magnitude) expended throughout the life-cycle of a building material come with associated greenhouse gas emissions (GHGs)—including carbon dioxide, often referred to as "carbon" in the context of buildings. Table 2.4 summarizes information on the emitted (commonly called embodied) carbon associated with prevalent construction materials. The embodied carbon estimates shown in Table 2.4 refer to carbon dioxide equivalent (CO_2e), defined as the concentration of CO_2 that would cause the same climate impact as emissions of other greenhouse gases (e.g., methane, perfluorocarbons, and refrigerants). The tabulated embodied carbon value for any given building material is complicated by variations in raw resource availability (more difficulty in extracting materials requires more energy), variations in distance from raw resource to manufacturing locations, and variations in the fuels used (and their efficiency of use) in the refining or fabricating processes. Embodied carbon values for materials are complicated; carbon emitted during the material life-cycle is a negative environmental attribute, yet carbon sequestered during a material's life-cycle is a positive attribute.

Environmental product declarations (EPDs) are standardized documents that provide summarized information about the life-cycle environmental impacts of a building product or material, including its embodied carbon. In accordance with ISO (International Standards Organization) standards, Type III EPDs require verification by a qualified third party and are characterized by their comprehensive and comparative life-cycle assessment (LCA) data, making them particularly useful to building design professionals. Type III EPDs will adhere to product category rules (PCRs), which are standardized guidelines that define how to conduct an LCA for a specific product group. EPDs typically break down life-cycle assessment information into distinct stages:

- product (raw material acquisition), stages A1–A3;
- construction (installation), stages A4–A5;
- use (including maintenance and repair), stages B1–B7;
- end of life (including disposal), stages C1–C4;
- beyond the product life-cycle, stage D.

These life-cycle assessment stages are defined by standard *EN 15978 : 2011* (CEN, 2011). Developed by the European Committee for Standardization (CEN), this document defines a calculation method, based on

TABLE 2.4 Embodied Carbon Benchmarks for Materials Assessment

Materials	Benchmark GWP[a] ($kgCO_2e$/metric ton unless otherwise stated)
Asphalt	72.6
Concrete (strength class, PSI)	
≤2499	277
3000	318
4000	352
5000	382
6000	407
≥7200	402
Cement	858
Concrete Masonry Units	290
Glass	1401
Steel	
Rebar (fabricated)	850
Rebar (unfabricated)	760
Hollow Structural Sections (fabricated)	1898
Hollow Structural Sections from Electric Arc Furnaces (unfabricated)	1652
Hot-Rolled Sections (fabricated)	1163
Hot-Rolled Sections (unfabricated)	869
Cold-Formed and Galvanized (stud, track, framing, decking, etc.)	2408

Source: Adapted from GSA, 2023.

[a]Table exhibits the "better than average" values from the GSA data.

LCA and other quantified environmental information, to gauge the environmental performance of a building and its materiality (which includes building products). *EN 15978:2011* also specifies the protocol for reporting and communicating the outcome of the environmental performance assessment. *EN 15978:2011* describes all stages of a building's life-cycle and is based on data obtained from a variety of information resources. The assessment includes all building-related construction products, processes, and services used over the life of the structure. Figure 2.7 clarifies the various LCA stages of a building in accordance with *EN 15978:2011*. Note that the standard includes both embodied and operational impacts as it includes operational energy (B6) and operational water (B7). Moreover, stage "D" affords an opportunity to incorporate the environmental benefits of improved circularity through reuse, recovery, and recycling.

The stages captured by an EPD will determine the scope of the document. "Cradle-to-gate" EPDs only summarize stages A1 through A3; whereas "cradle-to-grave" EPDs cover all stages from A1 through C4. Though it is not defined as such by *EN 15978:2011*, conceptually, one could approach the notion of "cradle-to-cradle" to cover stages A1 through D. The information in Table 2.4 is an example of cradle-to-gate data that may be considered for benchmarking embodied carbon from the materials exhibited.

Consider two alternatives for exterior wall cladding: wood siding and aluminum siding. Wood contains embodied energy and sequestered carbon from a renewable resource—the sun. It takes the energy of human beings and chain saws to cut the trees and fuel to haul them, perhaps 100 mi [160 km], to a mill. At the mill, more energy is invested as logs become lumber, and still more energy is used as lumber becomes finished siding—which is then transported to a construction site. Aluminum begins as bauxite, which requires energy to mine, then more energy to ship great distances to smelters, which use large quantities of electricity in the refining process. Cheap electricity (as in the Pacific Northwest, with its once-surplus hydropower) attracts bauxite that is mined thousands of miles away. Aluminum ingots are formed at the smelter, then they are shipped—again, sometimes thousands of miles—to factories that make products such as siding; the products are then transported to a construction site. For a given surface area of finished siding, the aluminum alternative represents about 100 times as much embodied energy as the wood. The comparison of carbon emissions is more extreme since wood is (by most accounts) a net sequesterer of carbon. This comparison, however, is not the end of the story—a designer must consider the impacts of these two siding materials on building energy consumption and enclosure maintenance and replacement needs and schedules.

(b) Recycled or Virgin Material

It seems paradoxical that, while the world's population is increasing and its raw materials are decreasing,

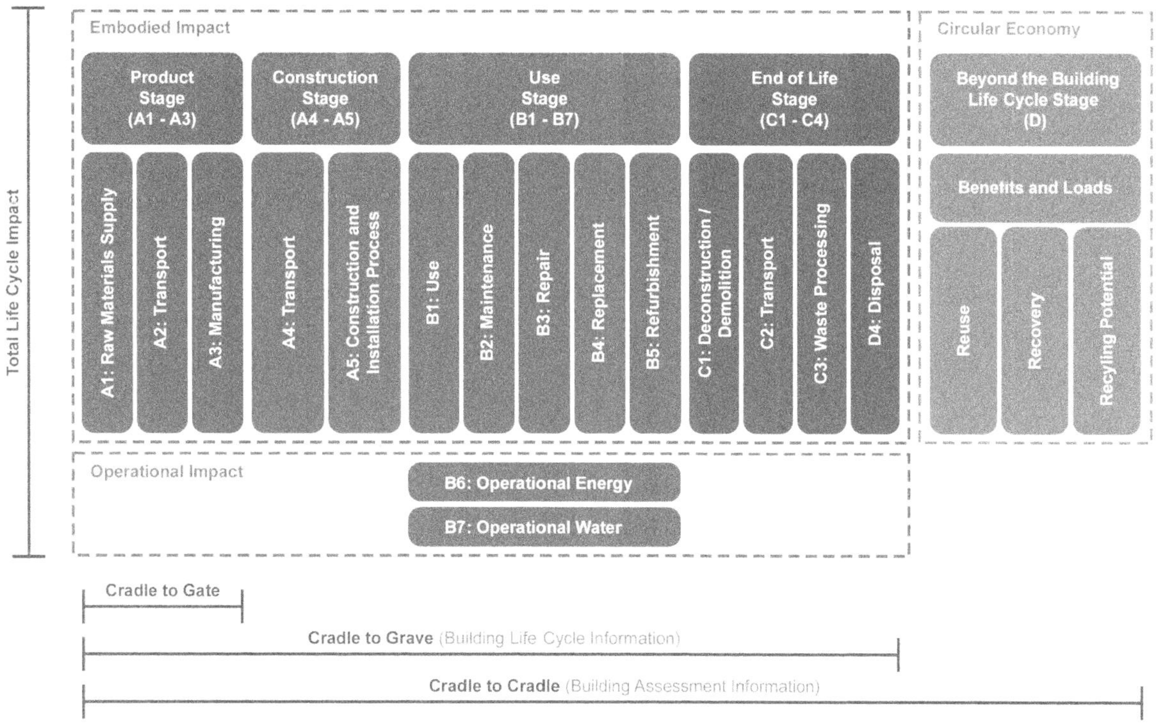

Figure 2.7 *Building life-cycle stages as defined in EN 15978 : 2011 along with conceptual depictions of "cradle-to-gate," "cradle-to-grave," and "cradle-to-cradle" scopes and embodied versus operational impacts overlain for clarity. (© Daniel Overbey; all rights reserved.)*

labor costs are growing so much faster than the costs of raw materials. One consequence is that labor-intensive practices become less economically attractive. Recycling is such a labor-intensive practice, whether at the scale of a household, an office building, or an entire industry.

Building construction, renovation, and demolition involve many opportunities for recycling. According to the U.S. Environmental Protection Agency (EPA), in 2018, the United States generated approximately 600 million tons [544.3 million Mg] of construction and demolition (C&D) debris. Just over 75% (455 million tons [412.8 million Mg]) of the C&D debris was directed to "next use"—a designation that the EPA uses to indicate that the debris was directed toward a market or process where it can be repurposed or recycled into new products or materials. Next use may include fuel, manufactured products, aggregate, compost and mulch, or soil amendment (EPA, 2024).

As more of a demolished building is recycled, more of the energy embodied in its material is recovered, fewer virgin materials are required for some other project, and carbon emissions from material manufacturing are reduced. At present, the recovery of usable materials from demolition is limited because the cost of labor is high, and the cost of energy and new products is comparatively low. In many instances, it may be easier, quicker, and cheaper to reduce a building to rubble and haul it to a landfill than to recycle its components. As landfill capacity becomes increasingly scarce, design regulations concerning recycled material use can be expected.

An Atlantic City, New Jersey, project was able to recycle 90% of its demolition waste. Of a total of 1583 tons [1400 Mg] of demolition waste, only 152 tons [140 Mg] were nonrecyclable. Concrete and masonry became crushed aggregate for road building. Glass became "glasphalt" embedded in road surfaces as reflectors. Wood waste became mulch. At Fort Ord in California, four buildings totaling about 11,000 ft^2 [1022 m^2] were dismantled rather than being demolished, saving roofing boards, framing lumber, and tongue-and-groove wood flooring. Unpainted drywall was reclaimed for composting.

Construction recycling opportunities include crushed wallboard as a replacement for lime in agriculture, carpet ground up for use as attic insulation, plate glass crushed for use in manufacturing glass fiber insulation, and pulverized wood as a composting aid at sewage sludge treatment facilities. Old acoustic ceiling tiles can become part of the slurry from which new acoustic tiles are made. Building materials are now increasingly being made from recycled materials: reinforcing bars from ferrous scrap metal; cellulose insulation from newsprint; parking lot bumper strips, fence posts, and park benches from recycled plastics; and nonstructural concrete from incinerator ash. Even plastic yogurt containers, complete with scraps of aluminum foil, are made into a terrazzo-like floor tile.

Architect Pliny Fisk, codirector of the Austin, Texas-based Center for Maximum Potential Building Systems (CMPBS), has developed a wide array of such applications. His "Advanced Green Builder" home near Austin displays several applications of "ashcrete," made with coal fly ash and bottom ash, producing a 97% recycled-content concrete. This is used as ferro-cement for columns and beams and is foamed for hollow wall infill. Numerous other innovative recycled-material applications are showcased in the home.

The most effective form of recycling involves reuse of a building or building shell. The next most effective form of recycling involves the reuse of a building component as is. A residential demolition-by-hand salvage project in Portland, Oregon, recovered doors, windows, bathroom fixtures, framing lumber, plywood, siding, flooring, and bricks.

As with water and renewable energy sources, much of the current interest in lowering the embodied energy (and thus carbon) content of construction materials and increasing the recycling and reuse of materials is attributable to the evolution of green building design efforts. The impact of building materials on occupant health and well-being is also an evolving area of concern and interest. There is a direct link between the selection and maintenance of building materials and indoor air quality (see Chapter 5).

Although no building should provide an unhealthy environment for its occupants, the term "healthy building" is generally used to denote a building that provides exemplary interior conditions for users with serious allergy (or other) conditions that are exacerbated by indoor pollutants. An advanced discussion of these types of buildings is beyond the scope of this book, but there are several readily available texts that can provide fundamental information on design for healthy buildings. Many building products with substantially reduced heath impacts have been developed and are in the market.

2.5 RESOURCE CHALLENGES

The buildings we design today are very likely, over their lifetimes, to experience major changes in the way they are used and in their sources of energy. Policies that favor electrification are already stirring interest and consternation among designers. The societal value of water and embodied materials will also change. The perspective on resources held by future generations is highly unlikely to be that which we hold today. This poses some intriguing challenges for the architectural and engineering professions.

(a) Design for Building Recycling

Designing for the recycling of buildings is a two-part balancing act. First, the design team should provide enough flexibility to prolong the useful life of a building by enabling it to adapt easily to changed usage. Flexibility, however, can be expensive to implement physically and can result in a bland "sameness." The latter characteristic is easier for the designer to engage than the former. Second, the design solution can allow for demounting of parts so that the structure can remain safely intact, while reusable materials and components are removed. This, however, can result in heavier buildings in which floor systems are not structurally integrated with beams. This approach also discourages integration of mechanical and structural systems. Furthermore, a demountable building may be especially subject to energy leaks, such as from cracks around self-contained components of the façade.

Some initial guidelines for recyclable buildings are as follows:

1. Design the structure to be separable from everything else and to be easily disassembled. Extensive remodeling is then possible without major structural modifications, and at the end of a building's life, elements of its structure can be reused elsewhere.
2. Design for "breathing room" where possible: between a building and its neighbors or between major spaces within a building. Some expansion is thus possible without relocation. This could include designing the columns and footings to support an extra floor or two for vertical expansion.
3. Maximize the use of on-site (natural) forces such as sun and wind. The less sophisticated the mechanical and electrical equipment, the less evident will be the obsolescence of such equipment with the passing of time.
4. Use materials and components distinctly: avoid combinations that make recycling of these elements difficult. A steel or plastic pipe embedded in a concrete slab is neither easily repaired nor easily recycled; some assembly "sandwiches" (manufactured building panels) do not allow constituent metals, plastics, and other components to be separated for reuse at the end of the panel's life.

Although maximum savings of embodied energy and emitted carbon can be realized when a building component is reused as is, even a "downuse" of building materials will save energy and resources compared to manufacturing from virgin materials.

(b) Design for Energy Transition

Two more challenges to designers arise:

1. *To design buildings not only to save energy, but also so that they can eventually be weaned away from dependence on nonrenewable fuels.* A transition away from electricity from the utility grid, to site-generated photovoltaic or fuel-cell electricity, seems easy enough given planning for appropriate building orientation, collection surfaces, and/or equipment spaces.
2. *To use energy wisely.* This means expecting only a fair share of locally available renewable resources, recognizing that such resources are limited even though they are continuously available. For example, in a high-density setting, it may be tempting to erect a large solar collector to intercept sunlight that would otherwise be utilized by a neighboring building. This temptation grows stronger as a building is designed to rely more heavily upon the sun. The concept of a solar envelope to protect each building's fair share is discussed in Chapter 3.

(c) Design for Transportation

There are clearly links among design decisions at regional and neighborhood scales (urban planning and subdivision design), transportation, and resulting energy use for commuting, shopping, and recreation. This scale of design, however, is beyond the scope of this book. The link between transportation systems and building mechanical/electrical equipment may seem obscure, but consider the impact on buildings of the automobile and its internal combustion engine. Fresh air intakes at street level face significant pollution from engine exhausts. Parking lots below buildings compete for space with heavy mechanical and electrical equipment such as boilers, chillers, ice storage tanks, and switchgear, and greatly complicate vertical transportation design. Sloped parking floors make future space use for other purposes quite difficult. Ventilating parking levels to remove fumes from automobiles requires large fans and energy to run them.

In a likely future of electrically powered vehicles, photovoltaic arrays over parking areas can provide electricity for a building and recharge the batteries of parked cars throughout the day (Figure 2.8). In an alternative future of hydrogen fuel cell-powered vehicles, engine discharge consists simply of water. No fumes are emitted from such cars, saving parking exhaust fan space and energy. To the extent that cars become smaller, or are replaced by public transit or bicycles, significant space may be reclaimed from parking for other uses. When entire buildings are now built as single-purpose parking garages, skillful design

Figure 2.8 *Building-integrated photovoltaics (BIPV) provide shelter, shading, and power for a fueling station/convenience store in Eugene, Oregon. Note the green roof on the store and the biofuel pumps. (Photo by Nathan Majeski.)*

for reuse might allow such buildings to be renovated to serve new functions rather than being demolished.

(d) Design for the Information Age

Controls for mechanical and electrical systems have become increasingly sophisticated, thanks to developments in information systems and electronics. With the advent of smart houses, intelligent buildings, and IoT appliances, it is now possible to regulate an array of building systems collectively and across great distances to optimize performance and minimize resource consumption. For a building with some zones requiring heating while other zones need cooling, some zones with available daylight and others without, an automatic central control system can, without human intervention, integrate the flow of fresh air, radiation through movable shading devices, and intensity of electric lighting to achieve maximum use of on-site renewable energy.

Data-driven control systems promise optimized operations and potentially enormous energy conservation achievements while using very small amounts of energy themselves. They also require one of the least space-consuming distribution systems, or distribution trees, of all building service systems, especially compared to air ducts and plumbing pipes. In return for such positive characteristics, building designers must recognize that developments in information technology are so rapid that the nature of these systems is likely to undergo frequent and dramatic change. Information system distribution spaces may be quite small, but they must be highly accessible. Where information can be transmitted without wires or cables, the impact on building service space demands is even smaller. Adaptable information systems can make more feasible the renovation, rather than demolition, of older buildings for new tenants. These potentials notwithstanding, the opportunity for occupants to play an active role in the use and control of their environments should not be ignored. In addition, the need to secure computerized controls against malicious hacking must be considered.

2.6 HOW ARE WE DOING?

From the preceding discussion and that in Chapter 1, it might seem that, environmentally, the building professions are doing well. Minimum standards for energy efficiency and plumbing fixture water consumption affect virtually every North American building design. Similar standards are also common internationally. There is growing interest in green buildings, generally fueled by private sector and government owners seeking to set an example. This interest is moving the center of gravity of design efforts beyond the just-acceptable minimum requirements of codes and

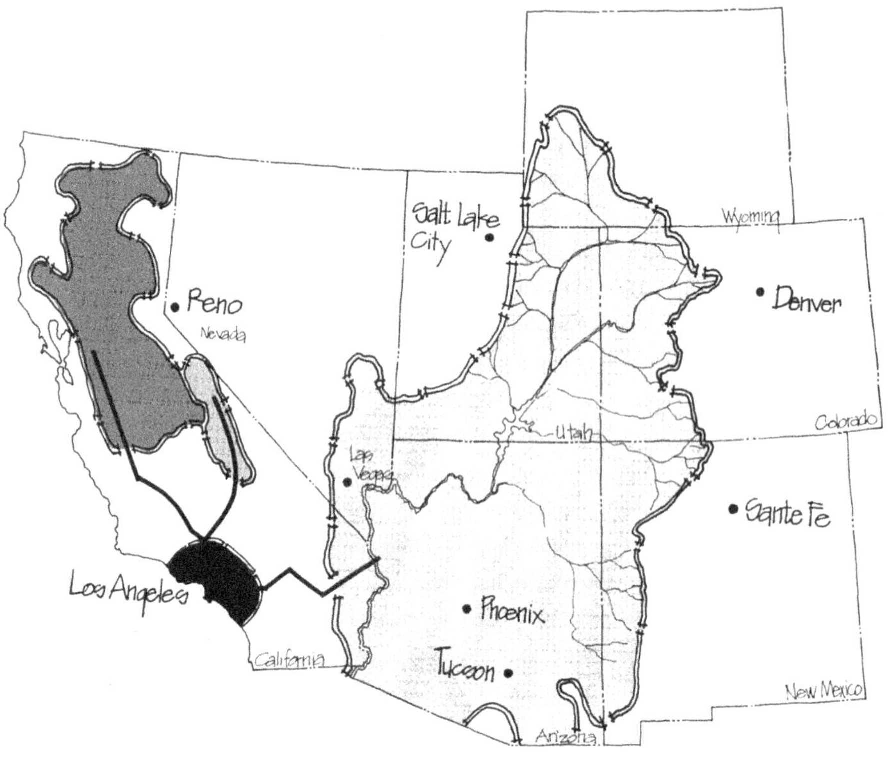

Figure 2.9 The effective watershed of the greater Los Angeles area. The area needed to provide water to this metropolitan area (its water footprint) is vastly greater than the politically defined city limits. (From Lyle, 1999.)

standards. Concern for energy consumption, renewable energy use, water resources and quality, and materials resources and consumption is an integral element of the green building movement. Per capita energy and water use in the United States appears to be stable and/or decreasing.

From the perspective of the previous century, today's building designs (even the worst) are arguably more resource-efficient and respectful of the environment. The question is: From the perspective of tomorrow, is today's good design good enough? The answer in one context is, unfortunately, no. That context is the environmental footprint. Environmental footprints are a concept promoted by Rees and Wackernagel (1995) that plots the gross resource demands of a geographic area as a footprint on the planet. Figure 2.9 provides an illustration of the environmental footprint concept applied specifically to water resources. The entity in question may be a city, state/province, or country. If the footprint is larger than the geographic boundaries of an entity, then the area is stepping on someone else's environmental toes. Such a city, state/province, or country needs more environmental resources (generally equating to land area) to support itself than are available—thereby surviving through imports from other places. All is well if there are other places with surpluses; all is not well when surpluses diminish or disappear.

Table 2.5 shows estimated ecological footprints for several countries. Some countries are substantially overstepping their ecological boundaries, while others are able to help support that overstep because of their less consumptive lifestyle or lower population density. Continuing worldwide population growth, however, makes the footprint balance tenuous. Table 2.6

TABLE 2.5 Ecological Footprints for Selected Countries

Country	2019 Population (millions)	Biocapacity (gha[a]/person)	Ecological Footprint (gha/person)	Biocapacity Reserve (+) or Deficit (−) (gha)
Australia	25.4	11.5	6.1	5.4
Austria	8.9	2.9	5.8	−2.9
Bangladesh	165.5	0.3	0.7	−0.4
Brazil	211.8	8.3	2.6	5.7
Canada	37.5	14.5	7.9	6.6
China	1453.8	0.8	3.5	−2.7
Egypt	105.6	0.3	1.6	−1.3
Germany	83.1	1.6	4.7	−3.1
India	1383.1	0.4	1.1	−0.7
Nepal	28.8	0.4	0.9	−0.5
United Kingdom	66.8	1.1	3.9	−2.8
United States	334.3	3.7	7.8	−4.1
WORLD	7765.0	1.6	2.6	−1.0

Source: Global Footprint Network, 2023.
[a]gha stands for global hectare (a productivity weighted land measure); a hectare is an SI measure of land equal to 2.5 acres.

TABLE 2.6 Per Capita Energy and Water Use and CO_2e Emissions for Selected Countries

Country	2019 Population	Energy Use[a]	Water Use[b]	CO_2e Emissions[c]
Australia	25,357,200	238.5	724.7	18.6
Austria	8,879,940	173.9	413	7.5
Bangladesh	165,516,000	10.5	231	0.7
Brazil	211,783,008	60.1	369.7	2.3
Canada	37,522,600	388.7	1113	16.7
China	1,453,799,936	101.8	425	7.7
Egypt	105,619,000	35.7	910.6	2.4
Germany	83,148,104	160.1	410.5	8.4
India	1,383,110,016	24.2	602.3	1.9
United States	334,320,000	286.2	1543	16.4
WORLD	7,764,950,016	75.6	583	5.0

Source: Adapted from Energy Institute, 2024; Global Footprint Network, 2025.
[a]Units are primary energy consumption per capita, given in gigajoule (GJ) per person per year (1 gigajoule = 947,817 Btu). Data are for 2019.
[b]Units are cubic meters (1 cubic meter = 35.31 cubic feet) of water withdrawals for agricultural, industrial, and municipal purposes per person per year. Data are for 2015.
[c]Units are carbon dioxide equivalent emissions per capita per year from energy, process emissions, methane, and flaring. Data are for 2019.

provides similar environmental benchmarking for the same countries with respect to energy and water use and carbon dioxide (CO_2) emissions. CO_2 has become a key environmental metric in the United Kingdom and parts of Europe (Roaf, 2004). In contrast, energy is still entrenched as the key environmental metric in the United States.

The ecological footprint for the "world" presented in Table 2.5 should lead to serious reflection among those who claim to be producing sustainable design solutions. One unfortunate offspring of the otherwise positive interest in green design is a seemingly endless stream of one-upmanship that glibly promotes "sustainable" this and "sustainable" that—including buildings (virtually impossible in today's economic climate) or communities (possible, but rare today). The term sustainable has lost almost any meaning through incessant misuse. This is unfortunate if one believes the story of the ecological footprint—sustainability is essentially keeping the Earth's footprint on the planet we inhabit. This seems a really good idea, as it is the only planet we have right now.

2.7 CASE STUDY—DESIGN PROCESS AND ENVIRONMENTAL RESOURCES

The Bullitt Center, Seattle, Washington

Project Basics

- Location: Seattle, Washington, USA
- Latitude: 47.6°N; longitude: 47.6°E; elevation: 20 ft [6.1 m]
- Heating degree days: 4908 base 65 °F [2727 base 18.3 °C]; cooling degree days: 2120 base 50 °F [1177 base 10 °C]; annual precipitation: 37 in [941 mm]
- Building type: New construction; commercial class A office space
- Size: 50,000 ft^2 [4645 m^2]; six occupied stories
- Completed: 2013 (Earth Day)
- Client: Bullitt Foundation
- Design team: Miller Hull Partnership, PAE Consulting Engineers (and consultants)

Background. The Bullitt Center in Seattle, Washington, was intended to be the largest commercial building to achieve net-zero energy and water, and thus serve as a model for other commercial developments to viably meet future environmental challenges. The Bullitt Center offers competitive market rate class A office space and houses tenants that are leaders in sustainable business and education, such as the Cascadia Region Green Building Council, the International Living Futures Institute, the University of Washington's Integrated Design Laboratory, and the Bullitt Foundation headquarters.

Context. The Bullitt Foundation's mission focuses on sustainable efforts in business, technology, urban issues, ecosystem services and planning, and civic engagement. The vision for the new Bullitt Center was championed by its president, Denis Hayes, to establish a market rate commercial office building in downtown Seattle that produces as much energy as it uses, is a healthy place to work, and can be an educational beacon for progressive planning and ecological design. The Bullitt Foundation planned this through an ambitious process to meet the 2030 Challenge and the Living Building Challenge.

Design Intent. The Bullitt Center serves as a "living laboratory" that visually conveys the benefits of sustainable design and energy efficiency by engaging users of the building with passive systems. The primary strategy for achieving energy efficiency and water independence was to integrate architectural and engineering designs that focus on a high-quality construction using locally sourced building materials, new technology, and an emphasis on reducing plug loads in the building through energy limits on tenants.

The architecture and building systems of the Bullitt Center were designed to engage occupants so that they consider their energy consumption. Plate-glass walls display electrical and plumbing equipment, while a generous rooftop photovoltaic array cantilevers well beyond the enclosure, calling attention to the building as shelter from weather and generator of power. The Bullitt Center siting provides improved accessibility to walking, biking, and public transit by offering proximity to an adjacent park. The project's ability to be a neighborhood resource led to discussions with the City of Seattle, Department of Parks and Recreation, Seattle Parks Foundation, and neighbors on how to create a vibrant public space adjacent to the Bullitt Center and improve a pedestrian crossing. Collaboration with the City of Seattle led to a reevaluation of building ordinances that prevent sustainable design. The collaboration led to an initiative called the Living Building Ordinance (No. 123206) that sets a goal of motivating the

Figure 2.10 *Street façade view of the Bullitt Center, showing stairway, adjacent park, and cantilevered photovoltaic array. (© Miller Hull Partnership; used with permission.)*

construction of 12 Living Buildings in the Seattle area over the next three years.

Integrated Design Process. The desire for a simple architectural design with integrated building systems required an iterative performance-driven design process on the part of the owners, engineers, architect, contractor, and consultants.

The first step included assembling a team of design specialists at a two-day workshop that defined the design problems, criteria, and vision for the project. The Miller Hull Partnership worked with Solar Design Associates to create a schematic drawing that responded to the vision of the project and accommodated the major target of achieving an energy utilization index (EUI) of 20, while generating an

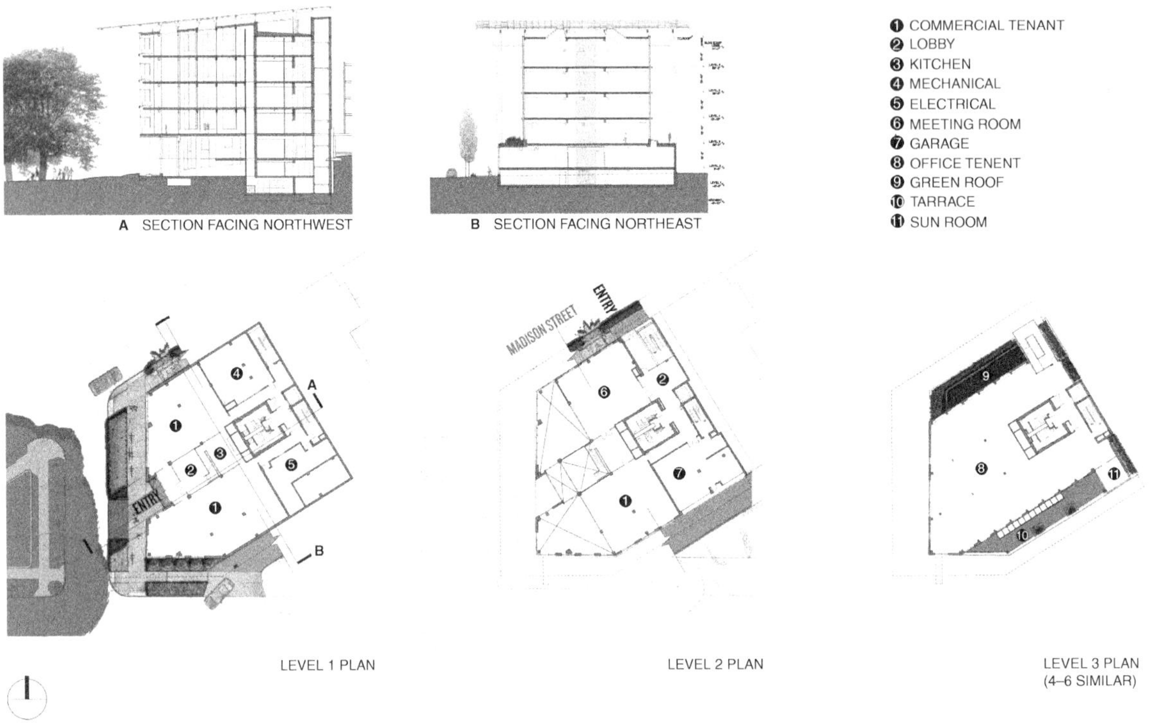

Figure 2.11 *The Bullitt Center sits on a tight urban site, with six stories above grade. (© Miller Hull Partnership; used with permission.)*

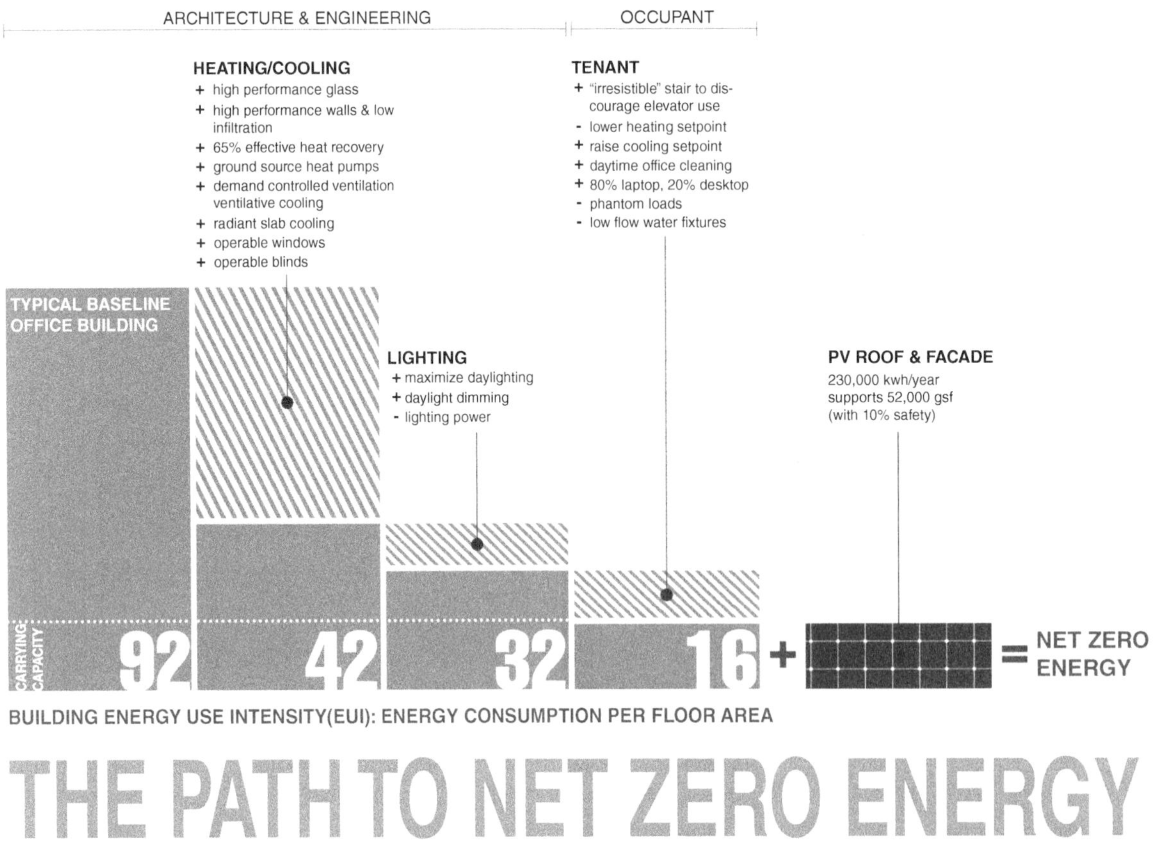

Figure 2.12 *Path to net-zero energy from a baseline building and load reductions through heating, cooling, lighting, occupants (behavior and tenant contracts), and energy generated on site. (© Miller Hull Partnership; used with permission.)*

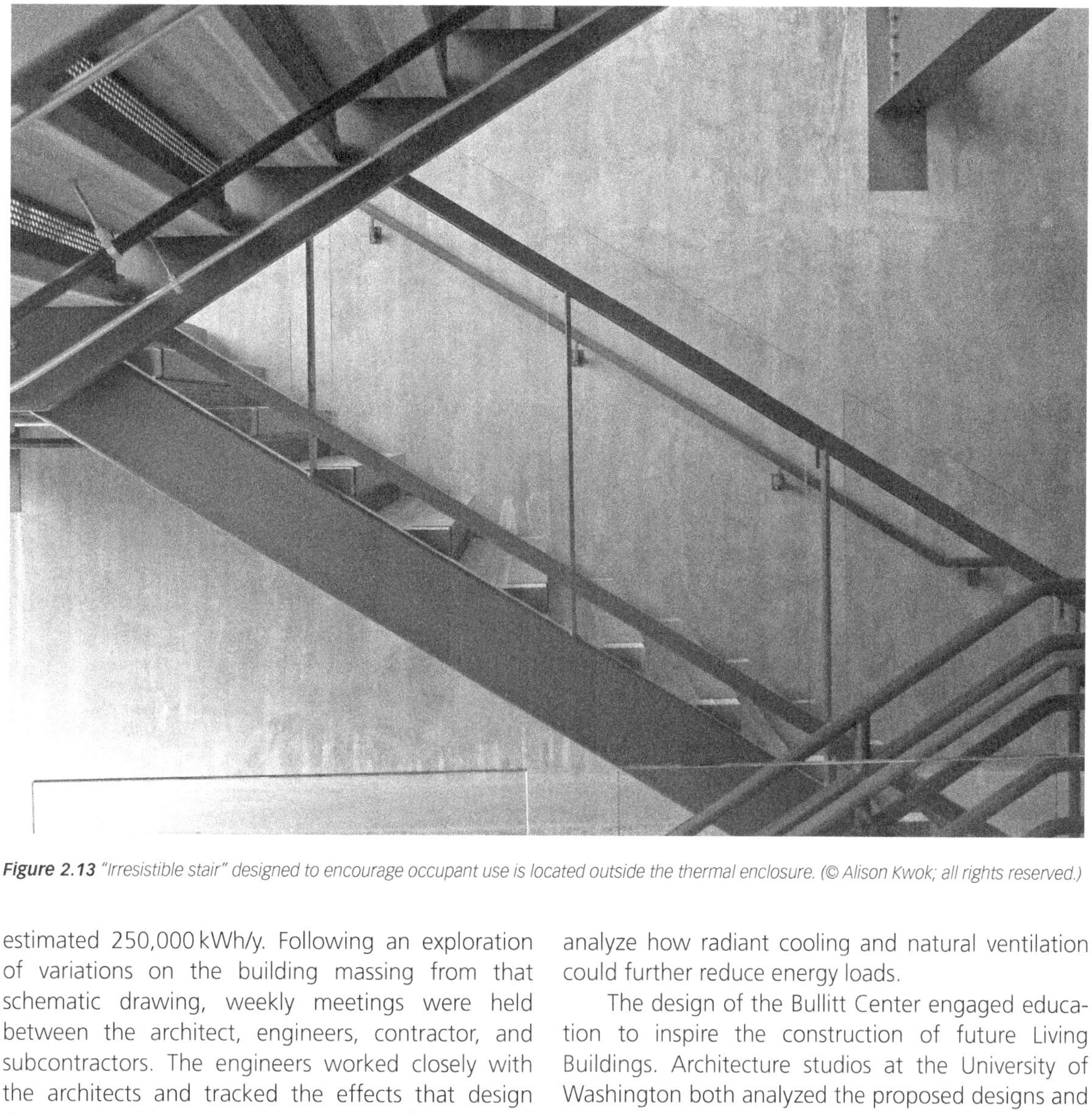

Figure 2.13 "Irresistible stair" designed to encourage occupant use is located outside the thermal enclosure. (© Alison Kwok; all rights reserved.)

estimated 250,000 kWh/y. Following an exploration of variations on the building massing from that schematic drawing, weekly meetings were held between the architect, engineers, contractor, and subcontractors. The engineers worked closely with the architects and tracked the effects that design changes would have on the energy performance of the building. The design team developed relationships with local contractors who were fabricating and installing high-performance building components, such as a curtainwall system designed and engineered in Germany.

Digital tools were used to optimize the systems in the building. The CAD modeling software Rhino combined with Grasshopper (parametric software) was used to quickly analyze iterations of roof forms and solar panel orientations. The panel power production was estimated with a variety of tools and methods. Analysis of daylighting performance was made possible by digital software such as Ecotect and collaboration with the Integrated Design Lab at the University of Washington. PAE Consulting used the thermal and airflow analysis software Bentley Tas v 9.1.4 combined with eQuest energy modeling to analyze how radiant cooling and natural ventilation could further reduce energy loads.

The design of the Bullitt Center engaged education to inspire the construction of future Living Buildings. Architecture studios at the University of Washington both analyzed the proposed designs and designed alternative schemes for the site. Throughout the construction process, students visited the site to learn about the challenges of designing and constructing a Living Building.

Criteria. The Bullitt Center sought Living Building Challenge certification that required the site and building to provide all of its own power, water, and waste management. Certification involved addressing seven categories (termed "petals" in reference to a flower).

1. Site: The building provides amenities for pedestrian, bicycle, and transit access.
2. Water: A concrete cistern under the site stores rainwater for reuse in the building. The system is designed to accommodate all water needs on site. Wastewater is treated on site.

Figure 2.14 *The elevator is inconspicuously located in the building so as not to draw attention to itself as people enter the building. (© Alison Kwok; all rights reserved.)*

3. Energy: Energy conservation is achieved by the high-performance enclosure, daylighting, and imposed energy budgets for occupants. These measures reduce energy loads to the point where all energy needs can be met by a solar photovoltaic array on a net annual basis.
4. Health: The inviting staircase, access to bike racks, and open floor plan with operable windows and access to daylight motivate a social and active lifestyle.
5. Materials: "Red List" hazardous materials, including polyvinyl chloride (PVC), cadmium, lead, mercury, and hormone-mimicking substances were not included in construction, to the extent feasible. All wood was certified to Forest Stewardship Council (FSC) standards.
6. Equity: The open floor plan of the building and tall floor-to-ceiling windows with operable sashes can provide fresh air and daylight to all occupants.
7. Beauty: Exposed structural timbers, access to views, native trees and shrubs, and a well-crafted building work together to create a comfortable and inspirational work environment.

Key Design Features

Building Massing Orientation. Six stories with 70-ft [21.3-m]-deep floor plates and a central service core are surrounded by operable floor-to-ceiling windows that daylight and ventilate the spaces. A glazed "irresistible" stair, situated outside the conditioned enclosure of the building, serves as the entry to the building, such that occupants and visitors are encouraged to take the stairs (rather than the elevator) and enjoy views to downtown Seattle.

Solar Array. A photovoltaic (PV) array of 575 panels, each at 425 watts per panel, produces a total of 242 kW under rated conditions and was projected to generate 230,000 kWh in a typical year. A steel frame holds the array, which cantilevers beyond the perimeter of the building. The photovoltaic

(a)

(b)

Figure 2.15 *(a, b) Tenant space features flexible, open-office plan with shared services at the inner core. (© Alison Kwok; all rights reserved.)*

array shelters a mechanical area on the roof and houses 19 inverters that convert DC power to AC power (bullittcenter.org/news/blog/solar-at-the-bullitt-center).

High-Performance Enclosure. The enclosure features a triple-glazed curtain wall system that was engineered in Germany but fabricated locally. Super-insulated wall elements were designed to eliminate thermal bridging and greatly reduce air infiltration. The design team took care to select a fluid-applied, weather-resistive air barrier that was produced within 300 mi of the Bullitt Center and did not include any materials from the Red List.

Heating and Cooling. Little heating and cooling are required, due to the high-performing enclosure, and loads are met with a ground-source heat pump fed

(a) (b)

Figure 2.16 *(a) Phoenix composters (made in Montana) in the basement of the building combine waste from the toilets with wood shavings and a small amount of water, causing aerobic decomposition. (© Alison Kwok; all rights reserved.) (b) Close-up of water and liquid composter. (© Miller Hull Partnership LLP.; used with permission.)*

Figure 2.17 *500-gal [1893 L] "day tank" contains rainwater that has gone through a purification system. The building will only use ~300 gal [1136 L] per day, so this tank serves as a buffer for the purification system that produces about 3–5 gal [11–19 L] of water per minute. (© Alison Kwok; all rights reserved.)*

by 26 wells. Heating and cooling are provided to the spaces via water piping in the concrete floors. Operable windows both cool and provide fresh air to the building.

Ventilation. An air-to-air heat exchanger, located on the roof and sheltered by the photovoltaic panels, efficiently brings in outdoor air that is distributed through a dedicated outdoor air system (DOAS). When cooling is desired, automated actuators on the windows allow natural ventilation to be the first cooling strategy, and the automation allows for both nighttime and daytime ventilation.

Transportation. Site planning and building circulation feature innovative ideas that elevate the experience of place through support of access to city transit stops, bike paths, and bike racks. The main circulation stair rewards users with views, convenience, and social interaction. The stairs are made more appealing by competing with a slow elevator that requires key card access.

Water Use and Collection. The project was designed to use rainwater to satisfy all of its water needs,

Figure 2.18 Walking toward Madison Street, near the alley behind the Bullitt Center. (© Miller Hull Partnership LLP.; used with permission.)

including a filtration system that treats rainwater to provide potable water for drinking, cooking, and showers. (Seattle ordinance, at the time of design, did not allow the use of rainwater for potable uses.) A 56,000-gal [211,980 L] rainwater cistern collects water from the roof, while graywater is filtered and recycled for nonpotable use on the green roof.

Waste Management. Foam-based composting toilets reduce water demand on the site, while anaerobic bacteria process human waste in 10 composting tanks located in the basement.

Long Life-Cycle. Life-cycle design considerations include a timber structure that is predicted to last 250 years, a durable 50-year building enclosure, and a photovoltaic array technology that will be competitive for 25 years. Construction costs for the project were 20% higher than comparable types of projects.

Site Sustainability. A green roof, bioswales, and a constructed wetland in the adjacent park will filter and treat all water that falls onto the site. Collection, reuse, evapotranspiration, and infiltration of graywater into the ground will limit the impact that the Bullitt Center has on non-source-point pollution of water that makes its way to Puget Sound.

Reduced Plug Loads. A limit of 0.8 W/ft^2 [8.6 W/m^2] was set for all office equipment, including computers, copiers, scanners, and the like. Typical office plug loads can be 1.5 W/ft^2 [16.1 W/m^2] or greater. Occupants will have to select energy-efficient technology to operate within the plug load limit.

Post-occupancy Evaluation. In compliance with the Living Building Challenge, the Bullitt Foundation was equipped to collect detailed performance information regarding energy, water, and waste via monitoring systems for at least one year—thus, allowing results to be compared with the predicted energy study conducted by PAE Consulting, who created the eQUEST energy model. Monitors that display the building energy use are on view in the lobby of the building.

After its first decade of operations, the Bullitt Center has achieved 30% net positive energy as its PV system has generated 2,475,021 kWh of electricity, which was 551,481 kWh more than the building required for all purposes.

However, it is prudent to acknowledge that experimentation with building systems may yield failures. For instance, after seven years of operation, the Bullitt Center's composting toilet system was removed in 2021. It was replaced with a highly water-efficient vacuum flush system connected to the municipal sanitary sewer system. In a white paper released by the Bullitt Foundation, several pragmatic lessons learned were offered to help ensure the success of future composting toilet installations (Kahn, 2021).

Project Data

Energy usage as predicted by DOE-2 software:

Predicted building energy use for HVAC (heating, cooling, pumping, fans): 3.96 kBtu/ft^2 per year [12.5 kWh/m^2 per year]

Predicted energy utilization index (PEUI) (total building consumption): 16 kBtu/ft^2 per year [50.5 kWh/m^2/y] (gross floor area); 21 kBtu/ft^2 [66.2 kWh/m^2 per year] (conditioned floor area)

Awards/Certifications

- First heavy-timber office building in Seattle since the early twentieth century
- First commercial building in the U.S. to earn Project Certification from the Forest Stewardship Council (FSC)

- 2012 Design & Build with FSC Award
- Project Certification: Living Building
- 2011 Eco-Structure: Evergreen Award, On the Boards category
- 2012 AIA Seattle: What Makes It Green Award, Mixed-Use category
- 2012 Forest Stewardship Council (FSC): Designing and Building with FSC Lumber Award
- 2013 Architizer.com: A+ Awards, Special Mention, Architecture and Sustainability categories
- 2013 *Metal Architecture* magazine: Metal Design Awards, Sustainable Design category

FOR FURTHER INFORMATION

The Bullitt Foundation places high value on making the design, construction, and operations of the Bullitt Center as transparent and accessible as possible. Many reports and analyses are available on the websites of the design team members.

References and Resources

Brand, S. 1994. *How Buildings Learn: What Happens after They're Built.* New York: Viking Press.

Brower, D. 1990. *Cool Energy*. Cambridge, MA: Union of Concerned Scientists.

Bullitt Foundation. 2025. "Bullitt Center." bullittcenter.org

CEN. 2011. *EN 15978:2011: Sustainability of Construction Works—Assessment of Environmental Performance of Buildings—Calculation Method*. Brussels: European Committee for Standardization.

DOE. 2012. *2011 Buildings Energy Data Book.* Washington, DC: U.S. Department of Energy.

EIA. 2012. *Annual Energy Outlook 2012*. Washington, DC: U.S. Energy Information Administration. www.eia.gov/outlooks/aeo

EIA. 2023. *Annual Energy Outlook 2023*. Washington, DC: U.S. Energy Information Administration. www.eia.gov/outlooks/aeo

EIA. April 2024. *Monthly Energy Review*. Washington, DC: U.S. Energy Information Administration. www.eia.gov/totalenergy/Data/monthly

Elly, T. 2011. "The Bullitt Center Bulk Airflow Analysis." Blog, October 4, 2011. pae-engineers.com/2011/10/04/the-bullitt-center-bulk-airflow-analysis

Energy Institute. 2024. *Statistical Review of World Energy 2024*, 73rd edn. London: Energy Institute. www.energyinst.org

EPA. 2024. *Sustainable Management of Construction and Demolition Materials*. Washington, DC: U.S. Environmental Protection Agency. epa.gov/smm/sustainable-management-construction-and-demolition-materials

FAO. n.d. *AQUASTAT Dissemination System*. Rome: Food and Agriculture Organization of the United Nations. data.apps.fao.org/aquastat

Fisher, J. 1974. *The Energy Crisis in Perspective*. Hoboken, NJ: John Wiley & Sons.

Global Footprint Network. 2023. *National Footprint and Biocapacity Accounts, 2023 Public Data Package*. (Data year 2019). Oakland, CA: Global Footprint Network. www.footprintnetwork.org/licenses/public-data-package-free

Global Footprint Network. 2025. *Ecological Footprint Explorer*. Oakland, CA: Global Footprint Network. data.footprintnetwork.org

GSA. 2023. *Interim IRA Low Embodied Carbon Material Requirements*. Washington, DC: U.S. General Services Administration. gsa.gov/real-estate/gsa-properties/inflation-reduction-act/lec-program-details/material-requirements

Hammond, G. P. and C. I. Jones. 2008. "Embodied Energy and Carbon in Construction Materials." *Proceedings of the Institution of Civil Engineers* 161(2): 87–98.

Hannon, B., R. Stein, B. Segal, and D. Serber. 1977. *Energy Use for Building Construction*. Champaign-Urbana, IL: Center for Advanced Computation, University of Illinois.

Hayes, D. 1977. *Rays of Hope*. New York: W. W. Norton & Co.

Hayes, D., et al. n.d. "Bullitt Center Energy Performance." bullittcenter.org/learning/ccsdc-energy-performance-report

ILFI (International Living Future Institute). living-future.org

Irurah, D. 1997. "An Embodied Energy Algorithm for Energy Conservation in Building Construction as Applied to South Africa." Ph.D. thesis, University of Pretoria.

Joslin, J., et al. 1993. *The Waste Papers: Analysis and Discussion of the Potential for Salvage and Re-Use of Construction Materials from Residential Demolition*. Portland, OR: Metro Solid Waste Department Publications.

Kahn, B. 2021. *Composing Toilets: Lessons Learned*. Seattle, WA: Bullitt Foundation. bullittcenter.org/2021/03/16/composting-toilets-lessons-learned

Kenny, J. F., N. L. Barber, S. S. Hutson, K. S. Linsey, J. K. Lovelace, and M. A. Maupin. 2009. *Estimated Use of Water in the United States in 2005 (USGS Circular 1344)*. Reston, VA: U.S. Geological Survey. pubs.usgs.gov/circ/1344

Lowry, W. 1988. *Atmospheric Ecology for Designers and Planners*. McMinnville, OR: Peavine Publications.

Lyle, J. T. 1999. *Design for Human Ecosystems: Landscape, Land Use, and Natural Resources*. Washington, DC: Island Press.

Meyers, R. 1983. *Handbook of Energy Technology and Economics*. Hoboken, NJ: John Wiley & Sons.

Rees, W., and M. Wackernagel. 1995. *Our Ecological Footprint: Reducing Human Impact on the Earth*. Gabriola Island, BC: New Society Publishers.

Roaf, S. 2004. *Closing the Loop: Benchmarks for Sustainable Buildings*. London: RIBA Enterprises, Ltd.

Stein, R., et al. 1981. *Handbook of Energy Use for Building Construction*. Washington, DC: U.S. Department of Energy.

UNEP. 2019. *The Inventory of Carbon and Energy (ICE) Database*. Global Alliance for Buildings and Construction. Nairobi: United Nations Environment Programme.

USGS. 2016. *Water Resources of the United States*. Reston, VA: U.S. Geological Survey. water.usgs.gov/index.html

USGS. 2017. *Estimated Use of Water in the United States in 2015 (USGS Circular 1441)*. Reston, VA: U.S. Geological Survey.

CHAPTER 3

Sites

SITE ANALYSIS SHOULD PRECEDE SITE PLANNING. The purpose of a site analysis is to understand the character of a given site. Such an analysis usually includes the collection of information on climate, solar access, availability of utilities, noise sources, zoning, views, traffic and pedestrian patterns, soil conditions, and the like. In some cases, long-term statistical data are available as an information resource (such as for climate, solar position, and utility services); for other variables (such as noise, views, and pedestrian density in an urban area), there is usually no existing database, and all information must be collected directly by the designer.

To be useful and successful, a site analysis must do more than simply catalog information; it must place value on the collected information in the context of a proposed project and its design intent. For a given building, is solar radiation a desirable resource or a problem to be solved? Is wind a usable design element or an environmental force to be avoided? Understanding which resources are available for inclusion in a design solution, and which natural forces are potential problems to be mitigated by design, is the essence of site analysis—and a necessary precursor to green design.

A designer's early site-planning decisions will, at a later date, influence available options for a building's climate control and lighting systems and affect a building's overall consumption of energy. Site planning must be informed by site analysis. When the site is seen as a collection of resources (sun, wind, water, and plants), and also as part of the environment we all share, buildings that greatly reduce dependence upon nonrenewable fuels can be developed. This can be accomplished without limiting the availability of local energy resources for neighboring buildings. In addition to saving energy, the thoughtful consideration of on-site resources can create outdoor spaces that are especially pleasant to be in. Such spaces can direct winter sun to a glass wall while blocking the wind, or funnel summer breezes through shading to an open window. Site planning is greatly influenced by economic considerations, zoning regulations, and adjacent developments, all of which can interfere with the design of a site to best utilize the sun, the sky, and the wind. Integration of these concerns at the site-planning stage is the first step in adapting a building to its climate. This chapter looks briefly at some aspects of site–climate interactions.

3.1 CLIMATES

Climate is a long-term statistically derived picture of weather. Weather is what happened today or yesterday, while climate has historically provided a picture of what happened over the past 10, 15, or 20 years. Our most familiar names for climates describe their most severe season, as shown in Figure 3.1. This is a convenient means of description, but it can be misleading for designers. "Cold" climates can have very hot, sometimes humid, summer days; hot-arid climates can have bitterly cold winter conditions. Before designing buildings that will interact with exterior conditions to provide indoor comfort, we should know in some detail the character of those conditions. This is not necessarily easy, as there are 8760 hours in a year and several climatic variables of interest for each hour (temperature, relative humidity, solar radiation, wind speed, etc.)—and each year is different. A graphical

Mechanical and Electrical Equipment for Buildings, Fourteenth Edition. Walter Grondzik, Daniel Overbey, and Alison Kwok.

Companion website: www.wiley.com/go/meeb14e

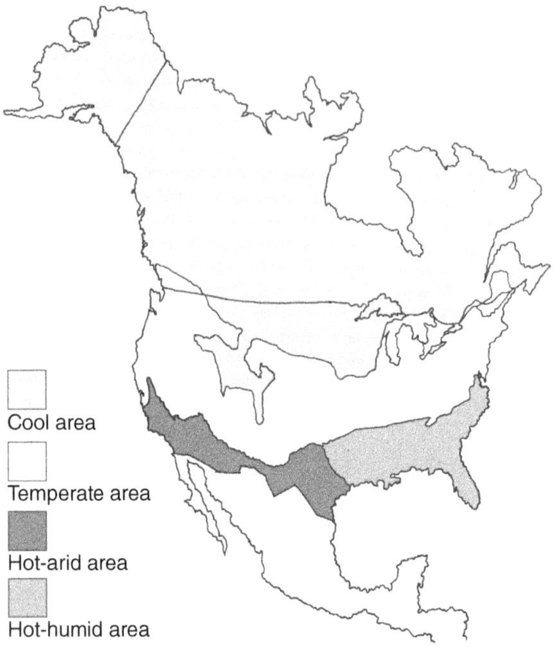

Figure 3.1 *Regional climate zones of the North American continent. (Redrawn by Tyler Mavichien from Olgyay, 1963 / with permission of Princeton University Press.)*

means of portraying these numerous variables is helpful. A very useful set of graphic analysis tools for climates was developed by Olgyay (1963) under the name bioclimatic design. Bioclimatic design links comfort and climate. Chapter 6 looks at thermal comfort, but a discussion of Olgyay's approach will introduce a few key considerations.

Figure 3.2 shows "timetables" developed by Olgyay for cities representative of two of the four climate zones shown in Figure 3.1. These graphical views of thermal environmental needs are plotted across an entire year (the horizontal axis) for all hours of the day (the vertical axis). Those portions of the year within the bounds of the bold shading line require that a person be shaded to achieve thermal comfort. The darker-colored (overheated) zones are bounded by an isoline that corresponds to a temperature of about 78–82°F [26–28°C], depending upon relative humidity. At higher temperatures, there is a need for air motion as well as shade to remain comfortable outdoors. Areas farther within the overheated zone correspond to thermal comfort needs for 300 and 700 fpm [1.5 and 3.6 m/s] air velocities. An occupancy schedule for a proposed building can be overlaid on such a timetable, providing insights into climate/building/user interactions.

The timetable for New York City (Figure 3.2*a*) shows that during about one-fifth of the year people require shade for outdoor comfort, with most of the hours from mid-July to mid-August falling in the overheated period. Little relief can be expected at night in summer. Perhaps one-third of the year is "too cold," with temperatures too low for shirtsleeve comfort outdoors and requiring more than 300 Btu/h ft^2 [960 W/m^2] of radiation, an unlikely solar bounty during New York winters.

The timetable for Miami, Florida (Figure 3.2*b*), shows that about four-fifths of the year requires shade for outdoor comfort, with about three-fifths falling in the overheated period. During a large portion of the year, people need over 700 fpm [3.6 m/s] wind speed for thermal comfort, with these hours occurring both during the day and at night. There is also a need for wind to counteract high vapor pressure (humidity), independently of high temperature. In Miami, wind is clearly important to achieving comfort in any building without mechanical cooling. There is no "too cold" time period in Miami.

The dry bulb hourly and monthly output from *Climate Consultant* (Figures 3.2*c*, 3.2*d*) shows similar visual patterns (and dozens of other meaningful displays), produced by a free graphics-based software program that will accept input for weather stations around the world. Of these two climates, the New York timetable suggests an emphasis on space heating but with a need for cooling as well. In Miami, the focus is upon shading, air motion, and cooling. Several building design strategies can utilize the ambient climate as a source or sink for heating and cooling.

3.2 CLIMATE DATA

Weather data (i.e., temperature, precipitation, solar radiation, cloud cover) come from a variety of standardized and quality-controlled measurements gathered by a global network of weather stations that adhere to guidelines and standards set by the World Meteorological Organization (WMO). Most countries have their own national meteorological service, such as the National Weather Service (NWS) in the United States, the Met Office in the United Kingdom, or Environment and Climate Change Canada. These national services adhere to WMO standards while also considering local conditions and requirements. The National Renewable Energy Laboratory (NREL) offers weather data for over 3000 locations worldwide at energyplus.net.

Climate normals are average values of weather data recorded over a 30-year period. These multidecade averages may serve as a baseline to compare weather patterns against "normal" or "typical" long-term climate conditions. Climate normals can help establish a baseline for planning agricultural activities, water management, or infrastructure development by providing expected climatic conditions. If available in

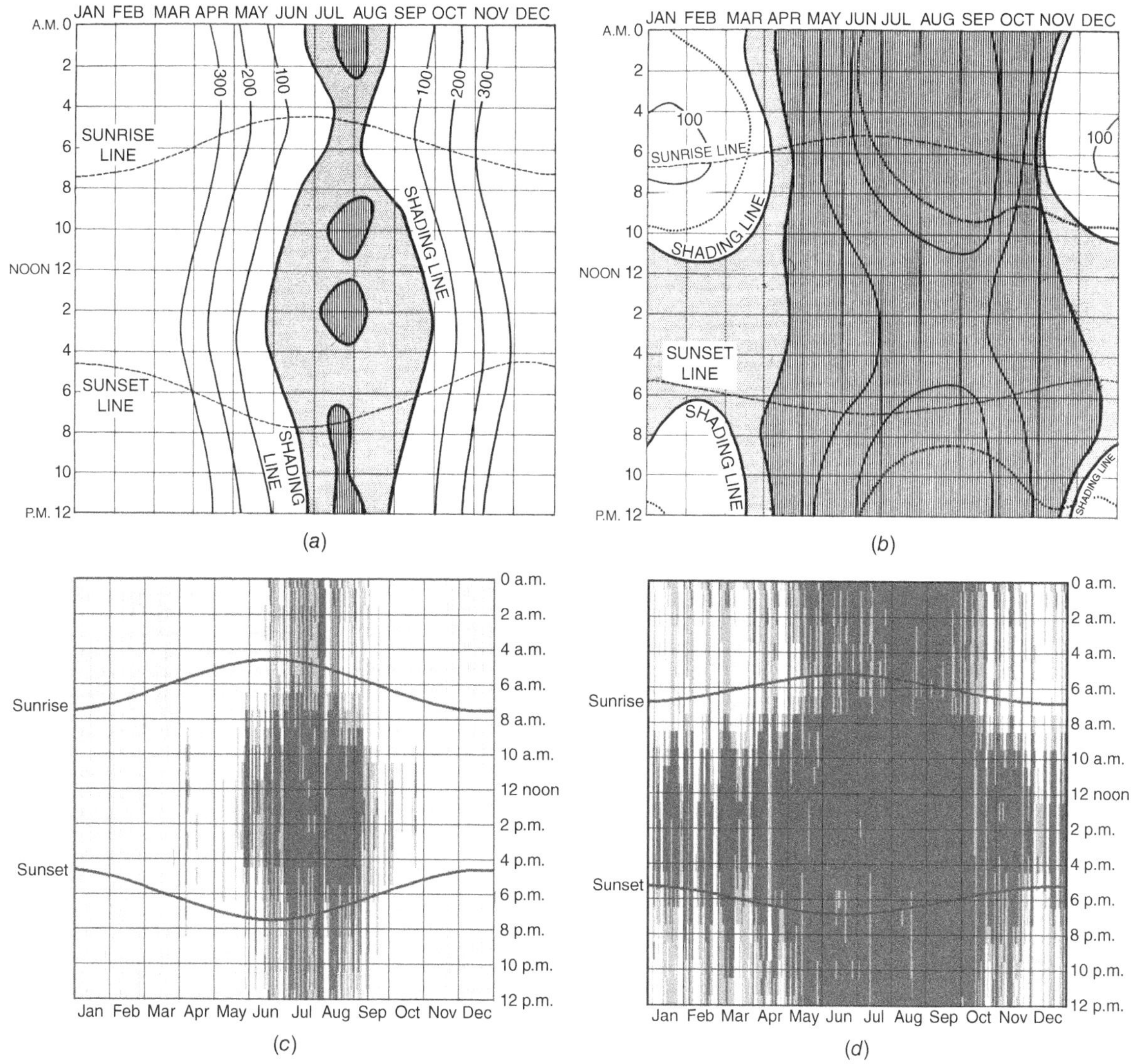

Figure 3.2 *Timetables of climatic needs for (a) New York City and (b) Miami, cities representative of two of Olgyay's North American regional climate zones. The shaded regions represent overheated zones; the isolines outside of the shaded areas represent solar radiation intensity needed to remain comfortable outdoors without wind. (Olgyay, 1963 / with permission of Princeton University Press.) Hourly and monthly dry bulb temperature representations of the same cities (c) New York City and (d) Miami using Climate Consultant. The visual patterns and details help the user to characterize the climate more readily than with data tables. (Milne / with permission of Regents of the University of California.)*

compatible file formats (e.g., EPP, CLM, WEA, PVSyst, DDY, RAIN, and STAT), climate normals can also be utilized in building energy modeling (BEM) to assist with high-performance design analysis.

Climate, however, is always changing. There is really no such thing as "normal." As such, the WMO recommends updating climate normals every decade to reflect more recent climate characteristics. The theory is couched in the notion that, given the natural fluctuations in the climate, a 30-year average is sufficient to determine a reasonable baseline condition. This is debatable given the rate at which the climate appears to be shifting (Figure 3.3). According to the National Oceanic and Atmospheric Administration (NOAA), the Earth's temperature has risen by 0.14 F° [0.08 C°] per decade since 1880, but the rate of warming since 1981 is more than twice that at 0.32 F° [0.18 C°] per decade.

Despite such evidence of changing climate, in much of the world (including the United States), it is still common for building design teams to reference historical (i.e., past) climate data for energy analyses. Often, designers will utilize Typical Meteorological Year (TMY) datasets for building energy modeling exercises. Produced by NREL, each iteration of TMY datasets (i.e., TMY, TMY2, and TMY3) references historic weather data from specific timeframes (the more recent TMY3 files are built from weather data collected between 1995 and 2005) and, because of continued changes to the climate, are potentially inappropriate to help building project teams design for the

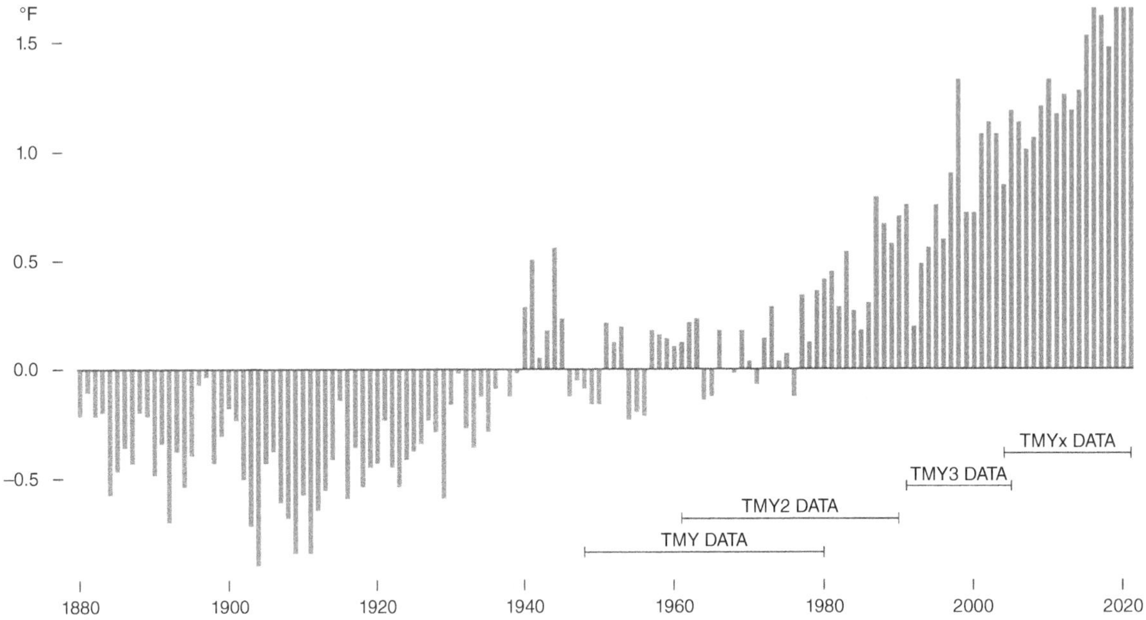

Figure 3.3 *Global average temperature compared to the middle of the twentieth century and TMY dataset year ranges. Data depict July global land and ocean temperature deviations. Note that years covered by TMYx datasets can vary. Climate data from NOAA National Centers for Environmental information, Climate at a Glance: Global Time Series, August 2022. (From NOAA, 2025b. Drawing © Daniel Overbey; all rights reserved.)*

future. Rhetorically, why would designers use the past 30 years of climate data to design for the upcoming 60 years of a building's life-cycle?

While TMY3 is currently the default within most building energy modeling platforms, climate.onebuilding.org offers TMYx files (the "x" distinguishes this file type from the official NREL iterations). While some TMY datasets may be decades old, the most recent TMYx datasets use meteorological data from 2007 through 2021.

Increasingly, there is advocacy for utilizing weather files that anticipate how the climate may change in future. For years, Weathershift, a service provided by Arup and Argos Analytics, has offered proprietary fTMY weather files (the "f" references "future") for purchase. Now, the U.S. Department of Energy (DOE) is providing publicly accessible fTMY files produced by the Oak Ridge National Laboratory (ORNL) through a rigorous and transparent development process that uses a combination of dynamic and statistical downscaling. The fTMY files from ORNL are downscaled from global climate models to specific geographical locations in a manner that takes localized, topographic-dependent weather variability into account.

Given the uncertainty of climate projection data and the outdated timeframes utilized by NREL's TMY weather files, professionals may reasonably contend that using multiple fTMY files for energy modeling exercises can better futureproof our buildings. Running multiple scenarios may be essential because the future is, of course, uncertain. In its Fifth Assessment Report (AR5), the Intergovernmental Panel on Climate Change (IPCC) defined several future "pathways" regarding global greenhouse gas emission projections based on the degree to which the global economy may (or may not) decarbonize. Called Representative Concentration Pathways (RCPs), these projections focus primarily on GHG concentrations and their direct impact on climate models (IPCC, 2014). However, the RCPs did not explicitly consider the social, economic, and technological pathways that lead to these emissions scenarios—which, by extension, limited the IPCC's ability to explore different mitigation or adaptation strategies. In the more recent Sixth Assessment Report (AR6), the IPCC offered a more comprehensive framework that incorporates socioeconomic factors called Shared Socioeconomic Pathways (SSPs). SSPs are not a replacement for RCPs but are used alongside them. SSPs allow researchers to establish a more nuanced and integrated understanding of the relationships between socioeconomic factors and climate outcomes (IPCC, 2021).

By running building simulations using fTMY files under different SSP/RCP scenarios, project teams can better understand and assess the potential vulnerabilities with their building design and/or systems configurations. It is important to note that using weather files derived from downscaled climate projection data is not yet standard practice and it may take time for model codes and standards to fully adopt such practices—let alone the time it will take for jurisdictional uptake.

See Appendix B for selected historic climatic conditions for the United States, Canada, and Mexico.

3.3 CLIMATES WITHIN CLIMATES

The climate at a particular site can be quite different from data indicated by the nearest available climate data file location. This is particularly evident when visiting a building site where a neighboring hill blocks wind or winter sun, or an adjacent lake cools summer breezes or adds a damp chill to the winter air. Such local variations constitute microclimates, which feature one or more characteristic distinctly different from the conditions prevailing in the larger macroclimate. The characteristics of a microclimate are influenced by the interaction of the local site conditions with the macroclimate. Table 3.1 lists representative site and climate design variables.

Most urban sites possess a subclimate that differs from the conditions of the surrounding suburban or rural countryside. Probably the best-known urban climate feature is the heat island. Urban heat island effects are summarized in Table 3.2 and Figure 3.4. Designers should note that city climatological stations are often located at nonurban sites, such as an outlying airport, masking heat island effects.

The primary reason for a city's relative year-around warmth is its concentration of heat sources: air conditioner condensers, furnace flues, electric lighting, and internal combustion engines in vehicles. This urban heat production is shown in Table 3.3 and Figure 3.5. It appears that cities and industrial regions of the world release less internal heat per capita as people live and work closer together—although the heat density (temperature) is greater. Commercial and industrial cities release more heat for a given population density, and tropical cities release less, while still conforming to the pattern that greater compactness permits less energy use per capita.

Rain that falls on a city can be an effective evaporative cooling mechanism, as water evaporates from wet surfaces. Yet streets and buildings are usually designed to shed water quickly and thoroughly, sending rainwater into storm sewers instead of letting it evaporate slowly in the wind and sun. As a result, evaporative cooling from these surfaces is minimized.

TABLE 3.1 Examples of Site and Climate Variables

Site Characteristics	Climate Characteristics
Soil type	Solar radiation
Ground surface	Air temperature
Topography	Humidity
Vegetation	Rainfall
Water bodies/flows	Air speed and direction
Views	Air quality
Human effects (heat, noise, etc.)	Snowfall

TABLE 3.2 Average Changes in Climate Caused by Urbanization[a]

Effect	Comparison with Rural Environment
Contaminants	
Condensation nuclei and particulates	10 times more
Gaseous admixtures	5–25 times more
Cloudiness	
Cloud cover	5–10% more
Fog, winter	100% more
Fog, summer	30% more
Precipitation[b]	
Totals	5–10% more
Days with less than 2 in. [5 mm]	10% more
Snowfall	5% less
Relative humidity	
Winter	2% less
Summer	8% less
Solar radiation	
Global	15–20% less
Ultraviolet, winter	30% less
Ultraviolet, summer	5% less
Sunshine duration	5–15% less
Temperature	
Annual mean	0.9–1.8°F [0.5–1.0°C] higher
Winter minima (average)	1.8–2.6°F [1–2°C] higher
Heating degree days	10% fewer
Wind speed	
Annual mean	20–30% lower
Extreme gusts	10–20% lower
Calms	5–20% more

Source: From Landsberg, 1970.

[a]These effects vary from city to city and from day to day.

[b]Research since 1970 has shown that it is not at all certain that urbanization causes increases in precipitation within a city.

A city also changes the overall cooling action of the wind by channeling it into narrow streets. A geometry of high vertical walls and narrow streets also increases summer heat collection in cities as the high sun is reflected downward to be absorbed and then reradiated by the often rock-like streets and building surfaces. In winter, however, this geometry puts most urban surfaces at a solar disadvantage because the low sun strikes only the upper portion of south-facing walls. A reduction in radiant heat loss at night, caused by this lack of access to the sky, is a key element in the formation of the urban heat island, as summarized in Figure 3.6. Sky access is discussed in more detail in Section 3.7.

A subtler urban climate influence is contaminated air. Small particles in the air can keep some sunlight from reaching the city yet can also help to keep the city's heat from radiating outward. These particles also form additional nuclei for fog droplets. Table 3.2 suggests that fog may occur in a city in winter up to twice as often as in the surrounding countryside. Trees and greenery, which can act as crude filters of airborne dust, are not as available in a city.

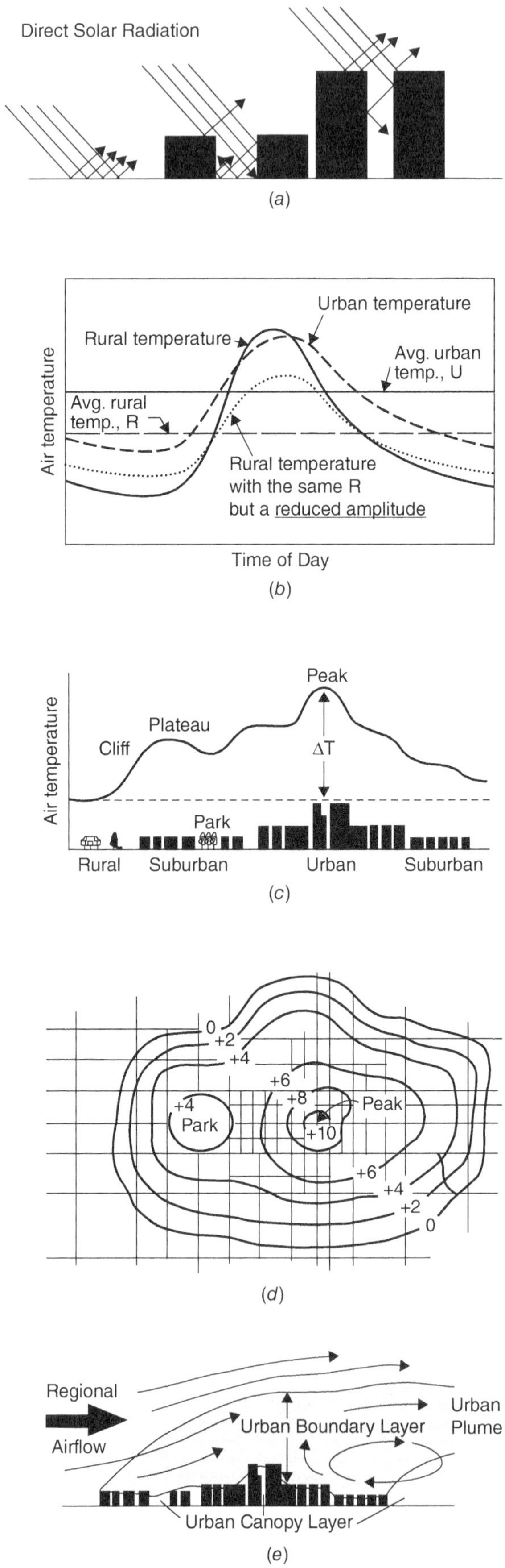

Figure 3.4 *Urban heat island: a densely occupied area with a temperature distinctly higher than that of the surrounding rural area. (a) Direct solar radiation is likely to be reflected within the city, thereby increasing solar heat gain in urban areas. (b) Temperature records at a rural site (solid line) and in the center of a city (dashed line) during a typical night and day. The city's heat-conducting materials and thin cloud of polluted air acting alone would not change the average air temperature but would reduce the day–night difference (the dotted line). In addition, the heat from increased solar gain and city-specific heat sources (cars, buildings) warms the air at all hours, producing the observed urban record (dashed line). (c) Idealized profile of the air temperature difference between urban and rural areas at times of peak differences—calm, clear nights. (d) Based upon (c), typical isotherms (lines of equal temperature) provide a "contour map" of the urban heat island. (e) An urban heat island can affect the "downstream" countryside. (Lowry, 1988 / with permission of Van Nostrand Reinhold.)*

The city thus modifies its climate from that of its surroundings. In winter, the city's factories, vehicles, surface materials, and geometry combine to increase temperature and reduce the amount of energy needed for heating buildings. The typical means of providing needed winter heating (fossil fuel-burning heating equipment and power plants) contribute airborne particles and cause urban fog. Solar-assisted heating would diminish this pollution—and with less air pollution, more sun would reach a solar collector. The greater the population density, however, the more difficult it is to access winter sun, especially at latitudes farther from the equator, where winter heating is most needed. High buildings readily block low-altitude sun.

In summer, a city's internal heat makes things worse. Air conditioners add their own process heat to the building heat that they pour into the air; their electric load can strain power plants, in turn adding more waste heat. Again, solar energy can help, this time by using photovoltaic (PV) modules to generate electricity to run the air conditioners. High summer sun angles make solar access reasonably easy to obtain. The stronger the solar resource, the more electricity will be produced. PV arrays are currently high in first cost but are noiseless and emit no exhaust gases.

While considering how sun, wind, and other climate elements can be utilized on a site for the benefit of a building's occupants, it is important to remember the need to protect the access of others to these shared resources. In "The Tragedy of the Commons" (1968), Garrett Hardin writes about the example of publicly owned meadows shared by many herders. Each herder realizes that his/her personal wealth will increase as animals are added to his/her herd, so all herders increase their livestock holdings. The carrying capacity of the meadow, however, does not increase; overgrazing occurs, and, as a result, the commons become unable to support any animals at all. The following discussions of sun, air, and water resources available on a site are influenced both by the "private" needs of a building

TABLE 3.3 Heat Generated within Cities

City or Region[a]		Population (millions)	Area mile² [km²]	Population Density 10³/ mile² [10³/km²]	Energy Use per Capita[b] (kW/person)	Energy Use per Unit Area[b] (W/m²)
			Cities and Industrialized Regions			
3	Fairbanks[c]	0.045	30.8 [80]	1 6 [0.6]	10.91	6.55
11	Vancouver	0.6	42.8 [111]	14.0 [5.4]	3.55	19.2
12	Brussels	1.3	62.9 [163]	20.7 [8.0]	3.5	28.0
14	West Berlin	2.3	90.7 [235]	25.4 [9.8]	2.14	21.0
9	St. Louis	0.75	96.5 [250]	7.8 [3.0]	5.3	15.9
10	Munich	0.9	116 [300]	7.8 [3.0]	3.0	9.0
5	New Jersey suburbs of NYC	4.7	2355 [6100]	2.1 [0.8]	9.1	7.3
1	Los Angeles County	7.0	3860 [10,000]	0.2 [0.07]	10.5	0.74
4	Nordrhein–Westfalen[d]	16.9	13,045 [33,800]	1.3 [0.5]	8.0	4.0
2	BosNyWash[e]	33.0	33,582 [87,000]	1.0 [0.38]	11.2	4.3
			Commercial/Industrial Cities			
15	Sheffield	0.5	18.5 [48]	26.9 [10.4]	1.83	19.0
17	Manhattan	1.7	22.8 [59]	74.6 [28.8]	5.52	159.0
16	Montreal	1.1	30.1 [78]	36.5 [14.1]	7.02	99.0
13	Budapest	1.3	43.6 [113]	29.8 [11.5]	3.74	43.0
8	Cincinnati (summer)	0.6	85.7 [222]	7.0 [2.7]	9.3	25.1
7	Hamburg	1.83	295 [763]	6.2 [2.4]	5.3	12.7
6	Chicago	3.5	711 [1842]	4.9 [1.9]	27.2	51.7
			Tropical			
18	Hong Kong	3.9	40.5 [105]	96.1 [37.1]	0.88	32.6
19	Singapore	2.1	54.0 [140]	38.9 [15.0]	0.81	12.2

Source: Lowry and Lowry, 1995.

[a]The numbers that precede city names refer to Figure 3.5.

[b]0.317 W/m² = 1 Btu/ft². These columns were not converted to I-P units. The units shown are as presented in the source, but technically energy should be expressed in kW-h while power is in W.

[c]Data for Fairbanks, Alaska, include a sparsely populated incorporated area surrounding the central business district.

[d]Nordrhein–Westfalen is the heavily urbanized and industrialized region in the lower Rhine valley near Düsseldorf–Dortmund.

[e]BosNyWash is a term often used by urbanists to refer to the megalopolis that stretches from Boston through New York to Washington, DC.

and the "public" needs for resource availability, which should remain accessible to all.

3.4 BUILDINGS AND SITES

Buildings are temporary occupants of their sites. The construction of a building usually produces rapid and dramatic changes to the biological systems on a site, to the microclimate, and often to the surface geology. Buildings are guests, sites are hosts; a fundamental design question is: How can the two most productively coexist?

A site offers a building earth for support, as a potential heat source and heat sink, and for the growth of plants where building density permits. Sounds on a site depend upon the context, urban or rural. Water is somewhere below a site, flows across it, falls on it as rain, and perhaps collects on its surface. Wind moves erratically across a site. Solar energy arrives in diurnal and seasonal cycles. Long-term patterns of sun, wind, and water are steady and generally predictable, although great variations can occur over shorter time spans (daily, monthly, or annually). At some level, an ecosystem of life is already established. See Chapter 1 for Malcolm Wells's take on this web of life.

During the construction process, a building evolves, bringing with it people and vehicles, a flow of materials into and out of the site, sounds of activity, and imported utility services such as electricity, water, and natural gas. The building offers the site electric light by night, a continuous flow of heat (to or from), a radical change in water flows (both at and outside of the building enclosure), and liquid and solid wastes (or nutrients, depending upon one's viewpoint). In our society, the waste outflows are most often whisked off for treatment "elsewhere."

When a site is larger than its building, both may be designed to improve building performance and user experiences. Large sites offer smaller buildings the opportunity to accept, filter, or block sun, wind, sound, and rainwater in order to make effective use of on-site resources. Vegetation, ground forms, and orientation, as well as roofs, walls, and floors, can play a role in this interaction.

When a building fully fills its site, opportunities for on-site resource use are more limited. Because buildings that completely fill sites are often located in densely built-up areas, less usable wind is available around the building walls, and solar radiation may be blocked. The built-upon earth is less able to act as a heat sink or absorb water. The building roof often

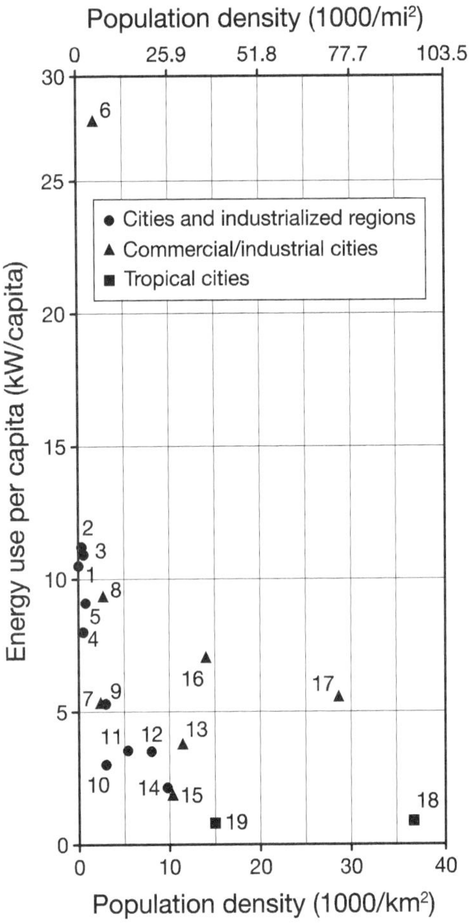

Figure 3.5 Population density and energy use per capita for 19 cities and regions. Numbers refer to locations in Table 3.3. The heat island effect is influenced by both density and energy use. (Adapted from Lowry and Lowry, 1995. Redrawn by Alisa Kwok.)

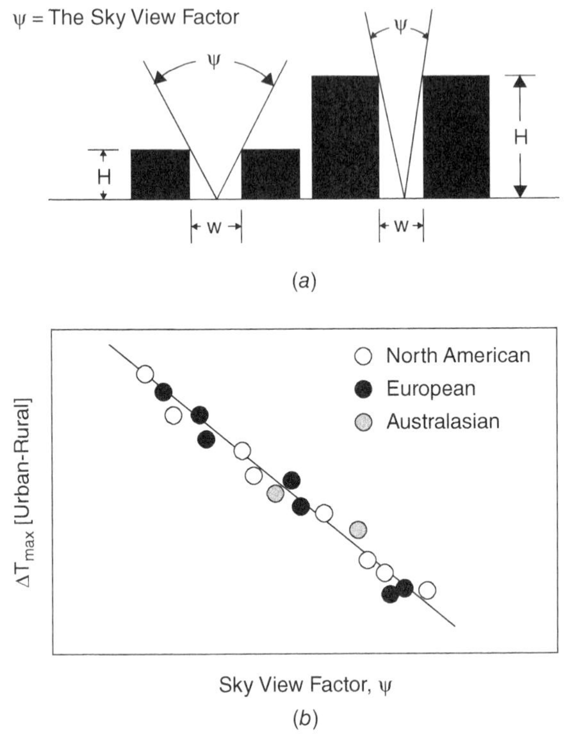

Figure 3.6 The urban heat island effect is particularly strong on calm, clear nights. (a) With a greatly reduced "sky view factor" (Ψ) to the cold night sky, the walls and floors of urban canyons (the right part of the sketch) cannot lose heat as readily as can the open countryside or less dense suburban areas (the left part of the sketch). (b) The narrower the Ψ, the more pronounced is the effect (ΔT) of the urban heat island in cities throughout the world. (From Lowry, 1988.)

provides the major opportunity to receive sun and rain, and discharge heat to the environment. It may offer the only opportunity to grow plants on the site. In dense urban areas, the future roofscape will be dominated by climate–building interactions: solar collection, rain collection, heat rejection, water treatment, and gardens. Compared to today's urban roofscape dominated by horizontal black or gray surfaces behind parapet walls, this represents a dramatic design opportunity for aesthetic, social, and technical change. Such a change is already under way in some cities (such as Chicago and Portland, Oregon, with their green roof initiatives).

3.5 ANALYZING THE SITE

After identifying the resources that exist on and around a site, a designer can decide how best to integrate these resources into the building design, while making the building and site a successful addition to the larger patterns of their surroundings. Schematic site plans should be used as an inventory of interactions. Diagrams can test design arrangements that relate rooms or functions to their surroundings in plan view. Sun and wind conditions (in both summer and winter), noise sources, and water runoff patterns are often included in a site analysis plan. It is particularly important to identify microclimates on the site, the places that have special characteristics differing from the regional climate—from what one may refer to as the macroclimate (e.g., long-term climate data recorded at a weather station). Microclimates can present opportunities for an expanded comfort zone: more sun in cool periods, more wind in warm periods. Microclimates can sometimes provide building sites where less energy is consumed because the winter is warmer or the summer cooler. Microclimates may also present problem areas to be avoided for buildings or outdoor activity, or where special design measures need to be taken to correct their difficulties.

Microclimates on a site are not limited to those visible on the surface (and presented in plan view). Consequently, both vertical and horizontal site analyses are needed. Conditions of privacy and accessibility, view, heat, light, air motion, sound, and water all change with height above (or below) the surface (Figure 3.7*a*). To minimize the energy required to

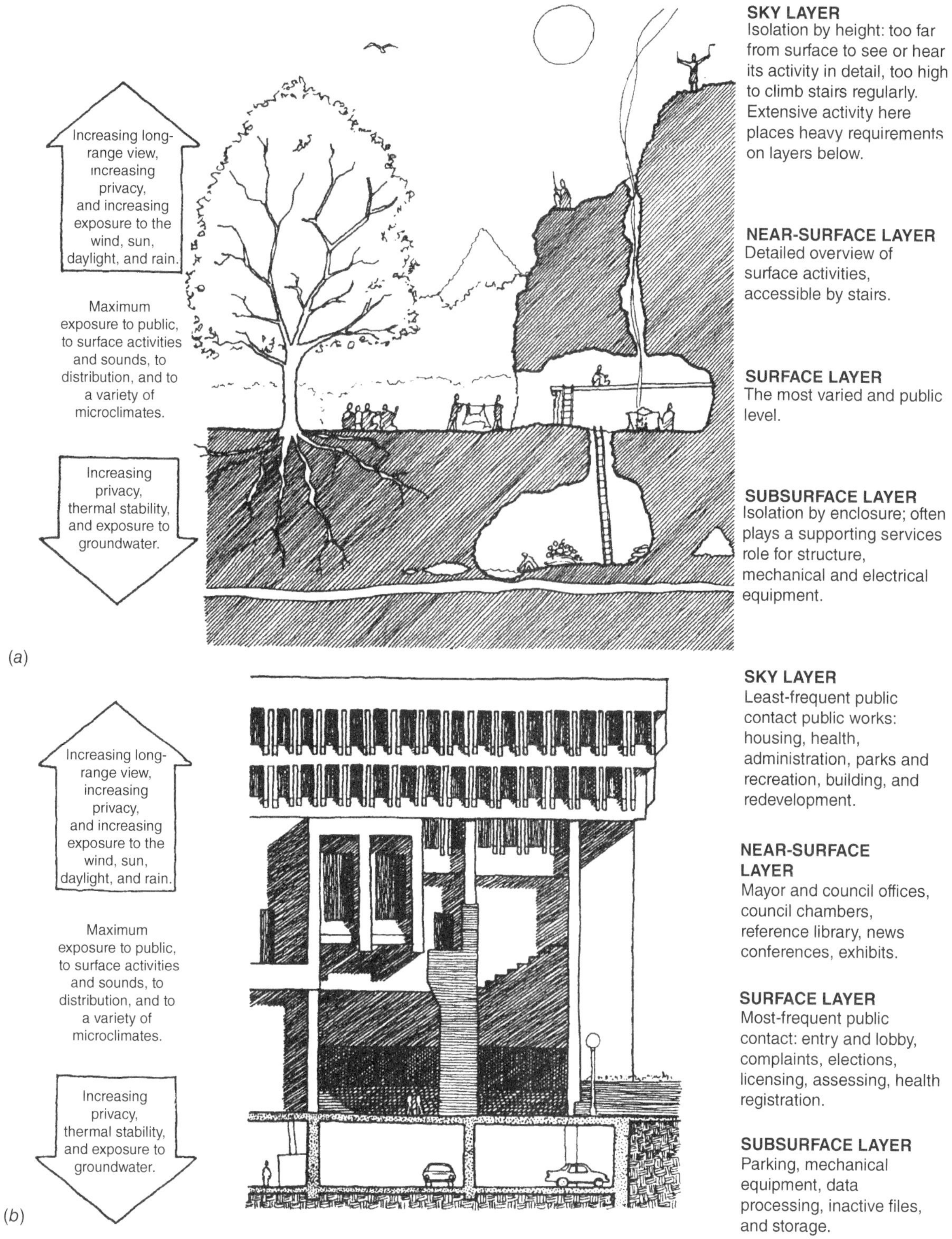

Figure 3.7 *(a) Characteristics of vertical layers of a site. (b) Vertical layers and form: Boston City Hall, 1969. (Kallman, McKinnell and Knowles, Architects.)*

construct and operate buildings and to better integrate buildings with their surroundings, programmatic functions should integrate with the characteristics of the most appropriate layer of the site. A lecture hall, requiring an acoustically isolated and closely controlled thermal environment, is a good candidate for the subsurface layer. Spaces that will most benefit from views and daylight should occupy levels that best provide such resources.

One building whose form responds to these layers is Boston City Hall (Figure 3.7*b*). Activities with the most frequent public interaction are located near the surface, such as the skylit, high-ceilinged lobby, whereas special ceremonial functions are elevated to distinctive forms in the near-surface layer above. City offices with less frequent public contact occupy several floors in the sky layer, where daylight is plentiful. Storage and mechanical functions, as well as parking, are in the subsurface layer. An aesthetic equivalent of this vertical layering is architect Louis Sullivan's concept of a façade as "base, body, and capital."

3.6 SITE DESIGN STRATEGIES

Sites can be utilized to assist in heating, lighting, cooling, and noise control for buildings. A range of site strategies were compared with their seasonal roles and effects by Watson and Labs (1983). An extract from this extensive comparison is shown in Figure 3.8. Several of these strategies are discussed in more detail later in this chapter. The graphics in this figure suggest a focus upon housing but also apply to any occupant situation that can utilize an extended comfort zone.

A house designed by Frank Lloyd Wright in the 1940s (Figure 3.9) is a great example of how the information in Figures 3.7 and 3.8 can be skillfully applied in the planning of a building that works with its site. The direct solar heat gain through south-facing windows in winter makes this an early example of passive solar heating. (Active solar heating, by contrast, includes collector devices and a tank of water or a bed of rocks for heat storage.) This house, known as the Solar Hemicycle, was built in 1948 near Madison, Wisconsin, which lies between the cool and temperate climate zones. Winter heating is the dominant thermal influence in this area.

The house stands on a rise in the prairie, particularly vulnerable to winter winds. During construction, earth was scooped from in front of the south face of the house and bermed against the entire curved north wall, almost to the roof level. Only a narrow strip of second-floor windows separates the berm from the roof. This berm and curve combination gives winter wind protection from the northeast to the northwest and provides further insulation for the north wall. The north wall is made of stone, which absorbs and stores winter solar radiation (heat) that comes in through the floor-to-ceiling, southeast-to-southwest-facing glass. The concrete slab-on-grade first floor also stores solar heat in winter. The impression that this house and site are sun collectors is heightened by an entrance tunnel in the northeast end of the house that leads from the parking area through the berm and onto the

	Winter				Summer				
	Minimize Conductive Heat Flow	Minimize External Heat Flow	Promote Solar Gain	Minimize Infiltration	Delay Periodic Heat Flow	Promote Ventilation	Minimize Solar Gain	Promote Radiant Cooling	Promote Evaporative Cooling
Windbreaks		●		●					
Plants and water							●		●
Indoor/outdoor rooms			●			●			
Earth sheltering	●	●		●	●		●		●
Solar walls and windows			●						
Thermal envelope	●			●	●			●	
Sun shading							●	●	
Natural ventilation						●			

Figure 3.8 *Generic bioclimatic site design concepts and building strategies. (From ASES, 1981.)*

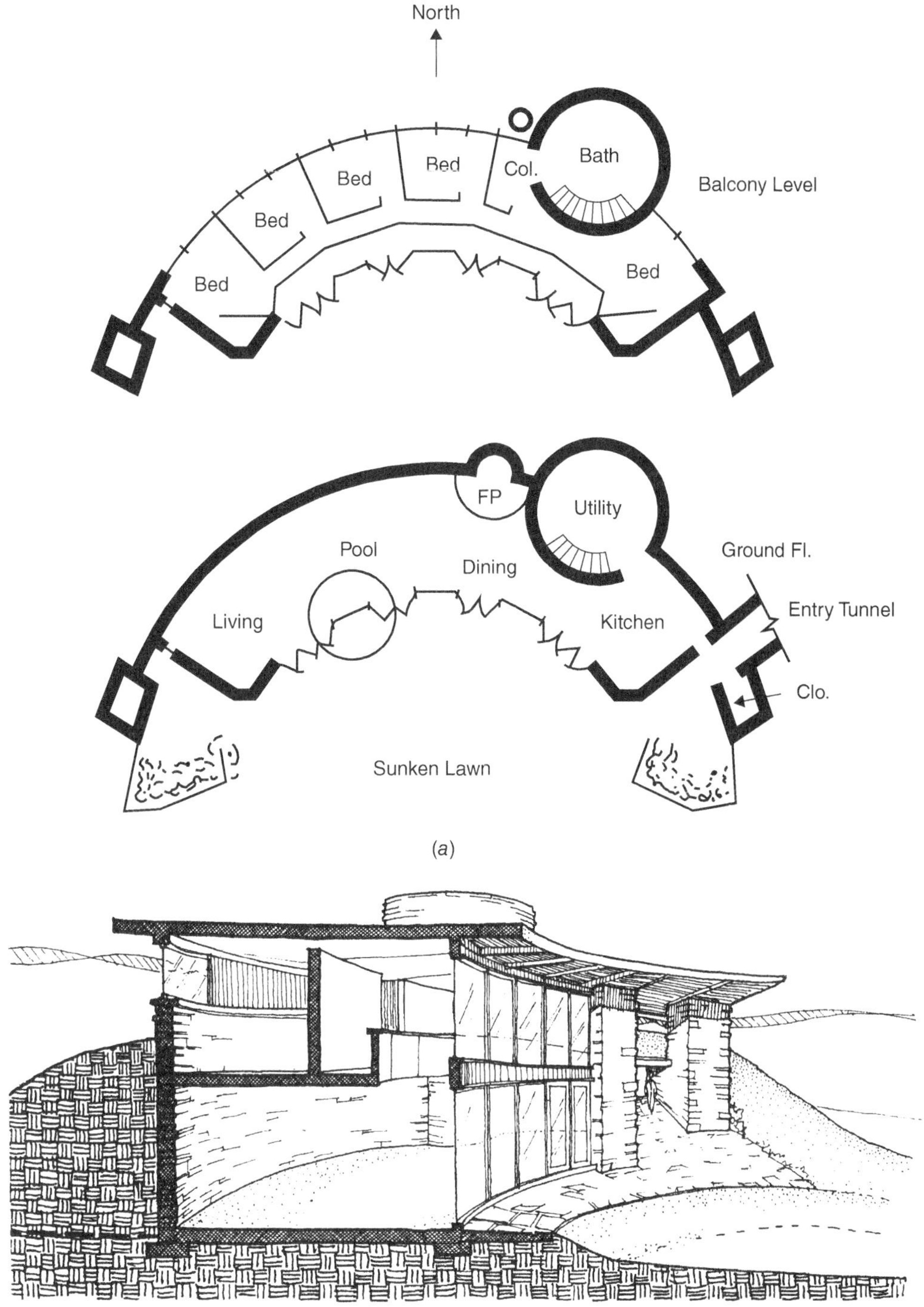

Figure 3.9 *An early passive solar-heated home, Frank Lloyd Wright's Solar Hemicycle (Jacobs House II) near Madison, Wisconsin. The house was designed in the early 1940s and built in 1948. (a) Floor plans. (b) Section-perspective, looking east toward the entry tunnel in the berm wall.*

sunny, protected south terrace overlooking a sunken "sun-trap" circular lawn. Passage from the cold north side through an even colder tunnel to the sunny south side transitions a visitor into the solar-heated interior.

The Solar Hemicycle house plan is shorter in the north-south and longer in the east-west direction (about a 1:3 ratio). This elongation is typical of passive solar designs, which can store and use winter solar heat gain, and thus profit from having long south-facing glass walls to act as passive collectors. Had these large windows been well insulated at night and the roof and walls insulated to current standards, this house would be a very up-to-date case study of passive solar heating.

In the Solar Hemicycle house, protection from summer overheating is provided by an overhang along the south glass walls, as well as by the shaded and cool thermal mass of the north stone/berm wall combination. The high windows in both the north and south walls allow warm air to rise and escape as required for temperature control.

3.7 DIRECT SUN AND DAYLIGHT

The Earth receives only a very small percentage of the sun's daily energy output. This small portion, however, is critically important to life on Earth. Almost all of today's energy sources have the sun as an ancestor. Fossil fuels are solar energy that has been concentrated by time and geologic conversion. Biomass, hydropower, and wind power represent shorter-term concentrations of solar energy. Even without concentration, the renewable solar resource is often more than adequate for heating and lighting a building.

The amount of solar energy available at any site varies both seasonally and daily. Typically, the nearer the sun approaches a position directly above a site, the more solar energy reaches a horizontal surface. Fortunately, solar radiation's embodied heat energy is not its only resource; direct sun is our most intense light source—too intense, in fact, for most building situations. Indirect sunlight, however, as available on an overcast day or from the clear north sky, is a wonderfully diffuse and readily utilized light source.

(a) Access to Light and Sun

The value of daylight (and fresh air) to buildings has long been recognized in zoning laws, which typically require that minimum distances (setbacks) be maintained between a building and the property line in lower-density areas. Height restrictions often accompany these setbacks, defining the maximum buildable volume that a building can fill (Figure 3.10). As space density increases and buildings become taller, daylight is diminished as it trickles down between buildings; in response, the maximum buildable volume becomes narrower as a building rises.

When direct sun in winter is desirable at ground level on all sites, the buildable volume must be sharply reduced in height due to the low angle of the winter sun in northern latitudes. Protection of this most-restricted buildable envelope, called the solar envelope, is at present rarely mandated, but a few cities have enacted solar access ordinances for residential zones. The most restrictive feature of the solar envelope is the low slope of its northern face, usually corresponding to the altitude of the sun (angle above the horizontal) for about two hours before and after noon on December 21. This feature allows four hours of access to direct sun for a neighbor on the site just to the north, on even the shortest day.

A brief history of daylight and solar access regulations and an overview of current design guidelines are given in DeKay (1992). Figure 3.11 shows the application of six variations of regulations on an east-west elongated urban block at 40°N latitude. The east-west streets are wider than the north-south streets. Further development of the solar envelope concept is shown in Figures 3.12 and 3.13, based upon the work of Knowles (1981). As suggested, maximum density of development is closely related to maintaining winter solar access.

For daylight access, building surfaces can be almost as important as building geometry. The importance of light reflected from vertical and horizontal

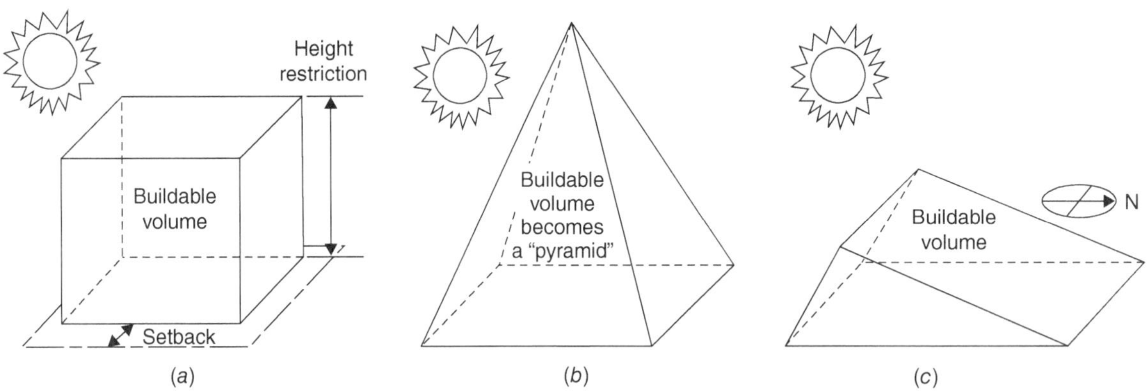

Figure 3.10 Protecting access to light and solar radiation. Three regulatory approaches that compromise between private optima (e.g., maximum rentable floor space) and public optima (e.g., daylight at street level). (a) Simple daylight access, residential and low-rise commercial areas. (b) Daylight access in high-density areas. (c) Access to direct sun for winter heating.

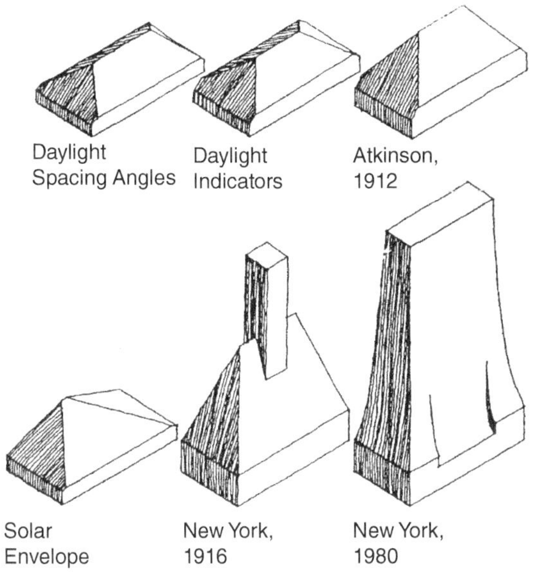

Figure 3.11 *Several approaches to defining maximum allowable building envelopes for daylight access. These envelopes are applied to a 200-ft × 400-ft [61-m × 122-m] block at 40°N latitude. The east-west streets (along the longer side) are 65 ft [20 m] wide; the north-south streets are 45 ft [14 m] wide. In this case, "daylight spacing angles" and "daylight indicators" produce nearly identical envelopes. (From DeKay, 1992.)*

surfaces is apparent in daylighting analyses. Lighter-colored surfaces result in more daylight, especially in crowded urban conditions, where a view of the sky from windows is not common. An increase in daylight for all surfaces on an urban street will result if the building and site surfaces have light colors. This can conflict with the use of planting for shading and evaporative cooling, although light-hued groundcovers are available. Although daylight diffusely reflected from building and ground surfaces is a potential benefit, harsh specular reflections of direct solar radiation from mirror-like surfaces are often unwelcome. See Section 3.7(f).

(b) Charting the Sun

During site analysis, one of the earliest tasks is to determine when direct solar radiation would reach a proposed space on a site—such as a classroom, playground, office, deck, or courtyard. Information on solar geometry is presented in Chapter 11. Numerical

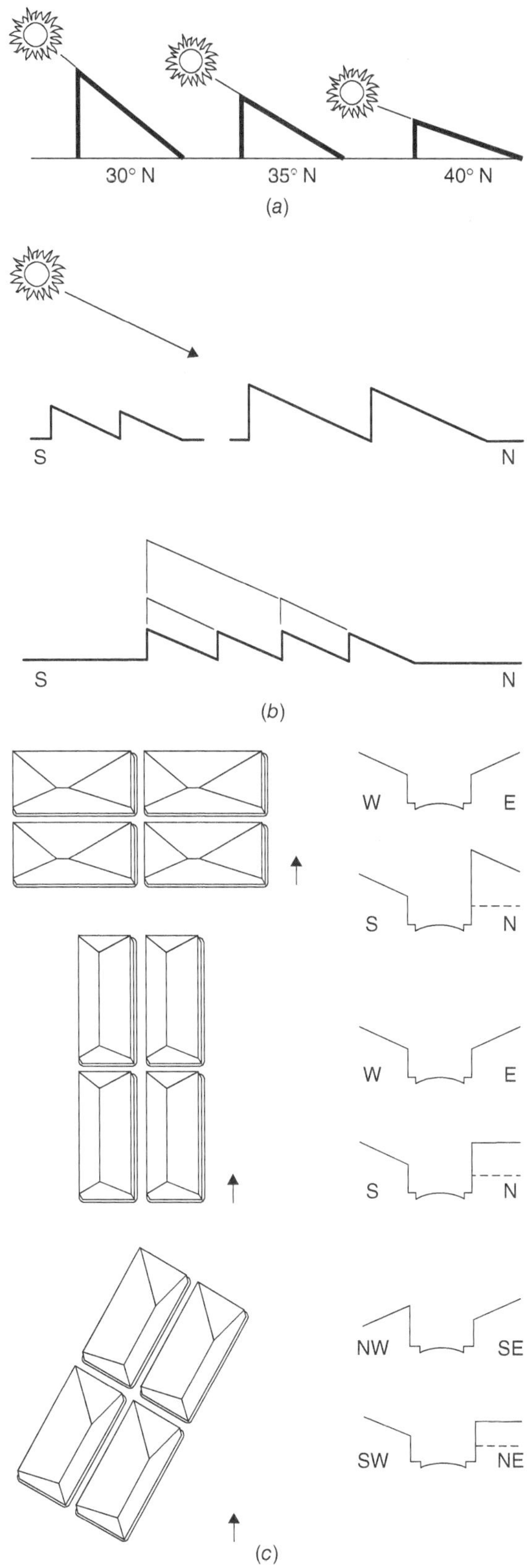

Figure 3.12 *These solar envelopes are refinements of the solar access "pyramid" of Figure 3.10. (a) The slope of the solar envelope changes with latitude. (b) The larger the site, the greater the buildable volume of the solar envelope. (c) Solar envelopes for various individual site orientations. (Knowles, 1981 / with permission of MIT Press.)*

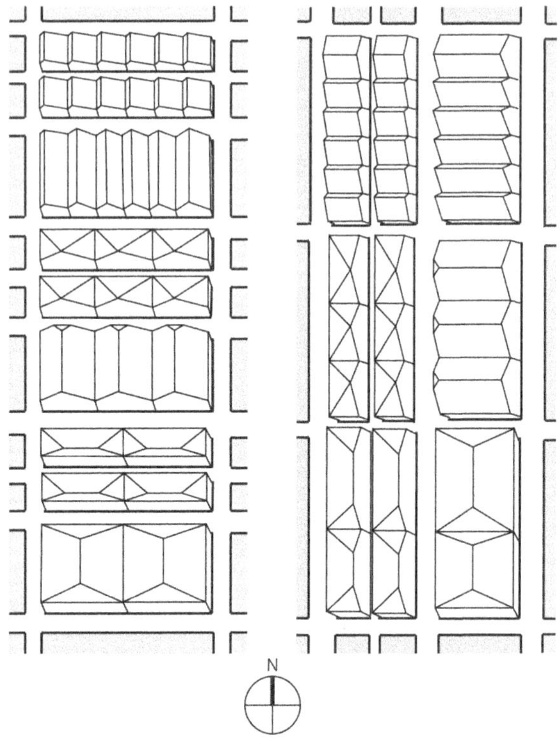

Figure 3.13 Solar envelopes for east-west elongated blocks (left) and for north-south elongated blocks (right). (From Knowles, 1981. Redrawn by Nathan Majeski.)

data on sun position and intensity are found in Appendices C and D. Details of how a building can utilize the sun are found in later chapters dealing with solar space heating, solar water heating, and daylighting. Effective design for solar utilization, however, begins with an understanding of the basics.

The chart in Figure 3.14 shows the sun's position for 40°N latitude and also shows, for the six hours of greatest insolation, what percentage of clear-day insolation is gained each hour by an unshaded south-facing vertical window. To convert these percentages to actual heat gains, see the numerous tables in Appendix C with data for clear days at a given latitude and Table C.19 (showing January and July only) for an average day by climate station. The emphasis on a southern orientation (in the Northern hemisphere) calls attention to the useful fact that such an orientation receives more sun in winter and less sun in summer than any other orientation.

(c) The "Band of Sun"

A building section drawing can be a powerful site analysis tool. As shown in Figure 3.15, a north-south section can be drawn through a proposed building and its site. For three days of the year (each solstice and the equinox), the solar altitude angle at noon is determined. This allows the band of sun that would strike the building and site at the noon altitude to be drawn. The portion of this direct solar radiation resource that is used by the building and site can be easily visualized. The portion of the band that is unused in each season illustrates the potential for redesign to better utilize the solar resource. This analysis helps to show how wintertime solar utilization depends upon lower sun altitudes, and summertime uses depend upon higher altitude angles.

(d) Skylines and Winter Sun

The skyline as seen from a given position on a site can be charted to determine access to direct solar radiation at any time of the year (Figure 3.16). This type of obstruction analysis should precede the placing of any solar

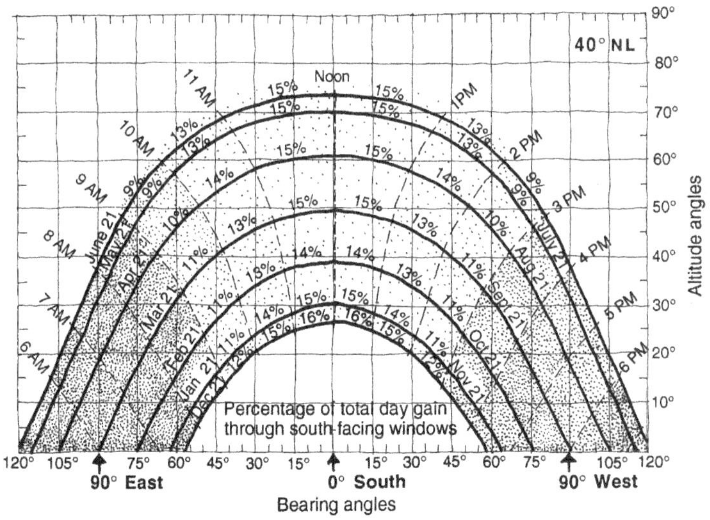

Figure 3.14 Sun chart for 40°N latitude shows the approximate percentage of clear-day insolation for south-facing windows for each of the six maximum hours of sun each month. (From Mazria and Winitsky, 1976.)

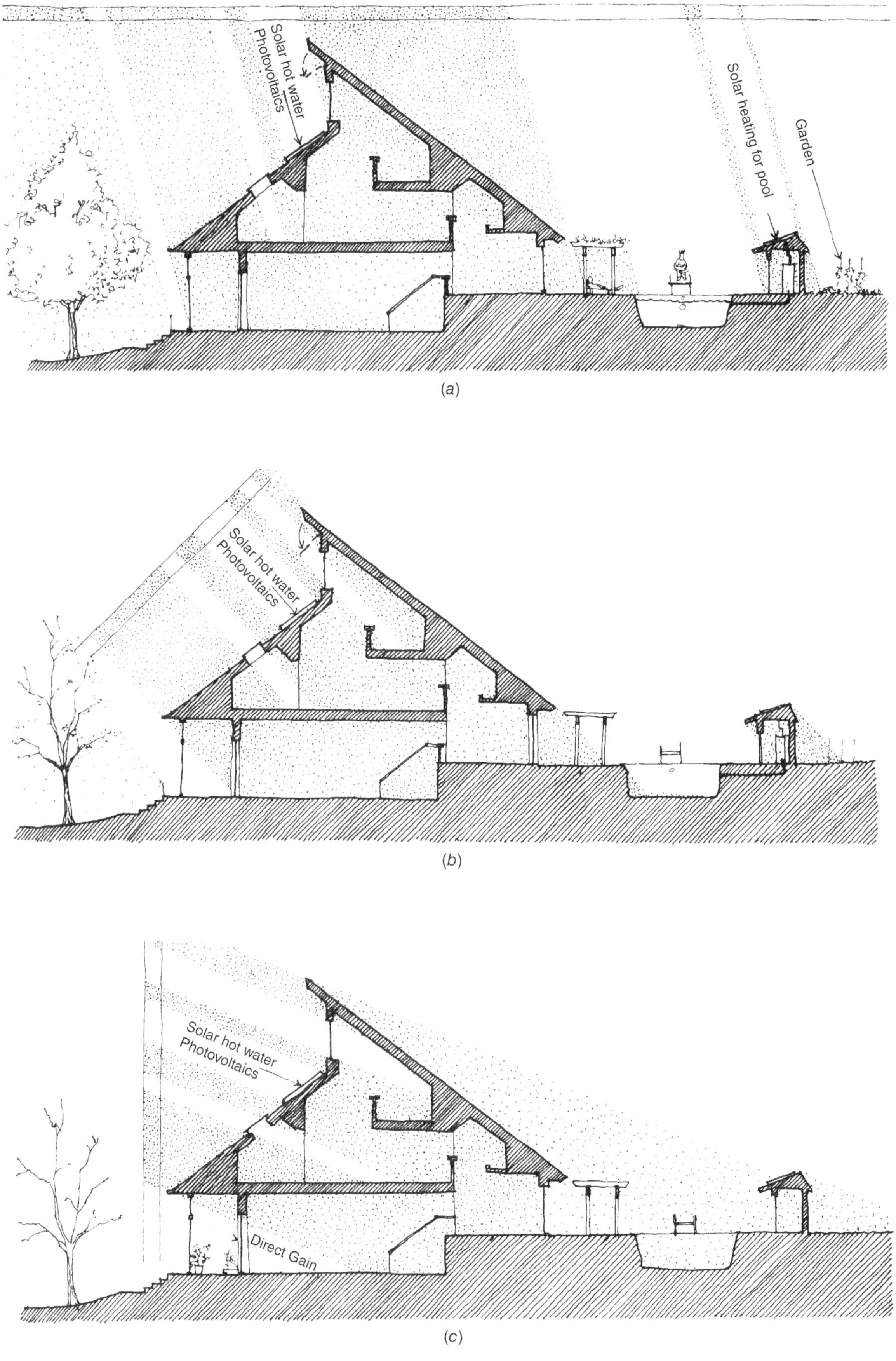

Figure 3.15 *The band of sun available to a proposed building at solar noon is charted on a north-south section. (a) The summer solstice, where optimum collecting surfaces are at near-horizontal tilt angles. (b) The equinox. (c) The winter solstice, where optimum collecting surfaces are at near-vertical south-facing tilt angles.*

collector, whether it is a passive south-facing window or an active manufactured collector (PV or solar thermal). Obstructions that occur within the six "best" collection hours (9:00 a.m. to 3:00 p.m.) are particularly serious and should be minimized when siting a collector.

Conversely, consideration of neighbors' access to direct solar radiation is appropriate. This can be checked by charting the skyline somewhere along the northern boundary of the site. The outline of the proposed building design should be included on this skyline. If the building form being considered casts shade onto the adjacent site, it can perhaps be modified as necessary to preserve solar access for the neighbor.

The location evaluated in Figure 3.16 is a good choice for summer solar usage, such as a swimming pool heater or a solar collector for domestic water heating, which would work well during the six months from March through September; obstructions within this period are few. It is a quite poor location for winter solar collection, however, allowing only about one-fifth of the potential solar gain in the November to January period.

(e) Sun and Shadows: Model Techniques

The graphic site analysis techniques just described have a limitation: Each graph applies only to one particular location on a site. To study multiple locations, multiple graphs must be constructed. In contrast, a three-dimensional model used in conjunction with a sun peg chart (Figure 3.17) can provide three-dimensional patterns of sun penetration as they change over time, for any position, indoors or out, in/on the model. Physical models are initially a bit time-consuming to build but can reduce analysis time when considering alternative locations on a site, alternative window and space combinations, alternative shading devices, and so on. Perhaps most important to a designer, models suggest three-dimensional solutions, because a volume is being tested rather than merely a plan or section. If sun obstructions exist far from a site (such as shading by nearby mountains), it might be better to rely on graphs rather than trying to build a huge model.

A sunpeg chart, correctly attached to a model of any scale, allows a designer to quickly determine exact sun penetration and shadow patterns at many times for any date. Sunpeg plots are important analog tools for a site-conscious designer throughout the design process.

In Figure 3.17, a model is observed at a sun position corresponding to 3:00 p.m. on December 21. At the afternoon end of the six-hour period of best solar collection, sun still fills the south-side circulation space and even enters spaces beyond. It can also be inferred that passing from these spaces into the circulation space might involve considerable glare because the sun is so low in the sky. In such a case, vertical fins inside the south glazing would help intercept and diffuse direct solar radiation while still permitting solar heat gain.

Computer modeling (simulation) can also be used to explore solar access and shading issues for a site. Whether to use a physical model or a digital model depends upon designer preferences and design process workflows.

(f) Controlling Solar Reflections

The use of highly reflective (or "mirror") glass to reduce heat gain in office buildings has increased the frequency of annoying solar reflections from buildings (Figure 3.18). Large areas of non-mirror glazing, such as found on passive solar-heated or glass-box buildings, can also cause reflection problems. Glass and plastic glazing materials become more reflective (less transmissive) as the angle of incidence of solar radiation moves away from a line normal (perpendicular) to the glazing. Thus, the intensity of reflection is greatest (and the transmission of solar gain is least) when the sun's rays are nearly parallel to a surface.

Several high-profile incidents of unacceptable reflections have been reported over the past few decades, including the Walt Disney Concert Hall (Los Angeles, Figure 3.19) and the Vdara Hotel (Las Vegas)—both of which required post-construction (i.e., expensive) modifications to ameliorate problems caused by the reflections.

Because the most intense reflections occur when the sun's rays are nearly parallel to a wall, such reflections can be blocked or intercepted through responsible design. For example, foliage (Figure 3.20) can intercept reflected solar radiation from a south-facing window without blocking solar collection during the best hours. Another common approach is to use external projections around windows, as in Figure 3.21. Scale models and sunpeg charts or digital models are good ways to explore façade reflections. Using mirrored surfaces to represent reflective glass on a model will make resulting reflections evident even when complex geometries are involved.

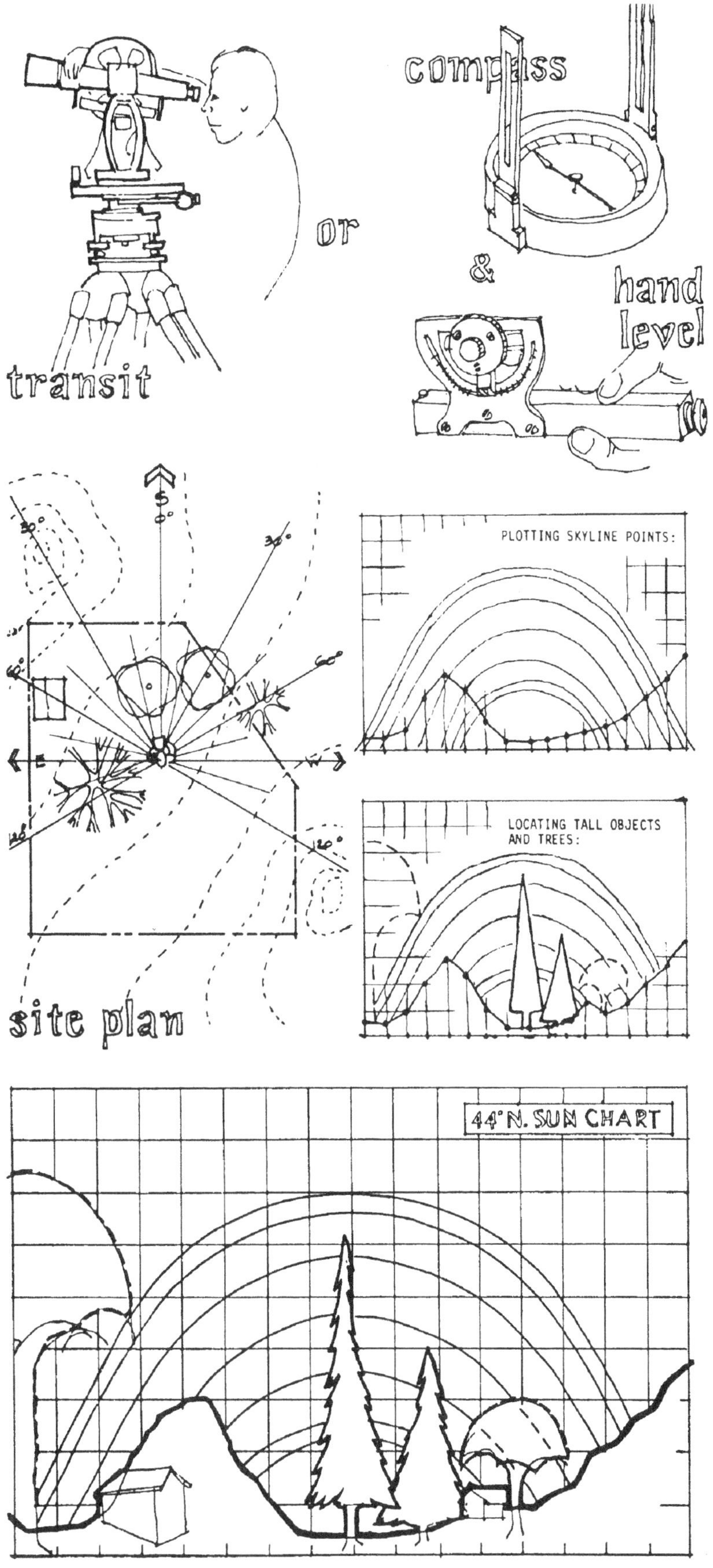

Figure 3.16 *Charting the skyline from a specific site position. (From Mazria and Winitsky, 1976.)*

Figure 3.17 *A model of a small building with a glazed open-frame circulation space on the south side is observed at the sun's position at 3:00 p.m. on December 21 using a sunpeg chart. (Photo by Tyler Mavichien; © Alison Kwok; all rights reserved.)*

Figure 3.19 *After construction, modifications were made to the highly polished stainless steel exterior of the Walt Disney Concert Hall in Los Angeles, California (by Frank Gehry), to reduce reflections impacting the neighboring condominiums; offending surfaces now have a matte finish. (© Karen Tse; used with permission.)*

Figure 3.18 *Mirror-glass windows in an office building (left) in San Francisco, California, cast strong reflections on the north- and west-facing walls of an older building next door. Although this reflected radiation/heat might occasionally be welcome in winter, the resulting glare can be intense. In summer, the older building is particularly disadvantaged by additional heat loads on its enclosure. (© Alison Kwok; all rights reserved.)*

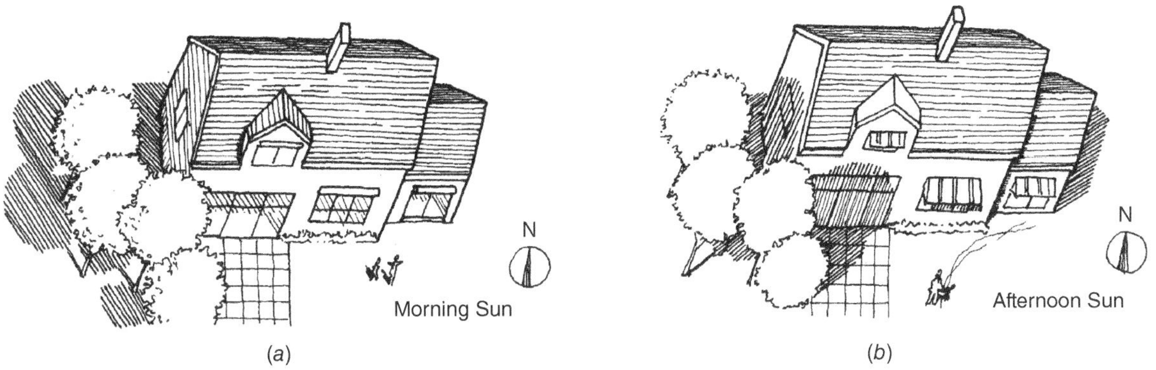

Figure 3.20 Selective protection from reflections. (a) The trees standing west of this south window wall do not interfere with solar access during the best hours for solar collection (around noon) nor do they prevent early morning sun from entering the windows. Any reflections of the early morning sun are intercepted by the trees before they can annoy those in nearby buildings. (b) The late afternoon sun is blocked by the trees before either solar gain or reflections can occur.

Figure 3.21 The "eggcrate" shading devices shown on the southeast corner of an office building in Nepal reduce solar heat gain by blocking acute sun angles from either side of the window. (© Ayush Vaidya; used with permission.)

3.8 SOUND AND AIRFLOW

Sound and airflow are discussed together because they are so difficult to separate in a site analysis. Many buildings that could be opened for ventilation or cooling breezes rely instead on forced ventilation because of noise that would accompany breezes through an open window. Pollution is another potential deterrent to natural ventilation. Almost any object or device that reduces noise will also reduce the velocity of a breeze, as is true of most filtering devices used to remove dust particles.

(a) Noise

Any unwanted sound is termed noise. The urban building in Figure 3.22 is unusual both in its potential for wind-driven ventilation—air moves freely below it as well as around it—and in the extraordinary intensity of airborne traffic noise.

Two key characteristics of cities contribute to increased noise at street level. These are hard surfaces that reflect (rather than absorb) sound and parallel walls that intensify sound by inter-reflection rather than dissipating it. Increasing horizontal and vertical distances from urban noise sources affect outdoor noise levels in various ways, as shown in the graphs of Figure 3.23. Unfortunately, obtaining useful distance between noise source and receiver in dense urban areas is quite difficult.

Although building surfaces are generally made of hard materials for durability, softer and multiplaned materials (such as plants) are desirable from an urban noise-reduction viewpoint. Their impact on measured sound levels may be slight, but visually softer surfaces reinforce a perception of acoustically softer environments (much as the sound of running water reinforces our perception of cool environments). Fountains are especially useful sources of masking sound; they can be kept flowing when a noise persists, and they can enhance the cooling function of natural ventilation via evaporative cooling, especially in drier climates. Not all spaces benefit from masking sound; where a single sound source is expected to predominate over all others, masking sound will interfere and become unwanted "masking noise."

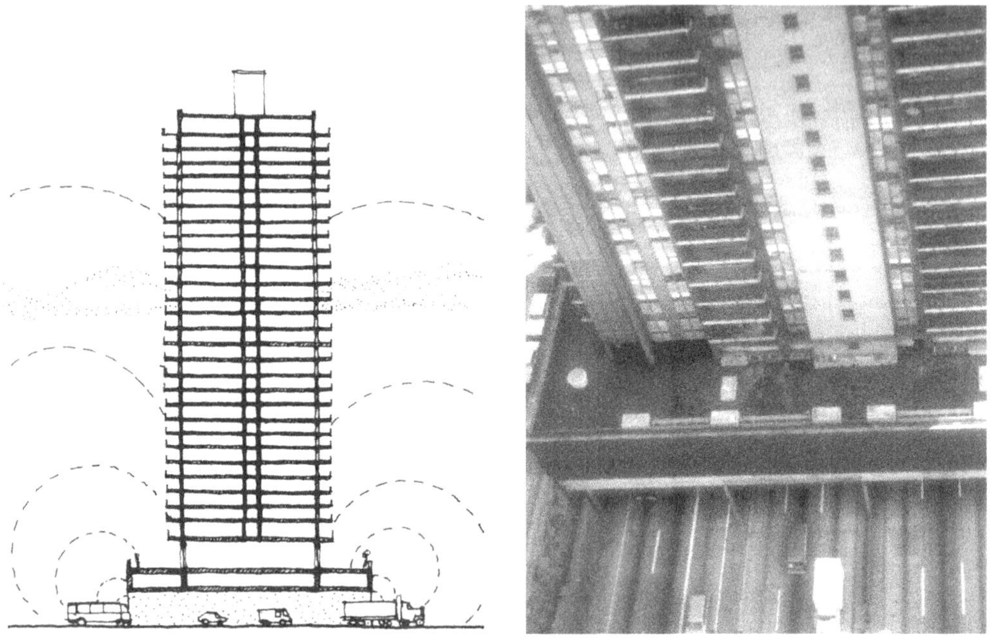

(a) (b)

Figure 3.22 Apartment buildings in series straddle the approach ramps to New York City's George Washington Bridge. (a) Section along the freeway. (Cohen et al. with permission of Elsevier.) (b) Looking down to the freeway. These buildings were the scene of a study linking noise levels with reading disabilities for occupants of the apartments. (From Cohen et al., 1973.)

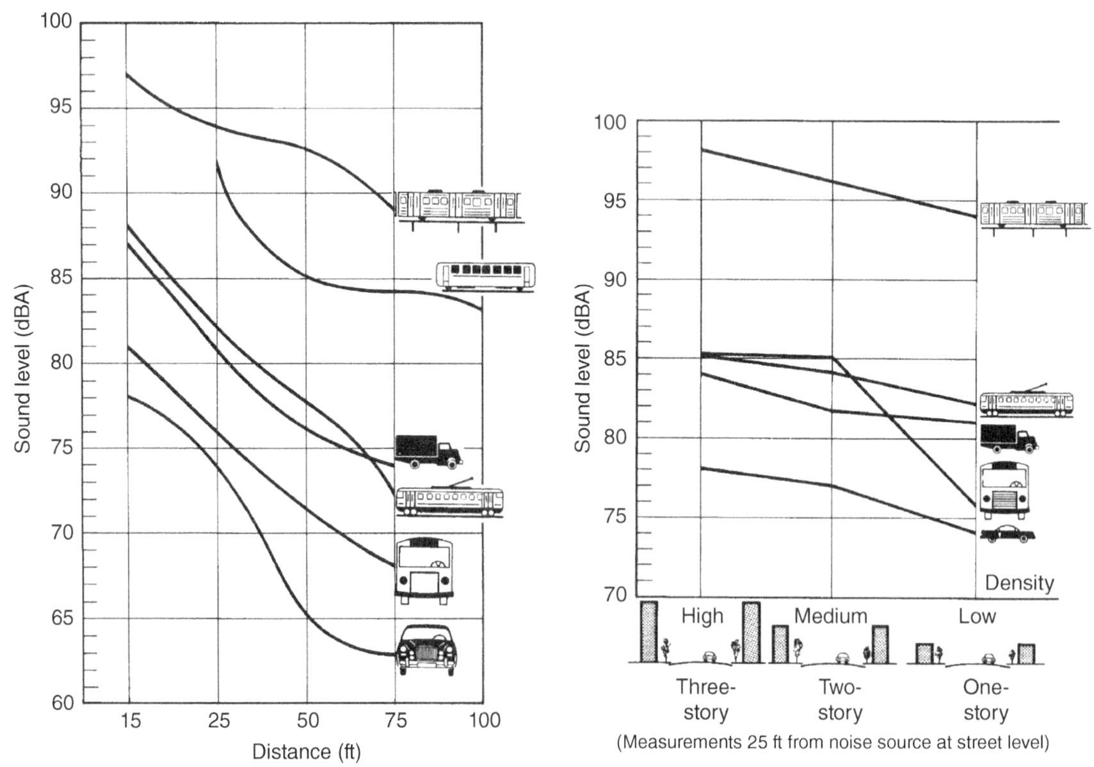

(a) (b)

Figure 3.23 Predicting noise levels outdoors. (a) Distance as a factor influencing sound pressure level. (b) Building height as a factor in noise propagation. (From Bragdon, 1971.)

Where site conditions allow, barriers to street noise can be installed. These cast sound shadows on a site (Figure 3.24). Such barriers may do little to reduce noise reaching upper windows, but activities near the ground can be given much lower noise levels, especially near the barrier. Many cities now require such barriers between new housing developments and highways or railroads.

Building mechanical equipment is a common urban noise source. Many noise complaints involving buildings are caused by air-conditioning equipment. When densely packed buildings are forced to rely upon mechanical cooling, such closeness makes the noise produced by the systems difficult to avoid. Noise is generated both by the air-conditioning compressor and by the large quantities of outdoor air that must be moved through outdoor condenser units. Attempts to surround condensing equipment with noise shields can hinder efficient operation, since a noise barrier is usually a barrier to airflow. Fortunately, energy-efficient equipment tends to be quieter, although not silent.

In residential neighborhoods, more spacing between buildings might be expected to lessen mechanical equipment noise problems. Yet the much lower ambient (or background) sound level in residential areas is one of their more appealing characteristics, and an intruding compressor on a hot summer night can irritate neighbors who would like to enjoy cool—and quiet—night breezes.

(b) Air Pollution

Building construction and operation contribute to air pollution. Air pollution contributes to problems of local and global importance. Key pollutants and their effects are summarized in Table 3.4. These effects include acidification, eutrophication, smog, a general decrease in ambient air quality, ozone depletion, and global warming.

An enhanced greenhouse effect (Figure 3.25) fueled by human-caused (anthropomorphic) emissions of a variety of gases is bringing about substantial global climate change. The naturally occurring and generally beneficial greenhouse effect is being amplified because gases that block the outgoing flow of long-wave radiation (heat) from the Earth's surface are accumulating in the atmosphere. Energy production and use (especially of fossil fuels) are contributing heavily to these greenhouse gas emissions, which include carbon dioxide, methane, nitrous oxide, and chlorofluorocarbons (CFCs). Collectively, the effects of these climate-modifying gases are termed global warming potential (GWP).

Another ongoing threat from pollution is stratospheric ozone depletion. Ozone in the atmosphere acts as a block to incoming ultraviolet radiation (a component of solar radiation). A reduction in this protective shield can allow increased ultraviolet radiation to be received at the Earth's surface, with potentially devastating consequences to ecosystems.

Building designers can mitigate these negative trends in several ways. First, by designing for greater energy efficiency and conservation and by utilizing renewable energy sources for building systems, they can greatly reduce the air pollution caused by electric power plants and by burning fuels on site. Chapters 12, 13, 14, 15, and 16 are concerned largely with this approach. Second, designers can specify materials and equipment that, through their manufacture or operation, lessen air pollution. This involves avoiding fuel combustion (coal, oil, trash, wood, and natural gas, roughly in descending order of air pollution threat). This approach also encourages selection of refrigeration equipment that uses environmentally friendly refrigerants. Insulation and furniture products made using CFC-free processes should be given preference. Many of these mitigation practices are mandated by codes, but a designer can always go further and seek out more environmentally benign products.

Local sources of air pollution must be minimized, and, as far as possible, such pollution be prevented from entering buildings. Combustion gases pose a threat, especially where vehicles interact with buildings (Figure 3.26). Idling truck motors at loading docks or automobile motors at drive-up service windows can threaten building occupants through openings such as doors and windows. Mechanical system fresh air intakes are particularly vulnerable, because outdoor air is intentionally drawn into them. Consider neighboring buildings. Will the thoughtless location of some activity on your site threaten a neighbor's fresh air?

(c) Wind Control

The focus of this discussion is normal ambient wind conditions—not extreme wind events. For most buildings and climates, wind (like sun) changes from resource to detriment with the changing of seasons. In many locations, wind also changes its prevailing direction with the seasons. Beneficial control of wind often means developing winter-wind-sheltered areas while designing to increase summer wind speeds. From spring through fall, outdoor spaces might benefit from a difficult-to-accomplish daily transition from site-as-barrier to site-as-connector—resulting in less wind during cool morning hours, more during warm afternoon hours. Fortunately, winds are generally weakest in the early morning and strongest in

(a)

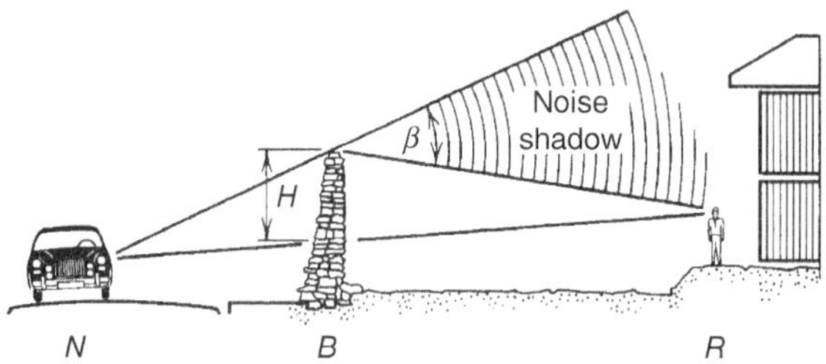

(b)

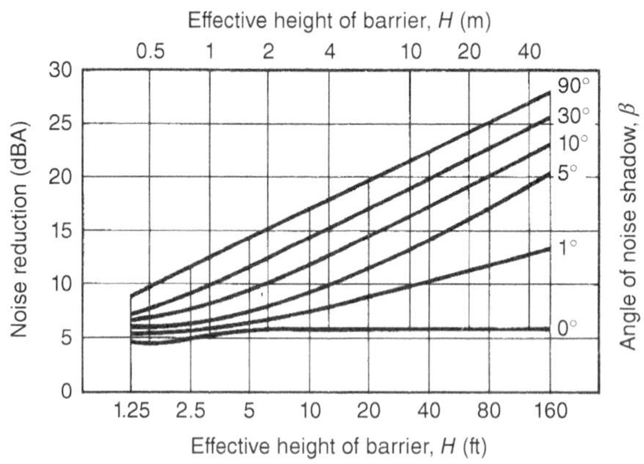

(c)

Figure 3.24 *Outdoor noise barriers. (a) A noise barrier abutting a highway in central Oregon. (Photo by Nathan Majeski.) (b) To determine the approximate noise reduction (in decibels) due to an outdoor barrier, construct a section locating the noise source (N), the solid barrier (B), and the receiver's location (R). On this section, determine the effective height (H) of the barrier and the diffraction angle (β) with the resulting "sound shadow" (noted as "noise shadow"—a less preferred term, as noise is subjective). Enter graph (c) with H and β; where the lines intersect determines the noise reduction (in dBA, left axis). A reduction of 10 dBA is perceived as half as loud as the original source. Note the perceptible noise reduction from simply breaking the line of sight (β = 1°). (b and c from Doelle, 1972.)*

TABLE 3.4 Air Pollution: Sources and Effects

Gas or Pollutant	Sources	Effects
Carbon monoxide (CO)	Gasoline-powered vehicles, industry using oil and gas, building heating using oil and gas, biomass burning	Enters human bloodstream rapidly, causing nervous system dysfunction and death at high concentrations; interferes with self-cleansing of atmosphere
Carbon dioxide (CO_2)	Fossil-fuel combustion, deforestation	Contributes to greenhouse effect
Methane (CH_4)	Rice fields, cattle, landfills, fossil-fuel production	Contributes to greenhouse effect
Sulfur oxides such as sulfur dioxide (SO_2) and sulfur trioxide	Industry using coal and oil; heating using coal and oil; power plants using coal, oil, and gas; ore smelting	Acid rain, damaging plants and attacking building skin materials; irritates human respiratory tract and complicates cardiovascular disease; decreases visibility in atmosphere
Nitrogen oxides (NO_X), such as nitric oxide (NO) and nitrogen dioxide (NO_2)	Gasoline-powered vehicles, building heating using oil and gas, industry and power plants, biomass burning	Acid rain, damaging plants and attacking building skin materials; irritates human eyes, nose, and upper respiratory tract; triggers development of smog; decreases visibility in atmosphere
Nitrous oxide (N_2O)	Nitrogenous fertilizers, deforestation, biomass burning	Contributes to greenhouse effect
Hydrocarbons (compounds of hydrogen and carbon)	Petroleum-powered vehicles, petroleum refineries, general burning	Promotes smog; toxic to human beings at high concentrations
Chlorofluorocarbons	Aerosol sprays, refrigerants, foams	Contributes to greenhouse effect and stratospheric ozone depletion
Particulates (liquid or solid particles smaller than 500 μm)	Vehicle exhausts, industry, building heating, general burning, spore- and pollen-bearing vegetation	Promotes precipitation formation; some are toxic to human beings; some pollens and spores cause allergic reactions in human beings

Sources: Adapted from Marsh, 1991, and Graedel and Crutzen, 1989.

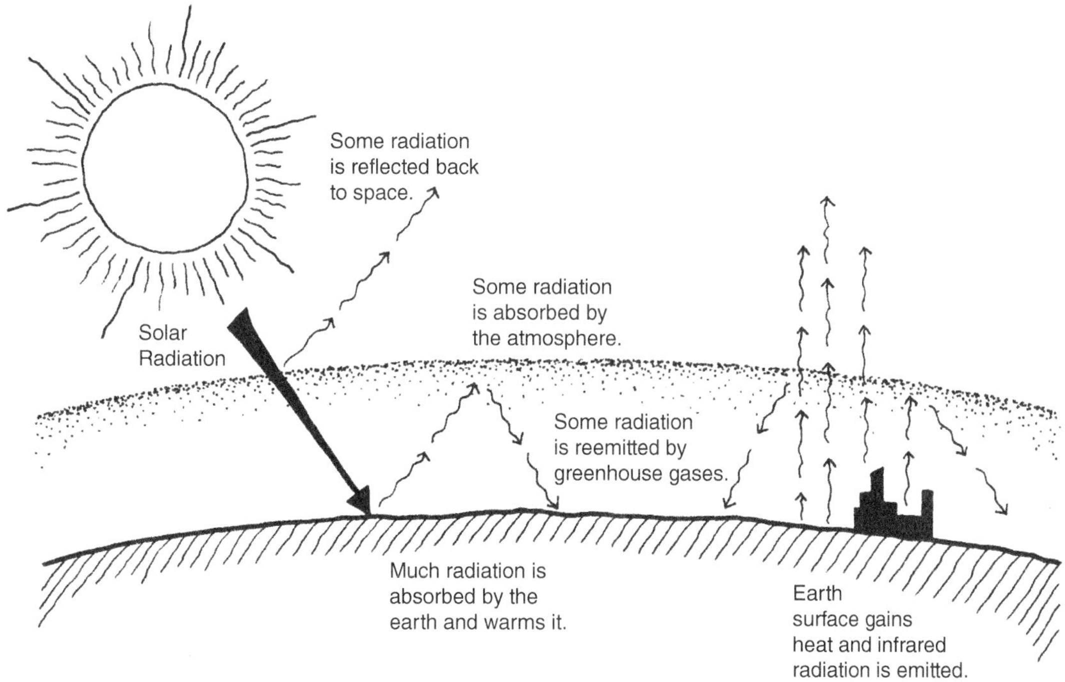

Figure 3.25 *The greenhouse effect traps heat in the Earth's upper atmosphere. Clouds and particles in the atmosphere reflect about one-fourth of incoming solar radiation while blocking about two-thirds of the heat that the Earth would otherwise lose to outer space. Historically, the atmosphere kept the Earth about 33°C [60°F] warmer than it would be without this heat-trapping process. Increases in greenhouse gas concentrations reflect more incoming solar radiation but block even more outgoing radiation, resulting in global warming and regional changes in climate. (Drawing by Amanda Clegg.)*

Figure 3.26 *Reactive protection of a poorly-located outdoor air intake. A loading dock near the intake was a source of indoor air pollution from truck motor exhaust, prompting the installation of a warning sign.*

the afternoon because of the sun's heating of the surrounding environment.

Generalized patterns of wind flow around thin windbreaks and thicker buildings (Figure 3.27) can help a designer understand where reduced and increased airflows may occur. These patterns, however, are much more complicated than they may appear and are highly influenced by objects upstream, to the sides, and downstream of the wind-directing object being analyzed. Wind-tunnel tests using scale models are far more reliable design tools than these generalized patterns. Unfortunately, such tests are expensive and still fraught with opportunities for misprediction. Notwithstanding these considerations, a site can and should be analyzed as accurately as possible for site-specific seasonal wind patterns. Readily accessible computational fluid dynamics (CFD) software

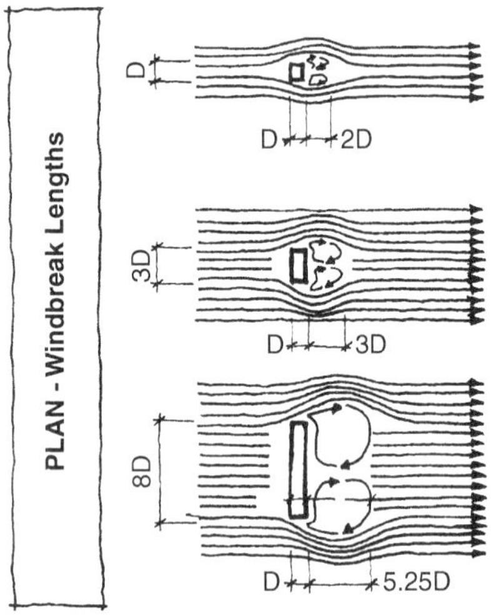

(*a*)

Wind Speed Reduction (%)

Density of Belt	Average Over First: 50 Yd	100 Yd	150 Yd	300 Yd
Very Open	18	24	25	18
Open	54	46	37	20
Medium	60	56	48	28
Dense	66	55	44	25
Very Dense	66	48	37	20

(*b*)

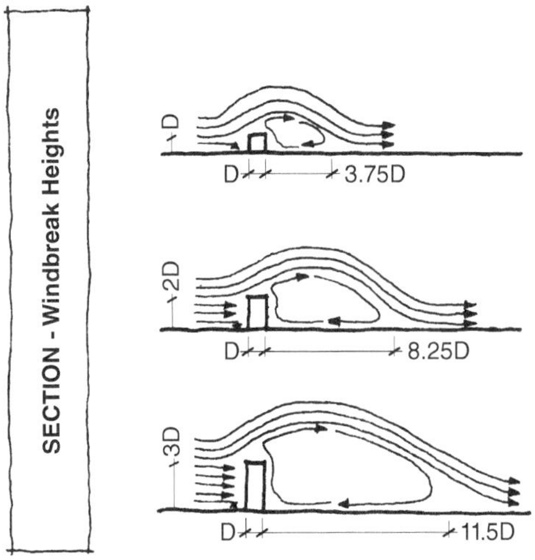

(*c*)

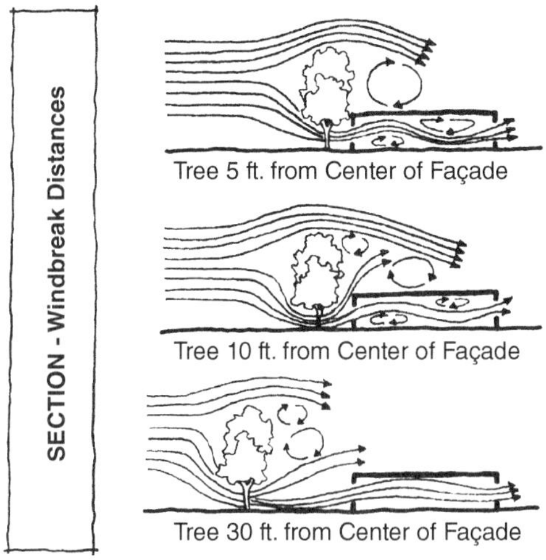

(*d*)

Figure 3.27 *Approximate patterns of wind around objects. (a) Effects of different barrier lengths (widths). (b) Reduction in wind speed due to windbreak density. (c) Effects of different barrier heights. (d) Wind flow through trees and buildings. (Redrawn by Jonathan Meendering / with permission of AIA, 1981.)*

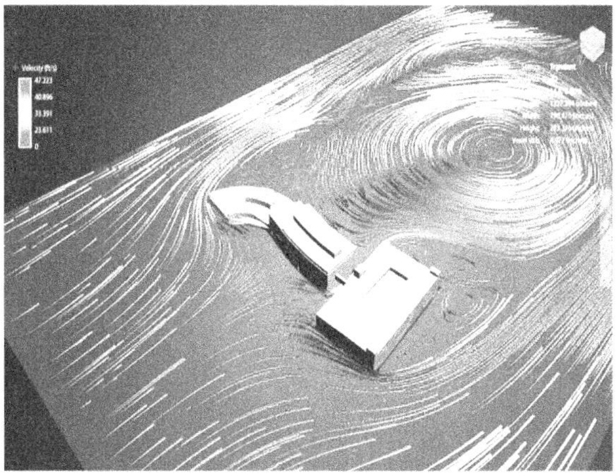

Figure 3.28 *Visualization of wind patterns around a simple 3D building model using Autodesk CFD. (Image by Daniel Overbey; © Browning Day; all rights reserved.)*

should soon make this analysis effort less challenging (Figure 3.28).

Before it reaches an obstacle, wind slows, builds up a positive pressure, and turns upward or sideways. As it passes the obstacle, it increases speed, and a reduced pressure occurs at the sides of and behind the obstacle. Such pressure differences, flow patterns, and the size and shape of the wind-protected areas behind the obstacle can be used to beneficially control air motion, both inside and outside a building. Wind ultimately returns to its original flow pattern after encountering an obstacle such as a windbreak or a building.

Windbreaks (such as fences or plants) are commonly used to protect outdoor areas. Figure 3.29 shows the relative reduction in wind velocity at the level of a windbreak as wind approaches and then passes. The distances (horizontal axis) are in multiples of the height of the windbreak. Note that the densest windbreak produces the greatest reduction in wind speed behind it—but that the wind recovers its full velocity closer to such a barrier, compared to a less-dense windbreak. Thus, the denser the windbreak, the greater the reduction in wind speed but the smaller the area so affected.

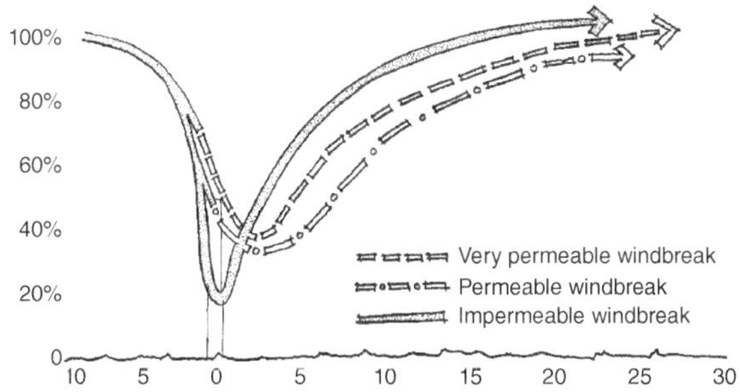

Figure 3.29 *Wind speed reduction behind windbreaks of varying permeability. Solid (impermeable) barriers produce the lowest wind speeds, but these are effective for the shortest distance beyond the windbreak. Units of distance equal heights of windbreak. (Brown, and Gillespie, 1995. Redrawn by Erik Winter.)*

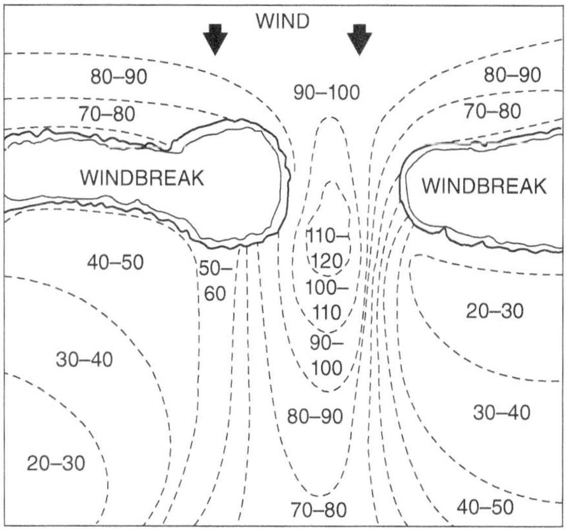

Figure 3.30 *Wind speeds accelerate through a gap in a windbreak. Numbers indicate the percentage of the incoming (unaffected) wind speed. (From Caborn, 1957. Cited in McPherson, 1984.)*

Increased wind speed can be produced by a gap in a windbreak. Figure 3.30 shows this effect. Although a gap is a threat to protection from unwanted winter winds, it is an opportunity for enhancement of desirable summer winds. Given constant prevailing summer wind directions, a windbreak gap could provide a small area of a site with above-average air motion. The Strata SE1 in Figure 3.31 is a 43-story skyscraper, which houses 408 apartments. The building features three wind turbines rated at 19 kW each that were designed to produce 50 MWh of electricity per year. Recent reports, however, found that vibration from the wind turbines disturbed occupants and led to them being shut down. Thus, acoustic mitigation should be included in such an application.

Wind flow around buildings is a complex matter; nevertheless, some general patterns for shelter areas are shown in Figure 3.27. Figures 3.32 and 3.33, on the other hand, show wind behavior likely to be encountered with typical building arrangements. Some patterns seen in the situations shown in Figure 3.33 include:

- with the bar effect (*a*), the downward-spinning wind behind a building can reach 1.4 times the ambient wind speed
- with the venturi effect (*b*) and few obstructions upwind or downwind from the narrow neck of a building, wind through the neck can reach 1.3 times the ambient speed, up to heights of 100 ft

Figure 3.31 *Strata SE1, London (by BFLS, now Flanagan Lawrence) was designed to produce 8% of its total estimated energy consumption. (© Tisha Egashira; used with permission.)*

[30 m], and 1.6 times the ambient speed at about 165 ft [50 m] in height

- the gap effect (*c*) begins to occur with perpendicular winds and buildings of more than 5 stories (50 ft [15 m]) in height; by 7 stories, wind speeds 1.2 times the ambient can occur through the gaps; by 60 stories, gap wind speeds can be 1.5 times the ambient
- for higher buildings, note the corner effect (*d*), where increased wind speeds occur at the corners (localized within a radius from the corner equal to the width, d, of the building); where height is 50 ft [15 m], wind speed can reach 1.2 times the ambient; for heights above 115 ft [35 m], wind speed can be 1.5 times the ambient. Where two towers approach each other, increased wind around corners and between the towers can be as high as 2.2 times the ambient speed for towers 330 ft [100 m] high
- the wake effect (*e*) refers to increased wind speed and turbulence within the wake of buildings, which can be especially serious for towers at heights from 16 to 30 stories, where wind speeds can reach 1.4–2.2 times the ambient

Beranek (1980) discussed methods for charting the shelter areas in such building groups. Localized effects of wind turbulence between buildings can have a particularly powerful impact on entryways. People leaving the calm of a building lobby are often unprepared for the speed and turbulence of channeled winds just beyond the doorway.

Wind flow for the cooling of buildings is discussed in subsection (d) and in detail in Chapter 14.

(d) Ventilation and Cooling

Outdoor air is purposefully brought into buildings for two distinct reasons. Ventilation (using the correct

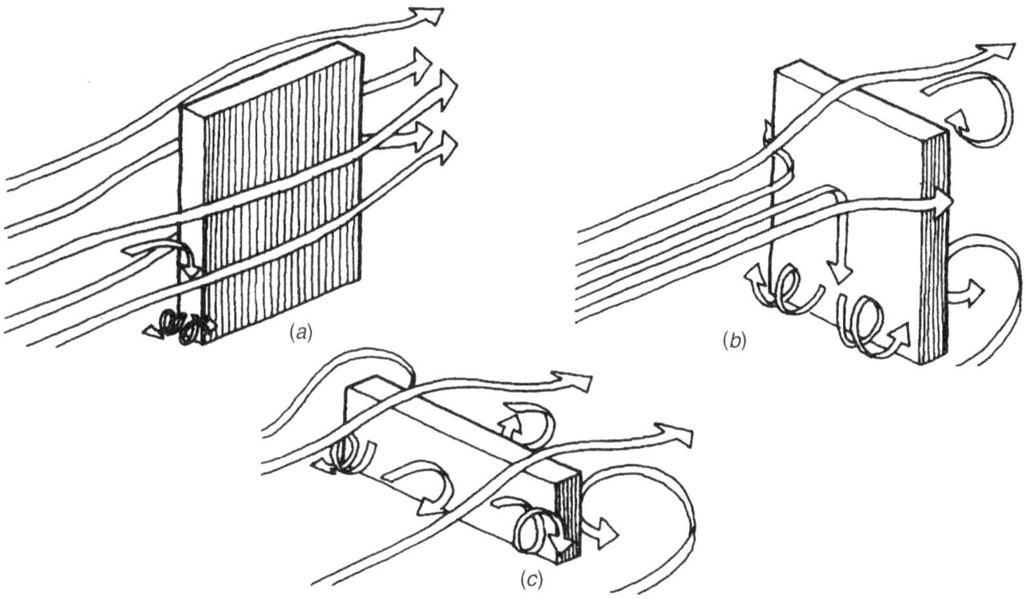

Figure 3.32 *Wind patterns around single buildings. (a) Tall, slender buildings: height greater than 2.5 times the width. (b) Tall, rather wide buildings; height between 2.5 and 0.6 times the width. (c) Long buildings; height less than 0.6 times the width. (From Beranek, 1980.)*

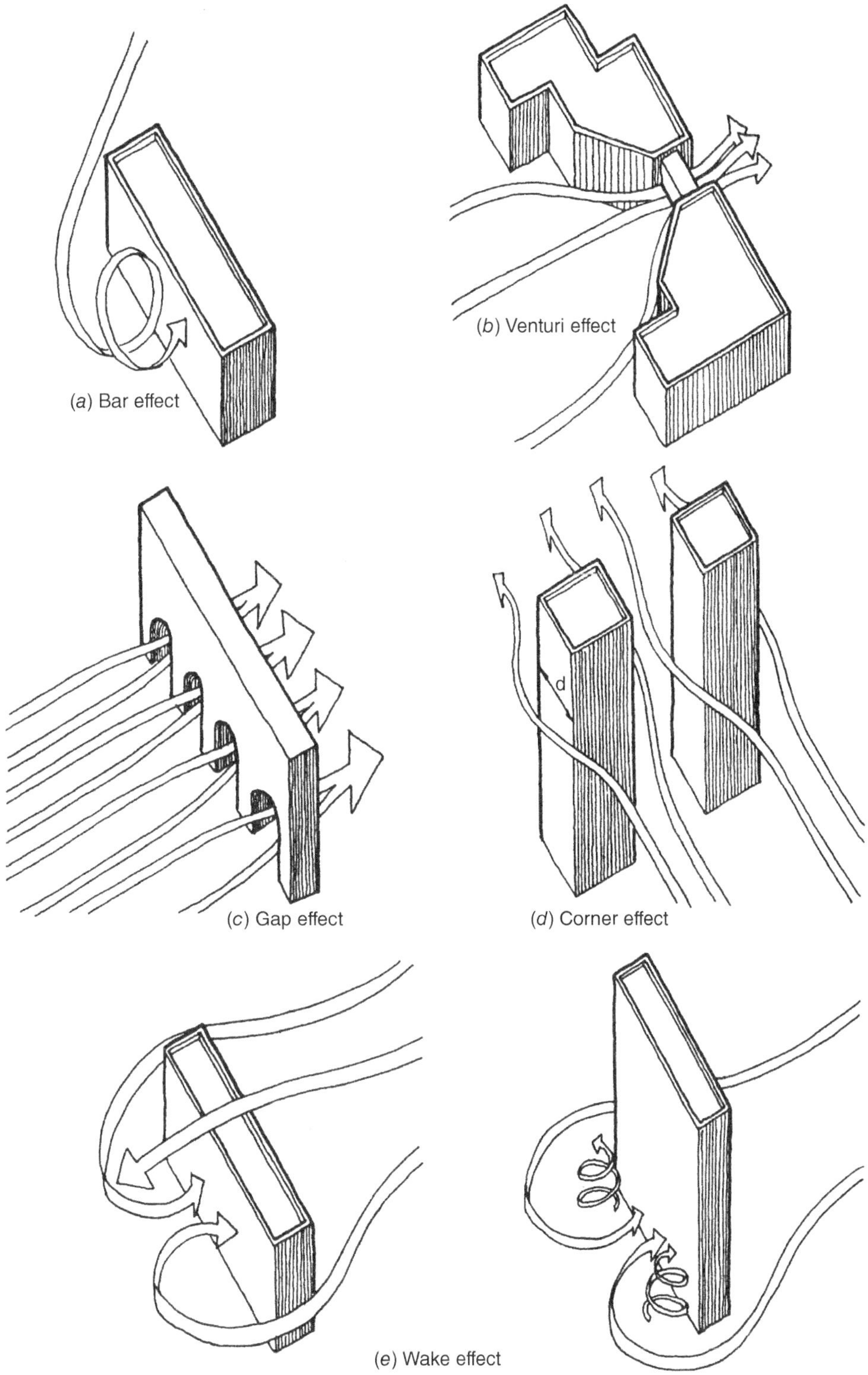

Figure 3.33 *Wind patterns among building clusters (see text for quantification). (From Gandemer, 1977.)*

design terminology) is the delivery of fresh outdoor air to building interiors to replenish the oxygen used by people and to help carry away a variety of indoor pollutants. Ventilation is desirable year-round and is the most common way to provide acceptable indoor air quality. Recommended minimum rates of ventilation airflow are found in Appendix F. Airflow for passive cooling (informally termed ventilation) replaces uncomfortably warm indoor air with cooler outdoor air. Passive cooling by cross-ventilation or stack ventilation

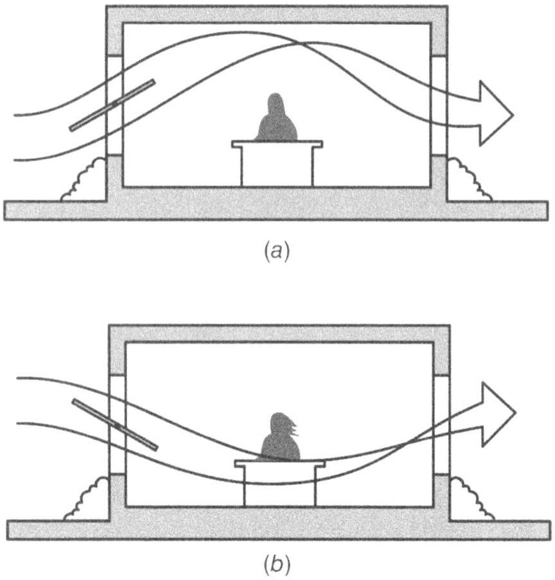

Figure 3.34 *Ventilation with and without direct occupant cooling. The size and position of a window will influence the flow of air within a space. (a) Indirect ventilation: the window directs breezes upward, removing hot air at the ceiling. Airflow has minimal contact with occupants. (b) Direct ventilation: the window directs breezes toward the floor and across occupants and provides a direct people-cooling effect from air motion. Both approaches provide air flow for indoor air quality. (Drawn by Alisa Kwok.)*

is a seasonal opportunity limited to times when the outdoor air temperature is lower than the indoor air temperature, and the outdoor humidity is at or below that desired indoors. When outdoor air temperatures are about the same as interior temperatures, ventilation might still be useful to increase interior air motion, thus extending the comfort zone, as discussed in Chapter 6.

Although airflows for indoor air quality control and for cooling are both commonly (and confusingly) referred to as "ventilation," they are not identical design responses. Building codes/standards typically rigidly define ventilation requirements for air quality. This is not the case for ventilation for cooling. Airflows for effective passive cooling can far exceed airflow rates for control of air quality. The influence of cooling ventilation requirements on building siting and window size and placement is considerable. Figure 3.34 illustrates the differing approaches to these two uses for airflow and suggests that window position and human occupants be considered together in design.

Two interesting buildings designed to use the wind appear in Figures 3.35 and 3.36. Portland Community College in Newberg, Oregon, is a 13,500-ft^2 [1254-m^2] facility designed as the first net-zero-energy higher education building in Oregon (and the second in the United States). The second is a chapel in a cemetery in hot, humid Houston, Texas. The architects intended that this shaded, open-air space act as a respite from the

Figure 3.35 *(a) Natural ventilation and passive cooling strategies articulated by the ventilation stacks at the Portland Community College Newberg Center in Oregon. (Photo © Nic Lehoux; used with permission.) (b) Small ventilation turbines in each stack help to draw fresh air into louvers along the building's perimeter and exhaust air through the top of the stack. (Drawing © Hennebery Eddy Architects; used with permission.)*

sealed, air-conditioned buildings and cars prevalent in Houston. In this building, a funeral becomes a time and place for a reunion with physical phenomena; breezes and shade trees provide cooling, rainwater collected in a pool reflects skylight, and a narrow shaft of direct

(a)

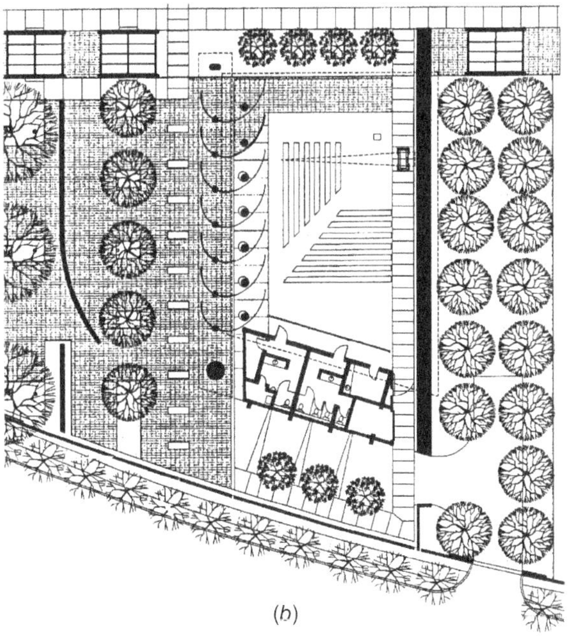

(b)

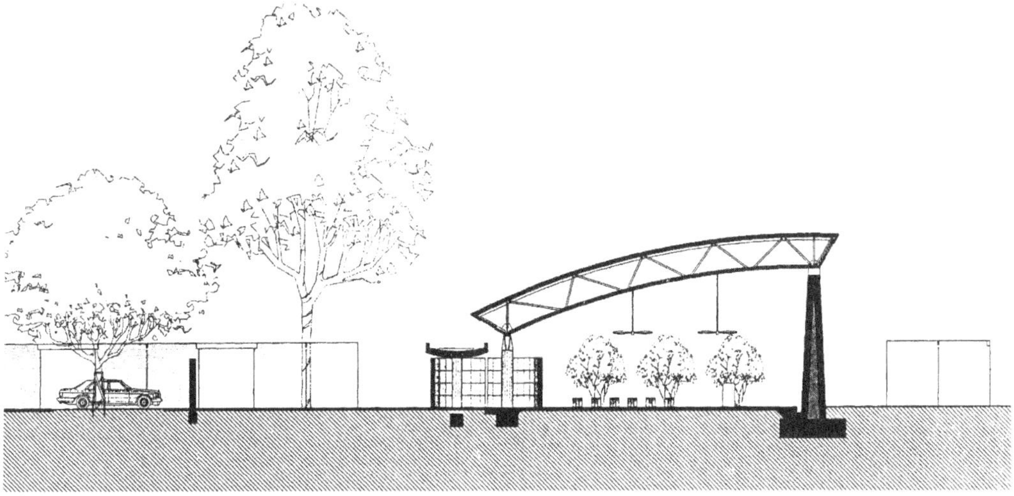

(c)

Figure 3.36 *The Beth Israel Chapel and Memorial Garden, Houston, Texas. (a) View from the west. The oversized gutter delivers rainwater to a pond that reflects daylight. Trees on islands in the pond provide evening shade. (b) Plan. Curved walls visually separate the open south courtyard from the roofed chapel but allow breezes to pass. A narrow triangular roof opening allows a shaft of direct sunlight to fall along the interior north wall, marking the passing of time. (c) Section, south to north. The curved roof sheds rainwater to a gutter above the curved walls. Suspended ceiling fans can augment air motion. (Photo by Timothy Hursley. Courtesy of Solomon Inc., Architecture and Urban Design, San Francisco.)*

sun marks its daily path along the rough stone interior north wall.

A building that relies on prevailing winds for cooling must be sited with attention to wind direction. Relatively unobstructed access to breezes is an important design consideration. As seen previously, obstacles upstream from intake openings or downstream near an outlet can substantially reduce the speed—and thereby the direct cooling effect—of the wind. Wind can cool people in hot weather (primarily by increasing evaporation from the skin), yet it can become an irritant at higher speeds. The Beaufort scale, originally devised to help mariners, provides a means to estimate and communicate wind speed (Table 3.5). Manual controls for openings are a necessary part of a natural ventilation system. Finally, properly sized and placed openings, with an unobstructed path for airflow through the building, must be provided.

(e) Wind, Daylight, and Sun

The design constraints of both daylighting and wind-driven (cross-)ventilation tend to limit building width.

TABLE 3.5 Beaufort Scale (Lower Speeds Only)

Beaufort Number	Speed mph	Speed knots	Description	Effects Outdoors On Land	Effects Outdoors Over Water
0	0–1	0–1	Calm	Calm; smoke rises vertically	Sea like a mirror
1	1–3	1–3	Light air	Direction of wind shown by smoke drift but not by wind vanes	Ripples with the appearance of scales are formed but without foam crests
2	4–7	4–6	Light breeze	Wind felt on face; leaves rustle; ordinary vanes moved by wind	Small wavelets, still short, but more pronounced. Crests have a glassy appearance and do not break.
3	8–12	7–10	Gentle breeze	Leaves and small twigs in constant motion; wind extends light flag	Large wavelets. Crests begin to break. Foam of glassy appearance. Perhaps scattered white horses.
4	13–18	11–16	Moderate breeze	Raises dust and loose paper; small branches are moved	Small waves, becoming larger; fairly frequent white horses

Source: National Weather Service (NWS), n.d.

Large, multistory hotels are an example; they often fill an entire block, are arranged around a central atrium, and have a floor plan that is only two hotel rooms and one corridor deep. Multistory office buildings used to have a similar form, but increasing urban density and reliance on electric lighting and mechanical cooling have changed design responses considerably (Figure 3.37). A trend back to arranging office workspaces close to daylight openings is occurring in response to green design concerns; in appropriate climates, buildings are also trending (although slowly) toward including operable windows for fresh air.

Wind and sun as design considerations tend to be most influential in opposite seasons. Fresh air, however, is necessary year-round; thus, even in winter a supply of tempered fresh air is desirable. Figure 3.38 captures some issues surrounding the design of a clerestory—whether it should face to the north or to the south for a given prevailing wind pattern.

3.9 RAIN AND GROUNDWATER

Most buildings interact with four forms of water. Rainwater and groundwater involve exterior interactions minimally impacted by building codes, whereas potable water and wastewater involve interior services tightly controlled by codes.

Rain, like solar energy, is a generally diffuse, intermittent, and often seasonal resource. It may be collected on site and used as a source of water for building services, where other water resources are scarce or of poor quality. Rain, like sun, has an influence on building design: Heavy rains and pitched roofs have long been found in the same locales. Overhangs may extend farther beyond walls exposed to storm winds; gutter and downspout details can become a design feature, as shown in the chapel in Figure 3.36 and in several other examples throughout the book. A design solution that reflects the combined influences of daylight, wind, and rain is shown in Figure 3.39.

Once on the surface of a site, rainwater will flow downhill in a sheetlike path, and structures that obstruct this path must make provisions to accommodate the flow. Shallow ditches called swales are frequently used to direct rainwater flow. The orientation of a building relative to the slope of a site can also affect surface water diversion, as shown in Figure 3.40.

Surface water can be used to advantage in thermal, acoustic, and daylighting roles. Hot, dry breezes that pass over water surfaces (and especially through misty sprays above ponds) gain substantial moisture while undergoing a drop in dry-bulb temperature because of the evaporative cooling process. Such conditioned air can provide improved comfort in hot, dry conditions. If a water feature also provides the sound of running water, it can serve to mask noises such as traffic or conversations in adjacent rooms.

Surface water plays a more complex role in daylighting due to the reflection characteristics of water. The surface of water is highly reflective to light striking at low angles of incidence; for example, the reflections of the setting sun from the small pond will illuminate the ceiling and east wall of the chapel shown in Figure 3.36. On the coast, reflected sunlight from an infinitely huge ocean or lake surface can throw a blinding sheet of light toward buildings on the shore. Conversely, sunlight near noon in summer strikes a water body at high angles of incidence—nearly perpendicular at southern U.S. latitudes. Water is highly absorptive and transmissive to radiation at these angles, thereby reflecting relatively little light. Therefore, water bodies east, south, and west of buildings can provide increased reflected light on sunny winter days and somewhat decreased reflected light in summer, relative to alternative surface materials. On heavily

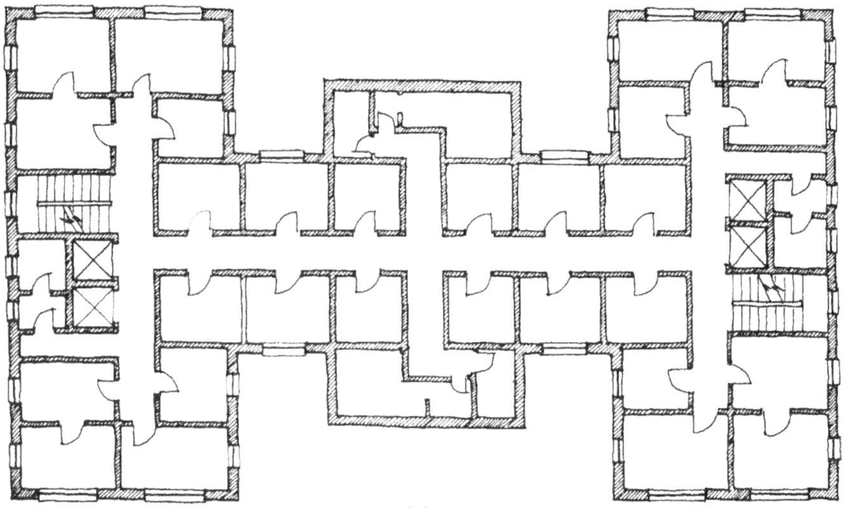

(a)

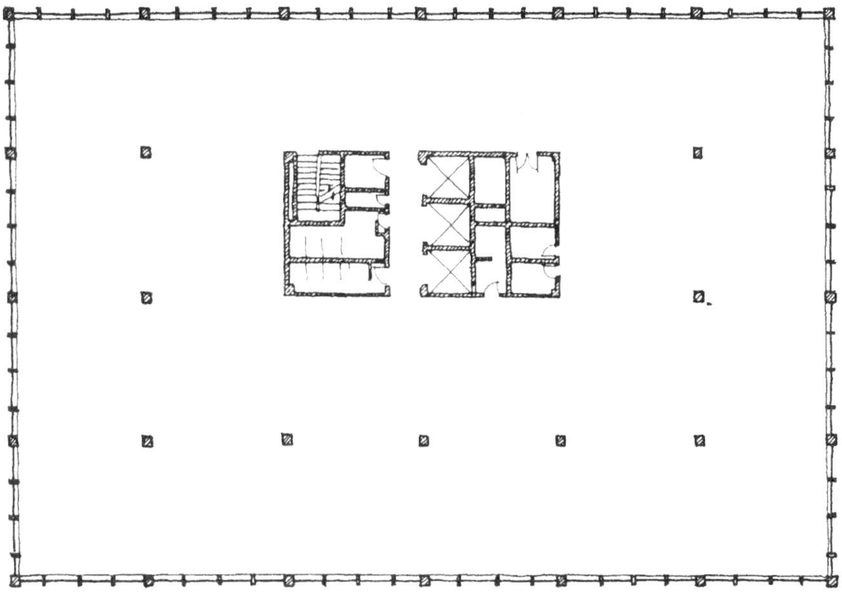

(b)

Figure 3.37 *In contrast to office building plan (a), which provides daylight and natural ventilation in each office, office building plan (b) receives mechanically cooled and filtered air, is less subject to exterior noise, typically provides constant light and temperature throughout, and provides for more rentable floor space on its site. Plan b also allows less daylight to reach the street level, consumes much more electricity (though probably less heating fuel), and thus contributes more waste heat (and possibly noise from mechanical equipment) to its surroundings year-round.*

overcast days, however, when the sky is uniformly gray (and the least quantity, but most glare-free quality, of daylight is available), water surfaces will not be particularly helpful in directing daylight.

The reflection of sunlight from moving water tends to be in sparkling patches of always-changing patterns. This can act as a fascinating design feature, or it can be annoying (perceived either as a welcome distraction or as glare). Reflected from a matte-finish ceiling, such dancing light might be welcome; reflected directly into someone's eyes or onto a work surface, it easily becomes a problem. What one viewer sees as sparkle may be seen as glare by another.

Groundwater is generally avoided by designers whenever possible, as it is a threat to foundations and belowground spaces. This concern must be an integral part of site planning. In urban areas, dryer soil conditions are often the norm as ponds are drained, streams are contained, and hard surfaces impede percolation of water into the soil.

Groundwater has interesting thermal potential as a heat sink, providing a place to discharge building heat in summer and providing heat for winter building heating. The quantity of groundwater available to act as a heat sink varies with geographic location and subsurface conditions, and little macroscale

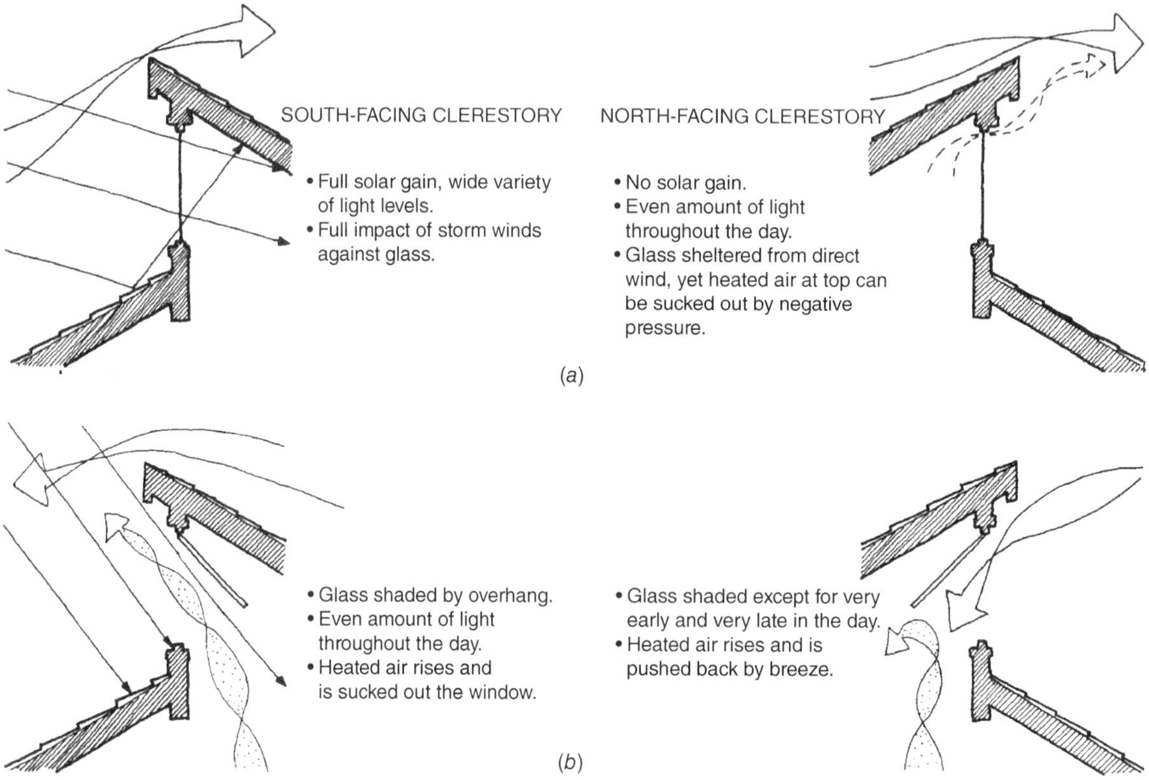

Figure 3.38 *Some relative advantages of north versus south orientation for a clerestory window/shed roof combination. (a) Winter, with low sun and southerly storm winds. (b) Summer, with high sun and northerly breezes. These wind directions are prevalent in the Pacific Northwest.*

Figure 3.39 *Integrated design is expressed by this gutter detail in the 2005 Cornell Solar Decathlon House. (© Nicholas Rajkovich; used with permission.)*

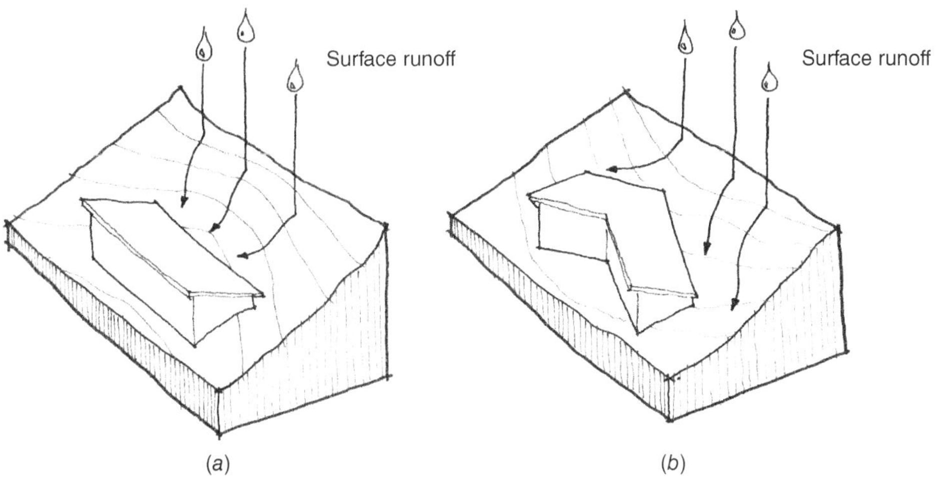

Figure 3.40 *Rain as surface flow. (a) Where buildings intercept surface water, provisions for diversion are necessary. A building sited as in (b) needs less elaborate provisions, as the form itself is a diverter. (Drawing by Dain Carlson.)*

information is available to indicate its potential on a given site. Ground-source heat pump systems using soil and/or groundwater as a heat sink/source are common as high-efficiency system options. As with any site resource, ambient groundwater and soil temperatures can be changed through use over time, but by how much and how quickly are difficult to predict. Groundwater pollution is another risk, through increased handling of groundwater or from refrigerants in the heat pump equipment. As a result of such concerns, many localities now strictly regulate the use of groundwater as a heat sink/source. In some locales, geothermal—referring to high-temperature water or steam—resources may be tapped for building use. Geothermal is a distinctly different concept than ground source.

The ready ability of water to conduct and store heat, which leads to its common use as a heat sink and heat-transfer medium, also makes it a thermal nemesis when heat containment is the design objective. Heat containment is a serious concern for hot water tanks and rock bins in solar-heated buildings (as well as underground wall and floor surfaces in heating-dominated climates). There is little point in storing heat from solar collectors in an underground tank if surrounding groundwater is allowed to strip the tank of its heat.

Snow has special site design implications: It stores water for delayed runoff, provides a blanket of thermal insulation, absorbs sound, and reflects more solar radiation than almost any other naturally occurring surface. Wind patterns can deposit snow much more thickly in some places on site than in others, for better or worse. Snow can hamper the movement of dynamic façade devices such as awnings or thermal shutters, can collect to excess on external light shelves, and can create disabling glare if it reflects low winter sun into view windows.

3.10 VEGETATION

Plants on a building site can play several roles. They affect the solar absorptivity and emissivity of the Earth's surface; they are part of both the food and water cycles (affecting surface permeability to rainfall); by day they turn carbon dioxide into oxygen; they can provide organic matter suitable for building materials; and they help people mark time both by growth and by change with the seasons. Our associations with plants are mostly pleasant ones, and they contribute to enjoyment of the places where we encounter them.

Plants are also of immediate practical value to the building designer because they can enhance privacy, slow the winter wind, reduce glare from strong sunlight, and/or prevent summer sun from entering and overheating buildings. In the last role, plants are particularly noteworthy, because they enhance a feeling of coolness when breezes rustle or sway their leaves. Most importantly, they respond more to cycles of outdoor temperature than to the cycles of sun position. Unlike fixed-position shading devices on buildings, plants provide their deepest shade in the hottest weather.

Shading from plants is conceptually different from shading from fixed architectural elements. This is illustrated in Figure 3.41, which shows the rigid predictability of shading from an architectural element. An overhang for south-facing windows at the Springs Preserve becomes a learning tool through educational markers denoting the position of the sun at

(a) (b)

Figure 3.41 Solar angle educational display (a) for a roof overhang at Springs Preserve, Las Vegas, Nevada. (b) Detail displaying the shade provided during the summer solstice. (© Alison Kwok; all rights reserved.)

various times of the year. Maximum shade is provided at noon on the summer solstice. A typical approach to fixed sunscreen design in a U.S. temperate zone is to shade at least half of a south window in a residence (or all of a window in an office with internal heat loads) from March 21 to September 21 (vernal equinox to autumnal equinox). Yet, March is (on average) a colder month than September; March 21 is the last official day of winter, whereas September 21 is the last day of summer. Indeed, average monthly temperatures for June and September can be quite similar. Solar radiation is more welcome in early spring than in early fall, yet sun position is identical at these times, thus, the benefit of appropriate vegetative shading.

In contrast to fixed sunscreens, deciduous plants do most of their shading from the middle of June to early October, giving windows access to solar radiation throughout much of the spring (Figures 3.42 and 3.43).

Deciduous trees have a potential disadvantage for solar heating, in that certain species (such as sweet gum and some oaks) hold on to their leaves well into the heating season, a tendency increased by fertilizing or irrigating near the tree. Other trees have a dense branch structure, blocking a surprisingly

Figure 3.42 A deciduous tree as a naturally "smart" shading device. (Courtesy Tyler Mavichien.)

(*a*) (*b*) (*c*) (*d*)

Figure 3.43 *Deciduous vines, temperature, and sun position. The sun's path through the sky is identical in late May (a) and late July (b). Similarly paired—but lower—sun paths occur in late November (c) and late January (d). This deciduous vine responds more to the temperature of its Oregon climate than to the sun's position, which makes it particularly useful as a sun control device. For pest control and wall longevity, it is best to keep vines on a trellis rather than directly on the wall surface. (From Reynolds, 1976.)*

TABLE 3.6 Deciduous Trees for Summer Shading and Winter Solar Collection

		Transmissivity[a] Range (%)				Mature Height	
Botanical Name	**Common Name**	**Summer**	**Winter**	**Foliation[b]**	**Defoliation[c]**	**m**	**ft**
Acer platanoides	Norway maple	5–14	60–75	E	M	15–25	50–82
Acer rubrum	Red maple	8–22	63–82	M	E	20–35	65–115
Acer saccharinum	Silver maple	10–28	60–87	M	M	20–35	65–115
Acer saccharum	Sugar maple	16–27	60–80	M	E	20–35	65–115
Aesculus hippocastanum	Horse chestnut	8–27	73	M	L	22–30	72–98
Amelanchier canadensis	Serviceberry	20–25	57	L	M		
Betula pendula	European birch	14–24	48–88	M	M–L	15–30	50–98
Cercis canadensis[d]	Red bud	62	74	L	M	12	40
Carya ovata	Shagbark hickory	15–28	66			24–30	78–98
Catalpa speciosa	Western catalpa	24–30	52–83	L		18–30	60–98
Cornus florida[d]	Dogwood	43	53	L	E	11	36
Fagus sylvatica	European birch	7–15	83	L	L	18–30	60–98
Fraxinus pennsylvanica	Green ash	10–29	70–71	M–L	M	18–25	60–82
Gleditsia tricanthos inermis	Honey locust	25–50	50–85	M	E	20–30	65–98
Juglans nigra	Black walnut	9	55–72	L	E–M	23–45	75–148
Liquidambar styraciflua[d]	Sweet gum	33	47	M	L	24	78
Liriodendron tulipifera	Tulip tree	10	69–78	M–L	M	27–45	88–148
Platanus acerifolia	London plane tree	11–17	46–64	M–L		30–35	98–115
Populus detoides	Cottonwood	10–20	68	E	M	23–30	75–98
Populus tremuloides	Trembling aspen	20–33		E		12–15	40–50
Quercus alba	White oak	13–38					
Quercus palustris[d]	Pin oak	45	47	L	L	23	75
Quercus rubra	Red oak	12–23	70–81		M	23–30	75–98
Robinia pseudoacacia[d]	Black locust	38	40	L	E	21	69
Tilia cordata	Littleleaf linden	7–22	46–70	L	E	18–21	60–69
Ulmus americana	American elm	13	63–89	M	M	18–24	60–78

Source: Brown and Gillespie, 1995.

[a]Transmissivity to solar radiation; estimates vary with instruments used by various researchers.

[b]Foliation: E = early (before April 30); M = middle (May 1–15); L = late (after May 15).

[c]Defoliation: E = early (before November 1); M = middle (November 1–30); L = late (after November 30).

[d]From Montgomery (1987). Transmissivity % = (100% – blockage percentage).

high percentage of solar radiation even when bare (Table 3.6). Agricultural extension services in most areas can provide information on the hardiness and average dates of defoliation for various tree species. For solar collection, avoiding trees or large shrubs within the area shown in Figure 3.44 is recommended. In this protected zone, a deciduous vine on a trellis above a south-facing window is a better shading strategy.

Many large buildings have a relatively short heating season because they have substantial and consistent internal heat gains. For such internal-load (heat-dominated) buildings, a tree or vine that foliates early in spring, defoliates late in fall, and has low transmissivity in summer is advantageous. The choice of trees, shrubs, and vines to improve a site and/or building microclimate can best be made after a month-by-month analysis of specific building heating and cooling needs.

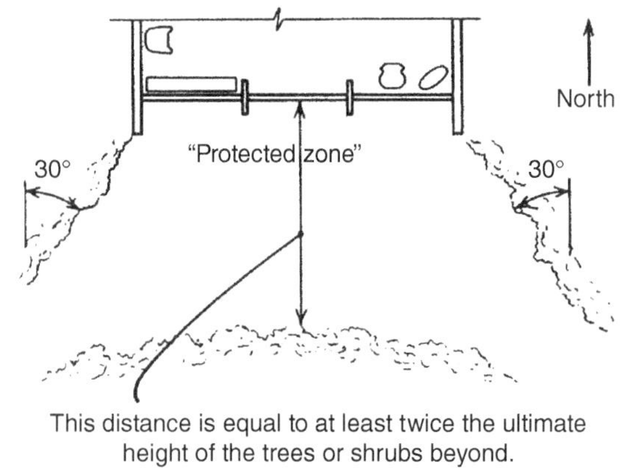

Figure 3.44 *Protecting access to winter sun, given a lawn or terrace of limited size to the south of solar collecting surfaces. Coniferous and even deciduous plants within the "protected zone" should be avoided unless they are very low growing or are a reliably early defoliating species (see Table 3.6). Summer sun protection for such south-facing windows is best provided by flexible architectural controls such as awnings or hanging screens.*

3.11 CASE STUDY—SITE AND RESOURCE DESIGN

Aldo Leopold Legacy Center, Baraboo, Wisconsin

PROJECT BASICS

- Location: Baraboo, Wisconsin, USA
- Latitude: 43.52°N; longitude: 89.46°W; elevation: 974 ft [297 m]
- Heating degree days: 7673 base 65°F [4263 base 18.3°C]; cooling degree days: 2389 base 50°F [1327 base 10°C] at Madison airport; annual precipitation 33.8 in. [858 mm]
- Building type: New construction; offices, meeting rooms, exhibit area
- Building area: 13,000 ft^2 [1208 m^2]; 8644 ft^2 [803 m^2] conditioned space
- Completed Spring 2007
- Client: Aldo Leopold Foundation
- Design team: The Kubala Washatko Architects, Inc. and consultants

Background. The Aldo Leopold Legacy Center is a Leadership in Energy and Environmental Design (LEED) Platinum-certified project, the first such project ever earning 61 (of 69 possible) points and the first to earn a point for being a net-zero energy, carbon-neutral building (Figures 3.45 and 3.46). The site and its resources were integral components in achieving this aggressive design goal. This will be further discussed in the following sections to emphasize the opportunities and resources that are on-site, free, and ready to be captured. The Aldo Leopold Foundation's (ALF) mission is to uphold the land ethic, and this unwavering goal helped to motivate the design team to exemplify the founding principles of the Foundation through innovation and sustainability. The Aldo Leopold Legacy Center utilizes several passive strategies and is constructed with local materials that were transported an average of 100 miles per pound [359 km per kg].

Context. The Aldo Leopold Legacy Center (ALLC) is an organization dedicated to promoting Aldo Leopold's land ethic. Leopold, author of the *Sand County*

Figure 3.45 *The Aldo Leopold Legacy Center in Baraboo, Wisconsin (north of Madison), which attained LEED Platinum certification. (© The Kubala Washatko Architects, Inc.; used with permission.)*

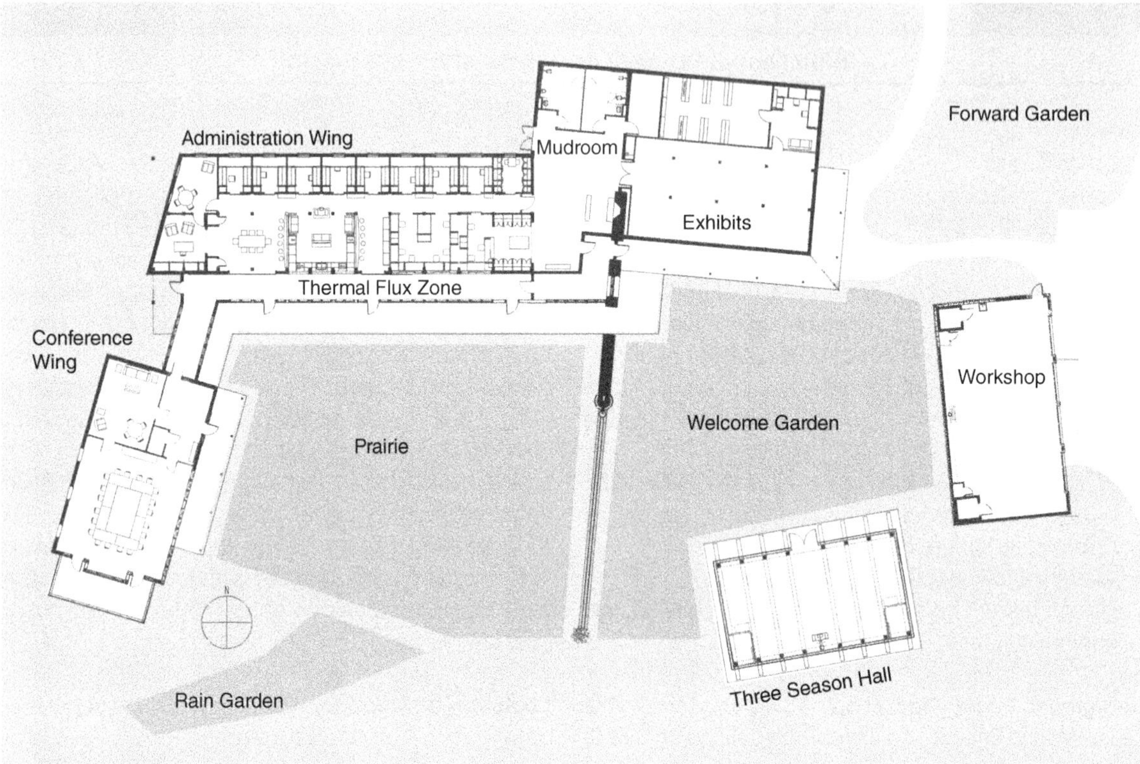

Figure 3.46 *Site plan of the Aldo Leopold Legacy Center showing elongated buildings and southern orientation around a courtyard. (© The Kubala Washatko Architects, Inc.; used with permission.)*

Almanac, developed a belief that the term "community" should expand to include elements such as the "soils, waters, plants and animals, or collectively: the land." His writings set the stage for the modern conservation movement and particularly our current tendencies toward ecological thinking. An original building on the site, called the "Leopold shack," served as a retreat as well as a center for ecological research. Formerly a chicken coop, the shack was built primarily of found materials, used passive strategies for daylighting and ventilation, and required no electricity. Design of the new center retained the same ethic of conservation present in the shack, even with increased occupancy, energy demands, and building functions.

Design Intent. The ALLC is a direct reflection of Aldo Leopold's values. The leading principles (intent) behind the design included:

- respect Aldo Leopold's land ethic in all aspects of design
- all occupied zones are perimeter zones designed to use daylight and natural ventilation
- minimize electric lighting use during occupancy
- minimize HVAC systems use during occupancy
- minimize motive power in all piping and ductwork systems
- achieve a building energy demand goal of 17 kBtu per ft^2 per year [53.6 kWh per m^2 per year]
- generate energy on site from solar radiation and biomass
- achieve carbon-neutral building operation
- utilize on-site wood resources in the building

The center was built on land that was previously used for housing; hence, no new land was built upon. Ninety-thousand board feet [27,400 board meters] of wood harvested from nearby forests, including many red and white pines planted by Aldo Leopold himself in the 1930s, were used in the building. The smaller trees were kept "in the round," which allowed the sapwood, the strongest part of the wood, to remain intact. Each tree selectively harvested from the site was measured and its use within the building accounted for as part of the design process. Natural ventilation, daylighting, and passive solar heating strategies helped the center reach the carbon-neutral performance benchmark, although active systems are needed to meet the occupants' comfort needs during very cold and during hot and humid conditions (Figures 3.47–3.50). Energy sources for the building include on-site photovoltaic panels, solar hot water collectors, and wood-burning heat sources. Radiant floors meet heating and

Figure 3.47 *Administrative center of the Aldo Leopold Legacy Center, with clerestories and daylight zoning throughout the building—strategies that helped to reduce the use of electric lighting. (© The Kubala Washatko Architects, Inc.; used with permission.)*

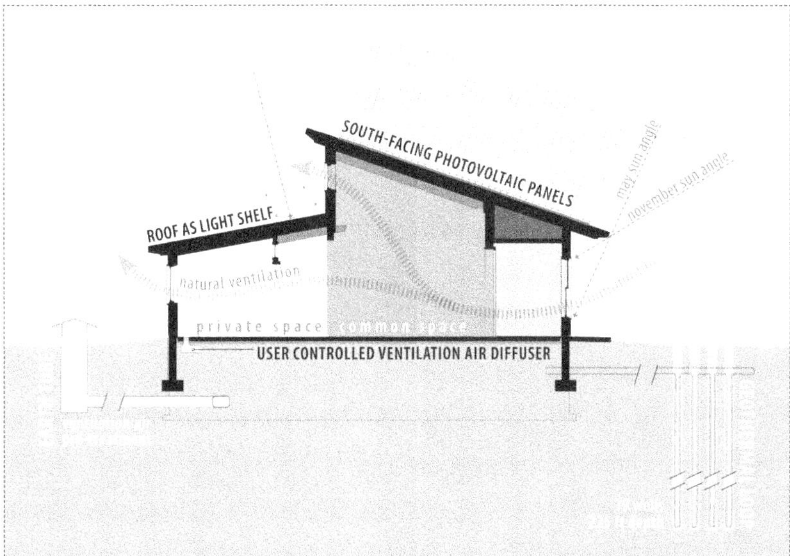

Figure 3.48 *Building section showing strategies for lighting, heating, cooling, and ventilation. (© The Kubala Washatko Architects, Inc.; used with permission.)*

cooling loads in the offices and exhibit space. Required ventilation air is tempered by earth tubes. Additional energy, if needed, is purchased from wind power providers. On a net annual basis, the ALLC is designed to produce 110% of the energy required for its operation.

Design Criteria and Validation. The project achieved a LEED Platinum rating, with an exceptionally high number of certification points. The project specifically earned a point in the Innovation and Design category for achieving carbon neutrality in building operation. To develop an appropriate energy use goal

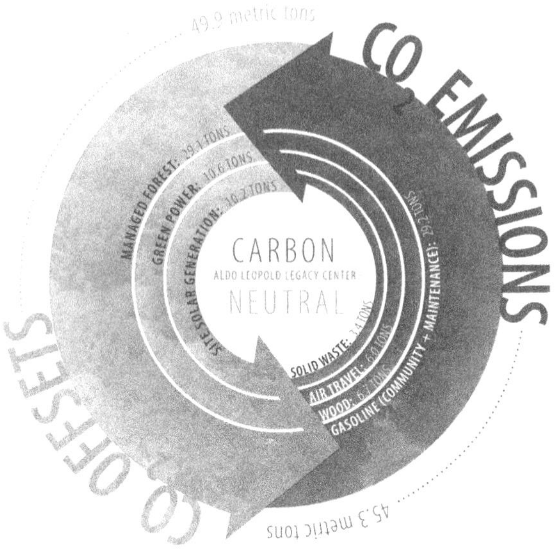

Figure 3.49 A carbon emissions diagram showing carbon neutrality: an annual balance of emissions versus offsets. (© The Kubala Washatko Architects, Inc.; used with permission. Redrawn by Sharon Alitema.)

(*a*)

(*b*)

Figure 3.50 (a) Trees selectively harvested from the site. (b) Accounting for each tree and placement of each board within the building. (© The Kubala Washatko Architects, Inc.; used with permission.)

for the building, the design team researched energy consumption and production values for other buildings of similar size and function and also calculated the amount of solar radiation available on the site, to match the potential for solar energy use with building needs. The design team set an energy use intensity (EUI) goal of 17 kBtu per ft^2 per year [53.6 kWh per m^2 per year] based on the performance of Woods Hole Research Center and the Schlitz Audubon Nature Center (both high-performing buildings). Using the program size at that time, 10,000 ft^2 [929 m^2], the design team established a renewable energy production requirement of 50,000 kWh, which would require roughly 3000 ft^2 [280 m^2] of PV collectors.

Post-Occupancy Validation Methods. Energy modeling using the TRNSYS software package integrated with the CONTAM program was performed during the preliminary design phases. Performance monitoring has occurred since the building was completed. The building uses two energy sources: electricity and wood. Electricity is monitored by two installed meters that measure electricity produced on site (surplus sent to the grid) and electricity consumed by the building.

The original design model suggested that the building would generate somewhat more energy from on-site renewable energy than the building would require for operation. In its first year of operation, the building required approximately 4300 Btu per square foot per year [13.57 kWh/m^2] more than the renewable energy generation (Figure 3.51). A colder than normal winter, record annual snowfalls, snow cover on the PV array in January and February, higher plug loads than estimated, and some inverter issues in 2008 and 2011 are presumed to have caused the first year's performance gap and other slight shortfalls.

For the years 2008–2010 (Figure 3.51), the simulated annual net energy flow is 0.51 kWh/ft^2 [5.4 kWh/m^2] per year. The actual net positive flow to the grid in years 2009, 2010, and 2012 was about 0.77 kWh/ft^2 [8.3 kWh/m^2] per year, which means the building did well! Wood consumption has averaged half a cord of wood per year, contributing an equivalent EUI of 1.06 kBtu/ft^2 [3.34 kWh/m^2] per year (Utzinger and Swenson, 2012).

Although the building falls short of achieving net-zero energy goals, the Foundation far surpasses carbon neutrality through the managed forests, which remove 110.1 metric tons of carbon dioxide from the atmosphere per year, more than offsetting the 46.99 metric tons of carbon dioxide emitted by the organization.

Performance Data. Information available to date suggests that the design team succeeded in upholding

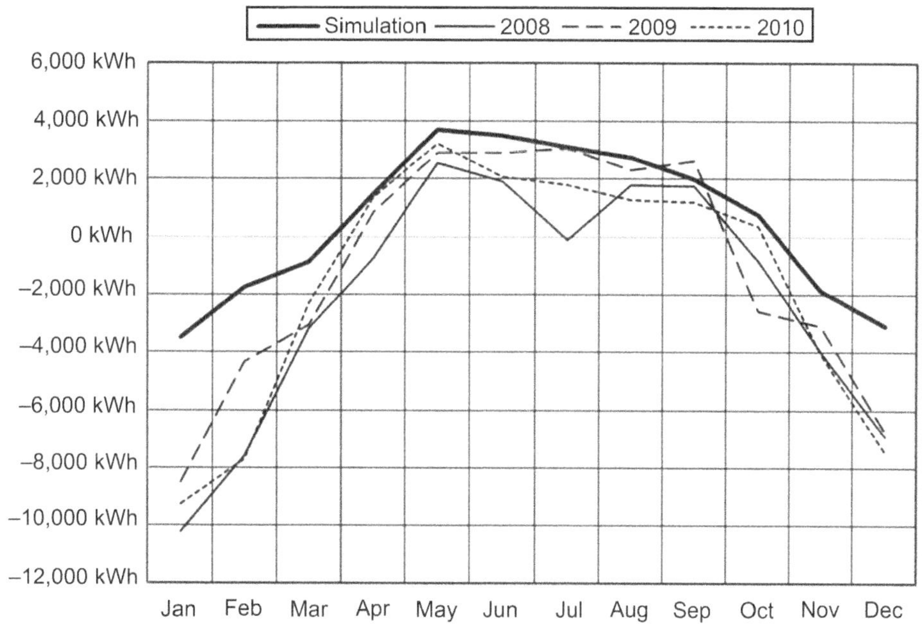

Figure 3.51 *Simulated and actual monthly electrical energy use. (© D. Michael Utzinger; used with permission.)*

the Leopold legacy by utilizing on-site energy and resources:

- The building achieved a LEED Platinum rating.
- Total emissions: 45.3 metric tons of CO_2/yr.
- Offset from renewable energies: −20.8 metric tons of CO_2/yr.
- Forest sequestration: −29.1 metric tons of CO_2/yr.
- Net result: −4.6 metric tons of CO_2/yr.
- Half of all wood used in the building, 90,000 board feet [27,400 board meters], was harvested on site and milled 70 mi [112 km] away.
- The building materials on average only traveled 100 miles per pound [359 km per kg] to the site.
- More than 95% of construction waste was recycled.
- The project was designed to meet the goals of the 2030 Challenge.
- The building uses 70% less energy than a typical building of its size.

The building was named an AIA COTE Top Ten Building for 2008.

For Further Information

Aldo Leopold Foundation. www.aldoleopold.org

Boehland, J. 2008. "Building on Aldo Leopold's Legacy." *Green Source* (April): 80–85.

Chapa, J. 2007. "First LEED Platinum Carbon Neutral Building!" Inhabitat. inhabitat.com/first-leed-platinum-carbon-neutral-building

Kubala Washatko Architects, Inc. www.tkwa.com

Utzinger, M. and S. Swenson. 2012. "The Aldo Leopold Legacy Center: Expanding the Definition of 'Community' in Carbon Management." *Planning Theory and Practice* 13(1): 156–166.

References and Resources

AIA. 2013. *Top Ten Metrics: Portland Community College Newberg Center*. Washington, DC: American Institute of Architects. www.aiatopten.org/node/75

ASES. 1981. *Passive Cooling*. Boulder, CO: American Solar Energy Society.

Beranek, W. J. 1980. "General Rules for the Determination of Wind Environment," in *Wind Engineering: Proceedings of the Fifth International Conference* (Vol. 1, pp. 225–234), J. E. Cermak (ed.). Elmsford, NY: Pergamon Press.

Bragdon, C. 1971. *Noise Pollution: The Unquiet Crisis*. Philadelphia, PA: University of Pennsylvania Press.

Brown, R. D. and T. J. Gillespie. 1995. *Microclimatic Landscape Design: Creating Thermal Comfort and Energy Efficiency*. New York: John Wiley & Sons.

Caborn, J. M. 1957. *Shelterbelts and Microclimate*. Edinburgh: H.M. Stationery Office.

Caulkins, M. 2012. *The Sustainable Sites Handbook: A Complete Guide to the Principles, Strategies, and Best Practices for Sustainable Landscapes*. Hoboken, NJ: John Wiley & Sons.

Cohen, S., D. C. Glass, and J. E. Singer. 1973. "Apartment Noise, Auditory Discrimination, and Reading Ability in Children." *Journal of Experimental Social Psychology* 9(5): 407–422.

DeKay, M. 1992. "A Comparative Review of Daylight Planning Tools and a Rule-of-Thumb for Street Width to Building Height Ratio," in *Proceedings of the 17th National Passive Solar Conference*. Boulder, CO: American Solar Energy Society.

DOE. 2024. *EnergyPlus*. Washington, DC: U.S. Department of Energy. energyplus.net

Doelle, L. 1972. *Environmental Acoustics*. New York: McGraw-Hill.

Gandemer, J. 1977. "Wind Environment Around Buildings: Aerodynamic Concepts," in *Wind Effects on Buildings and Structures*, K. J. Eaton (ed.). New York: Cambridge University Press.

Graedel, T. E. and P. J. Crutzen. 1989. "The Changing Atmosphere." *Scientific American* 261(3): 58–69.

Hardin, G. 1968. "The Tragedy of the Commons." *Science* 162(3859): 1243–1248.

IPCC. 2014. *Climate Change 2014: Synthesis Report*. Contribution of Working Groups I, II and III to the Fifth Assessment Report of the IPCC, R. K. Pachauri, and L. A. Meyer (eds.). Geneva: Intergovernmental Panel on Climate Change. https://www.ipcc.ch/report/ar5/syr/

IPCC. 2021. *Climate Change 2021: Sixth Assessment Report*. Contribution of Working Groups I, II and III to the Sixth Assessment Report of the IPCC, H. Lee, and J. Romero (eds.). Geneva: Intergovernmental Panel on Climate Change. www.ipcc.ch/assessment-report/ar6/

Knowles, R. 1981. *Sun, Rhythm, Form*. Cambridge, MA: MIT Press.

Landsberg, H. E. 1970. "Climates and Urban Planning," in *Urban Climates* (WMO Technical Note 108). Geneva: World Meteorological Organization.

Lawrie, L. K. and D. B. Crawley. 2022. "*Development of Global Typical Meteorological Years (TMYx)*." climate.onebuilding.org/

Lowry, W. P. 1988. *Atmospheric Ecology for Designers and Planners*. New York: Van Nostrand Reinhold.

Lowry, W. P. and P. P. Lowry. 1995. *Fundamentals of Biometeorology, Vol. 2: The Biological Environment*. McMinnville, OR: Peavine Publications.

Marsh, W. M. 1991. *Landscape Planning, Environmental Application*, 2nd edn. New York: John Wiley & Sons.

Mazria, E. and D. Winitsky. 1976. *Solar Guide and Calculator*. Eugene, OR: University of Oregon.

McPherson, E. G. (ed.). 1984. *Energy-Conserving Site Design*. Washington, DC: American Society of Landscape Architects.

Milne, M. n.d. *Energy Design Tools*. Los Angeles, CA: University of California. www.energy-design-tools.aud.ucla.edu

Montgomery, D. 1987. "Landscaping as a Passive Solar Strategy." *Passive Solar Journal* 4(1): 78–108.

National Weather Service (NWS), National Oceanic and Atmospheric Administration (NOAA). n.d. "Beaufort Wind Scale." www.weather.gov/mfl/beaufort

NOAA. 2025a. "Assessing the Global Climate in 2024," in *National Centers for Environmental Information*. Asheville, NC: National Oceanic and Atmospheric Administration. www.ncei.noaa.gov/news/global-climate-202413

NOAA. 2025b. "Climate at a Glance: Global Time Series," in *National Centers for Environmental Information*. Asheville, NC: National Oceanic and Atmospheric Administration. www.ncei.noaa.gov/access/monitoring/climate-at-a-glance/global/time-series

Olgyay, V. 1963. *Design with Climate: Bioclimatic Approach to Architectural Regionalism*. Princeton, NJ: Princeton University Press.

Reynolds, J. S. 1976. *Solar Energy for Pacific Northwest Buildings*. Eugene, OR: University of Oregon.

Watson, D. and K. Labs. 1983. *Climatic Design*. New York: McGraw-Hill.

PART II

DESIGN FUNDAMENTALS

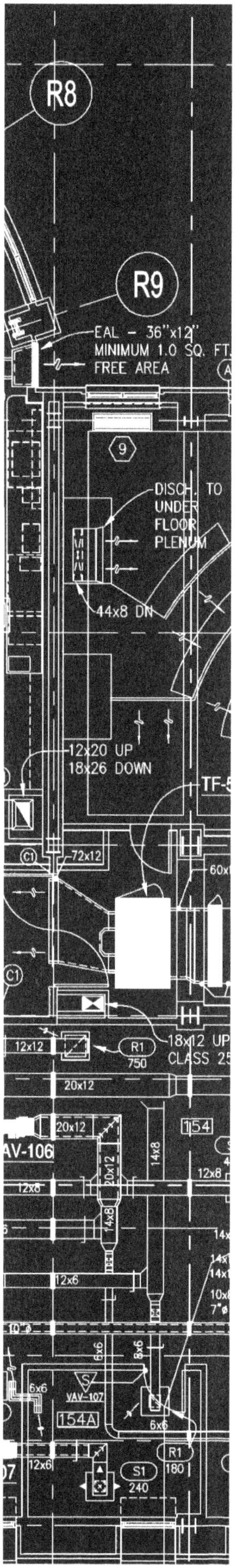

The chapters that comprise Part II lay the foundation for decisions that will be made from the earliest conceptual design through design development. The topics addressed herein—health and wellness, indoor air quality, thermal comfort, visual comfort, and acoustical comfort—can be a source of design inspiration or a challenge. Chapters 4–8 provide context and content for the development of a useful and viable owner's project requirements document, which is arguably the key to a successful building project.

Chapter 4 discusses health and wellness. These outcomes have always been in the background during the design process but are moving toward more explicit design expectations and engagement (propelled in part by the COVID-19 pandemic). Chapter 5 introduces indoor air quality, which is a major concern of building occupants and designers and a focus of green building design efforts. Chapter 6 outlines the basics of human thermal comfort, the nature of conditions that are generally deemed comfortable, and the design implications of a physiologically-driven occupant perception. Chapter 7 presents the fundamentals of light, its reception by the human eye, and visual comfort under various building conditions. Chapter 8 discusses the physics of sound, noise, and vibration and how the human ear is impacted—setting the stage for designing to deliver excellent acoustical conditions.

CHAPTER 4

Health and Wellness

4.1 INTRODUCTION

Ongoing developments have led to this new chapter. Perhaps most public is the emergence of the *WELL Building Standard* as a voluntary certification platform with a clear focus on both health and wellness in the built environment (IWBI, 2020). Less public is the ongoing evolution of the American Society of Heating, Refrigerating and Air-Conditioning (ASHRAE) position on health as a design outcome of interest. The formal thinking of this key professional organization is in flux, with the development of a position document on "Human Health and Wellness in the Built Environment" underway as this chapter is written.

The term health refers to a multifaceted and dynamic state encompassing various dimensions of an individual's physical, mental, and social condition. The World Health Organization (WHO) defines health as "a state of complete physical, mental and social well-being and not merely the absence of disease or infirmity" (WHO, 2025). An important implication of this definition for building design is that human health is more than just the absence of illnesses, disorders, or disabilities. Buildings must do more than just avoid causing health problems; they should engender well-being.

Wellness is a broad concept. It encompasses an individual's proactive pursuit of activities and lifestyle choices that lead to improved physical, mental, and social outcomes. Much like the notion of "comfort," wellness is subjective and can be evaluated in terms of an individual's satisfaction or sense of fulfillment across the various dimensions of life.

The term "healthy buildings" is used to denote structures and systems that endeavor to enhance the physiological, psychological, and sociological standing of their occupants through their materiality and performance attributes (Figure 4.1). The related term "active design" refers to the development of built environments that encourage physical activity, access to recreational amenities, walkable communities, nonvehicular mobility infrastructure, and universal access.

Designers cannot guarantee that building occupants will be healthy or well—but must surely make the effort to ensure that the environments building users encounter support health and encourage wellness.

4.2 HEALTHY BUILDINGS

(a) General Determinants of Health

The contribution of human health determinants to premature deaths have been categorized as follows (McGinnis et al., 2002):

- behavioral patterns: 40%
- genetic predisposition: 30%
- social circumstances: 15%
- healthcare: 10%
- environmental exposure: 5%

Stated differently, our lifestyles and surroundings account for approximately 45% of premature deaths. Second only to smoking, obesity and inactivity account for nearly twice the number of deaths in the

Mechanical and Electrical Equipment for Buildings, Fourteenth Edition. Walter Grondzik, Daniel Overbey, and Alison Kwok.

Companion website: www.wiley.com/go/meeb14e

Figure 4.1 The Living Building Challenge-certified Kendeda Building for Innovative Sustainable Design at the Georgia Institute of Technology integrates healthy building design strategies, active design elements, and biophilic principles.

United States as drugs, alcohol, guns, motor vehicles, and sexual behavior combined (Pilar et al., 2020). Moreover, humans have become an indoor species. According to the United States Environmental Protection Agency (EPA), people spend approximately 90% of their time indoors (EPA, 2022). While many countries have instituted a comprehensive regulatory structure monitoring outdoor environmental quality, there is nothing more impactful to our health than our indoor environments. Through concern for healthy buildings and active design, our structures and their systems can have a profound impact on human health and wellness.

(b) Psychological and Physiological Benefits

The terms "psychological" and "physiological" are often used in the context of healthy buildings. Psychological pertains to the psyche (think "mind"), which encompasses cognitive abilities, emotions, personality characteristics, and subconscious processes. Physiological refers to physical and biological processes (think "body"). It is important to distinguish between the two. Potential psychological benefits from healthy buildings may include stress reduction, enhanced cognitive function, and improved mood, whereas potential physiological benefits may include enhanced sleep quality due to lighting schemes, fewer respiratory issues because of clearer air, and decreased risk of chronic diseases through increased physical activity encouraged by building features.

(c) Sociological Benefits

When the term "sociological" is used in the context of healthy buildings, it typically refers to benefits that enhance how communities interact and how social norms can influence behavioral responses. If our built environment fosters greater engagement among users, social relationships and institutional bodies can benefit. An office environment developed with healthy building strategies and active design principles—in addition to well-documented benefits that include increased productivity, enhanced cognitive function, and improved moods—can lead to stronger relational networks that benefit users and institutions alike by encouraging user engagement, enhanced social interaction, and collaboration.

(d) Economic Benefits

Buildings in a Post-Pandemic World. With a market value well over US $1 trillion, commercial real estate is among the most valuable asset classes in the United States. Yet, in many cities, our streets feel empty, storefronts look vacant, and offices feel abandoned. The COVID-19 pandemic was an accelerant for hybrid and remote workplace trends and now employers are searching for ways to attract workers back to the office and investors are looking for a competitive edge.

Spike in Demand for Healthy Buildings. One potential avenue of commercial real estate market differentiation may be healthy buildings. A recent survey by the United Nations Environment Programme Finance Initiative, BentallGreenOak, and the Center for Active Design solicited feedback from a range of enterprises with combined US $5.75 trillion in global assets revealed an "extensive" post-pandemic spike in demand for "heathy buildings" (Murray et al., 2021). A staggering 89% of respondents described the current demand for healthy buildings as either "moderate" or "strong." The data also indicated that healthy building may yield a 4.4–7.0% increase in rental premiums.

Healthy Buildings Unlock People Power. Increased market demand is only the tip of the iceberg. Emerging data offer a compelling economic case for health buildings, including:

- 26% higher cognitive function scores and better sleep (Allen et al., 2016; MacNaughton et al., 2017)
- 1.6 fewer days of absenteeism due to illness each year (Myatt et al., 2002)
- 2–10% productivity gains by workers (Fisk, 2000; Wargocki et al., 2000; Olesen, 2007)

Healthy Buildings and Personnel Costs. Consider these data in the context of the "3-30-300™ rule" of real estate, which was popularized by the international facilities management company, Jones Lang LaSalle (JLL, 2016). The numbers represent the magnitude of typical commercial real estate costs per unit floor area per year for utility costs associated with energy ($3), rent and related operating costs ($30), and employee salary and benefit costs ($300). By this rule-of-thumb, a modest 2% increase in staff productivity from a healthy building would equate to an annual savings of $6 per square foot [$65 m^2]—double what a company will spend on energy costs. This is not an argument against energy efficiency, but rather an argument for clearly considering and articulating desired building outcomes—including health and wellness.

4.3 CONSIDERATIONS FOR HEALTHY BUILDINGS

This section presents various considerations for healthy buildings. Like other generalized design concepts, such as "high-performance" or "resilience," the term "healthy" encompasses a wide range of considerations in building design and construction.

(a) Air Quality

Indoor air quality (IAQ) plays a pivotal role in establishing healthy building environments, as it significantly influences the physiological health and cognitive performance of occupants. Healthy buildings should take measures to safeguard occupants against both indoor and outdoor contamination sources that can impact IAQ. See Chapter 5 for extensive discussion of IAQ concerns and solutions.

- *Ventilation:* Ventilation in the context of healthy buildings refers to the circulation of clean air (typically outdoor air or highly filtered recirculated air) as a means of diluting contaminants in the indoor air. Ventilation may be accomplished passively or actively. Ventilation for IAQ is distinct from ventilation for building cooling—although ventilation for cooling and for air quality may occur simultaneously. See Chapter 5 for more extensive discussion of ventilation.
- *Enhanced ventilation: ASHRAE Standard 62.1: Ventilation and Acceptable Indoor Air Quality* and *ASHRAE Standard 62.2: Ventilation and Acceptable Indoor Air Quality in Residential Buildings* establish baseline requirements for ventilation system design, including minimum ventilation rates and other measures intended to deliver acceptable IAQ. These standards are also discussed in Chapter 5. Enhanced ventilation, however, is discussed in this section because of its strong connection to healthy building design.

Increased ventilation air delivery to indoor spaces can increase oxygenation, reduce carbon dioxide (CO_2) levels, and lower the concentration of air pollutants—all of which can increase wellness. Increased ventilation is relative to the minimum ventilation rate required by prevailing standards (such as *ASHRAE Standard 62.1*). An enhanced ventilation strategy can help achieve the following benefits.

Increased Cognitive Function. There is growing evidence suggesting a correlation between increased ventilation and improved cognitive function. Ongoing research at Harvard University examining links between IAQ and cognitive function—known collectively as The COGfx Study—found that doubling the ventilation rate from 20 cfm [9.4 L/s] per person to 40 cfm [18.9 L/s] per person significantly improved cognitive function. By doubling the ventilation rate, the indoor CO_2 levels dropped from 950 ppm to 600 ppm and the concentration of VOCs (volatile organic compounds) dropped by at least 90% (MacNaughton, et al., 2015, 2017).

Reduced Disease Transmission. Many disease carriers, such as bacteria and viruses, are largely spread through the air. Coupled with an effective filtration strategy, increased ventilation can reduce the concentration of such airborne pathogens; thus, lowering the risk and rate of transmission.

Better Humidity Control. In mechanically conditioned buildings, certain spaces can experience increased humidity due to a variety of factors including outdoor air infiltration, vaporization of localized moisture exposure, or certain types of equipment and/or activities within the spaces. In such instances, increased ventilation can help disperse and stabilize humidity levels inside a building. Effectively managed relative humidity can lead to increased thermal comfort and mitigate mold and mildew growth.

Dilution and Removal of Organic and Inorganic Gases. An indoor air volume may contain a combination of organic and inorganic gaseous compounds. Organic gases are produced by biological processes and consist of carbon atoms bonded to hydrogen atoms. Examples include benzene, formaldehyde, and toluene. Inorganic gases do not contain carbon–hydrogen bonds but may still be naturally occurring, such as carbon dioxide, carbon monoxide, nitrogen dioxide, and ozone.

Indoor air may contain harmful volatile organic compounds emitted from paint, furniture, and cleaning products. Such gases are composed of very small particulates—thus, a robust filtration strategy will be required to reduce the concentration of such contaminants. In many instances, however, the source of these gases may be within the space itself.

Through continuous air quality monitoring and increased ventilation when/where warranted, both organic and inorganic gas concentrations can be managed by introducing increased volumes of filtered and/or outdoor when necessary. For example, enhanced IAQ parameters may set a maximum concentration threshold of 500 $\mu g/m^3$ for total volatile organic compounds (TVOCs), 10 $\mu g/m^3$ for carbon monoxide, and 100 $\mu g/m^3$ for ozone.

Whole-Building Flush-Out. Because organic and inorganic gases often emit most acutely from indoor pollution sources soon after installation, a whole-building flush-out is a common best practice to mitigate the concentration of these contaminants. A flush-out procedure involves ventilating a building by running its heating, ventilation, and air-conditioning (HVAC) system continuously, supplying 100% outside air at a specified rate to dilute or effectively eliminate construction-related indoor pollutant concentrations. This process is usually conducted before occupancy and after all construction and finishing work, including painting and carpeting, is complete. Prior to occupancy, HVAC air filters are replaced.

Air Testing. Verification of appropriate IAQ, including levels of particulate matter and TVOCs (total volatile organic compounds), can be ascertained from on-site testing of air samples from strategic locations throughout a building. Continuous monitoring through permanently installed sensors is becoming a common testing practice (Figure 4.2). Current best practice, including criteria advanced by Leadership in Energy and Environmental Design (LEED) and WELL Building Standard, entails TVOCs (total volatile organic compounds) testing the air within occupied indoor spaces for particulate matter and inorganic gases under ventilation conditions typical for occupancy. Thresholds for particulate matter typically address $PM_{2.5}$ and PM_{10} (two sizes of particulate matter of common concern); VOCs of interest include benzene, formaldehyde, and toluene; inorganic gases include carbon monoxide, carbon dioxide, and ozone. Continuous radon monitors are also recommended where such gases may be present.

Figure 4.2 Kaiterra's smart air quality monitors are designed for residential and commercial workplace applications and feature modular sensor designs for easy calibration, real-time data analytics, and seamless integration with building management systems. (© Kaiterra; used with permission.)

(b) Water Quality

The supply of adequate water quantities to buildings under appropriate delivery and safety standards has been a historical concern of building MEP (mechanical, electrical, and plumbing) systems. More recently concern for water conservation has become a focus of green design efforts. Even more recently, water quality within buildings has come on the radar of designers as a health and wellness factor. Water quality will generally be determined by the following properties.

Physical Characteristics. Physical impurities such as sediments, debris, and suspended solids can affect water quality. While physical contaminants may not directly cause disease, they can impair water clarity, taste, and odor. Additionally, particulate matter in water within certain temperature ranges may harbor microorganisms and foster their growth, which can impact human health. Physical characteristics such as turbidity, color, taste, temperature, and foamability can be altered through water treatment systems.

Chemical Characteristics. Chemical pollutants in water can pose a human health risk. These contaminants may include heavy metals, pesticides, herbicides, industrial biproducts, pharmaceuticals, or even an overabundance of certain nutrients. Chronic exposure to such impurities through supply water or water features can lead to a variety of health issues, including neurological disorders, reproductive disorders, immune system dysfunction, developmental abnormalities, hormonal imbalances, and cancer. The presence of chemical impurities in water can impact the alkalinity, hardness, and toxicity of water.

Biological Characteristics. Potable water sources should be kept as free as possible from disease-producing organisms such as bacteria, viruses, and protozoa. Such pathogens can cause humans to contract a range of diseases such as cholera, typhoid fever, hepatitis A, and giardiasis. Contaminated water sources can lead to outbreaks of illnesses, particularly in areas with inadequate sanitation and water treatment facilities. Biological hazards can be mitigated by careful water sourcing, protection, monitoring, and treatment.

Radiological Characteristics. Radioactive substances (e.g., uranium, radium, or radon) can contaminate water sources through natural processes or human activities, such as nuclear power generation or mining. Exposure of water sources to radioactive contaminants can pose dire health risks, including an increased probability of cancer, genetic mutations, and organ damage. Monitoring and regulation of radiological contaminants in water sources are vital to public safety and can help avert long-term human health consequences.

These defining characteristics of water quality are prone to interplay. Certain physical characteristics (e.g., temperature) can exacerbate thermal degradation and leaching of chemicals through polymer-based water storage and conveyance systems. The presence of nutrient pollution (e.g., nitrogen and phosphorus) can foster toxic algal blooms. A comprehensive approach to water quality is prudent. The characteristics of water quality as well as filtration, disinfection, and other system-related treatments for water supply, are addressed more thoroughly in Chapter 21.

(c) Moisture

The section above dealt with water provided for domestic building purposes—including consumption. This section on moisture looks at water more broadly, including in the vapor state. Poor moisture management—whether from bulk water, moisture diffusion, or moisture in the air—can lead to threats to human health and wellness.

Bulk Water. Water leaks—from mismanaged rain, leaky plumbing systems, or uncontrolled condensate from mechanical systems—can lead to the rot, decay, or corrosion of building materials that are meant to stay dry. This can lead to mold and mildew growth, which can cause contaminants to enter the building's air mass—posing a direct human health risk via degraded air quality. Proper architectural detailing should anticipate bulk water intrusion and provide assemblies and systems that mitigate the human health threats stemming from bulk water. Interior leaks should also be anticipated and addressed through design.

Moisture Diffusion. Moisture easily and frequently changes phase between solid, liquid, and gas. As such, moisture can migrate through building enclosure assemblies in several ways. If mismanaged, water vapor can meet its dew point and condense—leading to bulk water collection and the aforementioned human health concerns. Even within well-conceived building enclosure solutions, envelope transitions can be fraught with conditions conducive to potential moisture intrusion. Envelope transitions are areas in a building's enclosure where different materials and/or assemblies come together. Examples include the roof meeting walls, windows or doors intersecting with walls, or walls meeting foundations. The design and construction of these transitions are critical because they naturally consist of irregularities, which can be susceptible to moisture-related problems if not properly managed. Proper moisture control throughout all enclosure assemblies and transitions can mitigate the human health threats of moisture diffusion.

Relative Humidity. Relative humidity is a key factor in thermal comfort. There are also direct human health implications of poor indoor air humidity control. Excessively high relative humidity can facilitate the propagation of certain types of mites and fungi in addition to exacerbating chemical interactions (e.g., material off-gassing of VOCs). Low relative humidity can lead to increased rates of respiratory infections and ozone production. Either humidity extreme can promote bacterial growth, virus transmission, allergic rhinitis, and asthma symptoms. More information about thermal comfort and the interplay of heat and moisture can be found in Chapter 6. Read more about moisture control fundamentals in Chapter 9, which presents the concept of control layers for building enclosures.

(d) Materials

The creation, delivery, installation, and removal of every item of a building come with finite and measurable environmental and human health impacts—whether positive or negative. Simply put, materials matter. From paint that emits VOCs to plastic pipes that react with chlorinated water to furniture that contains brominated flame retardants, every building product and material has an impact on human health.

Invariably, every item procured and installed requires material and energy resources.

Healthy buildings aim to reduce human exposure, whether direct or indirect, to chemicals that may adversely impact human health. There are two basic considerations through which healthier building products and materials may be pursued: disclosure and optimization.

Building Product Disclosures. The first strategy is to increase awareness of environmental and human health risks posed by products. Such awareness may be conveyed to all stakeholders of a building through product disclosures. The transparency offered by these disclosures enables comparisons between similar choices of materials for a building and can foster better-informed decision making. The two most common types of building product disclosure documents are environmental product declarations and health product declarations.

Environmental Product Declarations. An environmental product declaration is a building product disclosure document that summarizes the results of a life-cycle assessment (LCA) and conveys a product's environmental impacts. EPDs are often based on internationally recognized standards and utilize adopted product category rules (PCRs) to offer a transparent, consistent, and comprehensive summary across a number of specific environmental impact categories.

There are three basic types of EPDs:

Type I EPD constitutes a third-party verified "eco-label" that indicates a product has met certain environmental impact criteria.

Type II EPDs are self-declared by a manufacturer and therefore unverified.

Type III EPDs are independently verified by a third party. Type III EPDs may address a generic product (providing industry-wide characteristics) or may be product-specific. In either case, Type III EPDs are registered documents that present transparent and comparable data about the overall environmental performance of a product, including factors such as energy consumption, global warming potential, and resource depletion.

EPDs are increasingly used in various industries, especially in construction and building materials, to support more environmentally responsible design and procurement practices. EPDs do not, however, explicitly address human health impacts.

Health Product Declarations. A health product declaration is similar in intent to an EPD; however, rather than reporting on building product or material environmental impacts, it outlines human health concerns. HPDs provide a thorough accounting of potential chemicals of concern—including any "red list" chemicals. Developed by the nonprofit Health Product Declaration Collaborative (HPDC), the protocol for developing an HPD involves a thorough assessment of a product's ingredients and the human health concerns associated with those ingredients. Manufacturers are encouraged to report on every ingredient down to 1000 ppm (or even 100 ppm for more comprehensive declarations). This level of specificity helps in identifying and minimizing the use of materials with known or suspected health hazards.

As with EPDs, not all HPDs are third-party verified. Manufacturers have the option to self-declare their HDPs or seek a third-party verification to enhance the credibility of the information disclosed in the HPD. Third-party verification requires an independent review by an HPDC-approved verifier.

Material Ingredient Reporting. Beyond EPDs and HPDs, other commonly encountered building product and material ingredient reporting programs and frameworks include:

- ANSI/BIFMA e3 Furniture Sustainability Standard
- Cradle to Cradle Certified Products Standard
- Declare
- Global GreenTag
- Living Product Challenge
- NSF/ANSI 336: Sustainability Assessment for Commercial Furnishings Fabric
- UL Solutions Sustainable Product Certification

The Red Lists. The International Living Future Institute's (ILFI) "Red List" represents the "worst in class" materials, chemicals, and elements known—or highly suspected—to pose a significant risk to environmental and human health. Regularly updated, the list is compiled using various criteria, including the presence of toxins, carcinogens, mutagens, and other harmful substances. Originally developed in 2006 in support of the Living Building Challenge (LBC) certification program, the LBC Red List contains a wide range of substances that have been associated with negative human health outcomes including certain polymers, flame retardants, phthalates, and heavy metals.

Though well-known and frequently referenced, ILFI's LBC Red List is but one of several generally termed "red lists" curated by different entities in the building design and construction space. The intent behind all the so-called red lists is the same: to provide a useful resource for building project teams that will allow them to avoid building products and materials that contain ingredients that can pose a significant environmental and/or human health risk.

The Precautionary Principle. "Vorsorgeprinzip" is a German term that roughly translates to "precautionary principle" in English. It is a concept that dates back at least to the 1970s which advocates for preventive action in the face of uncertainty regarding health and safety risk. A 1998 consensus statement stated it most clearly, "when an activity raises threats of harm to human health or the environment, precautionary measures should be taken even if some cause-and-effect relationships are not fully established scientifically." As it relates to healthy building design, the Precautionary Principle goes beyond the Red List in that if a building product or material is only suspected of posing an environmental and/or human health risk—call it a "watch list" item—then it should be avoided. In an abundance of caution, some building design and construction professionals might take a hardline stance against utilizing certain products and materials pursuant to the Precautionary Principle.

(e) Thermal Conditions

ASHRAE Standard 55: Thermal Environmental Conditions for Human Occupancy outlines conditions that will provide thermal comfort for most occupants, and it specifies combinations of indoor climate variables and personal factors that will positively influence comfort perceptions. A key to designing for better thermal "health" is realizing that the concept of comfort is subjective. What feels comfortable to one individual might be too warm or too cool for another. This variability in responses is influenced by environmental factors (i.e., ambient air temperature, air speed, relative humidity, and mean radiant temperature) and personal factors (i.e., metabolism and clothing) and an individual's psyche.

Thermal comfort is one of the more critical factors influencing overall satisfaction in buildings, impacting individual motivation, alertness, focus, and mood. Its influence on the integumentary, endocrine, and respiratory systems also underscores that our thermal environment impacts our health and wellness.

Healthy building considerations related to thermal comfort may include the following elements.

Air Temperature Control. Maintaining an appropriate ambient air temperature range is crucial for a healthful environment. A healthy building strategy might entail controls that allow for more personalized settings by occupants.

Humidity Management. Controlling the humidity within a building is essential for comfort; however, humidity also impacts health. Improper humidity can induce thermal stress in occupants. Sustained high or low relative humidity can foster certain types of bacterial growth, virus transmission, the spread of fungi and mites, elevated risk of respiratory infection, exacerbate allergic reactions, promulgate chemical interactions, or even produce ozone.

Radiant Surfaces. The mean radiant temperature of a space can have considerable sway over the thermal comfort of occupants. Though often uniform and consistent with the interior air temperature, the mean radiant temperature of a space can be affected by the ability of surfaces to absorb or emit radiant heat, the extent to which the surface is exposed to occupants (i.e., view factor), and the temperatures of other surrounding surfaces. Cold windows, uninsulated walls, and improperly sized radiant heating elements can be detrimental to thermal comfort—and thus to health. A healthy building will be designed with mindful consideration of occupant interactions with surface and component temperatures.

Reduced Infiltration. An airtight building enclosure mitigates sensible and latent heat exchange through unintended airflow. This will save energy; it will also improve thermal comfort by reducing "draftiness" in indoor spaces. In fact, infiltration has implications on air temperature control, humidity management, air quality, and noise pollution. Infiltration (uncontrolled air flow through the building enclosure) may be a transport mechanism for exterior pollutants.

Personalized Control. Personalized thermal controls in buildings are intended to cater to the individual comfort needs of occupants, thus recognizing that people have different thermal preferences. There is a wide range of potential strategies to provide such control, including localized ventilation controls, personal heating or ventilation devices, furniture that features integrated heating and cooling, smart building systems with Internet of Things (IoT) integration, and (of course) the ability to operate windows, shades, and blinds. Another form of personalized thermal control is the ability for occupants to relocate within a facility that offers a variety of conditions. The concept of "free address" allows users of a building to choose a space based on activity and personal preference regarding comfort conditions (including thermal, lighting, acoustic, and ergonomic). Personalized controls have the potential to leverage the underlying concept of adaptive thermal comfort to allow a range of indoor thermal conditions to improve the overall wellness of occupants.

See Chapter 6 for a more detailed discussion of thermal comfort. Generally, an occupant who is thermally comfortable will not experience thermal conditions detrimental to good health.

(f) Acoustic Conditions

Within any environment, human beings experience sounds in different ways. From delight to disturbance, from sounds that we barely notice on a conscious level to those we cannot ignore if we tried, from sounds that amplify our ability to interpret an environment to those that "mask" a space such that we will not be distributed by unwanted sound. A dynamic interplay of sounds is continuously around us as our acoustic environment is constantly changing.

Acoustic comfort can be defined as the subjective positive perception of sound in an environment. The way a human will experience sound in a space is fluid, complex, and individual. Acoustic comfort is an opinion of a person experiencing a given environment. Acoustic comfort encompasses the following considerations.

Sound Pressure Level. Exposure to high sound pressure levels (SPLs) can result in the perception of a noisy environment, causing stress, annoyance, and potential hearing damage over time. High SPLs can also interfere with sleep quality, communication, and have been shown to correlate with adverse cardiovascular effects and mental health impacts.

Reverberation Time. Even if sound pressure levels are appropriate, sounds can persist in spaces for too long. The time it takes for sound to decay in a space can be measured and expressed as reverberation time. An inappropriate persistence of sound can create a noisy and distracting environment, making communication difficult, which can lead to stress, fatigue, and decreased productivity.

Noise Transmission. "Noise" is defined as unwanted sound. Noise from HVAC systems, machinery, vehicular traffic, or neighboring spaces can be transmitted to a space and disrupt activities, causing a range of adverse human health impacts from the psychological (e.g., stress and anxiety due to overheard just-intelligible communication) to physiological (e.g., sleep deprivation, tinnitus, and hypertension) effects.

Emerging data are showing that acoustic comfort is important to human health and wellness. For instance, noise stressors have been shown to negatively impact productivity in the workplace (Mak and Lui, 2012). Global workplace productivity studies have consistently associated lack of concentration with noisy environments (Hedge, 1982; Navai and Veitch, 2003; Kaarlela-Tuomaala et al., 2009; Frontczak et al., 2012; Keus van de Poll et al., 2015). Research is also revealing potential physiological concerns from poor acoustic environments. A cross-national review study on children and traffic/aircraft noise indicated modest negative physiological health impacts at sound pressure levels of 40 dB(A), while levels between 40 dB and 50 dB were associated with strained concentration and impaired memory (Miedema, 2007; Babisch, 2014). Moreover, levels of 55 dB and above were associated with an increased risk of certain severe health impacts, including ischemic heart disease (Van Kempen and Babisch, 2012).

Healthy buildings can bolster occupant health and wellness through strategies aimed at improving acoustic comfort, including (but not limited to):

- designing exterior and/or interior building assemblies that aptly manage sound transmission
- noise control through sound-blocking or absorbing materials
- management of reverberation time through a combination of sound absorptive and/or reflective materials within a space
- sound-masking by which human conversation can be obscured to provide privacy
- creating "loud" and "quiet" zones for occupants such that they can choose a preferred acoustic environment

See Chapter 8 for a more detailed discussion of acoustic comfort. Generally, an occupant who is acoustically comfortable will not experience acoustic conditions detrimental to good health.

(g) Visual Conditions

The visual environment can impact visual comfort, circadian rhythms, and the mental health of building occupants. Lighting (daylighting and electric lighting) plays a key role in establishing the visual environment.

Visual comfort in any environment is influenced by multiple factors, including light levels, glare, and color rendering. Each factor contributes to how a person perceives and interacts with their surroundings, impacting not only the visual experience, but also mood, productivity, and overall wellness.

Light Levels. Proper light levels (both illuminance and luminance) are crucial for performing tasks effectively and comfortably. Insufficient light can cause eye strain and fatigue, making it difficult to focus on tasks. Excessive light can be just as detrimental, leading to discomfort and a waste of energy. Ideal light levels will vary depending on the activity taking place and any particular needs of the users of a space.

Glare. Glare occurs when there is a bright light source or excessive brightness contrast within the field of vision, such as when a bright window is in direct view or its image is reflected from a surface. Glare can cause discomfort, impair visibility, and lead to eye strain.

Color Rendering. This aspect of lighting pertains to how well colors are represented. Our collective and technical benchmark for color rendering is daylight. For instance, electric lighting sources with a high color rendering index (CRI) can accurately reproduce the colors of objects, relative to daylight, which is

Figure 4.3 The LEED Platinum certified Infosys Technology & Innovation Hub in Indianapolis, Indiana, achieved every credit in the Indoor Environmental Quality category under LEED v4/4.1 BD + C, including optimized daylight and verified IAQ through air testing. (Image courtesy of Browning Day; © Al Ensley Photography.)

essential in areas where color differentiation is important, such as retail spaces. Good color rendering can contribute to an environment's aesthetic appeal, but it can also affect psychological and physiological health. In building design, color rendering is determined by light source wavelength.

Balancing light levels, glare, and color rendering is essential for visual comfort, and thus for health and wellness (Figure 4.3). See Chapter 7 for further information on light, color, and visual comfort.

Circadian Rhythms. Lighting in healthy buildings should help to regulate circadian rhythms. The term refers to the natural, internal processes that regulate the diurnal sleep–wake cycle. The word "circadian" is a derivative of the Latin words "circa" (around) and "diem" (day). Some refer to this as a "biological clock." Maintaining a regular circadian rhythm is an important aspect of human health.

Circadian rhythms are kept on a regular cycle by various cues from the environment, including light. The human body's response to light is facilitated by intrinsically photosensitive retinal ganglion cells (ipRGCs). Essentially, ipRGCs are the eyes' non-image-forming photoreceptors. They relay ambient light levels and wavelength to the suprachiasmatic nucleus (SCN) in the brain through the retinohypothalamic tract—this is the main clock that acts as an oscillator to synchronize clocks in peripheral tissues and organs. High intensity light will promote bodily alertness, while a lack of light will cue the body to reduce energy expenditures and prepare for rest.

ipRGCs are especially sensitive to blue light. Daylight will prompt the brain to become more awake and alert. This is why staring at a computer screen before sleeping might not be a good idea for those with sleeping difficulties.

Throughout a 24-hour period, the body responds to a number of zeitgebers—the external cues that align

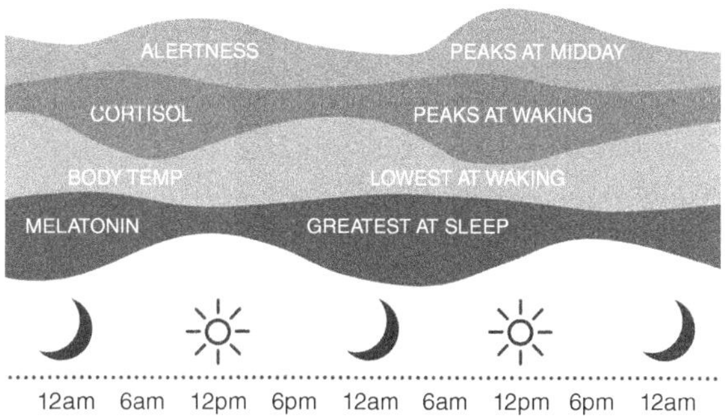

Figure 4.4 The human body's circadian rhythms follow a diurnal cycle in response to a number of zeitgebers, the most important of which is light. (Courtesy of Daniel Overbey.)

physiological functions to the daily solar cycle. Light is one of the most important zeitgebers. Light stimulates the ipRGCs, which trigger the suprachiasmatic nucleus to prompt the pineal gland to secrete less melatonin—the "darkness hormone" whose levels acts a driver for sleep (Figure 4.4).

Psychological Benefits of Lighting. Lighting can have a profound effect on the psychological health of building occupants. Lighting that considers human health can improve sleep quality, enhance mood, increase productivity and alertness, improve cognitive performance, and reduce Seasonal Affective Disorder (SAD). The Mayo Clinic defines this disorder as "a type of depression that's related to changes in seasons," which "starts in the fall and continues into the winter months, sapping [a person's] energy and making [them] feel moody" (Mayo Clinic, 2025). SAD is well-documented and characterized by recurrent episodes of depression that occur at the same time each year. Full-spectrum light has been demonstrated to hold antidepressant properties and can be used therapeutically in clinical settings (Wirz-Justice et al., 2013). Blue or blue-enriched light (which is easily achieved with access to daylight) can improve a person's mood (Glickman et al., 2006; Anderson et al., 2009).

(h) Mind

Mental health is an essential aspect of overall human health. It should not be defined simply as the absence of a diagnosed adverse psychological condition. Rather, mental health is a state of wellbeing that exists on a complex continuum, experienced differently by every individual. The World Health Organization (WHO) defines mental health as "a state of mental well-being that enables people to cope with the stresses of life, realize their abilities, learn well and work well, and contribute to their community" (WHO, 2022).

Mental health is influenced by the dynamic interplay of psychological, physiological, socioeconomic, and environmental factors. In this regard, our mental health is in constant flux—and at certain times, anybody may experience poor mental health. Approximately 1 in 5 adults will experience a common mental health condition, such as anxiety, depression, or substance abuse, over a 12-month period (NAMI, 2025); and over 30% of adults will experience a mental health condition during their lifetime (Steel et al., 2014). According to the WHO, around 1 in 7 adolescents will develop a mental disorder (WHO, 2024).

Just as our environment can impact our physical health, it can impact our mental state-of-mind. Considering that humans now spend most of their time indoors, our buildings can serve as tools to improve mental wellness.

Safety and Security. Healthy buildings must be safe and secure. Such buildings must meet or exceed code requirements for fire safety, provide appropriate emergency egress, deliver effective light and sound levels, monitor for hazards (such as high levels of carbon monoxide), and be designed around a holistic emergency action plan with systems and protocols for communication with building occupants.

There are psychological and physiological benefits to providing occupants with a personal sense of safety and security. Enhanced confidence in building safety and security can reduce anxiety and increase perceptions of comfort. Reduced production of stress hormones (e.g., cortisol) can positively affect various bodily functions, including sleep and immune response.

The Biophilic Hypothesis. In the decades after World War II, building designs trended toward emphasizing functionality, industrial advancements, and a modern aesthetic. Combined with low-cost fossil fuel energy resources and the advent of engineered systems, our structures made electric lighting and mechanical heating, ventilation, and air-conditioning (HVAC) commonplace. To occupants, a thermostat and fine-tuned indoor environments became the expectation. The emergence of mechanical and electrical equipment for buildings coincided with a relegation of nature and its dynamic conditions to the "outdoors."

As buildings increasingly separated humans from regular direct contact with nature, the Biophilic Hypothesis emerged—with roots in both evolutionary biology and psychology—which acknowledges an innate human affinity for nature. The term biophilia was popularized by biologist Edward O. Wilson in his 1984 book, *Biophilia*. The notion that people benefit from connections to nature is, however, not new. Chinese Taoists recognized the health benefits of gardens and greenhouses nearly 2000 years ago (Louv, 2005).

The 14 Patterns of Biophilic Design. Ever since Wilson's publication, architects and interior designers have explored the concept of biophilia. At first, biophilia was pursued in a rudimentary manner through basic design strategies focused on daylight, quality views of nature, and bringing vegetation indoors. In 2014, Bill Browning and co-authors from Terrapin Bright Green published *14 Patterns of Biophilic Design*, which outlined a variety of ways to incorporate biophilic design into architecture and urban design (Table 4.1). The 14 Patterns (now numbering 15) certainly included bringing nature into interior spaces through visual connections to vegetation, water features, and dynamic daylight; however, the publication expanded the biophilic design repertoire to include natural analogues such as biomorphic forms and visual complexity along with spatial experiences that evoke a sense of refuge, prospect, mystery—even peril—because of the unique psychological and physiological responses the human body can have to such experiences.

TABLE 4.1 The 14 Patterns of Biophilic Design

Nature in the Space Patterns	
1	Visual Connection with Nature
2	Non-Visual Connection with Nature
3	Non-Rhythmic Sensory Stimuli
4	Thermal & Airflow Variability
5	Presence of Water
6	Dynamic & Diffuse Light
7	Connection with Natural Systems
Natural Analogues Patterns	
8	Biomorphic Forms & Patterns
9	Material Connection with Nature
10	Complexity & Order
Nature of the Space Patterns	
11	Prospect
12	Refuge
13	Mystery
14	Risk/Peril
15	Awe

Source: From Browning et al., 2024.

The premise behind integrating biophilic design features is that healthy buildings should provide not only a variety of thermal, acoustic, and lighting conditions to enhance human health but also a variety of experiential and natural stimuli in order to reap the mental health benefits of reconnecting occupants to nature. Emerging research supports this premise. Recent studies suggest that green spaces and natural settings can provide a restorative effect as they have been associated with a variety of positive health outcomes (Burls, 2007; Van den Berg et al., 2010). Moreover, biophilic elements such as connections to nature through quality views, indoor elements that evoke nature, and direct access to the outdoors have been associated with various mental health benefits including improvement in mood, stress, and overall health (Ulrich, 2001; Browning et al., 2024). Exposure to nature has also been associated with improved cognitive functioning, specifically attention and working memory (Mayer et al., 2009; Tyrväinen et al., 2014; Gidlow et al., 2016).

Biophilia Versus Biomimicry. Biophilia and biomimicry are distinct concepts that are often erroneously interchanged. Whereas biophilia is the innate human affinity for connections to nature, biomimicry is the practice of "mimicking" biological solutions to address human-centric challenges. The concept of biomimicry was coined and popularized by biologist Janine Benyus in her 1997 book, *Biomimicry: Innovation Inspired by Nature*. Biomimicry can be defined as the interdisciplinary practice of observing and emulating nature's strategies, structures, and functions to develop technologies, materials, and methodologies.

These two terms can easily become conflated because certain building solutions can be both biophilic and biomimic in nature. This frequently occurs with structural solutions. Buildings that utilize geodesic domes, hyperbolic paraboloid structures, catenary arches, fractal growth patterns, the Fibonacci sequence, or harmonic proportions may offer the appeal of biophilia by mimicking the effective solutions of the world's oldest and most open-source design system—nature. In essence, biomimicry can be realized in buildings in three distinct ways:

- A building's structure and/or façade may be directly inspired by nature.
  Example: DP Architects and Michael Wilford & Partners' design of the Esplanade—Theaters on the Bay in Singapore utilizes a curved form comprised of an elaborate "spiky" skin inspired by the casing of the Durian fruit. The kinetic shading structure synchronizes with the movement of the Sun to protect the building from the tropical heat.
- A building's structure, façade, and/or systems may be inspired by the working or functioning of nature.
  Example: Mick Pearce, the architect of Eastgate Centre in Harare, Zimbabwe, studied the cooling chimneys and tunnels of termite dens to create a stack ventilation effect throughout the structure.
- A building's structure, façade, and/or systems may draw on nature to create a self-sufficient environment.
  Example: The Eden Project in Cornwall, United Kingdom, is a botanical garden and environmental education center housed in iconic biome structures comprised of steel and ETFE (ethylene tetrafluoroethylene) panels arranged in geodesic domes comprised of hexagonal panels. These enclosures modulate external climate conditions to mimic different climates within different domes—which house a wide variety of plant species from around the world.

(i) Active Design

The concept of "active design" has emerged in architecture and planning in response to increasing public concerns regarding obesity, diabetes, heart disease, and other chronic adverse health conditions exacerbated by a predominantly sedentary lifestyle, active design seeks to encourage increased physical activity among occupants through a range of strategies, including the following.

Enhanced Accessibility. The concept of "universal access" is centered around ensuring that all people—regardless of abilities—have equal access to spaces and services throughout a building or site. This requires attention to both physical environmental elements (e.g., ramps, elevators, restroom configurations) and informational access (physical and digital), which should accommodate visual, auditory, motor, and cognitive impairments.

Encourage Physical Movement. By integrating design features such as a prominently placed staircase, clearly demarcated walking paths with varied points of interest (e.g., views, art installations), and convenient proximity and access to public space and amenities, occupants may be more likely to take on increased physical activity—and benefit from the health outcomes.

Promote Active Travel. This can be accomplished by establishing facilities that encourage people to seek alternative modes of transportation other than automobiles. Such alternative modes are categorically referred to as "active travel" and can include cycling, running, walking—even emerging modes such as e-bikes, e-scooters, scooters, cargo bikes, recumbent bikes, and skateboards. Active travel can be supported through effective active travel facilities, conducive infrastructure (e.g., a variety of modes, access to routes and wayfinding, proper lighting, and security), suitable amenities (e.g., short-term and long-term storage options, shower and changing rooms, drying/airing provisions), maintenance and repair stations, clear signage, and "future-proofing" through a strategic degree of overprovision. The notion is that these active travel options become attractive to users as more cost-effective and efficient ways to travel that also contribute to improved health.

Community Connectivity. A successful active design solution must consider points of respite and destination. Strategic programmatic, spatial, and/or land-use layouts can create walkable communities and provide proximity to amenities such that the convenience of active travel results in improved human health through regular physical activity, environmental benefits through reductions in carbon-based transportation, and increased local economic support.

Integrate Biophilic Design. Active design solutions will often seek to reap the human health benefits of biophilic design features, which can include direct and indirect access to natural elements, forms, and patterns. Biophilic design may utilize vegetation, water features, natural materials, daylight and views, and various other forms of sensory stimuli (Figure 4.5).

By encouraging people to adopt a more active lifestyle and engage their communities, active design has the potential to yield several compounded benefits, including:

- *Improved physical and mental health:* Active design directly combats the risk of chronic health conditions that stem from a predominantly sedentary lifestyle. Moreover, physical activity, access to daylight, and interactions with nature have all been shown to help improve mental health.
- *Positive environmental impacts:* By encouraging walking and active travel, transportation-related greenhouse gas (GHG) emissions will be reduced.
- *Economic benefits:* In addition to lower healthcare costs and increased productivity, active design promotes local activity, which results in increased commerce.

Figure 4.5 *The design of Irsay Family YMCA at CityWay integrated a range of biophilic and active design elements. Pictured here are prominent stairways and a living wall within a daylit concourse offering ample views of downtown Indianapolis, Indiana. (Image courtesy of Browning Day; © Douglas Adams Photography.)*

The interrelationship between the mind and body is increasingly recognized. Mental and physical health impact each other. Healthy buildings have the unique opportunity to improve human health by enhancing both psychological and physiological well-being. By engaging active design principles, any building can improve human health.

4.4 DESIGN GUIDANCE

(a) Best Practices and Design Standards for Healthy Buildings

As noted throughout this chapter, a growing body of research is demonstrating the impact that building design decisions can have on human health and wellness. Beyond its self-defined ethical mandate to uphold the health, safety, and welfare of building occupants and the public, the American Institute of Architects (AIA) has issued a position statement advocating for the design of healthier built environments. Moreover, "design for well-being" is one of the 10 principles explicitly defined by the AIA Framework for Design Excellence (AIA, 2023).

Beyond the design phase, projects must verify performance outcomes to fully realize the benefits available to healthy buildings. Careful attention must also be given to building operations. ASHRAE emphasizes the importance of providing healthy buildings, particularly in the context of IAQ and the control of infectious aerosols. The *Indoor Air Quality Guide: Best Practices for Design, Construction and Commissioning* (ASHRAE, 2009) offers a comprehensive summary of best practices for all aspects of IAQ related to building design and construction. The guide includes recommendations for design tools, maintainability, suggestions for demonstrating the value of IAQ to owners, and a wide range of strategies for achieving critical IAQ objectives.

Since the COVID-19 pandemic, ASHRAE has sought to advance design practice regarding infectious disease control. One of the organization's key healthy buildings documents is *ASHRAE Standard 241-2023: Control of Infectious Aerosols*, which emerged from discussions with the White House COVID-19 Response Team about the need for new and better IAQ standards. The document sets minimum requirements to reduce the transmission of diseases through fine infectious aerosols. *ASHRAE Standard 241* builds upon pre-existing ventilation and IAQ standards and assumes that buildings meet the applicable requirements, including ventilation rates, as described in:

- *ASHRAE Standard 62.1-2022: Ventilation and Acceptable Indoor Air Quality.* This standard is focused on ventilation and acceptable IAQ in commercial and institutional buildings.
- *ASHRAE Standard 62.2-2022: Ventilation and Acceptable Indoor Air Quality in Residential Buildings.* This standard is similar to 62.1, but it is scoped for residential buildings.
- *ANSI/ASHRAE/ASHE Standard 170-2021: Ventilation of Health Care Facilities.* This standard addresses ventilation in healthcare facilities.

ASHRAE Standard 241 expands upon these standards to include steps to further dilute, remove, and inactivate airborne pathogens. The document underscores the critical role of effective design, installation, maintenance, and operation of ventilation controls in mitigating infectious aerosol transmission. Standard 241 includes provisions for a Building Readiness Plan (BRP) that allows HVAC systems to switch from normal mode to infection risk management mode as needed. The standard also places emphasis on assessing and validating the functionality of HVAC systems in existing buildings, ensuring they provide basic ventilation rates and can deliver increased clean air when necessary.

ASHRAE's approach to healthy buildings is not limited to infectious aerosol control; other standards address aspects of healthy buildings:

- *ASHRAE Standard 55-2023: Thermal Environmental Conditions for Human Occupancy.* This standard outlines the thermal environmental conditions that are most acceptable for human thermal comfort.
- *ASHRAE Standard 188-2021: Legionellosis: Risk Management for Building Water Systems.* This standard addresses risk management in building water systems—with a specific emphasis on Legionellosis.
- *ASHRAE Guideline 10-2023: Interactions Affecting the Achievement of Acceptable Indoor Environments.* This guideline provides design recommendations regarding the physical and psychological interplay of building environmental factors—with an evolving concern for health and wellness as an issue of concern.

(b) Certification Systems for Healthy Buildings

There are several design frameworks in the market that can offer building project teams a "road map" for healthy buildings and active design, including:

- *ActiveScore:* This rating system assesses, certifies, and enhances active travel facilities and services in buildings. ActiveScore maintains an eponymous certification system that sets benchmarks for active travel standards and can be leveraged on a building project to better ensure the successful implementation of nonvehicular mobility assets.
- *LEED:* Developed by the U.S. Green Building Council (USGBC), the *Leadership in Energy and Environmental Design* (LEED) rating system features credits that promote walkable communities, active travel, and design and construction best practices that support improved occupant health and wellness.
- *Fitwel:* Developed by the U.S. Centers for Disease Control and Prevention (CDC) and the General Services Administration (GSA) and operated by the Center for Active Design (CfAD), the Fitwel certification system evaluates the health and wellness attributes of buildings and communities. The system includes strategies that directly align with active design principles related to walkability, proximity to non-automobile transit, bicycle facilities, and enhanced accessibility.
- *The WELL Building Standard:* Developed by the International WELL Building Institute (IWBI), the *WELL Building Standard* is a performance-based system for verifying, certifying, and monitoring features of a building that impact human health and wellness, in the categories of air, water, nourishment, light, fitness, comfort, and mind (Figure 4.6). Active design

Figure 4.6 *The Center for Sustainable Landscapes (CSL) at Phipps Conservatory and Botanical Garden in Pittsburgh, Pennsylvania, achieved LEED Platinum, WELL Platinum, and Fitwel 3 Star certifications. (Courtesy of Daniel Overbey.)*

DESIGN FUNDAMENTALS

principles and biophilic considerations are embedded throughout the standard.

(c) Healthy Buildings Require a Holistic Approach

The benefits of healthy buildings should be integral to a holistic high-performance building design strategy as this can ensure that the health and wellness of occupants are prioritized alongside other competing project imperatives. A holistic approach to healthy buildings will include building systems that enhance IAQ, improve water quality, prioritize mindful materials, and maximize thermal, lighting and acoustic comfort for its occupants. Moreover, healthy buildings will leverage opportunities to prompt occupants to be more active and appeal to their affinity for connections with nature.

References and Resources

ActiveScore. www.activetravelscore.us

AIA. 2023. *AIA Framework for Design Excellence*. Washington, DC: American Institute of Architects. content.aia.org/sites/default/files/2023-06/Framework_for_Design_Excellence_June_2023.pdf

Allen, J. G., P. MacNaughton, U. Satish, S. Santanam, J. Vallarino, and J. D. Spengler. 2016. "Associations of Cognitive Function Scores with Carbon Dioxide, Ventilation, and Volatile Organic Compound Exposures in Office Workers: A Controlled Exposure Study of Green and Conventional Office Environments." *Environ Health Perspectives* 124(6): 805–812.

Anderson, J. L., C. A. Glod, J. Dai, Y. Cao, and S. W. Lockley. 2009. "Lux vs. Wavelength in Light Treatment of Seasonal Affective Disorder." *Acta Psychiatrica Scandinavica* 120(3): 203–212.

ASHRAE. 2009. *The Indoor Air Quality Guide: Best Practices for Design, Construction and Commissioning*. Atlanta, GA: American Society of Heating, Refrigerating and Air-Conditioning Engineers, Inc.

ASHRAE. 2021a. *ANSI/ASHRAE/ASHE Standard 170-2021: Ventilation of Health Care Facilities*. Atlanta, GA: American Society of Heating, Refrigerating and Air-Conditioning Engineers, Inc.

ASHRAE. 2021b. *ASHRAE Standard 188-2021: Legionellosis: Risk Management for Building Water Systems*. Atlanta, GA: American Society of Heating, Refrigerating and Air-Conditioning Engineers, Inc.

ASHRAE. 2022a. *ASHRAE Standard 62.1-2022: Ventilation for Acceptable Indoor Air Quality*. Atlanta, GA: American Society of Heating, Refrigerating and Air-Conditioning Engineers, Inc.

ASHRAE. 2022b. *ASHRAE Standard 62.2-2022: Ventilation and Acceptable Indoor Air Quality in Residential Buildings*. Atlanta, GA: American Society of Heating, Refrigerating and Air-Conditioning Engineers, Inc.

ASHRAE. 2023a. *ASHRAE Guideline 10-2023: Interactions Affecting the Achievement of Acceptable Indoor Environments*. Atlanta, GA: American Society of Heating, Refrigerating and Air-Conditioning Engineers, Inc.

ASHRAE. 2023b. *ASHRAE Standard 55-2023: Thermal Environmental Conditions for Human Occupancy*. Atlanta, GA: American Society of Heating, Refrigerating and Air-Conditioning Engineers, Inc.

ASHRAE. 2023c. *ASHRAE Standard 241-2023: Control of Infectious Aerosols*. Atlanta, GA: American Society of Heating, Refrigerating and Air-Conditioning Engineers, Inc.

Babisch, W. 2014. "Updated Exposure-response Relationship between Road Traffic Noise and Coronary Heart Diseases: A Meta-analysis." *Noise and Health* 16(68): 1–9.

Benyus, J. 1997. *Biomimicry: Innovation Inspired by Nature*. New York: HarperCollins.

Browning, W. D., C. O. Ryan, and J. O. Clancy. 2024. *14 Patterns of Biophilic Design*, 10th anniversary edition. New York: Terrapin Bright Green, LLC. www.terrapinbg.com/report/14-patterns

Burls, A. 2007. "People and Green Spaces: Promoting Public Health and Mental Well-being through Ecotherapy." *Journal of Public Mental Health* 6(3): 24–39.

EPA. 2022. *Indoor Air Quality*. Washington, DC: U.S. Environmental Protection Agency. www.epa.gov/report-environment/indoor-air-quality

Fisk, W. J. 2000. "Health and Productivity Gains from Better Indoor Environments and Their Relationship with Building Energy Efficiency." *Annual Review of Energy and the Environment* 25: 537–566.

Fitwel. www.fitwel.org

Frontczak, M., S. Schiavon, J. Goins, E. Arens, H. Zhang, and P. Wargocki. 2012. "Quantitative Relationships between Occupant Satisfaction and Satisfaction Aspects of Indoor Environmental Quality and Building Design." *Indoor Air* 22(2): 119–131.

Gidlow, C. J., M. V. Jones, G. Hurst, et al. 2016. "Where to Put Your Best Foot Forward: Psycho-physiological Responses to Walking in Natural and Urban Environments." *Journal of Environmental Psychology* 45: 22–29.

Glickman, G., B. Byrne, C. Pineda, W. W. Hauck, and G. C. Brainard. 2006. "Light Therapy for Seasonal Affective Disorder with Blue Narrow-band Light-emitting Diodes (LEDs)." *Biological Psychiatry* 59(6): 502–507.

Hedge, A. 1982. "The Open-plan Office: A Systematic Investigation of Employee Reactions to Their Work Environment." *Environment and Behavior* 14(5): 519–542.

IWBI. 2020. *WELL Building Standard version 2*. New York: International WELL Building Institute. v2. wellcertified.com/

JJL. 2016. *Commentary: A Surprising Way to Cut Real Estate Costs*. Jones Lang LaSalle. www.jll.ca/en/trends-and-insights/workplace/a-surprising-way-to-cut-real-estate-costs

Kaarlela-Tuomaala, A., R. Helenuis, E. Keskinen, and V. Hongisto. 2009. "Effects of Acoustic Environment on Work in Private Office Rooms and Open-plan Offices—Longitudinal Study during Relocation." *Ergonomics* 52(11): 1423–1444.

Keus van de Poll, M., J. Carlsson, J. Marsh, et al. 2015. "Unmasking the Effects of Masking on Performance: The Potential of Multiple-voice Masking in the Office Environment." *Journal of the Acoustical Society of America* 138(2): 807–816.

Louv, R. 2005. *Last Child in the Woods: Saving Our Children from Nature-Deficit Disorder*. Chapel Hill, NC: Algonquin Books.

MacNaughton, P., J. Pegues, U. Satish, S. Santanam, J. D. Spengler, and J. G. Allen. 2015. "Economic, Environmental and Health Implications of Enhanced Ventilation in Office Buildings." *International Journal of Environmental Research and Public Health* 12(11).

MacNaughton, P., U. Satish, J. G. C. Laurent, S. Flanigan, J. Vallarino, B. Coull, J. D. Spengler, and J. G. Allen. 2017. "The Impact of Working in a Green Certified Building on Cognitive Function and Health." *Building and Environment* 114: 178–186.

Mak, C. and Y. Lui. 2012. "The Effect of Sound on Office Productivity." *Building Services Engineering Research and Technology* 33(3): 339–345.

Mayer, F. S., C. M. Frantz, E. Bruehlman-Senecal, and K. Dolliver. 2009. "Why Is Nature Beneficial?: The Role of Connectedness to Nature." *Environment and Behavior* 41(5): 607–643.

Mayo Clinic. 2025. Seasonal Affective Disorder (SAD). www.mayoclinic.org/diseases-conditions/seasonal-affective-disorder/symptoms-causes/syc-20364651

McGinnis, J. M., P. Williams-Russo, and J. R. Knickman. 2002. "The Case for More Active Policy Attention to Health Promotion." *Health Affairs (Millwood)* 21(2): 78–93.

Miedema, H. M. E. 2007. "Annoyance Caused by Environmental Noise: Elements for Evidence-based Noise Policies." *Journal of Social Issues* 63(1): 41–57.

Murray, A., R. Ladha, C. Iacoucci, A. Roberts, et al. 2021. *A New Investor Consensus: The Rising Demand for Healthy Buildings*. Geneva: United Nations Environment Programme Finance Initiative, BentallGreenOak, and the Center for Active Design.

Myatt, T. A., J. Staudenmayer, K. Adams, M. Walters, S. N. Rudnick, and D. K. Milton. 2002. "A Study of Indoor Carbon Dioxide Levels and Sick Leave Among Office Workers." *Environmental Health* 1(3).

NAMI. 2025. *Mental Health by the Numbers*. Arlington, VA: National Alliance on Mental Health. www.nami.org/about-mental-illness/mental-health-by-the-numbers

Navai, M. and J. Veitch. 2003. "Acoustic Satisfaction in Open-plan Offices: Review and Recommendations." Ottawa, Canada: *Institute for Research in Construction*, pp. 1–22.

Olesen, B. W. 2007. "Indoor Environment Health-Comfort and Productivity." *Proceedings of Danske Køledage* pp. 40–50.

Pilar, M. R., A. A. Eyler, S. Moreland-Russell, and R. C. Brownson. 2020. "Actual Causes of Death in Relation to Media, Policy, and Funding Attention: Examining Public Health Priorities." *Frontiers in Public Health* 8, article 279.

Steel, Z., C. Marnane, C. Iranpour, et al. 2014. "The Global Prevalence of Common Mental Disorders: A Systematic Review and Meta-analysis 1980–2013." *International Journal of Epidemiology* 43(2): 476–493.

Tyrväinen, L., A. Ojala, K. Korpela, T. Lanki, Y. Tsunetsugu, and T. Kagawa. 2014. "The Influence of Urban Green Environments on Stress Relief Measures: A Field Experiment." *Journal of Environmental Psychology* 38: 1–9.

Ulrich, R. 2001. "Effects of Healthcare Environmental Design on Medical Outcomes." Design and Health: Proceedings of the Second International Conference on Health and Design, Stockholm, SE.

USGBC. *LEED Rating System*. Washington, DC: U.S. Green Building Council. www.usgbc.org/leed

Van den Berg, A. E., J. Maas, R. A. Verheij, and P. P. Groenewegen. 2010. "Green Space as a Buffer Between Stressful Life Events and Health." *Social Science & Medicine* 70(8): 1203–1210.

Van Kempen, E. and W. Babisch. 2012. "The Quantitative Relationship between Road Traffic Noise and Hypertension: A Meta-analysis." *Journal of Hypertension* 30(6): 1075–1086.

Wargocki, P., et al. 2000. "The Effects of Outdoor Air Supply Rate in an Office on Perceived Air Quality, Sick Building Syndrome (SBS) Symptoms and Productivity." *Indoor Air* 10(4): 222–236.

WHO. 2022. *Mental Health*. Geneva: World Health Organization. www.who.int/news-room/fact-sheets/detail/mental-health-strengthening-our-response

WHO. 2024. *Mental Health of Adolescents*. Geneva: World Health Organization. www.who.int/news-room/fact-sheets/detail/adolescent-mental-health

WHO. 2025. *Health and Well-Being*. Geneva: World Health Organization. www.who.int/data/gho/data/major-themes/health-and-well-being

Wilson, E. O. 1984. *Biophilia*. Cambridge, MA: Harvard University Press.

Wirz-Justice, A., F. Benedetti, and M. Terman. 2013. *Chronotherapeutics for Affective Disorders: A Clinician's Manual for Light and Wake Therapy*, 2nd edn. Basel, Switzerland: Karger Medical and Scientific Publishers.

CHAPTER 5

Indoor Air Quality

SINCE THE ADVENT OF MODERN HEATING, VENTILATION, AND AIR CONDITIONING (HVAC) SYSTEMS, control of indoor air quality has generally been integrated into the design of heating and cooling comfort systems. In centuries past, however, dedicated ventilation systems provided outdoor air to buildings, even at the residential scale. Banham (1984) describes both large and small buildings where outdoor air was deliberately introduced to improve air quality. In the home that Dr. John Hayward built for his family in 1867, the Octagon in Liverpool, England (Figure 5.1), outdoor air was brought into the basement, slowed down to precipitate some of the air-borne particulates, then heated to help it rise throughout the four-story building. Ceiling vents just above gaslights drew "vitiated" air from each room, which was then vented to a "foul air chamber" in the attic. From there, using a combined siphon and stack effect, the exhaust air entered a large shaft. Powered at its low point by ever-present heat from the kitchen cooking range, the shaft drew the foul air down, then routed it up a very high chimney to discharge.

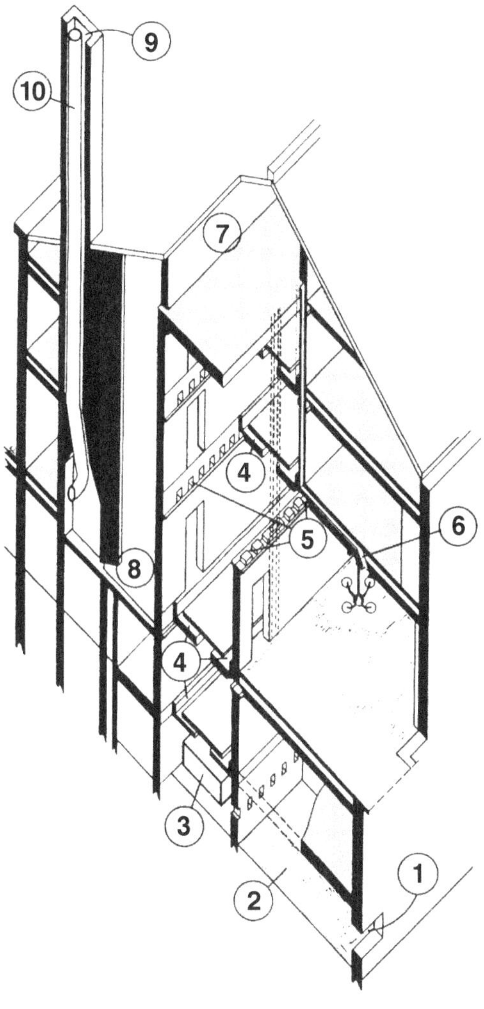

Figure 5.1 *Fresh air intake and stale air exhaust at the Octagon, Liverpool, England, 1867. Dirty air from outside (coal-fired industries and buildings were common) and inside (gas lamps) produced an IAQ likely considerably worse than today's norms. 1. Fresh air intake. 2. Settling chamber in the basement. 3. Heating coils. 4. Air passages in lobby floors. 5. Air passages in the cornice. 6. Extract above a gas lamp. 7. Foul air chamber. 8. Foul air duct. 9. Foul air chimney. 10. Flue from a kitchen range. (From Banham, 1984.)*

Mechanical and Electrical Equipment for Buildings, Fourteenth Edition. Walter Grondzik, Daniel Overbey, and Alison Kwok.

Companion website: www.wiley.com/go/meeb14e

5.1 INTRODUCTION

Several factors have brought indoor air quality (IAQ) to the forefront of building design considerations. A primary concern is health impacts from chemicals in our environment produced by a vast array of air pollutants, coming from synthetic products permanently installed in buildings, from indoor equipment, and from products used for cleaning and maintenance.

The effects of poor indoor air quality are often attributed to an increase in hours spent indoors and buildings that were built too tight. Over the past decades, an increasingly large percentage of people's time is spent indoors, in tightly controlled environments, as the service-based economy overtakes the previous manufacturing-based (and before that, agricultural-based) economy. The lingering memory of the oil embargo of 1973 raised the world's awareness of finite energy sources, which produced a powerful move toward energy-conserving buildings. This focus encouraged designers to limit the introduction of outdoor air volumes that required energy for cooling in summer and heating in winter.

Over time, the design profession's understanding of and approach to regulating IAQ have evolved. The International Living Future Institute (ILFI), the International WELL Building Institute (IWBI), the American Society of Heating, Refrigerating and Air-Conditioning Engineers (ASHRAE), and the United States Green Building Council (USGBC) are a few notable organizations that inform design professionals of the importance of acceptable interior conditions and support design for better indoor environments—including indoor air quality—through the development of standards, initiatives, and certifications.

Numerous factors and variables contribute to the quality of air in a building. Thus, beyond just complying with codes and standards, building designers face a challenging task when dealing with indoor air quality. Controlling indoor air quality is as important to building occupants as controlling for thermal, visual, or acoustical comfort, but the health implications of IAQ substantially differentiate it from design for comfort.

ASHRAE—through *Standard 62.1-2022*—defines acceptable indoor air quality as: "air in which there are no known contaminants at harmful concentrations as determined by cognizant authorities and with which a substantial majority (80% or more) of the people exposed do not express dissatisfaction." There are two distinct components to this definition—a comfort response and a health benchmark. Meeting the requirements for healthy (medically acceptable) air usually (but not in all cases) will meet the requirements for comfortable (occupant-acceptable) air. The opposite is not true.

Providing acceptable indoor air quality involves four major considerations, three of which depend largely on the building design team:

1. Limiting pollution at the source (choosing materials and equipment carefully and isolating unavoidable emissions of pollutants).
2. Providing for an adequate supply of "fresh" air (outdoor and/or recirculated air).
3. Filtering outdoor and recirculated air as necessary.
4. Maintaining the building and its equipment in a clean condition (done post occupancy).

These considerations can be condensed into three diverse IAQ approaches. These are listed in order of prevalence of use.

1. *Ventilation* (whereby cleaner air is introduced to a space to replace less clean air); the cleaner air is typically outdoor air; ventilation may be accomplished passively or actively.
2. *Filtration* (whereby a device of some type is used to remove pollutants from building air); the device may be local (in an occupied room) or central (serving several or all occupied spaces); filtration is an active strategy.
3. *Source control* (whereby pollutants are prevented from entering the building air); this can be an active strategy (as in mechanical kitchen exhaust) or a passive strategy (as in good material choices).

Each of these approaches is discussed in greater detail in the following sections of this chapter. The designer can further provide for improved IAQ by carefully locating a building on its site and zoning the spaces to isolate pollutant sources. Control systems can assist in ensuring that appropriate air quality is actually being provided to building users. The building commissioning process can help tie these design approaches together across the phases of building design, construction, and usage.

5.2 AIR CONTAMINANTS

Potential air contaminants (or pollutants) are numerous, can present a range of possible impacts for humans, and originate both within and from outside a building. This section presents common air contaminants organized by most likely source (interior or exterior). The impacts of these contaminants are discussed in the following section.

(a) Outdoor Air Contaminants and Sources

Outdoor air pollutants can come from a variety of sources—and can enter buildings through a variety of mechanisms. Common outdoor air contaminants include the following.

Vehicular Emissions. Combustion engines release a range of pollutants, including nitrogen oxides, carbon monoxide, hydrocarbons, and forms of particulate matter. Poorly located indoor air intakes and inadequate air filtration may allow such contaminants to permeate indoor environments.

Industrial Emissions. From power plants to large-scale mining activities to factories, industrial processes result in a range of environmental pollutants, including sulfur dioxide, nitrogen dioxide, carbon monoxide, hydrocarbons, and varying types of particulate matter. Chemicals and solvents in industrial and commercial processes can release volatile organic compounds (VOCs) and other harmful compounds such as synthetic per- and polyfluoroalkyl substances (PFAS) and airborne microplastics.

Agricultural Activities. Livestock and crops require significant resources—including the application of fertilizers and pesticides—which can lead to airborne particulates. Agricultural processes also result in methane, ammonia, and certain particulate contamination being released into the environment.

Construction and Demolition Activities. When we construct (or demolish) buildings, roads, and infrastructure, a variety of hazardous particulates become airborne. Such activities often involve solvents and materials that can emit VOCs.

Natural Events. Wildfires, earthquakes, volcanic eruptions, and other natural events can expel acute levels of combustion byproducts, particulates, and other types of emissions into the atmosphere. In some instances, such pollution can reach the upper atmosphere where air currents may carry it a considerable distance.

Organic Processes. Particulates (such as dust, pollen, mold spores, bacteria, and viruses) generated by organic processes can naturally become airborne in outdoor environments.

Landfills and Waste Disposal. Decomposing waste in landfills and other collections of refuse can release methane along with a range of other harmful compounds—some of which may include significant odors.

Exhaust from Buildings. When air is expelled from heating, ventilation, and air-conditioning (HVAC) systems, plumbing vents, kitchen or bathroom exhausts, and direct-exhaust indoor pollution sources, these "indoor sources" can become outdoor pollution for adjacent buildings, depending upon where and how outdoor air is drawn into the adjacent building.

Soil Gases. Certain soil gases, including methane (usually odiferous) and some pesticides, can enter buildings. Of particular concern is radon, which is a radioactive gas that decays rapidly, releasing radiation at each stage of decay. It is colorless and odorless, and thus undetectable by human senses. If we inhale radon, radiation released in the lungs can cause lung cancer. In many buildings with high levels of radon, the problem has been traced to exposure to soil. Radon penetrates through floor and wall cracks and openings around plumbing pipes; thus, below-grade spaces are particularly at risk. Penetrations of below-grade walls and floors should be both minimized and well-sealed; under-slab ventilation may be appropriate (or mandated) in areas of high radon risk (Figure 5.2).

Outdoor air contaminants can be mitigated through a robust IAQ management strategy that includes superior air tightness of the building enclosure (including in subgrade spaces and at foundations in order to address radon), mindful location of supply air intakes, robust air filtration, and pedestrian entryway systems (e.g., grates, grilles, rollout mats).

(b) Indoor Air Contaminants and Sources

Specific sources of indoor air contamination will be unique to a particular building. There are, however, many contaminants that are essentially universal. Common sources of indoor air contaminants include the following.

Mold and Mildew. Certain types of fungus can thrive in moist indoor environments (provided by damp-surface materials and/or high relative humidity of the air). While these fungi naturally exist in nature, exposure to mold and mildew concentrations in indoor environments can pose a range of health issues—especially to people with allergies, asthma, or weakened immune systems. In some instances, exposure to certain types of mold can cause serious health issues.

Environmental Tobacco Smoke. Secondhand smoke from cigarettes, cigars, or pipes (termed environmental tobacco smoke) introduces a complex mixture of chemicals and toxic substances into the air. In addition to being a respiratory irritant, environmental tobacco smoke contains carcinogens, nicotine, and carbon monoxide. Environmental tobacco smoke has been associated with negative effects on the immune system, fetal development, and respiratory infections.

Material Off-Gassing. Interior finishes and furniture will often emit high levels of VOCs when first installed and newly exposed to an indoor environment. Commonly used pressed-wood products, glues,

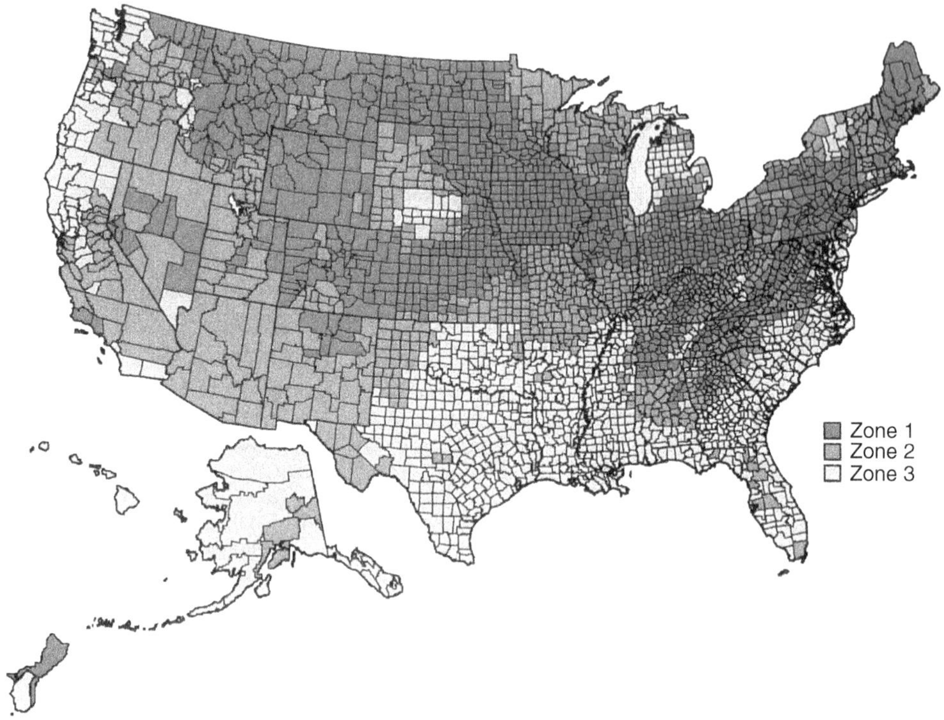

Figure 5.2 *Relative risks of high radon levels in the soil by county in the United States. The darker the zone, the higher the risk; zone 1, for example, presents more risk than zone 2. (EPA, 1996.)*

insulation, and fabrics can release formaldehyde—a carcinogenic gas—into an indoor environment where high concentrations can pose health risks.

Household Chemicals. Certain cleaning agents, pesticides, and other chemicals used to maintain a sanitary indoor environment release a range of pollutants into the air.

Dust and Allergens. Indoor environments contain many sources of dust—including skin cells from people and pets, fibers from clothing and upholstery, soil and pollen brought in on shoes and clothing, and airborne byproducts from indoor activities such as cooking. Dust (fine particles of solids) can be suspended in the air and travel considerable distances in an indoor environment. Most dust is harmless but in some instances it can cause respiratory issues in individuals with sensitivities and allergies.

Any substance in the air that triggers an allergic reaction from a person when inhaled is considered an airborne allergen, or aeroallergen, for that individual. Common allergens include pollen from plants, mold spores, pet dander, and insect particles Allergic reactions tend to be sneezing, runny or stuffy nose, itchy eyes, and respiratory issues. Some individuals may have severe allergic reactions that require immediate medical attention.

Combustion Byproducts. Many buildings contain indoor combustion equipment (such as stoves, heaters, fireplaces, and water-heaters). Indoor combustion sources can introduce harmful gases such as carbon monoxide and nitrogen dioxide into a building.

(c) Air Contaminants and Human Responses

In addition to source or origin, air pollutants can be described by how humans react to them.

Odors. One of the most immediate indicators of IAQ problems is odor. People are sensitive to odors over an extraordinary range, whereas equipment to detect and classify odors is woefully lacking. Odors are perceived most strongly on initial encounter; then "fatigue" occurs and perception fades. Thus, visitors are more likely to detect odors than are the long-term inhabitants of a space. Odors may be simply unpleasant, with psychological consequences, or may be indicators of a more serious IAQ problem with physiological consequences. When an unfamiliar odor is detected, our reactions are positive, neutral, or negative, depending upon whether we perceive a threat or get enjoyment from the odor.

Sometimes odors are directly traceable to a source, such as cooking aromas. In office situations, however, the odor environment is usually complex. A typical office-odor blend may include contributions from body odors, grooming products (perfumes, colognes), copy machines, food products, cleaning solutions, and outgassing from materials. In a decreasing number of locales, tobacco smoke may also be present. This complexity produces an interesting reaction; people tend to become less sensitive to each of the component odors, with a resulting overall masking. However, an architecturally imposed masking approach—the deliberate introduction of a "perfume" to cover offending odors—is rarely successful. Conversely, as the indoor environment is cleansed of multiple odors, people become more sensitive to the one or two odors that remain.

Irritants. Unlike odors, which are immediately perceived and fade with prolonged exposure, irritants are often imperceptible at first but cause increasing distress over time. Symptoms of irritants include itching or burning eyes, sneezing, coughing, dry nose and throat, sore throat, and tightness of the chest. Most irritants are present in the form of particles and gas dispersoids (Table 5.1).

Sources of irritants typically include the building itself and the equipment and occupants within. New and newly renovated buildings are particularly prone to problems from outgassing of paints, adhesives, sealants, office furniture, carpeting, and vinyl wall coverings. Volatile organic compounds (VOCs) are chemicals containing carbon molecules that are volatile; that is, they off-gas or evaporate from material surfaces at room temperatures. The VOC list is long: methane, ethane, methylene chloride, trichloroethane, chlorofluorocarbons (CFCs), hydrochlorofluorocarbons (HCFCs), hydrofluorocarbons (HFCs), formaldehyde, and hydrocarbons such as styrene, benzene, and alcohols. All are now found frequently in new buildings.

Long-term occupancy brings other irritants. Ozone, valuable in the upper atmosphere but a smog component below, is produced by copy machines, high-voltage electrical equipment, and—ironically—electrostatic air cleaners. Mineral fibrous particles can be produced by the breakdown of duct liner/insulation and fireproofing. Hydrocarbon compounds come from copy machines and copy papers. Tobacco smoke is a mixture of gases and fine particles that can be especially irritating to many individuals. Low humidity can exacerbate problems with irritants, producing symptoms like those from chemicals. Carpet shampooing yields organic solvents and ammonia; nighttime cleaning done under reduced or nonexistent ventilation is especially problematic. In contrast, night maintenance with increased ventilation rates—as with cooling by night ventilation of thermal mass—can reduce this threat.

ASHRAE Standard 62.1 defines acceptable indoor air quality in terms of odors and irritants (although using different terminology). As with odors, the impacts of irritants can be reduced by source control, ventilation, and filtration. Filters for the removal of irritants usually consist of particulate filters; less common are gaseous-removal filters, air washers, and electronic air cleaners.

(d) Contaminant Impacts

Contaminants in the air can have numerous and varied negative impacts on humans. Contaminants likely to be encountered in buildings have been demonstrated to cause these potential health effects (EPA, 2025):

- hypersensitivity pneumonitis
- allergic rhinitis
- some types of asthma
- fatigue
- chest pain
- angina
- reduced brain function
- impaired vision and coordination
- headaches
- dizziness
- confusion
- nausea
- flu-like symptoms
- eye, nose, and throat irritation
- aggravation of coronary and respiratory disease symptoms
- increased bronchial reactivity
- decreased lung function
- increased risk of respiratory infections
- damage to the central nervous system and kidney
- increased risk of cancer
- fatalities

The severity of contaminant impacts depends upon the concentration of the contaminant, the length of exposure, interactions that occur between various contaminants, ambient relative humidity, and the initial state of health of those who are exposed. These combinations of variables lead to a complex design situation. Prudent risk management suggests caution and due consideration for readily available design guidance regarding indoor air quality

Table 5.2 summarizes common indoor air contaminants and contaminant-specific design solutions. Effectively reducing indoor air contaminants, however, requires a comprehensive systems response that includes proper air filtration; localized direct-exhaust at certain indoor pollution sources; humidity and moisture management; specifications for finishes; fixtures and furnishings devoid of harmful emissions; elimination of combustion-based equipment and

TABLE 5.1 Characteristics of Particles and Particle Dispersoids

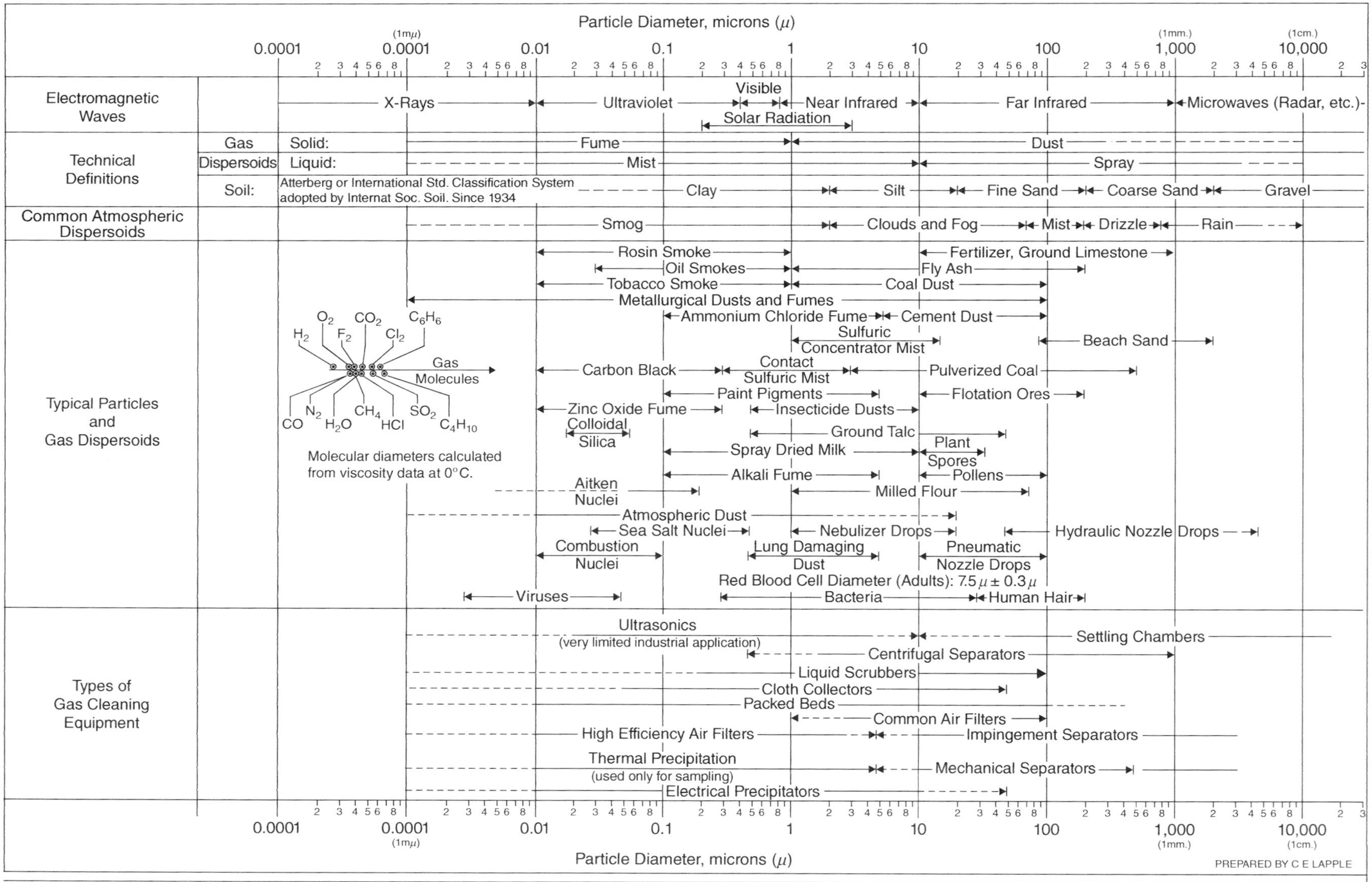

Source: With permission of SRI International.

TABLE 5.2 Common Indoor Air Contaminants

Contaminant	Sources	Effects	Control Strategies
Carbon dioxide (CO_2)	Human respiration	Minor discomfort at high concentrations; "stuffiness"	CO_2 is a good indicator of the ventilation rate in tightly enclosed spaces or where occupancy is high
Carbon monoxide (CO)	Incomplete combustion: furnaces, stoves, fireplaces; motor vehicle exhaust	Headaches, dizziness, sleepiness, muscle weakness, potentially lethal	Sealed combustion burners, adequate combustion air, safe exhaust flues
Nitrogen oxides	High-temperature combustion	Irritation, possible immune suppression	Safe exhaust flues, sealed combustion burners
Sulfur oxides	Combustion fuels containing sulfur (oil, coal)	Potential irritant, burning eyes, reduces lung function	Alternative fuels, safe exhaust flues, sealed combustion burners
Polynuclear aromatic hydrocarbons	Smoking, combustion of wood or coal, barbecuing, burnt food	Irritants and carcinogens	Prohibit smoking, lower temperature in cooking, use clean fuels, burn wood in enclosed firebox with adequate oxygen supply
Ozone	Laser printers, photocopiers, small motors, electronic air cleaners	Inflammation of bronchia, wheezing and shortness of breath, dizziness, asthma attacks	Remove sources or exhaust at source, maintain electronic air cleaners
Volatile organic compounds (VOCs)—formaldehyde	Particle board, interior laminated panels, glues, fabric treatments, paints	Burning eyes and nose, skin rash, shortness of breath, headaches, nausea, dizziness, fatigue	Use alternative materials, seal particle board if used, ventilate
Volatile organic compounds (VOCs)—others	Paints, solvents, carpets, soft plastics, adhesives, caulking, softwoods, paper products, cleaning and maintenance products	Intoxication, burning eyes and nose, shortness of breath, headaches, nausea, dizziness, loss of judgment, panic	Use alternative materials, age materials before installing, ventilate
Lead	Pre-1970s paint, pre-1985 pipes and solder, dust and soil near roads (residue from leaded gas)	Neurotoxic, especially if ingested by young children; learning disabilities, nausea, trembling, numbness of extremities	Identify and remove or seal old paint, replace pipes and solder, avoid foods grown by roadside
Pesticide residues[a]	Treated basements and foundations, treated ceiling and wall cavities, treated cabinets and closets, treated soil outside foundation	Neurotoxic or long-term risk of liver, kidney, and other diseases, including cancers	Identification and removal by expert if history known, sealing in pesticide if possible
Asbestos fiber	Pre-1975 steam pipe and duct insulation, furnace and furnace parts, pre-1980 reinforced vinyl floor tile, and fiber cement shingles and siding	Long-term cancer risk from inhaling fibers	Leave material undisturbed, get expert identification and removal if required, seal with special sealant and cover with sheet metal if not crumbling
Mineral and glass fiber	Thermal insulation, pipe insulation, fire-resistant acoustic tile and fabrics	Potential irritant, burning eyes, itching skin, long-term risk of lung damage and cancer	Handle only with respirator and gloves, seal and enclose, do not disturb in place
Fungus particles, dust mites	Grow in basements, damp carpet, bedding, fabrics, walls and ceilings, closets	Very allergenic, burning eyes and nose, sneezing, skin rash, congestion, and shortness of breath	Keep surfaces dry and clean, cover bedding and upholstery with barrier cloth, ventilate, use borax treatments to retard fungus
Hazardous bacteria (e.g., Legionella)	Standing warm water, untreated hot tubs, air-conditioning drain pans, humidifier reservoirs	Severe respiratory illness, potentially lethal	Prevent standing water, clean and treat tubs and reservoirs
Radon gas	Natural radioactivity in soils	Increased lifetime lung cancer risk	Seal foundation and floor drains, ventilate subsoil
Methane and other soil gases	Decomposing garbage in landfills, leaking sewage lines, toxic waste	Possibly explosive or toxic, nuisance odors	Know site history before building, remove soil if necessary, seal foundation and floor drains, ventilate subsoil
Excess moisture[b]	Cooking (heating open liquids), washing, exhaling	Increases growth of fungi, bacteria, and dust mites	Exhaust ventilation at source; dehumidification

Source: Adapted from Rousseau and Wasley, 1997.

[a] Pre-1980s treatments are more likely to leave residues and be toxic.

[b] Too little moisture also adversely affects health; be careful to not confuse the indoor air quality effects of moisture with its thermal comfort effects.

appliances; along with a holistic and well-executed operations and maintenance (O+M) strategy.

5.3 IAQ DESIGN STRATEGIES

(a) Source Control

Dealing with indoor pollution at its source, by choosing materials and equipment with care, rather than increasing ventilation rates (and related energy consumption) is the most rational IAQ strategy. Although the most logical starting point for design for indoor air quality, source control alone will seldom be sufficient for acceptable IAQ.

Material Selections. Many contaminants enter the indoor air from the materials that bound building spaces (floor, wall, and ceiling finishes and their substrates) and from furnishings and equipment installed in a building. Conceptually, building materials place occupants in proximity to pollutants. Since these materials are the result of conscious decisions and specifications, they can (in addition to all else they are expected to do) be selected to be indoor-air friendly. Resources set up in support of green and healthy building certification efforts will provide guidance and insights into wise material selections relative to indoor air quality.

The American Institute of Architects (AIA) prepared a *Prescription for Healthier Building Materials* (AIA, 2018), which provides a conceptual framework for environmentally responsible material choices. The Parsons School of Design has developed an online resource (titled *Material Collections*) to assist in the selection of healthier building materials (Parsons, 2025).

Multiple Chemical Sensitivity. This is an unusual (and sometimes controversial) condition, also known as environmental illness. The controversy arises because the causes of this condition are poorly understood and likely involve numerous factors. When causes are mysterious, establishing targeted remedies becomes difficult. People with multiple chemical sensitivity are likely to see the provision of outstanding IAQ as a primary intent of the design process. People with this condition avoid environments with any known environmental risk factors.

Ecology House (in San Rafael, California) is an apartment complex for people with multiple chemical sensitivity. Its construction avoided plywood, using Douglas fir sheathing instead; the floors are tile instead of wall-to-wall carpet; cabinets are metal, not plywood or oriented strand board; the heating system is radiant hot water, not forced air. Barbecues and fireplaces are absent, painted surfaces are minimized, and any window coverings are alternatives to curtains. There is even an "airing room," where, for example, newspapers can be hung before they are read in order to evaporate ink odors.

Localized Exhaust. Source control for point sources (as opposed to area sources, such as carpets) tends to involve localized exhaust systems. The concept is simple—capture and remove pollutants before they can disperse and spread throughout a space or building. Point sources that generate contaminants in a residential setting include cooking devices, water closets, showers, and bathtubs—and, increasingly, home office equipment. Targets for localized exhaust in non residential buildings include bathrooms, janitor closets, process equipment (many and varied), and some storage areas. Chapter 16 provides more information on exhaust fans.

(b) Ventilation

The ventilation design question focuses upon determining how much outdoor air is necessary to produce acceptable indoor air quality. This question is less easy to answer than might be expected, even with codes and standards that define acceptable ventilation rates. The definition of an "appropriate" ventilation rate—if above and beyond code minimum requirements—will be project-specific (in some cases even room-specific) and be established as part of the owner's project requirements (OPR).

Minimum Ventilation. Although very small amounts of outdoor air will provide enough oxygen, and although human body odor control is usually achievable at a rate of 6–9 cfm (cubic feet per minute) [3–4.5 L/s] of outdoor air per occupant, ventilation is expected to do more than just provide oxygen and control odors. Defining minimum outdoor air supply rates has proven to be a controversial task. Minimum ventilation rate recommendations have changed substantially over time. Nevertheless, the 2022 edition of *ASHRAE Standard 62.1* establishes current consensus and code-adopted minimum ventilation rates based on the summation of an occupancy component and a building component that recognizes contaminant sources specific to certain types of building uses. *ASHRAE Standard 62.2* provides similar requirements for residential-scale buildings. These recommendations are summarized in Appendix F, Tables F.1 and F.2 respectively. Equation 5.1 expresses the current ASHRAE approach to ventilation values mathematically.

$$V_{bz} = (R_p \times P_z) + (R_a \times A_z) \qquad \text{(Eq. 5.1)}$$

where

V_{bz} = outdoor airflow required in the breathing zone, cfm [L/s]

R_p = people outdoor air rate (see Appendix F, Table F.1)

P_z = zone population, the number of people in the ventilation zone during use
R_a = area outdoor air rate (see Appendix F, Table F.1)
A_z = zone floor area, the net occupiable floor area of the ventilation zone, ft² [m²]

For example, a 2000-ft² open office space with 12 workers would require 5 cfm/person [2.5 L/s] and 0.06 cfm/ft² [l/s m²]. This would result in a minimum outdoor air ventilation rate of 140 cfm [66 L/s]. ASHRAE specifies that such ventilation air be delivered to the breathing zone of occupants so that it is effective.

Although standardized (and codified), it is notable that some designers feel that the current ASHRAE outdoor air requirements are generally too high, while others believe that they are too low.

The concept of replacance is associated with the design of ventilation systems. Table 5.3 shows that at the rate of one air change per hour (ACH) of outdoor air, an indoor space would have only 63% "new" air after one hour; about eight hours at this rate is required for all the "old" air to be exhausted. There is, then, a difference between the fresh air input rate (ACH) and the replacance—the fraction of air molecules at one specified time that was not in the indoor space at an earlier reference time. This relationship, along with details of air pollutants (and of heat exchanger design for energy conservation), is discussed in Shurcliff (1981).

Enhanced Ventilation. Whether natural or mechanically induced, ventilation is vital for safe and healthy indoor environments. Code-minimum ventilation rates are established by *ASHRAE Standards 62.1* and *62.2*. These rates are intended to provide "acceptable" indoor air quality. There is a growing body of evidence that suggests that higher than minimum ventilation rates can improve indoor air quality and health outcomes. The Leadership in Energy and Environmental Design (LEED) green building certification program (USGBC) and the *WELL Building Standard* (IWBI, 2020) both provide recommendations for enhanced ventilation rates.

Ventilation Approaches. There are several viable approaches to providing acceptable or enhanced ventilation.

Natural or Mechanical Ventilation. Ventilation airflows can be provided using passive (natural) approaches—such as cross ventilation or stack ventilation. Natural ventilation for indoor air quality is a recognized option in *ASHRAE Standard 62.1 and 62.2.* Numerous architectural design features can support natural ventilation. See Chapter 14 for further discussion of ventilation for passive cooling. The principles behind designing airflow for ventilative cooling and ventilative air quality are similar—but the objectives are not the same. Mechanical ventilation using HVAC system components is very common in many parts of the world.

Continuous Versus Intermittent Ventilation. Ventilation air can be provided to spaces continuously or intermittently. Continuous supply is provided independently of specific spatial conditions. It is basically always on (except perhaps for unoccupied periods). Continuous ventilation is the most common approach in most non residential buildings. Intermittent ventilation is basically on-demand ventilation. Air flow is activated by some event within a space. In a residence, this event may be operation of a bathroom exhaust fan. In a commercial building space, the event may be a signal from an IAQ sensor installed in the space. This approach is also called demand-controlled ventilation. See Section (e) for further information on IAQ sensors.

Dedicated Outdoor Air Systems. Historically, mechanical ventilation in larger buildings has been integrated into comfort control HVAC systems. Ventilation is provided by the same system that provides heating or cooling. This sets up a potential conflict between comfort needs and air quality

TABLE 5.3 Air Replacance Related to Air Changes per Hour (ACH)

Time from Start of Run (h)	Rate of Air Input (ACH) ⟶	Replacance (%)								
		0.06	0.12	0.25	0.5	1	2	4	8	16
1/16		0.4	0.8	1.6	3.1	6.1	11.7	22.1	39.3	63.2
1/8		0.8	1.6	3.1	6.1	11.7	22.1	39.3	63.2	86.5
1/4		1.6	3.1	6.1	11.7	22.1	39.3	63.2	86.5	98.2
1/2		3.1	6.1	11.7	22.1	39.3	63.2	86.5	98.2	99.9
1		6.1	11.7	22.1	39.3	63.2	86.5	98.2	99.9	100
2		11.7	22.1	39.3	63.2	86.5	98.2	99.9	100	100
4		22.1	39.3	63.2	86.5	90.2	99.9	100	100	100
8		39.3	63.2	86.5	98.2	99.9	100	100	100	100
16		63.2	86.5	98.2	99.9	100	100	100	100	100

Source: From Shurcliff, 1981.

needs. Separating the ventilation function from the comfort function can improve air quality outcomes. This separated approach is called a dedicated outdoor air system (DOAS). See Chapter 16 for more details on DOAS.

(c) Filtration

Ventilation alone is not necessarily enough to ensure appropriate indoor air quality. The ventilation air must also be reasonably free of particulate matter. The size of particles of concern will range considerably. Best practice, however, addresses two classes of particulate matter: $PM_{2.5}$ and PM_{10}. $PM_{2.5}$ refers to "particulate matter" with a diameter less than 2.5 micrometers (μm). For perspective, $PM_{2.5}$ particles are about 3% of the diameter of a human hair. These particles can be inhaled deep into the lungs and can even enter the bloodstream (a typical coronavirus is 0.13 μm, significantly smaller than 2.5).

PM_{10} particulate matter has a diameter less than 10 μm. Dust and smoke often contain particles in this range. PM_{10} particles are still a fraction of the size of a human hair and not much larger than red blood cells. As such, PM_{10} particles are easily inhaled and can cause health issues. Inevitably, some particulate matter will be carried into indoor spaces with ventilation air. If the annual average outdoor $PM_{2.5}$ level is 35 micrograms (μg) per cubic meter (μg/m^3), then an indoor air design target might be 15–25 μg/m^3. For PM_{10}, a good indoor concentration target may be 50 μg/m^3 or lower.

Air filters with a high MERV (Minimum Efficiency Reporting Value) rating, as defined by ASHRAE, or meeting the U.S. Department of Energy's HEPA ("high efficiency particulate air") standard can mitigate particulate matter concentrations in indoor environments. A wide variety of particulate filters are available for use with building air handling equipment. See Chapter 16 for detailed information on filter types.

(d) Ultraviolet (UV) Radiation

Since early in the twentieth century, ultraviolet (UV) radiation in the C band (200–280 nm wavelength) has been used to kill harmful microorganisms, but under tightly controlled conditions. Now there are UV lamp units that work within HVAC systems, promising to control fungi, prevent the development and spread of bacteria, and reduce the spread of viruses. These devices take up very little space (they are placed within ductwork) and generate no ozone or other chemicals. The germicidal output of these lamps is somewhat higher at room temperatures (and above) than at lower temperatures. They are even more effective when installed in a UV-reflective duct interior. Aluminum seems to be the best, commonly available UV reflector. UV lamp life is 5000–7500 hours. As an additional benefit, cooling coils and drain pans stay cleaner in a UV-treated air system.

The National Renewable Energy Laboratory has explored the prospects of using UV radiation, helping to develop a process that bombards polluted air with UV radiation in the presence of special catalysts. Pollutants including cigarette smoke, formaldehyde, and toluene are quickly broken down into molecules of water and carbon dioxide.

(e) IAQ Monitoring and Control

The success of ventilation strategies has typically been inferred from compliance with codes, standards, or guidelines. Following consensus requirements was deemed to provide acceptable indoor air quality. Until recently, it was not common to measure contaminants in buildings as part of normal operations. The advent of lower cost sensors and the trend toward monitoring and verification across the building industry have begun to change this pattern.

Adequate ventilation should maintain carbon dioxide (CO_2) concentrations under 1000 ppm (ppm). As a frame of reference, typical carbon dioxide levels in outdoor air are generally around 400 ppm (although persistently increasing because of atmospheric carbonization). Prolonged exposure to concentrations over 2000 ppm may lead to significant negative health effects. Demand response ventilation systems can be utilized to continuously monitor CO_2 concentrations in indoor spaces to ensure healthy rates of ventilation air supply while also balancing energy performance.

Many devices for monitoring air quality are now available, some of which can control the operation of IAQ-related equipment. One of the oldest and simplest devices measures the concentration of CO_2. Wherever there are indoor concentrations of people, elevated levels of CO_2 can be expected. Thus, CO_2 becomes a kind of "canary in the coal mine," an early indicator of pollutant buildups due to occupancy.

In the theater at the Bedales School in Hampshire, England, the operation of exhaust air dampers in the cupola is controlled by a building management system. Under ordinary conditions, damper opening is regulated by information from the CO_2 monitor. This can be overridden by indoor and outdoor temperatures (summer night ventilation is encouraged), air velocity monitors below the seats (too much draft could produce discomfort), the presence of wind and rain (openings limited), or a fire alarm (damper fully open to aid in smoke extraction).

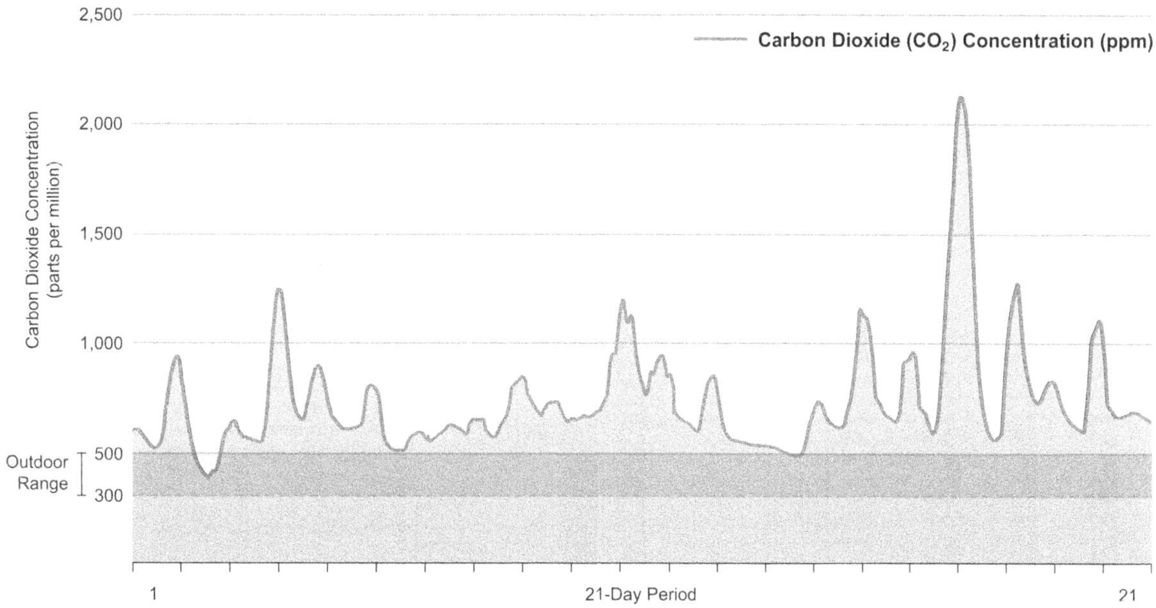

Figure 5.3 Continuous carbon dioxide (CO_2) monitoring over a 21-day period in a newly constructed high-performance home. The spike in CO_2 concentration levels during the third week coincided with a series of open-house events in which the home had a much denser occupancy. (Courtesy of Daniel Overbey.)

A sample of CO_2 monitoring results in a residential setting is shown in Figure 5.3.

Where contaminants are expected from both materials and human beings (typically the case), monitors may be installed to detect carbon monoxide, or combinations of VOCs, or fuels such as propane, butane, or natural gas—or even depletion of oxygen. These monitors can be installed as stand-alone alarms or with additional relays that activate equipment. They are about the size of a programmable thermostat; the mounting height depends upon which gas is to be monitored.

Such devices can regulate ventilation heat exchangers, such as an energy recovery ventilator (ERV) (see Chapter 16). This could be especially useful during unoccupied periods, holidays, and weekends. Many commercial/institutional HVAC systems are shut off during such periods; VOCs from finishes and furnishings continue to accumulate, however. Periodic flushing from ERVs, controlled by VOC monitors, will help maintain acceptable air quality and could eliminate (or greatly reduce) the need for a Monday morning pre-flush.

(f) Individual Space Air Cleaning

Energy conservation considerations have reduced the air circulation rate in many central air-handling systems; moving less air reduces the energy used by fans. One result can be low ventilation air distribution efficiency, which causes poorly mixed air within the occupied spaces. With local (individual space) air-filtering equipment, both a high circulation rate and proper air mixing are achievable. Each such unit has its own fan that can operate either with or without the central HVAC fan. An example of an independent electrostatic air filter is shown in Figure 5.4.

Air circulation through these filters should occur at rates between 6 and 10 air changes per hour. The air is then ducted to diffusers, hence circulating in a sweeping pattern across a space to return air intakes on the opposite side.

A variation on electrostatic air cleaning is shown in Figure 5.5. In this equipment, a mixture of outdoor and indoor air is passed through a complex electrical field produced by both high-voltage DC and high-frequency AC. This supply air emerges with greatly reduced submicron particles that have coagulated into larger particles more easily carried by air currents. As this air moves through the occupied space, it picks up submicron particles and is returned to the equipment (some is exhausted), where it is filtered to remove the larger particles.

Portable air cleaners abound. One elaborate model combines UV germicidal radiation with carbon/oxidizing media and HEPA. It is a counter-height rolling device about 15 in. [380 mm] square that emits cleansed air from the top. It claims to purify air in a space of about 12,000 ft^3 [340 m^3]. Use of a portable air cleaner will involve architectural design decisions regarding space use and visual appropriateness.

Figure 5.4 A stand-alone electrostatic air filter operates independently of the central HVAC system; no ductwork is required. Airflow rates vary; the slower the flow, the more efficient the filtering. Flush-mount (recessed in suspended ceiling), ceiling surface-mount, wall-mount, and portable models are available. (Courtesy of Tectronic Products Co. Inc.)

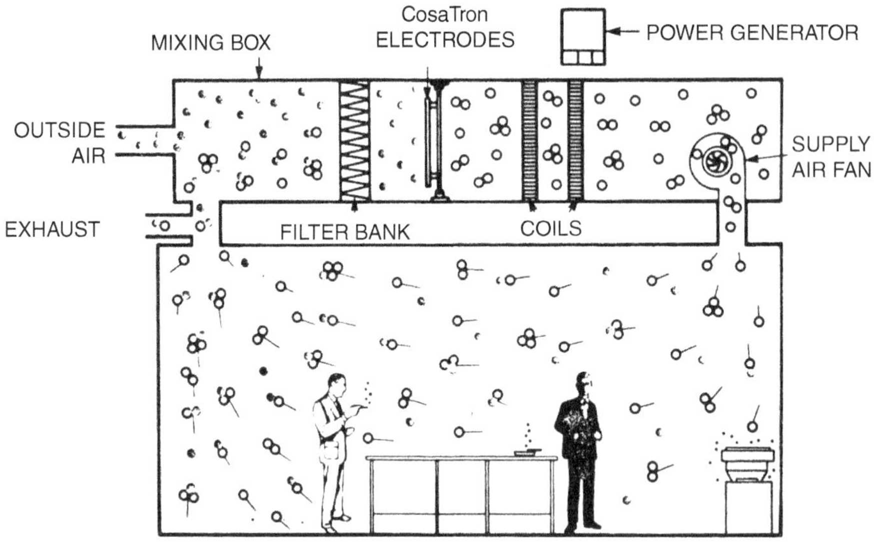

Figure 5.5 This air-cleaning system uses a combination of high-voltage DC and high-frequency AC. The supply air emerges with greatly reduced submicron particles, which have coagulated into larger particles more easily carried by air currents. As this air moves through the occupied space, it picks up submicron particles and is either exhausted or returned to the equipment, where it is filtered to remove the larger particles. (Courtesy of CRS Industries Inc.)

Although specialized smoking rooms in otherwise smoke-free airports have declined in recent years, some airports still provide such spaces (Figure 5.6). These rooms are constructed to provide separate ventilation systems to exhaust particulates in secondhand smoke.

(g) Zoning for IAQ

After pollution control is implemented at the source (equipment exhausts, prohibiting smoking, careful material choices, etc.), all remaining unavoidable pollutant sources should be identified. Then, during early design, more IAQ-sensitive building areas should be isolated from the key contaminators. This is sometimes difficult, as in "open offices," where walls are unwelcome but copying machines are essential. In such cases, one approach is to erect as much of a barrier as possible around an offending device, then task ventilate to remove locally contaminated air immediately. Sometimes air pollution sources also produce noise, in which case the argument for a more complete barrier may become more compelling.

When indoor pollution sources can be isolated in designated spaces (e.g., garages, custodial closets, laundry areas, copying and printing rooms) interior cross-contamination prevention becomes critical. Sufficient exhaust spaces where hazardous gases or chemicals may be present to create negative pressure with respect to adjacent spaces when the doors to the room are closed. Additionally, provide self-closing doors and floor-to-floor partitions or a hard-lid ceiling.

Many healthcare and laboratory buildings have designated "clean" and "dirty" zones, including separate circulation pathways. Differential air pressures

Ramsey, C. G. and H. R. Sleeper. 2000. *Architectural Graphic Standards*, 10th edn. New York: John Wiley & Sons.

Ramsey, C. G. and H. R. Sleeper. 2016. *Architectural Graphic Standards*, 12th edn. New York: John Wiley & Sons.

Rousseau, D. and J. Wasley. 1997. *Healthy by Design*. Point Roberts, WA: Hartley and Marks.

Shurcliff, W. A. 1981. *Air-to-Air Heat Exchangers for Houses*. Andover, MA: Brick House.

USGBC. *LEED Rating System*. Washington, DC: U.S. Green Building Council. www.usgbc.org/leed

CHAPTER 6

Thermal Comfort

ONE OF THE EARLIEST REASONS FOR BUILDING was to create shelter from often inhospitable climates and to attempt to obtain thermal comfort. This chapter addresses the interrelationships among human beings, ambient conditions, and conditions within the built environment. It begins by discussing heat flow to and from the body and then thermal comfort. Visual and acoustical comfort are dealt with later in the book; however, as will be seen, these important aspects of human response are not nearly as well developed from a design perspective as thermal comfort. Other aspects of comfort, such as ergonomics and our responses to spatial geometries, are beyond the scope of the book.

6.1 THE BODY AND HEAT

Across the full range of diverse daily activities, the human body is continuously generating heat as a byproduct of metabolism. The conversion of energy inputs (food) to work output is not terribly efficient, resulting in waste heat production. Because the body's core temperature must stay within a narrow range to support health, we nearly always need to lose heat to our environment. The rate at which we produce heat changes frequently (with activity level), as does the surrounding environment's ability to accept heat. This is a dynamic relationship.

When we are in a building, three key layers interface between our core body conditions and the ambient environment. These layers regulate our body's heat loss. The first layer is our skin; the second layer is our clothing; the third layer is a building enclosure. The relative importance of these layers varies with the thermal conditions we confront, and with our desire to select clothing ensembles that respond to other than thermal demands.

(a) Metabolism

The rate at which we generate heat (the metabolic rate) depends mostly upon our muscular activity, partly upon what we eat and drink (and when), and partly on where we are in our normal daily cycle. Our heat production is measured in metabolic (met) units. One met is defined as 50 kcal/h/m^2 [equal to 18.4 Btu/h ft^2 or 58.2 W/m^2]. One met is the energy produced per unit of surface area by a seated person at rest. Under these conditions, the total heat produced by a normal adult is about 360 Btu/h [106 W]. The more active we are, the more heat we produce. Table 6.1 lists metabolic heat production for a range of activities; Figure 6.1 illustrates various metabolic activity levels.

There are several interactions between our skin and the rest of our body. These include the sense of touch, the circulation of blood, and the exchange of water vapor. The touch sensations include pressure and pain as well as heat and cold. The experiences of heat and cold are produced by contact with surfaces, by immersion in air, and by radiation. These sensations detect changes in the environment and signal impending needs to adjust bodily heat regulation, a process that is controlled by the portion of the brain called the hypothalamus.

The hypothalamus triggers changes in our blood circulation patterns in response to signals from our skin and changes in our core body temperature. If the body temperature is dropping (we are becoming cold), the

Mechanical and Electrical Equipment for Buildings, Fourteenth Edition. Walter Grondzik, Daniel Overbey, and Alison Kwok.

Companion website: www.wiley.com/go/meeb14e

TABLE 6.1 Metabolic Rates for Typical Tasks

	Metabolic Rate[a]		
Activity	Met Units[b]	Btu/h ft²	W/m²
Resting			
Sleeping	0.7	13	40
Reclining	0.8	15	45
Seated, quiet	1.0	18	60
Standing, relaxed	1.2	22	70
Walking (on the level)			
2 mph [0.9 m/s]	2.0	37	115
3 mph [1.2 m/s]	2.6	48	150
4 mph [1.8 m/s]	3.8	70	220
Office activities			
Reading, seated	1.0	18	60
Writing	1.0	18	60
Typing	1.1	20	65
Filing, seated	1.2	22	70
Filing, standing	1.4	26	80
Walking about	1.7	31	100
Lifting, packing	2.1	39	120
Driving/flying			
Car	1.0–2.0	18–37	60–115
Aircraft, routine	1.2	22	70
Aircraft, instrument landing	1.8	33	105
Aircraft, combat	2.4	44	140
Heavy vehicle	3.2	59	185
Miscellaneous occupational activities			
Cooking	1.6–2.0	29–37	95–115
Housecleaning	2.0–3.4	37–63	115–200
Seated, heavy limb movement	2.2	41	130
Handling 110-lb [50-kg] bags	4.0	74	235
Pick and shovel work	4.0–4.8	74–88	235–280
Machine work			
Sawing (table saw)	1.8	33	105
Light (electrical industry)	2.0–2.4	37–44	115–140
Heavy	4.0	74	235
Miscellaneous leisure activities			
Dancing, social	2.4–4.4	44–81	140–255
Calisthenics/exercise	3.0–4.0	55–74	175–235
Tennis, singles	3.6–4.0	66–74	210–270
Basketball	5.0–7.6	90–140	290–440
Wrestling, competitive	7.0–8.7	130–160	410–505

[a]For average adult with a body surface area of 19.6 ft² [1.8 m²]. For whole-body average heat production, see also Table G.5.

[b]One met = 18.4 Btu/h ft² = 58.2 W/m².

rate of heat loss from the body needs to be reduced. This is accomplished through a decrease in the flow of blood from the core toward the surface of the skin. This decrease is called vasoconstriction. Blood carries heat around the body, and reduced flow to the extremities under cold conditions reduces heat loss. Vasoconstriction is triggered in part by temperature (cold) signals from our skin. Under this condition, our sweat glands also force less water to the skin surface, which reduces evaporation and its associated heat loss.

Note the implications of this arrangement. We strive (without thinking) to maintain, at all costs, a nearly constant core temperature for our vital organs. The core organs take thermal precedence over the less vital extremities, including the arms and legs and then the fingers and toes. The farther from our central body mass (fingers and toes) and the greater the exposed surface area (ears), the more and faster temperature will drop in cold conditions. This intentional variation of temperatures is an example of zoning, a principle that will be further discussed in connection with the design of several building systems. The most variable bodily thermal zone is our skin surface.

When cold conditions worsen, we get "goosebumps," symptoms of an unsuccessful attempt to increase insulation by fluffing up our (now generally missing) body hair. Because we cannot reduce heat loss this way, we soon increase our metabolic rate to produce more heat to protect the core organs. Essentially, we burn more fuel, by involuntary shivering, muscular tension, or increased muscular activity. At the point where shivering incapacitates us, we may experience a metabolic rate of 6 met. Before reaching this uncomfortable and unhealthy point, we have learned to seek help from our second and then our third skins—clothing and buildings.

The opposite occurs when we are too hot: first, blood flow toward the skin surface increases (vasodilation), triggered primarily by warm signals from our core. The sweat glands greatly increase the secretion of water and salt to the skin surfaces. This increases heat loss by evaporation (although salt accumulations impede evaporation by lowering the vapor pressure of water).

(b) Heat Flow

Once blood and water transport surplus heat from the core to the skin surface, we have four ways to pass it to the environment: (1) convection (air molecules contact our body, absorb heat, and move away by buoyancy), (2) conduction (heat is transferred when we touch cooler surfaces), (3) radiation (when our skin surface is hotter than other surfaces "seen" but not touched, heat is transferred to these cooler surfaces, and vice versa when surfaces are hotter than our skin surface), and (4) evaporation (an evaporating liquid removes large quantities of heat from the surface it is leaving). The amount of heat we lose by each of these four methods depends upon the interactions among our metabolism, our clothing, and our environment. Figure 6.2 illustrates the typical situation of a person at rest as environmental conditions change.

As surrounding air and surface temperatures approach our skin temperature, the effectiveness of convection, conduction, and radiation decreases. Evaporation then becomes essential, and access to dry, moving air is greatly appreciated. As air and surface temperatures fall, evaporation decreases while convection, conduction, and particularly radiation increase. Under the normally comfortable temperature of 70 °F [18 °C], the proportions of body heat loss are as shown in Table 6.2.

(*a*) 0.7 Met—Sleeping

(*b*) 1.4 Met—Standing, Modelling

(*c*) 2.1 Met—Leaning, Modelling

(*d*) 3.4 Met—Heavy Lifting

Figure 6.1 *An architecture student demonstrates various activity levels: (a) sleeping at the studio desk (0.7 met); (b) standing, working on a model (1.4 met); (c) standing/leaning, intently working on a physical model (2.1 met); and (d) heavy lifting of textbooks (3.4 met).*

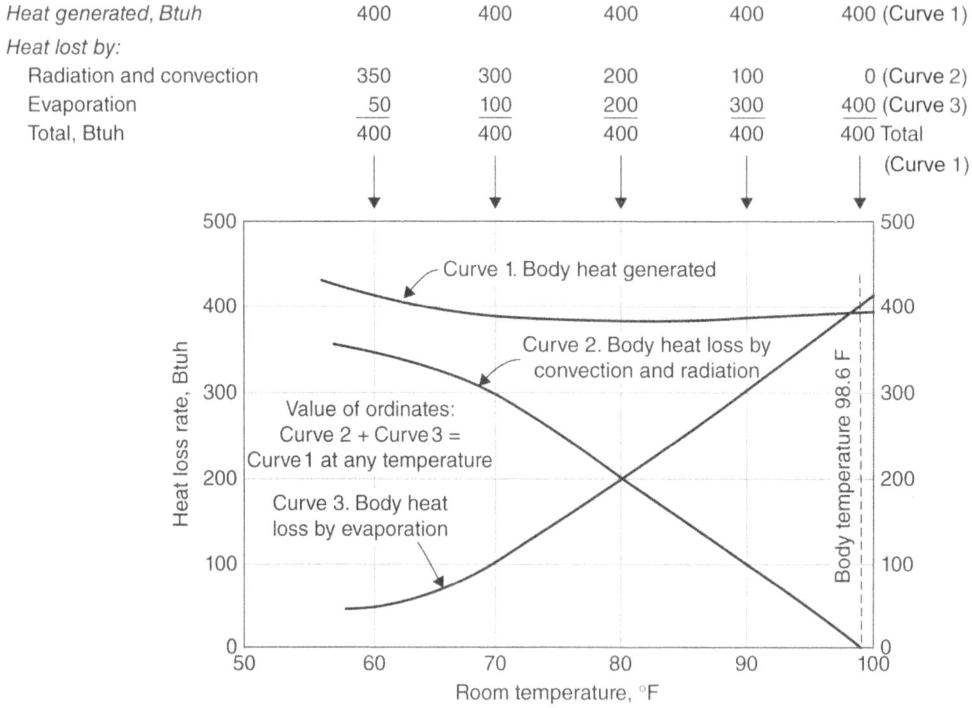

Figure 6.2 *Heat generated and lost (approximate) by a person at rest (with 45% relative humidity).*

TABLE 6.2 Approximate Heat Loss Distribution under Comfort Conditions

Component	(%)
Radiation, convection, and conduction	72
Evaporation	
From skin surface	15
From lungs (exhaled air)	7
Warming of air inhaled to lungs	3
Heat expelled in feces and urine	3

(c) Clothing

Most clothing acts as an insulating layer and is particularly effective at retarding radiation, convection, and conduction. As ambient air and surface temperatures fall well below our body's temperature, our clothing provides a second skin and increased insulation. In hot and humid environments, our bodies need air moving against the skin to remove heat, while clothing provides protection from the sun's radiant heat. In hot and arid environments, clothing layers can prevent our bodies from losing too much water, while also provide shading to our skin.

The insulation value provided by clothing is measured in clo units, where 1 clo is equal to $0.88\ \text{ft}^2\ \text{h F/Btu}$ [$0.155\ \text{m}^2\ \text{K/W}$]. The total clo value of a person's attire may be estimated by simply adding the clo of each item as listed in Table 6.3; this total will be a

TABLE 6.3 Typical Insulation Values for Clothing Ensembles

Ensemble Description[a]	clo[b]
Walking shorts, short-sleeve shirt	0.36
Trousers, short-sleeve shirt	0.57
Trousers, long-sleeve shirt	0.61
Same as above, plus suit jacket	0.96
Same as above, plus vest and T-shirt	1.14
Trousers, long-sleeve shirt, long-sleeve sweater, T-shirt	1.01
Same as above, plus suit jacket and long underwear bottoms	1.30
Sweatpants, sweatshirt	0.74
Long-sleeve pajama top, long pajama trousers, short 3/4-sleeve robe, slippers (no socks)	0.96
Knee-length skirt, short-sleeve shirt, pantyhose, sandals	0.54
Knee-length skirt, long-sleeve shirt, full slip, pantyhose	0.67
Knee-length skirt, long-sleeve shirt, half-slip, pantyhose, long-sleeve sweater	1.10
Same as above; replace sweater with suit jacket	1.04
Ankle-length skirt, long-sleeve shirt, suit jacket, pantyhose	1.10
Long-sleeve coveralls, T-shirt	0.72
Overalls, long-sleeve shirt, T-shirt	0.89
Insulated coveralls, long-sleeve thermal underwear, long underwear bottoms	1.37

Source: © ASHRAE, www.ashrae.org. 2021 ASHRAE Handbook—Fundamentals, Chapter 9, Table 7.

[a] All ensembles include shoes and briefs or panties. All ensembles except those with pantyhose include socks, unless otherwise noted.

[b] One clo = $0.88\ \text{ft}^2\ \text{h °F/Btu} = 0.155\ \text{m}^2\ \text{K/W}$.

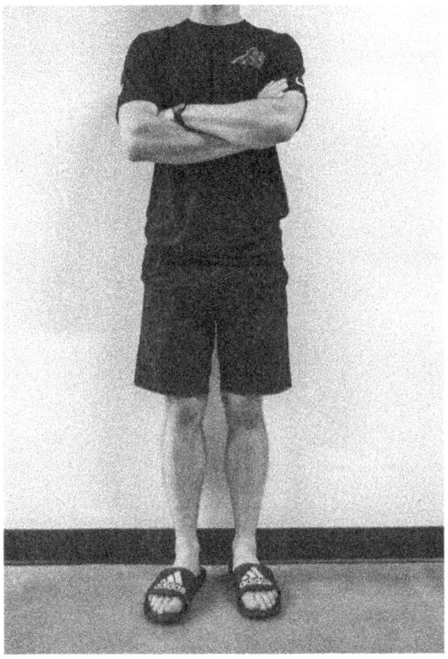

(*a*) 0.5 clo

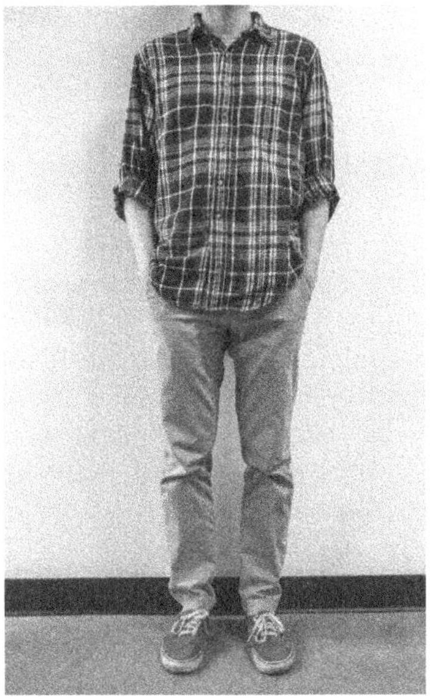

(*b*) 1.0 clo

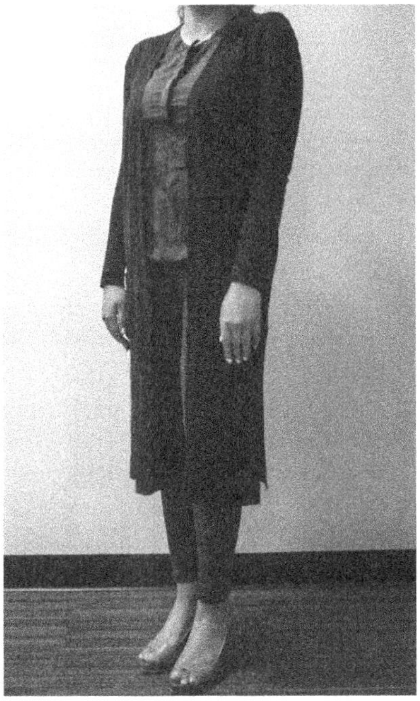

(*c*) 1.25 clo

(*d*) 2.8 clo

Figure 6.3 *Approximate insulating value of clothing: (a) shorts and T-shirt (0.5 clo); (b) trousers, long-sleeve shirt, T-shirt (1.0 clo); (c) skirt, half-slip, leggings, long-sleeve shirt, long-sleeve sweater (1.25) clo; (d) insulated trousers, long-sleeve thermal underwear, down jacket, long underwear bottoms, wool scarf (2.8 clo).*

bit higher than the actual clo value of the ensemble. The total clo can also be estimated by assuming 0.15 clo/lb [0.35 clo/kg] of clothing weight. Sometimes the position of clothing is as important as its clo value: consider the role of socks as they separate our feet from contact with a cold floor. Figure 6.3 shows examples of clothing ensembles and their clo values.

Our second skin (clothing) is just as likely as our third (building) skin to be more dominated by considerations of style than of thermal regulation; we cannot always count on our clothing—or buildings—to increase our thermal comfort.

6.2 PSYCHROMETRY

The air around us contains energy—specifically, thermal energy (heat). It exists in two forms: sensible heat and latent heat.

If the air in a room were completely dry, devoid of any moisture whatsoever, then its thermal energy could be measured using a conventional thermometer to assess the dry-bulb (DB) temperature. This is the type of "dry" heat that the human body can sense, thus the term sensible heat. The thermal energy added to air by the glowing coil of an electric range is an example of sensible heat.

Air mass in terrestrial environments, however, always contains some amount of moisture—even if it is minuscule. This moisture in the air embodies the "wet" heat that the air mass takes on as water undergoes a phase change from liquid to vapor via evaporation or boiling. This is referred to as latent heat because it is not able to be sensed directly. For example, consider the water vapor added to an air volume by a tea kettle boiling on a stove. The boiling point of water is 212 °F [100 °C]. Thermal energy is required to change the water into vapor, but during the phase change the temperature remains the same. If it did not heat up the moisture, then where did this energy go? It is contained in the air volume as latent heat.

Unlike sensible heat, latent heat is not directly "sensed" by a conventional thermometer with a dry-bulb. Rather, latent heat can be detected by comparing the DB temperature with the temperature of air measured using a thermometer with a wetted bulb moving rapidly through the air in order to promote evaporation. This is known as wet-bulb (WB) temperature.

Heat and moisture interact in air volumes in ways that are very consequential to building performance and human health. In winter, condensation within insulation due to falling air temperatures can be disastrous. In summer, adding moisture to hot, dry air can lower its DB temperature while raising its humidity to more comfortable levels.

These heat and moisture interactions in an air volume are complex. As air temperature rises, its capacity to hold moisture also rises, and the warmer air becomes less dense. These combined interactions are described by psychrometrics, the study of air and water vapor. Fortunately, these interactions can be combined within a single chart—see Appendix G, Figures G.1 (I-P units) and G.2 (SI units).

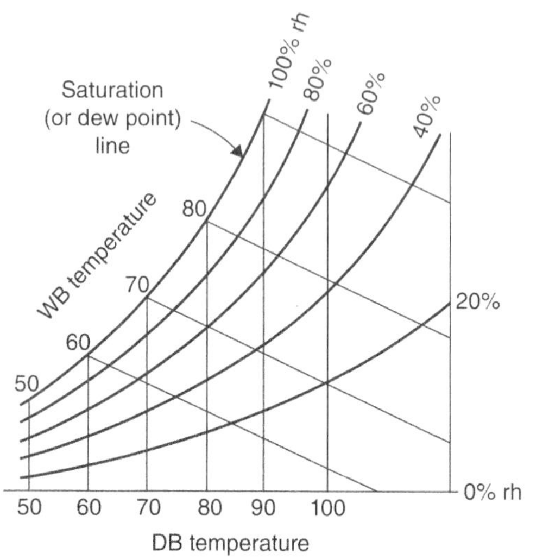

Figure 6.4 Some basic components of the psychrometric chart: DB and WB temperatures and RH.

Though the psychrometric chart may at first appear esoteric and complex, it is essentially a two-axis chart of sensible heat (i.e., DB temperature) and latent heat (i.e., humidity ratio) as depicted in Figure 6.4. Any air–vapor mix can be plotted on this chart; however, every air volume has an absolute maximum humidity ratio that can exist at a given DB temperature. This is called the saturation point. If the saturation point is plotted for all dry-bulb temperatures, one would define the saturation line on the psychrometric chart. An air–vapor mix cannot exist beyond the saturation line; therefore, the portion of the psychrometric chart is removed, giving the chart its unique curved profile. The saturation line is also referred to as the dew point (DP) because dew forms (water vapor condenses) when saturated air touches any surface at or below the air's dew point temperature. This saturation is sometimes undesirable, as within walls or roofs, or on ceiling, air duct, or glass surfaces. However, it is often desirable, as on air-conditioner coils, where the resulting reduction of the moisture content in the air is deliberate.

The psychrometric chart may be used to graph a wide variety of processes, which are summarized in Figure 6.5. To better understand these processes, we may consider informational overlays to the basic chart of Figure 6.4. The first addition is the relative humidity (RH), which indicates the humidity ratio of the air–vapor mix expressed as a percent of the total saturation point for that temperature.

The humidity ratio indicates the amount of moisture by weight within a given weight of dry air. Air treatment processes that travel to the right along the horizontal axis of constant humidity ratio (Figure 6.6) are the familiar processes of simple heating (air passing

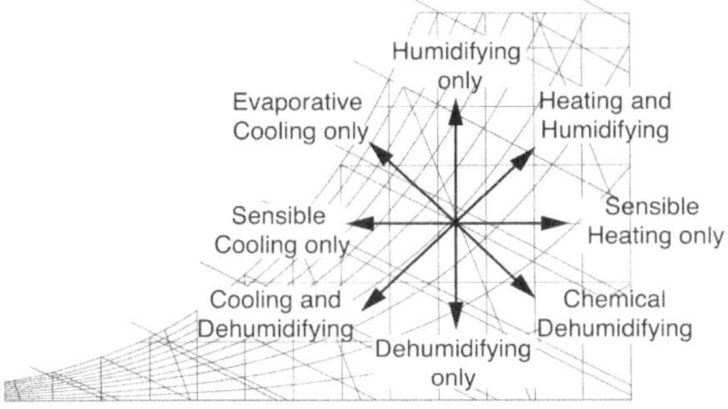

Figure 6.5 *Climatic-conditioning processes expressed on the psychrometric chart. (Adapted from Givoni and Milne, 1979; redrawn by Sharon Alitema.)*

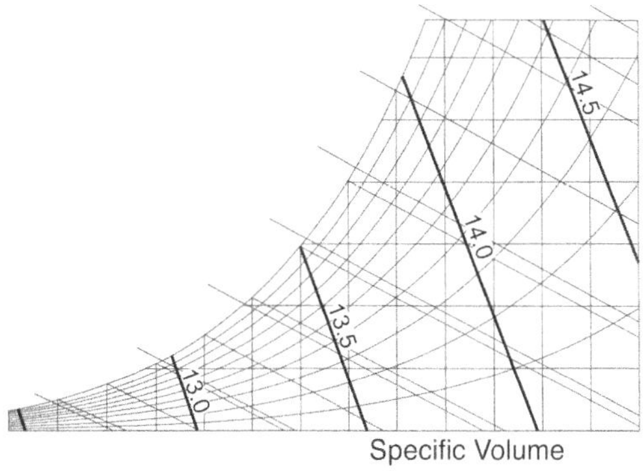

Figure 6.7 *Specific volume on the psychrometric chart: I-P units (pictured) are ft³/lb dry air; SI units are m³/kg dry air. (Drawn by Sharon Alitema.)*

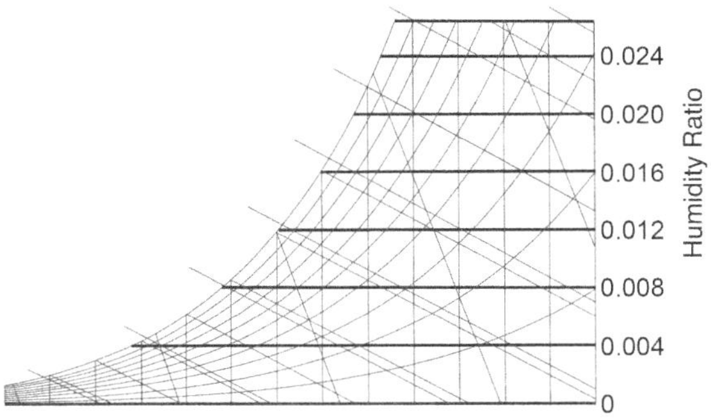

Figure 6.6 *Humidity ratio on the psychrometric chart: I-P units (pictured) are lb moisture/lb of dry air; SI units are kg moisture/kg dry air. (Drawn by Sharon Alitema.)*

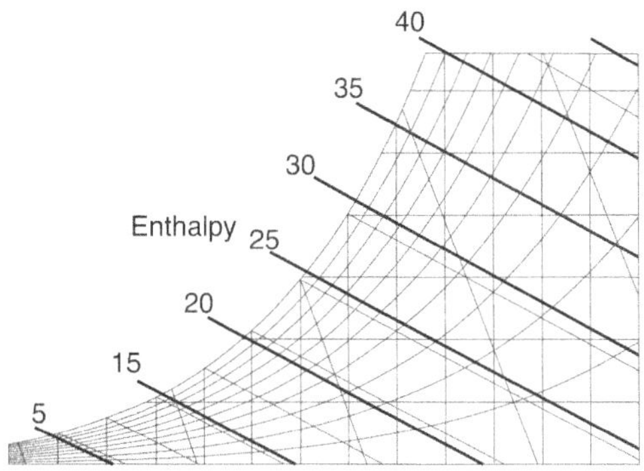

Figure 6.8 *Enthalpy on the psychrometric chart: I-P units (pictured) are Btu/lb; SI units are kJ/kg. (Drawn by Sharon Alitema.)*

through the heating coil of a furnace or through a solar collector) and simple (sensible) cooling (air passing through the cooling coil of an air conditioner before saturation). If sensible heat is removed such that the air–vapor mix is pulled left of the saturation line, condensation will occur, reducing the humidity ratio of the air mass in the process. The relative humidity curves intuitively convey the proportion of total potential saturation based on the DB temperature and humidity ratio of an air volume.

The next addition shows how the density of air varies as its temperature and moisture content vary. These lines are those of specific volume, the reciprocal of density, a useful quantity in air-conditioning calculations and helpful in understanding the stack effect in passive design. The specific volume is given in ft^3/lb [m^3/kg] of dry air. From these lines in Figure 6.7, it can be seen that a pound of hot air is larger (has more volume) than a pound of cold air. This larger volume per unit of weight increases buoyancy; thus, hot air rises, whereas cold air sinks.

The next addition (Figure 6.8) involves enthalpy, the sum of the sensible and latent heat content of an air–moisture mixture relative to the sum of the sensible and latent heat in air at 0 °F [0 °C in SI units] at standard atmospheric pressure. Enthalpy units are Btu/lb [kJ/kg] of dry air. Enthalpy lines are almost parallel to those of WB temperature. Perhaps the most familiar process to travel along the lines of constant enthalpy is evaporative cooling, whereby increased moisture and lower air DB temperature are obtained without changing the enthalpy (total heat content) of the air. There is indeed a drop in sensible heat as the temperature drops, but this is matched by an increase in latent heat as the moisture content increases. The opposite process is chemical (desiccant) dehumidifying, whereby decreased moisture content is obtained at the price of increased air DB temperature; again, no change in enthalpy (total heat) occurs.

The psychrometric chart is essentially a composite of these layers of information. Figure 6.9 depicts the psychrometric chart in I-P with some layers of information omitted for visual clarity. As previously mentioned, a comprehensive version of the psychrometric chart better suited for use in design is offered in Appendix G with I-P and SI units.

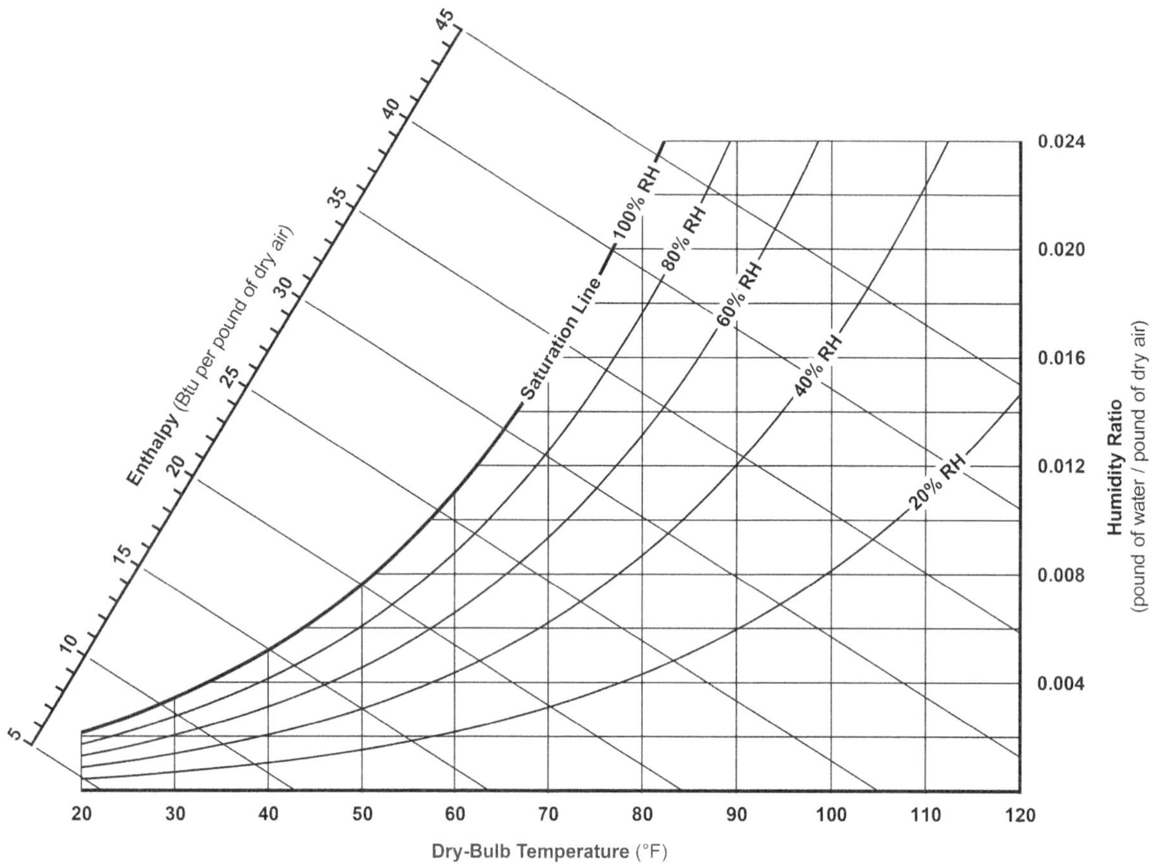

Figure 6.9 *The psychrometric chart in I-P units. See Appendix G, Figure G.2 for a version of the chart with SI units. (Drawn by Daniel Overbey.)*

6.3 THERMAL COMFORT DETERMINANTS

In our everyday experiences, thermal comfort is simply a lack of discomfort—of being unaware that we are exchanging heat with our environment. *ASHRAE Standard 55-2023: Thermal Environmental Conditions for Human Occupancy* technically defines thermal comfort as "that condition of mind which expresses satisfaction with the thermal environment and is assessed by subjective evaluation."

Three conditions combine to affect our perceptions of thermal comfort: personal factors, measurable environmental factors, and psychological factors. Personal factors are typically under our control and can be quantified. These include activity level (metabolism) and clothing (as an insulation layer), as well as various behavioral adaptations such as migration to a more comfortable place, or drinking or eating warm or cold foods. Measurable environmental factors are the primary tools of the designer: air temperature, radiant temperature, air motion, and humidity. Psychological factors are also familiar designers' tools, but are difficult to quantify when designing for comfort: occupant attitudes, previous experiences, and expectations. Several of these factors are often overlooked as we strive to meet the numerical (therefore calculable) physical criteria for thermal comfort. If our design intent is to make people comfortable, however, all aspects of buildings are a means to that end. And—all factors can contribute to or mitigate against thermal comfort depending upon circumstances.

Consider a courtyard in a hot, dry climate (Figure 6.10). Its fountain suggests coolness by the color and texture of its water. The running water provides splashing sounds and sparkles of light and may generate some air motion. Vines provide shade, and their leaves sway in the slightest breeze, visual evidence of some air motion. Blossoms of flowering plants yield a cool fragrance that blends with the aroma of moistened surfaces in the otherwise hot, dry surroundings. The measured microclimate conditions of the courtyard may be but a slight improvement over the microclimate beyond, but it can be perceived as refreshingly cool.

Next consider a fireplace in a cold climate (Figure 6.11). The fire's color is intensely warm, and it dances and casts a flickering light; it crackles, and it yields a smoky aroma (not appreciated by all). Few textures seem hotter than that of glowing coals. Yet a fireplace might add but slightly to the measured

Figure 6.10 *Psychological indicators of coolness in a courtyard in Savannah, Georgia, include running water and shade from trees that move with the breeze. The senses of sight, hearing, touch, smell, and taste all may be involved in a perception of coolness. (© Alison Kwok; all rights reserved.)*

Figure 6.11 *The open fire seems the very spirit of warmth, despite the heat lost by large quantities of exhaust air going up the chimney. The senses of sight, hearing, touch, and smell may be involved in a perception of warmth, wrapped in this red-brick environment. (Courtesy of the architect and photographer, Edward Allen.)*

temperature of a room because it draws cold air into the room to replace combustion air that passes up the chimney. We seek a place near the fire for measurable radiant warmth but also for the psychological comfort the fire offers.

In our society, designers are encouraged to comply with standards and codes. *ASHRAE Standard* 55 conveys information on appropriate values of quantifiable thermal comfort factors. These have been extensively tested in laboratories—but *Standard* 55 does not explicitly address qualitative factors. A holistic view of designing for comfort considers numbers as a foundational guide; common sense and a designer's own thermal experience play important roles as well. Lisa Heschong's influential work *Thermal Delight in Architecture* (1979) provides an excellent expression of this point of view.

Ultimately, our buildings are expected to demonstrate success in providing the measurable environmental factors that will result in comfort. Building environments and materials respond to the interactions of environmental factors via various mechanisms of heat transfer. Table 6.4 shows how air and surface temperatures, air motion, and humidity are related to heat transfer.

These comparisons are of primary importance in hot conditions, where humidity will influence the effectiveness of the heat transfer mechanisms. For example, humidity is critical in hot, dry environments dominated by evaporative heat loss. In cold conditions, heat loss by convection, radiation, and conduction is dominant, so humidity is less influential. Comfort studies have shown that skin temperature is an important factor in cold conditions, whereas skin "wettedness" (percentage covered by water) is most important in hot conditions.

(a) ASHRAE Thermal Comfort Standard

Extensive study of human responses to thermal conditions shows that people will express a perception of comfort across a range of environmental conditions. This range has come to be known as the thermal comfort zone. The general concept of a comfort zone is diagrammed in Figure 6.12. Because of the logical link to psychrometrics (moist air conditions), thermal comfort zones are plotted on the psychrometric chart framework. Unfortunately, these charts are two-dimensional and can only directly map two of the six primary quantitative comfort factors. Luckily, the psychrometric chart variables of dry bulb temperature and relative humidity are highly relevant comfort factors.

TABLE 6.4 Environmental Factors Influencing Heat Transfer

Heat Transferred by	Primarily Dependent upon
Conduction	Surface temperature
Convection	Air temperature, air motion, humidity
Radiation	Surface temperature, orientation to the body
Evaporation	Humidity, air motion, air temperature

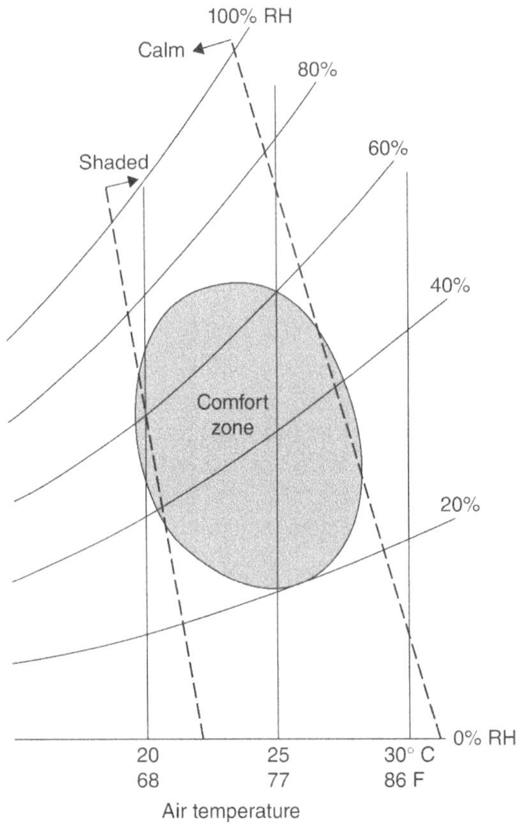

Figure 6.12 *Stylized comfort zone defined by relative humidity and air temperature.*

A human thermal response, perceived broadly as thermal comfort or discomfort, is shaped primarily by metabolic rate (met), clothing insulation (clo), dry bulb (DB) air temperature, mean radiant temperature (MRT), air speed, and relative humidity (RH). The first two factors are personal variables of the occupants; the remaining four are characteristics of the thermal environment. The six variables can be measured directly, derived from other measured variables (using a psychrometric chart), or imputed from collected or observed data.

It is important to recognize that "thermal comfort" is an opinion. Human comfort is substantially influenced by the aforementioned six factors—but it is an opinion, nevertheless. This opinion is tempered by the psychological state and experiences of the individual. Two people in the same room may (and often do) have differing perceptions of thermal comfort. For this reason, designers should recognize that a single ideal condition for human thermal comfort does not exist. Buildings should be designed to support the perception of comfort, but they cannot be guaranteed to deliver comfort.

Two internationally recognized standards provide design support for achieving thermally comfortable building conditions. *ASHRAE Standard* 55 is the predominant design resource in North America. *ISO Standard 7730* (ISO, 2005) is commonly used outside North America. These documents provide similar guidance. This chapter describes the ASHRAE view of thermal comfort. There is no building code in North America that requires thermal comfort, thus, minimum performance is generally governed by voluntary requirements.

ASHRAE Standard 55 offers a methodology for determining the likelihood that a majority of occupants will find the conditions of a certain space to be thermally comfortable. The methodology is based on the predicted mean vote (PMV) index. The PMV index, frequently attributed to the work of Danish professor Povl Ole Fanger, is a seven-point numerical scale that expresses the thermal sensation experienced by individuals:

+3	hot
+2	warm
+1	slightly warm
0	neutral
−1	slightly cool
−2	cool
−3	cold

Fanger (1982) used the PMV index to predict the average value of thermal sensation votes from groups of individuals via self-reported perceptions. He determined a correlation between the PMV and the predicted percentage of dissatisfied people (PPD) for a given interior environment (Equation 6.1):

$$PPD = 100 - 95 \exp[-(0.03{,}353\ PMV^4 + 0.2179\ PMV^2)] \quad \text{(Eq. 6.1)}$$

This correlation is graphically depicted in Figure 6.13. Note that even a slightly uncomfortable interior environment (say, a PMV of 1.0) will prompt a considerable percentage of occupants to deem the space unsatisfactory. An interesting feature of this equation is the minimum PPD of 5% for a thermally neutral mean vote (PMV = 0). This suggests that in any thermal comfort survey, at least 1 out of every 20 individuals will be dissatisfied.

ASHRAE Standard 55 uses the relationship shown in Figure 6.12 to establish a quantitative prediction of the percentage of thermally dissatisfied people.

Figure 6.14 shows an example graphical ASHRAE thermal comfort zone, where two shaded boxes indicate acceptable ranges of operative temperature (T_{op}) and

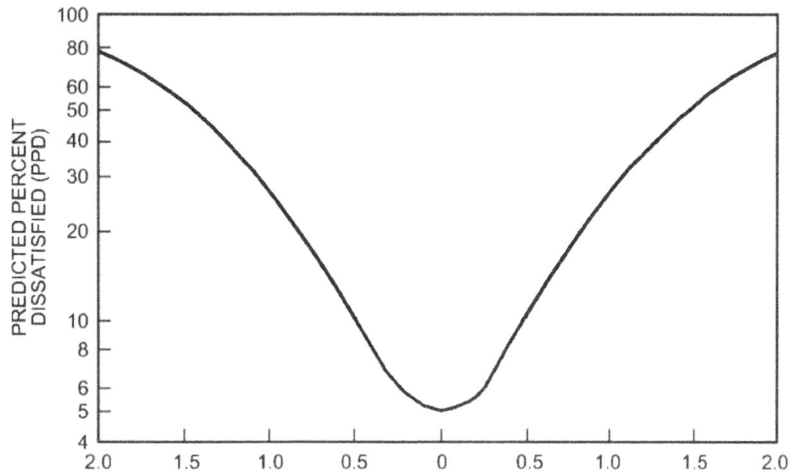

Figure 6.13 *Predicted percentage of dissatisfied (PPD) as a function of predicted mean vote (PMV). (© ASHRAE, www.ashrae.org. 2021 ASHRAE Handbook—Fundamentals, Chapter 9, Figure 16.)*

relative humidity (RH) for sedentary (1.3 met) persons wearing typical clothing (0.65 and 1.0 clo). Two zones (boxes) are shown—one for 0.65 clo of clothing insulation (assumed worn when the outdoor environment is warm) and one for 1.0 clo of insulation (for when the outdoor environment is cool). Other comfort zone charts exist for different values of assumed variables (such as air speed or activity). When using these charts, it is critical that assumptions appropriate to the space under consideration are made. Notes on Figure 6.14 suggest the effect of different activity, clothing, and radiant conditions on the position of the comfort zone. ASHRAE-endorsed software permits the plotting of comfort zone charts for any combination of factors.

The thermal comfort variables related to the psychrometric chart are defined as follows.

Dry-bulb (DB) temperature: DB temperature is the ambient air temperature as measured by a standard thermometer, thermocouple, or resistance temperature device. DB temperature can be used in combination with globe temperature and air velocity to calculate mean radiant temperature (see below). Temperature affects comfort in several ways and, in combination with the other variables described in this section, is a key determinant of our energy balance, thermal sensation, and perception of air quality.

Operative temperature: This is the average of the DB temperature and the mean radiant temperature. Using operative temperature as the horizontal axis of the comfort chart allows three variables to be directly represented on a diagram with x and y axes.

Wet-bulb (WB) temperature: WB temperature is measured by a thermometer with a wetted bulb rotated rapidly in the air to cause evaporation of its moisture. The sling psychrometer shown in Figure 6.15 is an analog WB measuring device. In dry air the moisture on the bulb readily evaporates and draws heat out of the thermometer as it does so. This produces a lower temperature reading, called the WB depression (the difference between the DB and the WB temperatures). A large depression is indicative of low relative humidity. Less evaporation, as when the air is moisture-laden, results in a small WB depression and indicates high RH. At 100% RH, DB and WB temperatures are equal.

Because WB temperature is easily measured and can be overlaid on a psychrometric chart as a series of diagonal lines, it is often the preferred latent (moisture-related) indicator for design purposes. Humidity ratio, the vertical axis on the psychrometric chart is much more challenging to directly measure. Humidity ratio, however, is absolute and independent of dry bulb temperature.

Relative humidity (RH): RH may be measured directly or derived from the DB and WB temperatures. RH is the ratio of the actual density of water vapor in air to the maximum density of water vapor that such air could contain, at the same temperature, if it were 100% saturated. RH appears on the psychrometric chart as a series of curved lines—the upper of which indicates 100% RH (saturated) air.

Mean radiant temperature (MRT): The temperatures of surrounding surfaces influence human comfort. The MRT at a particular location in a space is defined as the hypothetical uniform temperature of a space with which radiant heat flow to/from the human body would equal the radiant heat transfer in the actual nonuniform enclosure (ASHRAE, 2023). MRT as a mathematical construct is a calculated variable and cannot be

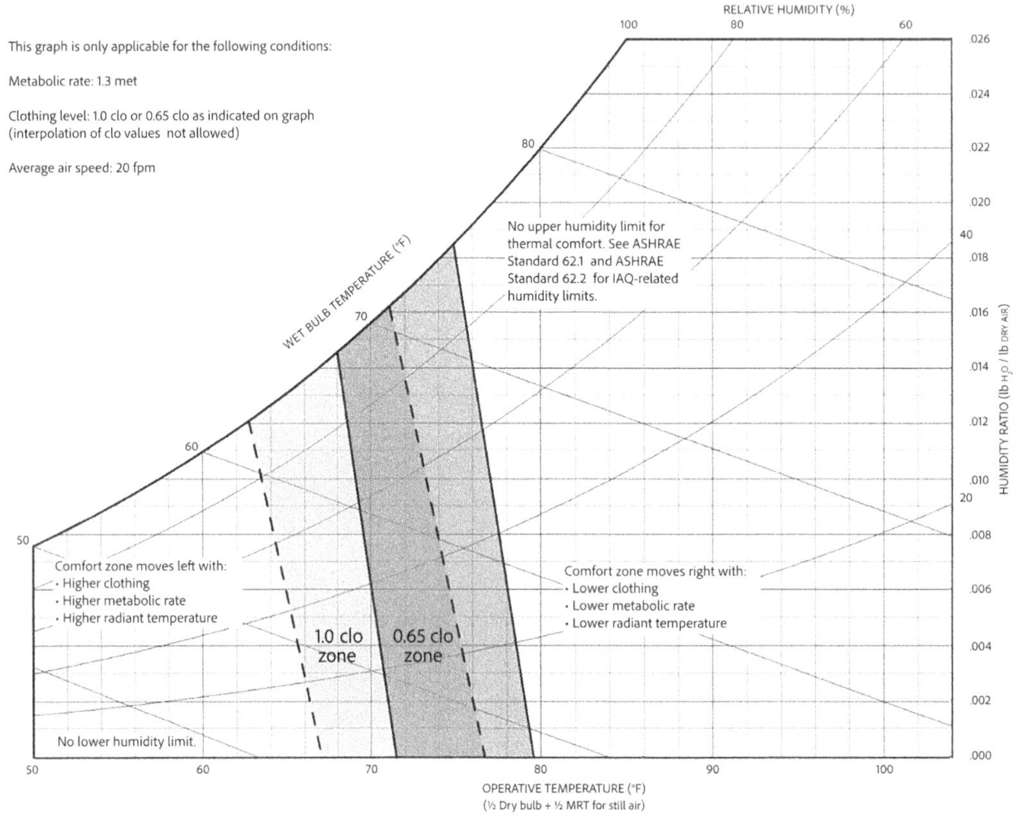

(a)

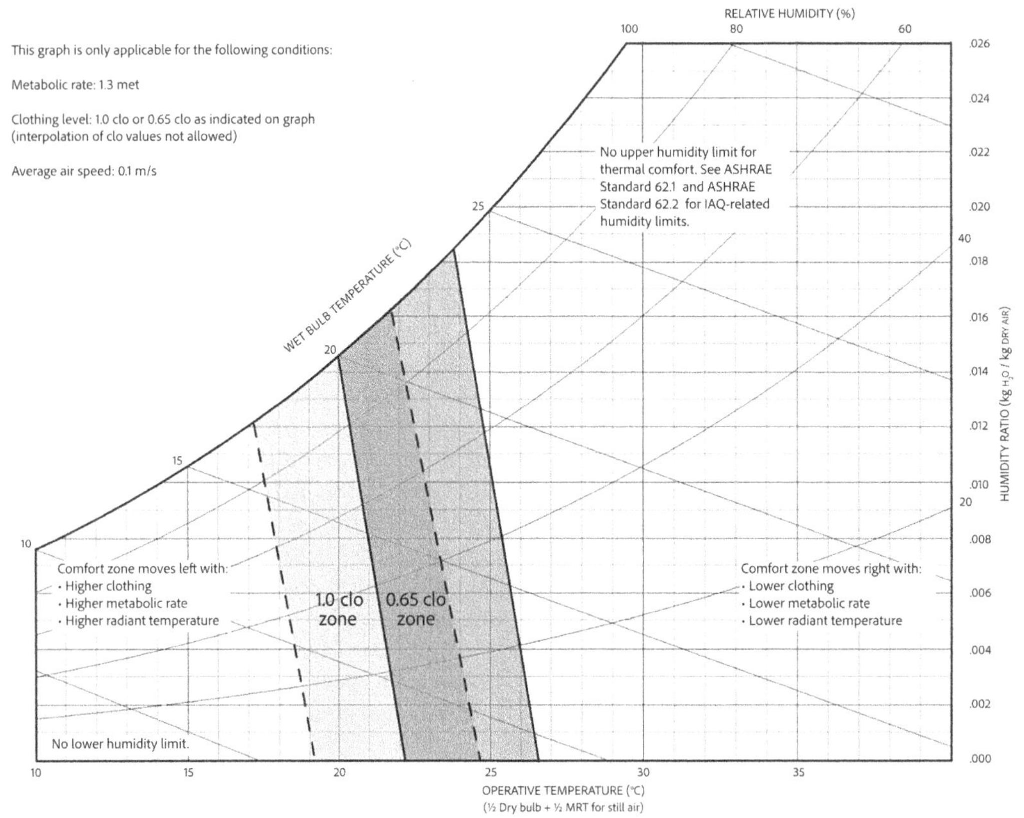

(b)

Figure 6.14 *ANSI/ASHRAE Standard 55–2023 comfort zone example for commonly encountered building conditions. I-P Units (a) and SI units (b). (Reprinted with permission; © ASHRAE, www.ashrae.org. ANSI/ASHRAE Standard 55–2023: Thermal Environmental Conditions for Human Occupancy.)*

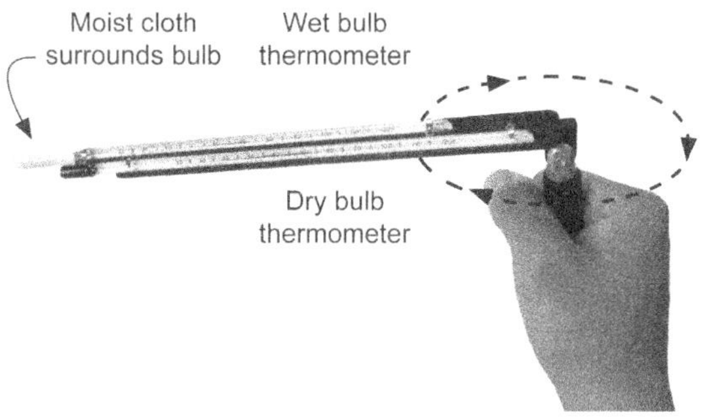

Figure 6.15 Sling psychrometer and its usage. Air motion encourages evaporation from the moist cloth, lowering the wet-bulb temperature reading, whereas the dry-bulb temperature reading directly reflects the surrounding air temperature.

directly measured. Surface temperatures, however, can be directly measured with a variety of devices.

There are several ways to determine mean radiant temperature. In a built space, a globe thermometer can measure globe temperature. This device can be constructed using a hollow sphere (e.g., a ping-pong ball) with a thermocouple or thermometer bulb at its center. Equation 6.2 describes how to determine MRT from globe temperature (ASHRAE, 2021). A graphic method of calculating MRT from surface temperatures is offered in Figure 6.16. This method can also be

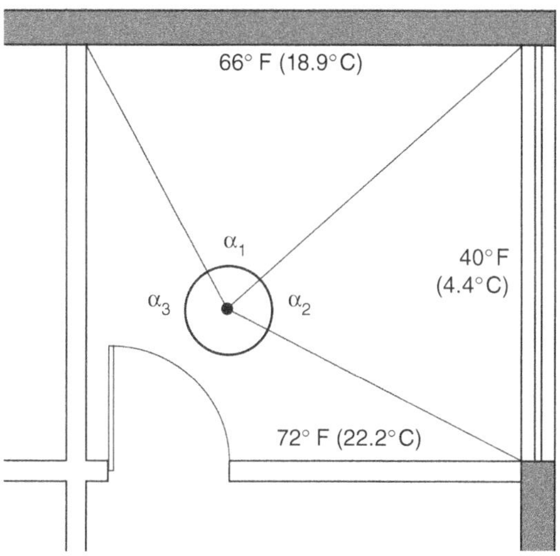

Figure 6.16 Graphic method of calculation for mean radiant temperature (MRT) using the geometry of the room relative to a particular location. Using Equation 6.3, the surface temperature of the wall or window is multiplied by the exposure angle relative to the occupant's location (the center of the circle). This process is repeated in section so that each surface is included twice. (Drawn by Tyler Mavichien; © Alison Kwok, all rights reserved.)

used during design, with estimated surface temperatures replacing measured values. Means of estimating surface temperatures are discussed in a later chapter.

I-P units: $\bar{t}_r$

$$\left[(t_g + 460)^4 + \frac{4.74 \times 10^7 V_a^{0.6}}{\varepsilon D^{0.4}}(t_g - t_a)\right]^{0.25} - 460$$

SI units: $\bar{t}_r$

$$\left[(t_g + 273)^4 + \frac{1.10 \times 10^8 V_a^{0.6}}{\varepsilon D^{0.4}}(t_g - t_a)\right]^{0.25} - 273 \quad \text{(Eq. 6.2)}$$

where

$\bar{t}_r$ = mean radiant temperature, °F [°C]
t_a = air temperature, °F [°C]
t_g = globe temperature, °F [°C]
V_a = air velocity, fpm [m/s]
D = globe diameter, ft [m]
ε = globe emissivity (dimensionless)

Another calculation approach involves the geometry of the room relative to a specific point and uses Equation 6.3 (Egan, 1975):

I-P or SI units

$$\bar{t}_r = \frac{\sum t\alpha}{360} = \frac{t_1\alpha_1 + t_2\alpha_2 + t_3\alpha_3 + \cdots t_n\alpha_n}{360} \quad \text{(Eq. 6.3)}$$

where

$\bar{t}_r$ = mean radiant temperature, °F [°C]
α = surface exposure angle (relative to the occupant) in degrees
t = surface temperature, °F [°C]

(b) Adaptive Thermal Comfort

Dry-bulb air temperature by itself is not an adequate thermal comfort indicator. This is especially so in passively solar-heated or passively cooled spaces, where radiant temperature or air motion may be more influential than air temperature. Occupants of naturally ventilated buildings, for example, seem to find comfortable a combination of higher temperatures and RH than do occupants of sealed, air-conditioned buildings. Figure 6.17 shows a much higher incidence of these warmer, more humid environments in naturally ventilated classrooms in Hawaii than in similar air-conditioned classrooms. Yet, the majority of students in these Hawaiian schools voted these

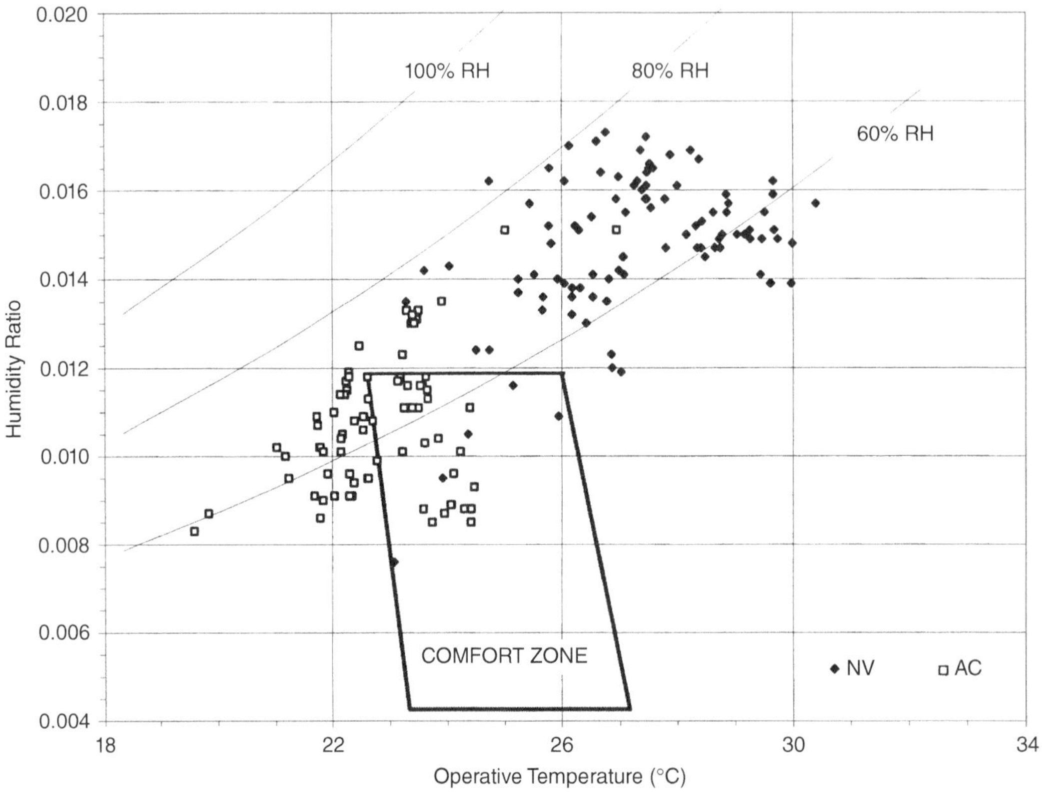

Figure 6.17 *Natural ventilation and comfort. Measured conditions compared to the ASHRAE comfort zone (warm season) for naturally ventilated and air-conditioned classrooms in Hawaii. The majority of the occupants voted these conditions acceptable. A higher temperature-and-humidity comfort zone for naturally ventilated buildings is supported by studies such as this. (Kwok, 1998. Reprinted with permission of ASHRAE, from the ASHRAE Transactions, 1998, Vol. 104, Number 1.)*

conditions as acceptable. These observations were made in both hot and cool seasons (Kwok, 1998).

Does long-term acclimatization (i.e., the physiological and psychological adaptation of a person to changes in their thermal environment) influence the sensation of thermal comfort? Givoni (1998) advocated for a higher-temperature, higher-humidity limit for "hot-developing" countries (Figure 6.18), based upon a combination of traditional and economic factors.

An adaptive model of comfort recognizes that acclimatization as well as personal actions can influence comfort perceptions. *ASHRAE Standard* 55 thus provides an optional method for determining acceptable thermal conditions for naturally ventilated spaces. Using this adaptive method to determine acceptable thermal conditions in naturally ventilated spaces one can find allowable indoor operative temperatures from Figure 6.19. This chart (Figure 6.19) includes two sets of operative temperature ranges—for 80% and for 90% acceptability limits, when the prevailing outdoor temperature is greater than 50 °F [10 °C] and less than 92.3 °F [33.5 °C]. Occupants are assumed to be near sedentary with metabolic rates of 1.0–1.3 met and will adjust their clothing choices to match conditions.

Humphreys and Nicol (1998) elaborate on this approach, with adaptive occupant behaviors, summarized in Table 6.5. They present a formula to determine the indoor comfort temperature, T_n, relative to an exponentially weighted running average of outdoor temperature, T_{rmo}, and applicable to free-running buildings (those without mechanically—that is, narrowly—controlled indoor temperatures):

$$T_n = 0.534\ T_{rmo} + 12.9\ (°C) \qquad \text{(Eq. 6.4)}$$

where T_{rmo} is in °C and represents the mean outdoor air temperature during a period of hot weather. Note that relative humidity is not included in this relationship. Using this equation, with Phoenix, Arizona's average July–August temperature of 32.3 °C [90.2 °F] (to approximate an exponentially weighted running average):

$$\begin{aligned} T_n &= 0.534\,(32.3) + 12.9 \\ &= 17.25 + 12.9 \\ &= 30.1\ °C\ [86.2\ °F] \end{aligned}$$

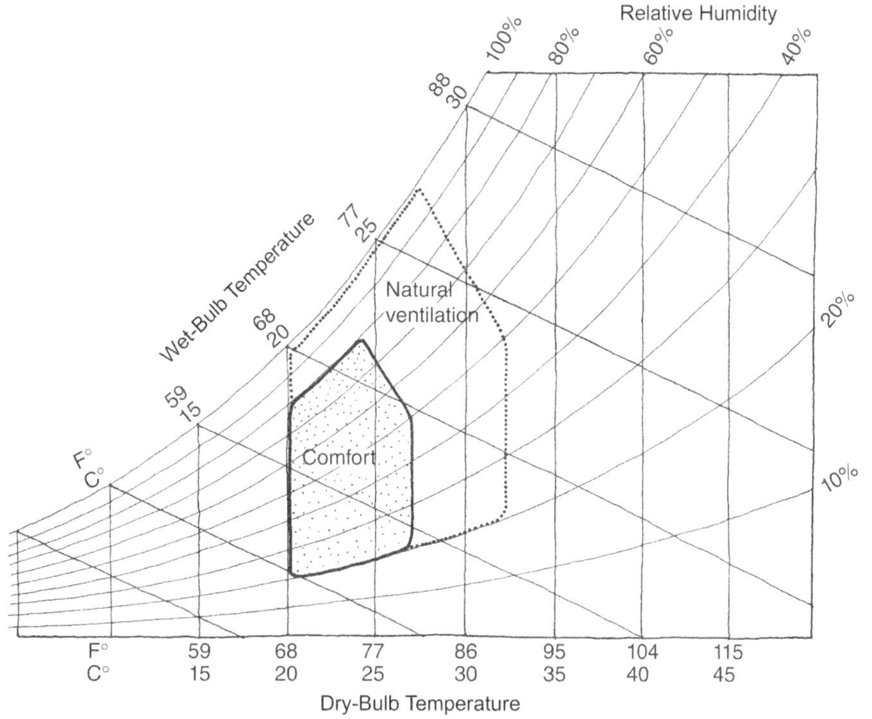

Figure 6.18 *Suggested limits of outdoor air temperature and humidity within which indoor comfort can be provided by natural ventilation. The assumed air speed is 2 m/s [400 fpm]. These higher limits for "hot-developing countries" assume cultural acclimatization. (Adapted from Givoni, 1998.)*

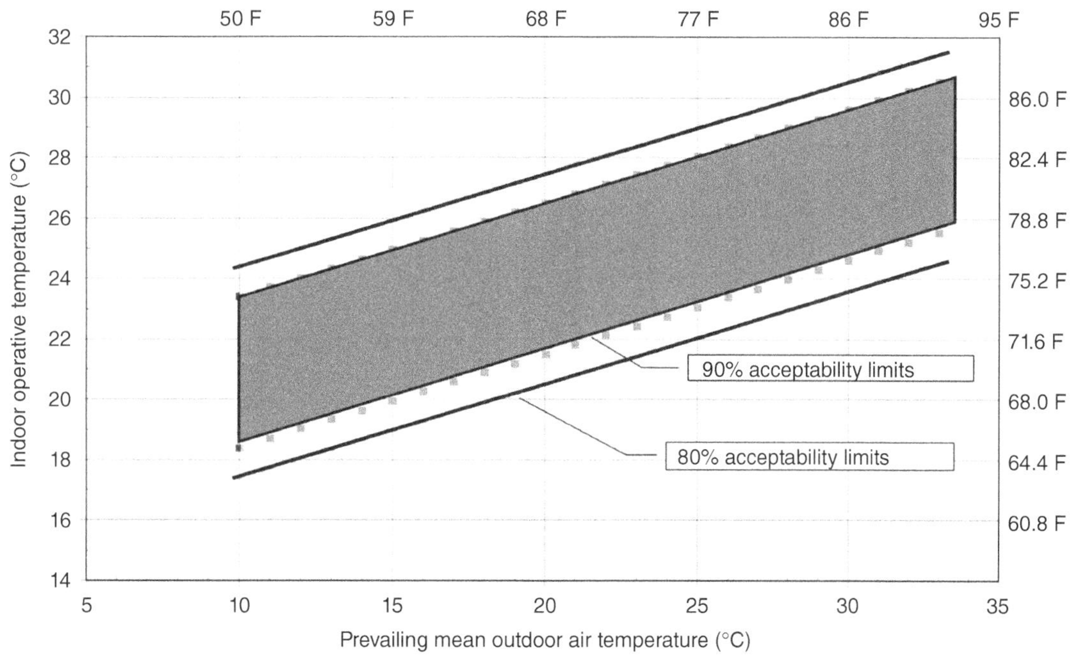

Figure 6.19 *Adaptive comfort. Acceptable operative temperature ranges for naturally ventilated spaces. (Reprinted with permission; © ASHRAE, www.ashrae.org. ANSI/ASHRAE Standard 55–2023: Thermal Environmental Conditions for Human Occupancy.)*

TABLE 6.5 Adaptive Behavior for Thermal Comfort

Response Category	Actions in Response to Cold	Actions in Response to Heat
Regulating the rate of internal heat generation	Increasing muscle tension and shivering	Reducing one's level of activity
	Drinking hot liquids	Drinking cold liquids
		Drinking hot liquids (to induce sweating)
		Eating less
		Adopting a siesta routine (matching activity to the thermal environment)
Regulating the rate of body heat loss	Vasoconstriction (reduces blood flow to the surface tissues)	Vasodilation (increases blood flow to the surface tissues)
	Curling or cuddling up (reduces exposed surface area)	Adopting an open posture (increases exposed surface area)
	Adding some clothing	Taking off some clothing
Regulating the thermal environment	Turning up the thermostat	Turning down the thermostat
	Lighting a fire	Switching on a fan or opening a window
	Complaining to management (so that someone else will raise the temperature)	Complaining to management (so that someone else will lower the temperature)
	Insulating a loft or wall cavities	Shading a window from the sun
	Improving windows or doors, weather-stripping (to reduce drafts)	
Selecting a different thermal environment	Finding a warmer spot (such as going to bed)	Finding a cooler spot
	Visiting a friend (with a warmer place)	Visiting a friend (with a cooler place)
	Visiting a heated public building	Visiting a cooled public building (or going swimming)
	Building a new home	Building a new home
	Emigrating: a long-term solution	Emigrating: a long-term solution
Modifying the body's physiological comfort conditions	Acclimatizing, letting the body and mind become more resistant to cold stress	Acclimatizing, letting the body and mind adjust so that heat is less stressful

Source: Adapted from Humphreys and Nicol, 1998.

6.4 THERMAL COMFORT DESIGN CRITERIA

The typical hard-copy comfort zone chart (such as shown in Figure 6.14) specifies comfort zone boundaries for specific operative temperature values, where the air speeds are not greater than 40 fpm [0.20 m/s], based upon stated clothing insulation values. Each chart is a snapshot representing a defined combination of the six key comfort factors—three of which are explicit and three of which are implicit. Using such charts requires the designer to mentally adjust for explicit variables that are not as defined on the chart and to account for the implicit (assumed) variables.

The ready availability of online design tools makes "fudging" a comfort chart to fit expected conditions unnecessary. The CBE Thermal Comfort Tool (Figure 6.20) allows a designer to enter values for all six comfort factors and get detailed information on predicted comfort outcomes, including a customized comfort chart. The Thermal Comfort Tool easily crunches the numbers—given some thoughtful input. A real benefit of this tool is the ability to "what-if" during design: to see the impact of better glazing (reduced summer MRT) or radiant cooling panels on occupant responses. Thermal comfort calculations can be exported into a format that is ready to support green building certification documentation (e.g., LEED or the WELL Building Standard) for thermal comfort credits. Some examples of design for thermal comfort follow.

Air speeds greater than 40 fpm [0.20 m/s] may be used to increase the upper operative temperature limit of the comfort zone in certain circumstances. For example, the borders of the warmer comfort zone (0.50 clo) may be raised 1 °F for each 30 fpm [1 °C for each 0.28 m/s] increase in air speed up to a limit of 82.5 °F at 160 fpm [28 °C at 0.8 m/s]. At this air speed, however, loose paper, hair, and other objects might start to be blown about (see Table 6.6). Visualizations of various air speeds are shown in Figure 6.21. Likewise, for each 0.1 clo decrease in clothing, a shift of 1 °F [0.6 °C] can occur at the border of the summer comfort zone.

The current version of *ASHRAE Standard* 55 prescribes no upper or lower comfort boundaries for relative humidity. This is a change from previous versions of the standard, where humidity ratios above 0.012 were flagged for caution. It is good practice to generally avoid low RH to avoid problems such as coughs, nosebleeds, static electricity, and dust mites resulting from excessively dry air. Avoiding high RH

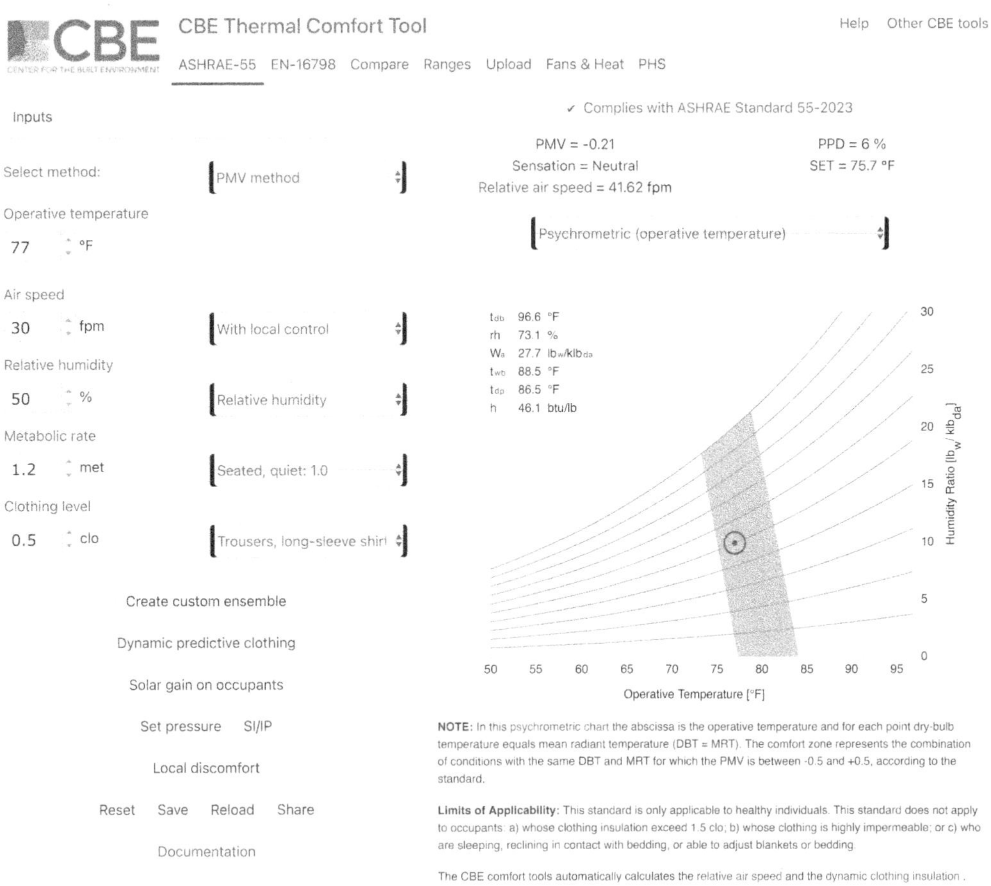

Figure 6.20 *CBE Thermal Comfort Tool. Left side shows user input fields. Top-right section contains the results of the calculations. Bottom-right section contains a visualization of thermal comfort conditions on a psychrometric chart, temperature-humidity chart, or the adaptive comfort chart. (Tartarini et al., 2020 / CC BY 4.0.)*

TABLE 6.6 Air Velocity and Comfort

Air Velocity	Possible Lower-Temperature (between 80 °F and 90 °F [26.7–32.2 °C]) Comfort Sensation; Higher Values Correspond to High-Humidity Areas	Probable Impact
Up to 50 fpm [0.25 m/s]	No change in comfort sensation	Unnoticed
50–100 fpm [0.25–0.51 m/s]	2–3 °F lower [1.1–1.7 °C]	Pleasant
100–200 fpm [0.51–1.02 m/s]	4–5 °F lower [2.2–2.8 °C]	Generally pleasant but causing a constant awareness of air movement
200–300 fpm [1.02–1.52 m/s]	5–7 °F lower [2.8–3.9 °C]	From slightly to annoyingly drafty
Above 300 fpm [1.52 m/s]	More than 5–7 °F lower [2.8–3.9 °C]	Requires corrective measures if work is to be efficient and health secured

Source: Adapted from Olgyay, 1963.

can help to keep skin "wittedness" within acceptable levels, as well as discourage the growth of mold and mildew.

As described previously, there is a substantive range of thermal conditions that will be deemed comfortable by the average building occupant or user. *ASHRAE Standard* 55 does not distinguish between psychrometric points that fall within the comfort zone—being in the center of the zone is not substantively different from being on the edge of the zone. There are, however, energy consumption and carbon emission differences for otherwise acceptable design conditions. As an example, in the summer a higher thermostat set point will use less active cooling than a lower set point—with the same comfort outcomes. Understanding this allows for informed design.

Arens et al. (1980) demonstrated a wider range of comfort conditions and considerably more tolerance for

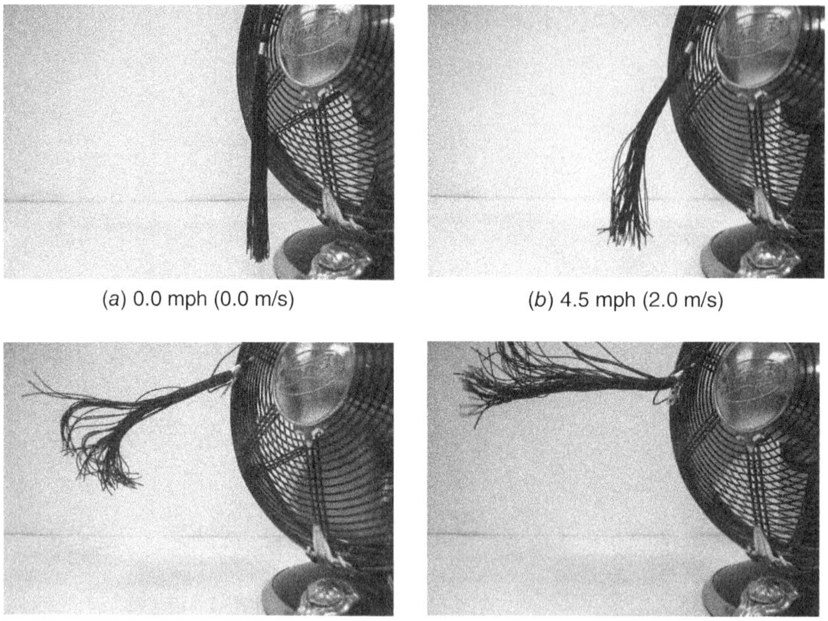

(*a*) 0.0 mph (0.0 m/s) (*b*) 4.5 mph (2.0 m/s)

(*c*) 7.8 mph (3.5 m/s) (*d*) 11.2 mph (5.0 m/s)

Figure 6.21 *Various air speeds are illustrated using a fan and graduation tassel.*

summer air motion than did ASHRAE. A summary of these results is shown graphically in Figure 6.18. Such graphs are called bioclimatic charts because of their interrelation of climate and human comfort factors. When the users of buildings are expected to be able to adjust to the wider temperature swings associated with environmentally responsive buildings, such guidelines may be used. Some differences in the basic assumptions of Arens from those of ASHRAE (2023) should be noted:

Activity: 1.3 met (ASHRAE, nominal 1.2 met)
Clothing winter: 0.8 clo (ASHRAE, 1.0 clo)
Clothing summer: 0.4 clo (ASHRAE, 0.5 clo)

Note that, rather than operative temperature, ordinary air temperature (DB temperature) is used for the horizontal axis in Figure 6.22. Adding radiant heat for conditions below the "shading line" can extend the comfort zone to lower temperatures; adding air motion above the "nearly still air" line can extend the comfort zone to higher temperatures. Radiant heat is expressed in Figure 6.22 using two quantities. The effective radiant field (ERF) is a measure of the net radiant heat flux to the body from all surfaces at temperatures other than air temperature. A more convenient quantity for designers may be the total solar radiation (or insolation) on a horizontal surface, termed I_{th}. Insolation values are readily available, and the added-radiation lines of Figure 6.18 represent I_{th} converted to its approximate radiant impact on a human body's surface area when the sun is at an altitude of 45°. Arens et al. (1980) give further details on the effects of solar radiation on comfort at other sun altitudes, activity levels, and clothing combinations.

The effect of air motion on thermal comfort is shown in Figure 6.22 for a range of velocities. People can easily tolerate air motion outdoors up to about 600 fpm [3 m/s], but indoor studies (Table 6.6) have indicated that about 400 fpm [2 m/s] is the maximum tolerable air speed. This limit applies to air motion caused by overhead fans. The comfort zones seen in Figure 6.22 constrain relative humidity above 70% or below 20% for the reasons cited earlier.

In addition to the use of environmental factors to modify the thermal comfort zone, personal comfort factors can be considered. Higher activity and more clothing, for example, would also help achieve comfort in cool situations. Similarly, under warmer air temperatures, comfort may be attainable by lowering the activity level and/or (within reason) removing clothing. There are limits to such adjustments in both directions, but the important point is that the basic comfort zone can be expanded by changing environmental variables and/or human behavior.

(a) Localized Comfort Considerations

How can comfort be ensured at each workstation in an office or wherever people spend a lot of time? The location of heating, cooling, and ventilating components is an important detail. The human body is most affected

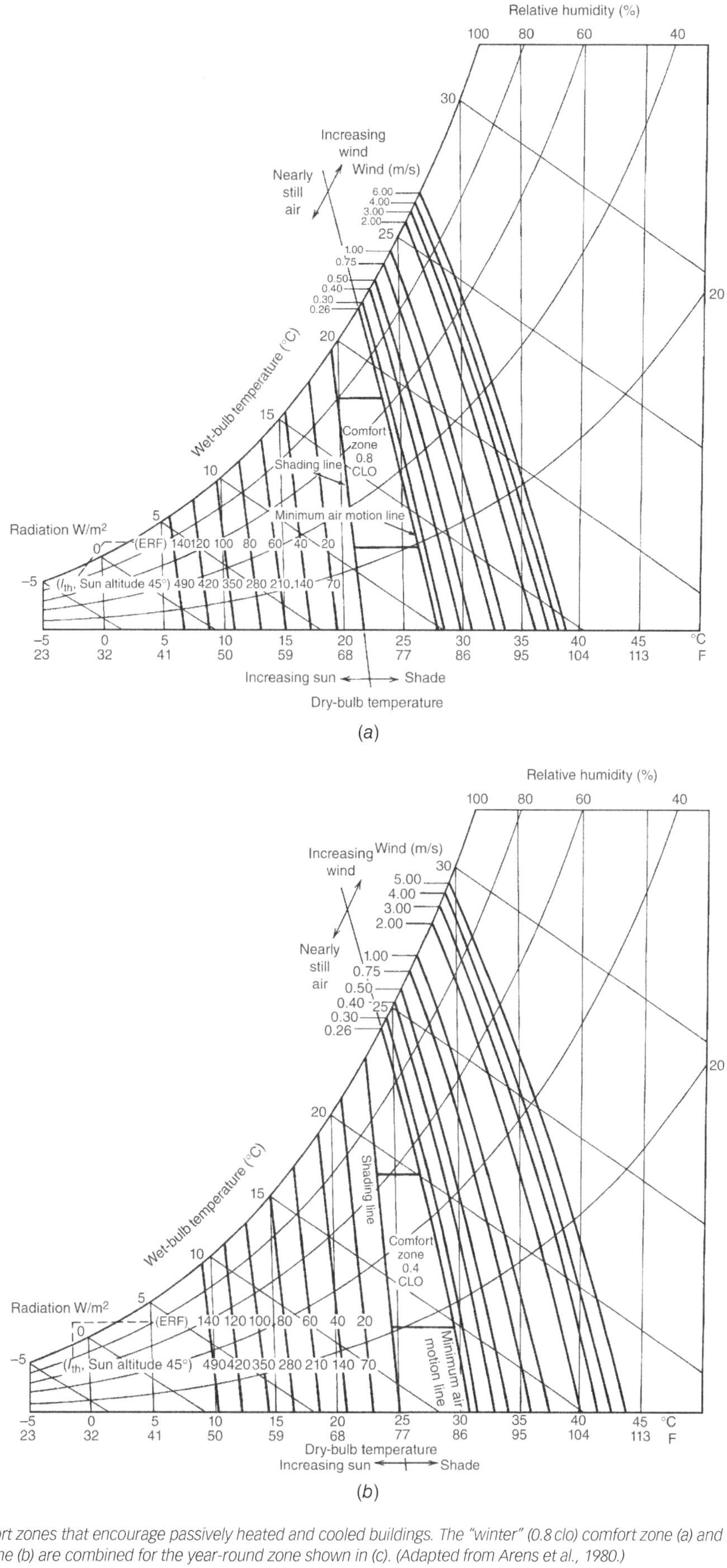

Figure 6.22 Comfort zones that encourage passively heated and cooled buildings. The "winter" (0.8 clo) comfort zone (a) and the "summer" (0.4 clo) comfort zone (b) are combined for the year-round zone shown in (c). (Adapted from Arens et al., 1980.)

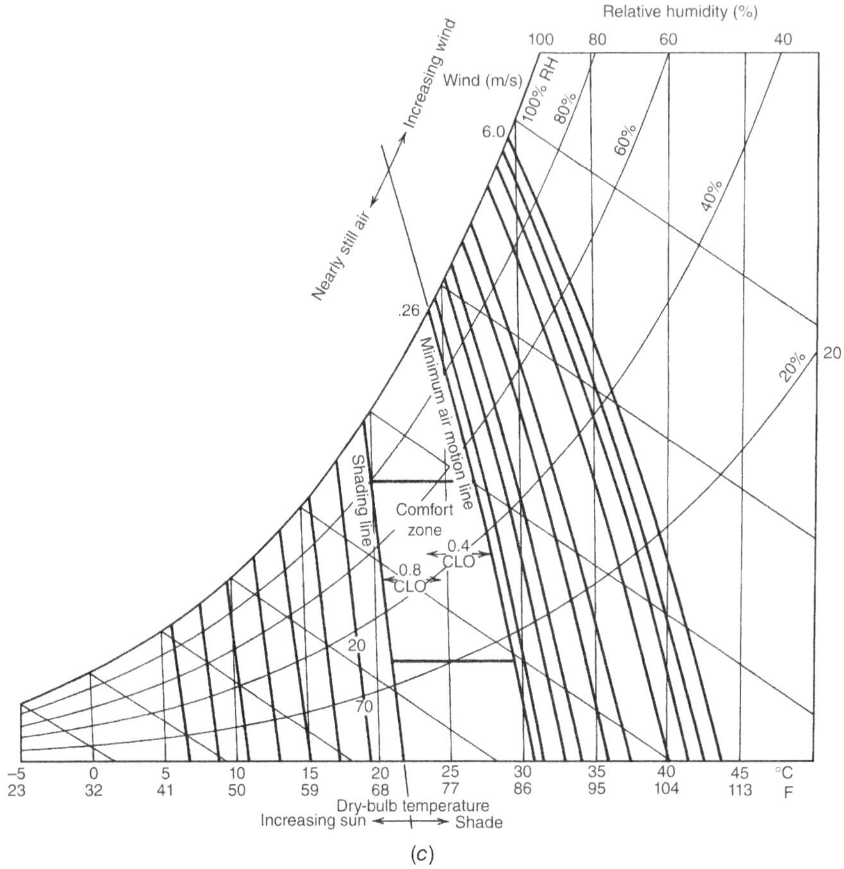

Figure 6.22 (continued)

by the thermal environment in its thermally sensitive places. Although we sweat and are sensitive to heat or cold over nearly all of our skin surface, we are most thermally sensitive in these places:

Heat receptors: fingertips, nose, elbows

Cold receptors: upper lip, nose, chin, chest, fingers

So, on a hot, humid day, cool air moving across the face is a particularly strong promise of comfort, whereas on a cold day, a burst of heat (such as radiant heat from a window, a heater, or a cup of coffee) to the face and fingers is quickly effective.

Regarding our sense of touch: Our fingertips are most sensitive to the rate at which heat is being conducted to colder objects or from hotter objects. The temperature of our skin at the fingertips under ordinary conditions is in the high 80s°F [high 20s°C], so our sense of touch works against many passively heated surfaces in winter, which feel cool even though they are warmer than room air temperature. These surfaces are made of materials that conduct heat rapidly so that they can soak up solar radiation without overheating the room. This high conductivity makes them eager to accept human warmth as well and persuades us, as we touch them, that they are cooler than, in fact, they are. Anyone who has walked barefoot from a rug to an unheated tile floor will sense that the tile is colder, even when both have exactly the same surface temperature. This same characteristic works for passively cooled surfaces in summer.

With the exception of mean radiant temperature, comfort standards typically assume that the thermal conditions within a space are homogeneous and static. Thus, dry bulb air temperature and relative humidity, for example, are presumed to be the same at all points in a room. This is not necessarily the case, and non-uniformity can lead to localized discomfort.

Unwanted local cooling of the body is called a draft. Drafts are a particularly serious concern in winter, yet excessive summer drafts can also be bothersome. Figure 6.23 shows the percentage of people dissatisfied due to a perceptible draft on their head, neck, shoulders, and back. Draft sensation depends upon air speed, activity, and clothing. These results (derived primarily from laboratory studies) apply to sedentary people wearing normal indoor clothing. Many air-conditioning systems deliver summer supply

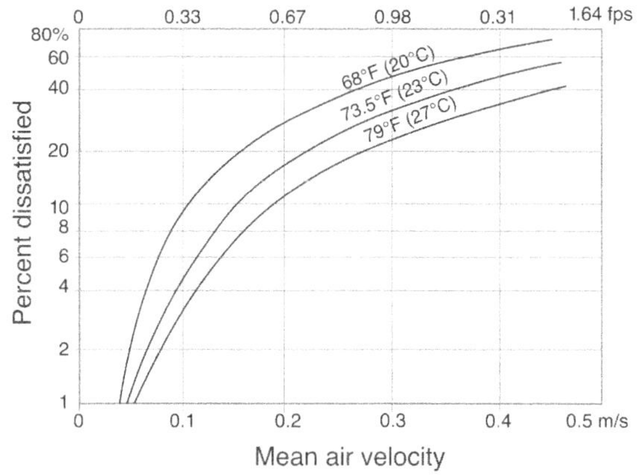

Figure 6.23 Percentage of people dissatisfied as a function of mean air velocity. Note the influence of air temperature: The lower the room air temperature, the greater the dissatisfaction. (© ASHRAE, www.ashrae.org. 2021 ASHRAE Handbook—Fundamentals, Chapter 9, Figure 10; SI units added.)

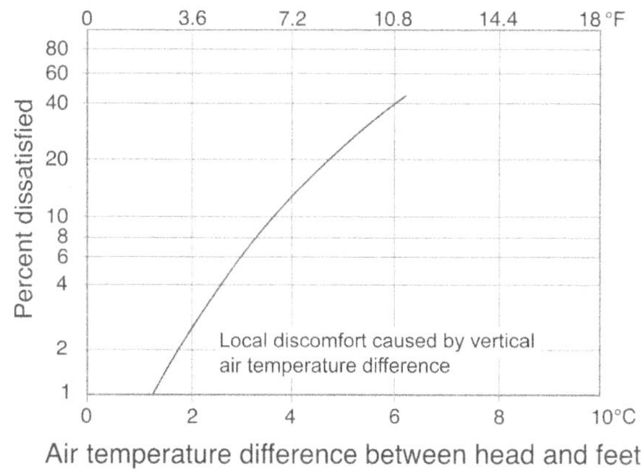

Figure 6.25 Percentage of people dissatisfied as a function of the vertical air temperature difference between the head (higher temperature) and ankles (lower temperature). A cold floor is deemed uncomfortable. (© ASHRAE, www.ashrae.org. 2021 ASHRAE Handbook—Fundamentals, Chapter 9, Figure 12.)

air at temperatures as low as 55 °F [13 °C], well below the lowest temperature shown here. If this air is not well mixed before striking a person, local discomfort is highly likely. During the heating season, 50 fpm [0.83 fps; 0.25 m/s] air speed at typical room temperatures may result in more than 40% dissatisfaction from drafts. However, ceiling fans producing the same air speed, to extend the comfort zone at higher temperatures (such as 80 °F [27 °C]), should cause less than 15% dissatisfaction.

Another local comfort concern is radiant asymmetry, the difference between the temperatures of two opposite surfaces as experienced by a body seated between them. Radiant asymmetry becomes an issue when one of the surfaces is markedly different in temperature from other surfaces within a space. All asymmetry is not perceptually equal. Figure 6.24 shows that more dissatisfaction arises from warm ceilings (typical of radiantly heated ceilings in winter) than is caused by warm and cold wall surfaces. Less serious are "cool walls" (typical of a cold window in winter). Less than 10% dissatisfaction resulted from "cool ceilings" (such as radiant cooling panels in summer) and even less from "warm walls" (typical of a sunny window in summer or a solar heated wall in winter).

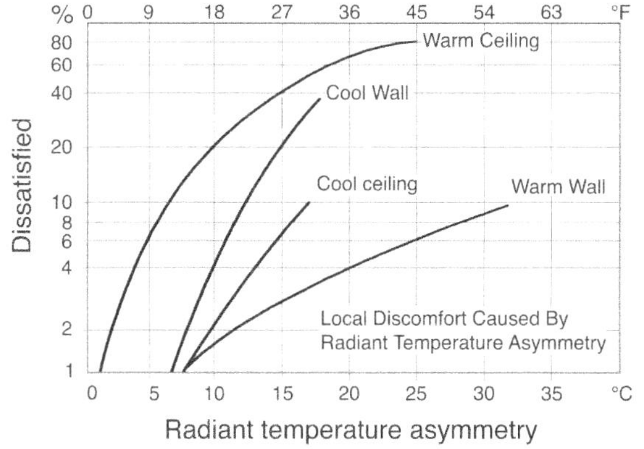

Figure 6.24 Percentage of people expressing discomfort due to radiant asymmetry. A warmer ceiling produces the highest discomfort; a warmer wall, the least discomfort. (© ASHRAE, www.ashrae.org. 2021 ASHRAE Handbook—Fundamentals, Chapter 9, Figure 10; SI units added.)

Vertical air temperature differences can also cause local discomfort. Because warm air is less dense than cold air, it rises; therefore, in most building spaces, air temperature is somewhat higher at the ceiling than at the floor, regardless of the season. Figure 6.25 shows experimental results of dissatisfaction with a vertical air temperature difference between the head at 43 in. [1.1 m] above the floor and the feet at 4 in. [0.1 m]. The air temperature difference between head and ankle should not exceed 5.4 °F [3 °C] to limit dissatisfaction to less than 10%. (When the temperature at head level is cooler than the floor, greater differences are tolerated.)

Floor surface temperature is another potential contributor to local discomfort. The design guidance on this situation is direct: floor temperature should be between 66.2 and 84.2 °F [19 and 29 °C]. These limits apply to situations where occupants are wearing shoes.

A bit more about the percentage of dissatisfied folks. The basic ASHRAE comfort zone boundaries are drawn to enclose conditions that will result in 90% satisfaction—when conditions are in fact uniform. The objective of the ASHRAE comfort standard is to deliver spaces with at least an 80% satisfaction rate. Thus, no more than 10% additional dissatisfaction can be tolerated. Awareness (and sometimes mitigation) of non-uniform conditions are critical to success.

(b) Evaluating Comfort in Occupied Buildings

Designing comfortable conditions that maintain the environment within the range specified by these standards involves assembling a complex combination of building enclosures, mechanical systems, and control systems. There is no reason to just assume that everything works as intended in the completed building. Post-occupancy evaluation is recommended. Luckily, "testing" thermal comfort is straightforward. Building users can simply be asked their opinion about thermal comfort—using a seven-point Likert-Scale questionnaire. The seven response options are as described in Section 6.3(a): hot, warm, slightly warm, neutral, slightly cool, cool, and cold. Reasonably low-cost measurement devices can be employed to determine the physical basis for unsatisfactory environments.

References and Resources

Arens, E., R. Gonzalez, L. Berglund, P. McNall, and L. Zeren. 1980. "A New Bioclimatic Chart for Passive Solar Design," in *Proceedings of the Fifth National Passive Solar Conference* (Vol. 5.2, pp. 1202–1206). Boulder, CO: American Section of the International Solar Energy Society.

ASHRAE. 2021. *2021 ASHRAE Handbook—Fundamentals.* Atlanta, GA: American Society of Heating, Refrigerating and Air-Conditioning Engineers.

ASHRAE. 2023. *ANSI/ASHRAE Standard 55-2023: Thermal Environmental Conditions for Human Occupancy.* Atlanta, GA: American Society of Heating, Refrigerating and Air-Conditioning Engineers.

Egan, M. D. 1975. *Concepts in Thermal Comfort.* Englewood Cliffs, NJ: Prentice-Hall.

Fanger, P. O. 1982. *Thermal Comfort: Analysis and Applications in Environmental Engineering.* Malabar, FL: R.E. Krieger Publishing Company.

Givoni, B. 1998. *Climate Considerations in Building and Urban Design.* New York: Van Nostrand Reinhold.

Givoni, B. and M. Milne. 1979. "Architectural Design Based on Climate," in *Energy Conservation in Building Design*, D. Watson (ed.). New York: McGraw-Hill.

Heschong, L. 1979. *Thermal Delight in Architecture.* Cambridge, MA: MIT Press.

Humphreys, M. and J. F. Nicol. 1998. "Understanding the Adaptive Approach to Thermal Comfort." *ASHRAE Transactions* 104(1): 991–1004.

ISO. 2005. *Standard 7730: Ergonomics of the Thermal Environment — Analytical Determination and Interpretation of Thermal Comfort using Calculation of the PMV and PPD Indices and Local Thermal Comfort Criteria.* www.iso.org/standard/39155.html

Kwok, A. 1998. "Thermal Comfort in Tropical Classrooms." *ASHRAE Transactions* 104(1): 1031–1047.

Olgyay, V. 1963. *Design with Climate: Bioclimatic Approach to Architectural Regionalism.* Princeton, NJ: Princeton University Press.

Tartarini, F., S. Schiavon, T. Cheung, and T. Hoyt. 2020. *CBE Thermal Comfort Tool: Online Tool for Thermal Comfort Calculations and Visualizations.* Berkeley, CA: Center for the Built Environment, University of California Berkeley. comfort.cbe.berkeley.edu

CHAPTER 7

Visual Comfort

7.1 LIGHT

THE ILLUMINATING ENGINEERING SOCIETY (IES) DEFINES light as visually evaluated radiant energy. Light is a unique form of electromagnetic energy that permits us to see. If light is considered a wave, like a radio wave or an alternating-current wave, it can be described by its frequency and wavelength. Figure 7.1 shows the position of light in the electromagnetic spectrum and its relation to other wave phenomena of various frequencies.

Figure 7.1 shows that the longest-wavelength light (red) has a much higher frequency than radio waves and radar, and that light constitutes only a very small part of the electromagnetic energy spectrum. What we call color is determined by wavelength—starting with the longest visible wavelengths (the lowest frequency, perceived as red) on through the shortest visible wavelengths (the highest frequency, perceived as violet). The visible spectrum (light) encompasses the spectrum of what we recognize as red, orange, yellow, green, blue, indigo, and violet (mnemonic Roy G Biv). Bordering the visible spectrum, on the low-frequency end, is infrared radiation; at the high-frequency end lies ultraviolet radiation. Both are invisible to human beings (but not to some animals). Neither infrared nor ultraviolet should be called light because these portions of the spectrum cannot be seen by humans.

When a light source produces energy in approximately equal quantities over the entire visible spectrum, the combination of wavelengths appears white. A source with energy output over only a small section of the spectrum produces a colored light characteristic of its generation process. Examples are the blue-green of a clear mercury lamp and the yellow of a sodium lamp.

Herein, all light will be considered white unless specifically noted. This is a reasonable design assumption because most (but not all) light sources generally appear white after a short viewer accommodation period. Most standard commercial light sources permit object colors to be easily and correctly identified. Only when sources differing widely in spectral content are viewed side by side can a variation in whiteness be noticed by the effect on colored objects and on neutral surfaces. Exceptions to this white light assumption, when they occur, are critical for a designer to understand.

In building design, light is defined relative to human occupants/users. Radiation that can be detected by a statistically typical young adult is deemed to be light. This is commonly understood to encompass wavelengths between 380 and 780 nanometers (nm). Other wavelengths are not visible to the general population and are not "light." There is no need to say "visible light" as only light is visible. Light from the sun/sky should be called daylight (never natural light); light from an electric lamp should be called electric light.

An understanding of the units and terminology associated with light in the built environment is critical to lighting design. The International System of Units (SI) is used throughout most of the world—a major exception being the United States. Increasingly, SI units for quantities of light are being seen in the United States, although the Inch-Pound (I-P) system still predominates. Throughout this chapter—consistent with other portions of this book—dual units are presented.

Mechanical and Electrical Equipment for Buildings, Fourteenth Edition. Walter Grondzik, Daniel Overbey, and Alison Kwok.

Published 2026 by John Wiley & Sons, Inc.
Companion website: www.wiley.com/go/meeb14e

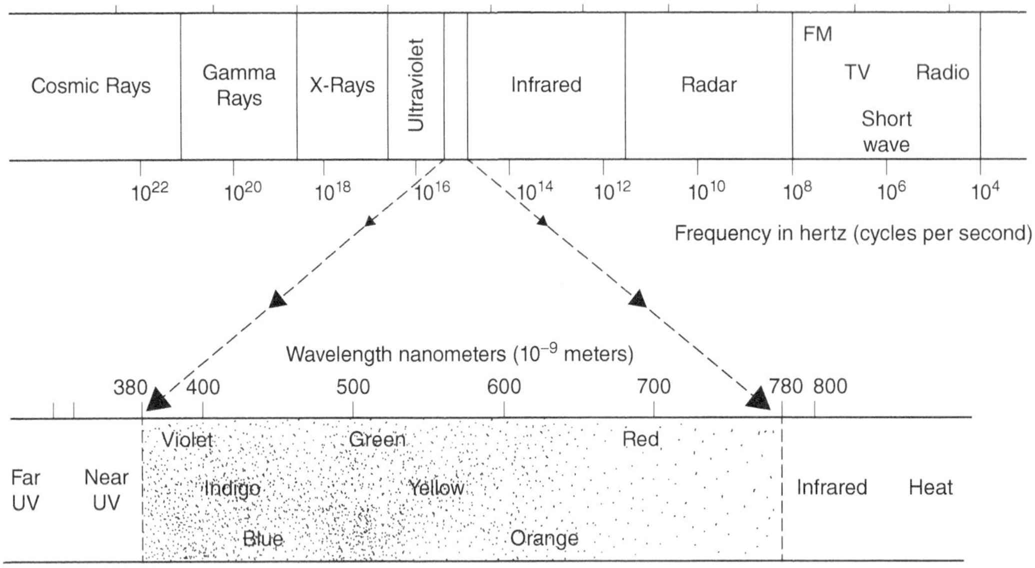

Figure 7.1 The electromagnetic spectrum.

(a) Luminous Flux

The unit of luminous flux, in both SI and I-P units, is the lumen (lm). The lumen, a measure of the quantity of light at some point in the environment, is analogous to water flow in a hydraulic system or current in an electrical system; it is normally represented by the Greek letter φ.

If we surround a 1-cd (candlepower) source that radiates light equally in all directions with a transparent sphere of 1 m [ft] radius (Figure 7.2), then by definition the luminous energy (flux) emanating from 1 m^2 [ft^2] of surface on the sphere is 1 lm. Because the surface area of such a sphere is 4π m^2 [ft^2], it follows that a source of 1 cd [candlepower] intensity produces 4π (12.57) lumen.

In physical terms, the lumen is a unit of power, like the watt. However, unlike the watt, which is a radiometric unit directly convertible to other power units such as Btu/h, the lumen is a measure of photometric power. This means light power as

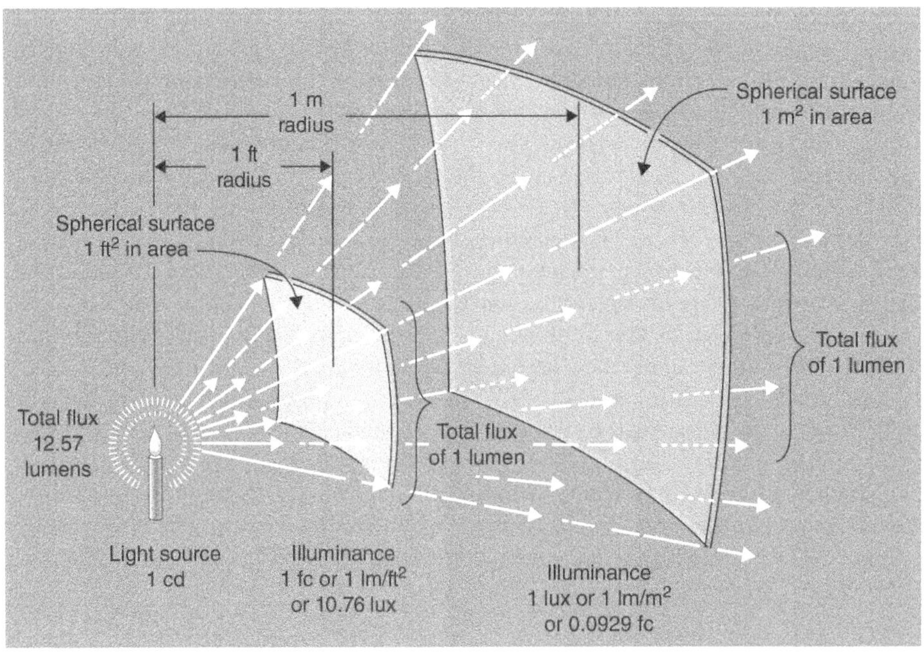

Figure 7.2 A source of 1-cd intensity produces 4π (12.57) lumens of light flux. Thus, each square foot [square meter] of spherical surface surrounding such a source receives 1 lm of light flux. This quantity of light flux produces an illuminance of 1 fc (lux) on the spherical surface.

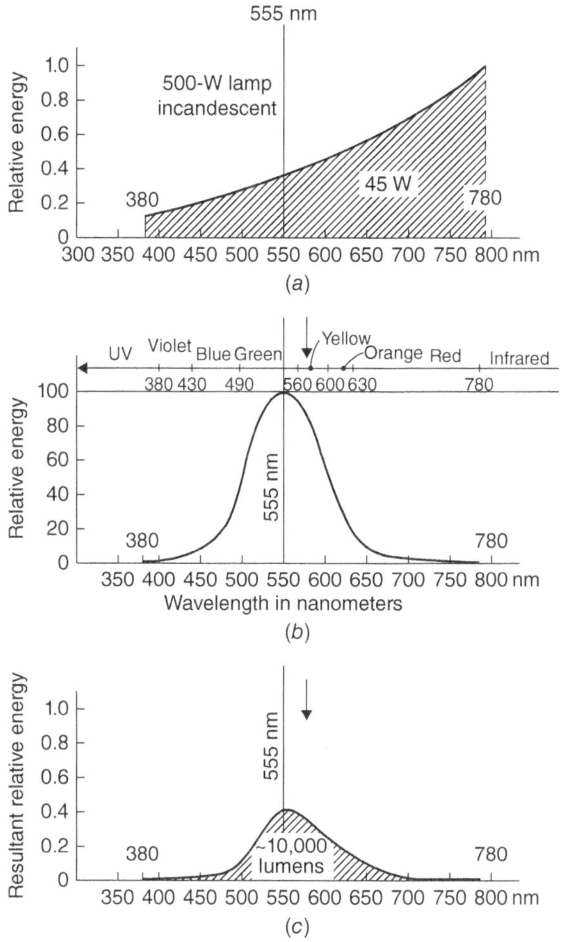

Figure 7.3 *Graphical demonstration of the method by which the unit of light flux is defined. (a) The spectrum of the light produced by a 500-W incandescent lamp amounts to approximately 45 W measured radiometrically. When engaged by the human eye, whose spectral sensitivity curve is given in (b), this light power is perceived as shown in (c). The new light power curve is expressed in lumens and indicates the quantity of light as perceived by the eye.*

perceived by the human eye and therefore as a function of human physiology. Put another way, lumens, or luminous flux, is the time rate of flow of perceived luminous energy. Because the visual response of the eye is frequency dependent, perceived light power is also frequency dependent—varying with the spectral content of the impinging light and the spectral sensitivity of the eye. Figure 7.3*a* shows the spectral content of the visible energy produced by a 500-W incandescent lamp. Measured radiometrically, it amounts to 45 W. However, when passed through a selective filter (Figure 7.3*b*), which is effectively what happens when the light is processed by the eye, the light power usefully received appears as in Figure 7.3*c*, and can no longer be measured in watts. Instead, we use the unit of eye-perceived, or photometric, power: the lumen. If the spectral content curve in Figure 7.3*a* were differently shaped, even if the total radiometrically measured power were the same, the resultant perceived power would be different.

Referring to Figure 7.3*b*, a correlation can be made between photometric and radiometric power at the point of maximum response of the eye, which occurs at 555 nm wavelength—1 nm is 10^{-9} m. One watt of monochromatic light at that wavelength produces 683 lm. However, because common light sources are not monochromatic but produce light across many parts of the spectrum, there is no single factor to convert between watts and lumens. Each source has its own luminous efficacy (lumens/watt) determined by its mode of light production and by its spectrum. For the 500-W lamp discussed above, the luminous efficacy is 10,000 lm/500 W, or 20 lumens per watt (lm/W). The term efficacy is used instead of efficiency in this situation as the output (lumens) and input (Watts) are not expressed in the same units (and thus do not cancel).

(b) Luminous Intensity

The SI unit of luminous intensity (I) is the candela (candlepower), abbreviated cd (cp). Luminous intensity represents luminous flux in a specific direction—with the direction being defined by the center of a solid angle measured in steradians. Thus, luminous intensity would be expressed as lumens per steradian—which is a candela. It turns out that an ordinary wax candle has a horizontal luminous intensity of approximately 1 cd, hence the name. The candela and candlepower have the same magnitude. Luminous intensity is solely a characteristic of the source; it is independent of the visual sense. Understanding luminous intensity can be important to point source lighting design efforts and to glare control.

(c) Illuminance

One lumen of luminous flux, uniformly incident on 1 m^2 [ft^2] of area, produces an illuminance (E) of 1 lx or 1 footcandle (fc). Illuminance is a measure of the density of light (i.e., lumens per unit area). If we consider a lightbulb as analogous to a fire sprinkler head, then the resulting coverage of water per unit floor area would be the illuminance. The SI unit of illuminance, lux, is smaller than the corresponding I-P unit, footcandles, by the ratio of square meters to square feet. That is, 10.764 lx = 1 fc. Conversely, footcandles multiplied by 10.764 yield lux. This relationship is shown in Figure 7.2. Mathematically:

$$\text{lux} = \frac{\text{lumens}}{\text{square meter area}}$$

$$\text{lx} = \frac{\text{lm}}{\text{m}^2} \qquad \text{(Eq. 7.1)}$$

and

$$\text{footcandles} = \frac{\text{lumens}}{\text{square foot area}}$$

$$\text{fc} = \frac{\text{lm}}{\text{ft}^2} \qquad \text{(Eq. 7.2)}$$

As an approximation (with 8% error):

$$10\ \text{lx} \cong 1\ \text{fc} \qquad \text{(Eq. 7.3)}$$

EXAMPLE 7.1 A fluorescent tube produces 3200 lm. What is the illuminance on the floor of a 3-m^2 room, assuming 60% overall delivery efficiency and uniform light distribution?

SOLUTION

Delivered lumens = 0.6 × 3200 = 1920 lm

$$\text{lx} = \frac{1920}{3 \times 3} = 213.3$$

$$\text{fc} = \frac{213.3}{10.76} = 19.8$$

Calculating footcandles directly (with 3 m = 9.84 ft):

$$\text{fc} = \frac{1920}{9.84 \times 9.84} = 19.8\ \text{fc}$$

By approximation:

$$\text{fc} \cong \text{lx}/10 = 21.3$$

Note that this calculation gives average illuminance in the space. ■

(d) Luminance, Exitance, and Brightness

An object can be visually perceived when light from the object (either emitted or reflected) enters the eye. The impression we receive regarding light coming from the object is termed brightness. Brightness is a subjective sensation and depends upon the luminance (L) of the object (an objective variable), but also upon the state of adaptation of the eye (see Section 7.2) and to some extent the mental state of the receiving person. This perception is generally referred to as subjective brightness or apparent brightness—or simply brightness. The measurable-by-instruments object luminosity is termed luminance. Luminance is normally defined in terms of intensity; it is the luminous intensity per unit of apparent (projected) area of a primary (light-emitting) or secondary (light-reflecting) object. Luminance is expressed in candela per surface area. The SI unit of luminance is candela per square meter (cd/m^2). The I-P unit of luminance is the footlambert ($1/\pi\ cd/ft^2$). Luminance has no meaningful mechanical or electrical analogy.

Conversion factors for SI and I-P lighting units are given in Table 7.1. The term luminance always refers to

TABLE 7.1 Common Lighting Units and Conversion Factors

Unit	Multiply	By	To Obtain
Illuminance	Lux (lx)	0.0929	Footcandle
	Footcandle (fc)	10.764	Lux
Luminance (L)	cd/m^2	0.2919	Footlambert
	cd/cm^2	10,000	cd/m^2
	$cd/in.^2$	1550	cd/m^2
	cd/ft^2	10.76	cd/m^2
	millilambert	3.183	cd/m^2
	Footlambert	3.4263	cd/m^2
Luminous Intensity (I)	Candela (cd)	1.0	Candlepower (cp)

the quantitative phenomenon, while brightness refers to the human perception of luminance.

Another concept the lighting designer may encounter is luminous exitance, or simply exitance. As the name implies, exitance describes the total luminous flux density leaving (exiting) a surface, irrespective of directivity or viewer position. For instance, if a 1 m^2 surface emits 1 lm, its luminous exitance is 1 lm per square meter (1 lm/m^2) or 0.093 lm/ft^2. A surface that is a perfect diffuser, whether in emitting light or reflecting light, is known as a Lambertian surface. The luminance of such a surface equals $1/\pi$ times its exitance. This relationship is useful as a design approximation. Although very few surfaces are truly Lambertian, many are approximately so.

Exitance is important in detailed photometric calculations—such as those involved in determining coefficients of utilization and surface luminance coefficients—and in detailed point illuminance calculations. These analyses are beyond the scope of this book because they are not usually performed by the lighting designer. Detailed point calculations are today almost universally performed by computer, and the necessary math is built into the software.

Because object luminance is a prime factor in visibility (and glare), it is important to understand basic luminance calculations. The eye does not differentiate between primary sources (those that generate/emit light) and secondary sources (those that derive their luminance from reflection or transmission); the differentiation, however, is important in calculation procedures. See Figure 7.4 for a graphic representation of the basic relationships.

EXAMPLE 7.2 Consider the luminance of a light-emitting surface.

1. Calculate the luminance of an historic A-19 standard inside-frosted, 100-W incandescent lightbulb with a maintained output of 1700 lm. Assume (for simplicity's sake) that the bulb is spherical.

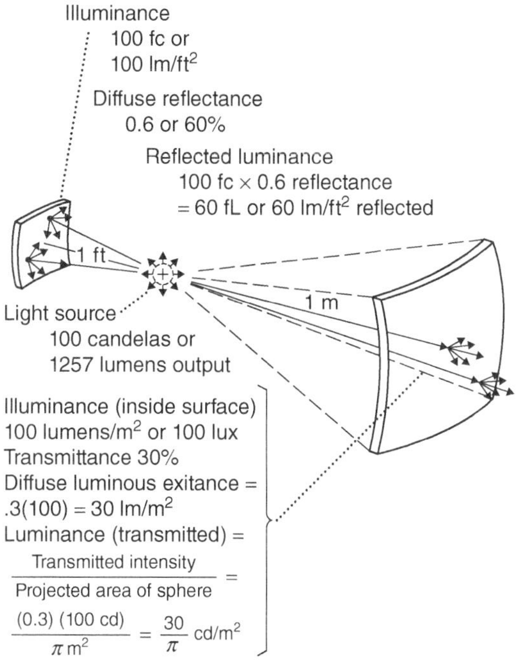

Figure 7.4 Luminance may be the result of light that is either reflected or transmitted. In the former case, it is calculated as the product of the incident lumens and the reflectance; in the latter case, it is calculated as the transmitted intensity divided by the projected area.

2. Assume that an opal glass globe of 8-in. [200-mm] diameter and a transmittance of 35% surrounds the bulb described in step 1. Calculate the luminance of the globe. Use SI units.

SOLUTION

1. Assume that the filament is a point source (an essentially valid assumption) and that the inside frosting of the glass does not reduce the output (it actually does so by about 1%). The inside frosting acts to convert the point source filament to a uniformly emitting globe. The definition of a point source tells us that 1 cd produces 4π lumens distributed spherically. Therefore:

$$\frac{1\ \text{cd}}{4\pi\ \text{lm}} = \frac{I\ \text{cd}}{1700\ \text{lm}}$$

The intensity, I, of the filament, and therefore also of the bulb (because the frosting is assumed not to reduce output), is:

$$I = 1700\ \text{lumens} \times \frac{1\ \text{cd}}{4\pi\ \text{lm}} = 135\ \text{cd}$$

An A-19 bulb has a diameter of 19/8 in. [60 mm]. The definition of luminance is:

$$L = \frac{I}{A}\ \frac{\text{(intensity)}}{\text{(projected area)}}$$

The projected area becomes approximately a circle with a diameter of 60 mm (assuming a spherical bulb), whose area equals $\pi\ r^2$ or $\pi\ (0.03\ \text{m})^2$. So:

$$L = \frac{135\ \text{cd}}{\pi\,(0.03\ \text{m})^2} = 47{,}750\frac{\text{cd}}{\text{m}^2}$$

This luminance is a potentially severe glare source if it is in the field of vision (depending on its distance from the eye; see Section 7.4(b)).

2. The intensity of the source (the bulb) is reduced by the 35% transmittance of the globe. The projected area of the globe is larger than that of the bulb; that is,

$$A\ \text{proj.} = \pi r^2 = \pi\frac{(0.2\ \text{m})^2}{(2)}$$

Thus, the luminance of the globe is:

$$L = \frac{135\ \text{cd}\ (0.35)}{\pi\ (0.10\ \text{m})^2} = 1500\frac{\text{cd}}{\text{m}^2}$$

This is no longer likely to be a potential source of direct glare.

Note two important principles demonstrated by this example:

1. Intensity is neither area- nor distance-dependent; it varies only with transmission factors.
2. Luminance varies with both intensity and area. ■

For example, if a white lamp whose inside coating reduces the lamp output by about 10% is used in lieu of an inside frost, both intensity and luminance would be reduced by 10%. However, if a clear glass lamp is used, the intensity would remain the same, while the luminance would increase inversely with the ratio of filament area to bulb area. Therefore, a clear 100-W lamp, whose filament has an area of perhaps 0.3 cm², would have a luminance of:

$$\frac{135\ \text{cd}}{0.3\ (0.0001)\ \text{cm}^2} = 4{,}500{,}000\ \text{cd/m}^2$$

which is so severe a glare source as to be disabling when in the near field of vision.

EXAMPLE 7.3 Calculate the luminance of a 34-W, T12, 4-ft white fluorescent lamp. Assume a viewing angle normal to the long axis of the lamp and that the lamp is a diffuse (Lambertian) emitter. Use SI units.

SOLUTION

This problem can be solved in two ways: by using the relationship between exitance and luminance or by calculating intensity.

By exitance/luminance: Use 2770 lm at 40% lamp life as an average output. The luminous length is 4 ft [1200 mm], and the diameter is 12/8 in. [38 mm]. The luminous surface area of the tube is then:

$$A = (L)(d)(\pi)$$

$$= (1200\ \text{mm})(38\ \text{mm})(\pi)$$

$$= (1.2\ \text{m})(0.038\ \text{m})(\pi)$$

$$= (0.0456)(\pi),\ \text{m}^2$$

$$\text{exitance} = \frac{\text{luminous flux}}{\text{area}}$$

DESIGN FUNDAMENTALS

$$\text{exitance} = \frac{2770 \text{ lm}}{0.0456\,\pi,\ \text{m}^2}$$

$$\text{luminance} = \frac{1}{\pi}(\text{exitance})$$

$$= \frac{2770}{0.0456\,\pi}$$

$$= 6155 \text{ cd/m}^2$$

By intensity: The equivalent spherical intensity of the lamp can be calculated by using the relationship that 1 cd produces 4π lumens. Therefore,

$$\text{equivalent intensity } I = \frac{2770 \text{ lm}}{4\pi}\text{cd}$$

The radius of a sphere of equivalent surface area to the fluorescent tube would be:

$$4\pi r^2 = (0.0456)(\pi),\ \text{m}^2$$

$$r^2 = \frac{0.0456}{4}\text{ m}^2 = 0.0114 \text{ m}^2$$

$$r = 0.1067 \text{ m}$$

The equivalent projected area of such a sphere is πr^2, or

$$A = 0.0114\,\pi$$

and its luminance is:

$$L = \frac{\text{intensity}}{\text{area}} = \frac{2770/4\,\pi,\ \text{cd}}{0.0114\pi,\ \text{m}^2}$$

$$= \frac{2770}{0.0456\,\pi^2}\frac{\text{cd}}{\text{m}^2}$$

$$= 6155 \text{ cd/m}^2$$

A luminance this high can constitute a glare problem, depending on the lamp's position in the field of view. Considering a 40-W lamp with 3200 lm output, the luminance would be higher in direct proportion to the ratio of output; that is,

$$L_{3200} = L_{2770} \times \frac{3200}{2770} = 6155\,\frac{3200}{2770}$$

$$= 7110 \text{ cd/m}^2$$

This is the origin of the frequently used average value of 7000 cd/m^2 as the luminance of a standard 48-in. [1200 mm] fluorescent tube. It is a borderline glare source. Lamps of similar size, but with higher lumen output create increasing potential for glare problems. ■

(e) Transmittance, Reflectance, and Absorptance

Lighting design is possible because light is predictable; that is, it obeys certain laws and exhibits certain fixed characteristics. Although some of these are so well known as to appear self-evident, a review is in order.

The luminous transmittance of a material (such as a pane of glass or a sheet of plastic) is a measure of its capability to transmit incident light. This quantity, known variously as transmittance, transmission factor, or coefficient of transmission, is the ratio of the total transmitted light to the total incident light. When light containing several spectral components passes through a material with selective absorption, overall transmittance is an average of the individual transmittances for the various wavelengths—and must be used cautiously. A piece of frosted glass and a piece of red glass may both have a 70% transmittance, but they affect the incident light differently. In general, transmission factors should be used only when referring to materials displaying nonselective absorption; that is, those that transmit the various component colors equally. A single sheet of clear glass has a transmittance of 80–90%, frosted glass 70–85%, and solid opal glass 15–40%. Light not transmitted is either absorbed or reflected. See Chapter 12 for typical transmittance values.

Similarly, the ratio of reflected light to incident light is variously called reflectance, reflectance factor, or reflectance coefficient. If half the light incident on a surface is bounced back, the reflectance is 50% (or 0.50). Light that is not reflected is either absorbed and/or transmitted. In all cases, reflection plus transmission plus absorption equals 100%. The extent of absorption, reflection, and transmission depends upon the type of material and the angle of light incidence; light impinging on a surface at grazing angles tends to be reflected rather than absorbed or transmitted (Figure 7.5). An example of almost total reflection from an opaque surface would be from a high-quality mirror, whereas almost complete absorption takes place on an object with a matte-finish black paint. The effect of the material surface finish on reflection is shown in Figure 7.6. See Chapter 12 for typical reflectance

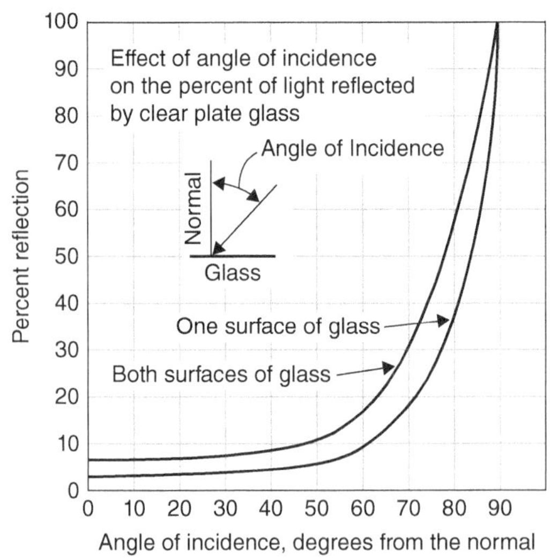

Figure 7.5 *Relation between angle of incidence and percentage of reflectance. This effect is important when considering the penetration of sunlight into interior spaces and, conversely, the exterior glare produced by reflection of the sun from building windows.*

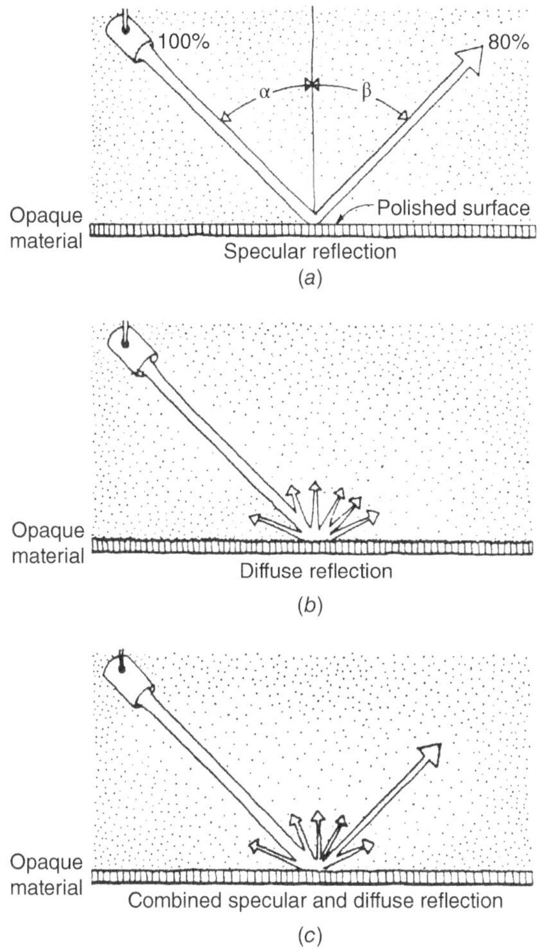

Figure 7.6 *Reflection characteristics. (a) In specular reflection, angle of incidence equals angle of reflection (α = β). Because 80% of light is reflected, reflectance is 80%; 20% of light is absorbed. (b) In diffuse reflection, incident light is spread in all directions by multiple reflections on the unpolished surface. Such surfaces appear equally bright from all viewing angles. (c) Most materials exhibit a combination of specular and diffuse reflection. Such a surface mirrors the source while producing a bright background.*

values. Reflectance measurement is discussed later in this chapter.

The reflection that occurs on a smooth surface such as polished glass or stone is called specular reflection, as in Figure 7.6*a*. If the surface is rough, multiple reflections take place on the many small surface projections, and the light is diffused, as in Figure 7.6*b*. Reflectance is a measure of total light reflected; it may be specular or diffuse, or a combination of both, as shown in Figure 7.6*c*.

Diffuse transmission takes place through any translucent material such as frosted glass, white glass, milky Plexiglas, tissue paper, and so on. Diffuse transmission is widely employed in lighting fixtures (luminaires) to better spread the light generated by the lamp within the fixture. Diffuse and nondiffuse transmission are illustrated in Figure 7.7*a* and 7.7*b*.

(f) Measuring Light

The design of lighting systems involves both quantitative and qualitative variables. The quantitative aspects have great value because they can be used in calculations, be specified as enforceable design criteria, and be measured in post-occupancy evaluations of built spaces. The benefit of the quantitative approach is that these elements are absolute. Qualitative lighting elements, on the other hand, are equally important—as they are how humans perceive the luminous environment. Building designers need to be aware of, and willfully engage, both sides of lighting. This section describes a number of the more commonly encountered quantitative lighting measurements.

Illuminance. Illuminance (E) is the density of light falling upon a surface. Field measurements of illuminance, necessary to confirm that design criteria have been met in a built project or for existing condition assessments, are commonly made with a portable illuminance meter. Two examples of illuminance meters are shown in Figures 7.8 and 7.9. These devices contain a photoelectric sensor connected to a microammeter via electronic control circuitry and display in lux or footcandles or both.

As shown in Figure 7.3, the human eye is not equally sensitive to the various wavelengths that comprise the visible spectrum. Maximum sensitivity at high illuminance is in the yellow–green area (555 nm wavelength), whereas sensitivity at the red and blue ends of the spectrum is quite low. This effect is so pronounced that 10 units of blue energy are required to produce the same visual effect as 1 unit of yellow–green. If a light meter is to be useful, its inherent sensor response, which is quite different from that of the human eye, must be corrected to correspond to the eye. For this reason, meters are "color-corrected." Illuminance meters must also be corrected for light incident at oblique angles that do not enter the meter due to reflection from the sensor surface and the shielding of the sensor by the meter housing. This correction is known as cosine correction. A good light meter will therefore be color- and cosine-corrected.

Digital illuminance meters may have considerable electronic circuitry—which provides functions such as automatic ranging, unit switching, reading hold, integration for flickering or time-varying sources, and connections for external data storage and transmission. Field illuminance measurements should be made using common sense. Horizontal measurements should be made at task height. Where no definite height is specified, readings are typically taken at 30 in. [750 mm] above the floor (approximately normal desk height). The meter must be held with the sensor parallel to the

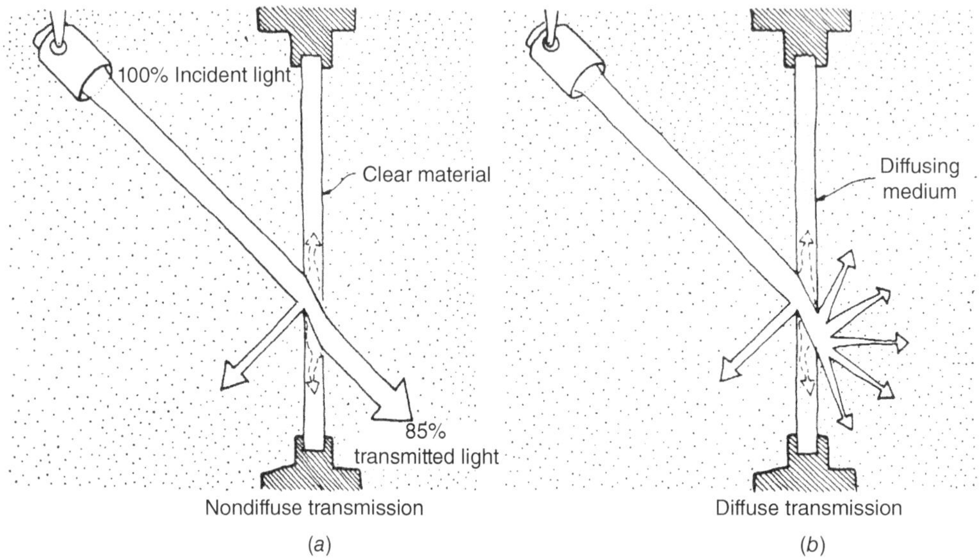

Figure 7.7 *Transmission characteristics. (a) In nondiffuse transmission, the light is refracted (bent) but emerges in the same beam as it enters. Clear materials such as glass, water, and certain plastics exhibit this type of transmission. In the instance illustrated, the transmittance is 85% (the remaining 15% is reflected and absorbed). The source of light is clearly visible through the transmitting medium. (b) With diffuse transmission, the source of light is not visible and, in the case of multiple sources, the diffusing surface exhibits generally uniform brightness if the spacing between the light sources does not exceed approximately 1½ times their distance from the material.*

Figure 7.8 *Electronic, digital, color-corrected, and cosine-corrected illuminance meter from Konica-Minolta.*

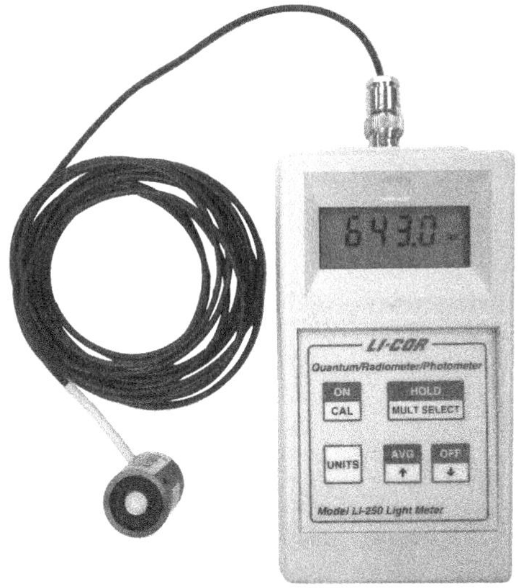

Figure 7.9 *Digital meter (from Li-Cor) has a variety of sensors that measure illuminance, solar irradiance, or photosynthetic radiation. Due to its small size, the illuminance sensor can be easily used for architectural model measurements.*

plane of the surface of concern. Thus, to measure wall illuminance, the meter must be held with the sensor parallel to the wall. If electric lighting readings are desired and the test is being conducted during daylight hours, readings should be taken with and without the electrical illumination, and the results subtracted. Detailed instructions for conducting field surveys are contained in the 10th edition of the *IES Lighting Handbook* (2011). Briefly, a survey of an existing indoor lighting installation should establish:

- type, rating, and age of sources
- type, design, and model of luminaires
- maintenance schedule and status

It should also measure:

- mounting height of luminaires
- spacing and pattern of luminaires

- reflectances of walls, floor, ceiling, and major items of furniture and equipment
- illuminance levels throughout the area, plus levels at all working plane elevations
- vertical surface illuminance at walls and other major vertical planes

It is commonly recommended that average illuminance be determined from a reasonable sample of individual readings (reasonable being defined by the context of the situation). This is not good advice. If three measurements are made and are found to be 100, 200, and 300 lx, the average is 200 lx. This may match the design criteria for the space, but some areas of the space have half as much light as desired, while some have 50% more. This is not representative of a good design. An individual experiences the illuminance at his or her work area—occupants do not experience average illuminance.

Luminance. Luminance (*L*) is more important to general appreciation of a visual scene, and to glare potentials, than is illuminance. We see luminance and contrast—not illuminance. Light, in transit and falling on a surface, is invisible. Illuminance measurements are more widely utilized than luminance measurements (often to the detriment of properly evaluating the adequacy of a lighting installation) due to two factors:

1. Illuminance meters are cheaper and simpler to use than luminance meters.
2. Quantitative lighting design recommendations often focus on illuminance. Verification that such design targets have been met is commonly used to confirm the success of a lighting design, even though there are a number of success elements above and beyond illuminance.

Several configurations of luminance meters are available, one of which is shown in Figure 7.10. Luminance meters of reasonable quality are substantially more expensive than a comparable illuminance meter. As a result, they are not as commonly encountered or used in the field. A lower-cost workaround for lack of a luminance meter is discussed below.

An approximation of the luminance of a reflecting surface or luminous source can be obtained using an illuminance meter of the type shown in Figure 7.8. For diffuse reflecting surfaces, the cell of the meter is placed against the surface and then slowly retracted 2–4 in. [50–100 mm] until a constant reading is obtained. The luminance, in footlamberts, is approximately 1.25 times the reading in footcandles—the 1.25 factor compensating for wide-angle light losses (reflected light not captured by the meter sensor).

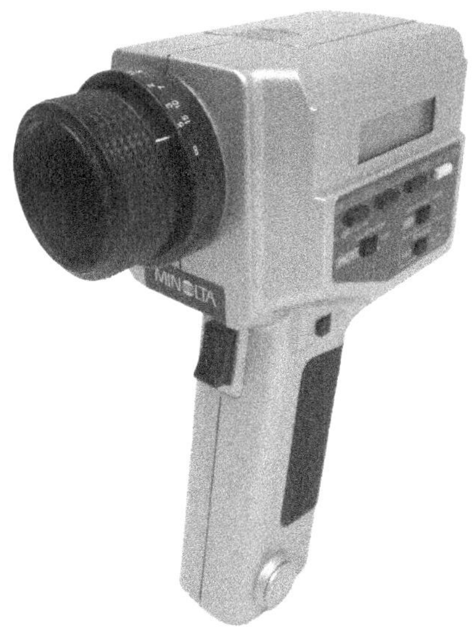

Figure 7.10 *Direct-reading, narrow-angle, spot-type luminance meter has an acceptance angle of 1°, a range of 0.001–299,000 cd/m^2 [0.001–87,000 fL], a variable response speed to permit measurement of flickering sources, and a comparison mode that permits direct luminance comparison of two sources. Results are displayed digitally.*

Figure 7.11 *When the cell of a direct-reading illuminance meter is held in contact with a luminous source, the surface luminance can be read directly or calculated.*

For a diffuse luminous source, the cell of an illuminance meter is placed directly against the surface (Figure 7.11); the source luminance in footlamberts is equal to the reading on the meter in footcandles because footlamberts = lumens per area = footcandles. When using a meter calibrated in lux, the readings must be divided by π (i.e., 3.13285 ···) to obtain the diffuse source luminance in cd/m^2.

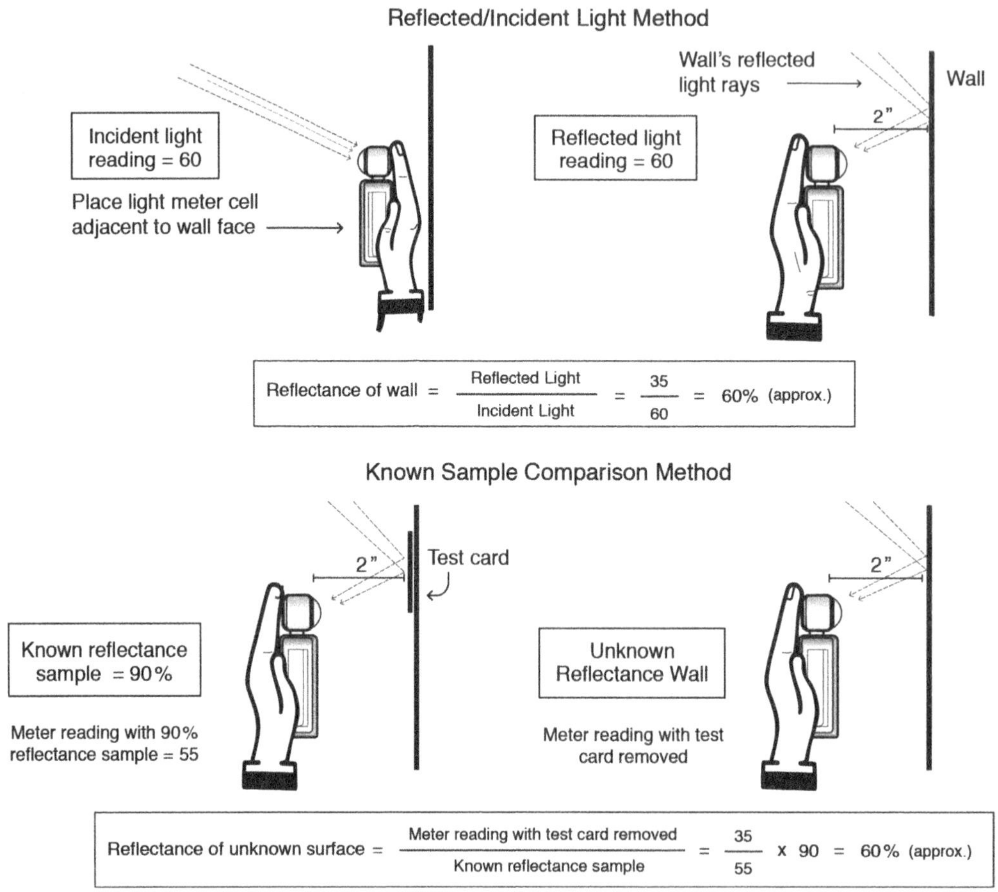

Figure 7.12 *Two simple methods of measuring the diffuse reflectance of a surface. (Drawn by Martin Lee.)*

Reflectance. It is often desirable to determine the reflectance of a given surface because luminance can then be readily computed (see Figure 7.4). Two methods of measuring diffuse (nonspecular) reflectance are shown in Figure 7.12: the known-sample method and the light-ratio method. If a sample of known reflectance is available, this method should be used because it provides more accurate results than the ratio method. The sample should be no smaller than 8 in. × 8 in. [200 mm × 200 mm]. See Chapter 12 for some typical reflectance values.

Those involved with lighting design should be generally aware of the reflectances of common materials and paint colors as well as the range of luminances encountered in typical spaces. This will enable the designer to better appreciate and visualize architectural variables that substantially affect the visual environment and visual comfort.

Luminous Intensity. Luminous intensity (I) cannot be measured directly in the field but must be derived from its illumination effects. The simplest way of doing this is to use the inverse square relationship developed in the following section. Measure the illuminance produced on a plane at right angles to a source at a known distance and apply Equation 7.6. For accurate measurement, the distance should be at least 5 and preferably 10 times the maximum dimension of the source because, for anything other than a point source, the equation is an approximation. The candela (candlepower) thus calculated is the luminous intensity in the direction being viewed.

Because luminous intensity is not uniform in all directions for anything except an ideal point source, and because a single intensity value for a source is sometimes desirable for design purposes, the average of a number of intensity measurement values taken from several directions is used. This average figure is called the mean spherical candlepower (mscp) and represents an equivalent point source that produces 4π lm for every candela. Thus, a 10-cd lamp exhibits an average intensity of 610 cd in all directions and produces 40π lm.

Inverse Square Law. By definition, a point source of 1-cd intensity produces an illuminance of 1 lx on the inside surface of a surrounding sphere of 1-m radius (r). Because the surface area of this sphere is 4π m^2,

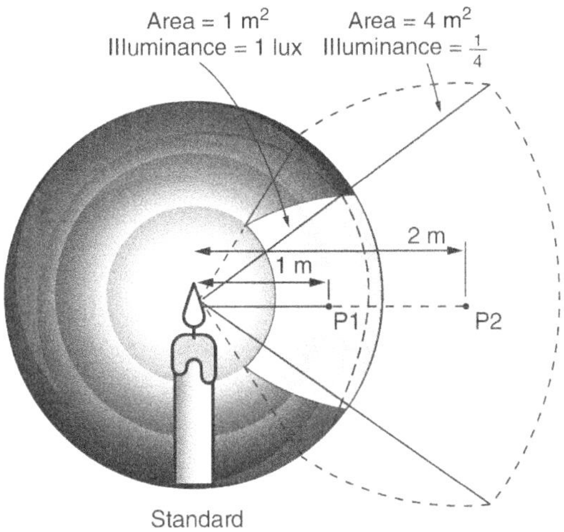

Figure 7.13 *Relationship among candelas, lumens, and lux defined with reference to a standard light source of 1 mean spherical cp (1 cd) located at the center of a sphere with a 1-m radius. (Drawn by Martin Lee.)*

a 1-cd source produces 4π lm of luminous flux. Now, assume a sphere of 2-m radius surrounding this same source (Figure 7.13). Because the same flux is spread over a larger area, the illuminance on the larger sphere is inversely proportional to the ratio of the sphere areas; that is:

$$\frac{\text{lux}_2}{\text{lux}_1} = \frac{\text{area}_1}{\text{area}_2} \qquad \text{(Eq. 7.4)}$$

or

$$\text{lux}_2 = \text{lux}_1 \times \frac{\text{area}_1}{\text{area}_2}$$

therefore,

$$\text{lux}_2 = \text{lux}_1 \times \frac{4\pi r_1^2}{4\pi r_2^2}$$

$$= \text{lux}_1 \times \frac{r_1^2}{r_2^2} \qquad \text{(Eq. 7.5)}$$

In other words, the illuminance is inversely proportional to the square of the distance from the source. In general terms,

$$\text{lux} = \frac{\text{cd intensity}}{\text{distance}^2} \qquad \text{(Eq. 7.6)}$$

where distance is expressed in m [ft]. This holds true for surfaces normal to a source.

This relationship can also readily be derived by using any solid angle and the area it intercepts, as in Figure 7.14. This figure shows that the area intercepted is proportional to the square of the distance from the source; therefore, the illuminance is inversely proportional, as stated previously.

Intensity Distribution Curves. If luminous intensity values for a light source are plotted on polar

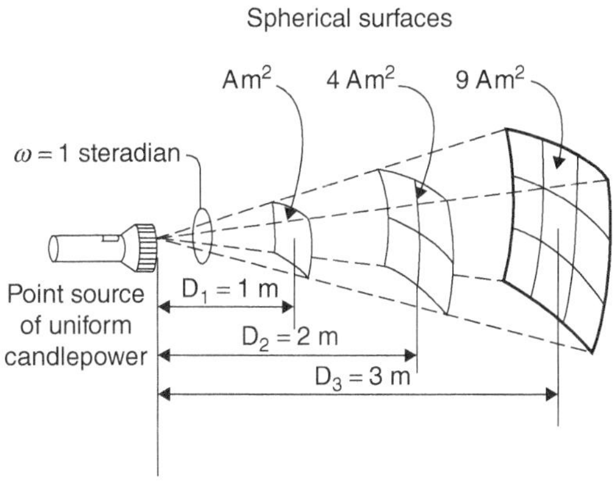

Figure 7.14 *Demonstration of inverse square law properties using a solid angle of unit size. Note that the surfaces are necessarily spherical because points on a planar surface are not equidistant from the source.*

coordinate axes, the resultant figure is called a candlepower distribution curve (CDC). Different sources will have different curve shapes and magnitudes. The procedure for making a CDC is straightforward. An illuminance meter is rotated around the source in a defined plane, illuminance is measured at multiple points, and luminous intensity (cd) is calculated. Alternatively, the illuminance meter can be fixed and the source rotated. If the source distribution is symmetrical, as shown in Figure 7.15, then only a single set of values is required, and the resultant plot is valid in all vertical planes through the source. Thus, for point-like sources only a single CDC is usually required. For a nonsymmetric source, such as a tube, CDC curves in several planes are required to define the fixture distribution characteristic. Normally, manufacturers will provide longitudinal and crosswise candlepower

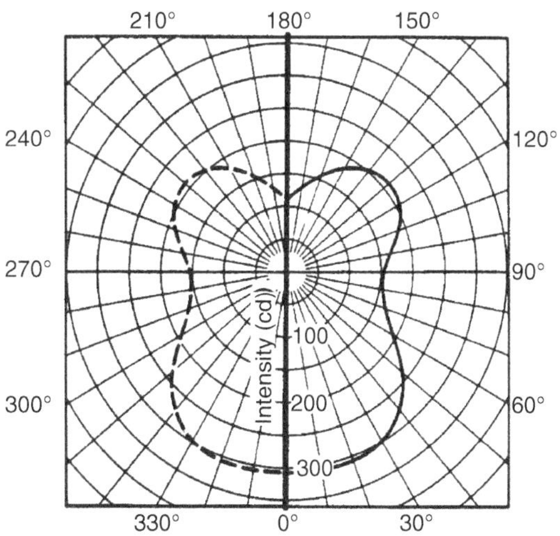

Figure 7.15 *Typical luminous intensity (cd) distribution curve for a general diffuse-type luminaire. Because the unit is symmetrical about its vertical axis, only one curve need be shown. Furthermore, only the right side of this curve need be shown, due to symmetry.*

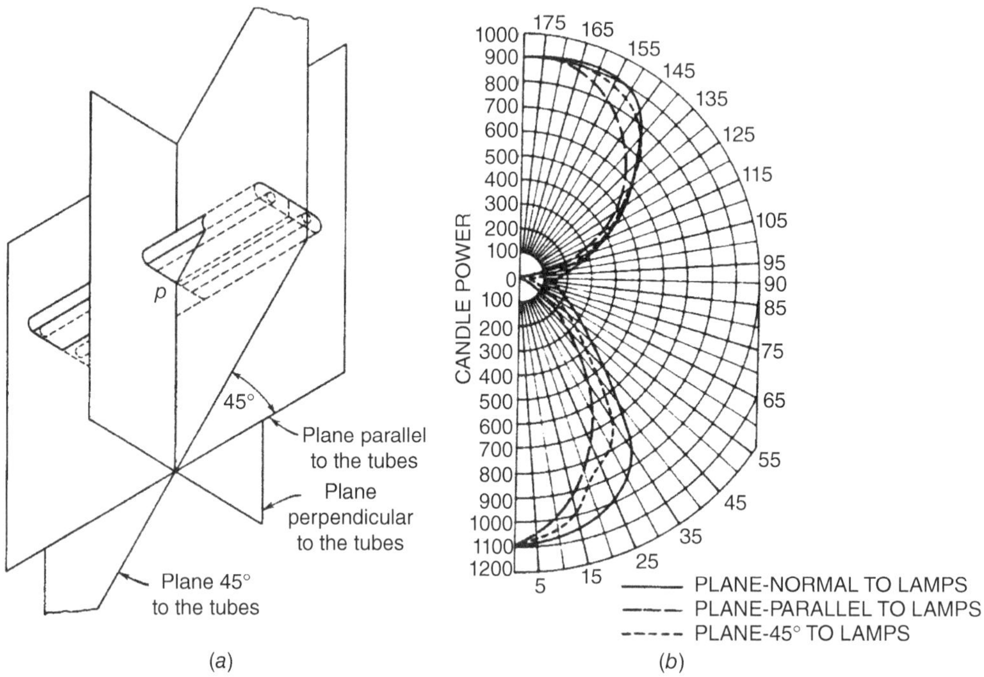

Figure 7.16 *(a) Due to the asymmetry of a fluorescent luminaire, intensity distribution curves in (at least) three planes are required. (b) Photometric distribution for this luminaire is symmetrical in each individual plane; therefore, only one side of a curve is required. By convention, the right side is used.*

distribution curves, plus a diagonal (45° plane) curve on request. This is illustrated in Figure 7.16, where curves for three planes are shown. Note that the area of a CDC curve is not a measure of total lumen output.

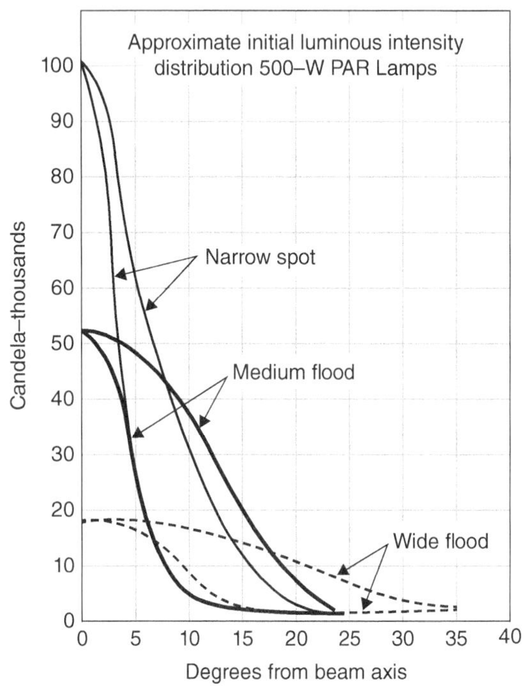

Figure 7.17 *Luminous intensity distribution curves plotted in rectangular coordinates. Note that candela values near the cutoff angles are easily read, which is not the case in polar plots.*

Most CDC plots are made on polar coordinates because such a plot clearly shows directions and magnitudes. Polar plots tend to crowd near the nadir, however, and accurate magnitude readings at the cutoff angle are difficult to make. For this reason, it is occasionally desirable to obtain a plot on rectangular coordinates. One such plot is shown in Figure 7.17. Intensity distribution curves are important to an understanding of lighting fixture performance, point-by-point illuminance calculations, and direct and reflected glare analysis.

7.2 VISION

Both light and the measurement of light are directly related to how the human eye receives and processes this special form of energy. The eye and vision are both integral to understanding light in the built environment. Figure 7.18 shows the structure of the eye and the parallel structure of a camera. Light passes through the pupil, the size of which is controlled by the iris, thereby controlling the amount of light entering the eye. The lens focuses the image conveyed by the light on the retina, from which the optic nerve conveys a visual message by electric impulse to the brain.

Light entering the eye is focused on the retina, which contains in all some 150 million light-sensitive cells of two types: rods and cones. The central portion of the eye, near the fovea, is an area of pinhead size

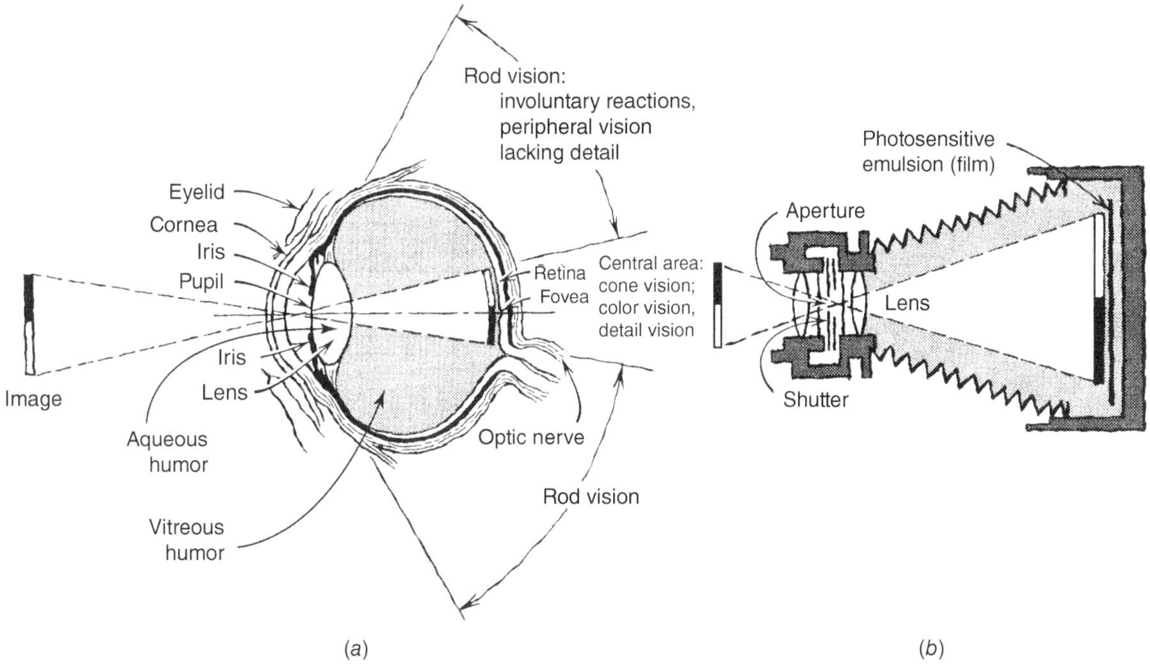

Figure 7.18 *The human eye (a) and a camera (b) operate on similar optic principles. The cornea acts as an outer refracting lens that introduces light into the iris. The iris and pupil control the extent of opening of the eye, and correspond roughly to an f-stop range of f2.1–f11. The lens, which acts as a perfectly smooth automatic zoom lens, can focus from about 2 in. [50 mm] to infinity.*

containing about 100,000 cone cells, which accounts for the extreme precision of foveal (center-focus) vision. The cones are responsible for the ability to discriminate detail; they also give us the capacity to perceive color and can detect luminances ranging from 3 to 1,000,000 cd/m^2. Proceeding outward from the fovea, rod cells are encountered. Rods can detect luminances from 1/1000 cd/m^2 to approximately 120 cd/m^2 and are extremely light-sensitive, having a response to light 1/10,000 that is required by cone cells. Rods lack color sensitivity, however, thus accounting for the fact that in dim light (rod vision), we have no color perception and all colors appear as varying shades of gray. Rods also lack detail-discrimination, making "night vision" quite coarse. Finally, rods are slower acting than cones and therefore have a low degree of flicker fusion; they are highly motion-sensitive. Because rods dominate at the outer portions of the retina, we are best able to detect movement when looking out of the "corner of the eye."

Figure 7.19 illustrates the angles related to the field of vision. Of interest is the extreme narrowness of the cone of central (foveal) vision, in which acute perception of detail takes place. This area is so small that your eye must refocus on each dot in a colon (:) if you wish to examine each one individually. Surrounding this central area is a cone of binocular vision of 30° half-angle, called the near field or surround, in which most of the coarser sight information is gathered. Beyond this cone, we have far-field and peripheral, primarily horizontal, monocular vision. The far-field and peripheral areas largely give us our subjective reactions of ambience.

We encounter a wide range of visual tasks during a typical day. Some of these tasks occur outdoors, and many are relatively simple and undemanding. For most of us, however, many visual tasks are conducted in buildings—and many of these indoor tasks are reasonably demanding. Visual acuity describes the ability to see detail (think of an eye test) and in many building spaces adequate visual acuity is important to the success of the space in use.

There are three main components to any seeing task: the task or object itself, the lighting conditions, and the observer. Several variables affect each of these three components. Based upon many investigations, they can be categorized as of primary or secondary importance. The variables (some controllable by design and some not) are the following :

Visual Acuity Factors

I. **The Task (or Object)**

Primary Factors

a. Size
b. Luminance
c. Contrast, including color contrast
d. Exposure time—needed or available

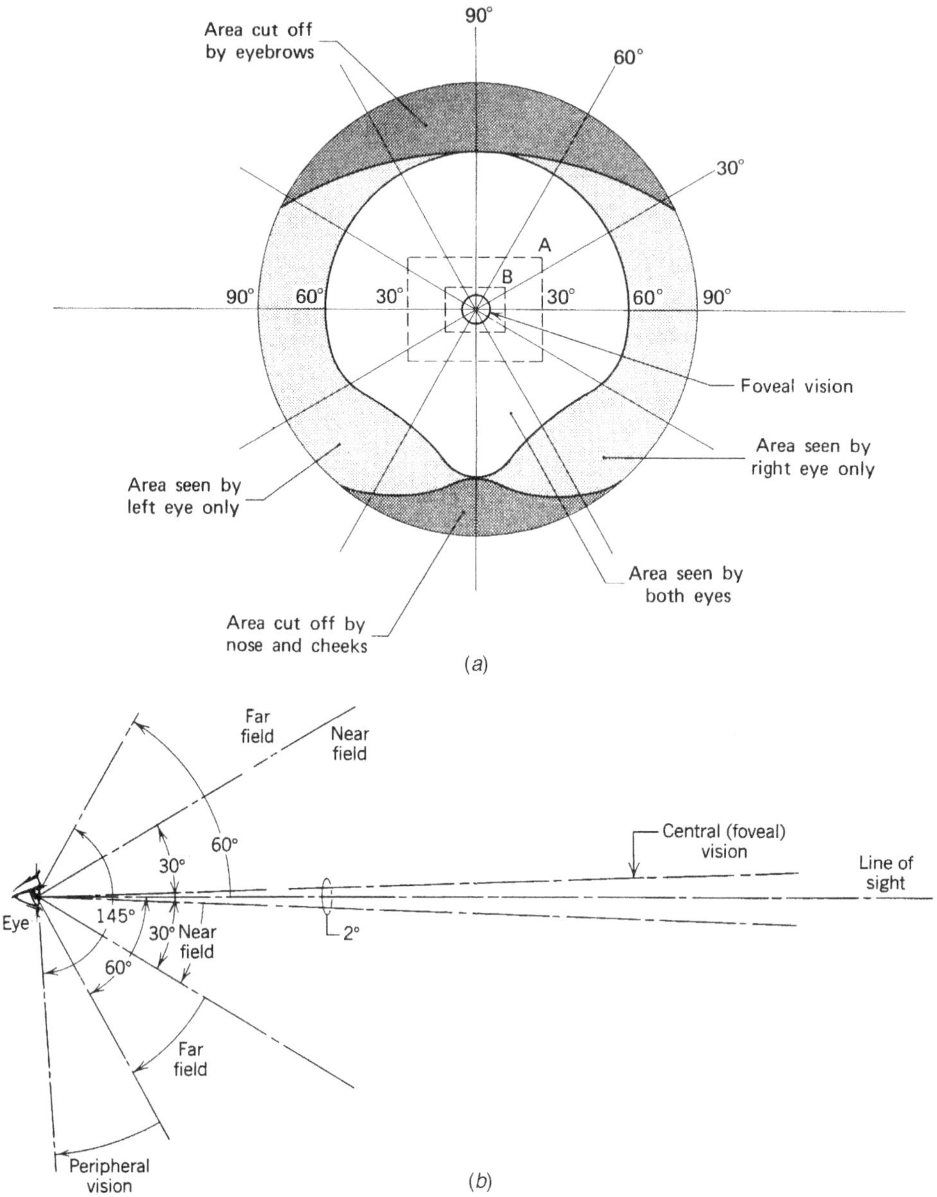

Figure 7.19 *The fields of vision of a normal pair of human eyes (a) and the subtended angles (b). Rectangles A and B superimposed on the field of vision in (a) represent a large magazine and a small book, respectively.*

Secondary Factors

e. Type of object—required mental activity; familiarity with the object (in reading, familiarity is so important as to become a primary factor)
f. Degree of accuracy required
g. Task—moving or stationary
h. Peripheral patterns

II. **The Lighting Conditions**

Primary Factors

a. Illuminance
b. Disability glare (blinding glare being rare indoors)
c. Discomfort glare

Secondary Factors

d. Luminance ratios
e. Luminance patterns
f. Chromaticity

III. **The Observer**

Primary Factors

a. Condition of the eyes (relative to both health and age)

b. Adaptation state
c. Fatigue state

Secondary Factors

d. Subjective impressions; psychological reactions

These factors are considered individually in the following discussions, but many are interrelated. Luminance (I*b*) and adaptation (III*b*) are the result of illumination (II*a*); subjective impressions (III*d*) are dependent on brightness patterns (II*e*) and chromaticity (II*f*); fatigue (III*c*) results from a combination of many of the factors, and so on.

It is common to find reference to the quantity and quality of light—dealing generally with lighting design elements that are objective and subjective, respectively. In the above list, the quantity of light connects directly to item II*a* and quality to items II*b* through II*f*.

Assuming a good lighting situation—one with low glare, acceptable luminance ratios, white light, and a normal pair of unfatigued eyes—visual acuity is primarily dependent upon items I*a* to I*d*, the interrelated effects of which have been revealed by numerous field tests. Remember that the seeing task for visual acuity involves foveal vision (i.e., focusing and concentrating on small-area detail). This is a vastly different task than normal reading, where the eye rapidly scans familiar images without focusing on details, and the brain immediately understands even if much of the information is missing. The task discussed next is quite different and could be compared to studying mathematical equations or reading an unfamiliar language or even proofreading. Such tasks require detailed examination of each symbol individually.

(a) Size of the Visual Object

Visual acuity is generally proportional to the relative size of the object being viewed, given fixed luminance, contrast, and exposure time. Because the parameter of concern is not absolute size but subtended visual angle, visual acuity can be increased by bringing the object nearer the eye (Figure 7.20). This assumes a pair of young eyes, because at ages above 40, the accommodation ability of the eye becomes limited and bringing the object closer may blur the focus.

(b) Luminance

The sensation of vision, as explained previously, is caused by light entering the eye. This light may be thought of as a group of convergent rays, each ray coming from a different point in space and therefore carrying different visual information. The composite of these rays constitutes the entire visual picture that the eye sees and the brain comprehends. The individual rays differ from each other in intensity and chromaticity, depending on the part of the viewed object from which they emanate. The intensity of these cones of light determines the luminance of the object being viewed.

The human eye can detect luminance over an astonishing range of more than 100 million to 1. This range is made possible by a visual process called

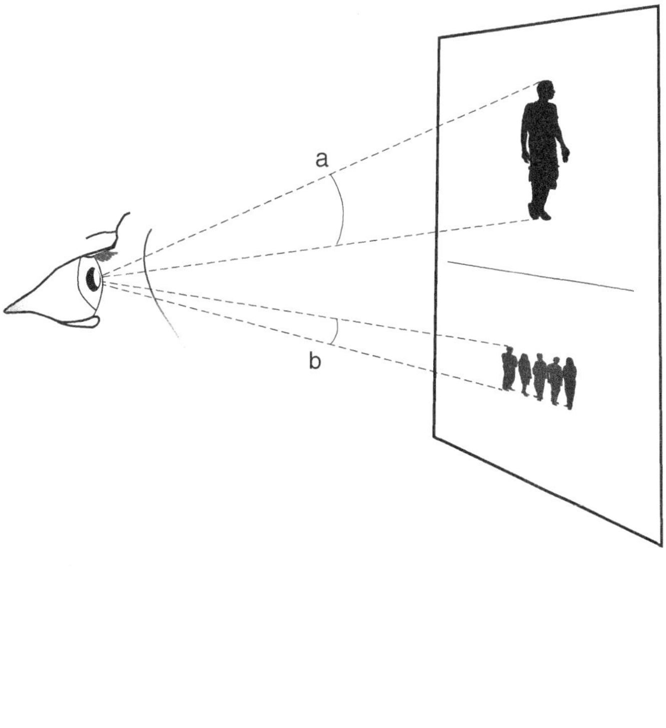

Figure 7.20 *Relationship between object size and visibility is demonstrated by comparison of subtended angles a and b. (Drawn by Martin Lee.)*

TABLE 7.2 Typical Luminance Values[a]

	Luminance	
Object	**cd/m²**	**Footlamberts**
Black glove on a cloudy night	0.0003	0.0001
Wall brightness in a well-lighted office	100	30
A sheet of white paper in an office	120	35
Green electroluminescent lamp	150	45
Asphalt paving—overcast day	1300	380
North sky	3500	1000
Moon, candle flame	4000–5000	1200–1500
Fluorescent tube	6000–8000	1800–2400
Kerosene flame	8500	2500
Hazy sky or fog	15,000	4400
Snow in sunlight	25,000	7300
100-W inside-frost incandescent lamp	50,000	14,600
Sun	2.3 E9	0.67 E9

[a]Values are rounded off.

TABLE 7.3 Example Luminances

Item	cd/m²	Footlamberts
Recommended road luminance, night	1–2	0.3–0.6
Minimum luminance for discernible chromatic distinctions	2–3	0.6–0.9
Clearly discernible human features	15–20	4.5–6.0
Preferred wall luminance	25–150	8–45
Preferred ceiling luminance	50–250	15–75
Preferred task luminance	100–500	30–150
Permissible luminaire luminance (depending on position)	1000–7000	300–2100

adaptation. Dark adaptation may take up to 2 minutes (for cone vision) to up to 40 minutes (for rod vision). Light adaptation (going from dark to light) involves a lesser time lag for both types of vision. The effects of adaptation on brightness are discussed in the following section. Tables 7.2 and 7.3 list some measured luminances of everyday visual tasks.

An interesting characteristic of light-level adaptation is a shift in the sensitivity curve of the eye (Figure 7.3*b*). For the light-adapted eye (described as photopic vision) maximum sensitivity occurs at 555 nm in the yellow–green region; for the dark-adapted eye (scotopic vision) sensitivity peaks at 520 nm in the blue–green region. This means that as luminance decreases, warm colors—yellow, orange, red—become grayed, and the blues and violets stand out. This phenomenon can be important in the design of restaurant lighting, where light levels generally vary inversely with restaurant quality. Very few foods are blue or violet.

In general, visual performance increases with object luminance. This pattern, however, is affected by the background against which an object is viewed and the consequent contrast in luminance between the object being viewed and its surroundings.

(c) Contrast and Adaptation

The discussion that follows assumes full-spectrum white light and ignores the effects of chromaticity, which is considered separately. Many researchers have concluded that contrast is the single most important factor in visual acuity. This is evident when we realize that, in fact, the acute eye sees only contrast. There is no detail to be seen when viewing a large, evenly lighted, monochromatic, diffuse-finish surface.

The effect of contrast (luminance ratio) on visibility depends on the nature of the visual task (more simply, exactly what it is that we are trying to see). The basic visual tasks are detail discrimination and detection under low contrast. Examples of the former are drafting, industrial product inspection, or something as simple as discriminating between the numbers 3, 6, and 8, which are similarly shaped. Detection under low contrast includes reading faint copy, sewing a black fabric with black thread, and the like.

Contrast is a dimensionless ratio, defined as:

$$C = \frac{L_T - L_B}{L_B} \text{ or } \frac{L_B - L_T}{L_B} \text{ or } \left|\frac{L_B - L_T}{L_B}\right| \qquad \text{(Eq. 7.7)}$$

where L_T and L_B are the luminance of the task and background, respectively, in consistent units. Thus, C varies from 0 (no contrast) to 1.0 (maximum contrast). In most situations, the illuminance on the task and background is the same. Therefore, because luminance is the product of illuminance (lux) and reflectance, contrast can also be expressed as:

$$C = \left|\frac{R_B - R_T}{R_B}\right| \qquad \text{(Eq. 7.8)}$$

where R_T and R_B are the reflectances of the task and the background, respectively. From this equation, we can conclude that contrast is generally independent of illuminance (ignoring specularity).

High contrast is the critical factor in visual appreciation of outline, silhouette, and size, which are the factors involved in the task of reading. Thus, reasonably large black-on-white print can be read with ease even in moonlight (around 0.1 lx illuminance) because the contrast is so high (94%). For a reading task—with high contrast (clear, legible print), visibility is essentially independent of illuminance above a certain minimum. Indeed, high illuminance values can be detrimental because they generally go hand in hand with high-luminance sources, and these can potentially lead to glare.

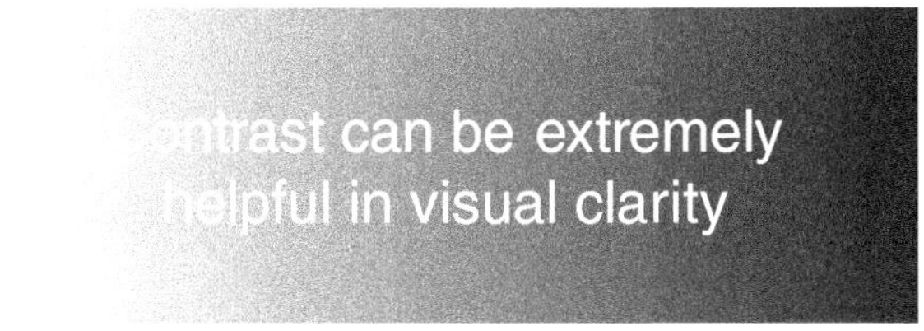

Figure 7.21 *High contrast is helpful when the seeing task involves detection of silhouette detail. (Martin Lee.)*

Referring to Figure 7.21, note that as the contrast between the letters and the background diminishes, the individual letters become harder to read. The left-hand letters require an illuminance of up to 1000 lx [90 fc], and that suffices only because we can imply the letters from the sentence. If the left-hand letters were instead an unknown, illuminance of a magnitude of 10,000 lx [930 fc] or more would be needed. These initial letters are an example of the second type of visual task mentioned previously—low-contrast poor copy requiring surface detail study.

The *h* in "helpful" in Figure 7.21 is printed with the same density as the *y* in "clarity." It exists, and with enough light, the negative effect of lack of contrast can be overcome. This is not so with copy from a used-up printer cartridge or a washed-out photocopy. There, the data simply do not exist, and increased illuminance only makes this fact more evident.

High background luminance makes an object look darker and therefore assists in outline detail discrimination. For this reason, black-on-white is desirable for reading. Conversely, high background luminance makes surface examination more difficult. A simple experiment demonstrates this effect. Stand near a window and hold your hand in front of you with the floor as background. The skin surface detail is perfectly clear—in rough proportion to its luminance. Now hold your hand up against the window with the daytime sky as background. The hand outline is clear, but the skin surface appears dark—the brighter the sky, the darker the skin surface. The reason for this is that the eye automatically adapts to the average luminance of the entire scene.

When taking a photograph of a dark object on a light background (such as a person in a snow scene) the camera aperture should be increased to obtain additional light to photograph the detail of the darker object (in doing so, the rest of the scene is overexposed). Because we cannot easily control the aperture of our eyes, the detrimental effect of high background luminance must be compensated for in another way—for example, by increasing the surface luminance of the visual task. This method is frequently employed in lighting design. Squinting can provide limited visual compensation; doing so reduces the field of vision and the overall scene brightness. For maximum visual acuity, the luminance of a surface-type task should be the same as, or slightly higher than, that of the background, but ratios of 3:1 are acceptable in most circumstances.

Another way of understanding this is to consider the adaptation characteristic of the human eye (Figure 7.22). The eye adapts to the brightness level of the overall scene and sees each object in the scene in the framework of that adaptation level. Thus, at an adaptation level of 1 fL [3.4 cd/m^2], a measured luminance ratio of 1:10 (horizontal scale on Figure 7.22) appears to be approximately 1:4 (vertical scale); that is, the apparent ratio is smaller than the actual one. The low

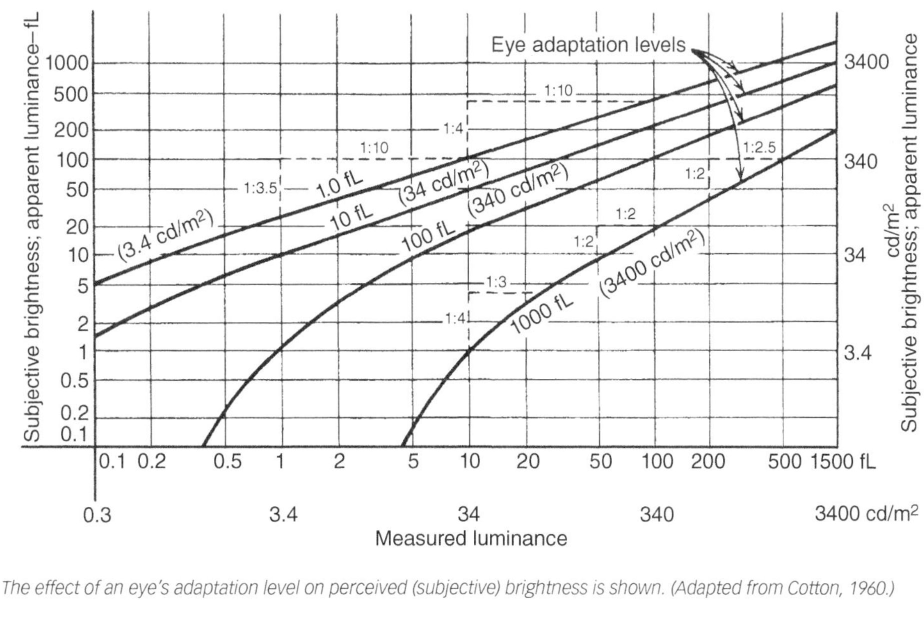

Figure 7.22 *The effect of an eye's adaptation level on perceived (subjective) brightness is shown. (Adapted from Cotton, 1960.)*

state (1 fL) [3.4 cd/m^2] of eye adaptation causes the eye to diminish the difference between high brightnesses. This effect decreases as the adaptation level rises, until at an adaptation level of 1000 fL [3400 cd/m^2; typical of daylight conditions], the apparent and actual ratios correspond; that is, smaller ratios are recognizable. Because visual acuity is the ability to distinguish between different levels of luminance, it follows that visual acuity increases with increased adaptation level.

The second important conclusion that can be drawn is that at high adaptation levels, apparent brightness is lower than actual brightness, and vice versa. Thus, a shadowed object near a window looks darker than it actually is; contrary to first expectation, it must be better lighted than a similar object further inside the room for equal visibility. That this effect (high-level adaptation) is primarily important in daylight situations is also apparent from the curves. At a 100-fL [340-cd/m^2] adaptation level, which is approximately that of a brightly lit interior space, apparent and actual luminance levels coincide. The reverse effect, resulting from low adaptation levels, can be very important in design situations where low lighting levels are found, such as theaters, lecture halls, restaurants, and storage spaces. Sources of light that would be entirely acceptable at a higher adaptation level can easily become an annoying glare under low level adaptation. Good examples (at night) are a flashlight or the blinding glare of an oncoming car's headlights.

The above discussion of contrast deliberately avoided discussion of object colors and the effect of color contrast on visual acuity for several reasons:

- Office- and school-type tasks (paperwork) are often black-on-white tasks.
- Most work tasks involving colored objects involve unsaturated colors, where the pronounced effects of color contrast are minimal.
- The effect of object color on visual acuity is very complex because it involves the color characteristic not only of the object, but also of the background and the surround plus the chromaticity of the illuminating light.

Nevertheless, several important object color phenomena that bear on visual acuity should be mentioned:

- The brightness of a colored (heterochromatic) object is greater than that of an achromatic object for the same luminance. This effect varies with hue and saturation. It is more pronounced with saturated colors than with those of low chroma, and more so in the blue–purple–red area than the yellow–green area.
- Colored objects on a dark-to-black background appear light and desaturated. Conversely, colors on a light-to-white background appear darker and more saturated.
- Adjacent complementary colors produce a pale-to-white border between them. This effect also appears when the task and background colors are complementary, and is most pronounced with saturated (high-chroma) colors.

These remarks address object color only. Most design situations assume white, full-spectrum light, such that neither object color nor visual acuity is negatively affected. The effect of color-unbalanced light sources (such as a high-pressure sodium lamp) on visual acuity is less clear—although it appears that any negative effects may be most pronounced with the elderly and persons with visual defects. This impact should be considered when designing with demonstrably nonwhite light sources.

(d) Exposure Time

Registering a meaningful visual image is not an instantaneous process; it requires a finite amount of time. Just as a photograph can be taken in dim light by using a longer exposure, so can the human eye better distinguish and discriminate fine detail in poor light given time (and neglecting eyestrain). Of course, the time needed depends on the type of task, but the principle of shorter time at higher illuminance, within limits, remains the same. This is particularly true when the object being viewed is not static but in motion.

The effect of viewing time, however, is not linear. In one experiment, increasing luminance by a factor of six reduced the seeing time by 50%, whereas a further sixfold increase in luminance reduced the time only another 20%. Thus, as in the situation of improved contrast with increasing background brightness, there is a case of diminishing returns.

With the parameter of time, as with other parameters of visual acuity, the same qualification applies. When dealing with material that does not require detail-discrimination, improved performance does not necessarily result from improved illuminance. Speed of reading and comprehension are substantially independent of illuminance above a minimum, but are very much dependent on the contrast quality of the material.

(e) Secondary Task-Related Factors

Consider the secondary factors I*e–h* in the list of visual acuity considerations presented above. These factors essentially concern the level of concentration required. Spray-painting a large metal object or packing fruit are very different from inspecting the painted object for defects or the fruit for bruises. The former tasks are largely mechanical and repetitious, whereas the latter tasks require continuous judgment based on visual information. Because both inspection tasks are frequently moving (in assembly-line-type work) and both involve penalties for inaccuracy (rejection at a later inspection point or complaints from a purchaser), the lighting required for these tasks must be several orders of magnitude better than that for related but largely mechanical tasks.

There are several potential ways that a worker might compensate for an unsatisfactory visual condition:

- Moving the work to a better viewing angle.
- Moving the head and eyes to a more comfortable position.
- Reducing the distance between the eyes and the task to the extent that the eyes can accommodate.
- Complaining about a poor-contrast task so that something is done about it (such as replacing the toner in a photocopier).
- Taking more time to perform the seeing task (which, if it affects production, frequently spurs management to make appropriate alterations in the work environment).

The last item in the visual acuity factor task-related list—peripheral patterns—deals with the visual surround rather than the immediate area of the work. Glare sources will be discussed separately, but there are other items outside the central field of vision that can disturb the viewer's concentration and therefore task performance. These include movement (of vehicles, machines, persons), to which peripheral vision is particularly sensitive; large variations in the brightness pattern of the background caused by such activities as periodic opening of an outside door or welding; and even nonvarying patterns that are disturbing because of their very nature, such as checkerboard light-dark patterns, or devices on which it is difficult to focus the eyes, such as crossed patterns of wires and bars. None of these items is strictly lighting-oriented, but are noted here to demonstrate that a good lighting design will include adjustments for anticipated field conditions.

(f) Observer-Related Factors

The lighting designer must be concerned not only with providing adequate illuminance and luminance, but with all the factors involved in the process of seeing. Many are beyond his or her control—including the state of the eye of the observer—but must be considered when addressing those factors that are controllable.

Excellent visual performance may not be synonymous with maximum comfort or minimum fatigue. The reverse may sometimes be true. Most experts agree that what is normally referred to as eyestrain is a condition of the eye muscles resulting from extensive and intensive eye use. Thus, excellent performance under excellent lighting conditions can still produce fatigue because of the demanding nature of a task. As discussed later in this chapter, discomfort glare or even excessive light can cause fatigue without affecting performance.

The visual performance of healthy eyes decreases with age. This reduction is demonstrated principally in two areas: an increase in minimum focusing distance caused by increasing lens rigidity and a decrease in light sensitivity caused by clouding of the cornea, lens, and vitreous humor. Both effects can be compensated for, the former with external lensing (eyeglasses) and the latter by increased task size, luminance, contrast, and/or exposure time, as explained above. The age of the observer is an important parameter in lighting design, as discussed below. Most lighting design standards were established to meet the needs of average young adults.

(g) The Aging Eye

Recent decades have seen an increase in human life expectancy in modern Western countries and the aging of the Baby Boomer generation, resulting in a sharp increase in the elderly population. In the United States, approximately 17% of the population is 65 years of age or older (U.S. Census Bureau, 2023)—a proportion expected to reach 22% by the year 2050. As a result, lighting designers must consider not only the special requirements of buildings specifically intended for use by the aged but also in general-use public buildings. A brief review of these special requirements is presented here.

Light enters the eye through the cornea, passes through the aqueous humor, and enters the lens through the pupil (Figure 7.18). After being focused by the lens, light continues through the vitreous humor and finally projects the viewed image, reversed, onto the retina. As the eye ages, numerous physiological changes may occur; some are usual and are therefore classified as normal; others, such as cataracts, are less common but are still considered by ophthalmologists to be an expected development. The nature of these changes and their influence on design are briefly described in the following subsections.

The Cornea. This initially perfectly clear outer lens tends to become cloudy, with corresponding reduction of visual clarity and acuity. This results in a need for more ambient light to overcome the reduction in light falling on the retina. The need for additional light for aging eyes is recognized in most modern lighting design guidelines.

The Lens. The lens, which begins life as a very lightly yellow tinted, flexible crystalline body, gradually thickens and darkens in tone. The thickening reduces flexibility, leading to an inability to focus on objects that are near (presbyopia or hyperopia). The yellowing reduces the overall light intensity in the eye and selectively filters the blue portion of the spectrum. The overall effect is again a requirement for additional light to achieve the desired visual acuity.

A second and more important degenerative phenomenon of the lens is its gradual clouding (increasing opacity). When such opacity is confined to the perimeter, its effect is negligible because vision is unaffected. When small, opaque areas appear within the visual axis of the lens, however, vision is affected in two ways:

1. The viewed image is dimmed and blurred due to opacities in the field of view.
2. Light entering the lens is scattered by interreflections from the opaque particles, resulting in a subjective impression of glare. This effect is particularly severe outdoors, where light enters the lens from all angles.

The net result of these reactions is a requirement for more ambient light, but an even more pressing requirement that sources of glare and peripheral light be eliminated. (People with this condition frequently wear eyeglasses and sunglasses with large, opaque side shields to block peripheral light and thus reduce glare.) Also, because short-wavelength (blue) light interreflects and scatters more readily than does long-wavelength (yellow–red) light, such persons are more comfortable with low-color-temperature lamps than with sources rich in the blue–green spectrum. (Ophthalmologists frequently prescribe yellow-tinted eyeglasses for people with this lens condition in order to filter out blue light.)

A less common, but still prevalent, condition of the aging lens is the development of fluorescent particles, called fluorigens, in the vision path. In the presence of UV radiation (as exists with daylight, fluorescent, and high-intensity discharge sources), these particles fluoresce—causing scatter, blur, and glare. The solution to this problem is a combination of yellow-tinted eyeglass lenses and a reduction of light from sources containing appreciable quantities of UV.

The Pupil. The pupil controls the amount of light entering the eye and is therefore intimately involved in the constantly changing accommodation level of the eye. The pupil muscles react more slowly as they age, thus lengthening accommodation time. Dark-to-light accommodation is very rapid in the young eye, and is barely noticed except for extreme changes, such as exiting a movie theater into sunlight. With an aging eye, the slower pupil response results in severe glare sensations with even small brightness changes.

The net result of all the normal changes taking place in the aging eye is a heightened sensitivity to glare, intolerance to the blue–ultraviolet end of the spectrum, and an overall requirement for higher illuminance levels. For the lighting designer, these needs translate into requirements for very careful selection and placement of luminaires, increased use of indirect lighting, and attention to the spectrum of the light sources used. Because some of these requirements are not only mutually incompatible but also contrary to energy-efficient design practices, it may be particularly difficult to satisfy all the requirements in spaces occupied by people with a wide range of ages. In spaces of this type, it may be wise to provide for the possibility of readily changing lighting conditions in a limited area to accommodate older occupants, keeping in mind the glare and color factors discussed previously.

(h) Subjective Impressions: Psychological Reactions

As is the case with thermal and acoustic comfort, visual comfort is also affected by the "X-factor" of one's mindset. As such, visual acuity can be impacted by psychological factors. Stress, anxiety, and fatigue can reduce attention and impair visual processing. Emotional stimuli can activate the autonomic nervous system, which can affect the eyes. Hormonal changes (e.g., cortisol levels), the brain's perceptual load, and memory recall can all influence one's ability to visualize and interpret visual stimuli, which can hamper or enhance one's ability to execute tasks that require good visual acuity.

7.3 COLOR

The use of color in design is a vast subject. Only those aspects of color that are fundamentally important to the design of lighting systems are presented herein. We perceive what we call color because of the wavelength mix of light received by the eye. Quantitatively, this wavelength mix can be very precisely described (as discussed below). Qualitatively, our descriptions of color are usually limited to the colors of the rainbow (e.g., red)—plus hybrids (red–orange) and creative adjectives (flaming red) often used for marketing purposes (paint, fabrics, carpets). The use of qualitative color descriptors is adequate for early design to ensure that parties to the process are in the same ballpark. Descriptions of colors in final design documents typically must be quantitative to ensure that what is truly wanted is actually installed. A specification that calls for green paint in a bathroom is not tight enough for most projects.

Colored light (other than white) is rarely used for general illumination—other than to create a special atmosphere. When a space is lighted with colored light, the eye adapts by a phenomenon known as color constancy so that it can, to a considerable degree (depending on the chromaticity of the light), recognize colors of objects despite the effect of the colored illuminant. Thus, even when wearing tinted sunglasses, we can still easily distinguish the color of objects. After only a short while, we no longer notice the green, yellow, amber, or other color cast caused by the tinted lenses. The eyes do, however, become more sensitive to the missing colors that would make up white light. This phenomenon can be used to make meat look redder in a display by using blue-rich, red-poor, cool white lighting in the remainder of the store.

A similar phenomenon occurs when the eye is exposed to a monochromatic scene, where the chromaticity is due to the coloration of the objects rather than the illumination. The eye in such a situation becomes sensitized to the complementary color; thus, if after looking at a green surface one shifts one's gaze to a white surface, one sees the complementary red color. Returning to our meat display, the use of green paint on the walls can enhance the redness of the meat. This effect in reverse also partly accounts for the extensive use of green for paints, linens, gowns, and so on, in operating rooms. The eyes of surgeons and nurses, when diverted from the redness of the surgical area, are more comfortable with a green background than with a white one.

By a process known as lateral adaptation, the apparent color of an object changes when the background color is changed. Thus, a green object looks somewhat blue–green on a yellow background because the eye is supplying the complementary color to yellow—that is, blue. Similarly, the same green object looks slightly yellow–green when on a blue background, due to the eye supplying the yellow.

The apparent brightness of a color is a function of its hue, in that light colors appear lighter than dark colors even when luminance is the same. Thus, spaces may be defined by color within an area of equal illumination. Also, all colors tend to appear less saturated—that is, they appear "washed out"—when illumination is high. Colors with high saturation (chroma) must be used in well-lit spaces if they are to be effective, although extensive use of saturated colors is generally best avoided.

Other well-known psychological effects of colors are the perceived coolness of blues and greens, and the perceived warmth of reds and yellows. Thus, cool colors might well be used in a display of winter wear, and warm colors in a display of summer wear. Red and yellow are also considered "advancing" colors because objects lit with them are perceived as advancing toward the observer—giving the appearance that they are becoming larger. The opposite effect is noted with blue and green, accounting for their being known as "receding" colors.

In an atmosphere designed to be calm and restful, greens should generally predominate in illuminant color, object color, or both—except in eating areas, which should be illuminated with reds and yellows because cool colors are generally unappetizing. Yellows and browns encourage motion sickness, whereas blues and greens tend toward the reverse. Warm, saturated colors encourage activity; conversely, cool, unsaturated colors are conducive to meditation. Cool colors also seem to shorten time passage and are often applied in areas of dull, repetitive work.

(a) Light Source Color

The term light source color is used to describe the wavelength characteristics of a source of light—independent of its interactions with objects in a space being illuminated. There are several means of expressing light source color, as discussed below.

Color Names. The simplest, and least accurate, means of describing the color of a light source is through names. For example, cool-white, warm-white, daylight, or deluxe warm-white. The use of such names is not regulated and is at best an approximate way of expressing design intent or criteria. Nevertheless, color names are often used to describe the output of electric lamps, and can be a convenient means of conveying a general preference for wavelength mix.

Color Temperature. Light sources are often designated by color temperature—for example, 3400 K for halogen lamps, 4200 K for certain fluorescent lamps. This nomenclature is derived from the fact that when a light-emitting body (specifically, a blackbody with essentially zero reflectivity) is progressively heated, it first glows deep red, then cherry red, then orange, until it finally becomes blue–white hot. The color of the light radiated from such an object is related to its temperature. Thus, a blackbody temperature scale can be used to express the color of a light source through its color temperature—that is, the temperature to which

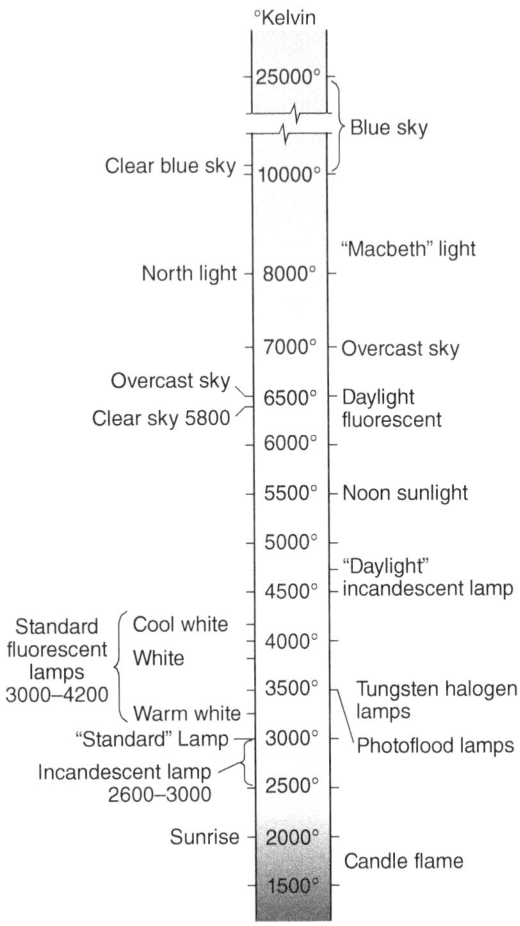

Figure 7.23 Approximate color temperatures of common illuminants.

a blackbody must be heated to radiate a light similar in color to the color of the source in question. Color temperature is measured in Kelvin (a scale with its zero point at −459.67 °F). Figure 7.23 shows the assigned color temperature of some common light sources.

Strictly speaking, a color temperature can be assigned only to a light source that produces light by heating a solid, such as an incandescent lamp. Light sources that operate on a different principle, such as gaseous discharge and LED lamps, are described using correlated color temperature (CCT). CCT is the temperature of a blackbody whose chromaticity most nearly matches that of the light source in question. Under CCT, there is no relationship between the lamp operating temperature and correlated color temperature.

Chromaticity. The CIE (International Commission on Illumination) chromaticity system is a globally accepted standard for designating illuminant color. Under chromaticity, the relative proportions of each of the three primary additive colors (red, green, and blue) required to produce a given illuminant color are determined. These three values are called the tristimulus values for that color and are designated by capital letters: X (red), Y (green), and Z (blue). See Figure 7.24 for an example of a chromaticity diagram. It is possible to show X, Y, and Z values on a two-dimensional diagram because $X + Y + Z = 1.0$—such that knowing any two of the chromaticity coordinates leads to the third.

The chromaticity diagram represents all colors that can be perceived by humans. Light with roughly equal X, Y, and Z coordinates is considered to be white light. Compared to color names, chromaticity provides a very precise means of specifying the color of a light source. A reasonably low-cost meter (similar in appearance to an illuminance meter) can be used to determine chromaticity coordinates in the field (Figure 7.25). Such a meter measures the *X* and *Y* coordinates of an illuminant, which can then be plotted on a standard CIE color diagram to graphically determine chromaticity. This is very useful in comparing illuminants to predict color responses.

Spectral Distribution. The relative energy content of light produced by a source can be plotted across the wavelengths of the visible spectrum. The result is a graph that shows the content of the light in a form similar to the way a mass spectrometer displays the chemical composition of a substance. As an example, the relative spectral energy distributions of a few common light sources have been plotted as a function of wavelength in Figure 7.26. Figure 7.26*a* and 7.26*c* shows the spectral distribution for two still common light sources—cool white and warm white fluorescent lamps, respectively. Note that the principal difference lies in the amount of blue in the spectrum. As a result, a blue object will be bright under cool white light, and dull (grayed) under warm white light. The color situation is more pronounced with a standard high-pressure sodium lamp (Figure 7.27*c*), compared to a clear metal-halide lamp (Figure 7.27*b*). A blue object will appear gray under the sodium lamp, whereas under the metal-halide lamp its blue color will show clearly. Spectral distribution plots can be useful in understanding where in the range of wavelengths energy from a lamp lies and thus likely color rendering effects. Graphs as shown in Figures 7.26 and 7.27 are available from electric lamps manufacturers, and these resources should be considered whenever rendition of colors is an important project requirement.

An interesting way to explore the effect of illuminant color is to expose a dull white surface to two illuminants—side by side, but separated by an opaque divider—to get an impression of the illuminant color. Then expose a series of colored samples—again, side by side—to see which colors are brightened and which are grayed by the light sources. The intensity of

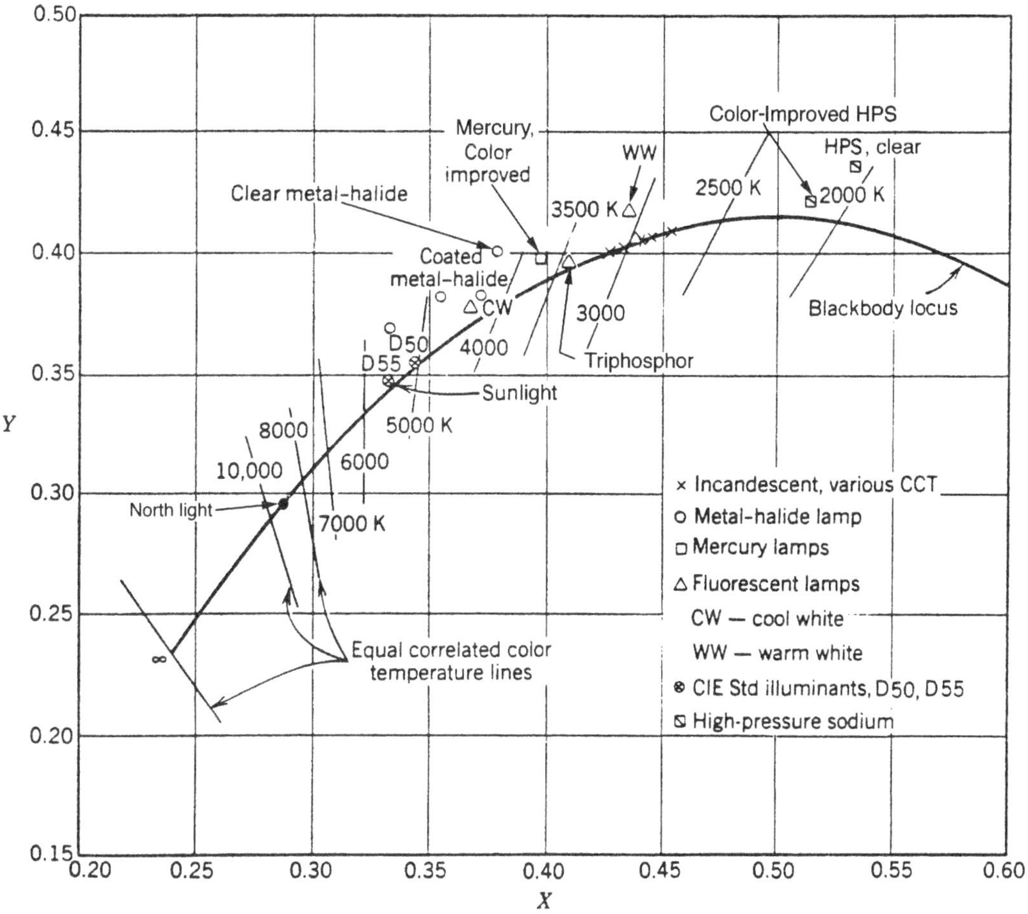

Figure 7.24 *Portion of a chromaticity diagram showing the relation of common illuminant chromaticities to that of the blackbody locus. Illuminants whose coordinates fall on the same line crossing the blackbody locus have the same CCT but may have entirely different component colors.*

illumination also influences the appearance of colors, and it must be considered in choosing object colors. As intensity is increased, reflection increases, particularly with pale tints (high value) that contain much white pigment and thus tend to wash out color. Therefore, with high-intensity lighting, saturation of colors should be high for true, brilliant color rendition.

In Figure 7.28, note in *b*, *c*, and *d* that the spectrum of a light source that produces light as a result of heating is continuous. Sunlight is equal in spectrum to a blackbody radiator at 5500 K; north light is equal to one at about 8000 K to 10,000 K; a 500-W incandescent lamp is approximately equal to one at 2850 K. If the spectrum of a blackbody radiator is plotted on a chromaticity diagram, its locus is a continuous curved line, as seen in Figure 7.24. The chromaticity of all true blackbody radiators falls exactly on this line, with the location depending on temperature. Daylight generally falls on this locus, although because of selective atmospheric absorption and other phenomena, it is slightly off. Incandescent lamp chromaticity is very close to this locus because it is also a heat-based radiator.

A source that produces light by means of individual phosphors can also have chromaticity on this locus if the phosphors are selected to produce a continuous spectrum like that of a blackbody radiator. Thus, we see in Figures 7.26 and 7.27 that triphosphor fluorescent lamps (Figure 7.26*d*) and metal-halide lamps (Figure 7.27*d*) have spectral components over the entire spectrum, yielding chromaticities fairly close to the blackbody locus. For other sources, the CCT is established by their chromaticity locus in relation to the diagonal lines crossing the blackbody locus, as seen in Figure 7.24. Each of these lines is isothermal—that is, all chromaticities on it have the same CCT. Thus, two sources with widely differing spectral content, and therefore object color rendering, can have the same CCT.

Color Rendering Index. Color rendering is defined as the degree to which perceived colors of objects, illuminated by a light source, conform to the colors of the same objects as illuminated by a reference source. The color rendering index (CRI) of a source is a two-part concept, comprising a color temperature

Figure 7.25 Handheld chromaticity meter. Developed specifically for LED, organic LED, or other electroluminescent sources, this meter can provide simultaneous readings of illuminance, color temperature, excitation purity, dominant wavelengths, and chromaticity. (Courtesy of Konica Minolta, Inc.)

that establishes the reference standard, and a number that indicates how closely the illuminant approaches the standard. The standard is always daylight at the defined color temperature. Therefore, the CRI of a lamp is really a measure of how closely it approximates daylight of the same color temperature. Two sources cannot be compared unless their color temperatures are equal or quite close. A CRI of 100 indicates an illuminant whose spectral content is equal to daylight of that temperature. CRIs for typical common lamps are given in Figures 7.26 through 7.28.

Table 7.4 lists color descriptors and effects for a few common light sources. Note especially the wide range of CRI values. CRI provides a quick, if not absolute, means of comparing the color rendering capabilities of light sources. It would be reasonable to find CRI targets as part of the project requirements for many space types.

All considerations of color—comparison, matching, and rendering—must remember that object color depends on the spectral energy distribution of the light source and that any change in spectral content changes an object's appearance. Two sources of the same color temperature—and, therefore, apparent source whiteness—can have quite different spectral content and subsequently render object colors differently. An example would be a 3000-K warm-white fluorescent tube versus an incandescent lamp (500-W photoflood) of approximately the same color temperature. Color temperature is an expression of dominant color, not spectral distribution.

(b) Color Names

As with light source color, names can be used to describe object color. In fact, the use of names to identify colors is more often and more creatively applied to objects than to sources. A quick look at the website for a paint company or automobile manufacturer will give

Fluorescent Lamps

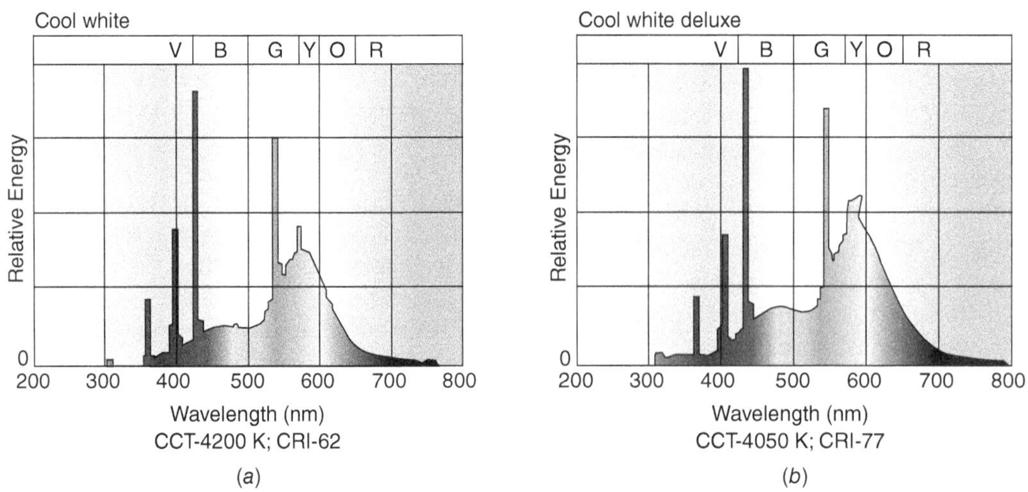

Figure 7.26 Spectral energy distribution of several fluorescent lamp types with their correlated color temperatures (CCT) and color rendering indices (CRI). The curves are not truly continuous, but consist of individual color lines; they are shown connected for simplicity. These graphs are generic and show patterns of wavelength distribution; consult manufacturers' data for the output spectrum of a specific lamp.

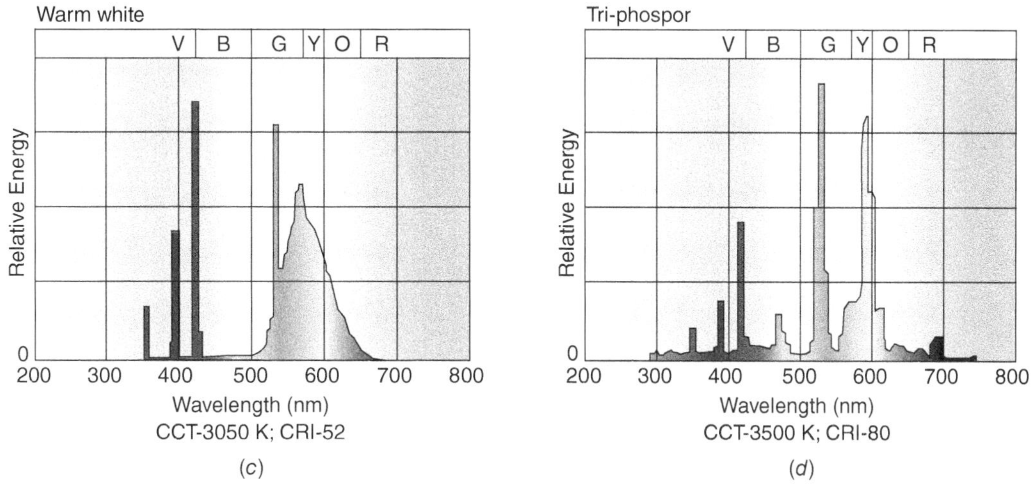

Figure 7.26 *(continued)*

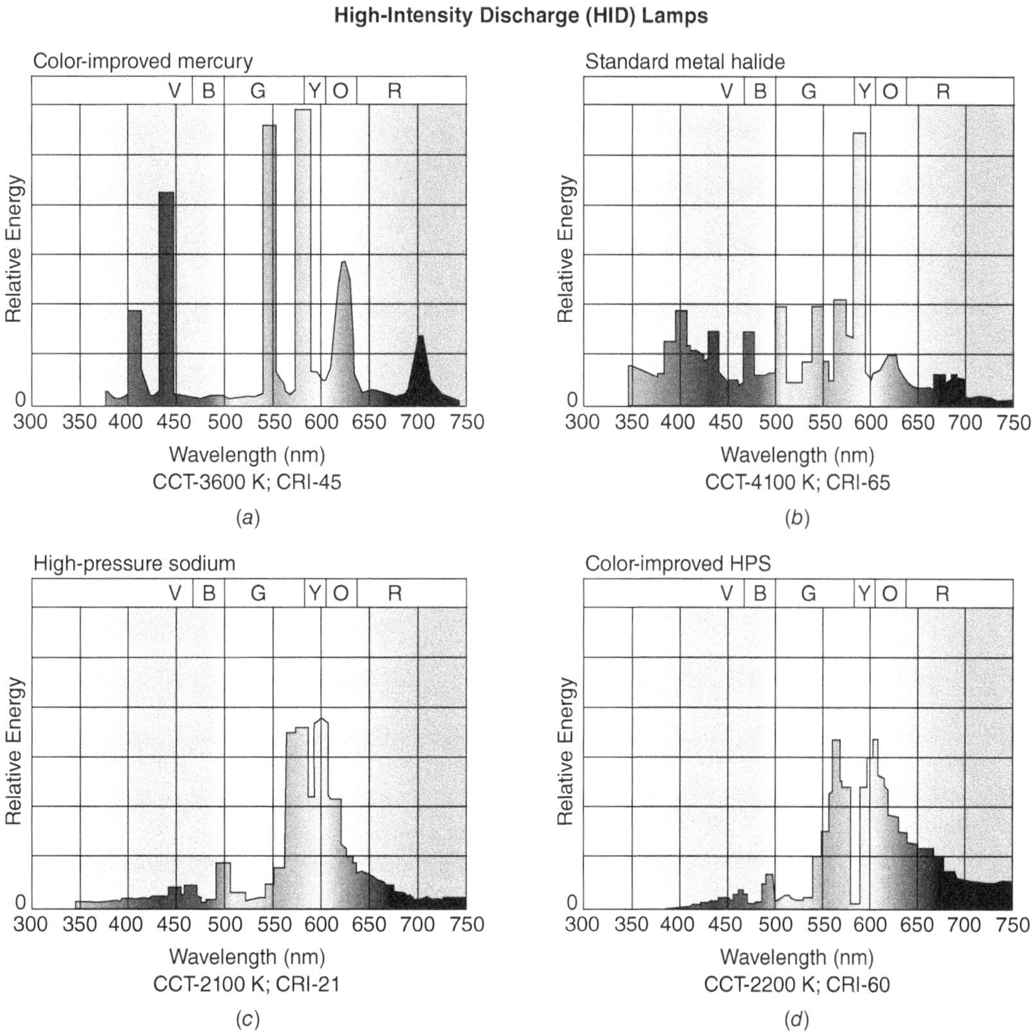

Figure 7.27 *Spectral energy distribution of typical high-intensity discharge (HID) lamps with their CCT and CRI. These graphs are generic and show patterns of wavelength distribution; consult manufacturers' data for the output spectrum of a specific lamp.*

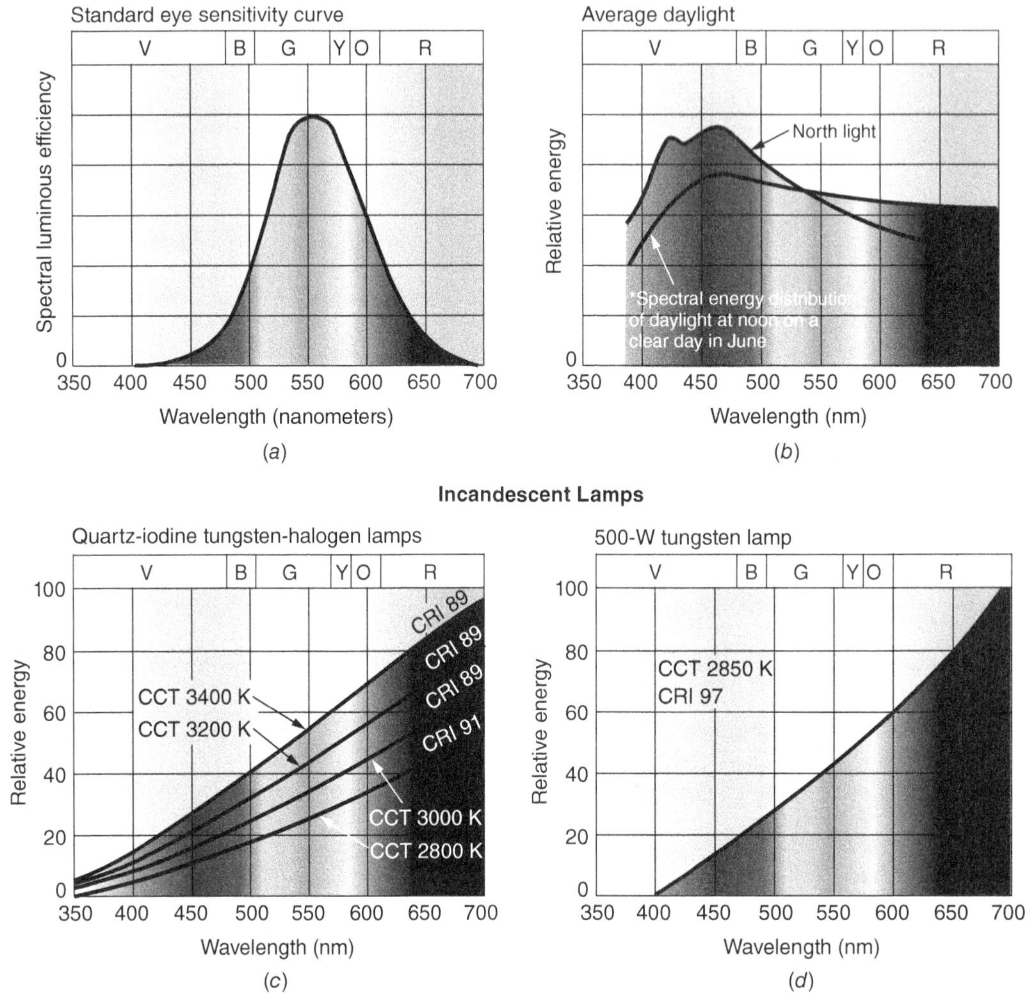

Figure 7.28 *(a) Standard photopic (cone-based) eye sensitivity curve. Note that maximum sensitivity occurs in the daylight range of 500–750 nm. (b) Spectral energy distribution of two specific types of daylight. North light, with a color temperature of 8000–10,000 K, peaks in the blue range, whereas noon daylight contains all spectral colors in roughly equal proportions. (c) Tungsten-halogen lamps are incandescent sources and therefore contain all spectral colors. As wattages increase, the color changes from orange-red to white, and the CRI drops slightly. (d) A simple filament-type incandescent lamp is very close to being a blackbody radiator (i.e., its actual temperature and CCT are almost the same). This accounts for its high CRI (97).*

a sense of the importance of color names in day-to-day commerce. As with light source color, the use of names is informal and nontechnical. On the other hand, when proprietary specifications are allowable as part of the project contract documents, there is no easier way to get the precise color of paint or finish intended than by product name. Names become more of a challenge when open bidding across manufacturers is required or desired.

In theory, chromaticity coordinates could be used as an accurate means of specifying object color—with light leaving an object being captured by a meter. This is not commonly done in architectural design.

(c) Object-Light Interactions

Light source color, discussed above, is a property of the source and is independent of the environment in which the source is used. Object color is not an independent property of an object—but represents the effect of a light source interacting with an object. The effect of light source on object color is well within the realm of architectural concern—which makes studio presentations of materials palettes with no reference to a lighting system problematic. We see only what light permits us to see.

In addition to providing sufficient illuminance and minimizing glare potential, a lighting designer must be concerned with the spectral content of light sources, because the object color experienced by an observer depends heavily upon the illuminant. The object color experienced by a person is the result of selective absorption and reflection of components of the illuminating light by the pigments of the object being viewed. An illuminant must contain the inherent color of the

TABLE 7.4 Effect of Illuminant on Object Colors

Lamp	CRI (Approximate)	CCT (K)	Whiteness	Colors Enhanced	Colors Grayed	Notes
Daylight						
Daylight	99+	5000		All colors		
Fluorescent						
Warm white	52	3050	Yellowish	Orange, yellow	Red, blue, green	
Cool white	62	4200	White	Yellow orange blue	Red	
Cool white deluxe	77	4050	White	Green orange yellow	Red	
Triphosphor	75	2800	Yellowish	Red orange	Deep red, blue	
	80	3000	Pale yellowish	Red orange green	Deep red	
High-pressure sodium						
Standard	21	2100	Yellowish	Yellow green	Red, blue	
Color-corrected	60	2200	Yellowish-white	Red green yellow	Blue	CRI decreases slightly over life
Incandescent						
Incandescent	99+	2900	Yellowish	Red, orange, yellow	Blue, green	
LED						
High-CRI LED	99	4990		All colors	Blue	
Mid-CRI LED	80	5000		Orange, green, blue	Red	
RGB LED	73	5090				
Mercury						
Clear	20	7000	Blue–green	Blue green	Red orange	Poor overall color rendering
Deluxe	45	3700	Pale purplish	Deep blue, red	Blue–green	Shift over life to greenish
Metal–halide						
Clear	65	4000	White	Blue green yellow	Red	May shift to pinkish over life
Clear	65	4000	White	Blue green yellow	Red	May shift to pinkish over life

object for us to see the object's actual color (let's say "actual" is the color that would be seen under daylight). In addition, the relative energy of an illuminant at a particular wavelength determines the saturation and brilliance with which we see a color.

Most architectural light sources emit several component color illuminants. When such composite light—for example, white light—falls on a surface other than black or white, selective absorption occurs. The component colors of the light are absorbed in different proportions so that the light reflected or transmitted is composed of a new combination of the same colors that first impinged on the surface. Thus, a white light reflected from a red wall acquires a red tint because the component colors of the white light other than red were absorbed in greater proportion than the red. When reflected, the red light takes prominence, thus giving the reflected light a red tint. This is illustrated in Figure 7.29*a*. Similarly, a white light passed through a piece of red glass emerges as a reddish light because the other spectral components are absorbed in much greater proportions than the red. This phenomenon is illustrated in Figure 7.29*b*. An individual object's pigmentation, by absorbing other colors of light, reflects or transmits to the eye its own hue in a greater concentration than existed in the incident light.

The color of the illuminant and the inherent color of the objects within a space constitute important facets of lighting quality. These two design variables, however, cannot be considered separately. The color of an object is its ability to modify the color of light incident on it by selective absorption—the color reflected or transmitted by the object is experienced by the eye as the color of the object. An object, therefore, is technically said to be "colorless" (not transparent) when it does not exhibit selective absorption—that is, when an object reflects and absorbs the various components of incident light nonselectively. Thus, white, black, and all shades of gray are colorless, neutral, achromatic—or more precisely, lacking in hue.

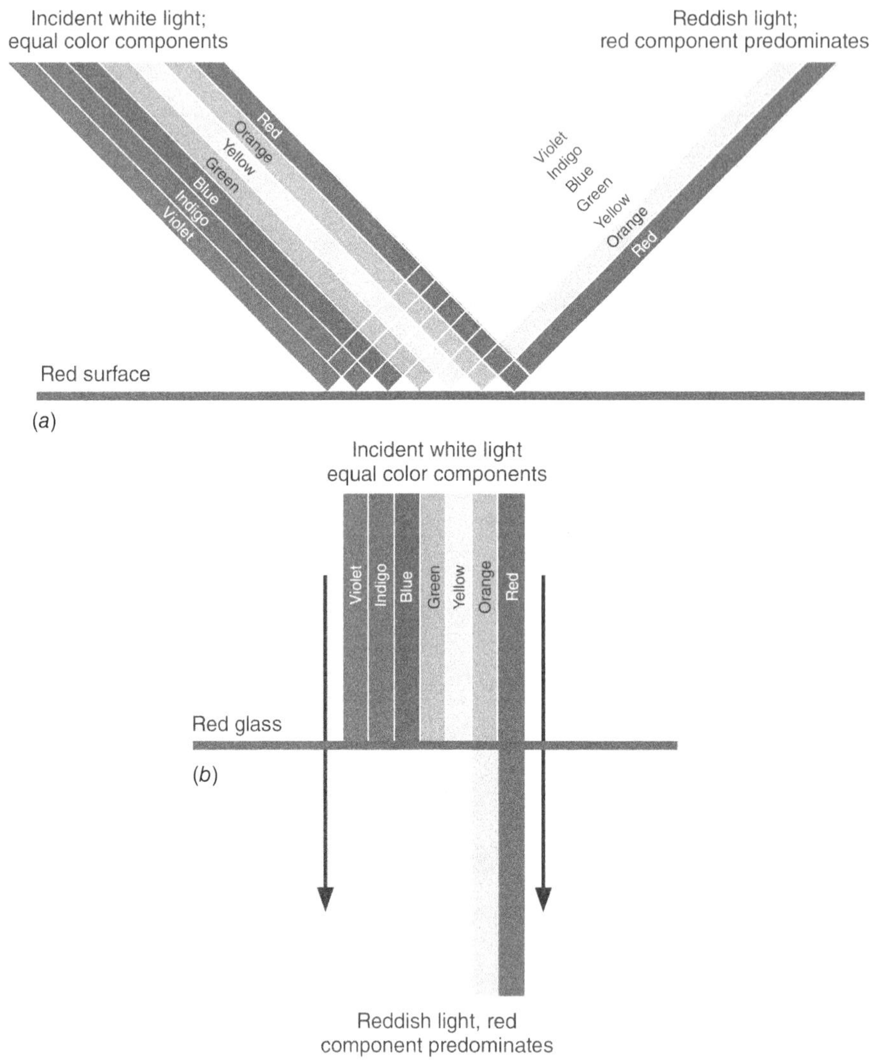

Figure 7.29 *Selective absorption of materials relative to (a) reflected light and (b) transmitted light.*

(d) Classification Schemes

Hue is defined as that attribute by which we recognize and therefore describe colors as red, yellow, green, blue, and so on. Just as it is possible to form a gradient from white to black with intermediate grays, it is possible to do the same with a hue.

The difference between the resultant colors of the same hue, when arranged in this manner, is called brilliance or value. White is the most brilliant of the neutral colors, and black the least; pink is a more brilliant red hue than ruby; and golden yellow is a more brilliant (lighter) yellow hue than raw umber.

Colors of the same hue and brilliance may still differ from each other in saturation, which is an indication of the vividness of the hue, or the difference of the color from gray. Thus, pure gray (black plus white) has no hue; as color is added, the saturation changes without changing the brilliance. The three characteristics, then, that define a particular color are: hue, brilliance, and saturation. Using these terms, "bay" can be defined as a color red–yellow in hue, of low brilliance and low saturation; whereas "carmine" can be defined as a color red in hue, of low brilliance and very high saturation.

Various systems of color classification with a focus on object color have been devised. The Munsell Color System and the Ostwald Color System are two that are commonly encountered. In the widely used Munsell Color System (Figure 7.30), brilliance is referred to as value and saturation as chroma; thus, a color is defined by hue, value, and chroma. The brilliance (value) of a pigment or coloration is related to its reflectance to white light—that is, the higher the brilliance or value, the higher the reflectance (as might be expected when one considers that white and black are the poles of brilliance). Chroma or saturation may be thought of as either the difference from gray or the purity of the

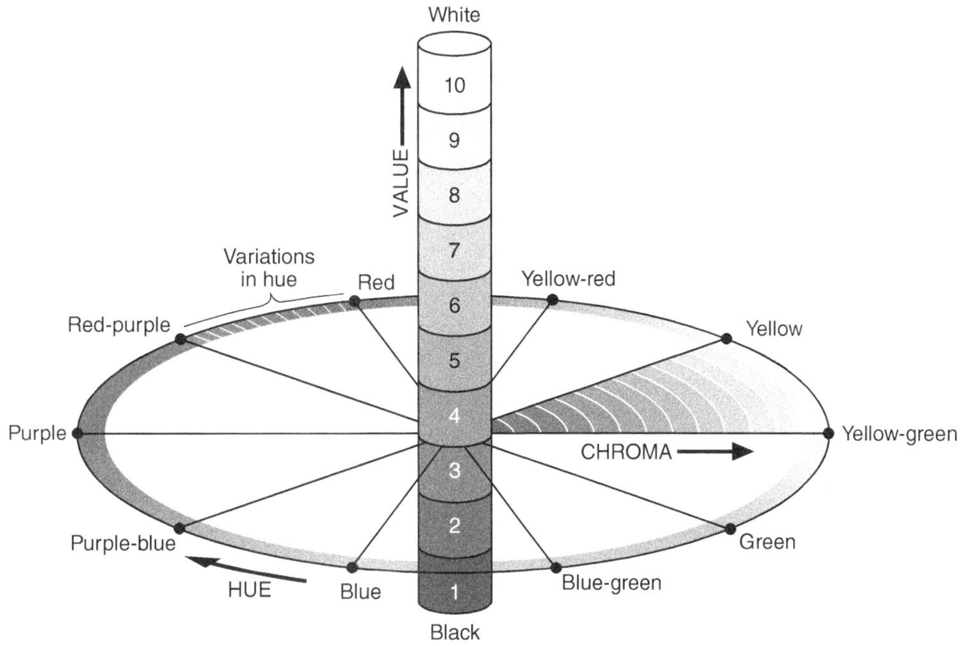

Figure 7.30 *The Munsell Color System defines a color by three characteristics: hue (color), chroma (saturation), and value (grayness).*

color. Spectral colors have 100% purity and therefore maximum chroma.

(e) Mixing Colors

When white is added to a pigment, it produces a tint; adding black produces a shade. When pigments are mixed to produce a particular color, the color that results is created by a subtractive process. Each pigment absorbs certain wavelengths of the illuminating light. With a mix of pigments, these absorptions combine to subtract various colors from the reflected light—leaving only those colors that will constitute the hue, value, and chroma of the pigment. This subtractive effect is also utilized when producing colors by filtering white light. Each filter selectively absorbs component colors, transmitting only the desired component. Thus, a red filter transmits only red, and so on (see Figure 7.29*b*). Conversely, when lights of the three primary colors—red, green, and blue—are combined, they form white by an additive process (Figure 7.31).

The additive and subtractive primary colors are complementary; they combine to give a white or neutral gray, respectively. Red and blue–green, blue and yellow, and green and magenta are complementary. Therefore, if a red object is illuminated with blue–green light, the object's color appears gray because the red pigment absorbs the blue–green and reflects nothing—hence, gray.

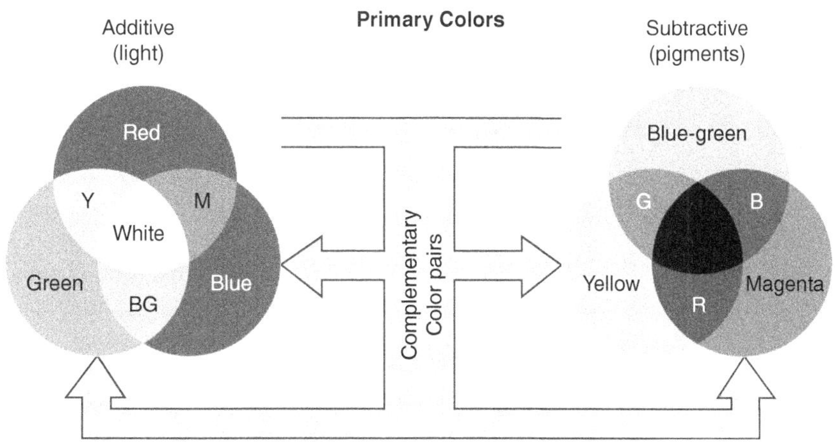

Figure 7.31 *Primary and complementary colors. Complementary color pairs are shown by arrows. Pigments form color by a subtractive process; colored lights form colors by an additive process.*

7.4 VISUAL COMFORT DETERMINANTS

(a) Illuminance

Illuminance, the measure of the amount of light incident on a surface per unit area, directly affects the lighting condition of an object or space by determining visibility, color perception, and the ability to discern fine details.

Chapter 18 offers design phase guidance regarding light delivery, light distribution, and illuminance targets for indoor spaces, including recommended illuminance categories from the Illuminating Engineering Society (IES).

(b) Glare

Excessive luminances and/or excessive luminance ratios in the field of vision are commonly referred to as glare. Glare, however, is a human perception—an opinion—regarding a particular aspect of the luminous environment. Glare is the potential subjective result of an objective condition (excessive luminance). Glare cannot be measured with any device; it can only be declared by an observer. As with many aspects of building design, we use quantifiable values during analysis as a surrogate for qualitative responses by users. Thus, luminance is a good predictor of glare.

Glare that is caused by a light source seen by the users of a space is termed direct glare; glare that is caused by the reflection of a light source in a viewed surface is known as reflected glare or, when the surface involves a visual task, as a veiling reflection (see Figure 7.32). The severity of glare is usually expressed through the classifications of discomfort, disability, and blinding glare. Blinding glare, with effects as per its name, should never be encountered in a building. Disability glare will measurably decrease visual acuity and should be absent from most buildings. Discomfort glare will not measurably decrease visual acuity, but does enter the consciousness or subconsciousness of an occupant—with a range of potential negative consequences. Factors affecting the severity of glare are the adaptation level of the eyes, the apprehended size of the glare source, luminance ratios, room size, surface finishes, and the size and location of lighting fixtures and windows. Light sources in the far-field and peripheral-vision areas beyond the central 90° cone are less troublesome as glare sources.

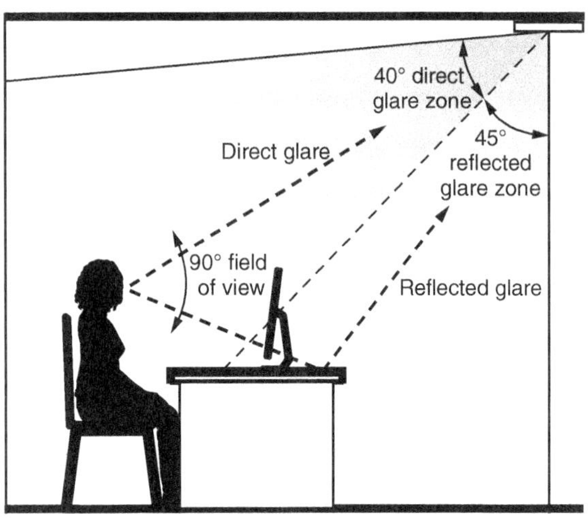

Figure 7.32 *Glare zones. Direct glare assumes a head-up position, whereas reflected glare assumes eyes down at a reading angle. (Drawn by Martin Lee.)*

Direct Glare. The factors that contribute to the perception of direct glare are the luminance, size, and position of a source within the field of vision, plus the adaptation of the eye. The discomfort resulting from direct glare stems from two conditions. First, the eye adapts rapidly to the average brightness of the overall visual scene; second, the eye is attracted to the highest luminance in that scene. (This fact is used effectively in merchandising displays.) Thus, if an area of high brightness, such as a window or a light fixture, is in the visual scene, and we are looking at an area of lower brightness, such as a work task, three visually disturbing things occur:

1. The eye adapts to a higher luminance level, thus effectively reducing the subjective brightness of the task—essentially making it harder to see what we are looking at (see Figure 7.22). This can be experienced by alternately blocking and unblocking a direct glare source while trying to perform a moderately difficult visual task—there will be an immediate improvement in visibility when the glare source is obscured.
2. The eye is drawn simultaneously in two directions: involuntarily to the source of high luminance, and intentionally to the object we are looking at. The resulting tension causes considerable visual discomfort.
3. Adaptation is continuously varying as the eye is drawn to the glare source and away again.

Glare potential is proportional to source luminance and its apprehended solid angle. Therefore, a small but bright source is usually not a visual problem, whereas a large, low-brightness source (such as a highly illuminated ceiling) may be. A small, bright source may add sparkle to the field of vision, which many observers may find to be a pleasant addition to a monotonous lighting environment. Although discomfort glare from a scene is cumulative (multiple sources can contribute), luminance of the sources is more important than the number of sources. For instance, if the luminance of several sources is halved, the reduction in glare potential is greater than is achieved by reducing the number of sources by half. In fact, the latter has little effect on discomfort glare.

Factors 2 and 3 are less evident. Glare potential decreases rapidly as a bright source is moved away from the direct line of vision; it depends on the source position in the field of view. The likelihood of direct discomfort glare from a source is inversely proportional to the background luminance (which establishes eye adaptation level). Thus, a ceiling fixture with a luminance of 4000 cd/m^2 [1167 fL] at a 65° view angle would likely constitute a source of discomfort glare to a user with eyes adapted to 150 cd/m^2 [44 fL]. The same fixture would not be objectionable in a daylight condition, where the eye might be adapted to 1500 cd/m^2 [438 fL]. A common example is automobile headlights, which at night present such a severe contrast as to constitute disabling or blinding glare—whereas in daylight, with its high eye adaptation level, headlights, although noticeable, are not usually disturbing.

Considering the dependence of direct glare on eye adaptation, a useful design recommendation is that the luminance of large sources should not exceed 2500 cd/m^2 [730 fL] and that of small sources should not exceed 7500 cd/m^2 [2190 fL]. The first value is roughly the luminance of blue sky; the second is roughly that of a fluorescent lamp. The terms "large" and "small" are relative—depending on the source distance from the observer as well as the actual dimensions of the source. The determining characteristic is apprehended size, or subtended visual angle, as shown in Figure 7.33.

The statistically likely cumulative effect of several direct glare sources (in an interior space) can be quantified by visual comfort probability (VCP). VCP is defined as the percentage of normal-vision observers who would not express direct-glare discomfort in that specific visual environment. IES has established (IES, 1966) a set of standard conditions for which VCP of sources can be calculated. These include a 1000-lx [93-fc] illuminance, representative room dimensions, fixture height, observer position, and a head-up field of view limited to 53° above and directly forward from the observer (Figure 7.33). The unified glare rating (UGR) is a International Commission on Illumination (CIE) design index that can be used for similar purposes.

Direct glare should not be a problem if all three of these conditions are satisfied:

1. The VCP is 70 or more.
2. The ratio of maximum-to-average luminaire luminance does not exceed 5:1 (preferably 3:1) at 45°, 55°, 65°, 75°, and 85° from the nadir, crosswise, and lengthwise.
3. Maximum luminaire luminances crosswise and lengthwise do not exceed the following:

Angle Above Nadir (degrees)	Maximum Luminance cd/m^2	fL
45	7710	2250
55	5500	1605
65	3860	1126
75	2570	750
85	1695	495

A typical set of manufacturer's luminance and VCP data is shown in Figure 7.34 for a ceiling-mounted fluorescent fixture with four 40WT12 lamps. Note that all VCP values are considerably above the minimum criterion of 70. If full VCP data of this type are not

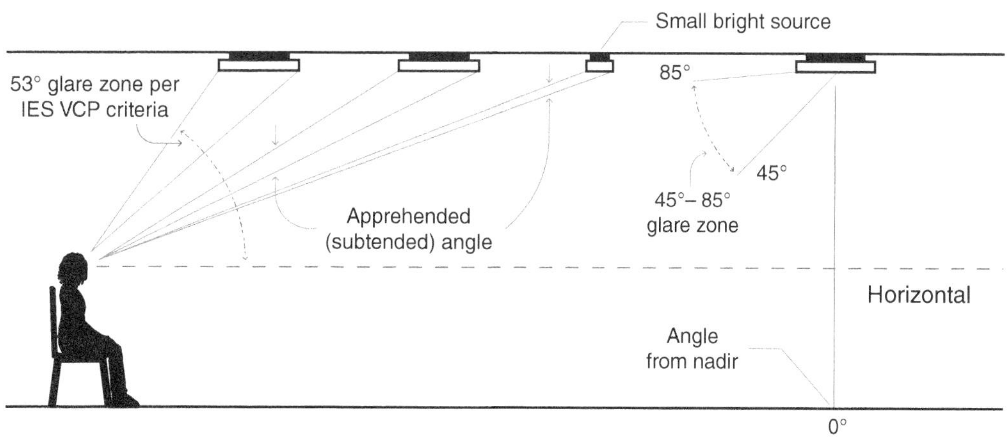

Figure 7.33 Evaluation of direct glare potential. The glare contribution of each source depends on its size (subtended solid angle), luminance, and location in the field of view. Note that the solid angle of a small source is such that even with high luminance, it is not generally objectionable. Such sources are normally called sparkle. Glare is much more likely when a bright source is seen against a darker background; therefore, light-colored ceilings and upper walls are recommended. (Drawn by Martin Lee.)

4–40-W Lamps
Prismatic Lens Diffuser

Average Luminance Data cd/m²

Vertical Angles	Across Axes	Along Axes
60°	2000	1750
65°	1060	1075
70°	500	560
75°	410	380
80°	480	343
85°	560	420

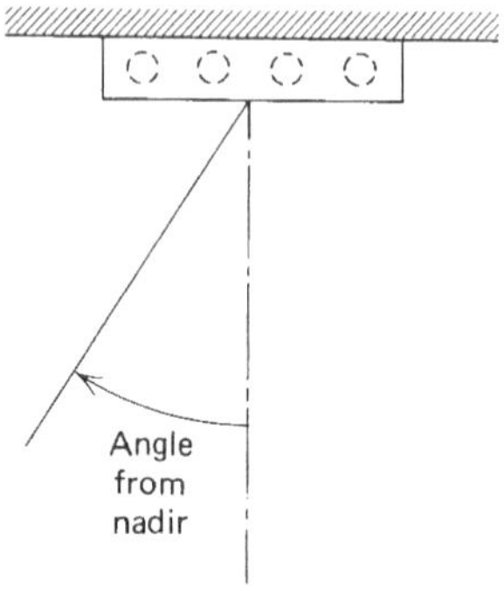

Reflectances:
Wall 50%
Ceiling cavity 80%
Floor cavity 20%
Work plane illumination: 1000 lux

IES Visual Comfort Probability Data

Room Size W (ft)	L (ft)	Luminaires Lengthwise, Ceiling Height (in ft) 8.5	10.0	13.0	Luminaires Crosswise, Ceiling Height (in ft) 8.5	10.0	13.0
20 ×	20	80	77	76	79	75	72
	30	80	78	76	79	76	73
	40	82	79	77	79	77	74
	60	80	80	78	79	77	75
30 ×	20	84	80	77	82	78	74
	30	83	80	77	81	79	74
	40	82	80	78	81	79	75
	60	82	80	78	80	79	75
	80	82	80	78	80	78	76

Figure 7.34 *A typical set of a manufacturer's published VCP and luminance data.*

available, they can be calculated with almost any lighting calculation program, given the luminaire luminance data.

Despite the usefulness of the VCP criterion in designing to reduce glare potential, it is inherently limited by its standard conditions, which are not easily applied to other situations. Thus:

- In small spaces, VCP has little significance.
- Tabulated VCP figures are given for the worst viewing position in the room. Because VCP varies dramatically with observer position, the VCP values given are always lower than the space's average VCP. (Then again, no individual space user can experience the average VCP. This issue speaks directly to how design criteria for buildings are established.)

Given these considerations (which tend to make the direct glare situation better in a built space than predicted by VCP calculations), a proposed lighting design with a VCP somewhat below 70 should not be immediately discarded. This is not a recommendation to accept bad design solutions, but rather to evaluate designs holistically using well-developed owner's project requirements that appropriately balance intended outcomes.

Reflected Glare. No formal definition distinguishes reflected glare from veiling reflection. Both terms refer to the introduction of luminance to the eye through the action of a reflection. Herein, veiling reflection will be used to describe a situation where a source reflection decreases the visibility of a specific task (such as reading a computer screen). Reflected glare will be used to describe the broader situation.

Designing to control reflected glare is a more complex problem than that of direct glare control. Reflected glare involves both a primary source (say, a light fixture) and a secondary source (such as a computer screen), which are connected by sometimes complex and varying geometry (Figure 7.35). Vision is made possible by light reflected from the majority of the objects we see. If a mirror replaced the object being viewed, we would see the source of illuminating light clearly (Figure 7.35*a*).

In commercial/institutional spaces, there are usually one or more lighting fixtures near the observer that furnish most of the light required to see. These principal sources are the main contributors to reflected

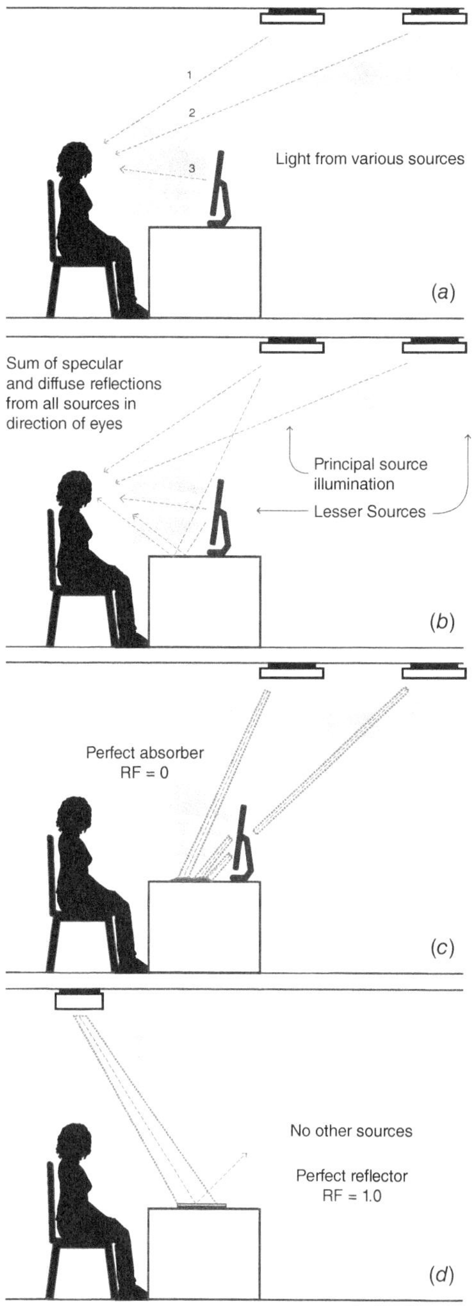

Figure 7.35 *(a) The nature of the seeing process requires that light from a source or sources be reflected by the task into the eye. (b) The light entering the eye is the sum of all the reflected light, specular and diffuse, from all sources in the direction of the eye. If the task is specular, all the sources will be seen reflected in the task. (c) A perfectly absorptive object is jet black because it reflects nothing. (d) A perfectly reflective object positioned as shown is also seen as black because geometrically it cannot reflect light into the eyes. (Drawn by Martin Lee.)*

glare. Other, more remote fixtures in the room are lesser sources (Figure 7.35*b*).

Reflected glare exists to the extent that light sources can be seen in the task being viewed. The reflection characteristics of the object being viewed are important to an understanding of this type of glare. If an object is perfectly absorbent—with a reflectance of 0%—it would appear completely black, as no light would be reflected into the eye (Figure 7.35*c*). If the object is perfectly specular, like a clean mirror, and there is no light source within the geometry of reflection, it too would appear black (Figure 7.35*d*). Reflected glare is due to task surface specularity, whereas object definition (the ability to see the task itself) is due to task surface diffuseness. Veiling reflections, which are caused by the mirroring of a source in the task, are proportional to source luminance and substantially independent of illuminance—the brighter the source, the more troublesome its reflection.

Glare sources within the geometry of reflected vision are shown in Figure 7.36, and the effects are shown in Figure 7.37. Figure 7.36 shows that although large sources are difficult to avoid, small sources can usually be easily avoided by a small change in the source-task-eye geometry (for instance, by moving the head or tilting the task). Table 7.5 lists a few sample reflectance figures to demonstrate that most materials exhibit both specular and diffuse reflectance. In Figure 7.36, it is important to note that most visual work is done in the zone of 20° to 40° from the vertical, below the eye, with a maximum at the 25° reading angle (Figure 7.38).

Contrast Reduction. The principal effect of the reflection of a light source in a visual object is a reduction in contrast between the object and its background—and hence, a reduction in the object's visibility. It is as if a bright veil were spread over the object being viewed, which accounts for the term veiling reflection. As the angle of the incident light approaches the viewing angle, the specularly reflected component of the light becomes more and more pronounced, and task contrast drops. This is illustrated in Figures 7.37 and 7.38. The worst situation occurs when the incident angle equals the viewing angle. When the specular reflectance of the task and background is high, as with a glass computer or tablet screen, an image of the source is superimposed on the object, making viewing impossible (Figure 7.38). Even with the highly specular finish of "slick" magazine paper, however, vision is still possible because of the very high contrast between black ink and white paper, although with much reduced clarity and considerable annoyance.

When considering specular and diffuse reflectance, the relationship for contrast given in Equation 7.7 must be rewritten as:

$$C = \frac{(L_{BD} + L_{BS}) - (L_{TD} + L_{TS})}{L_{BD} + L_{BS}} \qquad \text{(Eq. 7.9)}$$

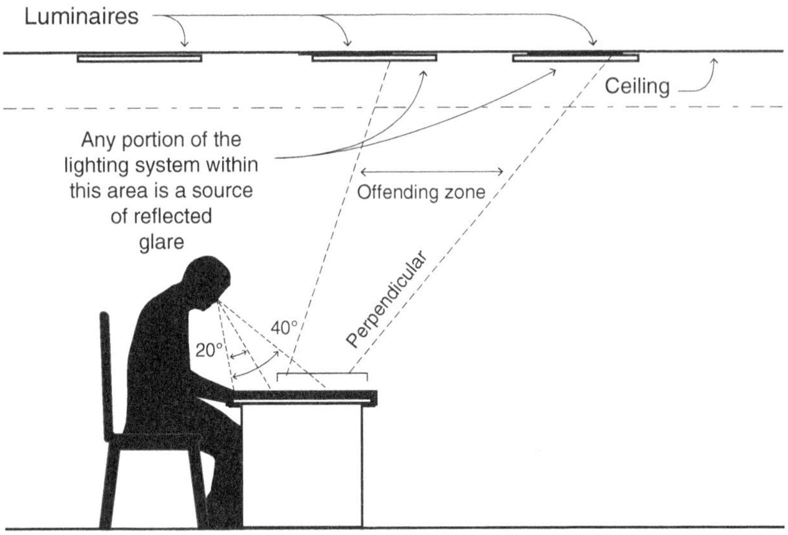

Figure 7.36 Geometry of reflected glare. Because normal desktop, head-down viewing angles vary from 20° to 40° from the vertical, the offending zone (wherein light sources are likely to be reflected from tasks) is the area on the ceiling corresponding to specular reflection between these two angles. Note that the higher the ceiling, the larger this area becomes. Also note that the offending zone will move as the table tilt changes. (Drawn by Martin Lee.)

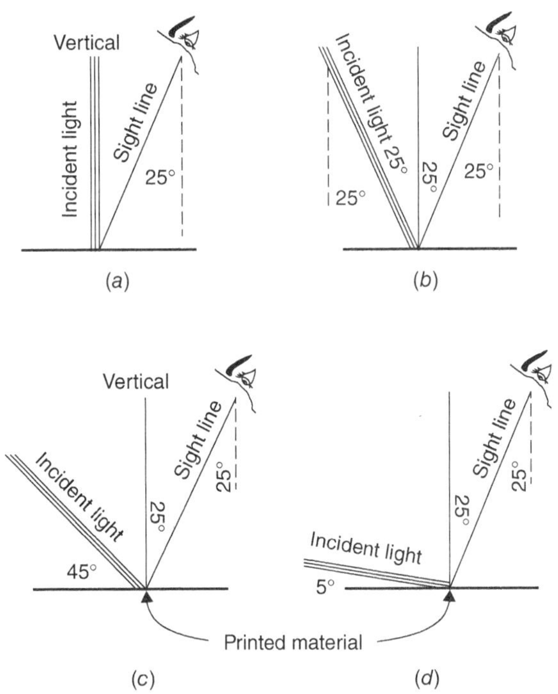

Figure 7.37 The usual viewing angle to a horizontal surface is between 20° and 40° from the vertical—25° is the most common viewing angle. With vertical incident light on a diffuse surface (such as a textbook page) (a), the print can appear dark and clear. When the angle of light incidence is equal to the viewing angle (b), a mirrorlike reflection can make the page illegible. Even with a diffuse paper, the print is light at best and almost invisible at worst. As the angle of incidence becomes larger (c), reflected glare decreases. When the incident light is at a very low angle (d), there is little reflected glare, and the print appears lighter.

TABLE 7.5 Typical Reflectances

	Reflectance	
Material	**Specular**	**Diffuse**
Matte black paper	0.0005	0.04
Matte white paper	0.0030	0.77
Newspaper	0.0065	0.68
Very glossy white photo paper	0.048	0.83
Metallic paper—copper	0.11	0.28
Dull black ink	0.006	0.045
Super gloss black ink	0.039	0.016

Source: Courtesy of the Illuminating Engineering Society (IES).

where L_B and L_T are background and task luminances caused by diffuse (D) and specular (S) reflectance; that is, L_{BS} is the background luminance due to specular reflectance, and so forth. Reworking the calculation of contrast by including specularity, the result is very different.

EXAMPLE 7.4 Assume an interior space lit to an average illuminance of 75 fc [750 lx] using bare bulb fluorescent fixtures (with luminance = 7000 cd/m^2 [2000 fL]). The task is drawing with ink on paper. Reflectances are:

	Specular	**Diffuse**
Ink	0.021	0.038
Paper	0.018	0.71

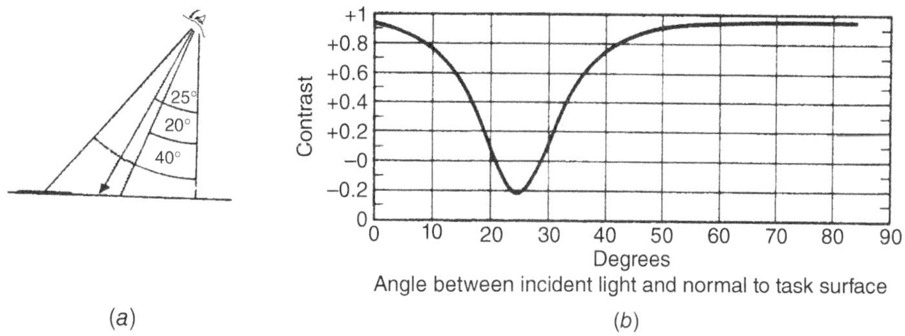

Figure 7.38 *(a) Typical viewing angles for a task on a horizontal surface. (b) Graph showing contrast reduction of a task with a specular background (such as a sheet of glossy paper) as a function of the angle between incident light and a line normal to the task surface. Viewing angle is assumed to be 25° from normal. Note that between 22° and 27° the contrast is negative. This indicates that background luminance exceeds that of the task, making the task essentially invisible. What will be visible is a reflected (mirrored) image of the source.*

Calculate the task contrast without and with reflection of the 2000 fL source on the work.

SOLUTION

1. Without specularity, using Equation 7.7 for diffuse reflection only:

$$C = \frac{R_B - R_T}{R_B} = \frac{0.71 - 0.038}{0.71} = 0.947$$

2. With specularity, using Equation 7.9,

$$C = \frac{(L_{BD} + L_{BS}) - (L_{TD} + L_{TS})}{L_{BD} + L_{BS}}$$

where

$$L_{BD} = 75\ \text{fc} \times 0.71 = 53.25$$

$$L_{BS} = 2000\ \text{fL} \times 0.018 = 36.0$$

$$L_{TD} = 75\ \text{fc} \times 0.038 = 2.85$$

$$L_{TS} = 2000\ \text{fL} \times 0.021 = 42.0$$

and

$$C = \frac{(53.25 + 36) - (2.85 + 42)}{53.25 + 36} = 0.497$$

which is roughly half of the previous contrast. If the contrast is normalized to the maximum contrast (as is usually done), the contrast reduction R can be expressed as:

$$R = 1 - \frac{C}{C_{MAX}} \qquad \text{(Eq. 7.10)}$$

Thus, in this case, contrast reduction would be:

$$R = 1 - \frac{0.497}{0.947} = 0.47 \quad \text{or} \quad 47\%$$

■

A similar calculation for black printed text on good white bond paper yields a contrast reduction from 94% to 77%, or $R = 17\%$. This suggests that with high task-to-background contrast, effective seeing is possible almost regardless of the lighting condition. This does not, however, relieve the lighting designer of the responsibility to provide a comfortable and effective lighting environment that minimizes veiling reflections. In general, a contrast reduction of more than 15% is undesirable.

Because both specular and diffuse reflectances frequently vary with the angle of view, and exact values are rarely available, accurate calculations are difficult. If a lighting system exists or a mock-up can be made, measurements of contrast reduction can be made accurately with a contrast/luminance meter of the type shown in Figure 7.39. In such an experiment, a standard contrast device that corresponds to a normal office task (e.g., reading black typeface on a white paper background) is positioned on the work surface, and exposed to ambient illumination. The task contrast is measured at the same angle at which it would normally be viewed. Contrast reduction is automatically calculated and displayed.

Equivalent Spherical Illumination. Another approach to addressing contrast reduction is to define a reference lighting system that is effectively free of veiling reflections, and compare an actual lighting system to this "ideal" system. This is the basis of the concept of equivalent spherical illumination (ESI).

A lighting system almost free of reflected glare will emulate daylight from a fully overcast sky. Figure 7.40 shows an integrating sphere that simulates such a condition. Light is introduced from the outside, split by a deflector, and evenly distributed throughout the sphere by the multiple reflections from the white-painted walls. The result is an evenly illuminated volume. When a task is introduced, the illumination falling on it is entirely uniform; there are no high-luminance sources reflected in it. The task is said to be spherically illuminated. The extent to which another illumination system can duplicate

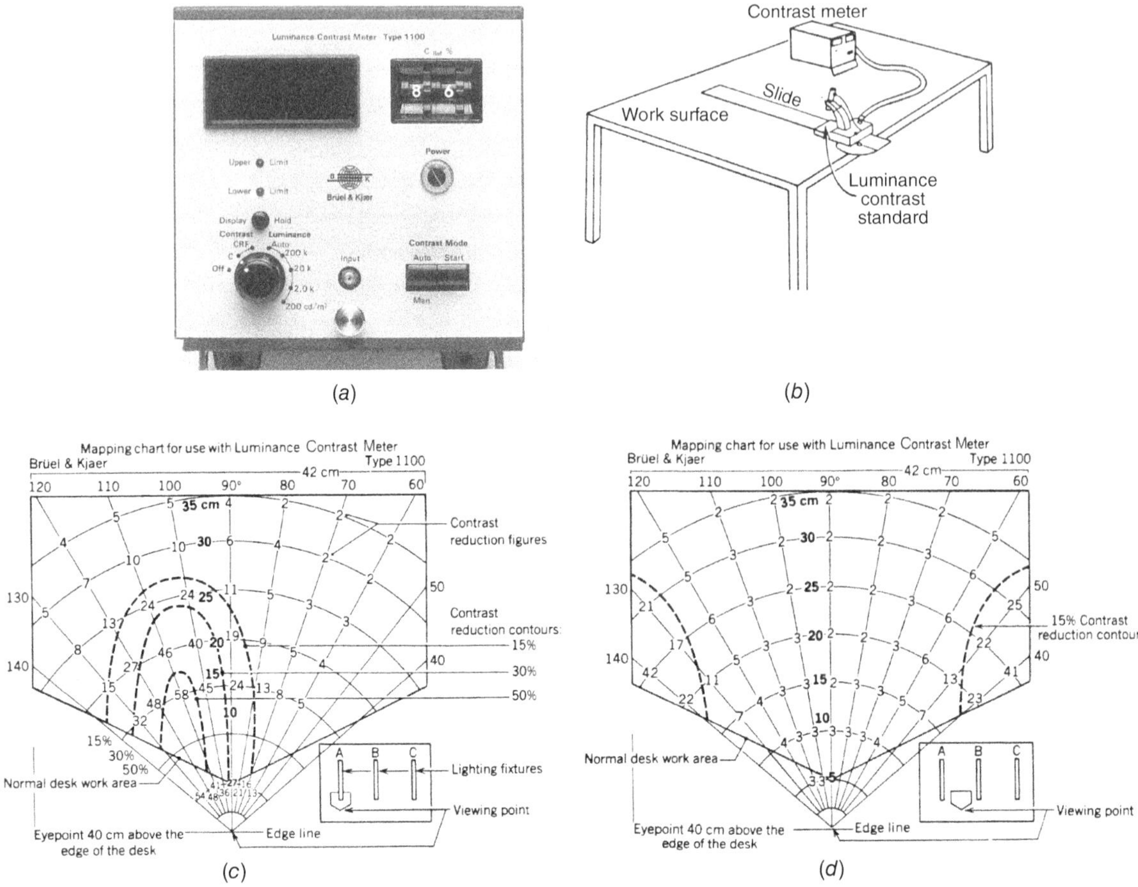

Figure 7.39 *Precision meter for measuring luminance and contrast. (a) Able to accommodate up to four sensor heads, the luminance contrast meter is suitable for measuring the luminance, luminance ratio, contrast, and contrast reduction. (b) A typical assembly will include a light source, a luminance lens attached to the sensor head, and a luminance contrast meter. Measurement outputs can be adjusted utilizing the touchscreen on the front of the meter. (c) With a lighting fixture directly in the offending zone—that is, above and in front of the viewer—a severe loss of contrast occurs over much of the work surface. (d) By shifting the relative position of the viewer and the source so that there is no source in the offending zone, contrast reduction is held to 3% to 4% over most of the work surface. (a: Courtesy of Gamma Scientific; b, c, d: Courtesy of Brüel & Kjær.)*

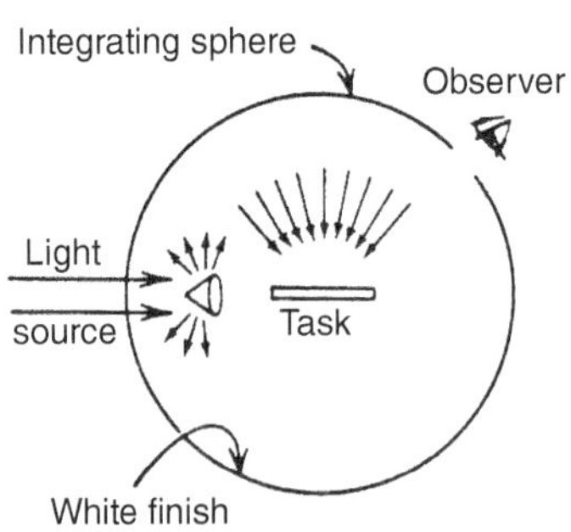

Figure 7.40 *Spherical illumination is produced by illuminating an object by diffuse reflection from the inside walls of an integrating sphere. The light source and observer are normally external.*

this glare-free environment is that system's ESI, representing the portion of its total illumination that is spherical—that is, diffuse and glare-free. ESI is determined by comparing the contrast rendition produced by the spherical and alternative systems.

An investigation of school lighting options (Figure 7.41) provided ESI results for four viewing positions in a classroom lit with ceiling-mounted continuous rows of 2-ft × 4-ft [0.6 by 0.6 m], 4-lamp, 40-W fluorescent fixtures with lens-type wraparound diffusers on 10-ft [3.0 m] centers. Note that:

- ESI depends entirely on the viewing position and viewing angle, other factors in the space being equal.
- In an ostensibly well-lit (215 fc [2314 lx]) position (M1), the glare-free illuminance is only 28 fc [301 lx]. This does not mean that visual work in this position is impossible. It does mean that in position M1, a pronounced veiling reflection exists on all specular objects. (Because of the size, orientation, and location of the glare source, this reflected glare is difficult to avoid.) It further means that a large amount of energy is being used (effectively wasted) to produce negative results.

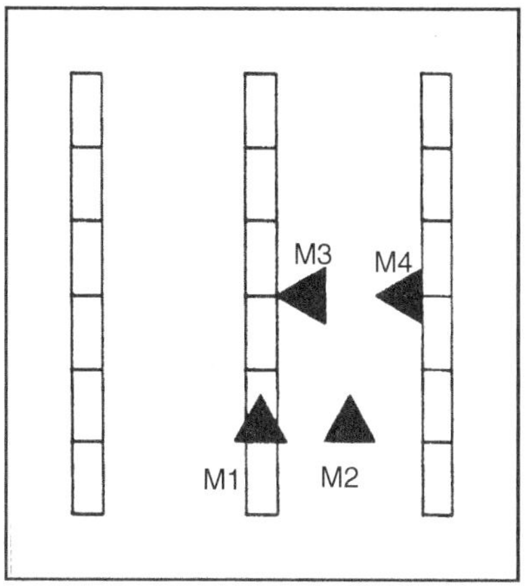

		Observer Position			
		M1	*M2*	*M3*	*M4*
TI	2L	108	92	125	118
	4L	215	185	250	235
CRF	2L	0.75	1.00	0.82	1.01
	4L	0.76	1.00	0.83	1.03
ESI	2L	17.8	91.9	31.5	127.8
	4L	28.4	185.3	58.1	308.3

TI–Task illuminance
2L–2 lamps (inside pair)
4L–4 lamps
CRF–Contrast rendition factors

Figure 7.41 *A test classroom illuminated by three widely spaced rows of four-lamp fixtures with lens-type wraparound diffusers. Observer positions are shown by arrows. The row of fixtures in front of position M4 is too far forward to be in the offending zone. (Sampson, 1970.)*

These results can be qualitatively anticipated by an examination of the observer positions in Figure 7.40. Positions M1 and M3 have bright sources in the offending zone—M1 more so than M3. M2 is an excellent position that receives light contributions from the two sides, its illuminance being lower than the others due to wide fixture row spacing. M4 is ideally placed; no glare sources are in the offending zone, and a row of fixtures is positioned behind it, which makes it geometrically impossible for the fixtures to act as a glare source. The ESI analysis gives quantitative value to qualitative judgment, which makes it a valuable design tool. Note particularly that the ESI results shown in Figure 7.41 correspond to the results of a similar test made with the contrast meter, as shown in the charts of Figure 7.39.

As with VCP criteria for direct glare, there are ameliorating factors with ESI that generally make a given lighting system better than these values would indicate. Some of these factors are:

- ESI is critically dependent on observer position and viewing angle. Although position is generally fixed by chair location, observers can and do change their viewing angle and head aspect to correct for glare situations.
- The nature of the task (i.e., its specularity) is assumed to be fixed and unique. In some situations, the task varies, and thus also the contrast. When tasks are constant, severe veiling reflections frequently lead to measures being taken that improve the task, the lighting, or both. This should not be taken as license to design systems that users will need to fix.
- The lighting distribution characteristic of the fixture involved is a critical factor in glare production. The characteristic of a wraparound lens diffuser is such that considerable light falls in the glare zone. Other diffuser characteristics yield different results.

The concept and use of ESI have been questioned in the lighting design professions for several technical reasons. The ESI procedure, however, does exactly what it is designed to do: point out locations with poor lighting geometry and identify luminaires with distribution characteristics unsuitable for the intended application. As stated in the *1981 IESNA Lighting Handbook* (from the heyday of ESI; and with a note that IESNA is now IES), it is "a tool in determining the effectiveness of controlling veiling reflections and as part of the evaluation of lighting systems."

Relative Visual Performance. Relative visual performance (RVP) has gained considerable acceptance as a tool for lighting system evaluation. It examines (via computer calculation) the effectiveness (relative to ideal visual performance) of a given visual environment in terms of task accomplishment (measured by speed and accuracy). Like ESI, it is based on luminance and contrast. Unlike ESI, it considers the performance of a task rather than simply contrast reduction. It seems to ignore discomfort from veiling reflections if the task can be performed effectively. A study computed RVP for several common lighting layouts with various types of lighting fixtures. The results uniformly gave values for RVP varying between 0.95 and 0.99—which suggests that using RVP as a sole criterion for glare situations is questionable. Lighting designers are encouraged to use lighting design software to calculate ESI and RVP for proposed lighting layouts and, where possible, to compare the results to post-occupancy evaluations of the performance of the installed system.

TABLE 7.6 Recommended Maximum Luminance Ratios[a]

To achieve a comfortable brightness balance, limit luminance ratios between areas *of* appreciable size as seen from normal viewing positions as follows:	
1 to 1/3rd	Between task and adjacent surroundings
1 to 1/10th	Between task and more remote darker surfaces
1 to 10	Between task and more remote lighter surfaces
20 to 1	Between luminaires (or fenestration) and surfaces adjacent to them
40 to 1	Anywhere within the normal field of view

[a]These ratios are recommended as maximums; lesser ratios are generally beneficial.

Luminance Ratios. Other factors being equal, visual performance increases with contrast—that is, with the difference in luminance between the object being viewed and its immediate surroundings. Conversely, however, the difference between the average luminance of the visual field (task) and the remainder of the field of vision should be low, to avoid the discomfort of large, rapid changes in eye adaptation level. As a rule, contrast is desirable in the object of view but undesirable in the wider surrounding field of view.

Providing reflectances of 50%, 30%, and 80% for walls, floors, and ceilings, respectively, and 35% for furniture, establishes a fairly high eye adaptation level so that direct glare (which results from excessive luminances in the field of view) is minimized. Maximum luminance ratio recommendations to achieve a comfortable environment are presented in Table 7.6. Effective visual performance is entirely possible in environments with much higher ratios—they are not as visually comfortable and may be fatiguing.

Careful control of the reflectances of the major surfaces in a room is necessary to achieve these recommended luminance ratios. The surface reflectance values given above are reasonable for commercial and educational spaces. The difference between a background with proper reflectance and one with excessive brightness ratios caused by low surrounding reflectances is shown in Figure 7.42.

Figure 7.42 *The reflected glare from luminaires disappears when a piece of light, diffuse linoleum is placed over the dark, polished desktop. Light-colored desktops with 35% to 50% reflectance result in task-to-background ratios within the 3:1 recommended range. Before the linoleum was placed, a reflection like the one seen on the right also existed on the left of the desk due to another luminaire. (Courtesy of the Illuminating Engineering Society.)*

7.5 VISUAL COMFORT DESIGN CRITERIA

This book historically looked at lighting design from a quantity and quality perspective. The quantity element was primarily illuminance (adequate for task completion). The quality element was mainly glare control (both direct and reflected). This was not an unreasonable approach, although it might suggest that adequate quantity of light is independent of acceptable quality of light. A more holistic view of the luminous environment would look at designing for visual comfort—which in any given building situation would include several variables of substantial importance to a successful space.

As opposed to the thermal environment, however, there is little resource material to help the design team define (or even engage) visual comfort as a design objective. There is no unified theory of visual comfort as there is for thermal comfort. A unified theory enables designers to identify the key design variables for comfort—but beyond that to understand how an increase in one variable might allow for a decrease in another variable. In the absence of a unifying theory, it is not possible to state that an increase in CRI of 5% will result in an increase in viewer satisfaction of 3%; or that an increase in contrast rendition will permit a known decrease in illuminance for equal occupant satisfaction. This makes design of the luminous environment more of a challenge than design of the thermal environment. There have been attempts to consolidate aspects of the luminous environment—most notably by Flynn and Mills (*Architectural Lighting Graphics*, 1962)—but these efforts have not led to easily actionable design guidance.

Neither the IES nor the CIE defines visual comfort. There is no IES or CIE standard on visual comfort. *ISO 16817:2017: Building Environment Design—Indoor Environment—Design Process for the Visual Environment* addresses visual comfort but does not define it. In the absence of an existing definition of visual comfort, it seems reasonable to appropriate the definition of thermal comfort. Thus, let's state that "visual comfort is that condition of mind that expresses satisfaction with the visual environment." This is a definition that allows for easy verification of visual comfort status during a post-occupancy evaluation, but provides little guidance to design efforts. Lack of a well-established linkage between design moves and occupant responses is not ideal, but also not unusual in building design.

Lacking a unified theory of visual comfort, it is recommended that lighting criteria be specified through the owner's project requirements (OPR) document and lighting design strategies be outlined through a project team's basis of design (BOD) document. Both of these documents are developed early in building project and are critical for driving high-performance outcomes—including with regard to lighting. The OPR details the ideas, concepts, and criteria determined by the owner to be important to the success of the project. The corresponding BOD, developed by the design team, clarifies how a building will accomplish the OPR, including system descriptions, indoor environmental quality criteria, design assumptions, and references to applicable codes, standards, regulations, and guidelines. Often, an owner will need input from design professionals to fully understand and specify certain nuances of lighting requirements within the OPR—but this should not be a problem on a well-structured project.

Typical lighting requirements for a commercial/institutional building might include:

- minimum illuminance levels (see Chapter 18)
- acceptable direct glare
- acceptable reflected glare
- appropriate control options
- appropriate color rendering
- appropriate energy efficiency
- aesthetic criteria
- spatial daylight autonomy (sDA) (see Chapter 12)
- annual sunlight exposure (ASE) (see Chapter 12)
- adequate concern for circadian rhythms (see Chapter 4)

The words "appropriate" and "acceptable" can become design criteria in the context of a specific project (owner, occupants, budget, greenness, etc.). Depending on the project, "adequate" may become "best practice" or "world-class" or "0.375." Guidance on setting criteria for several of these potential OPR can be found in lighting handbooks and high-performance building certification program requirements (e.g., LEED or the WELL Building Standard).

References and Resources

Cotton, H. 1960. *Principles of Illumination*. New York: John Wiley & Sons.

Flynn, J., and S. Mills. 1962. *Architectural Lighting Graphics*. New York: Van Nostrand Reinhold.

IES. 1966. "Outline of a Standard Procedure for Computing Visual Comfort Ratings for Interior Lighting." Committee Report. *Illuminating Engineering* 61: 643.

IES. 2011. *The Lighting Handbook*, 10th edn. New York: Illuminating Engineering Society.

ISO. 2017. *ISO 16817 : 2017: Building Environment Design—Indoor Environment—Design Process for the Visual Environment*. Geneva: International Organization for Standardization.

Lighting Design Lab. lightingdesignlab.com

Lighting Research Center. lrc.rpi.edu

Murdoch, J. B. 1985. *Illumination Engineering*. New York: Macmillan.

Ross and Baruzzini, Inc. 1975. "Energy Conservation Applied to Office Lighting" (Conservation Paper No. 18). Washington, DC: Federal Energy Administration.

Ryer, A. 1997. *The Light Measurement Handbook*. Peabody, MA: International Light Technologies. intl-lighttech.com/services/light-measurement-handbook

Sampson, F. K. 1970. *Contrast Rendition in School Lighting*. New York: Educational Facilities Laboratory.

U.S. Census Bureau. 2023. *2020 Census: 1 in 6 People in the United States Were 65 and Over*. Washington, DC: U.S. Department of Commerce. census.gov/library/stories/2023/05/2020-census-united-states-older-population-grew.html

CHAPTER 8

Acoustical Comfort

THE ACOUSTIC ENVIRONMENT PLAYS AN IMPORTANT role in the overall sense of comfort we experience (or not) in many of the spaces we occupy daily—both residential and commercial/institutional spaces. For several reasons, many design solutions shortchange the acoustical environment. This is partly due to the perceived complexity of architectural acoustics, and partly due to lack of coverage of the topic in many architectural education curricula. Good acoustics is not required by most building codes; nor is it a required element in most green building rating systems. Nevertheless, providing acceptable acoustical conditions is a fundamental part of good design practice. This can be accomplished by concern for acoustical comfort and development of an owner's project requirements that properly address such a concern.

8.1 SOUND, VIBRATION, NOISE

Sound can be looked at from several perspectives, depending upon the aspect of most interest or concern. It can be described as a physical wave, or as a mechanical vibration, or as a series of pressure variations in an elastic medium. For airborne sounds, the medium is air. For structure-borne sounds, the medium may be concrete, steel, wood, glass, or combinations of these materials. A more limited definition of sound, appropriate to architectural acoustics, is that it is an audible pressure variation. Sound is energy that a human can hear.

Although the terminology is a bit confusing (since sound is a mechanical vibration), in architecture a vibration is a pressure variation that can be felt but not heard. Vibration is tactilely sensed, not audibly sensed. Vibration control is a distinct subset of architectural acoustics.

Noise is unwanted sound—requiring qualitative evaluation on the part of the party hearing the sound. Noise is a perception of an individual occupant. It is analogous to glare. Noise cannot be measured by electromechanical instrumentation. It is not described by numerical units. Since most occupants do not want to feel a building or its furnishings moving (even if visually imperceptible), vibration can be considered noise in most architectural situations.

There are three key terms in architectural acoustics: sound, noise, and vibration. All are variants of sound, but have different means of reception or different impacts on the receiver. The discussion of "sound" characteristics that follows covers all three variants.

(a) Speed of Sound

Sound travels at different speeds, depending upon the medium. In air, at sea level, the speed of sound is 1130 fps [344 m/s]. This is 770 miles per hour [1239 kilometers per hour]—slow compared to light, which has a speed of 186,000 miles per second [299,338 kilometers per second]. Since sound travels not only through the air in a building but also through its structure and partitions, it is good to know the speed

Mechanical and Electrical Equipment for Buildings, Fourteenth Edition. Walter Grondzik, Daniel Overbey, and Alison Kwok.

Companion website: www.wiley.com/go/meeb14e

TABLE 8.1 Speed of Sound Propagation in Various Media

	Speed	
Medium	Meters per Second	Feet per Second
Air	344	1130
Water	1410	4625
Wood	3300	10,825
Brick	3600	11,800
Concrete	3700	12,100
Steel	4900	16,000
Glass	5000	16,400
Aluminum	5800	19,000

Note: These values are approximate, since the listed materials vary in density.

of sound in other media (Table 8.1). For architectural design, speed variations due to changes in temperature and altitude (atmospheric pressure) may be ignored, and for most calculations, 1130 fps [344 m/s] may be used as the speed of sound in air (usually within 3% error). From a practical standpoint, the speed of sound in air is slow enough that the travel time of a sound signal can be a key design issue.

(b) Wavelength

The wavelength of sound is defined as the distance between similar points (peaks or valleys) on successive waves—which is the distance sound travels in one cycle. The relationship between wavelength, frequency, and speed of sound is expressed as:

$$\lambda = \frac{c}{f} \qquad \text{(Eq. 8.1)}$$

where

λ = wavelength, ft [m]
c = velocity of sound, fps [m/s]
f = frequency of sound, Hz

Low-frequency sounds are characterized by long wavelengths, and high-frequency sounds by short wavelengths. Sounds with wavelengths ranging from 0.5 in. to 50 ft [12 mm to 15 m] can be heard by human beings. A simple nomograph (see Chapter 24) permits rapid determination of wavelength, given a sound frequency, and vice versa.

(c) Frequency

The number of times that a cycle of compression and rarefaction of air occurs in each unit of time is described as the frequency of a sound. For example, if there are 1000 such cycles per second (cps), the frequency is 1000 hertz (Hz)—where Hertz (after Heinrich Hertz, German physicist) is the unit name applied to this measurement. Thus, in Figure 8.1, higher frequencies would be shown by compressions and rarefactions that are closer together, and lower frequencies by those that are farther apart. In architectural acoustics, frequency is sometimes referred to using the musical term pitch. The higher a sound's frequency, the higher its pitch, and vice versa.

Sound frequency is integrally linked to speech and hearing (see Figure 8.2). The approximate frequency range of a healthy young person's hearing is 20–20,000 Hz. The human speaking voice has a range of approximately 100–600 Hz in fundamental frequencies, but harmonics (overtones) extend to approximately 7500 Hz. Most speech information is carried in the upper frequencies, whereas most of the acoustic energy exists in the lower frequencies. The critical frequency range for speech communication is 300–4000 Hz. Overtones outside these core frequencies, however, give the voice its characteristic sound and specific identity. Telephone and radio communication are accomplished using a considerably narrower frequency band that sacrifices some voice quality and intelligibility.

A sound composed of only one frequency is called a pure tone. Except for the sound generated by a tuning fork, few sounds are truly pure. Musical sounds (tones) are composed of a fundamental frequency and integral multiples of the fundamental frequency (harmonics). Most common sounds are complex combinations of frequencies. Figure 8.3 shows examples of pure tones, musical notes, and common sounds, while Figure 8.4 shows the frequency ranges of some common devices and phenomena.

(d) Octave Bands

The frequencies on the axis of Figure 8.4 stand in the ratio of 2:1 to each other—that is:

16.5:31.5 ···
31.5:63 ···
63:125 ···
125:250 ··· and so on.

Using musical terminology, they are one octave apart, which refers to the interval between one musical pitch and the next sequential pitch with double (or half) of its frequency. The name "octave" is derived from the Latin word "octavus," which means "eighth" in reference to the eight notes in the diatonic scale in Western music—in this sense, "octave" is a bit of a misnomer in

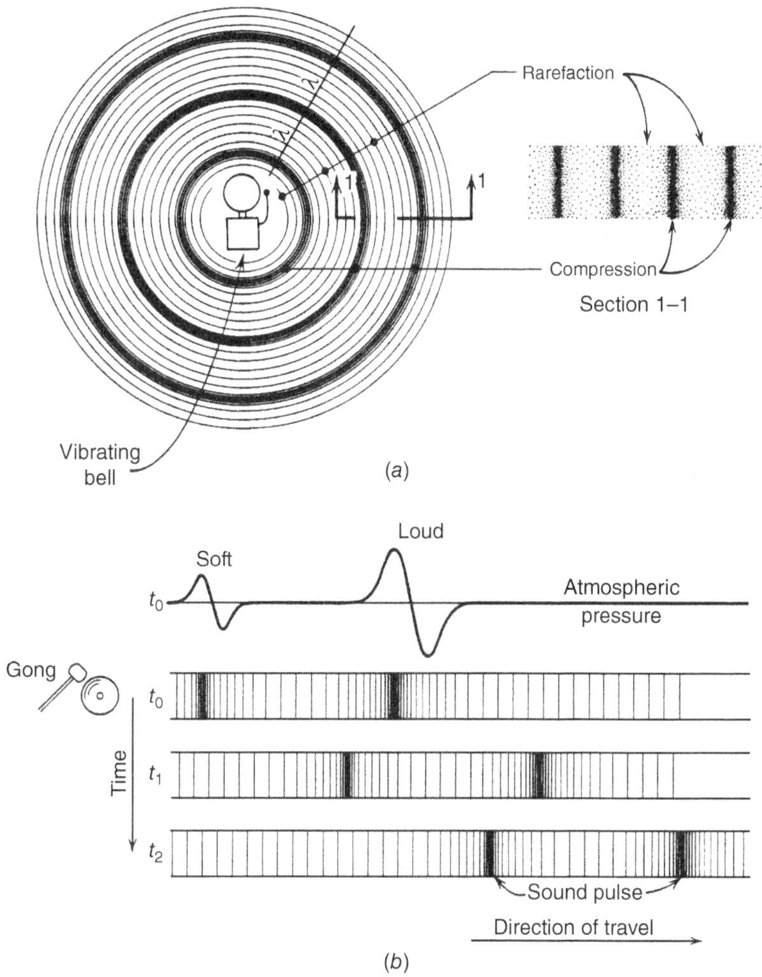

Figure 8.1 *Sound pressure waves. (a) The continuous vibration from the bell causes a series of compressions and rarefactions of the air to travel outward longitudinally from the source. Amplitude information is carried by pressure; that is, greater amplitude means greater compression and greater rarefaction. (Compression and rarefaction are shown diagrammatically as line density, although they are actually molecular phenomena, as shown in the upper drawing.) (b) Two single impulses of different magnitude (amplitude) traveling away from the source. Note how amplitude information is carried by the difference in pressure.*

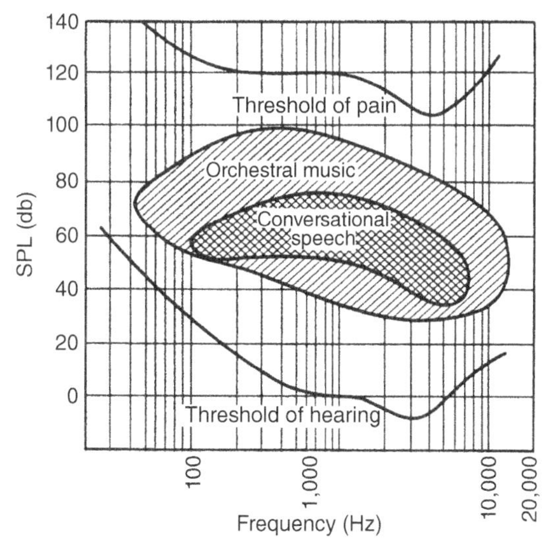

Figure 8.2 *The positions of speech and wide-range music in the aural field of the human ear. Speech is in the nominally linear response area of the ear, as is most music. Beyond these frequencies, the ear effectively attenuates the incoming signal.*

acoustic measurements because there are 11 standard octave bands, as depicted in Table 8.2. Each octave band has a center representative frequency. These frequencies are accepted internationally as the reference frequencies of octave bands used for acoustical analysis. For technical reasons, a geometric mean is used. Thus, 250 is the center frequency of an "octave band" ranging from $250/\sqrt{2}$ to $250\sqrt{2}$, with that particular octave being described as the 250-Hz octave. If a finer division is required for analysis purposes (unusual in architectural acoustics), ½-octave and ⅓-octave bands are used. Octave bands serve as a consistent way to divide the full range of frequencies in sound into smaller, more manageable sections (i.e., "bands").

Because frequency is so important to architectural acoustics, and because there are 19,980 discrete whole frequencies in the normal range of hearing, octave bands are used repeatedly during design as a way of capturing frequency-specific information about sounds without becoming buried in detail. Using a single

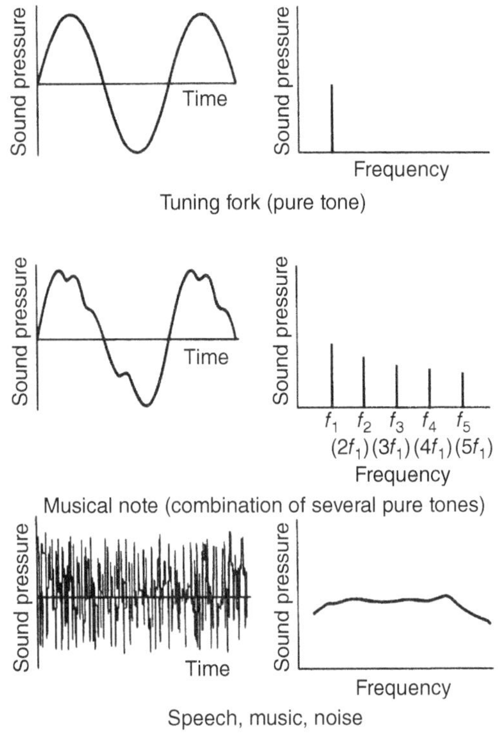

Figure 8.3 *Schematic representations of a pure tone, a musical note, and a more complex sound (such as speech or music), showing the variation of sound pressure with time and frequency.*

frequency to identify most sounds will be woefully inaccurate. Using an average of frequencies to describe a sound using a single number is a bad idea, as averaging discards critical information about the character of a sound. Using several octave band center frequencies to describe a sound is a good idea—with "several" sometimes being four, sometimes six, sometimes eight, depending upon the importance of the acoustical situation.

TABLE 8.2 Standard Octave Bands

Band	Center Frequency
1	16.5 Hz
2	31.5 Hz
3	63 Hz
4	125 Hz
5	250 Hz
6	500 Hz
7	1000 Hz (1 kHz)
8	2000 Hz (2 kHz)
9	4000 Hz (4 kHz)
10	8000 Hz (8 kHz)
11	16,000 Hz (16 kHz)

(e) Sound Magnitude

Expressing the magnitude of a sound signal is relatively complex because there are several interrelated metrics (and associated units) in common use and because of the great range of values involved in day-to-day acoustic situations. Sound magnitude is often equated with loudness, which is a subjective receiver-oriented response and not linearly related to the power of a sound (in watts). The physical magnitude of sound is variously described as follows:

Sound power (P): The energy emitted by a sound source per a unit of time, measured in watts (W). For example, the sound power of a speaker to define how much energy it outputs as sound.

Sound power level (PWL): A logarithmic measure of sound power, measured in decibels (dB). A decibel is a unit used to express the relative magnitude on a uniform scale from a reference magnitude.

Sound pressure (p): The localized pressure deviation from the average atmospheric pressure created by a sound wave, measured in pascals (Pa). This is a

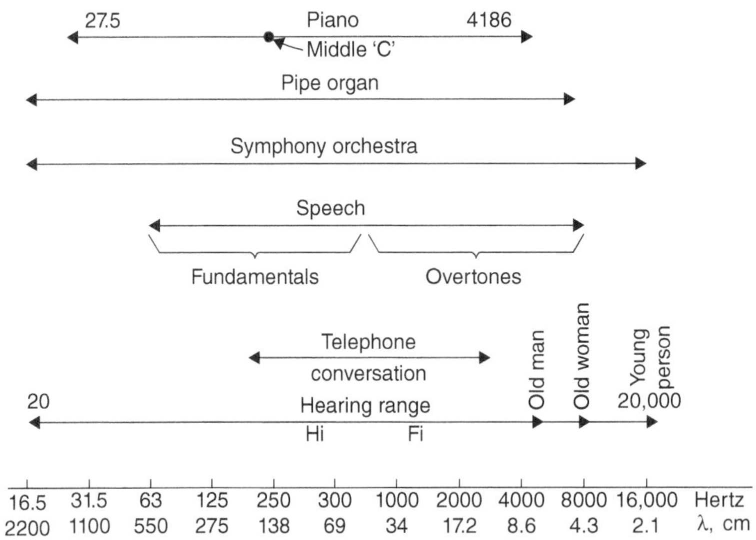

Figure 8.4 *Frequency ranges of common instruments. Wavelength (λ) is based upon a sound speed of 344 m/s [1130 fps].*

measure of the fluctuation in air pressure that your ear detects through a passing sound wave.

Sound pressure level (SPL): a logarithmic measure of the effective sound pressure of a sound compared to a reference, usually the threshold of hearing (20 micropascals, μPa, or one-millionth part of a pascal); measured in dB (again, an expression of relative magnitude).

Sound intensity (I): Sound power measured per unit area, or watts per square meter (W/m^2).

Sound intensity level (IL): A logarithmic measure of sound intensity relative to a reference intensity, measured in dB.

Each of these metrics has a place, and each differs from the others and from subjective loudness. To clearly understand these concepts—and it is imperative that they be understood—a comprehension of how we hear and how sound is propagated in free space is necessary. Thus, discussion of specific measures of sound magnitude will be found below, following a discussion of hearing.

(f) Sound Propagation

For simplicity, it is useful to view sound as a series of pressure variations. In air, these pressure variations take the form of periodic compressions and rarefactions. The bell in Figure 8.1 radiates energy in all directions more or less equally—that is, it creates a circular wavefront. As the material comprising the bell vibrates, it sets up vibrations of the same frequency in the air, which can best be visualized in a sectional view. The pressure changes containing the sound information travel in the same direction as the wavefront—longitudinally. Sound is therefore a longitudinal wave motion. This is unlike (for example) an AM radio signal, in which the wave travels longitudinally, but the information—that is, the modulation—is transverse. Sound is a mechanical wave, whereas light and electricity are electromagnetic waves.

8.2 HEARING

As noted above, the approximate frequency response of a healthy young person's hearing is 20–20,000 Hz. The upper limit decreases with age due to a process called presbycusis (Figure 8.5). The decrease is more pronounced in men than in women. Recognition of this phenomenon can be of importance in schools, since very high-pitched sounds that are inaudible to most adults can be a source of extreme annoyance to young students. For example, dentists report that high-speed drills and tooth-cleaning devices cause extreme auditory discomfort in many young patients. These devices produce sounds in the 15–20-kilohertz (kHz) range.

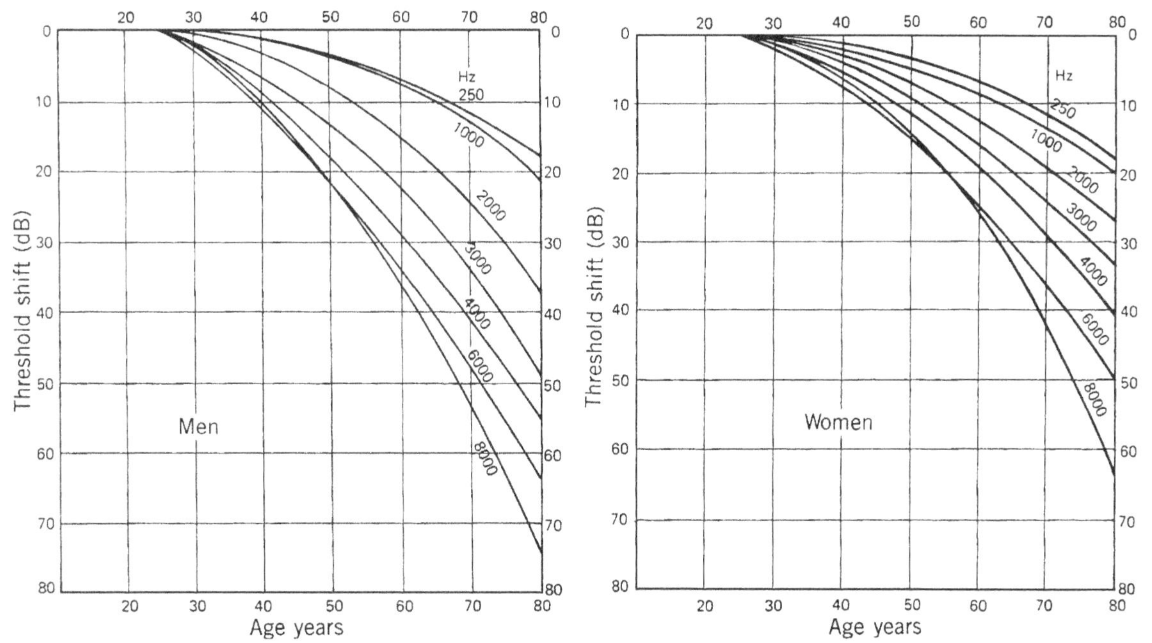

Figure 8.5 The average hearing threshold shift for men and women with increasing age, relative to a 25-year-old reference group. Note that, at age 60, the male threshold shift at upper speech frequencies (4000 Hz) is 30 dB, compared to 20 dB for women, a difference of 10 dB or one-half of the subjective loudness (as noted in Table 8.3). For men older than 50 and women older than 60, frequencies above 10 kHz are effectively inaudible, and even at normal speech frequencies of 1000–2000 Hz, subjective loudness is reduced to one-half of that of a 25-year-old listener. (Reproduced with permission from F.A. White. 1975.)

TABLE 8.3 Intensity Level Changes and Corresponding Subjective Loudness Changes

Change in Intensity Level (dB)	Subjective Change in Loudness
3	Barely perceptible
6[a]	Perceptible
7	Clearly perceptible
10	Twice (or half) as loud
20	Four times (or one-quarter) as loud

[a]Six decibels corresponds to the change encountered when the distance to the source in a free field is doubled (halved).

(a) The Ear

The outer ear is funnel-shaped and serves as a sound-gathering device for the auditory system (Figure 8.6). Sound energy travels through the auditory canal (outer ear) and sets in motion the components of the middle ear, comprising the eardrum, hammer, anvil, and stirrup. The stirrup acts as a piston to transmit vibrations into the fluid of the inner ear. The motion of this fluid causes movement of hair cells in the cochlea, which, in turn, stimulates nerves at the bases of the hairs. The nerves, in turn, transmit

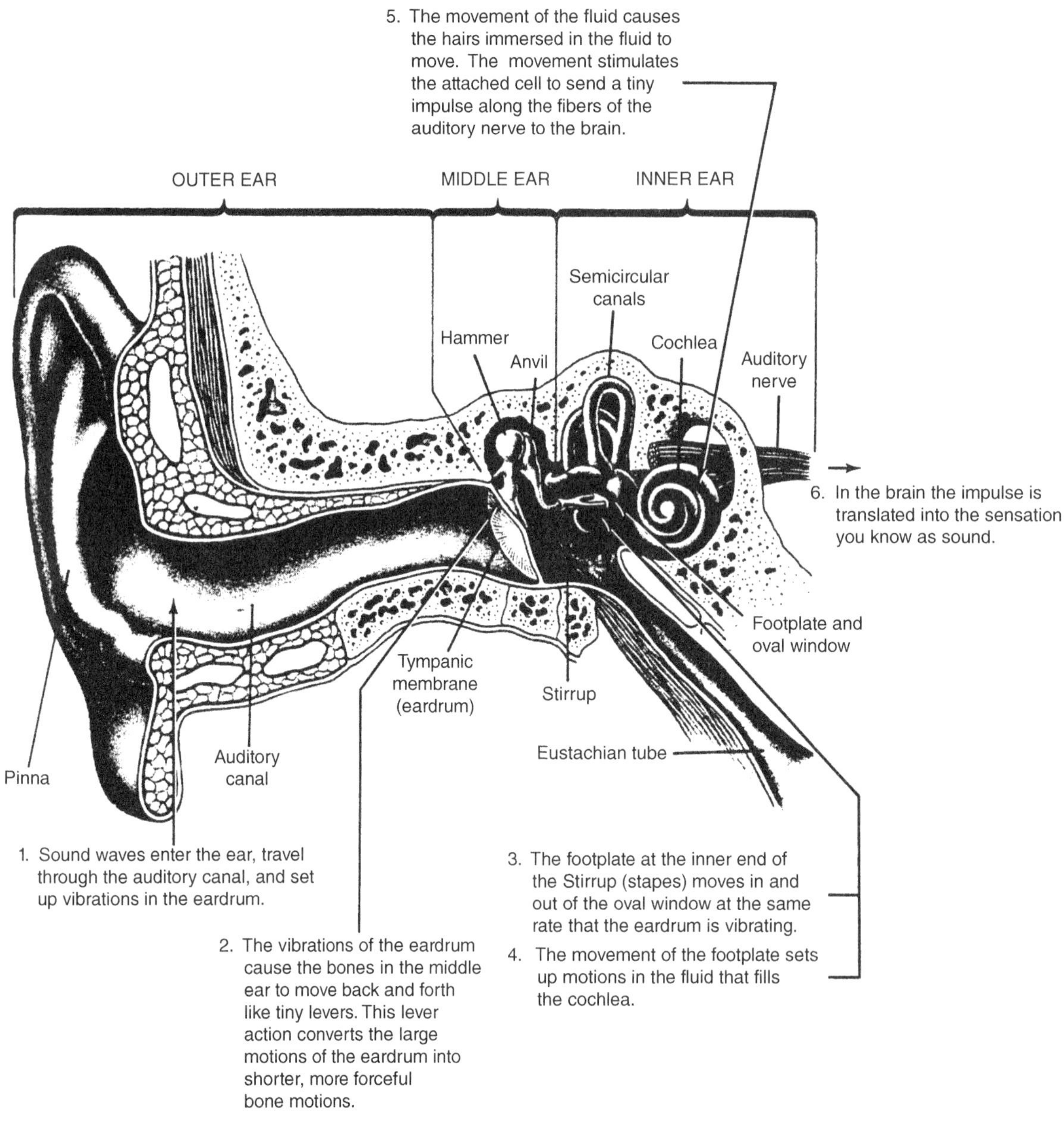

Figure 8.6 *Drawing of the human ear showing the principal parts and its function as a sound receptor. (Drawing reproduced with permission from F.A. White, 1975. Notes are reproduced from NBS, 1976.)*

electrical impulses along the eighth cranial nerve to the brain. These impulses we understand as sound.

It is often assumed that the ear ignores phase differences and combines frequencies. This may not always be the case, however, particularly when the frequencies are very far apart. For this reason, a single-number representation of a complex sound can be misleading and must be used with caution.

At the threshold of hearing (approximately 0 decibels; dB), the displacement of air molecules impinging on the eardrum, and the eardrum excursion, are approximately 1 angstrom unit (1 Å = 10^{-8} cm), which is approximately the diameter of an atom. Were the ear an order of magnitude more sensitive, it would hear thermal noise. The human ear thus operates close to the practical limit of sensitivity. At the other end of the magnitude range, the threshold of pain corresponds to a sound pressure level of 130 dB and to an eardrum motion of approximately 0.25 mm [0.01 in.]—truly an astonishing range.

Several "measures" of sound magnitude are encountered in architectural acoustics. Movement of the eardrum (and thus hearing) is caused directly by air pressure variations. Therefore, the physical variable generally of most interest to architectural acoustics is sound pressure (force density), and its derivative—sound pressure level—which is the ratio of a given sound pressure to a reference base level, expressed in decibels (dB).

(b) Equal Loudness Contours

The human ear is not uniformly sensitive over its frequency range of 20 Hz to 20 kHz. The 120–130-dB upper limit (pain threshold) occurs at all frequencies. At the lower limit, however, the 0-dB threshold occurs only at 1000 Hz. The ear is in fact most sensitive at 3000–4000 Hz, at which frequencies the threshold is about −5 dB (relatively speaking). This type of nonlinear response exists throughout the ear's hearing range. To determine the nature of this nonlinearity, many tests were conducted with pure tones of different frequencies, in which listeners were asked to equate the subjective loudness of signals. These test results produced a family of curves called equal loudness contours (sometimes called Fletcher–Munson equal loudness contours after two of the principal researchers). These curves (Figure 8.7) are internationally recognized and standardized, and are used as the reference for normal hearing response. They are also used to "weight" measuring devices, as explained later in this chapter. Note that by definition:

1. All points on a single contour have the same subjective sensation of loudness.

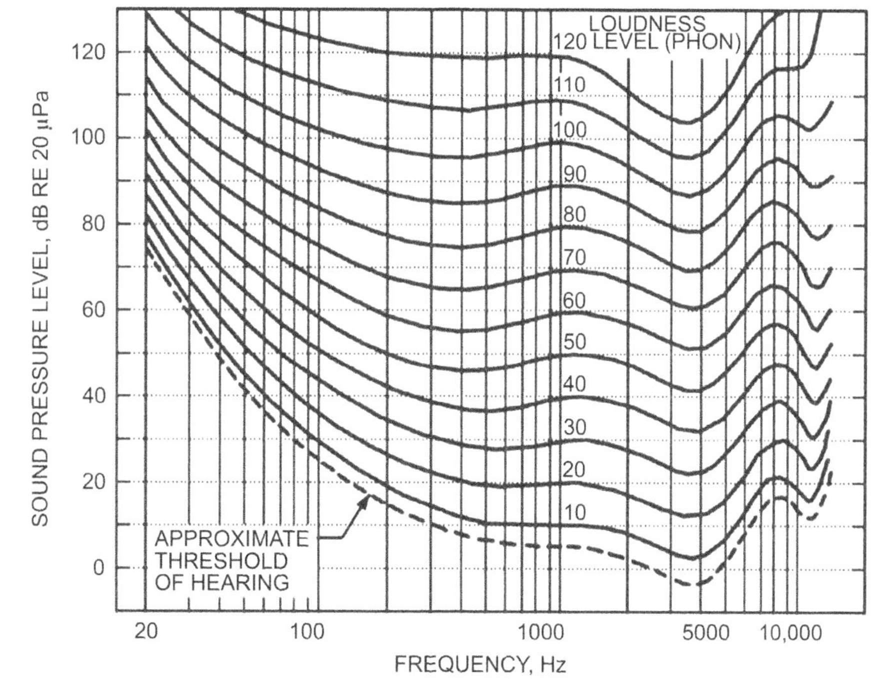

Figure 8.7 Free-field equal-loudness contours for pure tones. These curves are accurate for a listener with normal binaural hearing situated in the near field of a source producing pure tones directly ahead of the listener. The subjective (perceived) loudness of each contour is quantified in a unit called the phon. The phon value of each curve is shown in the center of the figure. Note that a sound pressure level of 60 dB corresponds to 30 phons at 50 Hz, 50 phons at 100 Hz, 63 phons at 500 Hz, 60 phons at 1 kHz, 68 phons at 4 kHz, and 60 phons at 6 kHz. This shows the relative flatness of the ear's response in the central frequency range and the sharp drop at low frequencies. (© ASHRAE, www.ashrae.org. 2021 ASHRAE Handbook—Fundamentals, Chapter 8, Figure 4.)

DESIGN FUNDAMENTALS

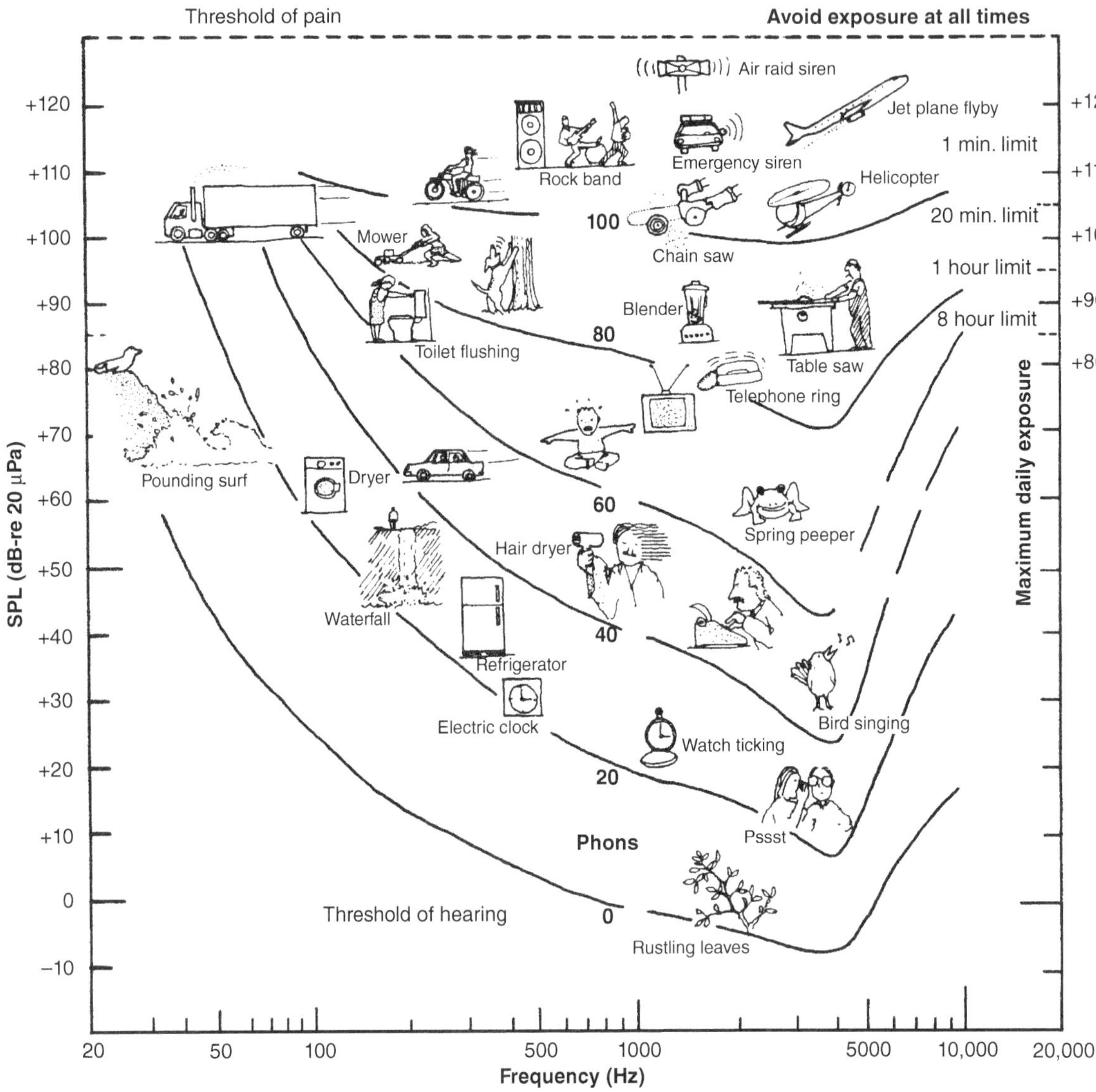

Figure 8.8 *Common sound sources plotted at their dominant frequencies and levels as typically heard by the observer. The equal-loudness contours (see Figure 8.7) show why certain sounds seem louder than others, despite pressure levels that would indicate the contrary. (Reproduced with modification from NBS, 1976.)*

2. The loudness level in phons (the number shown in the center of each curve in Figure 8.7) of the entire contour is defined by the decibel level of that contour at 1000 Hz. Stated differently, a phon is a unit used to measure the apparent loudness of a given sound that is equal in number to the decibels of a 1000-Hz reference sound that listeners would most likely perceive as equal in loudness.

The phon scale was constructed as a way of defining perceived sound (subjective loudness impressions) in terms of the ear's nonlinear response. Assigning a single number—using a unit called the phon—to an equal loudness contour makes it possible to compare subjective loudness impressions of two sounds, regardless of frequency or measured sound pressure level. Thus, in Figure 8.8, the subjective loudness of a whisper in the ear (20 phons) is the same as that of distant pounding surf, even though the SPL of the whisper is 20 dB and that of the surf is 80 dB. The 60-dB differential is due to the ear's sharp drop in sensitivity at low frequencies. However, because the phon scale is nonlinear in terms of perceived loudness (a given phon differential does not correspond to the same perceived loudness change throughout the phon scale) and because phons cannot be combined arithmetically to give a resultant subjective loudness (60 phons plus 50 phons is not 110 phons), the scale has found few uses. One common application is rating the noisiness of bathroom exhaust fans.

The equal loudness contours demonstrate some interesting phenomena:

- Sensitivity drops off sharply at low frequencies, particularly at low decibel levels. (For this reason, most stereo amplifiers provide automatic bass boost at low volume settings.)
- Maximum sensitivity occurs between 3 and 4 kHz—precisely the frequencies that convey the most information in human speech (see Figure 8.2).
- In a normal listening range of 45–85 dB, and in the most-often-encountered frequency range of 150 Hz to 6 kHz, the contour is substantially flat, that is, the ear's response is effectively linear in this zone. Only at extremes of sound level and frequency does nonlinearity occur.

The ear "averages" sounds over a minimum time increment; sound impulses of shorter duration will be perceived as quieter than they would if received as a steady-state sound. The minimum sound pulse length (i.e., the time required for the sound to achieve full loudness) varies between 50 and 200 ms (ms; i.e., 1/20–1/5 s) and depends to an extent upon the frequency content of the sound. The value of 70 ms (1/14 s) is frequently found in the literature. This threshold time becomes important when considering the effect of echoes (see Chapter 24).

(c) Masking

When two separate sound sources are received simultaneously, the perception of each is made more difficult by the presence of the other. This effect is known as masking, which is defined technically as the number of decibels by which the threshold of audibility of one sound is increased by the presence of another sound. The masking effect is greatest when two sounds are close in frequency or frequency content, since the ear has greater difficulty separating like frequencies. Also, a low frequency will mask a high frequency more than the reverse for the same decibel levels. With broad-frequency sounds, the masking effect is difficult to predict, since it depends in part upon how intently a listener is listening. Masking is an extremely important and useful technique for noise control, wherein background sound levels are deliberately manipulated to mask unwanted sounds. The background sounds typically used for this purpose are broadband and continuous in nature, such as "white" or "pink" noise, and are noninformation-bearing. It is important to distinguish a few common types of random noise. The *2021 ASHRAE Handbook—Fundamentals* offers the following definitions:

White noise *has a continuous frequency spectrum with equal energy per hertz over a specified frequency range. Because octave bands double in width for each successive band, for white noise the energy also doubles in each successive octave band. Thus white noise displayed on a 1/3 octave or octave band chart increases in level by 3 dB per octave.*

Pink noise *has a continuous frequency spectrum with equal energy per constant-percentage bandwidth, such as per octave or 1/3 octave band. Thus pink noise appears on a 1/3 octave or octave band chart as a horizontal line.*

Red noise *has a continuous frequency spectrum with octave band levels that decrease at a rate of 4–5 dB per octave with increasing frequency. Red noise is typical of noise from well-designed HVAC systems.*

Sound masking serves to depress (or "mask") the intelligibility of lower-magnitude, information-bearing sounds that would otherwise cause annoyance, and are particularly useful in open office situations where few noise control alternatives exist. Mechanical equipment sound (such as that from air diffusers and fans) can sometimes be used to good advantage as masking sound.

(d) Directivity

The exact mechanism by which the binaural aspect of hearing detects direction is not entirely understood. The brain can, however, sense the direction of a sound source in many environments. In enclosed spaces, echoes from surface reflections (reverberation) will blur most directivity, and any "stereo" information will be almost completely dependent upon high frequencies in the near field (the sound field that is close to the source and therefore minimally affected by reverberation). Since high frequencies (with short wavelengths) travel in a relatively straight line, they will reach a listener before their reflections, and in so doing will cue the hearer as to their origin. Low frequencies (with long wavelengths) are difficult to localize because their wavelengths are large compared to the dimension between our ears. Low frequencies, even in the near field, are readily mixed with their reflections, thus confusing the location of their origin.

(e) Discrimination

Frequency recognition is accomplished in the cochlea by the hair cells embedded in the basilar membrane. The ear can hear and recognize distinct frequencies, yet the brain has the ability either to hear individual frequencies or to combine them into a single more complex sound. Thus, when we hear a string quartet we can, generally at will, hear either the entire quartet or each instrument individually. With concentration (vision helps in this regard), a trained ear can pick out a single instrument in an orchestra of 120 pieces, even if there is more than one such instrument in the same section. Conductors do this regularly. Similarly, the ear can perform the feat known as the cocktail party effect, that is, pick out one voice among background sounds that may be substantially louder than the wanted signal. In effect, the brain in this case attenuates the unwanted sound signals. Normally, however, hearing does precisely the opposite. It combines sounds that are clearly distinct from each other in frequency and phase. The three tones in a chord struck on a piano are different in frequency and, if played as a very rapid triplet, are out of phase. Yet the ear combines them and hears a single sound, even though the maxima of the three tones do not occur simultaneously.

(f) Sound Sources

Building occupants encounter a wide range of sounds during a typical day. Some of these sounds are produced by human beings, others by mechanical equipment or natural phenomena. Some of the sounds are information-bearing, while others convey nothing intelligible. Speech and music are sounds of interest to building designers, as are those sounds generally deemed to be noise.

Speech. The ear's sensitivity (Figure 8.4) is highest in the speech frequency and normal energy range. Individual speech sounds vary in duration between 30 and 300 ms, and the ear normally perceives these individually and clearly. Speech is composed of phonemes, which are individual and distinctive sounds that allow one word to be distinguished from another (such as in mad, cad, had, hat). Certain phonemes exist in one language and not in another. Some phonemes carry more information than others, and architectural acoustics must be particularly careful to preserve and support these to maintain intelligibility. In English, consonants carry much more information than vowels, as can be seen by:

1. Writing a sentence:
   Most speech energy is concentrated in the one-hundred to six-hundred Hertz range.
2. Then, rewriting it without consonants:
   o ee eey i oeae i e oe-ue o i-ue e ae.
3. And then, without vowels:
   Mst spch nrg s cncntrtd n th n-hndrd t sx-hndrd Hrtz rng.

The male voice energy is centered at around 500 Hz, the female voice at around 900 Hz. Consonants, however, have most of their energy in the high frequencies. Phonemes such as *s* and *sh*, for example, have most of their energy above 2 kHz, and both are particularly important in conveying intelligible content.

Normal speech averages between 55 and 65 dBA sound pressure level at 3–4 ft [0.9–1.2 m] from the source, with a dynamic range from about 30 dBA for soft speech to somewhat above 65 dBA for loud speech (at the same distance). Extremes of speech sound pressure level are 10 dBA for a soft whisper and 80 dBA for a shout, but in both instances, intelligibility is sharply reduced because of lack of consonant power. In shouting, emphasis is placed on vowels, so that 70-dBA sound pressure level is generally accepted to be the upper limit of fully intelligible human speech. Singers who frequently exceed 90 dBA do so at great loss of intelligibility.

Another result of the high-frequency content of consonants, and therefore intelligibility, is directiveness. The higher the frequency, the greater a sound's directivity, and the less its diffraction (ability to be heard beyond a partial barrier). Therefore, intelligibility of speech is greatest directly in front of a speaker and least behind him or her. High-frequency tones are most easily absorbed and least diffracted.

Instrumental Music. Instrumental music is much broader in dynamic range and more complex in frequency than speech. It has no direct parallel to intelligibility. A person's "reception" of music is a combination of physiological and psychological phenomena.

Mechanical and Electrical Equipment. Mechanical and electrical equipment produce sounds are commonly adjudged to be noise. Such equipment may be located exterior to a building, within a building, or within a habitable space in a building.

Exterior Sound Sources. Exterior sounds (often deemed noises) also enter most buildings; these may include traffic sounds (cars, trains, planes), people speaking, and the sounds of commerce.

8.3 EXPRESSING ACOUSTIC VARIABLES

Six different quantitative measures of sound magnitude are commonly encountered in architectural acoustics. Three of these—sound power, sound pressure, and sound intensity—are absolute measures; the other three—sound power level, sound pressure level, and sound intensity level—are ratio values that compare an absolute measure to a baseline reference value. Use of the qualitative metric of loudness is also common. It is common practice to use SI units for all architectural acoustics metrics—even in the United States. I-P and SI units, however, are both presented herein.

(a) Sound Power

Sound power is an independent property of a sound source that quantifies the source's acoustical output. Sound power is constant for any given source operating under defined conditions (a certain level of speech effort or a load on a motor) and is not influenced by the nature of the surroundings into which the source is placed. The sound power (output) of a chiller or orchestra or orator is not changed by the distance to a receiver or the characteristics of a mechanical room, auditorium, or classroom.

Sound power is expressed in watts (of acoustical power) and varies widely from source to source. Some examples of sound power values are: a jet engine, 100,000 W; a symphony orchestra, 10 W; a loud radio, 0.1 W; normal speech, 0.000010 W. Note the wide range of values just in this sample: from 10^5 to 10^{-6} W. Sound power values for natural phenomena need to be obtained from an appropriate information source. Sound power data for manufactured equipment and devices can be obtained from the manufacturer. A manufacturer should be able to provide information on the sound power of a product under specified operating conditions. Manufacturers cannot provide accurate information on sound pressure levels (see below), as these are influenced by environmental variables beyond the manufacturer's control or knowledge.

(b) Sound Pressure

Sound pressure is the deviation from ambient air pressure that is caused by a sound wave. Sound pressure is instigated by the acoustic output (sound power) of a source, but is modified by the environment between the source and the receiver. Sound power is a characteristic of a source; sound pressure is the effect of a source as experienced at some specific location. Sound pressure must be referenced to a point in a space, because pressure will typically vary from location to location in a room of reasonable size—and may differ radically among points in a larger space.

TABLE 8.4 Some Representative Sound Pressures

Near a jet plane	200 Pa
The threshold of pain	20 Pa
A loud nightclub	2 Pa
Next to a highway	0.2 Pa
Normal speech	0.02 Pa

Sound pressure is expressed in pascals (Pa) [or microbars (μbar)]. Table 8.4 gives sound pressures for some common situations.

As with sound power, note the substantial range of values in these sample situations.

(c) Sound Intensity

The ear responds directly to pressure variations, but such pressures involve various energy densities. The threshold of hearing—that is, the minimum sound intensity (I) that a normal ear can detect—is 10^{-16} W/cm^2. The maximum sound intensity that the ear can accept without damage is approximately 10^{-3} W/cm^2. This gives an intensity range of 10^{13}, or 10 trillion to 1. Table 8.5 gives an idea of the physical significance of these numbers. Note that the maximum (painful) acoustic intensity in Table 8.5 is 0.001 W/cm^2 (or 10 W/m^2).

A point sound source of constant power radiating in free space—that is, at a location far from the effects of any reflecting surface—is represented in Figure 8.9. The sound intensity at any (defined) distance from a source is expressed as:

$$I = \frac{P}{A} \quad \text{(Eq. 8.2)}$$

where

I = sound (power) intensity, W/cm^2
P = acoustic power, W
A = area, cm^2

since the sound radiates freely in all directions.

TABLE 8.5 Comparison of Decimal, Exponential, and Logarithmic Statements of Various Acoustic Intensities

Intensity (W/cm^2)		Intensity Level,	
Decimal Notation	Exponential Notation	Logarithmic Notation (dB)	Examples
0.001	10^{-3}	130	Painful
0.0,001	10^{-4}	120	
0.00,001	10^{-5}	110	75-piece orchestra
0.000,001	10^{-6}	100	
0.0000,001	10^{-7}	90	Shouting at 5 ft [1.5 m]
0.00,000,0001	10^{-9}	70	Speech at 3 ft [0.9 m]
0.0000,000,0001	10^{-11}	50	Average office
0.000,000,000,0001	10^{-13}	30	Quiet, unoccupied office
0.0000,000,000,0001	10^{-14}	20	Rural ambient
0.000,000,000,000,001	10^{-15}	10	
0.0000,000,000,000,001	10^{-16}	0	Threshold of hearing

DESIGN FUNDAMENTALS

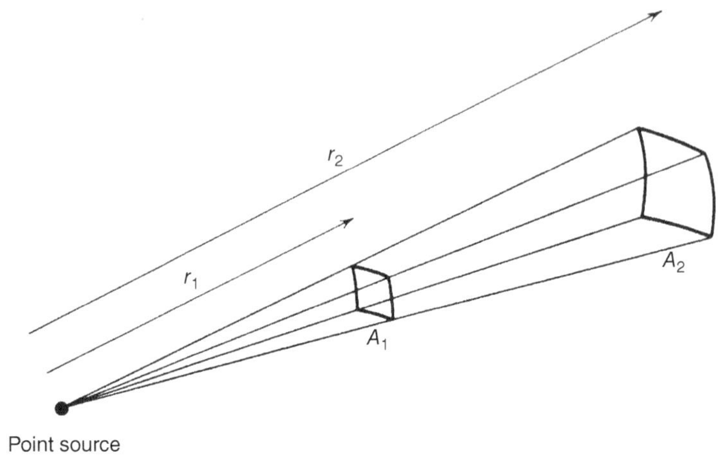

Figure 8.9 *The same total energy passes through areas A_1 and A_2. Since A is proportional to the square of r, the energy density or intensity is inversely proportional to r^2.*

$$I = \frac{P}{4\pi r^2} \text{W/cm}^2 \qquad \text{(Eq. 8.3)}$$

where

r = radius of an imaginary sphere enclosing the sound source

This reflects the inverse square law, which states that intensity is inversely proportional to the square of the distance from the source. In I-P units, this is:

$$I = \frac{P}{930 \times 4\pi r^2} \text{W/ft}^2 \qquad \text{(Eq. 8.4)}$$

since 1 $\text{ft}^2 = 930\,\text{cm}^2$. Using Equation 8.3 to determine the intensities I_1 and I_2 at distances r_1 and r_2 from point source P, we find that the intensities at distance r_1 and r_2 from the source stand in the ratio

$$\frac{I_1}{I_2} = \frac{r_2^2}{r_1^2} \qquad \text{(Eq. 8.5)}$$

Figure 8.10 shows how a sound pulse is attenuated by the action of distance in strength (but not in waveform) as it travels outward from a source.

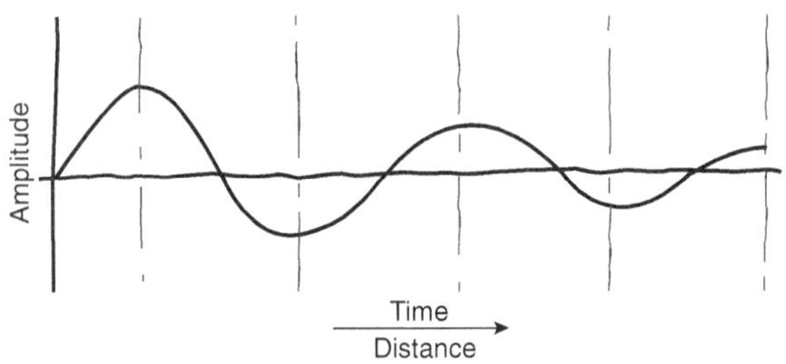

Figure 8.10 *Attenuation of a sound signal in air as it travels away from the source. The shape (information) remains constant when traveling in a nondispersive medium such as air. This is not the case with travel in solids, where different frequencies travel at different velocities, causing a wave-shape change with time and distance. Since velocity of propagation is constant, time and distance are linearly related and can therefore be plotted on the same axis. (Drawing by Jonathan Meendering; © Walter Grondzik; all rights reserved.)*

Note that it is traditional in architectural acoustics to express area in square centimeters, although the SI system uses square meters. Conversion data for units are given in Chapter 25. A list of symbols and abbreviations is also provided in Chapter 25.

The preceding derivation is based upon a point source—that is, a source that is small relative to the wavelength of the sound produced. This type of source produces spherical waves. Line sources, such as strings, produce cylindrical waves. Large vibrating surfaces, such as walls, produce plane waves. The importance of these distinctions will become clear in the discussion of sound barriers and diffraction in Chapter 25.

(d) The Decibel

Two problems arise immediately when dealing with quantities of the type encountered with sound power, pressure, and intensity. The numbers themselves are very small or very large. Furthermore, the human ear responds logarithmically, not arithmetically, to sound pressure (or intensity); that is, doubling the intensity of a sound does not double its loudness—such a change is barely perceptible. To address these problems, it would be much more convenient if there were a scale that:

1. Started at zero for the minimum sound pressure (or intensity) that can be heard.
2. Used whole numbers rather than powers of 10.
3. Had some fixed relationships between an arithmetic difference and a loudness change, say, 10 units equals a doubling (or halving) of loudness. On such a scale, the difference between 20 and 30, and between 60 and 70, would always be a doubling of loudness.

Such a scale exists. It is the decibel scale. When appended to power, pressure, or intensity, the word level indicates a quantity expressed relative to a base quantity—in decibels. Intensity level is thus the ratio between a given intensity and a base intensity. If intensity level is expressed as:

$$IL = 10 \log \frac{I}{I_0} \qquad \text{(Eq. 8.6)}$$

where

IL = intensity level, dB
I = intensity, W/cm^2
I_0 = base intensity (i.e., 10^{-16} W/cm^2, the threshold of hearing)
log = logarithm to base 10

then a scale is established that satisfies the three conditions presented above. The quantity IL is dimensionless, since it is a ratio between two numbers (whose units cancel). The resulting value is expressed in decibels, however, to clarify its status as a ratio quantity. This provides a convenient way to express the large range of sound magnitudes encountered. Table 8.5 shows the ease of using the logarithmic decibel scale, compared to either decimal notation or exponential notation. Table 8.3 gives a sample of subjective loudness changes expressed in decibels. Note that 10 dB represents a doubling of loudness, and 20 dB a quadrupling of loudness. The difference (Δ) between any two intensity levels can be expressed as:

$$\Delta IL = IL_2 - IL_1 = 10 \log \frac{I_2}{I_0} - 10 \log \frac{I_1}{I_0}$$

therefore,

$$\Delta IL = 10 \log \frac{I_2}{I_1} \text{dB} \qquad \text{(Eq. 8.7)}$$

A few examples using decibel notation and logarithmic calculations should help explain this useful system. By the way, the bel in decibel is in honor of Alexander Graham Bell (thus the capital *B* in dB).

EXAMPLE 8.1 Two sound sources (I_1 and I_2) produce intensity levels of 60 and 50 dB, respectively, at a point. When these sources are operating simultaneously, what is the total sound intensity level? Assume identical frequency content and random phasing—that is, the phase relationship between the two sources changes in a random manner.

SOLUTION

Note that this example deals with intensity level, not intensity; intensity itself has little significance in architectural acoustics. The technique involved in adding two sound intensity levels has three steps:

1. Convert both levels to actual intensity.

$$IL = 10 \log \frac{I}{I_0}$$

so

$$60 = 10 \log \frac{I_1}{10^{-16}}$$

or

$$6.0 = \log \frac{I_1}{10^{-16}}$$

Then, using the definition of a base 10 logarithm:

$$10^6 = \frac{I_1}{10^{-16}}$$

$$I_1 = (10^{-16})(10^6) = 10^{-10} \text{ W/cm}^2$$

By similar calculation,

$$I_2 = 10^{-11} \text{ W/cm}^2$$

2. Add the intensities arithmetically.

$$I_1 + I_2 = 10^{-10} + 10^{-11}$$
$$= (10 \times 10^{-11}) + 10^{-11}$$
$$I_{tot} = 11 \times 10^{-11} \text{ W/cm}^2$$

3. Reconvert to decibels. To find the intensity level (IL) corresponding to the combined or total intensity $I_1 + I_2$, simply apply Equation 8.6:

$$IL_{tot} = 10 \log \frac{I_{tot}}{I_0}$$
$$= 10 \log \frac{11 \times 10^{-11}}{10^{-16}}$$
$$= 10(\log 11 + \log 10^5)$$
$$= 10(1.04 + 5)$$
$$= 60.4 \text{ dB}$$

which is only a fraction larger than the original 60 dB (the intensity level of the stronger of the two sounds). This example shows that decibels cannot be added arithmetically. ■

EXAMPLE 8.2 Assume two sounds of 60 dB each. What is the combined sound intensity level in decibels?

SOLUTION

One method would be to calculate levels as in Example 8.1. A shorter method is to find the difference between the sum and either of the (equal) signals and add it to either individual signal. Using Equation 8.7:

$$\Delta IL = IL_{comb} - IL_1 = 10 \log \frac{I_{comb}}{I_1}$$
$$= 10 \log \frac{2I_1}{I_1}$$
$$= 10 \log 2$$
$$= 10(0.30)$$
$$= 3 \text{ dB}$$

This answer (which is independent of any particular level) yields the useful fact that doubling a signal's

DESIGN FUNDAMENTALS

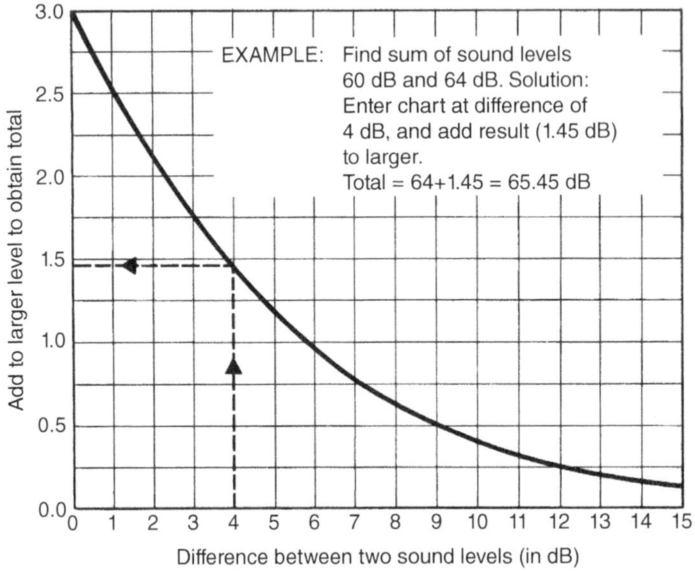

Figure 8.11 Chart for adding two uncorrelated sound pressure levels when both are expressed in decibels. (Reproduced with permission from E.B. Magrab, 1975.)

intensity increases the intensity level by 3 dB. In this case, the combined intensity level would be 60 dB + 3 dB, or 63 dB. Similarly, quadrupling a signal's intensity raises the resulting level by 6 dB. This is because

$$\begin{aligned}\Delta IL &= 10 \log \frac{4I}{I} \\ &= 10 \log 4 \\ &= 10\,(0.60) \\ &= 6\text{ dB}\end{aligned}$$

Therefore, quadrupling 60 dB gives 66 dB (or, alternatively, 60 dB + 60 dB = 63 dB and 63 dB + 63 dB = 66 dB). This finding is very useful when combining many identical sound levels, as in the following example. ■

EXAMPLE 8.3 A factory will contain 20 identical machines, each of which generates a sound intensity level of 80 dB at some point in the factory. What will be the combined sound intensity level? Ignore frequency content, phasing, and sound field distributions.

SOLUTION

$$\begin{aligned}\Delta IL &= IL_{tot} - IL_{single} \\ &= 10 \log \frac{I_{tot}}{I_{single}} \\ &= 10 \log \frac{20\, I_{single}}{I_{single}} \\ &= 10 \log 20 \\ &= 10\,(1.3) = 13\text{ dB}\end{aligned}$$

Therefore, the total sound intensity level will be:

$$IL_{tot} = 80\text{ dB} + 13\text{ dB} = 93\text{ dB}$$ ■

Using a chart (Figure 8.11) to combine the decibel levels of two sources eliminates the somewhat lengthy procedure detailed in Example 8.1. Referring to Table 8.3, note that the human ear is not responsive to fractional decibel changes; indeed, even a 3-dB change is barely perceptible. This being the case, it is recommended that the detailed calculations, and even the chart, be reserved for situations where a high degree of accuracy is required.

For everyday calculations, the following approximations may be used to combine the decibel levels of two sources:

- When the difference between two sources is 1 dB or less, add 3 dB to the higher decibel level to obtain the total.
- When the difference is 2 to 3 dB, add 2 dB.
- When the difference is 4–8 dB, add 1 dB.
- When the difference is 9 dB or more, ignore the lower-level source (add 0 to the higher).

The comparison in Table 8.6 of addition using these rules versus calculations shows that, at typical levels, the error resulting from simplification is always less than 1%.

Using the inverse square law from Equation 8.5, it is possible to determine the effect on sound intensity level of moving away from a sound source.

TABLE 8.6 Addition of Uncorrelated Sound Pressure Levels

dB Levels		Sum	
Lower	Higher	Approximate[a]	Accurate[b]
60	60	63	63.0
60	62	64	64.4
60	64	65	65.5
60	66	67	68.0
60	68	69	68.7
60	70	70	70.5

[a]See text for approximation rules.

[b]In architectural acoustics, decimal values of decibels are not warranted, and these values would be rounded off to a whole number.

EXAMPLE 8.4 Given a sound source that produces an intensity level IL at a distance d_1 from a source, what is the intensity level at twice the distance? At four times the distance?

A sound intensity distribution that follows the inverse square law (on which these calculations are based) requires that a source be in an open (unenclosed), obstruction-free space (e.g., outdoors). Also, intensity measurement I_1 must be taken at a sufficient distance d_1 from the source that a free field has developed. See Chapter 24 for a discussion of sound fields.

SOLUTION

From Equation 8.7 it is known that:

$$\Delta IL = \frac{I_2}{I_1}$$

and from Equation 8.5 that:

$$\frac{I_1}{I_2} = \frac{d_2^2}{d_1^2}$$

Therefore, since $d_2 = 2d_1$

$$\frac{I_2}{I_1} = \frac{(d_1)^2}{(2d_1)^2} = \frac{1}{4}$$

Substituting in Equation 8.7, we have:

$$\begin{aligned}\Delta IL &= 10 \log \frac{I_2}{I_1} \\ &= 10 \log {}^1\!/_4 \\ &= 10(-0.6) \\ &= -6\,\text{dB}\end{aligned}$$

which tells us that sound intensity level (not pressure) is reduced by 6 dB. Similarly, if the distance is quadrupled, it is reduced by 12 dB. ■

To summarize, the intensity level changes by 3 dB with every doubling or halving of power and changes by 6 dB with every doubling or halving of the distance from a point source. Figures 8.12 and 8.13 illustrate the latter relationship.

(e) Sound Power Level

Sound power levels may be derived from sound power values using the ratio manipulation described in Section 8.4. The reference sound power is 10^{-12} W. Sound power level is expressed in decibels.

(f) Sound Pressure Level

Sound pressure levels are derived from sound pressures using the ratio approach described in Section 8.4. The usual reference sound pressure corresponds to the threshold of hearing and is taken to be 20 μPa or 2×10^{-5} Pa (2×10^{-4} microbars [μbar]). As with intensity, this sound pressure reference establishes the 0 dB base for sound pressure level. See Table 8.4

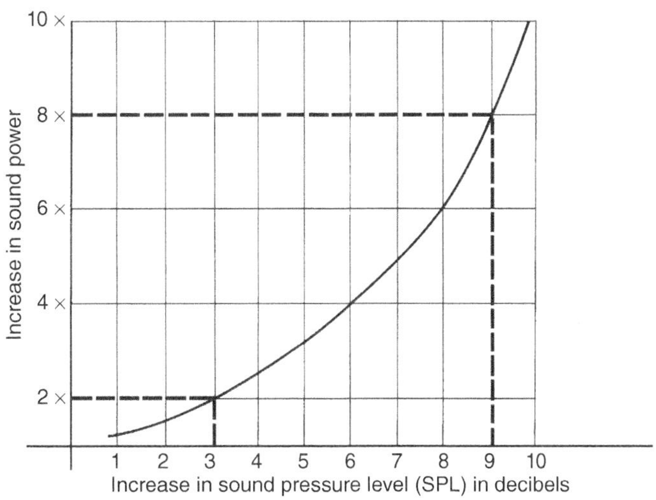

Figure 8.12 *Decibel level increase as a function of power increase.*

DESIGN FUNDAMENTALS

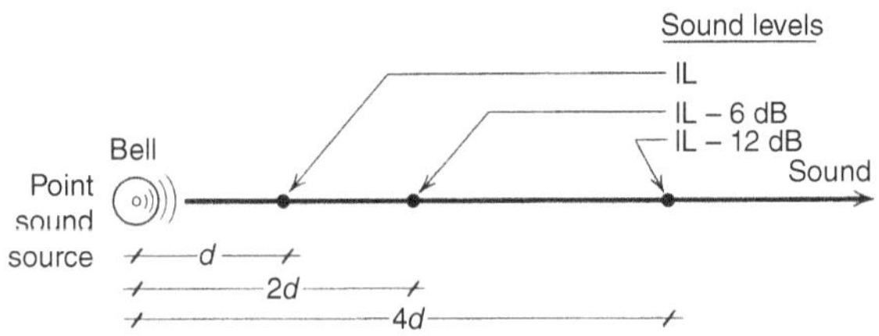

Figure 8.13 Sound intensity levels at varying distances from a source. Each doubling of distance reduces the intensity level by 6 dB. These relationships hold true only in a free field.

for representative values of sound pressure levels and corresponding subjective responses. Since the ear responds logarithmically to intensity and since pressure varies as the square root of intensity, we can write the expression

$$SPL = 10 \log \frac{p^2}{p_0^2}$$

or

$$SPL = 20 \log \frac{p}{p_0} \qquad \text{(Eq. 8.8)}$$

where

SPL = sound pressure level, dB
p = pressure, Pa or μbar
p_0 = reference base pressure, 20 μPa or 2×10^{-4} μbar

Since the 0-dB base corresponds to the threshold of hearing for both sound intensity level and sound pressure level, the decibel scales for sound pressure level and sound intensity level are equalized and the decibel values of the two can be used interchangeably. The actual intensity and the actual pressure corresponding to a particular decibel level, however, are completely different—in magnitude and units. For instance, 70 dB corresponds to 10^{-9} W/cm^2 intensity or to 0.063 Pa pressure. The common thread is that 70 dB corresponds approximately to a given sound magnitude. It is necessary to say "approximately" because assigning a single-number decibel level to a sound faces two difficulties:

1. Sound pressure level varies with time, except for a pure steady tone.
2. The different frequency components of common complex sounds vary in pressure level.

Two techniques are used to overcome this problem. If a sound has a dominant frequency, the sound level at that frequency can be used (see Figure 8.8). This would be the case for a relatively constant sound such as from a motor, fan, or pump. Sounds that vary widely in constituent level and frequency can be plotted on an octave band chart using maximum level for some minimum percentage of time (Figure 8.14). Where the position of the listener is not specified in the table, it is assumed to be at normal close distances: that is, 10–20 ft [3–6 m] from a train, 3–5 ft [0.9–1.5 m] from a radio, and the like.

As noted previously, the combined effect of two sounds depends upon their frequency content. The above examples assumed signals either of identical frequency and random phase or of a very-wide-frequency spectrum—so wide that phase phenomena are not significant. In architectural acoustics work, such an assumption is generally valid.

(g) Measuring Sound

Confirming that design criteria have been met requires a means of measuring sound in the field. A useful, and reasonably economical, measurement instrument is the integrating sound-level meter illustrated in Figure 8.15. To correlate meter readings

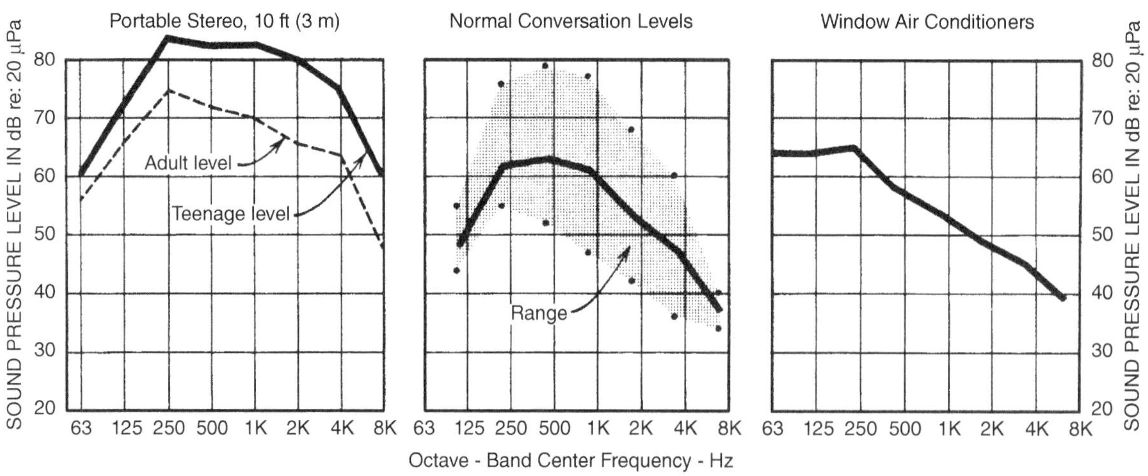

Figure 8.14 Sound pressure level curves of common noise sources plotted according to octave band frequency content. Values shown are averages of multiple measurements. (Reproduced from HUD, 1968.)

Figure 8.15 *General-purpose (type 2) integrating, data-logging sound level meter. This handheld unit will measure and record isolated, nonrepetitive sounds in addition to continuous, fluctuating, and impulsive sounds, and will give both instantaneous and equivalent (Leq) sound levels. The latter are used when an equivalent level of varying sound conditions over a selected measurement time is required. The meter will measure dBA, dBC, and linear with fast, slow, peak, or impulse response. When equipped with a filter set, the meter can be used for octave or ⅓-octave analysis. (© 2013 3M. All rights reserved.)*

with subjective loudness impressions, most such instruments that provide a single-number output are furnished with weighting curves (networks or factors), the characteristics of which are given in Figure 8.16 for the most common weightings: *A*, *B*, and *C*. The weighting curves are derived from mathematical functions applied to sound measurements to account for the frequency sensitivity of the human ear. The *A*-weighting (dBA) is the most commonly used measurement of sound pressure levels and reflects the sensitivity of human hearing. It corresponds to an inverted 40-phon contour and deemphasizes low frequencies (see Figure 8.7), as does the human ear. The *B*-weighting (dBB) and *C*-weighting (dBC) correspond to the 70-phon and 100-phon contours, respectively. In basic terms:

- dBA emphasizes midrange frequencies and deemphasizes low frequencies
- dBB provides less attenuation of lower frequencies than *A*-weighted
- dBC provides a flatter response across more frequencies and offers much less attenuation than the *A*-weighted

A completely linear, precise unweighted response is often also available—this is sometimes referred to as a Z-weighting (dBZ). The logic behind these weighting curves is embedded in the equal loudness curves of Figure 8.7—to have a meter that responds generally similar to a human ear. The original intention was to use the *A*-weighting at levels of up to 55 phons, the *B*-weighting to 85 phons, and the *C*-weightings at higher level sound measurement. It was found, however, that only the *A*-weighting corresponds well to subjective loudness impressions. As a result, the *B*-weighting has fallen into disuse, and the *A*-weighting

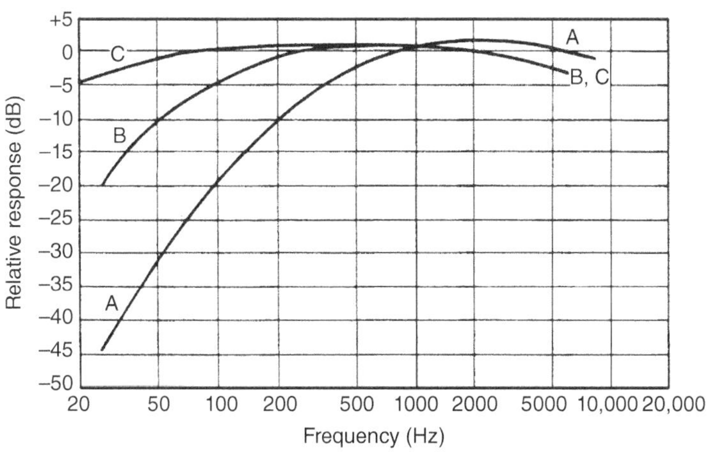

Figure 8.16 *Internationally standardized A, B, and C weighting curves.*

TABLE 8.7 Common Sound Pressure Levels and Subjective Impressions

Sound Pressure Level (dBA[a])	Typical Sound	Subjective Impression
150		(Short exposure can cause hearing loss)
140	Jet plane takeoff	
130	Artillery fire, riveting, machine gun	(Threshold of pain)
120	Siren at 100 ft (30 m), jet plane (passenger ramp), thunder, sonic boom	Deafening
110	Woodworking shop, hard-rock band, accelerating motorcycle	Sound can be felt (threshold of discomfort)
100	Subway (steel wheels), loud street noise, power lawnmower, outboard motor	
90	Noisy factory, unmuffled truck, train whistle, machine shop, kitchen blender, pneumatic jackhammer	Very loud, conversation difficult; ear protection required for sustained occupancy
80	Printing press, subway (rubber wheels), noisy office, supermarket, average factory	(Intolerable for phone use)
70	Average street noise, quiet typewriter, freight train at 100 ft (30 m), average radio, department store	Loud, noisy; voice must be raised to be understood
60	Noisy home, hotel lobby, average office, restaurant, normal conversation	
50	General office, hospital, quiet radio, average home, bank, quiet street	Usual background; normal conversation easily understood
40	Private office, quiet home	
30	Quiet conversation, broadcast studio	Noticeably quiet
20	Empty auditorium, whisper	
10	Rustling leaves, soundproof room, human breathing	Very quiet
0		Intolerably quiet Threshold of audibility

[a]dBA denotes an A-weighted sound pressure level.

is used today for all measurements, regardless of loudness. All sound pressure level measurements must be associated with the weighting network used (e.g., 50 dBA or 100 dBC).

More accurate measurements of complex sounds than are possible with a standard sound level meter are made with sophisticated instruments that measure intensity in octave bands and often plot the results, as per the graphs in Figure 8.14. Such measurements are necessary for accurate application of sound absorption and attenuation materials whose characteristics are nonlinear over the frequency spectrum.

Single-number dBA readings are known as overall levels and are useful as preliminary data and for broad-spectrum design. Table 8.7 lists common sound pressure levels as measured by the dBA scale. Such single-value numbers are useful to establish a reference base and for use in maximum noise exposure calculations, as discussed in the following section.

8.4 ACOUSTICAL COMFORT DETERMINANTS

(a) Noise

Noise is unwanted sound. It should be assumed that any sound can at some point be referred to as noise by someone—my "music" is your "noise."

Although noise and its control are the subject of Chapter 25, noise criteria (NC) and their development are discussed here as part of an overview of sound and hearing. There are two basic approaches to the negative effects of noise: a psychological-practical one, and a purely physiological one. The former is concerned with annoyance and disturbance to daily activities, including work, relaxation, and rest. The latter is concerned with the physical impact of noise on the body, including hearing loss.

(b) Annoyance

Research has developed accurate data on perceptions of loudness. Predicting the annoyance of a sound is more elusive—as it is even more subjective than noise, with at least six factors of concern. Tests have shown that in general, annoyance resulting from noise is:

1. proportional to the loudness of the noise
2. greater for high-frequency than low-frequency noise
3. greater for intermittent than continuous noise
4. greater for pure-tones than for broadband noise
5. greater for moving or unlocatable noise than for fixed-location noise
6. much greater for information-bearing noise than for nonsense noise

(c) Noise Criteria

To establish criteria for acceptable background noise (i.e., noise that is not too annoying to space users), some of these effects must be neglected for the sake of simplicity. They are considered in architectural design and in setting masking noise levels (see Chapter 25). Ignoring, for the time being, these factors from the list in the preceding section:

Factor 3—by assuming continuous sounds

Factor 4—by assuming broadband sound

Factor 5—by assuming the noise source is fixed in location

Factor 6—by assuming only general noise level, not actual content

Thus, the characteristics of sounds such as a barking dog (3), a whistle (4), a single passing vehicle (5), and intelligible sounds (6) are not considered when establishing conventional noise criteria.

To quantify acceptable background noise conditions required removing annoyance from the purely psychological arena by relating it to a physical variable. This was done by studying the effect of annoyance factors 1 and 2 on speech communication. A study by Leo Beranek resulted in two design concepts:

Articulation Index (AI): This is the ratio of correctly identified syllables to total syllables read. An AI greater than 0.5 was considered indicative (Beranek, 1988a) of a condition in which acceptable intelligibility could be expected for male voices.

Speech Interference Level (SIL): This is a simplified version of the AI, and was also devised by Beranek. It consists simply of the arithmetic average (in decibels) of the background sound pressure levels in the four octave bands centered on 500, 1000, 2000, and 4000 Hz, under which acceptable intelligibility could be expected, for a given voice effort, at a given distance between the speaker and listener.

Both AI and SIL are determined by reading a carefully selected set of phonetically balanced nonsense syllables to a test audience in the presence of different levels and compositions of background noise.

(d) Noise Criteria Curves

Beranek also developed the widely used NC curves shown in Figure 8.17. NC curves offer design teams a convenient way to specify "criteria" for background noise in a space. Design teams may set a specific NC curve for any given space. Each NC curve represents the SPL across the frequency spectrum for background noise that should not be exceeded in a specified environment. Higher decibel noise levels (i.e., higher SPL) are permitted at the lower frequencies since the human ear is less sensitive to sound in this frequency region. NC curves are based on the observation that most people prefer to speak at a level no greater than 22 dB above the background noise level. The NC curves were derived by combining the SILs (in decibels) with this behavioral phenomenon; they represent an SPL 22 dB higher than the SIL. The contours then represent the maximum continuous background noise likely to be considered acceptable in a specified environment, and generally correspond to background sound pressure levels found reasonably acceptable in selected commercial/institutional building environments. Designers should remember that the quantitative NC curves are based upon observations of building user behaviors and perceptions. A similar set of design targets called noise rating (NR) curves finds considerable application outside the United States. They are less stringent than NC curves in low frequencies but more stringent in high frequencies.

To apply the NC curves, the spectrum (from 63 to 8000 Hz) of a specific sound of concern is overlaid on a graph of the NC curves. The lowest NC curve that is not exceeded by any portion of the selected sound spectrum is the NC rating of that sound (which may be a noise to most occupants). Thus, specifying a maximum noise level of NC-30 for a space means that no portion of the SPL plot of any continuous background noise in the space may cross the NC-30 contour. A piece of equipment rated NC-35 has a spectrum completely below NC-35. A fan rated NC-53 means that at some frequency the fan exceeded NC-50 by 3 dB.

The NC rating of a sound (potential noise) usually falls between 5 and 10 dB below the measured dBA for the sound, depending upon the shape of its spectrum. The virtue of the NC curves is that they provide a single-number specification for sound across a wide frequency range—without losing important information in the frequency distribution. Their disadvantage is that they were derived for, and are most accurate with, speech conversation conducted against a backdrop of continuous, broadband noise. This is not the situation that is typically most troublesome in an office; indeed, continuous equipment noise may be helpful in masking unwanted speech. Nevertheless, NC (or RC, see below) curves remain the most commonly used design criteria for establishing acceptable continuous background sound levels.

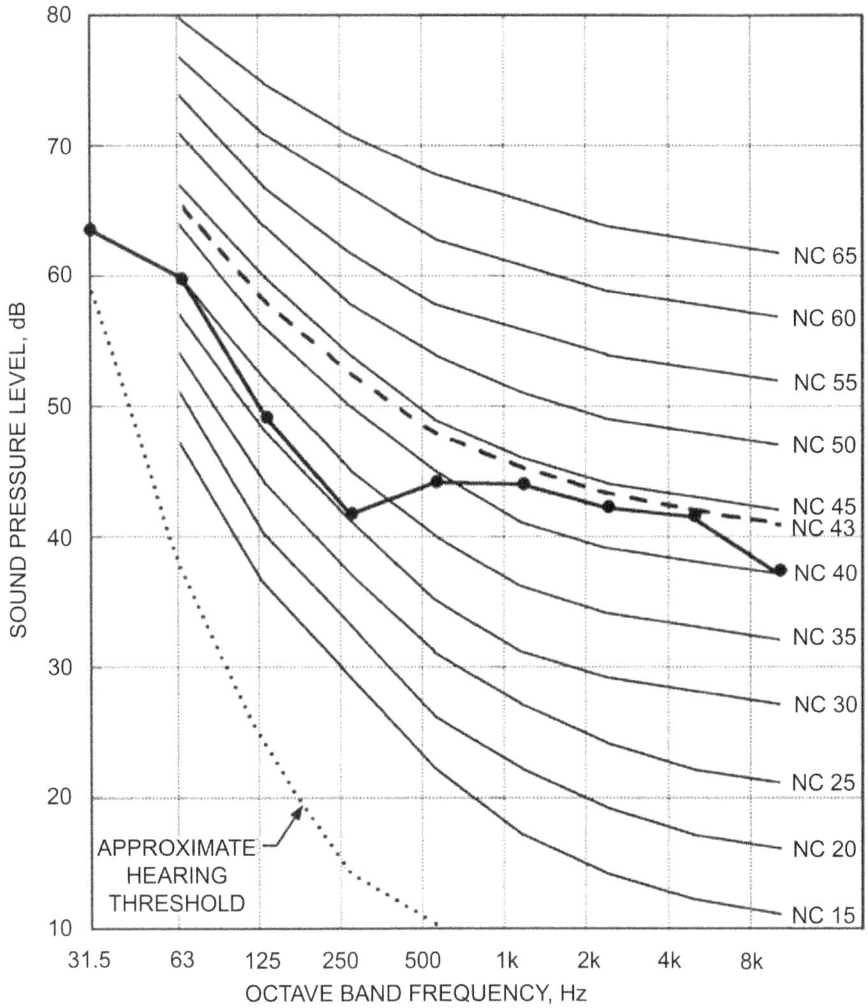

Figure 8.17 *NC (Noise Criteria) curves and sample spectrum (curve with symbols). The distribution of sound pressure level versus frequency shown in the heavy line with the dots would be judged to comply with the limits for an NC 43 rating. (© ASHRAE, www.ashrae.org. 2021 ASHRAE Handbook—Fundamentals, Chapter 8, Figure 7.)*

(e) Room Criteria Curves

Historically, NC curves have been commonly used as a quick and easy way to set and assess HVAC noise specification and general building acoustic criteria. However, there are shortcomings in the NC curves in that they do not consider the very low frequencies (16 and 31.5 Hz octave bands) and seem insufficiently stringent at frequencies above 2 kHz. To address these issues, a similar approach to setting design criteria was developed, called room criteria (RC) curves. These curves, shown in Figure 8.18, were adopted by ASHRAE as the suggested background sound target, in preference to NC curves. The curves were developed to evaluate the acceptability of mechanical system background sound for typical space types. RC curves differ from NC curves in several ways:

- They are straight lines.
- Their slope is constant at −5 dB per octave (determined from extensive tests, mostly in the range of 40–50 dB).
- Regions labeled A and B, as in Figure 8.18, address the problem of very low frequencies and high sound pressure level, an issue that is ignored in the NC criteria. This region deals with rumble and vibration that can cause extreme annoyance for many occupants.

Referring to Figure 8.18, the procedure for determining the RC value of a specific item of equipment with a known sound spectrum is as follows:

1. Calculate the arithmetic average of sound pressure levels in the 500-Hz, 1000-Hz, and 2000-Hz octave bands. This number is the RC value of that spectrum.

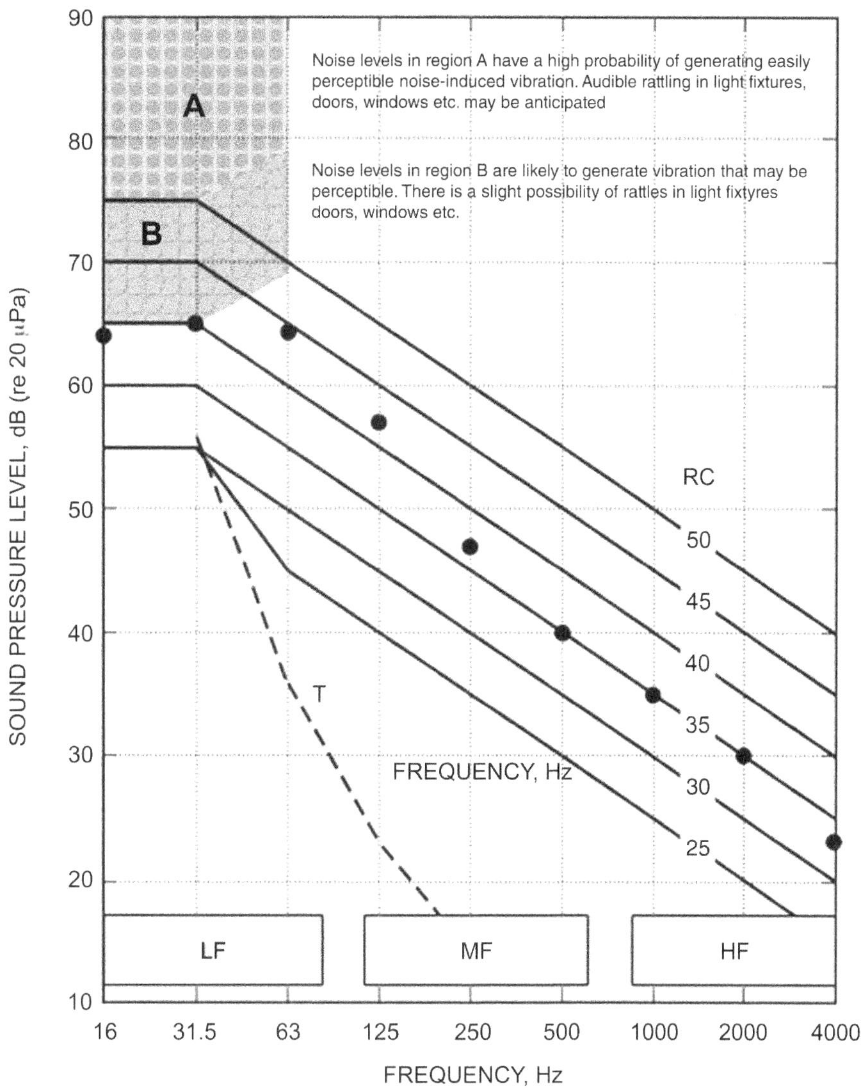

Figure 8.18 *RC curves. (© ASHRAE, www.ashrae.org. 2023 ASHRAE Handbook—HVAC Applications, Chapter 49, Figure 6.)*

2. At 1000 Hz, draw a straight line at a – 5 dB slope through this value of RC.
3. Plot the SPL values for the octave band center frequencies on the RC curve sheet and compare the plot to the RC line drawn in step 2.
4. Classify the equipment sound quality from this comparison as follows:

    a. *Neutral:* If the octave band data plotted in step 3 do not exceed the RC line drawn in step 2 by more than 5 dB at or below 500 Hz and more than 3 dB at or above 1000 Hz, then the sound is considered neutral (bland, uncharacteristic), and the designation *N* is placed after the RC level. A piece of mechanical equipment with an *N* designation is understood to have a neutral tone quality that most people would classify as unobtrusive and lacking specific character. ASHRAE design guidelines for HVAC system sound pressure levels are listed as RC(*N*).
    b. *Rumble:* If the octave band plot does exceed the RC line by more than 5 dB at 500 Hz or below, the spectrum is classified as "rumbly," and the descriptor *R* is appended to its RC level number.
    c. *Hiss:* If the octave band plot does exceed the RC line by more than 3 dB at or above 1000 Hz, it is classified as "hissy," and the descriptor *H* is appended to its RC level.
    d. *Vibration:* The shaded areas in Figure 8.18 labeled *A* and *B* represent high sound pressure levels in the 31.5-Hz and 63-Hz octave bands. In the *A* area, vibration will likely be felt in light construction and furniture, and rattling may

occur in loosely connected devices, cabinets, glassware, and the like. In the lower-energy *B* area, rattling is less likely but vibration may still be felt.

The result of this classification is to give the designer not only an average SPL for an item of equipment, but also a sense of subjective sound quality. This information should assist in determining the noise abatement measures that need to be taken.

ASHRAE updated the original RC curve concept to the *RC Mark II* method. This updated method is a bit more complex than the original approach. For further information, see the *2023 ASHRAE Handbook—HVAC Applications* (ASHRAE, 2023). The original NC curves have also been modified (by Beranek) and issued as balanced noise criteria (NCB) curves. This revision amended a failing of the NC approach by adding very-low-frequency coverage, and the NCB curves were straightened, so that they resemble RC curves above 125 Hz, except that the slope angle is −3.33 dB, compared to −5 dB for the RC curves. The slope at the higher frequencies eliminates the hiss that characterized equipment noise conforming to the original NC criteria. These adjustments yield a "balanced" neutral sound, which is reflected in the chart's name. Application of these curves to mechanical equipment noise is similar to that for RC curves, except that letter descriptors are not used. For a full discussion of the construction and application of NCB curves, see Beranek (1988b).

Developed to address the limitations of the NC curves, the RC curves provide a more detailed criteria by addressing both the sound pressure level and the quality of noise in a space. RC curves are often used for spaces (e.g., hospitals) where the quality of noise (e.g., rumble or hissing) is as critical as the overall noise level.

Given several competing options for criteria, selection of NC, RC, RC Mark II, or NCB to establish background sound criteria is project-specific and may be part of the development of the owner's project requirements. The approach that best meets the needs (and budget) of the client (tempered by design team professional judgment) should be chosen.

(f) High Noise Levels and Hearing Protection

Continuous exposure to high sound pressure (noise) levels causes a degree of temporary deafness in most people. Long periods of such exposure, even on an intermittent eight-hour workday basis, can produce permanent hearing impairment. Most experts place the safe eight-hour upper limit at 85 dBA. In addition, studies have indicated that continual exposure to noise levels as low as 75–85 dBA can produce or contribute to numerous physical and psychological ailments, including headache, digestive problems, tachycardia, high blood pressure, anxiety, and nervousness—an extensive palette of human illnesses. Since continuous noise exposure is most severe in industry, regulatory legislation in the United States has been directed at this area.

In 1969, the Walsh-Healey Public Contracts Act was passed; thereafter its provision for maximum permissible exposure to noise was incorporated into the Occupational Safety and Health Act. The associated regulatory agency is the Occupational Safety and Health Administration (OSHA). Key noise provisions of this act are reproduced in Figure 8.19. To avoid overly complex regulations, limitations on exposure are given as single-number dBA values. Since workers rarely remain in a single acoustic environment for eight hours, their total daily exposure, or time-weighted average (TWA) exposure, can be calculated from timed dBA measurements, using formulas and tables given by OSHA, and then compared to permissible levels. Alternatively, a dosimeter (Figure 8.20) can be used, which automatically integrates the noise to which a person is exposed over a given time period and reads out the permissible TWA exposure directly.

Typical industrial noise levels are given in Table 8.8. When permissible levels are exceeded, management must take steps to reduce the exposure, either by reducing the noise or by providing hearing protection. Characteristics of a few typical types of ear protectors are given in Figure 8.21. Photographs of three of the most common ear protectors in industrial use are shown in Figure 8.22.

OSHA does not deal extensively with impulse noises, except to state that they shall not exceed a 140-dB peak sound pressure level. Impulse noise is quite different from continuous noise, since effective noise levels depend on duration, and specification is difficult. Much work has been done in this area by the military, but this subject is outside of the realm of architectural acoustics.

(g) Vibration

Sound is heard; vibration is felt. If sound is defined as a pressure variation that is audibly received, then vibration is a tactilely received pressure variation. Vibrations of concern to building design have frequencies that

§ 1910.95 Occupational noise exposure.

(a) Protection against the effects of noise exposure shall be provided when the sound levels exceed those shown in Table G-16 when measured on the A scale of a standard sound level meter at slow response. When noise levels are determined by octave band analysis, the equivalent A-weighted sound level may be determined as follows:

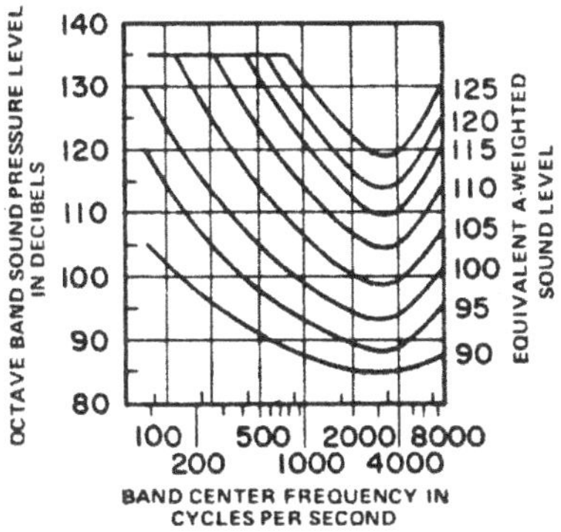

Equivalent sound level contours. Octave band sound pressure levels may be converted to the equivalent A-weighted sound level by plotting them on this graph and noting the A-weighted sound level corresponding to the point of highest penetration into the sound level contours. This equivalent A-weighted sound level, which may differ from the actual A-weighted sound level of the noise, is used to determine exposure limits from Table G-16.

[1910.95 amended at 39 FR 19468, June 3, 1974]

(b)(1) When employees are subjected to sound exceeding those listed in Table G-16, feasible administrative or engineering controls shall be utilized. If such controls fail to reduce sound levels within the levels of Table G-16, personal protective equipment shall be provided and used to reduce sound levels within the levels of the table.

(2) If the variations in noise level involve maxima at intervals of 1 second or less, it is to be considered continuous.

(3) In all cases where the sound levels exceed the values shown herein, a continuing, effective hearing conservation program shall be administered.

TABLE G-16—PERMISSIBLE NOISE EXPOSURES[1]

Duration per day, hours	Sound level dBA slow response
8	90
6	92
4	95
3	97
2	100
1½	102
1	105
½	110
¼ or less	115

[1] When the daily noise exposure is composed of two or more periods of noise exposure of different levels, their combined effect should be considered, rather than the individual effect of each. If the sum of the following fractions: $C_1/T_1 + C_2/T_2 + \cdots C_n/T_n$ exceeds unity, then, the mixed exposure should be considered to exceed the limit value. Cn indicates the total time of exposure at a specified noise level, and Tn indicates the total time of exposure permitted at that level.

[1910.95 Table G-16 amended at 39 FR 19468, June 3, 1974]

Exposure to impulsive or impact noise should not exceed 140 dB peak sound pressure level.

Figure 8.19 The OSHA standard for exposure to noise in the workplace (Standard Number 1910.95—Occupational noise exposure.). (From OSHA, July 1988; current as of May 2025.)

begin just below the range of human hearing at around 20 Hz. For someone with a hearing threshold above 20 Hz, the vibration frequency transition might be higher. The means of energy reception by the body is more important than specific frequency cutoffs. In general, vibration in a building is always an unwanted experience. Most mechanical and electrical equipment produces energy output that will be sensed as noise and also output that will be sensed as vibration. Various means of vibration control are presented in Chapter 25. As vibration is not received through the sense of hearing, vibration mitigation approaches are quite different from noise mitigation approaches.

8.5 ACOUSTICAL COMFORT DESIGN CRITERIA

As with the topic of visual comfort (see Chapter 7), it is reasonable to assume that acoustical comfort would be a design objective for a variety of building space types. This is (or should be) the case. Unfortunately, there is no

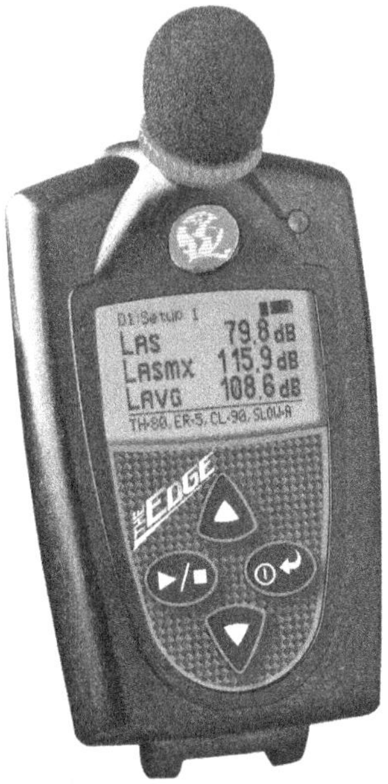

Figure 8.20 A noise dosimeter. The unit is carried by a person whose noise exposure is being tested, with the clip-mounted microphone at shoulder (ear) level. The meter will display, store, and calculate maximum, minimum, and peak levels of SPL, Leq, TWA, exposure levels, and dosage. Noise-level histories are stored and can be printed out as desired. The unit shown measures 5.5 × 2.8 × 1.4 in. [140 × 70 × 40 mm] and weighs 15.5 oz. [440 g]. (Courtesy of Quest Technologies.)

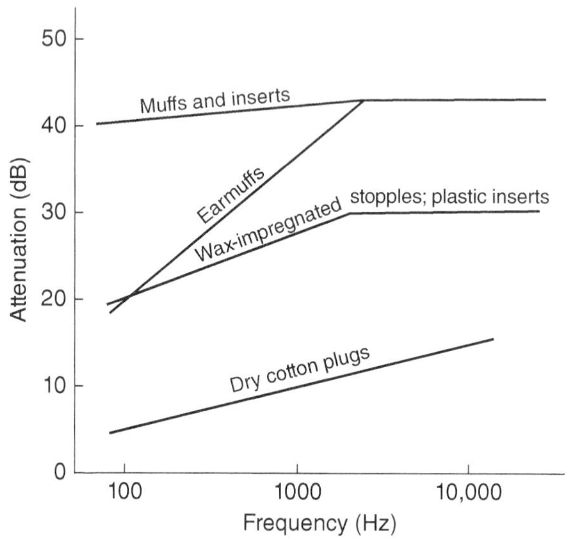

Figure 8.21 Sound attenuation characteristics typical of various types of ear protectors. For design purposes, specific attenuation characteristics must be used and the attenuation figures modified to reflect field usage. (From NBS, 1976.)

TABLE 8.8 Typical Industrial Sound Pressure ("Noise") Levels[a]

Equipment	dBA
Printing press plant (medium-sized automatic)	86
Heavy diesel-propelled vehicle (about 25 ft [7 m] away)	92
Heavy-duty grinder	93
Air compressor	94
Plastic chipper	96
Cutoff saw	97
Multiple spot welder	98
Turbine condenser	98
15-ft^3 [425-L] air compressor	100
Drive gear	103
Banging of steel plate	104
Magnetic drill press	106
Air chisel	106
Positive displacement blower	107
Air hammer	107
Vacuum pump	108
Jolt squeeze hammer	122

[a]These are approximate values for some typical generic equipment types and should not be used as design values.

formal definition of acoustical comfort, no clear understanding of which acoustical characteristics contribute to such comfort in a particular situation, and no unified theory of acoustical comfort. This makes designing for good acoustics more of a challenge than might otherwise be the case but is no excuse for bad acoustical design.

The owner's project requirements (OPR) for a building or a given space within a building can set specific criteria for acoustical design efforts. OPR consists of design issues, related design intents, and verifiable design targets (criteria). Typical acoustical design intents might include:

- adequate loudness
- appropriate reverberation
- acceptable articulation
- spatial flexibility
- acceptable background sound levels via NC or RC curves
- avoidance of anomalies (echo, creep, focusing, distortion)
- cultural appropriateness

Project context will determine whether "acceptable" articulation needs to be "outstanding" articulation. Project context will also guide establishment of design criteria that will work for the project and lead to owner and user satisfaction. Suggestions to address specific criteria will be found in later chapters of this book that deal with noise control and room acoustics.

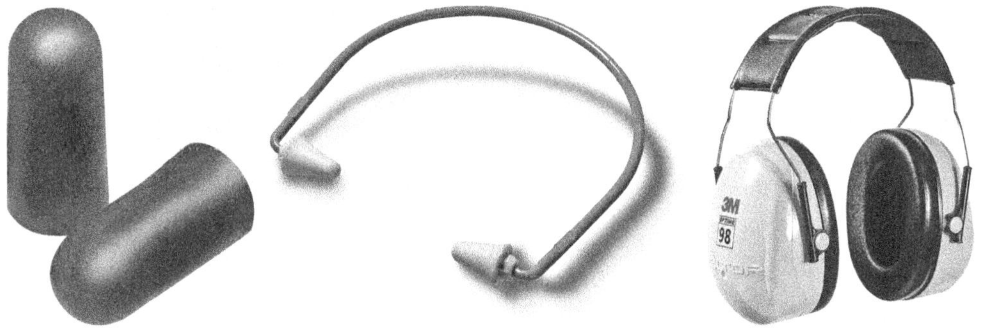

(a) (b) (c)

Figure 8.22 *Three common types of ear protection devices: (a) PVC foam ear plugs are rolled and squeezed to reduce their diameter before being placed in the ear canal, where they re-expand. They are generally discarded after a single use. The ear plugs have an EPA noise reduction rating (NRR) of 32 when tested to ANSI 53.19–1974; the National Institute for Occupational Safety and Health (NIOSH) recommends multiplying such NRR values by an adjustment factor to account for real-world conditions. (b) Ear protector for intermittent use consists of (replaceable) foam pads mounted on a flexible neck band. The foam does not require rolling. The unit is rated NRR 28. (c) Spring-type headband and foam-filled ear cushions provide an NRR rating of 30 for this earmuff design. (© 2013 3M. All rights reserved.)*

References and Resources

ASHRAE. 2021. *2021 ASHRAE Handbook—Fundamentals*. Atlanta, GA: American Society of Heating, Refrigerating and Air-Conditioning Engineers.

ASHRAE. 2023. *2023 ASHRAE Handbook—HVAC Applications*. Atlanta, GA: American Society of Heating, Refrigerating and Air-Conditioning Engineers.

Beranek, L. L. 1988a. *Acoustical Measurements*, rev. ed. Melville, NY: Acoustical Society of America.

Beranek, L. L. 1988b. *Noise and Vibration Control*, rev. ed. Washington, DC: Institute of Noise Control Engineering.

HUD. 1968. *A Guide to Airborne, Impact, and Structure Borne Noise–Control in Multifamily Dwellings*. Washington, DC: U.S. Department of Housing and Urban Development.

Magrab, E. B. 1975. *Environmental Noise Control*. Hoboken, NJ: John Wiley & Sons.

NBS. 1976. *Quieting: A Practical Guide to Noise Control* (Handbook 119). Washington, DC: National Bureau of Standards (National Institute of Standards and Technology).

OSHA. 1988. "Occupational Noise Exposure." Occupational Safety & Health Administration. Washington, DC: U.S. Department of Labor. www.osha.gov/SLTC/noisehearingconservation/index.html

White, F. A. 1975. *Our Acoustic Environment*. New York: John Wiley & Sons.

P A R T I I I

BUILDING ENCLOSURE

Part III focuses on conveying an understanding of the building enclosure as a mediator between ambient outdoor conditions (including both useful resources and potential problems) and an indoor environment that will support the health and comfort of building users. These indoor environments are provided by passive and/or active building systems. Appropriate enclosure design can make life easier for such systems and contribute greatly to energy efficiency and carbon emissions mitigation.

Chapter 9 addresses building enclosures as part of the larger building system and discusses the components of an enclosure (primarily from a hygrothermal and lighting perspective). Chapter 10 explores in detail the flow of heat, air, and moisture through building enclosures and the role of control layers in meeting design criteria. Chapter 11 discusses the principles and concepts of solar geometry and offers guidance on the design of shading devices.

CHAPTER 9

Building Enclosure and Systems

WHY PRESENT INFORMATION ABOUT BUILDING ENCLOSURES in a book about mechanical and electrical systems? Simply put, key MEP (mechanical, electrical, and plumbing) systems are tasked with providing services to close the gap between desired indoor conditions and the ambient outdoor environment that exists after intervention by the building enclosure. Further, desired building outcomes (such as thermal, visual, or acoustic comfort) are directly affected by the quality of the separation between outdoor and indoor conditions. It is not possible to design for indoor conditions without designing the building enclosure.

9.1 THE BUILDING ENCLOSURE

From a building science perspective, the exterior enclosure (or envelope) of a building consists of numerous materials and components that are assembled on-site to meet the intents of the owner and the design team. A building enclosure typically includes some prefabricated components (such as windows and doors) that are available off the shelf and have well-defined and tested performance characteristics. The typical enclosure also includes materials in a variety of forms (sheets, blocks, bulk products, membranes, etc.) that are site-assembled to meet design requirements. These components and materials may be assembled into commonly used configurations with generally understood performance, or into configurations unique to a given project and of uncertain performance. The job of enclosure design and analysis is to ensure that a proposed enclosure solution will meet the project's design intent and criteria (including building codes).

From a functional perspective, the enclosure of a building is not merely a two-dimensional exterior surface; it is an active three-dimensional transition space. Sun and daylight are admitted or rejected; breezes and sounds are channeled or deflected; and rain is repelled or collected. This transition space is where people indoors might experience something of what the outdoors is like at the moment (Figure 9.1), as well as where people outside can get a glimpse of the functions within.

Figure 9.1 *A south-facing office façade in Oregon forms a microclimate zone that buffers the transition between indoor and outdoor conditions. Groundcover plants are placed at eye level for seated occupants and deciduous vines overhead give a seasonally changing view to the outdoors through a façade that also admits winter sun, year-round daylight, and summer breeze. (Photo by Amanda Clegg.)*

Mechanical and Electrical Equipment for Buildings, Fourteenth Edition. Walter Grondzik, Daniel Overbey, and Alison Kwok.

Companion website: www.wiley.com/go/meeb14e

9.2 BUILDING ENCLOSURE DESIGN INTENT

(a) Enclosure Functions

Called by their familiar names, the basic components of building enclosure include windows, doors, floors, walls, and roofs. On closer inspection, windows can include skylights, clerestories, screens, shutters, drapes, blinds, diffusing glass, and reflecting glass—an array of components that determines how the enclosure does its job of making the transition between inside and outside. Norberg-Schulz (1965) suggests that a component can more fundamentally be thought of in terms of its design intent relative to the exchange of energies: as a connector, filter, switch, or barrier:

> In general, we define a connector as a means to establish a direct connection, a filter as a means to make the connection indirect (controlled), a switch as a regulating connector, and a barrier as a separating element ··· An opaque wall thus serves as a filter to heat and cold, and as a barrier to light. Doors and windows have the character of switches, because they can stop or connect at will.

In addition to these historic enclosure intents, a contemporary option is notable: a transformer. A transformer is intended to convert an environmental force (such as solar radiation) directly into a different and desirable energy form (such as electricity). A photovoltaic roof shingle is an example of a transformer (it does more than simply connect, filter, block, or switch).

The fundamental range of choices surrounding enclosure intent and components can be illustrated by two opposite design concepts: the open frame and the closed shell. In harsh climates (or where unwanted external influences such as noise or visual clutter abound), the designer frequently conceives of the building enclosure as a closed shell and proceeds to selectively punch holes in it to make limited contact with the outdoors. Such an approach might be called barrier-dominated. When external conditions are very close to the desired internal ones, the enclosure often begins as an open structural frame, with pieces of building enclosure selectively added to modify only a few outdoor forces. Such an approach might be called connector-dominated. A connector, filter, or barrier for one natural force may change its role for another force. Glazing, for example, may be a connector to light but a barrier to wind. The open-frame or the closed-shell approach to enclosure design, when combined with material availability and the influence of local culture, can produce a distinct regional architecture, as shown in Figure 9.2.

With a wide range of energy sources, building materials, and mechanical equipment available, it is possible to build connector-dominated buildings anywhere, regardless of the climate. The consequences of the resulting energy consumption can be severe. In contrast, with the advent of ubiquitous air conditioning, the most energy-efficient building—even in hot and humid climates—often becomes a barrier-dominated enclosure, often in direct contrast to the vernacular architecture of the region. The designer's use of connectors, filters, barriers, switches, and transformers is basic to the design of building enclosures.

(b) Switches

Switches are a designer's way of allowing an enclosure to respond in a variable manner and/or giving building occupants some control of their own environment. Seasons and spatial functions change; people are unpredictable. If the designer has carefully considered the range of choices that a switch will provide, successful user control of that switch is possible. Building enclosures plentifully supplied with switches become demonstrations that architecture can have dynamic and evolving form. Figure 9.3 shows a remarkably integrated switch-dominated enclosure in vernacular architecture.

Solar control devices are perhaps the most common and visible switches. Awnings block direct solar radiation at some times and admit it at others. Opaque drapes can expose all of a window to incoming daylight on dark days or block out daylight entirely when desired. Translucent curtains can change bright sun into a diffuse light or be drawn back to allow strong direct sun and shadow to be cast inside a room. Passive solar heating systems rely on switches to control incoming solar radiation on warmer days. Operable windows are a commonly used thermal switch. Ventilating switches may be separate from windows.

Visually, switches are a particularly promising source of three- and four-dimensional interest on a building's exterior. Considering the office buildings in Figure 9.4, on which façade would the daily and seasonal changes be most visually evident? If it is easier to imagine human beings working behind the windows of one building than of the other, might that suggest a more satisfactory work environment? Might more personal control of a window promote greater comfort and increased productivity? Climate-responsive buildings are likely to look different on each façade due to the changing solar geometry that is unique to each orientation as discussed in Chapter 11.

Switches encourage interaction between people and their environments. This is usually satisfying to the users, who are able to select a desired exposure to the

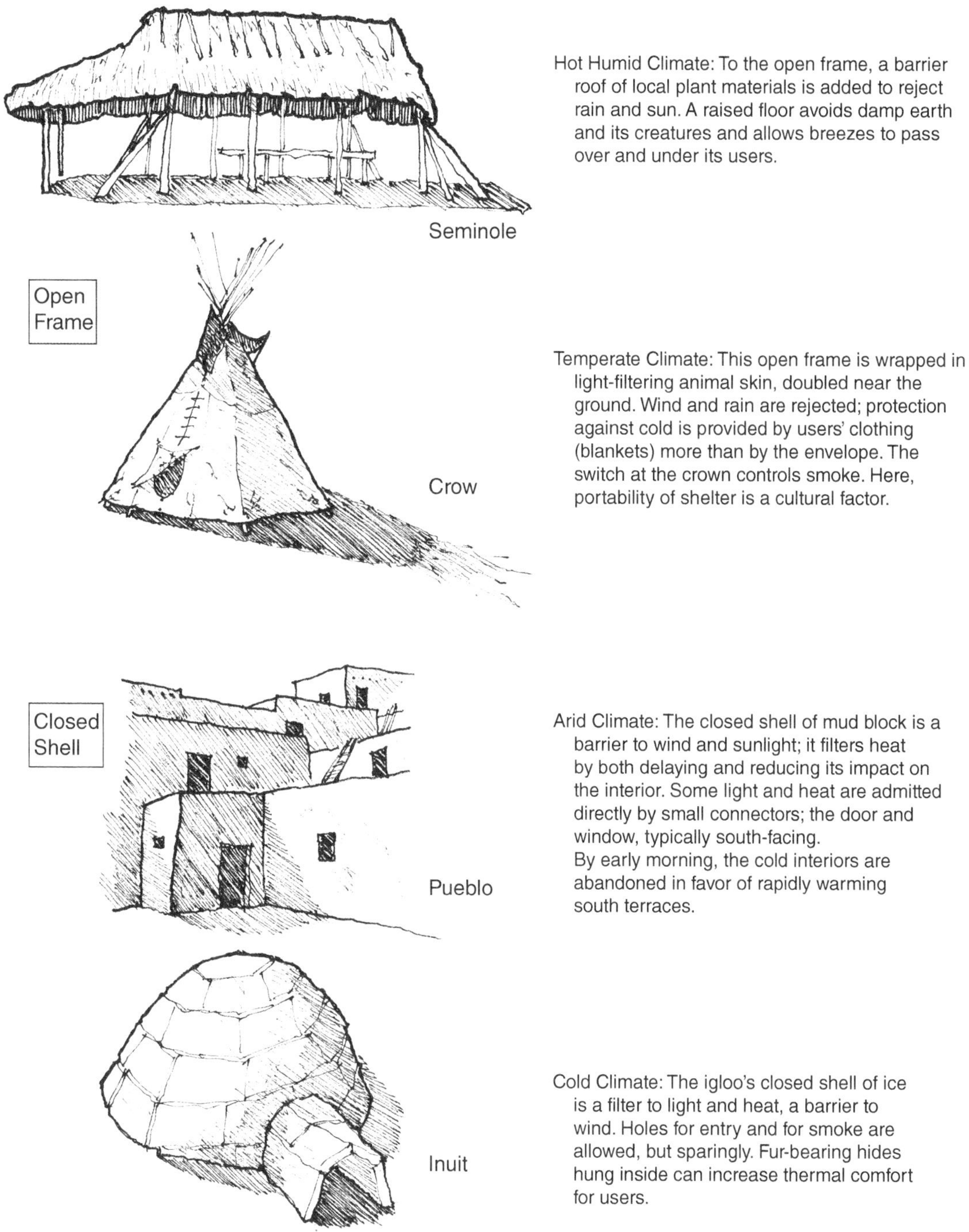

Figure 9.2 *Open-frame (Seminole and Crow) and closed-shell (Pueblo and Inuit) enclosure approaches are influenced by climate, materials, and culture. The influence of climate is dramatic, but material availability and cultural expectations also influence the enclosure design solutions in these examples. (Drawing by Dain Carlson.)*

climate at a given moment. Yet without automation, supervision, or training, switches can also be detrimental to system performance. Examples include a thermal shade in a passive solar heating system left closed on a sunny winter morning; a vent left open during the hottest hours in a high-mass, night-ventilated building; or an awning rolled up to expose a window to summer sun. People in buildings often use switches as they feel appropriate, as demonstrated in Figure 9.5. Air-conditioned buildings typically do away with user-operated ventilating switches (operable windows) so that the system will instead function with a closely and automatically controlled flow of filtered outdoor air. Thermally

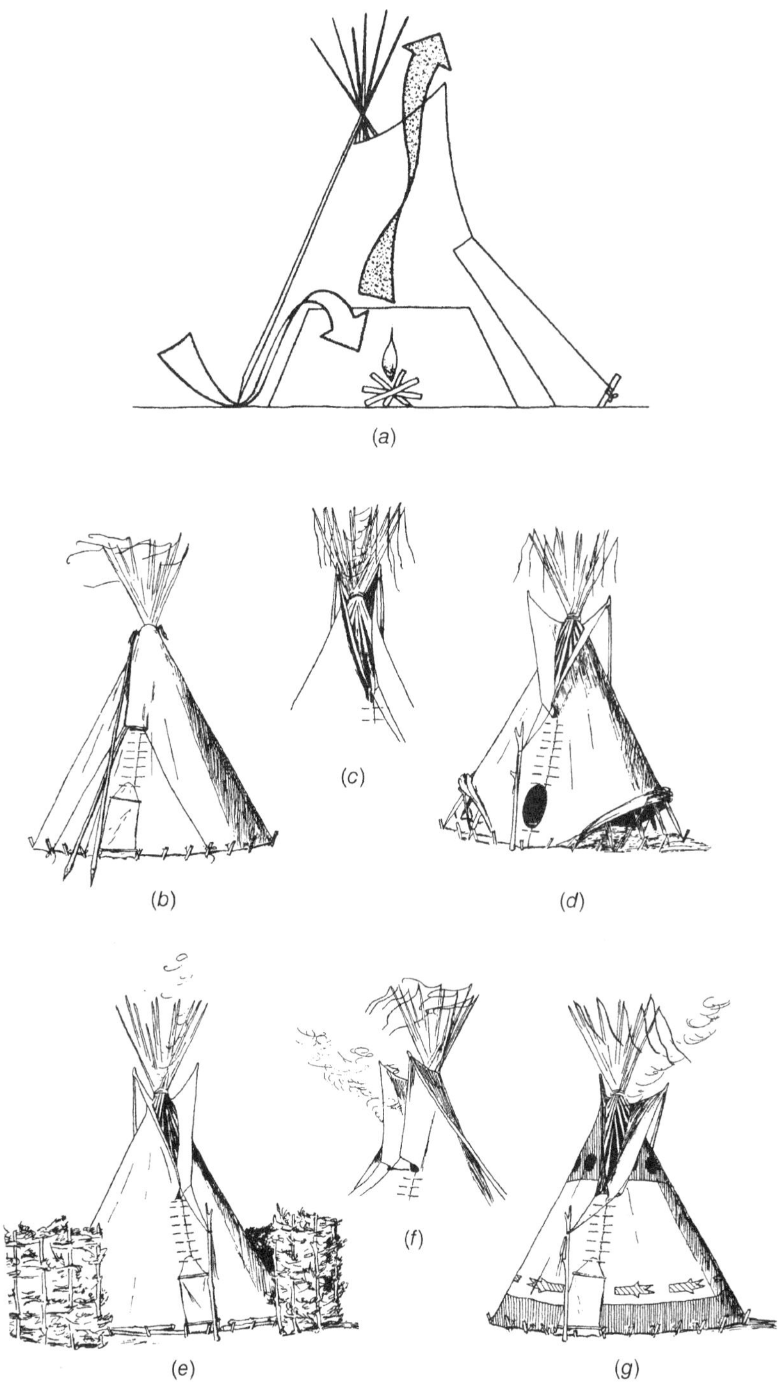

Figure 9.3 *Thermal switches on the Great Plains. (a) With an interior fire, smoke flaps at the top are adjusted to vent the smoke, and an interior liner forces the cold combustion air entering at ground level to first rise along the sides of the tipi, gaining some warmth, before it moves across the occupants on its way to the fire. Six adjustments to the lightweight, translucent, and portable tipi are shown. The tipis in these diagrams are facing east, with their backs to the prevailing westerly winds. (b) In severe rainstorms, the smoke flaps at the top can be closed. (c) For ordinary conditions of west wind, the smoke flaps block the wind, thus creating suction to draw out the smoke. (d) In hot weather, breezes are admitted under the cover at the ground. (e) In extremely cold weather, a temporary windbreak can be added. (f) For the unusual wind or (g) southwest wind, the smoke flaps are manipulated to block wind, again encouraging smoke draw-out, as in (c). (From Reginald and Gladys Laubin.)*

(a) (b) (c) (d)

Figure 9.4 *Sun control for office buildings. (a) Shading switches both inside the glass (blinds) and outside the glass (awnings) are evident, and the windows themselves are operable. Awning use varies with window location and occupant needs and desires (New England Merchants' Bank; Shepley Rutan & Coolidge, Architects; demolished 1966). (b) Mechanized external shades provide filters and switches (Bateson Building, Sacramento, CA; Sym Van der Ryn). (c) Three-dimensional filters (overhangs) dominate the south façade (right) of this building at Boston University; the adjacent west windows (where overhangs are less effective) have only internal switches (blinds) deployed in a variety of positions. (Law and Education Building; Sert, Jackson and Gourley, Architects). (d) A two-dimensional filter of reflective glass equally sheathes all faces of this office tower, sending reflected solar radiation to the neighborhood below. Variability is removed from the enclosure, and users are not involved in defining the outside–inside relationship (a, c, d: photos by Stephen Tang; b: © Alison Kwok; all rights reserved.)*

Figure 9.5 *Multiple window switches: shutters, curtains, inner shutters, operable windows, and planter boxes on the balconies overlooking Piazza dei Signori in Vicenza, Italy. (© Alison Kwok; all rights reserved.)*

efficient as this practice may be, it can also be a source of widespread dissatisfaction with air-conditioned spaces.

Sealed windows greatly curtail people's contact with sounds, smells, and breezes from the outdoors. This is frequently beneficial in urban areas, yet on beautiful days it can be very frustrating. A lack of switches can contribute to a feeling of helplessness regarding one's personal environment. Studies indicate that user satisfaction grows with increased control over the indoor environment (Brager et al., 2004)—whether that is with operable windows, dimmable light fixtures, access to the thermostat, etc. Even if the occupant does not operate these switches, knowing they have the ability to use them if needed appears to significantly increase user satisfaction.

Passive heating and cooling systems are especially reliant upon switches, hence on the knowledge and cooperation of their users. These users often must base their manipulations of thermal switches at one point in time on the effect that will be needed at a later time. This practice, called thermal sailing, is similar to the actions of outdoor workers in the far north, who learn to unbutton their coats in the cold early hours of the workday before they begin to sweat. Sweat would soak their clothing, with harmful results later in the day as temperatures fall rapidly near dusk.

Misjudgments in the design of passively solar-heated residences can result in extraordinarily high temperatures on a sunny day or uncomfortably cold nights without stored solar heat. For a building closely connected to its climate, the design of switches is also the design of an educational process for the users. The challenge is to involve, but not bind, the users in the management of their environment. Automated controls are a partial answer to this challenge; switches that are easy and understandable to use are another. In residences, occupants are more open to learning to use switches to control their environment. In large commercial and institutional buildings, however, it is not realistic to expect short-term occupants to learn the details of how the building functions. Therefore, automated controls are often required for these types of facilities.

9.3 MANAGEMENT OF CONTROL LAYERS

Best practices in building enclosure design are evolving. Historic practices of enclosure detailing are being adapted or modified to account for the increasing demand for high performance. Changes in design and detailing show several trends.

Control Layer Definition and Continuity. A control layer is a material or materials in an assembly tasked with addressing a specific design intent related to the response of an enclosure to an environmental force The forces of most concern include water vapor, air (wind), heat, and bulk water. The possible responses include the intents discussed above: barrier, connector, filter, switch, or transformer.

The materials selected to serve as control layers need to be well thought out in terms of desired performance and location within an assembly. The layers need to be continuous across the assembly—especially at junctions between materials and where there are abrupt changes in enclosure (such as at a window, door, or skylight). Modern, layered enclosure assemblies make the transfer of environmental forces complicated and difficult to predict. Yet poor control can lead to thermal discomfort, mold, rot, corrosion, mildew, material decay, biological growth, and even structural failure.

Sometimes one material can accomplish control of two or three environmental forces at once. For example, closed cell (or high-density) spray polyurethane foam can (if properly used) accomplish control of heat, air, vapor, and bulk water simultaneously. In other cases, multiple materials located at different places in an assembly may be necessary for appropriate control. Table 9.1 outlines common design concerns for heat and moisture transfer.

Control of bulk water is by far the most critical control layer function for most buildings. Preventing bulk water infiltration should never be compromised even in projects with the most ambitious goals in terms of energy-efficiency and comfort. Joe Lstiburek cogently suggests (Lstiburek, 2010) that the next most important layers are air, vapor, and then heat. The argument he makes is: "if you can't keep the rain out, don't waste your time on the air. If you can't keep the air out, don't waste your time on the vapor." The order of control layer importance does not expressly define the extent of necessary control—which is project-specific and specified by performance metrics listed in the far-right column of Table 9.1.

In addition to the focus on hygrothermal (water-heat) performance seen in Table 9.1 the concept of control layers can be usefully applied to other environmental elements—such as insects, sounds, or burglars. The question is: what material will adequately mitigate the environmental element of concern and where should the layer be placed within the enclosure assembly for best effect and assembly longevity? Development of a very detailed control layer diagram (a wall, roof, or floor section with attention to junctions and transitions) is a necessary step in getting this correct. An example control layer diagram is shown in Figure 9.6.

Thermal Bridges. A thermal bridge is a location in a construction assembly where higher (compared to the rest of the assembly) heat flow occurs due to a construction detail. It is a special control layer case. Examples of thermal bridges include wall and roof framing elements, fasteners, and window and door frames. Attention to detail is the key to reducing thermal bridges. Continuous insulation is a common solution for wall and roof assemblies. Mitigating thermal bridges reduces building energy use, reduces the risk of condensation, and increases thermal comfort (because thermal bridge elimination contributes to a more consistent mean radiant temperature within a space).

Figure 9.7 provides an example of a thermal bridge situation commonly found in frame wall construction. The framing (wood or, even more so, steel) has a lower thermal resistance (R-value) than the insulation that is interrupted by the framing. More heat flows through the framing than through the insulation. Similar thermal bridging can occur with cantilevered balconies or fasteners that hold insulation in place.

Several software programs can predict heat flow through nonhomogeneous construction assemblies and show the effects of thermal bridges. An example analysis (produced using the software *THERM*, see Appendix L) is provided in Figure 9.8.

TABLE 9.1 Control Layer Design Matrix

Environmental Force	Transfer Mechanism	Typical Control Strategy	Performance Metric
Sensible heat	Conduction, convection, radiation	Insulation[a]	R-value, U-factor
Air (wind)	Convection	Air barrier	Air permeability
Water vapor	Diffusion	Vapor retarder	Vapor permeability
Bulk water	Gravity, surface tension, capillary action, momentum, air pressure	Shingle laps, drip edges, air gaps, baffles, splines, waterproofing, WRB[b], roofing, flashing, etc.	Yes or no (typically)

[a] Radiative barriers and shading not included for simplicity.

[b] Water-resistive barrier.

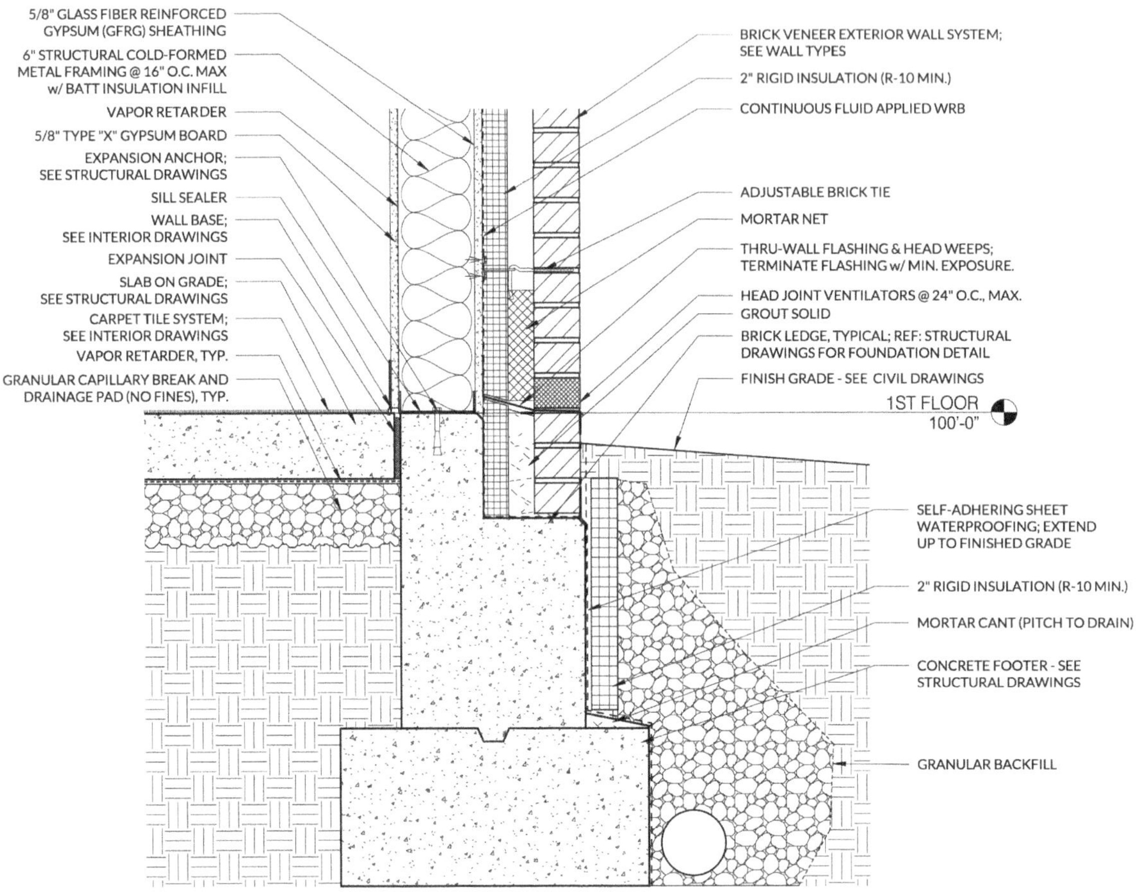

Figure 9.6 *Typical foundation detail demonstrating how standard best practices can largely achieve continuity of a building enclosure's various control layers. However, achieving complete continuity requires overcoming construction-related practicalities. A common example is exhibited here, by which there is a minor break in continuity of the thermal barrier where the wall assembly meets the foundation. (© Browning Day; all rights reserved.)*

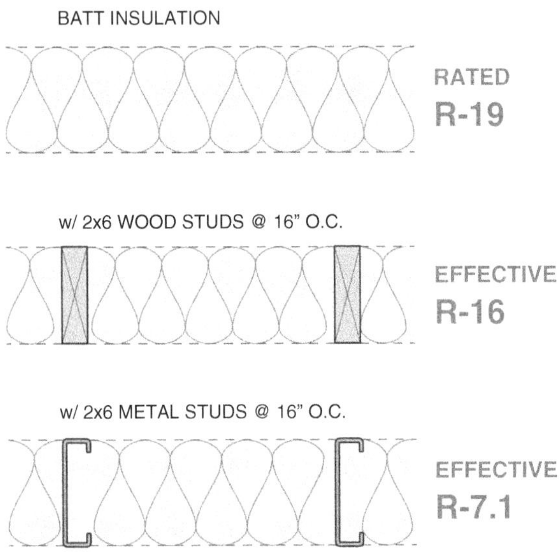

Figure 9.7 *Example of how thermal bridging can significantly reduce the effective R-value of a wall assembly's thermal barrier. (© Daniel Overbey; all rights reserved.)*

9.4 BUILDING ENCLOSURE DESIGN REQUIREMENTS

(a) Thermal Performance

Energy efficiency codes establish minimum thermal enclosure performance requirements for virtually all jurisdictions in North America and for virtually all building types. In the United States, code adoption is a local matter, and the specific energy code to be followed depends on the jurisdiction. Even so, the energy code most likely to be encountered for nonresidential buildings is derived from *ASHRAE/IES Standard 90.1: Energy Standard for Sites and Buildings Except Low-Rise Residential Buildings*. In California, *Title 24* is the statewide counterpart of *Standard 90.1*. Residential buildings are likely to fall under the requirements of the ICC's *International Energy Conservation Code (IECC)*, which is really a standard until adopted as law. ASHRAE also publishes a residential energy standard, *Standard 90.2:*

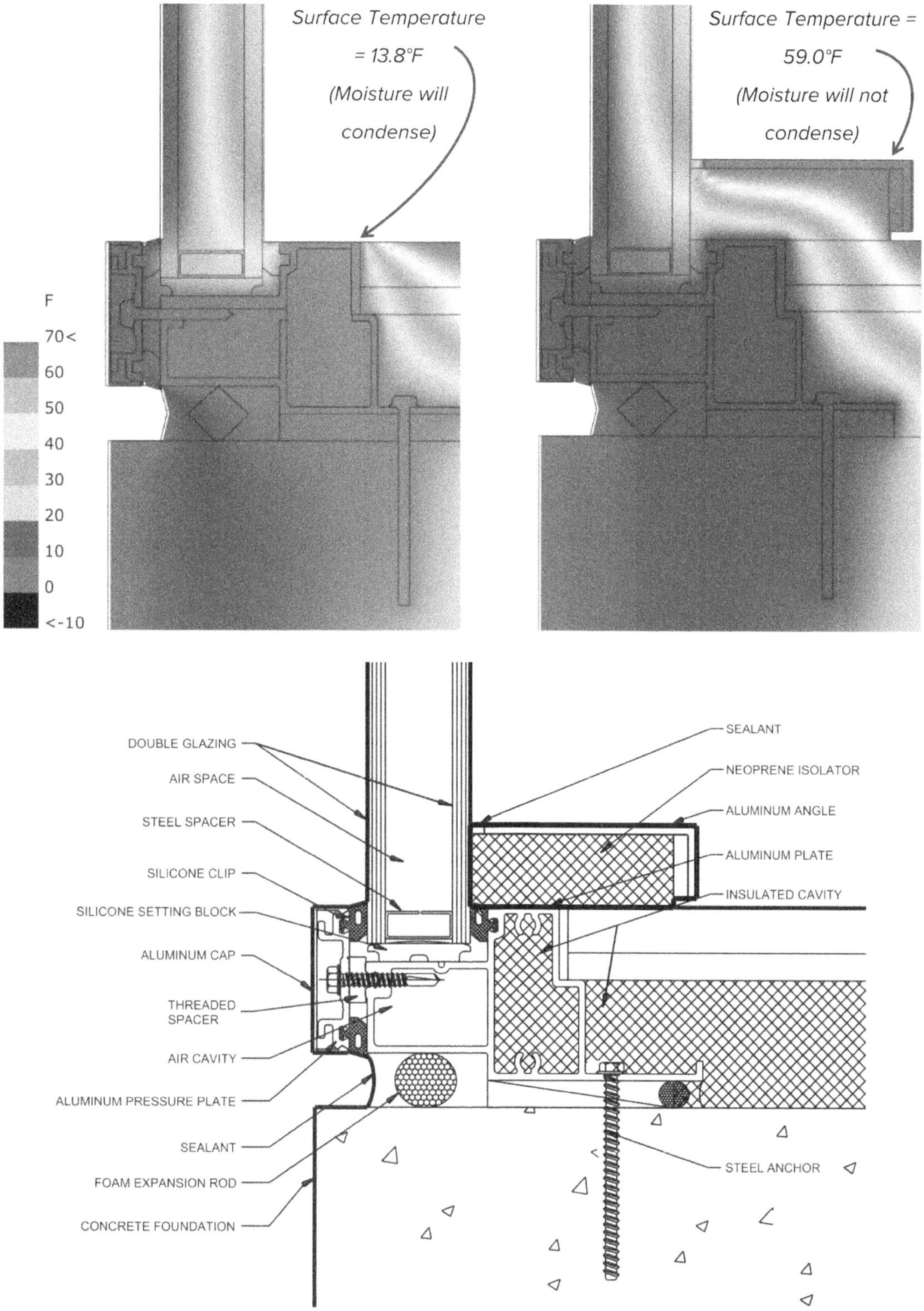

Figure 9.8 *Thermal analysis of heat flow patterns under different conditions at curtainwall-to-floor assembly transition. THERM was used to determine how to adapt an existing condition to mitigate an existing condensation risk. (© Browning Day; all rights reserved.)*

High-Performance Energy Design of Residential Buildings. Although energy codes provide efficiency requirements for more than just the enclosure (usually also dealing with HVAC equipment efficiency, hot water heating, and lighting), this discussion focuses upon enclosure requirements.

It is typical for energy codes to provide three paths to building enclosure compliance:

1. *Prescriptive option*, which provides precise statements of minimally acceptable component characteristics (e.g., a minimum R-value for wall insulation and a maximum U-factor for a window).
2. *Trade-off option*, where lack of adequate U-factor for some assembly (e.g., a wall) can be made up for by better performance in other assemblies (e.g., a roof).
3. *Energy budget option*, which is a performance path that sets a minimum level of overall energy performance that may be met using a wide range of solutions. Code compliance via a prescriptive option is usually straightforward and requires little analysis or creativity. The compliance values are well-vetted and offer a reasonable balance between performance and economics for most buildings. Following a performance option may require detailed analysis (by way of a computer program) but may open the door to innovation and creativity.

Building energy codes need to be placed in context. They establish a set of minimum requirements. Just barely meeting an energy code is not aspirational, it simply means that a building enclosure design proposal is not illegal. Today, doing better—perhaps much better—than the minimum required by code is becoming a more common design approach. See *Beyond-Code Design Guidance*.

Appendix H provides excerpts from two influential high-performance building design standards and guidelines. The excerpts include:

- a sample of the prescriptive enclosure requirements from *ASHRAE/IES Standard 90.1-2022: Energy Standard for Sites and Buildings Except Low-Rise Residential Buildings* (for nonresidential buildings)
- a sample of recommendations from ASHRAE's *50% Advanced Energy Design Guide for Small to Medium Office Buildings*

The various ASHRAE energy standards are continuously being updated. The intent of these excerpts is not to provide the most current information, but rather to show the general format by which such design requirements are conveyed and to compare the relative requirements.

EXAMPLE 9.1 A two-story medical office building is being designed for Indianapolis, Indiana. What are the maximum permissible U-factors (and other enclosure characteristics) for such a building in this location?

SOLUTION

Refer to Appendix B for climate data and to Appendix H, Table H.1, for sample provisions of *ASHRAE/IES Standard 90.1–2022*.

- Location: Indianapolis, Indiana
- HDD65: 5224 [SI: HDD18: 2902]
- CDD50: 3758 [SI: CDD10: 2088]
- Climate Zone (CZ): 4 (see the note at the end of this example)

HDD and CDD values, as well as CZ designations, can be obtained from *Standard 90.1* or another data resource.

These benchmark climate indicators (heating and cooling degree days: HDD and CDD, respectively) are used to establish which set of prescriptive requirements in *Standard 90.1* apply to a building in Indianapolis (or the map in *Standard 90.1* can be used for the same purpose). Within the applicable climate zone data set (only CZ 4 information is shown in Appendix H, Table H.1), note that one of the labeled columns of design values applies to a nonresidential building (for this example, an office), while the other columns apply to other building contexts.

- Roof: maximum U-factor: 0.032 [SI: 0.184], assuming a roof/attic construction typical of this type of building.
- Frame walls: If wood-framed, the maximum U-factor is 0.064 [SI: 0.36]; if steel-framed, the maximum U-factor is also 0.064 [SI: 0.36]. See Section 9.3 regarding the difference wood and steel studs can make in terms of an assembly's effective insulation value due to thermal bridging.

Looking at the above-grade wall data in Appendix H, Table H.1 shows that the assembly maximum U-factor varies with the type of construction (ranging from 0.060 to 0.104). This "flexibility" in performance requirements reflects the fact that *Standard 90.1* is a consensus document attempting to satisfy many constituencies, and that it is not based on scientific absolutes or idealized efficiencies.

- Floor (assuming wood-framed, over a crawl space): maximum U-factor: 0.033 [SI: 0.187]
- Opaque swinging doors: maximum U-factor: 0.370 [SI: 2.101]

Doors are a substantial weak link in the thermal enclosure resistance but are usually a small percentage of the overall wall area.

- Windows: If operable windows constitute 25% of the wall area (or even up to 40%), their maximum U-factor is 0.45 [SI: 2.56], and the maximum SHGC is 0.33.

Compare the minimum acceptable ("code-minimum") enclosure properties outlined here with the recommended properties given in Appendix H, Table H.2—which contains excerpts from the 50% *Advanced Energy Design Guide* for offices.

Note: ASHRAE's *50% Advanced Energy Design Guide* was published in 2011. At that time, the climate zone designations for U.S. counties matched *Standard 90.1*. Since then, however, changes to the U.S. climate have prompted ASHRAE to change the climate zone designations of a number of U.S. counties (including Marion County, where Indianapolis is located). Marion County was in Climate Zone 5 at the time the *50% Advanced Energy Design Guide* was published. However, in *ASHRAE/IES Standard 90.1-2022*, Marion County is in Climate Zone 4. For consistency, a contemporary application of the *50% Advanced Energy Design Guide* for Indianapolis should use the recommended properties for Climate Zone 4. This climate zone shift is emblematic of the reality of climate change. ■

There have been huge increases in installed component R-value (therefore, huge decreases in assembly U-factor) over the past decades. Sometimes this has occurred because of the use of increased insulation thickness (as demanded by evolving codes), sometimes because of substantially improved products (as with the emergence of high-performance windows). Improvements in code-minimum values for enclosure performance at times seem painfully slow to occur, especially given growing concern for climate change mitigation and green design initiatives. Luckily, alternatives to code-minimum design requirements abound.

(b) Beyond-Code Design Guidance

Publication of the *Advanced Energy Design Guides* (ASHRAE, various dates) was a welcome addition to the designer's tool kit. Although not mandatory, these guides (dealing with a range of building types and setting three levels of improved building energy efficiency) suggest easy-to-adopt ways to reduce building energy consumption—including improvements in enclosure performance. Although now dated, these *Guides* still provide useful information and recommendations for building enclosure design.

Passive building design certifications (offered by Phius and Passivhaus Institute) provide recommendations for enclosure design that will lead to ultra-low-energy buildings. These recommendations address high thermal resistance values, reduced thermal bridging, summer shading, and exceptional air tightness. The "passive" approach promoted by these organizations is related to—but not identical to passive heating or cooling systems discussed elsewhere in this book. Figure 9.9 provides a conceptual summary of "passive building" design moves. Chapters 13 and 14 provide details of "passive heating" and "passive cooling" respectively.

Building Science. There is also a resurgence of interest in building science—including the nuts and bolts of heat and moisture (hygrothermal) flows

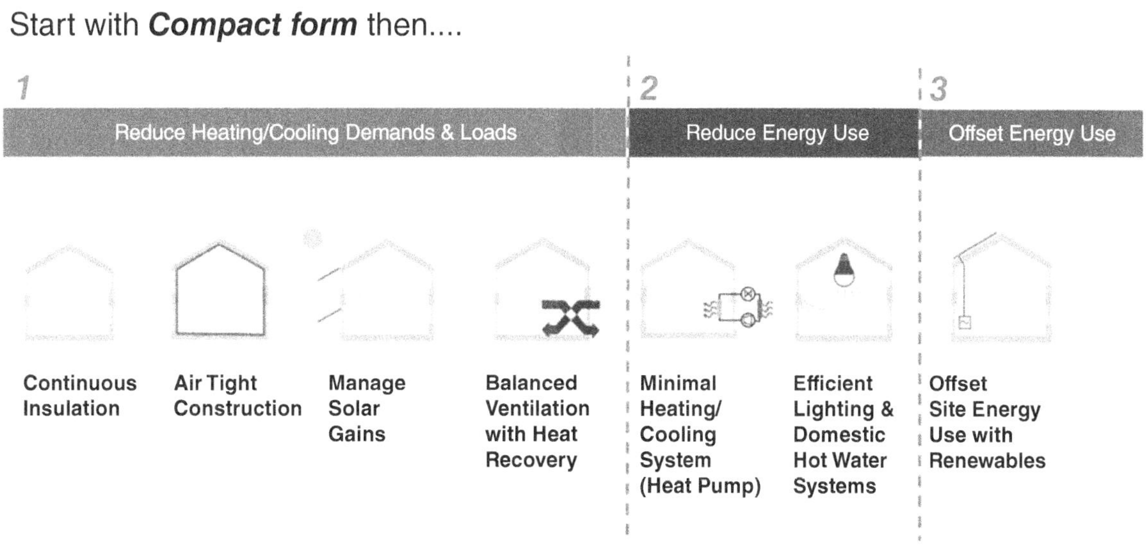

Figure 9.9 *Conceptual diagram of passive building design principles. (Reprinted with permission; © Phius, www.phius.org)*

through building enclosure elements. This interest is being driven by the efforts of several organizations and events, including:

- Society of Building Science Educators (SBSE)
- Building Science Education (DOE, 2025a)
- Phius (formerly Passive House Institute US)
- Race to Zero—reframed as the Solar Decathlon-Design Challenge and more recently as Buildings-NEXT—student design competition (DOE, 2025b)

9.5 BUILDING ENCLOSURE COMMISSIONING

Building enclosure commissioning (BECx) is a rapidly emerging means to improve enclosure performance outcomes. BECx goes beyond construction quality control to attempt to ensure (through predesign and design reviews, documentation evaluation and verification, and testing) that enclosure design, construction, and in-situ performance meet the owner's project requirements (OPR). The National Institute of Building Sciences (NIBS) published *NIBS Guideline 3: Exterior Enclosure Technical Requirements for the Commissioning Process* (NIBS, 2012), which, when originally developed in 2006, was intended to work in concert with *ASHRAE Guideline 0: The Commissioning Process* (ASHRAE, 2019). NIBS subsequently handed responsibility for enclosure commissioning over to ASTM. ASTM substantially reworked the NIBS process requirements and published them as *ASTM E2947: Standard Guide for Building Enclosure Commissioning* (ASTM, 2024) and the associated *ASTM 2813: Standard Practice for Building Enclosure Commissioning* (ASTM, 2021). The user of these ASTM documents should be aware that they are not fully aligned with ASHRAE commissioning documents—although the objectives of the process are the same. The International Organization for Standardization (ISO) also has published documentation on BECx—*ISO 21105-1: Performance of Buildings — Building Enclosure Thermal Performance Verification and Commissioning; Part 1: General Requirements* (ISO, 2019).

ASHRAE/IES Standard 90.1-2022 requires verification, testing, and commissioning of the energy performance of the building enclosure (ASHRAE, 2022c). Although somewhat vague in the specifics, this bringing of commissioning into mainstream code compliance is an important step. *LEED v5 for Building Design and Construction* (BD+C) requires compliance with *Standard 90.1*'s commissioning requirements for the building enclosure (USGBC, 2025). See further discussion of the commissioning process applied to HVAC systems in Chapter 16.

9.6 EMERGING TRENDS IN BUILDING ENCLOSURES

What might the next quarter-century bring, and how might the designer of today's building enclosures anticipate—or participate in—these advances?

Insulation Developments. These will likely include aerogel, gas-filled panels, powder-evacuated panels, and compact vacuum insulations. Some of these products are suitable for retrofit applications in existing construction. Aerogel is transparent and porous; it is one of the lightest solid materials. Silica aerogel can be foamed into cavities without ozone-depleting chlorofluorocarbons CFCs. In a 90% vacuum, silica aerogel has a resistance of R-20 per inch [SI: R-139 per meter]. By adding carbon to absorb infrared radiation, thermal resistance of R-32 per inch [SI: R-222 per meter] is possible.

Gas-filled panels are hermetically sealed (i.e., completely airtight) plastic bags enclosing honeycomb baffles of thin polymer films and low-conductivity gas (argon, krypton, or xenon). Relative resistance values of these insulating cores are: argon, R-5.2 per inch; krypton, R-13.4 per inch; and xenon, R-19.3 per inch [SI: R-36, 93, and 134 per meter, respectively]. Powder-evacuated panels contain a vacuum and compacted silica-based powder sealed within a multilayer gas barrier. Expensive to produce and threatened by punctures, this may be a development more suited for appliances than for buildings. Thermal resistance is about R-20 to R-25 per inch [SI: R-139 to R-173 per meter].

Windows. These enclosure components have undergone the most dramatic thermal improvement. Resistances for window units (glazing plus frame) have evolved from about R-1 [SI: R-0.18] to R-6.5 [SI: R-1.14] and higher. Prototypes as high as R-20 [SI: R-3.52] exist. Such improvement has been accomplished through a series of developments: a second layer of glass; a wider air space between glazing layers; tinted, reflective, photochromic, and low-emissivity coatings; low-conductivity gas between glazing panes; intermediate films between panes; operable blinds between panes; lower-conductivity glazing spacers and frames; and more airtight weather stripping.

Future commercially viable developments are likely to include smart windows with on-demand variable light transmission. For example, with electrochromic glazing, an applied electric field switches the window from a clear state to one with a deep coloration; intermediate states of coloration are possible. The most likely application is in cooling-dominated climates where control of glare, solar gain, and visual privacy are paramount. The net-zero energy Research Support Facility at the National Renewable Energy Laboratory in Golden, CO, employed an area of such glazing (NREL, 2012).

Windows do so many things in addition to providing weather protection—admitting daylight, allowing views outdoors and in, admitting breezes, and admitting warming winter sun—that their relative thermal weakness compared to their desirability continues to challenge building designers. Over the next quarter-century, windows may well continue to display the most dramatically improved resistance to heat flow of all the components of the enclosure. Carmody et al. (2007) provide an excellent overview of residential window technologies, while Carmody et al. (2003) provide a similar overview of commercial window technologies.

Roofs. Roofs are of particular interest relative to heat gains and losses because they are subject to extreme conditions. On a clear night, radiant losses to the sky can lower a roof's surface temperature below that of the outdoor air—beneficial in summer, but a concern in winter. In summer, elevated roof surface temperatures caused by absorption of solar radiation are dramatic and problematic. Information regarding the benefits of cool roofs and has been assembled and disseminated by the Cool Roof Rating Council (CRRC, 2025). Both reflective and green (or eco-roof) designs have been successfully employed as cool roofs. Requirements for cool roofs in several warm climate zones are incorporated in *ASHRAE Standard 90.1*, as is a requirement for cool walls in the warmest climate zone.

Infiltration. Windows historically represented the component with the greatest heat flow rate in a typical building enclosure. Today, the highest rate of heat flow is more likely to be from outdoor air infiltration (or deliberate ventilation). Limiting heat flow via unwanted air flow is an important aspect of enclosure design. The tightness of buildings has improved remarkably over the past 20 years, partly due to improved components (such as windows) and partly due to increased care in construction (involving sealants and air barriers). As a result of this radical improvement, building airtightness as a means of reducing energy usage is now in conflict with IAQ needs and concerns. Future improvements in this area will revolve around the use of heat recovery ventilators that permit reasonable ventilation rates without excessive energy penalties.

Transformers. Flat-panel photobioreactors (PBRs) are under development by Colt, ARUP, and SSC (Strategic Science Consult) for the Federal Ministry of Transport, Building and Urban Development, Germany. These panels cultivate microalgae as an energy resource and were demonstrated as an integrated system in a four-story residential building, BIQ House, in Hamburg, Germany (see Figure 9.10). The BIQ House has 129 flat-panel glass bioreactors (SolarLeafs) on the southwest and southeast façades, which facilitates photosynthesis. ARUP's Europe Research Leader, Jan Wurm, writes (Wurm, 2012):

> This passive-energy house generates biomass and heat as renewable energy resources while also having a dynamic shading device, thermal insulation and noise abatement.
>
> The microalgae circulate through the panels with water and nutrients, absorbing light and carbon and producing biomass. The part of the solar spectrum that isn't absorbed by the algae heats the water, and this solar thermal heat is removed so it can be used in the building or stored for when it's needed.

The photobioreactors are linked in a closed loop to the building's plant room where they are fed with carbon from combustion processes in the neighborhood. After harvesting, the algae are transformed into methane, and the heat generated is taken out of the system by heat exchangers. It is stored geothermally or fed back into the building using a heat pump, for heating and hot water supply. Some of the benefits include:

- generation of biomass as a renewable energy resource in the urban context; the calculated net energy gain is approximately 30 kWh per square meter [2.8 kWh/ft^2] per year
- additional generation of solar thermal heat of around 30 kWh per square meter [2.8 kWh/ft^2] per year
- absorption of carbon directly at the source of domestic or industrial emissions

Figure 9.10 *Photobioreactors in the BIQ House. (a) Southeast and southwest elevations of the world's first SolarLeaf-Building in Hamburg, Germany, with 200 m² [2153 ft²] of active SolarLeaf area. (b) Panels track solar radiation around a vertical axis. (c) SolarLeaf louver with hidden horizontal support system. (d) AirLift-System bubbles rising in the SolarLeaf louvers. (© Colt International, Arup Deutschland, SSC GmbH; used with permission.)*

Acknowledgments

Keith A. Simon, FAIA, is acknowledged as a significant contributor to the contents of this chapter.

References and Resources

ASHRAE. 2004–2012a. *Zero Energy Design Guides (for Three Building Types)*. Atlanta, GA: American Society of Heating, Refrigerating and Air-Conditioning Engineers.

ASHRAE. 2004–2012b. *30% Advanced Energy Design Guides (for Six Building Types)*. Atlanta, GA: American Society of Heating, Refrigerating and Air-Conditioning Engineers.

ASHRAE. 2011–2012. *50% Advanced Energy Design Guides (for Five Building Types)*. Atlanta, GA: American Society of Heating, Refrigerating and Air-Conditioning Engineers.

ASHRAE. 2018–2019. *Zero Energy Advanced Energy Design Guides* (multiple published components). Atlanta, GA: American Society of Heating, Refrigerating and Air-Conditioning Engineers.

ASHRAE. 2019. *ASHRAE Guideline 0-2019: The Commissioning Process*. Atlanta, GA: American Society of Heating, Refrigerating and Air-Conditioning Engineers.

ASHRAE. 2022a. *ANSI/ASHRAE Standard 62.1-2022: Ventilation for Acceptable Indoor Air Quality*. Atlanta, GA: American Society of Heating, Refrigerating and Air-Conditioning Engineers.

ASHRAE. 2022b. *ANSI/ASHRAE Standard 62.2-2022: Ventilation and Acceptable Indoor Air Quality in Low-Rise Residential Buildings*. Atlanta, GA: American Society of Heating, Refrigerating and Air-Conditioning Engineers.

ASHRAE. 2022c. *ANSI/ASHRAE/IESNA Standard 90.1-2022: Energy Standard for Sites and Buildings Except Low-Rise Residential Buildings*. Atlanta, GA: American Society of Heating, Refrigerating and Air-Conditioning Engineers.

ASHRAE. 2024. *ANSI/ASHRAE/IES Standard 90.2-2024: Energy-Efficient Design of Low-Rise Residential Buildings*. Atlanta, GA: American Society of Heating, Refrigerating and Air-Conditioning Engineers.

ASTM. 2021. *ASTM E2947-21a: Standard Guide for Building Enclosure Commissioning*. West Conshohocken, PA: ASTM International.

ASTM. 2024. *ASTM E2813-24: Standard Practice for Building Enclosure Commissioning*. West Conshohocken, PA: ASTM International.

Brager, G., G. Paliaga, and R. de Dear. 2004. "Operable Windows, Personal Control and Occupant Comfort." *ASHRAE Transactions* 100: 17–35.

Carmody, J., S. Selkowitz, and L. Heschong. 2007. *Residential Windows: A Guide to New Technologies and Energy Performance*, 3rd edn. New York: W. W. Norton and Company.

Carmody, J., S. Selkowitz, E. Lee, D. Arasteh, and T. Willmert. 2003. *Window Systems for High-Performance Buildings*. New York: W.W. Norton & Company.

CRRC (Cool Roof Rating Council). 2025. www.coolroofs.org

ICC. 2024. *2024 International Green Construction Code*. Washington, DC: International Code Council.

ISO. 2019. *ISO 21105-1:2019: Performance of Buildings—Building Enclosure Thermal Performance Verification and Commissioning*. Geneva: International Organization for Standardization.

Lstiburek, J. 2010. *BSI-001: The Perfect Wall*. Westford, MA: Building Science Corporation. buildingscience.com/documents/insights/bsi-001-the-perfect-wall

Lstiburek, J. 2011. *BSI-049: Confusion About Diffusion*. Westford, MA: Building Science Corporation. buildingscience.com/documents/insights/bsi-049-confusion-about-diffusion

Lstiburek, J. and J. Carmody. 1994. *Moisture Control Handbook: Principles and Practices for Residential and Small Commercial Buildings*. Hoboken, NJ: John Wiley & Sons.

NIBS. 2012. *NIBS Guideline 3-2012: Building Enclosure Commissioning Process*. Washington, DC: National Institute of Building Sciences. www.wbdg.org/FFC/NIBS/nibs_gl3.pdf

Norberg-Schulz, C. 1965. *Intentions in Architecture*. Cambridge, MA: MIT Press.

NREL. 2012. *The Design-Build Process for the Research Support Facility*. Golden, CO: National Renewable Energy Laboratory. www.nrel.gov/docs/fy12osti/51387.pdf

Passivhaus Institute. www.passivehouse.com

Phius. www.phius.org

SBSE (Society of Building Science Educators). www.sbse.org

USDOE. 2025a. "Building Science Education." bsesc.energy.gov/

USDOE. 2025b. *Solar Decathlon*. Washington, DC: U.S. Department of Energy. www.solardecathlon.gov

USGBC. 2025. *LEED v5*. Washington, DC: U.S. Green Building Council. http://www.usgbc.org/leed/v5

Wurm, J. 2012. "Bioreactive Façade with Photobioreactors for Cultivating Microalgae as an Energy Resource." *ARUP Thoughts*, May.

CHAPTER 10

Building Enclosure Heat Flow

BUILDING ENCLOSURE

UNDERSTANDING HEAT FLOW IS FOUNDATIONAL to all aspects of climate control. Chapter 6 addresses heat flows to and from the body that affect thermal comfort. Chapters 13, 14, and 15 deal in part with heat flows to, from, and within various elements of passive climate-control systems. This chapter deals with heat flows through building enclosures, both through the materials of the building skin and by way of infiltration of outdoor air that replaces conditioned indoor air. Basic concepts and calculations of heat flow are presented in this chapter, whereas applications of these concepts (passive solar heating, passive cooling, active HVAC system/equipment sizing, seasonal energy usage) are found in subsequent chapters. Numerous data tables that accompany this chapter are presented in Appendix E.

10.1 HEAT FLOW FUNDAMENTALS

The magnitude of heat flow through a building enclosure is a function of several variables. These include the surface area of the enclosure, the temperature difference across the enclosure, and the thermal properties of the materials that comprise the enclosure. This section of the chapter defines basic types of heat flow and addresses these thermal properties.

The flow of heat through the building enclosure is highly variable. Heat flow varies broadly by season (heat generally flows from a building in winter and to a building in summer) and also hourly (heat flows from hot to cold, from humid to dry, and air flows from high pressure to low pressure in response to constantly changing conditions). The precise path of heat flow can also vary (through the numerous materials and assemblies of the building enclosure or by way of outdoor air entering the interior through intentional and unintentional pathways). These complexities must be considered by a designer who intends to deliver thermal comfort in an energy-efficient and environmentally responsible manner. The following discussion of heat flow will focus first upon the components of building enclosure (opaque elements first, then transparent elements), followed by heat flow via air exchange through the enclosure.

In the 1970s, in the wake of global concerns about energy costs and availability, designers began placing increasing emphasis on the thermal performance of building enclosures. Buildings (and building performance) changed because of this focus. Tighter building enclosures resulted in decreased air leakage, leading to (still ongoing) concerns about indoor air quality and "sick building syndrome." Today's building code requirements likely strike a fair balance between enclosure energy efficiency and air quality requirements—although in the United States these two issues are addressed by separate codes/standards.

A typical new small North American building loses somewhat more winter heat via incoming fresh air than it loses through its enclosure. In summer it gains somewhat more heat through its enclosure than it does via incoming fresh air (although this is very climate-dependent). This pattern is due in part to the importance of solar radiation to enclosure heat gains in summer, and in part to the design outdoor–indoor temperature difference being much greater in winter than in summer. The handling of latent (humidity-based) loads also influences these

Mechanical and Electrical Equipment for Buildings, Fourteenth Edition. Walter Grondzik, Daniel Overbey, and Alison Kwok.

Companion website: www.wiley.com/go/meeb14e

load patterns. Larger buildings tend to be internal-load-dominated.

(a) Static versus Dynamic; Sensible versus Latent

Although the general principles remain the same, analysis of heat flow under dynamic (rapidly changing) conditions is more complex than under static or steady-state (generally stable over time) conditions. The effects of heat storage within materials become a greater concern under dynamic conditions than under static conditions. A static analysis requires consideration of fewer variables than a dynamic analysis and is therefore simpler. The key enclosure characteristic determining steady-state heat flow is thermal resistance. An analysis of heat flow under dynamic conditions involves more variables—including thermal capacitance. The following discussion begins with steady-state heat flow and then looks at dynamic situations.

Heat flows through an enclosure are of two forms—sensible heat and latent heat. Sensible heat flow results in a change in temperature. Latent heat flow results in a change in moisture content (typically humidity of the air). Total heat flow is the sum of sensible and latent heat flows. Materials react differently to sensible and latent heat flows; therefore, the discussion that follows begins with sensible heat flow and then deals with latent heat flows.

(b) Heat Flow Processes

Whenever an object is at a temperature different from its surroundings, heat flows from the hotter to the colder. Likewise, water vapor flows from areas of greater concentration to areas of lesser concentration. Buildings, like human beings, experience sensible heat loss to and from the environment in three principal ways. Convection refers to heat exchanged between a fluid (typically air) and a solid, with the motion of the fluid due to heating or cooling playing a critical role in the extent of heat transfer. Conduction refers to heat transferred directly from molecule to molecule, within or between materials, with proximity of molecules (material density) playing a critical role in the extent of heat transfer. Radiation refers to heat flows via electromagnetic waves from hotter surfaces to detached, colder ones—even across empty space and potentially great distances. Evaporation can also be involved in enclosure heat loss, carrying heat away from wet surfaces, but this is less influential for most building enclosures than for our bodies. Water vapor flows through enclosure assemblies and via air leakage are the principal means of latent heat gain (or loss).

A combination of sensible heat flow by convection, conduction, and radiation through some typical material assemblies is shown in Figure 10.1. Heat flow through the various components of a building enclosure assembly involves both heat flow through solids and heat flow through layers of air. Multiple air spaces and reflective surfaces are useful and inexpensive ways to reduce the flow of heat via radiation when temperature differences are large. Some of the most effective insulating materials, therefore, combine dead-air spaces and layers of reflective films.

(c) Thermal Properties of Components

Each material used in an enclosure has fundamental physical properties that determine how that material will enable or resist sensible heat flow. These key properties are described in the following sections.

Conductivity. Every material has a characteristic rate at which heat will flow through it. For homogeneous solids, this property is called conductivity (designated as k). In the inch-pound (I-P) system, it is the number of British thermal units per hour (Btu/h) that flow through 1 square foot (ft^2) of material that is 1 inch (in.) thick when the temperature difference across that material is 1°F (under conditions of steady heat flow). Thus, the I-P units of conductivity are Btu in./h ft^2 °F.

The Système International (SI) equivalent is the number of watts that flow through 1 square meter (m^2) of material 1 meter (m) thick when the temperature difference across that material is 1 K (equal to 1°C). Thus, the SI units of conductivity are W/m K or W/m °C. It is traditional to use °C (Celsius) in building design work rather than K (Kelvin).

Conductivity values are established by laboratory tests and are published as a basic property of homogeneous solids. Conductivity is an important factor in passive heating or cooling designs that depend heavily upon the rate at which heat is conducted into a material from its surface.

Conductance. Many solids such as common brick, wood siding, batt or board insulation, gypsum board, and so on are widely available in standard thicknesses. For such materials, it is useful to know the rate of heat flow for that standard thickness instead of the rate per inch. Conductance, designated as C, is the number of Btus per hour that flow through 1 ft^2 of a given thickness of material when the temperature difference is 1°F. The I-P units are Btu/h ft^2 °F. The SI units are W/m^2 K. Conductance is also used to describe the rate of heat flow through defined sizes of modular units of nonhomogeneous materials (such as a concrete masonry unit—composed of pockets of air surrounded by concrete).

BUILDING ENCLOSURE

TABLE 10.1 Energy, Power, and Heat Terminology[a]

Concept	Terminology	Symbol	Discussion
Energy			The term energy implies a cumulative perspective, such as the potential for work in a barrel of oil or the solar radiation collected during a heating season
	British thermal unit	Btu	The British thermal unit is the fundamental I-P unit of energy. 1 Btu is the amount of energy (heat) required to raise the temperature of 1 lb of water by 1°F. A burning wooden match releases approximately 1 Btu.
	Joule	J	The joule is the fundamental SI unit of energy. A joule is a newton-meter (a force of 1 N acting over a 1-m distance). A joule is a fairly small unit of energy, so the kilo-joule (kJ), equal to 1000 J, is commonly used in building design.
	Watt-hour	Wh	The watt-hour is a commonly used SI unit of energy (and also applies to I-P electricity consumption) 1 W = 1 J/s 1 Btu = 1.055 kJ 1 kJ = 0.9478 Btu
Power			The term power is used to describe the rate of energy usage, production, or flow. Power is always associated with a time frame (often an hour in building design).
	British thermal unit per hour	Btu/h or Btuh[b]	An I-P unit. Btu/h is used to express heat gains and losses and the design capacity of heating and cooling equipment.
	Watt	W	The watt is the SI unit of power and is used the same as Btu/h 1 Btu/h = 0.0293 W 1 W = 3.412 Btu/h
	Horsepower	HP	An I-P expression of power, typically used to describe the capacity (size) of certain types of HVAC equipment (such as motors) 1 HP = 2545 Btu/h 1 HP = 746 W
Heat flow			Heat flow is a particular form of power, and in building design is typically expressed in Btu/h or W. Several heat-flow-related properties of materials, with specific definitions and uses, are described herein. These properties are based upon a unit temperature difference.[c]
	Conductivity[d]	*k*	The rate of heat flow through a unit thickness of a homogeneous (such as wood or brick) material I-P units: Btu in./h ft^2 °F SI units: W/m K Btu in./h ft^2 °F × 0.1442 = W/m K
	Conductance	C	The rate of heat flow through a specific nonhomogeneous object (such as a concrete masonry unit) or a defined thickness of a homogeneous material I-P units: Btu/h ft^2 °F SI units: W/m^2 K Btu/h ft^2 °F × 5.678 = W/m^2 K
	Film or surface conductance	h	The conductance of an air film; same concept and units as conductance
	Overall coefficient of heat flow	U, U-factor, U-value	The rate of heat flow through an assembly (window, wall, etc.) bounded by air on both sides. Includes the effects of all materials, air films, and air spaces I-P units: Btu/h ft^2 °F SI units: W/m^2 K Btu/h ft^2 °F × 5.678 = W/m^2 K
	Resistance	R, R-value	A measure of resistance to heat flow; the reciprocal of conductivity (or conductance). Essentially the force required to cause a unit flow of heat. I-P units: h ft^2 °F/Btu SI units: m^2 K/W h ft^2 °F/Btu × 0.176 = m^2 K /W (I-P) R per in. × 6.93 = (SI) R per m
	Permeance	M	Although permeance deals directly with water vapor flow, it is related to heat flow through the need to add or remove heat to humidify or dehumidify a building I-P units: grains/ft^2 h in. Hg SI units: ng/m^2 s Pa grains/ft^2 h in. Hg × 56.7 = ng/m^2 s Pa

[a] A more extensive listing of I-P (inch-pound) and SI (Système International) units and conversions is provided in Appendix A.

[b] Units oddity: The alternative symbol "Btuh" is shorthand for Btu/h and does not imply a product of energy and time; on the other hand, Wh is a product (1 watt of power over a 1-hr period); furthermore, Btu (without a time unit) is a measure of energy, as is Wh (with a time unit).

[c] "Degrees temperature" vs. "temperature degrees": Please note that "degrees temperature" such as degrees Fahrenheit (°F) or degrees Celsius (°C) indicate an actual temperature reading (e.g., via a thermometer). However, "temperature degrees" such as Fahrenheit degrees (°F) or Celsius degrees (°C) indicate increments of temperature difference (e.g., the difference between two thermometer readings).

[d] Use of K vs. °C for thermal increments: With regard to SI units, ASHRAE (reflecting a consensus within the HVAC profession) uses degrees Celsius (°C) for specific temperatures (e.g., setpoints) and degrees Kelvin (K) for temperature increments. As such, ASHRAE uses units of W/m K to express conductivity (k) and this preference is adopted herein. Readers should note that, historically, ASHRAE used units of W/m °C to express conductivity (k). This dated unit for conductivity may appear in some legacy materials. However, increments of Kelvin and Celsius are identical. Consistent with ASHRAE, the use of Kelvin within SI units has been applied to other related terminology of heat flow.

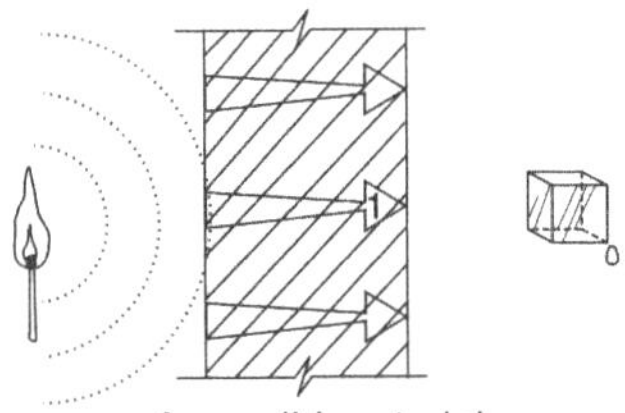

Any solid material wall, floor, or roof

A single solid material illustrates the transfer of heat from the warmer to the cooler side by conduction (1).

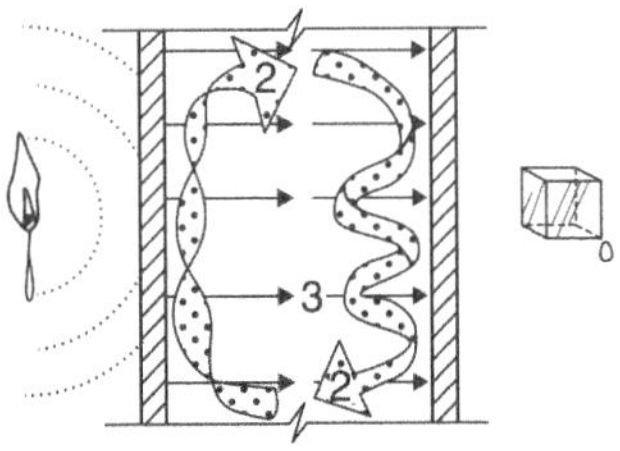

An air space in a wall

As air is warmed by the warmer side of the air space, it rises. As it falls down along the cooler side, it transfers heat to this surface (2). Radiant energy (3) is transferred from the warmer to the cooler surface. The rate depends upon the relative temperature of the surfaces and upon their emissive and absorptive qualities. Direction is always from the warmer to the cooler surface.

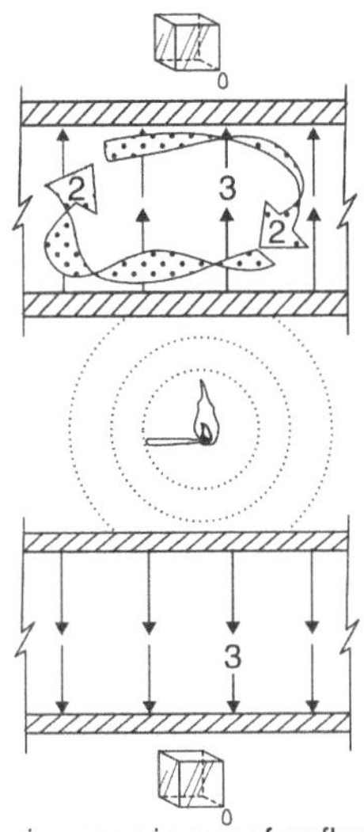

The convective action (2) in the air space of a roof is similar to that in a wall, although the height through which the air rises and falls is usually less. The radiant transfer is up in this case because its direction is always to the cooler surface.

When the higher temperature is at the top of a horizontal air space, the warm air is trapped at the top and, being less dense than the cooler air at the bottom, will not flow down to transfer its heat to the cooler surface. This results in little flow by convection. The radiant transfer in this case is down because that is the direction from the warmer surface to the cooler one.

An air space in a roof or floor

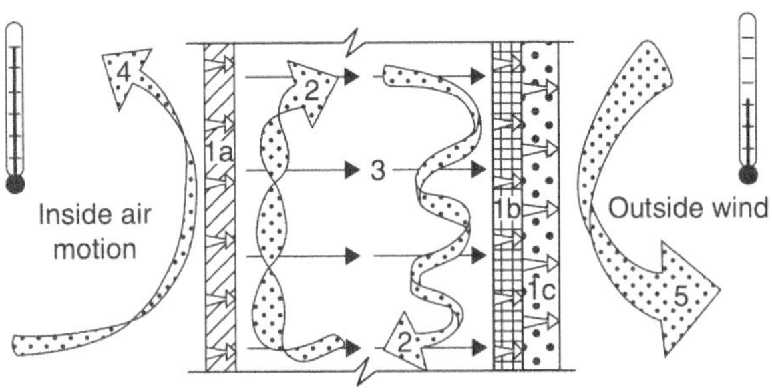

A composite assembly of building materials in place at the site

This example of a wall in place illustrates the several methods by which heat is lost through a composite assembly of materials. Conduction at varying rates in different materials is accounted for in 1a, 1b, and 1c. Convection currents (2) and radiation (3) carry the heat across the air space.

Figure 10.1 *Heat flow through materials, across air spaces, and through construction assemblies. Means of heat flow include conduction, convection, and radiation.*

Resistance. Conductivity and conductance are compared in Table 10.1 and Figure 10.2, which also include another useful property: resistance. Designated as R, resistance indicates how effective a material is as an insulator. The reciprocal of conductivity (or conductance), R is measured in hours needed for 1 Btu to flow through 1 ft^2 of a given thickness of a material when the temperature difference is 1°F. In the I-P system, the units are h ft^2 °F/Btu. SI units are m^2 K/W. Resistances and other important thermal properties are listed for many conventional building materials in Appendix E, Table E.1. Appendix E, Table E.2 provides

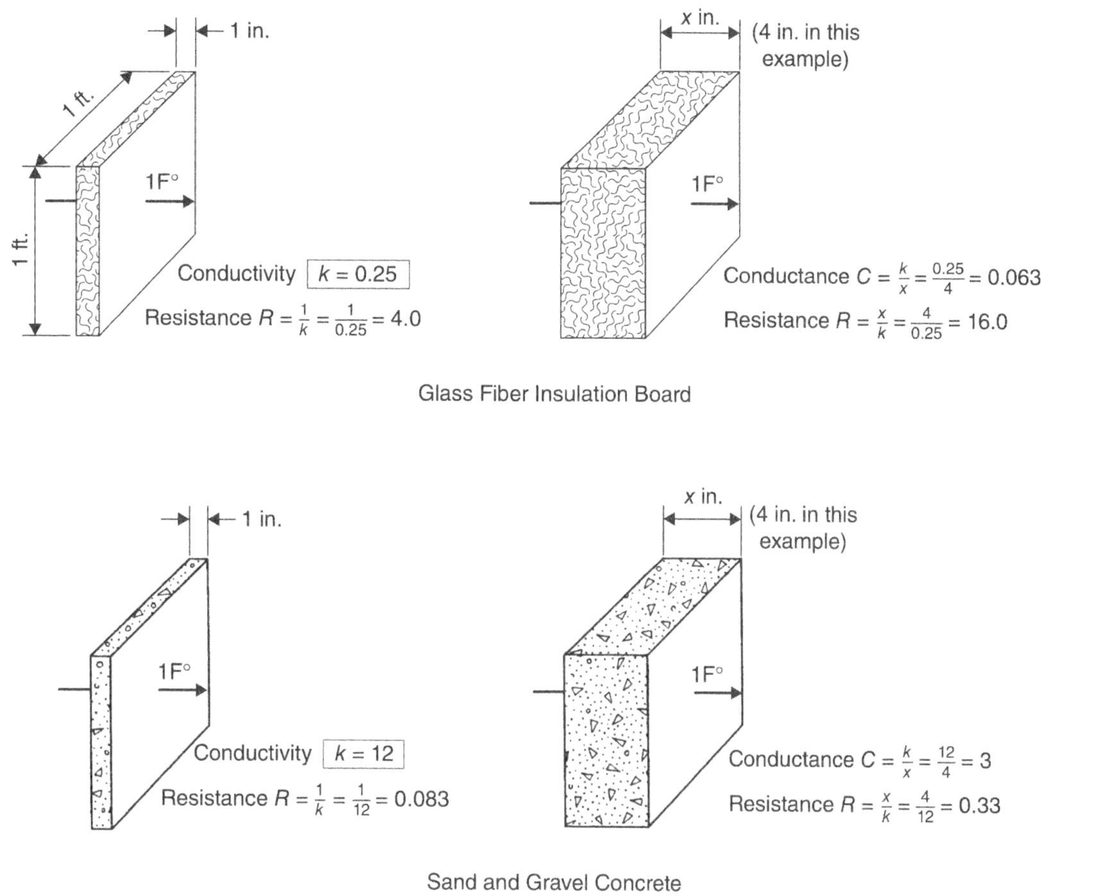

Figure 10.2 *Relationship between conductivity, conductance, and resistance for two typical materials. Glass fiber is a material of low conductivity (high resistance); concrete is a material of high conductivity (low resistance). Note: Standard unit of area is 1 ft^2 [1 m^2]; standard air temperature differential is 1°F [1°C].*

similar information for alternative (less commonly used and/or emerging) construction assemblies.

Resistance is especially useful when comparing insulating materials, because the greater the R-value, the more effective the insulator. For this purpose, R-value is sometimes listed as "per inch" of thickness, in which case the I-P units are h ft^2 °F/Btu-in. SI units are m K/W (indicating "per meter" of thickness across 1 m^2, even though the resulting canceling of units looks odd).

Emittance. Radiation heat transfer is greatly influenced by material surface characteristics. Shiny materials are much less able to radiate than commonly used rough building materials. This characteristic is called emittance, the ratio of the radiation emitted by a given material to that emitted by a blackbody at the same temperature. The impact of emittance (shiny vs. matte surfaces) is seen in Appendix E, Tables E.3, E.4, and E.5, which present properties of air layers and air spaces within construction assemblies. The lower the emittance, the lower the radiative heat exchange potential. For most materials, emittance is related to absorptance. A highly absorptive (low-reflectance) material will usually also have a high emittance. Selective surfaces (sometimes used in solar collectors) are highly absorptive yet have very low emittance.

10.2 SENSIBLE HEAT FLOW THROUGH OPAQUE ASSEMBLIES

Sensible heat is a form of energy that flows whenever there is a temperature difference. Sensible heat manifests itself as an internal energy of atomic vibration within all materials. Temperature is an indication of the extent of such vibration, essentially the "density" of heat within a material. Other forms of energy (such as solar radiation or sound) can be converted to heat and vice versa (within limits). Latent heat is heat that changes the state of (to evaporate or condense) water. Power refers to the instantaneous flow of energy (at a given time). Energy refers to power integrated over time. Table 10.1 lists commonly encountered terms related to energy, power, and heat flow. Some of the units used to quantify heat, power, and energy are quirky. Nevertheless, time spent understanding these units will be useful to building design efforts.

(a) Thermal Classifications of Materials

Architectural materials generally interact with heat either as insulators that retard the flow of heat (useful as thermal barriers) or as conductors that encourage heat flow (useful as thermal storage materials). It is common to find both insulators and conductors in the same construction assembly. For example, a wall can have an inner layer that is highly conductive and thermally massive (for thermal storage), an outer layer that is also highly conductive and thermally massive (for durability and weathering), and a highly insulative, low-mass material in between.

Insulations. Materials used for insulation fall into two basic categories:

1. Those to be installed between framing members—typically, between studs in walls and between trusses or beams in roofs.
2. Those designed to be used as continuous insulation outboard of sheathing on walls or above structural decking on roofs.

Metallic or metalized organic reflective membranes (which must face an air space to be effective) can also be used in wall or roof assemblies to retard the flow of heat.

Insulating materials are available in a wide variety of forms. These include loose fill (as might be poured between rafters on the floor of an attic); insulating cement, a loose material that is mixed with a binder and troweled onto a surface; and formed-in-place materials such as expanded pellets or liquid-fiber mixtures that are poured, frothed, sprayed, or blown into their final locations. Less form-fitting but more common are batts and blankets of flexible, semirigid insulation, with varying degrees of compressibility and adaptability to substrates. Rigid insulation, with little on-site adaptability, is applied in blocks, boards, or sheets and can be preformed to fit nonplanar surfaces such as pipes.

Foamed plastics are the most common type of rigid material used for continuous insulation. There are three basic types: polyisocyanurate (highest R-value per inch, least moisture resistance, least harmful manufacturing process), expanded polystyrene or "EPS," and extruded polystyrene or "XPS" (highest moisture resistance, but most harmful manufacturing process). All foamed plastics are flammable. Mineral wool is a non-flammable alternative that can be used for continuous insulation; it has a similar R-value to EPS, but is air- and vapor-permeable. While mineral wool loses R-value when wet, the material does not degrade when exposed to moisture and the R-value returns when it dries.

Exterior insulation and finish systems (EIFS) were controversial due to significant failures in the 1980s and 1990s when they were installed as barrier systems. The name EIFS still carries a stigma today because of that past. However, modern drainable EIFS systems have become popular as an easy way to achieve a high-performance enclosure that is very low-risk. Utilizing EPS insulation boards adhered to a water-resistive barrier (WRB), then covered with fabric-reinforced acrylic stucco finish, this construction method achieves slightly more than R-4 per inch of EPS thickness [SI: about R-28 per meter]. Because the system relies on adhesion between the layers instead of fasteners, there is far less thermal bridging and a reduction of degradation risk as fasteners are not punching holes in the WRB every 7 in. [180 mm] on center over the building façade, as is required with stucco lath and plaster systems.

Reflective materials are available in sheets and rolls of either single or multiple layers, sometimes as preformed shapes with integral air spaces. When used without attachment to blanket or batt insulation, a reflective layer is called a radiant barrier and is especially applicable to roofs in warmer climates. Radiant barriers are also useful in east- and west-facing walls in such climates. Details of applications in the southern United States are found in Melody (1987). Reflective layers, such as foil-facings, are also typically Class I vapor retarders and should be used with caution. If they are located incorrectly in an assembly they can trap moisture and reduce assembly drying potential.

A combination of dead (still) air spaces and reflective surfaces produces some of the most effective insulating products, especially when assembled of lightweight materials of low conductivity. Glass fiber, cellular glass, expanded styrenes (foamed plastics), and mineral fibers all enclose vast numbers of dead-air spaces per unit volume. When they are bonded to reflective films and properly installed (the shiny film facing a dead-air space), high resistance to heat flow is achieved.

For a summary of contemporary insulation materials, including environmental impacts and life-cycle considerations, see Wilkinson (1999). For a range of information on insulations, see Bynum (2001). Recent life-cycle environmental analyses have been critical of many common insulation materials. Oak Ridge National Laboratory and others have been researching the potential for bio-based insulation materials to reduce the upfront carbon associated with many current insulations. BuildingGreen published a guide that addresses the environmental impacts of thermal insulations (BuildingGreen, 2018).

Conductors. Materials used as conductors are typically dense, durable, and able to readily diffuse heat. Density, conductivity or conductance, and specific heat are important thermal properties for conductors. For a given material, the higher the numerical values of these three characteristics, the more successful that material's performance as a conductor. Appendix E, Table E.1 lists a variety of thermal properties for many common materials.

(b) Air Films and Air Spaces

Air films and spaces are interesting thermal components. Although they are void of material, they possess useful thermal properties and contribute substantially to the insulating capabilities of some construction assemblies. All above-grade enclosure assemblies include at least two air films (interior and exterior), and many common enclosure assemblies include a substantial air space among their components.

Air Films. At the exposed surface of a solid, heat transfer takes place both by convection and by radiation. Convection is highly dependent upon air motion, so air speed must be considered when estimating exterior surface convection. Also, because warm air rises and cold air falls, vertical surfaces that encourage natural convection will exchange heat faster than similar surfaces placed horizontally, unless the direction of the heat flow is upward through the horizontal layer, as illustrated in Figure 10.1.

When air motion along a surface is minimal, an insulating layer of air attaches itself to the surface. The resistance of this layer of still air along a vertical surface is equivalent to that of a thickness of ½-in. [12.7-mm] plywood. When this air layer is disturbed, however, its resistance drops quickly; with a 15-mph [6.7-m/s] wind, resistance drops to about one-quarter of the still-air value (see Appendix E, Table E.3). Similar drops in air film resistance occur when forced-air HVAC registers are located immediately above or below windows.

The insulation value provided by layers of air is often listed as conductance, the reciprocal of resistance. Surface conductances are designated h_i for interior air layers and h_o for exterior or outside air layers (sometimes the symbols f_i and f_o are used instead). Like other conductances, they are expressed in Btu/h ft^2 °F [in SI units, W/m^2 K]. The variations in surface resistances and conductances that are typical of building constructions are shown in Appendix E, Table E.3.

Air Spaces. An air space is a planar volume of air contained on two sides by elements (drywall, brick, insulation, etc.) of an enclosure assembly. Like the air films just discussed, air spaces contribute to the overall thermal resistance of a construction assembly. As indicated in Appendix E, Tables E.4 (I-P units) and E.5 (SI units), the resistance provided by an air space is a function of its width, position (vertical, horizontal, tilted), and surrounding surface emittances. To be effective in resisting heat flow, an air space must be relatively "dead"—without substantial air circulation—a consideration that limits the useful depth of an insulating air space. Because air spaces within a building assembly can promote convection and drying while at the same time facilitating drainage—air spaces are often a key component of high-performance assemblies.

(c) Composite Thermal Performance

U-factor. The variety of terms used to express thermal properties is potentially bewildering. These properties are, however, only components of a larger picture. Fortunately, there is one overall property that expresses the steady-state rate at which heat flows through architectural enclosures. This property is the U-factor (also referred to as the U-value). U is the overall coefficient of thermal transmittance, expressed in terms of Btu/h ft^2 °F [in SI units, W/m^2 K]. U-factors are commonly used to specify enclosure thermal design criteria, as presented in Appendix H. Energy codes and standards specify maximum U-factors for various components of the enclosure (or, for insulation alone, minimum R-values or maximum C-values). An example of such requirements for a small office is found in Chapter 9, Example 9.1. Because U-factors are so important and so often encountered, data for typical wall, floor, roof, door, and window constructions are presented in many design resources (see, for example, Appendix E, Tables E.6 through E.16).

U-factors are calculated for a particular element (roof, wall, etc.) by finding the resistance of each constituent part, including air films and air spaces, then summing these resistances to obtain a total resistance. The U-factor is the reciprocal of this total (Σ) resistance:

$$U = \frac{1}{\sum R} \qquad \text{(Eq. 10.1)}$$

This calculation procedure is shown in Example 10.1. "Precalculated" U-factors for many common constructions are found in Appendix E, Tables E.6 through E.16, typically based upon an outside wind speed of 15 mph [6.7 m/s], except for summer conditions as noted. Architectural creativity often leads to the design of unique enclosure assemblies for which precalculated U-factors cannot be found. U-factor values must then be developed by the design team.

EXAMPLE 10.1 What is the winter U-factor for the wall assembly shown in Figure 10.3?

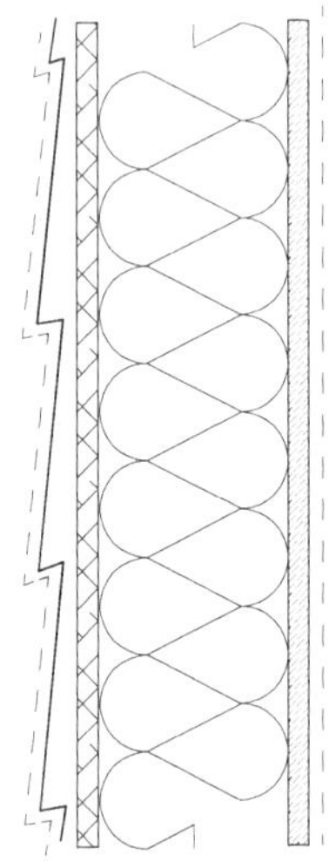

Figure 10.3 *Wall section used in Example 10.1 to illustrate the process for determining a U-factor (overall thermal transmittance). (Drawn by Sharon Alitema.)*

SOLUTION

Component	**R (I-P)**	**R (SI)**	**Data Source, Appendix E**
Inside air film	0.68	0.12	Table E.3
Gypsum board, 0.375 in. [10 mm]	0.34	0.063	Table E.1
Plastic film vapor retarder	nil	nil	Table E.1
Glass-fiber batt insulation, 0.61 lb/ft^3 [9.8 kg/m^3] density, 5.5 in. [140 mm]	19.64	3.49	Table E.1
Plywood sheathing, 29 lb/ft^3 [460 kg/m^3] density, 0.5 in. [12.7 mm]	0.79	0.14	Table E.1
Wood siding (drop, 8 in.), 1 in. [25 mm]	0.79	0.14	Table E.1
Outside air film	0.17	0.03	Table E.3
Total resistance (R)	**22.41**	**4.55**	

$$U\ (\text{I-P}) = 1/\Sigma R = 1/22.41 = 0.045$$

$$[U\ (\text{SI}) = 1/\Sigma R = 1/4.55 = 0.220]$$ ■

There are several important points to remember about U-factors. The U-factor is a sensible heat property—addressing heat flow resulting from a temperature difference but not addressing latent (moisture-related) heat flow. The U-factor is an overall coefficient of heat transfer and includes the effects of all elements in an assembly and all sensible modes of heat transfer (conduction, convection, and radiation). The term U-factor should be used only where heat flow is from air to air through an enclosure assembly. Air-to-ground heat flow (as through a slab-on-grade floor) is a different matter and is considered in Section 10.2(d).

Wall U-factors. Compared to other elements of the building enclosure, wall U-factors are quite straightforward. Ground contact, crawl spaces (as with floors), or intermediary attic spaces (as with roofs) are not present to complicate analysis. There is, however, the issue of thermal bridging. Where framing interrupts an insulation plane, there are two thermally different wall constructions, and a weighted average U-factor (reflecting an insulated wall area paired with a less-insulated wall area) must be found. Example 10.2 illustrates this procedure.

EXAMPLE 10.2 Considering thermal bridging, what is the winter U-factor for the wall assembly shown in Figure 10.4?

SOLUTION

For those portions of the wall where insulation is encountered:

Component	**R(I-P)**	**R (SI)**
(6) Inside air film, still	0.68	0.12
(5) Gypsum board, 0.5 in. [12.7 mm]	0.45	0.079
Plastic film vapor retarder	nil	nil
(4) Glass-fiber batt insulation, 0.61 lb/ft^3 [9.8 kg/m^3] density, 5.5 in. [140 mm]	19.64	3.49
(3) Vegetable fiber board, regular density, 0.5 in. [12.7 mm]	1.32	0.23
(2) Lapped wood siding, 0.5 in. [13 mm]	0.81	0.14
(1) Outside air film, 15 mph [24 km/h]	0.17	0.03
Total resistance through insulation	**23.07**	**4.09**

For those portions of the wall where framing is encountered:

Component	**R(I-P)**	**R(SI)**
(6) Inside air film, still	0.68	0.12
(5) Gypsum board, 0.5 in. [12 mm]	0.45	0.079
Plastic film vapor retarder	nil	nil
(4) Wood studs, Douglas fir, 34 lb/ft^3 [535 kg/m^3] density, 5.5 in. [140 mm]	5.79	0.99
(3) Vegetable fiber board, regular density, 0.5 in. [12.7 mm]	1.32	0.23
(2) Lapped wood siding, 0.5 in. [12 mm]	0.81	0.14
(1) Outside air film, 15 mph [24 km/h]	0.17	0.03
Total resistance through framing	**9.22**	**1.58**

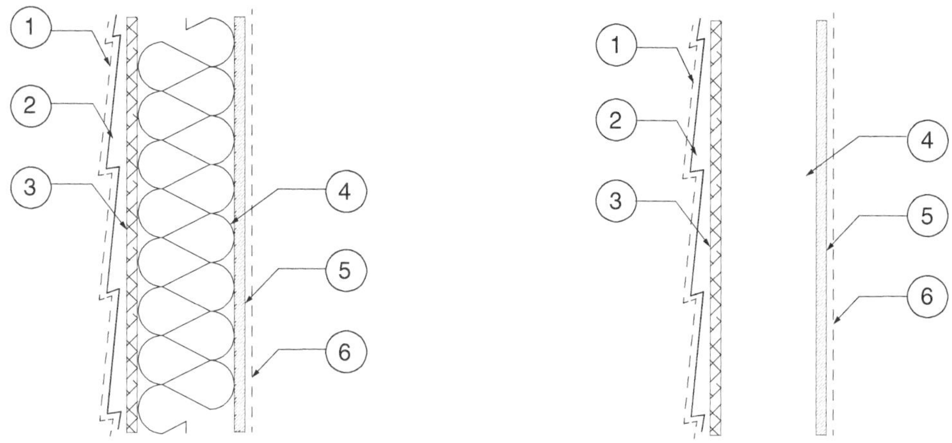

Figure 10.4 *Wall sections used in Example 10.2 to illustrate the process for determining a weighted-average U-factor to account for thermal bridging. (Drawn by Sharon Alitema.)*

If 12% of the surface area of the wall consists of framing (e.g., studs, sills, and plates), the area-weighted U-factor of the overall wall is as follows:

$$\text{I-P: } R_{insulated} = 23.07$$

$$\text{I-P: } U_{insulated} = (1/23.07) = 0.043$$

$$\text{I-P: } R_{frame} = 9.22$$

$$\text{I-P: } U_{frame} = (1/9.22) = 0.108$$

$$\text{I-P: } U_{weighted} = (0.88)(0.043) + (0.12)(0.108)$$

$$= 0.051$$

The predicted U-factor for the actual wall assembly is thus 0.051 (compared to 0.043 if framing is ignored).

$$\text{[SI: } R_{insulated} = 4.09$$

$$\text{SI: } U_{insulated} = (1/4.09) = 0.245$$

$$\text{SI: } R_{frame} = 1.58$$

$$\text{SI: } U_{frame} = (1/1.58) = 0.631$$

$$\text{SI: } U_{weighted} = (0.88)(0.245) + (0.12)(0.631)$$

$$= 0.291]$$

■

The effects of wood stud framing on wall thermal integrity are quite noticeable. With metal studs, the detrimental effects of thermal bridging are more serious; Appendix E, Table E.8 provides information on correction factors and resulting R-values. *ASHRAE/IES Standard 90.1: Energy Standard for Buildings Except Low-Rise Residential Buildings*, and *ASHRAE/IES Standard 90.2: Energy-Efficient Design of Low-Rise Residential Buildings*, present examples of whole-wall U-factors that account for thermal bridging. See Appendix H for excerpts from *ASHRAE/IES Standard 90.1-2022*.

To illustrate the effects of wall construction and variations in thermal bridging on performance, consider the following cases. From Appendix E, Table E.8, using the case of a nominal 4-in. [100-mm] steel-framed wall with R-15 [SI: R-2.64] cavity insulation, increasing the stud spacing from 16 in. on center to 24 in. on center [400–600 mm] increases (improves) the effective R-value from 6.4 to 7.8 [SI: R-1.13 to 1.37]—a nominal difference.

However, changing the framing material from steel to wood studs at 16 in. on center [400 mm] has a more significant impact. Consider a wall assembly option offered in Appendix E, Table E.6. If R-4 [SI: R-0.7] continuous insulation and the balance of other wall components (i.e., exterior air film, stucco, exterior gypsum board, interior gypsum board, and interior air film), were added to a nominal 4-in. [100-mm] steel framed wall with R-15 [SI: R-2.64] cavity insulation the effective R-value of the wall assembly is approximately R-11.49 [SI: R-2.04]. By only changing the framing from steel to wood, the effective R-value of the assembly increases (improves) to R-14.7 [SI: R-2.56]. The thermal bridging effect of material selections and construction details can make a difference. Mitigating thermal bridges is a required aspect of enclosure design for a passive house project. If a detailed calculation of thermal bridging effects is required, consult the *2021 ASHRAE Handbook—Fundamentals*, Chapter 25.

Structural insulated panels (SIPs). Several alternative assemblies can improve wall system thermal performance. Structural insulated panels are available in a wide variety of finishes, sizes, insulation types, and thicknesses. Because factory-built panels replace site-built framing, the savings in labor can be substantial. Thermal performance is considerably improved because no framing members penetrate the insulated core. The typical SIP consists of two structural surfaces (often oriented strand board, or OSB) that enclose a core of either expanded polystyrene (R-4 per inch; [SI: R-27.7 per meter]) or polyurethane foam (R-6.5 per inch; [SI: R-45 per meter]). The panels are joined with plywood splines that connect the structural surfaces of adjacent panels in a manner that reduces the thickness of, but does not interrupt, the insulated core. Another joining method, shiplap joints, maintains full-thickness core insulation. U-factors of framed walls and SIPs are compared in Appendix E, Table E.9.

Insulated Masonry. Insulated masonry systems can offer low U-factors while preserving a masonry exterior (for weathering and appearance) and a masonry interior (for thermal mass). Past insulation approaches (filling the hollow cores of concrete block or clay tile with insulation) left considerable thermal bridging through the solid masonry flanges, resulting in rather high U-factors. If one of the masonry surfaces is not intended to be exposed, then a continuous layer of insulation can be applied to that face, providing much lower whole-wall U-factors. In masonry cavity walls, the layer of insulation between wythes of block is usually penetrated by ties—with resulting thermal bridging. The most elementary improvement in this design uses ties with lower thermal conductivity, such as fiber composite materials in place of steel. Another approach is to use precast blocks that can be stacked integrally with preformed rigid insulation. To the extent that the insulation layer can be both thick and unbroken by ties, thermal bridging is greatly reduced. The following is a sample of manufacturers' whole-wall U-factors for medium-density (120 lb/ft^3) [1920 kg/m^3] concrete block insulation alternatives:

8-in. two-core concrete block, uninsulated	U = 0.39
[SI: 200-mm two-core concrete block, uninsulated]	U = 2.21
8-in. two-core concrete block with insulated cores	U = 0.19
[SI: 200-mm two-core concrete block with insulated cores]	U = 1.08
[SI: 200-mm proprietary two-wythe concrete block and core]	U = 0.62
8-in. proprietary two-wythe concrete block and core	U = 0.11

Poured-in-Place Concrete. Where poured-in-place concrete is not intended as a surface material, insulating concrete forms (ICFs) can be used to improve thermal performance. This system employs preformed rigid insulation as a formwork for poured concrete; the form/insulation remains in place after the wall is poured. Rigid insulation thus protects both faces of the structural wall; exterior and interior finishes are then applied to the insulation. As the concrete within the ICF is permanently covered by insulation on both sides, the benefits of thermal mass are essentially removed. Because concrete contains a very high upfront carbon cost, concrete should be utilized carefully and wisely to maximize its performance and environmental cost. For example, standard masonry construction with EIFS or another insulation and cladding system on the exterior is a high-performance combination that works in both hot and cold climates to provide thermal mass on the interior while at the same time providing insulation on the exterior.

Roof U-factors. In the simple case of a roof/ceiling separating a conditioned space from the outdoors, the U-factor is calculated as for walls. Where insulation is broken by framing, the effects of thermal bridging must be considered. Appendix E, Table E.8 shows correction factors to be applied to metal truss assemblies with insulation. Similar adjustments, although of lesser magnitude, are required with wood framing. *ASHRAE/IES Standard 90.1* presents several tables of whole-roof U-factors that account for thermal bridging. If a detailed calculation of thermal bridging is required, consult the *2021 ASHRAE Handbook—Fundamentals*, Chapter 25.

Roof insulation, however, is often placed entirely above the supporting structure, greatly reducing the effects of thermal bridging. There is a growing availability of above-the-roof-structure insulation materials, some penetrated by fastenings, others essentially unbroken. These options should be explored as a means of reducing roof heat loss and heat gain—with appropriate consideration of the environmental impacts of material selections. Polyisocyanurate is by far the most common roofing insulation when used above the structural decking. However, if an inverted roof membrane assembly (IRMA) is utilized (where the membrane is below the insulation), then polyisocyanurate cannot be used because of the increased exposure to moisture. XPS rigid foam insulation is more common in these applications.

An attic is a space having an average distance of 1 ft [0.3 m] or more between the ceiling and the underside of the roof. Where an insulated ceiling separates a space from an uninsulated attic, the simplest approach is to assume that the attic is at outdoor air temperature. This assumption likely overestimates the winter heat flow rate because the temperature in a vented attic may be higher than the outdoor air temperature—but it simplifies analysis. If an accurate estimate of attic temperature is required, a more complex procedure is used, as presented in the *2021 ASHRAE Handbook—Fundamentals*, Chapter 27. A vented attic is often designed to carry away moisture that may have migrated through the insulated ceiling. See Lstiburek and Carmody (1994) for detailed information on attic design practices. Recent analyses indicate that venting an attic space in hot and humid climates is counter productive because it brings in humid air, and the limited convection does little to promote drying (Lstiburek, 2023).

Floor U-factors. When the underside of a floor is exposed to air (as with a cantilever or crawl space), the U-factor is calculated as for walls and roofs. When insulation is placed within framing cavities, the effects of thermal bridging must be included. It is becoming increasingly common to provide a continuous layer of insulation below the framing cavity, thus greatly reducing the detrimental effects of thermal bridging. *ASHRAE/IES Standard 90.1* presents several tables of whole-floor U-factors that account for thermal bridging.

If a vented crawl space is utilized, the simplest procedure is to insulate the floor above the crawl space and assume that the crawl space is at outdoor temperature. This assumption overestimates the heat flow rate because a vented crawl space will rarely be as cold (or as warm) as outdoor air under winter (summer) design conditions. If a more accurate analysis of crawl space design temperatures is required, or if crawl space walls are insulated, a more complex procedure is used, as presented in the *2021 ASHRAE Handbook—Fundamentals*, Chapter 27. It is rarely energy-conserving, however, especially in colder climates, to insulate the walls of a crawl space instead of insulating the floor above. See Lstiburek and Carmody (1994) for detailed information on crawl space design practices.

In addition to insulation below raised floors, it is very important to stop the flow of air and vapor at these locations. Crawl spaces are often laden with moisture, biological growth, and animal feces, among other undesirable qualities. Therefore, maintaining air, vapor, and thermal control at the underside of a raised floor is critical for maintaining good indoor air quality and health. Closed-cell (or high-density) spray polyurethane foam is one option that can serve as a control layer for heat, air, and vapor at these locations with just one material.

Door U-factors. Appendix E, Table E.10 lists U-factors for solid (no glazing) doors in common use in North America. These precalculated values are convenient but should be used with care, as air film resistances are a relatively large factor in overall door thermal performance. Many doors are placed near forced-air supply registers or return air grilles, reducing the resistance of the indoor air film. Conversely, the exterior air film may be more effective than anticipated because doors are typically protected somewhat by overhangs, porches, and so on. The wind conditions assumed in Appendix E, Table E.10 are typical of winter (not summer) conditions. Where significant door glazing is involved, U-factors should be adjusted to account for the glazing performance, or data for a specific product should be used.

(d) Special Enclosure Heat Flow Conditions

The overall coefficient of heat transfer (U-factor) is a convenient property for analysis of above-grade enclosure assemblies. It is only applicable, however, to assemblies exposed to air on both the interior and exterior surfaces. In other situations, such as slab-on-grade floors and below-grade walls and floors, the U-factor is not a well-defined property because soil temperature varies considerably, based on depth and location, and it is not clear how much of the soil should be considered as part of the floor assembly for U-factor calculation. Typically, data derived from empirical studies are used in lieu of fundamental heat flow equations. Because the concept of the U-factor is so simple to understand and apply, much of the empirical data are presented in terms of equivalent U-factors.

Slab-on-Grade Floors. Heat flow dynamics become complicated when the lower surface of a floor is in direct contact with the ground. The ground is often at a temperature different from that of the outdoor air, and earth is more conductive than air. Testing has shown that heat flow in this complex situation is strongly related to slab perimeter length. Appendix E, Table E.11 shows the heat flow rate through a concrete slab-on-grade, given as F_2 units (per foot or meter of perimeter length) rather than U-factors (per unit area). Four edge insulation conditions are shown, with corresponding values of F_2 for three climate zones. Interpolation can be used to estimate F_2 for other climates (correlated to heating degree days). This procedure relates all heat loss from the floor to the length of the perimeter.

Slab-on-grade heat loss illustrates the principle of diminishing returns: An incrementally higher R-value does not produce an equally lower heat flow. An insulation truism is that the first few inches [millimeters] of insulation have a much greater impact on the rate of heat flow than do the last few inches [millimeters]. Appendix E, Table E.11 lists a maximum slab-edge R-value of 5.4 h ft^2 °F/Btu [SI: R-0.95]. If more slab-edge insulation seems desirable, consult Appendix E, Table E.12. Note how much more effective vertical insulation is at the exposed perimeter compared to insulation beneath the slab. The highest F_2 value listed in Appendix E, Table E.12 is for insulation at R-10 [SI: R-1.8] not only around the perimeter but also under the entire area of the floor slab. This lowest rate of heat flow is a reduction of just 25% from the performance provided by the same insulation employed only around the perimeter to a depth of 4 ft [1.2 m].

EXAMPLE 10.3 A warehouse will be built with a slab-on-grade floor. The warehouse is 80 ft [24.4 m] square in plan. What is the relationship between total slab insulation and rate of heat flow through the slab?

SOLUTION

From Appendix E, Table E.12, F_2 is related to total insulation as follows: building slab perimeter—four sides at 80 ft [24.4 m] each = 320 ft [97.5 m]; building slab area: 80 ft [24.4 m] × 80 ft [24.4 m] = 6400 ft² [595 m²].

I-P Units:

Insulation Approach	Total Volume of Slab Insulation Used (ft³)	F_2 (Btu/h °F)	F_2 × Perimeter = Rate of Heat Flow (Btu/h ft °F)	Incremental Reduction in Heat Flow (Btu/h °F)	Reduction per Unit (ft³) of Insulation
Uninsulated slab	None	0.73	234	None	—
R-5 (1 in.) vertical, 2 ft	(2 ft) (320 ft) (1/12 ft) = 53.3	0.58	186	48	0.90
R-10 (2 in.) vertical, 4 ft	(4 ft) (320 ft) (2/12 ft) = 213.3	0.48	154	32	0.15
R-10 (2 in.) fully insulated	(2 ft) (320 ft) (2/12 ft) + (6400 ft²) (2/12 ft) = 1173.3	0.36	115	38	0.03

SI Units:

Insulation Approach	Total Volume of Slab Insulation Used (m³)	F_2 (W/K)	F_2 × Perimeter = Rate of Heat Flow (W/m K)	Incremental Reduction in Heat Flow (W/K)	Reduction per Unit (m³) of Insulation
Uninsulated slab	None	1.26	123.0	None	—
R-0.9 (25 mm) vertical, 0.6 m	(0.6 m) (97.8 m) (0.025 m) = 1.5	1.00	97.5	25.4	17.0
R-1.8 (50 mm) vertical, 1.2 m	(1.2 m) (97.8 m) (0.051 m) = 6.0	0.83	81.0	16.6	2.8
R-1.8 (50 mm) fully insulated	(0.6 m) (97.8 m) (0.051 m) + (595 m²) (0.051 m) = 33.3	0.62	60.5	20.5	0.6

Thus, a minimal initial investment of 640 ft² [60 m²] of 1-in.- [25-mm]-thick insulation reduces the uninsulated slab heat flow by one-fifth (i.e., 186/234). However, a vastly greater investment of 7040 ft² [654 m²] of 2-in. [50 mm]-thick insulation (22 times as much insulation!) cuts the uninsulated slab heat flow only by one-half (i.e., 115/234). This exemplifies the diminishing return of increasing insulation values in thermal enclosures. ■

Basements. Heat flow through basement walls and floors is complicated by an increasing length of heat flow path with increasing depth, as shown in Figures 10.5*a* and 10.5*b*. A further complication is that the temperature of the earth is not equal to ambient air temperature and becomes more and more out of phase with air temperature at increasing depths. To obtain a "design temperature" for below-grade heat loss, first estimate the mean winter temperature of the site location. For example: using Appendix C, Tables C.19 and C.20, take the average of the ambient temperature (TA) for January and for the year. Then, from this mean winter temperature, subtract the value of the constant amplitude, Figure 10.5*c*. This gives a design temperature. Appendix E, Table E.13 is then used to determine the heat flow rate.

EXAMPLE 10.4 Calculate the design heat loss for a basement in Minneapolis, Minnesota, which is 28 ft [8.5 m] wide by 30 ft [9.1 m] long, sunk 6 ft [1.8 m] below grade. An insulation of R-10 [SI: R-1.8] is applied to this wall to a depth of 2 ft [0.6 m] below grade. An internal temperature of 70°F [21°C] is to be maintained.

SOLUTION

The soil-path design temperature is estimated from Appendix C, Tables C.19 [C.20]; for Minneapolis, the TA in January is 15.9°F [−9.0°C]; TA for the year is 47°F [8.3°C]; the average of these values is 31.5°F [−0.4°C]. From Figure 10.5c, the amplitude at Minneapolis is about 24°F degrees [13.3°C]. The design temperature at the ground surface is therefore 31.5 − 24 = 7.5°F [−0.4 − 13.3 = −13.7°C].

The heat loss through the basement wall is estimated using Appendix E, Table E.13, Part A, by determining the loss per degree of temperature difference per 1-ft- [0.3-m]-high strip of wall and then summing for the full height of the wall:

1st ft below grade (R-10 insulated) (0.080) × (1 ft)	= 0.080 Btu/h ft °F
[SI: 1st 0.3 m: per m height (0.458) × (0.305 m)	= 0.140 W/m K
2nd ft (use 2.6 ft value) below grade (R-1.8 insulated)	= 0.075
[SI: 2nd 0.3 m (use 0.79 m value): (0.427) (0.305)	= 0.130
3rd ft below grade (uninsulated)	= 0.273[a]
[SI: 3rd 0.3 m: (1.571) (0.305)	= 0.479[a]

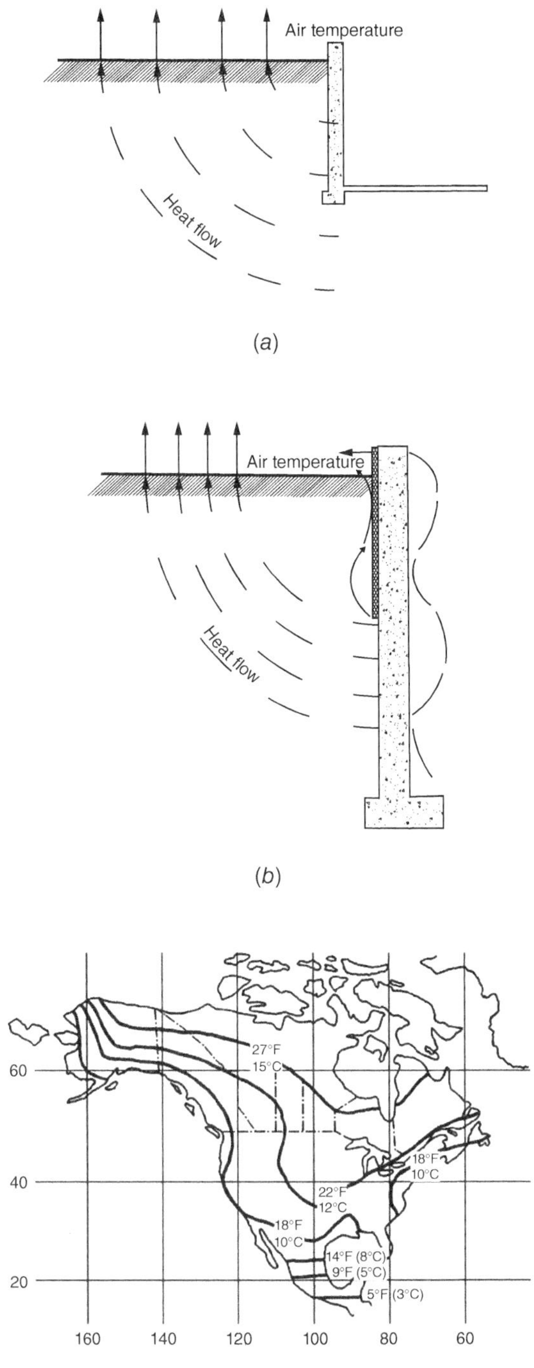

Figure 10.5 *Heat flow below grade. (a) Heat flow through basement walls and floors follows radial paths (that are generally perpendicular to soil isotherms). (b) Heat flow paths of differing lengths for a partially insulated basement wall. (c) Lines of constant amplitude of ground temperature; just below the surface, the ground temperature fluctuates around a mean annual temperature by this amplitude. (a: Reprinted with permission; © ASHRAE. 2017 ASHRAE Handbook—Fundamentals; b: Reprinted with permission of ASHRAE, from the 2001 ASHRAE Handbook—Fundamentals; redrawn by Sharon Alitema; the referral to an older version of the ASHRAE Handbook is intentional and provides access to historic reference information of ongoing interest; c: © ASHRAE, www.ashrae.org. 2021 ASHRAE Handbook—Fundamentals, Chapter 18, Figure 13; SI units added.)*

4th ft below grade (uninsulated)	= 0.235
[SI: 4th 0.3 m: (1.353) (0.305)	= 0.413]
5th ft below grade (uninsulated)	= 0.208
[SI: 5th 0.3 m: (1.195) (0.305)	= 0.364]
6th ft below grade (uninsulated)	= 0.187
[SI: 6th 0.3 m: (1.075) (0.305)	= 0.328]
Total per 6-ft height and 1-ft length of wall	= 1.058 Btu/h ft °F
[Total per 1.8-m height and 1-m length of wall	= 1.854 W/m K]
Basement perimeter = 2 (28 + 30 ft) = 116 ft	
[SI: 2 (8.5 + 9.1 m) = 35.2 m]	
Total wall heat loss = 1.058 Btu/h ft °F × 116 ft = 123 Btu/h °F	
[SI: 1.854 W/m K × 35.2 m = 65.3 W/K]	

[a]Although this is an increase in heat loss relative to the segment of wall directly above, it represents a design decision that reflects an understanding of the law of diminishing returns.

The heat loss through the basement floor is calculated using Appendix E, Table E.13, Part B.

Average heat loss per ft^2 of floor:

I-P: 0.030 Btu/h ft^2 °F (for 28-ft width and 6-ft depth)

[SI: 0.176 W/m^2 K (for 8-m width and 1.8-m depth)]

Floor area:

$$\text{I-P: } 28\,\text{ft} \times 30\,\text{ft} = 840\,\text{ft}^2$$

$$[\text{SI: } 8.5\,\text{m} \times 9.1\,\text{m} = 77.4\,\text{m}^2]$$

Total floor heat loss (U×A):

$$\text{I-P: } 0.030\,\text{Btu/h ft}^2\,°\text{F} \times 840\,\text{ft}^2 = 25.2\,\text{Btu/h}\,°\text{F}$$

$$[\text{SI: } 0.176\,\text{W/m}^2\,\text{K} \times 77.4\,\text{m}^2 = 13.6\,\text{W/K}]$$

Total heat loss (U) per degree of temperature difference for the basement below grade:

$$\text{I-P: } 123\,\text{Btu/h}\,°\text{F (wall area)} + 25.2\,\text{Btu/h}\,°\text{F (floor area)} = 148.2\,\text{Btu/h}\,°\text{F}$$

$$[\text{SI: } 65.3\,\text{W/K (wall area)} + 13.6\,\text{W/K (floor area)} = 78.9\,\text{W/K}]$$

Design temperature difference:

Internal temperature-design temperature at the ground

$$70°\text{F}\,[21°\text{C}] - 7.5°\text{F}\,[-13.7°\text{C}] = 62.5°\text{F}\,[34.7°\text{C}]$$

Design below grade heat loss:

$$\text{I-P: } 148.2\,\text{Btu/h}\,°\text{F} \times 62.5\,°\text{F} = 9262.5\,\text{Btu/h}$$

$$[\text{SI: } 78.9\,\text{W/K} \times 34.7°\text{C or K} = 2737.8\,\text{W/K}]$$ ■

(e) Predicting Surface Temperatures and Condensation

Enclosure surface temperatures, as well as the temperature at any point within a wall, roof, or floor assembly, can be predicted if the indoor and outdoor air temperatures and the thermal properties of the construction are known. (This process is more challenging for below-grade assemblies.) The variance in temperature through a cross-section of a construction assembly is called a thermal gradient. For any construction, the thermal gradient can be predicted by proportioning the collective thermal resistance at any point in the assembly to the overall difference in temperature across the assembly. This procedure is illustrated in Example 10.5.

EXAMPLE 10.5 Determine the thermal gradient, under winter design conditions, through the wall assembly shown in Figure 10.6. The interior air temperature is 68°F [20°C], and the exterior air temperature is 32°F [0°C].

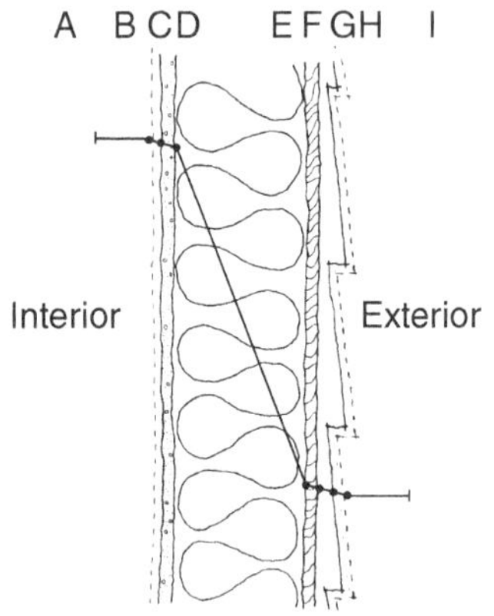

Figure 10.6 Wall section used with Example 10.5 to illustrate the procedure for calculating a thermal gradient through a construction assembly. (Drawing by Jonathan Meendering.)

SOLUTION

I-P Units: See the tabulation. Thermal resistance values are from Example 10.1; "reference point" is a shorthand means of describing a series of locations throughout the wall assembly, each on the exterior side of a component; the total temperature gradient spans (68 – 32) = 36°F:

Thermal Component	Component Resistance	Cumulative Resistance	Reference Point	Temperature Difference to Reference Point (°F)	Temperature at Reference Point
Room air	—	—	A	—	68.0
Inside air film	0.68	0.68	B	(0.68/21.58) × (36) = 1.1	66.9
Gypsum board	0.32	1.00	C	(1.00/21.58) × (36) = 1.7	66.3
Vapor retarder	nil	1.00	D	(1.00/21.58) × (36) = 1.7	66.3
Batt insulation	19.00	20.00	E	(20.00/21.58) × (36) = 33.4	34.6
Plywood sheathing	0.62	20.62	F	(20.62/21.58) × (36) = 34.4	33.6
Wood siding	0.79	21.41	G	(21.41/21.58) × (36) = 35.7	32.3
Outdoor air film	0.17	21.58	H	(21.58/21.58) × (36) = 36	32.0
Outdoor air	—	—	I	—	32.0

I-P Tabulation (continued)

SI Units: See the tabulation. Thermal resistance values are from Example 10.1; "reference point" is a shorthand means of describing a series of locations throughout the wall assembly, each on the exterior side of a component; the total temperature gradient spans (20 – 0) = 20°C:

Thermal Component	Component Resistance	Cumulative Resistance	Reference Point	Temperature Difference to Reference Point (°C)	Temperature at Reference Point (°C)
Room air	—	—	A	—	20.0
Inside air film	0.12	0.12	B	(0.12/3.81) × (20) = 0.6	19.4
Gypsum board	0.056	0.18	C	(0.18/3.81) × (20) = 0.9	19.1
Vapor retarder	nil	0.18	D	(0.18/3.81) × (20) = 0.9	19.1
Batt insulation	3.35	3.53	E	(3.53/3.81) × (20) = 18.5	1.5
Plywood sheathing	0.11	3.64	F	(3.64/3.81) × (20) = 19.1	0.9
Wood siding	0.14	3.78	G	(3.78/3.81) × (20) = 19.8	0.2
Outdoor air film	0.03	3.81	H	(3.81/3.81) × (20) = 20	0.0
Outdoor air	—	—	I	—	0.0

■

Understanding enclosure surface temperatures is fundamental to predicting thermal comfort. How cold in winter, or hot in summer, will an interior surface be, and how will this affect mean radiant temperature? Understanding temperature gradients within enclosures is critical to establishing where water vapor might condense within or on a wall, roof, or floor assembly; this is discussed further in Section 10.2(e).

Knowing whether a building assembly is likely to experience condensation at some moment in time is important—yet does not provide enough information to make solid design decisions. Even moisture-sensitive materials like paper-backed gypsum board or wood-based sheathing can handle a certain amount of bulk water exposure during the year. But if elevated relative humidity (even without condensation) is present at these moisture-sensitive materials (a rolling average of 80% relative humidity or higher over a 30-day period per *ASHRAE Standard 160: Criteria for Moisture-Control Design Analysis in Buildings*), then conditions favorable to mold growth may be present. Hygrothermal modeling software (e.g., Wärme Und Feuchte Instationär [WUFI] Pro by Fraunhofer) may be necessary to analyze complicated assemblies and ensure durability. Such simulations perform thousands of calculations to provide a picture of how an assembly will perform in a specific climate over a longer period of time, such as five years or more.

(f) Dynamic Thermal Effects

The preceding discussion of thermal properties of materials and assemblies focused upon static (steady-state) conditions. Such conditions are usually presumed for analysis of winter heat loss (assumed to occur under the darkness of night). A more complex situation exists when thermal conditions experience rapid changes—such as those brought about by solar radiation impinging upon various portions of the building enclosure. Dynamic conditions are usually presumed for analysis of summer heat gain (assumed to occur during daylight hours). The dynamic situation requires that the enclosure designer consider yet more thermal properties.

Under static conditions, heat flow is primarily a function of temperature difference (the driving force) and thermal resistance (the mitigating force). Under dynamic conditions, these two factors are still important, but heat storage in the enclosure assembly itself becomes a compounding issue. Heat storage is a function of the density of a material and its specific heat; the product of these two properties is known as thermal capacity. Thermal capacity can (but will not always) reduce heat flow via diversion to storage. Heat entering a wall assembly during the daytime, for example, can be stored within the wall for several hours until it flows back out to the cool night air—assuming appropriate weather conditions and adequate thermal capacity. As a minimum, the effect of heat entering a massive wall at, say, 10:00 a.m. will not be seen on the interior surface until a few hours later. This strategy works in climates with substantial diurnal temperature swings (hot days, cold nights) such as in the arid southwest.

Density. Density is the weight of a material per unit volume. In the I-P system, density is expressed as lb/ft^3; in the SI system, it is expressed as kg/m^3. For a fixed volume of material, greater density will permit the storage of more heat. Appendix E, Table E.1 provides density data for common building materials. Note that porosity is often the indirect corollary of density.

Specific Heat. Specific heat is a measure of the amount of heat required to raise the temperature of a given mass of material by 1°. In the I-P system, this is expressed as Btu/lb °F; in the SI system, it is expressed as kJ/kg K. It takes less energy input to raise the temperature of a low-specific-heat material than that of a high-specific-heat material. Appendix E, Table E.1 provides values of specific heat for common building materials.

Thermal Capacity. Thermal capacity is an indicator of the ability of a defined volume of material to store heat. The greater the thermal capacity of a material, the more heat it can store in a given volume per degree of temperature increase. Thermal capacity for a material is obtained by taking the product of density and specific heat; for example:

Thermal capacity for concrete:

$$\text{I-P: } (140\ \text{lb/ft}^3)(0.24\ \text{Btu/lb °F}) = 33.6\ \text{Btu/ft}^3\ \text{°F}$$

$$[\text{SI: } (2240\ \text{kg/m}^3)(1.0\ \text{kJ/kg K}) = 2240\ \text{kJ/m}^3\ \text{K}]$$

Thermal capacity of water:

$$\text{I-P: } (62\ \text{lb/ft}^3)(1.00\ \text{Btu/lb °F}) = 62.0\ \text{Btu/ft}^3\ \text{°F}$$

$$[\text{SI: } (992\ \text{kg/m}^3)(4.18\ \text{kJ/kg K}) = 4147\ \text{kJ/m}^3\ \text{K}]$$

Thermal capacity of air:

$$\text{I-P: } (0.075\ \text{lb/ft}^3)(0.24\ \text{Btu/lb °F}) = 0.02\ \text{Btu/ft}^3\ \text{°F}$$

$$[\text{SI: } (1.2\ \text{kg/m}^3)(1.0\ \text{kJ/kg K}) = 1.2\ \text{kJ/m}^3\ \text{K}]$$

Time Lag. Time lag is a measure of the delay in the flow of a pulse of heat through a material that results from thermal capacity. Units are hours. As an example, if the sun comes out from behind clouds and strikes the exterior surface of a mass wall at 10:00 a.m., the exterior surface temperature will rise quickly. It may be several hours, however, before this temperature "spike" is seen at the inside surface of the wall. The reason is that some heat is being stored in the wall material, and that amount of heat will not continue to flow through the wall. Time lags of several hours are possible with very heavy wall constructions.

Heat storage in a material is similar to the storage of electricity within a capacitor. Thus, the term "capacitive insulation." The design effects of capacitive insulation are difficult to analyze without computer assistance. As a result, most hand calculation methods incorporate capacitive effects through the use of surrogate variables that attempt to simplify calculations for the end user. All reasonable computer simulations of heat flow should intrinsically address capacitive effects.

10.3 LATENT HEAT FLOW THROUGH OPAQUE ASSEMBLIES

The information presented in Section 10.2 focused upon sensible heat flow through various elements of the building enclosure. Water can also move through building enclosure assemblies—in both liquid and vapor states. The focus here is upon water vapor movement (assuming that proper architectural design of the enclosure will control the movement of liquid water). Water vapor will often need to be handled by a climate control system through the use of energy (termed latent heat). In the summer (depending upon climate), moisture will often flow into an air-conditioned building, increasing humidity and requiring dehumidification—which is often accomplished through removal of the latent heat of condensation of the added moisture. In the winter, it is not unusual to add water vapor to the air in a building to keep the relative humidity from dropping too low. This is often accomplished by evaporating water by adding the latent heat of vaporization.

In some building types and climates, dealing with latent heat may be as big a problem as dealing with sensible heat. In hot and humid climates, a dedicated dehumidification system is often provided to meet demand for both high-performance and thermal comfort. Otherwise, the air conditioner is the only means of interior dehumidification and building occupants are known to lower the thermostat to ironically uncomfortably cold temperatures as their only means of lowering the interior humidity. This phenomenon can be observed from Texas to Florida as occupants are forced to wear sweaters indoors during the middle of sweltering hot and humid days.

(a) Moisture Control Fundamentals

A difference in vapor pressure is the driving force behind water vapor flow through components of a building enclosure assembly. Gaps in the enclosure (holes, cracks, etc.) can provide a route for airflow that carries water vapor. Vapor pressure difference is to latent heat flow as temperature difference is to sensible heat flow. The permeance of a construction material is the latent equivalent of sensible conductance. As with conductance, permeance refers to the bulk properties of a material, and permeability refers to unit thickness properties. The less permeable a material is, the greater the resistance to water vapor flow. Materials with low permeance are termed vapor retarders and are incorporated into enclosures as a means of reducing the flow of water vapor and the associated risk of condensation of the vapor within the assembly.

$$\text{I-P units of permeance (perms): gr/h ft}^2 \text{ in. Hg}$$

$$\text{SI units of permeance (perms): ng/s m}^2 \text{ Pa.}$$

$$\text{I-P units for vapor pressure: in. Hg}$$

$$\text{SI units for vapor pressure: Pa}$$

where

gr = grains of water, with 7000 grains per pound
in. Hg = refers to inches of mercury, a pressure measurement
ng = nanogram or 10^{-9} of a gram
Pa = pascals

Permeance ratings and material classifications are given in Table 10.2. The International Code Council (ICC) defines the three classifications of vapor retarders in the *International Building Code* (IBC) and the *International Residential Code* (IRC) by permeance (ICC, 2024a, 2024b). The permeance of building materials is determined by testing the materials in accordance with the water method (Procedure B) per *ASTM E96/E96M: Standard Test Methods for Gravimetric Determination of Water Vapor Transmission Rate of Materials* (ASTM, 2024).

A water-resistive barrier (WRB) membrane can vary from vapor impermeable to permeable depending upon the manufacturer and model. Membranes known as "smart" vapor retarders have the ability to adjust their permeability based upon the relative humidity of the ambient air.

TABLE 10.2 Vapor Permeance Ranges and Classifications

Permeance (perms)	Rating[a]	Definition	Material Examples
Less than or equal to 0.1	Class I Vapor Retarder	Impermeable	Glass, metal, waterproofing membranes
Greater than 0.1 and less than or equal to 1.0	Class II Vapor Retarder	Semi-Impermeable	Kraft paper, XPS at ½ in. [12 mm]
Greater than 1.0 and less than or equal to 10	Class III Vapor Retarder	Semi-Permeable	Gypsum board, brick, wood sheathing
Greater than 10.0	Vapor Permeable	Permeable	Fiberglass, siding, felt paper

[a]The term "vapor barrier" should not be used in building design. When encountered in the industry, it is often meant to be a Class I Vapor Retarder.

From an architectural design perspective, reducing water vapor flow is accomplished using very thin materials (membranes) that must be carefully specified and located to ensure that they work as intended. Although placement within an enclosure assembly is critical, vapor retarders take up virtually no space—in drastic contrast to the thickness requirements of sensible heat retarders (insulations). The specific location of a vapor retarder within a wall, roof, or floor cross section will vary with climate and construction type. The fundamental principle, however, is for the retarder to stop the flow of water vapor before the vapor can come in contact with a material at its dew point temperature within the assembly. Sometimes the vapor retarder is also the bulk water and/or air control layer. Sometimes the vapor retarder is a separate and distinct control layer within an assembly.

(b) Cold Climate Moisture Control

Most common building materials, including gypsum board, concrete, brick, wood, and glass fiber insulation, are easily permeated by moisture. Many surface/finish materials are also permeable. In cold climates, the winter outdoor air contains relatively little moisture, even though the relative humidity may be high. By contrast, indoor air contains much more moisture per unit of volume, despite its probably lower relative humidity. The resulting differential vapor pressure drives the flow of water vapor from high to low vapor pressure (typically from warmer to colder temperature).

The primary problem with water vapor flow in winter occurs when the temperature somewhere within a wall, floor, or roof drops low enough for water vapor passing through the assembly to condense. The temperature at which this occurs is the dew point temperature of the air which is acting as the source for the water vapor. The dew point temperature often occurs within the insulation layer. The insulation can then become wet and thereby less effective, because water conducts heat far better than the air pockets it has filled. If wet insulation compacts, the air pockets are then permanently lost. Worse yet, moisture damage can occur, such as dry rot in wood structural members. In practice, condensation is observed to occur on the "first condensing surface" encountered by the water vapor as it flows from high to low pressure through an assembly. This condensing surface is not necessarily coincident with the precise location of the dew point. For further information, see Building Science Corporation (Lstiburek, 2011).

The usual remedy for enclosure condensation problems in cold climates is a vapor retarder installed within the building enclosure assembly. Because very low permeability is desired, these retarders are commonly a plastic film installed with as few gaps and holes as possible. Because the moister air on the warm side of the enclosure is the source of the problem, the vapor retarder needs to be installed as close to the warm side as possible—typically, just behind the interior surface (gypsum board, wood flooring, etc.) before the dew point can be encountered by wandering water vapor. See Lstiburek and Carmody (1994) for details on moisture control in cold climates.

Another approach is to use vinyl wallpaper or vapor retarder paints on interior surfaces of the enclosure. These give less protection, however, than the around-the-corners wrap that can be achieved with properly installed plastic films. This performance disadvantage is shared by aluminum-foil-faced insulation; it is very effective thermally but less effective as a vapor retarder. Also, the aluminum foil must face an air space if it is to be thermally effective as a radiant barrier.

A substantial benefit of plastic films and Class I vapor retarders is that they also reduce airflow through construction—acting as an air barrier. Outdoor air is always infiltrating a building, gradually replacing the indoor air. This unintended flow of cold air becomes a problem when temperatures outside are very different from those inside, especially when strong winds force outdoor air indoors fast enough to produce noticeably cold drafts. Therefore, the combined moisture-blocking and infiltration-blocking characteristics of plastic film vapor retarders are sometimes beneficial. However, a continuous and robust dedicated air barrier (often the same material as the building WRB at the exterior side of the sheathing) is typically the most effective strategy to reduce air infiltration.

In cold climates, condensation on cold interior surfaces can also occur, especially at low-R windows (with little resistance to reduce the slope of the thermal gradient). Occupant annoyance or material damage can result. Although the air indoors may not be particularly humid (perhaps 30–50% RH), it often contains enough moisture to support condensation on cold surfaces. Condensation can be readily predicted (see Figure 10.6) and addressed during design. Historically, warm air was introduced near a cold glass surface to reduce condensation potential and increase mean radiant temperature. Strategically placed air registers/diffusers and baseboard radiators were commonly used tools. This approach carries an energy burden, however, as heat loss through the glazing is increased via an increased delta *t* and decreased R-value. Better glazing is recommended as an environmentally preferred solution.

(c) Hot, Humid Climate Moisture Control

In hot, humid conditions, cool surfaces are often encountered inside a building—for example, a radiant cooling panel, a chilled water pipe, or a supply air diffuser. If hot, humid air contacts such a surface, condensation can occur—with the moisture vapor in the air condensing to form visible droplets of water on the cool surface. The result can be mildly annoying if droplets of condensation fall on occupants, or serious if water stains occur and, eventually, mold grows on damp surfaces.

In hot, humid climates, the intent of the enclosure is to keep moisture in the warmer outdoor air from penetrating to the cooler (and usually less humid) interior. As in cold climates, the most effective strategy to reduce air infiltration is to utilize a continuous, robust air barrier (which is often the same material as the water-resistive barrier on the exterior face of the wall sheathing). No vapor retarder (such as plastic, vinyl wallpaper, or vapor retarder paints) should be used on interior surfaces in hot and humid climates as this can block water vapor flow (too late) and create a condensing surface. Numerous severe problems with water vapor damage to building interiors—and related mold and mildew growth, with associated indoor air quality problems—have been publicized, typically related to condensation of moisture on the exterior-facing (hidden) side of vinyl wall coverings used as an interior finish. See Lstiburek and Carmody (1994) for details on moisture control in hot and humid climates.

10.4 HEAT FLOW THROUGH TRANSPARENT/TRANSLUCENT ASSEMBLIES

Heat flow through windows and skylights requires special attention for several reasons. Despite dramatic improvements, these transparent/translucent enclosure components still usually have the lowest R-value (highest U-factor) of all components that constitute a building enclosure. Also, they are major contributors to the infiltration of outdoor air, which adds to winter heating or summer cooling loads. Perhaps thermally most important, they admit solar heat, for better (winter, in enclosure-load-dominated buildings) or for worse (summer, in all buildings in most climates). Windows and skylights are designed to allow solar radiation (heat) to pass with minimal resistance, unlike the opaque elements discussed in Section 10.2. Transparent/translucent devices also admit daylight to buildings and often provide desired ventilation. The variety of such components and the several important roles they play require special attention from designers.

Figure 10.7 *Certifying window thermal performance; a sample NFRC window label. (© National Fenestration Rating Council; used with permission.)*

A transparent material permits a generally undistorted view (as with clear glass); a translucent material permits at best a distorted view (as with glass blocks or milky plastic). Both types of material, however, permit some unimpeded transmission of radiation—as opposed to an opaque material/assembly that permits no direct solar radiation transfer.

A standardized approach to rating window performance characteristics has been developed in North America. This approach is coordinated by the National Fenestration Rating Council (NFRC). Each certified window or skylight bears an NFRC label verifying that the window has been independently rated. The label carries a brief description of the product, such as "Model #, Casement, Low-e = 0.2, 0.5″ gap, Argon Filled," and lists the following performance information: U-factor, solar heat gain coefficient (SHGC), visible light transmittance (VT), and air leakage. Figure 10.7 shows a sample NFRC label. Similar labels are available for site-assembled window units. Characteristics of some residential-sized windows are found in Appendix E, Table E.15.

The following sections provide information on fundamental thermal properties of transparent/translucent assemblies and approaches that can be taken to enhance the performance of such assemblies.

(a) U-factor

As with opaque enclosure components, heat flow due to temperature differences through windows and skylights is a function of the U-factor. The determination of U-factors for windows and skylights is complicated by significant differences in heat flow rates between the center-of-glass, edge-of-glass, and frame portions of a unit. The NFRC "U-factor" melds these into a single representative value for an entire window or skylight unit. The size of the air gap between glazing panes, the coatings on the glazing panes, the gas fill used between glazing panes, and the frame construction all influence the U-factor. The lower the U-factor, the lower the heat flow for a given temperature difference. Appendix E, Tables E.14, E.15, and E.16 list U-factors for various windows and skylights. Appendix H presents sample requirements for window U-factors and SHGC (solar heat gain coefficient) for nonresidential buildings (these have been excerpted from *ASHRAE/IES Standard 90.1-2022*; ASHRAE, 2022c).

(b) Solar Heat Gain Coefficient (SHGC) and Shading Coefficient (SC)

SHGC is also based upon the performance of the entire glazing unit, not just that of the glass portion. SHGC represents the percentage of solar radiation (across the spectrum) incident upon a given window or skylight assembly that ends up in a building as heat. It is a measure of the ability of a window to resist heat gain from solar radiation. SHGC can theoretically range from 0 to 1, with 1 representing no resistance and 0 representing total resistance. SHGC values for commercial products typically range from about 0.9 to 0.2. SHGC is dimensionless. A high SHGC (meaning poor resistance to radiant gain) is desirable for solar heating applications, whereas a lower SHGC (good resistance) is better for windows in buildings where cooling is the dominant thermal issue. SHGC depends upon the type of glass and the number of panes, as well as tinting, reflective coatings, and shading by the window or skylight frame. NFRC tests and lists SHGC values only for glazing units; ratings do not include auxiliary elements such as draperies, overhangs, trees, and the like.

Prior to the development of the SHGC approach, the term shading coefficient (SC) was universally used to quantify the same concept. Shading coefficients, however, considered only the glass portion of a glazing unit—specifically excluding the frame. Although theoretical SC values also range from 0 to 1, the basis of SC is different from that of SHGC. SC is the ratio of radiant heat gain through a given type of glass relative to 1/8-in. [3-mm]-thick single clear glass. SC is still a useful concept for expressing the effects of external or internal shading devices—which can provide valuable shading but are not included in SHGC values. Procedures for merging SC and SHGC values have not truly been formalized. A conversion estimate of (0.86) (SC) = SHGC is suggested. The cumulative effect of series shading devices is multiplicative:

$$SHGC_{overall} = (SHGC_{overhang})(SHGC_{window})(SHGC_{drapery})$$

A wide variety of window and skylight U-factor, SHGC, and VT data can be found in the *2021 ASHRAE Handbook—Fundamentals*, Chapter 15. Appendix E, Tables E.17 through E.24 list SHGC, SC, and/or VT values for various types of transparent/translucent products and shading approaches.

A small sample of U-factors for skylights is shown in Appendix E, Table E.16. This table illustrates the great thermal penalty paid when using smaller units; the "manufactured skylights" are 2 ft × 4 ft [600 mm × 1200 mm], whereas the "site-assembled glazing" unit size is 4 ft [1200 mm] square. For otherwise identical construction, the whole-component U-factors are almost twice as high for the smaller units. With these smaller units, a frame with better thermal performance is a smart investment.

(c) Visible Transmittance (VT) and Light-to-Solar Gain (LSG)

Visible transmittance (VT) is a thermal/optical property that represents the percentage of incident light (only the visible spectrum) at a normal angle of incidence that passes through a particular glazing. VT is dimensionless. The higher the visible transmittance, the greater the daylight transmission. VT is influenced by the color of the glass (clear glass has the highest VT) as well as by coatings and the number of panes. VT may be expressed relative to the glass portion only of a glazing unit or relative to the glass and frame. The appropriate expression will depend upon the nature of an analysis; in any case, noncomparable values should not be compared. All NFRC-certified VT values should be directly comparable.

It might seem intuitive that any glazing or coating that reduces SHGC (via lower radiation transmission) will also reduce VT (implying lower radiation transmission). This is not always the case, however, because SHGC deals with the full solar radiation spectrum (including light), whereas VT deals only with the visible (light) spectrum. Spectrally selective glazings and selective coatings can greatly reduce SHGC with little reduction in VT. The relationship between SHGC and VT is expressed as the light-to-solar gain (LSG) ratio, obtained by dividing the VT by the SHGC. The greater the LSG, the more suitable a glazing is for daylighting in hot climates (or wherever cooling is the dominant thermal condition). Values of LSG are not directly listed on NFRC labels, but can be derived. LSG values for common glazing units (including the effects of the frame) are included in Appendix E, Table E.15.

(d) Air Leakage (AL)

The air leakage (AL) rating indicates the rate of outdoor air infiltration between a new window and its frame measured under defined conditions—usually under pressure equivalent to that of a 25-mph [40-km/h] wind, with the window locked. High-performance windows may be tested/rated at even higher pressures. The NFRC will only certify windows that achieve an AL at or below 0.3. The best rating possible is 0.1 (if a product performs lower than 0.1, the reading will be rounded up for the NFRC label). As weather-stripping deteriorates with age, higher rates of infiltration may be expected with older and well-used windows. Design air leakage values are not required to be listed on NFRC labels but may be obtained from manufacturers' catalog data. AL data are expressed in units of cfm/ft^2 [L/s m^2].

(e) Condensation Resistance (CR) and Condensation Index (CI)

Historically, the NFRC also offered a condensation resistance (CR) rating; this was optional for certification (and therefore may not appear on a label). The CR rating is dimensionless and ranges from 0 to 100. The higher the CR rating, the better the product is at resisting condensation (i.e., less potential deposition of moisture). The CR rating was developed over two decades ago and has notable shortcomings. For instance, the CR rating cannot be used to indicate the potential of condensation forming on a window product based on indoor relative humidity, indoor dew point temperature, or exterior air temperatures during the coldest days of year. To remedy this, the NFRC approved the condensation index (CI), which avoids the shortcoming of the CR rating and permits users to choose products with reduced potential for condensation based on location. The CI rating is also dimensionless and ranges between 0 and 100, with the higher values indicating better performance.

Fenestration Product Enhancements. ASHRAE defines fenestration as: "all areas (including the frames) in the building enclosure that let in light, including windows, plastic panels, clerestories, skylights, glass doors that are more than one-half glass, and glass block walls." Thus, this term covers the transparent and translucent building elements being discussed in this section.

There are a number of well-established options for fenestration products that can improve their thermal and visual performance. Several of these are described in the subsections that follow.

Low-Emittance (Low-ε) Coatings. These coatings are typically applied to one glass surface facing into an air gap between multiple panes of a fenestration product. A low-ε coating blocks a great deal of the potential radiant heat transfer between the glazing panes, reducing the overall flow of heat and thus improving the U-factor. Indeed, one such coating is almost as effective as adding another layer of glazing. In Appendix E, Table E.15, compare the U-factor for Window 5 with the U-factors for Windows 7 and 12. An important added benefit of these films is their reduction of ultraviolet (UV) transmission, thus reducing fading of objects and surface finishes in rooms.

Two approaches to providing low-ε films are hard-coat (durable, less expensive, but less thermally effective) and soft-coat (better thermal performance but more expensive and subject to degradation by oxidation in the manufacturing stage).

Three common types of low-ε coatings are:

1. *High-transmission low-ε:* For passive solar heating applications, where a low U-factor is combined with a high SHGC; Window 7 in Appendix E, Table E.15 is an example. The coating is on the inner glazing, where it blocks outgoing infrared radiation that otherwise would be lost. Summer overheating can be avoided with external shading devices.
2. *Selective-transmission low-ε:* Where winter heating and summer cooling are both important, requiring low U-factor and low SHGC, but with a relatively high VT for daylighting. Window 9 in Appendix E, Table E.15 is an example, with an LSG ratio of 1.65. The coating is on the outer pane, where it blocks incoming infrared radiation, which as heat is then convected away by outdoor air.
3. *Low-transmission low-ε:* Where the sun is the enemy, low U-factor, low SHGC, and even low VT seem warranted. Window 10 in Appendix E, Table E.15 is an example, again with the coating on the outer pane, where it rejects solar heat gain. With a tinted exterior glazing, even lower SHGC and VT could result.

A note about low-ε glass terminology: some manufacturers market low-ε products with claims for performance enhancements (such as reduced short-wave transmission) that are not associated with the effects of a low-ε coating. This is not necessarily inaccurate but can be confusing. Emissivity affects the release of long-wave radiation from a surface. Only those effects associated with changing emissivity should be connected to the term "low-ε."

Selective Transmission Films. Heat flow due to radiation can be greatly reduced by the introduction

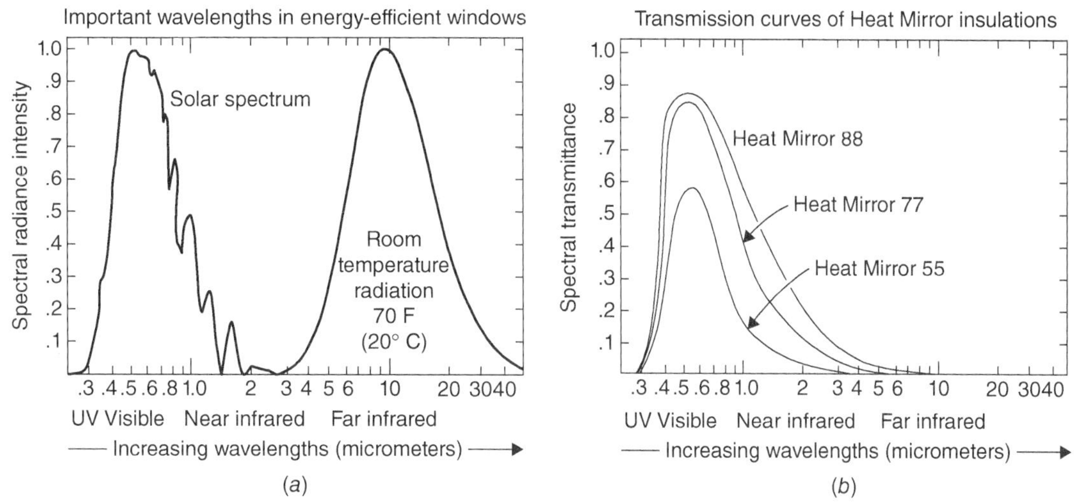

Figure 10.8 *Performance of selective transmitters. (a) Spectral characteristics of solar (short-wavelength) radiation and room-temperature (long-wavelength) radiation. (b) Transmission and reflection performance for several Heat Mirror selective transmitter films. Incoming solar radiation (both visible and near-infrared) is mostly transmitted, whereas heat radiated from room-temperature objects is reflected and thereby kept within heated spaces. (Courtesy of Southwall Corporation.)*

of a selective transmitter (or low-emittance) film somewhere within the glazing cavity. As shown in Figure 10.8, these films admit most of the incoming solar radiation in both the visible and near-infrared (short) wavelengths. Warm objects within a room emit far-infrared (long-wave) radiation. This long-wave radiation is reflected back into the room by the selective film. These selective films typically are available as separate sheets that can be inserted between sheets of glazing as a window is fabricated. As a separate sheet, a selective film could also be applied to existing windows, for instance, between storm windows and the primary windows they protect.

With continuing advances in glazing performance, an increasing range of selective transmission options is available. For example, products from Southwall Technologies range from Heat Mirror 77 (low reflectance, recommended for vertical glass) to Heat Mirror 22 (low transmission, recommended for sloping glass), and provide various colors. A clear Heat Mirror 88 has the lowest reflectance of all. Some skylight and greenhouse manufacturers offer this low-emittance product as a standard option.

Inert Gas Fills. Filling an enclosed space with argon or krypton has thermal benefits. These less-conductive (than air) gases greatly reduce heat transfer by convective currents within the gap between multiple glazing panes, producing lower U-factors. Because heat flow to or from the window surface is reduced, the inner surface of the glass is maintained at a temperature closer to that of the indoors. This provides greater comfort and less chance of condensation on the inside surface. To preserve this gas fill over the life of the window, the manufacturer must install a very reliable edge seal.

For argon, the optimum gap width is about 1/2 in. [12 mm]. When a thinner window is needed, more expensive krypton can be used with a gap of only 1/4 in. [6 mm]. Because the combination of inert gas and low-ε coatings is so effective at lowering the U-factor, most manufacturers offer them together (as with Windows 7 through 11 in Appendix E, Table E.15) rather than separately.

"Superwindows". When available high-performance glazing options are combined in one product, it is often called a "superwindow." The combination of multiple glazing panes and/or suspended films, coatings, inert gas fill, and a sealed/thermally broken frame yields a lower heat flow rate and a higher price. Window 11 in Appendix E, Table E.15 is a superwindow. An early example of a superwindow is shown in Figure 10.9: a double-glazed assembly with two selective films, yielding three gas-filled cavities. With a combination of superior thermal resistance and useful solar gain potential, superwindows can conceivably provide better heating-season thermal performance than an insulated opaque wall. The potential design consequences are enormous. Figure 10.10 compares the heating- and cooling-season performance, in a residential application, of several windows in three US locations with quite different climates.

(f) Shading

Perhaps the single most important energy-related component for energy-efficient buildings is a shading

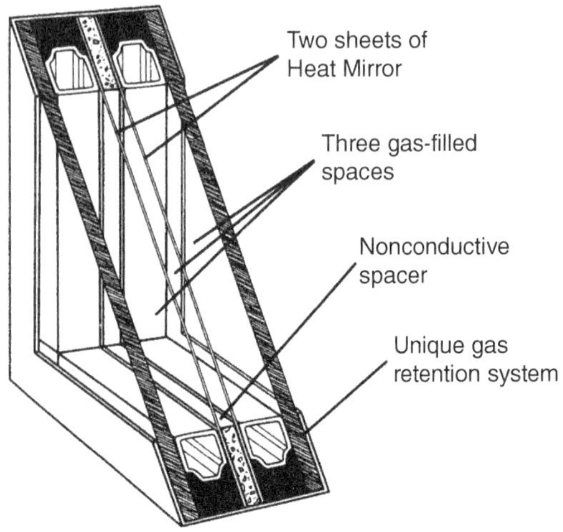

Figure 10.9 *The Superglass window system (an early superwindow) utilized two Heat Mirror films between two outer panes of glass. Three air spaces are thus created and filled with a nonreactive gas mixture to retard convection. Nonmetallic spacers reduce heat flow at window edges. The resulting window provided a center-of-glass insulating value of R-8.1 [SI: R-1.43]. Very low transmission of UV radiation is another characteristic. (Courtesy of Hurd Millwork Company, Medford, WI.)*

device. This is especially true for passive buildings of all types. Because solar-heated buildings will often experience overheating in hot weather, sunshading is critical for even for passively heated buildings. When correctly implemented, sunshading can reject most unwanted solar heat gain while helping distribute daylight deep into buildings, thus reducing internal heat gains caused by electric lighting.

If a building is arranged to intercept the intense rays of the sun before they pass through its transparent enclosure elements, instead of afterward, the building cooling load can often be cut in half. In general, effective external shading rejects about 80% of solar energy, whereas internal shading accepts about 80%. Outdoor louvers have a chance to cool off in the occasional breeze, but indoor draperies are part of a heat trap, and they constitute a system of hot-weather heating that causes discomfort for those who work near perimeter surfaces.

To properly reject direct sun yet allow for a view and daylight, many sunshades project out from the windows they protect. These exterior projections become highly visible elements of façades, and they tempt some designers to impose formal aesthetic criteria that can be damaging to the solar control functions. A frequent example is the application of the same sunshade geometry to all façades of a building (see Figure 3.21). Such sunshades, if not moveable, tend to help one façade but not the others. Where they are movable, as with awnings, such a façade-indifferent sunshade approach is not as serious a failing.

Fixed sunshading devices are very common, partly because they lack moving parts and controls that can be expensive and troublesome. They pose a dilemma in the spring and fall, however, because in order to block the sun on any elevation in September, they will also block it in March. For many buildings in temperate climates, March is a heating-need month, whereas September is a cooling-need month. A procedure for evaluating fixed shading device performance is presented in Chapter 11; some approximate shading effects, expressed as shading coefficients, are found in Appendix E, Tables E.20 and E.21.

Adjustable sunshading, once extremely common (it seemed that every 1930s shop and office window had an awning) but later considered old-fashioned, has made a comeback with an increased interest in passive systems. Some basic approaches to adjustable shading devices are shown in Appendix E, Table E.22. The first type shown, made of durable materials, has proved effective against both break-ins and hurricanes.

How are movable shading devices adjusted to the desired position? Most manufacturers offer three types of controls: manual, motorized, and automatic. Manual systems are cheap and relatively trouble-free, but they require thoughtful, timely action by the users of a building. So do motorized controls, but for adjusting large/heavy devices in hard-to-reach locations (clerestories, for example), motorized assistance is a practical necessity. Automatic systems have the advantages of freeing users from adjustment tasks and of taking into account the thermal needs of the building as a whole when setting the sunshade's position. With computerized controls now commonplace for large buildings, the added costs of controls for automatic sunshading can easily be incorporated into the overall cost of controls.

Interior shading devices are less effective than exterior devices but are far more commonly used. There are several reasons for this seeming contradiction. They are not subject to weathering or dirt accumulation and generally are easier for users to adjust. The designer who prefers a clean, apparently unchanging façade appearance will rely on interior devices—often at substantial energy and comfort cost. In Appendix E, Tables E.23 and E.24 list shading coefficients for several combinations of interior shading devices.

The overall performance of multiple shading devices/approaches can be estimated by taking the product of the shading coefficients for individual devices that are used in series. For example, the overall SC of a window with external shading with an SC of

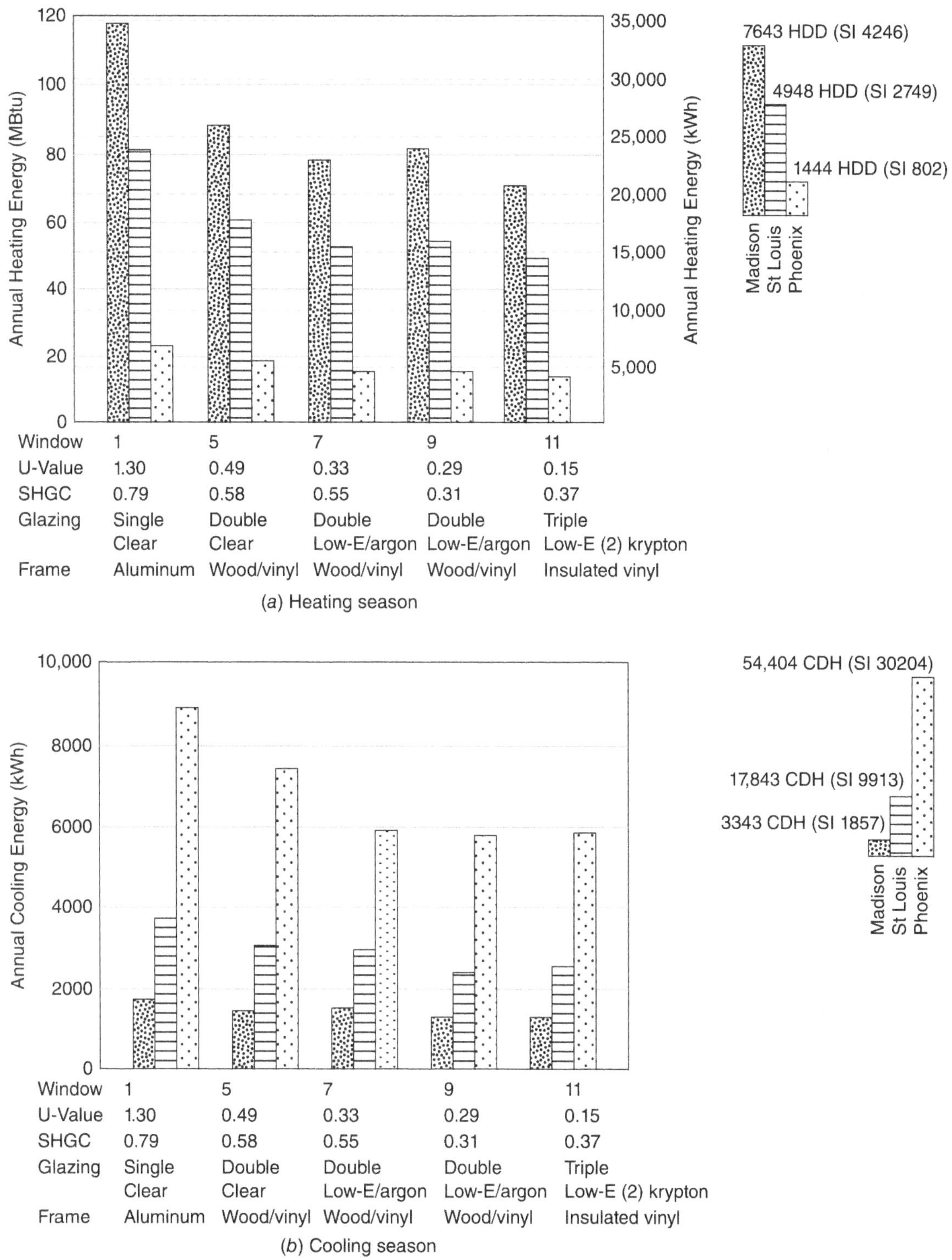

Figure 10.10 *(a, b) Seasonal energy performance comparison of various windows—assuming a typical 1540-ft^2 [143-m^2] residence with a glazing area of 15% of the floor area, equal glazing area on all four orientations, and no external or internal shading. Window numbers correspond to those in Appendix E, Table E.15. MBtu = million Btu; HDD = heating degree days, base 65°F [18.3°C]; CDH = cooling degree hours, base 74°F [23.3°C]. U-factors in this figure are I-P values; multiply these by 5.678 for SI U-factors. (Adapted from Carmody et al., 1996.)*

0.75, integral shading from the glazing unit of 0.65, and internal shading providing an SC of 0.7 is (0.75) (0.65) (0.7) = 0.34. In general, shading coefficients lower than 0.2 should be included in energy analyses with caution.

In hot climates, the single greatest source of heat gain comes from solar radiation. Therefore, shading of windows and shading of walls (via overhangs, etc.) are critical to improve performance. This point is so extreme in the southern United States that a fully shaded south-facing 1960s-era single-pane window actually outperforms, in terms of energy, a modern code-minimum insulated glazing unit that is not shaded. Overhangs, awnings, and some shading devices also have the added benefit of protecting the building from rain and associated moisture infiltration. Unfortunately, for stylistic reasons, many modern architects design—even in hot climates (see Figure 10.11)—buildings without overhangs (or any type of shading device) and therefore subject their buildings to the full brunt of radiative heat gains, thermal discomfort, visual glare, and at the same time unnecessarily increase the risk of water infiltration.

10.5 HEAT FLOW VIA AIR MOVEMENT

Outdoor air can enter a building by means of infiltration and/or ventilation. Infiltration is an unintended flow of outdoor air due to air leakage through the building enclosure. Ventilation is the deliberate, designed introduction of outdoor air. In either case, air at a different temperature and humidity interacts with the air inside a building. This interaction brings sensible and latent heat loads. Of the two means of air entry, infiltration is the more difficult to predict, since it is by definition unintended. Ventilation air quantities are intended and thus easily quantified. Although it is possible for infiltration and ventilation to occur simultaneously, usually it is reasonable to assume one of the following scenarios: only infiltration occurs (which is typical of many smaller buildings with no mechanical ventilation) or mainly ventilation occurs (which is typically required by code for larger buildings, and which usually pressurizes a building and blocks infiltration). Both infiltration and ventilation airflow rates are expressed in cubic feet per minute (cfm); the SI units are liters per second [L/s].

Figure 10.11 *Construction of a building in Austin, TX. Notice the absence of external shading devices. (Courtesy of Keith Simon.)*

(a) Infiltration

Air Barriers. Air barriers are materials, systems, and assemblies that stop the flow of air through walls, roofs, floors, and fenestration. Continuity of an air barrier is critical because just a few holes can dramatically short-circuit the entire system. Air barriers are often membranes that at the same time may act as the water-resistive barrier, waterproofing, and/or vapor retarder. An air barrier, however, does not need to be a membrane—an air barrier can be wood sheathing, gypsum sheathing, rigid foam insulation, or any other material that stops the flow of air. The key to ensuring airtight buildings is not so much picking the most airtight material, but rather ensuring that all transitions between air barrier systems, terminations, and penetrations are well sealed to ensure continuity. Examples of common air barrier discontinuities that require special detailing are the roof-to-wall transition, the base-of-wall termination at the foundation, soffits, loading docks, elevator and stair shafts, and elevated floors/balconies.

Estimating Infiltration. The main design problem with infiltration is estimating how much air will leak into a building once it is built. There are two primary means of calculating the infiltration airflow rate during design: the air-change method and the crack method.

The air-change method is very simple and quick but tends to overestimate. Air-change is a correlational method wherein observed performance of a set of existing buildings is matched (correlated) to key characteristics of a proposed building related to construction type and climate. Virtually no detailed information about a building or its spatial arrangement is required. Appendix F, Table F.3 lists estimated air changes per hour (ACH) as a function of these two characteristics. ACH is indicative of the "turnover" of air within a building or space. The greater the ACH, the greater the rate of outdoor airflow. A fairly tight building might permit 0.50 ACH or so; a recently built leaky building, 1.50 ACH. ACH can be converted to airflow rate as follows:

In I-P units:

$$V = \frac{(\text{ACH}) \times (\text{volume, ft}^3)}{60\ \text{min/h}} \qquad \text{(Eq. 10.2)}$$

where V is airflow rate in cfm.

In SI units:

$$V = \frac{(\text{ACH}) \times (\text{volume, m}^3)}{3600\ \text{sec}/h}$$

where V is in m^3/s. Although both m^3/s and L/s are used as SI units of airflow rate, L/s is more commonly encountered in building design work.

$$\text{L/s} = \text{m}^3/\text{s} \times 1000.$$

The crack method requires more effort and assumes that data on window and door construction and wind velocities are known. It also assumes that all infiltration under design conditions is due to cracks around doors and windows. To estimate the airflow rate, calculate the length of cracks on the windward exposure(s) only, and multiply this by the respective assembly air leakage rate(s) as follows:

In I-P units:

$$V = (\text{total lin. ft of window/door edge}) \times (\text{air leakage rate(s), cfm per lin. ft}) \qquad \text{(Eq. 10.3)}$$

where V is in cfm.

In SI units:

$$V = (\text{total lin. m of window/door edge}) \times (\text{air leakage rate(s), L/s per lin.m})$$

where V is in L/s.

See Appendix E, Table E.15 for generic window air leakage data or get NFRC or manufacturer's ratings for a specific fenestration product. If air leakage rates are unknown, Appendix F, Table F.4 shows how much leakage per unit length of crack should be assumed, for the windward exposure(s) only, to arrive at the total infiltration airflow rate. Multiply the total length of the crack by the infiltration rate from Part B of Appendix F, Table F.4.

It is possible to measure the actual infiltration of a constructed building using a fairly simple device called a blower door. A blower door is a high-powered fan that a trained professional temporarily mounts into the frame of an exterior doorway in a building. After calibrating the device, the fan pulls air out of the building, lowering the indoor air pressure. The higher outdoor air pressure then forces air in through all unsealed gaps, cracks, and openings. If conditions do not allow for lowering the pressure in the building, the fan may also be operated in reverse, with air pressure increased inside the building. If the results of a blower door test are unsatisfactory, the points of air infiltration can be located and sealed, and the blower door test repeated. Energy codes increasingly require that building air tightness be verified by blower door tests. As a result, local resources may be able to provide reasonably accurate values for air-change rates for buildings constructed using prevailing practices. Local experience data may provide better estimates of infiltration than the crack method.

There is a common misconception in the industry that one should let the enclosure of a building

TABLE 10.3 Prevailing Enclosure Air Tightness Values

Air Tightness Reference Condition	cfm/ft² @ 75 pa [L/s m²]
U.S. commercial average	1.55[a] [7.9]
2024 International Energy Conservation Code	0.35 [1.8]
U.S. Army Corps of Engineers	0.25 [1.3]
Canadian Standard	0.15 [0.8]
Phius	0.06 [0.3]

[a]Value per Emmerich and Persily (2005).

"breathe" by allowing unintended airflow. This is problematic, as unintended airflow often leads to humid air coming in and causing moisture accumulation in areas such as window corners or penetrations. Letting the walls "breathe" eventually leads to bad indoor air quality. By contrast, it is common in the high-performance building world to "build airtight and ventilate right." Passive buildings follow this credo and utilize continuous, robust air barriers to make the buildings extremely airtight, while providing balanced, controlled ventilation via mechanical systems resulting in dramatically improved indoor air quality and health. Prevailing building air tightness results are shown in Table 10.3. These values are for leakage through the plane of the building enclosure.

(b) Ventilation

It is important to provide some minimum amount of fresh air to indoor environments as a means of indoor air quality control. Odors and a sense of staleness can be uncomfortable, and dangerous buildups of pollutants such as formaldehyde and radon can occur within buildings. These pollutants are most effectively removed via fresh airflow and filtration (although with a notable energy penalty). In Appendix F, Tables F.1 and F.2 list recommended design outdoor airflow rates to produce acceptable indoor air quality (IAQ). A more thorough discussion of IAQ is found in Chapter 5.

Many building codes require a minimum outdoor airflow rate based upon either building population or floor area. These requirements are generally based upon *ASHRAE Standard 61.1* (for larger buildings) or *ASHRAE Standard 62.2* (for smaller residential buildings). This minimum is often (but not always) provided by mechanically induced ventilation. Since ventilation is intentional, estimating such airflow rates is straightforward.

With some simplification, when a building's required outdoor air quantity (Appendix F, Table F.1) is based upon population:

$$V = (\text{outdoor air rate per person}) \times (\text{number of people}) \quad \text{(Eq. 10.4)}$$

When the required outdoor airflow rate is based upon floor area:

$$V = (\text{outdoor air rate per unit floor area}) \times (\text{total floor area}) \quad \text{(Eq. 10.5)}$$

10.6 CALCULATING ENCLOSURE HEAT FLOWS

When the fundamentals of heat flow through a building enclosure are understood, calculations to determine the magnitude of such flows can be undertaken. Three analyses of heat flow are usually of primary interest:

1. Design heat loss based upon "worst hour" conditions, which is used to size heating systems.
2. Design heat gain (or cooling load) also based upon worst-hour conditions, which is used to size cooling systems.
3. Annualized heat flow, based upon year-long climate conditions, which is used to predict annual energy usage and costs and/or demonstrate compliance with energy standards.

Hand calculations can provide reasonably accurate estimates of design heat loss and gain—even though the use of computer programs for these analyses is very common. Although a rough estimate of annual energy usage can be made using hand calculations, it is much more likely that a computer program will be used for this purpose. Many such programs are readily available, although they generally have a moderately steep learning curve and require detailed input information. Regardless of whether a manual or computerized calculation approach is taken, understanding the concepts and issues involved is critical to good design decisions.

(a) Design Heat Loss

Calculation of design heat loss is an attempt to estimate what is reasonably the worst-likely hourly heat flow from a building to the surrounding environment. This value is used to size heating systems; the greater the design heat loss, the larger the required heating system capacity. The design heat loss is not the highest heat loss that can (or will) ever occur; rather, it is a statistically reasonable maximum heat loss based upon a chosen outdoor air temperature. By convention, design heat loss is assumed to occur at night (with no solar radiation), during the winter (coldest temperatures), with no occupants, lights, or equipment to offset heat lost through the enclosure. These assumptions lead to an easy analysis based upon steady-state conditions.

To calculate the design (the statistically "worst-hour") heat loss through a building enclosure, the following information is necessary:

1. The rate at which heat flows through each of the elements that make up the building enclosure—that is, the various assembly U-factors (or equivalents).
2. The area (or perimeter) of each of these assemblies.
3. The design temperature difference between indoors and outdoors.

With only the first two factors, a comparison of the thermal integrity of one building enclosure versus another, in any location, can be made. These two factors are an outcome of the design process. The third factor, temperature difference, is a function primarily of climate, but also of comfort criteria. Design temperatures for various geographic locations (climates) are listed in Appendix B.

The design sensible heat loss through all types of above-grade elements of the building enclosure can be calculated as follows:

$$q = (U)(A)(\Delta t) \qquad \text{(Eq. 10.6)}$$

applied to each element of the building enclosure (e.g., above-grade walls, roof, windows, doors), where

q = hourly heat loss through a specific enclosure component (by conduction, convection, and radiation) under design conditions, Btu/h [W]
U = U-factor for a given enclosure component, Btu/h ft^2 °F [W/m^2 K]
A = surface area of the enclosure component, ft^2 [m^2]
Δt = the design temperature difference between indoor and outdoor air (see Appendix B), °F [K]

(If a representative heat loss is desired, use average temperatures, such as from Appendix C or local climatological data.)

In addition to above-grade components, heat loss through at-grade and below-grade elements and via airflow must be estimated.

For slab-on-grade floors (see Example 10.3):

$$q = (F_2)(P)(\Delta t) \qquad \text{(Eq. 10.7)}$$

where

q = hourly sensible heat loss through the slab-on-grade floor under design conditions, Btu/h [W]
F_2 = heat loss coefficient for the floor slab/insulation configuration, Btu/h ft °F [W/m K]
P = perimeter of the floor slab, ft [m]
Δt = design temperature difference between indoor and outdoor air, °F [K]

For below-grade basement walls and floors (see Example 10.4):

$$q = (F)(A)(\Delta t) \qquad \text{(Eq. 10.8)}$$

summed across various depths-below-grade of the walls and floors, where

q = hourly sensible heat loss through the basement walls and floors under design conditions, Btu/h [W]
F = heat loss coefficient for the wall-floor/insulation configuration at a given depth below grade, Btu/h ft^2 °F [W/m^2 K]
A = area of the floor or wall at a given nominal depth below grade, ft^2 [m^2]
Δt = design temperature difference between indoor air and surface of the ground, °F [K]

For airflow due to infiltration or ventilation, design sensible heat loss is calculated as follows:

$$\text{I-P units: } q = (V)(1.1)(\Delta t) \qquad \text{(Eq. 10.9)}$$

$$\text{SI units: } q = (V)(1.2)(\Delta t)$$

where

q = sensible heat loss due to infiltration or ventilation, Btu/h [W]
V = outdoor airflow rate, cfm [L/s]
Δt = temperature difference between outdoor and indoor air, °F [K]
1.1 = (used with I-P units) a constant derived from the density of air at 0.075 lb/ft^3 under typical building conditions, multiplied by the specific heat of air (the heat required to raise 1 lb of air 1°F, which is 0.24 Btu/lb °F) and by 60 min/h; the units of this frequently encountered (but quirky) constant are Btu min/ft^3 h °F
1.2 = (used with SI units) a constant derived from the density of air at 1.20 kg/m^3 under typical building conditions, multiplied by the specific heat of air (1.0 kJ/kg K)

The total design sensible heat loss (q) for a building is the sum of all component losses (i.e., above-grade elements + above-grade elements + below-grade elements + airflow). All the above-grade enclosure elements could be lumped together (using an area-weighted U-factor) under the same equation (because assembly orientation, tilt, and transparency are of no consequence in the absence of sun). It is very useful, however, to identify the loss of each component (window, floor, roof, etc.) so that design improvements can focus upon those enclosure components with the biggest contribution to overall heat loss.

Design for energy efficiency in the building enclosure essentially consists of trying to reduce the value of any variable that occurs on the right-hand side of these design heat loss equations. No other design changes will affect design heat loss. This is the underheated period design palette:

- Reduce component U-factors (within the constraints of the budget and constructability).
- Reduce areas of components (within the constraints of the program and design intent).
- Reduce the design temperature difference (usually by lowering indoor air temperature, perhaps by microclimate improvement of outdoor conditions).
- Reduce airflow into the building (consistent with IAQ criteria).

In practice, design latent heat loss is often ignored. There are two reasons for this:

1. Indoor humidity (directly affected by latent loss) is often allowed to float during the winter.
2. There is no single system or piece of equipment that will handle both sensible and latent heat loss.

Even so, it is possible to estimate design latent heat loss for the various enclosure elements. The calculation of latent heat loss through above-grade components involves an equation similar to that used for sensible heat loss:

$$q = (M)(A)(\Delta p)(C) \quad \text{(Eq. 10.10)}$$

where

q = latent heat exchange due to vapor pressure difference, Btu/h [W]
M = permeance of the enclosure element, lb/h ft^2 in. Hg [ng/s m^2 Pa]
A = surface area of the enclosure component, ft^2 [m^2]
Δp = difference in vapor pressure between outdoor and indoor air, in. Hg [Pa]
C = an approximate value of energy content of the air (enthalpy) under typical indoor conditions, 1076 Btu/lb [2500 kJ/kg]

Because it is common to install an effective (impermeable or Class I) vapor retarder in cold climate enclosure assemblies to minimize condensation problems, latent enclosure heat loss is often assumed to be negligible. Because the on-grade and below-grade portions of a building are commonly waterproofed with very-low-permeability materials, these components are typically assumed to be impermeable and to support no latent heat loss. Substantial latent loss can occur, however, from ventilation or infiltration airflow.

The design latent heat loss due to the flow of outdoor air into a space is calculated as follows:

$$\text{I-P units: } q = (V)(4840)(\Delta W) \quad \text{(Eq. 10.11)}$$

$$\text{SI units: } q = (V)(3010)(\Delta W)$$

where

q = latent heat exchange due to ventilation, Btu/h [W]
V = outdoor airflow rate, cfm [L/s]
ΔW = difference in humidity ratio between outdoor and indoor air, lb of moisture/lb of dry air [kJ/kg]
4840 = (used with I-P units) a constant derived from the density of air (0.075 lb/ft^3) multiplied by the heat content of water vapor (1076 Btu/lb) and by 60 min/h; the density and heat content values noted are for conditions typical of indoor conditions
3010 = (used with SI units) a constant derived from the density of air (1.20 kg/m^3) multiplied by the heat content of water vapor (2500 kJ/kg) for typical indoor conditions

(b) Design Heat Gain

The calculation of design heat gain (more correctly called cooling load) is far more complex than the calculation of design heat loss. The simplifying assumptions made for winter conditions (nighttime with no solar radiation and ignoring occupancy-related heat sources) are untenable for a summer situation. There is sun during the "worst" summer hour, and the highest temperatures are usually in the afternoon, when a nonresidential building would be in full work mode with occupants, lighting, and equipment generating heat. In a nutshell, design cooling load calculations must deal with many more variables—including the compounding effects of solar radiation, which create the need for dynamic heat flow analysis. Things get very complicated.

There are hand-calculation methods for estimating design cooling load, and computer programs to do the same. It is important, however, to establish the conditions under which such calculations are conducted. In addition to the design variables that affect design heat loss (enclosure assembly U-factors, surface areas, temperature difference, and airflow rate), the following variables will affect design cooling load:

- the orientation of an assembly (north, south, etc.)
- the tilt of an assembly (vertical, horizontal, inclined)
- the surface reflectance of an assembly
- the thermal capacity of an assembly
- the SHGC of a transparent/translucent assembly
- shading for any enclosure component
- heat gain (sensible and latent) from occupants
- heat gain from lighting
- heat gain (sensible and latent) from equipment

Design decisions influence all these variables and therefore the resulting design cooling load.

This list of variables is again the palette from which energy-efficiency decisions may be selected. It is critical that this palette be fully appreciated early in the design process and not be delegated to a mechanical engineering consultant who may enter the design process too late to provide effective advice.

How does design heat gain differ from design cooling load? In the past, calculations for summer loads simply totaled all the heat flows at a given hour that entered or originated inside of a building enclosure. This total instantaneous heat flow is known as heat gain. It turns out, however, that some of this instantaneous gain does not immediately affect indoor air temperature—this is specifically true of radiation heat flows that are absorbed and stored by the internal mass of a building. Those heat gains that affect air temperature at the hour of interest are collectively called the cooling load. It is only this distinct subset of heat gains that must be handled (at the time) by a cooling system. Conceptually, cooling load is equal to instantaneous heat gain minus any part of that heat gain that is stored within the building, plus any previously stored gains that are now affecting air temperature (through release from storage). The distinction can be substantial, providing first cost savings in equipment capacity and life-cycle savings through improved operating efficiency.

Solar radiation plays an important role in building enclosure heat gain, and not just via transparent/translucent components. The impact of solar radiation on opaque enclosure assemblies can be substantial. The concept of sol-air temperature helps to illustrate this impact. Sol-air temperature is the apparent outdoor air temperature that would produce the same heat flow experienced under the combined effects of temperature difference (based upon actual outdoor air temperature) and radiation. Essentially, sol-air temperature lumps the heat flow caused by radiation absorbed and retained by a surface with the heat flow caused by the air-to-air temperature difference. This is a convenient concept, as it produces a delta-t value that can be plugged into the conventional heat gain equation. The simplified sol-air temperature formula is as follows:

I-P units:

$$t_e = t_o + \frac{\alpha I}{h_o} - 7\,°\text{F} \qquad \text{(Eq. 10.12)}$$

SI units:

$$t_e = t_o + \frac{\alpha I}{h_o} - 3.9°\text{C}$$

where

t_e = sol-air temperature
t_o = outdoor (ambient) dry-bulb temperature
α = absorptance of surface for solar radiation (for light-colored surfaces, usually assumed as 0.45; for dark-colored surfaces, usually assumed as 0.90; detailed values are tabulated in Chapter 16)
I = total solar radiation incident on the surface, Btu/h ft² [W/m²] (see Appendix C for solar heat gain factors, which are approximately equivalent to I for horizontal surfaces)
h_o = coefficient of heat transfer by long-wave radiation and convection at the surface; usually assumed as 3.0 Btu/h ft² °F [17.0 W/m² K]

EXAMPLE 10.6 What is the sol-air temperature for a horizontal white roof compared to a dark roof on a clear July 21 at noon, at 40°N latitude? Assume outdoor air temperature of 90°F [32°C].

SOLUTION

From Table C.3 of Appendix C, on July 21 at noon (12) at 40°N latitude, the solar heat gain factor on a horizontal (HOR) surface = 262 Btu/h ft² [827 W/m²].

$$\text{Sol-air temperature } t_e = t_o + \frac{\alpha I}{h_o} - 7°\text{F}$$

For a white roof:

$$t_e = 90°\text{F} + \frac{0.45 \times 262 \text{ Btu/h ft}^2}{3.0 \text{ Btu/h ft}^2 °\text{F}} - 7°\text{F}$$

$$t_e = 90°\text{F} + 39.3°\text{F} - 7°\text{F} = 133.3°\text{F}$$

For a dark roof:

$$t_e = 90°\text{F} + \frac{0.90 \times 262 \text{ Btu/h ft}^2}{3.0 \text{ Btu/h ft}^2 °\text{F}} - 7°\text{F}$$

$$t_e = 90°\text{F} + 78.6°\text{F} - 7\,°\text{F} = 161.6°\text{F}$$

In SI units:

$$t_e = t_o + \frac{\alpha I}{h_o} - 3.9°\text{C}$$

$$t_e = 32°\text{C} + \frac{0.45 \times 827 \text{ W/m}^2}{17 \text{ W/m}^2 \text{ K}} - 3.9°\text{C} = 50.0°\text{C}$$

$$t_e = 32°\text{C} + \frac{0.90 \times 827 \text{ W/m}^2}{17 \text{ W/m}^2 \text{ K}} - 3.9°\text{C} = 71.9°\text{C}$$

Solar radiation has a marked impact upon surface temperature (and resulting heat flow). Elevated temperatures in full sun will drive considerably more heat through both roofs compared to shaded conditions. The white roof has a Δt about 1.4 times greater in full sun than in shade. The dark roof has a Δt about 1.3 times greater than the white roof under the stated conditions. ■

Because cooling loads will most likely be calculated using computer simulation rather than hand calculations, cooling load equations are not presented herein. The enclosure designer, however, should be aware of the impacts of solar radiation and thermal mass on the energy performance of a building. Decisions about the enclosure will often be made before such decisions are subject to computational analysis.

10.7 CASE STUDY—HEAT FLOW AND ENCLOSURE DESIGN

Orchards at Orenco

PROJECT BASICS

- Location: Hillsboro, Oregon, USA
- Latitude: 45.53 N; longitude: 122.94 W; elevation: 194 ft [60 m]
- Heating degree days: 4750 base 65°F [2639 base 18.3°C]; 280 base 65°F [156 base 18.3°C]; annual precipitation: 43.5 in. [1105 mm]
- Building type: Multifamily, affordable housing
- Floor area: Phase I : 42,584 ft^2 [3956 m^2] conditioned floor area, 57 units; Phase 2 : 58 units
- Completed: Phase I 2015; Phase II 2016
- Design team: Owner: REACH Community Development; Ankrom Moisan Architects, Inc. (architect of record); William Wilson Architects (Design Architect); Walsh Construction Company (contractor); Phase I: Greenhammer/Dylan Lamar (Certified Passive House Consultant)

Background. This project demonstrates the successful implementation of affordable passive building strategies and their scalability to a large, multifamily, affordable housing project, comparing high-performance enclosure designs from two different housing certifications.

REACH Community Development, a nonprofit affordable housing provider sought to build a comprehensive, sustainable model for affordable living and purchased a vacant, 2-acre [8094 m^2] site in Hillsboro, Oregon, to build 150 units. Affordability goals included low rents and proximity to work, but also low monthly utility bills. The project was intended to be convenient, transit-oriented, and affordable. Completed in 2015, Phase I was certified with Phius as the largest multifamily passive house development in the United States (Figure 10.12).

Figure 10.12 Orchards at Orenco, Phase I, completed in 2015, opened as the largest certified passive house multifamily, affordable housing project in the US. (© Casey Braunger, Ankrom Moisan Architects, Inc.)

BUILDING ENCLOSURE

Phase II, completed in 2016 (Figure 10.13) was pre-certified under the Phius+2015 Standard. Phase III followed (Figure 10.14). The buildings were predicted to achieve nearly 90% energy reduction for space heating and 60–70% for overall energy use compared to a comparable Leadership in Energy and Environmental Design (LEED) certified building. REACH installed the Imagine Energy monitoring system to track and improve upon tenants' energy usage habits.

Context. During Phase I of Orchards at Orenco, the architects, contractors, and passive house consultants embraced careful adherence and intent to meet the stringent passive house requirements and criteria commonly achieved in Europe. The unique challenges for this phase were in the HVAC integration and reduction of thermal bridging in the enclosure. Many lessons were learned by the design team in this first project of this scale in the United States. Primary strategies used in this project included

Figure 10.13 Orchards at Orenco, Phase II, completed in 2017, was precertified to the Phius+2105 standard. (© Casey Braunger, Ankrom Moisan Architects, Inc.)

Figure 10.14 Orchards at Orenco plan showing three phases of development. (© Walker | Macy Landscape Architects.)

continuous insulation, no thermal bridging, airtight enclosure, balanced heat recovery ventilation, and optimized solar heat gains for heating during the heating season and minimized overheating during the cooling season. Just after completion of Phase I, the Phius+2015 climate-based North American Passive Building Standard was released; the Phase II project was designed and pre-certified to the criteria of this standard.

Design Criteria and Validation. The criteria used in Phases I and II are shown in Table 10.4. The Phius+2015 passive building standard prescribes climate-specific annual heating and cooling demand (consumption) and peak heating and cooling loads. The source energy demand is based upon occupancy, and air tightness targets are based upon area of the enclosure rather than enclosed volume.

Design Features

- *Superinsulation and hygrothermal performance*: Like the cutting-edge 1970s houses from which the passive house standard was derived, Phius-certified buildings feature highly insulated enclosures, with typical wall assembly R-values ranging from 35 to 55 Btu/h ft^2 °F [7–10 W/m^2 K] and roof assembly R-values ranging from 60 to 90 Btu/h ft^2 °F [10–17 W/m^2 K]. Unlike their 1970s counterparts, however, modern enclosure assemblies are carefully analyzed for their hygrothermal performance to assess the degree of vapor diffusion through the assembly.

The Phase 1 roof has 12 in. [305 mm] of polyiso insulation, approximately three times the amount required by code. A light color surface reflects solar radiation and helps moderate the building temperature. The foundation and above-grade enclosure are shown under construction in Figures 10.15*a* and 10.15*b*. The construction detail (Figure 10.16) shows the reinforced concrete perimeter foundation wrapped with 4 in. [102 mm] of EPS insulation and the exterior wall using 1.5 in. [38 mm] exterior insulation and 2 × 10 in. [38 × 235 mm] wall framing filled with blown-in fiberglass insulation. In Phase 2, the exterior wall uses dramatically less material and insulation—1 in. [25 mm] of mineral wool exterior insulation and 2 × 8 in. [38 × 184 mm] wall framing—due to the newly revised, climate-specific Phius+2015 criteria and careful cost-saving measures.

TABLE 10.4 Performance Metrics and Criteria for Orchards at Orenco Phase I

Metric	Single Performance Target[a]	Phius+2015 Passive Building Standard
Climate Zone	NA	ASHRAE Climate Zone 4C; Portland/Hillsboro
Annual Heating Demand[b]	≤4.75 kBtu/ft^2 yr [15 kWh/m^2 yr]	5.1 kBtu/ft^2-iCFA[c] yr 1.0 kBtu/ft^2-iCFA yr
Annual Cooling Demand		
Peak Heating Load	≤3.17 Btu/hr ft^2 [10 W/m^2]	3.7 Btu/ft^2-iCFA yr 3.9 Btu/ft^2-iCFA yr
Peak Cooling Load		
Source Energy Demand	≤38 kBtu/ft^2 yr [120 kWh/m^2 yr]	≤6200 kWh/yr/person
Airtightness	≤0.6 ACH_{50}	≤0.05[d] cfm_{50}/ft^2 [0.3 L/s m^2]

Source: Wright and Klingenberg, 2015.

[a]This is a single target standard that applies the same design requirements to all climate zones (one size fits all).

[b]Demand in this context refers to annual consumption (of energy for heating or cooling, respectively).

[c]iCFA stands for interior conditioned floor area.

[d]The area referred to in this requirement is surface area of the building enclosure (not floor area).

(*a*) (*b*)

Figure 10.15 *(a) Orchards at Orenco Phase I construction of foundation and (b) enclosure assembly. (© Walsh Construction Company.)*

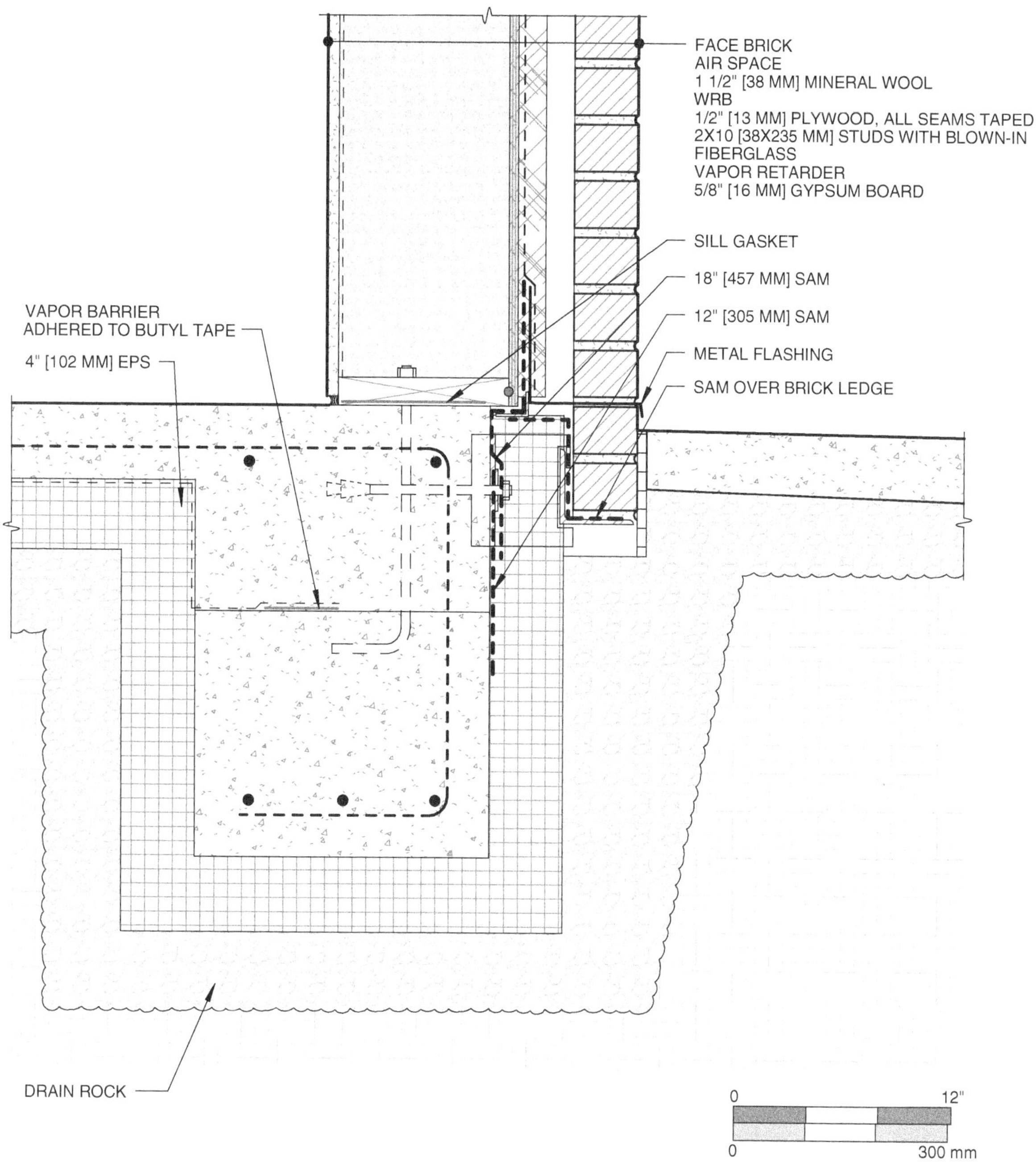

Figure 10.16 *An offset steel angle supports the brick wainscot with reduced thermal bridge effects. Wall construction uses 10 in. [13 mm] deep studs and two types of cladding: fiber cement panels and brick veneer. (Ankrom Moisan Architects, Inc., redrawn © Corner et al., 2018.)*

As the complexity of the enclosure assembly increases, so must the rigor of its hygrothermal analysis. Components must be thoughtfully organized to reduce the potential for vapor condensation and accumulation, which is a precursor to mold growth. At Orchards at Orenco, extensive hygrothermal modeling was used to verify the "vapor-open" nature of the assembly. While the assembly is both airtight and highly thermally resistive, moisture is allowed to dry to both sides.

- *Air sealing:* One of many energy-saving measures required of certified passive buildings is a continuous air barrier. Great care (and much tape) are required to ensure that the "airtight layer," which includes floor, walls, and ceiling, remains free of both intentional and unintentional holes.

Shown in Figure 10.16, the primary control layer for airtightness is formed by 0.5 in. [13 mm] plywood, carefully taped at its seams. The layer is placed

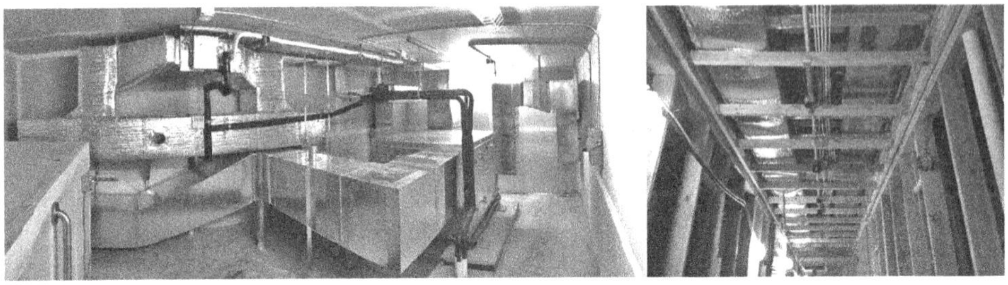

Figure 10.17 *Left: Penthouse location houses an energy recovery ventilator (ERV) and heat pump inline with the supply air trunk that carries fresh air to the apartments and common areas. Right: Ventilation air is routed through the space above the third-floor corridor. (© Walsh Construction Company.)*

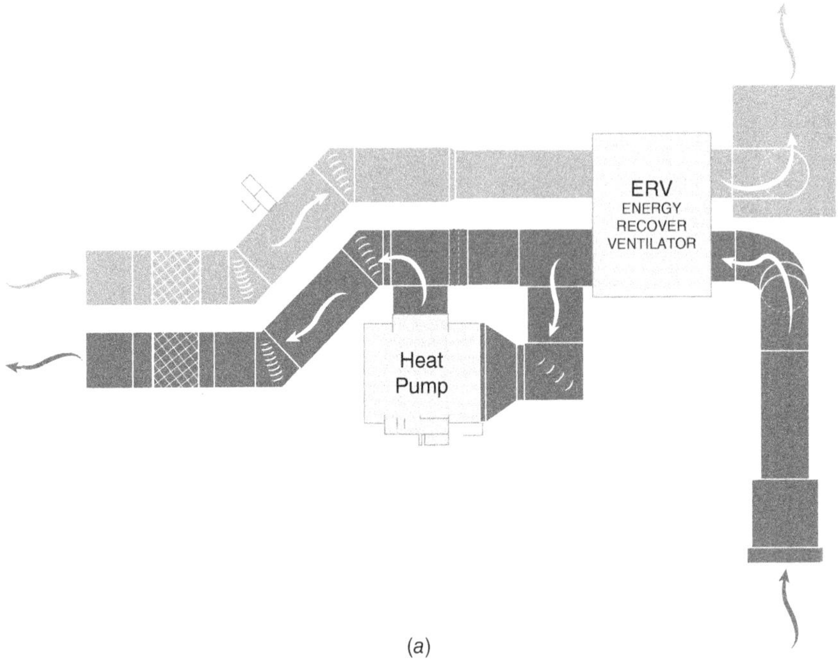

(*a*)

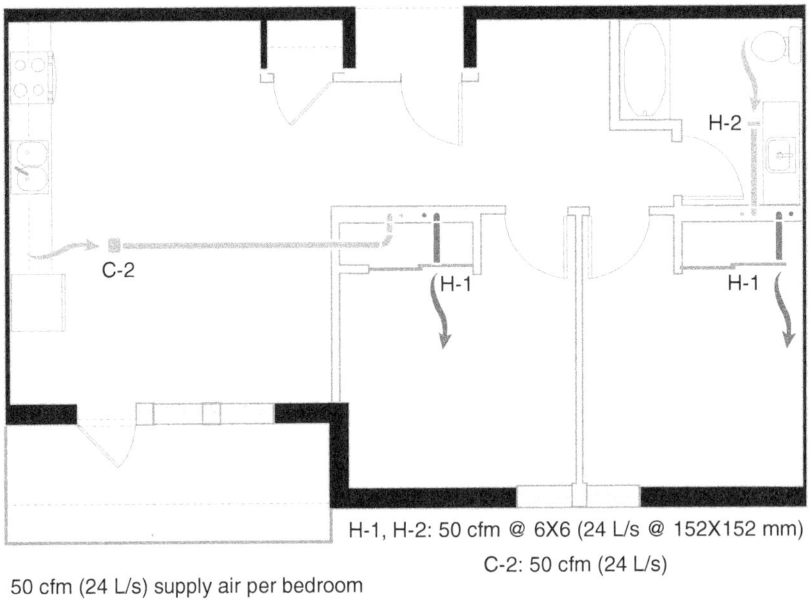

(*b*)

Figure 10.18 *(a) Heating and partial cooling using an ERV and heat pump; (b) mechanical layout for supply to bedrooms and exhaust from the bathroom in a typical 2-bedroom apartment. (© PAE Consulting Engineers, Inc.)*

between the 2 × 10 in. [38 × 235 mm] wall and the mineral wool. This strategic placement protects the airtight layer from weather and from occupant wear, and it minimizes necessary service punctures to the airtight layer, as the service cavity is placed in the innermost position. Meticulous detailing is provided at the wall intersections, and at penetrations such as windows, to ensure continuity of the air barrier system.

- *Shading devices:* Shading devices that allow the winter sun in, but block the summer sun were placed on façades. This was a particularly important element at the west- and south-facing façades. Exterior shading is one of the most critical components of HVAC design and this very simple addition to the façade will be experienced through low energy use and excellent thermal comfort.
- *Heat recovery ventilators:* In Phase I, three rooftop mechanical penthouses (Figure 10.17) house US-made heat recovery ventilators. Heat recovery ventilators use the energy from kitchen and bathroom exhaust air to warm the fresh incoming air which is supplied to the bedrooms. Each ventilator unit serves 18–20 apartments (see typical duct layout for the apartments in Figures 10.18*a* and 10.18*b*) and provides supply to some of the common areas. Rather than designing a fully centralized system, or a decentralized system with individual ventilator units in each apartment, this grouped approach provides high efficiency while simplifying distribution and likely cutting down on maintenance needs.
- *High-performance windows:* The team chose to use windows from the Loren Cook Company—a North American manufacturer—due in part to cost considerations, but also based to a significant degree on the mechanical contractor's prior experience with that manufacturer's products. The windows are PVC-fiberglass hybrid window frames with argon-filled, triple-pane glazing and operate with a tilt-turn, European-style movement.
- *Performance data:* Pre-certification was applied for to the current Phius standard. The project used a Phius + Rater and Earth Advantage, to perform on-site inspections and blower door testing. Final certification occurred at the end of construction after the final blower door test and commissioning of the heat recovery ventilators. An energy-tracking system (Figure 10.19) records the performance of each unit and displays the results in the lobby (Project 1203: The Orchards at Orenco). Numerous awards testify to the quality of this project, as seen in Figure 10.20.

Awards

- 2015 Oregon Opportunity Golden Hammer Award for Best Overall Project

Figure 10.19 *Display shows comparative energy use by unit and for the entire building, located near the mailboxes and lobby in the Orchards at Orenco, Phase I Project. (© Casey Braunger, Ankrom Moisan Architects, Inc.)*

Figure 10.20 Common space and kitchen areas are spacious and designed to balance daylighting and electric lighting. (© Casey Braunger, Ankrom Moisan Architects, Inc.)

- 2015 Phius Building Project Competition—Best Overall Project and the Best Affordable Housing Project awards
- 2015 *Portland Business Journal* Better Bricks "Sustainable Project of the Year"
- 2015 *World Architecture News* (WAN) Sustainable Buildings Awards
- Award of Merit Green Project: Orchards at Orenco, ENR Northwest, December 10, 2015 People's Choice category—First Place; Private Buildings Category Honorable Mention; High Performance Building Award—Top Choice. *Oregon Daily Journal of Commerce* Top Projects, May 5, 2016

FURTHER INFORMATION

Passivhaus Institute. www.passivehouse.com

Phius. www.phius.org

Project 1203: The Orchards at Orenco. www.phius.org/phius-certification-for-buildings-products/case-studies/orchards-at-orenco-phase-i

Steffen, M. 2014. "The Largest Passivhaus Building in the U.S." *Green Building Advisor*, November 13. greenbuildingadvisor.com/article/the-largest-passivhaus-building-in-the-u-s

Steffen, M. 2016. "Mechanical and Electrical Systems at the Orchards at Orenco Project," *Green Building Advisor*, August 17. greenbuildingadvisor.com/article/mechanical-and-electrical-systems-at-the-orchards-at-orenco-project

Wright, G. S. and K. Klingenberg. 2015. *Climate-Specific Passive Building Standards (DOE/GO-102015–4679)*. Washington, DC: U.S. Department of Energy. www.nrel.gov/docs/fy15osti/64278.pdf

ACKNOWLEDGMENTS

Keith A. Simon, FAIA, is acknowledged as a significant contributor to the contents of this chapter.

References and Resources

ASHRAE. 2001. *2001 ASHRAE Handbook—Fundamentals*. Atlanta, GA: American Society of Heating, Refrigerating and Air-Conditioning Engineers.

ASHRAE. 2017. *2017 ASHRAE Handbook—Fundamentals*. Atlanta, GA: American Society of Heating, Refrigerating and Air-Conditioning Engineers.

ASHRAE. 2021. *2021 ASHRAE Handbook—Fundamentals*. Atlanta, GA: American Society of Heating, Refrigerating and Air-Conditioning Engineers.

ASHRAE. 2022a. *ANSI/ASHRAE Standard 62.1-2022: Ventilation for Acceptable Indoor Air Quality*. Atlanta, GA: American Society of Heating, Refrigerating and Air-Conditioning Engineers.

ASHRAE. 2022b. *ANSI/ASHRAE Standard 62.2-2022: Ventilation and Acceptable Indoor Air Quality in Low-Rise Residential Buildings*. Atlanta, GA: American Society of Heating, Refrigerating and Air-Conditioning Engineers.

ASHRAE. 2022c. *ANSI/ASHRAE/IES Standard 90.1-2022: Energy Standard for Buildings Except Low-Rise Residential Buildings*. Atlanta, GA: American Society of Heating, Refrigerating and Air-Conditioning Engineers.

ASHRAE. 2024a. *ANSI/ASHRAE/IES Standard 90.2-2024: Energy-Efficient Design of Low-Rise Residential Buildings*. Atlanta, GA: American Society of Heating, Refrigerating and Air-Conditioning Engineers.

ASHRAE. 2024b. *ASHRAE Standard 160: Criteria for Moisture-Control Design Analysis in Buildings*. Atlanta, GA: American Society of Heating, Refrigerating and Air-Conditioning Engineers.

ASTM. 2024. *ASTM E96/E96M: Standard Test Methods for Gravimetric Determination of Water Vapor Transmission Rate of Materials*. West Conshohocken, PA: ASTM International.

BuildingGreen. 2018. "The BuildingGreen Guide to Thermal Insulation." www.buildinggreen.com/continuing-education/buildinggreen-guide-thermal-insulation

Bynum, R. T. Jr. 2001. *Insulation Handbook*. New York: McGraw-Hill.

Carmody, J., S. Selkowitz, and L. Heschong. 2007. *Residential Windows: A Guide to New Technologies and Energy Performance* (3rd ed.). New York: W. W. Norton and Company.

Carmody, J., S. Selkowitz, and L. Heschong. 1996. *Residential Windows: A Guide to New Technologies and Energy Performance* (2nd ed.). New York: W. W. Norton and Company.

Corner, D., J. Fillinger, and A. Kwok. 2018. *Passive House Details*. New York: Routledge.

Emmerich, S. J. and A. K. Persily. 2005. "Airtightness of Commercial Buildings in the U.S," in *Proceedings of the 26th IEA Conference of the Air Infiltration and Ventilation Centre*, (pp. 65–70). Brussels: International Energy Agency.

ICC. 2024a. *2024 International Building Code*. Washington, DC: International Code Council.

ICC. 2024b. *2024 International Energy Conservation Code*. Washington, DC: International Code Council.

ICC. 2024c. *2024 International Residential Code*. Washington, DC: International Code Council.

Lstiburek, J. 2011. *BSI-049: Confusion About Diffusion*. Westford, MA: Building Science Corporation. https://buildingscience.com/documents/insights/bsi-049-confusion-about-diffusion

Lstiburek, J. 2023. *BSI-141: Shakespeare Does Roofs*. Westford, MA: Building Science Corporation. https://buildingscience.com/documents/building-science-insights/bsi-141-shakespeare-does-roofs

Lstiburek, J. and J. Carmody. 1994. *Moisture Control Handbook: Principles and Practices for Residential and Small Commercial Buildings*. Hoboken, NJ: John Wiley & Sons.

Melody, I. 1987. *Radiant Barriers: A Question and Answer Primer (FSEC-EN-15)*. Cocoa, FL: Florida Solar Energy Center. publications.energyresearch.ucf.edu/wp-content/uploads/2018/06/FSEC-EN-15-87.pdf

NFRC. n.d. National Fenestration Rating Council. www.nfrc.org

NFRC. 2025. "NFRC Consumer Guide to Windows." National Fenestration Rating Council. efficientwindows.org

Wilkinson, G. 1999. "Beyond R-Value: Insulating for the Environment." *Environmental Design and Construction*, January–February.

CHAPTER 11

Solar Geometry and Shading Devices

BUILDING ENCLOSURE

AN UNDERSTANDING OF THE SUN'S POSITION during the day is critical to site planning, daylighting design, passive solar design, and controlling unwanted heat gains. In early civilizations, most cultures revered the sun as a deity and understood its patterns and importance in the cycle of life. Today, such understanding is just as important when considering the relationships between the environment and technology in the design of buildings. We can no longer afford the resources to sustain the construction and operation of buildings with poorly designed glazed areas habitable only through the use of active climate control systems. We can rely on the patterns of the sun as a constant in the design process and create environments that work with these patterns, climate, and well-considered design intents.

11.1 SOLAR POSITION

The sun is a giant star and the largest object in the solar system. The sun's energy is produced by nuclear fusion that occurs at temperatures in the range of 18–25 million °F (10–14 million °C). The energy from thermonuclear fusions at the sun's surface is released as electromagnetic radiation at approximately 10 million °F (5.6 million °C). The solar spectrum (Figure 11.1) consists of approximately 5% ultraviolet (shortwave) radiation (less than 350 nm), 46% light—what we see—(350–750 nm), and 49% infrared (longwave) radiation (greater than 750 nm). Though 93 million miles from Earth, the amount of the sun's radiation (heat energy) reaching the Earth's outer atmosphere is relatively stable and is called the solar constant—433 Btu/ft^2 h (1.37 kW/m^2). Some annual variation in the solar constant occurs because of the elliptical orbit of the Earth around the sun, but it is so small that it has little effect on building design.

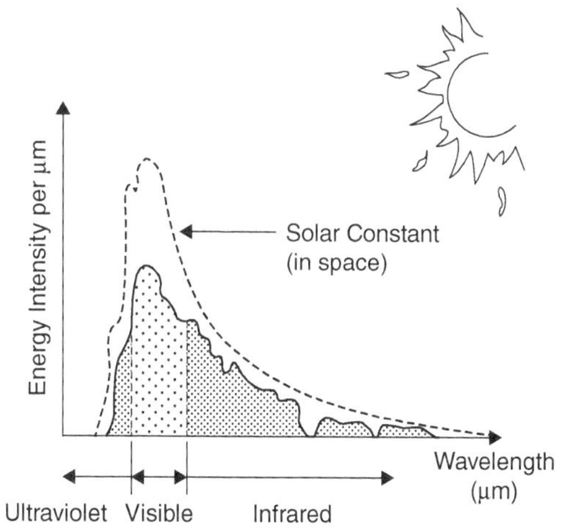

Figure 11.1 At the Earth's surface the solar radiation spectrum consists of ultraviolet radiation, visible radiation (light), and infrared radiation. (Drawing by Erik Winter; © Alison Kwok; all rights reserved.)

Mechanical and Electrical Equipment for Buildings, Fourteenth Edition. Walter Grondzik, Daniel Overbey, and Alison Kwok.

Companion website: www.wiley.com/go/meeb14e

The length of the radiation's path through the atmosphere is the most important factor in determining how much radiation reaches the surface of the Earth. This received radiation is termed insolation.

(a) Earth's Rotation and Tilt

The Earth's axis of rotation is tilted at 23.5° (precisely, 23.47°), also known as its declination. The Earth travels in the plane of the ecliptic (Figure 11.2) in its orbit around the sun. The tilt of the Earth is toward the sun (June +23.5°) or away from the sun (December −23.5°). This tilt is responsible for the seasons. In the Northern hemisphere in June, the sun's direct radiation is perpendicular to the Earth's surface; in December, there are fewer hours of sun, and the radiation passes through a longer, nonperpendicular, length of atmosphere to reach the Earth. The rotation of the Earth (once every 24 hours) and its tilt determine the length of atmosphere that solar radiation passes through, and consequently how much energy the Earth's surface receives.

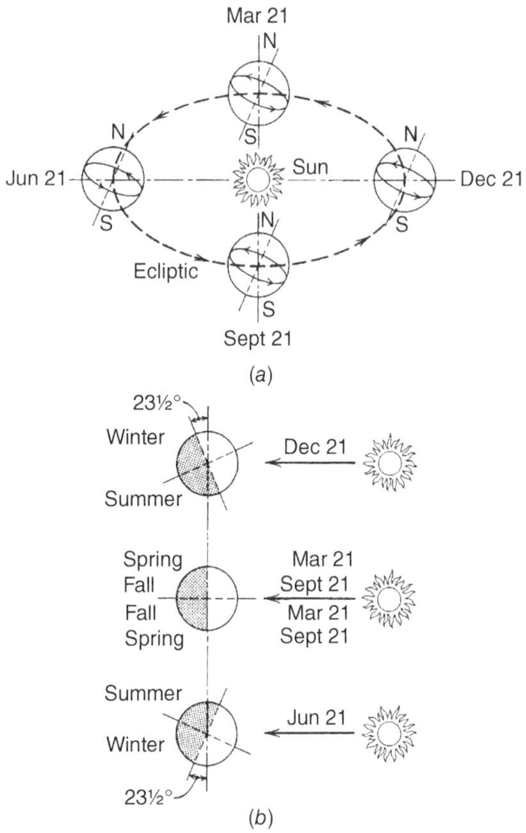

Figure 11.2 *(a) The ecliptic is the annual path of the Earth around the sun. (b) The tilt of the Earth's axis in the plane of the ecliptic results in the seasonal variations. In the northern hemisphere the Earth tilts away from the sun in December (−23.5°), resulting in winter's low sun altitude and cold weather. In June, the effect is reversed.*

(b) Altitude and Azimuth

The position of the sun at any instant with respect to an observer on the ground is defined by its altitude angle and its azimuth angle. The altitude of the sun is the angle between the horizon and the sun's position above the horizon. The altitude varies during the day, beginning and ending at 0° at sunrise and sunset and reaching a daily maximum at solar noon. The altitude angle at noon varies from day to day, reaching a yearly maximum on June 21 (the summer solstice), a minimum on December 21 (the winter solstice), and a point halfway between the two on the vernal and autumnal equinoxes (March 21 and September 21). The azimuth (also called the solar bearing angle) is the angle along the horizon between the projected position of the sun and true (solar) south. Note that the azimuth is referenced from south herein, whereas in other sources it is sometimes referenced from north; there is no clear consensus on this issue. Refer to Figure 11.3 and Appendix C, Tables C.11 to C.18 for solar position and insolation information.

Referring to the graphic relationships in Figure 11.2, the height of the sun in the sky (altitude angle) depends on the observer's position on the Earth (the latitude) and the seasonal changes (the tilt of the Earth). This can be expressed in Equation 11.1, as follows:

$$\text{altitude angle at solar noon} = 90° - \text{latitude} \pm \text{declination} \qquad \text{(Eq. 11.1)}$$

For example, the solar altitude for Minneapolis (44.9°N, approximated to ~45°) for various seasons at solar noon is:

Maximum (June 21) = 90° − ~45° + 23.5° = 68.5°

Equinox (Mar/Sept 21) = 90° − ~45° − 0° = 45°

Minimum (Dec 21) = 90° − ~45° − 23.5° = 21.5°

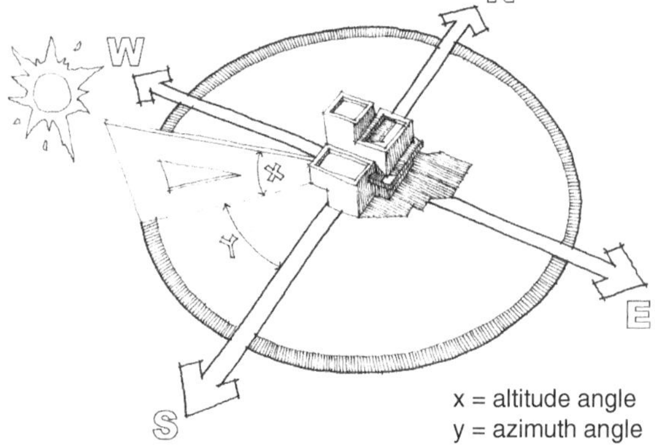

Figure 11.3 *Altitude and azimuth angles. (Drawing by Erik Winter; © Alison Kwok; all rights reserved.)*

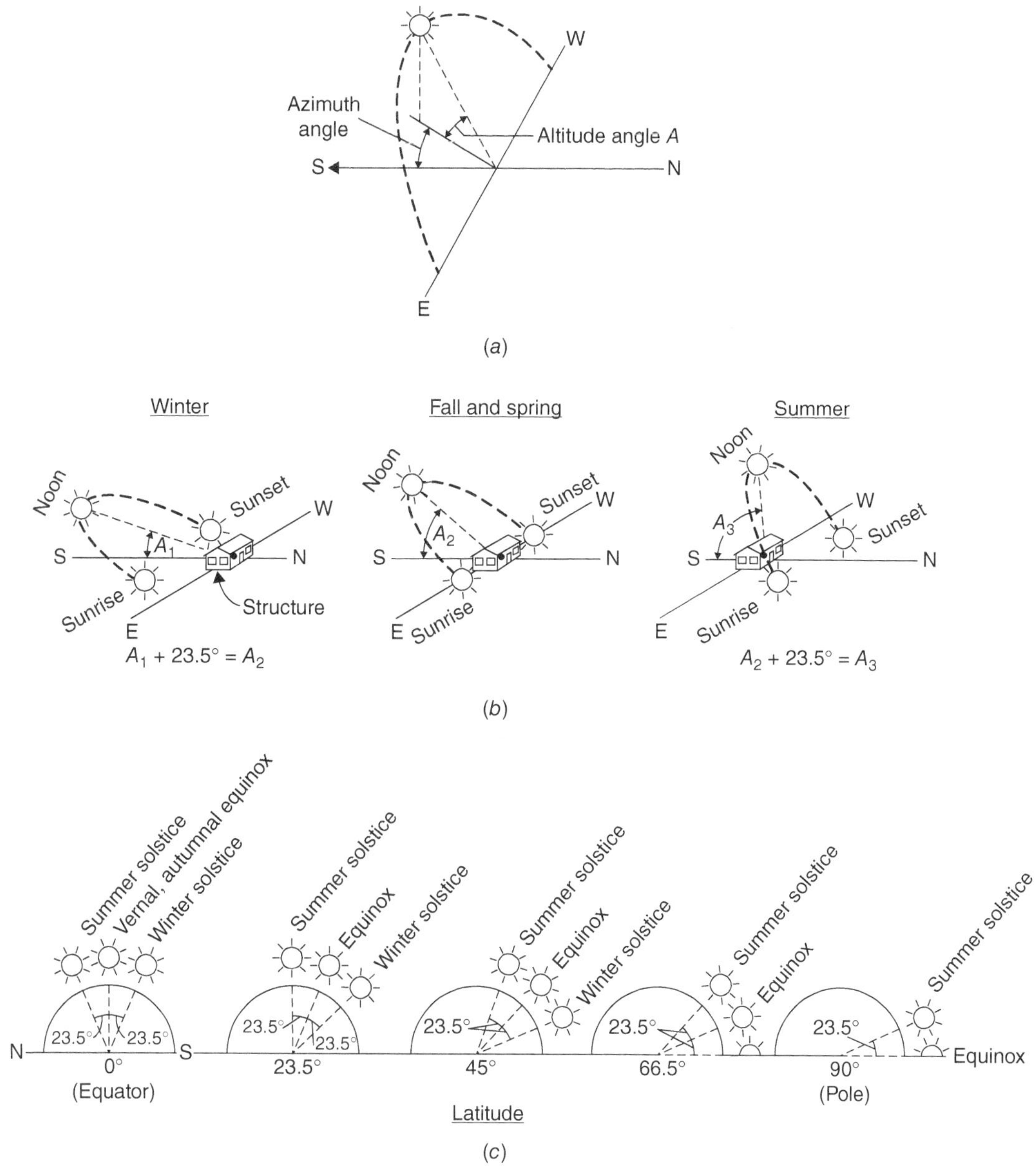

Figure 11.4 *(a) The position of the sun is expressed in terms of vertical angle above the horizon (altitude) and horizontal angle (azimuth, measured from the south). (b) Approximate position of the sun in each of the seasons at a mid-northern latitude (approximately 45°). Note that the altitude angle is maximum in summer, minimum in winter, and in between in spring and fall. (c) Maximum sun altitude at various latitudes for both solstices and equinoxes.*

These relations are shown graphically in Figure 11.4.

The azimuth angle during its path from sunrise to sunset changes with the season. On the equinoxes (March 21 and September 21), the sun rises due east and sets due west. In the northern hemisphere, the sun rises north of east and sets north of west between March 22 and September 20. This is shown graphically in Figure 11.5 for one specific latitude (32°N). Sunrise on June 21 is 28.5° north of east, and sunset is correspondingly 28.5° north of west. On December 21, sunrise and sunset are 28.5° south of east and west.

A designer can use three basic solar pattern concepts when designing with the sun:

1. The altitude of the sun is highest in summer, lowest in winter, and in between in spring and fall for all latitudes. The maximum difference in altitude angle between summer and winter solstices is about 47°.
2. The daily maximum altitude of the sun increases as a location approaches the equator (low latitudes, either north or south). The seasonal altitude variation, however, is the same for all latitudes (except at those extreme north and south latitudes

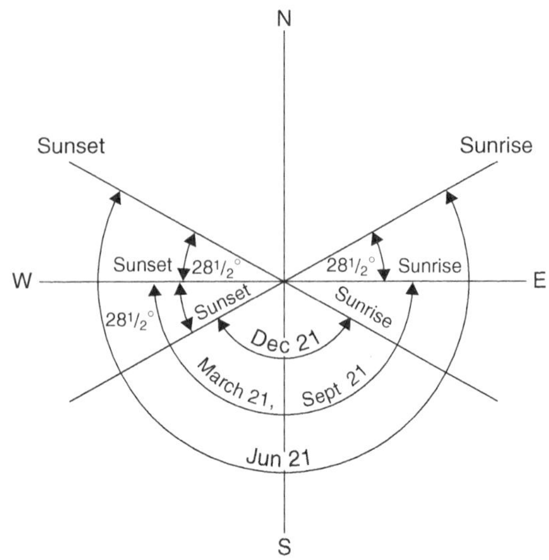

Figure 11.5 A diagram of the sun's azimuth path from sunrise to sunset projected on a horizontal plane for solstices and equinoxes at latitude 32°N (Jerusalem, Savannah, Tucson). The sun's azimuth angle, like its altitude, varies with the time of day and the date.

where the sun is above or below the horizon for extended time periods). This factor not only has a pronounced effect on the design and efficacy of sunshading devices but also affects exterior daylight illuminance levels.

3. The azimuth angle of the sun is dictated by the time of day and by the season. The sun by definition traverses the sky between sunrise and sunset and in the northern hemisphere rises north of east during the summer, due east on the equinoxes, and south of east during the winter. The principal significance of the azimuth angle is its role as an important consideration when orienting a building on a site, establishing building exposures, and analyzing shading angles.

11.2 SOLAR TIME

The basis of all timekeeping is the length of the solar day, that is, one full rotation of the Earth on its axis. The Earth's motion on the ecliptic is not uniform—it moves less rapidly when farther from the sun. As a result, the actual length of a solar day varies. Since it is impractical to use timepieces that need daily adjustment, we use an average, or mean solar day, as the basis for timekeeping. There are three reasons for the difference between solar time and clock time:

1. *Location within a time zone:* The 360° circumference of the Earth is divided into 24 one-hour time zones, each with a width of approximately 15° of longitude, which corresponds to approximately 1000 miles (1610 km). An observer located at any point other than directly on a standard time zone reference longitude (normally in the center of a time zone) must make a time correction for his or her distance from the reference longitude. (In some countries, a reference longitude is more than 30 minutes from time zone extremes. In such places, the time correction for longitude plus the equation of time can total well over an hour. This situation does not exist in the continental United States.)
2. *Equation of time:* The speed of the Earth in its orbit around the sun is nonuniform; the planet moves less rapidly when farther from the sun. As a result, the actual length of the solar day varies as defined by the equation of time (Figure 11.6*a*). The extent of this variation can be determined from a curve called the analemma (Figure 11.6*b*) or from tabulations in various sources. The analemma is the shape that results if the sun's position in the sky is recorded at the same time of day throughout the year.
3. *Daylight saving time (DST):* During DST, clocks are moved forward one hour to effectively give (from the morning) one more hour of daylight in the evening during the summer months. (Note: The official term is daylight *saving* time, not daylight savings time.) The original purpose of DST was to save electricity—a wartime measure that started in Europe in the midst of World War I. Different regions

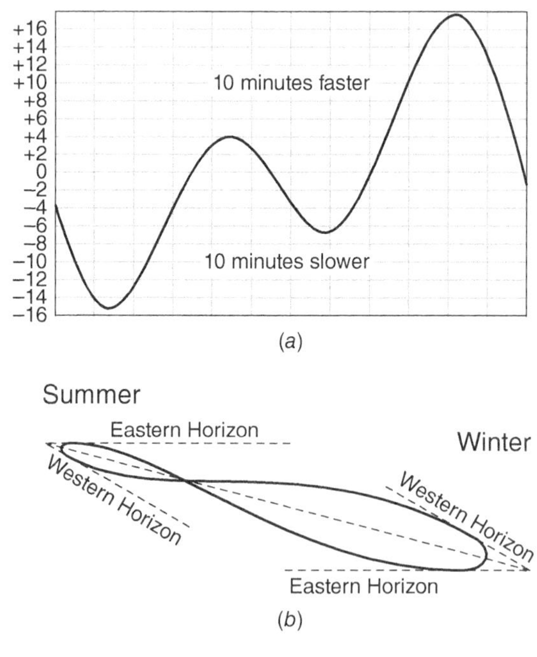

Figure 11.6 (a) Equation of time. (b) The analemma is a miniature almanac where the vertical coordinates of the path represent the sun's declination on a particular day of the year, while the horizontal coordinates tell how much the sun is ahead of or behind clock time on that day. (Drawings by Erik Winter; © Alison Kwok; all rights reserved.)

of the globe implement DST differently. In most of the United States (except Arizona and Hawaii), DST begins at 2:00 a.m. on the second Sunday of March and reverts back to standard time on the first Sunday in November. By comparison, Central European Summer Time (CEST) begins at 1:00 a.m. Universal Time (Greenwich Mean Time) on the last Sunday in March and ends the last Sunday in October.

Unless otherwise clearly noted, most solar position tables and charts (including sunpath diagrams) are based on solar time, not clock time. This must be considered when solar-driven loads are being added to clock-driven loads (e.g., the heat gain from the western sun is being added to the heat gain from office occupants to determine the total load). Such an addition must use coincident values, and solar and clock times are rarely coincident.

11.3 TRUE SOUTH AND MAGNETIC DEVIATION

Working from a "true south" or "solar south" orientation (opposite of "true north," location of the Earth's axis of rotation) is the starting point in the planning and design of a building, particularly for passive solar strategies. When a compass is used on a building site to find "north," the compass yields a reading toward magnetic north, determined by the Earth's internal magnetic field. This is usually (depending on the locale) not the same as true north. Figure 11.7 illustrates this phenomenon. To correct between magnetic north (compass) and true north (solar), a magnetic deviation (sometimes called magnetic declination) is applied to the compass reading. Depending on the location on the Earth's surface, the deviation may be as large as 50°—but in most locations this is not the case. In some locations, magnetic deviation is negligible (i.e., virtually 0°). The theoretical routes created along points

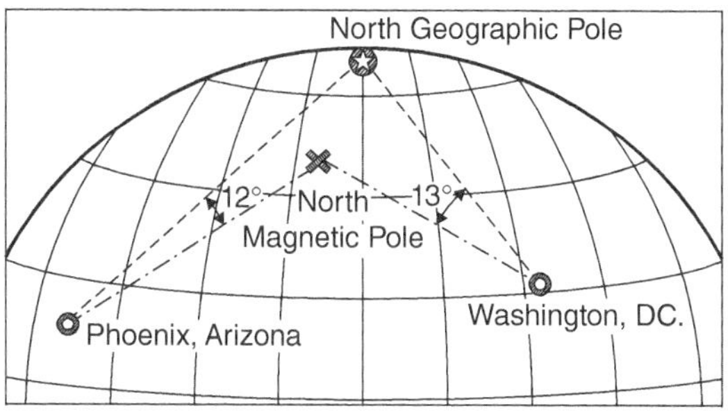

Figure 11.7 *The concept of magnetic deviation (from true north), illustrated for Phoenix, Arizona (12° easterly) and Washington, DC (13° westerly). (Drawing by Erik Winter; © Walter Grondzik; all rights reserved.)*

at which magnetic deviation is the same are called isogonic lines. The isogonic lines of 0° are referred to as agonic lines. It should be noted that the magnetic field is not uniform, stationary, or perfectly aligned with the Earth's poles; thus, magnetic deviations can change and have changed over time. The U.S. National Oceanic and Atmospheric Administration (NOAA) maintains an interactive public database of historical isogonic lines and magnetic poles going back to 1590.

11.4 SUNPATH PROJECTIONS

Design tools such as sunpath projection charts can be used to assess site conditions, orient a building, or provide information critical for passive solar design. Sunpath charts represent the sun's position throughout the year. There are two fundamental types of sunpath projections: vertical and horizontal. Horizontal projections are more commonly encountered. The basic idea behind the horizontal projection is shown in Figure 11.8. The sky vault projection is represented as a hemisphere that projects as a circle, the center of which is the zenith. The observer looks down onto the hemispherical sky vault and projects the sunpaths from the vault onto the two-dimensional projected circle. The basic difference between the various horizontal projections is in the spacing between circles of equal altitude. Vertical projection charts are also available.

Four of the more common sunpath projections are: (1) gnomonic, (2) equidistant, (3) rectilinear, and (4) stereographic.

1. The *gnomonic projection* (used with sundials and sunpeg charts) is derived from the sundial. The observer is at the center of the projection; therefore, low sun angles (sunrise, sunset) extend to infinity. For this reason, this method is rarely used for solar charts but is easily applied to building shadow studies. Sunpeg charts for specific latitudes (see Figure 11.9 and Appendix D, Section D.2) will show the exact position of sun penetration and shadow on a model of any scale, on any date, at any time of day between shortly after sunrise and shortly before sunset. Use of the sunpeg chart is explained in the following outline (reprinted by permission from Brown, Reynolds, and Ubbelohde, 1982) and shown in Figure 11.10.

   - *Find the sunpeg chart corresponding to the nearest latitude for a site.*
   - *Make a copy of this chart.* If the copier changes the chart size, the peg height line will also change; there should be no problem.

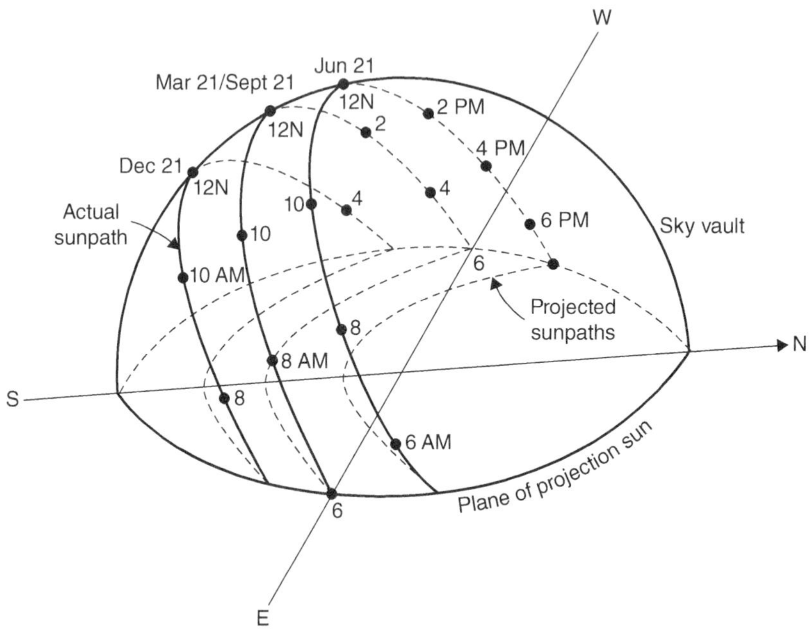

(a)

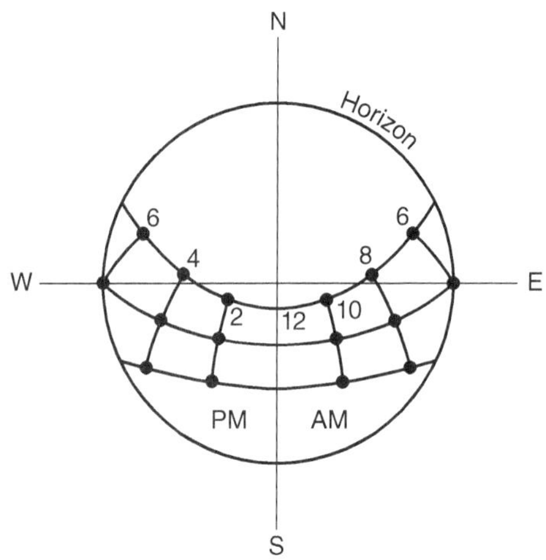

(b)

Figure 11.8 *(a) The sky vault is represented as a hemisphere, which projects as a circle (b) on a projection plane. The exact shape of the sunpath chart depends on the projection method. The hour points on the sunpath chart are connected to form hour lines that are useful for interpolation.*

- *Construct a "peg" (gnomon) whose finished height above the chart surface corresponds exactly to the "peg height" shown on the copy of the chart.* This peg must stand and remain perfectly vertical relative to the model and not be bumped out of vertical alignment.
- *Mount the copy of the chart on the model to be tested.* The chart must be perfectly horizontal over its entire surface, and the north arrow on the chart must correspond to true (solar) north on the model.
- *Choose a test time and date. Take the model out into direct sunlight.* (With a small model, a camera tripod might serve as a mount.) Tilt the model until the shadow of the peg points toward the intersection of the chosen time line and the chosen date curve. When the end of the peg's shadow touches this intersection, the model will show the same sun-shadow patterns as would occur on the time and date chosen.

2. The *equidistant projection* (horizontal, polar) is used almost exclusively in the United States because of the ready availability and wide acceptance of the sunpath charts provided in the Pilkington (formerly Libbey Owens Ford) Sun Angle Calculator.

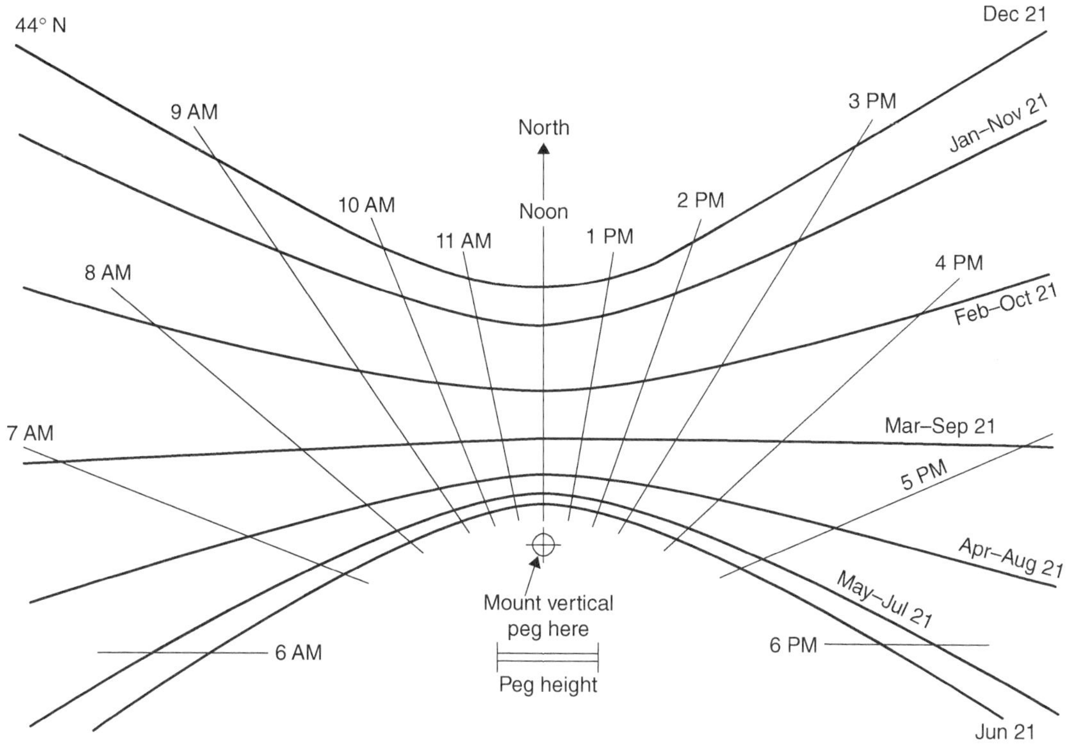

Figure 11.9 *Sunpeg chart for 44°N latitude.*

Figure 11.10 *A sunpeg chart attached to the base of a scale model to study solar access and shadow patterns in and about a building. (Photo by Jonathan Meendering; © Alison Kwok; all rights reserved.)*

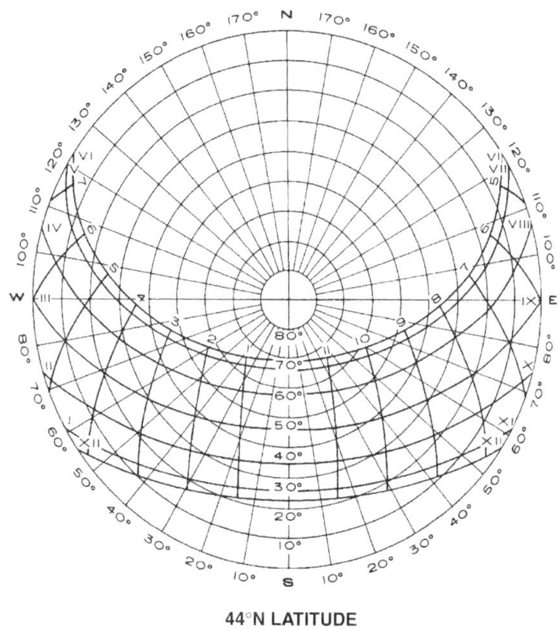

Figure 11.11 *Equidistant (horizontal, polar) sunpath diagram for 44°N latitude. (From Ramsey and Sleeper, 2016.)*

The observer moves around the skydome, like the sun's view to the Earth. Figure 11.11 shows a sample equidistant projection sunpath diagram. A full set of equidistant projections is provided in Appendix D, Section D.3. This projection method is equally applicable to all latitudes, and a latitude-specific sunpath is needed for a location under study. The Pilkington Sun Angle Calculator sunpath diagrams are available in steps of 4° of latitude between 24° and 52° latitudes. The Sun Angle Calculator can be purchased through the Society of Building Science Educators (SBSE).

3. The *rectilinear projection* (vertical, cylindrical) is a two-dimensional graph of the sun's position in Cartesian coordinates. Azimuth is plotted along the horizontal axis and altitude on the vertical axis.

The development of this type of chart is shown in Figure 11.12. These charts are graphs superimposed on the eye of an observer. At the vertical center of each chart, the observer is looking due south. The horizon (a horizontal plane at the observer's eye level) is the line at the bottom of the chart. A set of rectilinear charts for several latitudes (Mazria, 1979) is found in Appendix D, Section D.4.

These charts may also be used to easily determine the number of hours of greatest insolation, that is, what percentage of clear-day insolation is gained each hour by an unshaded south-facing vertical window (Figure 11.13). This emphasis on

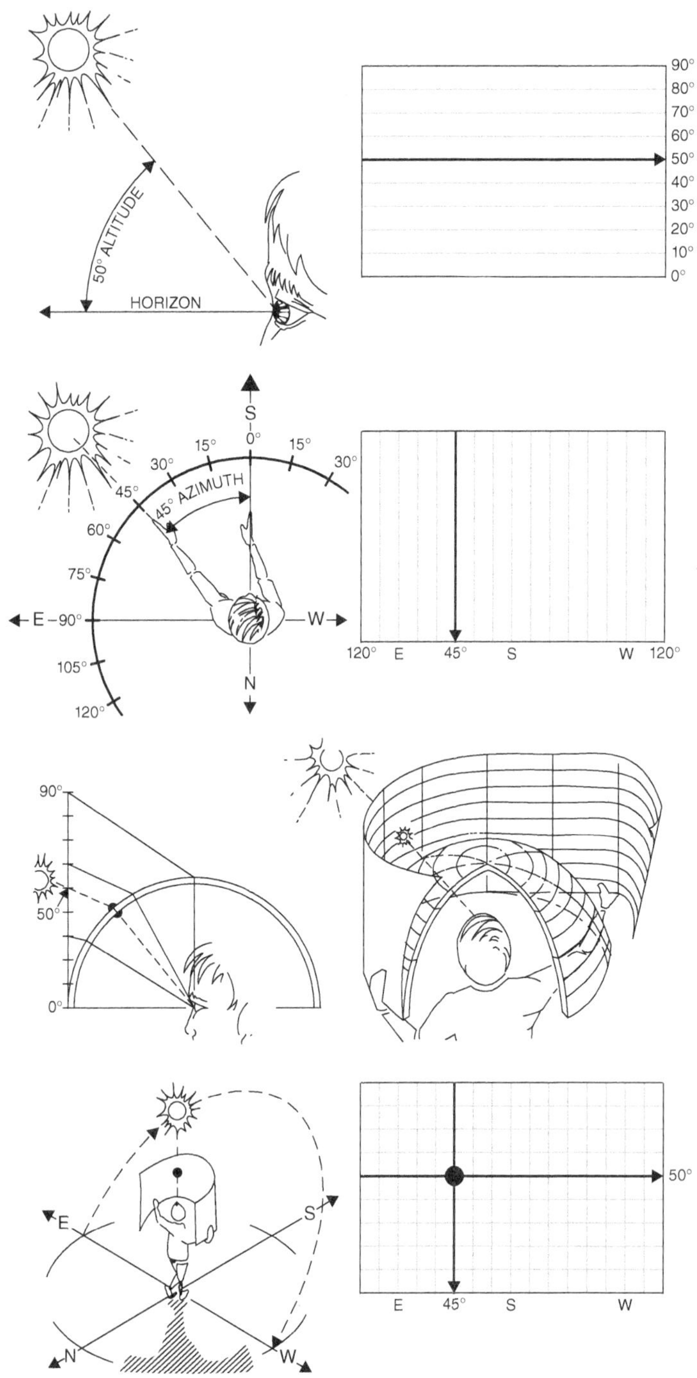

(*a*) Altitude

Two coordinates are used to locate the position of the sun in the sky. They are called the *altitude* and the *azimuth* (also called the bearing angle).

Solar altitude is the angle, above the horizon, of the position of the sun measured from the horizon. The horizontal lines on the chart represent altitude angles in 10° increments above the horizon.

(*b*) Azimuth

Solar azimuth is the angle, along the horizon, of the position of the sun measured to the east or west from true south.

(*c*) Skydome

The grid on the chart thus represents vertical and horizontal angles of the whole skydome. It is as if there was a clear dome around the observer, and then the chart was peeled off this dome and laid flat.

(*d*) Sun Position

With the altitude and azimuth angles, the sun can be located at any point in the sky.

Figure 11.12 *Illustrations (a) through (g) describe the development of the rectilinear sunpath chart. (h) U.S. map showing cities along similar latitudes. (From Mazria and Winitsky, 1976.)*

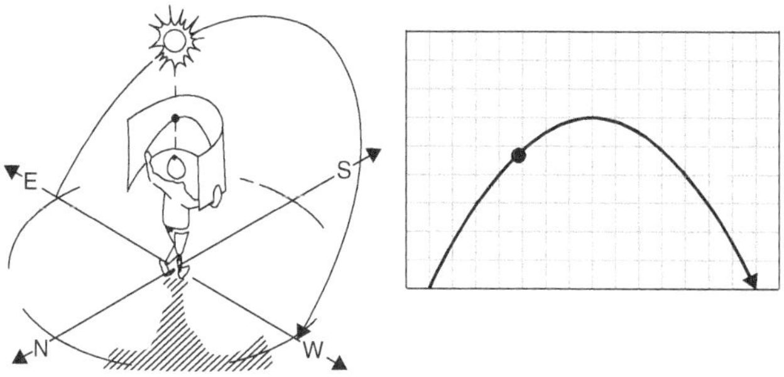

(*e*) Sun Path

By connecting the points of the location of the sun throughout various times of the day, the sun's path for that day can be drawn.

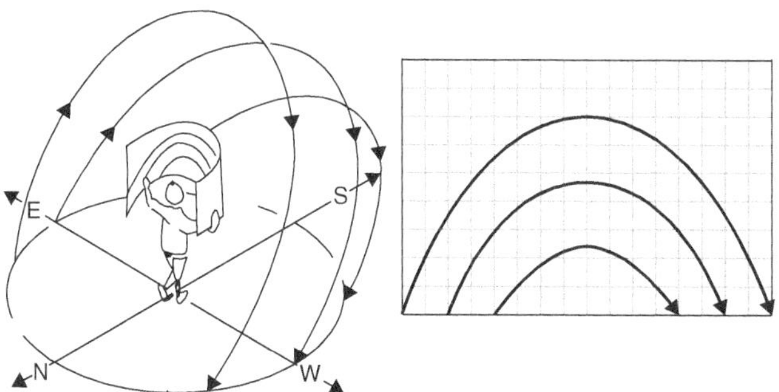

(*f*) Seasonal Paths

Thus, we can plot the sun's path for any day of the year. Those paths shown represent summer, fall/spring, and winter. The sun path is greatest during the summer when it reaches its highest altitude and rises and sets with the widest azimuth angle from true south. During winter, the sun is much lower in the sky and rises and sets with the narrowest azimuth angles from true south.

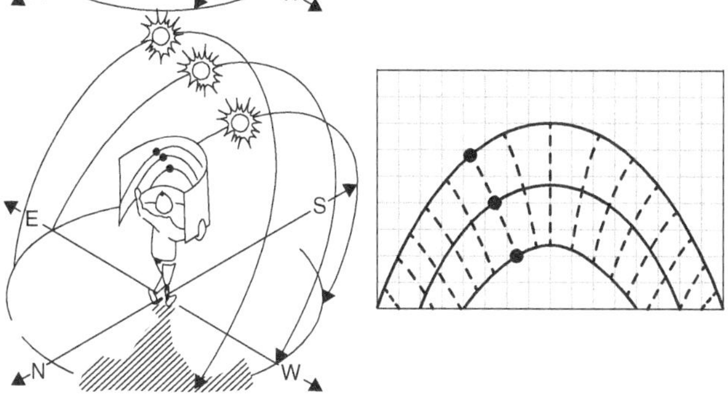

(*g*) Times of Day

Finally, if we connect the hour lines on each path, we get the heavy dotted line which represents the hours of the day. This completes the sun chart. NOTE: The times on the sun chart are for solar time. This may vary from standard time as much as an hour and fifteen minutes for different locations and different times of the year. This is adequate for most practical uses of the sun chart. It is important to remember to at least use standard time when using the charts.

(*h*)

Figure 11.12 *(continued)*

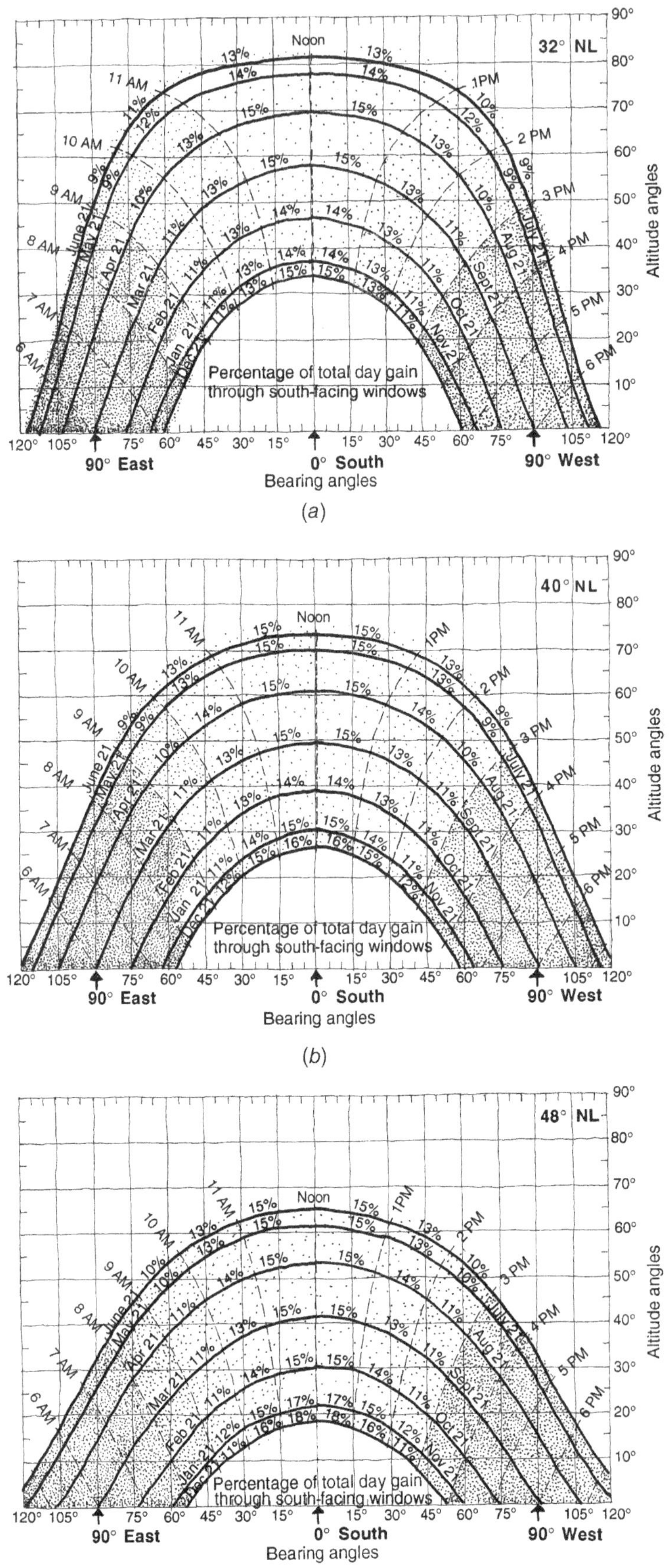

Figure 11.13 *Sunpath charts for three latitudes (32°, 40°, 48°) show the approximate percentage of clear-day insolation for south-facing windows for each of the maximum hours of sun each month.*

south orientation (in the Northern Hemisphere) calls attention to its useful characteristic of receiving more sun in winter and less sun in summer than any other orientation. To convert these percentages to actual heat gains, see Appendix C, Tables C.1 to C.10 (for clear days at a given latitude) or Tables C.19 and C.20 for average day data by climate station (January and July only).

4. The *stereographic projection* (circular, equal spacing) also represents the sun's changing position in the sky throughout the day and year. The format is like a fish-eye photograph of the sky pointed to the zenith directly above. The various paths of the sun throughout the year can be projected onto the sky, then "flattened." The principal advantage of this projection is that it is very simple to draw sunpaths at any scale and for any latitude without having to rely on published sunpaths. Thus, for any serious sunshading design, a designer can prepare a large-scale sunpath chart at the exact latitude of a project and do accurate graphical analysis. For a detailed description of the method of preparing these charts, see Szokolay (1980).

The design of sunshading devices necessitates projecting the three-dimensional solar path onto a two-dimensional surface. Stereographic and equidistant projections are the most popular for sunshading analysis.

11.5 SHADING

Shading windows from solar heat gain is a key design strategy for passive cooling and to reduce cooling loads on HVAC systems. Shading the opaque building envelope is also important, but since thermal resistance is usually greater through such elements than through glazing, the discussion in this section refers primarily to design strategies for shading windows oriented south, east, and west. Northern windows often also need shading devices, contrary to the myth that the north façade never receives direct beam radiation.

(a) Shading for Orientation

Because of the high altitude of the sun, the most effective shading device for south-facing windows during the summer is a horizontal overhang. On east- and west-facing windows, a horizontal overhang is somewhat effective when the sun is at high positions in the sky but is not effective at low-altitude angles. Various shading devices are illustrated in the diagrams and examples in Figures 11.14 through 11.18.

Direct solar gain through east- and west-facing windows can be an extraordinary heat gain liability and produce thermal and visual discomfort. During the early design stages, it is best to orient spaces to face north or south to avoid the east/west sun's low

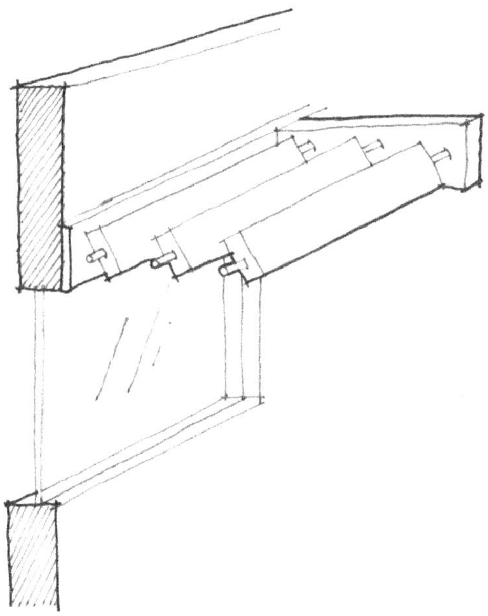

(a)

(b)

Figure 11.14 *(a) Overhang with horizontal louvers. (b) Horizontal louvers allow free air movement, as seen on an office building in Honolulu, Hawaii. (Photo by Alison Kwok; © Alison Kwok; all rights reserved.)*

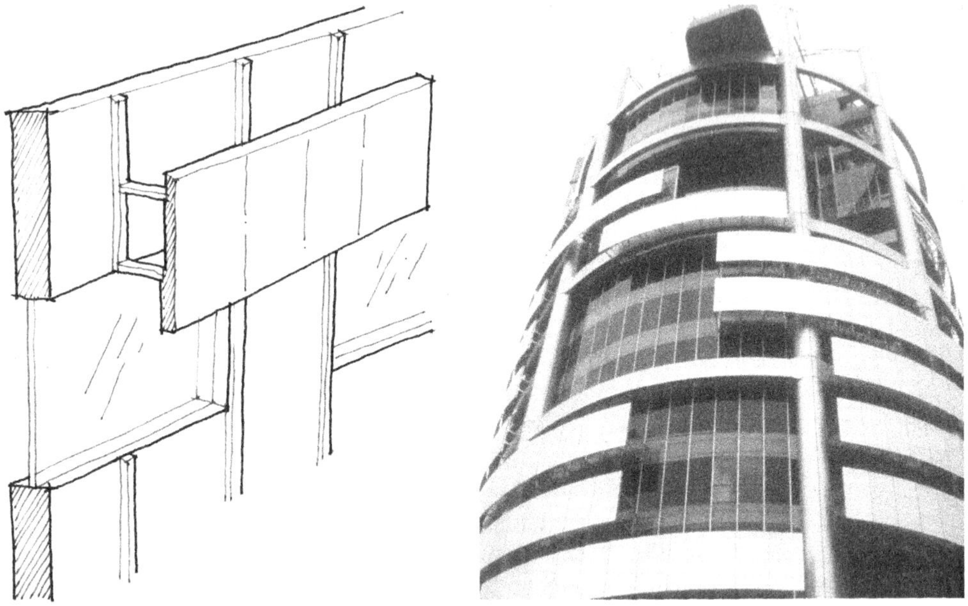

(*a*) (*b*)

Figure 11.15 *(a) Overhang using a series of vertical panels. (b) Application of horizontal overhangs at the IBM Tower in Kuala Lumpur, Malaysia. (Photo by Alison Kwok; © Alison Kwok; all rights reserved.)*

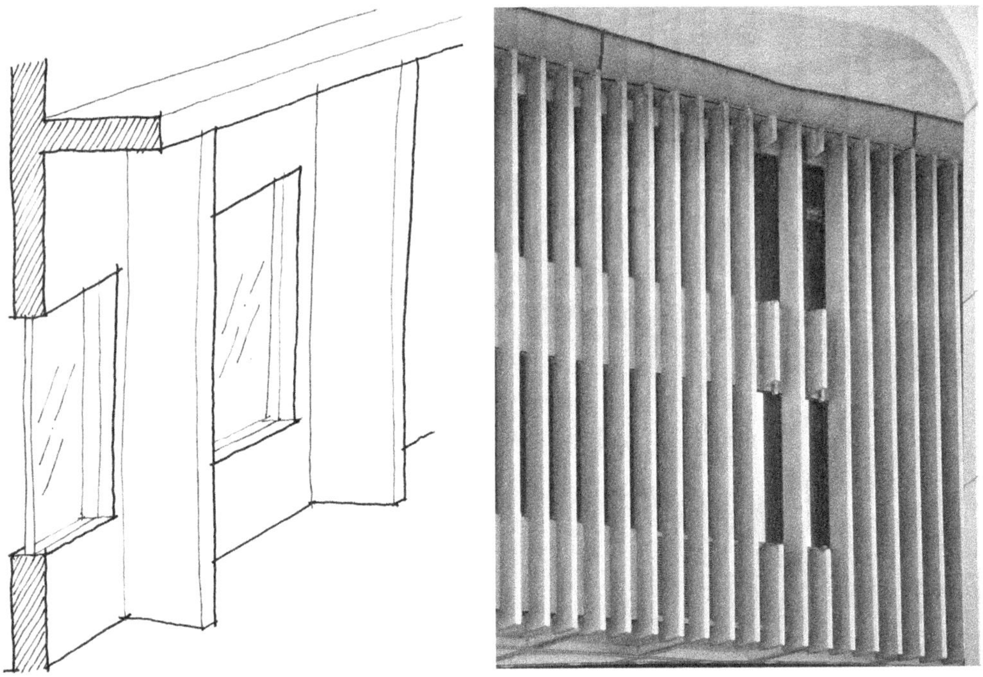

(*a*) (*b*)

Figure 11.16 *(a) Vertical fins. (b) Fixed vertical fins on a government building in Honolulu, Hawaii. (Photo © Alisa Kwok; used with permission.)*

angle. If this is not possible, vertical fins are an effective strategy for east and west orientations. Eggcrate shading devices (a combination of overhangs and fins) provide optimal shading, particularly in hot climates (Figure 11.19). North-facing windows receive direct solar radiation in the summer in the early morning and near sunset, when the altitude of the sun is very low. For shading on the north side at these times, vertical fins are most effective.

(b) Operable Shading Devices

Operable exterior shading devices are useful because they respond to daily and seasonal variations in solar and weather patterns in ways that fixed shading devices

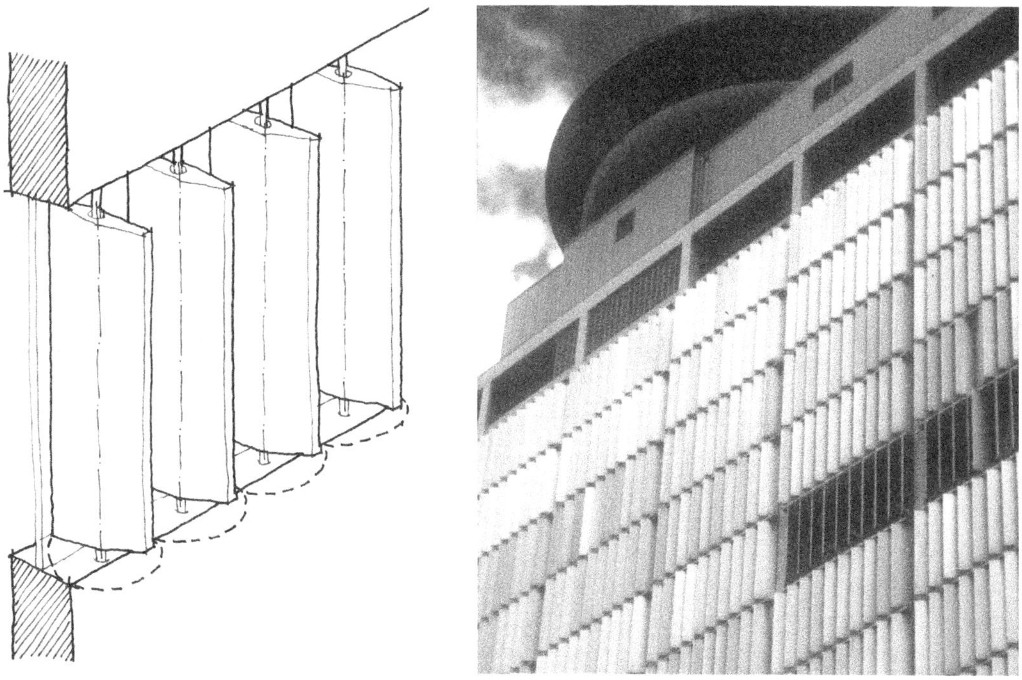

(a) (b)

Figure 11.17 *(a) Movable fins. (b) Movable fins are positioned according to the sun's position at the Ala Moana office building in Honolulu, Hawaii. These effective shading devices were removed when the building was remodeled to make it look more contemporary. (Photo by Alison Kwok; © Alison Kwok; all rights reserved.)*

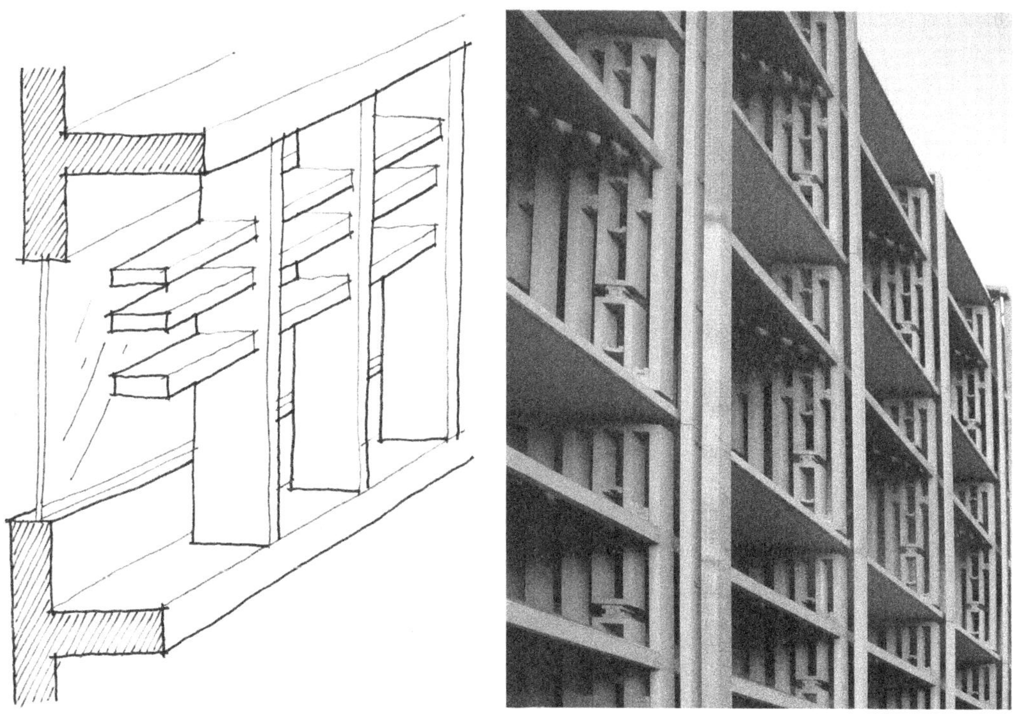

(a) (b)

Figure 11.18 *(a) Eggcrate shading device. (b) Modified eggcrate to allow air movement and lighter structure at the Board of Water Supply in Honolulu, Hawaii. (Photo © Alisa Kwok; used with permission.)*

(a)

(b)

Figure 11.19 Integration of eggcrate shading devices into the envelope at (a) the Hawaii Medical Services Association building in Honolulu, Hawaii, and (b) the University of Arizona library in Tucson. (© Alison Kwok; all rights reserved.)

simply cannot do. The operation of a movable shading device can be as simple as twice-a-year adjustment—for example, manually extending roller shades, awnings, rotating fins, or louvers at the beginning of summer and retracting the shade after the hot season has ended (in fall). These devices are very effective at blocking low sun angles from the east or west. More complex movable devices are typically on automated daily and seasonal programs. Although many facility managers are of the opinion that movable exterior shading devices require high maintenance and are prone to malfunctioning, the designer can apply appropriate technology to provide a low-maintenance solution.

Designers can also situate deciduous trees and vines at key locations around a building, which will act as natural shading devices with a natural cycle for shade in the summer and loss of leaves in the winter (Figure 11.20). Plants have the advantages of costing little, reducing glare, providing an attractive connection to nature, and reducing exterior surface (wall and ground) temperatures.

Figure 11.20 Vines grow on an exterior structure at the Finnish Embassy in Washington, DC. (© Erica Ling; used with permission.)

The City of Melbourne, Australia, has worked to become one of the world's greenest cities, initiating several programs to support commercial building owners' efforts to retrofit their properties with modern, energy-efficient technologies; to assist residents with recycling and water conservation; and to implement an urban tree plan, pilot programs for green roofs, water catchment, and green transport. Many buildings in the last two decades have developed unique shading devices while also promoting a harmonic coexistence (and interdependence) with nature. (Figures 11.21 through 11.26.)

11.6 SHADOW ANGLES AND SHADING MASKS

Altitude and azimuth angles are very useful in understanding solar position and sunpath diagrams, but are much less useful in defining the shadow angles cast by a projection on a wall exposed to the sun. More angles come into play with shading device design.

(a) Shadow Angles

When designing shading devices, the geometry of the device itself and its relationship to the face of a building produce a number of angles relative to the desired shadow being cast. Since there is no universal nomenclature for these relationships, the angles involved in the design of shading devices will be very carefully defined. Refer to Figure 11.27, which shows the shadow cast by a horizontal overhang on a wall exposed to sunlight. Note that the shadow is defined by two angles: the vertical shadow angle (VSA), which indicates the position of the leading edge of the shadow as defined from the leading edge of the overhang, and the horizontal shadow angle (HSA), which defines the leading edge of a shadow cast by a vertical element

(a) (b)

Figure 11.21 (a) The Forte Apartments building (Lend Lease) features solar control, rainwater catchment, vegetable gardens, and bike sharing, and is the world's largest cross-laminated timber building. (b) Overhang and fin design create dramatic shadows on the façade. (© Toshi Woudenberg; used with permission.)

(a) (b)

Figure 11.22 (a) Sandblasted glass disks make up the exterior shading skin of the Design Hub at the Royal Melbourne Institute of Technology (RMIT; Sean Godsell Architects). (b) The disks can rotate and meet climate and solar needs. (© Toshi Woudenberg; used with permission.)

(a) (b)

Figure 11.23 *(a) Colorful, jagged shading panels cover the exterior of the Pixel Building (Studio 505). (b) The fragmented fins are one of many strategies for this high-performance building, which is designed to capture precious rainwater and filter it through native grasses covering 75% of the roof's surface. Note also the vertical axis wind machines. (© Toshi Woudenberg; used with permission.)*

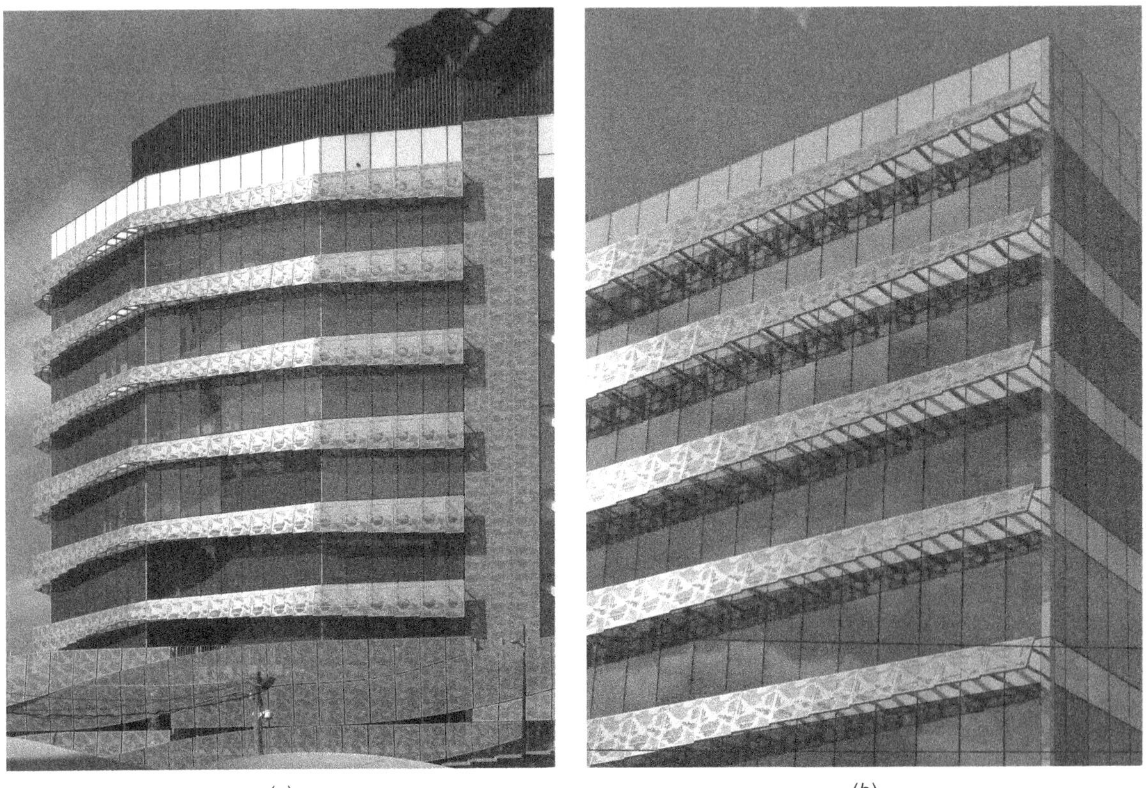

(a) (b)

Figure 11.24 *(a) Overhang shades undulate across the Melbourne Water Headquarters (Woods Bagot). (b) Sunshades are screens shading a substantial amount of glass. The building also features innovative water-conserving toilets. (© Toshi Woudenberg; used with permission.)*

(a) (b)

Figure 11.25 (a) A triangular pinwheel grid system on Federation Square (LAB Architecture Studio) employs a series of fractal shapes intended to be viewed as a dynamic façade. (b) Three cladding materials: zinc, glass, and sandstone tiles are used on the façade. (© Toshi Woudenberg; used with permission.)

(a) (b)

Figure 11.26 (a) A biomimicry concept was developed for Council House 2 (CH2; Melbourne City Council and DesignInc) as a study of mediation between the environment and its occupants in a 10-story office building. (b) Operable, recycled timber shutters shade against the low, direct rays from the west. (© Toshi Woudenberg; used with permission.)

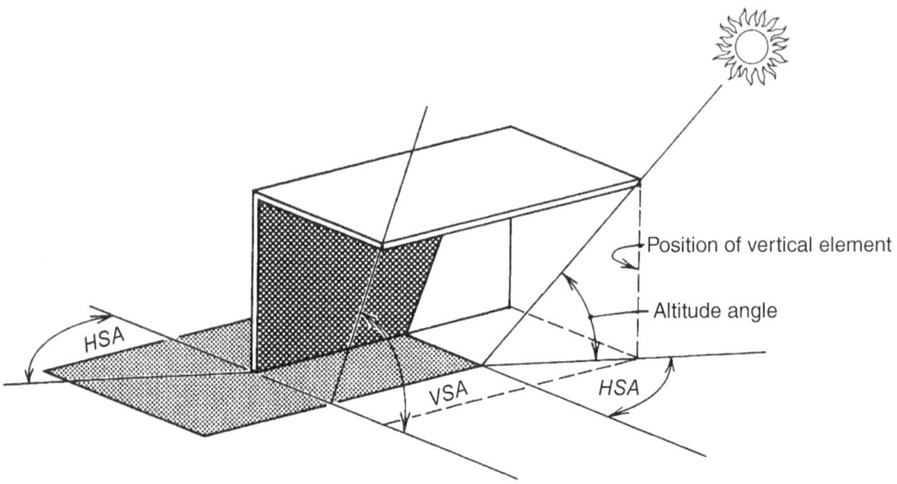

Figure 11.27 The shadow cast by a horizontal overhang is best defined by the vertical shadow angle (VSA) and the horizontal shadow angle (HSA). The VSA is also known as the profile angle in the United States.

(indicated by a dashed line) as defined with respect to that element's leading edge. The variables VSA and HSA are in use throughout the world, although in the United States the VSA is also known widely as the profile angle because it is so designated on the Pilkington Sun Angle Calculator.

Figure 11.28 shows the same information in slightly different form; line *DA* is the shadow cast by line *DE*, which is the intersection between a horizontal projection and a vertical projection. This line is particularly important in determining the required extent of a shading element, as will be shown in the following sections.

(b) Shading Masks

A shading mask is a sunpath chart (horizontal projection) that shows the shadow cast by a particular shading device, as shown in Figure 11.29. For a horizontal shading device, the leading edge of the shadow

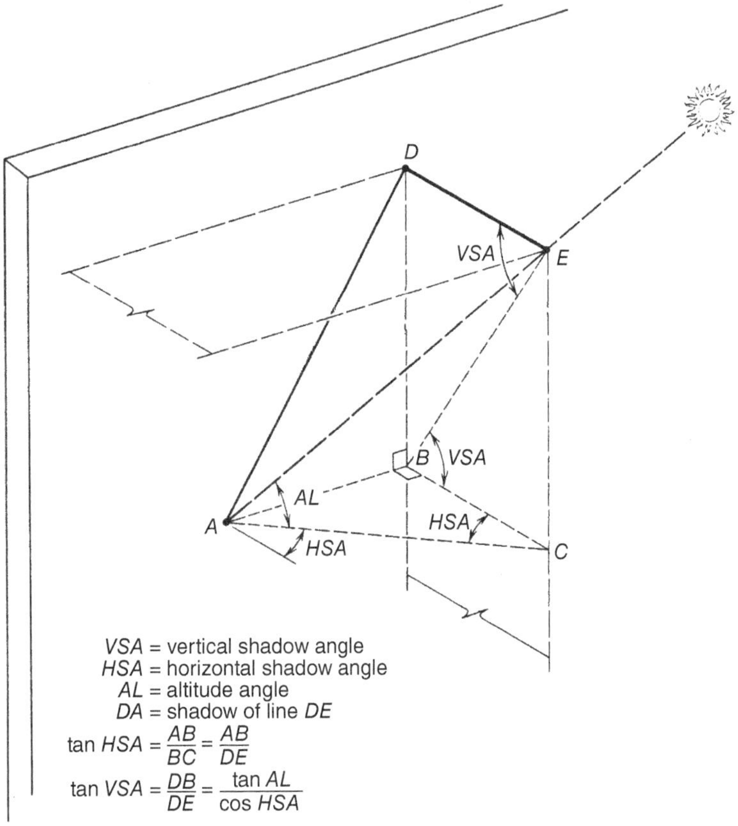

VSA = vertical shadow angle
HSA = horizontal shadow angle
AL = altitude angle
DA = shadow of line *DE*

$$\tan HSA = \frac{AB}{BC} = \frac{AB}{DE}$$

$$\tan VSA = \frac{DB}{DE} = \frac{\tan AL}{\cos HSA}$$

Figure 11.28 The shadow DA cast by the intersection line DE between a horizontal and a vertical shading element defines the edge of each of these shadows. DE can also be thought of as a pin, normal to and extending from the wall, casting a shadow, DA, on the wall. The location and size of line DA are best defined in terms of angles VSA and HSA.

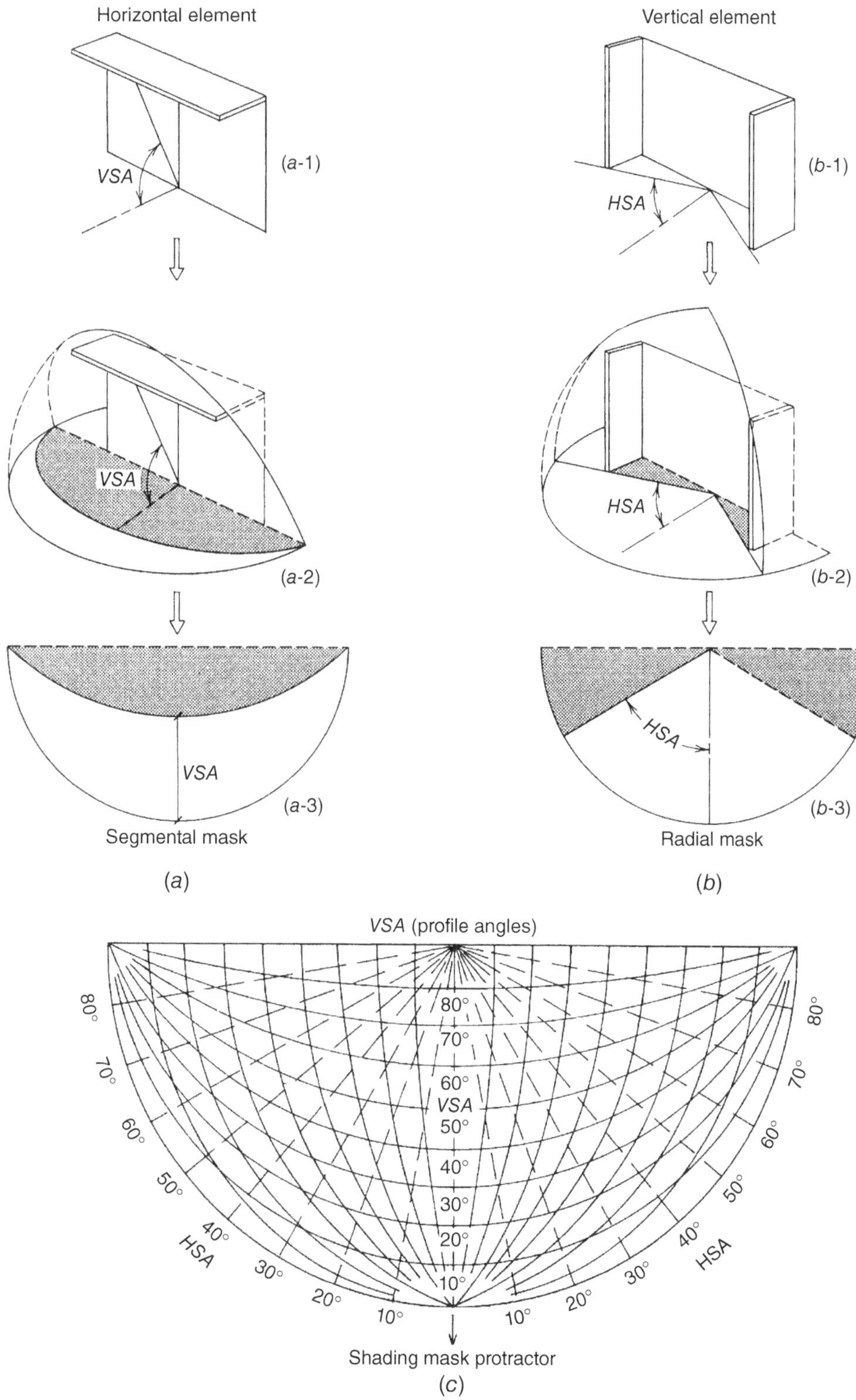

Figure 11.29 A shading mask is the horizontal projection of the shadow cast by the elements with the same VSA or HSA. Thus, any infinitely long horizontal element with the VSA shown in (a-1) will have a shading mask, as shown in (a-3). Similarly, any infinitely high vertical elements with the HSA shown in (b-1) will have the shading mask shown in (b-3). A protractor (c) is required to draw the segmental horizontal element mask properly.

cast by all horizontal elements with the same VSA (profile angle) projects as a segmental line, and when plotted on the sunpath chart, it is shown as a segmental mask. To draw this segment, a protractor is required. The Pilkington Sun Angle Calculator includes a profile angle protractor (an overlay to the sunpath chart) to draw the required segment. With no shading, there is no shadow, the VSA is 90°, and no segment exists. With a very deep overhang, the VSA approaches 0° and the unshaded area shrinks.

For vertical shading elements, such as fins (Figure 11.29*b*), the leading edge of vertical shading elements forms a shadow, shown as radii projecting from the center, which forms an angle HSA, from a line normal to the wall. These radial lines can be drawn with the assistance of an ordinary protractor or by use of the protractor in Figure 11.29*c*. The full segments and full radial lines of the shading masks are of infinitely long elements, which, of course, do not actually exist. Shading masks that appear in the literature (Olgyay and Olgyay, 1951; Ramsay and Sleeper, 2016) are frequently drawn as if for infinite elements, and the masks must be truncated for real-world shading design.

(c) Use of Shading Masks

A shading mask is drawn on some transparent medium (e.g., paper or plastic sheet) and laid on top of a sunpath diagram for the proper latitude drawn to the same scale (see Figure 11.30, which illustrates the placement of a shading mask). Its center point is placed on the center point of the sunpath diagram, and it is rotated until its facing direction (the direction normal to the wall) is aligned with the appropriate azimuth line on the sunpath diagram. Assuming that the shading mask has been correctly drawn to provide shading for the window (typically 50% or 100% coverage), the shaded hours for the window and device in question can then be read directly from the underlying sunpath diagram, as summarized in Table 11.1. (This is why the shading mask must be drawn on a transparent medium.)

It is extremely important to note that a horizontal overhang shading element that has the same width as a window can provide full shade for the window only when the sun is exactly opposite the window (when the solar-window azimuth is 0°). This situation occurs for only an instant (Figure 11.31). At all other times,

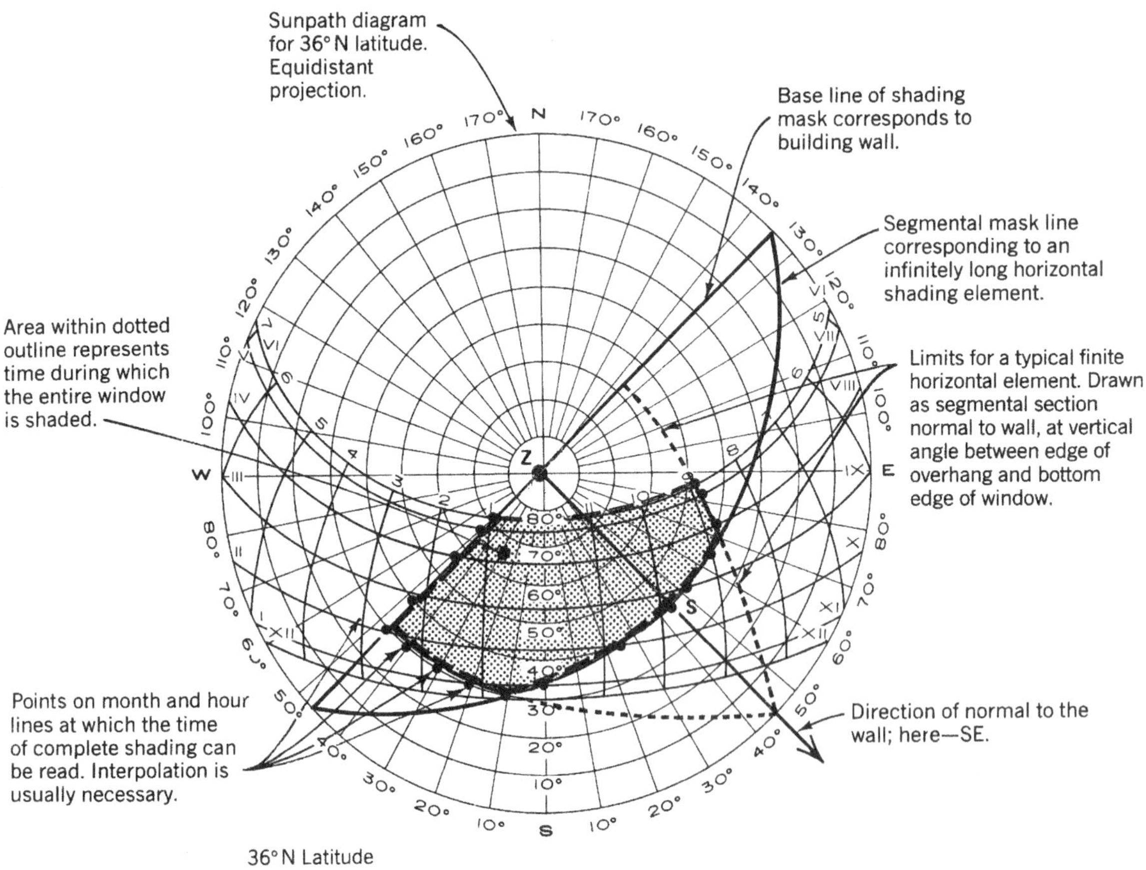

Figure 11.30 *A shading mask for a typical horizontal overhang on a window is laid over a sunpath diagram to permit determination of the shaded hours. The mask is drawn to provide 100% shading for the entire window. The hours during which the entire window is shaded are determined by the intersection of the mask perimeter with the date/hour lines of the sunpath diagram and are tabulated in Table 11.1. Only during the hours falling within the dotted area is a window fully shaded by an overhang element that is as wide as the window. The shading element shown here is not symmetrical about the center, indicating that an extension to the left was used to provide a larger shade period after noon. This is typical for east-facing windows and is reversed for western exposures.*

TABLE 11.1 Time of Day When Window Is Fully Shaded

	With Required Overhang Extensions		With Element Same Width as Window
Date	**From**	**To**	**Time**
21 June	0900	1240	1115
21 July/May	0850	1250	1100
21 Aug./Apr.	0845	1330	1035
21 Sept./Mar.	0940	1400	0955
21 Oct./Feb.	1040	1410	—
21 Nov./Jan.	1200	1300	—
21 Dec.	1230		—

some part of the window will be exposed to the sun. To provide full shading for more than an instant with a horizontal element and no vertical (side) elements, an overhang must extend beyond the sides of the window. The amount of such an extension can be determined both graphically and analytically. Since graphic solutions are amply treated in the literature (Harkness and Mehta, 1978; Lim et al., 1979; Cowan, 1980), the focus here is the analytic solution. In reviewing this discussion, keep in mind the requirement that shading masks for 100% coverage must be prepared so that the extremities of the opening are shaded.

(d) Designing Finite Horizontal Shading Devices

Although any percentage of window shading coverage is possible, most shading devices are designed to give either 50% or 100% coverage. Deciding upon shading coverage is the first step in design. The second step is to establish the required depth of the shading device (the distance it projects from the wall). Figure 11.32

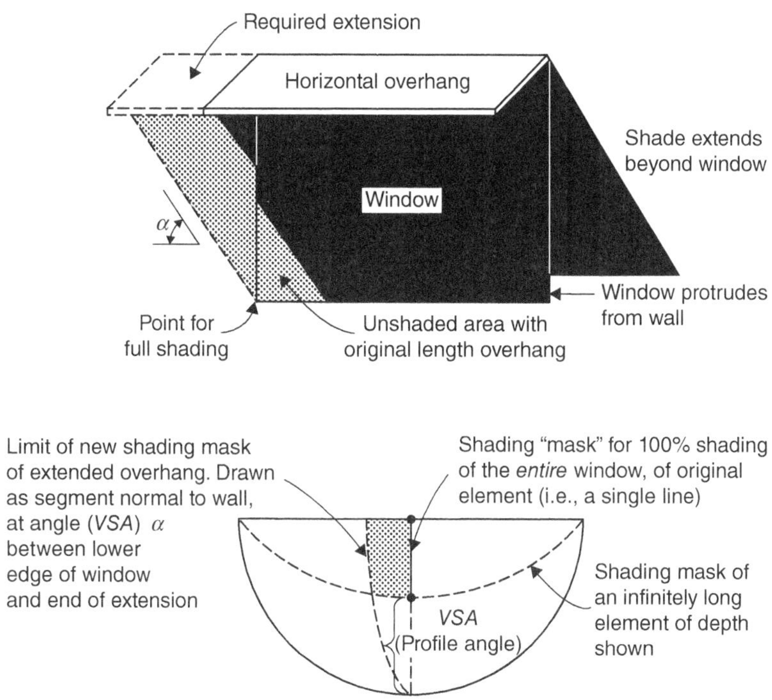

Figure 11.31 *A horizontal overhang with sufficient depth to provide 100% shading can provide full shading only when the sun is exactly opposite a window. At any other time, part of the window will be exposed to the sun. Full shading from low early morning or late afternoon sun is not possible with horizontal elements.*

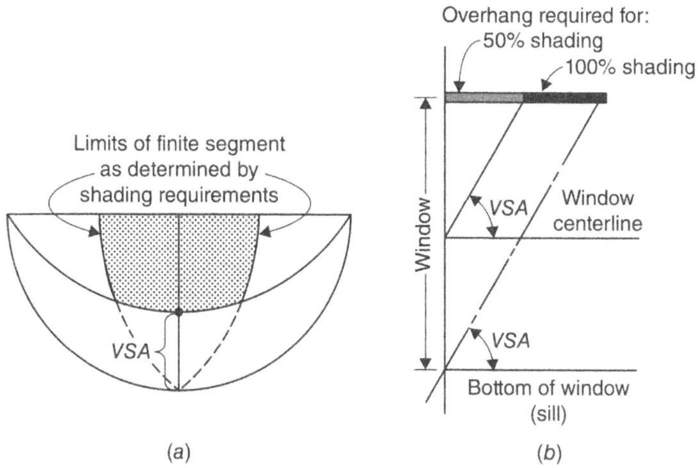

Figure 11.32 *One way to find the required depth of a horizontal overhang is to establish the required segmental shading mask (a) and read the VSA (profile angle) off the protractor that corresponds to the segment. Draw a wall section (b) to scale with the VSA. The VSA can be drawn for 50% shading coverage, 100% coverage, or any other value. The depth of the overhang can be measured directly from the section. Alternatively, if the overhang depth is known, the segment can be drawn.*

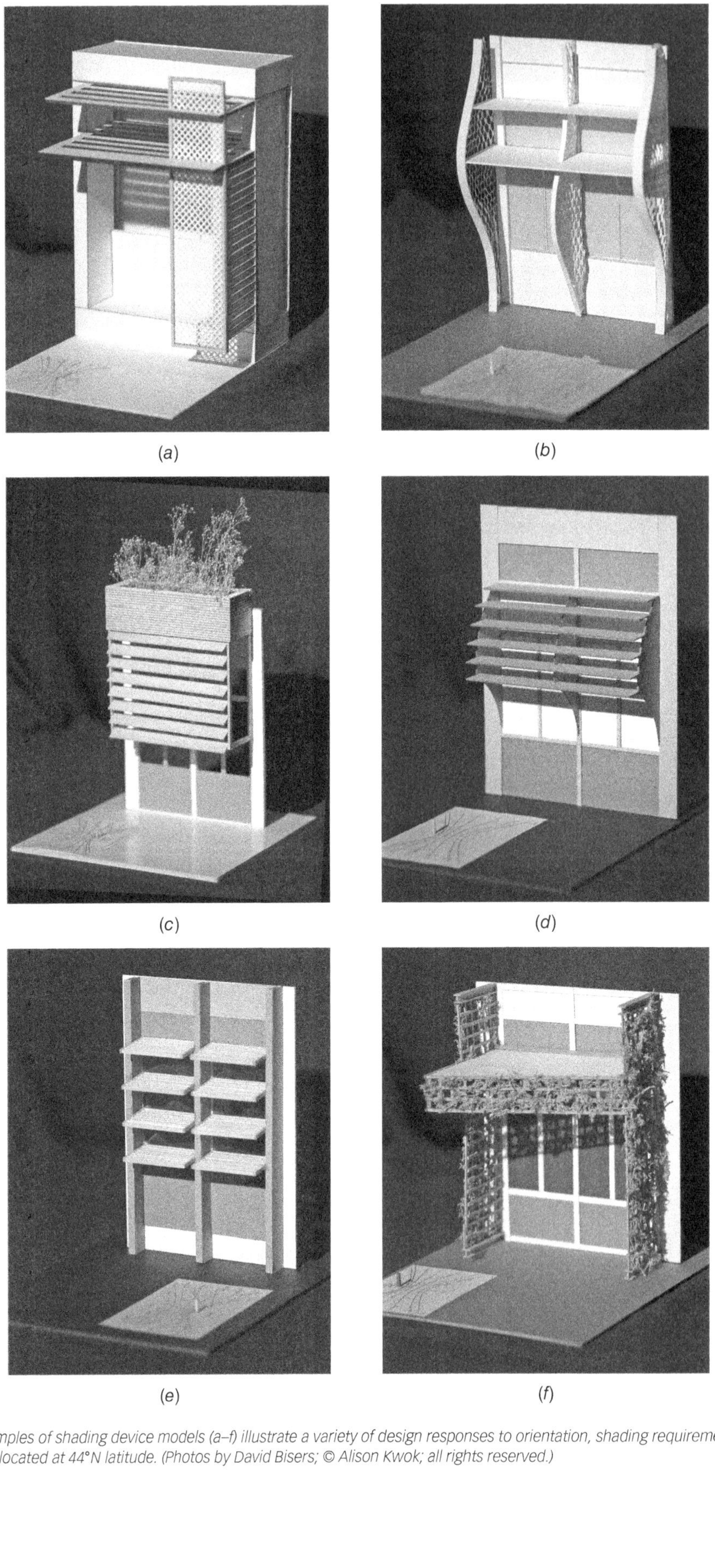

Figure 11.33 *Examples of shading device models (a–f) illustrate a variety of design responses to orientation, shading requirements, and view for a university building located at 44°N latitude. (Photos by David Bisers; © Alison Kwok; all rights reserved.)*

shows how the required depth and the corresponding segmental mask are determined. At times, because of window areas left unshaded by the original overhang, it is necessary to determine the required side extensions beyond the window's edges.

Due to the symmetry of solar motion about the ecliptic, the position of the sun is symmetrical on both sides of its maximum and minimum positions, that is, the solstices. Thus, the sun's position on May 21 is the same as on July 21, since both dates are one month from the summer solstice. As a result, any fixed shading device will give the same shade in the spring (before June 21) as in the summer (after June 21). Since in many locations spring is cool, desired late summer shading will produce spring shading that may not be desirable. Solutions to this apparent dilemma are either to use a movable or a variable-size fixed shading device or to compromise on the amount of shading—that is, to design for 50% shading for late summer (and early spring) and 100% shading for early summer (and late spring).

(e) Design Approaches

In addition to the approach just described, design might start with conceptual sketches of an appropriate shading device. As the sketches are developed into scaled drawings, a sectional drawing will show the geometry of the building. The VSA for 100% shading would be an angle from the windowsill to the outer edge of the overhang. Once the VSA is established, a shading mask can be created and overlaid onto the appropriate sunpath chart to determine whether there is adequate shading during specific times of the day and year. A more detailed description of this design process is found in Olgyay and Olgyay (1951).

Shading devices can greatly enrich the aesthetics of a building as well as improve performance. Examples of shading device models for various orientations (Figure 11.33) show a range of shading design solutions for a building in a temperate climate.

References and Resources

Brown, G. Z., J. S. Reynolds, and S. Ubbelohde. 1982. *Inside Out: Design Procedures for Passive Environmental Technologies*. Hoboken, NJ: John Wiley & Sons.

Cowan, H. J. (ed.). 1980. *Solar Energy Applications in the Design of Buildings*. London: Applied Science Publishers.

Gronbeck, C. 2009. "Sun Angle Calculator." Sustainable By Design. www.susdesign.com/sunangle

Harkness, E. and M. Mehta. 1978. *Solar Radiation Control in Buildings*. London: Applied Science Publishers.

Lim, B. P., K. R. Rao, K. Tharmaratnam, and A. M. Mattar. 1979. *Environmental Factors in the Design of Building Fenestration*. London: Applied Science Publishers.

Mazria, E. 1979. *The Passive Solar Energy Book: A Complete Guide to Passive Solar Home*. Emmaus, PA: Greenhouse and Building Design, Rodale Press.

Mazria, E. and D. Winitsky. 1976. *Solar Guide and Calculator*. Eugene, OR: Center for Environmental Research, University of Oregon.

NASS (North American Sundial Society). sundials.org

NOA (National Centers of Environmental Information). "Magnetic Declination," in . Washington, DC: U.S. National Oceanic and Atmospheric Administration. www.ncei.noaa.gov/products/magnetic-declination

Olgyay, A. and V. Olgyay. 1951. *Solar Control and Shading Devices*. Princeton, NJ: Princeton University Press.

Ramsey, C. G. and H. R. Sleeper. 2016. *Architectural Graphic Standards*, 12th edn. New York: AIA/John Wiley & Sons.

SBSE (Society of Building Science Educators). "Sun Angle Calculator." (The Pilkington Sun Angle Calculator is available through SBSE.). www.sbse.org/sun-angle-calculator

Szokolay, S. V. 1980. *Environmental Science Handbook*. New York: Halsted/John Wiley & Sons.

UO. 2024. *Sun Path Chart*. Eugene, OR: University of Oregon, Solar Radiation Monitoring Laboratory. solardata.uoregon.edu/SunPathChart.html

USNO. *Complete Sun and Moon Data for One Day*. Washington, DC: U.S. Naval Observatory. aa.usno.navy.mil/data/RS_OneDay

PART IV

PASSIVE ENVIRONMENTAL SYSTEMS

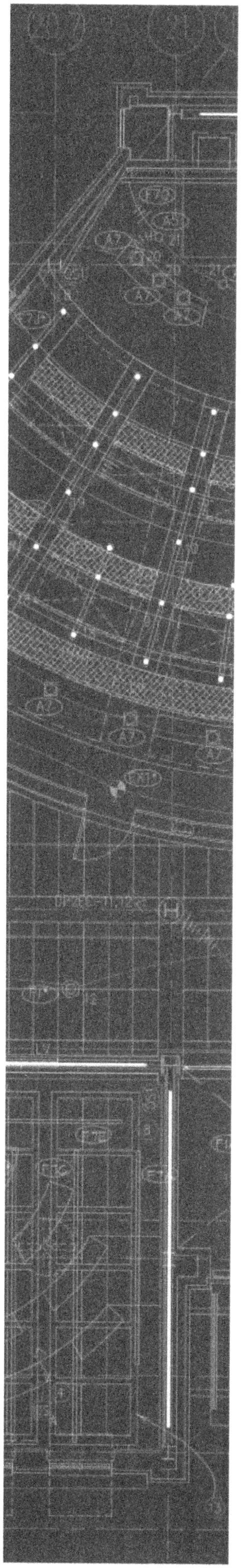

The passive environmental systems explored in Part IV focus on understanding the potentials of energy flows, building form, comfort, and function derived from ambient energy resources that are essentially free. Chapters 12–14 offer guidance for building form and the verification of schematic design proposals utilizing daylight, solar energy, wind, water, soil temperatures, and the night sky as resources. Simple guidelines and detailed calculations, along with related codes, standards, and recommended practices, are presented with sample exercises and solutions. Chapter 15 presents a fully realized project and supporting calculations for achieving passive design goals.

With building operations accounting for about 40% of global carbon emissions, we argue forcefully that the use of passive strategies to produce low-energy-use, low-carbon-emission buildings is the road to a sustainable future.

CHAPTER 12

Daylighting

PASSIVE ENVIRONMENTAL SYSTEMS

DESIGNING WITH DAYLIGHT IS BOTH AN ART AND A SCIENCE. The designer, in concert with the appropriate combination of building geometries, materials, and light produced by site conditions, can produce health, well-being, and visual delight (Figure 12.1), while also achieving an intended ambience and reducing dependency on electrical energy use. Although studies have demonstrated that daylighting improves indoor environmental quality for occupants, daylighting design is often mistakenly understood to mean that an abundance of light should fill a space. Successful design involves a careful balance and control of heat gain and loss. A variety of strategies are available to control and enhance daylight through shading devices, lightshelves, glazing, atria, courtyards, and material finishes (both interior and exterior).

12.1 THE DAYLIGHTING OPPORTUNITY

(a) Importance of Daylighting Design

Historically, the architectural form of buildings, placement of windows, and location of rooms were guided by the availability of daylight as the primary source of illumination. Daylight was the only source of abundant light for buildings, provided through deep, tall windows and thick walls, and perhaps replaced (although inadequately) in the evening by the flicker of a candle flame or an oil lamp. Electrical lighting began around 1870 with the development of commercially usable arc lamps, and was given greater impetus nine years later by Edison's first practical incandescent lamp. Subsequently, building form changed dramatically with the development of fluorescent lighting technologies, allowing interiors to be uniformly lit by electric lighting systems and to function at cooler temperatures (as opposed to the higher temperatures produced by heat-intensive incandescent lamps). With such advances, buildings such as offices, shopping centers, and factories could now operate during evening hours.

Designing with daylight can improve energy efficiency by minimizing the use of electricity for lighting as well as reducing associated heating and cooling loads. Daylighting is a critical design factor to those concerned about global warming, carbon emissions, and sustainable design—in addition to visual comfort.

Research has found daylight to be an important factor influencing human behavior, health, and productivity. Windows admitting daylight provide occupants with a view and a temporal connection with the outdoors. Daylight renders the environment in a vivid range of experiences and delight. It is important for basic visual requirements to be able to view tasks and to perceive space. How daylight is delivered is in the hands of the designer at the beginning stages of design. The option of ignoring daylight—in a world characterized by high energy costs and rapidly diminishing natural resources—is no longer viable. This chapter describes daylight strategies that can be used to increase occupant satisfaction, control glare, provide appropriate vertical and horizontal illumination, and address the potential for energy savings to enable the designer to create a proper visual environment.

The Illuminating Engineering Society (IES) defines daylight's role in facilitating visual performance, desirable energy and economic outcomes, and aesthetic

Mechanical and Electrical Equipment for Buildings, Fourteenth Edition. Walter Grondzik, Daniel Overbey, and Alison Kwok.

Companion website: www.wiley.com/go/meeb14e

(a)

(b)

Figure 12.1 (a) Visually appealing daylighting elements at Chiswick Park Office Complex, London, UK. (Richard Rogers Partnership). (b) Hong Kong International Airport, Hong Kong, China (Norman Foster). (© Alison Kwok; all rights reserved.)

intent, regarding a wide range of interior and exterior spaces in *IES LP-1-20: Lighting Practice: Designing Quality Lighting for People and Buildings* (IES, 2020a).

(b) Planning for Daylight Throughout Design

Designing buildings for daylighting is a complex systems integration process. Daylighting design begins with situating a building on its site and continues through each phase of design; making the best use of daylight continues throughout a building's occupancy. While overall design goals remain generally fixed throughout each design phase, there are key concerns associated with each of the phases. For example, in the conceptual design phase, building form, orientation, layout, and major apertures might be primary elements. Further into design development, there would be specification of materials and interior finishes, as well as zoning for integration with electric lighting and other services; control systems would be coordinated with occupancy schedules, and commissioning test procedures set in place. During occupancy, fine-tuning and maintenance of the system would occur, and a post-occupancy evaluation would be conducted, in order to determine satisfaction, visual comfort, and lighting system performance.

IES outlines the various elements of lighting design throughout the building design and construction process in *IES LP-7-20: Lighting Practice: The Lighting Design and Construction Process* (IES, 2020c).

(c) Energy Savings with Daylighting

To obtain lighting energy savings in a building, a project team should consider the six essential "ingredients" for daylighting design:

1. Assign performance parameters and plan interior spaces for proper access to daylight.
2. Manage sunlight in the vicinity of critical visual tasks.
3. Design spaces to maximize daylight autonomy and minimize glare.
4. Zone electric lighting for daylight-responsive control.
5. Provide for daylight-responsive control of electric lighting.
6. Provide for commissioning and maintenance of fenestrations and lighting control systems.

Energy savings from reduced electric lighting will be compromised if any of these factors is overlooked. If any one of the ingredients is missing, a daylight strategy may make little contribution to the illumination of the space. Each of these factors is discussed in this chapter.

A detailed breakdown of how to optimize daylighting for energy savings is offered by IES in the document *IES LP-3-20: Lighting Practice: Designing and Specifying Daylighting for Buildings* (IES, 2020b).

(d) Goals of Daylighting

Improved aesthetics, provision of human biological needs (circadian rhythms and visual relief), and reduction of electric lighting energy usage are the most important advantages of daylighting a building. Key goals in daylighting design are to provide sufficient illuminance, minimize the perception of glare, and provide for overall visual comfort.

12.2 HUMAN FACTORS IN DAYLIGHTING DESIGN

The following human-related factors (as opposed to the physical aspects of light) are described briefly to illustrate the importance of considering daylight, and especially these factors in the design of daylighting strategies for indoor spaces.

The human needs served by lighting in general are defined in *IES LP-1-20* (IES, 2020a). The human health, wellness, and performance benefits of daylighting specifically are described in *IES LP-3-20* (IES, 2020b).

Chapter 5 outlines human health and wellness considerations in more detail.

(a) Direct Daylight and Views

The presence of daylight and quality views generally improves visual and psychological comfort. Windows are the de facto design element to accomplish both. There is a common belief that if a window is placed in a wall, there will be sufficient view and daylight. The view function of a window, however, is very different from the daylighting function. The most preferred views from a window include the sky, the horizon, and the ground. In offices, people enjoy having windows in their work space because of the views. The functional advantage to a window is that people can look into the distance to reduce eye fatigue after doing close desk tasks. Depending upon the type of facility, the designer should be aware of special circumstances—for example, ensuring that bedridden occupants of care facilities have views from their vantage points, providing lower sills in facilities for children (depending upon safety), or accommodating people in wheelchairs by providing low-sill windows in bedrooms and other areas.

(b) Productivity and Satisfaction

Productivity is a complex issue that is difficult to isolate or attribute to a single parameter such as daylighting. The connection to the temporal qualities of daylight improves our psychological well-being and productivity. In studies of classrooms, windows, daylight, and performance, researchers found that students with more daylighting in their classrooms progressed faster on math and reading tests than students with less daylighting. Also, sources of glare can negatively impact students' learning, and the issues of control of windows, blinds, sun penetration, and acoustic conditions are important for teachers. In another study, retail stores were found to have a "daylight effect on increased monthly sales" (Heschong Mahone Group, 1999–2003).

(c) Visual Comfort and Glare

A well-daylit indoor environment can improve visual comfort, which can affect occupant performance, productivity, satisfaction, and well-being. Daylight in indoor spaces may offer:

- Improved general visibility through appropriate illumination.
- Enhanced task performance from the improved color and object visibility fostered by full-spectrum luminance.
- Maximized circadian benefits, which can have a direct impact on the users' moods and experiences, including relaxation, stimulation, and/or a sense of safety.

Simply providing daylight does not ensure visual comfort. Improperly managed daylight can lead to discomfort in the form of glare. Glare can be described as a personal sensation produced when luminances in one's visual field are greater than the luminance to which the person's eyes are adapted—which can cause annoyance, discomfort, or loss of visual performance or visibility. Glare is a difficult problem to overcome when balancing daylight and view. Any window (including north exposures) can produce problematic glare if the window is within the field of view. High contrast ratios between a window and adjacent surfaces can occur unless the window is designed to reduce luminance ratios through the use of sunshading devices, lightshelves, high-reflectance interior surfaces, light-colored window surrounds and mullions, and low-transmittance glazing (though such glazing will reduce light flux through the window). Furniture should be oriented to work with sidelighting (as opposed to having an occupant face a window).

(d) Daylight Controls

Daylight, whether diffuse light or direct sunlight, provides significant benefits associated with psychological well-being. On the other hand, there are potential problems—such as glare or substantial cooling loads—caused by uncontrolled quantities and qualities of light. Direct sunlight is not, however, always a liability. In nontask areas, a momentary sunny patch, a streak of sunlight against a wall, or a series of multiple shapes provides visual interest and dynamism to a space. Sunlight in task areas can be controlled in a number of ways:

- Provide exterior fixed shades that exclude sunlight for all sun positions.
- Use systems that diffuse the incident sunlight sufficiently to eliminate glare potential.
- Provide occupant-controlled adjustable shades.
- Provide automated controls that use sensors to balance daylight, control glare, and integrate with electric lighting systems to optimize lighting quality and energy performance.

The use of manual versus automated daylight controls is an aspect of indoor daylighting that must be carefully considered. User controllability of lighting (both daylight and electric lighting systems) can lead to an improved sense of visual comfort and satisfaction. However, automated controls provide the greatest opportunity for building performance optimization. The best approach may be a blended approach by which daylight controls are largely automated; however, users are provided with localized controls. Keep in mind that users will always find a way for "ad hoc" daylight controls if necessary (e.g., taping paper over windows for privacy or installing temporary curtains to mitigate glare)—and this can be counterproductive to building performance.

12.3 SITE STRATEGIES FOR DAYLIGHTING BUILDINGS

Optimal daylighting opportunities depend upon a building's position on a site relative to available daylight, horizon obstructions, orientation, and building form. The quality of daylight, its effects on illumination, and solar position for passive heating and

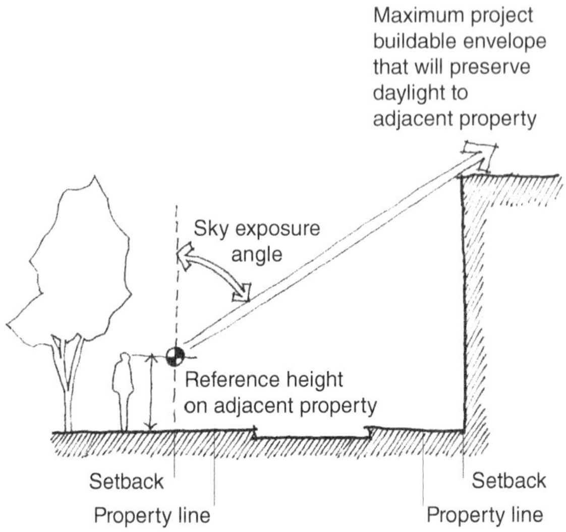

Figure 12.2 Protecting a site from obstructions to daylight.

control of cooling loads are of particular importance to the designer. Site obstructions such as neighboring buildings, trees, and landforms will determine the maximum available daylight on a site and the maximum project envelope that will preserve daylight access to adjacent properties (Figure 12.2).

(a) Orientation

In many locations, an ideal orientation for buildings is an elongated, narrow plan allowing the north and south façades of the building maximum exposure to more easily controllable daylight. From a daylighting standpoint, this is desirable because direct solar radiation received by the south façade is easier to control to prevent excess solar gain, is relatively uniform, and is necessary for passive solar heating strategies. The nearly constant diffuse skylight availability on the north façade is advantageous for uniform and soft daylighting. Figure 12.3 shows how orientation affects cooling, lighting, and heating energy for a building.

(b) Zoning

Zoning considerations are among the most important influences on building form and external design, along with the familiar aesthetic, social, legal, economic, and technical influences that combine in a tug of war familiar to the building designer.

The thermal and luminous zoning of a building recognizes that different envelopes and support systems may be required around and within the building. The more carefully zoning is considered in these early design stages, the lower the building's annual energy

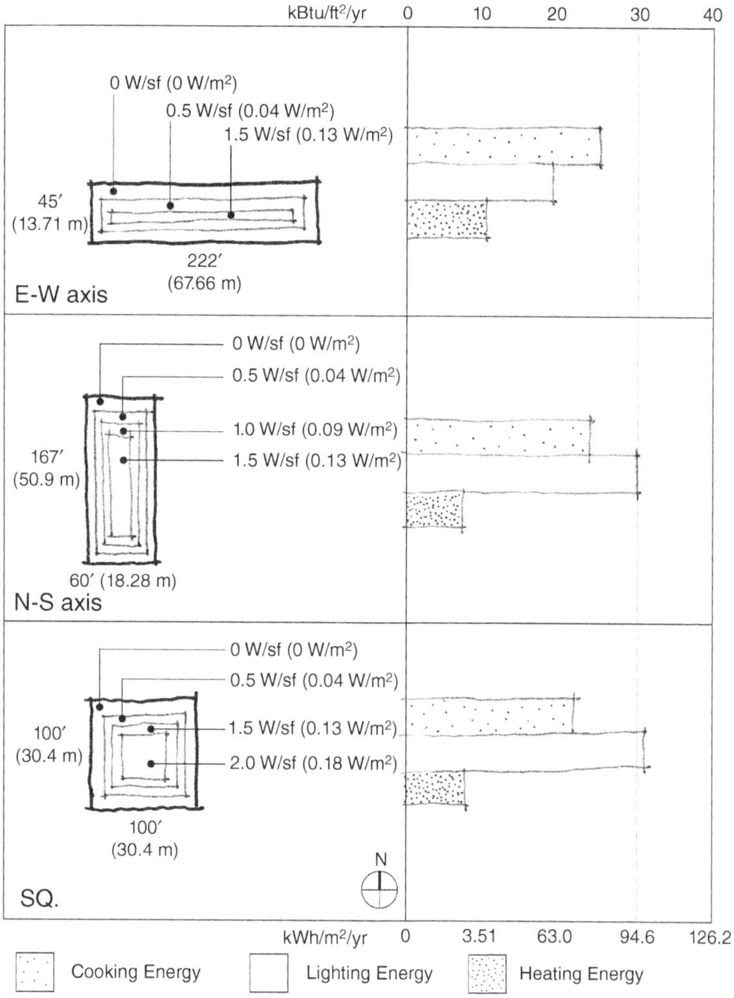

Figure 12.3 *Effect of building orientation on energy consumption.*

consumption will be because of enhanced lighting and thermal performance. Zoning is one of the first steps to passive heating and cooling design.

Zoning is most often influenced by the following visual and thermal factors:

1. *Function*: Particularly important because of the variations in internal heat gains between functions, function may also influence the zonal organization of a building, as in Figure 12.4. Comfort conditions may vary considerably between functions; air temperatures can be lower for a strenuous activity than for a sedentary activity, or heat tolerance may be greater for some activities (restaurant kitchens) than for others. Some functions thrive in daylight; others shun it. Some functions adversely affect the indoor air quality of other functions.
2. *Schedule*: Closely related to function, scheduling can influence both the envelope and the support system. An activity scheduled only between 9:00 a.m. and 4:00 p.m. can often be entirely daylit at a time when the outside temperatures are the warmest of the daily cycle. By contrast, an activity that takes place only from 9:00 p.m. to 4:00 a.m. will be entirely dependent on electric lighting, whose

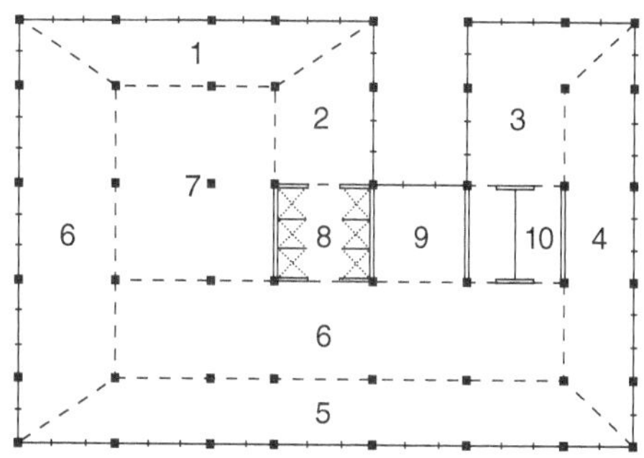

Figure 12.4 *Daylight zoning for a mixed-use building with perimeter and internal zones as well as varying types of use. Scheduling and/or internal load differences within any one of these zones could require division into additional zones.*

heat can be used to overcome the chill of the outside temperatures during these hours in winter. (In the summer, such heat can be flushed away with the cool outside night air in many U.S. locations.) Support systems are often divided by scheduling considerations: If one activity has operating hours different from those of the remainder of the building, a separate mechanical system is often provided. This saves energy, because large equipment will not be underused to provide heating or cooling for only one zone.

3. *Orientation*: The degree of exposure to daylight, direct sun, and wind is critically important to zoning. Consider the block-square office building floors (Figure 12.4) on a cold, sunny, and windy day. Perimeter spaces with direct sun through the windows may gain more heat than is lost and thus need cooling. This might be done by the opening of windows, but too much cold air (especially on the windy side of a building) may make occupants near the windows uncomfortable. Perimeter spaces without direct sun may have a net heat loss due to heat loss through glass, infiltration, and a lack of electric lighting (because daylight is adequate). These spaces will need heat from a mechanical support system. Interior (no-daylight) spaces are often overheated by electric lighting because they cannot lose heat. These spaces will need cooling from the support system.

(c) Form

At its simplest, form can be reduced to questions of tall or short, thick or thin. Figure 12.5 compares these form variations to their impacts on heating, cooling, and daylighting solutions. Thicker, taller buildings have more floor space away from climate influences; being electrically lit rather than daylit, they generate heat and need cooling all year. These buildings are called internal-load dominated (ILD). In contrast, thinner buildings—in which nearly all spaces have an exterior wall—need heating in cold weather and cooling in hot weather; electric lights by day are largely unnecessary. These buildings are called skin load-dominated (SLD).

The ultimate choice of building form is ascertained through a combination of programmatic needs, contextual constraints, and myriad design issues. In particular, design issues related to the potential for daylighting and the form's influence on heating and cooling loads have a direct impact on a building's energy performance. Because the question of daylighting versus electric lighting frequently is so influential in determining whether heating or cooling will be the dominant concern, we begin with daylighting.

Daylighting Issues

1. What will be the relative emphasis on sidelighting (characterized by uneven distribution and glare in the visual field but little glare on horizontal surfaces) and toplighting (the reverse characteristics)?
2. What role will direct sun play in daylighting? In winter, can solar heat without glare be admitted?
3. How can seasonal adjustments be made in the size of daylight openings?
4. To what extent will daily changes in daylighting control be necessary?
5. How can adequate daylight be admitted in an even way, such that unwelcome dark-appearing places are avoided?

Heating Issues

1. Can the sun be used to heat spaces? If so, how will south-wall design be affected?
2. How can openings in walls facing other directions be kept to a minimum without daylight being shut out? Where such openings are desirable, how low a fenestration U-factor can be justified?
3. What role will direct sun through south glass or skylights play in daylighting?
4. How can daylight be admitted but the chilling effects of large, cold glass surfaces be minimized?
5. How can incoming fresh air be warmed before it chills the people sitting near the fresh air opening?
6. Is there surplus heat elsewhere in the building that can be used to help warm perimeter spaces?

Cooling Issues

1. Will the strategy be to open the building to breeze or close the building for coolth retention, or to use a combination of these alternatives (open by night, closed by day)?
2. How can direct sun be kept out of the building? Can east and west windows be minimized and adequate daylight still be provided?
3. How can adequate daylight be admitted for winter conditions without overlighting (and thus overheating) for summer conditions?
4. When can cooling be provided by outdoor air rather than by a refrigeration cycle?
5. Can the operation of refrigeration machinery be concentrated during the coldest (nighttime) hours, when electric power is cheapest?
6. How can incoming fresh air be cooled before it warms the people sitting near the fresh air opening?
7. Can the structure of the building be used to absorb heat by day, then be flushed with night air in climates with cool nights?

Establishing an appropriate building form in the early stages of design is critical to daylighting

BUILDING TYPE	DAYLIGHT	HEATING	COOLING
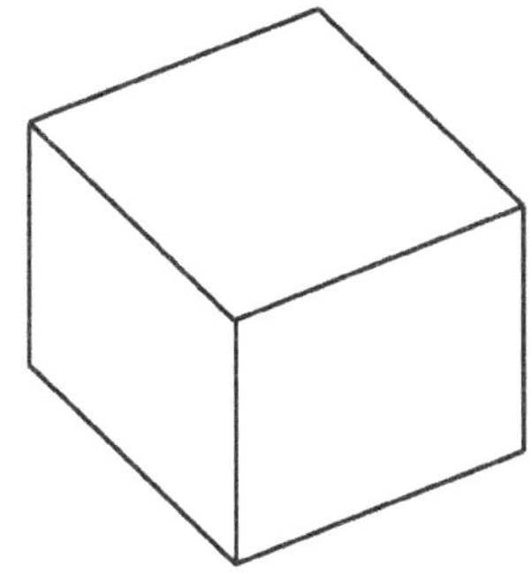 THICK TALL BUILDING	✓– Core is dark. Perimeter lighting is unidirectional	✓– Unbalanced solar gain makes perimeter zoning and control critical	✓– Large core requires mechanical cooling. Perimeter ventilates poorly
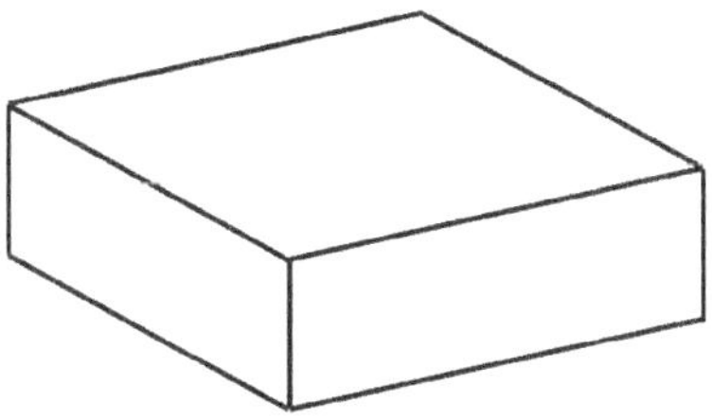 THICK SHORT BUILDING	✓ Toplights and perimeter windows can provide adequate daylighting	✓ Passive heating systems may be implemented although somewhat limited and unbalanced	✓– Passive cooling possible but costly. No cross ventilation
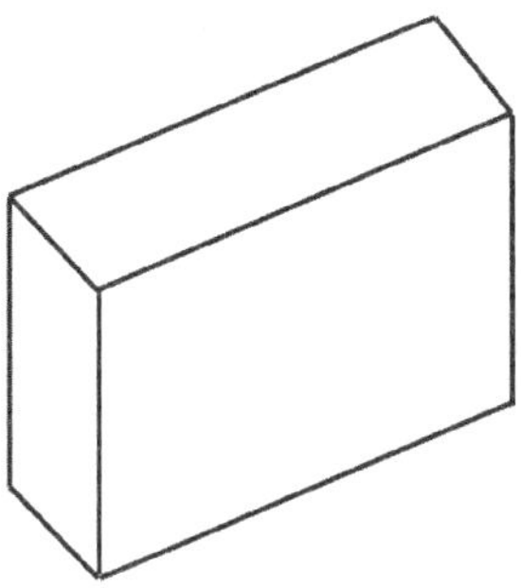 THIN BUILDING, TALL OR SHORT	✓+ Complete sidelighting for all spaces can be achieved	✓+ Passive heating and solar heat gain mitigation possible	✓+ Natural ventilation is easily accomplished
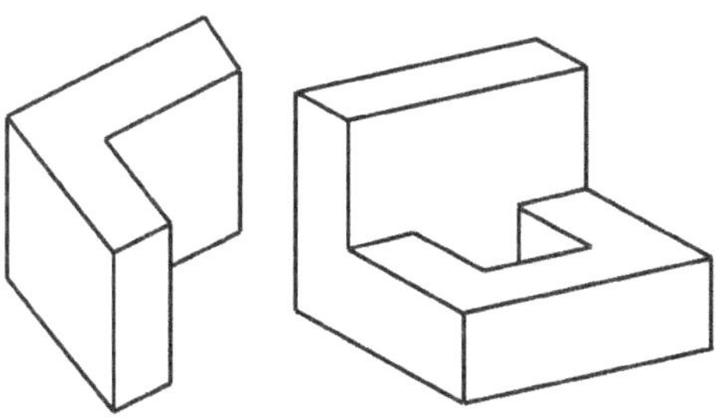 THIN COMBO BUILDING	✓ Complete daylighting for most spaces	✓+ Passive heating and solar heat gain mitigation possible	✓+ Natural ventilation is easily accomplished

Figure 12.5 *The effect of building form on environmental control strategies. The illustration shows how building layout affects cooling, daylighting, and heating opportunities.*

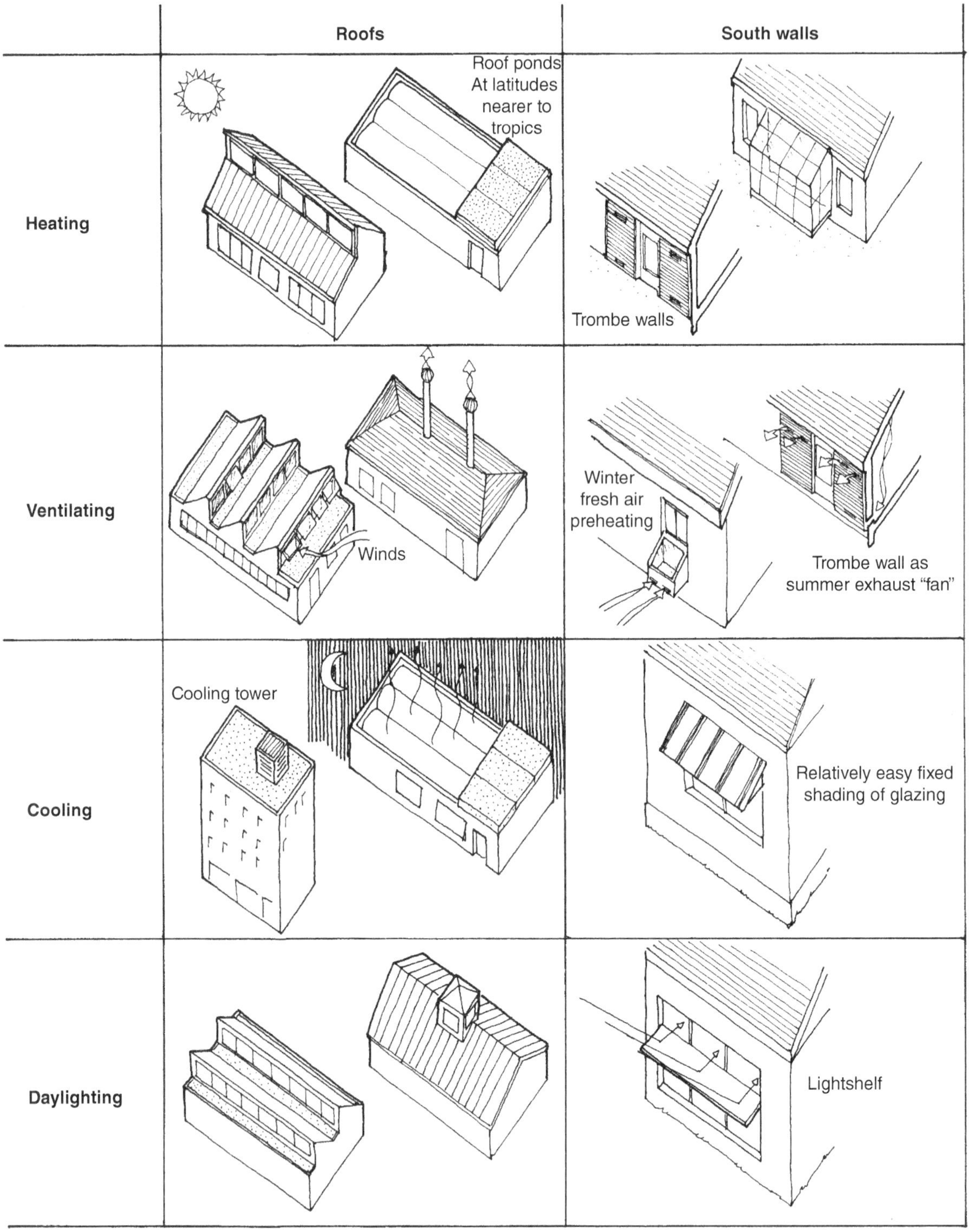

Figure 12.6 *The components of a building's enclosure can be used both to conserve energy and to admit on-site or natural energy sources.*

performance. The width of the long, narrow plan previously described will determine how much of the floor area will have access to usable daylight. Generally, a 15-ft [4.5-m]-wide perimeter zone can be completely daylit; the next 15- to 30-ft [4.5- to 9.0-m] area can be partially daylit; and beyond 30 ft [9 m] will be electrically lit; the total of which can be used to determine the width of a building.

(d) Enclosure

The next design step involves relating the climate to the design of the building's skin. Each skin element

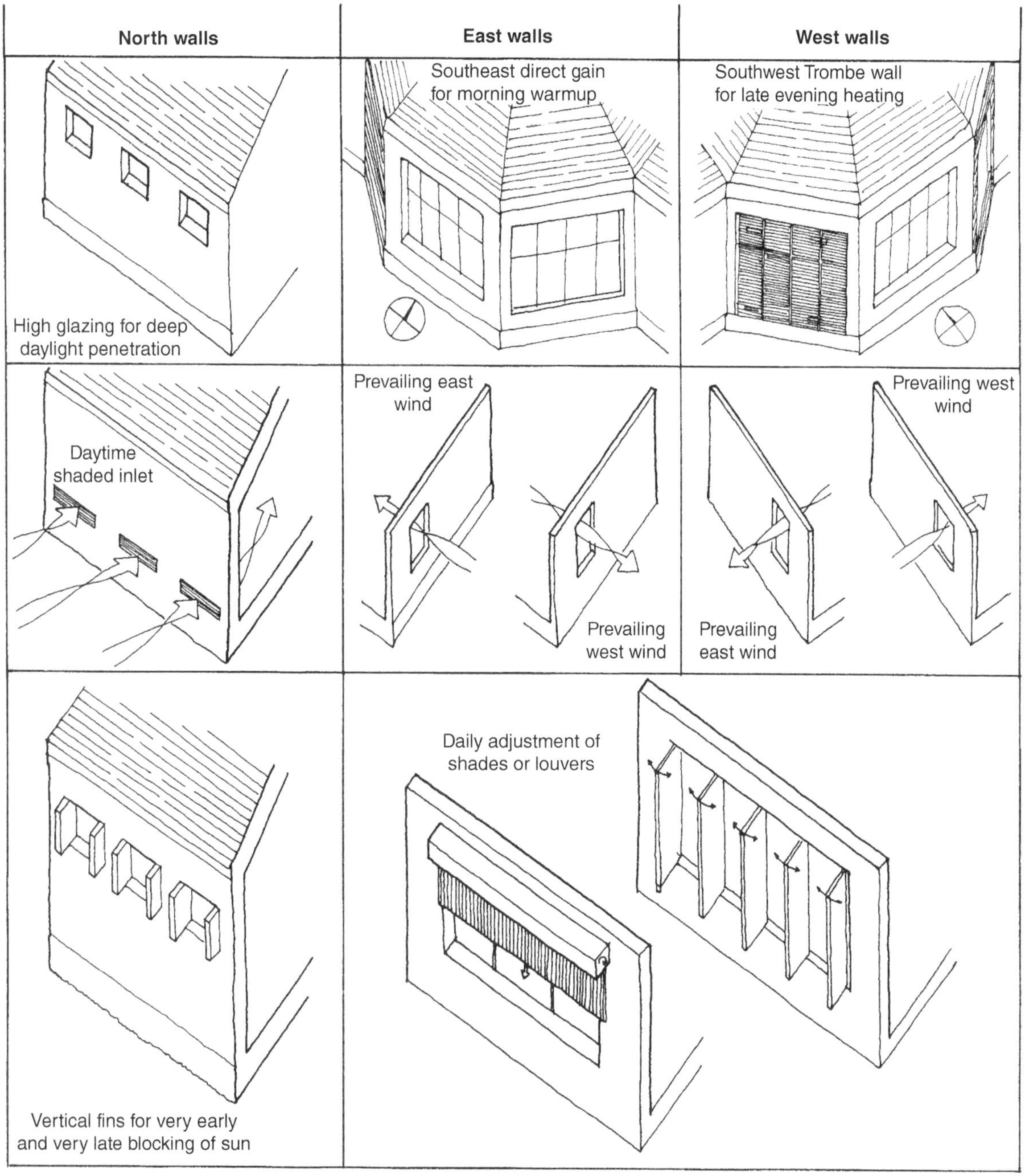

Figure 12.6 *(continued)*

provides an opportunity for thermal and luminous exchange between inside and outside; heating, cooling, ventilating, and daylighting devices can be mixed as needed. Figure 12.6 shows some of the most common of these devices for varying orientations. Sections in this chapter give numerical criteria for sizing these skin elements, with an emphasis on the use of on-site, renewable energy resources. These criteria may conflict with the size relationships that may be prescribed by codes and standards for conventional buildings.

(e) Fenestration

Codes and standards typically prescribe a relationship between floor area and fenestration area (residential buildings) or total wall area compared to the total fenestration area (nonresidential buildings). For example, the prescriptive compliance path in *ASHRAE/IES Standard 90.1: Energy Standard for Buildings Except Low-Rise Residential Buildings* calls for a maximum window-to-wall ratio (WWR) of 40%. Such prescriptions assume that a building will be designed conventionally, that is, to rely on imported energy for lighting, heating, and cooling. Thus, prescribed areas of fenestration tend to be rather small. If a designer wishes to rely on daylighting to a greater extent, then some proof of benefit will be needed.

Daylighting is accompanied by large glass areas, which increase a building's heating needs in winter; yet in such buildings, less heat from electric lighting is available to fill those needs. Another complication is that adequate daylight requires much larger glass areas under dim winter skies than under bright summer skies. If the glass area is sized for winter daylighting, then excessive daylight—and, along with it, excessive heat—might be admitted in summer. With proper controls, daylighting can reduce summer cooling loads, relative to electric lights, but it will usually increase winter heating loads. Where passive solar heating or surplus heat from another source is readily available, this trade-off is attractive.

One of the earliest and most difficult questions for the designer is how much fenestration is optimum for a building. A recent technical report by the National Renewable Energy Laboratory (NREL, 2020) indicated that an average office building energy use breakdown may be similar to:

14% space heating
10% refrigeration
9% space cooling
8% ventilation
8% lighting
4% water heating
47% plug and process loads (PPL)

The proportion of total building energy use attributed to PPL has increased in recent decades. However, the collective loads for lighting, heating, and space cooling constitute approximately a third of the overall energy requirement. The story is similar with residential electricity consumption, where lighting, air conditioning, and space heating comprise approximately 42% of total usage (EIA, 2015). As such, a successful daylighting strategy can greatly influence the overall energy requirements of a commercial and residential buildings.

12.4 APERTURE STRATEGIES: SIDELIGHTING

Sidelighting systems admit light from apertures in window walls, and light sweeps across the space from one or more sides. The distance to which usable daylight penetrates a space and falls onto a work plane is the variable that designers work with to provide for sufficient illuminance. Generally, sidelighting is best for desk tasks because there are no veiling reflections, provided there is proper orientation of the worker. A number of generalized sidelighting strategies provide good illuminance fairly far into a space and improve visual comfort. A complete discussion is provided in *Daylight in Buildings: A Source Book on Daylighting Systems and Components* (IEA, 2000). The variety of approaches and components available to the designer is extensive and includes the following:

- *Design for bilateral lighting*: Daylight within a space is generally most evenly distributed when a space is lit from two walls (bilateral lighting), as shown in Figure 12.7. Bilateral lighting from "opposite" walls produces the most evenly distributed lighting condition. Unilateral lighting (windows on one wall) can increase the potential for glare.

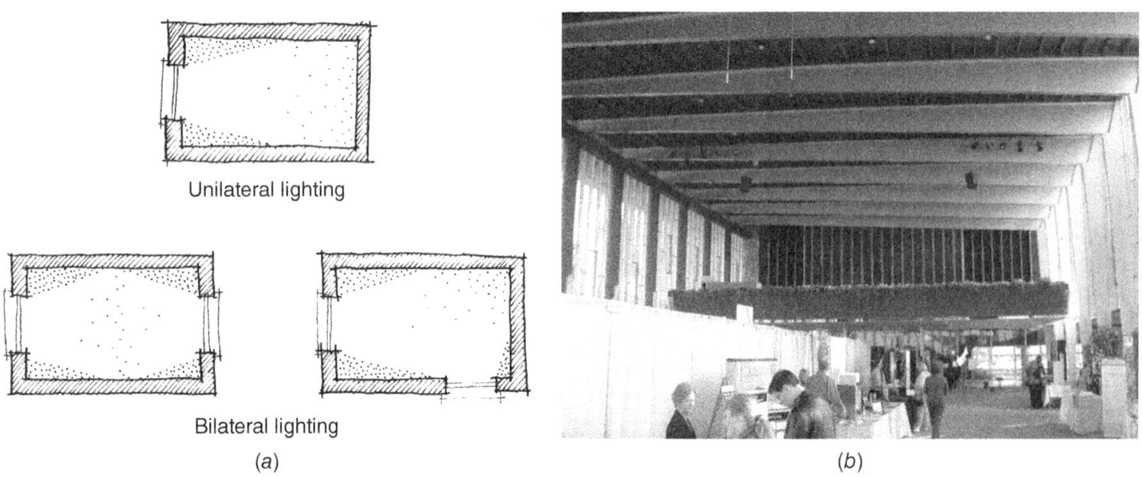

Figure 12.7 *(a) Plan diagrams of unilateral and bilateral daylighting. (b) Windows on two sides (a bilateral approach) at the Crystal Cathedral campus, Anaheim, California. (photo by Alison Kwok.)*

- *Place windows high on a wall*: In general, for a given window area, daylight will penetrate farther into a space and have more uniform distribution when windows are placed high on a window wall (Figure 12.8). If possible, raise the ceiling height to accommodate a higher window position. Use the ceiling as a reflecting surface by placing window heads as close as possible to the ceiling.
- *Use adjacent walls as reflectors*: Interior walls become reflectors when windows are placed adjacent to them, thus reducing the contrasting edge around the window (Figure 12.9). This arrangement can also bring visual delight if the reflecting wall is a light color, which will reveal patterns and colors from sunlight and reflect diffuse light farther into the space.
- *Splay the walls of an aperture*: This strategy is similar to the reflector strategy described previously, where light washes across a longer or rounded surface area around the window. When the edges of window openings are splayed (Figure 12.10) or rounded, these illuminated surfaces surrounding the window reduce contrast and are more visually comfortable, thereby reducing the potential for glare.
- *Provide daylight filters*: Daylight may be modified (either blocked or diffused) by a number of elements, which include trees, vines, and trellises (Figure 12.11) on the exterior of a building; filters for the interior of a building may include blinds, drapes, or translucent glazing.
- *Provide summer shading*: Depending upon passive solar heating and cooling design strategies, in some instances direct sunlight should be blocked before it enters a space at certain times of the year. Figures 12.12–12.14 show examples of exterior louvers (horizontal or vertical), overhangs, trellises, trees, and lightshelves that can block direct sunlight, reflect diffused sunlight into a space, and provide

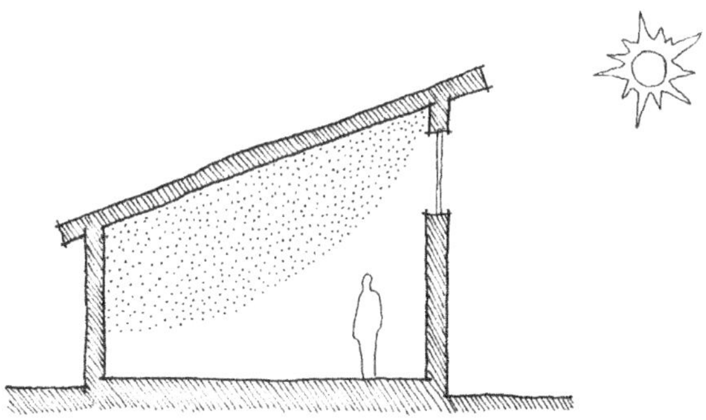

(a)

(b)

Figure 12.8 *(a) With higher windows, daylight extends farther into a space. (b) High windows in a classroom at the University of Oregon. (photo by Nathan Majeski.)*

(a)

(b)

Figure 12.9 *(a) Windows adjacent to a wall provide an additional reflecting surface. (b) Reading carrel adjacent to a window at the Graduate Theological Union Library, Berkeley, California. (photo by Alison Kwok.)*

(a) (b)

Figure 12.10 *(a) Splayed window and wall provide additional reflecting surfaces. (b) Splayed window opening to increase visual comfort and reduce glare potential at the 2011 Solar Decathlon EMPOWER House by the New School and Stevens Institute of Technology. (photo by Alison Kwok.)*

(a) (b)

Figure 12.11 *(a) Trellis at Westcave Environmental Center, Round Mountain, Texas. (b) Several layers (trees, shading, curtains) at a window can filter light and provide shade. (a: © Walter Grondzik; all rights reserved; b: photo by Nathan Majeski.)*

solar control. Light-colored materials or finishes on these components will reduce contrast between the element and the view or sky beyond. Lightshelves that also serve as a shading device can be designed as a horizontal (integrated or attached) component positioned inside and/or outside a window. Typically above eye level, they divide a window into a lower view portion and an upper area exclusively for daylight.

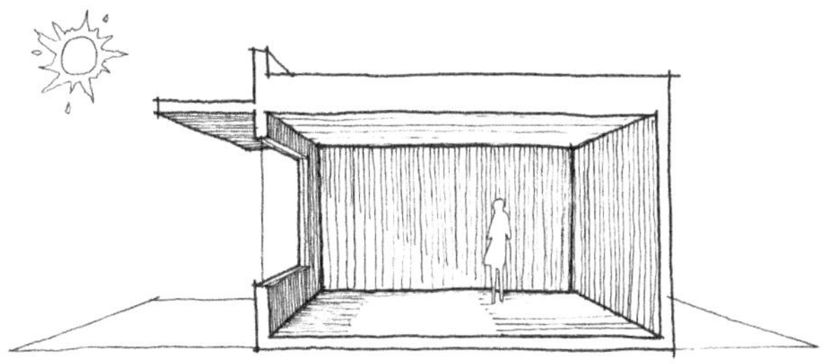

(a)

(b)

Figure 12.12 *(a) Horizontal overhangs block light but also act as a reflector for light from the ground plane. (b) Horizontal shading devices at Ash Creek Intermediate School, Monmouth, Oregon. (photo by Alison Kwok.)*

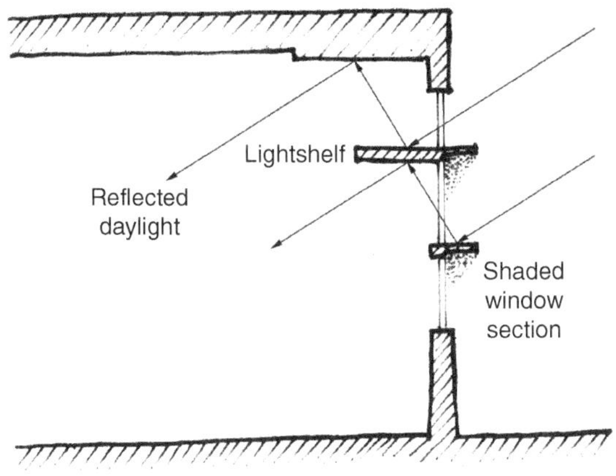

(a)

(b)

Figure 12.13 *(a) A lightshelf reduces the daylight factor near a window and increases it at greater depths. Shelf material (opaque, translucent) and angle of installation (horizontal, sloped up) markedly affect performance. (b) Classroom lightshelf, Allen Hall School of Journalism, Eugene, Oregon. (© Karen Tse; used with permission.)*

Figure 12.14 *Fixed horizontal louvers on the south façade at the Phoenix Public Library, Phoenix, Arizona. (© Walter Grondzik; all rights reserved.)*

12.5 APERTURE STRATEGIES: TOPLIGHTING

Skylights, roof monitors, and clerestories are suitable aperture strategies for the top floor of a building, particularly for interior locations of large floors that are far from perimeter windows. To prevent veiling reflections or direct glare situations, toplighting components should be placed away from the offending zone (areas with a direct view from an occupant), or a baffle or interior reflector should be used to diffuse and control daylight. Most of the strategies for sidelighting also apply to toplighting, several of which are discussed here.

- *Splay the "walls" of an aperture*: Splaying the sides of a skylight makes the skylight appear larger because

(a) (b)

Figure 12.15 (a) Splayed surfaces of a skylight provide areas for diffusely reflected light. (b) Conical skylights at Millesgården Museum and Sculpture Garden, Lidingö, Sweden (Everet Milles). (© Karen Tse; used with permission.)

light washes along a larger surface area and reflects diffuse light into the space (Figure 12.15). This strategy reduces the potential for glare similarly to the way splayed windows function.

- *Place toplights high in the space*: Higher ceilings with skylights allow more surface area for light to diffuse upon, virtually becoming a larger source (Figure 12.16). This strategy works well in spaces where the skylight is well above the field of view.
- *Use interior devices to block, baffle, or diffuse light*: Direct sunlight can be redirected by a reflector below a skylight, clerestory, or roof monitor that, depending upon the surface material, diffuses the light onto another surface within the space (Figures 12.17 and 12.18).

12.6 SPECIALIZED DAYLIGHTING STRATEGIES

A number of innovative daylighting systems can be categorized as experimental, yet they have tremendous potential. Some of these strategies include laser-cut or prismatic panels, fiber optics, solar tubes, and heliostats. More advanced systems use reflectors and lenses to introduce concentrated luminous energy into some type of light-conducting device. These may be fiber-optic bundles, prismatic light pipes, or some type of mirrored channel. The problem of heat rejection becomes more severe as the degree of solar energy concentration increases. Thus, light pipes are much less critical in this regard than are optical-fiber systems. The efficiency and economic feasibility of these systems are interdependent because of the materials used; the farther the light is transmitted, the lower the system's overall efficiency and the higher the cost to make the system technically feasible.

(a) Light Pipes

This term refers to several strategies: daylight pipes, electric light pipes, and fiber-optic pipes (Figure 12.19). The light pipe operates by collecting light through a heliostat; channeling daylight (or electric light) through a reflective tube made of prismatic glass, plastic film, and mirrors; and diffusing light at the end of the pipe. It is an exciting strategy because of the length it can transport light—65 ft [20 m]—from a single-point light source.

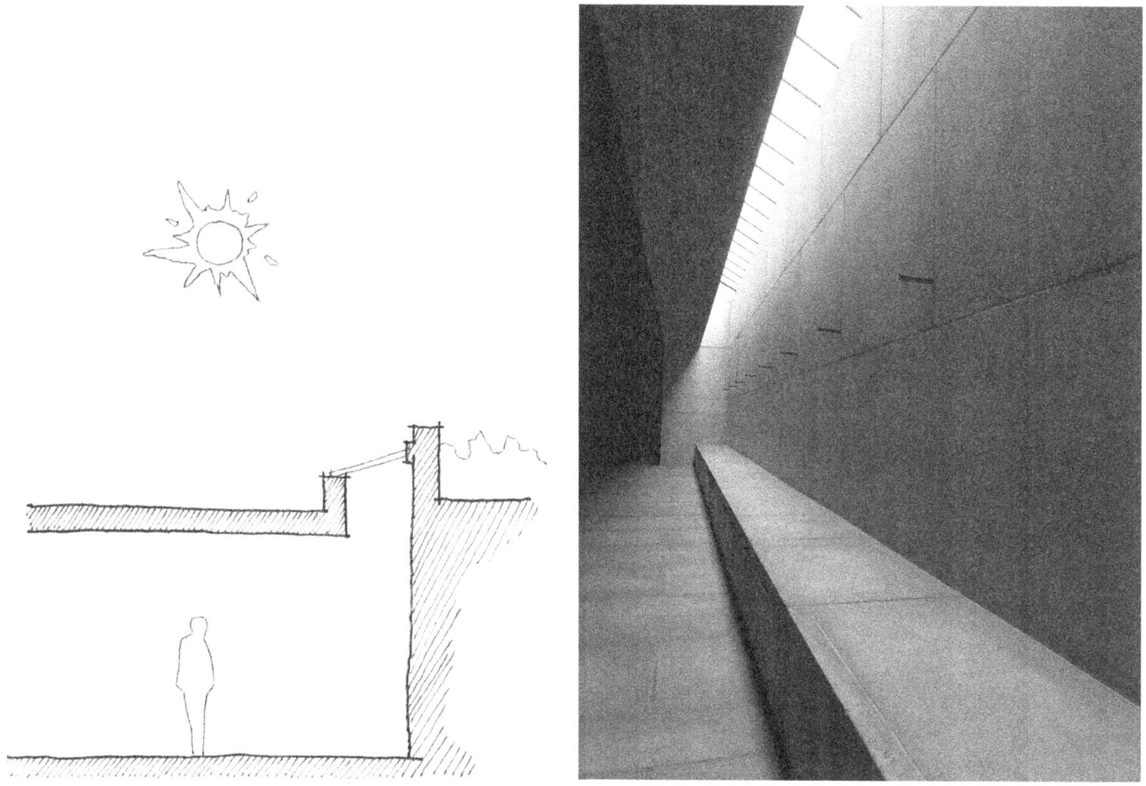

(a) (b)

Figure 12.16 *(a) A skylight near a north wall provides reflecting surfaces for uniform light distribution and reduces the potential for glare. (b) The linear toplight enables light to wash an interior concrete wall at the Chapel of the Holy Cross, Turku, Finland (Pekka Pitkänen). (© Karen Tse; used with permission.)*

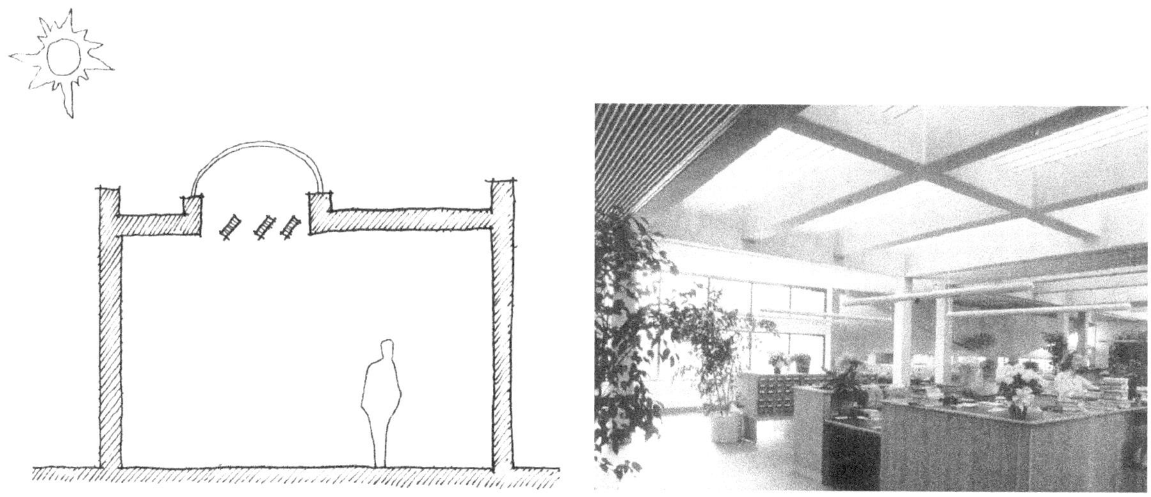

(a) (b)

Figure 12.17 *(a) Skylight with baffles that block direct solar radiation. (b) Baffled skylight daylighting design at Mt. Airy Public Library, Mt. Airy, North Carolina (Edward Mazria). (photo by Fuller Moore; © 2004 The Society of Building Science Educators; used with permission.)*

PASSIVE ENVIRON-
MENTAL SYSTEMS

Figure 12.18 Clerestory skylights with louvers at the U.S. Holocaust Memorial Museum in Washington, DC (Pei, Cobb, Freed & Partners). (© Alison Kwok; all rights reserved.)

(b) Tubular Skylights

These light shafts have highly reflective surfaces and are capped by a clear skylight (Figure 12.20). The amount of light transmitted and delivered varies with the diameter of the shaft. They are convenient and economical for supplemental illumination in hallways, closets, and areas without a need for a lot of control. The light quality is comparable to that of a ceiling-mounted fluorescent fixture (often indistinguishable).

12.7 BASIC CHARACTERISTICS OF LIGHT SOURCES

For daylighting design a good understanding of light sources, characteristics of light, sky conditions, and lighting "behavior" will rationalize the daylighting goals of the project. Technological developments in the lighting industry offer the designer a variety of energy-efficient and environmentally responsible sources and controls to fully integrate daylight and electric light into the design process (Figure 12.21).

Daylight sources may be categorized as *direct* (direct sunlight or diffuse skylight) or *indirect* (light reflected or modified from its primary source). Efficacy is a basic characteristic common to daylight (and electric light) sources—measured in lumens per watt (lm/W). *Efficacy* is the ratio of lumens provided to watts of heat produced by a light source (i.e., how much

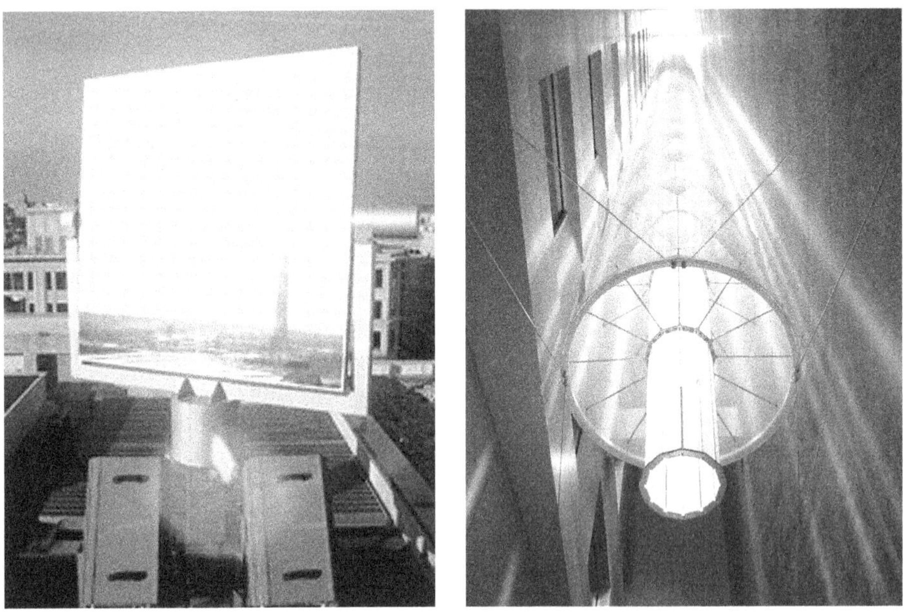

(a) (b)

Figure 12.19 (a) Heliostat on the rooftop of a building tracks the sun and directs light into an 8-ft [2.4-m]-wide atrium and down 14 floors. (b) The heliostatic light tube is a 120-ft [37-m]-long, 6-ft [1.8-m] diameter, 12-sided steel-and-aluminum frame—enclosing laminated glass panels and surrounded by fabric—within the atrium at the Morgan Lewis building in Washington, DC. (© Carpenter Norris Consulting; used with permission.)

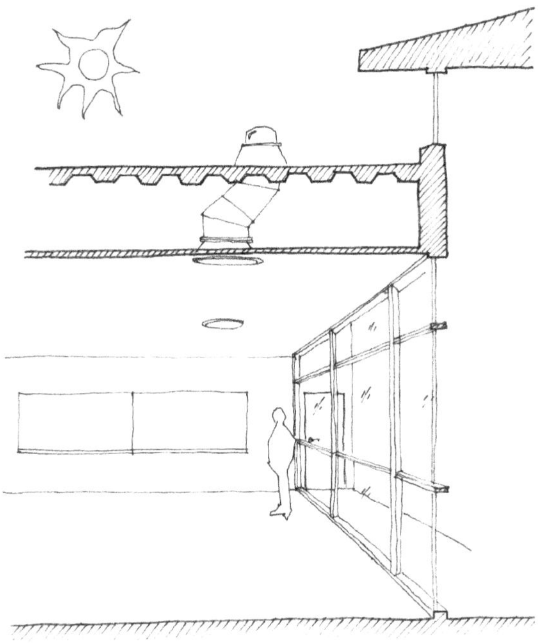

(a)

(b)

Figure 12.20 (a) Skylight through a roof structure. (b) Top of Solatube® skylights on the roof at Ash Creek Intermediate School in Monmouth, Oregon (Bora Architects). (© Alison Kwok; all rights reserved.)

(a)

(b)

Figure 12.21 (a) A daylit space; sunlight streams through a window at the Santa Anna Monastery in Santa Anna, Italy. (b) An electrically lit space—a wall sconce at the Westin Peachtree Plaza Hotel in Atlanta, GA. (© Alison Kwok; all rights reserved.)

light is provided per unit of heat produced). Table 12.1 lists efficacies of common light sources. Due to its high efficacy, diffuse daylight introduces less heat per lumen than electric sources, making use of daylight an attractive strategy for reducing cooling loads in buildings caused by lighting (assuming effective, balanced distribution and utilization of illumination).

The efficiency of a standard incandescent lamp in converting electrical energy to light is approximately 7%; the other 93% is released as heat. Compare this to fluorescent lamps, which are approximately 22% efficient. Although fluorescent lamps are a great improvement over incandescent lamps, the generally low efficacy of lighting in older buildings accounts for a significant proportion of their energy use (which could be well over 25%, may be even over 50%).

Selecting light sources for buildings—whether daylight or electric light (or, more likely, a combination of both)—involves simultaneous lighting and thermal considerations. Over the past three decades, light-emitting diode (LED) technology has advanced considerably and become much more economic. High-efficacy LED luminaires and LED lamps designed for direct swap-out with incandescent and compact

TABLE 12.1 Efficacy of Various Light Sources

Source	Efficacy (Lm/W)
Candle	0.1
Oil lamp	0.3
Original Edison lamp	1.4
1910 Edison lamp	4.5
Incandescent lamp (15–500 W)	8–22
Tungsten–halogen lamp (50–1500 W)	18–22
Mercury-vapor lamp (40–1000 W)[a]	32–63
Fluorescent lamp (15–215 W)[a]	35–80
OLED Luminaire[d]	43
Induction lamp[c]	48–70
Compact fluorescent lamp[b]	55–75
High-pressure sodium lamp (35–100 W)[a]	55–115
Metal-halide lamp (70–1500 W)[a]	80–125
Sulfur lamp[c]	90–100
LED 6″ downlight (warm white)[d]	64
LED PAR38 lamp (warm white)[d]	70
LED A19 lamp (dimmable, warm white)[d]	78
LED troffer 2′ × 4′ (warm white)[d]	94
LED high/low-bay fixture (warm white)[d]	102
LED T8 tube (neutral white)[d]	107
Direct sun (low altitude = 7.5°)	90
Direct sun (high altitude > 25°)	117
Direct sun (mean altitude)	100
Sky (clear)	150
Sky (average)	125
Global (average)	115
Maximum source efficacy predicted (in the year 2010)	150
Maximum theoretical limit of source efficacy	250 (approximate)

[a]Includes ballast losses (with electronic ballasts, lumens per watt become much higher). Losses vary between ballasts and manufacturers.

[b]With electronic ballasts.

[c]With a power supply.

[d]System-level efficacies for typical 2015 performance of solid-state light (SSL) sources listed for comparative purposes (noted examples are from the DOE's *Solid-State Lighting R&D Plan* (DOE, 2016)). Due to the continued improvement of the technology, the efficacy of LED lamps and luminaire types will vary.

PASSIVE ENVIRON-MENTAL SYSTEMS

fluorescent lamps (CFL) in certain types of luminaires have precipitated in significant reductions in lighting loads across the building sector. Many buildings today utilize only LEDs for lighting. Lighting retrofits often have considerable economic returns.

As a result of the ubiquitous utility of LEDs, electric lighting in American nonresidential buildings has dropped to around a tenth of the total electric energy utilized (but, again, this can be considerably more in existing buildings that have not recently undergone lighting retrofits). Any attempt to further reduce lighting loads must necessarily include integration of the cheapest (insofar as energy is concerned), most abundant, and, in many ways, most desirable form of lighting available—daylight. In selecting appropriate light sources for buildings, understanding the characteristics of the light sources will allow a designer to use them appropriately for energy efficiency and to provide visual and thermal comfort. For resource efficiency, a designer should first optimize daylight sources through building geometries and material finishes, and then design the electric lighting system to supplement and enhance illumination and effect.

12.8 SKY CONDITIONS

The most prominent characteristic of daylight is its variability. The source of all daylight is the *sun*. Exterior illumination, at a particular place and time, depends upon (a) solar position, which can be determined if the latitude, date, and time of day are given; (b) weather conditions (e.g., cloud cover, smog); and (c) effects of local terrain (natural and built obstructions and reflections). The position of the sun in the sky is expressed in terms of its altitude above the horizon and its azimuth angle. For all latitudes in the northern hemisphere, the sun's altitude is highest in summer, lowest in winter, and in between in spring and fall. Azimuth angle is defined as the sun's horizontal position angle, measured from the south. Solar position is absolutely predictable for any given time and location.

Cloud cover, unlike solar position, is only statistically predictable, on the basis of extensive U.S. National Weather Service observations at numerous weather stations throughout the United States. At locations other than those for which recorded data are available, an educated guess is necessary. Outside the United States, a designer must rely on locally available data, which are often difficult to obtain.

The third factor—local terrain and construction conditions that either reduce illumination by shadowing or increase it by reflection—can be considered only on a case-by-case and site-specific basis.

For manual calculation procedures, it is sufficient to establish four basic sky conditions. These are:

1. solid overcast sky
2. clear sky without sun (in the field of view)
3. clear sky with sun
4. partly cloudy sky

(a) Standard Overcast CIE Sky

This condition, which occurs for much of the year in northerly climates such as England, Scandinavia, and the Pacific Northwest, is called the CIE sky, because it was adopted by the Commission Internationale de l'Eclairage (CIE) as the standard design sky for daylighting calculations (CIE, 1970). The CIE sky has a nonuniform brightness distribution, increasing from horizon to zenith in approximately a 1:3 ratio. Sky luminance at any altitude angle above the horizon is defined in Equation 12.1 as

$$L_A = L_z \frac{1 + 2\sin A}{3} \quad \text{(Eq. 12.1)}$$

where

L_A = luminance at $A°$ above the horizon (in any direction)
L_Z = luminance at the zenith

Thus, at the horizon, where $A = 0$,

$$L_A = \frac{L_Z}{3}$$

The illuminance (density of light in lux) on unobstructed exterior horizontal and vertical surfaces produced by this luminance distribution has an approximate ratio of 2.5:1 (Figure 12.22*a*). This results from an integration of Equation 12.1 over the whole sky.

There is agreement among all sources that with an overcast sky, exterior horizontal illuminance varies directly with the sun's altitude, irrespective of azimuth. Various formulations for this relationship have been put forward. One formulation that gives good agreement with observations is by Krochman (1963), shown in Equation 12.2:

$$E_H = 300 + 21{,}000 \sin A \qquad \text{(Eq. 12.2)}$$

where

E_H = exterior illuminance (lux) on an unobstructed horizontal surface
A = solar altitude (degrees)

Solar altitude (and azimuth) for various times of day can be obtained from Appendix D, Table D.1. Figure 12.23 is a plot of year-round averages for both vertical and horizontal illuminance from an overcast sky as a function of solar altitude, based on U.S. National Weather Service observations.

It is interesting to compare the exterior horizontal illuminance obtained from the two sources given: Krochman's formula (Equation 12.2) and the observation-based data of Figure 12.23, for a few typical conditions. Solar altitude values were obtained from Appendix D, Table D.1.

Latitude:	38°
Solar Time:	10:00 a.m.
Dates:	Dec. 21, Mar./Sept. 21, June 21
Solar Altitude:	23°, 43°, 61°

	Equation 12.2	Figure 12.23
Dec. 21	8505 lx [790 fc]	8608 lx [800 fc]
Mar./Sept. 21	14,622 lx [1358 fc]	15,923 lx [1480 fc]
June 21	18,667 lx [1734 fc]	23,134 lx [2150 fc]

The degree of agreement is generally satisfactory and either source will yield suitable results for designers seeking to approximate available exterior horizontal illuminance for given day and time.

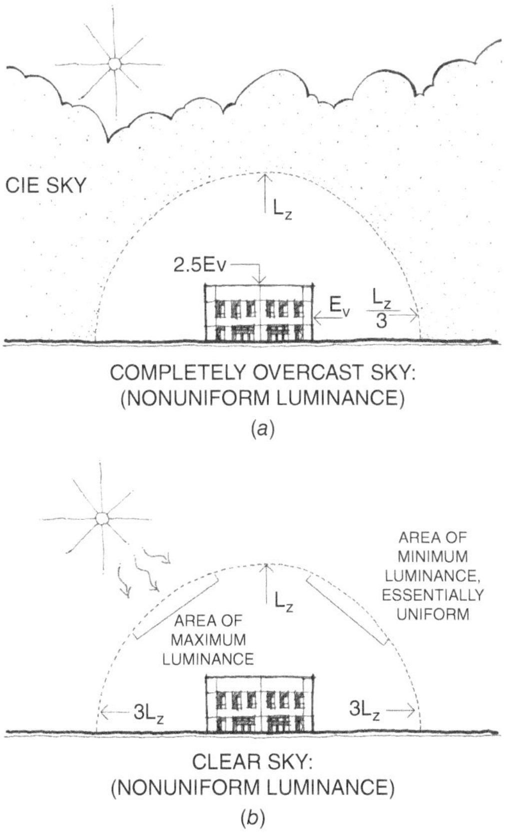

Figure 12.22 *(a) The completely overcast sky has a zenith luminance L_Z, which is three times the horizon luminance. With such a sky, illuminance on unobstructed exterior horizontal surfaces (E_H) is about 2½ times that on similar vertical surfaces (E_V). (b) The clear sky has the area of brightest luminance around the sun. The area opposite the sun is darkest and can be considered as essentially uniform at approximately 3500 cd/m² [1000 fL].*

(b) Clear Sky With and Without Sun

Horizontal Illuminance. Exterior horizontal illuminance on a clear (cloudless) day consists of two source components: diffuse illumination from the entire sky plus the much larger component of direct sunlight. As with overcast sky, various empirical formulas for both components have been proposed, and here too, all sources concur that the total illumination, diffuse skylight plus direct sunlight, varies directly with solar altitude.

Figure 12.24 gives values for both components of exterior horizontal illuminance based upon observations. The sky component only values are used to determine the shaded skylight illuminance or ground illuminance outside a shaded window—that is, a north-facing window, or an east/west window when the sun is on the opposite side of the building.

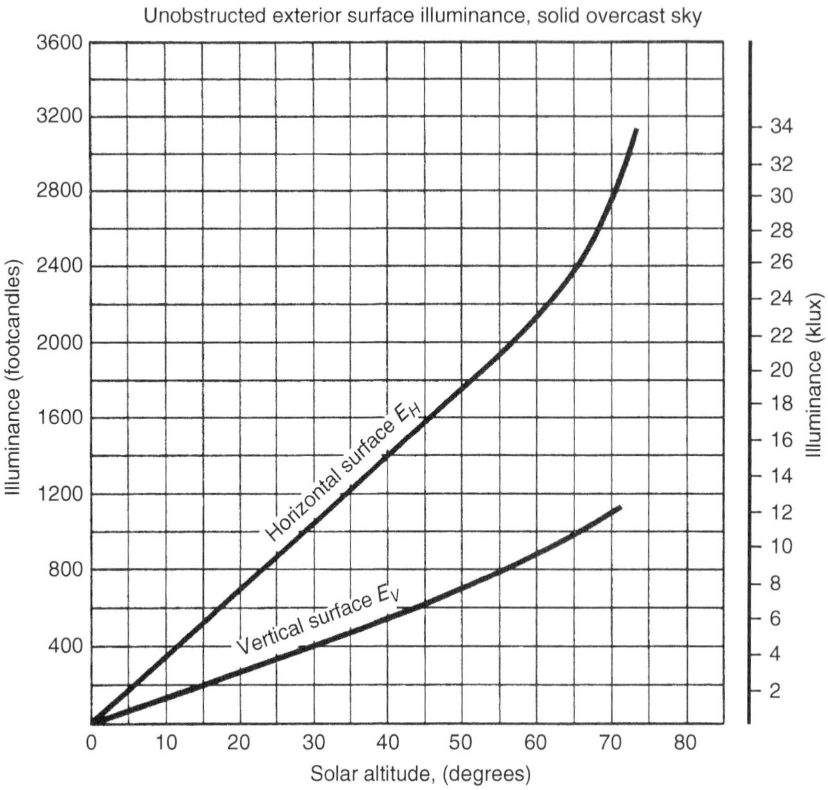

Figure 12.23 Curves giving unobstructed exterior surface illuminance directly from an overcast sky. (Data based on U.S. National Weather Service observations; courtesy of Libbey-Owens-Ford.)

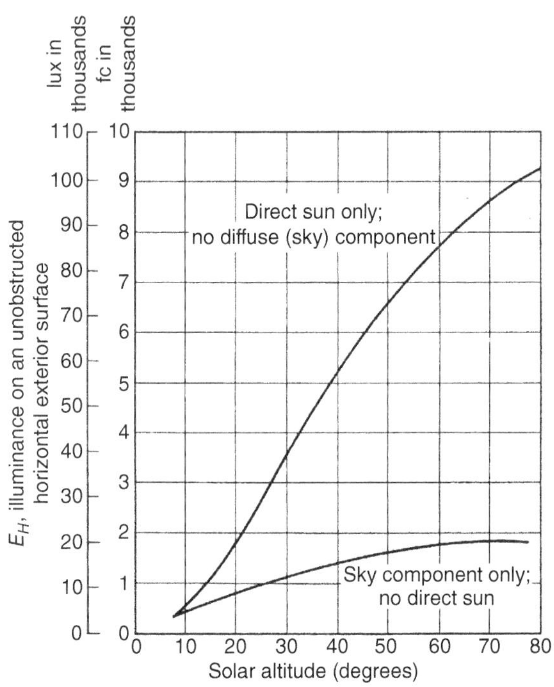

Figure 12.24 Components of the exterior horizontal illuminance on an unobstructed surface, from a clear sky, as a function of solar altitude. Total illuminance E_H is the sum of the two components. (Rennhackkamp / MDPI / CC BY 4.0.)

In determining ground illuminance, the values given in Figure 12.24 must be reduced somewhat, because they represent unobstructed horizontal illuminance; whereas the area outside a building window is partially obstructed from skylight by the building itself. If a building is so large that the ground outside the shaded window effectively receives diffuse radiation only from the half of the sky opposite from the sun, then an overall average value for E_H of 1000 fc [~10,000 lx] can be used. This is because the luminance of the half of the sky away from the sun varies from a minimum of approximately 300 fL [1031 cd/m^2] for the deep-blue patch directly opposite the sun to about 2000 fL [6874 cd/m^2] at the sides, giving an average half-sky luminance of about 1000 fL [~3400 cd/m^2]. This, in turn, gives a horizontal illuminance E_H of about 1000 fc [~10,000 lx] for the (diffuse) sky component (see Figure 12.22b).

Figure 12.24 also gives horizontal illuminance from the sun only, as a function of solar altitude. For example, using this figure, a solar altitude of 43° (Mar./Sept. 21 from the example above) would yield an E_H value of approximately 5850 fc [~63,000 lx]. This value, when combined with the proper portion of diffuse illuminance, as discussed previously, is useful in

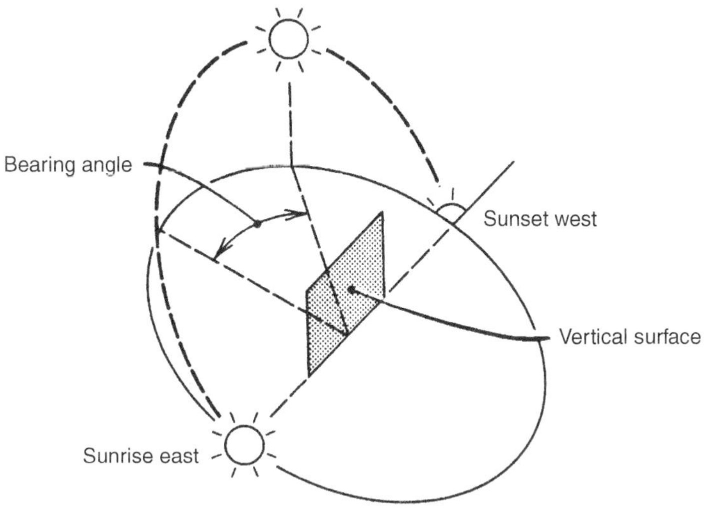

Figure 12.25 *The bearing angle of a vertical surface—the angle between a hypothetical vertical plane perpendicular to the surface (say, a window) and a hypothetical vertical plane containing the sun. Other sources refer to this angle as the window-to-sun azimuth angle or surface azimuth.*

approximating the ground illuminance outside a sunny building exposure or illuminance on an unshaded skylight. The light incident on an external reflector or lightshelf at a window can also be determined from these figures.

Vertical Surface Illuminance. Inasmuch as most daylighting is accomplished via vertical fenestration, vertical surface illumination is the major component of interior daylight. It is also important for determining the daylight contribution of vertical elements in skylights. There is no simple relationship between horizontal and vertical illuminance from a clear sky, as there is for an overcast sky, because the illumination on a vertical surface depends upon solar azimuth as well as altitude. More specifically, it depends upon the bearing angle (Figure 12.25), which is defined as the horizontal angle between a vertical plane containing the sun and a plane perpendicular to the vertical surface in question. A bearing angle of 0° indicates that the sun plane is perpendicular to the vertical surface. Like E_H, E_V (vertical illuminance) is divided into two components: sky only and direct sun only, as plotted in Figure 12.26 as a function of solar altitude and bearing angle. The sky only component is effectively for the half-sky because a vertical surface can be exposed to a maximum of only half of the full sky. Solar radiation data may be translated into illuminance by using average "efficiency" values for solar energy in units of lumens per watt of received radiation (Figure 12.26).

(c) Partly Cloudy Sky

The luminance of a partly cloudy sky cannot be expressed mathematically because of its infinite variability of conditions. However, statistical data on cloud cover are available from observations at many weather stations, and these data should be used in computer-calculated, hour-by-hour energy analysis programs. For the purpose of lighting design, it is important to note that the illumination from a partly cloudy sky is higher than that from a clear sky by 10% to 15% because of additional reflected sunlight from cloud edges. Several attempts have been made to account for this type of sky in terms of the effect on the daylight factor within a room, but none has received general acceptance.

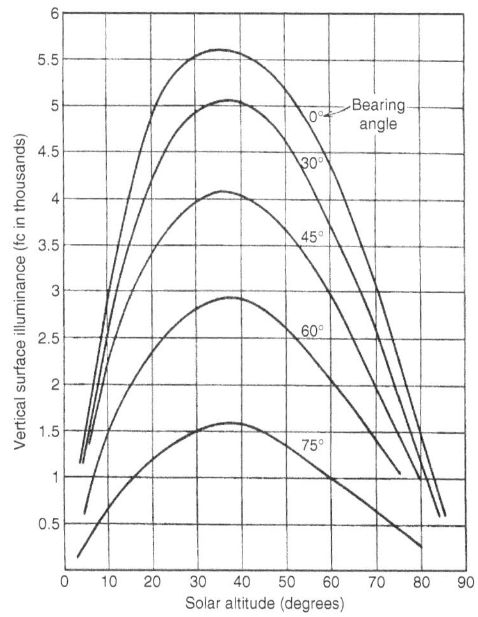

(*a*)

Figure 12.26 *(a) Vertical surface illuminance, year-long average, sun only, no sky contribution. (b) Vertical surface illuminance, clear summer sky, no sun contribution. (c) Vertical surface illuminance, clear sky, no sun contribution. (Courtesy of Libbey-Owens-Ford.)*

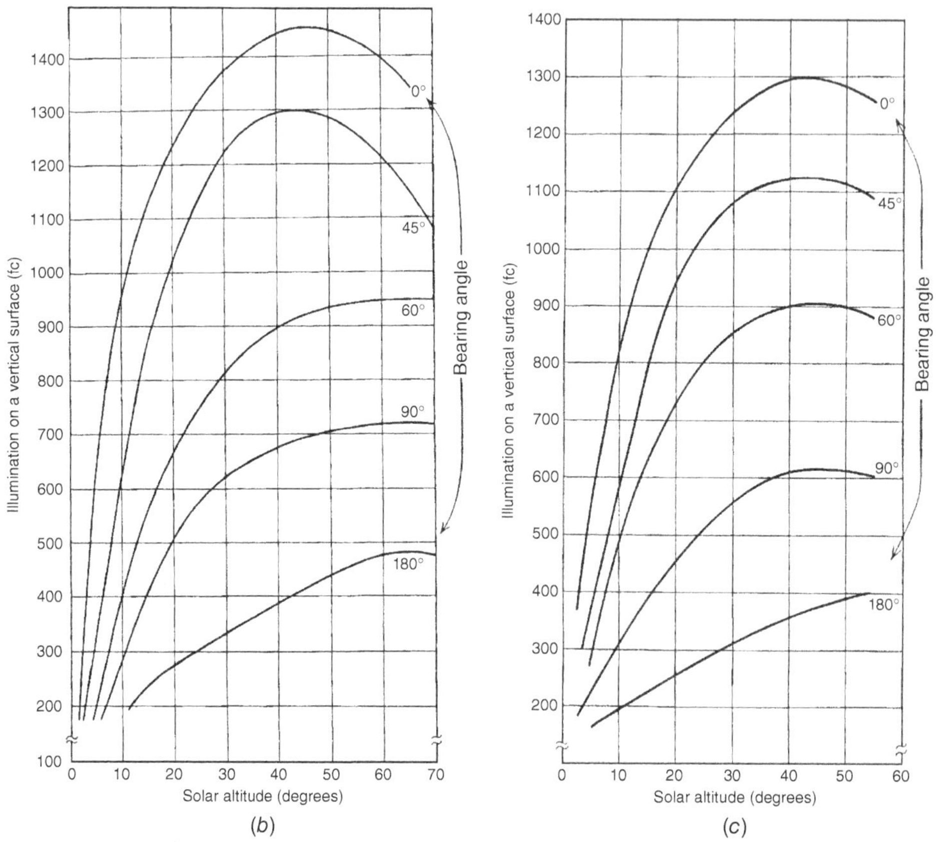

Figure 12.26 (continued)

12.9 PERFORMANCE METRICS FOR DAYLIGHTING

A number of metrics have been developed to assess daylighting performance. IES offers a comprehensive compilation in *IES LP-3-20, Lighting Practice: Designing and Specifying Daylighting for Buildings* (IES, 2020b). In this section, descriptions of daylighting performance metrics most frequently utilized by building design professionals are offered.

(a) Daylight Factor

When a building is designed to rely on daylighting, the daylight factor (DF) may be considered, which is expressed as the ratio of interior illuminance (E_i) to available outdoor illuminance (E_H) under overcast skies.

$$DF = \frac{E_I}{E_H} \times 100 \qquad \text{(Eq. 12.3)}$$

where

E_I = indoor illuminance, at a given point
E_H = unobstructed exterior horizontal illuminance

The daylight factor concept is applicable only where the sky luminance distribution is known or can reasonably be estimated. CIE defines an overcast sky and a clear sky whose luminance distributions are fixed for the purpose of calculations. Daylight factor cannot be used with skies with constantly changing luminance (partly cloudy and direct sun) because under such conditions the daylight factor at a given point also varies continuously, making the concept useless as a calculation tool for absolute daylight values.

The daylight factor as a means of expressing interior daylight illuminance is both absolute and relative. With a given sky luminance distribution, variations in daylight illuminance inside correspond exactly to variations outside (i.e., the daylight factor remains the same). This assumes a minimal effect from obstructions and ground reflections. Thus, the daylight factor allows determination of interior daylight distribution for varying fenestration, spatial arrangement, and building orientation.

The daylight factor is constant for a given space and window configuration. Interior illuminance can easily be calculated by knowing the daylight factors for locations in a given space and the exterior illuminance derived from sky luminance data. Daylight design analysis can use a combination of minimum exterior illuminance and corresponding minimum daylight factor requirements to predict daylight sufficiency under almost all exterior conditions.

(b) Spatial Daylight Autonomy

Though it has utility during the conceptual and schematic design phases, the daylight factor is considered a crude, limited, and oversimplified metric for advanced daylight analysis. The DF does not take into consideration location, climate, or building orientation. Nor does it convey the quality of light received or glare potential. As a result of these limitations, more comprehensive metrics have come into use. Spatial Daylight Autonomy and Annual Sunlight Exposure together provide more accurate predictions of daylight performance over the entire year. These metrics are used in green building codes, standards, and rating systems such as the *International Green Construction Code* (IgCC), the *ASHRAE/ICC/USGBC/IES Standard 189.1*, and the Leadership in Energy and Environmental Design (LEED) rating system.

IES LM-83-23, Approved Method: IES Spatial Daylight Autonomy (sDA) and Annual Sunlight Exposure (ASE) defines Spatial Daylight Autonomy (sDA) as a metric that "describes the sufficiency of ambient daylight levels in interior environments across the year, while also accounting for the operation of daylight management devices that limit direct sunlight penetration into the space" (IES, 2023). This metric conveys how much of an interior space receives sufficient daylight. For example, data assessing $sDA_{300/50\%}$ is conveying the percentage of regularly occupied floor area receiving at least 300 lx [28 fc] for at least 50% of the interior space's annual occupied hours. Project daylighting goals or requirements may specify a percent of floor area that must meet an average sDA value. For instance, LEED version 5 Building Design and Construction (LEED v5 BD + C) offers points for projects that can demonstrate (through a daylight simulation analysis) that an average $sDA_{300/50\%}$ value is achieved for at least 40% of the regularly occupied floor area (USGBC, 2025).

(c) Annual Sunlight Exposure

IES LM-83-23 defines Annual Sunlight Exposure (ASE) as a metric that "describes the risk of visual discomfort from direct sunlight penetration into an interior space." This metric conveys how much of an interior space could receive too much direct sunlight, which can cause visual discomfort (glare) or increase cooling loads. For example, data assessing $ASE_{1000,250}$ is conveying the percentage of floor area that receives at least 1000 lx [93 fc] for at least 250 occupied hours per year. Similar to sDA, project daylighting goals or requirements may specify a percent of floor area that must meet a target ASE value. For instance, the *2024 International Green Construction Code* (IgCC), in addition to a requirement that the area-weighted $sDA_{300/50\%}$ shall not be less than 40% of the total applicable floor area, specifies that $ASE_{1000,250}$ shall not exceed 20% of the applicable floor area (ICC, 2024).

(d) Continuous Daylight Autonomy

In a broader sense, *IES LP-3-20* describes the notion of Daylight Autonomy (DA) simply as "the percentage or fraction of the occupied time (or analysis period) throughout the year when a target (or criterion) illuminance value is met or exceeded at a specified point by daylight alone." sDA takes the point-based concept of DF and expands it to the percentage of room area that meets a given DA value.

IES LP-3-20 offers a related daylighting performance metric that is increasing in popularity among design professionals: Continuous Daylight Autonomy (cDA). This metric is as an extension of the point-based DA metric but asserts additional credit for any fraction of the target illuminance that is met by daylight. In application, this means that if a point in a space has a DA target of 300 lx, half credit would be given for any hour at 150 lx. cDA has notable utility in evaluating energy performance as it estimates dimming system savings for a control system that is programmed to maintain a specific illuminance at a critical task point illuminated by both daylight and electric lighting.

12.10 COMPONENTS OF DAYLIGHT

Understanding the components of daylight is important to the design of apertures and the selection of materials. Daylight illuminance in a building consists of three components (see Figure 12.27):

1. sky component (SC)
2. externally reflected component (ERC)
3. internally reflected components ($IRC_1 + IRC_2$)

DF is the sum of these three components, each calculated individually for each location being considered. DF is a ratio, but the value of a given DF is based upon contributions from these components:

$$DF = SC + ERC + IRC$$

The sky component (SC) is that portion of total daylight illuminance at a point received directly from the area of the sky visible through an aperture. As the SC represents received light, it takes into account

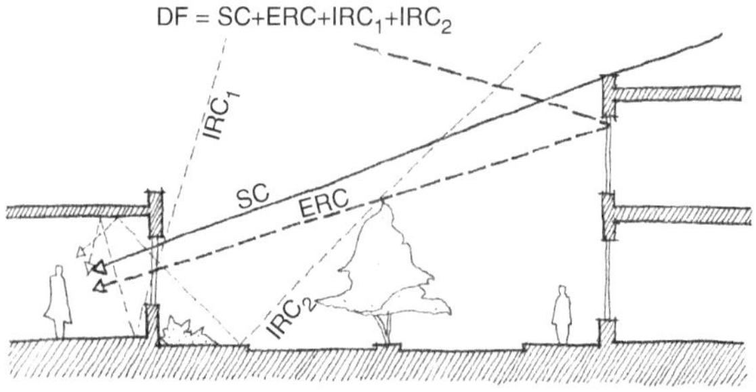

Figure 12.27 *Total daylight factor (DF) is composed of the SC, the ERC, and the IRC. The IRC, in turn, is subdivided into reflected skylight and reflected ground light components. Note that surfaces deep in the room are illuminated with re-reflected light.*

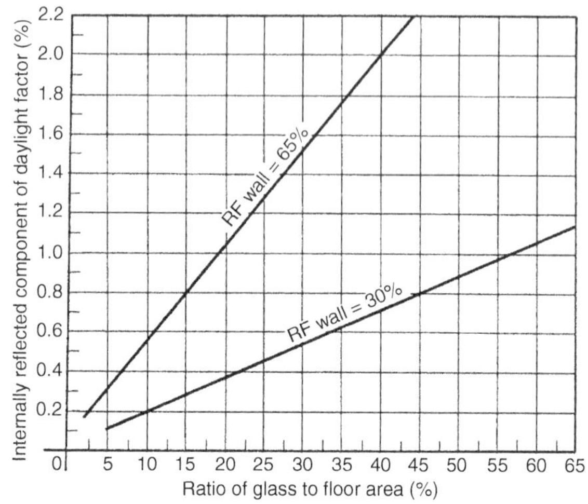

Figure 12.28 *Plot of the IRC of the daylight factor as a function of the amount of glazing, expressed as per the DF (i.e., as a percentage of exterior illuminance). As expected, the effect of a lighter wall finish becomes more pronounced as the fenestration area increases.*

reductions due to window obstructions (mullions, etc.) and losses in transmission; that is

$$SC = incident\ skylight - window\ losses$$

The externally reflected component (ERC) represents light reflected from exterior obstructions onto the point under consideration. This does not include ground-reflected light. ERC is of significance only in built-up areas (where there are structures opposite an aperture) and can be estimated as the portion of the SC for that area of obstructed sky, reduced by the percentage of the sky obstructed (RD) and the reflectance factor (RF) of the obstruction; that is

$$ERC = SC \times RD \times RF$$

Thus, if 25% of the sky is obstructed (RD) by a building with a 20% reflectance factor (RF), we have

$$ERC = SC \times 0.25 \times 0.20$$

For this particular example, then

$$ERC = 5\% \text{ of } SC$$

to be added to the remaining 75% of SC (again, 25% of the sky is obstructed; so the unobstructed SC value is reduced accordingly).

The internally reflected component (IRC) represents the daylight received at the point under consideration that has been reflected from interior surfaces. IRC is subdivided into two values: reflected skylight (IRC_1) and reflected ground light (IRC_2). IRC_2 is generally small, and $IRC \cong IRC_1$. IRC is, therefore, primarily dependent upon interior surface reflectances and upon the amount of window glazing, and becomes a large portion of the DF deep within an interior space (see Figure 12.28 illustrating IRC as a function of the amount of glazing and Table 12.2 relating IRC to room depth). The IRC is normally calculated using published interreflectance tables, because direct calculation is extremely complex. It is worth noting that the DF formula ($SC + ERC + IRC_1 + IRC_2$) underscores the influence on daylight performance of internal finishes and the reflectance of visible ground cover outside of an aperture.

Typical curves for both horizontal and vertical daylight factors for a room with a single (unilateral) sidelighting (windows on one side) are shown in Figure 12.29. While these curves communicate how the DFs vary by component and the degree to which proximity to the window impacts the DFs, these curves were developed by a longhand daylight-protractor-aided technique (Building Research Station, London). Any change in parameters, such as window dimensions or height above the work plane, ceiling height, surface reflectance, ground reflection, and obstructions, alters these curves and requires recalculation and replotting. Exact calculation of even a few variants for a space is a tedious and time-consuming procedure. As such, utilization of DF curves is not practical in most project-specific circumstances.

The following methods describe alternative approaches available to save time and increase accuracy (however, each is still limited by early project approximations and assumptions):

1. Use of simplifications, such as standard curves, tabular data, or the CIE method.
2. Use of a library of graphic light distribution plots with varying parameters.
3. Use of a less-laborious manual calculation procedure (one such technique is known commonly as the lumen method, as defined by the IES).
4. Use of computer simulation software.

TABLE 12.2 Effect of Wall Reflectance Factor (RF) on the Proportion of IRC in the DF

Distance from Window in ft [m]	30% Wall Reflectance		60% Wall Reflectance	
	Total DF	$\frac{IRC}{DF}$ (%)	Total DF	$\frac{IRC}{DF}$ (%)
0	30	1	31	3.5
5 [1.5]	16	1.9	17	6.5
10 [3.0]	5.5	5.5	6.3	16.9
15 [4.5]	2.1	14.3	2.9	37.9
20 [6.0]	1.3	23	2.1	52.4

Room data:
Room 24 ft × 28 ft [7.3 m × 8.5 m]; 70% ceiling reflectance.
Window on 28-ft [8.5-m] wall—one side only; 20% floor reflectance.
Window area = 20% of floor area.

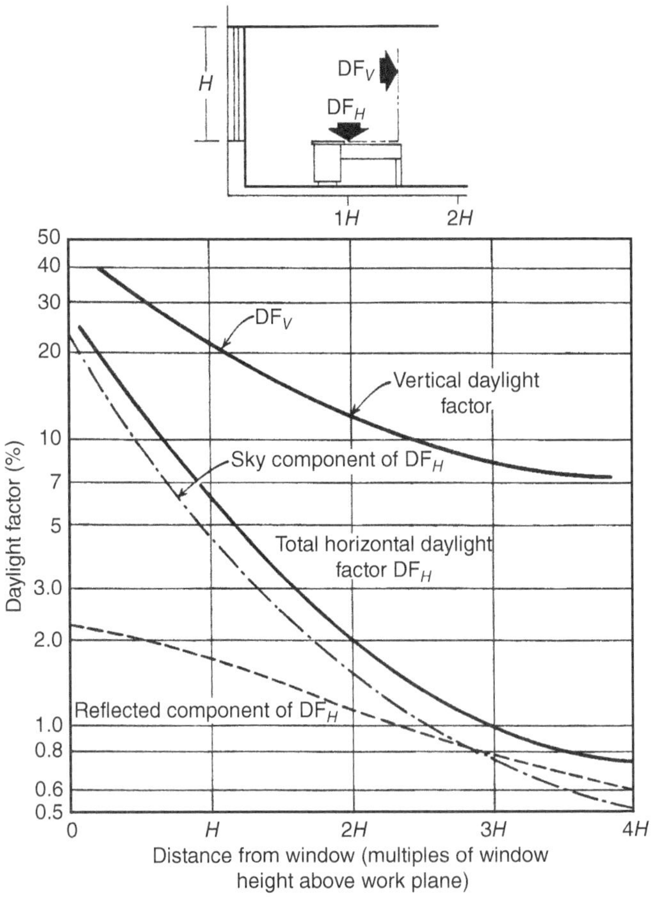

Figure 12.29 Typical daylight factor curves for horizontal (DF_H) and vertical (DF_V) illuminance for a room with large windows on one side only. Note that the SC represents almost the entire DF near the window, but its proportion reduces at greater depths. There, interreflected light constitutes 50% of the available daylight.

Though they may not be leaned upon for daylighting analysis, designers may use daylight factor criteria as a starting point for design goals, translating the DF values (such as those given in Table 12.3) into actual illuminances in footcandles [lux] and comparing calculated or simulated results to the recommended illuminance values. As an example, let us use the daylight factor method and consider two cities in the United States that have overcast skies for appreciable portions of the year: Columbus, Ohio (40°N latitude), and Seattle, Washington (48°N latitude). Table 12.4 compares illuminance values

TABLE 12.3 Recommended Daylight Factors

Task	DF (%)[a]
Ordinary seeing tasks, such as reading, filing, and easy office work	1.5–2.5
Moderately difficult tasks, such as prolonged reading, stenographic work, normal machine tool work	2.5–4.0
Difficult, prolonged tasks, such as drafting, proofreading poor copy, fine machine work, and fine inspection	4.0–8.0

Source: Millet et al., 1980.

[a]Use the smaller DF values for southern latitudes with plentiful winter daylight.

TABLE 12.4 Horizontal Illuminances (E_H) from Overcast Sky, at Selected Times, in Columbus, Ohio, and Seattle, Washington, Corresponding to the Recommended DF

Location	10 a.m. Solar Altitude[a]	Available Daylight, E_H fc [lux][b]	DF Recommendation (%)[c]	Illuminance, fc [lux] Calculation from DF	Illuminance, fc [lux][d] Recommendation
Columbus 40°N latitude	June 21 60°	2100 [22,500]	1.5–2.5	31–52 [338–563]	28–47 [300–500]
			2.5–4	52–84 [563–900]	47–70 [500–750]
			4–8	84–167 [900–1800]	70–93 [750–1000]
	Mar./Sept. 21 41°	1400 [15,500]	1.5–2.5	21–35 [225–375]	28–47 [300–500]
			2.5–4	35–56 [375–600]	47–70 [500–750]
			4–8	56–112 [600–1200]	70–93 [750–1000]
	Dec. 21 21°	700 [7500]	1.5–2.5	11–18 [113–188]	28–47 [300–500]
			2.5–4	18–28 [188–300]	47–70 [500–750]
			4–8	28–56 [300–600]	70–93 [750–1000]
Seattle 48°N latitude	June 21 56°	1950 [21,000]	1.5–2.5	29–49 [315–525]	28–47 [300–500]
			2.5–4	49–78 [525–840]	47–70 [500–750]
			4–8	78–156 [840–1680]	70–93 [750–1000]
	Mar./Sept. 21 36°	1220 [13,000]	1.5–2.5	18–30 [195–325]	28–47 [300–500]
			2.5–4	30–48 [325–520]	47–70 [500–750]
			4–8	48–97 [520–1040]	70–93 [750–1000]
	Dec. 21 14°	500 [5400]	1.5–2.5	8–13 [81–135]	28–47 [300–500]
			2.5–4	13–20 [135–216]	47–70 [500–750]
			4–8	20–49 [216–532]	70–93 [750–1000]

[a]From Appendix D, Table D.1.

[b]From Figure 12.23.

[c]From Table 12.3.

[d]From former CIBSE and IES recommendations retained here for this example: *CIBSE Code for Interior Lighting* (CIBSE, 1994). Reference 25; *CIBSE Lighting Guide LG3: The Visual Environment for Display Screen Use* (CIBSE, 2006); *The Lighting Handbook*, 9th ed. (IES, 2000).

(for DF_H, which is located at 1H feet from the window—see Figure 12.29) calculated by the DF method with those recommended by the IES and the Chartered Institution of Building Services Engineers (CIBSE). For most of the year (with the exception of winter), daylight provides all the light necessary for the tasks in Table 12.3. In this case, where the available exterior daylight in the winter is as low as 700 fc [7500 lx] in Columbus and 500 fc [5400 lx] in Seattle, supplemental electric lighting would be required for all interior areas beyond H feet (i.e., one window height from the window—see Figure 12.29).

The calculated daylight factor will vary throughout an interior space. In addition to the recommendations in Table 12.3, the ratio between the minimum and average daylight factor in a space, which relates to contrast ratios, should be no less than 30%:

$$\frac{DF_{min}}{DF_{avg}} \geq 0.3$$

The minimum daylight factor in any portion of a space should not drop below 0.5%, which is necessary for circulation.

12.11 GUIDELINES FOR PRELIMINARY DAYLIGHTING DESIGN

Guidelines offer designers a variety of broadly based "rules-of-thumb" that set simplified and actionable parameters informed by basic principles or phenomena. In the absence of calculations or simulations, guidelines can serve as a shorthand during rapid or real-time conceptual design work to help safeguard against egregiously poor design decisions. Based upon design experience and lighting research, the following daylighting design guidelines assume overcast sky conditions. Though they might serve to establish a starting point for design, beginning at the schematic design phase, such "rules-of-thumb" should give way to iterative performance analyses that incorporate other parameters such as sky conditions, orientation, building envelope characteristics, interior finish information, and more, using manual calculation methods, daylighting simulation programs, or physical models.

(a) The 2.5H Guideline

This longstanding guideline in the lighting design field (Figure 12.30) assumes that there will be sufficient work plane illuminance from a window up to a distance of 2.5 times the head height of the window above the work plane—assuming clear glazing, overcast skies, no major obstructions, and a total window width that is approximately half that of the exterior perimeter wall. The work plane is the plane on which most visual tasks take place, and on which illuminance is specified and measured. Unless noted otherwise, designers should assume the work plane to be 30 in. [0.76 m] above the finish floor.

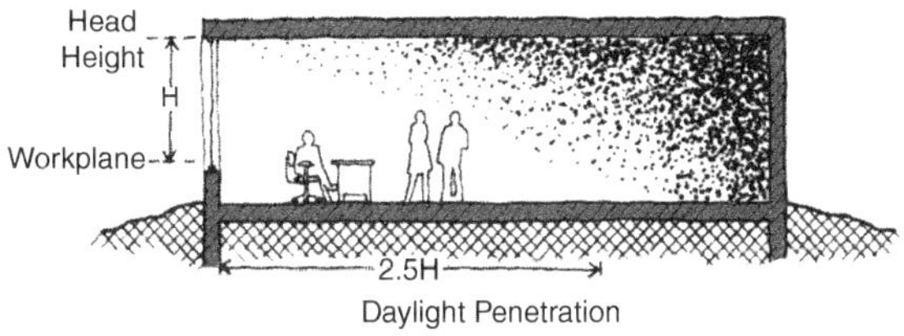

Figure 12.30 Section shows the 2.5H guideline, which assumes that sufficient daylight for the desk plane will be delivered at a depth 2.5 times the height of the window above the desk plane. (© Walter Grondzik; all rights reserved.)

(b) The 15/30 Guideline

This preliminary design guideline assumes that a 15 ft [4.6 m]-wide zone from a window wall (Figure 12.31) can be daylit sufficiently for office tasks. The next 15 ft [4.6 m] zone can be partially daylit and supplemented with electric lighting. Zones farther than 30 ft [9.1 m] from the window would receive very little daylight. In schematic design, these most-distant areas might ideally be allocated to circulation.

(c) The Sidelighting and Toplighting Daylight Factor Guideline

The size of windows, clerestories, or skylights may be estimated by using the simple formulas in Table 12.5, Parts A and B, which provide target daylight factor values. These design guidelines consider two factors: the height of the window in the wall and the window or skylight area compared to the floor area for each daylit space.

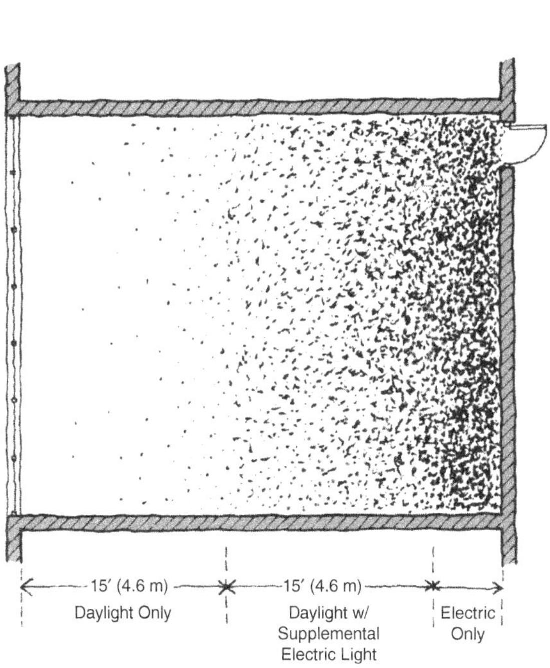

Figure 12.31 Plan shows the 15/30 guideline, which assumes that sufficient daylight will be delivered to the desk plane at a 15 ft [4.6 m] distance from the window wall. The 15 to 30 ft [4.6 to 9.1 m] daylight zone will need supplementary electric lighting, and the zone beyond 30 ft [9.1 m] will receive virtually no daylight. (© Walter Grondzik; all rights reserved.)

Table 12.5, Part C, shows design guidelines for buildings with an atrium that provides daylighting to surrounding offices. Because the lowest daylight factor will occur in offices on the lowest floor (deepest within the atrium), the designer might find the required atrium aspect ratio first for the lowest floor and then size the atrium on that basis. The atrium aspect ratio equals [(length × width)/height^2]. This design approach would provide a higher daylight factor in all the offices higher in the atrium. For a detailed discussion of the relationship of atrium size, rentable office floor area, and latitude, see DeKay (1992).

12.12 MANUAL CALCULATION METHODS

Because of the variability of daylight, the designer may provide a balance of illumination to save electric energy and reduce utility costs while addressing issues of glare, direct sunlight, and heat gain. The art and science of daylighting is largely about understanding how to control the admission of daylight into buildings. As such, in the following sections, three basic approaches to interior daylighting analysis are described:

- *Manual calculation methods* (Section 12.12) constitute "hand calculations" that address only minimum, maximum, and average conditions. The manual calculation methods are inexpensive but limited to simple spatial geometries.
- *Computer simulation programs* (Section 12.13) can produce detailed and realistic presentations in three-dimensional graphical form. Software is widely available, but its use is dependent upon cost and training, and the user must understand daylighting concepts and principles in order to interpret the results and to overcome the limitations of simulation.
- *Physical models* (Section 12.14) offer the designer an economical, realistic, and accurate alternative. Additionally, the intuitive understanding provided by scale models may increase the client's understanding of lighting phenomena.

Three manual calculation methods are offered in this section: the CIE method, the Graphic Daylighting Design Method (GDDM), and the IES lumen method of daylight calculations.

TABLE 12.5 Daylight Factor Design Estimates for Overcast Sky Conditions

Part A: Sidelighting[a,b]

$$DF_{avg} = 0.2\left(\frac{\text{window area}}{\text{floor area}}\right)$$

$$DF_{min} = 0.1\left(\frac{\text{window area}}{\text{floor area}}\right)$$

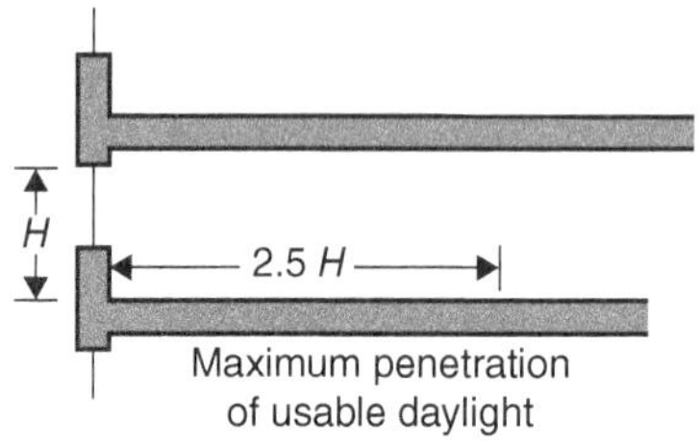

Part B: Toplighting[c]

Vertical monitors:

$$DF_{avg} = 0.2\left(\frac{\text{skylight glazing area}}{\text{floor area}}\right)$$

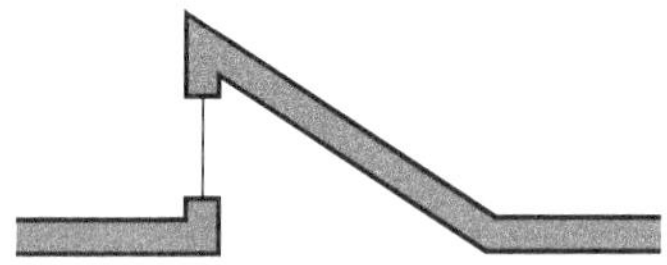

North-facing sawtooth:

$$DF_{avg} = 0.33\left(\frac{\text{skylight glazing area}}{\text{floor area}}\right)$$

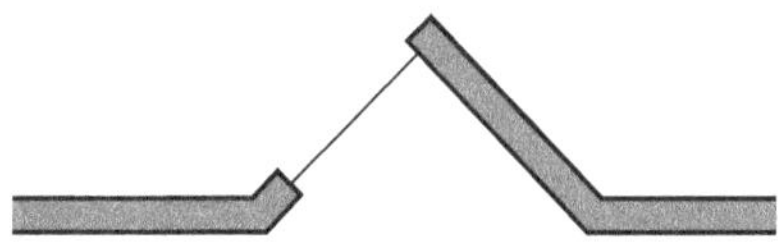

Horizontal skylights:

$$DF_{avg} = 0.5\left(\frac{\text{skylight glazing area}}{\text{floor area}}\right)$$

Part C. Buildings with atrium[d]

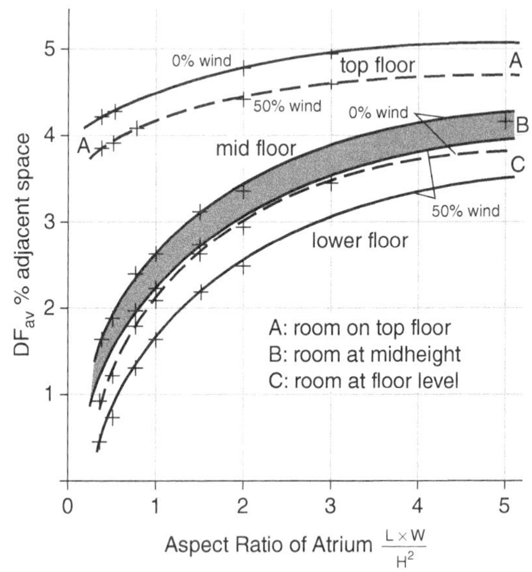

Source: Parts A and B: Millet et al., 1980; Part C: Brown and DeKay (2014).

[a] Assumes windows in one wall of a room with relatively light-colored surfaces.

[b] Window height/room depth relationships based on the works of R.G. Hopkinson (1966) and others at the British Research Station.

[c] Assumes an even distribution of such skylights in the roof so that an even distribution of light results in the room below: thus, only average DF, no minimum, is listed.

[d] Based on model tests of a square atrium with white walls open to the sky. "No windows" average atrium wall reflectance = 70%. "50% windows" average atrium wall reflectance = 40%. DF values are for an office of 9 m × 9 m × 3 m [30 ft × 30 ft × 10 ft]. The window opening to the atrium is 1.5 m high × 9 m long [5 ft high × 30 ft long], and the sill height is 0.85 m [2.8 ft].

(a) The CIE Method

This method resulted from a search for a simple, rapid, straightforward, and reasonably accurate daylighting calculation method that would yield reliable results without the time-consuming constructions and calculations necessitated by other manual methods. After a study of considerable length and intensity, the CIE adopted and adapted a system developed in Australia by Dr. Albert Dresler (1963). The current CIE method was published in *Daylight* (1970).

This system is based upon the daylight factor (DF) described previously as applied to the standard overcast CIE sky. Dresler developed a set of more than 100 curves covering rooms of varying proportions and fenestration. A typical set of curves is shown in Figure 12.32. The curves relate minimum daylight factor (at a point 2 ft [0.6 m] from the wall opposite a window) to the maximum permissible room depth, for given reflectances and a standard window design, thus establishing the room's proportions. Depth, or width, is the dimension at right angles to the window wall.

The curves imply that the number of design variables is so large and daylight itself is so variable that a simple routine method can be based only on minimal conditions for a given (selected) daylight duration. Therefore, the diagrams give the lowest level of daylight that can reliably be expected for a given percentage (percentile) of normal working hours in sidelighted rooms and the average level in toplighted spaces.

Advantages of the CIE method are:

1. It allows consideration of obstructions, exterior reflections, and interior reflections.
2. It is applicable to a very wide range of side and top fenestration designs.
3. The establishment of required room proportions is architecturally more useful than solving for specific dimensions.

Limitations of the CIE method are:

1. It assumes the standard overcast CIE sky and is inapplicable to clear-sky and direct-sun conditions.
2. It is inapplicable to spaces other than rectangular rooms.
3. It is unusable with sunshading devices or high-reflectance ground surfaces.
4. Results provide points of minimum, twice minimum, and four times minimum daylight only. Other points must be interpolated or extrapolated.
5. Window proportions and position in a wall are fixed.

Overall, the CIE system accomplishes what it intended. The limitations listed are inherent in any quick, simplified daylight calculation technique.

PASSIVE ENVIRONMENTAL SYSTEMS

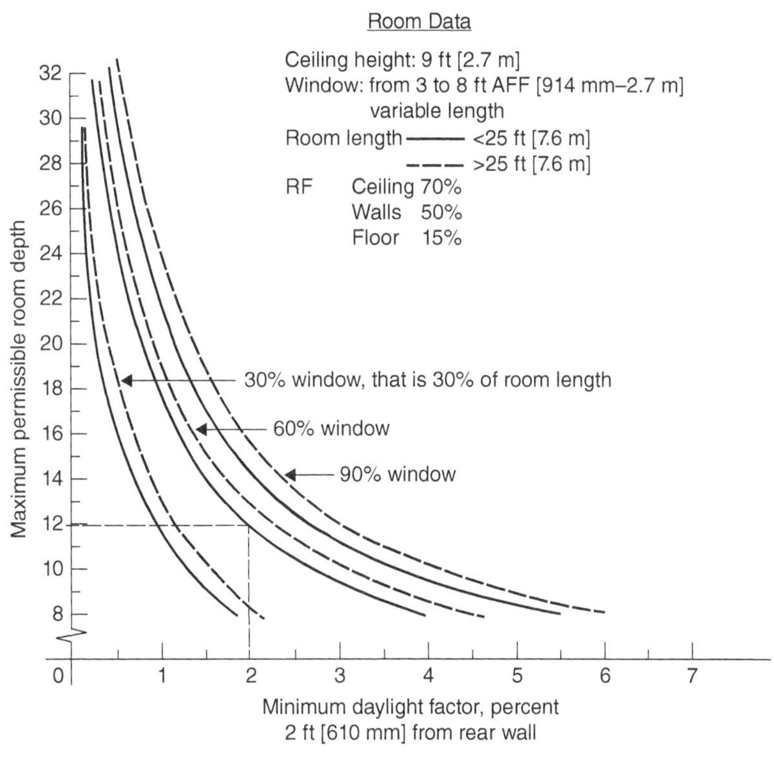

Figure 12.32 *Maximum room depth that will maintain a minimum daylight factor is proportional to window size. Thus, for a room less than 25 ft [7.6 m]-long with a 5-ft [1.5-m]-high window for 60% of the room's length, the depth cannot exceed 12 ft [3.7 m] if 2% DF is to be maintained at a point 2 ft [61 cm] from the rear wall. (From Dresler, 1963.)*

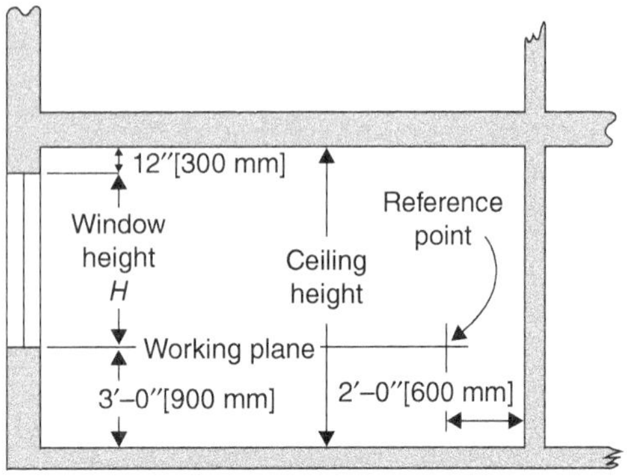

Section through a unilaterally lit room showing the assumed dimensions. These dimensions are the same for bilateral lighting except that the reference point is midway between the window walls.

(*a*)

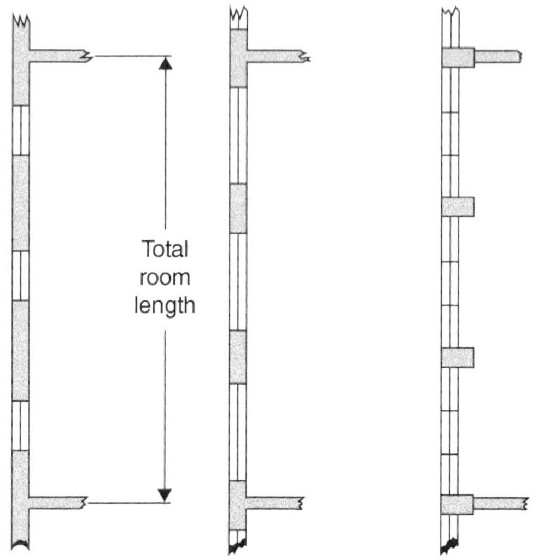

Plane of window walls showing window width expressed as a percentage of total room length.

(*b*)

Figure 12.33 *Sketches indicating the parameters of the CIE method. (a) A vertical section through a room with dimensional data relevant to this system. Note that the sill height has been selected to coincide with a work plane at 900 mm [3 ft]. The height of the work plane usually varies between 760 and 910 mm [30 and 36 in.], the former being more common in North America, the latter in Europe. A lower sill contributes only ground-reflected light at the work plane. Where the window sill is significantly above the work plane (i.e., short windows high on a wall), this analysis system is inapplicable. (b) Calculation size (length) of windows with respect to overall room length. (CIE, 1970; reproduced with permission.)*

The CIE system is usable in two modes:

1. Given complete architectural dimensional data, to find interior illuminance.
2. Given incomplete architectural dimensional data but required interior illuminance, to find maximum room depth and/or other room proportions that satisfy the illuminance requirement.

Mode 1 is simpler because it leads directly to an answer. For this reason, a designer should set the room length (window wall dimension) and percentage of fenestration of the window wall, leaving the room depth (perpendicular to window wall) as a variable. Alternatively, room length and depth may be set, with the percentage of fenestration as the variable. Ceiling height is usually fixed. See Figure 12.33 for sketches showing room parameters.

EXAMPLE 12.1 An example of the CIE method in Mode 1 uses a classroom in a single-story Seattle elementary school, 25 ft [7.6 m] length, 18 ft [5.5 m] depth, and with a 9.5 ft [3 m] ceiling height. It receives daylight unilaterally from three 6 ft [1.8 m] windows totaling 18 ft [5.5 m] in length (see the room sketch in Figure 12.34). Window glazing is wired glass having a transmittance of 80%. The school is situated in a dense residential area. Determine the portion of the year during which tasks requiring a minimum illuminance of 14 fc [150 lx] can be carried out by daylight throughout the room. Also, determine what illuminance levels can be maintained for 85% of daylight hours and at what distances from the window. (Note that 14 fc [150 lx] corresponds to a DF of 2% applied to an unobstructed outdoor horizontal illuminance (E_H) of 70 fc [7500 lx].)

SOLUTION

The latitude of Seattle is 47.6°N. The design condition for Seattle is solid overcast sky for 85% of the hours between 9:00 a.m. and 5:00 p.m. From Figure 12.35, the minimum external illuminance (i.e., unobstructed outdoor horizontal illuminance, E_H) during these hours is 669 fc [7200 lx].

STEP 1. Determine the room depth in terms of window height. Window height (H) is 5 ft [1.5 m] and the sill is 3.5 ft [1.0 m] (see elevation in Figure 12.34). In plan, the room depth is expressed as multiples of window height above sill level:

$$\frac{\text{18-ft room depth}}{\text{5-ft window height}} = 3.6H$$

STEP 2. Determine the window coverage of the window wall. This variable is expressed as a percentage of the total room length based upon the width of the glazing used in the room. It is assumed that the window head depth is 12 in. [305 mm] below the ceiling. This head depth was

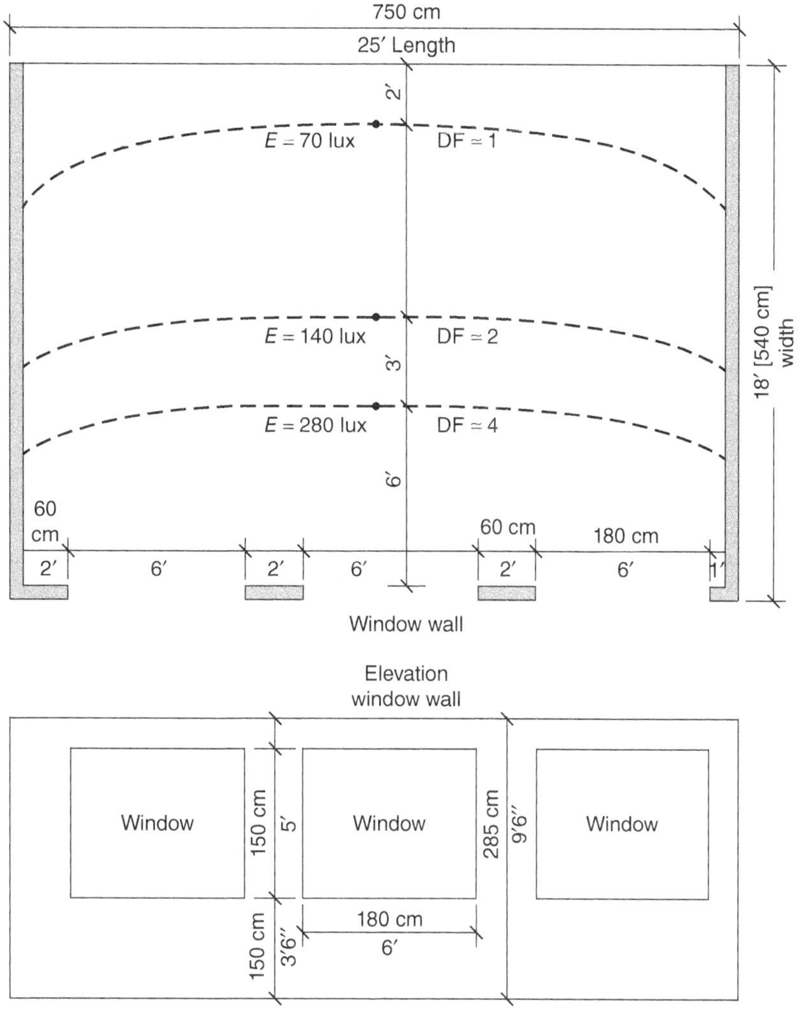

Figure 12.34 *Plan and window wall elevation of the Seattle classroom calculation example using the CIE method. The three daylight contours are estimated based upon the calculated center point. They represent the levels maintained for 85% of daylight hours. Levels twice as high are maintained for 60% of the daylight hours.*

selected as the representation of best practice "without being unduly optimistic."

$$\frac{3\times 6\,\text{ft}}{25\,\text{ft}} = 72\%$$

STEP 3. Determine the design daylight factor and the service daylight factor.

The design daylight factor (DF_{design}) is calculated at a point 2 ft [0.6 m] from the wall opposite the window—on the window centerline. This represents the minimum daylight factor that will occur in a room of a given depth (lower values that will occur closer to the wall and/or off the centerline being discounted). The room depth as a multiple of *H* is 3.6*H* (see step 1). The window coverage is 72% (see step 2). Recall the room length of 25 ft [7.6 m] and ceiling height of 9.5 ft [2.9 m]. An application of Figure 12.36 yields the design daylight factor:

$$DF_{design} = 1.3\%$$

Note that some interpolation between the solid curves is required. Also, even though the solid curves are more applicable to the room length than the dashed curves regarding room length, the interpolated value derived needs to be adjusted up slightly to account for the room length in this example.

The service daylight factor ($DF_{service}$) takes into account correction factors such as glazing transmission and dirt accumulation. The service daylight factor is a product of the design daylight factor and correction factors.

$$DF_{service} = DF_{design} \times \text{correction factors}$$

Assume the window glass has a diffuse transmittance of 80%. Also, account for this school being located in a built-up residential area and exhibiting a "clean" class of industry (regarding the potential for dirt accumulation on the window glass). Correction factors from Table 12.6 are:

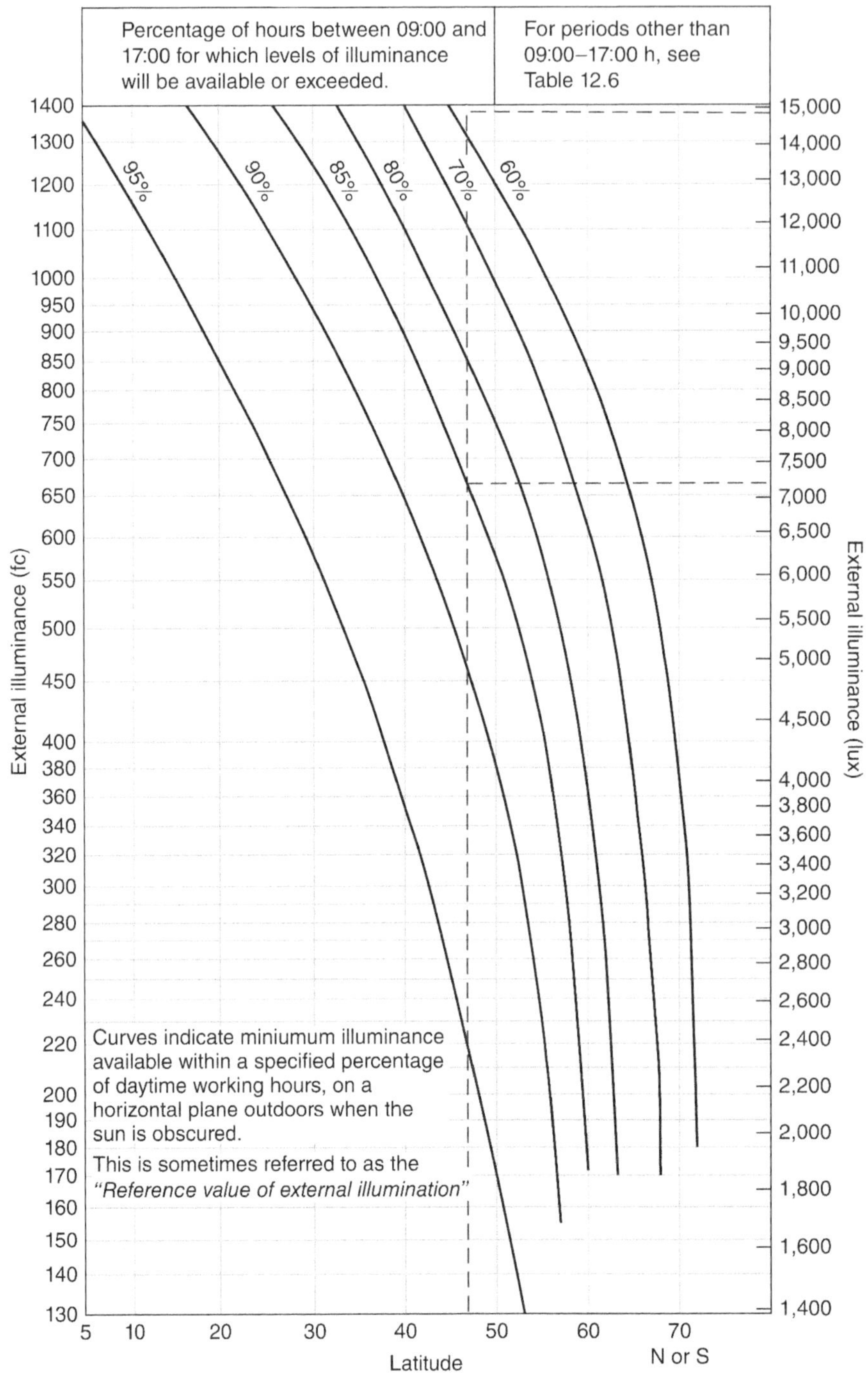

Figure 12.35 Minimum maintained external illuminance as a function of latitude for a given percentage of the normal working day. (CIE, 1970; reproduced with permission.)

correction factor to account for glass transmittance: 0.95; correction factor to account for dirt accumulation on glass: 0.8.

Therefore:

$$DF_{service} = 1.3\% \times 0.95 \times 0.8$$

$$DF_{service} \approx 1.0\%$$

For curiosity, consider the impact that a nearby structure could have on this example due to obstructing a portion of the sky. If the structure's angle of obstruction to the windows was 20°, then according to Figure 12.37, a correction factor of 0.65 would also need to be applied to step 3. This would take the service daylight factor down to 0.65% (i.e., 1.0% × 0.65).

Note that the terms "design" and "service" as used in this method have meanings that are not consistent with normal design practice. Design daylight factor typically would include the effects of building components such as glazing—and would be described

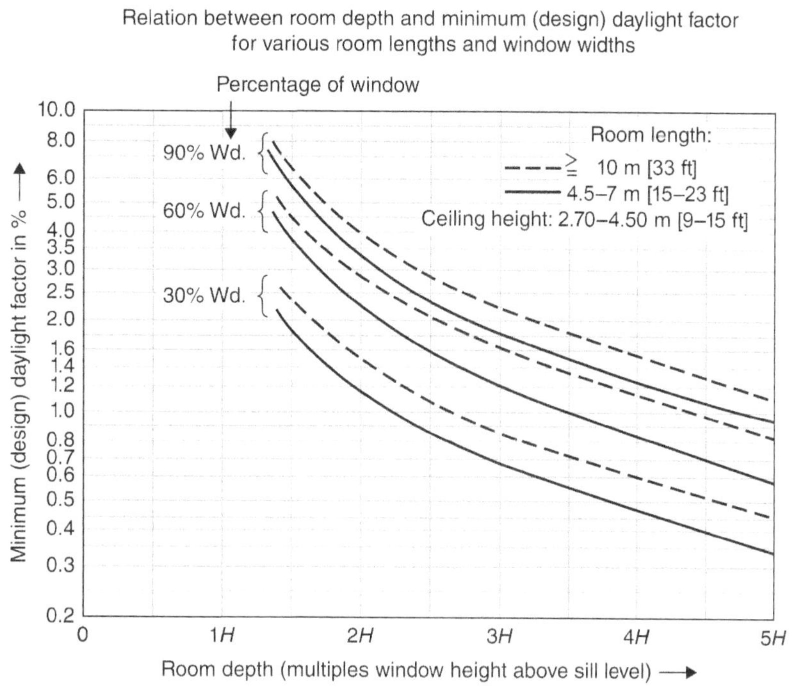

Figure 12.36 *Basic design diagram that relates minimum daylight factor to room depth. Inasmuch as room depth is expressed in terms of window height, the curves effectively relate minimum daylight factor (2 ft [610 mm] from the back wall) to room proportion. (CIE, 1970; reproduced with permission.)*

TABLE 12.6 Correction Factors to Be Used in CIE Daylight Calculations

Part A. Correction Factor to Account for Glass Transmittance	
Diffuse Transmittance of Glass (%)	**Correction Factor**
80	0.95
70	0.80
60	0.70
50	0.60
40	0.45
30	0.35

Part B. Correction Factors to Account for Dirt Accumulation on Glass				
		Angle of Slope (Measured to the Horizontal)		
Locality	**Class of Industry**	**90–75°**	**60–45°**	**30–0°**
Country or outer-suburban area	Clean	0.9	0.85	0.8
	Dirty	0.7	0.6	0.55
Built-up residential area	Clean	0.8	0.75	0.7
	Dirty	0.6	0.5	0.4
Built-up industrial area	Clean	0.7	0.6	0.55
	Dirty	0.5	0.35	0.25

Part C. Percentages to Use When Figure 12.35 Curves Are Applied to Periods Other Than 09:00–17:00						
Curve in Figure 12.35	**95%**	**90%**	**85%**	**80%**	**70%**	**60%**
Alternative Period	**Percentage of Alternative Period**					
07:00–15:00	95	90	85	80	70	60
08:00–16:00	100	100	95	85	70	60
07:00–17:00	95	85	75	65	55	45
06:00–18:00	75	70	65	60	50	40

Source: CIE, 1970; reproduced with permission.

PASSIVE ENVIRON-MENTAL SYSTEMS

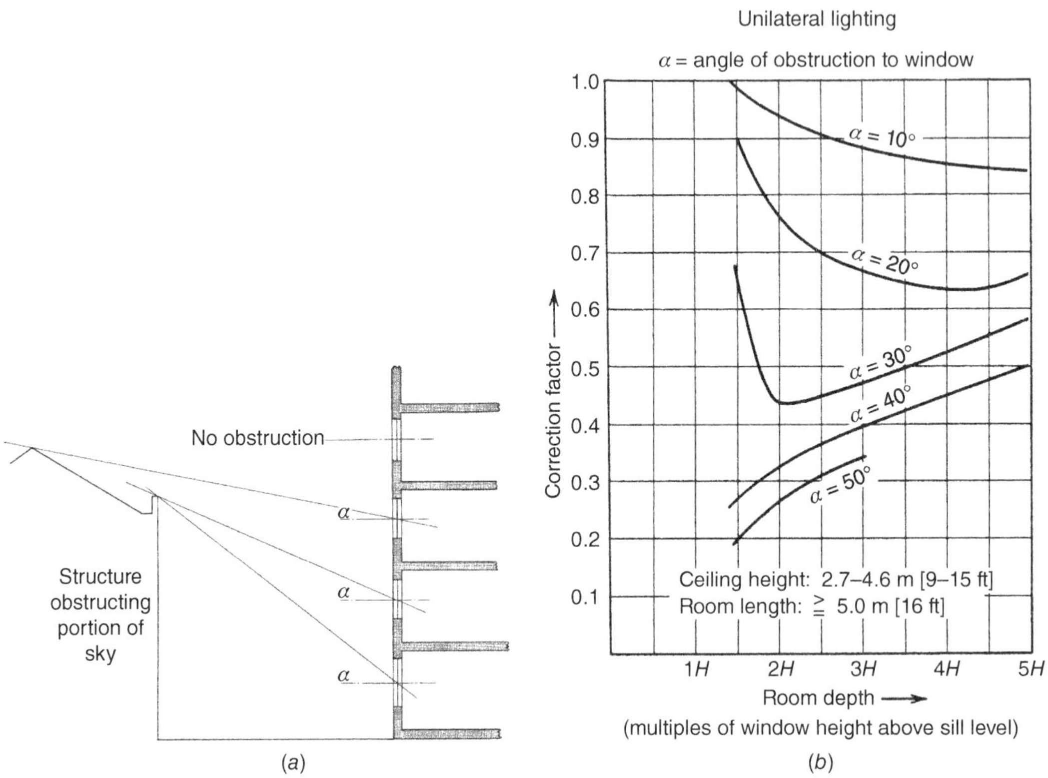

Figure 12.37 *(a) Angle of obstruction α of an external object. (b) Correlation factors to account for the influence of external obstructions on the minimum daylight factor. (CIE, 1970; reproduced with permission.)*

as "initial" daylight factor. Service daylight factor would be called "maintained" daylight factor and include those effects that would reduce daylight illuminance over time (such as dirt on glazings or reductions in surface reflectances). Maintained daylight factor should be equal to or exceed the designer's daylight factor criterion for the system to be successful in the long run. If the maintained daylight factor is under the design daylight factor, the daylight strategy may be challenged. In this example, the DF_{design} is 1.3% and the $DF_{service}$ is 1.0%—which is under the 1.3% target. This is not ideal, but the values are likely close enough to proceed, knowing that the classroom's daylighting may be insufficient at times.

STEP 4. Determine the required exterior illuminance. Recall that this example asks us to determine the portion of the year during which tasks requiring a minimum illuminance of 14 fc (150 lx) can be carried out by daylight throughout the room. Using the following formula, ascertain the required exterior illuminance on an unobstructed horizontal surface (E_H) needed to achieve the service daylight factor ($DF_{service}$):

$$E_H = \frac{\text{required interior illuminance}}{DF_{service}} \qquad \text{(Eq. 12.4)}$$

$$E_H = \frac{150\ lx\ min}{1.0} \times 100 = 15{,}000\ lx$$

STEP 5. Obtain the percentage of hours between 9:00 a.m. and 5:00 p.m. during which the required illuminance is maintained. From Figure 12.35 with the given conditions (roughly 48°N latitude, exterior illuminance 15,000 lx [1500 fc]), an illuminance of 150 lx [15 fc] will be maintained for less than 60% of the hours between 09:00 a.m. and 5:00 p.m. (for time periods other than this, see Table 12.6C). The level that is maintained for 85% of the hours is at an E_H of 7200 lx [669 fc]:

$$E_{min} = 7200\ lx \times DF_{service} \text{ of } 1.0\%$$

$$E_{min} = 7200\ lx \times 0.01$$

$$E_{min} = 72\ lx [6.9\ fc]$$

STEP 6. Determine the locations in the room that will receive adequate illuminance. Figure 12.38 shows the distance from the window at which the daylight factor is twice (a) or four times (b) the minimum, which is the $DF_{service}$ of 1.0%.

Using Figure 12.38*a*, the room depth is 3.6*H* (step 1). This point intersects with the α = 0° angle of obstruction line (step 3, recall no obstructions) at a 1.8*H* distance (1.8 × 5 ft = 9 ft [2.7 m]) from the window, resulting in a doubling of the minimum daylight factor. Recall E_{min} is 7200 lx (step 5).

$$E_{@1.5H} = 7200\ lx \times DF_{service} \text{ of } 1.0\% \times 2$$

$$E_{@1.5H} = 144\ lx$$

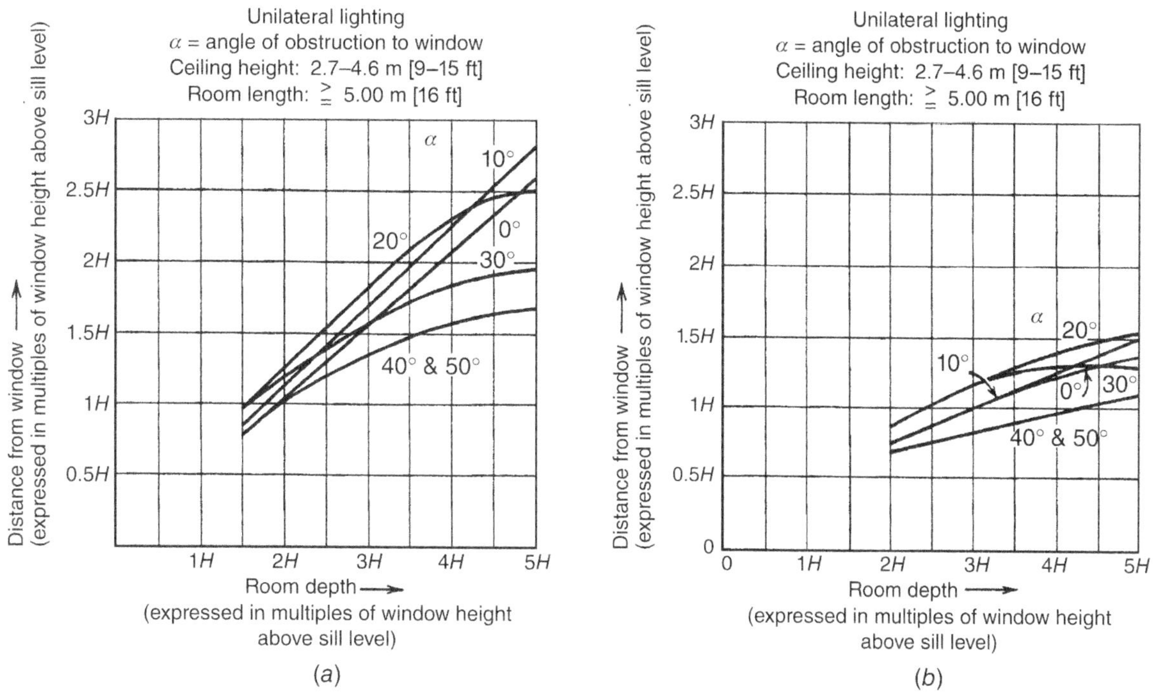

Figure 12.38 *(a) Distance from the window at which the daylight factor is twice the minimum daylight factor. (b) Distance from the window at which the daylight factor is four times the minimum daylight factor. (CIE, 1970; reproduced with permission.)*

Using Figure 12.38*b*, the room depth is 3.6*H*. This point intersects with the $\alpha = 0°$ angle of obstruction line at a 1.2*H* distance ($1.2 \times 5\,\text{ft} = 6\,\text{ft}$ [1.8 m]) from the window, resulting in a quadrupling of the minimum daylight factor.

$$E_{@1.2H} = 7200\ lx \times DF_{service} \text{ of } 1.0\% \times 4$$

$$E_{@1.2H} = 288\ lx$$

These calculated illuminance levels are accurate only at the centerline of the window wall. By visual estimate and extrapolation, rough contours can be drawn and are shown in Figure 12.39 (for comparison with the graphic method discussed in the next section). For bilateral sidelighting, toplighting (skylights, sawtooth roofs, and monitor roofs), and calculations for other α angles of obstruction, additional information is found in the CIE document (*CIE*, 1970). ■

In summary, the CIE method is relatively simple but provides only limited data on predicted performance. In this example, its exterior illuminance data (7250 lx from Figure 12.35) seem to agree well with a measured average value of 7200 lx for Seattle. From the rough contours shown in Figure 12.35, an integrated daylight and electric lighting strategy should be designed for areas farther from the window wall.

Reminder: The term design illuminance as used with the CIE method differs from common lighting system usage where design illuminance is used to identify the criteria [benchmark] illuminance established for a space or position; initial illuminance identifies the illuminance actually provided at system startup, and maintained illuminance is the illuminance provided after some defined time period.

(b) Graphic Daylighting Design Method

This method, which applies to overcast sky conditions and shows results as daylight factor (isolux) contours within a room (rather than individual daylight factors at specific points), was developed by Millet, Adams, and Bedrick (1980). The primary advantage of the graphic daylighting design method (GDDM) over the CIE method is that its results are a family of daylight factor contours that are more useful to a lighting designer than is numerical output. The disadvantages of this method are that it is not readily applicable to clear-sky conditions, and it requires that a designer acquire a "library" of 200 or so patterns that cover most design situations. An outline of the method is presented here.

A computer simulation program (UWLIGHT) developed the daylight distribution patterns resulting from either sidelights or skylights. To generalize the system, windows are identified by height-to-width proportion (*H*/*W*), and the positions of the isolux contours on the plan are determined by the ratio of the height of the sill above the work plane to the window height. The GDDM can account for high windows, clerestories, and other designs intended to introduce

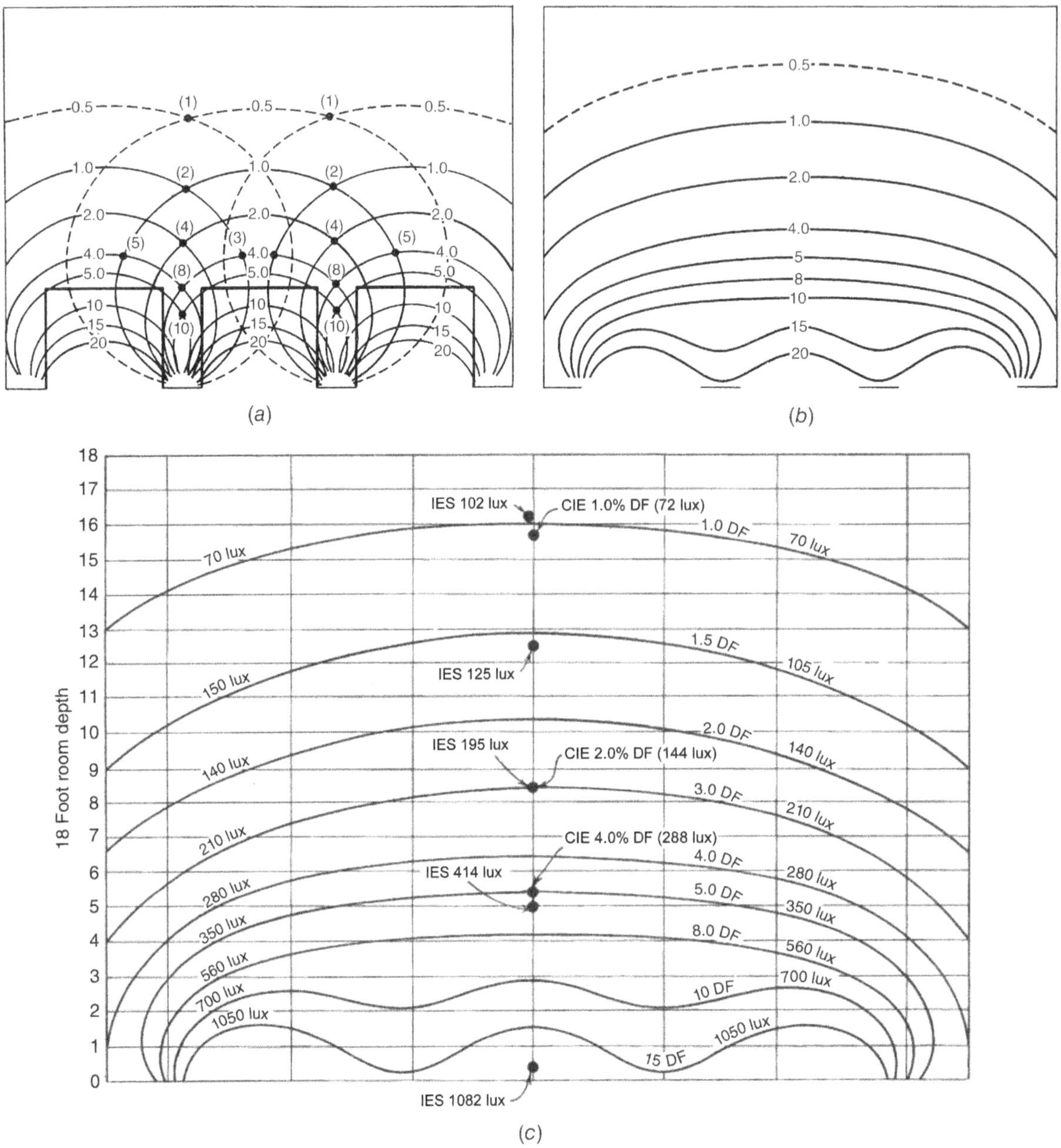

Figure 12.39 (a) Daylight contours for each window of Figure 12.34 are plotted on the floor plan of the room being studied. Numbers in parentheses are combined sky component (SC) values. (b) The isolux contours of (a) are combined to form new isolux contours that represent the total SC of daylight within the room. (c) The final isolux contours are calculated, including correction factors (accounting for internally reflected components [IRC] of daylight plus light reduction due to glazing). The numbers represent daylight factors. Note the variance between these contours and the points calculated by the CIE method. The five design points defined by the IES lumen method of daylight calculations are also shown. Compare the results on the room centerline (a location where comparison of all methods is possible); agreement is within engineering accuracy (see text discussion).

daylight deep into a space—something that the CIE method cannot do because it is restricted to a sill height at the work plane.

EXAMPLE 12.2 Figure 12.40 shows a typical isolux pattern for a window whose room height-to-width (*H/W*) ratio is 0.8 and whose sill is at the work plane. This particular window pattern was selected because it corresponds to the window in the previous example, Figure 12.34, enabling graphical comparison.

SOLUTION

Employing the GDDM method with the dimensions of the Seattle classroom used in the last section:

STEP 1. Determine the window proportion.

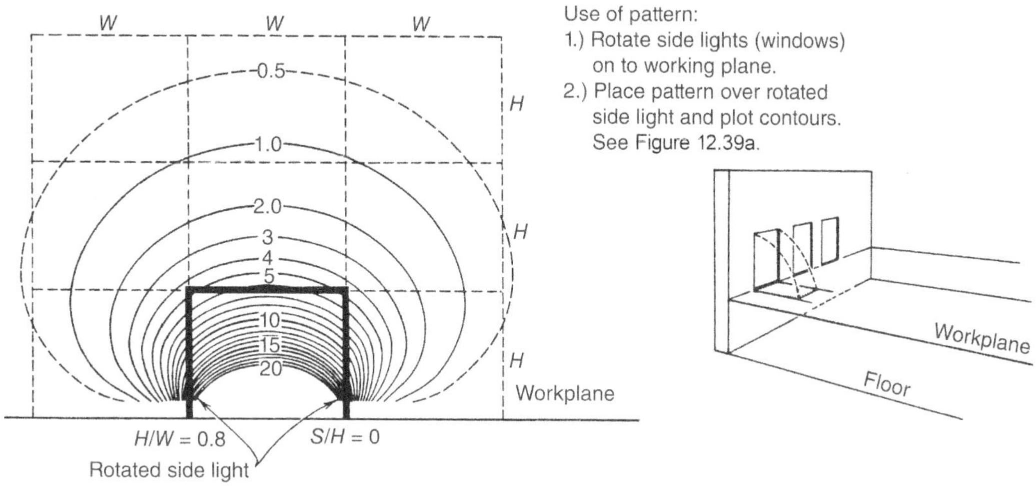

Figure 12.40 *Typical isolux contour map for a window with a height-to-width ratio of 0.8 and a sill at the work plane elevation. Numbers represent the sky component (SC) of daylight factors for an overcast sky condition. The rectangles are the window outlines rotated (projected) onto the work plane. See insert. (Millet et al., 1980.)*

Referring to Figure 12.34, the window proportion (height/width) is

$$\frac{H}{W} = \frac{5}{6} = 0.83$$

STEP 2. Select the appropriate window pattern. In this case, Figure 12.40 (selected from a library of isolux patterns developed by Millet, Adams, and Bedrick, 1980) is the closest pattern to match the example. $S/H = 0$ indicates that the pattern begins at the window wall, as shown.

STEP 3. Develop an isolux pattern for each window of the space. On a plan of the room, trace the isolux pattern for each window (Figure 12.39*a*). The patterns overlap because the windows are close together. Where contours meet, the daylight factors of the contours are added together, producing values for the new combined contours. The combined contours and their daylight factors are shown in Figure 12.39*b*.

STEP 4. Make corrections to the isolux pattern. The value of the combined contours is corrected to account for IRCs plus light reduction due to glazing. The final contours are shown in Figure 12.39*c*. Note that this diagram gives the designer a much more complete picture of the daylight contours than do the results of the CIE method. For comparison, the three calculated daylight factors from the CIE method are shown in Figure 12.39*c*. ■

(c) The IES Lumen Method of Daylight Calculations

IES developed the lumen method of daylight calculations (published as *RP-23-89: Recommended Practice for the Lumen Method of Daylight Calculations*; IES, 1989). Although inexpensive, like many manual methods, it is limited in application—in this case, to rectilinear spaces with flat ceilings. Today, IES favors daylight simulation programs (see Section 12.13) over this method—even at early design stages, as such simulations can incorporate the unique characteristics of a specific circumstance.

A trade-off between usability, learning curve, and cost is often made when selecting a daylighting design method; and while the IES lumen method of daylight calculations may not offer a high degree of utility to designers, the procedure may offer value in the absence of digital tools or for instructional purposes. Do not conflate the IES lumen method of daylight calculations as defined by *RP-23-89* with the IES lumen method for electric lighting as defined by IES (2020d) in *IES LS-6-20: Lighting Science: Calculation of Light and Its Effects*.

The calculation procedure for sidelighting is outlined in this section, as this is a more frequently encountered strategy than toplighting. The method, as fully described in *RP-23-89*, consists of four detailed steps. In the section that follows, the same notation and terms found in *RP-23-89* are used except for bearing angle, which is instead referred to in the IES procedure as solar window azimuth. The term bearing angle is commonly used in international sources.

Characteristics of the Method. The IES lumen method of daylight calculations is probably the most flexible manual technique available. It has the following major characteristics for sidelighting:

1. It takes into account reflected light from the ground and adjacent structures, as well as the reduction in skylight due to such structures.

2. It cannot accommodate direct sunlight, but, conversely, it readily accommodates the shading devices normally used to block direct insolation.
3. Provision is made for various types of glazing, as well as common window controls, such as horizontal and vertical blinds.
4. The principles of the zonal cavity calculation approach for interior lighting are applied. The window height determines the cavities (i.e., the floor cavity extends to the window sill height, and the ceiling cavity from the top of the window to the ceiling). The room cavity is therefore the window height.
5. The work plane is always at the sill height of the window. Where this is decidedly not the case (a difference of up to 1 ft [0.3 m] is usually negligible), such as when a clerestory or a floor-to-ceiling window is used, work plane illuminance can be calculated by superposition. For instance, with a clerestory, subtract a work-plane-to-clerestory sill height window from a work-plane-to-top-of-clerestory window to obtain the desired result. A degree of inaccuracy is unavoidable in the calculation when the work plane is above the sill.
6. Cavity reflectances are fixed (Figure 12.41) at 70%, 50%, and 30% for ceiling, room, and floor cavities, respectively.
7. The system calculates only five points in a room on the window centerline. As noted with reference to the three points calculated by the CIE method, this is not normally sufficient to give a picture of the interior daylight distribution.
8. The method is usable in only one mode—that is, given location and full dimensional data, daylighting can be calculated. It cannot readily be used to determine desirable room proportions, given the other data, as can the CIE method.

EXAMPLE 12.3 To again use the Seattle classroom as an example, here are the conditions of the problem:

- Location: Seattle, Washington
- Latitude: 47.6°N
- Room: 25 ft [7.6 m] long, 18 ft [5.5 m] deep, with a 9.5 ft [3 m] ceiling
- Window: height 5 ft [1.5 m] above a 30 in. [760 mm] work plane; total length, 18 ft [5.5 m]
- Glazing transmittance: 85%
- Net glazing percentage of the window area: 92%
- Ground reflectance: Not previously specified. Assume an extensive area of mixed grass, asphalt, and concrete walkways, with an average overall ground reflectance (RF_g) of 20% (see Table 12.10 on p. 40).

SOLUTION

Illuminance (E_i) will be found at the five points shown in Figure 12.41 for a spring day and a winter day at 10:00 a.m. and 2:00 p.m. Assume that the sky is overcast so that a direct comparison with other methods can be made. Reflectances are assumed to be 70% for the ceiling, 50% for the wall, and 30% for the floor to correspond to the IES method standard conditions.

STEP 1. Determine the vertical (E_{xvk} and E_{xvg}) and horizontal (E_{xhk}) illuminance components on the exterior of the vertical window area.

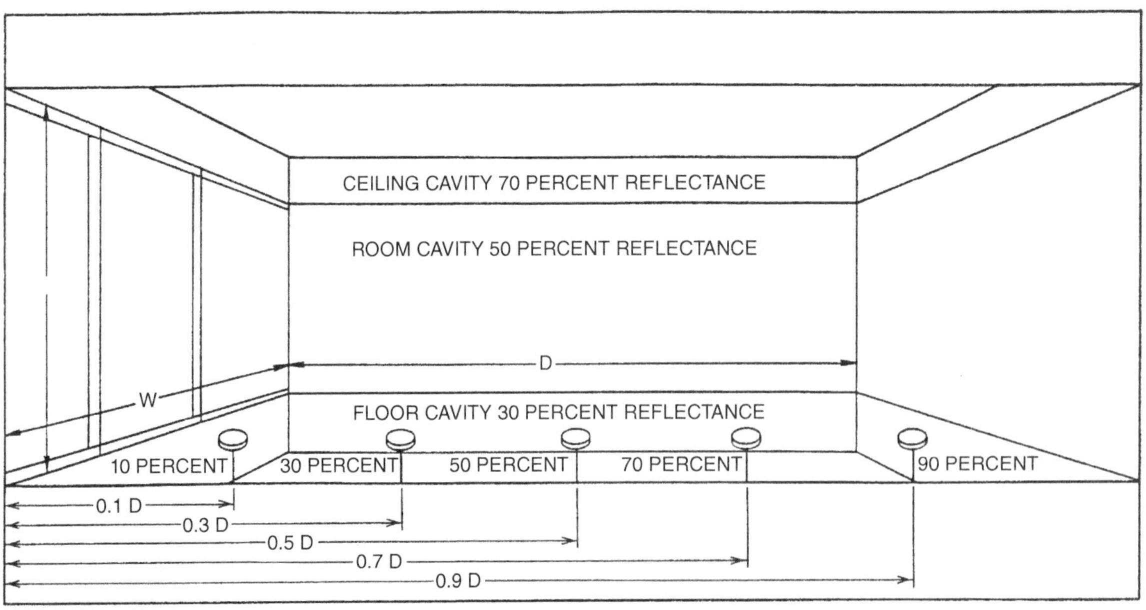

Figure 12.41 Standard conditions in a room for daylighting calculations: sidelighting. (Reproduced from IES RP-23-89: Recommended Practice for the Lumen Method of Daylight Calculations with permission of Illuminating Engineering Society.)

TABLE 12.7 Vertical and Horizontal Illuminance Values for Spring and Winter, Seattle, Washington

Solar Altitude (Appendix D, Section D.1)	Vertical Illuminance, E_{xvk} (Figure 12.23)	Vertical Illuminance from Ground Reflectance, E_{xvg} (Equation12.5)	Horizontal Illuminance from Full Sky, E_{xHk} (Figure 12.23)	Horizontal Illuminance from Half Sky, E_{xhk}
Dec. 21: 14°	200 fc [2150 lx]	50 fc [538 lx]	500 fc [5380 lx]	250 fc [2690 lx]
Mar. 21: 36°	500 fc [5380 lx]	124 fc [1334 lx]	1240 fc [13,340 lx]	620 fc [6670 lx]

First, using the solar altitude for the spring and winter day (Appendix D, Table D.1, solar data, use reference 48°N), determine the vertical window illuminance values and full sky horizontal illuminance values from Figure 12.23 (E_V to be used as E_{xvk}; E_H to be used as E_{xHk}) for a spring (March 21) and winter (December 21) day.

Then, divide the full sky illuminance (E_{xHk}) by two in order to determine the half sky illuminance (E_{xhk})—this is necessary because a vertical window can only "see" half of the sky.

In addition to the E_{xvk}, vertical illuminance on the exterior window would also include illuminance from ground reflectance (E_{xvg}). Using the ground reflectance value (RF_g), E_{xvg} is calculated as follows:

$$E_{xvg} = RF_g \times \frac{E_{xHk}}{2} \qquad \text{(Eq. 12.5)}$$

$$\text{Dec. 21: } E_{xvg} = 0.2 \times \frac{5380}{2} = 538\ lx$$

$$\text{Mar. 21: } E_{xvg} = 0.2 \times \frac{13{,}340}{2} = 1334\ lx$$

Values for E_{xvk}, E_{xvg}, E_{xHk}, and E_{xhk} are compiled in Table 12.7.

STEP 2. Determine net glazing transmittance (τ) of the window. A number of factors affect the transmittance of light through glazing. The net glazing transmittance is the product of:

- Glazing transmittance (Table 12.8 or manufacturer's data). Manufacturer's data should be consulted for the transmittance for specific glazing assemblies (i.e., insulated glass, triple-pane glass).
- Light loss factor (LLF) (Table 12.9), which represents the cleanliness of the window.
- Net glazing percentage of the window area. Net glazing transmittance should only account for the net glazing area (i.e., gross window area less mullions, glazing bars, and so on) and any other factor (such as insect screens) that would reduce actual transmittance.

The glazing transmittance in our example is given as 85% (Table 12.8), light loss factor is 0.9 (Table 12.9), and the net glazing percentage is given as 92%. The net transmittance (τ) of the window is a product of these factors:

$$\tau = \text{(glazing transmittance) (LLF)} \times \text{(net glazing percentage of the window area)}$$

$$\tau = 0.85 \times 0.9 \times 0.92 = 0.70$$

TABLE 12.8 Transmittance Data for Glass and Plastic Materials

Material	Approximate Transmittance (%)
Polished plate/float glass	80–90
Sheet glass	85–91
Heat-absorbing plate glass	70–80
Heat-absorbing sheet glass	70–85
Tinted polished plate	40–50
Figure glass	70–90
Corrugated glass	80–85
Glass block	60–80
Clear plastic sheet	80–92
Tinted plastic sheet	9–42
Colorless patterned plastic	80–90
White translucent plastic	10–80
Glass-fiber-reinforced plastic	5–80
Double glazed—two lights clear glass	77
Tinted plus clear	37–45
Reflective glass[a]	5–60

Source: Reproduced from IES RP-23-89: *Recommended Practice for the Lumen Method of Daylight Calculations* with permission of Illuminating Engineering Society.

[a]Includes single glass, double-glazed units, and laminated assemblies. Consult manufacturer's material for specific values.

TABLE 12.9 Typical Light Loss Factors for Daylighting Design

	Light Loss Factor per Glazing Position		
Location	Vertical	Sloped	Horizontal
Clean areas	0.9	0.8	0.7
Industrial areas	0.8	0.7	0.6
Very dirty areas	0.7	0.6	0.5

Source: Reproduced from IES RP-23-89: *Recommended Practice for the Lumen Method of Daylight Calculations* with permission of Illuminating Engineering Society.

STEP 3. Select coefficients of utilization for sky (CU_k) and ground (CU_g) at the five calculation locations (10%, 30%, 50%, 70%, and 90%) from Appendix C, Tables C.21–C.26 on the basis of room dimensions and the portion of the sky seen by the window.

To determine CU_k, first, determine the sky component (SC) seen by the window. It is determined by the ratio of vertical illuminance (E_{xvk}) to horizontal illuminance for half sky (E_{xhk}) at the window.

$$\text{Recall from Table 12.7 that } SC = \frac{E_{xvk}}{E_{xhk}}$$

$$\text{Dec. 21: } SC = \frac{2150}{2690} = 0.8$$

$$\text{Mar. 21: } SC = \frac{5380}{6670} = 0.81$$

PASSIVE ENVIRON-MENTAL SYSTEMS

TABLE 12.10 Reflectances of Building Materials and Outside Surfaces

Material	Reflectance (%)
Aluminum	85
Asphalt (free from dirt)	7
Bluestone, sandstone	18
Brick	
Light buff	48
Dark buff	40
Dark red glazed	30
Red	15
Yellow ochre	25
White	75
Cement	27
Chromium	65
Concrete	55
Copper	40
Earth (moist cultivated)	7
Granolite pavement	17
Glass	
Clear	7
Reflective	20–30
Tinted	7
Grass (dark green)	6
Gravel	13
Granite	40
Marble (white)	45
Macadam	18
Marble	45
Paint (White)	
New	75
Old	55
Plaster	
Smooth	80
Rough	40
Stippled	40
Slate (dark clay)	8
Snow	
New	74
Old	64
Vegetation (mean)	25

Source: Values compiled from Lam (1986), Stein and Reynolds (1992), and the Lighting Design Lab (© 2005; used with permission). Data are reaffirmed in *IES LP-3-20: Lighting Practice: Designing and Specifying Daylighting for Buildings* (IES, 2020b).

The E_{xvk}/E_{xhk} value 0.80 corresponds most closely to the E_{xvk}/E_{xhk} value of 0.75 in Appendix C, Table C.21.

Use the following values to find the coefficients of utilization for sky (CU_k):

$$\frac{\text{room depth}}{\text{window height}} = \frac{18}{5} = 3.6$$

(use 4; first column variable)

$$\frac{\text{window length}}{\text{window height}} = \frac{18}{5} = 3.6$$

(use 4; first row variable)

Next, find the coefficients of utilization for ground (CU_g), using Table C.26. Again, use 4 as the room depth/window height variable and 4 as the window length/window height variable.

The sky (CU_k) and ground (CU_g) component values are compiled in Table 12.11 for each of the five reference locations (from Appendix C, Tables C.21 and C.26).

TABLE 12.11 Coefficients of Utilization for Sky and Ground Components for Five Interior Locations

Location[a]	CU_k (Sky)	CU_g (Ground)
10	0.673	0.183
30	0.235	0.159
50	0.104	0.103
70	0.065	0.071
90	0.053	0.060

[a]Expressed as percent of room depth.

TABLE 12.12 Illuminance Values for Winter and Spring in Seattle, Washington, at Five Reference Locations

Location[a]	E_i Dec. 21, lx [fc]	E_i Mar. 21, lx [fc]
10	1082 [101]	2705 [251]
30	414 [38]	1033 [96]
50	195 [18]	488 [45]
70	125 [12]	311 [29]
90	102 [9.5]	256 [24]

[a]Expressed as percent of room depth.

STEP 4. Calculate the illuminances (E_i) for each of the five reference locations. (Table 12.12 tabulates the illuminance values from this calculation.) The basic equation for each location is:

$$E_i = \tau\,(E_{xvk} \times CU_k + E_{xvg} \times CU_g) \qquad \text{(Eq. 12.6)}$$

where

E_i = interior illuminance at a specific reference point
τ = net transmittance of the window
E_{xvk} = exterior vertical illuminance at the window from half of the sky (an unobstructed vertical window sees only half of the sky)
E_{xvg} = exterior vertical illuminance at the window from the ground
CU_k = coefficient of utilization for sky light
CU_g = coefficient of utilization for ground light

For example, at the 10% depth on December 21:

$$E_i = \tau\,(E_{xvk} \times CU_k + E_{xvg} \times CU_g)$$

$$E_i = 0.7\,(2150 \times 0.673 + 538 \times 0.183)$$

$$E_i = 1082 \text{ lx}$$

To compare the result of the IES method with the results of the CIE and GDDM methods, the IES and CIE values are plotted on the finalized GDDM isolux contours in Figure 12.39c. The values exhibited are for December 21. The IES method's December values are minimum values from the solstice (Dec. 21) and correspond most closely to the 85th percentile figures of the CIE and GDDM methods. The agreement among the three methods is excellent for the half of the room nearest the window. For the deeper half, where there is significant ground contribution, the IES method yields higher illuminance values because it considers the re-reflected ground light contribution.

The IES values for March 21 correspond most closely to the 50th to 60th percentile of Figure 12.35, that is, somewhat more than double the minimum figures—and here again, the agreement is good.

To demonstrate the use of the IES method for clear-sky conditions, we can work through a brief example where five IES point illuminances are calculated for the same Seattle classroom, on June 21 at 10:00 a.m. for clear-sky conditions. Assume that the window faces southwest (azimuth angle = 45° west of south).

The solar azimuth at 10:00 a.m. on June 21 at 48°N latitude is 55° (Appendix D, Table D.1). The bearing angle is therefore 45° + 55° or 100° (i.e., no direct sunlight enters the window, which is a necessary condition of the IES clear-sky method). Solar altitude is 56°.

SOLUTION

STEP 1. Determine the horizontal illuminance (refer to Figure 12.24):

Sun only:	80,000 lx [7432 fc]
Sky only:	18,000 lx [1672 fc] full sky (E_{xHk})
	9000 lx [836 fc] half sky (E_{xhk})

Determine the vertical illuminance (refer to Figure 12.26*b*):

E_{xvk}: At a solar altitude of 55° and a bearing angle of 100°, the vertical illuminance (E_{xvk}) is 7500 lx [697 fc].

E_{xvg}: From Equation 12.5, vertical illuminance on the exterior of the window resulting from ground light (E_{xvg}) is:

$$E_{xvg} = 0.2\left(\frac{80{,}000,sun + 18{,}000,sky}{2}\right)$$

$$E_{xvg} = 9800 \text{ lx } [827 \text{ fc}]$$

Note that even though the window does not receive direct sun in this example, the sun does have an influence on the ground-reflected component.

STEP 2. Determine the net glazing transmittance of the window:

$\tau = 0.70$, as previously

STEP 3. Select the coefficients of utilization (as was previously done for overcast conditions).

Solar altitude for June 21 at 10:00 a.m.: 56° (Appendix D, Table D.1)

$$SC = \frac{E_{xvk}}{E_{xhk}}$$

$$\text{June 21: } SC = \frac{7550}{9000} = 0.83$$

(therefore, utilize Appendix C, Table C.21 for SC = 0.75)

TABLE 12.13 Illuminance Values for Clear-Sky Conditions on June 21 at Five Reference Locations

Location[a]	E_i June 21, lx [fc]
10	4789 [445]
30	2324 [216]
50	1253 [116]
70	828 [77]
90	690 [64]

[a]Expressed as percent of room depth.

STEP 4. Calculate the illuminance (E_i), using Equation 12.6:

For example, at the 10% depth on June 21:

$$E_i = \tau\,(E_{xvk} \times CU_k + E_{xvg} \times CU_g)$$

$$E_i = 0.7\,(7500 \times CU_k + 8900 \times CU_g)$$

Table 12.13 tabulates the illuminances at the five reference locations. ■

12.13 DAYLIGHTING SIMULATION PROGRAMS

Historically, daylighting simulation programs were too expensive and complex to use on day-to-day designs; they were, as such, primarily utilized by lighting consultants or researchers. Today, there are a number of software tools, both publicly and commercially available, that allow designers to quickly assess key daylighting performance metrics, develop immersive visualizations of simulation outputs, and create realistic visual representations of daylighting results with varying degrees of accuracy.

Computer simulation programs for daylight design strategies can integrate seamlessly into a project team's workflow, incorporate dynamic design and environmental factors, and can do so in moments. Many of these programs use radiosity techniques. Radiosity-based renderings are produced by dividing all the surfaces in a scene into a mesh of small polygons. Each polygon takes on a different value of light absorption/reflection, depending upon its relationship to a light source and its surface parameters. These values simulate the light distribution throughout the scene.

The most well-established software "engine" for daylighting simulation programs is Radiance. Developed in the 1980s by Gregory J. Ward at the Lawrence Berkeley National Laboratory (LBNL), Radiance is open-source and has been validated through peer-reviewed research. Instead of radiosity techniques, Radiance utilizes a backward ray-tracing approach for its calculations. What ray-tracing tries to simulate is a "tracing" of light "rays" from the observer's eye back to the light source(s). Through a process called "ambient calculation," Radiance can process both direct and indirect illumination, which is critical for accurately simulating complex lighting scenarios.

As a standalone application, Radiance allows designers to quantitatively assess daylighting performance, gauge the potential for glare and visual discomfort, and output detailed visualizations (Figure 12.42). However, Radiance lacks a user-friendly interface and, as such, has historically remained largely inaccessible to most designers. LBNL sought to address this issue through the software tool, Desktop Radiance; but even that program proved too esoteric for many design professionals and the software is no longer updated. Yet, Radiance continues to be highly utilized as an "engine" behind several commercially available tools that offer more convenient and user-friendly access to Radiance's calculations. As validated, open-source software, Radiance has been integrated into many popular building performance simulation (BPS) platforms including (among others) ClimateStudio, Cove, OpenBuildings Energy Simulator, OpenStudio, and Sefaira (Figures 12.43, 12.44, and 12.45).

(a) (b)

Figure 12.42 *Rendering studies using Radiance. (a) Shading provided by the blinds at the New York Times building. (b) Veiling reflections on a computer monitor. (a: © Greg Ward and Judy Lai; b: Chas Erlich, Lawrence Berkeley National Laboratory; used with permission.)*

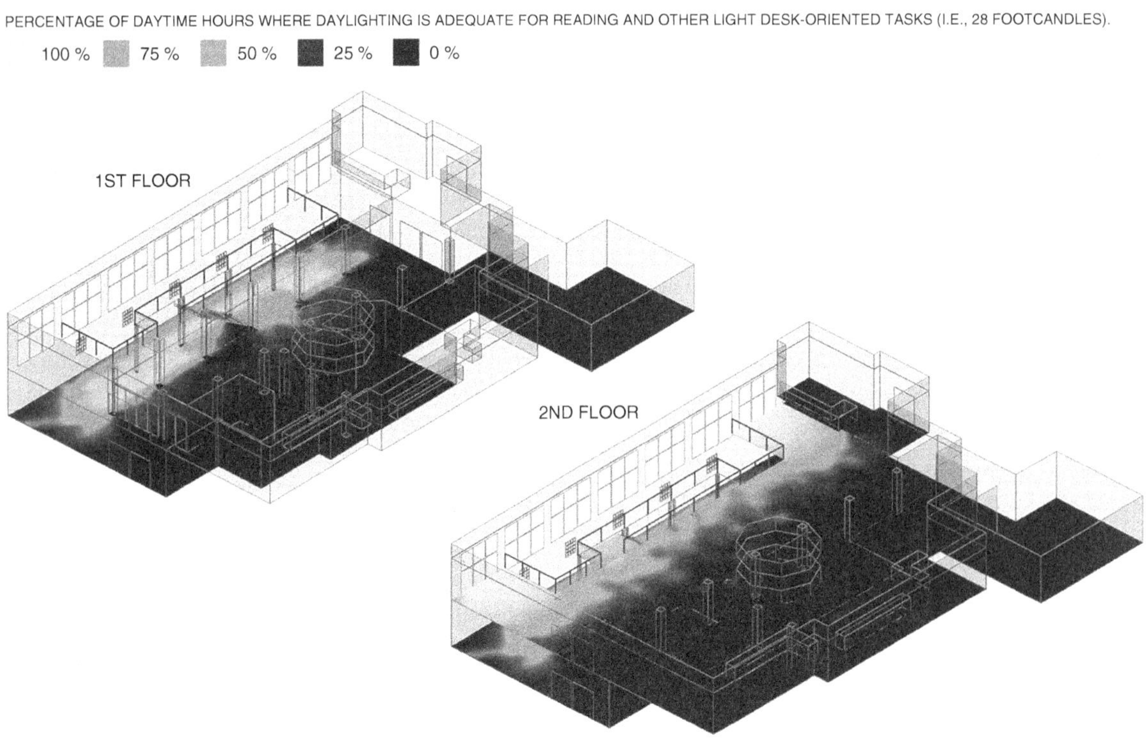

Figure 12.43 *Schematic daylighting study using Sefaira depicts spatial daylight autonomy (sDA) for a renovation project in Indianapolis, Indiana. (Illustrations courtesy of Daniel Overbey; © Browning Day.)*

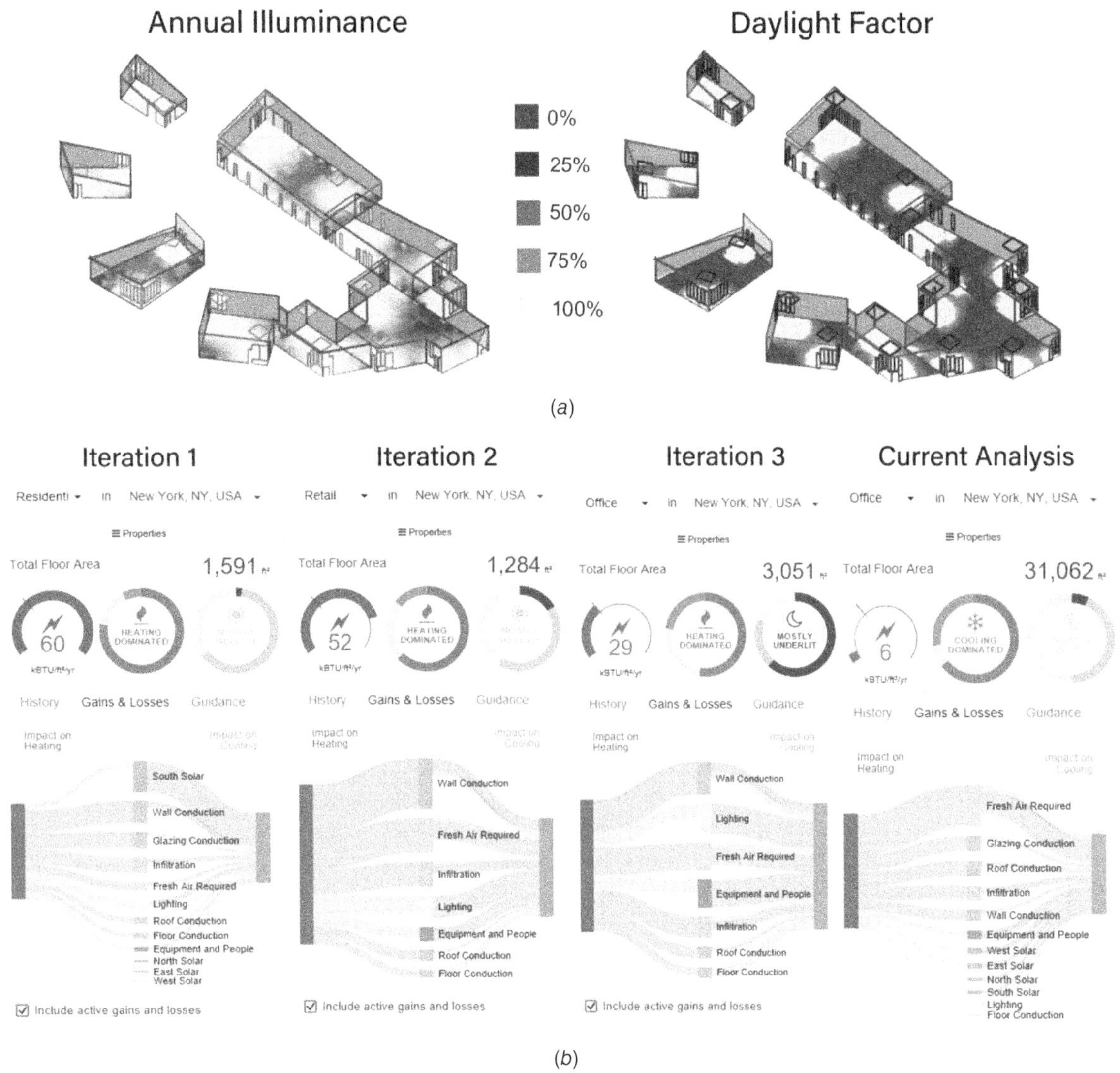

Figure 12.44 (a) Dynamic and diffuse daylighting analysis by targeting a range of illuminance levels for specific spaces and (b) estimating energy performance (EUI) for various parts of a studio program using Sefaira. (Illustrations by Megan York; used with permission.)

Figure 12.45 Renderings of daylight in a residence. (Images reproduced from Autodesk®.)

(a) (b)

Figure 12.46 *(a) Daylighting rendering for a middle school design in Las Vegas, Nevada, and (b) corresponding photometric plan using AGi32. (Courtesy of Daniel Overbey; © TSK Architects.)*

Algorithms for daylight simulation vary in output and accuracy, especially in complex and dynamic scenarios. As such, there is room for differentiation in the marketplace. Several commercially available daylight simulation programs utilize their own proprietary algorithms. AGi32 from Lighting Analysts, Inc., for example, offers its own lighting calculations and renderings of electric lighting and daylighting systems based on radiosity (Figure 12.46). This software is widely used in the lighting industry and includes a web-based interface that allows users to access manufacturers' photometric data and provides for easy updating. Autodesk Insight, IES Virtual Environment, Visual from Acuity Brands Lighting—each of these commercial software platforms uses their algorithms for daylighting simulation. For a description of these and other daylight simulation programs, see Appendix L.

12.14 PHYSICAL MODELING

Physical modeling is a useful and indispensable tool for the investigation of complex daylighting phenomena. Simple physical models can give both the designer and the client a visual understanding of a daylighted space. Physical models can duplicate the lighting phenomena that would occur in a full-scale space and, when placed under identical sky conditions, will yield accurate results relative to brightness, shadows, and daylight factor. By changing window design or orientation, adding lightshelves, reflectors, or shading devices, and/or modifying surface materials, a designer can quickly produce a three-dimensional visual image that displays qualitative and quantitative performance results for a proposed daylighting design.

The advantages of physical models include:

- the opportunity for accurate daylight measurements and for qualitative evaluation
- easy construction (for most designers)
- even crude models can yield critical information
- easy comparisons of various schemes (using interchangeable wall or ceiling elements)
- realistic visualization for clients

The principal disadvantage of using physical models is the need to expose them to the desired sky conditions. For example, waiting for suitable sky conditions in order to view a particular space under both overcast and clear sky conditions, or at different seasons of the year, or during different times of day, is not always practical.

Constructing scale models is relatively simple, using corrugated cardboard, mat board, and colored paper—mounted on a base for ease of manipulation (Figure 12.47). The model should be made modularly so that alternative design proposals can be interchanged. For example, to compare various skylight configurations, several replaceable roof configurations can be constructed. Model size depends upon the size of photometers used to measure interior illuminance, the size of the space, and the need to accommodate a camera viewport. Considering ease of construction and visualization opportunities, bigger is usually better—although larger models are often preceded by smaller/cruder study models.

Cardboard is an ideal material for daylighting models because it is opaque, unlike foam core board, which is translucent and transmits some light. Unintentional light leaks must be prevented, typically by sealing the joints of a model with black electrician's (or duct) tape or by using strips of black cardboard to close gaps. "Portholes" in one or both of the long sides of a model (approximately 2 in. [50 mm] in diameter) will accommodate visual inspection and insertion of a camera lens to photograph the distribution of light. Model surface reflectances (both interior surfaces and exterior surfaces that contribute to daylight

(a)

(b)

(c)

__Figure 12.47__ The effectiveness of even a crude model for daylighting studies. (a) A faculty office served as an exercise for a daylighting study. (b) The scale model for this office was constructed of cardboard. Important reflecting surfaces such as desk surfaces and windowsills were carefully modeled. (c) A quick modification was made to the model so that the rear of the office would receive more light. All photos were taken on site at midday on an overcast day. (© Alison Kwok; all rights reserved.)

distribution) should be the same as those proposed for the actual building (see Appendix C, Table C.27 for reflectances and mat board colors). Special care should be taken to accurately replicate details around daylight openings—the size and depth of mullions, the depth and reflectivity of the sill, louvers, shading devices,

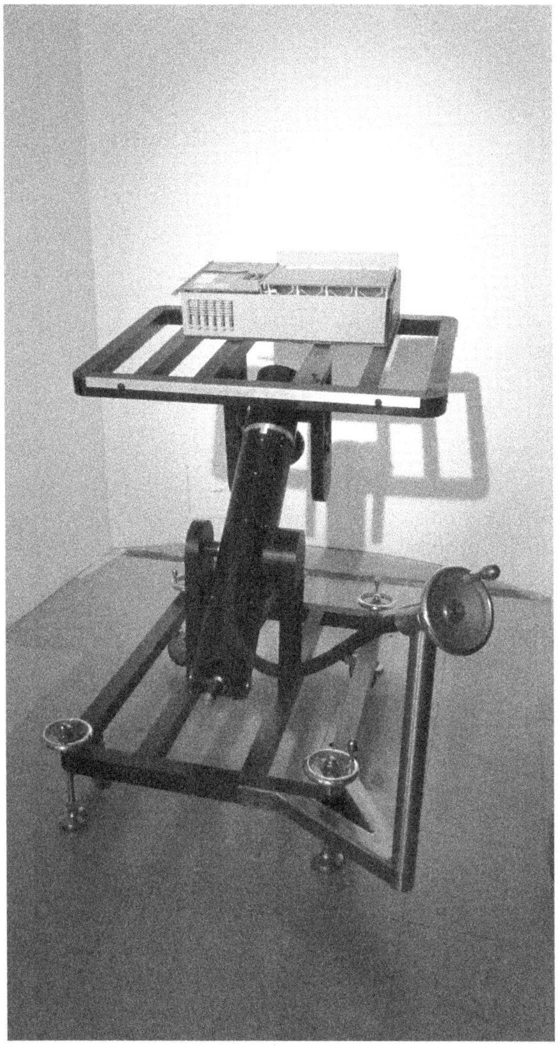

__Figure 12.48__ ShadowTracker heliodon at the Baker Lighting Lab at the University of Oregon. Automated adjustments for latitude, elevation, and tilt permit exposure of the model to desired clear sky solar conditions. (© Alison Kwok; all rights reserved.)

and surfaces just outside daylight openings. Any major furnishings that might have a significant impact upon light distribution should also be included.

Daylight model testing may be conducted under a real or an artificial sky and may also involve heliodon studies. Some considerations are as follows:

1. Use of a real sky with daylighting models is logical but often difficult to coordinate (as described earlier in this section).
2. Use of an artificial sky or mirror box. A carefully designed and controlled artificial sky dome or mirror box can duplicate overcast sky conditions with a high degree of accuracy and is ideal for testing physical models. Several such units exist in major universities and lighting laboratories around the world. Sky domes are usually illuminated by

(a)

(b)

Figure 12.49 (a) Shaded walkway at the Horseshoe Lodge, Beulah, Colorado. (b) Sidelighting in dormitory room of the summer camp. (Photos © Dan Bihn; used with permission.)

interior perimeter lamps with the model located in the center. A mirror box is essentially a room with a luminous ceiling (using fluorescent lamps) and mirrored walls to create a sky with an "infinite" horizon. For construction details, see Moore (1985).

3. Use of a heliodon. A heliodon, as shown in Figure 12.48, is a sophisticated device that allows the study of shading and solar access at a specific latitude and longitude, time of day, and time of year, using architectural scale models. It operates by rotating and tilting a building model with respect to the real sky or an "artificial sun" (a narrow-beam electric light source) until the desired solar altitude and azimuth are reached. Over the years, heliodons (sun machines, sun tables, helioluxes, sun emulators) have been built using a variety of configurations to simulate the sun's position relative to an architectural scale model. In all cases, the device establishes a geometric relationship for three variables: site location (latitude), solar declination (time of year), and the Earth's rotation (time of day). With an adjustment to any one of these variables, a heliodon can simulate sunlight penetration and shading for any combination of site location and time. Other types of heliodons keep the position of the model fixed and use a band of lamps that move along three axes to simulate the sun's position for different times and seasons.

12.15 RECAPPING DAYLIGHTING

This chapter began by describing the control of daylight in buildings as both an art and a science. Daylighting is a cornerstone of green design and a major contributor to good building energy performance, as well as occupant comfort, productivity, and health. Addressing daylighting early in schematic design is a critical step for successfully zoning activities and massing the building to improve the use of daylight and minimize the use of electric lighting. Skillfully employing strategies such as toplighting, sidelighting, lightshelves, and shading devices involves numerous qualitative judgments and quantitative calculations to achieve design intents and criteria. While it is important to conduct the necessary calculations, it is also important to recognize the beauty in the simplicity of the straightforward design of the past—as shown at the historic Horseshoe Lodge, built in 1939 (Figure 12.49).

12.16 CASE STUDY—DAYLIGHTING DESIGN

The Hive: Worcester Library and History Center, Worcester, England

Project Basics

- Location: Worcester, England, United Kingdom
- Latitude: 52.2°N; longitude: 2.2°W; elevation: 58 ft [17.7 m]
- Heating degree days: 5866 base 65°F [3259 base 18°C]; cooling degree days: 1355 base 50°F [752 base 10°C]; annual precipitation: 26.3 in. [669 mm]
- Building type: New construction; library
- Building area: 142,654 ft^2 [13,253 m^2] gross floor area divided into four levels
- Completed: January 2012
- Client: WLHC ProjectCo for Worcestershire County Council and the University of Worcester
- Design team: Feilden Clegg Bradley Studios (architects), Max Fordham (services), Hyder Consulting

Figure 12.50 Aerial view of the site near the Severn River; site diagram showing the massing of the building. (Photo by Simon Kirwan; © Simon Kirwan Photography; used with permission.)

Ltd. (structural); Galliford Try Construction Central (contractors) and specialist consultants

Background and Context

The Hive: Worcester Library and History Center is an integrated public and university library in Worcester, England (Figures 12.50 and 12.51). The city's existing library was no longer fit, and the university sought to improve its connection to the city. So the County Council put forward a proposal of both a public and university library, to meet the needs of both Worcester County Council and the University of Worcester. In 2007, a competition included a program for a new building that would achieve a 50% reduction in carbon emissions and fill 50% of its energy needs from renewable sources. The brief also called for the building to have a minimum environmental rating of "Excellent" by the Building Research Establishment Environmental Assessment Method (BREEAM) for buildings. In addition, the building must be able to adapt to the extreme environmental conditions expected in 2050.

Design Intent

Seven golden pyramidal roofs cover an irregular, pentagonal, four-story building plan. The roof profile originates from the surrounding ridgeline of the Malvern Hills and the traditional hop-kiln houses found in the region. The roof cones maximize daylight and provide natural ventilation through a central atrium and a series of voids strategically positioned in the building's various floor plates (Figures 12.52–12.54, and 12.55). The nearby Severn River is used as a heat sink for the cooling system in the library. Biomass from local sources provides the primary heating source.

Design Criteria and Validation

The Hive secured a BREEAM "Outstanding" rating with a score of 86.4% at the final post-construction review. The scheme also exceeds the 50% carbon reduction target, with renewables contributing almost 35% to the reduction. The building is designed to adapt to climate change predicted by the UK Climate Impacts Programme (UKCIP) to 2050. The building has an "A" rated Energy Performance Certificate (given by the government whenever a property is bought, sold, or rented).

Key Design Features

- *Daylighting*: Skylights with large chamfered surfaces maximize daylight deep into the atrium

Figure 12.51 *The Hive is in a historic part of the city of Worcester. It is designed to be an integrated public and university reference and loan library. (© Feilden Clegg Bradley Studios; used with permission.)*

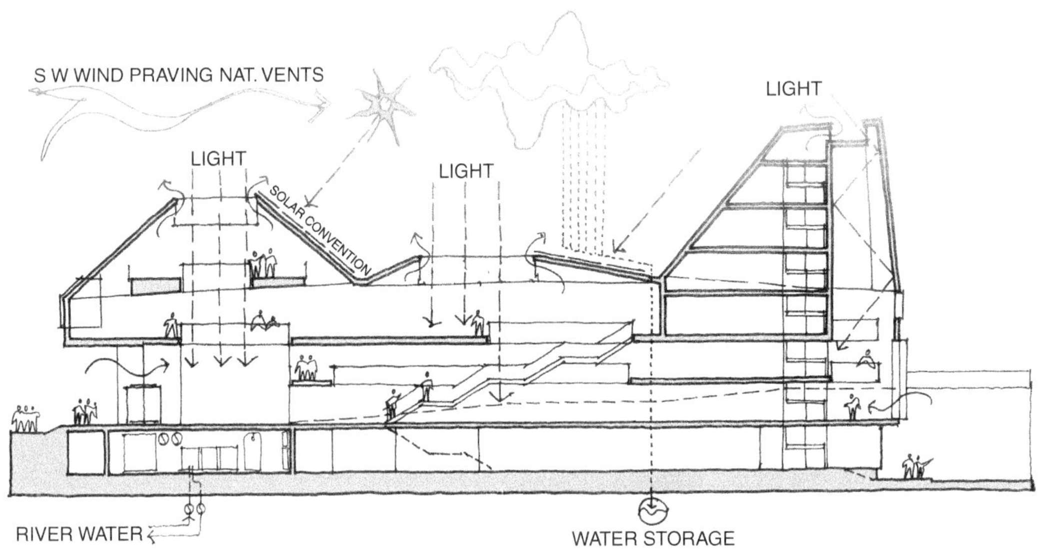

Figure 12.52 *Early design sketch showing environmental strategies. (© Feilden Clegg Bradley Studios; used with permission.)*

and down through voids cut into the floor plate (Figure 12.56). Sidelighting from large areas of shaded vertical glazing supplements the toplighting strategy and ensures an average daylight factor of 3%.

- *Natural ventilation*: The building is naturally ventilated by air entering through windows at the building perimeter and rising up by the stack effect through the main atrium, then exiting through the roof vents assisted by negative wind pressure. Baffles in the skylights create ventilation troughs that ensure the presence of negative pressure regardless of wind direction (Figure 12.57). Additional ventilation is delivered to the central

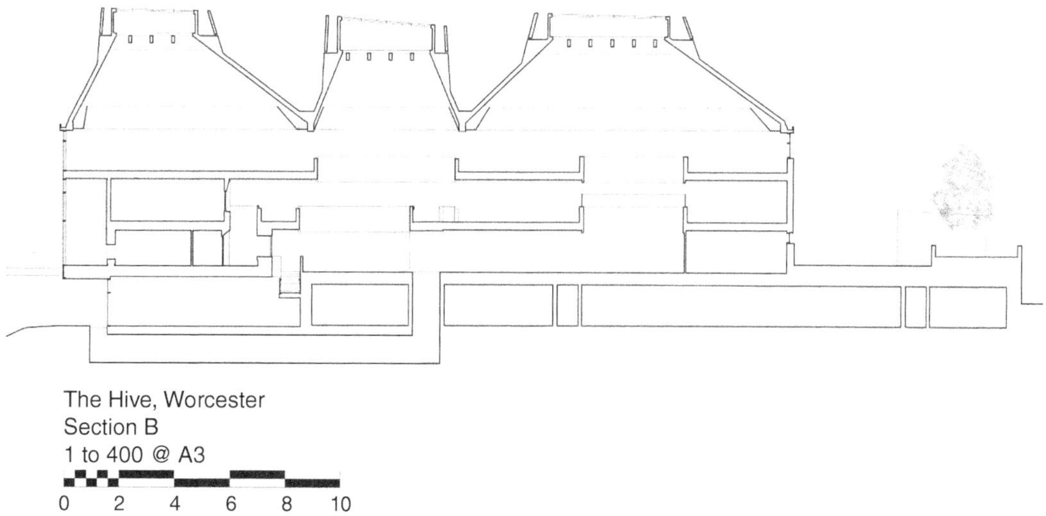

Figure 12.53 *Section drawing showing roof cowls. (© Feilden Clegg Bradley Studios; used with permission.)*

Figure 12.54 *Daylit atrium is controlled by the structural timber baffles; use of absorptive material around voids in the atrium provides a good acoustical environment. (Photo by Nick Hufton; © Hufton & Crow; used with permission.)*

atrium via a concrete duct; air travels from an exterior intake vent, under the basement to the base of the atrium, and rises into the atrium.

- *Thermal mass*: Thermal mass in the concrete soffits stabilizes temperature swings in the library by absorbing heat in the building. During the summer, the soffits are precooled by night ventilation that flushes the heat off the concrete and keeps the building cool during the day. During the winter, the soffits absorb daytime radiation and keep the building warm at night.
- *Heat sink*: The site's proximity to the Severn River allows it to be used as a heat sink. Water is pumped from the river to the building, passes through a heat exchanger and then into radiant coils in the concrete, and is then returned to the river. Piping within the concrete floor cools the structure of the building and provides radiant cooling to occupants. Engineers predicted that the cost of running the water pumps would be a third of the operation costs for conventional heating, ventilation, and air-conditioning (HVAC) equipment.

Figure 12.55 Rooftop cowls and skylight features. (© Hufton & Crow; used with permission.)

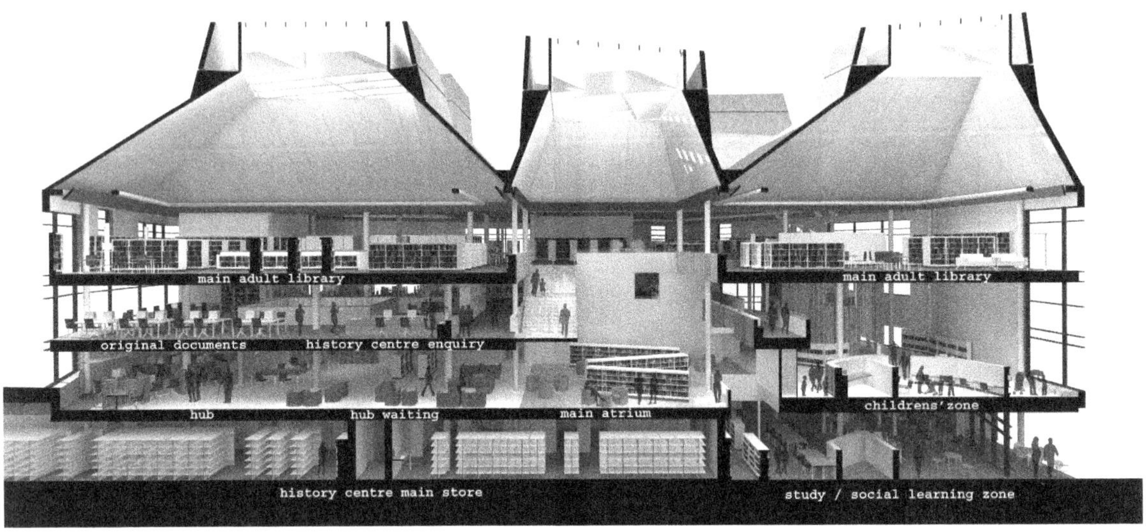

Figure 12.56 Section drawing of the building, showing the daylight strategy. Baffled skylights allow daylight to wash along the large surfaces, creating a "larger" aperture. (© Feilden Clegg Bradley Studios; used with permission.)

- *Rainwater harvesting*: Water from the rainwater harvesting system is used for flushing toilets and in the wash and silt trap, which forms part of the archeology service in the history center.
- *Alternative energy source*: Biomass from local wood chips provides peak heat to the building and further reduces demand on the energy grid.

Post-Occupancy Validation Methods and Performance Data

The public can monitor the building's performance using the display screen in the entrance foyer. Max Fordham was appointed to carry out additional post-occupancy monitoring and to work with facilities contractor SGP to improve the building performance.

Figure 12.57 *Section of the building showing the natural ventilation strategy. Air enters around the building perimeter, rises through atria by the stack effect, and exits via the roof cowls under negative wind pressure. (© Feilden Clegg Bradley Studios and Max Fordham LLP; used with permission.)*

Performance Data

- Building emissions rate: 3.57/lb CO_2/ft^2 per year [17.4/kg CO_2/m^2 per year].
- Predicted electricity consumption: 9.84 kWh/ft^2 per year [105 kWh/m^2 per year] including small power and IT use.
- Predicted fossil fuel consumption: 0.467 kWh/ft^2 per year [5 kWh/m^2 per year] not including biomass fuel, assumes gas usage for peak loads and backup only.
- Predicted renewable energy generation: 3.561 kWh/ft^2 per year [38 kWh/m^2 per year].
- Predicted water use: 668 gal [2530 L] per occupant per year; 396 gal [1500 L] potable water per occupant per year; 264 gal [1000 L] harvested rainwater per occupant per year.
- Rainwater is projected to supply 40% of the building's water needs.

Awards

- 2013 RICS Awards Regional Design and Innovation—Winner
- 2013 RICS Awards Regional Community Benefit—Winner
- 2013 RIBA Awards National—Winner
- 2013 RIBA Awards Regional Sustainability Award—Winner
- 2013 Building Awards: Sustainability Project of the Year—Winner
- 2013 Civic Trust Awards—Winner
- 2013 CIBSE Building Performance Awards: New Build Project of the Year (value above £5 million)—Winner
- 2012 PFI Partnership Awards: Best Sustainability in a Project
- 2012 Wood Awards: Commercial & Public Access: Shortlist
- 2012 Partnerships Awards: Best Community Project: Shortlist
- 2012 Partnerships Awards: Best Sustainability in a Project—Winner
- 2012 Partnerships Awards: Best Pathfinder Project: Shortlist
- 2012 Partnerships Awards: Best Local Government Project Team: Shortlist
- 2012 Partnerships Awards: Best Designed Project: Shortlist
- 2012 WAF Awards: Culture—Libraries: Shortlist
- 2009 Bentley Awards—Be Inspired: Infrastructure Best Practices Symposium and Awards: Innovation in Generative Design—First Prize

For Further Information

Feilden Clegg Bradley Studios. fcbstudios.com

Max Fordham LLP. maxfordham.com

Pearson, A. 2013. "Gold Standard." *CIBSE Journal* March. cibsejournal.com/archive/PDFs/CIBSE-Journal-2013-03.pdf

"The Hive/Feilden Clegg Bradley Studios," *ArchDaily* April 13, 2012. archdaily.com/225962

"The Hive by Feilden Clegg, Bradley Studios," *De Zeen Magazine*. April 12, 2012. dezeen.com/2012/04/12/the-hive-by-feilden-cleggbradley-studios/

University of Worcester. "The Hive: An Introduction to Worcester's New Library and History Centre." thehiveworcester.org/documents/the-hive-level-by-level.pdf

References and Resources

ASHRAE. 2022. *ANSI/ASHRAE/IESNA Standard 90.1-2022: Energy Standard for Buildings Except Low-Rise Residential Buildings*. Atlanta, GA: American Society of Heating, Refrigerating and Air-Conditioning Engineers.

Brown, G. Z. and M. DeKay. 2014. *Sun, Wind and Light: Architectural Design Strategies*, 3rd edn. Hoboken, NJ: John Wiley & Sons.

CIBSE. 1994. *Code for Interior Lighting*. London: Chartered Institution of Building Services Engineers.

CIBSE. 2006. *CIBSE Lighting Guide LG3: The Visual Environment for Display Screen Use*. London: Chartered Institution of Building Services Engineers.

CIE. 1970. *Daylight: International Recommendations for the Calculation of Natural Daylight* (Publication No. 16 E-3.2). Paris: International Commission on Illumination.

DeKay, M. 1992. "Volumetric Implications and a Rule-of-Thumb for Thickness of Atria Buildings," *In Proceedings of the 17th National Passive Solar Conference*. Boulder, CO: American Solar Energy Society.

DOE. 2016. *Solid-State Lighting R&D Plan. Office of Energy Efficiency and Renewable Energy*. Washington, DC: U.S. Department of Energy. energy.gov/sites/prod/files/2016/06/f32/ssl_rd-plan_%20jun2016_2.pdf

Dresler, A. 1963. *Daylight Design Diagrams*. Melbourne: Service Division, Commonwealth of Labor and National Service.

EIA. 2015. *2015 RECS Survey Data*. Washington, DC: Energy Information Administration. www.eia.gov/consumption/residential/data/2015/index.php

Heschong Mahone Group. 1999–2003. Executive summaries and reports (out of print).

Hopkinson, R. G., R. Pethebridge, and J. Longmore. 1966. *Daylighting*. London: Heinemann.

ICC. 2024. *2024 International Green Construction Code*. Washington, DC: International Code Council.

IEA. 2000. *Daylight in Buildings: A Source Book on Daylighting Systems and Components*. International Energy Agency, Solar Heating and Cooling Programme, Energy Conservation in Buildings and Community Systems. Berkeley, CA: Lawrence Berkeley National Laboratory.

IES. 1989. *IES RP-23-89: Recommended Practice for the Lumen Method of Daylight Calculations*. New York: Illuminating Engineering Society.

IES. 2000. *The Lighting Handbook*, 9th edn. New York: Illuminating Engineering Society.

IES. 2020a. *ANSI/IES LP-1-20: Lighting Practice: Designing Quality Lighting for People and Buildings*. New York: Illuminating Engineering Society.

IES. 2020b. *ANSI/IES LP-3-20: Lighting Practice: Designing and Specifying Daylighting for Buildings*. New York: Illuminating Engineering Society.

IES. 2020c. *ANSI/IES LP-7-20: Lighting Practice: The Lighting Design and Construction Process*. New York: Illuminating Engineering Society.

IES. 2020d. *ANSI/IES LS-6-20: Lighting Science: Calculation of Light and Its Effects*. New York: Illuminating Engineering Society.

IES. 2023. *ANSI/IES LM-83-23, Approved Method: IES Spatial Daylight Autonomy (sDA) and Annual Sunlight Exposure (ASE)*. New York: Illuminating Engineering Society.

Krochman, J. 1963. "Über die horizontal Beleuchtungsstärke der Tagesbeleuchtung." *Lichtechnik* 25(11).

Lam, W. 1986. *Sunlighting as Formgiver for Architecture*. New York: Van Nostrand Reinhold.

LBNL. "Windows and Daylighting." Lawrence Berkeley National Laboratory. windows.lbl.gov

Lighting Design Lab. lightingdesignlab.com

Millet, M. S., C. Adams, and J. R. Bedrick. 1980. "Graphic Daylighting Design Method: Including Clear Sky Conditions," in *Proceedings of the Annual Meeting of the American Section of the International Solar Energy Society, 5.2*. Amherst, MA.

Moore, F. 1985. *Concepts and Practice of Architectural Daylighting*. New York: Van Nostrand Reinhold.

Moore, F. 1993. *Environmental Control Systems*. New York: McGraw-Hill.

NIBS. *Whole Building Design Guide*. Washington, DC: National Institute of Building Sciences. wbdg.org

NREL. 2020. *Assessing the Reducing Plug and Process Loads in Office Buildings*. Golden, CO: National Renewable Energy Laboratory. nrel.gov/docs/fy20osti/76994.pdf

Reinhart, C. F., J. Mardaljevic, and Z. Rogers. 2006. "Dynamic Daylight Performance for Sustainable Building Design." *Leukos* 3(1): 7–31.

Rennhackkamp, W. M. H. 1967. "Sky Luminance Distribution in Warm Arid Climates," in *Proceedings of the 16th International Conference on Illumination*.

Washington, DC: National Building Research Institute.

Stein, B. and J. Reynolds. 1992. *Mechanical and Electrical Equipment for Buildings*, 8th edn. New York: John Wiley & Sons.

USGBC. 2025. *LEED v5*. Washington, DC: U.S. Green Building Council. www.usgbc.org/leed/v5

Van den Wymelenberg, K. and A. Mahić. 2016. "Annual Daylighting Performance Metrics Explained." *Architect Magazine*, April. architectmagazine.com/technology/lighting/annual-daylighting-performance-metrics-explained_

Wienold, J. and J. Christoffersen. 2005. "*Towards a New Daylight Glare Rating*." www.researchgate.net/publication/313608753_Towards_a_new_daylight_glare_rating

CHAPTER 13

Passive Heating

PASSIVE HEATING SYSTEMS COLLECT SOLAR ENERGY through glass, and store that energy in thermal mass, to provide warmth to habitable space (Figure 13.1). This approach is considered passive because it avoids the use of either purchased or site-generated energy to operate active systems (i.e., furnaces, boilers, fans, or pumps), supplying heat with extremely low operational energy requirements.

Space heating is currently the single greatest energy end-use in U.S. buildings, and because that energy is supplied primarily by fossil fuels, it is also the greatest source of building-related CO_2 emissions. The 16 quadrillion Btu [~5 trillion kWh] devoted to space heating amount to 38% of U.S. building-related energy consumption, provided primarily by natural gas (69%), fuel oil (8%), and electricity (13%) (EIA 2009, 2012). Similar proportions are found in the developed countries worldwide. Across the planet, space heating energy use is such a great contributor to greenhouse gas emissions that the Intergovernmental Panel on Climate Change has called for renewed efforts to avoid active space heating altogether (IPCC, 2014).

Currently, however, passive strategies provide less than 1% of all space heating in the United States (U.S. Census, 2016), with similarly low proportions in other countries; passive solar adoption has primarily been limited by low confidence among designers and risk-aversion among builders, especially in cloudy areas, combined with relatively low fuel prices (Garrett and Koontz, 2008). The rapid intensification of climate change concerns, however, is adding new urgency. In addition, recent progress in solar physics, materials, and soil science, controls engineering, and computational tools, combined with field verification, now allow passive solar performance to be predicted with greater reliability than has previously been possible. This chapter integrates these findings into a new design approach, allowing designers and builders to create passive solar heating systems that perform well in a diversity of climates.

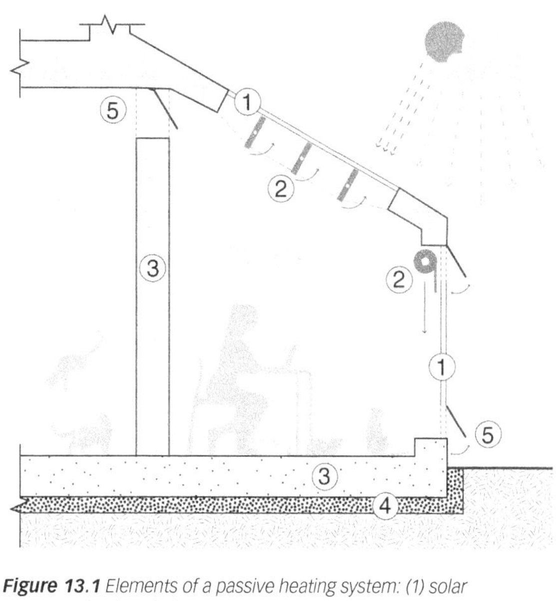

Figure 13.1 *Elements of a passive heating system: (1) solar energy-collecting glazing, (2) movable insulation and shading, (3) thermal storage mass, (4) separation of floor mass from the ground, and (5) vents to the interior and exterior for warm air passage and ventilation, respectively. (Drawing by Alexa Thornton; © Walter Grondzik; all rights reserved.)*

Mechanical and Electrical Equipment for Buildings, Fourteenth Edition. Walter Grondzik, Daniel Overbey, and Alison Kwok.

Companion website: www.wiley.com/go/meeb14e

13.1 BRIEF HISTORY

(a) Energy Transitions

Firewood shortages drove the earliest known experiments in passive solar heating, first in ancient Greece and China, later in Rome, and then in the Americas and Europe. As cities grew and people burned wood or charcoal for heat, forests vanished across vast areas of land, requiring people to seek wood in distant places and to invest in military force to do so. Wood and charcoal prices rose accordingly, and designers and builders gradually discovered that solar heating was inexpensive, effective, and enhanced by certain building practices. This situation was partly analogous to conditions today, in which climate change and diminishing supplies of readily extractable fossil fuels are promoting a new energy transition.

(b) China and Solar Geometry

Buildings in ancient Chinese settlements consistently faced south, as revealed by numerous archeological excavations. The northern Chinese city of Erlitou, settled about 4000 years ago, showed urban and building forms that were reproduced throughout China, with major streets aligned in an east-west, north-south grid, as well as ample courtyards that allowed primary living spaces to face south. Beijing, though settled thousands of years later, was given the same patterns (Figure 13.2). In addition, ancient lore associated the south with warmth and brightness, while the north was associated with dark, cold, and somber ideas; the southwest corner of a house was considered the seat of honor, and emperors were required to face south when addressing an audience to embrace enlightenment symbolically.

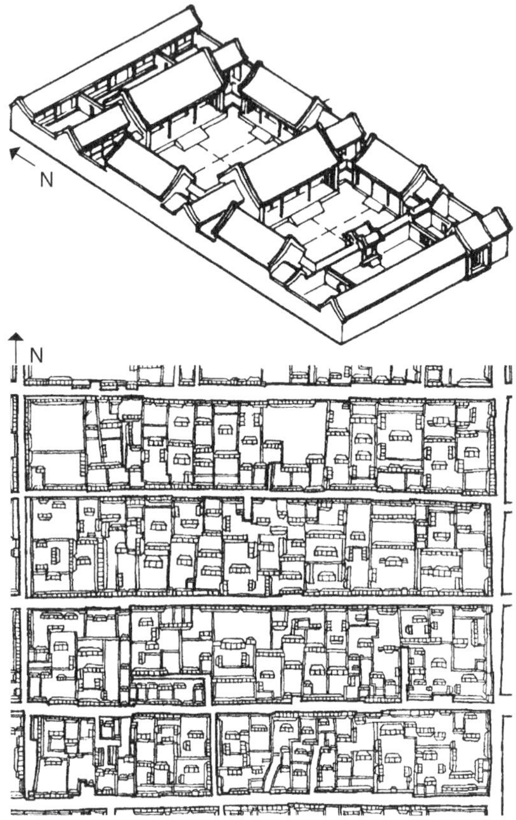

Figure 13.2 An ancient Chinese house (above), in which the main building faced south and was assured of solar access by a large courtyard, and ancient Beijing's grid layout (below), with broad streets running from east to west that provided the possibility of a southern orientation to each dwelling. (From Perlin, 2013.)

About 2000 years after Erlitou was settled, the Chinese invented systematic methods for tracking the sun's motion, allowing the solar position to indicate the time of day and year precisely. One of these was the gnomon (from the root *gno-*, "to know"), which uses a vertical post to cast a shadow (Figure 13.3). These shadow patterns revealed the correspondence between seasonal temperatures and solar position; the fifth-century BCE book *Zhouli* first documented the use of gnomons for building orientation. From the fifth century on, extensive deforestation around Chinese cities increased pressure to conserve firewood and charcoal, and in apparent response, other climate-responsive design elements were adopted: adobe and rammed earth were developed, allowing early experimentation to occur with thermal storage, and buildings were partly sunken into the ground for thermal buffering (Perlin, 2013).

(c) Greece and Solar Equality

Although Greece's sunny Mediterranean climate was milder than the northern Chinese continental climate, winters were still cool, deforestation was severe, and heat was greatly needed. Learning of Chinese gnomons through travelers, the Greeks adapted them into sundials and used them to develop new city and building forms. Dwellings were positioned at the northern ends of large courtyards, allowing low-altitude winter sun to reach them easily through colonnades, while floors and northern walls were built of thick adobe to store heat. The planned cities of Olynthus and Priene were arranged in grids, such that equal solar access was preserved for all buildings,

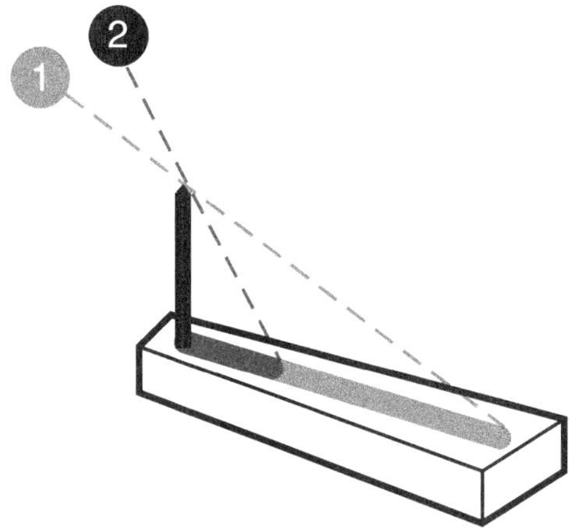

Figure 13.3 *Early passive solar heating design used gnomons with pegs that cast shadows, revealing the extent of solar penetration into south-facing dwelling spaces in winter (1) compared to summer (2). (Drawing by Alexa Thornton; © Walter Grondzik; all rights reserved.)*

Figure 13.4 *Section of a Roman heliocaminus (above), in which large mica or glass windows admitted solar energy and kept it from escaping, and ruins of the heliocaminus at Hadrian's Villa (below). The windows, as shown in the overlay, faced southeast, south, and southwest, so that the room would capture solar energy throughout the day during the winter. (From Perlin, 2013 / Reprinted with permission of New World Library.)*

and even irregular cities such as Delos maximized solar access to the extent possible. Strikingly, Greek practices valued solar rights for all citizens, regardless of wealth or rank, and by the fourth century BCE, solar orientation was regarded as a normal characteristic of Greek homes (Butti and Perlin, 1980; Perlin, 2013).

(d) Rome and Glass

By the first century BCE, Roman deforestation was also extreme. Guided by the teachings of Vitruvius and Greek settlers, Romans began to orient not only buildings but particular spaces toward the sun; winter dining rooms, for example, were to face the direction of sunsets for evening warmth. Romans also began to use thin sheets of mica, and later glass, to cover solar-collecting apertures, creating intensely warm rooms called *heliocamini* (sun furnaces) (Figure 13.4). Glass coverings were also used to enclose Roman baths and greenhouses, significantly improving passive solar performance. Solar rooms were privileges of only the wealthy—one poor poet remarked that he would be warmer as the guest of his patron's trees—but even among the relatively few affluent citizens, disputes over solar access arose, ultimately leading to the first legal guarantees of ongoing solar access for existing solar rooms (Butti and Perlin, 1980; Perlin, 2013).

(e) Anasazi and Multifamily Dwellings

In the high-desert climates of the Southwest United States, passive solar architecture showed greater seasonal distinction. The Sky City of Acoma, in present-day northern New Mexico, consisted of three long rows of dwellings running east to west along a plateau, with thick adobe walls facing south to intercept low-altitude winter sun. The roofs, in contrast, were built of layers of straw and adobe over pine timbers, providing insulation from, rather than storage of, high-altitude summer sun. The Anasazi dwellings at Mesa Verde also used massive south-facing walls for solar collection but sheltered them under rock overhangs for further protection from high-altitude sun. In both arrangements, the clustering of dwellings allowed deep interior spaces to remain cool, while south-facing exterior spaces remained warmer, echoing Roman practices of arranging living spaces to take advantage of interior microclimates (May and Reid, 2010; Perlin, 2013).

(f) European Horticulture

The use of glass for solar heating was virtually unknown in Europe for several centuries after the fall of Rome, but by the 1500s, political stability, scientific curiosity, and increasing wealth promoted new interest in horticulture. Growing seasons were short during the Little Ice Age, and gardeners discovered that sloping glass and movable insulation allowed them to cultivate flowers and vegetables for months longer than was possible outside. With time, greenhouses of the wealthy became lavish glass-enclosed gardens, or conservatories, for entertaining and leisure activities. Opening them to a home's primary spaces provided such warmth, light, and welcome humidity during the winter that they were adopted in smaller form by the middle class as enclosed balconies on south-facing façades. By the 1800s, however, conservatories were increasingly heated with newly available coal. When fuel rationing was instituted during World War I, conservatories were largely abandoned; the lesson of passive solar heating had temporarily been lost (Butti and Perlin, 1980).

(g) An American Revival

In 1931, the Royal Institute of British Architects developed the heliodon, a device that uses a sunpeg chart, a rotating platform, and a fixed light source to simulate the solar exposure of a model building at any hour of the year (see also Chapter 11). This device, combined with numerous encouraging performance studies, the availability of large, double-paned glass windows, and successful experimentation of architects such as George Fred Keck, led to an American revival. Supported by the wartime conservation ethic, thousands of passive solar homes were sold across the United States in the 1930s and 1940s as the manufacturers of prefabricated and mass-assembly homes adopted passive solar designs. By the late 1940s, however, falling fuel prices dampened interest in passive solar heating, and the increased first cost of homes with large areas of double glazing further discouraged investment; by 1950, new passive solar homes had all but vanished in the United States.

(h) The Balcomb Method

By the late 1960s, new fears of fuel shortages again spurred interest in passive solar heating. In response, the U.S. Department of Energy deployed a cadre of nuclear engineers at the Los Alamos National Laboratory, led by Douglas Balcomb, to develop quantitative methods for passive solar design and performance prediction for every climate in the United States. Within 10 years, this group had developed a completely new method for passive solar design based on empirical models of test cell performance in the high-desert climate of New Mexico; the results were presented in the *Passive Solar Design Handbook* (Balcomb et al., 1980) and *Passive Solar Heating Analysis* (Balcomb et al., 1984). This method established the Solar Savings Fraction (SSF) as a metric of passive solar potential, in a given climate, and supplied it with prescriptive guidelines for glazing and mass configuration. This method quickly dominated the passive solar design literature and is still cited in virtually all contemporary design guides and references (e.g., Chiras, 2002; Kachadorian, 2006; Bainbridge and Haggard, 2011; DeKay and Brown, 2013).

(i) Recent Progress

Because the "Balcomb method" was developed from field data collected in the sunny, high-desert climate of New Mexico, it could not fully predict the effects of cloud cover, moist soils, tree shading, and other factors important in cooler, wetter climates. Reports from Alaska (Seifert, 1981), the United Kingdom (summarized in Porteous, 2005), and the U.S. Department of Energy itself (Swisher and Duffy, 1983, Swisher, 1984) showed that passive solar heating system performance was at odds with the Balcomb method's predictions, and questions were raised regarding the validity of the SSF and the wisdom of the massing guidelines. However, more recent applied research at Ball State University and the University of Nevada, Las Vegas, concluded that despite major deviations from the massing guidelines used by Balcomb, his method was reasonably adequate for predicting the annual SSF for passively heated buildings in harsh winter climates with predominantly overcast sky conditions (Fernández-González, 2007; Overbey, 2007). Study into passive solar heating and associated fields has continued to advance, leading to the development of a contemporary method of assessing passive solar heating performance, as offered below.

13.2 THE PASSIVE HEATING OPPORTUNITY

The Balcomb method's SSF metric was defined as the energy consumption avoidable by use of passive solar rather than conventional heating, expressed as a percentage of the energy that an otherwise analogous baseline building would use (Balcomb et al., 1980). This emphasis on the percentage of energy saved, rather than the absolute amount of energy saved, drew strong resistance from researchers in the United Kingdom when it was introduced (MacGregor, 1980; Bartholomew, 1985). Given that the goal of passive heating is to avoid fossil fuel use, they argued, the absolute amount of energy saved was far more important than the percentage. Moreover, they found, percentages were misleading. An SSF of 90 was viewed as promising, while an SSF of 20 was seen as much less so, encouraging passive solar design in the former case and discouraging it in the latter. However, locations in which SSFs are low often have extensive heating needs, such that the absolute energy savings are potentially substantial. While avoiding 90% of a relatively low 10,000 kWh heating need saves 9000 kWh, for example, avoiding 20% of a 50,000 kWh heating need saves 10,000 kWh—greater absolute energy savings.

Instead of the SSF, which typically assigns high values to sunny climates with limited heating needs and low values to cold or cloudy climates with extensive heating needs, the passive heating opportunity herein described is expressed in terms of absolute solar resources, absolute heating needs, and the extent of their coincidence, resulting in absolute energy savings estimates. These quantities, rather than abstract percentages, are the values most relevant to passive heating design decisions.

(a) Solar Resources

A site's solar resource describes the solar energy available for capture, including both short-wave (ultraviolet and visible) and longer-wave (infrared) radiation; over 40% of the sun's energy arrives at the Earth's surface in the visible range (i.e., having wavelengths of 0.4–0.7 μm), while nearly 50% falls in the infrared (>0.7 μm) range (Figure 13.5). This distribution is significant because short- and long-wave radiation interact with glass differently: while most visible wavelengths are transmitted readily through clear glass, infrared wavelengths are primarily absorbed by glass (and window framing) and re-radiated or transferred by convection to the interior (Figure 13.6). The solar

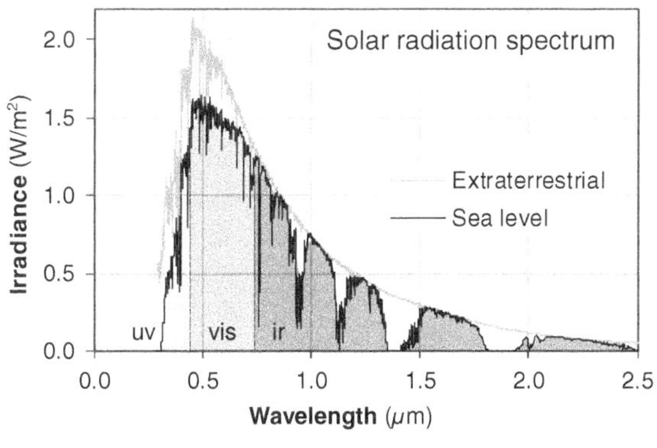

Figure 13.5 *Solar radiation intensity (irradiance) as a function of wavelength at the outer surface of the atmosphere (extraterrestrial) and on a 37° tilted surface at sea level. (Adapted from ASTM, 2023.)*

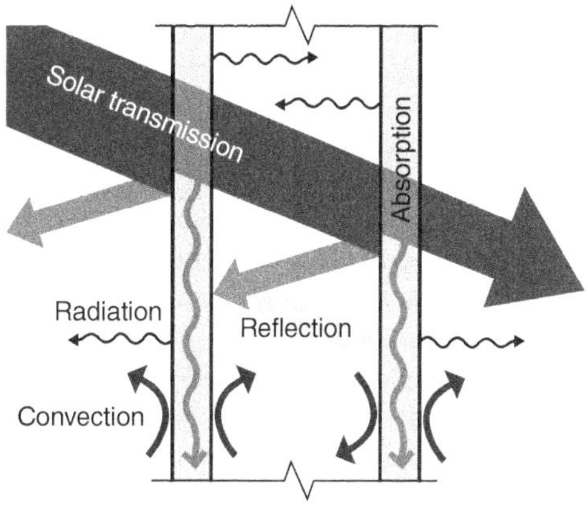

Figure 13.6 *Solar transmission through double glazing. Short-wave (ultraviolet and visible) solar energy is largely transmitted, with reflection increasing as the angle of incidence becomes more oblique. Longer-wave (infrared) energy is primarily reflected or absorbed and then partially transmitted to the interior by convection and radiation. (Drawing by Alexa Thornton; © Walter Grondzik; all rights reserved.)*

heat gain coefficient (SHGC) describes a given window assembly's ability to transfer this solar resource to the interior as a fraction of the total radiation incident on the exterior.

Solar resources vary with latitude, season, climate, elevation, and sky exposure; in other words, they tend to be greatest in equatorial regions, warm seasons, clear climates, high elevations, and on sites without significant shading. For passive heating design, however, only the solar resources that occur during the heating season are important. For this reason, sunny dry climates are not necessarily better suited to passive heating than cool cloudy ones; the potential solar heating benefit depends on the distribution of the solar resource with respect to a building's heating needs, as discussed below.

(b) Heating Season and Degree-Days

The heating season is the underheated period of the year, or that part of the year in which a given building requires heat. The extent of underheating is often expressed in terms of heating degree-days (HDD) with respect to a base temperature. The base temperature is an established temperature—assume 65°F [18.3°C] for a typically-heated house; lower for a solar-heated house and commercial buildings—against which the ambient outdoor air temperature is compared for heating (and cooling) degree-day calculations. The base temperature is equal to the average of all hourly balance point temperatures across a heating season. The balance point temperature (BPT) is the ambient outdoor air temperature at which the heat gains inside the building "balance" the heat losses through the building enclosure.

Though the balance point temperature is estimated and applied as a static temperature for the sake of calculation procedures, it is important to recognize that this estimate is based on specific heat gain conditions (e.g., number of people, lighting power density, equipment use, solar radiation), and that, in actuality, the balance point temperature will fluctuate throughout a typical daily cycle as conditions change (Moore, 2004).

Note the word "day": one HDD is defined as a 24-hour period (i.e., 1 day) in which the average outdoor temperature is one degree below a particular base temperature. For example, if a cool fall day has an average 24-hour temperature of 35°F [1.7°C], that single day has 30 HDD evaluated against a base temperature of 65°F, or 30 HDD_{65F}. In SI units, evaluated against the analogous base temperature of 18.3°C, it has 18.3 − 1.7 = 16.6 $HDD_{18.3C}$. The base temperature chosen depends on a building's enclosure construction and internal gains: while buildings built decades ago often need heat when the daily average temperature drops below 65°F [18.3°C], those with greater insulation, lower infiltration, and/or higher internal gains may (on average) not need heat until the average daily outdoor air temperature drops below 60°F [15.5°C], 55°F [12.8°C], or even lower.

In the early stages of design, the heating season may be approximated as the set of months with more than about 100 total HDD_{65F} [55 $HDD_{18.3C}$]. Since these months may be either concentrated into a few severe months or distributed over many mild months, heating seasons have not only annual magnitude but also length and intensity. The resulting shape (load profile) of a heating season has important implications for passive solar design. Consider, for example, several cities that each have about 4800 annual HDD_{65F} [~2700 $HDD_{18.3C}$], shown in Figure 13.7. While those in humid mid-latitude climates (e.g., Wilmington, Baltimore, or New York) concentrate their heating seasons into several intensely cold months, those with West Coast marine climates (e.g., Eugene or Seattle) distribute the heating season over a greater number of milder months. Such long heating seasons, typical of cloudy climates, extend the heating need into spring and fall months with many hours of solar radiation each day, giving them excellent passive heating opportunities despite their cloudiness.

(c) Intersection of Resource and Need

The passive solar heating opportunity is the extent of co-occurrence between the solar resource and the heating need, where the latter is determined by the heating season and building characteristics. An example is shown in the shaded portion of Figure 13.8 for a building in Seattle that requires about 2.5 kWh of heating energy for each HDD_{65F} [about 4.5 kWh for each $HDD_{18.3C}$], typical of an existing residence in the region. When the solar resource is less than the heating need, the entire solar resource is useful; when the solar resource exceeds the heating need, however, only the resource up to the level of need is useful. This constitutes the net solar heating resource (NSHR), or the solar resource available to a passive heating system.

The solar resource captured during operation is limited in practice by the glazing chosen and by site shading, of course, and each building's heating need depends on its enclosure construction, hours and intensity of occupancy, and internal gains. Nevertheless, the passive solar opportunity is readily quantifiable in terms of the NSHR for an individual building, giving a designer a clear standard against which to evaluate choices of solar-collecting glass, glazing tilt and

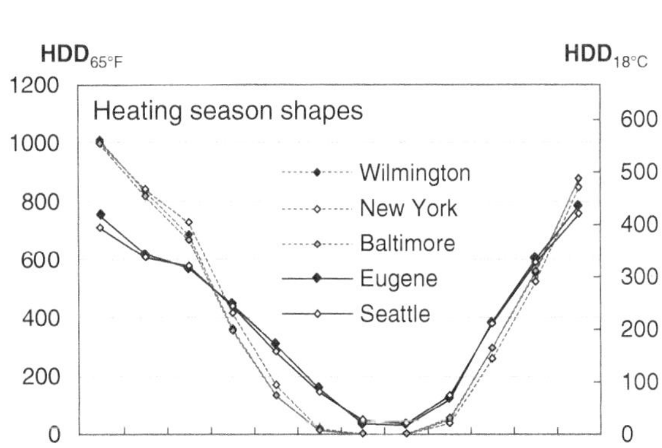

Figure 13.7 Contrasting heating season length and severity, as shown by monthly heating degree-days in Wilmington, DE; New York, NY; Baltimore, MD; Eugene, OR; and Seattle, WA, all with about 4800 annual HDD_{65F} [2650 $HDD_{18.3C}$]. (Adapted from Arguez et al., 2010.)

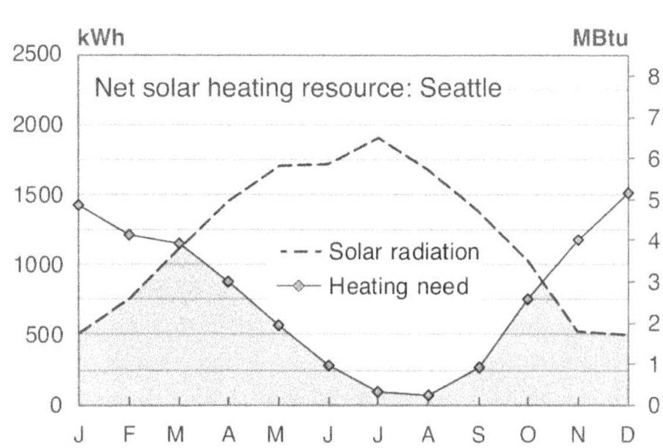

Figure 13.8 The net solar heating resource (shaded) is the extent of co-occurrence between the solar resource (dotted line) and the heating need (solid line); this example assumes a collector area of 108 ft^2 [10 m^2] at a collector tilt of 40° above horizontal, facing south, for a typical dwelling (EIA, 2009) in Seattle. (Adapted from NREL, 2007.)

orientation, thermal storage material and size, and movable insulation type and schedule.

13.3 DESIGN STRATEGIES

The four fundamental passive solar heating approaches are direct gain, in which solar energy is admitted directly to living space; indirect gain, in which solar energy is intercepted (at least in part) by thermal mass before passage through it to living space; isolated gain, in which solar energy enters a thermally separate, habitable space that stores and/or distributes the energy to living space; and flat collection, in which solar energy warms air in a shallow chamber for distribution to thermal storage and/or living space. Multiple systems may be combined, as well.

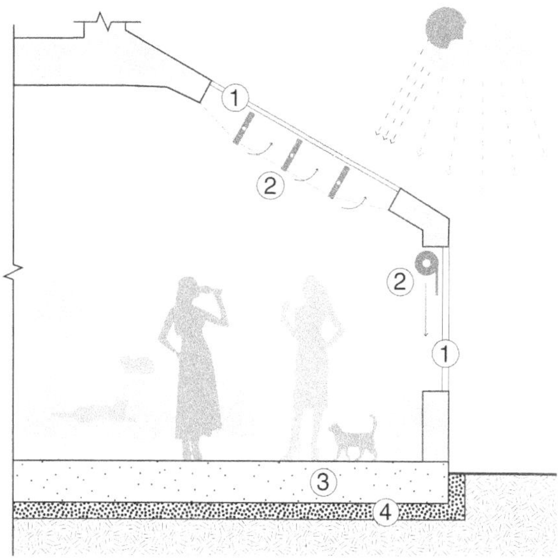

Figure 13.9 *Direct gain configuration, showing solar energy collection in living space through tilted and, optionally, vertical glazing (1) equipped with movable insulation and shading (2); thermal storage is provided by floor mass (3) that is thermally separated from the ground (4). (Drawing by Alexa Thornton; © Walter Grondzik; all rights reserved.)*

(a) Direct Gain

Direct-gain systems admit solar energy directly to living space through windows and/or skylights (Figure 13.9). Because of this broad definition, almost any building with windows or skylights experiences some direct-gain heating; if the heat provided is modest, the approach is described as solar tempering. For intentional solar collection, however, glazing must be sized and optimally tilted for the climate; as a result, a direct-gain space is typically bright and warm, creating either a welcome microclimate for basking, socializing, bathing, etc. or unwelcome glare for fine visual tasks.

Atria, for example, provide effective direct-gain heating or solar tempering in spaces designed for circulation and casual use, as in the Federal Center South Building in Seattle, Washington (Figure 13.10). In the Ogden Street passive solar renovation in Denver, Colorado, an atrium was created between two apartment buildings during their conversion to offices, forming an inviting entry and waiting space (Figure 13.11*b*). In clerestory configurations,

Figure 13.10 *Direct-gain atrium in the Federal Center South Building, Seattle, WA. (Courtesy of ZGF Architects.)*

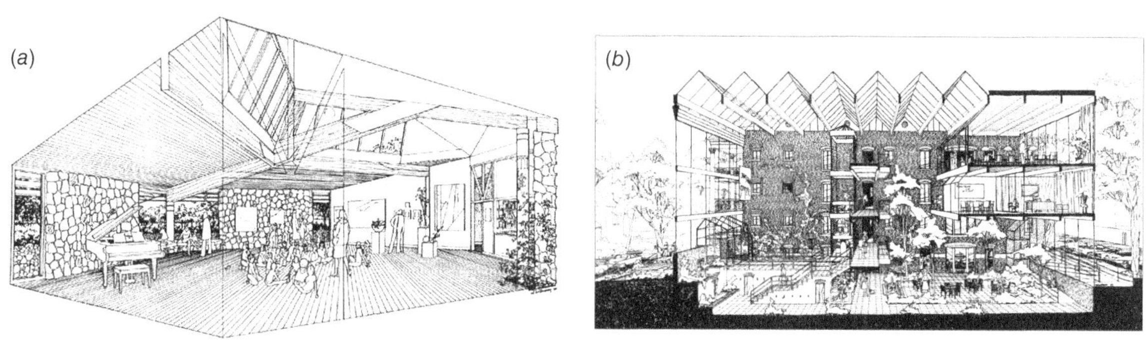

Figure 13.11 *Direct-gain configurations in (a) a passive solar addition to the Unitarian Church of Boulder, Colorado, showing a solar-collecting clerestory in the sanctuary that allows radiation to fall onto stone walls (right side of image), as well as two stone Trombe walls (left side of image), and (b) the Ogden Street apartments-to-office solar retrofit in Denver, Colorado, showing the central atrium with its glazed sawtooth roof. (© Dennis R. Holloway, Architect.)*

glazing may be positioned more vertically, allowing it to collect low-altitude winter sun in sunny winter climates and to direct it toward vertical thermal storage mass (Figures 13.11*a*, 13.12). Direct-gain spaces are also effective in residential applications, often using floor-based thermal mass (e.g., ceramic tile, brick, or concrete) to receive much of the incoming sun while the space is unoccupied, providing warm floors as well as warm air for occupants when they return in the evening (Figure 13.13).

At night, movable insulation can help minimize heat loss and the thermal discomfort that results from exposure to cold surfaces. In summer, direct-gain systems are well matched to summer cooling strategies

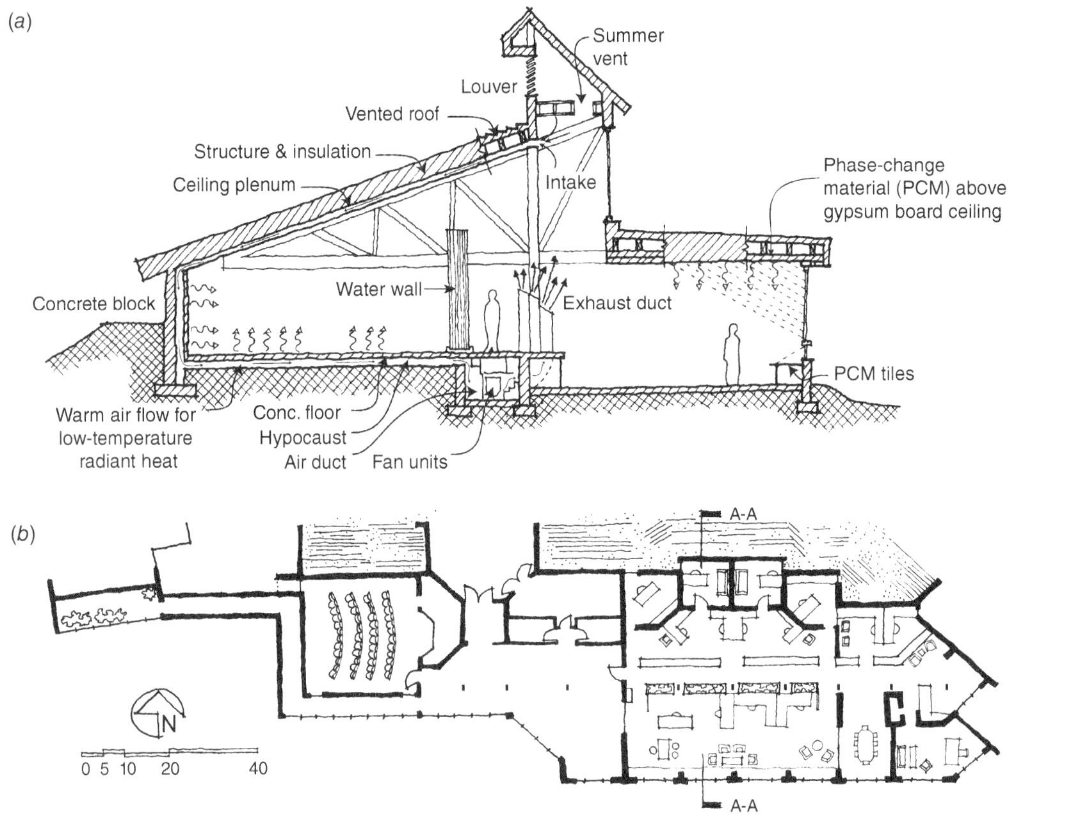

Figure 13.12 *The Conservation Center for the Society for the Protection of New Hampshire Forests, Concord, by Banwell White Arnold Hemberger & Partners, Architects. (a) Section with south glazing and thermal storage materials: translucent water tubes, masonry walls, and phase-change materials in ceiling and windowsill positions. The circulation of hot air that collects at the clerestory is also shown. In summer, the hot air is vented, and an awning shades the clerestory. (b) The plan is elongated on the east-west axis to maximize southern exposure. Direct gain serves the reception area, workroom, and offices, and a sunspace double-envelope combination warms the lecture room. A wood-fired boiler provides backup heat. (Drawings by Jonathan Meendering after original works by Banwell White Arnold Hemberger & Partners, Architects.)*

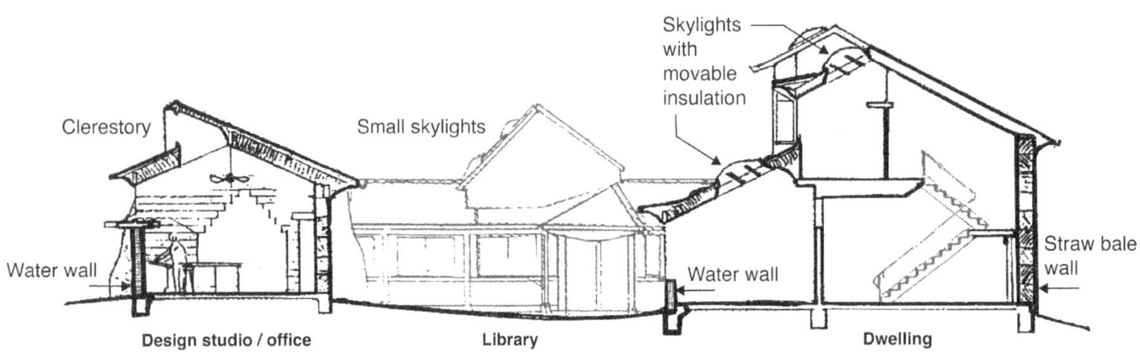

Figure 13.13 *North-south section of the San Luis Solar Group Complex near Santa Margarita, California, including a house and an architectural studio. Both buildings collect solar heat through clerestories and skylights equipped with movable insulation as well as water walls consisting of steel tanks behind double-glazed surfaces. (Drawings courtesy of the San Luis Solar Group, Santa Margarita, CA.)*

of shading, potentially accomplished by dual-purpose movable insulation, and night ventilation of thermal mass. Visually, direct-gain spaces may be prominent features on a building's sun-facing façades and roof surfaces, or they may be virtually invisible from the outside by relying largely on skylights.

(b) Indirect Gain

While massive exterior surfaces such as brick and concrete absorb solar heat effectively, solar energy retention is improved significantly by the addition of an external glass layer a short distance from the mass, creating an indirect-gain system (Figure 13.14). Glazed walls, also known as Trombe walls after the French engineer, Félix Trombe, who pioneered their development in the 1960s, are usually vertical, making them best suited to sunny winter climates with low sun angles; because of this, they feature prominently on the sun-facing façades of their buildings (Figure 13.15).

Trombe wall glazing may be backed by any massive material for heat storage; those backed by tanks or tubes of water are often known as water walls (Figure 13.16). Because a space exists between the glazing and mass, an indirect-gain system creates warm air as well as warm mass. If this warm air is also desirable for heating, the wall may be vented (Figure 13.17), admitting cool air at the bottom and removing warm air from the top; otherwise, the wall may be unvented. Other considerations include cleaning of the space between the mass and glass, facilitated either by ample width or by operable exterior glazing, and ability to keep the interior surface of the mass clear of hangings or furniture to facilitate radiant heat transfer.

Compared to direct-gain systems, indirect-gain systems diminish unregulated heat and glare in adjacent spaces while allowing clerestory windows (Figure 13.16) or even apertures in the wall itself (Figure 13.18) to provide daylight. Through choice of thermal storage material, thickness, and venting, they also allow the timing and intensity of heat delivery to be adjusted to occupant needs.

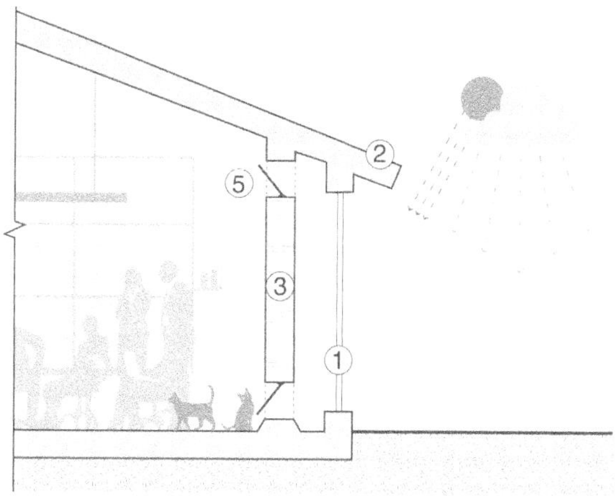

Figure 13.14 *Indirect-gain configuration, showing solar energy collection through vertical glazing (1), which may be selectively coated to increase solar absorption; upper portions often remain clear to provide daylight. Overhangs (2) provide shade from high sun angles; movable insulation is uncommon. When sun angles are low, solar energy is partly absorbed by wall mass (3), and warm air may be vented to the interior if desired (5). Element (4), separation of floor mass from the ground seen in other system diagrams, is not critical to this system configuration. (Drawing by Alexa Thornton. © Walter Grondzik; all rights reserved.)*

(c) Isolated Gain

Isolated-gain strategies are popular because they provide sunny habitable rooms, also known as sunspaces, that capture, store, and manage delivery of

(a) (b)

Figure 13.15 *Exterior (a) and interior (b) of the Trombe wall at the Zion National Park Visitor Center, Springdale, UT, designed by architects of the National Park Service in collaboration with scientists and engineers at the National Renewable Energy Laboratory. (Photos by Robb Williamson, NREL image 09256 and NREL image 09237.)*

Figure 13.16 A clear plastic water wall system in the 2009 University of Arizona Solar Decathlon entry transmits filtered daylight; note that as a wall becomes less opaque, less energy is stored. (© Alison Kwok; all rights reserved.)

heat to adjacent, thermally separate occupied spaces (Figure 13.19). Because of this separation, sunspaces may have considerable outdoor character, supporting a variety of plants and permitting wide temperature swings (Figure 13.20). Alternatively, however, they may easily be controlled to narrower, more comfortable limits through use of movable insulation, shading, adjustable thermal mass, and natural ventilation.

Numerous sunspace configurations are possible: a sunspace may share only one common wall with the primary building, in an attached configuration, as in Figure 13.21, or it may share two or three common walls, in a semi-enclosed configuration, such as that in Figure 13.22. In addition, the common wall between the sunspace and living space may be insulated or massive. A similarly diverse range of performance options is available, determined primarily by massing choices: low-mass sunspaces are used primarily to produce warm air for delivery through interior vents or windows to occupied spaces, while massive floors are used to provide evening warmth in the space itself; alternatively, internal water storage may be used for low-intensity heat delivery throughout the night.

Like direct-gain spaces, sunspaces require tilted glazing for optimal collection of winter sun, combined with movable insulation to diminish heat loss. Like indirect-gain systems, however, sunspaces often also have extensive vertical glazing; while this provides views and connection to the outdoors, it may also pose a thermal liability, especially in cloudy climates, and should be sized carefully.

Sunspaces must also be equipped with operable shading, natural ventilation, and, optionally, adjustable thermal mass (e.g., additional potted plants or water barrels) to avoid overheating on unusually warm or sunny days of the heating season. With these in place, sunspaces are also among the easiest passive heating systems to cool in the summer, since their overnight temperatures may drop significantly without compromising indoor comfort.

(d) Flat Collector

When warm air is desired early in the day, a flat collector is the system of choice. This is an isolated-gain system in the sense that the collection area is thermally

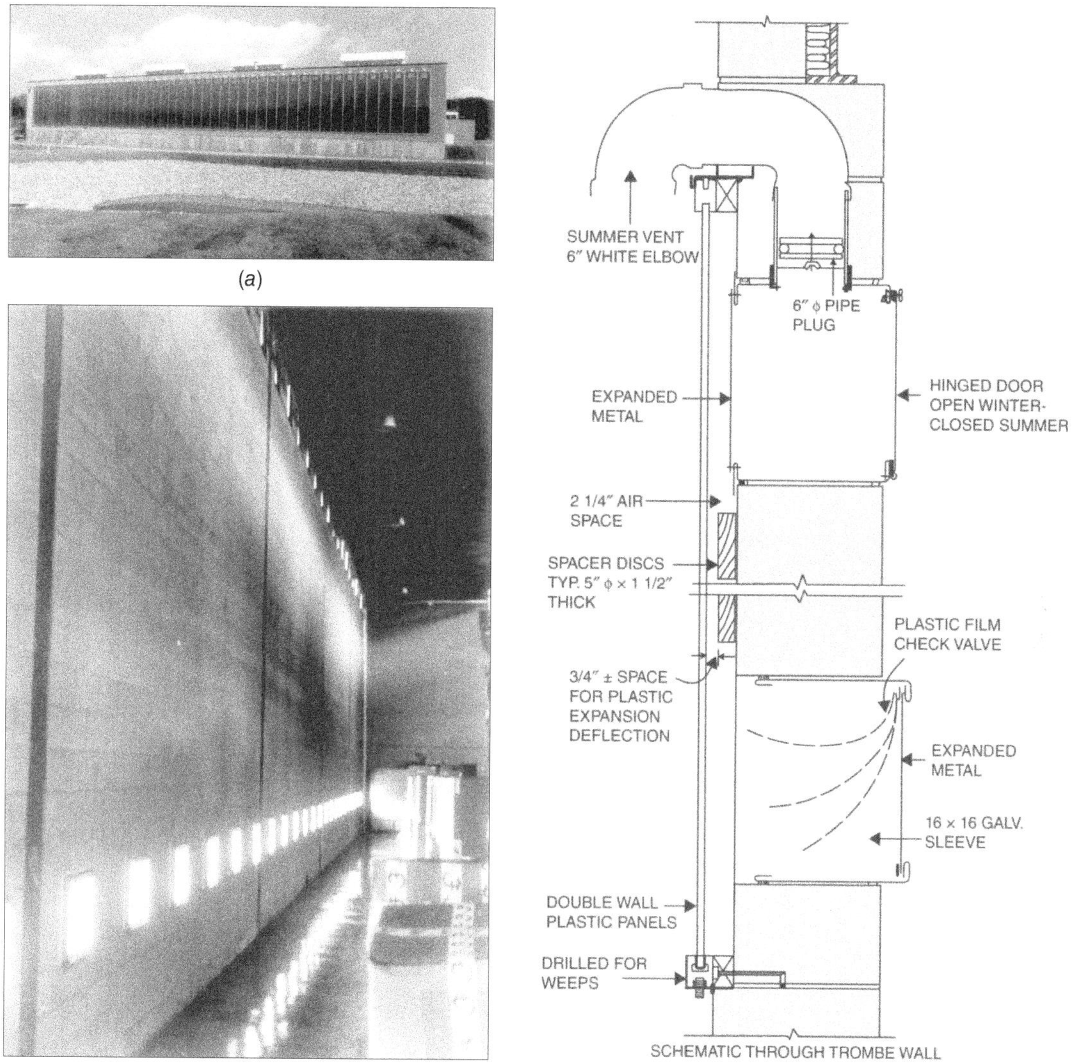

(a) (b) (c)

Figure 13.17 *(a) The Famolare shoe company warehouse in Brattleboro, Vermont, uses (b) a Trombe wall of 20 ft × 184 ft [6.1 m × 56.1 m] facing 10° east of south with lower and upper vents. (c) Switches at the bottom prevent flow reversal on cold nights, and at the top, provide building ventilation in summer. (a: Photo by Robert Perron; c: Reproduced from Architectural Record, November, 1979.)*

separate from living space, but it performs much like a vented Trombe wall. Occupying a substantial area of sloping roof or ground surface (Figures 13.23 and 13.24), a flat collector warms air in shallow chambers between glazing and dark-colored corrugated metal, causing the air to flow upward along the sloping surface. In a fully passive configuration, the channels open to living space at the top of the slope, such that warm air enters from above. Alternatively, in an active/passive hybrid configuration, fans draw the air downward and through thermal storage before entering the living space from below. Because the air volumes warmed are small, the temperatures reached can be quite high, allowing the fan energy required to be minimal. In either configuration, air may enter the collector from the outdoors, from living space, or from a combination of the two. A fully passive configuration that allows living space air to enter the collector, such that it circulates air from collector to living space continuously without fan assistance, is known as a thermosiphon (also spelled thermosyphon).

13.4 CONCEPTUAL DESIGN

A passive heating design process begins by defining the desired performance and experiential qualities of the system, considering winter weather patterns, and selecting a system appropriate to both. Once a system is selected, general rules and patterns are

(a) (b)

Figure 13.18 (a) A Trombe wall design for the Burger Residence in Lakeside, Oregon, completed in 2008, uses (b) circular openings to preserve lake views. (Courtesy of D. Holmes Chamberlin Jr., Architect LLC.)

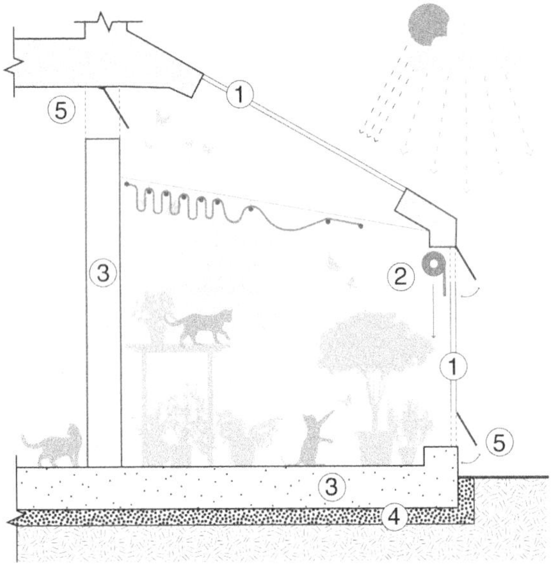

Figure 13.19 Isolated-gain configuration, showing solar collection through both tilted and vertical glazing (1), equipped with movable insulation and shading (2); mass provides thermal storage in a common wall shared with living space, in a floor, and/or in adjustable internal mass such as potted plants (3). Thermal separation from the ground (4) is also important, as is ventilation to both interior and exterior (5). (Drawing by Alexa Thornton. © Walter Grondzik; all rights reserved.)

used in conceptual design to size and configure the system's primary elements, namely solar-collecting glass, thermal storage mass, and movable insulation. Plans for shading, natural ventilation, and adjustable thermal mass (if desired) must be made at this stage as well, since passive heating systems overheat during warm months without them. Preliminary configurations are then refined in schematic design, through simple calculations, and in design development through computer simulations.

(a) Selecting a System

Initial system selection depends on the answers to several questions. In approximate order of importance, these are:

1. *Is a room with outdoor character desired?* If a bright, airy, and/or plant-filled space is appealing, an isolated-gain (sun-) space or an atrium is indicated; these are versatile systems, with sunspaces typically occupying south-facing façades and atria typically located internally. If space is at a premium, however, another configuration may be preferable.
2. *How tolerant will heat-needing spaces be of high daytime temperatures or glare?* If this tolerance is high, as in a room usually unoccupied during the day, a direct-gain system is eligible; if the tolerance is low, as in a workplace where glare would interfere with visual tasks, a Trombe wall or flat collector may be preferable.
3. *Is the winter sunny, and is the site unshaded?* In choosing between a Trombe wall and flat collector, the predominant angles of incident solar radiation

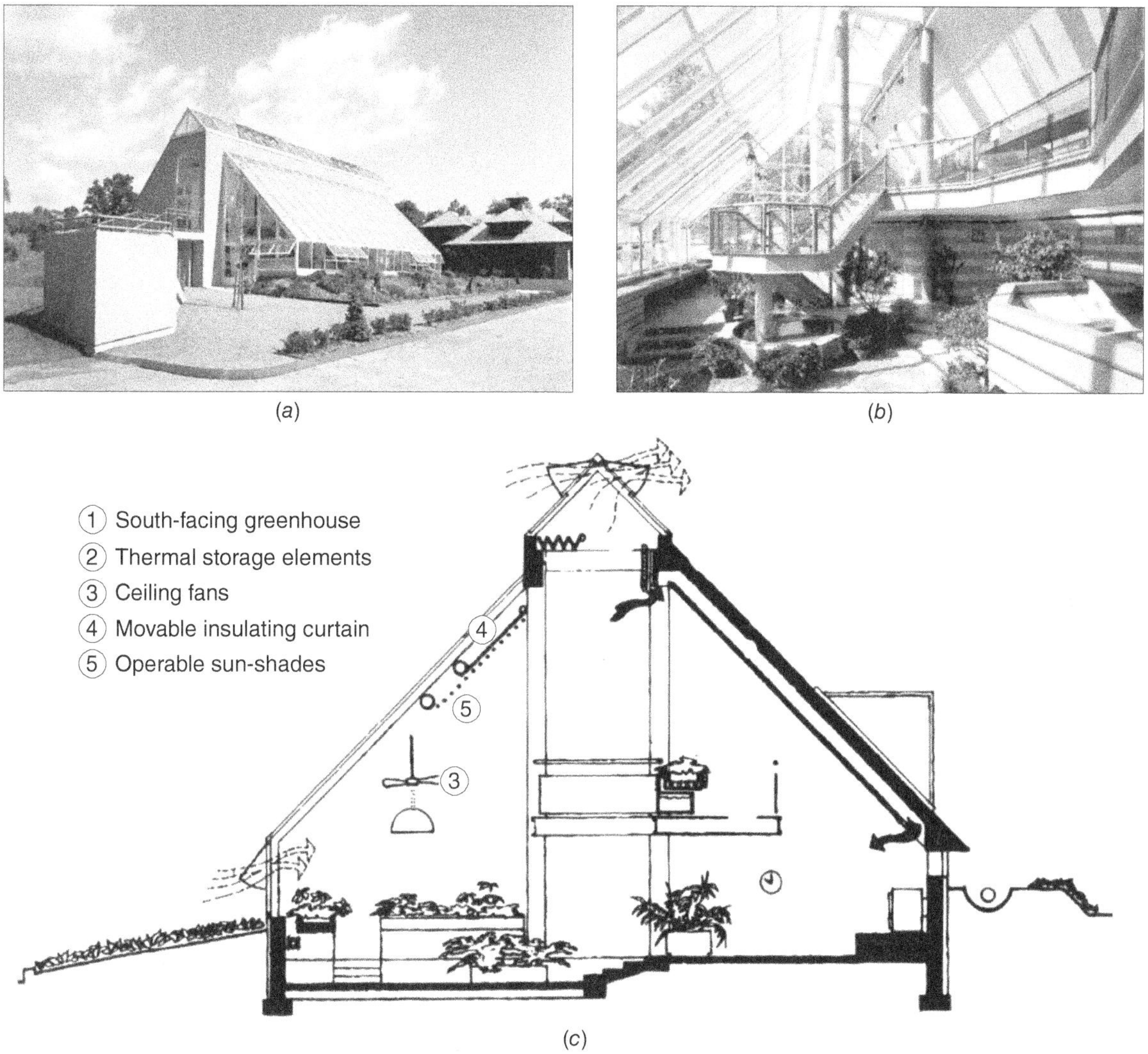

Figure 13.20 The New Canaan, Connecticut, Nature Center uses a greenhouse (1) to capture heat for the building, storing it in phase-change materials (2) and moving warm air through the building with fans (3); movable insulation (4) minimizes heat loss at night, while operable shades (5) minimize overheating during summer months. (Photos by Robert Perron. Courtesy of Donald Watson.)

are important: only sunny winter climates and unshaded sites provide the low sun angles required by vertical Trombe walls.

Questions 1–3 are summarized in Figure 13.25. However, the result obtained in this way may be overridden by the answer to a fourth question, namely: *What forms, materials, space types, and other elements would be best suited to the building?* Sunspaces are visually striking glazed elements on south façades, while Trombe walls resemble spandrel glass, and flat collectors resemble photovoltaic arrays. Direct-gain spaces may be as visually prominent as sunspaces, or they may be virtually hidden from outside view. If the most promising system from a performance perspective is ruled out by other considerations, the desired system may nevertheless be configured appropriately for the climate, and it may also be supplemented by an additional, more effective one.

(b) Estimating the Heating Need

In many cases, and particularly in extremely cold or cloudy winter climates, passive solar heating is not expected to meet the entire space-heating need; in other words, it is not expected to provide operative temperatures within thermal comfort zones during the coldest or cloudiest months. It is, however, expected to offset heating energy use during these months. For this reason, monthly heating needs for the entire heating season are used to estimate the optimal collector tilt and area.

Figure 13.21 The Class of 1959 Chapel at the Harvard Business School (a) features south-facing glazing and a sunken garden in the sunspace (b, d). The point of the sunspace is about 30° west of south (c). (a, b: © Thomas Collins; used with permission; c, d: courtesy of Moshe Safdie and Associates, Inc.)

Defining the Heating Season. For conceptual design estimates, the heating season is assumed to consist of months with more than about 100 HDD, using any base temperature, in Fahrenheit degrees [about 55 HDD in Celsius degrees]. If little information is available regarding the future building's balance point temperature, 65°F [18.3°C] is a conservative choice for passive heating design, accounting for the possibility that the system will be designed for an existing building and increasing the likelihood that subsequent revisions will reduce rather than increase system size.

"Normal" HDD over 30-year periods are tabulated by the National Atmospheric and Oceanic Administration (NOAA) for multiple base temperatures for hundreds of cities in North America; international weather services (e.g., Weather Underground, www.wunderground.com) compile HDD data as well.

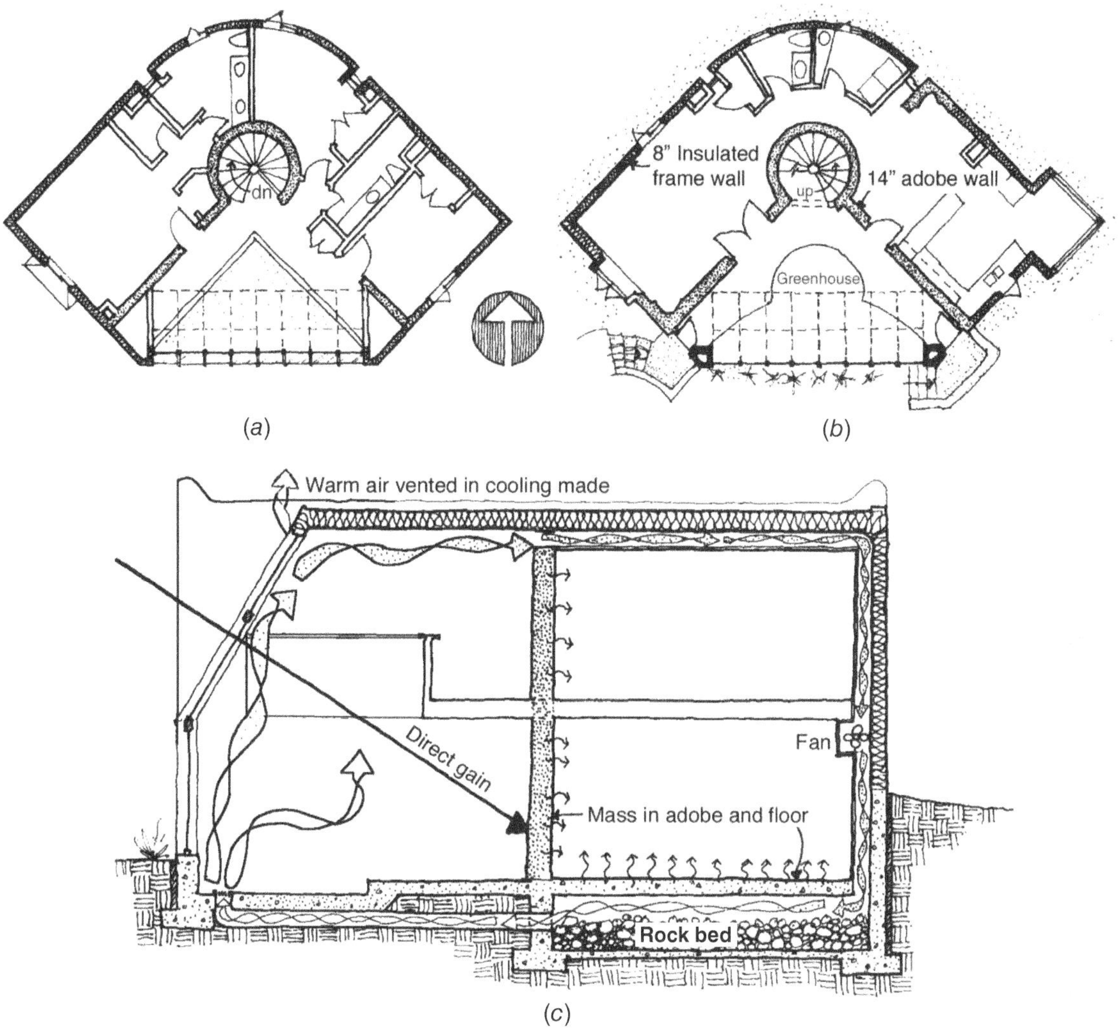

Figure 13.22 *Unit 1, First Village, an early and well-known example of sunspace passive solar heating. Living space encloses the sunspace's northeast and northwest sides on both ground (a) and second (b) floors. In heating mode (c), a mass common wall stores heat for delivery to interior spaces, and warm air is circulated through a rock bed below the main floor for further storage. In cooling mode, warm air is vented to the exterior. (Drawings by Michael Cockram, based on Stromberg and Woodall, 1977.)*

Monthly values for major U.S. cities are provided in Appendix I, Table I.1.

EXAMPLE 13.1 A small two-story barn in the cold continental climate of Token Creek, Wisconsin, is to be converted into a passive solar chamber music hall for an annual festival. What is its estimated heating season, given an estimated balance point temperature of 18.3°C [65°F]?

SOLUTION

Madison, WI, is the closest city to Token Creek in Appendix I, Table I.1, with monthly $HDD_{18.3C}$ values above 55 [HDD_{65F} values above 100] for all months except June, July, and August, giving the building a nine-month heating season (Figure 13.26). ■

Estimating Monthly Heating Needs. The next step is to estimate the building's monthly heating needs. If enclosure thermal resistance information and estimates of internal gains from people, lighting, and equipment can be made, these should be used to estimate total monthly heating needs according to the method described in schematic design (Section 13.5).

If little building-specific information is available, however, estimates may be obtained from regional space heating energy use values compiled in government or international databases, online building databases, or research publications. In the United States, the Energy Information Administration (EIA) compiles such data in Residential and Commercial Building Energy Consumption Surveys, or RECS and CBECS, respectively (EIA, 2009, 2012). Among existing residences in the United States, average space

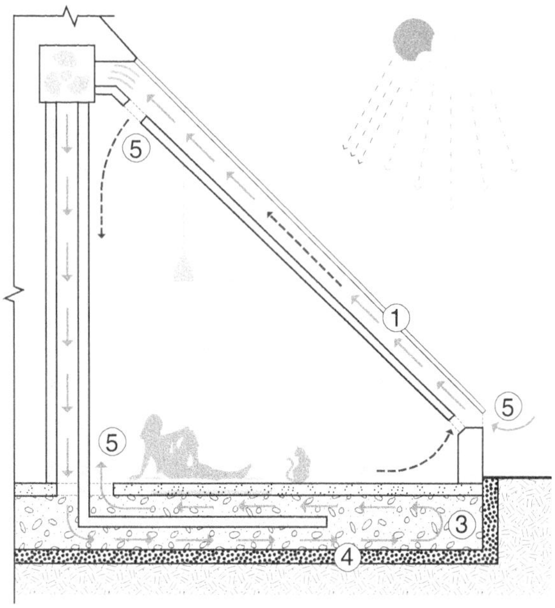

Figure 13.23 Flat collector configuration, showing solar collection through tilted glazing (1) that produces warm air, either to be introduced to living space directly or to be pulled downward by a fan through rock-bed thermal storage (3) that is thermally separated from underlying soil (4). Exterior and interior vent positions (5) determine whether interior air recirculates without fan assistance. (Drawing by Alexa Thornton. © Walter Grondzik; all rights reserved.)

Figure 13.24 Raven Run house near Lexington, KY, designed by Richard Levine and completed in 1975. Corrugated metal lies behind the collector's tilted glazing; a fan blows the hot air down to a basement rock bed for storage. (Courtesy of Allison Bailes III.)

heating energy use ranges from about 2 to 4 kWh per HDD_{65F} [~3.6–7.2 kWh per $HDD_{18.3C}$], with an average floor area of 1500 ft^2 [139 m^2] per dwelling; note that these are dwelling-specific rather than building-specific values, and that they include residences in multifamily buildings (Figure 13.27).

Nonresidential buildings show greater variability in space-heating needs due to their diversity of internal gains, but if no further information is available, CBECS values may be used to provide rough estimates in the continental United States; these range from about 16 to 26 kWh per HDD_{65F} [~29–72 kWh per $HDD_{18.3C}$] for an average building floor area of 15,000 ft^2 [1400 m^2] (Figure 13.27). If RECS, CBECS, or analogous values are used, multiplying the monthly HDD by the average space-heating energy use per HDD yields a monthly energy need.

EXAMPLE 13.2 Since the Token Creek festival barn will be equipped with a kitchen, office, and guest quarters, and its size is small, its monthly heating needs are expected to resemble those of a residence. What are these expected to be?

SOLUTION

The residential energy consumption survey map, Figure 13.27, shows that a typical residence in Madison uses 2.5 kWh for heating per HDD_{65F} [1.4 kWh per $HDD_{18.3C}$]. Multiplying this value by each month's HDD yields the monthly heating need, in energy units:

$$\text{monthly HDD} \times \text{energy use (kWh/HDD)}$$
$$= \text{monthly heating energy need (kWh)}$$

Plotting these results yields Figure 13.28. Values for months outside of the heating season should not be plotted. For this example, the values for June, July, and August are omitted. ■

(c) Planning the Solar-Collecting Glazing

A building's ability to collect solar energy depends on the tilt, area, orientation, and material of the solar-collecting glazing; in conceptual design, estimating the optimal tilt and area are the primary goals. At this stage, an orientation directly facing the equator is assumed, although fog patterns or site shading may later revise this assumption. Glazing material is similarly assumed, at this stage, to have a SHGC of 0.60; this value will also be refined in schematic design (Section 13.5).

Finding the Optimal Tilt. The optimal tilt is the angle above horizontal that receives the greatest incident solar radiation with respect to the building's space heating need, such that it can offset that need to the greatest extent. It is rarely the same as the optimal tilt for a photovoltaic array, which prioritizes sunny months that often do not coincide with the heating season.

While the Balcomb method, developed in a sunny climate, found that the vertically projected area of solar-collecting glazing predicted performance sufficiently well to ignore the glazing's tilt (Balcomb et al., 1980, 1984), work in cloudy climates has shown otherwise (Bartholomew, 1985; Porteous, 2005;

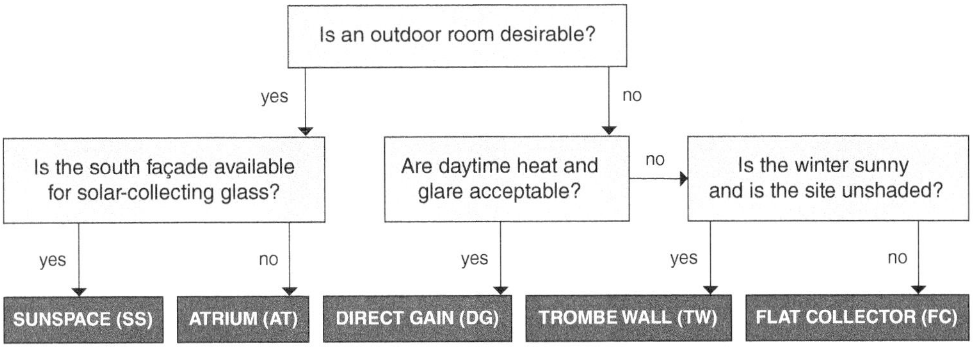

Figure 13.25 *Decision tree for passive solar heating system selection.*

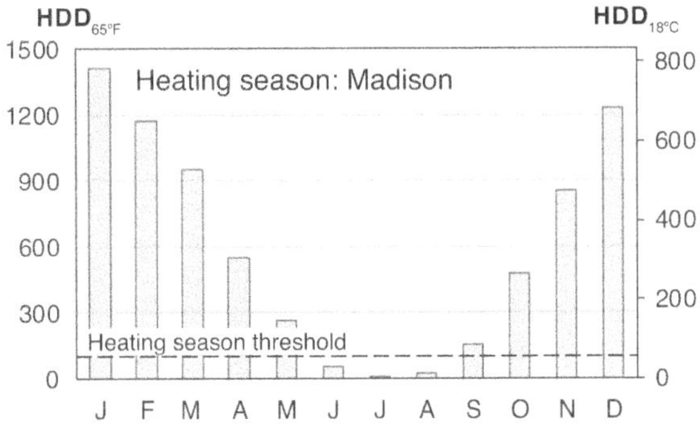

Figure 13.26 *Monthly heating degree-days, base 65°F [18.3°C], for Madison, WI (gray bars), and threshold above which a month is included in the heating season (dashed line).*

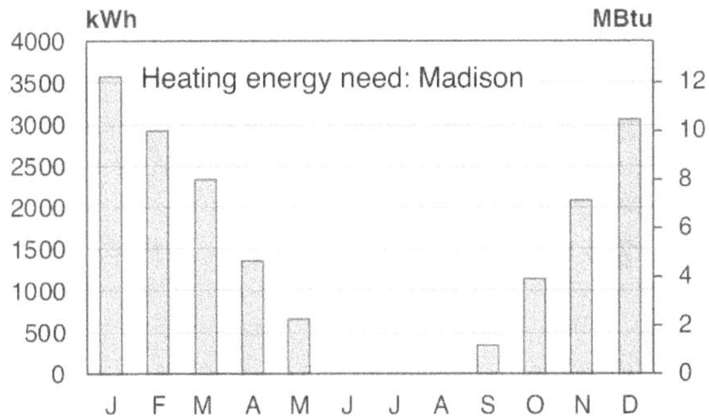

Figure 13.28 *Monthly estimated heating needs for a residence in Madison, WI, obtained from monthly HDD (Appendix I, Table I.1) and typical residential heating energy use per HDD in Wisconsin (Figure 13.27).*

Seifert, 1981; Rempel et al., 2013). Because clouds scatter both short- and long-wave solar radiation toward the top of the sky dome (Eymet et al., 2004; Demain et al., 2013; Perez et al., 1990), cloudy climates have apparent sun angles that are much higher in the sky, and much more broadly oriented, than the actual position of the sun. This is evident in plots of sky luminance under standard cloud conditions (Figure 13.29). As a result, the typical cloud cover during the heating season strongly influences the optimal tilt for solar glazing: climates with cloudy heating seasons require shallower tilts that allow glazing to collect solar radiation from the sky zenith, while climates with clear sunny heating seasons require steeper tilts that allow glazing to face the sun more directly.

Conceptual design-level estimation of the optimal tilt is accomplished by comparing a building's monthly

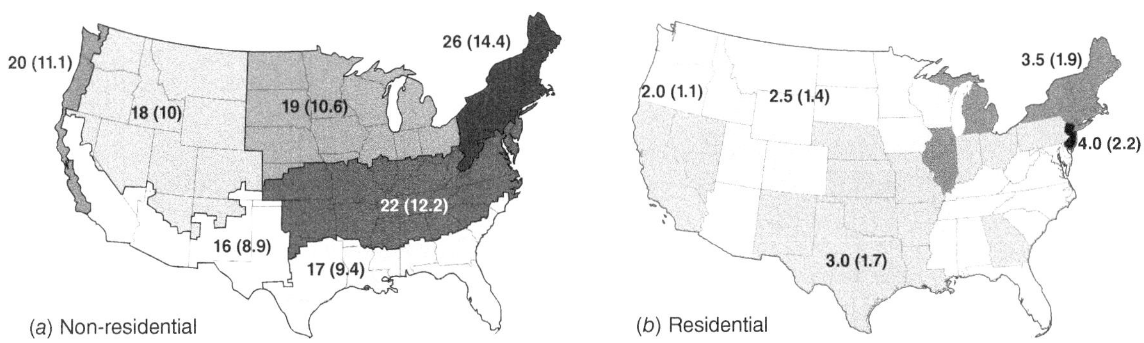

Figure 13.27 *Average space-heating energy use by residences in the continental United States, including apartments, of average area 1500 ft² [140 m²], and by nonresidential buildings of average area 15,000 ft² [1400 m²]. Labels show kWh/HDD_{65F} (kWh/$HDD_{18.3C}$); sector shapes reflect the data aggregation protocols of the 2012 Commercial Buildings Energy Consumption Survey (CBECS) (EIA, 2012) and the 2009 Residential Energy Consumption Survey (RECS) (EIA, 2009). (Drawings by Kelli Kimura.)*

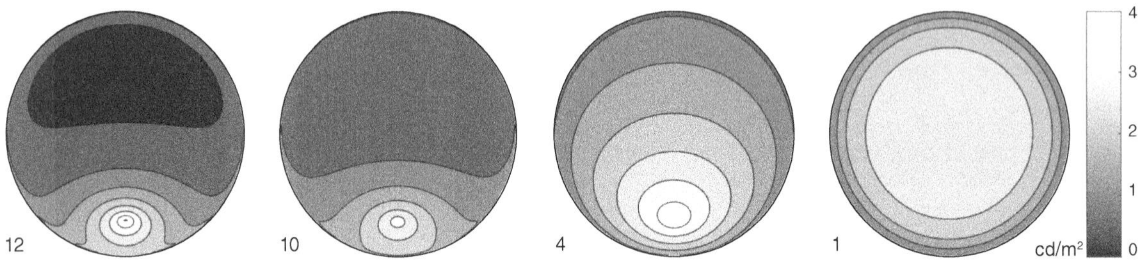

Figure 13.29 CIE Standard luminance distributions at solar noon, for a solar altitude of 38° above the horizon, for sky types (12) clear, (10) partly cloudy, (4) lightly overcast, and (1) heavily overcast, showing progressive brightening of the zenith and dimming of the solar position with cloud cover. (Adapted from the International Commission on Illumination. CIE, 2004. Images by Alan W. Rempel.)

heating need with monthly solar radiation values, for a range of tilt angles, and selecting the angle that provides the greatest solar radiation during the heating season. Tabulated values for incident solar radiation on surfaces of specific tilt are provided for numerous cities in Appendix I, Table I.1.

To estimate the optimal tilt, find the nearest city to the intended site in Appendix I, Table I.1. If the site is not in North America, a city with monthly HDD, latitude, and cloud cover patterns similar to those of the site may be used; the Köppen climate classification system is often useful in finding a suitable match (Peel et al., 2007). Alternatively, EnergyPlus (or another building energy-modeling tool that accounts for cloud cover effects) may be used with a local weather file to report the incident solar radiation on surfaces of various tilt.

Next, sum the incident solar radiation values over the months of the heating season for each tilt, and choose the tilt that receives the greatest solar radiation as the starting point for collector sizing.

EXAMPLE 13.3 What is the estimated optimal tilt for solar-collecting glazing in the Token Creek festival barn, given the heating season found previously?

SOLUTION

Figure 13.30 reproduces information shown for Madison, WI, in Appendix I, Table I.1. Summing these values over the heating season shows that a tilt of 40° receives the greatest solar radiation during the heating season, among the angles shown, and should therefore be close to the optimal tilt. ■

MADISON, WI		Solar radiation (kBtu/ft^2) on a surface of tilt:									
Mo	**D65**	**90°**	**80°**	**70°**	**60°**	**50°**	**40°**	**30°**	**20°**	**10°**	**0°**
Jan	1414	31	32	33	34	33	32	29	26	23	19
Feb	1172	30	33	35	35	35	35	33	30	27	24
Mar	945	31	35	38	41	42	43	42	41	39	36
Apr	549	29	35	39	43	46	48	48	48	47	44
May	247	29	36	42	47	52	56	58	59	59	57
Jun	43	27	34	41	47	52	57	60	62	62	61
Jul	8	29	36	42	48	53	57	60	61	62	60
Aug	17	31	38	43	48	52	55	56	56	55	53
Sep	137	33	38	41	44	46	47	46	45	42	38
Oct	461	32	35	37	38	39	38	36	34	30	26
Nov	834	26	27	28	28	28	27	25	23	20	16
Dec	1226	26	27	28	28	27	26	24	22	19	15
Total		267	298	322	339	348	**349**	342	327	305	275

MADISON, WI		Solar radiation (kWh/m^2) on a surface of tilt:									
Mo	**D18.3**	**90°**	**80°**	**70°**	**60°**	**50°**	**40°**	**30°**	**20°**	**10°**	**0°**
Jan	786	96	102	105	106	104	99	93	83	72	60
Feb	651	96	104	109	112	112	109	104	96	86	74
Mar	525	98	111	121	128	133	135	134	129	122	112
Apr	305	93	109	123	135	144	150	152	151	147	139
May	137	92	113	132	150	164	175	183	186	185	180
Jun	24	84	106	128	148	165	178	188	194	196	193
Jul	4	90	112	133	152	168	181	190	194	194	190
Aug	9	99	119	137	152	164	172	177	178	175	168
Sep	76	104	119	131	140	146	148	146	141	132	121
Oct	256	102	111	117	121	122	119	114	106	96	83
Nov	463	81	86	89	90	88	85	79	71	62	51
Dec	681	81	86	88	88	86	82	76	68	59	48
Total		842	940	1016	1070	1099	**1103**	1080	1032	961	868

Figure 13.30 Solar radiation incident on tilted surfaces, summed over the nine-month heating season of Madison, WI, showing that 40° is close to the optimal tilt. (Data from Appendix I, Table I.1.)

Estimating the Area. Once the optimal tilt is approximated, the next step is to estimate the area of solar-collecting glazing by examining the proportion of the heating need that each of several test areas is able to meet. Ultimately, the choice of solar-collecting area is subject to the designer's discretion: at a certain point, each additional square foot or square meter of glass intercepts less and less useful solar radiation, yet still requires movable insulation and shading proportional to its size. This is illustrated graphically in Figure 13.31, continuing the example of Madison, WI.

First, find the incident solar radiation, tabulated in Appendix I, Table I.1, on a range of tilted areas that are realistic for the building. If the solar-collecting glazing must have a particular, nonoptimal tilt, use the incident solar radiation on a surface of the nearest tabulated tilt for subsequent steps, with the understanding that performance will be somewhat diminished. In addition, multiply each of these values by 0.60 to approximate the SHGC of the glazing. Then, in any month with greater solar heat gain than heating need (e.g., late spring or early fall months), subtract the excess solar radiation (see Figure 13.8). Finally, plot the results alongside the monthly estimated heating need.

EXAMPLE 13.4 How much of the Token Creek festival barn's heating need can tilted glazing areas of 10, 20, 30, and 40 m^2 offset? Plot these results alongside monthly heating need values.

SOLUTION

First, use the monthly values for incident solar radiation (kWh) on a 1 m^2, 40°-tilted surface in Madison, Figure 13.30, to obtain monthly values for incident solar radiation on surfaces of 10, 20, 30, and 40 m^2. Then, multiply each of these values by 0.6 as a conservative estimate of solar heat gain. Third, subtract any solar gain in excess of the heating need. In this example, values for June, July, and August should be omitted. However, there are multiple months during the heating season where the solar gain is excessive. During those months, excess solar gain should be omitted as well. Plotting the results then yields Figure 13.31. ■

Examination of this graph shows that the smallest amount of glass shown, 10 m^2, is expected to offer 5635 kWh of useful incident solar radiation across the nine-month heating season (i.e., the summation of the monthly values from Figure 13.31).

Recall the monthly heating need in Example 13.2 (plotted in Figure 13.28). The total for the nine-month heating season is 17,463 kWh. Therefore, the smallest amount of glass shown, 10 m^2, is projected to meet about 32% of the estimated heating need.

The next 10 m^2, doubling the original area (or increasing it by 100%), meets another 23% of the heating need, approximately; note that no credit is given for solar collection in excess of the heating need. The third 10 m^2, tripling the original area, meets another 17% of the heating need yet requires as much movable insulation, summer shading, and attention to controls as the first 10 m^2. Quadrupling the initial area, finally, adds another 13%, gaining solar energy equal to ~85% of the heating need. This progression is shown in Table 13.1.

At this point, the designer should revisit the goals, ideals, and practical constraints of the project. In this example, the musicians viewed meeting 72% of the heating need as a worthy goal and planned for a 30 m^2 [324 ft^2] total collector area.

(d) Planning Thermal Storage Mass

Massive materials allow passive heating systems to store heat collected during the day for delivery during cooler hours or to cooler interior spaces. Numerous materials may be used for this purpose, and they may be configured in various ways to produce the heat delivery timing and intensity desired (e.g., Rempel and Rempel, 2013), providing greater flexibility to the designer than early design guides suggested (Balcomb et al., 1980, 1984). In conceptual design, characteristic patterns of thermal behavior are used to estimate the most promising materials, configurations, and sizes of thermal mass for the design intent and program; these will then be refined in schematic design using specific material properties.

Material Properties. A useful thermal storage material typically has four characteristics: (1) moderate-to-high solar absorptance, or ability to

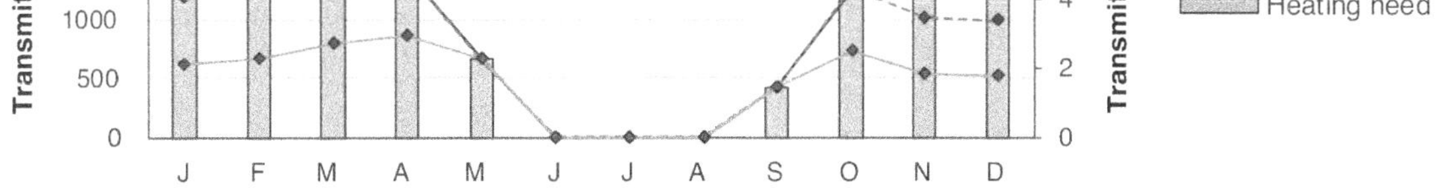

Figure 13.31 Rough-sizing the solar-collecting glazing for a system in Madison, WI, showing the estimated heating need (bars) compared to transmitted solar radiation through 10, 20, 30, and 40 m^2 glazing areas (lines) at the near-optimal tilt of 40° above horizontal. An equator-facing orientation and 60% solar transmission through glazing are assumed, typical of conceptual design.

TABLE 13.1 Progressive Benefits with Increasing Area of Solar-Collecting Glazing

Collector Area		Area Increase Over Previous			Useful Solar Radiation		Useful Solar Radiation Increase Over Previous			Heating Need Met	Increase in Heating Need Met Over Previous
m^2	ft^2	m^2	ft^2	+ %	kWh	MBtu	kWh	MBtu	%	%	+ %
10	108	—	—	—	5635	22.6	—	—	—	32	—
20	216	10	108	+100	9607	32.8	3972	10.2	70	55	+23
30	324	10	108	+50	12,600	43.0	2993	10.2	31	72	+17
40	432	10	108	+33	14,850	50.7	2250	7.7	18	85	+13

absorb incident solar radiation; (2) high heat capacity, or ability to store thermal energy; (3) moderate thermal conductivity, or ability to transfer heat through itself; and (4) high emittance, or ability to radiate stored energy. Solar absorptance and emittance are surface properties, and therefore affected by color, roughness, and coatings, while heat capacity and thermal conductivity are intrinsic properties of the material itself.

Solar absorptance, often symbolized by A, is a dimensionless quantity that describes the fraction of incident solar radiation, hemispherically distributed, absorbed by a material surface—values range from 0 (perfect mirror) to 1 (perfect absorber, or black body). Because transmitted solar radiation incident upon an indoor mass surface is not, in fact, uniformly hemispherically distributed, a material's solar absorptance value only approximates its actual absorption behavior, but this estimation is acceptable in conceptual design. Black, matte, and/or rough surfaces tend to have the highest solar absorptance values, while white, smooth, and/or polished surfaces have the lowest. Tabulated values for common materials are provided in Appendix I, Table I.2.

Heat capacity, often symbolized by C, quantifies the energy stored by one unit of material undergoing one unit of temperature increase (or, equivalently, the energy released by one unit of temperature decrease). Materials that require large energy inputs to become warm, such as concrete or stone, have high heat capacities, while those that become warm with small energy inputs, such as metals, have low heat capacities (see Appendix E, Tables E.1, E.25). Specific, or mass-based, heat capacities are expressed in Btu/lb °F [J/kg K], while volumetric heat capacities are expressed in Btu/ft^3 °F [J/L K]. To ensure that they occupy relatively small volumes of space, thermal storage materials should also have high density, or mass per unit volume (lb/ft^3 or kg/m^3), symbolized by ρ (rho).

Latent heat capacity quantifies the energy required to fully melt, or the energy released by fully freezing, a quantity of material, again expressed either specifically (Btu/lb °F [J/kg K]) or volumetrically (Btu/ft^3 °F [J/L K]). In any material, the transition between the solid phase and the liquid phase requires energy, as does the transition between liquid and vapor phases; this takes the form of kinetic energy among constituent molecules, freeing them either from the constraints of a solid lattice during melting, or from the transient attractive forces of a fluid during vaporization. During a phase change, therefore, energy uptake does not increase the overall temperature of the bulk material until the phase change is complete (Figure 13.32). The temperature at which a solid melts is called the melting temperature, or t_m; less intuitively, the energy needed to melt a unit mass of a material is called the latent heat of fusion, or L_f. Materials that melt at temperatures useful for thermal storage, described below, are known as phase-change materials (PCMs).

Thermal conductivity describes the rate of energy transfer through a material, in the form of molecular vibrations and collisions, induced by a temperature difference between two surfaces. Materials that move energy through themselves rapidly in response to relatively small temperature differences, such as metals, have high thermal conductivities, while those that move very little energy under such conditions, such as insulation, have low thermal conductivities. Thermal storage materials ideally have moderate values, such that they can move energy into their interiors effectively

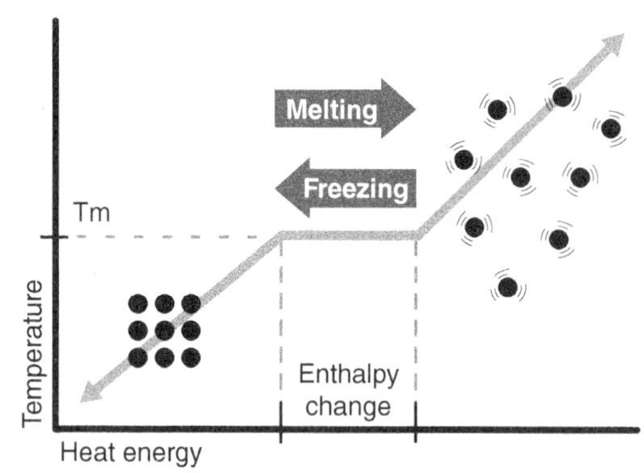

Figure 13.32 *Thermal storage by phase-change materials (PCMs), showing the relationship between temperature and energy. Below the melting temperature (t_m), a PCM is solid, and its temperature rises as it accepts energy. At the t_m, additional energy melts the solid but does not increase its temperature until the solid is fully melted. Above the t_m, additional energy raises the liquid's temperature. (Drawing by Alexa Thornton. © Walter Grondzik; all rights reserved.)*

PASSIVE ENVIRONMENTAL SYSTEMS

during solar collection without returning that energy too rapidly during cooler hours.

Fluids present a special case. In a fluid, most commonly water, material near the collecting surface expands as it warms, becoming less dense, and rises, allowing cooler material to take its place. This creates spontaneous mixing; as a result, the temperature in a fluid mass remains nearly uniform, maintaining cooler surfaces during collection and warmer surfaces during delivery than a solid mass could. This behavior is often estimated by using apparent thermal conductivity values that are several times greater than the value of the motionless material (Figure 13.33; Churchill and Chu, 1975).

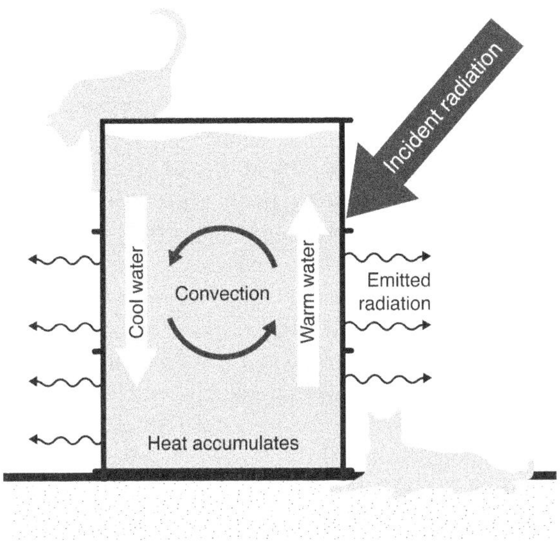

Figure 13.33 *Water expands as it warms during solar collection, causing a tank heated on one side to mix spontaneously. As a result, its surface temperature rises more slowly than a solid's would, facilitating greater energy gain. During cooling, in turn, its surface remains warmer, allowing it to deliver heat over longer periods of time than a comparable solid could. (Drawing by Alexa Thornton. © Walter Grondzik; all rights reserved.)*

Thermal conductivity, symbolized by k, is expressed in units of energy moved across a unit distance, during a unit time, for a unit temperature difference: W/m K [Btu/h ft °F].

Emittance, or the emissivity coefficient, symbolized by ε, describes the ability of a material to radiate stored energy—values range from 0.0 (no emission) to 1.0 (emission in perfect accord with its surface temperature, as in a black body). Most opaque building materials have emittance values of ~0.9, but a few, particularly metals such as aluminum, have much lower values; such low-emittance materials should be avoided in thermal storage, even as coatings on thermal mass surfaces.

Thermal Behavior. In solid materials, the process of heat storage and return follows the steps shown in Figure 13.34: the material first absorbs transmitted solar radiation according to its solar absorptance, while the remainder is reflected. If air in the collector space is warmer than the surface, energy may also be transferred to the mass by convection. After energy is absorbed, it travels through the material according to the material's thermal conductivity. Over time, the mass becomes warmer, as governed by its heat capacity and density. If the surroundings are cooler than the mass surface, however, as on a bright cold day, the mass may simultaneously absorb solar radiation and lose energy through radiation and convection. The warmer the mass surface becomes during collection, the more energy it tends to lose through these competing pathways.

PASSIVE ENVIRON-MENTAL SYSTEMS

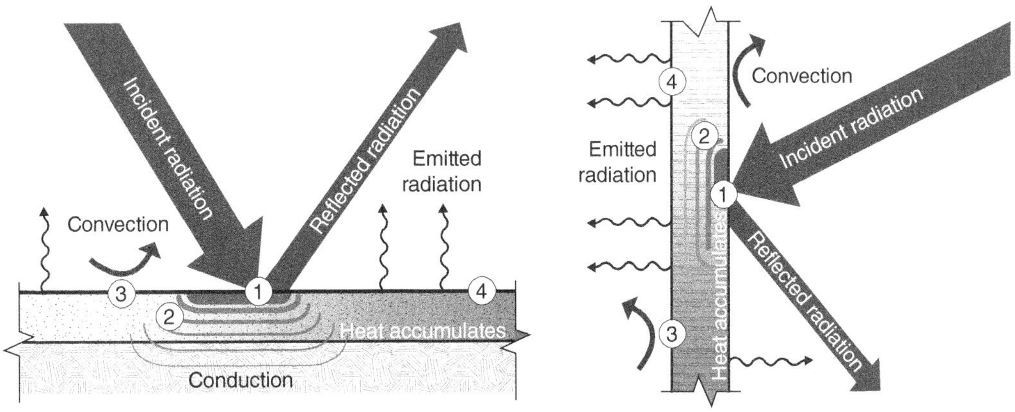

Figure 13.34 *Energy storage and release by solid materials in floors (left) and walls (right) involve the following processes: (1) absorption and reflection of solar radiation at the material surface, determined by its solar absorptance; (2) conduction of thermal energy into the interior, determined by the thermal conductivity, including further conduction into underlying soils in floor configurations, determined by soil thermal conductivity; (3) convection from warm surfaces, determined by convection coefficients; and (4) radiation from warm surfaces, determined by emissivity coefficients. (Drawing by Alexa Thornton. © Walter Grondzik; all rights reserved.)*

The temperature of a mass surface strongly influences both the extent of energy collection and the timing and intensity of heat delivery. During collection, mass that is more extensive relative to a given solar resource tends to remain cooler, allowing it to take up more total energy. Fluid mass, which cools its collection surface by mixing, is therefore able to take up more total energy than solid mass of comparable surface area and heat capacity, while phase-change materials (PCMs), which remain at their melting temperatures while they melt, are able to accept even more energy during their phase change.

During heat delivery, in turn, more extensive (and therefore cooler) mass delivers heat in lower intensities over longer periods, while modestly-sized (and therefore warmer) mass accomplishes more intense, shorter-term heat delivery (Figure 13.35).

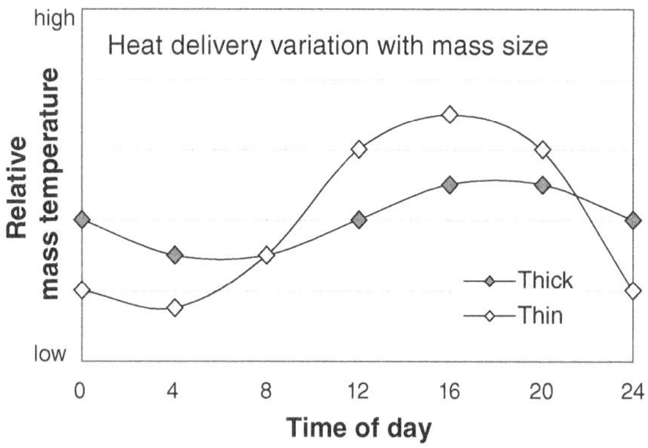

Figure 13.35 *Solid thermal mass that is thin relative to the solar resource becomes warmer during collection, delivers heat in a more intense pulse over shorter duration, and cools to a lower temperature overnight than mass that is thick relative to the solar resource.*

Materials, Configurations, and Rough Sizing. The most important question in thermal mass conceptual design is, simply, "when is heat needed most?" in order to determine whether daytime, evening, or all-night heat delivery should be given top priority. These categories may be envisioned as idealized heat flux patterns into and out of a thermal storage wall (Figure 13.36). When heat is needed during the day, delivery should begin soon after solar collection begins, with most heat delivered during daytime hours. When evening heat is most desirable, delivery should begin soon after collection ends and should be concentrated during evening hours. When heat is most desired during the coldest hours of the night, the ideal heat delivery is as great as possible during those hours, which equates in practice to nearly constant heat delivery at all hours.

Distinction among thermal mass design intents may seem surprising to some, since the Balcomb method and its derivatives advocated only one design intent: minimizing indoor temperature swings (Balcomb, 1978, Balcomb et al., 1980, 1984). This intent was an understandable expression of the desire to maintain comfortable indoor conditions in a climate of intensely sunny days and cold nights. The resulting guidelines, however, recommended extensive mass that released heat approximately continuously, as in Figure 13.36*c*. While this approach diminished temperature swings as intended, it was unable to concentrate heat delivery during daytime or evening hours when some program types needed it most. In climates with limited solar resources, it also provided too much thermal inertia for passive solar spaces to warm significantly at all, causing some researchers to abandon thermal mass altogether while others sought better

PASSIVE ENVIRON-MENTAL SYSTEMS

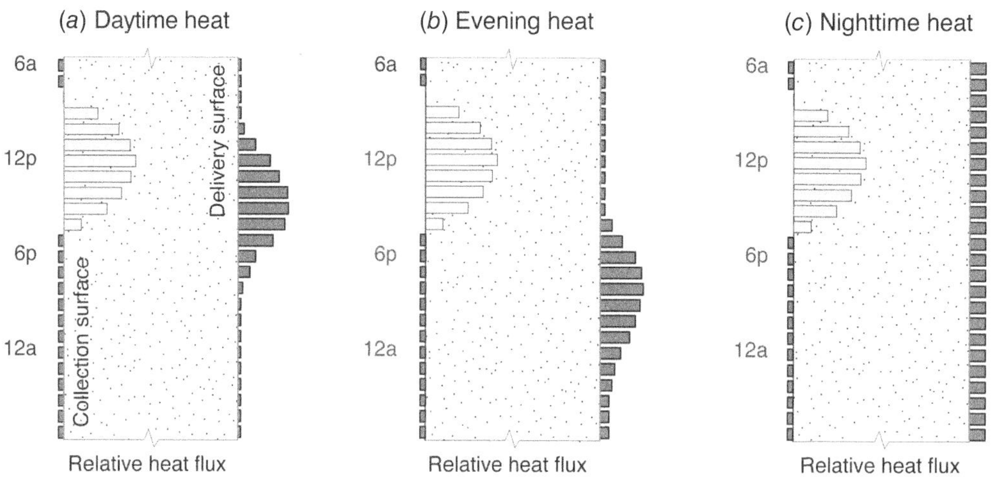

Figure 13.36 *Idealized heat flux patterns into and out of thermal storage wall sections, illustrating three example heat delivery priorities: (a) daytime, i.e., for workplaces, schools, and public buildings; (b) evening, i.e., for residences, and (c) nighttime, i.e., for maximum heat delivery during the coldest hours of the night. Solar energy uptake is shown with light bars; heat delivery is shown with dark bars. Horizontal extent represents relative heat flux. (Rempel and Rempel, 2013 / MDPI / CC BY 4.0.)*

approaches (e.g., Seifert, 1981; Swisher and Duffy, 1983; Swisher 1984; Rempel and Rempel, 2013; Rempel et al., 2016). Their evidence is incorporated into the conceptual and schematic design methods below, allowing thermal mass to be configured to deliver heat as desired in cloudy climates as well as sunny ones.

Once the heat delivery design intent is established, the next step is to identify a compatible material and configuration from the characteristic thermal behavior patterns described below, summarized in Figure 13.37 and Table 13.2.

When heat is desired as soon as possible after collection begins, air is the medium of choice. This air may be warmed in a flat collector, low-mass sunspace, atrium or other direct-gain space, or vented Trombe wall (Figure 13.37). Air is also the best medium for transferring heat to a location far from a solar collector. In each of these systems, warm air may be propelled by natural convection (i.e., motion arising from density differences between warm and cool air) or by fans. Thermal storage should be minimal, consisting of finishes no more extensive than a ceramic tile with a cement-board underlayment. Massive floors and walls should be omitted, and fish tanks, potted plants, and rainwater collection barrels minimized, though they may be usefully added in summer. Because thermal storage is minimal, overheating is likely if warm air removal stops during collection, requiring operable vents to the outdoors.

Water delivers heat less immediately than air, but a water wall can nevertheless provide a warm surface

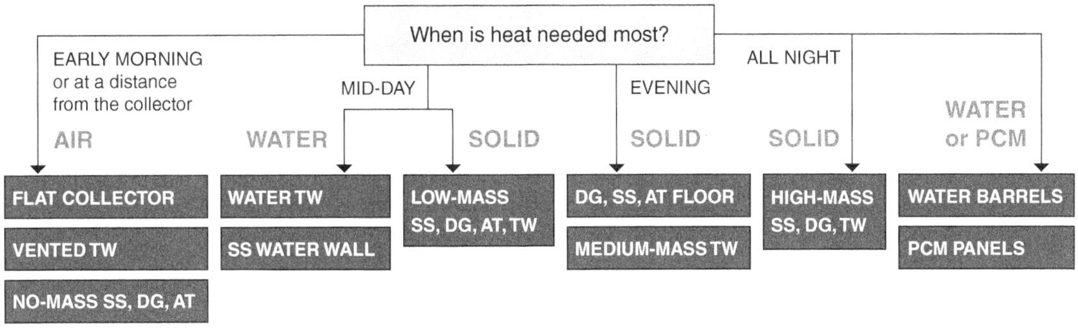

Figure 13.37 *Decision tree for thermal mass configuration in conceptual design. Note: TW = Trombe wall; SS = sunspace; DG = direct gain; AT = atrium; PCM = phase-change material. Low, medium, and high mass levels are defined in Table 13.2.*

TABLE 13.2 Thermal Mass Thickness Guidelines for Conceptual Design

	Sunny Winters				Cloudy Winters			
Desired Heat Delivery Time	**Morning**	**Mid-Day**	**Evening**	**Night**	**Morning**	**Mid-Day**	**Evening**	**Night**
Mass Configuration	**Thickness**				**Thickness**			
Air (No Mass)	None	None	—	—	None	None	—	—
Water Wall	—	Medium 4–6 in. [10–15 cm]	Med-high 6–8 in. [15–20 cm]	High 8–12 in. [20–30 cm]	—	—	—	—
Solid Wall	—	Low 2–4 in. [5–10 cm]	Medium 4–6 in. [10–15 cm]	High 8–12 in. [20–30 cm]	—	—	—	—
Solid Floor	—	—	Medium 4–6 in. [10–15 cm]	Med-high 6–8 in. [15–20 cm]	—	Very low 1–2 in. [2.5–5 cm]	Low 2–4 in. [2.5–7.5 cm]	Medium 4–6 in. [5–15 cm]
Water Barrels	—	—	—	High 8–12 in.[a] eq. [20–30 cm eq.]	—	—	Low 2–4 in. [2.5–7.5 cm]	Medium 4–6 in. eq. [5–15 cm eq.]
PCM Panels	—	—	Very low 1–2 in. [2.5–5 cm]	Low 2–4 in. [5–10 cm]	—	—	—	Very low 1–2 in. [2.5–5 cm]

Notes:

Symbol —: not recommended.

Values shown assume that mass areas are equal to solar-collecting glazing areas and are to be used only for planning purposes; optimal thicknesses depend on climate, site shading, and material thermal properties.

[a]eq.: equivalent.

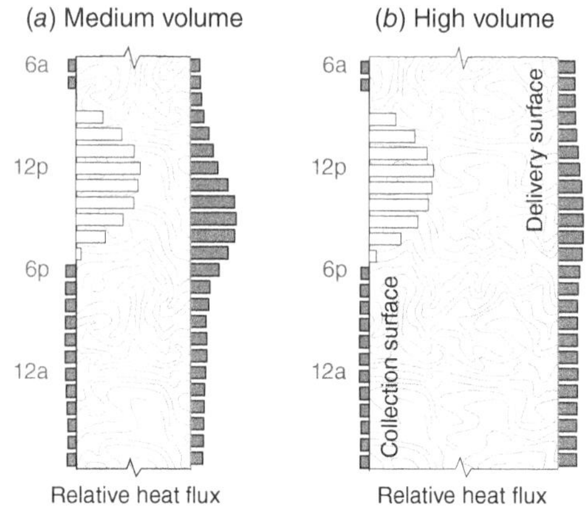

Figure 13.38 Effect of thickness on solar energy uptake and heat delivery by water Trombe walls in a sunny winter climate, showing (a) a medium-volume configuration appropriate for daytime heat delivery and (b) a high-volume configuration for heat delivery throughout the night. Solar energy uptake is shown in light bars, heat loss to the collecting space is shown in dark bars at the collection surface, and heat delivery is shown in dark bars at the delivery surface. Horizontal extent is proportional to heat flux, but uptake values are compressed, relative to delivery, for clarity. (Rempel and Rempel, 2013 / MDPI / CC BY 4.0.)

to an adjacent space soon after collection begins (Figure 13.38*a*). Water is ideal when late-morning to late-afternoon heat delivery is desired without glare, as in offices, schools, libraries, medical or gym facilities, retail spaces, and so on. In these applications, water may be configured as a Trombe wall or sunspace common wall (Figure 13.37), and the water itself may be contained in tubes, tanks, or any other vessel of the designer's invention (e.g., Figure 13.16). In these arrangements, water moves heat from one side to the other fairly quickly, with lower-volume walls accomplishing the most rapid transfer.

Because water can flow, its optimal size is more closely related to its volume than its area. In conceptual design, it is sufficient to plan for either a wall of medium volume (Figure 13.38*a*), to focus heat delivery during the day, or one of medium-high volume to extend heat delivery into the evening (Table 13.2).

For best performance, a water wall must be located in an area that receives direct solar radiation consistently. In a Trombe wall, the entire wall should have solar exposure; in a sunspace common wall, water should occupy lower portions of the wall to minimize shading by the roof. Like all vertical mass, water walls are best suited to sunny winter climates, allowing solar radiation to encounter their vertical surfaces from low angles.

Solid Trombe walls are another choice for daytime heat delivery in sunny climates (Figure 13.37). In contrast to water, solid mass cannot deliver heat rapidly through itself because its effective thermal conductivity is significantly lower, but it usually has greater density, allowing it to store more heat in a given volume. As in the case of water, thinner walls deliver heat sooner after collection begins and more intensely than thicker walls, because they become warmer, although the delivery typically begins later than that of a comparably-sized water wall. Conceptual design requires planning only for relatively thin, medium, or thick walls, focused on daytime, evening, or all-night heat delivery, respectively (Table 13.2), with the assumption that the mass area is comparable to the area of solar-collecting glazing. Again, sunny winter climates are required for effective use of vertical thermal mass.

In sunspaces, direct-gain spaces, and atria, floors are often convenient locations for mass because they are generally unshaded except near their edges and under furniture. In addition, materials like brick, adobe, concrete, tile, and stone become very warm at their collecting surfaces if sized modestly for the solar resource. This allows mass floors to deliver heat in a concentrated pulse of several hours as soon as solar collection ends, making them ideal for evening heat delivery. The thicker the mass, the cooler the surface remains, and the longer and less intense is the pulse of heat delivered, consistent with all mass applications (see Table 13.2).

Floor mass, however, is vulnerable to heat loss to underlying soil (see Figure 13.34). The belief that perimeter insulation is sufficient, found in some design guides, misinterprets a thermal storage floor as an unheated slab-on-grade in a conventionally conditioned space, a configuration thought to maintain the mass temperature near that of the underlying soil. In a passive solar heating system, in contrast, floor mass becomes 10s of degrees warmer than the underlying soil, just as a mechanically heated radiant floor would (e.g., Rempel et al., 2016). Thermal mass floors are therefore heated floors.

Moreover, soil thermal conductivity increases with soil moisture, causing it to be a consideration in all but the driest, most well-drained soils: field studies have established that conductive heat loss from mass floors to moist soils is a significant heat loss pathway for buildings in general (Deru, 2003; Janssen et al., 2004), as well as a problem for passive solar spaces in particular (Rempel et al., 2013; Rempel et al., 2016). To illustrate, consider the example of a 4 in. [10 cm] concrete floor in uninsulated, perimeter-insulated, and fully insulated configurations over a soil of moderate thermal conductivity (Figure 13.39). While collection patterns are similar, as shown by heat flux into the mass, the perimeter-insulated mass returns only about two-thirds of the heat returned by the fully insulated mass; related work has shown that a gravel layer works almost as well as insulation in mild climates and

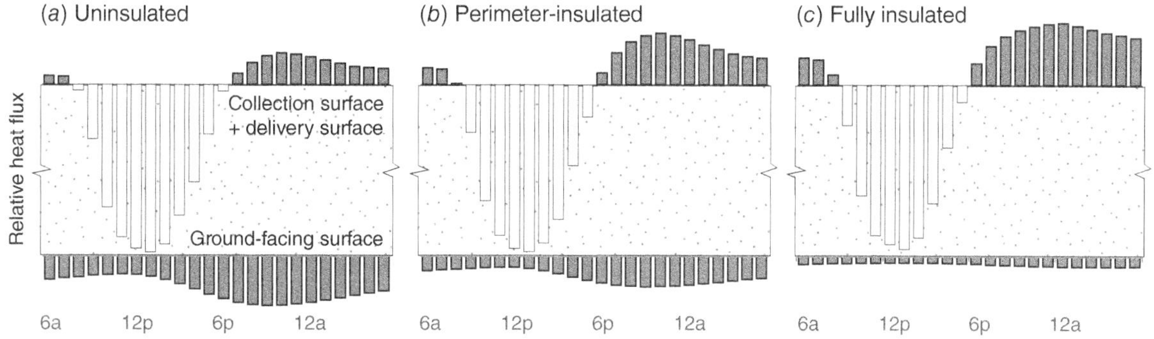

Figure 13.39 *Effect of insulation on heat loss to the ground from floor thermal mass in a direct-gain space in a sunny winter climate, comparing (a) uninsulated, (b) perimeter-insulated, and (c) fully insulated mass floors over loamy soils of moderate thermal conductivity (k = 0.4 Btu/h ft °F [0.7 W/m K]). Solar energy uptake is shown by light bars, evening heat delivery is shown by dark bars at the delivery surface, and heat loss to the ground is shown by dark bars at the ground-facing surface; vertical extent is proportional to heat flux. (Rempel and Rempel, 2013 / MDPI / CC BY 4.0.)*

requires lower embodied energy (Rempel et al., 2016). Thermal mass located in floors must therefore be well-drained, insulated, or otherwise thermally separated from moist soils.

Floor mass should be located under solar-collecting glass such that it intercepts as much of the direct solar radiation as possible. For best performance in cloudy winter climates, the floor mass should also be positioned to collect diffuse (i.e., high-angled) solar radiation. Evaluation of the relative benefits of various floor materials will be undertaken in design development (Section 13.6).

Because of its high heat capacity and ability to flow, water in larger volumes than discussed previously can deliver heat effectively throughout the night to a sunspace, atrium, or direct-gain space (Figures 13.38*b*, 13.40; Table 13.2). Compared to the leading alternative, PCMs, the volumes are greater, but costs are lower, and seasonal adjustment is more convenient. In conceptual design, it is sufficient to allow space for a high-volume wall or equivalent tanks or barrels; this quantity will be refined in schematic design.

Positioning of the mass for excellent solar exposure is again essential, as is the use of thermally conductive, nonreflective containers, such as metal or glass. Plastic containers are not recommended because their low thermal conductivities severely diminish both energy uptake and delivery. Again, if a vertical collection surface is planned, performance will be best in sunny climates.

PCMs neither take up energy as rapidly as water nor deliver it as intensely as solid mass. Instead, PCMs are ideal for extensive energy uptake as well as sustained, constant heat delivery when they are well-matched to the solar resource and space to be heated. The ideal PCM should first have a convenient melting temperature; this is ideally somewhat warmer (~10–20°F [~5–10°C]) than the desired room temperature, just as a mechanical radiator is kept warmer than the room it heats, so that the direction of heat delivery is into the space when heat is needed. This limits the options primarily to waxes and the hydrated versions of several salts (Appendix I, Table I.2; Cabeza et al., 2011). In addition, a substantial enthalpy change should accompany the phase change, so that substantial energy is stored. Third, the volume change must be manageable. Some PCMs form crystals that can puncture containers as they grow, and some expand significantly; in these cases, glass or metal containers are required to contain them.

Because of their substantial heat capacities near their melting temperatures (Figure 13.32), PCMs are typically effective in very small volumes. Since melting begins at the collecting surface, and since solid PCMs cannot flow, the thickness should be minimal (Table 13.2). An oversized PCM acts like a low-density solid, with lower heat capacity and lower thermal diffusivity than concrete, brick, or stone. An undersized PCM, in contrast, acts like water: it flows, it moves heat across itself rapidly, and it still delivers heat well while cooling within its liquid phase. As a result, PCMs are typically sized for the coolest months of the heating season, erring on the side of water-like behavior.

Like other massive materials, full solar exposure is necessary for good performance; PCM panels, tubes, or other containers should not be positioned close to an opaque roof or behind plants or furniture. Because PCMs work well in small volumes, their containers may be tilted or arranged horizontally to intercept diffuse solar radiation more directly. PCMs may also be used as the storage mass in Trombe walls, providing an exceptionally thin option. Micro-encapsulated PCMs within gypsum board-type products, however, are not sufficient to store heat from a passive solar system. Since PCM performance varies substantially by material, either experimentation and planned adjustability

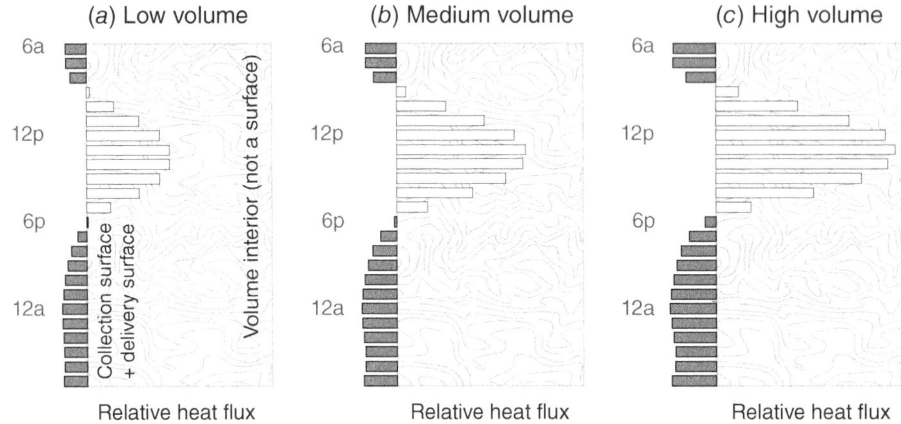

Figure 13.40 *Effect of water barrel volume on solar energy uptake and heat delivery patterns in a sunspace in a sunny winter climate, comparing (a) low volume (see Table 13.2) to (b) medium volume and (c) high volume configurations. Solar energy uptake is shown by light bars; heat delivery to the space is shown by dark bars. Horizontal extent is proportional to heat flux. (Rempel and Rempel, 2013 / MDPI / CC BY 4.0.)*

or thermal simulation modeling (Section 13.6) must be used for final sizing.

The ideal quantity of thermal mass for any passive solar heating system changes throughout the year: less is typically needed in colder months and more in warmer months. Because of this, particularly in climates with cloudy winters, permanent (i.e., floor or wall) mass may be sized for cooler months, with internal mass added during the warmer months. Fish tanks, potted plants, stone sculptures, and PCM bottles have been used effectively in this way (e.g., Rempel et al., 2013; Rempel et al., 2016).

(e) Planning Controls

The final step in conceptual design is to plan control elements: movable insulation, internal vents, exterior vents, and operable shading.

Once solar energy is captured and stored, a passive heating system next needs to prevent that energy from escaping. This effort focuses on movable insulation for the solar-collecting glazing, because that is the component most susceptible to heat loss. Even the most insulating window assemblies have U-factors of only about 0.15 Btu/h ft^2 °F [0.85 W/m^2 °C], while high-SHGC assemblies necessary for solar collection typically have U-factors of 0.3–0.65 Btu/h ft^2°F [1.7–3.6 W/m^2 °C] (Appendix E, Table E.15); both of these are far more thermally conductive than the opaque enclosure, which may be required to have a U-factor of 0.03–0.04 Btu/h ft^2 °F [0.17–0.23 W/m^2 K] or lower in some code jurisdictions where addressing heat loss is a priority (ICC, 2024).

Movable insulation is therefore essential for use when solar collection is negligible (i.e., between dusk and dawn and during rain and snow). This insulation may consist of panels or fabrics that roll, slide, or rotate into place, offering expressive opportunities as well as performance improvement. In conceptual design, the goal is to envision the movable insulation type(s) and mode(s) of action. Figure 13.41 shows a number of examples, and further options are described in Wright (1978) and Langdon (1980).

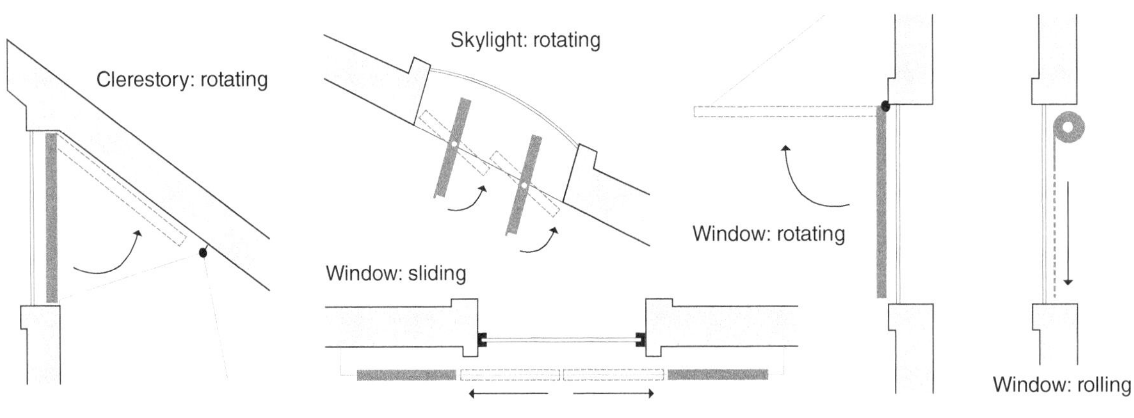

Figure 13.41 *Movable insulation examples for windows and skylights, including rotating, sliding, and rolling configurations. (Drawing by Alexa Thornton. © Walter Grondzik; all rights reserved.)*

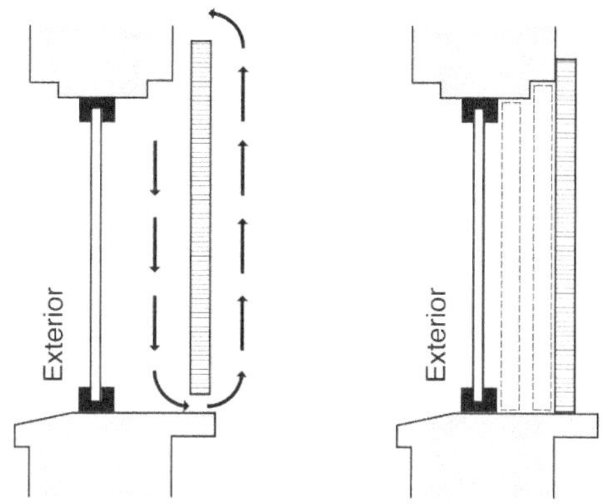

Figure 13.42 *Establishment of air currents behind movable insulation (left) must be prevented by effective edge seals (right), potentially using window trim to form secure connections. (Drawing by Alexa Thornton.*

The most important influence on movable insulation's effectiveness is not its U-factor, although that is important, but its edge sealing, which is necessary to prevent warm room air from flowing behind the insulation and losing heat to cold window surfaces (Figure 13.42). An edge seal may take the form of a pressure fitting between a panel's edge and the window frame, a channel that encases the insulation's edge, a rabbet into which a matching panel edge fits, a simple cord that holds a fabric edge in place, or any other mechanism that prevents room air from passing behind the insulation.

The physical mode of operation should be considered as well, even in conceptual design. While manual control is simpler and less expensive, it is also far less reliable, particularly in nonresidential buildings where occupants feel less freedom and/or less responsibility to operate building elements. Even in single-family homes, however, dedicated attention to movable insulation is required for good performance. Because automatic actuators are far less expensive than a solar-collecting window, and far more reliable, they are worth considering in most applications.

Plans should also be made for daytime shading, night ventilation, increased summer thermal mass, evaporative cooling in dry climates, and/or radiant cooling to night skies, to counteract overheating during warm or sunny days in the winter and to provide cooling during the summer. Mechanisms of operation for these elements should be planned as well, since consistent operation of movable elements greatly increases their effectiveness (e.g., Roetzel et al., 2010; Rempel and Remington, 2015).

13.5 SCHEMATIC DESIGN

In schematic design, the designer is assumed to have a specific site, a complete program, and preliminary enclosure performance goals, allowing the system choice, solar-collecting glazing, thermal mass, and control decisions made in conceptual design to be improved accordingly.

(a) Revising the Heating Need Estimate

The first step is to revise the expected monthly heating need in light of expectations for the future building enclosure. This revision may change the building's heating season and, as a result, the glazing tilt and area.

Estimating the Balance Point Temperature. Recall that a building's balance point temperature is the outdoor temperature at which heat gains to the building from solar radiation and internal sources (i.e., people, lights, and equipment) equal heat losses through the enclosure and infiltration, assuming the building is maintained at the desired indoor air temperature. To find this value for a particular building, for a particular time of the year (here, a winter month), the gains are therefore set equal to the losses:

$$q_w = q_b$$

where

q_w = sum of solar (q_s) + internal (q_i) gains under winter conditions
q_b = sum of enclosure (q_e) + infiltration (q_{inf}) losses at the balance point temperature
units = Btu/h ft² [W/m²]

Expanded, this becomes

$$q_s + q_i = q_e + q_{inf}$$

Solar Gains (q_s). In schematic design, solar heat gain through vertical double-glazed surfaces in sunny climates may be estimated from Appendix C, showing values at a number of latitudes and a range of orientations. In cloudy winter climates, however, solar gains on vertical glass should be neglected.

EXAMPLE 13.5 Estimate the hourly January solar gain (q_s) rate on vertical double-strength sheet (DSA) glass, at a latitude of 48° in the consistently sunny climate of Coeur d'Alene, Idaho, with the following orientations and areas:

North: 400 ft² [46.5 m²]
East: 50 ft² [4.6 m²]
West: 50 ft² [4.6 m²]

SOLUTION

Appendix C, Table C.4 shows solar heat gain estimates for DSA glass at 48° latitude. To find the average hourly heat gain rates, the gains for January need to be summed and averaged over 24 hours; note that the table assumes solar heat gains are symmetric around solar noon.

North: 400 $ft^2 \times$ (43 Btu/ft^2 half-day total × 2) × (1 d/24 h) = 1433 Btu/h
East: 50 $ft^2 \times$ (316 Btu/ft^2 AM half-day + 43 Btu/ft^2 PM half-day using the west value) × (1 d/24 h) = 748 Btu/h
West: 50 $ft^2 \times$ (43 Btu/ft^2 AM half-day + 316 Btu/ft^2 PM half-day using the east value) × (1 d/24 h) = 748 Btu/h
Total: 1433 + 748 + 748 = 2929 Btu/h

$$q_s = 2929 \text{ Btu/h } [860 \text{ W}]$$ ■

EXAMPLE 13.6 How would the solar heat gain estimate above change for Everett, WA?

SOLUTION

Appendix C, Table C.4 assumes clear conditions, with solar radiation originating from the direction of the sun, but Everett, WA, has a cloudy West Coast marine climate, in which much of the winter solar radiation emanates from the top of the sky dome (Figure 13.29). In climates such as this, a conservative estimate assumes that no significant solar gain on vertical glass occurs:

$$q_s = 0 \text{ Btu/h } [0 \text{ W}]$$

When calculating a winter balance point temperature, conductive heat gains through the opaque enclosure are assumed to be negligible. ■

Internal Gains (q_i). Next, internal gains may be estimated from Appendix G, Table G.1, showing approximate heat gains from people (q_{people}), lighting ($q_{lighting}$), and equipment ($q_{equipment}$).

$$q_i = q_{people} + q_{lighting} + q_{equipment}$$

Although this table specifies "summer" heat gains, those from people and equipment are fairly constant throughout the year; lighting tends to be higher in the winter, however, which has fewer hours of daylight. To account for this, a daylight factor at the low end of the planned range should be chosen.

EXAMPLE 13.7 Estimate the average hourly winter internal heat gain (q_i) rate in Coeur d'Alene, for each hour of a 24-hour day, for a 1500 ft^2 [140 m^2] household of two adults, assuming that the household is actively occupied for half of each day (i.e., 12 hours).

SOLUTION

From Appendix G, Table G.1 and its footnotes, estimated heat gains from people, lighting, and equipment are as follows:

People:

$$q_{people} = 230 \text{ Btu/h person} \times 2 \text{ people} \times 12 \text{ h/day} = 5520 \text{ Btu/day } [67.4 \text{ W}]$$

This value is per day, but we need q_{people} per hour.

$$q_{people} = 5520 \text{ Btu/day} \times 1 \text{ day/24 h} = 230 \text{ Btu/h } [67.4 \text{ W}]$$

Lighting: Assuming a low winter daylight factor, but energy-conscious occupants, we obtain an estimate of 2.7 Btu/h ft^2 of floor area.

$$q_{lighting} = 2.7 \text{ Btu/h ft}^2 \times 1500 \text{ ft}^2 \times 12 \text{ h/1 day} \times 1 \text{ day/24 h} = 2025 \text{ Btu/h } [593.5 \text{ W}]$$

Equipment: Assuming energy-conscious occupants, it is reasonable to estimate equipment use at the lower end of the range, or 1200 Btu/h. Since many appliances and other plug loads (e.g., refrigerators, dishwashers, electronics) consume energy independently of occupancy, however, this value should be used for all 24 hours.

$$q_{equipment} = 1200 \text{ Btu/h } [351.7 \text{ W}]$$

The total heat gain rate is therefore

$$q_i = q_{people} + q_{lighting} + q_{equipment} = (230 + 2025 + 1200) \text{ Btu/h} = 3455 \text{ Btu/h } [1013 \text{ W}]$$ ■

Enclosure Losses (q_e). Next, enclosure losses may be estimated from schematic design-level knowledge of enclosure U-factors using the familiar relationship:

$$q = U \times A \times \Delta t$$

where

q = heat loss through one enclosure element
U = thermal transmittance of the element
A = area of the element
Δt = temperature difference across the element

To account for enclosure heat losses of an entire building, then, it is necessary to consider the areas and U-factors of each unique aboveground enclosure construction. This may be expressed as:

$$q_e = \sum_j U_j \times A_j \times \Delta t$$

where

q_e = enclosure heat loss
$\sum_j$ = symbol indicating summation of j items
U_j = individual U-factors of enclosure constructions
A_j = individual areas of enclosure constructions
Δt = difference between the desired indoor air temperature (t_i) and the balance point temperature (t_b):

$$\Delta t = t_i — t_b$$

PASSIVE ENVIRONMENTAL SYSTEMS

Collecting these into one equation yields:

$$q_e = \sum_j U_j \times A_j \times (t_i - t_b)$$

or, equivalently, since the temperature difference is the same for each enclosure element:

$$q_e = (t_i - t_b) \sum_j U_j \times A_j$$

The balance point temperature (t_b) is the unknown quantity that will be found by the calculation. At this point, therefore, it is useful to tabulate the U-factors and areas of the building's total construction (UA_{total}), simplifying this equation to:

$$q_e = (t_i - t_b) \times UA_{total}$$

EXAMPLE 13.8 Given that the Coeur d'Alene residence has the following enclosure areas and U-factors, find the individual products of these quantities (UA) and sum them to find an expression for the temperature dependence of total enclosure heat loss.

Walls: $A = 3000\ ft^2\ [279\ m^2]$

$U = 0.05\ Btu/ft^2\ °F\ h\ [0.28\ W/m^2\ K]$

Roof: $A = 1500\ ft^2\ [139\ m^2]$

$U = 0.03\ Btu/ft^2\ °F\ h\ [0.17\ W/m^2\ K]$

Windows: $A = 500\ ft^2\ [46.5\ m^2]$

$U = 0.40\ Btu/ft^2\ °F\ h\ [2.30\ W/m^2\ K]$

SOLUTION

$$UA_{walls} = 3000\ ft^2 \times 0.05\ Btu/ft^2\ °F\ h = 150\ Btu/h\ °F$$

$$UA_{roof} = 1500\ ft^2 \times 0.03\ Btu/ft^2\ °F\ h = 45\ Btu/h\ °F$$

$$UA_{windows} = 500\ ft^2 \times 0.40\ Btu/ft^2\ °F\ h = 200\ Btu/h\ °F$$

$$UA_{total} = (150 + 45 + 200)\ Btu/h\ °F = 395\ Btu/h\ °F\ [209\ W/K]$$

Expression for I-P units:

$$q_e = (t_i - t_b) \times 395\ Btu/h\ °F$$

and, equivalently for SI units:

$$q_e = (t_i - t_b) \times 209\ W/K$$

■

The rate of warm air loss in a future building can be difficult to predict, but infiltration targets are often established in schematic design. A carefully detailed small building can easily achieve a rate of 1.0 air change per hour (ACH), and with fanatical attention to air barrier installation and caulking of cracks, rates below 0.33 ACH have been demonstrated. Once an infiltration rate is estimated, the resulting heat loss (q_{inf}) is estimated in the following general way:

infiltration heat loss rate (energy/time) =
volume of air lost per hour (volume/time) ×
density of air (mass/volume) ×
heat capacity of air (energy/mass-degree) ×
difference between indoor air and balance point temperatures (degree)

In condensed form, this expression is:

$$q_{inf} = ACH \times V_{air} \times \rho_{air} \times C_{air} \times (t_i - t_b)$$

where

V_{air} = volume of air in one air change, i.e., the building volume
ρ_{air} = density of air: $0.075\ lb/ft^3\ [\sim 1.2\ kg/m^3]$ at room temperature and atmospheric pressure
C_{air} = specific heat capacity of air: 0.24 Btu/lb °F [~1 kJ/kg K] at atmospheric pressure

As in the case of enclosure heat loss above, at this point, it is useful to simplify this expression with known information.

EXAMPLE 13.9 Given that the Coeur d'Alene residence has a volume of $30,000\ ft^3$ ($850\ m^3$) and infiltration of 0.75 air changes per hour, find an expression that shows the temperature dependence of infiltration heat loss.

SOLUTION

Substituting values into the relationship above:

$$q_{inf} = ACH \times V_{air} \times \rho_{air} \times C_{air} \times (t_i - t_b)$$
$$= (0.75\ vol/h) \times (30{,}000\ ft^3) \times (0.075\ lb/ft^3) \times (0.24\ Btu/lb\ °F) \times (t_i - t_b)$$

Expression for I-P units:

$$q_{inf} = (t_i - t_b) \times 405\ Btu/h\ °F$$

and, equivalently for SI units:

$$q_{inf} = (t_i - t_b) \times 212\ W/K$$

Two analogous expressions now exist for the Coeur d'Alene residence, showing the temperature-dependence of enclosure and infiltration heat loss, respectively:

$$q_e = (t_i - t_b) \times 395\ Btu/h\ °F \text{ (see Example 13.8)}$$

and

$$q_{inf} = (t_i - t_b) \times 405 \text{ Btu/h °F}$$

Combining these, we see:

$$q_e + q_{inf} = (t_i - t_b) \times (395 + 405) \text{ Btu/h °F}$$

$$= (t_i - t_b) \times (800) \text{ Btu/h °F}$$

Equivalently:

$$q_e + q_{inf} = (t_i - t_b) \times (209 + 212) \text{ W/K}$$

$$= (t_i - t_b) \times (421) \text{ W/K}$$

Recall that the balance point temperature (t_b) is the unknown of interest, and that:

$$q_e + q_{inf} = q_s + q_i$$

Also, recall from earlier examples in this chapter:

$$q_s = 2929 \text{ Btu/h} \quad \text{(see Example 13.5)}$$

$$q_i = 3455 \text{ Btu/h} \quad \text{(see Example 13.7)}$$

Therefore,

$$q_e + q_{inf} = q_s + q_i$$

$$= 2929 + 3455$$

$$= 6384 \text{ Btu/h [1871 W]}$$

From this, given the desired winter indoor air temperature (t_i) of 68°F [20°C], we can find the balance point temperature (t_b):

$$6384 \text{ Btu/h} = (68 \text{ °F} - t_b) \times 800 \text{ Btu/h °F}$$

$$6384 \text{ Btu/h} \div 800 \text{ Btu/h °F} = (68 \text{ °F} - t_b)$$

$$8.0 \text{ °F} = 68 \text{ °F} - t_b$$

$$t_b \approx 60 \text{ °F [15.6 °C]}$$

For a comparable residence in cloudy Everett, WA, however, winter heat gains on vertical glass are neglected, yielding a higher balance point temperature:

$$3455 \text{ Btu/h} \div 800 \text{ Bt/h °F} = (68 \text{ °F} - t_b)$$

$$4.3 \text{ °F} = 68 \text{ °F} - t_b$$

$$t_b \approx 64 \text{ °F [17.8 °C]}$$ ■

Revising the Heating Season Estimate. The balance point temperature estimated above may now be used to select a corresponding base temperature with which to gauge the heating season more accurately. While the Coeur d'Alene residence above would have had an estimated 9.5–10-month heating season with an estimated balance point of 65°F [18.3°C], in conceptual design, the revised balance point of 60°F [15.6°C] suggests that an 8-month heating season is more accurate (Figure 13.43).

Revising the Heating Need Estimate. The calculations used in the balance point estimation may now be used to revise the estimate of the monthly heating need.

EXAMPLE 13.10 Given a building in the sunny cold climate of Coeur d'Alene with the following characteristics, find and plot the monthly heating needs.

Balance point temperature (t_b) is ~60°F [15.6°C]

Total heat loss rate of ~800 Btu/h °F [421 W/K]

Total winter heat gain rate, estimated conservatively, of ~6400 Btu/h [1875 W]

SOLUTION

At a base temperature of 60°F, Coeur d'Alene has 916 January HDD. Since each HDD represents a 24-hour

Station: **COEUR D ALENE, ID US USC00101956**

Temperature (°F)																
Mean							Cooling Degree Days Base (above)						Heating Degree Days Base (above)			
Month	Daily Max	Daily Min	Mean	Long Term Max Std Dev	Long Term Min Std Dev	Long Term Avg Std Dev	55	57	60	65	70	72	55	57	60	65
01	35.5	25.4	30.5	3.3	3.7	3.4	0	0	0	0	0	0	761	823	916	1071
02	40.2	26.7	33.5	3.6	3.8	3.6	0	0	0	0	0	0	603	659	743	883
03	49.0	31.4	40.2	3.7	2.4	2.8	-7777	-7777	-7777	0	0	0	459	521	614	769
04	56.4	36.7	46.5	3.0	1.9	2.3	10	5	2	-7777	0	0	264	319	405	554
05	65.2	43.6	54.4	3.1	1.6	2.2	77	54	30	8	2	1	96	135	203	337
06	72.3	50.9	61.6	3.5	1.8	2.6	213	166	108	41	10	5	15	28	60	143
07	81.5	56.3	68.9	4.7	2.4	3.4	431	371	283	155	62	37	1	2	7	34
08	82.5	55.5	69.0	3.0	1.7	2.1	434	373	284	153	59	34	-7777	1	5	29
09	72.6	47.6	60.1	4.3	2.2	3.0	184	142	89	34	8	4	31	49	86	181
10	58.0	38.4	48.2	3.2	1.9	2.1	17	9	3	-7777	-7777	0	228	282	369	521
11	44.0	31.7	37.8	4.1	3.6	3.7	-7777	-7777	-7777	0	0	0	515	575	665	815
12	34.3	25.2	29.8	3.6	3.7	3.6	0	0	0	0	0	0	783	845	938	1093
Summary	57.6	39.1	48.4	3.6	2.6	2.9	1366	1120	799	391	141	81	3756	4239	5011	6430

-7777: a non-zero value that would round to zero

Figure 13.43 *Example report by the National Oceanic and Atmospheric Administration (NOAA) showing the normal heating degree-days at multiple base temperatures for Coeur d'Alene, Idaho.*

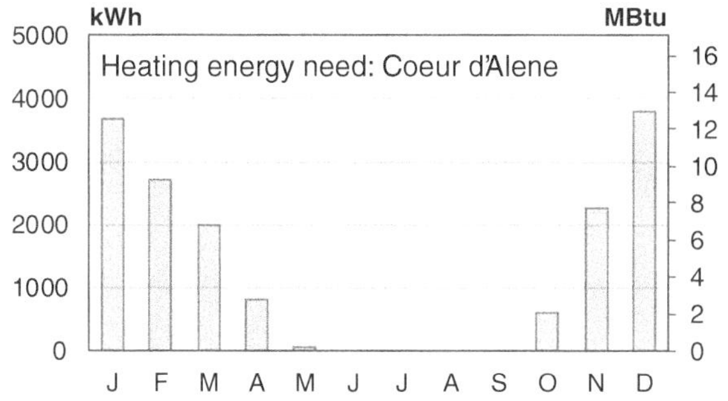

Figure 13.44 *Monthly heating energy needs found using enclosure and infiltration heat loss estimates, as well as window and internal gain estimates, for a 1500 ft² [139 m²] residence in Coeur d'Alene, ID.*

period in which the average outside air temperature is one degree below the base temperature (i.e., 24 heating degree-hours), monthly heating needs can be estimated as the product of hourly heat loss rate and heating degree-hours, offset by the hourly heat gain rate:

916 heating degree-days (HDD) in January (Figure 13.43)
× 24 heating degree-hours per heating degree-day
× 800 Btu/h °F
= 17,587,200 Btu
≈ 17.6 MBtu [~5.2 kWh] January heat loss
~6400 Btu/h average heat gain rate from $q_s + q_i$ (see Example 13.9)
× 24 hours per day
× 31 days per month of January
= 4,761,600 Btu
≈ 5 MBtu [~1.5 kWh] January heat gain
Net January heat loss ≈ January heating need − January heat gain (q_s + q_i) 17.6 − 5 ≈ 12.6 MBtu [~3.7 kWh]

Repeating this procedure for the other months produces a graph like that shown in Figure 13.44.

This revised heating need will ultimately be combined with a revised solar resource estimate, found next, to revise the solar-collecting glazing orientation, tilt, and area. The system choice may need to be reevaluated, as well. ■

(b) Revising the Solar Resource Estimate

Site shading can substantially diminish solar resource estimates made on the basis of climate alone in conceptual design. For this reason, an in-person survey of site shading is essential. Site topography, tree species, impending construction nearby, and other features are not fully evident in satellite images.

Solar Site Survey. For a solar site survey, the following materials are useful:

- 100-ft [30-m] tape measure
- field notebook; pencil or pen

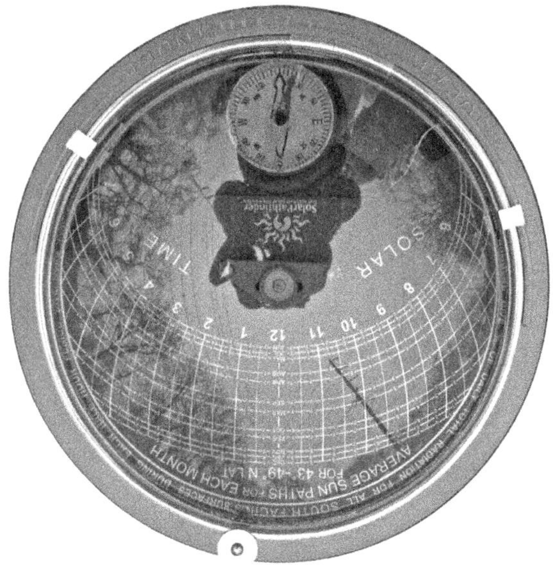

Figure 13.45 *Top view of a solar pathfinder positioned on a site with magnetic declination of 15° East. Note the projection of site tree images onto annual sun paths, particularly on the west. (Photo by Kelli Kimura.)*

- solar pathfinder (Figure 13.45)
- inclinometer (a.k.a. clinometer)
- 8-ft [2.5-m] step-ladder
- digital camera
- plastic bags

Plan to visit during the heating season so that deciduous tree branch density is evident. If a summer visit is unavoidable, tree identities may be used to estimate winter branch densities. Choose a day without rain or fog, if possible, since droplets on the pathfinder dome tend to obscure reflected images. Overcast days, however, are ideal.

1. *Magnetic declination:* Before the visit, look up the site's magnetic declination on a map such as the one in Figure 13.46. The National Oceanic and Atmospheric Administration (NOAA) has a number of resources publicly available (ngdc.noaa.gov/geomag/declination.shtml).
2. *Solar pathfinding:* At the site, locate one or more positions likely to become future centers of solar-collecting glazing. Set up the ladder at each position and position the pathfinder so that its top is at the approximate height and center of the future solar-collecting glass, assuming a first-floor space. If the planned space is higher, shading can be estimated from computer simulations in design development using other data collected at the site. Orient the pathfinder to solar north, taking into account the site's magnetic declination, and level it. Once the pathfinder is level and correctly oriented,

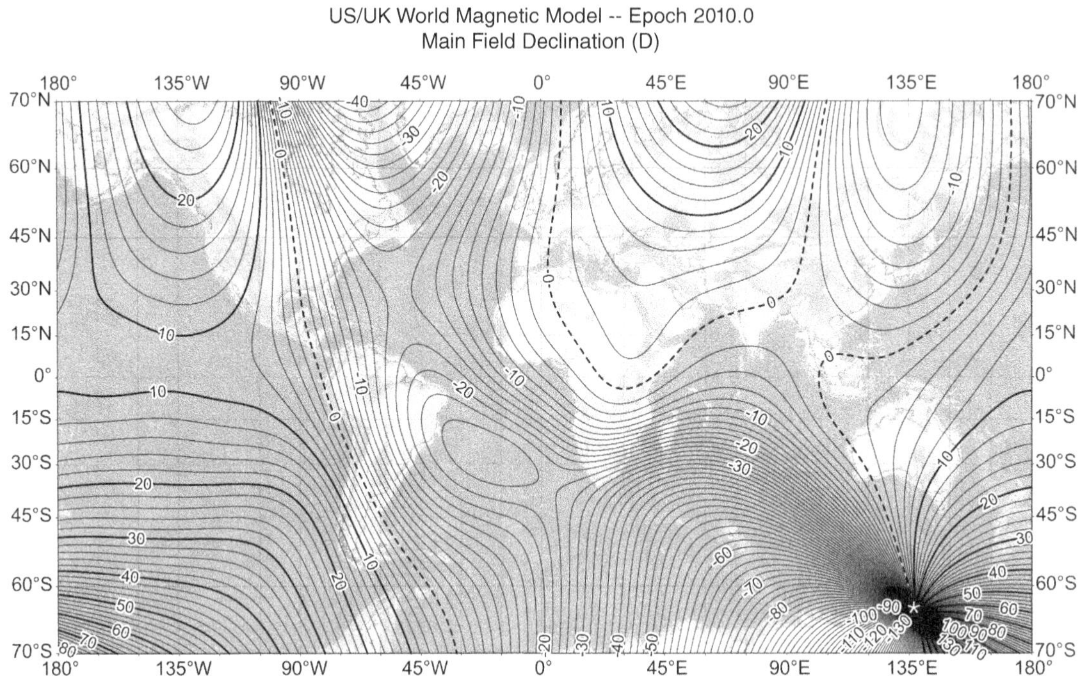

Figure 13.46 Global magnetic declination values at 10° intervals. (Reproduced from National Geophysical Data Center.)

stand on the north side and photograph it from directly above. Be sure the white sunpath lines are visible in the photograph; the result should resemble Figure 13.45.

3. *Photograph the horizon:* Next, from the same positions, fully document the surroundings with photographs at ~15° intervals from northeast to northwest, in the Northern hemisphere. Be sure the photographs show good contrast between tree branches and the sky; these will be used to estimate optical density.
4. *Document individual tree heights:* From one of the positions chosen above, measure the distance to each tree tall enough to appear in pathfinder photographs. Then, returning to that position, measure the angle to the top of each tree. To do this, hold the inclinometer at eye level and tilt it upward until the top edge of the inclinometer aligns with the line from your eye to the top of the tree. Ask a partner to read the angle between the horizontal and that line, or steady the tilted inclinometer against the ladder to allow you to read it yourself. A low or far-away tree typically results in a reading of 10–15°, while a close or tall tree typically gives a reading of 40–60°. These angles and distances will allow tree heights to be calculated, below.
5. *Identify important trees:* Next, gather enough information to identify each deciduous tree that appears on the pathfinder. Collect samples of leaves, twigs, flowers, and/or berries, and store these in plastic bags. In addition, closely photograph the bark and branching pattern of each tree. As you proceed, number each important tree on the site plan and label the bags respectively.
6. *Measure other site shading objects:* As for each tree, find the distance to each corner of each building and other structure that shades a potential location of solar-collecting glazing. Then, find the elevation angle of each upper corner with the inclinometer.

The next step is to find tree and building heights, tree shading density, and tree leaf-out and leaf-drop schedules for use in revising the solar resource estimates from conceptual design.

1. *Tree and building heights:* Consider the diagram in Figure 13.47. The distance from the central point to any given tree or building corner is "d," while the height of the tree or building corner is "h." The angle formed between the horizontal and the line to the top of the tree or building corner, in turn, is θ, and the tangent of θ is equal to x/*d*:

$$\tan\theta = x/d$$

Since θ and *d* are known, x may be found from:

$$x = d\tan\theta$$

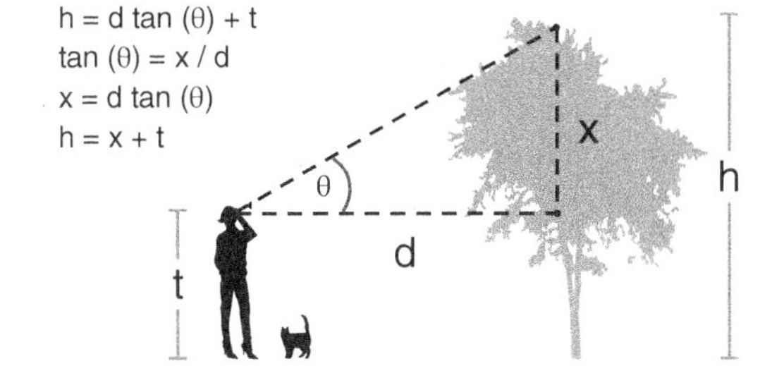

Figure 13.47 Procedure for calculating tree height from inclinometer measurements. (Drawing by Alexa Thornton. © Walter Grondzik; all rights reserved.)

EXAMPLE 13.11 If a tree is 50 ft [15.2 m] away (*d*) from a central location, and the sighted angle to the top of that tree (θ) is 40°, how tall is the tree? The eye of the person (*t*) is 5 ft [1.5 m] above the ground.

SOLUTION

Using the equation above,

$$h = d\ \tan\theta + t$$
$$= 50\ \text{ft} \times \tan(40°) + 5\ \text{ft}$$
$$\approx 47\ \text{ft}$$

The tree height is 42 ft [12.8 m] above the eye of the person sighting, or 47 ft [14.3 m] total. ■

2. *Tree optical density:* Deciduous tree branches typically have densities of 20–70%, and evergreen trees may be nearly 100% opaque. To estimate the optical densities of individual or groups of trees from photographs, use one of the digital densitometers found in photo-editing programs, subtracting the sky density value from that of a representative area that encompasses the branches of interest.
3. *Deciduous tree leaf-out and leaf-drop dates:* Using the collected leaves, twigs, fruits, flowers, etc., as well as bark and branching-pattern photographs, consult an extensive tree identification guide to identify each tree. The Arbor Day Foundation provides guides tailored to parts of the United States, for example. Once each tree is identified, consult regional phenology guides to find typical leaf-out and leaf-drop dates. In the United States, the extension services of land-grant universities are often good sources of such reports. If the site visit occurred in the summer, use these identities to find photographs of typical winter branch density.

Using Site Survey Results. The site shading information found above is now ready for use in revising monthly solar resource estimates and establishing the optimal orientation for the passive heating system. These steps are accomplished graphically by overlaying pathfinder images onto sunpath representations of incident solar radiation and adjusting the solar radiation values accordingly. See Chapter 11 to review solar geometry, sunpath diagrams, and shading masks if necessary.

First, select the sunpath diagram for the latitude closest to that of the site (Appendix D), or create one with the University of Oregon Solar Radiation

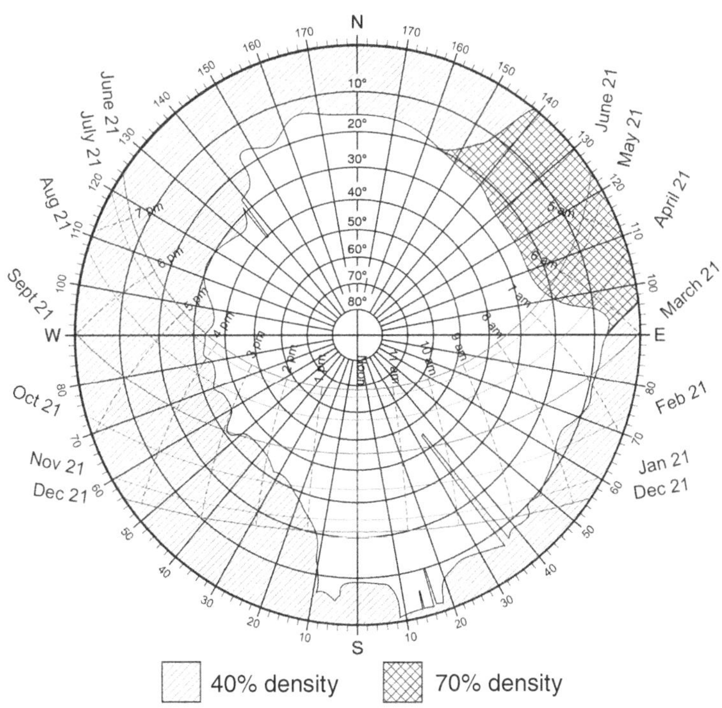

Figure 13.48 Shading on a site in Albany, OR, redrawn onto a 44°N sunpath diagram, showing relative optical densities of tree groups. (Drawing by Kelli Kimura with contributions from Kayla Bundy and Hieu Vo.)

Monitoring Laboratory online application. Next, for each position of interest on the site, prepare two blank diagrams: one for January through June (i.e., spring), and the other for June through December (i.e., fall). Then, redraw the relevant pathfinder image(s) onto the corresponding sunpath diagrams as density regions, assuming leafless states, to create shading masks (Figure 13.48).

In the second step, find typical hourly values for solar radiation each month, incident upon a surface of the near-optimal tilt found in conceptual design or, for Trombe walls, on vertical surfaces. One convenient method is to create a custom plot option within the University of California, Los Angeles' weather-visualization application Climate Consultant, using the weather file nearest the site to approximate solar radiation values on a surface of specified tilt. While these values are not as accurate as those provided by advanced solar radiation models, particularly for cloudy climates, they are acceptable for schematic design. Alternatively, a simulator such as EnergyPlus that incorporates such solar models may be used to establish more accurate values.

Plot these hourly values onto the corresponding spring and fall shading masks, as in Figure 13.49, to reveal the approximate percentage of solar radiation obstructed by shading in each hour of the year.

The resulting plots may show that a substantial fraction of the solar resource is obscured by site shading. Because most glazing reflects an increasing proportion of solar radiation as the angle of incidence becomes more oblique, solar-collecting glazing should face the direction of greatest solar resource. If much of the useful (i.e., heating-season) solar radiation is concentrated in a non-south direction, repeat the estimation of hourly solar radiation for more favorable orientations, and replot them over the shading masks, to reveal the orientation least compromised by shading. In the example shown (Figure 13.49), a south-southeast orientation was found to have greater solar exposure during the heating season than a due-south orientation.

In conceptual design, the useful solar heating resource estimate was based on monthly values of solar radiation incident upon unshaded, optimally tilted surfaces oriented due south, with accurate representation of the effects of cloud cover on apparent sun angles. Neither site shading nor hourly solar radiation values were considered.

Due to the shading information, the solar resource estimate should now be revised. The most straightforward procedure is to multiply each hourly solar radiation value (e.g., 10 W/m^2) by the corresponding shading density (e.g., 50%) to yield a new estimate (e.g., 5 W/m^2). This procedure is reliable for sunny winter climates, in which direct solar radiation emerging from the direction of the sun provides most of the useful solar resource. In cloudy winter climates, however, the effect of site shading is diminished by the fact that much of the solar radiation emanates from the top of the sky dome. In cloudy climates, it is therefore necessary to estimate the fraction of solar radiation that is direct (i.e., not diffuse) in each month, and to reduce only that fraction by site shading.

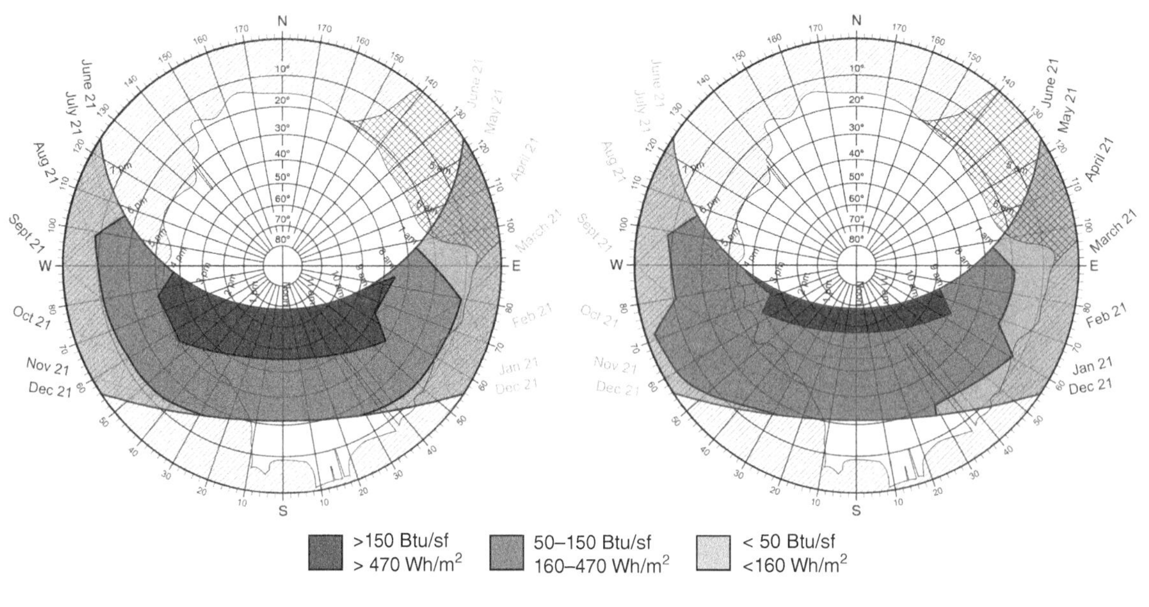

Figure 13.49 Plots of typical hourly solar radiation on a surface of 40° tilt in Albany, OR (44°N latitude), for the months of fall (left) and spring (right), obtained from the Corvallis TMY3 weather file (NREL, 2007), viewed in Climate Consultant 6.0. (Drawings by Kelli Kimura with contributions by Kayla Bundy and Hieu Vo.)

EXAMPLE 13.12 At 2 p.m. in February, the total solar radiation on an unshaded 45° tilted surface in Eugene, OR, is estimated by Climate Consultant to be ~300 W/m^2 [95 Btu/h ft^2] using the Eugene TMY3 weather file. However, a site survey shows that the location planned for solar-collecting glazing is 60% shaded by the bare branches of deciduous trees at that hour. The trees extend about 35° above the horizon. What is the revised solar radiation for this hour?

SOLUTION

Since Eugene has a cloudy winter climate, the first step is to find the proportion of direct solar radiation in February. Climate Consultant shows this to be about 50%. The other 50%, which is diffuse, is assumed to emerge from apparent angles high in the sky dome and therefore to be unaffected by the shade of low trees. The revised solar radiation, for that hour, is therefore:

300 W/m^2 total radiation × 50% direct radiation

= 150 W/m^2 direct radiation before shading by trees

= 150 W/m^2 direct radiation

× 40% transmitted through trees

= 60 W/m^2 transmitted direct radiation

60 W/m^2 transmitted direct radiation

+ 150 W/m^2 diffuse radiation

= 210 W/m^2 [66.5 Btu/ft^2 h]

Although the location of interest is 60% shaded at that hour, the solar radiation incident on the tilted surface is only reduced by about 30%, because half of it originates at high apparent sun angles. ■

A spreadsheet should then be used to replicate this procedure for each typical daytime hour of each month; Climate Consultant, CBE Clima Tool, or another climate file visualization tool may be used to export typical hourly total, direct, and diffuse solar radiation values for each month on a surface of the desired tilt. These new values for the available solar resource, which account for site shading, will then be used to revise the glazing tilt and area.

Reconsidering Vertical Elements. Once the impact of site shading is clear, the use of vertical glazing and/or mass surfaces should be reconsidered. Vertical surfaces require access to low-angled solar radiation during the heating season. Sites in populated areas, however, are typically shaded to altitudes of 10–20° above the horizon by other buildings, with higher-altitude shading common in dense urban areas. In forested areas, shading to 40–50° above the horizon is not uncommon. To the extent that site shading is present, vertical surfaces require greater areas to provide the desired solar gain. If site shading appears to be significant, a system with tilted glazing is recommended, either to replace or to supplement the desired wall system(s). Refer to the system choice decision tree, Figure 13.25, may be useful at this point.

(c) Revising the Solar-Collecting Glazing

The next steps are to choose a glass or composite material, based on optical and thermal properties, and to use these properties with the revised solar resource estimate to resize the solar glazing area.

Choosing a Glazing Material. In choosing solar-collecting glazing, the goal is to admit as much solar heat as possible while limiting heat loss to the greatest extent possible. In other words, the ideal glazing has a SHGC that is as high as possible, but a thermal transmittance, or U-factor, that is as low as possible. Since high SHGC values tend to accompany high U-factors, and vice versa, this choice requires a balance between the two factors that depends on the severity of the outdoor temperatures during solar collection. Milder and cloudier climates tend to benefit from the highest possible SHGC at the expense of a higher U-factor, to the extent that single glazing is sometimes optimal (with night insulation, of course), while cold sunny snowy climates tend to benefit from lower U-factors, despite lower SHGCs.

In schematic design, an approximate glazing choice may be made by comparing performance among representative glazing assemblies, as shown in Appendix E, Table E.15.

EXAMPLE 13.13 Compare the February performance of the glazing types below for buildings in four contrasting climates: (1) the mild cloudy climate of San Francisco, CA; (2) the cooler, cloudier climate of Portland, OR; (3) the cold sunny alpine climate of Bend, OR; and (4) the subarctic climate of Fairbanks, AK. Assume that night insulation, applied for 14 h/day, causes differences between the glazing options to be negligible at night, and that building interiors are maintained at 68°F.

Single clear	SHGC = 0.79 U = 1.30 Btu/ft^2 °F [7.4 W/m^2 K]
Double clear	SHGC = 0.65 U = 0.64 Btu/ft^2 °F [3.6 W/m^2 K]
Double low-ε	SHGC = 0.55 U = 0.33 Btu/ft^2 °F [1.9 W/m^2 K]
Triple low-ε	SHGC = 0.37 U = 0.15 Btu/ft^2 °F [0.85 W/m^2 K]

SOLUTION

First, create a table similar to Table 13.3.

PASSIVE ENVIRONMENTAL SYSTEMS

TABLE 13.3 Choosing a Glazing Material

Part A. I-P Units

Glazing	SHGC	U-Factor	T_{in}	Solar Rad	T_{out}	Heat Gain[a]	Heat Loss[b]	Net Gain[c]	Solar Rad	T_{out}	Heat Gain[a]	Heat Loss[b]	Net Gain[c]
		Btu/h ft² °F	°F	Btu/d ft²	°F	Btu/d ft²	Btu/d ft²	Btu/d ft²	Btu/d ft²	°F	Btu/d ft²	Btu/d ft²	Btu/d ft²
				San Francisco, CA					**Portland, OR**				
Single	0.79	1.3	68	1250	55	988	169	819	750	48	593	260	333
Double clear	0.65	0.64	68	1250	55	813	83	729	750	48	488	128	360
Double low-ε	0.55	0.33	68	1250	55	688	43	645	750	48	413	66	347
Triple low-ε	0.37	0.15	68	1250	55	463	20	443	750	48	278	30	248
				Bend, OR					**Fairbanks, AK**				
Single	0.79	1.3	68	1000	32	790	468	322	500	10	395	754	−359
Double clear	0.65	0.64	68	1000	32	650	230	420	500	10	325	371	−46
Double low-ε	0.55	0.33	68	1000	32	550	119	431	500	10	275	191	84
Triple low-ε	0.37	0.15	68	1000	32	370	54	316	500	10	185	87	98

Part B. SI Units

Glazing	SHGC	U-Factor	T_{in}	Solar Rad	T_{out}	Heat Gain[a]	Heat Loss[b]	Net Gain[c]	Solar Rad	T_{out}	Heat Gain[a]	Heat Loss[b]	Net Gain[c]
		W/m²K	°C	kWh/d m²	°C	kWh/d m²	kWh/d m²	kWh/d m²	kWh/d m²	°C	kWh/d m²	kWh/d m²	kWh/d m²
				San Francisco, CA					**Portland, OR**				
Single	0.79	7.4	20	3.94	12.8	3.11	0.53	2.58	2.36	8.9	1.86	0.82	1.05
Double clear	0.65	3.6	20	3.94	12.8	2.56	0.26	2.30	2.36	8.9	1.53	0.40	1.13
Double low-ε	0.55	1.9	20	3.94	12.8	2.17	0.13	2.03	2.36	8.9	1.30	0.21	1.09
Triple low-ε	0.37	0.85	20	3.94	12.8	1.46	0.06	1.40	2.36	8.9	0.87	0.09	0.78
				Bend, OR					**Fairbanks, AK**				
Single	0.79	7.4	20	3.15	0.0	2.49	1.48	1.01	1.58	−12.2	1.25	2.38	−1.13
Double clear	0.65	3.6	20	3.15	0.0	2.05	0.73	1.32	1.58	−12.2	1.03	1.17	−0.14
Double low-ε	0.55	1.9	20	3.15	0.0	1.73	0.37	1.36	1.58	−12.2	0.87	0.60	0.27
Triple low-ε	0.37	0.85	20	3.15	0.0	1.17	0.17	1.00	1.58	−12.2	0.58	0.27	0.31

[a] Heat gain = Solar rad × SHGC.

[b] Heat loss = U-factor × 10 h/day × (T_{in} − T_{out}).

[c] Net gain = Heat gain−Heat loss.

Next, find the average February daytime temperatures for each city from weather files or historical weather data.

Then, estimate February values for incident solar radiation on surfaces of optimal tilt from Appendix I, Table I.1.

Finally, calculate the daily solar gain through a unit area by multiplying the incident solar radiation by the SHGC, and calculate daily heat loss by multiplying the U-factor (in units of energy per hour, per area, per degree) by the number of uninsulated hours per day (here, 10 hours) and the difference between indoor (T_{in}) and average outdoor daytime temperatures (T_{out}).

The sum of daily heat gain and daily heat loss, for each glazing type, then reveals the glazing type that achieves the greatest net solar gain under these conditions. Results are shown in Table 13.3. ■

This example shows that single clear glass is the best choice only in the mildest climate, San Francisco, because relatively warm daytime temperatures cause heat loss to be less of a concern than maximizing solar heat gain. Double-clear glass performs best in Portland, Oregon, with its cooler and rainier winter climate, while double low-ε glass is optimal in the sunnier, colder climate of Bend, Oregon; in Bend, the cold favors a lower U-factor, while the sunniness permits a lower SHGC. Of these cities, only the subarctic climate of Fairbanks, Alaska, benefits most from triple low-ε glazing.

Revising Glazing Tilt and Area. At this point, the combination of new enclosure information, site shading information, and glazing material choice may be used to revise the glazing tilt, area, and orientation. It is now useful to collect the following:

1. revised monthly heating season and need (Section 13.5a)
2. revised orientation for solar-collecting glass (Section 13.5b)
3. revised monthly solar resource (Section 13.5b)
4. glazing material (Section 13.5c)

If the optimal orientation is south, use the revised heating season to find a new optimal tilt from solar radiation data in Appendix I, Table I.1, following the conceptual design process (Section 13.4).

If the optimal orientation is significantly non-south, however, accurate solar radiation models such as those that generated Appendix I, Table I.1 will be necessary to find the optimal tilt in design development (Section 13.6). In the meantime, use Climate Consultant or another suitable weather visualizer to estimate the optimal tilt at the desired non-south orientation, with the understanding that in cloudy winter climates, the tilt obtained in this manner will be steeper than is truly optimal.

In conceptual design, an estimate of 0.6 was made for the glazing SHGC, since double-clear glazing performs well in passive heating in many climates. Now, however, the glazing chosen above should be used to refine this value. Using the monthly incident solar radiation upon surfaces of the revised tilt, and the estimated SHGC of the glazing material, calculate a new monthly transmitted solar resource, per unit area, to prepare for resizing.

Next, using the revised monthly heating need, as well as the revised solar gain through a unit area of glazing, repeat the process in conceptual design (Section 13.4) to revise the collector area.

(d) Revising Thermal Mass

The size of permanent thermal mass must now be adjusted to match the revised solar heat gain. If seasonal addition of water tanks or potted plants is planned, permanent mass should be sized for cooler months, before these masses are added.

Confirming Mass Configuration. Just as sunniness and solar exposure inform glazing design, the desired timing and intensity of heat delivery, derived from the owner's project requirements, guide thermal mass design. If the heat delivery design intent has changed, review the mass decision tree, Figure 13.37, as well as the general patterns of heat delivery in Section 13.4, to realign the planned mass configuration with the new design intent.

Estimating Temperature Boundaries. The rate of heat delivery from thermal mass, like all spontaneous heat flow, depends on the temperature difference between the heat source (the mass) and the heat sink (the space to be heated). Hotter sources deliver heat with greater intensity over a shorter duration, as discussed previously. As a result, designing mass to operate through a desired temperature range provides an effective preliminary sizing estimate.

The upper boundary is governed by the desired timing of heat delivery, while the lower boundary is governed either by the nighttime thermostat setpoint, if the mass is in contact with a conditioned space, or (approximately) by the average daily high outside air temperature, if it is only in contact with unconditioned space.

Mass intended for daytime heat delivery must be sized to reach fairly high temperatures, facilitating rapid heat transfer to interior spaces. In practice, peak temperatures of ~120°F [49°C] are consistent with such delivery, though individual optima vary (Table 13.4). For evening heat delivery, peak temperatures should be somewhat lower, in the range of ~100°F [38°C], while heat delivery throughout the night requires sizing for even lower peak temperatures (~85°F [30°C]).

Estimating Size Based on Heat Capacity. Once the temperature range is estimated in °F or °C, the mass (lb [kg]) may be approximately sized using the specific heat capacity (Btu/h ft^2 °F [J/kg °C]) and the absorbed solar radiation (Btu [J]), assuming that the mass releases about as much heat as it stores on a typical day. Since the absorbed radiation varies during the heating season, this estimate may need to be repeated for multiple months.

TABLE 13.4 Mass Temperature Boundaries

Time of Heat Delivery	Upper Temperature	Lower Temperature
Daytime	~120°F [49°C]	65°F [18.3°C] or avg. max OSAT[a]
Evening	~100°F [38°C]	
Nighttime	~85°F [30°C]	

[a]Average maximum outside air temperature (OSAT) for the dates of interest, for unconditioned spaces.

EXAMPLE 13.14 A direct-gain space at a ski lodge in Bennington, VT, needs thermal mass appropriate for heat delivery on late winter afternoons when skiers gather to eat, drink, and swap stories. The solar-collecting glass consists of skylights totaling 50 m^2 [540 ft^2], facing south at optimal tilt, with solar access unobstructed by site shading. The month of greatest interest is March, when the space tends to overheat, and the transmitted solar radiation on a typical March day is ~200 kWh [680 kBtu]. The mass is planned as a fired clay brick floor, insulated from below, also of 50 m^2 [540 ft^2], positioned under the skylights so as to receive about half of the transmitted radiation. How thick should the floor be?

SOLUTION

Appendix E, Table E.1 shows that fired clay brick has a density of 1920 kg/m^3 (120 lb/ft^3), while Appendix E, Table E.25 shows that it has a specific heat capacity ranging from 790 to 800 J/kg K [~0.19 Btu/lb °F] and solar absorptance values ranging from 0.5 to 0.7.

The radiation that encounters the floor is therefore:

$$\sim 200\ \text{kWh} \times 0.5 \approx 100\ \text{kWh}$$

and one day's absorbed solar radiation, using an intermediate value for the bricks, is approximately:

$$0.6 \times \sim 100\ \text{kWh} \approx 60\ \text{kWh}$$

Converting these units to kJ yields:

$$60\ \text{kWh} \times 3600\ \text{kJ/kWh} = 216{,}000\ \text{kJ}$$

The desired upper temperature for late afternoon heat is estimated from Table 13.4 to be partway between daytime and evening values, giving a temperature range estimate of:

$$45\text{–}18^\circ\text{C} = 27^\circ\text{C} = 27\ \text{K}$$

The heat capacity is then used to reveal the mass quantity required to limit 216,000 kJ of energy to a temperature rise of 27 K:

$$216{,}000\ \text{kJ} \times 1\ \text{kg K}/0.79\ \text{kJ} \div 27\ \text{K} \approx 10{,}000\ \text{kg}$$
$$[\text{or} \sim 22{,}000\ \text{lb}]$$

Next, density is used to convert mass to volume:

$$10{,}000\ \text{kg} \times 1\ \text{m}^3/1920\ \text{kg} \approx 5.2\ \text{m}^3$$

And, finally, dividing volume by area yields thickness:

$$5.2\ \text{m}^3/50\ \text{m}^3 \approx 0.1\ \text{m thick}\ [\sim 4\ \text{in.}]$$ ■

This sizing addresses only the needs of March, and as a result, this mass would be oversized for December–February, in that it would not often reach the desired peak temperature. However, thermal mass is easily inactivated by floor coverings, which shade the surface; in this case, seasonal rugs could be used to allow most of the transmitted solar energy to warm the lodge air in the colder months. An equally valid choice would be to size the mass for February, resulting in a thinner floor, and to add internal mass in March.

In this example, the mass area was comparable to the area of solar-collecting glass. In general, this is a good starting point; since solid mass distributes heat through itself relatively slowly, an area this large allows it to be sufficiently thin that the entire volume is useful. The balance between area and thickness will be explored further in design development (Section 13.6).

Evaluating Ground Contact. If floor mass is planned, it is essential to determine the local soil type and soil moisture regime so that potential heat loss to soil may be evaluated. Numerous soil moisture maps exist online, such as that shown in Figure 13.50. Moderate, high, and very high soil moistures (i.e., ustic, udic, or perudic moisture regimes) give soils high thermal conductivity; in addition, fine-grained (e.g., clayey or silty) soils tend to transport moisture laterally through capillary action, reaching floor undersides (Janssen et al., 2004). Unless floor thermal mass contacts a sandy soil in a xeric or aridic soil moisture regime, it must be well-drained with gravel or fully insulated from the soil below.

(e) Revising Controls

Movable Insulation. In conceptual design, plans were made for the general form and operation of movable insulation; the task of schematic design is now to choose the material, envision the edge-sealing details, and plan the control mechanisms.

A movable insulation material needs not only to insulate but also to be conveniently operable, easily sealed at the edges, and durable. In schematic design, the most important decision is whether curtain (i.e., flexible) or panel (i.e., rigid) materials provide the best combination of U-factor, operability, and sealability for the project, with the understanding that edge sealing is as, if not more, important than U-factor.

The schemes shown previously (Figure 13.41) can accommodate rigid panels of 2 in. [5 cm] or more, providing U-factors of ~0.1 Btu/h ft^2 °F [0.57 W/m^2 °C] or lower. Rigid panels are also readily edge-sealed; assuming these seals are excellent, the balance between cost (i.e., thickness) and benefit (i.e., heat loss avoided) may be evaluated with the familiar equation

$$q = U \times A \times \Delta t$$

during the months of the heating season. Rigid panels are usually the most effective options for cold climates. In milder climates, curtains or rollable fabrics may be preferred, depending on the area of solar-collecting glazing and maximum acceptable heat loss rates.

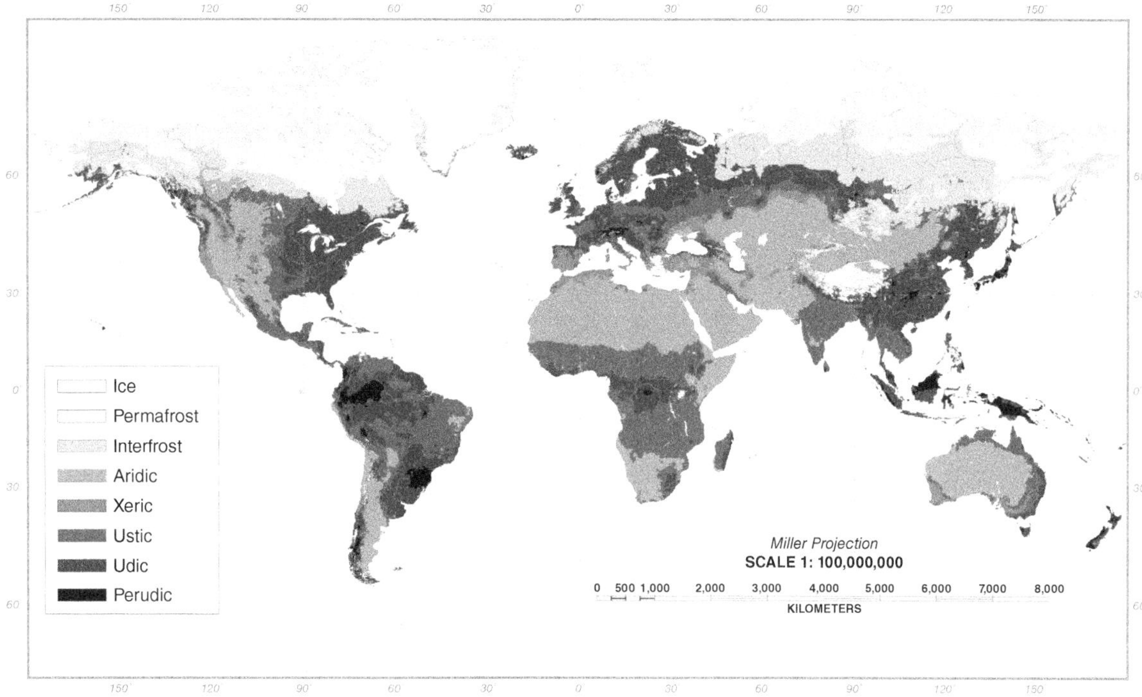

Figure 13.50 *Map of global soil moisture regimes. (U.S. Department of Agriculture.)*

Schematic design is also the time to envision edge-sealing and mechanism details, which are inextricably linked. Details should include all the elements necessary for the movable insulation to be retracted, stored, and deployed: slots, channels, tracks, rollers, pivoting elements, rabbets, bevels, troughs, anchors, pulleys, springs, and so on. Detailing movable insulation is a custom effort, since few systems are fully commercially available, and time devoted to creating a convenient, effective system will be rewarded in performance.

Occupants are notoriously unreliable in operating passive system elements, including shading and vents as well as movable insulation (e.g., Roetzel et al., 2010). In part, this is because status changes in passive systems are subtle: the switch from net heat gain to net heat loss by solar-collecting glazing, for example, is not readily apparent to a person. For this reason, seasonal schedules must always be included in movable insulation design. Ideally, schedules should also be equipped with a means of alerting occupants to the need for operation (i.e., signal lights, push notifications), and automatic control should be considered if the movable insulation is difficult to operate manually or if occupant availability is in doubt.

One convenient, reliable schedule type simply specifies morning retraction and evening deployment hours, at monthly resolution, switching mid-month in November and March for daylight saving time. Since dawn typically coincides with some of the day's coldest hours, while meaningful solar collection ends well before dark, movable insulation should typically be deployed at sunset and retracted about an hour after sunrise (e.g., Figure 13.51). On days with several hours of rain or any snow, however, the insulation should remain in place all day, since moisture accelerates conduction through glazing and little solar energy is available on such days.

More involved schedules may rely on temperature and/or heat flux sensors, weather predictions, and/or customized model predictions to determine hours of deployment and retraction; these are worth further

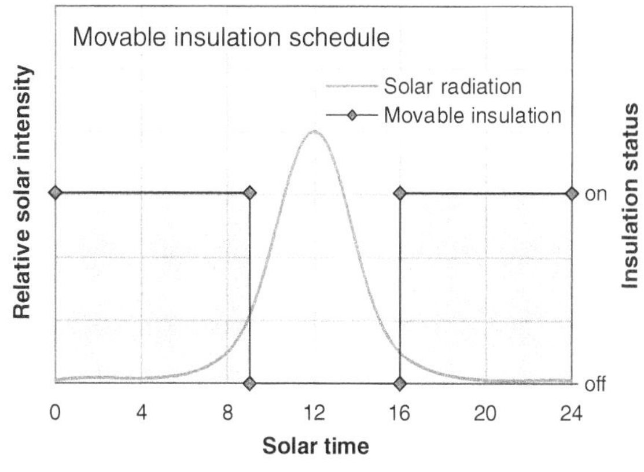

Figure 13.51 *General winter movable insulation schedule compared to solar radiation intensity (parabolic line), in which insulation is retracted about an hour after sunrise and deployed near dusk.*

PASSIVE ENVIRONMENTAL SYSTEMS

exploration when the highest possible performance is desired, as described briefly in design development (Section 13.6).

Interior Vents. If air warmed by flat-panel collectors, sunspaces, vented Trombe walls, or direct-gain systems is to be moved through living spaces, interior vent locations and operation should be planned alongside spatial layouts in schematic design. Distances between the collector and receiving space(s) should be shortened wherever possible to diminish the need for fan energy, if fans are planned, and heat-needing spaces should be positioned above warm air sources where possible to take advantage of buoyancy.

Mechanisms for opening and closing internal vents should also be envisioned, as should schedules for vent opening and, if necessary, fan operation when air is sufficiently warm. In low- or no-mass systems, it may be possible to open vents as early as an hour or two after sunrise on sunny days; in higher-mass systems, the ideal hour is more difficult to predict in a general way, and thermometers in the collecting space may be needed to signal operation. The choice of manual versus automatic operation should also be made during schematic design, with the understanding that automatic operation is far more reliable with all but the most diligent occupants. If automatic operation is cost-prohibitive, signals for manual operation should be considered as an inexpensive but effective alternative.

Cooling. In schematic design, passive cooling measures (e.g., operable daytime shading, night ventilation, evaporative cooling in dry climates, radiative cooling to cold night skies, and adjustable thermal mass) should be considered in light of heating elements to find synergies and limit the number of operable elements where possible. Shading may be accomplished by movable insulation in some cases, for example; similarly, exterior vents may be positioned and scheduled to provide safety valves to avoid winter overheating as well as to provide cooling on summer nights. At the end of this chapter, three case studies provide further examples of well-integrated control systems.

13.6 DESIGN DEVELOPMENT

In design development, the remaining tasks are to refine glazing tilt and area, mass area and thickness, and controls for movable insulation, vents, and shading, to ensure expected performance. At this stage, building energy modeling is highly recommended to address the considerations below. If computer modeling is impossible, however, the best approach is to reduce the quantity of permanent thermal mass, plan for additional internal mass in warmer months, and add air and mass surface temperature sensors. Once the system is built, monitoring these over time will reveal the extent to which mass should be added in any given season and whether control schedules are effective.

(a) General Considerations

Time-Dependent Behavior. The effectiveness of solar-collecting glazing and thermal storage mass changes hourly, monthly, and seasonally, due in part to the ever-changing position of the sun with respect to the site. Since solar transmission through glazing depends significantly on the angle of incidence, and since the estimates above combined multiple incidence angles into weighted averages, those estimates were necessarily simplified. An additional contribution to time-dependent behavior is the relatively slow rate with which thermal mass distributes energy throughout itself. While thermal mass is expected to be nearly thermally uniform by the beginning of collection each day, it is rarely expected to be at thermal equilibrium with its surroundings, and its rates of heat gain and loss are sensitive to the surrounding air and the surface temperatures that were simplified in the above procedures. All of these processes, however, may be taken into account with appropriate simulation programs.

Thermal Comfort. Providing thermal comfort during certain hours is a central goal of any passive heating system, whether it is operating alone or in conjunction with an active system. Since thermal comfort in a space depends approximately equally on air and mean radiant temperatures when air velocities are low, and since radiant temperatures depend on continually changing mass temperatures, it is difficult to predict the hours in which thermal comfort will be achieved through heat gain and loss calculations alone. Again, however, appropriate simulations can perform the necessary calculations rapidly.

Air-Based Heat Delivery. Even the most massive passive heating systems deliver part of their heat in the form of warm air, and low-mass systems use air delivery exclusively. To minimize fan use for moving warm air, it is extremely useful to understand likely air movement patterns arising from natural convection. Bulk air movement depends, however, on the same air and surface temperatures that affect thermal comfort, again requiring detailed hourly calculations that are most conveniently performed by simulation software.

(b) Simulation Guidance

Given the considerations above, building energy modeling is extremely helpful in optimizing thermal mass size, glazing area, tilt, and movable insulation parameters. While energy modeling may seem daunting, it is readily undertaken with the open-source tool EnergyPlus, which has been shown to simulate passive system performance well in field studies (e.g., Rempel et al., 2013; Rempel et al., 2016). This specific recommendation is made because many building energy modeling tools simplify representation of diffuse radiation on tilted surfaces, angular dependence of solar transmission through glass, and time-dependent behavior of thermal mass too extensively to provide meaningful results. The treatment of thermal mass is particularly important, because passive systems often rely on thermal mass to carry warmth or coolth through several subsequent hours. For computational speed, however, many simulations reset all model temperatures to a user-input "design" value each hour, obscuring the true mass temperature. Any software used for passive system simulation must therefore forward model temperatures from one hour to the next, in addition to simulating cloud scattering of solar radiation and representing the angular dependence of solar transmission accurately.

Geometry. In passive heating simulations, the geometry of solar-collecting zones must usually be simplified to avoid L-shaped or otherwise concave spaces or surfaces, which complicate solar reflection calculations and often result in errors.

Solar Transmission. For speed, most software assumes that all solar radiation is diffuse once it passes through glazing, which dramatically affects predicted thermal mass behavior. Even EnergyPlus defaults to this assumption; the alternative, which projects solar radiation through glazing onto the surface beyond and allows subsequent reflection, must therefore be selected by the user. In addition, the simulation of window assembly thermal performance can be greatly improved by the use of the application WINDOW, developed by Lawrence Berkeley National Laboratory, to represent angle-dependent optical properties for reference by EnergyPlus.

Ground Configuration. To represent heat loss to the ground accurately, the thermal properties of the underlying soil, sand, and/or gravel must be modeled explicitly to a depth of ~20–50 cm [~8–20 in.]. The default algorithms of many programs rely on weather files for soil thermal properties, but unfortunately, the widely used TMY3 files contain a value for soil thermal diffusivity (a composite of density, thermal conductivity, and heat capacity) that is ~1/1000 of those reported in soil science literature (Rempel and Rempel, 2013). Because of this, accepting ground default properties allows underlying soil to behave as an insulating layer.

Site Shading. Tree shapes may be simplified, but they must be given accurate positions, heights, and transmittance schedules, using data from the solar site survey, for effective shading simulation. Buildings and structures that shade the site should be modeled as fully opaque, which is usually the default.

Output Variables. The most useful outputs from a passive heating model are the total solar transmission through glazing, collector space and mass temperatures, mass surface heat fluxes, and receiving space operative temperatures and comfort conditions. Cloud cover and outside air temperatures are also useful for comparison with these values. If the temperatures of the space(s) receiving heat are controlled by a thermostat, space heating energy use is also of interest. In addition, reporting the open/closed status of each movable element is essential to reveal whether operation is occurring as intended.

Evaluation. When schematic design parameters are represented fully, including infiltration, movable insulation, edge sealing, ground configuration, and the like, monthly outputs should be examined carefully. If model temperatures seem unexpectedly high, infiltration may be too low; if they seem unexpectedly low, movable insulation may not be operating or its edges may not be sealed. When these issues are resolved, investigating the timing and duration of heat delivery, daily mass temperature patterns, and the solar heat gain in each month should reveal areas for improvement.

Experiments. To accomplish the desired improvements, relevant parameters should be varied systematically and individually. While it may be tempting to vary two or more parameters simultaneously, doing so obscures the effect of varying any one alone and ultimately requires more effort. Varied parameters should include, in approximate order of initial variation: glazing orientation (if unresolved in schematic design), tilt, and area; glazing assembly; thermal mass material, thickness, and area; and movable insulation material, edge sealing, and schedule. Several iterations may be needed to reveal the parameters that achieve the desired heating performance. When heating is satisfactory, cooling performance should be examined and improved as necessary.

Troubleshooting. When perplexing errors arise, the first step is to search for the error message in software documentation and online. Next, online organizations such as the Building Simulation User Group may be consulted, and if the software offers user support, as EnergyPlus does, questions may be directed to the corresponding helpdesk.

Accuracy. Building energy models are well known to misrepresent actual performance. Researchers have found several consistent reasons for this. In most cases, however, unwarranted assumptions were made in the modeling; the built version diverged from the model in important ways; and/or occupants behaved unexpectedly. The risks of unwarranted modeling assumptions and occupant reliability have been addressed above, and the importance of avoiding major discrepancies between design and construction is clear. As long as care is taken to represent each key element of a passive system accurately in both model and built form, simulation results should provide an excellent indication of future performance.

(c) Software Tools

Weather File Visualizers. Climate Consultant from the University of California at Los Angeles and Data Viewer (DView) from the National Renewable Energy Laboratory are suitable applications for visualizing weather data from a range of file formats. Simplified calculations are used to estimate solar radiation on surfaces of specific tilt and orientation that may be useful in schematic design, although they should not be relied upon to size solar-collecting glazing, and custom graph creation and data export are supported. DView, in turn, allows multiple weather files to be viewed simultaneously, facilitating comparisons that are sometimes necessary to choose the file that best represents a particular site. The CBE Clima Tool from the Center for the Built Environment at the University of California, Berkeley, offers easy access to, and visualization of, EnergyPlus (EPW) weather files.

Appendix L offers additional information on climate data analysis tools and resources.

EnergyPlus Suite. Among the research-grade, open-source building energy simulators currently available, EnergyPlus has received the most extensive development for passive system simulation. The underlying algorithms are published in the EnergyPlus Engineering Reference, confirming the use of advanced solar radiation models, temperature forwarding from hour to hour, and so on, and in combination with WINDOW, it has reproduced field performance of passive solar heating and cooling systems successfully.

Within the suite, Euclid is used for thermal zone and shading object input (bigladdersoftware.com/projects/euclid); the IDF Editor is used to edit EnergyPlus input files; WINDOW (windows.lbl.gov/software/window) is used to calculate window and skylight assembly properties for angle-dependent solar transmission; and xESOView (sourceforge.net) is used to visualize simulation output. In addition, automated optimization of model parameters is possible using GenOpt (simulationresearch.lbl.gov/GO).

Although the EnergyPlus user interface is minimal, graphic user interfaces (e.g., Sefaira, Cove, DesignBuilder) simplify some functionality. For passive heating systems, however, simulation in native EnergyPlus is not difficult.

Other Building Performance Simulation Tools. Software tools enable designers and students alike to incorporate passive solar heating performance to a limited degree. At the time of writing, most commonly utilized building performance simulation software does not support accurate passive solar heating simulation, in that one or more of the essential capabilities is omitted; necessary user inputs are omitted; and/or assumptions relevant to passive heating are not fully documented in user guides or calculation references. For these reasons, proprietary tools are not currently recommended for passive solar heating simulation.

Appendix L offers additional information on building performance simulation tools and resources.

13.7 CASE STUDY 1—BEDDINGTON ZERO ENERGY DEVELOPMENT, SUTTON, UK

Project Basics

- Passive heating and cooling strategies: Isolated gain (sunspaces) with movable insulation, thermal mass, and natural ventilation; enclosure superinsulation and low infiltration
- Location: Hackbridge, Sutton, UK; latitude 51°N; longitude: 8°W
- Climate: West coast marine; 5960 HDD_{65F} [3311 HDD_{18C}]; 1,411 CDD_{50F} [784 CDD_{10C}]; 23 in. [580 mm] precipitation annually
- Building type: Multifamily residential; mixed use
- Client: Peabody Trust
- Design team: Bill Dunster Architects/ZEDFactory, Bioregional Development Group, Ove Arup & Partners
- Completion date: 2002

Design Intent The Beddington Zero Energy Development, or BedZED (Figure 13.52), is a mixed-use eco-village designed under the One Planet Living principles of the Bioregional Development Group: that people should be able to live within an ecological

Figure 13.52 *South-facing units showing sunspaces with tilted and vertical glass (some with integrated photovoltaics), operable windows, and movable insulation. (© Alison Kwok; all rights reserved.)*

footprint small enough to equal no more than one "Earth's worth" of resources. While the average UK citizen's lifestyle required 15.3 acres [6.19 ha] at the time of design, equivalent to a need for 3.3 Earths for all of humanity, the BedZED goal was to facilitate living within a footprint of only 4.7 acres [1.9 ha], requiring only one Earth.

Units were designed to maintain comfortable temperatures in the winter, when occupied, through a combination of passive solar heating and heat gains from people, lights, and equipment. When unoccupied, air warmed over each unit's domestic hot water cylinder is blown into the corridors by low-energy fans; the water, in turn, is heated by the community's wood chip-fired combined heat and power plant.

Design Elements To reach its ambitious goals, BedZED's green design strategies include transportation, water use, and renewable energy production in addition to passive heating and cooling (Figure 13.53). The passive heating strategies, however, are among the most notable, since they represent an exemplary integration of passive solar heating and passive house enclosure construction, allowing conventional mechanical heating systems to be eliminated altogether.

To maximize solar heat gain, the homes have conservatory-like sunspaces with large, double-glazed, high-SHGC façades; these are oriented 15° east of south, consistent with typical afternoon cloudiness in the London area. Exterior sunspace glazing is insulated with thick curtains, and sunspaces are thermally separated from living spaces by double-glazed windows and doors as well as additional curtains in the living spaces. Elsewhere, heat loss is minimized with superinsulated air-tight construction, including triple-glazed windows in non-solar-collecting areas.

Occupant control of operable elements is central to passive heating success: once the sun warms the sunspace air to a comfortable temperature, on a given day, the interior doors and windows are opened to allow warm air to flow in; post-occupancy surveys have shown that even on the coldest days, sunspaces heat living spaces effectively when the sun is shining. When the sunspace becomes uncomfortably cool (late in the day, at night, or when the sky is heavily overcast), it is thermally separated from the rest of the home by closing these doors and windows. Summer operation is equally important: when sunspace air

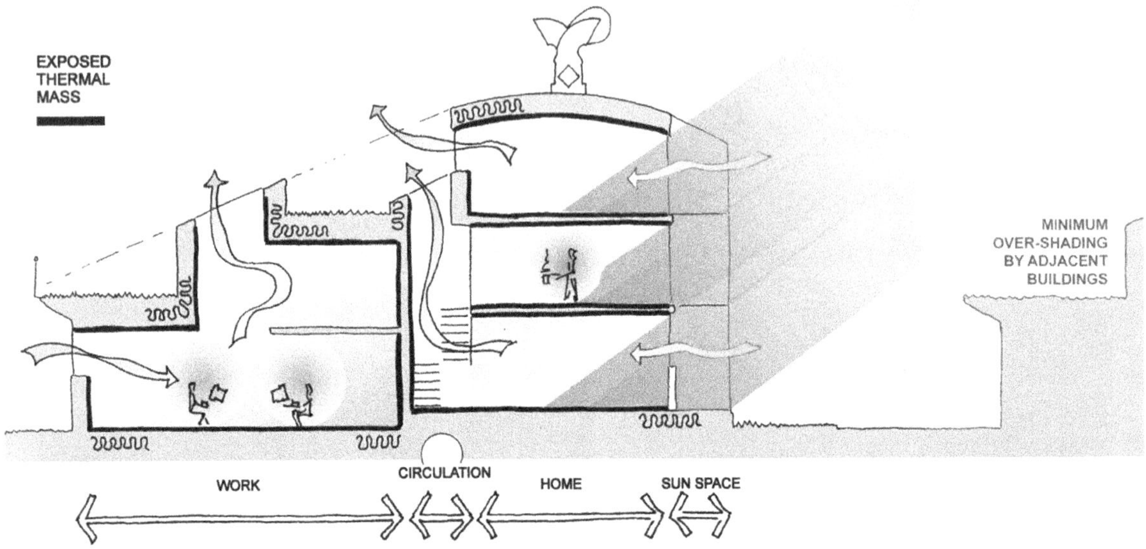

Figure 13.53 Sunspaces on south façades collect solar energy with double-clear glazing, warming sunspace air and interior thermal mass; operable windows and doors between sunspaces and living spaces allow warm air to circulate. Heat loss is minimized by movable insulation at night, as well as by superinsulated and low-infiltration enclosure construction in non-sunspace areas. Thermal mass and exterior sunspace windows provide passive cooling in the summer. (© Chris Twinn/Arup; see Twinn, 2003.)

becomes uncomfortably hot, exterior windows are opened at night to allow cooling, and sunspace enclosures may be partly removed as well to avoid overheating (Schoon, 2016).

Performance BedZED performance has been monitored closely through meter reading, waste audits, and post-occupancy surveys (Twinn, 2003; Hodge and Haltrecht, 2009). Results for the combination of space heating and domestic hot water heating have been encouraging, showing energy savings of ~80% on a per-dwelling basis and ~77% on a per-person basis compared to the local average usage of 22.5 kWh/person-day for space heating and hot water (Figure 13.54).

Although the original design included only south-facing dwellings and north-facing workspaces, some workspaces were later converted to two-bedroom dwellings, providing a useful test of passive solar heating effectiveness. In 2007, monitoring showed that these north-facing units required nearly

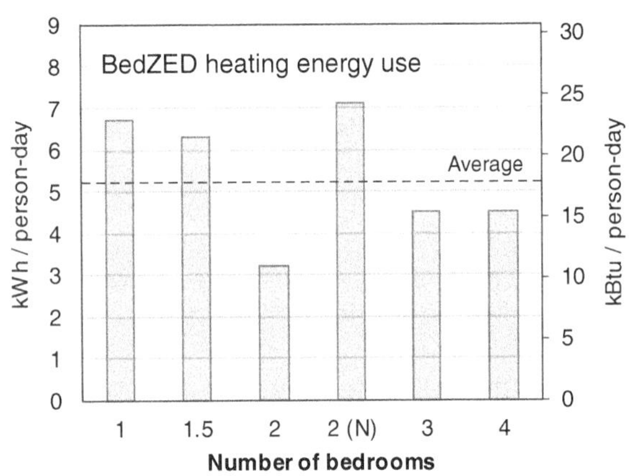

Figure 13.54 Space heating and hot water energy use in BedZED dwellings in 2007. The average of 5.2 kWh/person-day represents a 77% reduction from the city average of 22.5 kWh/person-day; the difference of 3.9 kWh/person-day between south-facing and north-facing two-bedroom dwellings (i.e., 2 and 2(N)) shows the contribution of passive solar heating. (Adapted from Hodge and Haltrecht, 2007, 2009).

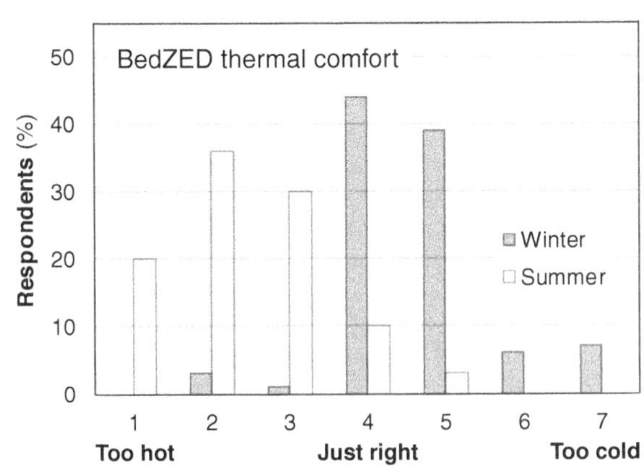

Figure 13.55 Thermal comfort reported in 2007 by BedZED occupants in winter (dark gray) and summer (light gray). While winter conditions were largely perceived as "just right" or "slightly cool," most respondents rated summer conditions as "very warm" or "too hot." (Adapted from Hodge and Haltrecht, 2007, 2009).

twice as much energy for space heating and hot water as the south-facing units, suggesting that passive solar heating is contributing significant heat, even in this cloudy climate (Figure 13.54). Post-occupancy surveys further attested to the heating system effectiveness, with over 80% of respondents reporting that winter conditions were "just right" or "slightly cool" (Figure 13.55). In contrast, over half of respondents reported summer conditions as "very warm" or "too hot," suggesting that the passive cooling elements (operable sunspace exterior windows and movable insulation/shading between the sunspace and dwelling units) may not have been operated consistently and/or with optimal timing. This observation reinforces the importance of passive cooling design for summer control of passive heating systems.

FOR FURTHER INFORMATION

bioregional.com/bedzed

zedfactory.com/bedzed

13.8 CASE STUDY 2—GREGORY BATESON STATE OFFICE BUILDING, SACRAMENTO, CALIFORNIA

Project Basics

- Passive heating and cooling strategies: Direct gain (atrium) with thermal mass, shading, and natural ventilation
- Location: Sacramento, CA, USA; latitude 38.6°N; longitude 121.5°W
- Climate: Mediterranean; 2700 HDD_{65F} [1500 HDD_{18C}]; 4360 CDD_{50F} [2422 CDD_{10C}]; 18.5 in. [470 mm] precipitation annually
- Building type: Office
- Client: State of California
- Design team leaders: Sim van der Ryn and Peter Calthorpe
- Completion date: 1981

Design Intent The architects' primary intent was to create a climate-responsive building that would illustrate the work exploring interrelatedness of individuals, societies, and ecosystems by the anthropologist and cyberneticist Gregory Bateson.

In the mid-1970s, when design began, the State of California had recently adopted new energy-efficiency standards for buildings in response to the energy crisis at the time. Architects were experimenting widely with new green design strategies, and van der Ryn, a pioneer of this movement, welcomed the opportunity to take these ideas to a larger public. The Bateson Building has subsequently been described as "the first large-scale building to embody sustainable architecture" and "a showcase for ecological design," as well as "the unlikely intersection of counterculture with systems control theory" (Sadler, 2008, 2016; Louw, 2013).

As part of its ecological mission, existing within greater systems of people, water, and the planet, the building established three guiding principles: (1) energy efficiency, with the goal of reducing total building energy use to 20% of that expected for a comparable building; (2) support for social interaction; and (3) attentiveness to system integration, reflecting pursuits of interdependence, adaptability, and self-reliance (Sadler, 2016).

Design Elements Responding to these principles, the architects organized the building around a large unconditioned direct-gain atrium, with a glazed sawtooth roof and operable shading, to preheat or precool outdoor air as needed before delivery to offices (Figure 13.56). The atrium was also designed to encourage casual visits by passers-by and social interaction among occupants, as well as to form a central hub for automated daylighting, ventilation, heating, and cooling controls.

Sacramento's climate is Mediterranean, with a sunny, cool heating season of October through April and hot sunny summers; as a result, effective solar controls and ample thermal mass were essential to avoid summer overheating. Nighttime temperatures typically dip below 60°F [15.5°C] in the summer, however, providing a useful cooling resource even in the hottest months.

In heating mode, vertical louvers on the tilted south-facing glazing rotate automatically to allow solar energy collection. This energy is then stored partly in the atrium paver floor and exposed concrete structure; in addition, warm atrium air was originally circulated through a rock bed under the atrium for additional storage; the bed could then use this heat to warm atrium air during cool morning and evening hours before delivery to offices (Figure 13.57). In this way, the bed served as a thermal buffer for fresh outside air throughout the day. Concerns regarding

Figure 13.56 Bateson Building atrium in heating mode, with south-facing skylight louvers open; the multi-story vertical tubes hold fans that draw warm air downward. (© Alison Kwok; all rights reserved.)

pathogen growth led to the rock bed's disconnection several years after opening, but its restoration has been considered recently as part of an energy systems update.

In cooling mode, the clerestory louvers automatically close, and colorful canvas exterior blinds automatically descend over the east, west, and south windows when solar gain reaches threshold levels. Originally, the rock bed also participated; cool outside air was circulated through it at night, discharging the previous day's warmth, and during the day, misters further cooled the exiting air before delivery to offices.

Performance An 80% reduction of whole-building energy use with respect to similar buildings was an explicit design goal. To achieve this, extensive computational simulation was conducted to find thresholds of temperature and solar gain, above and below which shading status and rock-bed fan operation should be adjusted to provide the desired indoor conditions. Although monitoring reports confirmed that summer indoor air temperatures consistently remained within the intended range of 72–77°F [22.2–25°C], with winter values between 72°F and 74°F [22.2–23.3°C] (State of California, 2002), a post-occupancy survey showed that slightly over half of the occupants were dissatisfied with their level of thermal control (Louw, 2013). While this was disappointing, it was also consistent with reports investigating other office spaces at the time (e.g., Schiller et al., 1988), suggesting that the inability of

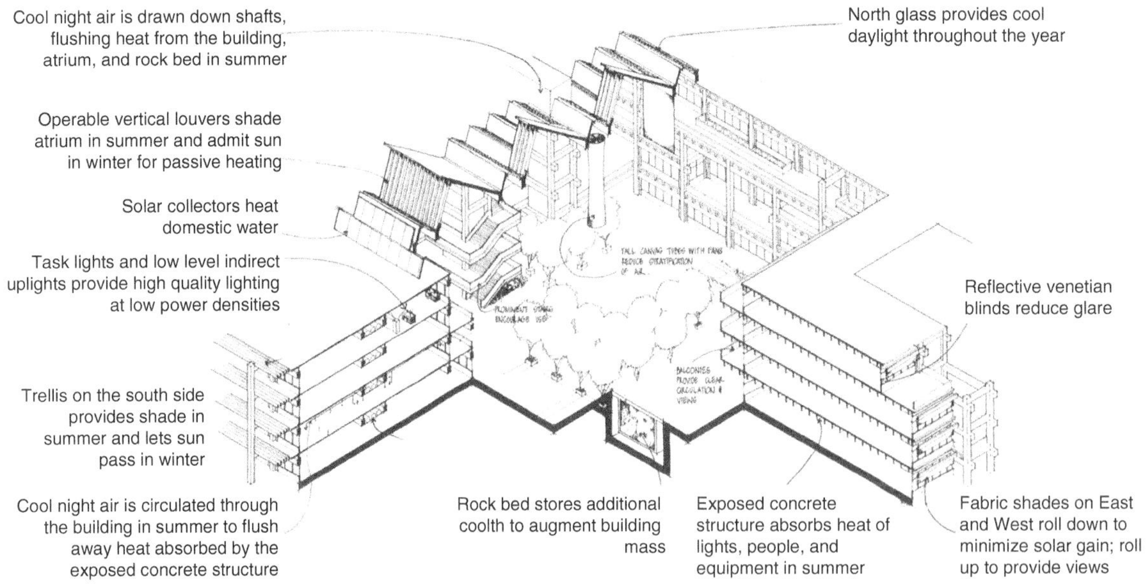

Figure 13.57 Bateson Building systems, showing vertical louvers regulating solar gain through the sawtooth roof; massive interior construction supplemented by rock bed thermal storage; air circulation pathways; and external shading devices for summer use. (Reproduced with permission of the California State Archives.)

the Bateson system to create individually-customized conditions, rather than underperformance of the system as a whole, was primarily responsible.

This is an important lesson for passive heating designers: high-mass systems, with their significant thermal inertia, require additional measures to provide customized conditions; in addition, large-scale systems with automatic controls may leave occupants feeling powerless. The integration of personally controllable windows and window shades, as well as the freedom to adjust clothing and metabolic activity and to work during the most comfortable hours of the day, have since proven to increase thermal satisfaction in passively conditioned spaces (e.g., Brager et al., 2004).

The post-occupancy survey also showed that the atrium was well liked as a social space, drawing people in with a "sidewalk cafe-like feel" and ample daylight as well as greenery and people-watching opportunities. This testifies to the potential value of direct-gain spaces as semi-outdoor rooms and gathering places where bright sunlight and high contrast are acceptable.

When the rock bed was disconnected from the atrium air circulation, followed by an increase in code ventilation requirements from 5 to 15 cfm [2.5–7.0 L/s] per occupant in offices, the ambitious energy-reduction goals for the building were viewed as so unattainable that a comprehensive study of its passive solar heating performance was never undertaken. Nevertheless, a large-scale renovation of the Bateson Building is being designed (as of 2024) with the intent of revisiting the rock bed operation and maximizing the contribution of passive systems using contemporary modeling tools (Taylor, 2016).

FOR FURTHER INFORMATION

critical-sustainabilities.ucsc.edu/bateson-building

simvanderryn.com/bateson

13.9 CASE STUDY 3—BLUE RIDGE PARKWAY DESTINATION CENTER, ASHEVILLE, NORTH CAROLINA

Project Basics

- Passive heating and cooling strategies: Indirect gain (vented Trombe wall); natural ventilation; night ventilation of thermal mass
- Location: Asheville, NC, USA; latitude 35.6°N, longitude 82.6°W
- Climate: Highland humid subtropical; 4308 HDD_{65F} [2393 HDD_{18C}]; 3365 CDD_{50F} [1869 CDD_{10C}]; 31 in. [787 mm] rainfall annually
- Building type: Information and orientation services, exhibition spaces, theater, retail
- Client: National Park Service, Southeast Region
- Design team: Lord, Aeck, & Sargent Inc. and consultants
- Completion date: 2007

Design Intent The Blue Ridge Parkway Destination Center, in mountainous western North Carolina, was designed for the National Park Service to provide ecological information about the Shenandoah and Great Smoky Mountains National Parks (Figure 13.58). To embody its educational and environmental missions, it embraces the protection, preservation, and promotion of regional cultural and natural heritages, including the minimization of its environmental impact. Specific goals included:

- reduction of heating and cooling loads through daylighting, passive solar heating, increased enclosure insulation, natural ventilation, shading, and a green roof
- provision of additional conditioning through high-efficiency radiant heating and energy recovery ventilation
- increased adaptive comfort opportunities through occupant control of passive systems

Design Elements The primary passive solar element is a sawtooth Trombe wall, with the solar-collecting faces oriented toward true south (Figure 13.59); this configuration lengthens the solar-collecting surface and allows the building entrance to face incoming visitors from the southwest (Figure 13.60). The concrete mass of the Trombe wall provides structure, as well, with each interior surface creating exhibit space. The wall is vented with two sets of operable louvers: one allows warm air to enter the exhibit space in winter, while the other allows warm air to escape to the outdoors in summer (Figure 13.61).

Asheville is in the Appalachian Mountains, and its climate is a cooler highland variant of the humid subtropical climate of the U.S. southeast, with

Figure 13.58 Blue Ridge Parkway Destination Center near Asheville, North Carolina. (© Evan Danchenko; used with permission.)

(a) (b)

Figure 13.59 South-facing (a) and north-facing (b) sides of the sawtooth façade. (© Evan Danchenko; used with permission.)

short warm humid summers and long cool humid winters; the heating season lasts about eight months, from October through May. Asheville's heating need is therefore greater than lower-elevation cities in the region, while its solar resources are comparable, providing a good opportunity for passive heating.

In combination with other green design elements, particularly occupancy sensors, daylighting controls, natural ventilation, and high-efficiency mechanical

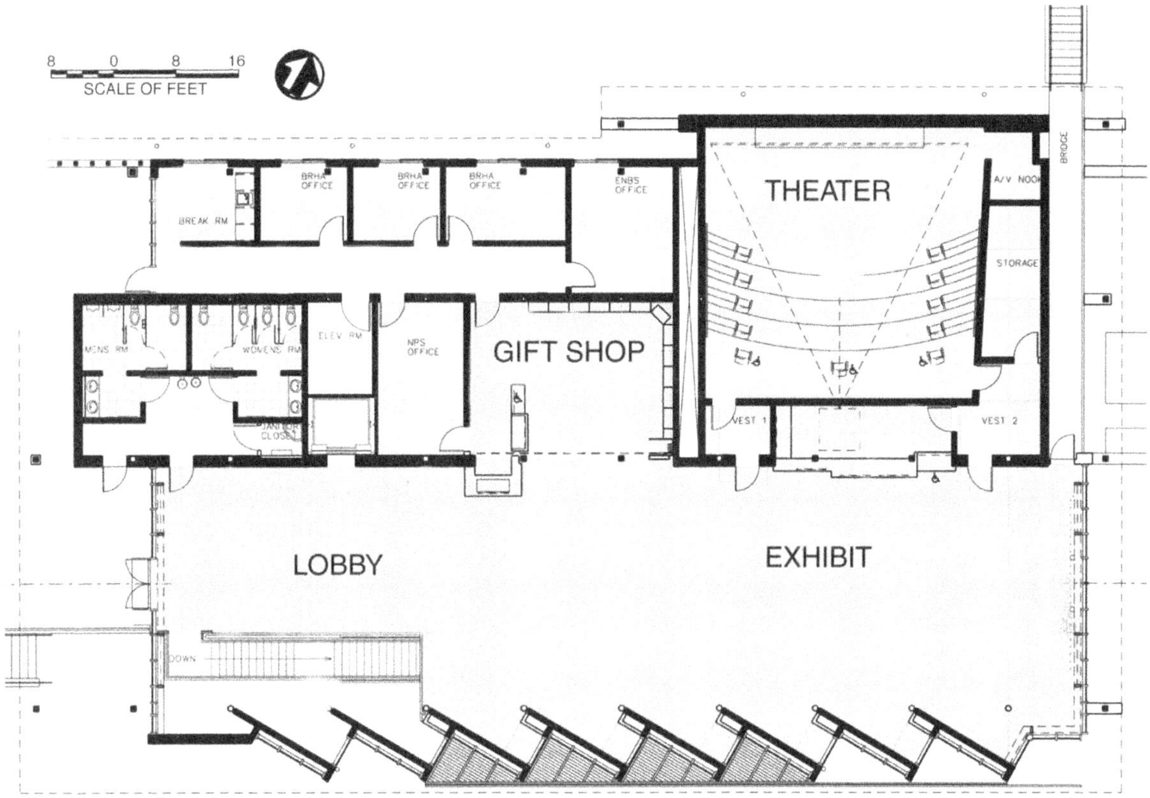

Figure 13.60 *The sawtooth Trombe wall allows each solar-collecting element to face true south while allowing the entrance to face visitors arriving from the southwest; the interior side of each Trombe wall element provides display space. (© Lord, Aeck, & Sargent Inc., used with permission.)*

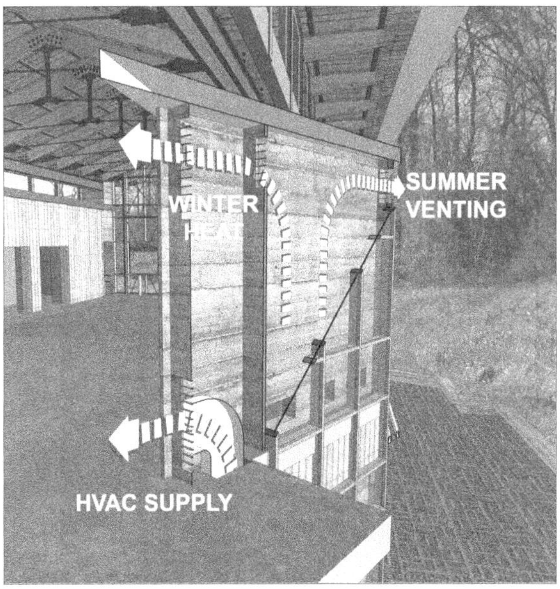

Figure 13.61 *Each Trombe wall panel has operable louvers that allow warm air to enter the building in winter, supplemented by actively-heated air when necessary, and to exit to the outdoors in summer. (© Lord, Aeck, & Sargent Inc., used with permission.)*

systems, the design team projected that the building would use 75% less energy than a minimally code-compliant building.

Model Predictions Preliminary building energy model simulations predicted that, in the absence of a heating, ventilation, and air-conditioning (HVAC) system, the Trombe wall would provide temperatures ~8°F [4.4°C] warmer than those of a standard high-mass building with optimized orientation and ~30°F [17°C] above outside air temperatures. Computational fluid dynamics (CFD) analysis by Pennsylvania State University's Applied Research Laboratory similarly predicted that the Trombe wall would warm the building 30–40°F [17–22°C] above the outside air temperatures in winter. The USGBC awarded this design the maximum number of points possible (10) in the category of optimized energy performance on the basis of these predictions, contributing to its Leadership in Energy and Environmental Design (LEED) Gold certification in 2009.

During the first winter of occupancy, the architects and a team of students from Georgia Institute of Technology (Georgia Tech) conducted a thorough post-occupancy evaluation, including monitoring of the Trombe wall inlet, outlet, and outside air temperatures. On sunny days, the Trombe wall outlet air averaged 35–40°F [19–22°C] warmer than outdoor air during the day and 30–35°F [17–19°C] warmer at night, correlating well with model predictions.

The excellent performance of this system illustrates several important themes regarding Trombe wall applications. First, obtaining a clear southern exposure is essential; in the western North Carolina forest, this was accomplished by the choice of an elevated site and by extending the solar-collecting glass to the full double height of the exhibit space (Figure 13.59). Second, the choice of a vented wall allows substantial heat to be delivered during the day, while the building is occupied, in the form of warm air; the role of the mass is then to maintain building warmth overnight. Third, the provision of controls not only to vent warm air to the interior but also to the exterior, protects the system from overheating in the winter and provides a mechanism for night cooling of mass during the summer, explicitly acknowledging the importance of integrating passive cooling elements into passive heating systems.

FOR FURTHER INFORMATION

Lord, Aeck, & Sargeant. lordaecksargent.com

Moffit, D. 2008. "Appalachian Suncatcher." *Architecture Week* 35: 381.

Sami, V. and J. Gassman. 2006. "A Simultaneous Modelling Methodology to Analyze Passive Solar Performance of Trombe Walls." *Proceedings of the 23rd Conference on Passive and Low Energy Architecture*, Geneva.

Acknowledgments

Special thanks to the following individual, who made significant contributions to this chapter: Alexandra R. Rempel, Ph.D., University of Oregon.

PASSIVE ENVIRON-MENTAL SYSTEMS

References and Resources

Arbor Day Foundation. 2025. "Tree Identification & Guides." arborday.org/tree-identification

Arguez, A., I. Durre, S. Applequist, M. Squires, R. Vose, X. Yin, and R. Bilotta. 2010. *NOAA's U.S. Climate Normals (1981–2010)*. Washington, DC: National Centers for Environmental Information: National Oceanic and Atmospheric Administration.

ASTM. 2023. *ASTM G173-23: Standard Tables for Reference Solar Spectral Irradiances: Direct Normal and Hemispherical on 37° Tilted Surface*. West Conshohocken, PA: ASTM International.

Bainbridge, D. and K. Haggard. 2011. *Passive Solar Architecture*. White River Junction, VT: Chelsea Green Publishing.

Balcomb, J. D. 1978. *Designing Passive Solar Buildings to Reduce Temperature Swings* (No. LA-UR-78-1316). Los Alamos, NM: Los Alamos Scientific Laboratory.

Balcomb, J. D., D. Barley, R. McFarland, J. Perry, W. Wray, and S. Noll. 1980. *Passive Solar Design Handbook, Vol. 2: Passive Solar Design Analysis*. Washington, DC: U.S. Department of Energy.

Balcomb, J. D., R. W. Jones, R. D. McFarland, and W. O. Wray. 1984. *Passive Solar Heating Analysis*. Washington, DC: American Society of Heating, Refrigerating, and Air-Conditioning Engineers and the U.S. Department of Energy.

Bartholomew, D. M. L. 1985. "Possibilities for Passive Solar House Design in Scotland." *International Journal of Ambient Energy* 6: 147–158.

Brager, G. S., G. Paliaga, and R. De Dear. 2004. "Operable Windows, Personal Control, and Occupant Comfort." *ASHRAE Transactions* 110: 17–35.

Butti, K. and J. Perlin. 1980. *A Golden Thread: 2500 Years of Solar Architecture and Technology*. Palo Alto, CA: Cheshire Books.

Cabeza, L. F., A. Castell, C. D. Barreneche, A. De Gracia, and A. I. Fernández. 2011. "Materials Used as PCM in Thermal Energy Storage in Buildings: A Review." *Renewable and Sustainable Energy Reviews* 15: 1675–1695.

Chiras, D. D. 2002. *The Solar House: Passive Heating and Cooling*. White River Junction, VT: Chelsea Green Publishing.

Churchill, S. W. and H. H. Chu. 1975. "Correlating Equations for Laminar and Turbulent Free Convection from a Vertical Plate." *International Journal of Heat and Mass Transfer* 18: 1323–1329.

CIE. 2004. ISO 15469:2004: Spatial Distribution of Daylight—CIE Standard General Sky (prepared as CIE S 011/E:2003 and adopted by ISO). International Commission on Illumination. Geneva: International Organization for Standardization.

DeKay, M. and G. Z. Brown. 2013. *Sun, Wind, and Light: Architectural Design Strategies*. Hoboken, NJ: John Wiley & Sons.

Demain, C., M. Journée, and C. Bertrand. 2013. "Evaluation of Different Models to Estimate the Global Solar Radiation on Inclined Surfaces." *Renewable Energy* 50: 710–721.

Deru, M. 2003. "Model for Ground-Coupled Heat and Moisture Transfer from Buildings" (No. NREL/TP-550–33954). Golden, CO: National Renewable Energy Laboratory.

EIA. 2009. "2009 RECS Survey Data." Washington, DC: Energy Information Administration. www.eia.gov/consumption/residential/data/2009

EIA. 2012. "2012 CBECS Survey Data." Washington, DC: Energy Information Administration. www.eia.gov/consumption/commercial/data/2012

Elman, B. and M. Kern (eds.). 2009. *Statecraft and Classical Learning: The Rituals of Zhou in East Asian History*. Leiden: Brill.

Eymet, V., J. L. Dufresne, P. Ricchiazzi, R. Fournier, and S. Blanco. 2004. "Long-Wave Radiative Analysis of Cloudy Scattering Atmospheres Using a Net Exchange Formulation." *Atmospheric Research* 72: 239–261.

Fernández-González, A. 2007. "Analysis of the Thermal Performance and Comfort Conditions Produced by Five Different Passive Solar Heating Strategies in the United States Midwest." *Solar Energy* 81: 581–593.

Garrett, V. and T. M. Koontz. 2008. "Breaking the Cycle: Producer and Consumer Perspectives on the Non-adoption of Passive Solar Housing in the US." *Energy Policy* 36: 1551–1566.

Hodge, J. and J. Haltrecht. 2007. *BedZED Monitoring Report 2007*. London: BioRegional Development Group.

Hodge, J. and J. Haltrecht. 2009. *BedZED Seven Years On*. London: BioRegional Development Group.

ICC. 2024. *2024 International Green Construction Code*. Washington, DC: International Code Council.

IPCC. 2014 in *Climate change 2014: Synthesis Report. Contribution of Working Groups I, II and III to the Fifth Assessment Report of the IPCC*, R. K. Pachauri, and L. A. Meyer (eds.). Geneva: Intergovernmental Panel on Climate Change. ipcc.ch/report/ar5

Janssen, H., J. Carmeliet, and H. Hens. 2004. "The Influence of Soil Moisture Transfer on Building Heat Loss via the Ground." *Building and Environment* 39: 825–836.

Kachadorian, J. 2006. *The Passive Solar House: Using Solar Design to Cool and Heat Your Home*. White River Junction, VT: Chelsea Green Publishing.

Langdon, W. 1980. *Movable Insulation*. Emmaus, PA: Rodale Press.

Louw, E. 2013. "Hanging Gardens of Sacramento." *Architectural Review* 233: 75—85.

Macgregor, A. W. K. 1980. "The Potential for Solar Space Heating in Scotland." *International Journal of Ambient Energy* 1: 149–1153.

May, J. and A. Reid. 2010. *Buildings Without Architects: A Global Guide to Everyday Architecture*. New York: Rizzoli.

Moore, F. 2004. *Environmental Control Systems: Heating Cooling Lighting*. New York: McGraw-Hill, Inc.

NREL. 2007. "National Solar Radiation Database (NSRDB), 1991–2005 Update: Typical Meteorological Year 3." Golden, CO: National Renewable Energy Laboratory.

Overbey, D. J. 2007. "Validation of the Load Collector Ratio (LCR) Method for Predicting the Thermal Performance from Five Passive Solar Test Rooms Using Measured Data," in *UNLV Retrospective Theses & Dissertations*. Las Vegas, NV: University of Nevada, Las Vegas.

Peel, M. C., B. L. Finlayson, and T. A. McMahon. 2007. "Updated World Map of the Köppen-Geiger Climate Classification." *Hydrology and Earth System Sciences Discussions* 4: 439–473.

Perez, R., P. Ineichen, R. Seals, J. Michalsky, and R. Stewart. 1990. "Modeling Daylight Availability and Irradiance Components from Direct and Global Irradiance." *Solar Energy* 44: 271–289.

Perlin, J. 2013. *Let It Shine: The 6,000-Year Story of Solar Energy*. Novato, CA: New World Library.

Porteous, C. 2005. *Solar Architecture in Cool Climates*. London: Routledge.

Rempel, A. R., and S. J. Remington. 2015. "Optimization of Passive Cooling Control Thresholds with GenOpt and EnergyPlus," in *2015 Proceedings of the Symposium on Simulation for Architecture & Urban Design* (pp. 111–118), H. Samuelson, S. Bhooshan, and R. Goldstein (eds.). Vista, CA: Society for Computer Simulation International.

Rempel, A. R. and A. W. Rempel. 2013. "Rocks, Clays, Water, and Salts: Highly Durable, Infinitely Rechargeable, Eminently Controllable Thermal Batteries for Buildings." *Geosciences* [Special Issue: Geoscience of the Built Environment] 3: 63–101.

Rempel, A. R., A. W. Rempel, K. V. Cashman, K. N. Gates, C. J. Page, and B. Shaw. 2013. "Interpretation of Passive Solar Field Data with EnergyPlus Models: Un-conventional Wisdom from Four Sunspaces in Eugene, Oregon." *Building and Environment* 60: 158–172.

Rempel, A. R., A. W. Rempel, K. R. Gates, and B. Shaw. 2016. "Climate-Responsive Thermal Mass Design for Pacific Northwest Sunspaces." *Renewable Energy* 85: 981–993.

Roetzel, A., A. Tsangrassoulis, U. Dietrich, and S. Busching. 2010. "A Review of Occupant Control on Natural Ventilation." *Renewable and Sustainable Energy Reviews* 14: 1001–1013.

Sadler, S. 2008. "An Architecture of the Whole." *Journal of Architectural Education* 61: 108–129.

Sadler, S. 2016. "The Bateson Building, Sacramento, California, 1977–81, and the Design of a New Age State." *Journal of the Society of Architectural Historians* 75: 469–489.

Schiller, G., E. A. Arens, F. Bauman, C. Benton, M. Fountain, and T. Doherty. 1988. "A Field Study of Thermal Environments and Comfort in Office Buildings." *ASHRAE Transactions* 94: 280–308.

Schoon, N. 2016. *The BedZED Story*. London: Bioregional Development Group.

Seifert, R. D. 1981. *A Solar Design Manual for Alaska* (No. AK-RD-82–01). Fairbanks, AK: Institute of Water Resources, University of Alaska.

State of California. 2002. "Summary of Ambient Indoor Air Quality Investigations: Measurements of Temperature, Humidity, Oxygen, and Carbon Dioxide." Sacramento, CA: State of California.

Stromberg, R. and S. Woodall. 1977. "Passive Solar Buildings: A Compilation of Data and Results" (Report Number: SAND-77–1204). Albuquerque, NM: Sandia National Laboratory.

Swisher, J. 1984. "Passive Solar Performance: Summary of 1982–1983 Class B Results." Golden CO: Technical Information Branch, Solar Energy Research Institute: U.S. Department of Energy.

Swisher, J. N. and J. J. Duffy. 1983. *"Measured Performance of 50 Passive Solar Residences in the United States"* (No. SERI/TP-254–2020; Conf-830929–9). Golden, CO: Solar Energy Research Institute /Winchester, MA: Duffy (JJ) and Associates.

Taylor, M. 2016. *The Administration's Sacramento Office Building Construction Strategy*. Sacramento, CA: Legislative Analyst Office.

Twinn, C. 2003. "BedZED." *Arup Journal* 38: 10–16.

University of Oregon. 2024. "Sun Path Chart." Solar Radiation Monitoring Laboratory. Eugene, OR: University of Oregon. solardata.uoregon.edu/SunPathChart.html

U.S. Census Bureau. 2016. "Annual Characteristics of New Housing: Heat Systems." www.census.gov/construction/chars

Wright, D. 1978. *Natural Solar Architecture*. New York: Van Nostrand Reinhold.

CHAPTER 14

Passive Cooling

DEPENDING ON THE LOCATION AND SEASON, human beings have always searched for suitable ways to improve or stabilize the conditions for comfort. Some say that we have been most successful in controlling environments that require heat and maintaining warmth during cold seasons. However, across much of the planet, passive cooling strategies have been practiced for thousands of years and remain part of the vernacular culture of building in many countries.

Controlling heat that a building gains from its environment is the first step in passive cooling design (as it is with passive heating). Passive cooling strategies can literally introduce "coolness" into a building without mechanical assistance—essentially exhausting the building of unwanted heat gains without any mechanical fans or energy.

The information in this chapter is organized from general to specific. General guidelines are given as a starting point for preliminary passive cooling sizing for typical buildings; these include determining heat gains, appropriate opening sizes, and thermal mass. Designers can make quick approximations to determine if the design is on the right track. As the building takes shape, more detailed information becomes useful and a section with more detailed calculation procedures will allow the designer to select a passive cooling approach with some confidence in its applicability and the need for auxiliary space cooling. A number of example exercises and solutions for the various strategies will demonstrate how to start making these schematic design calculations.

14.1 BRIEF HISTORY

Just as the Greeks and the Chinese practiced solar design planning for proper solar orientation and passive heating, there was also an understanding of shading and evaporative cooling (fountains, pools, water streams, and vegetation) in hot, dry conditions to accomplish cooling effects in and about buildings. Earth-sheltered dwellings used the massive construction to dampen temperature swings, block direct solar radiation, and provide "coolth" in desert climates. In hot, humid climates, where cross ventilation is desirable, it becomes advantageous to have large, shaded verandahs, shading devices, large openings to admit breezes, and structures raised above the ground to allow breezes to flow beneath.

With more affordable air-conditioning systems, hotter cities (urban heat islands), and people's changing expectations about comfort, often the passive approach is abandoned in favor of a total dependence on mechanical equipment. The green building movement and improved building techniques have reconsidered passive cooling principles as part of new integrated, high-performance designs. The downdraft evaporative cooltower at the Global Ecology Research Center at Stanford University (Figure 14.1) uses simple technology, with a mister (equivalent to a wetted pad) at the top where hot, dry air enters. As water evaporates inside the tower, the air temperature drops (the moisture content of the air increases), and the denser air drops to the opening in the base, exiting in this case to the lobby of the building.

Mechanical and Electrical Equipment for Buildings, Fourteenth Edition. Walter Grondzik, Daniel Overbey, and Alison Kwok.

Companion website: www.wiley.com/go/meeb14e

Figure 14.1 Downdraft evaporative cooltower at the Global Ecology Research Center at Stanford University, Menlo Park, California. (© Alison Kwok; all rights reserved.)

14.2 DESIGN STRATEGIES FOR COOLING

Taking into consideration site conditions such as location, orientation, wind, land massing, vegetation, and microclimate, a designer will develop screens, shading, wingwalls, building exposure, openings, and other features to follow four basic approaches to control the heat gains in a building: (1) natural ventilation cooling (cross and stack ventilation), (2) high-mass cooling (roofponds, earth contact), (3) night ventilation of thermal mass, and (4) evaporative cooling. All take advantage of the concept of "heat sinks," whereby heat will move to a cooler area.

The various approaches to passive cooling in temperate climates are best assessed on the building bioclimatic chart (Figure 14.2) developed by Milne and Givoni (1979). The chart allows the designer to visualize the suitability of the four approaches to passive cooling under summer conditions.

Average monthly climate data (as in Table 14.1) are now available from software and various websites (see Appendix L). Such climate data can be plotted on this chart. If they do not exceed the "boundaries" of a strategy zone, then a match is indicated between the climate and that strategy. Although the edges of each strategy zone are drawn as lines, these boundaries are broader and more vague than the lines suggest. When a climate surpasses the boundaries of all four strategy zones, conventional air conditioning is almost certainly desirable.

To plot the average monthly conditions for a climate on this strategy chart, assemble information such as that shown in Table 14.1. This is available for every climatological station maintained by the National Oceanic and Atmospheric Administration (NOAA) of the U.S. Department of Commerce. These climate summaries, often called Local Climatological Data (LCD), are available from the National Centers for Environmental Information (NOAA, 2025). It should be noted that an updated compilation of summaries, referred to as Local Climatological Data Version 2 (LCDv2), has been released. LCDv2 supersedes the original LCD (i.e., LCDv1).

In addition to a typical hot summer day, it is useful to check for the design condition, the statistically relevant condition for which engineers would design a building's mechanical cooling equipment. The design condition shows the climate near its worst, while avoiding the anomalous conditions of extreme temperatures listed in the NOAA data. Summer (and winter) design data are available in Appendix B (along with a discussion of the nature of the data).

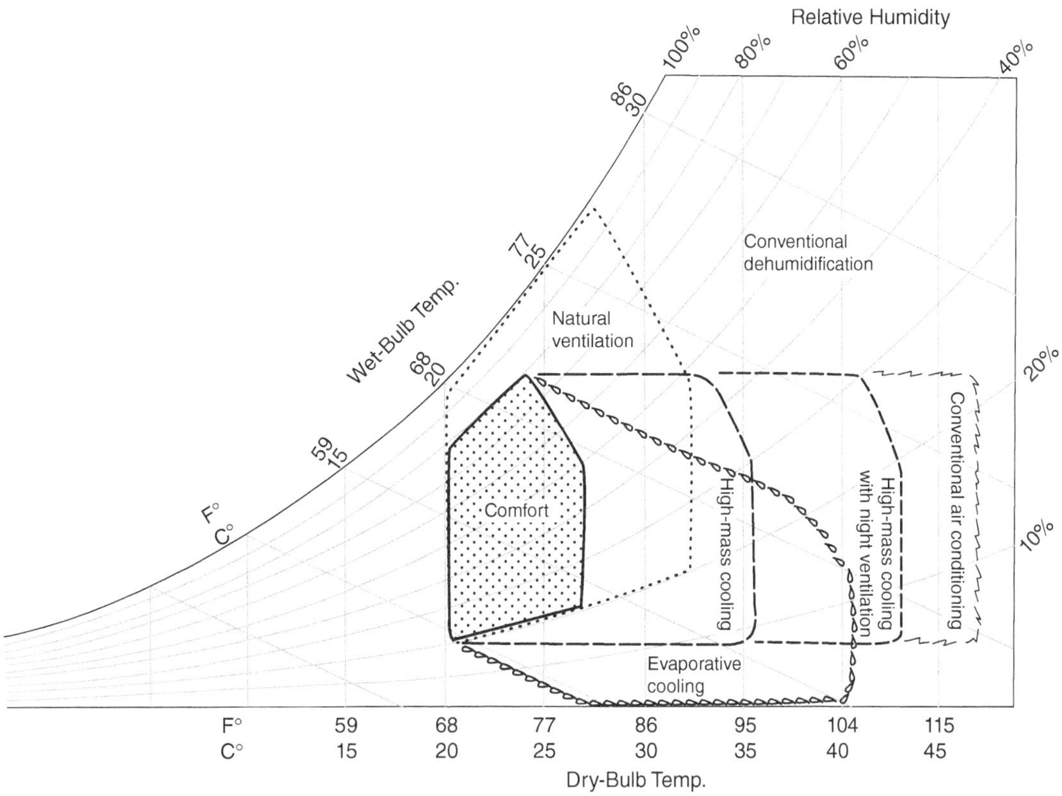

Figure 14.2 *Passive cooling design strategies by climate. After plotting the outdoor climate data on this chart, consider how that information would affect a design strategy. (Adapted from Milne and Givoni, 1979.)*

TABLE 14.1 Normal Data from Annual Summary of Local Climatological Data for Dodge City, Kansas[a]

	Daily Temperatures, °F [°C]				Relative Humidity (%) at					
	Normal		Extreme[b]		Hour				Wind	
Month	**Max**	**Min**	**Max**	**Min**	**00**	**06**	**12**	**18**	**Mean Speed, mph [km/h]**	**Prevailing Direction**
June	86.0 [30.0]	61.4 [16.3]	108 [42.2]	41 [5.0]	63	75	44	38	14.4 [23.2]	South
July	91.4 [33.0]	66.9 [19.4]	109 [42.8]	47 [8.3]	66	78	46	42	12.9 [20.8]	South
August	90.4 [32.4]	65.7 [18.7]	107 [41.7]	47 [8.3]	71	80	50	46	12.7 [20.5]	South

[a]Records as of 1979.

[b]Extremes do not occur on the same day or series of days.

EXAMPLE 14.1 To determine the viable passive strategies for Dodge City, Kansas, plot the typical day of the hottest month on the bioclimatic chart.

SOLUTION

Inspection of Table 14.1 shows that the hottest month is July; however, August is close, and sun angles are lower (more gain through windows). To chart the typical August day, find the approximate relative humidity (RH) for the coldest and hottest hours. Because RH is listed only at four times, first select the highest RH, occurring here at 6:00 a.m.; this will coincide approximately with the coldest hour. Thus, one end of the linear climate plot for August in Dodge City will be at the combination of 80% RH and 65.7°F [18.7°C]. The other end of the plot will be at the lowest RH and the hottest hour, in this case at 6:00 p.m., a combination of 46% RH and 90.4°F [32.4°C]. The line between these points is shown in Figure 14.3.

Figure 14.3 shows that one passive cooling strategy—high mass with night ventilation—appears clearly adequate to meet the needs of the typical August day in Dodge City. (Using this strategy, the low-temperature end of the line is also important and should be below the comfort zone for best results.) Another strategy, high mass, appears just barely adequate. Two other strategies, natural ventilation and evaporative cooling, appear inadequate because the highest temperature/lowest RH combination falls outside their boundaries—slightly so for natural ventilation, but greatly so for evaporation.

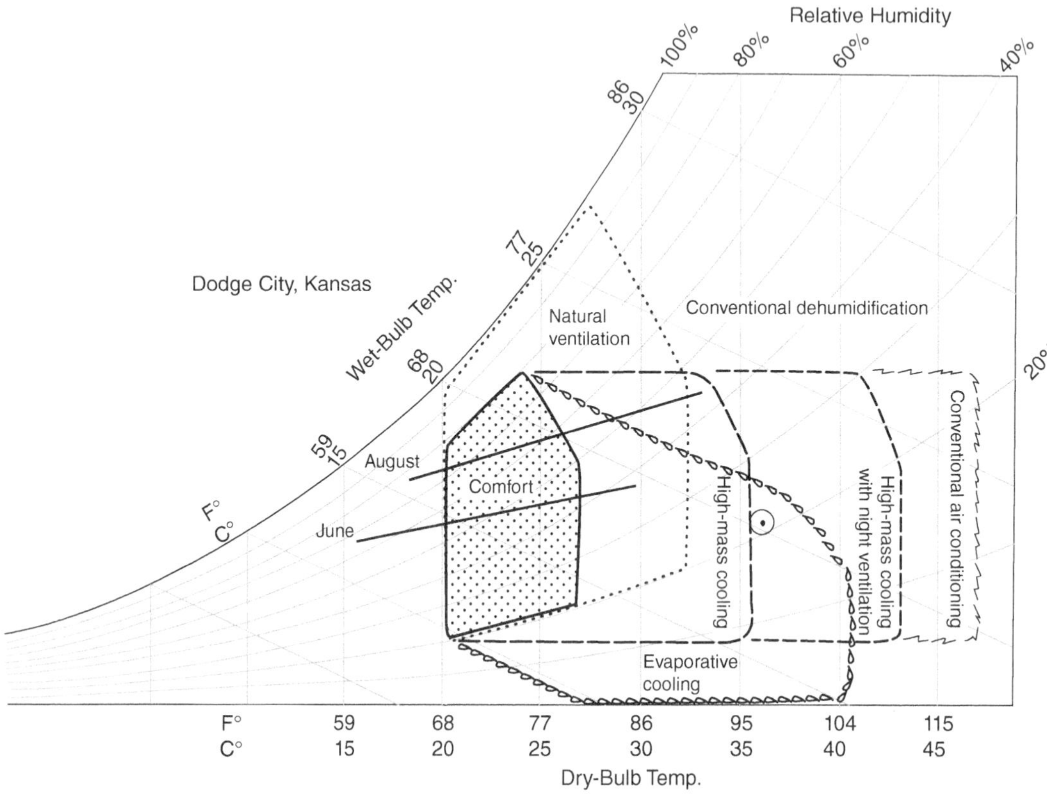

Figure 14.3 *Hot-month daily ranges and the summer design condition for Dodge City, Kansas, superimposed on the design strategy chart of Figure 14.2.*

Using design climate data corresponding to the values in Table 14.1, the summer DB and mean coincident WB temperature is 97/69°F [36.1/20.6°C]; this combination is shown as a circle around the point in Figure 14.3. The 97°F [36.1°C] DB temperature is plotted on the corresponding vertical line; the 69°F [20.6°C] WB temperature is found by first locating the 69°F [20.6°C] DB vertical line and following that line up to 100% RH, where 69°F [20.6°C] DB and WB are coincident. Then follow downward to the right along the constant WB line to its intersection with 97°F [36.1°C] DB.

The design condition point falls just outside the zone of high-mass cooling and just within the zone of evaporative cooling. Because the design condition falls well within the zone of high mass with night ventilation, this strategy is thus confirmed as clearly the best of the passive cooling strategies for Dodge City. ■

(a) Natural Ventilation Cooling

This is the most evident strategy suggested by the comfort charts presented earlier, in which higher air temperatures were offset by increased air motion. It may be the only passive strategy available in humid, hot climates in which temperatures are only slightly lower by night than by day. Buildings should be very open to breezes while simultaneously closed to direct sun. The materials of these buildings may be thermally lightweight as well, because the temperature of the night air is not cool enough to remove much stored daytime heat. Very high humidity may be avoided only by sealing and air-conditioning buildings.

Natural ventilation has two variations: cross ventilation and stack ventilation. Cross ventilation is driven by wind and is accomplished with windows. It relies upon rather narrow plans with large ventilation openings on either side. Thus, it is naturally compatible with daylighting. Stack ventilation depends upon very low openings to admit outdoor air and very high openings to exhaust air; it is driven by the principle that hot air rises. Stack ventilation is generally weaker than cross ventilation—except when there is no wind at all.

(b) High-Mass Cooling

This strategy is most successful in locations with warm, dry summers, where the extremes of hot days can be tempered by the still-cool thermal mass of a building. Cool nights then slowly drain away the heat that such mass accumulates during the day. The thermal mass can be located in floors, walls, or roofs but will need a sink to which it can reject its heat by night. The roof has the advantage of radiating to the cold night sky, but it should be protected from exposure to sun by day. The masonry courtyard-type buildings of the

Mediterranean are vernacular examples of this passive cooling strategy; their courtyard floors and roofs can be protected with movable shading devices (*toldos*) by day, then opened to night-sky radiation.

Roofponds (also termed "roof ponds") are a form of high-mass cooling for one- and two-story buildings. Because they require only the roof to be massive, they allow for considerable design freedom in walls and fenestration. Where cooling is the only objective, this approach uses water that is stored between the metal ceiling and the roof insulation; by night, the water is pumped (and/or sprayed) over the exposed roof surface and allowed to trickle back through the insulation to the storage pond. At lower latitudes (with high winter sun altitudes), roofponds can be used for passive solar heating as well. This strategy uses sliding panels of insulation over bags of water; the panels slide open on winter days to collect sun, and open on summer nights to radiate heat to the sky.

Another variation on high-mass cooling depends upon earth contact. The earth acts as a heat sink, keeping walls and floors (even roofs when earth-covered) cool. However, if the earth is allowed to continue to act as a heat sink in winter, heating needs could be greatly increased. Thus, a strategy for summer contact and winter isolation might be appropriate.

(c) High-Mass Cooling with Night Ventilation

This hot-dry summer design strategy must use outdoor air at subcomfortable nighttime temperatures to flush away heat stored during the daytime. The fewer the subcomfortable night hours, the greater the area of thermally massive surface that must be provided to store the day's heat. Also, because there are more hours of daylight and fewer of nighttime, the ventilation must occur quickly and thoroughly, probably using fans. (Nightly wind velocities are typically lower than daytime velocities, because summer wind is often driven by regional solar overheating of the ground.) The building switches from a thermally closed condition by day (to exclude sun and hot outdoor air) to an open condition at night (to allow ventilation to cool the mass). Note: Nighttime temperatures must be cooler than the comfort zone temperature if this strategy is to be effective. Figure 14.4 shows the need to inspect both ends of the typical day's bioclimatic plot for this strategy.

This cooling strategy is highly compatible with passive solar heating strategies that rely on large areas of thermal mass, such as direct gain. It is suitable for large, high buildings, particularly those that have concrete (thermally massive) structural systems.

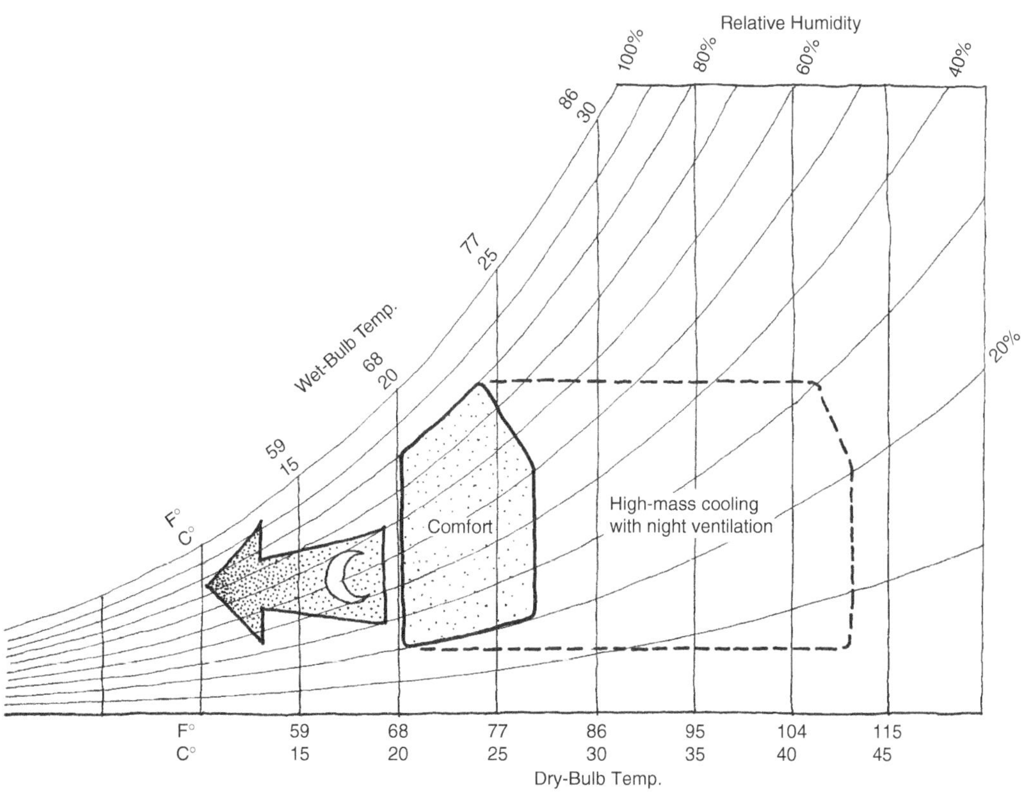

Figure 14.4 *Strategy chart segment for night ventilation of thermal mass. This strategy is most successful when cool nighttime temperatures are available, preferably below comfort zone temperatures for several hours.*

(d) Evaporative Cooling

This design strategy relies on the principle that when moisture is added to air, relative humidity increases while dry-bulb temperature decreases. (On the bioclimatic chart, this pattern exactly follows the constant wet-bulb line, upward and to the left.) In conditions that are more uncomfortably dry than uncomfortably hot, lower air temperature in exchange for higher humidity is a win-win proposition. However, large quantities of both water and outdoor air are needed; fan-driven evaporative coolers are the most common way to provide this kind of cooling.

Evaporative cooling imposes few constraints on designers because the equipment resembles conventional HVAC systems. Rather high indoor air velocities and their associated sounds are typical of these systems, and the aroma of the wetted material of the cooler is often noticeable. The coolest air will be in the vicinity of the air inlet to the space, and the warmest air at the outlet from the space.

14.3 SUMMER HEAT GAIN GUIDELINES

Estimating the extent of the heat gains in a building is the first step to designing a passive cooling approach. Passive cooling guidelines and calculations are expressed in heat to be removed per unit of floor area, so it is necessary to gather data and building information to do the calculations. Sometimes cooling loads are more closely related to individual building characteristics than to climate; for example, sunshading and internal heat gains are particularly influential on cooling loads. An underlying assumption in all passive design and energy performance calculations is the optimization of the building envelope to minimize heat gains. A comparison of these guidelines with more detailed analysis procedures is given in Table 14.2.

Note that "open" buildings, such as those that are naturally ventilated, do not have heat gains from infiltration, because they are assumed to maintain internal temperatures that are slightly above exterior temperatures. However, these buildings do experience

TABLE 14.2 Comparing Passive Cooling Design Guidelines with Detailed Calculation Procedures

Design Guidelines (Section 14.4)	Detailed Calculations (Section 14.8)
Part A. Cross Ventilation	
• Assume 3°F [1.6°C] Δt • Assume window is oriented to wind	• Use actual Δt • Use actual window orientation to wind
Part B. Stack Ventilation	
• Assume 3°F [1.6°C] Δt	• Use actual Δt
Part C. Night Ventilation of Thermal Mass	
• Assume ratio of mass area/floor area • Assume cooling during hour of maximum Δt • Assume maximum Δt for natural ventilation estimation • Find total cooling and required air flow rate	• Use actual exposed mass area • Use actual mass heat capacity • Use actual hourly chart of air and mass temperatures
Part D. Evaporative Cooling (Active)	
• Assume 2.67 cfm/ft^2 floor area [13.6 L/s m^2] • Assume 83°F [28.3°C] exhaust air • Find allowable Δt as air passes through indoors • Then cfm = (Btu/h)/(1.1)(Δt) [L/s = W/(1.2)(Δt)]	• Use actual outdoor temperature for analysis • Determine actual indoor air temperature
Part E. Downdraft Cooltowers	
• Find approximate exit air temperature • Find approximate flow rate • Then Btu/h = (cfm)(1.1)(Δt) [W = (L/s)(1.2)(Δt)]	• Use actual outdoor temperature and wet-bulb depression • Find actual exit air (supply) temperature • Find actual exit airflow rate • Then Btu/h = (cfm)(1.1)(Δt) [W = (L/s)(1.2)(Δt)]
Part F. Roofponds	
• Assume pond maximum temperature of 80°F [26.7°C] • Estimate pond minimum temperature • Assume 30% gain through roof insulation • Assume pond depth from 3 in. to 6 in. [75–150 mm] • Find pond depth and area	• Use actual outdoor temperature, resulting pond temperature • Use actual heat gain through roof insulation • Consider actual hours of internal heat gains • Find pond depth and area • Determine size of backup cooling
Part G. Earth Tubes	
• Assume 65°F [18.3°C] soil temperature • Assume soil conductivity • Does not specify depth • Assume 85°F [29.4°C] outdoor air • Assume 500 fpm [2.5 m/s] velocity • Choose diameter and length to match cooling load	• Actual underground temperature; assume resulting tube temperature is within 4°F [2.2°C] • Actual soil conductivity • Actual depth • Actual outdoor air temperature • Assume 500 fpm [2.5 m/s] velocity • Calculate actual cooling capacity

heat gains through windows, walls, and roofs due to solar impacts on these surfaces. For "closed" buildings, heat gain from infiltration or ventilation must be added, because these structures maintain internal temperatures lower than outside temperatures.

14.4 PASSIVE COOLING GUIDELINES

The design guidelines for passive cooling are much newer and less tested than those for passive heating. It is important to first check the match between the climate and the cooling strategy, as was done earlier in this chapter, before applying these guidelines to a building design.

(a) Cross Ventilation

One of the oldest cooling methods known, this strategy provides plentiful fresh air but maintains a building at temperatures slightly above those outdoors. The cross-ventilation inlet (window) areas, expressed as a percentage of the total floor area, are related to wind speed and resulting heat removal in Figure 14.5. Remember that an equal (or greater) area of outlet openings must also be provided. Furthermore, any internal obstructions (such as partitions) must have a total area of openings at least equal to this required inlet area. The assumptions about wind direction and indoor-outdoor temperature differences that were used to produce these guidelines are explained in Figure 14.5. For more detailed wind speed and direction information, see Section 14.8.

The Δt of 3°F [1.6°C] used in Figure 14.5 is deliberately kept small to encourage open strategies in milder summer climates. Thus, an interior temperature of 83°F [28.3°C], which is comfortable with sufficient air motion, lower RH, and comfortable surface temperatures, would be obtainable with an outside temperature of 80°F [26.7°C]. However, a greater Δt is often appropriate—for example, for spring or fall cooling of office buildings, or for summer cooling of factories or kitchens where internal temperatures may remain in the low 90s [+/− 32°C]. In such cases, find the percentage of inlet area required; then multiply by the ratio

$$\frac{3°\text{F [or } 1.6°\text{C]}}{\text{actual } \Delta t}$$

to obtain the required cross-ventilation areas for any specific temperature difference.

(b) Stack Ventilation

This is another historically useful strategy, which, like cross ventilation, provides plentiful fresh air but maintains a building at temperatures slightly above outdoor temperatures. The stack inlet areas, expressed as percentage of total floor area, are related to stack height and the resulting heat removal in Figure 14.6. Remember that an equal (or greater) area of stack outlet

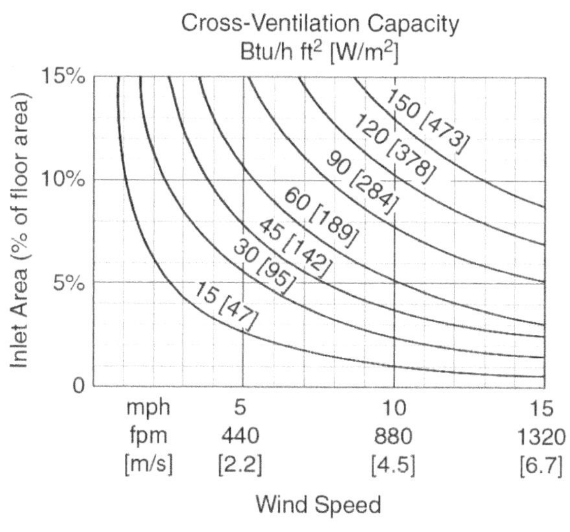

Figure 14.5 Cross-ventilation design guidelines for heat removed per unit floor area and relationship of inlet openings and wind speed. The total inlet opening area is expressed as a percentage of the total floor area served by cross ventilation. Note: Outlet areas must be at least equal to inlet areas. This graph is based on an internal temperature 3°F [1.7°C] above the exterior temperature and assumes that the wind is not quite perpendicular to the window openings, for a wind effectiveness factor of 0.4 (see Section 14.8). (Drawn by Tyler Mavichien with revisions by Evelyn Dinh.)

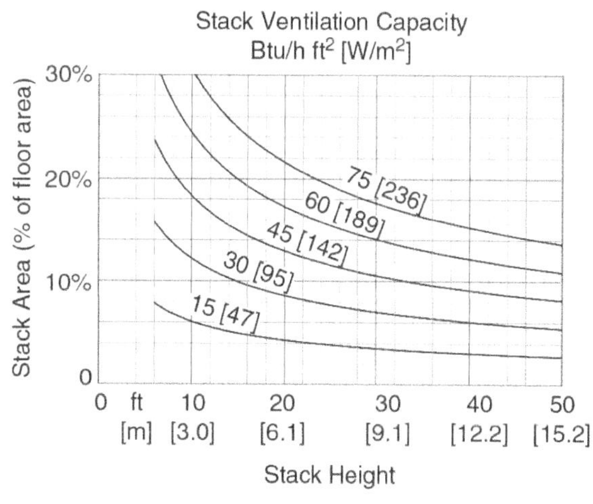

Figure 14.6 Stack ventilation design guidelines for heat removed per unit floor area and the relationship of stack height and inlet openings. The area of stack inlet/throat/outlet is expressed as a percentage of the total floor area served by stack ventilation. Note: Outlet areas and the stack throat area must be at least equal to inlet areas. This graph is based on an internal temperature of 83°F and an exterior temperature 80°F, for a 3°F differential [28.3°C internal, 26.7°C external, for a 1.6°C differential]. (Drawn by Sara Tepfer with revisions by Evelyn Dinh; © Walter Grondzik and Alison Kwok; all rights reserved.)

openings, as well as at least an equal cross-sectional area through the vertical stack, is also required. Again, consider internal obstructions: partitions must have a total of openings at least equal to this required inlet area. The assumptions about indoor-outdoor temperature differences that were used to produce this guideline are explained in the figure.

Adjustment of the Δt for this guideline is similar to the procedure for cross ventilation. It requires multiplication of the required percentage of stack area by the ratio

$$\frac{3°\text{F [or } 1.6°\text{C]}}{\text{actual } \Delta t}$$

to obtain the required stack ventilation areas for any other specific temperature difference.

Until the advent of air conditioning, these methods of cross ventilation and stack ventilation were used in virtually all commercial buildings. In some climates with mild summers, newer buildings are turning to these older methods. Figure 14.7 shows the Thoreau Center for Sustainability in San Francisco's Presidio National Park. An older military hospital was renovated for offices for nonprofit environmental organizations. To maximize daylighting and passive cooling, the original operable windows are utilized. Interior partitions are kept low enough to facilitate cross ventilation, with integral indirect electric lighting using the high white ceiling. The section indicates that both cross ventilation and the stack effect are anticipated. However, the stack openings to the attic and, subsequently, through the roof are less than 1% of the total floor area. The most likely benefit of the stack effect here is to keep the attic at cooler summer temperatures (with air drawn in at the eave vents, not flowing outward as shown).

(c) Night Ventilation of Thermal Mass

This strategy maintains a building at temperatures lower than those outside by day and flushes the building with plentiful fresh air by night. The sizing procedure is shown in Figure 14.8, where climate data are related to two representative types of thermally massive building. For each type, the graphs show the daily Btu per square foot of floor area that can be stored. The "average" mass building is represented by a building with an exposed concrete floor 4 in. [100 mm] thick (or an exposed ceiling of equivalent construction). There is one-unit area of exposed mass for each unit floor area. The "high" mass building is similar to the typical passively solar-heated, direct-gain building or to a multistory building with an exposed

(a)

Figure 14.7 The Thoreau Center for Sustainability occupies renovated hospital buildings (a) in San Francisco's Presidio National Park. (b) Relatively narrow floor plans facilitate daylighting and natural ventilation. (c) Interior at the center corridor. (d) Section illustrates daylighting and cross ventilation. Very small stack openings relative to the floor area served limit the stack effect to a minor role. (Drawings courtesy of Tanner Leddy Maytum Stacy Architects, San Francisco.)

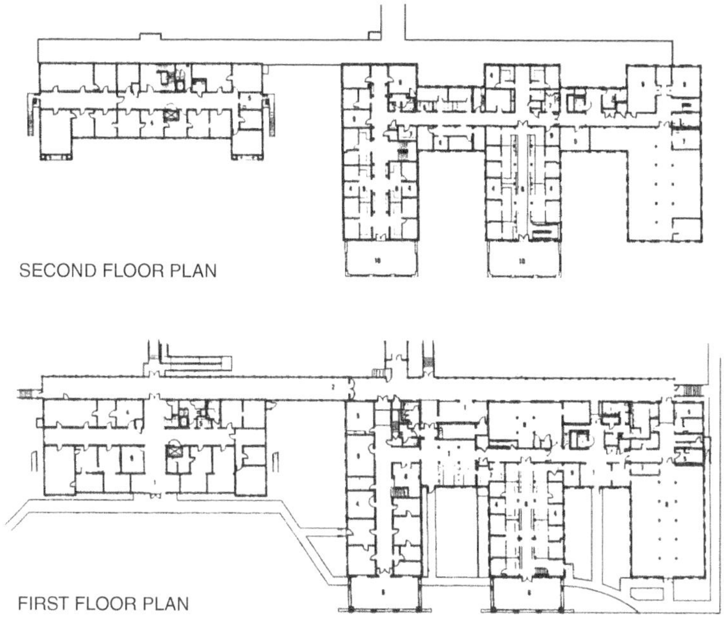

(*b*)

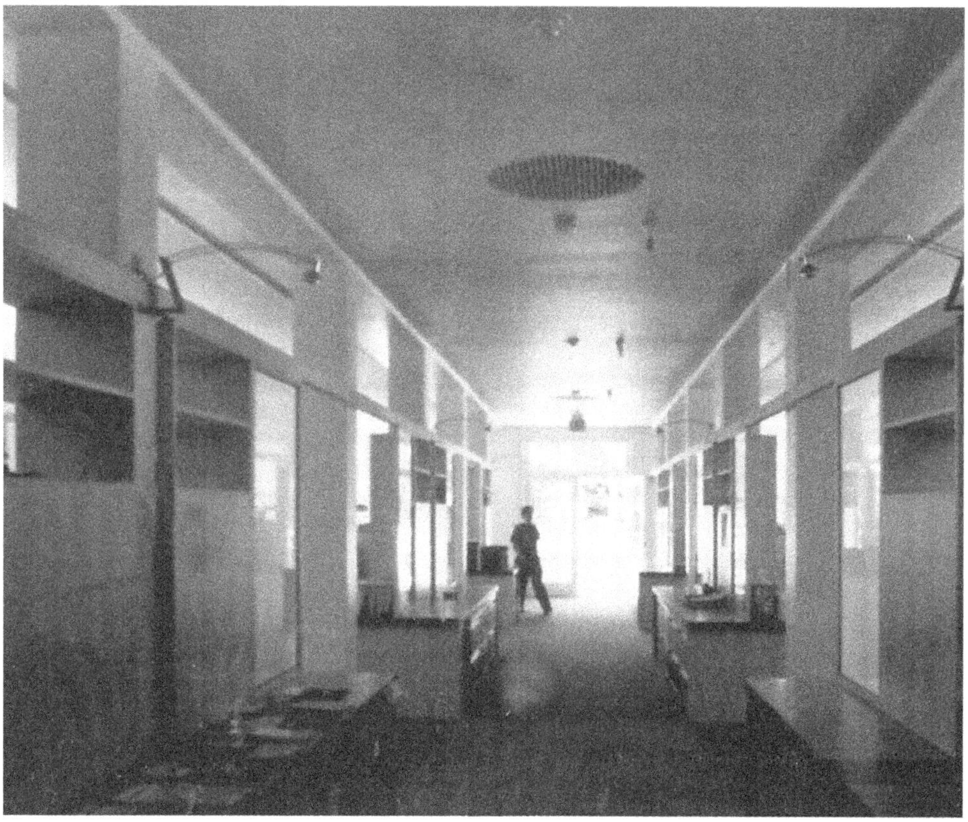

(*c*)

Figure 14.7 *(continued)*

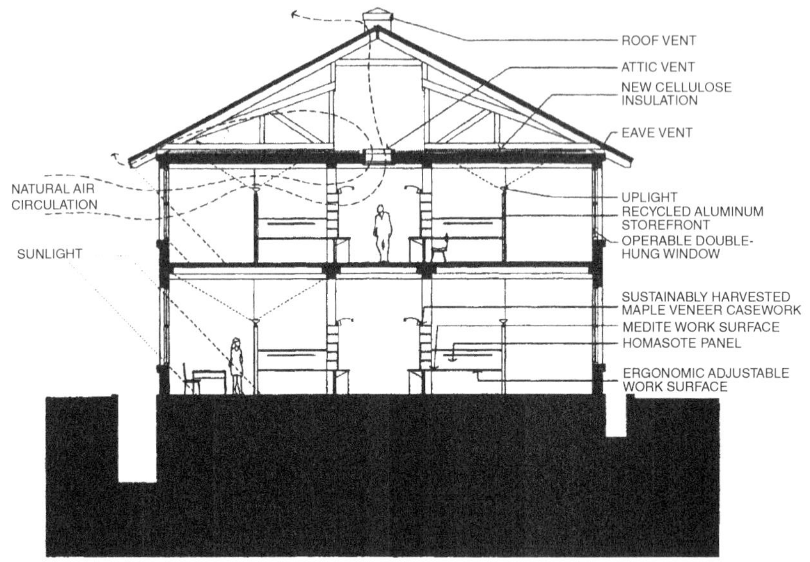

TYPICAL SECTION AT BUILDING 1013

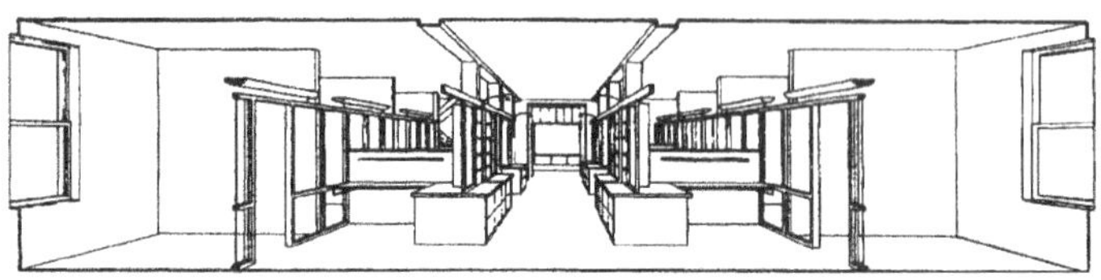

PERSPECTIVE AT BUILDING 1013

(*d*)

Figure 14.7 *(continued)*

concrete structure, in which both sides of the floor slab are available for thermal storage, or an equivalent exposed mass area in walls, and so on. For this building, there are two units of exposed mass for each unit of floor area.

The climate data needed for Figure 14.8, the maximum summer design dry-bulb (DB) temperature and the mean daily range (MDR), are given in Appendix B, Tables B.1 through B.5. These data also allow calculation of the minimum summer design DB temperature; that is

maximum design DB − MDR = *minimum* design DB

This minimum DB temperature is of interest here because the thermal mass of the building will be lowered toward (but not quite to) it during night ventilation. For high daily range climates, the lowest temperature obtained by the thermal mass will be about one-fourth (1/4) of the mean daily range above the minimum air temperature (for lower daily range climates—30°F [16.7°C] or less—one-fifth of the mean daily range).

EXAMPLE 14.2 Compare the potential for night ventilation of mass in Sacramento, California, and Oklahoma City, Oklahoma.

SOLUTION

Sacramento, California:

Maximum design DB = 97.5°F (Appendix B, Table B.1) [36.4°C (Appendix B, Table B.3)]

Mean daily range = 32.6°F (Appendix B, Table B.1) [18.1°C (Appendix B, Table B.3)]

Minimum design DB = 97.5 − 32.6 = 64.9°F [18.3°C]

Approximate lowest mass temperature

= 1/4 of MDR + minimum design DB

= 8.2 + 64.9

= 73.1°F [22.8°C]

Oklahoma City, Oklahoma:

Maximum design dry-bulb = 97.6°F [36.5°C]

Mean daily range = 21.6°F [12.0°C]

Minimum design dry-bulb = 97.6 − 21.6 = 76.0°F [24.5°C]

Approximate lowest mass temperature

= 1/5 of MDR + minimum design DB

= 4.3 + 76.0

= 80.3°F [26.8°C]

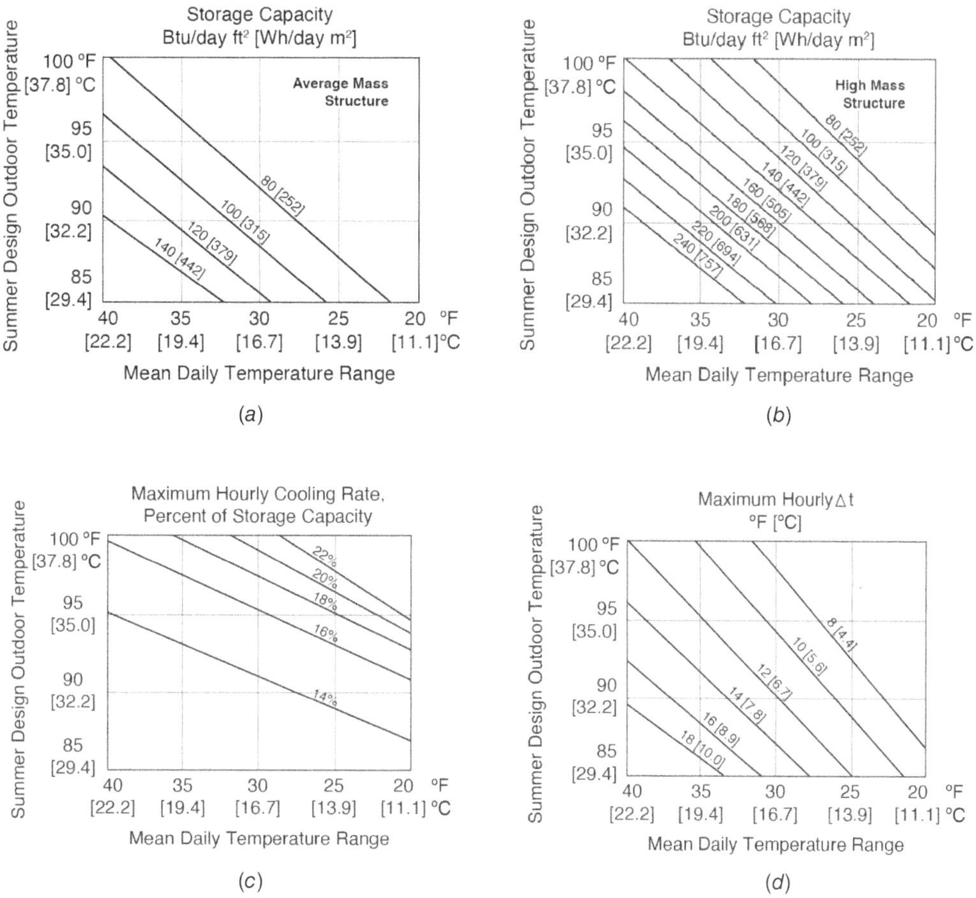

Figure 14.8 *Night ventilation of thermal mass design guidelines. Parts (a) and (b) show Btu/day stored and removed per ft² of floor area [W/day per m² of floor area]. Enter these graphs with your location's summer design outdoor DB temperature and the MDR (mean daily range). (a) "Average mass" structure has unit area of mass surface exposed per unit area of floor area, where the mass is a 4-in. [100-mm] ordinary-density concrete slab. (b) "High-mass structure" has two unit areas of mass surface exposed per unit area of floor area, where the mass is a 3-in. [75-mm] ordinary-density concrete slab (or both sides of a 6-in. [150-mm] concrete wall or slab exposed). Both average- and high-mass structures go into "open" mode when the outdoor temperature drops below 80°F [26.7°C]; therefore, the highest mass temperature is assumed to be 80°F [26.7°C]. Part (c) shows the approximate percentage of the daily total heat stored that is removed during the nighttime hour of maximum cooling. Part (d) shows the Δt between the outdoor air and inside mass that exists during that hour of maximum cooling. Required ventilation rates can then be determined from (c) and (d).*

From these data, it appears that Sacramento, with a comfortably low 73.1°F [22.8°C] mass temperature, is a likely site for night ventilation to remove heat stored in the building's mass. However, Oklahoma City's 80.3°F [26.8°C] lowest mass temperature is at the middle of the comfort zone, suggesting a climate in which this passive strategy will be harder to achieve. In either location, performance studies can best be carried out with the details of this mass cooling procedure, found in Section 14.8.

We can also compare the daily heat absorbed by the structure. For a high-mass building, using Figure 14.8*b*:

Sacramento (Summer Design DB = 97.5°F [36.39°C], MDR = 32.6°F [0.33°C]) storage is around 120 Btu/ft² [379 W/m²] day.

Oklahoma City (Summer Design DB = 97.6°F [36.44°C], MDR = 21.6°F [−5.78°C]) storage is much less than 80 Btu/ft² [252 W/m²] day. ■

Before this thermal storage information can be fully utilized, the number of hours for which the building must be "closed" must be determined. During these hours, heat will be stored in the structure's mass, up to the maximum storage indicated in Figure 14.8*a* (for an average mass structure) or Figure 14.8*b* (for a high-mass structure). Typically, these buildings will go into the thermally closed mode (allowing minimum ventilation only) at about 6:00 a.m. for a day with 100°F [37.8°C] maximum or about 8:00 a.m. for a day with 85°F [29.4°C] maximum, remaining in the closed mode until the outdoor temperature drops below 80°F [26.7°C]. (To approximate this hour, assume that the midpoint outdoor temperature, between daily high and daily low, occurs at around 10:00 p.m.) Thus, the typical office building remains thermally closed during the eight or nine hours of summer occupancy.

All the heat generated during the closed mode—eight to nine daytime hours—must be stored in the structure:

Heat to be stored per ft^2 [m^2] of floor area = hours (h) occupied in "closed" mode × heat gain of floor (Btu/h ft^2 [W/m^2]) (from Appendix G, Table G.1, Part D). For instance, consider a passively cooled office building. Assuming a nine-hour typical working day (including noon hour) with an average approximated heat rate of 13 Btu/h ft^2 (i.e., gains from people and equipment—Part A, from electric lighting—Part B, through the enclosure—Part C, and from infiltration/ventilation—Part E), the daily heat gain in the building would be:

$$9\,\text{h} \times 13\,\text{Btu/h ft}^2 = 117\,\text{Btu/ft}^2 \text{ per day}$$

If applied to Example 14.2, a high-mass structure in Sacramento could handle this amount of heat gain; not so in Oklahoma City.

Consider the administration building for the Fetzer Winery, which sits among grapevines in northern California's Mendocino Valley (Figure 14.9). The summer design DB is approximately 96°F [35.6°C] with an MDR 30°F [16.7°C]. As an average-mass structure, this building could store and remove about 60 Btu/ft^2 day [190 W/m^2 day]; as a high-mass structure, it could store and remove about 100 Btu/ft^2 day [315 W/m^2 day]. Neither amount would be suitable to handle 117 Btu/ft^2 per day using thermal mass alone. Measures would need to be taken to reduce the daily heat gain in the building.

Having found the amount of heat that can be stored each day, the designer must then solve the problem of removing the heat by night. Either natural or forced ventilation can be used, but in many locations wind speed is very low on summer nights. The ventilation rate is roughly determined by the "best hour" of cooling during the night; that is, the hour during which the temperature difference (Δt) between inside mass and outdoor air is greatest, and hence the most heat is removed. This "best hour" information can be found in Figure 14.8*c* and Figure 14.8*d*. Using Sacramento and Oklahoma City as examples:

Sacramento:

Summer design DB = 97.5°F [36.4°C]

Mean daily range = 32.6°F [18.1°C]

Figure 14.8*c*: maximum hourly cooling rate = 17% of storage capacity

(i.e., heat stored that is removed during the nighttime hour of maximum cooling)

Figure 14.8*d*: maximum hourly Δt = 10°F [5.6°C]

(i.e., the Δt between the outdoor air and inside mass that exists during that hour of maximum cooling)

Oklahoma City:

Summer design DB = 97.6°F [36.5°C]

Mean daily range = 21.6°F [12.0°C]

Figure 14.8*c*: maximum hourly cooling rate = 23% of storage capacity

Figure 14.8*d*: maximum hourly Δt = 6°F [3.3°C]

Then, the cross-ventilation or stack ventilation "rules-of-thumb" can be used to size the openings. See Table 14.2 for a comparison of passive cooling design guidelines and more detailed calculation procedures for night ventilation of thermal mass.

(d) Evaporative Cooling

The most common evaporative cooler is not strictly passive cooling, because it depends on a fan to force large quantities of outdoor air through a wet filter, thereby lowering the air temperature and raising the relative humidity before delivering the air to the space to be cooled. In hot and arid climates, the energy used by the fan in evaporative systems is less than the energy needed to achieve compression-refrigeration cooling. Although this process requires water, it does not use refrigerants that may pose a threat to the Earth's environment.

Before using the following guidelines, be sure to check the passive cooling strategy chart (Figure 14.2) to determine whether evaporative cooling is appropriate for your climate. It is unlikely that evaporative cooling will be helpful in the humid southeastern United States.

The rates of evaporative cooling presented in Figure 14.10 are based upon a rather high airflow of 2.67 cfm per ft^2 of floor area [13.56 L/s m^2]. (In conventional cooling, an airflow rate closer to 1 cfm per ft^2 [5 L/s m^2] is more common.) With this amount of air motion, a highest indoor air temperature of 83°F [28.3°C] is assumed, so the evaporatively cooled air, after absorbing the heat from the space, exits at 83°F [28.3°C]. (Note: This may still be well below the outdoor temperature—might it do some useful cooling?) To use Figure 14.10, first find in Appendix B, Table B.1 the summer design DB and mean coincident WB temperatures for your location. Enter the graph at these two data points, and at their intersection find the approximate amount of heat in Btu/h ft^2 [W/m^2] that evaporative cooling can remove. As WB temperatures surpass 68°F [20.0°C] (represented by dotted lines in the graph), indoor conditions become increasingly humid, producing almost certain discomfort with an indoor WB temperature of 75°F [23.9°C].

A more thorough method for evaluating evaporative cooling potential is presented in Section 14.8, where various airflow rates and temperatures of supply air and exit air can be examined. See Table 14.2 for a comparison of this guideline and the more thorough method.

EXAMPLE 14.3 Evaluate the potential for evaporative cooling for a 3000-ft^2 [279 m^2] retail store in Denver, Colorado. Because of large electric lighting and equipment display loads, the approximate heat gain is 30 Btu/h ft^2 [95 W/m^2].

SOLUTION

From Appendix B, Table B.1, Denver has a summer design dry-bulb temperature of 92.3°F [33.5°C], with a mean coincident wet-bulb temperature of 59.7°F [15.4°C]. Checking back with Figure 14.2, these design conditions fall well within the climate served by evaporative cooling. Next, enter Figure 14.10 at the points of 92.3°F [33.5°C] DB and 59.7°F [15.4°C] WB; the intersection of these data lines shows that under such conditions, heat gains

(a)

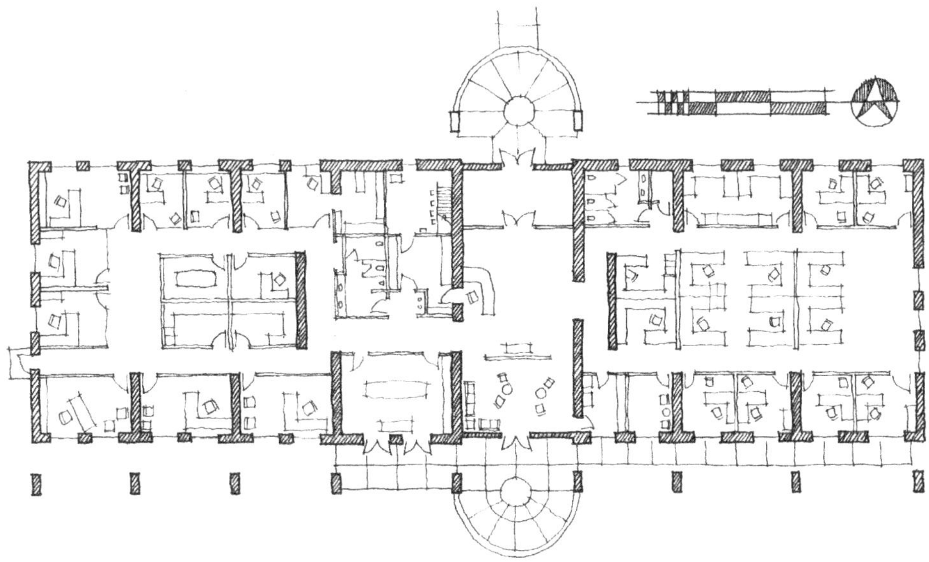

(b)

Figure 14.9 *The Fetzer Winery administration building (a) in its rural Mendocino County, California, setting. The south-facing clerestory and a deciduous vine and trellis are the more visible parts of the daylighting, passive solar heating, and passive cooling strategies. (b) Plan shows the relation of workstations to north and south daylight. (c) Section with ducts for the night air flush as well as for conventional heating and cooling. (d) North-side dormers are for night air intake and exhaust. (Parts b and c redrawn by Dain Carlson from original drawings provided courtesy of Valley Architects.)*

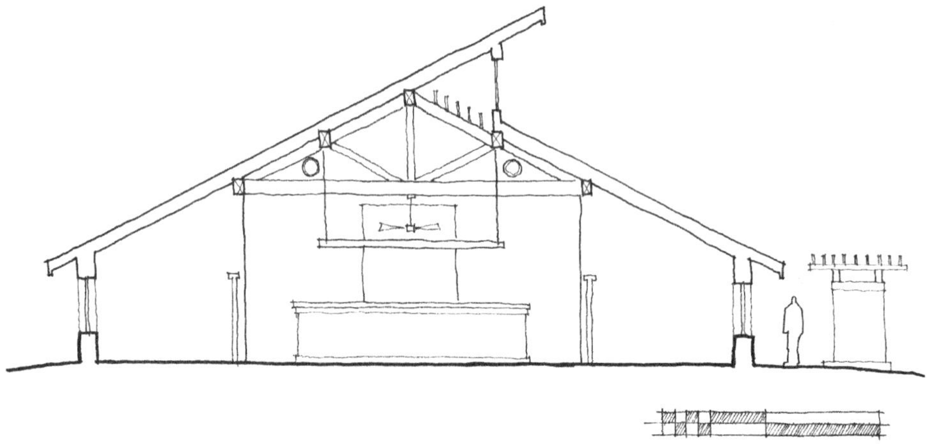

(*c*)

(*d*)

Figure 14.9 *(continued)*

can be removed by evaporative cooling at a rate of about 38 Btu/h ft^2 [120 W/m^2] of floor area.

$$35 \text{ Btu/h ft}^2 \ [110 \text{ W/m}^2] \text{ capacity}$$

$$>30 \text{ Btu/h ft}^2 \ [95 \text{ W/m}^2] \text{ approximate heat gains}$$

Therefore, evaporative coolers can meet the need. Because Figure 14.10 assumes an airflow rate of 2.67 cfm per ft^2 [13.56 L/s per m^2], the cooler's approximate size will be:

$$3000 \text{ ft}^2 \times 2.67 \text{ cfm/ft}^2 = 8010 \text{ cfm}$$

$$[279 \text{ m}^2 \times 13.56 \text{ L/s per m}^2 = 3783 \text{ L/s}]$$

The evaporative cooler size could be reduced by the ratio of cooling capacity to needed heat removal:

$$8000 \text{ cfm} \times \frac{30 \text{ Btu/h ft}^2 \text{ need}}{35 \text{ Btu/h ft}^2 \text{ capacity}} = 6857 \text{ cfm}$$

$$[\text{corresponding to } 3236 \text{ L/s}]$$

Sticking with the larger capacity unit, however, will allow for a lower indoor air exit temperature than the 83°F [28.3°C] assumed in Figure 14.10, so the larger capacity is a safer approximation. ■

(e) Downdraft Cooltowers

A more passive approach to evaporative cooling appears in the downdraft cooltower. This approach was originally developed by researchers at the University of Arizona's Environmental Research Laboratory (ERL) in Tucson, Arizona, as an improvement to the typical wind tower or wind catcher. Their proposal was to shift the direct evaporative cooling process from the lower portion of a wind tower (tower outlet) to its upper portion (tower inlet). As ambient air around the inlet gets evaporatively cooled, it becomes heavier

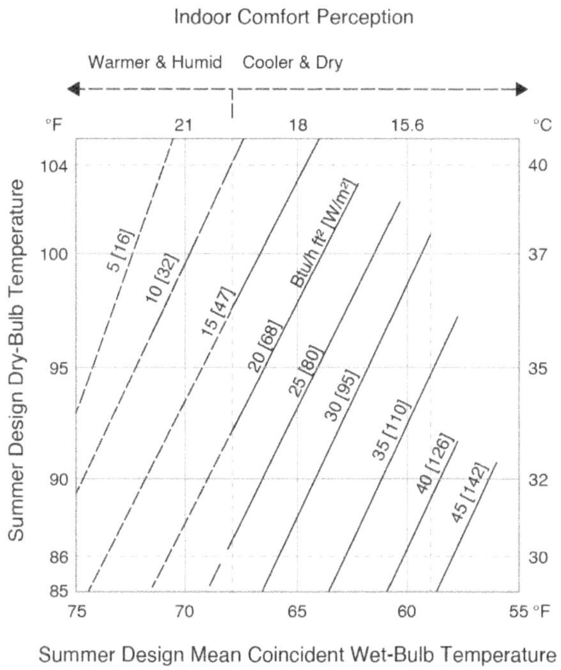

Figure 14.10 *Fan-forced evaporative cooling design guideline for Btu/h removed per ft² of floor area [W per m²]. A forced airflow rate of 2.67 cfm per ft² of floor area [13.56 L/s per m² of floor area] is assumed, as is an exiting (exhaust) air temperature of 83°F [28.3°C]. The dashed lines toward the left side of the graph represent increasingly humid indoor air, which may cause discomfort.*

and more humid, causing it to drop inside the tower structure by gravity and generate air motion without necessarily relying on wind speeds or fans; hence, the name downdraft evaporative cooltower.

The effectiveness of this approach was demonstrated in the University of Arizona experimental residential building shown in Figure 14.11, which was built near the Tucson Airport. Significant temperature drops, up to 30°F [16.67°C], were witnessed. This building used a downdraft cooltower with wetted pads on all four faces at the top. Hot, dry air is cooled as it passes through the pads, dropping to the base of the tower and then into the house. Analysis by Givoni (1994) indicated that such a tower's delivery of wetter, cooler air is almost independent of wind speed and also is not dependent on the second tower at the opposite end of this building, a solar chimney through which air from the house is discharged.

Other evaporative cooling methods that have been used to create a downdraft include showerheads and misting nozzles. With showerheads, large drops of water are sprayed vertically downward. The momentum of the falling water is transmitted to the air stream, creating an airflow down the tower shaft. Unevaporated water is collected at the bottom of the shaft and recirculated back to the showerhead. The advantage of this method over others is that it works with any type of water, including brackish or seawater. This method has been implemented in the downdraft cooltower within the enclosed courtyard of the Blaustein Institutes for Desert Research (Figure 14.12) as well as the four downdraft cooltowers at DPR Construction's Phoenix Regional Headquarters (Figure 14.13). In contrast to showerheads, misting nozzles introduce a very fine mist of water into the air stream and can be designed to ensure full evaporation of water, and thus provide the opportunity for occupants to stand underneath the tower while maximizing cooling output. Many projects, including the Global Ecology Research Center in Figure 14.1, have used this method. Studies have found this method to be the most efficient in terms of cooling output to water consumption as well as the most cost-effective among evaporative cooling methods.

To avoid recurring issues with evaporative cooling, such as accumulation of unevaporated water at the tower base and high humidity levels, some projects have used sensible cooling methods to create the downdraft.

Figure 14.11 *An experimental residence in Tucson, Arizona, is passively cooled by a downdraft evaporative cooltower (at the right in the photograph). Hot, dry air passes through the wetted pads at the tower's top and undergoes a drop in dry-bulb temperature that causes it to fall, entering the residence at the bottom of the tower. The tower at the left is a solar chimney to help increase the flow of warmed exhaust air.*

Figure 14.12 The downdraft cooltower shaft and outlet at the enclosed courtyard of the Blaustein Institutes for Desert Research in the Negev Desert, Israel. (© Wolfgang Motzafi-Haller.)

Figure 14.13 The four downdraft cooltowers along the eastern façade of the DPR Phoenix Regional Headquarters use a combination of showerheads and misters to create the downdraft. (© DPR Construction.)

This has been achieved by installing cooling coils at the tower inlet, such as the downdraft cooltowers along the perimeter of the Conrad N. Hilton Foundation building in Agoura Hills, California (Mead et al., 2013). Multi-stage hybrid systems, which combine evaporative and sensible cooling methods, have been implemented to expand applicability of downdraft cooltowers to a wider range of climate conditions. However, such systems are usually hosted in an atrium enclosure rather than in a tower enclosure. An example of this is the hybrid system in the atrium space at the Malta Stock Exchange in Valletta, Malta (Ford et al., 2020).

The enclosure of a downdraft cooltower has three main variables that impact conditions of the air exiting the tower. These are the tower inlet, the tower shaft, and the tower outlet (Figure 14.14). Ambient air enters the tower through its inlet, which sits on top of the tower shaft. In most of the examples found in the literature, the form of the inlet is a cube with openings on all four surfaces perpendicular to the ground as well as diagonal baffles on the inside. The tower shaft's effective height, which is the distance from the bottom of the inlet to the top of the outlet, mainly impacts air velocity at the tower outlet since most of the cooling happens within the upper 5 ft [1.5 m] of the tower shaft. It is recommended that the shaft height is between double and triple the length of the tower edge in cross-section. The downdraft cooltower at the University of Arizona

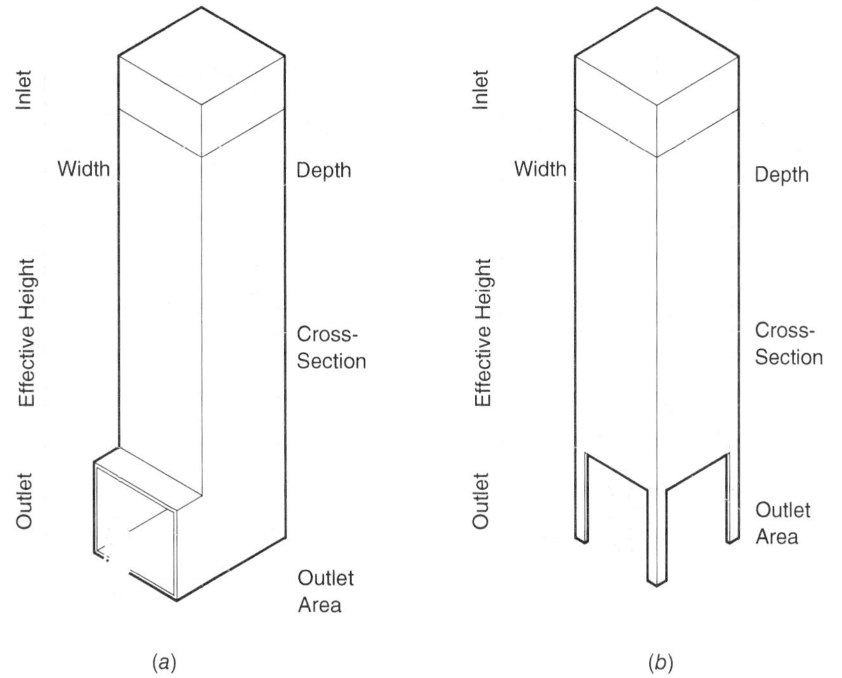

Figure 14.14 *Downdraft cooltowers enclosure major variables. (a) Side discharge cooltower. (b) Vertical discharge cooltower. (© Omar Al-Hassawi.)*

experimental building had an edge length of 6 ft [1.8 m] and an effective height of 15 ft [4.5 m]. The direction of discharge at the outlet, into the space being cooled, is typically to the side as cooled air makes a 90-degree bend into the space it is attached to (Figure 14.14*a*). Vertical discharge is another common approach to supply cooled air, usually used with towers that implement misting nozzles and wetted pads as these can ensure the full evaporation of water (Figure 14.14*b*). It is recommended that the area of the tower inlet openings, tower shaft cross-section, and tower outlet openings are approximately equal.

When using a downdraft cooltower to cool an outdoor space, it is necessary to provide some form of side and top enclosure to shade the space as well as retain the cooled air being supplied by the tower and prevent it from escaping (Figure 14.15). This could be achieved by surrounding the space with vertical elements for spaces located above ground level (Figure 14.15*a*) such as the downdraft cooltower at the Masdar Institute (Figure 14.16). Another approach would be to place the outdoor space below ground level, which results in a pit to contain the cooled air supplied by the tower (Figure 14.15*b*). A proposal developed by researchers from Arizona State University and the University of Arizona to provide passive cooling to ASU's Hayden Library sunken courtyard (Figure 14.17) is an illustration of this approach.

Temperature and flow rate of air delivered to a building by a downdraft cooltower can be approximated from Figure 14.18, but such data are based on very little experimental work. A well-regarded building using this principle, as well as Trombe wall passive solar heating, is the Zion National Park Visitor Center (Figure 14.19).

EXAMPLE 14.4 A 3000-ft^2 [279-m^2] retail store in Las Vegas, Nevada, has a summer heat gain of 14 Btu/h ft^2 [44.2 W/m^2]. How tall should a downdraft cooltower be to provide the needed cooling under summer design conditions?

SOLUTION

From Appendix B, Table B.1, Las Vegas has a summer DB = 106.7°F [41.5°C] with a mean coincident WB = 66.4°F [19.1°C]. This represents a wet-bulb depression (DB − WB) of 40.3°F [22.4°C]. From Figure 14.12*a*, at 104.6°F [40.33°C] DB and 38°F [3.33°C] (DB − WB), the air exiting a downdraft cooltower will be about 70°F [21°C].

Next, determine the amount of supply air at 70°F [21.11°C] needed to maintain this building at an indoor temperature of 82°F [27.8°C] (probably a reasonable indoor temperature when it is over 106°F outdoors, given the air motion provided by this downdraft cooltower).

$$\text{cfm} = (\text{Btu/h})/(1.1)(\Delta t)$$

$$= (3000\ \text{ft}^2 \times 14\ \text{Btu Btu/h ft}^2)/(1.1)(82°\text{F} - 70°\text{F})$$

$$= 3182\ \text{cfm}$$

At the Las Vegas design condition of 106.7°F [41.5°C] DB and 40.3°F [22.4°C] wet-bulb depression,

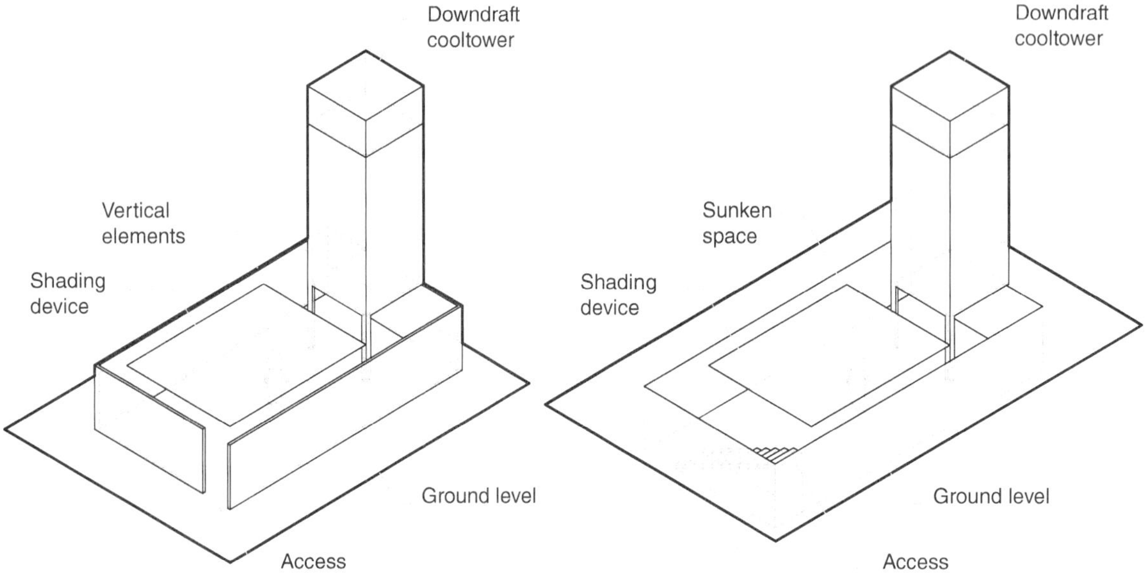

(a) (b)

Figure 14.15 Downdraft cooltower configuration for cooling an outdoor space. (a) Configuration when outdoor space above ground. (b) Configuration when outdoor space below ground. (© Omar Al-Hassawi.)

Figure 14.16 The downdraft cooltower within the courtyard space at the Masdar Institute in Abu Dhabi, United Arab Emirates. (© Nigel Young/Foster + Partners.)

(a) (b)

Figure 14.17 Proposed downdraft cooltowers coupled with operable shading for cooling at Arizona State University Hayden Library's sunken courtyard. During summer days (a), shading is deployed to retain the cooled air supplied by the cooltowers. During summer nights (b), shading is retracted to allow for heat absorbed during the day to radiate to the night sky. (© Omar Al-Hassawi.)

from Figure 14.18c it appears that a cooltower of 40 ft [12 m] in height, with a total wetted pad area of 32 ft^2 [3 m^2], will provide about 3300 cfm [1.42 m^3/s or 1558 L/s], somewhat more than the minimum just calculated. ■

(f) Roofponds

Roofponds are a promising strategy that has rarely been implemented, possibly due to water phobia among architects and clients alike. Yet roofponds demonstrate the most stable interior temperatures of any of these techniques, needing electricity only to open and close the sliding insulation panels on the roof. Roofponds sized for cooling will likely be nearly equal in area to the floors of the buildings they cool. Average pond depth is between 3 and 6 in. [75 and 150 mm]. A more detailed and precise procedure for determining pond area is presented in Section 14.8, but here is a quick approximation to size roofpond areas:

- Set pond maximum temperature: 80°F [26.7°C]; this temperature is based upon the assumption that any higher pond temperature would fail to produce cooling for the building interior.
- Assume pond minimum temperature is the minimum nighttime DB air temperature; = maximum daytime temperature−mean daily range (from Appendix B, Table B.1).
- Establish pond Δt = pond maximum temperature − pond minimum temperature.
- Estimate allowable daily building heat that can be stored in pond; Btu/day ft^2 [kWh/day m^2] = (0.7) (pond Δt, °F) (pond depth, ft) (62.5 lb/ft^3 water) (1.0 Btu/lb°F), with depth not to exceed 0.75 ft [0.23 m] and assuming that 70% of the pond's daily heat gain is from the building below and 30% is through the insulated panels above the pond [in SI units, this relationship is: (0.7) (pond Δt, °C) (pond depth, m) (1000 kg/m^3) (4.18 kJ/kg K)].
- Determine the required "size" of the pond (ft^2 [m^2] of pond per ft^2 [m^2] of floor area).

$$= \frac{\text{building sensible heat gain per day}}{\text{pond allowable heat stored per day}}$$

See Table 14.2 for a comparison of the design guideline and the more detailed calculations for roofponds.

EXAMPLE 14.5 An office building in Albuquerque, New Mexico, is to be cooled by a roofpond 8 in. (0.67 ft) [0.203 m] deep. An hourly heat gain of 15 Btu/h ft^2 [47 W/m^2] is assumed (excluding heat gain through the roof).

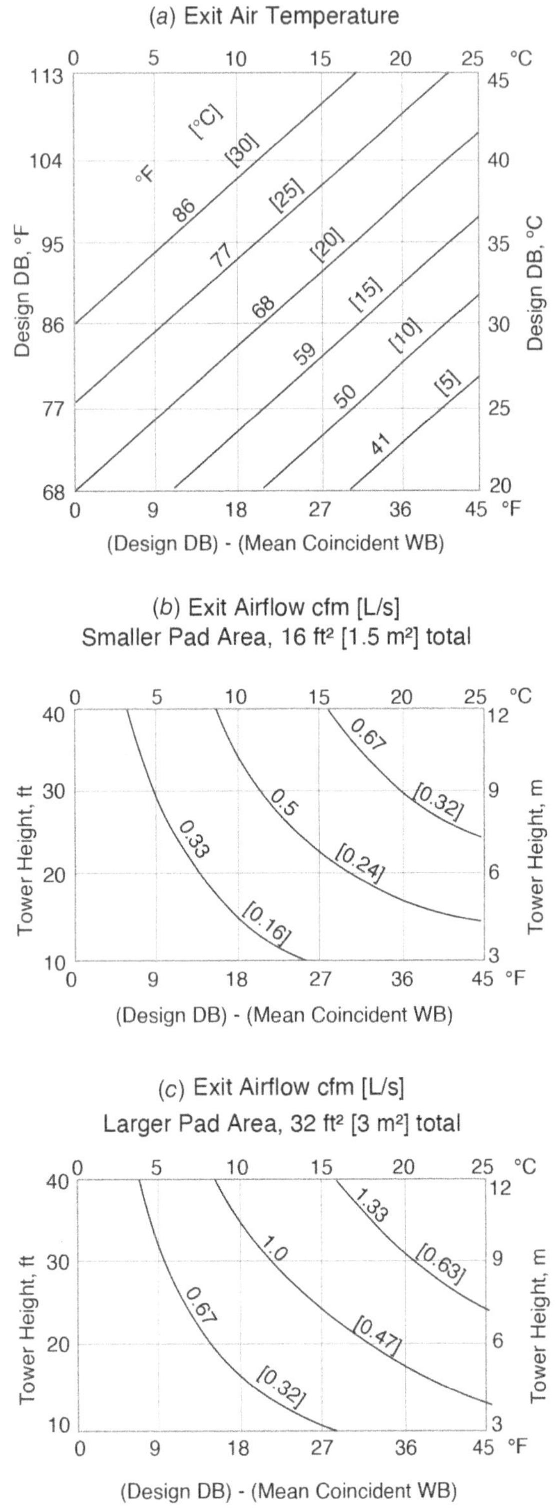

Figure 14.18 *Design guidelines for downdraft cooltowers. (a) The temperature of the air as it leaves the cooltower (enters the building) depends on the outdoor dry-bulb (DB) temperature, as well as the difference between the dry-bulb and wet-bulb (DB-WB) temperatures. The flow rate of the exit air depends on the tower height, as well as the difference between the dry-bulb and wet-bulb (DB-WB) temperatures. (b) Flow rate for a smaller area of wetted pads. (c) Flow rate for a larger area of wetted pads. (Adapted from Givoni, 1994.)*

SOLUTION

For Albuquerque:

maximum design temperature = 93.4°F (Appendix B, Table B.1) [34.1°C (Appendix B, Table B.3)]

mean daily range = 22.9°F (Appendix B, Table B.1) [12.7°C (Appendix B, Table B.3)]

minimum (night) temperature = 93.4 − 22.9 = 70.5°F [21.4°C]

Therefore:

pond Δt = 80°F [26.7°C] maximum −70.5°F [21.4°C] minimum = 9.5F° [5.3C°]

pond storage capacity for building heat:

I-P units:

$$= 0.7\,(9.5\text{F}°)\,(0.67\text{ ft})\,(62.5\text{ lb/ft}^3)\,(1\text{ Btu/lb °F})$$

$$= 279\text{ Btu/day ft}^2$$

SI units:

$$= 0.7\,(5.3\text{C}°)\,(0.203\text{ m})\,(1000\text{ kg/m}^3)\,(4.18\text{ kJ/kg K})$$

$$= 3148\text{ Wh/day m}^2$$

The required pond size, then, is:

$$\frac{\text{hourly gains: }15\text{ Btu/h ft}^2 \times 9\text{ h/day}}{\text{storage capacity: }343\text{ Btu/day ft}^2}$$

$$= 0.48\text{ ft}^2/\text{ft}^2\text{ floor area }[0.13\text{ m}^2/\text{m}^2]$$

Therefore, a 7-in. [178-mm] pond covering 72% of the one-story building's floor area will be approximately large enough to cool the building. (If the entire ceiling is desirable as a cooled surface, then a slightly shallower pond of greater area could be used. However, the sliding insulation panels must be stacked somewhere beyond the edge of the roofpond.) ■

We revisit this building as Example 14.10, in Section 14.8.

The residence in Figure 14.20 is one of the first roofpond structures. Located in Atascadero, California, at about 35°N latitude, it has a remarkable history of thermal stability, providing both passive solar heating in winter and passive cooling in summer. Roofpond inventor Harold Hay donated the building to California State Polytechnic at San Luis Obispo in the late 1990s. To emphasize the pond as the solar heating device, this house has almost no south-facing glass (most windows face east toward a small lake). The sliding insulation panels are stacked when open (by day in winter, by night in summer) over the carport on the north end of the building.

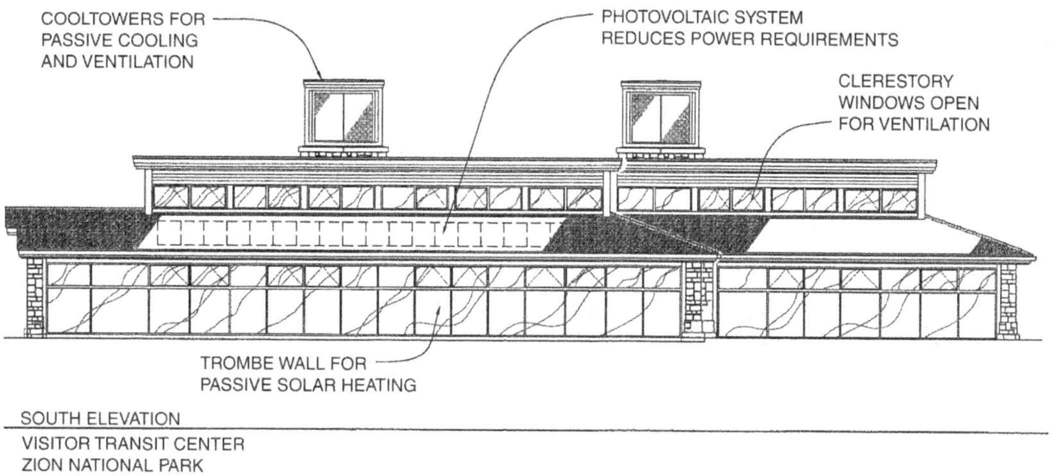

(*a*)

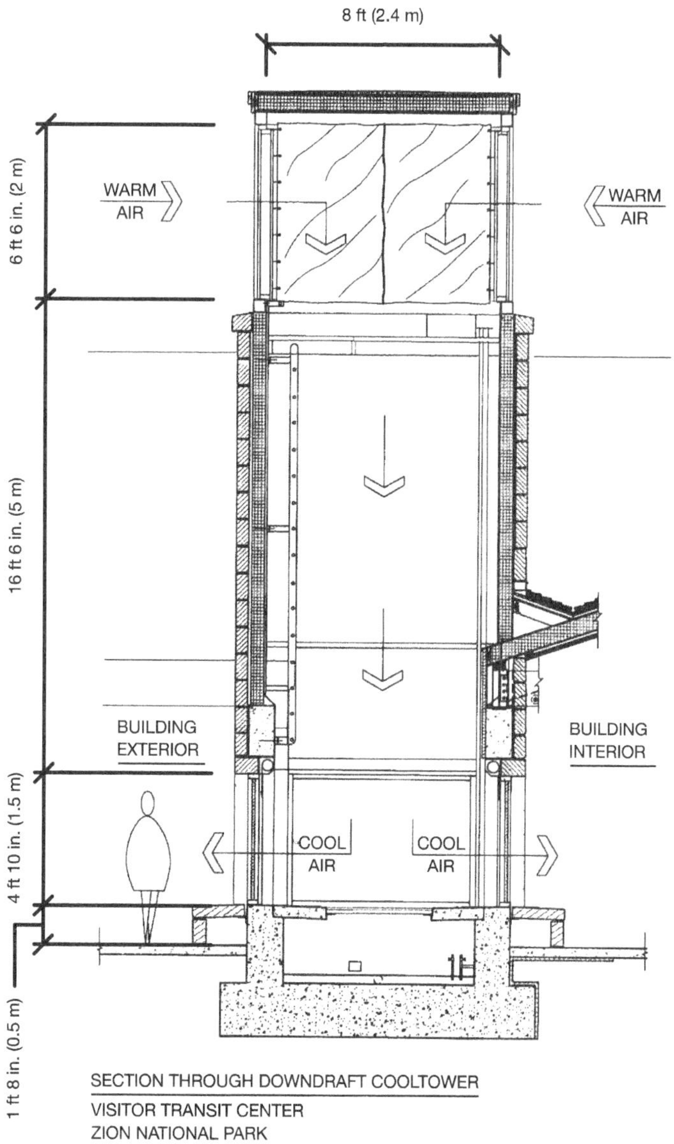

(*b*)

Figure 14.19 *The Visitor Center at Zion National Park (Utah) features both cooling by downdraft cooltowers and solar heating (a) by Trombe walls and a clerestory. Because many of the building's functions can be performed outside, each cooltower (b) has both an indoor and an outdoor outlet, either of which can be closed off when not needed. Air enters through wet media pads at the top of the tower (c). An X-baffle then diverts the air downward. The building also features photovoltaics (PV) and water harvesting. (Courtesy of the National Park Service, Denver Service Center.)*

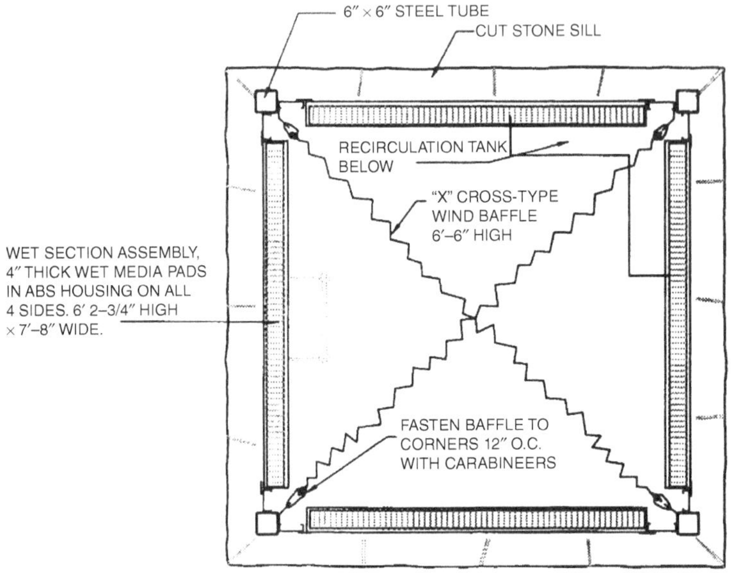

(c)

Figure 14.19 (continued)

(a)

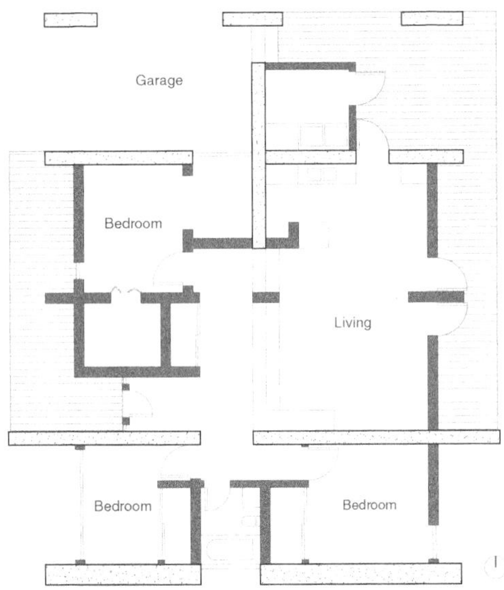

(b)

(c)

(d)

Figure 14.20 *One of the first roofpond buildings, this Atascadero, California, residence is now a research building for California State Polytechnic University at San Luis Obispo. (a) The exterior from the west. (b) Plan. South exposure is minimal; the roof is the surface of interest for heating and cooling. (c) The roof insulating panels in the closed position (as on a summer day). (d) The roof insulating panels in the open position (as on a summer night). Water is contained within plastic bags laid directly on the metal deck ceiling. (e) Sections showing summer day roofpond insulation and night exposure. (f) Sections showing winter day roofpond exposure and night insulation. (g) Year-long record of the indoor temperature range compared to the outdoor range. (Drawings based in part on Stromberg and Woodall, 1977.)*

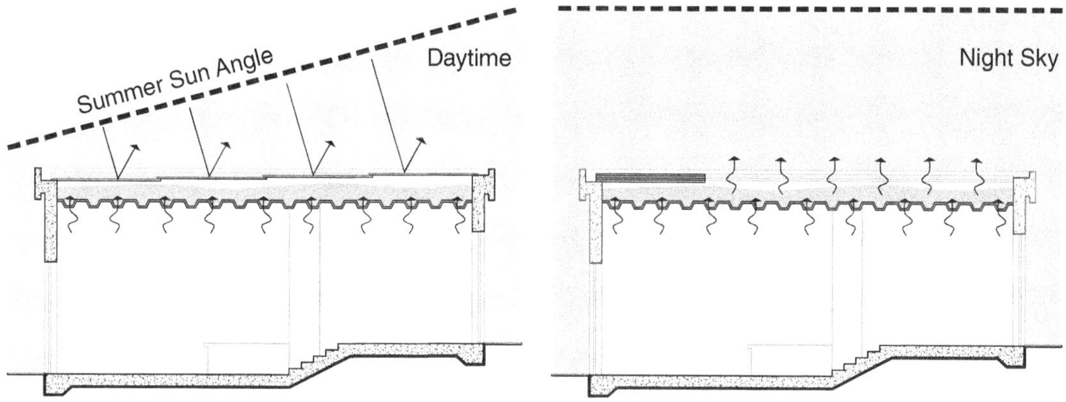

Roof Pond Cycle in Summer

(*e*)

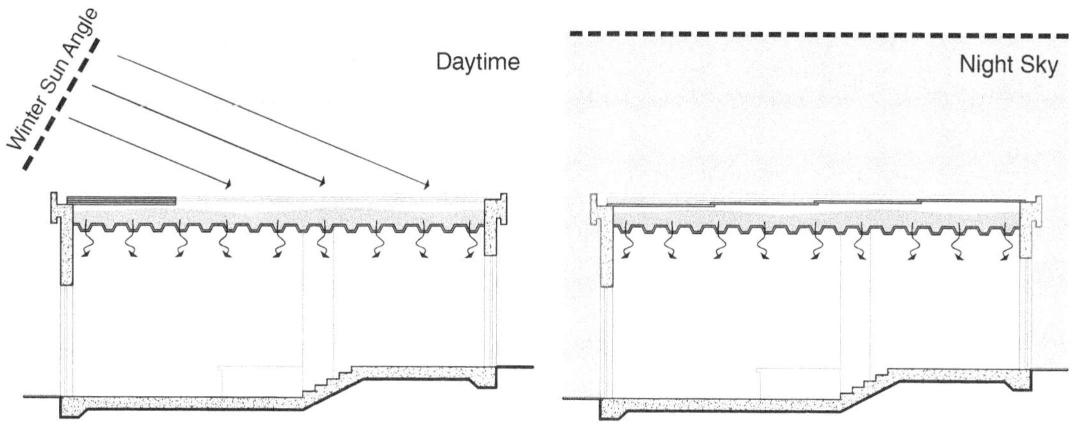

Roof Pond Cycle in Winter

(*f*)

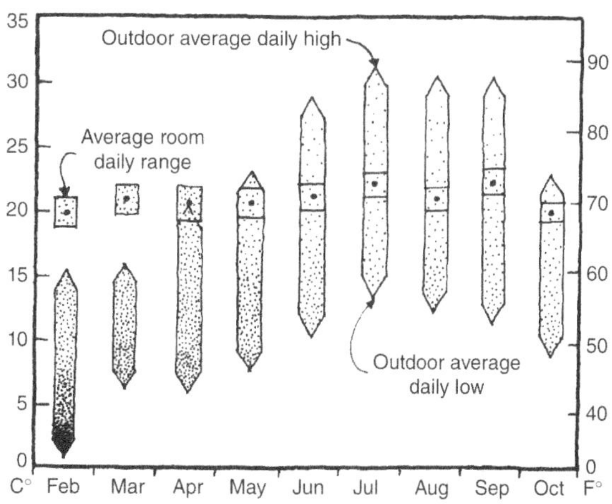

(*g*)

Figure 14.20 *(continued)*

(g) Earth Tubes

These provide a way to cool outdoor air before it enters a building. A fan is used to force sufficient quantities of air through these long tubes. Because earth tubes need to be well underground as well as rather long in order to cool outdoor air, it is rarely economical to install enough earth tubes to completely meet a building's need for cooling. If long trenches are needed for another purpose (underground water lines, for example), an earth tube is more feasible. Where earth tubes are considered, the component of building heat gain that is represented by cooling fresh air (Appendix G, Table G.1, Part E) can sometimes be mitigated using this strategy.

The estimate of cooling for the earth tubes described in Figure 14.21 is based on Abrams (1986), who assumed a soil conductivity equal to that of heavy, damp earth. One of the most influential variables in earth tube performance is soil conductivity, so Figure 14.22 (also from Abrams, 1986) compares the cooling performance of 6-in. [150-mm]-diameter earth tubes when soil conductivity changes. Note that this design guideline (developed using I-P units) presents Btu/h cooling of the outdoor air in the tube rather than Btu/h ft^2 (for a typical ft^2 of building floor area). Therefore, two adjustments must be made. First, find the Btu/h cooling load of the outdoor air alone; second, adjust for the difference between hotter air outdoors and cooler air desired within the building.

See Section 14.8 for a more detailed calculation procedure for earth tubes in which tube depth, soil temperature, and soil conductivity are variable.

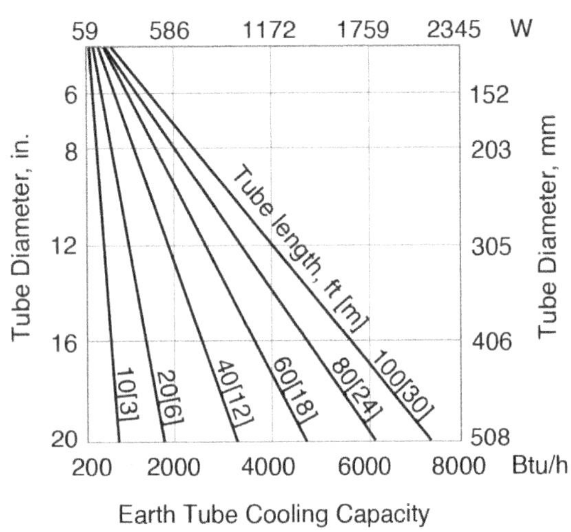

Figure 14.21 *Cooling capacity for airflow through an earth tube as a function of tube diameter and length. These cooling capacities assume heavy, damp soil with a conductivity (k) of 9.0 Btu in./h ft² °F [0.144 W/m K] and a soil temperature around the tube of 65 °F [18.3 °C]. Outdoor air entering the tube is 85 °F [29.4 °C] and flows at a speed of 500 fpm [2.5 m/s]. (Adapted from Abrams, 1986.)*

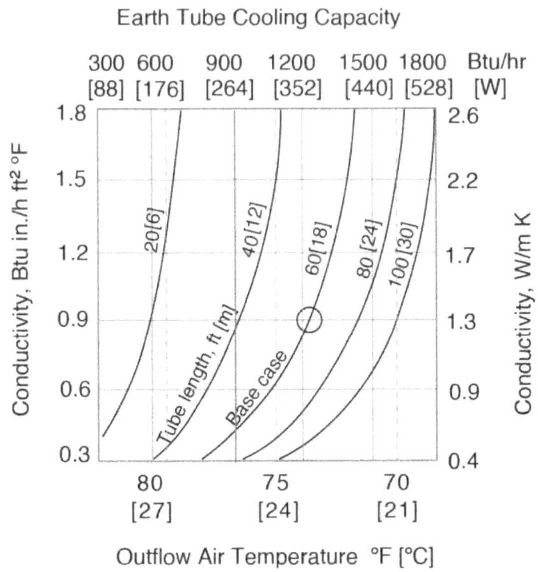

Figure 14.22 *Six-inch-diameter [150 mm] earth tube cooling capacities as soil conductivity varies. The base case is a 6-in.-diameter [150 mm] earth tube 60 ft [18.3 m] long, shown in Figure 14.21 to deliver 1200 Btu/h [352 W] of cooling at soil conductivity of 0.75 Btu/h °F ft [0.11 W/mK], with other conditions as specified in Figure 14.21. (Adapted from Abrams, 1986.)*

EXAMPLE 14.6 A partially underground nature center of 4000 ft^2 [372 m^2] in northern Pennsylvania, design DB = 87°F [65°C], has an hourly heat gain of 9.3 Btu/h ft^2 [29.3 W/m^2], of which 2.2 Btu/h ft^2 [6.9 W/m^2] are due to required outdoor air ventilation. If earth tubes are used only for cooling the fresh air, how many tubes at what length will be needed?

SOLUTION

First, the cooling needed from the earth tubes is (4000 ft^2) (2.2 Btu/h ft^2) = 8800 Btu/h [2580 W]. Next, adjust this figure to account for actual project temperature conditions. If we accept 82°F [27.8°C] indoors (the walls are earth-sheltered and therefore provide some radiant cooling), then the total earth tube cooling needed is:

(cooling need from the earth tubes)

× (design DB/target indoor DB)

(8800 Btu/h) × (87°F/82°F) = 9337 Btu/h [2736 W]

From Figure 14.21, assuming heavy, damp soil, a 65 ft-long [19.8 m] tube 20 in. [500 mm] in diameter will deliver about 5000 Btu/h [1465 W]; two such tubes will deliver about 10,000 Btu/h [2931 W] and exceed the target of 9337 Btu/h [2736 W]. ■

See the more detailed calculation procedures that are given later in Section 14.8. We return to this situation in Example 14.11.

14.5 COORDINATING DAYLIGHTING, PASSIVE SOLAR HEATING, AND COOLING

This section offers guidance on a quick reconciliation of the preliminary investigations (to this point) and presents a "check" of enclosure component U-factors and fenestration/floor area ratios as the design moves through various codes, standards, and design guidelines. Depending on the location, building type, and the client's objectives, a very different mix of glass, insulation, and thermal mass could result from prescriptive conservation approaches compared to a passive solar heating focus or any of several approaches to passive cooling.

When prescriptive standards (such as those presented in Appendix H, Table H.1) appear to preclude either passive solar heating or passive cooling strategies in a building, the designer should remember that there are procedures for either window area trade-offs (in which more insulation in walls or roofs might allow more glass area) or methods to compare whole-building annual energy consumption that can demonstrate the superiority of using well-designed passive strategies over a simple conventional prescribed approach.

EXAMPLE 14.7 Consider a one-story residence in Billings, Montana. How does the fenestration area prescribed for code compliance compare with desirable fenestration areas for daylight, solar heating, and ventilation cooling?

SOLUTION

A residential energy code states that windows that are ≥15% of the floor area must have a maximum U-factor of 0.36 [SI: 2.0] and a maximum SHGC of 0.7. From Appendix E, Table E.15, windows #7 through #12 provide these values for U-factor and SHGC.

We now need to assume the floor area and dimensions in order to proceed. A plan 32 ft [9.8 m] wide by 64 ft [19.5 m] long (with long walls facing north/south) yields a floor area of 2048 ft^2 [190 m^2]. Most windows can be placed in the long walls, and on long roof surfaces, for both daylight distribution and cross ventilation. However, the south wall and roof will need the most glazing area to facilitate solar heating. Tentatively, assign window areas as follows:

south window area	= 8% of floor area
north window area	= 5% of floor area
east and west window area	= 2% (1% each of floor area)
total	= 15%

Daylight: For sidelighting, referencing Chapter 10, Table 10.5, $DF_{avg} = 0.2$ (window area/floor area), where all of the floor area is within 2.5*H* of the window wall. If we assume that the ceiling height at the wall is 8 ft [2.4 m] with a window $H = 7.5$ ft [2.3 m], then the daylight area is 2.5 × 7.5 = 18.75 ft [5.7 m] deep. Light from opposite exterior walls produces the maximum daylight width of 2 × 18.75 = 37.5 ft [11.4 m]. With our stated 32-ft [9.8-m] room width, all floor area is within the daylight zone. For our total of 15% of floor area in fenestration, $DF_{avg} = 0.2 \times 0.15 = 0.03$, or 3%. This seems to be sufficient according to the target DFs in Table 10.3.

Passive Solar Heating: Windows #7 through #12 (Appendix E, Table E.15) are under consideration. Since Window #7 has the highest SHGC of these options at 0.55, it would be the best choice for passive heating. In accordance with Chapter 13, Section 13.4, and using data from Appendix I, Table I.1, we estimate that Billings has a 9-month heating season (i.e., has more than 100 HDD from September through May), giving solar-collecting glazing an optimal tilt of about 50°; that value is found by summing the incident solar radiation on each tilt for the duration of the heating season (i.e., 50° yields 389 kBtu/ft^2, more than any other tilt listed; 90° yield 309 kBtu/ft^2). If 5%, or about 100 ft^2 [9.3 m^2] of the south-facing glazing is placed in tilted skylights, while the remaining 3% (about 60 ft^2 [5.6 m^2]) is allocated to vertical windows unshaded by overhangs, the useful solar gain would be approximately:

$$\text{tilted} = 100\text{ ft}^2 \times (388\text{ kBtu/ft}^2\text{ per year}) \times 0.55\text{ SHGC}$$
$$= 21,340\text{ kBtu/year}$$

$$\text{vertical} = 60\text{ ft}^2 \times (309\text{ kBtu/ft}^2\text{ per year}) \times 0.55\text{ SHGC}$$
$$= 10,197\text{ kBtu/year}$$

$$\text{total useful solar gain} = 21,340 + 10,197$$
$$= 31,537\text{ kBtu [9243 kWh per year]}$$

The next step is to compare this solar gain to the estimated residential heating need in Billings. From Appendix I, Table I.1, we see that Billings typically has 6631 HDD_{65F} [3683 HDD_{18C}] during the nine-month heating season, and from Chapter 13, Figure 13.27 we see that a typical residence in Billings uses about 2.5 kWh of space heating for each HDD_{65F}, or about:

$$6631\ HDD_{65F} \times 2.5\text{ kWh}/HDD_{65F}$$
$$= 16,578\text{ kWh annually}$$

As a rough estimate, therefore, this quantity, type, and configuration of glazing is expected to meet a little more than half (~56%) of the heating need (9243 kWh per year) of this Billings residence. This may be acceptable; if not, the designer can:

- reallocate window area so that essentially all of the glazing faces south (with daylight and cross-ventilation penalties)
- allocate more of the south-facing vertical window area to skylights

- increase the total window area to more than 15% of the floor area, with correspondingly better fenestration U-factors required (from applicable energy code, up to 21% of floor area would require U = 0.29 [SI: 1.65])

Cross Ventilation: First, we need a quick estimate of heat gains. From Appendix G, Table G.1:

A. Internal gains from people and equipment: Assume 4 people at 230 Btu/h [67 W] plus 1200 Btu/h [352 W] from appliances:

$$(4 \times 230) + 1200$$
$$= 2120\ \text{Btu/h}\ [621\ \text{W}]$$
$$2120\ \text{Btu/h}/2048\ \text{ft}^2\ \text{floor area}$$
$$= 1.04\ \text{Btu/h ft}^2\ [3.3\ \text{W/m}^2]$$

B. Electric lighting gain: Assume with 3% DF a gain of about 2.0 Btu/h ft^2 [6.3 W/m^2].

C. Gain through enclosure: Billings's design temperature (Table B.1) is 88.4°F [31.3°C]. Use 90°F [32°C] reference in Table G.1.

Gains through externally shaded window:

$$(0.15) \times 16 = 2.4\ \text{Btu/h ft}^2\ [7.6\ \text{W/m}^2]$$

Gains through opaque walls:

Total vertical area = [(2 × 32 ft) + (2 × 64 ft)] = 1536 ft^2 [143 m^2]

Of this area, the window area = (0.15 × 2048) = 307 ft^2 [28.5 m^2]

Therefore, the opaque wall area = 1536–307 = 1229 ft^2 [114 m^2]

To meet the energy code, frame walls must have a U-factor no greater than 0.073 [SI: 0.4], whereas masonry walls may have a maximum U-factor of 0.08 [SI: 0.45].

Therefore, gains through opaque walls:

$$(1229\ \text{ft}^2/2048\ \text{ft}^2) \times 0.08 \times 15$$
$$= 0.72\ \text{Btu/h ft}^2\ [2.3\ \text{W/m}^2]$$

Gain through roof: Assume roof area equal to floor area (2048 ft^2), and to comply with the energy code, a maximum U-factor of 0.033 [SI: 0.19].

Therefore, gain through roof:

$$(1.0) \times 0.033 \times 35 = 1.16\ \text{Btu/h ft}^2\ [3.7\ \text{W/m}^2]$$

D. For an "open building" using cross ventilation:

$$\text{Total heat gains} = 1.04 + 2.0 + 2.4 + 0.72 + 1.16$$
$$= 7.32\ \text{Btu/h ft}^2\ [23.1\ \text{W/m}^2]$$

We can assume a total operable area of perhaps 90% of the glazed area using fully openable casement windows. With 8% floor area in south windows and 5% in north windows, the maximum available ventilation area is 90% of the smaller percentage; or 0.9 × 5% of floor area constituting inlet = 4.5%. From Figure 14.5, it appears that this rate of heat gain—7.32 Btu/h ft^2 [23.1 W/m^2]—can be removed through this inlet area by a wind velocity of only about 2 mph [3.2 km/h].

Summary. It is recommended to increase the south glass area (and the overall ratio of window/floor area) and use windows with lower U-factors, with summer external shading. The floor and many interior wall surfaces will be faced with thermally massive materials. ■

14.6 APPROXIMATE METHOD FOR CALCULATING HEAT GAIN (COOLING LOAD)

Unlike the calculations for winter worst-hour heat loss, which simply assume nighttime conditions and few if any internal gains, summer worst-hour heat gain calculations are very complex. During the summer, worst-hour heat gain is assumed to occur during the daytime under active occupancy conditions, so solar gains and internal gains from lights, people, and equipment must be included. Summer calculations are complicated further by the fact that the hourly change in summer load can be very great, both from changing sun position and from changing internal loads. Also, in summer the thermal mass of the building becomes influential, delaying the impact of the radiant component of heat gains from all sources.

Before making any detailed calculations for passive cooling performance, consider a simplified heat gain procedure that has been developed for residential buildings; with some risk and some judgment, it can also be used for a quick approximation of the conditions in commercial buildings. This simplified method was devised to be used with buildings that, like residences, share the following conditions:

1. Are occupied and air-conditioned (internal temperatures closely controlled) for 24 hours per day. (Separate calculations can be done for weekdays and weekends.)
2. Derive much of their gains through the building envelope and ventilation rather than internally.
3. Can accept the effects of undersized cooling equipment, with the result that unusually hot weather means noticeably higher indoor temperatures (and that interior temperatures will vary during a typical summer day).

Because many commercial buildings do not have these characteristics, this method should not be applied if very accurate estimates are desired for such buildings. However, the method is rapid, if risky, and though different methods are offered by ASHRAE, we include this method as an easy way to gain understanding of the necessary inputs to more complex calculations and software programs.

(a) Gains Through Roof and Walls

The sensible heat gains through opaque parts of a building's enclosure are calculated with the equation:

$$q = U \times A \times \text{DETD}$$

where

U = U-factors for summer
A = area of the roof or wall
DETD = design equivalent temperature difference

Design equivalent temperature differences (DETD) are listed for broad categories of construction in Appendix G, Table G.2. The values are based on an average indoor temperature of 75°F [23.8°C], and the outdoor conditions listed. Note that for lightweight wall construction (and doors) the DETD varies by orientation. A means of correcting DETD for other temperatures is as follows.

Where the design temperature difference (outdoor design temperature minus assumed indoor temperature of 75°F [23.8°C]) is not an even increment of 5°F [2.8°C], the equivalent temperature difference should be corrected 1°F [0.5°C] for each 1°F [0.5°C] difference from the tabulated values.

For rapid approximation, however, select the DETD directly from the table for the conditions nearest to your building/climate combination. The design temperature for your climate is listed in Appendix B.

Note that Appendix G, Table G.2 lists considerably higher DETD values for roofs than for any other component. This is due to the sol-air temperature, the effective temperature of outdoor air just above a solar-heated surface. When the sun strikes the surface, it adds its heat to that from the outdoor air. The darker the surface, the greater the resulting effective (sol-air) temperature. This elevated outdoor temperature thus increases the Δt through the roof, resulting in greater heat gains. The sol-air formula was discussed in Chapter 9. Sol-air temperature is discussed in the *2021 ASHRAE Handbook—Fundamentals*, Chapter 18 (ASHRAE, 2021). Uninsulated or poorly insulated roofs of darker colors are particularly impacted by sol-air temperature.

(b) Gains Through Glass

The quick way to approximate these gains is:

$$q = A \times \text{DCLF}$$

where

A = area of glass
DCLF = design cooling load factor

Design cooling load factors (DCLF) values are listed in Appendix G, Table G.3 and include the U-factors as well as the equivalent temperature differences. The values do not correspond to the worst-hour gains; they were obtained by averaging the hours from 5:30 a.m. to 6:30 p.m. at both 30°N and 40°N latitudes. Again, the DCLF values were based on an indoor temperature of 75°F [23.8°C] and outdoor temperatures as listed in Table G.3. (For rapid approximation, ignore the corrections procedure shown in note (a).) Glass protected by exterior shading devices that exclude all direct sun may be considered equal to the values listed in Appendix G, Table G.3 for north glass shaded by awnings.

(c) Gains from Outdoor Air

In residences, outdoor air often enters by infiltration. In many other buildings, codes require the deliberate introduction of outdoor air by mechanical ventilation. Whichever way outdoor air enters a building in summer, the sensible heat gain can be calculated by either

$$q_{\text{infiltration}} = (A_{\text{exposed}})\,(\text{infiltration factor})$$

where A_{exposed} is the total area of exposed wall surface, including windows and doors, and the infiltration factor is found from Appendix G, Table G.4, or

$$q_{\text{mechanical ventilation}} = (Q)\,(\text{ventilation factor})$$

where Q is the volume of outdoor air (cfm [L/s]; see Chapter 5), and the ventilation factor is found from Appendix G, Table G.4. Latent loads will be addressed later.

(d) Gains from People

Only the sensible gains are tabulated, because a simple overall factor for latent gains is included later. The rate of heat gain from people in various activities is shown in Appendix G, Table G.5; values in the "Sensible Heat" column are generally used. (For residences, sensible heat gain per occupant is often assumed at 230 Btu/h [67 W]):

$$q_{\text{people}} = (\text{number of occupants}) \times (\text{sensible gain per occupant})$$

(e) Gains from Lighting

The power supplied to electric lamps (those that normally are on while cooling equipment is functioning) can be added directly to the sensible heat gain. Be sure to include ballast heat gains along with fluorescent lamps, usually done by taking from 1.12 to 1.2 times the nameplate wattage of such lamps (use the lower figure with energy-efficient ballasts).

(f) Gains from Equipment

In residences, a standard assumption is that 1200–1600 Btu/h [350–470 W] of sensible heat gain is produced by appliances. (Other residential heat loads are assumed to be vented.) For other buildings, see the gains of specific pieces of equipment given in Appendix G, Tables G.6 and G.7.

(g) Latent Heat Gains

Latent heat gains vary a great deal with the type of occupancy, but this simple method assumes that the latent gains are closely associated with outdoor air infiltration. Figure 14.23 is a method to estimate additional latent heat as a percentage of total sensible gains. The design DB and mean coincident WB temperatures from Appendix B, as well as the relative tightness of building construction, determine the latent percentage of total sensible gain. A minimum of 10% and a probable maximum of 30% additional latent are recommended limits.

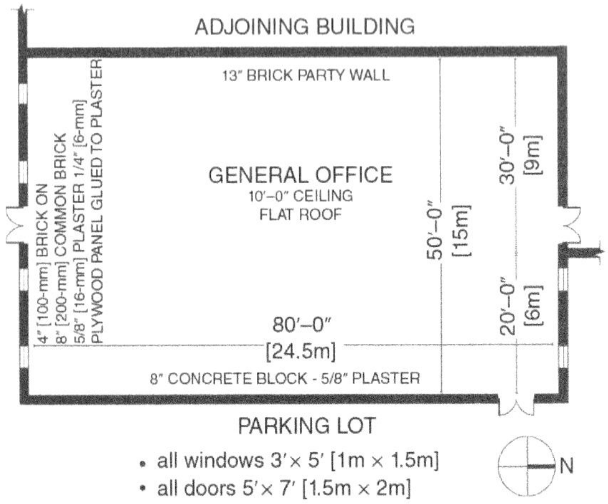

Figure 14.24 *Plan of the office building for which heat gain calculations are shown in Example 14.8 (and in Table 14.3). (Drawn by Tyler Mavichien with revisions by Evelyn Dinh.)*

EXAMPLE 14.8 A 4000 ft² [371.6 m²] one-story office building (Figure 14.24) is located in the eastern United States near 40°N latitude. The adjoining buildings on the north and west are not conditioned, and their inside air temperatures are, for simplicity, assumed equal to the outdoor air temperature at any time of day. This is an unusual building both for its uninsulated walls and for very high lighting loads. What are the heat gains, hence the cooling load?

Roof construction: 4.5-in. [115-mm] flat roof deck of 2-in. [50-mm] gypsum slab on metal roof deck, 2-in. [50-mm] rigid above-deck roof insulation, surfaced with two layers of mopped felt vapor-seal built-up

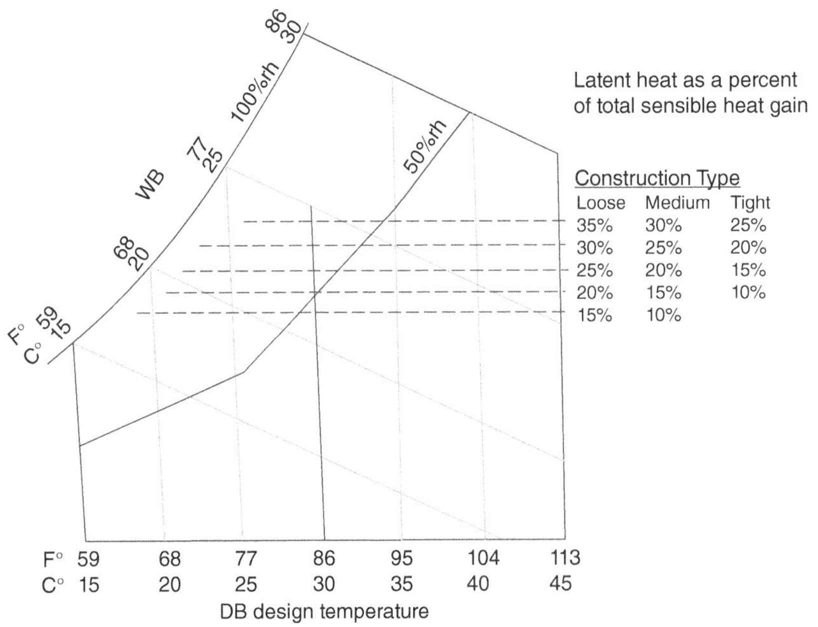

Figure 14.23 *Estimating latent heat gain as a percentage of total sensible heat gain, assuming that these additional latent gains are closely associated with outdoor air infiltration. Relative tightness of building construction is included as a variable.*

PASSIVE ENVIRONMENTAL SYSTEMS

roofing having dark-colored gravel surface, no false ceiling.

Summer U-factor = 0.09 Btu/h ft²°F [0.51 W/m² °C]

South wall construction: 4-in. [100-mm] face brick, 8-in. [200-mm] common brick, 0.625-in. [16-mm] plaster, 0.25-in. [6-mm] plywood panel glued on plaster.

Summer U-factor = 0.24 Btu/h ft²°F [1.36 W/m² °C]

West wall and adjoining north party wall construction: 13-in. [325-mm] solid brick, no plaster.

Interior partition U-factor

= 0.25 Btu/h ft²°F [1.40 W/m² °C]

North exposed wall and east wall construction: 8-in. [200-mm] concrete block and 0.625-in. [16-mm] plaster.

Summer U-factor = 0.48 Btu/h ft² °F [2.73 W/m² °C]

Floor construction: 4-in. [100-mm] concrete slab-on-grade

Fenestration: 3-ft × 5-ft [1-m × 1.5-m] non-operable windows of regular plate glass with light-colored venetian blinds.

Summer U-factor = 0.81 Btu/h ft²°F [4.6 W/m² °C]

Front doors: Two, 2.5 × 7 ft [1.5 m × 2 m]

Side doors: Two, 2.5 × 7 ft [1.5 m × 2 m]

Rear doors: Two, 2.5 × 7 ft [1.5 m × 2 m], interior

Door construction: Light-colored 1.75-in. [45-mm] steel door with solid urethane core and thermal break.

Summer U-factor

= 0.18 Btu/h ft² °F [1.08 W/m² °C]

U-factors for all doors and outside walls were calculated assuming a wind speed of 7.5 mph [12 km/h]. For party and inside walls, still air was assumed.)

Outdoor design conditions: Dry-bulb temperature, 94°F [35°C]; mean daily range, 20°F [11°C]; wet-bulb temperature, 77°F [25°C].

Indoor design conditions: Dry-bulb temperature, 75°F [24°C]; wet-bulb temperature, 62.5°F [18°C]

Occupancy: 85 office workers from 8:00 a.m. to 5:00 p.m. (nine hours)

Lighting: 17,500 W fluorescent total from the 8:00 a.m. to 5:00 p.m. workday; plus 4000 W tungsten, continuous; totaling 21,000 W/day

Equipment: This example does not include any office equipment.

Ventilation: The ventilation rate is 15 cfm [7 L/s] per person, for a total of 1275 cfm [595 L/s].

The conditioning equipment is located in the adjoining building to the north.

Determine the sensible, latent, and total space cooling load at design conditions.

SOLUTION

Before beginning the calculations, estimate the heat gain using the design guideline procedure of Appendix G, Table G.1.

People (omitting equipment in this example):

$$2.3 \text{ Btu/h ft}^2 \times 2^a = 4.6 \text{ Btu/h ft}^2 \ [14.5 \text{ W/m}^2]$$

Lighting[b]:

$$21{,}500 \text{ W}/4000 \text{ ft}^2 = 5.4 \text{ W/ft}^2 \times 3.41 \text{ Btu/h per W}$$

$$= 18.4 \text{ Btu/h ft}^2 \ [58.0 \text{ W/m}^2]$$

Enclosure gains:

Windows:	90 ft²/4000 ft² × 18[c] = 0.4 Btu/h ft²
Walls:	
South:	405 ft²/4000 ft² × 0.24 × 19[c] – 0.5 Btu/h ft²
East, north exposed:	935 ft²/4000 ft² × 0.48 × 19[c] = 2.1 Btu/h ft²
Party:	1065 ft²/4000 ft² × 0.25 × 19[c] = 1.3 Btu/h ft²
Roof:	4000 ft²/4000 ft² × 0.09 × 39[c] = 3.5 Btu/h ft²
Ventilation:	1275 cfm/4000 ft² × 21[c] = 6.7 Btu/h ft²
Envelope total:	14.5 Btu/h ft² [45.8 W/m²]
Total gains:	37.5 Btu/h ft² [118.3 W/m²]

Notes:

[a] 4000 ft²/85 people = 47 ft² per person, more than double the heat gain at a lower density of 100 ft² per person that is assumed in this approximate procedure, so increase people gain by a factor of 2.

[b] This is considerably in excess of the lighting loads upon which this approximate procedure is based. Without a detailed knowledge of the actual installed lighting loads, the design guideline would have suggested a lower load, as follows:

$$\text{DF} = 0.2 \times \frac{90 \text{ ft}^2 \text{ window area}}{4000 \text{ ft}^2 \text{ floor area}} = 0.5\% \text{ DF}$$

which is <1, therefore suggesting a load of only 5.1 Btu/h ft² (16.1 W/m²).

[c] Using Appendix G, Table G.1, Part C, interpolate for 95°F [35°C], between 90°F [32.2°C] and 100°F [37.8°C] outdoor temperature.

Cooling load due to heat gain through building enclosure: The heat gains through roof, exposed walls, and doors shown in Table 14.3 were calculated by:

$$q = U \times A \times \text{DETD}$$

for which DETD values are taken from Appendix G, Table G.2. In line with the approximate (and therefore rapid) nature of this calculation method, no corrections were made to DETD values to adjust for either outside or inside design temperatures differing from those listed, and actual gains were rounded to the nearest 10. An outside design temperature of 95°F [35°C] was used. The climate's daily temperature range is assumed medium (i.e., reference "M" in Appendix G, Table G.2). The roof is dark (typical of commercial buildings in air-polluted areas). To keep

TABLE 14.3 Cooling Load Through Building Enclosure (Example 14.8)

		I-P Units						SI Units					
Element	**Reference, Appendix G**	**Net Area (ft^2)**	**U-factor (Btu/h ft² °F)**	**Δt (°F)**	**DETD (°F)**	**DCLF**	**Cooling Load (Btu/h)**	**Net Area (m^2)**	**U-factor (W/m² K)**	**Δt (°C)**	**DETD (°C)**	**DCLF**	**Cooling Load (W)**
Roof	Table G.2	4000	0.09		44.0		**15,840**	372	0.51		24.4		**4620**
South wall	Table G.2	405[a]	0.24		16.3		**1580**	37.6[a]	1.36		9.1		**470**
East wall	Table G.2	765[a]	0.48		16.3		**5990**	71.1[a]	2.7		9.1		**1750**
North wall, exposed	Table G.2	170[a]	0.48		16.3		**1330**	15.8[a]	2.7		9.1		**390**
Party walls		1065[a]	0.25	20[b]			**5330**	98.9[a]	1.41	11.2[b]			**1560**
Doors:													
S	Table G.2	35	0.19		23.6		**160**	3.25	1.08		13.1		**50**
N	Table G.2	35	0.19	20[b]			**130**	3.25	1.08	11.2[b]			**40**
E	Table G.2	35	0.19		23.6		**160**	3.25	1.08		13.1		**50**
Windows:													
S	Table G.3	60	(0.81)[c]			31	**1860**	5.6	(4.6)[c]			97.9	**550**
N	Table G.3	30	(0.81)[c]			23	**690**	2.8	(4.6)[c]			72.6	**200**
						Total:	**33,070**					Total:	**9680**

[a]Calculated from gross wall area less windows and door areas.

[b]Design temperature difference, indoor to outdoor.

[c]DCLF for glass includes the U-factor.

things simple, the temperature difference through the party walls was made equal to that on which the table was based; in this case, 95 − 75 = 20°F [35 − 23.8 = 11.2°C].

Cooling load due to heat gain through glass: These heat gains were calculated by:

$$q = A \times \text{DCLF}$$

for which the DCLF values (which include the U-factor) were taken from Appendix G, Table G.3. Again, no corrections were made, and gains were rounded to the nearest 10.

Cooling load due to heat gain from lighting, people, and equipment: The rate of heat gain from the lighting can be approximated simply by taking the energy input (including that required for fluorescent ballasts). For fluorescent lamps, 17,500 W × 1.2 ballast factor:

$$= 21{,}000\ \text{W}\ [71{,}650\ \text{Btu/h}]$$

For tungsten lamps:

$$= 4000\ \text{W}\ [13{,}650\ \text{Btu/h}]$$

The sensible gains from people can be determined from Appendix G, Table G.5. For office work, assume 75 W [250 Btu/h]:

$$85\ \text{people} \times 75\ \text{W} \approx 6380\ \text{W}[71{,}650\ \text{Btu/h}]$$

Heat gains from office equipment are ignored in this example. (See Appendix G, Tables G.1 and G.6 for typical ranges of such gains.) The sensible gains from lighting, people, and equipment thus total 106,550 Btu/h [31,380 W].

Cooling load due to ventilation or infiltration: Because this is a commercial building incorporating deliberate introduction of outdoor air, infiltration will be ignored. This example assumes ventilation at 15 cfm [7 L/s] per person. Total mechanical ventilation is 1275 cfm [595 L/s].

Sensible heat gains due to mechanical ventilation (from Appendix G, Table G.4):

$$22.0\ \text{Btu/h per cfm} \times 1275\ \text{cfm} = 28{,}050\ \text{Btu/h}$$

$$[13.6\ \text{W per L/s} \times 595\ \text{L/s} = 8092\ \text{W}]$$

Latent heat gains: The climate description suggests a typical northeastern US climate at 40°N latitude. Newark, New Jersey (91.1°F DB, 72.8°F WB [32.8°C, 22.7°C]), and Philadelphia, Pennsylvania (90.8°F DB, 73.9°F WB [32.7°C, 23.3°C]), are examples. From Figure 14.23, estimate latent gain as about 20% of the total sensible load.

	W	**Btu/h**
Sensible gains, enclosure	9680	33,070
Sensible gains, lights and people	31,380	106,550
Sensible gains, ventilation	8090	28,050
Total sensible heat gains	**49,150**	**167,670**
+ latent gains (20%, this example)	9830	33,530
Total latent and sensible heat gains	**58,980**	**201,200**

■

14.7 DETAILED HOURLY HEAT GAIN (COOLING LOAD) CALCULATIONS

Ordinarily, the approximate method of calculating the peak cooling load (discussed in Section 14.6) is sufficient for a designer's preliminary estimates of cooling equipment capacity for a small building. However, more detailed procedures are used by engineers to actually size equipment and to assess the peak-load impact of various design options such as shading devices.

Three detailed methods for calculating heat gains and equipment cooling loads are described in Chapter 28 of the *1997 ASHRAE Handbook—Fundamentals* (ASHRAE, 1997):

1. the transfer function method (TFM)
2. the total equivalent temperature differential, time-averaging method (TETD/TA)
3. a one-step method, cooling load temperature difference/cooling load factor (CLTD/CLF)

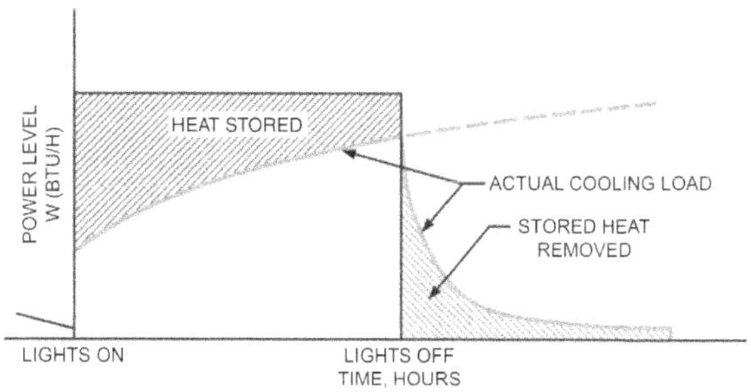

Figure 14.25 *Thermal storage effect on the cooling load caused by electric lighting. (Author/with permission of ASHRAE, www.ashrae.org, 1997. This citation to an older version of the Handbook is intentional and provides access to historic reference information of ongoing interest.)*

Both TFM and TETD/TA yield hourly values over a 24-hour day. They account for thermal storage by building mass, which can significantly shift the impact of instantaneous heat gains on the actual cooling load for the HVAC equipment, as shown in Figure 14.25. By comparison, CLTD/CLF yields a one-hour value that also accounts for the effects of thermal mass.

Though these methods are rarely used by architects, a summary of the results of the calculations by each method for the office building of Example 14.8 is shown in Table 14.4.

The quick and uncomplicated heat gain methods shown in Table G.1 and Section 14.6 served the simple building in Example 14.8 rather well. Had the design guideline (Appendix G, Table G.1) been used to size a cooling system, it would have matched the TFM worst-hour sensible gains, but it was only about 84% of the other methods' worst-hour sensible gains. Latent gain estimates would have been necessary as well. The approximate method (Section 14.6) underpredicted latent gains, but the total (latent and sensible) gains of this quick, by-hand method ranged from 83% to 95% of the computer-based methods' results.

As a quick approximation for estimating the overall equipment size or passive cooling approach, these by-hand methods are adequate for smaller and simpler buildings. When utilizing these quicker methods, keep in mind that some design judgment may require subdividing the calculations or even changing some of the multipliers to reflect unusual conditions.

TABLE 14.4 Comparison of Methods of Heat Gain Calculation (Example 14.8)

I-P Units							SI Units					
Maximum Hourly Gain (Btu/h)			24-h Average Gain (Btu/h ft²)				Maximum Hourly Gain (W)			24-h Average Gain (W/m²)		
Sensible	Latent	Total	Sensible	Latent	Total	Method	Sensible	Latent	Total	Sensible	Latent	Total
37.5/ft²						Design guideline (Appendix G, Table G.1)	118.3/m²					
167,670 (41.9/ft²)	33,530	201,200				Approximate[a] (Section 14.6)	49,150 (131/m²)	9830	58,980			
149,623 (37.4/ft²)	61,168	210,791	21.3	14.5	33.8	TFM[b] (worst hour 4:00 p.m.)	41,415 (110/m²)	15,040	56,455	68.3	32.0	100.3
179,140 (44.8/ft²)	63,208	242,348				CLTD/CLF[b] (for 4:00 p.m. only)	53,499 (143/m²)	15,040	68,489			
169,082 (42.3/ft²)	61,168	230,250	21.3	14.5	33.8	TETD/TA[b] (worst hour 4:00 p.m.)	50,096 (134/m²)	15,040	65,136	68.3	32.0	100.3

[a]See Table 14.3 and Example 14.8 for calculations.

[b]Previously extracted, with permission of ASHRAE (www.ashrae.org), from Chapter 28 in ASHRAE, 1997. This table is maintained for the purpose of this example; recently ASHRAE has developed a more complex calculation.

A designer's checklist of considerations for heat gain should include:

1. characteristics of the building enclosure: materials, sizes, external surface colors, and shapes
2. building location and orientation, as well as the extent of external shading of the building by trees or adjacent structures
3. outdoor design conditions
4. indoor design conditions: DB, WB, and ventilation rate
5. the schedule of lighting, occupancy, equipment, and any other processes that contribute to internal heat gain
6. thermal zoning requirements

14.8 DETAILED CALCULATIONS: PASSIVE COOLING PERFORMANCE

When outdoor air is 85°F [30°C] or below, it is marginally possible to cool buildings by simple ventilation, maintaining conditions indoors that are within the adaptive thermal comfort zone. The procedures in this section go beyond the quicker design guidelines that appeared in Section 14.4 and allow the designer to better adjust a preliminary design to the program and the climate.

In the many climates in which the outdoor temperature is above 85°F [30°C] for a large number of working-day hours, buildings that are closed to the hot exterior during those hours are practical. Between 80° and 85°F [27° and 30°C], outdoor air can keep people within the comfort zone if it moves across the body fast enough. In areas with reliable winds, open buildings are feasible up to 85°F [30°C]. Between 55° and 69°F [13° and 21°C], outdoor air is cool enough to use in place of, or to greatly supplement, mechanically cooled air if the humidity is sufficiently low. Passive cooling is even easier under these conditions.

To introduce these more detailed methods, consider the United Kingdom Pavilion (now dismantled) at the 1992 Seville World Expo (Figure 14.26). This was an unusually large passive building (213 ft × 105 ft × 82 ft high [65 m × 32 m × 25 m high]), with an unusually wide collection of passive cooling techniques, inspired by the traditional cooling methods of hot, dry southern Spain: shading, water, and mass. The expo was an April to October event, with the greatest comfort challenge in July–August, when average daily temperatures range from about 64° to 102°F [18° to 39°C].

Shading began with the flat roof, where standing wave-form racks carried both translucent fabric shades and PV cells facing south. Outside the south façade, projecting sail-like "fly-sheets" kept off the direct sun. The east and west façades used different strategies.

Water became a theme of the exhibit, beginning with the wavelike silhouette on the roof and featured as a continuous sheet of water streaming down the east façade, dripping into a pool. The water was pumped, with some power (about 50%) provided by the PV cells on the roof racks. Waiting visitors standing in line outside were treated to the sight and sound of this controlled waterfall, then crossed a bridge over the pond, flanked by spray fountains, to enter the Pavilion. The pond continued below the east wall, exposed to the interior. Water also played a much less evident role on the west façade.

Mass was a particular challenge, because the building was designed as a lightweight structure to be dismantled and largely reused elsewhere. The west wall consisted of white metal containers filled with water to reflect the greater part of the direct sun and store the remainder during the day, radiating by night to cooler outdoor air.

Thermal zoning was a vital part of this concept, recognizing that people were entering from very hot, unshaded conditions and that most visitors would then move through rather quickly. Passive cooling was given the task of keeping visitors cooler than the outside; mechanical refrigeration was responsible for maintaining even lower temperatures only where people gathered over extended periods of time, as in the "pods" that contained the restaurant and the cafeteria. Gardner and Hadden (1992) describe how the typical July visitor left a sunny outdoor environment at about 100°F [38°C] for the cooler interior at about 90°F [32°C] supplemented by radiant coolth from the huge east glass wall at a maximum of 75°F [24°C] thanks to evaporation from the running water. Exhaust air from the pods lowered temperatures in their vicinity to about 82°F [28°C]. The pods themselves, maintained at about 73°F [23°C], were entered only after the visitor underwent a lengthy transition through the pavilion from the hot outdoors.

To some extent, both cross ventilation and stack ventilation were also involved, despite the outdoor heat; the hottest air indoors rose to the ceiling and was replaced by outdoor air that was cooled slightly as it passed through the watery gap between the east glass wall and the pond. Very high vents at the top of both east and west walls admitted wind to sweep across the ceiling and remove the hottest air.

(a) Cross Ventilation

Once again, remember that cross-ventilation cooling works only when the outside is cooler than the inside.

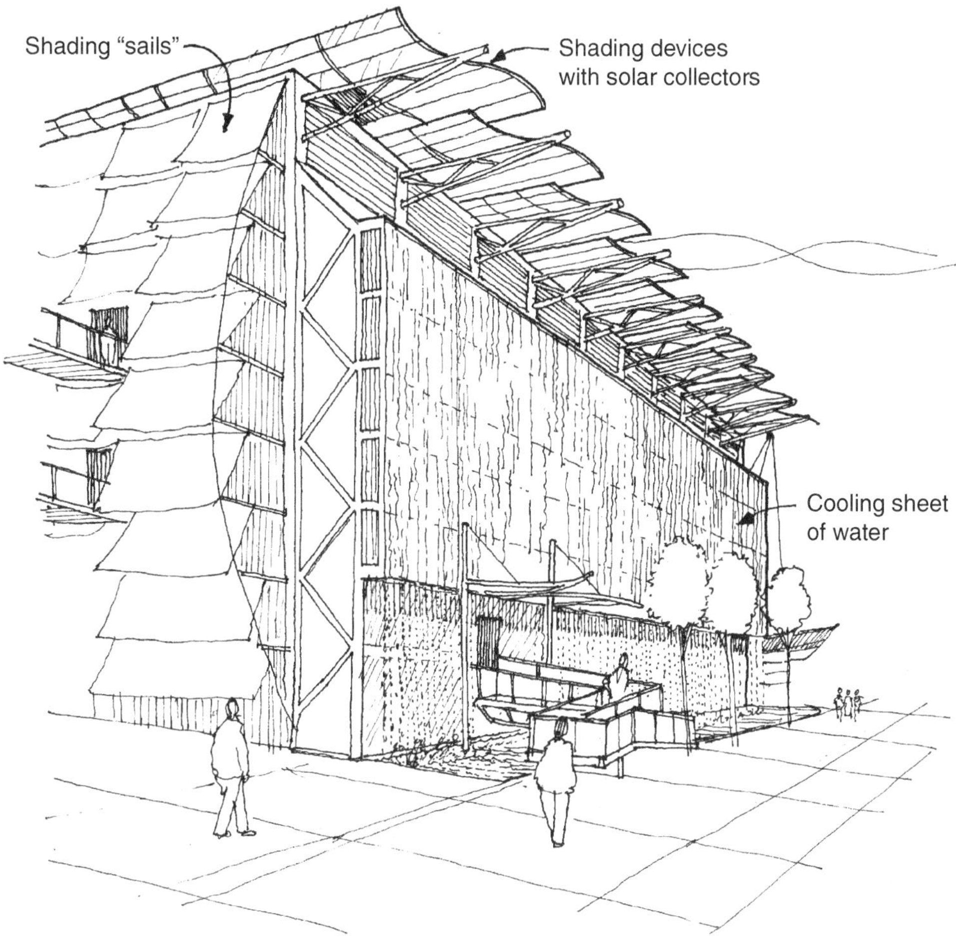

(a)

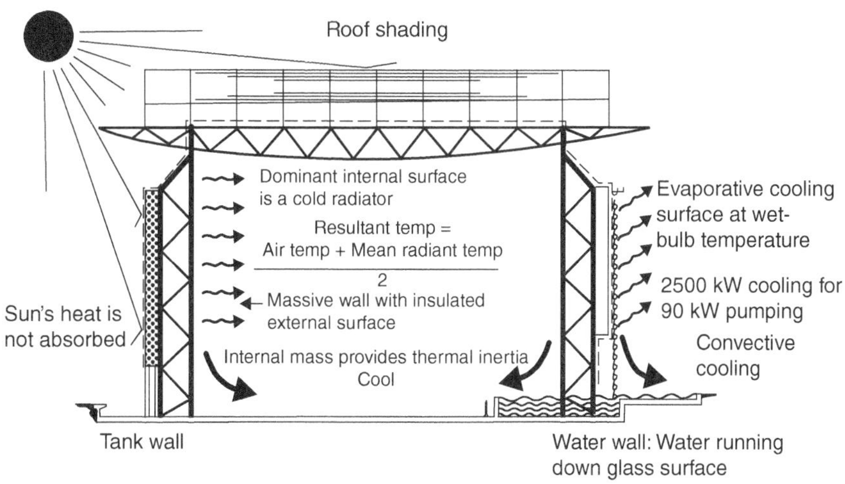

(b)

Figure 14.26 *The United Kingdom Pavilion at the Seville (Spain) World Expo, 1992. (a) Waveform shading devices on the roof also carry PV cells, whose electricity helps pump water from the pool that flows continuously over the east glass wall. (b) Section (east–west) with a combination of cooling strategies. (c) Section (east–west) with estimated maximum temperatures. (Sections are courtesy of Ove Arup Partnership, London.)*

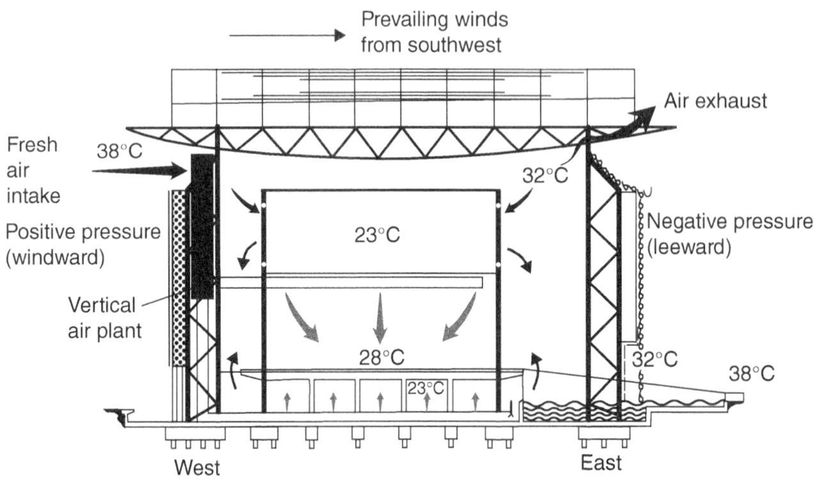

(c)

Figure 14.26 (continued)

Otherwise, why bring in air that is warmer to try to cool a space? For ventilation using windows, the quantity of outdoor air admitted is termed V. The flow of air from inlet windows through the building to outlet windows should have a minimum of obstacles.

In I-P units:

$$V = C_\upsilon A \upsilon$$

where

V = volume flow rate of air, cfm
A = area of operable windows on inlet side or sides, ft²
C_υ = effectiveness factor (dimensionless) that adjusts for different wind orientations: 0.5–0.6 for winds perpendicular to the window openings; 0.25–0.35 for wind diagonal to the window openings
υ = velocity of wind, fpm (= mph × 88)

In SI units:

$$V = 1000\, C_\upsilon A \upsilon$$

where

V is in L/s (= m³/s multiplied by 1000)
A is in m²
υ is in m/s

The sensible heat removed by this flow of outdoor air through indoor spaces was presented in Chapter 10:

In I-P units:

$$q_\upsilon = (V)(1.1)(\Delta t)$$

where

q_υ = sensible heat exchange due to ventilation (Btu/h)
V = volume flow rate, in cfm of outdoor air introduced (ft³/min)
Δt = temperature difference between outdoors and indoors (°F)
1.1 = a constant derived from the density of air under typical building conditions (0.075 lb/ft³) multiplied by the specific heat of air (0.24 Btu/lb °F) and by 60 min/h; units are Btu min/ft³ h °F

In SI units:

$$q_\upsilon = (V)(1.2)(\Delta t)$$

where

q_υ is in W
V is in L/s
Δt is in K
1.2 = a constant derived from the density of air under typical building conditions (1.2 kg/m³) multiplied by the specific heat of air (1.0 kJ/kg K); units are kJ/m³ K

A further modification to these cooling formulas is usually necessary because wind data are typically taken at airports at a height of about 30 ft [10 m] above open, unobstructed ground. Rarely will buildings that rely on cross ventilation be so favorably situated. Figure 14.27 shows correction factors to be applied to such airport wind velocities for variations in terrain and height above the ground. The dimensionless correction factor should then be applied to the quantity V, usually resulting in a reduced volume flow rate.

A more detailed procedure for calculating cross ventilation in residences given in Chandra et al. (1986) accounts for such factors as neighboring buildings and terrain, height of opening above grade, insect screens, and framing members within openings. To analyze

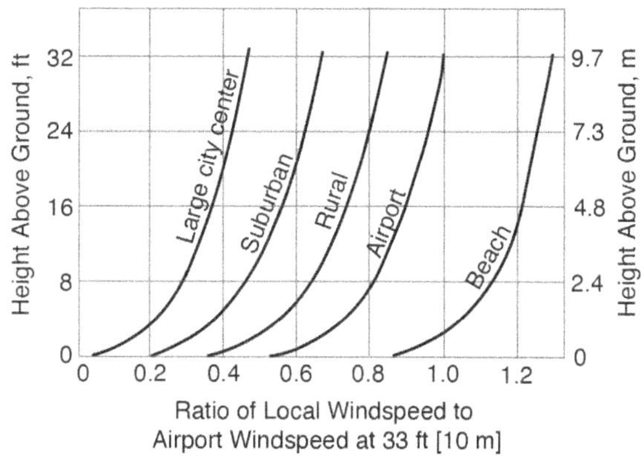

Figure 14.27 *Wind speed variation with height for various terrains. (From Chandra et al., 1986.)*

natural ventilation in even more detail, wind tunnel tests or computation fluid dynamic (CFD) analyses are more likely to be useful than are detailed hand calculations.

(b) Stack Ventilation

This is another reminder that stack ventilation cooling works only when the outside is cooler than the inside. The stack effect needs several conditions: warmer air indoors that can enter the bottom of the stack, cooler air outdoors, and low inlets to admit that cooler outdoor air to the building. This cooler outdoor air picks up heat from the building and enters the bottom of the stack. Within the stack, this now-warm air rises, because it is less dense—therefore lighter—than the cooler outdoor air that surrounds the top of the stack.

The following equation applies when there is no significant resistance to airflow within the building. To calculate ventilation cooling by the stack effect:

In I-P units:

$$V = 60\,KA\sqrt{2gh(t_i - t_o)/t_i}$$

where

V = airflow rate, cfm
= discharge coefficient for opening; assumed to equal 0.65 for multiple inlet openings
A = in square feet, the smaller of either total free area of inlet or outlet openings or horizontal cross-sectional area (throat area) of the stack
g = gravitational constant, 32.2 ft/s²
h = height of the stack from inlet to outlet, ft
t_i = temperature indoors ($>t_o$), °R
t_o = temperature outdoors ($< t_i$), °R

Note: °R = °F + 459.67

In SI units:

$$V = KA\sqrt{gh(t_i - t_o)/t_i}$$

where

V = airflow rate, m^3/s (× 1000 = L/s)
K = discharge coefficient for opening, assumed to equal 0.65 for multiple inlet openings
A = in m^2, the smaller of total free area of inlet or outlet openings or horizontal cross-sectional area (throat area) of the stack
g = gravitational constant, 9.81 m/s^2
h = height of stack from inlet to outlet, m
t_i = temperature indoors ($>t_o$), K
t_o = temperature outdoors ($< t_i$), K

Note: K = °C + 273.15

The height of a stack and its cross-sectional area present an unusual opportunity for building form. The Inland Revenue Centre at Nottingham, England (Figure 14.28), took advantage of this opportunity to provide cylindrical stair towers at corners (or at ends) of its buildings as stack ventilators. These are three- and four-story buildings, with a width of 44.6 ft [13.6 m] to encourage both daylight penetration and natural ventilation. The tower stacks serve all floors except the top. The fabric-covered top of each tower is 23 ft [7 m] above the highest floor served and can be raised up to 3.3 ft [1 m] or lowered (closed) to vary the flow of exhaust air. The stacks are enclosed in glass block so that solar gain can assist the heating—and thus the flow—of air by day within the stack. Berry et al. (1995) report that the airflow increased from 4.8 to 6.2 ACH with the tower operating in solar mode.

Each lower floor has a 43 ft² [4 m²] opening to each stack, with the (openings) doors held open magnetically unless fire alarms close them. This rather small opening to the stack (relative to the floor area served) is the limiting factor in the outward flow of heated air; to supplement the supply of outdoor air, small three-speed fans are installed in the floor below each window to draw in fresh air, then distribute it through a wide floor grille. The triple-glazed windows are also operable, both as sliding "doors" and as tilt-ins from the top.

This fan-assisted stack effect is used on summer nights to help precool the building for the following day; the lower floors feature very large areas of exposed structural mass in their precast concrete ceilings.

The top floor is a lighter-weight structure, with the stack effect culminating in a ridge vent. Monitoring reveals that this floor is usually a few degrees warmer than the massive night-ventilated lower floors.

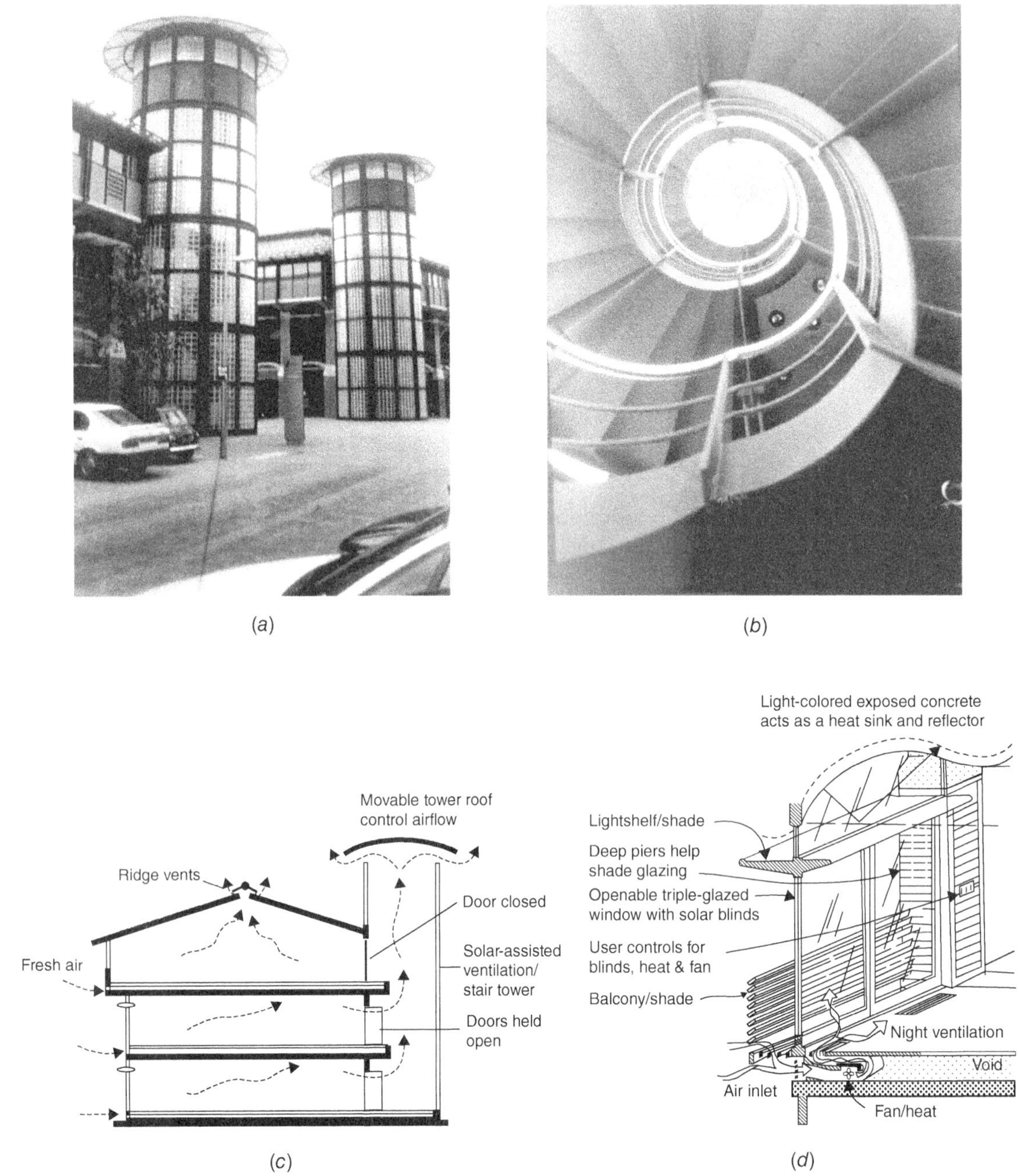

Figure 14.28 The Inland Revenue Centre at Nottingham, England. (a) Cylindrical stair towers at the corners and ends of three- and four-story office blocks act as stacks; solar gain through glass helps increase the flow of exhaust air. (b) Tops of the towers are fabric and can be raised to increase the stack aperture and therefore increase the flow. (c) Section of a three-story wing with a stack ventilation stair tower. (d) Section perspective with fan-assisted intakes below each window and an exposed precast concrete ceiling that helps store heat for night ventilation cooling. (a, b courtesy of Vaughan Reynolds; c, d courtesy of Ove Arup Partnership, London.)

(c) Night Ventilation of Thermal Mass

Before doing these detailed calculations, be sure that:

1. You have checked your summer climate against the passive cooling design strategies (Figure 14.2) and that night ventilation of thermal mass is appropriate.
2. Your building and climate have been checked for approximate performance based on the design guidelines (Figure 14.8) for this cooling strategy.

Also have in mind a positive night ventilation strategy; Figure 14.29 shows an example of a forced-air system integral with a concrete joist-and-girder structural system. The following procedure is adapted

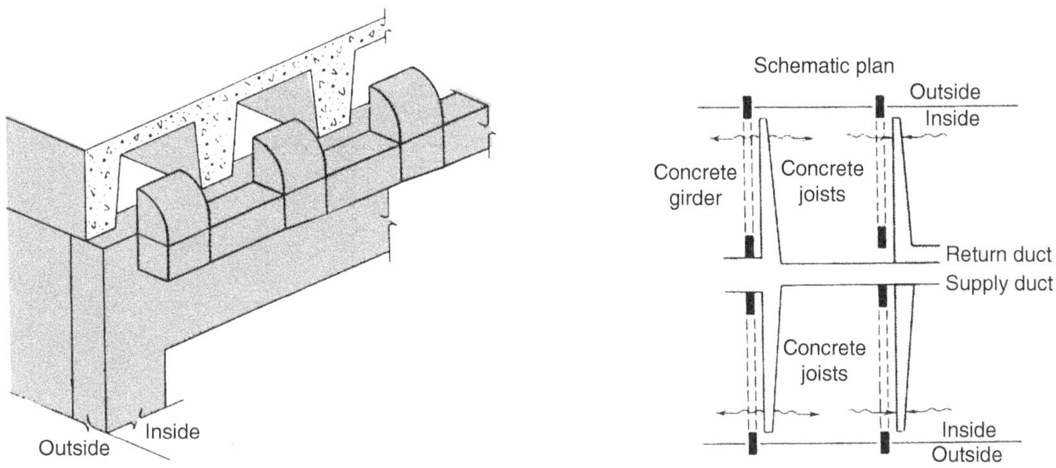

Figure 14.29 *Example of ductwork for night ventilation using the thermal mass of a concrete structural system.*

from one developed by Karen Crowther for a workshop at the Fifth National Passive Solar Conference in Amherst, Massachusetts, as presented by Miller (1980).

STEP 1. In column II of Table 14.5, list the hourly outdoor temperatures for the design condition (these may be approximated from the summer DB temperature and mean daily range). This will give the worst-day performance. (To get average-day performance, list average hourly temperatures, which are available from local weather service records.) You need not list temperatures above 80°F [27°C] because outdoor air will not be used for cooling above that temperature. (The information assembled in Table 14.5 and subsequent steps using I-P units can also be developed using SI units.)

TABLE 14.5 Night Ventilation of Thermal Mass

Calculation Procedure			
A. Mass surface area (step 3)	______ ft^2		
B. Mass heat capacity (step 4)	______ Btu/°F		
C. Floor area (supplementary cooling, step 5), and total building volume	______ ft^2 ______ ft^3		
(I) Hour	**(II) Outside Air Temperature (°F) (step 1)**	**(III) Cooling (Btu)**	**(IV) Mass Temperature (°F)**
8:00 p.m.	______	______	______
9:00 p.m.	______	______	______
10:00 p.m.	______	______	______
11:00 p.m.	______	______	______
Midnight	______	______	______
1:00 a.m.	______	______	______
2:00 a.m.	______	______	______
3:00 a.m.	______	______	______
4:00 a.m.	______	______	______
5:00 a.m.	______	______	______
6:00 a.m.	______	______	______
7:00 a.m.	______	______	______
8:00 a.m.	______	______	______
9:00 a.m.	______	______	______
D. Total mass cooling		______ Btu	______ °F
E. Final mass temperature		______ °F	
F. Supplementary cooling (see steps 5 and 11)		______ Btu	
G. Total cooling, D + F		______ Btu	
H. 24-hr heat (step 2)		______ Btu	
I. Flow rate required for night ventilation		______ cfh	or ______ ACH

STEP 2. Calculate the 24-hour heat gain for the building in Btu. Find the sum of all the hourly heat gains through the envelope, the minimum ventilation while "closed," and the internal gains while operating. List on line H of Table 14.5.

STEP 3. Find the total area of the thermal mass surface that is exposed (no rugs, etc.) both to the space to be cooled and to moving night air during the ventilation ("open") cycle; list on line A of Table 14.5. Note: The larger the mass area exposed, the better the performance. This is why direct-gain solar-heated buildings, with plentiful exposed mass, often make such good candidates for night ventilation cooling. Two additional comments on mass surface: (1) Place it where people "see" it so that it can readily receive their radiant heat, and (2) keep direct sun off the mass (and out of the building) during the cooling season.

STEP 4. Find the mass heat capacity for the entire space to be cooled:

$$\text{heat capacity} = \text{mass volume} \times \text{density} \times \text{specific heat}$$

Appendix E, Table E.1 lists both density and specific heat for most common building materials; Table 14.6 shows a quick way to get mass heat capacities for the most common thermal mass materials. Enter this total mass heat capacity on line B of Table 14.5.

STEP 5. Calculate the total building volume. For "supplementary" cooling due to surfaces other than those of the principal thermal mass, list the space's floor area. This step should be taken only for spaces with a significant amount of roof, wall, or floor area in addition to the thermal mass areas counted in step 3. For example:

- If the space has exposed concrete ceilings, walls, and floors—all counted already as thermal mass—skip this step.
- If all thermal mass is in the form of free-standing water containers, enter the entire floor area.
- If the entire ceiling is thermal mass, but the walls or the floors are not, enter half of the floor area.
- If the entire ceiling is thermal mass, but the floor is not, and there are few or no walls (e.g., open office plan), enter one-third to one-fourth of the floor area.

TABLE 14.6 Common Mass Heat Capacities

I-P Units ft³ × (Btu/ft³ °F) = Btu/°F		SI Units m³ × (kJ/m³K) = kJ/K
Volume × (62.4)	Water	Volume × (4181)
Volume × (22.5)	Ordinary concrete	Volume × (1507)
Volume × (18.7)	Masonry, grout-filled	Volume × (1253)
Volume × (15.6)	Brick	Volume × (1045)

Source: Adapted from Miller, 1980.

STEP 6. Complete column III hour-by-hour after determining the mass temperature in column IV for the preceding hour:

$$\text{cooling Btu/h} = \left[\begin{array}{c}\text{previous hour mass temp., °F (col. IV)}\end{array} - \begin{array}{c}\text{outside temp., °F (col. II)}\end{array}\right] \times \begin{array}{c}\text{mass surface area, ft}^2\text{ (line A)}\end{array} \times \begin{array}{c}\text{surface conductance, Btu/h ft}^2\text{ °F}\end{array}$$

For the first hour, assume that the mass temperature is 80°F. The surface conductance is usually assumed as 1.0 Btu/h ft² °F. (See Appendix E, Table E.3 for other surface conductances under various conditions, many of which are considerably more than 1.0.)

STEP 7. Complete column IV, hour by hour, after calculating the cooling (Btu/h) for column III:

$$\text{mass temp.} = \text{previous hour mass temp. (col. IV)} - \frac{\text{cooling Btu/h (col. III)}}{\text{mass heat capacity, Btu/°F (line B)}}$$

STEP 8. Continue this hourly process using columns III and IV until the falling temperature of the mass equals the rising temperature of the outdoor air. At that point, continuing with plentiful ventilation will only rob the mass of its coolth; the building therefore switches to (thermally) closed mode, with minimal ventilation.

STEP 9. Add all the hourly cooling Btu/h values (column III) to obtain the total mass cooling in Btu (line D).

STEP 10. Note the final mass temperature from column IV. This is probably at least 5°F above the lowest air temperature of the night (column II). If this lowest mass temperature is significantly higher, consider redesigning for more exposed mass surface area.

STEP 11. If supplementary cooling is appropriate (see step 5), calculate it as follows:

$$\text{supplementary cooling, Btu} = [80°\text{F} - \text{final mass temp. (line E)}] \times 2.25 \times [\text{floor area (line C)}]$$

The factor 2.25 assumes a modest role for the other, less thermally massive surfaces. Enter this supplementary cooling on line F.

PASSIVE ENVIRONMENTAL SYSTEMS

STEP 12. Obtain total cooling by adding lines D and F; enter the total on line G.

STEP 13. In step 2, you entered the 24-hour heat gain for the building on line H. Compare this required cooling to the total cooling provided on line G. If you have not provided enough cooling and the final mass temperature is more than 7°F [−13.8°C] above the lowest nighttime outdoor temperature, consider redistributing the building mass over a wider surface (e.g., 3000 ft^2 of 4-in. [101.6 mm] slab rather than 2000 ft^2 of 6-in. [152.4 mm] slab) and trying again. If you do not have enough cooling, and the final mass temperature is 5–7°F [−15°C to −13.8°C] above the lowest nighttime outdoor temperature, consider providing both more mass and more surface area (e.g., 3000 ft^2 of 4-in. [101.6 mm] slab rather than 2000 ft^2 of 4-in. [101.6 mm] slab) and trying again.

STEP 14. Determine and enter on line I the approximate flow required for night-ventilating air. Use the following formula:

$$\text{cfh} = \frac{\text{Btu/h}}{0.018\,\Delta t}$$

where

cfh (ft^3/h) = the minimum required flow rate of night air

Btu/h = the cooling Btu for the hour of maximum cooling during the night (column III)

Δt = the temperature difference between the final mass temperature (column IV) and the outdoor air (column II) for that same hour of maximum cooling

It is often useful to express this night ventilation flow rate in terms of ACH:

$$\text{ACH} = \frac{\text{cfh required}}{\text{building volume (ft}^3\text{)}}$$

Another night ventilation example is the California Highway Patrol offices at Gilroy, California, shown in Figure 14.30. This one-story building sits near a freeway interchange south of San José. In this building, the nighttime air is drawn in as far from the freeway as possible, taken underground a short distance (with just a bit of earth tube cooling), then distributed throughout the central corridor. At either end of a continuous skylight over the central corridor, large fans exhaust air by night. Corridor walls are concrete block, and the floor is a concrete slab on grade. The cool night air may carry a seasonal scent; Gilroy calls itself the garlic capital of the world.

(d) Fan-Assisted Evaporative Cooling

Before beginning the following calculations, be sure that:

1. You have checked your summer climate against the passive cooling design strategies (Figure 14.2), and the evaporative cooling strategy is appropriate.
2. Your building and climate have been checked for approximate performance based on the design guidelines for evaporative cooling (Section 14.4).

First, find the total sensible heat gain in Btu/h that is to be removed from your building by evaporative cooling. Generally, this is calculated at your climate's summer design DB and mean coincident WB temperatures (Appendix B, Table B.1). The psychrometric chart is then used to plot the progress of evaporatively cooled air, as shown in Figure 14.31. (The complete psychrometric chart is given in Appendix G, Figure G.1, using I-P units and Appendix G, Figure G.2, using SI units.)

STEP 1. Determine the outdoor air conditions. Enter the psychrometric chart at the summer design DB and mean coincident WB temperatures for your climate (point *A* in Figure 14.31). As the air is blown through the evaporative cooler, it proceeds along the constant WB line toward saturation or 100% RH (from point *A* toward and beyond point *B* in Figure 14.31).

STEP 2. Determine supply air temperature. The most efficient evaporative cooler will be able to cool and humidify the outdoor air to a point about three-fourths of the total distance on the chart from entering air conditions to saturation (the three-fourths point is reached at point *B* in Figure 14.31); a more reasonable operating assumption may be two-thirds of that total distance.

STEP 3. Determine Δt indoors. At conditions of point *B*, the air leaves the evaporative cooler and enters the building. It immediately begins to pick up sensible heat (moving from point *B* toward point *C* in Figure 14.31). When the air has picked up enough heat to reach a marginally warm temperature, it is exhausted to the outdoors (point *C* in Figure 14.31). Note, however, that this exhaust air is still well below outdoor temperatures and, therefore, is capable of some cooling just outside the building.

STEP 4. Calculate heat removed and airflow rate indoors. The heat removed from the building by this airflow depends on two factors:

1. The DB temperature difference (Δt) between the supply air and the exhaust air (point *B* to point *C* in Figure 14.31).

(*a*)

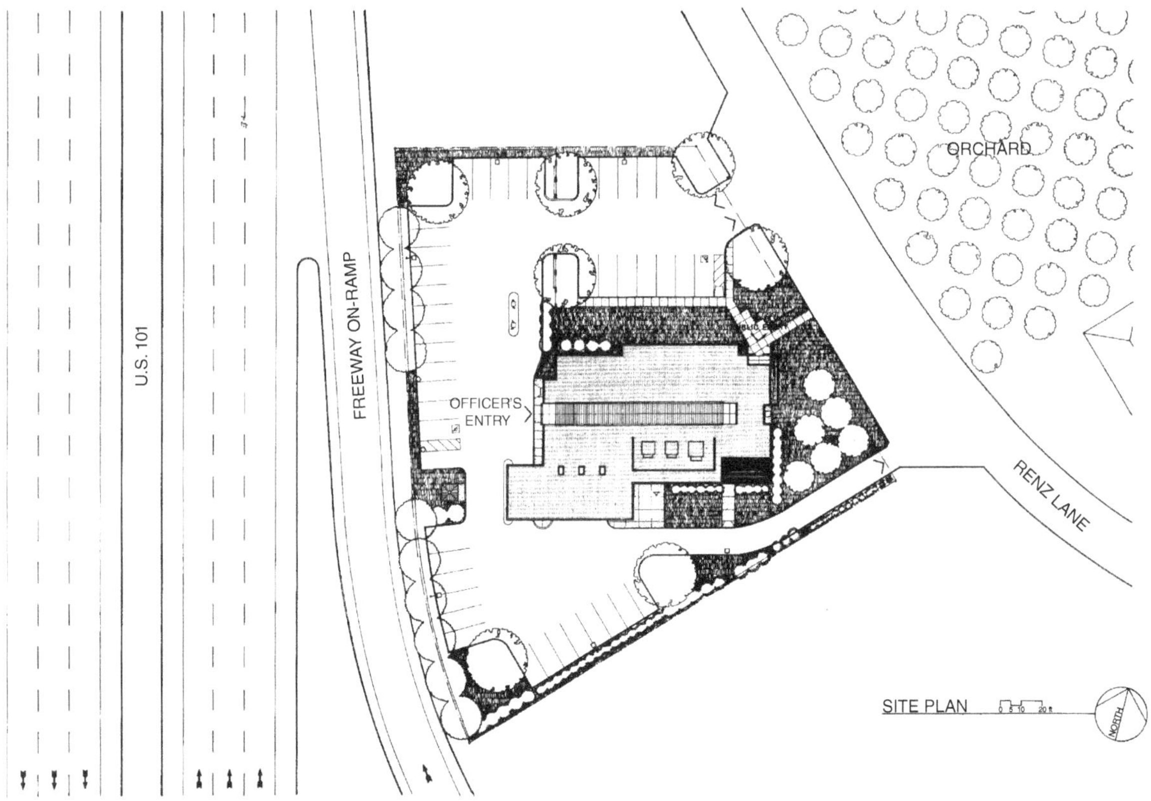

(*b*)

Figure 14.30 *The California Highway Patrol building near Gilroy, California (a) is flushed with night air for cooling. (b) Site plan. (c) Axonometric shows intake air through a very short earth tube; night flush exhausts are at either end of the central skylit corridor with thermally massive walls and floor. (d) Looking up at one-night flush exhaust duct. (e) Exterior with night flush exhaust grille. (Courtesy of the Colyer / Freeman Group.)*

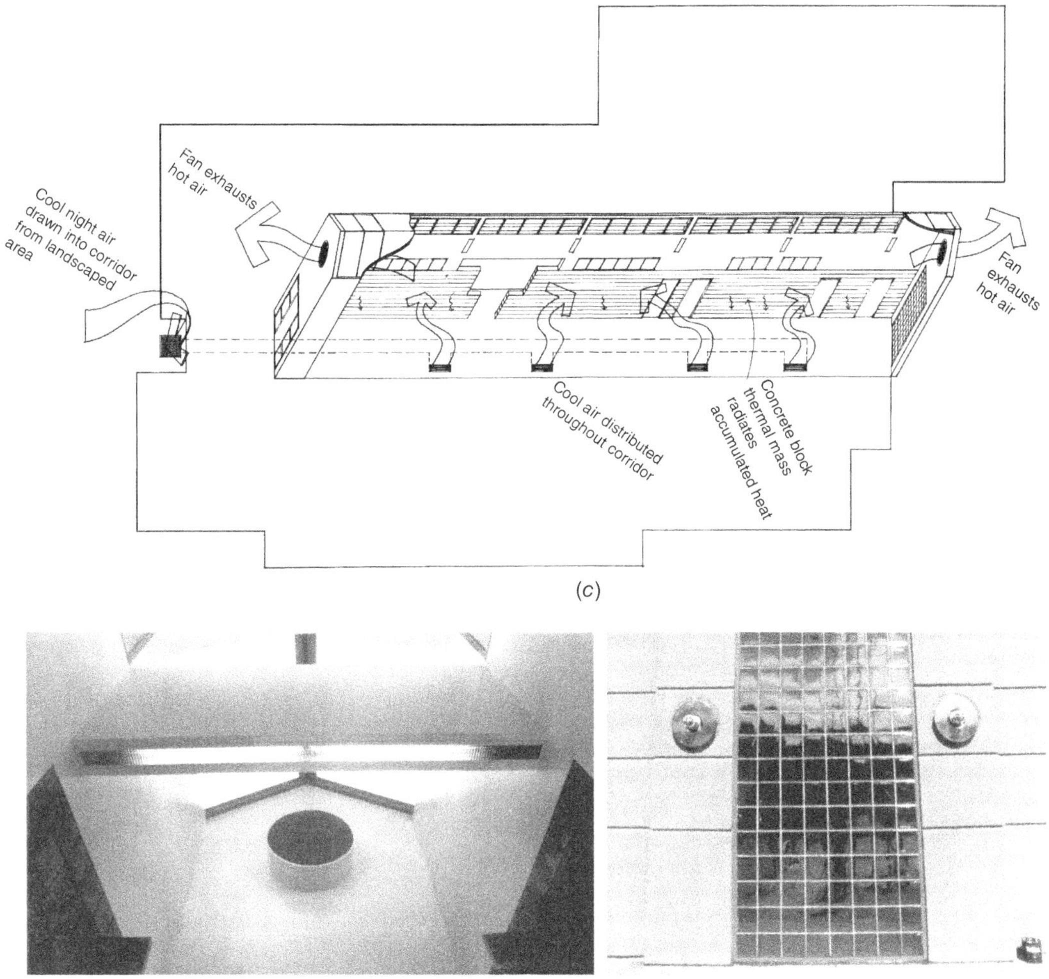

(c)

(d)

(e)

Figure 14.30 (continued)

2. The airflow rate, usually expressed in cubic feet per minute [L/s].

$$\text{Btu/h removed} = (\text{cfm airflow}) \times (1.1\ \text{Btu min/ft}^3\ {}^\circ\text{F h}) \times (\Delta t, {}^\circ\text{F})$$

$$[\text{W removed} = (\text{L/s}) \times (1.2) \times (\Delta t, {}^\circ\text{C})]$$

The designer can vary the airflow rate by choosing larger or smaller evaporative coolers; any point less than three-fourths of the total distance from outdoor air to saturation may also be chosen.

EXAMPLE 14.9 A 4000-ft^2 [279-m^2] retail store several miles outside of Tucson, Arizona, has been calculated to have a total sensible heat gain of 100,000 Btu/h [29,310 W] at summer design conditions (105°F DB, 66°F WB [40.6°C, 18.9°C] for this location, from local climate data). How many evaporative coolers are needed to remove heat gains?

SOLUTION

STEP 1. Enter Figure 14.32 at 105 DB, 66 WB (or equivalent SI values), shown as point *A*.

STEP 2. Measure along the constant 66 WB line to saturation and determine that three-fourths of this distance: 76 DB, 66 WB (point *B*).

STEP 3. Δt indoors depends on the assumed exhaust air DB temperature; assume 83°F in this example (point *C* corresponding to 83 DB, 68 WB).

$$\Delta t_{\text{indoor}} = 83^\circ\text{F (assumed exhaust DB)} - 76\ (\text{DB at point B}) = 7^\circ\text{F}$$

STEP 4. Calculate heat removed and airflow rate indoors.

$$100{,}000\ \text{Btu/h heat removed} = (\text{cfm airflow}) \times (1.1\ \text{Btu min/ft}^3\ {}^\circ\text{F h}) \times (7^\circ\text{F})$$

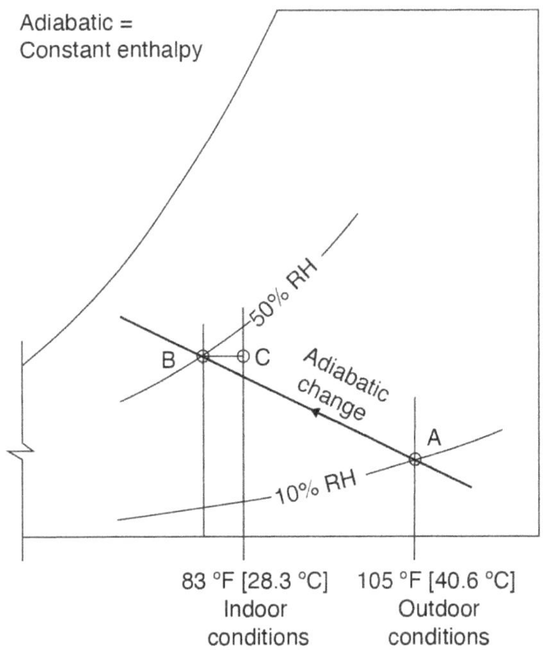

Figure 14.31 *Once-through cycle of outdoor air through a fan-forced evaporative cooler. Values are transcribed from the psychrometric chart (Appendix G, Figure G.1). Hot, dry air can be humidified adiabatically (no change in total heat content) from point A to point B, reaching indoor conditions that fall within, or close to, the optimum comfort zone shown in Chapter 6. This is accomplished without the use of the high-energy-consuming compressive refrigeration cycle. As the humidified air picks up sensible heat from the space, it moves from point B to point C.*

Rearranging the equation yields:

$$\text{cfm airflow} = \frac{100{,}000\ \text{Btu/h heat removed}}{(1.1\ \text{Btu min/ft}^3\ {}^\circ\text{F h}) \times (7^\circ\text{F})}$$
$$= 12{,}987\ \text{cfm}\ [6130\ \text{L/s}]$$

Note that this is a rate of 12,987 cfm/4000 ft^2 or 3.3 cfm/ft^2 of floor area [16.8 L/s m^2].

This airflow rate can be provided by one larger or several smaller cooling units, depending on possible zoning of the building. For instance, two evaporative coolers at 7000 cfm [3303 L/s] each would meet the requirements for this retail store; one might serve a smaller-area zone in front with show windows and a lot of solar gain, while another might serve a larger-area interior zone with less solar gain. ■

(e) Downdraft Cooltowers

One of the first North American downdraft cooltowers was shown in Figure 14.11, along with estimated cooling design guidelines in Figure 14.18. To more closely estimate outlet conditions and cooling capacity of such a passive evaporative device with calculations, some complex equations are involved. These equations were used to develop cooltower computer-based performance prediction tools (Thompson et al., 1994). Another set of complex equations were those presented by Givoni (1994), based on his research related to downdraft cooltowers that utilize showerheads for evaporative cooling. These equations were used to develop the EnergyPlus module. The dataset reported by Givoni (1994) on data collected from three different climate zones was used to partially validate this module (Kang and Strand, 2009). For a shorter but less-exact approach, consider the following formula proposed by Givoni (1993). This was validated using the results from the Tucson tower and is suitable for predicting outlet conditions from towers of similar size.

$$t_{exit} = \text{DB} - 0.87(\text{DB} - \text{WB})$$

where

t_{exit} = dry-bulb temperature of air entering the space (at cooltower outlet)
DB = dry-bulb temperature, outdoors
WB = coincident wet-bulb temperature, outdoors

This simple equation does not take into account the influences of both wind (driving more air through the evaporative cooling pads) and the optional solar chimney exhaust tower (seen in Figure 14.12).

Cooling achieved by the tower depends both on the temperature of this exit air and on its flow rate. Givoni (1993) proposes, in SI units:

$$V = 0.033\,A_{evap}\sqrt{H(\text{DB} - \text{WB})}$$

where

V = flow, in m^3/s (× 1000 = L/s)
A_{evap} = total area of wetted pads, m^2, at cooltower inlet
H = overall height of tower, m
DB = dry-bulb temperature, outdoors
WB = coincident wet-bulb temperature, outdoors
0.033 = accounts for pressure drop through the tower and the building

To calculate air velocity at tower outlet, simply divide flow rate by area of opening at outlet. Knowing both the temperature and the flow rate, the cooling capacity achieved is:

$$W = 1.2V[t_{int} - t_{exit}]$$

where

W = watts removed from space by cooltower
V = flow from cooltower, in m^3/s (× 1000 = L/s)
t_{int} = dry-bulb temperature to be maintained indoors, °C
t_{exit} = dry-bulb temperature of air entering the space (at cooltower outlet), °C

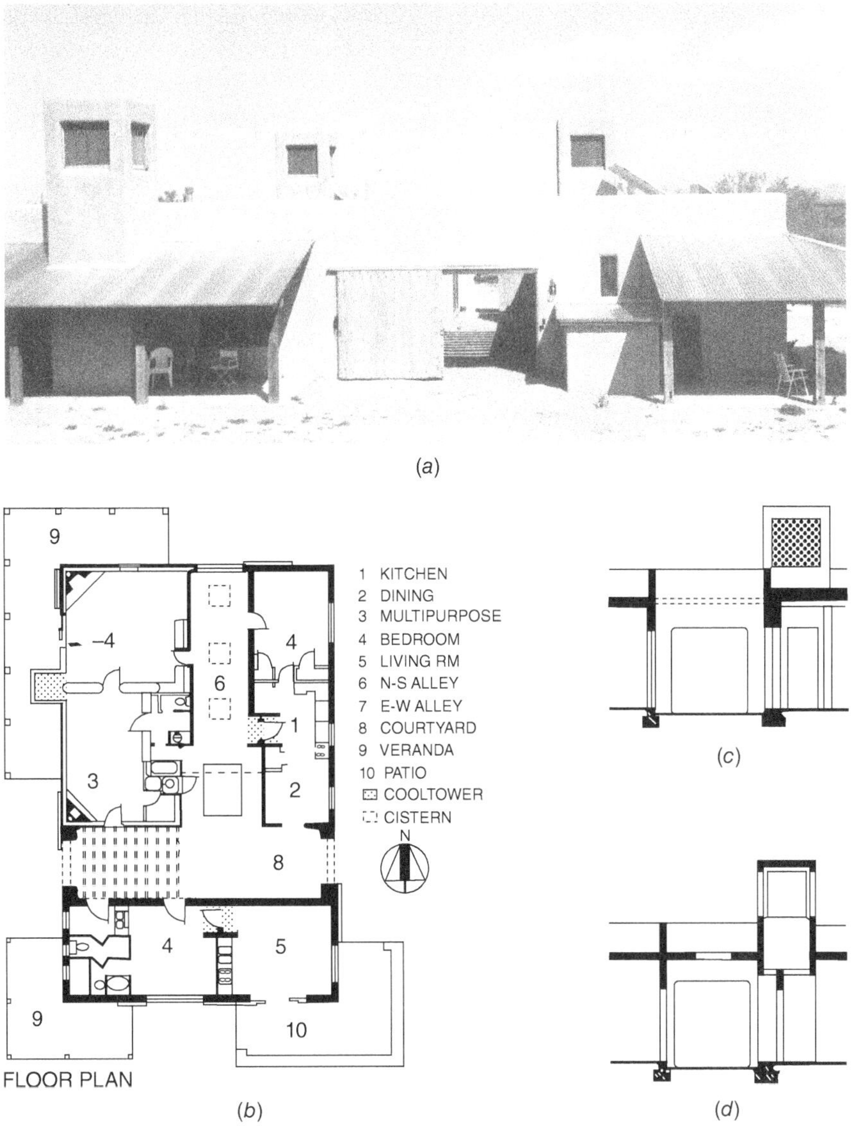

Figure 14.32 *Downdraft cooltowers dominate the silhouette of the Agua Blanca ranch in a remote desert near Tucson, Arizona. (a) West side, with one of the large rolling barn doors that allow the courtyard/outdoor spaces to be connected with cooler breezes or to be protected from hot winds. Air from the downdraft cooltowers, after use within the units, can be discharged to the courtyard, while (under the hottest conditions) it is still cooler than outdoor air. This helps to form a "cool pool" of air in the protected courtyard. (b) Plan of three units, each with a downdraft cooltower, around a T-shaped courtyard/covered passage outdoor space. (c) East–west section through the courtyard. (d) North–south section through one of the three downdraft cooltowers. (Courtesy of Martin Yoklic, Environmental Research Laboratory, University of Arizona.)*

In I-P units, this becomes:

$$V = 2.7\,A_{\text{evap}}\sqrt{H(\text{DB} - \text{WB})}$$

where

V = flow, in cfm
A_{evap} = total area of wetted pads, ft^2, at cooltower inlet
H = overall height of tower, ft
DB = dry-bulb temperature, outdoors, °F
WB = coincident wet-bulb temperature, outdoors, °F
2.7 = accounts for pressure drop throughout the tower and the building

To calculate air velocity at tower outlet, simply divide flow rate by area of opening at outlet. Knowing both the temperature and the flow rate, the cooling capacity achieved is:

$$\text{Btu/h} = 1.1V[t_{\text{int}} - t_{\text{exit}}]$$

where

Btu/h = heat removed from space by cooltower
V = flow from cooltower, in cfm
t_{int} = dry-bulb temperature maintained indoors, °F
t_{exit} = dry-bulb temperature of air entering space (at cooltower outlet), °F

In a remote desert west of Tucson, Arizona, the Agua Blanca ranch uses several downdraft cooltowers to serve a building around a T-shaped courtyard (Figure 14.32). This off-grid building complex uses PV for lighting and water pumping, propane for a refrigerator, and one wall heater and stand-by generator, and collects rainwater in a cistern below the courtyard. Wood stoves and fireplaces supplement direct solar gain in winter. The courtyard connects to the surrounding ranch by three passages, one covered with a roof and small open skylights, one covered with a deciduous vine on a trellis, and one open to the sky. Large rolling barn doors are located at the outer end of each passage; these provide both security and the ability to trap cool air from the cooltowers (or via the indoor spaces), thus increasing hot-weather comfort outdoors.

Three cooltowers, 22 ft [6 m] high, each with a total evaporative pad area of 64 ft² (5.9 m²), serve three units around the courtyard. At the outlet of each cooltower, the cool airflow can be controlled: shared or sent to either of two locations, or closed off. One of the cooltowers can deliver air directly to the covered passage off the courtyard.

Design simulations (Yoklic and Layseca, 1998) for a typical June in Tucson predict a flow from each cooltower of about 5000 cfm [2360 L/s], maintaining the indoor temperature at 74°F [23.3°C]. In these quite small buildings, this translates into an ACH of about 60, considering the volume to include a height up to 6 ft [1.8 m], above which stratification of warm air negates circulation through the upper portion of the space.

These computer-generated design simulations are summarized in Table 14.7. The results differ somewhat from the results of Givoni's (1994) equations, suggesting that perhaps in this case:

$$t_{exit}\,\text{DB} = 0.70\,(\text{DB} - \text{WB})$$

TABLE 14.7 Downdraft Cooltower Performance Hourly Simulations[a]

I-P Units								SI Units							
Ambient (°F)			Tower Outflow					Ambient (°C)			Tower Outflow				
DB	WB	RH (%)	DB	RH (%)	Velocity (fpm)	Flow Rate (cfm)	Water Consumed (gal/h)	DB	WB	RH (%)	DB	RH (%)	Velocity (m/s)	Flow Rate[c] (m³/s)	Water Consumed (L/h)
Part A. Tower Effective Height[b] = 10 ft [3 m]															
82.0	55.9	18.2	61.6	72.0	170	4254	9.8	27.8	13.3	18.2	16.4	72.0	0.86	2.00	37.1
87.1	59.0	18.0	65.1	71.0	175	4384	10.8	30.6	15.0	18.0	18.4	71.0	0.89	2.07	40.9
93.0	62.1	16.7	68.9	69.1	183	4572	12.3	33.9	16.7	16.7	20.5	69.1	0.93	2.16	46.6
98.1	64.0	15.0	71.8	67.0	190	4752	13.9	36.7	17.8	15.0	22.1	67.0	0.97	2.24	52.6
102.0	64.0	11.4	72.9	63.2	199	4987	16.1	38.9	17.8	11.4	22.7	63.2	1.01	2.35	60.9
Part B. Tower Effective Height[b] = 15 ft [4.6 m]															
82.0	55.9	18.2	62.2	69.6	205	5133	11.5	27.8	13.3	18.2	16.8	69.6	1.04	2.42	43.5
87.1	59.0	18.0	65.7	68.7	212	5289	12.7	30.6	15.0	18.0	18.7	68.7	1.08	2.5	48.1
93.0	62.1	16.7	69.6	66.7	221	5516	14.4	33.9	16.7	16.7	20.9	66.7	1.12	2.6	54.5
98.1	64.0	15.0	72.5	64.5	229	5734	16.3	36.7	17.8	15.0	22.5	64.5	1.16	2.71	61.7
102.0	64.0	11.4	73.7	60.5	241	6017	18.9	38.9	17.8	11.4	23.2	60.5	1.22	2.84	71.5
Part C. Tower Effective Height[b] = 20 ft [6.1 m]															
82.0	55.9	18.2	62.6	68.0	235	5864	12.9	27.8	13.3	18.2	17.0	68.0	1.19	2.77	48.8
87.1	59.0	18.0	66.2	67.0	242	6043	14.2	30.6	15.0	18.0	19.0	67.0	1.23	2.85	53.7
93.0	62.1	16.7	70.1	65.0	252	6302	16.1	33.9	16.7	16.7	21.2	65.0	1.28	2.97	60.9
98.1	64.0	15.0	73.0	62.7	262	6550	18.2	36.7	17.8	15.0	22.8	62.7	1.33	3.09	68.9
102.0	64.0	11.4	74.3	58.7	275	6874	21.1	38.9	17.8	11.4	23.5	58.7	1.40	3.24	79.9

Source: Excerpted from Yoklic and Layseca, 1998. Additional simulations courtesy of Martin Yoklic. SI conversions by the authors of this book.

[a] Tower has a total evaporative pad area of 64 ft² [5.95 m²]; pad thickness is 4 in. [100 mm]; tower outlet area is 25 ft² [2.32 m²]; building outlet area is 50 ft² [4.65 m²]. Ambient data are selected from Tucson, Arizona, typical June.

[b] Effective height is measured from the bottom of the evaporative pads to the top of the outlet to the building.

[c] Flow rate (m³/s) × 1000 = L/s.

and that the flow rate (in SI units) may be closer to:

$$V = 0.04\,A_{evap}\sqrt{2z(DB - WB)}$$

where

z = tower shaft effective height, which is the distance from the bottom of the inlet to the top of the outlet (refer to Figure 14.14)

(f) Roofpond Cooling

Before doing the following detailed calculations, be sure to (1) check the summer climate against the passive cooling design strategies (Figure 14.2) to confirm that either the evaporative or the high thermal mass strategy is appropriate, and (2) check the building and climate for approximate performance based upon the design guidelines. Note, however, that the design guidelines are based only on summer night DB temperature. With evaporative cooling (such as a light spray onto the surface of the roofpond containers), better performance can be expected, as shown by the following calculation procedure, based on Fleischhacker et al. (1982). This procedure can be used to check the size and depth required for an acceptable building roofpond. It assumes an optimum pond depth of 4 in. [100 mm] but allows for other depths as well. The procedure is presented in I-P units, but it can be conducted in SI units by appropriate conversions.

STEP 1. First, assemble the following data on the site climate (shown in I-P units for this example):

Maximum DB temperature (Appendix B)
Mean daily range (Appendix B)
Minimum DB temperature (= maximum DB − mean daily range)
Design WB (2%) (Appendix B)
Average maximum RH for July (from local climatological data)
July average temperature (TA July in Appendix C)

From these data, determine two further characteristics for the climate of the site:

1. Minimum WB temperature
2. Average July operating hours for residential air conditioning

   Minimum WB temperature can be determined from the psychrometric chart (shown in Appendix G, Figures G.1 and G.2). Enter the chart with the minimum DB and move vertically along the constant DB line until reaching the average maximum RH for July. At that intersection, refer to the diagonal WB temperature lines, from which the minimum WB temperature can be determined (Figure 14.33).

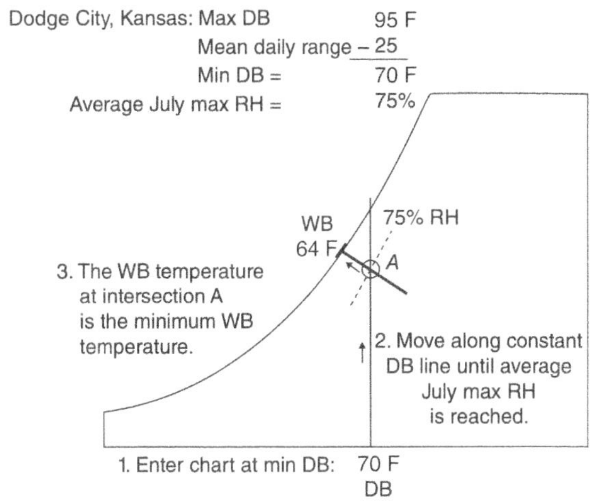

Figure 14.33 *Finding the minimum (nightly) WB temperature, summer design conditions, for Dodge City, Kansas.*

Average July operating hours, *N*, can be estimated from Figure 14.34. Enter at July TA for the site climate. The climate's *N* will fall somewhere between the maximum *N* and minimum *N* lines; as a guide to estimating *N* for the climate, compare the climate's mean daily range and design WB to those of the cities shown. The higher the mean daily range and the lower the design WB, the lower the *N*.

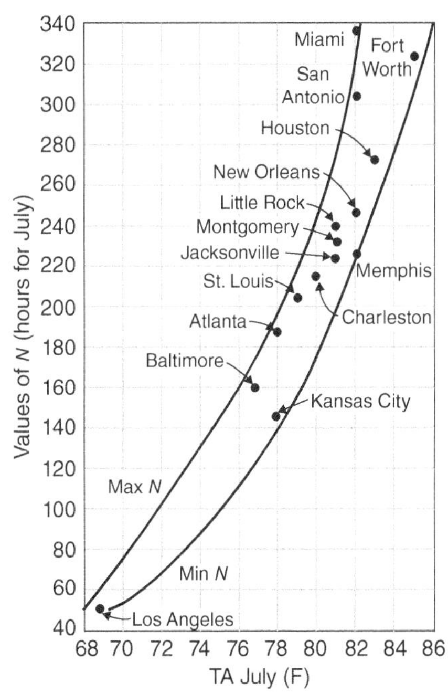

Figure 14.34 *Range of N, average July operating hours, for residential air conditioners. (Adapted from Fleischhacker et al., 1982.)*

TABLE 14.8 Roofpond Design Data

Part A. Values of Factor "F"

Location	Design DB/Mean Coincident WB °F [°C]	MDR °F [°C]	"F," Based on Interior Air Motion: None	Fans: 115 fpm [0.6 m/s]
Miami, Florida	90/77 [32.2/25]	15 [8.3]	2.0	1.0
San Antonio, Texas	97/73 [36.1/22.8]	19 [10.6]	3.0	1.5
Phoenix, Arizona	107/71 [41.7/21.7]	27 [15]	4.0	2.0

Part B. Corrections to Minimum (DB) Pond Temperature, °F [°C]

Pond Depth in. [mm]				Night Internal Load Btu/h ft² [W/m²]				Pond Portion Exposed to Sky			
2 [50]	4 [100]	6 [150]	10 [250]	0 [0]	2 [6.3]	4 [12.6]	8 [25.2]	⅓	½	⅔	Fully
−1.5 [−0.8]	+0 [+0]	+0.7 [+0.4]	+1.0 [+0.6]	+0 [+0]	+0 [+0]	+0.8 [+0.4]	+2.5 [+1.4]	+3.3 [+1.8]	+2.1 [+1.2]	+1.2 [+0.7]	+0 [+0]

Source: From Fleischhacker et al., 1982. SI units added by the authors of this book.

STEP 2. Calculate the peak hourly heat gain but exclude internal gains. (The method given in Section 14.6 is appropriate for this roofpond-sizing procedure.)

STEP 3. Approximate the daily total heat (excluding internal heat) to be stored in the roofpond, or Q_E, in Btu:

$$Q_E = \frac{(\text{Btu/h, peak hourly gain}) \times (N, \text{hours for July})}{31, \text{days in July}}$$

STEP 4. Determine the rate of the internal gain, in Btu/h, while the building is occupied.

STEP 5. Approximate the daily total internal heat to be stored in the roofpond, Q_I, in Btu:

$$Q_I = (\text{hourly internal gains, in Btu/h}) \times (\text{daily hours of occupation})$$

STEP 6. Calculate the daily heat gain directly to the roofpond through its insulated covers, Q_P, in Btu. This equation assumes R-16 (SI: R-2.8) insulating panels with white upper surfaces and foil faces on the under surface.

$$Q_P = (0.4)\,(A_C)\,((4)\,(DB_{max}) - DB_{min} - 200)$$

where A_c is the horizontal surface area of pond in ft² and temperatures are in °F.

STEP 7. Consider whether fans will be used to stir the room air below the roofpond (to help the heat exchange between the pond and the room, as well as to provide comfort), and determine the value to be used for h, the overall heat transfer coefficient. A corrugated steel deck ceiling is assumed.

Air Velocity	fpm [m/s]	0 [0]	44 [0.22]	73 [0.37]	115 [0.60]
Value of h		1.25	1.47	1.53	1.70

STEP 8. Determine the highest comfortable internal air temperature, t_{op}, from the comfort criteria provided in Chapter 6, Figure 6.16.

STEP 9. Calculate the maximum allowable internal air temperature, t_{imax}, which will be higher than t_{op} by a factor F that is related to your climate's characteristics (see Table 14.8, Part A, for values of F).

$$t_{imax} = t_{op} + F$$

STEP 10. Determine the allowable temperature swing of the roofpond.

a. Calculate the maximum pond temperature, *max. pond t*,

$$\text{max.}\textit{pond } t = \frac{\begin{pmatrix}\text{peak hourly heat gain} \\ \text{including internal heat gains, Btu/h}\end{pmatrix}}{h(A_c)}$$

where t_{imax} is from step 9, h is from step 7, and A_c is the horizontal surface area of the pond in square feet.

b. Calculate the minimum pond temperature, *min. pond t*, which depends on whether evaporative cooling will be used to help lower the pond's temperature and on several characteristics of the pond and the building.

1. For a dry pond surface:

$$\text{min. } \textit{pond } t_{dry} = DB_{min} + 1.5\,°\text{F} \pm \text{corrections } °\text{F}$$

(if any, from Table 14.8, Part B)

2. For a wet pond surface:

$$\text{min.}\textit{pond } t_{wet} = DB_{min} - \frac{DB_{min} - WB_{min}}{2}$$

c. Find the pond temperature swing Δt_p:

$$\Delta t_{p\,dry} = \text{max. } \textit{pond } t - \text{min. } \textit{pond } t_{dry}$$

$$\Delta t_{p\,wet} = \text{max. } \textit{pond } t - \text{min. } \textit{pond } t_{wet}$$

(If neither *min. pond* t_{dry} nor *min. pond* t_{wet} is lower than *max. pond t*, a roofpond cannot be used for cooling.)

STEP 11. Determine the required pond depth D_{rpd} (in inches).

$$D_{rpd} = \frac{(0.19)\,(Q_E + Q_I + Q_P)}{(\Delta t_p)\,(A_c)}$$

where Q_E, Q_I, and Q_P are the daily pond heat gains from steps 3, 5, and 6; Δt_p is from step 10; and A_c is the horizontal surface area of the pond in square feet.

Note: If D_{rqd} is less than 4 in. [100 mm], consider reducing the pond size and recalculating. If D_{rqd} is much more than 4 in. [100 mm], consider a larger pond area, more air motion indoors, the use of a wet pond surface to more closely approach the optimum 4 in. depth, or see optional step 12.

STEP 12. (optional) Auxiliary mechanical air-conditioning may offer a more economical alternative than increased pond size. The size, in tons (12,000 Btu/h) of air conditioning required, is determined as follows:

a. desired D_{rqd} is 4 in. [100 mm] optimum.

$$\Delta t_p = \left(\frac{0.19}{D}\right)\frac{(Q_E + Q_I + Q_P)}{A_c}$$

b. so, *max. pond t* = *min. pond t* + Δt_p

c. and tons of air-conditioning at peak total hourly heat gain

$$= \frac{(\text{Btu/h} - (h(A_c)))\ \ (t_{imax} - \text{max. } pond\ t)}{12{,}000}$$

where the peak total hourly heat gain includes internal gains, h is from step 7, A_c is the horizontal surface area of the pond in square feet, t_{imax} is from step 9, and *max. pond t* is from step 10.

EXAMPLE 14.10 (Extended from Example 14.5.)

What size should the roofpond be for the Albuquerque office building for which we previously predicted that a 4-in.-deep [100 mm] pond equal to three-fourths of the building's floor area would be sufficient? Recall that the one-story office is about 4000 ft² [372 m²] in area. Try a 4-in.-deep [100 mm] pond of 3000 ft² [279 m²].

SOLUTION

STEP 1. Albuquerque, design data:

DB_{max} is 93.4°F (Appendix, Table B.1) [34.1°C (Table B.3)]
MDR is 22.9°F (Appendix, Table B.1) [12.7°C (Table B.3)]
DB_{min} is 93.4 − 22.9 = 70.5°F [21.4°C]

July RH maximum (nighttime) is approximately 50% (local data).

July TA is 79.2°F [26.2°C] (Appendix C).

WB_{min} is from Appendix G, Figure G.1; at the intersection of the DB_{min} (70.5°F) and 50% RH, this is 58.8°F [14.9°C].

July operating hours N for this dry, high area should be close to the minimum for TA = 79.2°F [26.2°C]; from Figure 14.34, this is about 160 hours.

STEP 2. Determine the peak hourly gain. Earlier in this exercise, a total gain of 15 Btu/h ft² [47.3 W/m²] was assumed as an average. Assume that 9 Btu/h ft² [28.4 W/m²] of this total represents the load from electric lighting, people, and equipment, and that 6 Btu/h ft² [18.9 W/m²] is due to enclosure and ventilation gains. (No heat gains to the interior are assumed through the roofpond.)

$$\text{peak hourly gain} = 6\ \text{Btu/h ft}^2 \times 4000\ \text{ft}^2$$
$$= 24{,}000\ \text{Btu/h}\ [7035\ \text{W}]$$

STEP 3.

$$Q_E = \frac{(24{,}000\ \text{Btu/h}) \times (160\ \text{h})}{31\ \text{days}}$$
$$= 123{,}871\ \text{Btu}\ [36.3\ \text{kWh}]\ \text{per day}$$

STEP 4.

$$\text{internal gains} = 9\ \text{Btu/h ft}^2 \times 4000\ \text{ft}^2$$
$$= 36{,}000\ \text{Btu/h}$$

STEP 5.

$$Q_I = (36{,}000\ \text{Btu/h})\ (9\ \text{hours of operation})$$
$$= 324{,}000\ \text{Btu}\ [95\ \text{kWh}]\ \text{per day}$$

STEP 6.

$$Q_P = 0.4\,(3000\ \text{ft}^2)\,(4 \times 93.4°\text{F} - 70.5°\text{F} - 200)$$
$$= 1200 \times (373.6 - 70.5 - 200)$$
$$= 123{,}720\ \text{Btu}\ [36.3\ \text{kWh}]\ \text{per day}$$

STEP 7. Fans will be used, at 115 fpm [0.6 m/s], for added comfort as well as for heat transfer. Therefore, $h = 1.7$.

STEP 8. From Chapter 6, Figure 6.22*b*, with 115 fpm [0.6 m/s] air speed, it appears that 83°F [28.3°C] is within the comfort zone. So, $t_{op} = 83°\text{F}$.

STEP 9. $t_{imax} = t_{op} + F$. In Table 14.8, Part A, Albuquerque's conditions appear to be somewhere between those of San Antonio and those of Phoenix, so interpolate F to be 1.75.

$$t_{imax} = 83 + 1.7$$
$$= 84.75°\text{F}\ [29.3°\text{C}]$$

STEP 10.

(a) *max.pond t*

$$= 84.75°\text{F} - \frac{(24{,}000\ \text{Btu/h} + 36{,}000\ \text{Btu/h})}{(1.7 \times 3000\ \text{ft}^2)}$$

PASSIVE ENVIRONMENTAL SYSTEMS

$= 84.75 - 11.75$

$= 73.0°\text{F}\ [22.8°\text{C}]$

(b) Assume a dry pond surface. Because this is a 4 in.-deep [100 mm] pond that is fully exposed to sky, with no night internal load, there is no correction factor from Table 14.8, Part B.

$$\text{min. } pond\ t_{dry} = \text{DB}_{min} + 1.5°\text{F}$$
$$= 70.5°\text{F} + 1.5°\text{F}$$
$$= 72.0°\text{F}\ [22.2°\text{C}]$$

For a wet pond,

$$\text{min. } pond\ t_{wet} = \text{DB}_{min} - (\text{DB}_{min} - \text{WB}_{min})/2$$
$$= 70.5°\text{F} - (70.5°\text{F} - 58.8°\text{F})/2$$
$$= 64.7°\text{F}\ [18.2°\text{C}]$$

(c) The pond temperature swing is therefore:

$$\Delta t_{p\ dry} = 73.0 - 72.0 = 1.0°\text{F}\ [0.6°\text{C}]$$
$$\Delta t_{p\ wet} = 73.0 - 64.7 = 8.3°\text{F}\ [4.6°\text{C}]$$

STEP 11. Determine the required dry pond depth (D_{rqd}) in inches.

$$D_{rqd\ dry} = \frac{(0.19)(Q_E + Q_I + Q_P)}{(\Delta t_{p\ dry})(A_c)}$$

$$D_{rqd\ dry} = \frac{(0.19)\left(\begin{array}{c}123{,}871\ \text{Btu} + 324{,}000\ \text{Btu}\\ + 123{,}720\ \text{Btu}\end{array}\right)}{(1.0°\text{F})\,(3000\ \text{ft}^2)}$$

$$D_{rqd\ dry} = 36.2\ \text{in.}\ [919\ \text{mm}].$$

For a wet pond,

$$D_{rqd\ wet} = \frac{(0.19)\left(\begin{array}{c}123{,}871\ \text{Btu} + 324{,}000\ \text{Btu}\\ + 123{,}720\ \text{Btu}\end{array}\right)}{(8.3°\text{F})\,(3000\ \text{ft}^2)}$$

$$D_{rqd\ wet} = 4.4\ \text{in.}\ [112\ \text{mm}]$$ ■

It appears that a nominal 5-in. [127-mm] deep roofpond with a wet surface of somewhat less than 3000 ft^2 [279 m^2] would be sufficient for this building in this climate. However, a dry roofpond of the same depth would be insufficient. Moreover, the required 36 in. [919 mm] depth for a dry roofpond would be deemed not feasible in many design instances due to the weight of the water at such a depth. Note: climate change affecting Albuquerque has made a dry roofpond (previously feasible) much less feasible for the same building situation.

Once the type, area, and the depth of roofponds have been determined, questions arise about the architectural integration of such large horizontal surfaces and their relative emphasis on heating or cooling. Figure 14.35 shows variations in the treatment of containers and the insulating panels to match the most important thermal role of roofponds; Figure 14.36 shows three typical approaches to the placement of roofponds on buildings. A valuable source of information about the container bags, sliding insulation, roof decks, and other components needed for roofponds is the *California Passive Solar Handbook* by Phillip Niles and Ken Haggard, written for the California Energy Commission in 1980 (Niles and Haggard, 1980). Commercially available products include the Skytherm® system, developed and pioneered by Harold R. Hay.

A variation of the roofpond (applying primarily to the cooling mode) is to move the water rather than move the insulation. Water-impervious insulation (such as Styrofoam) is secured in place above the metal ceiling at a distance sufficient to be a reservoir for the cooling water. At night, this water is pumped onto the top of the insulation, where it cools by radiation and evaporation, then trickles slowly back into the reservoir. This variation was tested at the School of Engineering Technology, University of Nebraska at Omaha campus. In winter, the reservoir of water above the metal ceiling acted as thermal mass for a DG passive system. However, cold winter rain must be excluded from that reservoir.

(g) Earth Tubes

The use of the earth as a heat sink, in our culture at any rate, is still in an early development stage. The most

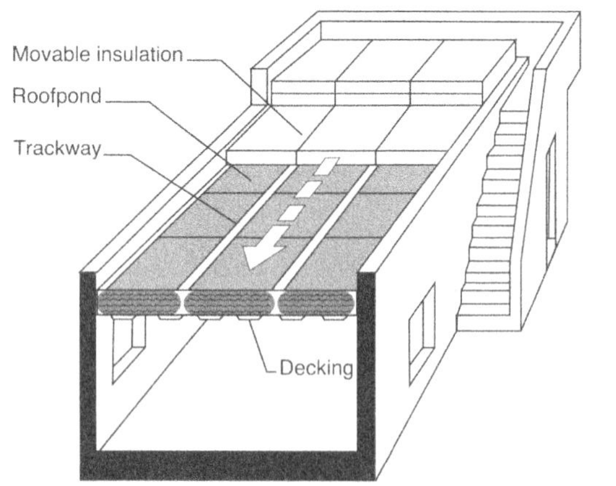

(a)

Figure 14.35 *(a) Key components of roofpond systems with movable insulation. (b–i) Variations on roofponds for optimization of heating or cooling performance. (Drawn by Sharon Alitema, after Niles and Haggard, 1980.)*

PASSIVE ENVIRONMENTAL SYSTEMS

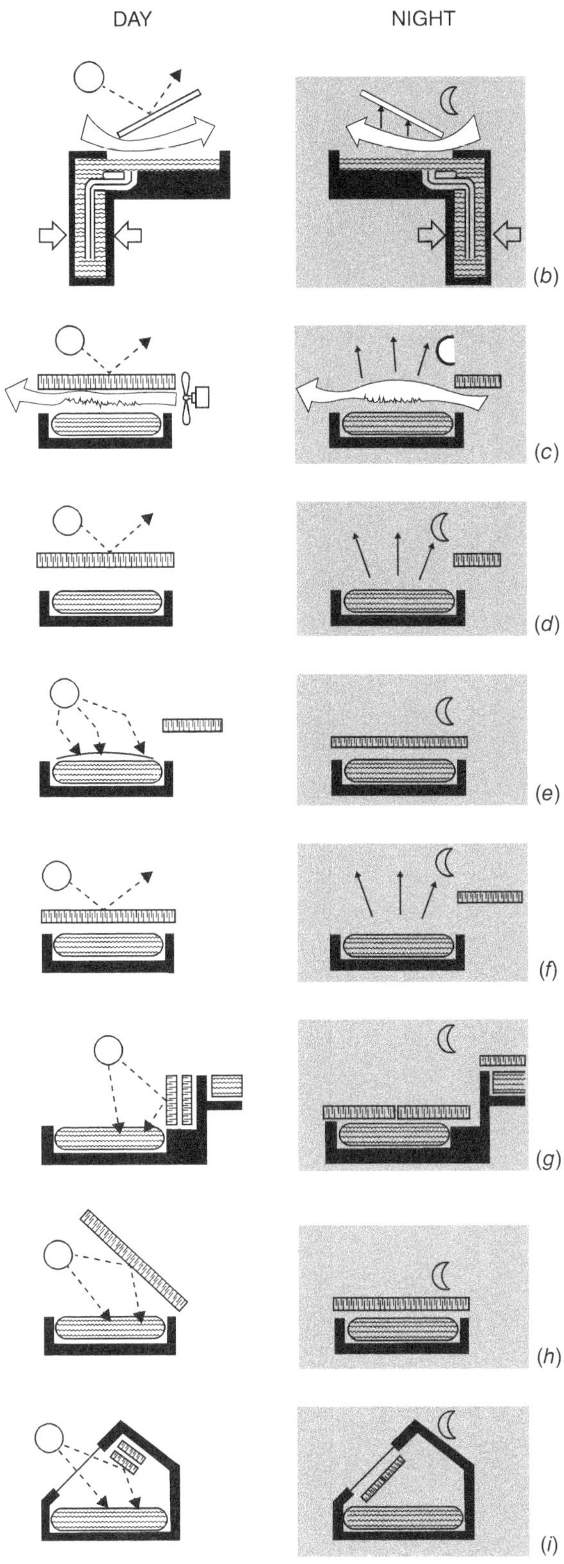

Figure 14.35 *(continued)*

direct application would be an underground building with uninsulated concrete walls set against the soil. The problem, of course, is that winter heat loss will likely exceed summer loss; if solar heat can be admitted (e.g., through skylights) to counterbalance the increased winter loss, uninsulated walls sized for the desired summer loss become more attractive. (However, except in dry climates, condensation on walls may pose a seasonal or even year-round problem.) A more thorough discussion of this option can be found in Watson and Labs (1983).

The long-term potential of the earth as a heat sink is lessened by the fact that soils are relatively slow heat conductors. For example, we ignore winter heat loss through concrete slabs to the earth below, calculating instead the heat losses through the slab's exposed perimeter.

Earth tubes (as discussed under design guidelines in Section 14.4) are devices for cooling the incoming ventilating air through earth contact before it enters the building. A small amount of air (perhaps equal to the minimum fresh air requirements—see Appendix F, Tables F.1 and F.2) is usually brought in through several tubes. These tubes are usually 8–20 in. [200–500 mm] in diameter, are buried at a depth of 5–10 ft [1.5–3 m], and are up to 200 ft [61 m] long. Appendix G, Table G.8 summarizes several tested summertime applications.

The lowest temperature to which air can be cooled in such tubes will approach ground temperature; the more slowly the air moves through the tube, the more time there will be for cooling. The air temperature might come within 4°F [about 2°C] of the soil temperature under good conditions.

To approximate soil temperature (in I-P units) around the tube:

1. Determine t_{GW}, the groundwater temperature (in °F) (assumed equal to deep-underground earth temperature) from the map in Figure 14.37.
2. Determine t_{amp}, the average amplitude of surface temperature (in °F), using the map in Figure 10.5*c*.
3. Determine CF, the amplitude correction factor, depending upon whether the soil outside the earth tube is dry, average, or wet during the cooling period. The amplitudes, or seasonal variations in earth temperature, diminish to near zero at identifiable depths:

	CF
Dry soil	14 ft [4.25 m]
Average soil	18 ft [5.5 m]
Wet soil	22 ft [6.7 m]

4. Determine D, the depth (ft) at which the earth tube will be installed.

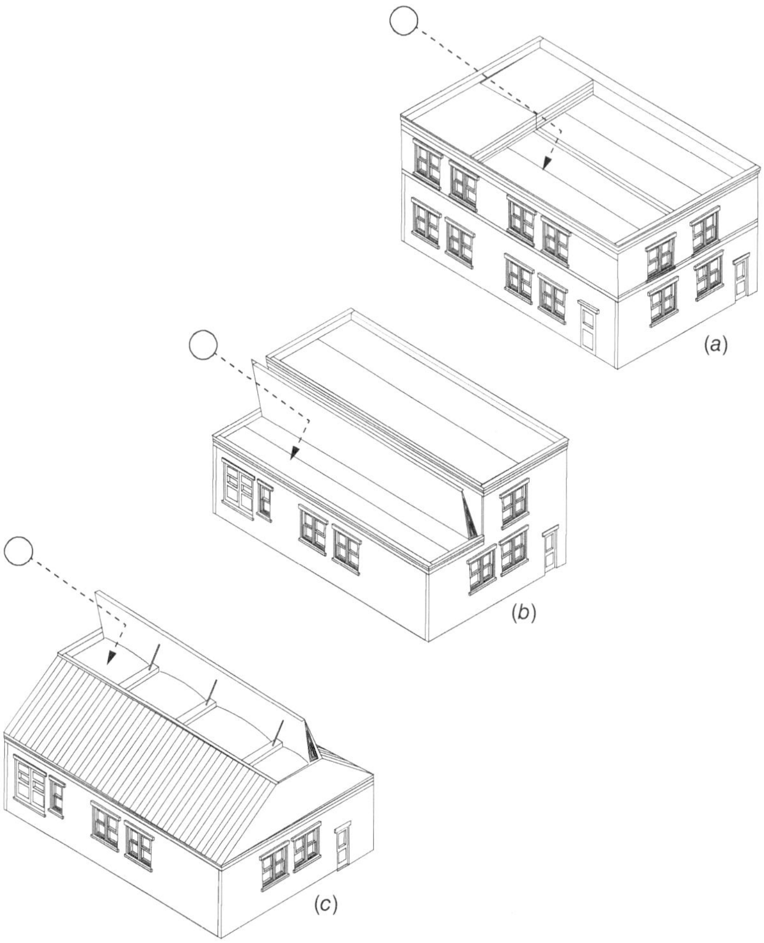

Figure 14.36 *Three examples of roofpond integration with building form. (a) Insulating panels slide open to stack over exterior (or untempered) space. See Figure 14.14 for this system in use. (b) Hinged insulating panels with reflective undersurfaces can act as reflectors to increase winter insolation. However, some interference with summer night sky radiation can result, because the pond surface cannot "see" the entire sky dome. (c) Bifold insulating panels also act as reflectors, interfering less with diffuse solar radiation and allowing somewhat more exposure to the night sky than (b). (Drawn by Sharon Alitema, after Niles and Haggard, 1980.)*

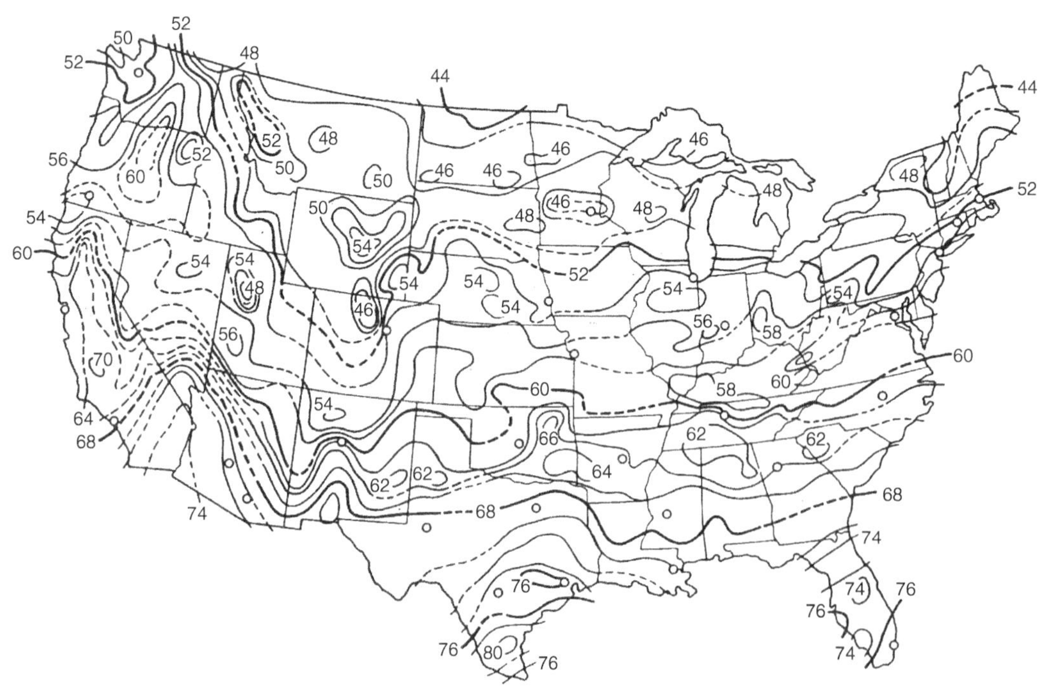

Figure 14.37 *Distribution of well-water temperature (°F) in the United States, adapted from the National Ground Water Association. These temperatures may be assumed to be the yearly average ground temperatures. For seasonal temperature change amplitude, see Chapter 10, Figure 10.5. (From Bowen et al., 1981.)*

5. Determine t_{SG}, the late summer ground temperature (°F) around the earth tube:

$$t_{SG} = (t_{GW}) + (t_{AMP})(1 - (D/CF))$$

(Note that when the tube is very deep and D ≥ CF, then $t_{SG} = t_{GW}$.)

6. Determine the temperature difference that makes cooling possible:

$$\Delta t = \text{desired indoor temperature} - t_{SG}$$

7. Determine the long-term cooling rate:

Btu/h ft of tube length

$$= (\Delta t)(C)(\text{area of tube surface, ft}^2/\text{ft of length})$$

where C is a factor that accounts for the long-term effects of heat flow from the tube to the soil:

In I-P units:

$C = 0.11$ for dry soil

(soil conductivity k = 0.25 Btu/h ft °F)

$C = 0.28$ for average soil (k = 0.75)

$C = 0.44$ for wet soil (k = 1.5)

In SI units:

$C = 0.62$ for dry soil

$C = 1.59$ for average soil

$C = 2.50$ for wet soil

This represents an attempt to simplify a complex problem. The airflow rate is ignored, even though, with very slow flow, there is more time for heat exchange in the tube. The longer the tube is flooded with hot air, the higher will be the earth temperature. For more detailed information about this strategy and other approaches to earth tubes, see Givoni (1994); C.E. Francis, "Earth Cooling Tubes, Case Studies of Three Midwest Installations," in Bowen et al. (1981); Abrams (1986); and Brown et al. (1992).

EXAMPLE 14.11 (Extended from Example 14.6.)

A partially underground nature center of 4000 ft² [372 m²] in northern Pennsylvania has an hourly heat gain of 9.3 Btu/h ft² [29.3 W/m²], of which 2.2 Btu/h ft² [6.9 W/m²] are due to required outdoor air ventilation. A long trench at a depth of 6 ft [1.8 m] in heavy, damp soil must be dug for a utility connection. If one earth tube, used only for cooling the outdoor ventilation air, is installed at the bottom of the trench, what length and diameter will be needed?

SOLUTION

The total cooling needed from the earth tube is:

$$(4000\ \text{ft}^2)(2.2\ \text{Btu/h ft}^2) = 8800\ \text{Btu/h}\ [2580\ \text{W}]$$

(In Example 14.6, using design guidelines, assuming heavy damp soil, a 65 ft-long tube [19.8 m], 20 in. [508 mm] in diameter promised to deliver about 5000 Btu/h [1465 W]; two such tubes will deliver about 10,000 Btu/h [2931 W].)

1. Determine t_{GW}: From Figure 14.37, northern Pennsylvania has an average underground temperature of 52°F [11.1°C].
2. Determine t_{AMP}: The seasonal amplitude (from Chapter 10, Figure 10.5) is about 18°F [10°C].
3. Determine CF: In wet soil, this amplitude extends to 22 ft [6.7 m].
4. Determine D: The given trench depth is 6 ft [1.8 m].
5. Determine t_{SG}: The late summer ground temperature (working in I-P units) around the earth tube:

$$\begin{aligned} t_{SG} &= (t_{GW}) + (t_{AMP})(1 - (D/CF)) \\ &= 52 + 18(1 - (6/22)) \\ &= 52 + 13 \\ &= \text{about } 65°\text{F}\ [18.3°\text{C}] \text{ at most} \end{aligned}$$

6. Determine the temperature difference that makes cooling possible:

$$\Delta t = \text{desired indoor temperature} - t_{SG}$$

In this case, we could accept a maximum 82°F [27.8°C] indoor air temperature (the walls will be cooler, thanks to their underground location):

$$\begin{aligned} \Delta t &= \text{desired indoor temperature} - t_{SG} \\ &= 82 - 65 \\ &= 17°\text{F}\ [9.5°\text{C}] \end{aligned}$$

7. Determine (again using I-P units) the long-term cooling rate: Btu/h ft of tube length

$$= (\Delta t)(C)(A)$$

where A is area of tube surface, ft²/ft of length

Here, C = about 0.44 Btu/h ft² °F for wet soil.

Try first the 20-in. diameter [508-mm] tube from the earlier design guidelines:

Btu/h ft of tube length

$$\begin{aligned} &= (\Delta t)(C)(\text{area of tube surface, ft}^2/\text{ft of length}) \\ &= (17°\text{F})(0.44)((20\ \text{in.} \times \pi \times 1\ \text{ft length})/(12\ \text{in./ft})) \\ &= 39\ \text{Btu/h ft} \end{aligned}$$

To deliver the full 8800 Btu/h [2580 W], the tube would need to be about 8800/39 = 225 ft [69 m] in length. The design guideline promised considerably more, in this case, than could be delivered.

In early summer, the earth temperature around the tube would be at its yearly average temperature of 52°F [11.1°C]. With this larger Δt, 82 − 52 = 30°F [16.7°C], the performance improves:

Btu/h ft of tube length

$$= (\Delta t)(C)(\text{area of tube surface, ft}^2/\text{ft of length})$$

$= (30°F)(0.44)((20\ in. \times \pi \times 1\ ft\ length)/(12\ in./ft))$

$= 69$ Btu/h ft

and the tube length under those conditions would be about 8000/69 = 127 ft [39 m]. ■

(h) Passive Cooling Summary

As a summary example of a passively cooled and solar-heated building, consider the Visitor Center for the Antelope Valley California Poppy Reserve, about 85 miles [135 km] northeast of Los Angeles (Figure 14.38). Set in a high desert where winter temperatures sometimes fall below 20°F [−7 °C], the building is 100% passively solar heated by a combination of direct-gain and Trombe wall strategies, interior thermal shades, and thermally massive ceiling, walls, and floor. An earth-sheltered, thermally massive building was chosen for its suitability to the environmentally sensitive site, as well as for its year-round thermal advantage in a desert climate. On summer days when air temperatures frequently exceed 100°F, a combination of an earth tube and evaporative cooling provides a modest supply of fresh air without overheating. Continuous power ventilation of the building at night draws the cool outdoor air through the tube, then over the thermally massive surfaces to provide added cooling capacity for the following day. The 2100 ft^2 [195 m^2] building also is served by an 8 kW wind generator and has its own well.

EXAMPLE 14.12 Use the psychrometric chart to investigate the cooling process at the Antelope Valley Visitor Center. The earth tube is 24 in. [610 mm] in diameter and 150 ft [45.7 m] long, at an average depth of 6 ft [1.8 m].

SOLUTION

First, determine the design conditions. For Palmdale, assume the design condition is:

Design DB = 104°F

Design WB = 68°F

MDR = 37°F

Thus, a nightly low of 104 − 37 = 67°F at design conditions [40/20°C, with MDR of 20.6°C]. Enter the psychrometric analysis (Figure 14.39) at the convenient and slightly more severe conditions of 105°F DB, 70 WB noted as point A [40.6 and 21.1°C]. Note that RH = about 17% in this desert location.

Next, examine the earth tube conditions:

1. Determine t_{GW}: From Figure 14.37, the deep-earth temperature is about 64°F [17.8°C].
2. Determine t_{AMP}: The seasonal amplitude (from Chapter 10, Figure 10.5) is about 19°F [10.6°C].
3. Determine CF: In dry (desert) soil, this amplitude extends to 14 ft [4.3 m].
4. Determine D: The given trench depth is 6 ft [1.8 m].
5. Determine t_{SG}: The late summer ground temperature around the earth tube:

$$t_{SG} = (t_{GW}) + (t_{AMP})(1 - (D/CF))$$

$$= 64 + 19(1 - (6/14))$$

$$= 64 + 10.9$$

$$= \text{about } 75°F\ [23.9°C] \text{ at most}$$

Note that during the April poppy season, the earth temperature is likely to be at the yearly average of 64°F [17.8°C].

The air in the earth tube will be about 4°F [2.2°C] above this late-summer temperature, or 74.9 + 4 = approximately 79°F [26.1°C]. (This may not sound very cool, but compare it to 104°F [40°C] outdoors.) To follow what happens to the air in the tube, move horizontally on the psychrometric chart in Figure 14.39, from point *A* toward point *B* at 79°F DB [26.1°C]: at 79°F DB and about 62°F WB, the RH in the air has risen to about 34%.

At this point, the air enters the building through a small fountain consisting of water droplets. To chart this evaporative cooling, move from point *B* along the constant 62°F [16.7°C] WB line. If this fountain is a really good heat exchanger, we might achieve 80% of the distance from point *B* to saturation; call that point *C* (about 66°F DB, 62°F WB [18.9°C DB, 16.7°C WB], RH now = 80%). This combination (of pessimistic air/earth temperatures and optimistic evaporative cooling expectations) is actually below the summer comfort zone!

The air now begins to pick up heat from the building's interior. With few sources of latent heat, we follow sensible heat gain horizontally to the right; if we are to maintain an interior temperature of, say, 82°F [27.8°C]—with air motion and cool underground surfaces—the air can be heated to point *D*: 82°F DB, 67°F WB [27.8°C DB, 19.4°C WB], RH = about 47%. In this condition, it is discharged through the stack ventilator at the west end of the building. (Note that it is much cooler than the outdoor air at point *A*, even at the point of discharge.)

But does this theoretical process actually occur? We must complete the calculation process:

6. Determine the temperature difference that makes cooling possible. For the psychrometric equivalent process,

$$\Delta t = \text{outdoor temperature} - t_{SG}$$

$$= 105°F - 75°F = 30°F$$

$$[40.6°C - 23.9°C = 16.7°C]$$

7. Determine the long-term cooling rate:

Btu/h ft of tube length

$$= (\Delta t)(C)(\text{area of tube surface, ft}^2/\text{ft of length})$$

(a)

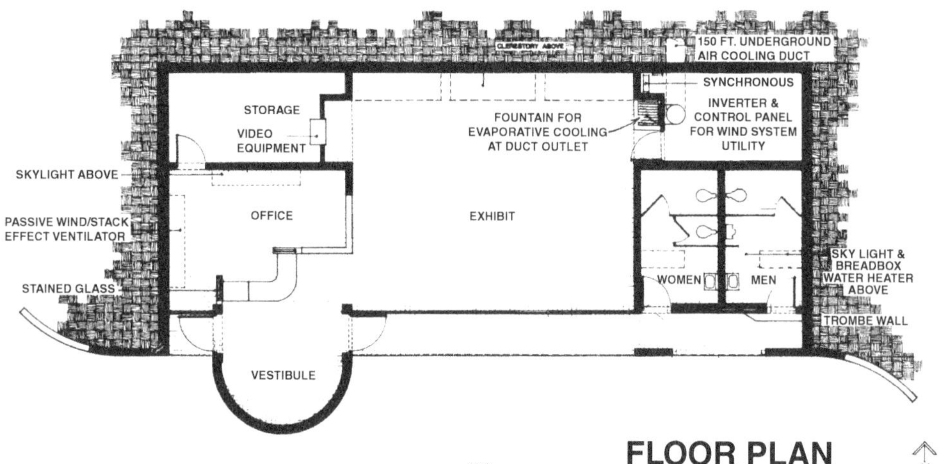

(b)

Figure 14.38 *Visitor Center for the Antelope Valley California Poppy Reserve near Palmdale, California. (a) The Visitor Center is set into the field of poppies; an 8-kW wind generator and the earth tube intake can be seen on the hillside above. (b) Plan shows earth sheltering for an elongated east–west axis, served both by direct-gain and Trombe wall solar heating and by earth tube/evaporative cooling. (c) Solar energy enters mostly through the south glass; the glass-block skylights and north clerestory are mostly for even distribution of daylighting. At rest rooms, a vented Trombe wall is used. Thermal shades on the interior protect the south glass and north clerestory on winter nights; a roll-down exterior sunscreen greatly reduces direct solar gain through south glass in summer. Concrete block walls and concrete roof and floor provide thermal mass, as useful for summer cooling as for winter solar heating. (d) In summer, the desert air first passes through an earth tube (assisted by a fan), then is evaporatively cooled within the building, in sight of visitors. This cooled air picks up the building's heat and is then exhausted from the stack ventilator. (Courtesy of the Colyer/Freeman Group.)*

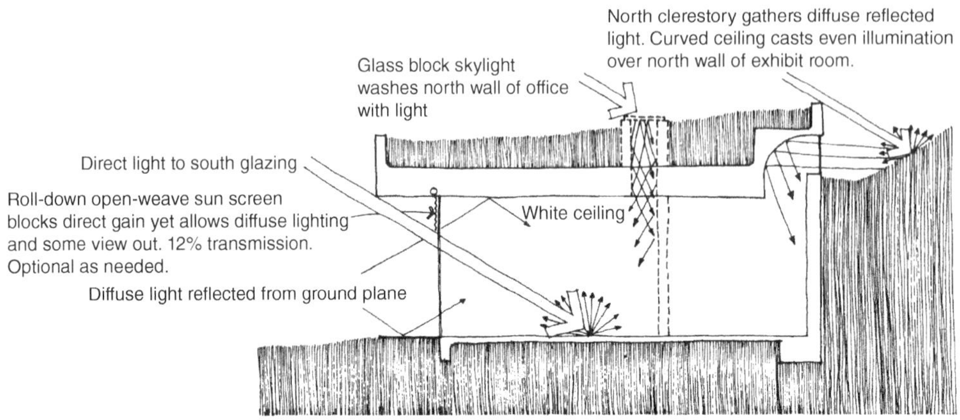

(c)

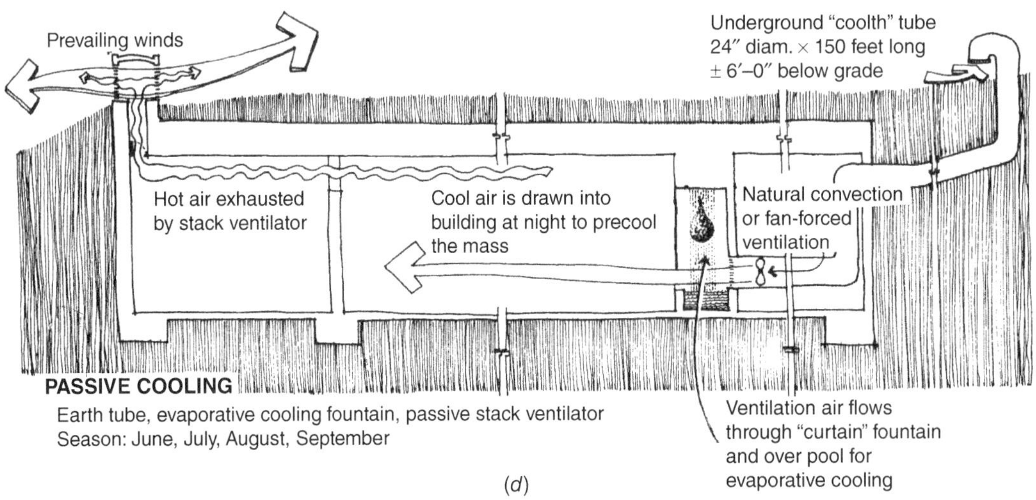

(d)

Figure 14.38 *(continued)*

Here, $C = 0.11$ for dry soil.

$$= (30°\text{F})\,(0.11)\,((24\text{ in.} \times \pi \times 1\text{ ft length})/(12\text{ in./ft}))$$

$$= 20.7\text{ Btu/h ft [equivalent to } 19.9\text{ W/m]}$$

Note that this rate is similar to that of the tubes in Appendix G, Table G.8, most of which are smaller in diameter but are located in moist soil. This tube will therefore cool outdoor air at the rate of

(20.7 Btu/h ft) × (150 ft) = 3105 Btu/h under summer design conditions [19.9 W/m × 45.7 m = 910 W].

If the air in the tube is to drop from 105°F [40.6°C] to the 79°F [26.1°C] assumed at point *B* in Figure 14.40, at what rate must it move?

$$\text{Btu/h} = \text{flow rate, cfm} \times 1.1 \times \Delta t$$

$$3105 = \text{cfm} \times 1.1 \times (105 - 79)$$

$$[\text{W} = \text{L/s} \times 1.2 \times \Delta t]$$

The flow rate, 3105/28.1 = 111 cfm [52.4 L/s], is substantially less than the flow rates typical of Appendix G, Table G.8. Checking the velocity in this 24-in.-diameter [610-mm] tube,

$$\text{ft/min} = \text{cfm/ft}^2 \text{ of cross-section}$$

$$\text{velocity, ft/min} = 111\text{ cfm}/\pi\,(1\text{ ft radius})^2$$

$$= \text{about } 35\text{ ft/min } [0.18\text{ m/s}]$$

again, substantially less than the velocities typical of Appendix G, Table G.8.

It is likely that much more outdoor air will be pulled through the tube, thus entering the building at a higher temperature. Then evaporative cooling through the fountain will be necessary to deliver air that is within the comfort zone.

For example, assume that for the cooling rate of 20.7 Btu/ft of length (based upon 105°F outside air, 75°F t_{SG}) and resulting total cooling of 3105 Btu/h, and assume the fan provides a modest flow rate of 500 cfm. Then

$$3105\text{ Btu/h} = 500\text{ cfm} \times 1.1 \times \Delta t$$

$$\Delta t = 5.65°\text{F}$$

$$[910\text{ W} = 236\text{ L/s} \times 1.2 \times \Delta t]$$

$$[\Delta t = 3.2°\text{C}]$$

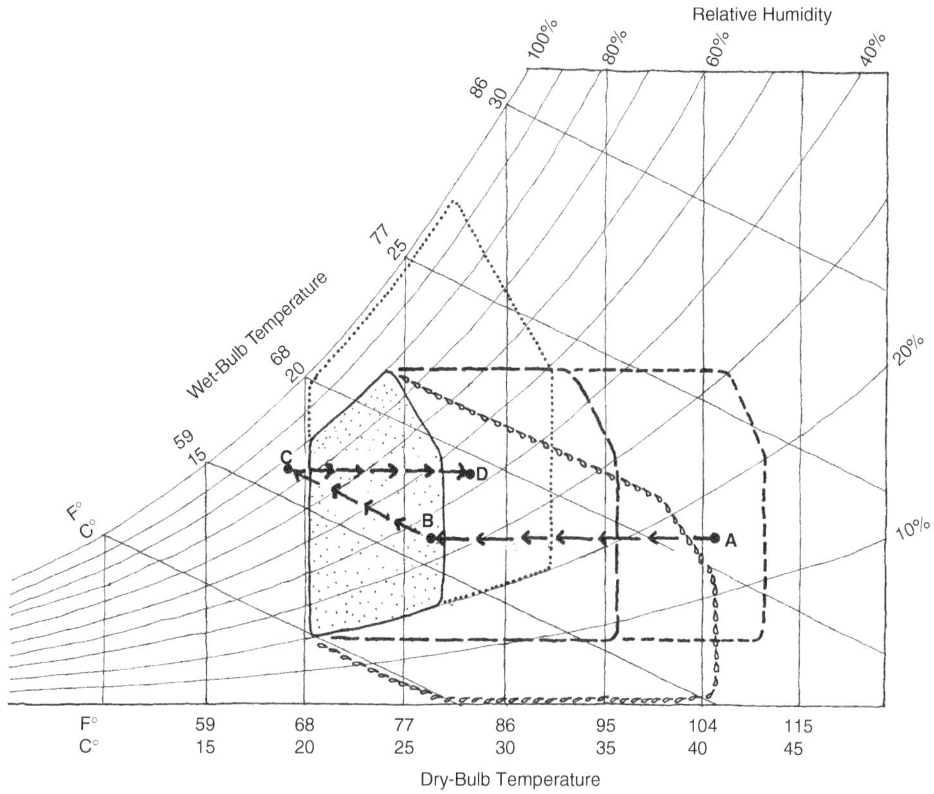

Figure 14.39 *Following the path of passively cooled air at the Antelope Valley Visitor Center. Under summer design conditions, air enters the earth tube at point A. It is sensibly cooled within the tube, perhaps to the extent shown at point B. It then enters the building through a small fountain, perhaps being evaporatively cooled as far as point C. As the cooled air then picks up sensible heat from the building, the air heads toward point D, where it is exhausted through the stack ventilator.*

The outdoor air then leaves the tube (new point *B*) at about 104°F − 5.65°F≈ 99°F DB and 68°F WB [37°C DB, 20°C WB]. With successful exposure to evaporative cooling in the fountain, the air might reach (new point C) 75°F DB, 68°F WB [23.9°C DB, 20°C WB]. This is quite humid, but within the comfort zone and likely welcome in this desert environment. It then could absorb some heat from the building before it is exhausted. ■

References and Resources

Abrams, D. W. 1986. *Low Energy Cooling*. New York: Van Nostrand Reinhold.

ASHRAE. 1997. *ASHRAE Handbook—Fundamentals*. Atlanta, GA: American Society of Heating, Refrigerating and Air-Conditioning Engineers.

ASHRAE. 2021. *ASHRAE Handbook—Fundamentals*. Atlanta, GA: American Society of Heating, Refrigerating and Air-Conditioning Engineers.

Berry, B., J. Bown, J. Thornton, J. Turzynski, and M. Walton. 1995. "The New Inland Revenue Centre, Nottingham." *Arup Journal* 30(4): 11–17.

Bowen, A., E. Clark, and K. Labs. 1981. *Passive Cooling*. Boulder, CO: American Solar Energy Society.

Brown, G. Z. and M. DeKay. 2014. *Sun, Wind & Light*, 3rd edn. New York: John Wiley & Sons.

Brown, G. Z., B. Haglund, J. Loveland, J. Reynolds, and S. Ubbelohde. 1992. *InsideOut: Design Procedures for Passive Environmental Technologies*, 2nd edn. New York: John Wiley & Sons.

Chandra, S., P. W. Fairey, and M. Houston. 1986. *Cooling with Ventilation*. Golden, CO: Solar Energy Research Institute.

Cunningham, W.A. and T.L. Thompson. 1986. Passive Cooling with Natural Draft Cooling Towers in Combination with Solar Chimneys. *Proceedings of PLEA 1986 Conference, Pecs, Hungary*. PLEA (Passive and Low Energy Architecture) Association.

Fleischhacker, P., D. Bentley, and G. Clark. 1982. "A Simple Verified Methodology for Thermal Design of Roof Pond Cooled Buildings," in *Proceedings of the 7th National Passive Solar Conference*. Boulder, CO: American Solar Energy Society.

Ford, B., R. Schiano-Phan, and J. Vallejo. 2020. *The Architecture of Natural Cooling*, 2nd edn. New York: Routledge.

Gardner, I. and D. Hadden. 1992. "UK Pavilion, Expo '92, Seville'." *Arup Journal* 27(3): 3–7.

Givoni, B. 1993. "Semiempirical Model of a Building with a Passive Evaporative Cool Tower." *Solar Energy* 50(5): 425–434.

Givoni, B. 1994. *Passive and Low Energy Cooling of Buildings*. New York: Van Nostrand Reinhold.

Johnson, T. 1981. *Solar Architecture: The Direct Gain Approach*. New York: McGraw-Hill.

Kang, D. and R.K. Strand. 2009. Simulation of Passive Down-Draught Evaporative Cooling (PDEC) Systems in ENERGYPLUS. *Proceedings of the Eleventh International IBPSA Conference, Glasgow*. International Building Performance Simulation Association.

Mazria, E. 1979. *The Passive Solar Energy Book*. Emmaus, PA: Rodale Press.

Mead, D., B. Baptista, and A. Corney. 2013. Net Zero Energy Through Passive Downdraft Ventilation: The Design & Operation of the Conrad N. Hilton Foundation. *Proceedings of BESS-SB13 CALIFORNIA: Advancing Towards Net Zero*, Pomona, CA.

Miller, H. (ed.). 1980. *Passive Cooling Handbook: Proceedings of the Passive Cooling Workshop, Held in Conjunction with the Fifth National Passive Solar Conference*. Boulder, CO: American Solar Energy Society.

Milne, M. and B. Givoni. 1979. "Architectural Design Based on Climate," in *Energy Conservation Through Building Design*, D. Watson (ed.). New York: McGraw-Hill.

Moore, F. 2004. *Environmental Control Systems: Heating, Cooling, Lighting*. New York: McGraw-Hill, Inc.

Niles, P. and K. Haggard. 1980. *California Passive Solar Handbook*. Sacramento, CA: California Energy Commission.

NOAA. 2025. Local Climatological Data (LCD). *National Centers for Environmental Information*. Asheville, NC: National Oceanic and Atmospheric Administration. www.ncei.noaa.gov/products/land-based-station/local-climatological-data

Olgyay, V. 1963. *Design with Climate: Bioclimatic Approach to Architectural Regionalism*. Princeton, NJ: Princeton University Press.

Santamouris, M. 2007. *Advances in Passive Cooling*. London: Earthscan.

SERI. 1981. *A New Prosperity: The SERI Solar Conservation Study*. Andover, MA: Brick House Publishing.

Stromberg, R. P. and S. O. Woodall. 1977. *Passive Solar Buildings: A Compilation of Data and Results, SAND-77-1204*. Albuquerque, NM: Sandia Labs.

Thompson, T., N. Chalfoun, and M. Yoklic. 1994. "Estimating the Performance of Natural Draft Evaporative Coolers." *Energy Conversion Management* 35(11): 909–915.

Utzinger, M. and J.H. Wasley. n.d. Building Balance Point. Vital Signs Curriculum Project Resource Module. webpages.uidaho.edu/arch463Bruce/LabExercises/balance_point_big.pdf

Watson, D. and K. Labs. 1983. *Climatic Design: Energy-Efficient Building Principles and Practices*. New York: McGraw-Hill.

Wittwer, V., W. Platzer, and M. Rommell. 1991. "Transparent Insulation." *Solar Today* 5(January–February): 20–22.

Yoklic, M. and M. Layseca. 1998. Earth Construction, Courtyards, Verandahs, and Cooltowers: An Arizona Ranch House with Middle East Vernacular. *Proceedings of the 23rd National Passive Solar Conference*, Boulder, CO: American Solar Energy Society.

CHAPTER 15

Integrating Passive Systems

CHAPTERS 1 THROUGH 14 PROVIDE SOME PREPARATION for design: a perspective on energy, water, and material resources; an understanding of human comfort and indoor air quality; an analysis of the climate resources available on site; a list of the general design strategies appropriate to various climates; a discussion of the basics of heat transfer calculations; components of the building enclosure; daylighting design; and passive heating and passive cooling guidelines. Hence, we now know much about the variables outside of a building, the desired conditions inside, and the individual components of a building's skin. In this chapter, all these variables are integrated into the process of designing for passive heating and cooling with the important related factor of daylighting.

This chapter is intended as a demonstration of the decision-making design process for the passive performance of a small office building. The project takes the user through the steps and calculations of the principles and concepts described in the previous chapters—serving as an example (with guidance) for any project at the schematic planning stages of design.

The decision-making process for this example is outlined in Table 15.1, which shows which chapters, guidelines, and strategies to employ for buildings that use renewable energy versus buildings that use imported energy.

15.1 ORGANIZING THE DESIGN PROBLEM

Renewable on-site energy sources for lighting, heating, and cooling allow a building to function with less need for imported energy resources. As building designs are shaped by the use of on-site sources, they become more regional in appearance. This works against the one-design-fits-all approach common to chain retailing (such as fast-food outlets or big box retail stores), but works for the subtle integration of a building into its regional climate and culture, while reducing the economic and environmental costs of imported energy.

How should the building enclosure respond to the sometimes-conflicting needs for heating, cooling, and daylighting? Design strategies were related to comfort and climate in Chapter 14, where it became evident that internal heat gains can shift the appropriate design strategies for a building toward cooling, perhaps eliminating heating needs entirely. Typically, much of this internal heat is provided by electric lighting. Daylighting can replace electric lighting for most of the typical working day in most building types—*if* the building is designed to allow daylight to reach most of the interior.

In summer, properly shaded daylighting openings contribute less heat gain than the electric lighting that the daylight is replacing. This reduces the need for both electricity and cooling. In winter, some solar gain through daylight openings assists with heating (especially with south-facing windows), but the daylight still replaces electric lighting that would otherwise be an additional source of heat. Thus, compared to a conventional building of the same size and function, the daylit building is likely to use more heating energy (unless passive solar heating is emphasized), less electricity for lighting, and less cooling energy.

A matrix of design strategies (Figure 15.1) covered in Chapter 12 (Daylighting), Chapter 13 (Passive

Mechanical and Electrical Equipment for Buildings, Fourteenth Edition. Walter Grondzik, Daniel Overbey, and Alison Kwok.

Companion website: www.wiley.com/go/meeb14e

TABLE 15.1 Decision-Making Process for Integrating Passive Systems

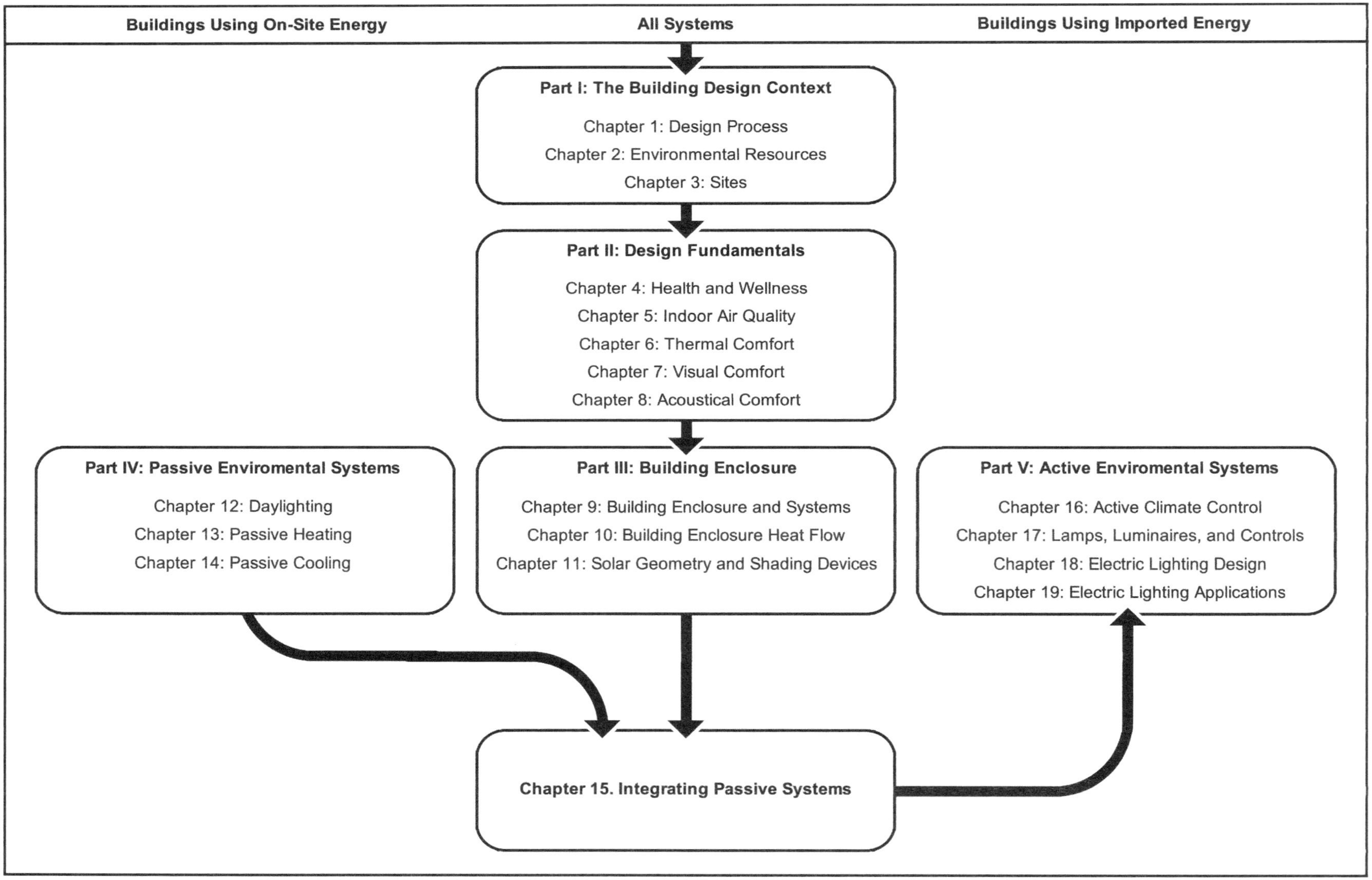

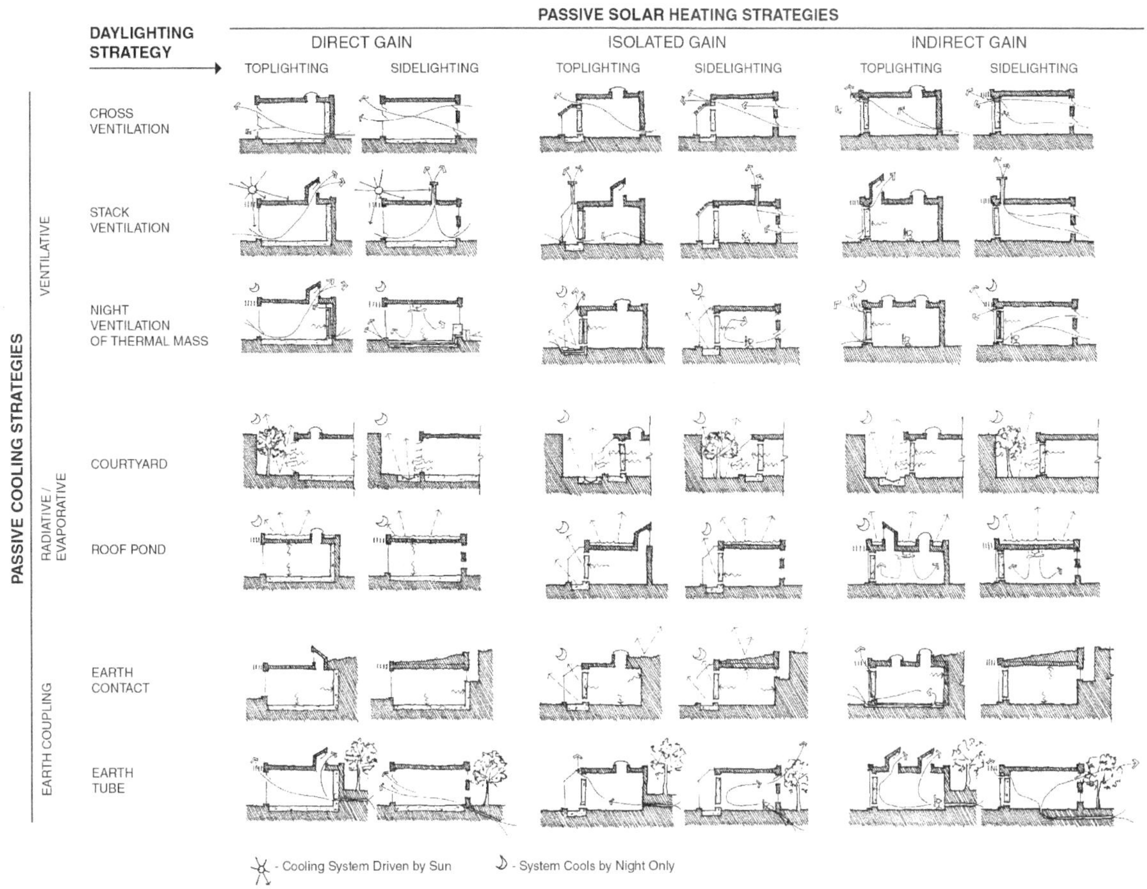

Figure 15.1 *Combining daylight, passive cooling, and passive solar heating opportunities. These schematic sections of simple one-room buildings remind designers of the potential for site-based, renewable resources and how they might interact. (Drawn by Sharon Shoshani and Alan Rutherford.)*

Heating), and Chapter 14 (Passive Cooling) illustrates some opportunities for a simple one-space building across a range of on-site energy sources. Ventilative cooling strategies depend upon the air as a heat sink; evaporative/radiative cooling strategies depend upon both sky and air as heat sinks; earth coupling cooling strategies depend upon the earth as a heat sink.

15.2 EXAMPLE DESIGN PROJECT

The following example will illustrate how to apply daylighting, passive heating, and passive cooling guidelines to the Emerald People's Utility District (EPUD) office building (Figure 15.2), located near Eugene, Oregon. This 24,000-ft^2 [2230-m^2] building is located in western Oregon's wet-winter, dry-summer climate, typical of the U.S. Pacific Northwest. Energy conservation and daylighting considerations were major goals in this building's design and were integrated with solar heating, night ventilation of mass, structure, and acoustics.

EXAMPLE 15.1 This sample problem is an ongoing illustration. Applications of various design guidelines appear in the context of this example throughout this chapter. Table 15.2 presents the type of information about a building that will be required in order to apply these design guidelines.

Part A. Daylighting Design

Part B. Overall Heat Loss

Part C. Approximate Passive Solar (Winter) Heat Gain

Part D. Approximate Summer Heat Gain

Part E. Cross-Ventilation Guidelines

Part F. Night Ventilation of Thermal Mass Guidelines

Part G. January Balance Point Temperature

Part H. Passive Heating Contribution

Part I. Thermal Storage Mass for Passive Heating

Part J. Detailed Night-Cooling of Mass Calculation ■

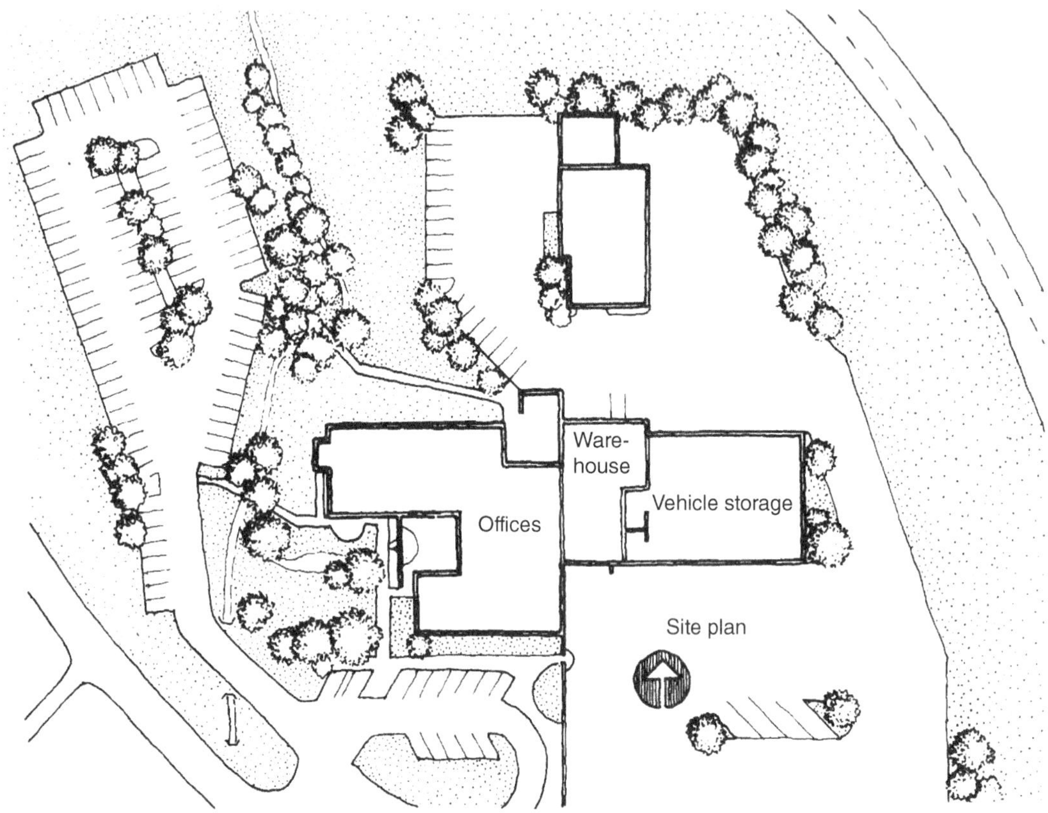

(a)

Figure 15.2 The office building for the Emerald People's Utility District (EPUD) near Eugene, Oregon, is elongated east–west for maximum window areas facing north–south. (a) The site plan, (b) the ground-floor plan, (c) the second-floor plan, and (d) a section through a two-story wing. The cutaway axonometric (e) of a typical open-office two-story bay shows air flush through hollow cores of exposed precast concrete slabs, windows and shading, and suspended sound-absorbing baffles. The deciduous vines on trellises at the south façade change the building's appearance by season. (f) Diagrammatic daylighting section of the office building for the Emerald People's Utility District near Eugene, Oregon. Lightshelves (g) reduce the contrast between a lot of daylight near windows and too little daylight farther in. The resulting windows (on north façade) are T-shaped (h), with more glass area above the lightshelves and less glass below. (i) At the summer solstice, the shading stripes will soften as vines leaf out with age; at the fall equinox, the leaves will still shade, whereas at the spring equinox (j) the branches will be bare to allow sun to reach the windows; (k) at the winter solstice, most of the window is exposed to the sun. (Drawing c is courtesy of Virginia Cartwright. Courtesy of Equinox Design, Inc., and WEGroup, PC, Architects, Eugene.)

EXAMPLE 15.1 PART A. DAYLIGHTING DESIGN Daylighting design for this office building began with design diagrams (Figure 15.3) and an assessment of daylighting potential (Figure 15.4). After it was established that daylighting was available during almost all normal working hours, the building plan was organized so that almost all windows faced either south or north to avoid the problems of low-altitude sun (year-round glare and summer heat gain) that accompany east- and west-facing windows. Then the building section was designed so that the height of the windows (H) was related to the depth of the floor plan served by those windows (2.5H).

After a generous target $DF_{avg} = 4.0\%$ was chosen from Chapter 12, Table 12.3, the next step was to size the windows and clerestories (as skylights), using the daylight factor guidelines from Chapter 12, Table 12.5. Applicable formulas for both sidelighting and for vertical monitor skylights:

$$DF_{av} = 0.2\,\frac{\text{window (or skylight) area}}{\text{floor area}}$$

Applied to the entire typical bay,

$$DF_{avg}$$

$$= 0.2\,\frac{97\ \text{ft}^2\text{north} + 97\ \text{ft}^2\text{south} + 132\ \text{ft}^2\text{clerestory}}{1440\ \text{ft}^2}$$

$$= 0.2\,\frac{326\ \text{ft}^2}{1440\ \text{ft}^2}$$

$$= 0.045\ \text{(or 4.5\%)}$$

slightly above the target DF_{avg} of 4.0%.

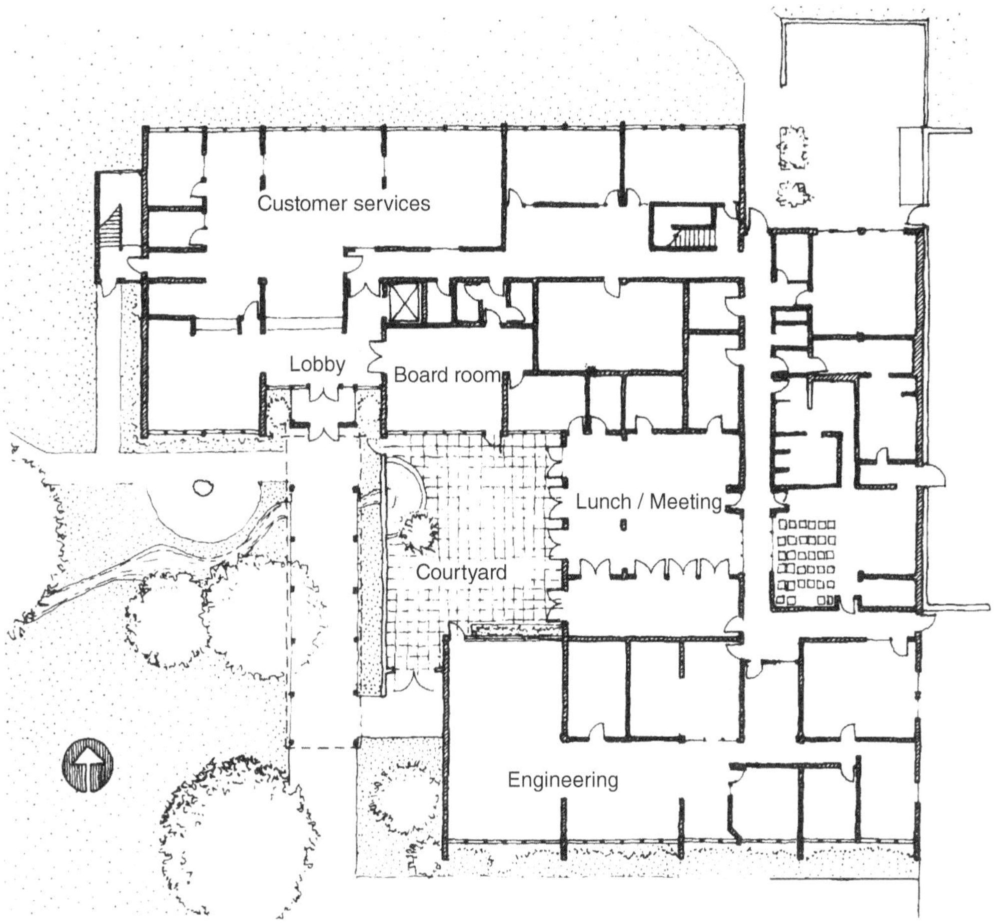

(*b*)

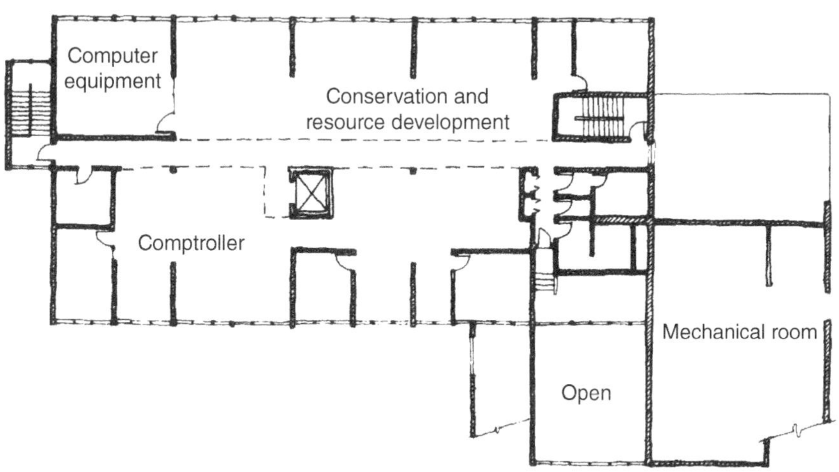

(*c*)

Figure 15.2 *(continued)*

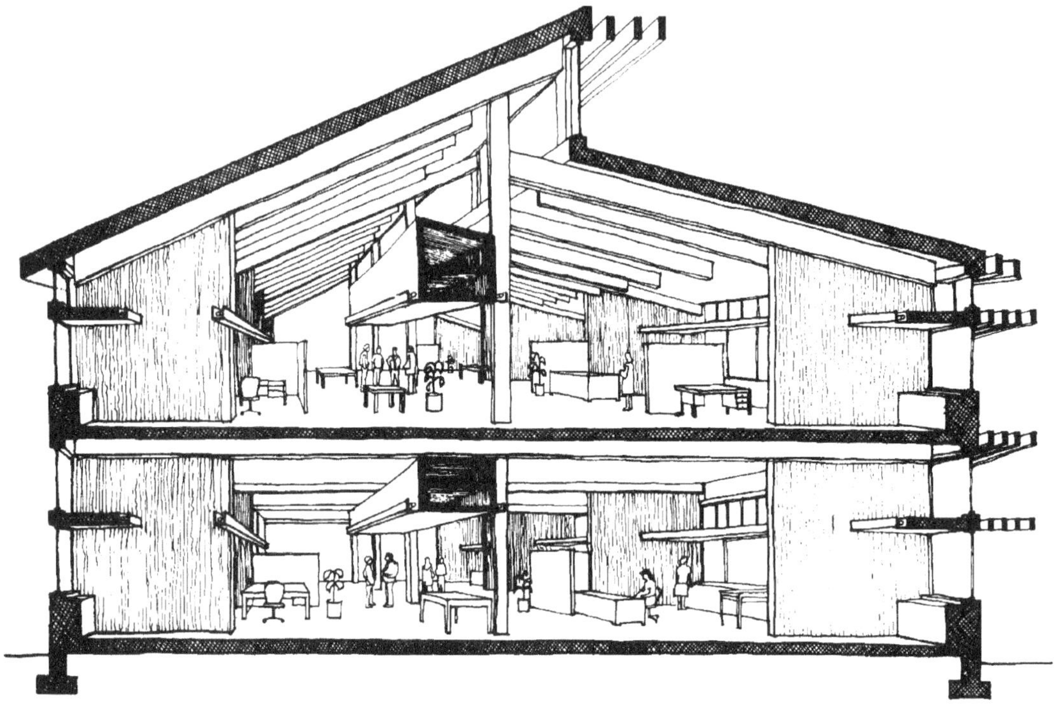

(*d*)

Figure 15.2 *(continued)*

Note that in this example, DF_{min} from the sidelighting occurs near the center of the building, at about the point (on the second floor) where the most light is available from the skylight. Therefore, relatively even daylighting distribution is expected on the upper floor. This is helped by the use of lightshelves and the T-shaped windows shown in Figures 15.2*g* and *h*.

The seasonal window-size question (more needed in winter, less in summer) was answered in this example by the use of deciduous vines outside the south windows. In winter and cool spring, the vines are bare of leaves. Warm weather brings leafy shade lasting well into the warm fall (Figures 15.2*i* and *j*).

The target daylight factors listed in Chapter 12, Table 12.3, provide sufficient light during most of the daylight hours on overcast winter days. The relationship between office working hours, daylight, and outdoor temperature is explored in the climatic timetables of Figure 15.4. Much more light will be available on summer days—probably more light than is needed, bringing heat along with it. When sizing windows and skylights, remember that controlling direct sun is necessary and that less opening area is needed in summer than in winter. ■

EXAMPLE 15.1 PART B. OVERALL HEAT LOSS Find the overall rate of heat loss (i.e., winter heat loss rate), in Btu/DD ft^2.

SOLUTION

This office building is passively solar heated, so the overall rate of heat loss, in Btu/DD ft^2, is determined by using the total $U \times A$ of the non-south-glass enclosure. From Table 15.2, this is 42 + 25 + 72 = 139 Btu/h °F.

To this must be added the effects of ventilation. From the same table, outdoor air is shown to be supplied at the rate of:

$$20 \text{ cfm/person} \times 8 \text{ persons} = 160 \text{ cfm}$$

which can also be expressed as 160 cfm × 60 min/h = 9600 cfh. Comparing this hourly rate to the volume of the typical bay, we have:

$$\frac{9600 \text{ cfh}}{22,000 \text{ ft}^3} = 0.44 \text{ ACH}$$

The $U \times A$ calculation for ventilation can be approximated as follows:

$$0.44 \text{ ACH} \times 22,000 \text{ ft}^3 \times 0.0765 \text{ lb/ft}^3 \times 0.24 \text{ Btu/lb}$$
$$= 178 \text{ Btu/h°F}$$

where

$$0.0765 \text{ lb/ft}^3 = \text{density of air}$$
$$0.24 \text{ Btu/lb} = \text{specific heat of air}$$

and the total rate of heat loss is:

$$\frac{(139 \text{ Btu/h°F} + 178 \text{ Btu/h°F}) \times 24 \text{ h/day}}{1440 \text{ ft}^2}$$
$$= 5.3 \text{ Btu/DD}_{65F}\text{ft}^2$$

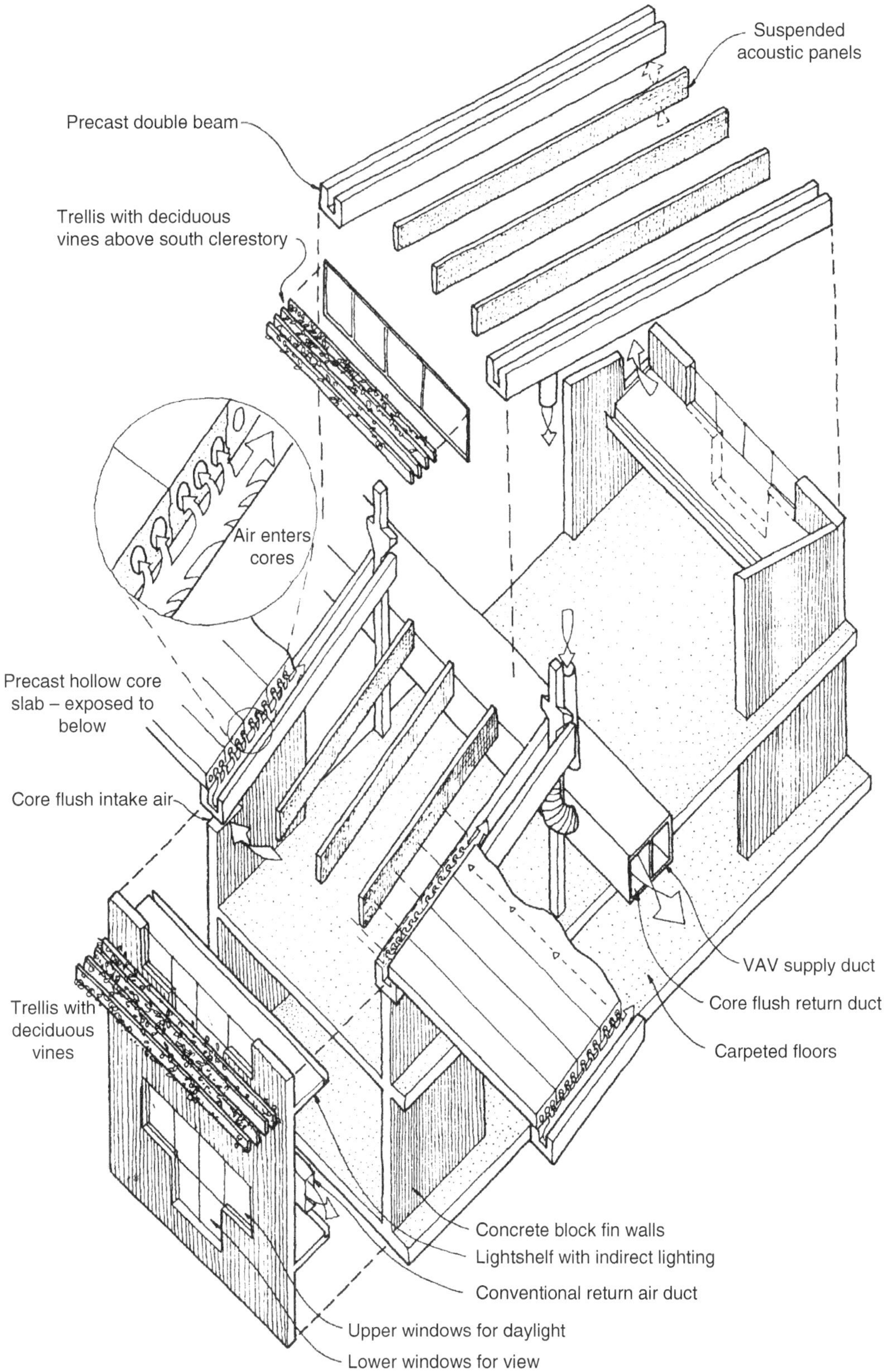

(*e*)

Figure 15.2 *(continued)*

PASSIVE ENVIRONMENTAL SYSTEMS

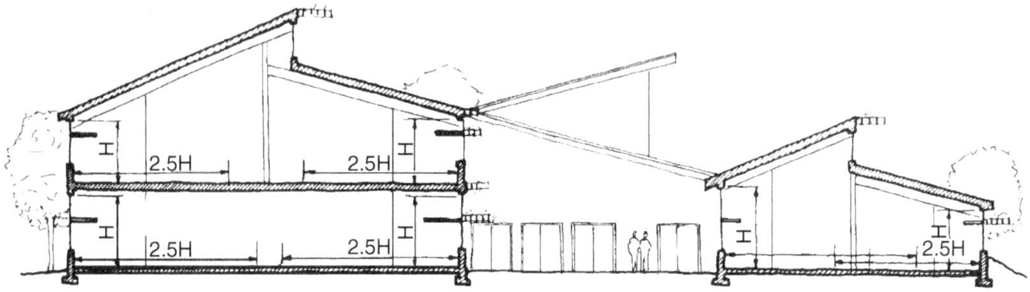

(*f*) Daylighting cross-section (south to right)

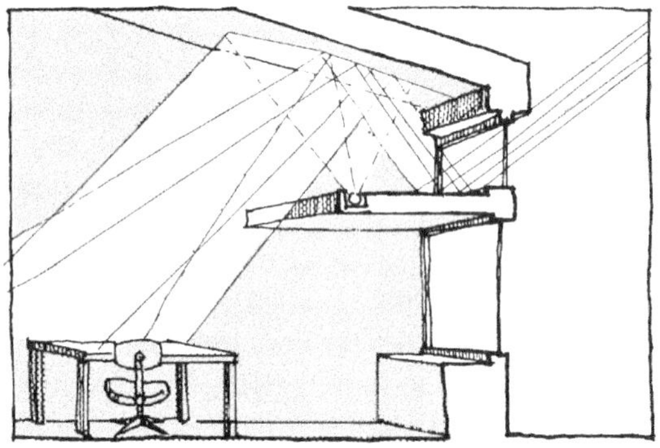

(*g*) Lightshelves

(*h*) North façade

Figure 15.2 *(continued)*

(*i*) South façade: summer solstice

(*j*) South façade: spring equinox

(*k*) South façade: winter solstice

Figure 15.2 *(continued)*

TABLE 15.2 Design Data for a Typical Second-Floor Bay of the Office Building

Component of Enclosure	U-Factor (Btu/h ft² °F)	Area (ft²)	U × A (Btu/h °F)
Floor area (basis for guidelines)		1440	
Exposed to Outside			
Roof[a]	0.028	1512	42
Exterior opaque walls[b] (north and south)	0.084	296	25
North window[c]	0.740	97	72
Total for passive solar guidelines[d] UA_{ns}			139
South window[c]	0.740	97	
South clerestory[c]	0.740	132	
Total for passive solar guidelines A_p		229	
Interior Thermal Mass			
Roof underside		1512	
Bearing walls, beams, and columns		720	
Total for passive heating and cooling		2232	
Ventilation			
Volume of enclosed space: 22,000 ft³			
Occupancy: 8 people			
Ventilation rate: 20 cfm per person			
Openable window area: 16 ft² on north, 16 ft² on south			

[a] R-40 over precast slab. Assume corrected R-35; therefore, U = 0.28.

[b] A 4-in. concrete block outside, 1-in. airspace, R-19 batt with metal studs at 24 in. and gypsum inside; approximate corrected R-11.9; therefore, U = 0.084.

[c] Clear double-glazed, 3/8-in. airspace, nonthermal break aluminum, U = 0.74.

[d] Passive solar heating calculations do not directly include losses from south-facing glass. Also, there is no floor heat loss or gain, because this is a second-floor space.

The Eugene, Oregon, climate has 4629 HDD_{65F} [2571 HDD_{18C}] annually, according to Appendix I, Table I.1 (note that totals may vary slightly from Appendix B, Tables B.1 and B.3 due to rounding). The majority of the HDD occur during a 10-month heating season (September through June)—recall that a heating season may be defined as the totality of months with more than about 100 HDD (heating degree-days) using any base temperature in Fahrenheit degrees [Celsius degrees]. Multiplied by the rate of heat loss found in Exercise 15.1, Part B, we estimate the annual heat loss as:

$$4629\ HDD65_{65F}/yr \times 5.3\ Btu/DD_{65F}\ ft^2 \times 1440\ ft^2$$

$$= 35.3\ MBtu/yr\ [10{,}354\ kWh/yr]$$ ■

EXAMPLE 15.1 PART C. APPROXIMATE PASSIVE SOLAR (WINTER) HEAT GAIN For one typical bay of the building, estimate the useful passive solar heat gain in each month of the heating season.

SOLUTION

Table 15.2 shows that each bay holds 97 ft² [9.0 m²] of vertical (i.e., 90°) south-facing clear double glazing; from Appendix E, Table E.15, we estimate that this glass has a Solar Heat Gain Coefficient (SHGC) of 0.65. Next, from Appendix I, Table I.1, we find the monthly incident solar radiation on vertical glass in Eugene. Approximations of useful solar gain necessarily include heat loss as well; in Exercise 15.1, Part B, a winter heat loss rate of approximately 5.2 $Btu/HDD_{65F}\ ft^2$ [0.03 $kWh/HDD_{18.3C}\ m^2$] was found, and Appendix I, Table I.1, shows monthly HDD values. With this information, we can estimate the useful solar heat gain for each month as follows:

September heat loss:

$$101\ HDD_{65F}/month \times 5.3\ Btu/DD_{65F}\ ft^2 \times 1440\ ft^2$$

$$= 0.8\ MBtu/month\ [226\ kWh/month]$$

September solar heat gain:

$$36\ kBtu/ft^2\ per\ month \times 97\ ft^2 \times 0.65\ SHGC$$

$$= 2.3\ MBtu/month\ [665\ kWh/month]$$

Since the passive solar gain appears to be less than the heat loss, the entire solar gain is considered to be useful, pending information on internal heat gains. Continuing this process, we find:

Monthly heat loss:

September	0.8 MBtu [226 kWh]
October	2.8 MBtu [819 kWh]
November	4.4 MBtu [1300 kWh]
December	5.8 MBtu [1691 kWh]
January	5.6 MBtu [1633 kWh]
February	4.7 MBtu [1367 kWh]
March	4.3 MBtu [1262 kWh]
April	3.4 MBtu [984 kWh]
May	2.2 MBtu [646 kWh]
June	1.1 MBtu [327 kWh]

Total ~ 35.0 MBtu [~10,254 kWh]

PASSIVE ENVIRON-MENTAL SYSTEMS

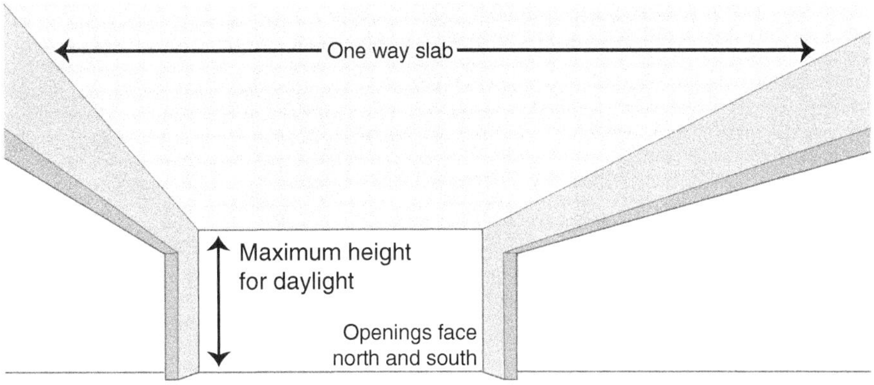

(*a*) Structure

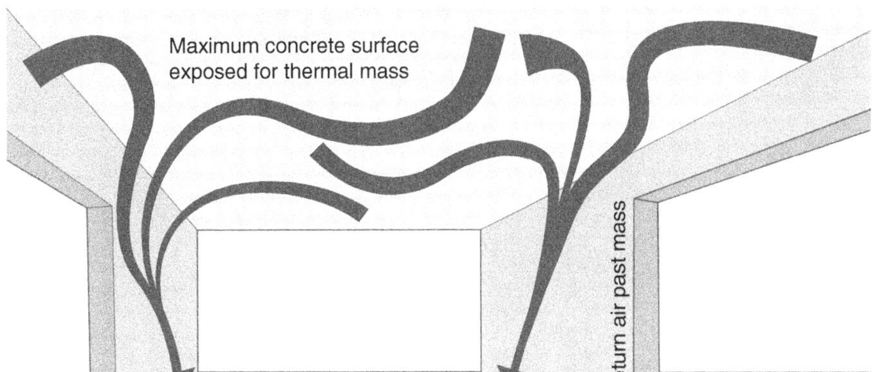

(*b*) Solar heating and night ventilation of mass

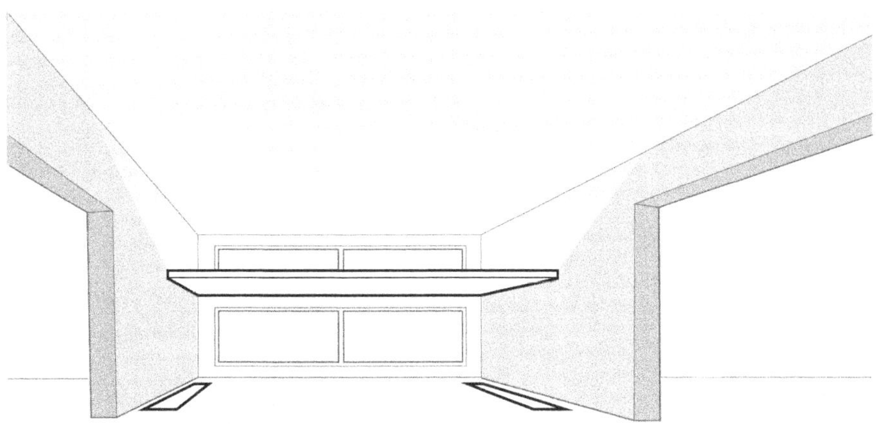

(*c*) Lightshelf for even daylight

Figure 15.3 *Design diagrams explore opportunities in the EPUD building. (a) Choosing a one-way structure allows windows to be placed as high as possible for deeper daylight penetration. (b) Maximizing exposed interior surfaces that are thermally massive facilitates both winter solar heat storage and summer night cooling. Carpet was installed at the insistence of the client. (c) Shaping the window with additional area above a lightshelf helps reduce the visual contrast near the window. (d) Indirect electric lighting is facilitated when higher ceilings are available; less glare on computer screens results. (e) With so much exposed mass surface, sound absorption is a must. This solution continues to expose the concrete ceiling and does not cast shadows from incoming daylight. (Drawn by Alexa Thornton; after work by John S. Reynolds.)*

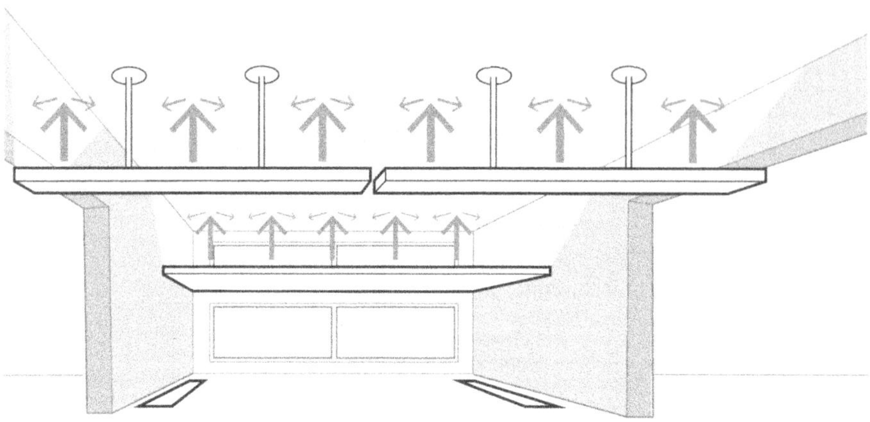

(*d*) Indirect electric lighting

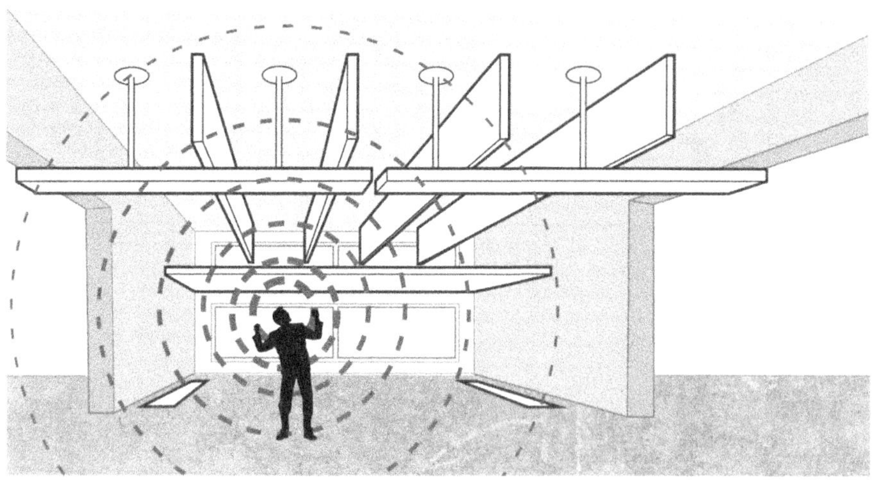

(*e*) Sound absorption

Figure 15.3 *(continued)*

Monthly solar heat gain:

September	2.3 MBtu [665 kWh]
October	2.0 MBtu [591 kWh]
November	1.1 MBtu [314 kWh]
December	0.9 MBtu [277 kWh]
January	1.2 MBtu [351 kWh]
February	1.4 MBtu [407 kWh]
March	1.8 MBtu [517 kWh]
April	1.7 MBtu [499 kWh]
May	1.7 MBtu [499 kWh]
June	1.7 MBtu [499 kWh]

Total ~ 15.8 MBtu [~4620 kWh]

During the heating season, the total heat loss is estimated to be ~35.0 MBtu [~10,254 kWh], while the useful solar gain is about ~15.8 MBtu [~4620 kWh], equaling approximately 45% of the heat loss. A greater proportion of that loss could have been offset through the use of tilted glazing: since the client chose not to include movable insulation, tilted glazing would have provided the greatest solar gain per area of glass, minimizing the area of uninsulated glass vulnerable to heat loss at night. ■

EXAMPLE 15.1 PART D. APPROXIMATE SUMMER HEAT GAIN Find the approximate heat gains from people, lighting, enclosure, and ventilation.

SOLUTION

A. People and equipment (using the high end for U.S. office buildings), select from Appendix G, Table G.1, Part A:

$$3.4\ \text{Btu/h ft}^2$$

B. Electric lighting (with a 4% DF from Example 15.1, Part A; this procedure assumes that most electric

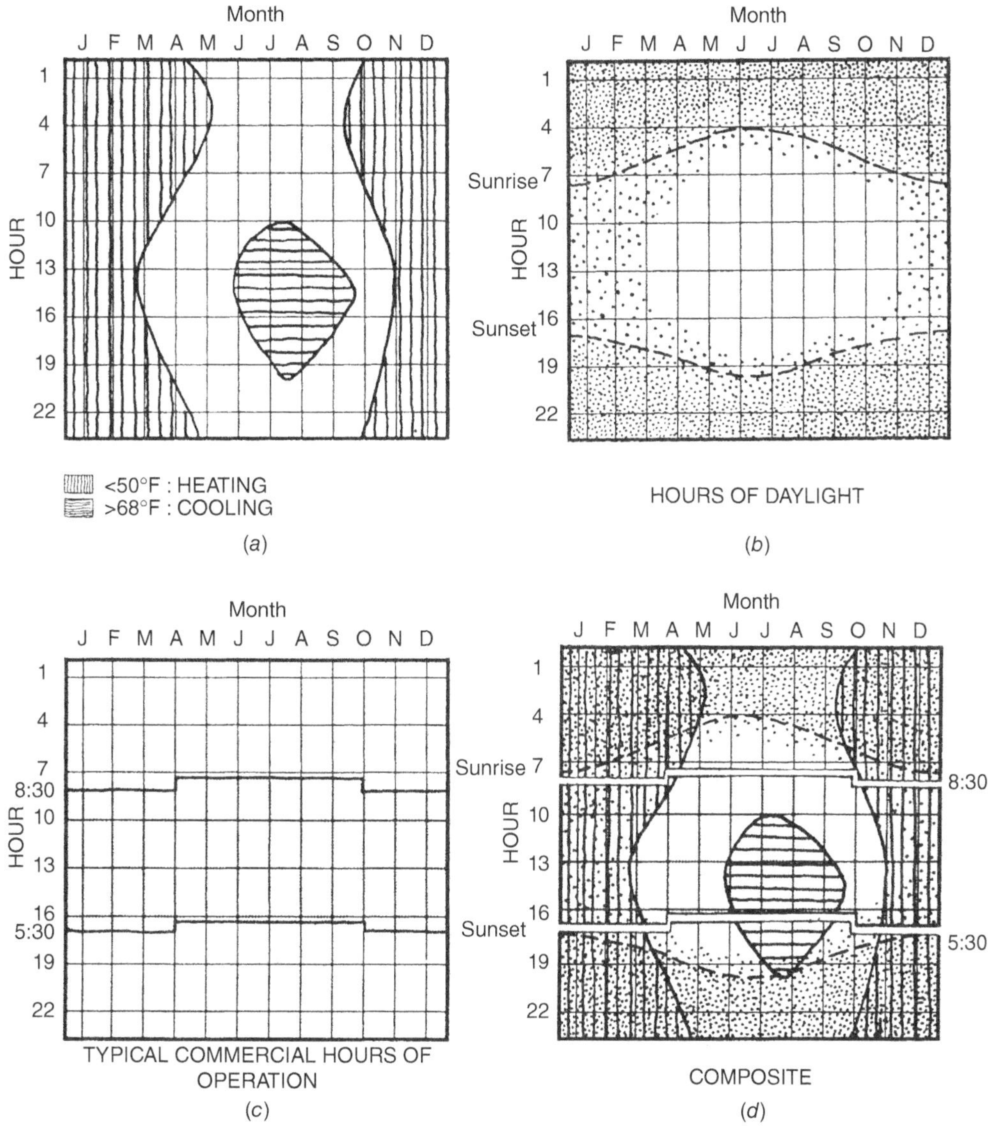

Figure 15.4 *A series of climate timetables was prepared while designing the office building for the EPUD building. (a) Average outdoor temperatures show little daily variation in winter but a large daily range in summer. (b) Daylight peaks around noon near the summer solstice. (c) Work schedule shows the impact of summer's daylight saving time. (d) A composite timetable indicates that daylight is available during almost all working hours and that half of the hottest hours occur after the building closes on summer afternoons—thanks to daylight saving time. Cool nighttime temperatures make night ventilation of thermal mass an attractive cooling option. (Courtesy of Equinox Design, Inc., and WEGroup, PC, Architects.)*

lighting is off when daylight is available), select from Appendix G, Table G.1, Part B:

$$0.5\ \text{Btu/h ft}^2$$

C. Enclosure (Eugene's design dry-bulb (DDB) temperature from Appendix B, Table B.1 is 88.3°F):

South glazing (window and clerestory), shaded by vines; apply Appendix G, Table G.1, Part C (reference the most appropriate temperature offered, 90°F):

$$\frac{(97\ \text{ft}^2 + 132\ \text{ft}^2)\ \times 16\ \text{Btu/h ft}^2}{1440\ \text{ft}^2} = 2.54\ \text{Btu/h ft}^2$$

North glazing, unshaded; apply Appendix G, Table G.3, Part B (reference regular double glass, venetian blinds):

$$\frac{97\ \text{ft}^2 \times 14\ \text{Btu/h ft}^2}{1440\ \text{ft}^2} = 0.94\ \text{Btu/h ft}^2$$

For the walls and roof, recall the formula from Chapter 12 that incorporates the DETD (design equivalent temperature difference); apply Appendix G, Table G.1 (reference 85°F):

$$q = U \times A \times \text{DETD}$$

DETD values are offered in Appendix G, Table G.2. Assume typical (lightweight) masonry block construction for the walls and a light-colored built-up (lightweight) roof assembly.

Walls:

$$\frac{296\ \text{ft}^2 \times 0.084\ \text{Btu/h ft}^2\text{°F} \times 10.3}{1440\ \text{ft}^2} = 0.18\ \text{Btu/h ft}^2$$

Roof:

$$\frac{1512\ \text{ft}^2 \times 0.028\ \text{Btu/h ft}^2\text{°F} \times 30}{1440\ \text{ft}^2} = 0.88\ \text{Btu/h ft}^2$$

D. Total heat gains, internal and through building skin:

$$3.4 + 0.5 + 2.54 + 0.94 + 0.18 + 0.88 = 8.44\ \text{Btu/h ft}^2$$

E. Heat gains from ventilation, apply from Appendix G, Table G.1, Part E:

$$\frac{(20\ \text{cfm/person} \times 8\ \text{people}) \times 16\ \text{Btu/h ft}^2}{1440\ \text{ft}^2} = 1.78\ \text{Btu/h ft}^2$$

Total approximate heat gain rate:

$$8.44 + 1.78 = 10.22\ \text{Btu/h ft}^2$$

■

EXAMPLE 15.1 PART E. CROSS-VENTILATION GUIDELINES Find the required inlet areas for cross-ventilation of a typical bay.

SOLUTION

The monthly average wind velocities for Eugene, Oregon, may be obtained from *ASHRAE Standard 169*, NOAA Local Climatological Data (LCD), or another source. For July, the hottest month, the average wind velocity is about 7 mph (typically from a nearly due-north direction). From Table 15.2, the inlet area (equal to outlet area also in this case) is 16 ft^2. The inlet area, as a percentage of floor area, is therefore:

$$16\ \text{ft}^2 / 1440\ \text{ft}^2 = 1\%$$

Entering Figure 14.5 in Chapter 14 with 7-mph wind velocity and 1% inlet area, we find that somewhat less than 15 Btu/h ft^2 heat gain can be removed by cross-ventilation.

In Example 15.1, Part D, we approximated a heat gain (excluding ventilation) of about 8.44 Btu/h ft^2. Therefore, under average July conditions, the window area is probably adequate to remove heat gains and maintain an indoor temperature 3°F *above* the outdoor temperature (as determined with Figure 14.5 in Chapter 14).

Eugene's average July daily high is about 80°F. This would produce an interior daily high of about 83°F. Such a temperature would require somewhat more than minimum air motion in order for the occupants to feel comfortable. At Eugene's summer design temperature of about 88°F, the interior temperature would be about 91°F. For these reasons, this building was designed to be night ventilated, relying on thermal mass rather than on cross-ventilation for hot-day comfort.

In milder weather, these windows can achieve substantial cooling because the temperature difference, Δt, is more than the 3°F assumed in Figure 14.5 in Chapter 14. For example, applying the equation such that the temperature indoors is 75°F and outdoors is 65°F; the fully opened windows with the 7-mph wind would result in a cooling rate of:

$$\frac{\text{About } 15\ \text{Btu/h ft}^2 \times 10\text{°F}}{3\text{°F}} = 50\ \text{Btu/h ft}^2$$

which is considerably more than the actual gain of 11 Btu/h ft^2. Therefore, under such conditions, the windows are only partially opened. ■

EXAMPLE 15.1 PART F. NIGHT VENTILATION OF THERMAL MASS GUIDELINES Find the potential for night ventilation of mass.

SOLUTION

For Eugene, Oregon, from Appendix B, Table B.1, the summer design temperature is 88.3°F and the MDR is 30.2°F. From Table 15.2, the typical bay has a ratio of exposed internal thermal mass/floor area of:

$$\frac{1512\ \text{ft}^2 + 720\ \text{ft}^2}{1440\ \text{ft}^2} = 1.55$$

which places this example halfway between average mass (ratio 1.0) and high mass (ratio 2.0). Entering Figure 14.8*a* in Chapter 14, with Eugene's climate data, we find that an average-mass building can store about 100 Btu/ft^2 day, whereas a high-mass building (Chapter 14, Figure 14.8*b*) can store about 180 Btu/ft^2 day. Applying generalized interpolation with consideration of the considerable amount of thermal mass exhibited in this building, we will proceed with an assumption of a thermal storage capacity of about 170 Btu/ft^2 day.

Assuming a nine-hour typical working day (including the noon hour), at the average approximate heat gain rate of 10.22 Btu/h ft^2 (Example 15.1, Part D), the daily heat gain in this building is:

$$9\ \text{h} \times 10.22\ \text{Btu/h ft}^2 = 92.0\ \text{Btu/ft}^2\ \text{day}$$

This average daily heat gain is considerably less than the approximated storage capacity of 170 Btu/ft^2 day, so the building should perform satisfactorily with this method of cooling.

Although this building was designed for nighttime forced ventilation because of summer night calm, we will check to see whether the operable windows are adequate for natural, night cross-ventilation. In Chapter 14, Figure 14.8*c*, we find that during the hour of maximum night cooling, slightly less than 14% of the total stored heat gains must be removed (however, with 14% as the minimum percent exhibited in the figure, we will assume this percentage):

$$14\% \times 92.0\ \text{Btu/ft}^2\ \text{day} = 12.9\ \text{Btu/h ft}^2$$

From Figure 14.8*d* in Chapter 14, the resulting maximum hourly Δt is about 13°F.

In Example 15.1, Part E, we determined that less than 15 Btu/h ft^2 would be removed by cross-ventilation under average July conditions at a 3°F Δt. Because during this hour of maximum cooling the Δt is 18°F, cross-ventilation could remove approximately:

$$\frac{13°\text{F} \times 15\ \text{Btu/h ft}^2}{3°\text{F}} = 65\ \text{Btu/h ft}^2$$

The daily heat gain was approximated at 92 Btu/h ft^2; therefore, at roughly 65 Btu/h ft^2, cooling by cross-ventilation at night may be challenging at certain times of the cooling season. Also, Eugene's summer nighttime average wind velocity is nearly zero. For these reasons, forced ventilation should be considered. This also allows a thorough flushing of all interior mass surfaces with cool air.

Recall that hourly design sensible heat loss due to ventilation (q_v) can be estimated as follows:

$$q_v = (V)(1.1)(\Delta t)$$

where 1.1 is a constant derived from the density of air at 0.075 lb/ft^3 under typical building conditions, multiplied by the specific heat of air (the heat required to raise 1 lb of air 1°F, which is 0.24 Btu/lb °F) and by 60 min/h. The units of this frequently encountered but quirky constant are Btu min/ft^3 h °F. (Recall that for SI units, this constant is 1.2.)

This formula can be adapted in order to estimate the required rate of forced ventilation, in cubic feet per hour (cfh), during the hour of maximum cooling as follows:

$$V = \frac{q_v \times 60\ \text{min/h}}{(1.1\ \text{Btu min/h ft}^3\ °\text{F})(\Delta t)}$$

where

q_v = approximated from the 14% maximum hourly cooling rate determined in Figure 14.8*c* during the hour of maximum night cooling at 12.9 Btu/h ft^2

V = being solved in cubic feet per hour (cfh) rather than cubic feet per minute (cfm)

Δt = approximated at 13°F using Figure 14.8*d*

Expressed per square foot floor area in this building,

$$12.9\ \text{Btu/h ft}^2 = (V\text{cfh})(0.018\ \text{Btu/ft}^3\,°\text{F})(13\ °\text{F h})$$

which can be solved for Vcfh,

$$V\text{cfh} = \frac{12.9\ \text{Btu/h ft}^2}{(0.018\ \text{Btu/ft}^3\,°\text{F})(13°\text{F})}$$

$$= 55.1\ \text{cfh outdoor air per ft}^2\ \text{of floor area}$$

This rate of forced outdoor air ventilation produces the following air change (AC) with average ceiling height at 15.3 ft:

$$\frac{55.1\ \text{ft}^3/\text{h per ft}^2\ \text{floor area}}{15.3\ \text{ft}^3\ \text{per ft}^2\ \text{for 1 AC}} = 3.6\ \text{ACH}$$

a rate that would not only help expel heat gains by the next morning but also improve indoor air quality through considerable air flush-out. Table 15.2 identifies the occupancy as 8 persons and the ventilation rate as 20 cfm/person. As such:

$$20\ \text{cfm per person} \times 8\ \text{persons} = 160\ \text{cfm}$$

$$160\ \text{cfm} \times 60\ \text{min/h} = 9600\ \text{cfh}$$

$$9600\ \text{cfh}/1440\ \text{ft}^2 = 6.7\ \text{cfh/ft}^2\ \text{of ventilation}$$

Compare this minimum 6.7 cfh/ft^2 for indoor air quality (IAQ) ventilation versus the forced ventilation rate of 55.1 cfh/ft^2 in the example for cooling—nearly an 8:1 ratio. 55.1 cfh/ft^2 is considerably greater than that required for minimum outdoor air per person by day. This promises an attractive indoor air quality by the next morning. ■

EXAMPLE 15.1 PART G. JANUARY BALANCE POINT TEMPERATURE Calculate the January balance point temperature for a typical bay. Assume that the average indoor January temperature over a 24-hour period is an energy-conserving 68°F.

SOLUTION

The (summer) internal gains of 0.5 Btu/h ft^2 from Example 15.1, Part D, may need some modifications for a January condition. In darker winter, assume that the gain from electric lighting is closer to the 2 < DF < 4 value of 2.0 Btu/h ft^2. The total hourly internal gains during occupancy are then 3.4 Btu/h ft^2 (people and equipment) + 2.0 Btu/h ft^2 (electric lighting) = 5.4 Btu/h ft^2. The daily average internal gains are as follows:

$$\frac{5.4\ \text{Btu/h ft}^2 \times 9\ \text{h}}{24\ \text{h}} = 2.03\ \text{Btu/h ft}^2$$

The January solar gain, VS (vertical surface), from Appendix C, Table C.19, for Salem, Oregon, is 471 Btu/ft^2 day per ft^2 of south glass:

$$\frac{471\ \text{Btu/h ft}^2 \times (97\ \text{ft}^2 + 132\ \text{ft}^2)\ \text{south glass}}{24\ \text{h} \times 1440\ \text{ft}^2}$$

$$= 3.12\ \text{Btu/h ft}^2$$

Therefore,

$$Q_i = 2.03 + 3.12 = 5.15\ \text{Btu/h ft}^2$$

UA for the typical bay (from Table 15.2 and Example 15.1, Part B) includes, for this calculation, all elements of the building enclosure:

Roof: 42

Exterior opaque walls: 25

PASSIVE ENVIRONMENTAL SYSTEMS

North window: 72
South window: 0.74 × 97 = 72
South clerestory: 0.74 × 132 = 98
Ventilation: 178

$$(42 + 25 + 72 + 72 + 98 + 178) = 487 \text{ Btu/h°F}$$

expressed per square foot of floor area:

$$487 \text{ Btu/h°F}/1440 \text{ ft}^2 = 0.34 \text{ Btu/h°F ft}^2$$

Therefore,

$$t_b = 68°\text{F} - \frac{5.15 \text{ Btu/h ft}^2}{0.34 \text{ Btu/h°F ft}^2}$$
$$= 52.9°\text{F}$$

Based on this information, the January balance point temperature for a typical bay is 52.9°F. ■

EXAMPLE 15.1 PART H. PASSIVE HEATING CONTRIBUTION How much does passive heating contribute to the annual heating need?

SOLUTION

The estimation of internal gains in Exercise 15.1, Part G now allows us to revise the heating need, as well as the portion of that need met by passive solar heating, in each month. In Exercise 15.1, Part C, we estimated a January solar gain of 1.2 MBtu [351 kWh], representing a small fraction of January's 5.6 MBtu [1633 kWh] heat loss. Internal gains, however, are now understood to offset a significant fraction of that heat loss:

$$2.03 \text{ Btu/h ft}^2 \text{ (from Part G)} \times 24 \text{ h/d} \times 31 \text{d/mo}$$
$$\times 1440 \text{ ft}^2 = 2.2 \text{ MBtu } [645 \text{ kWh}]$$

As a result, January has a net heating need of:

$$5.6 \text{ MBtu (heat loss)} - 2.2 \text{ MBtu (internal gains)}$$
$$= 3.4 \text{ MBtu } [996 \text{ kWh}]$$

The January passive solar gains of 1.2 MBtu [350 kWh] are now expected to meet over a third of the heating need. Repeating this process for the remaining months shows that an increasing proportion of the need is met in months with longer days and warmer temperatures as monthly heat loss declines and solar gain increases. ■

EXAMPLE 15.1 PART I. THERMAL STORAGE MASS FOR PASSIVE HEATING How does the actual mass in this building compare to the amount needed to store energy for passive heating?

SOLUTION

Thermal stability, resulting from virtually constant (or all-night) heat delivery, was the design intent for the concrete thermal mass, which was intended to operate through a daily temperature range of less than 10°F [5.5°C] and to store energy not only from solar radiation but also from people, lights, and equipment. As seen in Exercise 15.1, Part G, internal gains were, in fact, expected to provide the majority of energy stored.

To evaluate the building storage capacity, we first find that the concrete has a specific heat capacity of around 0.20 Btu/lb°F [0.84 kJ/kg K] at a density of ~140 lb/ft^3 [2240 kg/m^3], from Appendix E, Table E.1. Note that the material primarily occupies ceiling and interior wall surfaces, removed from direct solar exposure, and that floors are carpeted. Because of this, the mass must acquire most of its energy from warm air, through natural convection, rather than from absorption of solar radiation incident upon the mass surface. Natural convection typically involves relatively low rates of heat transfer, so based upon experience we estimate that the mass will be able to take up at most 35% of the daily solar + internal heat gain. Consider this an assumption for the purposes of this example. With this information, we can then estimate the volume and thickness of concrete consistent with the design intent.

For the month of January:

$$(1.2 \text{ MBtu solar gains} + 2.2 \text{ MBtu internal gains}) \times 35\%$$
$$= 1.2 \text{ MBtu of energy to be stored during the month}$$
$$(1.2 \text{ MBtu/mo}) \times (1 \text{ mo/31 days}) \times (1000 \text{ kBtu/1 MBtu})$$
$$\approx 39 \text{ kBtu/day of energy to be stored}$$
$$39{,}000 \text{ Btu} \times \frac{(1 \text{ lb°F}/0.20 \text{ Btu specific heat capacity of concrete})}{10°\text{F}}$$
$$= 19{,}500 \text{ lb of concrete needed}$$
$$19{,}500 \text{ lb} \times (1 \text{ ft}^3/140 \text{ lb density of concrete})$$
$$= 139 \text{ ft}^3 \text{ of concrete volume}$$
$$139 \text{ ft}^3 \text{ concrete}/0.3 \text{ ft thickness}$$
$$\approx 464 \text{ ft}^2 \text{ } [43 \text{ m}^2] \text{ of 4 in. thick concrete to be exposed in the space}$$

With actual exposed thermal mass in a typical bay of 2232 ft^2 [207 m^2] (Table 15.2), or nearly five times the area estimated above, we see that the mass is sized to provide exceptional thermal stability, with daily temperature variation in the material of only a few degrees. The tradeoff for such extensive massing, of course, is a building that tends to remain cool during occupied hours without the operation of supplemental active heating. Indeed, as described in Section 15.3, avoiding active cooling was a central goal of the project and one that drove the decision to create a high-thermal-mass building. ■

EXAMPLE 15.1 PART J. DETAILED NIGHT-COOLING OF MASS CALCULATION How much heat will be removed, and what is the final thermal mass temperature in a typical bay? Assume that only the exposed surfaces of the thermal mass participate in night flush cooling. (In the actual building, the interior surfaces of the continuous cores in the precast floor slabs are also flushed with night air, yielding added cooling capacity.)

SOLUTION

STEP 1. Again, from Appendix B, Table B.1, the summer design dry-bulb (DDB) temperature is 88.3°F and the mean daily range (MDR) is 30.2F°. These two values can be used to approximate the hourly distribution of outdoor air temperature throughout a day that exhibits both the summer DDB temperature and the MDR by applying the following equation and the applicable fraction value from Table 15.3:

$$\text{Hourly outdoor air temperature}$$
$$= \text{DDB} - (\text{MDR} \times \text{fraction})$$

For example, at 8 p.m. (hour of the day, h = 20) the hourly outdoor air temperature would be:

$$88.3 - (30.2 \times 0.50) = 73.2°\text{F}$$

Using this process, hourly outdoor air temperatures have been calculated and applied to column II of Table 15.4 with the minimum daily temperature assumed at 4:00 a.m. and the maximum at 4:00 p.m. It is important to note that the fraction values from Table 15.3 represent a mathematical response to the premise that daily temperature variation is driven by heat from the sun. As such, the sine curve profile derived from the tabulated hourly values are specified in terms of solar time.

STEP 2. The 24-hour heat gain is estimated from Example 15.1, Part D. During the 9-hour workday, the heat gains total 10.22 Btu/h ft^2. During the unoccupied hours, there are negligible gains from electric lighting, people and equipment, windows, or ventilation. However, the opaque enclosure gains remain because the enclosure stores heat and releases it gradually. These unoccupied gains (from Table 15.2) total 0.26 Btu/h ft^2 from the walls, plus 1.03 Btu/h ft^2 from the roof, totaling 1.29 Btu/h ft^2.

Occupied:

$$9\,\text{h} \times 10.22\,\text{Btu/ft}^2 = 92.0\,\text{Btu/ft}^2$$

Unoccupied:

$$(24 - 9)\,\text{h} \times 1.29\,\text{Btu/h ft}^2 = 19.4\,\text{Btu/ft}^2$$

$$\text{Daily gains} = 92.0 + 19.4 = 111.4\,\text{Btu/day ft}^2$$

For the entire typical bay:

$$111.4\,\text{Btu/day ft}^2 \times 1440\,\text{ft}^2$$
$$= 160{,}416\,\text{Btu/day (enter on line H, Table 15.4)}$$

STEP 3. The area of exposed mass, from Table 15.2, totals 2232 ft^2. The concrete floor is covered by carpet and is therefore assumed to be disconnected from thermal storage for this space. The wallboard interior surfaces of the exterior walls are also ignored.

STEP 4. The mass heat capacity is estimated, assuming that all elements are solid concrete at a density of 140 lb/ft^3 and a specific heat of 0.20 Btu/lb°F. The bearing walls are 8 in. thick, exposed on both sides, with concrete fill in the cores of all concrete blocks. The beams are 8 in. thick, but their inner faces are unexposed (they form the air duct for the night flush system). The precast ceiling slabs are 10 in. thick, but much of this thickness consists of air-filled cores; therefore, only the lower 4 in. (0.33 ft) are assumed to participate as "exposed" mass.

$$\text{mass heat capacity} = (\text{mass surface area})$$
$$\times (\text{mass thickness}) \times (\text{density} \times \text{specific heat})$$
$$= 2232\,\text{ft}^2 \times 0.33\,\text{ft} \times 28.0\,\text{Btu/ft}^{3}°\text{F}$$
$$= 20{,}624\,\text{Btu/°F (enter on line B, Table 15.4)}$$

STEP 5. Little supplementary cooling is assumed, although the carpeted floor provides some (albeit insulated) mass, as does the wallboard surface of the exterior walls. Assuming that only one-fourth of the floor area contributes to cooling performance:

$$0.25 \times 1440\,\text{ft}^2 = 360\,\text{ft}^2\ \text{(enter on line C, Table 15.4)}$$

STEPS 6–10. See Table 15.4.

STEP 11. Determine supplementary cooling:

$$(80°\text{F} - \text{final mass temperature, line E, Table 15.4})$$
$$\times 2.25 \times (\text{floor area, line C, Table 15.4})$$
$$(80°\text{F} - 64.7°\text{F}) \times 2.25 \times 360\,\text{ft}^2$$
$$= 12{,}393\,\text{Btu (enter on line F, Table 15.4)}$$

Recall that the factor 2.25 assumes a modest role for the other, less thermally massive, surfaces.

STEP 12. See Table 15.4.

TABLE 15.3 Fraction of Daily Temperature Range

Time (h)	Fraction	Time (h)	Fraction	Time (h)	Fraction
1	0.88	9	0.55	17	0.14
2	0.92	10	0.38	18	0.24
3	0.95	11	0.23	19	0.39
4	0.98	12	0.13	20	0.50
5	1.00	13	0.05	21	0.59
6	0.98	14	0.00	22	0.68
7	0.91	15	0.00	23	0.75
8	0.74	16	0.06	24	0.82

Source: © ASHRAE, www.ashrae.org. 2021 ASHRAE Handbook—Fundamentals, Chapter 14, Table 6.

TABLE 15.4 Oregon Office Building (Figure 15.2) Performance: Cooling by Night Ventilation of Thermal Mass for a Typical Second-Floor Bay (use Table 14.5 in Chapter 14, and procedure to fill in values)

A. Mass surface area (Step 3)	2232 ft^2
B. Mass heat capacity (Step 4)	20,624 Btu/°F
C. Floor area (supplementary cooling) (Step 5)	360 ft^2
Total building (bay) volume (Table 15.2)	22,900 ft^3

(I) Hour	(II) Outside Air Temperature (°F)	(III) Cooling[a] (Btu/h)		(IV) Mass Temperature (°F)
PM				
8	73	No heat removed		73
9	70	(73 − 70) 2232 (line A)	= 6696	$70 - \frac{6696}{20,624\ (\textit{line B})} = 72.7$
10	68	(72.7 − 68) 2232	= 10,490	$72.7 - \frac{10,490}{20,624} = 72.2$
11	66	(72.2 − 66) 2232	= 13,838	$72.2 - \frac{13,838}{20,624} = 71.5$
12	64	(71.5 − 64) 2232	= 16,740	$71.5 - \frac{16,740}{20,624} = 70.7$
AM				
1	62	(70.7 − 62) 2232	= 19,418	$70.7 - \frac{19,418}{20,624} = 69.7$
2	61	(69.7 − 61) 2232	= 19,418	$69.7 - \frac{19,418}{20,624} = 68.8$
3	60	(68.8 − 60) 2232	= 19,642	$68.8 - \frac{19,642}{20,624} = 67.8$
4	58	(67.8 − 58) 2232	= 21,874[b]	$67.8 - \frac{21,874}{20,624} = 66.8$
5	59	(66.8 − 59) 2232	= 17,410	$66.8 - \frac{17,410}{20,624} = 65.9$
6	59	(65.9 − 59) 2232	= 15,401[c]	$65.9 - \frac{15,401}{20,624} = 65.2$
7	61	(65.2 − 61) 2232	= 9374	$65.2 - w\,\frac{9374}{20,624} = 64.7$
8	66	Stop flush; mass temperature (64.7) is now below outdoor temperature (66)		

D. Total mass cooling (sum of III)	170,302 Btu
E. Final mass temperature (ref. IV)	64.7°F
F. Supplementary cooling (Step 11)	12,393 Btu
G. Total cooling (line D + line F)	182,695 Btu
H. 24-h heat gain (Step 2)	160,416 Btu
I. Flow rate required for night ventilation (Step 14)	5.4 ACH

[a] A surface conductance of 1.0 Btu/h ft^2°F is assumed in this calculation.

[b] Hour of maximum cooling.

[c] At this point, enough heat has been removed to meet the typical bay's design-day heat gain.

STEP 13. The building has more than adequate cooling capacity (~185 kBtu/day, line G, Table 15.4) to meet the 85°F outdoor design condition (~160 kBtu/day, line H, Table 15.4); on typical summer days (recall from Exercise 15.1, Part E that Eugene's average July daily high is about 80°F), there will be even more excess mass capacity. (A more conservative and very detailed analysis, assuming more electric lighting left on, extremely high temperatures, and a less-favorable coefficient of heat transfer between mass and air, led to the decision to also flush the precast cores at night.)

Because of this excess capacity, it appears advisable to begin the night flush a few hours later, when the Δt between indoors and outdoors is greater and therefore more cooling is achieved per hour of fan operation. (Repeat the procedure, beginning the flush at midnight with a Δt of 80–68 = 12°F, to see how similar results can be obtained with 3 fewer hours of fan operation.) The disadvantage is that the interior will remain hotter for a longer period in the evening, which could cause discomfort for a late worker. Alternatively, the flush could be stopped after 5:00 a.m., by which time the stored heat has been removed.

STEP 14. The airflow required will be checked at 4:00 a.m., the hour of maximum cooling:

$$V = \frac{21,874\ \text{Btu/h}}{0.018\ \text{Btu/ft}^3\,°\text{F h} \times (67.8 - 58°\text{F})} = 124,000\ \text{cfh}$$

$$\text{ACH} = \frac{124,000\ \text{cfh}}{22,900\ \text{ft}^3} = 5.4\ \text{(enter on line I, Table 15.4)}$$

which is notably more than the 3.6 ACH predicted by the design guideline in Exercise 15.1, Part F. ■

15.3 PROJECT PERFORMANCE

The advantages of this passive cooling approach compared to conventional mechanical cooling can be seen in a comparison of several August monthly records for the conventional-cooling compressors and the night-flush fans. Figure 15.5 shows two Augusts of operation (1989, 1990) when the building operated as a night-ventilated structure. In a third August (1991), due to a building-wide computer control changeover, the building operated only as a conventionally cooled structure. The sun's impact on the building was thought to be relatively similar for these three months, but outdoor temperature varied as shown by the CDD values. Not only was as much electricity required to cool the conventionally operated building in 1991, despite lower temperatures than the previous August, but all the fan energy use and almost all the compressor energy use occurred during the daytime hours of high consumption. In many locations, utilities charge more for electricity used during the daytime. With night ventilation, the fans deliver huge quantities of fresh

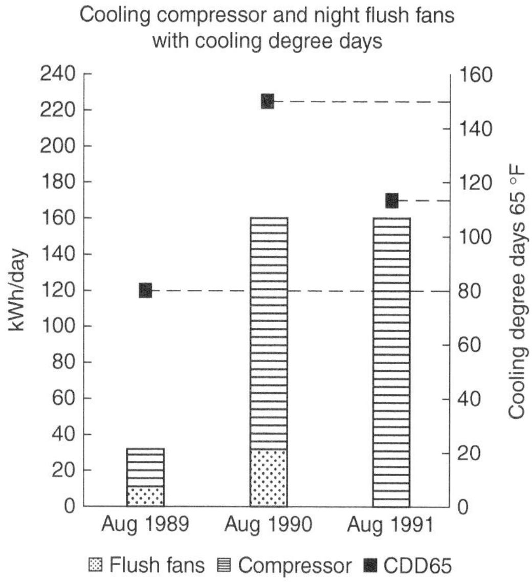

Figure 15.5 *Comparing night ventilation of mass with conventional compressive cooling. The Emerald People's Utility District offices are monitored for end use of electricity. In August 1991 (117 CDD65 [65 CDD18]), with compressive cooling only, 160 kWh/day of electricity was used, nearly all of it during periods of heavy use (daytime). Compare this to the hotter August 1990 (150 CDD65 [83 CDD18]), when 126 kWh/day was used for compressive cooling by day, plus 34 kWh/day for night ventilation fan-forced cooling. (From Ashley and Reynolds, 1994.)*

air off-peak by night, unlike the conventional daytime operation that recycles some air to save energy.

15.4 PROJECT SUMMARY

As seen in this example, combining strategies is not a linear process. Daylight influences form; form changes the extent of solar access; choice of materials and construction methods impact heat flow; strategically placed thermal mass impacts when a space will become too cold or too warm—all of this in the context of achieving thermal, visual, and acoustical comfort. Passively cooled or heated spaces generally rely on the occupants to adapt to wider thermal fluctuations in the environment, as opposed to conditions in mechanically controlled spaces. The designer must balance thermal, visual, and acoustic conditions that contribute to our health and well-being.

Thermal comfort is represented by a fairly unified theory in which the interrelationships of environmental and personal variables are defined and quantified; however, visual and acoustical comfort variables are not so well developed or researched. The visual and acoustical comfort pictures are complex: If one variable changes, the impact on other variables is unclear, and there are no guarantees of comfort. For example, if daylight and electric light are integrated, is glare necessarily eliminated by adequate illuminance and color rendering? Visual comfort can also mean providing a connection to the outdoors and visual stimulation through the use of view windows. Acoustic comfort in an open office plan may require enclosed conference spaces for meetings, or acoustical paneling to baffle ambient sounds and conversations.

A focus on visual and acoustical comfort is no less important to building design, but the information available to the designer is not as well integrated or standardized. The designer can begin to understand the complexities of these situations by examining issues individually and then integrating concerns via the design process.

15.5 CASE STUDY—DESIGNING FOR PASSIVE HEATING AND COOLING

Manitoba Hydro Place, Winnipeg, Manitoba, Canada

PROJECT BASICS

- Location: Winnipeg, Manitoba, Canada
- Latitude: 49.9°N; longitude: 97.1°W; elevation: 740 ft [225.6 m] above sea level
- Heating degree days: 10,350 base 65°F [5750 base 18.3°C]; cooling degree days: 1819 base 50°F [1011 base 10.0°C] for Winnipeg, MB; annual precipitation: 19 in. [483 mm]
- Building type: New construction; commercial offices and interpretive center
- Size: 823,535 ft^2 [76,508 m^2]
- Date completed: 2009
- Client: Manitoba Hydro
- Design Team: KPMB Architects (design architects), Smith Carter Architects and Engineers (architects of record), Prairie Architects Inc. (advocate architects), Transsolar (Energy/Climate Engineers)

Background. Using high-performance, low-impact, active and passive green systems, while simultaneously maintaining high standards in design quality and human comfort, KPMB Architects designed Manitoba Hydro Place as an iconic prototype challenging the potentials of sustainable architecture. Concern for multifunctional passive systems, high-performance active systems, intelligent building management networks, and green management policies, among other factors, allowed Manitoba Hydro Place to achieve a high level of sustainable performance in many areas. Design principles, client values, and integrated as well as interdisciplinary design processes were important to the success of the project.

Context. Producing most of its electricity through renewable sources, Manitoba Hydro is the province of Manitoba's primary energy utility company. Manitoba Hydro Place, the company's headquarters, was built on a remediated site in a dense part of downtown Winnipeg. The structure, designed to support the company's current as well as predicted future needs, was to be the largest office building in Winnipeg at the time of construction. Despite its size, Manitoba Hydro set the goal that the building should achieve at least 60% energy efficiency, while still maintaining user comfort and design quality, when weighed against a comparable office building.

Design Intent. The new headquarters was designed not only to serve Manitoba Hydro's programmatic needs, but also to function as a model for low-impact, high-performance, sustainable architecture. The design was to use a mix of passive, active, and intelligent systems, combined with specific green policies and urban integration. The building was to respond to the climate, the city and downtown, user habits, local vegetation, soils, views, solar orientation, and functional requirements (Figures 15.6 and 15.7).

The leading principles during the design process were as follows:

- Create a model prototype for low-impact, high-performance architecture.
- Achieve energy saving goals and cost effectiveness, and attain at least Leadership in Energy and Environmental Design (LEED) Gold certification.
- Use mix of passive/active systems.
- Respond to climate.
- Create a workplace that is functional, connective, and supportive of a sustainable lifestyle.
- Support/stimulate urban revitalization of downtown.
- Integrate within the existing transportation system of downtown Winnipeg.

Design Criteria and Validation. The end goal of the building was to achieve energy and water conservation, while also establishing an integrated hub that promoted sustainable living on a number of scales. An interdisciplinary and integrated design process was key to testing and optimizing the project's aims. Aside from the architectural actors involved, specialist engineers and consultants from a wide range of disciplines

Figure 15.6 *Aerial view of north elevation showing solar chimney. (© Maris Mezulis; used with permission.)*

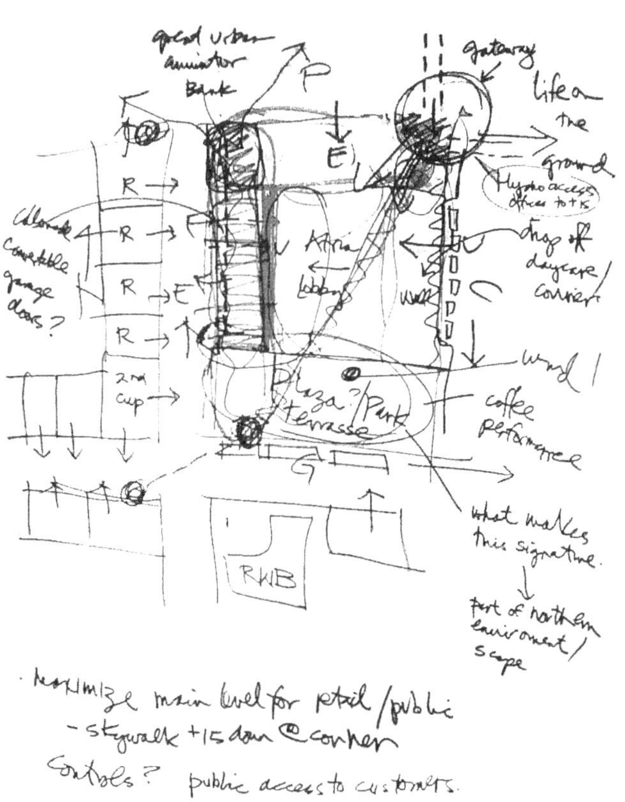

Figure 15.7 *Conceptual sketch developed from the first charrette. (© KPMB Architects; used with permission.)*

Figure 15.8 Street view of Manitoba Hydro Place, showing the tower above the podium and the solar chimney rising above the office tower roof. (© Maris Mezulis; used with permission.)

(*a*) (*b*)

Figure 15.9 (a) One of three six-story south atrium wintergardens. (b) Six-story atrium. (© Maris Mezulis; used with permission.)

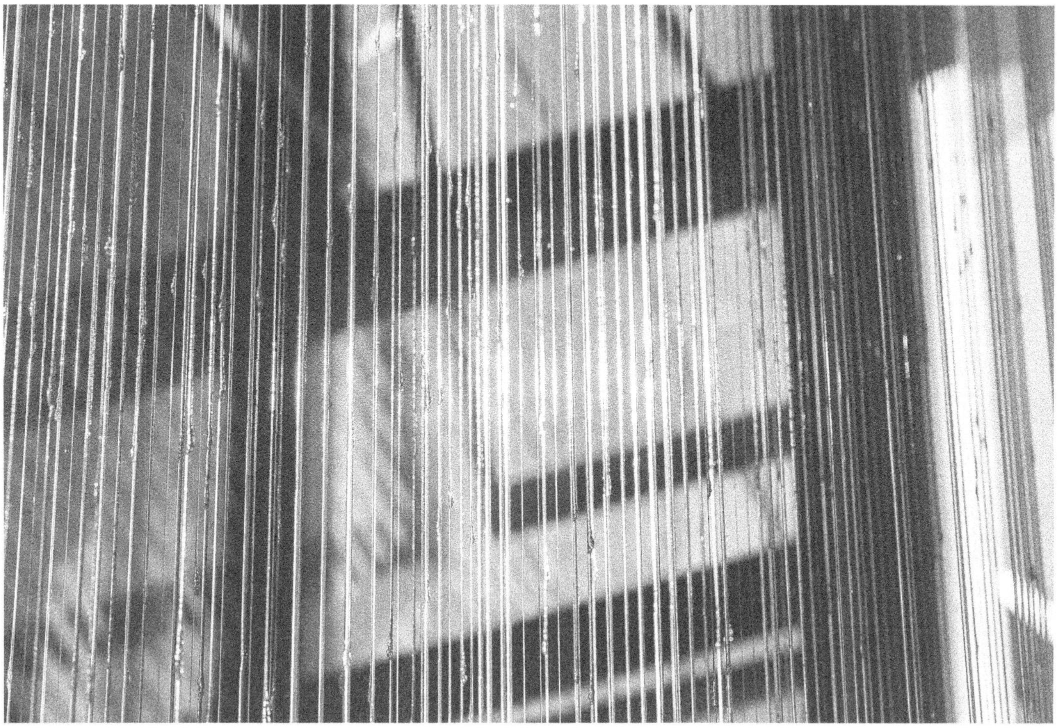

Figure 15.10 *Detail of wintergarden water element. (© Maris Mezulis; used with permission.)*

Figure 15.11 *Solar chimney detail. (© Maris Mezulis; used with permission.)*

were called upon—including structural, mechanical, electrical, geothermal, building enclosure, microclimate, traffic/parking, water, and geotechnical consultants, among others. A combination of computer simulations and real-model tests was conducted in order to verify the capacities and capabilities of the passive and active systems to be employed.

Post-Occupancy Validation Methods. Manitoba Hydro established a monitoring system in order to verify the building's performance over a two-year period following full occupancy. The building uses an intelligent building management system in order to regulate its ventilation, heating, and lighting systems. For its success in achieving its goals, the company credits the ground-source heating/cooling system, solar chimneys, wintergardens, automated shading systems, high-performance and intelligent lighting systems, ventilation systems, and management policies, among other factors (Figures 15.8 through 15.12).

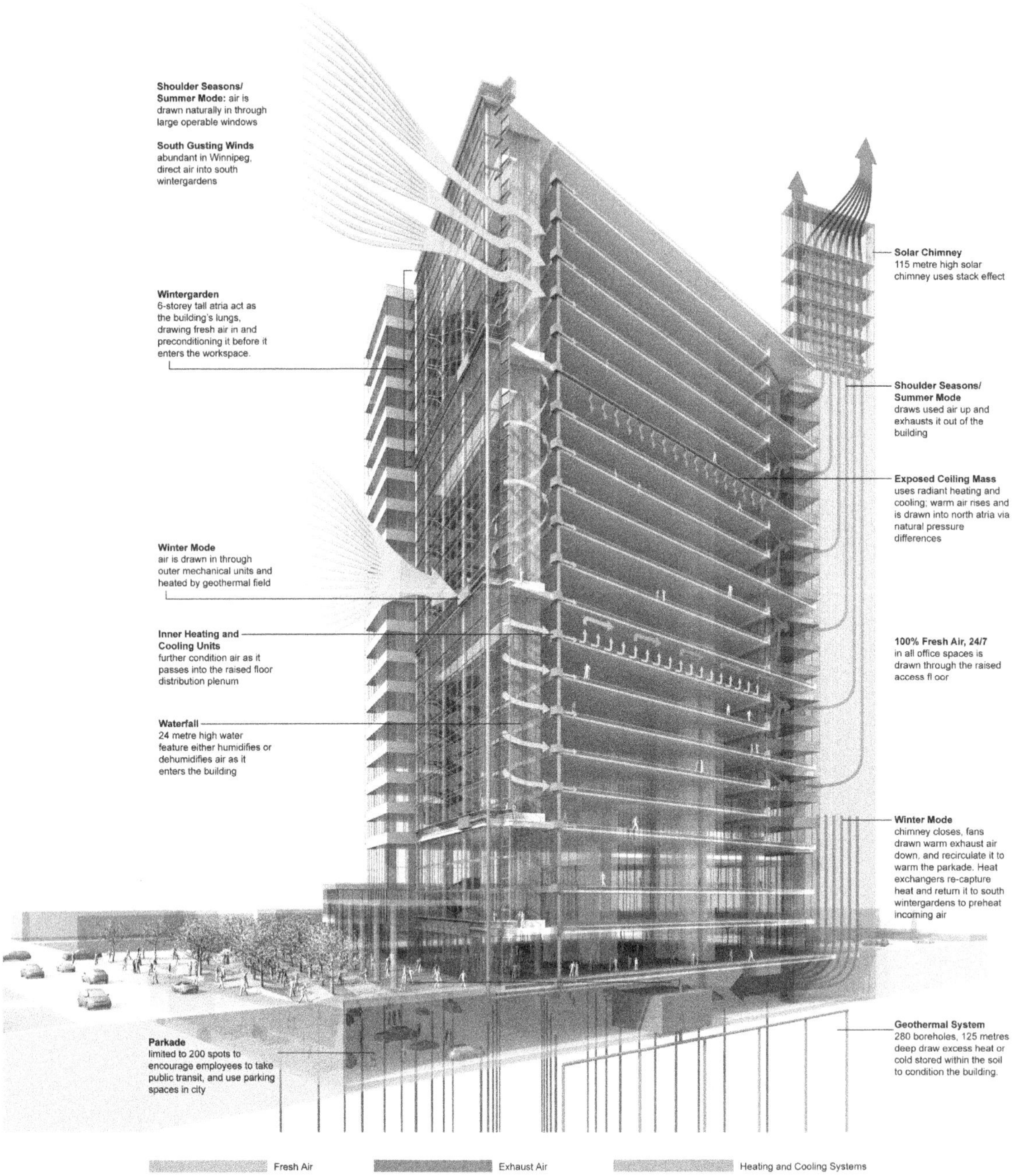

Figure 15.12 *Natural airflows through the building. (Bryan Christie Design; © KPMB Architects; used with permission.)*

PASSIVE ENVIRONMENTAL SYSTEMS

Design Process Specifics. The Integrated design process utilized in the design and construction of Manitoba Hydro Place was based on the C-2000 Program for Advanced Commercial Buildings, designed by a branch of the Canadian government (Natural Resources Canada). Key steps in the process were as follows.

An integrated design team was formed, composed of the client, architects, engineers, consultants, cost estimators, and contractors. Manitoba Hydro worked with KPMB in order to form this team.

- Before the design of the project began, a project charter was formed, establishing the primary goals (see Design Intent) to be achieved.
- One year was set aside in order to form a critical design concept for the new building, involving charrettes and workshops, among other activities.
- During the design development phase, bi-weekly meetings and charrettes were also held, for discussing and resolving the issues brought up by the interdisciplinary actors involved.
- The contracting and bidding phases were coordinated earlier in the process; through this method, material/labor costs and schedules were confirmed far ahead of time, allowing the interdisciplinary design team to avoid any surprises in terms of material costs or labor supply during actual construction.

Performance Data. Information available to date suggests substantial success in a range of critical sustainability fields—particularly in the areas of energy conservation, water-use reduction, and material use.

- Manitoba Hydro was awarded LEED Platinum certification in May of 2012.
- The building achieved an energy cost savings of 70% when weighed against a comparable office tower constructed according to the building standards specified by the Model National Energy Code.
- Through the use of aerator faucets, dual-flush toilets, and waterless urinals, water use was reduced by 43%.
- The building uses a 100% fresh air ventilation system.
- Throughout the construction phases, 85% of wastes were recycled or salvaged; over 23% of new materials contained recycled content; and 33% of new construction materials were extracted within an 800-km distance from the site, or transported via railway within a 2400-km distance from the site.
- Manitoba Hydro provides 16 vehicles, 11 of which are hybrids, for employee use.
- A reduction in heat island effect was achieved through the implementation of reflective roofing or green roof systems.

The building was named one of the AIA/COTE Top Ten Green Projects in 2010; it was given the Urban Land Institute Global Award for Excellence in 2012; it also received the Best Tall Building Award in the Americas category via the Council for Tall Buildings and Urban Habitat in 2009.

FOR FURTHER INFORMATION

Canada Newswire. 2012. "Manitoba Hydro Sets New Record with LEED Platinum," May 25, 2012. prnewswire.com/news-releases/manitoba-hydro-place-sets-new-record-with-leed-platinum-154070225.html

KPMB Architects. kpmb.com

Natural Resources Canada. nrcan.gc.ca/home

Prairie Architects, "Manitoba Hydro Place: Eco Facts." prairiearchitects.ca/manitoba-hydro-place

Transsolar Energietechnik GmbH. transsolar.com

PASSIVE ENVIRONMENTAL SYSTEMS

References and Resources

Ashley, R. and J. Reynolds. 1994. "Overall and Zonal Energy End Use in an Energy Conscious Office Building." *Solar Energy* 52(1): 75–83.

ASHRAE. 2021. *2021 ASHRAE Handbook—Fundamentals*. Atlanta, GA: American Society of Heating, Refrigerating and Air-Conditioning Engineers.

Emerald People's Utility District. epud.org

University of Oregon. 1999. "EPUD REVISITED: A Post Occupancy Analysis of Thermal and Daylight Design Implementation." Vital Signs case study by University of Oregon ARCH 407/507. seminar. darkwing.uoregon.edu/~akwok/VSCS/epud.pdf

PART V

ACTIVE ENVIRONMENTAL SYSTEMS

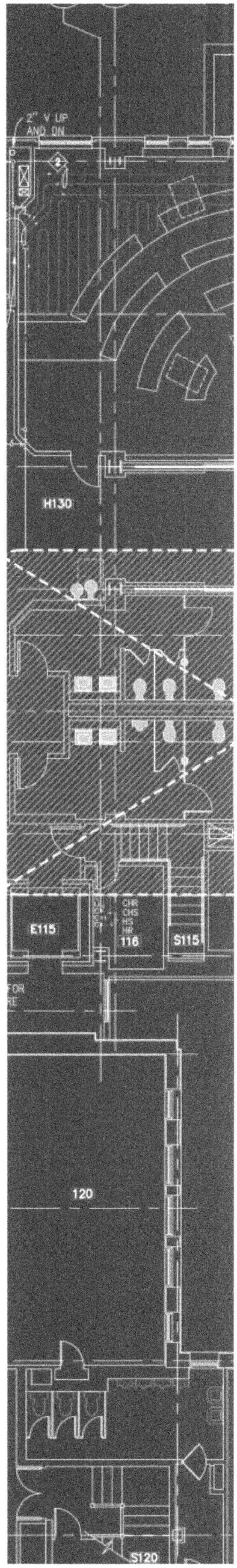

In Part IV (Passive Environmental Systems), daylighting, passive solar heating, and passive cooling systems were discussed as effective, energy-efficient environmental control approaches. In many buildings, due to climate or function, there will be a need to supplement passive strategies with active systems. The chapters comprising Part V: Active Environmental Systems address heating and cooling (HVAC) systems, electric lighting fundamentals and design, and water and waste systems. These active systems are typically designed by consultants, use specialized components, and require purchased energy (or water) resources. Information to support informed decision-making regarding components and systems is presented in this Part with example problems and guidelines. The interconnectivity of the thermal, luminous, and aqueous environments will vary depending upon the climatic conditions of the site/region, occupancy, function, behavioral activities and attitudes, and the quality of the building materials and enclosure. It is critical that these systems be designed to conserve resources.

Chapter 16, admittedly a long chapter, discusses active climate control (HVAC) systems for buildings of all scales. The main functional elements (source, distribution, delivery, and control) of HVAC systems are addressed, both conceptually and practically through the use of equipment diagrams and photographs. Two case studies illustrate the application of HVAC systems at different scales.

Chapters 17, 18, and 19 deal with electric lighting systems. Chapter 17 presents basic information on system components: lamps, luminaires, and controls. Chapter 18 deals with the quantitative and qualitative analysis of a lighting system relative to the owner's project requirements. Chapter 19 presents information on the application of lighting systems to a variety of occupancy types.

Chapters 20, 21, and 22 address water and its use in buildings. Chapter 20 discusses the basics of water and its roles in building design. Chapter 21 provides detailed information on the design of water supply systems and components. Chapter 22 presents information regarding the systems used to convey liquid wastes and stormwater from a building to appropriate disposal.

Chapter 23 deals with solid wastes. Although this chapter fits logically in this Part of the book, many solid waste solutions will be architectural in nature—and thus generally passive within the confines of a building.

CHAPTER 16

Active Climate Control

ACTIVE CLIMATE CONTROL SYSTEMS are thermal comfort and indoor air quality modification systems that:

- require the use of purchased energy
- involve numerous single-purpose components
- frequently are only lightly integrated into the overall building fabric
- typically are designed by an engineering consultant rather than an architect

Active climate control systems are, conceptually, the opposite of a passive systems approach to climate control. Although this book exhibits a preference for passive solutions, the use of active systems is very commonly dictated by realities of climate or building function. A designer of buildings virtually anywhere in the world should be familiar with active systems (for climate control, lighting, vertical circulation, fire protection, and sanitation)—they are a part of modern design.

The acronym HVAC is typically used to describe an active climate control system. HVAC stands for heating, ventilation, and air conditioning. Heating is a process whereby the temperature of something (such as the air in a building) is increased. Ventilation is a process that moves or circulates air. It would be easy to assume that the AC part of HVAC refers to "and cooling." This would be incorrect—AC stands for air conditioning, which has very specific process requirements. An air-conditioning system should be able to:

- control air temperature (by heating and/or cooling as required)
- control the humidity of air (perhaps bi directionally)
- control air distribution (speed and direction)
- control indoor air quality (IAQ)

If it sounds as if the term AC generally covers the expectations of the term HVAC, this is true. AC is commonly used as a synonym for HVAC and vice versa. The term HVAC will be used in this chapter, for consistency with usage in the design professions. Occasionally, a project may demand only a heating system, or only a ventilation system, or only a cooling system—more typically, however, today's projects demand (or at least expect) air conditioning as provided by an HVAC system. Ideally, the capabilities of the passive climate control systems discussed in Part IV (i.e., Chapters 12–14) would be considered first. Even when a passive system cannot fully provide for project requirements, such an approach might be used during part of the year or in some part of a building to mitigate the energy and carbon impacts that come with active systems.

16.1 INTRODUCTION

The process of acquiring a viable HVAC system parallels the process of acquiring a building and generally involves decisions and actions taken during the predesign, design, construction, and occupancy phases of a project. The following list describes some of the key aspects of HVAC system acquisition in the order that would be typical for a new construction project. The list presumes that the commissioning process is in place (as would be expected for any high-performance project).

Mechanical and Electrical Equipment for Buildings, Fourteenth Edition. Walter Grondzik, Daniel Overbey, and Alison Kwok.

Companion website: www.wiley.com/go/meeb14e

- Engage the commissioning provider (CxP) as part of the project team from the earliest phases of design.
- Establish HVAC-related owner's project requirements (OPR: design issues/intents/criteria; including code/standard compliance). Note that the commissioning process may also include the building enclosure, this is called building enclosure commissioning (BECx).
- Establish zoning requirements.
- Draft the basis of design (BOD), which should determine a preliminary system selection based on the OPR and zoning requirements. Regularly update the BOD in response to subsequent development.
- Calculate the design heating/cooling loads.
- Select the appropriate source equipment (to meet loads, intent, and context).
- Select the appropriate distribution approach (to meet intent and fit context).
- Coordinate the HVAC components with other building systems.
- Rough-size equipment (fans, pumps, valves, dampers, pipes, ducts, condensers, air-handlers, tanks, etc.).
- Run energy analyses to optimize equipment selections and system assemblies.
- Final-size equipment based upon optimization studies.
- Coordinate the final individual equipment selections into a cohesive whole.
- Develop the appropriate control logic and strategies.
- Develop commissioning test protocols and checklists.
- Witness systems installation and verifications.
- Develop systems manuals for the owner.
- Prepare a current facilities requirements (CFR) and operations and maintenance (O+M) plan that contains the information necessary to operate the building effectively and efficiently.
- Provide benchmark (new system) performance data for the owner.
- Assist in the initial operations to maintain the owner's project requirements.

The scope and extent of the actions described here will vary from project to project. They generally apply to small and simple (single-family residence) as well as large and complex (teaching hospital) projects, but specific actions will be adapted to fit the project scale, schedule, and budget. For smaller buildings, an HVAC system selection/design may be done by the architect alone (or with assistance from a mechanical contractor). For larger, more complex buildings, consulting engineers will be involved (perhaps along with other specialists such as fire protection engineers or laboratory consultants). At whatever scale, a high-performance outcome will normally be most easily achieved through an integrated design process. Several of the earlier actions in this list (specifically, developing the owner's project requirements and thermal zoning) are arguably best done by the project architect (who will likely spend the most time with the client). Other actions will be completed by a consulting engineer, yet others by a contractor, and others by the owner's operating personnel. A process that taps into the experiences and expertise of these diverse players in a timely manner simply makes sense. With inspired teamwork, the integration of HVAC services can enhance building form, as many examples in this chapter show.

Earlier chapters in this book deal with foundational considerations that feed into HVAC systems design. Chapter 1 addresses the design process and the many influences on that complex activity (such as design intent and codes/standards) that may impact HVAC system and equipment selection. Chapter 2 deals with environmental resources, which are a consideration in all aspects of building design. Chapter 3 addresses sites and their resources, which will affect building heating/cooling loads and the selection of appropriate systems solutions. Chapter 4 clarifies the role of buildings with regard to human health and wellness. Chapter 5 provides background information on indoor air quality. Along with comfort, acceptable IAQ will be an intended outcome for most projects. Chapters 6 through 8 deal with comfort—thermal, visual, and acoustical, respectively. Specific to HVAC systems, thermal performance will usually be a major element of an owner's project requirements for virtually all building types. Chapters 9 and 10 present the fundamentals of building thermal enclosure design and performance—which will affect the loads that an HVAC system must mitigate. Chapters 11 and 12 introduce solar access, shading, and daylighting, which can have profound impacts on a building's HVAC system. To reemphasize, Chapters 13 and 14 present passive heating and cooling systems as preferred starting points for climate control. Chapter 15 discusses integration of passive heating, cooling, and lighting systems.

A potential problem faced by one attempting to understand HVAC systems is their seeming complexity. HVAC system types are numerous, their components are more numerous, and the expected system outcomes can vary widely. This is unavoidable as such diversity permits these systems to work in a wide range of contexts while providing some flexibility in selection. Fewer options would ease understanding—but would be unacceptable to the design professions, which thrive on choice. This chapter presents aspects of HVAC systems that will facilitate introductory engagement with important concepts and systems.

16.2 HISTORY AND CONTEXT

One of the most basic functions of a building is to provide shelter from weather. In the words of James Marston Fitch: "to interpose itself between people and the natural environment . . . to remove the gross environmental load from their shoulders" (Fitch and Bobenhausen, 1999). Historically, this role was accomplished through passive systems—for heating, cooling, and daylighting. In a carefully designed building, the roofs, walls, windows, and interior surfaces alone can maintain comfortable interior temperatures for most of the year in most North American climates. With appropriate scheduling, the most uncomfortable hours within buildings can often be avoided; for example, the siesta avoids the hottest afternoon hours within stores and office buildings. Several aspects of comfort and climate, however, pose difficult challenges for ordinary building forms and materials. Under these circumstances, when passive systems were the only available solutions, human comfort and well-being suffered. Active systems provide another option.

A building surface influences comfort primarily through its surface temperature; secondarily, a surface can modify air temperature (as when cool air moves across a warm surface). As important as these two determinants of comfort are, they are often not sufficient by themselves. In cooling situations, air motion and relative humidity are significant comfort determinants. For most building occupancies, air quality is an important issue under both heating and cooling conditions.

Building form can work with climate to produce air motion for cooling, although the higher air speeds that can extend the human comfort zone may be difficult to provide without mechanical assistance. Relative humidity is most readily controlled by mechanical or chemical means. Building form and materials may be able to keep spaces surprisingly cool, but without dehumidification, surfaces in many North American summer climates can become clammy and covered with mold. Further, it is difficult to filter air for IAQ purposes without a fan to force the air through the filtering medium.

Thus, whereas air and surface temperatures can often be successfully manipulated by passive means (a combination of building form, surface material, and informed user response), the comfort determinants of air motion and relative humidity, and the health and comfort considerations of air quality, often require mechanical systems. As the control of air properties—motion, moisture, pollutant content—becomes more critical to project success, the designer becomes more likely to respond with a sealed building, excluding outdoor air except through carefully controlled mechanical equipment intakes. In the recent past, this exclusion of outdoor air has often been accompanied by the exclusion of daylight, of view, of solar heat on cold days—in sum, by a general rejection of all aspects of the exterior environment. As designers come to terms with the role of mechanical systems in high-performance buildings, they should also clarify the role of these systems in reaching net-zero energy or carbon-neutral outcomes. Passive systems can be historic, but have limitations. Active systems are more recent and seem to have few if any limits on capabilities, but do have serious environmental impacts.

Depending upon how one interprets the definition of an active system, it is reasonably fair to claim that active heating systems have been in use for a rather long time. Roman hypocaust systems exhibit basic characteristics that mirror today's central hot air heating systems. Such sophistication was generally lost for almost two millennia until scientifically designed central heating systems reappeared in the 1800s. Rationally designed central ventilation systems also appeared in the early to mid-1800s. Design methods and component analyses were documented in early heating/ventilating texts (see, e.g. the timeline from Haberl et al., 2012). The cooling part of an active climate control system is a much more recent development. Early forays into central cooling occurred in the mid-1800s, but true diffusion of active cooling into the building sector came only after World War II in the 1940s. By the 1960s, approximately 13% of homes in the United States had HVAC systems—a figure that grew to over 50% by the 1980s. Today, close to 90% of all homes in the United States have air conditioning. The uptake of HVAC systems post-WWII made settlement in harsh climate regions much more viable, and cities such as Las Vegas, Nevada, saw a population boom.

As active climate control systems became commonplace and found increasing use in a range of building types, they also became more complex and more specialized. This has resulted in a differentiation and compartmentalization of design responsibilities that mirror the situation in many other aspects of design, and modern life in general. Current terminology describes this as working in separated silos or towers. As suggested in Chapter 1, a relay-race-like, nonintegrated design process seldom serves the long-term interests of the owner and places roadblocks in the path to high-performance building outcomes. A fully integrated design process is strongly recommended. If this is not yet feasible on a specific project, then the commissioning process is suggested as a bridging mechanism. The ongoing collaborative work of the commissioning team (composed of diverse project participants) can provide some of the benefits of a fully integrated process—and is well documented and well accepted in many parts of the world. It is unreasonable for important aspects of building design to remain a mystery or appear to be magic to other design team participants.

16.3 DESIGN CRITERIA

The role of codes, standards, and guidelines in the design of an HVAC system is quirky, but consistent with the pattern seen in other areas of building design. In the United States, there is rarely a code mandate that requires that an HVAC system be provided for a building. There may be requirements here and there that dictate that building occupants not be allowed to freeze in winter (we seem less concerned as a society about summer heatstroke), but there is no known code requirement that U.S. buildings be thermally comfortable. Increasing adoption of *ASHRAE Standard 62.1* is moving the profession toward provision of code-mandated minimum air quality. Once a decision has been made to provide an HVAC system, however, all sorts of codes, standards, and guidelines come into play. These may address the design and installation of specific components or equipment (such as boilers, flue stacks, or window air-conditioning units), the arrangement and installation of systems (primarily from the perspective of life safety and energy), and the overall performance of a system (mainly its energy performance).

It would be unreasonable to attempt to list in this chapter all the design guidance documents that would apply (in the United States) to an HVAC system and its components. If the geographic perspective is widened to systems designed worldwide, the number of applicable standards and guidelines mushrooms to easily exceed a thousand documents. Chapter 40 of the *2021 ASHRAE Handbook—Fundamentals* provides an extensive and up-to-date listing of HVAC-applicable codes and standards sorted by subject. Some of the more important and commonly encountered standards and guidelines include:

- *ASHRAE Guideline 0–2019: The Commissioning Process for New Buildings and New Systems*
- *ASHRAE Guideline 1.1–2025: Application of the Commissioning Process to New HVAC&R Systems*
- *ASHRAE Standard 90.1–2022: Energy Standard for Sites and Buildings Except Low-Rise Residential Buildings*
- *ASHRAE Standard 62.1–2022: Ventilation for Acceptable Indoor Air Quality*
- *ASHRAE Standard 62.2–2022: Ventilation and Acceptable Indoor Air Quality in Low-Rise Residential Buildings*
- *ASHRAE Standard 55–2023: Thermal Environmental Conditions for Human Occupancy*
- USGBC: *LEED* v5 rating systems for various building types
- NFPA: Numerous fire protection standards (see Part VII of this book)

Please note that the editions of the documents listed here are subject to change as each is updated regularly.

16.4 FUNDAMENTALS

A number of fundamental principles underlie the selection and design of an HVAC system. The most basic of these considerations are summarized herein.

(a) Design Intent and the Commissioning Process

The design of any building system must start with a clear understanding of the outcomes required and/or expected by the client, by users/occupants, and by society. Moreover, there should be an established process engaged to ensure that this design intent is achieved on a project. This is especially true for complex HVAC and refrigeration (i.e., HVAC&R) systems. Society's expectations are typically expressed through building codes (in particular, those dealing with energy and indoor air quality). The aspirations of the client and subsequent building users are best expressed through a comprehensive and cohesive statement of what the owner expects regarding a project's building systems performance. The design intent as outlined by this statement should be adopted and pursued through an HVAC&R commissioning process.

Commissioning (Cx) can be defined as a process of verifying and documenting that a facility and all of its systems and assemblies are planned, designed, installed, tested, operated, and maintained to meet the owner's project requirements (OPR). The OPR is a formalized document that identifies the ideas, concepts, and criteria determined by the project owner to be important to the success of the project. A complete OPR may address thermal comfort requirements for all occupied spaces, indoor air quality requirements for all occupied spaces, minimum acceptable energy performance, environmental preferences for constituent components, first cost, life-cycle costs, reliability, maintainability, building space and volume demands, aesthetics, and other considerations.

The project team will receive the OPR and should respond by developing a basis of design document (BOD), which summarizes the information necessary to accomplish the OPR, including system descriptions, indoor environmental quality criteria, design assumptions, and references to applicable codes, standards, regulations, and guidelines. If the OPR defines the "what" of a project, the BOD defines the "how."

The commissioning process for a facility should adhere to a defined set of guidelines. *ASHRAE Guideline 0-2019: The Commissioning Process for New Buildings and New Systems* offers an outline of best practices for applying whole-building commissioning to a project. Guideline 0 is aimed at defining a uniform, integrated, and consistent commissioning process. *ASHRAE Guideline 1.1-2025: Application of the Commissioning Process to New HVAC&R Systems Application of the Commissioning Process to New HVAC&R Systems* contains both general and technical requirements in support of the commissioning process for HVAC&R systems and assemblies. From the outset of a project, the commissioning provider (CxP) will require both the OPR and BOD with an understanding that these documents are subject to change as a project is developed. The commissioning process will verify that the latest version of the BOD is adhered to and better ensure the likelihood that the HVAC system(s) proposed, designed, and installed will deliver on the performance targets established in the latest OPR.

The Lawrence Berkeley National Laboratory determined that roughly 90% of buildings have HVAC controls and systems that do not function properly in the first year of occupancy. Commissioning addresses this shortcoming squarely by ensuring that building systems are ultimately installed and function in accordance with the design intent. In short, the commissioning process better ensures that the owner gets what they think they are paying for with regard to building systems. Selecting an HVAC system in the absence of clear requirements, selection criteria, and a commissioning process is not designing—it is wishful thinking.

Key design criteria may include the following:

- *Temperature setpoints (for heating and cooling):* Setpoints establish the range of dry-bulb temperatures which the HVAC system will attempt to maintain for each specified zone during occupied hours.
- *Temperature setbacks (for heating and cooling):* Setbacks establish the range of dry-bulb temperatures which the HVAC system will maintain for each specified zone when the building is unoccupied or when less stringent temperature control is otherwise required. Setbacks can save energy by allowing an HVAC system to accept a wider range of temperatures than would otherwise be provided.
- *Relative humidity setpoints and setbacks:* Similar to temperature setpoints and setbacks.
- *Annual diversity factors:* Broken down by day and further by hour throughout the entire year, these basically define the occupancy profile for the building and/or thermal zones. A percent-based diversity schedule can establish patterns by which the HVAC system can adjust capacity based on the number of people expected to be present at certain times along with the corresponding lighting and equipment power loads needed to serve the occupant load.
- *Outside air rate per person.*
- *Outside air rate per unit area.*
- *Air changes per hour* (ACH).

(b) Thermodynamics Laws

The laws of thermodynamics apply to the design of HVAC systems and are worthy of a quick summary and interpretation.

- *Zeroth Law:* Two systems that are in thermal equilibrium with a third system are in thermal equilibrium with each other. This law helps to define the notion of temperature. Heat will only flow from a higher to a lower temperature.
- *First Law:* Heat and work are forms of energy transfer. Energy is always conserved; thus, all energy associated with a system must be accounted for. Perpetual motion is impossible, and efficiencies greater than 100% are impossible. All macro energy forms within a building devolve to heat.
- *Second Law:* A system attempts to move toward a state of thermodynamic equilibrium, which increases entropy (disorder). It takes energy to order (and maintain order within) a system, which will otherwise naturally move toward disorder.
- *Third Law:* The entropy of a system approaches a constant value as the temperature approaches zero; this is not particularly germane to HVAC system design.

(c) Important "e" Terms

Several terms that start with the letter "e" are important to consider when designing an HVAC system. These terms include:

- *Energy:* Energy is the ability to do work. In HVAC system design, the amount of energy required to accomplish a defined work task is often of substantial interest to the design team. Power is an instantaneous snapshot of work (over a very short time increment); energy is the integration of power over time (per hour, per day, per year). Energy requirements may be analyzed at the site boundary of a project (effectively at the electric or gas meter), or they may be analyzed at the source of the particular energy supply (such as at the electrical power plant). These two system analysis boundaries will yield very different results. The appropriate system boundary is presently a matter of serious contention.

- *Effectiveness:* Effectiveness is a qualitative descriptor that describes success in meeting defined objectives. An HVAC system that delivers the OPR would be properly termed effective.
- *Efficiency:* Efficiency is a defined ratio (output/input) that is a quantitative assessment of the ability of a system to provide some effect relative to the resource cost of doing so. An HVAC system that provides a lot of useful heat while requiring just slightly more heat as input would be termed efficient. A system can be efficient while being ineffective. Conversely, a system can be effective while being inefficient. A high-performance building seeks systems that are simultaneously effective and efficient.
- *Enthalpy:* Enthalpy is a measure of the total (sensible and latent) energy content of a sample of air. See Chapter 6 for further information.
- *Entropy:* Entropy is a measure of the disorder of a system. Without counterbalancing energy inputs, most building systems lose order over time (thus increasing in entropy). A hot-water pipe loses energy when not in use. A layer of paint degrades over time. Water must be pumped to overcome the effects of gravity. Buildings continually require the input of resources for operation and maintenance.
- *Exergy:* Exergy is sometimes referred to as "available energy" and can be defined as the amount of work theoretically available by bringing a system into equilibrium with its surroundings through a reversible process. Exergy is a measure of the utility of an energy exchange and it is generally not well understood. In the future we may evaluate the exergy of competing HVAC system solutions to judge their appropriateness. As an example, the exergy of using a nuclear power plant to heat water by 50 degrees is quite different from the exergy involved when using a passive solar thermal system to accomplish the same task.
- In addition, one "c" term is also of interest—*conservation:* Conservation implies the use of less of something (energy, water, materials, cash). Conservation has both a noble history and detractors. In some circles, conservation is seen as unnecessarily doing with less. Conservation is a necessary concept when efficiency makes no sense. For example, discussing the efficiency of a showerhead is irrational (output equals input). A water-conserving showerhead is a logical way to address this situation—where there is an attempt to provide equal effectiveness using less water.

(d) Applied Psychrometrics

Psychrometrics (introduced in Chapter 6) is the study and manipulation of the properties of moist air. Adjusting the properties of air (which, in a building, always contains water vapor) is precisely what the thermal side of an air-conditioning system is intended to do. HVAC systems are psychrometric systems, and a fundamental understanding of moist air properties and their modification is critical to an understanding of climate control systems. Psychrometrics, which can be addressed through equations, data tables, and/or a diagram called the psychrometric chart, is a means of understanding the relationships among various properties of air. The psychrometric chart (Appendix G, Figures G.1 and G.2) is commonly used as an HVAC system design tool.

As defined in Chapter 6, there are eight distinct psychrometric processes. These are related to commonly encountered HVAC systems in the following discussion.

Sensible heating: Under this process, room air temperature is increased with no change in absolute humidity. Relative humidity will decrease as the air is heated, but there is no change in air moisture content. This process is very common and represents the outcome of the typical building heating system, which is purely sensible and has no inherent capability to affect latent loads. Hot air heating, baseboard radiation, and solar heating systems are all sensible heating systems.

Heating and humidifying: This process, which moves up and to the right on the psychrometric chart, is often desired in smaller buildings located in cold climates. There is no single HVAC device that will produce this effect; sensible heating equipment must be paired with a humidification device.

Heating and dehumidification: This process, which moves down and to the right on the psychrometric chart, is not a commonly desired or encountered process. No single HVAC device will intentionally produce this effect (although it is possible to accomplish it).

Humidification: This process adds moisture to the air without intentionally changing air temperature. A device called a humidifier can accomplish this effect (although rarely will this alone be adequate to produce thermally comfortable conditions).

Dehumidification: This process removes moisture from the air without intentionally changing air temperature. A device called a dehumidifier can accomplish this effect. This may occasionally (in benign climates) be the only effect required to produce thermally comfortable conditions.

Sensible cooling: This process reduces air temperature without a change in absolute humidity. Some HVAC systems will produce sensible cooling through a part of their operating range. Other HVAC systems (such as a radiant cooling system) are specifically designed to operate as sensible cooling devices. Most active cooling systems, however, produce sensible and latent cooling effects.

Cooling and dehumidification: This combined sensible and latent cooling process is commonly desired in a wide range of building situations and is produced by the common vapor compression cooling process. A cooling coil is brought to a temperature below the dew point of the room air such that moisture condenses and leaves the air while the air is reduced in dry-bulb temperature.

Evaporative cooling: This process, which involves simultaneous sensible cooling and latent heating (cooling and humidification), is an intriguing HVAC option. The evaporative cooling process occurs along a line of near-constant enthalpy; sensible heat is exchanged for latent heat at little net energy cost. Where climate conditions will support this process, direct evaporative cooling can be a very energy-efficient space-cooling option. Indirect evaporative cooling can be employed where high RH resulting from a direct evaporative cooling process would be objectionable.

An example showing the use of the psychrometric chart to analyze conditions that occur during typical HVAC system operation may be helpful.

EXAMPLE 16.1 Find the total heat to be removed, and thus the required cooling capacity, for a dance hall. The design conditions are:

Room conditions (summer):	75°F DB [24°C], 50% RH
Number of occupants:	80 people
Activity:	Dancing
Size of room:	10,000 ft² [930 m²]
Ventilation provided:	35 cfm [18 L/s] per person
Outdoor air conditions:	90°F DB, 75°F WB [32.2 and 23.9°C]

Non-occupant heat gains: 67,600 Btu/h [19,814 W]

Heat Gains (with 80 people dancing, see Appendix G, Table G.5)	Sensible Heat, SH (Btu/h)	Latent Heat, LH (Btu/h)
80 @ 305 Btu/h	24,400	
80 @ 545 Btu/h		43,600
Total enclosure, lighting, equipment, etc.	67,600 (per above)	None
	Room sensible heat (RSH)	Room latent heat (RLH)
	= 92,000	= 43,600

SOLUTION

First, determine the portion of the heat gain that is due to sensible heat gain, called the sensible heat factor (SHF). Using I-P units

$$\text{SHF} = \frac{\text{RSH}}{\text{RSH} + \text{RLH}} = \frac{92{,}000}{135{,}600} = 0.68$$

On the psychrometric chart (simplified in Figure 16.1), draw a line between the fixed "bull's-eye" target (80°F DB, 50% RH) and the value of 0.68 on the SHF scale at the upper-right edge of the chart (Figure 16.1*a*). This is called the SHF line, and its slope is of interest.

Point *A* (Figure 16.1*a*) is the condition of the air within the dance hall as it is returned to an air-handling unit (AHU) for reprocessing: 75°F DB, 50% RH (62.5°F WB). The system designer must decide how much cooler the supply air should be than the room air in order to provide capacity to absorb the heat load of the dance hall. To avoid uncomfortable drafts, the supply air temperature is usually 20°F (or less) below the room air temperature. In this case, a supply-to-room Δt of 20°F is chosen. The quantity of air required to sensibly cool the room is found as:

$$\text{cfm} = \frac{\text{RSH}}{(1.1)(\Delta t)} = \frac{92{,}000\ \text{Btu/h}}{(1.1)(20°\text{F})} = 4182\ \text{cfm}$$

This above equation is a reformulation of the following ventilation load equation:

$$q = (\text{cfm})\,(1.1)\,(\Delta t)$$

The portion of the required supply air (4182 cfm) that should be outdoor air is found as follows (reference Appendix F, Table F.1):

$$(20\ \text{cfm/person})(80\ \text{people}) + (0.06\ \text{cfm/ft}^2)(20{,}000\ \text{ft}^2) = 2800\ \text{cfm}$$

So the percentage of outdoor air is:

$$\frac{2800}{4182} = 67\%$$

Several important process points can now be plotted on the psychrometric chart. Point *B* (Figure 16.1*a*) is the condition of the supply air entering the room; it will be 20°F below 75°F, which places point *B* somewhere on the 55° DB line. To determine exactly where, a dashed line is drawn through point *A*, parallel to the previously plotted SHF line, and extended until it crosses the vertical 55° DB line. This crossing point is point *B*, and occurs at 55° DB, 51.3° WB; enthalpy (h_B) = 21.0 Btu/lb.

Point *D* (Figure 16.1*b*) is the condition of the outdoor air, given (at design conditions) as 90°F DB, 75°F WB. Point *C* represents the mixture of 67% outdoor air and 33% return air that is brought to the AHU for treatment and redistribution to the dance hall. Connecting points *A* (return air) and *D* (outdoor air) and marking 67% of the distance from *A* to *D* gives point *C*. This occurs at 85°F DB, 71.3°F WB; enthalpy (h_C) = 35.2 Btu/lb.

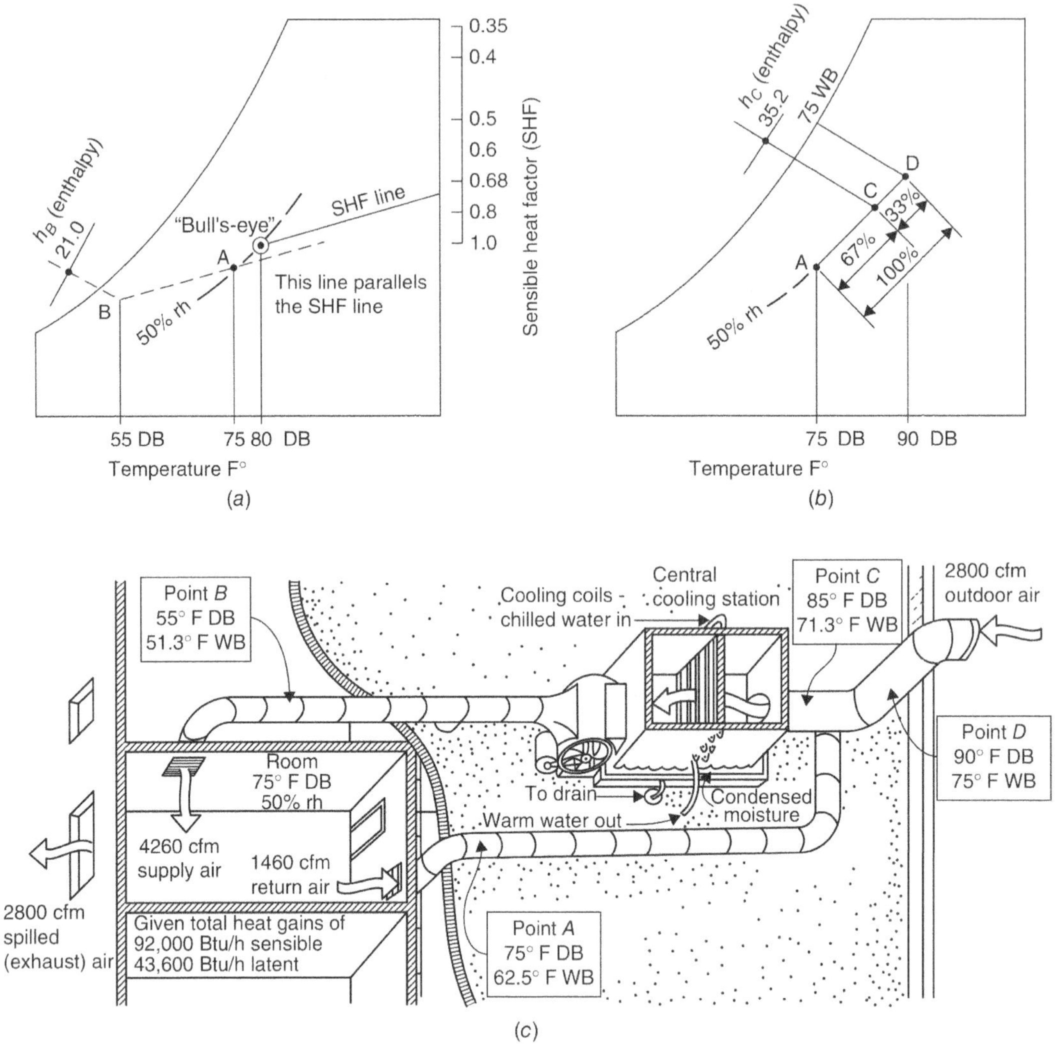

Figure 16.1 *Sizing cooling equipment using the psychrometric chart. (a) Finding the conditions for the supply air. [SI values are: 12.8, 23.9, and 26.7°C DB.] (b) Finding the conditions for the return air–outdoor air mixture. [SI values are: 23.9 and 32.2°C DB; 23.9°C WB.] (c) Points A, B, C, and D are representative conditions within the cooling cycle. [SI values are: Point A 23.9/16.9°C; Point B 12.8/10.7°C; Point C 29.4/21.8°C; Point D 32.2/23.9°C; outdoor air 1321 L/s; exhaust air 1321 L/s; supply air 2010 L/s; return air 689 L/s; room temperature 23.9°C.]*

The cooling equipment must remove the grand total heat (GTH), which constitutes sensible and latent heat, according to this formula:

$$\text{GTH} = 4.5 \times \text{cfm}\,(h_C - h_B)$$

where 4.5 is a constant derived from 60 min/h × 0.075 lb/ft^3 average air density. So, in this example:

$$\text{GTH} = 4.5 \times 4182\ \text{cfm}\,(35.2\ \text{Btu/lb} - 21.0\ \text{Btu/lb})$$

$$= 267{,}230\ \text{Btu/h}$$

the capacity of a refrigeration unit is typically specified in tons, where 1 ton = 12,000 Btu/h. The refrigeration required for this example is:

$$\text{refrigeration capacity} = \frac{267{,}230\ \text{Btu/h}}{12{,}000\ \text{Btu/h per ton}}$$

$$= 22.3\ \text{tons}$$

Note: If the minimum outdoor airflow rate were reduced from 2800 cfm to, say, 2000 cfm, then the percentage of outdoor air becomes 2000/4182 = 48%. Point *C* then moves to about 82°F DB, 68.8°F WB, at which point h_C = about 33 Btu/lb.

The required refrigeration capacity then becomes:

$$\text{refrigeration capacity} = \frac{(4.5)(4182)(33 - 21)}{12{,}000}$$

$$= 18.8\ \text{tons}$$

This change in outdoor air rate would provide a first cost saving for reduced equipment size and energy savings over the life of the dance hall. IAQ, however, might suffer. This type of trade-off currently confronts any designer balancing the competing priorities of energy conservation and human health. In all cases, code requirements will be met. ■

A focus on the psychrometrics of an active cooling process will further help explain how this tool is used. The work to be done by active air-conditioning equipment is indicated by the total change in enthalpy that must occur as the air is treated by the equipment. The psychrometric chart is used to mathematically determine air-conditioner performance requirements.

First, consider the problem of determining the total change in enthalpy. Assume outdoor conditions of 90°F DB and 76°F WB, with desired indoor conditions at 75°F DB and 50% RH. Using Appendix G, Figure G.1, the indoor conditions equate to 62.7°F WB. In the simple (but relatively rare) case of cooling 100% outdoor air, the total heat to be removed is determined as shown in Figure 16.2. How much total heat will be removed in conditioning air from 90°F DB and 76°F WB to 75°F DB and 62.7°F WB?

For every pound of "dry" air (based on the weight of the air alone) that is cooled and dehumidified, 39.6 − 28.3 = 11.3 Btu must be extracted. Similarly, for every pound of air so treated, 0.0162 − 0.0093 = 0.0069 lb of condensed moisture must be disposed of.

The actual cooling process is more complex, as shown in Figure 16.3. The conditioned air must be introduced to the space at both a lower DB temperature and lower RH than the desired indoor conditions, so that the supply air can "soak up" heat and moisture, then leave through the return air grilles at the desired indoor conditions (75°F DB and 50% RH). Thus, another set of conditions, of lower DB and RH, will be established. These conditions depend upon the rates at which air is introduced and the amounts of heat and moisture to be removed by the air passing through the space.

Within the cooling equipment, the outdoor air follows a complex path; the lines labeled 1, 2, and 3 in Figure 16.3 trace the sensible cooling and dehumidifying steps. Outdoor air is cooled without a loss of moisture (step 1) until it reaches saturation (its dew point). The cooling then continues; the air continues to lose sensible heat (step 2) and now loses moisture as well, extracting even more heat. When step 3 begins (known as a reheating process), the less-moist and lower-temperature air is heated to 65°F and 40% RH—slightly below the conditions to be maintained in the space. The changes in the heat and moisture content of the air at the various stages are measured along the enthalpy scale.

Deliberately adding heat to just-cooled (actually, intentionally overcooled) air sounds wasteful—and it is. The reheat process, however, is also a very effective thermal control mechanism. Heat-recovery processes can be incorporated into cooling equipment so that the reheat is provided by heat that was removed and would

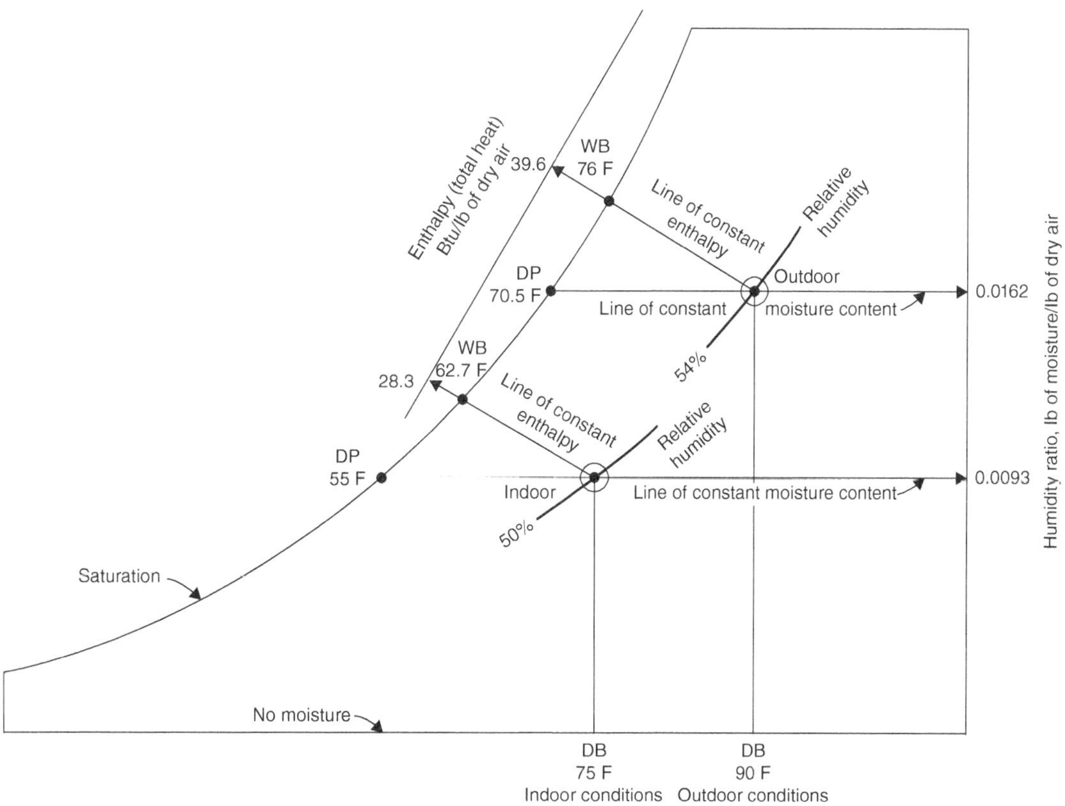

Figure 16.2 *Use of the psychrometric chart to determine the change in enthalpy between given outdoor and indoor air conditions.*

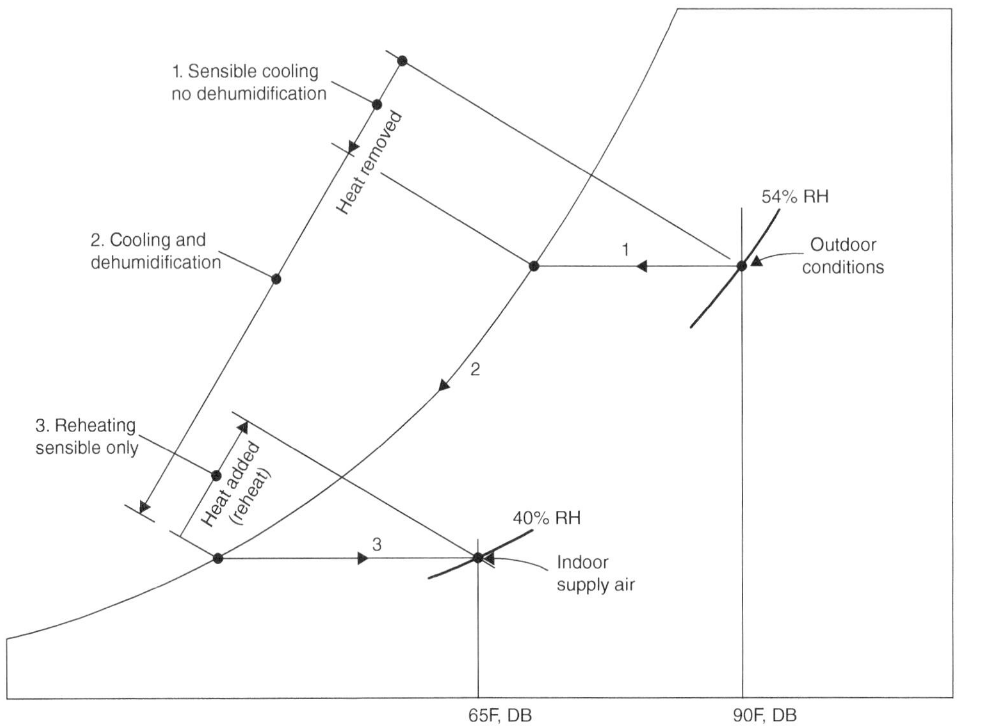

Figure 16.3 *Process of cooling and dehumidifying outdoor air, summer conditions.*

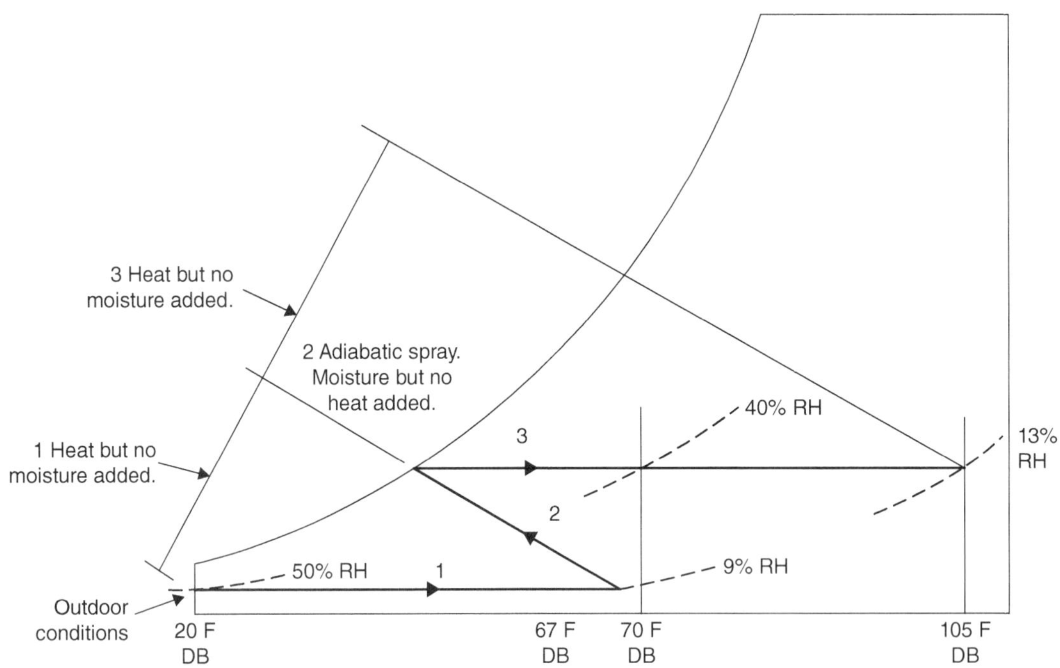

Figure 16.4 *Process of heating and humidifying outdoor air, winter conditions.*

otherwise be wasted. Heat-recovery technology is introduced throughout this chapter, but specific equipment is addressed in Sections 16.14 and 16.20.

Shifting focus to heating, the process of heating outdoor air in winter is charted in Figure 16.4. Air at low temperatures in winter often has a humidity ratio so low that this moisture content would be unacceptably low when the outdoor air is warmed (step 1). Therefore, moisture must be added. Often this is accomplished through injection of a warm water spray. In

Figure 16.4, however, use of an adiabatic spray is illustrated in step 2—adiabatic, meaning no change in enthalpy. Then, in step 3, the saturated air is warmed to a higher temperature (105°F in this case) than the desired indoor conditions of 70°F DB and 40% RH. This supply air then transfers its heat to the space. Note that this process starts at the desired indoor conditions and works backward (along step 3) to saturation, then downward along the constant enthalpy line (step 2) until reaching the line of constant humidity ratio of the warming outdoor air—in this case, 67°F DB. As with cooling, the changes in heat content can be read on the enthalpy scale.

(e) Metrics for HVAC System Performance

A wide variety of metrics (measures) for the expression of HVAC system and equipment performance will be encountered. In many cases there will be a mandated minimum performance threshold that is embedded in a building code. In the United States, for example, minimum acceptable performance values for most common types of source equipment will be found in *ASHRAE/IES Standard 90.1* (ASHRAE, 2022a). These values are usually the same minimum performance limits promulgated by the U.S. government. In almost all design situations, higher-performing equipment will be available from manufacturers. Such equipment will normally have a higher first cost than the code-minimum equipment—but the additional first cost may often be offset by energy savings and reduced life-cycle costs.

No matter the specific metric being considered, the performance of any given type of equipment will vary as its operating parameters vary. Listed equipment performance (and capacity) ratings are typically based upon a consensus standard that defines the parameters under which equipment will be rated. Such conditions are not necessarily the peak conditions that will be experienced in an actual building situation, so caution needs to be exercised when considering metrics such as efficiency or coefficient of performance (COP). In addition, most equipment operates less efficiently at partial load, and such reduced performance will affect annual energy consumption and payback analyses.

The following are some commonly encountered performance metrics that will affect HVAC system selection and design for buildings. Where applicable, the descriptions are consistent with ASHRAE's current definitions.

Annual fuel utilization efficiency (AFUE): The ratio of annual output energy to annual input energy, which includes any off-season pilot input loss. AFUE is a measure of how efficient a piece of equipment (e.g., a furnace or boiler) is in converting the energy from fuel to heat over the course of a typical year. An AFUE of 90% indicates that 90% of the energy input from fuel becomes useful energy output.

Coefficient of performance (COP): A measure of efficiency for HVAC systems based on their output. COP is defined in a slightly different way, depending upon the task. For cooling, COP is the ratio of the rate of heat removal to the rate of energy input in consistent units, for a complete cooling system (or factory-assembled equipment), as tested under a nationally recognized standard or designated operating conditions. For heating (i.e., a heat pump), COP is the ratio of the rate of heat delivered to the rate of energy input in consistent units, for a complete heat pump system as tested under designated operating conditions. Unless applicable and indicated under certain conditions, auxiliary heat is not included in this definition. The greater the COP, the more efficient the HVAC system is. A system with a COP of greater than 1 is delivering more energy than is required to run it (the COP of a ground-source heat pump may be in the 3–5 range).

Efficiency: Efficiency is the ratio of system output to system input when both values are presented in consistent units; it is expressed as a decimal value or percentage. Unless otherwise noted, efficiency should be taken to mean instantaneous efficiency at some defined point in time and under some specific set of operating conditions.

Energy: Energy is the capacity for doing work. It takes a number of forms that may be transformed from one into another such as thermal (heat), mechanical (work), electrical, and chemical.

Energy efficiency ratio (EER): The ratio of net equipment cooling capacity, in Btu/h, to the total rate of electric input in watts under designated operating conditions. (If consistent units were used, this ratio would be COP.)

Energy use intensity (EUI): Also "energy utilization index." An indicator of total annual building energy usage normalized per unit floor area: Btu/ft^2 yr [W/m^2 yr]; for a typical building, HVAC energy is a large part of EUI.

Heating seasonal performance factor (HSPF): The total heating output of a heat pump during its normal annual usage period for heating (in Btu), divided by the total electric energy input during the same period (in watt-hours, Wh).

Integrated part load value (IPLV): A single-number figure of merit, incorporating part-load EER or COP, expressing part-load efficiency for air-conditioning and heat pump equipment, based on weighted operation at various load capacities for the equipment.

Seasonal coefficient of performance–cooling ($SCOP_C$): The total cooling output of an air conditioner during its normal annual usage period for cooling divided by the total electric energy input during the same period in consistent units (analogous to SEER but in SI or other consistent units).

Seasonal coefficient of performance–heating ($SCOP_H$): The total heating output of a heat pump during its normal annual usage period for heating divided by the total electric energy input during the same period in consistent units (analogous to HSPF but in SI or other consistent units).

Seasonal energy efficiency ratio (SEER): Indicator of the total cooling output of an air conditioner during its normal annual usage period for cooling (in Btu) divided by the total electric energy input during the same period (in watts, W).

(f) Thermal Zoning

A thermal zone is an area of a building that must be provided with separate control if thermal comfort expectations are to be met. Thermal zones are very often a key basis for HVAC system selection, and zoning decisions can play a critical role in occupant comfort responses. Theoretically, the design team wants to provide just enough zones to meet the owner's project requirements (with perhaps some flexibility for anticipated future needs). Fewer zones will result in some level of discomfort. More zones than necessary will increase the first cost of the project.

Thermal zoning must be established prior to the selection of a climate control system. The concept of thermal zones applies equally well to passive climate control systems as to active climate control systems. A zone may be a room in a building, may consist of multiple rooms, or may be a portion of a room (a room may have more than one zone). The intent of zoning is to set up a control scenario that can respond to changing room loads and maintain thermally desirable conditions. Zone control in an HVAC system is most commonly initiated by a thermostat—a temperature-sensing device used to control air temperature within a thermal zone. As noted in Chapter 6, thermal comfort is affected by variables other than dry-bulb air temperature—nevertheless, temperature by itself is a de facto control variable in most buildings. Successful zoning allows the HVAC control system to provide conditions amenable to thermal comfort in the face of changing heating and cooling loads. In a passive system, zone control might be exercised by opening and closing windows or draperies.

Thermal zoning decisions are typically driven by differences in the timing of loads from one room to another. Timing (scheduling) is the driver—not differences in the magnitude of loads. As an example, an east-facing office must be zoned separately from a west-facing office due to solar radiation patterns. If the two offices are on the same zone (with one thermostat), the occupants of the office without a thermostat (without control capability) will be consistently uncomfortable. The same would be true of a classroom placed on the same zone as an office. If the thermostat is in the office, the system cannot respond to changes in classroom occupancy (empty, half-full, full).

To repeat, thermal zoning must precede system selection—so that a system with adequate zoning capability for the project context is selected. The minimum number of thermal zones for a conventionally designed multipurpose, mid-rise building is shown in Figure 16.5. More than the minimum number of zones could be required due to differences in load scheduling within a zone, such as between offices and stores or between offices and conference rooms. The apartments are also shown with the minimum five zones (based on solar orientation); an emphasis, however, on individual ownership of units and the variation in usage patterns typical of residential occupancies often results in a decision to provide as many zones as there are apartment units. Otherwise, the hard decision of which apartment gets a thermostat, and which does not, will need to be addressed.

It may be premature to call it a trend, but there has been a lot of interest in Europe in personal control systems for ventilation and thermal comfort. Under this approach, each workstation is set up as a separate

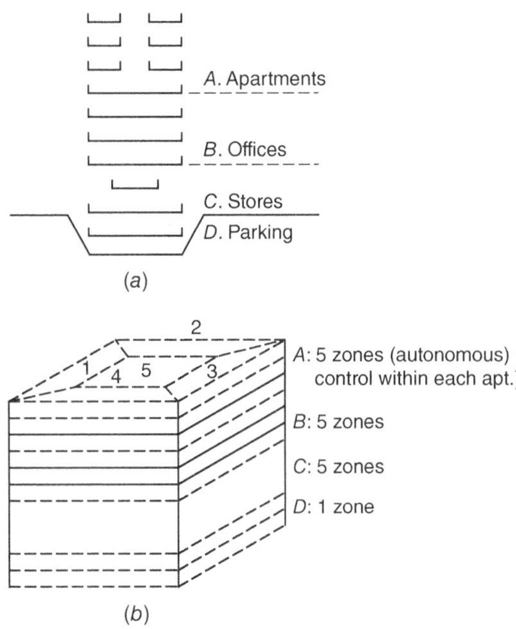

Figure 16.5 *A reasonable minimum number of thermal zones for a large multipurpose building.*

thermal zone whose conditions are controlled by the occupant. This approach is logical; although likely expensive—not just for the control capability, but also for the distribution delivery elements. There are also limits upon building types where this would make sense (each classroom seat in a school would be hard to individually zone). On the other hand, comfort, energy efficiency (perhaps), and diversity of conditions would be fostered when this zoning arrangement is possible.

Once the basics of system zone determination have been established, a design philosophy question is in order. How similar should the interior environments of buildings be? This question encompasses not only thermal experiences, but visual and acoustical ones as well.

The advantages of uniformity are most evident in ease of design and construction that, through mass production and speed, often brings lower first costs. In the case of an office, uniformity of ceiling heights, light fixture placement, grille locations, and so on promote flexibility in varying arrangements that can extend a building's usable life span. There are, however, at least four types of offices that may need to be interchangeable within such generic space. The typical enclosed office has the privacy of four walls and a door. The bullpen office has repeated, identical workstations, with low dividers at about the height of the desk surface. The uniform open-plan office resembles the bullpen, but with higher divider partitions for added privacy. The free-form open-plan office has some individually designed workstations with divider partitions of varying heights (sometimes reflecting the varying status of workers). In the bullpen and uniform open-plan office, the resulting uniformity is not always attractive to users, and diversity is often encouraged at a more personal level—with office furnishings, for example. A more thorough approach to diversity can provide stimulus to the user who spends many hours away from the variability of the exterior climate.

If offices must be uniform with respect to ceilings, lighting, air distribution, and even size, the corridors that connect them and the lounges or other supporting service spaces can deliberately be made different. Diversity requires a complete and detailed design of places; it gives the builder a more complex and interesting task; and it can provide orientation and interest to the users. The attractiveness of diversity is evident in most collections of retail shops, in which light and sound—and sometimes heat and aroma—are used to distinguish one shop from the next.

Diversity in the thermal conditions to be maintained, such as warmer offices and cooler circulation spaces in the winter, can be used to enhance the comfort of the office users. Designers have long recognized that a space can be made to seem brighter and higher if it is preceded by a dark, low transition space. Thermal comfort impressions can be manipulated similarly. Less than comfortable conditions in circulation spaces or other less-critical zones not only make the key functional spaces seem more comfortable by contrast, but also save significant amounts of energy over the life of a building. Furthermore, allowing diversity in conditions can make passive strategies more feasible.

A large-scale demonstration of diversity in thermal zones is shown in Figure 16.6. Passive solar heating can make a significant contribution, even through a shallow-sloped, single-glazed cover in cloudy Glasgow, Scotland, largely because the mall area and leisure areas are allowed a much wider thermal range than would be permitted in stores and offices. The overcast skies are quite suitable for daylight, and the addition of summer sunshading makes natural ventilation (through the stack effect, assisted by fans) possible during the cool summers. U.S. Pacific Northwest climate conditions are similar.

Designers should also be mindful of occasional sub-zone requirements. A sub-zone is an area within a larger thermal zone that requires additional or more refined control. Sub-zones are typically used when different areas within the same thermal zone have different heating or cooling requirements. For example, a thermal zone might be served by a constant air volume (CAV) system; however, it may include a sub-zone along a largely glazed perimeter that requires reheat coils to counteract exacerbated heat loss during winter conditions.

(g) Preliminary Space Requirements for HVAC

Detailed information regarding the floor area and distribution volume requirements for an HVAC system will become available during the design development phase of a project. Unfortunately, architectural design decisions regarding these space requirements were made much earlier—probably during the conceptual design phase as floor plans were first developed and accepted by the owner. Providing too much space for HVAC artifacts is not architecturally desirable, but a boon to the HVAC designer. Providing too little space for HVAC equipment, distribution, and maintenance is a serious problem that will likely compromise system performance and longevity. Having HVAC expertise available to the design team early in the design process can go a long way toward right-sizing space allocations. This can be accomplished through an integrated design process or (somewhat less so) via a well-implemented commissioning process. In the absence of such early expertise, precedent buildings can be used to inform space decisions; or, a design guidance tool (such as seen in Figures 16.7 or 16.8) can be employed.

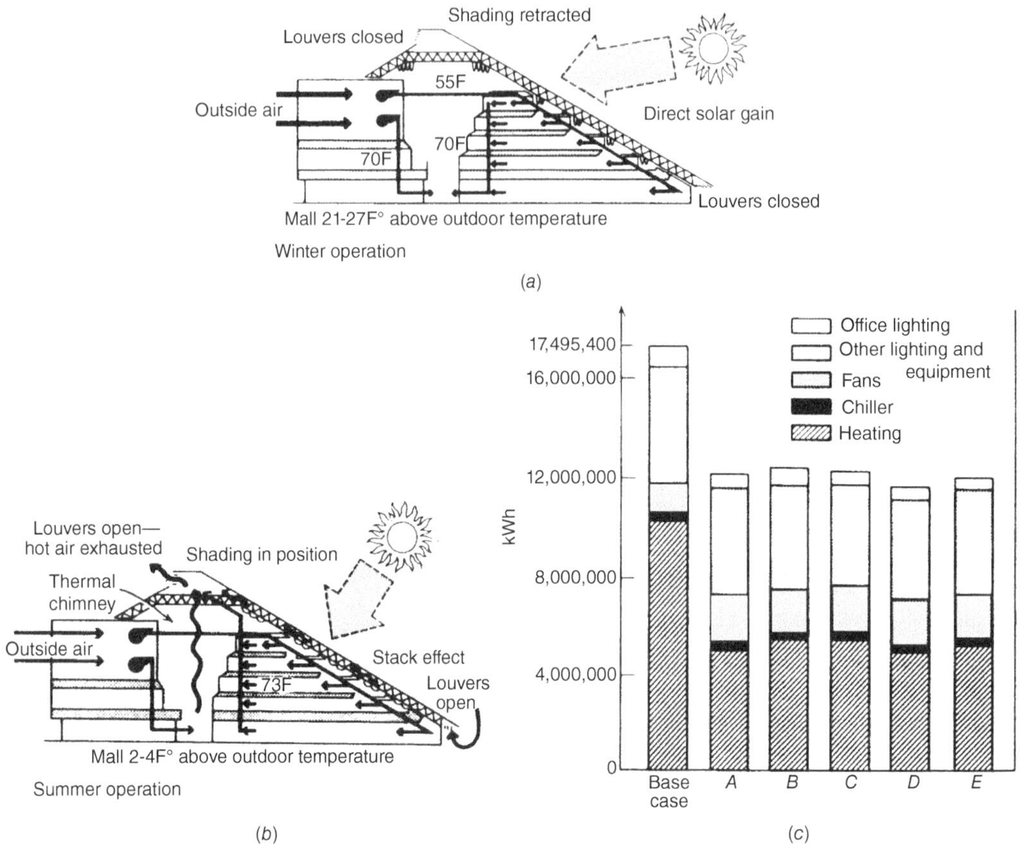

Figure 16.6 *St. Enoch's Square, Glasgow, Scotland: a proposal to use passive solar heating, daylighting, and natural ventilation. Reiach & Hall and GMW Partnership, architects (joint venture); Cosentini Associates, energy consultants; Princeton Energy Group, daylighting consultants.*
(a) Schematic section showing winter operation; the mall temperature varies around 63°F [17°C] during operating hours, while offices are kept near 70°F [21°C]. (b) Schematic section showing summer operation; the mall temperature varies from about 68–74°F [20–23°C] during operating hours.
(c) Estimates of annual energy consumption for a conventional-design base case and several alternative configurations. Note the significantly lower heating energy requirements, resulting in part from the lower winter temperatures allowed for the less-critical zones, such as the mall and the leisure areas in configurations A to E.

There is no single correct HVAC system for most building types, no single correct way to arrange an HVAC system, and no single correct way to place the system into a building. Fortunately, many decisions that need to be made to properly integrate an HVAC system into a building are contextual, rather than technical, issues. The project architect usually knows more about project context than any other design team member. Some of the issues to be considered early in the HVAC placement process are whether the system will be a local (requiring no central space allocation), central, or district (where source equipment is remote from the building) system and what the most logical distribution of major equipment might be.

An important decision is whether to spatially group or separate the HVAC source equipment and the HVAC distribution (air-handling) equipment (see Figure 16.83). A single central mechanical room in a good location might serve an entire building, with area and height sufficient for both heating/cooling and air-handling equipment. If separated, a large space for heating/cooling equipment is typically located in the basement or the penthouse, with a fan room on each floor (or alternate floors).

A central mechanical (plant, source equipment) room should be located so as to minimize distribution component runs to fan rooms, minimize electrical distribution runs from the building service, allow for easy installation and replacement of large and heavy equipment, allow for easy connection to cooling towers/air-cooled condensers, simplify noise control, minimize structural systems costs, and (as necessary) simplify provision of fuel storage. Mechanical rooms need relatively high ceilings; 12 ft [3.7 m] clear is a typical minimum.

A fan room should be logically located relative to the areas being served and have easy access to the exterior for connection to an outdoor air intake. Central locations minimize the distribution tree size and length (100 ft [30 m] might be considered a maximum run for energy efficiency); access to the exterior facilitates the use of outdoor air for both IAQ and economizer

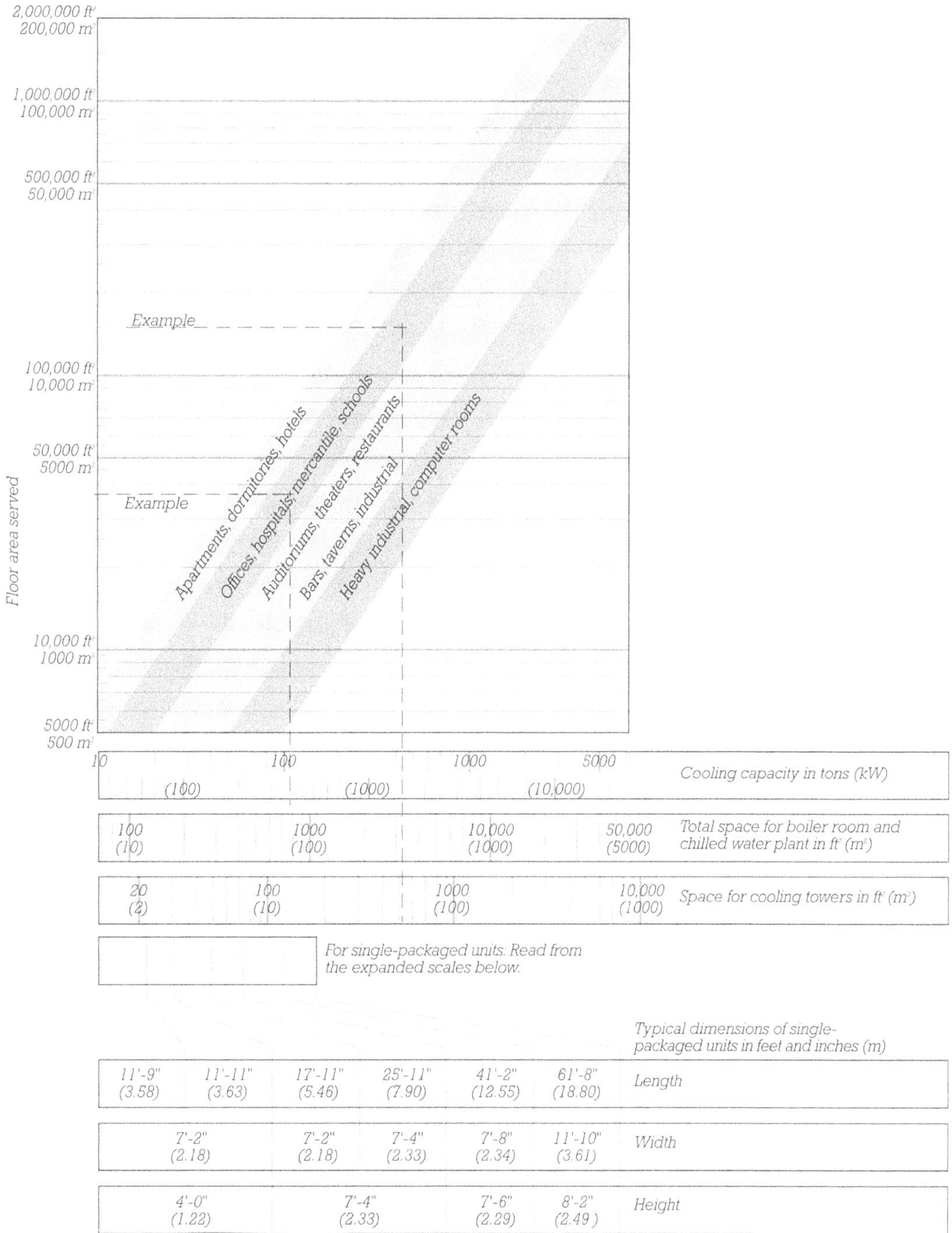

Figure 16.7 *Approximate space requirements for major heating and cooling equipment. (From Allen and Iano, 2022.)*

cycle purposes and simplifies equipment installation (or removal during remodeling). Fan rooms do not require the same ceiling height as a central mechanical room; 10 ft [3.1 m] clear is often acceptable—although getting ductwork into and out of the room may be as volumetrically demanding as placing the air-handling unit. Although AHUs are typically not nearly as noisy as chillers, placement of a fan room to facilitate duct-borne noise control makes economic sense.

Figures 16.7 and 16.8 present approximate space requirements for conventional mechanical systems. For more detailed information on equipment room space requirements, see Figures 16.13 and 16.41.

The charts presented in Figures 16.7 and 16.8 are intended to give a quick approximation of areas;

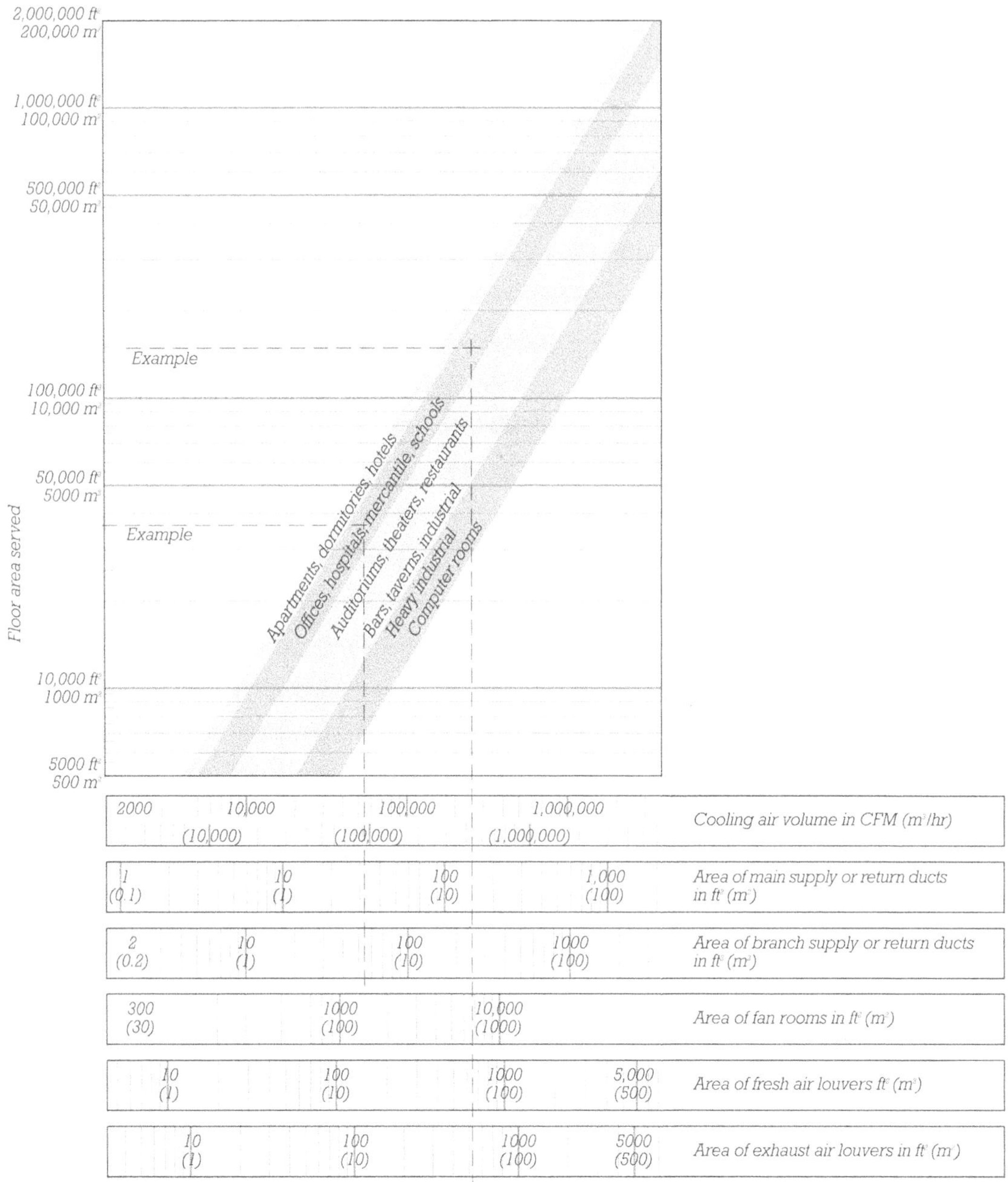

Figure 16.8 *Approximate space requirements for air-handling equipment. (From Allen and Iano, 2022.)*

more detailed procedures for sizing boilers, chillers, fan rooms, and air ducts are presented later in this chapter. The width of the diagonal "line" drawn to represent each building type in Figures 16.7 and 16.8 allows some opportunity for consideration of building energy efficiency. An efficient building will fall toward the left of a given building-type-line width. These preliminary estimates for equipment sizes are likely excessive for a substantially high-performance building (such as one based upon the recommendations in the *Advanced Energy Design Guides* from ASHRAE). For buildings with large heat gains or losses, however, these graphs may slightly undersize the areas needed.

EXAMPLE 16.2 Estimate the central equipment room floor space needed to serve an energy code-compliant office building in Portland, Oregon. The total floor area is 200,000 ft^2 [18,581 m^2].

SOLUTION

Enter Figure 16.7, which addresses approximate space requirements for major heating and cooling equipment. Consider an office building floor area of 200,000 ft^2.

TABLE 16.1 Preliminary Estimates for Cooling Capacity

Type of Occupancy	Floor Area Served ft²/ton	m²/ton	Assumptions
General occupancy:	350–550	32.5–51	400 cfm/ton [189 L/s per ton]
Perimeter spaces			1.5 cfm/ft² [0.7 L/s per m²]
Interior spaces			0.6 cfm/ft² [0.3 L/s per m²]
Offices	500	46.5	200 ft²/person [18.6 m²/person] 3.5 W/ft² [38 W/m²] for lights, equipment
High-rise Apartments:			
North-facing	1000	93	
Other orientations	500	46.5	
Hospitals	333	31	1000 ft²/bed [93 m²/bed]
Shopping centers:	400	37	
Department stores			2 W/ft² [21.5 W/m²]
Specialty stores			5 W/ft² [53.8 W/m²]
Hotels	350	32.5	
Restaurants	150	14	
Central Plants:			
Urban districts	380	35.5	
College campuses	320	29.5	
Commercial centers	475	44	
Residential centers	500	46.5	

Source: Grondzik (2007).
Note: 1 ton = 12,000 Btu/h [3.517 kW].

Note the dashed line indicating "Example" which is placed at approximately the 200,000 ft² mark. The chart indicates:

- cooling capacity: 440 tons [1547 kW]
- total space for boiler room and chilled water plant: 3300 ft² [307 m²]
- space for cooling towers: 520 ft² [48 m²]

Note that the size of the office building exceeds the general capacity that can be served by a single-packaged unit.

Enter Figure 16.8, which addresses approximate space requirements for air-handling equipment. The chart indicates:

- cooling air volume: 265,000 ft³/min [450,238 m³/h]
- combined area of main supply ducts: 160 ft² [15 m²]
- combined area of main return ducts: 160 ft² [15 m²]
- combined area of branch supply ducts: 260 ft² [24 m²]
- combined area of branch return ducts: 260 ft² [24 m²]
- total area needed for fan room(s): 6900 ft² [641 m²]
- total area of fresh air louvers: 650 ft² [60 m²]
- total area of exhaust air louvers: 520 ft² [48 m²]

Now, approximate the largest duct sizes. Consider that this large, multi-story office building has five stories at 40,000 ft² [4716 m²] each. Each floor has five bays, each at 8000 ft² [743 m²]. This building is served by a central HVAC system. The mechanical room is located on the east side of the building, so both the supply and return air are served from the east end of the building by primary 160 ft² [15 m²] ducts. Figure 16.8 indicates the following for the 40,000 ft² [4716 m²] area of each building floor (note the dashed line indicating "Example" which is located at approximately the 40,000 ft² mark):

- cooling air volume: 55,000 ft³/min [93,446 m³/h]
- combined area of branch supply ducts: 55 ft² [5.1 m²]
- combined area of branch return ducts: 55 ft² [5.1 m²]

The supply and return ducts must be approximately 55 ft² [5.1 m²] in total, meaning that several smaller ducts may serve various portions of the floor.

The information derived from Figures 16.7 and 16.8 may be considered conservative, which serves a code-compliant project well. At a conceptual or schematic design stage, these approximations provide a great starting point for a design team. High-performance building enclosure (well-insulated, shaded, and daylit) and HVAC system solutions may result in reduced system sizes compared to the figures obtained here. ■

The capacity of cooling equipment is commonly rated in tons of refrigeration. Rough estimates of cooling capacity versus floor area for several building types are provided in Table 16.1.

For a thermally well-designed detached residence, a design guideline of 1000 ft² of floor area/ton [26 m²/kW] is suggested.

HVAC COMPONENTS

An HVAC system is composed of a number of components (Figure 16.9 and Table 16.2)—each with a

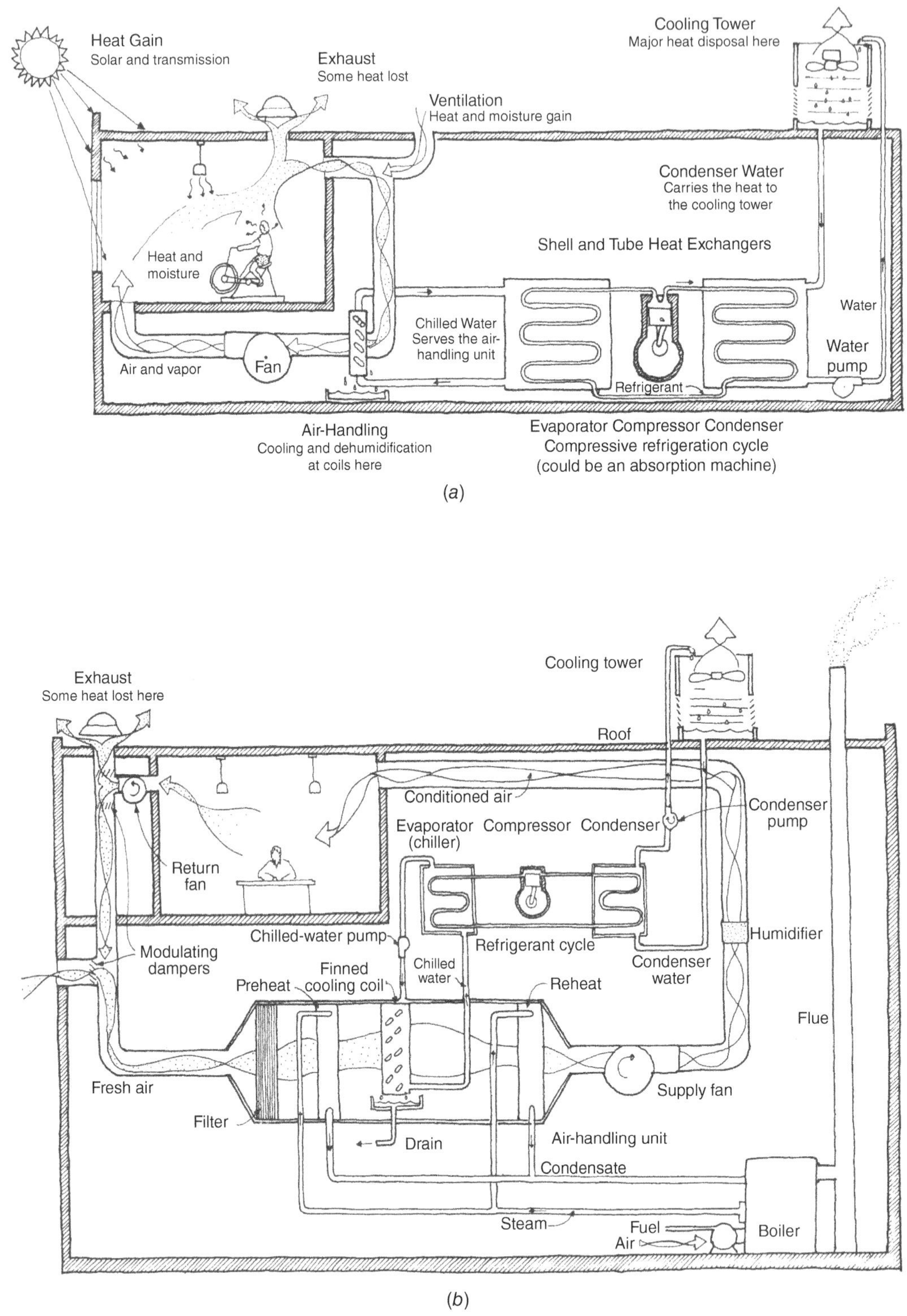

Figure 16.9 Some basic components of HVAC central equipment. (a) A simplified diagram of a cooling cycle, in which chilled water is circulated to air-handling coils, and heat is disposed of through a cooling tower. (b) Schematic diagram of major components of central equipment for both heating and cooling.

TABLE 16.2 Basic HVAC System Tasks and Components

Intake Exhaust	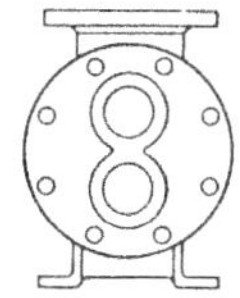 Production/Motion Movers, converters, processors	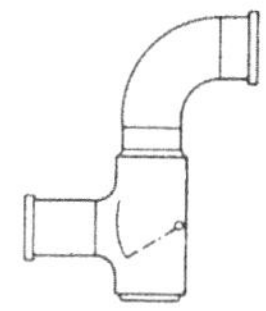 Distribution Supply and return trees, delivery and control components	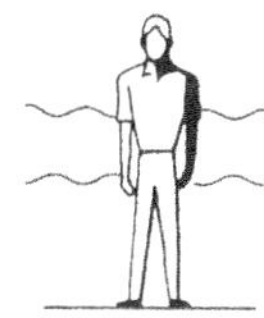 Results
Heat: Fuel Combustion Air → / ← Heat CO_2	Boilers Furnaces Pumps Fans Filters Heat pumps	Pipes Ducts Electricity conduits Diffusers Grilles Radiators Thermostats Valves, dampers	Warm air or surfaces Air motion often controlled Humidity control sometimes needed
Cool: Air Water Fuel → / ← Air Vapor Water Heat CO_2	Evaporative coolers Heat pumps Chillers and cooling towers Coils Pumps Fans Filters	Pipes Ducts Diffusers Grilles Radiators Thermostats Valves Dampers	Cool air or surfaces Air motion usually controlled Humidity control usually provided
Vent: Air → / ← Air	Fans Filters	Ducts Diffusers Grilles Switches Dampers	Fresh air Air motion usually controlled Air quality control often needed

Source: Class notes developed by G.Z. Brown, University of Oregon.

specific function. Because of the potential complexity of many common HVAC arrangements, looking at conceptual building blocks for these systems may assist in demystifying the options. Each of these categories of components will be looked at in turn prior to exploring the numerous HVAC systems that may be considered for building applications. Four main categories of functional elements will be encountered in the typical HVAC system:

Source components: These produce heating effect and/or cooling effect. Source components for heating and cooling will be addressed in turn.

Distribution components: These circulate the heating/cooling effects from the source(s) to the various conditioned zones. In a local system this component category is minor or nonexistent.

Delivery components: These introduce the heating/cooling effect into the various spaces being conditioned.

Control components: These provide for beneficial operation of a system, such as on–off functionality, temperature control, energy efficiency, freeze protection, fire response, etc.

16.5 SOURCE COMPONENTS: HEATING

An HVAC system that will offset building heat losses must have a heat source. Selection of a heat source is a high-level decision that should generate some design reflection. There are four conceptually (and physically) different means of introducing heat into a building. They have different architectural design impacts and environmental concerns/benefits. The four basic approaches mushroom into dozens of specific equipment options—and these options expand into more dozens of specific equipment offerings.

On-site combustion: A fuel of some type (natural gas, oil, propane, firewood, coal) is burned on site, usually within the building; heat is produced as an outcome of the combustion process. The fuel may be delivered to the building site upon demand (natural gas) or be delivered in bulk for on-site storage (fuel oil). Combustion air will need to be provided at the combustion location and combustion gases exhausted from the building. With the exception of biomass, heating fuels are nonrenewable and all produce carbon emissions. Efficiencies of combustion equipment can vary substantially, but modern equipment (such as a condensing boiler or furnace) tends to perform at or above 95% efficiency. Hot water or hot air can be produced by combustion sources.

Electric resistance: Electricity is passed through an electrical resistance element to produce heat; the electricity may come from an off-site (via a utility) or on-site source (such as photovoltaic or wind). Electric resistance heating is typically cited as being 100% efficient—but this is a site efficiency (similar to the 95%+ value noted for on-site combustion). Electricity is generally delivered to a building on demand (with storage only required for some approaches to on-site generation). No combustion air or exhaust for products of combustion is necessary. Hot air or hot water can be produced by electric resistance. The distinction between site and source efficiency is of particular interest for electric resistance (typically involving a 1:3 magnitude ratio).

Heat transfer: Finding some heat already on site and moving this heat to a place where it is more useful is the essence of heat transfer. An air-to-air heat exchanger (whereby heat in exhaust air is transferred to incoming fresh air) is an example of this approach; a heat pump (discussed in more detail in Section 16.8) is another example of this approach. A heat-transfer system may use electricity as a driving force or be self-powered. Heat transfer does not require the introduction of "new" energy and can be very energy-effective; a ground-source heat pump COPs can reach 5 or 6, and air-source units should be greater than 1.5 (denoting more than 100% "efficiency"). Heat pumps and air-to-air heat exchangers are both commonly found in high-performance buildings. In general, hot water or hot air may be produced—but this choice is limited by the specific transfer equipment being considered.

Energy capture: This approach is similar to heat transfer, but an energy form other than heat is tapped into and converted to heat for use in a building. Solar energy is by far the most common such energy resource (although wind energy is also covered by this approach). Energy capture sources tend to be carbon-free and also economically free for the taking. Hot water or hot air may be readily produced.

16.6 HEATING EQUIPMENT

Several examples of heat source equipment and applications are provided by way of introduction to this important part of an HVAC system.

(a) Boilers

Boilers produce hot water or steam; the heat source may be electric resistance or on-site combustion. Boilers provide the heating effect required to elevate water temperature to a point where it can be used for building heating (or to evaporate water to steam). There are a number of different types of boilers on the market—attesting to a wide range of application contexts and system demands. The type of boiler selected for a project depends upon the size of the building heating load, the heating fuels available (and desired), the desired efficiency of operation, and whether single or modular boilers are the most reasonable. Boiler sizes are commonly expressed either in Btu/h [kW] of net output or in (gross) boiler horsepower.

In I-P units:

$$\text{boiler horsepower} = \frac{\text{heating load, Btu/h}}{\text{\% boiler efficiency}} \times 33,470 \text{ Btu/h per horsepower}$$

In SI units:

$$\text{boiler horsepower} = \frac{\text{heating load, kW}}{\text{\% boiler efficiency}} \times 9.81 \text{ kW per horsepower}$$

Efficiency depends partly on the number of passes that the hot gases (from on-site combustion) make through the water—the more passes, the higher the efficiency. It also depends on burner efficiency and on regular maintenance. Finally, efficiency is highest when boiler equipment is operating near its capacity. Figure 16.10 compares typical boiler types, including two- and three-pass boilers.

Fire tube boilers: The hot gases of the fire are taken through tubes that are surrounded by the water to be heated. Fire tube boilers can be either dryback or wetback. Dryback designs have chambers outside the vessel to take combustion gases from the furnace to the tank. Wetback designs have water-cooled chambers that conduct the combustion gases. Firebox boilers place the boiler shell on top of the combustion chamber. Scotch marine boilers feature multiple passes of the combustion gas through tubes.

Water tube boilers: The water to be heated is taken through tubes that are surrounded by the boiler's fire. They hold less water than the fire tube models and so respond faster and can generate steam (where desired) at higher pressures.

Cast-iron boilers: Often used in residential and light-commercial applications, these boilers operate at lower pressure and lower efficiency. They do have the advantage of being modular.

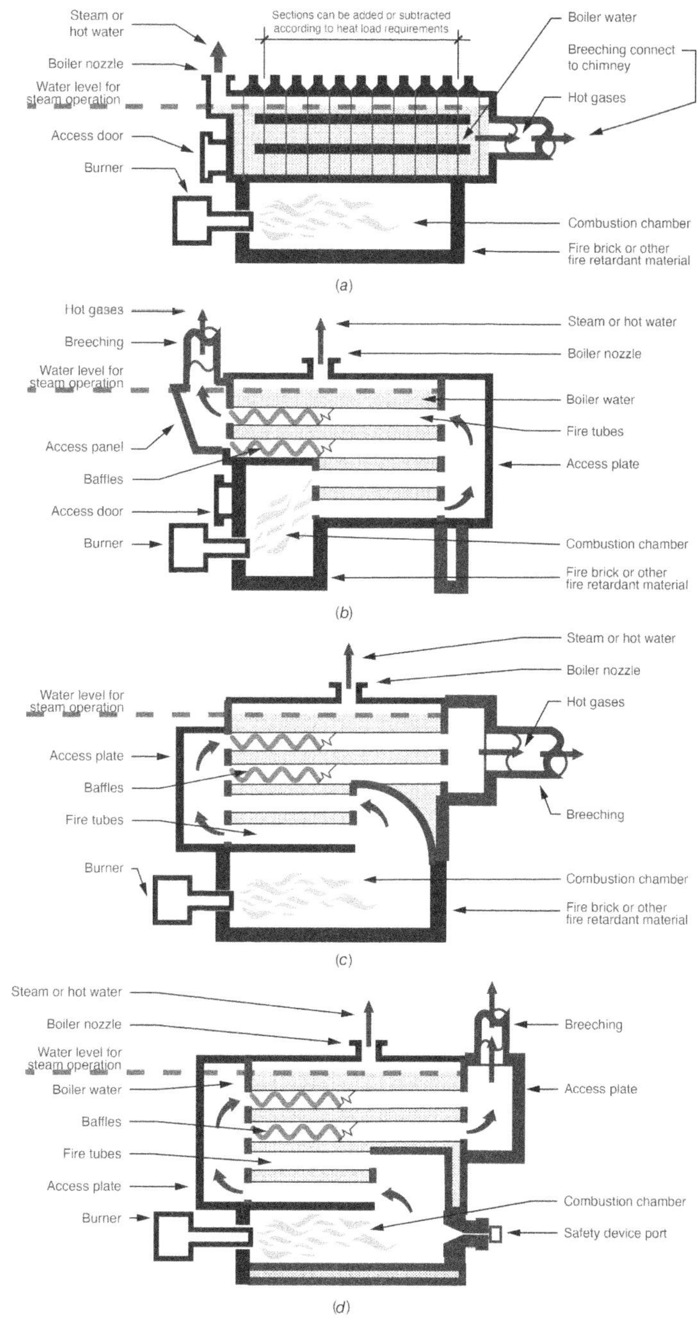

Figure 16.10 *Comparison of boiler types. (a) Cast-iron sectional type. (b) Two-pass fire tube. (c) Three-pass fire tube. (d) Three-pass wetback Scotch marine. (a: Redrawn by Alexa Thornton; adapted from AIA: From Ramsey and Sleeper, 2016; b, c, d: Redrawn by Alexa Thornton; from AIA: From Ramsey and Sleeper, 1994.)*

In addition to basic boiler construction, there are choices for burner types (depending on the fuel used), burner controls, and boiler feedwater systems. See Chapter 32 in the *2024 ASHRAE Handbook—HVAC Systems and Equipment* (ASHRAE, 2024) for details.

(b) Boiler Installation Considerations

Fuel-burning boilers need flues for exhaust gases, fresh air for combustion, and have mandatory air pollution control equipment and controls. The exhaust gas is usually first taken horizontally from the boiler; this horizontal enclosure, or flue, is called the breeching. The vertical flue section is called the stack. Guidelines for sizes and arrangements of breeching and stacks are shown in Figure 16.11. Local codes determine the quantity of air required for combustion; local or national air pollution authorities set pollution control requirements. As a general rule, combustion air can be supplied through a duct to the boiler at an average velocity of 1000 fpm [5.1 m/s]. The duct should be large enough to carry at least 2 cfm [1 L/s] per boiler horsepower. Furthermore, ventilation air to the boiler room should be provided; preferably, the inlet and outlet should be on opposite sides of the room. Minimum sizes: enough for 2 cfm [1 L/s] per boiler horsepower at a velocity of about 500 fpm [2.5 m/s].

Electric boilers may be used when and where electricity costs are competitive with the cost of combustion fuels. Both hot water and steam electric boilers are available. The advantage of electric boilers is the elimination of a need for combustion air, a flue, and air pollution mitigation at the building site. The disadvantages are the use of a high-grade energy source for a relatively low-grade task and the pollution impact at the electric generating plant. In order to reduce the possibility of high electric demand charges, a large number of control steps may be desirable.

Space requirements and arrangements for boilers are summarized in Figure 16.12, which is predicated on multiple boilers. Note that clear space within the room must be provided so that the tubes of the boiler can be pulled out when they must be replaced. A viable strategy for eventually replacing an aging boiler must be considered, and adequate access provided.

When space demands are a vexing design consideration, smaller-dimension boilers—compact boilers (Figure 16.13)—with high thermal efficiencies are available. In addition to their space-saving footprint, these boilers feature a variety of venting options that make them easily adaptable to smaller equipment rooms.

Two other types of boilers are discussed in the following material: steam boilers and modular boilers. Modular boilers are often preferred for energy conservation. Steam boilers are used, as may be apparent, to produce steam. Steam was very commonly used as a heat-transfer medium in older buildings and is still used for many district heating systems. The main rationale for steam climate control systems is the high thermal capacity of steam (changing the state of a substance is an effective way to embody energy in a fixed mass of substance). Steam systems, however, require greater attention to safety and maintenance than hot water systems—and are on the decline as an option of choice for single-building systems. Steam may still be encountered for space-heating purposes, and is still necessary in some building types for tasks such as sterilization.

High-Output, Package-Type Steel Boiler. One or several such boilers may be installed in a large building that uses steam as a primary heating medium. Direct use of steam for climate control can be seen in Figure 16.9b, supplying preheat and reheat coils and also a humidifying unit. The relative lightness of this boiler type, compared to the older styles with heavy masonry bases (boiler settings), makes it suitable for use on an upper floor of a tall building. Figure 16.86 shows two such boilers on the 13th floor of the Fox Plaza Building.

A building that uses primary steam boilers or receives steam from a district heating system may have secondary hot water distribution circuits. In such a situation, a steam to hot water converter (Figure 16.14) is used as a heat exchanger. In the system shown in Figure 16.86, there is a downfeed steam supply from the two boilers on the 13th floor to two such converters, one for hot water heating in the apartments and one below the garage ceiling for hot water heating in the commercial area. A converter may also be used to transfer heat from steam to domestic (potable) water.

Modular Boilers. The primary advantage of modular boilers (Figure 16.15) is efficiency. Boilers achieve maximum efficiency when they are operated continuously at their full-rated fuel input. Large-capacity boilers operate this way only under design conditions, which occur, at most, 5% of the time during a normal winter. Most of the time, large boilers in building applications are running at part load and reduced efficiency. In a modular boiler design, each boiler section is run independently. Therefore, only one section need be fired for the mildest heating needs; as the weather gets colder, more sections are gradually added. Because each section operates continuously at full-rated fuel input, efficiency is greatly increased (Figure 16.16). Each module, being rather small, requires little time to reach a useful temperature and (unlike a larger single boiler) does not waste a lot of heat as it cools down. Thus, modular boilers usually produce a 15% to 20% fuel savings for the heating

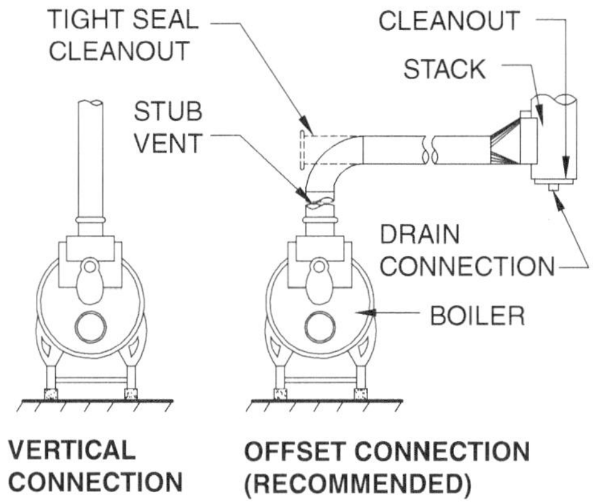

(a)

STACK DIAMETER–MULTIPLE BOILERS, COMMON BREECHING AND STACK

MINIMUM STACK DIAMETER (IN.)						
NO. OF BOILERS	2		3		4	
LENGTH OF STACK	100′	200′	100′	200′	100′	200′
BOILER HORSEPOWER						
25–40	11	12	13	14	14	16
50–60	13	14	15	16	17	18
70–100	16	17	19	20	21	23
125–200	21	22	24	26	28	30
250–350	26	28	32	34	34	40
400–600	32	34	38	40	42	46

(b)

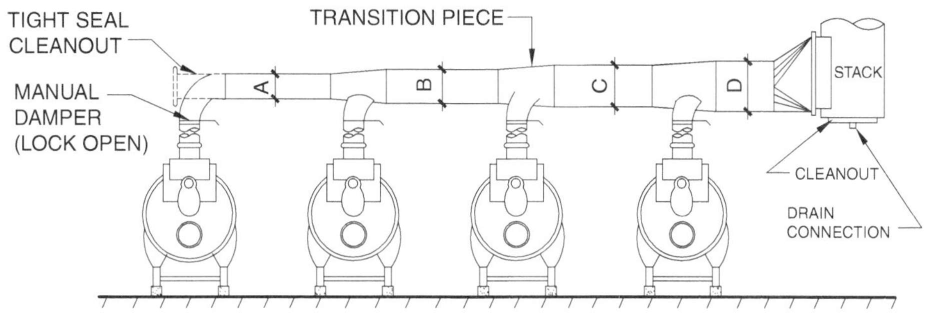

(c)

MULTIPLE BOILERS WITH COMMON BREECHING

MINIMUM BREECHING DIAMETER (IN.)				
NUMBER OF BOILERS	1	2	3	4
BOILER HORSE POWER	A	B	C	D
15–20	6	8	9	9
25–40	8	10	11	12
50–60	10	12	14	15
70–100	12	15	17	18
125–200	16	20	22	24
250–350	20	25	28	30
400–600	24	30	33	36
700–800	24	34	38	42

(d)

Figure 16.11 *Breeching and stack size guidelines for fossil-fuel-fired boilers. (From AIA: Ramsey and Sleeper, 2016.)*

season relative to single boilers. Their other advantages include ease of maintenance (one module can be cleaned while others carry the heating load) and small size (allowing for easy installation and replacement in existing buildings).

Modular boilers also eliminate the initial cost of oversizing heating equipment. In cold climates, conventional boiler systems often use two or three large boilers to ensure that heat is available if one boiler fails. When two such boilers are used, it is common practice to size each boiler at two-thirds of the total heating load; an oversize of one-third results. When three such boilers are used, it is common practice to size each boiler at 40% of the total heating load; an oversize of 20% results. However, when at least five modular boilers are used, oversizing can be eliminated because the failure of a single module will not have a crippling impact on the overall heat output.

Gas-fired pulse boilers are a smaller and more energy-efficient choice for modular boilers. Pulse boilers utilize a series of 60–70 small explosions per second, making the hot flue gases pulse as they pass

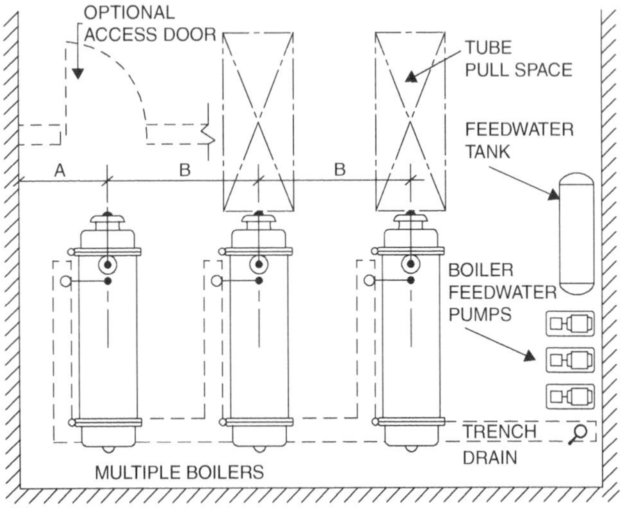

BOILER ROOM SPACE REQUIREMENTS

BOILER HP	15–40	50–100	125–200	250–350	400–800
Dimension A	5′–9″	6′–6″	6′–10″	7′–9″	8′–6″
Dimension B	7′–5″	8′–9″	9′–7″	11′–9″	14′–3″

Figure 16.12 *Boiler room space requirements. Dimension A includes an aisle of 3 ft 6 in. [1 m] between the boiler and the wall. Dimension B between the boilers includes an aisle of at least 3 ft 6 in. [1 m], up to 5 ft [1.5 m] for the largest boilers. (Ramsey and Sleeper, 2000 / with permission of John Wiley & Sons.)*

through the fire tube. This makes for very efficient heat transfer. Pulse boilers are available up to about 300,000 Btu/h [88 kW] capacity.

Pulse boilers operate with lower water temperatures so that water vapor in the flue gas can condense and drain. This change of state (or phase change) liberates additional heat, allowing these pulse boilers to achieve efficiencies up to 90%. They exhaust moist air, not hot fumes, so flues can be small-diameter plastic pipe rather than large-diameter, heat-resistant materials. The upcoming discussion regarding the appurtenances that come with on-site combustion may help place these comments in perspective.

The boilers in the preceding discussion tend to be used in larger building applications. Residential applications will tend to employ the following boiler types:

Oil-fired steel boiler: A refractory chamber receives the hot flame of the oil fire. Combustion continues within the chamber and the fire tubes. Smoke leaves through the breeching at the rear. Water, outside the chamber, receives the heat generated in the combustion chamber. If a domestic hot water coil is included, a larger-capacity boiler is selected to handle this additional load. An aquastat (a temperature-sensing device used to control water temperature—effectively, it is a "water thermostat") turns on the burner whenever the boiler water cools below a designated setpoint, thereby maintaining a reservoir of hot water ready for heating the building.

Gas-fired cast-iron hot water boiler: Cast-iron sections contain water that is heated by hot gases rising through these sections. Output is related to the number of sections. Improved efficiency is gained through use of a heat extractor in the flue. With induced draft combustion, a condensing unit in the flue, and electronic ignition instead of a pilot light, up to 90% AFUE is attainable (Figure 16.17). The American Gas Association (AGA) sets standards for gas-fired equipment.

Oil-fired, cast-iron hot water boiler: Primary and secondary air for combustion may be regulated at the burner unit. Flame enters the refractory chamber and continues around the outside of the water-filled cast-iron sections.

As fuels burn to produce heat, they require oxygen to support the combustion process. Because oxygen constitutes only about one-fifth of the volume of air, reasonably large rates of airflow are required. The air should be drawn from outdoors at a position close to the fuel burner or (preferably) led to the burner location by a duct. Combustion air should not be drawn from the general building volume. Doing so is a waste of energy, and contemporary "tight" construction inhibits replacement airflow through the building enclosure, creating a dangerous condition.

Prefabricated chimneys (Figure 16.18) have replaced bulkier and heavier field-built masonry chimneys. They offer a number of advantages and can be easily supported by a normal building structure. Further, high-efficiency boilers and furnaces manage to remove so much heat from the exhaust gases that smaller flues operating at much lower temperatures result. These relatively small pipes can be vented through a wall to the exterior. Eliminating a chimney

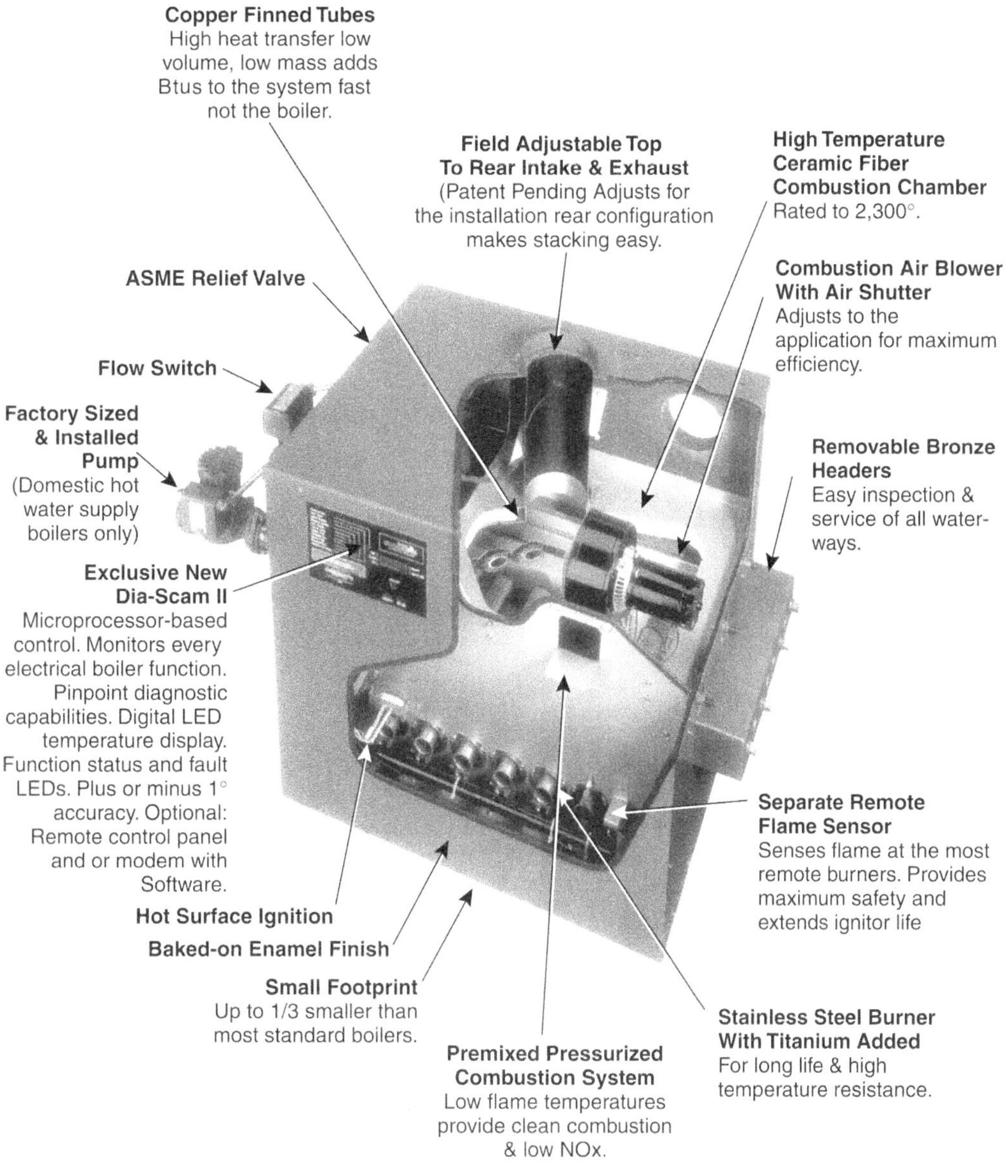

Figure 16.13 *A Burkay Genesis hot water boiler, fueled by either natural gas or propane, is available in ratings from 200,000 to 750,000 Btu/h [58,620 to 219,825 W]. All units are 30 in. high × 24 in. deep [762 mm × 610 mm]; the smallest boiler is 23 in. [584 mm] wide, and the largest is 57 in. [1454 mm] wide. The copper heat exchanger has an 83.7% thermal efficiency rating, and a variety of venting options are available. (Courtesy of A.O. Smith Water Products Company.)*

has lessened the impact on the building design of combustion-based boilers and furnaces.

For older or less-efficient fuel-burning equipment, it is important that chimneys carrying high-temperature flue gases be safely isolated from combustible construction to prevent the possibility of fire. The size of a flue will depend upon the capacity of the boiler or furnace. Flue height had traditionally been 35–40 ft [11–12 m]. Providing a draft, for which chimney height was an important consideration, is now accomplished by fans. Draft hoods above gas burners prevent downdraft from blowing out the flames.

The storage space to be provided for fuel oil or propane depends upon the proximity of the supplier and the space available at the building site. For oil, when more than 275 gal [1040 L] was stored, it was previously common practice to use an outside tank buried in the ground. This practice eventually led to leaking tanks and contaminated soil and groundwater. Several factors have converged to discourage the use of oil in newer buildings: a cleaner and more efficient alternative (natural gas), few basements in new residential construction where oil tanks might be placed, and a dislike of unsightly aboveground oil tanks.

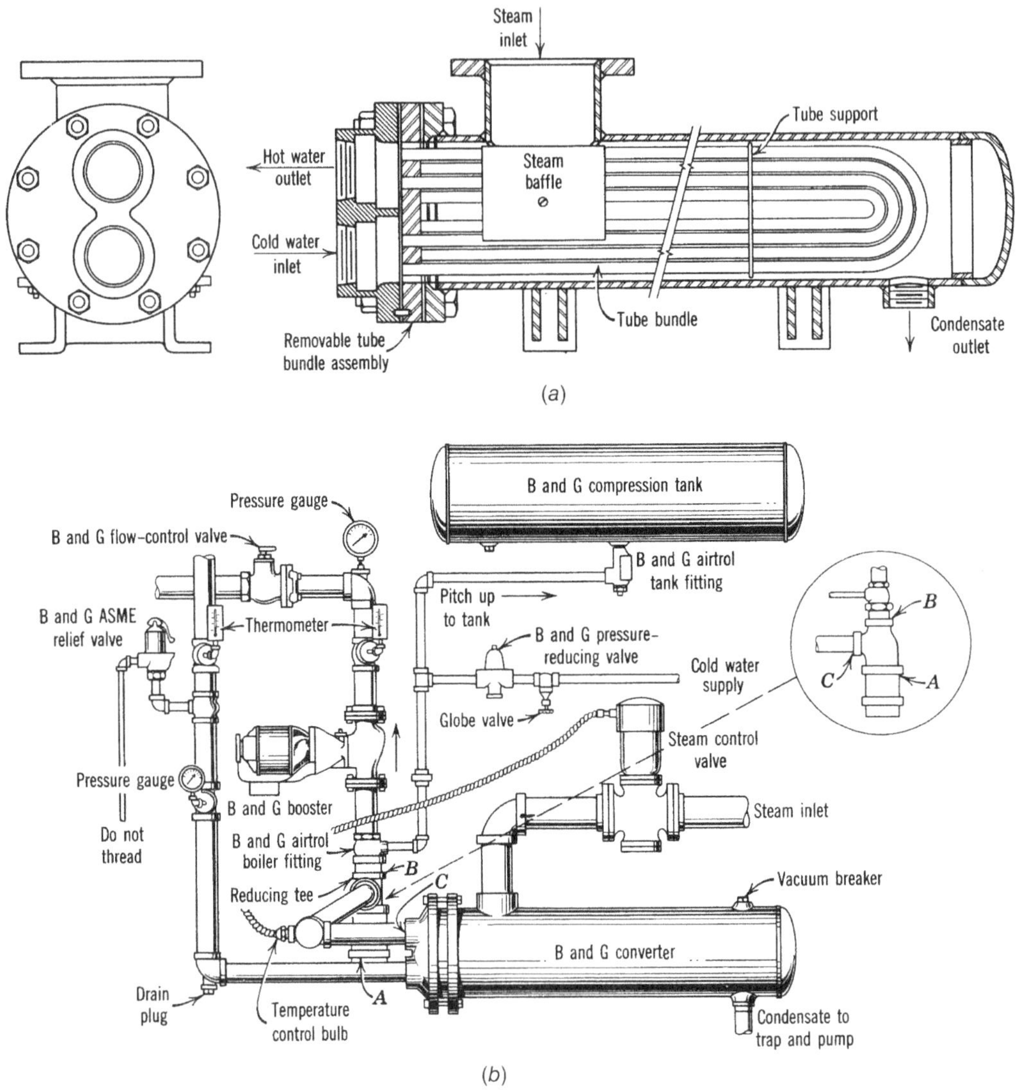

Figure 16.14 *Conversion unit that transfers heat from steam to hot water. (a) Section illustrating the principle of heat transfer from steam to water. (b) A converter connected to the steam supply and equipped with all devices necessary for a complete hot water heating system. (Courtesy of ITT Bell and Gossett.)*

Hot water boilers are rated according to heating capacity using several metrics:

- *Gross rating* expresses the boiler input, in Btu/h [W], required by the boiler (found on the equipment plates).
- *Net rating* expresses the boiler output, in Btu/h [W], and is determined as: gross rating × efficiency (also found on the equipment plates).
- *Boiler capacity* expresses the useful heat output when the boiler is operating under steady-state conditions, often expressed in Btu/h [W] or MMBtu/h (1 million Btu/h [293 kW]). The term "useful heat" assumes that the boiler is within the heated enclosure of the building; thus, any heat that escapes through the boiler walls is available to help heat the building. Hot water applications will use lower boiler capacities. A small apartment or office building may require a boiler capacity less than 1 MMBtu/h [293 kW]. Many commercial sector boilers have capacities ranging up to 10 MMBtu/h [2931 kW] range. Higher capacities will be required for steam applications. A district heating plant may need boilers with capacities up to 100 MMBtu/h [29,307 kW]. Boilers in the industrial sector may up to/beyond 1000 MMBtu/h [293,071 kW]. Power generation at the utility scale may require boilers with capacities nearing 10,000 MMBtu/h [2,930,710 kW].
- *Firing rate* indicates the amount of fuel that a boiler burns over a specific period, typically measured in Btu/h, kW, or ft^3 of gas per hour. This metric is used to gauge how quickly a boiler can produce heat. Depending on the boiler type, it may exhibit a fixed

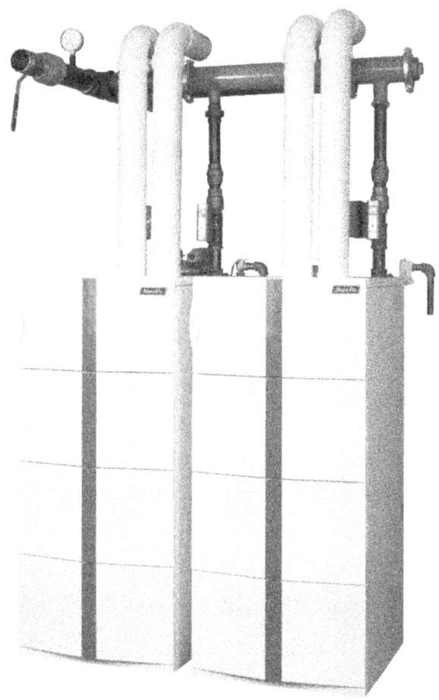

(*a*)

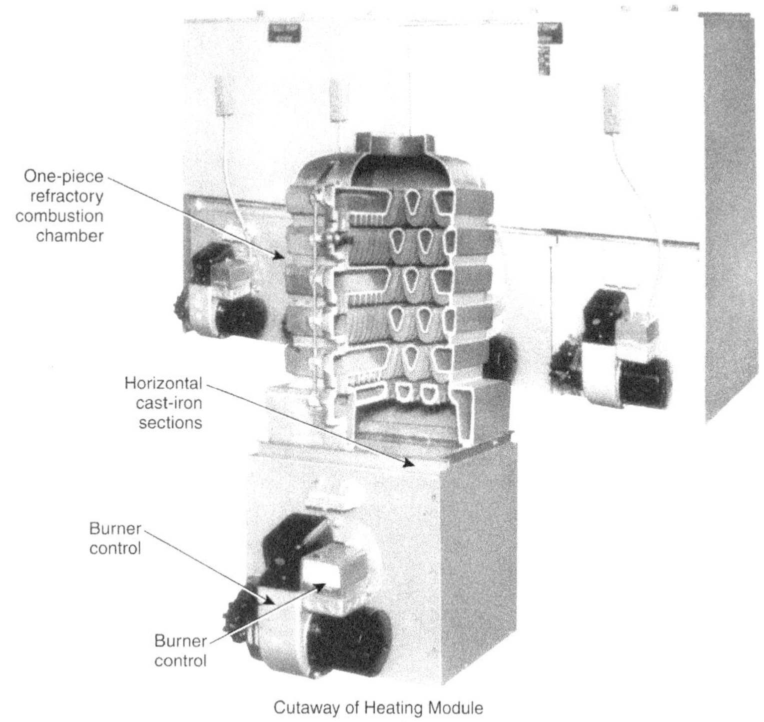

(*b*)

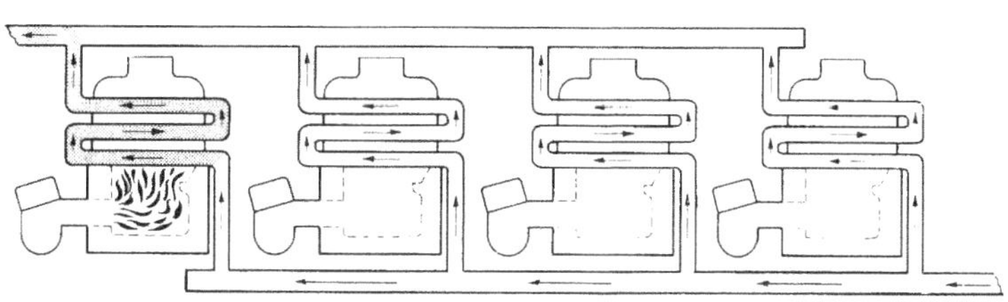

SCHEMATIC FLOW DIAGRAM

(*c*)

Figure 16.15 *Modular boilers. (a) A bank of two modules—with a total input of 1.5 million Btu/h [439 kW]. (b) Details of a module (20 × 32 × 48 in. H [510 × 812 × 1220 mm]) with a 385,000 Btu/h [113 kW] input. (c) Schematic showing multiple boiler flow conditions in mild weather, with only one module in operation. (a: Courtesy of Slant/Fin.)*

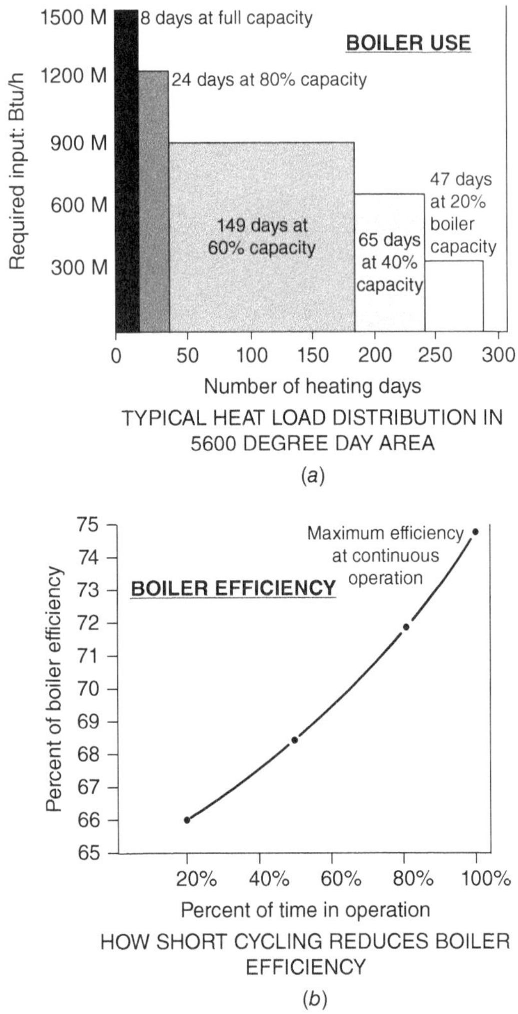

Figure 16.16 *One large boiler versus many smaller ones. (a) Boilers rarely operate at full capacity; instead, they respond to part loads most of the time. (b) Under part load conditions, a boiler will often short-cycle, which on a single large boiler could drop the annual efficiency into the 66% to 75% range.*

(constant) firing rate or modular firing rate that can adjust based on demand.

- *Annual fuel utilization efficiency* (AFUE), as previously noted in this chapter, is used to express boiler performance.
- *Net I = B = R ratings* (a designation of the Institute of Boiler and Radiator Manufacturers) are published by the Hydronics Institute Division of the Gas Appliance Manufacturers Association (GAMA). (The term "hydronic" refers to the use of water as a heat-transfer medium.) The net I = B = R rating load is lower than the heating capacity rating, because it consists only of the heating to be delivered to the spaces and excludes the heat loss of the boiler itself.

Select a boiler whose rating matches the calculated design heat loss of the building; too small a boiler results in lower indoor temperatures at design conditions; too large a boiler costs more and is a waste of space. If using AFUE to select an efficient boiler, ensure that the assumptions about "inches of water draft" and percentage CO_2 are similar for the boilers being compared. Minimum acceptable AFUEs are specified in *ASHRAE/IES Standard 90.1* (ASHRAE, 2022a). ENERGY STAR-certified boilers have AFUE ratings of 87% or greater for oil boilers and 90% or greater for gas boilers.

(c) Furnaces

Furnaces provide the heating effect required to elevate air temperature to a point where it can be used for building heating. A furnace produces hot air that can be distributed to spaces via a central ductwork system; local furnace configurations without ductwork are also available. The heat source for a furnace is most commonly on-site combustion (using natural gas, propane, or fuel oil) or electric resistance. A combustion furnace will require outdoor air to support the combustion process, a flue to vent products of combustion to the outdoors, and perhaps fuel storage capabilities (if the fuel used is not an on-demand utility), see Figure 16.18. Heat transfer can also be used as the heat source for a furnace via a coil connected to a heat pump system. Energy capture is also a possible heat source (perhaps from an air-based solar thermal collector).

The term "furnace" is generally used to describe residential-scale equipment. In larger building applications, an air-handling unit will perform the same basic functions as a furnace. In addition to coupling a heat source to circulating air, a furnace will contain a fan to provide the motive force to circulate that air, and a filter to clean the air passing through the unit. Controls will also be provided to ensure safe and effective operation. These components are all included in a packaged sheet-metal enclosure. Vertical and horizontal configurations (Figure 16.19) are available to provide flexibility in installation.

Furnaces have become much more efficient over the past decades, thanks to forced-draft chimneys and improved heat exchangers, as shown in Figure 16.20. For ENERGY STAR-certified residential models, expect AFUE ratings of up to 98% with gas furnaces and 87% for oil furnaces. Compare these figures to 65% or so for older furnaces. AFUE ratings for high-efficiency furnaces are based on an isolated combustion system that requires that all combustion air be drawn from outdoors (instead of from the occupied volume of a building).

Modern high-efficiency gas and oil furnaces should be equipped with an electronically commutated (EC) fan motor and an air leakage (Q_{leak}) rate equal to or less than 2.0%. This is the airflow that the fan requires to maintain pressure after accounting for the

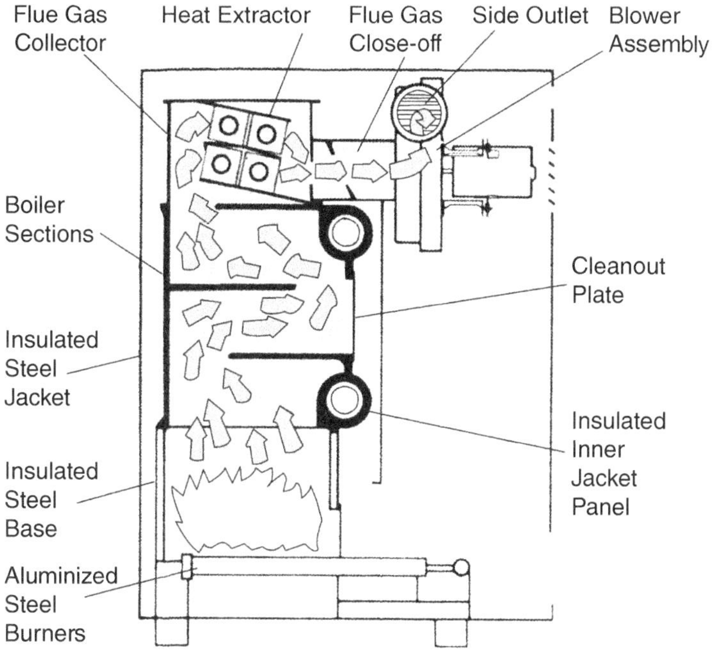

Figure 16.17 *Gas-fired cast-iron sectional boiler for hot water heating. Very high operating efficiency is possible; no chimney is required, as low-temperature exhaust gases can be vented through a wall to the exterior.*

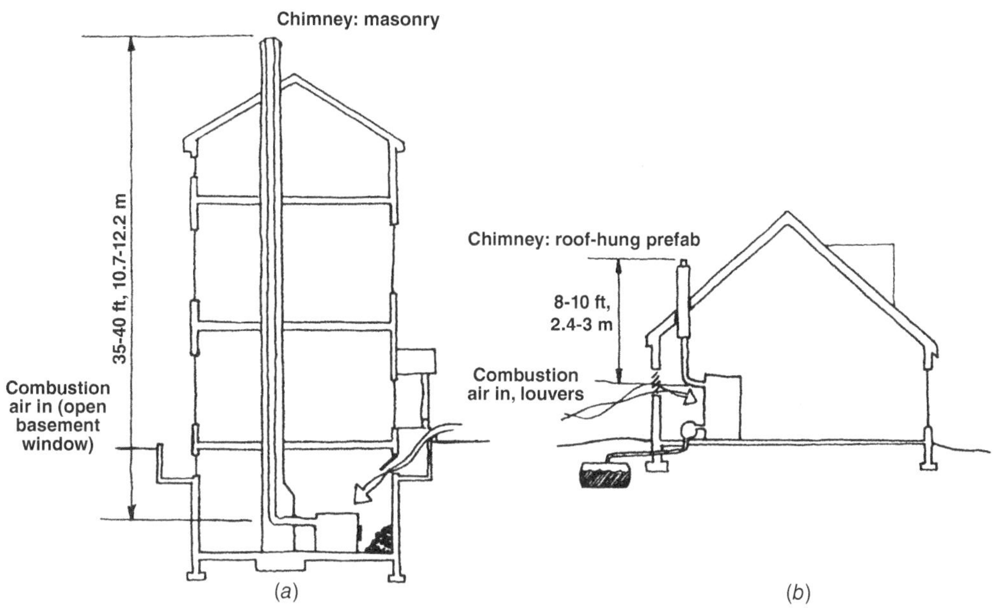

Figure 16.18 *The historic need for 40-ft [12-m] chimneys (a) has been eliminated by controlled draft in burners (b) and by the use of high-efficiency heating equipment that can be directly vented to the exterior through prefabricated chimneys. Note the buried fuel oil tank, an architectural consideration and potential environmental concern.*

air that "leaks" out of the furnace through cracks, joints, and various penetrations in the furnace cabinet (i.e., furnaces are not air-tight).

(d) Wood-Burning Appliances

After passive solar heating, the most ancient method of heating is the radiant effect of fire. With each improvement from campfire to fireplace to wood stove, more of the heat from the fuel was captured for use rather than wasted to the outdoors (Figure 16.21). Although many people enjoy the sight, sound, and smell of an open fireplace, a tightly enclosed wood stove with a catalytic combustor is a substantially more efficient and less polluting approach to heating. Indoor as well as outdoor air pollution is a serious issue with fireplaces and stoves. Combustion generates carbon monoxide, breathable particulates, and, at times, nitrogen dioxide. Wood smoke can cause nose and throat irritation, it can remain in the lungs, and it can trigger asthmatic attacks. Keeping a clean chimney, burning small, hot fires rather than large, smoky ones, using seasoned wood, and ensuring adequate

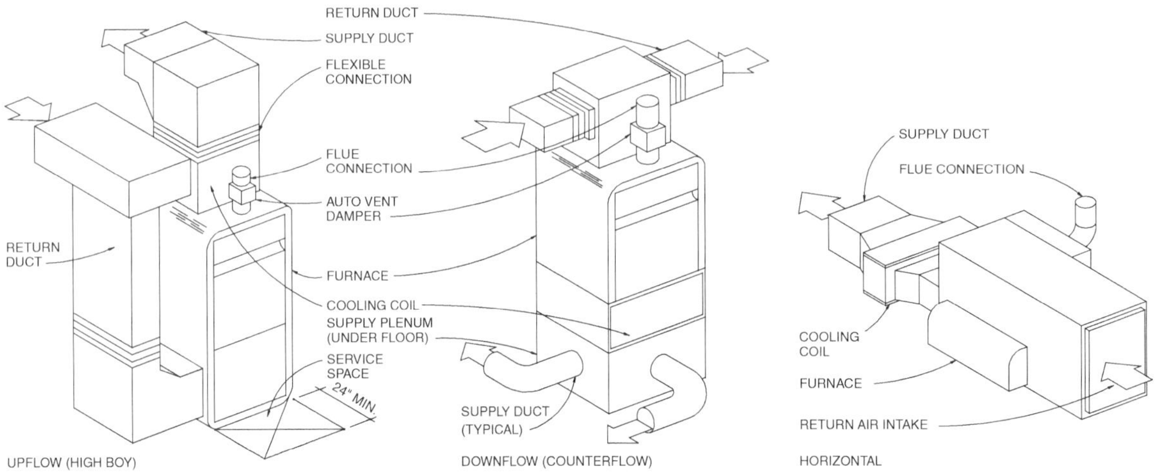

Figure 16.19 *Typical furnace types. (From Ramsey and Sleeper, 2000.)*

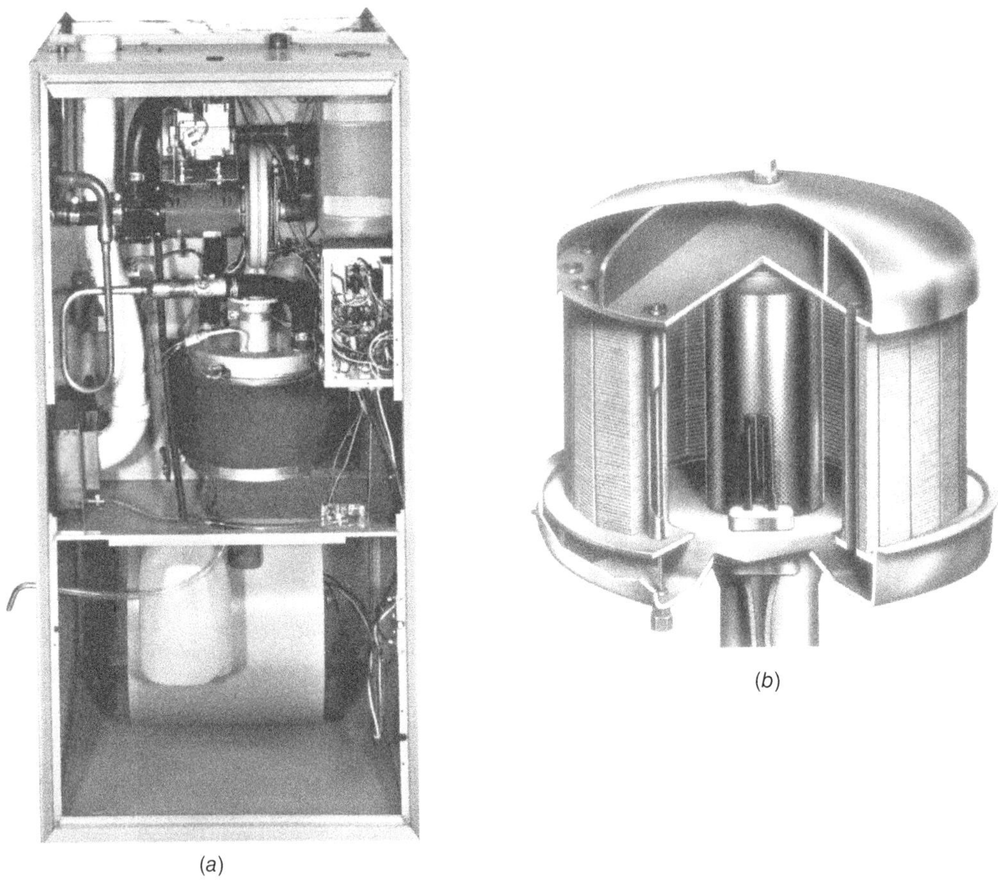

Figure 16.20 *Furnaces with high operating efficiencies are available. (a) An Amana gas furnace. (b) The small, high-efficiency heat exchanger utilized by this furnace, which recovers heat from exhaust gases. (Courtesy of Amana Refrigeration, Inc.)*

ventilation to the wood-burning device are strategies to minimize pollution and health and fire risk.

Open fireplaces may be lovely to look at, but the amount of air exhausted up the chimney can quickly cause more heat loss than heat gain from the fire. The colder the outdoor air, the greater the net heat loss. Masonry mass around fireplaces can store and release some heat; for energy conservation, the mass should be surrounded by the building rather than located on an exterior wall. Building codes typically require a fireplace to have a tight-fitting damper, firebox doors, and a source of outdoor combustion air feeding the firebox.

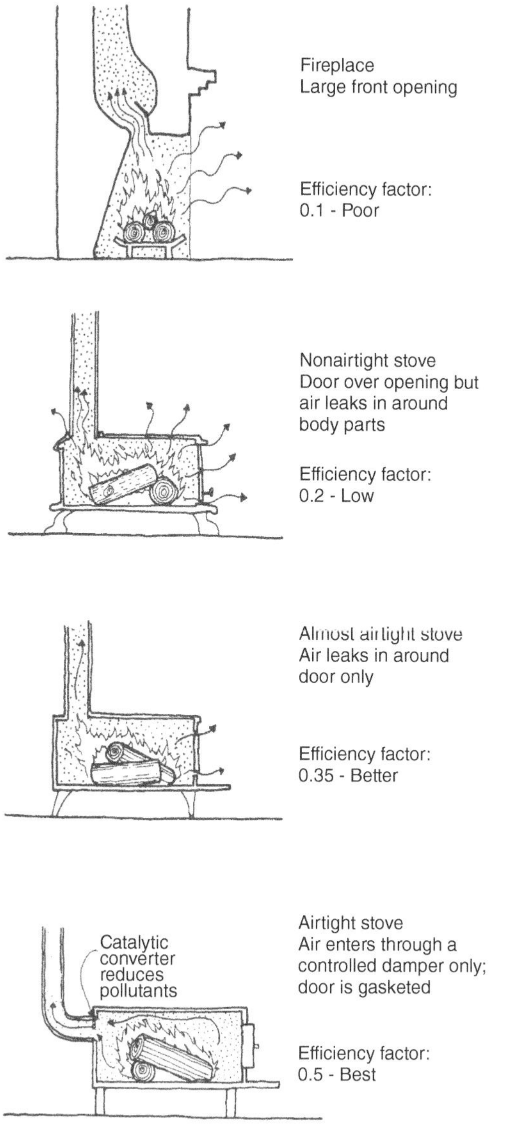

Figure 16.21 *Wood-burning devices have substantially increased in efficiency since the time of the open fireplace. (Copyright © 1978 by Alternative Sources of Energy, Issue 35.)*

Wood stoves are available in a wide variety of styles and are made of several materials. The heating capacities of such stoves are often difficult to determine. Manufacturers rarely specify the Btu/h output, which depends upon the density, moisture content, and burn time of the wood fuel. Wood that has been split, loosely stacked, and protected from rain for at least six months should achieve a moisture content of about 20% by weight, thus becoming "seasoned." The following sizing procedure assumes no more than this 20% moisture content. For more details, see Issue 35 of *Alternative Sources of Energy* magazine (1978).

The formula for the hourly heat output to a room from a wood stove is:

$$\text{Btu/h} = \frac{(V)(E)(D)(7000)}{T}$$

where

V = useful (loadable) volume of the stove (ft^3)
E = percent efficiency, expressed as a decimal (<1.0); see Figure 16.21
D = density of the wood fuel; see Table 16.3
T = burn time (hours) for a complete load of firewood; usually assumed as 8 hours at a minimum
7000 = Btu/lb of firewood, 20% moisture content

So-called "bone-dry" wood can be assumed to have 8600 Btu/lb [20,000 kJ/kg]. Higher-density wood will release more thermal energy through combustion on a per volume basis than lighter-density wood. However, lighter density wood will release thermal energy more quickly as it burns faster. Note that a reduction in burn time increases the heat output: when the air supply to the stove is increased, the fire burns hotter, consuming the wood more quickly. To meet the design heat loss (worst reasonable condition) for a room, burn times of 8–10 hours should be assumed. Stoves rarely need relighting with a 10-hour burn time.

Pellet stoves were first introduced in 1984 and have several desirable characteristics. The pellets are made from densified quality sawdust, a manufacturing by-product. The form and content of this fuel produce a highly efficient burn with less pollution emitted. The fuel is cleaner and takes less storage space than cordwood; an electric auger automatically feeds fuel into the burnplace to maintain a fire. From 10,000 to 50,000 Btu/h [2930 to 14,650 W] can be produced, depending on the model and operating settings.

There are two types of "efficiencies" used to characterize the performance of wood (and pellet) stoves. Combustion efficiency represents a calculated determination (in percent) of how well a wood stove can convert the wood into useable heat. It does not reflect how much of the usable heat produced is transferred

TABLE 16.3 Approximate Average Wood Density

Type	Density lb/ft^3 [kg/m^3][a]
Shagbark hickory	40.5 [648]
White oak	37.4 [598]
Red oak	36.2 [579]
Beech	36.2 [579]
Sugar maple	34.9 [558]
Yellow birch	34.3 [548]
White ash	33.7 [539]
Black walnut	31.2 [499]
American elm	28.7 [459]
Spruce	25.6 [410]
Hemlock	23.7 [379]
Aspen	23.1 [370]
Red cedar	18.7 [299]
White pine	17.5 [280]

Source: Alternative Sources of Energy, Issue 35, © 1978. Reproduced by permission.

[a]These values are approximate; they vary a great deal. SI units added by the authors of this book.

to the home. Compare this measurement to the overall efficiency, which is the percentage of heat that is transferred to a space to be heated when a load of fuel (e.g., wood, pellets) is burned. The actual efficiency of the device will vary, depending on many factors, such as firebox volume, device type, wood moisture content, user operation, and installation details (e.g., outside piping, chimney height). The overall efficiency of wood stoves, which will typically range between 50% and 90%, will be lower and is a better gauge of the amount of heat that is delivered to the home than combustion efficiency. Often manufacturers will list combustion rather than overall efficiency since it is the higher of the two. For example, an advertisement for a wood stove may tout that the device is "85% efficient" and although not clarified, the figure is the combustion efficiency. The overall efficiency may be 10% to 20% lower.

Wood stoves are frequently used as the sole active heat source for an entire building, such as a residence or a small commercial building that is passively solar heated. Because radiation is the dominant form of heat output, the areas that "see" the stove get most of the benefit. Circulating stoves, however, convert a larger portion of their heat to convection, which produces a layer of hot air at ceiling level. By providing a path between rooms at ceiling height, this hot air can slowly spread throughout a building; it can also easily find its way upstairs, because warm air rises. The more thermally massive the ceiling construction, the longer it can store and reradiate the heat from the stove.

The flue leading from a wood stove carries very hot gases that are a potential source of heat (and pollution). The flue can be exposed to a space, making radiant heat available, or simple heat exchangers can be constructed (such as for the preheating of domestic hot water). More elaborate heat recovery devices—for boiler flue heat recovery—are commercially available.

Catalytic combustors reduce the air pollution that results from burning wood. These devices are honeycomb-shaped, chemically treated disks as much as 6 in. in diameter and 3 in. thick [150 mm in diameter, 75 mm thick]. They are either inserted into the flue or built into the stove itself. When wood smoke passes through the combustor, it reacts with the chemical treatment and ignites at a much lower temperature; this causes combustion products to burn that otherwise would have gone up the flue. The result is more heat produced, less creosote buildup in the flue, and fewer pollutants in the atmosphere. Like the catalytic converters in autos, these devices impose limits on the fuel: plastic, colored newsprint, metallic substances, and sulfur are ruinous to combustors; this simply means that the stove must be used as a wood burner, not a trash incinerator.

Wood stoves have a larger impact on building design than do most other heating devices. Noncombustible materials must be placed below and around a stove or a minimum clearance to ordinary combustible building materials must be provided. Furniture arrangements and circulation paths must be designed with the very hot stove surfaces in mind. Hot spots occur near the stove; cold spots occur whenever visual access to the stove is blocked. Thermally massive materials near the stove are advantageous in leveling the large temperature swings that can accompany the on–off cycle of the stove; this affinity for thermal mass has made the wood stove a popular choice for backup heat in passively solar-heated buildings. Finally, the amount of space required for wood storage should not be overlooked; recall the impact of the wood storage space on the house shown in Chapter 2. A covered, well-ventilated, easily accessible, and quite large space is desirable.

Masonry heaters overcome many of the disadvantages of a metal wood stove. Their footprint is rather small compared to their height; typically, they are used to heat an entire building (such as a residence). An inner vertical firebox supports a hot, clean burn, resulting in efficient combustion; combustion gases then flow downward in outer chambers, transferring heat to exterior masonry surfaces. In Finnish masonry heaters, this is termed contraflow (Figure 16.22). Cool

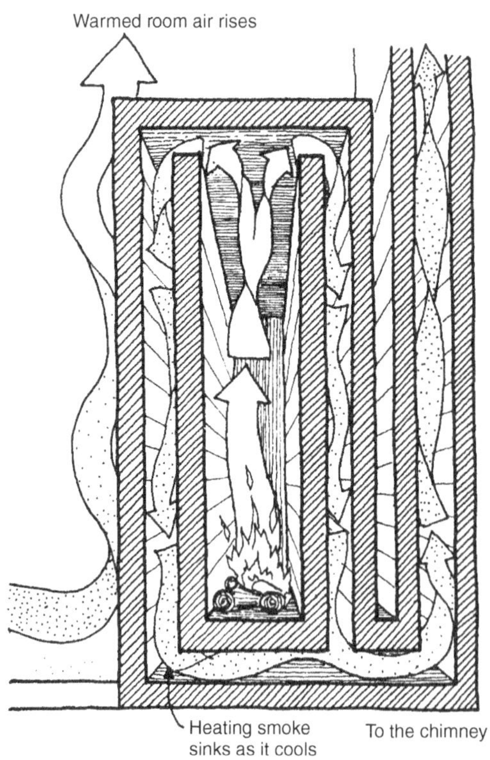

Figure 16.22 *The Finnish contraflow masonry heater.*

air at the floor of the room flows upward as it is heated by contact with this masonry; the temperature difference between masonry and air remains fairly constant, with the highest temperatures at the top. Heat is gentle and even; dangerously hot surfaces are avoided. Fires may be built at 6:00 p.m., and combustion is completed by bedtime; the heat continues to radiate all night, but no fire is burning while people sleep. Research at Finland's Tampere University of Technology has resulted in very effective masonry heater designs, described in Barden and Hyytiäinen (1993).

(e) Infrared Heaters

Infrared heaters are often seen in semi-outdoor locations such as loading docks, repair shops, and higher-end bus shelters. They are fired with either natural gas or propane, or powered by electricity. When vented, they can be used in more traditional environments such as the retail store in a remodeled warehouse shown in Figure 16.23. Their advantage is that they heat surfaces first rather than air, so that comfort is obtained without the need for high air temperatures. When high rates of air exchange are expected, high-intensity radiant heaters are often used.

Radiant heaters should be sized by the surface temperature change they produce. Many manufacturers specify this surface Δt for specified mounting heights and angles relative to the surface of interest.

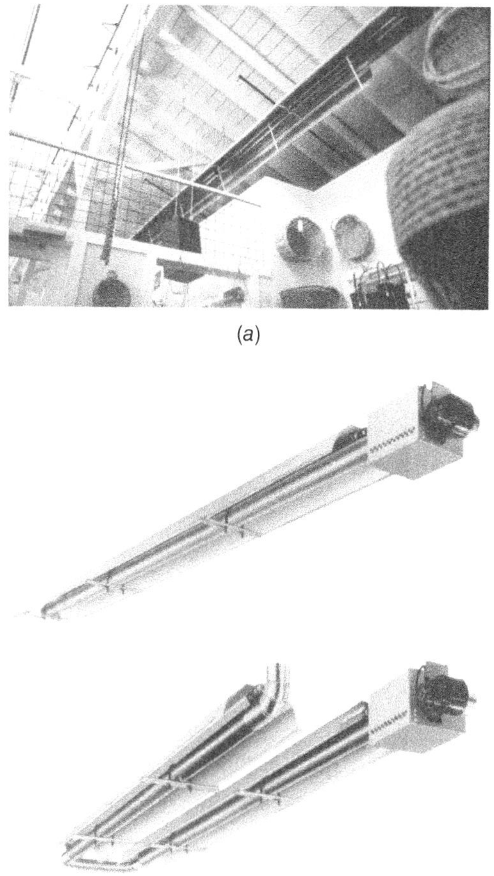

(a) (b)

Figure 16.23 A high-intensity infrared heater adds to the historic atmosphere of this retail store in an old warehouse (a). Exposed mechanical equipment includes the chains that operate the clerestory windows. The vented gas heater has an adjustable reflector that enables its radiant heat to be directed. (b) Vented gas high-intensity infrared heaters are available in both straight-line and U-shaped units. (a: Clark-Ditton Architects; b: Courtesy of Solaronics, Inc.)

16.7 SOURCE COMPONENTS: COOLING

An HVAC system that will be asked to provide cooling to offset building heat gains needs a "coolth" source. There are three conceptually (and physically) different means of introducing coolth into a building:

- vapor compression refrigeration
- absorption refrigeration
- evaporative cooling

These three basic approaches mushroom into dozens of specific equipment options—and these options expand into more dozens of specific equipment offerings.

Vapor compression refrigeration: The vapor compression cycle is a mechanical–electrical circuit in which a refrigerant is circulated under temperature conditions that allow it to pick up heat from within a building and dump heat to the outdoor environment (under weather conditions that will not permit heat to passively flow from in-to-out). Vapor compression is by far the most commonly used and encountered means of producing a cooling effect for an HVAC system. All vapor compression systems require an externally placed heat rejection unit (a condenser). The capacity of vapor compression equipment covers a wide range (from small to huge); the specifics of equipment types is also diverse. The vapor compression cycle can be used directly to cool air or to produce chilled water. Capacity is expressed in tons [kW] of cooling (reminder: 1 ton is equal to 12,000 Btu/h [3.517 kW]) and "efficiency" via COP, which will be above 2.0 and can be as high as 6.0 or more.

Absorption refrigeration: The absorption refrigeration cycle is diagrammatically similar to the vapor compression cycle—but employs a chemical refrigerant flow process driven by heat (versus mechanical compression). Absorption refrigeration equipment will typically have a lower COP than comparable vapor compression equipment but offers a unique advantage in that it may be driven by waste heat, solar hot water, or natural gas. Absorption equipment today is typically used to produce chilled water in moderately large capacity applications. The equipment can be quieter than comparable vapor compression equipment. Expect a COP between 0.6 and 1.2, perhaps even higher for certain enhanced systems.

Evaporative cooling: In an HVAC system, an evaporative cooling effect will be produced by equipment called an evaporative cooler. This approach to coolth production can be very energy-efficient (operating as it does along a line of approximately constant enthalpy). Rather than a refrigeration cycle, evaporative cooling utilizes the natural cooling effect of water evaporation and does not need much more than a fan and a water pump to operate. As such, its COP can range from 10 to 40. With variations in equipment type, evaporative cooling can be used to directly cool air or to cool water. The performance of an evaporative cooler is more dependent upon climate than that of either vapor compression or absorption refrigeration (low relative humidity is generally desirable, although indirect evaporative cooling can extend the climate range of applications). Where feasible, evaporative cooling has interesting potential as a low-GWP (global warming potential) high-efficiency coolth source.

Several examples of cooling equipment and applications are provided in Section 16.8 by way of introduction to this part of an HVAC system.

16.8 COOLING EQUIPMENT

The fundamental question that faces the designer of an active cooling system is how to get heat to flow from a cooler indoor environment to a hotter outdoor environment—which is the situation faced during the overheated portions of the year. Rather than violate the laws of thermodynamics, a refrigeration cycle sets up a manufactured intermediary thermal system that exhibits desirable temperature regimes that will support this "uphill" heat flow. This intermediating system comes with a cost—in terms of equipment, energy, and environmental pollution—but it allows us to be comfortable under conditions that 100 years ago would have been most uncomfortable. The two refrigerant-based cooling processes commonly used in buildings are the vapor compression and the absorption refrigeration cycles. The vapor compression cycle can be adapted to a huge range of capacities and a variety of configurations—such as direct expansion (DX) or chiller arrangements. Heat rejection to the outdoors may be accomplished through an air-cooled condenser or via a cooling tower. Vapor compression cooling is used in at least 95% of HVAC systems.

(a) Vapor Compression Refrigeration

Figure 16.24 illustrates a water-cooled (i.e., uses water to absorb and carry away the heat from the closed refrigerant cycle) vapor compression cycle. The system acts as a thermal bridge between one water system (chilled water) and another water system (condenser water). Heat is removed from a building via the chilled water and rejected from the building via the condenser water—thus cooling the building. The cycle operates by cyclical liquefaction (converting a gas into a liquid) and evaporation (liquid to gas) of a refrigerant, during which processes the refrigerant releases and absorbs heat, respectively.

Refrigerants are the heart of a refrigeration cycle. To liquefy a gaseous refrigerant (Figure 16.24), it is first compressed to a high-pressure vapor (by which the

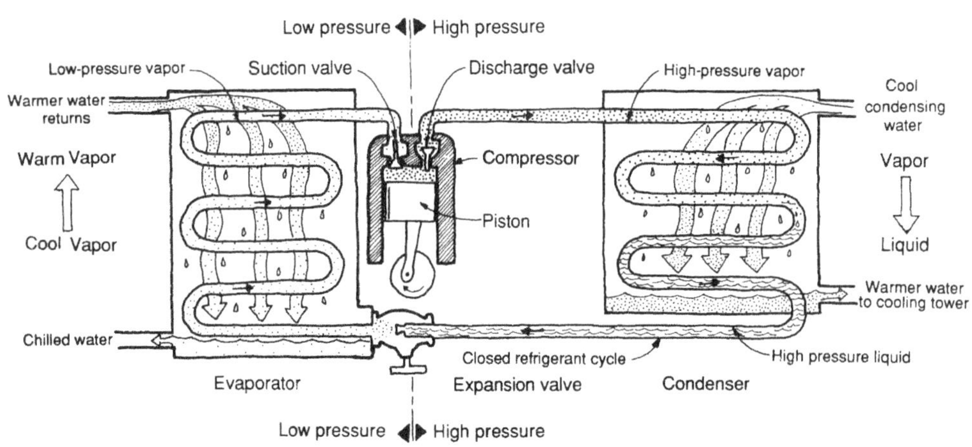

Figure 16.24 *Schematic arrangement of a vapor compression refrigeration cycle, providing chilled water for a building cooling system.*

molecules within the vapor are forced closer together and the temperature of the vapor increases) via the aptly named compressor; then, by means of a heat sink, sensible and latent heat is extracted from the refrigerant, which condenses it back into a liquid within the condenser. The resulting high-pressure liquified refrigerant has a potential for heat absorption again, but only after it is passed through an expansion valve where it is depressurized and allowed to expand back to lower-pressure gaseous form within the evaporator. During this change of state, the refrigerant must take on latent heat by reducing its own temperature and subsequently drawing heat out of the building via the specified cooled medium (i.e., water or air).

A refrigeration cycle is a heat-circulating system. Heat is "pumped" out of a building through the refrigeration cycle. This pumping action was originally conceived as a unidirectional process. By reverse cycle arrangements of the vapor compression components, however, heat flow in the other direction (from the outdoors to the indoors) can be accomplished. A cooling-only cycle is called a refrigeration system. A bi-directional (heating and cooling) cycle is called a heat pump. The compressive refrigeration cycle can be used to transfer heat between almost any media; Figure 16.24 illustrates a water–water cycle, Figure 16.45 shows a heat pump in both air–air (i.e., heat is extracted via air; heat is delivered via air) and water–air (i.e., heat is extracted via water; heat is delivered via air) applications.

Compressors. A reciprocating (piston-type) compressor shown in Figure 16.24 is one option. The compressor can instead be one of several other types: rotary, scroll, screw, or centrifugal, each with characteristics suitable to particular applications. For instance, rotary compressors are typically used in small/compact refrigeration units; reciprocating compressors are commonly used in small-to-medium refrigeration systems such as residential refrigerators or commercial freezers; scroll compressors are frequently used in heat pumps; and screw and centrifugal compressors apply to medium-to-large HVAC systems and chillers.

Chillers. Chillers are basically large-scale refrigeration units used to remove heat from a fluid medium and then the chilled fluid is used to distribute cooling for various processes (such as a central air-handling unit for thermal comfort purposes). Please note that there are both vapor compression chillers and absorption chillers, defined in terms of the refrigeration cycle the equipment utilizes. This section pertains to vapor compression chillers (the following section addresses absorption chillers).

Based upon their method of heat rejection, there are three types of vapor compression chillers:

1. *Air-cooled chillers* by which air is used to cool the refrigerant in the condenser.
2. *Water-cooled chillers* by which treated water is used to cool the refrigerant in the condenser.
3. *Heat pump chillers* which can reverse the vapor compression refrigeration cycle, effectively offering cooling and heating modes.

Chiller type selection will hinge on project-specific priorities. Considerations of initial cost, performance efficiency, maintenance costs, service life, and refrigerant environmental impact should all be assessed and compared.

A project may require one or two chillers—or it may require several chillers, constituting a chiller plant. Chillers or chiller plants can be located in dedicated mechanical rooms (Figure 16.25) or outdoors, including atop roofs (Figure 16.26) and perhaps combined with other components as part of a mechanical penthouse. Chiller space requirements can be considerable, and they may require setbacks for access and ventilation (if air-cooled).

Vapor compression chillers are commonly specified in terms of their compressors. As such, large refrigeration units are usually centrifugal chillers (Figure 16.27), with compressors that can be driven either by an electric motor or a turbine driven by steam or gas. When a steam-driven turbine is used, the exhaust steam is often used to drive an auxiliary absorption cycle machine. The coupling of these two devices makes an efficient combination (and the steam plant that supplies them in summer can supply heating in winter). Centrifugal chillers usually require about 1 hp/ton (equivalent to 0.57 kW, or 10 ft^3 gas, or about 15 lb of steam per ton). These chillers usually use a cooling tower as the heat rejection device.

Dual-condenser chillers (Figure 16.28) can reject heat to a cooling tower (via a heat rejection condenser) or to building heating uses (via a heat recovery condenser). As the name implies, a dual-condenser chiller is equipped with two condensers rather than one. Typically, one condenser is utilized to capture/repurpose waste heat (i.e., heat recovery) and the other for disposal of excess heat when recovery is not required or cannot be utilized (i.e., heat rejection). This two-condenser feature allows the system to switch between or simultaneously use two methods of heat rejection, typically air-cooled and water-cooled condensers, which can enhance operational flexibility, energy efficiency, and backup functionality (should a condenser fail).

Medium-capacity chillers may use either twin screws or a scroll compressor. The screw compressor

Figure 16.25 Chiller plant comprised of four individual chillers, located in a dedicated mechanical room. (Courtesy of Trane. © Trane. All rights reserved.)

Figure 16.26 A roof-mounted air-cooled chiller plant. (Courtesy of Trane. © Trane. All rights reserved.)

(Figure 16.29) has a pair of helical screws; as they rotate, they mesh and thus compress the volume of the refrigerant gas. They are small and quiet, with little vibration. The scroll compressor (Figure 16.30) uses two inter-fitting spiral-shaped scrolls. The refrigerant gas is compressed as one scroll rotates against the other fixed scroll. Quiet and low-maintenance, they are more efficient than reciprocating compressors.

Reciprocating chillers (Figure 16.31) are small-capacity vapor compression machines. Usually electrically driven, they are often combined with an air-cooled heat rejection loop rather than a cooling tower.

Chilled water is usually supplied at between 40°F and 48°F [4°C and 9°C]. When the chilled water is supplied cold and returns much warmer, the large rise in temperature reduces the initial size (cost) of equipment and increases its efficiency (thereby reducing the operating cost as well). Water treatment may be needed for chilled water to control corrosion or scaling (cooling tower water may similarly need water treatment to control corrosion and scaling—and biological growth, such as algae and bacteria).

Typical cooling capacities and space requirements for chillers are shown in Figure 16.32—with dimensions as indicated. Each refrigeration machine in this illustration requires two pumps—one for the chilled water (supplied to the building loads) and one for condenser water (that circulates to a cooling tower). Typically, space is provided for future chiller additions, which may be required by building expansion and/or by higher internal gains from as-yet-uninstalled plug loads. Improved-efficiency chillers may replace older ones when energy costs and environmental regulations become compelling. Adequate clearance and access to the equipment room are a major design issue.

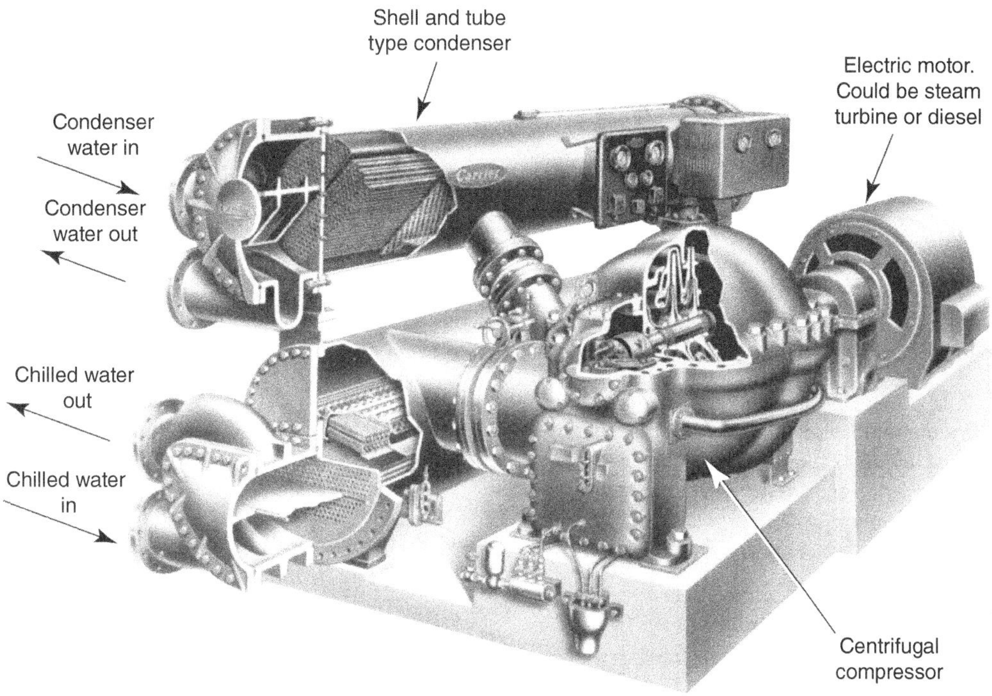

(a)

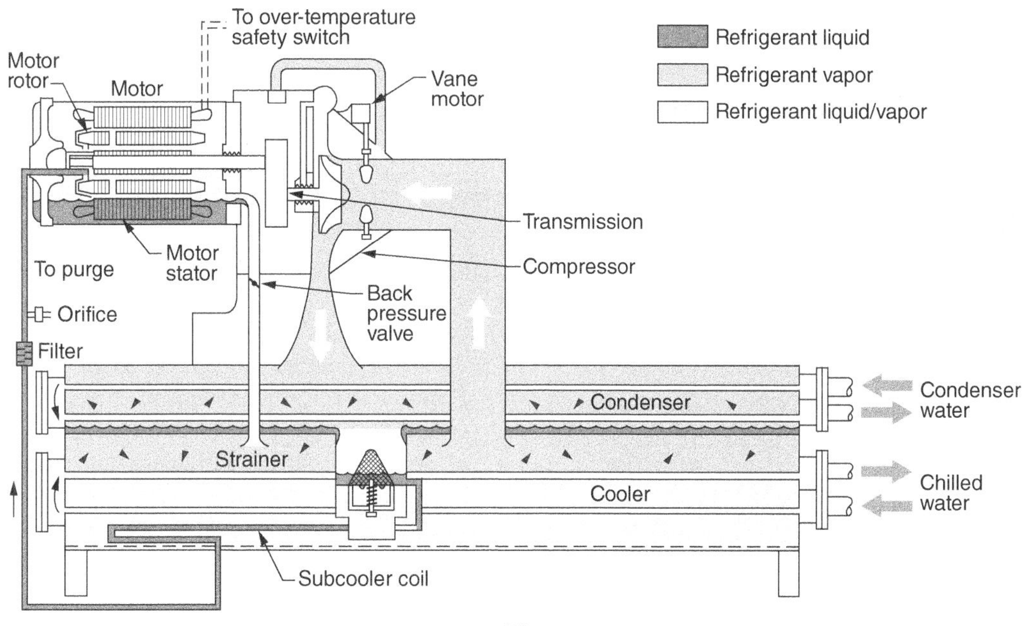

(b)

Figure 16.27 *(a) A large-capacity centrifugal chiller—using the vapor compression refrigeration cycle. (b) Centrifugal chiller with a flooded cooler and condenser within a single outer shell. This low-pressure unit typically produces 100–400 tons [350–1400 kW] of cooling. Typical dimensions are: 14 ft L × 5 ft W × 8 ft H [4.3 × 1.5 × 2.4 m]; weight 16,000 lb [7260 kg]. (a: Courtesy of the Carrier Corporation; b: Ramsey and Sleeper, 1981.)*

Refrigerants. Refrigerants are chemical compounds specifically designed for the purpose of heat transfer in HVAC systems. As they move through the vapor compression or absorption refrigeration cycle, refrigerants can exist in a liquid and gaseous state. Refrigerants are engineered to facilitate temperature–pressure relationships such that they can operate at a condensing temperature above the outdoor air temperature when used in an air-cooled configuration, and at an evaporator temperature below room air temperature when in direct expansion configuration—with a reasonable expenditure

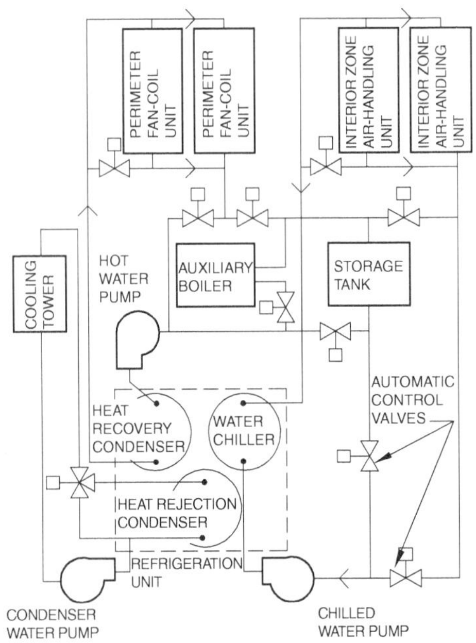

Figure 16.28 Dual-condenser chiller. The heat removed from the chilled water loop is either rejected to the cooling tower or recovered for building heating. (From Ramsey and Sleeper, 2007.)

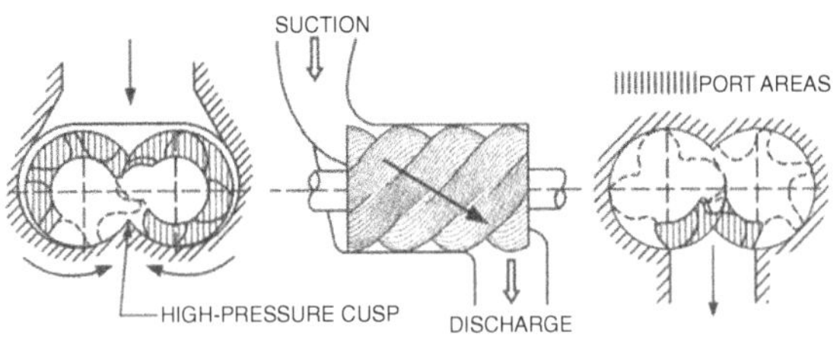

Figure 16.29 A screw, or helical, compressor is a quieter, smaller machine with little vibration. (© ASHRAE, www.ashrae.org. 2024 ASHRAE Handbook—HVAC Systems and Equipment, Chapter 38, Figure 32.)

of energy for pressurization. Refrigerants also exhibit freezing points that are well below the temperatures encountered in typical HVAC applications to ensure that they remain in a fluid state during operation.

A rather large number of compounds have been used as refrigerants for various applications. In addition to desirable temperature–pressure characteristics, it would be ideal if a refrigerant were nontoxic, low-cost, long-lived, not too viscous, and environmentally benign. This last characteristic is taking on increasing importance to the HVAC industry. Most refrigerants encountered in new and existing HVAC systems will fall into one of the following categories:

Chlorofluorocarbons (CFCs): The refrigerants most commonly used through the 1990s were CFCs—compounds made up of chlorine, fluorine, and carbon. CFCs were a serious threat to the atmosphere because their chlorine content depleted the ozone layer when released into the atmosphere. This galvanized international action to phase out CFC production and consumption, which culminated in the signing of the Montreal Protocol in 1987. As such, production of CFC refrigerants was banned in various countries, including the United States, in the mid-1990s. Fears that replacement refrigerants would be less efficient have largely disappeared; combinations of better chillers and new refrigerants produce energy savings. The energy aspect of refrigeration is intriguing. CO_2 is released during the production of the electricity that will power chillers and other refrigeration machines. There is a quandary if lower-global warming potential (GWP) refrigerants are also less efficient and therefore require higher energy consumption. This discussion assumes continued power generation from fossil fuels; if renewable power sources (wind, photovoltaic) are used to power cooling equipment, then this aspect of climate change can be mitigated.

Hydrochlorofluorocarbons (HCFCs): These refrigerants are an interim replacement for CFCs. HCFC compounds include hydrogen in addition to chlorine, fluorine, and carbon. As such, they are still a threat to the atmosphere, including ozone depletion, but not as severe as CFCs. In accordance with international agreements, such as the Montreal Protocol, HCFCs are being phased out as well. The EU completely banned HCFCs as of 2015; while new production and import of most HCFCs were phased out in the United States as of 2020.

Hydrofluorcarbons (HFCs): A longer-term replacement option for CFCs and HCFCs involves HFCs, which do not contain chlorine or bromine and therefore do not deplete the ozone layer of the atmosphere. However, as their name suggests, they are composed of hydrogen, fluorine, and carbon, the latter two of which cause these refrigerant gases to exhibit a significant negative environmental impact.

Hydrocarbons (HCs): Yet another possibility are natural hydrocarbons (HC). In general, comparing these two alternatives, HFCs have higher global warming potential and long atmospheric lifetimes, but they have low toxicity and are nonflammable. HCs have negligible global warming potential and short atmospheric lifetimes, but are flammable and explosive. Another alternative is ammonia (refrigerant R-717), used in the early years of vapor compression refrigeration but discontinued because of its acute toxicity and flammability. It is now returning in some refrigeration cycles because it does not contribute to global warming when it escapes to the atmosphere. Carbon dioxide as a refrigerant (R-744) is also seeing a resurgence in interest due

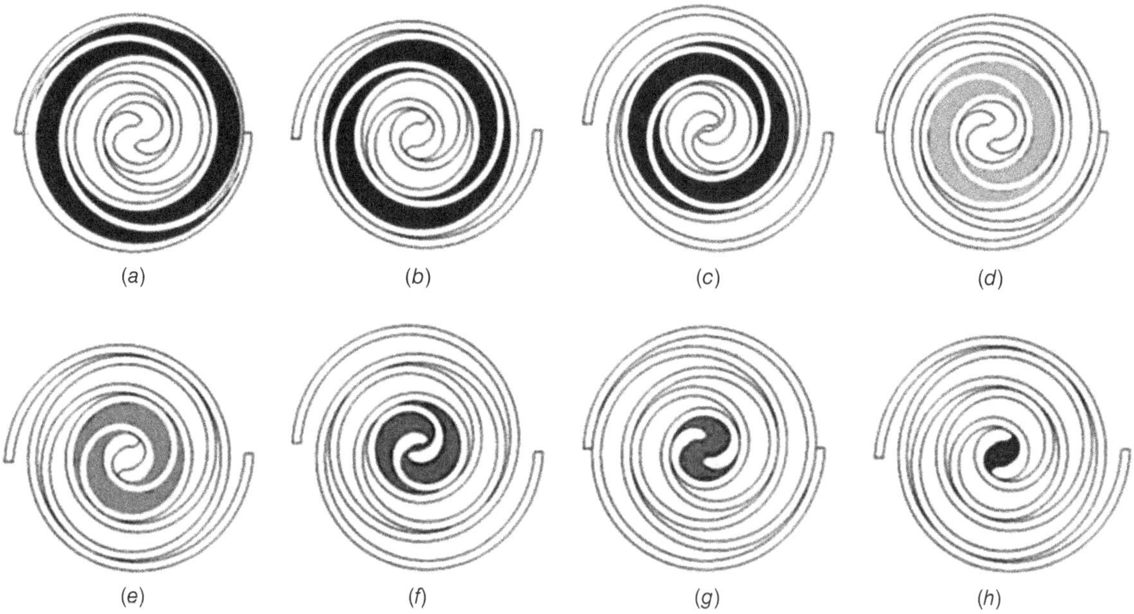

Figure 16.30 *The scroll compressor rotates one scroll element against another, with a quiet and efficient compression of the refrigerant. (© ASHRAE, www.ashrae.org. 2023a ASHRAE Handbook—HVAC Systems and Equipment, Chapter 38, Figure 45.)*

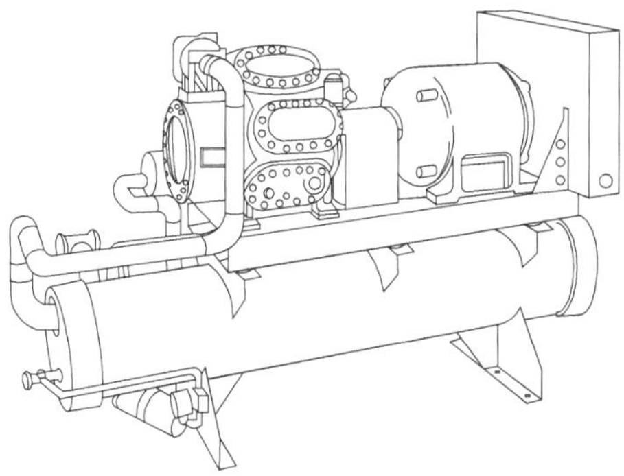

Figure 16.31 *A reciprocating chiller—a small-capacity machine that uses the compressive refrigeration cycle. This type of chiller typically produces less than 200 tons [700 kW] of cooling. Such a machine might be around 8 ft L × 3 ft W × 5 ft H [2.4 × 0.9 × 1.5 m] and weigh 3500 lb [1590 kg]. (From Ramsey and Sleeper, 2016.)*

to its minimal global warming potential. Research will ideally sort this situation out. For a comprehensive review of CFC, HCFC, and HC refrigerants, see Calm (1994).

Refrigerant blends: Driven by the need to replace environmentally harmful refrigerants, the HVAC industry has developed refrigerant blends that effectively balance performance, safety, and environmental impact. Some blends have been developed by commercial enterprises and have trademarked common names. However, all refrigerants used in HVAC systems and appliances are indexed. The variety of naming conventions can become confusing when reading the literature. CFC-11 may be referred to as R-11. HCFC-22 may go by R-22 or its trademarked name "Freon," which has now become a shorthand reference term.

Refrigerants continue to improve, but still pose an environmental threat because they can escape from equipment during operation or repair as gases that mix into the atmosphere. The two key threats are stratospheric ozone depletion and climate change. The potential of a refrigerant to harm the ozone layer is expressed by its ozone depletion potential (ODP) value. Global warming potential (GWP) is a measure of the ability of a refrigerant to effect climate change. Carbon dioxide (CO_2) has been assigned a GWP of 1. As such, it serves as the yardstick against which all other greenhouse gases (GHGs) are compared.

There are thousands of GHGs and most have a much higher GWP than CO_2. For instance, methane (CH_4) has a GWP of 28 and nitrous oxide (N_2O) has a GWP of 273. As potent as these other gases are, there is far more CO_2 in the atmosphere (even though its proportion of the atmosphere is only around 0.04% and rising). Moreover, the building design and construction industry has galvanized decarbonization efforts around using CO_2 as a proxy—and with the shorthand misnomer of "carbon." As such, the total GWP of any activity, including utilization of refrigerants, is estimated and defined in terms of carbon dioxide equivalent (CO_2e).

Once refrigerants are exposed to the atmosphere, they vaporize and persist as a gas. Some of these gases are highly reactive and exhibit a very brief atmospheric

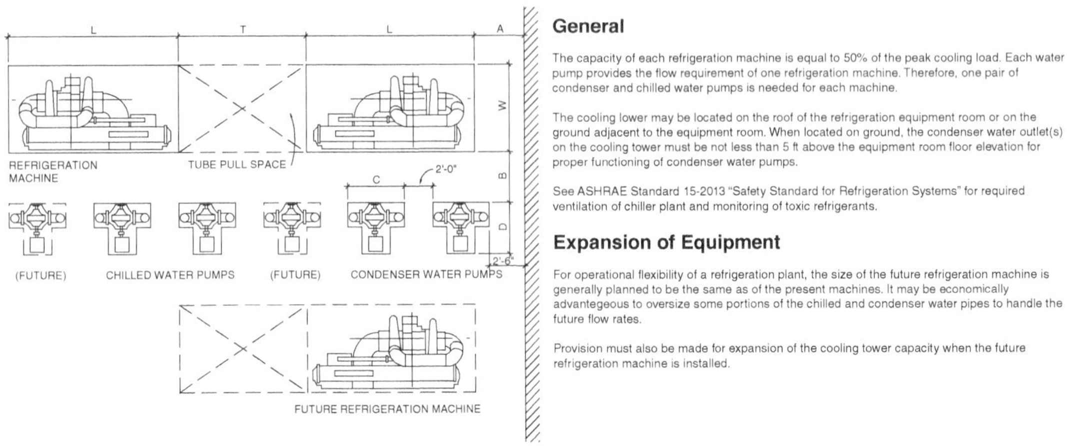

Figure 16.32 *Chiller room space requirements. Each refrigeration machine is served by two pumps (chilled water and condenser water). (From Ramsey and Sleeper, 2000.)*

lifetime (e.g., vaporized HC-1270 has one of the shortest atmospheric lifetimes at 0.001 years) while other gases can persist for thousands of years (e.g., vaporized PFC-116, which may linger for 10,000 years). For this reason, when comparing gases, GWPs must be averaged over the same number of years. That number appears as a subscript. GWP_{20} is the heat-trapping capacity of a gas when compared with CO_2, averaged over 20 years. Conventionally, a 100-year timeframe is assumed (i.e., GWP_{100}). If the literature simply states "GWP," assume 100 years.

Table 16.4 offers the ODP and GWP_{100} of some of the most frequently encountered refrigerants in buildings and their related systems, including appliances. These figures are approximations and subject to change—including when research prompts a change to how the CO_2 baseline of 1 is assessed. While these figures offer reasonable assumptions, design teams should assess the refrigerant properties for any given instance in practice. Considerable efforts are under way both to reduce the likelihood of escaping refrigerant and to develop more efficient non-CFC refrigerants.

Enhanced Refrigerant Management. Most HVAC systems are manufactured packages. As such, there is somewhat limited room for building designers to influence decision making on refrigerants (e.g., one cannot just simply ask/pay for a better refrigerant). However, the collective pursuit by the building design and construction industry to decarbonize the built environment has precipitated the emerging practice of conducting carbon assessments on building design options in terms of carbon dioxide equivalent emissions (CO_2e). These assessments typically range from 25 to 60 years and encompass a building's embodied carbon (including site hardscape), its operational carbon (which will vary by location due to regional energy resources), the transportation carbon (from building users, including visitors), and the carbon emissions resulting from refrigerant leaks over the time period of the carbon assessment. Recall the note above on the misnomer of using the term "carbon" in this context; though it is admittedly convenient as a shorthand and therefore uses across the industry and in this book.

Emissions from refrigerant leaks are an example of what is referred to as fugitive emissions. The word "fugitive" is used because these emissions are not specifically considered and calculated during the design of a building or its mechanical systems—rather, such emissions result from the miscellaneous leakage or discharge of vaporized refrigerants from pressurized equipment or facilities. Fluorinated gases (which constitute most refrigerants in the building sector) have a high GWP, making refrigerant leaks a substantial contributor to rising CO_2e concentrations in the atmosphere. Enhanced refrigerant management will play an even bigger role in the near future as other CO_2e contributions are curtailed or eliminated. Estimates suggest that practices to avoid leaks from refrigerants and safely dispose of refrigerants at the end of life can substantially reduce emissions. Project Drawdown, the nonprofit think tank founded by environmentalist Paul Hawken in 2014, consistently cites enhanced refrigerant management among the highest potential impact solutions across all sectors and scales toward reducing global carbon emissions (Accuardi et al., 2022).

Refrigerant Charge. The specific amount of refrigerant required by an HVAC system is called the refrigerant charge. There are myriad factors that will affect the amount of refrigerant charge for any given system. A conceptual rule-of-thumb estimate for commercial chillers is 2–3 lb [0.9–1.4 kg] of refrigerant per ton of cooling capacity. However, this number will change with system type (e.g., the extent of piping to be filled). For instance, a complex heat pump solution,

TABLE 16.4 Refrigerant Environmental Properties[a]

Refrigerant	Common Name[b]	ODP	GWP_{100}
CFC-11 (or R-11)		1.00	5506[c]
CFC-12		0.73	11,200[c]
CFC-13		1.00	16,200[c]
CFC-113		0.81	6520[c]
CFC-114		0.50	9430[c]
CFC-115		0.26	9600[c]
HCFC-22	Freon, Chlorodifluoromethane	0.034	1960[c]
HCFC-123		0.01	90.4[c]
HCFC-124		0.02	597[c]
HCFC-142b		0.057	2300[c]
HCFO-1233zd(E)		0.00034	3.88[c]
HE-E170		0	1
HFC-23		0	14,600[c]
HFC-32	Difluoromethane	0	771[c]
HFC-125	Pentafluoroethane	0	3740[c]
HFC-134a	1,1,1,2-Tetrafluoroethane	0	1530[c]
HFC-143a		0	5810[c]
HFC-152a		0	164[c]
HFC-227ea		0	3600[c]
HFC-236fa		0	8690[c]
HFC-245fa		0	962[c]
HFO-1234yf	2,3,3,3-Tetrafluoropropene	0	0.501[c]
HFO-1234ze(E)		0	1.37[c]
HFO-1336mzz(Z)		0	2.08[c]
PFC-116		0	12,400[c]
PFC-218		0	9290[c]
PFC-C-318	Octafluorocyclobutane	0	10,200[c]
HC-290	Propane	0	5
HC-600	Isobutane, Butane	0	4
HC-600a		0	~20
HC-601a		0	~20
HC-1270	Propene, Propylene	0	1.8
R-717	Ammonia	0	0
R-744	Carbon Dioxide	0	1[e]
R-407C[d]		0	1620
R-410A[d]	Puron	0	1920
R-449A[d]		0	1280
R-452A[d]		0	1950

Source: © ASHRAE, www.ashrae.org. 2021 ASHRAE Handbook—Fundamentals, Chapter 29. Adapted by permission.

[a]Content is from Chapter 29, Table 3B unless noted otherwise.

[b]Common names of some refrigerants added by the authors for clarity.

[c]Updated information from the 6th Assessment Report (AR6) of the Intergovernmental Panel on Climate Change (IPCC, 2023).

[d]Data are from Chapter 29, Table 4.

[e]Chapter 29, Table 3B offers an additional parenthetical GWP_{100} value stipulated for reporting under the Kyoto Protocol. This value has been omitted.

such as a VRF system, will require a relatively large amount of refrigerant to fill the extensive refrigerant piping in the system. As such, a VRF system may require 4 to 6 lb [1.8–2.7 kg] per ton of cooling capacity. Awareness is key. It would behoove a project team to ascertain the specific refrigerant charge for any proposed HVAC system.

Refrigerant Leakage Rates. All HVAC systems leak refrigerant over time. System type will affect the leakage rate among many other factors, but as a practical matter: more pipes and/or more refrigerant means more leakage. A conventional approach to approximating refrigerant leakage is to assume a 5% annual leakage rate. However, with best management practices regarding enhanced refrigerant management, including commissioning and proper maintenance, teams may assume a 2.5% annual leakage rate. Guidance regarding broad-based refrigerant leakage rates can vary. The Chartered Institute of Building Services Engineers (CIBSE) has produced a technical memorandum, *CIBSE TM65*, that suggests the following annual leakage rates (CIBSE, 2021):

Packaged heat pump or chiller, where no refrigerant is managed on site (type 1)	2%
Heat pump or chiller where some works on refrigerant piping is carried out on site (type 2)	4%
VRF systems where substantial refrigerant piping is installed and filled on site (type 3)	6%

Table 16.5 offers a list of typical annual refrigerant leakage rates by equipment type, as recommended in *ASHRAE Standard 228-2023* (ASHRAE, 2023b) The leakage rates offered by CIBSE and ASHRAE are broad estimates for assessment purposes. Actual leakage rates will vary based on specific equipment and the quality of installation, operation, and maintenance.

Armed with the refrigerant charge and the leakage rate, the annual CO_2 equivalent emissions contributed by a building's refrigerants may be calculated as follows:

$$\begin{aligned}\text{Annual } CO_2 \text{ equivalent emissions (kg } CO_2\text{e)} &= \text{(total amount of refrigerant, kg)}\\ &\times \text{(annual leakage rate)}\\ &\times \text{(GWP of the refrigerant)} \qquad \text{(Eq. 16.1)}\end{aligned}$$

An interesting consideration here is the accelerated uptake in all-electric buildings as a means to

TABLE 16.5 Annual Refrigerant Leakage Rates by Equipment Type

Equipment Type	Typical Annual Leakage Rate of Refrigerant Mass Charge Per Year (%)
Supermarket refrigeration	30
Commercial condensing units	15
Water chillers	5
Hermetic units with no field installed refrigerant piping	1
Rooftop unit air conditioner	6
Residential heat pump and air conditioner	2
Variable refrigerant flow air conditioner	10
Other refrigeration	2
Other air conditioning	2

Source: © ASHRAE, www.ashrae.org. ANSI/ASHRAE Standard 228-2023: Standard Method of Evaluating Zero Energy and Zero Carbon Building Performance.

reduce operational carbon emissions as the utility grid decarbonizes by transitioning to renewable energy resources. Heat pump systems offer economically viable all-electric HVAC options for many residential and nonresidential buildings, but these systems involve refrigerants. A holistic carbon assessment must consider both operational carbon emissions and contributions from refrigerants. See Example 16.3 for concept-level assessment of the annual operational and refrigerant CO_2e contributions for two different HVAC system options.

EXAMPLE 16.3 A client has asked your design team for a 25-year carbon assessment. As part of the assessment, the team needs to approximate and compare the current annual CO_2e emissions contributions from building energy consumption and refrigerant leakage. Consider a five-story, 53,000 ft^2 [4924 m^2], office building in Boston, Massachusetts. The base HVAC system consists of:

- variable air volume (VAV) system
- central air-handling units
- three @ 100-ton air-cooled chillers: 4.0 COP
- chiller plant cooling design capacity: 300 tons
- refrigerant used: R-410A (a blend of R-32 and R-125)
- natural gas-fired boilers: 90% AFUE
- heating capacity: 2500 Mbtu/h
- refrigerant charge: 2.5 lb per ton
- annual refrigerant leakage rate: 5%

Determine the annual operational and refrigerant carbon emissions of the building using the base HVAC system. Then, compare the base system to an all-electric option with better energy performance. This alternate HVAC system consists of:

- variable refrigerant flow (VRF) heat pump system
- dedicated outdoor air system (DOAS) for ventilation
- 4.5 COP
- cooling design capacity: 300 tons
- refrigerant used: R-410A (same as base design)
- refrigerant charge: 5 lb per ton
- annual refrigerant leakage rate: 5% (same as base design)

SOLUTION

STEP 1. Annual operational carbon emissions, base HVAC system. Using building performance modeling, the predicted energy use intensity (pEUI) of the building using the base VAV system was determined to be 44 kBtu/ft^2 [499 MJ/m^2] per year.

Based on the current energy resource mix of the regional electricity grid along with the natural gas emissions contribution, the operation of the base building will result in the following operational carbon emissions:

I-P units:
433,411 lb CO_2e per year

SI units:
196,592 kg CO_2e per year

STEP 2. Annual refrigerant carbon emissions, base HVAC system. Recall Equation Eq. 16.1:

$$\text{Annual } CO_2 \text{ equivalent emissions (kg } CO_2\text{e)} = (\text{total amount of refrigerant, kg}) \times (\text{annual leakage rate}) \times (\text{GWP of the refrigerant})$$

First, determine the refrigerant charge:

$$(2.5\ \text{lb/ton}) \times (0.4536\ \text{kg/lb}) \times (300\ \text{tons}) = 340.2\ \text{kg of refrigerant}$$

Second, recall the annual leakage rate of 5%.

Third, reference Table 16.4. The GWP for R-410A is 1920. Therefore:

$$\text{Annual } CO_2 \text{equivalent emissions (kg } CO_2\text{e)} = 340.2\ \text{kg} \times 0.05 \times 1920 = 32{,}659\ \text{kg } CO_2\text{e/yr}$$

In summary:

Annual operational carbon emissions:	196,592 kg CO_2e/yr
Annual refrigerant carbon emissions:	32,659 kg CO_2e/yr
Combined annual greenhouse gas emissions:	**229,251 kg CO_2e/yr**

This example indicates that with the base VAV system, refrigerant emissions constitute approximately 14% of the combined annual greenhouse gas emissions—a significant proportion.

STEP 3: Annual operational carbon emissions, alternate HVAC system.

Using building performance modeling, the predicted energy use intensity (pEUI) of the building option using the alternate VRF system was determined to be 41 kBtu/ft^2 [465 MJ/m^2] per year, a 7% reduction in energy use versus the base option.

Considering the current energy resource mix of the regional electricity grid, the operation of the alternate building will result in the following operational carbon emissions:

I-P units:
369,882 lb CO_2e per year

SI units:

167,776 kg CO_2e per year

This represents a 17% reduction in emissions versus using a VAV system.

Annual refrigerant carbon emissions, alternate HVAC system.

First, determine the refrigerant charge:

$$(5.0\ \text{lb/ton}) \times (0.4536\ \text{kg/lb}) \times (300\ \text{tons})$$

$$= 680.4\ \text{kg of refrigerant}$$

Second, assume the same annual leakage rate of 5%.

Third, recall the GWP for R-410A is 1920. Therefore:

$$\text{Annual } CO_2 \text{equivalent emissions (kg } CO_2\text{e)}$$

$$= 680.4\ \text{kg} \times 0.05 \times 1920$$

$$= 65{,}318\ \text{kg } CO_2\text{e/yr}$$

In summary:

Annual operational carbon emissions:	167,776 kg CO_2e/yr
Annual refrigerant carbon emissions:	65,318 kg CO_2e/yr
Combined annual greenhouse gas emissions:	**233,094 kg CO_2e/yr**

This example indicates that with the alternate VRF system, refrigerant emissions constitute approximately 28% of the combined annual carbon emissions—well over a third of the operational emissions! Perhaps the more bothersome outcome is that despite choosing a system that reduced operational greenhouse gas emissions by 17%, the VRF system actually increased the combined annual emissions by 1.7% due to the more extensive refrigerant charge requirement and the proportional increase in leakage.

The silver lining for the VRF system is that it is all-electric. As Boston's regional electrical grid advances as part of the transition to mostly renewable energy resources, the VRF system is likely to see the annual operational carbon emissions decrease dramatically over a 25-year carbon assessment projection. Herein lies a design challenge. A rational analysis of projected carbon emissions demands consideration of likely future conditions. Future conditions are unknown, although they may follow predictable patterns. Some risk is involved with all attempts at risk mitigation. ■

Refrigerant End-of-Life Recovery Rates. During demolition/deconstruction, there is an opportunity to recover refrigerants. Regulation often requires the recovery to be carried out using environmentally preferred best practices. As such, it is a safe assumption in developed countries that the majority of refrigerant will be safely and effectively recovered. CIBSE TM65 offers the following assumed recovery rates:

Package heat pump or chiller, where no refrigerant is managed on site (type 1)	99%
Heat pump or chiller where some work on refrigerant piping is carried out on site (type 2)	98%
VRF systems where large amounts of refrigerant piping are installed and filled on site (type 3)	97%

Direct Expansion (DX) Systems. The preceding discussion of vapor compression refrigeration cycles focused mostly on chillers—which provide cooling effect through the medium of chilled water circulated to loads located within a building. A DX refrigeration cycle provides a cooling effect through the action of room air passing across a coil that contains circulating refrigerant. In other words, the refrigerant "directly" cools or heats the air that is supplied to the indoor space being conditioned. A DX system is simpler than a chiller system but has application limitations. In general, the evaporator coil (cooling coil with refrigerant) can be located only 100 ft [30 m] or so from the compressor of the system. DX systems tend to be close-coupled, which may be a limitation in larger buildings. DX refrigeration systems are very commonly used in smaller-scale buildings. Figure 16.33 illustrates a basic DX system—this diagram will form the basis for ensuing discussion of the heat pump system. In essence, there are three basic types of DX systems:

- DX unitary systems
- split systems (including mini-splits)
- VRF systems

DX Unitary Systems. This is a ubiquitous simple solution for low-rise nonresidential structures (particularly "big box" retail stores, schools, and other buildings where dedicated space for HVAC equipment is limited). In a DX unitary system, the compressor, condenser, evaporator, and fan-powered air-handing are all contained in a single cabinet or "unit" (sometimes referred to as a "package"). Often placed on the roof, these systems are commonly referred to as rooftop units (RTUs). These units can also be attached to the side of a building. Packaged terminal air conditioner (PTAC) units are compact, self-contained DX systems that can be installed in a wall and discharge cooled air directly into the space (these are commonly used to condition hotel rooms).

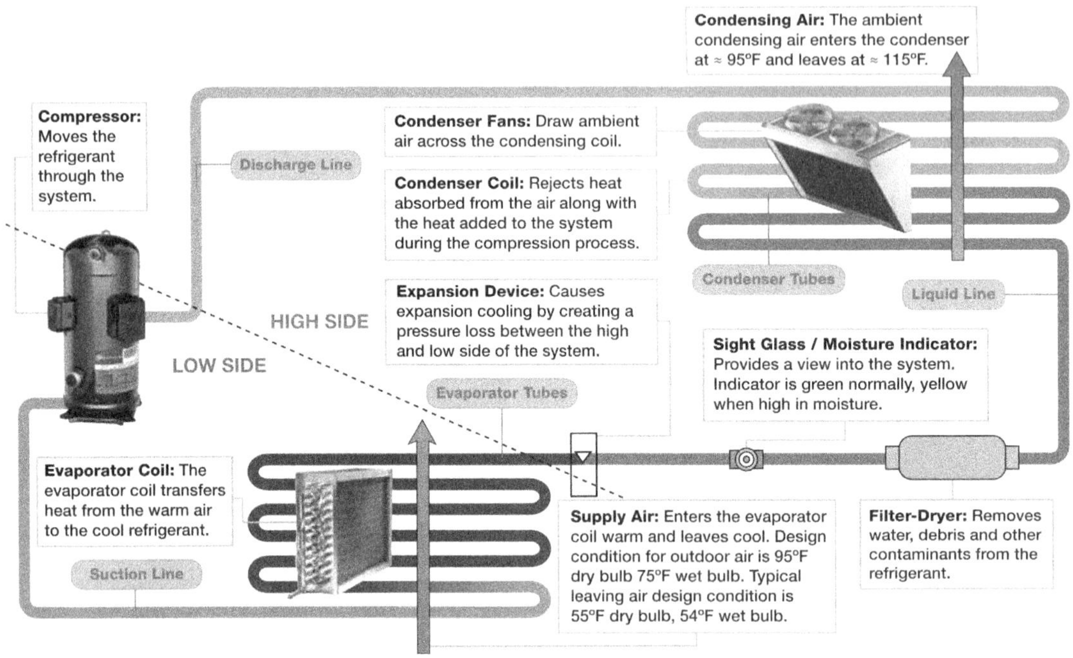

Figure 16.33 *Schematic diagram showing a basic DX refrigeration system. (Courtesy of Greenheck Fan Corp.)*

Adding a heating function to a DX unitary system may be considered. In a DX system, indoor air is directly cooled by the refrigerant; however, these unitary systems may include supplemental heating via electric resistance coils (think of a giant blow-dryer) housed inside the unit. This allows this otherwise cooling-only system to provide some degree of heating when needed. A heat pump is a twist on the DX cooling solution, but by being capable of reversing the refrigeration cycle, these systems can also transfer thermal energy (heat) from the outdoors (even in cold weather) and expel it indoors—meaning the system can operate in both a heating and cooling mode. Heat pumps are discussed later in this chapter.

Split Systems. Common in both residential and commercial buildings, a split system is similar to a unitary system; however, the components are "split" between an indoor and outdoor unit. The indoor unit contains the evaporator coil, an air-handing component (which may or may not introduce outdoor air, depending on the system), and a blower to facilitate air distribution. The outdoor unit contains the condenser coil, the compressor (which can be loud), and a fan or pump to facilitate heat exchange with the air, water, or ground. Refrigerant lines run between the two units.

Ductless Mini-Splits. These are a smaller-scale ("mini") variation of a split system that can condition air in a single room or relatively small thermal zone without ductwork (Figure 16.34). These systems consist of an indoor (evaporator) and an outdoor (compressor/condenser) unit like any other split system; but ductless mini-splits do not provide ventilation. Rather, the indoor unit cools the air in a zone and recirculates it. Even if the indoor unit features an air filter (as some do), fresh air to the zone will need to be provided by a different means. As such, it is often necessary to couple ductless mini-splits with a ventilation system that might include exhaust fans, a dedicated outdoor air system (DOAS), energy recovery ventilators (ERVs), or heat recovery ventilators (HRVs). See Section 16.14 for more information about ERVs and HRVs.

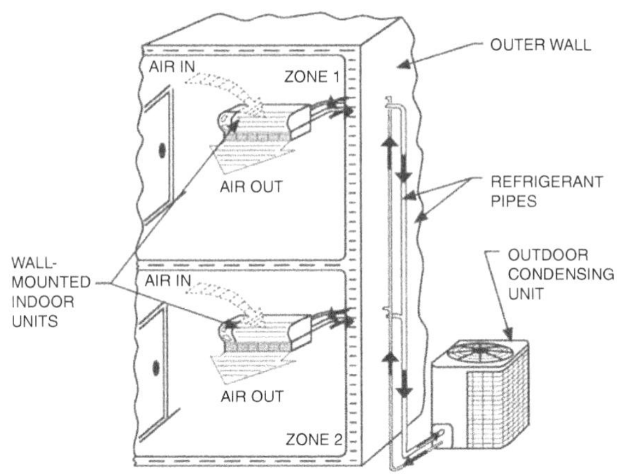

Figure 16.34 *Ductless mini-split HVAC system. (© ASHRAE, www.ashrae.org. 2024 ASHRAE Handbook—HVAC Systems and Equipment, Chapter 48, Figure 3.)*

Ductless mini-splits can easily serve large areas in open spaces with few obstructions. A multi-zone

ACTIVE ENVIRONMENTAL SYSTEMS

mini-split system is one in which the capacity of the outdoor unit can serve multiple indoor units, allowing each unit to serve its respective zone independently. The COP of ductless mini-splits will often range between 3.0 and 5.0—but advances are pushing systems beyond this range.

Variable Refrigerant Flow (VRF) Systems. The typical DX refrigeration cycle has one evaporator (cooling) coil paired with one compressor/condenser. In a VRF system, one compressor/condenser unit is connected to multiple evaporators (Figures 16.35, 16.36, and 16.37). Each evaporator unit can be individually controlled (each is a thermal zone). This arrangement is similar to that found in ductless mini-split systems, where one outdoor unit can be used with several evaporator units. Conventional split-systems turn on and off (like single-evaporator DX packaged systems) to maintain room set-point temperature. VRF systems, on the other hand, continuously modulate the amount of refrigerant being sent to each evaporator in response to zone loads.

By operating at varying speeds, VRF units work only at the needed capacity, which is how they consume less energy than on/off systems, even if they run more frequently. Like mini-splits, VRF systems are conducive to small-footprint configurations and require little or no intrusive ductwork. VRF systems are all-electric, which makes them an attractive option for project teams pursuing 100% electric HVAC solutions as part of a building's decarbonization strategy. Their COP typically ranges between 3.0 and 5.0—and even higher with some systems.

Modern VRF systems are commonly available in heat pump or heat recovery configurations. A heat pump arrangement can provide either heating or cooling (refer to the information on heat pumps later in this chapter). A heat recovery arrangement uses a branch circuit (BC) controller and a two-phase refrigerant from the outdoor unit by which a gas-liquid separator inside the BC controller sends gas refrigerant to certain indoor units for heating and liquid refrigerant to certain indoor units for cooling—effectively permitting coincident heating and cooling from one compressor/condenser unit and greatly increasing zoning flexibility and system performance. For detailed information on VRF systems, see Chapter 18 of the *2024 ASHRAE Handbook—HVAC Systems and Equipment* (ASHRAE, 2024).

(b) Absorption Refrigeration

This process (illustrated in Figure 16.38) uses water as the primary refrigerant and lithium bromide (a salt solution) as the absorber. As such, the GWP of absorption refrigeration is 0. However, a significant amount of thermal energy (heat) is required to drive this chemical process. Because of this relatively larger energy requirement, absorption refrigeration is less

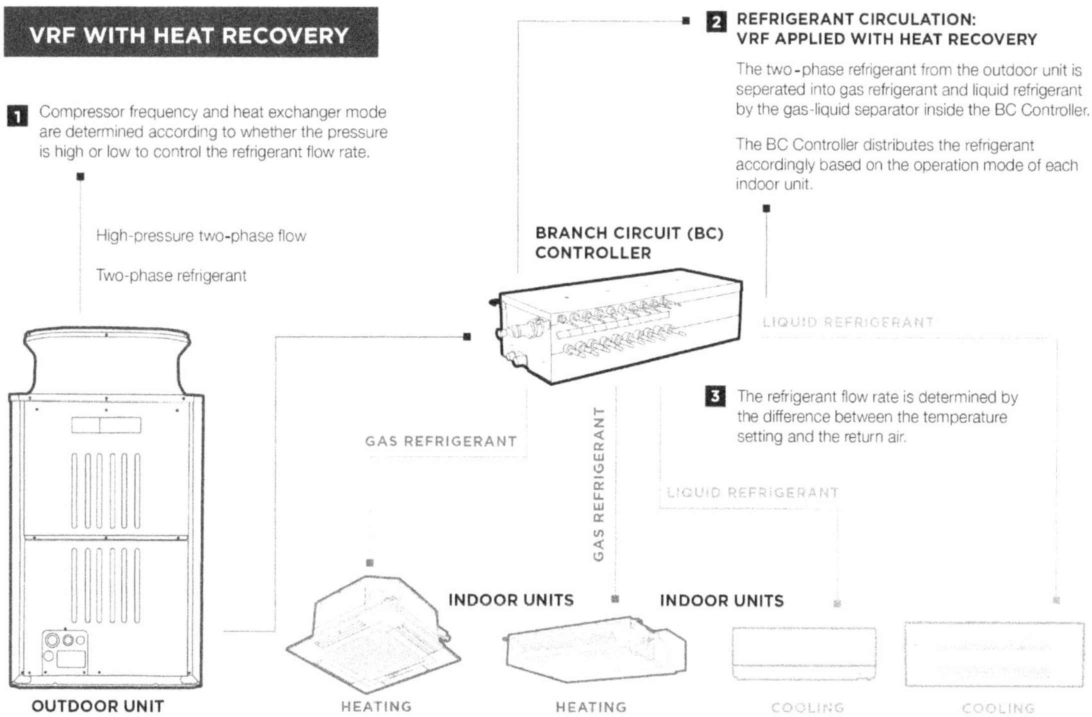

Figure 16.35 *Schematic diagram showing a variable refrigerant flow (VRF) system with heat recovery. (Courtesy of Trane. © Trane. All rights reserved.)*

Figure 16.36 *Outdoor units of a variable refrigerant flow (VRF) system. (Courtesy of Trane. © Trane. All rights reserved.)*

efficient (has a lower COP) than the vapor compression refrigeration cycle and requires about twice the heat rejection capacity. The input heat required to undertake salt solution regeneration may be provided by solar energy or by relatively high-temperature waste heat from an industrial-type process or a fuel cell. Because high-quality energy (electricity) is replaced by lower-quality heat, the absorption cycle can enjoy an exergy advantage over the vapor compression cycle, even though it is inherently less efficient.

There are several variations on the absorption cycle. When a combustion fuel such as natural gas is used to provide heat for the generator, the process is termed direct fired. When a non-combustion heat source (such as waste heat) is used, the process is termed indirect fired. Figure 16.38 shows a relatively simple single-effect absorption cycle that uses one heat exchanger between the strong and weak salt solutions. A double-effect absorption cycle (powered by steam in Figure 16.39) adds a second generator and condenser that operate at a higher temperature. It approximately doubles the COP of the single-effect cycle. A triple-effect cycle, in turn, provides a 50% COP improvement over the double-effect cycle.

A single-effect, indirect-fired absorption chiller (Figure 16.40) is attractive where central steam or high-temperature water (from solar collectors, waste heat from an industrial process, or outflow from a fuel cell, etc.) is available. Absorption equipment is less efficient than vapor compression refrigeration cycle equipment, although a cheap or even free heat source to power the cycle can rapidly overcome efficiency disadvantages. Natural gas and absorption refrigeration have often proven cost-effective alternatives to electrically driven vapor compression systems. Absorption machines have fewer moving parts (and therefore require less maintenance) and are generally quieter than vapor compression equipment of similar capacities. They can be environmentally attractive because they do not use CFCs or HCFCs and because they require far less electricity to operate. The higher waste heat output of about 31,000 Btu/ton [9299 kJ/kW], compared to at most 15,000 Btu/ton [4499 kJ/kW]

Figure 16.37 *An indoor terminal unit; part of a variable refrigerant flow (VRF) system. (Courtesy of Trane. © Trane. All rights reserved.)*

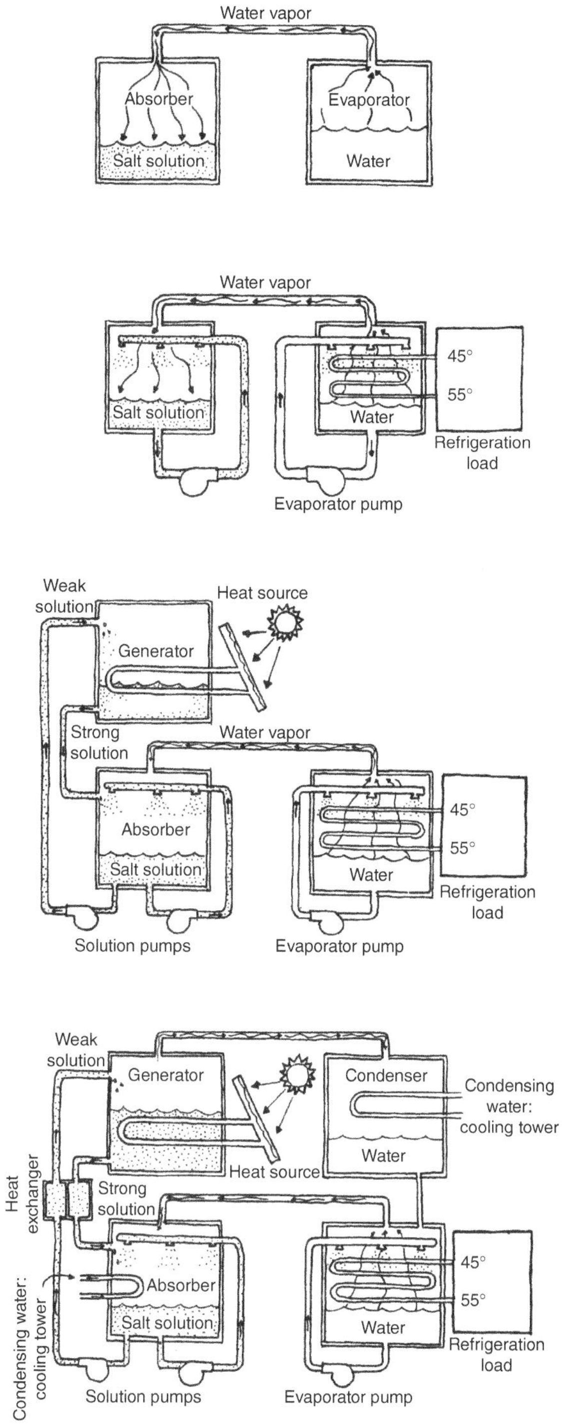

1 *Evaporator and absorber*
Consider two connected, closed tanks with a salt solution (lithium bromide) in one and water in the other. Just as common table salt absorbs water on a damp day, the salt solution in the absorber soaks up some of the water in the evaporator. The water remaining is thereby cooled by evaporation.

2 *Evaporator coil and pump added*
This refrigeration effect is utilized by putting a coil in the evaporator tank. Water from this tank is pumped to a spray header which wets the coil. The spray's evaporation chills water in the coil as it circulates to the refrigeration load. Solution pumped to spray in absorber raises efficiency.

3 *Solution pumps and generator added*
In an actual operating cycle, the salt solution is continuously absorbing water vapor. To keep the salt solution at the proper concentration, part of it is pumped directly to a generator where excess water vapor is boiled off. The reconcentrated salt solution is returned to the absorber tank, where it mixes with the solution sprayed to absorber in step 2.

4 *Condenser and heat exchanger added*
Water vapor boiled off from the weak solution is condensed and returned to the evaporator. A heat exchanger uses the hot, concentrated salt solution, leaving the generator to preheat the cooler, weak solution coming from the absorber. Finally, condensing water circulating through the absorber and condenser coils removes the waste heat.

Figure 16.38 *A single-effect absorption refrigeration cycle. The refrigeration capacity from this cycle (step 4) is delivered by a chilled water supply system. The heat source for the generator can be direct fired (natural gas) or indirect fired (steam, hot water, waste heat, or solar energy). (Adapted courtesy of the Carrier Corporation.)*

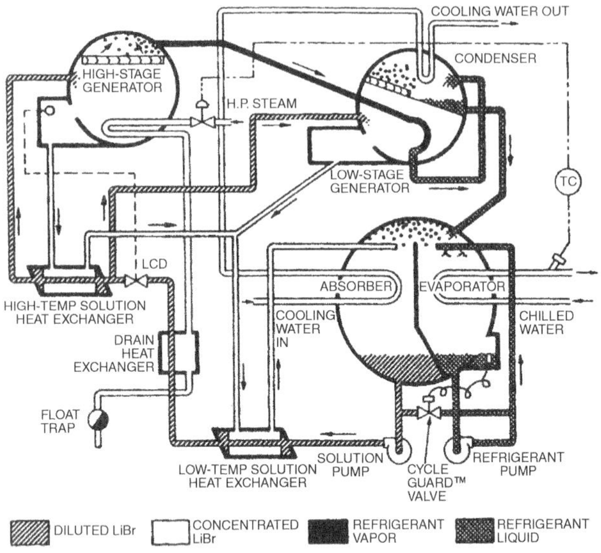

Figure 16.39 *Indirect-fired double-effect absorption cycle. Lithium bromide and water are used as refrigerants. (Courtesy of the Carrier Corporation.)*

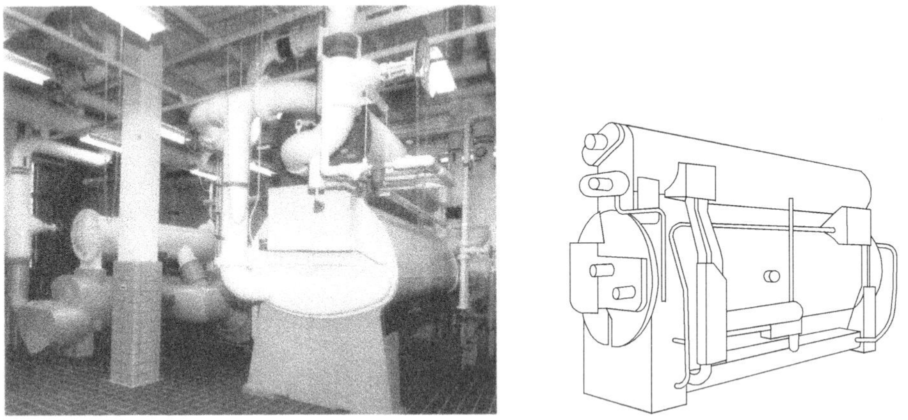

(*a*) (*b*)

Figure 16.40 *(a) An absorption chiller driven by heat. (b) Two-stage absorption chiller utilizing steam, producing 200–800 tons (700–2800 kW) of cooling. (a: Courtesy of Ingersoll-Rand, the Carrier Corporation; b: From Ramsey and Sleeper, 2016.)*

for vapor compression refrigeration cycle equipment, is a design consideration. Absorption refrigeration is almost always employed in a chiller configuration—to produce chilled water.

(c) Air-Cooled Condensers

Small-capacity refrigeration systems often employ an air-cooled condenser by which heat removed from a building is dumped to the outdoor environment. These types of condensers pass refrigerant gas through a coil over which outdoor air flows. The temperature of the refrigerant is higher than that of the outdoor air so that heat flows from the refrigerant to the air. This heat loss allows the refrigerant to change state or condense (from the gaseous phase to the liquid phase) while lowering its heat content.

The most common coil type is a fin-and-tube heat exchanger, otherwise known as a finned tube arrangement (a metal tube with an extended surface in the form of fins, usually placed behind a metal enclosure). The heat exchange is sensible (involving only temperature difference). Propeller fans are very common in small- to medium-capacity condensers and are employed to create a forced convection heat exchange (which accelerates the overall rate of heat exchange). Vertical and horizontal airflow patterns may be found in various equipment models. Airflow quantities of 600–1200 cfm/ton [80–160 L/s per kW] are common, and fan power requirements generally range from 0.1 to 0.2 hp/ton [20–40 W/kW]. Figures 16.92 and 16.110 show the role of an air-cooled condenser in a typical refrigeration cycle.

(d) Water-Cooled Condensers

Cooling Towers. Chillers often (but not always) employ a water-cooled condenser connected to a cooling tower to reject the heat that is removed from the chilled water system via condenser piping. The cooling tower receives warm water coming from the condenser and rejects it to the atmosphere. For a typical cooling tower, which may be categorized as an open-circuit cooling tower, this heat rejection is typically accomplished by use of evaporation by casting a water steam. This creates the steam plume that many visually associate with the towers.

The place of a cooling tower within the overall refrigeration system layout is shown in Figure 16.9*b*. More specific detail regarding cooling tower sizes and types is given in Figures 16.41 and 16.42. Cooling towers are designed to maximize the surface area contact between the outdoor air and the condenser water in order to maximize evaporation (and thus heat transfer). In cross-flow towers, fans move air horizontally through water droplets and wet layers of fill (or packing); in counterflow towers (prevalent in larger buildings), fans move the air up as the water moves down.

Cooling towers create a special—and unpleasant—microclimate. They move huge quantities of outdoor air (up to 300 cfm per ton [142 L/s per kW]), which they make considerably more humid. In cold weather, they can produce fog. They are typically noisy—a natural consequence of forced-air motion. The condensing water flows are about 2.8 gpm [0.18 L/s] per ton of vapor compression refrigeration and about 3.5 gpm [0.22 L/s] per ton of absorption refrigeration.

Between 1.6 and 2.0 gph [1.7 and 2.1 mL/s] of water leaves a typical cooling tower as vapor. This water must be replaced, which is done automatically—this is called makeup water. The steady evaporation and exposure to the outdoor environment under hot and humid conditions spell trouble for the condensing water: Controls for scaling, corrosion, and bacterial and algae growth are especially important. Ozone treatment systems have the advantage of reliable biological control and leave no chemical residue. Since the discovery of the link between Legionnaires' disease and cooling towers, biological control has assumed greater importance.

The water vapor that discharges from a cooling tower should be kept from the vicinity of fresh air intakes, and away from neighboring buildings or parked cars, where feasible. The footprint requirements for a cooling tower can be approximated from Figure 16.41. Alternatively, during schematic design, consider a typical space requirement of 1/500 of the building gross floor area (for towers up to 8 ft [2.4 m] high) or 1/400 of the building gross floor area for higher towers.

Although it is tempting to try to block the noise produced by a cooling tower with solid barriers, it is critical that noise control not interfere with air circulation. The manufacturer's recommended clearances to solid objects adjacent to cooling towers must be consulted before a tower is enclosed in any way. The roof is thus a favorite location for cooling towers, where the wind can disperse the water vapor, noise and odor are remote from the street, and aesthetics are less of a concern. A cooling tower can, however, become a design feature; near downtown Denver, the cooling tower for a large performing arts complex sits in a forlorn stretch of grass bordered by arterial streets and away from pedestrians (Figure 16.43). Its plume adds visual interest as it twists ghostlike above the equipment.

When fouling of the condensing water system cannot be tolerated, an alternative approach using a closed-circuit evaporative cooling tower may be considered. Its schematic operation is described in Figure 16.44. Usually used to cool refrigerant directly, it can also be used to cool condenser water, as well as being used in water-loop heat pump systems. The refrigerant or condenser water is protected within an always-closed loop, while a separate body of water is recirculated through the cooler, with steady evaporation and its attendant problems. This type of device requires much less makeup water than a cooling tower. Similarly, a closed-circuit air-cooled cooling tower may be used to conduct heat transfer with an always-closed loop through air-cooled heat exchangers, which completely eliminates the need for makeup water. However, there are notable cost and performance differentials between different cooling tower solutions.

(e) Heat Pumps

Discussion of heat pumps brings knowledge of the vapor compression refrigeration cycle and condensing equipment into active play. A heat pump uses the refrigeration cycle to both heat and cool; in an HVAC system it is thus both a heat source component and a coolth source component. As shown in Figure 16.45, heat is transferred ("pumped") from indoors to outdoors in summer (Figure 16.45*a*) and from outdoors to indoors in winter (Figure 16.45*b*). Heat pumps can be configured in air–air, air–water, and water–water arrangements, providing substantial flexibility for a building designer.

The outdoor "source" of heat exchange matters and serves as a basis to define heat pumps:

Air	When the units use outdoor air as the source of heat exchange, the units are termed air-source heat pumps (ASHP)
Water	If water is the source, then water-source heat pumps (WSHP)
Ground	If the ground is the source, then ground-source heat pumps (GSHP)

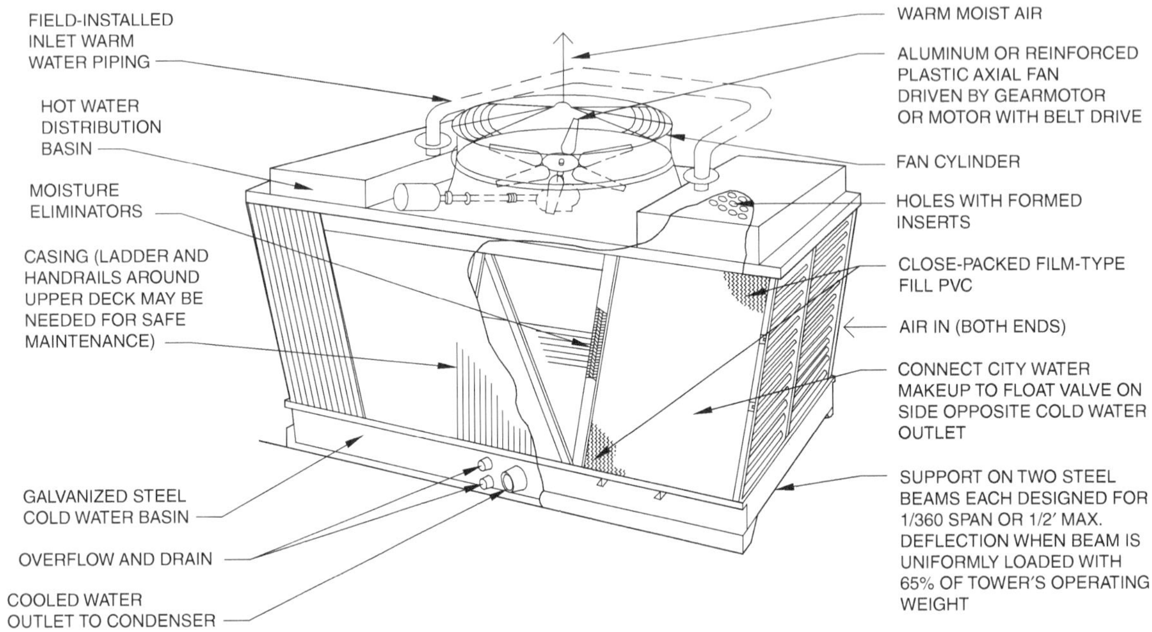

(*a*)

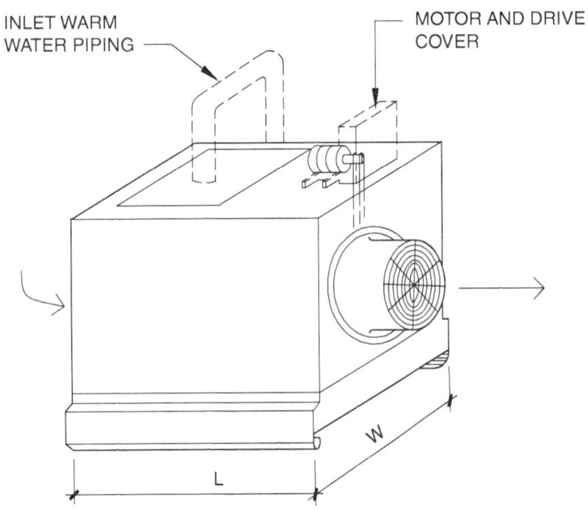

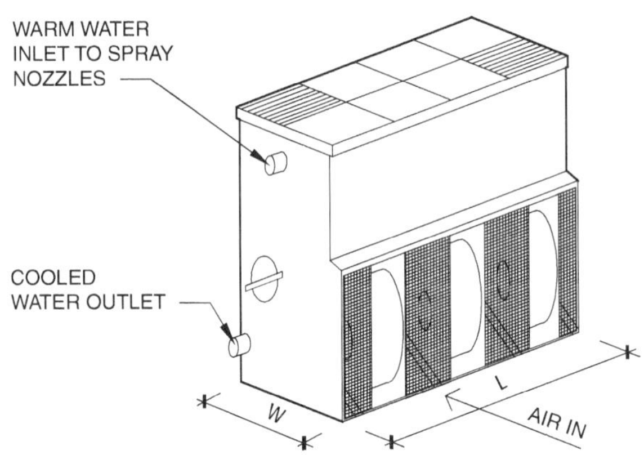

COOLING TONS	OVERALL DIMENSIONS (IN.) L	W	HT.	OPERATING WEIGHT (LB)	MOTOR (HP)
5	69	33	60	940	1/4
25	75	45	80	1600	1
50	84	64	92	2500	3
100	93	100	92	4200	5
150	100	144	112	8000	7-1/2

(*b*)

COOLING TONS	OVERALL DIMENSIONS (IN.) L	W	HT.	OPERATING WEIGHT (LB)	MOTOR (HP)
20	36	36	78	950	2
50	72	36	96	1700	7-1/2
150	144	56	122	4800	20
400	140	118	192	14,000	50

(*c*)

Figure 16.41 *Cooling towers for large buildings. (a) Cutaway view of a large-capacity (200–700 ton [700–2460 kW]) cross-flow induced-draft package cooling tower. (b) Size ranges for cross-flow induced-draft package cooling towers. (c) Size ranges for counterflow induced-draft package cooling towers. (From Ramsey and Sleeper, 2016.)*

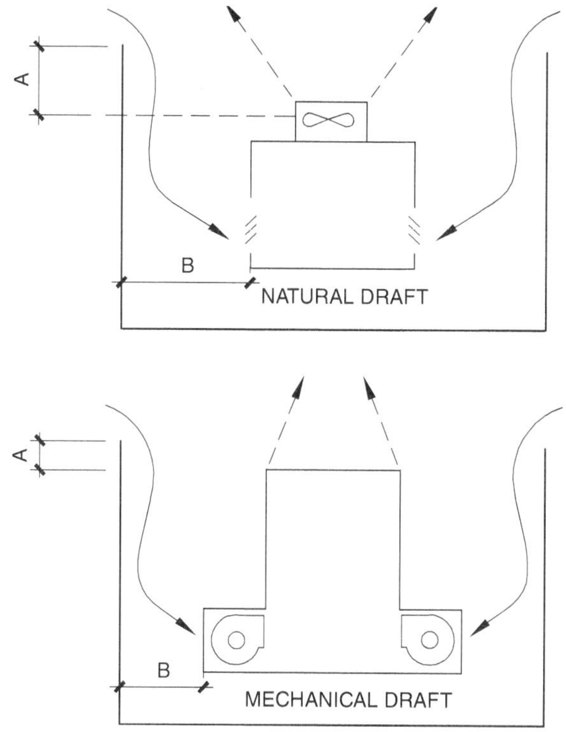

Figure 16.42 *The more clearance for airflow, the better the operation of a cooling tower. A = maximum height of enclosure above the tower outlet; minimize this dimension. B should be as large as possible, especially if the walls have no air openings. Consult manufacturers for preferred dimensions. (From Ramsey and Sleeper, 2016.)*

The ground is the most effective and stable heat exchange source. However, GSHPs are far more expensive to install than ASHPs due to the need to bury closed loop refrigerant lines in the ground. As expanded upon below, both WSHPs and GSHPs typically use closed-loop systems for heat exchange; however, WSHPs are also available in open-loop configurations, drawing the water source directly into the heat pump.

Single-Package Heat Pumps. In a single-package (also called unitary) system (Figure 16.46a), only one piece of equipment is involved. For instance, a single-package ASHP moves heat between an outdoor air stream and an indoor air stream; although kept separate, both streams pass through a single enclosure assembly. Typically applied to smaller buildings, the most common type of single package heat pump is an ASHP—an air–air heat pump—as shown in Figures 16.45 and 16.46. Single-package WSHPs and GSHPs are available.

Single-package heat pumps are commonly located on roofs, where they have ready access to outdoor air (the device's heat sink and heat source), and where their noise is less likely to annoy—provided they are adequately isolated from the building structure. This approach is shown in the daylit, passively solar-heated Mount Airy, North Carolina, library (Figure 16.47). This 14,000-ft^2 [1300-m^2] building also has a solar preheating system for its hot water. The five air–air heat pumps utilize economizer cycles (up to 100% outdoor air when temperatures are favorable). The average annual building energy consumption has been monitored at about 17,000 Btu/ft^2 [53,635 W/m^2]—approximately one-third of nearby buildings with similar functions.

Split System Heat Pumps. A heat pump with both outdoor and indoor components is called a split system. Again, the most common type of split system is an AHSP—an air-to-air heat pump (Figure 16.46b). A split system moves heat via a refrigerant loop between the outdoor unit (which also contains the compressor), and the indoor unit. The indoor unit has provisions for the treatment and circulation of indoor air and usually contains a backup heating coil for cold weather conditions that reduce the unit's heat-pumping capacity. WSHPs and GSHPs are available as split systems.

Split system heat pumps are popular because the noise of the compressor and the outdoor air fan are removed from the interior, and the size of the indoor unit can be quite small. As the distance between the indoor and outdoor units increases, so does the strain on the refrigerant loop. The indoor element is often mounted either high on the wall or on the ceiling. Indoor units are available with automatically changing louvers; when in the cooling mode, cool air is delivered along the ceiling, from where it sinks to the level of occupancy; cold air blowing directly on people is avoided. In the heating mode, the louvers shift to direct hot air steeply downward.

Coefficient of Performance (COP). One of the primary attractions of a heat pump is that in the heating mode it can deliver more energy than it consumes (as electricity). Although electricity is required to run the cycle, the heat pump draws "free" heat from a source such as outdoor air. The total heat delivered to the building is substantially more than the heat equivalent of the electricity required to run the cycle. The COP is a measure of this heat advantage, which was defined in Section 16.4. The COP varies by heat pump type:

- The lowest first-cost ASHP systems typically have COPs between 2.4 and 4.0.HS
- WSHP systems may have COPs between 3.0 and 5.0.
- The most expensive option, GSHPs, usually have the best performance with COPs ranging from 3.5 to 5.5.

Because performance for a given piece of equipment changes with varying outdoor conditions and with indoor loads, specific performance metrics have been developed to allow project teams to better assess and compare heat pumps (as well as other

Figure 16.43 A plume of mist hovers ghostlike above a cooling tower in full public view near the Denver performing arts complex.

air-conditioning equipment). These performance indicators are presented in Section 16.4(e), with the following additional comments.

- *Heating seasonal performance factor (HSPF2):* The testing procedure for determining HSPF has recently been revised, resulting in newer product data listing "HSPF2" ratings, which should be considered a more accurate metric to gauge the unit's heating efficiency. One should expect an HSPS2 rating of at least 6.7 for single-package heat pumps and 7.5 for split system heat pumps.
- *Seasonal energy efficiency ratio (SEER2):* Product data may list "SEER2." This is a revised version of the SEER metric, which accounts for updated testing procedures that better reflect real-world conditions. Because the updated testing conditions are more rigorous, a SEER2 rating is generally lower than an equivalent SEER rating for the same equipment. One should expect a SEER2 rating of at least 13.4 for

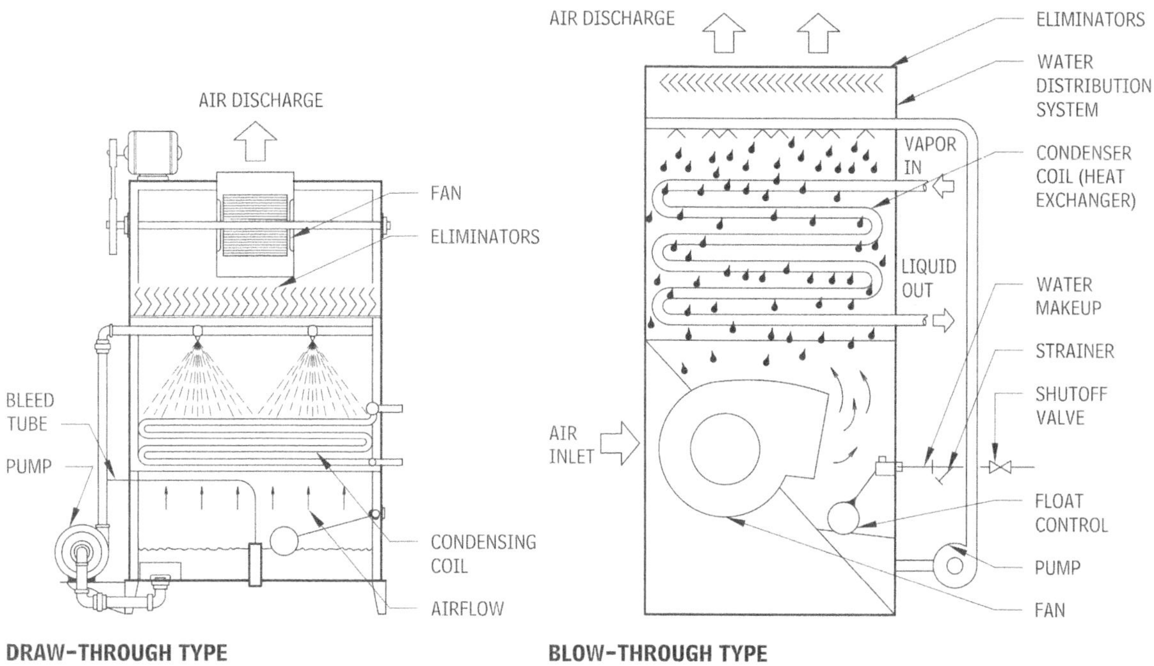

Figure 16.44 *Closed-circuit evaporative coolers cool a condensing water system while protecting it from contact with outdoor air. A self-contained water system is circulated through the evaporative cooler; steady evaporation losses are replaced by makeup water. (From Ramsey and Sleeper, 2016.)*

single-package heat pumps and 16.3 for split system heat pumps.

- *Energy efficiency ratio (EER2):* Product data may list "EER2." This is a revised version of the EER metric. Look for an EER2 rating of at least 13.4 on single-packaged units and 16.3 or higher on split systems.

Considering HSPS2, SEER2, and EER2 as a trio can help project teams establish and ensure minimum performance criteria. For example, the energy performance criteria for a WSHP split system may state:

$$\geq 7.8 \text{ HSPF2}/\geq 15.2 \text{ SEER2}/\geq 11.7 \text{ EER2}$$

A WSHP split system that has a HSPF2 rating of 8.0, a SEER2 rating of 16.0, and an EER2 rating of 12.0 would comply with these criteria.

As might be expected from a device that draws heat from the outdoor air in winter, there are limitations to a heat pump's heating performance. As outdoor temperatures approach 32°F [0°C], the COP drops, and the outdoor coil tends to ice over. Backup electric resistance heating must then be used, which ends the efficiency advantage that made the heat pump attractive. See Figure 16.48 for a demonstration of falling performance with falling temperatures. Because of this characteristic, air-to-air heat pumps are not commonly used in cold-winter climates. Air-to–air heat pumps are a questionable backup system for a passively solar-heated building in a cold climate, because backup sources are typically needed only in the coldest weather. The development of low-ambient (temperature) heat pumps is reducing the system's traditional cold weather disadvantage.

Heat pumps that transfer heat from water sources (such as wells or solar-heated storage tanks) or from the ground perform much more dependably and efficiently in cold weather compared to extracting heat from an air source. Water–air heat pumps (shown in Figure 16.45*c*) and solar collectors make an effective team. As a high-COP heat pump removes heat from the solar storage tank (for delivery to the indoor air), the resulting lower temperature of the solar-heated water increases the solar collector's performance. Assume that on a cold, partly sunny day, the collector is fed water from the tank at 90°F [32°C] and can raise its temperature by 4°F [2.2°C] to 94°F [35°C]. This improvement is slight because of the rather high heat loss that a 94°F [35°C] collector experiences when surrounded by cold air. If, however, the collector was fed 59°F [15°C] water, its heat loss would be greatly reduced. The heat that the collector does not lose to the cold air can be invested in the 59°F [15°C] water, which will thus leave at a considerably higher temperature of $59 + 4 = 63$°F [$15 + 2.2 = 17.2$°C]. Thus, more solar energy is collected, and more is available for transfer to the building via the heat pump. The heat pump, in turn, can heat the 59°F [15°C] tank water to much higher temperatures to serve the building's heating needs.

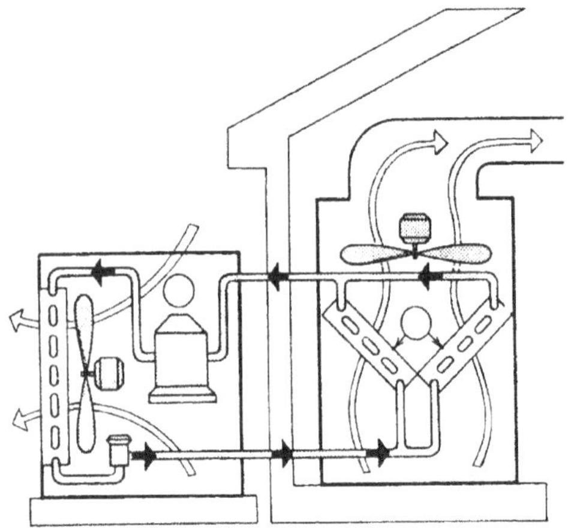

(a)

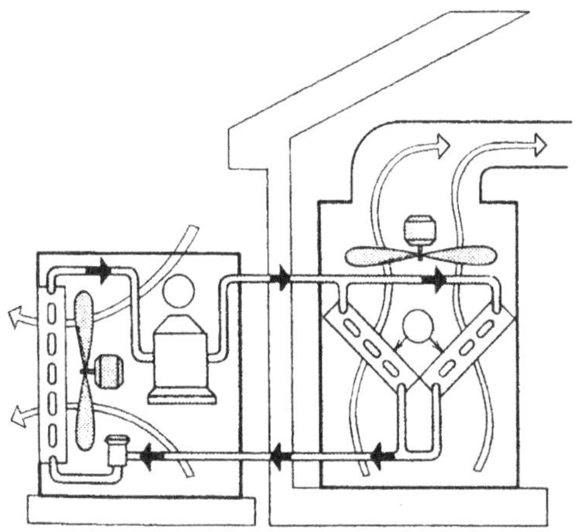

(b)

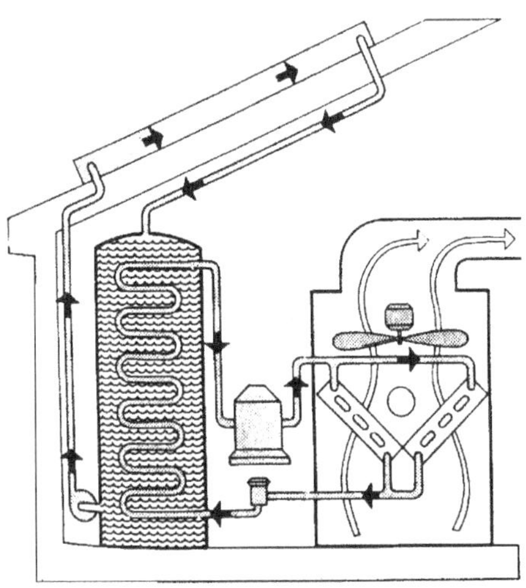

(c)

Figure 16.45 *Heat pump applications. A simple air–air heat pump provides space cooling (a) or space heating (b). Teamed with a solar collector and water storage tank (c), a heat pump can deliver usefully warm temperatures in the air stream while increasing the collector efficiency by lowering the storage tank temperature. (Popular Science; 1978 / Times Mirror Magazines, Inc.)*

Most heat pumps utilize electricity to drive the vapor compression refrigeration cycle; there are, however, absorption cycle heat pumps that utilize natural gas. The GAX heat pump (championed for some time by Oak Ridge National Laboratory and shown in Figure 16.49) not only uses natural gas for heating and cooling, it has no CFCs or HCFCs and functions at outdoor temperatures much lower than those of an electric air–air heat pump. Eventually, solar energy may be used to drive the absorption cycle. Solar-driven refrigeration is a particularly elegant blend of energy source and task—the hotter the sun, the greater the cooling capacity.

Geothermal Versus Geo-Exchange Systems. Ground source heat pumps (GSHPs) are often—but inaccurately—referred to as "geothermal" heat pumps. Geothermal systems extract the natural thermal energy (i.e., heat) stored deep within the Earth; while ground-source heat exchange, sometimes referred to as geo-exchange, is a specific technology that uses a heat pump to transfer (or "exchange") thermal energy from/to the ground nearer to the surface of the Earth to heat and/or cool a building.

GSHPs offer versatility through several configurations. They are often used to provide domestic hot water (DHW) in addition to heating and cooling ventilation air. An environmentally safe heat exchange fluid is circulated through a loop installed underground (or in a pond or lake), taking heat from the soil in winter and discharging heat to the soil in summer. The loop is often constructed of high-density polyethylene (HDPE). Soil temperatures below the surface are more stable year-round than outdoor air temperatures; tapping into this more benign environment raises the system COP relative to that of air–air heat pumps. The outdoor portion of this system is almost completely out of sight, with no maintenance or weathering of exterior equipment. Noise is confined to the compressor, which can be placed in a small indoor mechanical room.

Some common GSHP configurations are shown in Figure 16.50. In the closed systems (Figure 16.50a–c), the flow rate is typically 2–3 gpm/ton [0.3–0.5 mL/J] of refrigeration, with lower flows in the open-loop systems.

Horizontal closed-loop GSHP: This solution (Figure 16.50*a*) requires trenches 3–6 ft [1–2 m] deep; typically, 400–650 ft [120–200 m] of pipe are installed per ton (12,000 Btu/h or 3.5 kW) of heating and cooling capacity. To squeeze more pipe length into a trench, a "slinky" coiled pipe is sometimes used. The trenches can be placed below parking lots or lawns and gardens.

(a)

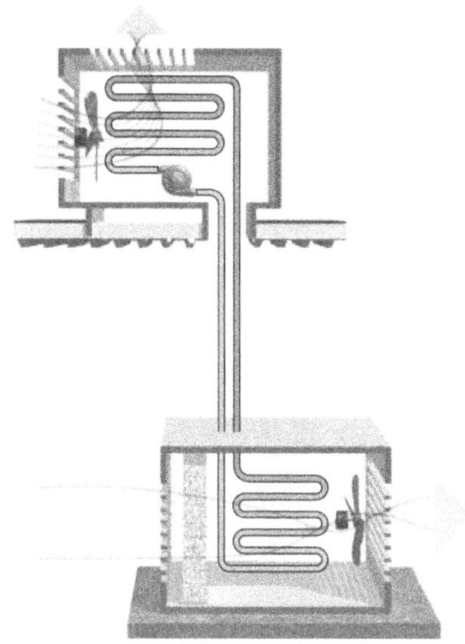

(b)

Figure 16.46 *The package unit (a) and the split system (b) are popular applications of air–air heat pumps. The package unit and the outdoor unit of the split system are typically placed either on the ground or on a roof (as shown here).*

Vertical closed-loop GSHP: This solution (Figure 16.50*b*) is particularly applicable where the site area is limited. Vertical holes are bored from 150 to 450 ft [46 to 137 m] deep. Each hole contains a single full-depth loop and is backfilled (or grouted) after the loop is installed. Because the temperature is more stable at greater depths, less pipe length is required than for horizontal loops. The distance between boreholes varies from a minimum of 15 ft [4.6 m] with high water table and low building cooling loads to as much as 25 ft [7.6 m] for buildings with high cooling loads. A minimum distance of 20 ft [6 m] is usually recommended.

Closed-loop WSHP: A pond- or lake-source closed-loop WSHP (Figure 16.50*c*) is sometimes used when a building is close to an adequately large body of water. The loop is submerged, and the surrounding water conducts heat far more rapidly than does soil. The resulting shorter loop length, and the low cost of placing the coils in water, can make this attractive. However, the water level in the pond should never drop below a minimum of 8 ft [2.5 m] and must have sufficient surface area for heat exchange.

Open-loop WSHP: The groundwater-source open-loop WSHP ("pump and dump") is suitable only where groundwater is plentiful; it may be prohibited by local codes and environmental regulations. One variant of this system (Figure 16.50*d*) takes water from one well, passes it through a heat exchanger within the building, then discharges it to a second well. Another variant (Figure 16.50*e*) takes water from the bottom and discharges back into the top of the same (standing) well, typically 6 in. [150 mm] in diameter and as deep as 1500 ft [460 m].

The Wildlife Center of Virginia at Waynesboro is a teaching–research hospital. Its 5700-ft^2 [530-m^2] floor area is served by four GSHPs (two at 4 tons, two at 5 tons) connected to 11,350 ft [3460 m] of underground horizontal pipe, laid in a "slinky" configuration and thus fitting within about 2500 ft [760 m] of trench. The trench was dug around existing trees and placed under future roadways in a forest setting. Energy simulations estimated that annual heating, cooling, and hot water use would be about 35,000 kWh. Had air–to-air heat pumps and an electric water heater been used instead, the estimate was 66,000 kWh—resulting in a 47% yearly savings advantage for the ground-source system.

GSHPs are often used in retrofits, especially for schools where site areas are plentiful, or historic structures where small mechanical systems are highly desirable. The Daniel Boone High School near Johnson City, Tennessee, was built in 1971 with a two-pipe chilled water system and electric resistance heat. An ASHRAE Technology Award was given to recognize this 160,000-ft^2 [14,864-m^2] school, which was retrofitted with a ground-source vertical closed-loop heat exchanger fed by 320 boreholes that are 150 ft [46 m] deep. Each borehole loop is 300 ft [91 m] of ¾-in. [19-mm] polyethylene pipe. The boreholes are arranged in sections of 20 holes each. The holes are

(a)

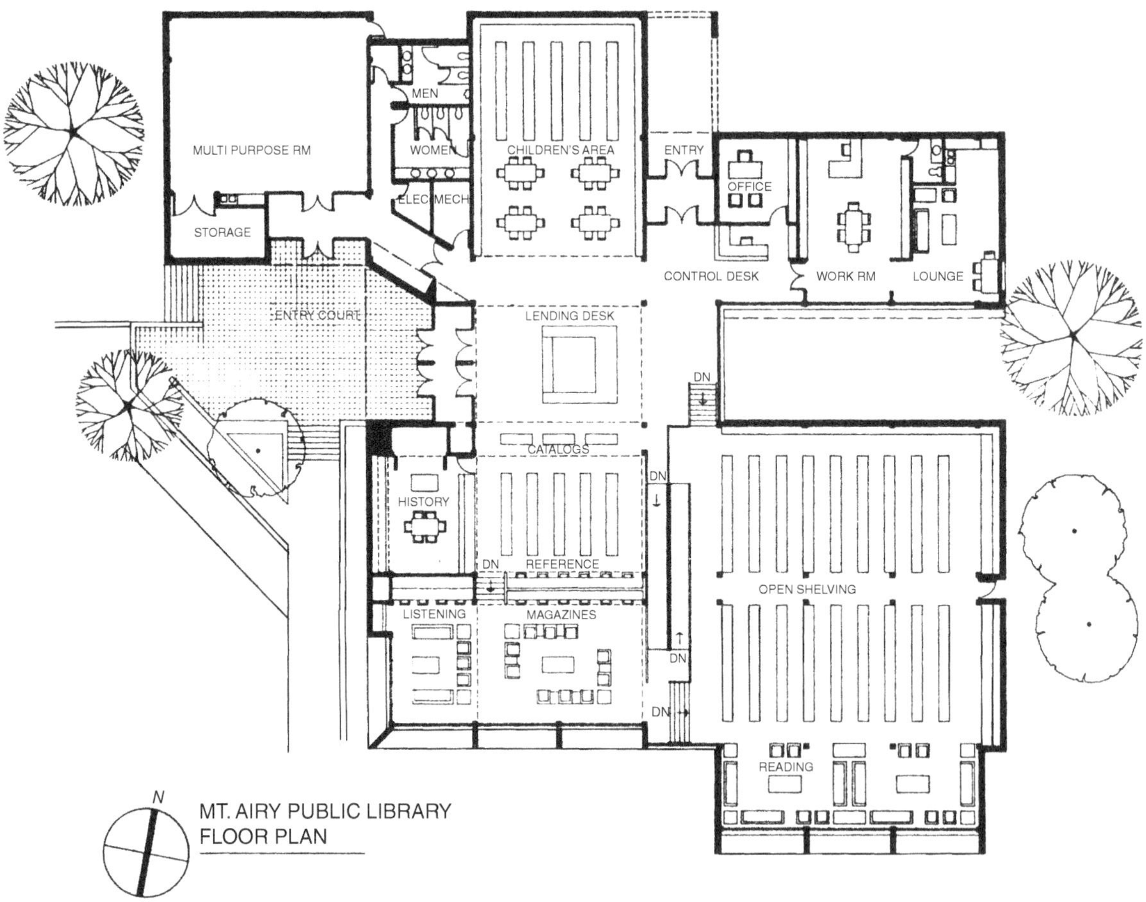

(b)

Figure 16.47 *Mount Airy Library, Mount Airy, North Carolina. Mazria/Schiff & Associates, Architects. (a) View from the southwest. (Photo by Gordon H. Schenck, Jr.) (b) Plan. (c, d) North–south sections relating sunlight and natural heat flow. Individual package heat pumps are set on the flat-roof sections.*

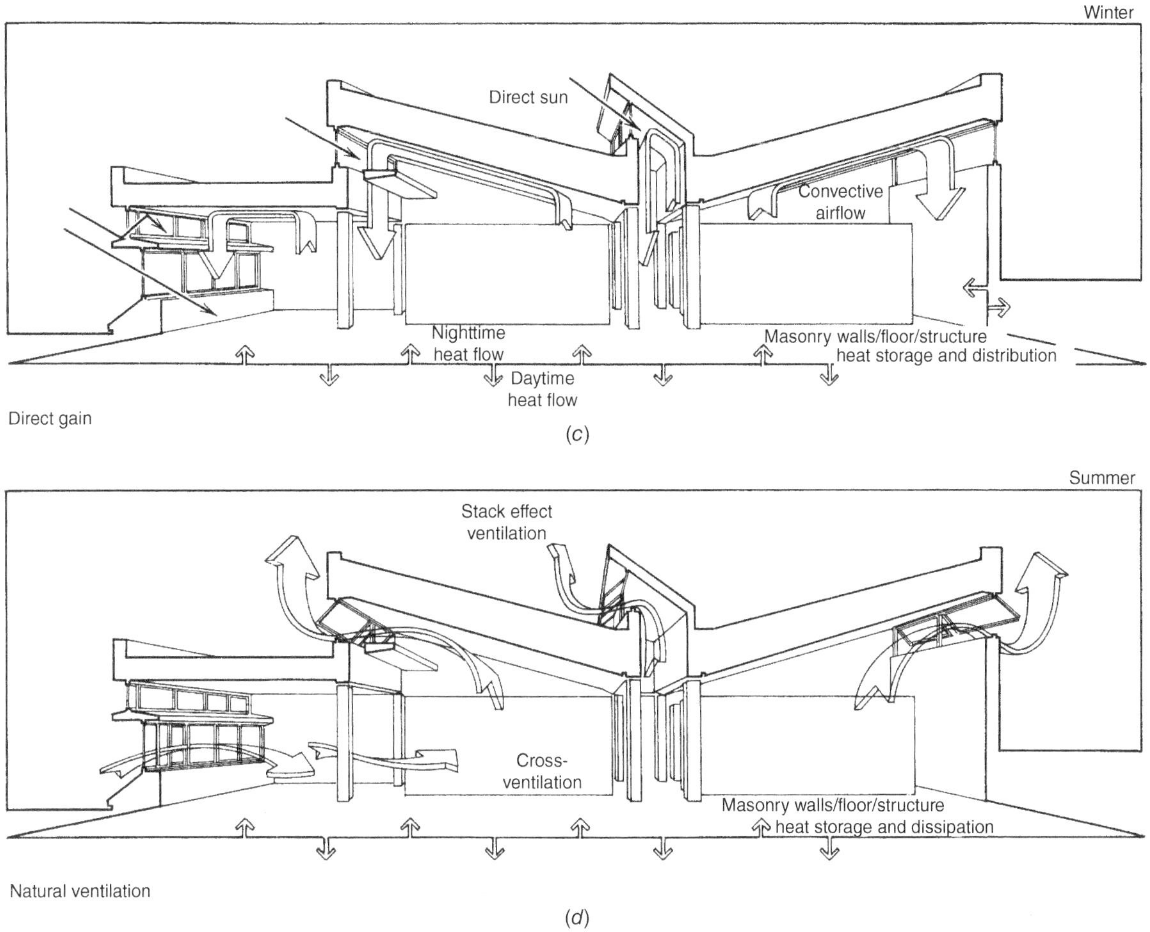

Figure 16.47 *(continued)*

15 ft [5 m] on center. The 20-hole sections are separated by 20 ft [6 m]. The loops all connect to an 8-in. [203-mm] supply and same-size return line to a new heat exchanger within the existing mechanical room. Within the building, a new water loop heat pump system is installed, one heat pump in each zone.

The Daniel Boone High School uses an interior closed water loop that connects all the heat pumps serving all thermal zones. This system facilitates heating in one zone while another zone is being cooled, because simultaneous heat "deposits" and heat "withdrawals" actually help the system to function most efficiently. Figure 16.51 diagrams a typical water loop system, where a supplementary heat source (such as a boiler for cold weather) and a supplementary heat rejector (such as a cooling tower for hot weather), are installed to maintain usable water temperatures within the loop across an annual cycle. A large office building in Pittsburgh that used such a system is shown in Figure 16.52.

Water loop heat pumps: This is a special case application of WSHPs. Interconnected heat pumps (water-to-air) draw heat from a common water circulation loop (in heating mode) or discharge heat to the loop (when in cooling mode). For a large building in cold weather, heat removed from the interior zones is used to warm the perimeter zones. The loop's temperature may range between 65°F and 90°F [18°C and 32°C]. In hot weather, a cooling tower disposes of excess heat (not required for heating any zone), whereas in cold weather a central boiler adds needed heat to the water loop. The loop is sized to carry 2–3 gpm [0.13–0.19 L/s] per ton, where the total tonnage equals the sum of the capacity of all the individual units (often greater than the actual load).

This loop approach is diagrammed in Figure 16.51. In the 175,000-ft^2 [16,260-m^2] Comstock Center in Pittsburgh (Figure 16.52), there were six to eight small heat pumps on each of the 10 floors; these units were located above the suspended ceiling. Their connecting water loop doubled as the building's wet-pipe sprinkler system supply; this was possible because the heat pumps kept the loop between

ACTIVE ENVIRONMENTAL SYSTEMS

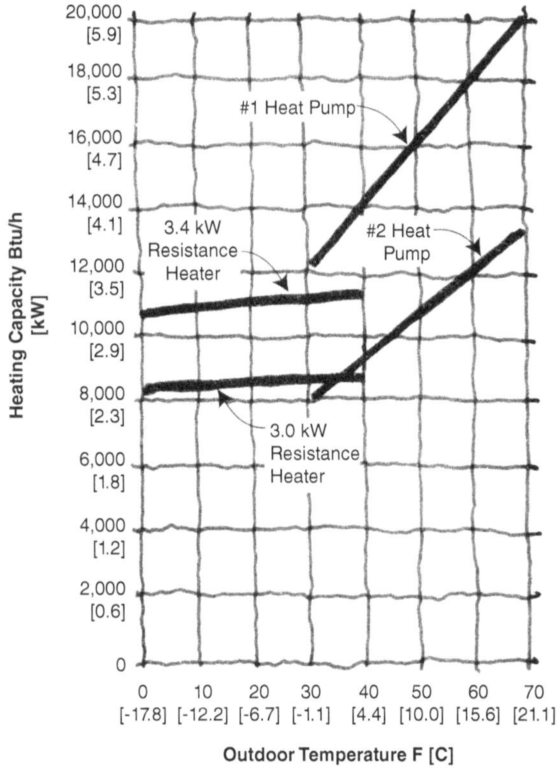

Figure 16.48 Typical air–air heat pump operating characteristics. Heat delivery falls with falling outdoor temperature. (Redrawn by Jonathan Meendering.)

65°F and 85°F [18°C and 29°C]. Because no additional water distribution piping was needed (beyond the sprinkler piping), there was a substantial first-cost saving, and relatively little building volume was consumed for HVAC distribution elements. A 23,000-gal [87,060-L] water storage tank allowed daytime heat, rejected to the loop by the heat pumps, to be stored and recalled for nighttime heating. When necessary, two small boilers maintained the 65°F minimum temperature in the water loop or an evaporative condenser held the maximum temperature at 85°F.

In addition to the innovative use of the sprinkler piping loop, the Comstock Center used air-extract windows (see Figure 16.59) to control infiltration and to moderate the perimeter zone temperatures. After return air passed up inside this window and arrived in the ceiling plenum, it was mixed with fresh air, then tempered and recirculated by the air side of the heat pumps. The Comstock Center also utilized a large daylighting atrium, whose temperature control was provided largely by exhaust air from the offices; the stack effect was utilized to provide natural ventilation on its west face in summer conditions.

Hotels and motels are often good candidates for water loop heat pumps, because some rooms receive solar radiation while others are shaded; some are occupied, some unoccupied; and there are substantial domestic hot water needs. All these factors combine to make a heat-sharing water loop attractive.

(f) Evaporative Coolers

The third currently-viable active source of coolth is evaporative cooling. This approach to cooling involves a basic psychrometric process (see Section 16.4). Evaporative cooling may be achieved through passive means, via a hybrid system (mostly passive with a bit of

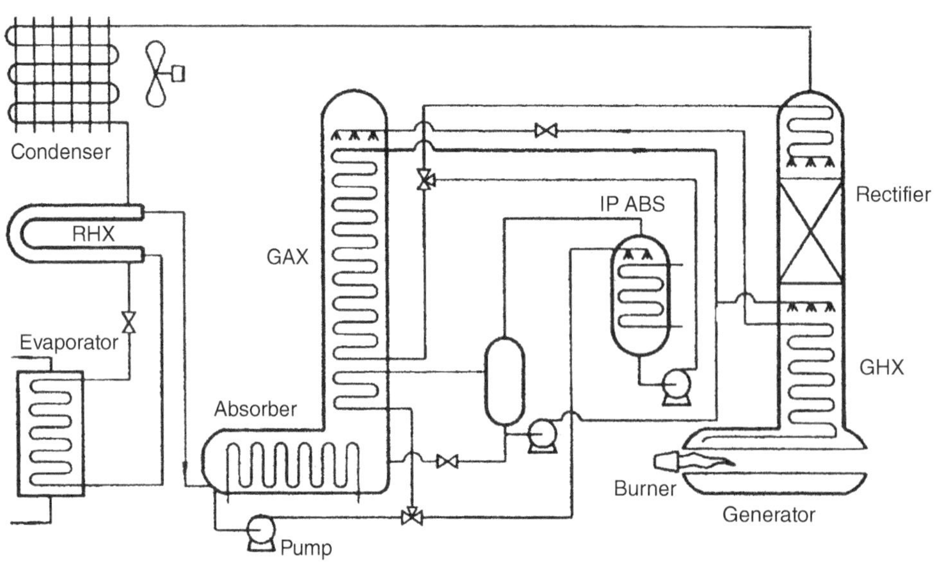

Figure 16.49 An air-cooled advanced GAX (generator-absorber heat exchange) heat pump uses aqua ammonia as the absorbent and is operational down to −10°F [−23°C]. The target cooling COP is 0.95; heating COP is 1.55. Designed for light commercial applications, it ranges from 5 to 25 tons [17.6 to 87.9 kW] capacity. (RHX = refrigerant heat exchanger; IP ABS = intermediate pressure absorber; GHX generator heat exchanger.) (Courtesy of Energy Concepts Company.)

(*a*) Horizontal ground source closed-loop heat pump

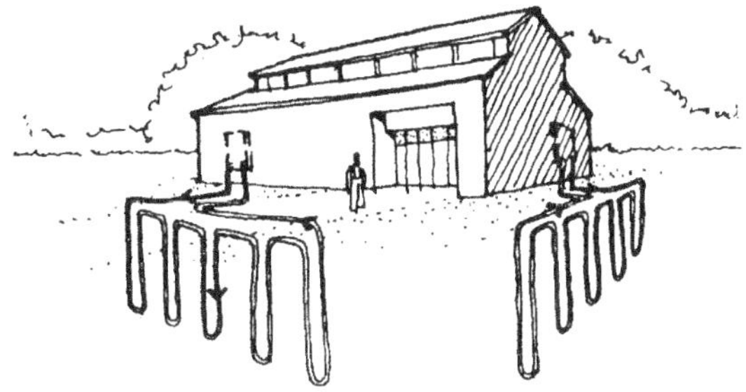

(*b*) Vertical ground source closed-loop heat pump

(*c*) Pond or lake closed-loop heat pump

(*d*) Groundwater-source heat pump

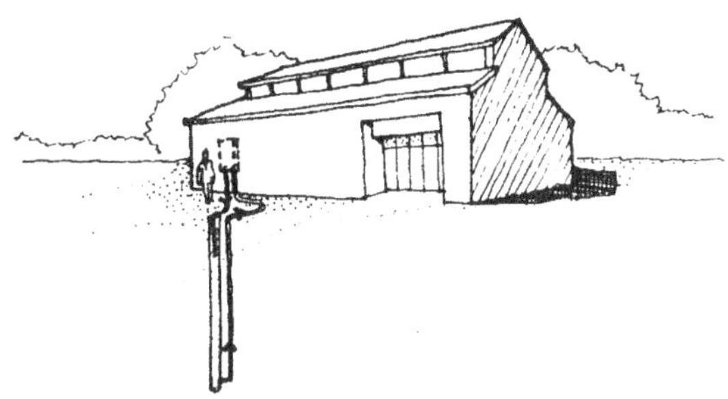

(*e*) Standing column groundwater-source heat pump

Figure 16.50 *Configurations of ground- and water-source heat pumps. (a) Horizontal closed loop GSHP laid in trenches. (b) Vertical closed loop GSHP placed in boreholes. (c) Pond- or lake-source closed-loop WSHP. (d) Groundwater-source open-loop WSHP taking water from one well and discharging it to another. (e) Standing column groundwater-source heat pump taking water from, then discharging to, the same well.*

help from active components), or by means of an active system. Active evaporative cooling employs equipment called an evaporative cooler. A range of evaporative cooling systems is discussed in Section 16.18.

(g) Dehumidifiers and Humidifiers

A reminder that humidity affects thermal comfort and indoor air quality. Humidity is thus a factor in occupant health, wellness, and satisfaction. For a sedentary person, a 30% change in relative humidity (RH) will produce about the same comfort sensation as a 2°F [1°C] change in temperature. Higher humidity adds to hot air discomfort, making evaporation more difficult, increasing skin wettedness, and increasing friction between skin and clothing or furniture surfaces. As RH exceeds 60%, problems with IAQ increase due to mold and mildew growth. Lower humidity produces cooler and drier sensations, but too-low humidity can irritate the skin.

For spaces that need only dehumidification (latent cooling) rather than conventional (sensible and latent) cooling, refrigerant dehumidifiers are commercially available (Figure 16.53). Their advantage over desiccant dehumidifiers is that the air temperature remains essentially unchanged during dehumidification (desiccant dehumidifiers raise the temperature of the dried air). These active devices consume energy to run the refrigeration cycle and this energy is added, as waste heat, to the space. Accumulated condensed water must periodically be removed from these units and, if untended, could become a source of disease. Most

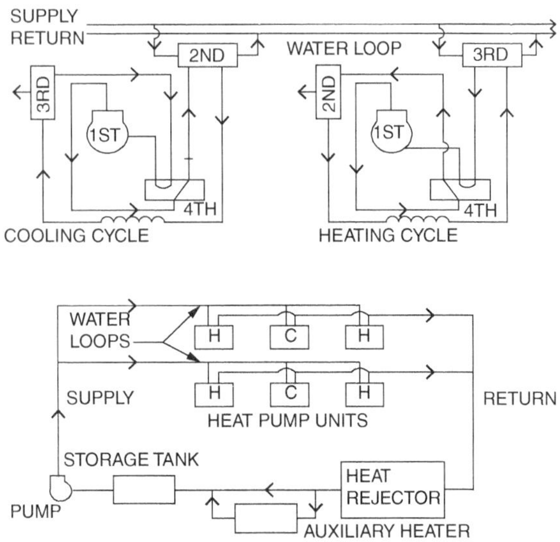

Figure 16.51 *Water-source heat pumps. (a) An individual water–air heat pump either deposits heat into the loop (while cooling) or withdraws heat (while heating). This system is particularly well suited to buildings where simultaneous heating and cooling needs occur. (b) Supplementary heat sources (boilers) and heat rejectors (cooling towers) are usually provided. (From Ramsey and Sleeper, 2016.)*

refrigerant dehumidifiers encounter operating difficulties at air temperatures below 65°F [18°C], at which point frost forms on their cooling coils. This could cause problems in a tightly enclosed residence in winter.

Task humidifiers are widely available and are often used to relieve symptoms of respiratory illnesses. Again, the problem of bacterial and mold growth in water reservoirs is of concern; some products add an ultraviolet (UV) lamp to counteract this threat.

(h) Desiccant Dehumidifiers

Desiccant dehumidifying systems (Figure 16.54) are potentially attractive because they use no refrigerants (and thus have low-global-warming potential) and reduce humidity without overcooling or heating the air being conditioned. Desiccants (such as silica gel, activated alumina, or synthetic polymers) in an active system must be heated to drive out the moisture they remove from the incoming air. This process is called regeneration and at present natural gas is typically used to regenerate, but solar energy is a promising substitute, given its plentiful availability. Waste heat from other mechanical systems may also be used. In a passive desiccant system, heat from a building's exhaust air is enough to release and vent the moisture removed from incoming air. Passive desiccant systems, however, are not quite ready for the marketplace.

Research on materials suitable for desiccant cooling and solar-driven regeneration should produce improvements in the types available as well as their performance. For an extended discussion of both solid- and liquid-based desiccant systems, see Chapter 24 of the *ASHRAE Handbook—HVAC Systems and Equipment* (ASHRAE, 2024).

16.9 DISTRIBUTION COMPONENTS: AIR

Distribution components connect source components to delivery components in a central HVAC system. (Local systems are close-coupled and have little need for distribution; district systems function as central systems at the building scale.) The building designer faces a fundamental decision regarding the medium that will be used to convey heat and/or coolth effect around a building. Should this medium be air or water? Both will work, both have been used effectively in the past, and both are competing for attention in high-performance buildings—see, for example, the discussion producing an *ASHRAE Journal* article comparing VAV (air) to chilled beam (water) systems (Stein and Taylor, 2013). Air can be directly introduced into a room, can inherently address indoor air quality concerns, but has very low thermal capacity—requiring large conduits (ducts). Water has the opposite characteristics; it is thermally efficient—permitting small conduits (pipes)—but cannot be allowed to flow directly into a room, and does not inherently deal with air quality.

Examples of air-based and water-based systems (including hybrid air–water systems) are discussed in Sections 16.18 and 16.19. This section looks at the equipment artifacts associated with air as a distribution medium.

(a) Fans

The purpose of a fan is to circulate (move) air. Buildings may employ stand-alone fans (such as a ceiling fan or industrial exhaust fan), fans that are packaged as a product (such as residential-scale bathroom exhaust fans), fans that are inserted into ductwork (such as an exhaust or return air fan), and fans that are part of a larger equipment item (such as an air-handling unit). Fans may be found in hybrid climate control systems and most certainly will be found in most HVAC systems.

Ceiling Fans. Before the advent of HVAC systems, cooling effect was commonly achieved with simple air motion provided by fans. The summer thermal comfort chart shown in Chapter 6, Figure 6.18 encourages increased air motion as a way to extend comfort into air temperatures in the mid-80s °F [roughly 29°C]. As a general rule, people will perceive a 1°F decrease in air temperature for every 15 fpm increase in air

(a)

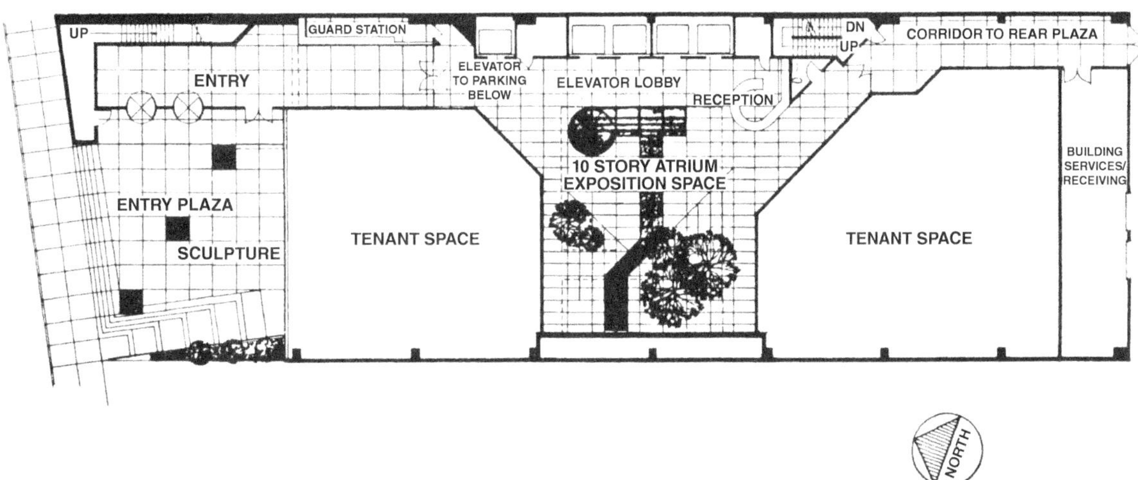

(b)

Figure 16.52 *The Comstock Center, Pittsburgh, Burt Hill Kosar Rittelmann Associates, architects. (a) The north and west faces. (b) Ground-floor plan. (c) Water loop heat pumps were linked via the sprinkler system on each floor. This may also be called a tri-water system. (d) Extract-air windows (see Figure 16.59) controlled infiltration and moderated the perimeter zone temperatures. (e) The central daylighting atrium was tempered by exhaust air from the offices; exhaust air then left via the stack effect. (f) The stack effect also controlled summer overheating.*

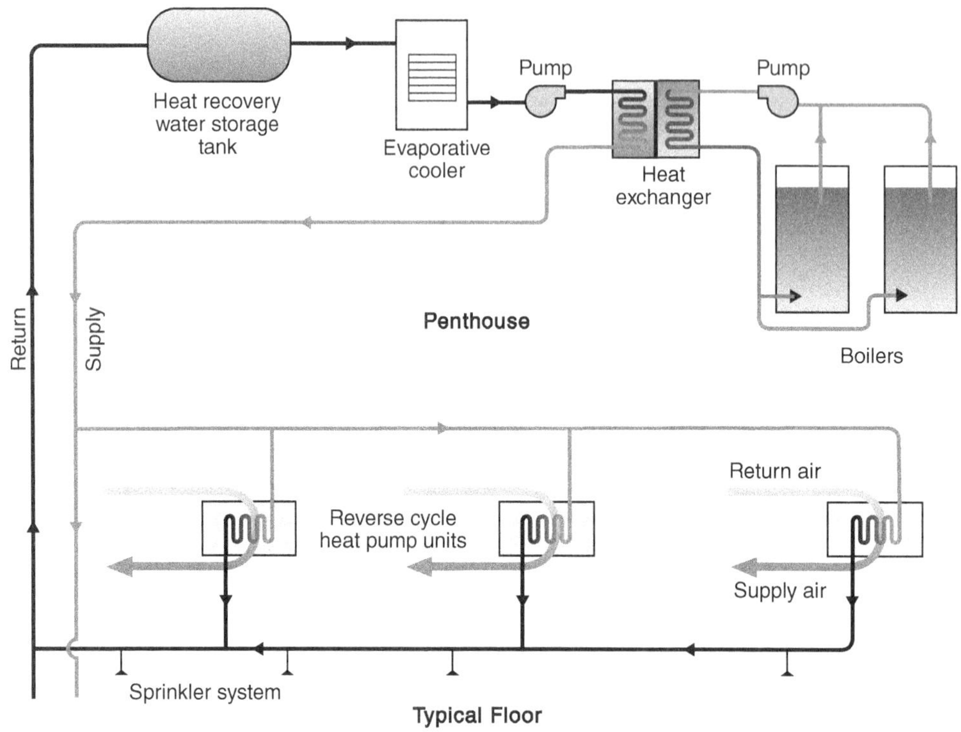

(c)

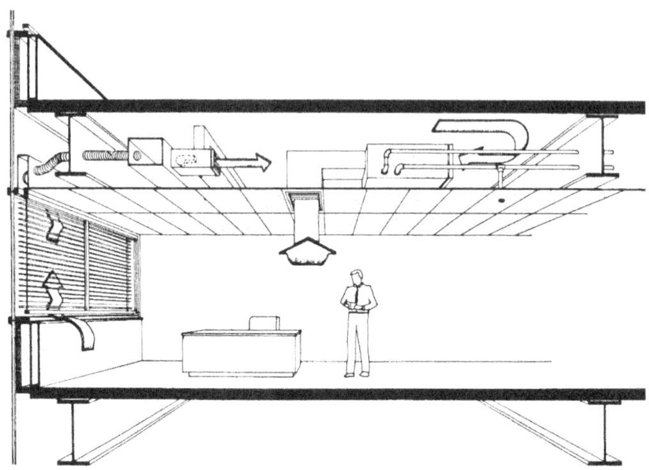

(d)

Figure 16.52 *(continued)*

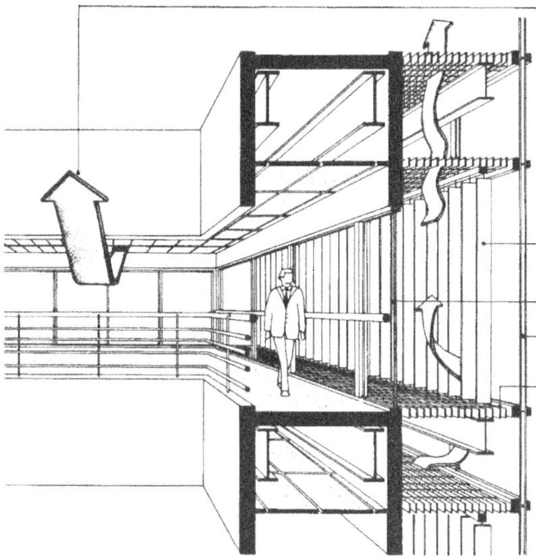

(*e*)

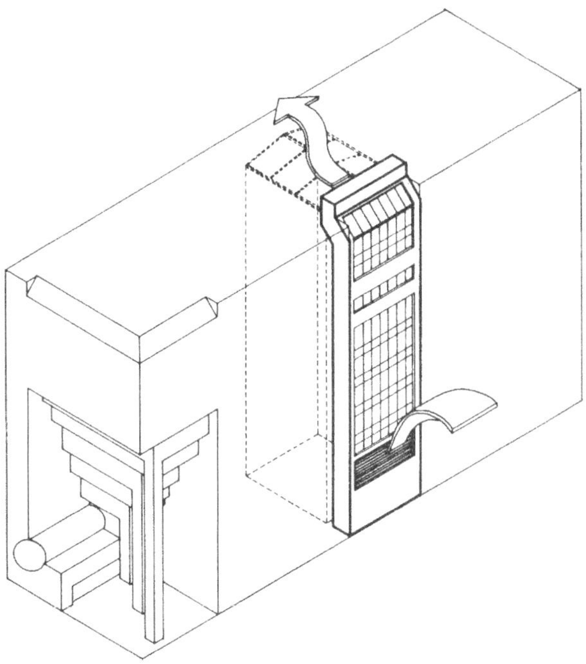

(*f*)

Figure 16.52 *(continued)*

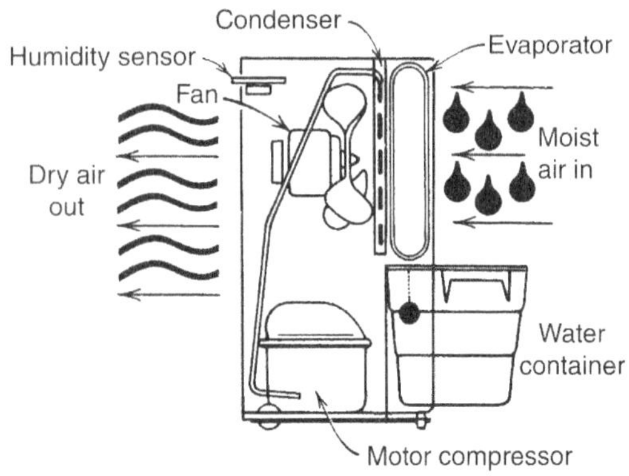

Typical Performance Data

Watts	290	450	465	630
Pints per day[a]	14	20	25	30
Unit dimensions (in.)				
Height	20 1/2	20 1/2	21 1/2	21 1/2
Width, 11 3/4				
Depth, 16 3/4				
"Average" room size served	20′ × 32′	35′ × 40′	40′ × 48′	40′ × 60′

[a]For air at 80°F, 60% RH.

Figure 16.53 *Refrigerant dehumidifiers typically are installed as free-standing units in smaller buildings.*

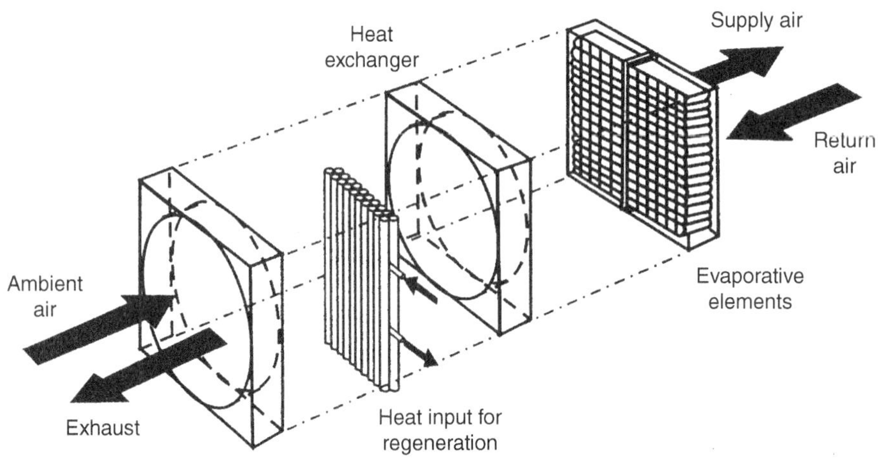

Figure 16.54 *Desiccant dehumidification typically utilizes a rotating wheel impregnated with either silica gel or another water-absorptive material. The wheel removes moisture from incoming outdoor air and transfers some of it to the outgoing exhaust air stream. The remaining moisture must be driven from the wheel by a heat source. (National Renewable Energy Laboratory.)*

speed at the body [about a 1°C decrease for every 1 m/s increase]. Ceiling fans are often installed to run at slow speed to destratify warm air at the ceiling in winter. They can be run at higher speed in summer to provide added comfort through increased air motion. The air motion produced by ceiling fans will vary with the fan height above the floor, the number of fans in a space, and the fan power, speed, and blade size. Figure 16.55 shows expected air speeds with one 48-in. [1220-mm] ceiling fan in a typical residential living room.

Whole-House Fans. As the name implies, these types of fans are used to extract warm indoor air through the attic or roof of a building (typically smaller structures, particularly housing units), which then pulls cooler air from open windows and other openings. When used effectively, whole house fans circulate air in an entire home. Certain climate conditions are more favorable for whole house fans. Typically, they are used at night or early in the morning. Whole house fans are most common in mild to moderate climates where the outdoor air temperature drops below the indoor air temperature by several degrees. However, whole house fans can be effective in more severe climates when effectively integrated. In the hot summer climate

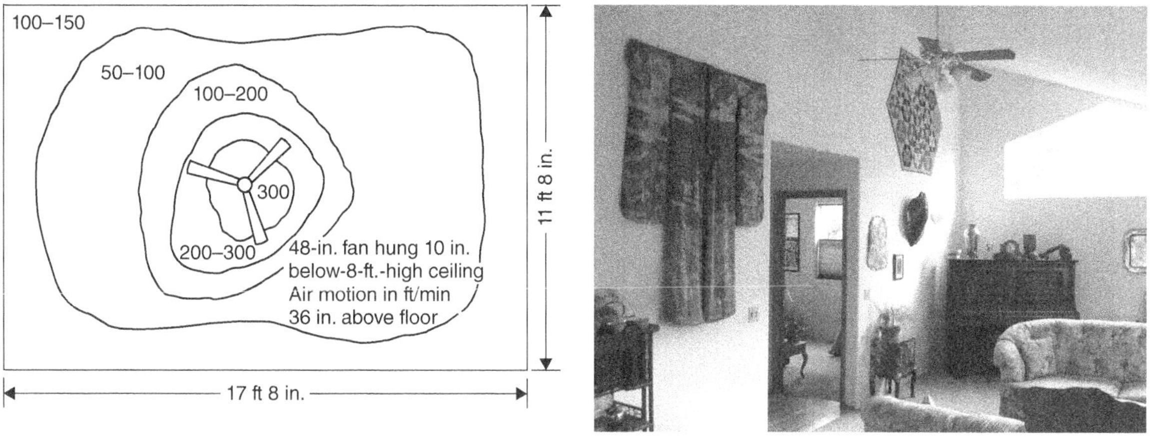

(a) (b)

Figure 16.55 *Ceiling fans are useful for both heating and cooling. In winter, at slow speeds, they destratify warm air at the ceiling. In summer (a) they extend the comfort zone by providing increased air motion. The illustrated room size is typical of residential living rooms. (b) A ceiling fan is a visual feature in this Oregon residence. (a: With permission of ASHRAE, 2011 (www.ashrae.org). This citation to an older version of the Handbook is intentional and provides access to historic reference information of ongoing interest; b: Photo by Nathan Majeski.)*

of Davis, California, an experimental house (sponsored by Pacific Gas & Electric and designed by the Rocky Mountain Institute) eliminated a conventional cooling system through a series of alternative approaches. Fans play a major role: In addition to ceiling fans, a whole-house fan removes the hottest air from the central hallway, exhausting to the ventilated attic. Increased thermal mass (tile floors, double drywall), an attic radiant barrier, and low-ε, gas-filled windows are also used.

Other Fans. Manufacturers of packaged HVAC equipment will normally select a fan type and capacity that best suit their product objectives. HVAC engineers, however, will need to select stand-alone fans and may be able to select from among fan options within packaged equipment selections. Key operational parameters are airflow and static pressure; for a given fan product, these are related through a fan curve. Analysis of fan curves allows a designer to ensure the selection of a fan that will perform well across the range of conditions it will likely encounter in use.

The Fan Affinity Laws. Designers should observe the fan affinity laws, which can be simplified as follows:

- Fan airflow is directly proportional to air speed.
- Static pressure is proportional to the fan speed squared.
- Fan energy consumption is proportional to the fan speed cubed.

These laws show that fan power is related to fan airflow via a cubic power relationship. Reductions in airflow result in substantial reductions in power draw—this effect is important to consider in high-performance buildings.

Fan types likely to be encountered in HVAC systems include:

Panel (or propeller): A basic air-moving device. The motor is generally mounted in the center of the propellers. It is not designed to work against substantial static pressure (as will result from pushing air through ductwork, filters, coils). This is a noisy type of fan, most commonly used for the V (ventilation) in HVAC.

Centrifugal: Centrifugal fans are the workhorses of HVAC system air moving. They are often used in air-handling units. With an airfoil blade wheel, a centrifugal fan has high efficiency over a wide operating range and is reasonably quiet. A major change in pressure can result from only minor changes in the volume of airflow.

Fixed pitch vane axial: Common for industrial applications. This fan is capable of working against somewhat more static pressure than a panel fan.

Vane axial adjustable pitch: The blade pitch is adjustable occasionally (not continuously) to provide for system balancing or seasonal changes in airflow volume. This fan requires less space than a centrifugal fan and can work against higher static pressures.

Vane axial controllable pitch: The automatically controlled (real-time) pitch adjustment can respond to changes in system temperature, airflow, pressure, and so on, depending upon the sensors employed. This fan is commonly used in VAV (variable air volume) systems.

Variable Frequency Drives (VFDs). These are electronic devices used to control the speed of certain electric motors in HVAC systems by adjusting the frequency and voltage supplied to the motors. By adjusting the speed of the electric motors, VFDs can allow control systems to monitor regulate the output of HVAC components, such as fans, pumps, and compressors, which can deliver energy savings. VFDs are becoming the norm in all new HVAC systems. For detailed information on fan applications, see Chapter 21 of the *2024 ASHRAE Handbook—HVAC Systems and Equipment* (ASHRAE, 2024).

(b) Air-Handling Units

An AHU is a packaged component that is found in both all-air and air–water HVAC systems. Although AHUs are typically selected from a manufacturer's catalog and delivered to the site as a unit, large HVAC systems may use built-up AHUs that are custom-designed and assembled for a given project. Depending upon project needs, an AHU will normally include a fan (to circulate air through the system), a filter (to clean the air and protect the fan and coils), and a cooling coil (when cooling is required). Heating coils are also commonly included with AHUs for a number of system types.

Numerous accessory components are available to meet the needs of specific projects. These may include a humidifier unit (to add moisture to the air, usually of interest during the winter in cold climates); a mixing box (to thoroughly mix outdoor and return air streams for better system performance); plenums of various types (to facilitate transitions in airflow); access doors and panels (to ease maintenance); preheat coils (to temper winter outdoor air intake in a cold climate so as to prevent freezing of the main heating/cooling coils); various types of dampers; and economizer sections (to permit use of unconditioned outdoor air as system supply air).

Air Handling via a Furnace. In many homes, the furnace equipment will contain both a heating function and a blower. The blower distributes air round the entire house via ductwork. When heating is required, the furnace's heating component will activate. In cooling mode, air-conditioning equipment (e.g., heat pump, evaporative cooler) will cool an evaporator coil and the blower will move supply air through the coil to cool the air and on through the ductwork. If a heat pump is used for cooling, the same coil can be used (in the reverse cycle) for more energy-efficient heating under functional conditions. In this case, the furnace's gas- or electric-based heating component will be activated for auxiliary heating during colder outdoor conditions.

Air-Handling Units (AHUs). The air handling units in large buildings can be huge. See Figure 16.56

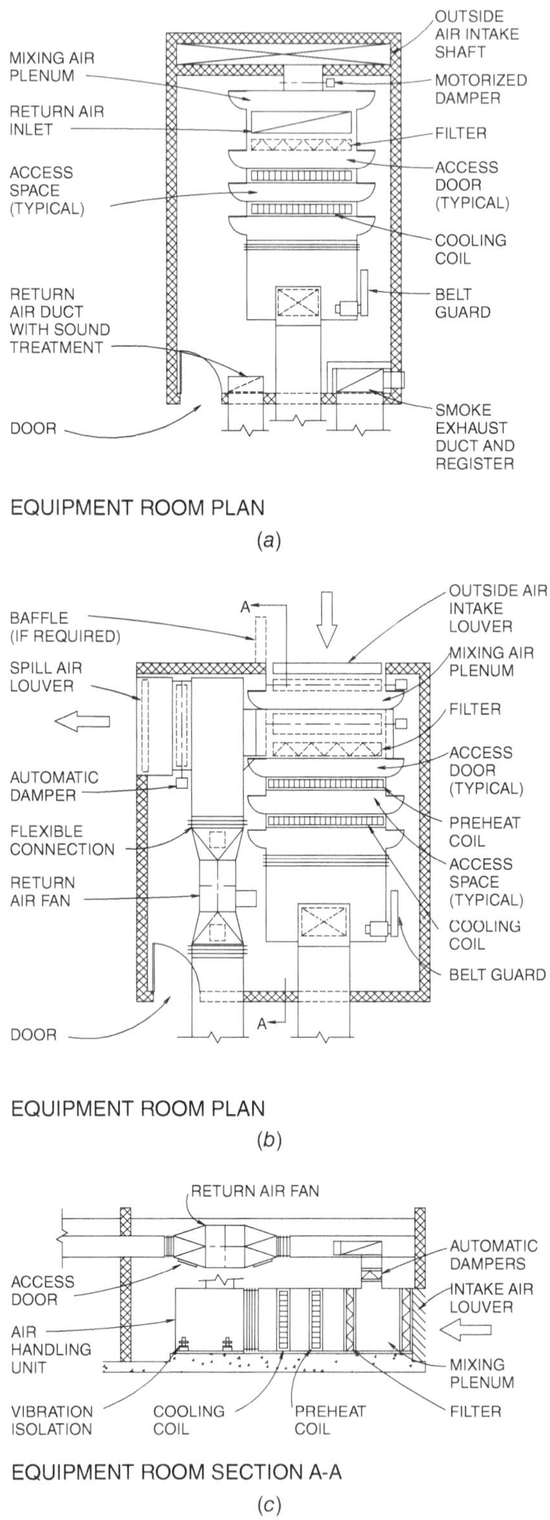

Figure 16.56 *Some variations on the wide range of HVAC equipment room arrangements. (a) Plan with floor-mounted air-handling unit. Outdoor air is drawn from an adjacent shaft; return air ducts are above the equipment, and a remote exhaust fan expels air that is not recirculated. (b) Plan of air-handling unit on an exterior wall. Outdoor air is taken through the wall, and a return air fan is overhead. (c) Section through plan b. (From Ramsey and Sleeper, 2000.)*

TABLE 16.6 Air-Handling Equipment Rooms

PART A. I-P UNITS						
	Approximate Overall Dimensions of Supply Air Dimensions (ft-in.)			Recommended Room Dimensions (ft-in.)		
Range (cfm)	Width ×	Height ×	Length	Width ×	Height ×	Length
1000–1800	4′-9″	2′-9″	14′-9″	12′-6″	9′-0″	18′-9″
1801–3000	5′-0″	3′-6″	16′-0″	13′-9″	9′-0″	20′-0″
3001–4000	6′-9″	4′-6″	16′-0″	17′-6″	9′-0″	20′-0″
4001–6000	7′-6″	4′-6″	16′-9″	18′-0″	9′-0″	20′-9″
6001–7000	7′-6″	4′ 9″	18′-3″	18′-6″	9′-6″	22′-3″
7001–9000	8′-0″	5′-0″	18′-9″	19′-0″	10′-0″	22′-9″
9001–12,000	10′-0″	5′-6″	21′-0″	23′-0″	11′-0″	25′-0″
12,001–16,000	10′-3″	6′-0″	22′-0″	13′-6″	12′-6″	26′-0″
16,001–19,000	10′-6″	6′-6″	23′-9″	24′-0″	13′-0″	27′-9″
19,001–22,000	11′-9″	7′-3″	25′-0″	26′-9″	15′-0″	29′-0″
22,001–27,000	11′-9″	8′-6″	26′-0″	27′-0″	16′-0″	30′-0″
27,001–32,000	13′-0″	9′-9″	27′-9″	29′-0″	18′-0″	31′-9″
PART B. SI UNITS						
	Approximate Overall Dimensions of Supply Air Equipment (m)			Recommended Room Dimensions (m)		
Range (L/s)	Width ×	Height ×	Length	Width ×	Height ×	Length
470–849	1.4	0.8	4.5	3.8	2.7	5.7
850–1415	1.5	1.1	4.9	4.2	2.7	6.1
1416–1887	2.1	1.3	4.9	5.3	2.7	6.1
1888–2831	2.3	1.3	5.1	5.5	2.7	6.3
2832–3303	2.4	1.4	5.6	5.6	2.9	6.8
3304–4247	2.4	1.5	5.7	5.8	3.0	6.9
4248–5662	3.0	1.7	6.4	7.0	3.4	7.6
5663–7550	3.1	1.8	6.7	7.2	3.8	7.9
7551–8966	3.2	2.0	7.2	7.3	4.0	8.5
8967–10,381	3.6	2.2	7.6	8.1	4.6	8.8
10,381–12,741	3.6	2.6	7.9	8.2	4.9	9.1
12,742–15,100	4.0	3.0	8.5	8.8	5.5	9.7

Source: From Ramsey and Sleeper, 2000. SI conversions added by the authors of this book.

and Table 16.6 for typical installation layouts and dimensions. Extremely large or sprawling buildings may require multiple AHUs. Some buildings may require multiple AHUs to serve separated air streams (i.e., ventilation zones)—such is the case with certain types of laboratories or zoological facilities. As with all mechanical and electrical equipment, providing adequate access for regular maintenance, refurbishment, and ultimate replacement is good design practice. Inability to easily reach or work on an AHU can negatively impact its performance and energy efficiency. Manufacturers provide a lot of variety in the AHU product lines: This includes a wide range of cooling, heating, and airflow capacities; units for exterior or for interior installation; and packaged units or built-up units. Whether to place an air-handling unit inside the building envelope (usually in a fan or mechanical room) or outside the envelope (usually on the roof) is worthy of serious consideration—as the decision can substantively impact energy efficiency and maintainability. Chapter 4 of the 2024 ASHRAE *Handbook—HVAC Systems and Equipment* (ASHRAE, 2024) provides details on building air-handling systems.

(c) Ductwork

Ductwork is the collective name for the air distribution conduits used in HVAC systems. Ductwork may be steel (or other metal), rigid glass fiber, flexible, or cloth. The first three materials are by far the most common. Duct shapes may be round, square, rectangular, or flat oval. In large buildings, ductwork can be physically large and spatially extensive. Coordination of ductwork with other building elements can be a challenge. Ductwork will be large in an all-air HVAC system, substantially smaller in an air–water system, and essentially nonexistent in an all-water or local system. See Chapter 19 of the *2024 ASHRAE Handbook—HVAC Systems and Equipment* (ASHRAE, 2024) for details on duct construction. Duct design (sizing and layout) is presented in Chapter 21 of the *2021 ASHRAE Handbook—Fundamentals* (ASHRAE, 2021).

Ductwork may contain supply air (in transit from an AHU to the terminal and/or delivery devices within the zones); return air (being recycled from a zone back to the AHU); outdoor air (being brought into a building to assist with IAQ); exhaust air (being channeled out of a building for IAQ or pressure-balance purposes); and/or relief air (return air that is not currently needed for system balance and is being conducted to the outdoors). An all-air system will typically require parallel but separate supply and return air ducts going to/from the various zones. At any point in a building, the supply and return ducts are roughly equal in area. In an air–water HVAC system, there is normally no return air (and thus, no return air ductwork). When there is return air, it is often channeled back to the AHU via a plenum (enclosed architectural volume) rather than a self-contained conduit.

Design considerations for ductwork include:

Pressure loss: The pressure loss (or drop) in a ductwork system can exert a major influence on system energy efficiency. Resistance to airflow in a ductwork system must be overcome by energy input via a fan. Key areas of pressure loss in a system include air filters, cooling and heating coils, ductwork and fittings, terminal devices, and delivery devices. Of these, ductwork losses are most amenable to reduction through integrated architectural and engineering design. High-velocity airflow will increase friction losses, long ductwork will increase friction losses, and convoluted ductwork will increase friction losses. From an architectural design perspective, ductwork should be kept reasonably short by selecting good fan room locations, and space/volume constraints that result from poor planning should not force the use of too-small ducts that result in high-velocity airflow. *ASHRAE/IES Standard 90.1* sets upper limits on fan system horsepower that are directly related to system airflow rate.

Insulation: Supply air ductwork may need to be thermally insulated (as much to prevent condensation as to reduce heat gains to the supply air); thermal insulation becomes critical when ductwork is run outside of the conditioned enclosure of the building (a practice that is generally frowned upon—but addressed by *ASHRAE/IES Standard 90.1*). Internal insulation (placed inside the ductwork) can act as both thermal insulation and sound-absorbing liner; external insulation will have no effect on duct-borne noise transmission. External insulation for condensation control will require a vapor retarder. The thickness of insulation must be considered when squeezing ductwork into tight places.

Noise control: Control of duct-borne noise is often a challenge in many building types. The unbroken path established to lead air from the air-handling unit to the occupied space transmits sound as readily as air. Mitigation of duct-borne noise is a technical design concern addressed in Chapter 25 of this book. Additionally, Chapter 49 of the *2023 ASHRAE Handbook—HVAC Applications* (ASHRAE, 2023a) provides detailed information on this topic. It is important to note that very logical noise control design moves (such as adding complexity to the ductwork layout) may act counter to energy-efficiency design moves.

Fire protection: Ductwork that penetrates fire-rated assemblies will need to be outfitted with fire dampers. Unnecessary damper installations (caused by uncoordinated design) can add to the cost of a building. See Chapter 26 for more information about fire protection. Additionally, note that the National Fire Protection Association (NFPA) is the definitive North American source for fire protection requirements related to HVAC systems.

Indoor air quality (IAQ): Building IAQ should be considered as ductwork systems are designed and constructed. Ductwork is generally dark, may be dusty, and may be humid. These are ideal environmental conditions for the development of mold and mildew. The initial cleanliness of ductwork and the ability to maintain such cleanliness over time should be considered by the design team. See Chapter 5 for more information about IAQ. Additionally, the *Indoor Air Quality Guide* from ASHRAE (2009) provides design, construction, and operation information that addresses ductwork and IAQ.

Approximate duct sizing: It is often very useful to have an estimate of duct size early in the design process. Duct sizes can influence floor-to-floor heights, structural element dimensions, and the area to be allocated for vertical chases. Approximate sizing can be easily accomplished. Figure 16.8 provides a graphic tool by which to estimate duct sizes based on building floor area. For preliminary estimates, assuming an airflow of 1 cfm per ft^2 [5 L/s per m^2] of conditioned floor area (adjustable up or down to accommodate dense occupancies/unusual loads or energy-efficient design) and a velocity of 1000 fpm [5 m/s] (also adjustable as appropriate) allows simple rough sizing of main and branch ducts. The following provides an example of this type of approximation.

In order to approximate the minimum required cross-sectional area of a duct (A), the following must first be determined:

- Quantity of air to be distributed through the duct element of interest within a certain period of time. This value can be found from Figure 16.8, from an estimation of the room air change rate (commonly expressed in air changes per hour), or from an airflow/unit area estimation.

- Express this volume of air distribution in cfm [L/s].
- Maximum velocity (fpm [L/s]) for airflow within the duct; this can come from project design criteria or be set arbitrarily (such as 1000 fpm [5 m/s] for larger ducts).

The approximate minimum required cross-sectional area of duct (A) is then:

$$A = \frac{VOL}{VEL} \times C \times FA \qquad \text{(Eq. 16.2)}$$

where

A = area of duct, ft² [m²]
VOL = volume of airflow, cfm [L/s]
VEL = velocity of airflow, fpm [m/s]
C = conversion factor, 1.0 for I-P [1000 for SI]
FA = friction allowance, decimal (see sample values below)

Typical frictional allowances are as follows:

Round ducts:

1.0 (may be neglected)

Nearly square ducts (ratio of width-to-depth, 1:1):

Small (<1000 cfm [470 L/s]) = 1.10

Large (>1000 cfm [470 L/s]) = 1.05

Thin rectangular ducts (ratio of width-to-depth, 1:5) have an allowance of 1.25.

Remember that providing only the minimum duct area will carry a penalty of increased noise and pressure loss. Extremely flattened duct cross-sections (with width much greater than depth) may be architecturally appealing but will increase pressure losses due to increased surface area.

EXAMPLE 16.4 A 150 ft² [13.9 m²] office is served by a small, nearly square, supply duct. The duct will provide 225 cfm [106 L/s] of air at a velocity of 30 fpm [0.15 m/s]. What is the approximate minimum required cross-sectional area of the duct (A)?

SOLUTION

Apply Equation 16.2 using the appropriate frictional allowances:

I-P units:

$$A = \frac{225\ \text{cfm}}{30\ \text{fpm}} \times 1.0 \times 1.10 = 8.2\ \text{ft}^2$$

SI units:

$$A = \frac{106\ \text{L/s}}{0.15\ \text{m/s}} \times 1000 \times 1.10 = 0.76\ \text{m}^2$$

For this small office space, the appropriate minimum required ductwork cross-sectional area as calculated is: 8.2 ft² [0.76 m²].

Since this duct was assumed to be nearly square (1:1), consider duct dimensions as the square root of the cross-sectional area: 2 ft-10 ½ in. × 2 ft-10 ½ in. [0.87 m × 0.87 m].

Note this is a minimum area. A slightly upsized 3 ft × 3 ft [0.91 m × 0.91 m] duct would be recommended. ■

Ductwork Accessories. A few accessories will be included in a ductwork system to facilitate control, improve performance, and/or increase safety. These devices might include the following:

- *Fire dampers*, to close off a duct with a fire rated barrier in the event of a fire.
- *Control dampers*, to change or redirect airflow.
- *Smoke detectors*, to activate a fire alarm system.
- *Temperature sensors*, to activate a thermostat.
- *Pressure sensors*, to ensure proper airflow.
- *Ultraviolet germicidal irradiation (UVGI) devices*, for inactivation of biological contaminants (e.g., viruses).
- *Photocatalytic oxidation (PCO) air cleaning system*, which uses UV radiation and a titanium dioxide catalyst to break down biological contaminants, volatile organic compounds (VOCs), odors, and more.

Consider how such devices might be accessed for maintenance and/or calibration once the building has been closed in.

16.10 DISTRIBUTION COMPONENTS: WATER

When water is employed as a distribution medium in a central or district HVAC system, duct sizes will be substantially reduced (in air–water systems) or ductwork will be eliminated (in all-water systems). This outcome is a result of the much, much higher thermal capacity of water versus air (by about a 3000:1 ratio). The piping systems that convey heat or coolth in an HVAC system are pressurized, of reasonably small diameter, and usually easily coordinated with other building systems. HVAC piping will usually be insulated—both to reduce unwanted heat loss/gain and to prevent condensation on chilled water pipes.

(a) Pumps

Pumps are used to circulate water through the various piping networks associated with building HVAC systems (including chilled water, heating water, and condenser water systems). Pumps are briefly discussed in Chapter 21. Centrifugal pumps are commonly used in HVAC systems. Pump performance hinges upon flow rate and head (pressure to be overcome); pump curves allow analysis of pump performance in the context of building performance (including operation at partial loads). Reducing pumping power draw and energy consumption are goals for a high-performance building. Chapter 44 of the *2024 ASHRAE Handbook—HVAC Systems and Equipment* (ASHRAE, 2024) provides detailed information on centrifugal pumps.

(b) Piping

Steel pipe and copper tubing are the default materials for HVAC piping systems. Pipe segments will be connected using any of several methods suitable for the pipe type and application. Fittings will be used as required to change direction, transition pipe sizes, and connect to equipment. Pipes will be insulated when heat loss or gain will affect system performance and energy efficiency. *ASHRAE/IES Standard 90.1* provides requirements for thermal insulation. Chapter 22 of the *2024 ASHRAE Handbook—HVAC Systems and Equipment* (ASHRAE, 2024) provides detailed information on pipes and fittings.

Accessories. Piping system accessories include valves, temperature sensors, pressure sensors, strainers, test ports, and the like. Valves are an integral aspect of water distribution. They may provide manual on-off control for isolating equipment during maintenance, or they may operate under automatic control to meet the control sequences established by the HVAC designer. Chapter 46 of the *2024 ASHRAE Handbook—HVAC Systems and Equipment* (ASHRAE, 2024) provides detailed information on valves.

(c) Steam

As discussed in Section 16.6, steam is seeing reduced use as a heat-transfer medium within buildings. Although robust and efficient, steam systems also demand a fair amount of design skill and an infusion of tender, loving care while in operation. Vaporized water is produced by a boiler and distributed to loads via a piping system. At the loads, steam is condensed and the condensate (water) returned to the boiler. Steam traps are installed to ensure that only condensate returns to the boiler. Chapter 11 of the *2024 ASHRAE Handbook—HVAC Systems and Equipment* (ASHRAE, 2024) waxes eloquent on steam systems.

16.11 AIR DELIVERY

The delivery of air to the various rooms within a building is conceptually simple, but with a fair amount of detail behind the simplicity. Because air is being delivered to another air volume, there is no need to interface between different media. Air could in theory just be dumped into a room to heat or cool it. Thermal and acoustical comfort demand greater sophistication. In general, devices termed diffusers, registers, or grilles will be used to introduce supply air into a space and/or allow return air to leave a space. Diffusers are the most refined of these devices; a grille is the least refined. Key design concerns related to air delivery include throw (spatial coverage, for supply devices), airflow capacity, noise production, and aesthetics (which seems to be often overlooked).

(a) Diffusers/Registers/Grilles

A *diffuser* is defined as a circular, square, rectangular, or linear air distribution outlet, generally located in the ceiling, and comprised of deflecting members discharging supply air in various directions and planes and arranged to promote mixing of primary air with secondary room air.

A *register* is defined as a combination grille and damper assembly over an air opening.

A *grille* is a louvered or perforated covering for an opening in an air passage which can be located in a wall, ceiling, or floor.

Supplying air from the ceiling of a space is so widespread a practice that many lighting fixtures are designed to either serve as diffusers for supply air or as intakes for return air. As return air fixtures, they are especially effective because they remove much of the heat from electric lighting before it can contribute a heat gain to the office space below. This reduces the quantity of supply air that must be provided (although the heat from lights must still be handled by the refrigeration system).

With enclosed individual offices, providing a supply diffuser and return grille can ensure good air circulation. As open-plan office space proliferated, so did concerns about whether the workspace cubicles were being adequately served with conditioned air. Air diffusers and grilles (Figure 16.57) on the ceilings are uncluttered by furniture and independent of rearrangements of workstation cubicles. Canadian research has shown that the most important variable is the quantity of airflow, not the location of the diffusers relative to

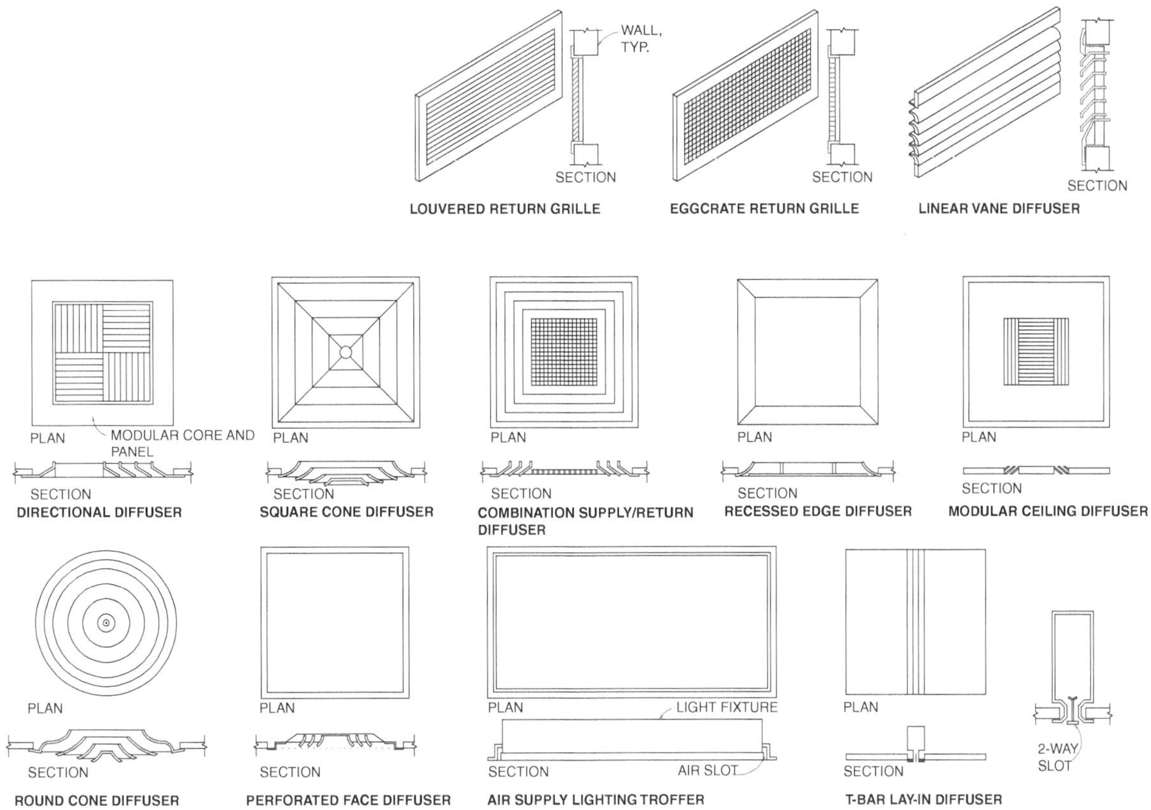

Figure 16.57 *Common air distribution outlets. (From Ramsey and Sleeper, 2016.)*

workstations or whether the cubicle partitions have a small gap where they meet the floor.

Are higher ceilings better for air distribution? There seems to be a trade-off between more vertical clear space (between the ceiling and the cubicle partitions) that would allow a wider range of air distribution and the increased distance from the actual occupant who should be receiving the conditioned air. An increased distance could allow colder supply air to be distributed, resulting in smaller ducts, and could also allow higher velocity of supply air, further reducing duct size. Higher ceilings allow for deeper daylight penetration and represent a larger "pool" of air that is slower to become polluted by occupants or office equipment.

When noise from diffusers is a critical concern, as in broadcast and recording studios, pinhole-perforated diffusers (Figure 16.58) can supply large quantities of low-velocity air. These types of spaces pose an especially challenging combination of high heat gains (from lights) and low background noise requirements. Ceiling distribution leaves the floor unencumbered for functional flexibility.

For a discussion of underfloor air supply and distribution, see Section 16.20.

(b) Alternative Air Delivery Systems

The supply/return systems discussed to this point have focused on the ceiling and the floor. Supply from the ceiling with return through the floor is another possibility when using these two planes for air distribution. Supply air gets to the occupied zone later, and individual control is much more difficult, but floor debris falls into a return, not a supply, plenum.

An alternative delivery approach with return at the perimeter is shown in Figure 16.59. It is variously called an air-extract window, an air curtain window, or a climate window. Developed in Scandinavia, this is a triple-glazed window that passes room air between a typical outer double-glazed window and an inner single pane. The inner pane thus is kept at very nearly the same temperature as the room air, which greatly increases comfort (via the mean radiant temperature [MRT] effect) near windows on very cold (or very hot) days. Venetian blinds are often inserted in this

Figure 16.58 *Pinhole-perforated diffusers (above a lighting grid) provide large quantities of low-velocity air to the television studio at the Community Media Center of Santa Rosa, California. The background noise from more typical HVAC systems would be picked up by sound recording equipment, which would be unacceptable. (© TLCD Architecture.)*

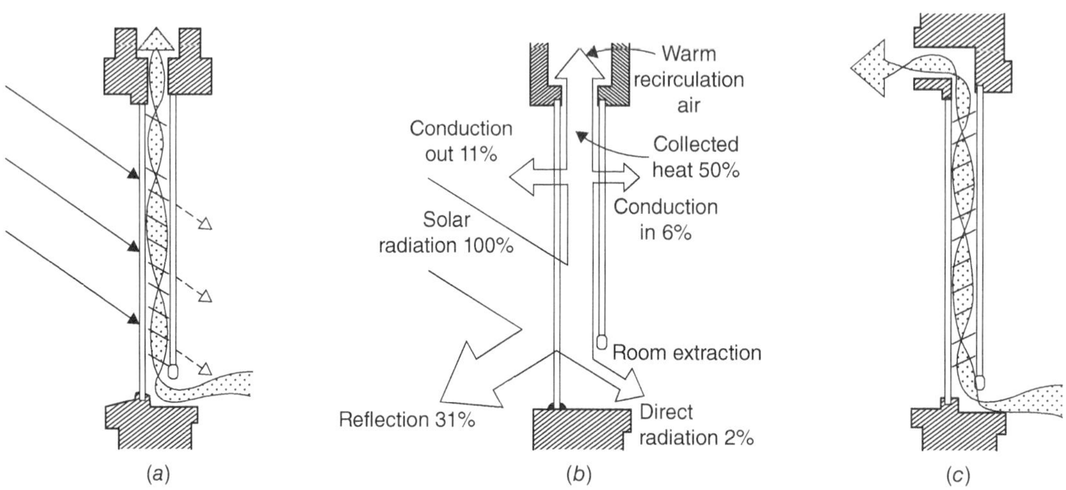

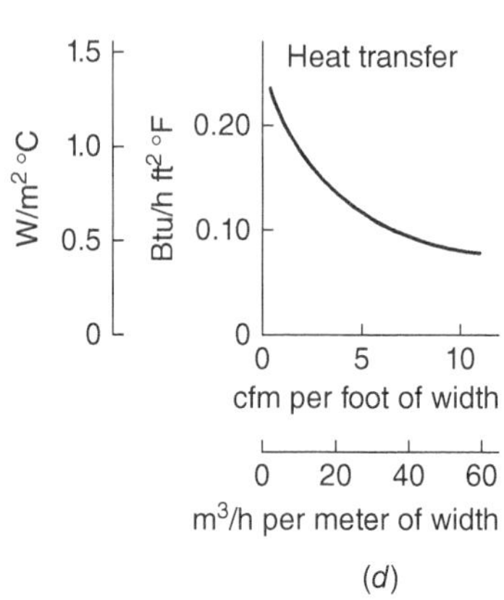

Figure 16.59 *Air-extract window. (a) Cross-section showing the window as a solar collector. (b) Distribution of solar heat in the collector mode. (c) The window as a "solar chimney" exhausting hot air. (d) The U-factor of these windows varies with the rate of airflow within the glazing. (From Aitken and Seppanen.)*

cavity, where they can intercept direct solar radiation and redirect light toward the ceiling. The solar heat intercepted by the blind is carried off by the room air into a plenum above the ceiling, where the air can be either exhausted or recirculated and its heat content either reclaimed or rejected. The U-factor of these windows is dependent upon the rate of airflow within the glazing panes (Figure 16.59*d*); typical flow rates are 4–6 cfm per foot [6–9 L/s per meter] of window width.

The Ocosta Junior/Senior High School in Westport, Washington (Figure 16.60), used these windows on the south façade. The classrooms had photocells that switched off some light fixtures nearest the windows when daylight is adequate, and water loop heat pumps helped to transfer passive solar heat from the south side to the sunless north-facing rooms. Excess heat was stored in an underground water tank. The compact plan, highly insulated envelope, and thermal mass capacity within the school reduced the need for a boiler to supplement the winter temperature of the water loop. During consistently warm weather, the building management system exhausted the air from the windows to the outdoors. For an example of such windows in a larger U.S. building, see Figure 16.52, which shows the Comstock Center.

Air supply openings in vertical risers are relatively rare but at times function effectively. The Frederik Meijer Conservatory (Figure 16.61) near Grand Rapids, Michigan, is a huge indoor tropical rain forest. This glass structure is more than 65 ft [20 m] high; it has more glass area than floor area—great for a high daylight factor, but imagine the heating loads. High humidity must be maintained, yet water condensation on leaves is undesirable. A strong horizontal airflow could be helpful to the plants, serving as wind would develop a stronger root structure.

A perimeter tunnel serves as an air supply duct, delivering air evenly around the glass walls, heated by finned tubes located at planting level (which allows them to be largely hidden by foliage). The tunnel insulates the planting beds from the extremes of the exterior, maintaining constant ground temperatures year-round. A separate "wind" system continuously feeds air to vertical ducts with several diffusers (Figure 16.61*b*), making leaves and branches sway.

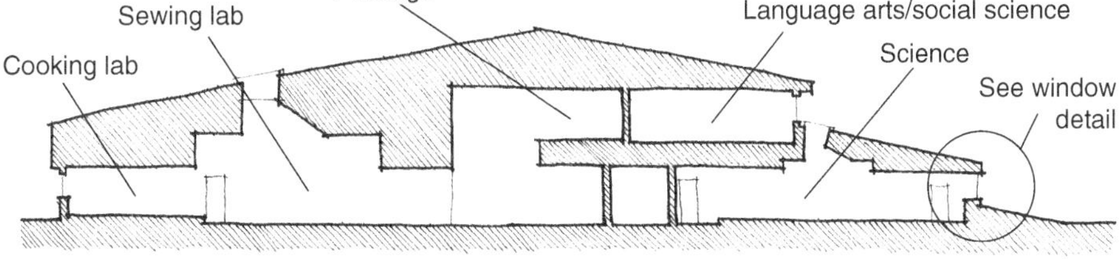

(*a*)

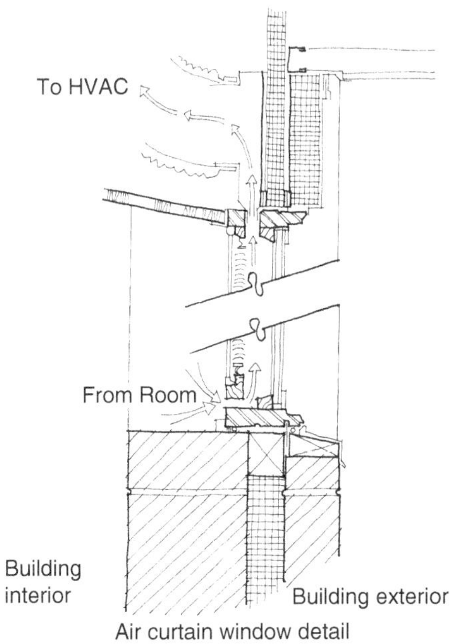

(*b*)

Figure 16.60 *The Ocosta Junior/Senior High School at Westport, Washington, used air extract windows to help distribute solar energy from the south side (to the right in section a) to the north classrooms. (b) Detail through the south-facing windows. Photocells controlled electric lighting during daylight hours; venetian blinds directed sunlight toward the ceiling and deeper into the classroom. (Courtesy of Burr Lawrence Rising + Bates Architects. Redrawn by Erik Winter.)*

Figure 16.61 *The Frederik Meijer Conservatory near Grand Rapids maintains a tropical environment in Michigan's winter. (a) A perimeter tunnel delivers air evenly around the walls, heated by finned tubes just above the planting level. (b) A separate "wind" system provides a continuous breeze from diffusers in the vertical ducts to strengthen plant root systems. (c) A fogging system maintains tropical humidity while providing evaporative cooling under summer conditions. (Photos by Roger Van Vleck; courtesy of Fishbeck, Thompson, Carr & Huber.)*

Humidity is provided by a high-pressure water system that creates fog (very small water particles). This is located high in the space (but also near waterfalls and a stream for special effects). This is also an evaporative cooling system to counteract summer overheating. Peak airflow is needed only for cooling; return air is taken through a large opening hidden behind a waterfall. About one-third of the peak supply air is returned; the remainder is exhausted at the top of the structure through operable openings high in the glass roof. A building management system controls the roof vents, the fogging system, the finned tube heating system, rolling shading screens, and the heating/cooling equipment (located in a basement), all in response to changing conditions surrounding this enormous glass house.

16.12 WATER DELIVERY

Delivering water to spaces conditioned by an air–water or all-water HVAC system is a much more difficult proposition than delivering air. Water cannot be directly introduced into a space. The heating or cooling effect carried by the water must be indirectly coupled to the space through the use of a water-to-air heat exchanger of some sort. Numerous such devices are available to meet the demands of a range of building types and spatial arrangements. A common characteristic of all such devices is their need to consume some amount of room volume (with a few key exceptions—such as recessed convector units or radiant hydronic panels).

The water-to-air heat exchangers that might be used as delivery devices in an HVAC system include the following:

Hydronic baseboard radiators: The term "hydronic" denotes heat transfer via a fluid medium, typically water (thus "hydro"). This type of heat exchange device primarily uses convective heat transfer. Baseboard radiators are essentially comprised of sections of finned tube with decorative covers. They come in a wide range of capacities, architectural styles, and are readily available. Room air that comes into contact with the finned tube (a mini coil) is heated by indirect contact with hot water, rises by convection, and distributes heat to the space. Air circulation is natural, and no filtering of circulated air occurs. Baseboard radiators are located near the intersection of a wall and the floor, usually on an exterior wall (where heat loss will be concentrated); it is common to see baseboard units located under windows (as a means of offsetting heat loss and increasing glass surface temperature—to improve MRT and resist condensation). Because of their placement near the floor, baseboard units are heating-only devices.

Valence units: This particular type of heat exchanger, which has never really caught on, has been used in a few energy-efficient buildings where only a small heating or cooling load exists. The underlying concept is that once freed from being place-bound to heat-loss locations, the intersection between a wall and the ceiling (where a window valance may be installed) may provide for more flexible use of space. There are a variety of valence unit types (e.g., hydronic, DX) and they may be used for heating and/or cooling.

Convectors: Convectors are large assemblages of finned radiation. They can provide a high-output heat source. Flush and recessed cabinet styles are available, natural and forced-convection (with fan) options are possible. Convectors are commonly seen in entry vestibules and similar locations with high heat loss. Located in the lower half of a space, natural convection convectors are typically used for heating; forced convection units might be used for heating and/or cooling.

Unit heaters: This is a heating coil coupled with a fan. The combination of a high-capacity heat exchanger (coil) with high airflow results in a high-capacity device that can be used to heat relatively large spaces. These are commonly used in industrial-type applications.

Hydronic radiant panels: Water tubing embedded in a floor or ceiling can be used as a heat exchanger—and is termed a hydronic radiant panel. These panels usually extend the full length and width of a space (although this is not necessary), providing a reasonably low-density heat source that can be very comfortable through its primarily radiant heat exchange. Production installation systems are readily available. Radiant panels may be used for either heating or cooling—although condensation (dew point) control is critical in a cooling application.

Chilled beams: A chilled beam (a misnomer) is a manufactured device that is located at ceiling level that uses chilled water flowing through finned tube arrangements to provide radiant cooling to a space; heating is also possible. There are two types of chilled beams:

- *Passive chilled beams (PCB)* which operate through simple convection without mechanically induced airflow.
- *Active chilled beams (ACB)* which feature mechanically induced airflow supplied at a low velocity. ACBs account for most chilled beam installations as their performance is markedly better than PCBs under most circumstances. Chilled beams are an emerging trend and have generated some controversy as well as buzz in the United States.

Induction units: This is a cabinet (historically installed at floor level) that includes a heating/cooling coil, air filter, and condensate drain pan. Air is provided to the induction unit from a central air-handling unit, along with water from a boiler and/or chiller. As the name implies, the primary air supply (from the AHU) is used to "induce" a secondary flow of room air across the coil. In North America, induction units have been largely supplanted by fan-coil units.

Fan-coil units (FCU): This is a cabinet with a heating/cooling coil, fan, and air filter. Unlike an induction

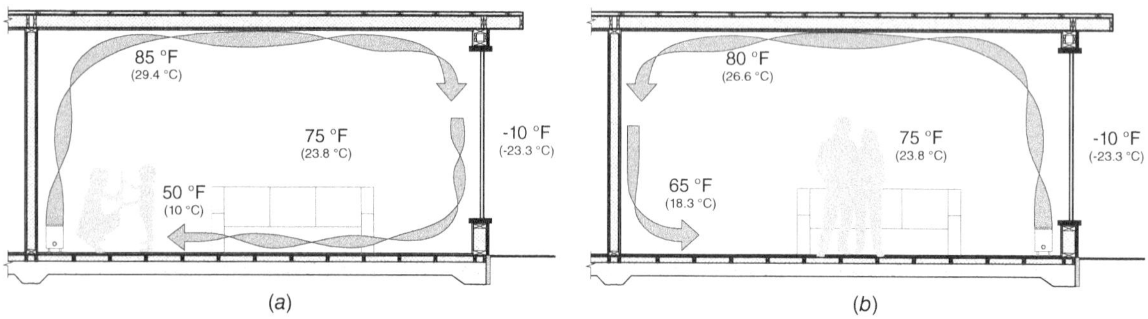

Figure 16.62 *Locating a heat source near an interior wall (a) encourages a cold draft along the floor in winter. Below a window (b) it evens the temperature throughout the room but also causes more heat loss through the window. (Drawn by Achyuthan Ramaswamy.)*

unit, an FCU does not introduce supply air from a central system. Instead, an FCU only circulates existing room air by the action of a fan. A wide range of capacities, arrangements, and architectural styles are available (including vertical, horizontal, and stackable configurations).

A designer will need to address the question: Where should water-based delivery devices be placed within a space? The relative comfort provided by radiant ceilings or floors is discussed in Chapter 6. When a smaller delivery device is selected, location becomes more critical. Figure 16.62 shows why designers usually locate heat sources below windows, despite the fact that warmer temperatures just inside an exterior wall will drive more heat through the wall in cold weather (since heat loss = $U \times A \times \Delta t$). As windows and wall assemblies provide better thermal resistance, their interior surface temperatures rise, and the need for heat at the perimeter is reduced. Indeed, with superinsulated enclosure components (such as found in Passive Building construction), the need for heating virtually disappears because internal gains from the sun, lighting, appliances, and occupants can almost heat the space.

16.13 AIR FILTERS

Air filters will be found in all HVAC systems that use air as a heating/cooling distribution and delivery mechanism. An air filter typically plays two roles in an HVAC system:

1. acting as an indoor air quality (IAQ) mitigation component
2. protecting HVAC equipment (such as fans and coils) from environmental dirt

When located within a terminal device (such as a fan-coil unit), filter quality and performance are usually aimed at equipment protection. Filters in central equipment (such as an air-handling unit) are usually selected to meet the demands of indoor air quality, which are more challenging than basic equipment protection. Air filters are also commonly part of ventilation systems with a supply air component (such as an HRV/ERV system) and may be employed as stand-alone devices for local IAQ control.

(a) Filter Performance

Air filters should be selected to accomplish the job for which they are intended. This seems to go without saying—but it reinforces the need to understand the owner's project requirements, meet minimum code requirements, and comply with the requirements of any voluntary building certifications being sought. Air filters are available in a wide variety of types. Figure 16.63 lists filter groups and testing methods with associated filter efficiencies; Table 16.7 summarizes filter characteristics. Current residential and commercial design efforts tend to use MERV ratings (see below). Filter groups I, II, and III are widely used in building ventilation systems. The highest-efficiency filters—high-efficiency particulate air (HEPA) and ultra-low penetration air (ULPA)—are more commonly utilized in IAQ-demanding environments. Not all pollutants can be readily removed by filters.

ASHRAE Standard 52.2–2017: Method of Testing General Ventilation Air-Cleaning Devices for Removal Efficiency by Particle Size, combines two previous standards to offer a test procedure for evaluating the performance of air-cleaning devices as a function of particle size. The standard addresses three air-cleaner performance characteristics:

1. ability to remove particles from the air
2. total dust-holding capacity with arrestance (i.e., the filter's ability to capture and retain larger particles)
3. resistance to airflow

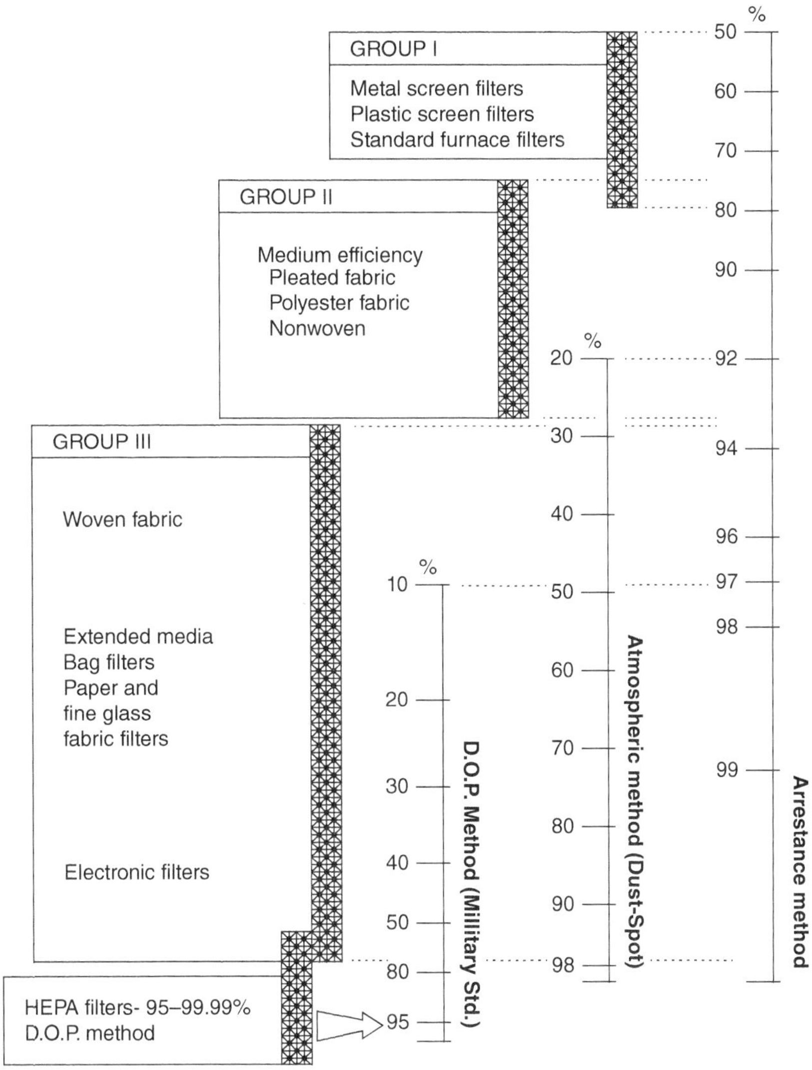

Figure 16.63 *Air filter efficiencies. Methods of determining efficiency vary. The Arrestance Method can be used across all groups. The Atmospheric (Dust-Spot) Method (now obsolete) was used for medium- and high-performance filters. The DOP Method is a military standard. (From Rousseau and Wasley, 1997.)*

TABLE 16.7 Air Filter Characteristics

	Percent Efficiency Range			
Media and Type	**Atmospheric Dust**	**Small Particles**	**Dust-Holding Capacity**	**Airflow Resistance, in. Water [Pa][a]**
Dry panel, throwaway	15–30	NA	Excellent	0.1–0.5 [25–125]
Viscous panel, throwaway	20–35	NA	Good	0.1–0.5 [25–125]
Dry panel, cleanable	15–20	NA	Superior	0.08–0.5 [20–125]
Viscous panel, cleanable	15–25	NA	Superior	0.08–0.5 [20–125]
Mat panel, renewable	10–90	0–60	Good to superior	0.15–1.0 [37–250]
Roll mat, renewable	10–90	0–55	Good to superior	0.15–0.65 [37–162]
Roll oil bath	15–25	NA	Superior	0.3 0.5 [75–125]
Close pleat mat panel	NA	85–95	Varies	0.4–1.0 [100–250]
High-efficiency particulate	NA	95–99.9	Varies	1.0–3.0 [250–745]
Membrane	NA	to 100	NA	NA
Electrostatic with mat	80–98	NA	Varies	0.15–1.25 [37–310]

Source: From Ramsey and Sleeper, 2000.

[a]Higher airflow resistance values will require increased fan energy.

TABLE 16.8 MERV Rating Application Guidelines

Standard 52.2 Minimum Efficiency Reporting Value (MERV)	Typical Controlled Contaminant	Typical Applications and Limitations	Typical Air Filter/Cleaner Type
16 15 14 13	**0.30–1.0 μm Particle Size** All bacteria Most tobacco smoke Droplet nuclei (sneeze) Cooking oil Most smoke Insecticide dust Copier toner Most face powder Most paint pigments	Hospital inpatient care General surgery Smoking lounges Superior commercial buildings	**Bag Filters** Nonsupported (flexible) microfine fiberglass or synthetic media. 300–900 nm (12–36 in.) deep, 6–12 pockets. **Box Filters** Rigid-style cartridge filters 150–300 mm (6–12 in.) deep may use lofted (air laid) or paper (wet laid) media
12 11 10 9	**1.0–3.0 μm Particle Size** Legionella Humidifier dust Lead dust Milled flour Coal dust Auto emissions Nebulizer drops Welding fumes	Superior residential Better commercial buildings Hospital laboratories	**Bag Filters** Nonsupported (flexible) microfine fiberglass or synthetic media. 300–900 nm (12–36 in.) deep, 6–12 pockets. **Box Filters** Rigid-style cartridge filters 150–300 mm (6–12 in.) deep may use lofted (air laid) or paper (wet laid) media
8 7 6 5	**3.0–10 μm Particle Size** Mold Spores Hair spray Fabric protector Dusting aids Cement dust Pudding mix Snuff Powdered milk	Commercial buildings Better residential Industrial workplaces Paint booth inlet air	**Pleated Filters** Disposable, extended surface, 25–125 mm (1–5 in.) thick with cotton-polyester blend media, cardboard frame **Cartridge Filters** Graded density viscous coated cube or pocket filters, synthetic media **Throwaway** Disposable synthetic media panel filters
4 3 2 1	**>10.0 μm Particle Size** Pollen Spanish moss Dust mites Sanding dust Spray paint dust Textile fibers Carpet fibers	Minimum filtration Residential Window air conditioners	**Throwaway** Disposable fiberglass or synthetic panel filters **Washable** Aluminum mesh, latex coated animal hair, or foam rubber panel filters **Electrostatic** Self-charging (passive) woven polycarbonate panel filter

Source: © ASHRAE, www.ashrae.org. ANSI/ASHRAE Standard 52.2-2017: Method of Testing General Ventilation Air-Cleaning Devices for Removal Efficiency by Particle Size.

Note: A MERV for other than HEPA/ULPA filters also includes a test airflow rate, but it is not shown here because it has no significance for the purposes of this table.

The prevalent approach to expressing air filter performance involves the use of Minimum Efficiency Reporting Values (MERV) ratings. The higher the MERV rating, the better the filter performance. *ASHRAE Standard 52.2-2017* (ASHRAE, 2017) offers guidance for selecting appropriate MERV-rated air filters for typical applications. These recommendations are summarized in Table 16.8.

(b) Filter Types

Particulate Filters. Particulate air pollutants are diverse, as described in Table 16.8. Large particulates are the easiest to remove, but much smaller respirable particles pose a greater threat to health. Particulate filters are very common and come in several types. Panel filters (Figure 16.64*a*) typically come with HVAC equipment and function mainly to protect fans from large particles of lint or dust. Because panel filters are relatively crude, they are not really considered to be air-cleaning equipment. Media filters are much finer, using pleated filter paper within a frame. They function both by straining and impaction. The larger particles are strained out by the closely spaced filter fibers, while some of the smaller particles that would otherwise pass through are pushed into the fibers by air turbulence. Particulate filters need regular maintenance, especially media filters, which can become blocked and cause damage to HVAC equipment and increased energy consumption if not replaced frequently enough.

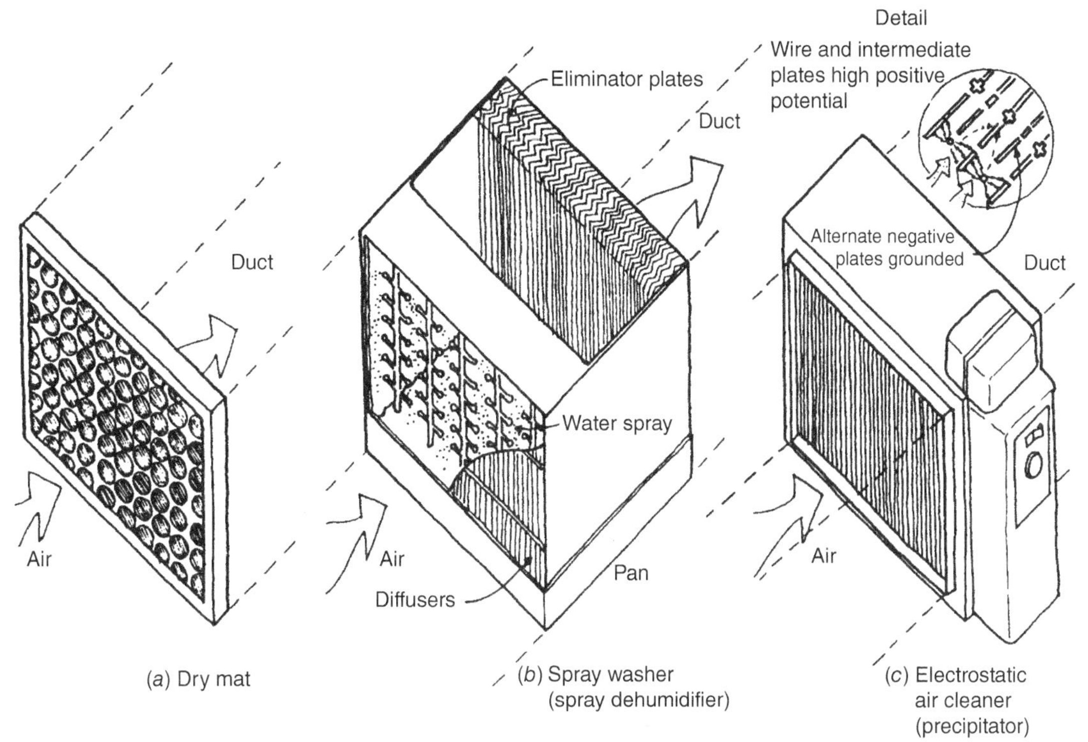

Figure 16.64 *Three diverse air filter types: (a) dry mat (i.e., panel filters); (b) spray washer (i.e., air washers); and (c) electrostatic precipitators. Not shown are pleated and roll types.*

Media filters of high quality (typically at least 6 in. [150 mm] deep) are expected to perform at an efficiency of 90% (minimum) for a six-month minimum life-cycle.

Adsorption Filters. These filters are used for gaseous contaminant removal and vary in response to the pollutant of concern. Activated-charcoal filters are the most common of these types; they absorb materials with high molecular weights but allow those of lower weights to pass. Odors are typically removed from indoor air by activated carbon filters. Other adsorption filters use porous pellets impregnated with active chemicals such as potassium permanganate; the chemicals react with contaminants, reducing their harmful effects. Adsorption filters must be regularly regenerated or replaced. High-quality adsorption filters contain gas adsorbers and/or oxidizers with sufficient capacity to remain active over a full-service cycle of 6 months at 24 hours per day, 7 days per week. Air velocity should be controlled to allow air to remain in the filter for about 0.06 second.

Air Washers. These types of filters (Figure 16.64*b*) are typically used in industrial applications. The moisture involved can be problematic relative to indoor air quality if these devices are not well maintained.

Electronic Air Cleaners. These include electrostatic precipitators (Figure 16.64*c*), which have the advantage of relatively low maintenance—but can pose health concerns due to ozone production. Static electricity is produced in the self-charging filter medium by flowing air; larger particles then cling to the filter. The more humid the air and/or higher the air speed, the lower the filtering efficiency. In a charged media filter, an electrostatic field is created by applying a high DC voltage to the dielectric material of the filter. Many particles are not captured, however, due to the weakness of the field. A two-stage electronic air cleaner first passes dirty air between ionizing wires of a high-voltage power supply. Electrons are stripped from the particulate contaminants, leaving them positively charged. Then these ionized particles pass between collector plates that are closely spaced and oppositely charged. The particles are simultaneously repelled by the positive plates and attracted to the negative plates, where they are collected.

(c) Locating Air Filters

The panel filters provided with HVAC equipment are usually located upstream from the unit fan. High-efficiency particle and adsorption filtering systems should be located downstream from the HVAC cooling coils and drain pans to ensure that any microbiological contaminants from those wet surfaces are removed rather than being distributed throughout the building.

In buildings where a filter system for indoor air quality is integrated with a central HVAC system, the HVAC system should be capable of continuously circulating air at the rate of 6–10 air changes per hour and of operating against the considerable static pressure that results from high-efficiency filters.

It is fairly common practice to protect high-efficiency filters with lower-efficiency pre-filters. With proper maintenance this can reduce the replacement cost of the more expensive high-MERV filters and reduce energy use by keeping high-efficiency filters from becoming clogged as quickly. Pay careful attention to the fan pressure capability of packaged devices such as an ERV/HRV if high-MERV filters are intended to be used.

16.14 VENTILATION EQUIPMENT

The focus of this section is active ventilation systems employed primarily to improve indoor air quality. Passive ventilation systems with a focus on cooling, often based upon cross or stack ventilation, are covered in Chapter 14. This section will start, however, with discussion of windows (a passive means of ventilation) as a basic building block of a ventilation strategy. This will be followed by stack ventilation (often arranged as a hybrid design strategy with special purpose equipment). Lastly, fully active ventilation systems will be presented.

(a) Windows

Operable windows are one of the oldest and most common enclosure "switches" of all. Ventilation through windows (and operable skylights) is influenced by the status of the window. If wind strikes the window in its closed position, it will be deflected. In the open position, a window will admit airflow. The direction of the wind approaching a window is broadly variable throughout the year, as is the wind speed. Both direction and speed are statistically predictable. Window status (open/closed/partially open) is typically a function of occupant decisions. Thus, the success of a window-based ventilation system is a function of several variables that can interact in complex ways.

Design intent for ventilation is crucial to success—is ventilation intended to improve IAQ, to cool occupants, or to do both? For ventilation for air quality control (without intentional cooling), it is usually desirable to keep wind away from people; for comfort cooling at ambient temperatures above the standard comfort zone, wind across the body is helpful. For these reasons, a window that can be opened in a variety of positions can be useful; some examples are shown in Figure 16.65.

Some windows offer more free area of opening than others of the same size. Figure 16.66 compares some common window types. The pattern of incoming flow is also highly influenced by the way in which a window opens. The outward flow is somewhat affected as well. Insect screens will reduce the flow of air. Details of windows and screens are discussed in Chandra et al. (1986).

Perhaps the best aspect of operable windows is that they can give building occupants some control over the flow of outdoor air. Such control, however, can be problematic in that windows will not usually be opened for IAQ purposes if doing so decreases thermal comfort (as will be the case during cold weather). Another concern is that they rarely offer any means of filtering incoming air. They also can confound attempts by a

(a) (b)

Figure 16.65 A window that can open in more than one position can enhance ventilation performance. (a) The window tilts from the top, directing the incoming air toward the ceiling; fresh air does not directly encounter workers within the space. (b) The window swings inward, allowing incoming air to move across people, enhancing warm-weather cooling. (Courtesy of Three Rivers Aluminum Company, Inc.)

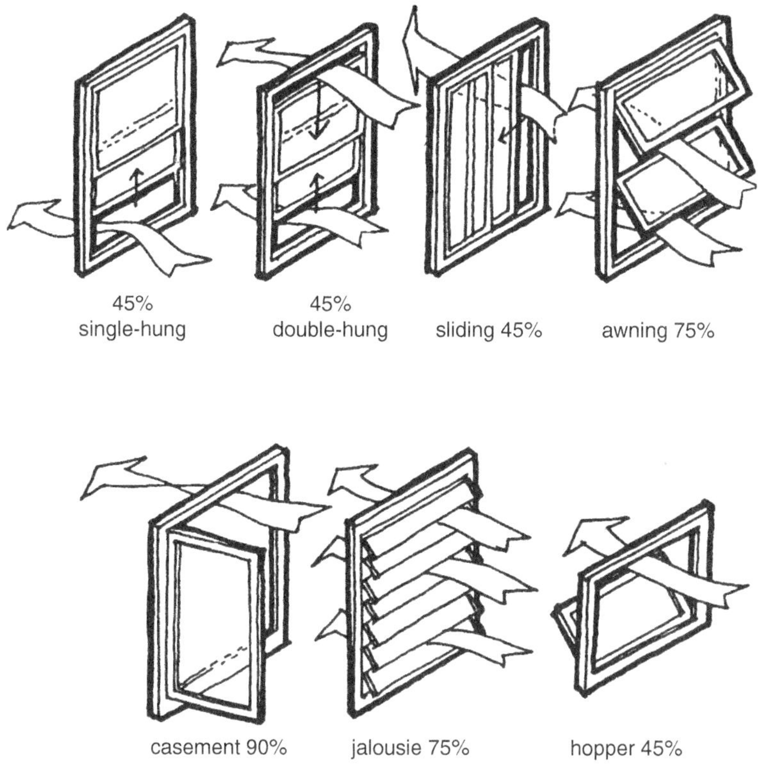

Figure 16.66 *The percentage of actual openable area varies with the window type. (Moore, 1994.)*

central HVAC system to regulate airflow and building pressures. Sometimes they admit air (windward side); at other times, they exhaust it (leeward side). Note that the EPA Campus (see Chapter 5) elected to install fixed, not operable, windows.

(b) Stack Effect

Windows may be used as apertures for both cross ventilation and stack ventilation. Windows work well in the presence of wind—in cross-ventilation arrangements. Estimating cross-ventilation airflow is discussed in Chapter 14. In calm conditions, however, windows may still admit—or exhaust—air due to the stack effect. The taller the building, the more pronounced this effect.

Operable windows in very tall buildings have been shunned by designers until recently. The Commerzbank in Frankfurt, Germany, is a 56-story tower. Each office's exterior window (Figure 16.67) is operable in temperate weather. The occupant associated with each window decides the degree of openness; a lock-out override is controlled by a building management system (BMS). The full-height office window's outer pane consists of fixed single-glazed safety glass, with 5-in. [125-mm] ventilation slots across the top and bottom. These serve the 8-in. [200-mm] wide cavity between the outer and inner windowpanes. The inner

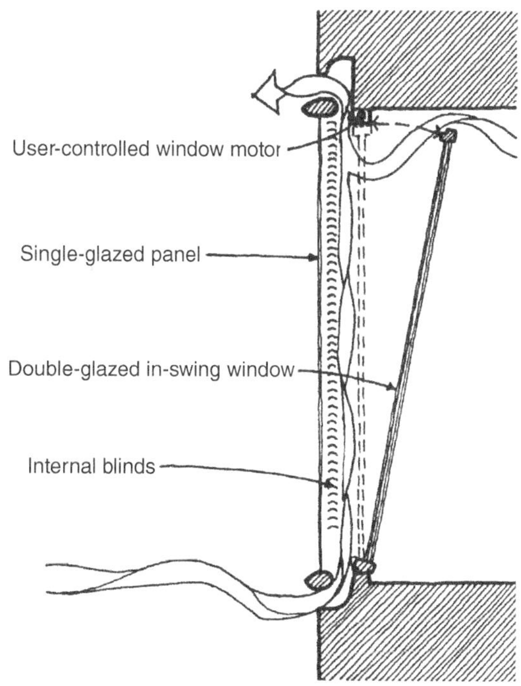

Figure 16.67 *Operable windows in the 56-story Commerzbank tower are tilted in from the top by occupants, while a fixed outer pane with continuous venting slots at the top and bottom keeps air currents under control. Shown here in exhaust mode, the window can also supply fresh air to an office. Under adverse outdoor conditions, the building management system (BMS) locks the inner window in the closed position.*

ACTIVE ENVIRONMENTAL SYSTEMS

window is double-glazed, hinged at floor level, and has a motor-operated tilt-in mechanism at the top. Motorized blinds for solar shading are located within the cavity between the windowpanes. A central atrium provides a stack effect so that an open window is usually a source of incoming rather than exhaust air, although the top and bottom slots allow for a slight stack effect at each window.

Estimating airflow due to the stack effect is discussed in Chapter 14. Devices can be used to enhance the stack effect by creating suction when wind blows across the top of a stack. Probably the most common and readily available are wind-gravity or turbine ventilators (Figure 16.68); typical performance characteristics are listed in Table 16.9. There are other types of stack-topping devices. Figure 16.69 compares volumetric airflow results for a turbine and several other ventilators with those for a simple open stack.

(c) Subslab Ventilation

As discussed in Chapter 5, in some areas the ground contains radon (or other soil gases). In addition, buildings constructed on brownfield sites or former landfills could be threatened by dangerous soil gases. Even benign sites can be threatened by methane gas from a leaking sewer line. A subslab depressurization system can mitigate soil gas intrusion. Such a system may be passive (relying upon stack-effect-driven airflow) or active (relying upon fan-induced airflow). A depressurization system involves installing at least one 4-in. [100-mm] vertical pipe open at both ends. (The number of pipes is a function of ground-contact floor area.) The lower end is set into a layer of clean, crushed rock at least 4 in. [100 mm] thick that lies immediately below the floor slab. The object is to allow air within this rock layer to enter the lower open end of the pipe. The slab is poured and carefully sealed around the pipe, which runs upward (via interior walls) through the roof—where radon and other soil gases are released into the ambient air (rather than leaking into the building). A temperature difference and pipe height drive passive subslab ventilation. A small continuously operating fan drives active subslab ventilation.

Figure 16.68 Two wind-gravity (turbine) ventilators accent the roofline of the addition to Cole Hall at the University of New Hampshire. These devices draw hot exhaust air from the auditorium; cool outdoor air is admitted through low windows and shutters in the auditorium's north wall. (Courtesy of Banwell White Arnold Hemberger & Partners, Inc.)

(d) Exhaust Fans

Exhaust fans are used to remove air that is odorous and/or excessively humid before it can spread more widely. They are an active form of source control, most commonly employed in bathrooms, kitchens, or process areas, to create a negatively pressured area that limits the spread of pollutants to the broader building. Exhaust fans may be stand-alone devices, part of a small simple ducted system, or part of a building-wide complex system. Thus, they can be local or central.

In buildings with heating systems lacking intentional air circulation (such as radiant heating), exhaust fans are often the only built-in devices for moving air. Local fans are often noisy, which is sometimes useful for covering other bathroom noises, but also noisy enough to discourage their use over long periods of time. In stand-alone applications there is seldom much thought given to where replacement air will be drawn from or concern about where exhaust air will be discharged. Although dumping exhaust into attics, basements, or crawl spaces is prohibited by building code.

Exhaust is required by building code. *ASHRAE Standard 62.2–2022, Ventilation and Acceptable Indoor Air Quality in Residential Buildings* (ASHRAE, 2022b), requires that a mechanical exhaust system be installed in each kitchen and bathroom. These exhausts can be either demand-controlled (used as needed) or continuous. Table 16.10 summarizes these exhaust requirements. The intakes for these exhausts should be placed as close to the source of polluted air as possible so as not to drag undesirable air around the room before it leaves. In kitchens, this location is directly above the range (to capture grease, odors, and water vapor); in bathrooms, in the ceiling above the toilet and shower areas to remove moist air. Directly "above the range" is not advisable for exhausts that connect to an ERV/HRV due to potential for clogging of the devices with grease. Some exhaust options are shown in Figure 16.70.

There are three conceptual approaches to building ventilation (and thus to exhaust). The supply-only

TABLE 16.9 Sample Turbine Ventilator Performance

Part A. I-P Units

Outdoor Wind Speed: mph		2			4			6			8			10		
ΔT, indoors-outdoors °F		10	20	30	10	20	30	10	20	30	10	20	30	10	20	30
Turbine Throat Diameter: in.	Height Above Intake: ft	Exhaust Capacity (cfm)														
6	10	114	125	130	210	221	226	314	325	330	426	437	442	534	545	550
	20	122	135	144	218	231	240	322	335	344	434	447	456	542	555	564
	30	129	144	156	225	240	252	329	344	356	441	456	468	549	564	576
	40	135	152	166	231	248	262	335	352	366	447	464	478	555	572	586
10	10	209	222	274	370	383	435	545	558	610	728	741	793	915	928	980
	20	234	269	301	395	430	462	570	605	637	753	788	820	940	975	1007
	30	254	301	328	415	462	489	590	637	664	773	820	847	960	1007	1034
	40	269	318	355	430	479	516	605	654	691	788	837	874	975	1024	1061
14	10	333	383	422	558	608	647	804	854	893	1062	1112	1151	1324	1374	1413
	20	376	444	496	601	669	721	847	915	967	1105	1173	1225	1367	1435	1487
	30	413	496	560	638	721	785	884	967	1031	1142	1225	1289	1404	1487	1551
	40	444	539	614	669	764	839	915	1010	1085	1173	1268	1343	1435	1530	1605

Part B. SI Units

Outdoor Wind Speed: m/s		0.9			1.8			2.7			3.6			4.5		
ΔT, indoors-outdoors °C		5.6	11.1	16.7	5.6	11.1	16.7	5.6	11.1	16.7	5.6	11.1	16.7	5.6	11.1	16.7
Turbine Throat Diameter: mm	Height Above Intake: m	Exhaust Capacity (L/s)														
150	3	54	59	61	99	104	107	148	153	156	201	206	209	252	257	260
	6	58	64	68	103	109	113	152	158	162	205	211	215	256	262	266
	9	61	68	74	106	113	119	155	162	168	208	215	221	259	266	272
	12	64	72	78	109	117	124	158	166	173	211	219	226	262	270	277
250	3	99	105	129	175	181	205	257	263	288	344	350	374	432	438	463
	6	110	127	142	186	203	218	269	286	301	355	372	387	444	460	475
	9	120	142	155	196	218	231	278	301	313	365	387	400	453	475	488
	12	127	150	168	203	226	244	286	309	326	372	395	412	460	483	501
350	3	157	181	199	263	287	305	379	403	421	501	525	543	625	648	667
	6	177	210	234	284	316	340	400	432	456	522	554	578	645	677	702
	9	195	234	264	301	340	370	417	456	487	539	578	608	663	702	732
	12	210	254	290	316	361	396	432	477	512	554	598	634	677	722	757

Source: Extracted from information provided courtesy of Western Ventilating Equipment, Inc., Los Angeles. SI units were appended by the authors of this book (they are not part of the original source).

approach is not viable as an exhaust strategy as it generally pressurizes a building such that exhaust leaves at random uncontrolled locations not necessarily useful for source control. An exhaust-only strategy is commonly employed but encounters problems in tightly sealed buildings (where getting adequate makeup air to replace the exhausted air cannot be assured just through infiltration) and where fuel-burning appliances are installed (which can compete with occupants for oxygen). Codes place specific requirements on makeup air for such appliances. The clear trend in modern building design is toward balanced ventilation systems. These systems have intentional exhaust and intake airflows and can be readily arranged for heat recovery.

ASHRAE Standard 62.1: Ventilation and Acceptable Indoor Air Quality (ASHRAE, 2022c) prescribes minimum exhaust airflow rates for several space types in buildings not covered by ASHRAE *Standard 62.2* (ASHRAE, 2022b). A random extract (not complete) of these requirements includes:

- art classrooms: 0.70 cfm per ft^2 [3.5 L/s per m^2]
- barber shops: 0.50 cfm per ft^2 [2.5 L/s per m^2]
- educational science laboratories: 1.0 cfm per ft^2 [5.0 L/s per m^2]

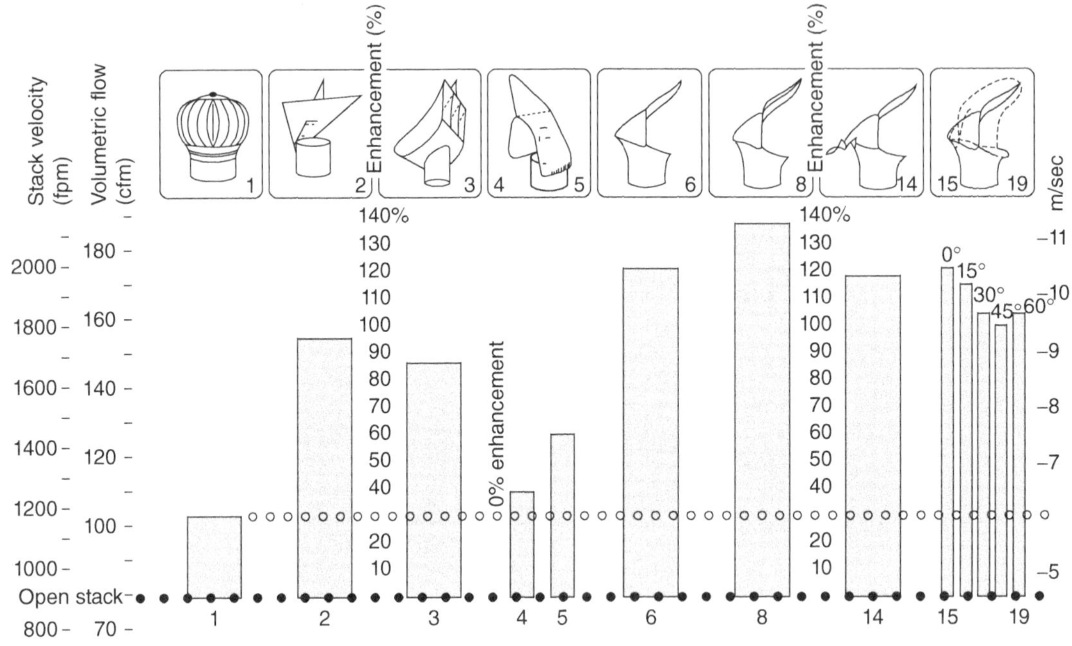

Figure 16.69 Performance of some passive ventilators compared to a simple open stack. The popular turbine ventilator is identified as #1 but is outperformed by many other devices. (From Schubert and Hahn, 1983.)

TABLE 16.10 ASHRAE-Required Exhaust Flow Rates for Residences

Space Type	Demand Controlled cfm [L/s]	Continuous cfm [L/s]
Bathroom	50 [25]	20 [10]
Enclosed Kitchen	Vented range hood: 100 [50] Other arrangements: 300 [150] or 5 ACH	5 ACH (of kitchen spatial volume)
Open Kitchen	Vented range hood: 100 [50] Other arrangements: 300 [150] or 5 ACH	---

Source: Extracted with permission; © ASHRAE, www.ashrae.org. ANSI / ASHRAE Standard 62.2-2022: Ventilation and Acceptable Indoor Air Quality in Residential Buildings.

- janitor closets, trash/recycling rooms: 1.0 cfm per ft^2 [5.0 L/s per m^2]
- parking garages: 0.75 cfm per ft^2 [3.7 L/s per m^2]

(e) Pre-Conditioning Ventilation Air

Fresh air brought into a building can improve IAQ, but at the expense of energy conservation. Summer or winter, outdoor air conditions usually deviate from desired indoor conditions, necessitating that heating or cooling be applied to the outdoor airflow. This sets up a tension between design intents: sometimes expressed as IAQ versus energy. This tension need not lead to compromise. Several low-energy strategies to mitigate this problem are available.

Double-Envelope Enclosure. The office building in Figure 16.71 was an early example of double-envelope building enclosures—a trend that became popular in Europe. This office building was surrounded by a 4-ft-wide [1.2-m] cavity between the inner and outer glass surfaces. Air within this cavity was heated, both by the sun and by indoor heat sources, and rose out of a damper-controlled opening. This building did not utilize its passively heated air for ventilation, but demonstrated the possibility of such an approach. Contemporary double-envelope buildings use a range of often elaborate control mechanisms to improve overall building performance.

Transpired Solar Collectors. The use of a south-facing wall as a winter preheating device with an (unglazed) transpired collector (available as the product SolarWall®) is illustrated in Figure 16.72. Aluminum sheeting, specially finished for excellent solar absorption and penetrated by thousands of tiny holes, is the exterior surface. Behind this is a cavity kept under negative pressure by a fan. Outdoor air is drawn through the holes and into the cavity, where it is heated by the dark outer surface and drawn up the cavity to the fan and then sent on to the space. Insulation and an interior surface complete the south wall assembly. Heat loss from the building through the south wall is mitigated by the solar-heated inflowing outside air. In summer, a fan draws air directly from the outside, bypassing the solar cavity. The holes at the top of the wall serve as outlets for the stack effect produced by the solar gains through the outer surface.

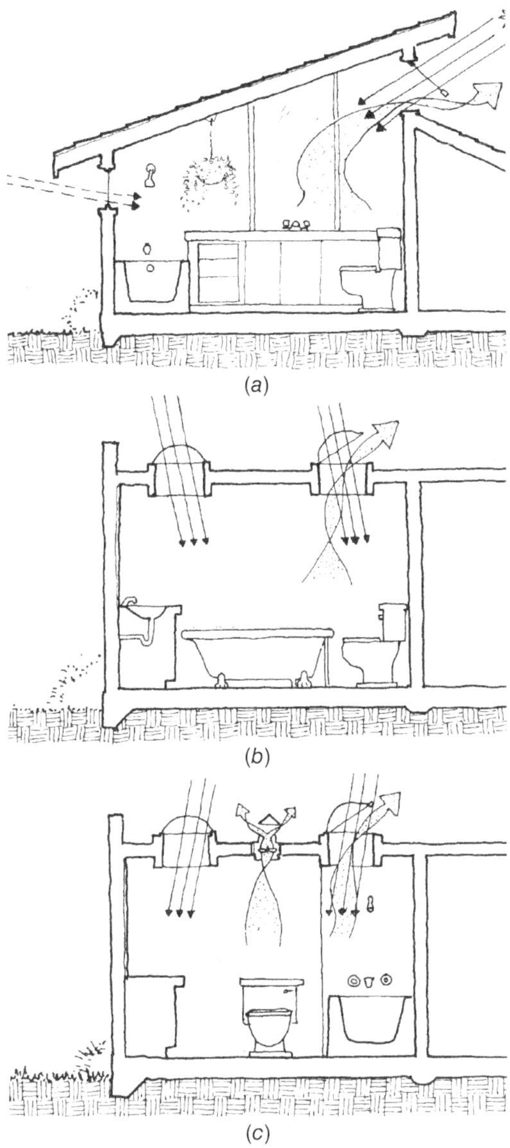

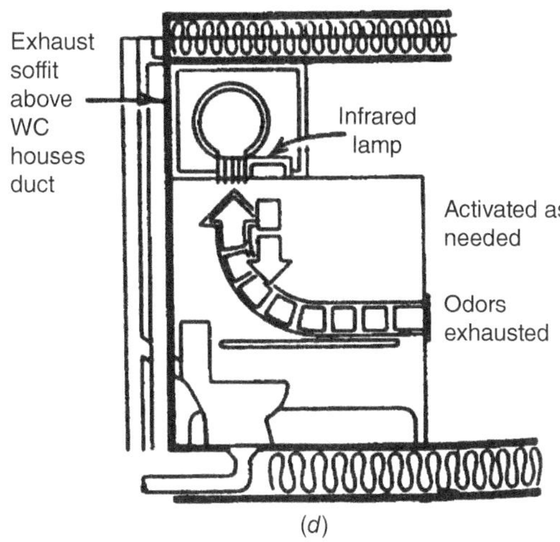

Figure 16.70 *Options for exhausting air from bathrooms: (a) Daylight enters from the sides; incoming air from the window picks up moisture near the ceiling and exits through a high clerestory. (b) Two skylights admit daylight; one exhausts moist air at the ceiling. Makeup air will need to flow from the adjacent rooms. (c) Adding an exhaust fan will ensure negative pressure in the bathroom, isolating moisture and odors from the adjacent rooms. (d) In public bathrooms, exhaust directly above toilets (water closets, WC) is desirable. In cold weather, small radiant heat lamps with timer switches could add heat.*

There are numerous installations around the world; one of the largest, at 108,000 ft^2 [10,034 m^2], is for an aircraft manufacturer in Quebec, Canada. For a design procedure, see NREL (1988).

Stand-Alone Equipment. In some situations, equipment (as illustrated in Figure 16.73) other than heat exchangers can be used to heat and/or cool large quantities of makeup air. Especially common in factories or laboratory buildings with high exhaust air requirements, these devices often supplement the building heating/cooling system, which deals primarily with heat gains/losses through the building enclosure. In warm, dry climates, direct evaporative coolers can be used to temper makeup air because they function with 100% outdoor airflow. In hot and more humid climates, indirect evaporative cooling can help lower the temperature of makeup air. In the Figure 16.74 example, outdoor air is assumed at 104°F [40°C] and 10% RH (Point A). Two streams of outdoor air are involved. An evaporative cooler cools one air stream along a constant wet-bulb temperature line to Point B, where it is now 70% RH but considerably cooler at 73°F [23°C]: in the comfort zone, but humid. At this point, it enters an air-to-air heat exchanger. The second outdoor air stream enters the other side of the heat exchanger, again at Point A conditions. As the two streams exchange heat, they move toward the same temperature: The evaporatively cooled stream moves from Point B to Point C, about 86°F [30°C] and 42% RH and is then exhausted. On the other side, outdoor air moves from Point A to Point D, about 91°F [33°C] and 16% RH. This second air stream is indirectly evaporatively cooled. Although it is still well above the comfort zone, it can then be either cooled by typical refrigerant-based systems or evaporatively cooled until it reaches the comfort zone.

Heat Pipe. This device, as shown in Figure 16.75*a*, transfers sensible heat between adjacent air streams. Within the heat pipe, a charge of refrigerant spends its life alternately evaporating, condensing, and migrating by capillary action through a porous wick. Because the only thing that moves is the refrigerant and it is self-contained, no maintenance and a long life are likely. Efficiency ranges from 50% to 70%; modular

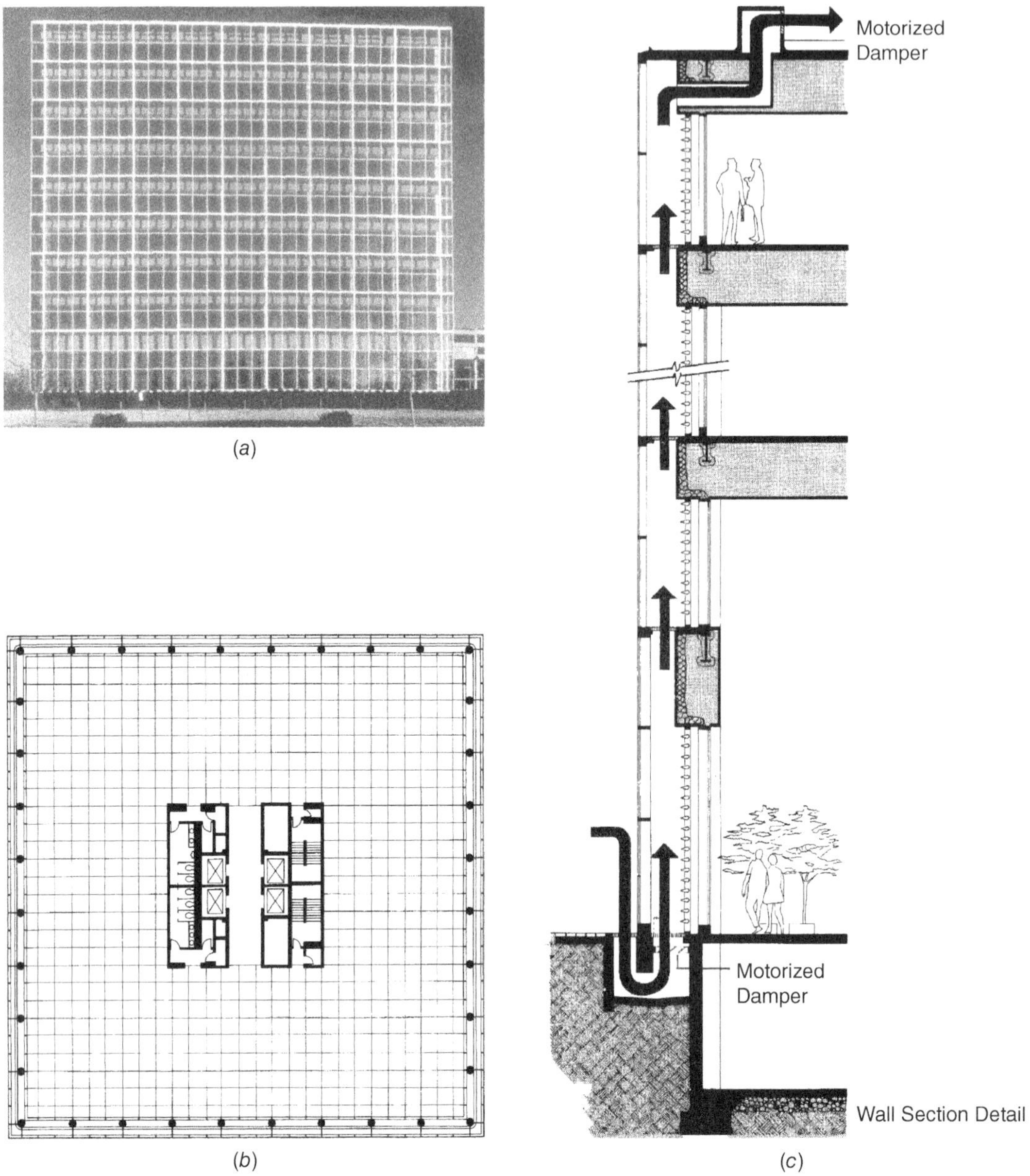

Figure 16.71 The Occidental Chemical Corporate Office Building, Niagara Falls, New York. (a) Fully exposed to sun and wind, the building appears as a conventional curtainwall box. (b) An ordinary-looking plan with a central core and suspended ceilings. (c) On all four sides, a substantial cavity allows for maintenance of movable daylight louvers and window washing, as well as for ample stack effect airflow. Although cavity airflow is released at the roof, it could be utilized during winter as tempered fresh air to the interior. (Architects: Cannon Design, Inc.)

sizes are available to 54 in. × 138 in. × 8 rows deep [1.4 × 3.5 m].

Heat pipes can assist in the dehumidification and cooling of incoming air. Conventional refrigerant-based cooling is an energy-intensive process that often overcools the air to remove (condense) water; the overcooled air is then reheated to make the temperature acceptable. A heat pipe precools the air (subtracting heat) before it enters a cooling coil, then warms it (adding heat) after the cooling coil. No external energy input is required. A heat pipe configured for this task is shown in Figure 16.75*b*.

(f) Energy Recovery Ventilators/Heat Recovery Ventilators

A heat exchanger is a device that facilitates the transfer of sensible heat between two fluids (e.g., air or water) without mixing of the adjacent fluids. Such a device can be used to recover heat from outgoing exhaust air and

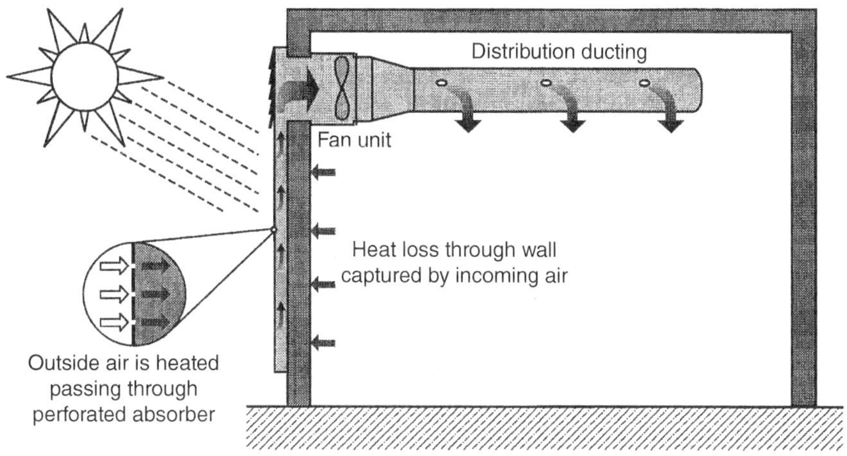

(a)

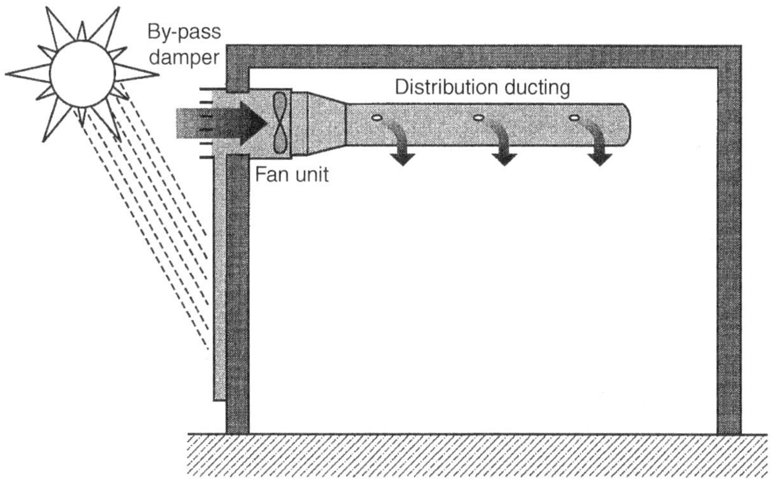

(b)

Figure 16.72 *Winter preheating of fresh air on a south wall using an unglazed transpired collector. (a) Tiny holes in the aluminum skin admit air to a cavity outside the wall's insulation, where it is preheated by sun on the dark aluminum, as well as by heat escaping through the insulation. A fan draws this air to a supply plenum, after which it is distributed to the space. In summer (b), the fan draws directly from the outside, and the collector is self-ventilated by the stack effect. (Courtesy of the National Renewable Energy Laboratory.)*

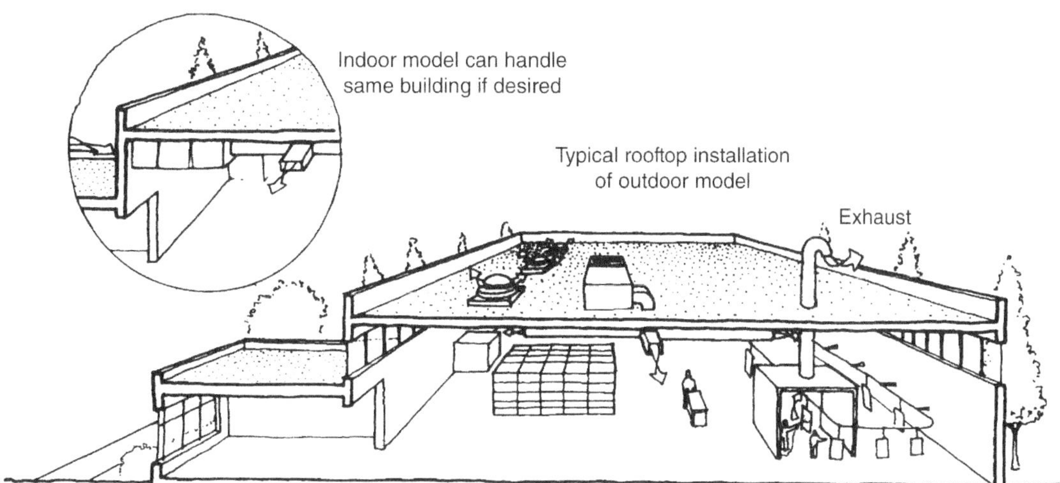

Figure 16.73 *Makeup air heaters/coolers. Distributed on a factory roof, these units prevent negative indoor air pressures by providing makeup air to replace the exhaust from processes. Often, a separate heating/cooling system will be used to deal with heat gains or losses through the building enclosure.*

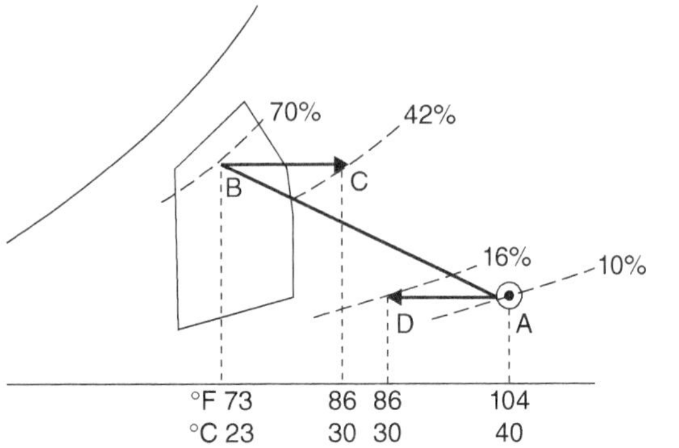

Figure 16.74 *Indirect evaporative cooling can precool ventilation air. The psychrometrics of the process.*

transfer the heat to incoming outdoor air. As the air tightness of buildings increases and fewer air changes per hour (ACH) occur from infiltration (unintended air leaks), mechanical (versus passive) ventilation becomes more attractive as a means of counteracting indoor air pollution. When a heat exchanger is used, it is possible to maintain an adequate supply of fresh air without severe energy consumption consequences. Figure 16.76 illustrates the basic principle of a simple air-to-air heat exchanger for residential-scale buildings.

These devices are termed heat recovery ventilators (HRVs) if they exchange only sensible heat (temperature). They are termed energy recovery ventilators (ERVs) if they exchange sensible and latent heat (temperature and moisture). ERVs and HRVs are both viable technologies; one will be selected versus the other based upon economics—usually driven primarily by climate and secondarily by the cost of energy. Note again that the outgoing and incoming air streams must be adjacent for both types of devices. This has architectural design implications.

Commercially available heat exchangers can extract anywhere from 50% to 90% of the heat from exhaust air. The lower the volume of airflow, the higher the efficiency. Table 16.11 shows representative sizes, airflows, and efficiencies for such devices. General best practice suggests that ERV/HRV devices be employed as stand-alone ventilation devices—although optional installation approaches have been used. Figure 16.76*a* diagrams the concept of an ERV/HRV; Figure 16.76*b* is an example of a typical residential-scale device.

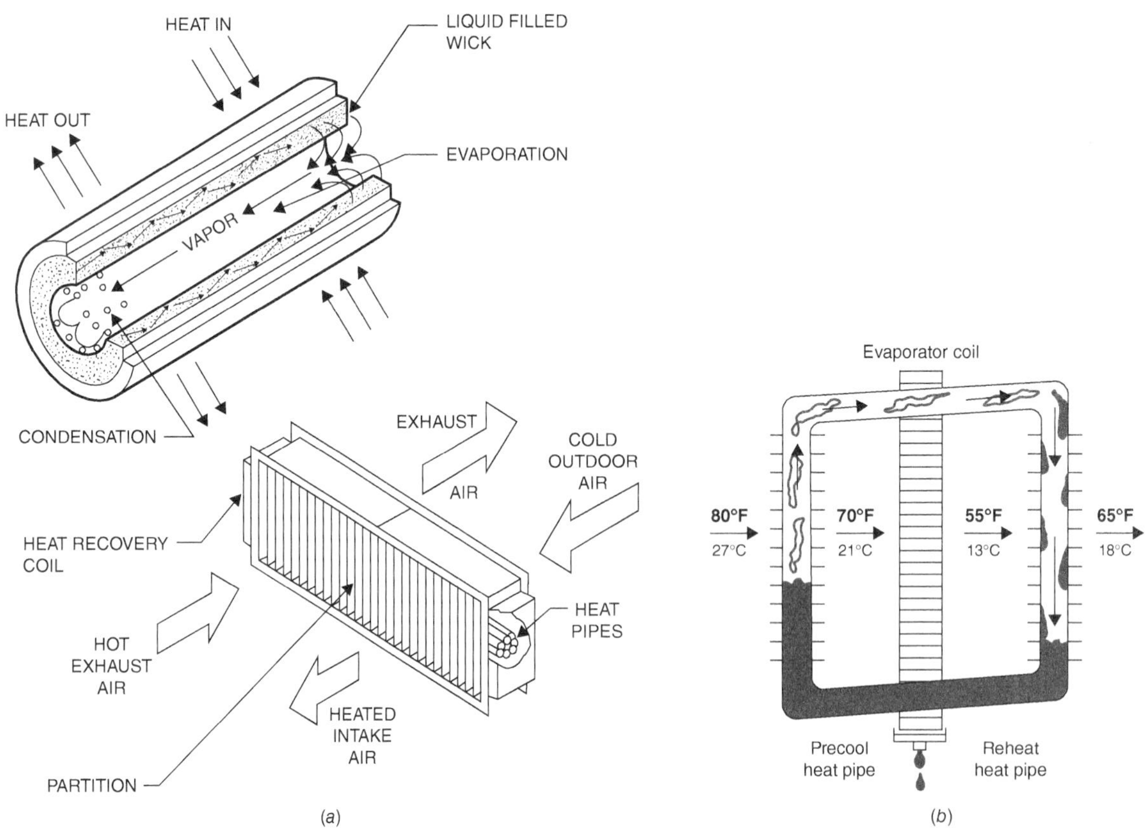

Figure 16.75 *A heat pipe is a self-contained device with no moving parts. (a) It silently transfers sensible heat between adjacent outdoor-air intake and exhaust air streams. (b) The heat pipe can increase moisture removal by first precooling incoming hot air before it reaches the cooling coils. Then the heat is returned to the cooled air, bringing it to a temperature-humidity combination acceptable for supply air to a space. (a: Ramsey and Sleeper, 2016 / with permission of John Wiley & Sons.; b: Courtesy of Heat Pipe Technologies, Inc., and Environmental Building News.)*

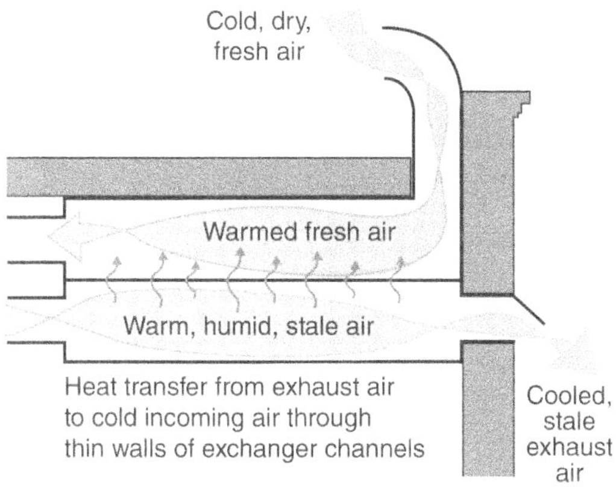

(*a*)

(*b*)

Figure 16.76 *Air-to-air heat exchangers are particularly beneficial in providing energy-efficient balanced ventilation. (a) The basic principle of operation. (b) An ERV installed at Ball State University's Alley House duplex in Indianapolis, Indiana, winner of the 2023 Solar Decathlon Build Challenge. (b: © Tom Collins, used with permission.)*

Some design considerations for air-to-air heat exchangers:

1. Avoid using them on exhaust air streams that are contaminated with grease, lint, or excessive moisture (through cooking and clothes drying in particular) because clogging, frosting, and fire-hazard problems can develop.
2. In cold weather conditions, a built-in defroster, which will consume energy, will be needed.
3. Carefully locate the outdoor air intake. Keep this intake as far as possible from the exhaust air outlet, to avoid drawing contaminated air back into the building. Keep the intake away from pollution sources such as vehicle exhaust, furnace flues, dryer and exhaust fan vents, and plumbing vents.

A student housing complex in Greensboro, North Carolina, utilizes ERVs to exhaust air from bathrooms (Figure 16.77). Each floor of the three-story building has four apartments, each with 1078-ft^2 [100-m^2] floor area and 8-ft [2-m] ceilings. Individual ERVs accept air from the two small bathrooms in the apartment. Control is accomplished by switches in the bathrooms. Prefiltered outdoor air is drawn into the ERV, exchanging heat with the outgoing exhaust air. This fresh air is then fed directly into the air handler for each apartment heat pump, located adjacent to the ERV. Thus, the bathrooms are under negative pressure relative to the rest of the apartment. The outdoor air intake is located high on the wall, the exhaust outlet is lower. Intake and outlet locations are separated by a minimum of 8 ft [2 m].

(g) Energy Transfer Wheels

Energy transfer wheels (Figure 16.78) are similar to ERVs in that they transfer both latent and sensible heat. These devices, however, provide greater capacities than ERVs and will be used in larger commercial/institutional buildings. In cold weather, they recover both sensible and latent heat from exhaust air; in warm weather, they cool and dehumidify incoming outdoor air. Seals and laminar flow of air through the wheels prevent mixing of the exhaust and incoming air streams. As a further precaution, each sector of the wheel is briefly purged, using outdoor air to blow away any residual effects of the exhaust air on the wheel surfaces. Carryover of undesirable exhaust air qualities is between 4% and 8% without purging and less than 1% with purging. Efficiency ranges from 70% to 80%, and available sizes range up to 144 in. [3.6 m] in diameter. Table 16.12 shows some representative sizes, airflows, and efficiencies.

16.15 CONTROLS

Controls and their underlying control logic are a critical part of an HVAC system. Most HVAC systems are actuated and regulated by automatic controls. Most control systems in larger buildings are computerized. The most noticeable HVAC control function might be to maintain desired thermal comfort conditions. HVAC controls are just now starting to focus on

TABLE 16.11 Representative Heat Exchanger Data for Smaller Buildings

Air-to-Air Sensible Heat Exchangers				
Approximate Size[a] *L* × *W* × *H*		Airflow Range		Temperature Recovery at 32°F [0°C] (%)
in	mm	cfm	L/s	
21 × 13 × 25	535 × 330 × 630	50–120	25–60	70
34 × 17 × 21	865 × 435 × 530	75–225	35–110	80
53 × 17 × 21	1335 × 435 × 530	75–225	35–110	93
47 × 20 × 37	1195 × 505 × 940	350–700	165–330	65

Source: Conservation Energy Systems, Inc., Saskatoon and Minneapolis.

[a]Access to one face (L × W) is needed for servicing.

Note: Higher airflow resistance values will require increased fan energy.

independently maintaining desired indoor air quality. Controls increase energy efficiency by promoting improved operation. They act as safety devices, limiting or overriding mechanical and electrical equipment. Automatic controls can also serve to eliminate human forgetfulness and bias. Then again, controls should act in favor of occupant satisfaction.

Although precise control of temperature and humidity everywhere in a building may be a tempting thought, controls can usually maintain only a range of conditions, not a specific setpoint. This range is influenced by the accuracy of sensors (cost often is an object) and the deadband of the control device (it is not desirable for conditions to change quickly and/or in very small increments). In addition, temperature variations will occur vertically within a space—the taller the space, the greater the variation. Variations between horizontal positions within a zone are also

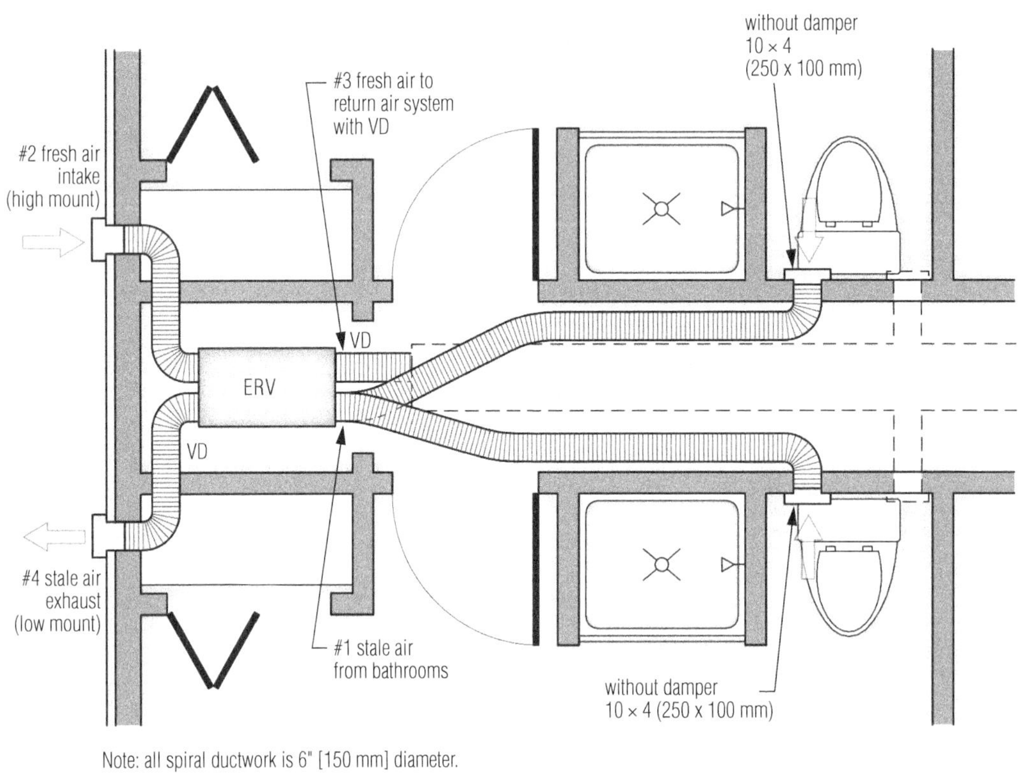

(*a*)

Figure 16.77 *Energy recovery ventilator (ERV) serves two small bathrooms (a) shown in plan in a student housing complex in Greensboro, North Carolina. The demand-controlled ERV (b) is activated by switches in either bathroom. Air is drawn out beside the water closet and exchanges heat (at about 85% efficiency) with incoming outdoor air. This tempered outdoor air is then mixed with some return air and fed directly to the indoor unit of a heat pump (c). The supply air is then fed to other rooms, ensuring a negative pressure in the bathroom while the ERV operates. Each of the 16 apartments has its own ERV and heat pump. (Design by Harry John Boody.)*

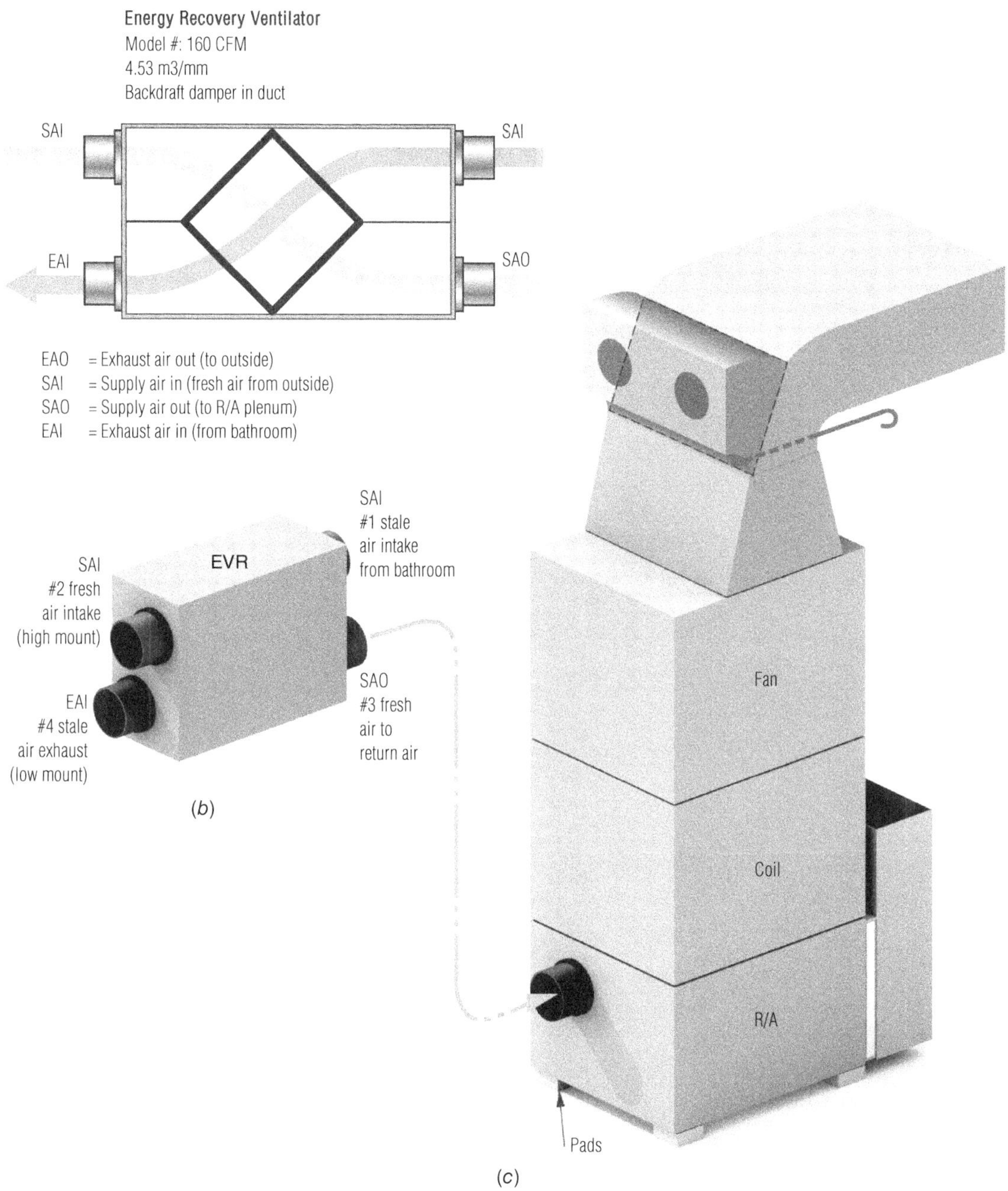

Figure 16.77 *(continued)*

highly likely, especially where one or more walls are exterior and where different rooms are combined in a zone. Given all these considerations, in noncritical applications, a control system generally tries to keep a space from becoming uncomfortable.

Individual control components can be classified as follows:

- *sensors*, devices which can detect various conditions including air temperature, relative humidity, bulk moisture presence, indoor air quality, illumination levels, and sound pressure levels
- *monitors*, similar to sensors in terms of detection, but these devices can regularly record conditions and electronically transmit the information
- *controllers*, which analyze input data and initiate action
- *actuators*, which implement a control action (such as opening a damper or modulating a valve)
- *limit and safety controls*, which function only infrequently, preventing damage to equipment or buildings
- *accessories*, a miscellaneous collection of special devices

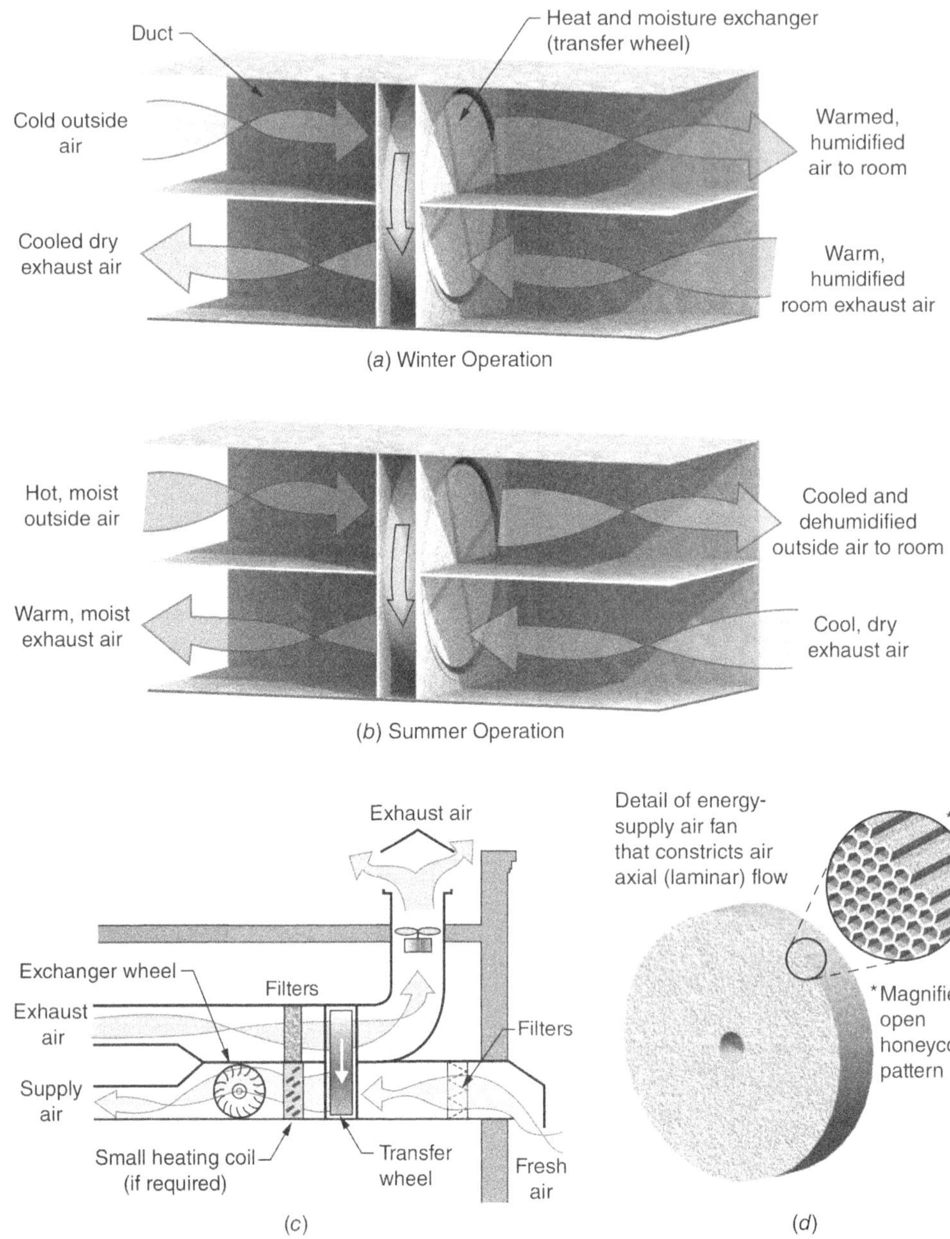

Figure 16.78 Energy transfer wheels. The wheel surface is impregnated with lithium chloride (or similar material), which absorbs moisture and transfers it to the other air stream. The wheel delivers moist air in winter (a) and dry air in summer (b). Cross-section (c) through the wheel and the two air streams it serves—with a closeup (d) of honeycomb construction. Exhaust air may be filtered to help keep the wheel clean. (Courtesy of Air Xchange, Inc.)

Control systems can be classified by power source: electric (both analog and direct digital control); pneumatic (in which compressed air is the motivating force); and self-contained, including "passive" controls such as those motivated by thermal expansion of liquids or metals. Control system devices can be classified by the motion of the controller: two-position systems are of the simple on–off type; multiposition systems have several variations of the on position, and are commonly used for separate operation of more than one machine; floating controls can assume any position in the range between minimum and maximum; central logic control systems can be programmed to integrate the many aspects of building control into one decision-making unit, and are now the prevalent approach to building management.

Control diagrams for two common HVAC applications are shown in Figure 16.79. In single-duct VAV systems (Figure 16.79*a*) with a constantly varying airflow rate, the fan must be regulated so as to maintain the minimum pressure (and therefore, flow) needed at the most demanding outlet. This outlet may be either

TABLE 16.12 Representative Energy Transfer Wheel Information

Air-to-Air Sensible + Latent Energy Wheels				
Approximate Size[a] L × W × H		**Airflow Range**		**Efficiency**
in.	**mm**	**cfm**	**L/s**	**(%)**
46 × 32 × 24	1170 × 815 × 610	500–1000	235–470	84–82
54 × 48 × 34	1375 × 1220 × 865	600–1600	285–760	87–80
		1600–2300	760–1090	75–70
67 × 67 × 43	1705 × 1,705 × 1095	1750–3250	830–1535	85–80
		3000–4500	1420–2125	76–71
124 × 84 × 60[b]	3150 × 2135 × 1525	3500–6500	1660–3070	85–80
		6000–9000	2835–4250	76–71

Source: Greenheck Fan Corporation, Schofield, WI.

[a]Several arrangements (interior and rooftop) are available; dimensions do not include service clearance access, supply hoods, or exhaust hoods.

[b]This unit contains two wheels.

Note: higher airflow resistance values will require increased fan energy.

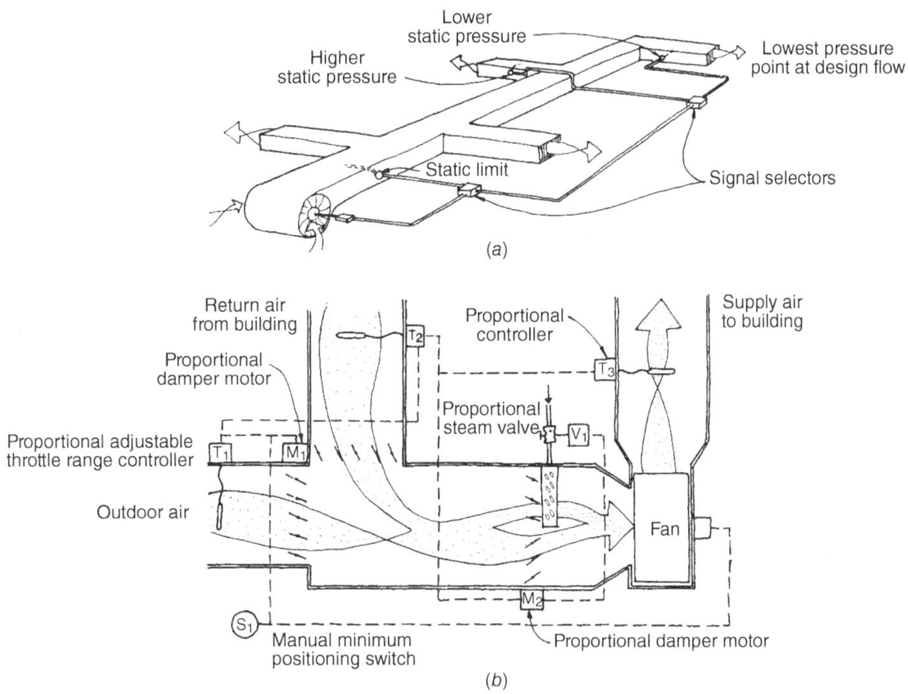

Figure 16.79 *Controls for some common HVAC applications. (a) Single-duct VAV. (b) Economizer cycle.*

the one most remote from the fan or the one needing the greatest flow (because it has the highest load) at the moment. Economizer cycles (Figure 16.79*b*) compare outdoor conditions to indoor conditions, and vary the proportion of fresh (outdoor) air to return air in order to provide "free" cooling (see Figure 16.146).

It is commonplace today for buildings to at least be installed with building-level meters for energy and water. Meter readings can be manual or automated. Often, a facility management team may benefit from more granular information offered by submetering—which, in regard to energy is sometimes referred to as advanced energy metering. Best practice with advanced energy metering is to track any individual energy end use that represents 10% or more of the total energy consumption of a building. Energy meters serve to monitor consumption and production. As such, they are more valuable if permanently installed and recording at intervals of one hour or less—with the ability to digitally transmit information.

Most of today's large buildings are regulated by centralized control systems, usually called building management systems (BMS). The BMS system collects data from various building meters, sensors, and equipment, processes the data, and issues information as well as commands to regulate various automated building systems and equipment based on predefined criteria and user inputs. The system is comprised of integrated hardware and software.

The goal of a BMS is a productive, cost-effective environment achieved by balancing the interrelationships among the building's structure, systems, services, and operation. The HVAC system is centrally regulated and interconnected with lighting, electric power (such as load shedding), elevators, service hot water, access control and security, telecommunications, and information management. Such systems not only maintain comfort with energy efficiency, they also sound the alarm about malfunctions, can learn from past experience, and keep records of system performance. In practice, the use of the term building automation system (BAS) is largely a case of semantics as both BMS and BAS are used loosely and interchangeably. However, some contend that "management" denotes more advanced building systems integration and interoperability. In the arena of HVAC systems, the terms energy management system (EMS) and building energy management system (BEMS) are also used. In any event, the advanced, integrated, centralized aggregation of hourly (or even more frequent) real-time performance data from energy-related building systems—especially HVAC systems—is essential for successful advanced building monitorization, diagnostics, early warning of system malfunction, and controls-based ongoing commissioning. See Chapter 32 for additional information on building automation.

Placement of certain control components, such as sensors and monitors, is critical. For instance, thermostat sensors should be located away from a primary heating and cooling sources to provide a more representative record of the average spatial condition. Similarly, IAQ monitors should be placed in the breathing zone. Additionally, the numbers of sensors and monitors are important and may be influenced by the number of occupants, the number and types of spaces, and/or code or green building certification requirements. Some monitoring systems have in-unit memory, while others are tied into a BMS system. Sensors and monitors need to be properly calibrated to ensure the validity of their measurements. These devices will exhibit "sensor drift" over time and, as such, a calibration schedule should be developed as part of a facilities management plan. Some devices offer the option of remote calibration through access via the BMS; some automatically recalibrate based on historical recorded data.

Extensive integration and automation of building control systems are made possible through direct digital control (DDC), which can be applied to a wide variety of controlled elements. DDC is a controls solution in which digital devices (e.g., microprocessors, controllers) are used to directly manage HVAC systems operations. DDC provides "dynamic control" in the case of HVAC, with the ability to anticipate time-based changes in heat flow patterns or in occupancy schedules. These actions depend on direct digital microcontrollers located on each piece of regulated equipment throughout the building. Three building types show applications of this comprehensive automated control opportunity.

Laboratories tend to present especially difficult HVAC control problems due to their many fume hoods. Conditioned air is provided from a central HVAC system, often by a VAV supply. Whenever a fume hood is exhausting air (frequently in huge quantities), the VAV supply and return systems are affected. Complicating this relationship is the nature of the laboratory work: when an experiment could be damaged by outside contaminants, the lab should be positively pressurized to minimize infiltration (and the VAV must therefore provide a net air flow of slightly more air than the fume hood exhausts). However, when experiments involve diseased, toxic, or other hazardous substances, the lab must be negatively pressurized (and the VAV must supply a net of slightly less air than the fume hood exhausts). DDCs tied to a central system can balance energy efficiency, lab worker safety, and safety for the non-lab environment.

Hotel rooms can incur serious energy waste from heating/cooling an unoccupied room or a room with open windows. With DDCs interconnected with the registration desk, an "unoccupied" mode of operation can be remotely established, with bare-minimum heating or cooling. When the room is occupied, the heating/cooling supply can be throttled back whenever a window is open. Also, a "purge" mode can enable a new arrival (or the front desk, in anticipation) to select a greatly increased flow of outdoor air for a limited time period to dilute odors. Chapter 32 presents more detailed information on intelligent buildings—for which some of these control ideas are precursors.

Offices might be provided not only with DDC for the VAV supply units but also with DDCs for an interconnected network involving a ventilating window (preventing simultaneous open windows and HVAC delivery), daylight reflectors (mini-light-shelves), venetian blinds, a radiant heater valve, an electric light switch, and an insulating shade. A control panel or "dashboard" can give each worker the opportunity to interact with the central control in operating these devices.

Office buildings that use passive strategies can also benefit from a BMS. The British Research Establishment building for fire research, "Building 16" (Figure 16.80), depends upon cross ventilation and stack ventilation for its cooling, and upon movable louvers on the south windows for sunshading and daylight distribution. A common network links this building's control systems; each worker has a monitor-based controller with a graphic user interface (GUI) that regulates lighting and

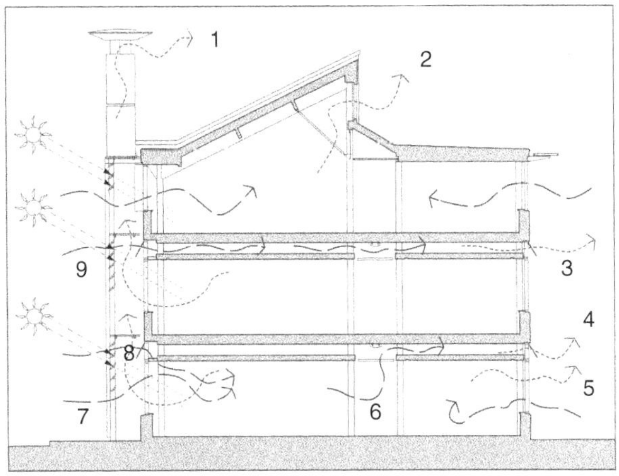

Office cross-section

(a)

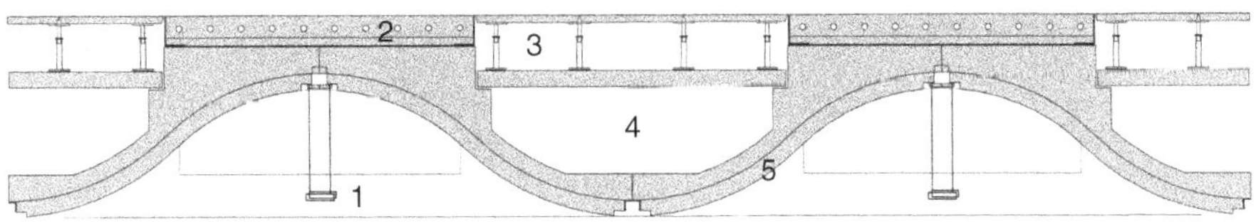

Floor construction

(b)

Figure 16.80 *Natural ventilation and daylight strategies dominate the north–south section (a) of the British Research Establishment Building 16. (1) Stack ventilation (hot, calm); (2) clerestory BMS-controlled ventilation; (3) night ventilation through slab, BMS-controlled; (4) cross-ventilation bypass over enclosed offices; (5) enclosed office single-sided ventilation; (6) corridor cross-over zone; (7) manually operated lower-level windows; (8) high-level BMS-controlled windows; (9) motorized external shading louvers, also BMS-controlled. (b) East–west section detail of precast floor structure. (1) Luminaire with integral photosensors; (2) heated/cooled screed using a ground-source; (3) raised access floor for wiring; (4) cross-ventilation duct (night ventilation and/or cross-over ventilation for enclosed offices); (5) waveform precast concrete ceiling with poured-in-place topping slab. (Courtesy of Feilden Clegg Architects.)*

can override the programmed settings of the nearest high-level windows (for ventilation) and south-window louvers (for sun control and daylighting). This innovative office building uses a ground-source heat pump (vertical field type) to supplement solar and internal winter heat gains or to provide supplementary cooling to the night ventilation system. Note the provisions for cross ventilation for individually enclosed offices on the ground floor, utilizing a cavity above the ceiling to carry ventilation air to an exhaust on the other side of the building.

Rapid changes in centralized control systems continue to occur. Initially, DDC systems were developed by individual companies, with little or no opportunity for communication between systems. Such proprietary systems left designers frustrated by the inability to specify a wide variety of products that could all be controlled by a single BMS. In the late 1990s, three competing communications protocols emerged that promised integration of products and systems from different manufacturers. BACnet (Building Automation and Control network) was developed by ASHRAE as a nonproprietary communication standard. ASHRAE continues to advance BACnet through *ASHRAE Standard 135-2020: BACnet—A Data Communication Protocol for Building Automation and Control Networks* (ASHRAE, 2020).

LonMark, a systems communications standard developed by the Echelon Corporation, is based on a proprietary communications protocol called LonTalk. This protocol integrates data from a comprehensive network platform called LonWork. Support for LonMark has waned compared to BACnet due to the closed-system nature of LonMark. A third protocol, Modbus, has been around since the 1970s when it was developed by Modicon, Inc. It is an open communications protocol primarily used to facilitate the interoperability of electronic equipment in industrial applications. The market has more recently seen an uptake of open communications protocols across the building sector. Open control system architecture is an approach that is intended to let components from several vendors interoperate over a BACnet-adapted Ethernet LAN (local area network). This makes possible a system combining BACnet, LonMark, and proprietary subsystems (Figure 16.81).

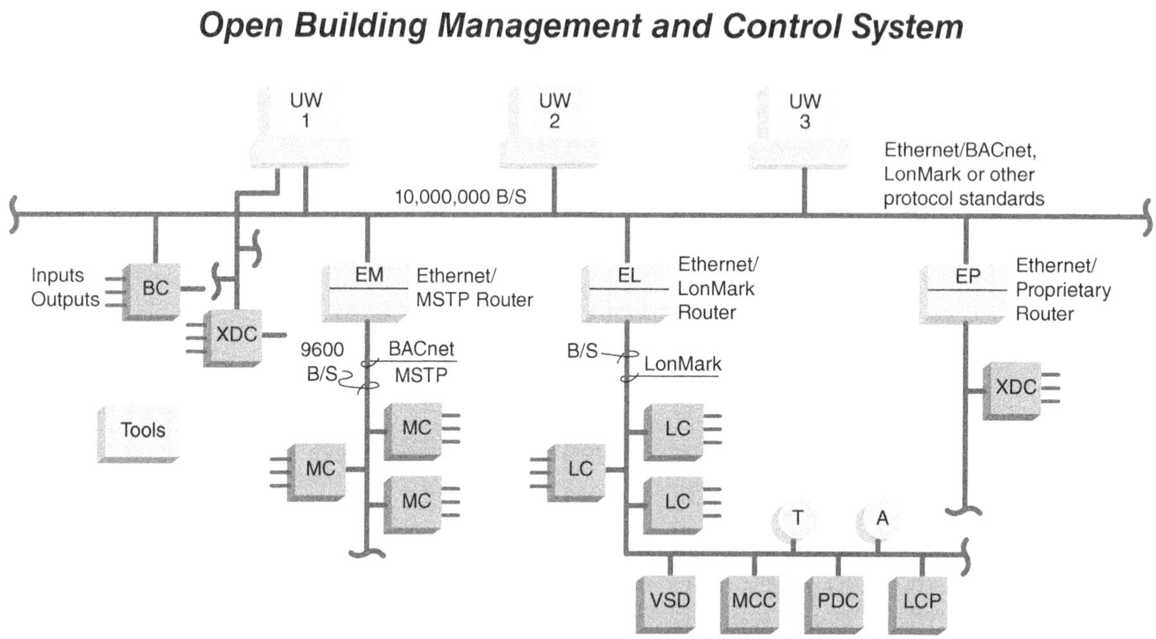

A	- Intelligent actuator	MCC	- LonMark-compatible motor control center
BN	- BACnet native controller that connects directly on the Ethernet/BACnet bus	MSTP	- Master/Servant Token Passing protocol
EL	- Ethernet/BACnet-to-LonMark router	UW	- User workstation with BMCS and BACnet software and Ethernet card
EM	- Ethernet/BACnet-to-BACnet Master/Servant Token Passing (MSTP) protocol router	PDC	- LonMark-compatible power distribution center
EP	- Ethernet/BACnet to a proprietary protocol router	T	- Intelligent temperature sensor
LC	- LonMark-compatible controller	Tools	- Programming, maintenance, and diagnostic tools required
LCP	- LonMark-compatible lighting control panel	VSD	- LonMark-compatible variable speed drive
MC	- BACnet/MSTP compatable controller	XDC	- Proprietary digital controller

Figure 16.81 *Building management system (BMS) that is open to various protocol standards. (Courtesy of Honeywell, Inc.)*

Wireless control systems are an emerging trend. A wireless system has both pluses and minuses. A big plus is the lack of need for power and data wiring at a hundred or multiple hundreds of locations in a building. This can reduce the cost of control systems and make retrofit installations feasible. Minuses include the need to change batteries on hundreds of devices scattered around a building in often inaccessible locations, and problems with wireless signal reception in some buildings. It is likely that the pluses will eventually outweigh the minuses, and wireless control systems will become more common. Security of wireless controls against hacking, however, is a potentially serious design consideration.

Other, more esoteric, control possibilities (such as self-diagnosing equipment monitors, nanocontrols, and neural networks) lie on the horizon. Neural networks involve automation systems that are capable of learning from use. They can predict usage patterns, adjusting operations in advance without needing specific commands from occupants. When the building use pattern is highly predictable, as with many retail and commercial occupancies, these self-programming systems can learn very quickly how to anticipate needs while conserving energy.

Advanced HVAC optimization solutions are rapidly emerging. Such systems leverage artificial intelligence (AI), machine learning, and Internet of Things (IoT) technologies to improve building performance both in terms of energy resource requirements and occupant comfort. Such systems aggregate building systems data and then apply the data to AI algorithms by which better performance can be achieved by incorporating usage trends, current and future weather conditions, and future performance projections.

The emergence of advanced digital controls and building management platforms has ushered in demand response systems and programs. Demand response (DR) is an energy management strategy through which a building's electricity consumption can be adjusted in response to supply conditions. Most BMS solutions can be programmed to integrate

DR capabilities by which the energy usage of certain building functions can be automatically adjusted in response to signals from the local utility. This will require the building operator to formally enroll into a DR program through the utility and for the BMS' DR functionality to be activated (i.e., just because a DR capability is present does not mean the building systems will automatically adjust to signals from the utility). The typical purpose of DR programs and systems is to enable the electrical utility to shave utility-scale peak loads—the maximum electrical power demand that a utility must supply. A smaller utility-scale peak load means the utility can function at a lesser capacity (and therefore a lesser frequency of activating fossil fuel-based auxiliary energy resources), which maximizes options for maximizing the utility's use of renewable energy resources.

Residences are seeing increasing use of remote control and automation (in many cases promoted by utilities or cable companies). Residential building management systems are now capable of regulating far more than temperature. They can activate systems in advance of the occupants' arrival and provide real-time status reports on building conditions. Door and window locks, security cameras, lighting, and appliances are potential actors in a comprehensive control system.

In residences, usage patterns are varied and difficult to predict. Experiments at the University of Colorado revealed that, even so, patterns may be more predictable than was first thought (Mozer, 1998). The Adaptive Control of Home Environments (ACHE) system has two objectives: to anticipate occupants' needs and to save energy. Lighting, air temperature, ventilation, and water heating are controlled so that just enough energy, just when needed, is provided. Lighting would be set to the minimum required levels, hot water maintained at the minimum temperature that would meet demands, only occupied rooms would be kept at a comfortable temperature, and so on.

The control framework then compares energy costs to "misery" costs. Minimum settings and their time patterns are developed as the building is occupied over time. Whenever a preset minimum is overruled by an occupant, the system learns and readjusts accordingly; however, it occasionally tests lower settings to be sure that energy conservation is not being unduly sacrificed to ingrained past practices.

HVAC SYSTEMS

With an understanding of fundamentals and components in place, it is reasonable to look at examples of actual HVAC systems and their application in buildings. Considering the wide variety in possible design objectives and the large number of individual components that may be assembled into an HVAC system, it should be clear that not every possible system will be addressed in the coming pages. It is also good to remember that all but local HVAC systems are field-assembled from a large number of components that are provided by a variety of suppliers and manufacturers. For example, the HVAC system in a single-family residence may use a heat pump from manufacturer "A," ductwork from fabricator "B," piping from hardware store "C," an energy recovery ventilator from manufacturer "D," and so on. Most HVAC systems are truly unique—often resembling other systems, but never quite replicating them. A lot of things can go wrong in such a scenario, so the commissioning process is highly recommended.

16.16 HVAC SYSTEMS: TAXONOMY

(a) System Scales

Starting from the philosophy that a sense of taxonomy can help with the understanding of complex systems, a classification of HVAC systems is presented. HVAC systems are commonly classified by scale as:

- *Local:* A local system is usually intended to serve one zone (with the zone usually consisting of one or at most a few rooms). The system is self-contained (the source, distribution, delivery, and control components are close-coupled and generally within one package). The system is usually located within the space being served (with implications for aesthetics and flexibility). The system is typically of small capacity and small size (with impacts on efficiency). The system is not commonly controlled from a centralized location (this can be a positive or a negative consideration). A window air-conditioning unit is an example of a local HVAC system.
- *Central:* A central system serves multiple zones from one location (or one zone from a remote location). A distribution system is required to transport heating/cooling effect from its place of origin (such as a mechanical room) to the system zones. System scope can vary greatly, from a single-family residence to thousands of square feet (square meters) of office or laboratory. A building may be served by one central system or by multiple central systems. A variable air volume (VAV) system is an example of a central system.
- *District:* A district system serves multiple buildings. Normally district systems provide heating and/or cooling effect for campuses or special utility

zones. The buildings being served usually have their own central HVAC systems. Economies of scale are possible with the large-capacity equipment typical of a district system (this can include bulk purchase of fuels or electricity, customized operating control sequences, outstanding maintenance quality, and highly trained operators). Ball State University's ground-source heat pump system (Ball State University, 2013) is an innovative example of a district system.

Long-distance steam distribution has been used for more than a century; the development of high-temperature water (HTW) and chilled water distribution among buildings is more recent. Offering many advantages, high-temperature, high-pressure hot water, and district chilled water systems are widely used on military bases, airports, and for groups of buildings such as hospital complexes and college campuses. Increasing efforts are being made to install new district heating/cooling networks served by existing fossil-fueled electricity generating plants. Such plants waste more than half of their fossil fuel input, and district heating/cooling could utilize much of that waste.

Water will not flash into steam if kept at sufficiently high pressure. It may then be circulated by pumps through supply and return mains and through branches to heat exchangers, which operate conventional low-pressure hot water systems, generate steam, and perform numerous other thermal tasks. Pressures are on the order of 400 psig (pounds per square inch gauge [2800 kPa (kilopascal)] and temperatures are about 300°F [150°C]. During its circuit, the water will sometimes lose up to 150°F [83°C] and 60 psig [414 kPa] in pressure. The section shown in Figure 16.82 illustrates a common arrangement.

High-temperature water has a number of advantages over steam for certain installations. It uses a two-pipe distribution system, and the temperature drop in the supply main is often as little as 10°F [5.5°C]. With reasonably high water velocities, main pipes can be reduced to almost half the size of those required for steam distribution, with no need for steam traps and pressure-reducing valves. The pipes need not pitch to low points (to accommodate condensation), as is the case with steam, but can follow the contours of the ground. Although installation costs are greater, operational costs are less than those for steam. Feed water treatment is negligible and corrosion is minimal. Expansion and insulation issues are the same as in other subterranean systems. Large sweep-type loops accommodate expansion between fixed points, and underground piping is embedded in thermally effective insulative fill.

District chilled water systems also offer advantages. Large central chillers are likely candidates to use waste heat in a non-CFC absorption refrigeration cycle. Natural coolth sources are possible; the Toronto (Canada) District Heating Corporation uses Lake Ontario water, drawn from a 1.6-mile [2.6-km] intake at a depth of 200 ft [61 m] with a year-round temperature of 40°F [4.5°C]. Passed through a heat exchanger (with the district chilled water) and then treated, the lake water then joins the city water supply.

With district heating/cooling, all source components are located together in a remote district plant. This frees other buildings from the space requirements and visible impacts of stacks, boilers, fuel storage, water chillers, and cooling towers; the associated heat, humidity, foul air, and noise are as remote as the district plant. When such a system serves commercial customers, the heated or chilled water is metered. When it is self-owned (such as on a college campus), it is usually not metered; this can become a problem when efforts to identify building energy waste and subsequent savings are undertaken.

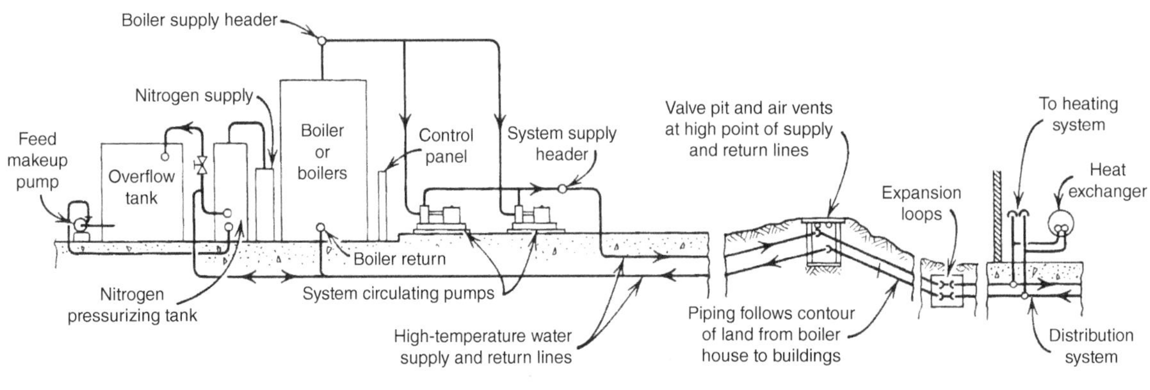

Figure 16.82 *Typical arrangement of a high-temperature water system. (Reprinted from Lieberg, 1963 / Industrial Press; courtesy of author Owen S. Lieberg, consulting engineer.)*

(b) Distribution Media

Distribution of the heating/cooling effect is (except for steam) accomplished using either water or air—or water and air. There are thus three distinct classifications of central HVAC systems:

All-air: In an all-air system, the heating/cooling effect is distributed from the source(s) to the spaces via heated or cooled air transported in ductwork; water is not used to transfer heat to/from the conditioned zones. The primary benefit of an all-air system is that air is used to modify the condition of air (sounds circular, but this approach is direct and logical); the main issue to be confronted in some building projects is the spatial volume that must be allocated to the ductwork. Conditioned air is delivered to the various spaces through diffusers/registers. An all-air HVAC system (of some configuration) should be able to readily meet the owner's project requirements for thermal comfort, IAQ, and energy efficiency.

Air–water: In an air–water system, the bulk of the heating/cooling effect is distributed from the source(s) to the spaces via hot or cold water transported in pipes. Air is also supplied to the spaces from a centralized unit—typically only enough air to ensure desired IAQ; this is often roughly 10% of the airflow seen in an all-air system. This air can also transport some heat/coolth. The primary benefit of an air–water system is reduced demand for distribution volume (piping is smaller than ductwork for equal heat transport). A concern in some building projects is the placement of water-to-air heat exchangers (the delivery devices) within the occupied spaces. An air–water HVAC system (of some configuration) should be able to readily meet the owner's project requirements for thermal comfort, IAQ, and energy efficiency.

All-water: The heating/cooling effect is distributed from the source(s) to the spaces via heated or cooled water transported via piping and introduced to the spaces via heat exchange delivery devices. Air is not used to transfer heat to/from the conditioned zones and is not supplied to the spaces by the HVAC system (air may be introduced to the spaces independently, e.g., by passive means). The primary benefit of an all-water system is that the spatial volume required for distribution will be the minimum possible (ducts are not used at all); a critical issue to be confronted when considering an all-water HVAC system is its ability to meet the owner's project requirements for IAQ.

16.17 HVAC SYSTEMS: ANATOMY

(a) Building Scale and Type

Larger buildings are generally more complex than smaller buildings. This also holds true for HVAC systems. This following discussion of HVAC system anatomy will begin with smaller-scale building considerations (buildings with, say, one to five thermal zones); larger-scale building situations will then follow.

Smaller buildings are typically skin-load dominated (SLD)—sometimes referred to as envelope (or enclosure) dominated—by which the climate (rather than internal loads) heavily influences the degree to which heating or cooling is the major design concern. In some climates, only heating is needed; a building can "keep itself cool" during hot weather without mechanical assistance. In other climates, only cooling is needed. In still other climates, both heating and cooling are required. It should be noted that "required" is in the eye of the client as well as in the climatic record—expectations for systems can drive decision making as much as numerical analysis.

A skin-load-dominated building may have such differing but simultaneous needs that a room-by-room solution for heating, ventilation, and cooling is desirable. Such a situation would suggest adoption of a *local* HVAC system approach. Consider the building with both north- and south-facing spaces on a cold, sunny winter day; one side gets ample solar heat, while the other side needs additional heat. An advantage of local systems is their ability to respond quickly to individual room (zone) needs.

A central HVAC system also has advantages: The equipment is contained within its own space rather than taking up space within each room, and maintenance can be carried out without disrupting activities within the occupied rooms.

Section 16.4 discussed equipment space requirements for large buildings. Determining the approximate size of small-building HVAC equipment during schematic design is also useful. Once design heating and/or cooling loads are known, manufacturers' catalogs can be consulted for the dimensions of appropriate heating and cooling equipment. A critical decision in sizing HVAC equipment regards the design temperatures: What are reasonable outdoor and indoor temperatures that would inform the selection of the system—and how to consider climate change?

The preliminary determination of required cooling capacity is not so straightforward. A very approximate early estimate can be obtained, however, from the estimated hourly gains listed in Appendix G. This type of estimate is likely to be lower than that obtained using the peak heat gain hour, for which cooling equipment is often sized. The ton is a commonly encountered unit of cooling capacity. One ton is the equivalent useful cooling effect of a ton of ice; this is 12,000 Btu/h [3516 W].

Larger buildings are typically internal-load dominated (ILD) by which their greater density tends to result in larger lighting, people, and equipment/appliances loads that dictate the overall mix between heating and cooling requirements. Interior areas of a building often have no thermal connection to the outdoor environment; all loads are internal. All buildings, however, will have perimeter areas that interface with the outdoor environment through the building enclosure. A design challenge is to ensure that an HVAC system can respond to such varying needs—both across space and across time. The following discussion is aimed primarily at such larger buildings.

Table 16.2 describes the basic components of any HVAC system—and reinforces the concepts of source, distribution, delivery (and controls). Three common tasks (heating, cooling, and ventilation) are accomplished. Intakes and exhausts accompany each task. Although the final choice of the HVAC system should follow a detailed analysis of the owner's project requirements, some basic concepts (introduced previously, and expanded herein) will underlie system choices.

Central systems require one or several large mechanical spaces (often located in a basement and/or on a roof), sizable distribution trees, and sophisticated control systems. The noise, heat, and other environmental conditions of such mechanical rooms can be controlled fairly easily, because the conditions are concentrated in a few locations not occupied on a regular basis. Similarly, maintenance is easy to perform without interrupting normal activities, although breakdowns in central equipment can paralyze an entire building. Air quality can be improved by locating air intakes high above street-level pollution and by regular maintenance of centralized air filters. Longer equipment life can be expected with regular maintenance. Energy efficiency can be served by the recovery of waste heat by-products. Also, there are many ways to provide for the differing thermal needs of the many zones that will be served by a central system. One important drawback of a central system is the potential size and length of the distribution tree(s) required to carry centralized services to many distributed spaces. Another potential drawback is scheduling of operations caused by differences in zone usage. Energy may be wasted when an entire HVAC system must be activated to serve one zone (such as a computer server space) when all other zones can be dormant.

Local systems may be attractive even for large buildings as scheduling differences between zones multiply. Also, pronounced differences in other factors—for example, function (with resulting comfort expectations) or placement within a building—can lead to the selection of a local system (usually, multiple local systems). Large, centralized equipment spaces are not required with local systems; rather, source equipment is distributed throughout a building (or over the roof and/or surrounding grounds). Dispersal of equipment minimizes the size of distribution trees (or eliminates them altogether) and greatly simplifies control systems. Moreover, a system breakdown will affect only a small portion of a building. Noise and other effects of local equipment, however, can cause discomfort in occupied spaces, and make quality maintenance challenging (because access to so many separate locations may be disruptive or restricted). Good air quality depends upon the regular cleaning of air filters, which is more difficult when they are scattered around a building, and are often within occupied spaces. The potential for energy conservation with local systems is promising, mainly because system operation and control are local and personal, but there is little opportunity to capture waste heat as a resource.

A central system approach does not imply that all spaces must be served from a single location. Though such an approach—as illustrated by Figures 16.83*a*, *c*, and *d*—is common, several central locations/systems may be used within a building. This latter approach is shown in Figure 16.83*b*, where a central boiler/chiller space is remotely located, and fan rooms are located on each floor. This greatly reduces the size of the bulky air distribution tree; although the distribution tree for hot and chilled water is extensive, the pipes are of much smaller diameter and therefore relatively easily accommodated and coordinated. The use of a central equipment room makes energy recovery systems more feasible.

(b) Zones and Systems

When selecting systems from among the wide variety available, it is helpful to match zone characteristics and system characteristics. Among the considerations are zone placement (close to or away from the building skin), zone thermal loads, the comfort expectations related to the zone activities, the space available for system components within the zone, and the life-cycle costs of various system alternatives.

Zone placement will sometimes preclude local systems, which depend on easy access to outdoor air, both for fresh air and for a heat source or sink. Local systems for interior (away-from-skin) zones can be awkward. Relationships between zone placement and building forms are shown in Figure 16.84.

The thermal loads on each zone determine the extent to which heating or cooling is the dominant

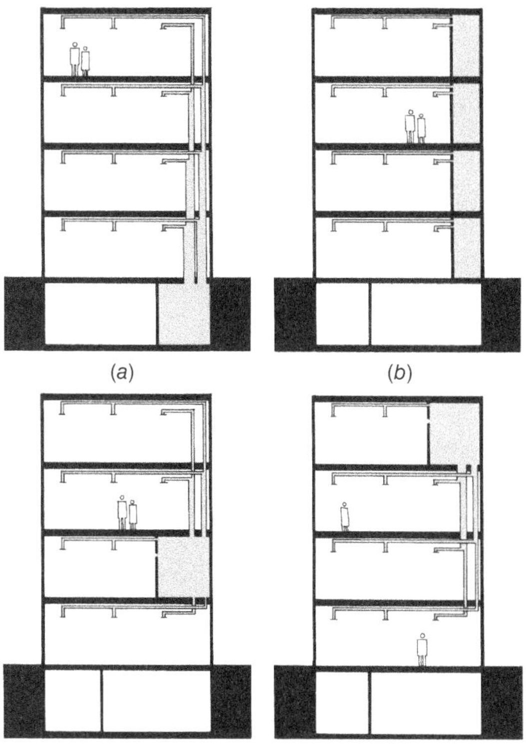

Figure 16.83 *Fan rooms (gray spaces) can either be combined with or separated from boiler/chiller "plant" rooms. (a) Common location for a central combined equipment room. (b) Increasingly common arrangement of a small fan room on each floor, with a plant room in the basement. (c) An intermediate floor may be able to provide space for a central fan room, while heavier and noisier equipment remains in the basement. (d) With a top-floor central fan room, plant equipment may be located either on the roof or in a mechanical penthouse, or it may remain in the basement. (From Allen and Iano, 2017.)*

problem—which, in turn, can influence the choice of system. A zone with little cooling load and low moisture production may be well served by a simple system providing fresh air plus heating, with no humidity control. Zones that require cooling will usually also require more complete control of air motion and relative humidity. Although it is risky to generalize about which comfort determinants are most important (given the differences between activities and between individuals), it can generally be assumed that comfort and thermal tasks are related. This relationship is summarized in Table 16.13. The choice of system may be based partly on whether a system can provide appropriate control of the more important comfort determinants.

Distribution Trees. Central heating/cooling systems produce a heating and cooling effect in one place, then distribute the effect to building spaces according to their respective needs. A tree is a metaphor for the means for distributing heating and cooling: The "roots" are the source machines that produce heat and coolth, the "trunk" is the main duct or pipe from the source equipment, and the "branches" are the many smaller ducts or pipes that lead to zones and individual spaces.

Questions regarding distribution trees might include: How many? What kind? Where? A building can have one giant distribution tree, several medium-sized trees, or a veritable orchard of much smaller trees. At one extreme, a large mechanical room is the scene of all heating and cooling production; leading from this room is a very large trunk duct with perhaps hundreds of branches. At the other extreme, each zone has its own mechanical equipment (such as a rooftop heat pump), with short trunks and relatively few branches on each tree.

What kind of distribution tree? Basically, the options are air (ductwork) or water (piping). Air distribution trees are bulky and therefore likely to have major visual impacts unless they are concealed above ceilings, below floors, or within vertical chases. Water distribution trees consume much less space (a given volume of water carries vastly more heat than does the same volume of air at the same temperature) and can be easily integrated within structural members such as columns. Air and water trees can both be noise sources.

How does the distribution tree relate to the building? If placed on the exterior, it can lend a three-dimensional organizational structure to a façade. Exterior distribution trees consume less enclosed and conditioned floor space but require expensive cladding and are subject to considerable heat losses and gains, which could increase energy usage. Interior trees are often housed in shafts and grouped with other continuous vertical spaces, such as elevator shafts and stairways. If the choice is an exterior distribution tree, its potential contribution to façade performance should be considered and architecturally integrated. For example, ductwork might act as a sunshade or as a lightshelf or as a visual organizing element.

HVAC system choice will be influenced by the amount of space the system requires. In some cases, it is easy to provide small equipment rooms at regular intervals throughout a building, such that little or nothing in the way of a distribution tree will be required. In other cases, a network of distribution trees and large central equipment spaces may be easier to accommodate. Figure 16.85 shows a matrix relating the influences on distribution trees of centralized/dispersed versus air/water systems.

To carry the tree analogy to its logical conclusion, consider the "leaves" as the delivery elements: the points of interchange between the distribution network and the spaces being served. Delivery device selection can substantially affect the aesthetics and usability of a space. Consider a bulky device such as a fan-coil unit

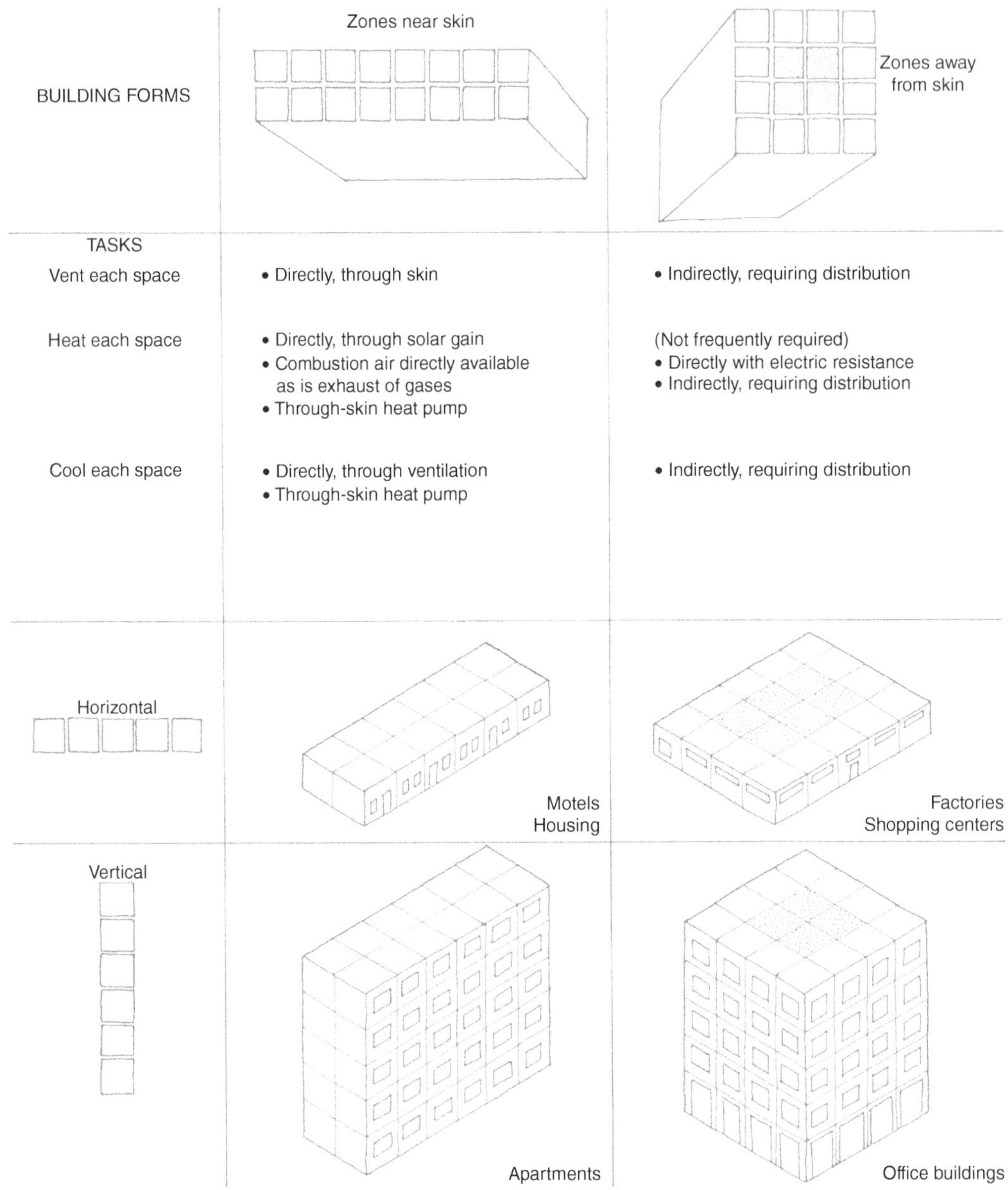

Figure 16.84 *Zone placement and building form are related to heating, cooling, and ventilation tasks; some applications take on typical building forms. (From class notes developed by G.Z. Brown, University of Oregon.)*

TABLE 16.13 Linking Thermal Comfort and Thermal Control

	For Heating of Spaces	For Cooling of Spaces
More important	Surface, air temperatures	Air motion
↓	Air motion	Relative humidity
Less important	Relative humidity	Air, surface temperatures

on an exterior wall below a window versus a perforated ceiling diffuser that may be essentially invisible.

The issue of distribution trees relates to the selection of an HVAC system and its ability to provide appropriate zoning. A simplified procedure for matching zone requirements and system capabilities is shown in Table 16.14. Using this process, the basic 16 zones are translated into three local systems and three central systems: one all-air, one air–water, and one all-water.

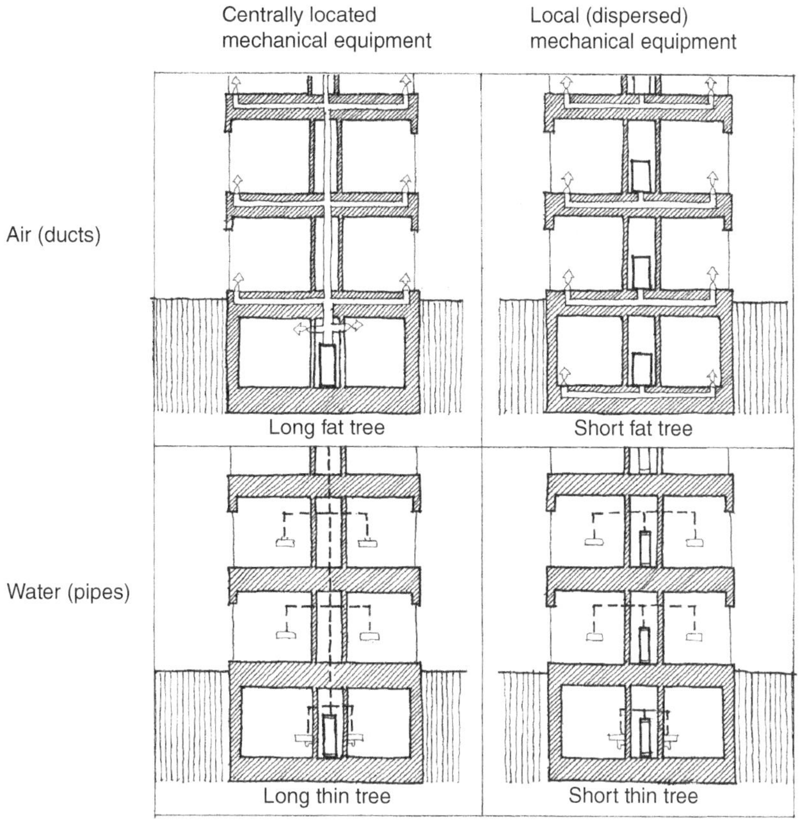

Figure 16.85 *Matrix of distribution trees. (© 1998 John S. Reynolds; drawing by Michael Cockram.)*

The Fox Plaza Building in San Francisco, which illustrates many such matches between systems and zones and resulting distribution trees, is shown in Figure 16.86. This project includes four major building types in one structure:

1. underground parking garage
2. commercial center at ground level, including a bank, a women's specialty store, and other commercial establishments
3. ten floors of offices
4. sixteen floors of apartments

The mechanical room/level is located between the office portion of the building and the apartments above. The distribution trees—HVAC, electrical, and so on—are thus directed both upward and downward, resulting in two shorter trees rather than one longer tree. The spatial requirements of offices and those of apartments are quite different; thus, the floor-to-floor heights, window treatment, and HVAC, electrical, elevators, and other services are different. The placement of the mechanical level between the offices and the apartments also provides for a definite visual separation between the two functions.

Quite unusual is the placement of the steam boilers on the 13th floor instead of in a more conventional basement location. Only a small amount of auxiliary equipment is located on the roof and in a small portion of the garage. Residential areas have hot-water heating (residential cooling being rarely needed in San Francisco), offices have dual-duct, high-velocity heating/cooling, and commercial (ground-floor) tenants are supplied with hot and chilled water for individual climate control requirements.

As noted, the designers of the Fox Plaza Building selected an intermediate location for the heating and cooling source equipment—as a means of separating floors of apartments from floors of offices. Other typical locations for central equipment are in the basement (where machine noise is most easily isolated, utilities are easily accessed, and machine weight is little problem) and on the roof, where access to air as a sink for rejecting heat is easiest of all, and floor-height is unlimited. Very tall buildings may require several intermediate mechanical floors.

TABLE 16.14 Procedure for Matching Zones and Systems

General Description

A multipurpose building is situated in a cold winter/mild summer climate.

Apartments are on upper floors surrounding an open-air central court; they have adequate daylight and cross ventilation. Floor-to-floor heights are low.

Offices are rented to various tenants. Exterior offices have high ceilings to facilitate daylighting, low internal gains, and restricted clearance for horizontal ductwork. Interior offices have lower ceilings; vertical chase space is limited because it reduces rentable area.

Shops are located around the perimeter on the high-ceilinged ground floor; some smaller shops are located on the mezzanine and ground floors in the interior zone. Space for vertical chases is severely limited on these highly rentable floors.

The parking area is below ground, surrounded by air/light wells. Floor heights are very limited to reduce ramp length.

Activities (Program)	Apartments	Computer Center	Offices		Shops		Restaurant	Parking
Schedule	24 h	24 h	9 h		9 h		12 h	
Placement	Exterior	Exterior for access	Exterior	Interior	Interior	Exterior	Exterior for access	Entire floor
Internal gains	Vary	High	Low	Medium	High	Medium	High plus moisture	Low with exhaust gas
Dominant HVAC task(s)	Heat-ventilate	Cool	Heat/cool	Cool	Cool	Cool	Cool	Ventilate
HVAC space available								
Vertical	Medium	Medium	Medium	Tight	Tight	Tight	Tight	Medium
Horizontal	Tight	Medium	Tight	Medium	Tight	Ample	Ample	Tight
System Choices								
Local		A					B	C
Central								
All-air			(D)	D	D	(D)		
Air and water			E			E		
All-water		F						

Summary

A. The computer center's unique operating schedule and high internal gains usually require a separate system equipped with humidity controls to protect the equipment. Some heat recovery for use in E and F seems possible.

B. The restaurant's special problems of heat, moisture, and aroma, as well as its schedule, require a separate system.

C. The parking area needs only plenty of outdoor air; it requires no thermal tie with the other zones.

D. The cooling-only loads of the interior zones are best served by all-air systems offering control of humidity and air quality. Vertical chase size is tight, however, and high-velocity air distribution may be required. The exterior zones could also be served by all-air systems, but the need for heating, plus the likelihood of outdoor air infiltration and the tight clearance for horizontal ductwork, suggest that the system for exterior zones should be independent of the system for interior zones.

E. Quick changes from heating to cooling are best handled by systems with little thermal inertia; some centralized control of air quality is offered by air–water systems.

F. A central all-water system offers energy conservation advantages, recovering waste heat from system D (and potentially from A). Outdoor air is easily and cheaply handled on a local basis, which can also provide cooling.

Mechanical Space

Probably best located on the top office floor or on a floor of its own between the offices and apartments. Distribution tree sizes will thereby be minimized on the high-rent ground floor.

(c) HVAC Systems Integration

Concealment or Exposure. The pipes, ducts, and conduits that move necessary resources around a building are often hidden in enclosed spaces unseen by anyone except contractors and repair people. The advantages of concealment include reduced noise broadcasting, fewer complex surfaces requiring cleaning, less care necessary in construction (leaks, not looks, are important), and more control over the appearance of the interior ceiling and wall surfaces. Although maintenance access to such hidden distribution elements is more difficult, various types of readily removable access panels are available, particularly for use with suspended ceilings.

The exposure of distribution networks, however, can provide an honest and direct source of visual interest. Exposure in corridors and service areas and concealment in offices constitutes an approach used in many office buildings. Flexibility is usually encouraged by exposure; changes can be easily made. One of the more spectacular examples of exposed mechanical (and structural) systems is shown in Figure 16.87—a building resulting from a design competition for a museum of modern art, reference library, center for industrial design, center for music and acoustic research, and supporting services in downtown Paris.

Mechanical–Structural Integration or Separation. The similarity of two technical support systems—structures and environmental controls—has intrigued designers ever since mechanical systems began to require substantial volume for distribution. As the complexity and size of mechanical distribution systems were increasing with technological development (typically, more air is required to cool a space than to heat it because of a lower Δt), the increasing strength

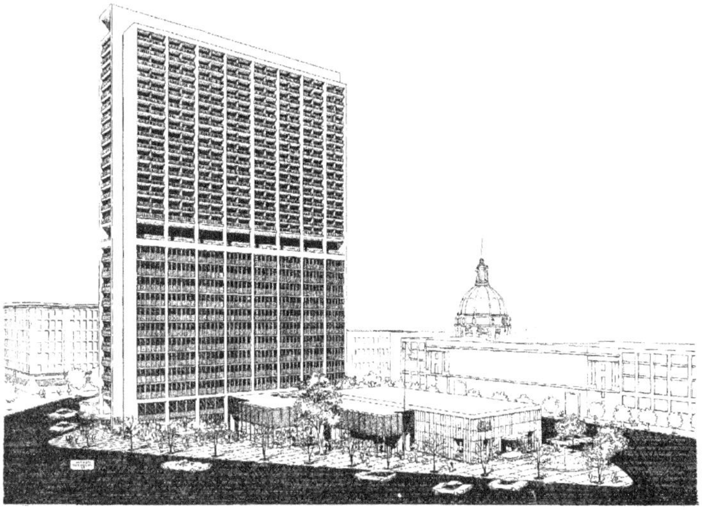

(a)

(b)

Figure 16.86 The Fox Plaza Building, San Francisco. Victor Gruen Associates, Inc., architects and engineers. (a) Elongated façade facing northeast shows the 16 floors of apartments above, the 10 floors of offices below, and the 13th-floor mechanical room—which hosts chillers and pumps for condenser water and chilled water, as well as boilers and converters (steam to hot water) for the fan-coil units in the residential floors above and hot water coils in the office floors below. Air-handling units for the offices are also located here, with air downfed through high-velocity ducts. Residences are heated; offices are heated and cooled. The cooling tower and the domestic hot water generator–storage units for the residential floors are located on the roof. (b) Photo showing clear vertical distinction between occupancy types. (a: Courtesy of Progressive Architecture; b: © Lee Eckert; used with permission.)

of materials was reducing the size of structural system components. Uncluttered floor areas between more widely spaced columns became desirable for spatial layout flexibility. With mechanical system elements located at or within columns, floor areas remained clear, thus giving mechanical–structural integration further impetus. Locating air-handling units on the roof further encouraged the merging of mechanical and structural systems. One system was growing while the other diminished (Figure 16.88). Thus, a fixed-column dimension, consisting mostly of the structure at the base and air duct at the top, became possible.

The functions of these two systems, however, differ widely. Compared to the dynamic air, water, and

Figure 16.87 A view of the mechanical support systems for the Centre Georges Pompidou, Paris. Piano + Rogers, architects. (© Terri Meyer Boake; used with permission.)

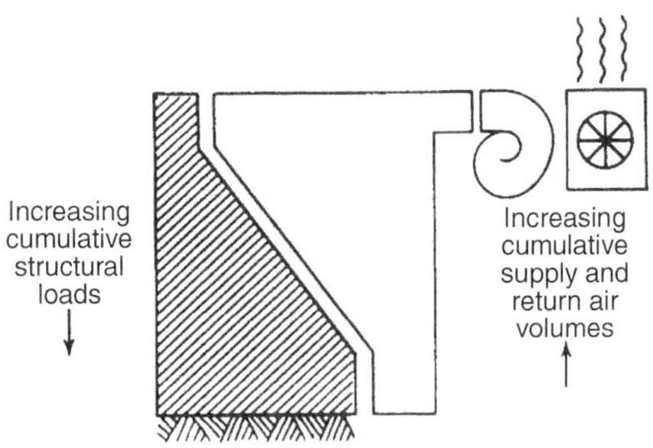

Figure 16.88 With rooftop central air handling, supply and return air duct sizes decrease as they approach the ground. Conversely, the structural element size increases toward the ground.

electrical distribution systems, the structural system is static—gravity never ceases. The moving parts in mechanical systems need maintenance far more frequently than the connections of structural components. Changes in occupancy can mean enormous changes in mechanical–electrical systems, requiring entirely different equipment; structural changes of such magnitude usually occur only at demolition. Mechanical systems invite user adjustment; structural systems rarely do.

Thus, although it is possible to wrap mechanical distribution elements in a structural envelope, it is of questionable long-term value, given the differing life spans and characteristics of these systems. The probability of future change suggests that the mechanical system be easily accessible and generally not be embedded in structural elements.

Distribution Tree Placement Options. These options are summarized in Figure 16.89. Vertical placement decisions are important because they affect floor space allocations, influencing the flexibility of spatial layout and the availability of usable (or rentable) floor space. Horizontal placement decisions affect ceiling height—a particular concern in daylighting design and sometimes a critical factor when building height limits are imposed yet maximum usable floor space is desired. (In Washington, DC, for instance, no building can rise higher than the Capitol.) Vertical and horizontal distribution at the edges of a building can have a dramatic impact on building appearance.

The history of HVAC distribution within high-rise buildings is one of trends and countertrends. Initially, multistory buildings relied upon daylighting and cross ventilation, so a thin, relatively high-ceiling plan with substantial perimeter was favored. Envelope loads dominated the heating and cooling needs, so perimeter distribution trees (with steam or hot water and of quite small diameter) were generally used. As electric lighting (and thus the need for air conditioning) increased, so did the depth of floor plans; large internal areas needed a lot of cooled air. Central boilers, chillers, and fan rooms were the norm. Thus, bulky air distribution trees appeared. At about the same time, the glass curtain wall and its slick, two-dimensional look of modernity became fashionable. Air distribution trees on the façade were so visually intrusive that they were pushed to the core, where cooling needs were relatively steady. The thin glass-enclosed perimeter, however, experienced extreme heating and cooling loads. Reaching these perimeter areas from vertical distribution trees at the core required larger cavities above suspended ceilings. This pushed the ceiling in the offices down to keep floor-to-floor spacing economical. The result was vast office areas that were visually dull, low-ceilinged, and without daylight.

Current countertrends include decentralized air handling, with small fan rooms on each floor. Vertical air distribution trees are shrinking, horizontal ones becoming more common. At the same time, daylighting is pushing office ceilings higher—as is a preference for indirect lighting and its compatibility with digital display equipment. Night cooling utilizing thermal mass encourages the exposure of concrete

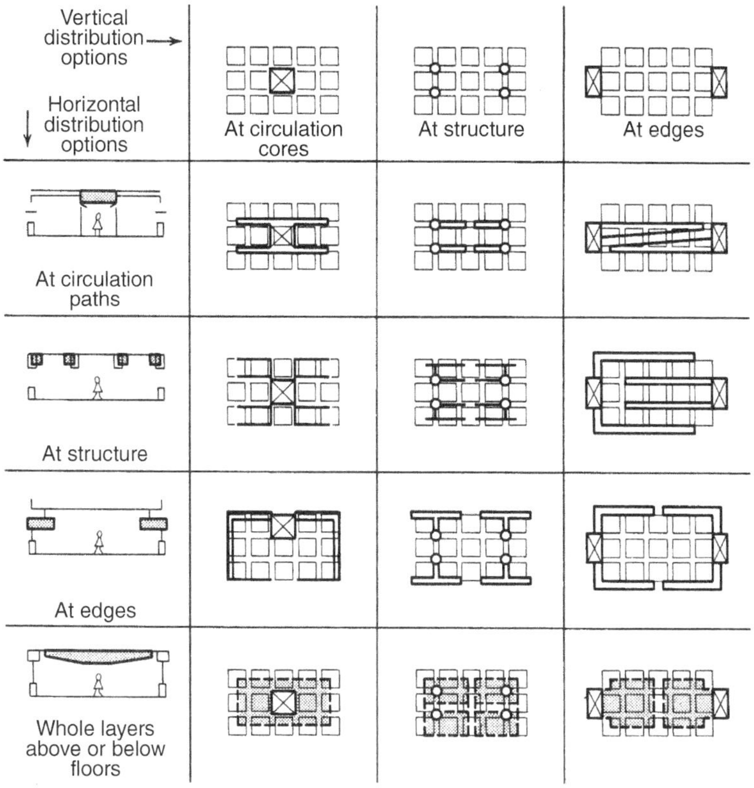

Figure 16.89 Distribution tree placement options: vertical (with impact on the plan) and horizontal (with impact on the section). (From class notes developed by G. Z. Brown, University of Oregon.)

structure and favors raised-floor air supply/ventilation systems. A renewed interest in sun control (driven, perhaps, by a quest for high-performance outcomes) is encouraging three-dimensional façades; these are slowly replacing two-dimensional reflective glass façades (which merely redirect solar radiation toward someone else). With increased three-dimensionality at the façade, perimeter distribution trees are once again conceivable.

It is logical to place at the perimeter the parts of the system that deal with the effects of sun, shade, and temperature change—leaving at the core a separate network to handle the more stable interior areas. The disadvantages of perimeter distribution include the potential for higher construction costs and the possibility of increased heat gains/losses to/from the distribution systems.

Vertical distribution within internal circulation cores is very common, as it maximizes plan flexibility for the rest of the floor and does not disturb the prized floor areas nearest windows. One centralized vertical distribution trunk, however, will require large horizontal branches near the core. With this choice, early thought must be given to horizontal placement decisions.

An unusual example of vertical air distribution at the core is shown in Figure 16.90. The unique features of the Fox Plaza office building, Los Angeles, include both fan rooms on each floor and a large central vertical air shaft. This air shaft begins at the bottom as a fresh air intake to each floor and tapers to become, at the top, an exhaust air outlet from each floor. The stack effect is utilized to help supply fresh air and to exhaust stale air from this large building, with help from small fans at each floor.

Vertical distribution integrated with structure creates some intriguing possibilities where the idea of integrating structure and HVAC is suitable. Multiple HVAC trees are implied (because there will be multiple columns with which they may be integrated); thus, the horizontal branches tend to be small. These branches, however, often join a vertical trunk at the same place where critical column-to-beam structural connections need to be made; interference is common and can be costly to correct. Vertical distribution at the edges is potentially dramatic in form but costly to enclose (if outside) or wasteful of prime floor space (if inside).

Horizontal distribution above corridors is very common, since reduced headroom in a circulation space is more acceptable than in a main functional area. Furthermore, corridors tend to be placed away from windows, so lower ceilings do not interfere with daylight penetration. Because corridors connect nearly all spaces, supply air access to the spaces is also provided by above-corridor ductwork. Exposure of services above corridors can heighten the contrast between such serving spaces and the uncluttered, higher-ceilinged offices that are served. Horizontal distribution at the

(a)

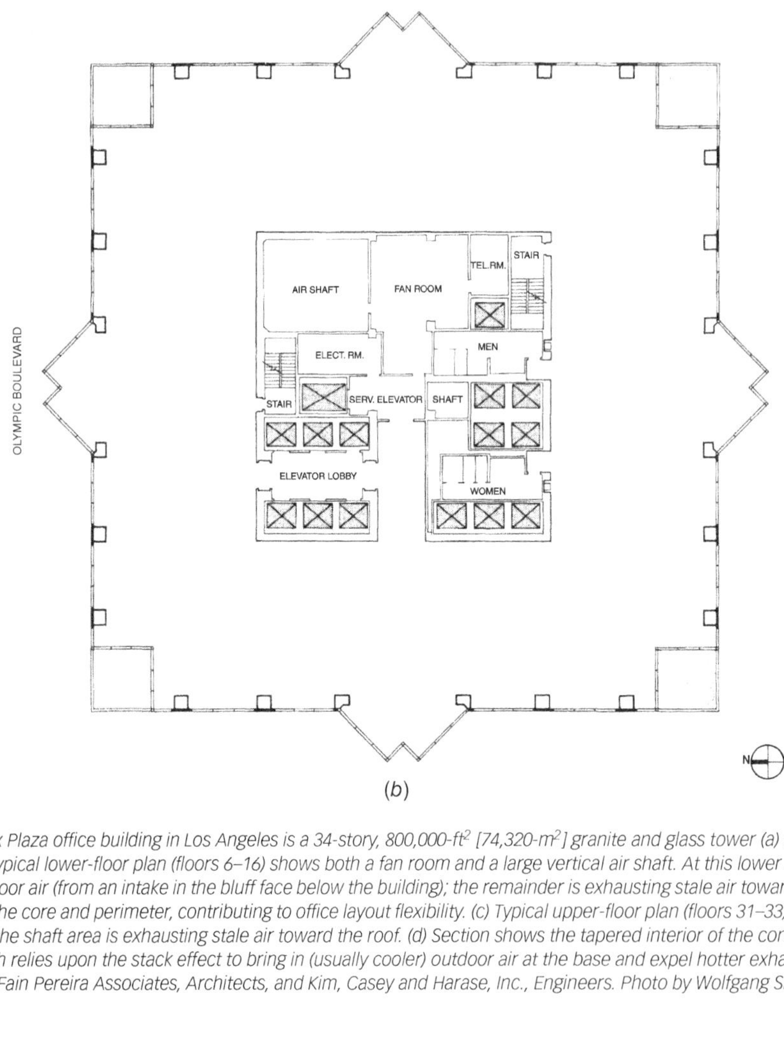

(b)

Figure 16.90 *The Fox Plaza office building in Los Angeles is a 34-story, 800,000-ft² [74,320-m²] granite and glass tower (a) with an unusual vertical distribution tree. (b) Typical lower-floor plan (floors 6–16) shows both a fan room and a large vertical air shaft. At this lower level, most of the shaft area is supplying outdoor air (from an intake in the bluff face below the building); the remainder is exhausting stale air toward the roof. Note the lack of columns between the core and perimeter, contributing to office layout flexibility. (c) Typical upper-floor plan (floors 31–33) shows fewer elevators; by this level, most of the shaft area is exhausting stale air toward the roof. (d) Section shows the tapered interior of the constant-cross-section central air shaft, which relies upon the stack effect to bring in (usually cooler) outdoor air at the base and expel hotter exhaust air at the top. (Courtesy of Johnson Fain Pereira Associates, Architects, and Kim, Casey and Harase, Inc., Engineers. Photo by Wolfgang Simon.)*

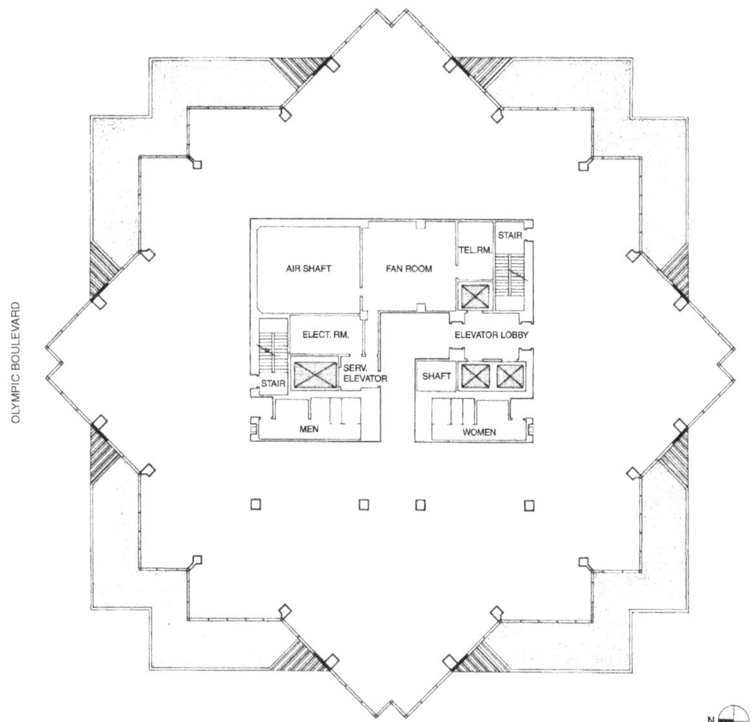

(c)

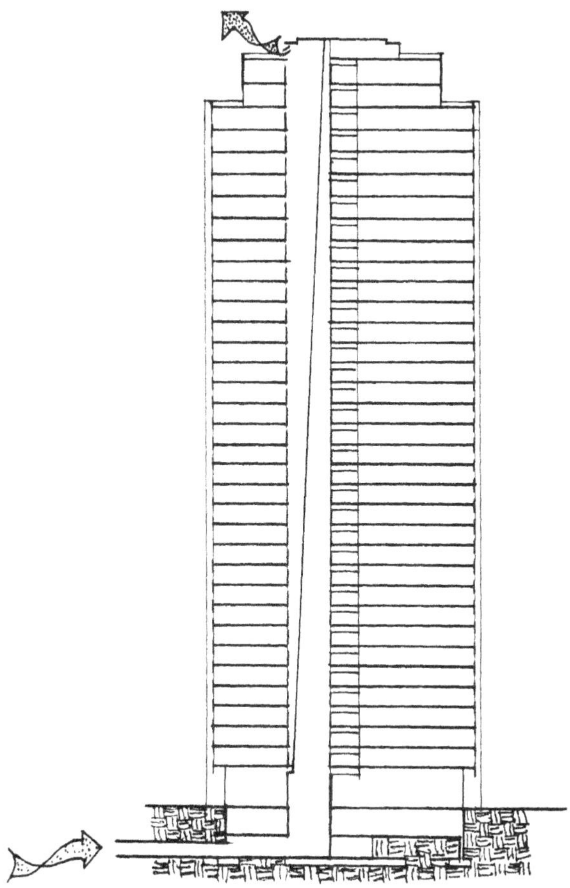

(d)

Figure 16.90 *(continued)*

structure is sometimes chosen, particularly where U-shaped beams or box beams provide ready channels for HVAC distribution. The penetration of horizontal structure members by continuous service runs must, however, be coordinated. Horizontal distribution at the edges can be integrated usefully with sunshading devices and lightshelves; it can also provide a spandrel element that contrasts with the window strips. Horizontal distribution within whole layers below floors (or above ceilings) is often utilized, becoming increasingly common with underfloor air distribution UFAD systems.

An example of supply at the edge for both vertical and horizontal distribution is found in the International Building in San Francisco (Figure 16.91). Vertical shafts are prominently exposed at the corners; these shafts carry supply and return ducts serving the four perimeter air-conditioning zones. Air-handling equipment and a 750-ton [2640-kW] chiller plant are located on the floors just below the terrace level (least desirable for rental). Each corner duct branches to serve two zones. Pressure reduction and blending are handled by equipment located above the hung ceiling; from these points, air flows to strip diffusers directly above the glass on the four sides of the building. Abundant control zones offer comfort to personnel in each area.

Interior zones on each floor are supplied by a riser duct in the building core, which branches at each floor to a loop just outside the line of elevators. The loop serves ceiling diffusers. Between the perimeter loop and the interior loop, another loop collects air for return to the central station air-handler (second and third floors). The return loops on the 11th to 21st floors are picked up by external return risers on alternate exterior corners. From the 10th floor down, the loops are picked up (as shown in Figure 16.91) by an interior return riser that extends down through the core in front of the blank faces of the high-rise elevators. To provide a clear space between the elevator banks on the main floor (fourth or terrace), the two core duct risers are offset at the ceiling of that story.

In summary, perimeter-area air for all stories is supplied through corner ducts. Central-area air for all stories is supplied through a core duct. All return air above the 10th floor is carried down through the return ducts at the other two corners. Return air from the 10th floor and below is carried down by a return duct in the core.

16.18 HVAC SYSTEMS FOR SMALL BUILDINGS

Several examples of HVAC systems applied to smaller buildings (or larger buildings employing local systems) are provided in this section. These systems do not represent all possible small-building HVAC arrangements—by a long shot. Designers are amazingly creative and come up with many, many solutions to similar problems. HVAC system selection is also heavily driven by context—fuel availability and cost, owner desires, local norms, and the like.

(a) Local Systems

Gas-Fired Baseboard Heaters. Using natural gas or propane, these devices might be employed where only space heating is required. The units heat by convection and by radiation (as is also the case for electric baseboard units). At a steady-state efficiency of 80%, these gas-fired units use a lower-grade resource (compared to electricity) to accomplish this simple thermal task. They are vented directly (using a built-in fan) to the outside and, therefore, must be installed on or near an exterior wall; vents are 1 ½ in. diameter, with a maximum length of 19 in. [38 mm in diameter, 483 mm in length]. With a cross-section of 9 in. high by 5 in. deep [230 mm × 130 mm], a 48-in. [1220-mm] length will deliver 5800 Btu/h [1700 W]; a 72-in. [1830-mm] length will deliver 9400 Btu/h [2755 W]. This is a local system.

Unitary Air Conditioners. The device shown in Figure 16.92 is perhaps the most seen piece of mechanical equipment in the United States. Usually perched in windows in full view of passersby, these packaged air conditioners noisily remind us that many of our buildings still are not centrally cooled (active cooling was considered a luxury until long after World War II).

Built-in, through-wall air conditioners offer a low-first-cost way to provide thermal zoning for individual apartments, motel rooms, and the like. As a local system of low capacity, these units are generally not as energy efficient as would be the case with larger central equipment. However, if turned on only when cooling is needed (i.e., when people are present), they can provide substantial savings over a larger always-on system. Properly operated local systems can be energy efficient over time—the key word is "properly."

Local Air-to-Air Heat Pumps. The use of individual (local) units is especially common in building types with all-perimeter spaces with varying orientations, numerous thermal zones, and a penchant for personal "ownership" of space. Motels are a prime example. In Figure 16.93, separate air–air heat pumps serve each motel room; at best, their constant noise helps mask the intermittent sounds from the adjacent parking lot/circulation space. Each customer controls his or her own climate (for better or worse, from an energy standpoint). If a central water loop replaced outdoor air as the heat source/sink, energy use and

(a)

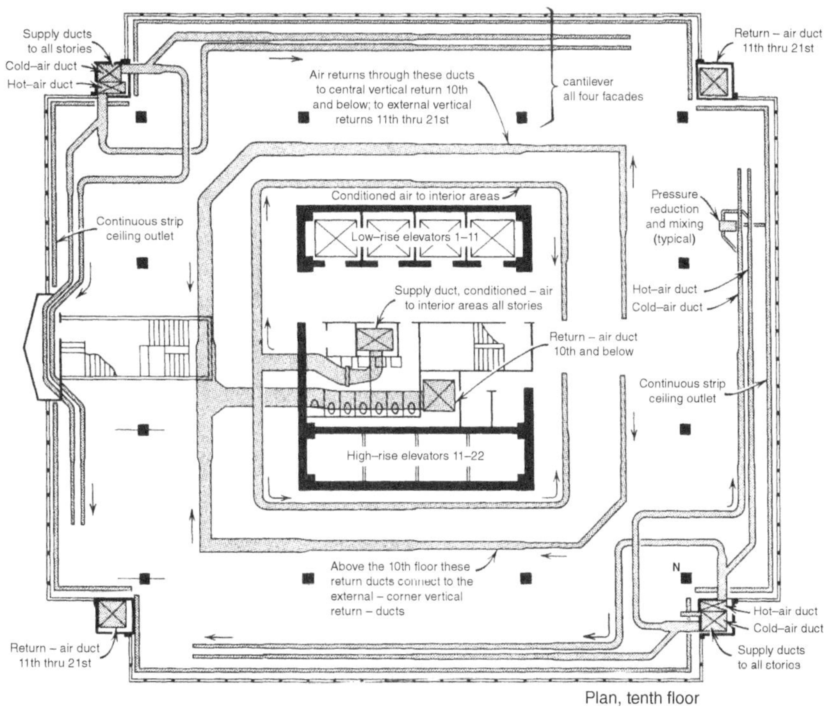

(b)

Figure 16.91 *The International Building, San Francisco. Anshen and Allen, architects; Eagelson Engineers (Charles Krieger, E.E.), mechanical designers. (a) Photo of one of the four corner main duct enclosures. (b) Tenth-floor plan. Column bay spacing is 24 ft, 6 in. [7.5 m], with a 16-ft [4.9-m] cantilever on all four façades. The major supply ducts (both hot and cold) to all 21 floors are in two opposite corners. Each of these supply distribution trees serves two adjacent sides of perimeter offices. The conditioned air is supplied from a third-floor mechanical space. In the opposite two corners, return air from the upper 11 floors is collected and taken down to the mechanical space. The remainder of the return air is taken down through the core. (Courtesy of Progressive Architecture.)*

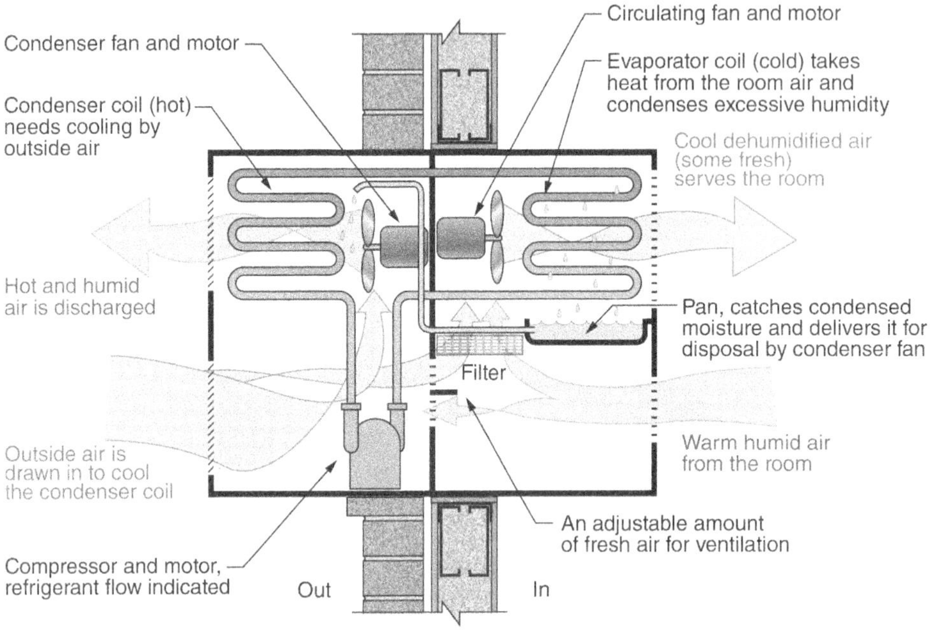

Figure 16.92 *Diagram of a typical through-the-wall air-conditioning unit. The unit is self-contained, requiring access to outdoor air and electricity to power the compressor and the two fans. Typical capacities are 1–2 tons (12,000–24,000 Btu/h [3.5–7 kW]).*

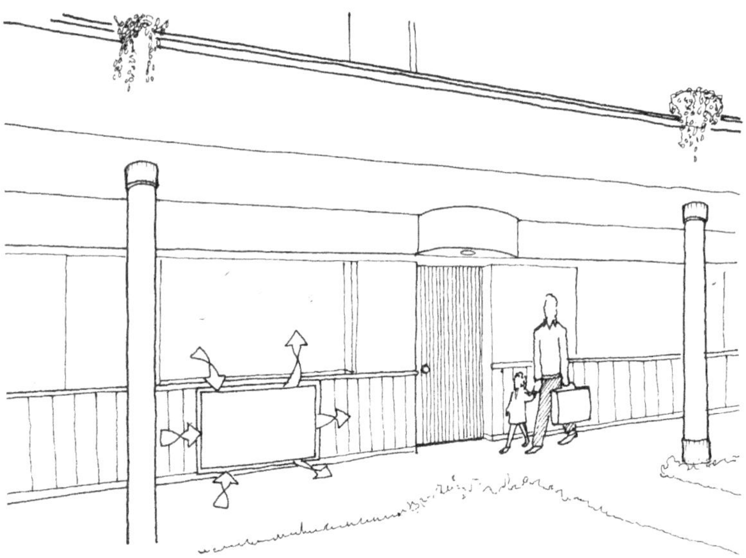

(a)

(b)

Figure 16.93 *Air-to-air heat pumps serve a motel. (a) A panel set just forward of the exterior wall allows the heat pump to intake and exhaust outdoor air around the panel edges. When cooling, the unit discharges warm air; when heating, cool air. (b) Interior cabinet contains the heat pump. Room air is taken in as shown, then discharged (after cooling or heating) upward across the glass surface—the likely point of maximum heat loss or gain.*

life-cycle cost would go down, although the first cost would increase.

Electric Resistance Heaters. These common devices have some impressive pluses; they are very low first cost and easily provide individual thermostatic control (making each room a separate heating zone). On the other hand, they bear the disadvantage of using high-quality energy to do a simple thermal task (resulting in poor exergy performance). Although electric resistance heating can legitimately be said to have an efficiency of 100% on site, the losses inherent to electrical generation and distribution drop the efficiency to around 35% when viewed from the perspective of the source. Because this is a local system, good operations habits can save energy if unused rooms remain unheated. A few of the many types of electric resistance heaters are shown in Figure 16.94. As in the case of metal wood stoves, surfaces can sometimes reach high temperatures, requiring care in the location of heaters relative to furniture placement, draperies, traffic flow, and children. Electric heaters are sized by their capacity in kilowatts (1 kW = 3413 Btu/h). The maximum allowable power density is around 250 W per linear foot [820 W per linear meter] of heater.

Radiant Ceiling with Electric Resistance. A ceiling can be constructed to include electric resistance via wiring (Figure 16.95). Because the ceiling is not touched by the occupants, it can be safely heated to a rather high temperature. The primary disadvantage of ceiling-based heat is that hot air stratifies just below the ceiling, so that passive air motion is discouraged. Remember also that Chapter 6 suggests more discomfort with a warmer ceiling. With wiring hidden within the ceiling surface, unwary occupants can damage wires while installing hooks or light fixtures.

(b) Central Systems

Hot Water Baseboard and Radiator Systems. Hot water heating circuits that serve baseboards or radiators come in four principal arrangements.

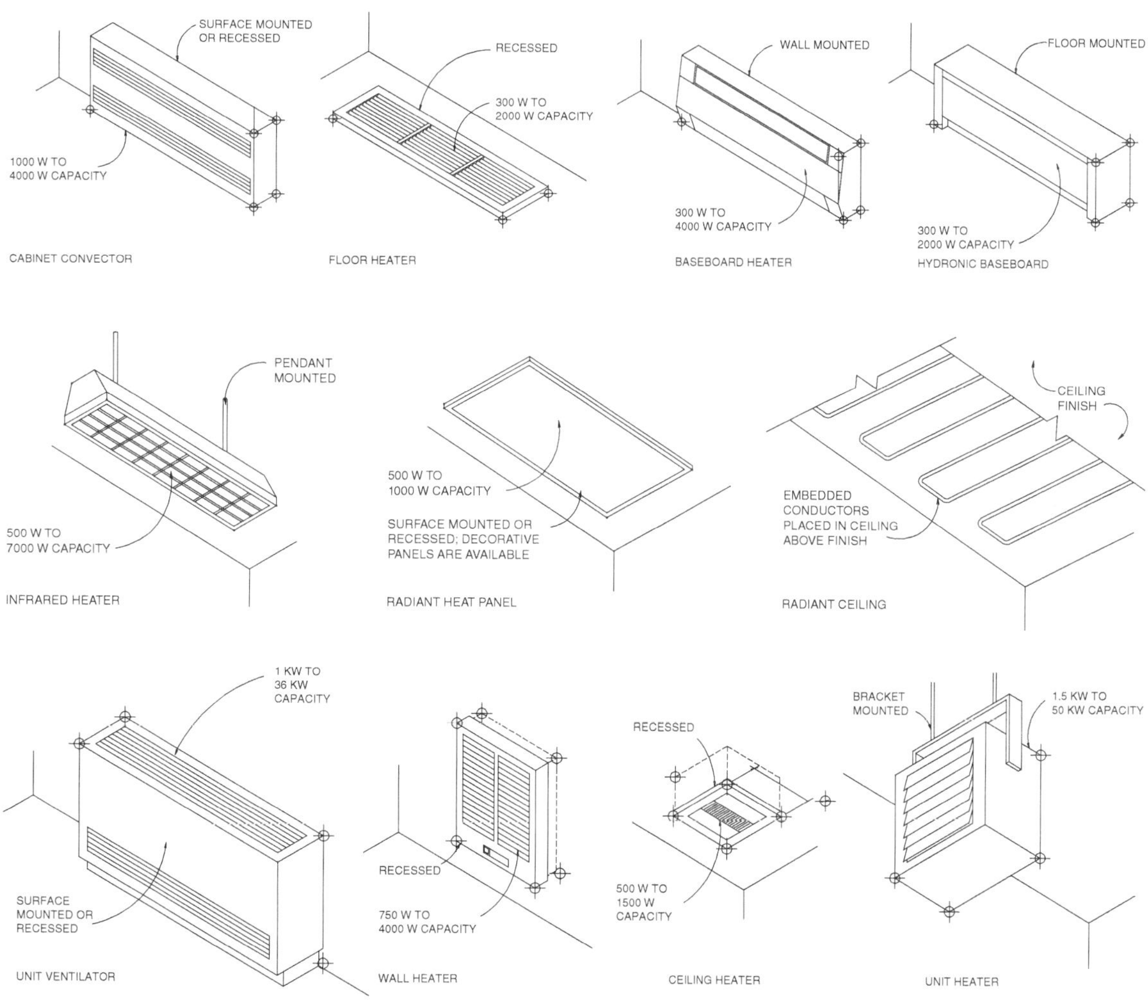

Figure 16.94 *Varieties of electric resistance heating units. (From Ramsey and Sleeper, 2000.)*

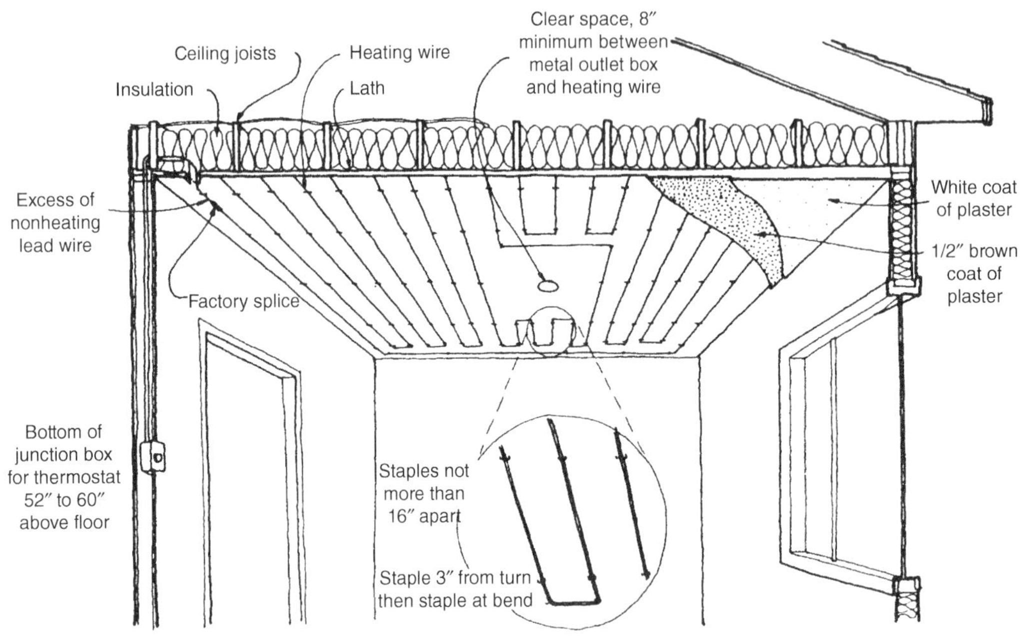

Figure 16.95 *Radiant heating electrical cable is installed prior to completion of a plaster ceiling.*

Figure 16.96*a* shows a series loop system, usually run at the building perimeter. The water flows to and through each baseboard or radiator in turn. As a result of this series flow, the water at the end of the circuit is a little cooler than at the start—but because the water temperature drop in a residential hot water system seldom exceeds 20°F [11°C], the average water temperature can usually be used to select the baseboard or other elements. Valves at each heating element are not possible, because any valve could shut off the entire loop. Control is by damper adjustment at each baseboard, which reduces the natural convection of air over the fins. This is a one-zone central system—all elements are on, or off, together. There is no general rule about the maximum allowable length of a water circuit, but for long runs, pipe size can be increased or several loops used in parallel to create more than one thermal zone.

The one-pipe system shown in Figure 16.96*b* and Figure 16.97 is a very popular option. Special fittings act to divert part of the water flow into each baseboard. A valve may be installed at each unit to allow for reduced heat output or for a complete shutoff to conserve energy—an advantage that the series loop system does not provide. The one-pipe system uses a little more piping and thus is not as economical to install as the series loop system, in which piping is minimal. Again, the supply water temperature will be lower at the end of the run than at the beginning.

The two-pipe reverse–return layout shown in Figure 16.96*c* provides the same supply water temperature to each baseboard or radiator, because (unlike a series system) supply water is not cooled by passing through a previous baseboard or by the addition of return water (as in a one-pipe system). Equal friction loss, resulting in equal flow, is achieved through all baseboards (numbers 1–5) by reversing the return instead of running it directly back to the boiler. This equality is accomplished by having equal flow-lengths of water pass through any baseboard unit together with its lengths of supply-and-return piping. More pipe is required for this system than for the systems shown in Figure 16.96*a* or Figure 16.96*b*.

Figure 16.96*d* shows an arrangement that is not usually favored because the path of water through baseboard number 1 is much shorter than that through the others, especially number 5. Baseboard number 5 could easily be undesirably cool, because it is short-circuited by the others.

Control of pipe expansion requires expansion joints in long runs of pipe and clearance around all pipes passing through walls and floors. Each time a hydronic system changes from room temperature to its heated condition, the piping will undergo the expansion shown in Table 16.15, assuming a 70°F [21°C] initial temperature.

An air cushion tank, compression tank, or expansion tank is a closed tank containing air and is usually located above the boiler. When water in a hydronic system is heated, it expands, compressing the air in the tank. This allows for operation in the usual range of system temperatures, including temperatures above the usual boiling point of water, without the frequent opening of the pressure relief valve. One type of air cushion tank, called a diaphragm tank, separates the air and water with an inert, flexible material; this prevents reabsorption of air by the water. For a

THERMAL CONTROL

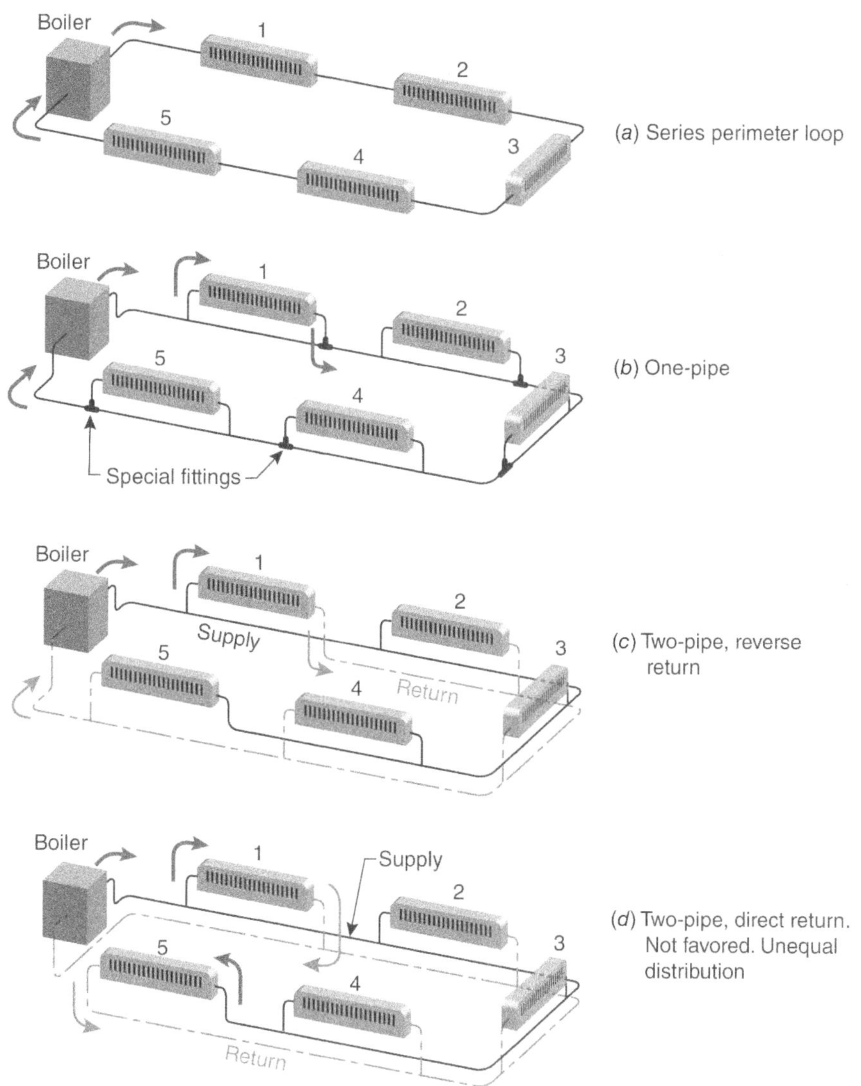

Figure 16.96 *Diagrams of hydronic distribution options (a, b, c, d), seen in plan view. Baseboard convector delivery devices are suggested here. Controls are not shown.*

conventional (unpressurized) air cushion tank, allow 1 gallon of tank capacity for every 5000 Btu/h [1 L for every 385 W] of total system heat capacity. For a pressurized tank (at least 8 lb/in^2 [55 kPa]), allow 1 gallon of tank capacity for each 7000 Btu/h [1 L for each 540 W], or see the manufacturer's recommendations.

Air vents and water drains are part of the water distribution system. Except for the necessary air cushion in the upper part of the compression tank above the boiler, air must not be allowed to accumulate at high points in the piping or at the baseboard or convector branches. Air vents at all high points relieve possible air pockets that would otherwise make the system air-bound and inoperative.

If a system is drained and left idle in a cold building, water trapped at low points can freeze and burst the piping or fittings. Operable drain valves must be provided at such locations and at the bottom of the boiler, as shown in Figure 16.98.

Hydronic and electrical controls allow for automatic system operation, described in Figure 16.98. There are two options for system control:

1. As in Figure 16.98, the thermostat controls the circulating pump and the boiler. In colder weather, the system operates almost continuously, and the average temperature in the system gradually rises.
2. The thermostat controls only the boiler, and the circulating pump operates continuously. This uses more energy for the pump but minimizes system temperature variation and thus the possibility of expansion noises.

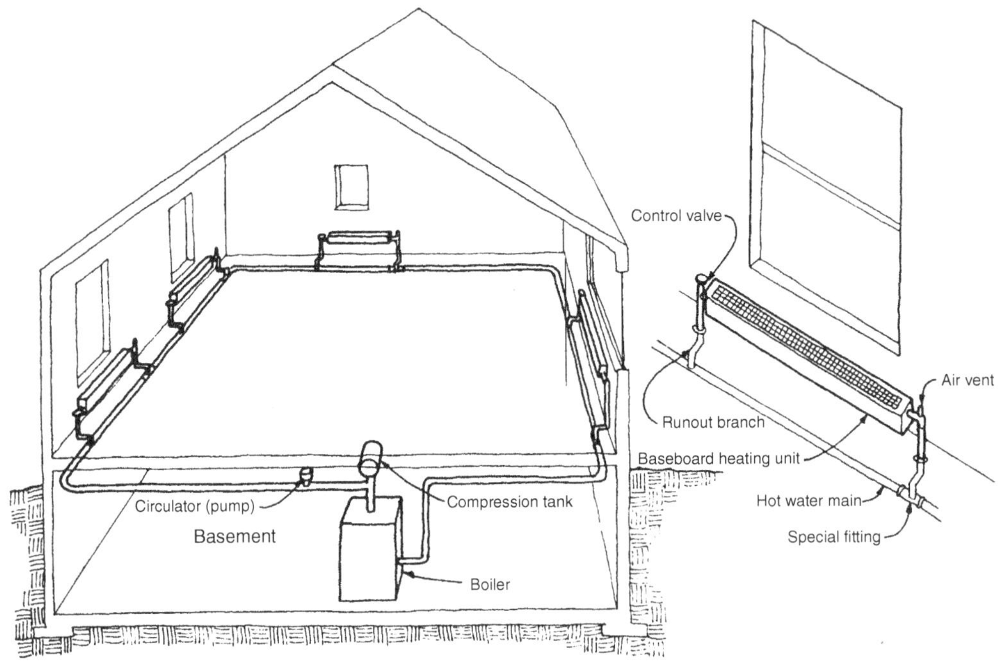

Figure 16.97 One-pipe hydronic system.

TABLE 16.15 Expansion in Copper Tube and Iron Pipe

	Pipe Expansion, in. per 100 ft [mm per 100 m]	
Water Temperature, °F [°C]	Iron Pipe	Copper Tubing
160 [71]	0.7 [58]	1.0 [83]
180 [82]	0.9 [75]	1.3 [108]
200 [93]	1.0 [83]	1.5 [125]
220 [104]	1.2 [100]	1.7 [142]

Makeup water is added as required, the air level in the tank is regulated by the air control fittings, and the circulator and burner operate as controlled by the aquastat and thermostat. If air vents in the piping are not automatic, they will require periodic manual "bleeding" of unwanted air. Circulating pumps are used to overcome the friction of flow in the piping and fittings and to deliver water at a rate sufficient to offset the hourly heat loss of the building. Pipe insulation is required whenever the pipes are outside the heated enclosure of the building; *ASHRAE Standards 90.1* and *90.2* specify insulation appropriate to the design water temperature.

Sizing of a Hydronic Heating System. Calculations for the sizing of a water distribution system are based upon consideration of the required flow and the desired friction loss in the piping. For an open-loop domestic water supply system, another factor is the vertical distance the water must be raised. In a closed-loop heating system, however, the weight of the cooler water returning to the boiler essentially counterbalances the weight of the hot water being raised upward.

The key to pipe sizing is the overall required flow rate. Ordinarily, the temperature drop that occurs as the hot supply water transfers heat to the space (through a convector) is about 20°F [11°C] in residences; in commercial applications, 30, 40, or 50°F [17, 22, or 28°C] temperature drops are also common, as recommended by the manufacturer of the baseboard or convector. Because the entire design building heat loss is overcome by this system, flow rate can be calculated by:

$$FR = \frac{Q}{(\Delta t)(C)} \qquad \text{(Eq. 16.3)}$$

where

FR = flow rate, gpm [L/s]
Q = design heat loss, Btu/h [W]
Δt = supply versus return water temperature difference, °F [°C]
C = conversion factor
(60 min/h)(8 lb/gal)(1 Btu/lb °F)
[(1 kg/L)(4180 W sec/kg °C)]

Friction loss through piping and fittings can be accounted for using the procedures shown in

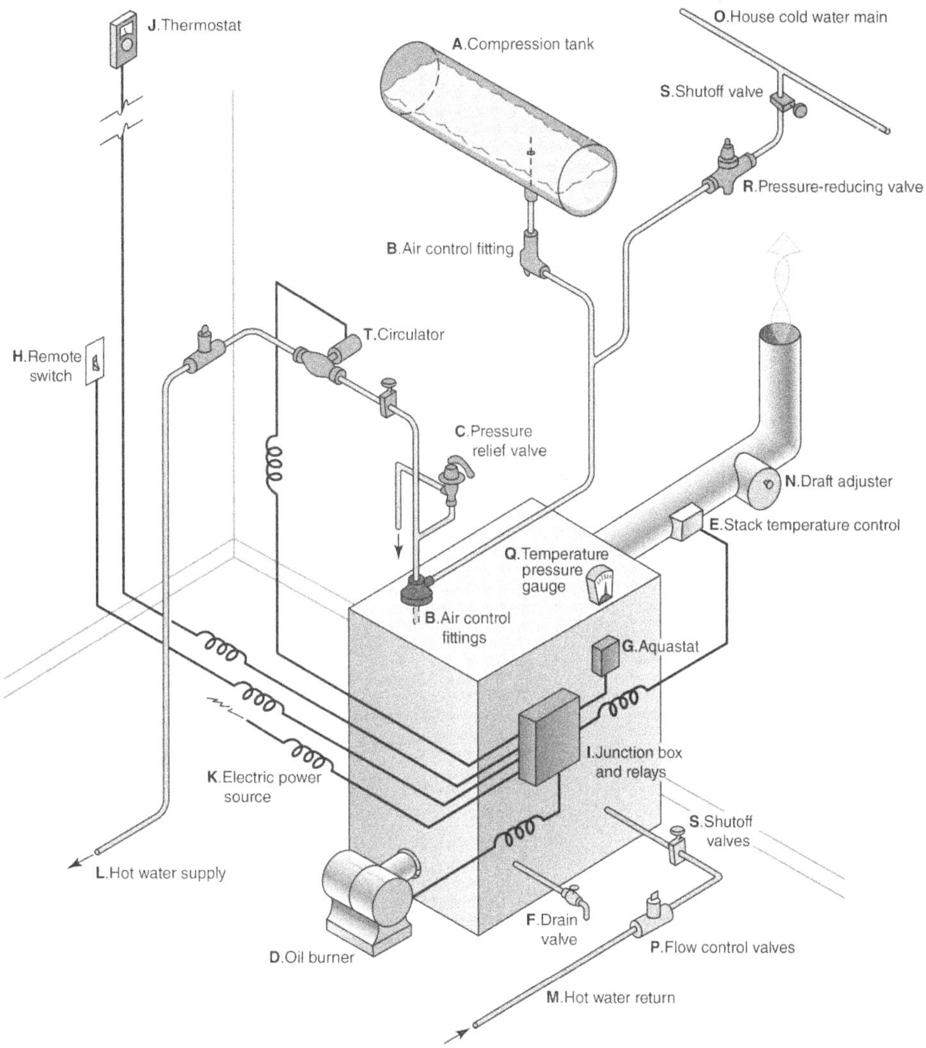

(A) *Compression tank*. Accommodates the expansion of the water in the system.
(B) *Air control fittings*. Vent out unwanted air in the boiler and maintain the level in the compression tank.
(C) *Pressure relief valve*. Usually set for 30 psi. Initial cold pressure about 12 psi. Relieves excessive system pressure.
(D) *Oil burner*. Responds to aquastat or thermostat.
(E) *Stack temperature control*. Senses stack temperature and stops oil injection if ignition has not occurred.
(F) *Drain valve*. At low point in the water system.
(G) *Aquastat*. Maintains temperature of boiler water by starting the oil burner when temperature of water drops below the aquastat's setting. Sometimes set at about 180 F.
(H) *Remote switch*. At a safe distance from the boiler so that the plant can be turned off in case of trouble during which the boiler cannot be approached.
(I) *Junction box and relays*. General control center.
(J) *Thermostat*. When the room temperature drops below its setting, it turns on both the oil fire and the circulating pump.
(K) *Electrical power source*. Operates from a separate individual circuit at the power panel.
(L) *Hot water supply*. Copper tubing to convectors or baseboards.
(M) *Hot water return*. Copper tubing from convectors or baseboards.
(N) *Draft adjuster*. Regulates the draft (combustion air) over the fire.
(O) *House cold water main*. From which water is fed automatically into boiler.
(P) *Flow control valves*. Prevent casual flow of water by gravity when the circulator is not running.
(Q) *Temperature pressure gauge*. Indicates water temperature and pressure. Sometimes supplemented by immersion thermometers in supply and return mains.
(R) *Pressure-reducing valve*. Admits water into the system when the pressure there drops below about 12 psi. Has a built-in check valve to prevent backflow of boiler water into the water main.
(S) *Shutoff valves*. Normally open. Can be closed to isolate the system and permit servicing of components.
(T) *Circulator*. Centrifugal circulating pump that moves the water through the tubing and heating elements.

Figure 16.98 *An oil-fired boiler and its hydronic and electrical controls.*

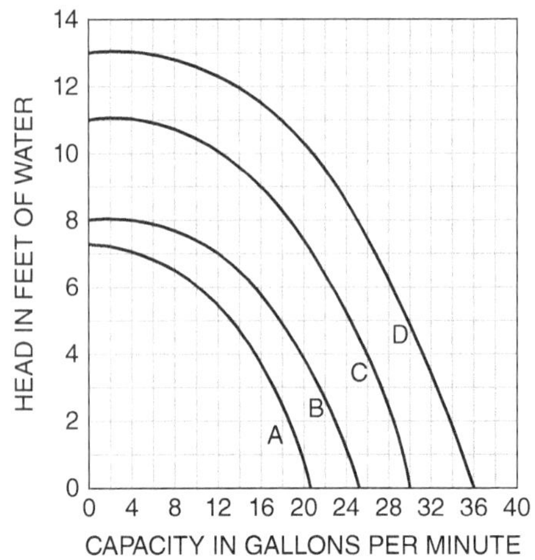

Figure 16.99 *Typical pump performance curves for four pumps used in hydronic heating systems. One foot of water = 2.99 kPa; 1 gpm = 0.0631 L/s. (Courtesy of the Hydronics Institute Division of GAMA, 1996.)*

Chapter 21, and the main supply and return pipes can be sized. The same procedure can be applied to branches, proportioned to the heat they must deliver. The total friction to be overcome in the most distant run is then converted from total psi to feet of head, with each foot of head = 0.433 psi (the pressure exerted by a foot-high column of water). For SI units, 1 ft of water = 2.99 kPa.

A pump can be selected with the friction loss of the system expressed in feet of water (or head) and the flow rate established. Figure 16.99 shows typical performance curves for four pumps. The designer enters the curve with a desired flow rate, then selects the pump with a head capacity greater than or equal to the head required. Pump performance curves are provided by the manufacturer.

The critical design choice, however, is not pipe size. It is relatively easy to install small-diameter pipes within typical building constructions. Rather, the critical choice is the hot water supply temperature: The higher this temperature, the smaller the convector units that deliver heat to each space. Higher temperatures, however, can endanger occupants, who may suffer burns if they touch exposed parts of the convectors or the piping. Higher temperatures also can lead to steam within the boiler or distribution tree—although the system is under pressure and the boiling point is therefore greater than 212°F [100°C]. These systems are not designed to accommodate steam, and serious injury can sometimes result. A safe choice for average water distribution temperature is 180°F [82°C], even though this results in larger convectors. A slightly lower annual average energy consumption should accompany a lower distribution temperature.

Baseboard convectors will be selected from manufacturers' data tables, such as the one shown in Table 16.16. Two common baseboard types are RC (usually cast iron with a water-backed front surface and an extended rear heating surface) and finned tube.

EXAMPLE 16.5 What length of baseboard convector (Table 16.16) is necessary along a 20-ft-long [6-m-long] living room wall? From calculations, the living room heat loss is 9000 Btu/h [2635 W]. The average water temperature is 180°F [82°C], and water flow is about 1 gpm [0.6 L/s].

SOLUTION

With 180°F supply water and 1 gpm flow rate, Table 16.16 shows that 580 Btu/h will be delivered for each linear foot of baseboard convector; 9000 Btu/h ÷ 580 Btu/h ft = 15.5 ft. Choose a baseboard combination of 16-ft overall length, consisting of two 8-ft (finished length) pieces. The total active finned length will be about 15.5 ft. ■

Hydronic Radiant Floors. This heat delivery approach has several comfort advantages over radiant ceilings. The system components are largely the same as with a baseboard or radiator system, except that coils of pipes replace individual radiators or baseboards. A balancing valve should be installed on each pipe coil. Where uninsulated spaces underlie floors (or are located above radiant ceilings), special attention should be paid to adequate insulation; a radiant panel will generate especially high temperatures, and thus a very high Δt through the floor.

Thanks to better insulation, modern buildings often require radiant panels that are smaller in area than the floor (or ceiling) area available. In a conventional radiant panel system, the panel is placed nearest the exterior walls, where the heat loss is greatest. In a solar-heated building, this is not so clear. If the panel heats the floor surface just inside south-facing glass, how much warming will be left to the sun? A preheated slab will absorb much less solar radiation. However, in cloudy cold weather, the area closest to the south glass could become uncomfortably cool.

In the past, copper tubing was widely used for the heating coils. This typically required a number of connections within a coil, and each connection represented a potential point of failure over the life of the panel. Today, coils are typically one-piece and are made of plastics such as cross-linked polyethylene tubing. When the floor is concrete or other cast-in-place material, the coil is either directly embedded in the slab (tied down to resist floating during the pour) or stapled to an underfloor, over which the slab is poured. Radiant

TABLE 16.16 Hydronic Baseboard Convectors

Water Supply					
Temperature		Flow Rate[a]		Heat Delivered per Unit Length[b]	
°C	°F	L/s	gpm	Watts/m	Btu/h ft
60	140	0.06	1	307	320
		0.25	4	326	340
66	150	0.06	1	365	380
		0.25	4	384	400
71	160	0.06	1	432	450
		0.25	4	461	480
77	170	0.06	1	490	510
		0.25	4	518	540
82	180	0.06	1	557	580
		0.25	4	586	610
88	190	0.06	1	614	640
		0.25	4	653	680
93	200	0.06	1	682	710
		0.25	4	720	750
99	210	0.06	1	739	770
		0.25	4	778	810

Source: Slant/Fin Ltd. Mississauga, Ontario.

[a]Use flow rate of 0.6 L/s [1 gpm] unless flow rate is known to be ≥0.25 L/s [4 gpm].
Pressure drop is as follows: flow 0.6 L/s [1 gpm] = 3.76 mm/m [0.045 in./ft].
flow 0.25 L/s [4 gpm] = 42 mm/m [0.5 in./ft].

[b]Baseboard finished length will be 76 mm [3 in.] longer than the length required to meet the heating need.

floors involving coils placed underneath wood floors are popular.

Rugs or carpets over radiant floors are a mixed blessing; they are soft on the feet but interfere with the exchange of heat. Special undercarpet pads can facilitate heat transfer; higher water temperatures can be used in the coils, because skin contact with the floor is prevented by the carpet.

Panel (radiant ceiling or floor) heating design usually depends on a water temperature of 120°F [49°C] for floors and 140°F [60°C] for ceilings. An uncarpeted concrete floor slab using ¾-in.-diameter pipe or tube on 12-in. centers [20-mm diameter on 300-mm centers], with an average water temperature of 120°F [49°C], will deliver 50 Btu/h ft^2 [158 W/m^2] of floor panel (as seen in Example 16.6). A ceiling panel with nominal ⅜-in. tube on 6-in. centers [10-mm diameter on 150-mm centers], with an average water temperature of 140°F [60°C], will deliver 60 Btu/h ft^2 [189 W/m^2] of ceiling panel.

To determine the required area of heated panel, divide the room's design heat loss by the rate of panel heat delivery. Although radiant ceilings deliver more heat per unit area, they also discourage air motion because the warmest air rises to lie against the warmest surface. In contrast, at a radiant floor the coolest air drops to contact the warmest surface, is then warmed, and rises to be cooled at the ceiling, drops to the floor, and repeats this cycle continually.

Each radiant panel contains one or more coils. In floor panels, each coil typically delivers 10,000 Btu/h [2930 W] (see Example 16.6) and should be no longer than about 200 linear feet [60 m]. In ceiling panels (with smaller-diameter tubes), each coil delivers around 3000 Btu/h [880 W] and should be no longer than about 100 linear feet [30 m].

Sizing of the main pipes and pump is based upon the longest circuit, measured along the length of the supply pipe from the boiler to the coil and back along the return line to the boiler. Two guidelines apply:

1. Do not include the length of the coil itself in this circuit length.
2. No section of a floor panel main pipe should be less than ¾ in. [20 mm] in diameter.

EXAMPLE 16.6 Determine the panel area and coils required for a radiant floor in a 15 ft × 25 ft (375 ft^2) living room with a design heat loss of 12,000 Btu/h [4.6 m × 7.6 m (34.8 m^2) with a design heat loss of 3514 W].

SOLUTION

The required panel area is:
In I-P units:

$$\frac{12{,}000\ \text{Btu/h}}{50\ \text{Btu/h ft}^2} = 240\ \text{ft}^2 \text{ of floor panel}$$

In SI units:

$$\frac{3517\ \text{W}}{158\ \text{W/m}^2} = 22.3\ \text{m}^2 \text{ of floor panel}$$

The room has 375 ft^2 [34.8 m^2] of available floor area. The heated panel is usually placed along the exterior walls, where the room heat loss is greatest (unless solar heat through south windows is involved). The number of coils in the panel is:

In I-P units:

$$\frac{12{,}000\ \text{Btu/h}}{10{,}000\ \text{Btu/h per coil}} = 1.2\ \text{coils}$$

In SI units:

$$\frac{3517\ \text{W}}{2931\ \text{W per coil}} = 1.2\ \text{coils}$$

It may be acceptable to use one coil in this panel; two coils will be conservative (overdesigned), but may be preferred. ■

Hydronic System Zoning. Figure 16.100 shows that zoning is relatively easy to accomplish with hydronic systems. The building shown in the schematic diagram has three separately heated areas—the first, second, and third floors. Each floor can be heated to different temperatures as called for by thermostats in the separate apartments. For example, if only the thermostat serving the second floor (zone B) calls for heat, it turns on pump B. Flow-control valves B open, admitting hot water from the boiler to piping circuit B. Flow-control valves A and C remain closed, preventing flow in circuits A and C. Any or all of the zones may operate at one time. The boiler keeps a supply of hot water continually ready to supply any zone on demand. This is achieved by an aquastat immersed in the boiler water. When the boiler water drops below the prescribed temperature, it turns on the firing device, such as an oil burner or a gas burner, which brings the water up to the temperature setpoint.

Two famous residences in the U.S. Midwest utilize hydronic systems. Frank Lloyd Wright's Robie House in Chicago has wall radiators integrated below the north windows in the living room. Underfloor radiators with grilles in the floor were to be provided below the full-height south windows, but were apparently never installed. The boiler sits in a basement room. Mies van der Rohe's Farnsworth House in Fox River, Illinois preserves its four walls of ceiling-to-floor glass by concealing radiant heating pipes in the floor slab. The boiler sits within a central utility "closet."

Modern radiators are designed to reflect the sleek and simple lines of contemporary architecture and are getting new exposure with some colorful products (Figure 16.101). The Mayer Art Center at Phillips Exeter Academy in New Hampshire (Figure 16.102) features new exposed radiators in older buildings. These radiators are based on simple components (typically, 2 3/4 in. [70 mm] wide) that can be combined in many heights and widths, inviting the designer to feature them rather than hide them in metal cabinets.

Warm Air Heating Systems. About a century ago, these systems began to supersede open fireplaces in residences. A typical early system involved an iron furnace that stood in the middle of the basement and was hand-fired with coal. Surrounding it was a sheet-metal enclosure. An opening in its side near the bottom admitted cool combustion air that gravitated to the basement. A short duct from the top of the enclosure delivered the warm air by stack effect to a large grille in the middle of the floor of the parlor. Other rooms, including those in upper stories, shared a little of this warmth when doors were left open. The system functioned much like a local system.

By the middle of the twentieth century, very gradual changes culminated in systems essentially like the ones illustrated in Figures 16.19 and 16.103. The improvements included:

- automatic firing of oil or gas burners
- operational and safety controls
- air ducted to and from each room
- blowers to replace stack effect
- air filters
- adjustable delivery registers

By the 1960s, basements were beginning to disappear, and subslab perimeter systems became popular for basement-less houses (Figure 16.104). The heat source was located centrally and fully within the insulated enclosure of the house; any heat escaping from the unit simply helped heat the building. In general, air was delivered from below, upward across windows, to be taken back at a central high-return grille.

When electricity was used for space heating instead of oil and gas, provisions for combustion air, chimneys, and fuel storage were unnecessary. Horizontal electric furnaces began to appear in shallow attics or above furred ceilings. Air was delivered down from ceilings across windows and taken back through door grilles and open plenum space. Electric resistance furnaces use more electricity than heat pumps to provide the same heating capacity, so heat pumps have largely supplanted electric resistance furnaces (where climate permits).

As energy-efficient design gained strength, insulated windows and well-insulated roofs, walls, and floors lessened the need for space heating. From a central furnace or heat pump, short ducts could deliver warm air to the inner side of each room, because providing heat at well-insulated windows was less essential. Air was returned to the furnace through grilles in doors and at the furnace or heat pump enclosure.

It is possible to clean both the recirculated air and the outdoor air by means of filters and (as appropriate) special air-cleaning equipment. Air may be circulated in nonheating seasons to reduce stuffiness. Outdoor

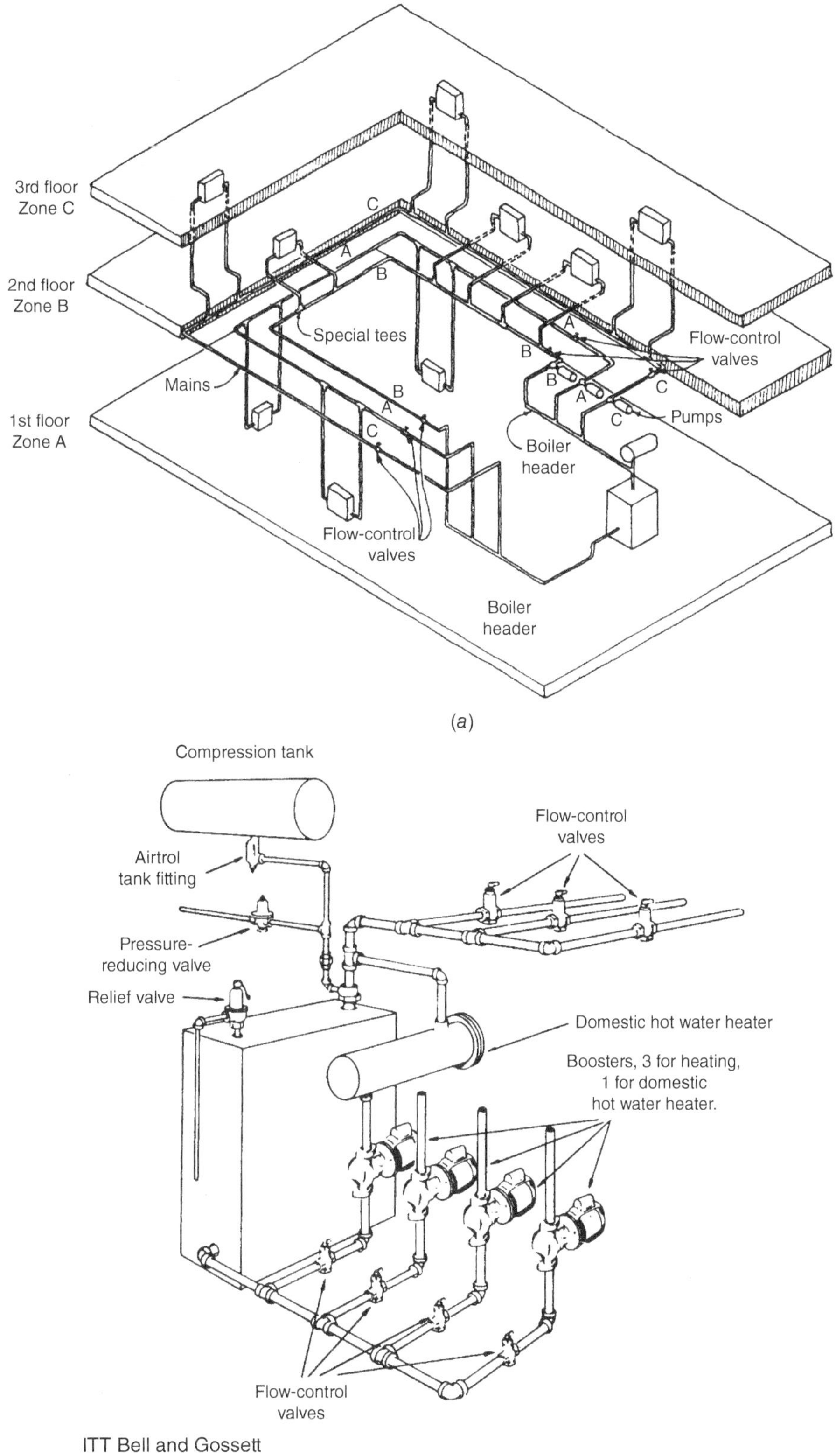

Figure 16.100 *Three-zone, multicircuit, one-pipe system. (a) Each convector has connections to the one pipe. (b) Boiler, piping, and water controls suitable for this three-zone, one-pipe system. Each one-pipe circuit should be provided with two flow-control valves and a circulator (also called a booster or pump) on the supply (or return) pipe. (Courtesy of ITT Bell and Gossett.)*

ACTIVE ENVIRONMENTAL SYSTEMS

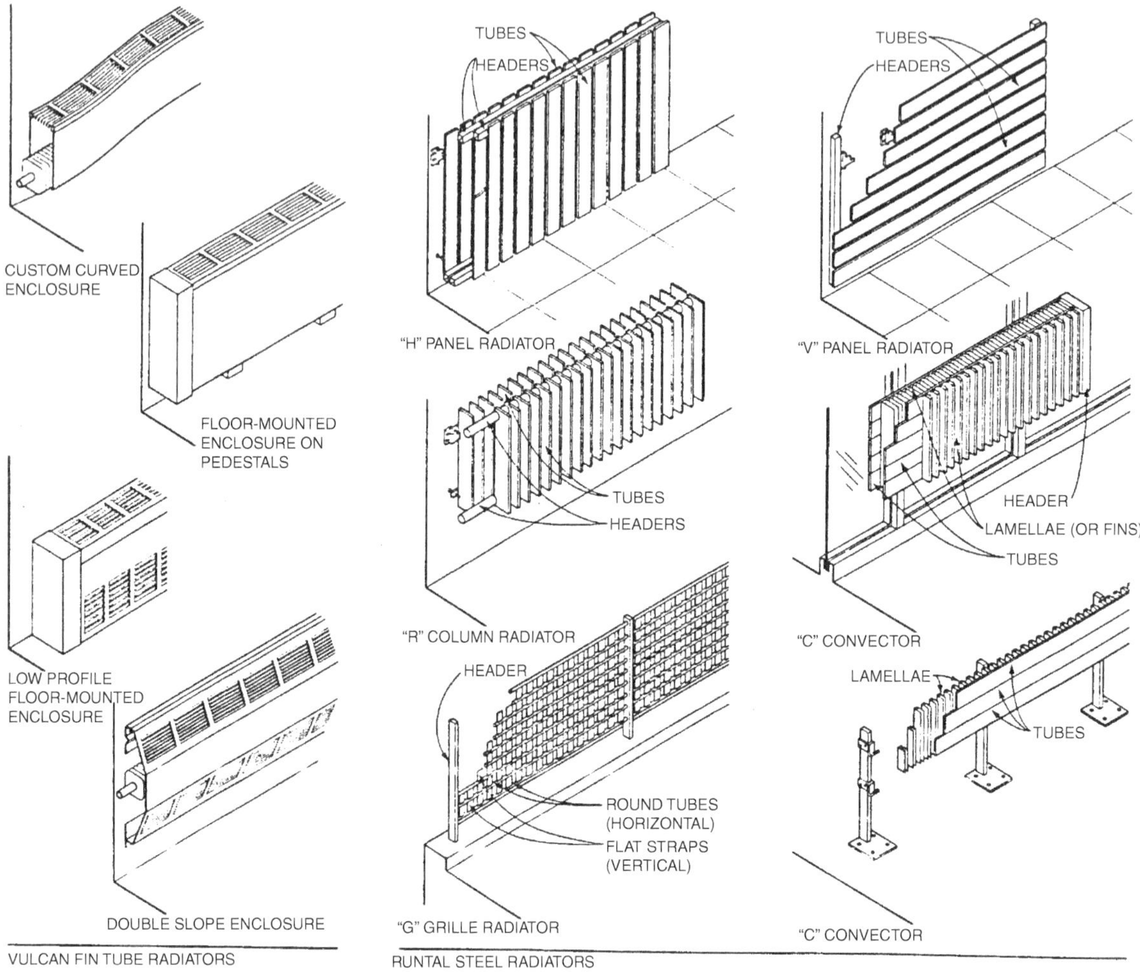

Figure 16.101 *Hot water radiators are available in bright colors and are based on simple components. (From Ramsey and Sleeper, 2000.)*

air may be introduced to reduce odors and to make up for air exhausted by fans in kitchens, laundries, and bathrooms. Central cooling can be retrofit into an air system if ducts are designed to accommodate this; cooling often requires greater rates of air circulation than does heating. Humidification can be provided by a humidifier located in the air stream, and if cooling is included in the design, dehumidification can be accomplished in summer. For both heating and cooling, a common arrangement is to place supply registers in the floor, below areas of glass. This is important for winter operation. With adequate attention to supply register placement, return grilles can be located so as to minimize return air ductwork. High return grilles pick up warmer air for reheating at the equipment. In many systems, air circulates at all times and is warmed or cooled as required.

Figure 16.105 shows the relationship between a furnace, the duct distribution tree, and some elements of the spaces they serve. Ducts also can be lined with sound-absorbing material to mitigate noise transfer—but beware of lining materials that support mold and mildew growth. The following noise-reduction suggestions are recommended:

- Do not place the blower too close to a return grille.
- Select quiet motors and cushioned mountings.
- Do not permit connection or contact of conduits or water piping with the blower housing.
- Use a flexible connection between furnace and ductwork.

In the absence of project-specific criteria, duct sizes may be selected based upon permissible air velocity in the duct segment (or delivery device) of interest. Table 16.17 provides recommended maximum velocities related to achievement of acoustical design objectives. These velocities are not based upon maximizing system energy efficiency (which would generally lead to lower velocities).

Figure 16.102 Hot water radiators fit below a window (a) at the Mayer Art Center, Phillips Exeter Academy, Exeter, New Hampshire. Amsler Hagenah MacLean, Architects. (b) Radiator formed by flat tubes. (a: by Alex Beatty; b: Courtesy of Runtal/North American Energy Systems.)

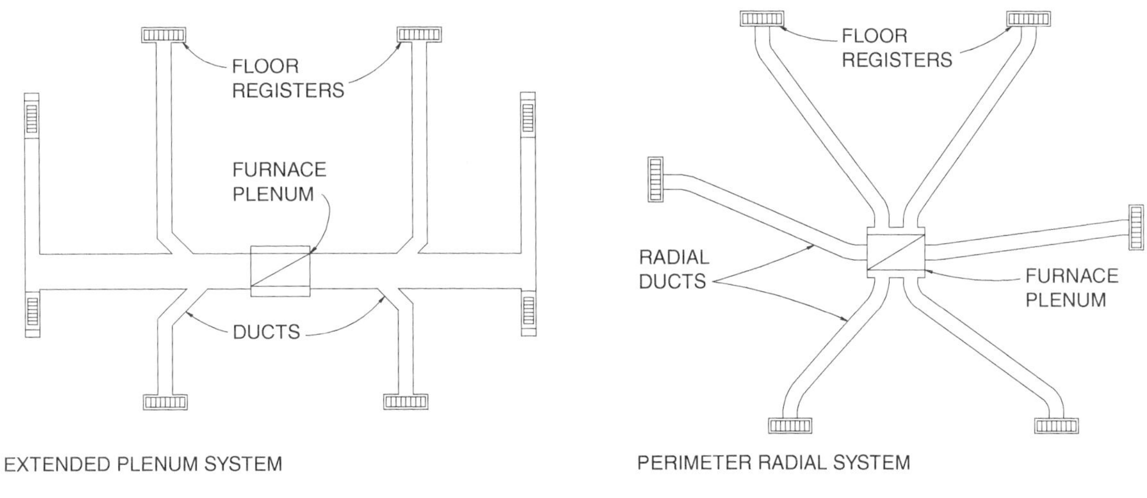

Figure 16.103 Typical duct distribution arrangements from residential furnaces. (From Ramsey and Sleeper, 2016.)

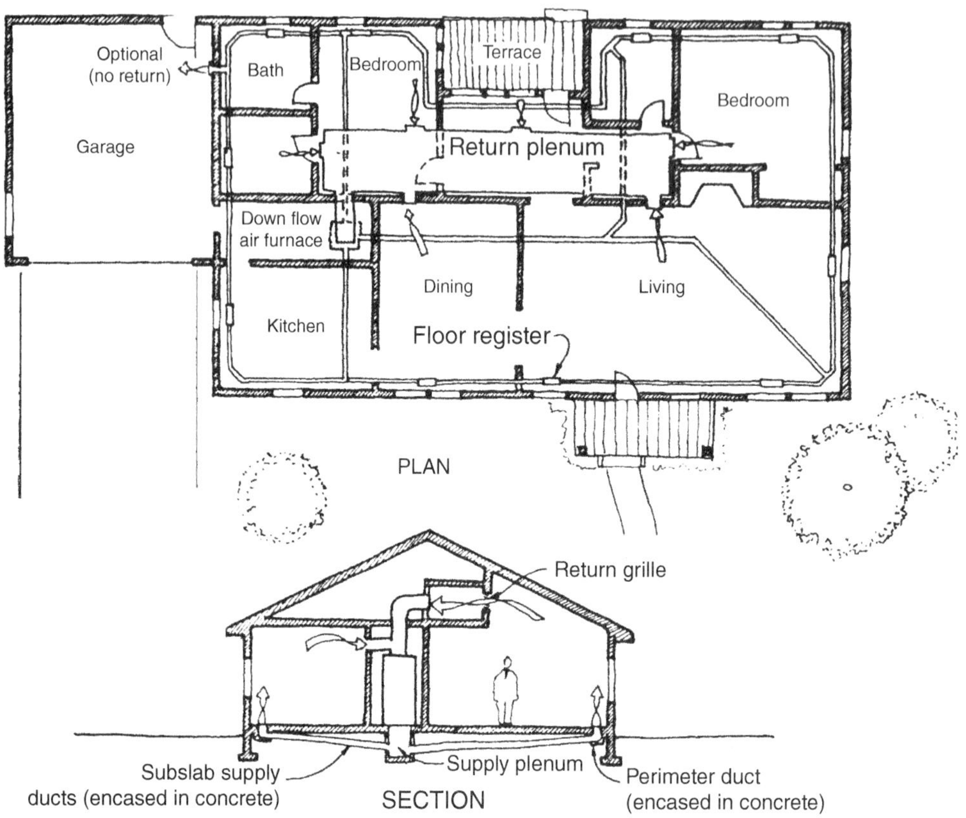

Figure 16.104 *Forced-warm-air, perimeter loop system, adaptable for a cooling coil at the furnace. No return air is taken from the kitchen, baths, or garage.*

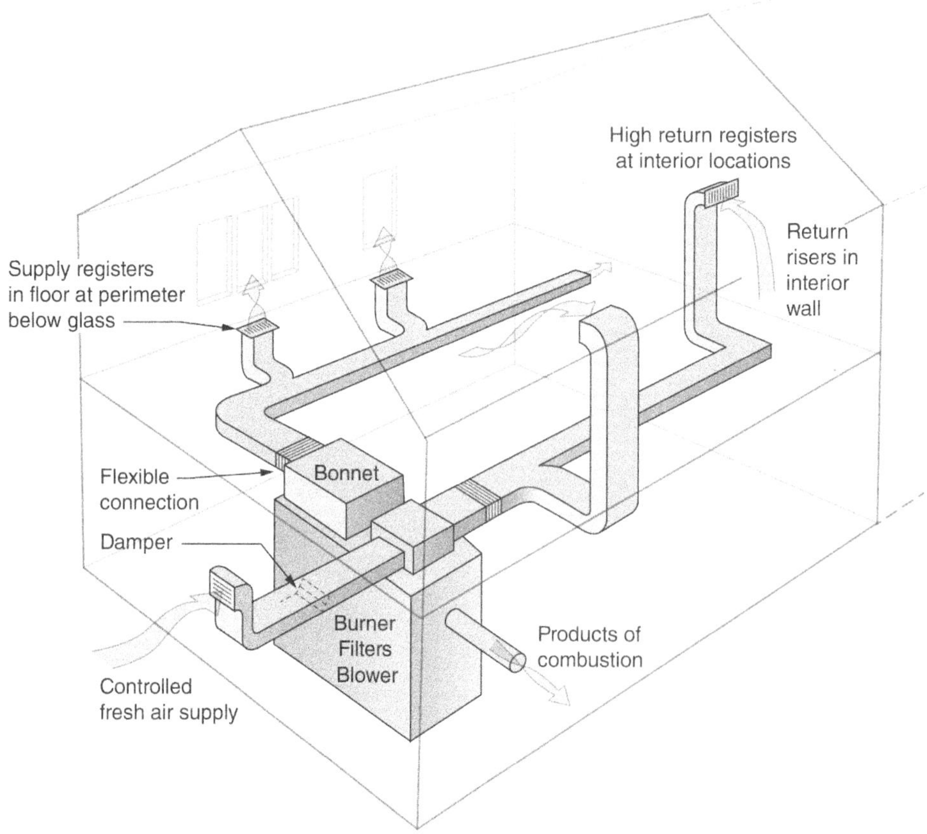

Figure 16.105 *Conventional warm-air furnace and ductwork. Low supply registers under windows and high return grilles at interior walls help ensure distribution within the spaces.*

TABLE 16.17 Maximum Air Velocities for Ducts and Grilles (Based on Noise Criteria)

PART A. Main Ducts[a]

		Maximum Airflow Velocity			
		fpm		m/s	
Duct Location	**Design RC(N)**	**Rectangular**	**Circular**	**Rectangular**	**Circular**
In shaft or above drywall ceiling	45	3500	5000	17.8	25.4
	35	2500	3500	12.7	17.8
	25	1700	2500	8.6	12.7
Above suspended acoustic ceiling	45	2500	4500	12.7	22.9
	35	1750	3000	8.9	15.2
	25	1200	2000	6.1	10.2
Duct within occupied space	45	2000	3900	10.2	19.8
	35	1450	2600	7.4	13.2
	25	950	1700	4.8	8.6

Part B. Face Velocities at Supply and Return Openings[c]

		Maximum Airflow Velocity	
Type of Opening	**Design RC(N)[b]**	**fpm**	**m/s**
Supply air outlet	45	625	3.2
	40	560	2.8
	35	500	2.5
	30	425	2.2
	25	350	1.8
Return air opening	45	750	3.8
	40	675	3.4
	35	600	3
	30	500	2.5
	25	425	2.2

Part C. Design Guidelines for HVAC-Related Background Sound in Rooms[d]

		Octave Band Analysis	Approximate Overall Sound Pressure Level	
Room Types		**NC/RC**	**dBA**	**dBC**
Residences, apartments, condominiums	Living areas	30	35	60
	Bathrooms, kitchens, utility rooms	35	40	60
Hotels/motels	Individual rooms or suites	30	35	60
	Meeting/banquet rooms	30	35	60
	Corridors and lobbies	40	45	65
	Service/support areas	40	45	65
Office buildings	Executive and private offices	30	35	60
	Conference rooms	30	35	60
	Teleconference rooms	25	30	55
	Open-plan offices	40	45	65
	Corridors and lobbies	40	45	65
Courtrooms	Unamplified speech	30	35	60
	Amplified speech	35	40	60
Performing arts spaces	Drama theaters, concert and recital halls	20	25	50
	Music teaching studios	25	30	55
	Music practice rooms	30	35	60
Hospitals and clinics	Patient rooms	30	35	60
	Wards	35	40	60
	Operating and procedure rooms	35	40	60
	Corridors and lobbies	40	45	65
Laboratories	Testing/research with minimal speech communication	45	50	70
	Extensive phone use and speech communication	40	45	65
	Group teaching	35	40	60
Churches, mosques, synagogues	General assembly with critical music programs	25	30	55

TABLE 16.17 (*Continued*)

Part C. Design Guidelines for HVAC-Related Background Sound in Rooms[d]				
		Octave Band Analysis	Approximate Overall Sound Pressure Level	
Room Types		**NC/RC**	**dBA**	**dBC**
Schools	Classrooms	30	35	60
	Large lecture rooms with speech amplification	30	35	60
	Large lecture rooms without speech amplification	25	30	55
	Transitional/collaborative areas	35	45	60
Libraries		30	35	60
Indoor stadiums, gymnasiums	Gymnasiums and natatoriums	45	50	70
	Phys. Ed. areas (per *ANSI S12.60* Part 4)	35	40	60
	Large-seating-capacity spaces with speech amplification	45	50	70

Source: © ASHRAE, www.ashrae.org. 2023 ASHRAE Handbook—HVAC Applications, Chapter 49, Table 1.

[a]Branch duct velocities should be about 80% of those listed, whereas velocities in final runouts to outlets should be 50% or less. Elbows and other fittings can increase noise substantially, so airflows should be reduced accordingly.

[b]RC(N) = Room criteria, neutral equipment sound quality. See Chapter 8 of this book for more information regarding room criteria.

[c]These are "free" opening velocities. Diffusers and grilles can increase sound levels, in which case these velocities should be reduced accordingly.

[d]Ideally, HVAC-related background noise should have the following characteristics:

- Balanced contributions from all parts of the sound spectrum with no predominant frequency bands of noise.
- No audible tones such as hum or whine (ASHRAE research project RP-1707 provides some metrics to evaluate annoyance due to prominent tones).
- No fluctuations in level such as throbbing or pulsing.

dBA and dBC: A- and C-Weighted Sound Level. The A-weighted sound level (described in Chapter 8 of the *2021 ASHRAE Handbook-Fundamentals*) has been used for more than 60 years as a single-number measure of the relative loudness of noise, especially for outdoor environmental noise standards. The rating is expressed as a number followed by dBA (e.g., 40 dBA).

A-weighted sound levels can be measured with simple sound level meters. The ratings correlate fairly well with human judgments of relative loudness but take no account of spectral balance or sound quality. Thus, two different spectra can result in the same numeric value, but have quite different subjective qualities.

Along with dBA, there is also a C-weighted sound level, denoted as dBC, which is more sensitive to low-frequency sound contributions to the overall sound level than is dBA.

When the quantity dBC—dBA is large (e.g., greater than 25 dB), significant low-frequency sound is present. It is recommended that when specifying background sound levels in dBA, the dBC is also included in the specification and does not exceed the dBA reading by more than 20 dB.

EXAMPLE 16.7 The main duct in a low-velocity, warm air system for a residence delivers 1600 cfm [755 L/s]. It is located above a drywall ceiling and has a rectangular cross-section. Select a size for this duct.

SOLUTION

Table 16.17 suggests that for a residence, an RC of 30 is generally appropriate. The homeowner desires less HVAC-related background sound than typical. As such, set the target RC = 25; thus, the maximum acceptable air velocity is 1700 fpm [8.63 m/s]. The minimum area of duct in square inches [square meters] would be:

In I-P units:

$$\frac{(1600\text{ cfm})(144\text{ in.}^2\text{per ft}^2)}{1700\text{ fpm}} = 135\text{ in.}^2$$

In SI units:

$$\frac{(755\text{ L/s})(0.001\text{ m}^3\text{ per L})}{8.6\text{ m/s}} = 0.09\text{ m}^2$$

A 10-in. × 14-in. (140-in.^{2}) [0.25-m × 0.36-m (0.09 m^2)] duct is acceptable; a larger duct would produce less noise and require less fan power to overcome friction. ■

Figure 16.106 illustrates "The Ductulator," a device that simplifies duct sizing. At a glance, it shows many duct cross-sectional configurations that will satisfy the combined requirements of friction, airflow, and air velocity.

Dampers and other accessories are necessary to balance an air system and adjust it to the desires of the occupants (Figure 16.107). Splitter dampers are used where branch ducts leave the larger trunk ducts. The flow of each riser can be controlled by an adjustable damper in the basement at the foot of the riser. Labels should indicate the rooms served. Some codes require dampers of fire-resistant material actuated by fusible links to prevent the possible spread of fire through a duct system. Figure 16.107*d* shows how turning vanes can be used to improve airflow at sharp turns. Turning vanes reduce friction within the ductwork, thus reducing the total static head (Figure 16.108) against which the supply fan must work.

Supply registers (Figure 16.109) should be equipped with dampers, and their vanes should be arranged to disperse the air and to reduce its velocity as it enters a room. This is commonly done by providing vanes that divert the air, half to the right and half to the

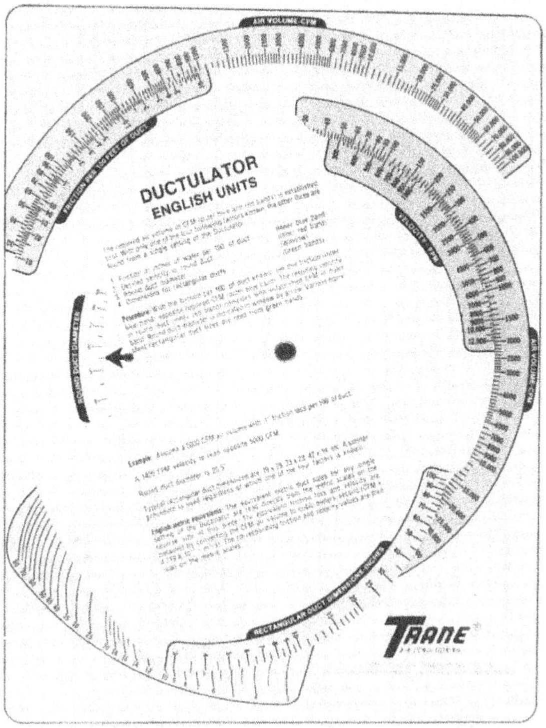

Figure 16.106 The Ductulator duct-sizing device is available from the Trane Company. The designer selects any two factors (e.g., friction and airflow volume), and the device displays all other factors (e.g., air velocity, required diameter of a round duct, or combinations of equivalent rectangular-duct cross sections).

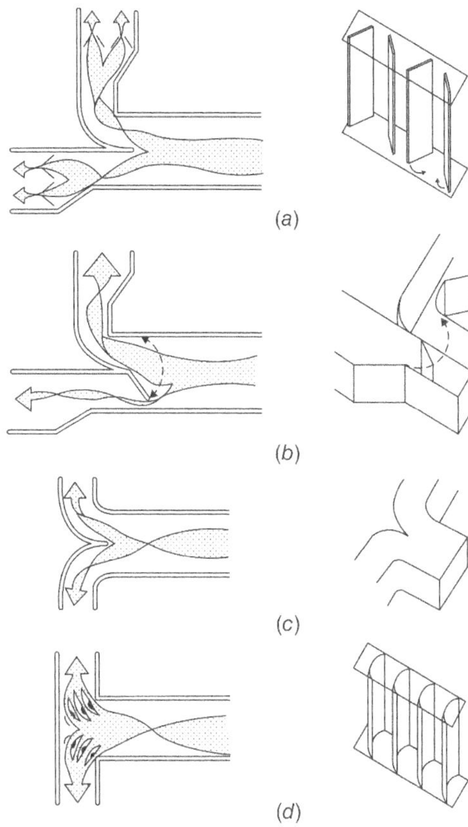

Figure 16.107 Airflow controls in ducts. (a) Air adjustment by opposed-blade dampers. (b) Air adjustment by a splitter damper. (c) Conventional turns in ducts. (d) Right-angle turns with turning vanes—a more compact method.

left. When a supply register is placed in the corner of a room, the vanes should deflect all the air away from the corner. Return grilles are usually of the slotted type in walls and of the grid type in floors. All registers and grilles should be made tight at the duct connection. See Tables 16.17 and 16.18 for a selection of registers based upon output and recommended face velocity.

Cooling Coils in Furnaces. Active cooling is a much more recent development than heating; in its early days, cooling was generally adopted as a retrofit to existing heating systems. Adding a cooling coil to a warm air furnace provided air conditioning (as opposed to heating). This conceptual approach to residential air conditioning persists. Figure 16.110 shows a schematic diagram of an air-to-air, direct expansion refrigeration system and how this can be coupled to a typical furnace. Figure 16.111 shows another popular arrangement in which airflow through the furnace and coils is horizontal.

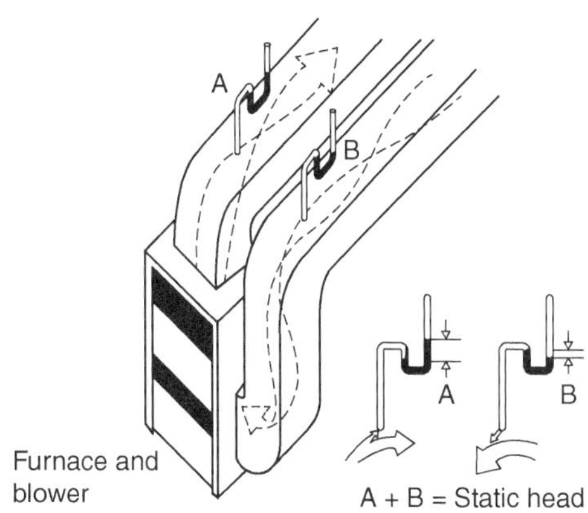

Figure 16.108 The static head is the pressure, measured in inches of water (or Pascals), required to overcome friction loss in an air distribution system.

Hydronic Perimeter and Air. This system combines a perimeter hot water heating pipe with an overhead air-handling system. A boiler with a tankless coil provides domestic hot water and also supplies a perimeter loop and a coil in an air-handling unit supplying a ductwork system. The total heating load is met by a combination of radiant heat generated by the perimeter loop and heated air from the overhead air distribution system.

As installed in the Levittown standardized houses, the perimeter loop consists of ½- or ¾-in. [13- or 19-mm] tubing embedded 4 in. [100 mm] below the

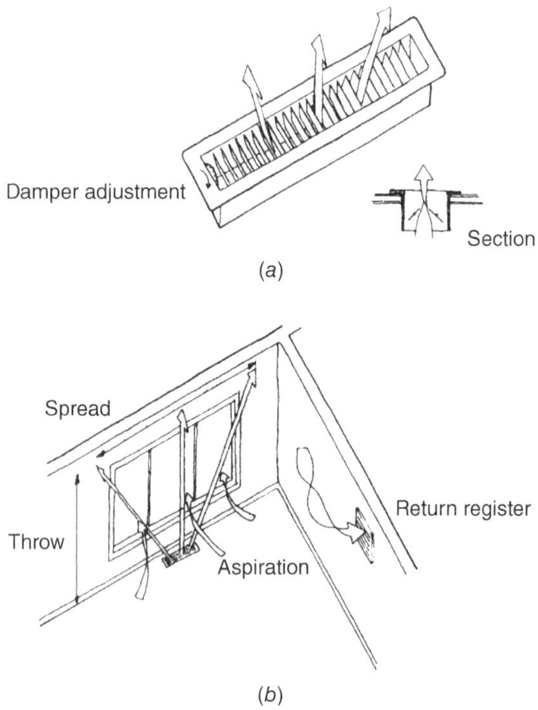

Figure 16.109 *The typical floor register (a) is 2¼ in. × 12 in. [60 mm × 300 mm]. It has diverting vanes for spread and an adjustable damper. See Table 16.18 for characteristics. (b) Spread and throw; cooler room air is induced by aspiration to join the stream of warm air, resulting in a pleasant air stream that crosses the room.*

top of the floor slab to overcome the cold slab effect. It has the capacity to maintain a 35°F [19°C] differential between the indoor and outdoor temperatures at the perimeter.

The air-handling unit and overhead duct system, incorporating supply outlets in each room and a central return, is used throughout the year. Its cooling coil is connected to an adjacent outdoor condensing unit (Figure 16.112). Because the heating load is shared by the slab loop and by the warm air system, winter indoor temperature remains rather constant. The air can be distributed at roughly 120°F [49°C], which is 20°F [11°C] less than with a conventionally ducted system.

(c) Evaporative Cooling Systems

Evaporative cooling systems are applicable (under the right climate and functional conditions) to both small and large buildings. Several of these systems are presented here. A few of these systems are hybrid (requiring some operational assistance from mechanical/electrical equipment); evaporative coolers (both direct and indirect) are best classified as active systems.

Evaporative Cooling—Misting. The net effect of evaporative cooling is no total change in the heat content (enthalpy) of the treated air; DB temperature is lowered, but there is an increase in RH. One of the most direct approaches to evaporative cooling is a misting or fogging system whereby a fine spray of water droplets is blown into the air. A common application is for small outdoor areas—team benches at football stadiums, refreshment pavilions, outdoor seating at a restaurant. Large spaces in hot, dry climates can also benefit; Figure 16.113 shows mist descending from a diffuser in Las Vegas, Nevada. A large space using fogging (in a Michigan conservatory) is shown in Figure 16.61.

Evaporative Cooling—Roof Spray. In the past, when roofs were poorly insulated, roof sprays were a rather common way to reduce heat gains. New variations on this approach promise potential energy savings.

Roof color is the first design consideration. White or near-white roofs are the first step toward energy savings through control of sol-air temperature. Emissivity is also involved; the higher the emissivity, the more readily a roof surface reradiates its heat to the sky. Table 16.19 compares the solar absorptance, albedo (overall solar reflectance), and emittance of

TABLE 16.18 Typical Residential Forced-Air Register, 2¼ × 12 in. [60 × 300 mm]

Part A. I-P Units								
Heating (Btu/h)	3045	4565	6090	7610	9515	11,415	13,320	15,220
Cooling (Btu/h)	855	1280	1710	2135	2670	3200	3735	4270
cfm	40	60	80	100	125	150	175	200
Vertical throw (ft)	3	4	5	6	8	10	12	14
Vertical spread (ft)	6	8	10	11	14	17	22	25
Face velocity (fpm)	280	420	565	705	880	1050	1230	1400
Part B. SI Units								
Heating (W)	890	1340	1780	2230	2790	3340	3900	4460
Cooling (W)	250	375	500	625	780	940	1090	1250
L/s	19	28	38	48	59	71	83	95
Vertical throw (m)	0.9	1.2	1.5	1.8	2.4	3.0	3.7	4.3
Vertical spread (m)	1.8	2.4	3.0	3.4	4.3	5.2	6.7	7.6
Face velocity (m/s)	1.4	2.1	2.9	3.6	4.5	5.3	6.2	7.1

Source: Lima Register Company.
SI conversions added by the authors of this book.

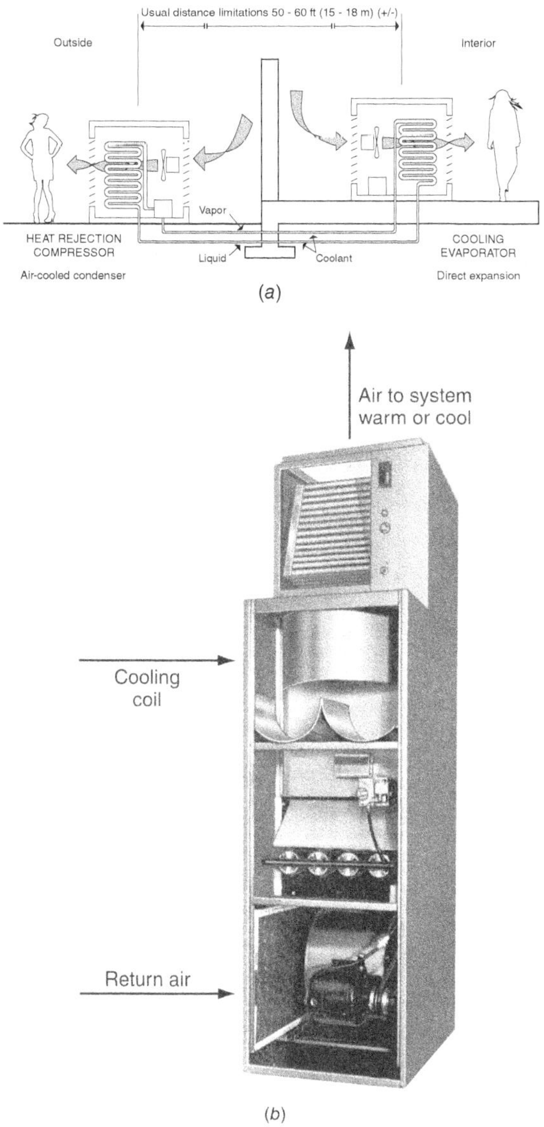

Figure 16.110 *The compressive refrigeration cycle (a) with an exterior air-cooled heat rejection unit and an interior cooling coil suitable for placement in a central air handler. (b) Cooling/heating residential air-handling unit. Cutaway view shows an upflow air pattern. At the lower left is the return air intake and a filter; adjacent are the fan and motor. Above these components are the gas-burning elements. At the top is the direct-expansion cooling coil. (a: Drawn by Achyuthan Ramaswamy; b: Courtesy of American Furnace Division, Singer Company.)*

some common roofing materials. A combination of high albedo and high emittance resists solar heat gain most effectively. This combination can also reduce heat island effect in urban areas. The concept of solar reflectance index (SRI) was developed to allow quick comparisons between roofing products. The SRI scale ranges from 0 (approximately 5% albedo and 90% emittance; roughly equal to black asphalt shingles) to 100 at 80% albedo and 90% emittance ("in the ballpark" with T-EPDM).

An approach called the night roof spray thermal storage system (NRSTS) cools water on a roof at night, using both night sky radiation and evaporation. The water is then stored for use the next day in building cooling. The water can be stored either on the roof (below floating insulation, above a structural ceiling) or in a tank below the roof. When stored on the roof, it becomes a variant of the roof pond cooling system. When stored in a tank, the water can be circulated through a cooling coil to precool air before the air is further cooled by chilled water provided by a chiller. Three variations are shown in Figure 16.114.

At a Nogales, Arizona, border patrol station, a 6500-ft^2 [604-m^2] flat white roof was retrofitted with an NRSTS, utilizing a 10,000-gal [37,850-L] above-ground tank. Over the summer of 1997, it averaged a cooling capacity of 250 Btu/ft^2 per day. Water use at various sites has averaged 4–5 gal/h [15–19 L/h] per 1000 ft^2 [per 93 m^2] of roof area. Details are presented in Bourne and Hoeschele (1998). A similar system was employed at the award-winning building housing the Department of Global Ecology at Stanford University.

Evaporative Coolers. These devices are a move toward mechanical implementations of evaporative cooling (as opposed to the fundamentally architectural approaches described previously). Evaporative coolers (Figure 16.115) are also affectionately termed swamp coolers and desert coolers and are familiar devices in hot, arid climates. They are also used in other climates for special high-heat applications such as restaurant kitchens. They require a small amount of electricity to run a fan and an input of water. The psychrometric process of evaporative cooling is explained in Section 16.4.

The typical evaporative cooler (as in Figure 16.115) needs full access to outdoor air and is thus often set on a roof, although through-the-wall units are also available. Large quantities of dry, hot outdoor air are blown through pads kept moist by recirculating and makeup water. The "cooled" air is delivered to the indoor space(s). In addition to the sensible cooling produced by the process itself, an additional cooling effect is often provided by the impact of air motion on bodily heat loss.

Air introduced into the indoor space must be exhausted in order for the system to operate properly. By skillfully selecting the room through which the air is exhausted, a designer can route cool air as desired along any chosen path from unit to relief opening. The closer to the inlet, the cooler the air; the closer to the exhaust, the warmer the air.

Indirect Evaporative Cooling. The preceding discussion relates to direct evaporative cooling (wherein processed outdoor air is directly introduced to a space).

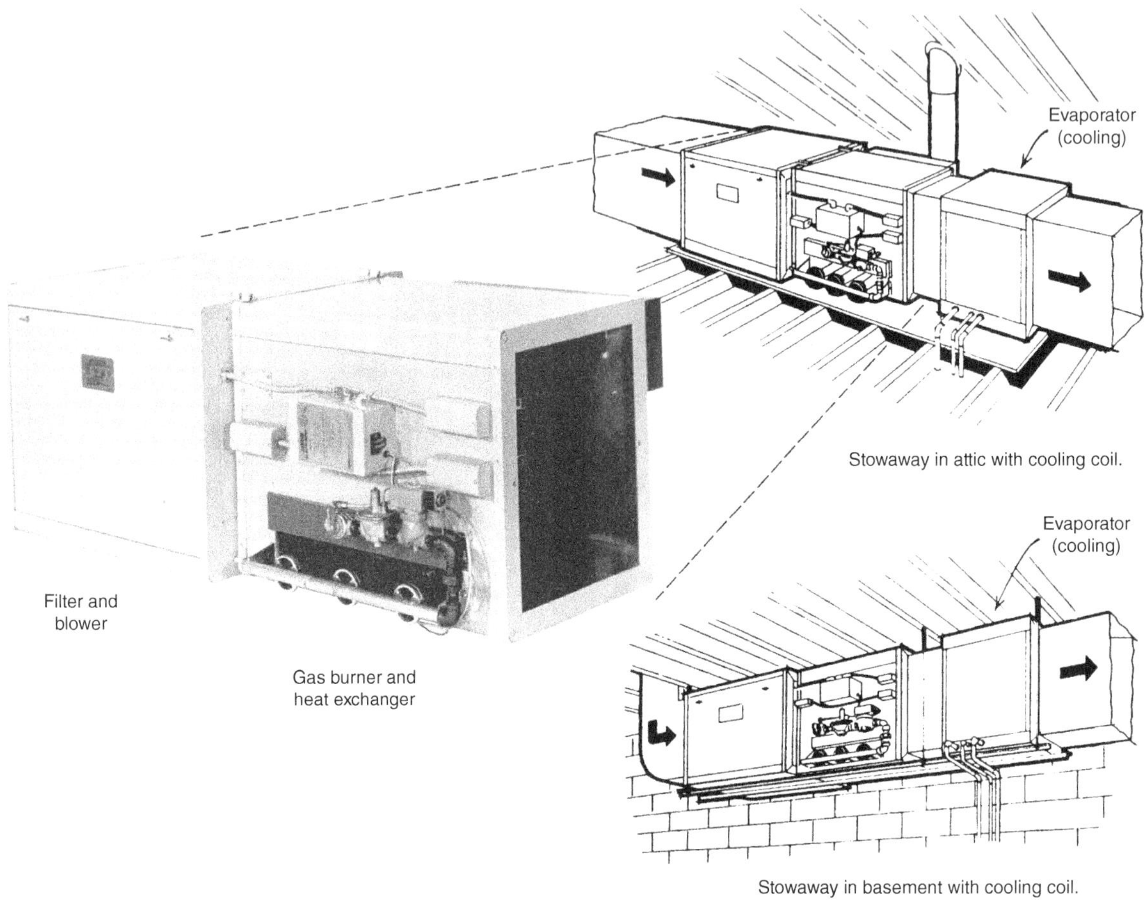

Figure 16.111 *Compact horizontal flow combinations for forced-air heating and cooling. The small pipe between the two refrigerant pipes of the cooling unit is a water drain that carries away the condensed moisture removed from the air. The heating unit requires gas and flue/exhaust. Refrigerant pipes connect to the outdoor compressor–condenser unit.*

Unfortunately, some areas have such hot and/or humid summer daytime conditions that a direct evaporative approach cannot produce thermally comfortable indoor conditions. An indirect process can produce better indoor comfort conditions. A psychrometric process diagram illustrating this is shown in Chapter 5, Figure 5.74.

One of several possible variations is shown in Figures 16.116 and 16.117. Warm, rather dry night outdoor air is evaporatively cooled and fed through a rock bed. The air temperature is low enough to cool the rock bed, and its RH is moderate. At the same time, the house is cooled by a second direct evaporative cooling unit. Figure 16.117 traces the process by day. Extremely hot, dry outdoor air (A) is drawn into the rock bed, where it is cooled by contact (D). It can then be passed through an evaporative cooler to achieve a better combination of RH and DB temperature (E). After picking up both sensible and latent heat, the air is exhausted (at approximately temperature F). Note that at condition F, it is still much cooler than outdoor air.

By comparison, simple direct evaporative cooling by day would have produced indoor supply air too hot and humid for comfort (B). Again, upon exhaust (C), it is still cooler than outdoor air.

Indirect evaporative cooling was combined with a direct refrigerant system in an innovative tent structure over a department store in the San Francisco Bay area (Figure 16.118). Two layers of fiberglass, Teflon-coated fabric, separated by an average of 12 in. [300 mm], were supported by a network of cables hung from eight masts. The tent roof covered about 70,000 ft^2 [6500 m^2] of sales floor; its 7% translucency to sunlight provided 450–550 footcandles [4800–5900 lux] of daylight. This greatly reduced the need for electric lighting, although some electric lamps, clipped onto exposed fire sprinkler pipes, were used as accent lighting. About 3.5 W/ft^2 [37.7 W/m^2] of solar gain, mostly in the form of diffused daylight, penetrated the tent cover. When this solar gain was combined with heat gains from people and electric accent lighting, a cooling load was always generated

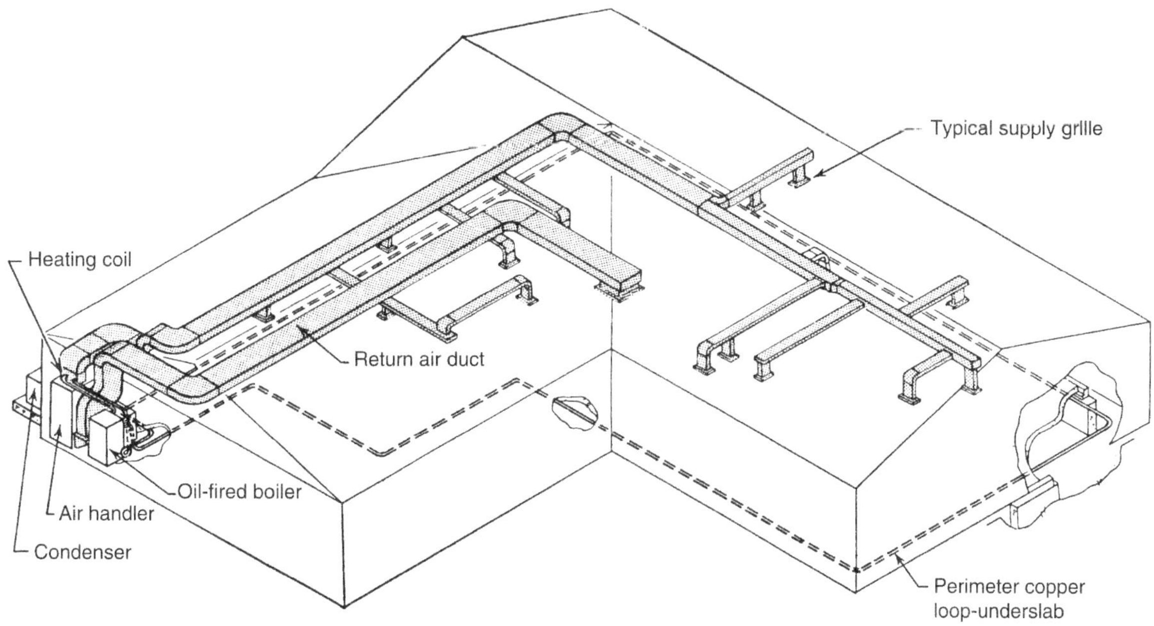

(*a*)

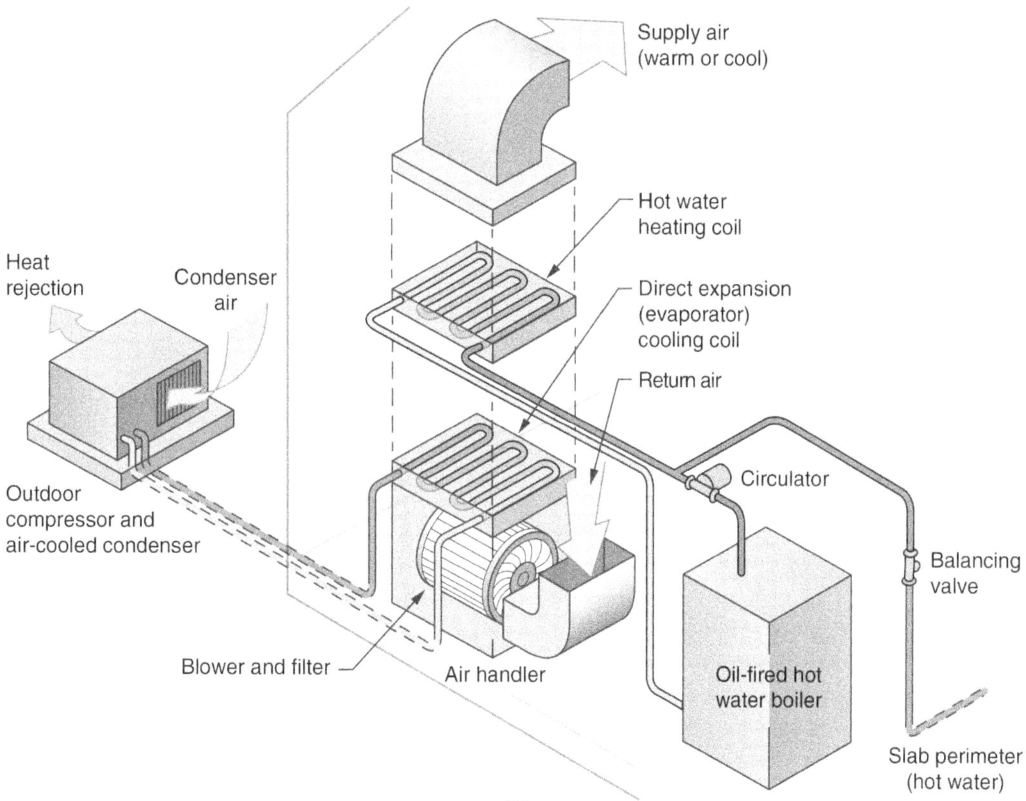

(*b*)

Figure 16.112 *Combination hydronic and forced-air system (a) using a perimeter hot water loop in the floor slab and an overhead air supply. (b) Close-up of the heating and cooling components. The boiler heats both the slab perimeter and the coils in the air stream. (Courtesy of Levitt and Sons. Design by mechanical engineer John Liebl.)*

Figure 16.113 Mist is sprayed from a diffusing system in a building located in Las Vegas, Nevada. It provides psychological reinforcement of the evaporative cooling effect during a hot, dry summer. (© Ayush Vaidya; used with permission.)

TABLE 16.19 Solar Performance of Roofing Materials

Material	Absorptance (%)	Albedo (%)	Emittance (%)
White asphalt shingles	79	21	91
Black asphalt shingles	95	5	91
White granular-surface bitumen	74	26	92
Red clay tile	67	33	90
Red concrete tile	82	18	91
Unpainted concrete tile	75	25	90
White concrete tile	27	73	90
Galvanized steel (unpainted)	39	61	4
Aluminum	39	61	25
Siliconized white polyester over metal	41	59	85
Kynar white over metal	33	67	85
Gray EPDM	77	23	87
White EPDM	31	69	87
T-EPDM	19	81	92
Hypalon	24	76	91

Source: Lawrence Berkeley National Laboratory.

(even in San Francisco Bay's mild climate, which rarely falls below 45°F [7°C]). The roof's relatively high U-factor was thus advantageous in helping the building lose heat. Because the San Francisco area overheats even more rarely, such low resistance to heat flow was not believed to be seriously disadvantageous in summer.

To remove the heat gain, four sets of direct refrigerant equipment were provided at the tent perimeter on grade. These fed into a perimeter plenum, from which the entire store was supplied with cool air. The exhaust air was collected at the center and returned to help with cooling. Indirect evaporative cooling (also called sensible evaporative refrigeration) units were used to cool air (see Figure 16.118*b*). In this application, two units worked in tandem. Air to be cooled (supply air) entered the heat exchanger of the first unit, which was cooled by evaporative cooling of outdoor air. During peak temperature periods, the supply air would only be somewhat cooled by this process; sufficiently low temperatures for use in the interior were obtained by passing the air through the second unit's heat exchanger. This second unit was cooled by evaporative cooling of the exhaust air from the store, which was cooler than outdoor air under summer conditions. Thus, the exhaust air did some useful work before exiting. This two-stage, indirect evaporative cooling process allowed the supply air to be cooled without raising the RH. This would not be the case if direct evaporative cooling had been used.

16.19 HVAC SYSTEMS FOR LARGE BUILDINGS

Large buildings typically have so many thermal zones, and there are so many ways to move heat/coolth from one place to another and to introduce heat/coolth into a space, that literally hundreds of HVAC system variations have been devised over the years. Several of the more typical systems are introduced in this section. Figure 16.119 provides a schematic representation of a range of HVAC system approaches—which, as they increase in complexity and flexibility, include increasing numbers of heat-transfer loops.

Direct expansion (DX) systems (Section 16.8) are typically used in smaller-scale buildings, but are also applicable to some larger building situations. Big box retail stores, for example, typically use large rooftop packaged air-conditioning units with minimum ductwork. These are essentially local systems—each serving a section of the store floor area (even though the store is open-plan). DX systems were discussed in Section 16.8(a).

Large-building HVAC systems in general tend to be central systems. As noted under taxonomy (Section 16.16), central HVAC systems fall into three broad categories based upon the distribution medium chosen:

1. all-air systems
2. air–water systems
3. all-water systems

Common implementations of these systems will be presented.

(a) All-Air Systems

Several variations of all-air systems are either commonly used in current building designs or are of historical interest. Because air is the only heat-transfer

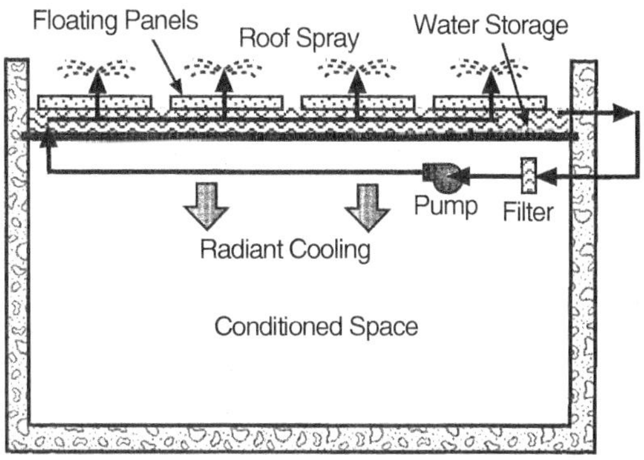

(a)

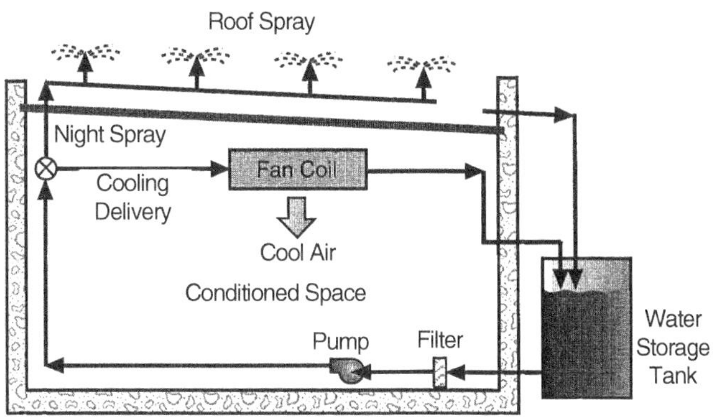

(b)

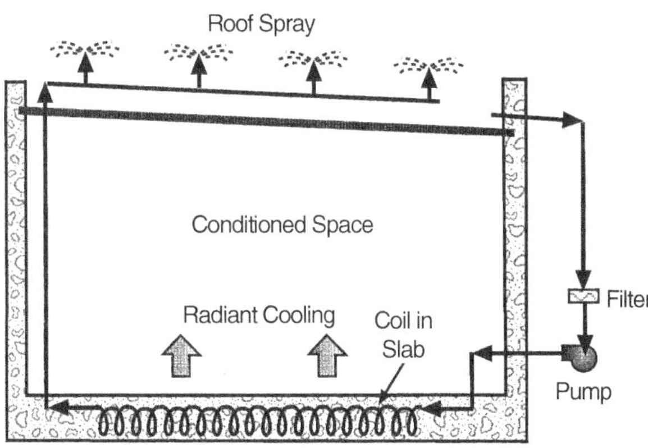

(c)

Figure 16.114 *Night roof spray thermal storage systems (NRSTS). (a) This version approximates the performance of the roof pond. (b) Remote water storage allows use of the cooled water at any time. (c) The floor slab is used as thermal storage in this version. (Technical Installation Review, 1997 / U.S. Department of Energy.)*

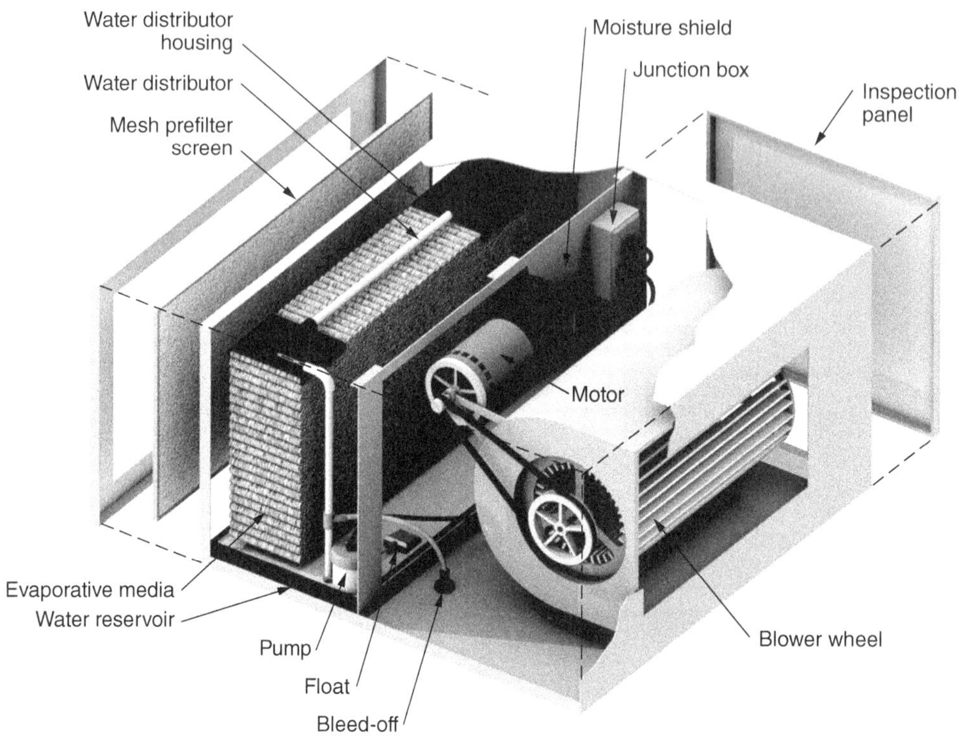

Figure 16.115 The UltraCool evaporative cooler has a low profile and uses internal components of plastic or stainless steel to reduce corrosion. Typical dimensions for a 4000-cfm [1890-L/s] unit are 35 in. H × 42 in. W × 48 in. D [890 × 1070 × 1220 mm]. (Courtesy of Champion Cooler Corporation.)

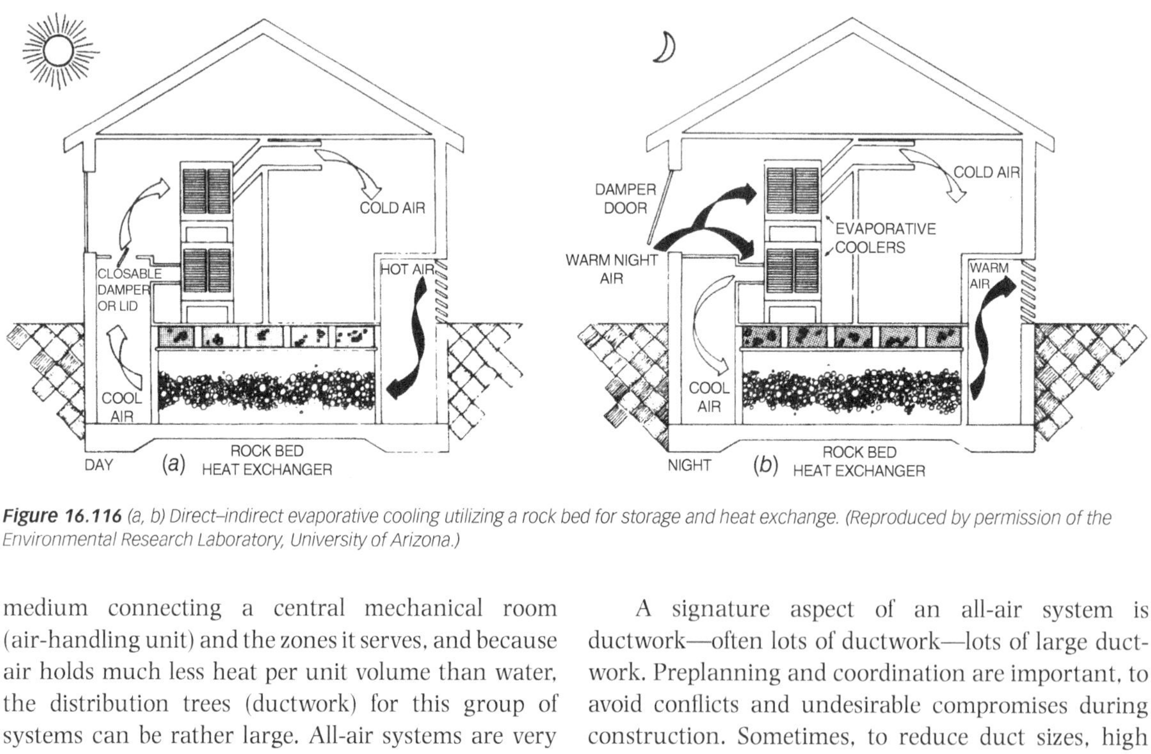

Figure 16.116 (a, b) Direct–indirect evaporative cooling utilizing a rock bed for storage and heat exchange. (Reproduced by permission of the Environmental Research Laboratory, University of Arizona.)

medium connecting a central mechanical room (air-handling unit) and the zones it serves, and because air holds much less heat per unit volume than water, the distribution trees (ductwork) for this group of systems can be rather large. All-air systems are very often the HVAC system type of choice because they can produce comfortable results while effectively regulating IAQ, temperature, and humidity at reasonable first cost and with acceptable energy efficiency (for most systems).

A signature aspect of an all-air system is ductwork—often lots of ductwork—lots of large ductwork. Preplanning and coordination are important, to avoid conflicts and undesirable compromises during construction. Sometimes, to reduce duct sizes, high supply air velocities are used. High velocity generates more airflow noise and results in higher friction losses (resulting in higher energy use by fans)—high velocity air distribution should be used only sparingly, where space limitations are real and extreme.

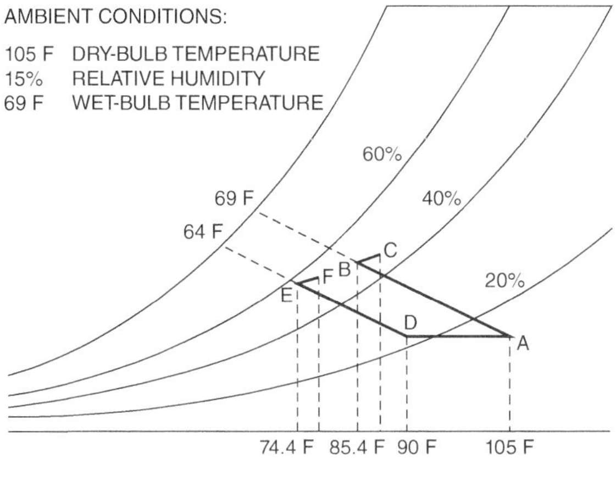

Figure 16.117 *Comparing direct (A–B–C) and indirect evaporative cooling (A–D–E–F) on the psychrometric chart.*

- *Room pressurization:* An all-air system allows for manipulation of pressure relationships throughout a building:
- *Positive pressure:* This involves an indoor space with higher air pressure relative to adjacent areas. This can be accomplished by providing a greater supply airflow rate than the exhaust airflow rate. Positive pressure will have the effect of pushing air through cracks or openings in the space enclosure at a magnitude necessary to achieve equal pressurization. Positive pressure is typical in occupied spaces.
- *Negative pressure:* This involves an indoor space with lower air pressure relative to adjacent areas. This can be accomplished by exhausting air at a greater rate than the supply airflow. Negative pressure will pull

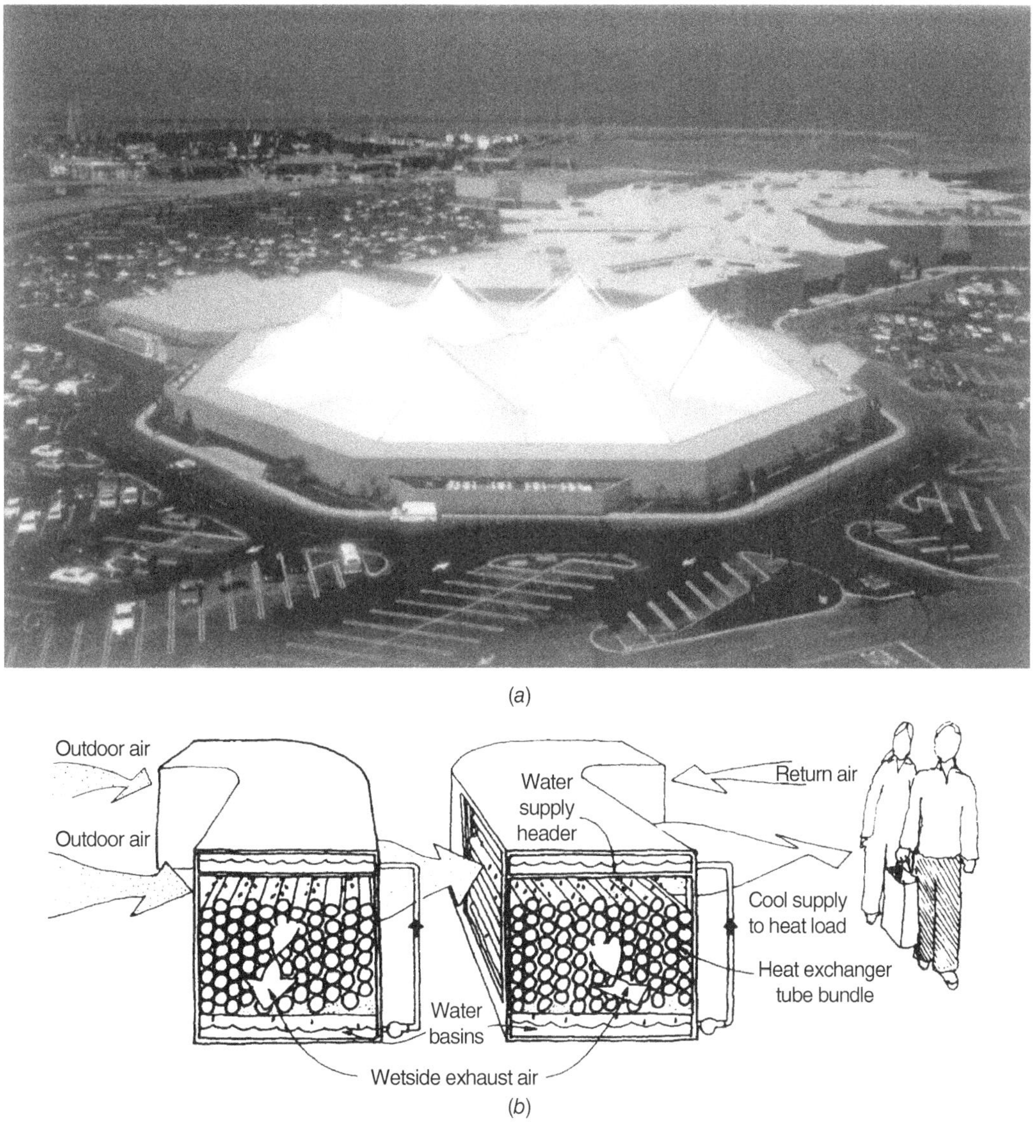

Figure 16.118 *Bullock's Department Store, San Mateo, California. (a) The eight-masted white fabric roof, highly reflective to ward off solar radiation, transmitted about 7% of daylight to provide ample diffuse daylight for sales areas. (b) Cooling was provided by a two-stage, indirect evaporative cooling system, which used much less energy than a conventional vapor compression refrigeration. (a: Steve Proehl. Environmental Planning and Research, Inc., architects; Giampolo and Associates, Inc., mechanical and electrical engineers.)*

ACTIVE ENVIRONMENTAL SYSTEMS

air through cracks or openings in the space enclosure at a magnitude necessary to achieve equal pressurization. Negative pressure requirements may be established in odorous or excessively humid locations (kitchens, bathrooms, pet shops, janitor closets, etc.).

Success in maintaining required room pressurization (positive or negative) requires sufficient airtightness within both the space (pay particular attention to doors and ceiling assemblies) and the ductwork serving the space. Pressure differences can be used to set up directional airflow patterns that help prevent the spread of contaminated air and can help manage smoke during a fire event (see Chapter 26). Indoor air quality control should not be a problem with an all-air system—although ensuring delivery of adequate outdoor air to all zones and under part-load operating conditions may require some creative thought.

Constant Air Volume (CAV) System. This is a category of HVAC systems that deliver a consistent—"constant"—flow of supply air as heating and cooling of the air is adjusted based on the needs of the zone(s) being served.

Single-Zone All-Air System. This is a CAV system controlled by a single thermostat (thus, "single-zone") (see Figure 16.120). A single supply duct leaves the AHU; it will divide into branches as necessary to deliver air to diffusers located in the zone. Temperature control (in response to changing loads) is accomplished by a modulating valve on the heating and/or cooling coils. The system can provide heating or cooling. Single-zone applications are surprisingly common across a range of building scales. A single-family residence is typically designed as a single thermal zone. A large warehouse might require 30 AHUs units, with each unit acting independently as a single-zone system. A high-rise building may require

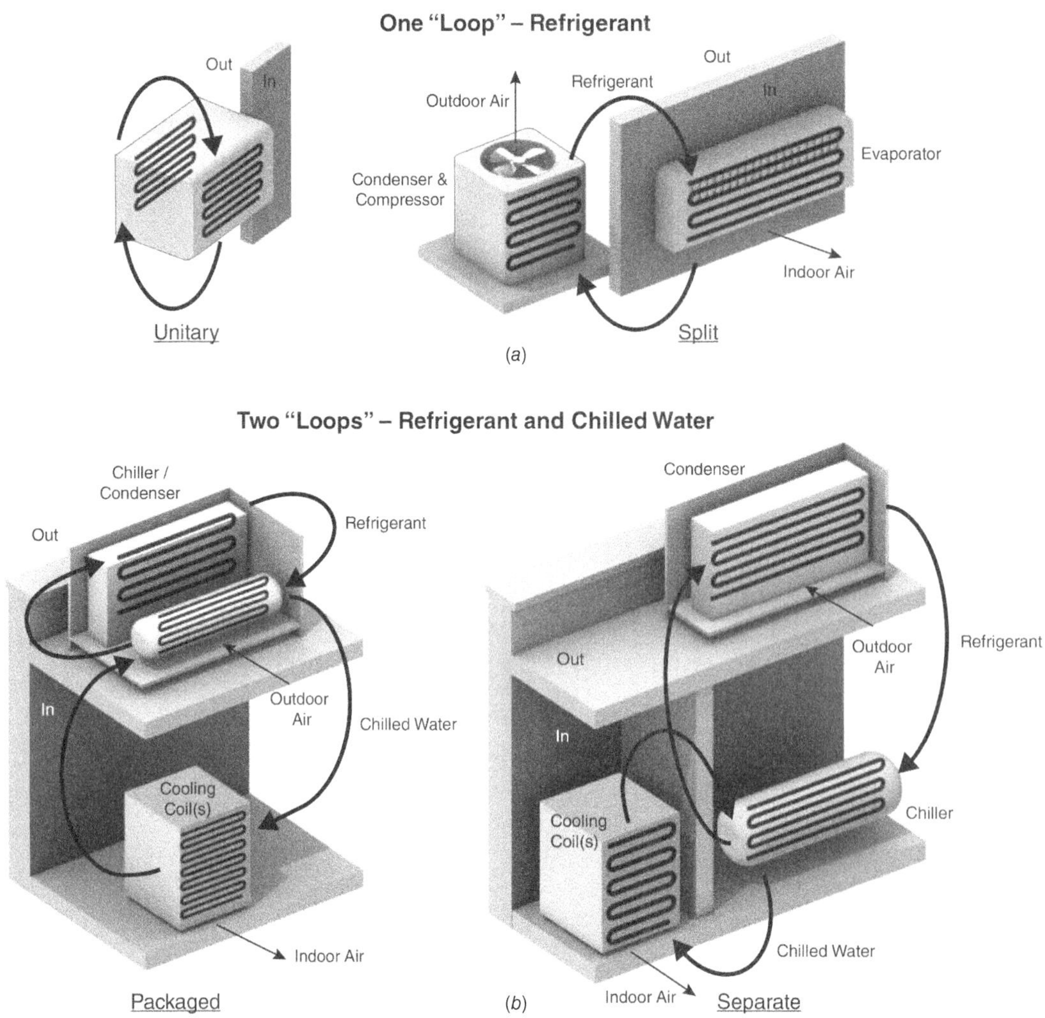

Figure 16.119 (a, b, c) Schematic representations of HVAC heat-transfer loops; the simplest are the least architecturally flexible; the more complex are very flexible.

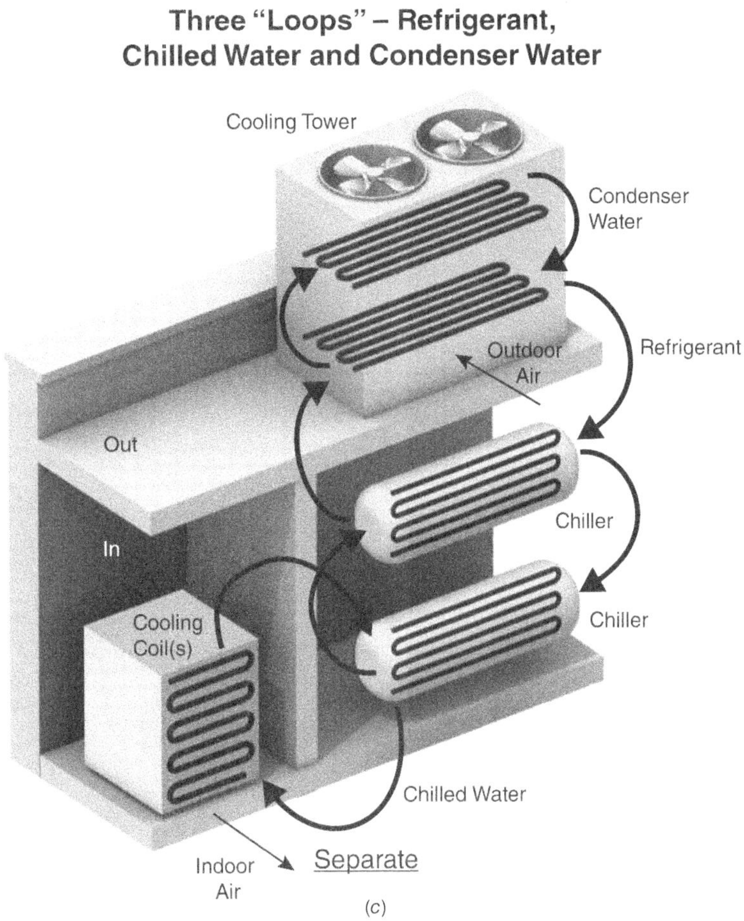

(c)

Figure 16.119 (continued)

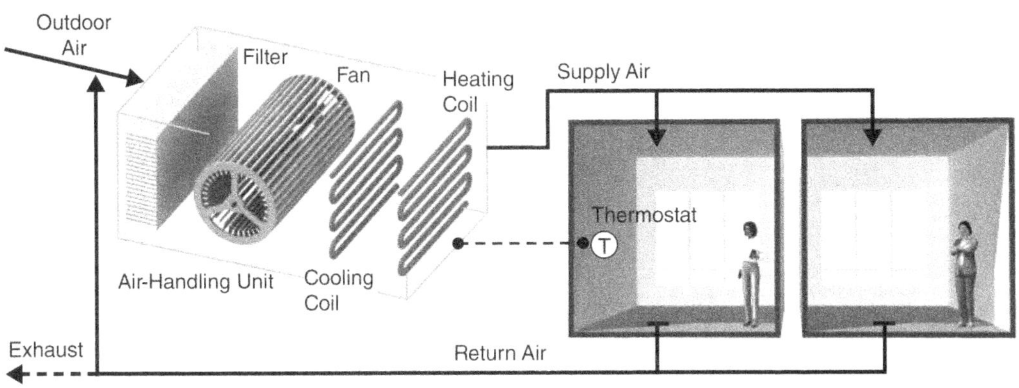

Figure 16.120 *Schematic diagram of a single-zone all-air HVAC system. Components are representative of a typical system installation. Additional components (such as a preheat coil or humidifier) may be required for some projects/climates. Various heat and coolth sources (such as a DX or chilled water cooling coil, hot water or electric heating coil) may be used as appropriate.*

several zones per floor, but the majority of the floor area might be an open-plan office that acts as a single zone. The zone capabilities of a system do not need to apply to an entire building, but only to the portion of a building being served by a particular air-handling unit. Because thermal control is exercised at the AHU (versus near the occupied space), this system is not very flexible with respect to adaptation to space use changes over time.

Single-Zone Terminal Reheat All-Air System. This is a CAV with a single-duct system controlled by multiple thermostats (Figure 16.121). The thermostats control reheat coils located near the various system zones. As many zones as necessary may be installed

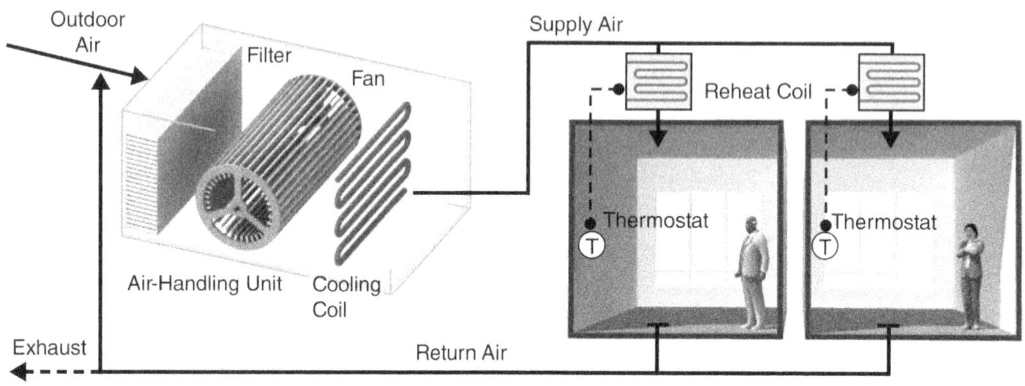

Figure 16.121 *Schematic diagram of a single-zone terminal reheat all-air HVAC system. Components are representative of a typical system installation. Various heat sources (such as a hot water or electric heating coil) may be used, as appropriate.*

(limited only by space constraints). No heating coil is required in the AHU (since air for each zone passes through a heating coil). Heating may be provided for one zone at the same time that another zone is being cooled. The system is termed a reheat system because all the air supplied by the air-handling unit is cooled to a lowest common denominator temperature that will handle the load in the "worst" cooling zone. Air supplied to any zone with "less-than-worst" loads is heated by the reheat coil (Figure 16.122). If this approach sounds energy-wasteful, it is. Constant-volume terminal reheat systems are prohibited by most energy codes. The system is shown here because of its historical significance (they were once very common) and because the reheat approach will show up with a different system type (e.g., VAV). If it weren't for the energy waste, this would be a very popular and effective system—providing for acceptable IAQ, with good temperature and humidity control, flexible and adaptable (zones can be easily added or removed), and with reasonable first cost. Several other all-air HVAC system types are in essence attempts to replicate the good points of a terminal reheat system while avoiding the energy waste.

Multizone All-Air System. This CAV system (Figure 16.123) mixes hot and cold air streams to provide appropriate-temperature supply air for each of several zones. Zone control is exercised at the air-handling unit (through the operation of mixing dampers). Once air has been mixed to meet zone loads, it must be conveyed to the zone separately (without further mixing with air for other zones). Thus, a separate supply air duct must run from the AHU to each zone. Because of the multiple supply ducts, such systems will rarely exceed ten zones per AHU. Even with that (physically imposed) limit, total distribution tree volume can grow to astonishing size. Careful spatial coordination is a must. A single return air stream collects air from all zones (this is the case for all the all-air systems). Use of a "bypass" deck, allowing return air to provide a tempering effect (instead of heated air) can improve energy efficiency. Leakage between zones at the decks of hot and cold coils is common, however,

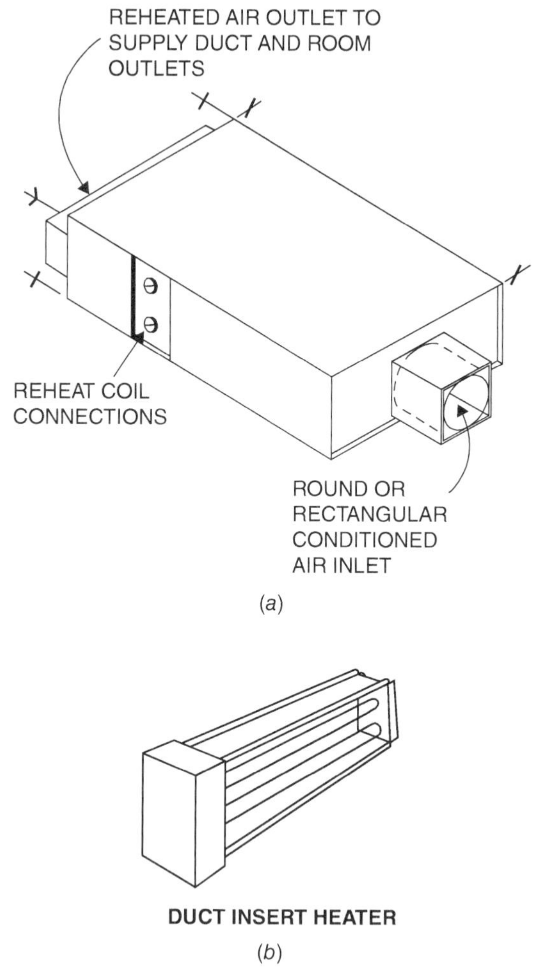

Figure 16.122 *Reheat system terminal-control options. (a) A terminal box where velocity and pressure are reduced while air is heated. (b) Electric heating coil designed for insertion within a duct. (From Ramsey and Sleeper, 1994.)*

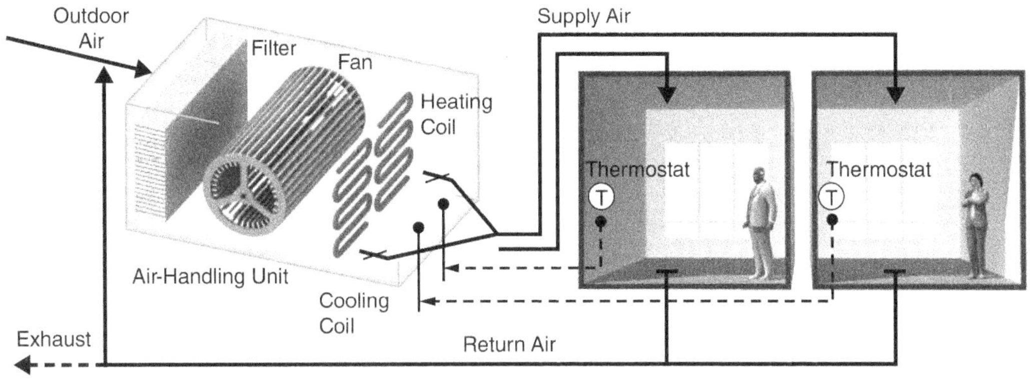

Figure 16.123 *Schematic diagram of a multizone all-air HVAC system. Components are representative of a typical system installation. Note that each zone requires a separate supply air duct.*

which decreases energy efficiency. This type of system is likely to be used in medium-sized buildings (or portions of buildings) where there is no expectation of flexibility for future zoning changes. Adding or deleting a zone requires replacing the supply air ducts. The system name is multizone (not "multiple zone").

Dual-Duct All-Air System. This CAV system (Figure 16.124) is conceptually similar to a multizone system in terms of control approach. Two air streams ("hot" and "cold") are mixed under control of a zone thermostat in order to produce supply air that will meet the cooling/heating needs of the zone. The air streams, however, are mixed not at the AHU but in a mixing box located near each zone. There is no limit to the number of zones provided, changes in zoning are easily made without system-wide disruption, and the system is reasonably energy efficient. Two supply ducts run throughout the building—impacting system first cost and spatial coordination difficulty. Dual-duct systems are rarely selected now, except where the pressure control benefits of a constant-volume system are compelling. The building spatial volume required to route the system's three (two supply, one return) full-sized air distribution trees is hard to justify, given that a VAV system can provide acceptable comfort for most spaces using a single supply duct. San Francisco's International Building, seen in Figure 16.91, employs a dual-duct system.

The mixing boxes (terminals; Figure 16.125) in dual-duct systems are similar to those used in other terminal-control all-air systems but are generally larger for the same airflow capacity. This adds still more to the system impact on building volume. Although most dual-duct systems are constant volume, they can be configured as VAV when the reduction in airflow is no more than 50% below the maximum. For details on other dual-duct variations, see Grondzik (2007).

Variable Air Volume (VAV) System. This system (in a single-duct configuration, Figure 16.126) is currently the most popular large-building HVAC system. Its single supply duct requires less building volume for distribution than do multiple-duct (sometimes called "multi-duct") systems, and control of space conditions through variation of air volume flow rate (rather than of air temperature, as with a constant-volume system) saves energy (reductions in airflow greatly reduce fan power). Because of the

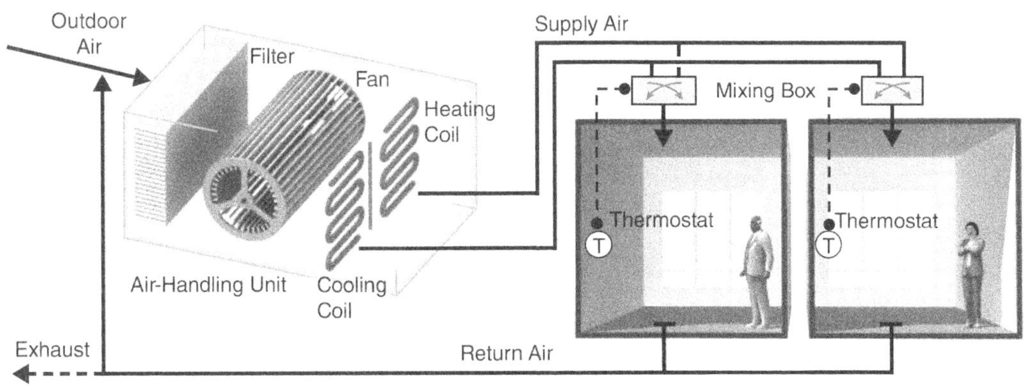

Figure 16.124 *Schematic diagram of a dual-duct all-air HVAC system. Components are representative of a typical system installation. Note that two supply air ducts will run throughout a building.*

ACTIVE ENVIRON-MENTAL SYSTEMS

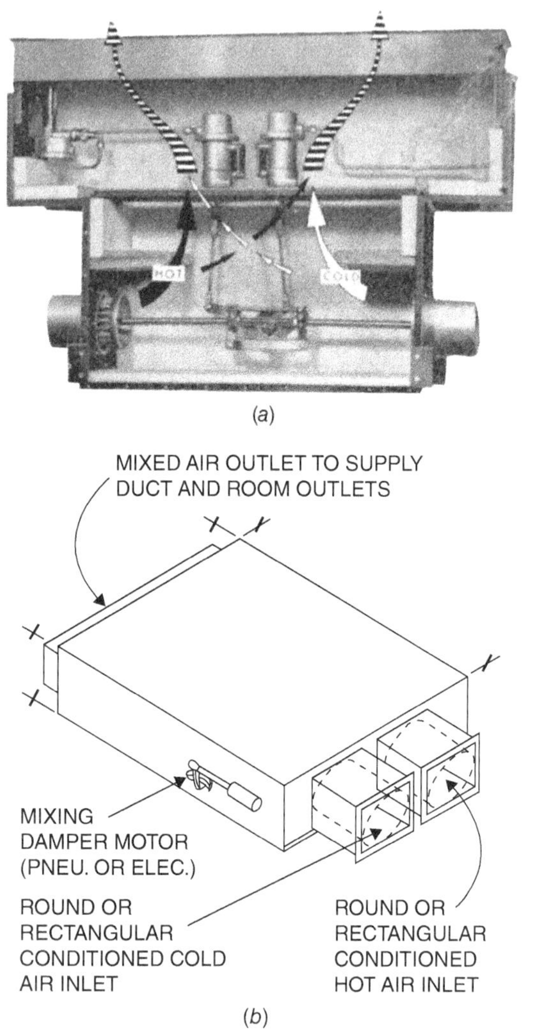

Figure 16.125 *High-velocity, dual-duct terminal providing mixing and attenuation (pressure and sound reduction). (a) Controlled from a thermostat, this unit blends the supply air streams and delivers air to the zone at the selected temperature. (b) Typical mixing box dimensions are: 10 by 50 by 30 in. [250 × 1270 × 760 mm] for a 400 cfm [190 L/s] unit to 18 by 60 by 66 in. [460 × 1520 × 1680 mm] for a 5000 cfm [2360 L/s] unit. (a: Courtesy of Anemostat; b: From Ramsey and Sleeper, 1994.)*

nature of the control logic (basically, a reduced load means a reduced airflow), VAV air-handling systems tend to be cooling-only systems. Space heating needs are commonly picked up by a supplemental heating system (such as hydronic baseboard radiators or reheat coils at the VAV boxes). Automatic controls (linked to each zone's thermostat) adjust the supply air volume admitted to each zone through the action of a VAV box (terminal) located near the zone. The control logic is based upon the equation for heat transfer via air:

heat transfer = airflow rate × Δt × conversion factor

A CAV system varies Δt to address varying loads; a VAV system varies airflow rate. Same result (almost); different means.

Design of a successful VAV system demands attention to space load dynamics. A space with little load variation over time (such as an open-plan office space) will experience little change in airflow over time—and little energy savings from fan power reductions. A space with more dynamic load patterns will result in greater fan power savings, but the details of part-load control, supplemental heating, and IAQ maintenance become more intriguing. When a VAV box starts to reach full closure, little outdoor air will be delivered to the space—this may (or may not) affect IAQ acceptability. Success is in the details.

With a VAV system, provision must be made for at least code-minimum fresh air supplies during occupied periods, sometimes resulting in overcooling or wasteful reheating. The changing airflow in a VAV system may result in changing noise production (which varies with airflow), and variable noise sources are inherently more noticeable than steady sources.

VAV terminal boxes are often placed above a suspended ceiling or below a raised floor adjacent to the zone being served. A typical VAV terminal box is shown in Figure 16.127. This terminal box serves not only to vary the quantity of air, but also to attenuate noise and reduce the speed of air before entry to the

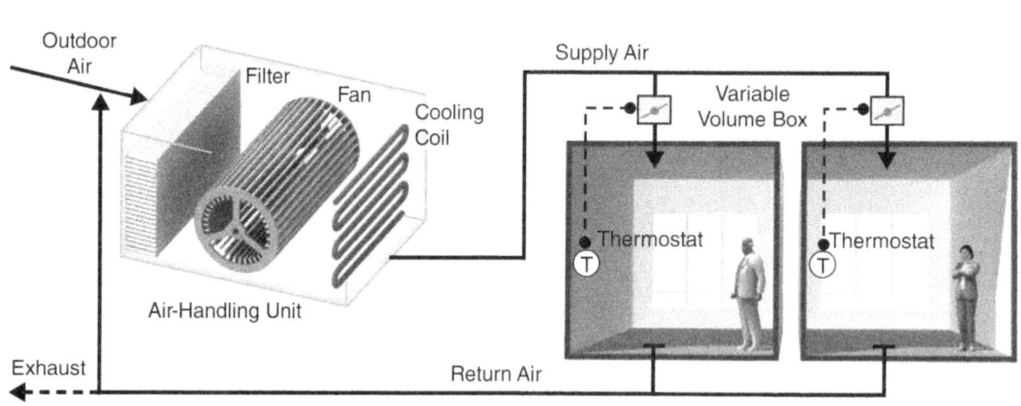

Figure 16.126 *Schematic diagram of a VAV all-air HVAC system. Components are representative of a typical system installation.*

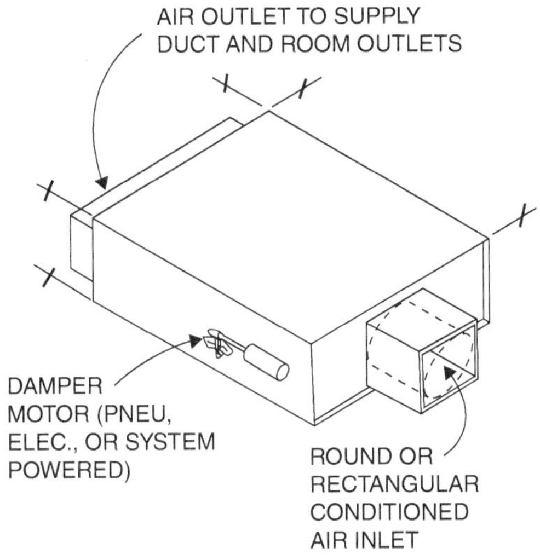

Figure 16.127 *The simplest type of VAV terminal, which pinches back the volume of incoming air as cooling load decreases. Boxes can range in size: 8–18 in. H by 24–67 in. L by 14–54 in. W [200–460 × 610–1700 × 360–1370 mm]. (From Ramsey and Sleeper, 1994.)*

space. High-velocity distribution is common in main ducts because it reduces the size of the distribution tree.

The Utah Department of Natural Resources in Salt Lake City (Figure 16.128) utilizes a VAV system throughout its three-story, 105,000-ft^2 [9755-m^2] office building. Air is supplied from linear ceiling diffusers located between the rows of indirect luminaires parallel to the windows. Return air is taken into the suspended ceiling plenum. Highly insulated walls, windows, and roof lessen the need for perimeter heating, although a supplementary warm water heating system is built in below the windows to offset perimeter heat losses. The lower 1 ft [300 mm] is a slot where room air is drawn in by convection, and a perforated metal grille at the top forms the windowsill.

The building is elongated east–west for best daylight access and control and winter solar gain; lightshelves even out the daylight penetration. With almost no east- or west-facing glass, the building relies on 4200 ft^2 [390 m^2] of south glass and 3800 ft^2 [353 m^2] of north glass to serve its 125-ft [38-m] width. Supplementary indirect lighting is controlled by photocells and dimming ballasts.

Utah's hot, dry summers permit a four-stage approach to cooling. First is an economizer cycle at lower temperatures; next, direct evaporative cooling is used; then, indirect evaporative cooling (see Figure 16.117), utilizing an oversized cooling tower; and finally (estimated at about 10 days per year), a conventional chiller.

An example of a floor-by-floor VAV system (Figure 16.129) is a 1-million-ft^2 [92,900-m^2] 28-floor Chicago office building designed by Skidmore, Owings & Merrill. This mid-rise approach to office towers utilizes three "stacked" atriums to relieve the monotony of the wide interior floors. Another result is lower structural and energy costs per unit floor area relative to conventional high-rise structures. The cubelike shape of the building exposes less skin area (38% of which is in insulating glass) to Chicago's cold winters.

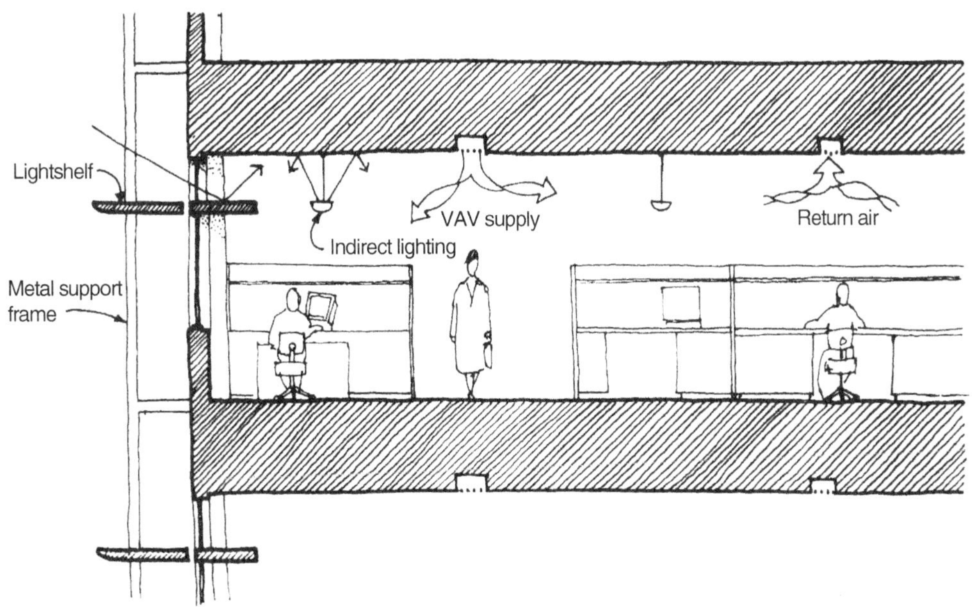

Figure 16.128 *The Utah Department of Natural Resources building features lightshelves, direct and indirect evaporative cooling, and an economizer cycle. Supply air is distributed by a VAV system with ceiling diffusers located between luminaires. (Based upon the design by Gillies Stransky Brems Smith Architects.)*

(a)

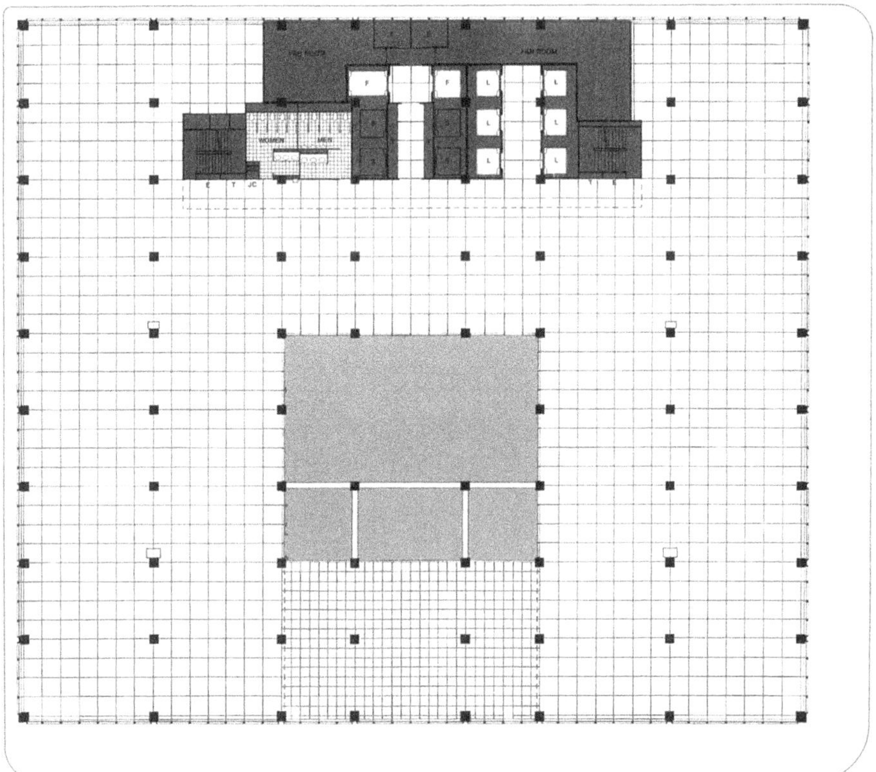

(b)

Figure 16.129 *Floor-by-floor fan rooms supply a VAV system for this Chicago office building. (a) Exterior view of 33 West Monroe. (Photo by Merrick, Hedrick-Blessing.) (b) Plan of a typical lower floor; two fan rooms occupy the space along the exterior wall. (c) Section perspective showing stacked atria. (Courtesy of Skidmore, Owings & Merrill, Architects-Engineers.)*

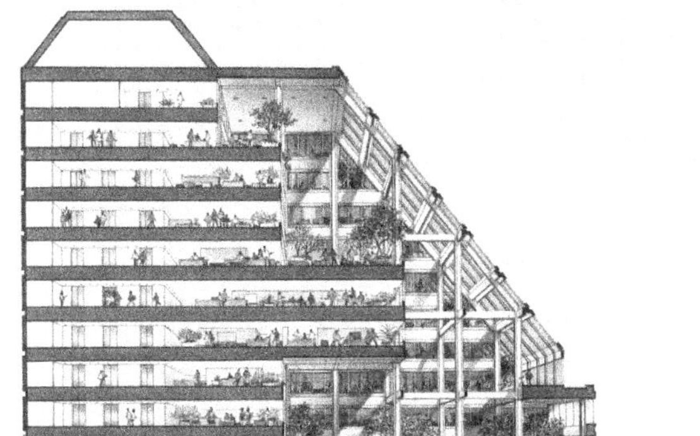

(c)

Figure 16.129 *(continued)*

To accommodate the differing schedules and comfort needs of a variety of tenants, each floor is provided with two VAV supply air handlers that can be operated at night and on weekends, independently of the rest of the building. Each floor's mechanical core has one exterior wall (on an alley) to facilitate fresh air intake/stale air exhaust. The perimeter heating system is electric resistance fin radiation; an economizer cycle provides cooling with outdoor air below 55°F [13°C].

Several variations on the basic VAV system have emerged, some of which respond to the problem of minimum airflows for rooms with little thermal load or to the desire to serve both interior and perimeter areas with the same HVAC system.

- *Fan-powered VAV terminals:* Under low loads, cool air from the AHU is reduced to the minimum flow required for acceptable indoor air quality, while the unit's built-in fan draws additional "supply" air from a ceiling (or floor) plenum, heating it as required to meet space conditions. This variation allows individual terminal units to heat when the main supply system is cooling; it might logically serve perimeter zones.
- *Minimum closure VAV terminals:* To maintain minimum fresh air flows, VAV terminals are usually set so that they cannot be entirely closed off, ensuring some outdoor airflow at all times while the system is operating. If this minimum airflow, entering at low velocity, does not provide the desired air motion and mixing within the room, VAV terminals can be equipped with fans, which are activated as needed in response to decreasing supply air volume. These self-contained fans mix room air with centrally supplied air to provide an air stream of the right temperature and velocity to maintain comfort.
- *Induction VAV terminals:* An induction-type VAV terminal induces plenum air (heated by space loads) to mix with the incoming cool supply air stream.
- *Reheat VAV terminals:* Large space heating loads often require the use of reheat terminals supplied by hot water or electric resistance. This type of reheat application is more energy- efficient than a standard constant-volume reheat system because a much smaller volume of air is first cooled, and then reheated. Water or electric coils can be incorporated into the VAV terminal or into the ductwork between the terminal and the room diffusers.
- *Return air recirculation percentage:* Central all-air systems circulate supply air and—in order to foster airflow, air mixing, and space pressurization—return air. Reviewing the schematic diagrams in this chapter may lead one to question why it is necessary to send any return air back to the AHU. The simple answer is energy conservation. The return air may be less-than-fresh, but has also been conditioned. A very practical means of saving energy is to send a portion of the return air back to the AHU to be filtered, mixed, and redistributed. The exhaust or return air recirculation percentage may range from 0% (i.e., 100% OA) to around 80% recirculated air.

(b) Air–Water Systems

Several variations of air–water systems are fairly regularly used in current building designs. Because both air and water are used as heat-transfer media, these systems tend to be more complex than an all-air or all-water system. The logic behind an air–water system is that it brings in the best of both media. Air distribution allows for mitigation of indoor air pollutants through delivery of outdoor air. Water distribution allows for a substantial reduction in the size of distribution conduits. Air–water systems are fully functional HVAC systems that can produce comfortable results while effectively regulating IAQ, temperature, and humidity at reasonable first cost and with acceptable energy efficiency.

Often the amount of supply air delivered from the central AHU is the minimum required to accomplish IAQ control. This is generally (and very roughly) around 10% of the airflow required in an all-air system—thus the smaller ductwork sizes. Return air is minimized or eliminated in an air–water system since it is a 100% outdoor air system by definition. The bulk of space heating and cooling loads are handled by the water side of the system. This includes latent cooling and the resultant condensation of water that is removed from the room air (and must be disposed of). The central supply air may be delivered to the zones in a neutral condition (near room temperature) or be delivered heated or cooled to assist with space conditioning.

A signature aspect of an air–water system is an in-space delivery device—even though these devices cover a range of configurations. Preplanning and coordination are important to avoid conflicts and undesirable compromises regarding placement of delivery devices during construction. Maintenance of these delivery devices will occur in occupied spaces. Unit filters will be changed in the spaces (not in a central fan room). Fan-powered systems (such as a fan-coil) may present acoustical challenges since the fan is located in an occupied room.

Exhaust air may be gathered in a centralized duct system, making heat recovery possible. Or (as a cheaper alternative) air can be exhausted locally to avoid the construction of yet another distribution tree. If the water-side system permits only heating or only cooling, then a two-pipe system will work. If simultaneous heating and cooling are necessary to meet the owner's project requirements, then a four-pipe system is necessary. Three-pipe systems are a lower-first-cost alternative, allowing simultaneous heating and cooling (from two supply pipes with a single return pipe), but they waste energy by mixing hot return and cold return water flows in one return pipe. They are no longer permitted in most locales.

Induction System. Induction terminals (the delivery device) are typically located either below perimeter windows (Figure 16.130) or above a suspended ceiling. The need to dispose of condensation from the cooling coil and clean the unit filters makes under-window locations much easier to maintain, even

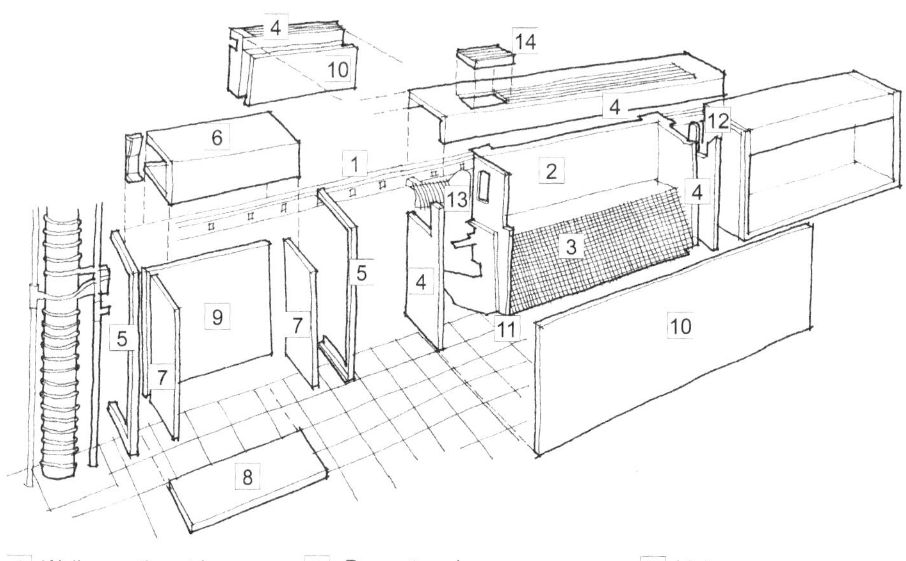

Figure 16.130 *Assembly of a high-velocity induction unit. Conditioned 100% outdoor air is brought in through a high-velocity duct to provide ventilation and to induce the circulation of room air. Air is attenuated and silenced in the chamber (2), and then, through jets in the front of the plenum, it induces the flow of room air, which is heated or cooled at the coil (3). The lint screen "filter" (11) requires periodic maintenance for proper airflow and IAQ performance. (Courtesy of Carrier Corporation. Redrawn by Erik Winter.)*

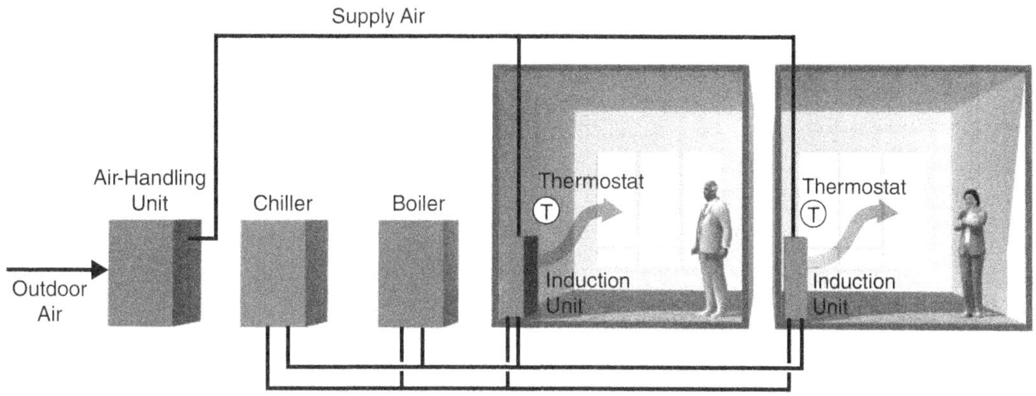

Figure 16.131 *Schematic diagram of an induction air–water HVAC system.*

if they intrude on the floor area available for occupancy. In a two-pipe system, either hot or cold water—not both—is available to deal with room loads under command of a thermostat linked to each terminal. Because cool air is distributed for most of the year, a two-pipe system is largely in heating mode in colder climates. Switch-over for two-pipe systems is a big deal—and a source of frequent occupant complaints. Switch-over is a slow process and is often scheduled without regard to actual weather conditions. In a four-pipe system, the simultaneous availability of hot and chilled water makes it possible to switch instantly from heating to cooling for excellent thermal control.

A schematic diagram of an induction system is shown in Figure 16.131. A high-velocity (and high-pressure), constant-volume air supply is delivered to each terminal from a central AHU. At the terminal it is forced through an opening in such a way that air within the room (secondary air) is induced to join the incoming jet of supply air. The terminal mixes 20% to 40% supply (fresh) air with the 80% to 60% room air. The combined air stream passes over finned tubes for heating or cooling and is introduced into the space. Thermostats control the induction unit output by controlling either the flow of the water or the flow of secondary air.

The CBS Tower in New York City (Figure 16.132) uses a high-velocity induction system for its perimeter zones. The triangular black granite-faced exterior columns give the façade a three-dimensionality quite unlike the slick glass boxes of its contemporaries (it was designed by Eero Saarinen and built in 1962). These thick columns enclose the perimeter distribution trees; every other column contains a high-velocity air supply; in between, every fourth column contains an HVAC water supply and every alternate fourth column a water return. The constant column sizes belie the thicker air, thinner water distribution trees. The interior zone of this tower is served by a VAV system; return air from both perimeter and interior zones is collected at the core.

Fan-Coil System with Supplemental Air. This system (Figure 16.133) is conceptually similar to the preceding induction system. The air and water sides of the system, however, are delivered to the spaces individually—water through a fan-coil device and air through diffusers. A fan, rather than high-velocity primary air, is used to circulate room air through the fan-coil unit. Fan-coil units are commonly employed in a variety of building types and are available in a range of configurations to accommodate such usage. Vertical, horizontal, and stackable units are available (Figure 16.134).

Fan-coil units are inherently noisy due to the operation of the fan, so careful attention to sound ratings is required when background sound levels must be low. In offices, a relatively high background sound level can help to maintain acoustic privacy at the workstation. The sound of the fan may also provide perverse reassurance that some kind of air-conditioning system is at work.

Radiant Panels with Supplemental Air. Radiant heating systems have a large following of comfortable users; radiant cooling systems are fairly commonly used in Europe and are getting a closer look in the United States. Radiant panels combined with a central air supply are another form of air–water HVAC system (Figure 16.135). Warm surfaces can be used to offset enclosure heat losses. Cool surfaces can be used to offset internal heat gains and/or enclosure heat gains. The ceiling is often favored as a radiant panel location, because it is uncluttered by furniture, carpet, tackboards, and the other items that regularly cover floors and walls.

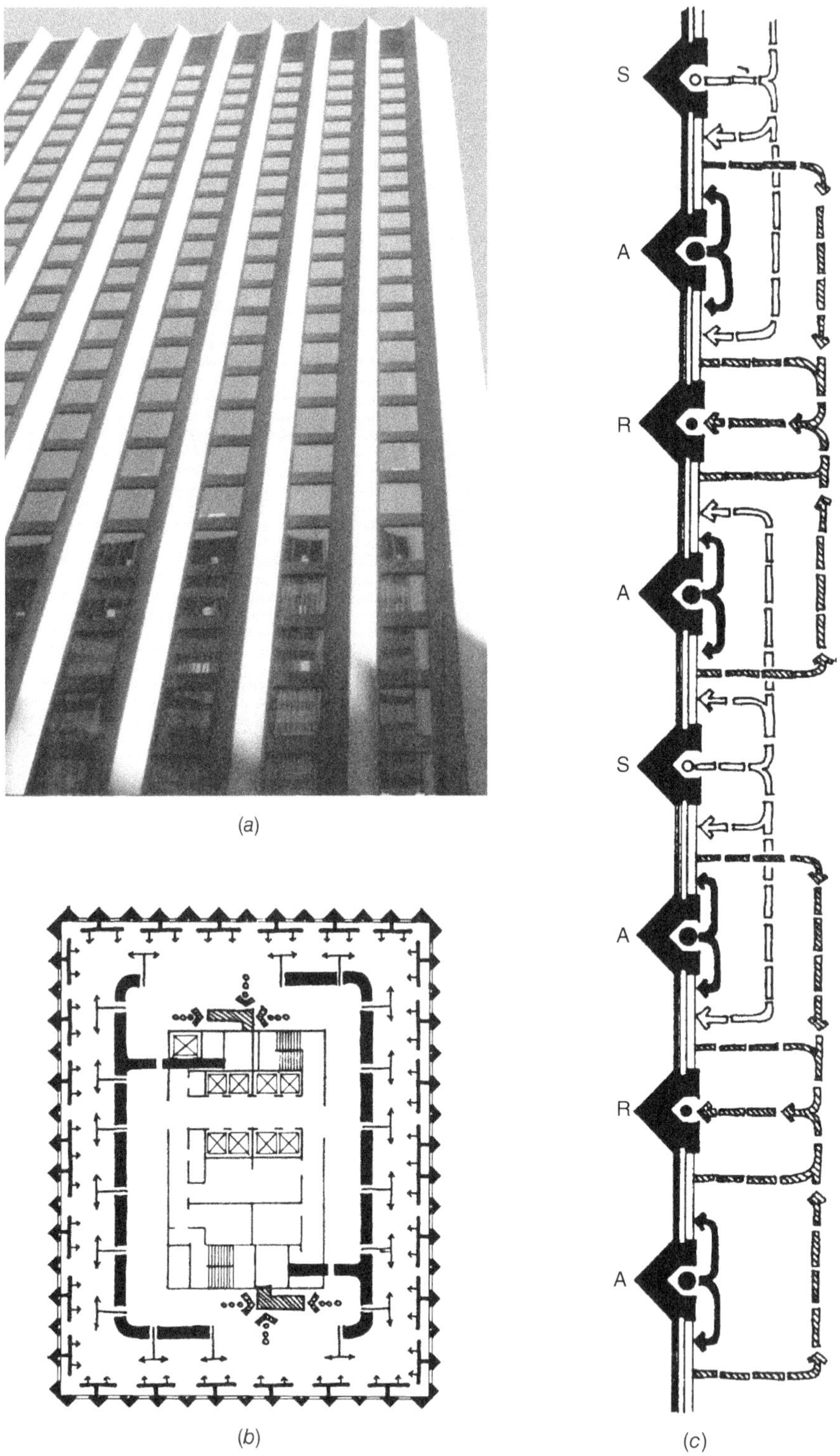

Figure 16.132 *The identical triangular exterior columns of the CBS Tower in New York City (a) enclose a variety of distribution trees. (b) Plan of a typical floor; perimeter zones are served by an induction air–water system, interior zones by a VAV all-air system. All return air is collected at the core. (c) At the perimeter, column type A contains an air supply tree feeding the induction units on either side of the column. Column type S contains the supply water tree, hot in winter and cold in summer. Column type R contains the return water tree. (From Guise, 1985.)*

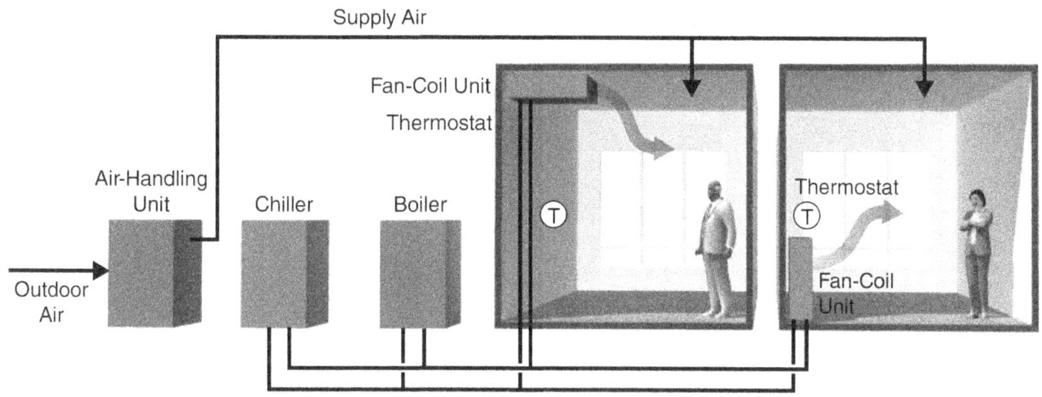

Figure 16.133 *Schematic diagram of a fan-coil air–water HVAC system.*

A building that uses radiant panels (in offices) and fan-coil units (at entries) is shown in Figure 16.136. Green on the Grand is a 23,465-ft^2 [2180-m^2] two-story office complex in Kitchener, Ontario. Its design combines a very highly insulated enclosure, daylighting, water conservation, a gas absorption (non-CFC, non-HCFC) chiller/heater with a heat rejection pond, a separate IAQ system, and heat recovery of exhaust air. It also features a dedicated outdoor air system for ventilation and a hydronic system for heating and cooling. The hydronic system uses fan-coils at the entryways and radiant panels at ceilings of office spaces. Fan-coils are able to handle the extremes of entryway heat losses and gains. The radiant panels in the offices are used for both heating and cooling, but are sized based upon cooling loads and cover 30% of the ceiling area. Tenants were given their choice of two designs: steel panels suspended below a drywall ceiling (painted to match the ceiling color) or extruded aluminum panels fit into a suspended ceiling system. To prevent condensation on these radiant panels during cooling mode, the ventilation air is dehumidified.

The ventilation system supplies air low on the walls and exhausts high on the walls, an informal application of displacement ventilation. There are two rates of outdoor air delivery: 20 cfm [10 L/s] per person and) 40 cfm [20 L/s] per person—when economizer cycles operate or when more fresh air or more cooling is required. The exhaust air flows through two heat exchangers: The first reheats outdoor air that has been deeply cooled for dehumidification; the second preconditions the incoming outdoor air at the point of entry. The latter heat exchanger is a rotary-wheel ERV capable of transferring both heat and moisture.

Several approaches to radiant cooling (and heating) panels, independent of a dehumidified air supply, are shown in Figure 16.137. These approaches assume a minimal floor/ceiling thickness, which could contribute either to reducing the floor-to-floor height or to greater ceiling height (for daylight penetration or displacement ventilation). If a raised floor is used above these concrete slabs, even higher percentage heat gains come from the ceiling.

Radiant floors are sometimes used for cooling, as in the Bleshman Regional Day School (Figure 16.138). The designers of this school for the disabled recognized that the majority of its users would spend much of their time quite close to the floor and that colder air near the floor could be uncomfortable, especially during New Jersey winters. The entire floor is warmed by supply ventilation air in winter; the air enters just below windows to counteract the downdraft off cold glass surfaces. In summer, the cool supply air first cools the floor and then cools the air in front of the warm glass. The concrete cellular "air floor" provides thermal mass that helps maintain steady temperatures. Heated or cooled air is provided by rooftop air-source heat pumps, one of which is provided for each cluster of three to six classrooms.

(c) All-Water Systems

As the name implies, an all-water system distributes heating and cooling effect from a central location to the various zones using only water. No air supply is provided by a central system. Although requiring the least amount of space for distribution of any central HVAC system, lack of a structured approach to supply air makes this type of system highly suspect with respect to indoor air quality. In most nonresidential applications, an all-water system (without a means of air supply) cannot meet the ASHRAE definition of an air-conditioning system. All-water systems must be considered with caution, even though they may be appealing in retrofit projects, historic renovations, or new projects if bad planning leaves no room for distribution elements.

A schematic example of an all-water system is shown in Figure 16.139. There is no air-handling unit or ductwork tree. IAQ is dealt with locally—by means of

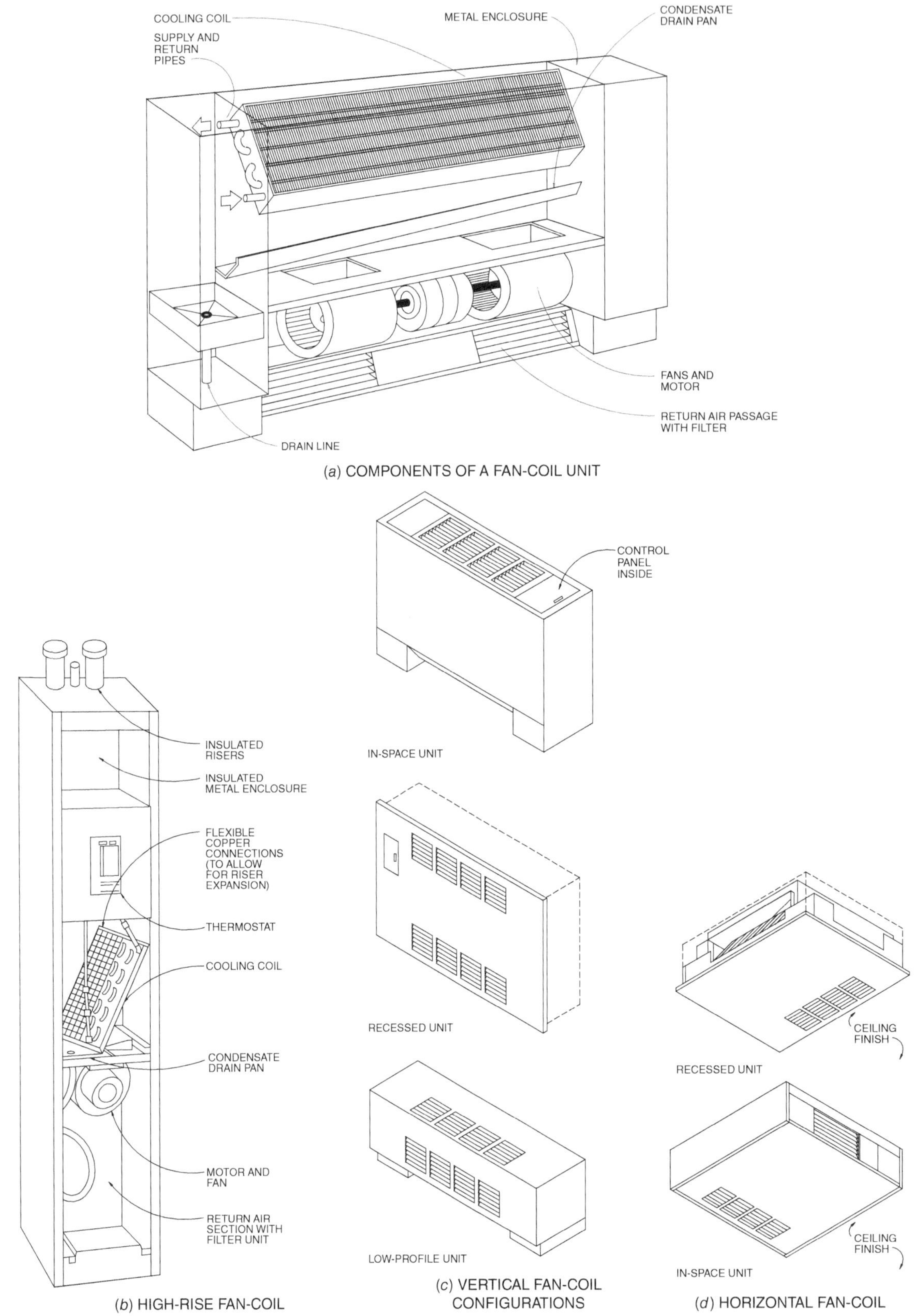

Figure 16.134 *(a) The innards of a standard below-window fan-coil unit without outdoor air connection. Widths can range from 2.5 to 7 ft [0.8 to 2.1 m], height is around 26 in. [660 mm], and depth between 9 and 12 in. [230–305 mm]. (b) High-rise unit for corners or cabinet locations. These can be stacked to reduce connecting piping runs and simplify condensate collection. (c) Vertical units for below windows. Low-profile unit height is around 14 in. [360 mm]. (d) Horizontal units for above-ceiling locations. (a, b: From Ramsey and Sleeper, 2016. c, d: From Ramsey and Sleeper, 2000.)*

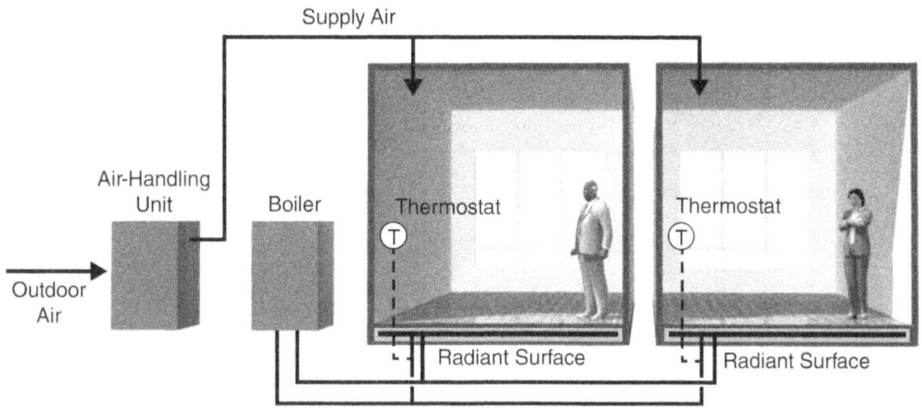

Figure 16.135 *Schematic diagram of an air–water HVAC system employing radiant panels and supplemental air supply.*

infiltration, or through windows, or by use of a delivery device that connects to the outdoor air via a penetration through the building wall (termed a unit ventilator). Any of the many water-based terminal devices may be used with an all-water system—radiant panels, fan-coil units, valence units, baseboard radiators, and so forth.

Provision of outdoor air (as with an all-water system) may take the simple form of operable windows. An expressive variation on this approach appears in the reading rooms and staff workrooms for the Seeley G. Mudd Library at Yale University (Figure 16.140). Limestone spandrels are curved in to allow outdoor air to enter these perimeter rooms just below the windows, where hot water finned tube radiators are available to provide heating as needed. The incoming outdoor air replaces exhaust air, which flows out of the upper operable windows. The remainder of the building has conventional forced-air heating and cooling.

16.20 HIGH-PERFORMANCE HVAC SYSTEMS AND EQUIPMENT

During the second decade of the twenty-first century, several trends appear to be bubbling in the world of HVAC design. These are being driven by a desire to produce higher-performance buildings—deep green projects, net-zero energy projects, carbon-neutral projects. One trend that seems unavoidable if net-zero energy buildings are to become a reality is a willingness to let active systems partner with passive systems. No matter how efficient an active HVAC system, an appropriate passive system will use less energy (with that being renewable energy, to boot). One of the more direct paths to net-zero energy is not to reduce energy use for heating and cooling, but rather to eliminate such energy use. Building automation systems make the integration of active and passive systems easier to manage. There is still a lot to learn, however, about system and occupant responses to rapidly changing thermal conditions (as in a mixed-mode cooling system—one that blends air conditioning and natural ventilation). Effectiveness is as important as efficiency.

(a) GSHPs at Large Scales

Ground-source heat pumps (GSHPs) are being used quite frequently in high-performance buildings. These are discussed in Section 16.8. All else being equal, changing to a GSHP system from any other active HVAC source approach should be able to reduce heating/cooling energy use by around 50%. This is a serious reduction in energy use and is attention-getting.

The technology is not new to the building sector; moreover, GSHPs can be used in dense urban environments. A late 1990s building just east of Central Park, in New York City, utilizes two wells 1500 ft [457 m] deep; all but the top 50 ft [15 m] of the wells are lined by bedrock. Heat is taken from (or discharged to) water, which is pumped from one well and discharged to the other—a "groundwater-source" system. The average year-round temperature in these deep wells (in an intensely urban area) is around 56°F [13°C], about the temperature at which chilled air is delivered to a space in summer. Going to such depths is perhaps the only ground-source option in densely built-up areas.

GSHPs are also increasingly applied to expansive residential projects. The Cambridge (Massachusetts) Cohousing Project (Figure 16.141) is also located in a densely settled urban neighborhood. The 41 living units feature passive solar heating, underground parking (this preserves some 20,000 ft^2 [1860 m^2] of the surface for open green space), and centralized heating/cooling utilizing ground-source heat pumps with numerous thermal zones. In this urban setting, the relative quiet of the indoor heat pumps is a welcome amenity.

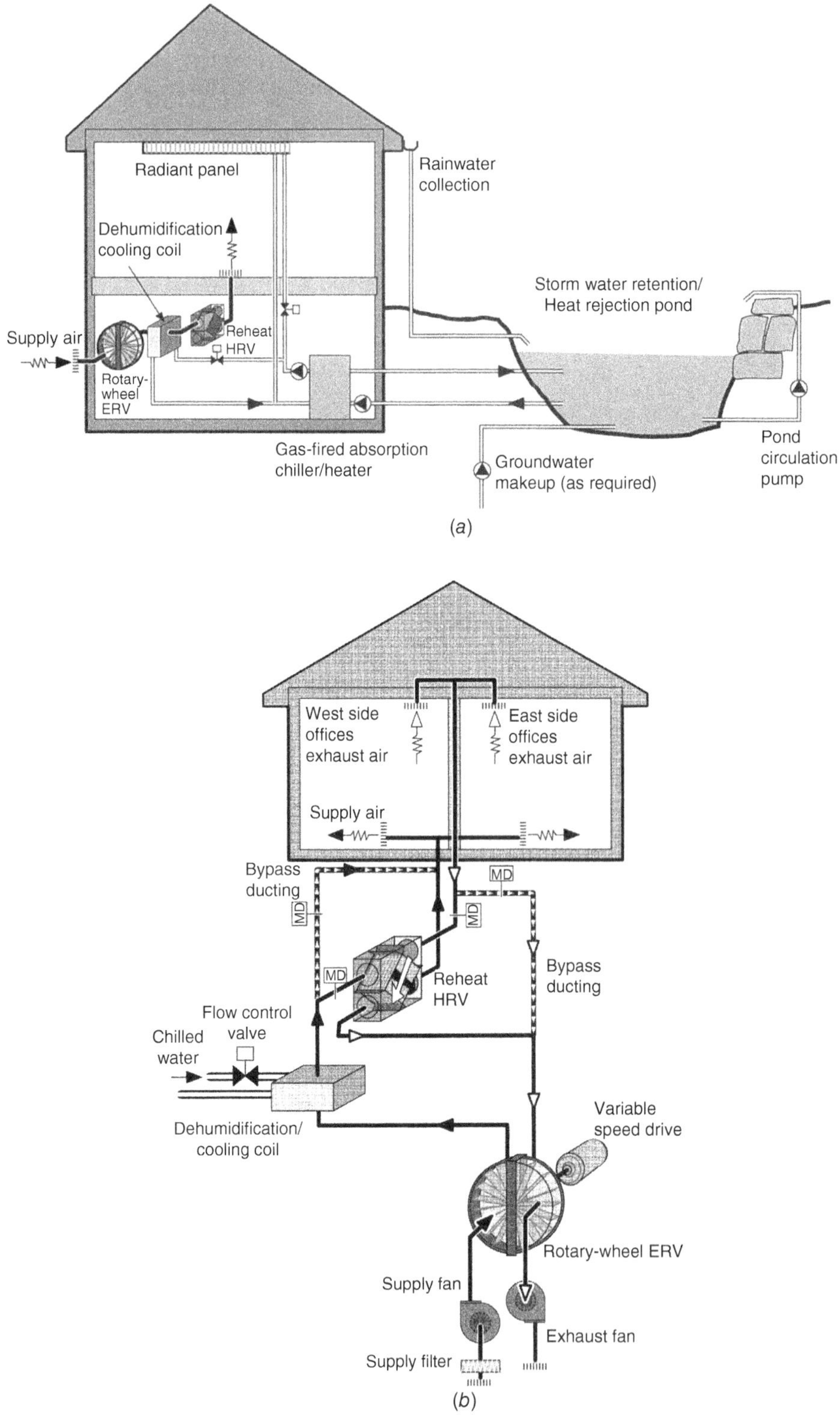

Figure 16.136 *An innovative Canadian office building, Green on the Grand at Kitchener, Ontario. The cooling system combines a radiant panel system (a) with a summer ventilation dehumidification system (b). The building also features daylighting, a gas absorption chiller/heater, rainwater storage pond heat rejection, and two-stage heat recovery. (Courtesy of Enermodal Engineering, Ltd.)*

ACTIVE ENVIRONMENTAL SYSTEMS

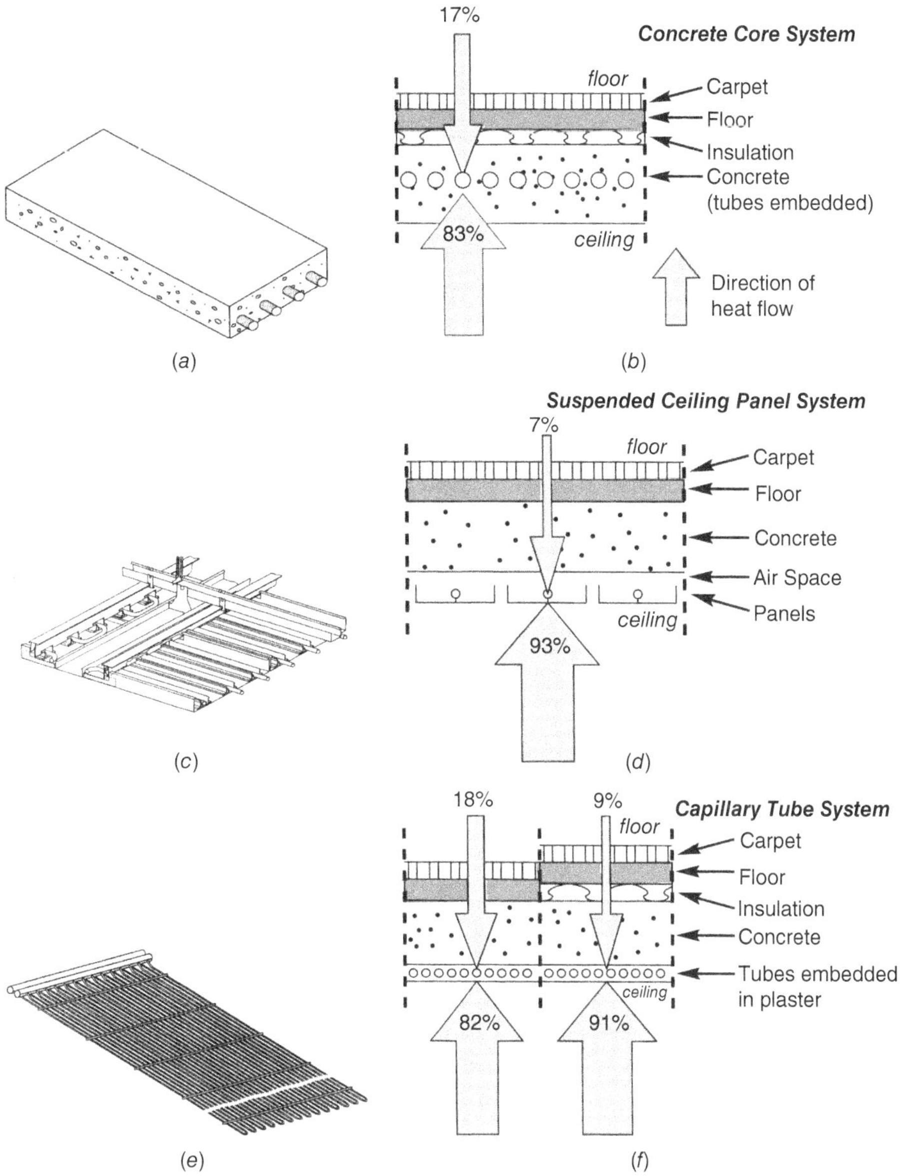

Figure 16.137 *Approaches to radiant cooling systems. (a) Concrete core system: water is circulated through plastic tubes embedded in the concrete floor/ceiling slab. (b) With carpet, pad, and insulation on the floor, most of the heat radiates to the ceiling. (c) Panel system, usually using aluminum facing, connected to metal tubes. (d) Panel system has the highest percentage of heat to the ceiling. (e) Cooling grid made of capillary tubes embedded in ceiling plaster (or in gypsum board or mounted on ceiling panels). This provides the most even surface temperature distribution. (f) A high percentage of heat is radiated to the cooled ceiling, depending on the use of insulation below the carpet and pad. (Lawrence Berkeley Laboratory.)*

Of particular note, however, is the application of GSHPs at the scale of a district energy system (DES). Ball State University in Muncie, Indiana, replaced coal-fired boilers (on-site combustion) and dedicated chillers (vapor compression refrigeration) that served as the source devices for campus climate control with a campus-wide (district) ground-source heat pump system. This major retrofit produced substantial environmental benefits and also provides ongoing energy (and cost) savings. The system uses two widely separated district plants, closed loop vertical bore holes, and underground distribution loops. See the case study of this system in Section 16.22.

(b) Underfloor Air Distribution

Underfloor air distribution (UFAD) systems are being promoted as an energy-efficient and air-quality-effective means of distribution for an HVAC system. The jury is still out on the ability of these systems to deliver on the promised benefits, but it is likely that they will become more common. ASHRAE released a revised

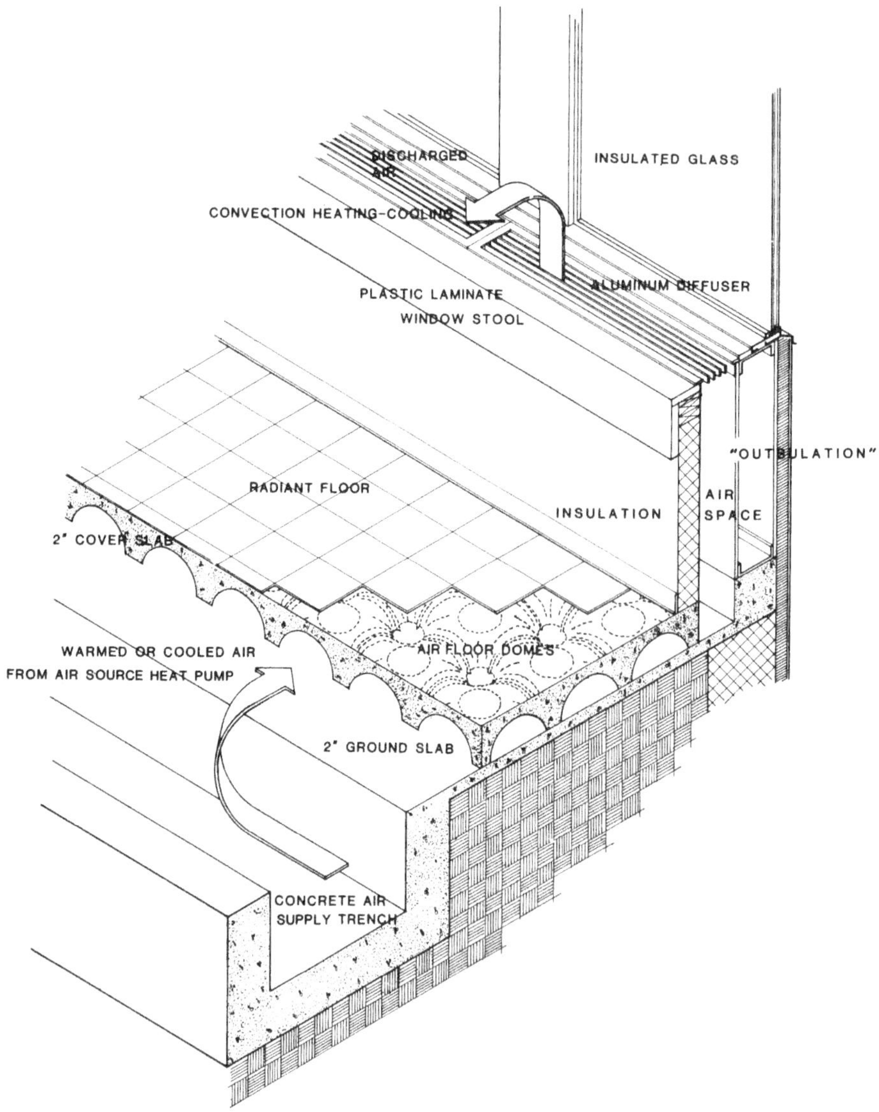

Figure 16.138 *Underfloor air distribution using a cellular concrete air floor serves the Bleshman Regional Day School in Paramus, New Jersey. (Courtesy of Rothe-Johnson, architects.)*

design guide for UFAD systems (ASHRAE, 2013). Some background on underfloor air distribution may help explain this distribution/delivery option.

As open-plan office spaces have grown larger in area and their ceilings higher (for daylight penetration and indirect lighting), and as concerns about IAQ have deepened, the floor has gained in popularity as an air source location. Underfloor air supply introduces supply air (cooled to just below the design room temperature) through floor outlets, at a low velocity. Air supply rates are typically greater than those used with ceiling supply systems. This high-volume, low-Δt, low-velocity combination requires greatly increased duct sizes, essentially provided by using the plenum formed by a raised floor system as a duct. The supply air rises quickly to the occupied zone, drawn upward by a stack effect driven by heat gains from lighting, people, and office equipment, and by supply-air pressure. The air continues to the ceiling, where it is collected and returned to the air-handling unit. With the freshest and coolest air at the floor and the hottest and most stale air at the ceiling, this system promises better IAQ and thermal comfort than would be available from ceiling supply–return systems. Furthermore, adjustable outlets at the floor are easily reached by occupants.

The raised floor that forms the upper boundary of the plenum is typically a 2×2 ft [600×600 mm] modular construction. The module elements that

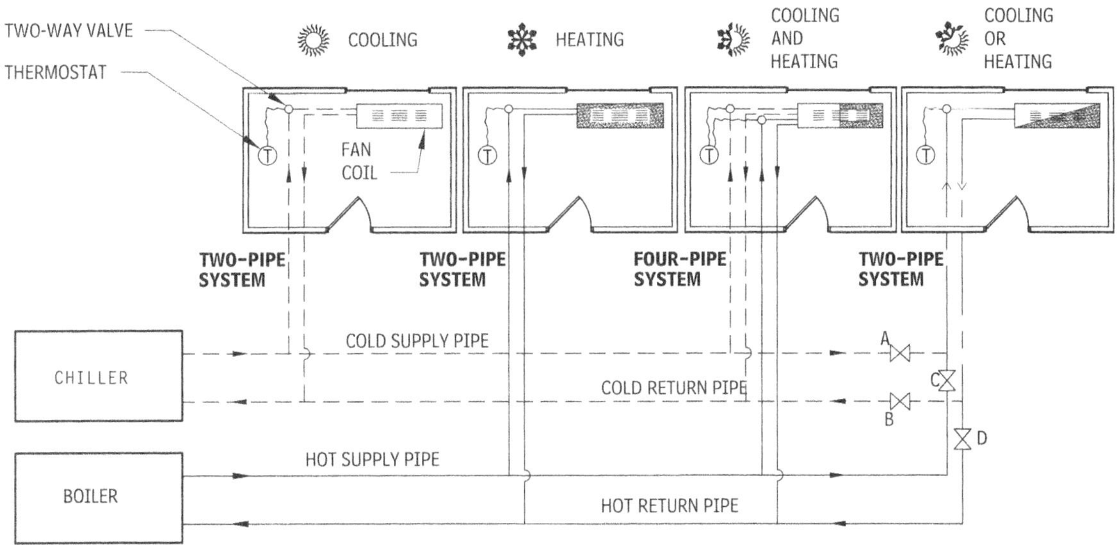

Figure 16.139 *Schematic diagram of an all-water HVAC system (with a four-pipe distribution arrangement). (From Ramsey and Sleeper, 2016.)*

contain diffusers often also contain an outlet for power and data. If the diffuser is placed off-center within the module, each plan rotation of a module moves the outlet slightly, expanding the variability of the floor openings relative to furniture placement. Some codes restrict the height of the raised floor plenum; many codes require specially coated wiring within the plenum.

The overall floor-to-floor height needs to be great enough to accommodate both the raised floor and a higher office ceiling (at least 9 ft [2.7 m] to accommodate warm air stratification). However, elimination of supply air ductwork can also eliminate the need for a ceiling cavity. This tends to expose the lighting, fire sprinkler, and public address systems below the structural ceiling, although the latter can be distributed in the raised floor cavity above, puncturing the structural ceiling as required.

The structural slab below the raised floor can be of concrete, and this thermal mass can be night-ventilated to store coolth to assist with cooling for the next day. The Inland Revenue Building incorporates a night-ventilated raised-floor area within its precast waveform floor sections.

Library Square in Vancouver, British Columbia (Figure 16.142), utilizes a raised floor to serve interior zones with book stacks. The library has seven floors, totals 390,000 ft^2 [36,320 m^2], and exposes the concrete structure as a finished ceiling. The floor-to-floor height is 16.4 ft [5 m]. The underfloor supply is fed by VAV boxes and floor fan units (FFU) that blend low-temperature supply air (about one-third) with locally filtered return air (about two-thirds) to provide 63°F [17°C] air to the pressurized plenum. Flush to the floor, high-induction swirl floor diffusers (8 in. [200 mm] in diameter) produce an upward air motion, taking the supply air directly into the occupied zone (the first 6.5 ft [2 m] of height on each floor). The stack effect caused by heat from people, computers, and lighting moves hot air to the stratification zone above. People-generated pollutants tend to rise to the hot, stale air zone, then into return air slots in the precast concrete ceiling. Return air is taken away in insulated ducts within the supply air plenum of the floor above. The plenum height is an unusually generous 2 ft [600 mm]; it also contains the sprinkler distribution and all wiring systems. The warm, dry summer climate supports a night ventilation system, flushing the plenum and storing coolth in the exposed concrete structure. This has reduced the need to use a provided ice-storage system.

(c) Displacement Ventilation

Displacement ventilation (DV) is an HVAC strategy by which air is supplied at a low velocity and at only a few degrees cooler than the zone's thermostat setpoint. Supply air is typically delivered through diffusers located near the floor, perhaps under furniture or seating. Return air is pulled from outlets located in or

(a)

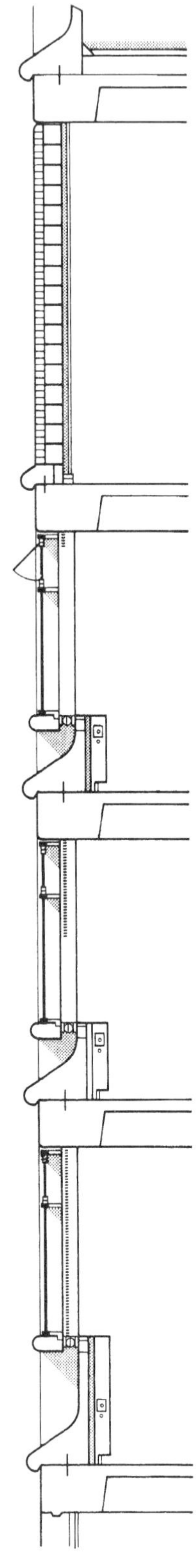

(b)

Figure 16.140 *Local outdoor air intake with finned tube radiation. The Seeley G. Mudd Library at Yale University; Roth and Moore, architects. (a) Exterior, showing curved limestone spandrels, along which flows incoming outdoor air. (b) Section showing the fresh air intake, finned tube radiation, and upper operable sash for exhaust air. This system is used for the smaller reading rooms and staff workrooms at the building perimeter. (a: Photo by Steve Rosenthal.)*

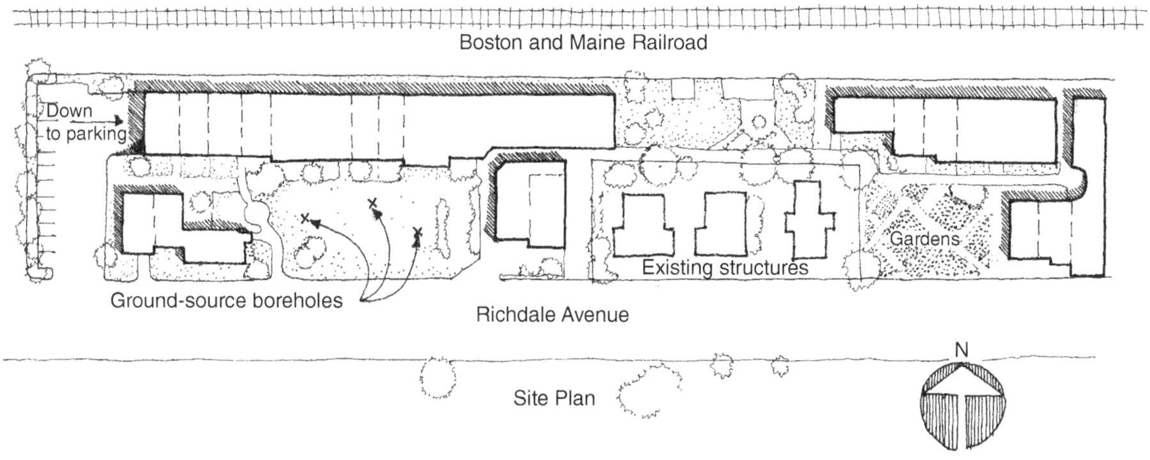

(a)

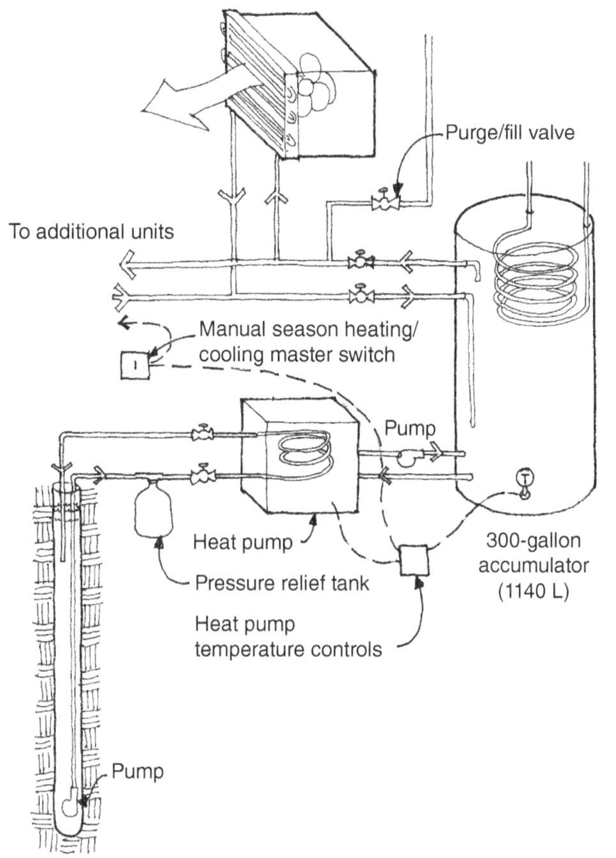

(b)

Figure 16.141 *Cambridge (Massachusetts) Cohousing development takes advantage of three boreholes on its urban site (a) to provide central heating and cooling (via heat pumps), serving individual residential fan-coil units. The heat pump also preheats water (b) for the DHW (domestic hot water) system. The residences are sited to provide winter solar access, and an outdoor air intake is located on the side sheltered from adjacent railroad tracks. (Courtesy of Building Science Engineering, Harvard, MA.)*

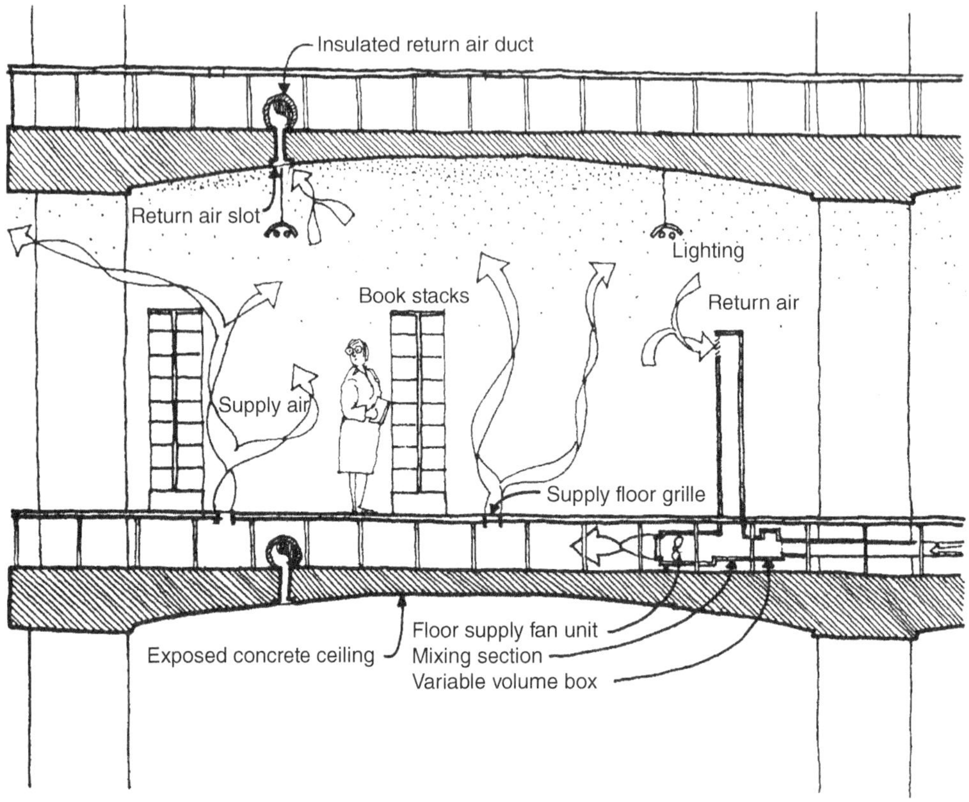

Figure 16.142 *Library Square, Vancouver, British Columbia, supplies conditioned air to the central book stacks via an underfloor plenum. (Section adapted from information supplied by Blair McCarry, Keen Engineering.)*

near the ceiling. By delivering air lower and removing air from the higher region of a space, more pronounced thermal stratification occurs than in a system where air is both supplied and returned from above the breathing zone. Versus ceiling-based delivery, DV systems can be more energy-efficient, comfortable, and quieter.

(d) Demand-Controlled Ventilation

Demand-controlled ventilation (DCV) uses occupancy monitors to automatically estimate the occupancy in a zone (or subzone) and adjust ventilation accordingly. The concept is based on the notion that less conditioned supply air is necessary in underoccupied spaces and greater ventilation is needed during times of increased occupancy. Carbon dioxide (CO_2) monitors are typically used because CO_2 concentrations typically correlate fairly well with human occupancy. The CO_2 monitors are integrated into a BMS that automates the HVAC operation for the thermal zone being monitored. DCV systems have the potential to save energy, improve thermal comfort, and better ensure healthy IAQ.

(e) Dedicated Outdoor Air Systems

Dedicated outdoor air systems (DOAS) are designed to essentially decouple the control of indoor air quality from the control of the thermal environment. This allows the two key objectives of an HVAC system to be designed independently and without conflicts. Although the idea has merit and has been developed in a number of projects, this is more of a potential than an actual trend. Increased first costs associated with two systems (versus one) appear to be slowing adoptions.

(f) Chilled Beam with DOAS

Radiant panels with supplemental air as an HVAC strategy is nothing new. However, chilled beams with DOAS are an emerging design trend. At the 56-story Commerzbank in Frankfurt, Germany, chilled ceilings serve all office floors in conjunction with a ventilation system. Chilled beams are a variation on radiant panels; DOAS are a variant on supplemental air systems. Combined, they are proving to be an effective strategy. Chilled beams (Figure 16.143) are an approach to HVAC delivery that has substantial traction in Europe but only modest uptake in North America. Although often viewed as simply a form of radiant cooling device, chilled beams can also deliver cooling effect via convective heat transfer. Passive chilled beams are all-water terminals that exchange heat via radiation and convection; active chilled beams are air–water terminals with convective transfer predominating.

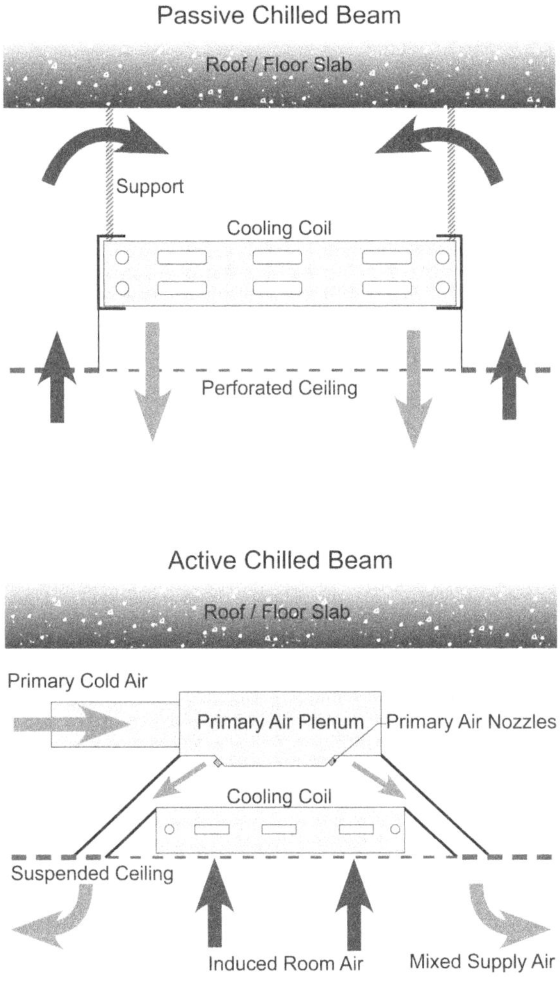

Figure 16.143 *Active and passive chilled beams. (Drawn by Tyler Mavichien; © Walter Grondzik; all rights reserved.)*

16.21 SPECIALTY HEAT TRANSFER EQUIPMENT

One major advantage of central systems with central equipment rooms is the opportunity they present for energy-efficiency design moves. Regular maintenance is simplified when all equipment lives in a generous space with good environmental conditions; regular maintenance brings increased efficiency of operation. The opportunity for heat transfer between various systems and pieces of equipment increases greatly when the equipment is co-located and the systems can easily be connected. A few examples of heat transfer and/or heat reclaim opportunities follow.

(a) Heat Reclaim

Boiler flue economizers (Figure 16.144) work by passing the hot gases in a boiler's stack through a

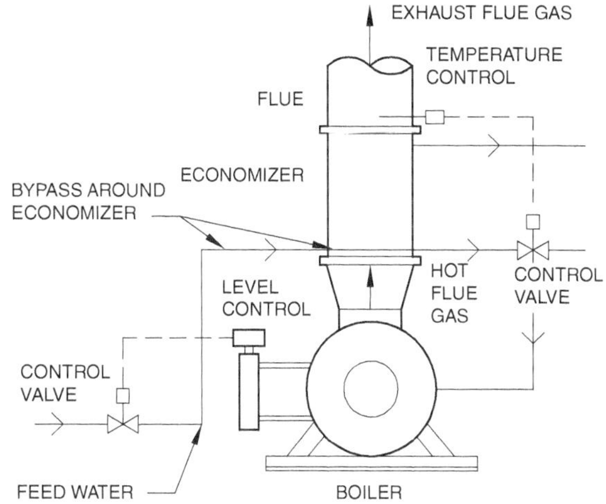

Figure 16.144 *Heat recovery for boilers. Flue gas entering at 500°F [260°C] leaves the "economizer" at 325°F [163°C], a temperature still high enough to prevent condensation in the stack. The recovered heat is applied to incoming boiler water, raising its temperature from 200 to 248°F [93 to 120°C]. (From Ramsey and Sleeper, 2016.)*

heat exchanger; the collected heat is used to preheat incoming boiler water.

Runaround coils (Figure 16.145) can be used to transfer heat between intake and exhaust air streams even when these two streams are rather far apart. The circulating heat-transfer fluid usually contains antifreeze and it provides simple sensible heat transfer, with no restrictions on locating air exhaust and air intake. No contamination of intake air is caused by this arrangement. The efficiency of such coils runs between 50% and 70%, and they are available in modular sizes up to 20,000 cfm [9440 L/s].

Energy-Recovery Ventilation Systems. These devices offer a means of siphoning useful "waste heat" from a return air stream before it is expelled from a

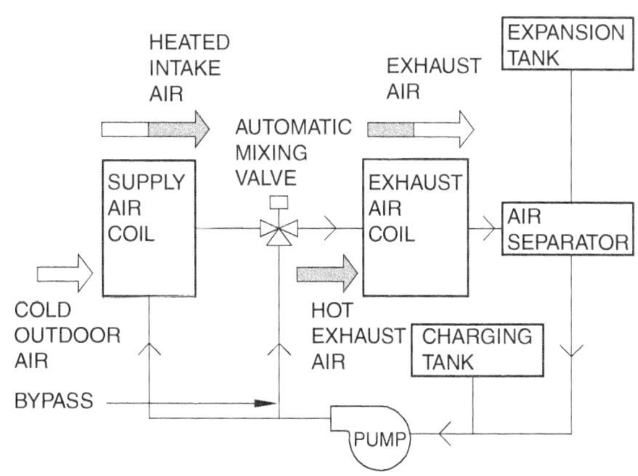

Figure 16.145 *Runaround coils for heat transfer between outdoor air and exhaust air are used where the air streams are in separate locations. (From Ramsey and Sleeper, 2016.)*

building. The devices transfer heat from an exiting air stream to an incoming air stream. Substantial energy savings can accrue from the use of such devices. These systems can be scaled to fit virtually any size of building. There are two basic types of energy-recovery ventilation systems:

- *Heat-recovery ventilators (HRVs):* These use fan assistance to foster a transfer of heat between return and supply air through a heat exchanger. The term "heat recovery" denotes sensible heat being recovered from return air before it is expelled. These are becoming common today in high-efficiency homes.
- *Energy-recovery ventilators (ERVs): These* similarly use fan assistance and a heat exchanger; however, the term "energy recovery" denotes both sensible and latent heat (through use of a desiccant-coated exchange medium) being recovered from the return air. ERVs are increasingly common in high-performance buildings of various scales.

Thermal Wheels. The term "thermal wheel" is often used as a catch-all term for both heat wheels and enthalpy wheels, encompassing the general concept of rotary heat exchangers. Sometimes also called energy recovery wheels, these devices transfer thermal energy, but they do not mix air streams:

- *Heat wheels* (or *sensible heat wheels*) transfer sensible heat.
- *Enthalpy wheels* (or *energy wheels*) also use a desiccant, allowing transfer of both sensible and latent heat.

Thermal wheels are typically installed as part of a central HVAC system in larger structures, such as commercial office buildings, health care facilities, schools, hospitality venues, and industrial facilities.

(b) Economizer Cycle

An economizer cycle (Figure 16.146) is a control strategy that uses cool outdoor air, as available, as supply air. The economizer cycle can be thought of as an active system mechanical substitution for an open window; when it is cool enough, 100% outdoor air can be provided as supply air, and no refrigeration is needed. When the outdoor air temperature is higher than the desired supply air temperature but lower than the return air temperature, 100% outdoor air is still brought in, but refrigeration is used to reduce its temperature. Above the return air temperature, outdoor air is reduced to the volume required for maintaining acceptable indoor air quality.

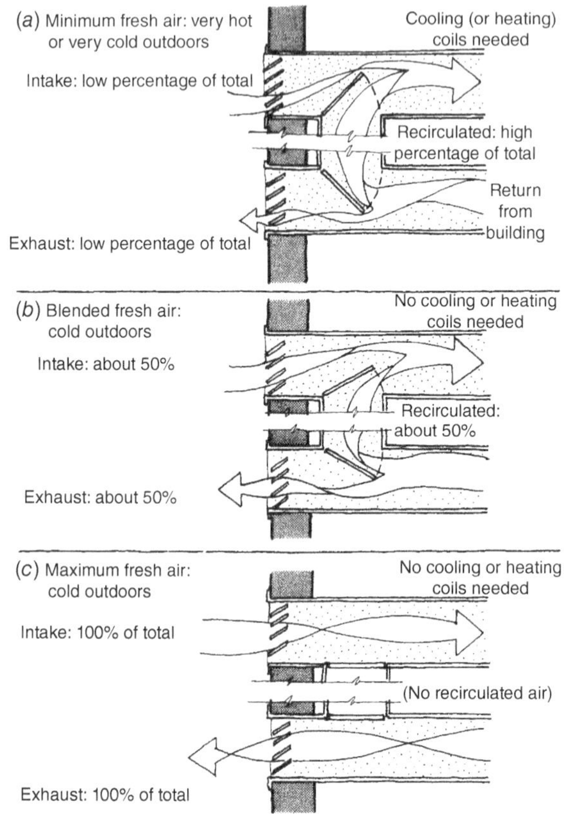

Figure 16.146 *An economizer cycle controls the relationships among outdoor, exhaust, and recirculated air. (a) When outdoor air is too hot or too cold, the economizer cycle is inactive, and IAQ-minimum outdoor air is introduced. (b) As very cold outdoor air gets warmer, it can be blended with recirculated air, and neither heating nor cooling coils are needed. (c) When outdoor air is cool, it can completely displace the need for mechanical cooling.*

Sensible and latent heat content affects the viability of using outdoor air in an economizer cycle. Although this discussion refers to temperature comparisons for simplicity, comparisons of enthalpy should, in fact, be used.

Relative to opening windows, an economizer cycle has several advantages: energy-optimizing automatic controls, filtering of the outdoor air, tempering of outdoor air to avoid unpleasant drafts, an orderly diffusion of outdoor air throughout a building (even locations distant from windows), and the opportunity for noise control. Its disadvantages are the loss of personal control that windows offer and thus loss of awareness of exterior–interior interactions.

Economizer cycles are available as an option on most direct expansion (DX) air-handling units (such as packaged rooftop units) and are typically installed in large-building central air supply systems. Buildings with high internal gains (internal load dominated) are particularly good targets for economizer cycles because they need cooling even when the outdoor temperature is chilly. Economizer cycles lend themselves readily to a cooling strategy of night ventilation of thermally massive structures because they have a built-in option for 100% outdoor air.

(c) Energy Storage

We commonly experience daily changes from warmer to colder. Central heat storage equipment in large buildings can take advantage of this cycle to increase operating efficiency, reduce resource consumption, and significantly reduce electricity demand charges. Some electric utilities offer incentives to install thermal storage in order to reduce the peak load strain on their generating facilities.

Water storage tanks, as shown in Figure 16.147, are a common approach to thermal storage. On a typical winter's day, the total internal heat generated by a large building can be somewhat greater than its total need for heating in the perimeter zones. Instead of being thrown away in exhaust air, this surplus heat can be captured and stored in large water tanks, from which it can be withdrawn and used on cold winter nights and weekends. In the summer, chillers can work

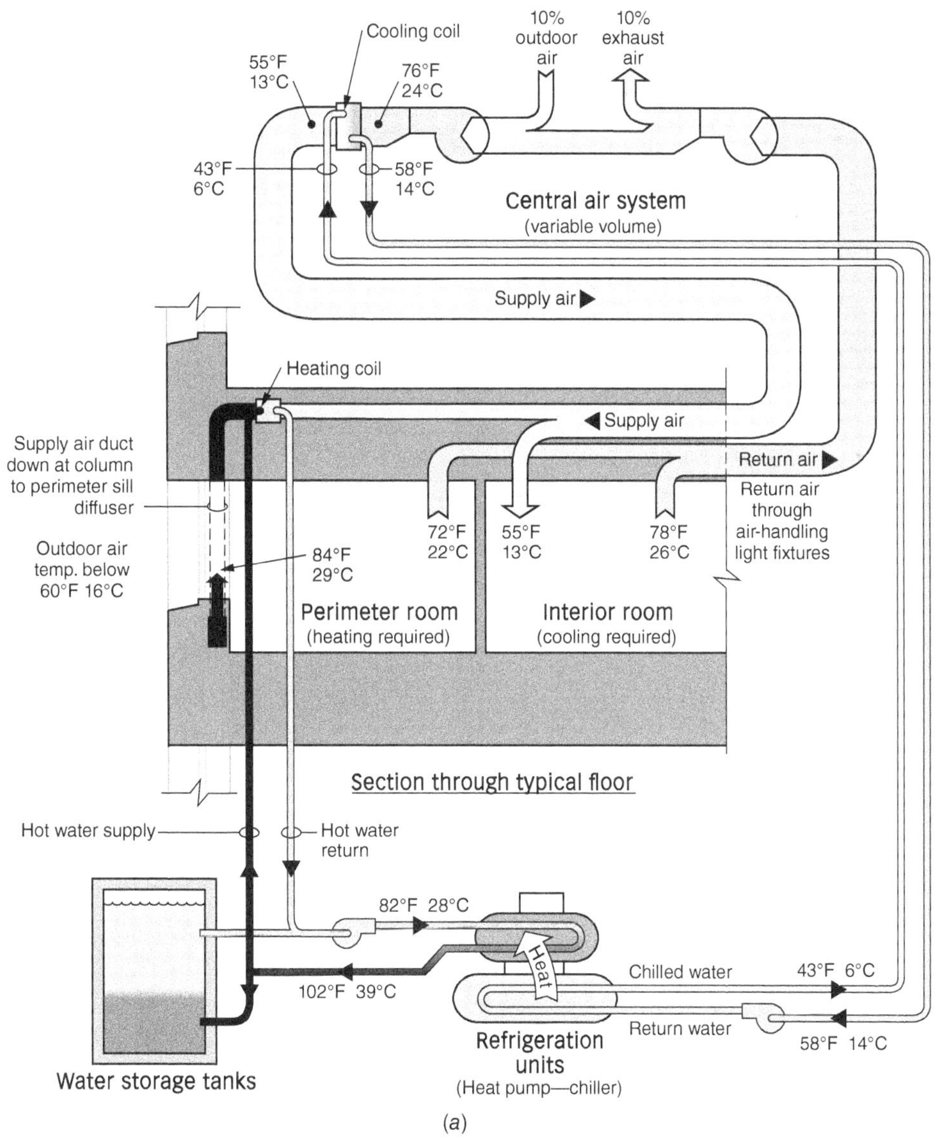

Figure 16.147 *Water tank heat storage. The 870,000-ft² [80,825-m²] transportation office building (Park Plaza, Boston), population 2000, uses a three-compartment insulated concrete tank storing 750,000 gal [2,838,990 L] of water. (a) At an outdoor air temperature of 40–50°F [4.4–10°C], surplus internal heat is stored in the tanks rather than rejected in the exhaust air. (b) By the time the outdoor air is about 50°F [10°C], the tanks are fully charged; up to an outdoor temperature of 60°F [15.5°C], the economizer cycle provides energy-conserving cooling. (c) At about 60°F [15.5°C] chillers must operate, but working all night when the outdoor air is cooler enables them to work less by day. Their nightly production is stored as cold water, available to help with the following day's peak. Smaller machines and more efficient operation are the result. (Courtesy of Shooshanian Engineering Associates, Inc.)*

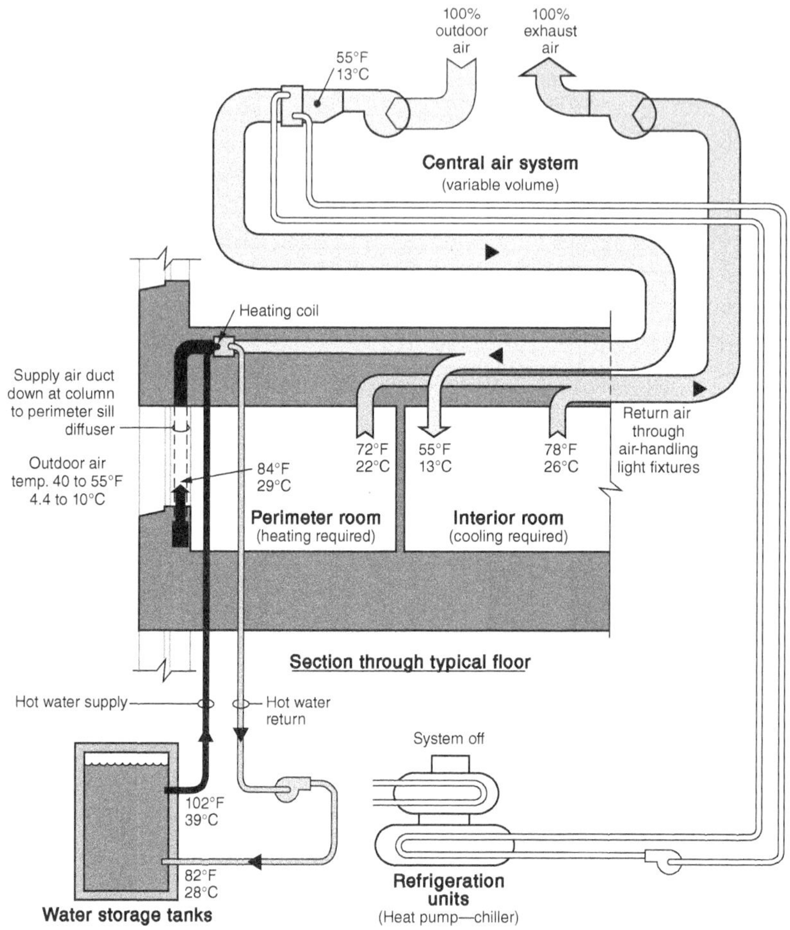

(*b*)

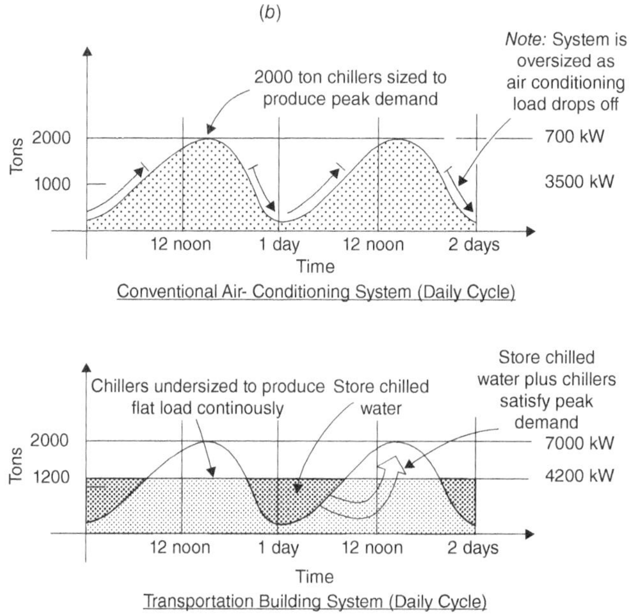

Note: Charging of storage tanks with cold water during off hours provides added cooling media for peak air-conditioning needs.

(*c*)

Figure 16.147 *(continued)*

at night, when efficiency is higher because cool outdoor air helps the refrigeration cycle more easily reject its heat. Storing the coolth thus produced means that less work needs to be done by chillers during the next day's peak load time, when electric rates are highest and operating efficiency is lowest.

- Design guidelines for water storage are 0.5–1 gal/ft² [20–40 L/m²] of conditioned space. Large, heavy tanks are frequently located in basements or underground parking facilities. Water tanks require that additional floor space be devoted to equipment, although they may allow somewhat smaller chillers to be installed. They may also contribute to lower fire insurance premiums because considerable water is stored and ready to use for fire protection.
- Ice storage tanks, as shown in Figure 16.148, are another thermal storage approach. Because ice storage takes advantage of the latent heat of fusion (143.5 Btu/lb of water at 32°F [334 kJ/kg at 0°C]), such units can store far more cooling effect in a given-size tank than can water. Design guidelines for ice storage are 0.13–0.25 gal/ft² [5–10 L/m²] of conditioned space. A comparison of the relative sizes of storage tanks needed for an 18-story office building is shown in Table 16.20. With ice storage there is less undesired mixing of hot and cold water within the storage tank.

Ice can be produced in several ways. A common method is to form ice as a layer around a pipe that carries refrigerant in a closed circuit through the tank. Control of the thickness of this layer is important, because if too little ice is made at night, too little cooling will be available the following day. In another method, ice is formed and thawed by circulating a brine through coils in cylindrical water tanks. Other methods form ice on plates, then harvest it in an insulated bin.

The Iowa Public Service Building (Figure 16.149) uses ice storage in six tanks that occupy about half of the floor space in the ground floor mechanical room. This installation serves a five-story, 167,000-ft² [15,515-m²] utility office building that utilizes solar collectors supplemented by small backup boilers. Six ice-making machines serve a 75,000-gal [283,900-L] ice storage pit with 90 million Btu [26,355 kWh] capacity. Winter heat exchange opportunities include reject heat from the ice-making machines, heat from the centralized toilet exhaust air, and heat from return air taken through luminaires. Solar collectors on the roof preheat the ventilation air, which is fed into the ceiling plenum at each floor. Fan-coil units then draw from this fresh air supply.

Water and ice storage tanks are sometimes located at the top of a building, especially when related mechanical equipment is there also. Despite the weight of such tanks, they can become a structural advantage. The tuned mass damper method of reducing lateral

TABLE 16.20 Cooling Storage Comparison

Assume an 18-story office building: 375,000 gross ft² [34,840 m²], 800 tons [2800 kW] peak load, 6250 ton-h/d [21,908 kWh/d] at design, 75 million Btu [21,900 kWh] storage needed.

	Water Storage	Ice Storage
Btu/lb[a] [kJ/kg]	15 [34.9]	164 [381.5]
Pounds storage	5,000,000	457,000
[kilograms]	[2,267,960]	[207,290]
Gallons	599,500	54,800
[liters]	[2,269,300]	[207,435]
Storage efficiency	0.90	1.0
Percent ice		66%
Net gallons	666,000	83,030
[liters]	[252,100]	[314,295]
Cubic feet	89,046	11,101
[cubic meters]	[2522]	[314]
Floor area of tank, approx. 8 ft deep [2.5 m deep]	80 by 150 ft [24 by 46 m]	30 by 50 ft [9 by 15 m]
Storage ratio, water to ice	8 : 1	

Source: Reproduced by permission from *Specifying Engineer*, January 1983. SI conversions added by the authors of this book.

[a]Btu/lb based on Δt = 15°F for water storage and on Δt = 20.5°F in the melted ice water (added to 143.5 Btu/lb at fusion).

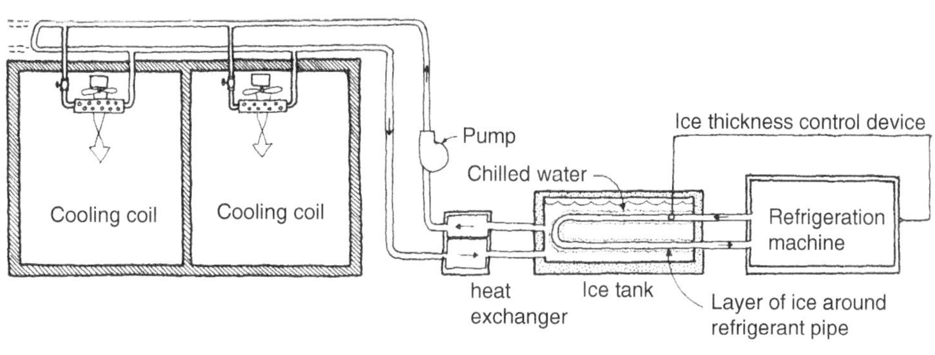

Figure 16.148 *Ice tank heat storage. Chilled water (supplied by a refrigeration machine) in an ice tank forms a layer of ice around the refrigerant pipe. As chilled water at about 35°F [1.7°C] is removed from the tank, it enters a heat exchanger so that its temperature is closer to 45°F [7.2°C] for distribution to cooling coils throughout the building.*

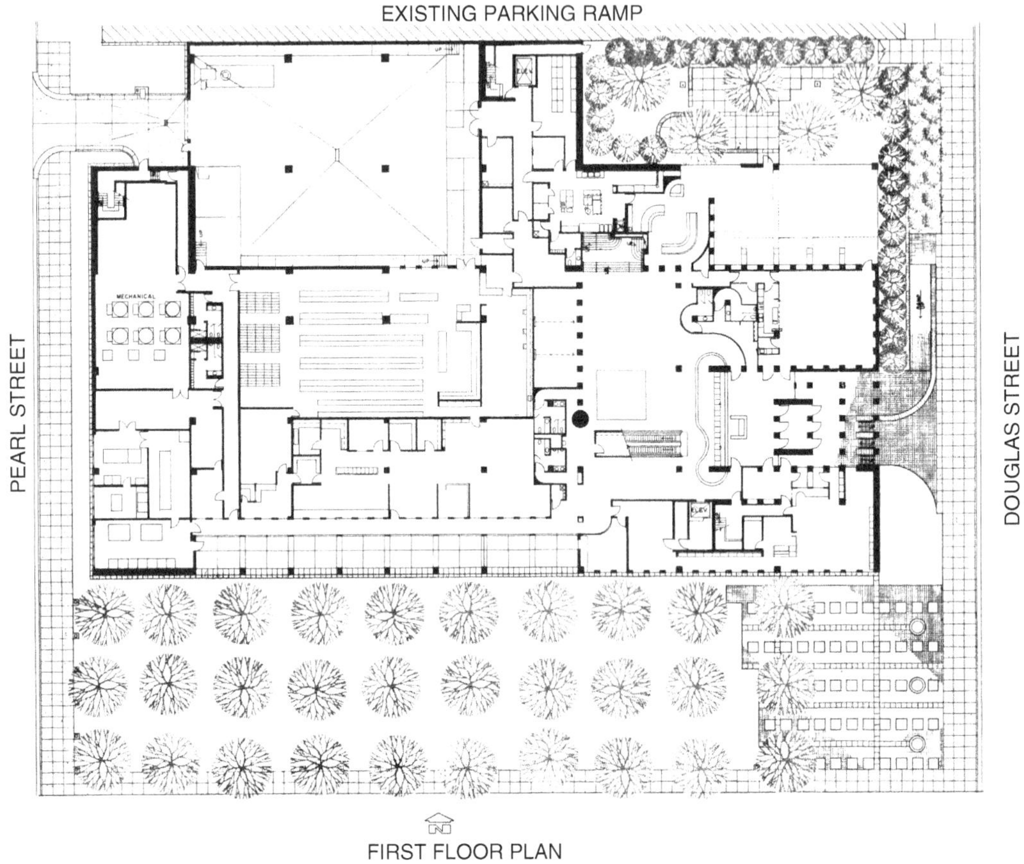

Figure 16.149 *The first (ground-level) floor of the Iowa Public Service offices (Sioux City) contains six ice machines within the mechanical space adjacent to the loading dock. (Courtesy of Rossetti Associates and Foss, Engelstad, Heil Associates, joint venture architects.)*

vibration (or sway) in high-rise buildings utilizes a heavy, moving mass at the top; when the building begins to sway, the mass is moved in the opposite direction. In the Crystal Tower in Osaka, Japan, this mass is provided by ice storage tanks. This 515-ft [157-m], 37-story building has an ice thermal storage volume of 25,400 ft^3 [720 m^3] divided into nine tanks. Six of the ice storage tanks provide the movable structural mass, suspended from roof girders, weighing 540 tons [489,880 kg], including steel framing. The chiller and condenser are also in the equipment penthouse.

A multi-pronged approach to heat storage is illustrated in the building featured in Figure 16.150. In England, the Hyndburn Borough Council decided that their new headquarters building should set an example as a "zero energy" building: That is, over a typical year, it should generate as much energy as it imports. The 38,750-ft^2 [3600-m^2] building is elongated east–west and boasts an ambitious section that combines daylighting, photovoltaics (PV), a well-insulated enclosure, and even rainwater collection to use for flushing toilets. Windpower adds to electricity generation. Summer cooling is by ventilation (assisted by night ventilation of mass) through a raised "Termodeck" system. This is a prefabricated hollow-core slab through which cool night air circulates, storing coolth to assist the next day's hottest hours. The raised floor then provides a distribution plenum, with a design rate of four air changes per hour ACH.

For more extreme future summer cooling, but now mostly for winter heating (beyond that provided by passive solar heating), an adjacent reservoir acts as a thermal sink. A "lake closed source heat pump" will serve the mechanical ventilation system. The lake is south of the building, providing some local evaporative cooling in summer and inviting a very climate-oriented south façade, behind which are open-office areas. The north façade faces the city and is more traditionally institutional in character, as are the individual offices behind it.

(d) Cogeneration

Electricity has repeatedly been called a high-quality resource in reference to the many, many tasks that it can be applied to—but also with regard to the high temperatures needed to produce electricity by conventional (fossil-fuel) means and to the large amount of waste heat (often twice the heat value of electricity) produced in the process. Cogeneration (also called total energy) represents an attempt to recover some of the otherwise wasted low-temperature heat that accompanies the generation of electricity by steam turbines. Industrial cogeneration facilities are especially cost-effective in the pulp and paper, petroleum, and chemical industries. Building-scale cogeneration is a less apparent energy bargain.

Where conditions are favorable, electricity for power and lighting can be generated economically by a system that also supplies the building with heating in winter and cooling in summer. Such a system utilizes a fuel such as gas or oil and is often supplemental to the local electric utility company. Although first costs are greater than for more conventional energy systems, the savings in annual operating costs can sometimes pay for the installation cost in a reasonable time. Operational savings continue thereafter. This approach was developed largely in the 1960s and is used in hundreds of commercial and industrial buildings and on university campuses. Cogeneration is particularly attractive for district heating/cooling plants.

Before 1900, nearly all large buildings and groups of buildings were supplied with direct-current electricity generated on or near the premises. The motive power was usually a steam-driven reciprocating engine with belt connections to direct-current generators. Direct current cannot be conventionally transformed to different voltages and was generated and distributed at the voltage used in the building. At these relatively low voltages, power loss in the distribution system is great, and distance adds greatly to the loss. This tended to favor local generation of power.

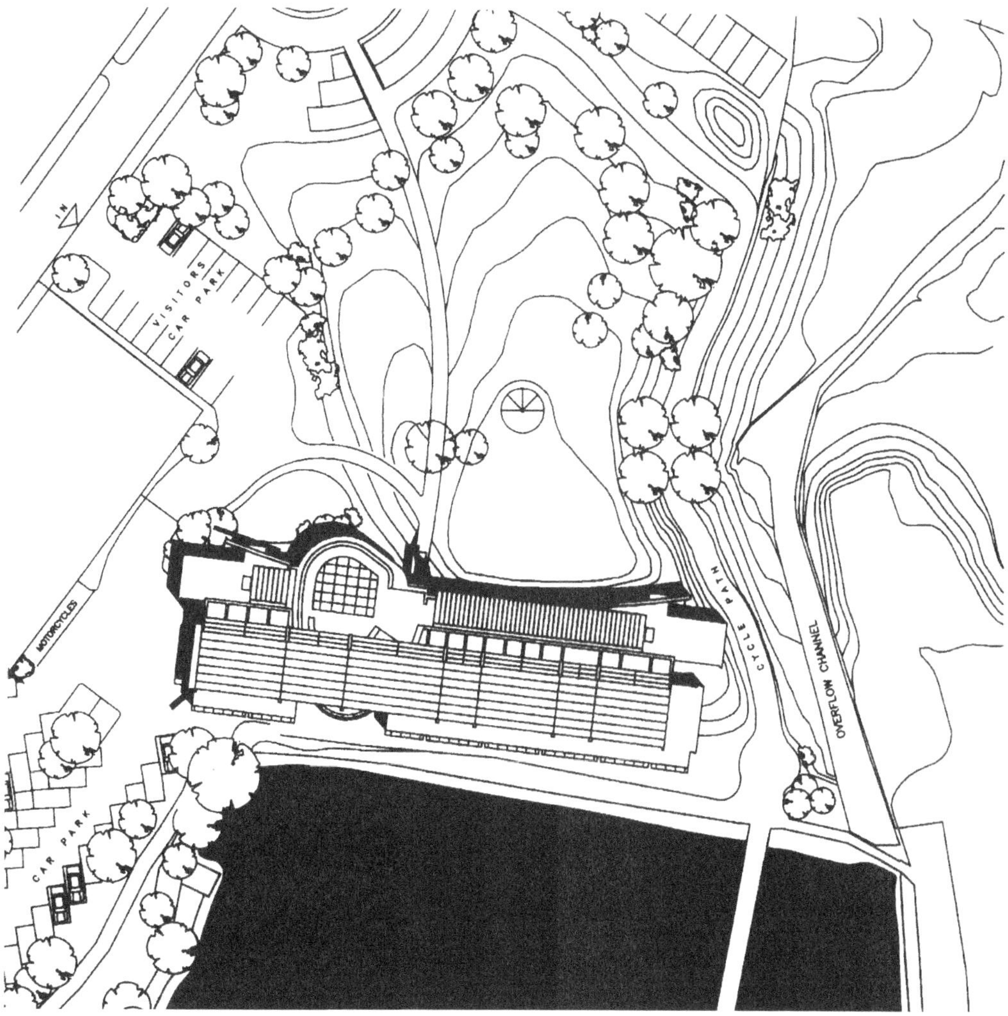

(a)

Figure 16.150 *The Hyndburn Borough Council headquarters faces south toward a reservoir (a) that provides an evaporatively cooled microclimate and also acts as a heat source/sink for a water–water heat pump. (b) As suggested in the diagrammatic section, daylighting, passive solar heating, PVs, and a well-insulated shell are featured. Termodeck is a precast hollow-core slab that stores coolth on summer nights and precools ventilation air. The supply air rises from the plenum created by the raised floor. (Courtesy of Jestico + Whiles + Associates, architects.)*

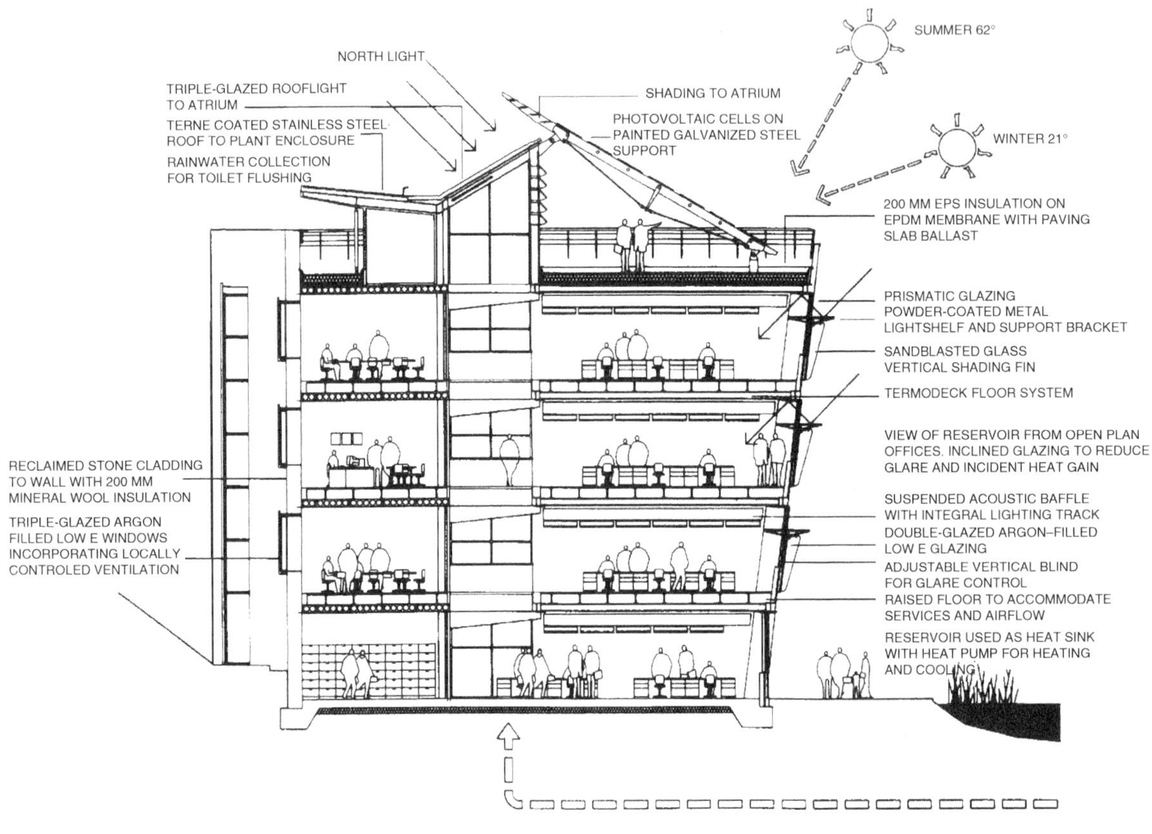

(*b*) DIAGRAMATIC SECTION

Figure 16.150 *(continued)*

With the development and use of alternating-current machines, utility companies were able to establish central power stations from which electricity at high voltage could be transmitted great distances to the users. It was transformed down to domestic voltages on the building site (or nearby). Because, at high voltage, power losses are fairly low, this system became universal. During the 1920s and 1930s, owners removed their private power generators and accepted utility service, with its savings in operating expense.

Back to the present. For cogeneration to be successful, there should be a reasonably steady demand for the power being generated and also for the heat being recovered. Lighting, computers, business machines, and other devices can create nearly constant demand for power throughout the year. Similarly, the exhaust heat recovery from the engines or turbines that power the generators is in demand for either heating or absorption refrigeration during most of the year.

Figure 16.151 shows two principal approaches to total energy. In both systems—one using a turbine and the other a reciprocating engine to operate the electric power generator—heat is reclaimed to produce steam or hot water. The steam or hot water is then used for heating or, via an absorption chiller, for cooling. When a turbine is used, the fuels are natural gas or fuel oil. The heat is recovered by passing the hot turbine exhaust through a waste heat boiler to produce steam. Fuel for a reciprocating engine is natural gas or diesel fuel. Both the jacket cooling water and the hot engine exhaust are passed through heat exchangers that utilize the heat to produce steam or hot water for heating or cooling. In both systems, an auxiliary boiler, fired directly by gas or oil, stands ready to help maintain a balance in the system.

The Harbortown apartment and townhouse complex in Detroit, Michigan, utilizes a year-round natural-gas-fired primary energy and cogeneration/waste recovery system (Figure 16.152). Each apartment has one or two stackable, upright fan-coil units using either chilled or heated water. This water is provided by two direct-fired chiller–boilers: Each incorporates a natural gas boiler and a two-stage absorption chiller in the same unit, taking less space in the mechanical room. The natural-gas-driven cogeneration electric plant serves rather constant loads such as corridor and outdoor lighting. The waste heat from the generator is passed to a storage tank, where it preheats domestic hot water; this meets about 87% of the hot water energy usage.

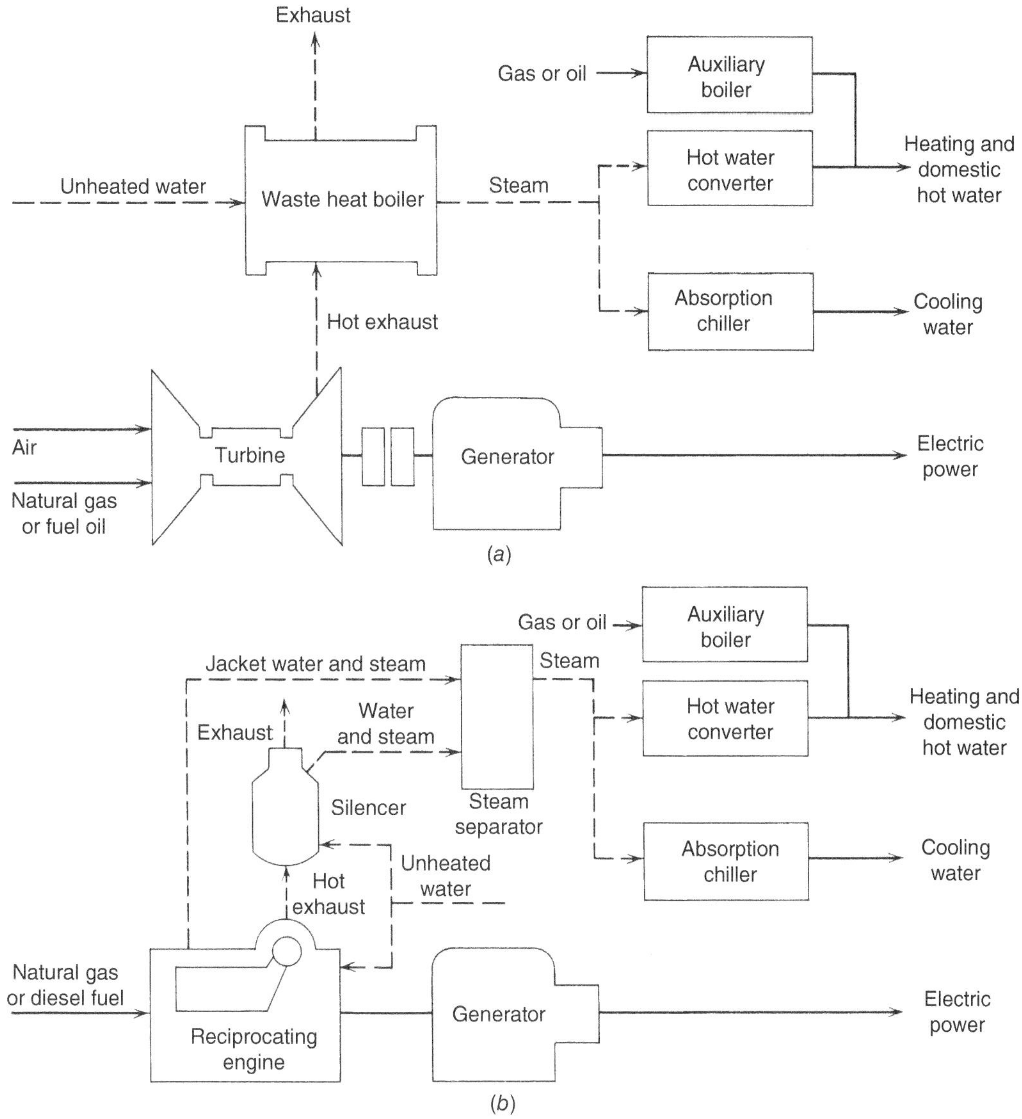

Figure 16.151 *Cogeneration (total energy) systems. (a) Using a turbine. (b) Using a reciprocating engine. (Reproduced by permission from* Total Energy, *Educational Facilities Laboratories.)*

(e) Fuel Cells

A fuel cell generates direct-current electricity by converting the chemical energy of hydrogen and oxygen into electricity and heat. Because no combustion is involved, nitrogen oxides and carbon monoxide are nearly eliminated as pollutants. For a building-based fuel cell power plant, several components are added. First, a fuel processor (or fuel reformer) to prepare a hydrogen-rich stream of fuel, then the fuel cell stack, then a power conditioner (or inverter) to convert the direct-current fuel cell output to alternating current.

Fuel cells are not, by themselves, much more efficient than combustion processes; they are about 40% efficient for power production. When combined with a way to use the 60% waste heat, however, they climb toward 90% efficiency, with very few harmful emissions.

Hydrogen is the ultimate fuel; it needs no processor and produces the least environmental impacts from the fuel cell. Other hydrocarbon fuels, however, can feed fuel cells: natural gas (most common for building-scale fuel cell power plants), propane, methanol, and, with additional pretreatment, methane from landfills and anaerobic digester gas from sewage treatment plants. Even with these less hydrogen-rich fuels, the emissions from fuel cells are well below those of combustion processes; the fuel processor emits carbon dioxide and a trace of carbon monoxide.

Proton-exchange membrane (PEM) fuel cells (Figure 16.153) are promising for transportation when

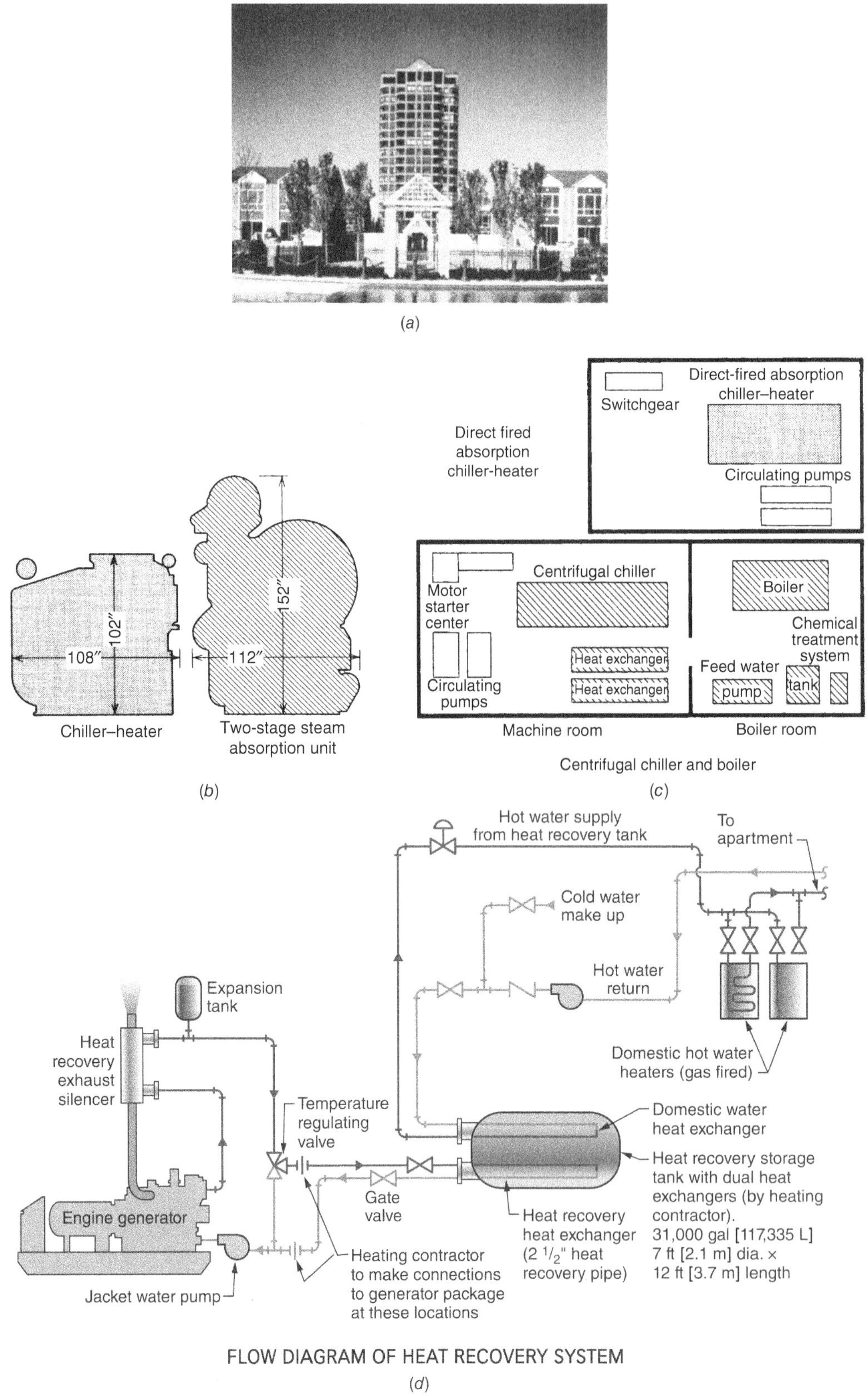

Figure 16.152 *Harbortown is a 120,000-ft² [11,150-m²] apartment and townhouse complex (a) in Detroit that uses natural gas as a primary fuel year-round. (Photo © 1989, William Kildow.) Chiller–heaters (Hitachi) combine a boiler and a two-stage absorption chiller in one compact unit. (b) Comparison of headspace and (c) floor space required by chiller–heaters versus conventional units of similar capacities. (d) Cogeneration plant sends waste heat to preheat domestic hot water. (Courtesy of Skidmore, Owings & Merrill, Architects-Engineers, Chicago.)*

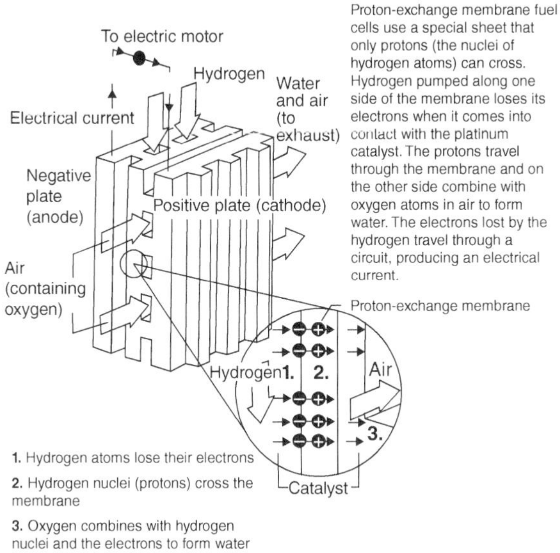

Figure 16.153 *Schematic diagram of a proton-exchange membrane (PEM) fuel cell with a description of its operation. (Courtesy of* Nucleus, *Fall 1994; © Union of Concerned Scientists, Cambridge, MA.)*

zero-emission vehicles are the objective and hydrogen is available as a fuel. This process has building applications as well. It works at a temperature below the boiling point of water. Relative to other fuel cell types, PEM has a smaller size, lighter weight, and lower noise levels. The PEM fuel cell stack involves several subsystems, typically including water purifiers and pumps, air compressors, and heat rejection components (coolant pumps and a radiator). This complicates performance at subfreezing temperatures and adds maintenance requirements. But the smaller size promises potential for small-building applications.

The phosphoric acid fuel cell power plant (Figure 16.154) is at work in (or outside) a number of buildings and municipalities around the world. The ONSI Corporation's 200-kW fuel cell power plant measures 10 ft × 18 ft, is 10 ft high [3 m × 5.5 m, 3 m high], weighs 40,000 lb [18,144 kg], and needs a clearance of 8 ft [2.5 m] around the module for maintenance. It also needs a cooling module (or use of a

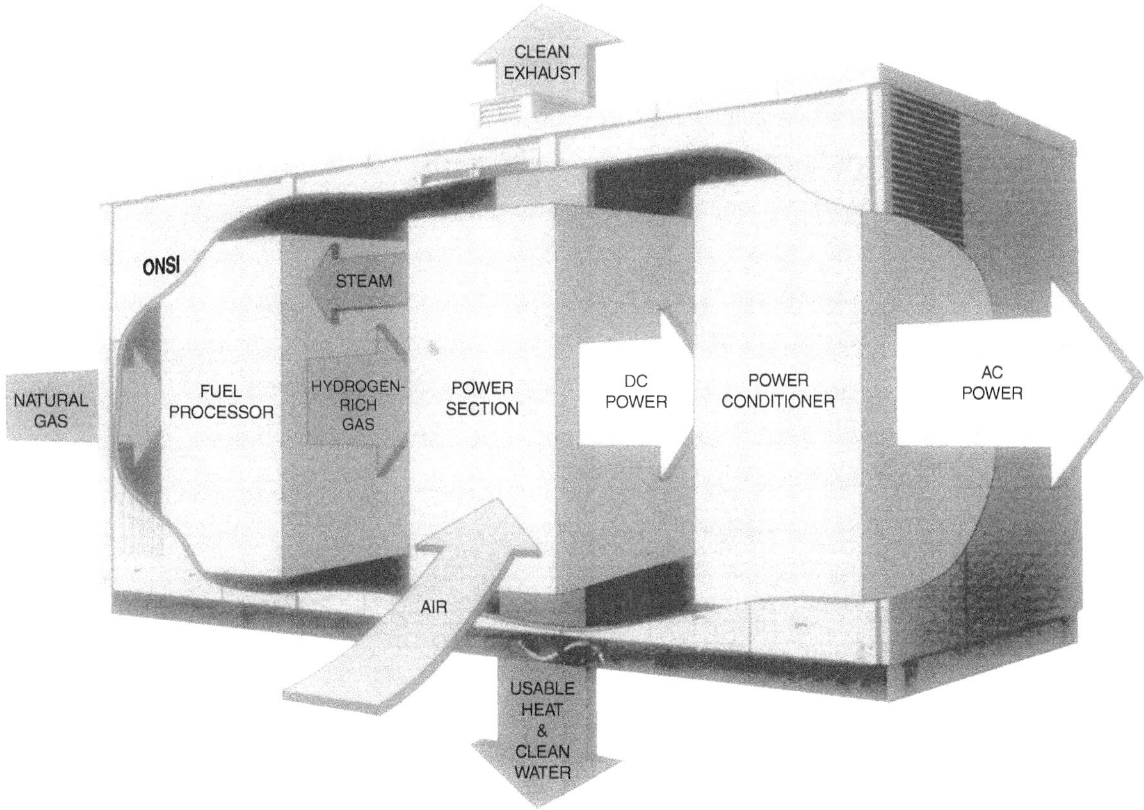

Figure 16.154 *A fuel cell power plant using natural gas (or other hydrocarbon fuel). The fuel processor (reformer) converts the natural gas to a hydrogen-rich stream that enters the fuel cell power section (or stack). The output is electricity (DC), water, carbon dioxide, and considerable heat. Some heat is used by the fuel processor and some is rejected, but most is usable in a building. A power conditioner (inverter) converts DC to AC power. (Courtesy of ONSI Corporation.)*

building cooling tower) for discharge of heat in excess of that recovered for building use. The acid electrolyte in the fuel cell stack works at about 390°F [200°C]. At its rated power, it also produces 700,000 Btu/h [205 kW] to heat a water stream to 140°F [60°C]. The rated sound level is 62 dBA at 30 ft [9 m] from the module.

The Durst Organization's high-rise office building at 4 Times Square in New York City (Figure 16.155) is a demonstration of several interesting energy-efficiency strategies: PV cells integral to the façade, direct-fired (natural gas) absorption chillers, increased outdoor air for IAQ (maybe not so energy efficient), a dedicated exhaust air shaft (for polluting activities), and waste chutes to facilitate recycling, among other features. It also utilizes two 200-kW fuel cell power packages fed by natural gas. The electricity is about 80% destined for external lighting by night; huge electrical signs are a prominent part of the Times Square façades. By day, about 80% of the power goes to the building's base load. The hot water will be used in winter for perimeter heating. In summer, the water is wasted. Although there was hope that one of the absorption chillers could be fed by this water, a variety of considerations precluded this: The fuel cells are located on

(a)

(b)

__Figure 16.155__ The 48-story office building at 4 Times Square, New York City (a), (b), (c). Many environmentally friendly features are included, such as façade-integrated PV and two fuel cell power plants like those shown in Figure 16.154. Fed by natural gas, the fuel cells are expected to work without maintenance interruptions for up to 5 years and provide considerable hot water as well as electricity. (The fuel cells are actually installed at the fourth-floor level rather than at the penthouse.) (a: Rustycale / Wikimedia Commons / Public Domain; b: © Gabrielle de Briey, used with permission; c: courtesy of Fox and Fowle, Architects, P.C.; Consentini Associates, Mechanical Engineers; the Durst Organization, Developer.)

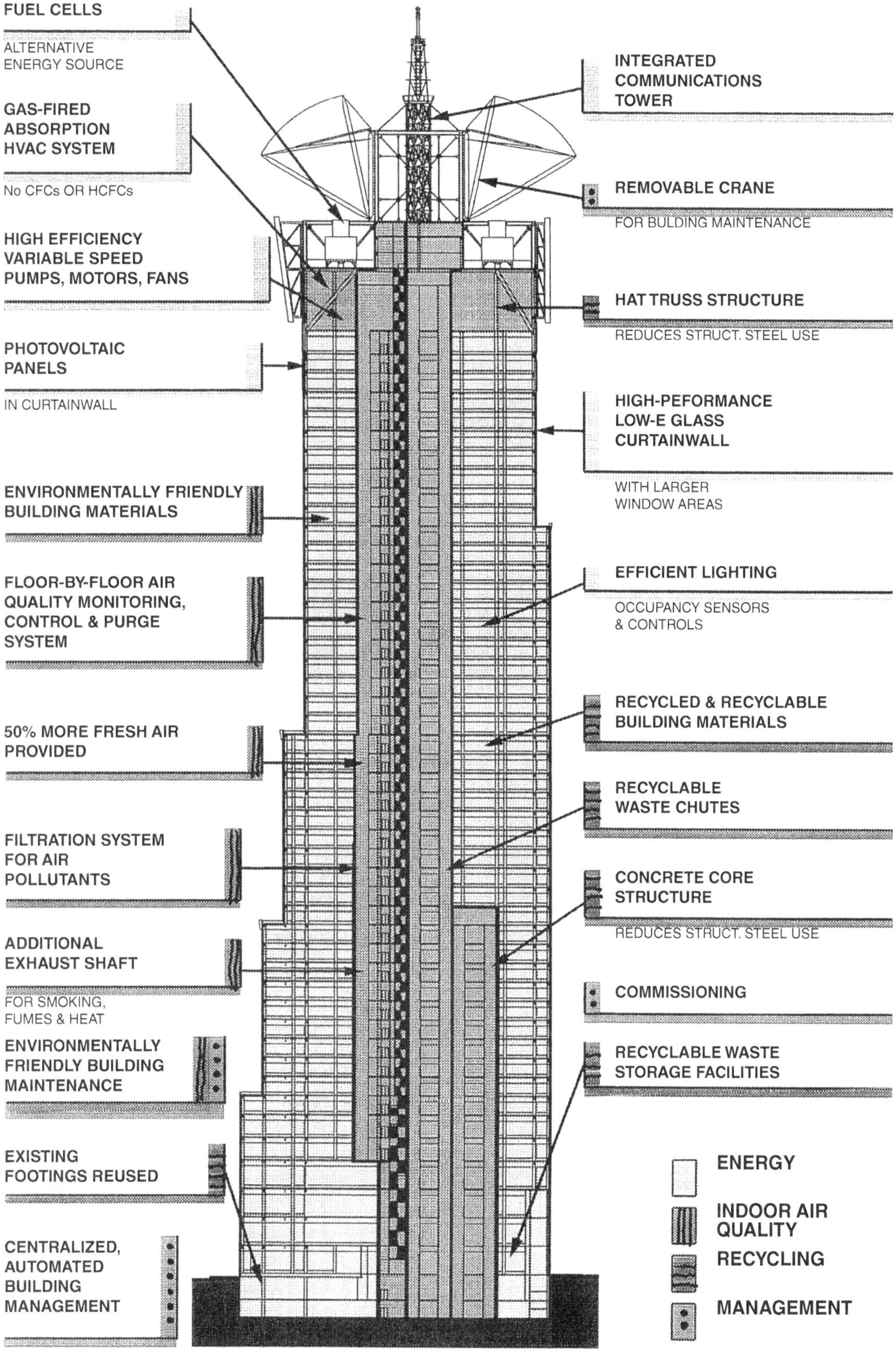

(*c*)

Figure 16.155 *(continued)*

the fourth floor (accessible to maintenance and close to the huge signs), whereas the chillers are located in the penthouse.

The 200-kW fuel cell in Figure 16.154 is usually fed by natural gas, but other examples include hydrogen fuel (Hamburg, Germany: electricity to the grid, hot water to heat an apartment building), anaerobic digester gas (Yonkers, New York: wastewater treatment plant; electricity and hot water used within the facility), and landfill gas (Groton, Connecticut: closed landfill; electricity to the grid). Plentiful hot water is a by-product of the fuel cell; it seems that finding a viable year-round use for this water is often a problem.

The HVAC systems in the Solaire apartments in New York City (Figure 16.156) bring fresh air into individual apartments and exceeded the New York City energy-efficiency standards by more than 35% by using variable speed fans, gas-fired absorption chillers and heaters, occupancy sensors, and optimized indoor air quality, using fresh and filtered air to all units and little or no volatile organic compounds associated with the interior finishes. The base system for both buildings is a gas absorption chiller system with a four-pipe fan-coil for delivery.

Figure 16.156 *The Solaire (Clarke Pelli Clarke Architects) was the first LEED Gold multifamily high-rise, in 2003. (© Photo by Tyler Duncan; used with permission.)*

16.22 CASE STUDY 1—ACTIVE CLIMATE CONTROL SYSTEMS

Ball State University Ground Source Heat Pump System, Muncie, Indiana

Project Basics

- Location: Muncie, Indiana, USA
- Latitude: 40.21°N; longitude: 85.41°W; elevation: 967 ft [295 m]
- Heating degree days: 5642 base 65°F [3176 base 18.5°C]; cooling degree days: 3666 base 50°F [2038 base 10°C] at Muncie airport; annual precipitation 40.1 in. [1019 mm]
- 731-acre university campus, 47 buildings
- Building area: 5.5 million ft^2 [510,967 m^2] of conditioned space
- Completed Spring 2015
- Client: Ball State University
- Design team: MEP Associates, Ball State Facilities Planning and Management

Background. Ball State University's ground-source heat pump (GSHP) system is the largest of its type in the United States. The $82.9 million, multi-phase project allowed the campus to replace its aging heating and cooling infrastructure, cut the campus carbon emissions by about half, save millions of dollars on annual energy costs, and meet the Presidents' Climate Commitment sustainability goals. The university saw the project as a way to advance its sustainability agenda by showing other institutions ways of overcoming barriers to ground-source heat pump implementation. As an educational institution, Ball State also saw the project as a way to engage faculty, students, and staff in greening-the-campus efforts.

Context. Ball State University began evaluating options for replacing its coal- and natural-gas-fired district energy system boilers in 2004. The four boilers, installed between 1941 and 1955, were at the end of their rated lifetimes and would soon require additional pollution controls for emissions compliance. Dissatisfied with the idea of replacing the equipment with new fossil-fuel-burning equipment, the university began looking at cost-effective ways to heat and cool the campus while achieving carbon neutrality goals set by the administration. Intrigued by U.S. Department of Energy research on ground-source heat pump systems, facilities planners began evaluating the viability of using the technology for the university's district heating and cooling needs. District heating/cooling systems are fairly common, as are ground-source heat pump installations. The combination of these technologies is not so common. Thus, Ball State took an informed risk when it constructed the nation's largest ground-source district heating and cooling system.

Design Intent. Ball State's GSHP district heating and cooling system solved practical infrastructure replacement needs and is a direct reflection of the university's commitment to carbon emissions reductions, sustainability, and reduced environmental impact. The design intent included:

- cutting carbon emissions by half (by 80,000 tons [72,575,000 kg] CO_2 per year)
- servicing the same campus buildings as before with the new system
- using the ground surface above the wells for parking and athletics
- reducing campus operations/energy costs
- improving regional air quality by eliminating the release of particulate matter from coal boilers
- minimizing system construction impacts to ongoing campus operations
- minimizing the area of ground required for the well fields
- minimizing environmental impact to underground aquifers

Construction. The project was completed in two phases. Phase 1 began in 2009 and was completed in 2012. This phase included the construction of two well fields, a north energy station, and connecting campus buildings on the north side of campus to the new system. Phase 2 began in 2011 and was completed in 2015. Figure 16.157 shows the two construction phases. The coal boilers were disconnected in 2016. The total project cost was $83 million with $5 million coming from a U.S. Department of Energy grant. Several campus buildings were not connected to the system because they are slated for demolition, and their replacements will be connected to the system at a later date.

Key Design Features

The Ball State GSHP system has four main components: boreholes, energy stations, hot and cold water distribution loops, and campus building interfaces.

Boreholes. Sometimes called "geothermal wells," boreholes are a defining feature of a large-scale GSHP system. To access stable ground temperatures that are unaffected by seasonal weather patterns, ground-source systems must go deep into the earth. Ball State has more than 3600 boreholes located in three separate well fields spread across campus. Each borehole is 6 in. [152 mm] in diameter and approximately 400 ft [122 m] deep on the north side of campus and 500 ft [152 m] deep on the south side of campus (Figure 16.158). A closed loop of 1.25 in. [32 mm] diameter high-density polyethylene (HDPE) tubing is inserted into each hole. Over 1000 mi [1609 km] of tubing were used in total. The holes are then filled with a grout to promote improved thermal conduction between the ground and the tubing. Boreholes are spaced 15 ft [4.6 m] apart, and the tubing from each hole is connected to a main manifold connected to a heat exchanger at the energy station (Figures 16.159, 16.160, and 16.161). Fresh water (no antifreeze is necessary) from the energy station is run through the tubes where it sheds heat to the ground in summer (cooling season) and absorbs heat from the ground in the winter (heating season). The tubing is closed loop, so there is no direct contact between the water and the ground. At the ground surface, the borehole well heads are covered and the wellfields returned to other campus uses.

Energy Stations. In a district energy system (DES), heat and coolth are generated at one or more central energy stations (Figure 16.162) and distributed to individual buildings. Ball State has two energy stations, one at the south end and one at the north end of campus. Each station employs two heat pump chillers that use a vapor compression refrigeration cycle and circulate R134A refrigerant. Heat is exchanged with the closed loop wells (Figures 16.165 and 16.166) to pull heat from the ground or to shed heat into the ground. This heat exchange (in conjunction with heat pumps) allows for simultaneous production of cold water (42°F [6°C]) and hot water (150°F [66°C]) for heating and cooling buildings. Each energy station can produce 38,000,000 Btu/h [11,137 kW] of heating and 2500 tons [8790 kW] of cooling.

District Loops. Hot and cold water generated by the heat pump chillers is sent to campus buildings via hot and cold water supply and return loops (Figure 16.163) in existing tunnels supplemented by new direct-burial piping runs. The original heat distribution to buildings at Ball State was via steam, but this piping network had to be replaced with hot water piping to work with the new GSHP system.

Figure 16.157 *Campus map illustrating the two construction phases, connected buildings, district energy stations, and well fields. (Reproduced with permission from Ball State University.)*

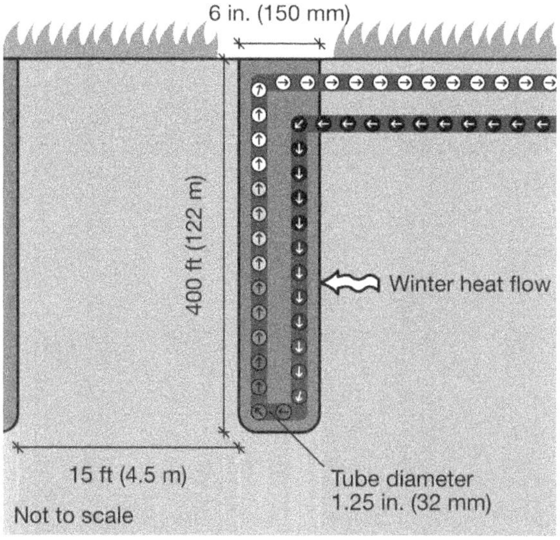

Figure 16.158 *Diagram illustrating the heat exchange process when water from the heat pump chillers is pumped through the closed-loop boreholes. (Drawing by Alisa Kwok after information provided by Ball State University.)*

Figure 16.160 *Borehole wellhead showing the 6 in. [152 mm] diameter HDPE closed-loop tubing. (Reproduced with permission from Ball State University.)*

Figure 16.159 *Borehole construction photo at the south well field. (Reproduced with permission from Ball State University.)*

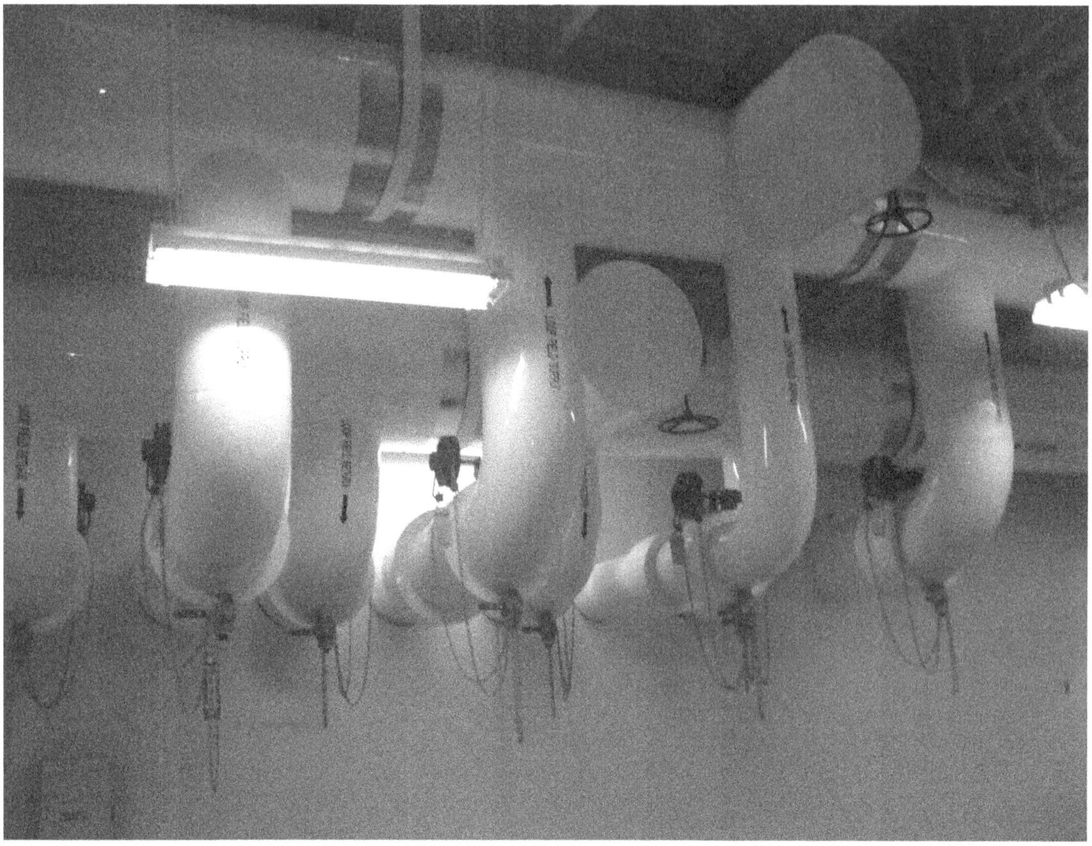

Figure 16.161 *Supply and return well field water headers at the energy stations. (Reproduced with permission from Ball State University.)*

Figure 16.162 *One of four 2500-ton [8790 kW] heat pump chillers used at the district energy stations on campus. (Reproduced with permission from Ball State University.)*

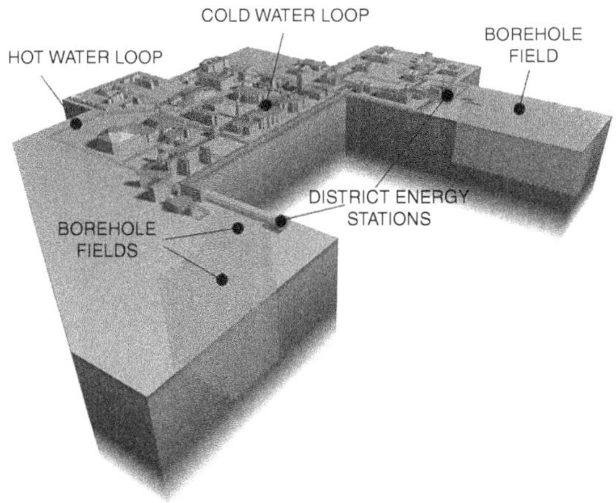

Figure 16.163 District hot and cold water loops connecting buildings to the central district energy stations. (Reproduced with permission from Ball State University.)

Building Interfaces. Heat and coolth from the district loops are exchanged with mechanical equipment at each campus building, which includes air handling, hydronic heating, and domestic hot water equipment. Pumps, variable frequency drives, and heat exchange coils at buildings were replaced or installed as necessary. Each building is connected to a building management system (BMS) that is monitored at the central energy stations (Figures 16.164–16.166).

Performance Data

Information available to date suggests that the Ball State GSHP system is performing well and achieving the original design intents:

- Energy use intensity (EUI) reduction from 175 kBtu/ft^2/yr [552 kWh/m^2/yr] to an estimated 105 kBtu/ft^2/yr [331 kWh/m^2/yr] for the campus as a whole.

Figure 16.164 One of two district energy stations constructed for the project. (Reproduced with permission of Ball State University.)

Figure 16.165 The south well field is a soccer field and is located next to one of the campus energy stations. (© Walter Grondzik; used with permission.)

Figure 16.166 Additional well field, adjacent to the soccer field. (© Walter Grondzik; used with permission.)

- Substantially increased coefficient of performance (COP): an improvement for heating from 0.62 (old coal and natural gas system) to roughly 7.0 (combined heating and cooling with GSHP system); less radical improvement in cooling system COP was also seen (in change from cooling tower heat rejection to ground-based heat rejection).
- Elimination of 36,000 tons [32,600,000 kg] of coal burned annually with associated greenhouse gas reductions and air quality improvements.
- Average $2 million per year saved in energy costs.

FOR FURTHER INFORMATION

How a Pioneering University Hopes to Cut Its Carbon Footprint by Half. *New York Times*, May 29, 2009. archive.nytimes.com/www.nytimes.com/cwire/2009/05/29/29climatewire-how-a-pioneering-university-hopes-to-cut-its-96239.html

Ball State U. to Install Nation's Largest Geothermal System. *Chronicle of Higher Education*, May 4, 2009. chronicle.com/blogs/buildings/ball-state-u-to-install-nations-largest-geothermal-system/5497

16.23 CASE STUDY 2—ACTIVE CLIMATE CONTROL SYSTEMS

The Mercy Corps Global Headquarters Building, Portland, Oregon

Project Basics

- Location: Portland, Oregon, USA
- Latitude: 45.5°N, longitude: 122.6°W
- Heating degree days: 4695 base 65°F [2608 base 18.3°C]; cooling degree days: 551 base 65°F [305 base 10°C]; annual precipitation: 37.1 in. [940 mm]
- Building type: Addition to and renovation of Packer-Scott Building; commercial office space
- Size: 82,800 ft^2 [7618 m^2] gross floor area
- Completed: September 2009
- Client: Mercy Corps
- Design team: Hacker (formerly THA Architecture), Walker Macy (landscape), ABHT Structural Engineers, Glumac (HVAC), David Evans & Associates (civil), American Heating Inc. (HVAC contractor), Thermal Supply (HVAC distributor), Walsh Construction Company (contractor)

Background. Mercy Corps developed their global headquarters as a renovation of the existing Packer-Scott Building (built in 1892) with a new 40,000 ft^2 [3726 m^2] addition (Figures 16.167 and 16.168). The site, located in the economically depressed Old Town area, provided an opportunity not only to unite Mercy Corps headquarters staff but also to provide a catalyst for positive change in the economically challenged neighborhood. Their new headquarters exemplifies the organization's sustainable, community-focused approach.

Mercy Corps' mission is to foster sustainable economic development through collaboration and self-empowerment. The headquarters also includes the Action Center on the ground floor—a window to the world—featuring interactive exhibits that educate visitors about the constantly changing nature of relief and development work and provide suggestions for concrete ways people can act now to help. Also on the ground floor, community education and meeting facilities further extend the headquarters' role within the community.

Context. Mercy Corps, a global humanitarian organization, is dedicated to building physical and human resources in communities that experience poverty, hunger, and war. The collaborative working culture of Mercy Corps synchronized with an integrated design process involving numerous workshops with the contractor, architect, engineers, and design consultants. The integrated design process was part of a critical set of steps leading to building a LEED Platinum building that required highly integrated HVAC, passive, electrical, and structural systems.

Design Criteria and Validation. LEED Platinum was targeted as a goal that would reflect the organization's commitment to sustainable development. Energy was the primary driver of design decisions, motivated by the goal of achieving all ten of the LEED Energy and Atmosphere credits. The motivation to reduce water consumption and adapt the Packer-Scott Building played a major role in determining design criteria and parameters for the new addition.

Design Approaches. From the early design phases, the team sought to implement sustainable design strategies. The reuse of the existing building minimized the

Figure 16.167 *The east façade opens to the adjacent riverside park. (© Jeff Amram, Hacker; used with permission.)*

Figure 16.168 *The historic Packer-Scott Building (built in 1892) offers important contextual and environmental control system cues such as punched window openings on the southern elevation. The new addition has a low window-to-wall ratio and helps to control solar gain in summer and heat loss in winter. (© Jeff Amram, Hacker; used with permission.)*

need for new materials and carbon-intensive construction, and reflected Mercy Corps' values of resourcefulness and sensitivity to culture. An open floor plan helps support a collaborative office culture, while allowing for passive strategies such as daylighting and stack ventilation to achieve improved energy performance.

Design Features

HVAC. A heat-pump system with variable refrigerant flow (VRF) was selected for heating and cooling because of its low initial cost, quietness, adaptability, and ability to provide the necessary thermal zones in the addition and in the renovation of the Packer-Scott Building. There are five VRF loops in the building.

The variable refrigerant flow system was critical to achieving energy performance goals. The layout of the heat exchange systems and the orientation of the building façades have a big impact on energy use. Due to the circulation layout and the nature of the building (an addition onto a renovation), the HVAC system was set up to allow for radiant gains in the south-facing spaces to be transferred to the north side of the building through the VRF loops.

Ventilation. A dedicated outdoor air system (DOAS) works in parallel with the VRF heat pumps. Passive cooling via ventilation is available when conditions are appropriate. The passive ventilation system is coordinated with the active DOAS system. Pathways for stack ventilation and cross ventilation are accommodated by an open vertical circulation core and by operable windows on the north and south enclosures (Figures 16.169 and 16.170), Passive cooling in the building has both automatic and manual aspects. It is mechanically actuated and user-operated, allowing for convenience and occupant control. A CO_2 sensor automatically opens the windows of the roof monitor to provide fresh air when needed, and occupants can control their own comfort and air quality through operable windows.

The ventilation system allows for a simplified sequence of control for the morning warmup. Prior to building occupancy and VRF system startup, the DOAS is converted to a recirculation air system and

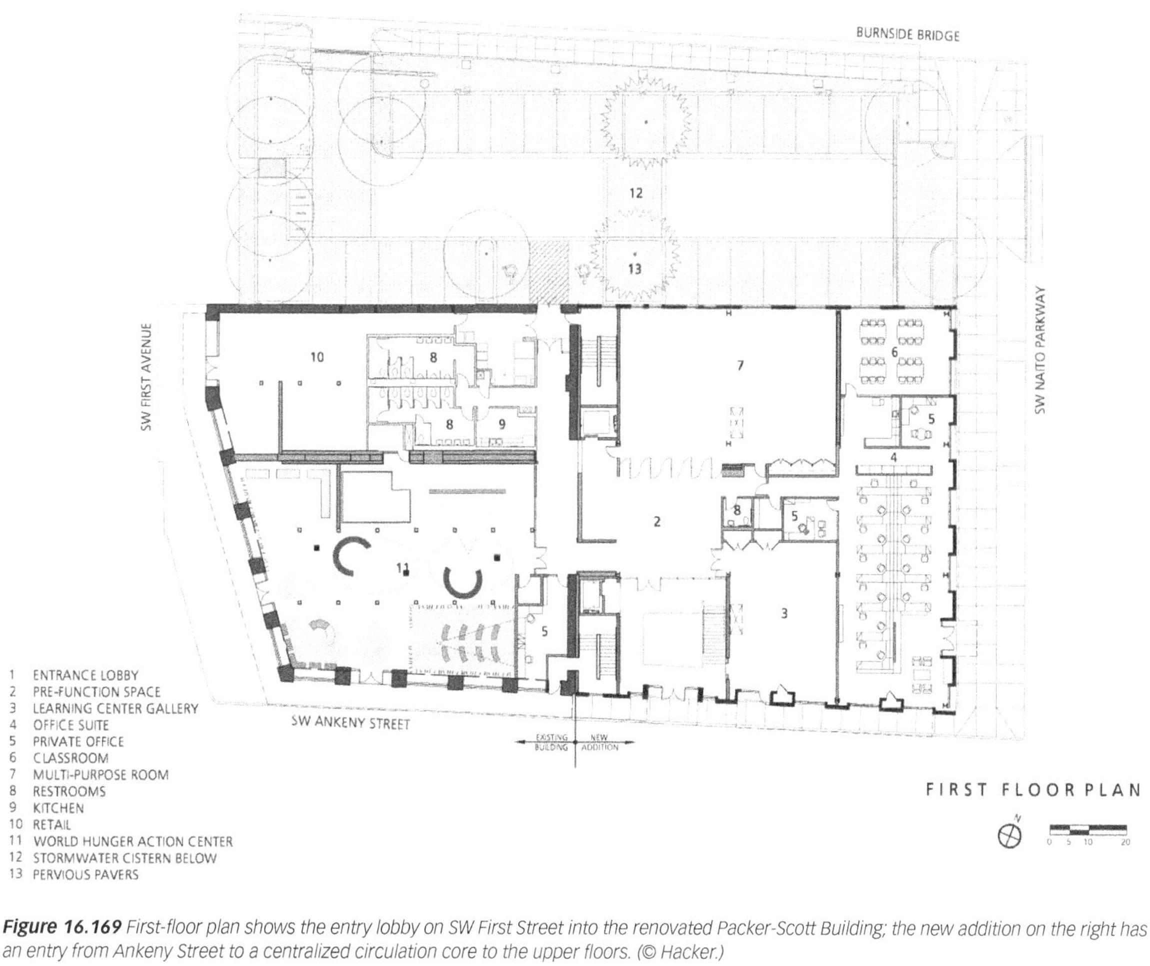

__Figure 16.169__ First-floor plan shows the entry lobby on SW First Street into the renovated Packer-Scott Building; the new addition on the right has an entry from Ankeny Street to a centralized circulation core to the upper floors. (© Hacker.)

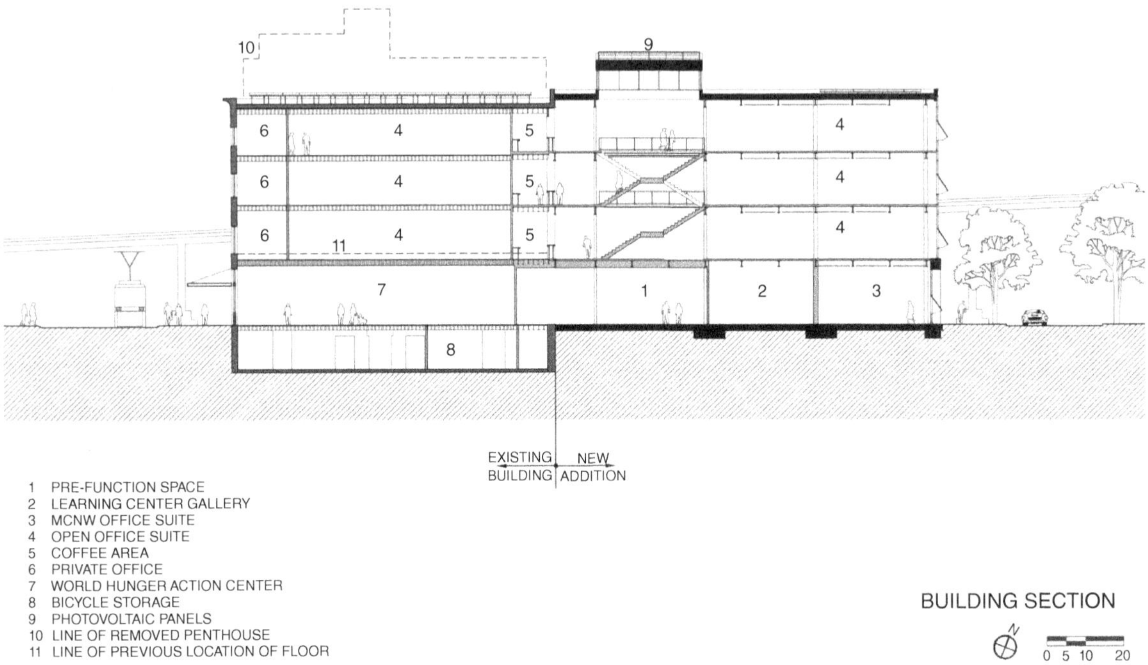

Figure 16.170 *Section showing central circulation core between the second and third floors that allows for stack ventilation through a motorized monitor above. (© Hacker.)*

uses low-cost natural gas to warm up the building from the overnight setback temperatures.

In the same manner, a nighttime flush sequence is implemented during the summer. The design team chose parallel DOAS systems over series DOAS systems for the ease of sequencing.

Occupant Control. Operable windows on all sides of the building allow occupants to open windows. In addition, occupants have a weather data dashboard on their computers to monitor outside conditions—with the idea of improving both comfort and energy performance in the building through informed decision making.

Daylighting. The east–west building massing optimizes solar access and daylighting of office space (Figures 16.171, 16.172, and 16.173). The new addition maintains a similar glazing-to-opaque-wall ratio as the existing historic building's punched window strategy. On the south side, trees provide a first level of solar control that provides light in the winter and shade in the summer. Photovoltaic integrated glass sunscreens shade southern-exposed glass and provide carbon-free power. The glazing strategy for the building includes: (1) a fairly low window-to-wall area ratio, (2) reasonably good window U-factors, (3) reasonably low solar heat gain coefficients, and (4) upgrading window performance in the retrofit portion of the building.

Figure 16.171 *The daylit entry provides a pathway for stack ventilation from the entry up through the central stair core. (© Lara Swimmer, Hacker; used with permission.)*

Figure 16.172 Clerestory windows of the roof monitor above the central stairs are mechanically actuated by CO_2 sensors. (© Lara Swimmer, Hacker; used with permission.)

Enclosure. The upgrades included adding continuous interior insulation of R-10, brick repair, and repointing of the existing masonry enclosure to improve its performance as an air barrier. The wall assembly design in turn drove the window improvements, which included upgrading windows' U-factors.

The enclosure assembly was designed to limit thermal bridging, provide a good air barrier system, and control solar radiation through windows. To limit heat transfer through the structure, slab, ceiling, floor plates, and steel frame, all the thermal control layers of the assembly are located on the outside of the structure.

Water. Water conservation at the Mercy Corps includes the reduction of water use by filtering and retaining storm water on site. Low-flow toilets and faucets reduce water consumption by 44% from typical potable water use, while landscaping is designed with drought-resistant native plants to eliminate the need for irrigation. Storm water is filtered by a 38,000 ft^2 (3530 m^2) vegetated roof that also reduces the air temperature around the building and delays the flow of rainwater into the Willamette River.

Energy. The Mercy Corps building employs a range of passive strategies that run in parallel with a VRF HVAC system to reduce energy loads in the building. The building owner plans to install solar panels on the roof with the intent to further reduce energy dependence, in anticipation of reaching net zero by 2030. As the project progressed, the DOE2-based eQUEST energy model software analysis was updated as the team of architects, engineers, and contractors fine-tuned the performance of the enclosure, window placement, sizing, and operation and electric lighting controls. The building is performing well. There is less natural gas usage than anticipated, suggesting that the ventilation strategy in the winter (using the dedicated outdoor air system for morning startup) is working.

Post-Occupancy Validation Methods. After a year of operation, post-occupancy energy analysis revealed that the measured data were consistent with modeled data (Figure 16.174). Actual gas use is significantly lower than predicted, a result that can likely be attributed to the highly adaptable ventilation rates within the building. This finding further strengthens the energy-saving value of a DOAS.

Performance Data

- Actual EUI: 36,000 Btu/ft^2 [408,960 kJ/m^2] per year
- Electricity (from grid): 34,000 Btu/ft^2 [386, 240 kJ/m^2]
- Natural gas: 2000 Btu/ft^2 [22,720 kJ/m^2]
- Annual actual energy costs: $0.98/ft^2 [$10.55/m^2]
- Percent reduction from national average EUI for building type: 53%
- Lighting power density: 0.74 W/ft^2 [7.98 W/m^2]
- Waste: 87% of the existing historic Packer-Scott Building was reused; 95% of waste was diverted from the landfill through recycling or reuse
- More than 10% of the new materials used were extracted, processed, and manufactured regionally
- Water: Annual water use is reduced by 44%

Awards

- 2009 LEED-NC v2.2 Platinum
- 2011 "Good Design is Good Business Award"
- 2011 National Housing & Rehabilitation Association J. Timothy Anderson ("Timmy") Award for Best Historic Rehabilitation Involving New Construction
- 2012 AIA Committee on the Environment "Top Ten Green Projects"

Figure 16.173 Aerial view showing integration of technologies on the roof: eco-roof, racks ready for photovoltaics to later serve as shading for the eco-roof and energy experiments, and clerestories over atrium. (© Bruce Forster, Hacker; used with permission.)

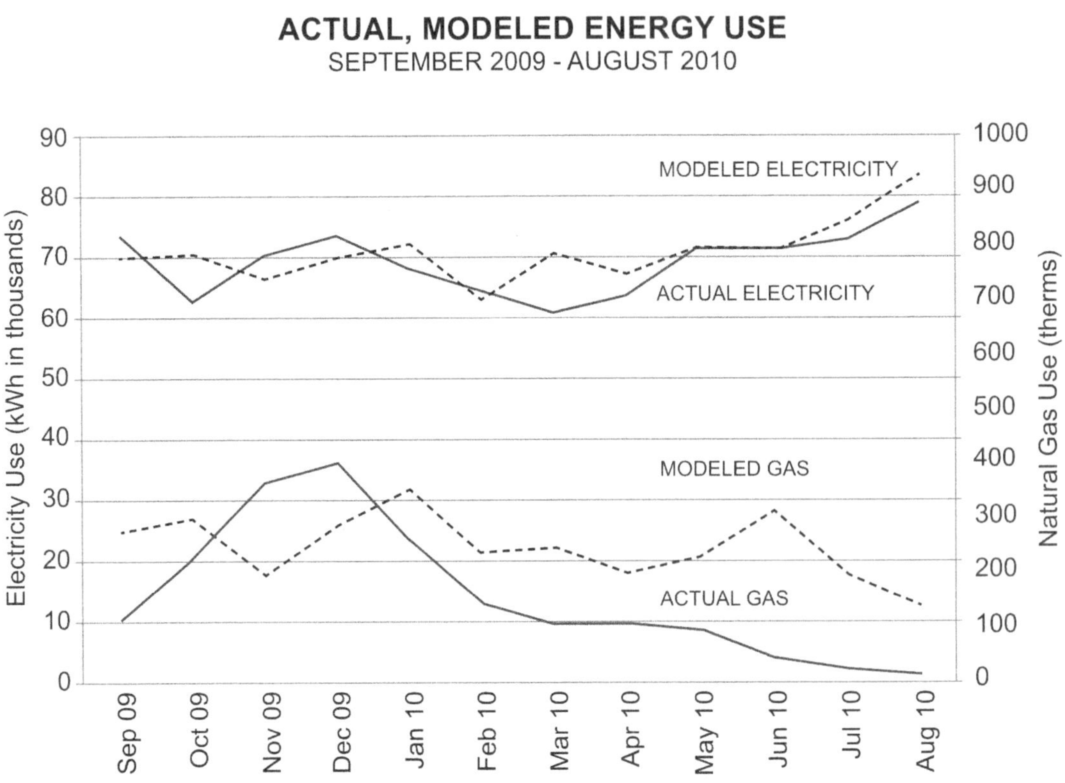

Figure 16.174 Modeled vs. actual energy use for one year. (© Glumac; redrawn with permission.)

FURTHER INFORMATION

Carinci, J. 2009. A Blend of Old and New at Mercy Corps' New Headquarters. *Daily Journal of Commerce*, August.

Dorn, C. and R. Schnare. 2013. "Building with a Mission," in *High Performance Buildings*. Atlanta, GA: American Society of Heating, Refrigerating and Air-Conditioning Engineers.

Eco-Structure Staff. 2012. "2012 AIA COTE Top Ten Green Project: Mercy Corps." *Architect*. www.architectmagazine.com/Awards/2012-aia-cote-top-ten-green-project-mercy-corps_o

Hacker. hackerarchitects.com/office/mercy-corps

Kolleeny, J. 2011. "A Catalyst for Change: The New Headquarters of Mercy Corps Accomplished Multiple Goals for Its Employees, for the Local Community, and for the Environment." GreenSource, May.

Mercy Corps. mercycorps.org

References and Resources

Accuardi, Z. et al. 2022. "Refrigerant Management." Project Drawdown. drawdown.org/solutions/refrigerant-management

Allen, E. and J. Iano. 1989. *The Architect's Studio Companion*, 1st edn. New York: John Wiley & Sons.

Allen, E. and J. Iano. 2017. *The Architect's Studio Companion*, 6th edn. Hoboken, NJ: John Wiley & Sons.

Allen, E. and J. Iano. 2022. *The Architect's Studio Companion*, 7th edn. Hoboken, NJ: John Wiley & Sons.

ASHRAE. 1997. *Handbook—Fundamentals*. Atlanta, GA: American Society of Heating, Refrigerating and Air-Conditioning Engineers.

ASHRAE. 2009. *Indoor Air Quality Guide, Best Practices for Design, Construction, and Commissioning*. Atlanta, GA: American Society of Heating, Refrigerating and Air-Conditioning Engineers.

ASHRAE. 2016. *UFAD Guide: Design, Construction and Operation of Underfloor Air Distribution Systems*. Atlanta, GA: American Society of Heating, Refrigerating and Air-Conditioning Engineers, Inc.

ASHRAE. 2017. *ANSI/ASHRAE Standard 52.2-2017: Method of Testing General Ventilation Air-Cleaning Devices for Removal Efficiency by Particle Size*. Atlanta, GA: American Society of Heating, Refrigerating and Air-Conditioning Engineers.

ASHRAE. 2018a. *Residential Indoor Air Quality Guide: Best Practices for Home Design, Construction, Operation, and Maintenance*. Atlanta, GA: American Society of Heating, Refrigerating and Air-Conditioning Engineers.

ASHRAE. 2018b. *ANSI/ASHRAE Standard 90.2-2018: Energy-Efficient Design of Low-Rise Residential Buildings*. Atlanta, GA: American Society of Heating, Refrigerating and Air-Conditioning Engineers.

ASHRAE. 2020. *ANSI/ASHRAE Standard 135-2020: BACnet - A Data Communication Protocol for Building Automation and Control Networks*. Atlanta, GA: American Society of Heating, Refrigerating and Air-Conditioning Engineers.

ASHRAE. 2021. *2021 ASHRAE Handbook—Fundamentals*. Atlanta, GA: American Society of Heating, Refrigerating and Air-Conditioning Engineers.

ASHRAE. 2022a. *ANSI/ASHRAE/IES Standard 90.1-2022: Energy Standard for Sites and Buildings Except Low-Rise Residential Buildings*. Atlanta, GA: American Society of Heating, Refrigerating and Air-Conditioning Engineers.

ASHRAE. 2022b. *ANSI/ASHRAE Standard 62.2-2022: Ventilation and Acceptable Indoor Air Quality in Residential Buildings*. Atlanta, GA: American Society of Heating, Refrigerating and Air-Conditioning Engineers.

ASHRAE. 2022c. *Standard 62.1: Ventilation and Acceptable Indoor Air Quality*. Atlanta, GA: American Society of Heating, Refrigerating and Air-Conditioning Engineers.

ASHRAE. 2023a. *2023 ASHRAE Handbook—HVAC Applications*. Atlanta, GA: American Society of Heating, Refrigerating and Air-Conditioning Engineers.

ASHRAE. 2023b. *ANSI/ASHRAE Standard 228-2023: Standard Method of Evaluating Zero Net Energy and Zero Net Carbon Building Performance*. Atlanta, GA: American Society of Heating, Refrigerating and Air-Conditioning Engineers.

ASHRAE. 2024. *2024 ASHRAE Handbook—HVAC Systems and Equipment*. Atlanta, GA: American Society of Heating, Refrigerating and Air-Conditioning Engineers.

Ball State University. 2009. "Going Geothermal: University Plans to Install the Largest System in the Country." bsu.edu/news/press-center/archives/2009/2/going-geothermal-university-plans-to-install-the-largest-system-in-the-country

Ball State University. 2013. "Going Geothermal." bsu.edu/about/geothermal

Barden, A. A. and H. Hyytiäinen. 1993. *Finnish Fireplaces: Heart of the Home*, 2nd edn. Helsinki: Finnish Building Centre, Ltd.

Bourne, R. and M. Hoeschele. 1998. "Performance Results for a Night Roof Spray Storage Cooling System," in *Proceedings of the 23rd National Passive Solar Conference*. Boulder, CO: American Solar Energy Society.

Calm, J. E. 1994. "Refrigerant Safety." *ASHRAE Journal* 36(7): 17.

Chandra, S., P. W. Fairey, and M. Houston. 1986. *Cooling with Ventilation*. Golden, CO: Solar Energy Research Institute.

CIBSE. 2021. *TM65 Embodied Carbon In Building Services: A Calculation Methodology*. London: Chartered Institution of Building Services Engineers.

EPA. *List of ENERGY STAR Energy Efficient Products*. Washington, DC: U.S. Environmental Protection Agency. www.energystar.gov/products/products-list

Fitch, J. M. and W. Bobenhausen. 1999. *American Building: The Environmental Forces That Shape It*. New York: Oxford University Press.

Grondzik, W. (ed.). 2007. *Air-Conditioning System Design Manual*, 2nd edn. Atlanta, GA: American Society of Heating, Refrigerating and Air-Conditioning Engineers, Inc.

Guise, D. 1985. *Design and Technology in Architecture*. Hoboken, NJ: John Wiley & Sons, Inc.

Haberl, J. S., J. Baltazar, and C. Mao. 2012. "*Thermal Mass Modeling: How We Got to Where We Are Today*." Energy Systems Laboratory. College Station, TX: Texas A&M University. oaktrust.library.tamu.edu/bitstreams/74d03237-a71b-4e5a-8659-b9617f344b53/download

Hydronics Institute Division of GAMA. 1996. *Installation Guide for Residential Hydronic Heating Systems, 200*. Berkeley Heights, NJ: GAMA.

IPCC. 2023 in *Climate Change 2023: Synthesis Report. Contribution of Working Groups I, II and III to the Sixth Assessment Report of the IPCC*, H. Lee, and J. Romero (eds.). Geneva: Intergovernmental Panel on Climate Change. ipcc.ch/report/ar6/syr

Lieberg, O. S. 1963. *High-Temperature Water Systems*. New York: Industrial Press.

Moore, F. 1994. *Environmental Control Systems: Heating, Cooling, Lighting*. New York: McGraw-Hill Company.

Mozer, M. 1998. "The Neural Network House: An Environment That Adapts to Its Inhabitants," in *Proceedings of the American Association for Artificial Intelligence Spring Symposium on Intelligent Environments*. Palo Alto, CA: Stanford University Press.

NREL. 1988. *Federal Technology Alert: Transpired Collectors (Solar Preheaters for Outdoor Ventilation Air)*. Boulder, CO: National Renewable Energy Laboratory.

Ramsey, C. G. and H. R. Sleeper. 1981. *Architectural Graphic Standards*, 7th edn. New York: John Wiley & Sons, Inc.

Ramsey, C. G. and H. R. Sleeper. 1994. *Architectural Graphic Standards*, 9th edn. New York: John Wiley & Sons, Inc.

Ramsey, C. G. and H. R. Sleeper. 2000a. *Architectural Graphic Standards*, 10th edn (American Institute of Architects ed.). New York: John Wiley & Sons.

Ramsey, C. G. and H. R. Sleeper. 2007. *Architectural Graphic Standards*, 11th edn. New York: John Wiley & Son, Inc.

Ramsey, C. G. and H. R. Sleeper. 2011. "Architectural Graphic Standards," in *American Institute of Architects*, 11th edn. Hoboken, NJ: John Wiley & Sons.

Ramsey, C. G. and H. R. Sleeper. 2016. "Architectural Graphic Standards," in *American Institute of Architects*, 12th edn. Hoboken, NJ: John Wiley & Sons.

Stein, J. and S. Taylor. 2013. "Reheat versus Active Chilled Beams & DOAS." *ASHRAE Journal* 55(5): 18–32.

Rousseau, D. and J. Wasley. 1997. *Healthy by Design*. Point Roberts, WA: Hartley & Marks.

Schubert, R. P. and P. Hahn. 1983. *Progress in Passive Solar Energy Systems*. Boulder, CO: American Solar Energy Society.

CHAPTER 17

Lamps, Luminaires, and Controls

CHOOSING LIGHT SOURCES FOR BUILDINGS—WHETHER DAYLIGHT OR ELECTRIC LIGHT (or, more likely, a combination of both)—involves visual and thermal considerations. In 2020, about 6% of the total electricity consumed in U.S. homes was for lighting. As of 2018, about 17% of electricity consumption by U.S. commercial buildings was for lighting. Energy consumption from electric lighting sources has decreased as a percentage of total energy use over the last decade due to the development and promotion of more energy-efficient light sources such as light-emitting diodes (LEDs). The emergence of new laws and regulatory standards at the federal and individual state level has been instrumental in this transition. While daylight is often the most energy-efficient, abundant, and desirable form of lighting, it's not always possible to integrate sufficient daylight to meet indoor illumination requirements.

In designing appropriate lighting systems for buildings, understanding the characteristics of light sources will allow a designer to use them appropriately to provide visual and thermal comfort while maintaining a minimal energy footprint. For resource efficiency, a designer should first optimize daylighting, and then design the electric lighting system to supplement and enhance illumination, visual comfort, and architectural effect where desired. A lamp is the basic source in an electric lighting system. Lamps are almost always paired with a luminaire (sometimes called a lighting fixture). Luminaires contain one or more lamps (we are including LEDs in this broad definition) along with all the accessory components required for the product to be energized and work. Luminaires are almost always managed by controls. Thus, a chapter that addresses lamps, luminaires, and controls is a good starting point for a discussion of electric lighting.

This chapter begins with a brief historical overview of evolving lamp technologies before describing the characteristics of various electric light sources, including the limits and capabilities of each source. Luminaire components, characteristics, selection considerations, and ratings are then presented. The chapter concludes with a discussion of lighting controls.

ELECTRIC LAMPS

17.1 EVOLUTION OF ELECTRIC LAMPS

Since their emergence over 150 years ago, electric light sources have been under constant development, from the first arc and incandescent lamps of the mid-1800s to Philips' award-winning screw-in LED replacement that offers 85% energy savings over its 60 W incandescent equivalent. Improvements in electrical distribution have been a driving force in the development of many active environmental systems described in this book, but electric light sources—in tandem with air conditioning—have had a profound impact on modern and contemporary architecture.

The late nineteenth century saw a plethora of new electric lamp technologies, but the cost of operating these technologies and their associated electrical supply plants kept them out of reach for all but the wealthiest consumers. It wasn't until growing commercial competition in the early twentieth century resulted in

Mechanical and Electrical Equipment for Buildings, Fourteenth Edition. Walter Grondzik, Daniel Overbey, and Alison Kwok.

Companion website: www.wiley.com/go/meeb14e

the advancement of alternating current "AC" electricity, allowing for long-distance electrical transmission and the birth of our contemporary electrical grid. This growth in electrical supply allowed the lighting market to expand into commercial, mainstream residential, and industrial applications. With it came advances in the efficiency of sources, including the stability and longevity of filaments, glass coatings, and gases. The introduction of linear fluorescent lamps, originally developed in Europe, drove the expansion of commercial U.S. applications to allow for deeper building floor plates and longer periods of indoor occupation. Parallel advances in indoor climate control allowed fluorescent sources to replace perimeter daylight design strategies. By the middle of the twentieth century, commercial and industrial buildings relied almost exclusively on electric light sources to meet general illumination requirements.

Linear fluorescent lamps provided a market-viable lamp for wide-spread commercial use, but it wasn't until the 1973 oil crisis, that energy-efficient lamps came into high demand in residential markets. At this time, researchers at General Electric and Sylvania developed a compact fluorescent lamp composed of a bent fluorescent tube and small hidden ballast. The compact fluorescent lamp (CFL) was brought to market in the 1980s and while initially expensive, offered significantly longer lamp life and higher efficiency. Over the next decade, this technology continued to improve, offering efficient light at a fraction of the energy of incandescent sources.

The next disruptive wave to hit the electric lighting industry occurred in the early 2000s when energy-related legislation put increased pressure on manufacturers to meet more stringent efficiency targets. The updated Energy Policy Act of 2005 (originally adopted in 1992) and the Energy Independence and Security Act, passed in 2007, threatened to decommission the manufacturing of many incandescent sources and promote the development of energy-efficient alternatives. The "L Prize" was announced by the U.S. Department of Energy in 2008 and was intended to drive competition to produce high-efficiency solid-state lamps. In 2011, Philips won the first "L Prize," demonstrating that LED lamps could meet ambitious targets for efficiency, color rendering, and lamp longevity. While LED lamps have existed since the 1960s, this leap from novelty application to mainstream residential and commercial market viability provided the necessary market boost. Since 2008, LED lamps have fallen in price more than 90% and today's lamps are 5–10 times more efficient than conventional incandescent sources.

The next several sections introduce the most commonly used electric light sources, from legacy technologies like incandescent and gaseous discharge to solid-state lamps, which have rapidly transformed the market in recent years. Many of the most energy-efficient LED sources were designed to replace legacy lamps and use the bases and bulb shapes of their predecessors. To contextualize these solid-state technologies, this section will begin with an overview of legacy technologies. While many incandescent and gaseous discharge sources are no longer recommended due to energy considerations, they are still commonly found in existing buildings and provide important technological context for today's more efficient light sources.

17.2 THE INCANDESCENT FILAMENT LAMP

(a) Construction

The standard incandescent lamp has existed in some form or another since the mid-nineteenth century and was the first electric source to sustain emittance for long enough to be considered market-viable. Today's incandescent lamp consists of a tungsten filament inside a gas-filled, sealed glass envelope. Current passing through the high-resistance filament heats it to incandescence, producing light. Gradual evaporation of the filament causes the bulb to blacken and the filament to eventually rupture, leading to lamp failure. Incandescent lamps are available in many bulb and base types, with specific designs to fit a range of luminaire applications (Figure 17.1). To diffuse the light output, most bulbs are coated on the inside with either white silica or neodymium, providing light diffusion with a slight reduction to light output. While white silica produces a color temperature in the standard incandescent range, neodymium filters out more yellow light and produces a color spectrum slightly more in the daylight range. More extreme colored light is available from either coated bulbs or bulbs of colored glass.

The incandescent lamp base is the connection between the socket and the source of electric current. Most lamps are made with screw bases of various sizes, the most common being the medium screw base. General service lamps of 300 W and larger use a mogul screw base. When lamps are placed in precise reflectors or in lens systems where exact positioning of the filament is important, a special base is used. A range of the most common bases are illustrated in Figure 17.1*b*.

(b) Operating Characteristics

The life, output, and efficiency of an incandescent lamp are critically dependent on the voltage being supplied and can be impacted by even a small change in operating voltage. For example, operating a 120-V lamp

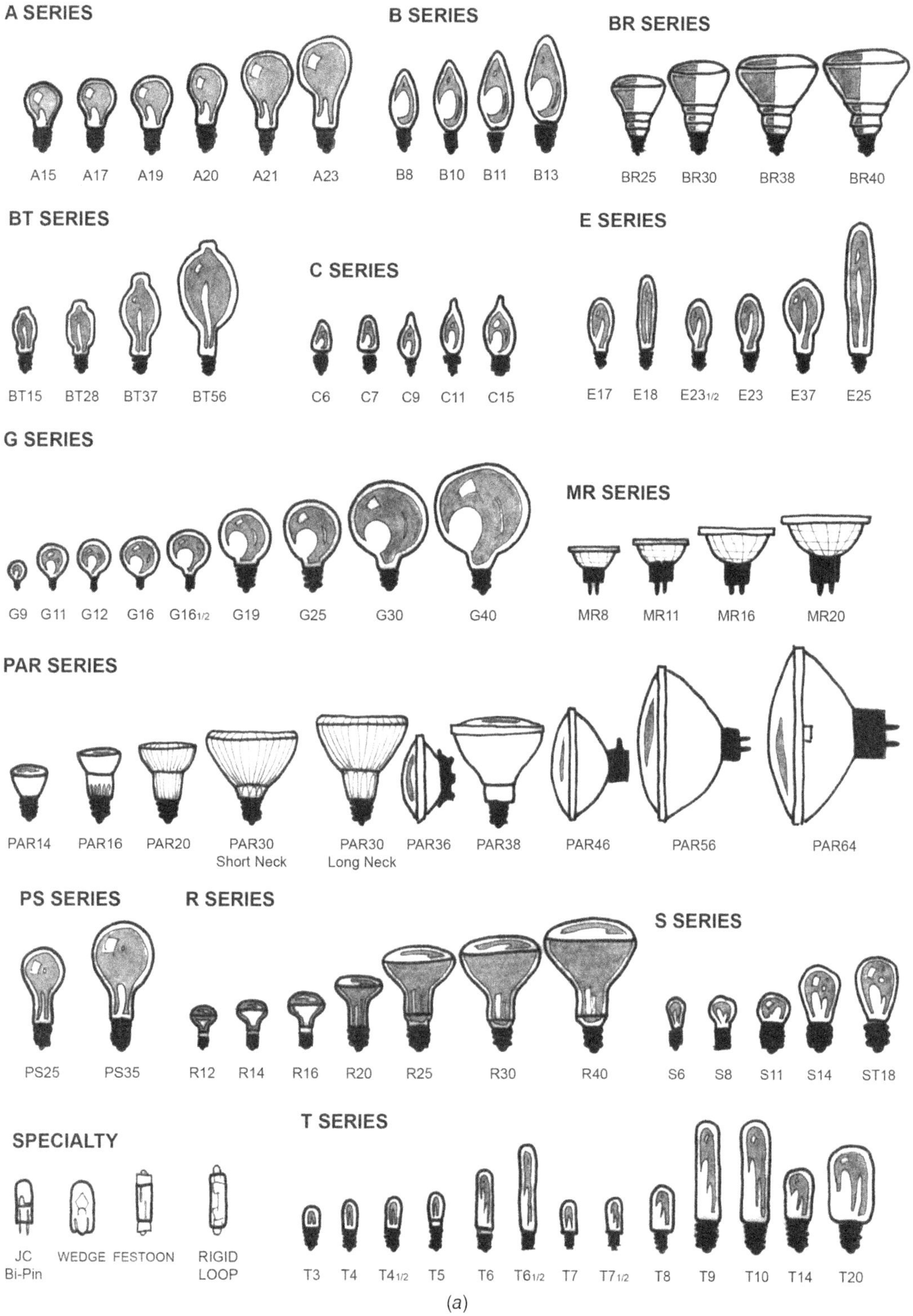

Figure 17.1 *(a) Common incandescent lamp bulb and (b) base types with associated nomenclature. Bulb nomenclature indicates type and size. The letter is an abbreviation of the shape, and the number is the diameter in eighths of an inch. An A-19 is a standard shaped bulb $^{19}/_{8}$ in. in diameter. A PAR-38 is a parabolic reflector lamp $^{38}/_{8}$ in. in diameter. (Drawing by Samantha Ruddy, © University of Oregon, Baker Lighting Lab.)*

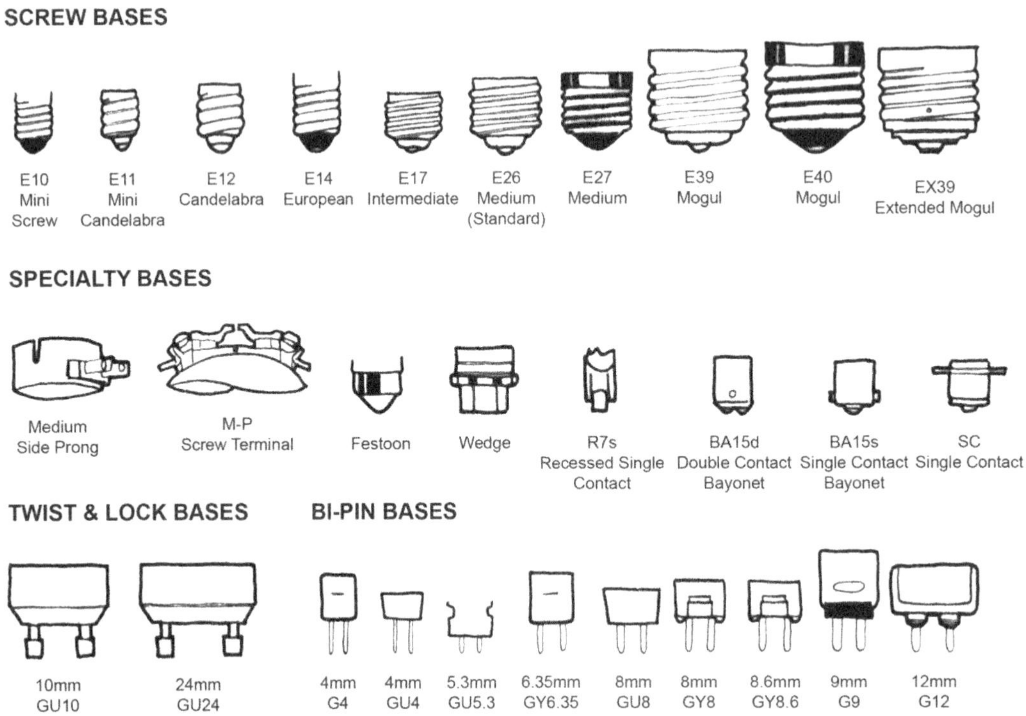

(*b*)

Figure 17.1 *(continued)*

TABLE 17.1 Comparison of Incandescent Operating Characteristics

Operation of Lamps	120-V lamp at 125 V (104.2%)	120-V lamp at 115 V (95.8%)
Amount of light (lumens)	16% more	15% less
Power consumption (watts)	7% more	7% less
Efficacy (lumens per watt)	8% higher	8% lower
Life (hours)	42% less	72% more

at 125 V or 115 V (Table 17.1) affects lumen output and lamp life. Because luminous efficacy (the number of lumens emitted for each watt of electricity used) is decreased by this procedure, a detailed life-cycle cost analysis (as per the principles described in Appendix J) should be made. Conversely, where lamps are replaced before burnout, using a group replacement system, and initial installation cost per lux and/or energy costs are high, lamps may be operated overvoltage, thereby increasing their output and efficacy but shortening their life. Before the emergence of cost-effective LED sources, this procedure was common in sports-lighting installations because of the high installation cost of tower-mounted floodlights, making it mandatory to extract the maximum light from each unit.

(c) Other Characteristics

1. *Lumen maintenance:* Light output decreases slowly with lamp life as an incandescent bulb blackens. Lamp position (vertical or horizontal) during operation and the resulting bulb temperatures affect this characteristic.
2. *Color:* Incandescent light has a large yellow to red component and is therefore highly flattering to the skin. The spectral content of the light produced by a heated source depends upon its temperature a high-wattage lamps are bluer, low-wattage lamps are yellower. Dimmed lamps give yellow–red light.
3. *Surroundings:* Generally, incandescent lamps are impervious to surrounding heat, cold, or humidity. Starting is unaffected by ambient conditions. Some bulbs, however, can be affected by wet conditions.
4. *Luminous efficacy:* Incandescent lamps produce light as a by-product of heat, where most of their output is in the infrared range of the electromagnetic spectrum. As a result, they are inherently inefficient, but luminous efficacy can be slightly improved with increased wattage.

(d) Summary

The principal advantages of incandescent lamps are low upfront cost; instant start and restart; simple and inexpensive dimming; lamp life independent of the number of starts; and skin-flattering color rendering. From a human factor perspective, the full-spectrum quality of incandescent light, with higher amounts of light in the red wavelengths, renders skin tones better than fluorescent sources, although advances in solid-state lamp technologies now offer a more energy-efficient alternative with many of these same benefits.

The principal disadvantages are low efficacy (see Chapter 10, Table 10.1), short lamp life, and critical voltage sensitivity. Low efficacy means more fixtures and larger heat gain than with more energy-efficient alternative sources such as fluorescent and LED. Short lamp life results in high lamp-replacement labor costs. Because of its poor energy characteristics, incandescent lamp use should be limited to the following applications:

- where use is infrequent
- where there is frequent short-duration use
- where minimum initial cost is essential
- where good color rendering and warm color temperatures are desired and use is infrequent

Specific lamp data for use during design development should be obtained from current manufacturers' literature.

17.3 SPECIAL INCANDESCENT LAMPS

In addition to the tungsten-halogen lamp, which is discussed separately, numerous special types of incandescent lamps are still available despite recent advances in solid-state technology. Some of the more well-known types are covered briefly below.

Rough service and vibration lamps are built to withstand rough handling and continuous vibration, respectively. Both conditions are extremely hard on general service lamp filaments. Neither of these types is intended for general use, as both have lower luminous efficacy than a general-service incandescent and neither can outperform an LED lamp.

Extended service lamps are designed for 2500-hour life. They were historically used in locations where maintenance is irregular and/or relamping is difficult. Such lamps are specifically designed for slightly higher voltage than that which is applied, and therefore efficacy is reduced. So-called long-life lamps, which are guaranteed to burn for two, three, or five years, are actually just lamps designed for higher voltages than those listed. As they normally sell at a high cost and are very inefficient, their use is seldom advisable.

(a) Reflector Lamps

These are made in "R," "BR," "ER," and "PAR" shapes (see Figure 17.1*a*). They contain a reflective coating on the inside of the glass envelope that gives the entire lamp accurate light-beam control. Many reflector lamp types are available in narrow or wide beam design, commonly called spot and flood, respectively. Basic reflector (R) lamps are generally made in soft glass envelopes for indoor use, whereas parabolic aluminized reflector (PAR) lamps are hard glass, suitable for exterior application. When using R and PAR lamps, the fixture acts principally as a lamp holder since beam control is built into the lamp. These lamps have an improved reflector design that increases their efficacy.

A number of incandescent reflector lamps are essentially obsolete in the United States because of ongoing rounds of energy legislation. Lamp efficiency and light output limits as shown in Table 17.2 are in place. All major manufacturers produce incandescent reflector lamps that meet current legislative requirements; among these are elliptical reflector (ER) and bulge reflector (BR) lamps that use a more efficient reflector design. Even the most energy-efficient reflector lamps can now be replaced by an LED option that produces an equivalent intensity, color rendering index (CRI), and correlated color temperature (CCT) with a significantly longer lamp life and higher lumens/watt rating.

(b) Energy-Saving Lamps

Major national energy legislation including the National Appliance Energy Conservation Act of 1987, the Energy Policy Act of 1992 (and its 2005 update), and the comprehensive energy legislation passed by the U.S. House of Representatives and the U.S. Senate in 2003 and again in 2007 (through the Energy Independence and Security Act) has created requirements to conserve lighting energy. The American Council for an Energy Efficient Economy (ACEEE), the U.S. Department of Energy (DOE), and the U.S. Environmental Protection Agency (EPA) have developed and support lighting energy efficiency programs, new research, and initiatives.

For example, the ENERGY STAR® program, established by the EPA in 1992 to improve and provide energy-efficient products (appliances, lighting, heating and cooling equipment) and practices, is aimed at reducing emissions from power plants, avoiding the need for new power plants, and reducing energy bills. The EPA continues to update its standard for ENERGY STAR-certified lamps, which must meet a prescribed set of standards for life expectancy, efficiency, color, and performance consistency. ENERGY STAR-certified light bulbs last 15 times longer, use up to 90% less

TABLE 17.2 Efficiency and Light Output Criteria for Lamps (per 2024 ENERGY STAR)

Performance Characteristics		Current Criteria	
		Color Rendering Index	Minimum Lamp Efficacy (initial lm/W)
Efficiency	Omnidirectional	<90	80
		≥90	70
	Directional	<90	70
		≥90	61
	Decorative		65

Light output requirements differ for different categories, based on the incumbent incandescent lamps. Most requirements are measured in lumens, but PAR, MR, and MRX bulbs have requirements in center beam candlepower based on equivalency claim and beam angle. This table is for omnidirectional shapes including A-line:

	Rated Wattage of the Referenced Incandescent Lamp (Watts)	Light Output (Lumens)
Light Output	25	250–449
	40	450–799
	60	800–1099
	75	1100–1599
	100	1600–1999
	125	2000–2549
	150	2550–3000
	200	3001–3999
	300	4000–6000

Source: EPA, 2024.

energy, and produce 70–90% less heat than traditional incandescent bulbs. A publicly available Light Bulb Calculator was developed by EPA and DOE to estimate the energy consumption and operating costs of various types of light bulbs along with the life-cycle savings one could achieve with ENERGY STAR-rated products.

All major manufacturers now produce a line of "energy-saving" incandescent lamps. Energy-efficient lamps are frequently known by trademarked names. They are rated at a lower wattage than that of the standard lamps they are intended to replace and are generally more efficient. Any additional first cost should be analyzed by a life-cycle cost analysis and the payback period calculated. The designer will find that low-wattage incandescent lamps are primarily useful for accent lighting in retrofit work as solid-state sources have become far more energy-efficient and should be considered before any legacy lamp technologies.

17.4 TUNGSTEN-HALOGEN (QUARTZ-IODINE) LAMPS

The tungsten-halogen lamp type, illustrated in Figure 17.2, is conceptually similar to the standard incandescent lamp in that it produces light by heating a filament. It differs in that a small amount of halogen gas (iodine or bromine) is added to the inert gas mixture that fills a small capsule, constructed of quartz glass, surrounding the filament within the bulb of the lamp. This results in retardation of filament evaporation, which is the usual cause of incandescent lamp failure, and thereby extends the lamp life (Figure 17.3).

Although the tungsten-halogen lamp has only slightly higher luminous efficacy than an equivalent standard incandescent lamp, it has the advantages of longer life, lower lamp lumen depreciation (98% output at 90% life), and a smaller bulb for a given wattage (see Figure 17.2). The last characteristic is due to the high temperature required by the halogen cycle, which in turn requires a compact, high-temperature filament. As a result, the lamp is effectively a point source, making it ideal for use in precision reflectors. Indeed, it is in this area that most of the developments in tungsten-halogen lamp technology have occurred.

Due to the lamp's high filament temperature, the bulb envelope is generally made of quartz or a special high-temperature glass, which can withstand high temperatures; this, in turn, gave rise to an alternative name, quartz-iodine, that is sometimes applied to this lamp type. Another result of high filament temperature is that the gas pressure inside the quartz envelope is elevated and these lamps have been known to rupture suddenly, spraying hot quartz fragments. As a consequence, manufacturers provide a cautionary notice with their lamps that includes directions to handle carefully, guard against abrasion and overvoltage operation, and not tamper with any shield or screen protecting the lamp. The shield can be a reflector cover, a fixture lens, a screen, or other device that will contain

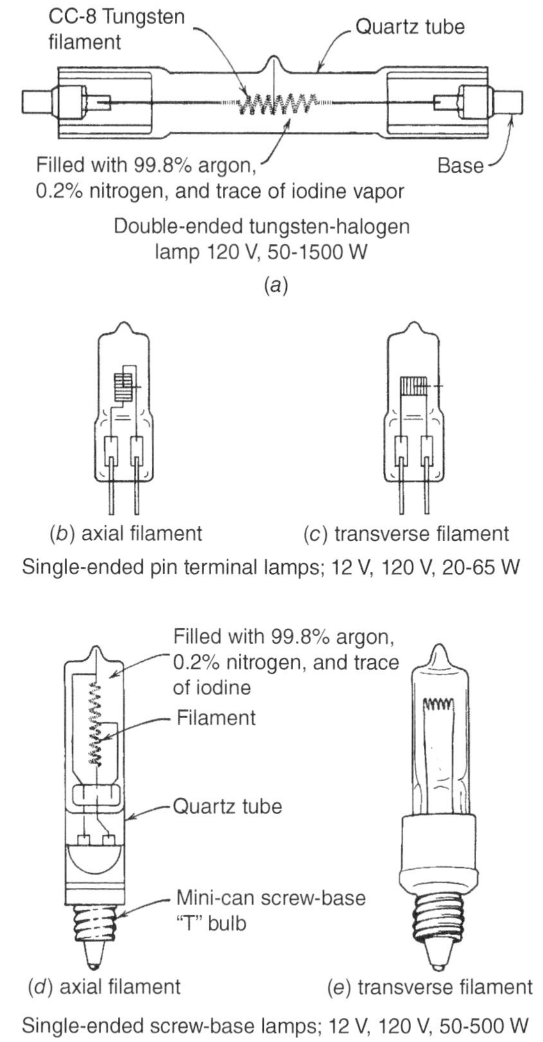

Figure 17.2 *Common tungsten-halogen lamps are available in a variety of designs.*

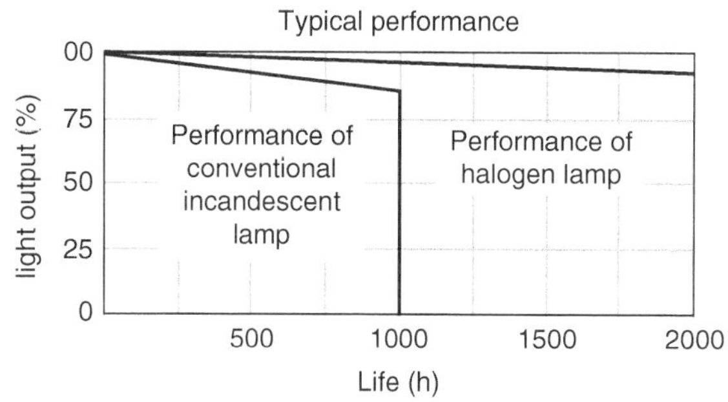

Figure 17.3 *The self-regenerative halogen cycle slows the evaporation of the tungsten filament, and consequently lowers light loss depreciation and lengthens the lamp life compared to a standard incandescent lamp.*

hot flying fragments in the event that a lamp shatters or high temperatures cause a fire hazard.

A number of standards organizations and professional societies have adopted and published cautionary notices, including the Illuminating Engineering Society (IES), the American National Standards Institute (ANSI), the Canadian Standards Association (CSA), and the International Electrotechnical Commission (IEC). Exceptions to these precautions exist when a lamp is protected by encapsulation inside a sealed envelope. Such encapsulated construction is now common in R, PAR, MR, and modified A-lamp shapes. These encapsulated lamps are intended for direct replacement of standard incandescent lamps of the same bulb shape.

Other halogen lamp characteristics are similar in all respects to those of the standard incandescent lamp. Color temperature ranges between 2000 K and 3400 K; spectral energy distribution is typical of blackbody radiation, and dimming characteristics are similar to those of standard incandescent lamps.

The basic tungsten-halogen lamp is a small gas-filled quartz tube, as shown in Figure 17.4. Because the lamp must be used with some sort of reflector, it is manufactured with different terminations to suit the fixture reflector or secondary lamp envelope in which it is placed. All the tungsten-halogen lamp types shown in Figure 17.4 can be used where lamp-only replacement is intended, such as in floodlights or reflector fixtures, by using appropriate bases, slide contacts for double-ended lamps, screw bases for screw-base lamps, and special ceramic pinhole bases for pin-type lamps.

(a) Encapsulated Lamps

These types of tungsten-halogen lamps (Figures 17.4 and 17.5) are sealed units intended for direct replacement of either a corresponding incandescent lamp—in the case of reflector units—or general-service incandescent lamps for reflectorless units. As a sealed unit, the halogen lamp is not replaceable, and the entire unit is discarded on burnout. Reflector units are available in a wide variety of beam patterns detailed in manufacturers' catalogs. Typical distribution data are given in Figure 17.6.

Reflector lamps are also available with a variety of filters: so-called "cool" lamps that direct much of the radiant heat out through the back of the lamp; high-efficiency units that reflect and concentrate radiant heat back onto the lamp filament; lamps with ultraviolet (UV) filters for use in displays of UV-sensitive objects; and others. For complete design information, see current manufacturers' catalogs.

(b) MR-16 Precision Reflector Units

Miniature single-ended 12-V lamps of 2-in. [50-mm] diameter (and smaller), with multifaceted dichroic

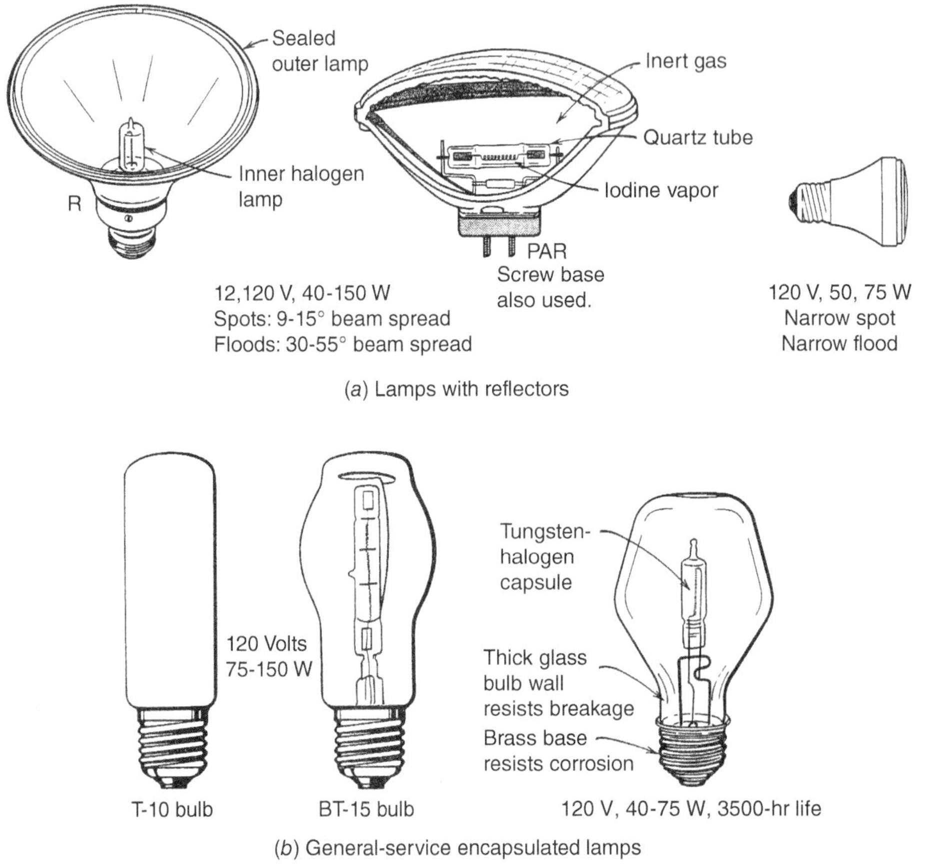

Figure 17.4 *Tungsten-halogen lamps can be mounted in a variety of ways inside an enclosing glass envelope. (a) Lamp is used either horizontally or vertically to the reflector or (b) used without a reflector in a protective glass envelope and an Edison medium screw base.*

Figure 17.5 *PAR halogen lamps with standard medium screw bases in sizes PAR 16, 20, 30, and 38 with a short or elongated lamp neck. Lamp wattages range from 45 to 90 W. (Courtesy of Philips Lighting Co.)*

(heat-ejecting) reflectors and a bi-pin base, found wide acceptance in all types of display and accent lighting applications before the advent of a viable LED alternative. They essentially constitute an entire lighting fixture, like R and PAR lamps, requiring only a base for electrification. The lamps are known by the generic name MR-16, after an early 2-in. [50-mm] diameter model, although each major manufacturer utilizes its own trade name. A multi-mirror-faceted dichroic reflector produces a "cool" precision light beam by ejecting approximately two-thirds of the lamp heat (long-wave radiation) through the back of the reflector. Luminaires must provide adequate means to dissipate this heat to avoid early lamp failure or creation of a fire hazard. The lamps are rated from 20 to 75 W. Beam characteristics of a few MR-16 lamps are given in Figure 17.7.

17.5 BALLASTS

Gaseous discharge lamps function by producing an ionized gas in a glass tube or container rather than heating a filament. Such lamps include fluorescent and high-intensity discharge (HID) lamps (mercury vapor, metal-halide, high-pressure sodium). When compared to incandescent sources, gaseous discharge lamps are

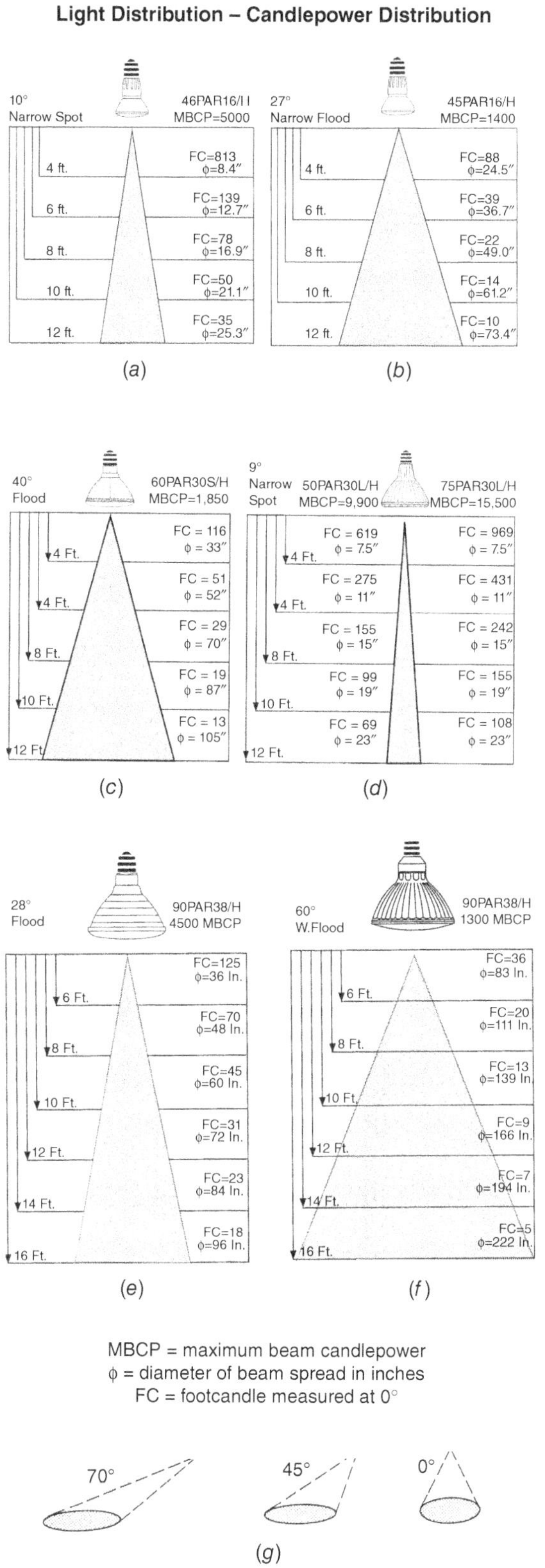

Figure 17.6 *Typical PAR halogen lamp data. The beam of the PAR lamp is conical in shape. Each type of PAR lamp has a distinct illumination pattern (a–f) that varies in size and light intensity, depending on the angle at which the lamp is aimed and its distance from the area illuminated. (g) The round lighting pattern changes to oval or elliptical when the lamp is aimed at an angle, making illuminance calculations much more complex. (Data in (a)–(f) courtesy of Philips Lighting Co.)*

known for their extended life and higher luminous efficacy, though solid-state sources such as LED and OLED now provide even higher luminous efficacies and extended lamp life.

All gaseous discharge lamps require a ballast to trigger the lamp with a high ignition voltage and to control the flow of electric current for proper operation. The ballasts discussed in this section primarily apply to fluorescent lamps. Refer to manufacturers' information for details regarding HID ballasts. Correct matching of the ballast to the lamp is critical to successful lamp operation.

A ballast has several primary functions:

- It supplies the voltage to break down the gas between the electrodes of arc lamps and initiate starting.
- It supplies the voltage and current to heat the electrodes to allow a low-voltage, high-current arc mode to develop (referred to as glow-to-arc transition).
- It supplies enough current to heat and evaporate the light-emitting components after an arc has been established; it provides enough sustaining voltage to maintain the arc during warmup and operation.
- It limits the lamp current once all the evaporable materials have reached thermal equilibrium.

Organizations involved with ballast standards and testing include:

- *American National Standards Institute (ANSI):* This oversees standards on a national level.
- *Electrical Testing Laboratories, Inc. (ETL):* A private, independent organization and recognized authority in measurement and testing of lamps and lighting equipment.
- *Underwriters Laboratories (UL):* An independent, nonprofit organization that certifies electrical products to ensure public safety from fire.

In the United States, ballasts should be UL labeled and ETL certified (the latter for a limited number of fluorescent ballasts). The UL label ensures intrinsic safety and ETL tests ballasts to determine that design standards have been met.

(a) Ballast Types

There are three basic types of ballasts: magnetic, electronic, and hybrid. Ballast technology greatly improved at the turn of the twenty-first century due to changes in energy policy developed by the U.S. DOE, state energy offices, the American Council for an Energy-Efficient Economy (ACEEE), the Alliance to Save Energy, the Natural Resources Defense Council (NRDC), and lighting manufacturers.

(a)

FOOT CANDLE CONES

TRU-AIM TITAN™

10° NSP

Distance from Source (in ft.)	Diameter (in ft.)	Center FC 20W	35W	50W	65W
0′					
3′	0.5	556	922	1278	1556
6′	1.0	139	231	319	389
9′	1.6	62	102	142	173
12′	2.1	35	58	80	97
15′	2.6	22	37	51	62

25° NFL

Distance from Source (in ft.)	Diameter (in ft.)	Center FC 50W	65W
0′			
3′	1.3	356	444
6′	2.7	89	111
9′	4.0	40	49
12′	5.3	22	28
15′	6.7	14	18

40° FL

Distance from Source (in ft.)	Diameter (in ft.)	Center FC 20W	35W	50W	65W
0′					
3′	2.2	78	139	222	233
6′	4.4	19	35	56	58
9′	6.6	9	15	25	26
12′	8.7	5	9	14	15
15′	10.9	3	6	9	9

60° VWFL

Distance from Source (in ft.)	Diameter (in ft.)	Center FC 20W	35W	50W	65W
0′					
3′	3.5	39	72	111	117
6′	6.9	10	18	28	29
9′	10.4	4	8	12	13
12′	13.9	2	5	7	7
15′	17.3	2	3	4	5

(b)

Figure 17.7 (a) MR-16 lamp; (b) Typical illumination cones for a MR-16 lamp. Abbreviations: NSP, narrow spot; NFL, narrow flood; FL, flood; VWFL, very wide flood. (Courtesy of Osram-Sylvania Products, Inc.)

Magnetic. Magnetic ballasts (core and coil) contain a magnetic core of several laminated steel plates wrapped with copper windings; they operate at line frequency (60 Hz in the United States). These ballasts (Figure 17.8) have become essentially obsolete, although they may still be found in older buildings.

Electronic. Solid-state electronic ballasts operate lamps at 20–60 kHz and have less power loss than magnetic ballasts. Electronic ballasts are lighter, are more energy-efficient, generate less heat, and are virtually silent. They are also available with dimming control (see below), which allows light output to be controlled between 1% and 100%.

Hybrid. Also called cathode disconnect ballasts, hybrid ballasts use a magnetic core-and-coil transformer and an electronic switch for the electrode

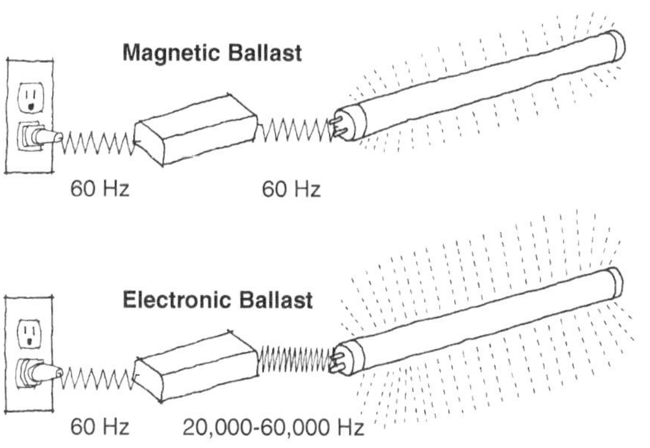

Figure 17.8 *Ballasts for fluorescent lamps have traditionally been of the electromagnetic type, operating at a frequency of 60 Hz. Electronic ballasts operate at frequencies of 20,000–60,000 Hz and cause lighting systems to convert electric power to light more efficiently than systems run by electromagnetic ballasts. (Drawing by Jonathan Meendering; © Alison Kwok; all rights reserved.)*

heating unit. Like magnetic ballasts, they operate at line frequency. The ballast disconnects the electrode heating unit after starting the lamp.

(b) Ballast Characteristics

Because of the considerable energy that is lost in inefficient ballasts, manufacturers and standards organizations have established criteria by which ballast energy efficiency can be judged. These characteristics allow comparisons of lighting system operation and performance parameters.

Ballast Factor. Ballast factor is the measured ability of a ballast to produce light from a connected lamp. It is the ratio of the light output of a lamp when operated on a tested ballast, to the light output of the same lamp when operated by a standard laboratory reference ballast (using ANSI test procedures). Ballasts with extremely high or low ballast factors can reduce lamp life because of inconsistencies in lamp current. *ANSI C82.11-2023: Lamp Ballasts - High Frequency Fluorescent Lamp Ballasts* covers high-frequency ballasts and includes guidance on minimum ballast factors for a certain number of ballast types. A ballast may have different ballast factors for different lamps—for example, one ballast factor for operating standard lamps, and another for operating energy-saving lamps. A lamp with a low ballast factor uses less energy, but also produces lower light output. A lamp with a high ballast factor uses more energy but provides higher light output. Energy savings with high ballast factors may be achieved by using lower-wattage lamps and fewer fixtures. Ballast factor is not a measure of energy efficiency. Although a lower ballast factor reduces lamp lumen output, it also consumes proportionately less input power.

Ballast Efficacy Factor. The ballast efficacy factor is the ratio of ballast factor (as a percentage) to power (in watts); it is expressed in lumens per watt for a given lamp–ballast combination. Comparisons using the ballast efficacy factor are valid only when comparing ballasts for equivalent systems in terms of lamp type and number of lamps.

Because the ballast factor is an indication of the amount of light produced by a ballast–lamp combination, and input watts an indication of power consumed, the ballast efficacy factor is an expression of lumens per watt for a given lighting system. This measurement is generally used to compare the efficiency of various lighting systems.

Power Factor. The power factor of a ballast is a measure of how effectively a ballast converts the voltage and current supplied by a power source into watts of usable power delivered to the lamp. In general, the power factor is determined from the ballast design and is considered high (if above 0.90), low (below 0.79), or "corrected" (0.80–0.90). The power factor addresses the effective use of power supplied to a ballast—it does not relate to the luminous output of a ballast–lamp combination.

(c) Ballast Performance

Heat. Ballast heat is usually transferred to the luminaire body by direct metal-to-metal contact (which must be unimpeded) and is then dissipated by radiation and convection from the fixture. The location and method of fixture installation affect the heat transfer from the fixture and, consequently, the ballast temperature. Operating temperature directly affects ballast life. At normal operating temperature, a ballast life of 12–15 years can be expected. Electronic ballasts will usually start a lamp at 50°F [24°C], but a special ballast is required for starting at temperatures down to 0°F [−18°C]. The more efficient (and thus cooler) operation of electronic ballasts reduces cooling loads.

Noise. All magnetic and some electronic ballasts make a humming sound that originates from electromagnetic action causing vibrations in the steel laminations of the core-and-coil assembly. Because electronic ballasts have a small (or no) core-and-coil assembly, they have the lowest noise output. Most electronic ballasts make almost no sound. Ballast noise, if any, may be amplified by (1) the method of mounting the ballast in the fixture, (2) loose parts in the fixture, and/or (3) ceilings, walls, floors, and hard furniture that reflect the noise. Ballasts are sound-rated by a letter, A through F, which indicates their performance in a space (not the actual sound they produce). A sound rating of A means an average ambient sound level between 20 and 24 dBA and designates the quietest ballast. While manufacturers all use the same ratings, there is no consistent standard that describes the method used to capture this measurement or how it is applied.

Flicker. Flicker is caused by extinguishment and reignition of the arc within a fluorescent tube and is visible only when a lamp is operated at a relatively low frequency (such as the 60 Hz typical of North American electrical systems) and where long persistence phosphors are thin or entirely absent (at the extreme ends of a lamp, for example). A magnetic ballast does not alter the incoming line frequency. Thus, the lamp voltage crosses zero 120 times each second, resulting in 120 light output oscillations per second. Flicker is not noticeable to most viewers, but there is evidence that flicker of this magnitude can cause adverse effects, such as eyestrain and headache. Most electronic ballasts operate at higher than line frequency, which reduces lamp flicker to an essentially imperceptible level. Manufacturers can specify the flicker percentage of a particular ballast working in conjunction with a given lamp type and phosphor composition.

Dimming Control. Dimming of electronically ballasted lamps is accomplished within the ballast itself. Electronic ballasts alter the power input to the lamps through a low-voltage signal into the ballast output circuit. High-power switching devices to condition the input power are not required. This arrangement allows control of one or more ballasts independent of the electrical distribution system.

17.6 FLUORESCENT LAMP CONSTRUCTION

Introduced in 1927, the fluorescent lamp quickly became the best-known and most widely used type of gaseous discharge lamp in the twentieth century. While not as energy-efficient as LED sources—and now being rapidly displaced by them—fluorescent lamps offered a replacement for incandescent sources with increased lamp life and low upfront cost. The typical linear fluorescent lamp comprises a cylindrical glass tube sealed at both ends and containing a mixture of an inert gas, generally argon, and low-pressure mercury vapor. Built into each end of the tube is a cathode that supplies electrons to start and maintain an electric arc, or gaseous discharge. Short-wave UV radiation, which is produced by the mercury vapor, is absorbed by phosphors coating the inside of the tube, causing a reaction that emits visible radiation (light). The particular mixture of phosphors used governs the quantity and spectral quality of the light output. Light from fluorescent sources radiates from a larger lamp surface area than is the case with incandescent sources. The light is diffuse, which is suitable for illuminating or washing large areas such as ceiling planes.

(a) Lamp Type Basics

While most new construction and interior renovations now specify LED sources, rapid-start and instant-start fluorescent lamps (having made preheat lamps obsolete) are still commonly found in existing buildings. Lamp families include linear and compact. Linear lamps are tubular in shape, with the most popular versions in the United States being T8 and T5 [26 mm and 16 mm] in standard and high output (HO) as well as the legacy T12 [38 mm] lamp. The nomenclature is such that "T" indicates "tubular" and the number corresponds with the diameter in eighths of an inch (e.g., T8 indicates a tubular lamp with an 8/8″ diameter, or 1-in.). Compact fluorescent lamps (CFLs) include tubular and helical shapes with a range of base connections. The descriptions that follow cover standard lamps and circuits. Special lamps, accessories, and circuits, including low-wattage lamps, triphosphor lamps, and special-shape lamps, are discussed separately.

(b) Preheat Lamps

Older fluorescent fixtures use a preheat technology that heats the fill gas in order to start the lamp; they use a mechanism called a starter. Fixtures with preheat lamps either have an automatic starter or require a manual starting action. The original T12 [38-mm] fluorescent lamp was a preheat design. Construction of a typical hot cathode lamp (used with both preheat and rapid-start types) is shown in Figure 17.9. All preheat lamps have bi-pin bases.

The lamp shown in Figure 17.9*f* utilizes a separate starter, a small cylindrical device that plugs into a preheat fixture. When the lamp circuit is closed, the starter energizes the cathodes; after a 2–5-second delay, it initiates a high-voltage arc across the lamp, causing it to start. Most starters are automatic, although in desk lamps, preheating may be accomplished by depressing a start button for a few seconds, and then releasing it. This closes the circuit and allows the heating current to flow; releasing the button causes the arc to strike. Preheat lamps are no longer the industry standard but are included here as a point of comparison.

(c) Rapid-Start Lamps

The most popular fluorescent lamp design still in use is the rapid-start lamp, shown in Figure 17.10. This design functions similarly to the traditional preheat lamp, but without a starter switch. Instead, the lamp's

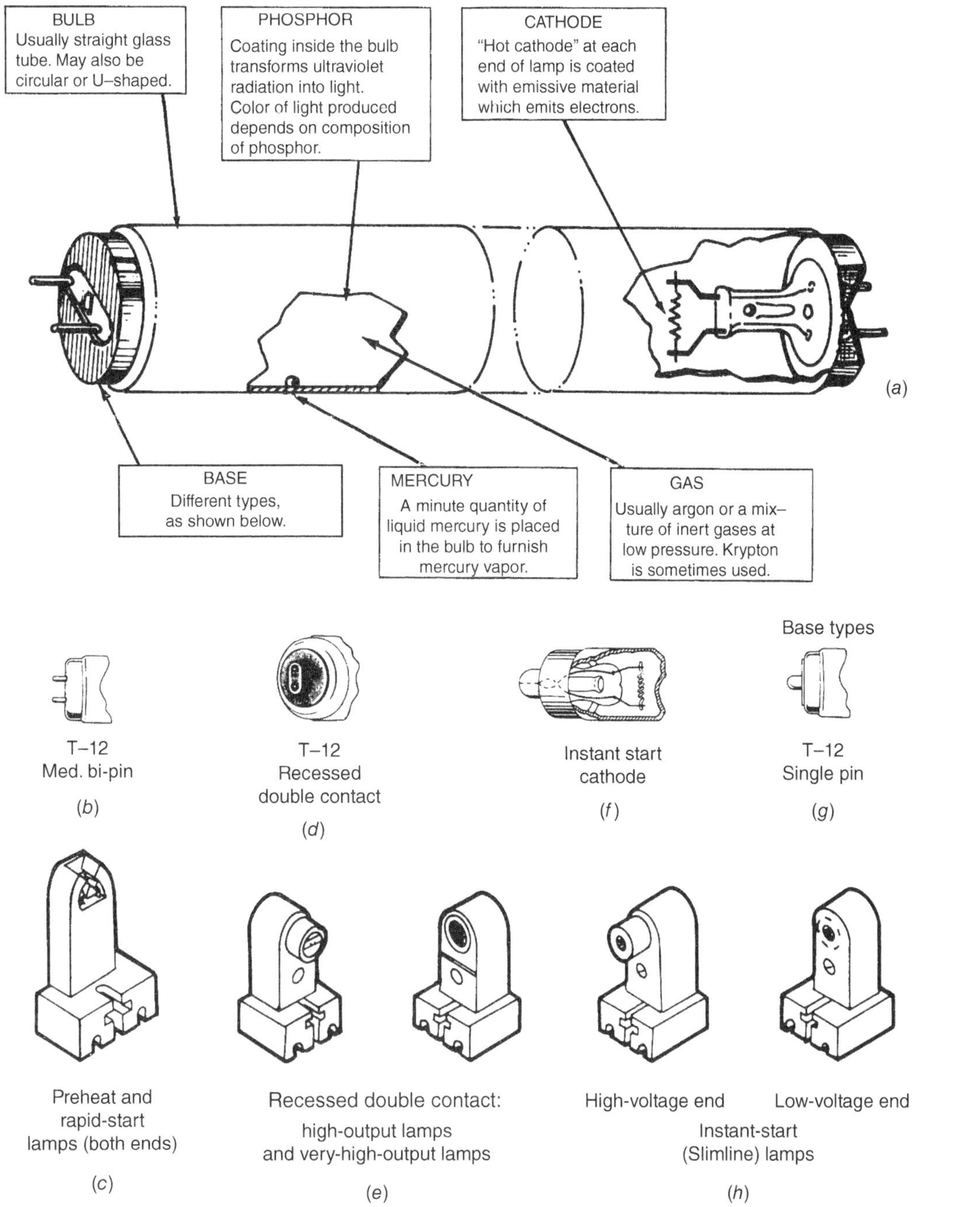

Figure 17.9 *Details of typical fluorescent lamps and associated lampholders. (a) Construction of preheat, rapid-start, bi-pin base lamp. This type of lamp has type (b) base and is held in type (c) lampholder. High output (HO) and very high output (VHO) rapid-start lamps use a recessed double-contact (dc) base (d) and lampholders (e). Instant-start lamps are similar in construction to (a) except with cathode construction (f), have a single-pin base (g), and use single-pin lamp holders (h), which are different at each end. (a: Courtesy of GTE/Sylvania, Inc.)*

ballast constantly channels current through both electrodes, eliminating the delay inherent in a preheat circuit. This current flow is configured so that there is a charge difference between the two electrodes, establishing a voltage across the tube. Most fluorescent fixtures with two or more lamps are rapid-start. When the lamp circuit is energized, the arc is struck immediately. No external starter is required. Because of the similarity of operation, rapid-start lamps will operate satisfactorily in a preheat circuit. The reverse is not true, because a preheat lamp requires more current to heat the cathode than the rapid-start ballast provides.

Most rapid-start T12 lamps operate at 430 mA. If the current is increased, the output of the lamp also increases. Two generic types of higher-output rapid-start lamps are available. One operates at 800 mA and is called simply high output (HO). The second, operating at 1500 mA (1.5 A), is called (by different

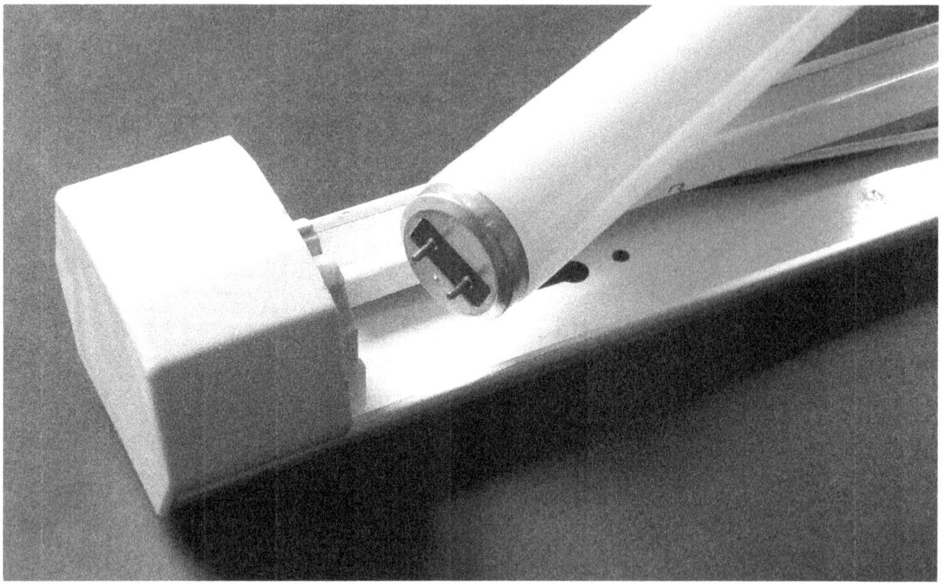

Figure 17.10 *Rapid-start fluorescent lamps have two pins that slide against two contact points in an electrical circuit. The ballast constantly channels current through both electrodes, creating a charge difference between the two electrodes and establishing a voltage across the tube. (Photo by Jonathan Meendering; © Alison Kwok; all rights reserved.)*

manufacturers) very high output (VHO), super high output (SHO), or simply the 1500-mA, rapid-start lamp. There is also a 1500-mA lamp that uses what looks like a dented or grooved glass tube. This lamp has a somewhat higher output than a standard VHO tube. All HO lamps use double-contact bases and special ballasts. Because of the serious heat problems involved, VHO lamps are frequently operated without enclosing fixtures. Conversely, HO and VHO lamps are frequently used in cold environments that would prevent proper operation of a standard output 430-mA lamp. Most HO and VHO lamps have slightly lower luminous efficacy than a standard 430-mA, rapid-start lamp and have a considerably shorter life. SHO lamps are specialized for maximum intensity in short runs. They are typically applied in specialty cases or niche circumstances where extremely high intensity is needed in compact installations—such as outdoor signage in very cold climates where rapid high light output is required.

(d) Instant-Start Fluorescent Lamps

Instant-start fluorescent lamps use a high-voltage transformer to apply a very high initial voltage to the cathodes. An excess of electrons on the cathode surface forces some electrons into the fill gas, which ionizes the gas. This creates an instant voltage difference between the cathodes, establishing an electric arc. These lamps have only a single pin at each end that also acts as a switch to break the ballast circuit when the lamp is removed, thus lessening the shock hazard. The lamps are generally operated in two lamp circuits at various currents; normal currents are 200 and 430 mA. The high voltage starting characteristic of instant-start circuits lowers the lamp life to about half that of a corresponding rapid start lamp. Instant starts have the advantage of being able to start at much lower ambient temperatures (below 50°F [10°C]) than rapid-start circuits. This starting characteristic makes the instant-start lamp and circuit particularly applicable to outdoor use.

17.7 FLUORESCENT LAMP LABELS

Standard fluorescent lamp labels are printed on the end of a lamp and use several identifying letters and numbers, as shown in Table 17.3. The typical labeling is in the form FSWCCC-TDD (each manufacturer has variations on this format). Depending upon the type

TABLE 17.3 Fluorescent Lamp Label Designations

Label	Explanation
F	Fluorescent lamp. "G" means germicidal shortwave UV lamp
S	Style—no letter indicates a normal straight tube; "C" means Circline
W	Nominal power in watts: 4, 5, 8, 12, 15, 20, 30, 40, etc.
CCC	Color. W = white, CW = cool white, WW = warm white, BL/BLB = black light, etc.
T	Tubular bulb
DD	Diameter of tube in eighths of an inch. T8 is 1 in. [nominal 25 mm], T5 is ⅝ in. [nominal 15 mm], etc.

ACTIVE ENVIRON-MENTAL SYSTEMS

of fluorescent lamp, designations for color rendering index and color temperatures are also included on the label.

17.8 FLUORESCENT LAMP TYPES

(a) T8 Lamps

When T8 lamps came onto the market, they afforded designers (and their clients) cost-effective and energy-efficient lighting systems that offered a high degree of flexibility in their application. T8 lamps are 1 in. [25 mm] in diameter; they are available in wattages of 17, 25, 32, and 40 at 2 ft, 3 ft, 4 ft, and 5 ft [600 mm, 900 mm, 1200 mm, and 1500 mm] lengths, respectively. Depending on the triphosphor coating used, T8 lamps come with a color rendering index (CRI) of 75, 82, or 85. Lamp manufacturers have developed a standard designation to indicate the color temperature of a lamp. For example, "30" indicates a 3000 K lamp. T8 lamps are available in 3000 K, 3500 K, 4100 K, and 5000 K color temperatures, designated "30," "35," "41," and "50" respectively.

(b) T5 Lamps

T5 lamp technology was developed in Europe and introduced to North America in 1996. Although it was expensive, the introduction of a T5 HO line in 1998 offered about twice the lumen output in the same length as its T8 counterpart, with an efficacy that was attractive in meeting project energy goals. The T5 was the first "metric" lamp introduced in the United States, yet it is commonly called T5 because of industry nomenclature. Standard and HO T5 lamps are available in 22 in., 34 in., 46 in., and 58 in. [560 mm, 864 mm, 1163 mm, and 1473 mm] lengths. The standard T5 and the T5 HO lamps are the same diameter and width. The 46 in. T5 (nominal 4 ft) is rated at 2900 lumens (lm), similar to the lumen per watt output of a T8 lamp (2950 lm). By contrast, the 46 in. T5 HO lamp is rated as high as 5000 lm, offering twice the maintained light output of a T8 lamp. Because of its smaller $^5/_8$-in. [15 mm] diameter construction, significantly less glass, mercury, and high-quality phosphors are needed for its construction. T5 lamps also allow a designer to use fewer lamps (and fixtures), thus providing certain savings on installation and long-term maintenance. The narrow lamp diameter provided an opportunity for the design of new fixtures and for use in low-profile, indirect luminaires. The color rendering quality of light from T5 lamps (CRI 85) is high. The potential for glare problems exists, however, which can be addressed by sophisticated shielding techniques. Utilizing T5 (and particularly T5 HO) lamps in direct delivery lighting installations requires special attention to glare control.

17.9 CHARACTERISTICS OF FLUORESCENT LAMP OPERATION

Five characteristics define the operation of a fluorescent lamp:

1. *Efficacy:* Light output per unit of power input.
2. *Unstable light output over time:* Output decreases as a lamp ages.
3. *Lamp life:* Average (statistically defined) lamp life expectancy.
4. *Temperature and humidity:* How a lamp responds to environmental operating conditions.
5. *Dimming:* Controlling light production of the lamp.

(a) Efficacy

The design efficacy (lumens per watt) of a fluorescent lamp depends upon the operating current and the phosphors utilized. Fluorescent lamp efficacy is further dependent upon the lamp length, ambient temperature, frequency of the electricity supply, and ballast operation. Wattage, in itself, is a meaningless quantity unless it is associated with a lumen output figure. Thus, energy-saving low-wattage or high-output lamps with their special ballasts are seldom the indicated choice in new design because their efficacy does not usually justify the cost premium.

(b) Lumen Maintenance

The lumen output of a fluorescent tube decreases rapidly during the first 100 hours of burning, and thereafter much more slowly. Phosphors deteriorate, typically blackening the ends of a lamp, thereby blocking some light. Most product catalogs list "initial lumens," which is the lamp output under laboratory conditions after 100 hours of burning, and "mean" or "design" lumens, which is lamp output at 40% of rated life. Unless otherwise controlled, the illuminance provided by these lamps gradually drops as a system ages until, somewhere in the middle of the effective life of the lamps, the intended "maintained" illuminance of a system is (temporarily) achieved.

(c) Lamp Life

The life of a standard fluorescent lamp is defined as the length of time an average lamp is expected to last.

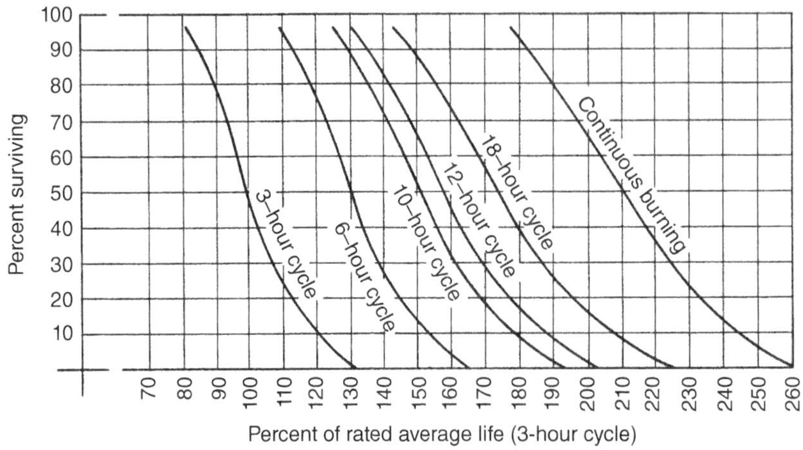

Figure 17.11 *Typical mortality curves of standard fluorescent lamps.*

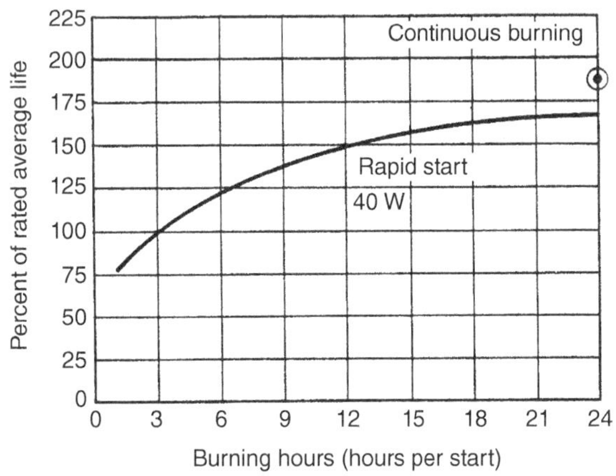

Figure 17.12 *Effect of burning hours on fluorescent lamp life. Note that at 3 burning hours per start, the average lamp life is 100% of the nominal catalog value.*

This value is dependent upon the burning hours per start. Life is expressed as "rated average life" in hours of operation. The life values listed in lamp catalogs are based upon a burning cycle of 3 hours per start (with 20 minutes of "off" status) and represent the average life of a group of lamps; that is, half of the lamps in any group will have burned out at that time. Typical lamp mortality curves are shown in Figure 17.11 and the effect of burning hours per start is shown in Figure 17.12. Average rated lamp life is not the same as the time at which lamps are typically replaced, which is usually well before 50% failures occur in a batch. Several factors affect fluorescent lamp life. Longer burning hours per start will extend lamp life. Lamp life is shortened by improper lamp current, improper voltage to the ballast, or improper cathode heating.

(d) Effect of Temperature and Humidity

Fluorescent lamps are affected by extremes in ambient temperature and by high humidity. Outside of the optimal operating temperature range, 41–77°F [5–25°C], there is a rapid drop in light output and difficulty in starting. High humidity causes electrical leakage along the lamp surface, lowering the starting voltage provided by the ballast. Lamps are precoated with silicone to break up moisture films and prevent such leakage.

The temperature of the coolest point on the wall of the lamp bulb determines a lamp's mercury-vapor pressure, which in turn determines the lamp lumen output, wattage, and color. Maximum output for standard lamps occurs at a bulb temperature of 104°F [40°C]. The bulb wall temperature itself is affected by room ambient temperature, airflow over the lamp (as with air-return fixtures), and the temperature of adjacent surfaces such as a ballast enclosure. Thus, catalog data on lamp output and wattage, based upon laboratory tests of bare tubes at 77°F [25°C] ambient temperature in still air, may be very far from actual field performance.

(e) Dimming

Dimming of a fluorescent lamp system can reduce energy consumption, correct over-lighting, balance illuminance through integration with daylighting, and allow flexibility when full lighting output is required. Unlike incandescent lamps, which can be dimmed with just a wall-mounted device, fluorescent lamps require dimming ballasts. The dimming range differs greatly among ballasts. With most electronic dimming ballasts, output can vary between full and a minimum of about 10% of full output. However, electronic, full-range dimming ballasts are also available for some lamp types and can operate lamps down to 1% of full lumen output.

A ballast can be configured so that (1) it receives a signal from a control device and, subsequently, (2) it changes the current flowing through a lamp, thereby achieving a gradual, controlled reduction in lamp output. The characteristics of the ballast circuitry affect the duration and extent of the change in current and subsequent lamp output.

Electronic dimming ballasts for fluorescent lamps are designed to respond to either an analog or digital signal to achieve the dimming effect. Because a dimming ballast must be able to communicate with connected controllers, the method becomes the basis for a protocol (common operating parameters adopted by all manufacturers of dimming ballasts and controllers that use that method). This ensures interchangeability between a ballast made by a particular manufacturer and various controllers made by controls manufacturers.

17.10 FEDERAL STANDARDS FOR FLUORESCENT LAMPS

A series of U.S. legislative Acts have mandated minimum standards for lamps in terms of efficacy and color rendering index. Cascading standards eliminated the manufacture and distribution of several major fluorescent lamp types with low luminous efficacies. Lamps with very good CRI and special service fluorescent lamps are excluded from the legislative bans.

17.11 UV FLUORESCENT LAMPS

Ultraviolet (UV) fluorescent lamps emit radiation in the ultraviolet spectrum, which includes all electromagnetic radiation with wavelengths in the range of 10–400 nanometers (nm). UV lamps include fluorescent black lights, fluorescent tanning and medical UV lamps, "RS" reflector ("floodlamp") sunlamps, and germicidal and EPROM (erasable programmable read-only memory) erasing lamps.

There are three common categories of UV radiation based on wavelength ranges:

- UVA refers to relatively longer wavelengths from 315 to 400 nm. Wavelengths from about 345 to 400 nm are used for "black light" effects (causing many fluorescent objects to glow). The UVA wavelengths border the visible spectrum, and these lamps are usually slightly visible if isolated from conventional lamps. Shorter UVA wavelengths from 315 to 345 nm are typically used for suntanning beds (along with a small amount from the UVB range).
- UVB refers to wavelengths from 280 to 315 nm. These wavelengths are more hazardous than UVA wavelengths and are largely responsible for sunburn.
- UVC refers to shorter UV wavelengths, usually from 200 to 280 nm. Wavelengths in this range, especially from the low 200 s to about 275 nm, are especially damaging to exposed biological cells. Such short-wave UV radiation is often used for germicidal purposes (either in industry or for indoor air quality mitigation).

Although UV lamps are not commonly used in architectural lighting, they are included in this section because they address a wide range of applications in industrial, technological, agricultural, laboratory, and medical settings.

17.12 COMPACT FLUORESCENT LAMPS

When compact fluorescent lamps (CFLs) first emerged on the market, they offered a comparable (in brightness), but energy-efficient alternative to incandescent lamps. Unlike standard fluorescent lamps, they can directly replace standard incandescent bulbs through a range of universal screw-base connections. Advances in solid-state sources now outperform CFLs in both efficacy and color rendering with the same range of screw- and specialty-base connections.

CFLs are simply folded fluorescent tubes with both ends terminating in a common base. Some compact fluorescent lamps have the tubes and ballast permanently connected with a screw-in medium base. Others have separate tubes and ballasts, allowing the tubes to be replaced without changing the ballast. As a result, an exhausted lamp is simply replaced in the existing ballast, resulting in considerable economy. A CFL produces a diffuse light distribution, unlike single-point incandescent lamps. This is an important factor to consider when replacing incandescent lamps with CFLs in high-ceiling applications.

CFLs are manufactured in a variety of styles or shapes: two, four, or six tubes; circular or spiral tubes (Figures 17.13 and 17.14). They are efficient at lower wattages and can produce light output equivalent to that of higher wattage incandescent lamps (e.g., a typical 60 W incandescent lamp with a 900-lm output could be replaced by a 13 W to 16 W CFL). The total surface area of the tube(s) determines how much light is produced. The efficacy of lamp–ballast combinations ranges from 50 to 70 lm/W, assuming an electronic ballast. Lamps with magnetic ballasts are available but are not favored because of excessive heat, weight, flicker, and reduced efficiency. Lamp colors are similar to those of straight fluorescent lamps (i.e., 3000 K, 3500 K, 4200 K, and 5000 K). All CFLs have a CRI of 80 or higher. Their life is about 6000–15,000 hours, which is 6–15 times longer than the typical 1000-hour

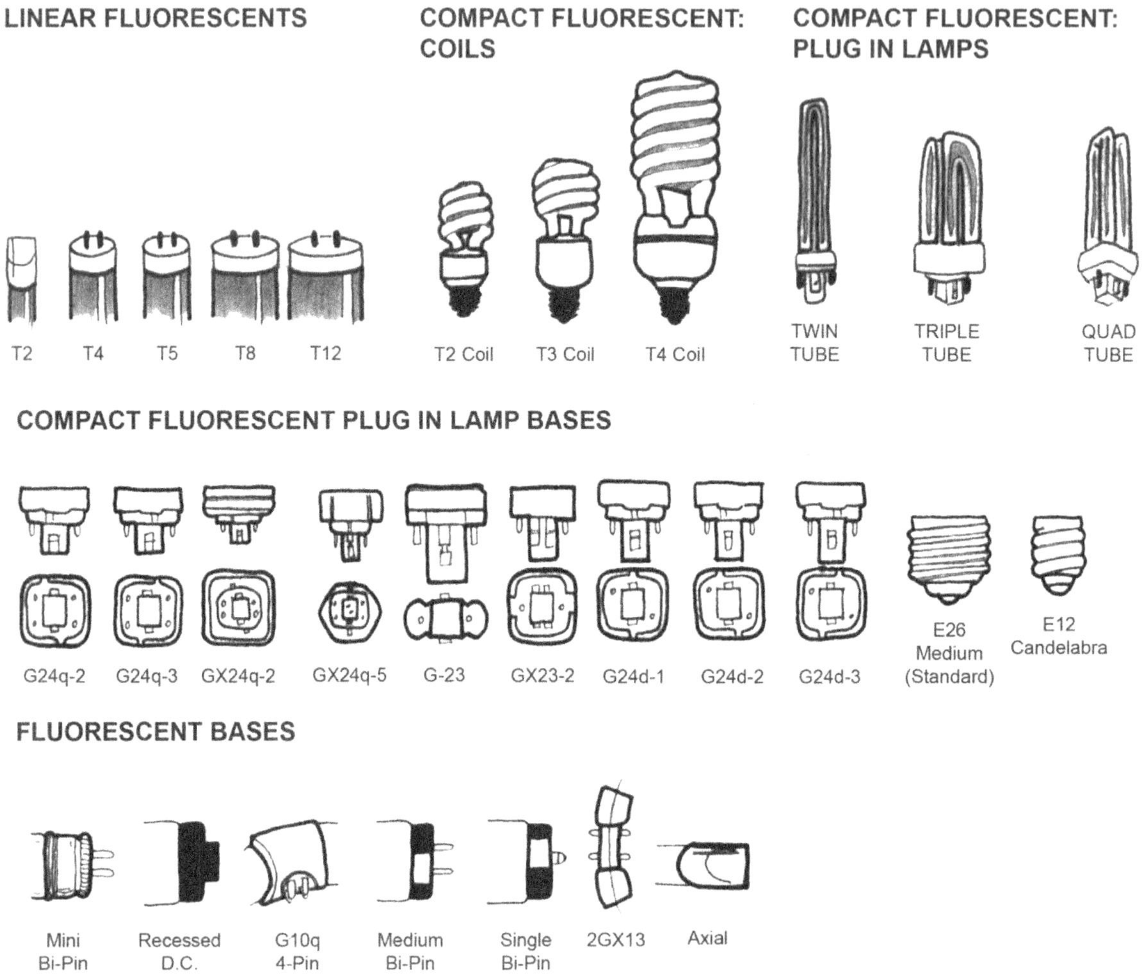

Figure 17.13 *A sample of linear and compact fluorescent lamp designs with associated lamp bases. (Drawing by Samantha Ruddy, © University of Oregon, Baker Lighting Lab.)*

life of an incandescent lamp. Table 17.4 compares the wattage of commonly available incandescent lamps to that of a CFL that provides similar light output.

A major advantage of using CFLs over traditional incandescent lamps is cost savings, as shown in Table 17.5. This analysis assumes that a lamp is on for 6 hours per day and that the electric rate is 8 cents per kWh.

17.13 HIGH-INTENSITY DISCHARGE LAMPS

High-intensity discharge (HID) lamps (Figure 17.15) produce light by discharging electricity through a high-pressure vapor. Lamps in this category include:

- mercury-vapor (CRI range 15–55)
- metal–halide (CRI range 65–80)
- high-pressure sodium (HPS) (CRI range 22–75)

These lamps are characterized by high luminous efficacy, defined warmup time, noticeable restrike time, and historically poor color rendering capabilities. Mercury-vapor lamps were the first commercially available HID lamps, and the early products produced a bluish green light. Today they are available with a color-corrected whiter light; because of inefficiency and potential hazards, however, such lamps are being replaced by newer and more efficient metal-halide and high-pressure sodium (HPS) lamps. Standard HPS lamps have the highest efficacy of all HID lamps, but they produce a yellowish light. HPS lamps that produce a whiter light are available, but their luminous efficacy is somewhat lower.

HID lamps were developed for use when high illuminance is required over large areas, and when energy efficiency and/or long life are desired. Typical applications include gymnasiums, large public areas, warehouses, outdoor activity areas, roadways, parking lots, and pathways.

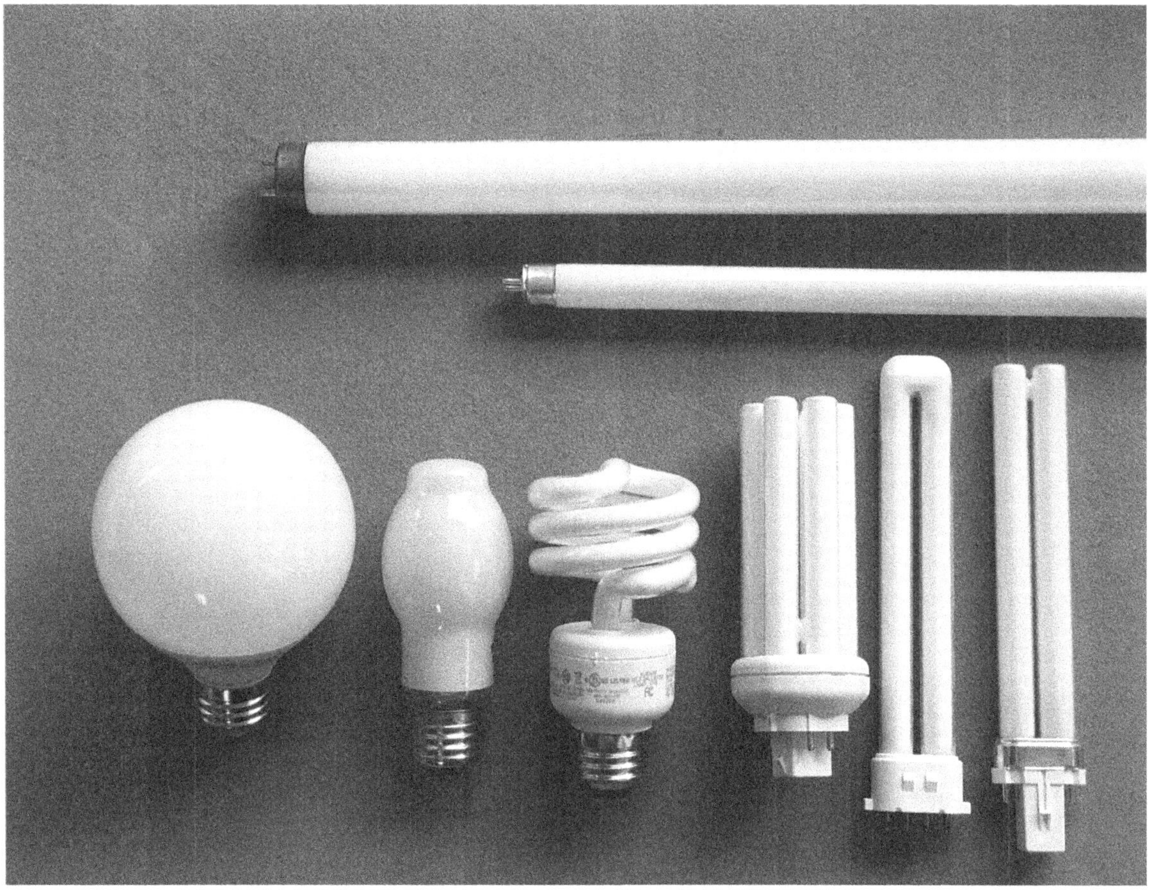

Figure 17.14 Collection of linear and compact fluorescent lamps. (© University of Oregon, Baker Lighting Lab.)

TABLE 17.4 Equivalent Wattage of Common Incandescent Lamps and Compact Fluorescent Lamps

Incandescent (W)	Compact Fluorescent (W)
50	9
60	15
75	20
100	25
120	28
150	39

Source: DOE, 2008.

TABLE 17.5 Cost Comparison for Operation of an Incandescent Lamp and a Compact Fluorescent Lamp

	60 W Traditional Incandescent	15 W CFL
Rated life (hours)	1000	10,000
Annual energy cost (@8¢/kWh for 6 h/d)	$4.80	$1.20
Annual energy savings (%)	—	75

Source: DOE, 2018.

17.14 MERCURY-VAPOR LAMPS

Invented in 1901, the mercury-vapor lamp was the first HID lamp to be developed, and for many years it was the only commercially available HID lamp. It has largely been supplanted by the metal-halide lamp because of the latter's better color rendering and luminous efficacy. Although installations with mercury-vapor lamps still exist, this technology is no longer specified for new buildings. For the same energy efficiency, metal-halide lamps have better color rendering properties and are often preferred for indoor applications. Mercury-vapor lamps, historically often used to light streets, gymnasiums, and sports arenas, must be maintained properly to be safe.

A mercury-vapor lamp operates by passing an electric arc through high-pressure mercury vapor contained in a quartz arc tube (Figure 17.16). This produces radiation in both the UV region (as in a low-pressure fluorescent lamp tube) and the visible region, principally in the blue–green band. This color mix is characteristic of a clear mercury lamp.

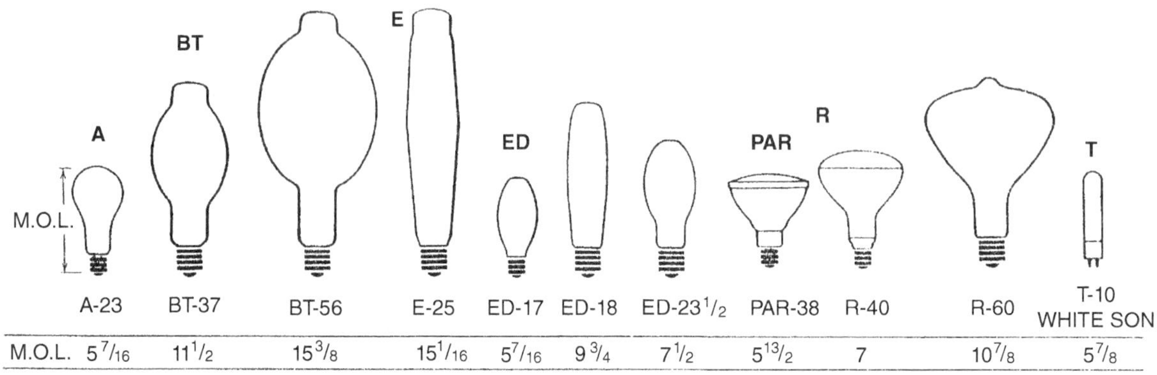

Figure 17.15 *Bulb shapes for typical HID lamps with their maximum overall length (M.O.L.).*

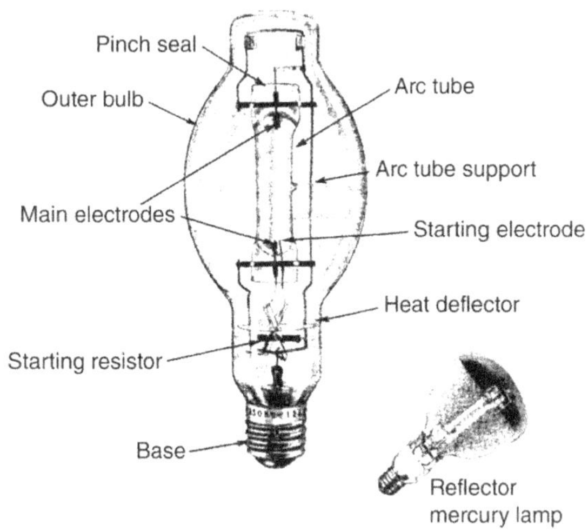

Figure 17.16 *Construction of a typical clear mercury-vapor lamp.*

(a) UV Radiation

A considerable portion of a mercury-vapor lamp's radiation spectrum is in the UV range. This does not normally constitute a hazard to people because the outer glass bulb (even a clear bulb) absorbs most of the UV radiation while transmitting light. If the outer bulb is broken, however, the quartz arc tube will continue to burn, and the unfiltered UV radiation becomes a safety hazard, particularly to the skin and eyes. As a result, manufacturers include a warning about this condition with all mercury lamps sold.

(b) Lamp Life

Mercury-vapor lamp life is extremely long, averaging 24,000 hours or more based upon 10 burning hours per start. Mercury-vapor lamps are not suitable for applications that are subject to constant switching. Their life is affected by ambient temperature, line voltage, and ballast design.

(c) Lumen Maintenance

Lumen maintenance performance depends upon the specific type of lamp and its burning position. Manufacturers publish data for each of their lamp types. In general, clear lamps have the best lumen maintenance, followed by color-improved and phosphor-coated units.

(d) Color Correction and Efficacy

Color correction is normally required because the blue–green light from a clear mercury-vapor lamp distorts almost all object colors. (These lamps are frequently used to illuminate outdoor gardens because the blue–green light enhances the green of trees and vegetation.) Color correction is achieved by adding phosphors to the inside of the outer bulb. The phosphors convert UV radiation to light exactly as in fluorescent lamps, and the glass acts as a filter to some of the blue–green radiation. The phosphors reradiate generally in the red band, which is entirely absent in a clear lamp. Depending upon the arc tube design and the phosphors used, the color of the emitted light can be corrected to make it acceptable for general indoor use. Lamps are available in clear, white, color-corrected, and deluxe white, in ascending order of color improvement.

(e) Ballasts and Lamp Starting

Ballasts are required, as with all gaseous discharge lamps, to start a mercury-vapor lamp and thereafter to control the arc. From 3 to 6 minutes are required for a lamp to reach full output because heat must be generated by electron flow to vaporize the mercury in the arc tube before the arc will strike. Once extinguished (turned off), a lamp must cool before restrike is possible. This restart delay amounts to several minutes, depending upon the ballast type, and is an important consideration in design, as a momentary power outage will extinguish all lamps, leaving an interior area in the dark.

The principal mercury-vapor ballast types are reactor, regulating, and electronic. Magnetic mercury ballasts are large, heavy, and quite noisy. Where this may be a problem, remote ballast mounting should be considered or lighter, quieter, and more expensive electronic ballasts used. Because lamp-operating characteristics depend heavily upon the type of ballast and because the choice of an appropriate ballast involves highly technical electrical considerations, selection should be left to an electrical engineer or lighting consultant.

(f) Self-Ballasted Lamps

These have been available for some years and consist of a screw-base color-corrected lamp with an internal resistive/reactive ballast. They have a CRI of 50, an efficacy of 20–25 lm/W, and a correlated color temperature (CCT) of 3500 K to 4000 K. Their great advantage is their long life, which can be used beneficially in applications involving: burning periods of 8–10 hours minimum, relative inaccessibility, limited space that precludes ballast installation, and indifferent color rendering requirements.

(g) Application

Mercury-vapor lamps were developed to withstand both indoor and outdoor use, with proper attention to performance characteristics. The most common exterior application is for parking lots. Indoor application is generally limited to mounting heights of 10 ft [3 m] AFF (above finish floor) or higher, to avoid direct glare potential and permit adequate floor area coverage. Their use in industrial spaces and stores was once common, but today use of metal-halide lamps is typical. Warehouses and non-color-sensitive industrial areas continue to use mercury-vapor lamps.

17.15 METAL-HALIDE LAMPS

The metal-halide lamp began its life in the early 1960s as a modified mercury-vapor lamp in which rare earth metal salts were added to improve performance. Major advances in miniaturization, color rendering, color temperature, and consistency—by the addition of halides such as thallium, indium, and sodium to the arc tube—resulted in changes in the output, efficacy, color, and life of the lamp. Metal-halide lamps have excellent color characteristics. Pulse-start metal-halide lamps utilize a glass arc tube to contain the arc. Pulse-start technology includes a new family of lamps (it has been used with high-pressure sodium lamps), and improves the start system, efficacy, and lumen maintenance, while also yielding shorter warmup and restrike times. The number of types and sizes is so large, and changes so rapid, that any abbreviated tabulation would be inadequate or misleading.

Ceramic metal-halide lamps were an industry standard for many years (up until the advent of LED alternatives), offering a high CRI (80–90), a color temperature of 3000 K or 4100 K, improved lumen maintenance, and stable color consistency. Typical metal-halide lamp characteristics and types are discussed in the following subsections. Always consult a manufacturer's catalog for current and accurate lamp data.

(a) Lamp Configurations

Construction details for a basic metal-halide lamp are shown in Figure 17.17 and are similar to those of its "parent" mercury-vapor lamp. In addition to this design (with a BT-shaped bulb), metal-halide lamps are manufactured in elliptical bulbs, PAR reflector lamps, and single- and double-ended tubular shapes (Figures 17.17 and 17.18). Figure 17.19 presents details of a PAR enclosure for a metal-halide lamp.

(b) Safety

Being essentially a modified mercury-vapor lamp, a metal-halide lamp carries the same safety warning as a mercury lamp. An additional warning, however, refers to the fact that metal-halide lamp arc tubes have a tendency to explode; therefore, the lamp must be used in an approved enclosing luminaire. All major manufacturers also make lamps with internal shields that will contain the flying pieces of a ruptured arc tube without damaging the outer bulb.

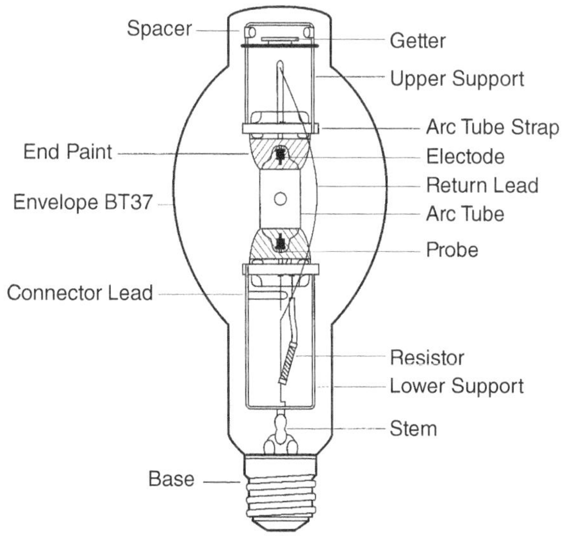

Figure 17.17 Construction details of a 400-W standard metal-halide lamp, which can be mounted either horizontally or vertically. (Courtesy of OSRAM Sylvania.)

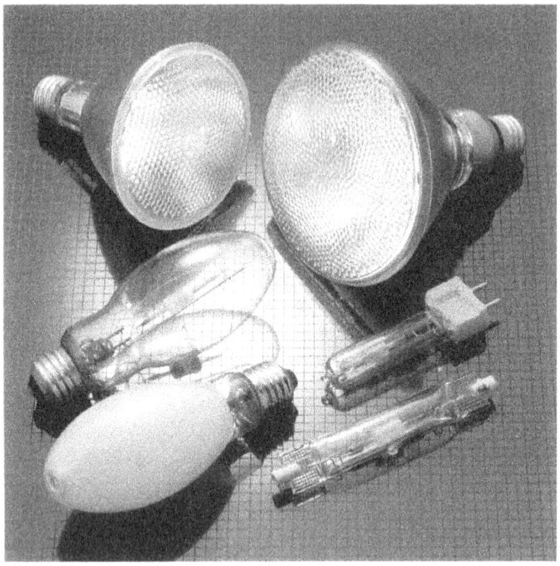

Figure 17.18 Various configurations of metal-halide lamps. Clockwise from the bottom left: phosphor-coated and clear elliptical bulbs; PAR 30 and 38 reflector lamps; single-ended and double-ended tubular lamps—all have ceramic arc tubes and a CRI greater than 80. (Courtesy of GE Lighting.)

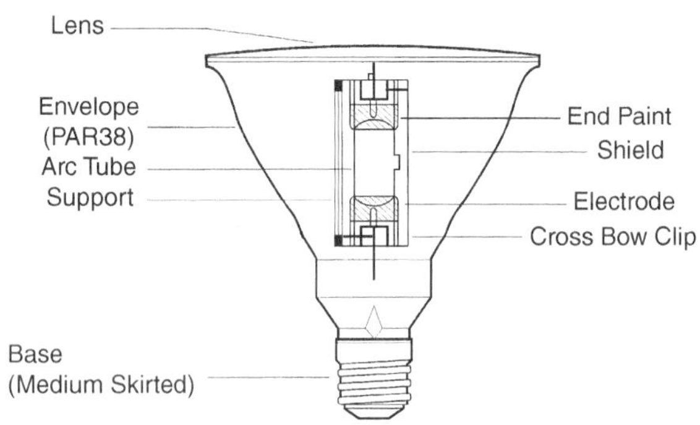

Figure 17.19 Construction details of a PAR enclosure for a metal-halide lamp. The lamp itself is tubular and constructed with a surrounding protective shield designed to contain arc tube fragments in the event of a violent rupture. (Courtesy of OSRAM Sylvania.)

(c) Designs, Shapes, and Ratings

Typical metal-halide lamp designs include:

- *Standard lamps*, available in ED, BT, and PAR shapes, in wattages from 50 to 1500 W, efficacies of 75 to 105 lm/W with magnetic ballasts, and slightly higher with electronic ballasts (efficacy also increases with wattage). Lamps are clear or phosphor-coated, with a CCT ranging from 3000 K to 4200 K and a CRI ranging from 65 to 85. Their life, at 10 burning hours per start, varies from 10,000 to 20,000 hours depending upon lamp type, size, and burning position.
- *Safety shielded lamps* with integral shields to contain a ruptured arc tube.
- *Self-extinguishing lamps* designed to shut down automatically upon a break in the outer glass envelope.
- *High-output lamps* designed for a specific burning position (as specified on each lamp). Output is 5% to 8% higher than that of standard lamps, but the color rendering is somewhat poorer, with a CRI range of 65–70.
- *Single-ended and double-ended tubular lamps* are characterized by a very high CRI (80–93); a somewhat shorter life than standard lamps, particularly for the single-ended units (6000–10,000 hours); and slightly lower efficacy (70–85 lm/W). They are intended for applications requiring very high color rendering.

(d) Operating Characteristics

Metal-halide lamps are not instant starting; they require approximately 2 to 3 minutes for initial startup and 8–10 minutes for restrike. (Tubular lamps require only about half of these times.) As a result, when they are used for indoor installations, a secondary instant-start source must be available. A number of manufacturers produce special hot-restrike ballasts that provide immediate restrike of lamps on restoration of power after an outage. Lamp output on restrike is inversely proportional to the duration of a power outage.

The spectrum of light produced by a metal-halide lamp changes as the lamp ages. The change is gradual, definite, but usually unnoticed and depends upon the particular design of lamp. Where color rendering is important, or where the lamp is used with other light sources, a designer should choose metal-halide lamps that are specially made for color stability. These lamps are designed not to vary in CCT more than 200 K over the lamp life. Finally, dimming or reduced output operation of metal-halide lamps is not normally recommended because of the very noticeable color shift that occurs when a lamp is dimmed.

(e) Lamp Ballasts

A metal-halide lamp will operate satisfactorily on a simple reactor ballast, although a separate ignitor is usually required to start the lamp. Such ballasts have a low power factor (about 50%), which is undesirable from the perspective of energy efficiency, wiring economy, electrical losses, and component heating. High-power-factor magnetic ballasts (pf > 90%) are also available. Magnetic ballasts are large, heavy, and often noisy. The last characteristic can be improved by the use of a potted (epoxy-filled) ballast. Electronic ballasts are also readily available, along with dimming and multilevel ballasts. As noted previously, dimming ballasts are not frequently used because of the large shift in lamp color that they cause.

17.16 HIGH-PRESSURE SODIUM LAMPS

Sodium-vapor lamps have been around since 1919. There are two varieties: high-pressure sodium and low-pressure sodium.

The basic construction of the high-pressure sodium (HPS or SON) lamp is illustrated in Figure 17.20, which shows schematic drawings of the design. Construction is quite different from that of a mercury-vapor or metal-halide lamp. The characteristic color of HPS lamps stems from the spectral absorption phenomenon of the sodium contained in the arc tube—with a resultant pronounced yellow-tinted light.

(a) Primary Characteristics of HPS Lamps

Standard HPS Lamps. The extremely low CRI (20–22) is not acceptable where any degree of color rendition is required, thus limiting the standard HPS lamp to use in exterior areas and for road lighting, where upfront cost is the primary concern and LED sources are not an option.

Color-Corrected HPS Lamps. Color can be improved considerably by increasing the pressure inside the arc tube. This causes some of the sodium in the arc to be reabsorbed, and the radiated light widens its spectrum into the red range (at the expense of efficacy and lamp life).

"White" HPS Lamps. A still greater increase in lamp pressure improves lamp color and yields a "whiter" lamp color in a limited wattage range. These low-wattage, reduced-efficacy, shortened-life lamps are normally operated with small, lightweight electronic ballasts.

Because of its high output and narrow linear arc tube, any HPS lamp is a potential glare source, and lamps of 150 W or higher must be either completely shielded or mounted at sufficient height (if in an open reflector) to be above the near field of vision. These glare prevention strategies also apply to metal-halide lamps.

If a diffusing coating is added to a sodium-vapor lamp, the entire glass envelope becomes the light-emitting source. This drastically reduces lamp luminance, and therefore glare potential, but also reduces output (and efficacy). Light distribution from a coated lamp in an open reflector is vastly improved, as can be seen in Figure 17.21.

(b) Other Operating Characteristics

In contrast to mercury-vapor and metal-halide lamps, high-pressure sodium lamps do not emit any appreciable UV radiation, do not tend to rupture violently, and can be installed in any position without affecting operating characteristics.

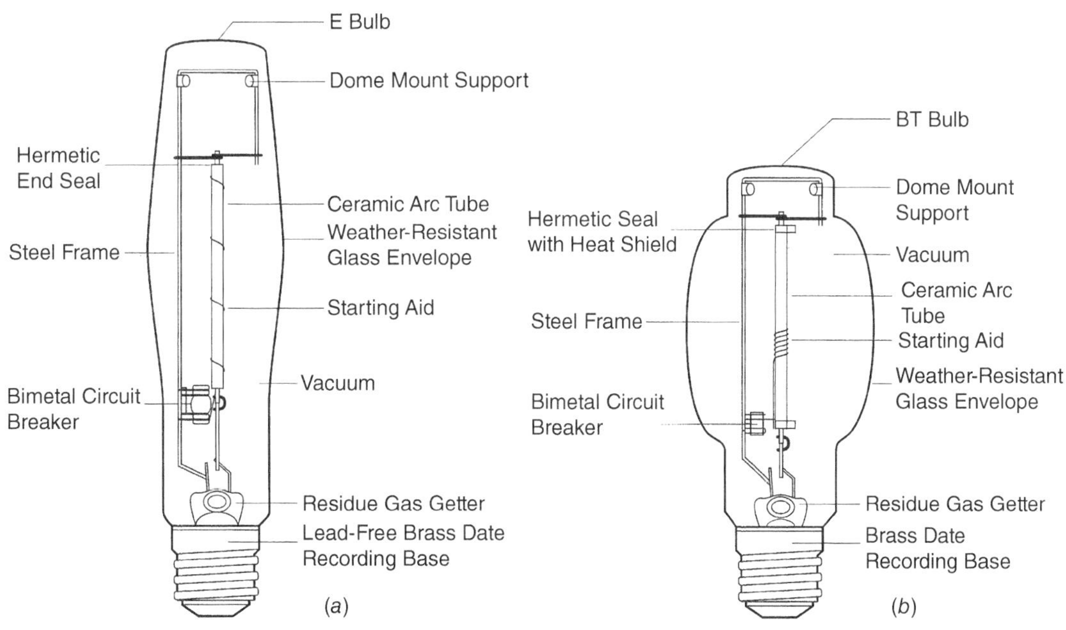

Figure 17.20 *Simplified drawings of the internal construction of two high-pressure sodium (SON) lamp designs. (a) Noncycling lamp (in an E-shaped glass bulb) designed to indicate by a special color when the lamp has reached the replacement stage. Unlike standard HPS lamps, this lamp will not cycle on and off at the end of its useful life. (b) Retrofit HPS (SON) lamp in a BT bulb, intended for direct replacement of an existing mercury-vapor lamp. This lamp operates efficiently on a mercury-lamp ballast. (Courtesy of OSRAM Sylvania.)*

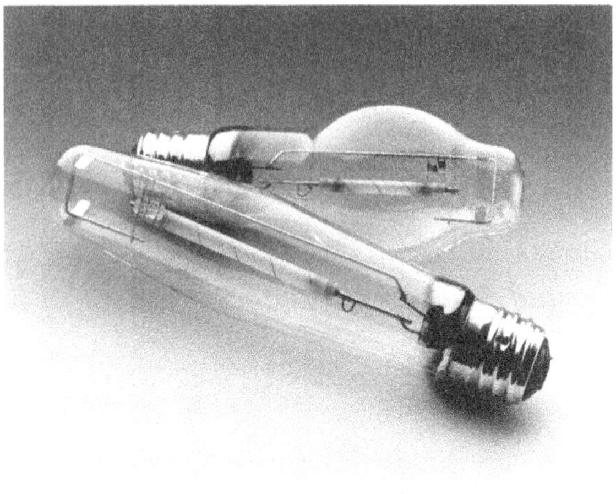

(a)

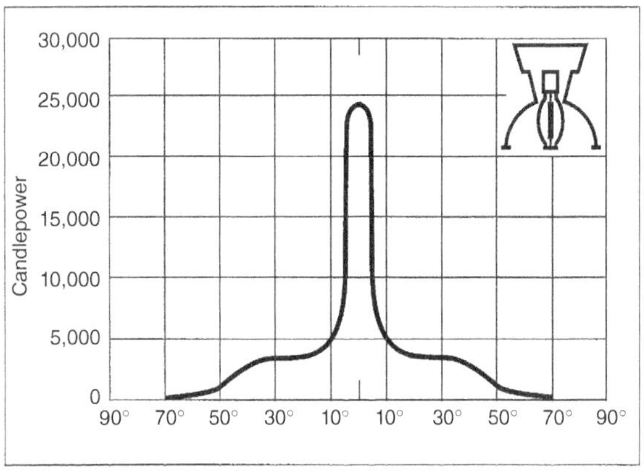

Clear HPS lamp operating at 12,000 lumens in open-bottomed fixture.

(b)

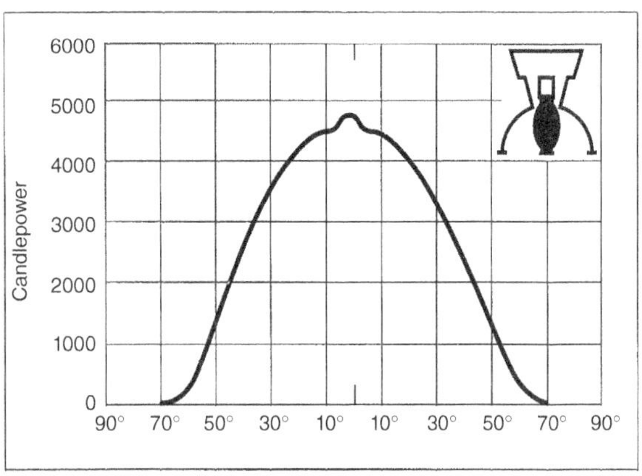

Coated HPS lamp operating at 12,000 lumens in open-bottomed fixture.

(c)

Figure 17.21 *(a) HPS retrofit and noncycling lamps. (b) The narrow linear arc tube of an HPS lamp is not suitable for use in an open reflector designed for a larger source such as a phosphor-coated mercury lamp. (c) A coated HPS lamp creates a large, diffuse source with improved light distribution. (Courtesy of OSRAM Sylvania.)*

HPS lamps require a ballast for ignition and arc control. Due to the extremely high voltage required for lamp ignition, both magnetic and electronic ballasts contain an electronic ignition circuit. Because of this, an HPS lamp must be used with a compatible ballast that carries the same ANSI designation as found on the outer glass bulb. If used with an incompatible ballast (and fixture), the lamp may rupture and constitute a serious safety hazard. As with metal-halide lamps, ballasts for HPS lamps are available that provide instantaneous restrike after a power interruption.

(c) Lamp Design Types

In addition to standard HPS lamps, including the color-corrected types, three additional special HPS lamps have been developed to solve specific problems. They are:

1. *Noncycling lamp:* As an HPS lamp ages, its arc voltage rises. Eventually, the ballast is unable to sustain the arc, and the lamp extinguishes. After cooling, the lamp lights to full brightness and soon thereafter extinguishes again. This on–off cycling is characteristic of an HPS lamp at the end of its life. To eliminate this condition, a special noncycling lamp was developed that uses very little sodium amalgam in the arc tube and is more environmentally friendly because of this reduced content. Photometric characteristics are similar to those of standard lamps.
2. *Standby lamps:* As noted previously, HPS lamps require a minute or more to restrike after being extinguished. A crowded public area plunged into complete darkness is a recipe for disaster; therefore, such areas must always be furnished with instant-on emergency lighting. Standby lamps have two arc tubes. When the lit one is extinguished due to a momentary power loss, the second (cool) arc tube immediately begins to glow—assuming that voltage has returned. It is not acceptable as emergency lighting as required by *NFPA 101: Life Safety Code*.
3. *Retrofit lamps:* These HPS lamps are designed as a direct replacement for mercury-vapor lamps of the same wattage. They are enclosed in BT-shaped bulbs of the same size as the lamp being replaced, and they operate properly on mercury-vapor lamp ballasts.

17.17 LOW-PRESSURE SODIUM LAMPS

The low-pressure sodium (LPS or SOX) lamp produces light characteristic of sodium's monochromatic saturated yellow color, making it inapplicable for general lighting. Because of its very high efficacy (more than 150 lm/W, including ballast loss—but with a CRI of 0), it can be applied where color rendition is not an important criterion, but energy efficiency is. Thus, LPS lamps are used for street, road, parking lot, and pathway lighting. LPS lamps are also used around astronomical observatories because the yellow light can be filtered out of a telescope.

17.18 SEMICONDUCTOR LIGHT-EMITTING DIODES (LED)

The term "solid-state lighting" currently encompasses light produced by electrified light-emitting semiconductors—otherwise known as light-emitting diodes (LEDs). Over the past two decades, LEDs have emerged not only as a viable option for architectural and urban scale lighting, but also as a preferred technology for general illumination, accent lighting, and high-intensity discharge applications. Due to energy-driven regulations at the state and federal levels, LEDs have led the market as a viable replacement for incandescent and gaseous-discharge sources.

LED technology is not new; it has been used since the 1960s in a wide range of applications, from medical instruments to bar coders, fiber optic communication, mobile electronic technologies, consumer appliances, automotive instrument panels, signal lights, courtesy lighting, and transport signaling (traffic, rail, aviation). LEDs were initially used for specialty applications in architectural illumination, such as signage, retail displays, emergency lighting (exit and emergency signs), and accent lighting for pathways. Since the proposed phase-out of energy-inefficient light sources through the federal Energy Independence and Security Act of 2007, the consumer lighting industry has taken large strides toward developing and marketing LED replacement options for many incandescent and fluorescent lamp sources. The energy and energy-cost savings are considerable. For example, replacing a 60 W incandescent bulb with a 9 W LED may result in about an 85% energy savings—potentially over a $70 net life-cycle savings in much of the contiguous United States.

LED lamps and luminaires now cover a broad range of application areas in conventional architectural and urban scale lighting. They are easy to install, last longer than incandescent or fluorescent lamps, are efficient (with high luminous efficacy values), and can provide good color rendering and even tunable color-changing capabilities.

LED lamps are essentially micro-light sources. A single A60 9.5 W LED lamp may have 30+ individual LEDs inside (Figure 17.22). Implementation of these

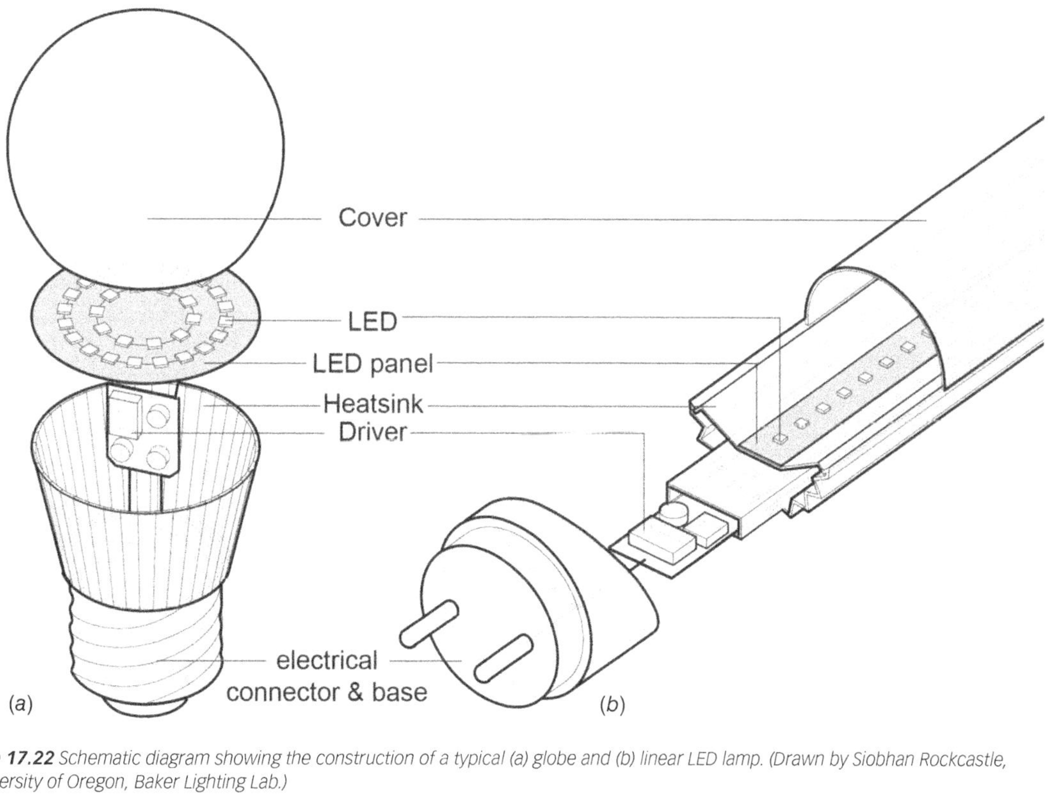

Figure 17.22 *Schematic diagram showing the construction of a typical (a) globe and (b) linear LED lamp. (Drawn by Siobhan Rockcastle, © University of Oregon, Baker Lighting Lab.)*

ACTIVE ENVIRONMENTAL SYSTEMS

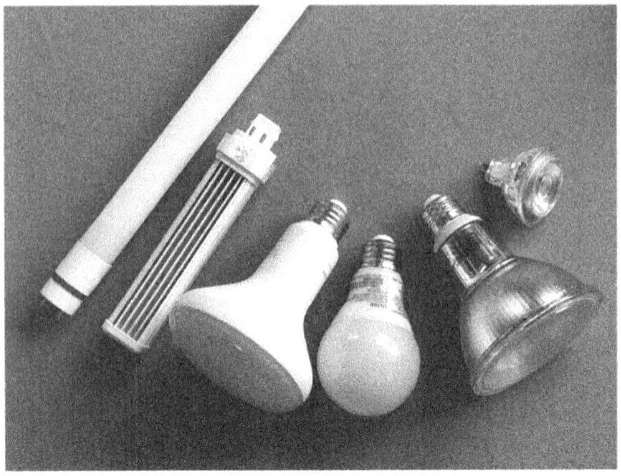

Figure 17.23 *A sample of LED light sources. (© University of Oregon, Baker Lighting Lab.)*

micro-light sources into mainstream architectural lighting applications requires the packaging of these point sources into composite lamps, many of which mimic the lamp sources they were designed to replace, with standard screw bases and shapes (Figure 17.23).

Manufacturers now bundle LEDs into a wide range of architectural lighting products that allow designers to use LED lamps for nearly all building lighting applications, including tubes, spots, HID replacement, bulbs, capsules, and candles.

Some positive characteristics of LEDs are:

- excellent longevity (25,000+ hours)
- excellent luminous efficacy
- high-quality color rendering
- variable color and color rendering capability within a single lamp
- directional and multi-directional light output
- lower environmental impact relative to mercury-based lamps

(a) Primary Characteristics of LED Lamps

LED Lamps and Tubes. Because many LED lamps were developed to replace existing electric sources, most are made to fit into existing luminaires designed for incandescent or gaseous discharge lamps. Lamps designed to replace incandescent, halogen, and compact fluorescent sources come with standard screw-base connections, such as medium and candelabra as well as miniature connections like G4 and GU5.3. Bulb types are available in a range of shapes including "A," "BR," and "PAR." Lamps designed to replace linear fluorescent sources are, like their predecessors, called "T" shapes. As most consumers and designers are accustomed to identifying lamps based on their wattage, many LED lamps contain both their actual wattage and the wattage of the lamp source they are replacing within their product title. For example, a LED T8 might be referred to as a "12 W (32 W) LED T8." Similarly, the LED replacement for a standard 60 W A19 lamp may be called a "7 W (60 W) A19 lamp." This makes it rather easy to identify the application and relative output of each lamp if you are familiar with the original source. Lumen output, however, is a more reliable metric for comparing electric light sources.

Operating Characteristics. Unlike other lamp technologies introduced in this chapter, LED sources are composed of a series of diodes and emit light directionally. LED lamps designed to replace omnidirectional technologies require clusters of diodes to replicate that light distribution. Whereas most electrical light sources run on alternating current (AC) electricity, LED sources run on direct current (DC) and require a driver to convert the alternating current electrical supply which is common in residential and commercial buildings. This driver is often embedded directly in the lamp as shown in Figure 17.22. The lifespan of LED sources is negatively impacted by high temperature, making heat dissipation critical for the thermal management of LED lamps. To prevent overheating from higher currents, LEDs are operated at 350 mA. Rather than increase the current to produce more light, multiple diodes are required to achieve a high-intensity light distribution.

White Light Most lighting applications in buildings require white light, but LEDs emit light in narrow wavelengths of red (R), green (G), and blue (B). To produce white light, red, green, and blue (collectively referred to as "RGB") diodes can be combined, but the CRI of this white light is typically low due to the narrow wavelength of each combined source. To produce LED lamps with a higher CRI, two strategies are used. The first strategy involves combining more than three diode colors to produce a "trichromatic" white light. The second strategy uses red, green, and blue LEDs in combination with a phosphor, which absorbs some light from the LEDs and allows them to fluoresce in different colors. This extended color range of the LEDs plus phosphor can produce CRIs up to 90, with a typical value of 82, which matches fluorescent sources. Using phosphors, LED lamps can achieve color temperatures across a broad range (from 2200 to 7000 K). This makes LEDs not only one of the most energy-efficient sources of white light, but also the longest-lasting and most versatile.

Dimming and Color Changing. The combination of individual diodes and phosphor coatings allows these sources to achieve a broad range of colors and intensities depending on which diodes are activated. LED lamps can be dimmed using a couple of different strategies. As LEDs require a constant current and voltage supply to operate efficiently, a simple analog dimming strategy (lowering the voltage to the lamp as is done with legacy lamps) does not work well. Reducing the current and voltage to an LED lamp can produce a dimming effect (until a given threshold where the lamp will fail), but it negatively impacts its efficacy and produces visible variations in the lamp's CCT.

Another strategy is called "pulse width modulation" (PWM), which causes the LED to flicker off and on at a rate that is undetectable to the human eye. By controlling the rate of flicker and the cycle of light output, it is possible to reduce the amount of light emitted from the lamp source by delaying its signal. Most LED drivers have been designed with a PWM generator, which controls the on and off cycle of the LED lamp. Because the frequency of this altered signal should not be visible, it is important to use a driver designed to work with the specific LED lamp assembly.

Lamp Life and Operating Efficiency. The principal advantages of LED sources are their efficacy and extended lamp life, which lead to significant life-cycle cost savings, as shown in Table 17.6. If we compare a 60 W incandescent to a 14 W CFL and a 12 W LED, the 14 W CFL has a payback period of only 0.3 years over the traditional 60 W incandescent (when operated 4 hours a day). The 12 W LED takes closer to a year to break even with the cost of purchasing and operating a 60 W incandescent, but the LED achieves the greatest life-cycle net cost savings. Despite the higher upfront costs of the lamp itself, the annual energy savings and extended lamp life (25,000 hours in this example) of the LED lamp mean a greater net cost savings over both the incandescent and CFL lamps. Other LED lamps can achieve even higher lamp life (up to 50,000 hours for a typical T8 LED). When compared to a typical T8 fluorescent lamp, the T8 LED is around 30% more efficient with a 60% longer lamp life.

It should be noted that integrated lamp products generally have lower luminous efficacies than products with replaceable lamps.

TABLE 17.6 Comparison of Bulb Life, Annual Electricity Cost, and Life-Cycle Energy Savings Between a 14 W CFL and a 12 W LED Lamp (Operated 4 Hours a Day)

	14 W CFL	12 W LED
Lamp Life	10,000 hours	25,000 hours
Annual Electricity Cost Savings	\$8.00	\$9.00
Life-cycle Energy \$ Saved (%)	~77%	~80%
Life-cycle Net Cost Savings[a]	\$49	\$110
Simple Payback Period[b]	0.3 years	1.0 years

Source: DOE, n.d.

[a] *Life-cycle Net Cost Savings refer to the life-cycle electricity cost savings plus the avoided cost for replacement, minus the purchase price of the lamp.*

[b] *Simple Payback is the time required to offset the additional cost of the LED lamp through accrued savings in energy (ignoring the time value of money).*

17.19 ORGANIC LIGHT-EMITTING DIODES (OLED)

Organic light-emitting diode (OLED) sources are a form of LED where the emissive surface is an organic compound or semi-conductor located between two electrodes (Figure 17.24). OLED sources were first introduced in digital displays such as computer monitors, television screens, and mobile phones, but are now being incorporated into architectural lighting applications. As a flexible emissive surface, OLED sources offer unique application potential.

Philips Lighting offers a brand called "Lumiblade," which contains a series of OLED panel products. Where normal LEDs are composed of a cluster or filament of diodes, OLEDs emit light through a flexible medium, giving them the ability to take a range of shapes depending on the intended light distribution and luminaire design.

As such, this technology does not conform to the "typical" range of lamp shapes and sizes seen with previous electric light sources. While currently more expensive than typical LED sources, OLED technologies are rapidly evolving and can be expected to match

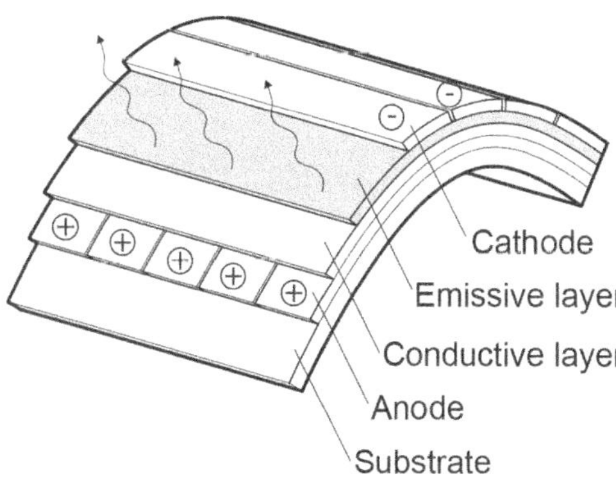

Figure 17.24 *OLED structure. (Drawn by Siobhan Rockcastle, © University of Oregon, Baker Lighting Lab.)*

and even exceed LEDs in terms of luminous efficacy, stability, dimmability, color change capability, and application versatility.

17.20 INDUCTION LAMPS

Induction lamps have been around in some form or another since Nikola Tesla demonstrated the technology in the late 1890s. Due to high upfront costs and the availability of other low-cost sources, this technology was not introduced to the market until the 1990s due to interest in energy-efficient solutions. An induction lamp is filled with low-pressure mercury vapor. When ionized by a high-frequency induction coil inside the lamp, the mercury vapor produces UV radiation, which then strikes a phosphor coating on the inside of the lamp, producing light. This is similar to the light-producing process used by standard fluorescent lamps; the difference is that the gas is ionized by an induction coil (rather than an electron stream)—thus the name induction lamp. Two such designs are shown in Figures 17.25 and 17.26.

Despite their high upfront cost, induction lamps offer low-maintenance operation, high efficacy (70+ lumens/watt), excellent color rendering (80 CRI+), a broad range of color temperatures (2700–6500 K), instant strike, dimming capability, high power factor, and excessively long life (100,000 hours+).

A self-contained induction lamp of lower wattage is illustrated in Figure 17.26. This lamp is built into a modified R-shaped envelope with a standard Edison base and is therefore readily applied as a direct replacement for a 100-W incandescent reflector lamp. The published performance values for the lamp are:

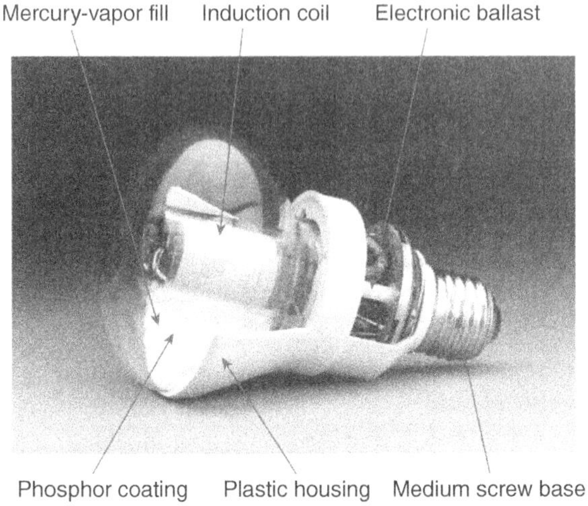

Figure 17.26 *Cutaway of the GE induction lamp showing the essential elements: induction coil, phosphor-coated bulb with mercury-vapor fill, and electronic ballast. This 23 W lamp in a modified R (reflector)-shaped bulb has a height just under 5 in. [127 mm] overall and a 3 in. [76 mm] maximum bulb diameter. (Courtesy of GE Lighting.)*

Total wattage	23
Initial lumens	1100
Efficacy	48 lm/W
CCT	3000 K
CRI	82
Life	10,000 hours
Lumen maintenance at 70% life	75%

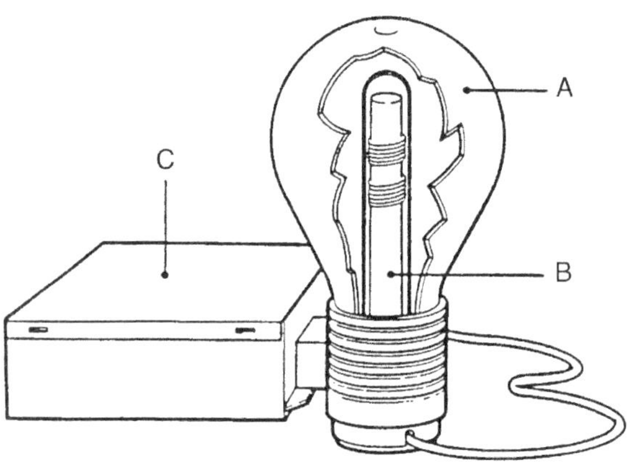

Figure 17.25 *Schematic diagram of an induction lamp (rated 85 W, including all losses) showing the operating principles. The lamp is 4.3 in. [109 mm] in diameter and 7.5 in. [191 mm] high overall. The generator (C) produces a high-frequency current, which circulates in the coil on the power coupler (B). This ionizes the mercury vapor inside the lamp (A), producing UV radiation. The UV radiation strikes the fluorescent coating inside the lamp, producing light. (Illustration courtesy of Philips Lighting Company.)*

17.21 SULFUR LAMPS

The principle of microwave energy excitation has been applied successfully to a lamp type that consists of a golf ball-sized globe filled with an inert gas and a few milligrams of sulfur. In contrast to the induction lamp described previously, no mercury vapor is used in this lamp, and the output is full-spectrum light with very little ultraviolet or infrared radiation. A prototype lamp was a 6-kW unit that emitted more than 400,000 lm (a lot of light). A more recent version of the lamp exhibited the following characteristics:

Total power input	1320 W
Initial lumens	130,000
Efficacy	101 lm/W
Diameter of globe	29 mm [1.1 in.]
CCT	5600 K
CRI	80
Life of lamp (estimated)	60,000 hours
Life of exciting magnetron	15,000 hours
Lumen maintenance	Very high
Color constancy with aging	Excellent

ACTIVE ENVIRONMENTAL SYSTEMS

Current sulfur lamp technology provides approximately the same output as a 1000 W HPS lamp, but with much better color characteristics. It also compares to about 1200 W equivalent of metal-halide lamp with similar efficacy, but, again, because of its full-spectrum radiation, it has superior color characteristics. Initial trials of the lamp as a driver for a "light pipe" were successful.

17.22 FIBER OPTICS (FO)

Although optical fibers (OF), or (more commonly) fiber optics (FO), have been available since the 1920s, practical applications (in the medical field) were not developed until the late 1950s and early 1960s. Bundled fibers used as a diagnostic tool can deliver light to remote regions of the body and carry coherent (understandable) images back to a doctor. In recent years, fiber optic systems have made their most significant advances in the communications field, wherein long-distance telephone cables and complex wire networks have been replaced by much simplified fiber optic installations. Architectural applications of fiber optics have included alternatives that directly replace recessed ceiling downlighting, track and display case lighting in museums, and lighting for pools/spas, supermarkets, and other commercial buildings.

Illuminators for fiber optic systems utilize a variety of lamp types. The primary advantages of using a fiber optic system are that no heat is produced where the light exits the fiber (permitting "cool" lighting installations), and no UV radiation is transmitted through the fiber.

LUMINAIRES

17.23 DESIGN CONSIDERATIONS

(a) General

A luminaire (lighting fixture) is the interface between a lamp and a building. The luminaire must be designed for the intended lamp type (LED, fluorescent, incandescent) and be designed for installation in the intended ceiling type (gypsum board, suspended acoustical tile, other). The aesthetics of the fixture are often of concern during the selection process.

(b) Luminaire Characteristics

The purpose of a luminaire or lighting fixture (the terms are synonymous) is twofold: physically, to hold, protect, and electrify the light source (lamp); and photometrically, to control the lamp output (i.e., to redirect the light produced by the lamp) because most common light sources emit light in all directions. How this distribution control is accomplished is well known to luminaire designers and manufacturers. The characteristics of the reflectors, baffles, lenses, and louvers that perform these functions are discussed in some detail in the following sections. However, the problem in luminaire selection is that the requirements are, in many respects, incompatible, and therefore a trade-off between various luminaire characteristics must be made. For instance, high-efficiency installations can entail high fixture luminance with resultant glare. Wall lighting from a ceiling fixture means high-angle luminaire output with resultant direct glare potential (see Chapter 7, Figures 7.23 and 7.24). Low-angle light means reduced direct glare potential but possible veiling reflections. A high shielding angle (>35°) means good visual comfort but reduced efficiency, and so on. It follows that:

- no single luminaire design is ideal for even a majority of applications
- to make an intelligent selection among the hundreds of lighting fixtures commercially available, it is necessary to understand both the specific requirements of the application and the light control characteristics of the luminaire being considered.

17.24 LUMINAIRE LIGHT DISTRIBUTION

The two intensity distribution curves shown in Figure 17.27*a* are actual test results from two 2-lamp, 1 ft × 4 ft [0.3 m × 1.2 m] semi-direct fluorescent fixtures with prismatic enclosures. The flat bottom of the curve in Figure 17.27*a* indicates even illumination over a wide area—therefore permitting a high spacing to mounting-height ratio (S/MH) of 1:4, while still providing uniform illuminance. The rounded bottom of the curve in Figure 17.27*b* indicates uneven illumination and closer required spacing for horizontal uniformity.

The straight sides of the curve in Figure 17.27*a* show a fairly sharp cutoff, and the small amount of light above 45° means high efficiency, probably insufficient wall lighting, barely adequate diffuseness, and very little direct glare potential but a distinct possibility of veiling reflections. Conversely, the curve in Figure 17.27*b* shows a large amount of horizontal illumination (above 45°) with resultant direct glare potential, diffuseness, and relative inefficiency, because horizontal light output is attenuated by multiple reflections before reaching the horizontal

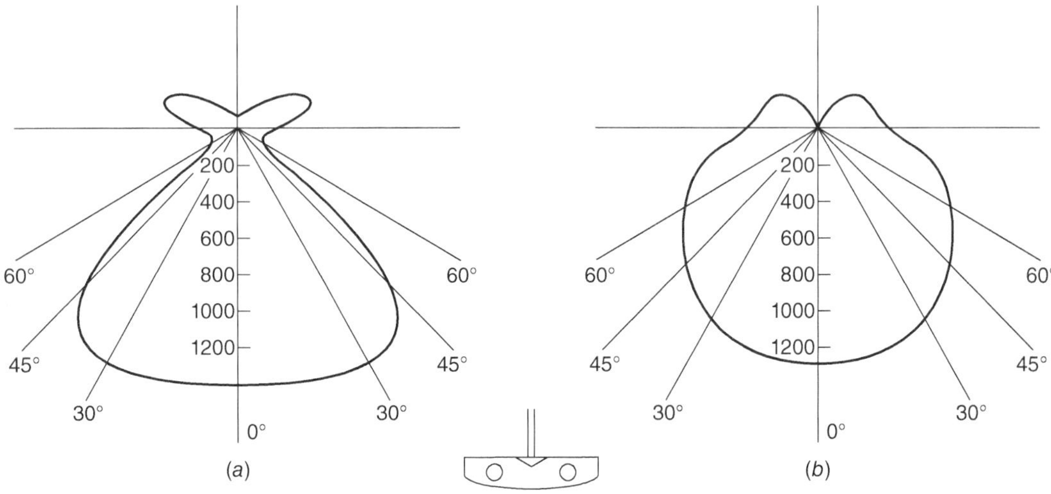

Figure 17.27 *Semi-direct fluorescent fixture crosswise distribution (two lamps, 32 W each, prismatic enclosure). Note the sharp cutoff and wide, horizontally even distribution of (a) in contrast to the diffuse, broad, and horizontally uneven distribution of (b).*

working plane. Here, however, low output below 45° minimizes reflected glare potential. The uplight component of luminaire (*a*) is directed outward to cover the ceiling and will not cause hot spots; the corresponding light from fixture (*b*) is concentrated above the fixture and gives uneven illuminance on the ceiling.

These conclusions were reached based on the following observations:

1. *Uniformity of illumination* requires that the intensity at angles above the nadir (0° from the vertical) be greater than the intensity at 0°, so that location-points distant from the fixture centerline obtain the same illumination as those below the fixture (because illuminance varies inversely with the square of distance). This is exactly the case with the flat-bottom characteristics of Figure 17.27*a*. Therefore, such fixtures can be spaced more widely than the units of Figure 17.27*b*.
2. *High efficiency* is achieved by directing the luminaire output to the work plane (i.e., from 0° to 45° from the vertical). Light above 45° is directed to the walls and reaches the working plane only after multiple attenuating reflections.
3. *Diffuseness* exists when light reaches the work plane from multiple directions. This requires that light be reflected from walls and ceilings to the work plane, which in turn requires luminaire light output above 45° from the vertical.
4. *Direct glare* is caused by light output at high angles (i.e., above 45° from the vertical). Direct glare from linear fluorescent fixtures can be minimized by placing the long axis parallel to the line of sight, because such fixtures normally have low endwise high-angle output.
5. *Reflected glare* is caused by reflection of low-angle output from the task. Therefore, fixtures with control means that limit output between 0° and 45° minimize the potential for veiling reflections (see Section 17.26b).
6. *Shielding* is a function of the shape of the fixture housing plus any additional lamp concealment means, such as louvers or baffles. The shielding angle is defined as the angle between a horizontal plane through the louvers or baffles and the inclined plane at which the lamp first becomes visible as one approaches the fixture (Figures 17.28 and 17.29). Cutoff angle is usually defined as being synonymous with the shielding angle. However, because some sources define it as the complement of the shielding angle, it is best to avoid the term and only use shielding angle.
7. *Ceiling illumination* is produced by light above the horizontal. As with light below the horizontal, a spread-out characteristic (Figure 17.27*a*) means good ceiling coverage, no hot spots, and, ultimately, good diffuseness. Concentrated uplight means a potential hot spot if the fixture suspension hanger is too short, and in any event, it yields uneven ceiling illumination.

Inspection of a luminaire intensity distribution curve can yield a large amount of information on the fixture's performance.

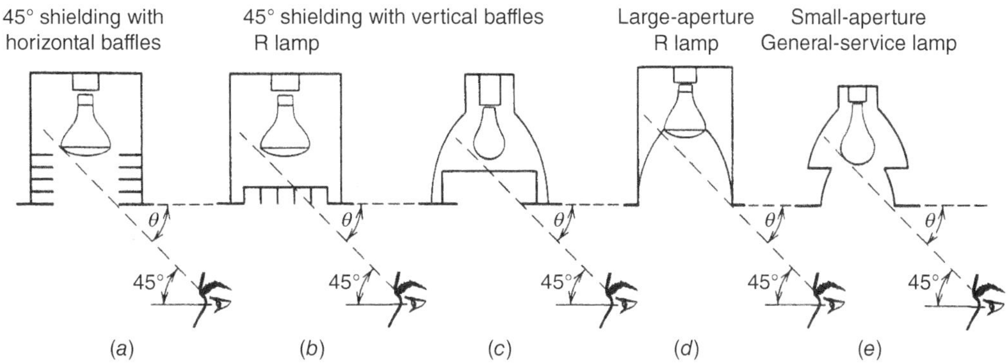

Figure 17.28 *Methods for shielding downlights using circular shields for vertically symmetrical sources such as incandescent and HID lamps. Halogen lamps mounted vertically in R lamp envelopes are symmetrical; horizontally mounted units in PAR or other reflectors are not. Baffled downlights (a–c) control unwanted high-angle light by cutoff as illustrated. Black baffles aid by absorbing light and appearing dark. Other colors give a ring of light at the baffled edge. Cones (d, e) control brightness by cutoff and by redirection of light due to their shape. They are either parabolic or elliptical. A light specular finish such as aluminum appears dull; a black specular finish appears unlighted. Black finishes require high-quality maintenance because dust shows as a bright reflection. CFL lamps in reflectors are not normally a serious direct glare concern and are considered to be vertically symmetrical. A shielding angle of 45° minimum is recommended for high-luminance lamps.*

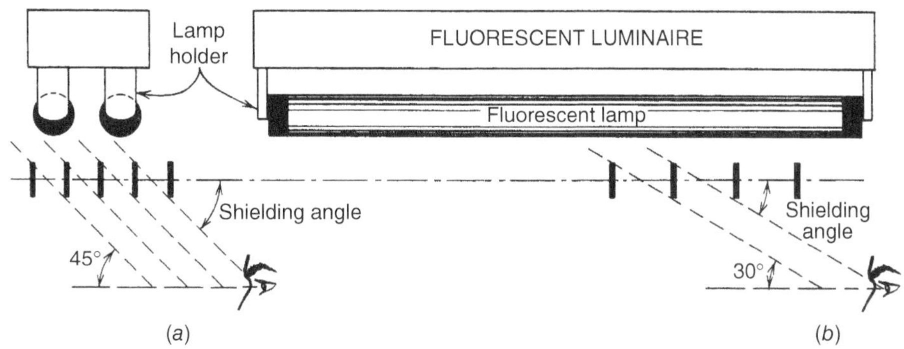

Figure 17.29 *Shielding of fluorescent lamps is less critical due to lower lamp luminance. For T12 lamps, 45° × 35° crosswise/lengthwise shielding as shown is excellent, and 35° × 30° is satisfactory. For T8 lamps, 45° × 35° should be used. Because fixture luminance is higher in the transverse direction (a) than lengthwise (b), a better cutoff angle is required. The shielding elements may be louvers or baffles. Opaque shielding elements provide better visual comfort than translucent plastic units.*

17.25 LUMINAIRE LIGHT CONTROL

(a) Shielding

Except where it is desired to use a bare lamp as a source of sparkle, such as in chandeliers and other decorative fixtures, all lamps in interior fixtures should be shielded from normal sight lines (i.e., sight lines in a head-up, eyes-straight-ahead position; see Figure 7.19). The reason is simple: bare lamps are so bright (see Table 7.2) that they usually constitute a source of direct or even disabling glare, depending on the apprehended angle (closeness to the eye and size of the lamp) and eye adaptation level. The range of permissible luminaire luminances (listed in Table 7.3) of 1000–7000 cd/m^2 depends upon apprehended angle and adaptation level.

As a rule, exposed incandescent lamps, 6 W and larger, are sources of direct glare and should be avoided. The upper direct-glare limit (7000 cd/m^2) corresponds to the luminance of a bare 34-W T12 fluorescent tube, which accounts for the fact that bare-lamp fluorescent fixtures are tolerated from a glare perspective. If such a fixture is relamped with a lamp whose luminance exceeds 10,000 cd/m^2, it becomes a source of annoying direct glare that actively impairs visual ability.

Shielding of lamps is accomplished with the fixture housing/reflector or with baffles and louvers, as mentioned previously (see Figure 17.30). Fluorescent fixtures require shielding most when placed crosswise to the line of sight, thus exposing the entire length of the lamp to the field of view. Such fixtures require longitudinal baffles, deep housings, or louvers to provide the necessary shielding. Alternatively, when at all possible, place fixtures with their long axis in the direction of sight lines.

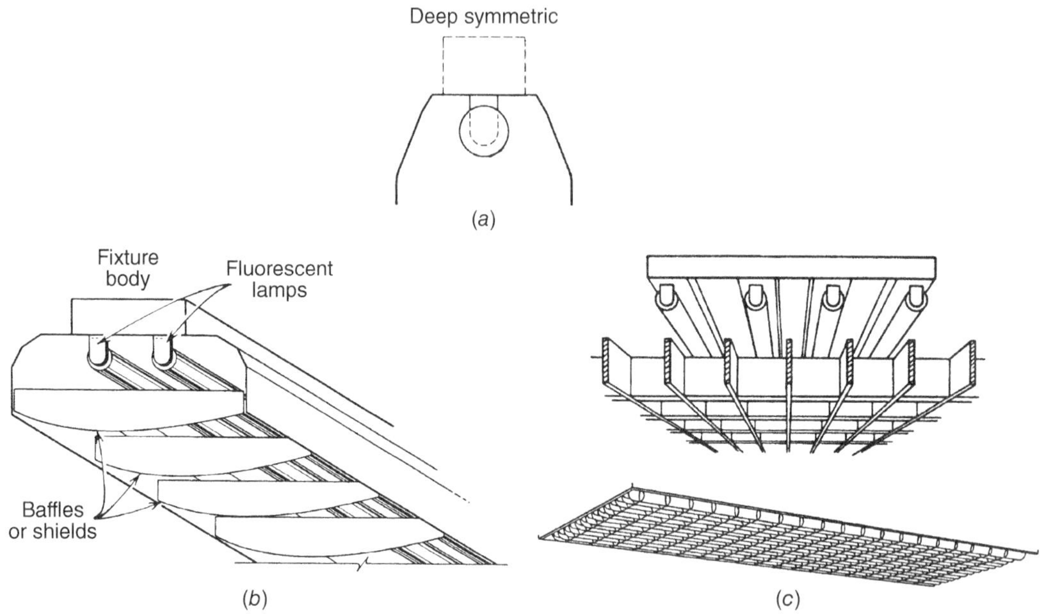

Figure 17.30 Shielding elements. (a) The most basic shielding element is the lamp reflector, which may double as the fixture body. (b) Shields perpendicular to the long axis of a linear fluorescent lamp are normally called baffles. They are less important than lengthwise shields (louvers) because endwise lamp luminance is lower than crosswise luminance. (c) Two-way shielding is most effective but seriously lessens luminaire efficiency. (From Stein, 1997 / with permission of John Wiley & Sons.)

(b) Reflectors

The basic shapes and beam patterns of luminaire reflectors are illustrated in Figures 17.31 through 17.33. A so-called pinhole downlight requires an elliptic reflector to focus the light while maintaining minimal fixture efficiency. Elliptic reflectors are large, and frequently the space above the ceiling is too restricted for their use. Lamps with integral elliptic reflectors can be utilized with a standard baffled reflector to achieve roughly the same effect.

(c) Reflector Materials

Until recently, reflector materials were of two types: white gloss paint for the portions of a luminaire interior that acted as a reflector, and anodized aluminum for the shaped reflectors of the types shown in Figures 17.31 through 17.33. The reflectances of both materials are approximately the same, varying between 0.84 and 0.88 when new and clean. Neither, however, is truly specular; the paint finish is primarily diffuse, whereas the aluminum is principally specular. Where shaped reflectors are not used, as in the case of a fluorescent troffer, the lack of specularity is immaterial because, at worst, the diffuseness will reduce luminaire output slightly by increasing the number of interreflections within the fixture body (Figure 17.34*a*). The idealized specular reflections for shaped reflectors shown in Figures 17.31 through 17.33 are just that; the reflectances are considerably more diffuse and become increasingly so with reflector aging and dirt accumulation.

Painted fixture body interiors lose their high reflectance by rapid aging due to elevated temperatures and accumulation of dust and dirt. This causes a decrease in overall reflectance, which is compensated for by initial overdesign. However, because a rectangular fixture body is not an accurately shaped reflector, the result of overall reflectance reduction is simply an overall reduction in output while maintaining the same photometric distribution characteristics.

Previous energy conservation programs and utility rebates led to the addition (via retrofits) of very-high-reflectance auxiliary reflectors to existing fluorescent troffers approximately as shown in Figure 17.34*b*. Unfortunately, many of the claims of highly increased efficiency were based on retrofitting aged, dirty luminaires, and the results were often very misleading. A reasonable estimate of the possible improvement in luminaire efficiency can be made by considering these facts:

- Approximately 40% of a lamp's output in an open luminaire is directed downward and is therefore completely independent of any reflector action.
- The difference in reflectance between a new, clean, painted surface and an old, dirty surface is, at most, 50%. That means that the maximum light loss of an open fixture due to poor maintenance is 50% of 60%

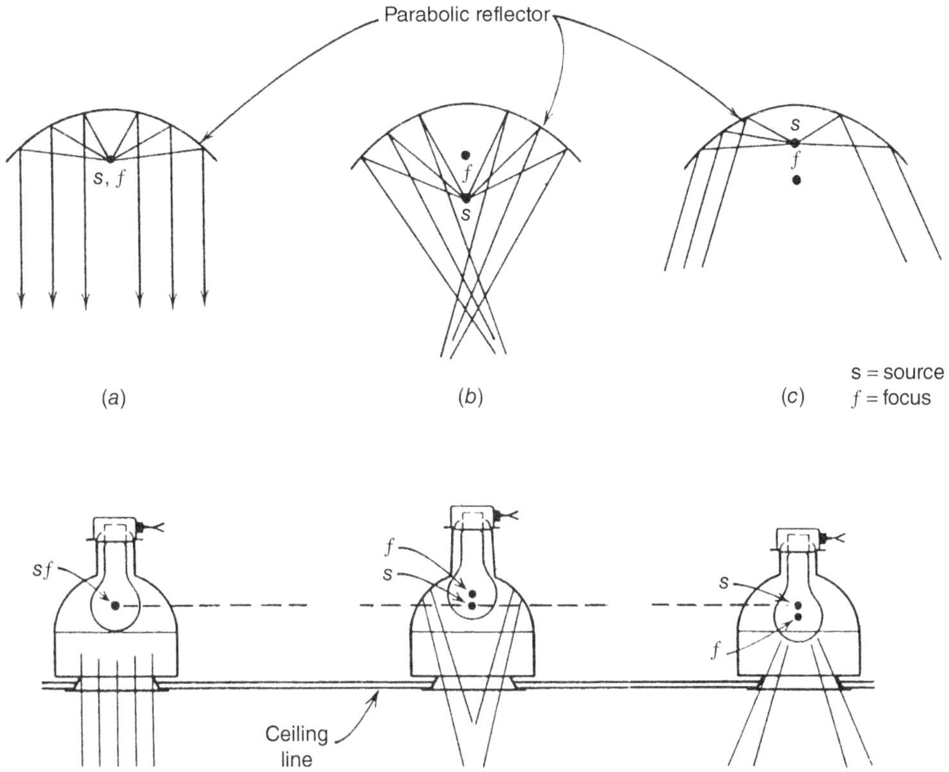

Figure 17.31 *Parabolic reflector action shown with the fixture below: (a) with the source at the focal point, rays are parallel; (b) with the source below the focal point, they converge; (c) with the source above the focal point, they diverge. This focusing action is illustrated by fixtures correspondingly designated. Note that type (c) requires a large ceiling opening to achieve even minimal efficiency.*

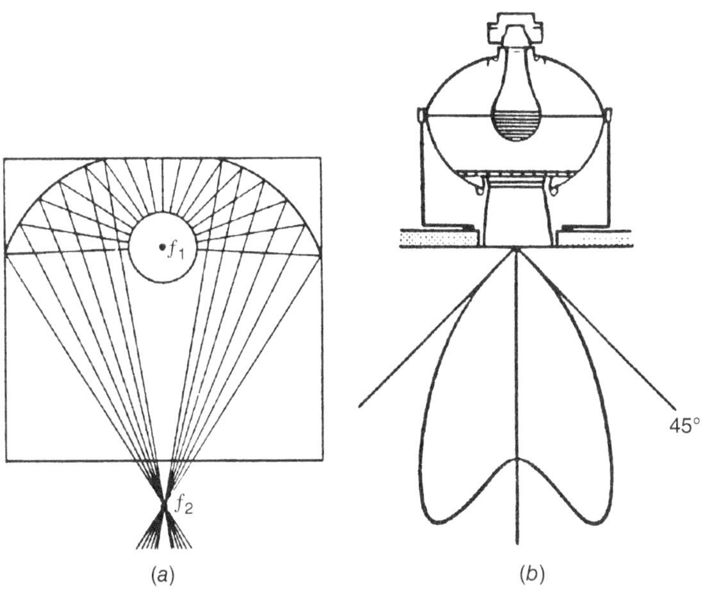

Figure 17.32 *(a) Action of an elliptical reflector section. With the light source at focal point f_1 the light converges at the other focal point, f_2. This effect is useful in fixture design, as in (b). By projecting light up only (using a silvered bowl lamp), the output light can be redirected through a constricted aperture at the other focal point, with little loss. This design is the basis of high-efficiency "pinhole" downlights.*

(the latter percentage being the maximum reflected light component), or 30% of the overall light output.

- The maximum reflectance of the best (and most expensive) silver reflectors is about 95%, comprising 93% specular and 2% diffuse. This is only 10% higher than the original minimum 85% paint reflectance. Therefore, at most, retrofit of a very dirty, old fixture with a high-quality (expensive) silver reflector improves performance by the 30% lost to dirt plus 10% of the 60% reflected light for a total of 36% maximum. Relamping also improves output, but that is not connected with fixture body reflectance.

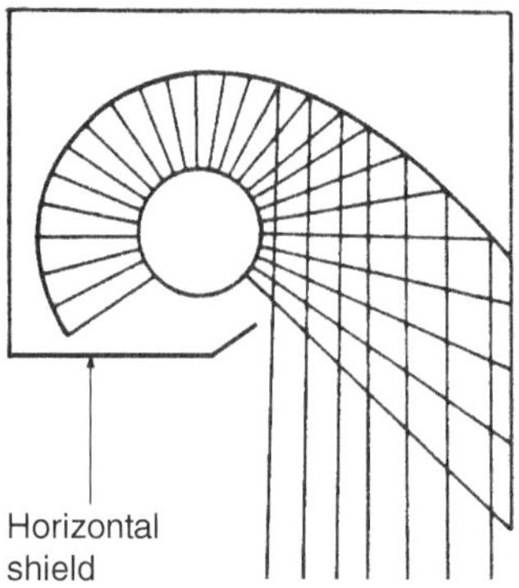

Figure 17.33 The extended section reflector allows the source to be concealed (shielded) while projecting its light directly down, but horizontally displaced, from the source.

- Simple cleaning of a very dirty fixture body restores 20% to 25% of the light loss. The remaining loss is due to aging of the paint. A cost analysis is required to determine whether the 10% to 15% differential in light output between simple cleaning and silvered reflector addition is economically feasible.

An important consideration in retrofit and new construction is the photometric characteristic of high-reflectance linear fluorescent luminaires. In general, they are shaped like the curve of Figure 17.27*b*. This means that light is concentrated downward, resulting in a requirement for closer luminaire spacing to obtain uniform illuminance. In new construction, this can be considered in design, although it can be a serious economic penalty. In retrofit work, it can result in unacceptable lack of illuminance uniformity, requiring additional luminaires and expensive relocation of existing units.

Another factor to be considered is the extent of maintenance required to keep "super" reflectors in pristine condition to achieve the 15% ± maximum output differential. To determine this, designers should request an aging test and inspect a previously retrofitted installation with ambient conditions and cleaning schedules like those of the area under consideration.

17.26 LUMINAIRE DIFFUSERS

Diffusers are elements of a luminaire placed between the lamp(s) and the illuminated space that function to disperse the light, control fixture brightness, redirect the light, and obscure (hide) and shield the lamps (and fixture components).

(a) Translucent Diffusers

Because these do not redirect the light but merely diffuse it, the distribution characteristic is circular, as seen in Figure 17.35*a*. Typical examples of this type are white opal glass, frosted glass, and white plastics such as Plexiglas, polystyrene, vinyl, and polycarbonates. The distribution is basically the same as it would be for bare lamps: lamp-hiding power is good; depending on the material, direct glare can be a problem; visual comfort probability (VCP) is poor; veiling reflection potential is high; the S/MH does not exceed 1.5; the fixture is generally inefficient; and wall illumination is good because of a large component of high-angle light (which reduces VCP). The net result of using this type of diffuser is a reduction in apparent lamp luminance due to a distribution of the lamp output over a larger diffusing area. Applications include corridors, stairwells, high-ceilinged spaces, and other areas without demanding visual tasks.

A polarizing flat plastic panel, when the product was initially introduced, held great promise because it decreased veiling reflection at an angle of 60°, but much less at other angles. Because most viewing is in the range 20–40°, using these panels did not result in any appreciable reduction in reflected glare in normal office work situations.

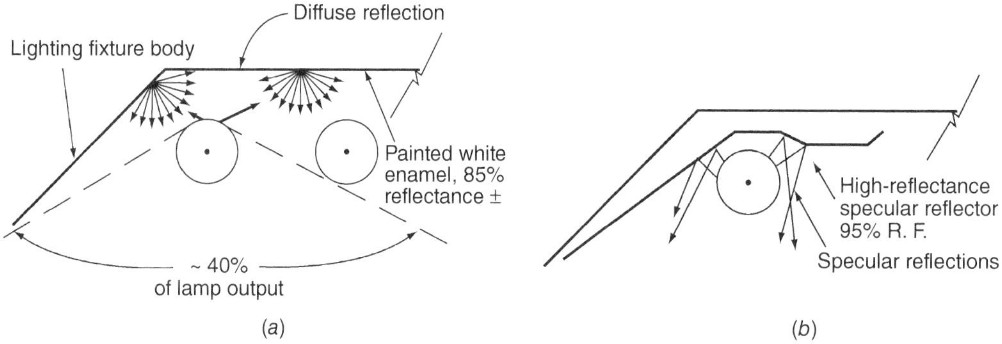

Figure 17.34 (a) Approximately 40% of lamp output in an open linear fluorescent fixture is unrestricted. The remainder leaves the fixture after one or more reflections. (b) A mirrored reflector narrows the distribution pattern of the luminaire by specular reflection and increases output somewhat (R.F. is reflectance factor; i.e., reflectance).

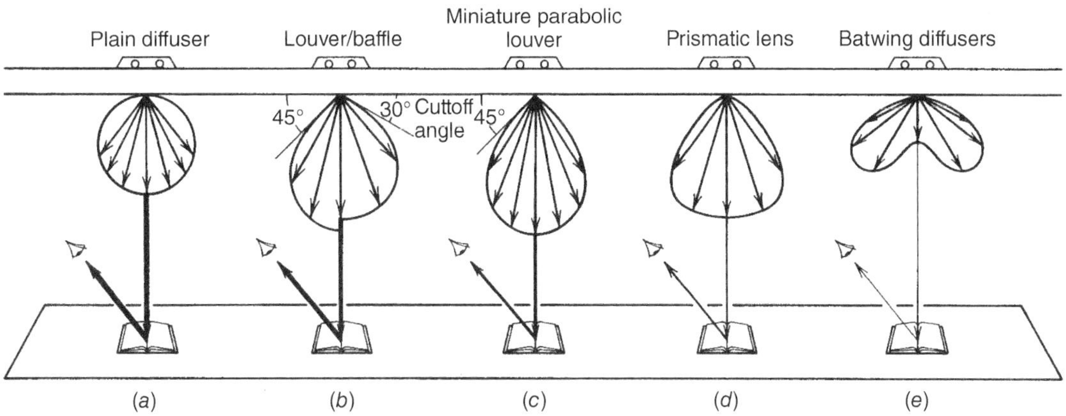

Figure 17.35 *Comparison of typical candlepower distribution curves for common linear or PL lamp fluorescent luminaire diffuser elements (for a full description of types a–d, see text). Note that for a given geometry of viewer and luminaire, the severity of veiling reflections depends entirely on the fixture's photometric characteristic. In the individual figures, the potential to produce reflected glare is indicated by the weight of the line representing fixture output and reflectance from the work task. The batwing distribution (e) concentrates its output in the 30–60° range, which minimizes both direct and reflected glare potential.*

(b) Louvers and Baffles

These are generally rectangular and are of metal or plastic; they serve primarily to shield the source (see Figures 17.28 and 17.29) and to diffuse the output, particularly when plastic translucent louvers are used. Candlepower distribution curves are shown in Figure 17.35*b*. The exact curve shape depends on the shielding angle, the design of the louver, and its finish. Louvers finished in specular aluminum or dark colors exhibit low direct glare potential. The large downward light component can cause serious veiling reflections. Overall fixture efficiency is average.

The S/MH ratio, which indicates the maximum spacing permissible for a given luminaire mounting height that yields uniform illumination, is fully explained in the next section. For this diffuser type, it does not exceed 1.5 and varies inversely with the shielding angle. This is because the basic circular distribution is changed to an egg shape by cutoff and redirection, reducing the high-angle light. Thus, a 45° shielding angle results in lower direct-glare potential but requires closer spacing.

A special design in this category is the miniature eggcrate parabolic wedge louver shown in Figure 17.36. These units redirect a large portion of the light directly downward, and because of this redirection and their specular finish, they appear completely dark—darker, indeed, than the unlighted portion of the ceiling when viewed obliquely. Fixtures using these louvers have low efficiency due to trapped light, with a maximum coefficient of utilization of about 0.5. VCP is very high, but veiling reflections can be troublesome; S/MH varies between 1.0 and 1.5. The shielding angle is usually 45°. A typical candlepower distribution curve is shown in Figure 17.35*c*. When these units are used, additional wall lighting is almost always required. The luminance of fixtures and surfaces can be easily checked in the field using a portable luminance meter, as in Figure 17.37.

Because of the low efficiency of the parabolic louver design shown in Figure 17.36*a*, caused by the wide light-trapping tops of the louvers, an improved version was developed with the tops shaped to reflect incident light efficiently. This improved design, which is shown in Figure 17.36*b*, increases luminaire efficiency by about 20%.

(c) Prismatic Lens

Many designs are available with varying distribution characteristics. Figures 17.27*a* and *d* can be taken as typical of this genre. They produce an efficient fixture (high coefficient of utilization), good diffusion, wide permissible spacing—an S/MH as high as 2.0—and low direct glare potential (high VCP). Veiling reflections can be troublesome, depending upon viewing angles and position (see Figure 7.41).

(d) Fresnel Lens

The action of this lens is like that of a reflector: lamp-hiding power is poor, but efficiency is high and visual comfort likelihood is good; S/MH is seldom more than 1.5 (Figure 17.38).

(e) Batwing Diffusers

The theory behind this type of diffuser is covered in Section 18.6h. A typical characteristic is shown in Figure 17.35*e*. There are two fluorescent luminaire designs that produce the batwing shape

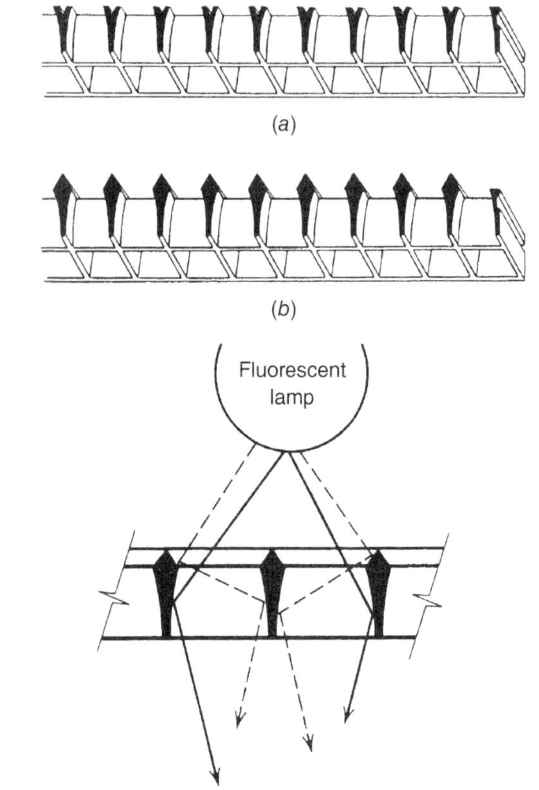

Figure 17.36 *(a) Section through a conventional, miniature parabolic wedge, eggcrate type of louver. These units give exceptionally low brightness when seen at a normal viewing angle. Most such units are made of aluminized plastic. Fixtures equipped with these louvers exhibit low overall efficiency due to the large amount of light trapped by the broad top of each parabolic wedge. (b) A modified wedge design uses a curved top on each wedge to redirect and utilize light striking the top. (c) Solid lines represent light rays redirected by the bottom curve, whereas dotted lines show light redirected by the top curve, which was lost in the design of (a). Typical louver cell dimensions are a ½-in. [12-mm] cube for design (a), with a consequent 45° shielding angle, and 5/6–3/4 in. [21–19 mm] square by ½ in. [12 mm] high for design (b), giving a 35–45° shielding angle.*

Figure 17.37 *Checking the luminance of a fixture with a luminance meter.*

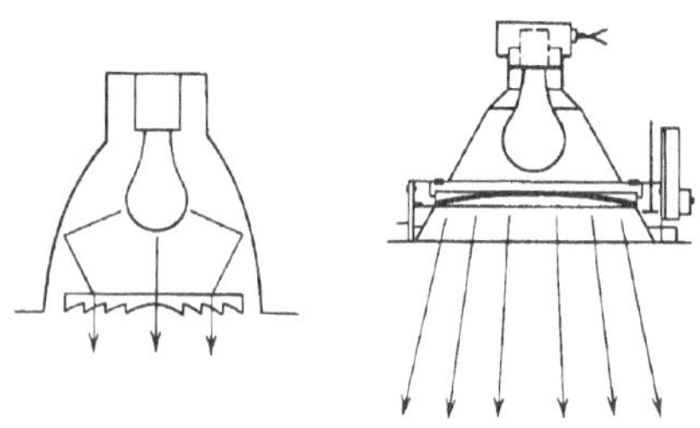

Figure 17.38 *Action of a Fresnel lens. With a Fresnel lens fixture, a smaller housing without a reflector can be used while still maintaining beam control. The lens performs the same function as a reflector, controlling the beam as a function of source placement. By utilizing a lens fixture, the curved reflector (a) can be largely eliminated, yielding a smaller fixture while maintaining accurate beam control. A common design (b) uses a regressed lens to provide shielding, although lens brightness is not normally objectionable.*

distribution characteristic—a prismatic lens and parabolic reflectors and baffles.

1. *Prismatic batwing diffusers:* These are either linear or radial; that is, they produce the batwing distribution either in one direction or in all directions. Typical characteristics of both types are shown in Figure 17.39. Note that the characteristic shape is more pronounced in the linear diffuser, which indicates better control of veiling reflections in that direction (usually crosswise). Fixtures equipped with these diffusers have good efficiency, low direct and reflected glare, and good diffusion. As with all enclosed ungasketed fixtures, the lens acts as a dust trap, necessitating frequent cleaning to maintain high output.

2. *Deep parabolic reflectors:* These luminaires (Figure 17.40) produce modified versions of the characteristic batwing distribution in the normal (crosswise) direction. Distribution in the parallel or lengthwise direction is circular (diffuse), indicating minimum beam control in that direction. These fixtures, like the batwing lens-type diffuser units, have high efficiency, high S/MH, low reflected glare, and low to very low surface brightness, making them usable in digital display areas. They are normally applied with the long axis in the direction of sight lines.

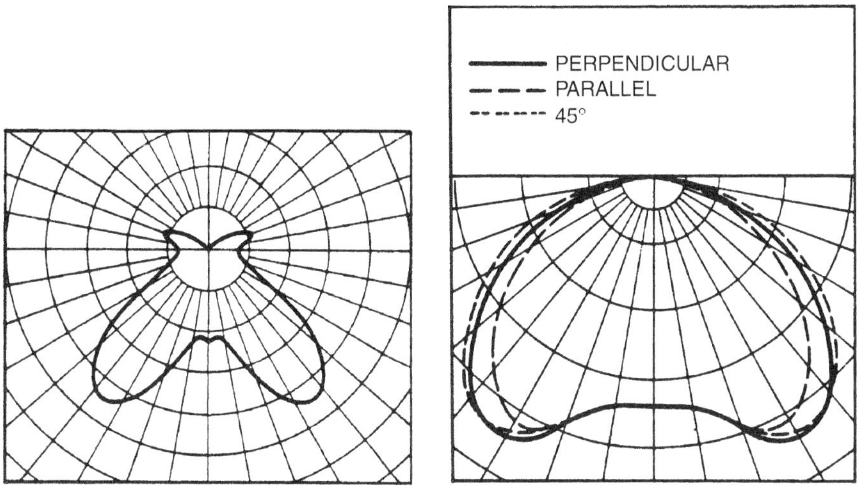

(a) (b)

Figure 17.39 *(a) Linear batwing distribution with extremely sharp cutoff in the upper and lower ranges. The curve is taken across the lamp axis for a single-lamp unit. (b) Distribution curves for a radial batwing distribution lens. Note that the perpendicular, parallel, and diagonal curves are almost identical. Zonal flux is maximum in the 30–60° range and drops off at both extremes, as desired.*

17.27 UNIFORMITY OF ILLUMINANCE

In a space intended to be lit uniformly with multiple, discrete, ceiling-mounted direct-distribution fixtures, a fixture spacing that gives acceptable uniformity of illumination must be used. A 1:1.3 ratio of maximum to minimum illuminance on the working plane is readily acceptable because such a difference is not easily noticed. For general background or circulation lighting, a ratio of up to 1.1.5 is acceptable. S/MH recommendations given by manufacturers are generally based upon a 1.1 illuminance ratio (Figure 17.41). Therefore, such S/MH recommendations may be somewhat exceeded without affecting occupant perceptions.

When a luminaire's distribution characteristic is symmetrical in all directions, as is generally the case with small-source lamps such as incandescent, compact fluorescent, and high-intensity discharge, only a single S/MH value is required for design. Most fluorescent fixtures, however, have an asymmetrical distribution and both crosswise and lengthwise S/MH ratios are required. The transverse (crosswise, perpendicular) ratio is almost always considerably higher than the longitudinal (parallel, endwise, lengthwise) ratio.

Luminaire manufacturers typically provide spacing criteria (SC), which are typically defined in terms of the spacing to mounting height ratios (S/MH), for specific light fixtures with direct lighting distribution. Adhering to these criteria better ensures that a lighting installation will provide the intended "even" uniformity of illuminance. The S/MH for a specific luminaire is determined by finding the distance between two test luminaires that yields the same illuminance on the working plane midway between the luminaires as directly under each one. This process deliberately ignores any contributions from other fixtures in a multifixture installation and from interreflections, accounting only for the direct component of illuminance from the two test fixtures. Therefore, in an actual installation with several rows of fixtures, the illuminance at point P_1 in Figure 17.42 is higher than the average by 20% to 30% because of the other fixtures and interreflections, and the illuminance at point P_2 is approximately equal to the room average. The illuminance levels along the walls, assuming a distance between the last row and the wall equal to one-half of the side-to-side spacing, range from 60% of average at point P_3 down to 50% at point P_4, assuming light-colored walls.

With dark walls, bookshelves, dark wood paneling, and the like, adjacent illuminance can drop to less than 50% of the average room illuminance. To counteract this negative effect, particularly when placement of furniture is such that visual tasks will occur near walls, the designer has three choices:

1. Reduce the distance between the last row of fixtures and the wall to a third or less of the row-to-row spacing. This also provides required wall lighting.
2. Provide some type of continuous perimeter lighting or wallwash units, both of which increase illuminance levels at the walls.
3. A combination of choices 1 and 2.

Because endwise illumination from linear fluorescent fixtures is considerably lower than crosswise

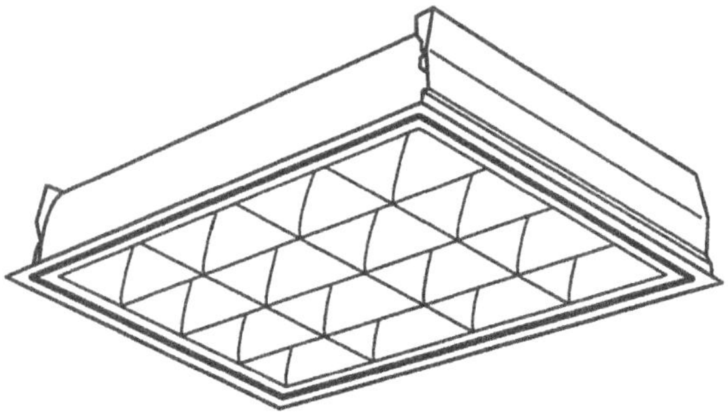

TOTAL LUMINAIRE EFFICIENCY =		66.0 %
TOTAL REFLECTANCE OF PAINT =		88.3 %
CIE TYPE – DIRECT		
PLANE	0-DEG	90-DEG
SPACING CRITERIA	1.2	1.6
SHIELDING ANGLES	36	43
PLANE	0-DEG	90-DEG
LUMINOUS LENGTH	21.250	21.250

LUMINANCE DATA IN FOOTLAMBERTS

ANGLE IN DEG	AVERAGE 0-DEG	AVERAGE 45-DEG	AVERAGE 90-DEG
45	2451.	3101.	1927.
55	494.	639.	55.
65	0.	0.	0.
75	0.	0.	0.
85	0.	0.	0.

Recessed

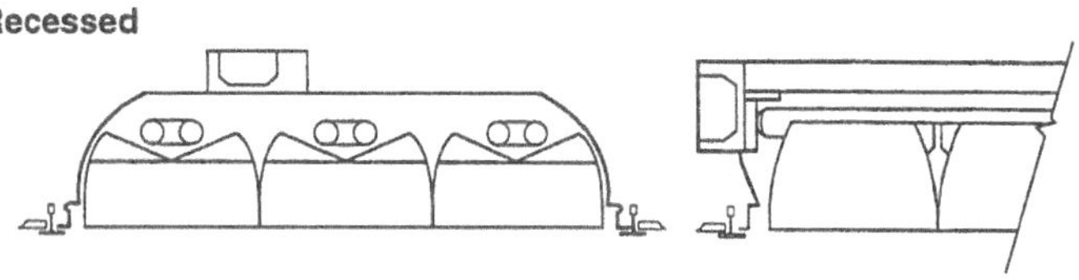

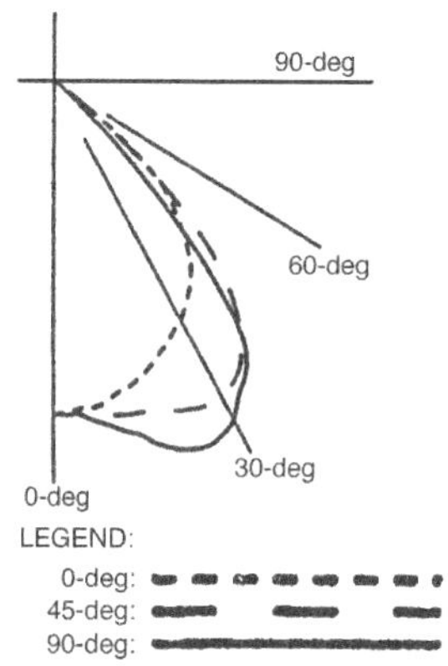

Visual Comfort Probability

Reflectance - 80, 50, 20
Work Plane Illumination - 100 FC@ 2.5 ft.
VCP

Room W	Room L	Luminaires Lengthwise Ht. 8.5	10.0	13.0	Luminaires Crosswise Ht. 8.5	10.0	13.0
20	20	68	81	77	90	90	87
20	40	69	81	77	91	90	84
30	30	69	81	77	91	90	84
30	60	69	82	78	91	90	84
40	40	69	82	78	91	90	84
40	60	70	82	78	91	90	84
60	30	69	82	78	91	90	84
60	40	70	82	78	91	90	84
60	60	70	82	78	91	91	84

Coefficients of Utilization

Coefficients of Utilization
Zonal Cavity Method — Floor Reflectance - .20

RC	80				50		
RW	70	50	30	10	50	30	10
1	75	73	71	69	69	67	66
2	71	67	65	62	64	62	60
3	67	62	59	56	59	57	54
4	63	57	54	50	55	52	50
5	59	53	49	46	51	48	45
6	55	49	44	41	47	43	41
7	51	44	40	37	43	39	36
8	48	41	36	33	39	35	33
9	44	37	32	29	36	32	29
10	41	34	29	26	33	29	26

Energy Data

LER: FP-42 **Energy Cost: $5.71***
Input Watts: 134 **BF: .90**

The above energy calculations were conducted using a specific lamp/ballast combination. Actual results may vary depending upon the lamp and ballast used. Lamp and ballast specifications are subject to change without notice.

*Comparative annual lighting energy cost per 1000 lumens based on 3000 hours and $0.08 per KWH.

***Figure 17.40** A 2-ft [610-mm] square, deep parabolic reflector luminaire designed for three F40, 22.5-in [572-mm] twin-tube CFL lamps, with a rated output of 3150 lm each. Total power input is 134 W and ballast factor (BF) is 0.9, giving a luminaire efficacy rating (LER) of 42. The typical modified batwing crosswise distribution and circular lengthwise distribution are clearly shown. Photometric data are also shown. (Courtesy of Columbia Lighting.)*

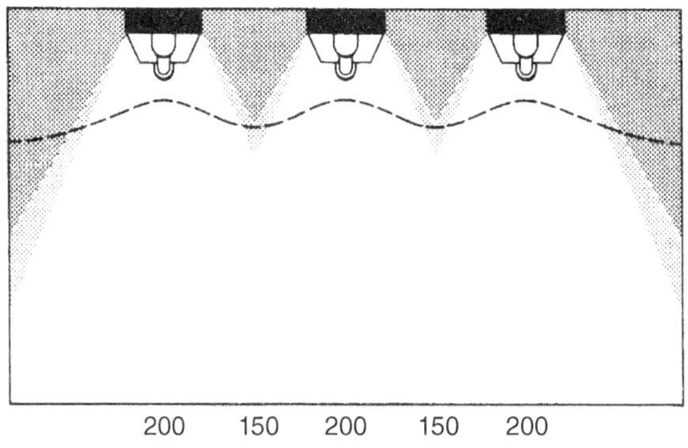

Figure 17.41 *The ratio of maximum to minimum illuminance should not exceed 1.3 in areas requiring uniform illumination.*

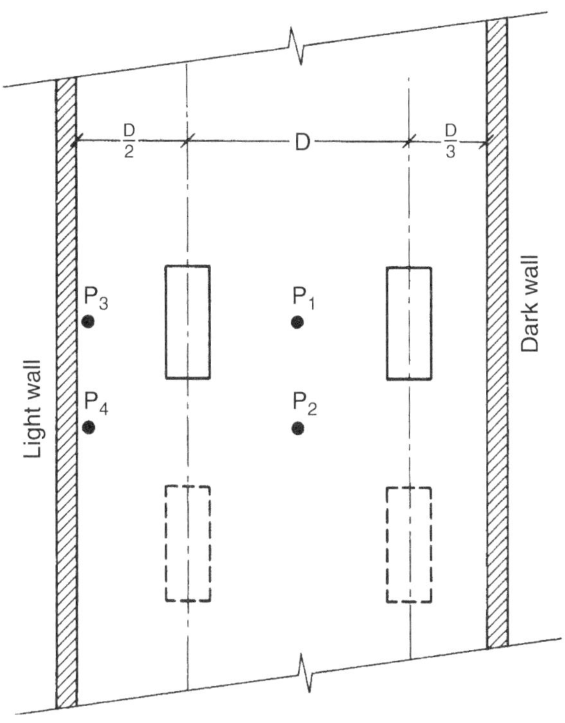

Figure 17.42 *The diagram shows lighting fixtures installed according to the manufacturer's recommended spacing criteria (ratio of spacing to mounting height above the work plane), with a row-to-row spacing, D, and a row-to-wall spacing, D/2, as shown at the left wall. Assuming a high-reflectance (light color) finish on the wall, illuminances P_3 and P_4 are at least one-half of the illuminance directly below a fixture. With a dark wall, as on the right, illuminance would fall below this value. Consequently, the right row of luminaires should be placed closer to the wall, as shown.*

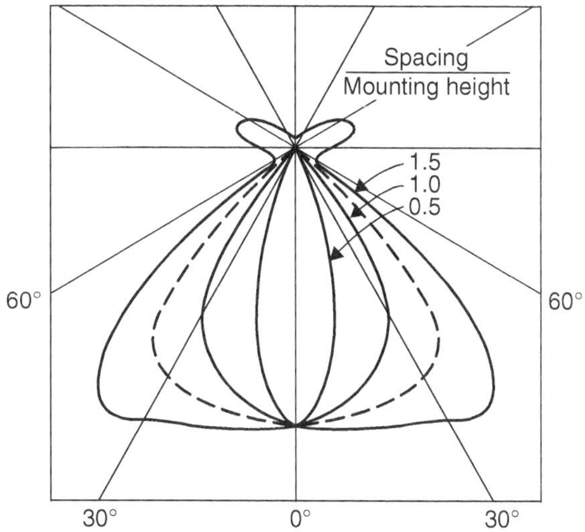

Figure 17.43 *Curves for approximating the ratio of fixture spacing to mounting height (S/MH) above the working plane for typical direct distribution luminaires. These curves were developed for point sources such as incandescent (and HID) but can also be applied to asymmetric distribution luminaires using fluorescent lamps. The practical permissible S/MH is somewhat higher than the curves indicate because this is a semi-direct distribution, and the ceiling light component permits wider spacing between units. (After Odle and Smith, from Illuminating Engineering, January 1963.)*

illumination, end walls have lower illuminance than side walls and greater illuminance variation. It is, therefore, particularly important to provide some additional illumination, as discussed previously, and to terminate fixture rows no more than 1 ft [0.3 m] from an end wall. This is even more important where visual tasks without supplementary task lighting occur at these walls.

It was mentioned previously that the fixture in Figure 17.27*a* had a high S/MH because of its flat-bottomed curve. This ratio, when not given by the manufacturer, may be approximated from Figure 17.43. A more accurate method of calculating maximum to minimum illuminance ratios is available in the 10th edition of the *IES Lighting Handbook* (2011a).

The foregoing discussion of illumination uniformity was concerned with uniformity on a horizontal work plane. Occasionally, it is necessary to know the degree of uniformity vertically—that is, on horizontal planes at different elevations directly below the fixtures. Four different lighting situations are normally encountered. They are point sources, such as point source downlights; line sources, such as continuous-row fluorescent fixtures; infinite sources, such as luminous ceilings—whether transilluminated or indirect; and parabolic reflector beams, such as from parabolic aluminized reflector (PAR) lamps. The vertical uniformity of each type is shown graphically in Figure 17.44.

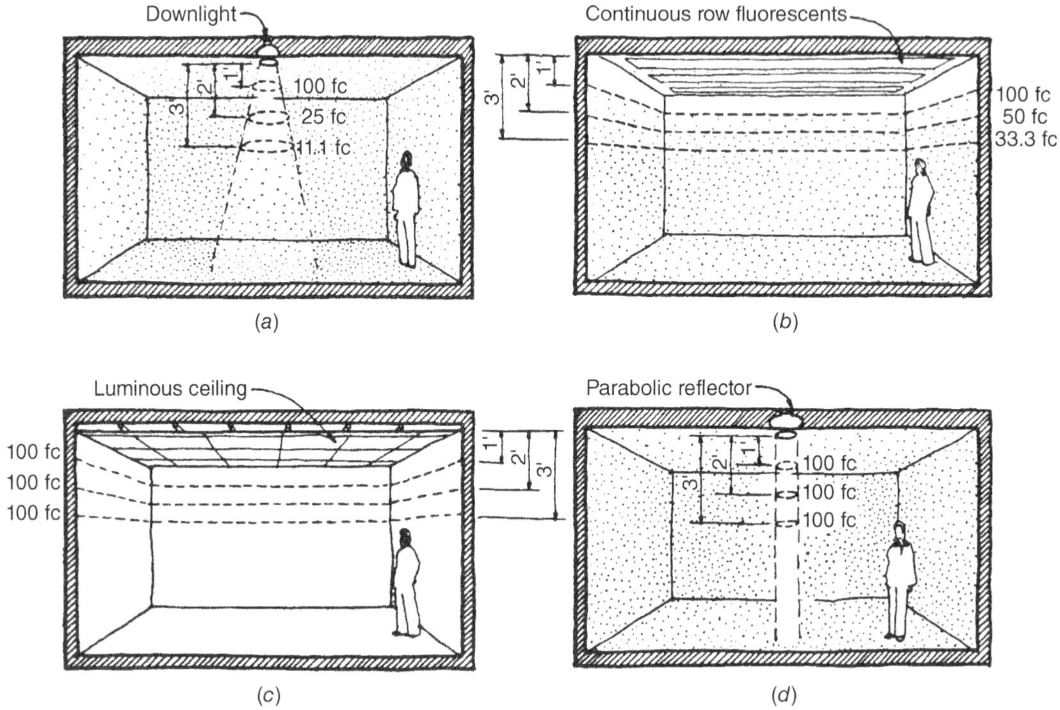

Figure 17.44 *Variation of illuminance vertically, directly below the fixtures, for different source types. (a, b) Illuminance directly below the fixture varies inversely with the square of the distance for a point source and inversely with the distance for a line source. (c, d) Illuminance remains constant at all distances from either an infinite (or nearly so) source or a parabolic reflector.*

17.28 LUMINAIRE MOUNTING HEIGHT

Luminaire mounting height is normally established before determining their spacing. For fixtures with an upward distribution component, a balance must be struck between low mounting (greater distance from ceiling), which affects ceiling brightness and gives good utilization of light, and a reluctance to dominate an area, particularly a large room, by using such a low mounting height that the apparent ceiling height is affected (Figure 17.45). General rules for mounting height are:

1. Indirect and semi-indirect luminaires should normally be suspended no less than 18 in. [460 mm] from the ceiling and preferably 24 to 36 in. [610 to 915 mm]. Single-lamp luminaires with a very wide distribution (inverted batwing) may be suspended as little as 12 in. [305 mm] from the ceiling. Manufacturers' recommendations should be sought and followed.
2. Direct-indirect and semi-direct fluorescent fixtures should be suspended not less than 12 in. [305 mm] for two-lamp units and 18 in. [460 mm] for three-lamp and four-lamp units.

The effect of mounting height (expressed as pendant length) on lighting system coefficient of utilization is given in Figure 17.46.

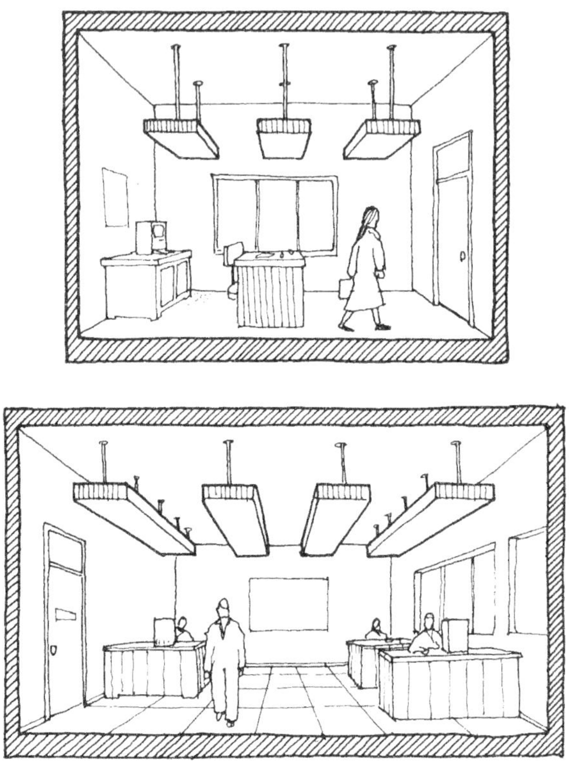

Figure 17.45 *Fixture mounting height may be lower in a small room than in a large room because of the illusion of lowness that is created in a large room.*

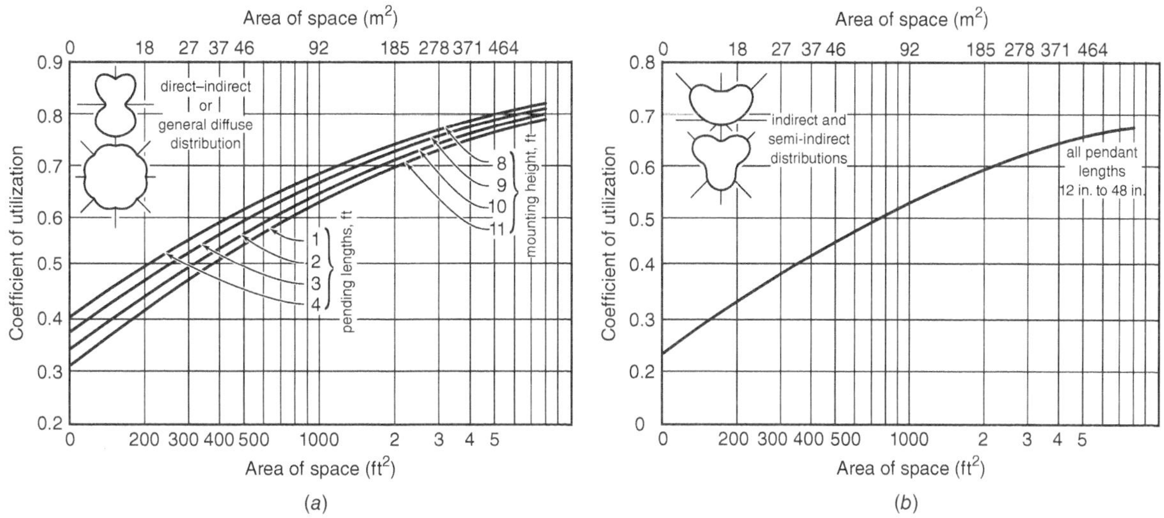

Figure 17.46 *Coefficient of utilization (CU) as a function of pendant length for various fixture distributions. With a substantial downward component, as in direct–indirect or general diffuse lighting (a), CU increases slowly as the fixture descends (pendant length increases). Maximum differentials occur in small rooms and can reach 20%. Where the ceiling is the light source, as in indirect and semi-indirect systems (b), the pendant length does not change the room illuminance. This curve can be used to estimate CU for indirect and semi-indirect luminaries in the absence of manufacturer's data. (Redrawn with SI units by Lisa Leal.)*

17.29 LIGHTING FIXTURE CONSTRUCTION

This and subsequent sections cover luminaire construction, installation, and assessment. Electric lighting fixtures typically constitute 25–30% of the electrical systems budget, which is normally around 4–5% of the overall building budget. The difference between a high-quality luminaire and an inferior one is often not apparent under casual inspection. Thus, care must be taken when specifying lighting fixtures and in reviewing luminaire submittals (shop drawings) and samples—to ensure that an appropriate quality fixture has been delivered and properly installed. The building commissioning process can assist in providing quality control for lighting installations.

All fixtures, if used properly, can give a sufficient quantity of light, but only a good quality unit will provide ease of installation and maintenance, lack of degradation, and expected life. Proper installation is key to ensuring mechanical rigidity and safety, electrical safety, freedom from excessive temperatures, and requisite accessibility of component parts and of the fixture outlet box. The following information is a combination of *NFPA 70*—the *National Electrical Code* (*NEC*)—minimum requirements and factors beyond these minima that may be important.

Although LED-based luminaires have diminished the fire safety threat otherwise posed by comparable luminaires with high operating temperatures due to their lamp types, it is important to ensure the incorporation of basic fire safety-related specification and installation criteria for any lighting fixture package:

1. All lighting fixtures should be wired and constructed to comply with local codes, *NEC*, Underwriters Laboratories (UL) standards for luminaires (such as *UL 1598*), and should bear the UL label where label service is available.
2. Extruded aluminum or composite materials are recommended for LED fixtures to optimize weight and improve thermal management. Non-LED fixtures should generally be constructed of at least 20-gauge (0.0359-in. [0.9-mm]) steel. Cast portions of fixtures should be no less than 1/16 in. [1.6 mm] thick.
3. Ensure proper thermal management. LED fixtures generate less heat but are more sensitive to it. As such, some LED fixtures may include integrated aluminum or copper heat sinks to better ensure longevity. LED fixtures often utilize highly reflective materials such as enhanced aluminum or reflective films optimized for LED applications. For non-LED fixtures, all metals should be coated. The final coat should be a baked-enamel white paint of at least 85% reflectance, except for anodized, aluminum, or silvered surfaces.
4. Although LED fixtures have relatively low operating temperatures far below code thresholds that pre-date mainstream adoption of LEDs, no point on the outside surface of any fixture should exceed 90°C [194°F] during continuous operation (unless otherwise explicitly exempted by the *NEC*).
5. All fixtures should be identified by a label carrying the manufacturer's name and address and the fixture catalog number. Consider digital tracking

in addition to physical labels as many manufacturers are now incorporating QR-coded and digital tags.

6. Ensure compatible diffuser materials. Both glass and plastic diffuser selection should incorporate the directional nature of the specific lamp type used in the luminaire to maximize light transmission while minimizing loss.
7. For all fixtures, select plastic diffusers that will resist the tendency to yellow over time due to the light spectrum emitted from the specific lamp type. For non-LED fixtures, plastic diffusers should be of the slow-burning or self-extinguishing type with a low smoke-density rating and low heat-distortion temperatures. The latter should be low enough that the diffuser distorts sufficiently to drop out of the fixture before reaching ignition temperature.
8. It is imperative that plastics used in air-handling lighting fixtures be of the noncombustible, low-smoke-density type. These requirements also apply to other nonmetallic components of such fixtures.
9. All plastic diffusers should be clearly marked with their composition material, trade name, and manufacturer's name and identification number. Results of ASTM combustion tests should be included with fixture submittals (shop drawings). The characteristics of many plastic diffusers change radically with age and with exposure to UV radiation.

A brief survey of the most common standard transparent and translucent lighting fixture diffusers follows.

1. *Glass:* Transparent to translucent, scratch-resistant, easily cleaned, nonflammable, available in all grades of impact resistance, non-yellowing, usable with all sources, unaffected by and effective in blocking UV, readily formed into desired patterns. Heavy and expensive.
2. *Acrylic (Plexiglas):* Clear to translucent, easily scratched, slow burning, low impact resistance, resistant to yellowing, and available in a special low-yellowing composition (UV grade). Usable indoors and outdoors. Not usable at the elevated temperatures (>90°C [194°F]) found in some HID applications. Does not readily embrittle, warp, or craze. Molds well.
3. *Polycarbonate:* Initially very clear and highly impact-resistant, but with a tendency to opacity and strength loss with age. Good scratch resistance and excellent thermal resistance. Usable indoors and outdoors with all sources. Readily molded to prismatic forms. Self-extinguishing (burning rate). Expensive.
4. *Polystyrene*: Usable only indoors because of rapid yellowing when exposed to exterior UV radiation. Slow burning, but smoke-generation properties problematic with some fire codes. Not usable in the long term because of discoloration, particularly with UV-producing sources. Good thermal resistance. Readily molded. Cheap. Not scratch-resistant.

As can readily be seen, no ideal diffuser material exists. A designer must select the material that best suits the use and budget, considering both initial and replacement costs as well as long-term optical properties. Ballasts should be mounted in fixtures with captive screws on the fixture body to allow ballast replacement without fixture removal.

All fixtures mounted outdoors, whether under canopies or directly exposed to the weather, should be constructed of appropriate weather-resistant materials and finishes, including gasketing to prevent entrance of water into wiring, and should be marked by the manufacturer as being "Suitable for Outdoor Use."

17.30 LIGHTING FIXTURE STRUCTURAL SUPPORT

Compared to their comparable predecessors, LED fixtures are relatively lighter and the need for heavy heat sinks and bulky housing has decreased with more contemporary design options. However, many type of lighting fixtures (especially highly decorative luminaires and/or those with non-LED lamps) may pose considerable structural loads. Although some codes allow certain fixtures weighing less than 40 lb [18 kg] to be mounted directly on the horizontal metal members of hung-ceiling systems, experience has shown that vibration, member deflection, routine maintenance operation on equipment in hung ceilings, and poor workmanship can cause such fixtures to fall, endangering life. It is, therefore, strongly recommended that all fixtures—surface, pendant, or recessed—whether mounted individually or in rows, be supported from the ceiling system support (purlins) or directly from the building structure, but in no case by the ceiling itself. This is particularly important in the case of an exposed "Z" spline ceiling system.

17.31 LUMINAIRE EFFICACY RATING

The concept of a luminaire efficacy rating has received variable traction over the years. The concept parallels that of lamp luminous efficacy—but applies to the complete light fixture (not just the lamp). Official recognition of a consensus standard for LER seems iffy at this time. NEMA has withdrawn two standards on the subject: *Standard LE 5-2001: Procedure for Determining Luminaire Efficacy Ratings for Fluorescent Luminaires* and *LE 6-2014: Procedure for Determining Target Efficacy Ratings for Commercial, Industrial, and Residential Luminaires*. Nevertheless, the concept of full-luminaire luminous effectiveness is valid and is noted here for reference.

The expression used to calculate a luminaire efficacy rating (LER) is:

$$LER = \frac{(EFF)(TRLL)(BF)}{LWI}$$

where

LER = luminaire efficacy rating, lumens per watt
EFF = percent of lamp lumens that exit luminaire, decimal
$TRLL$ = total rated lamp output, lumens
BF = ballast factor, decimal
LWI = total luminaire wattage input (including lamps and electronics)

LER can be summarized and simplified as: those lumens that leave a fixture / those watts that enter a fixture. This basic relationship applies to essentially any type of luminaire (including LED) and can be used as a figure of merit in lighting fixture selection. The United States Federal Energy Management Program (FEMP) in fact values LER as a primary selection criterion for parking lot lighting (DOE, n.d.).

17.32 BUG RATING

Outdoor lighting criteria may specify BUG ratings for a luminaire. This refers to a luminaire classification system that incorporates independent ratings of a luminaire's optical performance with regard to three light trespass criteria: backlight, uplight, and glare (thus, BUG).

The BUG method takes into consideration a luminaire's uplight shielding, glare shielding, and backlight shielding, and it limits lamp lumens to values appropriate for a given lighting zone (Figure 17.47). *IES TM-15-20*, Annex A offers tabulated BUG ratings based on zonal lumen calculations using data from IES-approved photometric testing procedures for outdoor luminaires. Based on the degree of backlight (i.e., backlight/trespass), uplight (i.e., uplight/skyglow), and glare (i.e., glare/offensive light) that occurs at certain angles, each criterion is given a rating between 0 and 5 (B0 to B5 for backlight, U0 to U5 for uplight, and G0 to G5 for glare).

BUG ratings supersede former "cutoff" ratings. The system is a convenient method to ensure such criteria are incorporated into outdoor lighting design. For example, a project owner may desire a light pollution reduction strategy in order to increase night sky access, improve nighttime visibility, and mitigate negative effects on wildlife in and around the site. The owner could set maximum uplight ratings for luminaires based on the BUG ratings (e.g., a luminaire uplight rating no greater than U2). Green building certification systems, such as Leadership in Energy and Environmental Design (LEED), may require teams to classify a project under a lighting zone (i.e., LZ0 through LZ4) as defined by the Illuminating Engineering Society and International Dark Sky Association (IES/IDA) *MLO-11: Model Lighting Ordinance with User's Guide* (IES, 2011) and then adhere to zone-specific maximum uplight ratings using the BUG method.

BUG ratings are commonly published by product manufacturers. However, the ratings are based on specific design conditions. If the position of a luminaire used on a project is not in accordance with the published data, then the BUG ratings may need to be recalculated. Software can be used to determine BUG ratings using the IES photometric file for a luminaire.

17.33 LIGHTING FIXTURE ASSESSMENT

Intense competition in the lighting products field necessitates close scrutiny of the characteristics of luminaires and all accessories. To compare the relative merits of similar lighting fixtures manufactured by different companies, complete test data plus a sample in a regular shipping carton from a normal manufacturing run are needed.

The following list should be used as a basic guide to required information, with additional items added according to job requirements:

1. *Photometric and design data:* Manufacturers should furnish complete test data, including candlepower distribution curve(s), coefficients of utilization, wall and ceiling luminance coefficients, luminance data from 45° to 85°, a table of VCP values, energy performance data including LER, and recommended SC (S/MH). These data should come from a reliable independent testing laboratory, not from the manufacturer's test facilities. In addition, many manufacturers either publish or make available upon request various design aids such as

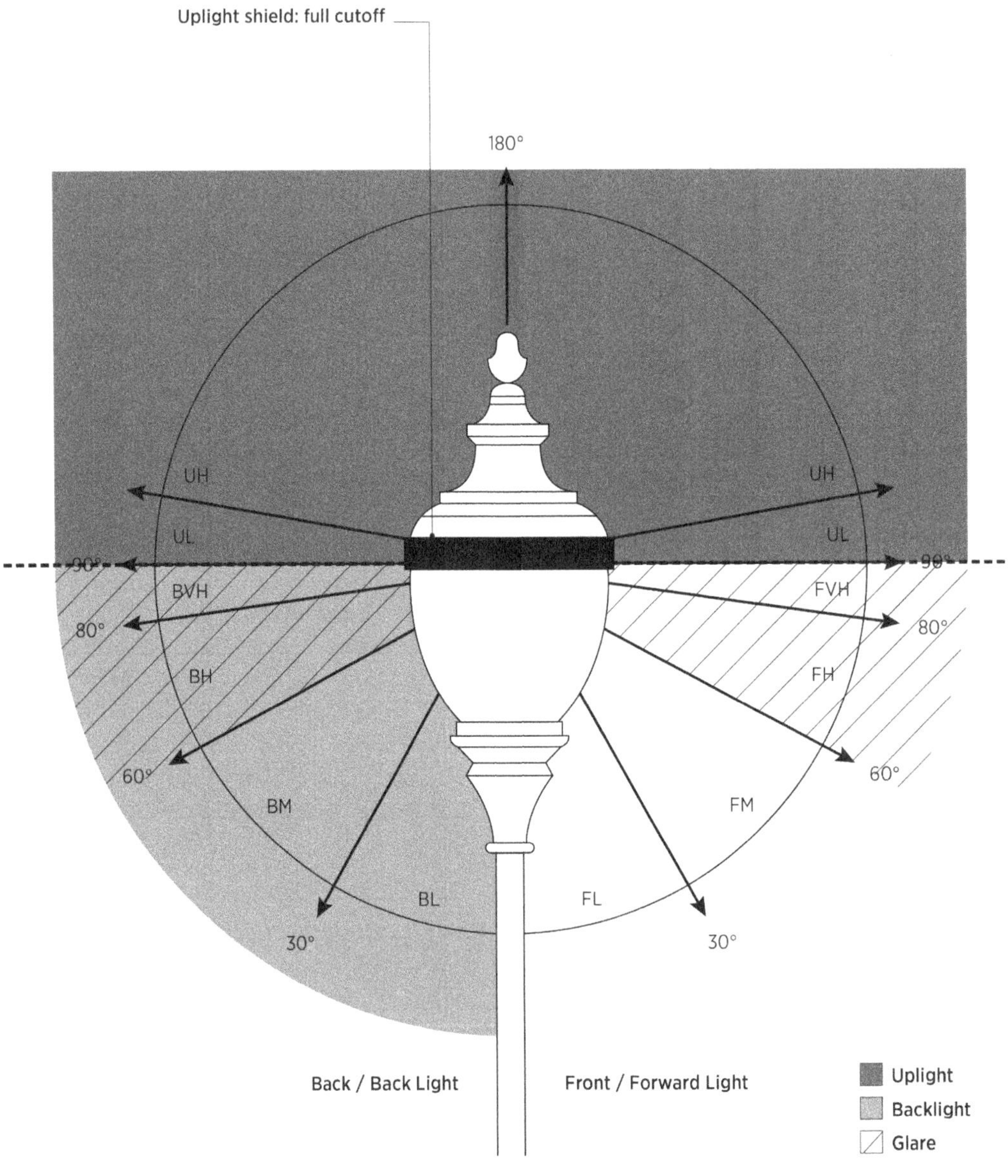

Figure 17.47 *In order to assess a luminaire's backlight, uplight, and glare (BUG) rating, its subzones depicted in this section diagram are individually rated on a scale from 0 to 5, going from lowest to highest luminous flux. The highest rating of a subzone determines the overall rating for that zone, and these readings are compiled into the BUG lighting classification. (Courtesy of U.S. Green Building Council.)*

isolux [isofootcandle] curves and point-by-point illuminance plots for different fixture layouts.

2. *Construction and installation:* The designer should check the sample for workmanship; rigidity; quality of materials and finish; and ease of installation, wiring, and leveling. Installation instruction sheets should be sufficiently detailed to be of use on the job site. Results of actual operating temperature tests in various installation modes should be included. Air-handling fixtures should be accompanied by heat-removal data, pressure-drop curves, air-diffusion data, and noise criteria (NC) values for different airflow rates.
3. *Maintenance:* Luminaires should be simply and quickly relampable, resistant to dirt collection, and simple to clean. Replacement parts must be readily available.

LIGHTING CONTROL

17.34 WHY LIGHTING CONTROL?

The term lighting control encompasses all the techniques by which a lighting system can be operated and includes both manual and automatic controls. A control strategy must be determined simultaneously with the lighting delivery design because the control scheme must be appropriate to the light source. In turn, the system's accessories and arrangement depend on the control scheme. For instance, if dimming a fluorescent light source is desired, the expected range of dimming determines the type of ballast, the ballast switching points, and the degree of dimming flexibility.

The primary purposes of lighting control are flexibility and economy. Flexibility to provide desired changes in illuminance, luminance, and brightness patterns; economy of both energy and costs. A properly designed lighting control system can reduce energy usage up to 60% over a simple on-off installation. In addition, other operating economies result from:

- reduced energy use
- reduced air-conditioning costs as a result of lower light-to-heat energy conversion
- longer lamp and ballast life due to lower operating temperatures and lower output
- lower labor costs due to control automation

Lighting in most new U.S. nonresidential construction is designed within the energy constraints of *ASHRAE/IES Standard 90.1*. This standard provides lighting power credits for automatic energy-conserving controls. These credits permit the effective connected load to be reduced by factors that increase with energy conservation effectiveness. Thus, for example, circuits with a simple on-off mode initiated by a daylight sensor have a lower power credit than daylight sensing with continuous dimming, because the latter is more energy economical (but more expensive initially).

To avoid confusion, particularly in view of overlapping and sometimes inaccurate terminology, it is necessary to differentiate between control functions, control devices, and control systems. For lighting, the only control functions are switching and dimming. The control devices are the means by which the switching and dimming functions are accomplished. These devices are numerous, ranging from a simple wall switch to time switches and dimmers of all sorts. Generally also included in this category are control initiation devices, such as occupancy sensors and photocells. The control system is the entire assembly of control and signal-initiating equipment together with all interconnections. Microprocessors and programmable controllers are also considered part of the control system. The system can be a stand-alone arrangement or, alternatively, as is the case in large facilities, it can be part of an energy management system (EMS) and/or a building automation system (BAS). The difference between an EMS and a BAS lies in the underlying objective—which is primarily energy-oriented for an EMS, and function-oriented for a BAS. BAS are defined in greater detail in Chapter 32. In the following sections, three key control criteria are considered: energy conservation, cost reduction, and operational flexibility.

17.35 LIGHTING CONTROL: SWITCHING

There are two basic control functions—switching and dimming. Switching is an on-off function. By selecting the number of lighting elements to be activated (or deactivated) by each switching action, the designer can establish the number of control levels. The more levels, the finer the control. Thus, in a space requiring several levels of uniform illuminance for different functions, there are many switching alternatives. Entire fixtures can be switched, but this adversely affects lighting uniformity. Using three-lamp fluorescent fixtures as an example, better uniformity and four levels of illuminance can be obtained by switching the ballasts (assuming appropriate ballast circuiting as described below):

All ballasts on	100% illumination
Two-lamp ballast on	66% illumination
Half of two-lamp ballast on	33% illumination
All ballasts off	0% illumination

The energy and flexibility benefits of such ballast control need to be assessed against the cost of the necessary ballasts and circuiting. Similar control possibilities, and cost considerations, are available for other lamp types.

Lamp switching has the advantage of light reduction in relatively small steps at low cost. A typical arrangement is shown in Figure 17.48. Use of split-wired two-lamp ballasts is advantageous from the cost and energy viewpoints. Even more uniform light reduction and finer control are possible with two-level ballasts (at increased cost). With a two-level system, each lamp can operate at either full or half output. Thus, the system could have 0%, 50%, and 100% output or, by combining alternate ballast switching with two-level ballasts, could have 0%, 17%, 33%, 50%, 67%, 83%, and 100% output options. However, if that range of control options is desirable, dimming is probably preferable. The choice depends on the type of space and on the project economics, as discussed later. An alternative method of using switching to achieve reduced lighting levels in discrete steps is to introduce impedance into the lighting circuit. This acts to reduce circuit current and light output. Such devices are readily available for control of fluorescent lamps (see Figure 17.52).

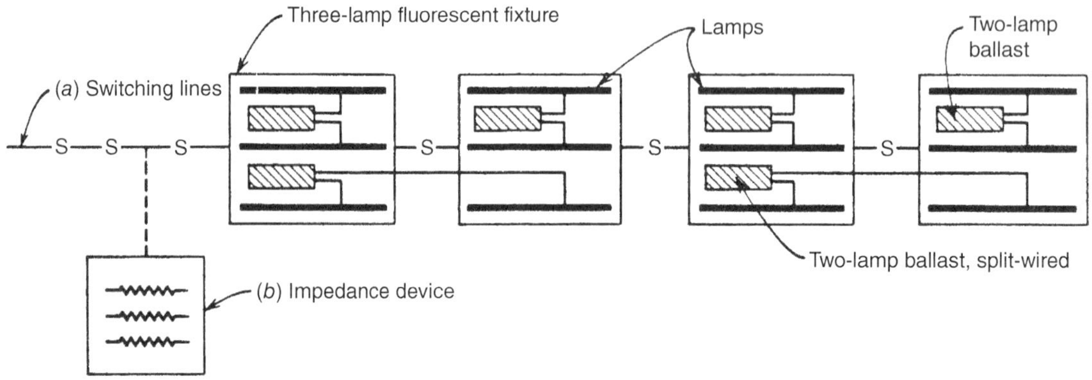

Figure 17.48 *Schematic diagram of switching arrangements to achieve multiple discreet lighting levels with three-lamp fluorescent lighting fixtures. Two-lamp ballasts are used in the interest of energy conservation and financial economy. In scheme (a) ballasts are switched, thus removing either one or two lamps from service. Finer control is achieved by using two-level ballasts or by introducing impedance (b) into the circuit, either in a block for an entire circuit or distributed in each fixture.*

Recognition of the fact that an increased number of control (switch) points makes possible finer control, and can reduce energy consumption if properly utilized, led to requirements in *Standard 90.1* relating to the number of control points in a space and their types. Many field studies have demonstrated conclusively that reliance on manual switching is not an effective long-term energy reduction strategy, regardless of the good intentions of the space occupants. Indeed, studies have shown that space "ownership" affects even the low level of energy savings possible with manual switching. Essentially, lighting in private offices, conference rooms, and small storage spaces may be switched off, whereas lighting in multi-occupancy and common-use spaces, such as libraries, large office spaces, break areas, and the like, may not be. As a result, *Standard 90.1* first specifies the minimum number of control points required in a space and then awards automatic switching or dimming a higher number of control credits.

The basic requirement is for one control point for every 450 ft^2 [42 m^2] or fraction thereof of enclosed lighted space, plus one control point for each task (or group of tasks) located in the space. Thus, a hypothetical classroom would require two control points for its 517 ft^2 [48 m^2] floor area plus one control point for all similar tasks grouped in the room, for a total of three control points. This requirement could be met with three wall switches, or one wall switch and an occupancy sensor, or an automatic dimming system most likely initiated by daylight sensors. The point is to encourage use of automatic controls, which is the only proven method of attaining significant energy use reductions.

17.36 LIGHTING CONTROL: DIMMING

The techniques and equipment required to dim each lamp type, as well as the effect on the color of the light

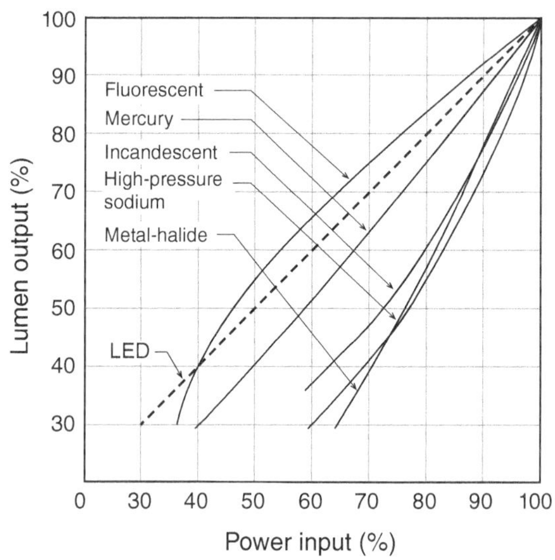

Figure 17.49 *Typical dimming curves for generic light source types. Note that fluorescent lighting is efficient and approximately linear down to 40% output. LED lamp dimming is similar to that of fluorescent. All other sources have reduced efficacy when dimmed. (Drawing revised by Martin Lee.)*

produced and on the lamp itself, are discussed in the sections above dealing with lamp types. Figure 17.49 shows typical lumen output versus power input curves for common light sources. Note that for fluorescent lamps, even with conventional magnetic ballasts, dimming down to approximately 40% of output is possible without reducing efficacy. This desirable characteristic can be exploited in control schemes where it is desired to change light output gradually without sacrificing efficiency. Due to efficacy reductions that occur below 40% output, an economical and efficient control scheme combines dimming and switching of multi-lamp fluorescent fixtures to yield an almost stepless output range of 13% to 100% output. Continuous dimming over a 10% to 100% range is practical with special individual magnetic dimming ballasts or with electronic ballasts. As discussed previously in this chapter, electronic ballasts are much more energy-efficient than magnetic ballasts and would be provided for all new installations, dimmed or not. For retrofit work, silicon-controlled rectifier (SCR) or triode for alternating current (TRIAC) dimmers give excellent results (down to 40% output) with existing magnetic core-and-coil ballasts.

17.37 LIGHTING CONTROL: CONTROL INITIATION

Control initiation is either manual, automatic, or a combined manual-automatic function. Combined control is usually in the form of an override function: for example, manual override of an automated procedure to address special or unusual situations with automatic reset to the automatic function to reestablish normal operation.

(a) Manual Control Initiation

Numerous studies dating back over 50 years have shown increased employee satisfaction when at least a degree of control over the working environment is provided to the employee. Such satisfaction is frequently accompanied by increased work output, at least in the short term. Unfortunately, manual control of lighting levels has been demonstrated to be energy-wasteful due to the tendency to leave luminaires at maximum output even when daylight is supplying more than enough light or when leaving a room for an extended period. Some energy use reduction in the latter instance is possible with the installation of time-out switches in spaces normally used for short periods, such as supply closets (see Chapter 28). For normal working spaces, however, even the installation of manual dimmers in private offices is not effective because of the need to go to the dimmer on the wall to readjust. Manual dimming in multiple-occupancy spaces is effective only in creating personnel dissatisfaction and friction.

There are practical remote-control dimming systems that can control single or multiple luminaires, making them applicable to all occupancies. Figure 17.50 illustrates the use of such a system in an open multiple-occupancy space. This enables an individual worker to adjust the output of those luminaires closest to his or her workstation without disturbing other employees. This adjustment, which is easily accomplished (see Figure 17.50*a*) by remote control, can alleviate direct and reflected glare experiences or can be a temporary expedient suitable to the task at hand, as, for instance, when working with a digital display. Figure 17.50*b–d* shows the system components for such control. Wiring of the system can be arranged so that a single receiver/dimmer can control up to 20 ballasts (or fixtures) or, conversely, multiple dimmers may be connected on a single circuit. The latter arrangement would be used in wiring a group of small single- or double-occupancy rooms. Such rooms are frequently wired with a wall-mounted dimmer, which can then be activated by the remote control, as shown in Figure 17.51.

(b) Automatic Control Initiation

Automatic controls are of two types: an open-circuit type and a closed-loop feedback type—these are also known as static and dynamic control, respectively. The former initiates a control function that is independent of the actual lighting situation, whereas the latter reacts to the condition of the lighting situation it controls via a feedback loop.

1. *Static control:* The most common type of open-circuit lighting control is the programmable time-base controller. These vary from small, relatively simple units that replace wall switches and fit into a common device box to the more sophisticated units shown in Chapter 28. These devices are available in myriad designs and capacities, but all perform the same basic function: (remote) control of loads and circuits on a preprogrammed time basis. The programming, in turn, is determined after analysis of operating schedules, task requirements, and field conditions. With "tight" programming, energy savings of up to 50% over an uncontrolled installation are possible.

   Because these devices act only on a time signal and are insensitive to actual field conditions, an override feature must be incorporated to permit accommodation of special conditions. Thus, if a timer is set to shut off a row of fixtures adjacent to windows between 10:00 a.m. and 3:00 p.m., local override must be provided to accommodate dark rainy days and the like. Similarly, if luminaires are turned off during nonworking hours, provision must be made for persons working overtime. The override arrangement can be entirely local, in which case it may lead to energy waste because it depends on local cancellation; it can be local with a time-out, which can be a nuisance to a person working for an extended period; or it can incorporate an override feedback link to the controller, usually operated by telephone lines. In general, programmable time controls are best applied to facilities with regular, repetitive schedules and few exceptional situations.

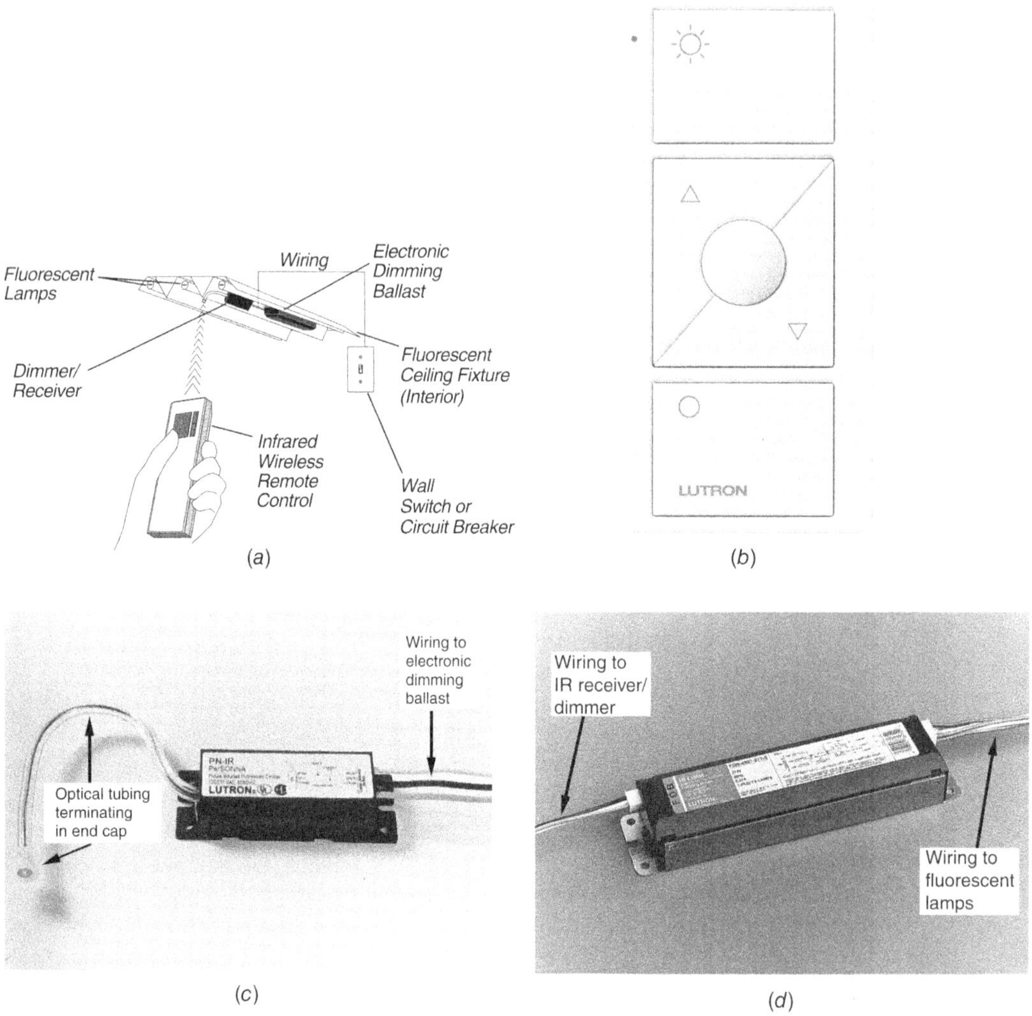

Figure 17.50 *(a) Ceiling-mounted fluorescent fixtures within an 8-ft [2.4-m] radius of the optical sensor can be dimmed by remote control and restored almost instantaneously. (b) The handheld IR remote-control device for continuous dimming and for dimming override for maximum or minimum light. (c) The IR-activated receiver/dimmer, which fits into a 4-in [100-mm] square standard outlet box, can be mounted inside the luminaire as in (a) or adjacent to it. (d) High-frequency electronic dimming ballast for one to four lamps can provide full-range dimming down to 1–5% of output or down to 10% of output, depending on the type. The optical tube that carries the IR signal must be exposed on the ceiling. It may be mounted on the underside of an open luminaire or close by. (Courtesy of Lutron Electronics Co.)*

2. *Dynamic control:* The second type of automatic control initiation responds to sensor-indicated field conditions via an information feedback loop. It is frequently referred to as dynamic control because the initiation of a control function depends not on a fixed parameter such as a specific time, but on real-time field parameters (as measured at that instant). In the case of lighting control, these parameters may be ambient illuminance, time, system kW demand, kWh usage during a specified time period, or space occupancy—singly or in combination—depending on the programming algorithm (i.e., how the controller microprocessor has been instructed). The control device in its entirety is called a programmable controller (see Chapter 28) which, in combination with the field sensors and the interconnecting wiring, constitutes the control system. Some systems are wireless, using high-frequency signals impressed on the power wiring system to transmit control signals. This arrangement is known as a power line carrier (PLC) system and is discussed in Chapter 28. Other systems are completely wireless, using radio frequencies and a system of wireless transmitters and receivers.

In addition to its CPU (central processing unit, or microprocessor), a programmable controller contains input/output (I/O) interfaces, memory, and means for programming (and reprogramming). An operational block diagram is shown in Chapter 27. The controller accepts not only the usual time-based signal function,

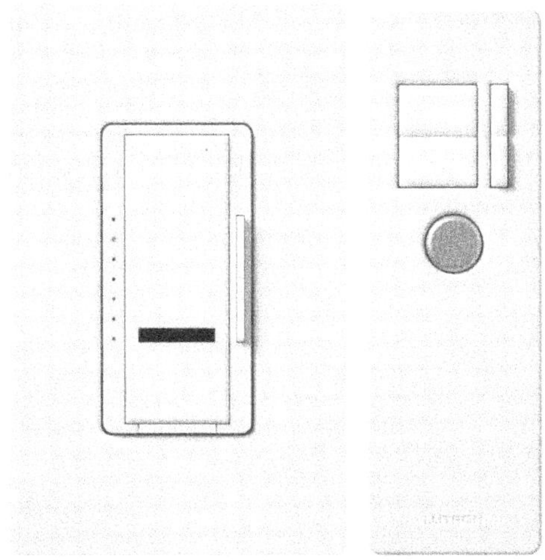

Figure 17.51 *Remote control (right) may be used to operate a wall-mounted dimmer (left). (Courtesy of Lutron Electronics Co.)*

but also information (feedback) from field devices via its I/O device. It then "processes" the signal in its CPU, which consists of logic and storage memory, and sends out a resultant processed control signal.

Large lighting control systems use a computer in lieu of a programmable controller and usually have additional control facilities, such as telephone interfaces and local control/relay/switching centers, which process local sensor input and control local lighting blocks (see Chapter 28). Identical systems are used for building automation, heating, ventilation, and air-conditioning control, energy management, and the like, the difference being the types of sensing devices and the control algorithms.

17.38 LIGHTING CONTROL STRATEGIES

A good lighting control system varies the light supplied to match the light required, in real time as the requirement varies; thus, overlighting and underlighting are avoided. In addition, the control system must be capable of permitting initial adjustments and of accommodating external, non-lighting constraints such as commands from a peak-demand controller. Common lighting system situations addressed by a control system follow.

(a) System "Tuning"

In every lighting installation there is a difference between the intended design and the installed result. This is due to unrealized assumptions, imprecisions in calculation, differences between specified and installed equipment, equipment location changes, and so on. Thus, a lighting system should be "tuned" in the field to attain the intended design results—prior to occupancy. This usually means reducing lighting levels in nontask areas because spill light is frequently sufficient for circulation, rough material handling, and the like. Field tuning can often result in an energy reduction of 20% to 30%, depending upon the control technique.

The fewer the number of light sources controlled, the more accurate the tuning and the larger the energy-saving potential. Lighting system retuning is required when the function of an entire space is changed, or when an area is altered by a furniture move or by a task change. An ancillary benefit of field tuning is glare reduction, which frequently improves task visibility. Tuning is a one-step task in the sense that, once accomplished, it does not require change unless the space function changes. Therefore, it should also be reversible to accommodate such changes. Tuning should not be confused with lumen maintenance, described in Section 17.38(d), which is a control strategy designed to compensate for normal system output decline and its corollary—initial system overdesign.

The tuning action most often required is a reduction in illuminance and this can be accomplished by:

1. making appropriate field modifications to adjustable fixtures (aiming, lamp position, etc.)
2. replacing installed lamps with lamps of lower wattage—changes that reduce light output reversibly
3. replacing installed fluorescent ballasts with low-current ballasts, thereby lowering output
4. adding current-limiting impedances to luminaires or lighting circuits, as previously mentioned (Figure 17.52)
5. ballast switching or use of multilevel ballasts (see Section 17.34)
6. dimming by adjustment of a potentiometer at individual fixtures that are so equipped

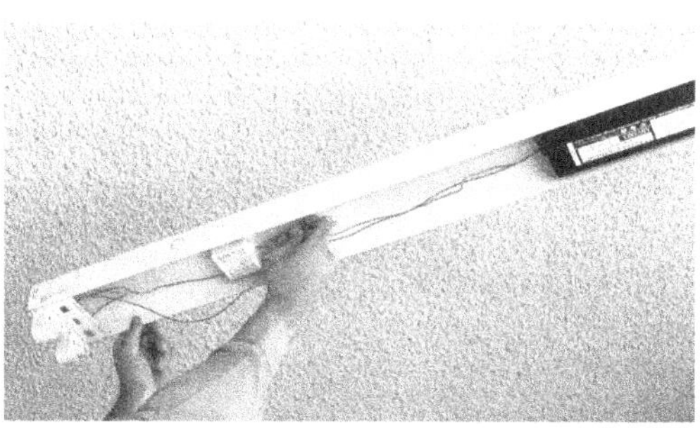

Figure 17.52 *This type of compact solid-state electronic circuit reduces ballast power input by approximately 30% and lamp output by about 28%, resulting in a net gain in efficacy. The power factor remains above 90%. Similar units are available for larger power decrease. They can be mounted on the lamp end or in the fixture channel, as shown. An ancillary benefit is cooler ballast operation, resulting in extended life. (Courtesy of Remtec Systems.)*

7. replacing standard wall switches with time-out units (see Chapter 27), programmable units, or dimmer units

Some of the suggestions listed above go beyond simple tuning and are suggestive of design considerations best addressed during design rather than during pre-occupancy. Including moderate cost tunable elements during the design process would be a benefit in many projects.

(b) Variable Time Schedule

No normal task area has a constant 24-hour, 365-day lighting requirement. In commercial and industrial spaces, work areas have regularly scheduled periods during which design illuminance is not required. These include coffee and lunch breaks, cleaning periods, shift changes, and unoccupied periods. Programmed time controls can readily save 10% to 25% of lighting energy use, compared to relying on occupants to manually operate controls. "Tight" program scheduling takes account of lunch hour and provides after-hours lighting only in those areas being cleaned rather than across whole floors. The payback period for the investment in control equipment varies between 1½ and 5 years. Note that static control that is insensitive to actual field conditions has only limited ability to save energy.

(c) Occupancy Sensing

Within a normal 9:00 a.m. to 5:00 p.m. working schedule, offices in commercial spaces are typically unoccupied for 30% to 60% of the time. The reasons are several—coffee breaks, conferences, off-site work assignments, illness, vacation, and reassignment to a different work location are a few. Occupancy sensors can operate relays to turn off lighting after a preset minimum period (about 10 minutes) or can dim the lighting to a minimum in areas such as corridors, which always require some light. Occupancy sensors can also turn off other energy consumers such as fan-coil units, fans, and plug loads. Reestablishment of the original lighting level can be instantaneous, delayed, or manual following action by the occupant. Another useful function of an occupancy sensor is to provide an automatic override in scheduled systems, thus both relieving the occupant of the necessity of using a manual override and limiting the energy use to actual occupancy time. Occupancy sensing can also light an occupant's way into a space and shut off the system after he/she has left.

Occupancy sensors (that can detect a human) are of three types—passive infrared (IR), ultrasonic, and a hybrid of both technologies. The IR sensor (Figure 17.53) reacts to the motion of a heat source within its range. It operates by creating a pattern of beams and activates when a heat source (such as a person) moves from one beam to another. It will not activate to a stationary heat source. Although the IR technology is quite sensitive, it has several disadvantages:

- Small movements may not be detected, as they may not cross from one beam (zone) to another. As a result, a person sitting quietly may not be detected and the sensor may shut down the lighting.
- Very slow movements may not be detected, even when they cross zones.
- The IR detector must "see" the heat source; a heat source blocked by furniture will not be detected.
- The beams have a discrete width and depth. Therefore, there may be "dead" spots under the beams if the units are not carefully selected to have adequate multilevel beams, or properly located to give the desired coverage.

Ultrasonic sensors emit energy in the 25–40-kHz range, which is well above the range of human hearing. The waves immediately fill a space by reflecting and rereflecting off all hard surfaces, establishing a pattern that is detected by the sensor. Any movement within the space disturbs this pattern and is immediately noted by the sensor. Ultrasonic sensors have distinct advantages over IR sensors in that they do not require a direct line-of-sight exposure to the movement and they can detect small movements. The latter characteristic, however, is also a disadvantage because the movement of curtains and even air movement can often trigger a sensor. It is frequently necessary to adjust (reduce) a sensor's sensitivity to avoid false sensing. Unfortunately, this also decreases its coverage.

Hybrid (dual-technology) sensors combine the characteristics of both sensor types by usually requiring both sensors to react to turn the lighting on. Once the lighting is on, a reaction in either sensor keeps the system on. In addition, sophisticated electronic circuitry "learns" a space's occupancy patterns and is programmed to react accordingly.

Placement of sensors is very important and should be tested before final installation. Studies have shown that reducing the minimum "on" period below 10 minutes is counterproductive, and frequently causes space occupants to shut off the sensors. Depending on their type and mounting position, sensors can cover a maximum area of 250–1000 ft^2 [23–93 m^2] per unit. The payback period for this equipment runs between six months and three years, depending on the type of space, the degree of control already

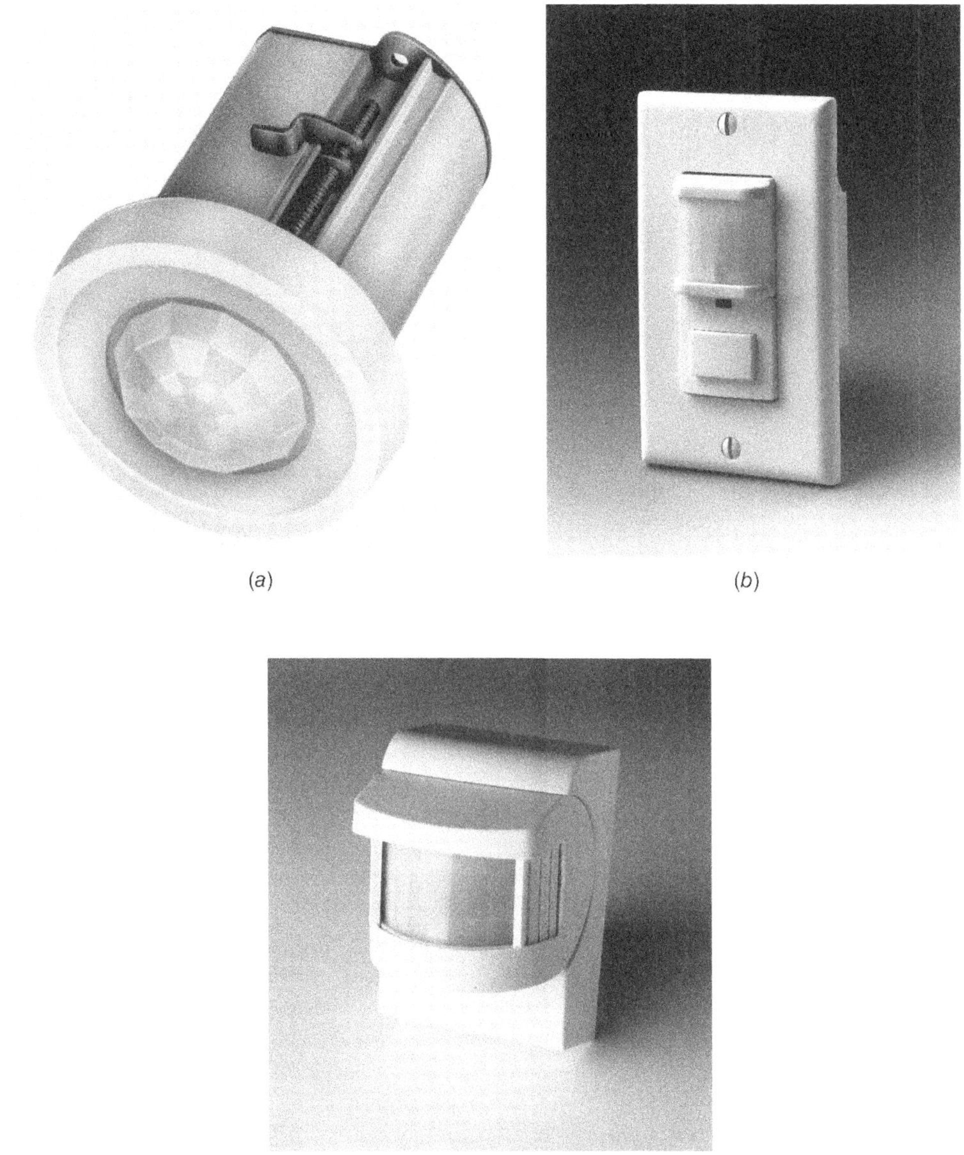

(a) (b) (c)

Figure 17.53 *Passive infrared (PIR) occupancy sensors. All sensors can be equipped with an adjustable override to prevent turning on lighting when ambient light is sufficient. All units have adjustable delayed-off timing and a flashing LED that indicates sensor operation. (a) Flush-mounted ceiling unit. The 360° circular pattern of the lens indicates omnidirectional coverage. The unit is approximately 4 in. [100 mm] in diameter. (b) Flush-mounted combination wall switch and PIR occupancy sensor. Operation of the switch overrides the sensor function. (c) Surface-mounted wall PIR occupancy sensor. The semicircular shape of the lens indicates wide horizontal coverage. (Courtesy of Leviton Manufacturing Co.)*

existing, and energy costs. Occupancy sensors are best applied in areas that are divided into individual rooms and workspaces. Sensors can be wall or ceiling mounted or mounted in a wall-outlet box in a combined sensor/wall-switch configuration. A few designs are shown in Figure 17.54. (See Section 30.2, which discusses the security application of motion detectors and illustrates their operation.)

ASHRAE/IES Standard 90.1 recognizes the great effectiveness of these devices by giving both connected power credit and a high control point equivalency.

(d) Lumen Maintenance

To maintain design illuminance across time, initial illuminance must be higher than the design illuminance to compensate for light losses. The extent of this overdesign is the reciprocal of the light loss factor (LLF). With a typical LLF of 0.60, the overdesign amounts to (1/0.60) or 166%. Assuming a linear light falloff over a two-year maintenance period, higher-than-necessary initial illuminance results in an average of 33% annual energy waste. Over the next two-year maintenance

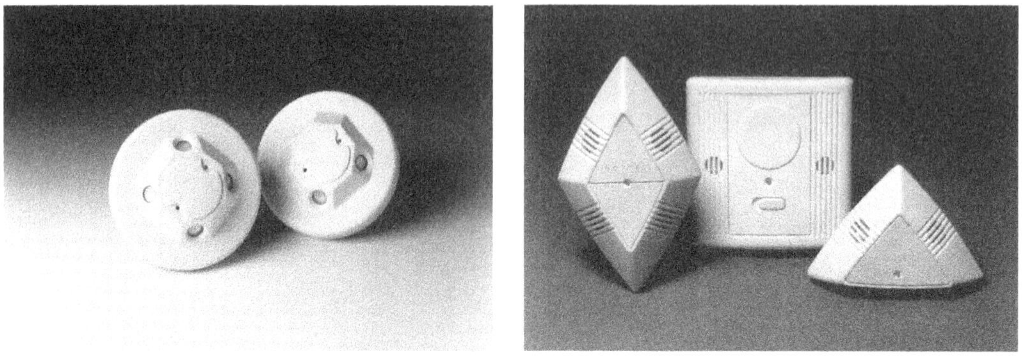

(a) (b)

Figure 17.54 *Ultrasonic occupancy detectors. (a) Ceiling-mounted sensors, bidirectional on the left and unidirectional on the right. (Courtesy of Leviton Manufacturing Co.) (b) Sensors in various designs, intended for differing applications. Left to right: Ceiling-mounted two-way sensor intended for large rooms of up to 2800 ft² [260 m²]. Dual-function sensor and wall switch designed for rooms of up to 300 ft² [28 m²]; wall mounted on a single- or double-gang wall box. One-way ceiling-mounted sensor designed for use in rooms of up to approximately 1250 ft² [116 m²]. (Courtesy of Novitas Inc.)*

cycle, the over-illuminance is slightly less due to a small increase in unrecoverable losses. Because the light delivery depreciation is a continuous and very gradual process over time, the most appropriate control strategy is one that reduces the initial overlighting by the required amount (as verified in the field) and then gradually restores it as the system ages. This control strategy, known as lumen maintenance, is accomplished by a dimming system operating in conjunction with local light sensors (photocells). The photocells measure ambient light, and, in response to their signals, the controller operates the dimming units to decrease (or increase) the light output. Depending on the size of the installation, the dimmers can either be dispersed or centralized. The modulating action of such a system over the maintenance period is shown in Figure 17.55. Lumen maintenance, manually enacted, gives only about a 10% energy use reduction; whereas with an automated system typically a 33% reduction can be delivered.

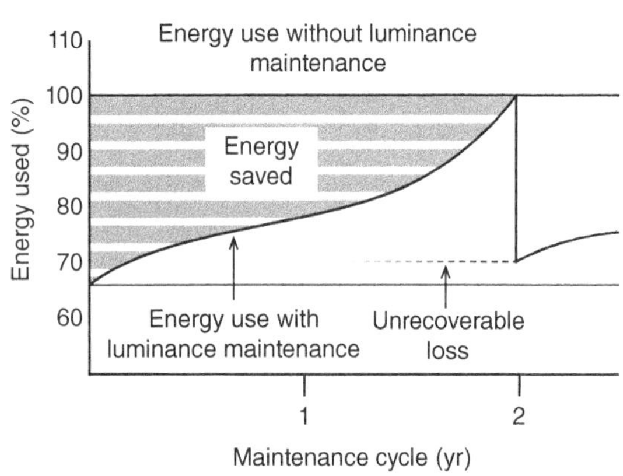

Figure 17.55 *Energy use pattern of a system that reduces the initial illuminance to compensate for necessary overdesign and gradually increases light output as system elements depreciate. In subsequent cycles the energy savings are reduced slightly because of unrecoverable light losses. (Redrawn by Martin Lee.)*

In a new installation, whether to use electronic ballasts and full-range dimming, or partial dimming, or a system of multilevel switching is an economic decision and depends upon many factors, one of the most important of which is the cost of energy. Often a combined control system is advisable. In interior zones, lumen maintenance reduction does not usually exceed 30% to 40%, and full-range dimming is not required. In perimeter zones, daylight often provides all the required illuminance, and full-range dimming, dimming plus switching, or a multilevel switching system is most logical. See the discussion of daylight compensation in Section 17.38*e* for more information. The payback period for a lumen maintenance installation of this type varies from 1 to 5 years. Shorter payback periods can be obtained by using multilevel switching rather than dimming because of its lower first cost.

An additional favorable effect of initial light reduction is the lengthening of effective lamp life and a reduction in the rate of lumen depreciation. When a lamp is operated at rated voltage, its lumen output drops during its life (according to the type of lamp). However, if lamps are operated at reduced output, the lamp life-cycle is greatly extended, lumen depreciation is reduced, and lamp energy consumption is linearly reduced.

Typical life extension (to the point where it is economical to replace the lamp) values are:

fluorescent	80
metal–halide	40
high-pressure sodium	20

These increases in lamp life are for interior zones; for perimeter zones with ambient daylight compensation, they are higher.

(e) Daylight Compensation

A control system arranged for continuous light compensation, as described for lumen maintenance purposes, might also automatically compensate for ambient daylight. The difference is that compensation for lumen maintenance due to light loss factors is a very gradual process of increasing output, whereas daylight compensation can be a minute-by-minute variation and generally in the direction of decreased electric lighting. Because of these possible rapid changes, switching systems are undesirable, as the constant on-off or level switching of lamps can be very annoying to occupants and bad for lamps.

Automatic dimming is the system of choice. In perimeter areas daylight often supplies all the required light, thus either full-range dimming or substantial partial dimming is necessary. Figure 17.56 shows the action of both types of dimming in a fluorescent installation with daylight compensation. The crucial design element in a daylight-compensating system is the establishment of zones (Figure 17.57). As a starting point, the size of the zones is established by determining the maximum room depth that receives at least half of its illuminance from daylight for several hours a day. A computer daylight and dimming cost study with perimeter zone depth as a variable is an effective design approach.

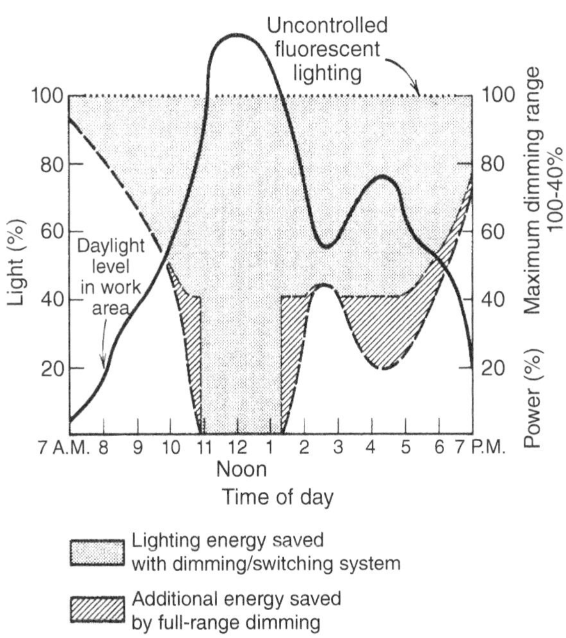

Figure 17.56 *Energy savings typical of a daylight-compensating lighting control. A full-range dimming system is more effective than one that dims down to only 40%, because daylight often supplies most of the required illuminance. An economic analysis is required to determine whether the additional cost of such a system is justified.*

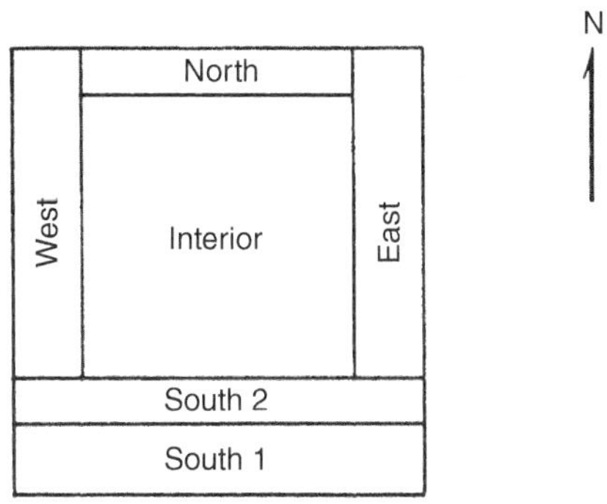

Figure 17.57 *Typical building plan showing approximate daylight perimeter zones. Exact delineation of zones depends upon latitude, climate, window design, and cost of electric energy.*

Placement of the control photocells depends on the control system. Where daylight compensation is desired in conjunction with lumen maintenance (see the preceding subsection), area photocells are desirable, as they give a feedback control signal for the specific area involved. Alternatively, a daylight-factor map of the space can be made—preferably after the installation is complete and the furniture in place—and photocells located at one select point, based on this map. Daylight factor at any interior point is constant. Therefore, by measuring actual daylight level at a point either immediately inside or outside the window, shielded from direct sun and from inside electric light (Figure 17.58), we can relate it to any area in the room by the ratio of daylight factors, and establish a switching value at which the lighting for that inside area is partially or fully switched.

In Figure 17.58, the photocells are mounted at point *A*. The ratio of daylight between points *A* and *B* is the ratio of daylight factors—that is, 20/10, or 2.0. Therefore, if 500 lx [47 fc] of daylight is required at point *B* before switching lights in that area, twice that

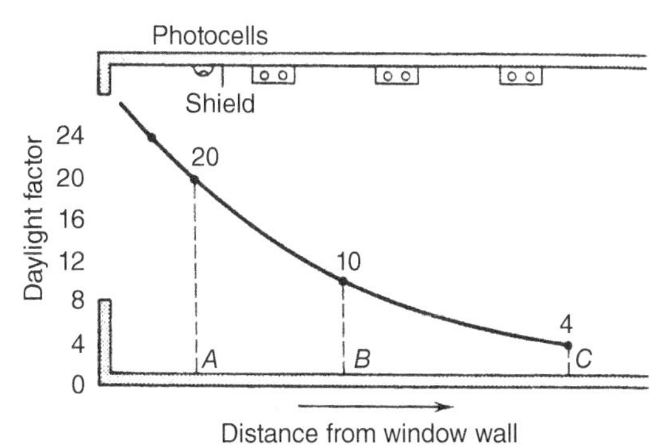

Figure 17.58 *Typical daylight factor curve plotted on a room section with unilateral sidelighting. The photocell control technique for daylight compensation described in the text assumes a constant daylight factor (DF) distribution indoors—that is, from a fixed-luminance sky and no direct sun.*

amount (2.0×500 lx) or 1000 lx [93 fc] is required at point *A*. Similarly, a 500-lx requirement at point *C* corresponds to (20/4)(500) or 2500 lx [230 fc] at point *A*. Therefore, switching can be initiated by two single-level photocells or a multilevel unit at point *A*, with settings at 1000 and 2500 lx [93 and 230 fc]. Other switching arrangements can be made on the same principle. Dimming initiation can also be arranged in this fashion.

Daylight compensation can reduce energy use in perimeter areas by up to 60%, depending on latitude, climate, depth of perimeter zone, hours of building use, initial power density, and so on. The amount saved for an entire building depends on the building configuration—that is, the ratio of perimeter area to total floor area. Payback time is usually in the range of three months to three years.

In summary, a well-designed lighting control system can provide energy conservation of up to 60%, extended lamp life, reduced cooling costs, extended ballast life, and reduced maintenance costs. When all these factors are considered, the investment payback period is always short and therefore financially attractive. An aspect of a centralized lighting control system, not discussed previously, is its use in connection with peak demand reduction efforts (see Chapter 27). When interconnected with a demand controller, this approach can reduce the electric lighting load in accordance with a predetermined preferential-load schedule and can achieve significant savings.

(f) Emerging Opportunities

As control system capabilities increase, many intriguing uses for lighting-based controls are emerging. Some of these applications include: remote diagnostics, demand control, color tuning, high-end tuning, plug load control, occupancy sensing, traffic sensing, and personal control. Several of these control options are inherent to the lamp/luminaire element itself; several take advantage of the fact the luminaires tend to be everywhere in and around a building. A common thread that underlies the expansion of how luminaires (and lamps) can be used and controlled is their networking and digitalization of signals.

References and Resources

ANSI. 2023. *ANSI C82.11-2023: Lamp Ballasts - High Frequency Fluorescent Lamp Ballasts*. New York: American National Standards Institute.

ASHRAE. 2022. *ANSI/ASHRAE/IES Standard 90.1-2022: Energy Standard for Buildings Except Low-Rise Residential Buildings*. Atlanta, GA: American Society of Heating, Refrigerating and Air-Conditioning Engineers.

DOE. 2008. "How Compact Fluorescents Compare with Incandescents." Washington, DC: U.S. Department of Energy.

DOE. 2018. "How Energy-Efficient Light Bulbs Compare with Traditional Incandescents." Washington, DC: U.S. Department of Energy.

DOE. n.d. "Federal Energy Management Program." Washington, DC: U.S. Department of Energy. energy.gov/femp/federal-energy-management-program

EIA. 2024. "How Much Electricity Is Used for Lighting in the United States?" Washington, DC: U.S. Energy Information Administration. eia.gov/tools/faqs/faq.php?id=99&t=3

Eley Associates. 1993. *Advanced Lighting Guidelines: 1993* (EPRI TR-101022s Rev. 1). Palo Alto, CA: Electric Power Research Institute.

EPA. 2024. "ENERGY STAR Light Bulb Key Product Criteria." Washington, DC: U.S. Environmental Protection Agency. energystar.gov/products/lighting_fans/light_bulbs/key_product_criteria

EPA. n.d.a. "ENERGY STAR Certified Light Fixtures." Washington, DC: U.S. Environmental Protection Agency.

EPA. n.d.b. "ENERGY STAR Product Finder." Washington, DC: U.S. Environmental Protection Agency.

IES. 2011a. *The Lighting Handbook*, 10th edn. New York: Illuminating Engineering Society.

IES. 2011b. *IES/IDA MLO-11: Model Lighting Ordinance with User's Guide*. New York: Illuminating Engineering Society.

IES. 2020. *ANSI/IES TM-15-20, Technical Memorandum: Luminaire Classification System for Outdoor Luminaires*. New York: Illuminating Engineering Society.

NFPA. 2022. *2023 NFPA 70: National Electrical Code (NEC)*. Quincy, MA: National Fire Protection Association.

Odle, H.A. and R.L. Smith. 1963. "Uniformity of Illumination in Fluorescent Applications." *Illuminating Engineering*, January 1963. pp. 34–38.

Stein, B. 1997. *Building Technology: Mechanical and Electrical Systems*, 2nd edn. New York: John Wiley & Sons.

UL. 2021. "UL 1598 Ed. 5–2021: Luminaires." Northbrook, IL: Underwriters Laboratories, Inc.

CHAPTER 18

Electric Lighting Design

CONTEXT

18.1 BIG PICTURE

The design of an electric lighting system—in its most basic sense—involves the selection of lamps and luminaires of adequate number that are placed in appropriate locations to meet the owner's project requirements (OPR). Conformance to the OPR for any proposed design solution can be evaluated by manual or computerized analysis methods. Prior to evaluation, any lighting proposal is simply an untested design hypothesis. Usually, the best of a range of potentially acceptable lighting solutions is determined by both technical and aesthetic considerations.

This sounds like a reasonably simple process, but, unfettered by constraints, it could easily involve hundreds of potential design options for a single space. Consider the math when four lamp types are paired with 15 luminaire types in 20 different spatial configurations (geometries and surface reflectances). Design constraints are found in the OPR and in the designer's preferences and biases. Precedents can effectively serve to narrow the range of options.

This chapter is centered on how a lighting design proposal may be evaluated during the design stage—prior to installation in a building. But it also addresses preparing for design during the pre-design stage and considerations related to the construction and occupancy phases of a project. Daylight is a highly preferred means of building illumination (see Chapter 12), yet most buildings will require an electric lighting system for use at night and to supplement daylighting. The electric lighting design process must acknowledge and link with daylighting efforts.

The fundamentals of light, light measurement, vision, and visual comfort are discussed in Chapter 7. Electric lamp types and their pluses and minuses are discussed in Chapter 17. Luminaires and controls are also discussed in Chapter 17. These lighting system components can, to some extent, be selected independently of a holistic system proposal—but must always be respected as part of a larger system.

Manual (hand) methods are the focus of discussion in this chapter. There are several reasons for this, even though manual methods are increasingly less likely to be used in an office. First, a manual method requires an explicit engagement with all important variables since they are unavoidably seen in the equations that represent the method. Second, computerized software is often proprietary and set up as a black box—the workings of the software are often opaque. Third, there are many lighting design software packages available, each with differences from the others. There are two key manual analysis methods: the zonal cavity method and the point-to-point method. Both are discussed herein.

18.2 PHILOSOPHY

Lighting design is a combination of art and science ("architecture" and "engineering"). There can be many solutions to the same lighting problem, all of which will satisfy the minimum project requirements, yet some will be dull and mundane, while others will display ingenuity and resourcefulness. The lighting

Mechanical and Electrical Equipment for Buildings, Fourteenth Edition. Walter Grondzik, Daniel Overbey, and Alison Kwok.

Companion website: www.wiley.com/go/meeb14e

designer should approach each problem with an open mind, bringing to it knowledge of current technology, as well as necessary background and experience. Rarely should an owner or designer be satisfied with a carbon copy system from a previous design.

Because of the large number of interrelated factors that affect the perception and performance of a lighting solution, no single proposed design can be claimed to be "correct." For this reason, it is not entirely desirable to solve a lighting problem using a step-by-step linear technique. Experience has shown, however, that this is a good approach for those who lack the experience necessary to envision an entire solution early and holistically. This chapter outlines a step-by-step process.

18.3 LIGHTING DESIGN OBJECTIVES

On one level, the objective of lighting design is to create an effective, efficient, and pleasing luminous environment. That these outcomes—encompassing the utilitarian and the aesthetic—are not contradictory is demonstrated by thousands of great lighting designs. Light can and should be regarded as an architectural material. Interestingly, many graduate architects have entered practice without ever designing an electric lighting system for any studio project—thus never experimenting with this valuable luminous material.

Some underlying architectural lighting design objectives are:

1. Illuminance should be adequate for effective viewing of the specific task involved; on a task-by-task basis.
2. Variation should be provided for, because this is desirable to avoid monotony and to create perspective effects; but this variation should respect acceptable luminance ratios.
3. Lighting equipment should be unobtrusive, but not necessarily invisible. Fixtures (luminaires) can be chosen and arranged in various ways to complement the other architectural elements or to create dominant or recessive architectural features or patterns. Fixtures may also be decorative and thus enhance the interior design. Accent lighting, directional lighting, and other highlighting techniques increase the utilitarian as well as the architectural quality of a space.
4. The lighting system must provide the proper luminous quality (glare control, color rendering, etc.) and be deemed appropriate by users.
5. The lighting outcomes must be efficient in terms of cost and energy use and respectful of resources. Cost should be life-cycle cost (acquisition plus operation); energy should consider both cost and resource consumption; resource-respectfulness should consider end-of-life and manufacturing.

On another level, the objective of lighting design is to meet the lighting-related criteria in the OPR for a building. Development of the OPR during pre-design is mandatory on a commissioned project—but good practice for any project. The purpose of the OPR is to delineate all issues of importance to project success, set intents for these issues, and establish criteria (design targets) for each intent. These targets are then used for evaluation during design, construction, and occupancy.

With these considerations in mind, a lighting design procedure can be laid out—keeping in mind that the suggested order of steps is not necessarily the same for each lighting problem. Since many lighting design factors are closely interrelated, it is often necessary to address several stages simultaneously before arriving at a decision.

The lighting design approach explained in this chapter is analytic; that is, the design procedure establishes requirements primarily in numerical form and then considers lamps, fixtures, fixture locations, and so on, to arrive at a design solution. In an alternative approach, frequently referred to as brightness design, the designer links surfaces on a proposed plan/section of a space with desired luminances established by a mental picture (or other means) and designs the lighting accordingly. This approach, which can be very effective, is highly intuitive and requires considerable prior experience on the part of the designer. For this reason, and because of the hands-on, trial-and-error fieldwork involved in the approach, it cannot be given adequate justice in a textbook and is not presented here.

PROCESS

18.4 OVERVIEW

Table 18.1 presents a simplified overview of a recommended lighting design procedure. This simplified overview does not presume a commissioned project, not does it assume that the integrated design process will be used. The "who is involved" list would be accentuated by the use of commissioning and integrated design.

18.5 PRE-DESIGN PHASE

A successful pre-design phase effort is essential to a successful lighting design. Project aspirations, constraints, targets, and biases must be enunciated

TABLE 18.1 Overview of the Lighting Design Process

Project Phase	Key Activities	Who Is Involved	Comments
Pre-Design	Develop lighting OPR	Client, users, architect, engineer, lighting designer	OPR (or equivalent) is essential to success
Design	Select approach Select components Analyze proposal Evaluate proposal and iterate as needed	Architect, engineer, lighting designer	A system is proposed, evaluated, tweaked, reevaluated, and then put into contract documents
Construction	Verify submittals Verify installation	Lighting designer	Site observation is beneficial
Occupancy	Tune systems Do a POE Assist owner	Lighting designer, owner, users	Close the design loop by confirming in the field that OPR are met

in pre-design. If they are not (if left to the design phase), then the lighting design will usually reflect only the wishes of the design team—and not those of the client. Historically pre-design has led to development of a program or brief that outlines project requirements. This is still a common approach. Unfortunately, many programs are at best rough outlines of owner expectations—often unsupported by specific design criteria.

The building commissioning process should begin a pre-design with the development of the OPR. The OPR will describe what success will look like in the occupied project—from the owner's perspective. In the absence of an OPR, the designer's perspective will be substituted. One of the strongest—and most challenging—elements of the commissioning process is OPR development. Preparation of a good OPR is best practice for any project—commissioned or not.

(a) Lighting OPR

The owner's lighting project requirements for a proposed building capture all the issues, intents, and criteria related to lighting that matter to the owner. Because an owner is typically not a trained designer, the design team will ensure that the owner's wishes are augmented by considerations of codes, standards, and current practices. Further discussion of numerous lighting design considerations is found in Section 18.6.

Lighting Design Issues. These are simply nouns that identify things related to lighting that should be considered during the design process. Typical lighting issues include first cost, operating cost, appropriateness, flexibility, energy efficiency, greenness, maintainability, appearance, durability, and the like.

Lighting Design Intent. The purpose of a lighting design intent is to convert each identified issue into a general direction to be pursued on the project. This may sound simplistic but qualifying an issue can lead to interesting discussions between a client (or user) and the design team—which become more interesting when quantified as criteria.

Examples of lighting intents (related to three of the issues noted above) are:

- low first cost (or perhaps regional-average first cost)
- code-minimum energy efficiency (or perhaps exceptional energy efficiency)
- easily maintained (or perhaps maintained with some difficulty)

Lighting Design Criteria. These convert the direction suggested by design intents into specific design targets (desired destinations). Criteria are usually quantitative and numeric but can be non-numeric—such as three photographic examples of luminaires that would be considered project-appropriate. Examples of criteria related to energy (note the range of possible targets) might include:

- code-minimum lighting power density (LPD)
- LPD 10% less than code minimum
- LPD 20% less than code minimum
- lamp luminous efficacy of 80 or more
- lamp luminous efficacy of 100 or more
- luminaire Target Efficacy Rating (TER) value of 40 or greater

(b) Integrated Process

An integrative design process entails a structured and intentional framework for building design by which all stakeholders—owners, architects, engineers, contractors, user groups, and others—are engaged at the earliest stage of the design process to ensure that project goals are clearly established and embedded into every phase of the project. Often, the process will yield discoveries and synergistic strategies by which project teams can optimize solutions and outcomes. The implementation of an integrated design process should bring more people to the table earlier than in the historic "siloed" type of design process. For instance, during the pre-design phase, it should be possible for the project lighting designer (let us assume a specialty consultant) to readily interact with (and seek advice from) the project architect, electrical engineer, interior designer, green building specialist, contractor, and the owner's operating personnel. This should lead to a more well-rounded lighting design outcome. The same would be true on a commissioned project, where the pre-design commissioning team would include the above players.

18.6 DESIGN PHASE

The "design" phase is typically divided into subphases—often including conceptual design, schematic design, and design development. The objective of conceptual design is to develop a design solution that will likely meet the client's needs and aspirations. Schematic design is typically used to test the feasibility of the conceptual design proposal by carrying out more detailed design explorations. In design development, contract documents (including drawings and specifications) are prepared. These contract documents, if faithfully executed by the contractor, will deliver the owner's project requirements (and more).

This discussion does not formally subdivide the design phase. Suffice it to say that design is iterative and evolves into ever more detailed exploration, analysis, and documentation. Some of the more common lighting design tasks are presented herein—not necessarily in the order in which they would be conducted on every project. For example, lighting design for one hypothetical classroom might start with selection of a lamp/luminaire, while design for another classroom might start with selection of a light distribution approach.

Surfaces other than the working plane must always be considered as part of the lighting system. The ratio of vertical illuminance to horizontal illuminance of a chosen lighting system determines wall luminance, which in turn greatly influences the overall impression of spatial brightness. The chromaticity of room lighting depends primarily on the lamp but secondarily on the luminaire and surface reflectances. A "white light" source can be tinted slightly by the use of a colored reflector in the luminaire; the effect on luminaire output of such a decision must be considered. In the case of semi-indirect and indirect lighting, the same effect can be accomplished by the use of colored ceiling and upper wall surfaces, which serve as secondary reflectors and become an actual luminous source for the room.

(a) Light Distribution Approach

This discussion is focused on electric lighting systems. The descriptions apply directly to electric lighting—but are also generally applicable to daylighting design efforts. There are four broad approaches to lighting a space in service to visual tasks more complex than simple orientation or spatial definition. These are described here as uniform illuminance, local illuminance, supplementary illuminance, and task-ambient illuminance.

Uniform Illuminance. This refers to a system designed to provide uniform, and generally diffuse, lighting throughout the area under consideration (a room or portion of a room). The methods by which this outcome can be accomplished are numerous. The key constraint is that the resulting illuminance on the working plane (sometimes called "work plane") must be reasonably uniform. Typically, the work plane is assumed to be 30 in. [76 cm] above the finished floor (AFF). This type of light distribution is appropriate for a variety of spaces—in particular, those where the design task may be located anywhere on the working plane (such as a classroom with movable desks or an open plan office). Lighting systems that deliver uniform illuminance tend to have a uniform physical presence.

Local Illuminance. This approach to lighting is described primarily in contrast to uniform lighting. With a local system, light is only delivered to a task of limited size located in a larger space. Such an approach would rarely be found acceptable in a building due to glare caused by excessive contrast. A desk lamp in an otherwise dark office is an example of local lighting—potentially very energy-efficient, while being visually ineffective due to discomfort.

Supplemental Illuminance. This illumination approach is used where some high-lux tasks are found in a space with generally lower illuminance demands. Uniform illuminance is provided for the space in general but is supplemented with additional light at the more demanding task locations. Track lighting for a bulletin board or pinup wall in a classroom is an example of supplemental illuminance. See Figure 18.1 for examples. Mixing additional light into a space already lit uniformly will normally reduce glare potential by reducing contrast.

Task-Ambient Illuminance. The term "task-ambient" means different things to different people. In this book it specifically refers to light delivered by a luminaire designed to provide both uniform and supplemental light components. Often, such a luminaire is part of the furniture layout of a space. Although not common, this lighting approach can be architecturally interesting (freeing the ceiling from luminaires) and may have some tax advantages.

These light distribution approaches are generally fully independent of lamp type or luminaire type—apart from task-ambient illuminance. Uniform illuminance, for example, can be provided by LED, fluorescent,

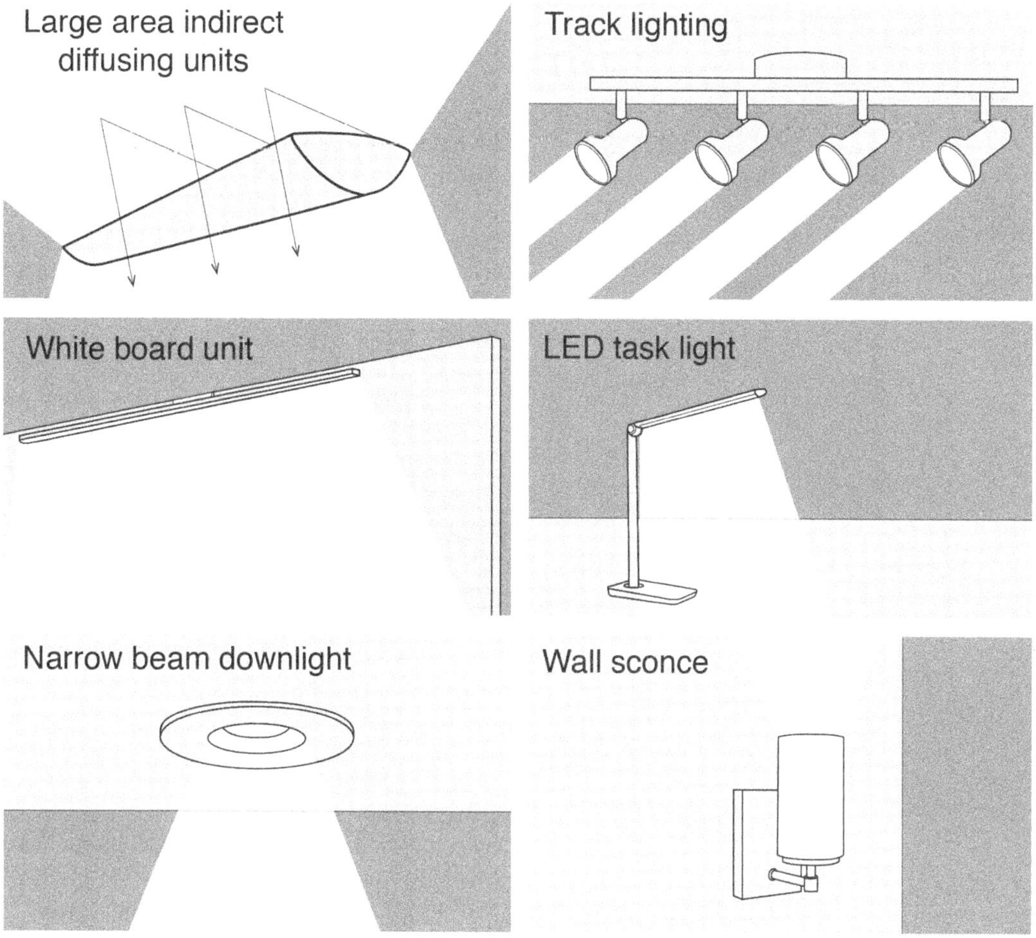

Figure 18.1 *Typical supplementary lighting units for incandescent, CFL, and linear fluorescent sources. (Drawn by Kelli Kimura; © Alison Kwok; all rights reserved.)*

metal-halide, or incandescent lamps installed in recessed, flush, or suspended luminaires.

(b) Light Delivery Options

There are six broad categories of luminaires based upon how lighting is delivered from the luminaire into the space being lit. These are: indirect, semi-indirect, direct–indirect, general diffuse, semi-direct, and direct. Each type is briefly discussed below.

Indirect Delivery. With this type of fixture, between 90% and 100% of the light output (see Figure 18.2*a*) is directed toward the ceiling and upper walls of the room (in other words, away from the typical working plane near the floor). The delivery is called indirect because practically all the light output will reach a horizontal working plane only via reflection from the ceiling and upper walls. The ceiling/upper wall planes become a secondary light source. If these surfaces have high-reflectance finishes, the room illumination will be highly diffuse (shadowless). Because the source must be suspended at least 12 in. [300 mm], and preferably 18 in. [450 mm] or more from the ceiling (depending upon the fixture) to avoid ceiling "hot spots," this system requires a minimum ceiling height of 9 ft, 6 in. [~3 m]. If luminaires are correctly spaced, the resulting illuminance is uniform, and both direct glare and reflected glare potentials are minimal.

To avoid excessive luminance difference between the luminaire bottom and its surrounding field (the ceiling plane), the luminaire can be translucent on the bottom, the sides, or both. Approximately 750 lux (lx) [75 footcandles (fc)] is the maximum horizontal-plane illuminance attainable without exceeding an overall ceiling luminance of about 2500 cd/m^2 [730 footlamberts (fL)]—a target limit for direct glare control. With practically no veiling reflections, this illuminance is sufficient for all but the most difficult tasks. The lack of shadow, the low source brightness, and the highly diffuse quality created by indirect lighting give a very quiet, cool ambience to a space, suitable for private offices, lounges, and waiting areas. Areas having specular visual tasks, such as an office with

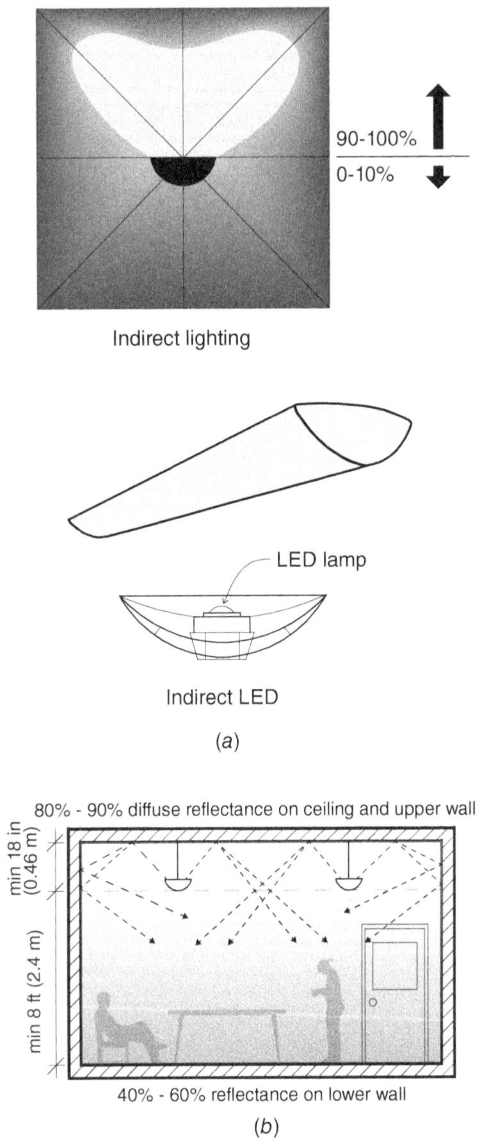

Figure 18.2 *Indirect lighting. (a) The luminaires deliver 90–100% of their output above their own horizontal plane. The ceiling and upper wall surfaces of the space are directly illuminated, and by reflection become large secondary sources that illuminate the space below. When properly designed, this type of installation yields a substantially uniform bright ceiling. (b) Indirect delivery of light to a space involves multiple reflections from multiple surfaces and, if properly designed, provides nearly uniform, glareless illumination in the room. This type of delivery is particularly useful in spaces with digital screens. (Drawn by Kelli Kimura.)*

digital screens, use this system to great advantage. In such spaces, indirect fixtures without luminous bottoms or sides should be specified.

When properly designed, particularly when the source of light is integrated with architectural coves (see Figure 18.2*b*), the ceiling has a floating sky quality, which is pleasant and can be used to give an impression of height in a low-ceilinged room. Another characteristic of indirect lighting is a loss of texture-read on vertical surfaces, which is typical of fully diffuse lighting.

Indirect lighting is inherently inefficient, because much of the useful light reaches the working plane only after at least two reflections—within the luminaire and off the ceiling/upper wall. However, to a considerable extent this inefficiency can be offset by the glare-free nature of the lighting—if the improved lighting quality allows for a reduction in target illuminance. Also, architectural intent may override concerns about reduced efficiency.

Semi-Indirect Delivery. With this system, between 60% and 90% of luminaire light output is directed to the ceiling and upper walls. This distribution is similar to that of indirect lighting, except that it is somewhat more efficient and allows higher levels of illumination without undesirable brightness contrast between the luminaire and its background, along with lower ceiling brightness. A typical fixture, illustrated in Figure 18.3, employs a translucent diffusing element through which the downward light component passes. The ceiling remains the principal light source, and the diffuse character of the lighting remains. Direct and reflected glare are both very low, as they tend to be with indirect lighting. With both indirect and semi-indirect delivery, it is often desirable to add accent lighting or downlighting to break the monotony inherent with these systems and establish visual points of interest.

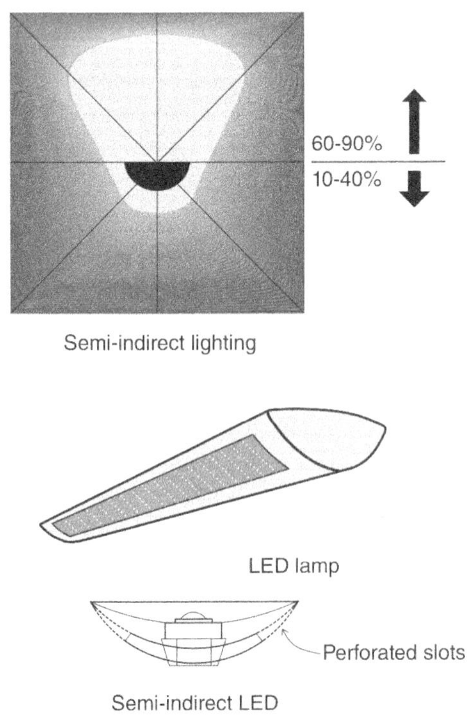

Figure 18.3 *Semi-indirect lighting. (Drawn by Kelli Kimura.)*

The light in both indirect and semi-indirect lighting systems undergoes several ceiling and wall reflections before reaching the horizontal working plane. Highly reflective surfaces are critical to system efficiency. The use of colored paints, particularly on the ceiling, can serve to tint the room illumination slightly via selective absorption—to either good or bad effect, depending upon the space and color in question.

Direct–Indirect Delivery. Direct–indirect delivery provides an approximately equal distribution of light upward and downward, resulting in a fairly bright ceiling and upper wall (Figure 18.4). For this reason, luminance ratios in the upper-vision zone are usually not a problem. As the ceiling is an important (although secondary) source of room illumination, diffuseness is substantial, with satisfactory vertical-plane illumination.

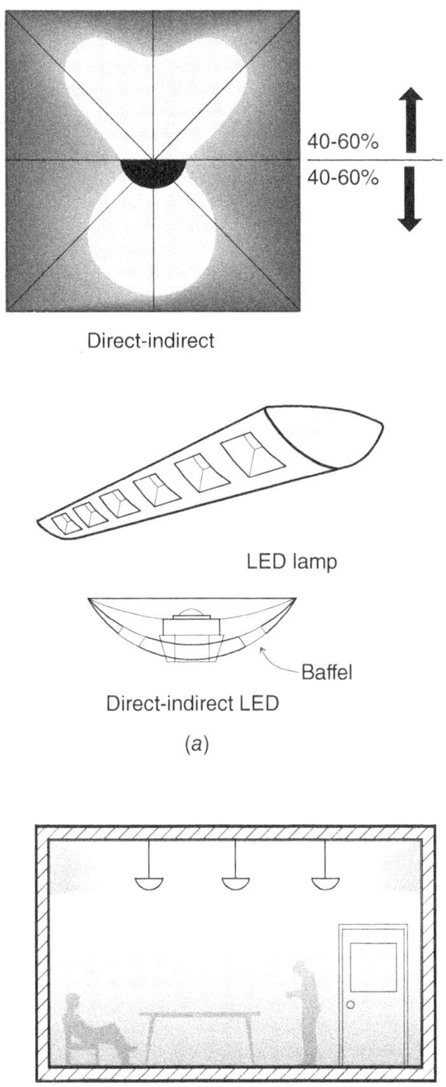

(b)

Figure 18.4 *(a) Direct–indirect lighting. Upper and lower room surfaces are luminous, but the center of walls is not because of the lack of horizontal light output from fixtures (b). Most of the light on the working plane comes directly from the luminaire. (Drawn by Kelli Kimura.)*

General Diffuse Delivery. General diffuse fixtures (Figure 18.5) distribute light in all directions, whereas direct–indirect fixtures have little or no horizontal component. Suspension distances for both types should be adequate (generally not less than 12 in. [305 mm]) to avoid excessive ceiling brightness and hot spots.

Because the impression of illumination depends, largely, on wall luminance (because this is the surface we see most often), a space with general diffuse delivery appears brighter than one with direct–indirect delivery due to the darker walls in the latter.

The potential for both direct and reflected glare can be minimized by avoiding excessively bright luminaires and giving attention to the positioning of sources and viewing angles. Furthermore, because a luminaire (like any other luminous source in the field of view whose luminance is higher than that of the average scene) draws the eyes' attention, care must be taken to limit its luminance and to avoid disturbing fixture patterns.

The efficiency of general diffuse and direct-indirect systems is generally good. Both are well applied in spaces requiring overall uniform illuminance at moderate levels, such as classrooms, standard office work spaces, and merchandising areas.

Semi-Direct Delivery. With this delivery, 60% to 90% of the luminaire output is directed downward (or toward the work plane), with the remaining upward component serving to illuminate the ceiling (Figure 18.6). If the ceiling has a high reflectance, this upward component is normally enough to minimize direct glare from the luminaires, depending on eye adaptation level. The degree of diffuseness of illumination depends largely on the reflectances of the room furnishings and of the floor. Shadowing should not be a problem when the upward component is at least 25% and ceiling reflectance at least 70%. With a lesser upward component, this is essentially a direct lighting system. Semi-direct light delivery is inherently efficient. Reflected glare can be controlled by the methods discussed in Section 18.6(h) that deal with reflected glare. With adequate wall illuminance, the quality of the light gives a pleasant working atmosphere. Semi-direct delivery is applicable to offices, classrooms, shops, and other work areas.

Direct Delivery. Essentially all the light from a direct luminaire is directed downward (toward the work plane). As a result, ceiling illuminance is entirely the result of light reflected from floor and room surfaces/furnishings. This delivery approach

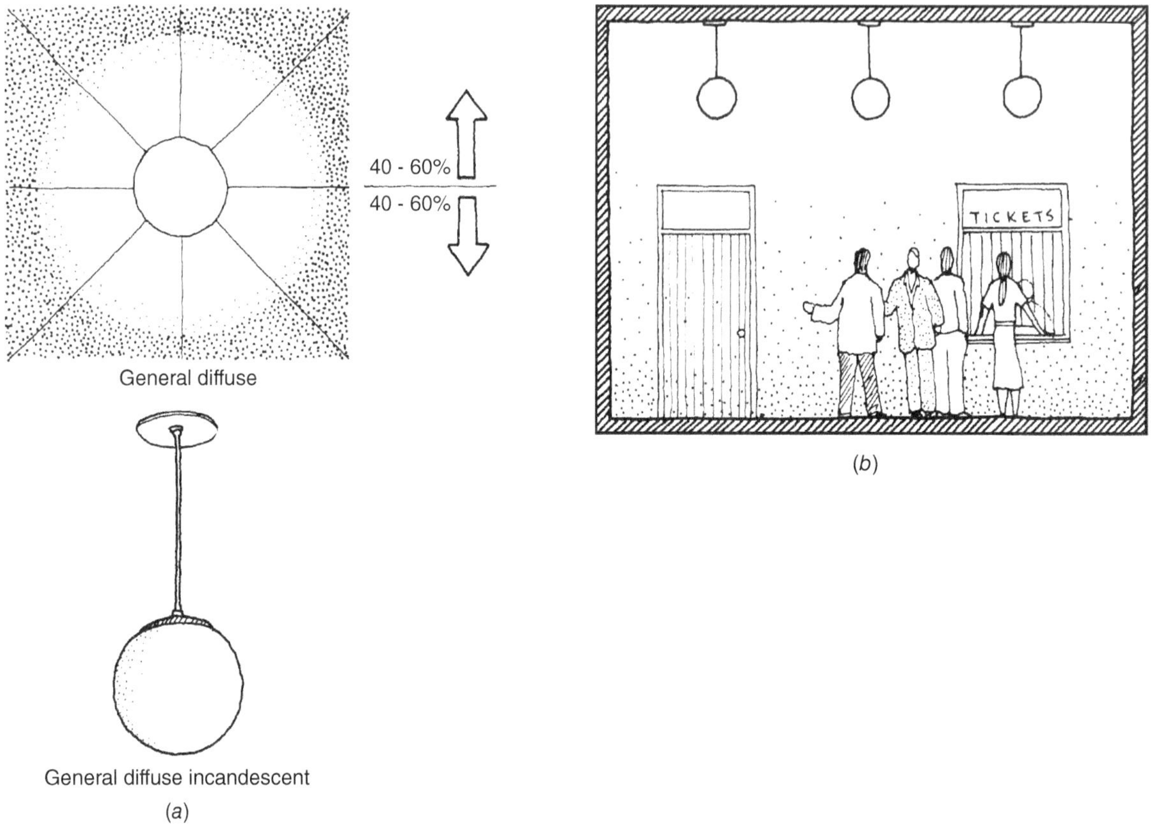

Figure 18.5 *(a) General diffuse lighting. (b) Note that all room surfaces are illuminated and become secondary sources, although those closest to the fixtures (ceiling, upper wall) are the brightest secondary sources. The primary source of illumination is the direct radiation from the fixture. The floor contribution to work plane illuminance is nominal due to its normally low reflectance.*

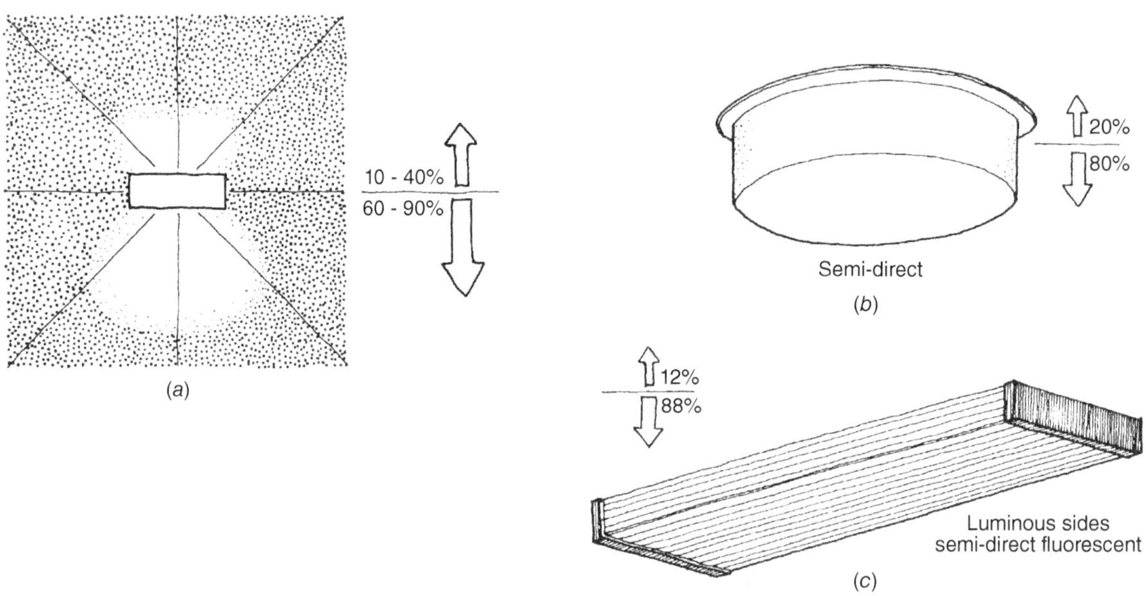

Figure 18.6 *Semi-direct light delivery provides its own ceiling brightness (a), with surface-mounted fixtures (b) or pendant/surface units (c) as illustrated. Other characteristics are like those for direct delivery.*

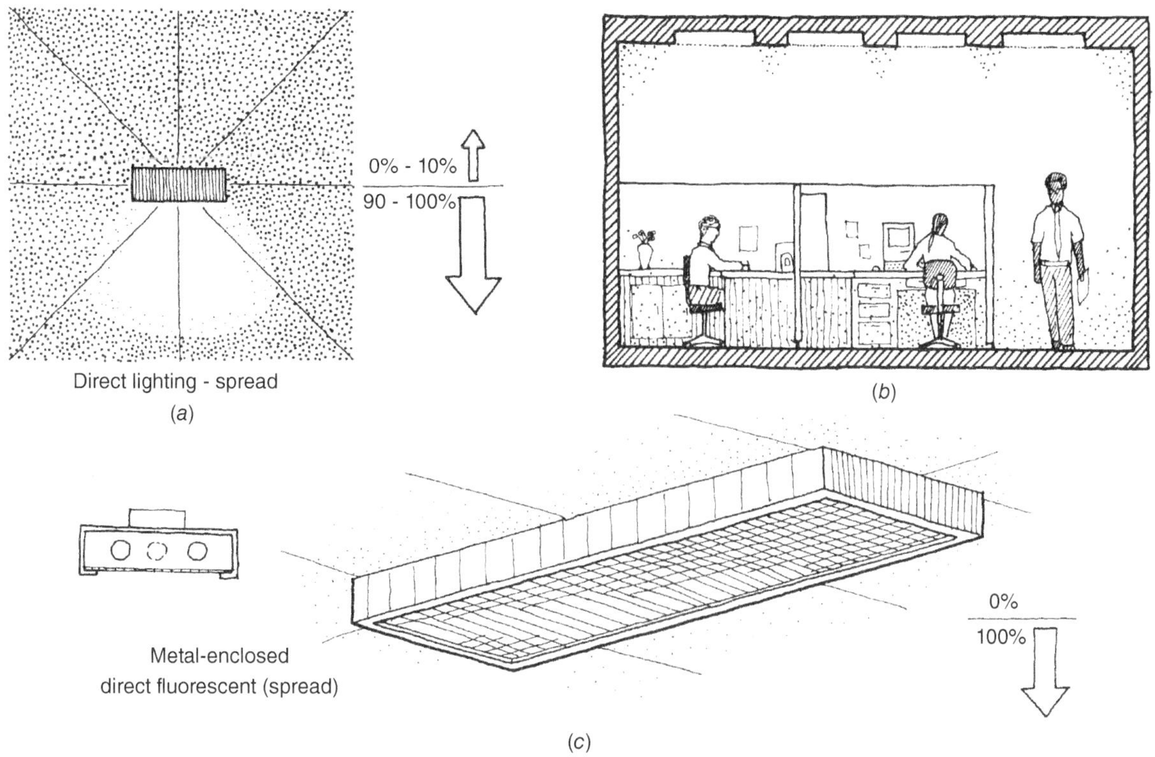

Figure 18.7 Spread-type direct light delivery (a) illuminates all room surfaces except the ceiling, which is only illuminated by reflection from the floor. Some diffuseness is evident. The most common fixture in this category is the direct fluorescent unit, either troffer-type, recessed in the ceiling (b), or surface-mounted (c).

requires a light, high-reflectance, diffuse floor—unless a dark ceiling is specifically intended as a way to lower apparent ceiling height, to correct a poorly proportioned room, or to hide unsightly piping or ductwork.

The effect of direct lighting depends greatly on whether the luminaire light distribution is spread out or concentrated (Figures 18.7 and 18.8). If spread, considerable diffusion of light results from reflections from the floor, furniture, and walls. The result is a working atmosphere with slightly darkened walls and ceiling. This type of light delivery, which is represented by recessed fluorescent troffers installed in a suspended ceiling, is common for general office lighting. The luminaires develop a ceiling surface pattern with light and dark areas. The quality of the entire system is generally not unpleasant. Although direct glare and veiling reflection issues may arise, they can be controlled using proper luminance ratios, the use of low-brightness fixtures, and thoughtful arrangement of viewing positions. When direct delivery fixtures are used in a uniform pattern, viewing-location flexibility disappears, and low-brightness luminaires and high ceiling reflectivity (or specialty diffusers, such as those with a batwing distribution) are necessary.

Direct lighting gives little vertical surface illuminance, requiring the addition of perimeter lighting in many space types (Figure 18.9).

Concentrating downlights create sharp shadows and a theatrical atmosphere that are not appropriate to an office-type commercial space. They can be used in restaurants and other areas where a feeling of privacy, generated by limited-area horizontal illuminance and minimal vertical-surface illuminance, is desired. A lighting fixture with a black cone, a baffle, or another device that is nonreflecting at the viewing angle, will appear dark even when operating. Installations providing high horizontal-surface illuminance, with no apparent source of brightness—such as those using black-cone downlights—tend to be disturbing to our normal bright-sun-and-sky orientation and should be used cautiously and only in limited areas. This same comment, but to a lesser extent, is applicable to very-low-brightness diffusers such as the parabolic wedge type (see Figure 17.36). Such fixtures at least have the redeeming characteristic of low reflected glare potential, which is not the case with downlights. Architecturally, the psychological impression of a space without a visible source of light is gloomy and cavelike. When very-low-brightness sources are used, as in

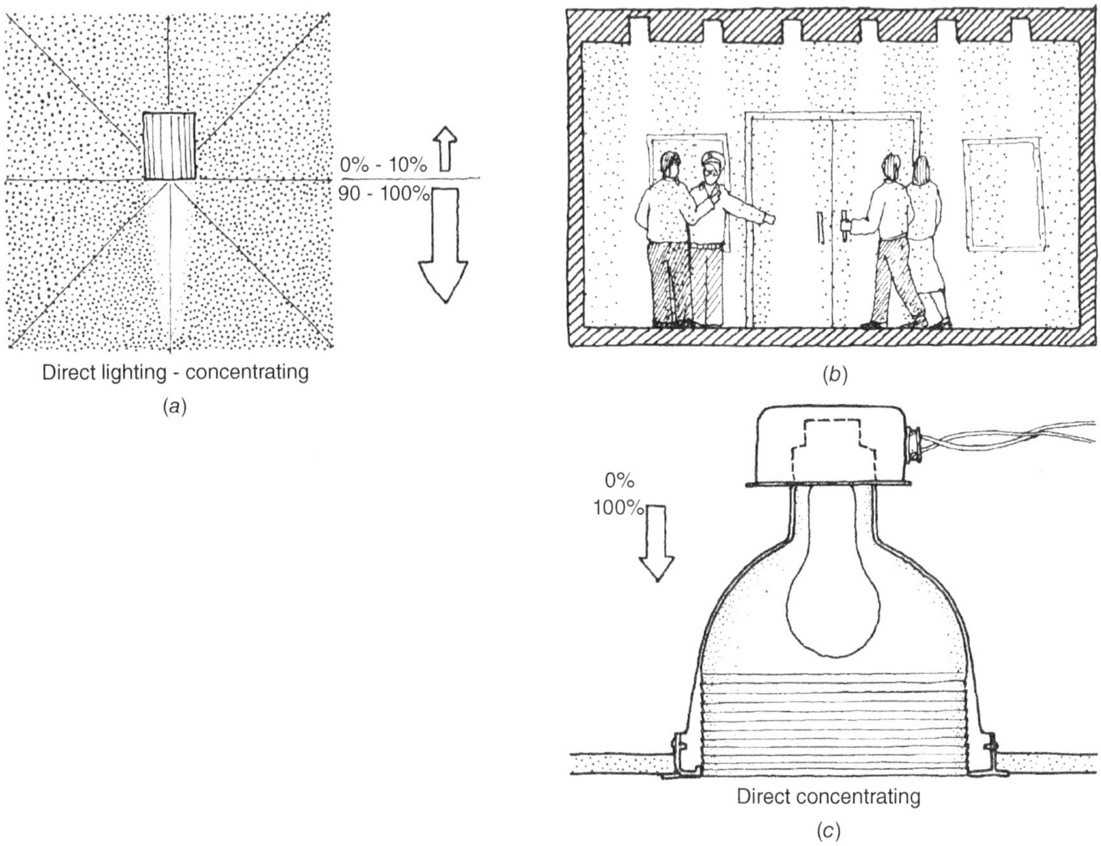

Figure 18.8 *(a) With concentrated direct delivery, the floor is the only luminous surface (b) other than the ceiling fixture. Diffuseness is absent. Walls are dark. LED/incandescent downlights (c) and, to a lesser extent, CFL downlights are of this type unless equipped with spread-type lenses.*

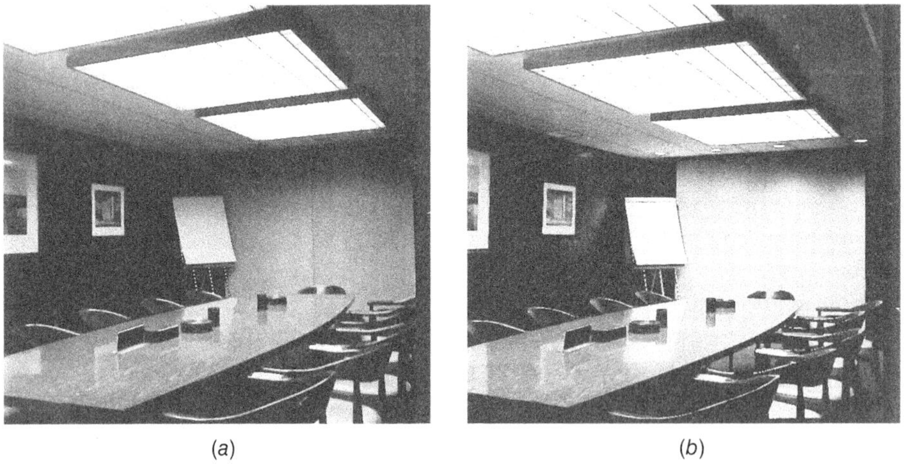

Figure 18.9 *Large-dimension lighting fixtures may be used in a low-ceilinged room if the apparent sizes of the units can be reduced. Here at a mounting height of 7.5 ft [2.3 m], 4 ft × 4 ft [1.2 × 1.2 m] units are acceptable because the lattice on the face of each unit gives the impression of reduced fixture size. Note also that the apparent illuminance in (b) is greater than in (a) although both are exactly equal on the table surface because of the wall wash in the background. The eye perceives vertical surface illuminance more readily than horizontal illuminance and retains the impression for the entire space.*

areas with lots of computer/TV screens, this negative impression can be alleviated by the addition of luminous surfaces in nonreflecting areas or points of light to add sparkle.

In summary, spread-type direct light delivery is suitable for uniform lighting, whereas concentrated direct light delivery (which reduces vertical illuminance) is appropriate for highlights, local and supplementary lighting, and private/intimate situations.

(c) Illuminance Targets

The lighting for many spaces is designed primarily—though ideally not exclusively—to support successful accomplishment of one or more defined visual tasks. This leads to a basic lighting design question: "How much light is needed for the visual task at hand?" This question is somewhat difficult to answer because recommendations for similar tasks vary by ratios as high as 10:1 among countries with developed technologies. Since the late 1980s, an attempt to reach a degree of international standardization has been underway.

The North American illuminance recommendations (promulgated by the Illuminating Engineering Society [IES]) were originally analytically developed by extrapolation from extensive laboratory tests. These tests determined the conditions under which small differences in contrast could be detected for specific degrees of accuracy, under variable task luminance, size, and exposure time. The idea behind the tests was that visual acuity could be defined as the ability to distinguish differences in contrast.

Because of social and legal pressure for energy-efficiency improvements, IES has taken steps toward the rationalization of its very influential illuminance guidelines. They include recognition of fatigue and task familiarity as factors in determining illuminance levels, in establishing lighting power budgets, and in setting energy standards that encourage the use of daylight and the use of supplemental lighting as the preferred technique wherever high levels of task lighting are required.

Illuminance and Luminance. Before discussing specific illuminance recommendations, it is good to understand the basis of their derivation, applicability, and shortcomings. As noted previously, most of the IES task illuminance recommendations are derived by extrapolation from threshold contrast visibility tests. Assuming uniform, diffuse task reflectance and uniform illuminance, it is simple to calculate required illuminance, since luminance is simply the product of illuminance and reflectance.

In SI units:

$$L = (E)(R/\pi)$$

where

L = luminance in cd/m^2
E = illuminance in lux
R = reflectance

In I-P units:

$$fL = (fc)(R)$$

Several reservations about this process of developing recommended task illuminances have been voiced by researchers. These will not be addressed here, but they suggest that the IES illuminance recommendations are best seen as guidelines and not as set-in-stone doctrine.

Visual task studies indicate that, assuming good contrast, required luminances for various categories of tasks are roughly as follows:

Category of Visual Task	Required Luminance cd/m^2 [fL]
Causal	10–20 [3–6]
Ordinary	20–100 [6–29]
Moderate	100–200 [29–58]
Difficult	200–400 [58–117]
Severe	Above 400 [117]

The dependence of required illuminance upon task reflectance (R) can be seen in the following tabulation, which shows quantitatively the illuminance requirements in the previous categories for tasks of radically different reflectance.

Category of Visual Task	Required E, lux[a] [fc]	
	R = 50%	R = 10%
Casual	62–125 [6–12]	300–625 [28–58]
Ordinary	125–625 [12–58]	625–3125 [58–290]
Moderate	625–1250 [58–116]	3125–6250 [290–580]
Difficult	1250–2500 [116–230]	6250–12,500 [580–1161]
Severe	>2500 [230]	>12,500 [1161]

[a]Lux and fc values are rounded.

This illustrates that a single illumination target may be inadequate for an area containing widely differing visual tasks. Note that a 10% reflectance makes all tasks difficult; and that casual seeing comprises only outline recognition.

Illuminance Criteria. The design targets for illuminance found in an OPR should come from a reputable source (association or organization) that provides guidance for a range of tasks. In the United States, this source is the Illuminating Engineering Society—which publishes consensus recommendations gathered from scientific research and experience, tempered by concern for available technology, economics, best practice, and energy efficiency. Building codes in the United States do not provide actionable guidance on adequate illuminance for anything beyond emergency egress. The recommendations presented in this book are expressly for North America. Many countries have their own published lighting standards. A specific project should follow the recommendations in force in the project locale.

Illuminance criteria should be used only in conjunction with other relevant lighting criteria (e.g., illuminance uniformity, surface modeling, color rendering, glare potential). Three factors will directly influence recommended illuminances: basic task characteristics, task importance, and observer characteristics (in the form of "visual age"). IES provides illuminance targets (Table 18.2) in a matrix organized with these characteristics in mind. The illuminance recommendations range from 0.5 to 20,000 lx [5 to 2000 fc] in lettered categories (A–Y) that progress from less demanding, primarily exterior lighting applications to highly demanding, primarily interior applications.

Table 18.2 also reflects a task's inherent potential contrast, size, and reflectance; and the importance of speed and accuracy. Those engaged with a task are estimated to be around 25 years old. IES states that these illuminance recommendations are to be doubled if more than 50% of the occupants are more than 65 years old.

For a space with several tasks of varying visual difficulty, the designer is expected to design the lighting and controls so that disparate tasks are accommodated without overlighting. Overlighting is common when a uniform illuminance system is based on the most light-intensive task in a space; being incredibly inefficient and resource-wasteful, intentional overlighting is strongly discouraged.

IES illuminance recommendations are not directly applicable to spaces in which a visual task is not the deciding factor in lighting design. Such installations include merchandising spaces, displays of all sorts, theatrical and artistic lighting, mood lighting, safety lighting, light used as part of an industrial process, and so on. Code requirements would supersede IES recommendations, but such conflict is rare.

IES maintains a wide variety of information resources to assist with lighting design. The society's *Lighting Library* (IES, 2024) is a curated five-collection series of standards that updates, expands, and replaces the 10th edition of *The Lighting Handbook* (IES, 2011) along with previous versions of IES standards. In instances where prior recommendations and/or manual calculation procedures are dispensed with in the latest IES resources due to diminishing utilization in practice, a specific past edition of *The Lighting Handbook* is referenced in this book as a historical resource for pedagogical purposes. Collectively, the IES' information resources offer a robust range of recommendations for illuminances and other lighting considerations for a wide range of applications and buildings such as schools, health care facilities, hospitality and entertainment venues, libraries, manufacturing plants, offices, residences, retail spaces, sports and recreation venues, transportation facilities, and places of worship. With qualitative and quantitative lighting criteria, a detailed user's guide offers the designer assistance in developing appropriate lighting solutions.

Lighting plays an important role in the functional use and general enjoyment of libraries. Table 18.3 includes a sample of the IES illuminance recommendations for a library: targets for horizontal surfaces (E_h) and vertical surfaces (E_v) are organized by task and visual ages of the users. As a lighting design progresses, thought must be given to aspects of the visual environment beyond illuminance. For example, a library circulation desk must be visible by patrons as well as provide visual interest for the general public; the library staff must be able to see certain areas of the library without experiencing glare. These kinds of refining details need to be considered.

(d) System Components

Lighting system components are the elements that enable design criteria (for illuminance, glare, energy, chemical pollution, aesthetics, cost, etc.). Lamps and luminaires are key system components and are discussed in detail in Chapter 17. Controls are another component, also discussed in Chapter 17. Beyond such hardware elements, the space itself is a major component in the operation and success of a lighting system. Space characteristics of importance to a lighting system include room dimensions and resultant proportions, surface and furnishing finishes (especially reflectance), and the geometric interrelationship between elements. The role of these components will be seen in the discussion of the zonal cavity method described in subsection *e* below.

(e) Uniform Illuminance Calculations

Lighting System Efficiency—Coefficient of Utilization. Some of the lumen output of a lamp is lost within a luminaire due to light absorption. The luminous efficiency of a fixture, expressed as a percentage, is the ratio of luminaire output lumens to lamp output lumens. This performance metric, while useful for cross-comparisons of like luminaires, is only

ACTIVE ENVIRONMENTAL SYSTEMS

TABLE 18.2 Recommended Illuminance Categories

Category		Recommended Illuminance Targets, lux [fc] — Visual Ages of Observers (Years) Where At Least Half Are <25	25–65	>65	Some Typical Application and Task Characteristics
INTERIOR and EXTERIOR applications	A[a]	0.5 [0.05]	1 [0.1]	2 [0.2]	Dark adapted situations
	B[a]	1 [0.1]	2 [0.2]	4 [0.4]	Basic convenience situations Very-low-activity situation
	C[a]	2 [0.2]	4 [0.4]	8 [0.7]	Slow-paced situations Low-density situations
	D[a]	3 [0.3]	6 [0.6]	12 [1.1]	Slow-to-moderate-paced situations
	E[a]	4 [0.4]	8 [0.7]	16 [1.5]	Moderate-to-high-density situations
	F[a]	5 [0.5]	10 [1]	20 [2]	Moderate-to-fast-paced situations
	G[a]	7.5 [0.7]	15 [1.4]	30 [3]	High-density situations
	H[a]	10 [1]	20 [2]	40 [4]	Some indoor very subdued circulation situations Some indoor social situations
INTERIOR and EXTERIOR	I[a]	15 [1.4]	30 [3]	60 [6]	Congested and significant outdoor intersections, important decision-points, gathering places, and key points of interest Some indoor social situations Some indoor commerce situations
INTERIOR and EXTERIOR applications	J[b]	20 [2]	40 [4]	80 [8]	Some outdoor commerce situations
	K[b]	25 [2.5]	50 [5]	100 [10]	Some indoor social situations
	L[b]	37.5 [3.5]	75 [7]	150 [15]	Some indoor commerce situations
	M[b]	50 [5]	100 [10]	200 [19]	
	N[b]	75 [7]	150 [15]	300 [30]	
	O[b]	100 [10]	200 [20]	400 [40]	
	P[c]	150 [15]	300 [30]	600 [60]	Some indoor social situations Some indoor education situations Some indoor commerce situations Some indoor sports situations
	Q[c]	200 [20]	400 [40]	800 [80]	Some indoor education situations
	R[c]	250 [25]	500 [50]	1000 [100]	Some indoor commerce situations
	S[c]	375 [37]	750 [75]	1500 [150]	Some indoor sports situations Some indoor industrial situations
	T[d]	500 [50]	1000 [100]	2000 [200]	Some sports situations
	U[d]	750 [75]	1500 [150]	3000 [280]	Some indoor commerce situations
	V[d]	1000 [100]	2000 [200]	4000 [380]	Some indoor industrial situations
	W[e]	1500 [150]	3000 [280]	6000 [560]	Some sports situations Some indoor industrial situations Some health care procedural situations
INTERIOR applications	X[e]	2500 [235]	5000 [470]	10,000 [930]	Some health care procedural situations
	Y[e]	5000 [470]	10,000 [930]	20,000 [1860]	

Source: IES, 2011. The I-P unit soft conversions were developed by the authors of this book.

[a]Orientation, relatively large-scale, physical (less-cognitive) tasks—visual performance is typically not work-related, but related to dark sedentary social situations, senses of safety and security, and casual circulation based on landscape, hardscape, architecture, and people as visual tasks.

[b]Common social activity and large and/or high-contrast tasks—visual performance involves higher-level assessment of landscape, hardscape, architecture, and people and can be work related.

[c]Common, relatively small-scale, more cognitive or fast-performance visual tasks—visual performance is typically daily life- and work-related, including much reading and writing of hard copies and electronic media consecutively and/or simultaneously.

[d]Small-scale, cognitive visual tasks—visual performance is work- or sports-related, close and distant fine inspection, very small detail, high-speed assessment and reaction.

[e]Unusual, extremely minute and/or life-sustaining cognitive tasks—visual performance is of the highest order in respective fields of health care, industrial, and sports.

TABLE 18.3 Library Facility Illuminance Recommendations

		Recommended Maintained Illuminance Targets, lux [fc]							
		Horizontal (E_h) Targets				Vertical (E_v) Targets			
			Visual Ages of Observers (Years) Where At Least Half Are				Visual Ages of Observers (Years) Where At Least Half Are		
Application and Tasks	**Notes**	**Category**	**<25**	**25–65**	**>65**	**Category**	**<25**	**25–65**	**>65**
Book Lending—Book Stacks									
General[a]	E_h @ floor of book stacks proper	O	100 [10]	200 [20]	400 [40]				
Shelving @ 2′–6″ [760 mm] AFF[a]	E_h and E_v @ front face of shelving	P	150 [15]	300 [30]	600 [60]	O	100 [10]	200 [20]	400 [40]
Lending Desk									
Self-service[a]	E_h @ 2′-6″ [760 mm] AFF; E_v @ 5′-0″ [1.5 m] AFF	P	150 [15]	300 [30]	600 [60]	M	50 [5]	100 [10]	200 [20]
Staffed[a]	E_h @ 2′-6″ [760 mm] AFF; E_v @ 5′-0″ [1.5 m] AFF	R	250 [25]	500 [50]	1000 [100]	O	100 [10]	200 [20]	400 [40]
Periodicals									
Shelving @ 1′-0″ [305 mm] AFF	E_v @ front face of shelving					M	50 [5]	100 [10]	200 [20]
Computer Center[a]	CSA/ISO Types I and II positive polarity screens. E_h @ 2′-6″ [760 mm]; E_v @ 4′-0″ [1.2 m] AFF	P	150 [15]	300 [30]	600 [60]	M	50 [5]	100 [10]	200 [20]
Reading Areas									
Grand Reading Room[a]	E_h @ 2′–6″ [760 mm] AFF; E_v @ 4′–0″ [1.2 m] AFF	R	250 [25]	500 [50]	1000 [100]	O	100 [10]	200 [20]	400 [40]
Stack Reading Areas[a]	E_h @ 2′–6″ [760 mm] AFF; E_v @ 4′–0″ [1.2 m] AFF	R	250 [25]	500 [50]	1000 [100]	M	50 [5]	100 [10]	200 [20]
Study Carrels[a]	E_h @ 2′–6″ [760 mm] AFF; E_v @ 4′–0″ [1.2 m] AFF	R	250 [25]	500 [50]	1000 [100]	O	100 [10]	200 [20]	400 [40]
Tables and Chairs[a]	E_h @ 2′–6″ [760 mm] AFF; E_v @ 4′–0″ [1.2 m] AFF	R	250 [25]	500 [50]	1000 [100]	O	100 [10]	200 [20]	400 [40]
Special Collections									
Archival Storage	E_h and E_v @ 3′–0″ [910 mm] AFF	P	150 [15]	300 [30]	600 [60]	M	50 [5]	100 [10]	200 [20]
Rare Books	E_h and E_v @ 3′–0″ [910 mm] AFF	P	150 [15]	300 [30]	600 [60]	M	50 [5]	100 [10]	200 [20]

Source: IES, 2011. The I-P unit soft conversions were developed by the authors of this book.

[a]Combination of daylighting and electric lighting strategies can be employed to achieve target values during daylight hours. Daylighting may require unconventional approaches.

a part of overall lighting system efficiency because a luminaire when installed in a specific room takes on a performance that depends on the characteristics of that room.

To illustrate, consider the case of a large room with a high, dark ceiling. Using a high-efficiency (say, 80%) indirect lighting fixture in such a room, most of the light directed upward would be absorbed by the ceiling, and the delivered illuminance on the working plane would be very low. If, however, the same room were illuminated with 50% efficiency direct-lighting fixtures utilizing the same lamps, the illuminance on the working plane would be considerably higher than in the first case.

Similarly, considering a small room with dark walls and ceiling, lighted either by diffuse lighting or direct lighting fixtures of the same wattage and efficiency, the horizontal-plane illumination is higher for the direct units because of the large loss of the horizontal and upward components of the diffuse lighting on the walls and ceiling.

Fixture efficiency alone is not sufficient to evaluate a lighting system; the overall luminous efficiency of a particular luminaire in a particular space is necessary to fully appreciate how well light source lumens are being used by a space. This value, since it describes the utilization of lamp/fixture output in a specific space, is known as the coefficient of utilization (CU). CU is defined as the ratio of lumens reaching the horizontal work plane to generated lumens. As each luminaire has a different CU for each different space in which it might be used, a system of standardization has evolved, utilizing room cavities (explained later) of certain proportions and with various surface reflectances. Coefficients of utilization are then computed and tabulated as shown in Table 18.4. The CU values given in this table are for generic fixture types only; in an actual job, luminaire data from manufacturers should be used. To summarize, CU is a performance factor that combines fixture efficiency and interroom distribution (as affected by room proportions, mounting height, and surface reflectances).

The Zonal Cavity Method. This method (also called the lumen method) is a hand calculation procedure for the analysis of average illuminance provided by a uniform illuminance lighting system. The purpose of the zonal cavity method is to determine how many lighting fixtures are required to deliver design illuminance to a space. One of the nice things about the zonal cavity method is that it lays out all the variables that affect the performance of a uniform lighting system—for inspection and consideration by the lighting designer. As opposed to black-box software-based analysis approaches, this hand calculation method is an open box. Understanding the design variables that impact system performance, and the general magnitude of such impacts, is very valuable to a lighting designer.

A number of calculation methods are available to address different design situations. Calculating the non-uniform illuminance provided by point, line, or area sources is covered later in this chapter. Surface luminance calculations are also discussed later in this chapter.

Before beginning a detailed description of the zonal cavity method, a general discussion of precision is in order. The precision of any calculation cannot exceed the precision of the input data. There is no point in working to three decimal places if the data driving the analysis are accurate to only one decimal place. The zonal cavity method of illuminance estimation is full of assumptions and estimates. Among these are the following:

1. It is assumed that the space is empty; this is not normally the case.
2. It is assumed that all surfaces are perfect diffusers; this is not the case.
3. All surface reflectances are estimates, usually with an accuracy of no more than ±10%.
4. Maintenance conditions are prognoses of future actions, at best ±10% accurate.
5. No allowance is made for deviation of the performance of an installed product from its design specification.

Any attempt to reduce the looseness of these approximations and assumptions would complicate the calculation process and would serve little purpose for this type of calculation. For this reason, the procedure presented herein introduces some approximations (which are justified) in the interest of simplification. These approximations are noted wherever used. This is not an argument for inaccuracy or sloppiness—but rather for realism.

Calculation of Horizontal Illuminance by the Zonal Cavity Method. The zonal cavity method is a hand calculation procedure for determining the average maintained and initial illuminance (in footcandles or lux) on a horizontal working plane in a room. The method presupposes that luminaires will be spaced such that uniformity of illumination is provided, so that an average calculation will have validity. The method is founded on the fact that 1 lx of illuminance is 1 lm incident on 1 square meter of area [1 footcandle incident on 1 square foot], that is,

$$\text{lux [or fc]} = \frac{\text{lumens}}{\text{area m}^2 \text{ [or ft}^2\text{]}} \qquad \text{(Eq. 18.1)}$$

TABLE 18.4 Coefficients of Utilization for Three Typical LED Luminaires

Luminaire Type	Photometric Distribution	RCC→ RW→ RFC→ RCR↓	80 50 20	80 30 20	80 10 20	70 50 20	70 30 20	70 10 20	50 50 20	50 30 20	50 10 20	30 50 20	30 30 20	30 10 20	10 50 20	10 30 20	10 10 20	WDRC↓
a. direct 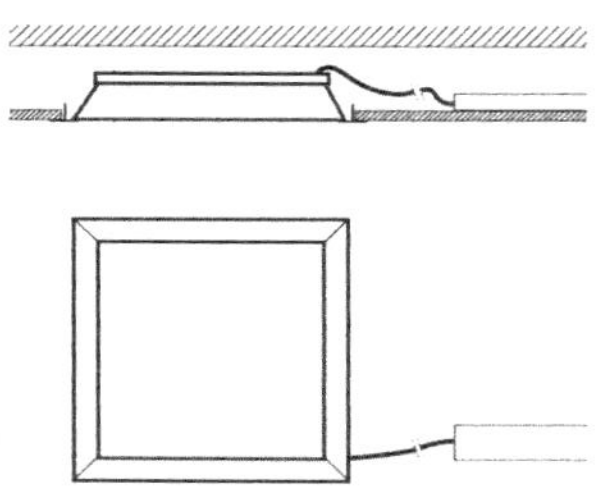	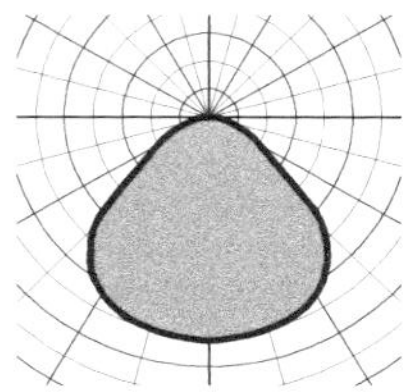	0	119	119	119	116	116	116	111	111	111	106	106	106	102	102	102	—
		1	107	103	100	105	101	99	101	98	96	97	95	93	93	91	90	0.301
		2	96	90	85	94	89	84	91	86	82	87	84	81	84	82	79	0.288
		3	86	79	74	85	78	73	82	76	72	79	75	71	77	73	69	0.270
		4	77	70	64	76	69	64	74	68	63	72	66	62	70	65	61	0.255
		5	69	61	55	68	61	55	66	60	55	64	59	54	63	58	54	0.242
		6	63	55	49	62	54	49	60	53	48	59	52	48	57	52	47	0.227
		7	57	49	43	56	48	43	54	47	42	53	47	42	52	46	42	0.214
		8	51	43	37	50	43	37	49	42	37	48	41	37	47	41	37	0.204
		9	46	38	32	45	38	32	44	37	32	43	37	32	42	36	32	0.194
		10	42	34	29	41	34	29	40	33	28	39	33	28	38	32	28	0.184
b. direct-indirect 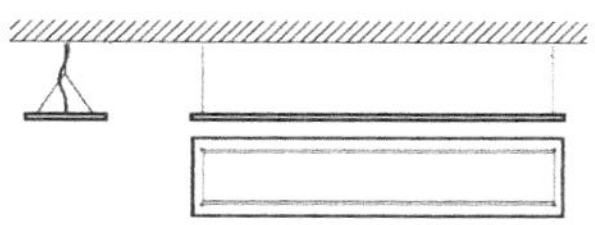	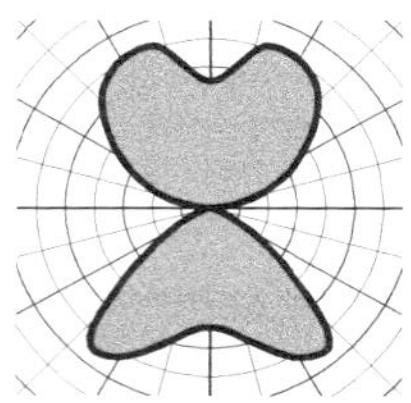	0	105	105	105	96	96	96	79	79	79	64	64	64	49	49	49	—
		1	93	89	86	85	82	79	70	68	66	57	55	54	44	43	43	0.138
		2	82	76	71	75	70	66	62	59	56	51	48	46	40	38	37	0.130
		3	72	65	60	66	61	56	55	51	48	45	42	40	36	34	32	0.121
		4	64	57	51	59	53	48	50	45	41	41	37	34	32	30	28	0.114
		5	57	49	44	53	46	41	44	39	35	36	33	30	29	26	24	0.108
		6	51	43	38	47	40	35	40	34	30	33	29	26	26	23	21	0.101
		7	46	38	32	42	35	30	36	30	26	29	25	22	23	21	18	0.095
		8	41	33	28	38	31	26	32	27	23	26	22	19	21	18	16	0.090
		9	37	29	24	34	27	23	29	23	20	24	20	17	19	16	14	0.086
		10	33	26	21	31	24	20	26	21	17	21	17	15	17	14	12	0.081
c. indirect 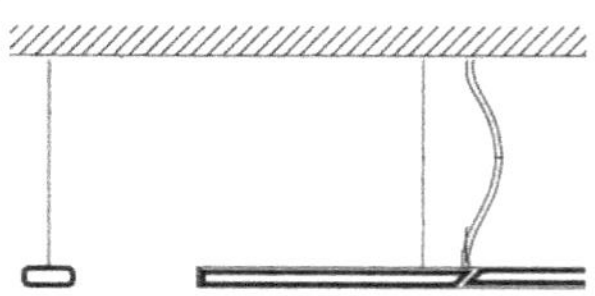	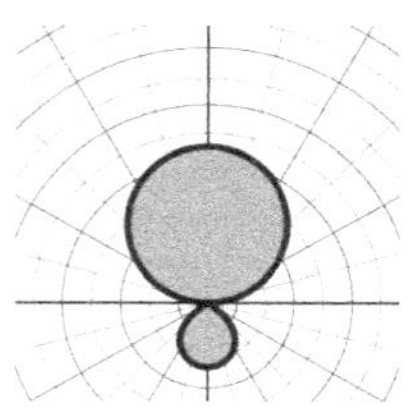	0	100	100	100	89	89	89	68	68	68	48	48	48	31	31	31	—
		1	88	85	82	78	76	73	60	58	57	43	42	41	28	27	27	0.058
		2	78	72	68	69	65	61	53	50	48	39	37	35	25	24	24	0.057
		3	69	62	57	61	56	51	48	44	41	35	33	31	23	22	21	0.054
		4	61	54	48	55	49	44	43	38	35	31	29	27	21	19	18	0.051
		5	54	47	41	49	42	38	38	34	30	28	25	23	19	17	16	0.049
		6	49	41	36	44	37	33	34	30	26	25	23	20	17	16	14	0.046
		7	44	36	31	39	33	29	31	26	23	23	20	18	16	14	13	0.044
		8	39	32	27	36	29	25	28	23	20	21	18	16	14	12	11	0.042
		9	36	29	24	32	26	22	25	21	18	19	16	14	13	11	10	0.041
		10	32	26	21	29	23	19	23	19	16	17	14	12	12	10	9	0.039

Abbreviations: RCC = effective reflectance of ceiling cavity; RW = average reflectance of wall surfaces; RFC = effective reflectance of floor cavity; RCR = room cavity ratio; WDRC = wall direct radiation coefficient.

These values and images were extracted by the authors from data generously provided by TRILUX gmbH & Co. for the following luminaires: type a-direct: ArimoS Sky; type b-direct-indirect: LuceoS H1-L; and type c-indirect: Parelia H-L.

As explained previously, the ratio of lumens reaching the working plane in a specific space to the lumens generated by the lamps in the lighting system is the coefficient of utilization, CU. As an equation,

$$\text{lumens falling on the working plane} = (\text{lamp lumens})\,(CU) \qquad \text{(Eq. 18.2)}$$

Therefore,

$$\text{illuminance } E = \frac{(\text{lamp lumens})\,(CU)}{\text{area}} \qquad \text{(Eq. 18.3)}$$

The coefficient of utilization is extracted from tables provided by the manufacturer of a selected luminaire by a technique known as the zonal cavity method. In the absence of project-specific CU data, a rough approximation can be made for LED fixtures by using the three generic fixture types in Table 18.4.

The illuminance value calculated by Equation 18.3 is the average initial illuminance. This initial illuminance is reduced by the effect of temperature and voltage deviations, dirt accumulation on luminaires and on room surfaces, lamp output depreciation, and other factors. These effects are collectively referred to as the system light loss factor (LLF):

$$\text{maintained } E = (\text{initial } E)\,(LLF) \qquad \text{(Eq. 18.4)}$$

The procedures used to estimate the various light loss factors, and establish an overall LLF, are explained in the following section.

Another equation for maintained illuminance (E) as calculated by the zonal cavity method is, therefore,

$$E = \frac{(\text{lamp lumens})\,(CU)\,(LLF)}{\text{area}} \qquad \text{(Eq. 18.5)}$$

where E is lux if area is expressed in square meters, or footcandles if the area is in square feet.

Lamp lumens is the total output of the lighting system proposed for the space and is equal to:

$$\text{lamp lumens} = (\text{number of fixtures})\,(\text{lamps per fixture}) \times (\text{initial lumens per lamp}) \qquad \text{(Eq. 18.6)}$$

The relationships can be restated (where # is the number of luminaires) as

$$E = \frac{(\#)\,(\text{lamps/luminaire})\,(\text{lumens/lamp})\,(CU)\,(LLF)}{\text{area}} \qquad \text{(Eq. 18.7)}$$

or, conversely, solving for the number of luminaires (#) required to achieve a target-maintained illuminance E:

$$\# = \frac{(E)(\text{area})}{(\text{lamps/luminaire})\,(\text{lumens/lamp})\,(CU)\,(LLF)} \qquad \text{(Eq. 18.8)}$$

For large areas, a sometimes-useful value is the area successfully illuminated per luminaire:

$$\text{area per luminaire} = \frac{(\text{lamps/luminaire})\,(\text{lumens/lamp})\,(CU)\,(LLF)}{E} \qquad \text{(Eq. 18.9)}$$

For instance, it may be more convenient to know that to maintain, say, 60 fc [650 lx] within a space with a given luminaire and lamp, that 70 ft^2 [6.5 m^2] per luminaire will work than it is to know that for an 18,000-ft^2 [1672-m^2] floor, 257 fixtures are necessary. The former value allows the establishment of a layout of a pattern, say, 7 ft by 10 ft [2.1 m by 3.0 m]; the latter value is too large to be immediately useful.

Estimation of Light Loss Factors. Light loss factor (LLF) is an interesting concept that seems to be uniquely applied to lighting system design. The basic idea is simple: system performance will deteriorate over time as the effects of entropy kick in. What makes estimating LLF for a project interesting is that these effects occur in the future—and are generally out of the direct control of the building design team. Whether an owner cleans luminaires every three years, six years, or never is not a design decision. Yet this owner choice will affect lighting system performance. One work-around for the disconnect between design and operations is the development of an owner's manual that describes design assumptions and recommends owner actions that would conform to those assumptions. Not attempting to make this connection would be professionally irresponsible.

Light loss factors for a given project typically include four recoverable and four nonrecoverable elements. The recoverable losses can be reversed or mitigated by maintenance; the nonrecoverable losses cannot. As discussed under lumen maintenance controls, recoverable losses increase over time (as, for example, luminaires collect dirt and lamps age). These slowly increasing loss factors decrease the amount of light that can be delivered to the task surface. Maintenance actions (such as lens cleaning or wall surface touchup) can reverse the reduction in light delivery and thus increase illuminance.

The total LLF for a lighting application is the product of the individual factors. This is an important consideration—three LLF values of 0.9 each (reasonably high values) collectively lead to a total LLF of 0.73 (which represents a 30% reduction in available lumens). The overview that follows includes approximations. For more precise data, see the 10th edition of *The Lighting Handbook* (IES, 2011).

Nonrecoverable light loss factors include:

(a) *Luminaire ambient temperature:* Light output from a lamp can vary from cataloged data when the lamp is operated at other than its design temperature. The extent of variation depends on the lamp type. With normal indoor installations (where human thermal comfort is a concern), use a value of 1.0 for this LLF—that is, there is no depreciation. For other ambient conditions, refer to technical data on the luminaire involved.
(b) *Voltage to luminaire:* LLF due to voltage occurs when the supplied voltage to a luminaire is different from the fixture's designed operating voltage. Over-voltage may cause the luminaire to initially exhibit more lumens, but it also tends to cause overheating, shorten the lifespan of components such as lamps and/or ballasts, and reduce the fixture's efficiency over time. Under-voltage can both reduce lumen output and hamper the fixture's efficiency over time. When a lamp operates at its rated voltage, use a value of 1.0 for this LLF.
(c) *Luminaire surface depreciation:* This LLF is proportional to age and depends upon the type of surface involved. The designer must estimate this factor based upon knowledge of the luminaire materials as specified for the project. In general, higher quality fixtures will have lower losses. This is a nonrecoverable factor (distinct from dirt depreciation) and represents a reduction in material reflectance or transmittance that can only be mitigated by replacement or resurfacing.
(d) *Other factors (components):* Losses due to system components may include ballast factor, ballast-lamp photometric factor, equipment operating factor, and lamp position (tilt) factor. Air troffers also introduce a thermal adjustment factor. In the absence of project-specific data for these component factors, use a combined value of 0.92.

In the absence of reliable data for any of the foregoing nonrecoverable factors, experience suggests using an overall nonrecoverable factor of 0.88 (representing the product of factors $a \times b \times c \times d$).

Recoverable light loss factors include:

(e) *Room surface dirt depreciation:* Lighting approaches that depend heavily on light reflected from room surfaces, such as indirect distribution systems, are more seriously affected by dirt than systems that deliver most of their light directly. Assuming a 24-month cleaning cycle and normal conditions of cleanliness, use the appropriate LLF value from the following list. Alter the chosen base factor for differential conditions such as infrequent maintenance and unusual cleanliness or dirtiness.

- direct lighting: 0.92 ± 5%
- semi-direct lighting: 0.87 ± 8%
- direct–indirect lighting: 0.82 ± 10%
- semi-indirect lighting: 0.77 ± 12%
- indirect lighting: 0.72 ± 17%

(f) *Lamp lumen depreciation (LLD):* This loss factor depends upon the type of lamp used and the facility's replacement schedule. Lamp lumen depreciation values are based upon lumen output at rated lamp life; the rated life of the lamp types listed below vary widely, thus these values may not truly represent apples-to-apples comparisons. Use the following values when project-specific data are unavailable:

	Group Replacement	Replacement on Burnout
Incandescent	0.94	0.88
Tungsten–halogen	0.98	0.94
Fluorescent	0.90	0.85
Mercury-vapor	0.82	0.74
Metal–halide	0.87	0.80
High-pressure sodium	0.94	0.88
LED	0.94	0.87

(g) *Lamp burnout factor:* This loss factor accounts for lamps that produce no output but have not been replaced (burnouts that remain in place). LLF depends upon building maintenance schedules and method of lamp replacement. The impact of a burnout also depends greatly upon the number and distribution of lamps in a space. Use the following values as a general rule:

- group replacement procedures: 1.0
- individual replacement on burnout: 0.95

(h) *Luminaire dirt depreciation (LDD):* The LDD factor depends upon luminaire design, dirt conditions in the space, and maintenance schedule. The luminaire maintenance category for a specific fixture is obtained from the manufacturer. The type of dirt condition is determined by considering the space and its functions. Assuming a 12-month cleaning schedule and normal room cleanliness, choose a base value from Figure 18.10 and adjust it to match the anticipated conditions of dirt and maintenance. The categories indicated correspond to those used by the IES.

Total LLF is the product of all eight depreciation factors:

$$LLF = (a)(b)(c)(d)(e)(f)(g)(h) \qquad \text{(Eq. 18.10)}$$

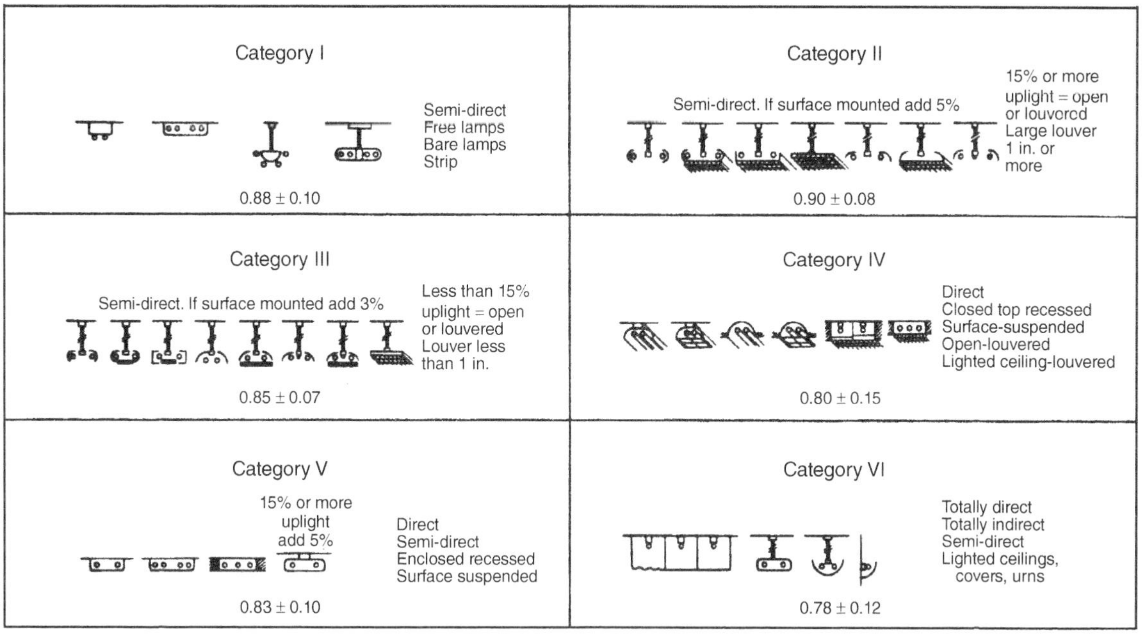

Figure 18.10 *The LDD factor is determined from the category of luminaire (which is an indication of its proneness to dirt accumulation) plus knowledge of the room ambient conditions.*

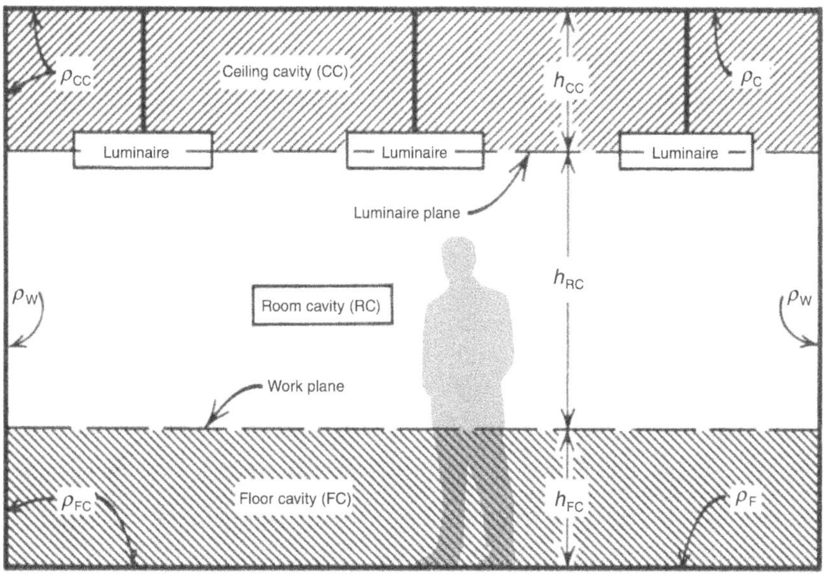

Legend:

ρ_C = ceiling reflectance
ρ_{CC} = ceiling cavity reflectance
ρ_W = wall reflectance
ρ_F = floor reflectance
ρ_{FC} = floor cavity reflectance
h = height in feet or meters
h_{RC} = height of room cavity (etc.)

Figure 18.11 *Room cavities as used in the zonal cavity method.*

For example, a fluorescent air troffer in a regularly maintained group-lamp-replacement, air-conditioned office might typically have an LLF of 0.8. The same fixture in the same setting, but where walls and the fixture are cleaned only when burned-out lamps are replaced, would typically have an LLF of 0.55. Thus, if in the first case the maintained illumination is *E* fc, in the second case it is 0.55/0.80 or 0.69 *E* fc, that is, a reduction of 31% because of poor maintenance.

When a detailed determination of LLF is not possible, use the factors described in the section "Zonal Cavity Calculation by Approximation" (they are somewhat conservative).

Determining the Coefficient of Utilization. The CU numerically connects a particular luminaire/lamp combination to a particular space by relating the luminaire's light distribution characteristic to the room dimensions and surface reflectances. To account for luminaire mounting height and its relationship to the working plane, the room is conceptually divided into three cavities:

- ceiling cavity (CC) above the fixture
- floor cavity (FC) below the working plane
- room cavity (RC) between these two

The cavities are graphically depicted in Figure 18.11. Knowing or assuming the major surface reflectances, effective reflectances of the floor and ceiling cavities can be obtained. An effective reflectance accounts for the effects of interreflections within a cavity and for differences in wall versus floor/ceiling reflectance values.

Using the above information, the applicable CU can be selected from tables (either Table 18.4 for a small sample of LED fixtures or from manufacturer's data) and the zonal cavity formula (Equation 18.8) applied to arrive at average illuminance. A step-by-step explanation of the method plus examples will demonstrate the procedure. The flow chart in Figure 18.12 and the calculation form in Figure 18.13 will be of great help in understanding the procedure.

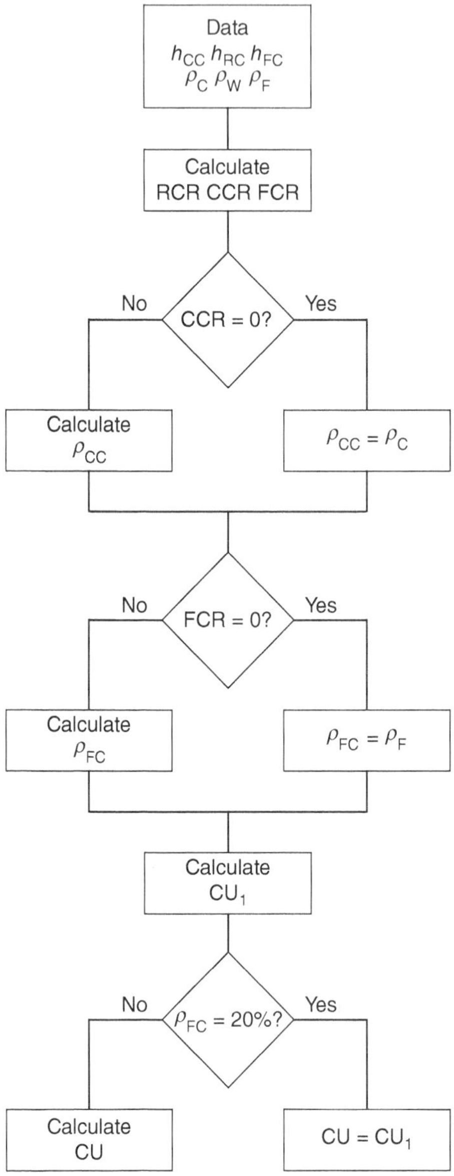

Figure 18.12 *Zonal cavity method flow chart.*

STEP 1. Establish dimensional data for the room. Room length, width, and height are determined by the architectural design process. In offices, schools, and many other occupancies, the working plane is typically 30 in. [760 mm] above finished floor (AFF). In shops, 42–48 in. [1066–1220 mm]; in a carpet store at floor level. The location of the working plane and placement of the lighting fixtures relative to the ceiling plane will determine the three *h* terms, representing the heights of the zonal cavities. Estimate the initial reflectances of major room surfaces (selecting reflectances that conform to those suggested by the values in Table 18.5 is fine for preliminary design). Fill in the information in a sketch, as in Figure 18.13.

STEP 2. See again Figure 18.13. Determine the three cavity ratios of the room by calculation. The basic expression for a cavity ratio (CR) is:

$$CR = (2.5)\left(\frac{\text{area of cavity wall}}{\text{area of work plane}}\right) \qquad \text{(Eq. 18.11)}$$

In a rectangular space, the area of a given cavity wall is $h \times (2l \times 2w)$ or $2h(l + w)$; therefore,

$$CR = \frac{(2.5) \times (2h(l + w))}{\text{area of work plane}} \qquad \text{(Eq. 18.12)}$$

or

$$CR = (5h)\left(\frac{l + w}{l \times w}\right) \qquad \text{(Eq. 18.13)}$$

GENERAL INFORMATION

1. Project identification: ____________________
(Give name of area and/or building and room number)

2. Average maintained illumination for design: ______ lux [footcandles]

Luminaire data:
3. Manufacturer: ____________________
4. Catalog number: ____________________

Lamp data:
5. Type and color: ____________________
6. Number per luminaire: ____________________
7. Total lumens per luminaire: ____________________

SELECTION OF COEFFICIENT OF UTILIZATION

8. Step 1: Fill in sketch at right.

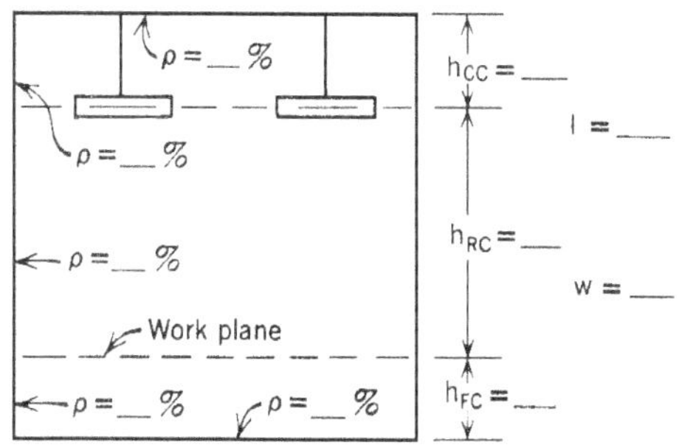

9. Step 2: Determine Cavity Ratios by formulas.

9a. Room cavity ratio, RCR = ____________________
9b. Ceiling cavity ratio, CCR = ____________________
9c. Floor cavity ratio, FCR = ____________________

10. Step 3: Obtain effective ceiling cavity reflectance (ρ_{CC}) from Table 18.5. ρ_{CC} = ________

11. Step 4: Obtain effective floor cavity reflectance (ρ_{FC}) from Table 18.5. ρ_{FC} = ________

12. Step 5: Obtain coefficient of utilization (CU) from manufacturer's data. CU = ________

SELECTION OF LIGHT LOSS FACTORS

Unrecoverable		Recoverable	
13. Luminaire ambient temperature	________	17. Room surface dirt depreciation	________
14. Voltage to luminaire	________	18. Lamp lumen depreciation	________
15. Luminaire surface depreciation	________	19. Lamp burnouts factor	________
16. Other factors **(components)**	________	20. Luminaire dirt depreciation LDD	________

21. Total light loss factor, LLF (product of individual factors above): ______

CALCULATIONS

(Average maintained illumination level)

$$\text{Number of luminaires} = \frac{(\text{Illuminance}) \times (\text{Area})}{(\text{Lumens per luminaire}) \times (\text{CU}) \times (\text{LLF})}$$

22. = ____________________ =

$$\text{Lux [footcandles]} = \frac{(\text{number of luminaires}) \times (\text{lumens per luminaire}) \times (\text{CU}) \times (\text{LLF})}{(\text{area})}$$

23. = ____________________ =

24. Calculated by: ____________________ Date: ____________

Figure 18.13 *Zonal cavity method calculation form. (Courtesy of Illuminating Engineering Society.)*

For other than rectangular rooms, the area can be calculated as required by geometry. For instance, in a circular room, the cavity wall area = $h \times 2\pi r$ and the work plane area is πr^2. Thus,

$$CR = \frac{(2.5)(h)(2\pi r)}{\pi r^2} = \frac{5h}{r} \qquad \text{(Eq. 18.14)}$$

For each of the cavities in a rectangular room, the cavity ratio is calculated as:

Room cavity ratio (RCR)

$$\text{RCR} = 5h_{\text{RC}}\left(\frac{l + w}{l \times w}\right) \qquad \text{(Eq. 18.15)}$$

Ceiling cavity ratio (CCR)

$$\text{CCR} = 5h_{\text{CC}}\left(\frac{l + w}{l \times w}\right) \qquad \text{(Eq. 18.16)}$$

Floor cavity ratio (FCR)

$$\text{FCR} = 5h_{\text{FC}}\left(\frac{l + w}{l \times w}\right) \qquad \text{(Eq. 18.17)}$$

Because the three cavity ratios for a given space are related, once one has been determined, the others can be obtained by adjusting for cavity height

$$\text{CCR} = \text{RCR}\left(\frac{h_{\text{CC}}}{h_{\text{RC}}}\right)$$

$$\text{FCR} = \text{RCR}\left(\frac{h_{\text{FC}}}{h_{\text{RC}}}\right) \qquad \text{(Eq. 18.18)}$$

$$FCR = RCR\frac{h_{\text{FC}}}{h_{\text{RC}}} \qquad \text{(Eq. 18.19)}$$

and

$$\text{CCR} = \text{FCR}\left(\frac{h_{\text{CC}}}{h_{\text{FC}}}\right) \qquad \text{(Eq. 18.20)}$$

STEP 3. See Table 18.5 and Figures 18.12 and 18.13. Obtain the effective ceiling cavity reflectance (ρ_{CC}) from Table 18.5 using the expected ceiling surface reflectance as ρ_C and the wall reflectance (ρ_W) from step 1. If the lighting fixtures are surface-mounted or recessed, then CCR = 0 (there is no cavity) and ρ_{CC} is equal to the reflectance of the ceiling surface plane.

Please note that, depending on the resource, reflectance values (ρ) may be offered as either a percentage (e.g., 80%) or as a decimal representing the same value (e.g., 0.80).

STEP 4. See Table 18.5 and Figures 18.12 and 18.13. Obtain the effective floor cavity reflectance ρ_{FC}, as described in step 3 for ρ_{CC}. If the floor is the working plane, FCR = 0 (there is no cavity) and ρ_{FC} is equal to the floor surface reflectance.

STEP 5. Select the applicable CU from the manufacturer's data—using room-specific values for ρ_{CC}, ρ_{FC}, ρ_W, and RCR. Interpolation may be necessary for project values that do not closely match the values given in the column headers in the CU table—for example, when RCR is 3.3 or ρ_W is 40%. CU correction factors for ρ_{FC} other than 20% (the standard value in

TABLE 18.5 Percent Effective Ceiling or Floor Cavity Reflectance (ρ_{CC}, ρ_{FC}) for Various Reflectance Combinations

Percent Ceiling ρ_c or Floor Reflectance ρ_F:		90				80				70			50			30				10		
Percent Wall Reflectance ρ_W:		**90**	**70**	**50**	**30**	**80**	**70**	**50**	**30**	**70**	**50**	**30**	**70**	**50**	**30**	**65**	**50**	**30**	**10**	**50**	**30**	**10**
Ceiling or Floor Cavity Ratios—CCR or FCR	0	90	90	90	90	80	80	80	80	70	70	70	50	50	50	30	30	30	30	10	10	10
	0.2	89	88	86	85	79	78	77	76	68	67	66	49	48	47	30	29	29	28	10	10	9
	0.4	88	86	83	81	78	76	74	72	67	65	63	48	46	45	30	29	27	26	11	10	9
	0.6	88	84	80	76	77	75	71	68	65	62	59	47	45	43	29	28	26	25	11	10	9
	0.8	87	82	77	73	75	73	69	65	64	60	56	47	43	41	29	27	25	23	11	10	8
	1.0	86	80	74	69	74	71	66	61	63	58	53	46	42	39	29	27	24	22	11	9	8
	1.2	86	78	72	65	73	70	64	58	61	56	50	45	41	37	29	26	23	20	12	9	7
	1.4	85	77	69	62	72	68	62	55	60	54	48	45	40	35	28	26	22	19	12	9	7
	1.6	85	75	66	59	71	67	60	53	59	52	45	44	39	33	28	25	21	18	12	9	7
	1.8	84	73	64	56	70	65	58	50	57	50	43	43	37	32	28	25	21	17	12	9	6
	2.0	83	72	62	53	69	64	56	48	56	48	41	43	37	30	28	24	20	16	12	9	6
	2.2	83	70	60	51	68	63	54	45	55	46	39	42	36	29	28	24	19	15	13	9	6
	2.4	82	68	58	48	67	61	52	43	54	45	37	42	35	27	28	24	19	14	13	9	6
	2.6	82	67	56	46	66	60	50	41	53	43	35	41	34	26	27	23	18	13	13	9	5
	2.8	81	66	54	44	66	59	48	39	52	42	33	41	33	25	27	23	18	13	13	9	5
	3.0	81	64	52	42	65	58	47	38	51	40	32	40	32	24	27	22	17	12	13	8	5
	3.5	79	61	48	37	63	55	43	33	48	38	29	39	30	22	26	22	16	11	13	8	5
	4.0	78	58	44	33	61	52	40	30	46	35	26	38	29	20	26	21	15	9	13	8	4
	4.5	77	55	41	30	59	50	37	27	45	33	24	37	27	19	25	20	14	8	14	8	4
	5.0	76	53	38	27	57	48	35	25	43	32	22	36	26	17	25	19	13	7	14	8	4

TABLE 18.6 CU Adjustment Factors for Effective Floor Cavity Reflectances Other Than 20% (Any Wall Reflectance)[a]

Room Cavity Ratio	Percent Effective Ceiling Cavity Reflectance, ρ_{CC} 80	70	50	10
1	1.08	1.06	1.04	1.01
2	1.06	1.05	1.03	1.01
3	1.04	1.04	1.03	1.01
4	1.03	1.03	1.02	1.01
5	1.03	1.02	1.02	1.01
6	1.02	1.02	1.02	1.01
7	1.02	1.02	1.01	1.01
8	1.02	1.02	1.01	1.01
9	1.01	1.01	1.01	1.01
10	1.01	1.01	1.01	1.01

Source: IES, 1993.

[a]For more precise data that address varying ρ_W values, consult the 8th edition of *The Lighting Handbook*.

Notes:

1. For 30% effective floor cavity reflectance, multiply the CU value for 20% floor cavity reflectance by the appropriate factor from the table.
2. For 10% effective floor cavity reflectance, divide the CU value for 20% floor cavity reflectance by the appropriate factor from the table.

CU tables) are given in Table 18.6. Be wary of extrapolating beyond the values given in published CU tables (for example, assuming a ρ_W of 75%); there is typically a practical reason for the range of values given in these tables. See Example 18.1.

STEP 6. Calculate the required number of fixtures to provide the design illuminance. If this number is impractical, adjust it (for example, 21.3 fixtures would likely become 22 fixtures) and recalculate the initial illuminance. Setting LLF to 1.0 allows calculation of initial illuminance; setting LLF to its actual project value will provide maintained illuminance. Maintained illuminance should equal or exceed design illuminance for a solution that meets the owner's project requirements.

Illustrative examples and shortcut methods are demonstrated in the following section. CU coefficients are listed in Table 18.4 for three generic LED luminaire types. Manufacturers can provide CU data for their specific products.

EXAMPLE 18.1 Figure 18.13 is a compact and useful worksheet for zonal cavity calculations—using this worksheet to track these examples is recommended. To simplify this example, it is worked using only SI units; the exact same process would apply with I-P units.

Given. Assume a classroom with dimensions of (W × L × H): 6 m × 8 m × 3.7 m, in an elementary school. Initial surface reflectances are: ceiling 80%, entire wall 50%, floor 20%. (Note that the Figure 18.13 worksheet can accommodate different reflectances for the upper, center, and lower wall sections.) The basic lighting objective is to provide adequate illuminance using LED fixtures. Assume a yearly maintenance cycle, lamp replacement at burnout, proper voltage and ballasts, and a medium-clean dirt condition.

SOLUTION

(a) *Illuminance target*. Refer to IES recommendations to establish an appropriate illuminance target. For this example, 500 lx will be assumed. Lines 1, 2, and 8 (the room sketch) of Figure 18.13 can be filled in at this point.

Note: The daylight contribution in many classrooms can be considerable because of the hours of space use. Daylight illuminance may frequently exceed 500 lx, but to cleanly demonstrate the zonal cavity calculation procedure, assume no daylight contribution. This will be the case if the classroom is used at night and/or during early morning/late afternoon hours. An appropriate control scheme would be used to reduce overlighting from the electric system when there is a daylight contribution.

(b) *Luminaire selection*. Classroom lighting precedents suggest the selection of luminaires that can provide:

- low direct glare potential (i.e., have a high visual comfort probability, VCP) because schoolchildren spend a large proportion of their time in a heads-up position
- low potential for veiling reflections because much of the seeing task involves high-reflectance materials, occasionally specular
- high efficiency to meet *ASHRAE/IES Standard 90.1* and the expectations of many clients
- minimal required maintenance in view of the limited operational resources available to many schools

Although the 3.7 m ceiling height is adequate to permit use of indirect lighting (e.g., luminaire *type c* in Table 18.4) with all of its distinct advantages, this approach is not chosen because:

- the luminaire maintenance category is of concern given the type of maintenance expected
- indirect lighting depends on a highly reflective ceiling, requiring yearly cleaning and repainting at intervals not exceeding five years; this is not generally the case in public schools

Considering the above, select luminaire *type a* from Table 18.4 as an initial informed trial fixture. This selection would meet requirements 1 through 4.

Although its distribution curve (see Table 18.4) shows no upward component, some fixtures of this basic design have slots in the top of the reflector to provide a 5% to 10% uplight component. In practice, such a unit would avoid an excessively dark ceiling and high luminance ratios between fixture and background. The mounting height is assumed as 2.7 m to permit an uplight component. The recommended maximum SC would be between 1.5 and 2.0. The work plane height is 750 mm.

(c) Calculations

STEP 1. The information that should appear in the sketch in Figure 18.13 is:

$h_{CC} = 1.0$ m	$h_{RC} = 1.95$ m
$h_{FC} = 0.75$ m	$l = 8$ m
$\rho_C = 80\%$	$w = 6$ m
$\rho_F = 20\%$	$\rho_w = 50\%$

STEP 2. Determine the three cavity ratios as follows:

$$\frac{l+w}{l \times w} = 0.29$$

$$RCR = 5h_{RC}\frac{l+w}{l \times w}$$

$$RCR = 5(1.95)(0.29) = 2.84$$

$$CRC = 5(1)(0.29) = 1.46$$

$$FCR = 5(0.75)(0.29) = 1.09$$

STEP 3. Obtain effective reflectances from Table 18.5. For ρ_{CC} use $\rho_C = 80\%$, $\rho_W = 50\%$, and CCR = 1.46.

Thus,

$$\rho_{CC} = 61\% \text{ (by interpolation)}$$

STEP 4. To determine ρ_{FC} use $\rho_F = 20\%$, $\rho_W = 50\%$, and FCR = 1.09.

Thus,

$$\rho_{FC} = 18\% \text{ (by interpolation)}$$

STEP 5. With a room cavity ratio of 2.84, the CU for the selected luminaire—*type a* (direct distribution)—can be extracted from Table 18.4 by interpolation.

$\rho_w = 0.50$		CU	
ρ_{CC} →	0.70	0.61	0.50
RCR			
2.0	0.94		0.91
2.84		?	
3.0	0.85		0.82

The CU = 0.85. No correction from Table 18.6 is required, as ρ_{FC} is close to 20%. At this point, lines 9 through 12 of Figure 18.13 can be filled in.

STEP 6. The LLF (see light loss factor subsection) is established by estimating items 13–20 in Figure 18.13. These are assumed as:

Items 13–16 (collectively)	0.88
Item 17	0.95
Item 18	0.85
Item 19	0.95
Item 20	0.80

Item 21: LLF = (0.88) (0.95) (0.85) (0.95) (0.80) = 0.54

STEP 7. Complete the illuminance calculation. A typical classroom normally consists of a student seating area and a teacher's area. The illuminance requirement is approximately the same for both, so the room can be treated as a single entity for visual task purposes. Assuming 7900 lm (at 4000 K and 80 CRI) per luminaire, the number N of luminaires required to provide design illuminance is:

$$N = (500 \text{ lx})(6 \times 8 \text{ m})/(7900 \text{ lm})(0.85)(0.54) = 6.6$$

The use of 6.6 fixtures is not possible, so the designer would need to decide whether six fixtures would be adequate (this would provide a maintained illuminance of 455 lx—substantially below the design target) or whether seven or eight fixtures should be used. Eight fixtures would simplify luminaire layout in the ceiling plane but result in substantially more illuminance than required; an eight-fixture layout might also better conform to the recommended SC limits for the luminaire. Perhaps a similar fixture with somewhat lower lumen output could be used. Lighting design does not stop with the fixture count.

The remaining lines of Figure 18.13 can now be completed. ■

EXAMPLE 18.2 Assume a large business office, with dimensions of (L × W × hung ceiling H) 60 ft × 100 ft × 8 ft. Initial reflectances (ceiling, wall, floor) are 0.80, 0.50, and 0.30. Provide uniform illuminance using LED fixtures. Lamps are replaced on a burnout basis, and the fixture is then cleaned.

SOLUTION

Because of the low ceiling height specified, a direct luminaire (such as *type a* in Table 18.4) will be required; this luminaire will be installed in the ceiling plane and will have only a direct distribution component. A section through such a fixture and its characteristic lumen distribution curve are shown in Table 18.4.

Calculations. This example is developed using only I-P units; the procedure would be the same using SI units. The working plane is taken as the desktop height (i.e., 30 in. or 2.5 ft).

STEP 1. $h_{CC} = 0$; $h_{RC} = 5.5$; $h_{FC} = 2.5$.

STEP 2. CCR = 0; RCR = 0.73; FCR = 0.33.

STEP 3. $\rho_W = 50\%$; $\rho_{CC} = 80\%$ (for a recessed or surface mounted fixture there is no ceiling cavity and the ceiling reflectance is ρ_{CC}).

STEP 4. From Table 18.5, $\rho_{FC} = 29\%$.

STEP 5. For a RCR of 0.73 and a ρ_{FC} of 20%, the CU can be found by extrapolation:

RCR	CU
0	119
0.73	?
1.0	107

For an RCR of 0.73, CU is interpolated from Table 18.4 as 110. Even though this value is mathematically correct, it is physically impossible. A luminaire cannot deliver more light than it produces. The problem in this case is that the CU tables for LED fixtures give quirky results at low RCR values due to the procedures used to develop the tables. As a workaround, it is suggested that a CU value of 95 (0.95) be used in this case—as a compromise between the data and reality.

STEP 6. From Table 18.6, the multiplier for $\rho_{FC} = 30\%$ (close enough to 29% to avoid interpolation) is 1.085.

STEP 7. Final CU = 1.085 (0.95) = 1.0 (rounded down in order not to mess with the laws of physics).

STEP 8. LLF following the steps in Figure 18.13:

1.0	0.92
1.0	0.85
0.9	0.95
0.92	0.88

LLF = (1.0) (1.0) (0.9) (0.92) (0.92) (0.85) (0.95) (0.88) = 0.54

STEP 9. Using the area per luminaire approach for this large space:

$$\text{area/luminaire} = \frac{(7900\text{ lm})(1.0\ CU)(0.54\ LLF)}{50\text{ fc design target}}$$

$$\text{area/luminaire} = 85\text{ ft}^2$$

An area of 85 ft^2 per fixture feels excessive and might lead to non-uniformity of illuminance. As in Example 18.1 (the school classroom), finding an LED luminaire with lower lumen output would increase the required number of fixtures and decrease the area served per fixture. ■

Zonal Cavity Calculation by Approximation. Even though zonal cavity calculations are reasonably straightforward, they can become tedious if more than a few rooms are involved. Two alternatives exist: to utilize one of the readily available computer programs (with an understanding of the impact of common design decisions on calculation outcomes), or simply to shorten the hand calculations by using reasonable approximations. Computer software is discussed in Appendix L. An effective computational method using approximations is demonstrated in this section, which is based upon a method developed by B.F. Jones.

A worksheet for illuminance calculation by approximation is shown in Figure 18.14. Assume that all rooms are square. To enable this assumption for a rectangular room, take one-third of the difference in dimensions and add this to the smaller plan dimension to obtain an equivalent width (W). The RCR for square rooms is:

$$RCR = \frac{10h_{RC}}{W}$$

Then assume:

1. $\rho_{CC} = 0.80$ for a large room, that is, equal to or larger than 30 × 30 ft [10 × 10 m].
   $\rho_{CC} = 0.70$ for a medium room, that is, between 30 × 30 ft [10 × 10 m] and 12 × 12 ft [4 × 4 m].
   $\rho_{CC} = 0.60$ for a small room, that is, equal to or smaller than 12 × 12 ft [4 × 4 m].
2. Assume that $\rho_{FC} = 0.20$.
3. Assume that LLF = 0.65 for good maintenance and dirt conditions, 0.55 for average conditions, and 0.45 for poor conditions.

EXAMPLE 18.3 Consider the classroom as in Example 18.1, but by approximation.

1. "Square" the room.

$$W_{SQ} = 6 + \left(\frac{8-6}{3}\right) = 6 + \left(\frac{2}{3}\right) = 6.67$$

2. Assume:

$$\rho_{CC} = 70;\ \rho_W = 50;\ \rho_{FC} = 20$$

3. Calculate RCR.

$$RCR = \frac{10 \times 1.95}{6.67} = \frac{19.5}{6.67} = 2.93$$

4. Obtain CU from Table 18.4, fixture *type a* (direct).

$$CU = 0.85\text{ (by visual inspection)}$$

5. Calculate required number of fixtures (N).

$$N = \frac{(500)(6 \times 8\text{ m})}{(7900)(0.85)(0.55)} = 6.5$$

Thus, the approximate result (6.5) is substantially the same as from the more detailed calculations (6.6). Next, consider the analysis in Example 18.1 by approximation. ■

EXAMPLE 18.4 Consider the office as in Example 18.2, but by approximation.

1. Determine W_{SQ}.

$$W_{SQ} = 60 + \frac{100-60}{3} = 73\text{ ft}$$

2. Establish ρ_{CC}.

$$\rho_{CC} = 80\%;\quad \rho_W = 50\%;\quad \rho_{FC} = 20\%$$

3. Determine RCR.

$$RCR = \frac{10 \times 5.5}{73} = 0.75$$

4. CU = 1.0 by inspection (eyeball interpolation); reduced to 0.95 as discussed in Example 18.2.
5. LLF = 0.55.
6. Determine the area per luminaire.

$$\text{area per luminaire} = \frac{(7900)(0.95)(0.55)}{50\text{ fc design target}} = 82.6\text{ ft}^2$$

This approximate result (82.6 ft^2) is within 5% of the more precise calculation result (85 ft^2). Thus, these simple approximations give answers sufficiently accurate for most uses.

In conclusion, the following statements apply to zonal cavity calculations:

1. For preliminary and routine calculations involving rectangular rooms, with typical reflectances, use the assumptions listed previously. A simplified calculation

Step 1: **GENERAL INFORMATION**

(a) Project identification: ______________________________
(Give name of area and/or building and room number)

(b) Average maintained illumination for design: ______ lux [footcandles]

Luminaire data: ______________
(c) Manufacturer: ______________
(d) Catalog number: ______________

Lamp data: ______________
(e) Type and color: ______________
(f) Number per luminaire: ______________
(g) Total lumens per luminaire: ______________

SELECTION OF COEFFICIENT OF UTILIZATION

Step 2: Determine the equivalent square-room RCR (room cavity ratio):

$$W_{sq} = W + \frac{L-W}{3}$$

$$RCR = \frac{10\ hrc}{W_{sq}}$$

Step 3: Determine ceiling cavity equivalent reflectance by equivalent square room size.

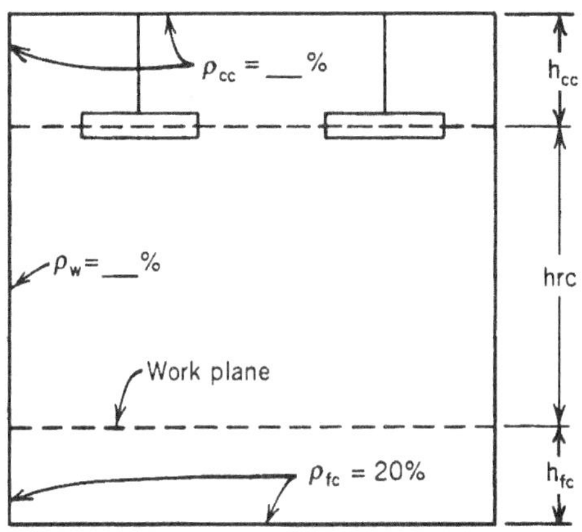

L = ______
W = ______

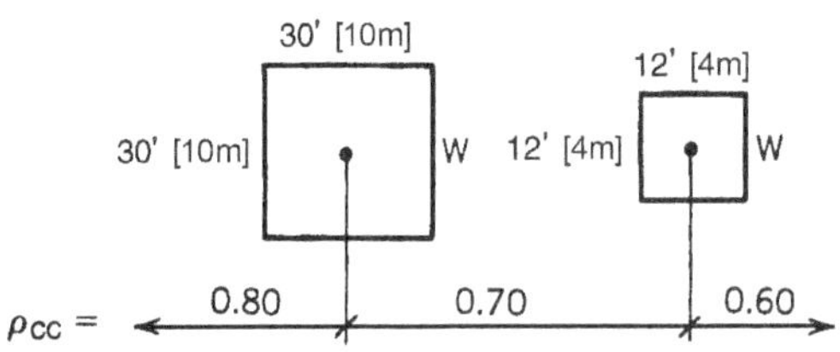

Step 4: Fill in sketch of room above.
Step 5: Assume floor cavity reflectance ρ_{fc} = 20%.
Step 6: Obtain coefficient of utilization CU from manufacturer's data: CU = ______

Step 7: Light loss factor LLF

Good conditions: 0.65
Average conditions: 0.55
Poor conditions: 0.45

Step 8: **CALCULATIONS**

$$\text{Number of luminaires} = \frac{\text{Recommended illuminance x area}}{\text{(Lumens per luminaire) x (CU) x (LLF)}}$$

OR

$$\text{Lux [footcandles]} = \frac{\text{(Number of luminaires) x (lumens per luminaire) x (CU) x (LLF)}}{\text{area m}^2\ [\text{ft}^2]}$$

Figure 18.14 *Worksheet for illuminance calculation using an approximate zonal cavity method. (Based on a method developed by B.F. Jones.)*

form is provided in Figure 18.14 to assist with this method.

2. For rooms where a high degree of accuracy is desired and actual reflectances are known, use the long method (Figure 18.13) with visual interpolation.
3. For rooms of unusual shape or rooms with special conditions, such as coffered ceilings, mixed-material walls, and partial height partitions, use computer software.
4. For spaces in which several different solutions are to be explored, use computer software. ■

Effect of Cavity Reflectances on Illuminance. The reflectances of the various room surfaces (cavities)

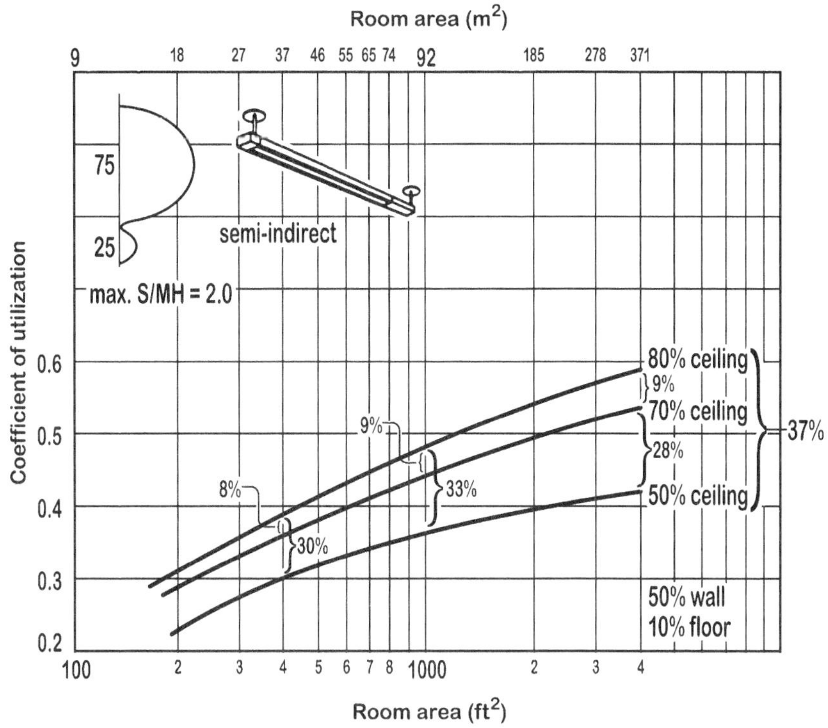

Figure 18.15 *Effect of surface reflectances on the CU of a luminaire with semi-indirect distribution. As expected, because the ceiling becomes the secondary light source, its reflectance has the most pronounced effect. With this particular fixture having a 25% downward component, the floor finish also has an appreciable effect, increasing the CU by an average of 10% for a 30% reflectance floor. The effect of wall reflectance increases as rooms become smaller and the proportion of wall surface becomes larger. The change in CU between a 30% and a 50% reflectance wall varies from 15% for a 400-ft² [37-m²] room to 5% for a 4000-ft² [372-m²] room. (Redrawn with SI units by Lisa Leal.)*

have a marked effect on CU because of the effect of reflection of light within a room. To demonstrate this graphically, see Figures 18.15, 18.16, and 18.17 for the impact of varying cavity reflectances on three common types of fixture distribution: semi-indirect, direct–indirect, and direct (spread). As expected, ceiling cavity reflectance has the most pronounced effect with fixtures with an indirect component; and floor reflectance the most impact with direct fixtures. Because lighting amounts to 3–5% of the total construction cost for many types of buildings, a 20% differential in lighting fixtures can amount to as much as 1% of the total cost of a facility. This amount would not only pay for higher-reflectance finishes and materials but would also reduce both initial and operating costs. These data clearly indicate the necessity for the lighting designer to have considerable influence on the selection of room materials and finishes, a situation that, unfortunately, does not usually occur.

Modular Lighting Design. Numerous buildings are designed around a modular system, resulting in a need for flexible lighting that can work with the module utilized. In such buildings, once the general lighting scheme and luminaire/lamp are established, it can be convenient to draw a family of curves for the fixture chosen, thereby facilitating the utilization of the modular unit in various spaces. "Area" may readily be replaced with multiples of modular areas, as shown in Figure 18.18.

Design Aids. This term refers to the various curves, charts, plots, or tables—either prepared by the designer or made available by luminaire manufacturers—intended to simplify and speed up lighting design when using a particular lighting fixture. The reliability of data so obtained depends entirely on the manufacturer involved—so let the user beware. Major manufacturers can also provide computer analysis charts and tables based upon proposed layouts submitted by a designer. When using these aids, the designer must fully understand the input data that was used for an analysis, as the fixture supplier is usually not a disinterested party. A brief description of common design aids follows.

Isolux charts: These charts, also called isofootcandle charts, are based on diagrams traditionally supplied

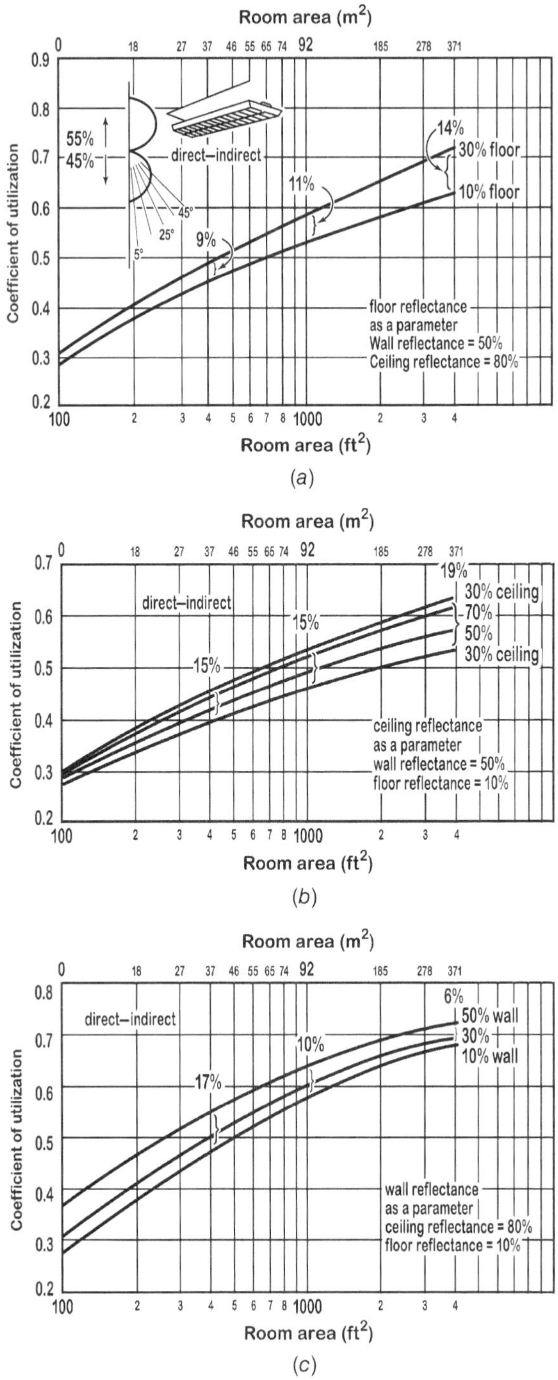

Figure 18.16 *Effect of surface reflectances (a: floor, b: ceiling, c: wall) on the CU of a luminaire with direct–indirect distribution. With such distribution, the effects of the ceiling and floor are most pronounced, with an appreciable wall effect only in small rooms. (Redrawn with SI units by Lisa Leal.)*

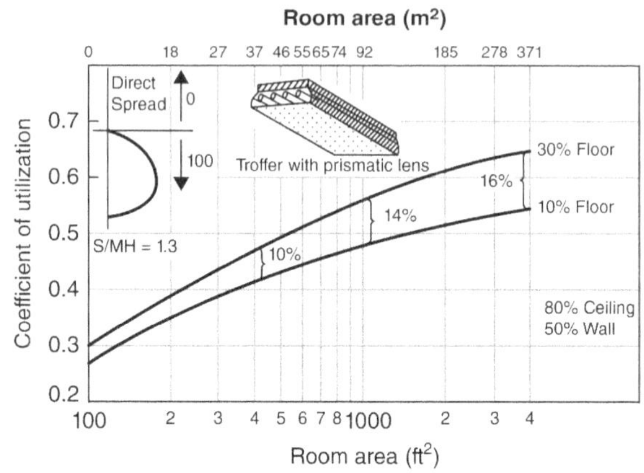

Figure 18.17 *Effect of surface reflectances on the CU of a luminaire with direct (spread) distribution. Floor finish is most important, with wall reflectance important only in small rooms. As these fixtures have no upward component, all ceiling illumination is derived from reflection. Thus, in a room with floor reflectance of less than 20%, ceiling finish has no effect on room illumination.*

by manufacturers of streetlights and floodlights, but are equally applicable to interior lighting. Their use is illustrated in Figure 18.19. The basic tool is an isolux diagram for a single luminaire. This information is either calculated, measured from a full-scale mock-up (the most accurate, if not the most practical method), or obtained from the manufacturer. Since the relative positions of the source and illuminated point are reversible—that is, if a source at *A* causes illuminance *E* at point *B*, then the source at *B* will cause the same illuminance *E* at point *A*—placing the center of the isolux chart at the point in question permits direct reading of the illuminance contribution of every other luminaire. It then remains simply to sum the individual contributions to obtain the (scalar) illuminance at the desired point. An example is shown in Figure 18.19.

Illuminance "cone" charts: When the light distribution of a direct downlight is symmetrical, as is generally the case, a cone (Figure 18.20) can be drawn showing the illuminance directly under the fixture at various distances. The inscribed circles (implied) are defined by maximum illuminance at the center and half of this illuminance at the edge. This type of diagram can be used in the same way as the isolux chart in the preceding section, except that only two illuminances are given—at the center and on the circumference.

Illuminance tables and charts: These take various forms, but all give specific illuminance data, numerically, for specific points. The values are obtained from a computer simulation or an actual test. Figure 18.21 shows the illuminance pattern on a wall produced by the wallwasher version of the downlight shown in Figure 18.20. The difference between the two fixtures is the addition of an interior reflector in the wallwasher version.

Computational Lighting Design. The use of computers in lighting design is a practical necessity if

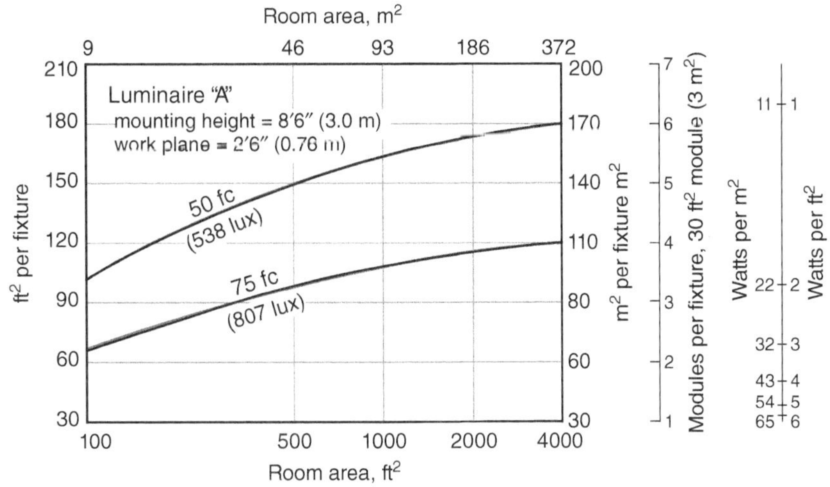

Figure 18.18 Modular design chart. For frequently used fixtures, this type of chart gives fixture requirements for various size rooms. As seen from the y-axes, the results can be translated into a number of modules and watts per unit area. (Redrawn by Martin Lee.)

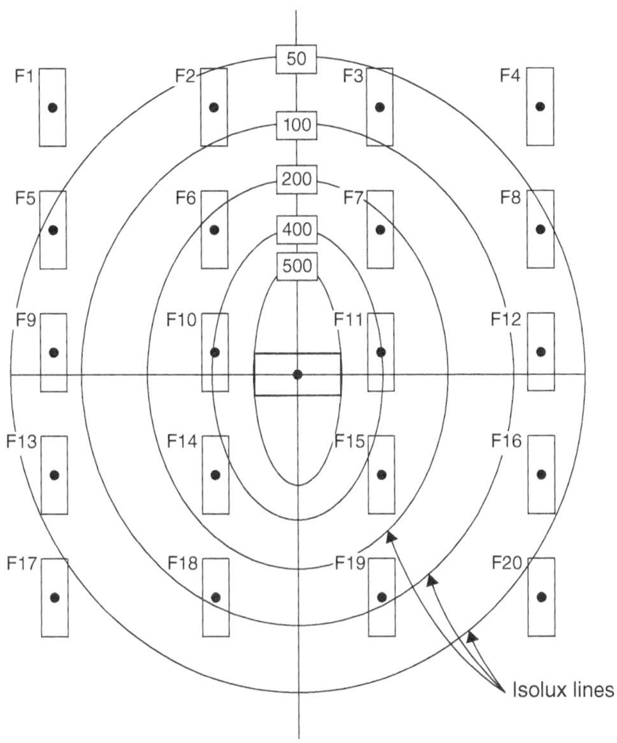

Luminaire	*Contribution (Lux)*
F2, F3	75 × 2
F5, F8	50 × 2
F6, F7	225 × 2
F9, F12	75 × 2
F10, F11	400 × 2
F13, F16	60 × 2
F14, F15	300 × 2
F18, F19	100 × 2
	2570 lux total

Figure 18.19 The ellipses represent isolux lines for a single luminaire at a given height above the work plane. They are centered on the point (the work area of a desk) for which the illuminance must be determined. The total illuminance at that point is the sum of the individual luminaire contributions. The center of the luminaire is the point of reference. Therefore, when two or more isolux lines pass through a fixture, its contribution is determined by the interpolated isolux line passing through its center. Note the symmetry around the vertical axis, necessitating a plot of only half of the ellipses. Note that the chart could be plotted with isofootcandle lines.

detailed results are required or iterative analyses are performed. Furthermore, a computer analysis can give useful point-by-point illuminance data plus projections of VCP and reflected glare for selected work locations and viewing directions.

In addition, computer analysis gives the designer a degree of flexibility not otherwise possible, in that:

1. The calculations are performed accurately and rapidly.
2. The designer is freed for other, less routine work.
3. The designer can change the parameters repeatedly without making the analysis excessively burdensome, as would be the case with hand calculations.

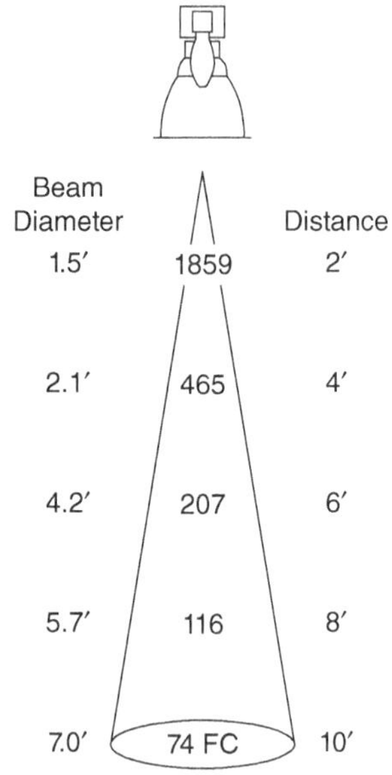

Figure 18.20 *For downlights with symmetrical circular distribution, a "cone of light," as shown, can be drawn. The illuminance at varying distances on the beam centerline directly below the luminaire is given in the center column. A circle with a circumference at which the illuminance is half of this maximum is drawn at each distance from the downlight (2 ft, 4 ft, etc.). The numbers in the left column show the diameter of this (beam) circle. Similar diagrams are available in SI units. (Courtesy of Zumtobel-STAFF.)*

This last characteristic gives a designer great flexibility. The ability to run a series of calculations for a pendant fixture with varying pendant lengths, or to change paint colors and reflectances for various surfaces, or to test different layout patterns, gives the designer the opportunity to optimize (at least partially) lighting decisions. In addition, a computer analysis can consider factors—such as first cost, energy use, operating costs, and impact on HVAC systems—whose interactive complexity puts them well beyond the realm of hand calculations. Several current lighting analysis programs are described in Appendix L.

(f) Nonuniform Illuminance Calculations

Calculating Illuminance from a Point Source. The basis of point source calculations is the inverse square law:

$$fc = \frac{cp}{D^2}$$

where *fc*, *cp*, and *D* are footcandle illuminance, candlepower intensity, and distance, respectively. Refer to Figure 18.22. The horizontal illuminance at a point *P* as shown in Figure 18.22 is:

$$\text{horizontal } E = \frac{cp}{D^2}\cos\theta$$

and the vertical illuminance at that same point is

$$\text{vertical } E = \frac{cp}{D^2}\sin\theta$$

However, because

$$\cos\theta = \frac{H}{D} \text{ and } \cos\theta = \frac{R}{D}$$

at point *P*

$$\text{horizontal illuminance} = \frac{cp}{H^2} \times \cos^3\theta$$

$$\text{vertical illuminance} = \frac{cp}{R^2} \times \sin^3\theta$$

Because the candlepower intensity in the direction of point *P* is taken from a candlepower distribution curve, and θ is known, these expressions are readily

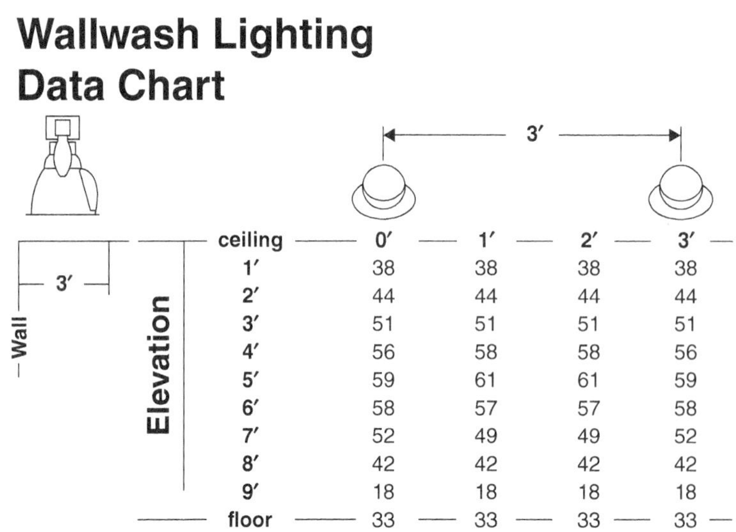

Elevation	0′	1′	2′	3′
ceiling				
1′	38	38	38	38
2′	44	44	44	44
3′	51	51	51	51
4′	56	58	58	56
5′	59	61	61	59
6′	58	57	57	58
7′	52	49	49	52
8′	42	42	42	42
9′	18	18	18	18
floor	33	33	33	33

Figure 18.21 *Addition of an interior reflector to the downlight of Figure 18.20 converts it to a dual-purpose downlight and wallwash unit. In this I-P unit example, the wall illuminances (in fc) produced by multiple units spaced 3 ft apart and ceiling-mounted 3 ft from the wall are given in the chart. Similar charts are available for other luminaire spacings (and in SI units). (Courtesy of Zumtobel-STAFF.)*

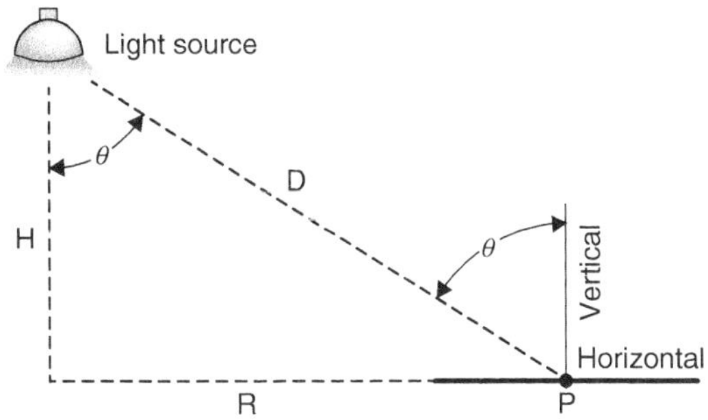

Horizontal illuminance at point P

$$E_H = \frac{CP \times \cos\theta}{D^2} = \frac{CP}{H^2}\cos^3\theta$$

Vertical illuminance at point P

$$E_V = \frac{CP \times \sin\theta}{D^2} = \frac{CP}{R^2}\sin^3\theta$$

Figure 18.22 *Relationship between intensity in candlepower (cp) and illuminance when the source can be considered a point source—that is, when the inverse square law applies. The source's major dimension must not exceed 0.2D to be considered a point source. Using feet yields fc; using meters yields lux. (Redrawn by Martin Lee.)*

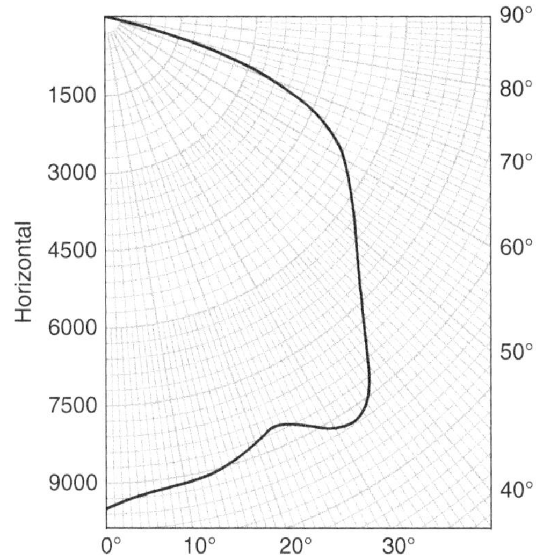

Figure 18.23 *Typical candlepower distribution plot.*

usable. Very few commercial light sources are actually point sources. However, when the maximum dimension of the source is less than five times the distance to point P, this method gives satisfactory results. Note that these equations can be used to calculate and plot isolux diagrams for point sources of the type shown in Figures 18.20 and 18.21.

EXAMPLE 18.5 Referring to Figure 18.22 and the candlepower distribution curve of Figure 18.23, find the horizontal and vertical illuminance at point *P*, which is 10 ft below and 12 ft horizontally distant from the source. The same analysis process would apply if using SI dimensions.

SOLUTION

$$H = 10\text{ ft} \quad R = 12\text{ ft}$$

$$\theta = \tan^{-1}\frac{12}{10} = 50°$$

$$\sin\theta = 0.766 \quad \cos\theta = 0.643$$

$$cp \text{ at } 50° = 6600 \text{ (from Figure 18.23)}$$

$$\text{horizontal illuminance: } \frac{6600}{10^2} \times (0.643)^3 = 17.5\text{ fc}$$

$$\text{vertical illuminance: } \frac{6600}{12^2} \times (0.766)^3 = 20.8\text{ fc}$$

■

Calculating Illuminance from Linear and Area Sources. When a source is too large to be considered a point source (as defined above), it is referred to as either a linear source or an area source. The direct component of the illuminance at a point, resulting from such sources, can be calculated by manual graphical or analytical means, both of which are based on an assumed distribution, generally Lambertian (diffuse). Since most lighting fixtures being used today do not have Lambertian characteristics (e.g., parabolic reflectors, prismatic diffusers), the results of such calculations are necessarily approximate. (Skylights and luminous ceilings do have a Lambertian distribution, and for these sources these calculation methods give reliable results.) In addition to the direct component of illuminance, a reflected component must be added that depends on the point's location in the room and the room characteristics. The calculations involve charts, diagrams, and tables. A full description of these methods can be found in the 10th edition of *The Lighting Handbook* (IES, 2011).

Because these manual methods are laborious and often not fully reliable, they are not presented here. The ready availability of computer software that can handle detailed input for a specific light source, without broad approximations, has made these manual procedures obsolete. When a point-by-point illuminance calculation is necessary, a computer analysis should be used. Alternatively, one of the graphic design aids described previously, based on a specific light source, can be used.

(g) Average Luminance Calculations

Knowing the horizontal illuminance (as calculated by any of the methods described above) and the reflectance of a horizontal, horizontal luminance can readily be calculated. However, the impression of brightness in the visual environment is affected more by vertical than by horizontal surface luminance. For this reason, it is important to be able to estimate average vertical surface (wall) luminance during design. Employing the same approach used to calculate average horizontal illuminance would be nice. It is also beneficial to know the average luminance of the ceiling cavity in a space as a means of evaluating the contrast between all objects, including luminaires, that have the ceiling cavity as background.

Straightforward calculation of both wall and ceiling cavity luminance (L_W and L_{CC}) is possible through the use of luminance coefficients that are similar in concept and application to coefficients of utilization. Such coefficients are listed in Table 18.7 for several generic fixture types. For actual project calculations, obtain luminance coefficients from luminaire manufacturers (or utilize computer software). Average luminance calculation procedures are parallel to those for illuminance.

Average initial wall luminance in cd/m^2 is calculated as:

$$L_W = \frac{\text{(lamp lumens)(wall luminance coefficient)}}{(\pi)(\text{floor area in m}^2)} \quad \text{(Eq. 18.21)}$$

Average initial ceiling cavity luminance in cd/m^2 is:

$$L_{CC} = \frac{\text{(lamp lumens)} \times \text{(ceiling cavity luminance coefficient)}}{(\pi)(\text{floor area in m}^2)} \quad \text{(Eq. 18.22)}$$

If area is expressed in square feet and π is omitted from these equations, L will be expressed in footlamberts.

To obtain maintained values, an LLF is introduced. It is calculated similarly to LLF for horizontal illuminance, except that room surface dirt loss values are as follow:

Lighting System	Wall Luminance (%)	Ceiling Luminance (%)
Direct	0.82610	0.75610
Semi-direct	0.8767	0.82610
Direct–indirect	0.9265	0.8568
Semi-indirect	0.8767	0.8867
Indirect	0.82610	0.9065

For ceiling-mounted or recessed luminaires, L_{CC} is the average luminance of the ceiling between luminaires. For pendant luminaires, the calculated L_{CC} is that of an imaginary plane at the height of the luminaires. L_{CC} is useful in determining the brightness ratio between a luminaire and the ceiling plane. The ceiling cavity, like the wall surface, is assumed to have perfect diffuseness—making luminance independent of viewing angle.

As an example, calculate the wall luminance of the office in Example 18.2. The photometric data in Table 18.4 do not include wall luminance coefficients because they are not normally published by manufacturers. However, based on other available data, a value of 0.22 for wall luminance is a good estimate given RCR, ρ_{CC}, and ρ_W values of 0.66, 80%, and 30%, respectively. Initial wall luminance is then:

$$L_W = \frac{3200 \times 0.22}{32\ \text{ft}^2} = 22\ \text{fL} = 75.5\ \text{cd/m}^2 \quad \text{(Eq. 18.23)}$$

This is within the preferred range of 25–150 cd/m^2 (see Table 7.3). In actuality, the average wall luminance would probably be higher because of the practice of placing the last row of luminaires quite close to the wall.

(h) Evaluating Glare Potential

Direct Glare. Control of direct glare (see Chapter 7) is an architectural design responsibility. The design solution is basic—avoid overly luminous (bright) light sources (typically a luminaire for electric lighting or a window/skylight for daylighting). For electric lighting systems, control of direct glare lies in the hands of well-designed products, well-selected by the lighting designer. Information on control of luminaire luminance is found in Chapter 17.

Reflected Glare. The causes of veiling reflections are well understood; unfortunately, this does not mean that solutions are simple. No known lighting method or material will eliminate veiling reflections. There are, however, several techniques to minimize contrast loss due to veiling reflections while maintaining adequate illuminance. These, discussed in more detail below, are:

- physically arranging sources, task, and observer so that reflected glare is minimized
- adjusting brightness (eye adaptation level) so that objectionable brightness is minimized
- designing the light source so that it causes minimal reflected glare
- changing the task quality

Physical Arrangement of Lighting System Elements. It can be challenging to arrange the lighting geometry to avoid sources of high luminance, undesirable angles of reflection, and specular tasks. This is often especially difficult in offices, which frequently have both horizontal and vertical work surfaces, the vertical being the specular surfaces of digital displays. This problem is widespread and can pervade the design of office lighting. Figures 7.39 and 7.41 show that in a space using multiple sources, particularly in continuous

TABLE 18.7 Wall Luminance Coefficients and Ceiling Cavity Luminance Coefficients for Typical Luminaires

To obtain a luminance coefficient, follow the procedure detailed at the beginning of Table 18.4 for finding a CU. More precise design data should be obtained from the manufacturers of the actual luminaires used.

	ρ_{CC}:	80			50			80		50	
	ρ_W:	50	30	10	50	30	10	50	30	50	30
Typical Luminaire	*RCR*	*Wall Luminance Coefficients for ρ_{FC} = 20*						*Ceiling Cavity Luminance Coefficient ρ_{FC} = 20*			
1	0							.42	.42	.25	.25
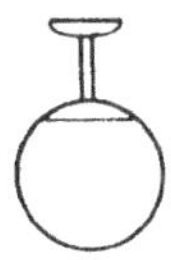	1	.32	.18	.06	.27	.15	.05	.42	.40	.25	.23
	2	.27	.15	.05	.23	.13	.04	.42	.38	.24	.23
	3	.24	.13	.04	.20	.11	.03	.41	.37	.24	.22
	4	.21	.11	.03	.17	.09	.03	.41	.36	.24	.22
	5	.19	.10	.03	.16	.08	.02	.40	.35	.24	.21
	6	.18	.09	.03	.14	.07	.02	.39	.34	.23	.21
	7	.16	.08	.02	.13	.07	.02	.39	.34	.23	.21
	8	.15	.07	.02	.12	.06	.02	.38	.34	.23	.20
Pendant diffusing sphere	9	.14	.07	.02	.12	.06	.02	.38	.33	.23	.20
with incandescent lamp	10	.13	.06	.02	.11	.05	.01	.37	.33	.22	.20
3	0							.15	.15	.09	.09
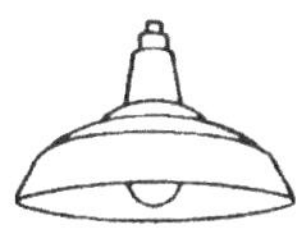	1	.23	.13	.04	.21	.12	.04	.15	.13	.09	.08
	2	.22	.12	.04	.21	.11	.04	.14	.11	.08	.07
	3	.21	.11	.03	.20	.11	.03	.13	.10	.08	.06
	4	.20	.10	.03	.19	.10	.03	.13	.08	.08	.05
	5	.19	.09	.03	.18	.09	.03	.12	.08	.07	.05
	6	.18	.09	.03	.17	.09	.02	.12	.07	.07	.04
	7	.17	.08	.02	.16	.08	.02	.11	.06	.07	.04
Porcelain-enameled	8	.16	.08	.02	.15	.07	.02	.11	.06	.06	.04
ventilated standard dome	9	.15	.07	.02	.14	.07	.02	.10	.05	.06	.03
with incandescent lamp	10	.14	.07	.02	.13	.07	.02	.10	.05	.06	.03
7	0							.08	.08	.04	.04
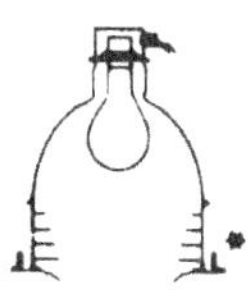	1	.06	.03	.01	.05	.03	.01	.08	.07	.04	.04
	2	.05	.03	.01	.05	.03	.01	.07	.06	.04	.04
	3	.05	.03	.01	.04	.02	.01	.06	.05	.04	.03
	4	.05	.02	.01	.04	.02	.01	.05	.04	.03	.03
	5	.05	.02	.01	.04	.02	.01	.05	.04	.03	.02
	6	.04	.02	.01	.04	.02	.01	.05	.03	.03	.02
Reflector downlight with	7	.04	.02	.01	.04	.02	.01	.04	.03	.03	.02
battles and inside-frosted	8	.04	.02	.01	.04	.02	.01	.04	.03	.02	.02
lamp	9	.04	.02	.01	.04	.02	.01	.04	.02	.02	.02
	10	.04	.02	.00	.04	.02	.01	.04	.02	.02	.01
28	0							.27	.27	.16	.16
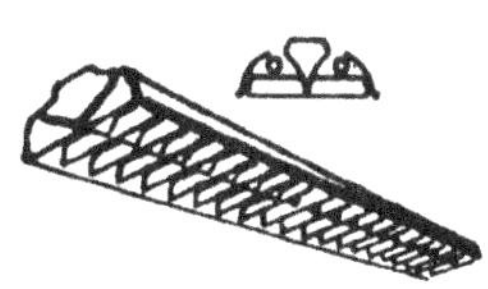	1	.17	.10	.03	.14	.08	.03	.26	.24	.15	.14
	2	.16	.09	.03	.14	.08	.02	.25	.23	.15	.14
	3	.15	.08	.02	.13	.07	.02	.24	.21	.14	.13
	4	.15	.08	.02	.13	.07	.02	.24	.21	.14	.12
	5	.14	.07	.02	.12	.06	.02	.23	.20	.14	.12
	6	.13	.07	.02	.11	.06	.02	.23	.19	.14	.12
	7	.13	.06	.02	.11	.06	.02	.23	.19	.13	.11
Diffuse aluminum reflector	8	.12	.06	.02	.11	.05	.02	.22	.18	.13	.11
with 35° crosswise and 35°	9	.12	.06	.02	.10	.05	.01	.22	.18	.13	.11
lengthwise shielding	10	.11	.05	.01	.10	.05	.01	.22	.18	.13	.11
33	0							.65	.65	.38	.38
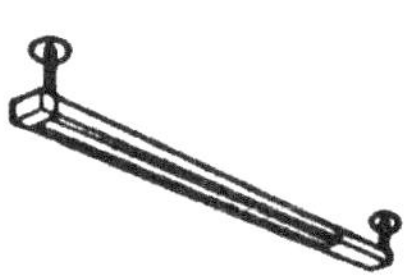	1	.20	.12	.04	.13	.08	.02	.65	.63	.38	.37
	2	.19	.10	.03	.12	.07	.02	.64	.61	.38	.37
	3	.17	.09	.03	.11	.06	.02	.64	.60	.37	.36
	4	.16	.08	.02	.11	.06	.02	.63	.59	.37	.36
	5	.15	.08	.02	.10	.05	.02	.63	.59	.37	.36
	6	.14	.07	.02	.09	.05	.01	.60	.58	.37	.35
	7	.13	.07	.02	.09	.04	.01	.62	.58	.37	.35
Luminous bottom-	8	.12	.06	.02	.08	.04	.01	.61	.57	.37	.35
suspended unit with extra-	9	.12	.06	.02	.08	.04	.01	.61	.57	.36	.35
high-output lamp	10	.11	.05	.01	.07	.04	.01	.61	.57	.36	.35

Source: Adapted from IES, 1993. Coefficients for fixtures 1, 3, 7, 28, 33, 35, and 42 have been rounded to two decimal places.

Notes:

1. Refer to the manufacturer's catalog data for more precise values when a specific luminaire is proposed for use.
2. Multiply coefficients by 1.05 for three lamps and by 1.1 for two lamps.

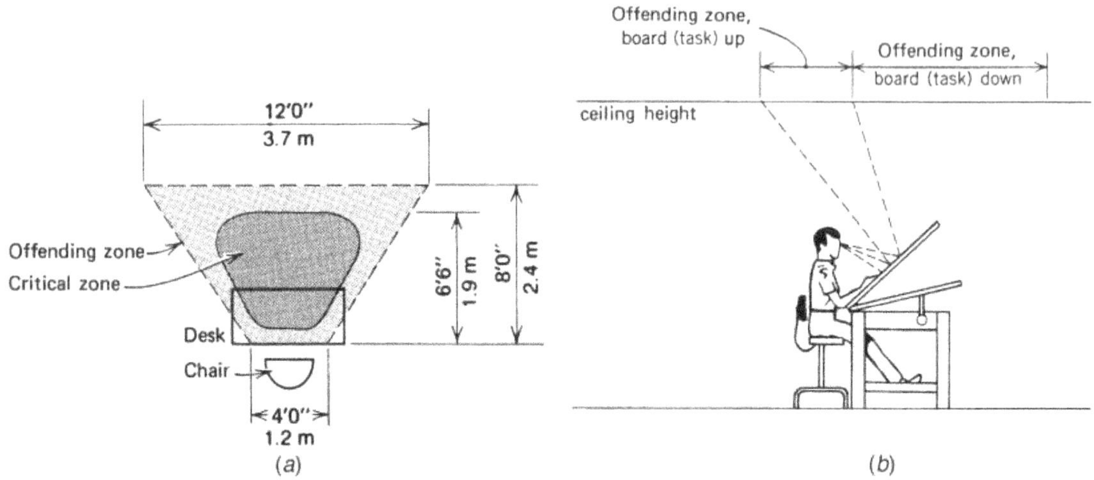

Figure 18.24 (a) If luminaires are kept out of the trapezoidal offending zone, contrast will be excellent. If the bulk of one or more luminaires projects into this zone, especially into the critical zone, contrast will drop sharply. The dimensions shown are for a flat desk 3 ft × 5 ft [0.91 m × 1.5 m] and a 9-ft [2.7-m] ceiling height. (b) The dependence of the glare zone on table tilt is illustrated. The offending zone becomes smaller as the table is raised, so that with a table near the vertical position, glare is all but eliminated. (Ross and Baruzzini, Inc., 1975.)

rows, placing the work between rows with the line of sight parallel to the long axis of the units is an effective technique. The offending zone for horizontal tasks depends upon the tilt of the desk. For a horizontal desk, the offending zone is forward of the desk, as in Figure 18.24*a*; with a tilted desktop, the offending zone may be behind the source, as in Figure 18.24*b*.

Developing a geometric solution to mitigate reflected glare potential presupposes a known (during the design phase) and fixed furniture layout. This may sometimes be the case but is not the norm. In the absence of such information, two alternatives are possible: a uniform lighting layout with furniture adjusted to work with it, or vice versa. A combination (compromise) of both is the most practical approach. High-luminous efficacy lighting systems have made ducted (air troffer) lighting fixtures less prevalent—thus properly designed, fixtures can be easily shifted. Such mobility is enhanced by the great flexibility of lighting fixtures fed from ceiling plug-in raceways. Figure 18.25 shows such a rearrangement, which results in saving five fixtures, a load reduction of 800 W, and an improvement in visibility.

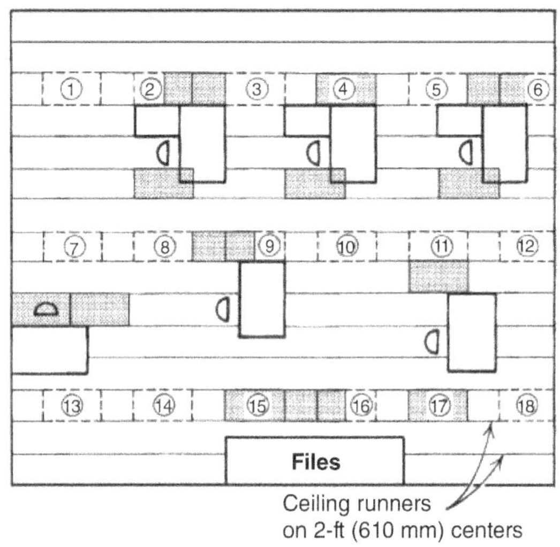

Figure 18.25 An original uniform fixture layout utilized three rows of six 2 ft × 4 ft [0.6 m × 1.2 m], four-lamp fixtures, giving a total load of 2880 W, a load density of 2.6 W/ft² [28 W/m²], and a uniform illuminance of approximately 90 (raw) fc [900 lx]. The original layout is shown dotted and numbered. A rearranged layout uses 13 fixtures (shown shaded) for a total of 2080 W, a load density of 1.9 W/ft² [20.5 W/m²], and more than 100 ESI fc [1000 lx] on each work surface. In addition, five fixtures are saved. Note: This level of illuminance is justified only for difficult visual tasks.

Control of Area Brightness and Eye Adaptation Level. Loss of visual contrast can be compensated for (and glare reduced) by increasing overall non-glare illumination. Doing so simply makes the task brighter in order to override the detrimental veiling reflection. The problem with this solution, however, is that a large increase in illuminance is required to overcome the glare. This increase can, in many instances, be most practically accomplished not by increasing overall room illumination—which comes with extremely high energy consumption—but by adding supplementary lighting arranged as to be free of reflected glare potential. Making the supplementary source's position adjustable, accomplishes three things:

1. Veiling reflections are overcome.
2. The high illuminance needed for exacting tasks is provided with minimum energy expenditure.
3. The user is granted geometric control with resultant better lamp placement, as well as psychological satisfaction, which generally reduces worker

complaints. (The best position is generally to the left and slightly forward of the task.)

The effectiveness of a supplemental desk lamp is demonstrated in Example 18.6, which revisits the situation described in Example 18.4.

EXAMPLE 18.6 Recalculate the contrast reduction of the ink-on-paper visual task of Example 18.4, assuming that a desk lamp raises the illuminance to 200 fc [2000 lx] and is positioned so as to be glare-free. Note: An adjustable lamp with 2 at 15-W fluorescent tubes would produce approximately that luminance on a desk.

SOLUTION

Contrast from Equation 6.9:

$$C = (L_{BD} + L_{BS}) - (L_{TD} + L_{TS})/(L_{BD} + L_{BS})$$

where

$$L_{BD} = 200\text{ fc} \times 0.71 = 142$$

$$L_{BS} = 200\text{ fL} \times 0.018 = 36$$

$$L_{TD} = 200\text{ fc} \times 0.038 = 7.6$$

$$L_{BD} = 2000\text{ fL} \times 0.021 = 42$$

and

$$C = (142 + 36) - (7.6 + 42)/(142 + 36) = 0.72$$

With this contrast (0.72), undesirable contrast reduction from a no-glare situation has been improved from the original 47% reduction (0.947 to 0.497) to a 24% reduction (0.947 to 0.72). However, because a contrast reduction of even 24% is unwanted, a change in task-source geometry or a change in source luminance would be required, assuming that the task itself must remain unchanged. ■

Control of Source Characteristics. The reflected light that causes loss of contrast is proportional to luminaire luminance at the viewing angle of concern, and therefore may be reduced by reducing the luminaire luminance at that angle. This can be accomplished in four ways:

1. *Dimming or switching lamps:* Reducing the total output of a fixture also reduces its output in the critical portion of the ceiling glare zone and can actually increase the ESI illuminance (by improving task contrast).
2. *Using luminaires with lower overall luminance:* Use a larger-area, low-output source (Figure 18.26) instead of a few small high-output sources. This will reduce the source luminance in the offending zone, while increasing the illumination contribution from outside the glare zone—resulting in better contrast for the same or lower illuminance. The disadvantage of this technique is increased lighting fixture cost.
3. *Using a luminaire as a primary source to illuminate a large, low-brightness secondary source:* To overcome the cost disadvantage of multiple low-output, low-luminance sources, the ceiling can become a secondary source illuminated by high-output indirect or semi-indirect fixtures. LED, fluorescent, or metal-halide lamps have high efficacy. Using this approach requires a high enough ceiling height to permit suspending the luminaire while avoiding "hot spots" on the ceiling. The minimum suspension length depends upon the luminaire characteristics and would normally be provided by the manufacturer. To ensure high system efficiency, the ceiling should be painted with a high-reflectivity matte white paint and kept clean. Results obtained from a semi-indirect installation using 1500-mA, very-high-output lamps are shown in Figure 18.27.
4. *Reducing the luminaire luminance only at the offending angles:* Because most horizontal task vision takes place between 20° and 40° from the vertical, any fixture that emits little or no light below 40° from the horizontal cannot produce a veiling reflection, regardless of its position in the field of view (Figure 18.28). As a result, luminaire manufacturers produce prismatic diffusers which have

(a) (b) (c)

Figure 18.26 *A concentration of light in the glare zone (a) produces the largest amount of reflected glare. As the number of light sources is increased (b) in the glare zone, and luminance is decreased, reflected glare is decreased. The least glare is from an all-luminous ceiling (c), which also has the lowest luminance.*

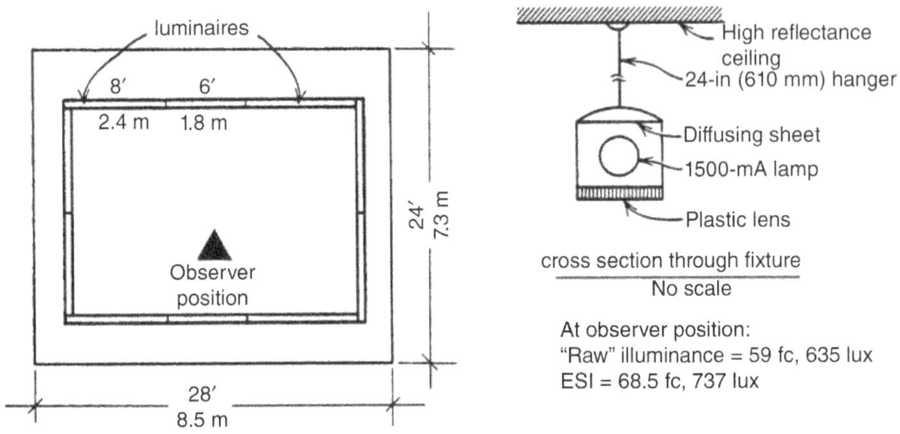

Figure 18.27 *With a high-reflectance diffuse finish ceiling, this semi-direct perimeter lighting installation yields a higher ESI illuminance than raw illuminance at the viewing location illustrated, indicating excellent contrast rendering. The plastic lens at the bottom of the fixture provides perceived light source luminance and avoids the impression of gloominess, despite the satisfactory overall luminance level. (From Sampson, 1970.)*

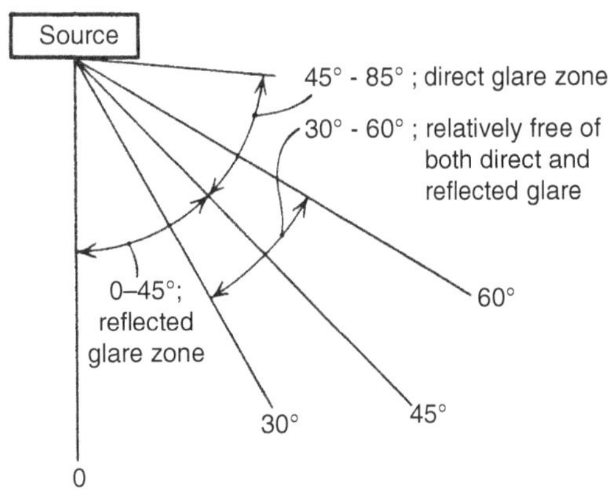

Figure 18.28 *Glare zones are 0° to 45° and 45° to 85° for reflected and direct glare, respectively. Therefore, a diffuser that emphasizes the 30° to 60° zone will be least objectionable on both counts.*

diminished output below 30° and above 60° in order to minimize both reflected and direct glare. Due to the characteristic shape of their intensity distribution curves, they are known as batwing diffusers. For observers positioned so that their sight lines are parallel to the longitudinal axis of the ceiling fixtures, lenses with linear (side-to-side) batwing characteristics perform well. If the observing position varies in aspect with respect to the fixture, a radial batwing curve (in all directions) is required. Note carefully that these diffusers have only limited usefulness in reducing reflected glare with specular vertical surfaces (digital display screens). Different types of diffusers and their characteristics are discussed in Chapter 17.

Changing the Task Quality. Reducing task specularity is at least as effective a means of reducing veiling reflections as changing the lighting system characteristics—if not more so. It is often cheaper, and always more energy-efficient, to upgrade the task (in the visibility sense) than to change the lighting system. Always cheaper, however, does not equate to always possible. Nevertheless, task contrast and specularity should be actively considered and recommendations for improvements made to the client in the pre-design phase when owner's project requirements are being adopted.

(i) Qualitative Design Considerations

Size and Pattern of Luminaires. A luminaire (or other luminous source) is often a point of visual attention because of its brightness. To the extent that luminaires are numerous, large, very bright, or arranged in striking patterns, attention is drawn to them and away from other surfaces. Color elements or accent lighting can also draw attention. Some examples can demonstrate the principles involved in design decisions regarding this issue.

Luminaire size should correlate with room size and ceiling height. Fixtures larger than 2 ft × 4 ft [600 mm × 1200 mm] should not be used with ceiling heights lower than 10 ft [3.1 m] unless the apparent size of the fixtures can be minimized through use of a surface pattern (see Figure 18.9). Transilluminated (luminous) ceilings (Figure 18.29) require a minimum height of 12 ft [3.7 m]. When they are installed below this level, particularly in large rooms, the effect is oppressive, as if the sky were lowered onto the viewer. To offset this effect, the use of colored, shaped, or dark panels is of some help. In place of a luminous ceiling, large-area, coffer-type

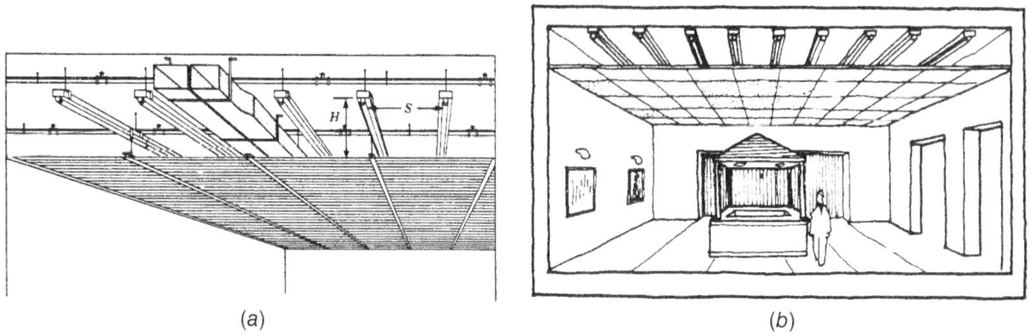

Figure 18.29 (a) The suspended diffusive material of a luminous ceiling—when properly designed—allows the designer to hide piping and ductwork, without affecting the ceiling's light distribution. As a rule, if uniform illuminance is desired, the spacing (S) of the strip fixtures should not exceed 1.5 times the height (H) between the fixtures and the diffusing element. (b) A luminous ceiling provides low-brightness, highly diffuse, uniform illuminance, generally exceeding 500 lx [45 fc]. Such a lighting approach is particularly useful for specular tasks where supplemental lighting is impractical. To relieve the monotony of large, unbroken expanses, as in (a), designers frequently use clearly defined diffuser panels, as shown.

fixtures can be utilized to give the impression of depth (Figure 18.30).

Regular fixture spacing is required to achieve uniform illuminance. A regular spacing can be used as a tool to accomplish an architectural purpose, as shown in Figure 18.31*a–d*. It is often recommended that the pattern of lighting ought to either reinforce an architectural form or remain neutral, rather than being at cross-purposes with a dominant architectural pattern—unless that is the desired effect. If a strong architectural element is absent, a dominant lighting pattern may be desirable. Conversely, a strong architectural element can either be reinforced (Figure 18.32*a*) or utilized as the base for a neutral lighting pattern (Figure 18.32*b*, *c*).

Continuous-row luminaire installations, aside from eliminating the dominant checkerboard effect of closely spaced individual luminaires, are also often cheaper. Coves and cornices give a ceiling a floating feeling or lightness. Geometric patterns can be used to add interest or to break the monotony of large areas, such as in department stores. Generally, downlights are not visually dominant, and regularity of placement is not essential (Figure 18.33). However, when downlights are surface-mounted (a generally inadvisable procedure), the result can be far from visually neutral (Figure 18.34). Nonuniform layouts with large fixtures can create problematic patterns since they are often too large to be neutral—their nonuniformity can create visual confusion (Figure 18.35). The only cure for this self-inflicted problem is to minimize the source brightness by using low-brightness luminaires.

In spaces where circulation is the primary "seeing task," but where there are isolated areas requiring greater illuminance (such as waiting areas in

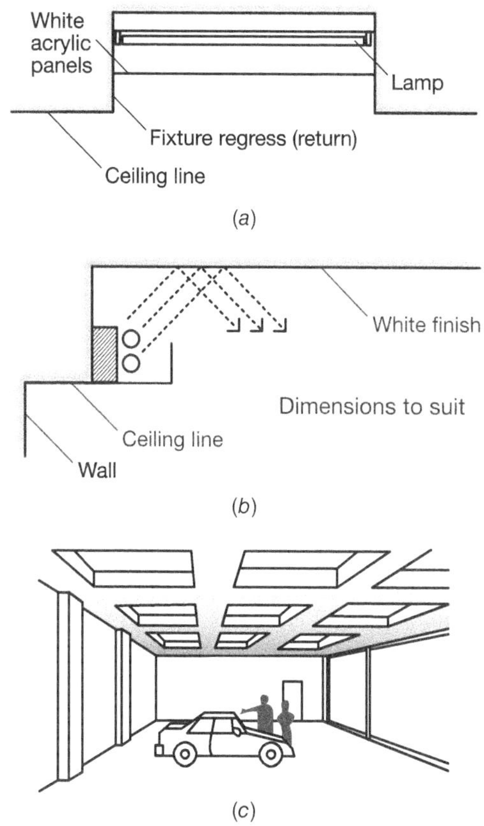

Figure 18.30 Coffer-type light sources come in both standardized (a) or custom-designed (b) sizes. Typical standardized coffers (a) are large direct lighting fixtures that are generally available in 4-ft [1.2-m] widths, and variable lengths. Custom-designed coffers (b) can be constructed in any shape or size. Both types of coffer can give an illusion of great depth and of a floating, illuminated surface. (c) Multiple coffers create a dominant architectural effect, and when used in conjunction with skylights, can furnish soft, glare-free illumination throughout the day and night. (Drawn by Alisa Kwok.)

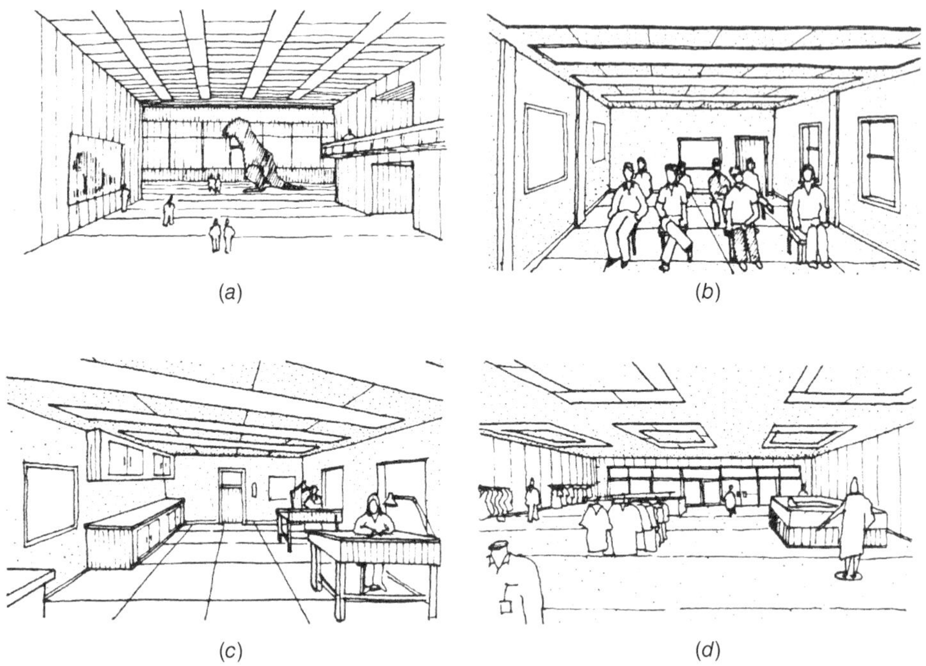

Figure 18.31 *(a) Longitudinal lines in the direction of the sight line increase apparent length, direct traffic flow, and decrease direct glare. (b) Lines perpendicular to the line of sight shorten and widen a space but also increase direct glare. (c) Diagonal lines minimize shadows and break rectangular patterns. (d) Rectangular pattern is architecturally dominant.*

transportation terminals), a perimeter lighting system with supplemental lighting can be a viable solution (Figure 18.36). The MAXXI: Museum of XXI Century Arts, in Rome, Italy, features elongated slot fixtures that reinforce movement through space and skylights with controllable louvers to illuminate the gallery space (Figure 18.37). Lighting patterns are often used as formal directional markers. This is particularly useful in transportation terminals, where signs only partially serve this purpose.

A frequently neglected consideration is the appearance of a luminaire when it is turned off. With proper daylighting and energy-efficient design, many fixtures can be turned off during the normal-use hours of a space. Low-brightness sources will change least in appearance between the on and off states. The visibility and appearance of luminaires in daylight, regardless of whether they are illuminated, from inside or outside a building, can be used as an advantage, as shown in Figures 18.38 and 18.39.

Patterns of Luminance. This term describes the patterns of light and shadow in a space that result from illumination provided by the lighting system. A single light source may produce sharp shadows, whereas a luminous ceiling or a completely indirect illumination system may produce almost completely diffuse light. Diffusion is the degree to which light is shadowless—and is a function of the number of directions from which light falls on a particular point and the difference in intensities of light from these directions.

Perfect diffusion, rarely obtainable (or desirable), results from equal intensities of light arriving from all directions—and yields no shadows. The only naturally occurring example of perfectly diffuse lighting is a daytime fog, which is a visually disturbing condition for most, suggesting that some directivity is desirable. Diffusion can be judged by the depth and sharpness of shadows. A room with well-diffused illumination, resulting from multiple sources and high room-surface reflectances, yields soft multiple shadows that do not obscure visual tasks. Because purely diffuse lighting is monotonous and not conducive to extended periods of effective work, some directional lighting is often introduced as a complement to diffuse general lighting—thus producing a more stimulating environment via variations in shadows and brightness. Where grazing angle reflections are critical to the examination of texture or surface imperfections, highly directional lighting is required. As seen in Figure 18.40, directional light is what creates shape and is precisely the characteristic best used to influence architectural space and form.

Various combinations of uplighting and downlighting can be implemented, with each arrangement producing its own shadows and modeling, and each having a unique quality. These patterns of light and dark and in-between produce an overall ambience, as well as the subjective perceptions and observations of

(a) (b) (c)

Figure 18.32 *Lighting designs for various spaces with high ceilings. The fixtures in (a) and (b) follow structural beams. (b) Floor reflection and daylight provide ceiling and wall illumination. (c) Lighting in this space was handled by recessing fixtures into the lattice ceiling pattern. Metal-halide HID lamps and tungsten-halogen lamps were used. (a: M.B. Warren; b: Photo by L. Reens; c: courtesy of GTE/Sylvania, Inc.)*

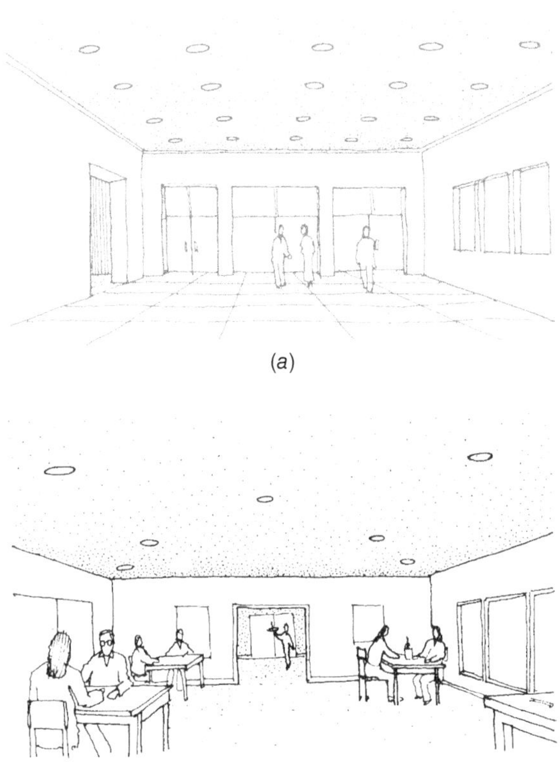

Figure 18.33 *Downlights are unobtrusive light sources. They can be spaced evenly throughout a room (a) or unevenly (b).*

Figure 18.34 *The array of large, cylindrical, surface-mounted downlights dominates the area's* appearance despite the high ceiling.

sociability/isolation, clarity/fuzziness, spaciousness/crowdedness, simplicity/clutter, formality/informality, boredom/excitement, definition/shapelessness, and so on.

Points of interest and visual excitement can be created by taking advantage of small, high-brightness sources (usually called sparkle). Lighting installations generally yield a sense of vividness or activity proportional to the level of illumination. This is not the case with very diffusely lighted areas, which, even at high illuminances, are tedious. Small LED units, a brightly lighted, rough-textured wall, and pendant fixtures with pierced reflectors are some of the techniques that can be used to create visual interest in a space.

Surfaces or points of high brightness can be used to draw visual attention. This inborn human response is used repeatedly to display merchandise. The following are typically encountered reactions:

- A 3:1 luminance ratio between object and surround will be noticed, but usually will not affect behavior or draw attention.

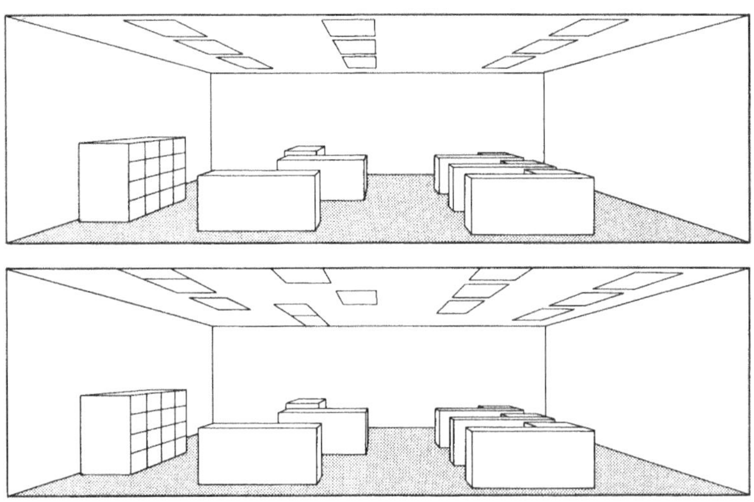
(a)
(b)

Figure 18.35 *(a) The layout of the lighting fixtures is economical and may provide uniform illuminance. (b) The nonuniform layout lacks integration with the functions below and may not provide the illuminance needed for the task locations.*

Figure 18.36 *Cornices, valances, and coves act as luminous ceiling borders. In large rooms, suspended coves achieve a uniform ceiling brightness gradient and, when designed with a downward component or combined with supplemental lighting as illustrated, can create a pleasant, intimate atmosphere.*

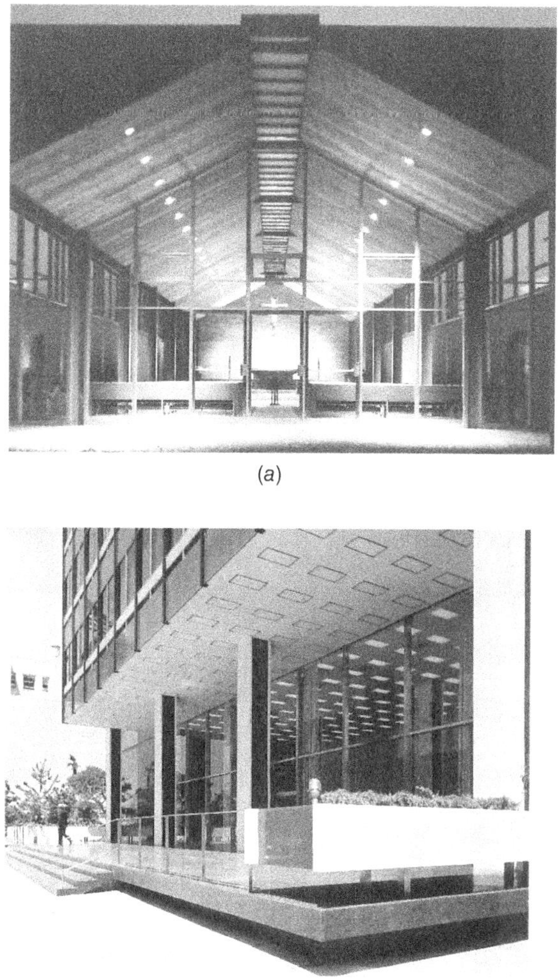

(*a*)

(*b*)

Figure 18.38 *(a) Lighting can be utilized as a medium to connect the inside and outside of a building. The simple maneuver of continuing the lighting pattern beyond the window visually connects the inside and outside spaces. Care must be taken to avoid a fixture placement that creates reflections in the glass. (b) Also, because fixtures are readily visible even when unlit during daylight hours, their pattern can be accentuated and the resultant outline continued from the inside to the outside as an architectural motif. (Courtesy of Welton Becket & Associates.)*

Figure 18.37 *Linear slot lighting emphasizes the curvilinear form of the gallery, further reinforcing a directional flow of traffic through the space. (© Donald Corner; used with permission.)*

- A 10:1 luminance ratio will attract attention and, if the object is interesting, hold it.
- A 50:1 luminance ratio or higher will highlight the illuminated object, often to the exclusion of all else in the field of view.

Because areas of high luminance draw the eye's attention, the individual brightness sources in the field of view produce a collective overall visual impression. If there is some form, order, or pattern to the sources (as with a pattern of lighting fixtures), then the overall impression is not disturbing—it can be thought of as visually harmonious. If, however, bright sources are in disarray, they produce discordance in the eye precisely as noise produces discordance in the ear. This visual "noise" is frequently referred to as visual clutter and can be very disturbing.

Other subjective reactions to lighting on which there is wide consensus are: (1) downlights (and color highlights) increase feelings of relaxation and comfort; (2) hidden-source indirect lighting and very-low-brightness lighting fixtures cause discomfort

Figure 18.39 Electric lighting fixtures can be seen through the glazing even during the daytime. (© Lee Eckert; used with permission.)

due to the viewer's inability to locate the source of the light.

(j) Other Evaluation Considerations

Codes, Standards, and Guidelines. Building codes are enforced by the AHJ (authority having jurisdiction; at a municipal, county, or state level). A lighting design must comply with all applicable code requirements. Standards and guidelines are typically brought into a project through the action of the owner (in the OPR) or design team (in project specifications). Several entities beyond the AHJ may provide input on lighting design, these include:

DOE—U.S. Department of Energy
GSA—U.S. General Services Administration
NFPA—National Fire Protection Association
ASHRAE—American Society of Heating, Refrigerating and Air-Conditioning Engineers
ICC—International Code Council
IES—Illuminating Engineering Society

Tasks. Identifying visual tasks is the key to selection of appropriate illuminance values. Factors to be considered beyond the general nature and location of each task are its repetitiveness, variability, who is expected to be performing the task (specifically the likely physical condition of the user), task duration, cost of errors, and special requirements. Several of these factors have been discussed under visual acuity in Chapter 7.

Multitasking. If there are several tasks to be performed at the same location and the most difficult one occurs infrequently, it may be reasonable to provide supplementary portable lighting or even to suggest moving the task to another location. If the most difficult task is the major task, the lighting design should be based on it and provision made for intensity reduction for less demanding work. As suggested by this discussion, illuminance within a

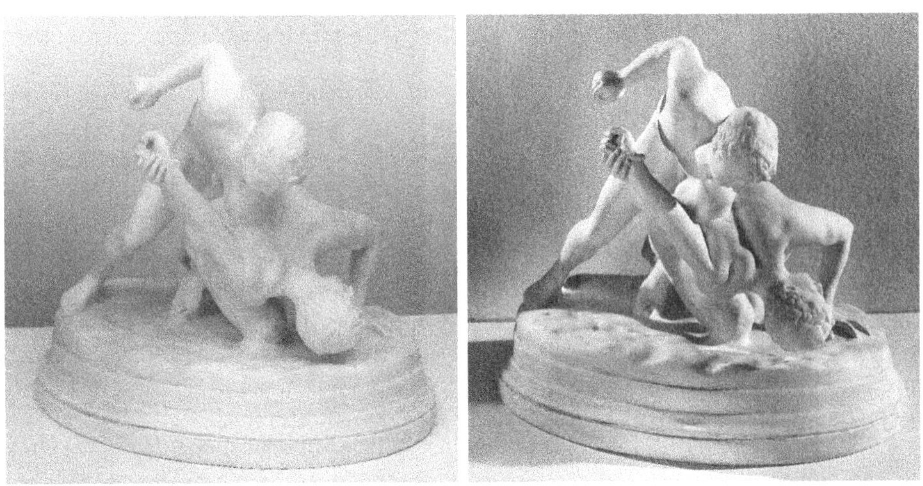

(a) (b)

Figure 18.40 Totally diffuse lighting (a) destroys texture, whereas a combination of diffuse and directional lighting (b) produces the required modeling shadows. (Courtesy of Holophane.)

building (or within a space) can vary with time and with location.

Variation in task difficulty is particularly common in spaces in public buildings. A school gym can be used for athletics, band concerts (despite the acoustics), testing, and town meetings—activities with disparate lighting requirements. In these and similar instances, it is common practice to treat the space as essentially four different spaces and to design a lighting system appropriate for each, while maximizing common equipment usage. Fortunately, many multipurpose spaces do not have severe seeing tasks.

The task variation referred to here is the variation that occurs in one very specific location, and is not to be confused with task variation across an area, however restricted. Thus, a small private office of, say, 8 ft × 8 ft [2.4 m × 2.4 m] has a desk, file cabinet, and circulation space, involving three tasks of differing but constant difficulty in one small space. The corresponding lighting solutions for these lighting zones should vary with the task severity.

Task Position. Recommended design targets represent the required illuminance on a defined task, whether horizontal, vertical, or slanted. Design often focuses on a 30-in. [760-mm] high horizontal work surface. It may be helpful in the early design stages to understand the ratio of horizontal-to-vertical illuminance for various lighting system approaches. This ratio is approximately:

Narrow distribution (direct and semi-direct)	3:1
Wide distribution (direct and semi-direct)	2.5:1
General diffuse (indirect)	1.5:1

Luminance Considerations. Recommended design illuminance values assume adherence to both recommended luminance ratios and surface reflectances. It is necessary to select—in conjunction with the interior designer—finishes and reflectances for surfaces within all illuminated spaces. If a dark wall finish with 10% reflectance is chosen for a private office, the lighting designer will need to compensate for this by additional wall lighting to maintain a recommended maximum 10:1 brightness ratio (see Table 7.6). Table 18.8 lists the reflectances of some common interior paint finishes.

Viewing Time. The length of time permitted for visual task accomplishment is important to establishing design task illuminance values in exacting work and production situations (data entry, quality assurance, etc.). Compared to moderately difficult tasks—requiring a luminance of, say, around 170 cd/m^2 [50 fL]—prolonged intensive or rapidly-changing tasks would require luminance (and thus usually illuminance) to be increased.

TABLE 18.8 Approximate Reflectances for Paints

Medium-Value Colors	Reflectance (%)
White	80–85
Light gray	45–70
Dark gray	20–25
Ivory white	70–80
Ivory	60–70
Pearl gray	70–75
Buff	40–70
Tan	30–50
Brown	20–40
Green	25–50
Olive	20–30
Azure blue	50–60
Sky blue	35–40
Pink	50–70
Cardinal red	20–25
Red	20–40

Alternatively, the quality of light could be improved by increasing daylighting or improving task contrast (as a means of increasing acuity without increasing raw illuminance).

Viewer Condition. Since the age and other specific characteristics of a worker/occupant are usually not known during design, a "standard" person is presumed to be the viewer. This standard person is usually a college-age student (a willing audience for scientific studies and statistical normalizations). Basic illuminance recommendations are often established with this standard user in mind. If there is likely to be a high percentage of older workers, as is the case in certain building space types or industries, illuminance should be increased. This compensates for the decreased ability of an aging eye.

Cost of Errors. This task consideration involves the economic trade-off between savings that result from improved visual acuity (work accuracy) and the cost of improved lighting to reach that higher accuracy. Performance can be brought close to perfection, but the cost of so doing increases more rapidly than the proportional increase in performance. Tasks where the costs of errors are a serious concern include inspection, proofreading, textile matching, very fine machining, and precision manufacturing. Lighting designed to meet normal recommendations is usually presumed to provide approximately 90% accuracy.

Special Requirements. These may include any nonstandard task lighting requirement, such as: need for a specific light-source color, directionality for shadowing, reflections needed for inspections, polarization, or control variations (as in a space with varied tasks or a varying daylight factor). The physical dimensions of a task itself often create special requirements. Designers get used to a small object in the horizontal plane because that is the normal office task. There are exceptions, however, such as an easel, a large machine, an inspection bench, or a cutting table. Consequently, these special requirements may arise:

1. *Large tasks:* With large tasks, the angle of view may vary from 20° to 70° from the vertical, resulting in radically changing reflection (glare) angles and reflection patterns.
2. *Three-dimensional tasks:* These tasks shadow themselves, particularly when containing undercuts and reveals. An architect's model shop presents such tasks. When it is necessary to see into an opening, an intense narrow light beam is required.
3. *Tools:* Tools cast shadows below and in front when lighted from above and behind. A fabric cutter must see ahead of and below the cutting machine.
4. *Nonhorizontal tasks:* These must be analyzed for the plane in which they stand. The ratio between horizontal and vertical illuminance can vary between 1.5:1 and 3:1, depending upon the lighting system. This variance can have a pronounced effect on the selection of a lighting system and its spatial arrangement.
5. *Task observed from various positions:* There are instances in which a fixed task is observed from several angles, such as a drawing in a conference room, a wall display, or a sculpture. Illumination design must be adequate for all viewing angles.

Costs. This is a particularly difficult item for a novice lighting designer because it requires experience in the field and an acquaintance with commercially available equipment. Also, the inevitable trade-offs between first cost and operating cost cannot be made intelligently unless the project cost structure is clearly understood. The following guidelines may be of assistance both in avoiding unpleasant surprises when a job is estimated and in preparing cost analyses:

1. Decide at the outset what cost criteria will be applied—that is, the relative importance of first cost, operating costs, annual owning costs, and collective life-cycle costs.
2. Trade-offs are necessary between first cost and operating costs. For example, high luminous efficacy lamps typically cost more than lower efficacy products (the same pattern is true of ballasts and luminaires), but can reduce operating costs.
3. Manufacturers' catalog items are always cheaper than custom items; off-the-shelf components can be estimated more readily and scheduled with more confidence.
4. Compare the annual owning costs of alternative systems or methods. Merging of these data into life-cycle cost comparisons is straightforward (see Appendix J).
5. The impact of lighting energy on the operating cost of the entire building must be studied and an apportionment of costs determined. The only practical means of accomplishing this is by using a computer simulation. Simulations can be readily adjusted to reflect the effect of a proposed lighting system on building costs and, in particular, on HVAC first cost and operating costs. It is incorrect to artificially separate design/cost of a lighting system from design/cost of the HVAC system, with which it intimately interacts.

The lower the lighting system energy use, the lower the building operating cost. Daylighting can radically impact the operation of an electric lighting system, and, thus, lighting energy use and cost. Only a well-integrated software analysis program can handle the complexities of daylight affecting electric lighting through control logic and the impact of those interactions on HVAC system operation, energy, and cost.

Lighting Power Budget. The requirement to establish a project lighting power budget in accordance with specified procedures has been long incorporated into the building codes of most U.S. states and many countries. The U.S. standard that establishes lighting power budgets is *ASHRAE/IES Standard 90.1: Energy Standard for Buildings Except Low-Rise Residential Buildings*, which is updated on a three-year cycle to coincide with the ICC's building model code update cycle.

The purpose of a lighting power budget is not to dictate a design solution. Instead, the purpose is to set a maximum permissible lighting power budget within which the designer is free to do as he/she deems most appropriate in the project context. Outcomes are set, not methods.

Table 18.9 provides examples of lighting power density (LPD) values (using the building area method) from five recent editions of *Standard 90.1*. The values show the differences in "expectations" between various space types and also show a clear trend of increasing stringency over a decade-long period.

Energy. Energy considerations must infuse every aspect of the design process. Some background is in order to place the subject of lighting system energy in proper perspective. Best current estimates indicate that lighting consumes approximately 25% of the electric power generated in the United States.

TABLE 18.9 Evolution of Lighting Power Density Values for Selected Space Types

Space Type	2022 LPD Limit	2019 LPD Limit	2016 LPD Limit	2013 LPD Limit	2010 LPD Limit	2007 LPD Limit	2001 LPD Limit
Dormitory	0.52 W/ft^2	0.53 W/ft^2	0.61 W/ft^2	0.57 W/ft^2	0.6 W/ft^2	1.0 W/ft^2	1.5 W/ft^2
Hotel	0.53 W/ft^2	0.56 W/ft^2	0.75 W/ft^2	0.87 W/ft^2	1.0 W/ft^2	1.0 W/ft^2	1.7 W/ft^2
Office	0.62 W/ft^2	0.64 W/ft^2	0.79 W/ft^2	0.82 W/ft^2	0.9 W/ft^2	1.0 W/ft^2	1.3 W/ft^2
Retail	0.78 W/ft^2	0.84 W/ft^2	1.06 W/ft^2	1.26 W/ft^2	1.4 W/ft^2	1.5 W/ft^2	1.9 W/ft^2

Sources: © ASHRAE, www.ashrae.org. 2022. ASHRAE Standard 90.1-2022, ASHRAE Standard 90.1-2019, ASHRAE Standard 90.1-2016, ASHRAE Standard 90.1-2013, ASHRAE Standard 90.1-2010, ASHRAE Standard 90.1-2007, and ASHRAE Standard 90.1-2001.

For LPD values in W/m^2, multiply the given values by 10.76.

In terms of resources, this amounts to approximately 4 million barrels (462,500 kL) of oil per day. The same sources indicate lighting energy usage percentages by occupancy as approximately:

Residential	20%
Industrial	20%
Retail	20%
Schools and offices	15%
Outdoor and other	25%

In commercial buildings, lighting consumes about 20–30% of electric energy; the percentage is greater in residences and lower in industrial facilities. A reduction of 40–50% in lighting energy is attainable through judicious design. Translated into resources, such a reduction can readily amount to more than 1 million barrels [115,630 kL] of oil per day. Few will deny that such a goal is well worth the effort. In a mechanically cooled building, every 1.0 W/ft^2 [10.7 W/m^2] reduction in lighting energy will result in at least a 1.25 W/ft^2 [13.5 W/m^2] energy savings. Offices and schools can be well lit with less than 1.5 W/ft^2 [18.1 W/m^2].

Resources to assist in reducing lighting energy consumption are numerous. For specific lighting system energy design targets and details, refer to *Standard 90.1*, the range of *Advanced Energy Design Guides* published by ASHRAE, and the LEED rating system by USGBC. For lighting effectiveness guidelines, refer to the *Lighting Library* from IES.

Conceptual approaches to energy-conscious lighting design include several models.

Design Lighting for the Expected Activity. It is energy-wasteful to light any surface to a higher illuminance than required. Nonuniform lighting is recommended where high illuminance levels are required for selected tasks in multitask spaces. One way to accomplish this for areas where an exact furniture layout is not available is to use readily movable fixtures. Providing overall high-lux illumination with a provision to switching to reduced lighting levels is not advisable because of the increased first cost and the psychological impetus to operate at maximum levels. Another solution is to use fixed luminaires for general low-lux lighting and supplementary lighting where needed. Also:

1. Place tasks with similar lighting requirements in the same general location.
2. Place the most severe seeing tasks at the best daylighting locations.
3. If possible, improve the quality of difficult visual tasks. This is more economical in terms of energy-use than providing additional light.
4. The advantages of nonuniform lighting increase as the space between work stations increases.
5. When using a task-based design approach, keep in mind that a nonuniform ceiling layout may give a chaotic appearance to a space. Therefore, the preferred approach is to use uniform ambient lighting and localized supplemental lighting.

Use Effective, High-Quality, Efficient, Low-Maintenance, Thermally Controlled Luminaires. "Effective" means providing useful light while minimizing direct glare. In cases where much of the viewer's time is spent in a head-up position, as in schools, or where the viewer can compensate for veiling reflections, the design should lean toward achieving high visual comfort probability. Where work and viewing position are fixed, and most of the viewer's time is spent head-down, the design should lean toward low reflected glare. The following should also be kept in mind:

1. A high-quality luminaire is made with permanent finishes such as Alzak® or multi-coat baked enamel or the high-quality permanent aluminum finishes currently available. This ensures that its performance after 8–10 years of service will be comparable to the original.
2. An energy-efficient luminaire is one with a high efficacy (LER/TER). Higher LER/TER (as with CU)

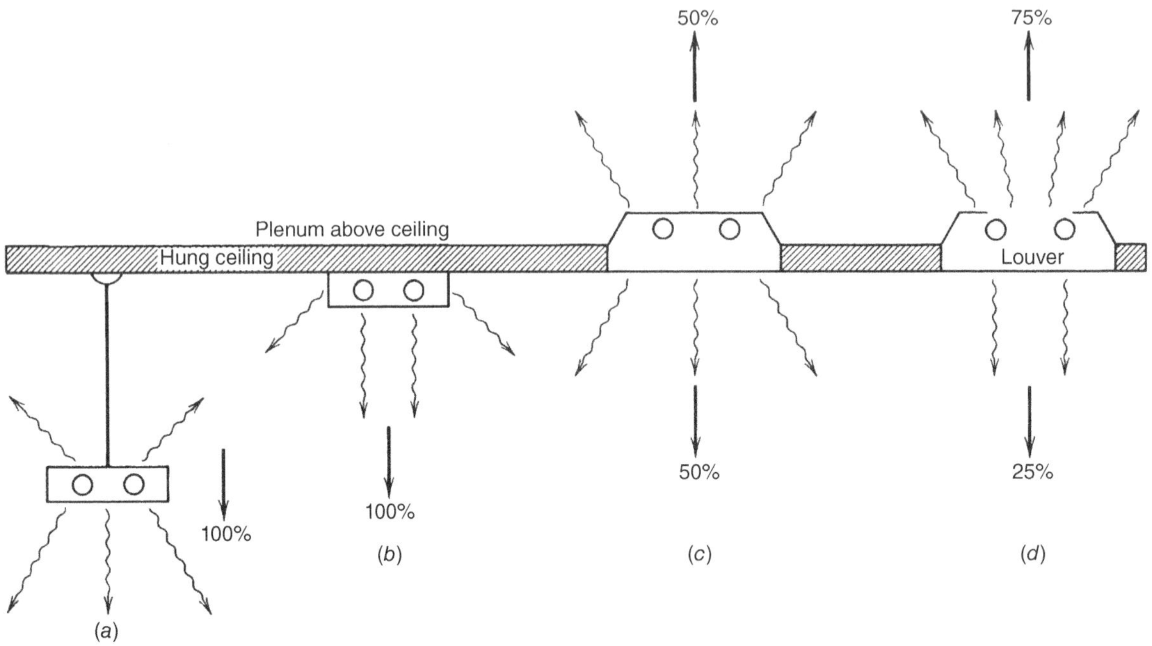

Figure 18.41 *Method of fixture installation controls the transfer of heat from a luminaire. (a) Suspended units transfer all heat to the space while remaining fairly cool. (b) Surface-mounted fixtures also transfer all heat to the space but, because of blocked transfer upward, run hot. (c) Completely enclosed recessed units transfer about 50% of their heat to the plenum. (d) Open-louvered, baffled units transfer about 75% of their heat. When they are ducted, heat transfer upward can be as high as 85%. (© ASHRAE, www.ashrae.org. 2022, ASHRAE/IES Standard 90.1-2022.)*

translates directly into effectiveness at lower energy cost.

3. A low-maintenance luminaire will remain clean for extended periods and is designed so that reflecting surfaces can be easily and rapidly cleaned without demounting. Enclosed fixtures should be gasketed. Nongasketed units collect and retain dust, thus causing rapid lumen output depreciation.
4. Relamping should be simple and rapid to encourage group relamping programs that are energy-efficient and cost-effective. A 20% increase in maintained light is possible if lamps are replaced at the end of their useful life—that is, when output is down to 70% of initial maintained lumens—and if fixtures are cleaned and maintained on a fixed schedule. No cost trade-off is generally involved because periodic maintenance and relamping are normally cheaper than one-at-a-time maintenance and burnout replacement, and yields 20% higher average lumen delivery. Fixtures in relatively inaccessible locations such as high ceilings must be designed for low maintenance, and maintenance should be on a fixed schedule. The power of a low LLF (light loss factor) to seriously reduce maintained illuminance shows the importance of reducing the impact of this variable to the lowest reasonable condition—first cost and energy costs both benefit.
5. A thermally controlled luminaire controls the heat generated by the light source (Figure 18.41). The importance of this effect depends upon the type of HVAC system, the lighting heat load, and the types of fixtures being considered. The regular use of electronic ballasts and high-efficacy lamps has reduced the impact of this issue over the past decade. Heat dissipation from LED fixtures, however, should follow manufacturer's recommendations.
6. Use efficient light sources and accessories. The ready availability of high efficacy, high color rendering index (CRI) lamps has made this far less problematic than previously. The primary cost trade-off for non-LED lamps is between high-CRI sources and low-CRI sources. The cost differential is not huge. LED lamps are no longer considered cutting edge, and most new commercial projects are using LED sources as the efficient source of choice.

Spill light and borrowed light are often neglected sources. For example, clerestory fenestration along corridors between offices can take advantage of borrowed office light to provide adequate corridor lighting.

Select an Appropriate Lighting System. There are six broad approaches to light delivery from a luminaire, each with its particular applicability. In addition, luminaires must be carefully located to provide uniformity of lighting, where applicable, and effective delivery of supplemental lighting. When using indirect luminaires, correct spacing and hanging distance will avoid ceiling "hot spots," which can cause direct and reflected glare. Properly designed indirect lighting allows flexible use of a space, in that users are not forced to face in any particular direction in order to avoid direct or reflected glare caused by fixtures. The mounting height of suspended fixtures must be coordinated with cavity sizes and finishes to decrease unnecessary light loss and maximize coefficient of utilization.

Use Daylight, and Use It Properly. Daylight must be considered as a basic light source that is subject to both weather variations and time variations. A three-shift industrial plant cannot use daylight on all shifts, but it can for at least one shift—lighting design should address this fact. Part of proper daylight design is control of window luminance, which can cause severe and even disabling glare. A corollary of excessive luminance is excessive heat gain—both of which are manageable with common manual and automatic control devices. Window control devices should be designed to reflect light back into a space at night to avoid light loss through transparent windows.

Use Energy-Efficient Lighting Control Strategies. The subject of lighting control, including manual and automatic switching, dimming, sensing, and intensity control, is covered in Chapter 15. Proper design of controls can reduce energy consumption over a noncontrolled installation by as much as 60% without reducing lighting effectiveness. Appropriate controls are mandated in *Standard 90.1*.

Use Light Finishes on Ceilings, Walls, Floors, and Furnishings. This point is repeatedly emphasized in this book. A brief summary of recommended reflectance ranges is:

Ceilings	80–92%
Walls	49–60%
Furniture, office machines, and equipment	25–45%
Floors	20–40%

In addition to producing higher illuminance levels in a room, high reflectances minimize uncomfortable luminance ratios between, for instance, a luminaire and upper wall, or task and background. Suggested maximum luminance ratios are as below. These are maximum values, and ratios above these limits should be accepted only with excellent justification.

Between task and near surround	3:1
Between task and immediate area	10:1
Between luminaires and their background	20:1
Anywhere in the normal field of view	40:1

Other Lighting Design Recommendations. Many of these ideas are from *Standard 90.1*:

1. Buildings shall have some form of automatic lighting shutoff control.
2. Where applicable, occupancy sensors and automatic daylight compensation controls should be used.
3. Separate spaces must have separate controls.
4. Enclosed rooms should have controls that allow for at least two lighting levels.
5. Where supplemental lighting is used, in order to avoid uncomfortable luminance ratios, the ambient level should not be lower than a third of the task level.
6. Accent lighting should not exceed five times the ambient lighting level. In merchandising areas where contrast between ambient and task levels is critical, ambient levels should be reduced as much as is practical.
7. Before specifying superreflective aluminum in luminaires, establish that the high reflectance will be maintained in the application being considered and that recommended luminaire maintenance procedures can and will be implemented.
8. Utilize low-energy/low-maintenance exit signage where permitted by local codes; such equipment is readily available.

The energy-efficiency benchmarks in *Standard 90.1* are updated on a regular basis. The current version should always be used for design. The accompanying *Standard 90.1 User's Manual* (ASHRAE, 2020), typically updated for each edition, is a valuable supporting document to assist with interpreting and implementing the document.

Other Considerations. Because the phenomenon of vision, as affected by lighting, is the core of lighting design, it is appropriate to present some miscellaneous observations that are not necessarily covered elsewhere in this chapter.

1. As shown in Figures 18.31*a*, *b*, the impression of room length and width can be emphasized by the direction of lines of lighting. An even wall wash of light can also be used to shorten and widen a hallway or corridor.

2. As seen in Figure 18.2, the character of a wall, in the line of sight, can affect the room's apparent size. Light and low-chroma, high-value paint hues can be used to expand a space; conversely, darker-hue paints and a lack of luminance can be used to contract it.
3. Vertical surface illuminance should be approximately 25–35% of horizontal illuminance for a space to appear dimensionally undistorted. Because high luminance attracts the eye, fixtures with sparkle draw the eye away from the walls and thereby shrink the space's horizontal dimensions.
4. As shown in Figures 18.30*c*, 18.31*c*, *d*, 18.33, and 18.34, luminaire patterns can be dominant (i.e., can become a focus of attention) by virtue of their size or arrangement. The same is true of wall-lighting patterns to an even greater extent, as walls are always in the direct line of sight. Therefore, the lighting designer should remember that wall lighting patterns can become dominant visual elements when they create meaningless scallops, spots, irregular gradients, points of sparkle, or focus on objects better left un-highlighted. Even though this is neither intended nor desired, it is common. Unscalloped, even-gradient wall illumination is readily accomplished with linear sources, with elliptical reflector lamps, or with luminaire reflectors, if proper luminaire spacing is utilized.
5. Concentrated pools of light, in a space with overall low ambient light, can create isolated pockets of illumination—often perceived as spatially differentiated areas. This can be advantageous in restaurants and work or school areas, where definition of individual "territories" in a single large space is desired.

(k) Construction Documents

The final product of the design phase of a lighting project is construction documents. These consist of the drawings and specifications necessary to ensure that the contractor can construct the lighting system as envisioned by the designer. The drawings are used to convey physical locations, quantities, and identifying characteristics of lighting system components. Reflected ceiling plans and details are common drawing elements. Specifications are used to describe lighting system component and installation quality.

18.7 CONSTRUCTION PHASE

During the construction phase of a project, the lighting system will be installed. It is common for the lighting designer to review submittals (shop drawings, cut sheets) and provide some site observation—both efforts to help ensure that what was designed is what is installed. On a commissioned project, lighting system submittals, installation, operation, and controls will be verified by the commissioning team—with the objective of assisting the owner in reaching his/her lighting OPR.

18.8 OCCUPANCY PHASE

Historically, owners were generally left to their own devices during the post-construction period (when the owner occupies and uses a building). This should not be the case. As a minimum, the lighting designer must prepare a user's manual to assist the owner with operation and maintenance of the lighting system. This manual might describe how daylighting compensation controls are intended to function or what maintenance schedule was assumed for cleaning of lighting fixtures or relamping—among other useful items of information.

Some lighting elements may need to be tuned to provide operational outcomes in accord with the OPR. Complex lighting controls almost always need to be adjusted in order to work properly.

Ideally, the owner would sponsor (or the designer would volunteer) a post-occupancy evaluation (POE) of the building lighting system. Visual comfort and energy performance are logical focuses for a POE.

References and Resources

ASHRAE. 2004–2012. *30% Advanced Energy Design Guides* (multiple published components). Atlanta, GA: American Society of Heating, Refrigerating and Air-Conditioning Engineers.

ASHRAE. 2011–2012. *50% Advanced Energy Design Guides* (multiple published components). Atlanta, GA: American Society of Heating, Refrigerating and Air-Conditioning Engineers.

ASHRAE. 2018–2019. *Zero Energy Advanced Energy Design Guides* (multiple published components). Atlanta, GA: American Society of Heating, Refrigerating and Air-Conditioning Engineers.

ASHRAE. 2020. *Standard 90.1 User's Manual*. Atlanta, GA: American Society of Heating, Refrigerating and Air-Conditioning Engineers.

ASHRAE. 2022. *ANSI/ASHRAE/IES Standard 90.1-2022: Energy Standard for Buildings Except Low-Rise Residential Buildings*. Atlanta, GA: American Society of Heating, Refrigerating and Air-Conditioning Engineers.

IES. 1993. *The Lighting Handbook*, 8th edn. New York: Illuminating Engineering Society.

IES. 2011. *The Lighting Handbook*, 10th edn. New York: Illuminating Engineering Society.

IES. 2024. *The Lighting Library*. New York: Illuminating Engineering Society. ies.org/standards/lighting-library

Ross and Baruzzini, Inc. 1975. *Energy Conservation Applied to Office Lighting*." (Conservation Paper (Vol. No. 18).) Washington, DC: Federal Energy Administration.

Sampson, F. K. 1970. *Contrast Rendition in School Lighting*. New York: Educational Facilities Laboratory. files.eric.ed.gov/fulltext/ED037943.pdf

USGBC. *LEED Rating System*. Washington, DC: U.S. Green Building Council. www.usgbc.org/leed

CHAPTER 19

Electric Lighting Applications

19.1 INTRODUCTION

Chapters 17 and 18 examined lighting fundamentals, sources, and design procedures. This chapter on lighting considers the application of lighting principles to specific situations, with a primary focus upon electric lighting systems. The facilities covered in some detail include residential, educational, commercial, institutional, and industrial occupancies. Each is examined from the viewpoint of its special requirements, and design approaches are suggested. The latter include lighting materials and sources as well as comparative economics and energy considerations. The chapter concludes with a consideration of special types of indoor lighting, plus a short section on exterior lighting.

RESIDENTIAL OCCUPANCIES

19.2 GENERAL INFORMATION

The buzzwords in modern residential construction are automation (convenience), environmental considerations (energy conservation, "green" design), and the home as a working area (home office, special communications considerations). These underlying considerations influence decisions on sources, control, energy use, and budget, as will be discussed later.

Residential lighting offers the lighting designer a great opportunity for originality and ingenuity because a residence combines more diverse functions and needs than almost any other building. Furthermore, it often requires that all work be done at minimal cost and that the results please persons with a range of tastes. The designer approaches the problems with a list of requirements, a perception of the space, and two basic tools: the lighting fixture and the architectural lighting element. There are numerous books and countless periodical articles devoted to residential lighting design. We recommend using *IES/ALA RP-11-20 Recommended Practice: Lighting for Interior and Exterior Residential Environments* (IES, 2020d) as a reference and useful guide.

19.3 ENERGY ISSUES

Although residences are generally excluded from the requirements of *ASHRAE/IES Standard 90.1-2022: Energy Standard for Sites and Buildings Except Low-Rise Residential Buildings* (ASHRAE, 2022), and lighting is not rigorously addressed by *ASHRAE/IES Standard 90.2-2024: High-Performance Energy Design of Residential Buildings* (ASHRAE, 2024), there may be local code requirements that deal with lighting energy efficiency. In any case, there are energy implications associated with residential lighting design decisions. General energy efficiency recommendations include:

1. Provide means for multiple light levels in all areas. A kitchen during food preparation does not have the same lighting requirements as a kitchen being entered for a late-night "refrigerator raid." Low-level lighting provisions should be made in all rooms, including bathrooms. To accomplish this, use high–low switches, simple dimmers, multilevel

Mechanical and Electrical Equipment for Buildings, Fourteenth Edition. Walter Grondzik, Daniel Overbey, and Alison Kwok.

Companion website: www.wiley.com/go/meeb14e

ballasts, and multilevel switching. An ancillary benefit is that ambience can be changed, thereby in multiuse rooms such as dining rooms, family and recreation rooms, and finished basements.
2. Provide local task lighting for areas where relatively difficult visual tasks are performed, such as the kitchen location where menus are planned and accounts are handled.
3. Provide dimming and switching for accent lighting.
4. Use programmable timers with photocell override for exterior lighting.
5. In large residences, consider low voltage or wireless control for ease of remote control and energy savings (see Section 19.7).
6. Use daylight in areas normally occupied during daylight hours, such as kitchens and living rooms. Consider skylights with built-in electric lighting for these areas.
7. Switch to an all-LED lighting package. A 100% LED lighting package could use up to 75% less energy than using luminaires with a mix of incandescent lamps and compact fluorescent lamps (CFLs).

Although automatic daylight compensation is generally not justified economically, area switching is and should be considered in all spaces, including work areas of children's rooms and home offices.

19.4 RESIDENTIAL LIGHTING SOURCES

Incandescent sources have traditionally found very wide use in residences (despite their inefficiency) because of their desirable characteristics, which include flattering skin color, low first cost, small size, focusability, and simple, economical dimmability. Contemporary LED sources can closely mimic the characteristics of incandescent sources for most residential circumstances. Especially with consideration of work in existing buildings, other sources with at least some of these characteristics and much higher efficiency than incandescent sources may be considered or encountered:

- Fluorescent sources with color temperatures between 3000 and 3500 K can be used in kitchens and other work areas. Where linear fluorescent lamps are not desirable, CFL can readily be used. Keep in mind, however, that the lamp life of all fluorescent sources is shortened by switching.
- Architectural elements such as coves and cornices can readily use CFL lamps.
- Tungsten-halogen lamps should be restricted to highlighting and specialty requirements. They are incandescent lamps and, in addition to having low efficacy, must be carefully applied because of their concentrated heat.
- Consider employing low-wattage (9–12 W) CFLs in frequently used corridors and stairwells, rather than incandescents. These small lamps can be left on for extended periods to provide the required low-level lighting without the constant switching and lamp replacement required for incandescents. This practice is even more practical in difficult access areas such as stairwells.
- For closets, pantries, and other small areas with frequently switched lighting, incandescent lamps remain the source of choice.
- For home offices and dual-purpose areas that serve for both work and recreation, design the lighting for each use individually, with maximum common use. Fluorescent sources are recommended for work purposes.
- Bathrooms can use 3000–3500 K fluorescent sources for general lighting, a separate incandescent source for short-time use, and low-brightness globe-shaped incandescents for mirror lighting.
- Although many manufacturers produce downlights and wallwashers for CFL sources, none that we have seen delivers the "punch" of incandescents. Where such brightness is not required, CFL downlights are a good choice when burned for at least three hours per use.
- High-intensity discharge (HID) sources are appropriate for all exterior lighting.
- But, again, LED lamps are becoming standard in all building types, especially new projects.

19.5 DESIGN SUGGESTIONS

The following are a few general design suggestions:

- Use a general/task lighting approach, with the levels recommended in Tables 19.1 and 19.2.
- Provide luminance ratios as in Figure 19.1.
- Provide general lighting sufficient for movement and casual seeing in all spaces. Hallways require little lighting; stairs require more. Illuminate stairs from directly above or ahead to create a shadow directly

TABLE 19.1 Illuminance Recommendations for General Lighting

Activity or Area	Average Lux
Conversation and relaxation	50–100[a]
Passage areas	50–100[a]
Areas other than kitchen	200–500
Kitchen	500–1000

[a]General lighting in these areas need not be uniform.

ACTIVE ENVIRON-MENTAL SYSTEMS

TABLE 19.2 Illuminance Recommendations for Specific Residential Visual Tasks[a]

Seeing Task	Typical North American Recommendation: Average Lux[b]	Other Authorities: Average Lux[b]
Dining	100–200	100–150
Grooming, makeup	200–500	500
Handcraft		
Ordinary seeing tasks	200–500	200–500
Difficult seeing tasks	500–1000	500–750
Critical seeing tasks	1000–2000	>1250
Kitchen Activities		
Food preparation and cleaning involving difficult seeing tasks	500–1000	750–1000
Serving and other noncritical tasks	200–500	200–300
Laundry tasks	200–500	100–300
Reading and Writing		
Handwriting, reproductions, and poor copies	500–1000	750
Books, magazines, and newspapers	200–500	300
Sewing, Hand or Machine		
Dark fabrics	1000–2000	>1250
Medium fabrics	500–1000	700–1000
Light fabrics	200–500	300–500
Table games	200–300	300

[a]Selection of illuminance within the given range is based on the criteria presented in Chapter 18.

[b]Divide by 10 to get footcandles. Due to the range of values, use of the exact 10.76 conversion value is unnecessary.

below the tread front. Lighting from the front eliminates shadows and can create a safety hazard.

- Do not avoid ceiling light sources, as is so frequently done. Wide-profile ceiling fixtures provide general lighting; switch-controlled table lamps do not.

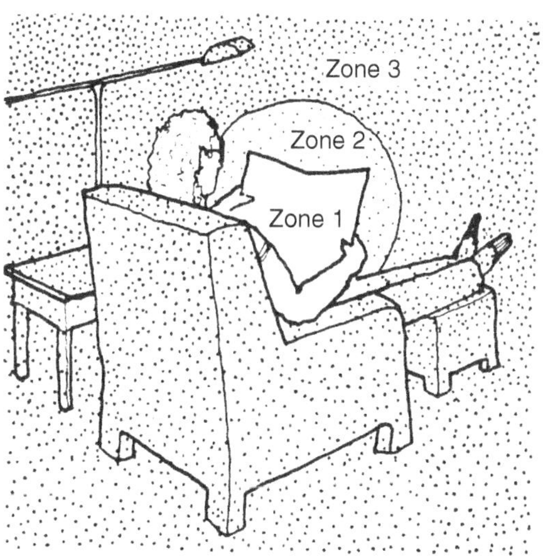

Zone 2 The immediate surroundings (area adjacent to the visual task)	
Desirable ratio	1/3 to equal to task*
Minimum acceptable ratio	1/5 to equal to task*

Zone 3 The general surroundings (not immediately adjacent to task)	
Desirable ratio	1/5 to 5 times task*
Minimum acceptable ratio	1/10 to 10 times task*

*Typical task luminance range is 40 to 120 cd/m^2 [12-35 fL] and seldom exceeds 200 cd/m^2 [60 fL]

Figure 19.1 *Seeing zones and recommended luminance ratios for residential visual tasks; Zone 1 is the task.*

- Lighting in areas specifically intended for use by older occupants, such as in-law units that are part of larger residences and residential buildings intended for older occupants, should take cognizance of the special illuminance recommendations in Chapter 18.

19.6 LUMINAIRES AND ARCHITECTURAL LIGHTING ELEMENTS

Guidelines that can assist the designer in selecting luminaires from the huge variety available commercially are as follows:

1. Utilize diffuse distribution for general lighting; narrow-distribution downlights for area and furniture accents; and narrow-distribution, ceiling-recessed incandescent wallwashers for accenting textured surfaces, such as brick.
2. Use built-in lighting to the extent possible, including architectural lighting elements. We believe that this demonstrates integrity of concept. For this reason, we recommend that the flexibility of track lighting be utilized for accent and task lighting but not for general lighting (Figure 19.2).
3. Private residences are the exception to the rule of selecting off-the-shelf items in preference to specials. The lighting should complement the architecture and furnishings, and frequently this can best be accomplished by original designs.

Architectural lighting elements include coves, cornices, valances, coffers, skylights, and other

Figure 19.2 *Track lighting is available with a wide variety of luminaire bodies. Those illustrated include a modular system to provide a practical approach to providing illumination in targeted areas. These feature 358-degree rotation and 90-degree tilt for precise aiming of halogen light sources to the object. In addition, a variety of single-circuit and multicircuit track designs are available. (Courtesy of Progress Lighting.)*

luminous constructions not normally constituting a lighting fixture. Although such units are normally less efficient than lighting fixtures, their use is often indicated by architectural considerations. Empirical design data are given in Figures 19.3 and 19.4.

When using fluorescent lamps in architectural lighting elements, dark spots between lamps can be avoided by placing lamps at a slight angle rather than end to end, thus enabling ends to overlap. Similar overlapping is readily accomplished with PL lamps as well. Reflectors increase efficiency of an installation. When used in coves, reflectors should be aimed 15–25° above the horizontal and field-adjusted for the best ceiling coverage. When two-lamp strips are used, they should be arranged vertically, as shown in Figure 19.4. Light output for a double-lamp installation rarely exceeds 1.75 times the single-lamp output.

19.7 RESIDENTIAL LIGHTING: CONTROL

In large residences, remote control of lighting becomes more of a necessity than a convenience from the viewpoints of control, safety, and status awareness. Completely wireless, radio-wave controlled lighting is particularly applicable to retrofit work because it requires only that control modules be substituted for existing switches. A conventional system might use a wall-mounted master at each entrance in the kitchen and in the home office (Figure 19.5), a bedside tabletop master in the master bedroom (Figure 19.6), and radio-operated wall switches and dimmers in all rooms.

EDUCATIONAL OCCUPANCIES

19.8 INSTITUTIONAL AND EDUCATIONAL BUILDINGS

The lighting requirements for some spaces in educational facilities coincide with the requirements for commercial (office) and institutional buildings. To that extent, the remarks herein are applicable there as well. Generally, educational buildings (excluding private colleges and universities) are maintained by operating funds obtained from taxes, and the budget is always tight. Therefore, all equipment in these public buildings must be extremely hardy, vandal-proof, as maintenance-free as possible, and low in energy consumption. Maintenance in such buildings is generally poor and is performed on a repair rather than preventive basis. With these factors in mind, the following remarks apply to lighting equipment:

1. Use sources with the highest possible efficacy. Remember that daylight has the highest efficacy, followed by high-pressure sodium (HPS), LED, fluorescent, and metal-halide sources.
2. Where specific color lamps are called for, such as 3500 K T8 and the like, this requirement should be permanently stenciled in large letters on the lighting fixture to ensure proper relamping.
3. Long-life sources should always be given preference because of their lower maintenance. Thus, corridor and stair lighting should use LED, fluorescent, or HID lamps. This is also important in locations where relamping is difficult, as in high-ceiling rooms such as gyms and assembly rooms.
4. In calculating illuminance, low light loss factor (LLF) values should be used to allow for aging of paints and dirt accumulation. Cleaning of lighting fixtures in schools is virtually unknown. An LLF value of 0.5 is reasonable. This being so, provision should be made to reduce initial overlighting or to provide for lumen maintenance.
5. Not all schools are air-conditioned. With the masking effect of air noise absent, careful control must be exercised over noise and vibration from ballasts, luminaire diffusers, and so on. Electronic ballasts are preferred if the construction budget tolerates the additional cost. Ballast noise increases with current rating. Therefore, 800-mA, high-output and 1500-mA, very-high-output lamps must be used with caution, particularly in locations that amplify sounds, or where low noise criteria (NC) obtain.

(*a*) Lighted Cornices

Cornices direct all their light downward to give dramatic interest to wall coverings, draperies, murals, etc. May also be used over windows where space above window does not permit valance lighting. Good for low-ceilinged rooms.

(*b*) Lighted Valances

Valances are always used at windows, usually with draperies. They provide up-light which reflects off ceiling for general room lighting and down-light for drapery accent. When closer to ceiling than 10 inches use closed top to eliminate annoying ceiling brightness.

(*c*) Lighted Coves

Coves direct all light to the ceiling. Should be used only with white or near-white ceilings. Cove lighting is soft and uniform but lacks punch or emphasis. Best used to supplement other lighting. Suitable for high-ceilinged rooms and for places where ceiling heights abruptly change.

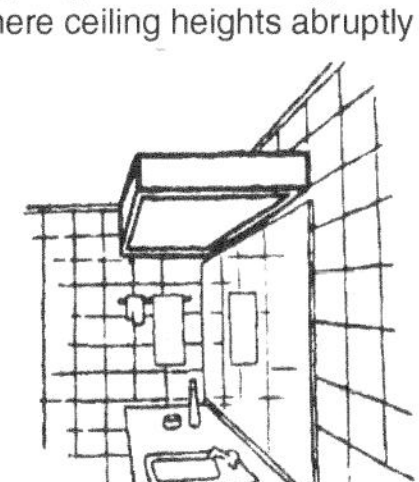

(*d*) Lighted Canopies

The canopy overhang is most applicable to bath or dressing room. It provides excellent general room illumination as well as light to the user's face.

(*e*) Lighted High Wall Brackets

High wall brackets provide both up and down light for general room lighting. Used on interior walls to balance window valance both architecturally and in lighting distribution. Mounting height determined by window or door height.

(*f*) Lighted Low Wall Brackets

Low brackets are used for special wall emphasis or for lighting specific tasks such as sink, range, reading in bed, etc. Mounting height is determined by eye height of users, from both seated and standing positions. Length should relate to nearby furniture groupings and room scale.

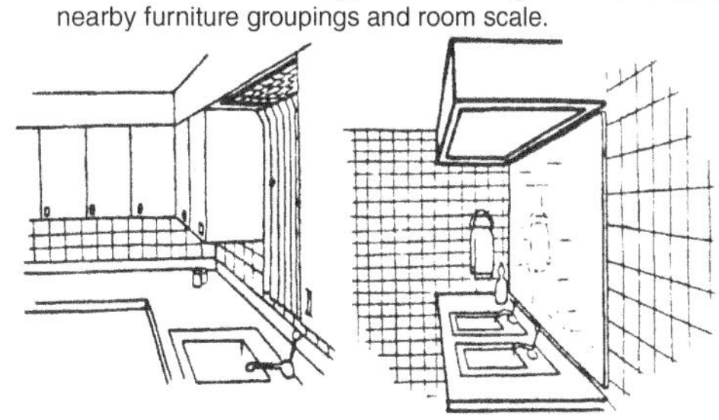

(*g*) Lighted Soffits

Soffits over work areas are designed to provide higher levels of light directly below. Usually they are easily installed in furred-down area over sink in kitchen. Also are excellent for niches over sofas, pianos, built-in desks, etc.

Bath or dressing room soffits are designed to light user's face. They are almost always used with large mirrors and counter-top lavatories. Length usually tied to size of mirror. Add luxury touch with attractively decorated bottom diffuser.

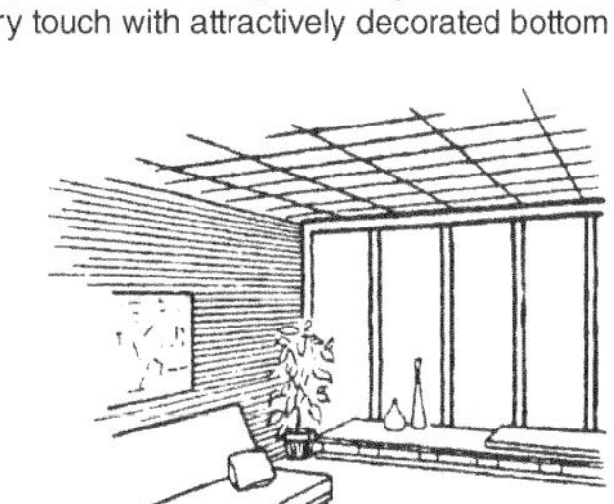

(*h*) Luminous Wall Panels

Luminous wall panels create pleasant vistas; are comfortable background for seeing tasks; add luxury touch in dining areas, family rooms, and as room dividers. Wide variety of decorative materials available for diffusing covers.

Figure 19.3 *Residential lighting elements. (Courtesy of IES.)*

(*a*) Typical Valance

This typical dimensional drawing applies only to commonly encountered window valance situations. Obviously, other window treatments could necessitate modifications in these critical dimensions; i.e., vertical blinds, double-track situations, curved bay windows, etc.

The same "job-tailored" variations can occur in the design of any type of structural lighting device. Therefore, no other dimensional drawings have been included here.

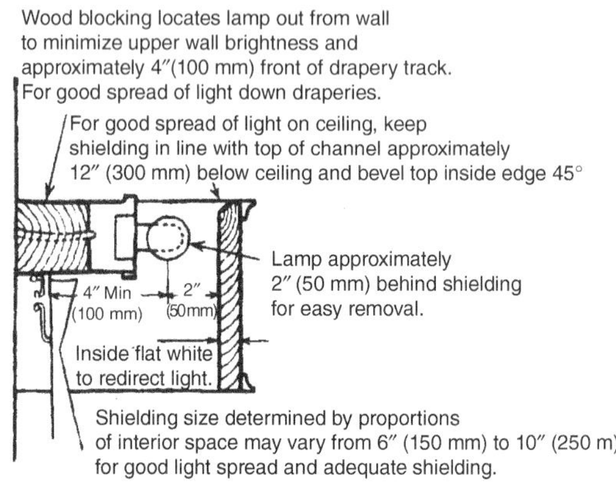

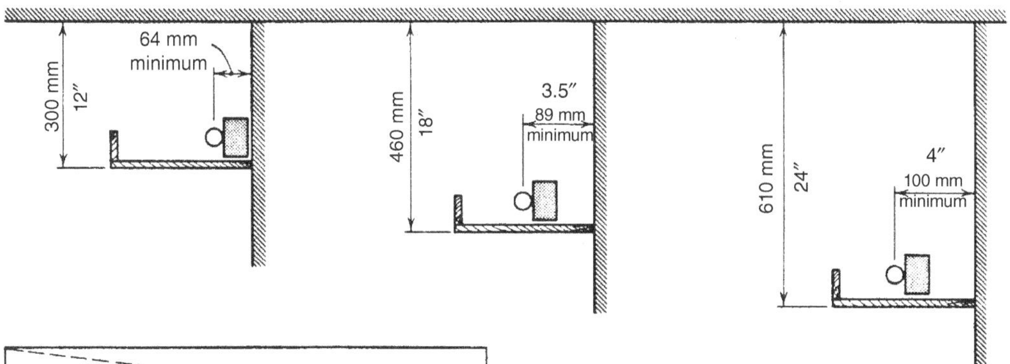

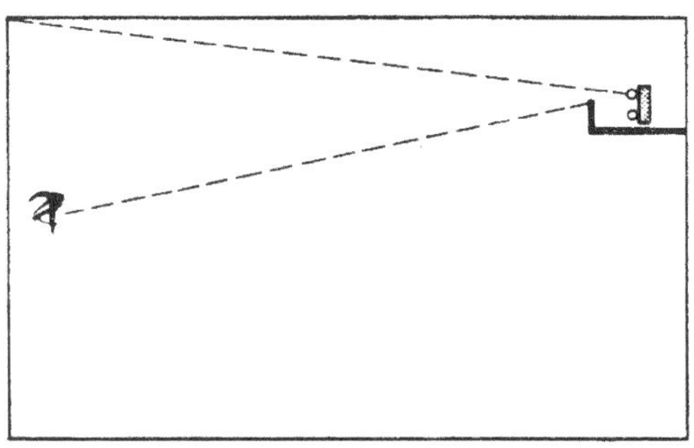

(*b*) and (*c*) Cove Installations

Proper cove proportions: Height of front lip of cove should shield cove from the eye yet expose entire ceiling to the lamp. Orientation of fluorescent strip as shown is preferable. Cove interiors should be painted with high reflectance matte-finish paint. *Westinghouse Lighting Handbook (Out of print)*

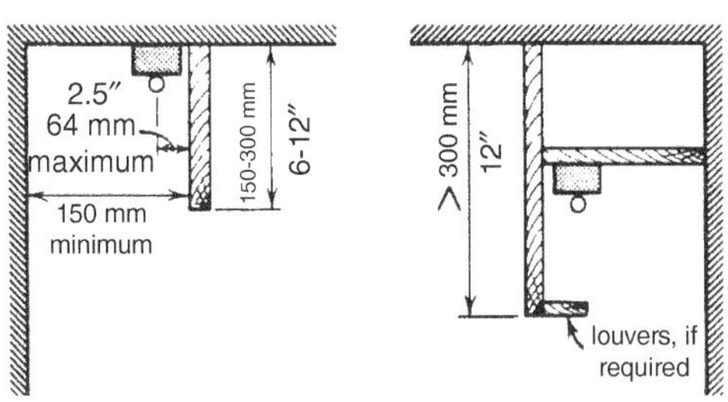

(*d*) and (*e*) Typical Cornices

Wallwashing equipment mounted in valances and cornices provides improved brightness ratios and may be used for lighting desks against walls or vertical illumination of walls and objects mounted thereon. *Westinghouse Lighting Handbook (Out of print)*

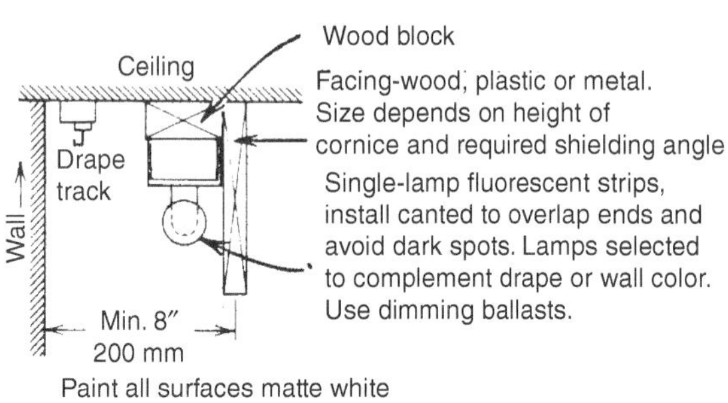

Figure 19.4 *Selected lighting elements: construction details.*

Figure 19.5 *Typical wall-mounted master station has five control buttons in addition to "All On" and "All Off" on every master. The control buttons can be arranged to control a single outlet or a group to establish lighting of a scene. Thus, the entry unit might have homecoming, extended absence, path lighting, nightlighting, and garage-lighting scenes, whereas the kitchen master may have lighting scenes for breakfast, dinner, formal and informal entertaining, and night. Each scene can be arranged to turn on, brighten or dim, and shut off selected lighting. (Courtesy of Lutron Electronics Co., Inc.)*

6. Lighting equipment must be designed for absolutely minimum maintenance, and those fixtures within easy reach should be vandal-proof. This means using captive screws, rust-preventive plated parts, captive-hinged diffusers whose cleaning requires only one person, ballast replacement without demounting fixtures (plug-in ballasts are available), nonyellowing plastics, and high-quality finish and assembly.

For recommended illuminances for the various visual tasks in an educational facility, see *IES RP-3-20 Recommended Practice: Lighting Education Facilities* (IES, 2020b). In the following sections, we discuss the lighting requirements of specific school building areas.

19.9 GENERAL CLASSROOMS

The classroom is the basic space in a school. Unlike the classic schoolroom with fixed seats, a single viewing

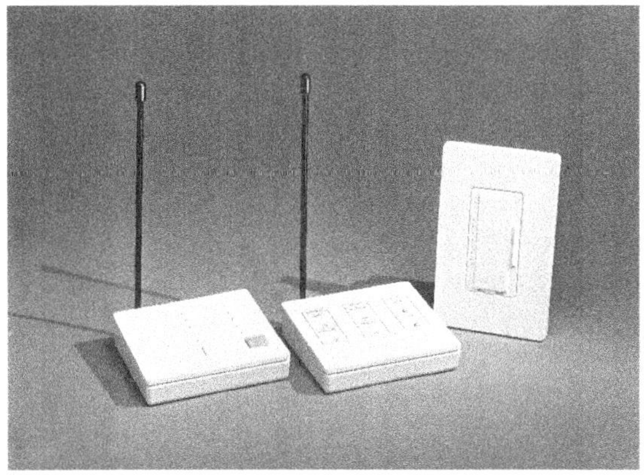

Figure 19.6 *Typical components of a wireless control system. Left: bedside master station has five scene buttons and 10 individual outlet controls in addition to "All On" and "All Off" buttons. Middle: repeater, which is used in large buildings to strengthen signals. Right: typical radio-controlled wall dimmer with local manual control. (Courtesy of Lutron Electronics Co., Inc.)*

direction, and a fixed teacher location, many modern classrooms, at all grade levels, utilize multiple-student groups, teacher mobility, multiple tasks in the same overall space, and movable seating arrangements. Such spaces require the following features:

1. Controls that permit subdivision of lighting, and level control within space groupings.
2. A lighting system with an appreciable indirect component and good diffusion to minimize the problem of veiling reflections due to viewing direction.
3. Low-brightness luminaires with high visual comfort probability (VCP) in all viewing directions, inasmuch as a considerable portion of the students' time is spent in a head-up position.

In addition to these special recommendations, some lighting recommendations applicable to all classrooms are as shown below:

4. Classrooms with digital displays require special lighting (see Section 19.19).
5. Mitigate lighting conditions that could cause veiling reflections on glossy surfaces, including computer screens, which can make it difficult to perform certain critical tasks, such as reading.
6. Generally, light sources with good color rendering should be specified. LEDs can achieve a broad spectrum of correlated color temperature (CCT) and high color rendering indices (CRI) and may be adjustable. Fluorescent luminaires should use T8 triphosphor lamps with a CCT of 3500–4100 K and high CRI. Electronic ballasts are preferred.

Incandescent sources should not be used. Daylight, to the maximum extent, is desirable as it provides full-spectrum luminance, supports energy conservation, and supports human health. Some form of daylight compensation may be mandatory by certain codes or standards.

7. In comparing the costs of alternative appropriate lighting systems, use life-cycle costing techniques only, because these facilities are nonprofit and long-term. See Appendix J for an explanation of the principles of economic analysis. Use a computer program to perform the economic analysis where possible. Do not compare costs on the basis of footcandles (lux) per dollar (i.e., by dividing maintained illuminance by cost), because this leads to a preference for higher illumination levels and, consequently, higher energy usage.
8. Illuminance recommendations are divided by application type (e.g., accenting, administration, auditoriums, building entries, classrooms, dormitories, reading and writing, transition spaces, and so on) and further divided by task (type of work), visual age of observers, and orientation of the task (horizontal or vertical).

Figure 19.7 shows a perimeter classroom lighting layout using a two-lamp T8 louvered luminaire and calculated initial horizontal and vertical illuminance values. A single-lamp version of the luminaire and its photometry are shown in Figure 19.8. Note the crosswise batwing distribution that assists in limiting reflected glare.

19.10 SPECIAL-PURPOSE CLASSROOMS

(a) Shops

In classroom-shops, where at least half of the people using the space are 25 years old or younger (visual age), the illuminance target for machine work, inspection, assembly, and woodwork is 500 lx. IES offers different recommended illuminance levels for various shops and tasks in educational facilities when the user group is between 25 and 65 years of age. Another aspect of lighting design in shops that does not exist to the same extent in other school environments is that of safety lighting (see Table 19.3, retained from an earlier IES report for its emphasis on safety).

(b) Music Rooms

The illuminance targets from IES again provide recommendations for both horizontal (300 lx at 1.22 m, 25–65 visual age of observer) and vertical surface illuminance (200 lx at 1.22 m, 25–65 visual age of observer). Specific applications will depend on the orientation of the music scores, which are normally held in the vertical position. Direct–indirect luminaires provide a high degree of diffusion that is helpful in this regard. For all spaces with vertical tasks, a computer

Figure 19.7 *Computer-generated image of a 24-ft-W × 28-ft-L [7.3-m by 8.5-m] classroom using a perimeter layout of 16 4-ft [1.2-m] luminaires in a 14-ft × 22-ft [4.3-m by 6.7-m] rectangle suspended 18 in. (457 mm) from the ceiling and 9 ft [2.7 m] above the finished floor (AFF). The luminaires use T8 triphosphor lamps and high-frequency electronic ballasts. Average horizontal luminance on the desks is 74 fc [800 lx], and average vertical luminance on the chalkboards is 39 fc [420 lx]. Room power density is 1.8 W/ft² [19 W/m²]. (Courtesy of Zumtobel-STAFF Lighting Co.)*

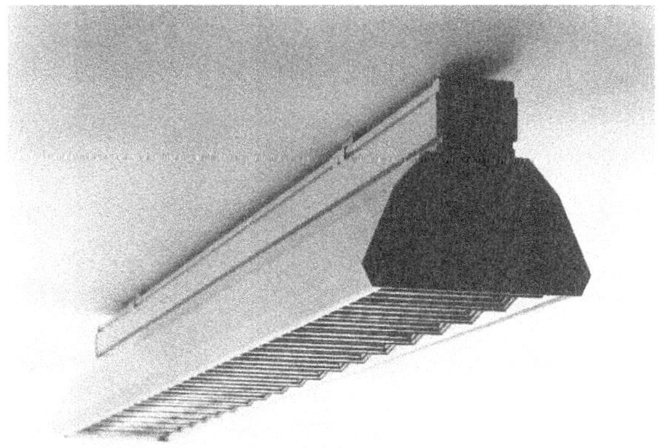

(a)

ZX-ST 1/32W T8
BIVERGENT STEPPED VANE LOUVER

Total Luminaire Efficiency 65%

0% Uplight	100% Downlight	
Spacing Criteria		
Lateral Plane	0	90
	1.2	1.5

TOTAL LAMP LUMENS = 2900
INPUT WATTS = 32

Candela Distribution

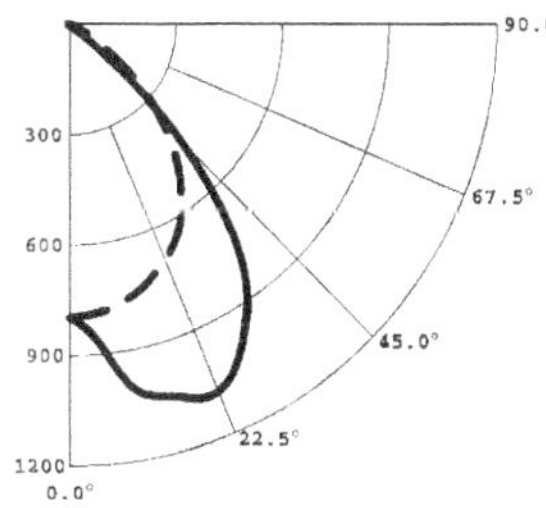

Vertical Angle	Horizontal Angle 0	45	90	Zonal Lumens
0	797	797	797	
5	790	846	861	82.2
15	748	929	1043	256.4
25	670	917	1059	412.3
35	541	851	861	486.6
45	363	590	425	400.2
55	170	245	29	184.6
65	53	42	4	45.9
75	20	13	2	15.5
85	6	0	1	3.6
90	1	0	0	

Luminance Data in Candela / Sq. Meter

Angle in Vertical	Average 0	Average 45	Average 90
45	2763	4491	3235
55	1595	2299	272
65	675	535	51
75	416	270	42
85	371	0	62

Coefficients of Utilization

Effective Floor Cavity Reflectance = 20%

pcc	0.8				0.7				0.5			0.3		
pw	0.7	0.5	0.3	0.1	0.7	0.5	0.3	0.1	0.5	0.3	0.1	0.5	0.3	0.1
0	78	78	78	78	76	76	76	76	73	73	73	69	69	69
1	73	71	69	67	71	69	67	66	67	65	64	64	63	62
2	68	64	61	58	67	63	60	57	61	58	56	59	56	55
3	63	58	54	50	62	57	53	50	55	52	49	53	51	48
4	59	52	48	44	57	52	47	44	50	46	43	49	45	43
5	55	48	43	39	53	47	42	39	46	42	39	45	41	38
6	51	44	39	35	50	43	38	35	42	38	35	41	37	34
7	48	40	35	31	46	39	35	31	38	34	31	38	34	31
8	44	37	32	28	43	36	31	28	35	31	28	35	31	28
9	42	34	29	26	41	33	29	26	33	28	26	32	28	25

(b)

Figure 19.8 *(a) This luminaire uses a very-high-reflectance silvered reflector and an unusual stepped mirrored baffle that provide low brightness for direct glare control, a crosswise batwing distribution for reflected glare control, and a sparkle on each baffle step for interest. (b) Photometrics of the luminaire shown in (a). (Courtesy of Zumtobel-STAFF Lighting Co.)*

TABLE 19.3 Illuminance Levels for Safety[a]

Hazard Requiring Visual Detection:	Slight	Slight	High	High
Normal Activity Level:	**Low**	**High**	**Low**	**High**
Areas	Normal classrooms Lounges Small offices Dorm rooms Washrooms	Shops Business classrooms Large offices Corridors Drafting rooms Lecture rooms Large classrooms Parking area Exterior walkways	Stairs Libraries Reading rooms Cafeteria Swimming pools Locker rooms Interior sports Bleachers	Boiler rooms Auditoriums Exits Exitways Laboratories Kitchens
Lux [footcandles]	5.4 [0.5]	11 [1.0]	22 [2.0]	5.4 [0.5]

Source: IES RP-3-88 Recommended Practice: Lighting for Educational Facilities (IES, 1988); reproduced with permission.

[a]Minimum illuminance for safety or personnel, absolute minimum at any time and at any location where safety is related to seeing conditions.

analysis of the design is recommended to determine compliance with recommendations.

(c) Art Rooms

The primary requirement here is for constant-color daylight. Thus, north windows and skylights are highly desirable. For electric lighting, because color is so important, high-CRI LED or fluorescent lamps are required. General illumination should be augmented by user-adjustable supplementary lighting. The illuminance targets are relatively higher for art and drafting room for both horizontal (500 lx at 0.76 m, 25–65 visual age of observer) and vertical surface illuminance

Figure 19.9 Art exhibition room illustrating good and bad lighting techniques. The upper wall fenestration is excellent for deep daylight penetration. Track lighting is ideal for display of art. The use of incandescent downlighting for general lighting is excessive and an eyesore. Also, the positioning of the track fixtures can create both direct and reflected glare problems and annoying shadows unless the sources are selected properly and ceiling height is at least 10 ft [3.1 m].

(300 lx at 1.22 m, 25–65 visual age of observer). If use of models is anticipated, adjustable accent fixtures are advisable. For display of artwork, adjustable wall illumination is required. Ceiling track-mounted units are an excellent choice (see Figures 19.2 and 19.9).

19.11 ASSEMBLY ROOMS, AUDITORIUMS, AND MULTIPURPOSE SPACES

The varied activities in these rooms make flexible lighting imperative. For performances, low-level dimmed incandescent lighting is required. Here, incandescent lamps are the recommended source because of the lower cost of dimming and the short burning periods. For assembly rooms, this can be augmented by architectural elements along walls and draperies and in the ceiling. For study, additional ceiling fluorescents or HID units can be switched on. The possible combinations are numerous; the different usages are the critical concern (Figures 19.10 and 19.11). Acoustic concerns are acute because of the low NC criteria. Thus, electronic ballasts for HID sources as well as fluorescents should be used.

Figure 19.10 Schools frequently utilize spaces for multiple functions. This space, typically used as a banquet room, doubles as an assembly room. High-intensity, recessed downlights in combination with large pendants provide sufficient light for both uses. (© Karen Tse; used with permission.)

An additional consideration is step lighting. These units should be mounted to the side, or in risers, to illuminate the tread, and particularly its leading edge. Stage lighting is too highly specialized to be discussed here. Some form of stage lighting is required, however,

Figure 19.11 Institutional cafeteria illuminated by cove lighting and daylight, accented with downlights and suspended pendants in a double-height space. Lighting is even, glare-free, soft in quality, and pleasant, yet of sufficient intensity to permit using the cafeteria as a working-meeting space. (© Karen Tse; used with permission.)

and the building designer is advised to consult a theater lighting designer.

19.12 GYMNASIUMS

Gyms present a situation similar to auditoriums in that they have widely varying usages. All fixtures should be sturdy and guarded. LED, phosphor-coated mercury, HPS, and high-CRI metal-halide are excellent choices for color, life, control, and efficacy. Multiple lighting levels should be available by switching or dimming. For dance and social events requiring low-level general lighting, other fixtures can be provided with long-life incandescent or tungsten-halogen lamps, which provide good color for low-intensity lighting as well as illumination during HID startup or restart after an outage. Illumination targets are typically based on the floor level. IES recommends 300 lx. All fixtures should be designed for relamping from the floor. Locker rooms should use guarded-strip fluorescents.

19.13 LECTURE HALLS

Lecture hall lighting is similar, with respect to sources and other considerations, to illumination for classrooms. However, lighting requirements may vary based on a wide range of tasks that might occur within the space. As such, adjustable-level lighting is necessary for demonstrations, videos, and the like—and auxiliary lighting for demonstration tables and chalkboards completes the design. High-ceiling installations can provide general lighting. Controls should be located at the demonstration table (Figure 19.12).

19.14 LABORATORIES

Laboratories differ from classrooms in that tables are fixed, bench surfaces are frequently very dark, many

Figure 19.12 Typical lecture room lighting utilizes 45° cutoff baffled parabolic reflector troffers for minimum direct and reflected glare, adjustable track lighting for demonstration table illumination, and an asymmetric reflector for chalkboard lighting. The large display area on the front wall of the room will not experience veiling reflections with the illustrated lighting arrangement.

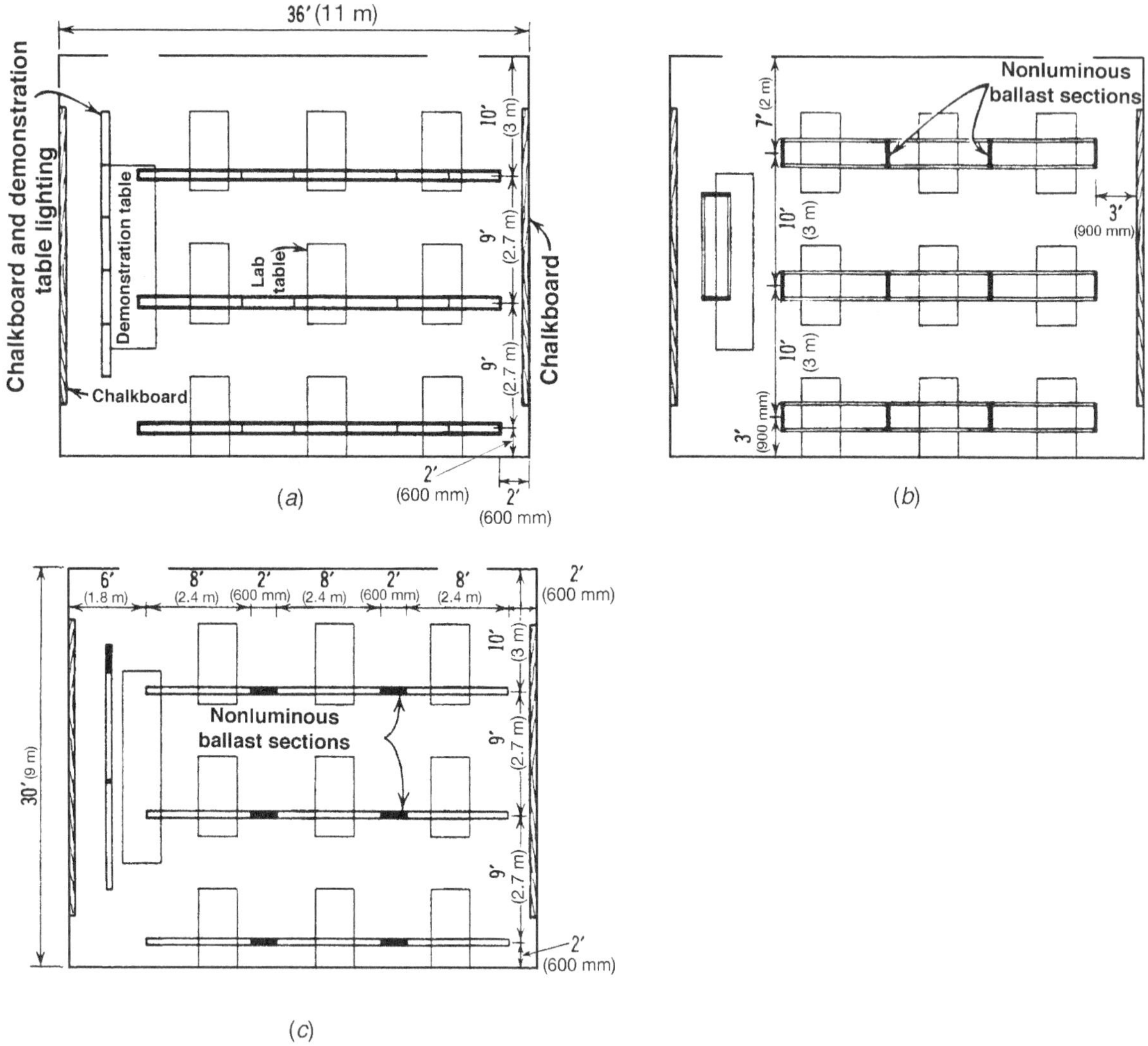

Figure 19.13 *Laboratory lighting schemes. Running fixtures across tables or in aisles is preferable to fixtures in the transverse direction, in terms of reflected glare. (a) Pendant direct–indirect units. (b, c) Variations of the single semi-direct high-output (HO) design.*

of the items used exhibit specular reflection, vertical surface illumination is important, and visual tasks are not normally prolonged or severe. With low ceilings, use direct fixtures with an uplight component run crosswise to the tables. Luminaires with a batwing distribution minimize reflected glare from specular equipment. If the ceiling height is sufficient, indirect lighting is highly desirable for the same reason. Indirect lighting also provides a high degree of diffuseness necessary for vertical surface illumination. See Figure 19.13 for suggested layouts.

19.15 LIBRARIES

Libraries comprise several different seeing tasks, each of which requires its own lighting solution.

(a) General Reading Room

Here two solutions are possible, and both are commonly used. In the first, general lighting is supplied over the entire area, which is sufficient for reading tasks. For this purpose, LED, fluorescent or metal-halide sources are normally applicable, the latter with ceiling heights of at least 10 ft (3 m). The long life and high efficacy of these sources are suited to the long burning hours found in libraries. The second and more energy-efficient solution involves low-level general lighting supplemented by local reading lighting on the tables or in carrels. This solution is consonant with task-lighting orientation and is preferred. IES recommends no less than 500 lx for task lighting reading rooms. Reading lighting should be user-adjustable, if possible, and arranged to avoid veiling reflections when not user-adjustable.

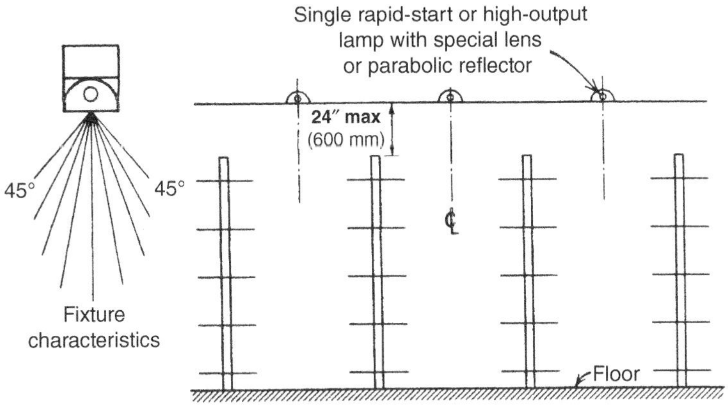

Figure 19.14 Stack lighting is best accomplished by fixtures with lenses specifically designed for the purpose. Fixtures with baffles and plastic diffusers generally do not give adequate vertical surface illumination.

Wherever HID sources are used, an instant-restart source must be available to supply minimal lighting after an outage. Many commercial HID luminaires contain a small tungsten-halogen source for this purpose. Ballast noise can be a problem in low-NC spaces such as libraries. Electronic ballasts for fluorescent and HID sources are available and should be employed.

(b) Stack Areas

Here the required vertical surface illumination is best supplied by specialty units designed for this purpose, mounted between stacks and no higher than 24 in. (0.6 m) above them, for best results (Figure 19.14).

For additional guidance on lighting criteria for libraries, see *IES RP-4-20 Recommended Practice: Lighting Library Spaces* (IES, 2020c).

19.16 SPECIAL AREAS

Most schools contain areas devoted to functions not covered previously. Some lighting recommendations for a few of these areas are as follows:

1. Spaces where color rendering and color matching are important, such as sewing rooms and textile and artwork spaces, must be lighted with particular attention to illuminant color. Consistent color rendering without disturbing metamerisms requires a continuous-spectrum source such as incandescent (including tungsten-halogen) and, of course, daylight.
2. Food service areas must be well lighted to emphasize cleanliness and food attractiveness. Color rendering of food is particularly important to enhance its appetizing appearance. Use LED or 3100 K fluorescent or metal-halide lamps in preparation and serving areas, and eating areas, but in the latter at lower illuminance levels.
3. Medical attention spaces require high-illuminance, good color rendering lighting for diagnosis and conventional office lighting for records and desk tasks.
4. Offices, storage spaces, industrial spaces, and outdoor facilities have the same requirements as similar spaces in other buildings and are covered in the discussion that follows.
5. Corridors and stairways in all types of buildings require special lighting. Corridors intended only for circulation need be lighted to only ±10 fc (100 lx) unless a specific seeing task requires higher levels—for example, bulletin boards and lockers (Figure 19.15). Lighting can also be used to give direction by longitudinal arrangement. Wall-mounted or recessed wall lighting is particularly effective in corridors, providing walking illumination plus lighting for posters, bulletins, and so on. Luminaires mounted across corridors, particularly when corridors are long, are effective in reducing the "tunnel" impression. Incandescent sources are not recommended because of their low efficacy, high maintenance, and frequency of relamping. LED, fluorescent and HID sources are also suggested for stairwells. Care must be exercised here, however, to avoid direct glare, which causes attention to shift from the stairs to the light source and may thereby create a hazard.

19.17 OTHER CONSIDERATIONS IN SCHOOL LIGHTING

(a) Controls

Because schools operate for the most part on fixed time schedules and during daylight hours, lighting controls of the preprogrammed time-base and daylight compensation types are readily applicable. Other energy-conserving control strategies can be utilized as applicable.

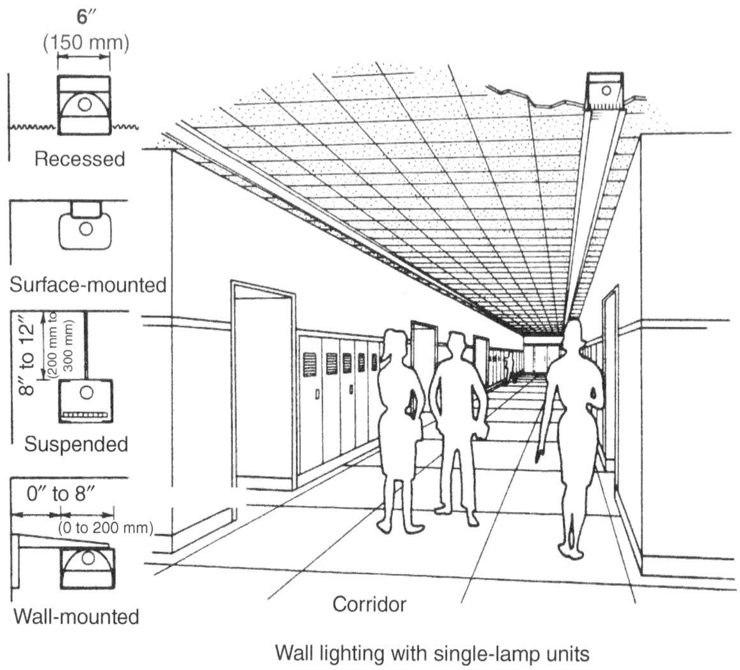

Figure 19.15 *Lighting of school corridors. High-reflectance walls, floor, and ceiling improve utilization of light and increase the feeling of cheerfulness. The lighting technique illustrated is appropriate for school corridors. The rows of luminaires at each sidewall illuminate bulletin boards, special displays, and the faces and interiors of lockers more effectively than do units centered in the ceiling.*

(b) Safety Lighting

See Table 19.3 for recommended illuminance levels. As pointed out, designers must be aware of the requirements of all jurisdictional codes.

(c) Emergency Lighting

Local, state, and National Fire Protection Association (NFPA) model codes establish minimum requirements. Exits must be clearly identified with lighted signs and a lighted path to these exits provided. As with safety lighting, it is not sufficient that average illuminance meets requirements. Illuminance in any given area must be free of large level differences, which can cause disabling glare, particularly in view of the relatively high adaptation levels of occupants' eyes immediately before a lighting outage.

COMMERCIAL OCCUPANCIES

19.18 GENERAL INFORMATION

The following information applies primarily to offices in commercial buildings and secondarily to similar spaces in other occupancies, such as educational and industrial buildings. In the latter cases, the general remarks applicable to facilities of those types take precedence. The special problems associated with digital displays are discussed in Section 19.19. Task-ambient (nonuniform) lighting is covered in Sections 19.20, 19.21, and 19.22. The reader is referred to *IES RP-1-22 Recommended Practice: Lighting Office Space* (IES, 2022) for a full discussion of office lighting.

(a) Light Sources

In the interest of energy economy and good color, LED, T8 3500 to 4000 K triphosphor linear lamps, or equivalent CFL units are recommended, along with high-frequency electronic ballasts. HID can be used in indirect installations with sufficient ceiling height (minimum 9.5 ft [2.9 m] clear). Color-corrected HPS and metal-halide units of high CRI are both suitable.

Source color must be coordinated with the color scheme of room surfaces and furnishings. In areas with a large daylight contribution, the source CCT should be at least 4000 K. Incandescents may be used for storage areas, closets, and other short-burning-period uses, although more efficient options exist. LED and tungsten-halogen track lighting is used to advantage to illuminate displays of all sorts.

(b) Illuminance Levels

These are discussed in Chapter 18 for the particular type of activity involved. See also *IES Standard*

RP-1-22. General recommended reflectances for room surfaces are:

Ceiling	80% minimum
Walls	50–70%
Partitions	50–70%
Floor	20–40%
Desktops, furniture	25–45%
Window blinds	40–60%

In open offices with workstation "cubicles," the typical half-height partitions block 30% to 80% of the light from ceiling luminaires, depending on the furniture arrangement. It is, therefore, all the more important that partition finishes (including fabrics) be light-colored. It is also desirable to have upper-wall sections painted to match the ceiling—that is, with a lighter finish than the remainder of the wall. This serves the dual function of increasing ceiling cavity brightness, particularly with suspended fixtures, and increasing vertical illumination due to reflection from this surface.

(c) Vertical Surface Illumination

This is required for many visual tasks in offices, such as those involving files, desk drawers, card files, and copy stands. Large area luminaires and a high degree of diffuseness are desirable. This is especially true in large offices where wall reflections are absent. Light-finished furniture surfaces, luminaires that yield an illuminated ceiling, and high-reflectance floors are also helpful.

19.19 AREAS WITH DIGITAL DISPLAYS

These devices, which include desktops, laptops, tablets, and so on, are standard office items. While some devices use matte, nonglare surfaces, there are still digital displays that have specular faces—creating a special problem for office lighting that can become the deciding consideration in selecting the lighting system. Simply stated, with such specular faces, the primary problem is to avoid reflection on the screen from any luminous source in the area, including luminaires, windows, illuminated walls, and even light-colored clothing. Any such reflection makes reading data on the screen difficult and sometimes impossible.

Generally, walls, floors, and furniture should be finished in low-chroma, low-value colors, with a maximum reflectance of 50%. Very dark and very light desktops are to be avoided because of excessive luminance ratios, and the latter also because of reflections. In general, the standard recommended maximum luminance ratios of 1:3 in the near field and 1:10 in the far field should be the design goal here as well.

19.20 OFFICE LIGHTING GUIDELINES

(a) Private Offices

In these spaces, a task-ambient approach is frequently appropriate because there is usually only one primary visual task location, with the remainder of the space devoted to circulation and storage. Often, sufficient illumination for the latter is supplied by spill light from the task area, particularly when the task fixture is a pendant unit with an uplight component, as seen in Figure 19.16. In large rooms, provide downlighting in sitting areas and some type of wall illumination using wallwashers, sconces, or a recessed perimeter unit to brighten the often dark wood-paneled walls. Accent lighting is required for pictures and other displays.

(b) General Offices

The two basic approaches to general office lighting are a uniform layout that provides task-level lighting in the entire area or a task-ambient

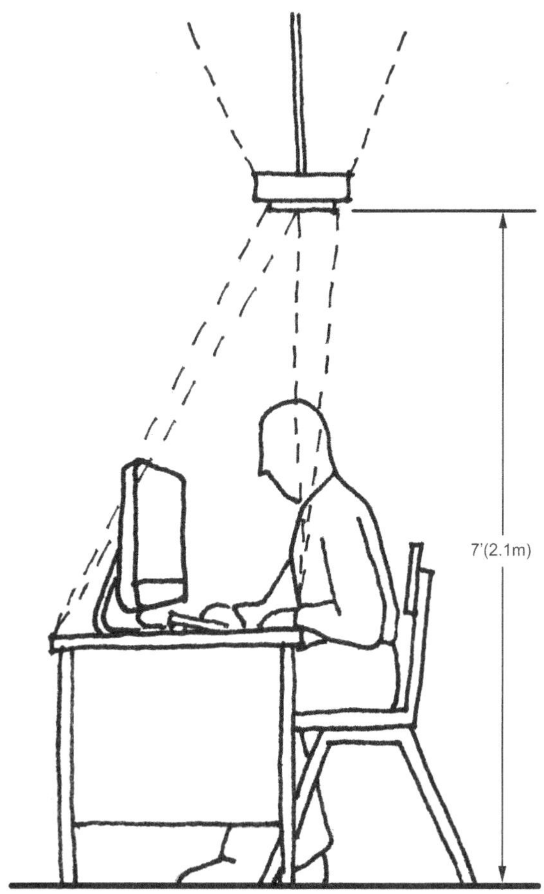

Figure 19.16 *Geometry of a pendant task-ambient fixture. This unit is intended to supply not only task illumination but also a measure of ambient lighting via its 20% uplight component. In a small room, this component is sufficient for general circulation. When installed as shown, veiling reflections are minimized and glare on the digital display is essentially eliminated.*

design. The former, discussed in this section, is most appropriate to speculative-type construction, where the furniture layout is unknown but a complete job is required. (Frequently, however, the construction contractor provides only sufficient electric power for lighting and miscellaneous power; the lighting for the space is designed after the space has been rented as "tenant" work. In such cases, a task-ambient design is possible because tenant work is tailored to the user's needs.)

A task-illuminance overall layout is wasteful of energy and increases building energy budgets. Therefore, a uniform layout, when used, should be designed so that levels can be lowered easily in areas not requiring task lighting. Furthermore, because no control of the geometry of direct and reflected glare is possible in such layouts, the luminaire selected must give minimum glare in all viewing directions. The system that best meets these requirements is one that uses a large number of low-brightness luminaires in a dense layout. This is a very expensive solution, however, unless the ceiling is either modular or coffered.

An interesting luminaire that creates a direct–indirect low-brightness unit in a troffer enclosure is shown in Figure 19.17. Where the viewing direction is established but the exact furniture layout is unknown, a uniform layout with low-brightness parabolic reflectors with a batwing characteristic can be used to advantage, although a batwing distribution is generally not sought because of high direct glare and strong veiling reflections. Luminaire selection criteria for digital display areas were given in the preceding section. Where digital displays are not the overriding consideration, the following criteria are helpful:

1. In terms of appearance, a 2 ft × 4 ft [0.6 m × 1.2 m] fixture is appropriate in a 2 ft × 2 ft [0.6 m × 0.6 m] tile ceiling; a 1 × 4 ft [0.3 m × 1.2 m] fixture is suitable in a 1 ft × 1 ft [0.3 m × 0.3 m] ceiling; and a square 2 ft× 2 ft [0.6 m × 0.6 m] fixture is suitable in all tile patterns.
2. Two-inch-deep parabolic louvers permit a shallow fixture, but lamp shielding is poor. A 3-in. [76-mm] minimum louver depth is suggested for all large-cell parabolic louver luminaires.
3. Three-lamp parabolic louver units, which are usually too bright for digital display areas, can be used to advantage. Two-lamp units give better uniformity and VCP at a premium price.
4. Two-foot-square deep cell parabolics with 31-W T8 U lamps or biaxial compact lamps are an excellent choice in terms of VCP, uniformity of illumination, and appearance.
5. Premium-quality specular parabolic louvers that eliminate hot (bright) spots are available.

(a) (b)

Figure 19.17 Luminaire construction giving a direct–indirect appearance (a) using a direct troffer construction (b). The medium-brightness luminaire interior reflects light, giving a deep, airy appearance. The translucent lamp enclosure maintains an acceptable luminance against the background (interior) reflector. The photometric characteristic is essentially circular in both longitudinal and transverse planes. (Courtesy of Zumtobel-STAFF Lighting Co.)

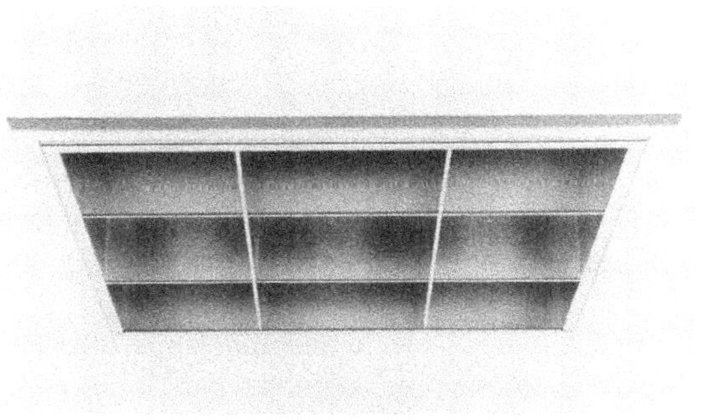

Figure 19.18 One-foot-square [305-mm-square], 9-cell luminaire with deep parabolic louvers. This unit is applicable to small offices and/or areas with 1 ft × 1 ft [305 mm × 305 mm] ceiling-tile arrangements. (Courtesy of Zumtobel-STAFF Lighting Co.)

6. In relatively small offices with 1 ft × 1 ft [0.3 m × 0.3 m] ceiling tiles, 1-ft^2 [nominal 0.09-m^2] troffers with deep miniature parabolic wedge louvers and 16-W U lamps or compact fluorescents are an excellent all-purpose lighting solution (Figure 19.18).
7. Direct–indirect shallow luminaires with a wide, inverted batwing uplight distribution can be pendant mounted as low as 7.5 ft [2.3 m] AFF.

(c) Office Lighting Equipment

Office lighting equipment is generally not treated roughly. Fixtures with touch latches, light hinges, and adjustable devices may be selected without fear of breakage or vandalism.

(d) Maintenance

In most offices, maintenance is provided on a trouble-call basis. Lamps are replaced on burnout, and the fixture is then cleaned. Because of the long life of LED, fluorescent, and HID sources, this generally means a three to five year cleaning cycle. An LLF of 0.65 is reasonable in air-conditioned spaces; a lower LLF is appropriate in open-window offices.

(e) Fenestration

When fenestration is absent, a lighted valance around the room is recommended. This removes the wall-ceiling line and partially compensates for the lack of windows. It also brightens the walls and increases illumination on desks placed adjacent to the walls.

(f) Control

The control strategy (see Chapter 17) should minimally provide (after tuning):

1. Daylight compensation.
2. The possibility of operating individual small groups of fixtures while the remainder are off, to permit off-hours work.
3. Path lighting through large spaces to permit traversing without turning on all lighting.
4. Careful scheduling with supervised local override.

With these general guidelines in mind, the following sections discuss specific topics in office lighting.

19.21 TASK-AMBIENT OFFICE LIGHTING USING CEILING-MOUNTED UNITS

Efficient office lighting, like other work-area lighting, is often predicated on a task-ambient design. This approach has the advantages of minimum contrast reduction at the task and minimum power use, the latter resulting in low operating cost. The method involves designing a uniform ceiling layout that provides low-level lighting for circulation and miscellaneous easy visual tasks, plus workstation-mounted task lighting. The latter can either be integral with the furniture (see the following section) or a separate, generally adjustable unit on or adjacent to the desk. The former is particularly useful for very severe seeing tasks when adjustability of angle and distance between light and work is vital. Integral furniture units are usually not adjustable, and are therefore limited to a maximum of about 70 footcandles [750 lx] on the task. Higher levels would generate excessive heat and, almost certainly, glare.

19.22 TASK-AMBIENT OFFICE LIGHTING USING FURNITURE-INTEGRATED LUMINAIRES

In lieu of ceiling-mounted luminaires to provide ambient lighting, indirect fixtures can be mounted on the top of furniture or can be free-standing. Task lighting would be integral with the office furniture. Advantages of this arrangement include the following:

1. The problem of furniture layout and layout changes is eliminated.
2. Initial construction cost is reduced.
3. Energy requirements are lowered because of short distances between light source and task.
4. Each occupant has on/off control of his or her task lighting, including, in some designs, positioning control.

5. Maintenance is greatly simplified because fixtures are readily accessible from the floor.
6. Floor-to-floor height can frequently be reduced.
7. Tax advantages normally accrue due to higher depreciation rates on furniture than on a building.

Disadvantages include the following:

1. Difficulty in dissipating heat and minimizing magnetic ballast noise due to proximity of sources to the user. For this reason, electronic ballasts should be used.
2. Veiling reflections are always present and can be severe.
3. Luminance ratios in the near and far surround may exceed the recommended levels.
4. There may be difficulty in lighting a free-standing open desk, because most of the fixture types are undercounter or sidewall mounted.
5. The concentrating nature of the lighting units may cause difficulty in evenly lighting large tables or L-shaped desk areas.
6. This arrangement is not readily applicable to automatic switching and dimming schemes.
7. For satisfactory operation with fixed task lighting, the desk must be usable by both left- and right-handed people. This requires a considerable degree of duplication, as shown by lamps on both sides of the drawing in Figure 19.19.

Figure 19.19 graphically shows the problem of local desk lighting. Experience suggests that only systems that permit user positioning adjustment of the light source have a high degree of user acceptability.

19.23 INTEGRATED AND MODULAR CEILINGS

The cost, appearance, and design-flexibility advantages of an integrated ceiling design over field-assembled and coordinated systems have long been known. As a result, ceiling systems with integrated lighting, acoustic control, and air-handling capabilities are commercially available in modular sizes (including 60-in. [1.5-m] square, 48-in. [1.2-m] square, and 30 in. × 60 in. [762 mm × 1524 mm]). Modules are made in flat and pyramidal shapes, the latter having several distinct advantages:

1. These modules are more interesting and aesthetically pleasing.
2. They offer more acoustic absorbency due to ceiling angles and a larger surface area.
3. The recessed center provides visual baffling, permitting the use of higher-brightness sources while maintaining high VCP.

Possible luminaire arrangements for both flat and pyramidal shapes are given in Figure 19.20. In addition to the design flexibility available, an electrified track can be integrated into the system runners to supply both the lighting fixtures and power poles.

19.24 LIGHTING AND AIR CONDITIONING

The reduction of lighting power densities to below 1 W/ft^2 [11 W/m^2] in all but special areas considerably reduces the impact of lighting-generated heat on a

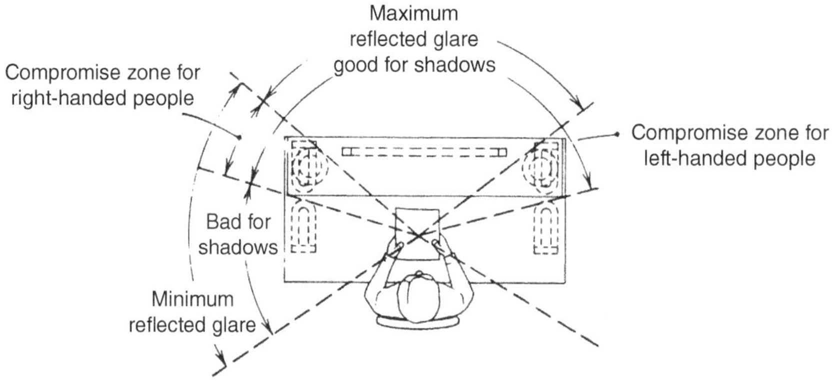

Figure 19.19 *Furniture-mounted task lighting can create severe reflected glare or hand shadows if the luminaire location is not selected carefully. Further constraints are location in elevation (for shielding, needs to be low; for good distribution, needs to be high), appearance considerations, and compatibility with a worker's physical movements. Light source position adjustability removes most user objections to this type of task lighting.*

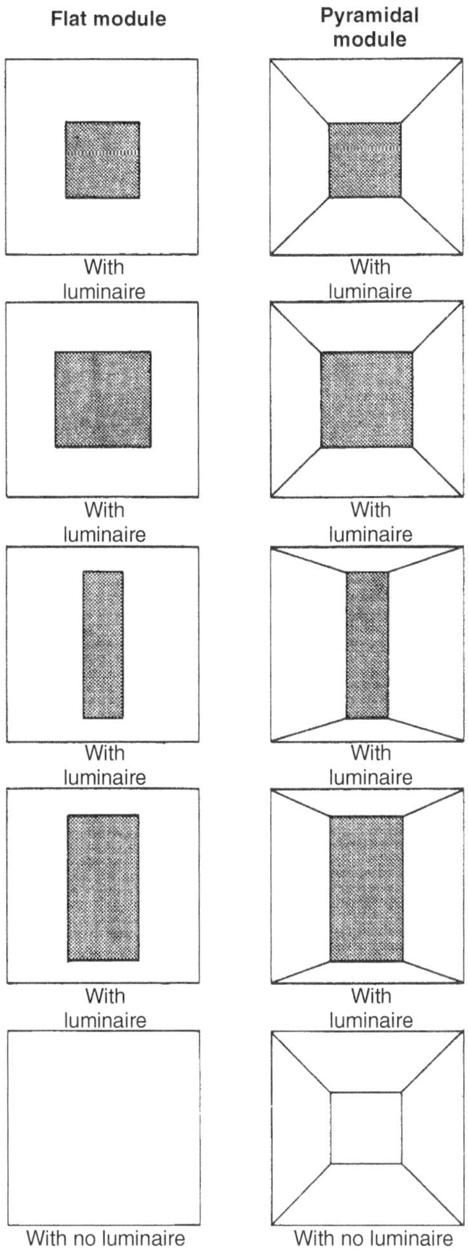

Figure 19.20 *Reflected ceiling plans showing various configurations of ceiling modules.*

building's heating, ventilation, and air-conditioning (HVAC) system. In non-air-conditioned buildings, the lighting heat contribution is partially applicable to building heating. Fixture efficiency is directly affected by its temperature. Fluorescent units operate at an optimum temperature of 77°F [25°C]. Temperatures above and below this decrease output and fixture efficiency. Thus, heat removal from units is desirable even at low lighting-energy levels. The most effective method of fixture heat removal is return air duct connection to the unit itself. This method is relatively expensive and immobilizes the fixture. Alternatively, the plenum can be exhausted with return air passing over the fixtures, picking up heat.

INDUSTRIAL LIGHTING

19.25 GENERAL INFORMATION

In industrial lighting, a primary design consideration is cost. Given acceptable standards of comfort and safety for the working staff, additional costs for lighting must be self-justifying economically. In one case, a good lighting installation was improved at considerable cost. Production jumped 15%, of which 3% was sufficient to amortize the cost of the lighting alteration. In another case, an outlay for new inspection lighting reduced product failures and proved economically sound. In a third, improved lighting reduced accidents, improved employee morale, and consequently improved production. The cases studied are far too numerous to mention; general principles are adduced instead.

19.26 LEVELS AND SOURCES

Illuminance requirements are detailed in *IES RP-7-21 Recommended Practice: Lighting Industrial Facilities* (IES, 2021). The guide focuses on establishing optimal lighting conditions to support the unique lighting needs of industrial spaces, which helps support safety, productivity, and visual comfort.

In general, with regard to industrial spaces, where levels higher than 50–75 fc [540–800 lx] are required, general illumination should be supplemented by task lighting. Industrial facilities lend themselves readily to daylighting because many are one-story structures. Thus, roof monitors, skylights, and clerestories are readily applicable and extremely desirable. However, inasmuch as industrial facilities are frequently sited in industrial areas with attendant heavy atmospheric soot and dirt, a frequent cleaning and maintenance program is necessary if the LLF is to be kept at reasonable levels. This observation is also applicable to indoor lighting systems.

Light sources for industrial applications should exhibit high efficacy, good lumen maintenance, and long life. Of the sources available today, LED, fluorescent, and HID lamps meet these criteria. Induction and sulfur lamps may join this group after additional development and field testing. Where color is not critical, HPS is a potential source. Adaptation to its warm yellow color is rapid, and if it is mixed with metal-halide or mercury sources, no problem should be encountered. HID lamps are easier to maintain, store, clean, and relamp than fluorescent lamps and have equal or better efficacy, but have the disadvantages of delay and lower output on restrike. Because of their relatively small size, HID sources are used in focusing reflectors that produce luminous intensity distribution characteristics designed for specific illumination objectives. Thus, HID sources are generally used for high-bay (>25 ft [7.6 m]) and medium-bay (15–25 ft [4.6–7.6 m]) installations. For low-bay lighting, appropriate LED fixtures, industrial reflector fluorescent luminaires, and low-bay HID reflectors are both applicable.

One of the most common industrial lighting tasks is warehouse aisle lighting that must provide adequate vertical surface illumination on racks on both sides of an aisle. The required candlepower (cp) distribution to perform this task efficiently is a modified batwing curve. This is more easily accomplished in low-bay lighting with a continuous row LED or fluorescent installation than with discrete HID units.

19.27 LUMINANCE RATIOS

For reasons explained in preceding sections, luminance ratios must be controlled. In many situations it is difficult to control the surrounding brightness. Ceilings, which frequently are covered with piping, ducts, and other equipment, should be light. Therefore, this mechanical equipment must be painted with matte, light unsaturated colors; maintenance and cleaning must be good; and fixtures should have an upward component of light to avoid more than a 20:1 ratio of task-to-ceiling luminance.

Use of bright saturated colors for general surface painting should be avoided because they draw attention and frequently have special significance. In addition to color-coded piping (banding is preferable), red frequently means fire equipment; green, first aid; orange, danger; and so on. White is also to be avoided, being excessively bright and susceptible to dirt. Recommended minimum reflectances are:

Ceiling	75–85%
Walls	40–60%
Equipment	25–45%
Floors	20%

19.28 INDUSTRIAL LIGHTING GLARE

The problem of direct glare can be acute in low-bay installations, and that of reflected glare in high-bay designs, when either uses a point source. One method of reducing direct glare is the use of low-brightness prismatic lens units with a black reflector behind the lens. Methods of minimizing veiling reflection from all sources were discussed previously.

19.29 INDUSTRIAL LIGHTING EQUIPMENT

The cost of maintenance increases with labor rates. For this reason, high-quality lighting equipment yields the lowest owning and life-cycle costs. For instance, the cost of replacing a ballast for HID lighting units frequently exceeds the cost of the ballast. It is thus more economical to utilize long-life, high-quality ballasts, particularly where luminaires are not readily accessible.

Other suggestions for lowering costs, both initial and operating, include using ventilated luminaires that tend to be self-cleaning by convection (Figure 19.21), in addition to giving the needed upward light component; using bus-mounted fixtures for rapid installation and repair; using lowering mechanisms on high-bay units, to avoid catwalk or platform relamping with a concomitant extremely high cost; using fixtures arranged for "stick" relamping from the floor in medium- and low-bay work; and generally incorporating modern equipment into the plant.

Proper maintenance is of paramount importance in industrial facilities because of the prevalence of dirt, vibration, and the overall wear-and-tear of the industrial environment. Maintenance includes cleaning, relamping, inspection, and preventive maintenance. Relamping on a burnout basis is extremely

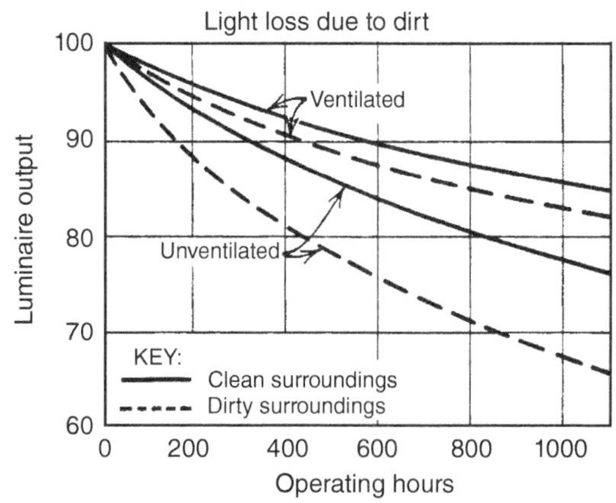

Figure 19.21 Graph demonstrating the advantage of ventilated fixtures.

uneconomical because of disruption of production and lowered production due to lumen depreciation before burnout. Relamping should be done on a planned group basis. Similarly, if the specific facility has a high dirt accumulation rate, cleaning must also be done on a planned group basis rather than only at relamping time.

Ballast noise, including the high levels of HID ballasts, is not usually a factor in industrial facilities because of high ambient noise. In relatively quiet installations and/or where fluorescent fixtures are mounted a short distance above a work bench—as in inspection and fine assembly—this is not true, and electronic ballasts are needed.

19.30 VERTICAL-SURFACE ILLUMINATION

In industrial facilities more than any other occupancy, the illumination of vertical surfaces is crucial. This is a result of the nature of the work: machines, storage, gauges, and so on, all require high vertical-surface illuminance (Figure 19.22). The illuminance on a vertical surface is the result of the horizontal component of the lighting. This is

$$\text{fc} = \frac{cd}{D^2}\sin\theta = \frac{cp \times R}{D^3} = \frac{cp \times \cos^2\theta \times \sin\theta}{H^2}$$

where

fc = illuminance (footcandles)
cd = luminous intensity (candlepower)
D = direct distance (feet), see Figure 19.22
H = vertical component (feet), see Figure 19.22
R = horizontal component (feet), see Figure 19.22

To maximize the horizontal component, we set to zero the derivative of fc with respect to θ. Thus

$$\frac{d\text{fc}}{d\theta} = \frac{cp}{H^2}(-2\cos\theta \times \sin^2\theta \times \cos^3\theta) = 0$$

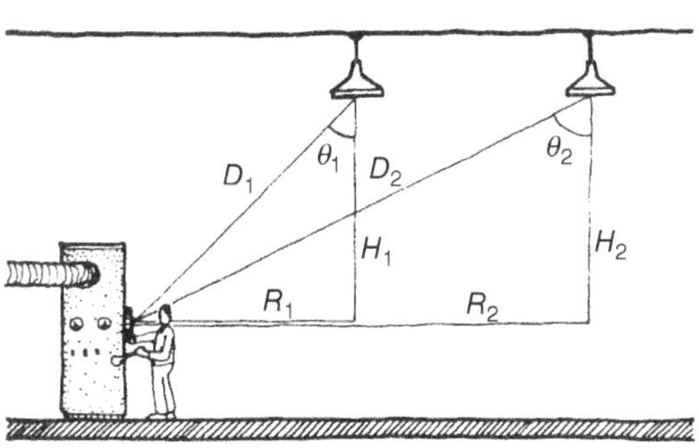

Figure 19.22 Vertical-surface illuminance is maximum where θ is approximately 35° (see derivation in text).

or

$$2\sin^2\theta = \cos^2\theta$$

$$\tan^2\theta = \frac{1}{2}$$

$$\tan\theta = 0.707 = \frac{R}{H}$$

$$\theta > 35^\circ$$

Therefore, maximum vertical illuminance (illumination resulting from the horizontal-lighting component) is obtained when the angle between the fixture's vertical axis and the work is approximately 35°. Hence, we should select a fixture whose candlepower distribution curve demonstrates a high value at that angle. Of course, the derivation is for a single location and fixture. For good vertical and angular illumination over a large area, arrange fixtures with considerable overlap.

SPECIAL LIGHTING APPLICATIONS

19.31 EMERGENCY LIGHTING

Emergency lighting is required when the normal lighting is extinguished, which can occur for any of three reasons:

1. general power failure
2. failure of the building's electrical system
3. interruption of current flow to a lighting unit, even as a result of inadvertent or accidental operation of a switch or circuit disconnect

As a result of the third reason, sensors must be installed at the most localized level—that is, at the lighting fixture (voltage sensor) or in the lighted space (photocell sensor).

(a) Codes and Standards

Because emergency lighting is a safety-related item, it is covered by various codes, several of which may have jurisdiction. In addition, there are widely accepted technical society and industry standards whose recommendations normally exceed the minima required by codes.

1. *NFPA 101: Life Safety Code* (NFPA, 2023a): This code defines the locations within specific types of structures requiring emergency lighting and specifies the level and duration of the lighting.
2. *NFPA 70: National Electrical Code* (*NEC*) (NFPA, 2022): This code deals with system arrangements for emergency light (and power) circuits, including

egress and exit lighting. It discusses power sources and system design.

3. *NFPA 99: Health Care Facilities Code* (NFPA, 2023b): This code deals with special emergency light and power arrangements for these facilities.
4. *OSHA regulations:* These are primarily safety-oriented and, in the area of emergency lighting, discuss primarily exit and egress lighting requirements.
5. *Industry standards:* These include the publications of the IES and the IEEE, in particular, *IEEE Standard 446: Recommended Practice for Emergency and Standby Power Systems for Industrial and Commercial Applications* (IEEE, 1995).

Because codes and standards are constantly being revised and updated, the designer for an actual project must determine which codes have jurisdiction, obtain current editions, and design to fulfill their requirements. The following material provides general information and focuses on good practice but is not intended to take the place of applicable construction and safety codes.

(b) Minimum Illumination Levels and Duration of Emergency Lighting

Most codes and authorities require a minimum average illuminance at floor level, throughout the means of egress, of 1.0 fc [10 lx]. This is considered sufficient, after eyes accommodate, to permit orderly egress. No point along the path of egress should have an illuminance of less than 0.1 fc [1 lx], and the maximum-to-minimum ratio of illuminances along the egress path should not exceed 40:1. The language of the codes is precise, and the consequences must be completely understood in order to properly design an emergency lighting system.

The *NFPA Life Safety Code* relates to life safety—that is, safe egress during an emergency that often involves fire. In such instances, smoke normally obstructs vision at eye level, and it is therefore extremely important that the required illumination be available at floor level. Ceiling-mounted emergency lighting may not illuminate the floor in smoky areas and may even worsen a situation by creating a bright, fog-like condition. It is therefore widely recommended that adequate egress lighting be provided at baseboard level. Some codes already mandate baseboard egress lighting. The requirement that the average minimum illuminance be 1.0 fc [10 lx] along with the mandated minimum of 0.1 fc [1 lx] at any point and a maximum 40:1 ratio of maximum to minimum illuminance, should effectively eliminate faulty design that utilizes several widely separated bright sources.

Other authorities recommend that escape routes be lighted to not less than 1% of their normal illuminance, but in no case less than 0.5 fc [5 lx]. Where the 1% of normal requirement recommendation falls below 1 fc [10 lx] (normal illuminance less than 100 fc [1076 lx]), we suggest the 1 fc [10 lx] minimum required by the *NFPA Life Safety Code*.

Duration of the 1 fc [10 lx] level of emergency lighting is normally specified to be a minimum of 90 minutes (for egress) and 0.6 fc [6 lx] thereafter. Facilities that cannot be evacuated quickly require higher levels for indefinite periods.

(c) Central Battery Emergency Lighting Systems

Generators are normally arranged to replace the utility service during outages, using the building's normal electrical distribution system, with the difference that only essential loads remain connected. This does not meet the requirement that emergency lighting be supplied on power failure, even if the fault is local (i.e., within the building).

Central battery systems are of three types:

1. *Systems supplying low-voltage DC (6–48 V) to dedicated emergency lighting units:* (See Figure 19.23*a*.) This type of central battery installation consists of a battery, charger, and switching equipment. The emergency lighting units are generally not part of the normal lighting system, consisting instead of dedicated incandescent fixtures. The system has the same limitation as a generator, and therefore is not applicable to most emergency lighting situations except when equipped with multiple downstream sensors. As such sensing would be highly uneconomical, this system is limited to a single (large) space with common outage sensing. This arrangement has the further disadvantage that the emergency lighting units may intrude on the building's decor and architecture, inasmuch as they are not part of the normal lighting system. Also, because the lighting units are isolated devices, it is difficult to obtain a satisfactory degree of emergency illumination uniformity, as explained previously. (See also Section 19.31*e*).
2. *Systems supplying 120-V DC:* (See Figure 19.23*b*.) This arrangement supplies line voltage DC to emergency lighting fixtures, which are also used as part of the normal system. HID and fluorescent luminaires are furnished with integral inverters to

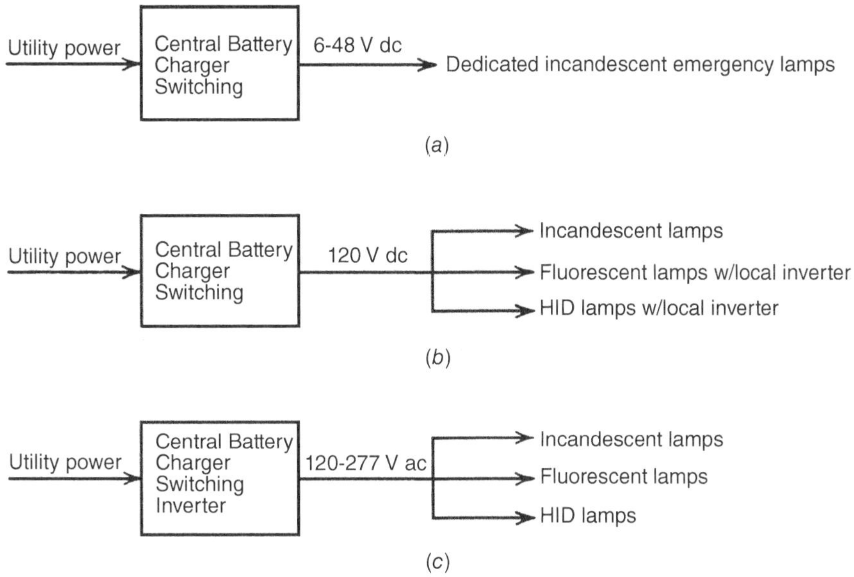

Figure 19.23 *Schematic diagrams of an emergency power system using central batteries. In (a) the emergency system is completely separate from the normal lighting system and utilizes dedicated lighting units. (b) This arrangement furnishes DC to the luminaire, where it is inverted to AC (except for incandescent lamps). The fixtures are utilized during both normal and emergency operation. As switching is instantaneous, HID lamps are not extinguished. (c) An arrangement that is similar to (b) except that a central inverter supplies AC continuously. Here, too, the lamps are used in both normal and emergency modes.*

change the incoming DC to AC. This arrangement has the advantage of a central, well-maintained battery supply, but the disadvantage of requiring a full DC distribution system.

3. *Systems supplying line voltage AC:* (See Figure 19.23*c*.) Here the central unit contains a battery, charger, central DC/AC inverter, and switching equipment. Such a system can supply power to incandescent, fluorescent, and HID systems in spaces where a single voltage-loss sensing point is sufficient. The emergency lighting units are part of the normal lighting system. This arrangement is most frequently used where a distributed system with local batteries (discussed later) would not be readily applicable due to the size of the load, as would be the case with high-wattage HID luminaires. A typical 10-kVA on-line UPS cabinet designed to supply emergency AC power to HID lamps measures approximately 40 in. × 50 in. × 20 in. [1.0 m × 1.3 m × 0.5 m] and weighs about 1000 lb [450 kg].

(d) Distributed (Local) Emergency Lighting Arrangements

As a result of the inability of a central system to respond to a localized outage, a frequently used emergency lighting arrangement is one in which the source of power, as well as all the voltage-sensing and voltage-switching equipment, are installed at the emergency lighting fixture, which is usually part of the normal lighting system. The emergency pack consists of a rechargeable battery and charger, voltage-sensing and voltage-switching equipment, and, in the case of fluorescent lamps, an electronic package (ballast) that operates the lamp at high frequency and generally at reduced output (Figure 19.24). These packages, which are designed to supply the code-required 90 minutes of emergency lighting, are maintenance-free for periods of

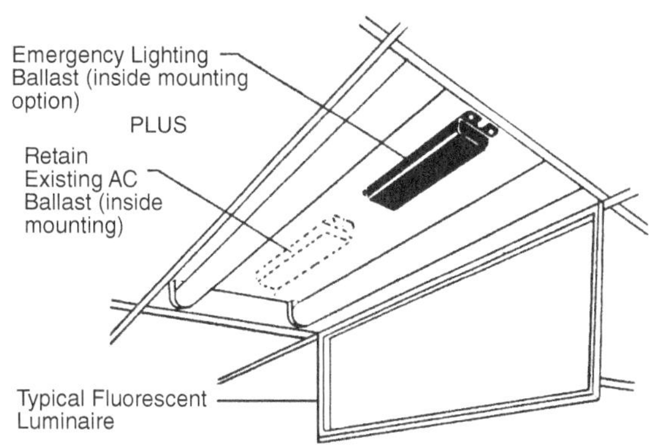

Figure 19.24 *The emergency lighting ballast package contains a 90-minute-capacity high-temperature battery, inverter, sensing and switching equipment, and electronic ballast circuitry to operate the lamp(s) at full or reduced output. The emergency ballast is energized on loss of normal power. The entire package measures 9½ in. × 2⅜ in. × 1½ in. [241 mm × 60 mm × 38 mm]. (Courtesy of Big Beam Emergency System, Inc.)*

(a) (b)

Figure 19.25 Commercial emergency lighting units. (a) This decorative unit contains a sealed, maintenance-free lead–acid battery, an automatic charger, and an automatic low-battery-voltage disconnect. The battery is designed to supply a minimum of 90 minutes of operation of two fully adjustable 5.4-W sealed beam lamps. The unit measures 16.5 in. × 5 in. × 5 in. [419 mm × 127 mm × 127 mm]. (b) Conventional design emergency unit contains a lead-acid or lead-calcium battery, which is housed with its required electronic control and safety equipment in a polycarbonate housing. The lamps are rated 7.2 W each. The entire unit measures approximately 15.75 in. × 11.5 in. × 4 in. [400 mm × 292 mm × 102 mm]; exact dimensions vary with battery type, voltage, and capacity. (Courtesy of Dual-Lite, a division of Hubbell Lighting, Inc.)

(a) (b)

Figure 19.26 Industrial-type emergency lighting units. (a) This unit has a maintenance-free, lead–acid battery. The cabinet conforms to NEMA Type 3R enclosure requirements. The unit can be arranged to supply an additional two remote 12-W sealed-beam heads. The unit measures 13 in. × 6.75 in. × 17 in. over the lamps [330 mm × 172 mm × 432 mm], and is provided with an external charge indicator light and a test switch. (b) This compact thermoplastic polycarbonate enclosure unit is sealed and gasketed for rugged industrial application. It measures approximately 10 in. × 11 in. × 16 in. [254 mm × 280 mm × 406 mm] and contains the usual electronics for control and charging of its lead-acid battery. A charge indicator and a test switch (to simulate power failure) are mounted on the side. (Courtesy of Dual-Lite, a division of Hubbell Lighting, Inc.)

up to five years. A second type of distributed emergency lighting unit is the familiar packaged unit with integral incandescent lamps. Several designs are shown in Figures 19.25 and 19.26.

(e) Emergency Lighting Design Considerations

The levels and duration discussed in Section 19.31*b* refer only to egress lighting. However, general emergency lighting is required to avoid distress and even panic that unfortunately may rapidly ensue. The required emergency illuminance in specific areas should be related to the area's normal illuminance and the degree of hazard in the area. Therefore, we suggest:

Exit area	5 fc [50 lx]
Stair	3.5–5 fc [35–50 lx]
Hazard areas, such as machinery room	2–5 fc [20–50 lx]
Other spaces	1.0 fc [10 lx]

These levels should be essentially uniform. When the illuminance in an interior space drops instantly from 30 to 100 fc [300–1000 lx] to 1–5 fc [10–50 lx], the eyes require up to five minutes to fully accommodate. During this long period, occupants are partially sightless—a condition that lends itself readily to panic. For this reason, bright, spotlight-type heads must be very carefully arranged. Otherwise, they can create disabling glare and distorting shadows and impede eye accommodation.

Although it is customary to furnish the required emergency illumination from ceiling- or wall-mounted fixtures, consideration must be given to the code requirement, as noted in Section 19.31*b*, that specified emergency lighting levels be maintained at floor level. Because heavy smoke can readily obscure light from overhead fixtures, it is advisable to install some (preferably directional) emergency lighting fixtures near floor level.

(f) Exit Lighting

Most codes require 5 fc [50 lx] on nonilluminated exit signs, internally illuminated signs with the same visibility as an externally illuminated sign, or self-luminous signs with a specified luminance (usually at least 0.21 cd/m^2). Some exit signs are equipped with a battery and controls that provide 1 ½ hours of illumination on loss of utility power. Others are arranged to illuminate the exit area beneath the sign (Figure 19.27), and still others are equipped with a flasher and/or an audible beeper that assists people in finding the exit in a light-obscuring, smoke-filled room. Finally, some exit signs are nonelectrical self-illuminating, requiring continuous illumination, and are part of the emergency lighting system.

Figure 19.27 *Typical application of an exit light with a built-in battery, charger, and controls. Note that the bottom of the unit is designed to illuminate the area immediately in front of the exit.*

19.32 FLOODLIGHTING

Floodlighting, both interior and exterior, is used extensively for the diverse locations listed in Table 19.4, in addition to the more common sports facility lighting, which is not listed. At the designer's disposal are a variety of sources with respect to output, color, life, efficacy, and wattage.

Although a detailed floodlighting design involves complex calculations beyond the scope of this book, it is often sufficient for the designer to utilize a watts-per-square-foot table such as Table 19.4 to determine the approximate floodlighting requirements. The designer should also consult the unit power allowances for exterior lighting listed in *ASHRAE/IES Standard 90.1*.

Thus, if one is concerned with lighting a self-service parking lot at a neighborhood shopping center, and metal-halide is selected, Table 19.4 tells us that approximately 0.055 W/ft^2 [0.6 W/m^2] will suffice. This is well within the limits of *Standard 90.1*. If the lot is 200 ft × 500 ft (100,000 ft^2), then 0.055 W/ft^2 × 100,000 ft^2 (5500 W) is required. This figure is a good first estimate. The actual power level depends upon the specific equipment and design. This may vary from the initial estimate by ±20%.

The arrangement and choice of equipment must be determined before the problem can be considered solved. Considerable assistance on this score can be obtained from either the lighting engineer involved or from representatives of equipment manufacturers.

Although most floodlight installations use a single type, the installation shown in Figure 19.28*b* used a combination of metal-halide and mercury to obtain the desired effect.

19.33 STREET LIGHTING

Although detailed outdoor lighting calculations and design considerations are beyond the scope of this book, a few remarks are in order. New installations now use LED sources almost exclusively. The low efficacy and

TABLE 19.4 Lighting Application Guide

Application	Minimum Footcandles [lux] Maintained[a]		Watts per Square Foot [W/m²][b] Generally Required							
			Tungsten-Halogen		Mercury		Metal-Halide		High-Pressure Sodium	
Automobile Parking										
Attendant parking	2 [20]		0.38 [3.8]		0.17 [1.7]		0.11 [1.1]		0.075 [0.75]	
Industrial lots	1 [10]		0.13–0.15 [1.3–1.5]		0.06–0.07 [0.6–0.7]		0.037–0.044 [0.4–0.5]		0.026–0.03 [0.26–0.3]	
Self-parking lots	1 [10]		0.13–0.15 [1.3–1.5]		0.06–0.07 [0.6–0.7]		0.037–0.044 [0.4–0.5]		0.026–0.03 [0.26–0.3]	
Shopping Centers										
Neighborhood	1 [10]		0.13–0.19 [1.3–1.9]		0.06–0.09 [0.6–0.9]		0.037–0.055 [0.4–0.6]		0.026–0.038 [0.3–0.4]	
Average commercial	2 [20]		0.26–0.3 [2.6–3.0]		0.12–0.135 [1.2–1.4]		0.075–0.087 [0.7–0.9]		0.052–0.06 [0.5–0.6]	
Heavy traffic	5 [50]		0.65 [6.5]		0.29 [2.9]		0.19 [1.9]		0.13 [1.3]	
Automobile Sales Lots										
Front row (front 20 ft [6 m])	50 [500]		10 [100]		4.5 [45]		2.9 [29]		2.0 [20]	
Remainder	10 [100]		1.5–1.8 [15–18]		0.68–0.81 [6.8–8.1]		0.44–0.52 [4.4–5.2]		0.3–0.36 [3.0–3.6]	
Building										
Construction	10 [100]		1.5–1.8 [15–18]		0.68–0.81 [6.8–8.1]		0.44–0.52 [4.4–5.2]		0.3–0.36 [3.0–3.6]	
Excavation	2 [20]		0.26–0.3 [2.6–3.0]		0.12–0.14 [1.2–1.4]		0.075–0.09 [0.7–0.9]		0.052–0.06 [0.5–0.6]	
	Adj. Area									
Buildings up to 50 ft [15 m] High	**Light**	**Dark**	**Light**	**Dark**	**Light**	**Dark**	**Light**	**Dark**	**Light**	**Dark**
Light surfaces	15 [150]	5 [50]	3.3 [33]	1.2 [12]	1.5 [15]	0.54 [5.4]	0.96 [10]	0.35 [3.5]	0.66 [7]	0.24 [2.4]
Medium light surfaces	20 [200]	10 [100]	4.3 [43]	2.2 [22]	1.94 [19]	1.0 [10]	1.25 [13]	0.64 [6]	0.86 [9]	0.44 [4.4]
Dark surfaces	50 [500]	20 [200]	10.0 [100]	4.3 [43]	4.5 [45]	1.94 [19]	2.9 [29]	1.2 [12]	2.0 [20]	0.86 [9]
	Adj. Area									
Billboards and Signs	**Light**	**Dark**	**Light**	**Dark**	**Light**	**Dark**	**Light**	**Dark**	**Light**	**Dark**
Good contrast	50 [500]	20 [200]	10 [100]	4.3 [43]	4.5 [45]	1.94 [19]	2.9 [29]	1.25 [13]	2.0 [20]	0.86 [8.6]
Poor contrast	100 [1000]	50 [500]	20 [200]	10 [100]	9.0 [90]	4.5 [45]	5.8 [58]	2.9 [29]	4.0 [40]	2.0 [20]
Protective Lighting										
Gates and vital area	5 [50]		1.2 [12]		0.54 [5.4]		0.35 [3.5]		0.24 [2.4]	
Building surrounds	1 [10]		0.15–0.19 [1.5–1.9]		0.07–0.09 [0.7–0.9]		0.044–0.055 [0.44–0.55]		0.03–0.04 [0.3–0.4]	
Roadways										
Along buildings	1 [10]		0.24 [2.4]		0.11 [1.1]		0.07 [0.7]		0.05 [0.5]	
Open areas	0.5 [5]		0.08–0.1 [0.8–1.0]		0.036–0.045 [0.36–0.45]		0.023–0.029 [0.23–0.29]		0.02 [0.2]	
Storage yards	20 [200]		3.6–0.3 [36–43]		1.6–1.94 [16–19]		1.04–1.25 [10.4–12.5]		0.72–0.86 [7.2–8.6]	
Storage yards (inactive)	1 [10]		0.15–0.19 [1.5–1.9]		0.07–0.09 [0.7–0.9]		0.044–0.055 [0.44–0.55]		0.03–0.04 [0.3–0.4]	
Shopping Centers										
Parking areas (attraction)	5 [50]		0.65 [6.5]		0.29 [2.9]		0.19 [1.9]		0.13 [1.3]	
Buildings (attraction)					(See Buildings)					
Used car lots					(See Automobile Parking)					

[a] All illuminance levels for ground area applications are horizontal values.

[b] SI conversions are approximate, using a factor of 10 (versus 10.76).

(a)

(b)

Figure 19.28 *(a) Floodlighted section of a wall surrounding the Old City of Jerusalem, Israel, adjacent to the Jaffa Gate. Light sources are 400-W HPS units giving an average illuminance level of 50 lx. A sodium source was chosen to enhance the yellow-red color of the stone. (b) Church of All Nations, Mount of Olives, Jerusalem, Israel. Floodlight sources are 250- and 400-W mercury and metal-halide units, giving an average illuminance of 70 lx. Sources were selected to complement the colors in the mosaic at the top of the façade. (Photos courtesy of City of Jerusalem and J. Stroumsa, Chief Engineer.)*

short life of incandescent sources and the bulkiness of linear fluorescents make them obsolete. Furthermore, effective outdoor lighting of streets and walkways can help reduce vandalism and crime, facilitate nighttime use, and add to an area's attractiveness. A walkway luminaire (with light pollution potential) is shown in Figure 19.29.

For detailed recommendation on outdoor lighting, please see *IES LP-11-20 Lighting Practice: Environmental Considerations for Outdoor Lighting* (IES, 2020a).

Figure 19.29 *A "lollypop" fixture, even if aesthetically pleasing to some, gives poor illumination downward (note the large collar). This type of luminaire causes light pollution. (Photo by Nathan Majeski.)*

19.34 LIGHT POLLUTION AND TRESPASS

A frequently neglected corollary to the principle of placing light where it is required is not to place light where it is not required. As such, among a range of considerations regarding outdoor lighting's potential effects on human health, animals, and plants, *LP-11-20* emphasizes the importance of minimizing light pollution and light trespass.

Light pollution is frequently defined as adverse or obtrusive effects of electric lighting that produces artificial sky glow. Sky glow is the brightening of the night sky due to the scattering and reflection of light from the molecular constituents of the atmosphere in the direction of an observer. Sky glow can arise from both natural and unnatural (i.e., electric lighting) sources. Persistent unnatural (sometimes called "artificial") sky glow can have an adverse effect on ecological systems and human health. Moreover, it constitutes distress not only to astronomers, but also to anyone who simply wishes to enjoy the beauty of a star-filled sky.

Light trespass is another form of obtrusive off-site light. It includes light intrusion on private property and through windows.

Lighting design to reduce light pollution and light trespass is an emerging best practice. Strategies might include:

- minimizing upward light emissions
- minimizing non-target lighting
- designs that do not over-light
- turning off or dimming exterior lighting during low or no use
- turning off interior high-rise office building lighting when not in use
- using no higher CCT than necessary

- conducting photometric assessments to assess and reduce light pollution
- specifying BUG (backlight, uplight, and glare) rated luminaires or fixtures with sharp cutoff beyond the illuminated area. Shields can be added to standard luminaires to accomplish this. See Chapter 17 for more on BUG ratings.
- after an installation is complete, inspecting it at night to determine whether any nuisance has been created

Further information on the subject of night pollution can be found in IES technical memorandum *IES TM-10-11 Technical Memorandum: Addressing Obtrusive Light (Urban Sky Glow and Light Trespass) in Conjunction with Roadway Lighting* (IES, 2011a), dealing with light trespass.

19.35 NON-VISUAL LIGHTING

Emerging practice in lighting design is placing an increased emphasis on what can be broadly defined as non-visual lighting. Non-visual lighting pertains to designed outcomes in lighting design that foster human health by maintaining circadian rhythms—that is, our internal clock that keeps the body's hormones and bodily processes on a diurnal cycle, even in continuous darkness. Circadian lighting may involve daylight applications, electric lighting applications, or an integration of both. With regard to electric lighting specifications, there are two key metrics to consider: equivalent melanopic lux (EML) and melanopic EDI.

Equivalent melanopic lux (EML) can be leveraged to gauge the potential biological effects of light on humans. As such, EML can be specified as part of a project's lighting criteria. In general, higher EML light sources (125–250) are appropriate for daytime activities, while sources with lower EML (>50) are better for nocturnal activities. Standards and green building certification systems, such as the WELL Building Standard v2 (IWBI, 2020), outline minimal EML for certain surfaces for specified periods of time. For example, lighting criteria could require at least 150 EML at a height at least 18 in. [458 mm] above the work plane for at least four hours beginning at noon at the latest.

Melanopic equivalent daylight illuminance (mel-EDI or m-EDI or MEDI) is another metric that can be utilized in circadian lighting design. Increasingly, m-EDI is being referenced in lieu of EML for non-visual lighting specifications. WELL v2 and other frameworks reference m-EDI. Melanopic EDI is similar to a traditional lux-based measurement for lighting, but it is adjusted to reflect the potential stimulation of melanopsin (a type of light-sensitive protein in certain cells of the human eye that are consequential in activating mode and biological "awakeness"). For example, lighting criteria might call for a minimum of 250 lx m-EDI of illuminance vertically at a specified "eye level" during certain daytime hours; but a maximum of 10 lx m-EDI for three hours before bedtime; and perhaps only a maximum of 1 lx m-EDI at night and only negligible illuminance in the blue spectrum as such wavelengths can especially stimulate melanopsin.

19.36 REMOTE-SOURCE LIGHTING

Every experienced lighting designer has at some time felt the need for a remote-source luminaire to fill a specific lighting need. Among the most common situations are:

- Display lighting for light- and heat-sensitive objects, such as old books, fabrics, drawings and paintings, and, in general, objects containing organic materials, dyes, and coloring. Of particular importance in this category is the acute sensitivity of such objects to UV radiation, which exists in daylight and light from most electrical sources.
- Installations where relamping is a major logistic and financial problem. These include high-ceiling auditoriums and public assembly spaces of all sorts, high-bay lighting areas and other difficult-access locations, clean rooms, spaces with security entry limitations, and rooms that cannot tolerate disturbances or interruptions, such as air-traffic control rooms, continuous process manufacturing control areas, and the like.
- Installations where lamp heat is highly undesirable and its removal is difficult, expensive, or both. Among these are store show windows, refrigerated showcases, and conditions where lamp heat is felt by the space occupant(s), such as halogen and other incandescent lighting at workstations and low-ceiling spaces.
- Installations where the presence of electrical wiring is undesirable, such as patient-controlled hospital bed lighting or light sources in devices used by children.
- Spaces classified by the *NFPA National Electrical Code (NEC)* as electrically hazardous. These areas typically use large, heavy, expensive, and inefficient explosion-proof lighting.
- Installations where the electric and magnetic fields produced by fluorescent and HID lighting fixtures are unacceptable.
- Applications where the light source must be very small and effectively invisible.

Two basic designs have been developed over the years to fill these lighting requirements: arrangements using optical fibers, and those using light guides of various types. Applications of these two techniques do not generally overlap, and they are therefore discussed separately in the following sections.

19.37 FIBER OPTIC LIGHTING

Fiber optics have been in use for years, principally in instruments designed to permit illumination and observation of essentially inaccessible locations. In medical instruments designed for intrusive applications, fiber optic (FO) cable has the additional advantages of absence of heat and elimination of electrical wiring, both of which could be at the very least uncomfortable and frequently dangerous when inserted into the body.

The physical principle by which an optical fiber conducts light is that of total internal reflection (TIR), illustrated in Figure 19.30. The fiber is constructed of an inner core of transparent material (silica, glass, optical plastics) and an outer coating (cladding) of another material of lower refractive index. Because of this difference in refractive indices, light rays within the acceptance cone angle (2θ) are reflected at the interface of the two materials and thus proceed almost unimpeded down the core by a series of 100% reflections. Although there is (theoretically) no loss at the reflection, the core material itself exhibits an optical impedance that varies with the type of material. Ultraclear glass fibers, used in communications, have losses as low as 1 dB per kilometer, whereas the best plastic optical fiber used today has losses of around 1 dB per meter—that is, several hundred times as high. However, because of the short distances involved, these plastic fibers are satisfactory for lighting work (1-dB loss is the equivalent of 8.6% loss per meter, or 2.6% loss per foot).

Air, with a refractive index of 1.0, can also be used as a "cladding" material (unclad core), resulting in a fiber that emits light over its entire length. This light is shown in Figure 19.30 as light "lost" in the cladding. In practice, a bare core is not used because it would be damaged at supports, and dirt accumulation on the surface would result in high losses. Side-light fibers (lateral mode) are constructed commercially of a core, cladding, and sheathing, all of which are transparent plastic materials with decreasing refractive index from core to sheath. Note also in the figure that the angle θ,

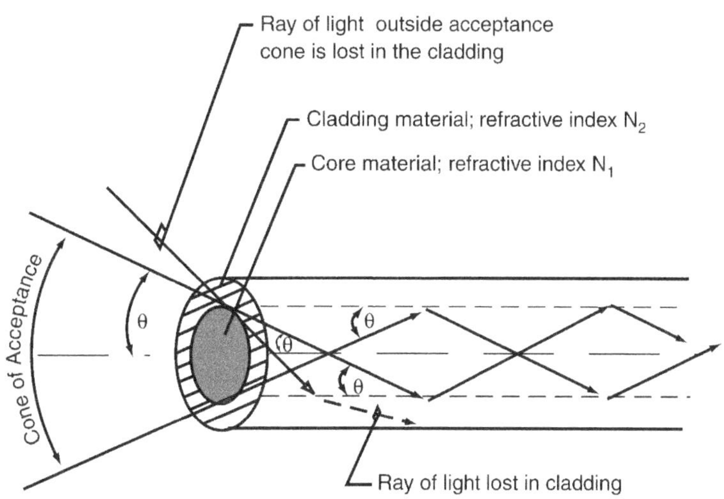

Core Index N_1	Cladding Index N_2	NA*	Cone Half Angle θ	Acceptance Cone (2θ)
1.4	1.3	0.52	31°	62°
1.4	1.0	0.98	78.5°	157°

*NA (Numerical Aperture) = $\sqrt{N_1^2 - N_2^2} = \sin\theta$

Figure 19.30 *Schematic representation of the phenomenon of total internal reflection (TIR) in fiber optics (FO). Light rays entering the end of an FO at an angle to the fiber centerline no greater than θ are totally reflected at the interface between core and cladding. Angle θ is determined by the difference between the refractive indices of the core and cladding materials and increases as that difference increases. Light rays entering within the acceptance core (solid angle 2θ) travel along the fiber; all other light rays are lost.*

TABLE 19.5 Typical Characteristics of Plastic Fiber Optic Cables

Type	Construction	Diameter Fiber/Core (mm)	Diameter OD (mm)	NA[a]	Cone Angle (Degrees)	Attenuation Percent per Meter
End-light Fiber bundle	See Illustration (*a*) below	0.15–2.0	6–18	0.5	60	4.7
Sidelight Fiber bundle		0.15–2.0	6–14	0.5	60	4.7
End-light Solid core	See Illustration (*b*) below	3–13	5–16	0.62	76	8.2
Sidelight Solid core		7–13	9–16	0.57	70	23.7

Note: Sheathing is opaque for end-light cables and transparent for sidelight cables.

[a]NA (Numerical Aperture) = $\sqrt{N_1^2 - N_2^2} = \sin\Theta$; see Figure 19.30.

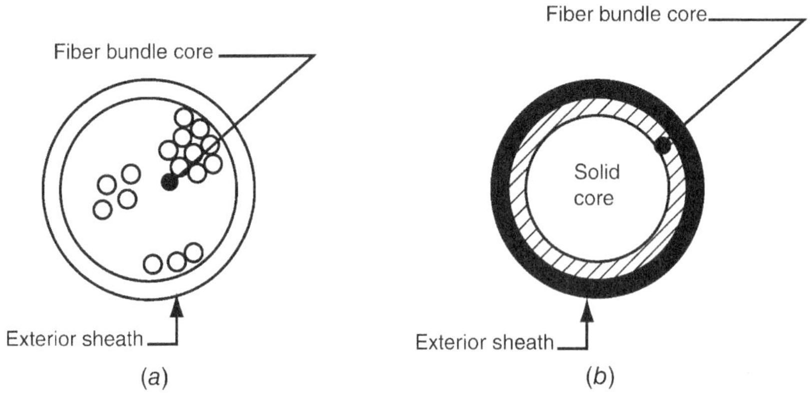

and therefore the size of the acceptance cone, increase with the difference in the refractive indices between cone and cladding. It is desirable to have an acceptance cone as large as possible (high numerical aperture) so that maximum light from the source (illuminator) can be introduced (coupled) into the fiber. Glass fibers are generally not used in lighting applications because of their high cost, small light-carrying capacity, and large bending radius. Table 19.5 shows the construction and major characteristics of typical lighting-use fiber optic cables.

19.38 FIBER OPTIC TERMINOLOGY

Because FO cable is becoming common in lighting work, a lighting designer should be familiar with FO terminology. A brief glossary of the most important terms follows.

Acceptance angle (also acceptance cone): The maximum solid interior angle of a cone that defines the spread of light rays that will enter a single fiber. Because this angle is determined only by the fiber and the cladding materials, it is also the cone of light ray acceptance of a fiber bundle.

Attenuation: The degree of light reduction as it travels along the fiber. Because attenuation varies with frequency, the light color exiting an optical fiber is different from that entering, and the change becomes more pronounced with increasing fiber length.

Axial mode (also end-light): The mode of conduction from one end of the fiber (cable) to the other, with no deliberate loss by emission over the cable length.

Core: The center of an optical fiber that carries the light.

Fiber optic port: A factory- or field-applied terminating connector that enables coupling of an FO cable to a light source or other item in the system. Glass FO is factory terminated; plastic fibers can be either.

Glass optical fiber: The type always used in communication work and seldom in lighting application. It is expensive, requires a large bending radius, and exhibits very low attenuation.

Illuminator: The light source in an FO lighting installation. It usually consists of a metal box containing the light source (usually metal-halide or halogen) and its accessories, plus required optical devices that collect and concentrate the light from the source and couple it optically to an optical port to which an optical cable is connected. The illuminator may also contain color filters, local and remote-control connections, and, almost always, a fan or blower for forced-air cooling of the source lamp.

Lateral mode (also side-light[ing]): A light conduction and distribution system whereby light is emitted over the entire length of the fiber, simulating neon tubing. This is accomplished by using transparent cladding and sheathing, with refractive indices selected to increase light rays entering (and leaving) the cladding.

Numerical aperture: A numerical metric based on the refractive indices of the core and cladding materials, indicating the angle of the acceptance cone (i.e., that portion of the light that is conducted by the fiber core).

Plastic optic fiber: The fiber material normally used in lighting work, usually a clear acrylic-type compound as the core, with various plastic compounds as cladding or sheathing, depending on the design and mode of operation. The fibers are classified by size as small core (up to 2 mm [0.08 in.] in diameter) and large core (up to 20 mm [0.78 in.] in diameter). Small-core fibers are manufactured in continuous and essentially unlimited lengths. Large-core fibers and sheathed fiber-bundle cables do not usually exceed 45 m [150 ft] in length.

Refractive index: The ability of a light-conducting (nonopaque) material to bend a light ray entering from another medium, expressed numerically.

Tail: A single optical fiber or a bundle of fibers extending from an illuminator to an output point. A single fiber separated from a bundle for a specific small, isolated illumination purpose (often decorative) is also referred to as a tail.

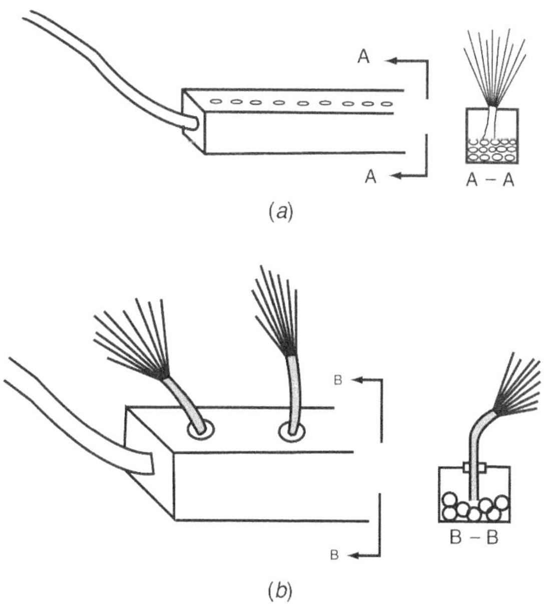

Figure 19.31 *Linear constructions using end-light fibers. (a) A light bar is simply a box containing multiple tails that are brought out of the box-type container at intervals, usually equal. Selecting fiber size, exit spacing, and illuminator size creates a low-intensity linear lighting fixture. (b) By using large fibers and/or bundles brought out in adjustable groups, the designer can readily construct a linear bar with adjustable multihead spotlights.*

19.39 FIBER OPTIC LIGHTING: ARRANGEMENTS AND APPLICATIONS

The development of large, efficient plastic optical fibers coupled with a concomitant price reduction has produced a lighting tool whose applications are limited only by the imagination of designers and manufacturers. In one form or another, optical fibers are applicable where:

- a single remote source can supply a large number of relatively small point-source lights
- burial of the light carrier (fiber bundle) in almost any substrate is desired or required
- the heat, UV content, and electrical fields associated with most sources are absent
- the presence of electrical wiring and its associated hazards are undesirable or unacceptable

A few of the many current configurations are described in the following subsections.

(a) Axial-Mode Linear Devices

In this arrangement, a bundle of fibers is placed in a longitudinal enclosure, and individual fibers or small groups of fibers are separated from the bundle(s) and brought out of the enclosure as a light-emitting point (Figure 19.31*a*). Most of these light bars are custom-made for a specific application, enabling the designer to specify all of the parameters, including spacing of points, intensity of each point, dimensions of the bar, individual fiber size and bundle size, color characteristics, and so on. Light bars are used in retail display lighting, accent lighting, directional devices, decorative applications, and other applications requiring a low-intensity, linear, and/or "sparkling" lighting device.

The essential design of a light bar can be adapted to the use of large-core fibers or multiple fiber bundles to produce a linear string of focusable, medium-intensity spotlights (Figure 19.31*b*). In lieu of the closely spaced tiny light points of Figure 19.31*a*, large fibers or bundles can be extracted. They can be used as light sources or can be coupled to optical plastic-fiber rods or other devices. These can then be "aimed" as desired to illuminate individual areas in a cabinet, store window, and the like. Here, too, the units are commonly custom-designed to perform a specific lighting task.

(b) Axial-Mode Discrete Sources

By using large groups of end-light FO bundles combined to produce a point light source, the designer can

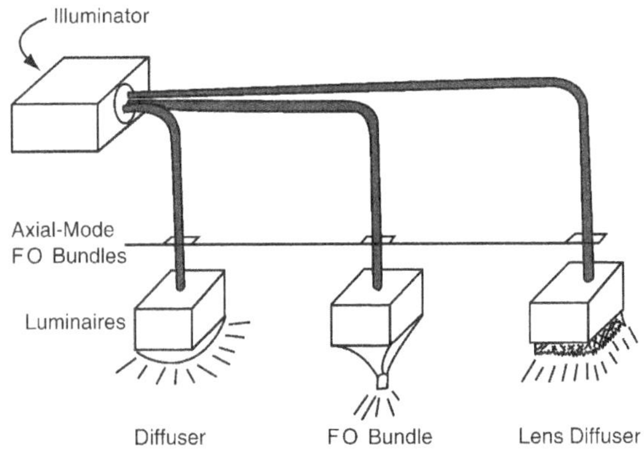

Figure 19.32 *Relatively large end-light FO bundles can be utilized as point sources in conventional types of lighting fixtures.*

construct semiconventional lighting fixtures with a variety of common diffusing elements (Figure 19.32). The advantages of these discrete sources compared to common electrically powered lighting fixtures are the same as those listed previously: absence of heat and electricity, and low maintenance and higher efficacy of high-output lamps in the illuminators. One disadvantage common to all plastic-fiber FO lighting is the necessity of keeping runs short due to the attenuation of the fibers. Lengths in excess of 20 ft [6.1 m] are inefficient and therefore uncommon.

(c) Lateral-Mode Fiber Optic Lighting

As explained previously, lateral-mode fibers emit light throughout their length. This makes them particularly suitable for linear lighting tasks such as illuminating stair nosings, path lighting, and all sorts of decorative trim lighting. Figure 19.33 shows a possible arrangement for under-shelf lighting, illumination under stair nosings, and the like. The practicability of burying parts of the cable in concrete is useful in giving the remaining lighted cable a disconnected, floating appearance. Color filters in the illuminator, when applied to multiple cable runs as in Figure 19.33, can produce dramatic outlining effects.

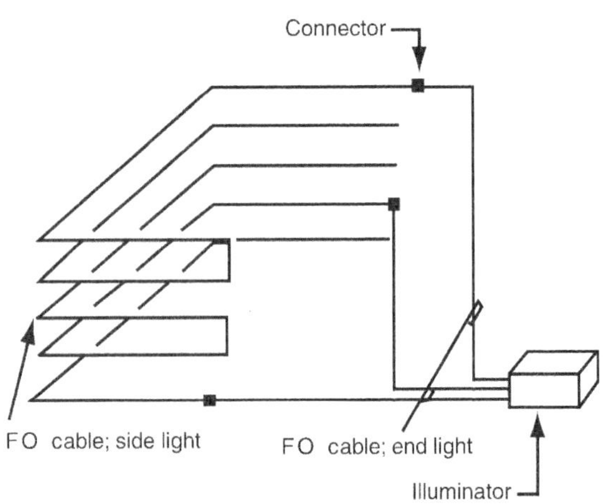

Figure 19.33 *Side-light-emitting FO cables are ideal for linear lighting tasks such as stair-edge illumination, under-shelf lighting, and outline lighting of all sorts. The illustrated 3-tail arrangement is representative of this type of FO lighting.*

19.40 HOLLOW LIGHT GUIDES

The idea of conducting light from one place to another apparently arose several millennia ago from an architectural desire to provide daylight deep in interior spaces. The interior court is one example of an architectural light guide that is only recognizable as such in multistory buildings. Figure 19.34 illustrates the use of an interior light shaft in a six-story building. Notice the rapid attenuation as daylight descends by multiple reflections, accounting for the increasing size of the windows at lower levels. Indeed, attenuation is the principal problem with light pipes and guides that depend upon multiple internal reflections to conduct light any appreciable distance. The best metallic mirrors available today specularly reflect only about

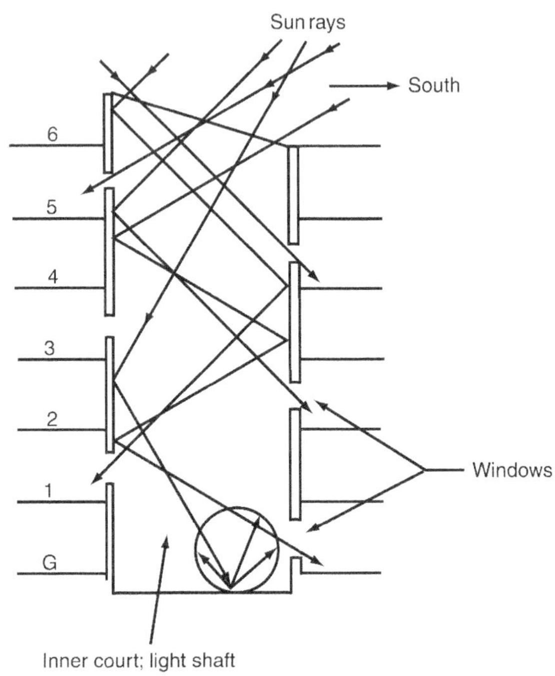

Figure 19.34 *An interior court or light well (shaft) with a reflective wall surface acts as a light guide to introduce daylight at lower floors. Buildings in the northern hemisphere are oriented as shown. Due to attenuation of re-reflected sunlight beams traveling down the shaft, lower-floor windows are larger than those at upper floors to capture more (of the attenuated) light. (From NASA Tech Brief LAR-12333.)*

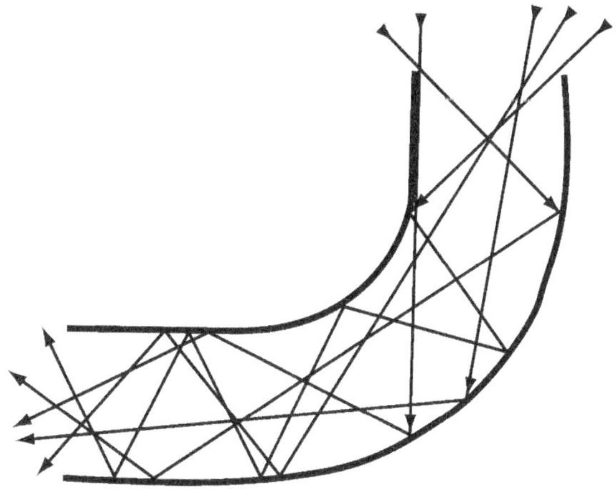

Figure 19.35 Curvature in a light guide increases the number of reflections that a light ray makes over the guide length, thereby increasing attenuation. (From NASA Tech Brief LAR-12333.)

95% of the impinging light. It is therefore relatively easy to calculate the number of 5% loss reflections that reduce the output light from a light pipe with random nondirectional input to an uneconomical level (where the "free" light received is more expensive due to construction and maintenance costs than alternative electric light).

Open-ended light pipes of this sort (Figure 19.35) are notoriously difficult to keep clean of dust and surface-reflectance degeneration, so that the 95% initial specular reflectance rapidly decreases to 85% semidiffuse reflection, thereby increasing ray re-reflection and consequently overall light attenuation. (Sealing the upper end of the pipe with a transparent medium is not advisable due to the rapid accumulation of dust and condensation.) In order to capture an appreciable amount of sunlight (sky light is insufficient), an open-ended pipe at roof level must be very large. This, in turn, increases construction and maintenance costs and decreases (in the case of a commercial building) the usable/rental area, making the light-guide idea generally impractical except for very short guides. Such commercially available units are essentially small skylights with an elongated, reflective collar.

One solution to the problem of rapid attenuation in a light guide caused by multiple interior wall reflections is to collimate the incoming light. This has the additional advantage of permitting reduction of the light guide's cross-sectional area. Such a system, however, requires a sun-tracking arrangement, plus mirrors and lenses in a collimating optical train. A typical arrangement is shown schematically in Figure 19.36. The increased efficiency of arrangements of this sort is often offset by losses in the optical train and increased equipment cost (although the light pipe itself is smaller and therefore cheaper). Systems of this type are in use today, with generally satisfactory results, although system economics is marginal except in arid, sunny climates.

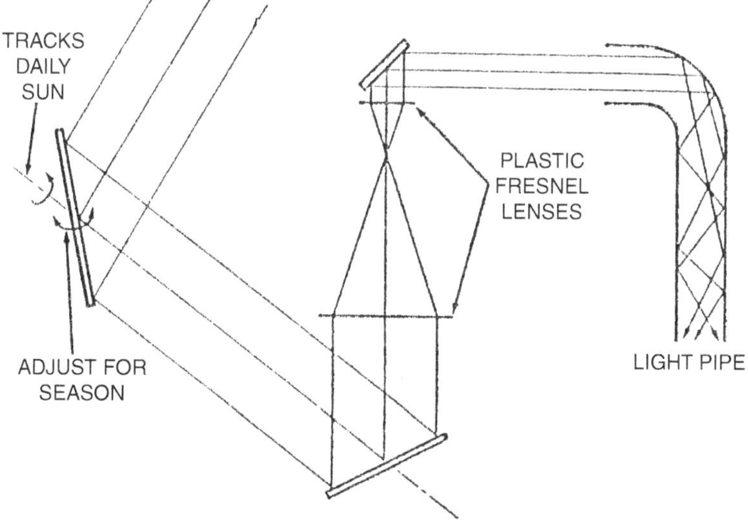

Figure 19.36 Concentration and collimation of sunlight can be accomplished with a sun-tracking mirror and an optical train of mirrors and lenses. Full tracking of the sun requires altitude and azimuth drives on the collection mirror. To reduce costs without severely reducing collection efficiency, a full azimuth drive tracks the sun from east to west daily, while the mirror tilt around a horizontal axis (altitude tracking) is adjusted only seasonally to the mid-season position. The sun's maximum altitude varies 23.5° from equinox to solstice. (From NASA Tech Brief LAR-12333.)

19.41 PRISMATIC LIGHT GUIDES

The first major improvement in hollow light guide design occurred in 1981, with the development and patenting of a prismatic rectangular hollow acrylic light guide (Whitehead et al., 1982). The patentable novelty of this device is that the facets of the prismatic exterior transparent acrylic walls of the pipe act as TIR "mirrors," thus preventing light from escaping from the transparent box guide. As a result, light is guided along the length of the device with very low losses, because reflection at the prismatic walls is theoretically perfect, and losses in the air space within the guide are very small. In practice, it was found that at each reflection about 2% of the light was lost by absorption and, because of microscopic imperfections in the prisms, about 6% escaped through the wall. This 6% "loss" converts the light guide into a long, rectangular lighting fixture emitting light uniformly over its entire length. Placing a mirror at the end of the guide causes all the light reaching the end to be reflected back into the guide, thereby increasing the light output over its length. Overall efficiency of this extended "lighting fixture" is high; this fact, combined with very low maintenance, produced a highly desirable lighting product. Its first application was in an electrically hazardous area previously illuminated by explosion-proof lighting fixtures (Figure 19.37).

19.42 PRISMATIC FILM LIGHT GUIDE

In 1988, the 3M Company developed a thin plastic prismatic film that utilizes the principle of TIR at the prism face. Tests have shown that this material loses only about 1% of the light at each reflection. As a result, tubular hollow light guides using this material can extend for several hundred feet if the input light is collimated (Figures 19.38 and 19.39). When used as an extended lighting fixture, this light guide exhibits good photometric characteristics.

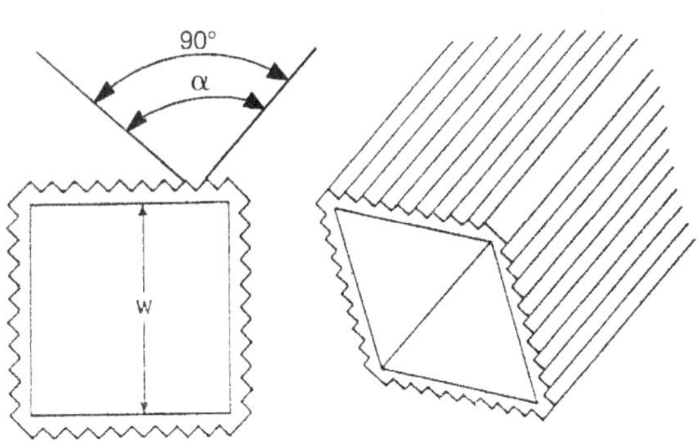

Figure 19.37 *Views of the hollow, clear acrylic, prismatic light guide developed by L. A. Whitehead. The cross-section measures 13-cm [≈5 in.] square, and the prism angle α is 90°. (Adapted from Whitehead et al., 1982.)*

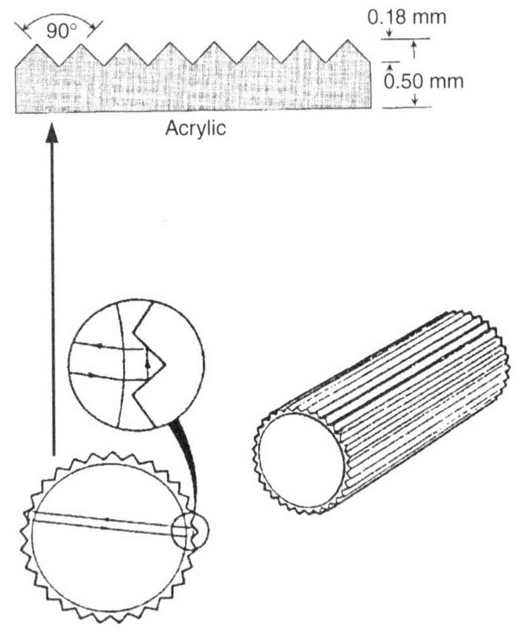

Figure 19.38 *Circular hollow light guide. If the prismatic plastic film is formed into a cylindrical shape as shown, it will act as a light guide with very low losses (0.18 mm = 0.007 in.; 0.5 mm = 0.02 in.). (Adapted from Whitehead et al., 1982.)*

Prismatic light pipes have all the advantages of FO lighting but fewer limitations, although in practice the two systems do not compete because they address entirely different needs. In addition to having all the advantages of remote-source FO lighting—including absence of heat, minimum UV radiation, absence of electrical wiring and the associated electromagnetic fields and electrical wiring hazards, and usability at any temperature, interior or exterior—prismatic light guides (and lighting fixtures) are constructed to handle very large quantities of light and do not produce color distortion (as do optical fibers). As a result, a luminous light pipe (light fixture) can be coupled to a very-high-output source such as an arc lamp, xenon lamp, or the relatively new sulfur lamp, and the pipe can be used to illuminate large areas, either interior and exterior. The economies in wiring, luminaires, installation, and maintenance help to offset the cost of the prismatic material.

Because of the very high light conduction efficiency of optical lighting film (OLF), using it as a lighting fixture requires special techniques and materials to extract light from the pipe. One method is to apply another prismatic material to the inside of the light pipe with a different prism angle that causes incident light to change direction and exit the pipe. When this material is placed judiciously, selected areas of the pipe circumference emit light, in effect, producing a longitudinal lighting fixture with the desired directionality. Typical lighting applications of prismatic film light-pipe fixtures are high-bay industrial and commercial installations, large exterior

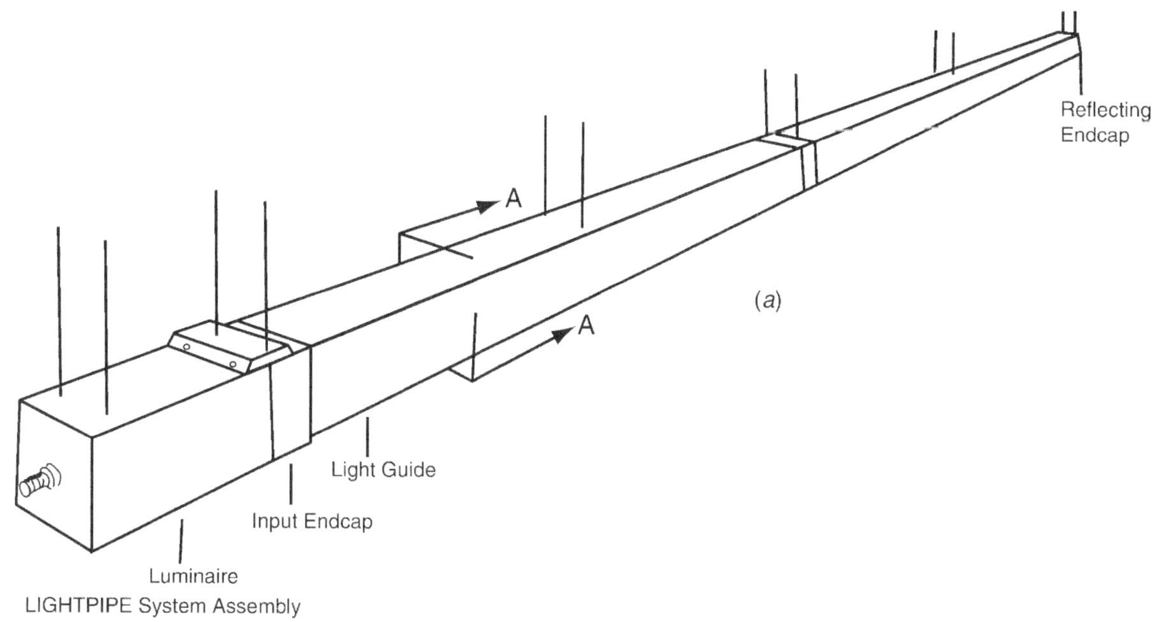

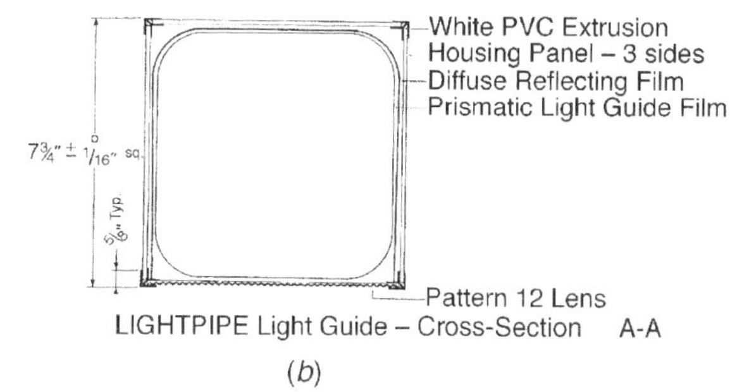

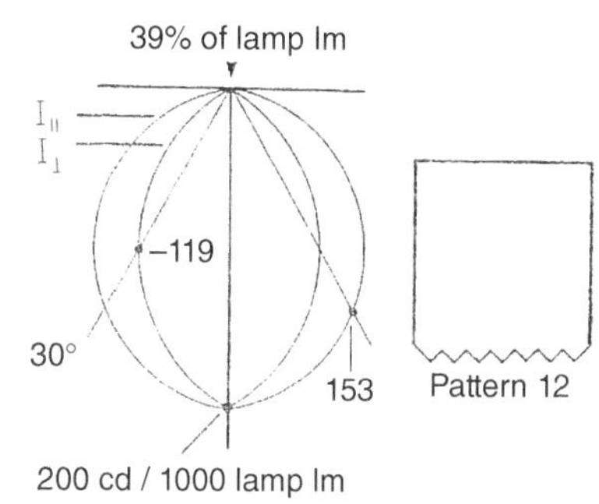

Pcc	80		50		30	
Pw	50	30	50	30	50	30
RCR						
0	.46	.46	.43	.43	.41	.41
1	.41	.40	.39	..38	.37	.37
2	.37	.35	.35	.33	.34	.32
3	.33	.31	.32	.30	.31	.29
4	.30	.27	.29	.26	.28	.26
5	.27	.24	.26	.23	.25	.23
6	.25	.22	.24	.21	.23	.21
7	.22	.19	.21	.19	.21	.19
8	.20	.17	.19	.17	.19	.17
9	.18	.15	.18	.15	.17	.15
10	.17	.14	.16	.14	.16	.13

CU Table – One Side Emitting

(c)

Figure 19.39 *Details of a rectangular light guide fixture. (a) The entire lighting unit comprises a "luminaire" at one end containing the light source [250- or 400-W metal-halide lamp] and its accessories connected to the hollow prismatic light guide. The end of the guide is sealed with a mirror that reflects the light back into the light guide. (b) The light guide itself is enclosed on three sides and equipped with a prismatic lens diffuser on the open (bottom) surface through which light is emitted. This specific design is usable to a length of 40 ft [12 m]. Longer units are equipped with light sources at both ends. (c) Photometric characteristics of the luminaire-light guide assembly. (Courtesy of TIR Systems, LTD.)*

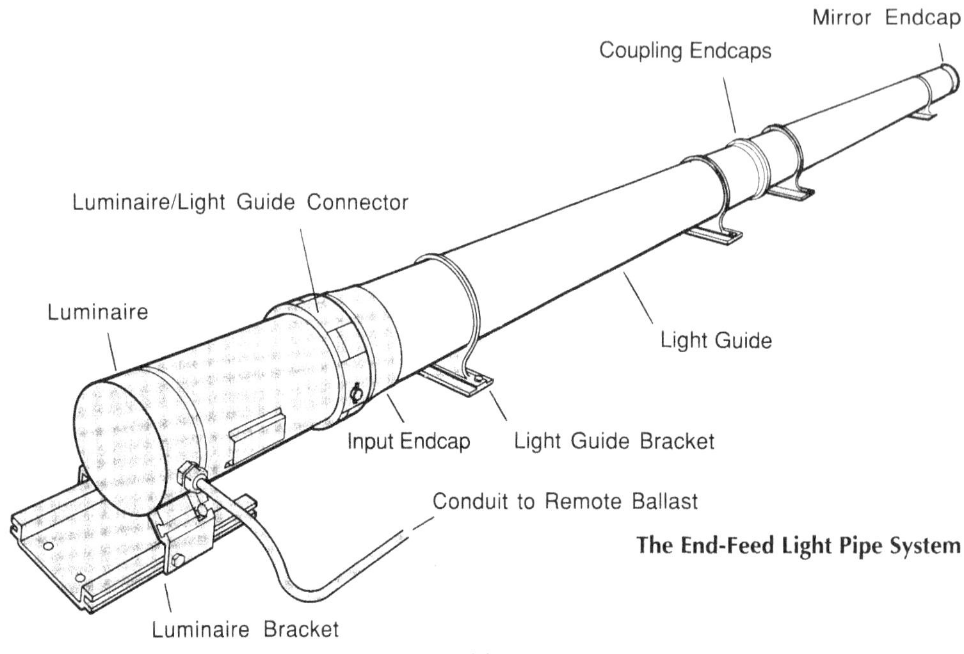

(*a*)

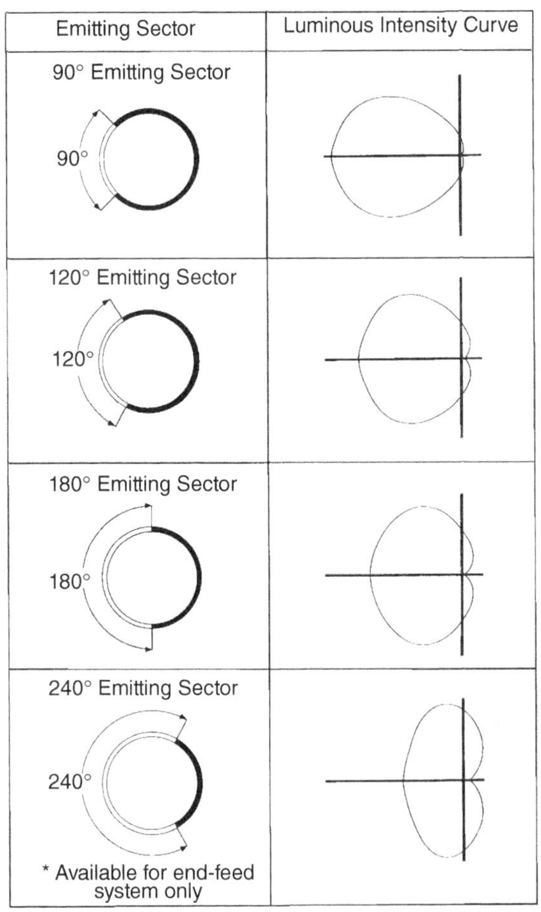

(*b*)

Figure 19.40 *Details of a 7-in. [178-mm] diameter hollow prismatic end-feed light-guide fixture. (a) The entire unit consists of an opaque "luminaire" section containing a 250-W metal-halide lamp connected to a 7-in. [178-mm] diameter acrylic tube up to 44 ft [13 m] in length containing prismatic optical film and terminating in a reflective (mirrored) endcap. (b) The portion of the tube circumference that emits light (uniformly) is controlled by placement of a diffuse reflecting film between the inner prismatic film and the outer transparent acrylic envelope (see Figure 19.39b). The reflective film is shown as the black portion of the tube circumference. (c) Photometric data for an end-feed system with a single 250-W metal-halide source. (Courtesy of TIR Systems, Ltd.)*

CU Tables — End-Feed System

ρ_{cc}	70		50	
ρ_w	50	30	50	30
RCR				
0	.34	.34	.32	.32
1	.30	.29	.29	.28
2	.26	.25	.25	.24
3	.24	.21	.23	.21
4	.21	.19	.20	.18
5	.19	.16	.18	.16
6	.17	.14	.16	.14
7	.15	.13	.15	.12

CU Table - 90° Emitting

ρ_{cc}	70		50	
ρ_w	50	30	50	30
RCR				
0	.36	.36	.34	.34
1	.30	.29	.28	.27
2	.26	.24	.24	.23
3	.23	.20	.21	.19
4	.20	.17	.19	.16
5	.17	.15	.17	.14
6	.15	.13	.15	.12
7	.14	.11	.13	.11

CU Table - 120° Emitting

ρ_{cc}	70		50	
ρ_w	50	30	50	30
RCR				
0	.39	.39	.36	.36
1	.33	.31	.30	.29
2	.28	.25	.26	.24
3	.24	.22	.23	.20
4	.21	.18	.20	.17
5	.19	.16	.18	.15
6	.17	.14	.16	.13
7	.15	.12	.14	.11

CU Table - 180° Emitting

ρ_{cc}	70			50	
ρ_w	50	30		50	30
RCR					
0	.39	.39	.39	.35	.35
1	.32	.30	.28	.29	.27
2	.27	.24	.22	.24	.12
3	.23	.20	.18	.21	.18
4	.20	.17	.15	.18	.16
5	.18	.14	.12	.16	.13
6	.16	.13	.10	.14	.11
7	.14	.11	.09	.13	.10

CU Table - 240° Emitting

Length of Run (feet)		10 ft.	20 ft.	30 ft.	40 ft.
Lumens/foot		500 lm/ft.	270 lm/ft.	165 lm/ft.	110lm/ft.
Peak Luminous Intensity (cd)\(Mean Exitance (lm/ft²))	90° emitting	1569\(1070)	1708\(583)	1569\(357)	1435\(245)
	120° emit	1143\(850)	1243\(462)	1143\(283)	1043\(194)
	180° emit	1091\(631)	1187\(343)	1091\(210)	998\(144)
	240° emit	696\(497)	757\(270)	696\(166)	N/A

Note: All values listed are based on the maintained output of a single T250 luminaire.

Photometric Data - End-Feed System

(c)

***Figure 19.40** (continued)*

signs, tunnel illumination, highway signs, continuous rail-type guidance illumination, and exterior architectural building lighting. Figure 19.40 shows a circular prismatic light guide designed for exterior use, with several configurations and the associated photometric data. Figure 19.41 shows an application of the prismatic light-guide fixture.

19.43 REMOTE-SOURCE STANDARDS AND NOMENCLATURE

Due to the relative newness of long, remote-source tubular lighting, industry-wide standards for said lighting do not exist, and even the nomenclature varies. Clear differentiation between devices intended primarily to carry light from one location to another and devices intended to illuminate along their length does not exist. Nor is terminology differentiated between light guides operating on the principle of internal reflection from specular surfaces and prismatic light guides. All these devices are variously and interchangeably referred to as:

- hollow light guide
- light pipe
- hollow light pipe

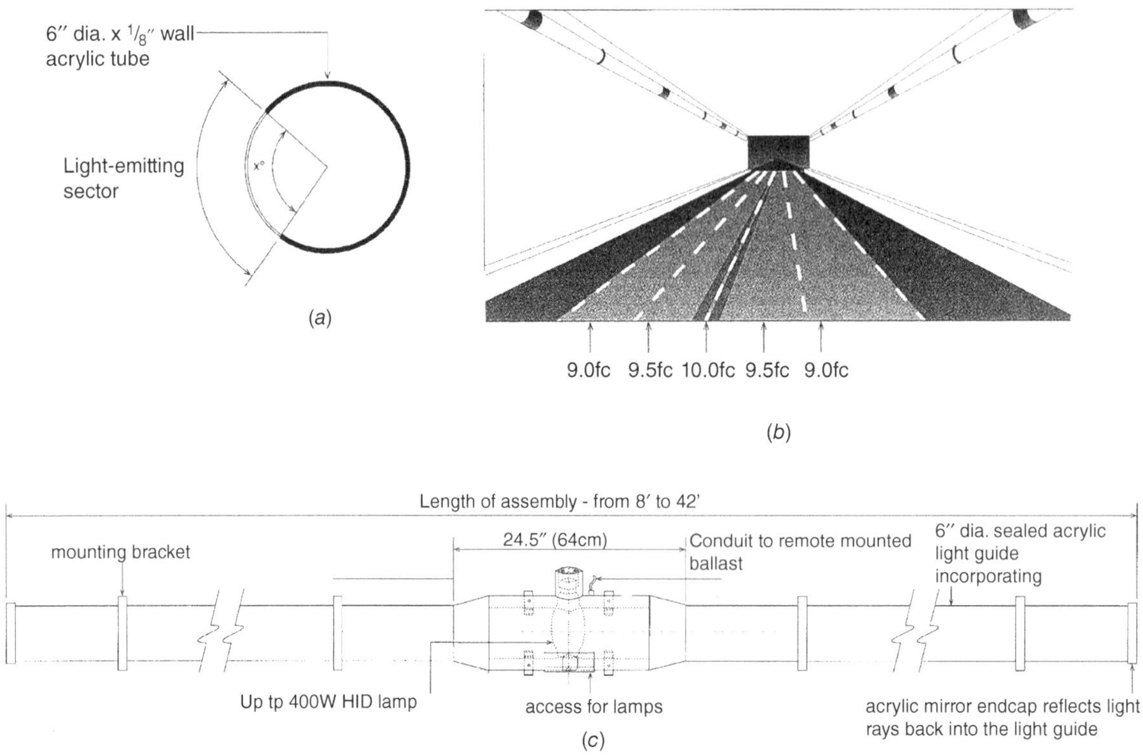

Figure 19.41 *Application of a prismatic tubular light guide to tunnel lighting. (a) Section demonstrating how light is emitted from the light guide. (b) Perspective drawing showing tubular lighting on both sides of the tunnel and the resultant uniform tunnel illuminance levels. (c) Drawing of one section of the center-feed light guide. (Courtesy of TIR Systems, Ltd.)*

- prismatic light guide
- prismatic light pipe
- hollow prismatic light guide
- remote-source hollow light guide

and various other names. See *NEMA LSD 4-1999: Glossary of Terms Pertaining to Remote Illumination Systems* (NEMA, 2004) for further information.

References and Resources

ASHRAE. 2022. *ANSI/ASHRAE/IES Standard 90.1-2022: Energy Standard for Buildings Except Low-Rise Residential Buildings*. Atlanta, GA: American Society of Heating, Refrigerating and Air-Conditioning Engineers.

ASHRAE. 2024. *ANSI/ASHRAE/IES Standard 90.2-2024: High-Performance Energy Design of Residential Buildings*. Atlanta, GA: American Society of Heating, Refrigerating and Air-Conditioning Engineers.

IEEE. 1995. *Standard 446: IEEE Recommended Practice for Emergency and Standby Power Systems for Industrial and Commercial Applications*. New York: Institute of Electrical and Electronics Engineers, Inc.

IES. 1988. *ANSI/IES/ALA RP-3-88 Recommended Practice: Lighting Educational Facilities*. New York: Illuminating Engineering Society.

IES. 2011a. *IES TM-10-11 Technical Memorandum: Addressing Obtrusive Light (Urban Sky Glow and Light Trespass) in Conjunction with Roadway Lighting*. New York: Illuminating Engineering Society.

IES. 2011b. *The Lighting Handbook* (10th ed.). New York: Illuminating Engineering Society.

IES. 2020a. *ANSI/IES LP-11-20 Lighting Practice: Environmental Considerations for Outdoor Lighting*. New York: Illuminating Engineering Society.

IES. 2020b. *ANSI/IES/ALA RP-3-20 Recommended Practice: Lighting Educational Facilities*. New York: Illuminating Engineering Society.

IES. 2020c. *ANSI/IES/ALA RP-4-20 Recommended Practice: Lighting Library Spaces*. New York: Illuminating Engineering Society.

IES. 2020d. *ANSI/IES/ALA RP-11-20 Recommended Practice: Lighting for Interior and Exterior Residential Environments*. New York: Illuminating Engineering Society.

IES. 2021. *ANSI/IES RP-7-21 Recommended Practice: Lighting Industrial Facilities*. New York: Illuminating Engineering Society.

IES. 2022. *ANSI/IES RP-1-22 Recommended Practice: Lighting Office Space*. New York: Illuminating Engineering Society.

IES. n.d. *The Lighting Library*. New York: Illuminating Engineering Society. ies.org/standards/lighting-library

International Dark-Sky Association. www.darksky.org

IWBI. 2020. *WELL Building Standard Version 2*. New York: International WELL Building Institute. v2. wellcertified.com

NASA. n.d. *NASA Tech Brief. PB80-974440 (LAR-12333): Optics for Natural Lighting*. Hampton, VA: Langley Research Center, National Aeronautics and Space Administration.

NEMA. 2004. *LSD-4-1999: Glossary of Terms Pertaining to Remote Illumination Systems*. Rosslyn, VA: National Electrical Manufacturers Association.

NFPA. 2022. *2023 NFPA 70: National Electrical Code (NEC)*. Quincy, MA: National Fire Protection Association.

NFPA. 2023a. *2024 NFPA 101: Life Safety Code*. Quincy, MA: National Fire Protection Association.

NFPA. 2023b. *2024 NFPA 99: Health Care Facilities Code*. Quincy, MA: National Fire Protection Association.

Whitehead, L. A., R. A. Nodwell, and F. L. Curzon. 1982. "New Efficient Light Guide for Interior Illumination." *Applied Optics* 21(15): 2755–2757.

C H A P T E R 20

Water and Design

"THE NEXT GREAT WORLD CRISIS WILL BE WATER SUPPLY." This prediction is becoming increasingly pertinent in the modern day, as many scales of conflict over water rights are coming to the forefront—downstream nation-states vocalizing threats of aggression against their upstream neighbors, municipalities engaging in legal battles over aquifer claims, and so on.

With a finite planetary water supply, pitted against an increasing population, seen in conjunction with an increasing per capita consumption of water, we see again (as with fossil fuels) the problem of limited resources meeting with growing demand. At least in this case, the amount of water is fixed, not diminishing. However, the problem of fair allocation remains (Figure 20.1). Countries have fought wars over oil; must they also wage war over water?

Agricultural and industrial use of water may dwarf that of buildings, but designers still have a role in this dilemma. We can use water both efficiently and aesthetically where it is appropriate, and avoid its use when possible.

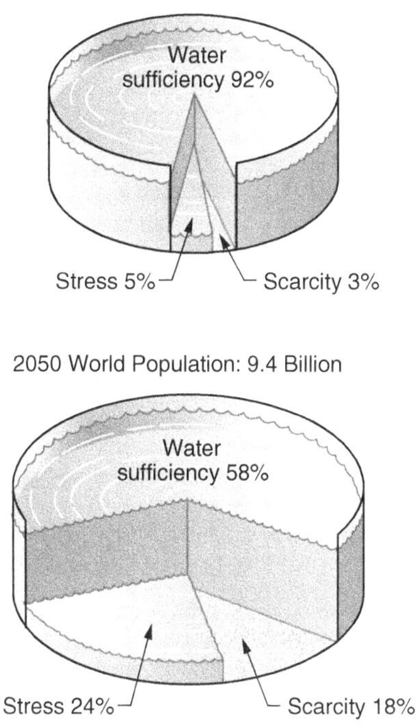

Figure 20.1 *Earth's past and projected annual per capita fresh water supply for all purposes. The relative size of the graphs reflects population growth. A "sufficient" supply is 1700 m^3 [449,000 gal] per person; the "stress" supply is between 1000 and 1700 m^3 [264,180, and 449,000 gal] per person; the "scarcity" supply is less than 1000 m^3 [264,180 gal] per person. Currently, worldwide water use is 70% irrigation, 20% commercial-industrial, and 10% residential. (Worldwatch, State of the World 1999; courtesy of Population Action International.)*

20.1 WATER IN ARCHITECTURE

Throughout history, in nearly all climates and cultures, the designer's major concern about water was how to keep it out of a building. Only since the end of the nineteenth century has a water supply within a building

ACTIVE ENVIRONMENTAL SYSTEMS

become commonplace in industrialized countries. In much of the world today, running water is still not available within most buildings. There are a variety of appropriate design responses to the supply, use, and return of this versatile commodity.

(a) Nourishment

Water is the most abundant chemical in our bodies as well as in our diet. From a larger perspective, though, the amount of pure (potable) water that we actually need for drinking and cooking is very small—only about 3 g/cd (gallons per capita per day) [11.4 L/cd] (liters per capita per day) in the United States. The most common supply system throughout history has been the central municipal fountain or well (Figure 20.2), whose technical importance to the community has often been emphasized by the aesthetics of both the fountain's overall sculptural composition and the elegance of detail in its water spouts, basins, and other elements. A fountain, centrally located within a municipality, serves not only as a geographic point from which the townspeople can access a vital resource, but also as a social hub wherein opportunities for conversations and interactions abound.

As potable water became available on demand within buildings, water-related social opportunities diminished. However, employees still converse around the drinking fountain—for example, when it is located in a place that invites lingering or relaxing.

(b) Cleansing and Hygiene

Water is a nearly ideal medium for the dissolution and transport of organic waste, and because of its high heat-storage capacity, it easily retains comfortable temperatures for bathing. Much larger quantities of water are used for cleaning than for nourishment: In the average U.S. home, about 14 g/cd [53 L/cd] are used for clothes washing and dishwashing, and another 21 g/cd [79.5 L/cd] are used for bathing and personal hygiene.

Figure 20.2 The fountain in San Lorenzo, Italy, serves as a social space and place of coolth. (© Donald Corner; used with permission.)

In the past, water for cleaning was carried to the home infrequently; the Saturday night (only) bath was typical well into the twentieth century in the United States. Bathing vessels were usually portable and sometimes were combined with other pieces of furniture (a couch that sat over a tub, a metal tub that folded up inside a tall wooden cabinet, etc.). Thus, a bathplace rather than a bathroom was the common design response.

Although the physical constraints of water carrying were important design influences on bathing, cultural attitudes were at least as strong an influence throughout history. Perhaps the most startling contrast was between the medieval concept of bathing as an almost sinful indulgence, and the earlier Roman attitude toward public baths as the social centers of cities.

Today bathing facilities are commonly designed to be used in a personal, private setting. There are also welcome opportunities to design more social bathing (recreational rather than cleansing) places, such as swimming pools, bathhouses, and hot tubs. The characteristics of the water supply (spouts, jets, cascades) can be matched with those of the water surface desired (mirror-smooth, gently flowing, rippled, rolling, foaming) in order to obtain the intended atmosphere.

(c) Ceremonial Uses

Water acquired a ceremonial significance that remains particularly evident in religious services. Examples of the ceremonial use of water include vessels containing holy water at entrances to Catholic churches, pools and fountains in the forecourts of mosques, full-immersion baptismal fonts at the altars of some Protestant churches, and the Jewish tradition of *tashlich*, whereby some shuls employ a fountain or pool for the purpose. The opportunities for aesthetic expression are particularly rich in these ceremonial applications.

(d) Transportation Uses

In stark contrast to its uses in nourishing, cleansing, and celebrating, water is used in our buildings principally to transport organic waste. The typical U.S. home uses 32 g/cd [121 L/cd] just to flush toilets. There is perhaps no more flagrant example of a mismatch in architecture than the high-grade resource of pure water being used for the low-grade task of carrying away a wad of toilet paper.

In the past, table scraps were commonly fed to animals or composted, and human waste was thrown out of windows (accompanied by warning cries) or deposited in holes below outhouses. Organic waste disposal was thus dependent on either portable vessels or special structures set apart from the typical building.

As water supplies were developed, water's advantages over the foul smell and inconvenience of these methods became irresistible. A typical sequence of events unfolded on Manhattan Island. In the 1700s, Manhattan was farm country that, like all other areas that later developed into large cities, had minimal water needs. Potable water was available in shallow wells and from some springs and streams. These sources were largely unaffected by the minor ground pollution from widely separated dry-pit privies (outhouses) that received human wastes. Paved city streets appeared in the 1800s, at which time the natural streams were enclosed in pipes called storm sewers. These pipes led the rainfall to the many waterways surrounding the island. Then in the late 1800s, flush toilets appeared. It seemed natural to connect the toilets to the already established storm sewers and to rename the pipeways combined sewers, which now carried both stormwater and so-called sanitary drainage to the rivers (sanitary for the building, but not for the rivers). Fast-flowing rivers were utilized as natural sewage treatment plants, and surprisingly, for many decades they did a fair job of keeping pollution reasonably in check. With the prospect of future sewage treatment plants, separate sanitary sewers were built. Also, there were some remaining (and some newly built) storm sewers that did not carry the wastes from toilets.

In cities where this confused pattern of sewer systems still exists (including most larger and older cities), it is now extremely difficult and expensive to sort out and reroute sewers so that only sanitary drainage goes to treatment plants and all storm drainage goes to waterways or into the ground. It seems particularly ironic that in most U.S. locations the rainwater that falls on a home's roof is adequate in both quality and quantity to supply a family's cleaning needs (21 + 14 = 35 g/cd) [132 L/cd]. In this chapter and in Chapter 21, these possibilities for rainwater are developed further.

As the human waste disposal place became a room within a building, the design issues grew more complex. Physically, there was a need for running water and for large-diameter pipes that sloped downward continuously from the toilet to a sewer or septic tank. As sewer gas became a recognized problem, an elaborate system of traps and vents became necessary. Again, cultural attitudes were also influential: How private an activity was this elimination from the body? To what extent could/should one plumbing fixture accommodate both body cleansing and waste elimination? If males insisted on standing rather than sitting while urinating, how could one devise a toilet that would also properly accommodate defecation, which requires a low seat?

(e) Cooling

Water has a remarkable cooling potential: It stores heat readily, removes large quantities of heat when it evaporates, and vaporizes readily at temperatures commonly found at the surface of human skin. In hot-dry climates, designers can place water surfaces (or sprays) upwind from the place to be cooled or resort to the evaporative coolers. Cooling towers are familiar components of large-building cooling systems.

Because all of us have experienced the physical cooling of the skin by water, we all carry psychological associations between water and cooling that can enhance our comfort on hot days. The sight of sunlight reflected on a water surface, with its characteristic "dancing" quality, connotes coolness, as does the sound of running or splashing water. Thus, even when water does not physically cool people, it can make an important psychological contribution to human comfort (Figure 20.3).

(f) Ornamental Uses

In almost any landscaping application, indoors or out, water becomes a center of interest. Our association of water with nourishing, cleansing, and cooling makes water a very powerful design element—a fact recognized by landscape designers throughout history. In arid regions, water is often used sparingly, in small, tightly controlled channels and at lower flow rates. The gardens of Islamic architecture in the Middle East are especially effective demonstrations of such design restraint. Where water is more plentiful, it has been used lavishly, as at the Villa d'Este in Tivoli, outside Rome, where much of a river's flow is diverted through the gardens.

Especially useful design characteristics of water include its reflectivity, which sets it apart from most plant and ground materials in a garden; its liquidity, which attracts attention to its motion and creates unique sounds wherever it is moved; and its life-sustaining potential, which allows the addition of both water plants and animals to a garden.

(g) Protective Uses

Every designer dreads water's ability to penetrate a roof and damage a building and its contents. However, we all depend on water as the best fire protection medium

Figure 20.3 Water as an aesthetic feature of a courtyard during the hot-arid season in Colima, Mexico. The sound and sight of running water add to a psychological impression of coolness.

available in most buildings. The vast quantities of water potentially required for firefighting must be delivered quickly—resulting in the necessity of large-diameter pipes regulated by very large valves. Because this system's distribution tree must be immediately apparent to firefighters, some degree of exposure is prudent. Despite its size and guarantee of at least partial exposure in public places, a fire protection water supply system is rarely treated as a visually integral design element. This mismatch of potential and actuality is discussed further in Chapter 25.

Another protective use of water has been as a means to control circulation; moats around castles may seem quaint today, but designers still sometimes use water as a means of directing traffic over a bridge to an entry.

20.2 THE HYDROLOGIC CYCLE

There is a finite quantity of water in the Earth and its atmosphere. The process whereby this water constantly circulates, powered by about one-fourth of the Earth's solar energy, is called the hydrologic cycle (Figure 20.4). More than 99% of this water is "inaccessible"—either because it is saltwater, or because it is frozen in glaciers and polar ice caps. The most accessible sources of water are precipitation and runoff.

Precipitation has the advantage of relative purity, although acid rain is a threat in many parts of the world, including much of the United States and Canada. Like solar energy, precipitation is a very large but very thinly spread resource; its capture is therefore likely to take place on an individual basis. It is predictable that until we experience a water crisis similar to the energy crisis that began in the 1970s, rainwater capture will remain a mostly untapped resource in the United States.

Runoff enjoys the advantage of a concentrated flow of water, which permits easy capture of large quantities. Its most serious disadvantage is the possibility of pollution—organic, chemical, and radioactive—depending on what is upstream from the point of capture. In some regions of North America, river water is reused 50 times on its way to the ocean. Further discussion of water sources and treatments occurs in Chapters 21 and 22.

That part of daily precipitation that neither evaporates nor joins the runoff becomes part of soil moisture. Much soil moisture is used by growing plants and is quickly transpired (evaporated) by the plant into the atmosphere. The remaining water works downward below the root zone of plants, eventually reaching a zone of saturation, where all voids in the ground are filled with water. This zone of saturation is called groundwater; the upper surface of groundwater is called the water table. Wells are commonly sunk to a point well below the water table so that the latter's seasonal fluctuations will not interrupt the well's access to groundwater.

20.3 BASIC PLANNING

After considering the relationship between a building and the roles that water plays within it, the designer must do basic sizing of the quantities of water needed, of the areas in which water will be used, and of the areas and equipment associated with water's return to the hydrologic cycle.

However, before discussing basic planning for the amounts of water used daily within buildings, and for the treatment of that water, we should note that water is also an important component of building construction. The production of 1 ton [907 kg] of bricks requires 580 gal [2200 L] of water; 1 ton of steel requires 43,600 gal [165,000 L]; 1 ton of plastic requires 348,750 gal [1.32 million L]. Water is one of the main ingredients of concrete; a typical 94-lb

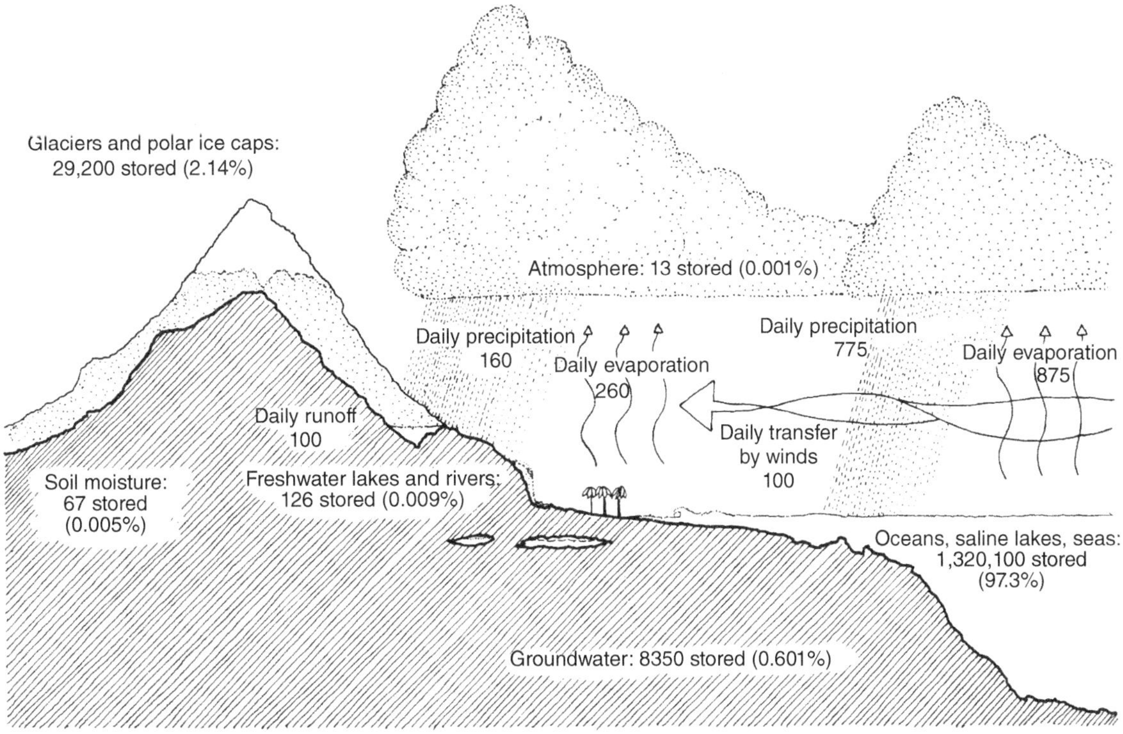

(*a*)

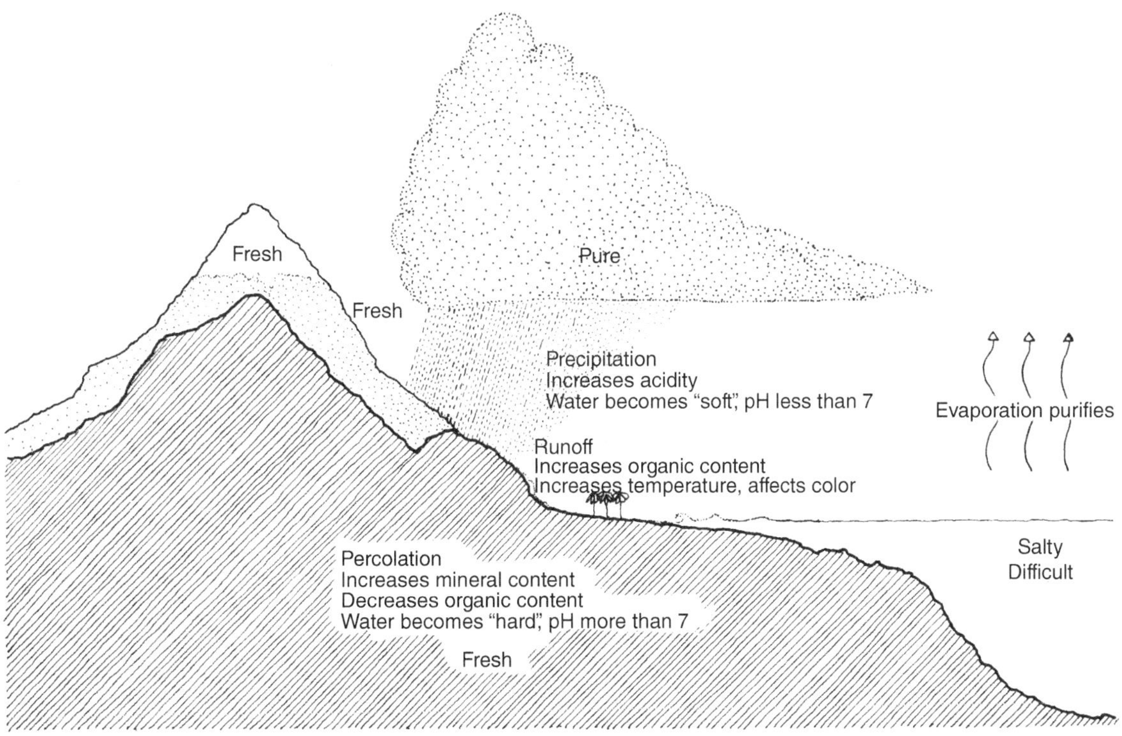

(*b*)

Figure 20.4 *(a) Hydrologic cycle. The figures given are in cubic kilometers. "Evaporation" includes transpiration from plants as well as evaporation from surfaces. "Precipitation" can be rain, hail, sleet, or snow. The vast majority of stored water is in the ocean. (b) Quality of water at various stages within this cycle.*

[42.6-kg] bag of cement requires about 6 gal [23 L] of water.

Another way in which buildings contribute to water consumption is through electricity use. Most power plants require very large quantities of water, which they quickly return to the hydrologic cycle warmer in temperature (and perhaps as vapor). A large nuclear power plant that utilizes a cooling tower can evaporate daily the approximate equivalent of transpiration from about 9 mi^2 [23 km^2] of forest.

(a) Water Supply

A water budget functions as a planning and operational management tool to help property owners quantify and allocate water supply needs. It is essentially a balance sheet of sorts which defines types and amounts of water supply and water usage sources. The task of estimating water needs is complicated by the conflict between current practice and conservation. Current practice tends toward the use of large amounts of water for very low-grade tasks. Conservation reserves high-quality water for high-grade tasks and emphasizes recycling as well as diminished overall usage of water.

Table 20.1 shows some common terms used in measuring the water supply. This book will rely primarily on gallons (1 gal = 128 fluid ounces) and liters (1 L = 33.8 fluid ounces).

Water supply must exceed water consumption requirements. As such, water supply can be estimated in a number of ways based on consumption patterns:

TABLE 20.1 Water Measurement Terms and Conversions

Quantity	
1 gal	= 8.33 lb
	= 231 in.^{3}
	= 0.134 ft^3
	= 3.785 L
1 liter (L)	= 0.0263 gal
1 cubic foot	= 7.48 gal
	= 62.4 lb
	= 0.028 m^3
1 cubic meter	= 1000 L
	= 35.32 ft^3
1 ton	= 240 gal
1 acre	= 0.4 ha
	= 40 km^2
1 acre-foot (a-f)	= 325,851 gal
	= 43,560 ft^3
	= 12 acre-inches (ac-in.)
1 million gallons (Mgal)	= 3.07 a-f
Flow	
1 cfs	= 1.98 a-f/day
	= 0.028 m^3/s
1 a-f/yr	= annual water supply for five people at 180 gal/cd
1000 gpm	= 2.23 cfs
	= 4.42 a-f/day
1 million gpd	= 694.4 gpm
	= 1.55 cfs
	= 1120 a-f/yr
Leaks	
Slow drip	= 170 gpd
	= 62,050 gal/yr
⅛-in. [3.2-mm] diameter stream	= 3600 gpd
	= 4 a-f/yr
Cost	
10¢ per 1000 gal	= 7.48¢ per 100 ft^3
	= \$32.59 per a-f
10¢ per 100 ft^3	= \$43.56 per a-f
	= \$13.40 per 1000 gal
10¢ per ton	= \$136 per a-f

Sources: Milne (1976) and Ferguson (1998).

Total Withdrawal. Water supply can be quantified on a total usage basis—or total withdrawal. For example, the USGS (2018) estimated that about 87% of the total U.S. population, approximately 283 million people, relied on public-supply water in 2015. The total withdrawal totaled 39,000 Mgal/d (million gallons per day).

Per Capita Per Day. Water supply requirements are often estimated in terms of gallons per capita per day [liters per capita per day]—abbreviated g/cd [L/cd]. This might also be conveyed as gallons [liters] per person per day. During conceptual or schematic design, water supply is often approximated in terms of gallons per capita per day [liters per capita per day]. Typical quantities matched with nourishing, cleansing, and other water supply requirements by building usage type in the United States are defined in Table 20.2. Within any given building usage type, a further breakdown is possible. For instance, the EPA (2024) estimates that the average U.S. citizen needs around 82 gal per day. Table 20.3 offers a breakdown from the U.S. Green Building Council (USGBC, 2019) for baselining current practice with regard to residential water supply needs in the U.S. By these figures, households require 50.2 gal [189.7 L] per capita per day—within the range of what Table 20.2 lists under "single family (per resident)."

Per Unit Per Day. For certain building types, it is instructive to consider water supply relative to functional units specific to the activities occurring within the building. According to the EPA, over 53,000 buildings (close to 20% of all buildings in their database) report water use data through the ENERGY STAR Portfolio Manager. Through this dataset, the EPA has ascertained indoor water usage data relative to certain business activities. For example, the median hospital needs 315 gal [1192 L] of water supply per bed per day; the median office requires 13 gal [49 L] per worker per day; and the median hotel requires 102 gal [386 L] per room per day (EPA, 2012).

TABLE 20.2 Planning Guide for Water Supply[a]

	Per Capita (as Listed) Daily Usage	
Building Usage	**Gallons**	**Liters**
Airports (per passenger)	3–5	11–19
Apartments, multiple-family (per resident)	60	227
Bath houses (per bather)	10	38
Camps		
Construction, semipermanent (per worker)	50	189
Day with no meals served (per camper)	15	57
Luxury (per camper)	100–150	378–568
Resorts, day and night, with limited plumbing (per camper)	50	189
Tourist, with central bath and toilet facilities (per person)	35	132
Cottages with seasonal occupancy (per resident)	50	189
Courts, tourist, with individual bath units (per person)	50	189
Clubs		
Country (per resident member)	100	378
Country (per nonresident member present)	25	95
Dwellings		
Boardinghouses (per boarder)	50	189
Additional kitchen requirements for nonresident boarders	10	38
Luxury (per person)	100–150	378–568
Multiple-family apartments (per resident)	40	151
Rooming houses (per resident)	60	227
Single family (per resident)	50–75	189–284
Estates (per resident)	100–150	378–568
Factories (per person per shift)	15–35	57–132
Highway rest area (per person)	5	19
Hotels with private baths (two persons per room)	60	227
Hotels without private baths (per person)	50	189
Institutions other than hospitals (per person)	75–125	284–473
Hospitals (per bed)	250–400	946–1514
Laundries, self-service (per washing)	50	189
Livestock (per animal)		
Cattle (drinking)	12	45
Dairy (drinking and servicing)	35	132
Goat (drinking)	2	8
Hog (drinking)	4	15
Horse (drinking)	12	45
Mule (drinking)	12	45
Sheep (drinking)	2	8
Steer (drinking)	12	45
Motels with bath, toilet, and kitchen facilities (per bed space)	50	189
With bed and toilet (per bed space)	40	151
Parks		
Overnight, with flush toilets (per camper)	25	95
Trailer, with individual bath units, no sewer connection (per trailer)	25	95
Trailer, with individual baths, connected to sewer (per person)	50	189
Picnic		
With bath houses, showers, and flush toilets (per picnicker)	20	76
With toilet facilities only (per picnicker)	10	38
Poultry		
Chickens (per 100)	5–10	19–38
Turkeys (per 100)	10–18	38–68
Restaurants		
With toilet facilities (per patron)	7–10	26–38
Without toilet facilities (per patron)	2.5–3	9–11
With bar/cocktail lounge (additional quantity per patron)	2	8
Schools		
Boarding (per pupil)	75–100	284–378
Day, with cafeteria, gymnasium, and showers (per pupil)	25	95
Day, with cafeteria but no gymnasiums or showers (per pupil)	20	76
Day, without cafeteria, gymnasiums, or showers (per pupil)	15	57
Service stations (per vehicle)	10	38
Stores (per toilet room)	400	1514
Swimming pools (per swimmer)	10	38
Theaters		
Drive-in (per car space)	5	19
Movie (per auditorium seat)	5	19
Workers		
Construction (per person per shift)	50	189
Day (school or office, per person per shift)	15	57

Source: Adapted from U.S. Environmental Protection Agency (1975).

[a]These values may be reduced as follows: with flow controls, up to 25% reduction; with water recycling, up to 50% reduction.

TABLE 20.3 Indoor Water Baseline Consumption (per Person per Day)

Fixture	Baseline Flush or Flow Rate		Estimated Fixture Usage	Estimated Water Usage	
Shower (per compartment)	2.5 gpm	9.5 Lpm	6.15 min	15.4 gal	58.4 L
Lavatory, kitchen faucet	2.2 gpm	8.3 Lpm	5.0 min	11 gal	41.5 L
Toilet	1.6 gpf	6 Lpf	5.05 flushes	8 gal	30.3 L
Clothes washer	9.5 WF	9.5 WF	0.37 cycles @ 3.5 ft^3 [@ 0.1 m^3]	15.1 gal	57.1 L
Dishwasher	6.5 gpc	24 Lpc	0.1 cycles	0.7 gal	2.4 L

Source: Adapted from U.S Green Building Council (USGBC, 2019).
gpm = gallons per minute
gpf = gallons per flush
WF = water factor
gpc = gallons per cycle
Lpf = liters per flush
Lpm = liters per minute
Lpc = liters per cycle.

Water Use Intensity (WUI). With increased frequency, WUI is being used to track a building property's water consumption much as EUI (energy use intensity) is used to measure energy consumption. Water use intensity basically refers to the rate at which water is used on a per building unit area basis. Measured in gallons per square foot per year (gal/ft^2/yr) or liters per square meter per year (L/m^2/yr), WUI is an indicator of how much water a building consumes per year during typical operation. Relative to a water budget or a comparable baseline, WUI can serve as a gauge of a building's water efficiency. Building performance simulation tools can help project teams develop a predicted water use intensity (pWUI), while metered data will reveal a building's actual WUI.

Due to the highly variable relative magnitude of outdoor water use compared to indoor water use, scoping should be clarified when assessing WUI figures. General WUI indicates a consideration of all water sources by a building and site excluding parking and irrigated areas. Indoor WUI considers only indoor water usage (i.e., showers, toilets, lavatories, kitchen equipment, and other indoor devices). If unspecified, assume WUI figures are general.

Like EUI, a building's WUI lends itself to comparison in order to gauge overall water use effectiveness. Table 20.4 offers WUI reference values from the EPA (2023) for certain property types.

Should water consumption data be utilized to quantify or otherwise assess water supply needs, it is notable that different utilities use different units for measuring water consumption. The most common units are the gallon but also centum cubic feet (CCF). The word "centum" is derived from the Roman word for hundred. As such, a CCF may also be identified as a hundred cubic feet (HFC). One CCF (visualize a 5-ft by 5-ft by 4-ft [1.5 by 1.5 by 1.2 m] container filled with water) is equal to 748 gal [2830 L] (EPA, 2024).

To gain an appreciation of how the rate of urban water usage has changed over time, consider the following figures (from Milne, 1976):

Imperial Rome	38 g/cd [144 L/cd]
London, 1912	40 g/cd [151 L/cd]
American cities just before World War II	115 g/cd [435 L/cd]
Los Angeles, mid-1970s	182 g/cd [689 L/cd]

Compare these usage rates to the "Basic Water Requirement" for four domestic needs—drinking, sanitation, bathing, and cooking—of about 13 g/cd [50 L/cd] proposed by Gleick (1998).

Although the historical trend has clearly been toward higher per capita use of water, a recent emphasis on conservation has resulted in significant changes in this pattern. Table 20.2 can be used to estimate the daily indoor usage of water in various facilities if current practices are anticipated. For a very rough approximation of conservation effects:

- Reduce Table 20.2 values by 25%, assuming simple conservation measures such as flow controls.
- Reduce Table 20.2 values by 50%, assuming partial recycling.

The amount of water used for flushing toilets within the building types listed in Table 20.2 shows the great potential for savings through conservation.

In urban areas, public water mains usually provide the necessary quantities of water at the pressures and rates of flow required to operate typical plumbing fixtures. For isolated buildings or those independent of public networks, the water supply can come from individual sources: wells, springs, cisterns, lakes, and so forth. For the minimum pressures and flow rates necessary from these sources (and/or the storage vessels associated with them), see Chapter 21.

TABLE 20.4 National Water Use Intensity (WUI) Reference Values for Select ENERGY STAR Portfolio Manager Property Types

Part A. I-P Units

Portfolio Manager Property Type	Number of Observations	WUI (gal/ft²/yr) Percentile 5th	25th	50th/ Median	75th	95th	Median Property Type-Specific Metrics
College/University	590	2.48	6.20	13.40	28.34	66.92	N/A
Fire Station	126	5.00	13.95	28.90	54.29	162.95	N/A
Hospital (General Medical & Surgical)	347	14.30	38.28	55.71	77.44	151.18	107,000 gal/hospital bed/yr
Hotel	1488	23.94	40.06	52.02	69.28	120.23	33,500 gal/guest room/yr
K-12 School (includes Preschool/Daycare)	1588	3.04	6.88	10.84	19.71	55.69	11,200 gal/worker/yr
Medical Office (includes Urgent Care/Clinic/Other Outpatient)	1177	5.73	12.86	23.40	38.82	84.71	N/A
Multifamily[a]	258	16.76	31.71	45.15	65.56	126.14	43,600 gal/unit/yr
Non-refrigerated Warehouse (includes Self-Storage Facility and Distribution Center)	1675	1.15	1.84	3.43	6.69	20.88	N/A
Office (includes Bank Branch/Financial Office/Other Offices)	9627	2.93	7.83	14.48	29.41	81.52	6020 gal/worker/yr
Retail Store	4382	1.41	2.84	5.24	8.99	28.59	N/A
Senior Living Community	1232	22.18	39.80	56.41	77.69	129.26	39,800 gal/resident/yr
Supermarket (includes Grocery Store, Wholesale Club and Supercenter)	432	9.63	21.03	31.00	45.58	78.42	N/A

Part B. SI Units[b]

Portfolio Manager Property Type	Number of Observations	WUI (L/m²/yr) Percentile 5th	25th	50th/ Median	75th	95th	Median Property Type-Specific Metrics
College/University	590	101	253	546	1155	2727	N/A
Fire Station	126	204	568	1178	2212	6640	N/A
Hospital (General Medical & Surgical)	347	583	1560	2270	3155	6160	405,039 L/hospital bed/yr
Hotel	1488	975	1632	2120	2823	4899	126,811 Liters/guest room/yr
K-12 School (includes Preschool/Daycare)	1588	124	280	442	803	2269	42,397 L/worker/yr
Medical Office (includes Urgent Care/Clinic/Other Outpatient)	1177	233	524	953	1582	3452	N/A
Multifamily[b]	258	683	1292	1840	2671	5140	165,044 L/unit/yr
Non-refrigerated Warehouse (includes Self-Storage Facility and Distribution Center)	1675	47	75	140	273	851	N/A
Office (includes Bank Branch/Financial Office/Other Offices)	9627	119	319	590	1198	3322	22,788 L/worker/yr
Retail Store	4382	57	116	214	366	1165	N/A
Senior Living Community	1232	904	1622	2298	3166	5267	150,659 L/resident/yr
Supermarket (includes Grocery Store, Wholesale Club and Supercenter)	432	392	857	1263	1857	3195	N/A

Source: Adapted from EPA (2023).

[a]Data from 2012 Fannie Mae Energy and Water Market Research Survey.

[b]SI unit conversions by the authors.

(b) Cisterns

Where rainwater is to be utilized, a rough approximation of catchment area and cistern storage volume is initially needed. This procedure is detailed later in this chapter. For now:

1. From Table 20.2, find the quantity of rainwater to be used daily:

$$\text{g/cd} \times \text{population} = \text{gpd}$$

$$[\text{L/cd} \times \text{population} = \text{L/d}]$$

2. Convert this quantity to the yearly need for water:

$$\text{gpd} \times 365 \text{ days} = \text{gal/yr}$$

$$[\text{L/d} \times 365 \text{ days} = \text{L/yr}]$$

3. Assume, conservatively, that a dry year will have two-thirds of the precipitation of an average year; this conservative value is the design precipitation. Average annual precipitation is available from National Oceanic and Atmospheric Administration (NOAA) annual summaries.

$$\text{average annual precipitation} \times (2/3)$$

$$= \text{design precipitation}$$

4. From Figure 20.5, determine the catchment area required.
5. Roughly size the cistern (storage) capacity by finding the longest dry period (consecutive days of negligible rainfall, from NOAA local climatological data):

$$\text{cistern capacity} = \text{gpd} \times \text{days of dry period}$$

6. Convert capacity to volume by the relationship:

$$1 \text{ ft}^3 \text{ stores } 7.48 \text{ gal of water}$$

$$[1 \text{ m}^3 \text{ stores } 1000 \text{ L of water}]$$

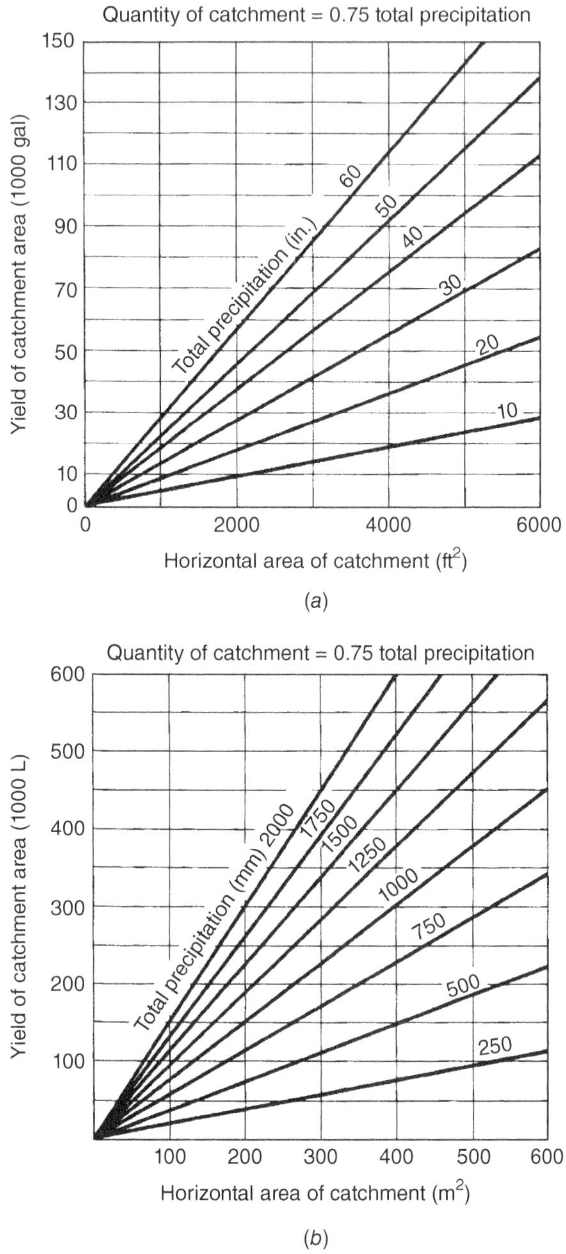

Figure 20.5 *Yields of rainfall catchment areas (roofs) in terms of total precipitation. In these graphs, 75% of the total precipitation is assumed to be catchable; the remainder is lost to evaporation or spillage. (a: I-P units from the U.S. EPA, 1975; b: SI units [1 m³ = 1000 L].)*

EXAMPLE 20.1 PART A A 20,000-ft² [1860-m²] one-story factory in central Oregon will use roof-collected rainwater to flush its toilets. Water-conserving 1.3 gal/flush [4.9 L/flush] toilets will serve 40 workers. The project team determined that the factory's workers would use 15 g/cd [57 L/cd]:

$$15 \text{ g/cd} \times 40 \text{ workers} = 600 \text{ gpd for all usages}$$

$$[57 \text{ L/cd} \times 40 \text{ workers} = 2280 \text{ L}]$$

Assuming the use of water-efficient fixtures and fittings, reduce this figure by 35%:

$$600 \text{ gpd} \times 0.65 = 390 \text{ gpd}$$

$$[2280 \text{ L/d} \times 0.65 = 1482 \text{ L/d}]$$

Approximate the proportion of water usage associated with the toilets. Assuming an average of 3.75 flushes per day per worker:

$$3.75 \text{ flushes/day} \times 1.3 \text{ gpf} \times 40 \text{ workers} = 195 \text{ gpd}$$

$$[3.75 \text{ flushes/day} \times 4.9 \text{ Lpf} \times 40 \text{ workers} = 735 \text{ L/d}]$$

The toilets are anticipated to account for roughly half of the per capita water consumption. Being a bit conservative, assume that up to 200 gpd [760 L/d] of rainwater will be utilized.

A. Catchment Area: The average annual rainfall in this central Oregon location is 41 in. [1042 mm].

Design (i.e., dry year) rainfall is:

$$2/3 \times 41 \text{ in.} = 27.3 \text{ in.}$$

$$[2/3 \times 1042 \text{ mm} = 695 \text{ mm}]$$

The approximate annual rainwater supply needed for the toilets is:

$$200 \text{ gpd} \times 365 \text{ days} = 73{,}000 \text{ gal}$$

$$[760 \text{ L/d} \times 365 \text{ days} = 277{,}400 \text{ L}]$$

The combination of 73,000 gal [277,400 L] and 27.3 in. [695 mm] is off the chart in Figure 20.5*a*, so divide the 73,000 gal [277,400 L] yearly need by two to obtain 36,500 gal [138,700 L] and then double the resulting catchment area. At 27.3 in. [695 mm] of precipitation, 36,500 gal [138,700 L] will be caught by 2800 ft^2 [260 m^2]. The catchment area for 73,000 gal [277,400 L] is, therefore, $2 \times 2800\ ft^2 = 5600\ ft^2$ [$2 \times 260\ m^2 = 520\ m^2$].

Therefore, 5600 ft^2 [520 m^2]—about 28% of the 20,000-ft^2 [1860-m^2] factory roof area—will suffice for a catchment area.

B. Cistern Capacity: The site location normally has very dry summers; average monthly rainfall is as follows:

May	2.1 in. [53 mm]
June	1.4 in. [36 mm]
July	0.4 in. [10 mm]
August	0.6 in. [15 mm]
September	1.5 in. [38 mm]
October	4.0 in. [102 mm]

The dry period, then, runs from mid-June to mid-September—about 90 days. Thus, capacity = 200 gpd × 90 days = 18,000 gal [760 L/d x 90 = 68,400 L]:

$$\text{volume} = (18{,}000 \text{ gal}/7.48 \text{ g/ft}^3) = 2406 \text{ ft}^3$$

$$[(68{,}400 \text{ L}/1000 \text{ L/m}^3) = 68 \text{ m}^3]$$

As one option, a 5 ft deep × 22 ft square = 2420 ft^3 [1.5 m × 6.7 m square = 67 m^3] would provide adequate capacity. A more detailed sizing procedure for this building's cistern is presented later in this chapter. ■

(c) Required Facilities

Another important early design question is: How many plumbing fixtures should be provided? Table 20.5 lists the minimum plumbing facilities required for various types and sizes of building occupancies. Note that local requirements may differ somewhat from this particular guide, which is taken from the International Code Council's *2024 International Plumbing Code (IPC)*. There are several plumbing codes in use in North America, and they sometimes disagree. Because all codes give minimum requirements, more generous provisions are sometimes appropriate. Chapter 21 discusses the design of the spaces in which these services are used.

Male/Female Fixture Ratios. The minimum required number of water closets and lavatories in Table 20.5 reflects the long-standing codification of providing separated restrooms for males and females. Section 403.1.1 of the *2024 IPC* states, "To determine the occupant load of each sex, the total occupant load shall be divided in half." It goes on to explain that any fractional numbers resulting from applying the fixture ratios depicted in Table 20.5 shall be rounded up to the next whole number. As such, the male/female separation of restroom facilities remains the default.

The 50/50 male-to-female split is codified in the *2024 IPC*; however, there are two notable exceptions in section 403.1.1. The first exception states, "The total occupant load shall not be required to be divided in half where approved statistical data indicate a distribution of the sexes of other than 50 percent of each sex." (The "approval" will need to come from the authority having jurisdiction with regard to the building project.)

All-Gender Restrooms. The second exception to the 50/50 male-to-female split recognizes and accommodates non-binary gender identification. The *2024 IPC* states, "Where multiple-user facilities are designed to serve all genders, the minimum fixture count shall be calculated 100 percent, based on the total occupant load. In such multiple-user facilities, each fixture type shall be in accordance with ICC A117.1." (*ICC A117.1: Accessible and Usable Buildings and Facilities* is a technical standard that provides guidance for building design solutions that accommodate persons with physical disabilities. Many jurisdictions have adopted editions of *ICC A117.1* as accessibility codes.) Recent advocacy for combined all-gender restroom solutions (otherwise referred to as gender-neutral or unisex restrooms) resulted in the ICC revising their model international codes (collectively referred to as "I-codes") to accommodate such solutions. ICC's work in this regard began when both the *2018 International Building Code (IBC)* and *2018 IPC* updated signage requirements for single-user restrooms, specifying that such facilities "shall be identified for use by either sex."

(d) Sewage

Where public sewers are available (as in most urban areas), the designer usually is not concerned with estimating the flow of sewage. However, where private or on-site sewage treatment is required, or where public sewage treatment facilities are overtaxed, total sewage flow is an early design concern. Table 20.6 lists these sewage flows by type of occupancy, again in g/cd [L/cd]. Note that sewage flow may differ from supply flow, especially where supply water is used for irrigation, for car washing, or in evaporative processes.

TABLE 20.5 Minimum Number of Plumbing Facilities[a]

Classification	Description	Water Closets, Urinals[b]		Lavatories		Bathtubs/ Showers	Drinking Fountains[c]	Other
		Male	Female	Male	Female			
Assembly	Theaters and other buildings for the performing arts and motion pictures[d]	1 per 125	1 per 65	1 per 200		—	1 per 500	1 service sink
	Nightclubs, bars, taverns, dance halls and buildings for similar purposes[d]	1 per 40	1 per 40	1 per 75		—	1 per 500	1 service sink
	Restaurants, banquet halls and food courts[d]	1 per 75	1 per 75	1 per 200		—	1 per 500	1 service sink
	Casino gaming areas	1 per 100 for the first 400 and 1 per 250 for the reminder exceeding 400	1 per 50 for the first 400 and 1 per 150 for the reminder exceeding 400	1 per 250 for the first 750 and 1 per 500 for the reminder exceeding 750		—	1 per 1000	1 service sink
	Auditoriums without permanent seating, art galleries, exhibition halls, museums, lecture halls, libraries, arcades and gymnasiums[d]	1 per 125	1 per 65	1 per 200		—	1 per 500	1 service sink
	Passenger terminals and transportation facilities[d]	1 per 500	1 per 500	1 per 750		—	1 per 1000	1 service sink
	Places of worship and other religious services[d]	1 per 150	1 per 75	1 per 200		—	1 per 1000	1 service sink
	Coliseums, arenas, skating rinks, pools and tennis courts for indoor sporting events and activities[e]	1 per 75 for the first 1500 and 1 per 120 for the remainder exceeding 1500	1 per 40 for the first 1520 and 1 per 60 for the remainder exceeding 1520	1 per 200	1 per 150	—	1 per 1000	1 service sink
	Stadiums, amusement parks, bleachers and grandstands for outdoor sporting events and activities[e]	1 per 75 for the first 1500 and 1 per 120 for the remainder exceeding 1500	1 per 40 for the first 1520 and 1 per 60 for the remainder exceeding 1520	1 per 200	1 per 150	—	1 per 1000	1 service sink
Business	Buildings for transaction of business, nonmedical professional services, other services involving merchandise, office buildings, banks, light industrial and similar uses	1 per 25 for the first 50 and 1 per 50 for the remainder exceeding 50		1 per 40 for the first 80 and 1 per 80 for the remainder exceeding 80		—	1 per 100	1 service sink[f]
	Ambulatory care facilities and outpatient clinics	1 per 25 for the first 50 and 1 per 50 for the remainder exceeding 50	1 per 25 for the first 50 and 1 per 50 for the remainder exceeding 50	1 per 50		—	1 per 100	1 service sink per floor

Educational	Educational facilities		1 per 50		1 per 50	—	1 per 100	1 service sink
Factory and industrial	Structures in which occupants are engaged in work fabricating, assembly or processing of products or materials		1 per 100		1 per 100	—	1 per 400	1 service sink
Institutional	Alcohol and drug centers[g] Congregate care facilities[g] Group homes[g] Halfway houses[g] Social rehabilitation facilities[g] Foster care facilities[g]		1 per 10 care recipients		1 per 10 care recipients	1 per 8 care recipients	—	—
	Assisted living and residential board and care facilities with care recipients who receive custodial care	Sleeping units for care recipients[h]	1 per 2 sleeping units		1 per 2 sleeping units	1 per 8 sleeping units	—	—
		Dwelling units for care recipients	1 per dwelling unit		1 per dwelling unit	1 per dwelling unit	—	1 kitchen sink per dwelling un t
		Employee facilities	1 per 60 care recipient units		1 per 60 care sleeping recipient units	—	1 per 100	1 service sink per floor
		Visitor facilities	1 per 75 care recipient units		1 per 75 care recipient sleeping units	—	—	—
	Nursing homes[g]	Sleeping units for care recipient[h]	1 per 2 care recipient sleeping units		1 per 2 care recipient sleeping units	1 per 8 care recipient sleeping units	—	—
		Employee facilities	1 per 60 care recipient units		1 per 60 care recipient sleeping units	—	1 per 100	1 service sink per floor
		Visitor facilities	1 per 75 care recipient units		1 per 75 care recipient sleeping units	—	—	—
	Hospitals	Sleeping units for care recipient	1 per care recipient sleeping unit		1 per care recipient sleeping unit	1 per 100 care recipient sleeping units	—	—
		Care recipient treatment areas	1 per 25 care recipient treatment rooms		1 per 50 care recipient treatment rooms	—	1 per 100	—
		Employee facilities	1 per 25 care recipient sleeping units or treatment room	1 per 25 care recipient sleeping units or treatment room	1 per 50 care recipient sleeping room or treatment room		1 per 100	1 service sink per floor
		Visitor facilities	1 per 75 care recipient sleeping units or treatment room	1 per 75 care recipient sleeping units or treatment room	1 per 50 care recipient sleeping room or treatment room		1 per 500	—

TABLE 20.5 Minimum Number of Plumbing Facilities[a] *(Continued)*

Classification	Description		Water Closets, Urinals[b]		Lavatories		Bathtubs/ Showers	Drinking Fountains[c]	Other
			Male	Female	Male	Female			
	Prisons[g]		1 per cell		1 per cell		1 per 15	1 per 100	1 service sink
	Reformatories, detention centers and correctional centers[g]	Cells	1 per 15		1 per 15		1 per 15	1 per 100	1 service sink
		Congregate living facilities	1 per 15		1 per 15		1 per 15	1 per 100	1 service sink
		Employees	1 per 25		1 per 35		—	1 per 100	—
	Adult day care and child day care		1 per 15		1 per 15		1	1 per 100	1 service sink
Mercantile	Retail stores, service stations, shops, salesrooms, markets and shopping centers		1 per 500		1 per 750		—	1 per 1000	1 service sink[f]
Residential	Hotels, motels, boarding houses (transient)		1 per dwelling or sleeping unit		1 per dwelling or sleeping unit		1 per dwelling or sleeping unit	—	1 service sink
	Dormitories, fraternities, sororities and boarding houses (not transient)		1 per 10		1 per 10		1 per 8	1 per 100	1 service sink
	Apartment house		1 per dwelling unit or sleeping unit		1 per dwelling unit or sleeping unit		1 per dwelling unit or sleeping unit	—	1 kitchen sink per dwelling unit; 1 automatic clothes washer connection per 20 dwelling units
	Congregate living facilities with 16 or fewer care recipients receiving custodial care		1 per 10		1 per 10		1 per 8	—	1 kitchen sink
	One- and two-family dwellings and lodging houses with five or fewer guestrooms		1 per dwelling unit		1 per dwelling unit		1 per dwelling unit	—	1 kitchen sink per dwelling unit; 1 automatic clothes washer connection per dwelling unit

Storage	Structures for the storage of goods, warehouses, storehouse and freight depots. Low and Moderate Hazard.	1 per 100	1 per 100	—	1 per 1000	1 service sink

[a]The fixtures shown are based on one fixture being the minimum required for the number of persons indicated or any fraction of the number of persons indicated. The number of occupants shall be determined by the *International Building Code*.

[b]In each bathroom or toilet room, urinals shall not be substituted for more than 67 percent of the required water closets for males according to the table in assembly and educational occupancies. Urinals shall not be substituted for more than 50 percent of the required water closets for males according to the table in all other occupancies.

[c]See Section 410 of the *2024 International Plumbing Code* for additional information and requirements. Drinking fountains shall not be required for an occupant load of 15 or fewer.

[d]The occupant load for seasonal outdoor seating and entertainment areas shall be included when determining the minimum number of facilities required.

[e]The required number and type of plumbing fixtures for indoor and outdoor public swimming pools shall be in accordance with Section 609 of the *International Swimming Pool and Spa Code*.

[f]For business and mercantile classifications with an occupant load of 15 or fewer, service sinks shall not be required.

[g]Toilet facilities for employees shall be separate from facilities for inmates or care recipients.

[h]A single-user toilet facility with one water closet and one lavatory serving not more than two adjacent care recipient sleeping units shall be permitted provided that each patient sleeping unit has direct access to the toilet room and provision for privacy for the toilet room user is provided.

TABLE 20.6 Estimated Sewage Flow Rates[a,b,c]

	"Unit"	
Type of Occupancy	**Gallons per Day**	**Liters per Day**
Airports (per employee)	15	57
Airports (per passenger)	5	19
Auto washers—check with equipment manufacturer	–	–
Bowling alleys—with snack bar only (per lane)	75	284
Campground—with central comfort station (per person)	35	132
Campground—with flush toilets—no showers (per person)	25	95
Camps (day)—no meals served (per person)	15	57
Camps (summer and seasonal camps)—(per person)	50	189
Churches—sanctuary (per seat)	5	19
Churches—with kitchen waste (per seat)	7	26
Dance halls—(per person)	5	19
Factories—no showers (per employee)	25	95
Factories—with showers (per employee)	35	132
Factories—with cafeteria (per employee)	5	19
Hospitals—(per bed)	250	946
Hospitals—kitchen waste only (per bed)	25	95
Hospitals—laundry waste only (per bed)	40	151
Hotels—no kitchen waste (per bed)	60	227
Institutions—resident (per person)	75	284
Nursing home—(per person)	125	473
Rest home—(per person)	125	473
Laundries—self-service with minimum 10 hours per day (per wash cycle)	50	189
Laundries—commercial, check with manufacturer's specifications	–	–
Motel (per bed space)	50	189
Motel—with kitchen (per bed space)	60	227
Offices—(per employee)	20	76
Parks—mobile homes (per space)	250	946
Parks (picnic) - with toilets only (per parking space)	20	76
Parks (recreational vehicles)—without water hook-up (per space)	75	284
Parks (recreational vehicles)—with water and sewer hook-up (per space)	100	379
Restaurants—cafeteria (per employee)	20	76
Restaurants—with toilet waste (per customer)	7	26
Restaurants—with kitchen waste (per meal)	6	23
Restaurants—with kitchen waste disposable service (per meal)	2	8
Restaurants—with garbage disposal (per meal)	1	4
Restaurants—with cocktail lounge (per customer)	2	8
School staff and office (per person)	20	76
Schools—elementary (per student)	15	57
Schools—intermediate and high (per student)	20	76
Schools—with gym and showers (per student)	5	19
Schools—with cafeteria (per student)	3	11
Schools (boarding)—total waste (per person)	100	379
Service station—with toilets for 1st bay	1000	3785
Service station—with toilets for each additional bay	500	1893
Stores—(per employee)	20	76
Stores—with public restrooms (per 10 square feet of floor space)	1	4
Swimming pools—public (per person)	10	38
Theaters—auditoriums (per seat)	5	19
Theaters—with drive-in (per space)	10	38

Source: © 2024 Uniform Plumbing Code. Reprinted with the permission of the International Association of Plumbing and Mechanical Officials. This copyright material and all points or statements in using this material have not been reviewed by IAPMO. The opinions expressed herein are not representations of fact from IAPMO.

Notes:

[a]Sewage disposal systems sized using the estimated waste/sewage flow rates shall be calculated as follows:

(a) Waste/sewage flow, up to 1500 gal per day [5678 L/day]
 Flow × 1.5 = septic tank size

(b) Waste/sewage flow, over 1500 gal per day [5678 L/day]
 Flow × 0.75 + 1125 = septic tank size

(c) Secondary system shall be sized for total flow per 24 hours.

[b]See Section H 201.1 of the *2024 Uniform Plumbing Code*.

[c]Because of the many variables encountered, it is not possible to set absolute values for waste/sewage flow rates for all situations. The designer should evaluate each situation and, where values in this table need modification; they should be made with the concurrence of the authority having jurisdiction.

Once the daily flow of sewage is established, some early guidelines are needed for determining the suitability of and the area required by treatment processes. One of the most common reasons for the rejection of a potential rural building site is a lack of suitability for sewage disposal. A geologic analysis of structural and sewage disposal potential is one of the first documents needed by the designer. Chapter 22 describes the sizing of some common treatment methods. At this earlier stage, the following design guidelines can be useful:

Septic Tank Drainfields In I-P units (minimum 750-ft^2 area for any system):

- For shallow trenches in poorly draining soil: drainfield area = total sewage flow in gpd × 3.6 ft^2/gal
- For deep trenches in well-draining soil: drainfield area = total sewage flow in gpd × 0.4 ft^2/gal

In SI units (minimum 70-m^2 area for any system):

- For shallow trenches in poor soil: drainfield area = L/day × 0.087 m^2/L
- For deep trenches in good soil: drainfield area = L/day × 0.01 m^2/L

These guidelines allow for an expansion area equal to the original size of the drainage field in case of field failure.

Mounds These are built-up leaching fields on top of the existing grade (see Chapter 22). For a single-family dwelling, allow for a 4-ft-high [1.2-m-high] mound whose bottom area is a square, 44 ft [13.4 m] on each side and whose sides slope at a 1:3 vertical-to-horizontal ratio.

Package Sewage Plant Drainfields In these, sewage is treated to a much greater extent than in septic tanks, and effluent is filtered:

- In poorly draining soil: drainfield area = total sewage flow in gpd × 0.49 ft^2/gal [L/day × 0.012 m^2/L]
- In well-draining soil: drainfield area = total sewage flow in gpd × 0.23 ft^2/gal [L/day × 0.006 m^2/L]

Sewage Lagoons These consist of at least two open treatment ponds (primary and secondary) and are sized on the basis of pounds of biochemical oxygen demand (BOD) rather than gallons of sewage flow. A typical assumption for estimating BOD is:

- 0.2 lb [91 g] BOD/person for ordinary domestic sewage
- 0.3 lb [136 g] BOD/person where garbage grinders or other devices contribute added organic material to domestic sewage

The total land area required for the two ponds can be estimated as:

- 20 lb BOD/acre [22 kg BOD/ha] for the (colder) northern United States
- 35 lb BOD/acre [39 kg BOD/ha] for the (drier, warmer) southern United States

(The primary pond is usually sized for 50 lb BOD/acre [56 kg BOD/ha].)

20.4 RAINWATER

There is a striking similarity between rainwater and solar energy. Both are essential for agriculture, and both have been well understood and utilized by farmers since agrarian societies first emerged. Both can be very beneficial to architecture and were utilized as needed by vernacular builders for centuries. Both fell out of favor with designers as plentiful supplies of pure, centrally treated water and concentrated, centrally controlled fuels became commonplace. Both are thinly, yet relatively evenly, spread over the world's population, so they are at least seasonally available to help meet nearly every building's needs for water and heat. Yet both are difficult to utilize in industrialized societies, because they require individual preoccupancy expenditures.

Consider the typical public water and electric utility in a U.S. city. It can raise the funds to build large water treatment plants, electricity-generating plants, and the network of pipes and wires that bring these commodities to every building. The utility's costs, including interest on its construction debts, will be passed on to its consumers on a monthly basis, along with a margin of profit that is usually controlled by state governments. Thus, our society has a well-established method for encouraging central suppliers of water and power.

Now consider the individual building owner. To build a cistern and a solar-heated building, she must borrow money at an interest rate higher than that which the utility pays, and both options cost more initially than simple connections to the utility's pipes and lines. Even though she is willing to flush her toilets with rainwater rather than with chlorinated and filtered potable water from the utility, and even though she is willing to heat with lower-grade solar energy rather than with higher-grade electricity, she must pay a substantial first-cost penalty—with interest—to do so.

Figure 20.6 This low-entropy kindergarten competition entry for Frankfurt, Germany, features sun, adjustable shade, and a roof that collects rainwater and leads it to an interior stone-filled catchment. The rainwater downspout can be seen in the center of the drawing; the stone-filled catchment is below the stair landing. The overflow then goes outside, through a layer of stones and into a belowground cistern. (Courtesy of Fernau & Hartman, Architects.)

The overall public good could be well served by a mixture of public networks of pure water and electricity, combined with individual cisterns and solar applications. The environmental benefits would be substantial: less water withdrawn from rivers, lakes, and underground aquifers; less energy and chemicals used to treat and deliver such water; less stormwater discharged to pollute rivers; less fuel used to generate electricity; and less environmental damage from power plants. Yet we continue to economically discourage the individual who uses the rain and the sun.

Architects Fernau and Hartman's design for an invited competition for a "low-entropy kindergarten" in Frankfurt, Germany (Figure 20.6), celebrates sun and rainwater as teaching elements. In their words, "The building reveals natural phenomena such as the play of light and shadow over a wall during the course of a year, the warmth of a wintergarden in January, the cooling shade from deciduous vines on a trellis overhead, water pouring off the roof into an interior stone catchment during a downpour."

20.5 COLLECTION AND STORAGE

In terms of both quality and quantity, rainwater is an attractive alternative. Figure 20.4*b* shows that rainwater is close to the purest state in the hydrologic cycle. Recently, it is true, air pollution has begun to threaten the quality of rainwater in some areas as acid rain became widespread in the northeastern section of the North American continent and in Europe. In some particularly air-polluted locations, lead poses a threat to rainwater quality. Also, on any catchment surface, dust and bird droppings are common pollutants that must be considered. Other factors that bear on rainwater quality are roofing materials and the form of the roof. An appropriate health authority should be consulted for a list of roofing materials that will have no toxic effects on rainwater capture. Steeper roofs are scoured by winds, and thus collect less dust and give cleaner runoffs. Devices to discourage roosting birds are strongly recommended, as are periodic checks of cistern water for bacteria. Fungicides (for moss control)

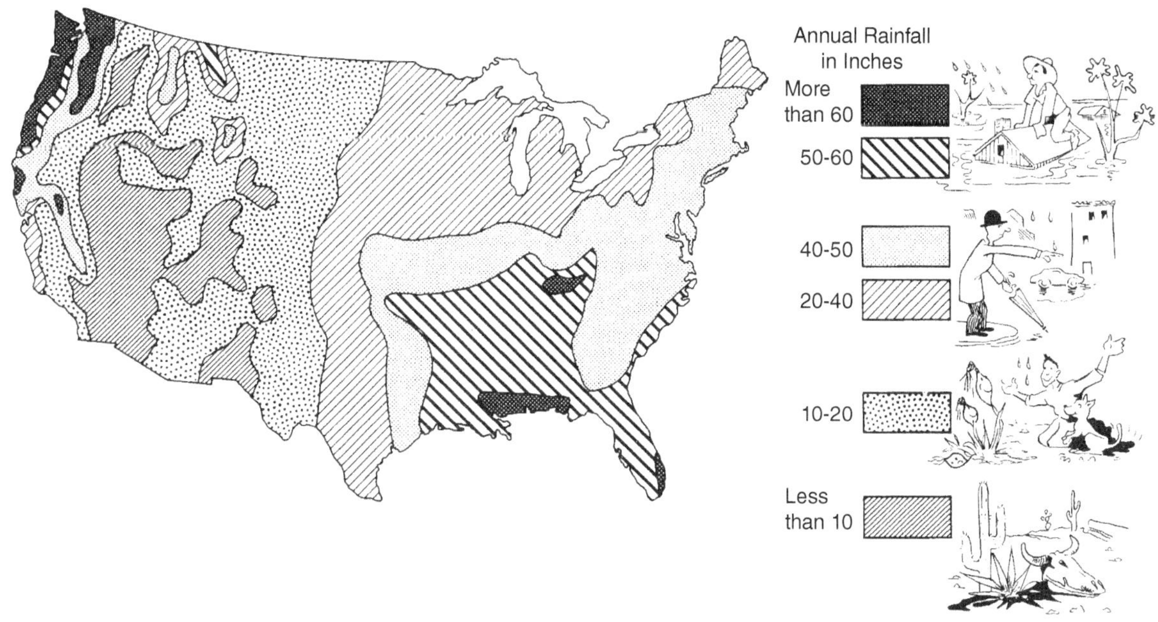

Figure 20.7 Annual rainfall in the United States. Cistern volume conversion factors: 1 ft^3 = 7.48 gal = 0.02832 m^3; 1 m^3 = 1000 L = 264.2 gal.

should be scrupulously avoided, as should roof coatings containing lead. For these reasons, urban rainwater commonly is not used for drinking and cooking. (Bottled water or on-site water distilling can supply potable water.) For the typical residence, however, that still leaves about 95% of indoor water usage that could be provided by rainwater. Also, in those cities that have very "hard" public water supplies, rainwater's "soft" characteristics make it particularly attractive.

The quantity of rainwater available in most U.S. locations could meet a high percentage of typical home or business needs. Milne (1976) pointed out that the 42 in. [1067 mm] of rain that falls annually on the streets and roofs of Manhattan Island could, if collected and stored, provide 148% of the residential needs of its 1.7 million inhabitants. For the typical U.S. suburban house, with a roof area of 1500 ft^2 [139 m^2], the annual catchment can be estimated by combining the rainfall quantities (Figure 20.7) with the resulting catchment yield (Figure 20.5). Even at a "dry" rate of only 20 in. [508 mm] annually, a 1500-ft^2 [130-m^2] roof would yield about 12,000 gal, or 33 gpd [45,424 L, or 124 L/d]—nearly enough to meet the clothes-cleaning and dishwashing needs of the family.

Unfiltered rainwater seems particularly well suited to the irrigation of small lawns or gardens, both because it lacks additives unneeded by plants, such as chlorine and sodium fluoride, and because it can reduce the user's demand on the public water supply on the hottest summer days. Cisterns located above an irrigated area have the advantage of replacing a pump with simple gravity flow. A rain barrel at the bottom of a downspout is an example of this approach—although one of limited capacity.

In many of the world's drier areas, small cisterns within a home are common. Such cisterns, which can be fed both by rainwater and by the public supply, are frequently used for all domestic purposes, including drinking. The presence of such a large water volume also can be advantageous in the event of fire. Sometimes these storage cisterns are required because the public supply is diminished or cut off at peak usage hours due to insufficient water main capacity. Figure 20.8 shows a cistern in a typical outdoor location, along with various options by which pollution from dirty roofs can be minimized.

When cisterns are taken seriously as water storage devices, their function and size can become strong design-form determinants. The country home of architect John Andrews in the dry ranchland of New South Wales, Australia, offers a particularly striking example of cisterns used as form-producing elements (Figure 20.9). These demonstrate the design image of "pregnant downspouts" storing water above the area where it will be used rather than belowground. The corner rainwater collectors also help deflect cooling summer breezes into the house along the diagonal walls at each corner. The central skylight then vents the breezes, along with the house's heat. From the corner cisterns, rainwater is pumped up to the central tower's storage tank. It then can be heated by solar collectors (to be installed on the sloping top of the

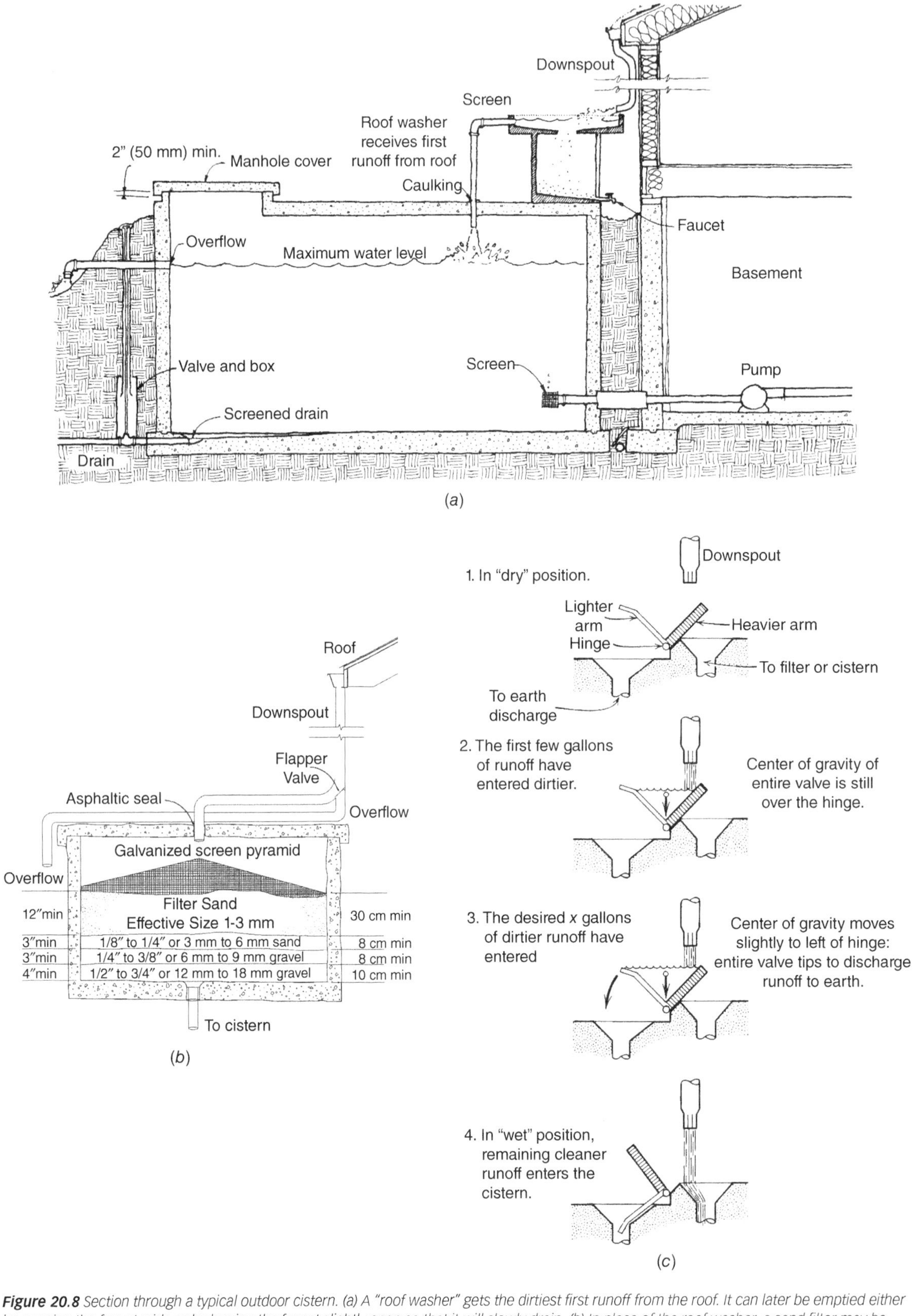

Figure 20.8 *Section through a typical outdoor cistern. (a) A "roof washer" gets the dirtiest first runoff from the roof. It can later be emptied either by opening the faucet wide or by leaving the faucet slightly open so that it will slowly drain. (b) In place of the roof washer, a sand filter may be used. The "flapper valve" is used rarely, to divert the first rainfall after a prolonged dry spell. After this, the valve is left in a position to divert all rainwater to the sand filter. (c) Another roof-washing option is a "tipping valve," which dumps the first few gallons of each rainfall. After each rainfall, the valve must be manually (or spring) reset to the "dry" position if it is to again intercept dirty water. In an extended wet period, it would probably be left in the "wet" position. (b: Redrawn with SI units by Nathan Majeski; based upon EPA, 1975.)*

(*a*)

(*b*)

Figure 20.9 *Country home of architect John Andrews, Eugowra, New South Wales, Australia. (a) View from the southwest showing corner cisterns, the central daylight/ventilation barrel vault, and the "energy tower" with a water storage tank. (b) View from the south showing the shading pergola to be covered by vines. (c) Floor plan showing the fireplace at the center and cisterns at all corners. (d) Section through the living area. (Courtesy of John Andrews International Pty, Ltd.)*

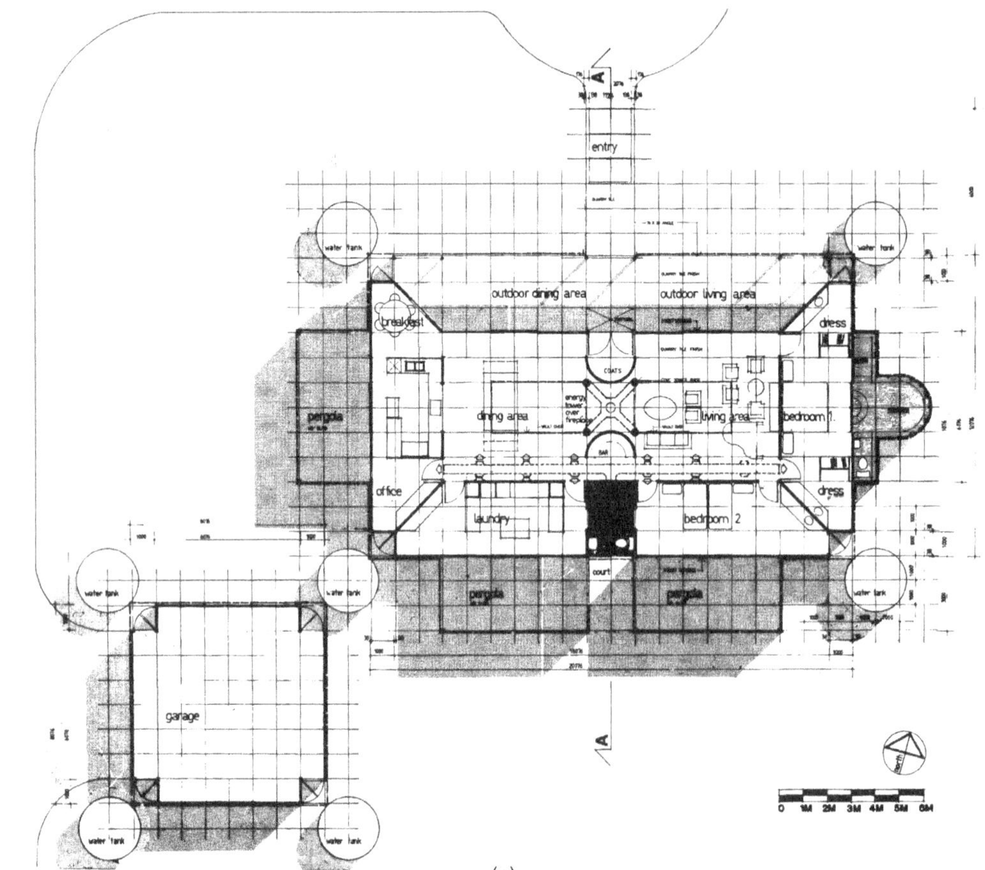

(*c*)

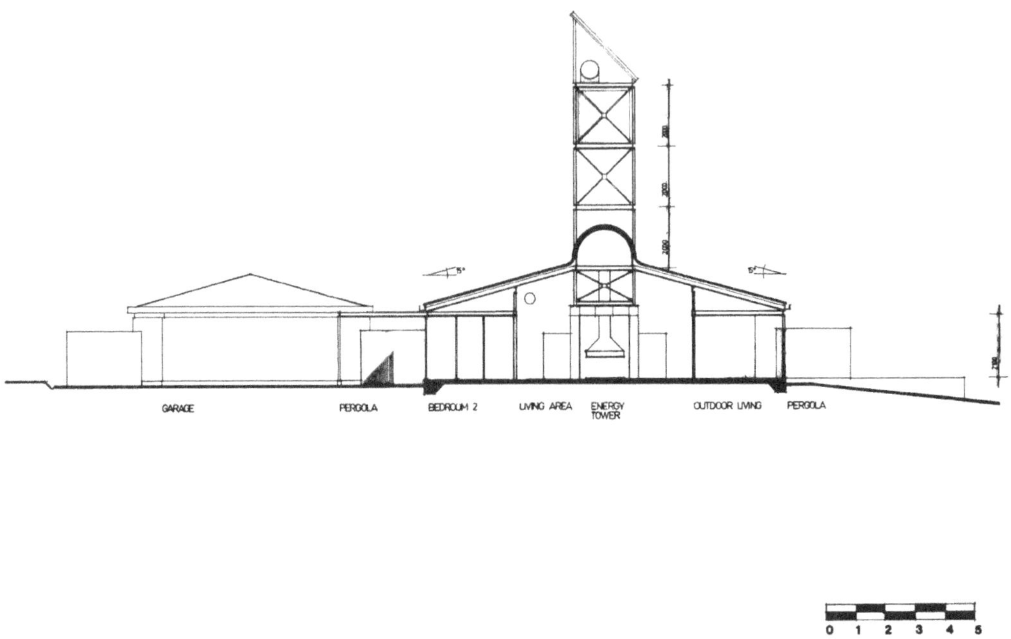

(*d*)

Figure 20.9 *(continued)*

tower), or used to supply the house's plumbing fixtures, or even used to sprinkle the metal roof surface, whose surface temperature could quickly be lowered by evaporation. Unevaporated roof water simply runs back into the cisterns. As a final design-with-water gesture, the shower (off bedroom 1) has been made into a true celebration place for cleansing and refreshing, with a sweeping view of the ranch.

In this passively solar-heated home, the living areas are placed on the elongated north side—the warmer side in the cold season for this southern hemisphere house. Passive cooling is aided by pergolas on the west and south, which are covered with vines for hot season shading of windows.

Sizing

The procedures described in Section 20.3 are for rough sizing of both the catchment area and the storage capacity for cisterns. When rainwater is to be a primary, as opposed to merely a supplementary, source, a closer look must be taken at rainfall deposits and user withdrawals from a cistern.

This procedure depends on the monthly average rainfall (from NOAA Local Climatological Data), the monthly water usage, and the catchment area yield (from Figure 20.5).

EXAMPLE 20.1 PART B Take a closer look at the cistern that was approximately sized in Part A. Daily usage for this cistern, to be used for toilet flushing in a central Oregon factory, was estimated at 200 gal [760 L] and the catchment area at 5600 ft^2 [520 m^2]. The cistern capacity was estimated at 18,000 gal [68,400 L]. Begin the process in the midst of the wettest months.

From Table 20.7, the following conclusions can be drawn:

1. For an 18,000-gal [68,400-L] cistern, when the end-of-December cumulative capacity is added to January's surplus, the cistern will be at capacity from November through April.
2. With no allowances for abnormally dry months, we could reduce the cistern size by about 3000 gal [11,355 L], the size of the smallest cumulative capacity in September.

TABLE 20.7 Rainfall Cistern Sizing Procedure

I Month and Rainfall[a],	in. [mm]	II Catchment Yield[b], gal [L]	III Usage[c], gal [L]	IV Net[d], gal [L]	V Cumulative Capacity[e], gal [L]	VI Cumulative Capacity Adjusted for Actual Size[f], gal [L]
January	6.9 [175]	18,630 [70,515]	6200 [23,476]	12,430 [47,048]	12,430 [47,048]	12,430 [47,048]
February	4.8 [122]	12,960 [49,054]	5600 [21,196]	7300 [27,631]	19,790[g] [74,905]	18,000 [68,130]
March	4.3 [109]	11,610 [43,944]	6200 [23,476]	5410 [20,477]	25,200[g] [95,382]	18,000 [68,130]
April	2.3 [58]	6210 [23,505]	6000 [22,710]	210 [795]	25,410[g] [96,177]	18,000 [68,130]
May	2.1 [53]	5610 [21,234]	6200 [23,476]	−530 [−2006]	24,880[g] [94,171]	17,470 [66,124]
June	1.4 [36]	3780 [14,307]	6000 [22,710]	−2220 [−8403]	22,660[g] [85,768]	15,250 [57,721]
July	0.4 [10]	1080 [4088]	6200 [23,476]	−5120 [−19,379]	17,540 [66,389]	10,130 [38,342]
August	0.6 [15]	1620 [6132]	6200 [23,476]	−4580 [−17,335]	12,960 [49,054]	5550 [21,007]
September	1.5 [38]	4050 [15,329]	6000 [22,710]	−1950 [−7381]	11,010 [41,673]	3600 [13,626]
October	4.0 [102]	10,800 [40,878]	6200 [23,476]	4600 [17,411]	15,610 [59,084]	8200 [31,037]
November	6.1 [155]	16,470 [62,339]	6000 [22,710]	10,470 [39,629]	26,080[g] [98,713]	18,000 [68,130]
December	6.9 [175]	18,630 [70,515]	6200 [23,476]	12,430 [47,048]	38,510[g] [145,760]	18,000 [68,130]

Source: Brown et al., 1982 / with permission of John Wiley & Sons.

[a]NOAA data for Salem, Oregon.

[b]From Figure 20.5, according to which 10 in. [254 mm] of precipitation yields about 27,000 gal [102,200 L] for this 5600-ft^2 [520-m^2] catchment area.

[c]The factory uses 200 gpd [760 Lpd] times days in each month.

[d]Col. II minus col. III.

[e]Values in col. IV, added month by month.

[f]Set at 18,000 gal [68,130 L] for the factory in Example 20.1.

[g]The cumulative capacity exceeds the actual storage capacity of 18,000 gal [68,130 L].

3. A larger cistern—one that could utilize everything from the catchment area—could be built. Its approximate size would be the year-end surplus of 38,510 gal [145,760 L] plus the maximum spring monthly cumulative capacity of 25,410 gal [96,177 L] in April.
4. Alternatively, the surplus could be devoted to additional usage of rainwater, beyond mere toilet flushing, from November through April. ■

20.6 RAINWATER AND SITE PLANNING

Prior to the spread of buildings, streets, roads, and paved parking lots to rural areas, water from rainfall and melting snow found its own route to natural destinations. Surface flow to creeks, streams, and rivers accounted for part of this drainage. Underground flow aided the general runoff. Outcropping of flowing groundwater created springs and artesian wells. Low, dished areas formed lakes that in turn overflowed to outlet streams. Flat areas sometimes developed into swamps or marshes.

At a time when there was a choice of locations for towns and villages, sites next to rivers were usually chosen. The waterways provided transportation, and water was supplied from the river or from adjacent wells. As streets and roads were built, slopes could be arranged whereby the rain falling on these areas and flowing onto them from roofs of buildings could run to the river. At interior parts of the country, high ground was favored for building sites and growing communities. Swampy or marshy ground would not be chosen, but did provide terminal locations for stormwater that ran off the high ground. In the course of this natural flow, much of the water was drawn by evaporation to the clouds. The rest, conforming to topographic river basins, continued to seek its way to larger bodies of water.

As building increased, desirable locations grew scarce. The possibility of selecting high, dry ground diminished. Large areas, formerly low and marshy, were filled in and buildings constructed, often on piles. From such locations stormwater could not be diverted to some adjacent lower area via drainage, or even recharged to the earth through dry wells. Moreover, extensive grids of paved streets and sidewalks in these level developments caught and held the water, resulting in considerable "ponding." Storm sewers had to be built and the water transported great distances, often having to be lifted at intermediate pumping stations before reaching its destination, which might have been a remote river.

This emphasis on the removal of stormwater has led to an expensive and elaborate system based on quick disposal of rainfall. By decreasing the time between precipitation and runoff, quick disposal increases the peak flows within such systems, thereby increasing flooding of rivers during storms but reducing the rivers' flow between storms. It also contributes pollutants to waterways that otherwise would have been filtered by the soil, as detailed in Table 20.8.

The overloading of storm sewers not only causes minor flooding, but also can influence building design. The designers of the New Orleans Convention and Exhibition Center, a building with 610,000 ft^2 [56,670 m^2] of roof area, spared the city's storm sewers from the impact of a 14-acre (5.7-ha) runoff by carrying the rainwater over the roofs of adjoining wharfs and discharging it directly into the Mississippi River.

TABLE 20.8 Some Constituents of Stream Water

Constituent	Source in Nature	Role in Natural Ecosystem	Source of Urban Excess	Role of Excess
Sediment	Banks of meandering channels	Maintain stream profile and energy gradient; store nutrients	Construction sites; eroding stream banks	Abrade fish gills; carry excess nutrients and chemicals in adsorption; block sunlight; cover gravel bottom habitats
Organic compounds	Decomposing organic matter	Store nutrients	Herbicides; pesticides; fertilizers	Deprive water of oxygen by decomposition
Nutrients	Decomposing organic matter	Support ecosystems	Organic compounds; organic litter; fertilizers; sewage; food waste	Unbalance ecosystem; produce algae blooms; by decomposition deprive water of oxygen
Trace metals	Mineral weathering	Support ecosystem	Cars; construction materials; all kinds of foreign chemicals	Reduce resistance to disease; reduce reproductive capacity; alter behavior
Chloride	Mineral weathering	Support ecosystems	Pavement deicing salts	Sterilize soil and reduce biotic growth
Bacteria	Native animals	Participate in ecosystems	Pet animals; dumpsters; trash handling areas	Cause risk of disease
Oil	Decomposing organic matter	Store nutrients	Cars	Deoxygenate water

Source: Adapted from Ferguson, 1998.

(a) (b)

Figure 20.10 *Bioswale at the John E. Jaqua Center for Student Athletes, University of Oregon, Eugene. (a) Upon completion, and (b) after two years of growth. (a: © Alison Kwok; all rights reserved; b: © Tyler Mavichien; used with permission.)*

As urban storm sewers reach capacity and suburban groundwater levels drop, designers have begun to emphasize (Figure 20.10) stormwater infiltration (or recharge of groundwater) rather than quick runoff. Three design strategies for encouraging such recharge have emerged: roofs that will retain water and slowly release it, porous pavement, and onsite infiltration of runoff.

(a) Roof Retention

If stormwater is to be sent to storm sewers or to soak into the ground, a slow flow from roofs will help by diminishing peak flows in sewers and giving soaked soil more time to absorb the runoff. A nearly flat roof with specially designed drains (see Figure 20.23) forms a temporary pond that permits slower discharge, yet eventually drains completely dry (to discourage mosquito breeding, etc.). For cisterns, however, sloped roofs should be used, as they stay cleaner than flat roofs. Another option is to store the "cistern" water on the roof itself (Figure 20.11). Another slow flow possibility is a sod roof that will retain water for longer periods than ponded flat roofs.

However achieved, a temporary pond on top of a building will clearly require additional structural reinforcement. Another problem could be posed by high winds blowing sheets of water onto people below. In summer, however, a flooded roof can provide a significant thermal advantage by greatly lowering daytime roof surface temperatures. Rainfall-retaining roofs are now required in some urban areas with overtaxed storm sewers. Sewer districts frequently charge for stormwater runoff; by ponding stormwater on site, such charges can be reduced or avoided.

(b) Porous Pavement

To retain rainfall on site, many builders use porous asphalt (Figure 20.12), porous concrete, "incremental" paving, or open-celled pavers (alternation of paving materials with grass or groundcover plants) for parking lots and roadways.

Porous concrete has been used for many years in building construction as a low-strength, high-porosity material that has some insulating properties (R-5 for 10-in thickness [RSI-0.88 for 254 mm]). Patented porous concrete pavement now in use in Florida has a compressive strength of more than 3800 psi [26,200 kPa] and a permeability of 2.3 gal of water per minute per square foot [94 L per minute per square meter]. (A 2500-psi [17,240 kPa] mix has

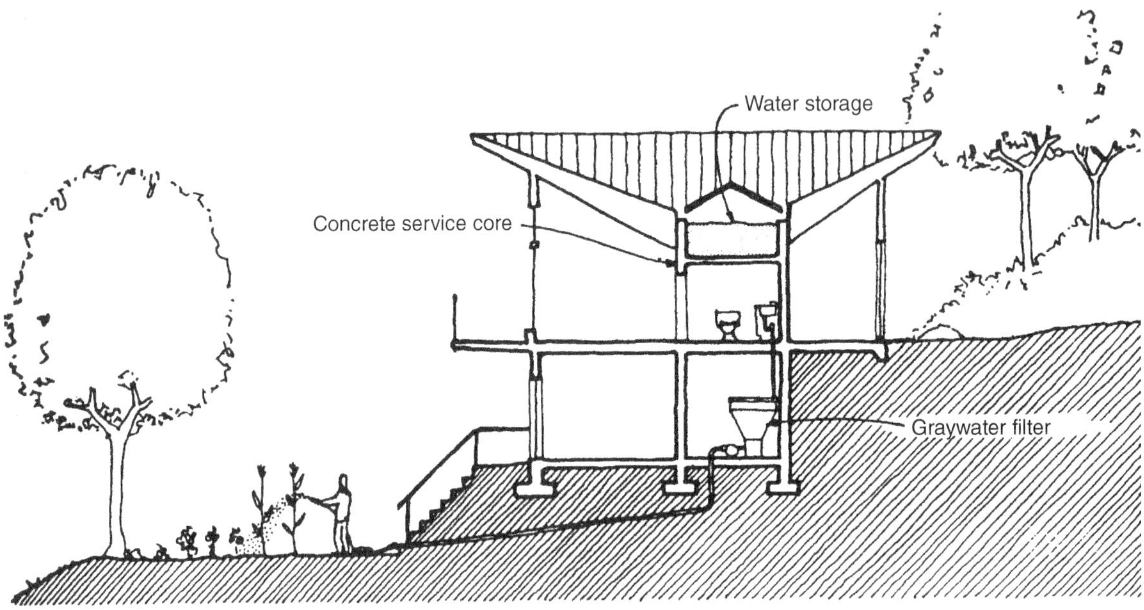

Figure 20.11 Retaining rainwater so that gravity flow can encourage its usage.

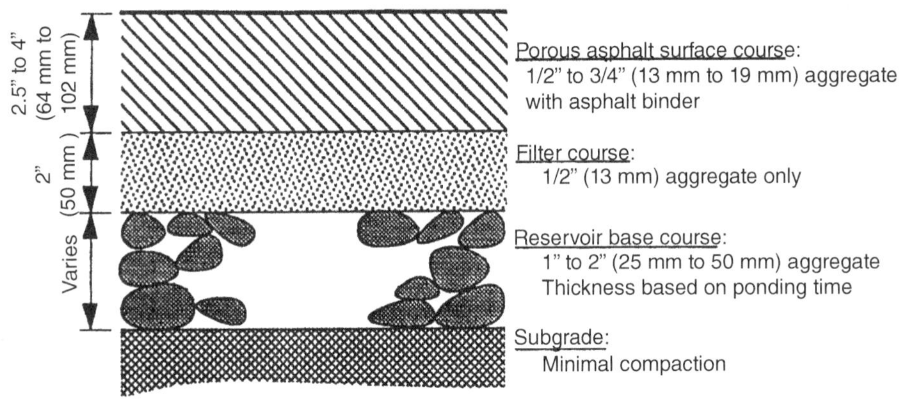

(*a*)

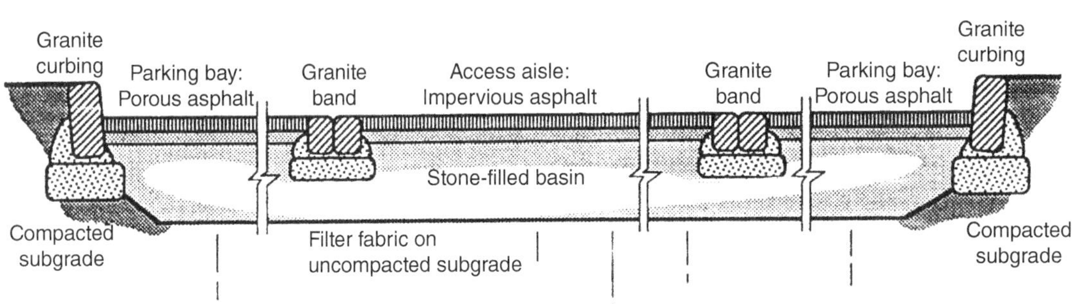

(*b*)

Figure 20.12 Porous paving encourages groundwater recharge rather than storm runoff. (a) Standard porous asphalt pavement as used in Rockville, Maryland. (b) Subsurface basin allows required retention ponds to serve as parking lots; this example is at the Morris Arboretum in Philadelphia. (From Ferguson, 1998 / with permission of John Wiley and Sons.)

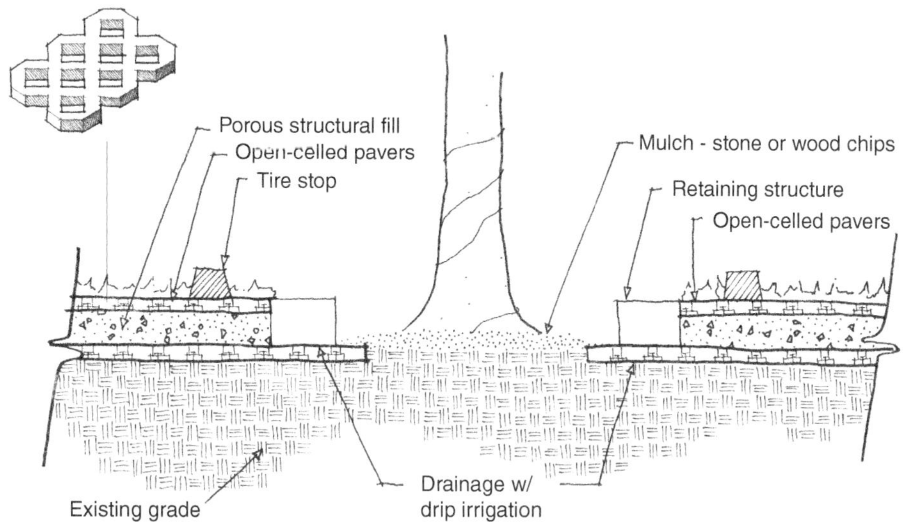

Figure 20.13 *Open-celled pavers on a porous fill allow grass or hardy ground cover to grow within the units. (Drawing by Erik Winter / with permission of Invisible Structures, Inc.)*

a permeability of 18.5 gal per minute per square foot [750 L/m per m^2].) In cold-weather areas, the freeze–thaw cycle could be destructive to porous concrete.

Incremental paving features many small, adjacent paving units; the many resulting joints allow water to pass through more readily than do solid, unbroken surfaces.

Open-celled pavers alternate small concrete or plastic paving units with grass or ground cover, as shown in Figure 20.13. These are particularly applicable to short-term parking, such as for a sports stadium, or for more remote areas of retail mall parking lots. They are also possible for roadway shoulders and emergency-vehicle-only access areas.

If stormwater is not retained, its runoff can be polluted to the extent that treatment is required. Using strips of planting for such treatment (phytoremediation) is discussed at the end of Chapter 22.

(c) Site Design for Recharging

This tactic is especially advisable for suburban-density residential developments in drier climates with absorptive soil (sand, gravel, etc.). The first option is to implement this at each house; design of entire subdivisions can also be influenced.

The simplest design approach is the gutterless sloped roof illustrated in Figure 20.14, which is applicable to one-story, basement-less homes with wide, overhanging roofs. A gravel-filled trench skirting the perimeter directly below the edge of the eaves catches the water flowing off the roof.

Some designers do not like the appearance of conventional gutters and leaders; other designers celebrate them (Figure 20.15). There are many ingenious ways to avoid or modify these elements, while still providing proper drainage. In many cases, however, gutters and leaders will be required, either to collect rainwater for cisterns or to control conditions around the perimeter of a house. Several options for stormwater recharge can be used with the gutter-leader combination.

A splash pan at the foot of each leader (Figure 20.16*a*) offers the simplest method. It will lead the water a few feet from the house but will accommodate only a relatively low rate of flow. A gravel-filled pipe is somewhat more effective (Figure 20.16*b*). When the soil is not very permeable (as, for instance, with clay), it is best to use a dry well with an extended area and many perforations through which the water can be discharged to the ground (Figure 20.16*c*).

Footing drains are often used to collect and lead away groundwater that accumulates around foundations. This reduces the likelihood of basement wall leakage. These drains are most necessary when higher ground near a building increases the flow of groundwater against underground walls. Figure 20.17 illustrates this and also shows how stormwater from drains and roofs may be led to a surface absorption area of rock and gravel, beyond a head wall, where the general storm drain outcrops. This method can be chosen where there is sufficient property area and slope. It has the advantages of easy maintenance and service. Also, one can observe whether it is functioning correctly.

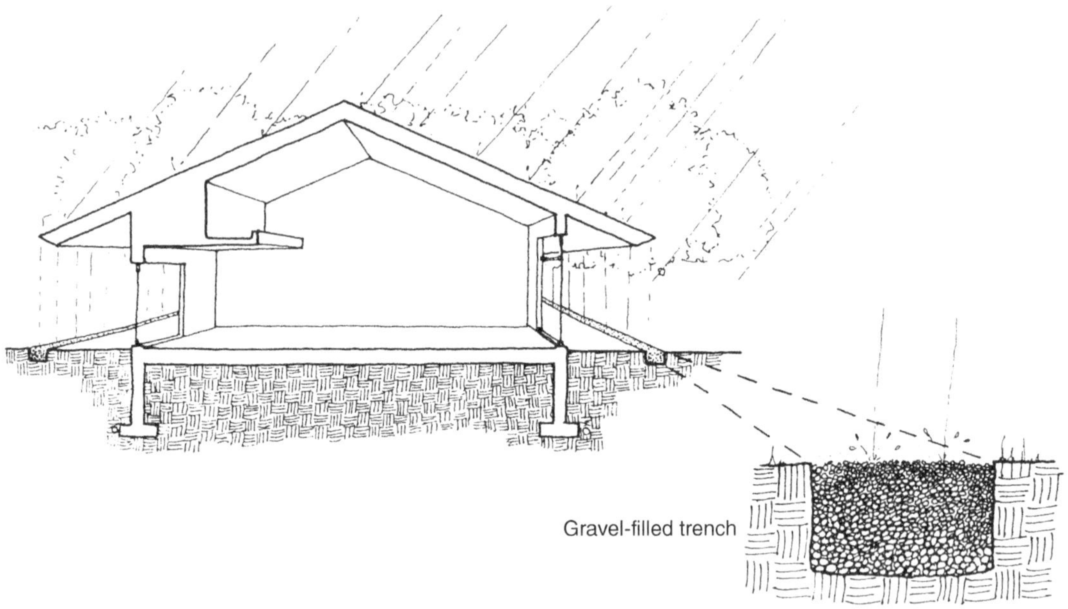

Figure 20.14 Gutters and leaders are not always essential, provided that doorways, walls, foundations, and landscaped lawns are not subjected to rain concentrations.

Figure 20.15 Rain gutters as an architectural focal point. The Desert Living Center at the Springs Preserve, Las Vegas, features a rain gutter that directs water from the trough of the butterfly roof to a cistern for irrigation. (© Alison Kwok; all rights reserved.)

In new developments, bioswales or recharge basins (Figure 20.18) are sometimes used to infiltrate stormwater into the ground. Water from numerous roofs, paved areas, and curb catch basins is collected and piped to an open, unpaved pit, where it sinks into the earth. This method is not recommended in areas of dense, impervious clay soil. A particularly effective example of this approach is offered by the community of passively solar-heated residences known as Village Homes, outside Davis, California. This area receives only about 20 in. [508 mm] of rain annually, so for this garden-oriented community, the recharge of groundwater was a far more attractive (and less expensive) option than loss of the rainwater to a storm sewer. Stormwater flows from leaders to dry, rockbed channels, along which are gardens and bicycle paths (Figure 20.19). Occasionally, small dams across these channels create temporary holding ponds in case the runoff has not yet soaked through the channel bottom. In the event of extraordinarily heavy rainfall, an inlet to the public storm sewer is available beyond the final holding pond; this inlet is needed approximately once every five years.

The National Wildflower Research Center (Figure 20.20) is built over an endangered aquifer, southwest of Austin, Texas. This complex of small buildings and larger gardens includes several cisterns, one of which forms the base of a central observation tower. Water gardens and irrigation are both served by the cisterns. The buildings' roofs are designed for rainwater collection, and a small channel (aqueduct) carries water over a wall to the outer cistern.

The planning of the landscape around buildings may closely follow such considerations as irrigation. The hydrozone concept of landscape planning is shown in Figure 20.21. To minimize water consumption,

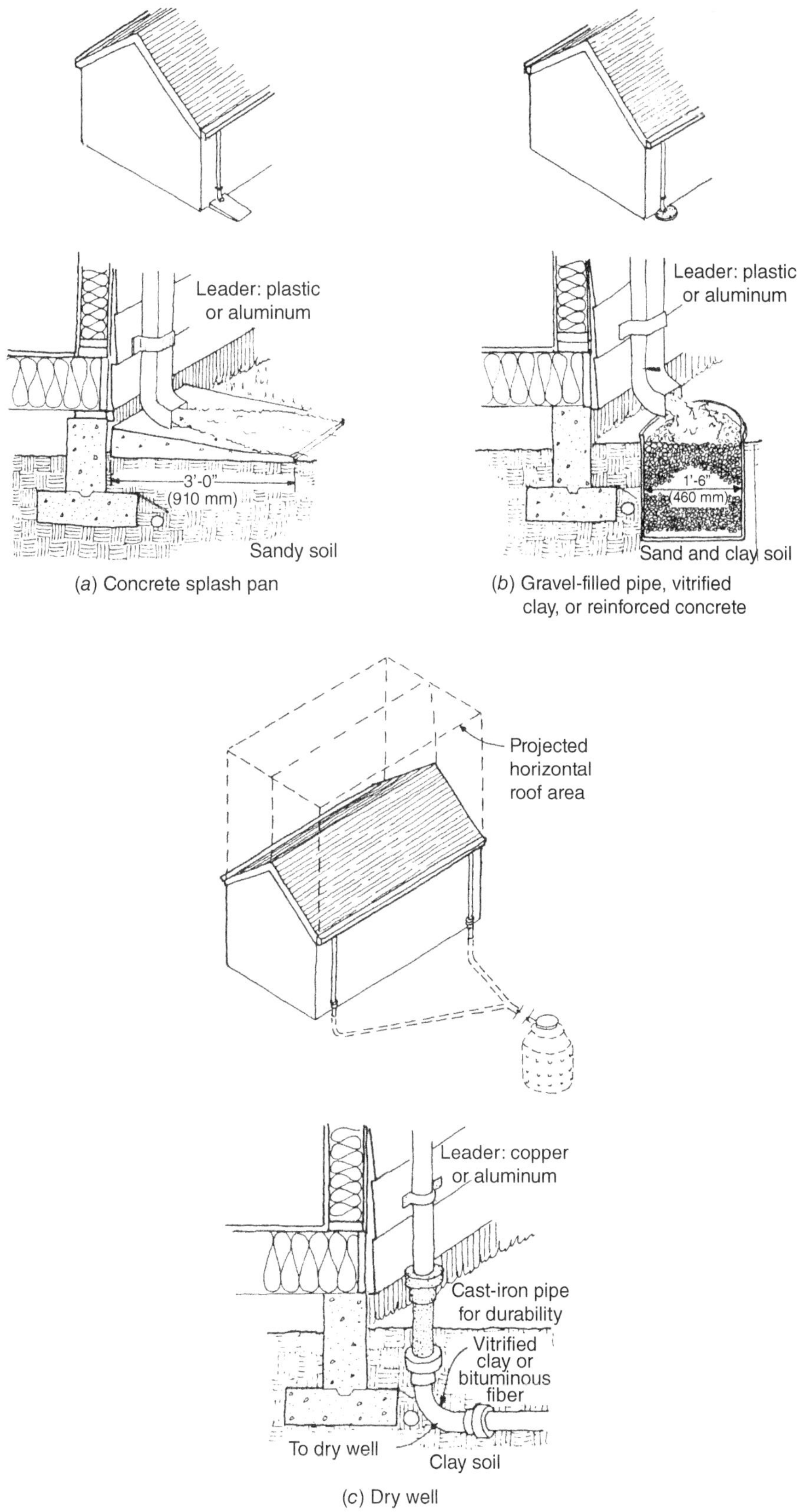

Figure 20.16 *Roof drainage for houses. Method (a) is suitable for low rates of flow introduced into very pervious soil. When denser soil is encountered, method (b) is used to get the water into the ground and thus avoid surface erosion. For heavy flow or to lead the water farther from the structure, method (c) may be used with one or several dry wells.*

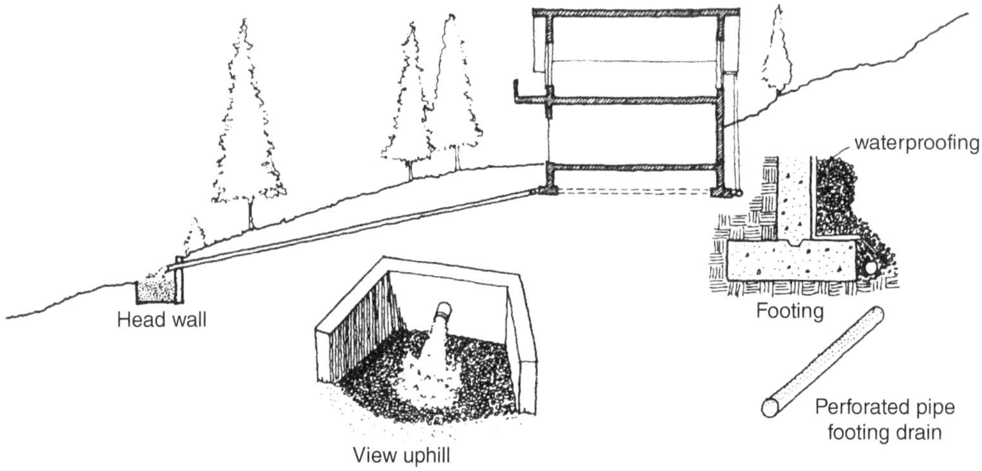

Figure 20.17 *Disposal of stormwater on the site but remote from the house or building. When a wall is against a hill, it is usually subjected to the pressure of groundwater during storms. Open-joint clay, plastic, or fiber tile accepts this water and carries it away. Footing drains are tight-joint clay tile or bituminous fiber pipe. Flow through stone and gravel returns the water to the earth. A head wall is appropriate in lieu of a dry well if the site permits.*

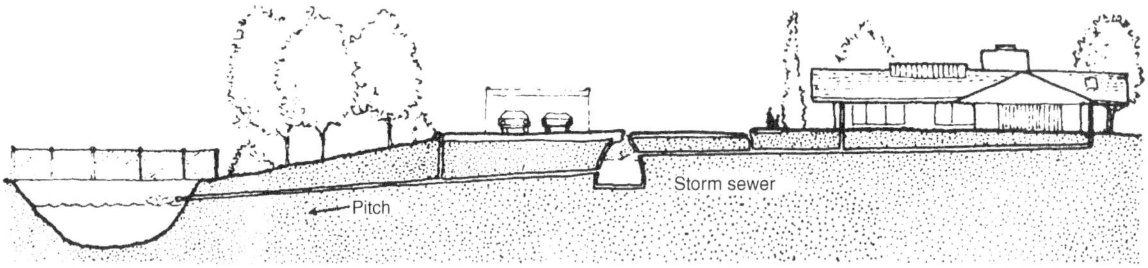

Figure 20.18 *Recharge basins in suburban communities. When topography, groundwater level, and porosity of the soil permit, developers are sometimes required to install systems that collect stormwater and carry it from catch basins at street curbs, the roofs of all houses, and paved areas to a recharge basin that receives the water and returns it to the ground. For the safety of children, a fence is sometimes required to prevent unauthorized access to the basin.*

exotic plant species are kept to a minimum and located near the house, where storm runoff and irrigating water are readily available. Native and adapted plantings, which can survive on normal rainfall, are utilized elsewhere. Landscape that requires no additional water is often called a xeriscape; see Chapter 21 for a discussion of irrigation systems and an example in California's Napa Valley.

20.7 COMPONENTS

The first design decision to be made for a stormwater system involves the establishment of "watersheds" on a building's roof. To what edges, or at what points, will runoff be directed? To what depths will it accumulate before it leaves the roof? Because the answers to these questions depend on the intensity of storms as well as on the roof's geometry, it is necessary to find the maximum hourly rainfall for each location. This figure is available from local building code officials, national climate data archives, or from Figure 20.22.

Gutters and leaders (downspouts) can be sized using Tables 20.9 and 20.10. The sizing of gutters and leaders depends both on the horizontal projected area of the roof, as shown in Figure 20.16*c*, and on the maximum hourly rainfall.

EXAMPLE 20.2 Select a gutter and two leaders for the front half of a house, as shown in Figure 20.16. The rainfall rate is 4 in./h [100 mm/h]. The projected roof area is 700 ft^2 [65 m^2], and the slope of the gutter (rising $^1/_{16}$ in. per foot of length [5.2 mm per meter]) is 0.5%.

SOLUTION

From Table 20.9, choose a semicircular gutter with a 6-in. [150-mm] diameter. Note that if a steeper gutter

(a)

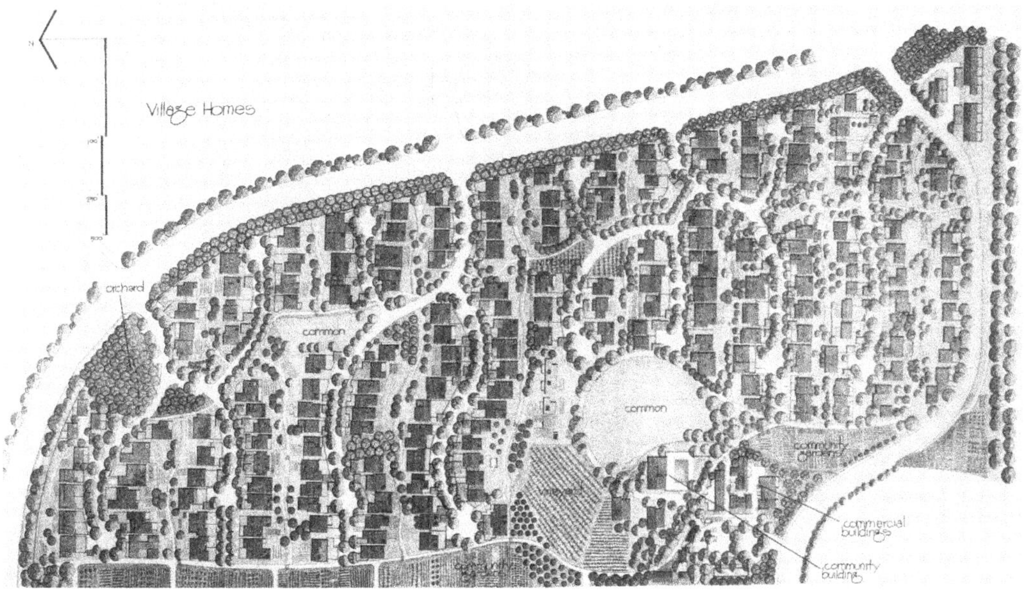

(b)

(c)

Figure 20.19 *Village Homes, a northern California subdivision of solar homes and stormwater recharge areas. (a) Photo illustrates the open drainage ways and pedestrian paths. (b) Plan shows the emphasis on bicycle paths, narrow streets, and widespread community-maintained garden space through which the recharge streambeds are led. (c) Site section explains the gradual drainage from leaders at houses to a recharge stream. (a: Photo by Alan Butler; c: From Corbett, 1981 / with permission of Rodale Press.)*

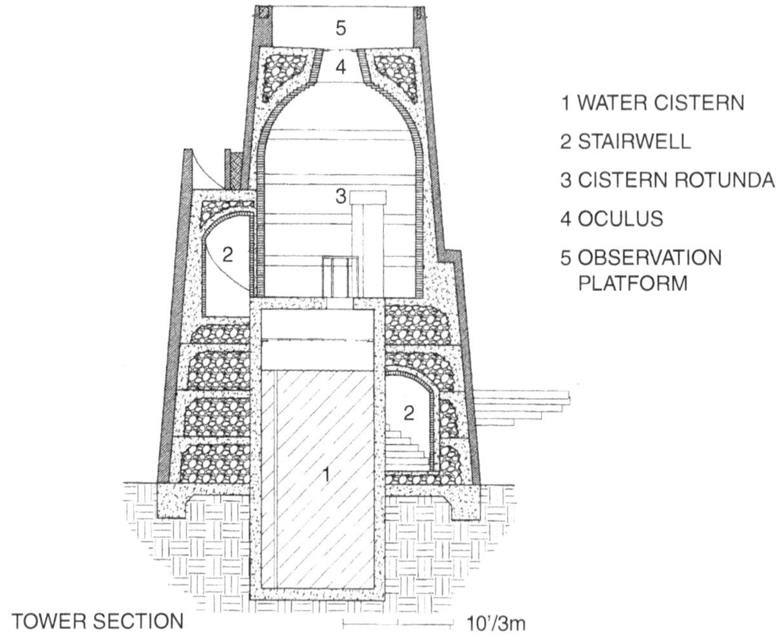

Figure 20.20 *Section through the observation tower/cistern at the Lady Bird Johnson Wildflower Research Center. (Courtesy of Overland Partners Architects.)*

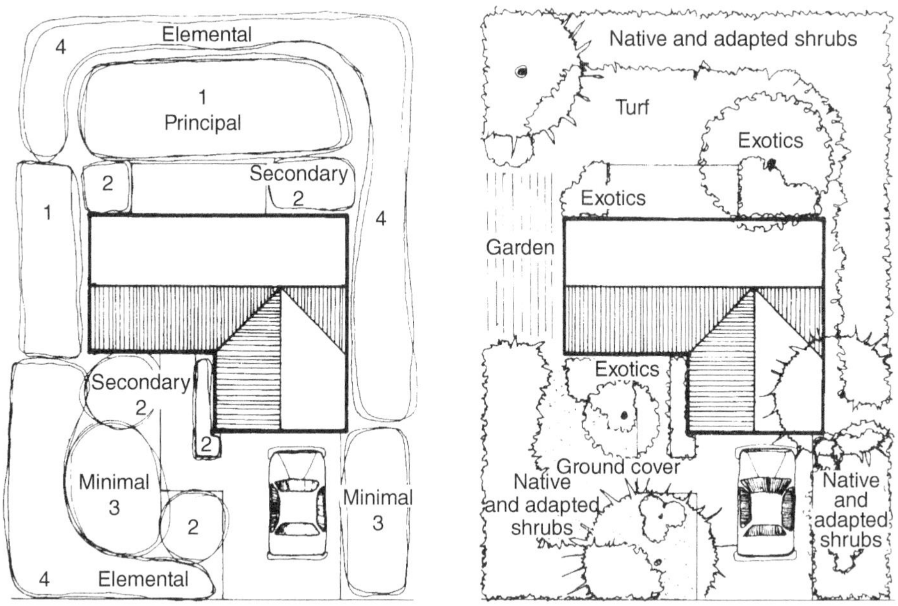

Figure 20.21 *The hydrozone concept of landscape planning restricts exotic planting to special, easily watered areas. Native and adapted plantings are used elsewhere. Zones are here shown on a typical suburban lot. (Adapted from McPherson, 1984.)*

(*a*)

Figure 20.22 *Maximum 100-year, 1-hour rainfall (in inches). (a) Eastern United States; (b) Central United States; (c) Western United States. (Note: Alaska, range 0.4–1.4 in. [10–36 mm]; Hawaii, range 1.5–8 in. [38–203 mm]) (Copyrighted excerpts from the 2024 International Plumbing Code. Copyright © 2023. International Code Council, Inc. All rights reserved. Reproduced with permission. www.ICCSAFE.org.)*

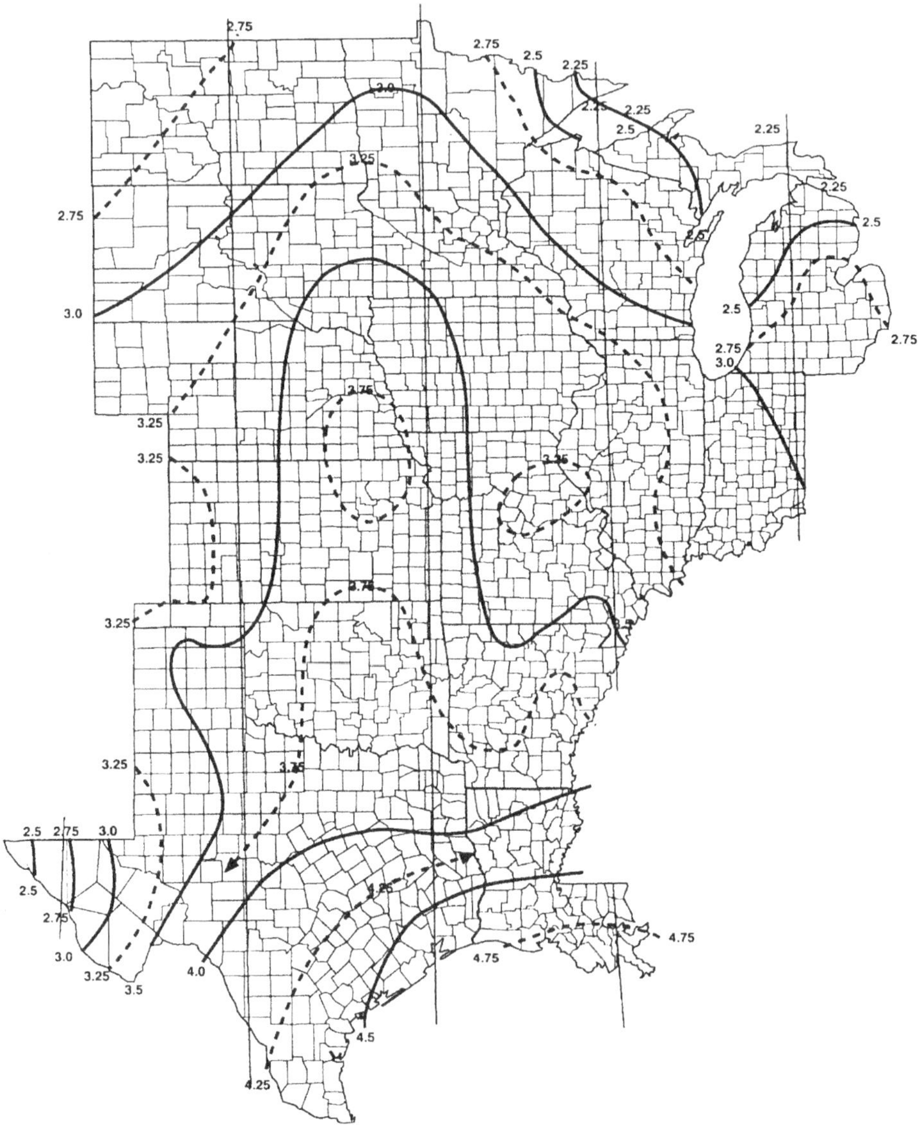

(*b*)

Figure 20.22 *(continued)*

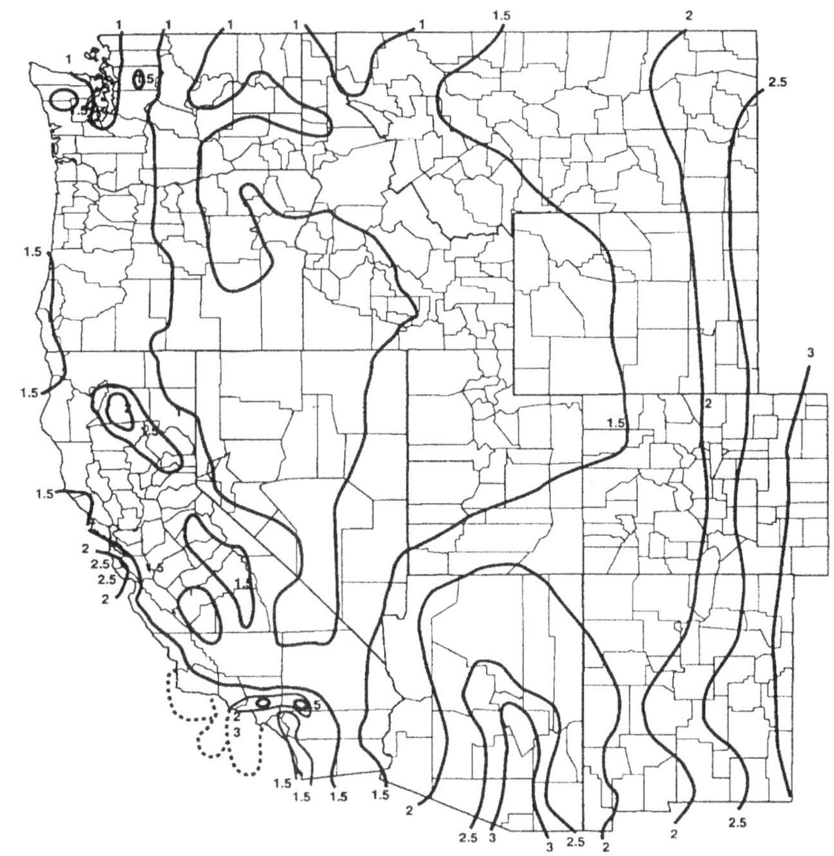

(c)

Figure 20.22 (continued)

TABLE 20.9 Size of Gutters

Part A. I-P Units (Showing Permissible Area of Drainage, ft²)					
	Maximum Rainfall (in. per h)				
Diameter of Gutter (in.) $^{1}/_{16}$ in./ft Slope	**2**	**3**	**4**	**5**	**6**
3	340	226	170	136	113
4	720	480	360	288	240
5	1250	834	625	500	416
6	1920	1280	960	768	640
7	2760	1840	1380	1100	918
8	3980	2655	1990	1590	1325
10	7200	4800	3600	2880	2400
	Maximum Rainfall (in. per h)				
Diameter of Gutter (in.) ⅛ in./ft Slope	**2**	**3**	**4**	**5**	**6**
3	480	320	240	192	160
4	1020	681	510	408	340
5	1760	1172	880	704	587
6	2720	1815	1360	1085	905
7	3900	2600	1950	1560	1300
8	5600	3740	2800	2240	1870
10	10,200	6800	5100	4080	3400

TABLE 20.9 (*Continued*)

Diameter of Gutter (in.) ¼ in./ft Slope	Maximum Rainfall (in. per h)				
	2	3	4	5	6
3	680	454	340	272	226
4	1440	960	720	576	480
5	2500	1668	1250	1000	834
6	3840	2560	1920	1536	1280
7	5520	3680	2760	2205	1840
8	7960	5310	3980	3180	2655
10	14,400	9600	7200	5750	4800
Diameter of Gutter (in.) ½ in./ft Slope	**Maximum Rainfall (in. per h)**				
	2	3	4	5	6
3	960	640	480	384	320
4	2040	1360	1020	816	680
5	3540	2360	1770	1415	1180
6	5540	3695	2770	2220	1850
7	7800	5200	3900	3120	2600
8	11,200	7460	5600	4480	3730
10	20,000	13,330	10,000	8000	6660
Part B. SI Units (Showing Permissible Area of Drainage, m²)					
Diameter of Gutter (mm) 5.2 mm/m Slope	**Maximum Rainfall (mm per h)**				
	50.8	76.2	101.6	127.0	152.4
76.2	31.6	21.0	15.8	12.6	10.5
101.6	66.9	44.6	33.4	26.8	22.3
127.0	116.1	77.5	58.1	46.5	38.7
152.4	178.4	119.1	89.2	71.4	59.5
177.8	256.4	170.9	128.2	102.2	85.3
203.2	369.7	246.7	184.9	147.7	123.1
254.0	668.9	445.9	334.4	267.6	223.0
Diameter of Gutter (mm) 10.4 mm/m Slope	**Maximum Rainfall (mm per h)**				
	50.8	76.2	101.6	127.0	152.4
76.2	44.6	29.7	22.3	17.8	14.9
101.6	94.8	63.3	47.4	37.9	31.6
127.0	163.5	108.9	81.8	65.4	54.5
152.4	252.7	168.6	126.3	100.8	84.1
177.8	362.3	241.5	181.2	144.9	120.8
203.2	520.2	347.5	260.1	208.1	173.7
254.0	947.6	631.7	473.8	379.0	315.9
Diameter of Gutter (mm) 20.9 mm/m Slope	**Maximum Rainfall (mm per h)**				
	50.8	76.2	101.6	127.0	152.4
76.2	63.2	42.2	31.6	25.3	21.0
101.6	133.8	89.2	66.9	53.5	44.6
127.0	232.3	155.0	116.1	92.9	77.5
152.4	356.7	237.8	178.4	142.7	118.9
177.8	512.8	341.9	256.4	204.9	170.9
203.2	739.5	493.3	369.4	295.4	246.7
254.0	133.8	891.8	668.9	534.2	445.9
Diameter of Gutter (mm) 41.7 mm/m Slope	**Maximum Rainfall (mm per h)**				
	50.8	76.2	101.6	127.0	152.4
76.2	89.2	59.5	44.6	35.7	29.7
101.6	189.5	126.3	94.8	75.8	63.2
127.0	328.9	219.2	164.4	131.5	109.6
152.4	514.7	343.2	257.3	206.2	171.9
177.8	724.6	483.1	362.3	289.9	241.4
203.2	1040.5	693.0	520.2	416.2	346.5
254.0	1858.0	1234.4	929.0	743.2	618.7

Source: © 2024 Uniform Plumbing Code. Reprinted with the permission of the International Association of Plumbing and Mechanical Officials. This copyright material and all points or statements in using this material have not been reviewed by IAPMO. The opinions expressed herein are not representations of fact from IAPMO.

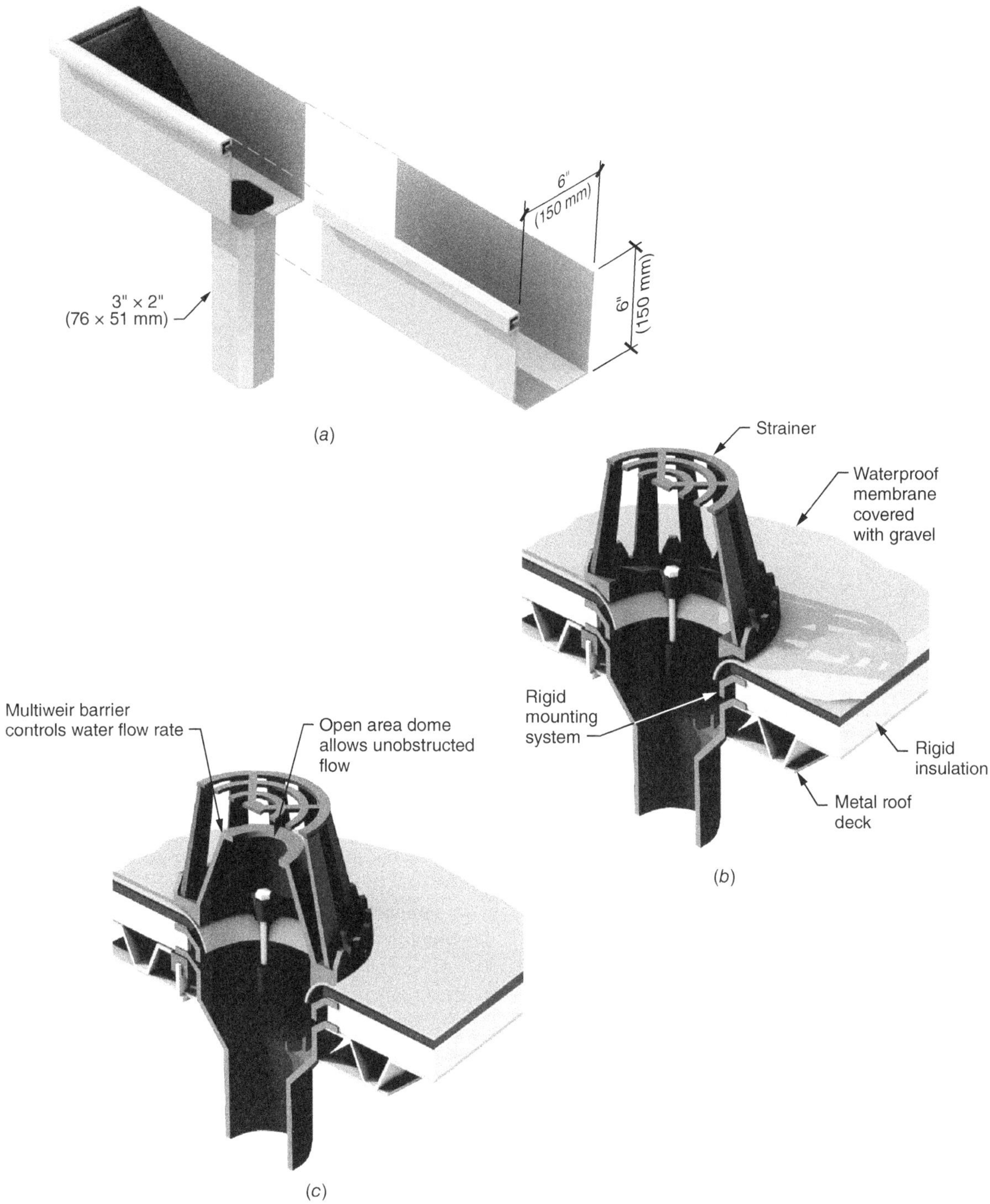

Figure 20.23 *Storm drainage components. (a) Conventional gutter and leader for houses; sizes vary by manufacturer. (b) Ordinary roof drain (c) Roof drain for controlled flow.*

TABLE 20.10 Sizing Roof Drains, Leaders, and Vertical Rainwater Piping[a, b]

Part A. I-P Units

Size of Drain, Leader, or Pipe (in.)	Flow (gpm[c])	Maximum Allowable Horizontal Projected Roof Areas (Square Feet) at Various Rainfall Rates											
		1 in./h	2 in./h	3 in./h	4 in./h	5 in./h	6 in./h	7 in./h	8 in./h	9 in./h	10 in./h	11 in./h	12 in./h
2	30	2880	1440	960	720	575	480	410	360	320	290	260	240
3	92	8800	4400	2930	2200	1760	1704	1260	1100	980	880	800	730
4	192	18,400	9200	6130	4600	3680	3070	2630	2300	2045	1840	1675	1530
5	360	34,600	17,300	11,530	8650	6920	5765	4945	4325	3845	3460	3145	2880
6	563	54,000	27,000	17,995	13,500	10,800	9000	7715	6750	6000	5400	4910	4500
8	1208	116,000	58,000	38,660	29,000	23,200	19,315	16,570	14,500	12,890	11,600	10,545	9600

Part B. SI Units

Size of Drain, Leader, or Pipe (mm)	Flow (L/s[c])	Maximum Allowable Horizontal Projected Roof Areas (Square Meters) at Various Rainfall Rates											
		25.4 mm/h	50.8 mm/h	76.2 mm/h	101.6 mm/h	127.0 mm/h	152.4 mm/h	177.8 mm/h	203.2 mm/h	228.6 mm/h	254.0 mm/h	279.4 mm/h	304.8 mm/h
51	1.9	182	91	61	45	36	30	26	23	20	18	16	15
76	5.8	555	278	185	139	111	93	80	69	62	56	50	46
102	12	1161	581	387	290	232	194	166	145	129	116	106	97
127	23	2183	1092	728	546	437	364	312	273	243	218	198	182
152	36	3407	1704	1135	852	681	568	487	426	379	341	310	284
203	76	7320	3660	2439	1830	1464	1219	1046	915	813	732	665	606

Source: © 2024 Uniform Plumbing Code. Reprinted with the permission of the International Association of Plumbing and Mechanical Officials. This copyright material and all points or statements in using this material have not been reviewed by IAPMO. The opinions expressed herein are not representations of fact from IAPMO.

Notes:

[a]For rainfall rates other than those listed, determine the allowable roof area by dividing the area given in the 1 in. per hour [25.4 mm/h] column by the desired rainfall rate.

[b]Vertical piping shall be round, square, or rectangular. Square pipe shall be sized to enclose its equivalent round pipe. Rectangular pipe shall have not less than the same cross-sectional area as its equivalent round pipe, except that the ratio of its side dimensions shall not exceed 3 to 1.

[c]Maximum discharge capacity, gpm [L/s] with approximately 1 ¾-inch (44 mm) head of water at the drain.

slope were designed, for instance, a ½-in. rise per foot of length (4% slope), only a 4-in. [100-mm] diameter gutter would be required.

Because two leaders will be used, each will drain 350 ft² [33 m²]. Table 20.10 shows that a 2-in. [50-mm] leader can be used. For this gutter-leader combination, specify the detail of Figure 20.23*a*. ■

Figure 20.24 *The whimsical transformation of a custom-designed rain gutter to an oversized spout serves as an architectural element at the Watering Can Theater at the Springs Preserve, Las Vegas. (© Alison Kwok; all rights reserved.)*

Storm gutters and leaders can have an important impact on a building's appearance (Figure 20.24). Alternatively, leaders can be set within buildings, and gutters can be built into a roof's surface to minimize the visual impact. Such decisions will affect ease of access and maintenance.

Where routing of stormwater inside a building is preferable, drains and leaders can be sized from Table 20.10, and horizontal piping can be sized from Table 20.11. Care should be taken to insulate such lines; cold rainwater inside pipes can cause condensation to form on the outside pipe surface, sometimes resulting in staining and other water damage.

EXAMPLE 20.3 Select the sizes for vertical conductors and horizontal storm drains for the building shown in Figure 20.25. The roof, balcony, and courtyard areas are as shown, the rainfall rate is 4 in./h [100 mm/h], and the pitch of horizontal drains is ¼-in. slope per 1 ft of run [2%; approximately 20 mm/m].

SOLUTION

The sizes selected and shown in Figure 20.25 may be verified in Tables 20.10 and 20.11. ■

TABLE 20.11 Sizing of Horizontal Rainwater Piping

Part A. I-P Units

Size of Pipe (in.)	Flow at 1/8 in./ft Slope (gpm)	Maximum Allowable Horizontal Projected Roof Areas (Square Feet) at Various Rainfall Rates: 1 in./h	2 in./h	3 in./h	4 in./h	5 in./h	6 in./h
3	34	3288	1644	1096	822	657	548
4	78	7520	3760	2506	1880	1504	1253
5	139	13,360	6680	4453	3340	2672	2227
6	222	21,400	10,700	7133	5350	4280	3566
8	478	46,000	23,000	15,330	11,500	9200	7670
10	860	82,800	41,400	27,600	20,700	16,580	13,800
12	1384	133,200	66,600	44,400	33,300	26,650	22,200
15	2473	238,000	119,000	79,333	59,500	47,600	39,650

Size of Pipe (in.)	Flow at 1/4 in./ft Slope (gpm)	Maximum Allowable Horizontal Projected Roof Areas (Square Feet) at Various Rainfall Rates: 1 in./h	2 in./h	3 in./h	4 in./h	5 in./h	6 in./h
3	48	4640	2320	1546	1160	928	773
4	110	10,600	5300	3533	2650	2120	1766
5	196	18,880	9440	6293	4720	3776	3146
6	314	30,200	15,100	10,066	7550	6040	5033
8	677	65,200	32,600	21,733	16,300	13,040	10,866
10	1214	116,800	58,400	38,950	29,200	23,350	19,450
12	1953	188,000	94,000	62,600	47,000	37,600	31,350
15	3491	336,000	168,000	112,000	84,000	67,250	56,000

Size of Pipe (in.)	Flow at 1/2 in./ft Slope (gpm)	Maximum Allowable Horizontal Projected Roof Areas (Square Feet) at Various Rainfall Rates: 1 in./h	2 in./h	3 in./h	4 in./h	5 in./h	6 in./h
3	68	6576	3288	2192	1644	1310	1096
4	156	15,040	7520	5010	3760	3010	2500
5	278	26,720	13,360	8900	6680	5320	4450
6	445	42,800	21,400	14,267	10,700	8580	7140
8	956	92,000	46,000	30,650	23,000	18,400	15,320
10	1721	165,600	82,800	55,200	41,400	33,150	27,600
12	2768	266,400	133,200	88,800	66,600	53,200	44,400
15	4946	476,000	238,000	158,700	119,000	95,200	79,300

Part B. SI Units

Size of Pipe (mm)	Flow at 10 mm/m Slope (L/s)	Maximum Allowable Horizontal Projected Roof Areas (Square Meters) at Various Rainfall Rates: 25 mm/h	50 mm/h	75 mm/h	100 mm/h	125 mm/h	150 mm/h
75	2.1	305	153	102	76	61	51
100	4.9	700	350	233	175	140	116
125	8.8	1241	621	414	310	248	207
150	14.0	1988	994	663	497	398	331
200	30.2	4273	2137	1424	1068	855	713
250	54.3	7692	3846	2564	1923	1540	1282
300	87.3	12,375	6187	4125	3094	2476	2062
375	156.0	22,110	11,055	7370	5528	4422	3683

Size of Pipe (mm)	Flow at 20 mm/m Slope (L/s)	Maximum Allowable Horizontal Projected Roof Areas (Square Meters) at Various Rainfall Rates: 25 mm/h	50 mm/h	75 mm/h	100 mm/h	125 mm/h	150 mm/h
75	3.0	431	216	144	108	86	72
100	6.9	985	492	328	246	197	164
125	12.4	1754	877	585	438	351	292
150	19.8	2806	1403	935	701	561	468
200	42.7	6057	3029	2019	1514	1211	1009
250	76.6	10,851	5425	3618	2713	2169	1807
300	123.2	17,465	8733	5816	4366	3493	2912
375	220.2	31,214	15,607	10,405	7804	6248	5202

TABLE 20.11 (*Continued*)

		Maximum Allowable Horizontal Projected Roof Areas (Square Meters) at Various Rainfall Rates					
Size of Pipe (mm)	**Flow at 40 mm/m Slope (L/s)**	**25 mm/h**	**50 mm/h**	**75 mm/h**	**100 mm/h**	**125 mm/h**	**150 mm/h**
75	4.3	611	305	204	153	122	102
100	9.8	1400	700	465	350	280	232
125	17.5	2482	1241	827	621	494	413
150	28.1	3976	1988	1325	994	797	663
200	60.3	8547	4273	2847	2137	1709	1423
250	108.6	15,390	7695	5128	3846	3080	2564
300	174.6	24,749	12,374	8250	6187	4942	4125
375	312.0	44,220	22,110	14,753	11,055	8853	7367

Source: © 2024 Uniform Plumbing Code. Reprinted with the permission of the International Association of Plumbing and Mechanical Officials. This copyright material and all points or statements in using this material have not been reviewed by IAPMO. The opinions expressed herein are not representations of fact from IAPMO.

Notes: The sizing data for horizontal piping are based on the pipes flowing full. For rainfall rates other than those listed, determine the allowable roof area by dividing the area given in the 1 in./hr [25 mm/hr] column by the desired rainfall rate.

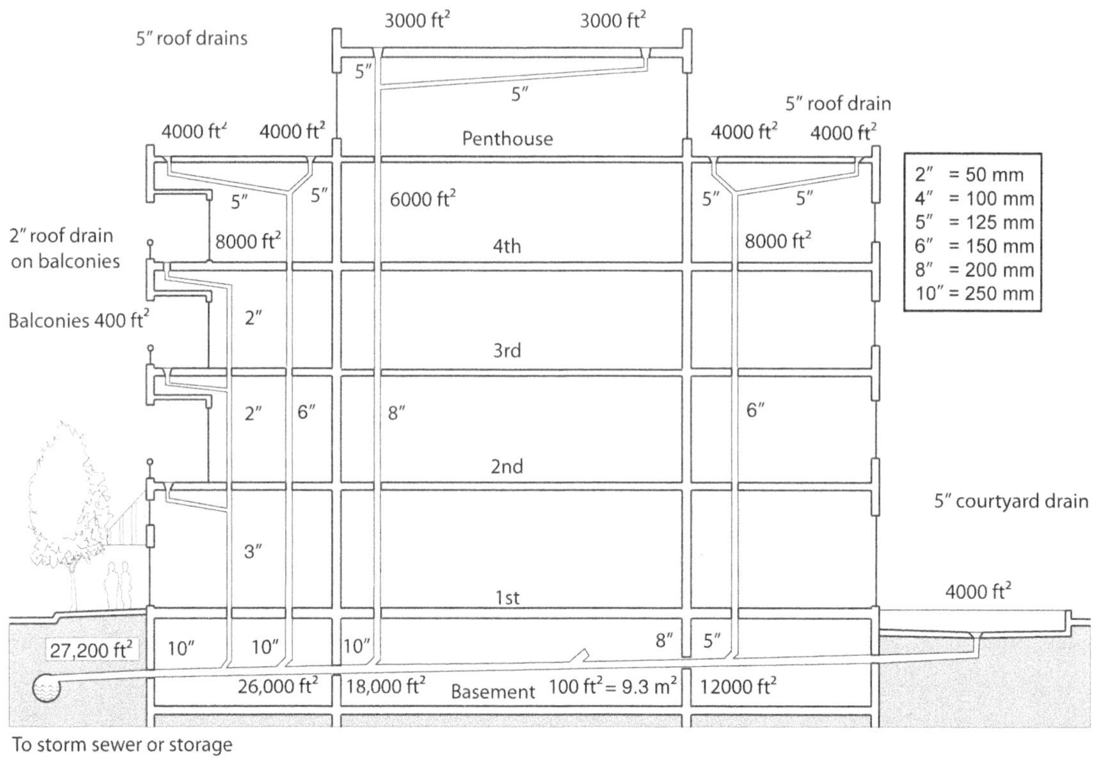

Figure 20.25 *Separate storm drainage. Areas drained and corresponding sizes of vertical leaders and horizontal drains are from Tables 20.10 and 20.11. Storm drain piping within a building requires insulating covering with a vapor retarder on the outside. This prevents condensation (sweating) on the pipes when, in winter, warm, moisture-laden air in the building could reach the pipe surface (which would be cold from carrying icy water) and condense there—leading to wet, dripping conditions on the pipes. Each roof has two drains in case one is temporarily blocked. (Drawing by Nathan Majeski.)*

20.8 CASE STUDY—WATER AND DESIGN

Lady Bird Johnson Wildflower Center, Austin, Texas

PROJECT BASICS

- Location: Austin, Texas, USA
- Latitude: 30.18°N; longitude: 97.89°W; elevation: 621 ft [189 m] for Austin city
- Heating degree days: 1688 base 65 °F [938 base 18.3 °C]; cooling degree days: 7171 base 50 °F [3984 base 10 °C]; annual precipitation: 30 in. [762 mm] per year
- Building type: New construction; offices, visitor's center, classrooms
- Building area: 54,000 ft^2 [5017 m^2]
- Completed 1995
- Client: LBJ Wildflower Center
- Design team: Overland Partners and consultants

Background. The Lady Bird Johnson Wildflower Center (Figure 20.26) is an award-winning project that exemplifies resource conservation through a careful union between the buildings and the surrounding landscape. Elements of the design process for water conservation and other strategies at the Wildflower Center are presented in the discussion that follows, in order to emphasize the importance of having an integrated design process when trying to achieve a wide array of goals—ranging from redeveloping the native landscape to showcasing the intrinsic beauty of local materials. In the case of the Wildflower Center, the client's values were important to the success of this project that aimed to manifest a 13-year campaign to bring research and resources together on behalf of native plants. It has become a place to learn about native plants and how to reduce water consumption while maintaining a beautiful outdoor habitat.

Initially named the National Wildflower Research Center, founded by Lady Bird Johnson and actress Helen Hayes in 1982, the center has expanded beyond its original intentions to incorporate education, progressive environmental movements, and the publication of data on landscaping with native plants in the United States. To accommodate its success and growth, the Wildflower Research Center moved to a new site (Figure 20.27) in 1995. In 2006, the Wildflower Center became an Organized Research Unit of the University of Texas at Austin's College of Natural Sciences and School of Architecture. Concern for water conservation led to many of the distinctive forms seen in the cisterns and waterways on the land. (The information that follows was provided by Overland Partners.)

Figure 20.26 *Lady Bird Johnson Wildflower Center in Austin, Texas; view to the Visitor's Center. (Photo by Jake Moore; © Alison Kwok; all rights reserved.)*

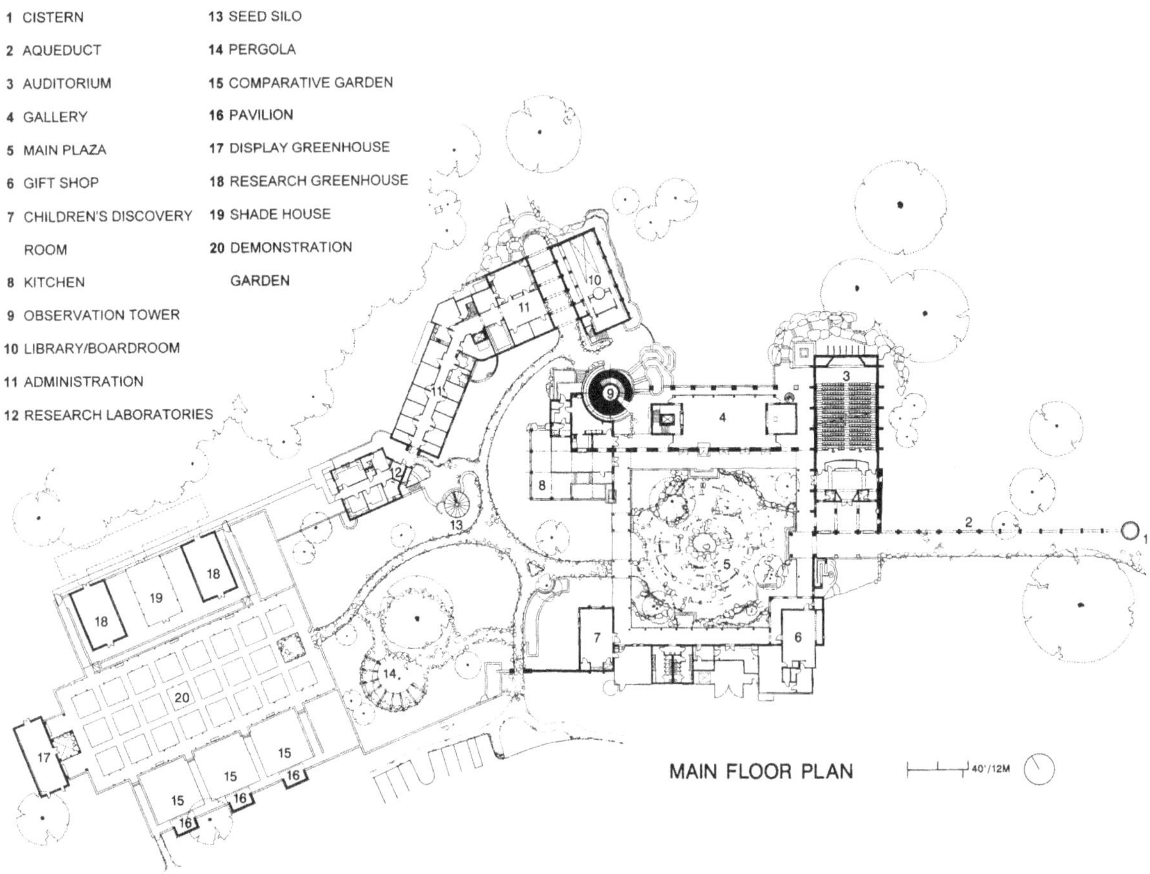

Figure 20.27 Site plan shows two cisterns, one at the far right (1), the other in the central observation tower (9). (Overland Partners Architects; used with permission.)

Context. The Lady Bird Johnson Wildflower Center is a research center designed to share information with the public by increasing awareness, and understanding, of local native plants and landscapes. In 1997, the center was renamed the Lady Bird Johnson Wildflower Center and its mission expanded to include educational opportunities for visitors. Within the last decade the center has played an active role in land restoration, plant conservation, the Millennium Seed Bank Project, and the creation of the Sustainable Sites Initiative with the aid of the American Society of Landscape Architects.

Design Intent. The mission of the Wildflower Center is to "increase the sustainable use and conservation of native wildflowers, plants and landscapes." Aligned with this, the design of the center teaches by example with particular attention being given to the visitor's experience. Rather than creating one monolithic building, the project comprises several smaller buildings wrapped by landscaping, demonstration gardens, and a main plaza. Framing views of the surrounding terrain, these buildings extend the notion that the center is more about the landscape than the buildings. A key element of the design was to eliminate the need to pump water from the endangered Edwards Aquifer. Many of the native plants help conserve water, since they require only that supplied by available rainfall. Runoff from the building rooftops drains into various cisterns (Figure 20.28) and storage tanks to be used later for irrigation. Some rooftops are connected with aqueducts (Figure 20.29) that allow the water to flow to a shared storage tank/cistern (Figure 20.30).

The leading principles behind the design were as follows:

- The building and programs should support "total resource conservation."
- Respect existing native wildflowers, landscaping, and plants.
- Harmonize built form with landscaping.
- Minimize the impact of the built environment.
- Enable visitor education through example.

Figure 20.28 Oculus of the water tower cistern. The tower serves as a focal point for views of the site. (Photo by Jake Moore; Alison Kwok.)

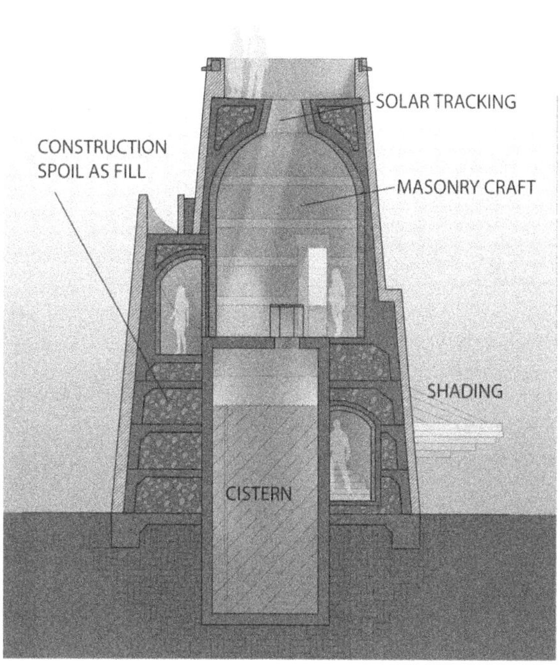

Figure 20.30 Section of the tower cistern. (Overland Partners Architects; used with permission.)

Figure 20.29 Entry cistern fed by aqueducts from the site's rainwater-harvesting system. (Photo © Alison Kwok; all rights reserved.)

- Eliminate the need to pump water from the endangered Edwards Aquifer.
- Respect and utilize the local vernacular, materiality, and craftsmanship.
- "Demonstrate an ecologically sensitive approach to the development of a site with fragile environmental conditions."
- Utilize and design sustainable technologies to aid in conservation efforts.

Design Criteria and Validation. The project was built in 1995, long before the Leadership in Energy and Environmental Design (LEED) certification process. In retrospect, the project complies with many of the principles that guide the LEED rating system. Sustainable design was desirable to maintain the center's founding philosophies and the ethics of the design team, which were the driving forces behind the environmental design decisions. The intentions of the project were to demonstrate an ecologically sensitive approach to the development of a site with fragile environmental conditions, and to integrate the buildings and programs (Figure 20.31) to support "total resource conservation," while showing the beauty of the native landscape and supporting over 400 native plant species.

Post-Occupancy Validation Methods The success of the Lady Bird Johnson Wildflower Center resulted in more than 20 international, national, state, and local awards for its facilities, programs, and leadership

Figure 20.31 Rainwater-harvesting system on local stone arches along the entry path surrounded by oak trees. (Photo © Alison Kwok; all rights reserved.)

in environmental design. It was calculated that the investment made in creating these facilities brought returns of more than $11 million in free media in the first two years. Based on the collected data, energy simulations were run using ENER-WIN—a software package used to analyze total building energy performance, developed in the Department of Architecture at Texas A&M University as part of the Vital Signs Curriculum Project. The overall energy performance of the building simulated by this software was calibrated to the actual building energy use based on utility bills for one year. The components of the energy usage were broken down and compared, to see the percent contribution to the HVAC load of the various design features—i.e., the use of daylighting strategies and a matching of HVAC and occupancy schedules.

Performance Data

- With an average rainfall of 30 in. [762 mm] per year, the whole rooftop system can collect approximately 300,000 gal [1,135,500 L] of rainwater per year.
- In total, the center collects water from 17,000 ft^2 [1579 m^2] of roof space and can store more than 60,000 gal [227,100 L] in on-site cisterns and storage tanks, providing approximately 10–15% of the center's yearly water needs for irrigation and landscaping.
- Approximately 10,600 gal of water per inch of rain [1580 L/mm] are collected for the whole center.
- Five cisterns are used in total, with an approximate capacity of 65,000 gal [246,025 L]: two cisterns at 20,000 gal [75,700 L] each, one at 5000 gal [18,925 L], the entry cistern at 12,000 gal [45,420 L], and the children's "little house" cistern holding 3000 gal [11,355 L].
- The entry collection cistern system is fed by 1167 ft^2 [108 m^2] of metal roof connected to a 12,000-gal [45,420-L] cistern. This cistern can collect 7000 gal per inch [1043 L/mm] of rain, or approximately 21,000 gal [79,485 L] per year.
- The children's area cistern collects water from 672 ft^2 [62 m^2] of roof, and has a 3000-gal [11,355-L] capacity. With approximately 420 gal per inch of rain [63 L/mm], it can collect 12,600 gal [47,691 L] of water per year.
- All cisterns are located aboveground except for the transfer station, which is underground.

Figure 20.32 Wood was reused from the site to create overhangs and trellises. (Photo by Jake Moore; © Alison Kwok; all rights reserved.)

- Recycled materials (Figure 20.32) included wood flooring and on-site non-biodegradable construction spoil used to fill the tower structure. Local stone, wood flooring, and site boulders were obtained from within a 150-mile [241-km] radius from site.

The project received the following awards and citations:

1995:

Texas Society of Architects—Honor Award

Austin Commercial Real Estate Society—Good Egg Award

Associated Builders and Contractors Awards—First Place for Excellence in Construction

1996:

Recycling and Education Awareness Citations by Keep Austin Beautiful

Good Steward Award by the National Arbor Foundation

American Society of Landscape Architects—San Antonio Design Honor Award

American Society of Landscape Architects—State of Texas Design Honor Award

American Society of Landscape Architects—National Design Honor Award

Associated General Contractors—San Antonio Design Honor Award

Associated General Contractors—Regional Honor Award

Associated General Contractors—Dallas Chapter Summit Award

Associated General Contractors—National Design Honor Award

American Institute of Architects—San Antonio Chapter Design Honor Award

1998:

Business Week/Architectural Record—Award Semi-finalist

2000:

American Institute of Architects Committee on the Environment—Earth Day 2000 Top Ten

FOR FURTHER INFORMATION

Lady Bird Johnson Wildflower Center, wildflower.org

Overland Partners Architects, overlandpartners.com

Rocky Mountain Institute, Green Developments CD, version 2.0, 2001

References and Resources

Brown, G.Z., J. Reynolds, and S. Ubbelohde. 1982. *Inside Out: Design Procedures for Passive Environmental Technologies*. Hoboken, NJ: John Wiley & Sons.

Corbett, M. 1981. *A Better Place to Live*. Emmaus, PA: Rodale Press.

EPA. 1975. *Manual of Individual Water Supply Systems*. Washington, DC: U.S. Environmental Protection Agency.

EPA. 2012. *Data Trends: Water Use Tracking*. Washington, DC: U.S. Environmental Protection Agency.

EPA. 2023. *U.S. Technical Reference: Water Use Intensity by Property Type*. Washington, DC: U.S. Environmental Protection Agency.

EPA. 2024. *Understanding Your Water Bill*. Washington, DC: U.S. Environmental Protection Agency. epa.gov/watersense/understanding-your-water-bill

Ferguson, B. 1998. *Introduction to Stormwater*. Hoboken, NJ: John Wiley & Sons.

Gleick, P. 1998. *The World's Water*. Washington, DC: Island Press.

IAPMO. 2023. *2024 Uniform Plumbing Code*. Walnut, CA: International Association of Plumbing and Mechanical Officials.

ICC. 2017. *ICC A117.1-2017: Standard for Accessible and Usable Buildings and Facilities*. Washington, DC: International Code Council.

ICC. 2023. *2024 International Plumbing Code*. Washington, DC: International Code Council.

McPherson, E.G. (ed.). 1984. *Energy-Conserving Site Design*. Washington, DC: American Society of Landscape Architects.

Milne, M. 1976. *Residential Water Conservation*. Springfield, VA: U.S. Office of Water Research and Technology, Department of Commerce, NTIS.

USGBC. 2019. *LEED v4 Reference Guide for Homes Design and Construction*. Washington, DC: U.S. Green Building Council. usgbc.org/leed

USGS. 2016. *Water Resources of the United States*. Reston, VA: U.S. Geological Survey. water.usgs.gov

USGS. 2018. *Estimated use of water in the United States in 2015*. Reston, VA: U.S. Geological Survey.

CHAPTER 21

Water Supply

WITH WATER, ONE OF THE DESIGNER'S FIRST CONCERNS is to match the quality of the water to the task it performs. As this becomes a more serious design issue, designers will provide for the recycling of water within and around buildings, as well as specify plumbing fixtures that use less water. Table 21.1 shows typical relationships between water quality and usage. Water recycling and conservation are discussed in detail in Chapter 22. In this chapter, we deal primarily with potable (drinkable) water, first with issues of water quality and then with the matter of ensuring an adequate supply of water throughout a building.

21.1 WATER QUALITY

A summary of water quality at the various stages of the hydrologic cycle was given in Chapter 20. As precipitation, water contains few impurities: Almost no bacterial content is present, and only small amounts of minerals and gases can be expected. Surfaces are needed to collect this nearly pure water—and from these surfaces, foreign substances can readily contaminate the water. These pollutants can affect water's physical (mostly organic), chemical (mostly inorganic), biological, or radiological characteristics. Both surface water and groundwater are subject to pollution.

The U.S. National Drinking Water Clearinghouse is a source of information on water supply and wastewater treatment systems. Its brief history of waterborne diseases worldwide begins with a major cholera epidemic that began in Calcutta, India, in 1817, killing thousands and spreading by 1832 to New York City, emptying the streets in panic over the disease. In 1854, a London physician demonstrated that local cases of cholera could be traced to one pump contaminated with sewage from a nearby house, but how the disease was transmitted remained a mystery. In summer 1859, London's Thames River, carrying combined storm and sanitary wastes, stank so badly that Parliament was suspended. In 1892, the bacterium causing cholera was identified during an epidemic in Hamburg, Germany, proving the relationship between contaminated water and the disease. In 1939, an outbreak of typhoid fever killed 60 people at an Illinois mental hospital; typhoid and gastroenteritis resulted from an accidental pumping of polluted river water into supply mains in Rochester, New York, in 1940. Another worldwide cholera epidemic began in Indonesia in 1961, eventually reaching Latin America by 1991. In 1993, an outbreak of cryptosporidiosis in the public water supply of Milwaukee, Wisconsin, shook the public's faith in water systems across the United States; that outbreak killed 104 and infected more than 400,000 people.

In response to this history, we disinfect huge quantities of water, expending energy and adding chemicals so that every drop of water entering North American buildings is potable. Again, the high-grade resource/low-grade use question arises; recycled water for lower-grade use is discussed in Chapter 22.

(a) Physical Characteristics

Some of the most noticeable aspects of water quality fall within this category. Water from surface sources

Mechanical and Electrical Equipment for Buildings, Fourteenth Edition. Walter Grondzik, Daniel Overbey, and Alison Kwok.

Companion website: www.wiley.com/go/meeb14e

TABLE 21.1 Water Use and Quality in Buildings

Use	Desired Quality
A. Consumed	
1. Drinking and cooking	Potable
2. Bathing	Potable
3. Laundering	Soft
4. Irrigation and watering of livestock	Unpolluted
5. Industrial processes	As required
6. Vapor to increase the relative humidity of air	
B. Circulated	
1. Hot water for heating 2. Chilled water for cooling 3. Condenser cooling water 4. Swimming pool water 5. Steam for heating, later condensed	*Note*: Makeup water should be soft or neutral and, for swimming, potable
C. Generally Static	
1. Water stored for fire protection	
2. Water in fire standpipes	
3. Water in sprinkler piping	
D. Controlled	
1. Vapor condensed to reduce relative humidity of air	

Note: For water uses in Section A, flow is often continuous. Section B comprises uses for which flow other than circulation is intermittent or at a relatively low rate, the water added to the systems being known as makeup water. Items C2 and C3 call for piping to provide adequate, though infrequent, flow in emergencies. Item D1 relates only to moisture condensed out of the air and involves no design for supply.

(roof runoff, streams, rivers, lakes, and ponds) is particularly subject to physical pollutants.

- *Turbidity*: This is the visible quality of being cloudy or opaque, is easy to see and thus is a likely source of dissatisfaction for the would-be consumer. It is caused by the presence of suspended material such as clay, silt, other inorganic material, plankton, or finely divided organic material. Even those materials that do not adversely affect health are usually aesthetically objectionable. There are three standard units of measuring turbidity, each differing by light source and methodology: nephelometric turbidity unit (NTU), formazin turbidity unit (FTU), and formazin nephelometric unit (FNU). Stringent water quality standards may require a threshold of 1.0 NTU, FTU, or FNU.
- *Color*: This is another visible alteration and is often caused by dissolved organic matter, as from decaying vegetation. Some inorganic materials also color water, as do microorganisms. Like turbidity, such color changes usually do not threaten health, but they are often psychologically undesirable.
- *Taste and odor:* These can be caused by organic compounds, inorganic salts, or dissolved gases. This condition can be treated only after a chemical analysis has identified which source is responsible.
- *Temperature:* This is another characteristic of psychological importance—we expect drinking water to be cool. In general, water supplied between 50°F and 60°F [10°C and 16°C] is preferred.
- *Foamability:* This is usually caused by concentrations of detergents. The foam itself does not pose a serious health threat, but it may indicate that other, more dangerous pollutants associated with domestic waste are also present. Because of increased foaming in water in the 1960s, today's detergents must use linear alkylate sulfonate (LAS), which biodegrades rapidly—except in the absence of oxygen. Because this lack of oxygen is characteristic of some septic tank drainage fields, foam in drinking water should be investigated promptly.

(b) Chemical Characteristics

Groundwater is particularly subject to chemical alteration because as it moves downward from the surface it slowly dissolves some minerals contained in rocks and soils. A chemical analysis (such as that given in Table 21.2) is usually required for individual water supply sources. Such an analysis will indicate: (1) the possible presence of harmful or objectionable substances, (2) the potential for corrosion within the water supply system, and (3) the tendency for the water to stain fixtures and clothing. Concentrations are expressed in milligrams per liter (mg/L), which is essentially equivalent to parts per million (ppm).

Some general terms commonly used to describe chemical characteristics of water are as follows:

- *Alkalinity:* Caused by bicarbonate, carbonate, or hydroxide components. Testing for these components of water's alkalinity is a key to determining which treatments to use.
- *Hardness:* A relative term (see Figure 21.1). Hard water inhibits the cleaning action of soaps and detergents, and it deposits scale on the inside of hot water

TABLE 21.2 Example of Chemical Analysis of Water

Quality		Parts per Million (ppm)[a]
Total hardness	$CaCO_3$	30
Calcium hardness	$CaCO_3$	20
Alkalinity (methyl orange)	$CaCO_3$	27
Alkalinity (phenolphthalein)	$CaCO_3$	0
Free carbon dioxide	CO_2	13.5
Chlorides	Cl	6
Sulfates	SO_4	4
Silica	SiO_2	19
Phosphates—normal	PO_4	0
Phosphates—total	PO_4	0.5
Iron—total	Fe	1.6
Total dissolved solids		66
Turbidity or sediment		Present

Source: Adapted from Olin Water Service for a private well in Virginia.

[a]Note that ppm and mg/L are essentially equivalent terms for concentration in water.

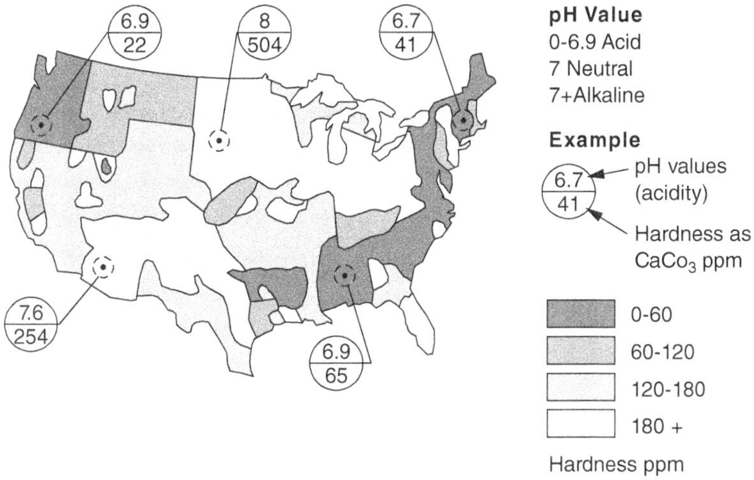

Figure 21.1 *Approximate chemical characteristics of groundwater across the United States. Treatment may be needed when the pH is less than 7.0 (acidity results in corrosion) or when hardness as $CaCO_3$ exceeds 65 ppm (ppm and mg/L are essentially identical measures). (Courtesy of Progressive Architecture.)*

pipes and cooking utensils, thus wasting heating fuel and making utensils unusable. Hardness, which is caused by calcium and magnesium salts, can be classified as temporary (carbonate) or permanent (noncarbonate). Temporary hardness is largely removed when the water is heated—it forms the scale just described. Permanent hardness cannot be removed by simple heating (see Section 21.4(c)).

- *pH:* A measure of the water's hydrogen ion concentration, as well as its relative acidity or alkalinity (Figure 21.1). A pH of 7 is neutral. Measurements below 7 indicate increasing acidity (and corrosiveness); water in its natural state can have a pH as low as 5.5, with 0 being the ultimate acidity. Measurements higher than 7 indicate increasing alkalinity. A pH as high as 9 can be found in water in its natural state, with 14 representing the ultimate alkalinity. The pH value is the starting point for determining treatments for corrosion control, chemical dosages, and disinfection.

Unintentional chemical additions to water supplies most commonly include the following elements:

- *Toxic substances:* These are occasionally present in water supplies. Local health authorities can provide information about acceptable concentrations of such substances as arsenic (As), barium (Ba), cadmium (Cd), chromium (Cr^{6+}), cyanides (CN), fluoride (F), lead (Pb), selenium (Se), and silver (Ag).
- *Fluoride:* Although limited amounts of fluoride are frequently added to water supplies to help prevent tooth decay, fluorides in excess of such optimum concentrations can produce mottling of teeth.
- *Lead:* This poses a dangerous threat, even in relatively small amounts, because it is a cumulative poison. Lead in water usually comes from lead piping (in older buildings or cities) or from corrosive water on lead-painted roofs.
- *Mercury:* This is a toxic pollutant that is released into the air and then falls into water supply sources.
- *Nickel:* This is a common element in water, but too much can be detrimental to water quality.

Unfortunately, the list of toxic chemicals is expanding, and as chemical waste dumps have been abandoned or mismanaged, groundwater has become contaminated. The U.S. Environmental Protection Agency (EPA) has estimated that 75% of both active and abandoned chemical waste dumps—some 51,000 in all—are leaking.

Per- and polyfluoroalkyl substances (PFAS) are a large group of inorganic synthetic chemicals that have been widely utilized in consumer and industrial products since the 1940s, PFAS exhibit strong carbon-fluorine bonds which make them resistant to degradation in nature or the human body. As a result, PFAS are bioaccumulative and thus referred to colloquially as "forever chemicals." PFAS persist in the human body and can be passed on to offspring through breastmilk or by crossing the placenta and reaching a developing fetus. According to the National Institute of Environmental Health Sciences (one of the federally funded centers that makes up the National Institute of Health), the EPA's current database of PFAS exceeds 15,000 different chemicals and the list continues to grow—making regulatory "bans" challenging (NIEHS, 2024).

The acronym "PFAS" is used as an umbrella term to address several different types of synthetic substances, including:

- *Perfluoroalkyl substances,* which do not contain a carbon–hydrogen bond.
- *Polyfluoroalkyl substances,* which do have some carbon–hydrogen bonds.
- *Long-chain PFAS,* which have eight or more carbon atoms (often termed "C-8"). Perfluorooctanoic acid

(PFOA) and perfluorooctanesulfonic acid (PFOS) are long-chain C-8 chemistries.

- *Short-chain PFAS* have fewer than eight carbon atoms and are sometimes labeled C-6, C-4, etc., depending on the number.
- *GenX*, a type of short-chain PFAS developed as an alternative to PFOA, which was eventually revealed to have similar toxicological effects to PFOA but can be even more mobile in the environment, making them often difficult to remove from water resources.

PFAS are increasingly prevalent in our water supply resources. Both the EPA and the U.S. Geological Survey (USGS) maintain publicly available interactive information resources that convey information about the monitoring and presence of PFAS in various water bodies and resources—including PFAS that persist in post-treatment tap water (EPA, 2024b; USGS, 2024).

The threat to our groundwater supplies is illustrated in Figure 21.2. Once polluted, aquifers are extremely difficult to clean. This is one reason why ground-source heat pumps are tightly regulated. In addition to the following list of inorganic chemicals, we are becoming aware of many new organic chemicals as well, some of which are suspected of causing 5–20% of US cancers.

- *Chlorides:* These can enter water as it passes through geologic deposits formed by marine sediment, or because of pollution from seawater, brine, or industrial or domestic wastes. A noticeable taste results from chloride in excess of 250 mg/L.
- *Copper:* This can enter water from natural copper deposits or from copper piping that contains corrosive water. Concentrations of copper in excess of 1.0 mg/L can produce an undesirable taste.
- *Iron:* This is frequently present in groundwater. Corrosive water in iron pipes will also add iron to water. At concentrations above 0.3 mg/L, iron can lend a brownish color to washed clothes and can affect the taste of the water.
- *Manganese:* This can both pose a physiological threat (it is a natural laxative) and produce color and taste effects similar to those produced by iron. The recommended limit is 0.05 mg/L.
- *Nitrates:* In high concentrations these pose a threat to infants, in whom they can cause "blue baby" disease. In shallow wells, nitrate concentrations can indicate seepage from deposits of livestock manure.
- *Pesticides*: These are a growing threat to water supplies, and are particularly common in wells near homes that have been treated for termite control. Avoid using pesticides near wells.
- *Sodium:* This is primarily dangerous for people with heart, kidney, or circulatory ailments. For a low-sodium diet, the sodium in drinking water should not exceed 20 mg/L. Salts spread on roadways for ice control can leach into the soil and enter groundwater. Note that some water softeners (discussed in Section 21.4(c)) can raise sodium concentrations in water.
- *Sulfates*: These have laxative effects, and can enter groundwater from natural deposits of Epsom salts (magnesium sulfate) or Glauber's salt (sodium sulfate). Concentrations should not exceed 250 mg/L.
- *Zinc:* Sometimes zinc enters groundwater in areas where it is found in abundance. Although not a health threat, it can cause an undesirable taste at concentrations above 5 mg/L.

(c) Biological Characteristics

Potable water should be kept as free as possible of disease-producing organisms—bacteria, protozoa, and viruses. These organisms are not easily identified; a thorough biological water test is complex and time-consuming. For this reason, the standard test is for

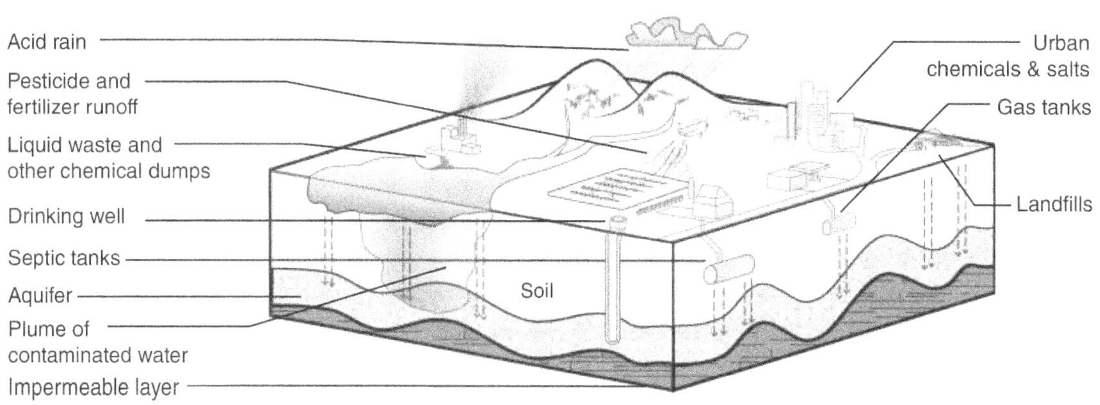

Figure 21.2 *How groundwater becomes contaminated. The "plume" formed by contaminants can often go undetected. (© Newsweek, Inc. 1982; all rights reserved. Redrawn by Martin Lee.)*

one kind of bacteria—the coliform group (*Escherichia coli*, better known as *E. coli*), which is always present in the fecal wastes of human beings (as well as those of many animals and birds) and which outnumbers all other disease-producing organisms in water. The recommended maximum concentration of coliform bacteria is one organism per 100 mL (about ½ cup) water. Stringent water quality standards may require that zero coliforms are detected in such a sample.

For biological activity to be kept to a minimum in drinking water, a water source should be chosen that does not normally support much plant or animal life—hence the popularity of groundwater rather than surface water as a source. In addition, the supply should be protected from subsequent biological contamination. Where cities depend on small lakes for water, human beings are frequently excluded from the watersheds. Organic fertilizers and nutrient minerals should also be kept out of the water supply to further discourage biological activity. For the same reason, stored water should be kept dark and at low temperatures. Finally, organisms (or their by-products) are commonly destroyed at treatment facilities.

(d) Radiological Characteristics

The mining of radioactive materials and the use of such materials in industry and power plants have produced radiological pollution in some water supplies. Because radiological effects are cumulative, concentrations of radioactive materials should be low. "Safe" minimum concentrations have continually been revised downward for other radiation exposures; it is advised to consult the local public health service for current recommendations.

(e) Setting Water Quality Parameters

Specific limits on the chemical composition of dispensed water supplies can vary widely by jurisdiction. Model codes, standards, and rating systems also establish certain water supply quality requirements. For instance, WELL Building Standard v2 (IWBI, 2020) sets the following health-based limits for drinking water (which must be verified by performance testing):

- arsenic ≤ 0.01 mg/L
- cadmium 0.003 mg/L
- chromium (total) ≤ 0.05 mg/L
- copper ≤ 2.0 mg/L
- fluoride ≤ 1.5 mg/L
- lead ≤ 0.01 mg/L
- mercury (total) ≤ 0.006 mg/L
- nickel ≤ 0.07 mg/L
- nitrate ≤ 50 mg/L as nitrate (11.0 mg/L as nitrogen)
- nitrite ≤ 3.0 mg/L as nitrite (0.9 mg/L as nitrogen)
- total chlorine ≤ 5 mg/L
- residual (free) chlorine does not exceed 4 mg/L
- concentration of total trihalomethanes (TTHM, sum of dibromochloromethane, bromodichloromethane, chloroform and bromoform) is 0.08 mg/L or less
- concentration of haloacetic acids (HAA5, sum of chloroacetic, dichloroacetic, trichloroacetic, bromoacetic, and dibromoacetic acids) is 0.06 mg/L or less

Additional water quality parameters may be set based on taste and/or odor rather than health. For instance, WELL v2 sets the following parameters for water delivered to projects for human consumption (which must also be verified by performance testing):

- aluminum ≤0.2 mg/L
- chloride ≤250 mg/L
- copper ≤ 1 mg/L
- manganese ≤ 0.05 mg/L
- iron ≤ 0.3 mg/L
- silver ≤ 0.1 mg/L
- sodium ≤ 270 mg/L
- sulfate ≤ 250 mg/L
- zinc ≤ 5 mg/L
- total dissolved solids (TDS) ≤ 500 mg/L
- free chlorine ≤ 1.25 mg/L

It is important to set appropriate, clear, and measurable parameters for water supply on any building project. Such information should be clearly outlined in the owner's project requirements (OPR). The water supplied to the building will exhibit a certain quality, which will dictate the local filtration and/or treatment required to meet the OPR's water quality parameters.

21.2 FILTRATION

In the preceding section, water pollution was broken down into physical, chemical, biological, and radiological categories. The various forms of treatment for such pollutants do not necessarily fall into the same categories, as one treatment may be effective for several different polluted conditions. A general look at common domestic water quality problems and treatments is provided in Table 21.3.

We begin with filtration, because so often it is the first treatment in a series; it is also one of the oldest and simplest methods. This very common treatment removes suspended particles, some bacteria, and color or taste by passing water through a permeable fabric or a porous bed of materials. The more common approaches are listed in the following subsections,

TABLE 21.3 Common Water Quality Problems and Treatment in Small Systems

Item	Cause	Bad Effect	Correction
Hardness	Calcium and magnesium salts from underground flow	Clogging of pipes by scale, burning out of boilers, and impaired laundering and food preparation	Ion-exchanger (zeolite process)
Corrosion	Acidity, entrained oxygen, and carbon dioxide (low pH)	Closing of iron pipe by rust, leaking connections, destruction of brass pipe	Raising the alkaline content (neutralizer)
Biological pollution	Contamination by organic matter or sewage	Disease	Chlorination by sodium hypochlorite or chlorine gas; or ozonation
Color	Iron and manganese	Discoloration of fixtures and laundry	Chlorination or ozonation and fine filtration
Taste and odor[a]	Organic matter	Unpleasantness	Filtration through activated carbon (purifier); aeration
Turbidity[a]	Silt or suspended matter picked up in surface or near-surface flow	Unpleasantness	Filtration

[a]These problems are not common in private systems that use deep wells.

beginning with filtering to remove suspended particles, and then moving on to more specialized applications for the removal of iron and/or manganese, tastes, and odors.

(a) Sedimentation

Before water enters a filter, this process removes some suspended matter simply by allowing time, the inactivity of the water, and gravity to do the work of settling out heavier suspended particles. Simple basins, ponds, or tanks constructed for this purpose are large enough to retain the water for at least 24 hours and are equipped with baffles to slow the water flow. To clean out the sediment, water is usually diverted to an identical second basin while the first is being cleaned.

(b) Coagulation

This process also removes suspended matter, along with some coloration. A chemical such as alum (hydrated aluminum sulfate) is added to water made turbulent by baffles or static mixers to distribute the chemicals evenly.

(c) Flocculation

The water is then held in a quiet condition in which the suspended particles will combine with the alum to form floc. These heavy particles then settle out in a process similar to sedimentation. Some adjustment of the pH may be necessary.

See the National Drinking Water Clearinghouse (1996a) "Tech Brief: Filtration," for details on the following methods of filtering.

(d) Slow Sand Filters

These are common in small-scale water supply systems (Figure 21.3; also see Chapter 20 for a rainwater application). Not suitable for water with high turbidity, they do not usually require coagulation/flocculation and may not even require sedimentation. Water should not be chlorinated before entering this filter, because it will interfere with the subsequent biological activity. These filters are able to remove up to 99.9% of *Giardia* cysts.

Slow sand filters are low-maintenance, easily constructed devices that should be cleaned as often as the turbidity of the water demands—from once a day to perhaps once a month. They are cleaned by removal and replacement of about the top 1 in. [25 mm] of sand, which has formed a layer of biological slime called (descriptively) the schmutzdecke, which traps small particles and degrades organic material in the water. This sand is then either washed for reuse or discarded.

The approximate rate of flow is slow, requiring a rather large surface area: 0.03–0.10 gpm per ft^2 [0.02–0.07 L/s per m^2] of filter bed surface; in other units, 40–140 gal per day per ft^2 [1630–5720 L per day per m^2] of filter bed surface. Overall thickness is usually 30 in. to 48 in. [762 mm to 1219 mm] of sand over 12 in. [304 mm] of gravel with an underdrain system.

In freezing temperatures, slow sand filters must be housed; if they develop an ice layer, this prevents cleaning.

(e) Diatomaceous Earth Filters

Also known as precoat or diatomite filters, these can be of either the vacuum or the pressure type. They rely on a layer of diatomaceous earth—a minimum of 1/8 in. [3 mm]-thick layer, placed on a septum or filter element (for *Giardia* removal, the thickness should be increased to about 1/5 in. [5 mm]). They are most suitable for water with low bacterial counts and low turbidity. Simple to operate and effective in removing cysts, algae, and asbestos, they require periodic attention to

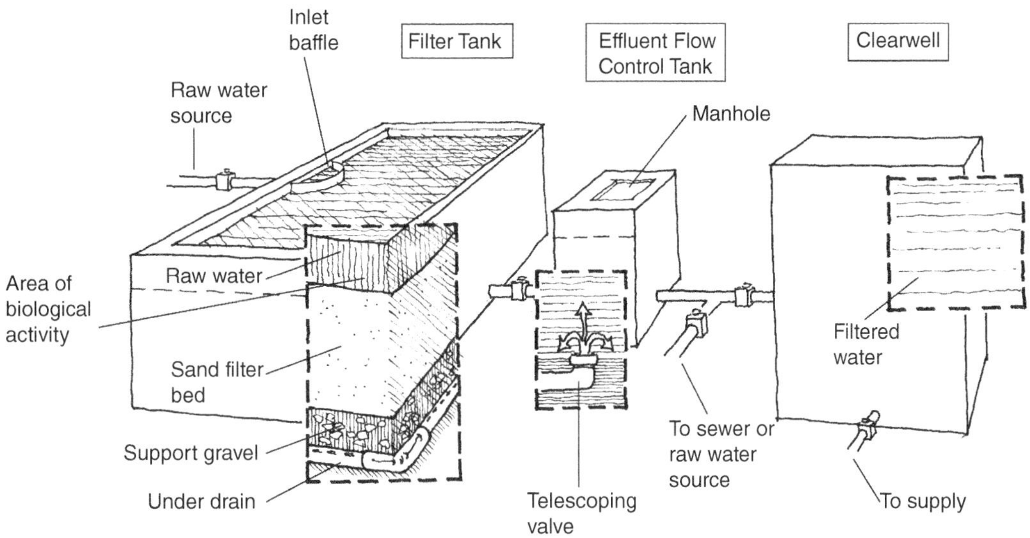

Figure 21.3 *Slow sand filter effective in removing cysts of Giardia. Water should not be chlorinated before it enters the filter. (From National Drinking Water Clearinghouse, 1996a. Redrawn by Dain Carlson.)*

remain effective, including backwashing every one to four days.

(f) Direct Filtration

Intended for water supplies of high quality and seasonally consistent flow, these systems omit sedimentation but should include coagulation for the most effective *Giardia* removal. These are often used with steel pressure tanks to maintain pressure in the water supply line.

Packaged filtration combines features such as chemical addition, flocculation, and sedimentation, along with filtration, in one compact unit. Most often used for small community water supplies, these systems treat surface water to remove turbidity, color, and coliform organisms.

(g) Membrane Filtration

Also called microfiltration or ultrafiltration, this rapidly developing technique can remove bacteria, *Giardia*, and some viruses. It does not require coagulation as pretreatment. Using hollow fiber or spiral-wrapped membranes, it is able to exclude all particles greater than 0.2 μm from the water stream. It is best used on water supplies of low turbidity because of fouling of the fibers or membranes.

Water is forced at high pressure through these filters. The contaminants trapped on the inflow side must be frequently removed by reversing the flow and flushing the waste; calcium and other persistent contaminants must be periodically removed with chemical cleaning.

Nanofiltration, using much smaller pores, is discussed in Section 21.3.

(h) Cartridge Filtration

Increasingly popular on lavatory faucets as well as on small supply systems, these systems are easy to operate and maintain. They require water of low turbidity and last longer when some prefiltering by more crude means is performed upstream. They can exclude particles of 0.2 μm (or even smaller). A disinfectant can prevent surface-fouling microbial growth on the cartridge filters; some periodic chemical cleaning will likely be required.

(i) Other Filters

Activated carbon filters are particularly effective for removing tastes and odors. The water is passed through granular carbon, which attracts large quantities of dissolved gases, soluble organics, and fine solids.

Porous stone, ceramic, or unglazed porcelain filters (also called Pasteur filters) are usually made in small sizes so that they can be attached to water faucets. They are used widely in some countries, such as Mexico, but poor maintenance or hairline cracks often lead to bacterial infiltration, complicating the filtration process. A more positive approach to the disinfection of drinking water thus is desirable.

21.3 DISINFECTION

Disinfection is the most important health-related water treatment, because it destroys microorganisms that can cause disease in human beings. Disinfection is required of water supply systems that rely on surface water or groundwater sources under the influence of surface water. Initially, primary disinfection achieves the desired level of microorganism kill (inactivation); then secondary disinfection maintains a disinfectant residual in the treated water that prevents microorganism regrowth.

Although chlorination has become the standard approach to removing harmful organisms from water, there are alternatives: nanofiltration, ultraviolet (UV) radiation (unsuitable for water with high turbidity because it cannot easily penetrate), bromine, iodine, ozone, and heat treatment, among others. Chlorine continues to disinfect after the initial application. It is this continuing secondary disinfection that has made it universally relied on, despite dangers such as that posed by deadly chlorine gas. Although chlorine affects the taste and odor of water, it is also effective in removing less desirable tastes and odors. Unfortunately, chlorine can react with organic materials in water to form halogenated by-products. Trihalomethanes (THMs) are chemical compounds that form when chlorine, or other halogen disinfectants, are used to treat water that contains organic matter. Haloacetic acids (HAAs) are a group of chemicals that form when disinfectants like chlorine or ozone react with organic matter in water. THMs and HAAs are types of disinfection by-products (DBPs), which are chemicals that form when chlorine is used to treat drinking water to prevent disease. DBPs are difficult to remove from water supplies. It is typically easier to either remove the organic materials before treatment, or to use another disinfectant strategy, than to try to remove these halogenated by-products after chlorine treatment.

See the National Drinking Water Clearinghouse (1996b) "Tech Brief: Disinfection," for details on the methods presented in Subsections 21.3(a) through 21.3(d).

(a) Chlorination

Factors that affect chlorine's ability to disinfect include the following:

1. *Chlorine concentration:* The higher the concentration, the faster and more complete the rate of disinfection.
2. *Contact time:* The longer the chlorine contacts the organisms in water, the more complete the disinfection. At a minimum, 0.4 mg/L of chlorine should contact water 30 minutes before use.
3. *Water temperature:* The higher the temperature during contact, the more complete the disinfection.
4. *pH:* The lower the pH, the more effective the disinfection.

There are three common forms in which chlorine is used to disinfect water supplies. Chlorine gas is stored in a cylinder (Figure 21.4) as a liquid under high pressure and is released (as a gas) by a regulator to an

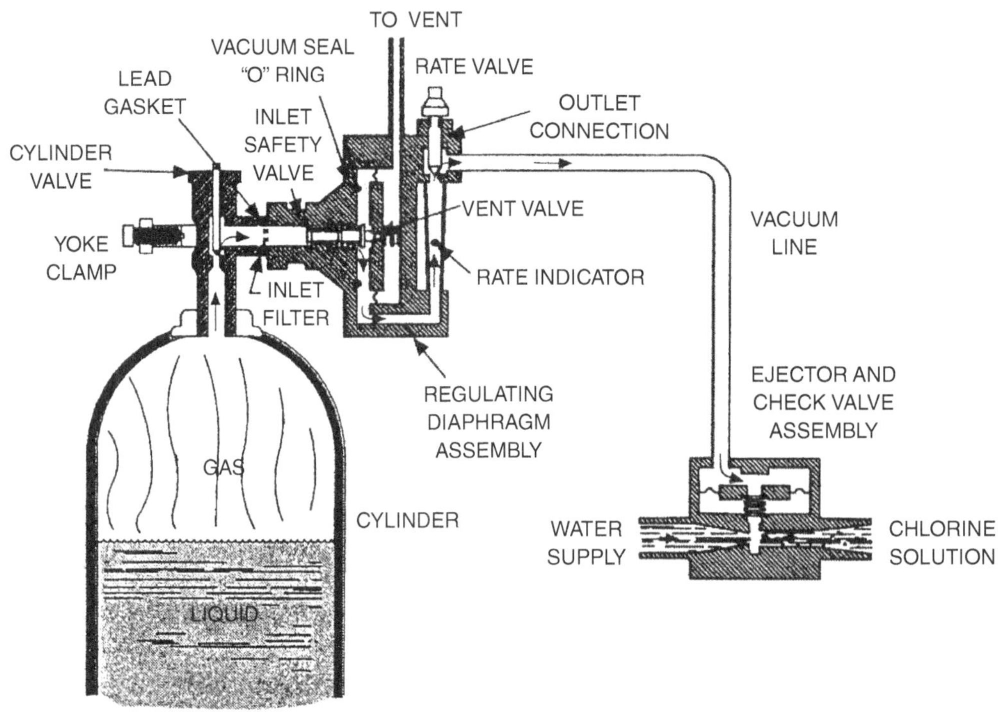

Figure 21.4 Cylinder-mounted chlorinator for automatic disinfection of supply water. (From National Drinking Water Clearinghouse, 1996b.)

injector attached to a water pipe or tank. The injector passes highly pressurized water through a venturi orifice, creating a vacuum that draws the chlorine into the water stream.

Sodium hypochlorite solution is easier to handle than deadly chlorine gas but is very corrosive and decomposes rather quickly. It should be stored in a cool, dark, dry area for no more than a month. Hypochlorinators automatically pump (or inject) a sodium hypochlorite solution into water. They are usually no larger than the pumps used in small water systems. Some hypochlorinators are specially designed for low and fluctuating water pressures or for use where electricity is not available.

Solid calcium hypochlorite is a white solid containing 65% available chlorine that dissolves easily in water. It is corrosive, with a strong odor, but is very stable and can be stored for up to a year. However, it readily absorbs moisture, forming chlorine gas; also, reactions between calcium hypochlorite and organic materials (wood, cloth, petroleum products) can generate enough heat to cause a fire or explosion. Again, hypochlorinators are used to deliver the disinfectant to water.

(b) Chloramine

This is generated on site by adding ammonia to water containing chlorine or when water containing ammonia is chlorinated. This is a weaker disinfectant against viruses or protozoa than the chlorination processes, but it produces fewer disinfection byproducts. It is most often used as a secondary rather than a primary disinfectant. Again, hypochlorinators are used to inject chlorine, after which ammonia is added.

(c) Ozonation

This was first used in full-scale drinking water treatment in 1906. It is a powerful oxidizing and disinfecting agent, destroying most bacteria, viruses, and other pathogenic organisms. It requires a shorter contact time and dosage than chlorine and leaves no chlorine taste. Ozone is formed by passing dry air (or pure oxygen) through a system of high-voltage electrodes. It is an unstable gas and must be generated on site. When ozone reacts with an organic, it produces oxygen and an oxidized form of the organic. Ozone not used in this process quickly decays to oxygen.

In the United States, ozone is commonly used in cooling tower water treatment (Figure 21.5), where its effectiveness against *Legionella pneumophila* is especially appreciated, as well as its control of algae and scale that can greatly reduce cooling efficiency. Ozone is also used in food processing, wastewater cleanup, smoke removal, swimming pools and spas, bottled water, and pulp and paper bleaching.

Equipment includes an ozone generator, a contactor, and a destruction unit, plus instrumentation and controls. Operation and maintenance are relatively complex; electricity accounts for 26–43% of

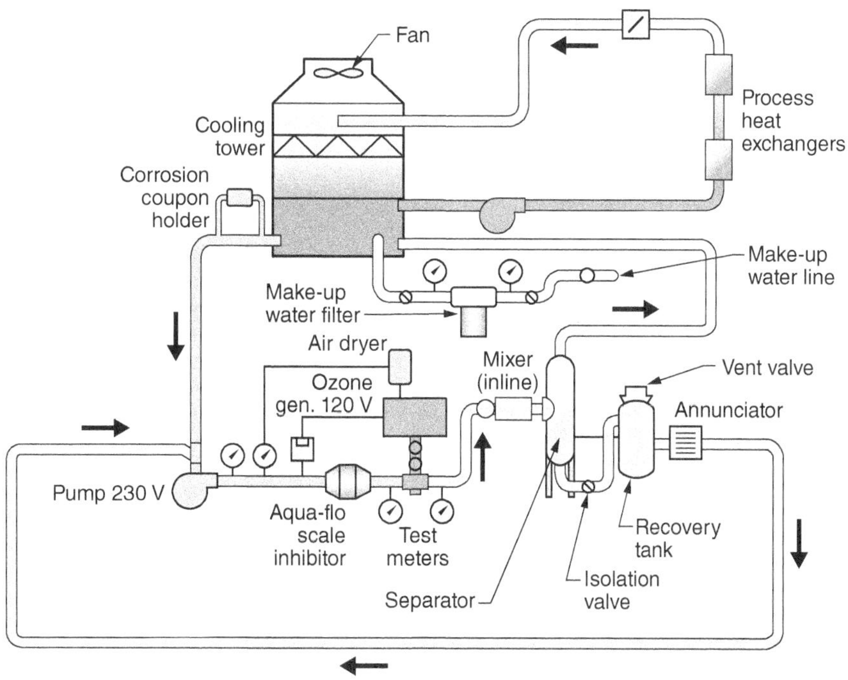

Figure 21.5 *Recycled cooling tower water is treated by an ozonator, a magnetic descaler, and a filtration system. This controls scale formation, algae and slime, corrosion, and sludge buildup. (Courtesy of Aqua-Flo, Inc., Baltimore.)*

the operating costs for small systems. Because it acts only as a primary disinfectant, a secondary disinfectant (often chlorine) is usually required.

(d) Ultraviolet Radiation

Special lamps are used within a reactor (Figure 21.6), whose radiation disrupts the genetic material of the cells of organisms, making them unable to reproduce. Effective against bacteria and viruses, UV radiation is known to inactivate *Giardia* or *Cryptosporidium* cysts. UV radiation is an effective primary disinfectant system, requiring a short contact time and without halogenated by-products. Yet again, a secondary disinfectant system is usually necessary. This system is not suitable for water that contains high levels of suspended solids, turbidity, color, or soluble organic matter.

(e) Nanofiltration

These filter membranes start with pore sizes of 0.2–0.3 μm and are then dipped into a polymer that leaves a thin film, decreasing the pore size to 1 nm. This pore size removes bacteria, viruses, pesticides, and organic material. It also gives the membranes an affinity for calcium, contributing to water softening. However, it also means that the membranes need periodic acid cleaning to remove the calcium deposits. Adding phosphates to nanofiltered water reduces its capacity to dissolve lead.

With such extremely small pore sizes, this process requires very high water pressures, in turn requiring energy. Yet again, a secondary disinfectant system is usually necessary.

21.4 OTHER WATER TREATMENTS

(a) Aeration (Oxidation)

This simple process can improve the taste and color of water and help remove iron and manganese. In aeration, as much of the water's surface as possible is exposed to air. The methods used are rich in aesthetic possibilities—the spraying of water into air, the fall of a turbulent stream of water over a spillway, and flowforms, sculptural waterfalls designed to carry water in a rhythmical, pulsating, figure-eight pattern. The Real Goods headquarters in Hopland, California, features these as a waterfall (Figure 21.7) within a recycled water irrigation system.

To guard against contamination, these processes are often enclosed; if exposed, they must be kept clean. For aeration within tanks, water is passed through a series of perforated plates in streams or droplets.

Aeration improves the flat taste of distilled water and cistern water by adding oxygen. It also oxidizes iron

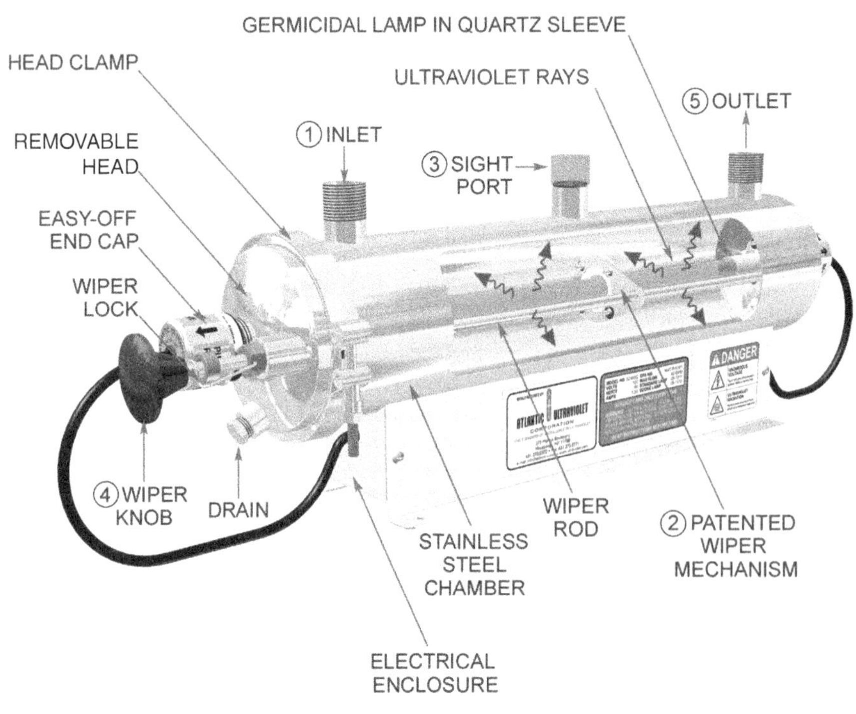

Figure 21.6 *Ultraviolet lamps in water purifiers eradicate harmful microorganisms. The dispersion of short radiation waves makes the water suitable for even the most stringent consumers, including hospitals, laboratories, and pharmaceutical manufacturing. (Courtesy of Atlantic Ultraviolet Corporation.)*

(a) (b)

Figure 21.7 The Real Goods Solar Living Center at Hopland, California, uses creative water recycling. Water from on-site irrigation ponds (a) is pumped by PV to a tank on a small hill, and (b) water is also dispersed into the air from the Agave Fountain for evaporative cooling. Landscape design by Land and Place. (© Alison Kwok, all rights reserved.)

or manganese, which then can more easily be removed by filtration. It also removes odors caused by hydrogen sulfide and algae.

Because aeration raises the level of dissolved oxygen in water, it should be avoided as a treatment when corrosion is a threat.

(b) Corrosion Control

It is important to control corrosion both to keep water systems operating freely and to prevent corrosive water from increasing the concentration of hazardous materials (as from copper pipes). Corrosion also imparts a taste and/or odor to water that is objectionable. Corrosion is a slow degradation of a metal by a flow of electric current from the metal to its surroundings. Some factors involved in corrosion control are:

1. *Acidity:* The more acid (low pH, less than 6.0), the more corrosive the water.
2. *Conductivity:* As dissolved mineral salts increase the water's conductivity, they encourage the flow of the electrical current of corrosion.
3. *Oxygen content:* Dissolved oxygen destroys the thin protective hydrogen film on immersed metals, thus promoting corrosion.
4. *Carbon dioxide content:* Carbon dioxide forms carbonic acid, which attacks metal surfaces.
5. *Water temperature:* Increased temperature increases corrosion.
6. *Lower flow rates:* Reduced turbulence means reduced erosion of the protective layers that form on the inner surfaces of pipes.

The products of corrosion often contribute to scale formation. Scale then lines surfaces, eventually clogging openings.

See the National Drinking Water Clearinghouse (1997a), "Tech Brief: Corrosion Control," for details on this subject.

Acid neutralizers can be installed on water supplies with low pH; their function is often combined with those of hypochlorinators. Typically, neutralizing solutions are mixtures of lime, soda ash, and water. However, pH adjustment should be made just before water delivery, after treatment processes such as coagulation and disinfection.

Corrosion inhibitors cause protective coatings to form on pipes. They are commonly fed into the water, as are other chemicals. Inorganic phosphates, sodium silicates, and mixtures of phosphates and silicates are the more commonly used additives.

Other corrosion control strategies include commercial pipe coatings/linings, installing dielectric or insulating unions (to avoid complications from dissimilar pipe metals), and avoiding metal piping and fixtures altogether.

(c) Softening

Water hardness is caused primarily by calcium and magnesium deposits; when they are removed, water will be soft. Where water hardness is mild enough to affect only laundering, cisterns may be used to collect soft rainwater to use for washing clothes. Where water hardness produces scale in pipes and water-heating appliances, and cisterns are not feasible, water-softening equipment is used. Demineralization of water is accomplished with one of three methods: ion exchange, reverse osmosis, or electrodialysis.

See the National Drinking Water Clearinghouse (1997b) "Tech Brief: Ion Exchange and Demineralization," for details on the following methods of treatment:

- *Ion exchange:* This is popular for small systems, and is effective not only with hardness ions but also with radionuclides. It can be used with fluctuating flow rates but requires pretreatment of most surface waters, and its waste is highly concentrated (requiring careful disposal). A large variety of resins are used for the exchange medium, each effective for specific contaminants.

  On the exchange medium's charged surface, one (contaminant) ion is exchanged for another (regenerant) ion. Eventually, saturation occurs; the contaminants are flushed and the medium is regenerated—once per day is the common shortest cycle. Sodium chloride is often used to regenerate the exchange medium, resulting in a rather high sodium residual—an undesirable development for people on low-sodium diets. Another regenerant material is potassium chloride.

  Equipment includes prefiltration, ion exchange, disinfection, storage, and distribution elements; in smaller systems, single-package units incorporate all of these processes.
- *Reverse osmosis (RO):* This, like ion exchange, is popular for small systems and can be used with fluctuating flow rates, but it requires pretreatment of most surface waters, and its waste is also highly concentrated. RO is effective not only with hardness ions but also with radium, natural organic substances, pesticides, and microbiological contaminants. RO units used in series can remove an even higher percentage of contaminants.

  An inert, semipermeable membrane has a high-pressure supply water on one side; as the pressure slowly forces water through this filtering membrane, most of the contaminants are removed. Water must be used to flush the membrane so that mineral buildup is avoided; this produces a brine requiring careful disposal. Membranes are available in varying types and pore sizes; they are prone to fouling.

  Commercial RO units are available in sizes ranging from a 1 gpm [3.9 L/m] water delivery rate (using two membranes and a 3-hp [2.2-kW] motor, requiring 4.5 gpm [17 L/m] feedwater for a 22% recovery rate) to a water delivery rate of 12.5 gpm [47.3 L/m] (using 12 membranes and a 15-hp [11.2-kW] motor, requiring 19.2 gpm [72.8 L/m] for a 65% recovery rate).
- *Electrodialysis:* This effectively removes fluoride and nitrate, and can also remove barium, cadmium, and selenium. It is relatively insensitive to levels of flow and of total dissolved solids but has an enormous appetite for water. From 10% to 80% of the total water supplied to an RO filtering unit is delivered to the user; the remainder is a waste stream. It also requires a higher level of pretreatment.

  In this process, membranes adjacent to the inflowing stream are charged (either positively or negatively), attracting counter-ions to these membranes. The membranes allow either positively or negatively charged ions to pass through; thus, the ions leave the inflow stream and enter the waste streams (on the other side of each membrane). High water pressure and a source of DC power are needed in this process.

  Membranes can become fouled when the pores are clogged by salt precipitation or blocked by suspended particulates. For either of these water conditions, pretreatment is essential. Reversing the charge on the membranes, electrodialysis reversal, helps to flush the attached ions from the membrane surface and can extend the time between membrane cleanings.

(d) Nuisance Control

Some organisms may not be injurious to health but can multiply so rapidly that piping or filters become clogged, or the water's appearance, odor, and taste are affected. This situation is most common with surface water sources, and it is within surface reservoirs that these treatments are most often applied. Algae growths, the most prevalent nuisance, can usually be controlled by applying copper sulfate (blue stone or blue vitriol) to the water body. Sudden and massive algae kills can have adverse impacts on other life forms within the water, because the decomposing algae rob the water of oxygen. As a further precaution, stored water should be shielded from sunlight whenever possible.

Cooling towers present an especially difficult water treatment problem. The murky water with high turbidity and a high bacterial count leads to clogged passages (thus inefficient performance), deteriorated surfaces, and the growth of potentially lethal bacteria (*Legionella pneumophila*). As a result, enormous quantities of water are commonly passed once through a cooling tower, and then dumped into storm sewers rather than being recycled in a closed system. (In 1989, San Francisco's City Hall was reported to be using 96,000 gal [363,000 L] per summer day for once-through cooling—in a year of drought.) To treat cooling tower water successfully, a method is needed for microbial control, removing organics, and precipitating inorganics; ozonation (see Figure 21.5) is a common answer to this widespread treatment problem.

(e) Fluoridation

A heated controversy continues over the addition of fluoride to drinking water. Despite misconceptions that the U.S. federal government mandates fluoridation, the decision to fluoridate a water supply is made by the state or local municipality. The advantage of fluoridation is that children who drink fluoridated water during the most active stages of tooth development have lower rates of tooth decay. Because everyone drinks water, all children (and adults) benefit, not just those who can afford fluoride pills or prescription toothpaste. Opponents of fluoridation suggest that because sugar is a leading cause of tooth decay in children, sugar, rather than water, should be fluoridated. Small water systems can be equipped with fluoridation units. However, fluoride levels in the water supply must be carefully monitored. In amounts above those used in water treatment, fluoride is toxic and can cause mottled teeth.

(f) Distillation

This is a simple, low-technology approach to purification that produces the equivalent of bottled water for drinking, cooking, and laboratory uses. In one process, it promises the removal of suspended solids, salts, bacteria, and (apparently) halogenated hydrocarbons. When water pollution is extreme, as in the case of sea (salt) water, distillation may be the best treatment. Water is heated to encourage evaporation. As the water turns to vapor, virtually all pollutants are left behind. When this vapor encounters a cooler surface, it condenses, and pure water (although flat in taste) can be collected from this surface.

Figure 21.8 *A solar still can be used to provide a small daily quantity of pure water; this installation serves a laboratory. (Courtesy of McCracken Solar Co.)*

Any heat source can be used in the distilling of water; as such, solar stills (Figure 21.8) are championed by advocates because the energy used is free. Solar distillation of cistern water is an autonomous approach. In semiarid, rather sunny climates, a solar still should produce about ½ gal per square foot [4 L/m^2] of collector surface area per day. This rate of production suggests that only the water used directly for drinking or other specialized purposes is usually feasible for distillation. Another factor to consider is that the cleaning of the still is generally accomplished by flushing it with twice as much water as was delivered. If not excessively brackish, this flush water could be used for irrigation or other non-potable-quality tasks.

21.5 WATER SOURCES

This section focuses on the equipment used to capture and store groundwater from wells. Other water sources are less often used for smaller systems, either because they require much more extensive treatment (surface water from lakes or rivers) or because they provide water intermittently (cisterns). A multistage treatment (flocculation, sedimentation, etc.) may be inappropriate for small water systems that receive only occasional maintenance. Cisterns were discussed in Chapter 20. Another increasingly important source of city water, the treated effluent of sewage treatment plants, is discussed in Chapter 22.

(a) Wells

Farms and remote housing developments usually have private water systems. In rural and suburban areas where the progress of building is faster than the development of municipal water supplies, private sources may also be sought. Driven or drilled wells are preferable; water from these sources usually has at least the advantages of purity, coolness, and freedom from turbidity, odor, and unpleasant taste—any of which may be encountered in well water, in addition to either acidity or hardness.

Bored wells, which are dug with earth augers, are usually less than 100 ft [30 m] deep. They are used when the earth to be bored through is boulder-free and will not cave in. The diameter range is 2–30 in. [50–760 mm]. The bored well is then cased with metal, vitrified tile, or concrete.

Driven wells are the simplest and usually the least expensive type. A steel drive-well point is fitted on the end of pipe sections and driven into the earth. The drive point is usually 1 1/4–2 in. [32–50 mm] in diameter. The materials and design of drive-well points vary according to the expected characteristics of the earth in which the well is driven. First, a pilot hole is bored (frequently with a simple hand auger), and the drive-well point and pipe sections are lowered into it. Then the well is driven to a point well below the water table.

Jetted wells require a source of water and a pressure pump. A washing-well point is supplied with water under pressure; this loosens the earth and allows the point and pipe to penetrate.

Drilled wells require more elaborate equipment of several types, depending upon the geology of the site. The percussion (or cable-tool) method involves the raising and dropping of a heavy drill bit and stem. Having thus been pulverized, the earth being drilled is mixed with water to form a slurry, which is periodically removed. As drilling proceeds, a casing is also lowered (except when drilling through rock).

Rotary drilling methods (either hydraulic or pneumatic) utilize a cutting bit at the lower end of a drill pipe; a drilling fluid (or pressurized air) is constantly pumped to the cutting bit to aid in the removal of particles of earth, which are then brought to the surface. After the drill pipe is withdrawn, a casing is lowered into position.

Another method is the down-the-hole pneumatic (air) hammer method, which combines the percussion effect with a rotary drill bit.

Local well drillers, who usually use the method most suited to the prevailing geology of a region, can offer useful advice about well construction methods. When clients plan to build in a remote location, the architect and engineer should advise them about water problems. Quality-corrective measures can always be taken and pumping equipment purchased, but the amount of water that can be obtained from the ground, as well as the depth and cost of wells, are important considerations. There are some problem areas where wells several hundred feet in depth will yield as little as 5 gpm [0.3 L/s] or nothing. The cost of drilling a number of exploratory wells may be excessive. Unfortunately, when such difficulties occur, there is often no easy solution. Discussions with neighboring owners, state and federal geologists, and local well drillers can all be helpful.

A low-yield well can be combined with storage tanks so that a pump can run all night, slowly filling the tanks to meet the next day's demands.

(b) Pumps

The Hydraulic Institute, a nonprofit trade association of pump manufacturers, has several recommendations for reducing the amount of electricity consumed by pumps. Some of these apply to small water supply systems.

1. Design systems with lower capacity and total head requirements. (Using larger pipe sizes and minimizing the elevation of tanks are examples.)
2. Avoid excessive capacity. It is typically less expensive to add pumping capacity later on if needs increase. Operating a smaller pump closer to its capacity saves energy compared to a larger pump operating well under its capacity.
3. Select the most efficient pump type and size, even if its first cost is greater; life-cycle costs are likely to be lower.
4. Use two (or more) smaller pumps instead of one large one so that excess pump capacity can be turned off.
5. Maintain pumps and system components to avoid efficiency loss.

Three common types of pumps used in well water supply are the positive displacement pump, the centrifugal pump, and the jet (or ejector) pump.

Positive Displacement Pumps. There are two principal types of positive displacement pumps. In a reciprocating pump, a plunger moves back and forth within a cylinder equipped with check valves. The cylinder is best located near or below the groundwater level. Water enters the cylinder through an initial check valve (which allows flow in only one direction). As the plunger moves toward this check valve, the water is forced through the second check valve, located within the plunger itself. Then, as the piston returns to its original position, the water is forced upward toward the surface.

A rotary pump has a helical or spiral rotor—a turning vertical shaft within a rubber sleeve. As the rotor turns, it traps water between it and the sleeve, thus forcing the water to the upper end of the rotor.

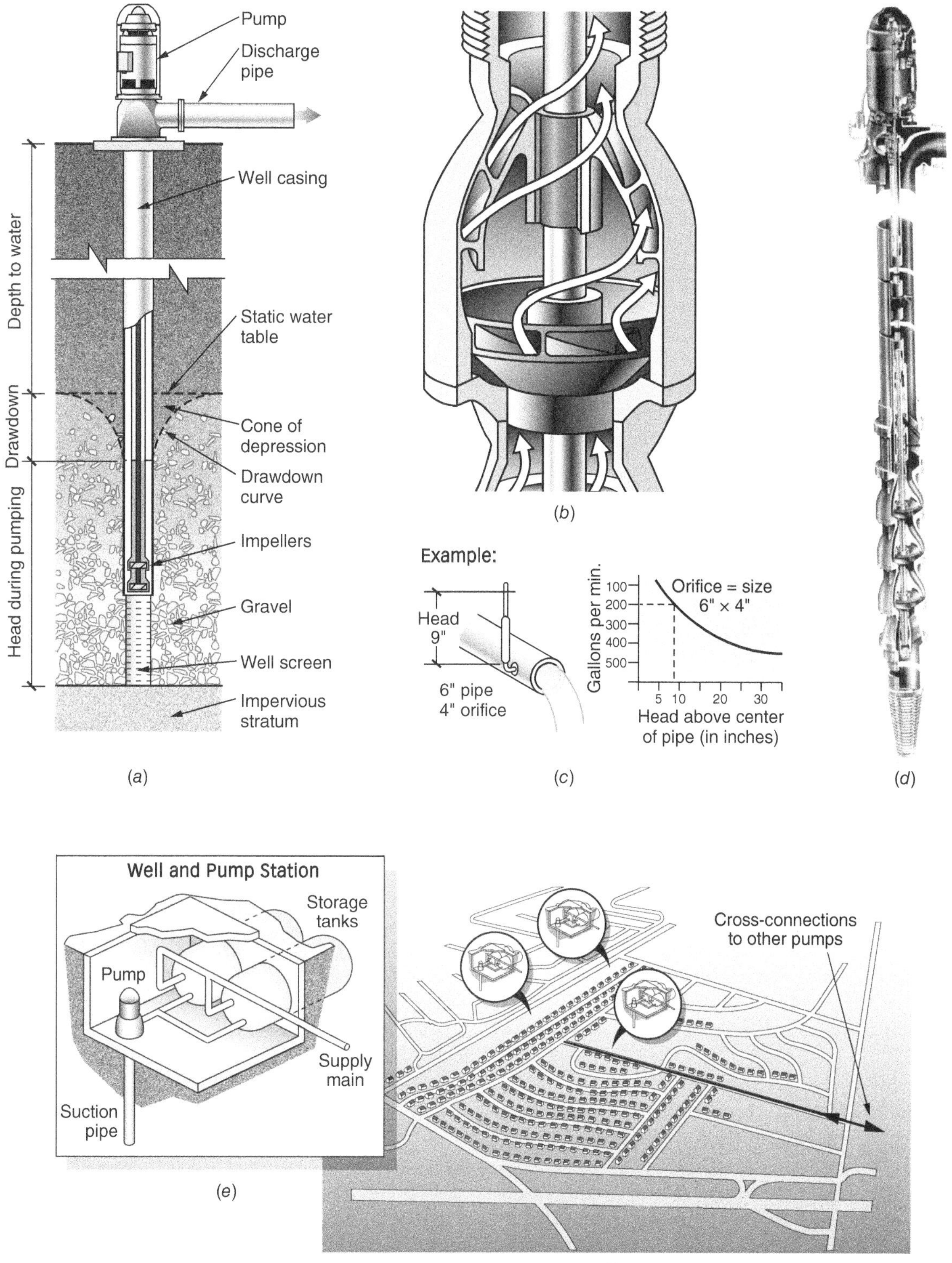

Figure 21.9 *(a) A turbine well pump. (b) Its operation. (c) Measurement of its capacity. (d) A Jacuzzi multistage lineshaft turbine well pump. (Courtesy of Jacuzzi Bros., Inc.) (e, f) Its use in supplying a small community with groundwater. Capacities of turbine pumps range from 50–16,000 gpm [3–1010 L/s]. (With permission of Progressive Architecture.)*

Centrifugal Pumps. This type of pump contains an impeller mounted on a rotating shaft. The rotating impeller increases the water's velocity while forcing the water into a casing, thus converting the water's velocity into higher pressure. Each impeller and its casing are called a stage; many stages can be combined in a multistage pump. The number of stages depends upon the pressure needed to operate the water supply system, as well as the height to which the water must be raised. The most common centrifugal pumps are those used in deep wells.

The turbine pump has a vertical turbine located below groundwater level and a driving motor located higher up, usually over the well casing at grade. A long shaft is thus required between the motor and the turbine. Substantial head clearances for this shaft's removal may be required. Figure 21.9 shows a turbine pump installation for a small community in Long Island, New York. The water is taken from the ground by multistage turbine pumps at depths of several hundred feet (100 ft = 30.5 m). It is delivered to submerged hydropneumatic tanks at a pressure of about 80 psi [about 550 kPa]. As water is required in the houses, the air under pressure in the upper part of the tanks forces water through the mains.

Submersible pumps are designed so that the motor can be submerged along with the turbine (Figure 21.10). The lengthy pump shaft is thus eliminated.

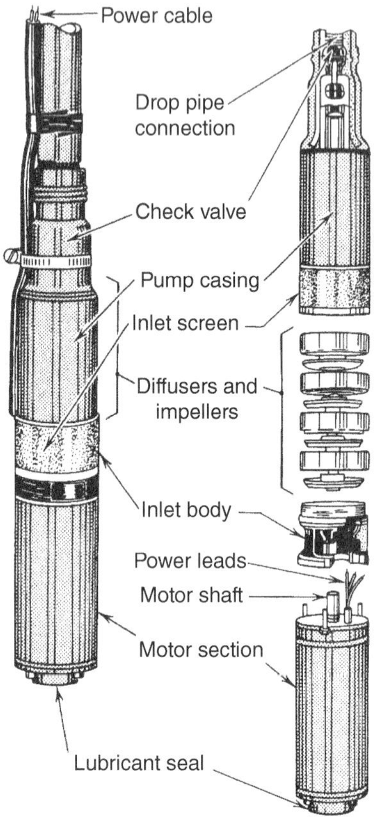

Figure 21.10 *Submersible pump (centrifugal type) in exploded view. (Adapted from U.S. Environmental Protection Agency, 1975.)*

Jet (or Ejector) Pumps. In a jet pump, a venturi tube is added to the centrifugal pump. A portion of the water that is discharged from a centrifugal pump at the wellhead is forced down to a nozzle and the venturi tube (Figure 21.11). The lower pressure within the venturi tube induces well water to flow in, and the velocity of the water from the nozzle pushes it up toward the centrifugal pump, which can then lift it more easily by suction.

Pump Selection. Pump variations and characteristics are summarized in Table 21.4. The type of pump selected depends upon many factors, including the rate of yield of a well, the daily flow (and maximum instantaneous flow rate) needed by the users, the size of the storage or pressure tank used, and the total operating pressure against which the pump works (including the height to which water must be raised within the well). First cost, maintenance, and reliability are also factors, as is the energy used by the pump. In cold climates, a pump and water supply system must be protected from freezing.

Of these factors, the two critical selection determinants are the flow rate (volume per minute or per hour to be delivered) and the total pressure (or head). The flow rate depends upon the number of fixtures to be served (Figure 21.12). The total pressure (Figure 21.13) includes the suction lift, static head, and friction loss, plus the pressure head. This relationship will be explained in detail in Section 21.11.

(c) Pressure Tanks

Serving also as water storage, these tanks are frequently used both to maintain constant pressure on a pump-supplied water system and to allow for temporary peaks in water supply rates that exceed the capacity of the pump. (Elevated tanks offer one alternative to pressure tanks, cisterns another—although the latter usually are not located high enough to provide pressure to the supply system.)

(a)

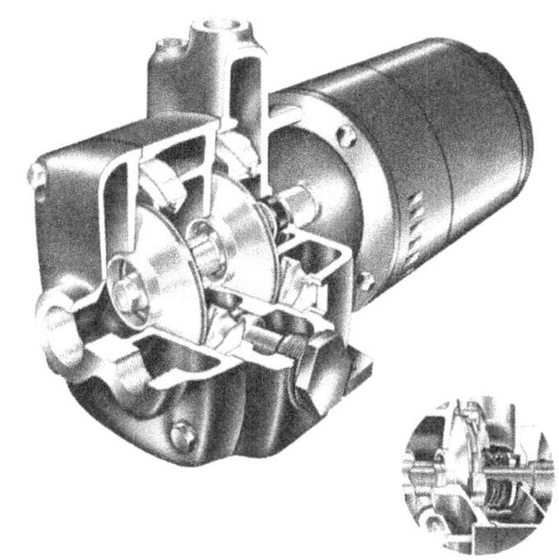

(c)

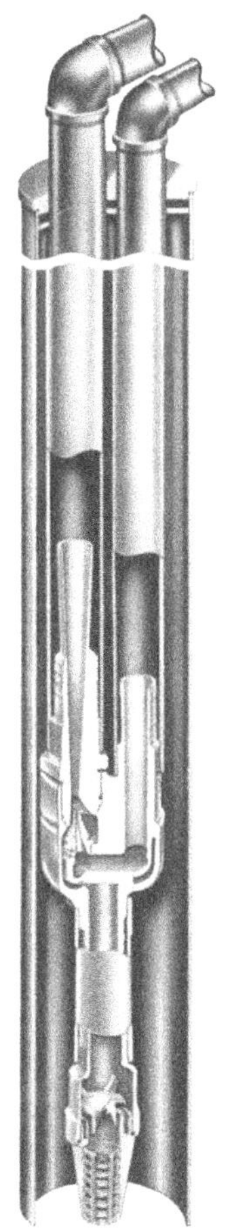

(b)

CHOOSE THE CORRECT JH FROM THESE CHARTS
DEEP WELL (Down to 120 feet)

The JH in the 1/2, 3/4, 1 and 1 1/2 horsepower rating. matched with the appropriate injector and pipe sizes. will pump this amount of water at the indicated lift or depth to the water in the well:

If suction lift or depth to water is	JH 1/2 HORSEPOWER	JH 3/4 HORSEPOWER	JH 1 HORSEPOWER	JH 1 1/2 HORSEPOWER
	Produces these gallons per hour between 20-50 lbs. discharge pressure		Produces rhese gallons per hour between 30-60 lbs. discharge pressure	
30 feet	795 G.P.H.	990 G.P.H.	1140 G.P.H.	1620 G.P.H.
40	680	875	1000	1470
50	575	735	875	1300
60	445	630	875	1200
70	360	495	620	920
80	310	385	530	820
90	255	315	435	700
100	220	275	340	590
110	195	240	295	540
120		205	250	480

(d)

Figure 21.11 *Details of a deep-well jet pump. (a) Photograph of a multistage jet pump housing and equipment. At the top is the on/off electrical switch activated by pressure settings. It controls the direct-connected electric motor to the left. Impellers are enclosed in the pump housing at the right. Circulating connections to and from the jet can be seen to the right; the pump discharge at the top. The pump can be set to operate up to 100 psi [690 kPa]. (b) Well casing and circulating lines. Jet element can be seen at the bottom of the left-hand (larger) pipe. (c) Cutaway section of the pump. (d) Pumping capacity in gph under various conditions and discharge pressure ranges of 20–50 psi and 30–60 psi. (e) Jet-type (also known as venturi or ejector) deep-well pump and storage tank for a house or small building (for well lifts greater than 25 ft [7.6 m]). Reduced pressure at (f), the jet nozzle, induces the flow of groundwater into the circulated flow. (Courtesy of Jacuzzi Bros.)*

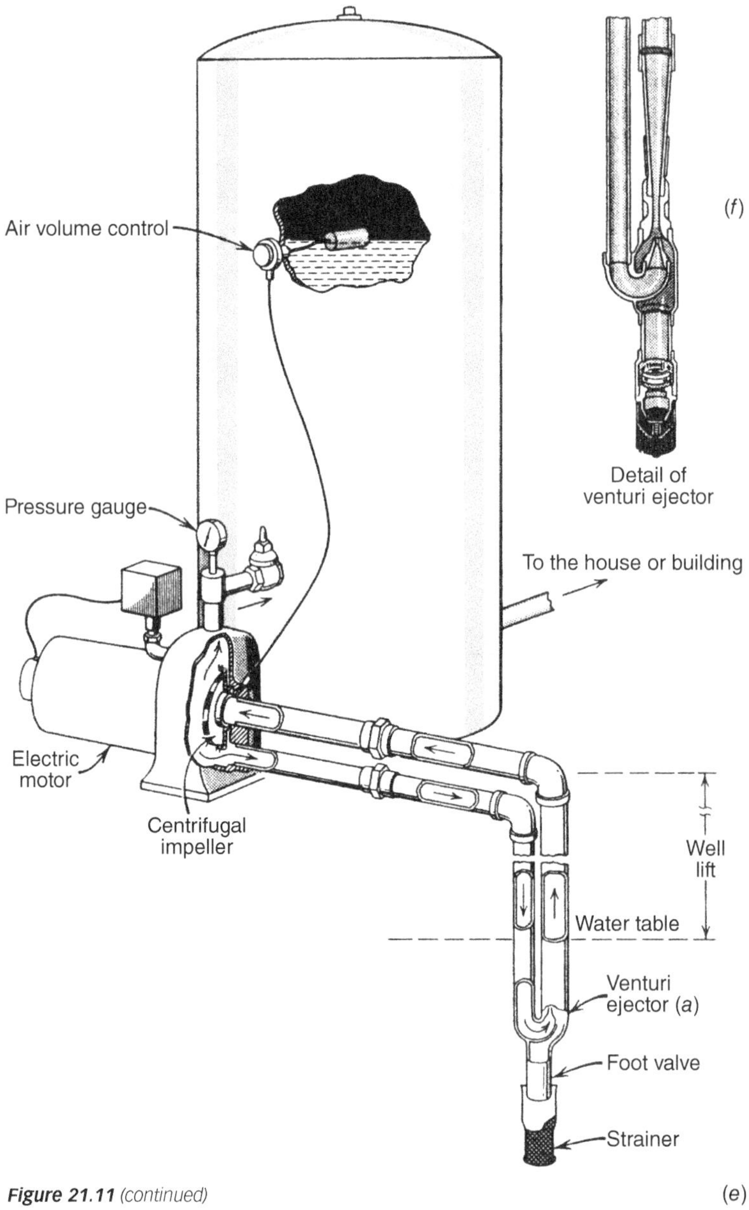

Figure 21.11 *(continued)*

Pressure tanks are often housed in outbuildings, along with the pump and any water treatment equipment (Figure 21.13). The temperature of the outbuilding must be kept above freezing, and its roof or walls should be removable to allow for replacement of parts over time. One type of pressure (storage) tank is shown in Figure 21.14.

The capacity of pressure tanks is usually small in comparison to the daily total water consumption; they provide short-term responses to peak flow demands. As a general rule, the pressure tank should be sized to deliver about 10 times the pump's capacity in gpm [L/m]. For a typical residence, allow 10–15 gal [38–57 L] tank capacity per person served.

For larger installations, the size of a pressure-storage tank can be calculated by:

$$Q = \frac{Qm}{1 - \frac{P_1}{P_2}}$$

where

Q = tank volume (gal [L])
Qm = 15 minutes of storage at peak usage rate (gal [L])
P_1 = minimum allowable operating pressure (psi [kPa]) plus atmospheric pressure (14.7 psi [101 kPa])
P_2 = maximum allowable operating pressure (psi [kPa]) plus atmospheric pressure (14.7 psi [101 kPa])

TABLE 21.4 Pumps for Water Supply

Type of Pump	Practical Suction Lift[a]	Usual Well-Pumping Depth	Usual Pressure Heads	Advantages	Disadvantages	Remarks
Positive Displacement						
1. Reciprocating						
a. Shallow well	22–25 ft [6.7–7.6 m]	22–25 ft [6.7–7.6 m]	100–200 ft [30–61 m]	1. Positive action 2. Discharge against variable heads 3. Pumps water containing sand and silt 4. Especially adapted to low capacity and high lifts	1. Pulsating discharge 2. Subject to vibration and noise 3. Maintenance cost may be high 4. May cause destructive pressure if operated against closed valve	1. Best suited for capacities of 5–25 gpm [0.3–1.6 L/s] against moderate to high heads 2. Adaptable to hand operation 3. Can be installed in very-small-diameter wells (2-in. [50-mm] casing) 4. Pump must be set directly over well (deep well only)
b. Deep well	22–25 ft [6.7–7.6 m]	Up to 600 ft [183 m]	Up to 600 [183 m] above cylinder			
2. Rotary						
a. Shallow well (gear type)	22 ft [6.7 m]	22 ft [6.7 m]	50–250 ft [15–76 m]	1. Positive action 2. Discharge constant under variable heads 3. Efficient operation	1. Subject to rapid wear if water contains sand or silt 2. Wear of gears reduces efficiency	
b. Deep well (helical rotary type)	Usually submerged	50–500 ft [15–152 m]	100–500 ft [30–152 m]	1. Same as shallow well rotary 2. Only one moving pump device in well	1. Same as shallow well rotary except no gear wear	1. A cutless rubber stator increases life of pump. Flexible drive coupling has been weak point in pump. Best adapted for low capacity and high heads.
Centrifugal						
1. Shallow well						
a. Straight centrifugal (single stage)	20 ft max. [6 m]	10–20 ft [3–6 m]	100–150 ft [30–46 m]	1. Smooth, even flow 2. Pumps water containing sand and silt 3. Pressure on system is even and free from shock 4. Low-starting torque 5. Usually reliable and good service life	1. Loses prime easily 2. Efficiency depends on operating under design heads and speed	1. Very efficient pump for capacities above 60 gpm [3.8 L/s] and heads up to about 150 ft [46 m]
b. Regenerative vane turbine type (single impeller)	28 ft max. [8.5 m]	28 ft [8.5 m]	100–200 ft [30–61 m]	1. Same as straight centrifugal except not suitable for pumping water containing sand or silt 2. They are self-priming	1. Same as straight centrifugal except maintains priming easily	1. Reduction in pressure with increased capacity not as severe as straight centrifugal

TABLE 21.4 (*Continued*)

Type of Pump	Practical Suction Lift[a]	Usual Well-Pumping Depth	Usual Pressure Heads	Advantages	Disadvantages	Remarks
			Centrifugal (Continued)			
2. Deep well						
a. Vertical line shaft turbine (multistage)	Impellers submerged	50–300 ft [15–91 m]	100–800 ft [30–245 m]	1. Same as shallow well turbine 2. All electrical components are accessible above ground	1. Efficiency depends on operating under design head and speed 2. Requires straight well large enough for turbine bowls and housing 3. Lubrication and alignment of shaft critical 4. Abrasion from sand	
b. Submersible turbine (multistage)	Pump and motor submerged	50–400 ft [15–122 m]	50–400 ft [15–122 m]	1. Same as shallow well turbine 2. Easy to frost-proof installation 3. Short pump shaft to motor 4. Quiet operation 5. Well straightness not critical	1. Repair to motor or pump requires pulling from well 2. Scaling of electrical equipment from water vapor critical 3. Abrasion from sand	1. 3500 RPM models, although popular because of smaller diameters or greater capacities, are more vulnerable to wear and failure from sand and other causes
			Jet			
1. Shallow well	15–20 ft [4.6–6.0 m] below ejector	Up to 15–20 ft [4.6–6.0 m] below ejector	80–150 ft [24–46 m]	1. High capacity at low heads 2. Simple in operation 3. Does not have to be installed over the well 4. No moving parts in the well	1. Capacity reduces as lift increases 2. Air in suction or return line will stop pumping	
2. Deep well	15–20 ft [4.6–6.0 m] below ejector	25–120 ft [7.6–37 m] 200 ft max. [61 m max]	80–150 ft [24–46 m]	1. Same as shallow well jet 2. Well straightness not critical	1. Same as shallow well jet 2. Lower efficiency, especially at greater lifts	1. The amount of water returned to ejector increases with increased lift—50% of total water pumped at 50-ft [15-m] lift and 75% at 100-ft [30-m] lift

Source: Adapted from U.S. Environmental Protection Agency, 1975.

SI units added by the authors of this book.

[a]Practical suction lift at sea level. Reduce lift 1 ft [0.3 m] for each 1000 ft [305 m] above sea level.

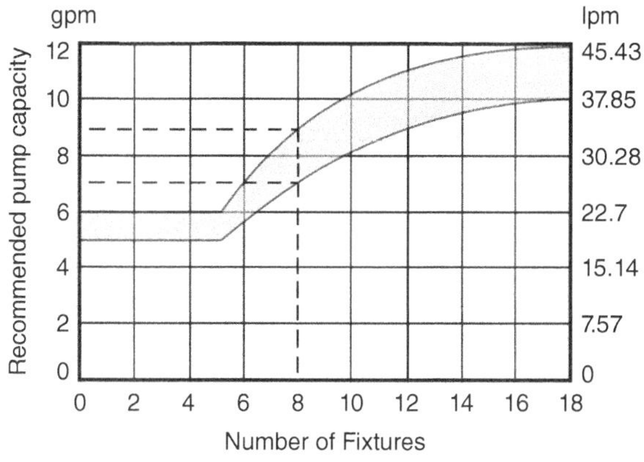

Figure 21.12 *The relationship between the recommended flow-rate capacity of a pump and the number of domestic plumbing fixtures it supplies. (Adapted from EPA, 1975. Redrawn with SI units by Nathan Majeski.)*

The ranges of allowable pressures are discussed in Section 21.9. An alternative tank-sizing procedure is shown in Table 21.5.

EXAMPLE 21.1 An office building in a remote location has a water supply system served by a pump and well. The peak demand is 50 gpm [3.15 L/s]. The fixtures will operate at a minimum of 50 psi [345 kPa]; a maximum of 70 psi [483 kPa] should not be exceeded. Therefore,

$$Qm = 15 \text{ min} \times 50 \text{ gpm} = 750 \text{ gal}$$

$$[Qm = 15 \text{ min} \times 60 \text{ s/min} \times 3.15 \text{ L/s} = 2835 \text{ L}]$$

and

$$Q = \frac{Qm}{1 - \frac{P_1}{P_2}}$$

$$Q = \frac{750 \text{ gal}}{1 - \frac{50 \text{ psi} + 14.7 \text{ psi}}{70 \text{ psi} + 14.7 \text{ psi}}} = 3125 \text{ gal}$$

$$\left[Q = \frac{2835 \text{ L}}{1 - \frac{345 \text{ kPA} + 101 \text{ kPA}}{483 \text{ kPA} + 101 \text{ kPA}}} = \text{apx.}12{,}000 \text{ L} \right]$$ ■

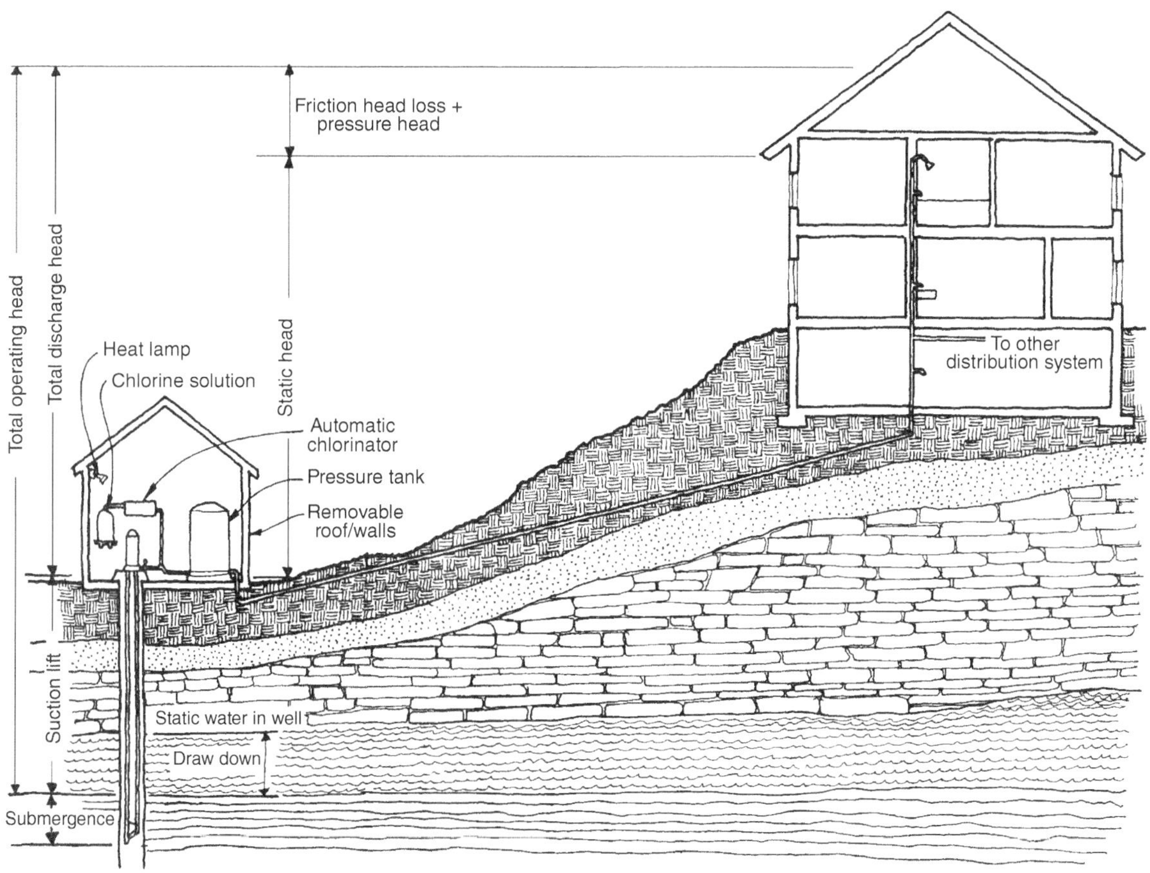

Figure 21.13 *The components of the total operating pressure (or head), a critical determinant of pump size. The pumphouse is usually a separate structure with water treatment and storage components. (Adapted from U.S. Environmental Protection Agency, 1975.)*

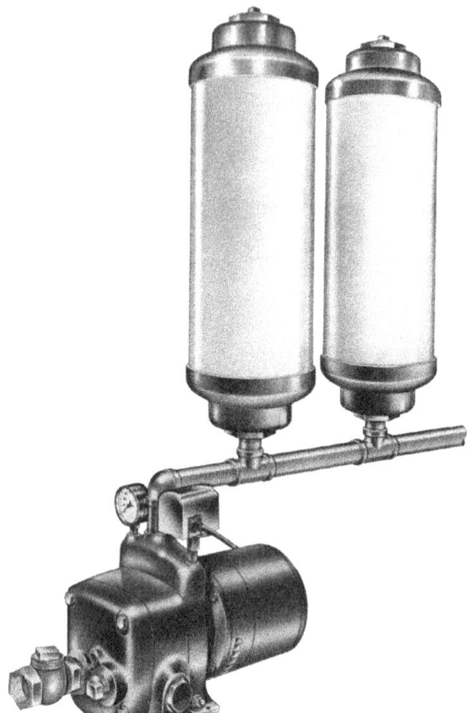

Figure 21.14 *Small pressure tanks are installed primarily to keep a water supply system at constant pressure. These Hydrocel models are small: 8½ in. [216 mm] in diameter, 27 in. [686 mm] long; they can be installed almost anywhere along the supply system. (Courtesy of Jacuzzi Bros.)*

The capacity of elevated tanks is usually equal to at least two days of average water usage. For firefighting or other special requirements, the capacity may have to be even greater.

A summary of the quality, treatment, and supply issues raised thus far is provided in the example of a rural estate shown in Figure 21.15 and Table 21.5. On this large estate, water was required for an estimated demand of 30,000 gpd [113,560 Lpd]. This was for domestic use only; irrigation was served by a separate installation pumping from a lake. Wells were dug for the domestic supply. It was quickly evident that, despite the great depths of the wells drilled, the available flow would be small. Four wells yielded a total rate of only 25 gpm [1.6 L/s].

Calculations for the amount of water to be stored to supplement this meager supply are shown in Table 21.5. The table shows that during the 14 daytime hours under conditions of peak demand, the well pumps would run continuously. Concurrently, an additional 9000 gal [34,065 L] would be drawn from the tanks. At night, the well pumps would run for six hours to restore the 9000 gal [34,065 L] drawn from the tanks during the day.

The operation of the system is illustrated in the diagram and notes of Figure 21.15*a*. The pressure of 75 psi [517 kPa] in the hydropneumatic tank is sufficient to raise the water to the greatest height in the distribution system, overcome friction in the piping, and leave a residual pressure available at each fixture of about 10–15 psi [69–103 kPa]. Excessive pressure can always be moderated by a valve in the branch supply of the fixture.

Pressure in this tank is ensured by the air compressor, activated by a pressure switch. A float switch starts the centrifugal pump to deliver water from the storage tanks as needed. The storage tanks, piped together and acting as a single reservoir, are fed by the four wells. The wells operate in unison, singly, or in groups, depending upon the level in the storage tanks. If the level drops rapidly, all well pumps can run. Although they are arranged to operate all day when needed, during periods of minimum demand only one or two may be called upon. Each has its own power supply, but controls emanate from panel *N*.

TABLE 21.5 Calculations to Establish Capacity of Water Storage Tanks at Kinloch (Figure 21.15)

1. Potable domestic water demand per day	30,000 gpd (113,550 L/d)
2. Assumed hours of use	7:00 a.m. to 9:00 p.m. (14 h)
3. Assumed hours of virtual nonuse	9:00 p.m. to 7:00 a.m. (10 h)
4. Well-yield rate per hour	25 gpm × 60 min = 1500 gph [1.6 L/s × 60 × 60 = 5679 L/h]
5. Total well yield in 14 h	1500 gph × 14 h = 21,000 gal [5679 L/h × 14 h = 79,506 L]
6. Total water needed in 14 h	30,000 gal [113,550 L]
7. Well yield in 14 h	21,000 gal [79,506 L] [see step 5]
8. Net-tank capacity required, minimum	30,000–21,000 = 9000 gal [113,550–79,506 = 34,044 L]
9. Net-tank capacity as designed with two tanks at 10,000 gal [37,850 L] each and one tank at 2000 gal [7570 L]	20,000 × 0.80 (80% full) = 16,000 gal [75,700 × 0.8 (80% full) = 60,560 L] 2000 gal × 0.70 (70% full) = 1400 gal [7570 L × 0.7 (70% full) = 5300 L] Total capacity = 17,400 gal [65,860 L]; which is OK, greater than 9000 gal [34,044 L] as per step 8
10. Well operation at night to restore 9000 gal [34,065 L] to tanks	9000 ÷ 1500 = 6 h of the 10 night hours available (34,065/5679 = 6 h)

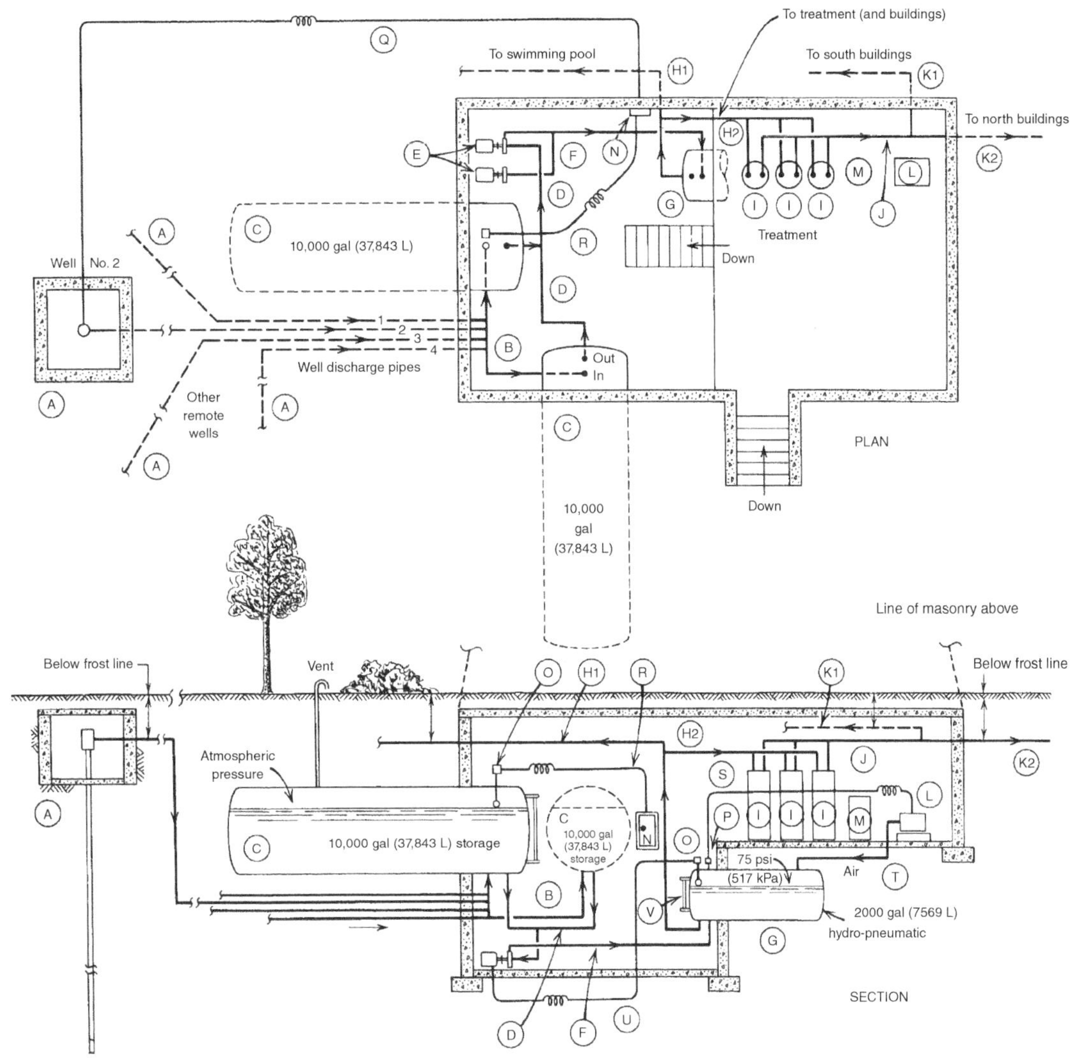

Legend

- (A) Wells
- (B) Well water header for four wells
- (C) Well water storage, vented tanks, (atmospheric pressure)
- (D) Pump supply pipe
- (E) Electric centrifugal pumps (one is a standby)
- (F) Pump-discharge to hydro-pneumatic tank
- (G) Hydro-pneumatic tank, 75 lbs per in.^{2} (517 kPa)
- (H1) (H2) Discharge pipes, pressurized well water
- (I) Water softeners (ion exchangers)
- (J) Treated well water under pressure
- (K1) (K2) Branch mains to buildings (treated well water)
- (L) Air compressor maintains 75 psi in tank
- (M) Sodium chloride for regeneration
- (N) Electrical control panel
- (O) Float switches
- (P) Pressure switch (actuates air compressor)
- (Q) Control wiring to well pumps
- (R) Control wiring, float switch to panel
- (S) Control wiring, pressure switch to compressor
- (T) Compressed air piping
- (U) Control wiring, float switch to centrifugal pump
- (V) Gauge glass indicates water level

This schematic diagram shows system components and connections. For clarity, details (valves, drains, checkvalves, etc.) are not shown. All control wiring, here simply indicated, operates through control panel (N). See text for operation.

(*a*)

Figure 21.15 *(a) Schematic diagram of the water control and distribution center for the Kinloch Estate. The plant is located on a hillside below the bathing and dressing pavilion. (b) One of two pavilions adjacent to the swimming pool that serves residents and guests at Kinloch. Below this pavilion is the control center. Exterior tubing to and from the control center is all below grade. (Courtesy of Bentel & Bentel, Architects.)*

(*b*)

Figure 21.15 *(continued)*

Swimming pools should be supplied with pure, potable water. Biological tests showed that this well water was safe. Thus, the fill line to the swimming pool is connected from this system rather than from the lake water supply used for irrigation. Because swimming pool water is separately recirculated and provided with its own purification treatment at a location adjacent to the pool, the supply line for makeup water to the pool would have infrequent and off-hour use. Water for domestic use in the buildings is passed through one of three ion exchangers to provide softening and to make the water more suitable for washing, bathing, and cooking. Periodically, the calcium precipitate can be flushed out and the tanks regenerated by sodium chloride.

21.6 HOT WATER SYSTEMS AND EQUIPMENT

There are many ways to provide the "domestic" or "service" hot water needs within a building—the hot water used not for space heating, but for bathing, clothes washing, dishwashing, and other related functions. Whereas much of the world either heats such water on a cookstove or does without it, North Americans enjoy an array of choices for domestic hot water (DHW) supply systems. In this section, rough supply estimates are considered first, then basic design choices, conventional heater tanks, solar water heaters, and finally, energy recovery heaters.

With today's better-insulated buildings, the demand for space heating can fall to a level about equal to DHW. As a result, a trend is developing toward using one rather larger water heater designed to meet the need both for space heating and for DHW. This saves both floor space and equipment cost, compared to the more conventional boiler for space heating and a separate water heater for DHW. Some restrictions apply; the *2024 International Plumbing Code* restricts the maximum potable hot water temperature to 140°F [60°C] and requires water in such a system to remain potable.

(a) Water Temperature

Higher water supply temperatures have several advantages but carry the risk of scalding. Although water becomes uncomfortably hot to the touch above 110°F [43°C], much higher temperatures are often used for some commercial processes, as shown in Table 21.6. When determining the temperature at which DHW is to be supplied, consider the following factors.

High temperatures do the following:

- Allow the installation of smaller storage tanks (less hot water is mixed with cold water to achieve a final usage temperature at the shower, sink, or lavatory) but require larger heating units.
- Can be achieved at the point of use (rather than in the tank) by heaters built into equipment such as dishwashers.
- Can cause scale to form on heating coils and within piping (above 140°F [60°C] in areas of hard water quality).
- May be required by code for some applications but limited by code for others.
- Limit the potential for growth of *Legionella pneumophila* bacteria (above 140°F [60°C]).

Lower temperatures do the following:

- Are less likely to cause burns but may not achieve desired sanitation.

TABLE 21.6 Representative Hot Water Temperatures

	Temperature	
Use	**°F**	**°C**
Lavatory		
Hand washing	105	40
Shaving	115	45
Showers and tubs	110	43
Therapeutic baths	95	35
Commercial and institutional laundry (based on fabric)	Up to 180	Up to 82
Residential dishwashing and laundry	140	60
Surgical scrubbing	110	43
Commercial Spray-type Dishwashing[a]		
Wash	150 minimum	65 minimum
Final rinse	180–195	82–90

Source: © ASHRAE, www.ashrae.org. 2023 ASHRAE Handbook—HVAC Applications, Chapter 51, Table 22; SI units added.

[a]For other types of commercial dishwashers, see ASHRAE (2023a).

- Mean less energy consumed, because storage and pipe heat losses are lower.
- Allow the installation of smaller heating units but require larger storage tanks.
- Make possible the use of lower-grade heat sources for DHW, such as solar energy or waste heat recovery. It has been approximated that around 13% of annual U.S. energy usage is for DHW, which requires the lowest-grade heat source of all the common usages.

Legionella pneumophila is a common aquatic bacteria found in both natural and building water systems as well as in some soils. This genus of bacteria can cause a range of illness when it is inhaled, aspirated, or directly introduced into the lungs of susceptible individuals, including a mild flu-like illness (Pontiac fever) and a type of pneumonia called Legionnaires' disease. ASHRAE offers a number of information resources to help design teams and facility managers mitigate the growth of Legionella and greatly reduce risk to human health.

ASHRAE offers three critical information resources that outline minimum risk-management requirements for the design, construction, commissioning, operation, maintenance, repair, replacement, and expansion of new and existing buildings and their associated potable and non-potable water systems and components—two of which are specific to *Legionellosis* (a blanket term used to describe Legionnaires' disease, Pontiac fever, and any other illness caused by exposure to Legionella bacteria):

- *ASHRAE Standard 514: Risk Management for Building Water Systems: Physical, Chemical, and Microbial Hazards*
- *ASHRAE Standard 188: Legionellosis: Risk Management for Building Water Systems*
- *ASHRAE Guideline 12: Managing the Risk of Legionellosis Associated with Building Water Systems*

(b) Heat Sources and Methods

These include familiar concentrated-energy, high-grade sources such as natural gas and electricity. Oil- and coal-fired boilers are frequently equipped with DHW coils as well. Buildings served by steam may use steam as a DHW heat source. Cogeneration provides heat for DHW, as in Chapter 16. Because of the relatively low-grade final temperatures needed, DHW can also be readily provided by wood-burning equipment, incinerators, solar energy equipment, heat pumps, and heat recovery devices (as in commercial ice-making machines whose discharged heat is contributed to DHW).

Heating Methods. There are two basic methods:

- *Direct heating:* This brings water into contact with directly heated surfaces: electric-resistance elements or other electrically warmed surfaces within tanks, or surfaces directly exposed to fire or hot gases.
- *Indirect heating:* This can be accomplished in several ways. Coils containing steam or fluids can be submerged within water tanks or (Figure 21.16) set within boilers, whose primary function usually is to provide space or industrial process heating. Alternatively, coils containing DHW can be placed outside a boiler but within a casing containing steam, hot exhaust gases, or very hot water.

Direct and indirect methods can be utilized in a variety of equipment:

- *Storage tank water heaters:* This is the type most commonly used for residential and small commercial purposes (see Section 21.6(h)).
- *Circulating storage water heaters:* In these the water is first heated by a coil, then circulated through a storage tank (as in some solar heaters—see Section 21.6(i)).
- *Tankless (instantaneous) heaters:* In these the water is very quickly raised to the desired temperature within a heating coil and immediately sent to the point of usage.

(c) Tankless Water Heaters

Also called instantaneous water heaters, these are available in larger sizes for central hot water systems (Figure 21.16) or in smaller sizes for wall mounting adjacent to remote plumbing fixtures that occasionally need hot water, or for isolated bathrooms, laundry rooms, and so forth. Decentralized tankless units can be as small as "instant hot water taps" for kitchen or bar sinks—electric-resistance heaters capable of generating perhaps 3 gph at up to 200°F [11 L/h at up to 93°C]. Bathroom groups can use either electric resistance or gas-fired units (Figure 21.17); electric heaters require very high amperage and only about one-third of the primary source energy is delivered as electricity; gas heaters must be vented and are most efficient without continuously burning pilots.

These tankless heaters can be very small; one unit of only 2.65 ft^3 [0.08 m^3] produces up to 4 gpm [0.25 L/s] with a temperature rise of 45°F [20°C]. With a greater temperature rise of 100°F [55.5°C], the flow rate drops to somewhat less than half. Because tankless heaters often consist of fairly long coils through which the water passes as it is heated, they may add

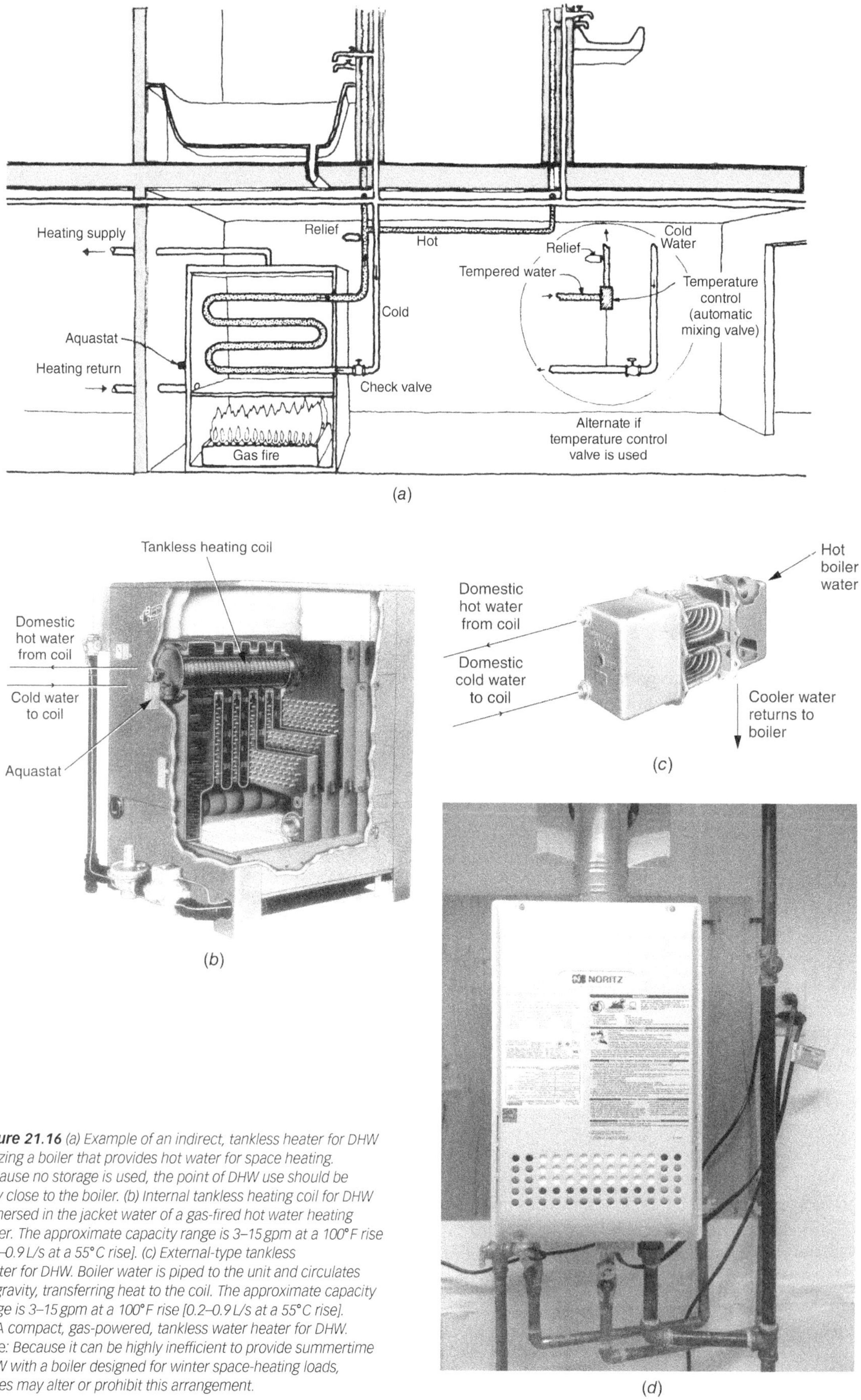

Figure 21.16 *(a) Example of an indirect, tankless heater for DHW utilizing a boiler that provides hot water for space heating. Because no storage is used, the point of DHW use should be very close to the boiler. (b) Internal tankless heating coil for DHW immersed in the jacket water of a gas-fired hot water heating boiler. The approximate capacity range is 3–15 gpm at a 100°F rise [0.2–0.9 L/s at a 55°C rise]. (c) External-type tankless heater for DHW. Boiler water is piped to the unit and circulates by gravity, transferring heat to the coil. The approximate capacity range is 3–15 gpm at a 100°F rise [0.2–0.9 L/s at a 55°C rise]. (d) A compact, gas-powered, tankless water heater for DHW.* Note: *Because it can be highly inefficient to provide summertime DHW with a boiler designed for winter space-heating loads, codes may alter or prohibit this arrangement.*

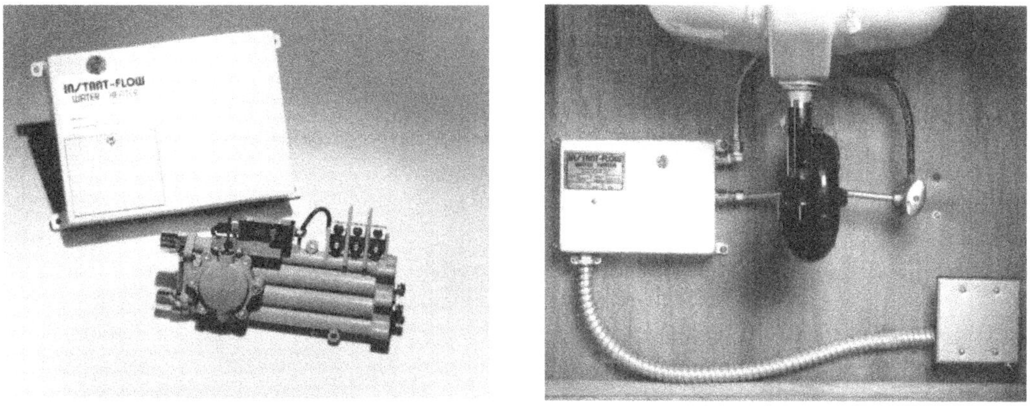

(a) (b)

Figure 21.17 Instantaneous or tankless water heater. (a) In this instant-flow water heater, a series of coils heats water as it flows through. (b) Installation below a lavatory. A wide range of sizes is available; the 4.6-kW unit raises water temperature by 31°F [17.2°C] at 1 gpm [0.06 L/s]; the 9-kW unit raises temperature by 61°F [33.9°C] at 1 gpm [0.06 L/s]. (Courtesy of Chronomite Laboratories, Inc.)

TABLE 21.7 Domestic Hot Water Consumption—Residences

	Hot Water Required per Use	
Clothes Washing Machine	**14-lb [6.4-kg] Machine**	**18-lb [8.2-kg] Machine**
Hot wash/hot rinse	38 gal [144 L]	48 gal [182 L]
Hot wash/warm rinse	28 gal [106 L]	36 gal [136 L]
Hot wash/cold rinse	19 gal [72 L]	24 gal [91 L]
Warm wash/cold rinse	10 gal [38 L]	12 gal [45 L]
Dishwashing	**Small**	**Large**
Dishwashing machine	10 gal [38 L]	15 gal [57 L]
Sink washing	4–8 gal [15–30 L]	
Personal Hygiene		
Tub bathing	12–30 gal [45–134 L]	
Wet shaving/hair washing	2–4 gal [8–15 L]	
Showering	2–6 gpm [13–38 L/s]	

Source: From Plante, 1983.

considerable friction to the total supply system design requirements (see Section 21.11).

To gain a better understanding of hot water used per activity, and the role of instantaneous heaters, see Table 21.7. Domestic water system design usually concentrates on the storage tank size (see Section 21.6(h)).

(d) Energy Factors

The U.S. Department of Energy (DOE) developed the energy factor (EF) as a standardized measure of annual overall efficiency. Standard storage tank water heaters that are gas-fired may reach an EF of 0.60–0.64. Because they have virtually none of the stand-by losses of storage tank heaters, gas-fired tankless water heaters offer EFs of up to 0.69 with continuous pilots and as high as 0.93 with electronic ignition (although this minor usage of electricity is not included in the EF rating).

Some gas-fired instantaneous heaters are designed to operate with battery-powered pilot light ignition. These are useful in remote installations with propane but without electricity.

(e) Central Versus Distributed Equipment

This choice can be particularly complex. In the United States, a central water-heater storage tank is standard equipment in homes and small stores. However, the growing use of solar water heating, along with awareness of the heat losses from water-heater storage tanks, has created new interest in combinations of central and distributed DHW.

Consider a large residential DHW application such as that shown in Figure 21.18. Here, two areas where hot water is used are separated by some 50 ft [15 m]. If the simple, centralized water-heater storage tank is used (Figure 21.18*a*), it would probably be placed nearest the maximum-use fixtures—dishwashers and clothes washers. (Also, floor space for the water-heater storage tank is usually more plentiful there.) Codes typically require that at least the first 8 ft [2.4 m] of the hot water pipe be insulated, as it leaves the water heater. However, each time hot water is needed in the remote bathroom serving the bedrooms, the previously heated water in at least 50 ft [15 m] of supply pipe will almost certainly have cooled—despite insulation even over its entire length—and must be drained off before hot water finally arrives. A ¾-in. diameter pipe, for example, will hold 4.6 gal [17.4 L] in 50 ft [15 m] of pipe; if the water heater is set at an energy-conserving temperature of 120°F [49°C] and the incoming city water is 50°F [10°C], the 4.6 gal [17.4 L] of wasted water will also waste the 2682 Btu [2830 kJ] invested during its heating.

Another way to conserve water is with a recirculating hot water system (Figure 21.18*b*). The primary disadvantages of this system are the increased heat loss through the hot water pipe (now kept at 120°F [49°C] for 24 hours per day rather than for just a few minutes) and the energy required to run the circulating pump (again for 24 hours daily). Codes typically require a conveniently located cutoff switch for the circulating pump and pipe insulation with a minimum k = 0.3 Btu in./h ft^2 °F [0.04 W/m K] over the entire length of the circulating hot water system.

A decentralizing option (Figure 21.18*c*) is to install two water-heater storage tanks, one for each group. The first cost will be greater, and more floor space required. The daily waste heat from two heaters is clearly a disadvantage that will at least partially offset the energy saved by eliminating the 50-ft [15-m] run of pipe. (Furthermore, in warm weather this waste heat will add to discomfort within the house.)

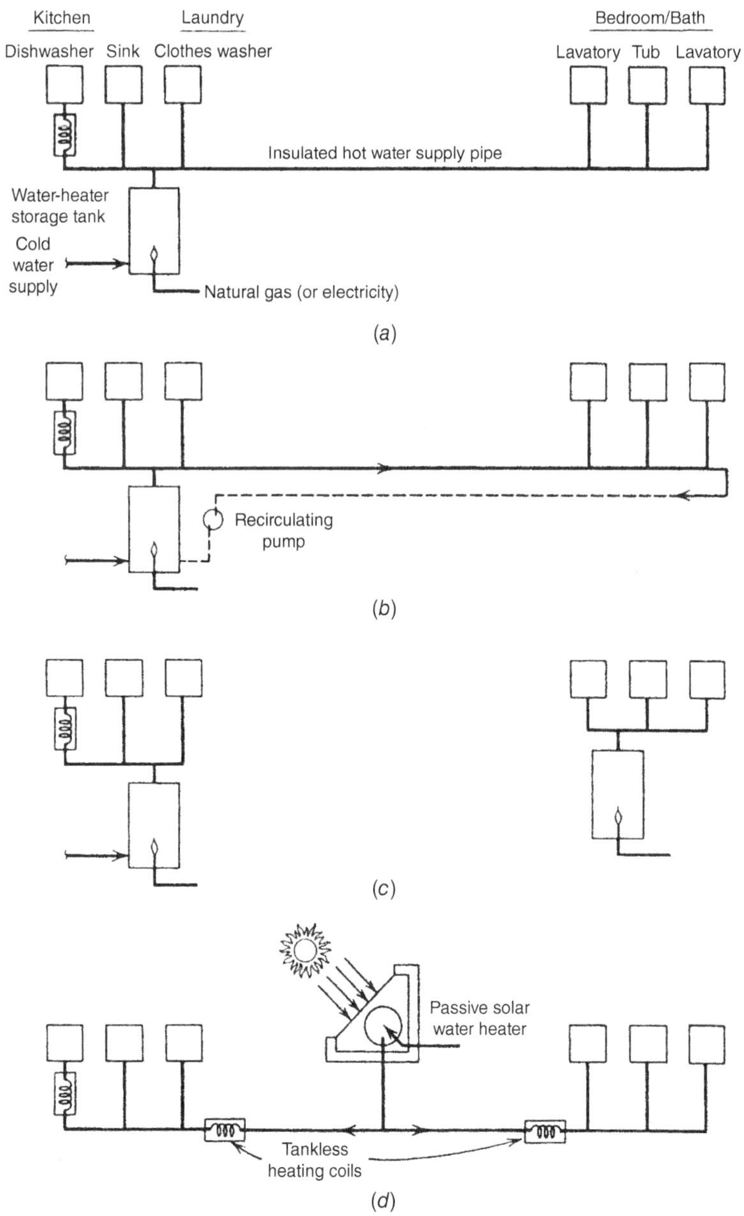

Figure 21.18 *Options for DHW in a larger residence. (a) Typical centralized water-heater storage tank; water and energy are wasted in the approximately 50-ft [15-m] run involved. (b) Recirculating pump added; this saves water but requires continuous energy to operate the pump and make up the pipe's heat losses. (c) Decentralized approach using two water-heater storage tanks; this roughly doubles the energy lost from storage but eliminates water and energy loss from the long pipe run. (d) Central solar water heater/decentralized tankless heating coil combination; it saves on water and energy but not on initial cost.*

A decentralized-centralized mix (Figure 21.18*d*) will save energy, water, and floor space, but it will almost certainly be costlier to install. A central solar water heater—either passive (shown) or active—brings the water up to warm (winter) or very hot (summer) temperatures. At each fixture group, a tankless heating coil instantaneously heats the water as needed. Almost no water is wasted, and the energy lost in the 50 ft [15 m] of pipe will be solar energy, not purchased energy from natural gas or electricity.

Another option is to omit the solar water heater and simply install two tankless heaters. The primary disadvantage of this setup is the need for larger-capacity heaters, whose instantaneous demand for energy could lead to electric cost penalties in areas where peak-load rates are high. Compared to central storage tanks, however, less heat is lost.

(f) Distribution Trees

The choice of a distribution tree for a central water heating system must take into account the gradual cooling of water within the distribution system once it has left the central heater. If the heat losses associated with constantly circulating hot water (looped trees) are preferable to the heat and water wastage associated with simple, single hot water distribution trees, such loop systems can be achieved in either of two ways.

Thermosiphon hot water circulation depends upon the fact that water expands and becomes lighter when heated, as can be seen in Figure 21.19*a*. If heat is applied to the lower loop of a glass tube, both ends of which have been inserted in an inverted bottle containing water, the water moves from *A* to *B* and rises through tube *BC* into the bottle. There it becomes cooled and drops through tube *DA* to *A*, is again heated, and rises in tube *BC*—thus completing the circulation. Because the movement depends upon the difference in weight between the two columns of water, the velocity and consequent effectiveness of the circulating system increase as both the temperature of the water and the height of the circuit increase. Hot water supply systems therefore usually consist of a heater with a storage tank, piping to carry the heated water to the farthest fixture, and a continuation of this piping to return the unused cooled water back to the heater. A constant circulation is thereby maintained, and hot water may be drawn at once from a fixture without first draining off through the faucet the cooled water that would be standing in the supply pipe if there were no recirculation. Because heat increases the corrosive action of water on metals, copper tubing or rated polyvinyl chloride (PVC) pipe is often chosen for use in hot water and hot

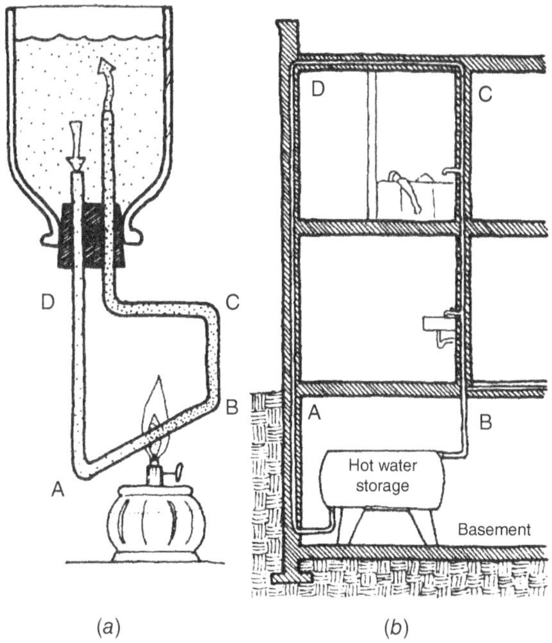

Figure 21.19 *(a) Principle of hot water circulation by gravity (thermosiphon). (b) Its application to hot water service. During periods of no demand, there is sufficient incidental cooling between C and D so that dense (less warm) water at A forces the lighter (hot) water at B to rise for speedy availability at each faucet.*

water circulating systems. Thermosiphon circulating systems are particularly effective in multistory buildings because of the beneficial effect of the increased circuit height on circulation.

Forced circulation of hot water is often utilized where a height advantage is unavailable. Low, long, rambling buildings, such as some large one-story residences, schools, and factories, lack the height needed to set up good hot water circulation by gravity. Also, flow is diminished by friction in long pipe runs. For such buildings, the forced-circulation scheme shown in Figure 21.20 offers one option. Three independent aquastats—devices that create an electrical signal when the water temperature drops—control this system. Aquastats *A*, *B*, and *C*, respectively, sense the temperatures of the water in the heater, in the tank, and at the end of the circulation-return main. As needed, they turn on the oil or gas burner, the tank-circulating pump, and the system-circulating pump. Fixtures remote from the tank are as close to hot water as the length of their hot water runout pipes. Water is usually available at full temperature in 5–10 seconds. Trial aquastat settings in °F could be *A* 180, *B* 160, *C* 120 [°C: *A* 82, *B* 71, *C* 49].

(g) Variable Storage Temperature

Energy-saving computer controls are available for hotels, motels, apartments, and larger commercial

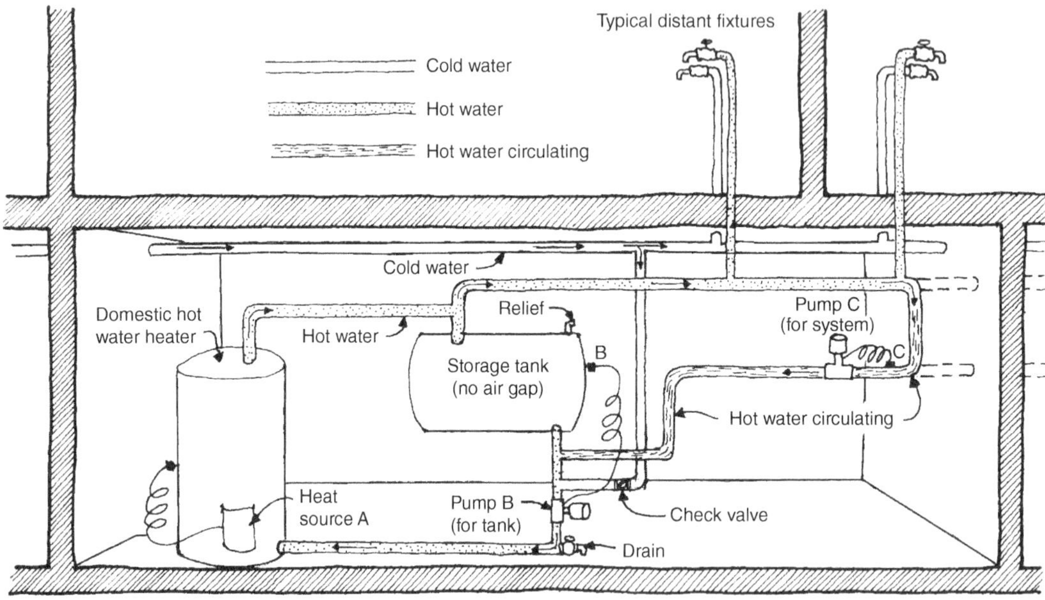

Figure 21.20 *Forced circulation of DHW for a long, low building. Note that the use of space-heating boilers for DHW may cause system inefficiencies in warm weather.*

buildings. These devices (Figure 21.21) vary the supply temperature of hot water so that the hottest water is supplied at the busiest hours. In hours of low usage, much lower supply temperatures mean that more hot water will be mixed with less cold water at showers, lavatories, and sinks. This increased hot water quantity poses no problem at off-peak hours, and the lower temperature means significant decreases in heat loss from the recirculating supply water system. A typical payback period is one year. Many of the devices store the typical daily patterns of usage in their memory banks and vary the supply temperature accordingly.

(h) Conventional Water Heater Selection

One of the most common appliances used in the United States is the water-heater storage tank, an energy-conserving model which is shown in Figure 21.22.

For residences, the capacity of hot water heating/storage equipment can be taken from the values in Table 21.8. Note that for the familiar tank-type direct water heaters (gas, electric, or oil), there is a stated relationship among storage size, rate of hourly heat input, rate of draw (demand) over one hour, and recovery rate. Example 21.2 uses Tables 21.8 and 21.9 for the sizing of a residential water heater.

EXAMPLE 21.2 Select a natural gas water heater for a five-bedroom house with three baths. The minimum requirements of HUD-FHA are shown in Table 21.8: 50-gal [190-L] storage, 47,000 Btu/h [1.38 kW] input, 90 gal [340 L] draw per hour, and 40 gph [151 L/h] recovery.

SOLUTION

Pull product data and select a model that meets the requirements. Figure 21.22 offers an example for our purposes here. Based on this manufacturer's product data, model BTH-120 may be selected as it exceeds all the minimums. ■

Estimation of the hot water demand for commercial and institutional buildings is not so simple, and design guidelines are less reliable. For larger buildings, there is a trade-off between quick recovery (high heating capacity) and storage size. Another variable is storage temperature, as discussed earlier in Section 21.6(a).

Small shops and very small office buildings can be treated like residences for hot water sizing. For larger buildings, the trade-off between heaters and storage tanks should be explored. To begin this process, consult Table 21.10, which shows the maximum hourly and daily demands and contrasts them to an average day (which should be used to estimate monthly energy consumption).

The relationship between heater size and tank size is graphed in Figure 21.23. The advantage of a larger heater is its smaller tank, which consumes less space and volume, weighs less, and probably has a lower first cost. The advantage of a larger tank is that its smaller heater tends to work steadily rather than in spurts; this lower, steadier demand for energy will lead to lower utility rates in the case of electric heating and thus to

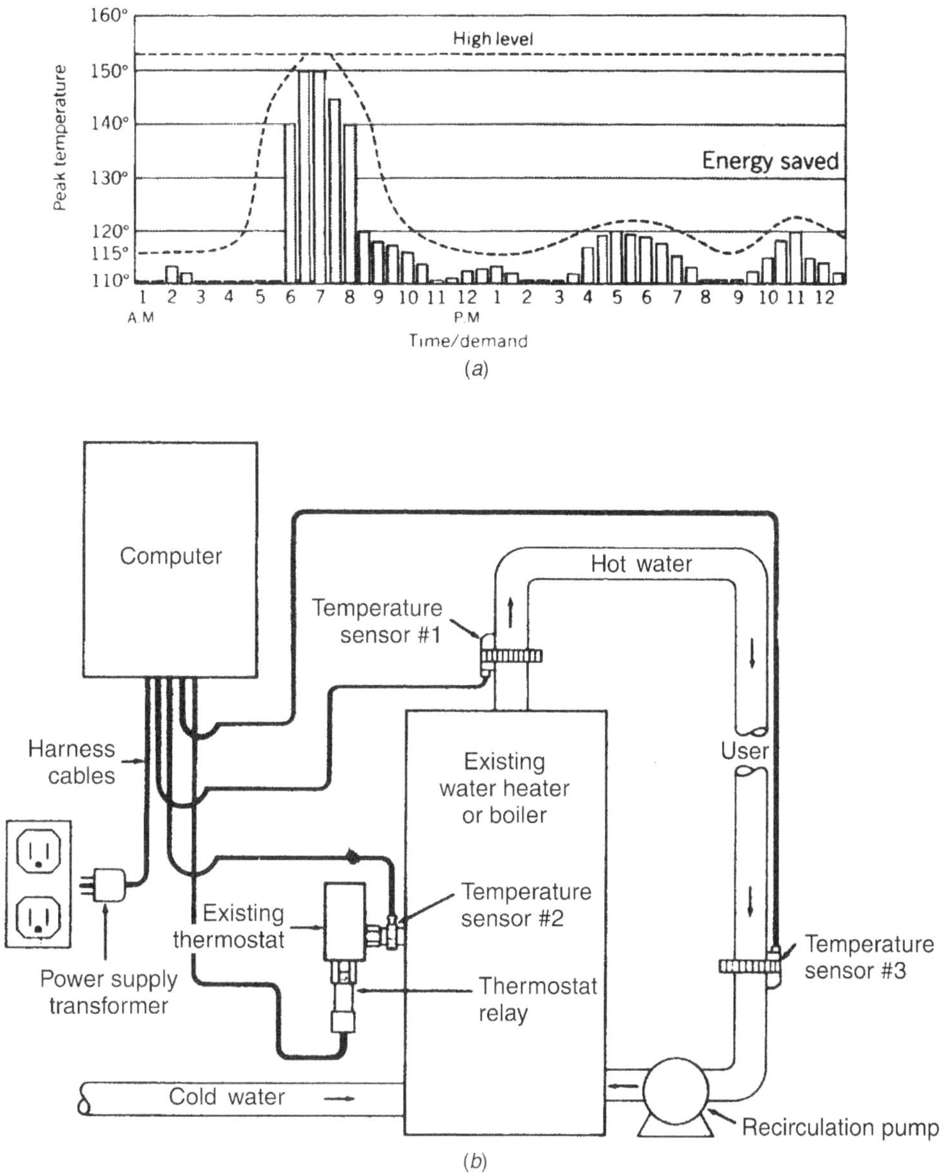

Figure 21.21 *Variable temperature DHW supply. (a) Energy savings are possible when the supply temperature of hot water is varied. Heat losses from supply pipes are greatly reduced for most of the typical day. A computer controls the supply temperature. (b) Schematic of computer-controlled, variable-temperature DHW system. (Courtesy of Fluidmaster.)*

money savings over the life of the system. When the heat source is solar energy or waste heat, a larger tank is usually best suited to the characteristics of the energy supply.

EXAMPLE 21.3 A women's dormitory housing 300 students, with a cafeteria serving 300 meals in one hour, is to be built. Find the required hot water storage size for two conditions: (1) assuming a minimum recovery rate for both dorm and cafeteria and (2) assuming a specified dorm recovery rate of 2.5 gph [9.5 L/h], which is half of the maximum hourly value given in Table 21.10, and a specified cafeteria recovery rate of 1.0 gph [3.8 L/h], which is two-thirds of the maximum hourly value given in Table 21.10.

SOLUTION

Condition 1—Minimum Recovery:
From Figure 21.23*a*, the minimum recovery rate for women's dormitories is 1.1 gph [4.2 L/h]. Therefore, for 300 students,

300 students × 1.1 gph = 330 gph

[300 students × 4.2 L/h = 1260 L/h] recovery

At this rate, again from Figure 21.23*a*, the minimum usable storage capacity is 12 gal/student [45 L/student] for women. Assume that 70% of the total capacity is

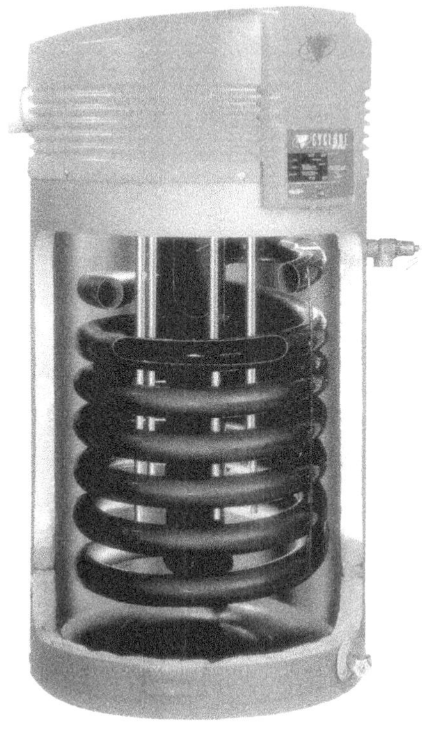

Cyclone XHE Data:

[All models are 27¾ × 30 in. (705 × 762 mm) with 3-in. (76-mm) PVC vents]

Model	Storage Capacity, gal (L)	Btuh Input/Output	Recovery (gal/h) at 94% Thermal Efficiency		Height	
			80F° Δt	100F° Δt	in.	mm
BTH-120	60 (227)	120,000/112,800	171	137	55½	(1410)
BTH-150	100 (378)	150,000/141,000	214	171	74½	(1892)

Figure 21.22 *"Cyclone XHE" water-heater storage tank using natural gas (or propane). This model achieves 94% thermal efficiency with a burner located at the top, forcing blower-driven air into a countercurrent contact with rising fuel. A precise mix of air and fuel at the point of ignition results in high combustion efficiency, and the swirling flame continues downward in a submerged central combustion chamber. The heated gas is then forced at high velocity through the spiral heat-exchanger coil, with a swirling action that maximizes heat exchange with the water. (Courtesy of the A.O. Smith Water Products Company.)*

usable capacity. This means that after 70% of the stored hot water is withdrawn, the remaining water has been cooled (by incoming water) to an unusably low temperature. Storage size must be increased by

$$\frac{100\%}{70\%} = 1.43$$

Thus:

12 gal/student × 300 students × 1.43 = 5148 gal

[45 L/student × 300 students × 1.43

= 19,305 L] storage

From Figure 21.23e, the minimum recovery rate for the cafeteria (serving full-meals, type A) is 0.45 gph [1.7 L/h]. For 300 meals,

300 × 0.45 = 135 gph

[300 × 1.7 = 510 L/h] recovery

At this rate, per the type A curve on Figure 21.23e, the minimum usable storage capacity is 7 gal/meal [26.5 L/meal]. Thus:

300 × 7 gal/meal × 1.43 = 3003 gal

[300 × 26.5 L/meal × 1.43 = 11,369 L] storage

Combining these requirements for dorm and cafeteria,

recovery = 330 + 135 = 465 gph

[1260 + 510 = 1770 L/h]

storage = 5148 + 3003 = 8151 gal

[19,305 + 11,369 = 30,674 L]

Condition 2—Faster Recovery:
At the specified dorm recovery rate of 2.5 gph [9.5 L/h],

300 × 2.5 = 750 gph

[300 × 9.5 = 2850 L/h]

TABLE 21.8 HUD-FHA Minimum Water Heater Capacities, Residential

Number of Bedrooms	1	2	3	2	3	4	5	3	4	5	6
Number of Baths	1–1.5			2–2.5				3–3.5			
Part A: I-P Units											
Gas[a]											
Storage, gal	20	30	30	30	40	40	50	40	50	50	50
Input, 1000 Btu/h	27	36	36	36	36	38	47	38	38	47	50
1-h draw, gal	43	60	60	60	70	72	90	72	82	90	92
Recovery, gph	23	30	30	30	30	32	40	32	32	40	42
Electric[a]											
Storage, gal	20	30	40	40	50	50	66	50	66	66	80
Input, kW	2.5	3.5	4.5	4.5	5.5	5.5	5.5	5.5	5.5	5.5	5.5
1-h draw, gal	30	44	58	58	72	72	88	72	88	88	102
Recovery, gph	10	14	18	18	22	22	22	22	22	22	22
Oil[a]											
Storage, gal	30	30	30	30	30	30	30	30	30	30	30
Input, kW	70	70	70	70	70	70	70	70	70	70	70
1-h draw, gal	89	89	89	89	89	89	89	89	89	89	89
Recovery, gph	59	59	59	59	59	59	59	59	59	59	59
Tank-type Indirect[b,c]											
I-W-H-rated draw, gal in 3 h, 100°F rise		40	66		66	66	66	66	66	66	66
Manufacturer-rated draw, gal in 3 h, 100°F rise		49	49		75	75	75	75	75	75	75
Tank capacity, gal		66	66		66	66	83	66	82	82	82
Tankless-type Indirect[c,d]											
I-W-H-rated draw, gpm, 100°F rise		2.75	2.75		3.25	3.25	3.75	3.75	3.75	3.75	3.75
Manufacturer-rated draw, gal in 5 min, 100°F rise		15	15		25	25	35	25	35	35	35
Part B: SI units											
Gas[a]											
Storage, L	76	114	114	114	150	150	190	150	190	190	190
Input, kW	7.9	10.5	10.5	10.5	10.5	11.1	13.8	11.1	11.1	13.8	14.6
1-h draw, L	163	227	227	227	265	273	341	273	311	341	350
Recovery, mL/s	24	32	32	32	32	36	42	34	34	42	44
Electric[a]											
Storage, L	76	114	150	150	190	190	250	190	250	250	300
Input, kW	2.5	3.5	4.5	4.5	5.5	5.5	5.5	5.5	5.5	5.5	5.5
1-h draw, L	114	167	220	220	273	273	334	273	334	334	387
Recovery, mL/s	10	15	19	19	23	23	23	23	23	23	23
Oil[a]											
Storage, L	114	114	114	114	114	114	114	114	114	114	114
Input, kW	20.5	20.5	20.5	20.5	20.5	20.5	20.5	20.5	20.5	20.5	20.5
1-h draw, L	337	337	337	337	337	337	337	337	337	337	337
Recovery, mL/s	62	62	62	62	62	62	62	62	62	62	62
Tank-type Indirect[f,g]											
I-W-H-rated draw, L in 3 h, 55 K rise		150	150		250	250	250	250	250	250	250
Manufacturer-rated draw, L in 3 h, 55 K		186	186		284	284	284	284	284	284	284
Tank capacity, L		250	250		250	250	310	250	310	310	310
Tankless-type Indirect[g,h]											
I-W-H-rated draw, mL/s, 55 K rise		170	170		200	200	240	200	240	240	240
Manufacturer-rated draw, L in 5 min, 55 K rise		57	57		95	95	133	95	133	133	133

Source: © ASHRAE, www.ashrae.org. 2023 ASHRAE Handbook—HVAC Applications, Chapter 51, Table 4.

[a]Storage capacity, input, and recovery requirements are typical and may vary with individual manufacturers. Any combination of these requirements to produce the stated one-hour draw will be satisfactory.

[b]Boiler-connected water heater capacities (180°F boiler water, internal, or external connection).

[c]Heater capacities and inputs are the minimum allowable. Variations in tank size are permitted when recovery is based on 4 gph/kW at 100°F rise for electrical, American Gas Association (AGA) recovery ratings for gas, and Institute of Boiler and Radiator Manufacturers (IBR) ratings for steam and hot water heaters.

[d]Boiler-connected heater capacities (200°F boiler water, internal, or external connection).

[e]Also for 1.5 baths and 4 bedrooms for indirect water heaters.

[f]Boiler-connected water heater capacities (82°C boiler water, internal, or external connection); I-W-H stands for indirect water heater.

[g]Heater capacities and inputs are the minimum allowable. Variations in tank size are permitted when recovery is based on 4.2 mL/s kW at 55°C rise for electrical, AGA recovery ratings for gas, and IBR ratings for steam and hot water heaters.

[h]Boiler-connected heater capacities (93°C boiler water, internal, or external connection).

TABLE 21.9 Procedure for Sizing a Residential Water-Heater Storage Tank

Characteristic	Data from Table 21.8 and HUD-FHA 5 Bedrooms, 3 Baths	Figure 21.22 Data: Cyclone XHE Model BTH-120
Storage, gal [L]	50 [190]	60 [227]
Input, Btu/h [kW]	47,000 [13.8]	120,000 [35.2]
1-h draw (= tank capacity + 1-h recovery)	90 [340]	231 at 80°F [875 at 44°C]
Recovery	40 [150]	171 [650]

and the minimum usable storage required is 5 gal/student [18.9 L/student]. Thus

$$300 \times 5 \times 1.43 = 2145 \text{ gal}$$

$$[300 \times 18.9 \times 1.43 = 8108 \text{ L}]$$

At the specified cafeteria recovery rate of 1.0 gph [3.8 L/h],

$$300 \times 1.0 = 300 \text{ gph}$$

$$[300 \times 3.8 = 1140 \text{ L/h}]$$

and the minimum usable storage required is 2 gal/meal [7.6 L/meal]. Thus

$$300 \times 2.0 \times 1.43 = 858 \text{ gal}$$

$$[300 \times 7.6 \times 1.43 = 3260 \text{ L}]$$

Combining these requirements for dorm and cafeteria,

$$\text{recovery} = 750 + 300 = 1050 \text{ gph}$$

$$[2850 + 1140 = 3990 \text{ L/h}]$$

$$\text{storage} = 2145 + 858 = 3003 \text{ gal}$$

$$[8108 + 3260 = 11{,}368 \text{ L}]$$

For this example, an increase in heater size of 225% for faster recovery allows the size of the tank to be reduced to only 37% of original size.

Note: This sizing procedure does not account for system losses, that is, heat lost from the storage tanks and from the hot water piping. Recovery capacities are usually increased because of these losses, which are simple to calculate when the U-factors of the tank and pipe insulation are known: Btu/h = $U \times A \times \Delta t$. ■

TABLE 21.10 Domestic Hot Water, Commercial/Institutional

Type of Building	Maximum Hour	Maximum Day	Average Day
Men's dormitories	3.8 gal [14.4 L]/student	22.0 gal [83.4 L]/student	13.1 gal [49.7 L]/student
Women's dormitories	5.0 gal [19 L]/student	26.5 gal [100 L]/student	12.3 gal [46.6 L]/student
Motels: No. of Units[a]			
20 or less	6.0 gal [23 L]/unit	35.0 gal [132.6 L]/unit	20.0 gal [75.8 L]/unit
60	5.0 gal [20 L]/unit	25.0 gal [94.8 L]/unit	14.0 gal [53.1 L]/unit
100 or more	4.0 gal [15 L]/unit	15.0 gal [56.8 L]/unit	10.0 gal [37.9 L]/unit
Nursing homes	4.5 gal [17 L]/bed	30.0 [114 L]/bed	18.4 gal [69.7 L]/bed
Office buildings	0.4 gal [1.5 L]/person	2.0 gal [7.6 L]/person	1.0 gal [3.8 L]/person
Food Service Establishments			
Type A—full meal restaurants and cafeterias	1.5 gal [5.7 L]/max meals/h	11.0 gal [41.7 L]/max meals/h	2.4 gal [9.1 L]/avg meals/day[b]
Type B—drive-ins, grilles, luncheonettes, sandwich and snack shops	0.7 gal [2.6 L]/max meals/h	6.0 gal [22.7 L]/max meals/h	0.7 gal [2.6 L]/avg meals/day[b]
Apartment Houses: No. of Apartments			
20 or less	12.0 gal [45.5 L]/apt.	80.0 gal [303.2 L]/apt.	42.0 gal [159.2 L]/apt.
50	10.0 gal [37.9 L]/apt.	73.0 gal [276.7 L]/apt.	40.0 gal [151.6 L]/apt.
75	8.5 gal [32.2 L]/apt.	66.0 gal [250 L]/apt.	38.0 gal [144 L]/apt.
100	7.0 gal [26.5 L]/apt.	60.0 gal [227.4 L]/apt.	37.0 gal [140.2 L]/apt.
200 or more	5.0 gal [19 L]	50.0 gal [195 L]/apt.	35.0 gal [132.7 L]/apt.
Elementary schools	0.6 gal [2.3 L]/student	1.5 gal [5.7 L]/student	0.6 gal [2.3 L]/student[b]
Junior and senior high schools	1.0 gal [3.8 L]/student	3.6 gal [13.6 L]/student	1.8 gal [6.8 L]/student[b]

Source: © ASHRAE, www.ashrae.org. 2023 ASHRAE Handbook—HVAC Applications, Chapter 51, Table 6; SI units added.

[a]Interpolate for intermediate values.

[b]Per day of operation.

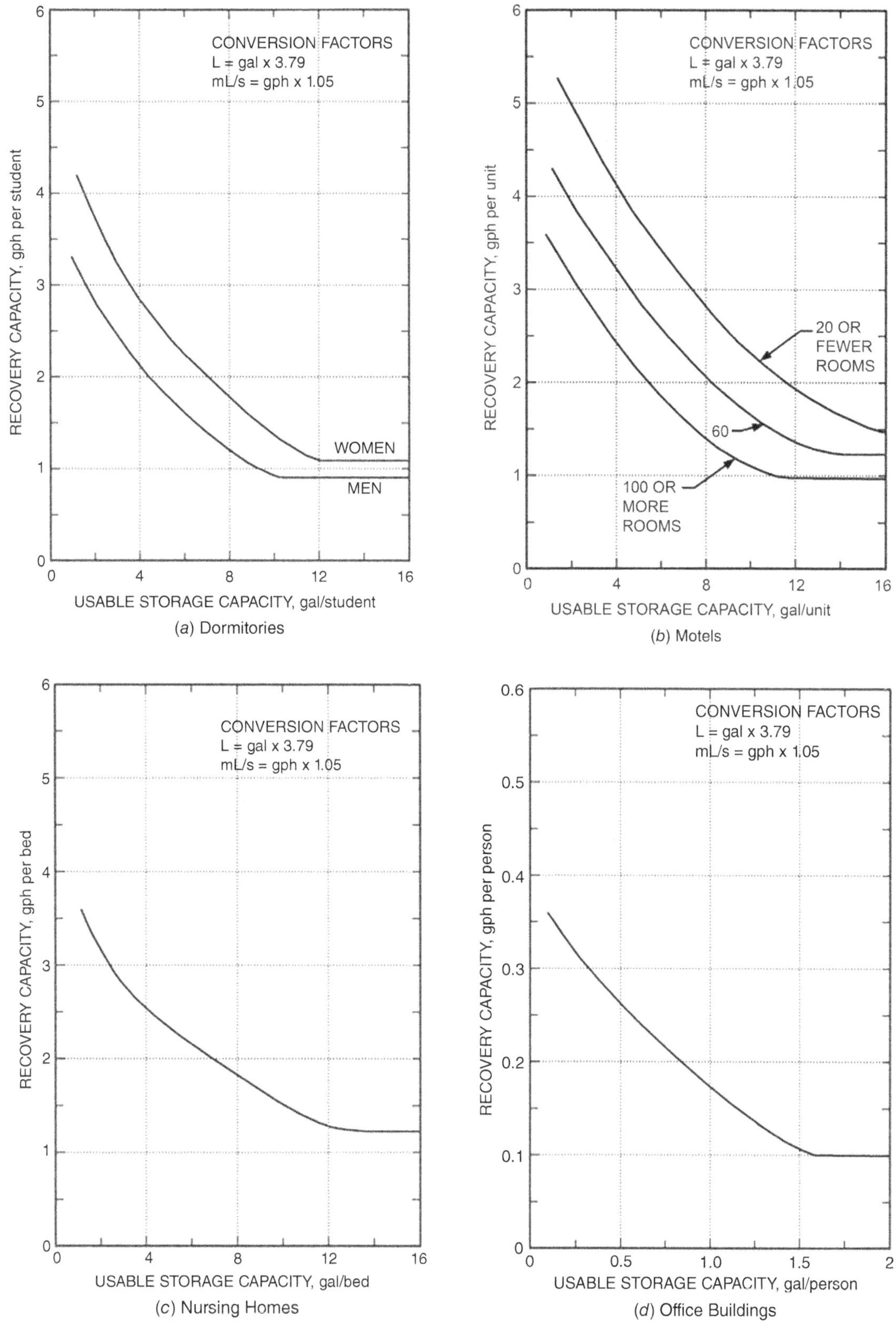

Figure 21.23 *Domestic hot water sizing; the trade-off between recovery (heater) capacity and storage capacity. Usable storage capacity is usually considered to be between 60% and 80% of the total tank capacity. For (b) Motels, and (f) Apartments, curves are based upon the number of dwelling units. For (e) Food service, Types A, B, see Table 21.10. (© ASHRAE, www.ashrae.org. 2023 ASHRAE Handbook—HVAC Applications, Chapter 51, Figures 25-32; SI notes added.)*

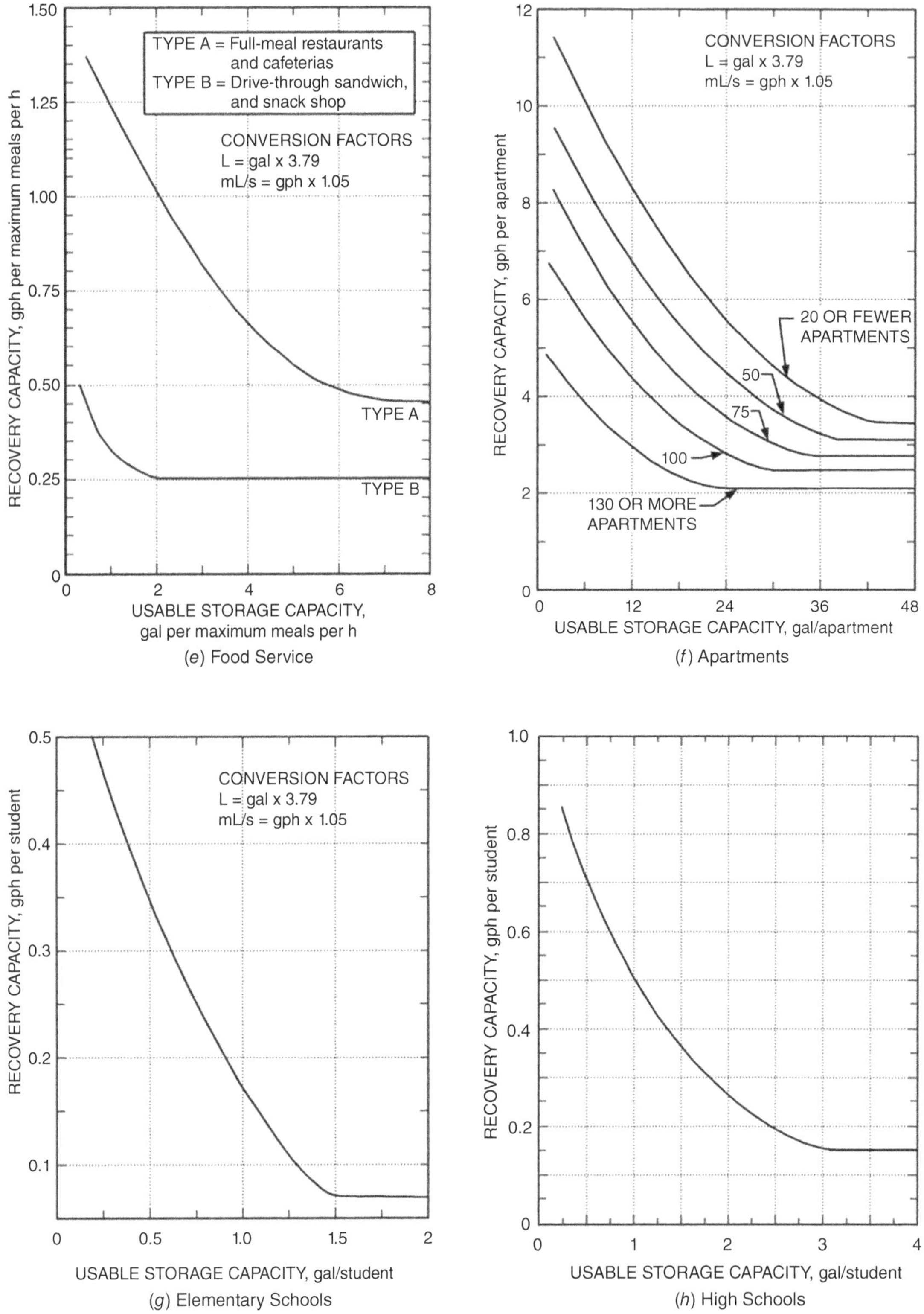

Figure 21.23 *(continued)*

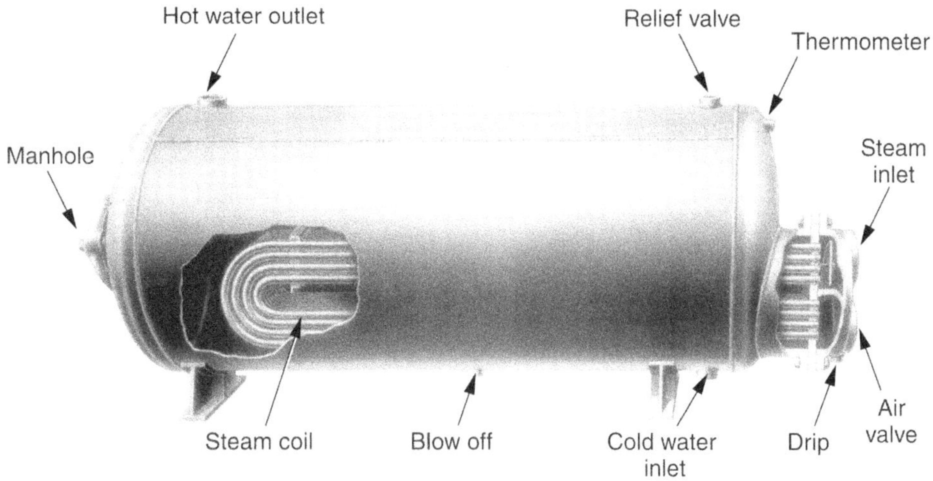

Figure 21.24 *Storage tank and heater for DHW for large-demand applications. A steam coil is submerged in the tank. Capacities from 100 to 10,000 gph [380 to 37,850 L/h], varying by length of coil, for a 140°F [78°C] temperature rise (from 40° to 180°F [4 to 82°C]).*

For these larger DHW applications, central steam is often used as a heat source. Figure 21.24 shows an indirect storage tank system. Such systems may cost more than tankless systems, but they are usually cheaper to operate, both because the peak prices for the instantaneous fuel demands of tankless heaters are avoided and because low-grade steam or waste high-temperature water sources can be utilized.

In any system that heats water under pressure, safety precautions are necessary. To minimize the dangers of excessive pressure (which can damage the system) and of superheated water (which can severely injure people), most codes require pressure and temperature (*P/T*) relief valves to be installed on top of all water heaters. Because these valves are designed to release hot water whenever a dangerous pressure or temperature is reached, they should be attached to a length of pipe that will conduct the released water to a drain.

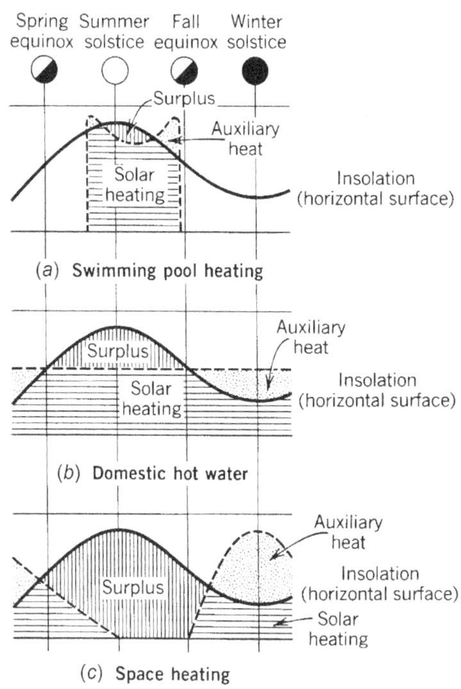

Figure 21.25 *Comparison of the pattern of solar energy supply (on a horizontal surface) to various patterns of heating needs.*

(i) Solar Water Heating

Solar energy is most attractive to a designer when it will do most of its work during the time that it is most available. On a seasonal basis (Figure 21.25), solar energy is especially attractive for outdoor swimming pool heating, as it will be used only in sunny, warm months. Throughout the United States, solar energy can easily meet most of the summertime demand for DHW as well. In the northern part of the country, a much smaller solar contribution can be expected in winter. An especially difficult problem is the winter space-heat mismatch between greatest need and least solar energy supply. For both solar DHW and space-heating systems, winter brings the added complication of the need to protect water in collectors against freezing.

Again, there are many ways in which solar energy can be used to heat water. Solar water-heating systems are usually classified by the means of fluid circulation (passive or active), the means by which heat from the collector piping is transferred to the DHW itself (direct, indirect, or closed-loop), and the means of protection against freezing.

Passive systems rely upon gravity for circulation, as explained in Section 21.6(f). Hence, the storage tank usually must be placed above the collector (Figure 21.26), and the number of bends in the system

Figure 21.26 *Solar water heaters are widely used. This house in Miami, Florida, incorporates a storage tank above the collector, which is enclosed in the chimney-like form. This permits passive (thermosiphon) circulation without a pump. (Photo by M. Steven Baker.)*

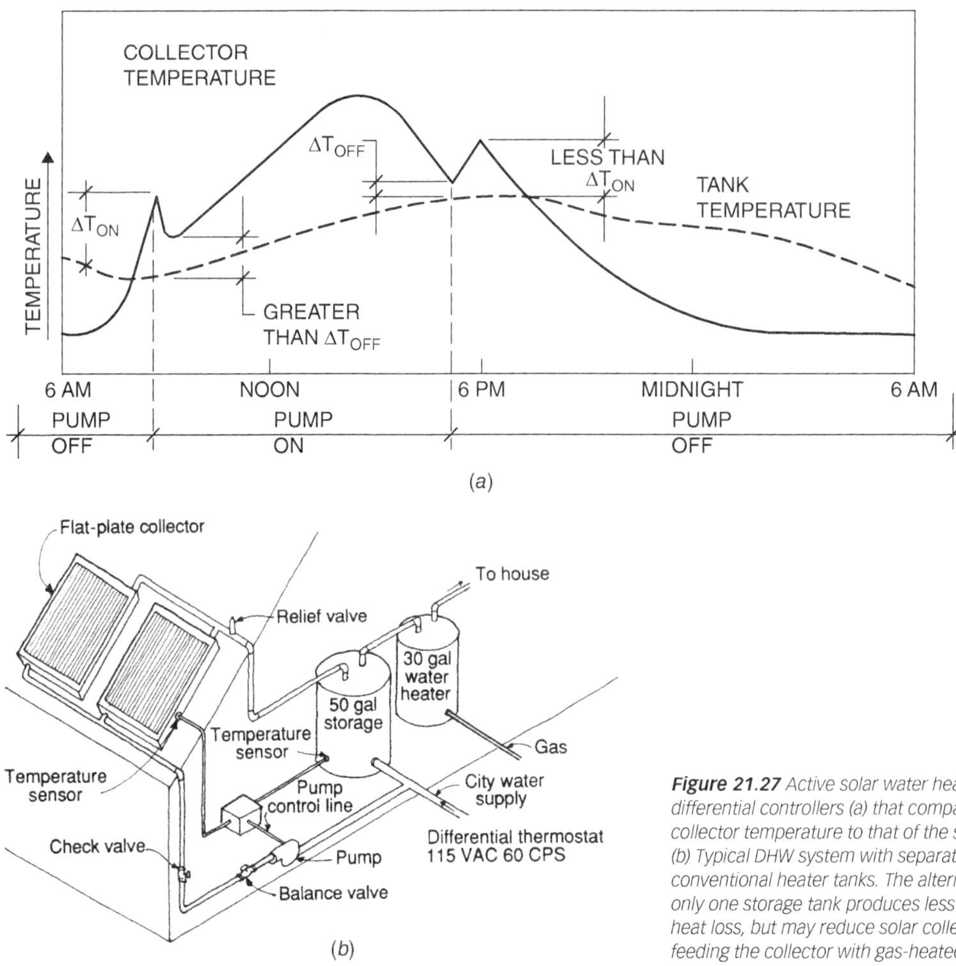

Figure 21.27 *Active solar water heating utilizes differential controllers (a) that compare the collector temperature to that of the storage tank. (b) Typical DHW system with separate solar and conventional heater tanks. The alternative of only one storage tank produces less standby heat loss, but may reduce solar collection by feeding the collector with gas-heated water.*

supply and return piping must be minimized to reduce friction. Heavy storage tanks located high in a building can cause structural problems. The advantages of this passive approach include lower cost of components, high mechanical reliability (no pump, etc.), and low operational costs.

Active systems use pumps to force the fluid into the collector. Although this setup allows the collector and storage to be located anywhere that is convenient for the designer, it introduces the complications of mechanical breakdown, increased maintenance, and the cost of the energy needed to run the pump. Active DHW systems are widespread in North America; see Figures 21.27 and 21.28 for typical systems.

Direct systems utilize only one fluid: The water to be heated for use in the building is circulated through the solar collector. Such a system has the advantages of simplicity and efficiency, as it does not require a separate fluid loop and the attendant piping complications and inefficiencies of heat exchange.

Indirect systems use a closed loop containing a fluid that circulates through the collector and storage tank. The fluid is not mixed with the DHW itself; rather, heat is passed from one fluid to the other through a heat exchanger. One advantage of this system is that it allows nonfreezing solutions to be used in the collector loop. It also allows collectors to be operated at low pressure rather than at the high pressures typical of urban public water systems. The choice of fluid is determined by its freezing and boiling points, its specific heat, and its level of toxicity.

As Table 21.11 shows, these design elements (passive and active, direct and indirect) are combined in a number of typical solar DHW systems. Brief descriptions of some of the more common systems follow.

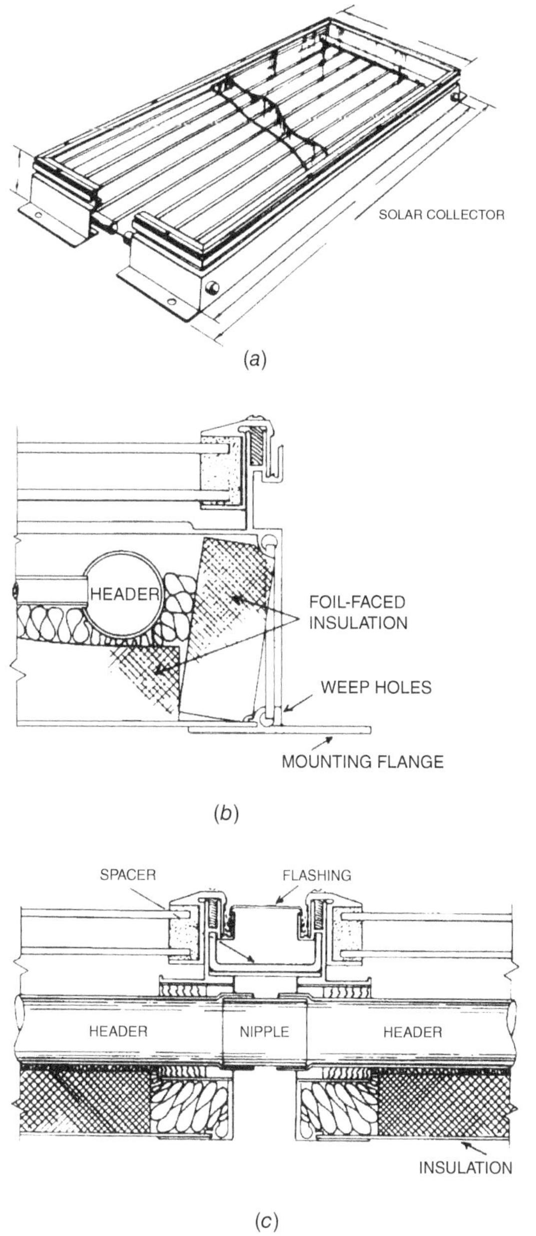

Figure 21.28 *(a) Cutaway section of a typical active solar collector. (b) Cross-section at the header. (c) Method of connecting manifolds of adjacent solar collectors.*

- *Batch systems:* In these, the simplest of all such systems, a black-painted storage tank is exposed to the sun within a glazed collector box (Figure 21.29). The price of simplicity is the inefficiency of exposing a tank of hot water to the cold nighttime conditions of the collector. Movable insulation, however inconvenient, can be used to reduce this problem. These heaters are also referred to as breadbox water heaters and integral passive solar water heaters. The batch system was used at the Antelope Valley California Poppy Reserve (Chapter 14), as shown in Figure 21.30. Because of their simplicity, they are favored by do-it-yourself builders, especially in mild winter climates.

  A more recent batch collector design utilizes eight copper tubes in a well-insulated, double-glazed (glass out, Teflon in) frame. The water flows in series, allowing the colder replacement water to enter at the bottom of the collector, with the hottest water at the top, ready for use. Insulation between the tubes helps maintain this temperature difference. Collectors are available in storage sizes of approximately 30, 35, 40, and 50 gal, weighing when filled 425–664 lb [114, 133, 152, and 190 L, weighing 193–301 kg].
- *Thermosiphon systems:* The sun acts as both the pump and the heat source for these systems (Figure 21.31). With no moving parts, maintenance needs are low.

TABLE 21.11 Solar Domestic Hot Water Systems

Type	Main Features	Advantages	Disadvantages
Batch system	Batch tank inside collector box Potable water within collector/tank	No external power Few components Collector/tank at any location	Seasonal; dependent on freezing locations Heat loss at night from storage
Thermosiphon system	Flat-plate liquid collectors Normally open loop but no pump or external power passive (DHW system) Storage tank higher than collector	No external power Few components High performance	Seasonal; dependent on freezing locations (if water is collector fluid) Needs structural support for high storage tank
Geyser pumping system	Flat-plate liquid collectors Methanol-water solution passively pumped by solar heat to a heat exchanger below	No freezing damage No mechanical or electric parts No liquid service	Temporarily stops operation at subzero temperatures, as solution freezes to a slush
Closed-loop freeze-resistant system	Flat-plate liquid collectors Closed loop of piping from collectors to storage tank Uses external energy (circulator and differential controller) Uses nonfreezing collector fluid Pressurized stone-lined storage tank	Can be used in coldest climates More and better-established competition Circulator; small consumption of external energy High performance	Liquid; service/maintenance required More components required
Drain-back system	Flat-plate liquid collectors Water is collector fluid (open loop) Potable water circulates through the heat exchanger in the storage tank (not through the collectors) Large heat exchanger Pitched headers	Can be used in the coldest climates No antifreeze used Most simple of active flat-plate systems (no valves)	Larger pump; larger consumption of external energy System must drain thoroughly Use of corrosion inhibitor recommended
Drain-down system	Flat-plate liquid collectors Potable water circulated through collectors Line pressure feeds collectors (open loop) Has automatic drainage valves Pitched headers	No heat exchanger or extra storage tank needed High performance	In some instances, a larger pump; larger consumption of external energy System must drain thoroughly No corrosion inhibitor possible Freeze danger with valve failure
Air-to-liquid system	Flat-plate air collectors Air-to-water heat exchanger Ductwork and blower Pipes and circulator. Larger collector area than liquid system	Won't freeze (dependent on exchanger location) Air leaks will not cause damage Integrates well with space heating	Hard to detect leaks More space required for ducts Blower and circulator required More carpentry involved Less efficient than other systems
Phase-change system	Flat-plate liquid collectors Freon 114 or R12[a] Storage tank higher than collectors (passive type) Closed-loop refrigerant-grade piping from collectors to storage tank (passive type) or to condenser (subambient type)	No external power (passive type) Can be mounted at any location (subambient type)	Very hard to detect leaks Special equipment to install More components required (subambient type)

Source: Plante, 1983.

Information on the geyser pumping and batch systems added by the authors of this book.

[a]Environmentally responsible refrigerants must be used in new systems.

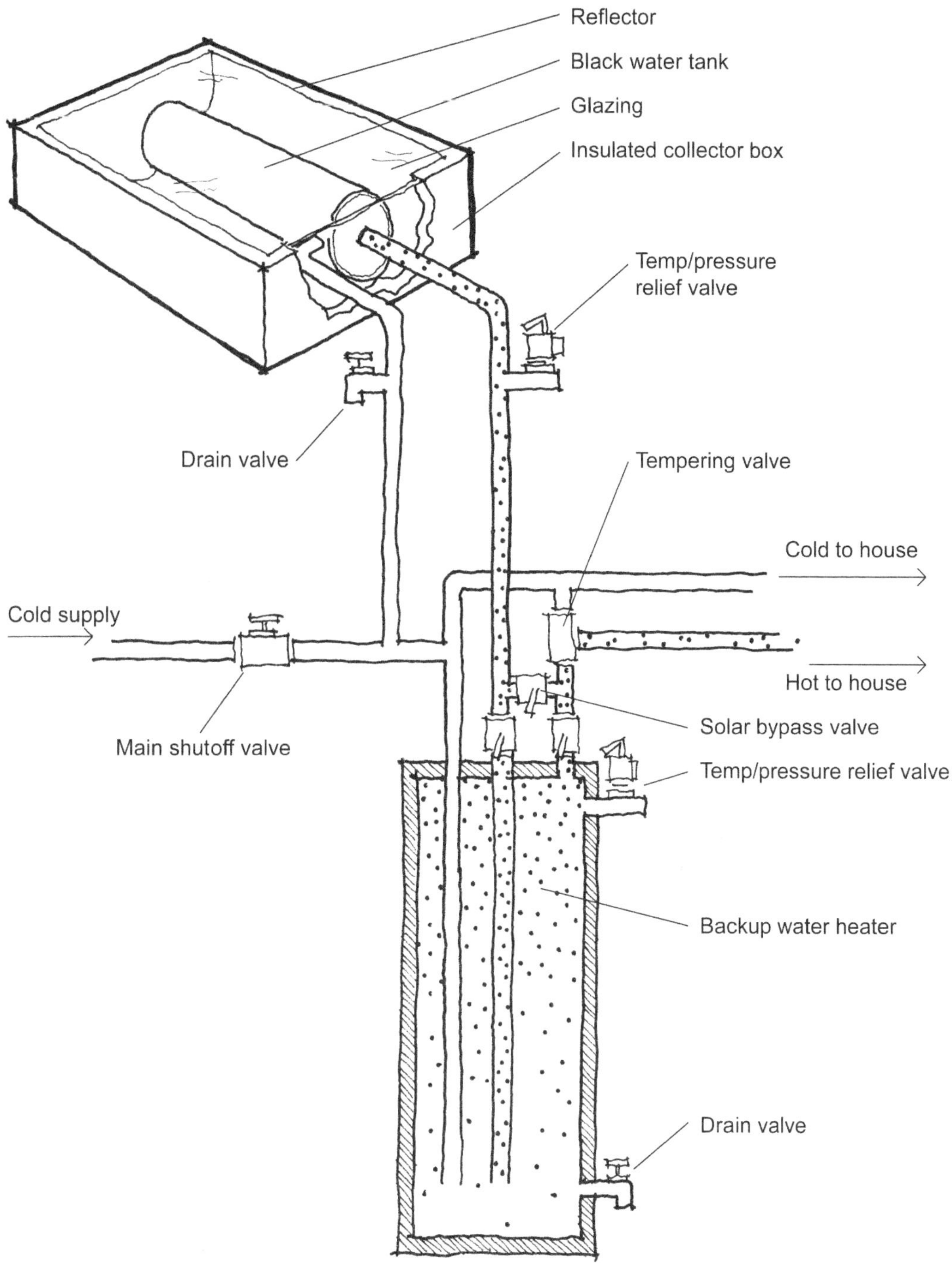

Figure 21.29 *Batch solar water-heating system. (Drawn by Dain Carlson; © Alison Kwok; all rights reserved.)*

The collector, however, must be lower than the tank, and piping must be kept as simple as possible. Because the coldest water remains in the lower collector at night, the hot water in the upper tank is not threatened with undue heat loss. However, freezing conditions pose a severe threat to the collector. Accordingly, indirect (closed-loop) systems containing a nonfreezing fluid are frequently used; phase-change systems are a promising cold-winter option for this type of passive system.

- *Closed-loop, freeze-resistant systems:* In addition to being used in thermosiphoning arrangements, these systems are commonly used in active systems (Figure 21.32). A small pump circulates nonfreezing fluid to the collector when there is sufficient sun. This process is governed by a differential controller.

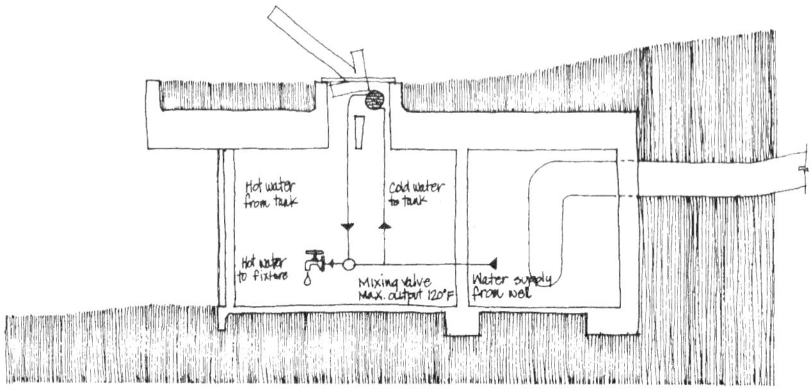

Figure 21.30 *A batch solar water heater supplies the hot water for the Antelope Valley California Poppy Reserve. (Courtesy of the Colyer/Freeman Group, Architects.)*

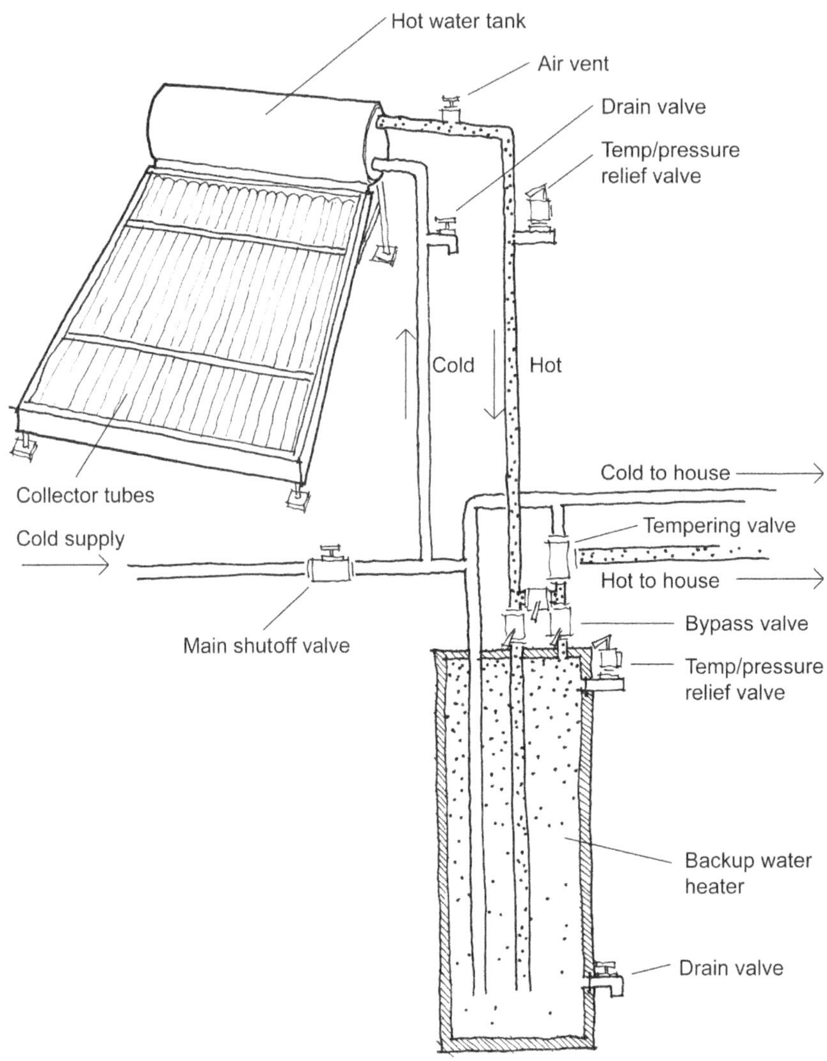

Figure 21.31 *Thermosiphon solar water-heating system. (Drawn by Dain Carlson; © Alison Kwok; all rights reserved.)*

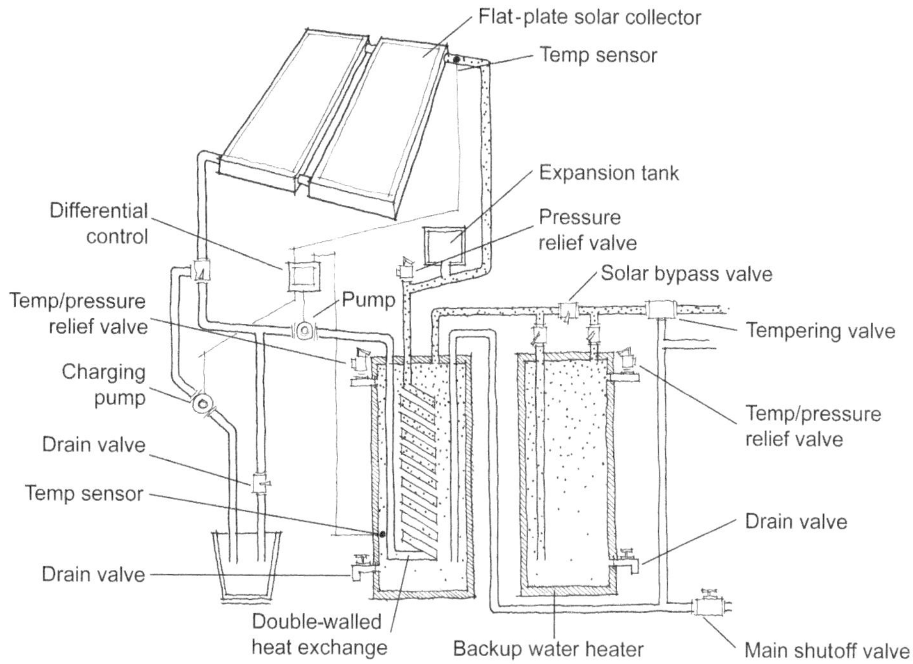

Figure 21.32 *Closed-loop, freeze-resistant solar water-heating system. (Drawn by Dain Carlson; © Alison Kwok; all rights reserved.)*

Alternatively, a photovoltaic-driven circulation pump can be used (see Chapter 31); when there is sufficient solar energy to activate the pump, there should be enough to also heat water in the collector.

The price of this freeze protection is the inefficiency of the heat exchanger, used between the collector fluid and the potable hot water. Some codes require a double-wall heat exchanger between any toxic nonfreezing fluid and the potable water, which further reduces efficiency.

- *Drain-back systems.* Although these systems (Figure 21.33*a*) use water as the fluid pumped from tank to collector, this water is not the potable hot water itself. Instead, the potable water passes through a heat exchanger in the solar storage tank. With the potable water kept out of the collector, the solar collector can operate at lower water pressure. This arrangement, however, requires a large heat exchanger, with attendant inefficiency. It also requires care in the design and installation of piping between the collector and tank so that the collector will drain thoroughly. When the controller senses that no solar energy can be gathered, it cuts off the pump, and water drains back into the tank. Therefore, the collector will be filled with air, not water, during all nighttime and cloudy-cold daytime hours. Corrosion inhibitors should be added to the collector/tank's water because the piping is frequently exposed to air.
- *Drain-down systems.* These are the only active systems that do not utilize heat exchangers (Figure 21.33*b*); DHW is circulated directly through the collector. Both higher pressure and higher efficiency result. In this system, the collector is usually filled with water that moves only when a differential controller activates the pump. Whenever the outside temperature drops near freezing, the controller activates solenoid valves and the water in the collector is drained down (dumped). In cold-winter areas, this process can result in several gallons of water wasted per winter day. Although the lack of a heat exchanger is attractive from the standpoint of cost savings and thermal efficiency, the set of electrically controlled solenoid valves is not foolproof. When malfunctions occur in pressurized systems, the loss of water can be enormous, and there is great potential for water damage to the building.
- *Air-to-liquid systems:* These systems use air collectors and rock storage beds; a heat exchanger transfers heat from the collector-heated air to the hot water. Much lower efficiencies result, and the ductwork is much more space-consuming than are pipes for water collectors. Leaks in air collectors or storage beds are very difficult to detect. However, air collectors are not damaged by freezing.
- *Phase-change systems:* In any of the aforementioned systems in which potable water is kept out of the collector, there are some missed opportunities—for instance, using the fluid within the collector as a mode of frost-resistance, taking advantage of the water's latent heat (the considerable amount of heat stored when fluid vaporizes and released when a fluid

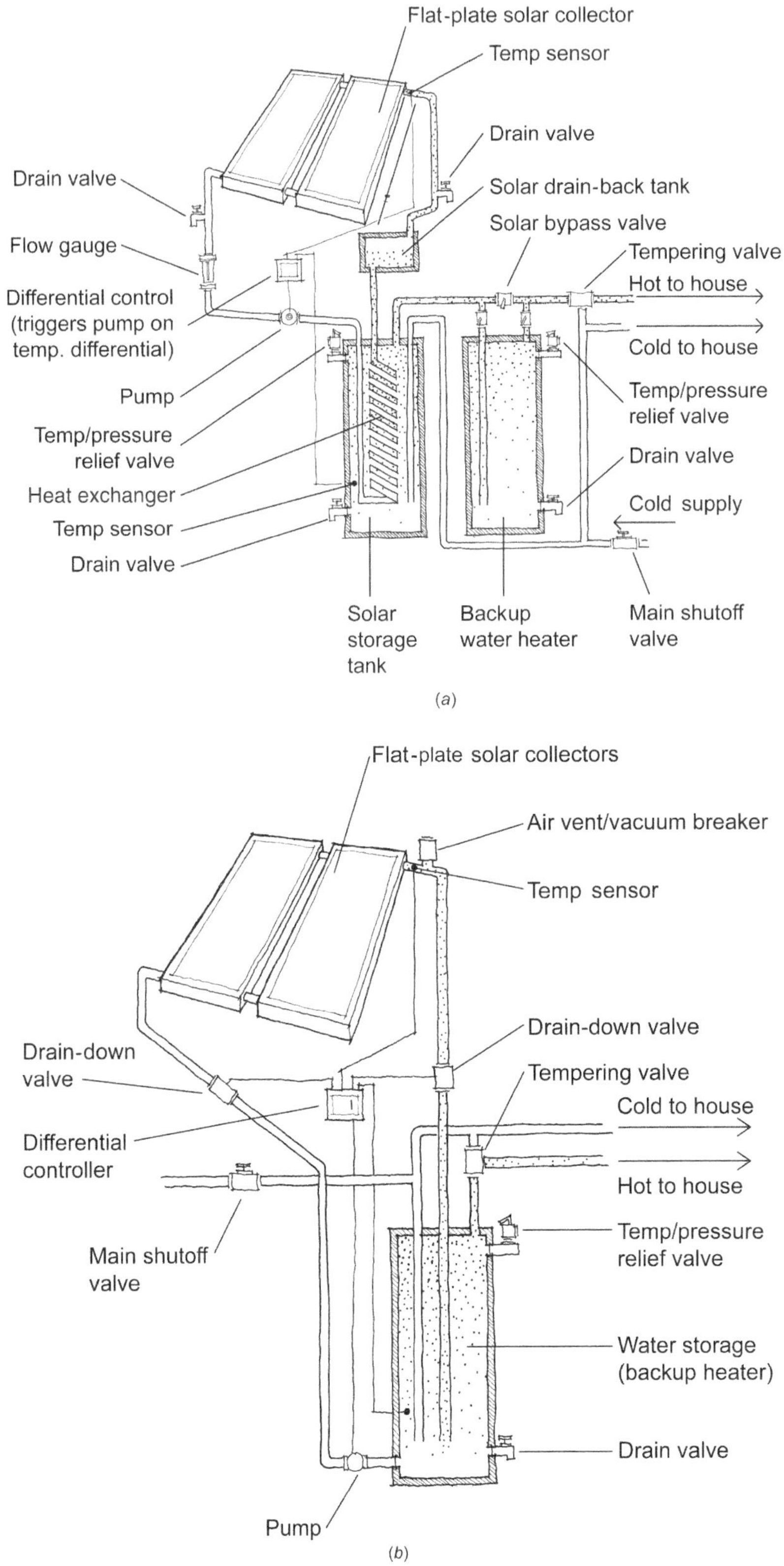

Figure 21.33 *Comparison of (a) a drain-back solar water-heating system with (b) a drain-down system. The drain-back system's performance can be increased by the addition of a check valve that allows the colder water at the bottom of the storage tank to circulate into the heat exchanger. The drain-down system is best used in mild winter climates with infrequent freezing temperatures, since the water in the collector is dumped with each freezing threat. (Drawings by Dain Carlson; © Alison Kwok; all rights reserved.)*

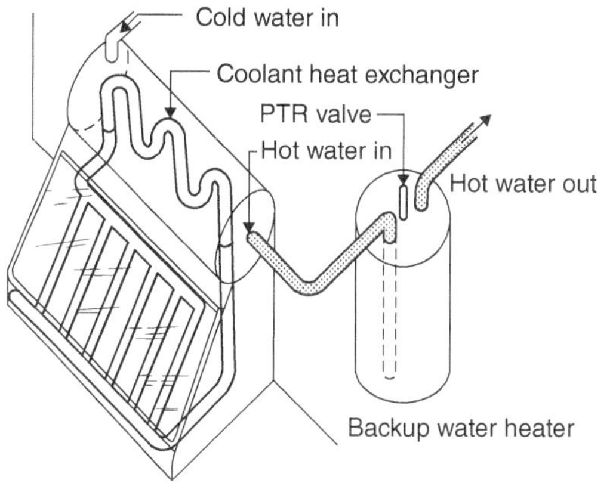

Figure 21.34 In a phase-change system, small heat exchangers can be built into the tops of thermosiphoning collectors; domestic hot water is drained down when the threat of freezing arises. (Adapted by permission from Solar Age, November 1983.)

condenses), and so on. The primary disadvantages of a phase-change system (Figure 21.34) are the high first cost, the difficulty in detecting leaks, and the resulting threat of chlorofluorocarbon or hydrochlorofluorocarbon refrigerant fluids leaking into the environment.

Passive approaches to phase-change materials were outlined in the discussion of thermosiphoning systems. Subambient approaches utilize a heat pump and can glean heat even from cold-cloudy ("subambient") conditions. They thus represent a form of solar-assisted heat pump, closely related to those discussed in Section 21.6(j).

The sizing of solar water heaters usually begins with simple design guidelines. For batch heaters, such a sizing guideline (Figure 21.35) is 0.45–0.65 ft^2 glazing per gallon of water stored [0.011–0.016 m^2 per liter].

For the flat-plate solar collectors used in all the solar DHW systems listed in this section, the sizing guidelines are:

- 12–25 ft^2 collector/person (residential) [1.1–2.3 m^2/person]
- optimum tilt (up from the horizontal) equal to latitude (or less)
- 1–1.5 gal of storage per square foot of collector area [41–61 L per square meter of collector area]

This collector-sizing guideline is for residential hot water usage, including cooking and bathing. For warmer climates with ample insolation, use the lower figure; this will supply about half of the hot water on an annual average basis. (For nonresidential DHW, adjust the sizing guideline by comparing the gallons of hot water per person per day used in residences to those used in the nonresidential function under design.)

A more detailed sizing procedure would consider the collector's expected heat contribution for some typical months. Such a procedure requires both climate data and assumptions about water temperatures and other values. As a result, detailed answers are best obtained from computer programs, such as those listed in Appendix L. For those interested in something more than sizing guidelines but less than computer programs, consider the following approach to collector sizing (adapted from Brown, Reynolds, and Ubbelohde, 1982):

STEP 1. *Select the collector tilt angle.* This can be done by consulting Appendix C, Tables C.11 to C.14 for I-P C.15 to C.18 for SI), which present clear-day insolation values for various tilt angles. Appendix C, Table C.19 for I-P (C.20 for SI) presents average insolation values on horizontal surfaces, which should also be checked. The tilt angle selected should be close to the optimum angle for the best month for total insolation.

STEP 2. *Check the collector efficiency.* This should be done at least twice: for the best insolation month and for the worst month. This step requires the following data:

Hourly insolation on tilted surface (Appendix C, Tables C.11 to C.14 for I-P [C.15 to C.18 for SI])
Outdoor average temperature (Appendix C, Table C.19 for I-P [C.20 for SI])

The hourly insolation values on tilted surfaces in Appendix C, Tables C.11 to C.14 [C.15 to C.18] are for clear days. For most locations, this should be adjusted for average conditions, which are shown for vertical (south) and horizontal surfaces in Appendix C, Table C.19 [C.20]. Because for most of North America the optimum yearly DHW tilt angle will be closer to horizontal than to vertical, a simple correction to insolation must be made:

Hourly average-day insolation on tilted surface:

$$= (\text{hourly clear} - \text{day insolation on tilted surface}) \times \left(\frac{\text{average} - \text{day total horizontal insolation}}{\text{clear} - \text{day total horizontal insolation}}\right)$$

This step also requires assumptions about the input temperature of the water supplied to the collector (T_i). In summer, this can be quite high—even higher than the thermostat setpoint temperature of the hot water tank. In winter, it is likely to be lower, by perhaps 10–20°F [5.5–11°C], than the thermostat setpoint temperature. Finally, this step requires a choice of collector type so that efficiency can be determined.

ACTIVE ENVIRONMENTAL SYSTEMS

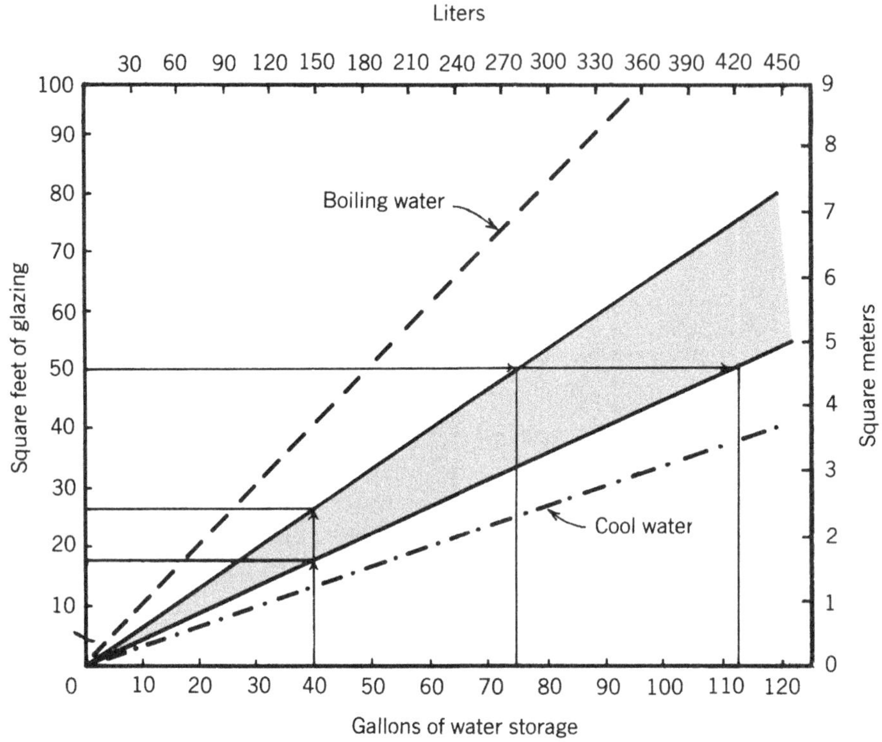

Figure 21.35 *Design guideline for the sizing of batch solar water heaters. In this example, a 40-gal [150-L] storage tank should be contained within a collector that has between 18 and 27 ft² [1.7 and 2.5 m²] of south-facing glazing. A 50-ft² [4.7-m²] collector box should contain a tank of between 75 and 113 gal [285 and 430 L]. (Adapted from Reif, 1983.)*

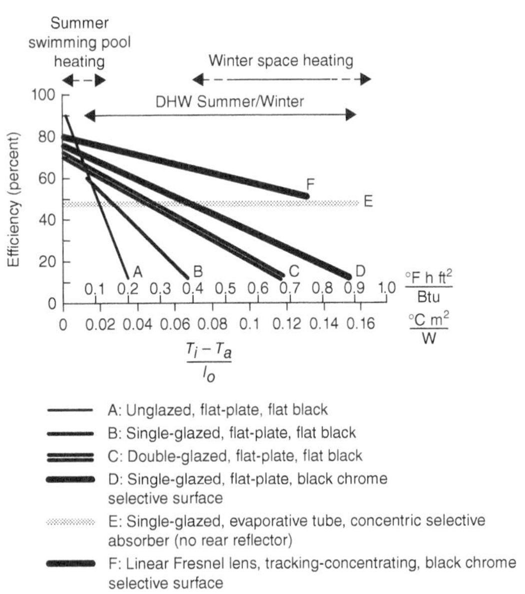

Figure 21.36 *Collector efficiencies depend upon insolation (I_o), water temperature entering the collector (T_i), and ambient (air) temperature (T_a). (Based on HUD, 1977.)*

Figure 21.36 shows the efficiency curves for a variety of solar collectors.

Note that the simplest of all, the unglazed flat black collectors, have the highest efficiency of all collectors at combinations of very low $T_i - T_a$ and very high I_o. These conditions are typical of summer days, on which the water needs to be heated only a few degrees. This corresponds to swimming pool applications and is one reason such collectors are so popular for that purpose; low cost is another reason.

For the opposite conditions—those of winter space heating—more elaborate collectors are appropriate. Evacuated-tube collectors have the advantage of nearly eliminating collector heat loss by convection. Fresnel lens and/or tracking-concentrating collectors increase the incoming solar energy per unit area, whereas heat loss increases only slightly (due to higher Δt). Both of these approaches are expensive, and tracking collectors require added maintenance for the tracking mechanism. Selective surfaces represent a simple, relatively cheap improvement over flat black collectors (compare *D* to *B*); they absorb solar energy just as well but emit radiant energy at a vastly lower rate, thus cutting collector heat loss by radiation.

STEP 3. *Approximate system efficiency.* Once the hot water leaves a collector, it must be led back to storage. In indirect systems, it must go through a heat exchanger. The tank will lose some heat; heat losses also occur as the cooled water is led back to the collector. Another "loss" to be considered is noncollection time. Estimates of collector performance are based on total daily insolation, yet the first few and last few hours of daylight generally will not bring enough insolation to warm the collector. Therefore, some reduction should be made in daily insolation totals when performance is estimated. All these factors can be accounted for by assuming a lower system efficiency. A perhaps optimistic assumption is that system efficiency = 0.8 × collector efficiency.

STEP 4. *Determine the total heat needed for DHW.* Estimates of the quantity of hot water needed were made in Section 21.6(h). Once the total gallons (or liters) per day are known and the desired storage temperature determined, the heat needed for the daily supply of DHW is easy to calculate.

In I-P units:

$$Q = 8.33\,(gpd)\,(t_s - t_g)$$

where

Q = daily heat need, Btu
8.33 = weight of water, pounds per gallon × Btu/lb °F
t_s = storage temperature, °F (see Table 21.7)
t_g = groundwater temperature, °F (see Chapter 14 for well-water temperatures)

In SI units:

$$Q = 1.16\,(L/d)\,(t_s - t_g)$$

where

Q = daily heat need, Wh
1.16 = the SI equivalent of the 8.33 factor
t_s = storage temperature, °C
t_g = groundwater temperature, °C

STEP 5. *Determine the desired percentage solar contribution to DHW.* There are several approaches to this problem. One common strategy is to have solar provide 100% of hot water needs in the best insolation month. A closely related strategy is to choose the yearly percentage of solar heat desired and then provide that percentage in the average month (such as March). Whatever strategy is chosen, a month and its percentage of solar heat must be identified.

STEP 6. *Size the collector.* This is done by combining the aforementioned steps as follows:

$$\text{collector area} = \frac{\text{daily heat need} \times \text{percent solar desired}}{\text{daily insolation} \times \text{system efficiency}}$$

This may be done for several months as a check on optimum collector size.

EXAMPLE 21.4 Determine the approximate size of solar collectors needed to serve the women's dormitory described in Example 21.3. Assume the location to be Springfield, Illinois.

STEP 1. *Tilt angle.* Refer to Appendix C (Table C.19 [C.20 for SI]). Springfield's latitude is 39.8°N. In Table C.19 [C.20], average insolation is given for July. Use this month as the best month.

From Appendix C, Table C.12 [C.16 for SI], the optimum tilt angle for July 21 at 40°N latitude is horizontal. However, at 30° tilt (latitude minus 10°), much better performance will be obtained in average months, such as March and September. For this example, choose a 32° tilt angle as Appendix C tabulates this angle.

STEP 2. *Collector efficiency.* For the best month (July), the best hourly clear-day insolation is 292 Btu/h per ft² [921 W/m²], at 32° tilt, in the hour centered at noon. To adjust this for average hourly value, calculate:

$$292 \times \frac{2058\ \text{Btu/day average horizontal}^{*}}{2230\ \text{Btu/day clear horizontal}^{**}} = 249\ \text{Btu/h per ft}^2$$

$$\left[921 \times \frac{6493\ \text{W/day average horizontal}^{*}}{7036\ \text{W/day clear horizontal}^{**}} = 787\ \text{W per m}^2\right]$$

*Table C.19 for I-P [C.20 for SI]
**Table C.12 for I-P [C.16 for SI]

Outdoor average temperature (TA) is approximately obtainable from Appendix C, Table C.19 [C.20 for SI]; TA for July is 76.1°F [24.5°C]. This is quite conservative, since it is the average daily temperature; the temperature at noon should be higher. From Appendix B, the mean daily range for Springfield is 19.4°F [10.8°C]. Thus, 76.1 + (19.4/2) = 85.8°F [24.5 + (10.8/2) = 29.9°C]. The thermostat setpoint temperature of the water storage tank is assumed to be 115°F [46°C]. This allows adequate hot water for most dormitory uses, although a higher temperature will be required for the cafeteria's dishwashing.

Assume, because it is summer, that T_i is at the setpoint of 115°F [46°C]. (In winter, it would be lower.) Therefore, the quantity $(T_i - T_a)/I_o$ can be calculated as

$$\frac{115°\text{F} - 85.8°\text{F}}{249\ \text{Btu/h ft}^2} = 0.117°\text{F h ft}^2/\text{Btu}$$

Enter Figure 21.36 with this number, and assume a single-glazed, flat-plate selective surface collector (type D). The efficiency will be about 70% under these conditions.

STEP 3. *System efficiency.* Use the design guideline:

$$0.8 \times (0.70\ \text{collector efficiency}) = 56\%$$

STEP 4. *Total heat needed for DHW.* The average usage in gpd (L/d) can be estimated from Table 21.10.

For women's dormitories, the rate is

$$12.3\ \text{gal/student} \times 300\ \text{students} = 3690\ \text{gal}$$

$$[46.6\ \text{L/student} \times 300 = 13{,}980\ \text{L}]$$

For the cafeteria, the average rate (type A) is 2.4 gal/meal × 300 students = 720 gal [9.1 L/meal × 300 = 2730 L]. Total volume needed: 3690 + 720 = 4410 gal/day [13,980 + 2730 = 16,710 L/day].

Groundwater temperature for Springfield (see Chapter 14) is about 56°F [13.3°C]. Assume that solar energy will be used to raise it to 115°F [46°C].

$$Q = 8.33\,(4410\ \text{gpd})\,(115°\text{F} - 56°\text{F}) = 2{,}167{,}383\ \text{Btu}$$

$$[Q = 1.16\,(16{,}710\ \text{L/d})\,(46°\text{C} - 13.3°\text{C}) = 633{,}844\ \text{Wh}]$$

STEP 5. *Desired percent solar.* For this best hour in July, assume a 100% solar contribution.

STEP 6. *Collector size.* Average daily insolation can be obtained from Appendix C as before:

Clear-day total, 32° tilt, July:

2230 Btu/day ft^2 [7036 W/day m^2]

(Appendix C, Table C.12 [C.16])

Clear-day total, horizontal, July:

2534 Btu/day ft^2 [7795 W/day m^2]

(Appendix C, Table C.12 [C.16])

Average-day total, horizontal, July:

2058 Btu/day ft^2 [6493 W/day m^2]

(Appendix C, Table C.19 [C.20])

Average July daily insolation (per unit area), at 30° tilt:

$$2230 \times 2058/2534 = 1811\ \text{Btu/day ft}^2$$

$$[7036 \times 6493/7795 = 5861\ \text{W/day m}^2]$$

$$\text{collector area} = \frac{(\text{load} \times \text{solar contribution})}{(\text{insolation} \times \text{system efficiency})}$$

$$\text{IP units} = \frac{2{,}167{,}383\ \text{Btu} \times 100\%}{1811\ \text{Btu/ft}^2 \times 56\%} = 2137\ \text{ft}^2$$

$$\left[\text{SI units} = \frac{633{,}844\ \text{Wh} \times 100\%}{5861\ \text{W/m}^2 \times 56\%} = 193\ \text{m}^2\right]$$

This size should be checked against the average-month performance and adjusted as desired. Note that this is a ratio of about 2137/300 = 7.1 ft^2 [193/300 = 0.6 m^2] of collector per student—lower than the general rule for typical residential DHW systems.

Note also that at the design guideline rate of 1.0–1.5 gal [3.8–5.7 L] of storage per square foot of collector, a tank size of ~2100 to ~3200 gal [~7949 to ~12,113 L] is indicated. In Example 21.3, the minimum recovery rate required a storage tank of 8151 gal [30,674 L], and the fastest recovery rate required 3003 gal [11,368 L]. Either seems sufficient for this size of collector array. ■

Swimming pool heating is an especially attractive application for solar energy. A common design guideline for collector sizing is

$$\text{collector area} = (0.5)\,(\text{pool surface area})$$

For summer ambient temperature operations, unglazed collectors are both the best-performing and the least expensive, as explained for Figure 21.36. Another important consideration is a pool cover, which will not only conserve the pool's heat but also reduce water lost by evaporation. On very hot and dry days, water losses can reach 100 gpd from a 20-ft × 40-ft pool [380 L/day from a 6-m × 12-m pool].

(j) Heat Pump Water Heaters

Air-water heat pumps are also used for DHW, as shown in Figure 21.37. Because these devices remove

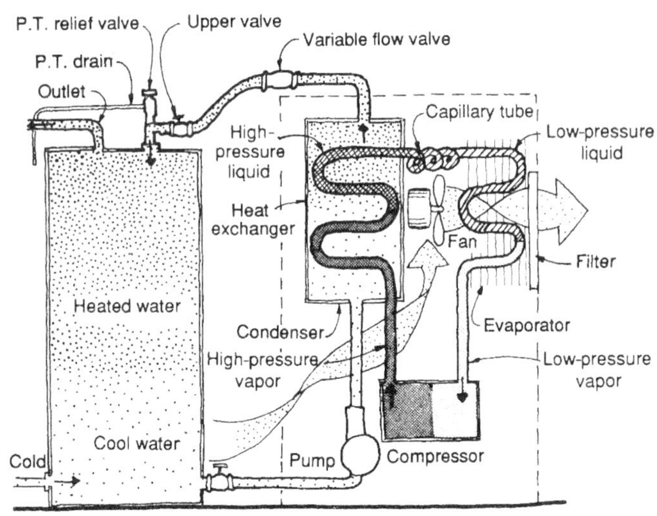

Figure 21.37 *Heat pump (air–water) used for heating domestic hot water. (Adapted by permission from Specifying Engineer, October 1983.)*

heat from the air, they are usually installed either in normally overheated spaces (such as restaurant kitchens) or in unheated spaces such as garages or basements. They can also remove heat from exhaust air streams. The spaces (or airflows) that contain these units will be cooled and dehumidified. Heat pumps require only a little more space than the simple hot-water storage tanks they serve. Some noise is created by the compressor and the fan that moves air across the evaporator.

Heat rejected from any refrigeration unit can be used for heating water via the heat pump. Applications include ice-making machines, refrigerated display units in grocery stores, walk-in freezers, and many others. Whenever constant refrigeration loads are present, there is an opportunity to utilize waste heat.

Graywater provides another opportunity for using a heat pump. Where filtered graywater is collected for recycling on site (as for irrigation systems), a water–water heat pump can heat a DHW storage tank as it lowers the graywater tank temperature. See Chapter 22 for graywater system design.

21.7 WATER-EFFICIENT FIXTURES AND FITTINGS

Building design and fixture choice can affect water and energy consumption over the life of a building. It is the users who turn the faucets on and off, but the designer who can encourage resource conservation through initial decisions, here centered largely in kitchens and bathrooms.

Historical (i.e., "conventional") fixtures and fittings encountered in existing buildings use significantly more water than current standards. The federally mandated minimum (i.e., "current") standard for the water efficiency of fixtures and fittings in the United States was set by the Energy Policy Act of 1992 (EPAact). See the information resources developed as part of U.S. Environmental Protection Agency's WaterSense program for information on the latest high-performance standards for water-efficient fixtures and fittings (EPA, 2024d). In the context of water supply systems, fixtures are permanently installed elements, such as toilets, sinks, or bathtubs; whereas fittings constitute elements that are affixed ("fitted") to plumbing-related components to control or guide water flow, such as faucets, shower heads, and valves.

Visible consumption is one strategy to encourage the users of fixtures to conserve water. We see, hear, feel, and sometimes taste water as it issues from a fixture. If we could see how much water is involved in each use, we could more readily alter our pattern of use to conserve water where appropriate. Rainwater storage tanks outside bathroom windows, used to flush toilets, might have a visible indicator of the water level; the lower this level, the more that conservation is an imperative. The water-level indicator might be as simple as visible condensation on the tank's outer surface. Strategies as simple as supplying water to lavatory sinks through small, transparent tanks of clean supply water placed above, might encourage occupants to use only a portion of the water already in the tank. Sufficient pressure at the fixture is achieved by such a tank's elevation above the fixture. The designer might also consider making the quantity of used (waste) water visible, but such water is often unsightly and the resulting quantity is "after the fact"—too late to conserve.

Audible consumption is another strategy, calling attention to the flow of supply (or waste) water. This is achieved most simply by slightly undersizing water supply pipes so that the water velocity becomes audible (at about 10 fps [3 m/s]). This strategy should be carefully considered; it is one thing to inform the water user, another to annoy. Audible flow is perhaps best justified on lines to exterior hose bibbs, where irrigation or car-washing hoses may inadvertently be left on.

Elimination of bodily wastes accounts for a significant portion of residential water usage (see Chapter 22), and bathing accounts for a significant portion of residential hot water usage (see Table 21.7). The choice of fixtures can have an impact on water conservation, and on energy conservation where hot water is involved. The bathroom is thus a site for potential resource conservation.

(a) Physiology, Psychology, and Fixtures

One of the more challenging aspects of bathroom design is the frequent conflict between physiological design criteria and such psychological influences as cultural attitudes about bathroom activities. Physiological criteria change as slowly as evolution. Cultural attitudes can change very rapidly, even within a generation. The bathroom supports two primary human activities: cleansing and elimination. Today's common attitudes about these closely linked activities are that one is "clean," the other "dirty." One might be discussed (although rarely) among friends; the other is simply not fit for conversation. This conflict in attitudes influences the design of our plumbing fixtures. The toilet is the fixture used for elimination; it is the logical place to provide for the cleansing of the perineal zone that should immediately follow elimination. (This is especially true of public toilets, where stalls close off toilets from lavatories.) Yet very few toilets presently incorporate a cleansing feature: the mixture of cleansing and elimination in the same fixture too often seems abhorrent.

(b) Lavatories

A key issue here is the contrast between running water and a standing water body. Lavatories are used primarily for the cleansing of hands, face, and teeth—activities done quickly with running water that is wasted directly rather than being collected. There are four broad types of faucet applications: private bathrooms, public restrooms, kitchens, and utility (think custodial, laundry, and the like). Most lavatories are designed as collection bowls—perhaps a reflection of the days when washbasins of standing water, drawn and heated elsewhere, were brought to the bathing place. Most lavatories also have fittings that project out over the sink. Although these fittings have slowly evolved into very sleekly designed objects, most of them still dump running water directly into the drain, are hard to use as drinking fountains, and can wound those who try to wash their hair in the lavatory.

In terms of faucet flow rates, projects teams may assume:

- conventional: 4–5 gpm [0.25–0.32 L/s]
- current standard: 2.2 gpm [0.14 L/s]
- WaterSense labeled: 1.5 gpm [0.09 L/s] maximum

In the United States, the federal efficiency standard requires both bathroom and restroom faucets to have a maximum flow rate of 2.2 gpm [0.14 L/s] at 60 pounds per square inch (psi) of water pressure. Currently, the EPA only offers WaterSense labels for private bathroom faucets; it does not have a specification that applies to public restroom, kitchen, or utility faucets. However, many models are available that meet or exceed the program's criteria. Newer low-flow faucets utilize a variety of devices to function as well (or better) with less water. Such devices include aerators (which add air bubbles to the stream, making it splash less and appear larger), flow restrictors, and mixing valves to control temperature. The lower flows achieved a range from 0.5 to 2.5 gpm [0.03 to 0.16 L/s]. Another promising development is the foot-operated faucet, which frees the hands from having to control water flow, thus saving a few seconds of flow during each lavatory usage. These devices could be particularly helpful at kitchen sinks, where extensive washing of objects takes place.

(c) Whole-Body Cleansing

The running/standing water contrast is even more clearly illustrated here. In cleansing, the ordinary sequence is wet, soap, scrub, and rinse. Standing water is good for wetting and very good for soaping/scrubbing; running water is superior for rinsing. The psychological aspect enters here in common attitudes toward tubs and showers. Tubs are often seen as places to relax in, to spend more time in, and perhaps to read in. Showers are viewed as quicker, "no-nonsense," stand-up places. Yet each could benefit from some features of the other.

Tubs should be designed so that the reclining body is supported at the back; this requires a contoured surface (Figure 21.38) rather than the ordinary straight-line design. It also requires braces for one's feet because otherwise the body will tend to float up and away from such a backrest. Tubs can be designed to accommodate persons of various leg lengths, as shown in Figure 21.38*b*. There must also be a seat to give most of the body a chance to be out of the water and to facilitate safe entry into and exit from the tub. Especially needed is a handheld shower for the final rinse; soapy standing water leaves a scummy film on both people and fixtures.

Showers may seem very efficient, but cleaning would be safer and more thorough if the bather could turn off the water and sit for at least part of the soap/scrub activity, especially for the lower legs and feet. Showers with integral seats are now common.

Another consideration is the location of water controls (fittings). The user should be able to reach them easily from outside the tub or shower without wetting his or her arm, but also should be able to manipulate them from within the tub/shower even if temporarily blinded by soap.

Shower heads (also commonly combined as one word: showerheads) have been notorious for encouraging prodigal water usage.

In terms of shower head flow rates, projects teams may assume:

- conventional: 5.5–8.0 gpm [0.35–0.50 L/s]
- current standard: 2.5 gpm [0.16 L/s]
- WaterSense labeled: 2.0 gpm [0.13 L/s] maximum

In the United States, the federal efficiency standard requires shower heads to have a maximum flow rate of 2.5 gpm [0.16 L/s] at 80 pounds per square inch (psi) of water pressure. Historically, flow rates around 6 gpm [0.4 L/s] and maximum rates of 12 gpm [0.7 L/s] were once common. As such, a short (five-minute) shower could consume as much as 60 gal [227 L] of water, much of it heated. Many codes now require a limitation on shower head flow; some with required flow rates below the U.S. federal standard of 2.5 gpm [0.2 L/s]. These flows can be designed into the shower head, or they can be achieved by cheap, simple flow restrictors in retrofit applications. Some utilities distribute

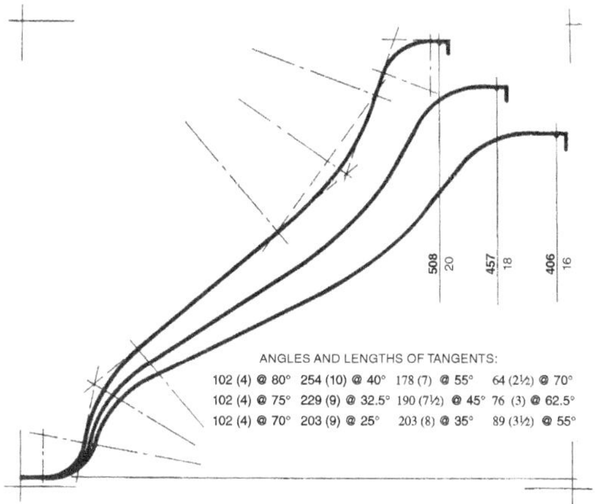

(*a*)

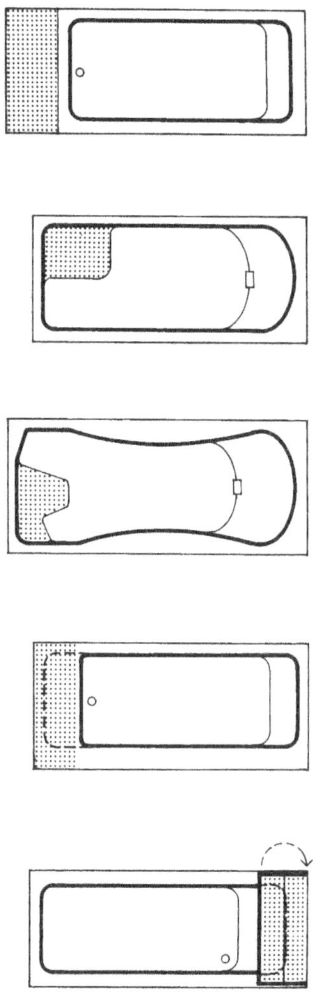

(*b*)

Figure 21.38 *Some considerations for tub design to facilitate relaxation during whole-body cleansing. (a) A contoured backrest allows comfortable reclining, with support for both the shoulder and lower back. Curve 2, with a median angle of 32.5°, promises comfort for most users. (b) Various tub plans, allowing long- and short-legged persons to brace their bodies against the contoured backrest. Raised seats are also provided to facilitate safe entry/exit. (From The Bathroom, new and expanded edition; © 1966, 1976 by Alexander Kira.)*

flow restrictors free of charge. Most bathers notice no difference, either in enjoyment or in cleansing, when using restricted-flow shower heads.

(d) Elimination

Several conflicts arise with toilets (which are also commonly referred to as water closets): the difficulty of combining cleansing with elimination, the issue of pure (high-grade) water for an impure (low-grade) purpose, and the conflict over the height of the toilet. A lower toilet is of benefit to most users, who will achieve far better bowel evacuation in a full squatting position. There are problems with low height. It is common for males to stand while urinating, and a lower toilet presents a more difficult target. This can have serious maintenance consequences unless males can be induced to sit, or a separate urinal is provided—an unlikely option for residential bathrooms. (Urinals are often considered to look institutional; they are also expensive.) Another problem with a lower toilet is that the elderly and some handicapped people will have difficulty getting on and off the seat (see Figure 21.48). Yet another problem arises from the fact that toilets are considered to be seats (for reading, toenail clipping, etc.). The low, squat-inducing toilet is decidedly not at a comfortable chair height. As a result, a conventional higher toilet is commonplace except for circumstances where the user group dictates otherwise (e.g., small children and groups with special needs). Low-cost ancillary devices can easily accommodate a more conducive squatting posture for users of conventional higher toilets. Toilet seats come in a range of different materials as well—some of which have lower thermal conductivity (and are therefore not as jarring when sat upon). Toilet seats can help promote better posture. For ordinary posture on conventional toilets, a physiologically sound contoured toilet seat is available (Figure 21.39). Note that some specialty toilets incorporate electric resistance heaters into the toilet seats to warm the small portion that contacts the body. In such cases, the toilet becomes a consumer of energy.

It is especially important that perineal cleansing be encouraged by the designers of bathrooms. The typical dry toilet tissue provided usually is not adequate for this task. In small, private bathrooms this problem is easily solved by placement of the toilet adjacent to the lavatory (or tub), where toilet tissue can be wetted with clean water. In public toilets separated by stalls, perineal cleansing requires either a clean water source built into the toilet or a separate bidet within the stall—a highly unlikely provision, given the extra floor space and added cost. A bidet is a bowl or receptable designed to be sat upon in order to receive spraying potable water for cleansing purposes. Cultural issues persist with bidets

Figure 21.39 Toilet seat for a conventional water closet, which encourages better posture for defecation and somewhat better access for perineal cleansing. (Courtesy of American Standard, Inc.)

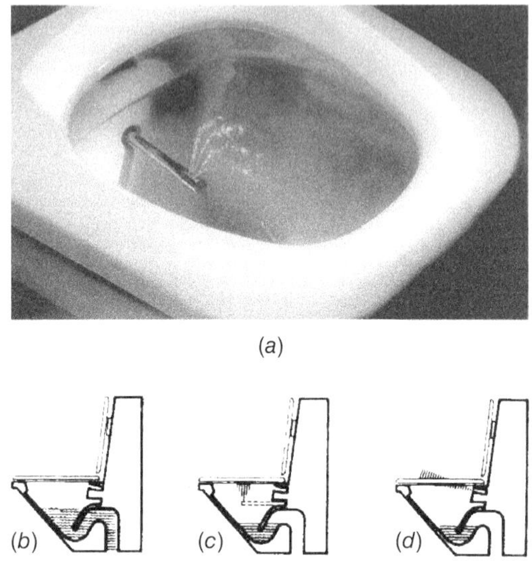

Figure 21.40 Provisions for perineal cleansing built into toilets. (a) An adjustable-intensity spray at warm (body) temperature continues for as long as a button is depressed in this Geberit Shower Toilet. (b) Section through the toilet before/after the cleansing function. (c) At the initiation of perineal cleansing, a pipe extends and emits a controlled spray of warm water. (d) After the pipe is retracted, a warm jet of air issues from just below the seat. (a: Courtesy of Geberit Manufacturing Inc.; b, c, d: From The Bathroom, new and expanded edition; © 1966, 1976 by Alexander Kira.)

as they are quite common in the private bathrooms of Europe but a rarity in North America.

Two common alternatives to provide integral cleansing capabilities at the toilet are: (1) to build a cleansing source into the toilet itself (Figure 21.40) or (2) to construct a toilet seat that provides for such cleansing. Several American manufacturers offer such seats, which can be easily adapted to existing toilets.

Increasingly, toilets are utilizing reclaimed or recycled water devoid of any fecal contamination, also called graywater, for waste expulsion. However, only high-grade potable water should be used for cleansing functions. Thus, using graywater in a toilet that also offers perineal cleansing complicates the high-grade water issue. Using graywater for cleansing would constitute a human health concern. In order to avoid unintentional mixing or switching of water supply sources, both the *International Plumbing Code* (IPC) and *Uniform Plumbing Code* (UPC) require all alternative (non-potable) water sources to be explicitly identified by purple colored pipes and bold, standardized lettering clarifying that the water source is not safe to drink (ICC, 2023; IAPMO, 2023a). Note that rainwater is not considered potable until it is sufficiently treated. Moreover, blackwater is a term used to define flushed water contaminated with urine, feces, and other biological waste. It can contain a range of pathogens and human contact constitutes an acute threat to human health and safety. As such, blackwater should never be used for water supply unless it is sufficiently treated to tertiary standards.

The urinal is an answer to the problem posed by lower toilets to males, who stand while urinating. Although urinals are culturally acceptable in public toilet rooms, they have never achieved a place in private bathrooms. They are, however, efficient relative to use of space and water efficiency. But make no mistake—in developed countries, toilets and urinals can account for nearly one-third of a building's water consumption. As such, many countries have established minimum standards for water efficiency with both types of fixtures. In the United States, the federal efficiency standard requires that urinals use no more than 1.0 gal [3.8 L] per flush; yet, in older buildings, inefficient "conventional" fixtures may still be encountered.

In terms of urinal flush rates, projects teams may assume:

- conventional: 1.0–5.0 gal [3.8–18.9 L] per flush
- current standard: 1.5 gal [5.7 L] per flush maximum
- WaterSense labeled: 0.5 gal [1.9 L] per flush maximum

For the toilet, there are four major categories of fixtures, depending upon the amount of water used per flush:

1. conventional: 3.5–7.0 gal [13.2–26.5 L] per flush
2. current standard: 1.6 gal [6.1 L] per flush maximum
3. high-efficiency: 1.5–0.38 gal [5.7–1.4 L] per flush
4. waterless: no water

(e) Conventional Toilets

Conventional toilets may no longer be legally installed in the United States due to their high-water consumption. They are still widely encountered in existing installations. In the common toilet, a sudden deluge of water removes human waste and simultaneously helps cleanse the toilet. Fast-moving water requires pressure and makes noise. The older flush-tank toilet (Figure 21.41*a*) stores a smaller quantity of water (about 2.5 gal [9.5 L]) at sufficient height above the toilet bowl to achieve fast flow. Although it uses minimal amounts of water, it is noisy, and its elevated tank has maximum visual impact. The more common (in North America) version is the two-piece flush-tank toilet with the tank bolted to the bowl (Figure 21.41*b*). With much less pressure available, this toilet requires 5–7 gal [19–26 L] per flush, but it is quieter. Shallow-trap models reduce the quantity needed to 3.5 gal [13.2 L]. The maintenance problem posed by the seam between tank and toilet is eliminated in the one-piece flush-tank toilet (Figure 21.41*c*). The cost of its low profile is an even greater need for water: 6–8 gal [23–30 L] per flush.

In addition to differing tank and bowl combinations, there are various types of flushing actions. Four common flush toilets are compared in Figure 21.42.

- *Washdown toilets:* These are no longer made in the United States, although many are still in use. This is the noisiest toilet and the most likely to become plugged because it has the smallest-diameter trap. It is usually found in the two-piece flush tank toilet; with an elevated tank, the flush required was only about 2.5 gal [9.5 L].
- *Siphon jet toilets:* These are in widespread use in North America, particularly in residences. A small priming jet hurries the bowl's contents into the trap and hastens the siphon action. With an elevated-tank (two-piece) toilet, this process requires a flush of about 3.75 gal [14.2 L]. With the more common close-coupled two-piece toilet, it requires a flush of 5–7 gal [19–26.5 L]. Siphon jet toilets are sometimes equipped with flush valves, which use less water (see the discussion of blowout toilets that follows).
- *Siphon vortex toilets:* These are especially suitable for low-velocity water (often as a result of low pressure). They are therefore also the quietest, making them a favorite wherever bathrooms are adjacent to sleeping areas or other acoustically sensitive spaces. The water enters the bowl off-center in such a way as to form a vortex; this swirling action cleans the sides of the bowl and the trap, helping the siphon action in emptying both bowl and trap. The one-piece flush tank toilets usually have the siphon vortex flushing action, which typically requires 6–8 gal [23–30 L].
- *Blowout toilets:* These combine very-high-velocity water and a simple trap to offer a noisy but very low-maintenance toilet dependent upon flush valves rather than tanks. They are very common in commercial and institutional toilet rooms, where large water supply lines and high pressures are available. The high velocity of the water lowers the quantity required from 3 to 4 gal [11.4 to 15.1 L] per flush.

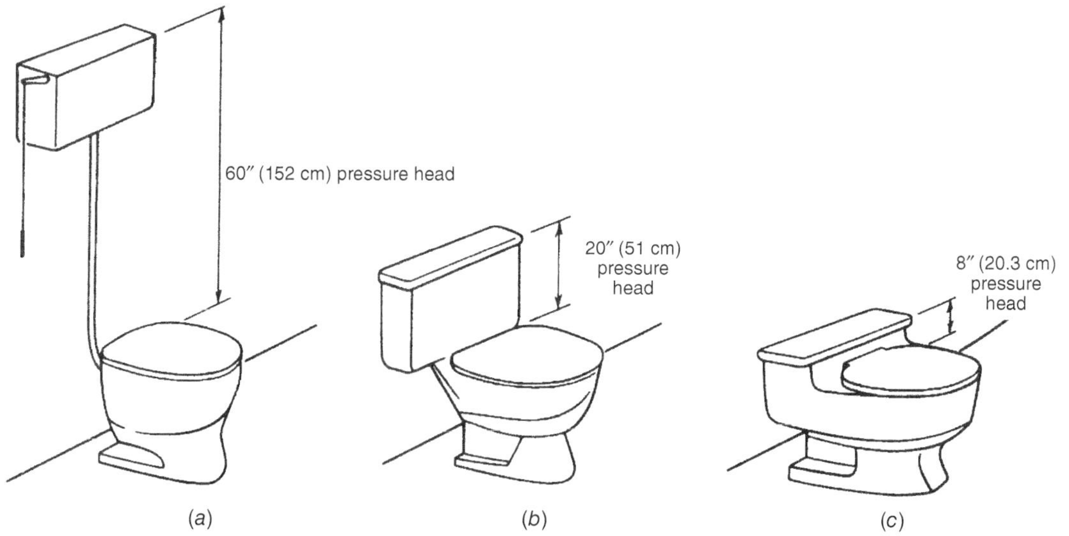

Figure 21.41 *Conventional flush tank toilets. (a) Two-piece flush tank toilet with an elevated wall-mounted tank and pull chain. (b) Two-piece flush tank toilet with the tank bolted to the bowl ("close-coupled"). (c) One-piece flush tank toilet with an integral tank. (From Milne, 1976.)*

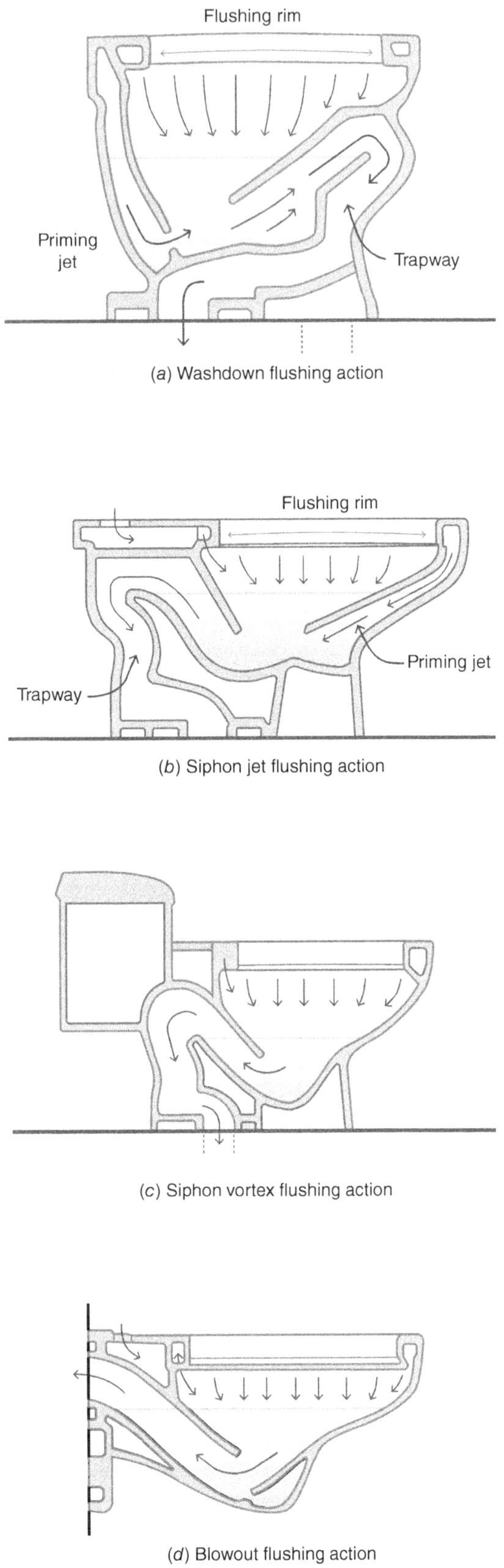

Figure 21.42 *Four common flushing actions that are built into toilets. (Drawn by Martin Lee after Milne, 1976.)*

Architects can easily specify toilets that require less water per flush and yet have very little impact on the user. With effort on the part of both users and architects, water can be recycled, or rainwater collected, to supply toilets with lower-quality water.

(f) Current Standard Toilets

In the United States, the federal efficiency standard requires no more than 1.6 gal [6.1 L] per flush. WaterSense labeled toilets use more than 1.28 gal [4.8 L] per flush maximum.

Toilets can be designed to use less water by varying several characteristics: the pressure of the water entering the bowl (generated either by the height of the tank above the bowl or by utilizing pressure from the water supply system or another source), the shape of the bowl, the way in which the bowl fills and empties, and the trap configuration.

Many toilets in the market today represent a compromise between the conventional "water waster" and the noisy operation and/or smaller water surface area that typifies so many low-consumption models. Standard toilet models that meet current minimum standards tend to use a conventional flushing action, and hence need enough height between the tank and the bowl to provide sufficient water pressure within the bowl during the flush.

A flush-valve (sometimes commonly referred to as a flushometer-valve, the name of the original valve-based invention patented by William E. Sloan in 1910) toilet is a tankless option very commonly encountered in nonresidential buildings. This type of toilet uses the building's water pressure to directly flush the toilet. Flush-value toilets are practical where durability, high-traffic use, and high-performance are priorities (e.g., high-traffic commercial spaces or public restrooms). Users activate a mechanism—using a handle, button, or sensor—that temporarily opens the flush-valve to release water. There are manual flush and automatic flush models of flush-valve toilets. Once activated, the diaphragm (or piston) lifts and allows pressurized water to flow into the toilet bowl. (Note: all conventional urinals use flush-valves.) Flush-valve toilets are typically noisy systems and may require significantly higher flow rates compared to tank-based conventional toilets.

In terms of flush-valve flow rates, projects teams may assume:

- conventional: 3.0–7.0 gal [11.4–26.5 L] per flush
- current standard: 1.6 gal [6.1 L] per flush maximum
- WaterSense labeled: 1.28 gal [4.8 L] per flush maximum

Note: WaterSense requires a 1.0 gal [3.8 L] per flush minimum.

(g) High-Efficiency Toilets

Many toilet models can function with far less water than current standard toilets. Collectively termed high-efficiency toilets (HETs), or low-flow toilets, these products will use 1.28 gal [4.8 L] per flush or less, depending on specific products, and can reduce toilet water consumption quite significantly. The terms ultra-high-efficiency toilets (UHETs) or ultra-low-flow toilets (ULFTs) are typically reserved for products that use less than 1.1 gal [4.2 L] per flush.

A heated controversy over low-flow toilets developed in the late 1990s due in part to steep price increases and consumer discontent over performance. It appears that regulation got ahead of technology for a time, resulting in a need for repeated low-consumption flushes to accomplish what one higher-consumption flush could do. To remove fecal matter, however, two 1.6-gal [6-L] flushes still approximately equal the old 3.5-gal [13.2-L] flush; to flush urine, one 1.6-gal [6-L] flush is adequate. Manufacturers are finding ways to change the entry point of water into the toilet bowl to increase swirl and thus waste-removal power.

Several of these toilets achieve water conservation by using a flushometer tank (Figure 21.43). Satisfactory flushing can be achieved with much less water by flushing with water entering at greater pressure. Instead of toilet tanks at ordinary air pressure, these tanks utilize water supply system pressure to compress air trapped within the tank. Water enters the bowl with much greater force with this combination of water under system pressure and compressed air. The flush is thorough, quick, and noisy. If the water supply pressure is greater than about 65 psi [448 kPa], some problems with excessive tank pressures may be expected. A pressure-reducing valve would be helpful in such cases.

Another water-saving technique is to redesign the bowl to hold less water. The siphon jet, gravity-fed toilet whose tank is shown in Figure 21.44 exposes a standing water surface of about $4\frac{1}{2} \times 6$ in. [114×152 mm], much smaller than that in Figure 21.43. Along with a very low 1.4-gal [5.3-L] flush, its trap is wide enough to pass a $2^1/_8$-in. [54-mm] ball.

Far lower water consumption is achieved when a central compressed air system is combined with water supply system pressure. The Microphor flush toilet (Figure 21.45) uses compressed air and 0.5 gal [1.9 L] of water per flush in a two-chambered toilet. When the flush lever is pressed, the water and waste in the bowl (upper chamber) are deposited into the secondary (lower) chamber, and 0.5 gal [1.9 L] of water (direct from the supply pipe) wash down the sides of the bowl to await the next user. The now-closed secondary chamber then is pressurized with compressed air, and its contents are deposited into a conventional sewer line.

Figure 21.43 Example of a 1.5-gal-flush [5.7-L], pressure-assisted, direct-fed siphon jet flush action flushometer tank, with a large 10 in. × 12 in. [250 mm × 300 mm] water surface area in the bowl: the Cadet Aquameter. (Courtesy of American Standard, Inc.)

Air compressors are needed, along with compressed air lines. A small compressor ($^1/_4$–$^1/_2$ hp [187–373 W]) with an accompanying air tank will operate up to three toilets. Although the compressor is noisy, the toilets themselves are no noisier than conventional toilets; they use the same plumbing lines. With such low flows, the designer may choose to increase the slope of waste lines; however, the 0.5-gal [1.9-L] quantity has proven sufficient to carry the waste in existing installations.

In contrast, the Envirovac flush toilet (Figure 21.46) uses a passive vacuum-assist technology and 0.38 gal [1.4 L] per flush. Because a central sewage tank (kept under vacuum by a pump) must be used, the drain line from the toilet to this tank may run horizontally without a slope or may even run vertically upward from the toilet. This can have significant architectural advantages when vertical clearance is tight, toilets are being added within existing structures, or toilets must be located below the level of the public sewer, as in marinas or basements. In tall buildings, significant savings in power for water supply pumping are achieved. Easier approval for building permits, reduced water hook-up fees, and lower monthly sewer charges can be substantial benefits. Furthermore, the central sewage holding tank can be flushed into the sewer at off-peak times—a benefit to the treatment

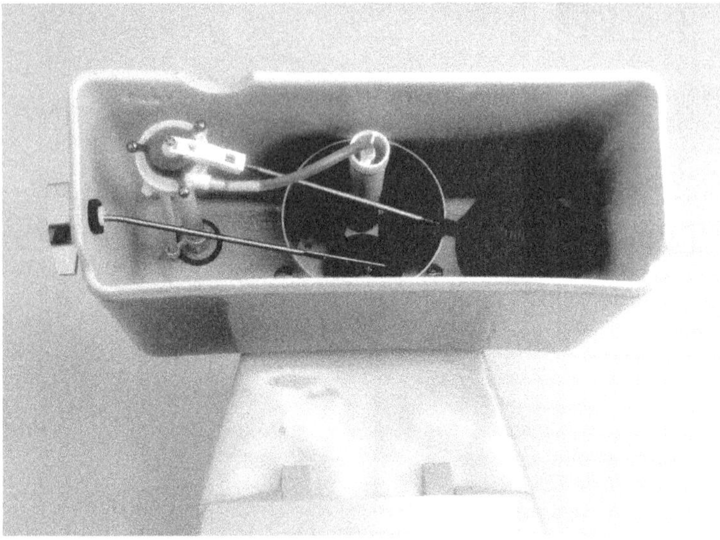

Figure 21.44 *A very-low-flow flush is achieved with a reservoir that meters about 1.4 gal [5.3 L] of water from the tank to the bowl during flushing. The standing water surface is about 4½ by 6 in. [115 by 150 mm]. The Ultra-One/G. (Courtesy of Eljer Industries.)*

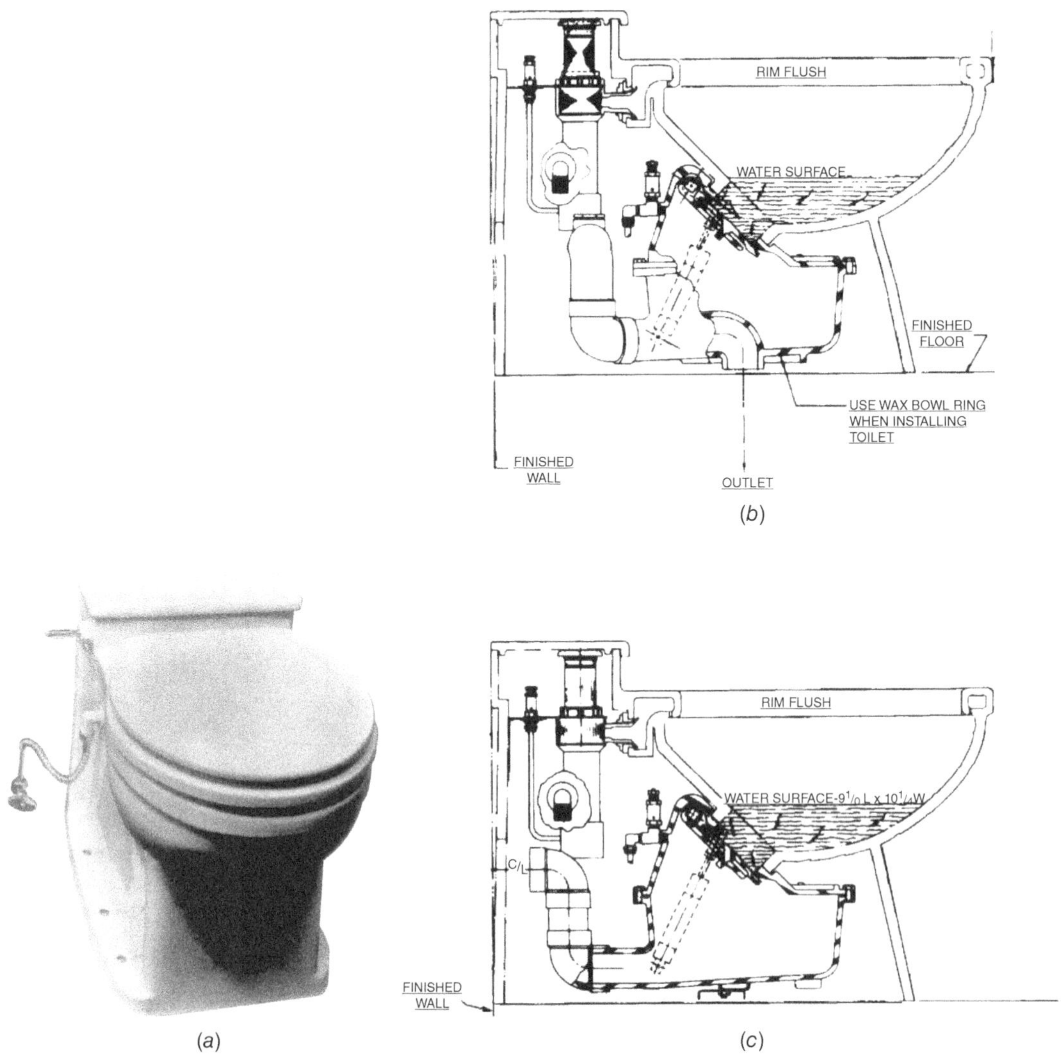

Figure 21.45 *(a) Compressed air is combined with a 0.5-gal [1.9-L] flush in the two-chamber Microphor toilet. (b) Section through the floor-discharge model. (c) Section through the wall-discharge model. (Courtesy of the Microphor Company.)*

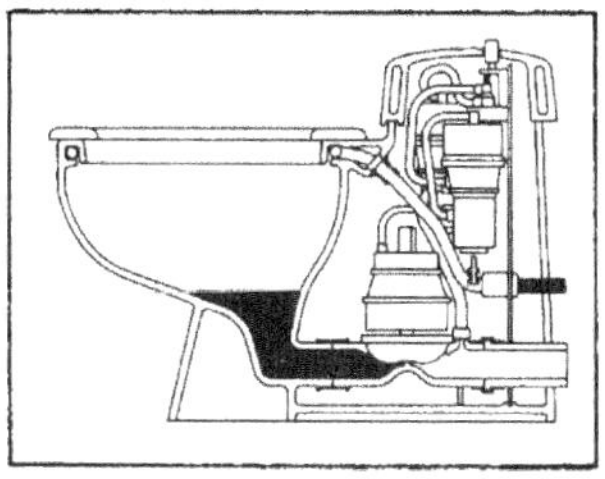

(a)

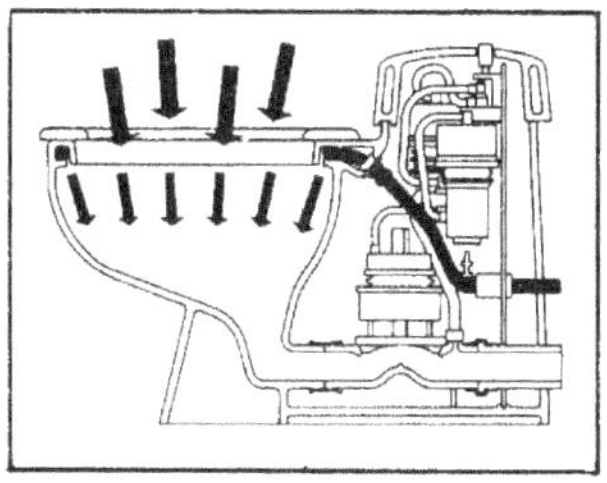

(b)

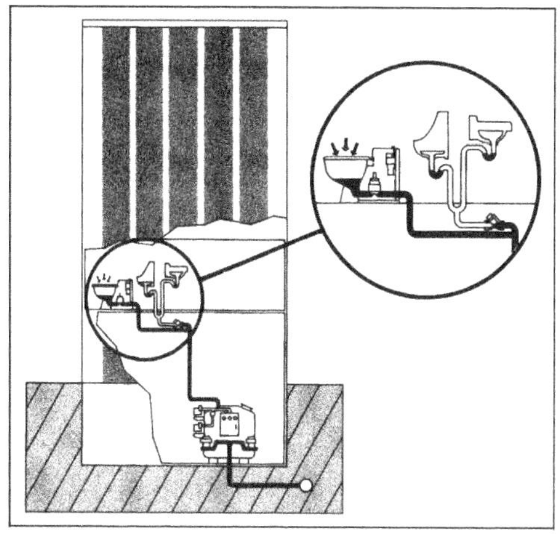

(c)

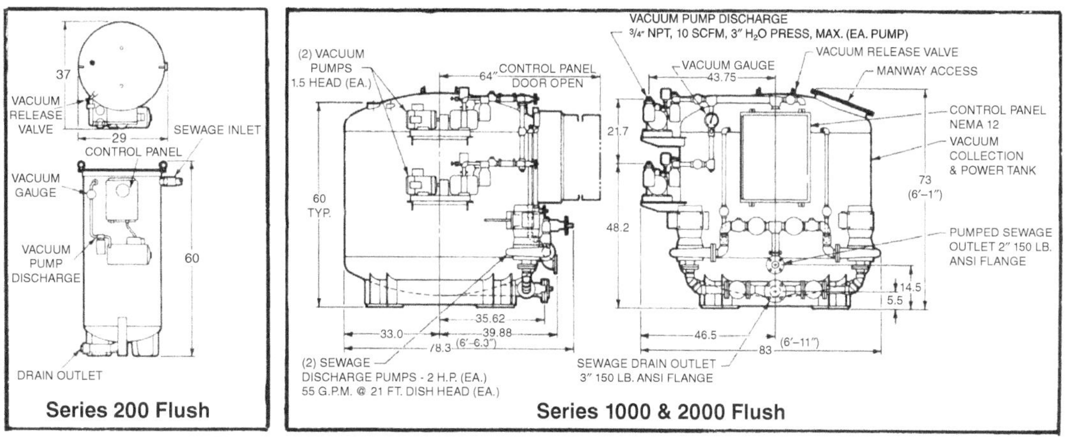

(d)

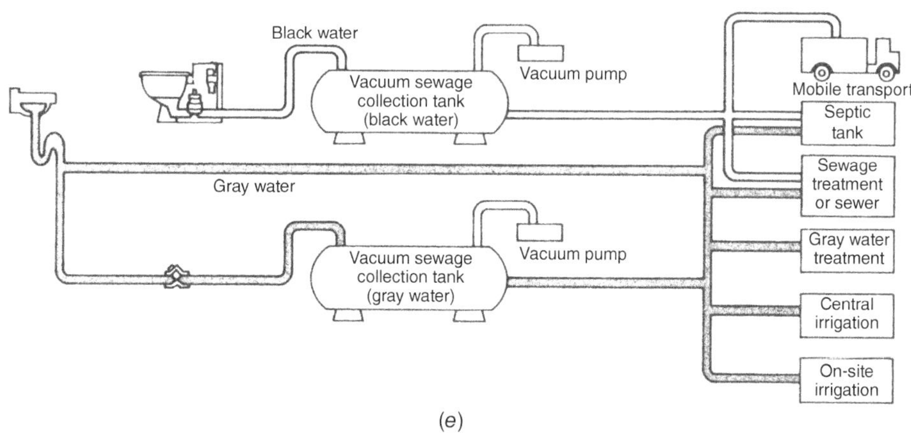

(e)

Figure 21.46 *The Envirovac system utilizes a vacuum and a 0.38-gal [1.4-L] flush. (a) Less than 0.38 gal [1.4 L] of water remain in the bowl until the toilet is used. (b) When the pushbutton flush is pressed, the discharge valve opens for about three seconds; a rush of air into the evacuated piping carries along sewage, odor, and airborne bacteria. Simultaneously, the washdown water flow begins. The flow continues for four seconds after the discharge valve closes, so that most of the 0.38-gal [1.4-L] flush remains in the bowl. (c) Toilets are directly connected to a central holding tank, kept under a vacuum by a pump. (d) Central tank/pump combinations can serve one house or a large building. (e) The separation of graywater and blackwater is another water-conserving opportunity. (Courtesy of Envirovac.)*

plant. These water-saving, architectural, and water treatment advantages must be weighed against the cost of the toilets and the tank/pump combinations, the space required for the tank/pump, the power used by the pump, and the possibility that a power failure will halt system operation.

Vacuum-assist systems can be installed for groups of buildings, such as subdivisions. Small pipe sizes, freedom from the need for continuous-sloping sewer lines, and conservation advantages for both water supply and treatment are the benefits. In addition, graywater (from kitchen, laundry, and bathing fixtures) can be kept separate from toilet water (blackwater) and thus made easier to recycle after moderate treatment.

(h) Flushing Controls

Water conservation is encouraged by the dual-flush (or dual-cycle) toilet, whose flushing mechanism allows a choice of fewer gallons for liquid wastes, more gallons for solid wastes. This simple mechanism, though beginning to spread in popularity here, is still more commonly seen outside the United States.

Another development is the automatic flush, common in public buildings and triggered by radiant heat from a body at the fixture or by light reflected off the user and back to the control. This "touchless" approach seems to promise more for hygiene than for water conservation, although it does prevent the flush valve from being held open too long by a careless user.

In general, plumbing fixtures will consume less water as the supply water pressure is reduced. For this reason, pressure-reducing valves are becoming popular as water conservation devices (when they would not otherwise be required to protect fixtures from overpressure). Installed on the supply line to a building, they can save water throughout the system.

Several newer low-consumption toilets use vertical flush valve sleeves, where the flush handle is in the center of the top of the toilet tank. The handle is lifted to initiate the flush. An advantage is that the less time the handle is raised, the less water enters the bowl. Again, less water may be admitted to flush away liquid waste, more water for solid waste.

(i) Waterless Toilets

Because these fixtures eliminate water entirely, they are described in Chapter 22.

(j) Appliances

Dishwashers, clothes washing machines, and other appliances are big users of water—and of energy to heat it.

In terms of dishwasher efficiency, projects teams may assume:

- conventional: 12–18 gal [45–68 L] per cycle
- current standard: 5.0 gal [18.9 L] per cycle
- ENERGY STAR: 3.2 gal [12.1 L] per cycle maximum

Older conventional residential dishwashers use 12–18 gal [45–68 L] per cycle, much of it heated well beyond the 120°F [48°C] typical of a household hot water supply. The current U.S. federal standard requires that dishwasher use not more than 5.0 gal [18.9 L] per cycle (gpc). WaterSense does not yet label dishwashers but recommends ENERGY STAR certified products, which are significantly more efficient. High-efficiency dishwashers will consume less water than typical "manual" dish washing using a kitchen faucet.

Conventional top-loading clothes washing machines use 40–55 gal [151–208 L] for full-sized loads. Front-loading machines, which are common in the market today, are typically more efficient than top-loading machines—using between 10 and 15 gal per washing cycle. High-efficiency (HE) top-loading and front-loading products will use much less energy and water than their standard counterparts. In the past, "suds saver" features allowed soapy, hot wash water to be reused. Many newer washers allow for a wider selection of water quantities and temperatures—a feature that can save considerable water and energy. Horizontal-axis (front-loading) washers have greatly reduced the need for hot water per wash cycle.

Given the wide range of clothes washers in the marketplace, comparing water efficiency between product options can be challenging. Capacities and performance vary by commercial and residential equipment and by top-loading and front-loading washers. On the residential side, standard washers have loading capacities at or greater than 1.6 ft^3 [0.05 m^3] and compact washers are below this threshold. It is common to see residential clothes washers with capacities between 3.5 and 5.0 cubic feet [0.10 and 0.14 m^3]. On the commercial side, top-loading washers can have capacities up to 4.0 cubic feet [0.11 m^3] and front-loading washers will typically be as high as 3.5 cubic feet [0.10 m^3].

The DOE maintains both energy and water consumption criteria for clothes washers (EPA, 2024a). Given the varying ranges of capacities, two specific metrics have been developed to help assist with comparing equipment energy and water performance. The integrated modified energy factor (IMEF) is a measure of the energy consumed per unit of capacity per average washing cycle (ft^3/kWh/cycle). The integrated water factor (IWF) is a measure of the water consumed per unit of capacity per average washing cycle (gal/ft^3/cycle).

As of 2018, the DOE's criteria for the maximum IWF of clothes washers are:

- residential, top-loading, standard: 6.5
- residential, top-loading, compact: 12.0
- residential, front-loading, standard: 4.7
- residential, front-loading, compact: 8.3
- commercial, top-loading: 8.8
- commercial, front-loading: 4.1

WaterSense does not yet label clothes washers but recommends ENERGY STAR certified products, which are significantly more efficient, as evident by the following maximum IWF criteria:

- residential, front-loading: 3.2
- residential, top-loading: 4.3
- commercial: 4.0

As such, an ENERGY STAR certified front-loading residential clothes washer with a capacity of 4.4 cubic feet and an IWF of 3.2 gal/ft^3/cycle will use 14 gal [53 L] per wash—significantly less than a conventional top-loading washer, which may use three times as much water per cycle.

For any appliance that requires considerable water consumption, it is important to assess the full range of available products and select equipment that will save both energy and water without sacrificing requisite performance.

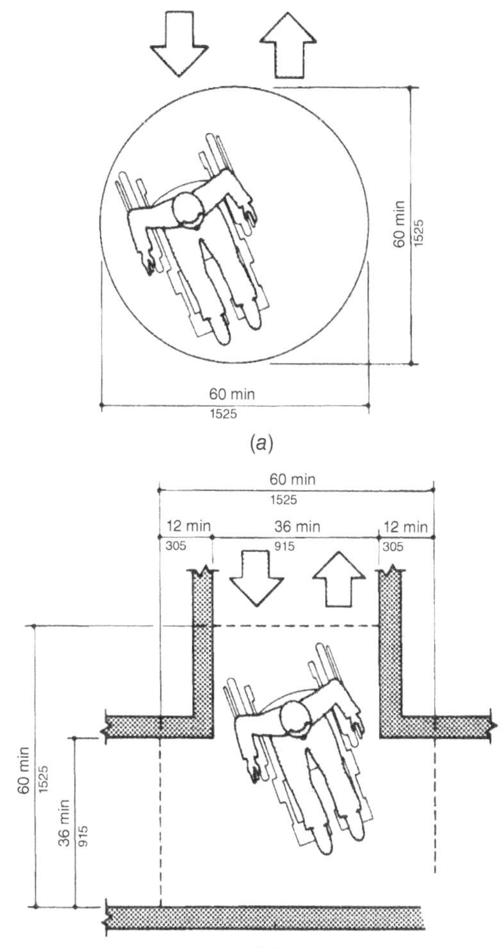

Figure 21.47 *Wheelchair turning space, minimum requirements. (a) A 60-in. [1.525-m]-diameter space. (b) T-shaped for 180° turns. Dashed lines indicate the minimum length of clear space required on each arm of the T-shaped space. (Reproduced from DOJ, 2010.)*

21.8 FIXTURE ACCESSIBILITY AND PRIVACY

After plumbing fixtures designed for the human physique and for water conservation have been chosen, they are placed in a space that has several unusual environmental needs. Much of the information in this section is reproduced from *ICC A117.1-2017: Accessible and Usable Buildings and Facilities* with permission from the International Code Council (ICC). Another useful guide is the *Americans with Disabilities Act (ADA) Accessibility Guidelines for Buildings and Facilities*, available from the U.S. Architectural and Transportation Barriers Compliance Board (U.S. Access Board, 2004).

(a) Accessibility

Designers must allow room for wheelchair maneuvering into and within toilet rooms. The minimum clear floor area should be a circle 60-in. [1.525 m] in diameter. Minimum clearances for wheelchair maneuvering are shown in Figure 21.47. Provisions for people in wheelchairs also influence the heights at which items such as light switches, electrical receptacles, paper towels, and water controls should be installed. In general, all objects to be reached by hand in toilet rooms should be placed more than 15 in. [380 mm] and less than 48 in. [1.22 m] above the floor.

Floor space requirements and mounting-height limitations also apply to the various plumbing fixtures in bathrooms, at least one of which (in each case) should be accessible to people in wheelchairs in most buildings. Requirements for drinking fountains (which often must be located outside toilet rooms), lavatories, bathtubs, and shower stalls include clearances beside the fixture (and clearance below lavatories) and grab bar locations; see *ICC A117.1* for details.

Accessible toilets (water closets) have particularly large floor space requirements, as shown in Figure 21.48. There is a conflict between the

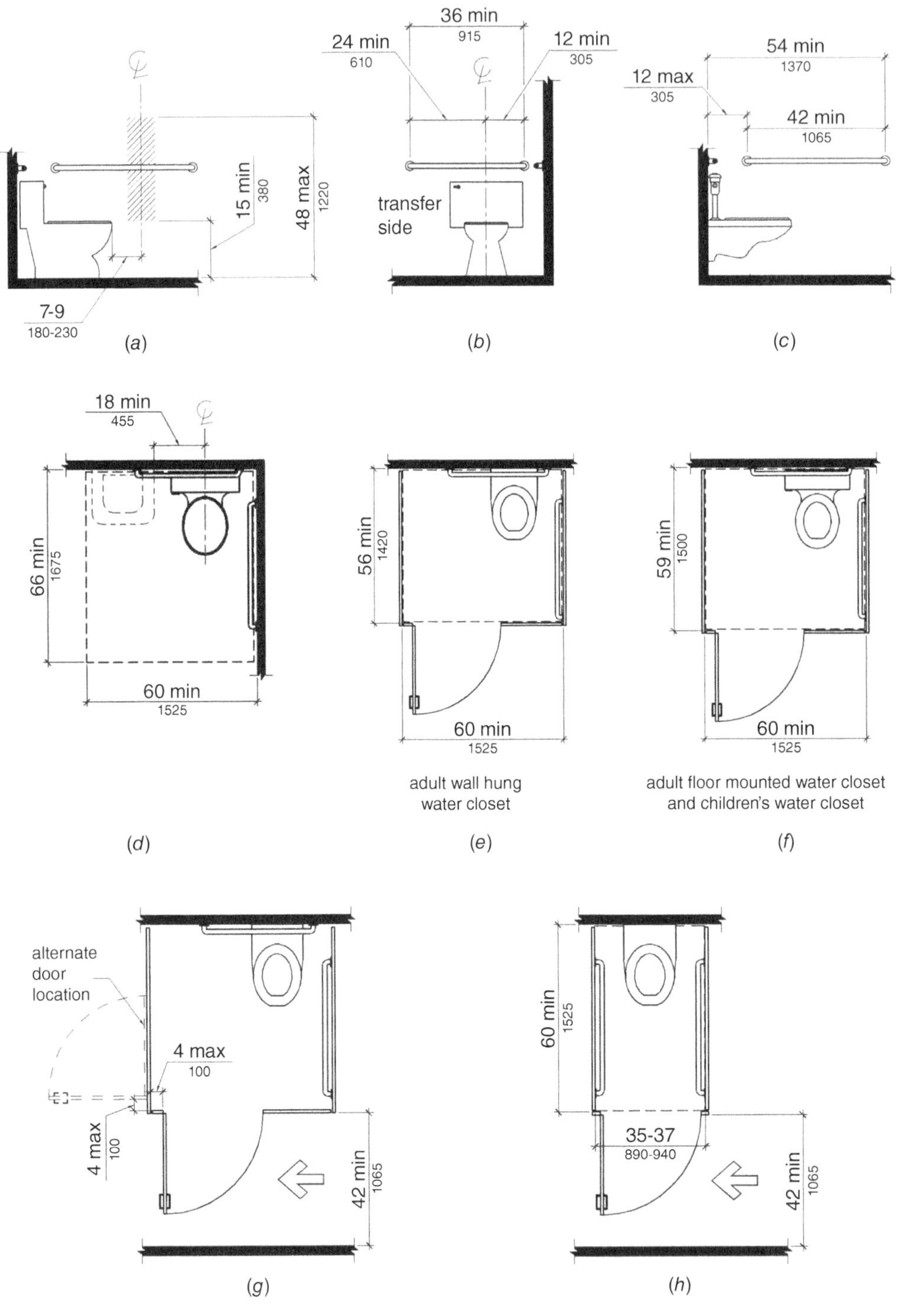

Figure 21.48 *Clearances and grab bar placements for toilets. (Reproduced from DOJ, 2010.)*

lower mounting height needed for proper defecation, the conventional height of 15 in. [380 mm], and the recommended height for handicapped users of 17–19 in. [430–485 mm]. The doors on stalls designated for wheelchair access should swing out rather than in. Provision for a wheelchair sitting beside the out-swinging door must also be made. No doors should swing into the clearance space required for any fixture.

Where six or more water closets (or water closets plus urinals) are located within the same toilet room, both the "standard stall" and one of the "alternate stalls" are required. This impacts the design of public toilet rooms.

Urinals for handicapped users should either be of the floor-mounted stall type or, if wall-hung, have an elongated rim at a maximum of 17 in. [430 mm]

above the floor. A clear floor space of 30 × 48 in. [760 × 1220 mm] is required in front of the urinal, and its flush controls should be no more than 44 in. [1.12 m] above the floor.

(b) Privacy

A desire for visual privacy is easily understood and at least minimally provided by partial-height partitions. However, acoustic considerations are another important aspect of bathroom design. Because of their nonabsorbent surfaces, bathrooms are frequently the most reverberant spaces in a building. They encourage the singer or the whistler and may even serve as practice rooms for the family musician. Yet they can also intimidate a person who prefers to be quiet, such as a guest using the toilet while the rest of the dinner party sits quietly in the adjacent room.

Two common design responses are isolation and masking sound. Bathrooms can be separated from acoustically sensitive spaces by closets or hallways. With careful attention to sufficiently massive construction and/or other construction details, the doors, walls, ceilings, and floors of bathrooms can be constructed so as to reduce the passage of sound to acceptably low levels. (These considerations are discussed in Chapters 22 and 23.) Sound isolation also depends upon attention to detail: no cracks around the bathroom door, no back-to-back electrical outlets between bathrooms and adjacent spaces, no air grilles into ducts that have other grilles nearby, and no other open windows near a bathroom window. If such goals are unattainable or if additional acoustic security is desired, masking sound can be added as needed within the bathroom. All too often, the masking noise is provided by repeated flushing of the toilet (itself an embarrassing sound for some) or the running of water in the lavatory. However, a noisy ventilating fan would serve the purpose, as would a music source such as a radio. An elegant, expensive touch would be a recirculating fountain gracing one corner; it might provide splashing sounds only when the light or fan is on.

The final consideration in bathroom acoustics is the sound of water moving within pipes. Where such pipes are exposed or barely concealed within, or firmly attached to, thin walls, perceptible sounds are highly likely. Where water noise is undesirable, pipes should be wrapped, resiliently mounted, and/or located in a less acoustically critical wall. But remember that running water can also be a conservation signal as well as a masking sound. Although many of us may have experienced this sudden acoustic signal that a bathroom is in use, few have experienced it quite like the occupants of the Aluminaire House on Long Island, cited by Reyner Banham (1969). Designed at the height of the modernist expression of mechanical services, this house featured a dining table cantilevered from the exposed waste stack serving the bathroom above.

21.9 WATER DISTRIBUTION

This section looks at ways to supply water throughout buildings at pressures sufficient to operate plumbing fixtures. Smaller buildings may be served simply by the pressure available in water mains (or pressure tanks fed by pumped wells). This is called upfeed distribution, because the water rises directly from mains to the plumbing fixtures. For taller buildings, several other options are available: pumped upfeed (in which pumps supply the additional pressure needed), hydropneumatic (in which pumps force water into sealed tanks, compressing the air within; this maintains the needed water pressure), and downfeed (in which pumps raise the water to storage tanks at the top of a building, and water then drops down to the plumbing fixtures).

In cities with municipal water supply systems, water is distributed through street mains at pressures varying at the main from about 50 psi [about 350 kPa] to about 70 psi [about 480 kPa]. For low-rise buildings of two or three stories, these pressures are adequate to act against the static pressure of water standing in the vertical piping, overcome the frictional resistance of water flow in the pipes, and still deliver water at the pressure required to operate plumbing fixtures. The flow pressure required at fixtures varies from 5 to 20 psi [35 to 138 kPa], depending on the type of fixture—for example, a basin faucet, shower head and faucet, or water closet. (Table 21.14 in Section 21.11 lists minimum flow rates and pressures for typical plumbing fixtures.)

(a) Static Pressure

The pressure exerted at the bottom of a stationary "head" of water is related directly to its height. One cubic foot of water weighs approximately 62.4 pounds. Consider a "cube" of water 1 ft square and 1 ft high—its weight (62.4 lb) rests on a bottom area of 1 ft^2 (144 in^2). The static pressure at the bottom is 62.4/144 = 0.433 psi [3 kPa]. Reciprocally, 1 psi of pressure will sustain a static (stationary) column of water 1/0.433 = 2.3 ft in height. [In SI units, 1 m of head = 10 kPa.] When fixture pressure and pressure lost in friction-of-flow in pipes are considered, the problem becomes more complex. Example 21.6 in Section 21.11 illustrates this problem in the calculation of pipe size. For upfeed and downfeed distribution, the relationship of heights and static pressure is one controlling design factor.

(b) Upfeed Distribution

In small, low buildings of moderate water demand, it is seldom difficult to achieve the proper flow pressure at fixtures by the use of an upfeed system. Pressure at the fixtures is usually more than required. When this causes an inconvenient splash, as at lavatory basins, a flow restrictor can be used in the faucet outlet.

Consider the typical upfeed system shown in Figure 21.49, beginning at the point where supply water enters the building. In cold climates, water in the service entry pipe must not freeze. The pipe must therefore be below the frost line of frozen ground. This could vary from 0 to 7 ft [0 to 2 m], depending upon the geographical location. The onset of winter in cold climates requires the closing and draining of pipes supplying the hose bibbs (and other external piping) by means of a stop-and-waste valve. Houses left unheated in cold-winter weather must be entirely free of water that could freeze and burst the pipes. Note the drain valve at every low point in the system. House shutoff controls are usually located at the main, at the curb, and within the house.

Meters have recently taken on a new role. Along with measuring the water quantity for which the occupant is to be charged, they now sometimes serve a restrictive function. During water shortages, they can be used to indicate water use in excess of established limits, beyond which fines are imposed and, in some cases, the water supply reduced by valves controlled by the water company.

In-house treatment is often performed to reduce water hardness that could clog piping and equipment or to neutralize acidity—a source of corrosion. During the short periods when the treatment tanks are valved off for backwashing or other servicing, the bypass shutoff valve is opened.

From this point on, the water continues under pressure to do the following:

1. Supply makeup water to the space-heating boiler, as required.
2. Supply water to and pressurize the cold water mains and branches, including the garden hose bibbs.
3. Supply water to and pressurize the domestic hot water system through the hot water heater, the hot water storage tank, and the mains, branches, and circulating lines.

The air-filled expansion chambers on cold water runouts absorb and reduce the shock of water hammer—the force exerted by decelerated flowing water that shakes and rattles pipes when faucets are shut off abruptly. On hot water runouts they perform

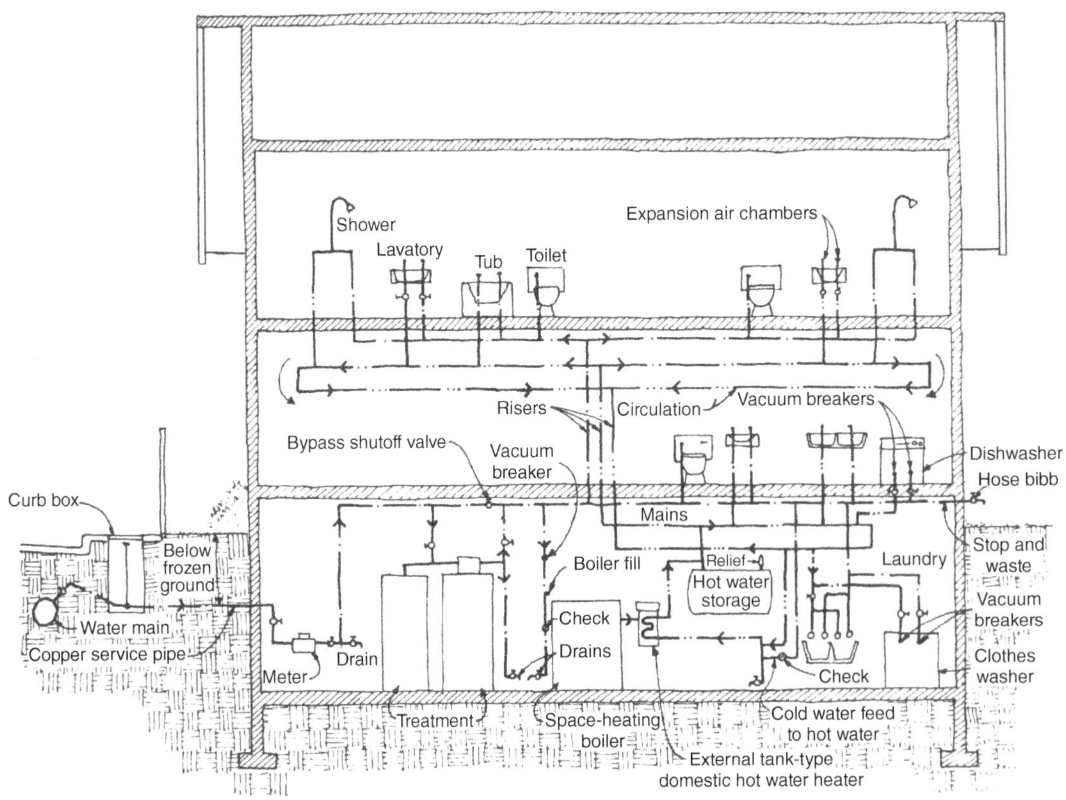

Figure 21.49 *Upfeed water distribution using pressure in street mains. A schematic section of the water services in a typical residence.*

the same function, as well as allowing for the expansion of the hot water as it increases in volume with increasing temperature. Vacuum breakers prevent backflow of polluted water into pipes carrying potable (hot or cold) water. Water from all fixtures and appliances, such as dishwashers, clothes washers, and boilers, is thus isolated.

Figure 21.49 shows all parts lying in a two-dimensional plane. In a real building, such a system is three-dimensional. For instance, economy of piping would suggest that, if possible, the kitchen and lavatory, as well as the two upstairs bathrooms, should be placed back-to-back.

(c) Principles of Downfeed Distribution

Water pumped directly from the street main (or from a basement "suction tank" filled by gravity from the main) is lifted to a roof-storage tank. In cold climates, the water in an outdoor tank is kept at temperatures above freezing by heating coils in the tank. The fact that water pressure increases with distance below the tank water level is suggested by the construction of the tank (shown in Figure 21.50). The iron rings, tensioned by adjustable threaded clamps, become more closely spaced toward the bottom of the tank, where the greater water pressure makes it increasingly difficult to restrain the vertical wooden staves of this cylindrical barrel.

Tanks like these add interest, though perhaps not beauty, to a building's silhouette. A view-blocking screen enclosing this tank would have to be about 24 ft [7.3 m] high. Architecturally, such a rectangular lump on the roof would not be much of an improvement. Yet in the century following the appearance of buildings such as shown in Figure 21.50, our technology has become more complex. Presently, for most high-rise buildings, including many 60 stories and more in height, an entire rooftop crowded with equipment and technical facilities is needed to serve the stories below (or the uppermost zone of a building). The items could include:

- water storage tanks
- two-story penthouses over elevator banks
- chimneys and/or flues
- numerous plumbing vents
- exhaust fans
- air-conditioning cooling towers
- cantilevered rolling rig to support a scaffold for exterior window washing
- perimeter track for a window-washing rig
- photovoltaic modules and/or solar thermal collectors

Thus, since the 1960s, tall buildings commonly have a band or screen two stories (or more) high above the structural roof. The best view locations are thus consigned to mechanical equipment. It might be said that it all began with the need for elevated water storage tanks.

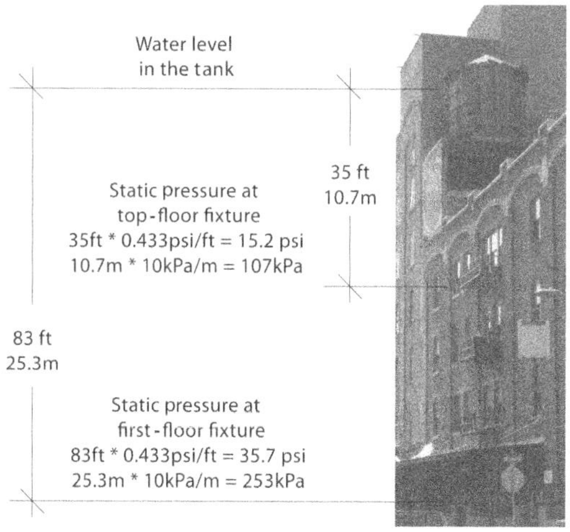

Figure 21.50 *Mid-rise building with gravity downfeed distribution. Although upfeed distribution from a street main is possible in buildings higher than two or three stories, when street main pressures drop below the usual 50 psi [345 kPa], combined with heavy use, very low-pressure results in fixtures in the upper stories. Therefore, water is pumped from the main to elevated roof tanks placed high enough to ensure reasonable pressure at the top story and ample pressure at the bottom of the downfeed run.*

(d) Tall Building Downfeed Distribution

Figure 21.51*a* shows a medium-rise building in which one elevated tank can serve all of the lower floors. For taller buildings, it is advisable to separate groups of floors into zones with a maximum height (for plumbing pressure limits) of about 150 ft [about 45 m]. This zone-height limitation is based upon the height-to-static pressure relationship. At the top of the zone (about 35 ft [10 m] below the storage tank), the minimum desirable pressure is probably at least 15 psi [103 kPa]. At the bottom of the zone, the maximum desirable pressure is perhaps 80 psi [552 kPa]; above this pressure, damage to fixtures might occur.

In I-P units:

$$80\text{ psi} - 15\text{ psi} = 65\text{ psi difference}$$

$$65\text{ psi difference} \times 2.3\text{ ft/psi} = \text{about } 150\text{ ft}$$

In SI units:

$$552\text{ kPa} - 103\text{ kPa} = 449\text{ kPa difference}$$

$$449\text{ kPa difference} \times 0.1\text{ m/kPa} = \text{about } 45\text{ m}$$

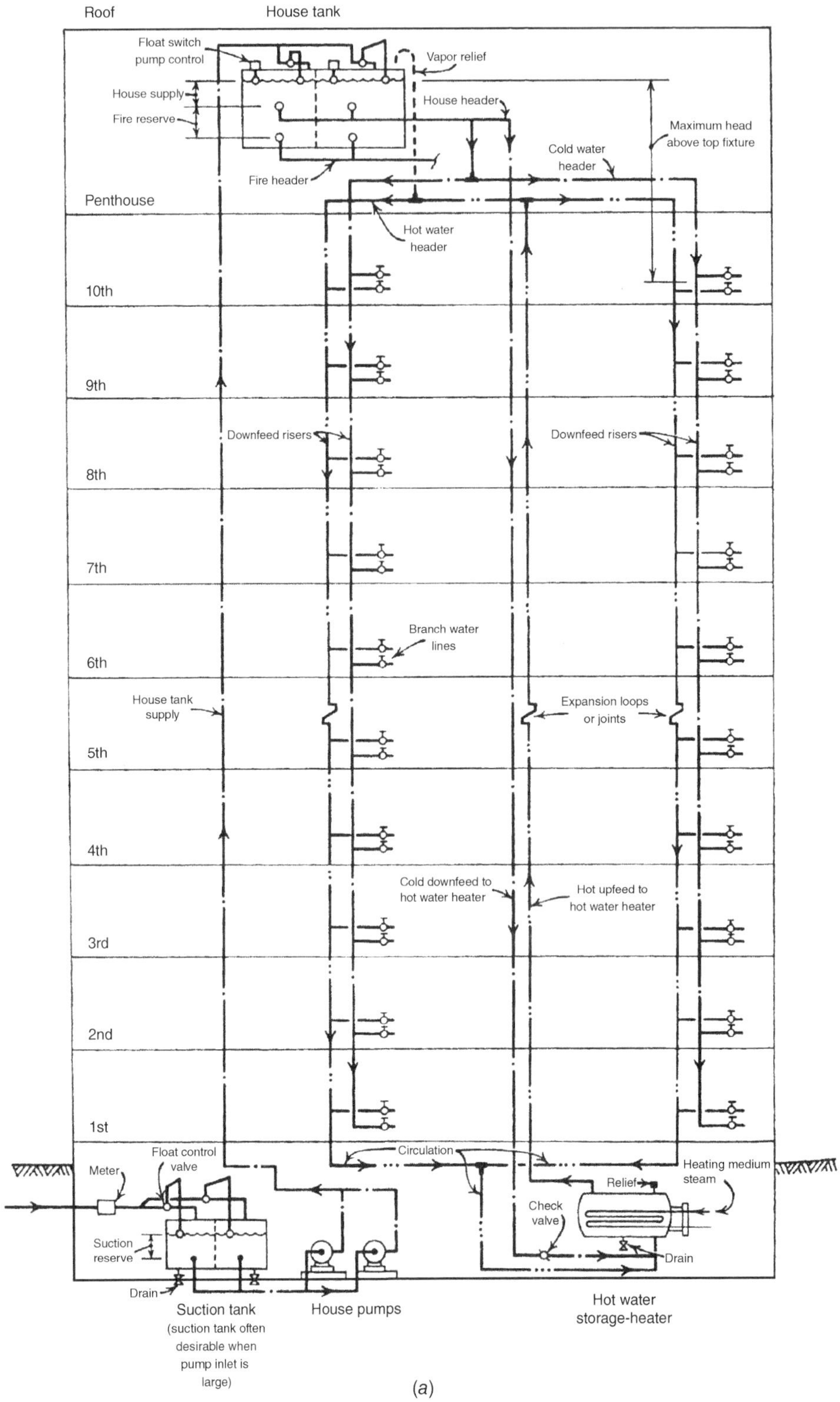

(a)

Figure 21.51 *(a) Downfeed water distribution, schematic section, part of the water services for a 10-story building. Hot water circulation moves from the hot upfeed in two directions at the hot water header, then down to the tank through the two downfeed hot water risers. For details of one type of steel house tank, a typical centrifugal house pump, expansion joints, and expansion in the hot water riser, see Figures 21.52 and 21.53 (b) Downfeed water distribution, schematic section, part of the water services for a much taller zoned building. Zone tanks include a fire reserve; standpipes are omitted from this drawing. For details of the steam-type domestic water heater, see Figure 21.24.*

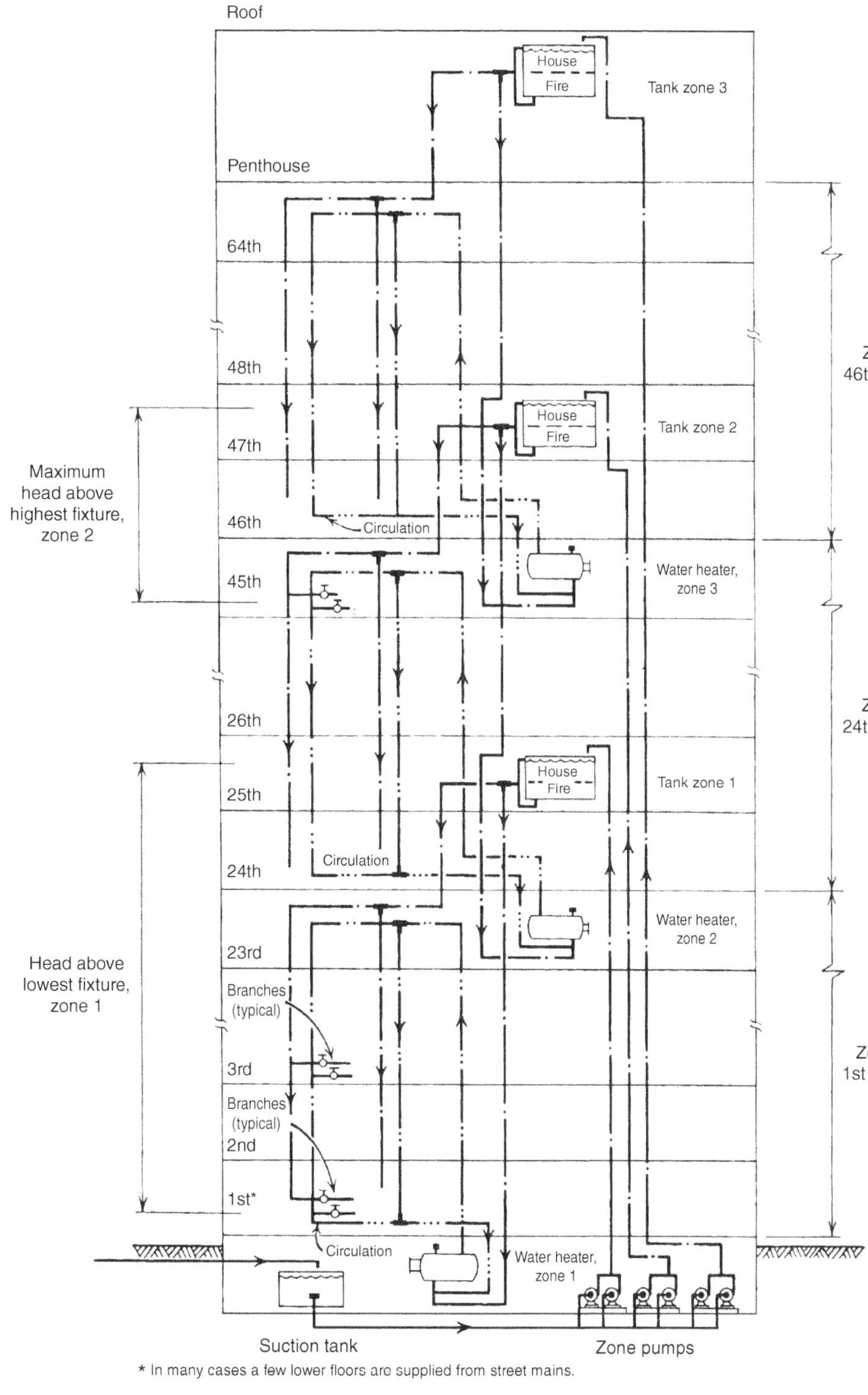

(b)

Figure 21.51 *(continued)*

With pressure-reducing valves at lower floors, however, these zones can be much higher, as shown in Figure 21.51*b*.

Consider the system described in Figure 21.51*a*, beginning with the elevated tank. The lower part of the tank often serves as a reserve space to hold a supply of water for a fire-extinguishing system. In this case, only the water in the upper part is available for use as domestic (service) water. The amount stored must be enough to supplement what the pump will deliver during the several daily hours of high demand that occur in most buildings. The pump then continues, often for several hours, to replenish the house supply that has become partially depleted during the busy period. The suction tank is a buffer between the system and the street mains. It usually holds enough reserve to allow the pumps to make up the periodic depletion in the house tank. It refills automatically by flow from the street main that, consequently, will not suffer as much of a drop in pressure as it would if it were connected directly to the suction side of the house pumps. Neighboring water users are protected from the adverse effects of sudden demands within adjacent large buildings (Figure 21.52).

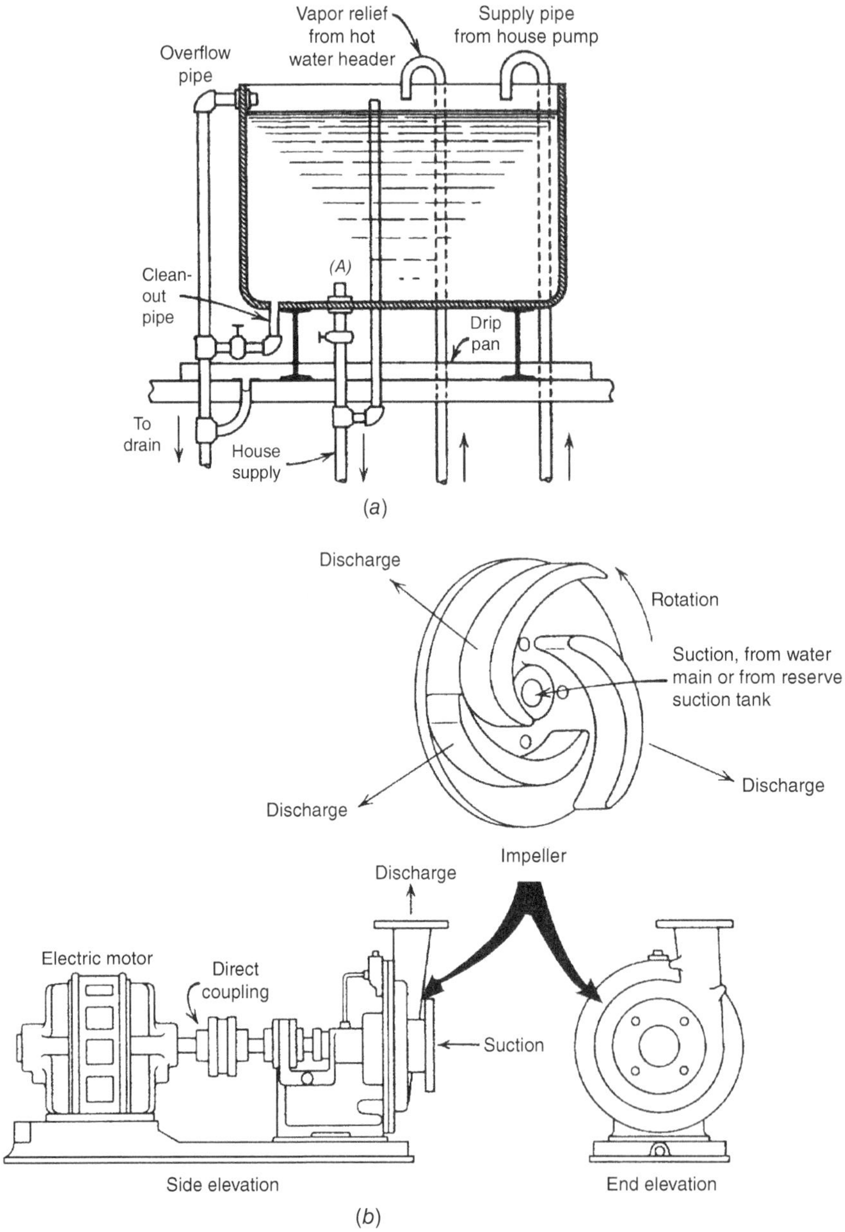

Figure 21.52 *(a) House tank in an elevated position for downfeed by gravity. Sediment in the tank is drawn off through the clean-out pipe and is prevented from entering the house water supply by the pipe projection at A. Water for fire reserve could be provided by additional piping or by a separate tank. (b) Centrifugal house pump, commonly used to fill an elevated house tank at the top story or at an intermediate mechanical floor. The electric motor responds to a float-switch pump control at the tank.*

House tanks and suction tanks are sometimes made of steel plate and divided vertically in half, each half having identical piping and controls. Hence, one-half of the tank can be cleaned out at a time of low demand without shutting down the entire system. One full-capacity pump is supplemented by an equivalent standby pump for alternate use. Because there is no suction lift below the pump, or any fixture pressure at the top of the house tank supply, the head against which the pump works is the sum of the distance from the suction tank water level to the top of the house tank and the feet of head equivalent to the friction loss in the tank supply pipe. For this kind of service, the vertical piping is on the order of 3 or 4 in. [76 or 102 mm] in diameter for large buildings. Sizes are established by formal calculations.

The house supply is fed by a short pipe from the house header to the cold water header that circles the top story and connects to many downfeed cold water risers. For simplicity, Figure 21.51*a* shows only two risers and also omits many valves and controls.

Figure 21.51*b* is even more simplified. The circulating hot water originates as cold water at the house tank header, then takes quite a long route, descending to the bottom of the hot water heater, rising to seek its own level at the hot water header, and becoming available there for hot water downfeed on demand. All of this occurs as flow below the general pressurizing effect of the house tank. In effect, when there is a cold- and hot-water demand on a story near the top of the building, the cold water makes a short trip down to the faucet, while the hot water goes through three vertical pipes instead of one.

This arrangement, with tank above and heater below, is used in multiple forms for very tall buildings. The zones are quite independent; the only common service is that provided by the general suction tank. With this zoning method, problems of pipe expansion, large pipe sizes, and high pressures in lower stories are minimized. Commonly, 2 1/2 stories, or about 35 ft [about 10 m], constitute the minimum pressure head above the top fixture served by any zone tank. The static pressure created at the fixture is thus 35 × 0.433 = 15 psi [10 × 10 = 100 kPa]. If, during flow, not too much pressure is lost in friction, flushometers (flush valves) can be placed at this level, although flush tanks, because of their lower pressure demand, must often be accepted. The opposite problem occurs at the bottom of the zones, where excessive pressures must be reduced at the fixtures. In zone 1 of Figure 21.51*b*, first-floor fixtures are below a head of 24 1/2 stories, or about 149 psi [1027 kPa] of static pressure. Pressure-reducing valves must be used, and fixture control valves must be throttled.

(e) Pipe and Tube Expansion

The range of temperature experienced by hot water supply piping, from the normal indoor air temperature of about 70°F [21°C] to that of service hot water (which often exceeds 160°F [71°C]), can be 90°F [50°C], with resulting pipe expansion (Table 21.12). This longitudinal elongation of pipe, though negligibly small in houses, can be appreciable in a tall building. Two methods of allowing freedom for this longitudinal motion in long runs of expanding hot water piping are shown in Figure 21.53*a*. These devices preclude the buildup of excessive stresses in the pipe material and the tendency of the pipes to buckle laterally.

EXAMPLE 21.5 A 20-story zone in a tall building has a height of 280 ft [85 m]. What will be the increase in length of a copper tube carrying "service hot water" (DHW) when its temperature increases from 70° to 160°F [21° to 71°C]?

SOLUTION

The difference in temperature is 90°F [50°C]. Approximate elongation per 100 ft for a 90°F increase is 1.01 in.

TABLE 21.12 Thermal Expansion of Pipes

Temperature Increase		Approximate Expansion of Piping Material in. per 100 ft [0.83 mm per meter] of Length					
°F	°C	Carbon and Carbon Moly Steel	Cast Iron	Copper	Brass and Bronze	Wrought Iron	Plastic
20	11.1	0.15	0.15	0.22	0.22	0.15	0.5
40	22.2	0.3	0.3	0.44	0.44	0.3	1.1
60	33.3	0.45	0.45	0.67	0.67	0.45	1.7
80	44.4	0.6	0.6	0.89	0.89	0.6	2.4
100	55.5	0.76	0.7	1.12	1.12	0.76	3.3
120	66.7	0.91	0.84	1.35	1.35	0.91	4.3
140	77.8	1.07	1.0	1.58	1.58	1.07	5.3
160	88.9	1.22	1.13	1.81	1.81	1.22	6.2
180	100	1.38	1.27	2.04	2.04	1.38	—
200	111.1	1.55	1.4	2.27	2.27	1.55	—

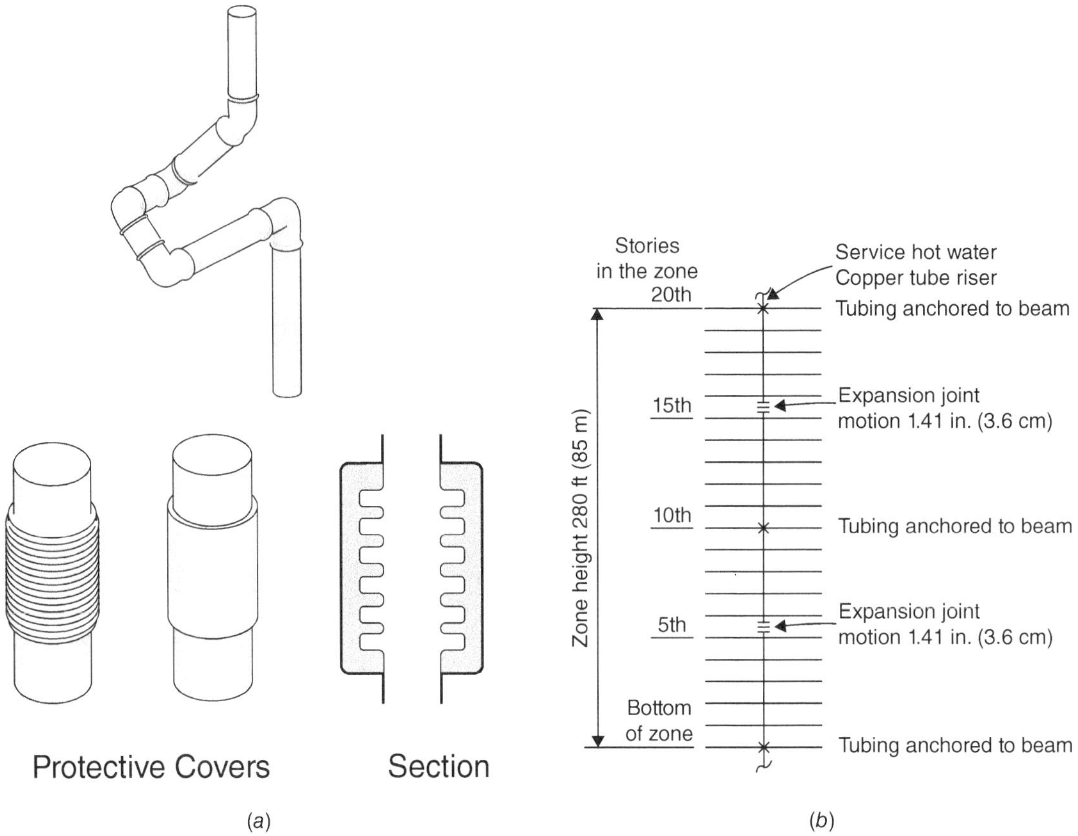

Figure 21.53 *(a) Accommodation for the expansion of hot water piping or tubing. Top: Expansion joint of pipe and fittings. Bottom: A manufactured product in elevation and section. (b) Suggested scheme for location of the points of anchorage and expansion for service hot water tubing in a 20-story zone (Example 21.5). (a: Drawn by Martin Lee.)*

(interpolating in Table 21.12).

$$\text{expansion} = 280 \text{ ft} \times (1.01 \text{ in.}/100 \text{ ft}) = 2.83 \text{ in.}$$

$$[85 \text{ m} \times 0.83 \text{ mm/m} = 71 \text{ mm}]$$

■

There are a number of ways of providing for this expansion. The one shown in Figure 21.53*b* accepts this motion at two locations, which would make the expansion in each case 1.41 in. Equidistant anchorage to fix the tubing is provided at the bottom, the 10th floor, and the 20th floor. The support of the vertical riser at floors other than those at which it is anchored could consist of clamps of the type illustrated later in Figure 21.58, supported on springs.

(f) Pumped Upfeed Distribution

This distribution system (Figure 21.54) is for medium-sized buildings—those too tall to rely on street main pressure but not so tall as to necessitate heavy storage tanks on the roof. The equipment associated with this system can deliver water at rates varying from those needed for two or three faucets to full building demand while maintaining at each outlet a pressure within 2 psi [14 kPa] of the design pressure for that outlet.

The installation shown in Figures 21.54*a, b* uses a triplex pump group. According to demand, one, two, or three pumps will operate. Because each pump is of the variable-speed type, virtually an infinite number of delivery rates can be achieved within the zero to maximum design range. The pumps operate in sequence. When a very small rate of demand occurs, the smallest or "jockey" pump starts in response to a low voltage impressed on its motor. All operations are triggered and adjusted by the pressure sensor at the base of the riser. The jockey pump continues to run until it has reached its maximum delivery rate, at which time the first of the larger pumps cuts in, joined by the other larger pump when required. Sequential operation of the three pumps, each increasing in delivery as called for by the sensor, meets the requirement for an increasing supply at nearly constant pressure.

Wear on the two large pumps is equalized by the assigning of "lead" and "lag" positions. For a period of 24 hours, one of the large pumps holds the lead position and starts after the jockey pump, giving the other large

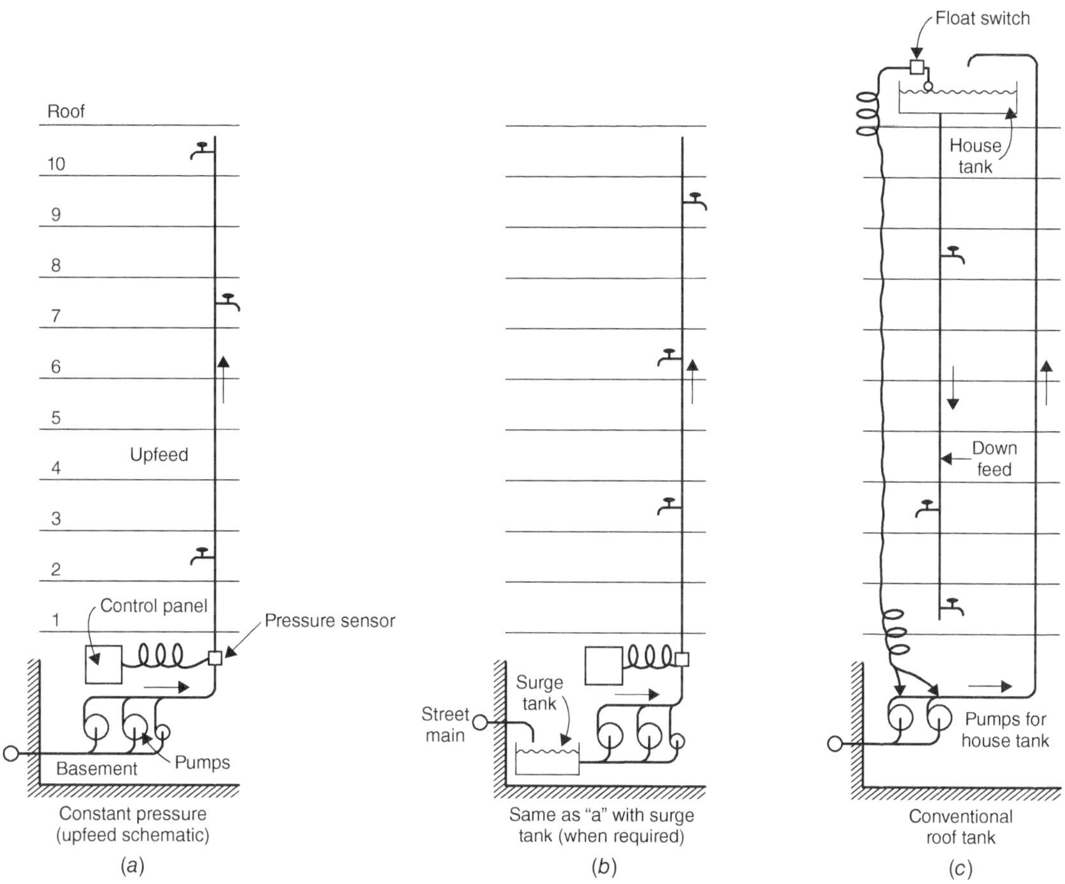

Figure 21.54 *Constant-pressure upfeed pumping (a, b) compared to gravity downfeed from a house tank on the roof (c). (With permission of Progressive Architecture.)*

pump a smaller burden. The next day the rested pump takes over the more active assignment. All of this occurs automatically.

At full operation, this triplex unit can put a suction demand on the street main that can seriously reduce the available water pressure in the neighborhood (Figure 21.54*a*). Therefore, utilities sometimes require that such a system draw from a surge tank, filled by casual flow from the street main, independent of the building requirements (Figure 21.54*b*). This requirement is often imposed when analysis indicates a maximum building demand in excess of 400 gpm [25 L/s].

The advantages of upfeed pumping are elimination of the house tank and the heavier structure required to transmit its weight down to the footings, and elimination of the necessary periodic cleaning of the tank. A shortcoming is the lack of reserve storage, which could cause a serious problem during an electrical power failure. However, minimum flow during this kind of emergency can be arranged if a diesel or emergency standby motor is available to drive one of the pumps.

21.10 PIPING, TUBING, FITTINGS, AND CONTROLS

(a) Piping, Tubing, and Fittings

The system of water-supply piping or tubing should efficiently fulfill its purpose, be easily maintained, and interfere as little as possible with architectural form and function. Except in basements, in utility rooms, and at points of access to controls, the piping system is usually concealed. Stud-and-joist construction can provide space for concealment, but in concrete or masonry buildings, vertical and horizontal furred spaces must often be provided.

Water supply piping is subject to corrosion over time. When pipe materials corrode, they first lose some carrying capacity (due to increased wall roughness and perhaps buildup of materials) and ultimately fail. Sediment from corrosion can adversely impact plumbing fixtures as well. Steel piping is particularly subject to corrosion. In the nonferrous group, red brass and copper tubing are effective in providing corrosion resistance. Copper tubing is less expensive than brass, assembles more easily, and is not subject

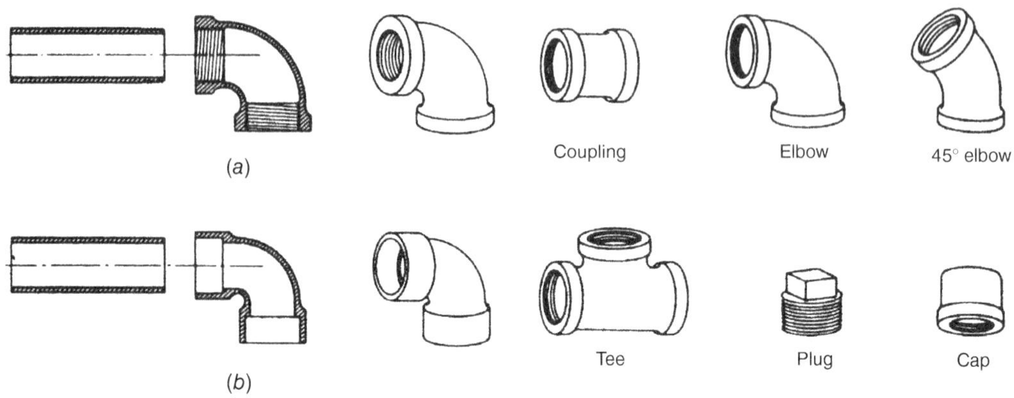

Figure 21.55 *Methods of connecting pipes and fittings, and tubes and fittings. (a) Threaded: for ferrous pipe and fittings and for iron pipe size (IPS) brass. (b) Soldered: for copper tubing and fittings. A sliding fit similar to that of (b) is used for the solvent weld of plastic connections.*

to dezincification (attack by acids on the zinc in brass). For use in handling aggressive waters, plastic is often a good choice. Like copper, it is light in weight and assembles with great ease.

For ferrous pipes and "iron pipe size" brass, threaded connections are used. The external, tapered thread on the pipe is covered with pipe compound and screws tight against the internal tapered thread of a coupling or other fitting. The solder-joint connection in copper depends on capillary attraction that draws the solder into a cylinder of clearance between the mating surfaces of tube and fitting. This occurs after the surfaces are polished and fluxed and the parts placed in final position. They are then heated, and molten solder is applied to the circular opening where the fitting edge surrounds the tube, with a small clearance. Solder is then drawn into the cylindrical connection. Solders are often tin-antimony alloys. This kind of joint permits the advantageous setting up of an entire tubing assembly without turning the parts (as in threaded installations) and before the soldering commences. For the same strength, copper tubing may have thinner walls than threaded pipes because no threads need to be cut into it. Its smooth interior surface offers low friction to flowing water. Although threaded and solder-joint connections are the most common in small work, there are many other types (Figures 21.55 and 21.56 and Table 21.13). Ferrous pipes in the larger sizes are often welded or connected by bolted flanges.

(b) Plastic Pipe

Most of the plastic pipes and fittings now produced are synthetic resins derived from such materials as coal and petroleum. These corrosion-resistant materials are widely used in water supply piping, fittings, and drainage systems (see Chapter 22). The National Sanitation Foundation (NSF) tests and certifies plastic pipe; the NSF seal must appear on pipes that are to carry potable water.

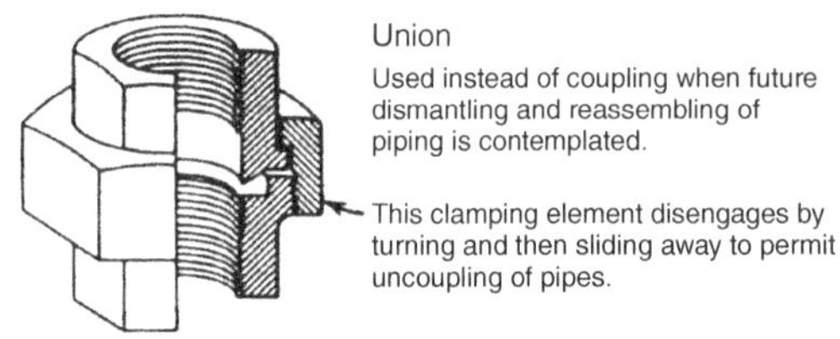

Figure 21.56 *Example of threaded pipe fitting (a union) for ferrous or brass pipe. These and all common fittings are also available for solder–joint (copper) or solvent–weld (plastic) connections and usually for transition from one system or material to another.*

Most of the materials used for piping are thermoplastics and will repeatedly soften under the application of heat. PVDC (polyvinylidene chloride) material can carry water at 180°F [82°C], but plastic pipe should not be subjected to temperatures higher than this. Expansion (see Table 21.12) is much greater than that of other piping materials for the same Δt and affects the piping design. Shock-resistant plastic is used in the plumbing systems of most mobile homes.

(c) Valves and Controls

It is usually desirable to valve every riser, the branches that serve bathrooms or kitchens, and the runouts to individual fixtures. This facilitates repair at any location with a minimum of shutdown within a system. Water treatment equipment will have a valved bypass (see Figure 21.57). Pumps and other devices that may need repair should be able to be disconnected by unions (Figure 21.56) after valves are closed.

A gate valve (Figure 21.57*a*), with a retractable leaf machined to seal tightly against two sloping metal surfaces when closed, offers the least resistance to water flow when open. It is usually chosen for locations where it is left completely open most of the time.

TABLE 21.13 Water Supply Piping Materials

Part A. Water Service Only[a]	
Material	**Connections[b]**
ABS (acrylonitrile butadiene styrene) plastic pipe	Mechanical with elastomeric seal (normally underground only); solvent cement; threaded joints
Asbestos–cement	Sleeve couplings of same material as pipe, sealed with elastomeric ring
Ductile iron water pipe	Mechanical joints (see manufacturer)
PE (polyethylene) plastic pipe and tubing	Flared joints (see manufacturer), heat fusion, mechanical joints (see manufacturer)
PE-AL-PE (polyethylene/aluminum/polyethylene) pipe, PVC (polyvinyl chloride) plastic pipe	Mechanical with elastomeric seal (normally underground only); solvent cement; threaded joints (may reduce pressure rating)
Part B. Water Service and Distribution[c]	
Material	**Connections**
CPVC (chlorinated polyvinyl chloride) plastic pipe (and tubing, indoors)	Mechanical (see manufacturer), solvent cement, or threaded joints (may reduce pressure rating)
Copper or copper-alloy pipe and tubing	Brazed, mechanical, soldered, threaded, or welded joints
Galvanized steel pipe[d]	Threaded or mechanical joints with an elastomeric seal
PB (polybutylene) plastic pipe and tubing	Flared joints, heat-fusion, or mechanical joints (see manufacturer)
PE-RT (polyethylene of raised temperature) plastic tubing	
PEX (cross-linked polyethylene) plastic tubing	
PEX-AL-PEX (cross-linked polyethylene/aluminum/cross-linked polyethylene) pipe	
PEX-AL-HDPE (cross-linked polyethylene/aluminum/high-density polyethylene)	
Stainless steel pipe	Mechanical, welded joints

Source: This table is a copyrighted excerpt from the 2024 International Plumbing Code. Copyright © 2023. International Code Council, Inc. All rights reserved. Reproduced with permission. www.ICCSAFE.org.

[a]Materials as listed in the *2024 International Plumbing Code* (ICC, 2023). Water service piping is for an outdoor, underground connection from the building to the main or supply tank.

[b]Joints between different piping materials require approved adapter fittings; dielectric or brass converter fittings required between copper (or copper alloy) and galvanized steel piping.

[c]Materials as listed in the *2024 International Plumbing Code*. Water distribution piping is within the building.

[d]Sediment from corrosion over the life of the pipe.

The compression-type globe valve (Figure 21.57*b*) is usually used for closing or throttling flow near a point of occasional use. Faucets are usually of the compression type, as are drain valves and hose connections. They are similar to an angle valve (Figure 21.57*d*). When it is necessary to prevent flow in a direction opposite to that which is planned, a check valve (Figure 21.57*c*) is introduced. The hinged leaf swings to permit flow in the direction of the arrow but closes against flow in the other direction.

(d) Pipe Support

Water piping systems are heavy because of their water content and need regular support (Figure 21.58). Vertical runs of piping should be supported at every story. Horizontal pipes should be supported at intervals of:

- 6 ft [1.829 m] for ½-in. [12-mm] pipe
- 8 ft [2.438 m] for ¾-in. or 1-in. [19- or 25-mm] pipe
- 10 ft [3.048 m] for 1 ¼-in. [32-mm] or larger pipe

Although somewhat counterintuitive, these recommendations reflect the ability of a larger pipe to span greater distances between supports. Horizontal copper tubing should be supported at closer spacing than steel. Adequate positioning of horizontal runs is important to ensure correct pitch and drainage. Hangers are adjustable for this purpose.

(e) Shock and Hot Water Expansion

Expansion chambers were shown diagrammatically in Figure 21.49, and water hammer was explained in Section 21.9(b). Expansion chambers are often made of capped lengths of vertical pipe about 2 ft [0.6 m] long at the fixture branches (Figure 21.59*a*). They trap air, which absorbs the impact of the water surge. A somewhat better (more controllable) device is a rechargeable air chamber (Figure 21.59*b*, *c*). By closing the valve and draining the water through the hose bibb while the petcock above is open to admit air, the chamber may be refilled with air. Closing the petcock and hose bibb and opening the valve complete the service operation and reconnect the device with the water system. Rechargeable chambers (instead of pipe extensions) are used on branch lines adjacent to groups of fixtures. Access for service must be provided. Perhaps the best device is the special shock absorber (Figure 21.59*d*).

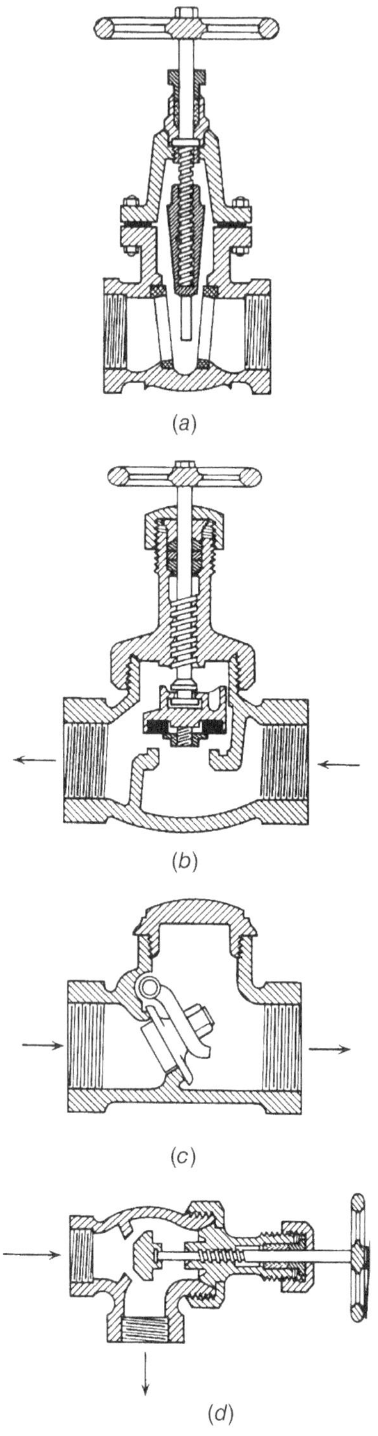

Figure 21.57 *Typical valves for water systems. (a) Gate valve. (b) Globe valve. (c) Check valve. (d) Angle valve.*

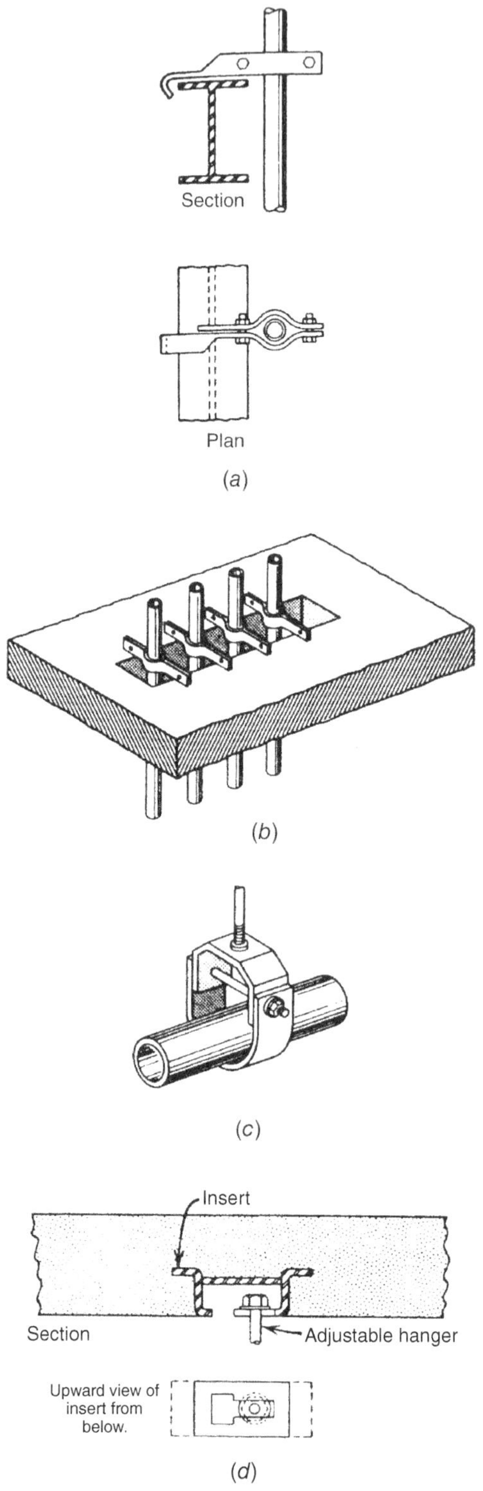

Figure 21.58 *Pipe supports. (a) Vertical riser supported at a steel beam. (b) Vertical riser group supported at a slot in a concrete slab. (c) Horizontal pipe hung from a slab above by an adjustable-length clevis hanger. (d) Typical metal insert in a concrete soffit to receive the hanger rod.*

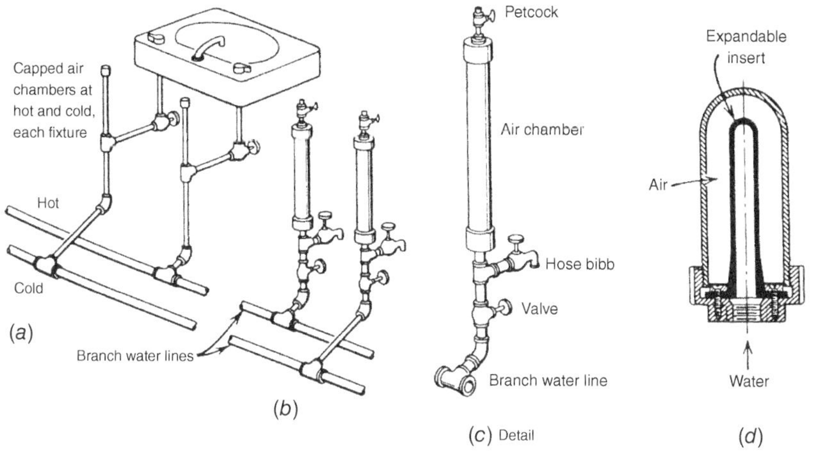

Figure 21.59 *Shock relief and expansion chambers. Air chambers cushion the shock of the water hammer when the fixture faucets are shut off abruptly. They also permit hot water to expand instead of periodically forcing open the hot water emergency pressure relief valve at the heater or tank. (a) Capped air chambers at each supply pipe of each fixture. (b, c) Rechargeable air chambers on hot and cold branch water lines (individual fixture chambers are omitted when these are used). (d) Special shock absorber.*

Air cushions also protect the water-heater tank's relief valve against frequent operation, with the resultant leakage of hot water, as hot water periodically expands and contracts in closed systems.

(f) Condensation or "Sweating"

The moisture that is always present in air can condense on the exterior surface of a cold pipe. Dropping off a pipe, it can create an unpleasant wet condition, disfigure finished surfaces, and provide conditions amenable to mold growth. Groundwater in some parts of North America is 50°F [10°C] and colder (see Chapter 14). A pipe carrying such water might have a surface temperature of about 60°F [15°C]. The psychrometric chart (Appendix G) indicates that at a summer air temperature of 85°F [29°C], condensation will occur on such a pipe when the relative humidity exceeds 40%. To avoid condensation damage, all cold-water piping and fittings should be insulated. Glass fiber ½–1 in. [12–25 mm] thick is commonly chosen for this purpose. A tight vapor retarder on the exterior surface of the insulation prevents water vapor from penetrating the insulation and reaching the cold surface. The insulation provides another important advantage: It retards heat flow from the warmer air to the water, thus preventing the water from becoming disagreeably warm.

(g) Heat Conservation

Pipes carrying domestic hot water should be insulated to conserve energy used to heat the water and to ensure a correct water temperature at the points of use. Parallel hot and cold water piping, even though insulated, should be separated to prevent heat exchange. The pipe insulation should have a $k = 0.22–0.28$ Btu in./h ft^2 °F, and its thickness should be a minimum of ½ in. (on pipes less than 1½-in. diameter) or 1 in. (on pipes 1½-in. diameter and above). In SI units, $k = 0.032–0.040$ W/m K, and the thickness should be a minimum of 1.3 mm (on pipes less than 25-mm diameter) or 2.5 mm (on pipes 25-mm diameter and above).

Storage tanks and water heaters are usually manufactured with integral insulation. Older devices, however, may have less insulation than today's energy concerns warrant. As a result, many older water heaters are retrofitted with added insulation. Some electric utilities find the savings from conservation so attractive (relative to the cost of building new generating facilities) that they provide water-heater wrapping as a free service to customers.

21.11 SIZING OF WATER PIPES

There must be sufficient pressure at fixtures to assure the user of a prompt and adequate flow of water. Municipal ordinances often state that the flow must be adequate to keep the fixtures clean and sanitary. These convenience and sanitation objectives result in prescribed pressures that must be maintained at the various fixtures to ensure the proper flow rates listed in Table 21.14.

Minimum fixture pressures vary from 4 to 20 psi [28 to 138 kPa] for fixtures other than hose bibbs. Because the pressure in street mains is usually about 50 psi [345 kPa], it is possible to ensure the minimum fixture pressure, provided that the water does not have to be lifted to too great a height and not too much pressure is lost by friction in distribution piping. Excessive

TABLE 21.14 Flow and Pressure to Typical Plumbing Fixtures

	Minimum		
Fixture Served	**Flow Rate gpm [L/s]**	**Pressure psi [kPa]**	**Maximum Flow Rate or Quantity[a]**
Bathtub	4 [0.25]	20 [138]	
Bidet	2 [0.13]	20 [138]	
Combination fixture	4 [0.25]	8 [55]	
Dishwasher, residential	2.75 [0.17]	8 [55]	
Drinking fountain	0.75 [0.05]	8 [55]	
Hose bibb	5 [0.32]	8 [55]	
Laundry tray	4 [0.25]	8 [55]	
Lavatory, private	2 [0.13]	8 [55]	2.2 gpm at 60 psi [0.14 L/s at 414 kPa]
Lavatory, public	2 [0.13]	8 [55]	0.5 gpm at 60 psi [0.03 L/s at 414 kPa]
Lavatory, public, metering or self-closing	2 [0.13]	8 [55]	0.25 gal [0.95 L] per metering cycle
Shower head[b,c]	3 [0.19]	8 [55]	2.0 gpm at 80 psi [0.13 L/s at 551 kPa]
Shower head, temperature controlled	3 [0.19]	20 [138]	2.0 gpm at 80 psi [0.13 L/s at 551 kPa]
Sink, residential	2.5 [0.16]	8 [55]	2.2 gpm at 60 psi [0.14 L/s at 414 kPa]
Sink, service	3 [0.19]	8 [55]	2.2 gpm at 60 psi [0.14 L/s at 414 kPa]
Urinal, valve	12 [0.76]	25 [172]	1.0 gal [3.8 L] per flushing cycle
Water closet, blow out, flushometer valve	25 [1.58]	45 [310]	1.6 gal [6 L] per flushing cycle
Water closet, siphonic, flushometer valve	25 [1.58]	35 [241]	1.6 gal [6 L] per flushing cycle
Water closet, tank, close coupled	3 [0.19]	20 [138]	1.6 gal [6 L] per flushing cycle
Water closet, tank, one piece	6 [0.38]	20 [138]	1.6 gal [6 L] per flushing cycle

Source: This table is a copyrighted excerpt from the 2024 International Plumbing Code.

[a]Consumption tolerances shall be determined from referenced standards.

[b]A hand-held shower spray is a shower head.

[c]Shower heads shall comply with all requirements for high-efficiency shower heads in *ASME A112.18.1-2020/CSA B125.1*.

friction results from piping that is too long in developed length (actual distance of water flow) or that interposes too many fittings (such as elbows and tees), or is too small in diameter.

The pressure losses in an upfeed system served by street main pressure are as follows. For A in the following chart, use the highest, most remote fixture from the main.

Minimum fixture flow pressure	A
Pressure lost because of height	+B
Pressure lost by friction in piping	+C
Pressure lost by flow through meter	+D
Total required street main pressure	=E

During design, items A, B, and E are known and are reasonably constant. A value for A can be found in Table 21.14. Street main pressure, E, is a characteristic of the local water supply and is obtained from the water utility. Item B, the pressure lost due to height, can be found by multiplying the height in feet by 0.433 (height in meters by 10) (see the discussion of static head in Section 21.9(a)). Item D, the pressure lost in flow through the water meter, depends upon flow and pipe size (Figure 21.60), neither of which is yet known. Therefore, the value of item D is estimated. (For residences and small commercial buildings, meter size rarely exceeds 2 in. [50 mm].) Later, it must be checked and a recalculation made if necessary. This leaves one unknown, the value of C, where:

$$C = E - A - B - D$$

Pipe size is based upon Figure 21.61. Pipe diameter is determined by the point of intersection of a horizontal line representing flow and a vertical line expressing friction loss. To select a pipe size, one needs to know the probable flow and the unit-friction loss in the pipe and fittings. The noise created by water flow also must be considered. Flow above 10 fps [3.1 m/s] is usually too noisy; flow above 6 fps [1.8 m/s] may be too noisy in acoustical-critical locations. To mitigate noisiness, a flow under 10 fps is recommended.

Flow can be found for any pipe by first listing all fixtures to be served by that pipe, assigning them water supply fixture units (WSFU, listed in Table 21.15), and finding the total WSFU. This total is then converted to the likely demand flow in gpm [L/s] from Figure 21.62. These curves show that actual flow does not increase in direct proportion to an increase in fixture units. The larger an installation, the less likely that many fixtures will be operating concurrently. This reflects the statistical concept known as diversity.

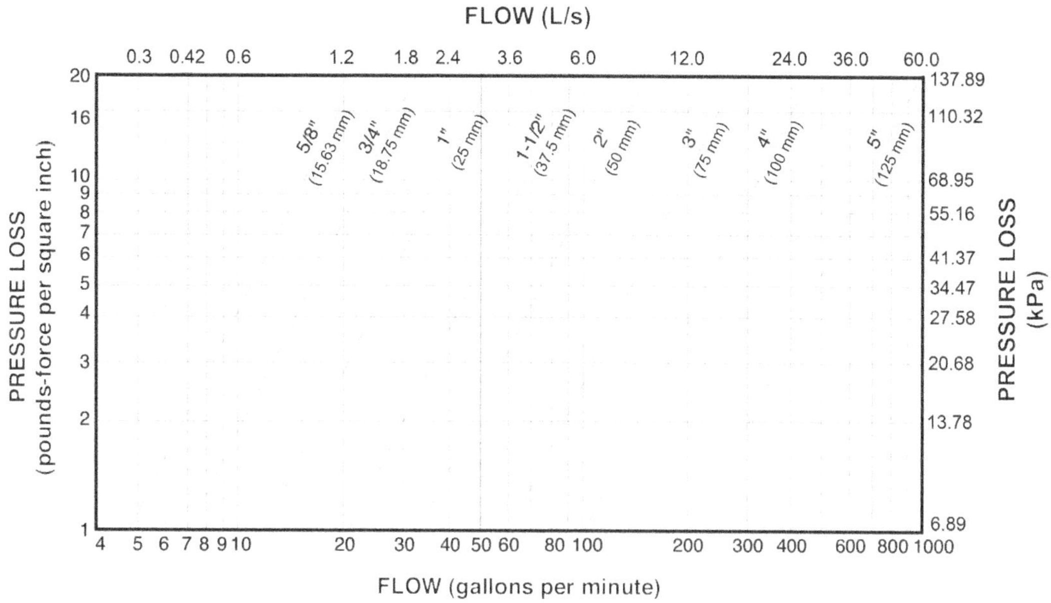

Figure 21.60 *Pressure losses in disk-type water meters. (© 2018 Uniform Plumbing Code; reprinted with the permission of the International Association of Plumbing and Mechanical Officials. This copyright material and all points or statements in using this material have not been reviewed by IAPMO. The opinions expressed herein are not representations of fact from IAPMO.)*

To establish the desired friction loss, divide value C (pressure lost by friction in piping) by the total equivalent length of the piping. This length is the sum of the developed length (total linear distance of water travel) and the length equivalent of the fittings. For example, Table 21.16 shows that, in a 1-in. [25-mm] diameter pipe run, a 90° ell causes a friction loss equivalent to that of 3 ft [0.9 m] of straight pipe. The number and style of fittings are estimated, and the size of fittings is assumed. This may be puzzling, but it is a common trial-and-error engineering procedure that sometimes requires several recalculations.

EXAMPLE 21.6 Using the following data—some values are from the assumptions referred to previously—find the proper size for a metered water supply main. (See Example 21.7 for SI units.)

Street main pressure (minimum)	50 psi
Height, topmost fixture above main	30 ft
Topmost fixture type	Water closet, tank, one piece (1.6 gal per flush)
Fixture units in the system	85 WSFU
Developed length of the piping (to the highest and most remote fixture)	100 ft
Pipe length equivalent of fittings (commonly estimated at 50% of the developed length)	50 ft
System uses predominantly	Flush tanks
Assumed peak demand	50 gpm
Assumed meter size	2 in.

SOLUTION

From the minimum street main pressure, subtract the sum of the fixture pressure, the static head, and the pressure lost in the meter. This sum is:

	psi
A: fixture pressure (Table 21.14)	−20
B: static head 30 ft × 0.433 psi/ft	−13
D: pressure loss in meter (estimated, Figure 21.60)	−2.3
Subtotal	−35.3
E: pressure in street main	50
(A + B + D)	−35.3
E−(A + B + D)	14.7

The pressure lost in 100 ft (developed length) of piping plus the 50 ft of piping equivalent for the pressure lost by friction in the fittings therefore can total 14.7 psi. Total equivalent length is 150 ft. This procedure ensures 20 psi at the critical fixture. The unit-friction loss, psi/100 ft of pipe, will be:

$$14.7 \text{ psi} \times 100 \text{ ft}/150 \text{ ft (total equivalent length)}$$
$$= 9.8 \text{ psi}/100 \text{ ft}$$

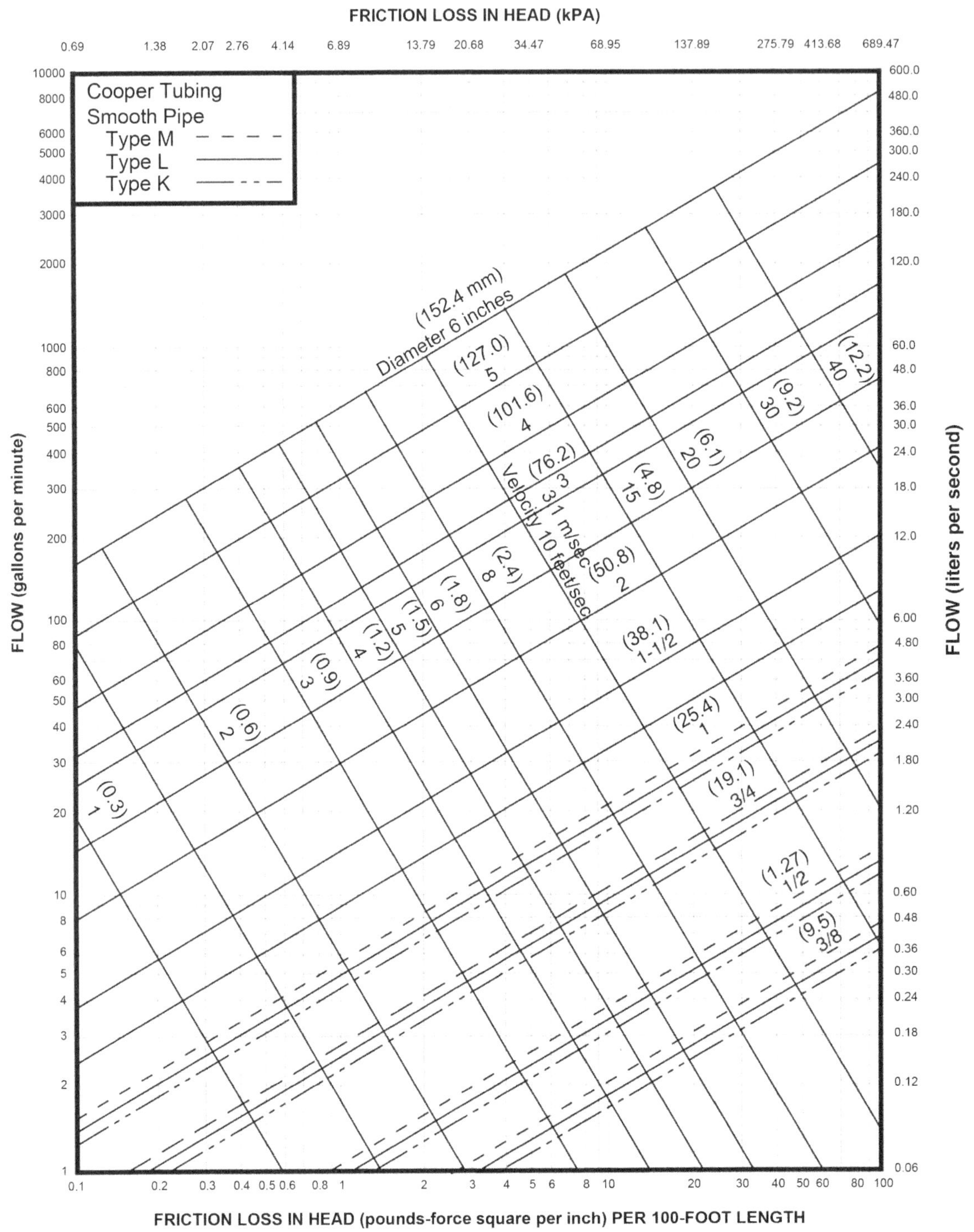

Figure 21.61 *Friction loss chart for smooth pipe. Velocity is shown to assist in noise control efforts: above 10 fps [3 m/s], moving water can be clearly heard within pipes. (© 2018 Uniform Plumbing Code; reprinted with the permission of the International Association of Plumbing and Mechanical Officials. This copyright material and all points or statements in using this material have not been reviewed by IAPMO. The opinions expressed herein are not representations of fact from IAPMO.)*

From Figure 21.62, curve 2, a flush-tank system with 85 WSFU will have a probable flow (diversified demand) of about 40 gpm. Given this information, enter Figure 21.64 horizontally at 40 gpm and vertically at 9.8 psi/100 ft. At the intersection of these lines, the pipe diameter and velocity are determined:

Diameter of pipe: between $1\,^1/_4$-in. and $1\,^1/_2$-in.

Velocity: ~9 fps (less than 10 fps, so OK regarding noisiness)

TABLE 21.15 Water Supply Fixture Units (WSFU)

			Load Values in WSFU		
Fixture	**Occupancy**	**Type of Supply Control**	**Cold**	**Hot**	**Total**
Bathroom group	Private	Flush tank	2.7	1.5	3.6
Bathroom group	Private	Flush valve	6	3	8
Bathtub	Private	Faucet	1	1	1.4
Bathtub	Public	Faucet	3	3	4
Bidet	Private	Faucet	1.5	1.5	2
Combination fixture	Private	Faucet	2.25	2.25	3
Dishwashing machine	Private	Automatic		1.4	1.4
Drinking fountain	Offices, etc.	⅜ in. [9.5 mm] valve	0.25		0.25
Kitchen sink	Private	Faucet	1	1	1.4
Kitchen sink	Hotel, restaurant	Faucet	3	3	4
Laundry trays (1–3)	Private	Faucet	1	1	1.4
Lavatory	Private	Faucet	0.5	0.5	0.7
Lavatory	Public	Faucet	1.5	1.5	2
Service sink	Offices, etc.	Faucet	2.25	2.25	3
Shower head	Public	Mixing valve	3	3	4
Shower head	Private	Mixing valve	1	1	1.4
Urinal	Public	1 in. [25 mm] flush valve	10		10
Urinal	Public	¾ in. [19 mm] flush valve	5		5
Urinal	Public	Flush tank	3		3
Washing machine, 8 lbs [3.6 kg]	Private	Automatic	1	1	1.4
Washing machine, 8 lbs [3.6 kg]	Public	Automatic	2.25	2.25	3
Washing machine, 15 lbs [6.8 kg]	Public	Automatic	3	3	4
Water closet	Private	Flush valve	6		6
Water closet	Private	Flush tank	2.2		2.2
Water closet	Public	Flush valve	10		10
Water closet	Public	Flush tank	5		5
Water closet	Public or private	Flushometer tank	2		2

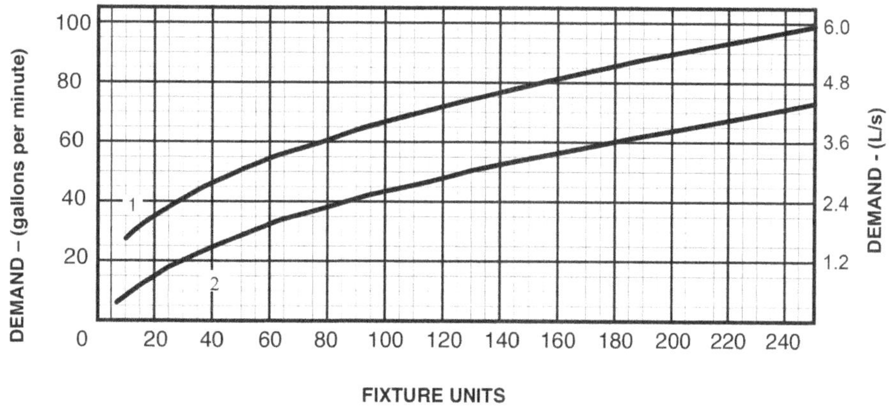

Figure 21.62 *Estimating curves for flow based upon total water supply fixture units. Curve 1 is for a system with predominantly flush valves; curve 2 is for a system with predominantly flush tanks. For fixture unit totals above 250, see the 2018 Uniform Plumbing Code. (© 2018 Uniform Plumbing Code; Reprinted with the permission of the International Association of Plumbing and Mechanical Officials. This copyright material and all points or statements in using this material have not been reviewed by IAPMO. The opinions expressed herein are not representations of fact from IAPMO.)*

Therefore, a 1½-in. diameter supply pipe will be chosen with a 2-in. water meter.

Now find the actual pressure loss in the 2-in. meter for a flow of 40 gpm (recall that the original assumption was 50 gpm). Figure 21.63 shows that this is about 1.5 psi. Because this is less than the 2.3 psi estimated, the pressure at the fixture will be slightly higher than the minimum needed. When a final system layout is made, the actual fittings are tabulated, and the equivalent length of fittings is found. If this differs greatly from the 50 ft estimated, a recalculation is made. ■

TABLE 21.16 Allowance in Equivalent Length of Pipe for Friction Loss in Valves and Threaded Fittings[a]

Part A. I-P Units							
	Equivalent Length of Pipe for Various Fittings						
Diameter of Fitting (in.)	90° Standard Ell (ft)	45° Standard Ell (ft)	Standard Tee 90° (ft)	Coupling or Straight Run of Tee (ft)	Gate Valve (ft)	Globe Valve (ft)	Angle Valve (ft)
3/8	1	0.6	1.5	0.3	0.2	8	4
1/2	2	1.2	3	0.6	0.4	15	8
3/4	2.5	1.5	4	0.8	0.5	20	12
1	3	1.8	5	0.9	0.6	25	15
1 1/4	4	2.4	6	1.2	0.8	35	18
1 1/2	5	3	7	1.5	1.0	45	22
2	7	4	10	2	1.3	55	28
2 1/2	8	5	12	2.5	1.6	65	34
3	10	6	15	3	2	80	40
3 1/2	12	7	18	3.6	2.4	100	50
4	14	8	21	4.0	2.7	125	55
5	17	10	25	5	3.3	140	70
6	20	12	30	6	4	165	80
Part B. SI Units							
	Equivalent Length of Pipe for Various Fittings						
Diameter of Fitting (mm)	90° Standard Elbow (mm)	45° Standard Elbow (mm)	Standard Tee 90° (mm)	Coupling or Straight Run of Tee (mm)	Gate Valve (mm)	Globe Valve (mm)	Angle Valve (mm)
9.5	305	183	457	91	61	2438	1219
12.7	610	366	914	183	122	4572	2438
19.1	762	457	1219	244	152	6096	3658
25.4	914	549	1524	274	183	7620	4572
32	1219	732	1829	366	244	10,668	5486
38	1524	914	2134	457	305	13,716	6706
51	2134	1219	3048	610	396	16,764	8534
64	2438	1524	3658	762	488	19,812	10,363
76	3048	1829	4572	914	610	24,384	12,192
102	4267	2438	6401	1219	823	38,100	16,764
127	5182	3048	7620	1524	1006	42,672	21,336
152	6096	3658	9144	1829	1219	50,292	24,384

Source: © 2024 Uniform Plumbing Code; Reprinted with the permission of the International Association of Plumbing and Mechanical Officials. This copyright material and all points or statements in using this material have not been reviewed by IAPMO. The opinions expressed herein are not representations of fact from IAPMO.

The SI conversions were appended by the authors of this book.

[a]Based on nonrecessed threaded fittings. Use one-half of these allowances for recessed threaded fittings or streamline solder fittings.

EXAMPLE 21.7 Solve the pipe sizing problem posed in Example 21.6 using SI units.

Street main pressure (minimum)	345 kPa
Height, topmost fixture above main	10 m
Topmost fixture type	Water closet, tank, one piece (6 L per flush)
Fixture units in the system	85 WSFU
Developed length of the piping (to the highest and most remote fixture)	30 m
Pipe length equivalent of fittings (commonly estimated at 50% of developed length)	15 m
System uses predominantly	Flush tanks
Assumed peak demand	3.2 L/s
Assumed meter size	50 mm

SOLUTION

From the minimum street main pressure, subtract the sum of the fixture pressure, the static head, and the pressure lost in the meter. This sum is:

A: fixture pressure (Table 21.14)	−138 kPa
B: static head 10 m × 10 kPa/m	−100 kPa
D: pressure loss in meter (estimated, Figure 21.60)	−16 kPa
Subtotal	−254 kPa
E: pressure in street main	345 kPa
(A + B + D)	−254 kPa
E−(A + B + D)	91 kPa

The pressure lost in 30 m (developed length) of piping plus the 15 m of piping equivalent for the pressure lost by friction in the fittings therefore can total 91 kPa. Total equivalent length is 45 m. This procedure ensures 103 kPa at the critical fixture. The unit-friction loss, kPa/100 m of pipe, will be:

91 kPa × 30 m/45 m (total equivalent length)

= 61 kPa/100 m

From Figure 21.62, curve 2, a flush-tank system with 85 WSFU will have a probable flow (diversified demand) of about 2.4 L/s. Given this information, enter Figure 21.64 horizontally at 2.4 L/s and vertically at 61 kPa/100 m. At the intersection of these lines, the pipe diameter and velocity are determined:

Diameter of pipe: between 31.8-mm and 38.1-mm

Velocity: ~2.7 m/s (less than 3 m/s, so OK regarding noisiness)

Therefore, a 38.1-mm diameter supply pipe will be chosen with a 51-mm water meter. ■

As in Example 21.6, the actual system pressures can be found using the exact pressure loss in a 51-mm water meter.

21.12 WE•STAND

The *Water Efficiency and Sanitation Standard* (WE•Stand) is an American National Standard developed by IAPMO to replace the earlier *Green Plumbing and Mechanical Code Supplement*. WE•Stand is noted here because one of its notable features is a spreadsheet-based Water Demand Calculator that provides an alternative to the Hunter Curve design approach described above.

This alternative is applicable to estimating water supply demand for residential buildings. The Water Demand Calculator addresses potable water demand for the building as well as cold and hot water branches and risers. IAPMO states that the Water Demand Calculator works in conjunction with any commonly accepted rules and procedures for sizing the water supply system. It claims that the Calculator reduces the overdesign (and thus oversizing) of pipes and results in reduced construction costs, improved water quality, greater energy efficiency, and faster hot water delivery.

In addition to the Water Demand Calculator, WE•Stand addresses:

- water efficiency and conservation
- systems for supply from alternative sources
- composting toilets
- onsite waste treatment systems

21.13 IRRIGATION

The highest design priority for landscape watering is to ensure optimum rainfall retention. In the past, provision for watering the landscape around buildings frequently was limited to the installation of hose bibbs at building exteriors. In many areas of North America, half of the residential water usage (see Chapter 22) is for outdoor purposes. Outdoor water use accounts for about 30% of average household water use nationwide. With increasing demands on a finite water supply, water-conserving irrigation equipment has become available.

Lawn sprinklers are relatively inefficient irrigating devices, as much of their water is lost to evaporation and runoff. It is commonly estimated that lawn sprinkling will provide ½ in. of water per hour per square foot of lawn (or 0.3 gph per ft^2) [0.005 L/s per m^2].

Landscape sprinkling is least efficient during the daytime, when hot sun and dry air combine to increase the rate of evaporation. However, nighttime sprinkling is rarely convenient for building custodians or homeowners. One solution to this problem is the use of sprinkler timing devices (Figure 21.63*a*), which control electronic valves and a network of underground supply pipes and sprinkler heads that are permanently installed within landscaped areas. These timing devices typically include controls governing the length of the watering cycle, the time at which the cycle begins (usually before sunrise, when relative humidity is highest), and the number of days between cycles. A "rainy day switch" is often included so that irrigation can be discontinued during rainy weather. Rain sensors can be used to shut off irrigation automatically (Figure 21.63*b*). Tighter control over the water–plant relationship can be obtained with tensiometers (Figure 21.63*c*), which monitor the moisture content of the soil at the depth of the plants' root zone. These can be installed so as to override the automatic timing device, thus watering more—or less—frequently, depending upon the plants' needs. Instead of sprinklers, bubblers (Figure 21.63*d*) can be installed, with very low flow rates and less evaporative water loss.

Drip irrigation takes an approach very different from that of the flooding method typical of sprinklers. From a network of plastic tubes, either just underground or on the surface, emitters (Figure 21.64) slowly and steadily drip water onto the ground surface at each needy plant. Most of this water soaks into the soil at a rate that is better for most plants than the sudden, short flooding of intermittent sprinkling. Two requirements are especially important for drip systems: The water must not contain materials that can clog the small holes of the emitters, and the pressure must be low. Pressure-reducing valves at the source of the drip system are advisable; if necessary, filters should also be installed there.

Drip irrigation is not a universal solution to landscape watering; it is best for individual plants such as

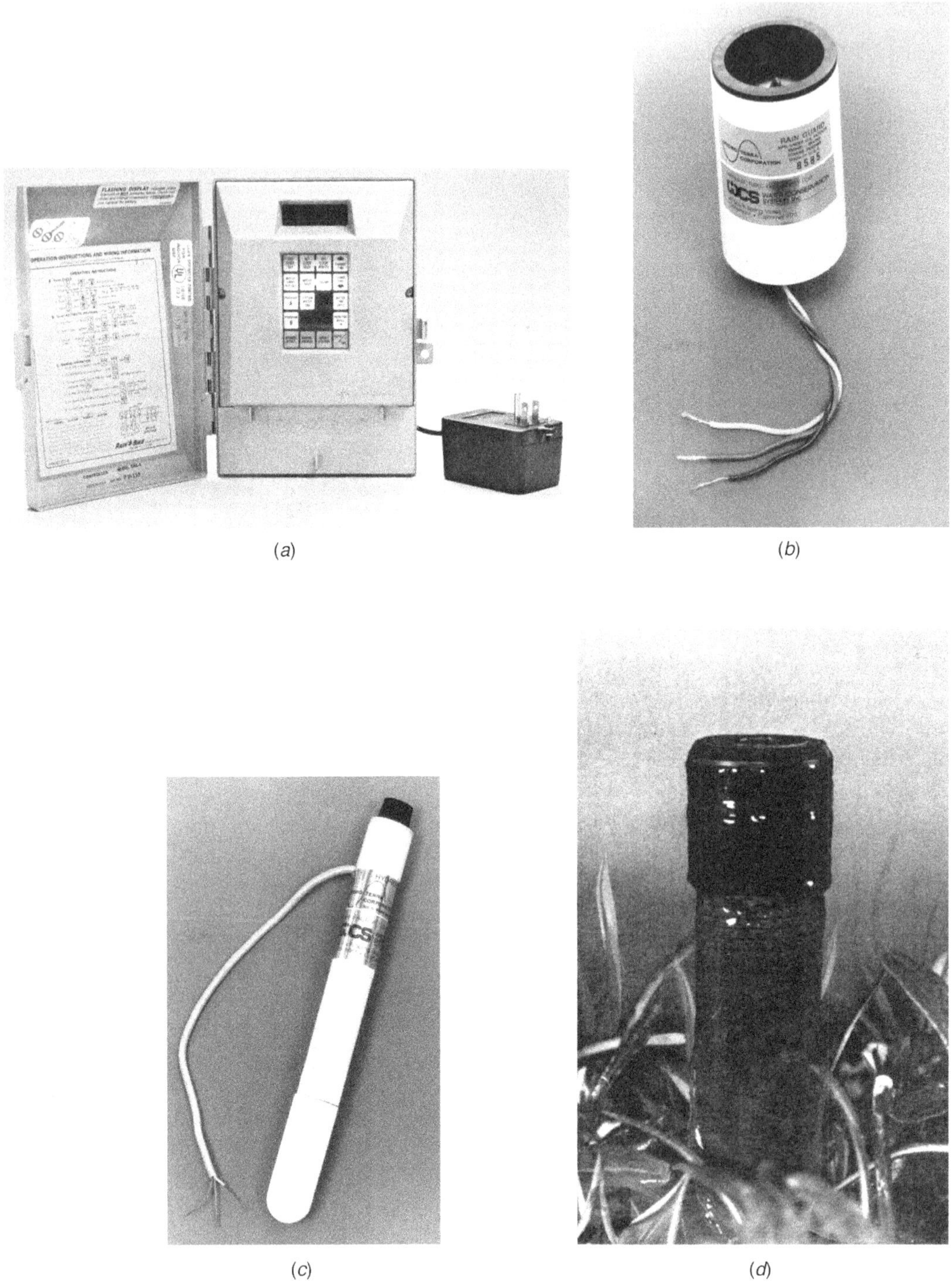

Figure 21.63 *Components for automatic landscape sprinkling systems. (a) Programmable controllers feature multiple stations, up to 14-day sequences, and memory provisions in case of power failure. (b) A solid-state rain sensor automatically prevents needless watering during rainfall. (c) A tensiometer controls irrigation by monitoring the moisture content of the soil. (d) Bubblers are low-flow substitutes for sprinklers—a step toward drip irrigation. These can be pressure-compensating to permit constant flow. (a, d: Courtesy of Rain Bird Sales, Inc.; b: Courtesy of Water Conservation Systems, Inc.)*

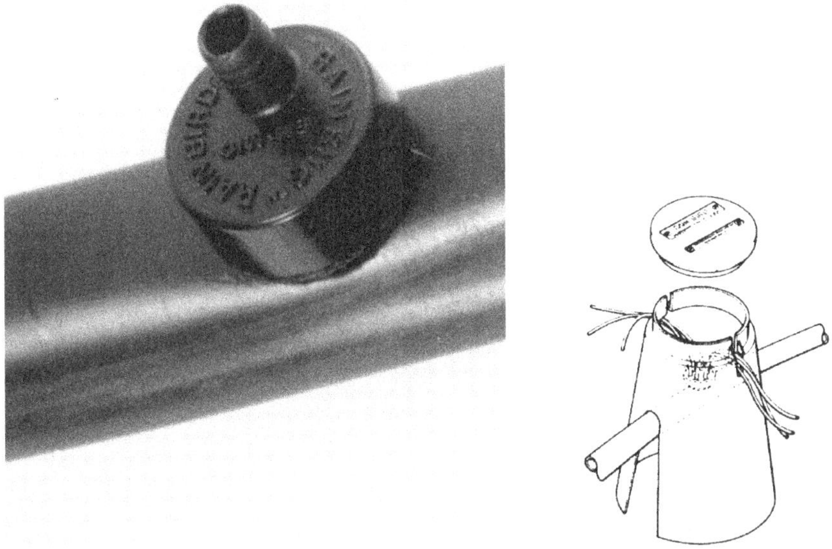

(a) (b)

Figure 21.64 *Equipment for drip irrigation. (a) On low-pressure lines, emitters are installed—one for each group of plants to be watered. The lines can be laid on the ground surface or just under the surface. (b) Simple emitter boxes allow easy access. (Courtesy of Rain Bird Sales, Inc.)*

shrubs and small trees but is difficult to apply to large lawns. Where appropriate, it can achieve significant water conservation compared to sprinklers. A detailed discussion of soil texture, moisture, and plant growth is provided in Lowry (1988).

Recycled or reclaimed water for irrigation, often provided by graywater systems and sometimes by stored rainwater, is gradually gaining acceptance in North American building codes. This topic is explored further in Chapter 22. Considered nonpotable, this water must sometimes be deliberately colored and/or be accompanied by warning signs at each outlet.

Of course, no irrigation is better that low irrigation. Site design and landscape plans that leverage indigenous flora can greatly reduce, or even eliminate, the need for permanent irrigation. A xeriscape is a landscape that is designed and developed in a manner which requires little or no irrigation. The concept is that vegetation that is native to the site is naturally suited to withstand the local climate, including the long-term rainfall patterns. Of course, both microclimate variabilities and the long-term effects of climate change may put certain stressors on even the most well-suited vegetation. It is also important to note that many landscape designs may require temporary irrigation during an establishment period (typically one to two years), even if the landscape solution is designed to not require permanent irrigation.

Irrigation, past and future, shaped the design of the Napa Valley Museum in California (Figure 21.65). An adjacent creek provides water behind a small dam. A pump house carries water to a central observation tower, where visitors see the history of irrigation displayed in the context of the Napa Valley. The path of water in this model landscape invites visitors to explore canal irrigation of the eighteenth and nineteenth centuries, as well as advanced drip irrigation and xeriscape plantings of the twentieth century and beyond. (Xeriscape planting, as a part of the hydrozone landscape concept, was shown in Chapter 20.)

WaterSense-labeled sprinkler bodies are required to meet the flow rate and outlet pressure requirements of the *ASABE/ICC 802: Landscape Irrigation Sprinkler and Emitter Standard* (ICC, 2020). As such, these products can help mitigate inefficient and ineffective outdoor water use. For instance, irrigation sprinklers are often installed in systems where the water pressure is higher than what is recommended for the sprinkler nozzle. This tends to lead to excessive flow rates, misting, fogging, and uneven coverage. WaterSense-labeled spray sprinkler bodies with integral pressure regulation can mitigate poor performance and wasted water by providing a constant flow at the sprinkler nozzle. Additionally, when a sprinkler body maintains a constant pressure near its optimal operating pressure, the connected nozzle is better able

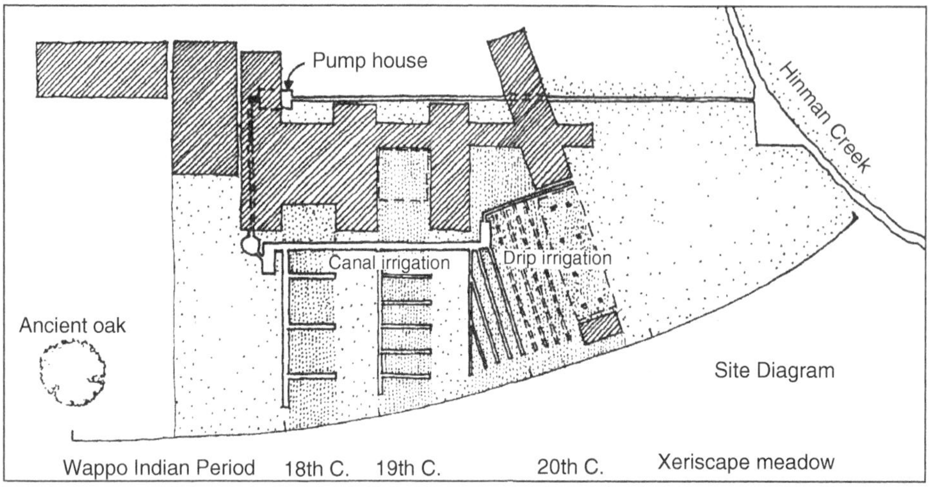

(a)

(b)

Figure 21.65 The Napa Valley Museum shows the history of irrigation in northern California. (a) Site plan. Water, piped from behind a small dam in the adjacent creek, flows through open canals, then provides drip irrigation, eventually discharging to a field of native plants. (b) The canals help guide the visitor through the exhibit gardens. (Fernau and Hartman, architects.)

to generate the proper amount of water spray and coverage for more uniform distribution of water across the landscape. For further information on planning for landscape water use, see the EPA's Water Budget Tool (EPA, 2024c)—part of the EPA's WaterSense program.

References and Resources

ASHRAE. 2021a. *ASHRAE Guideline 12: Managing the Risk of Legionellosis Associated with Building Water Systems*. Atlanta, GA: American Society of Heating, Refrigerating and Air-Conditioning Engineers.

ASHRAE. 2021b. *ANSI/ASHRAE Standard 188-2021: Legionellosis: Risk Management for Building Water Systems*. Atlanta, GA: American Society of Heating, Refrigerating and Air-Conditioning Engineers.

ASHRAE. 2023a. *2023 ASHRAE Handbook—HVAC Applications*. Atlanta, GA: American Society of Heating, Refrigerating and Air-Conditioning Engineers.

ASHRAE. 2023b. *ANSI/ASHRAE Standard 514-2023: Risk Management for Building Water Systems: Physical, Chemical, and Microbial Hazards*. Atlanta, GA: American Society of Heating, Refrigerating and Air-Conditioning Engineers.

Banham, R. 1969. *The Architecture of the Well-Tempered Environment*. London: Architectural Press.

Brown, G. Z., J. Reynolds, and S. Ubbelohde. 1982. *InsideOut: Design Procedures for Passive Environmental Technologies*. Hoboken, NJ: John Wiley & Sons.

DOJ. 2010. *2010 ADA Standards for Accessible Design*. Washington, DC: U.S. Department of Justice. ada.gov/law-and-regs/design-standards/2010-stds

EPA. 1975. *Manual of Individual Water Supply Systems*. Washington, DC: U.S. Environmental Protection Agency.

EPA. 2024a. *WaterSense at Work*. Washington, DC: U.S. Environmental Protection Agency. epa.gov/watersense/best-management-practices

EPA. 2024b. *PFAS Analytic Tools*. Washington, DC: U.S. Environmental Protection Agency. awsedap.epa.gov/public/extensions/PFAS_Tools

EPA. 2024c. *Water Budget Tool*. Washington, DC: U.S. Environmental Protection Agency. epa.gov/watersense/water-budget-tool

EPA. 2024d. *WaterSense*. Washington, DC: U.S. Environmental Protection Agency. epa.gov/watersense

HUD. 1977. *Intermediate Minimum Property Standards Supplement*. Washington, DC: U.S. Department of Housing and Urban Development.

IAPMO. 2023a. *2024 Uniform Plumbing Code*. Ontario, CA: International Association of Plumbing and Mechanical Officials.

IAPMO. 2023b. *WE•Stand: Water Efficiency and Sanitation Standard for the Built Environment*. Ontario, CA: International Association of Plumbing and Mechanical Officials.

ICC. 2017. *ICC A117.1-2017: Standard for Accessible and Usable Buildings and Facilities*. Washington, DC: International Code Council.

ICC. 2020. *ASABE/ICC 802: Landscape Irrigation Sprinkler and Emitter Standard*. Washington, DC: International Code Council.

ICC. 2023. *2024 International Plumbing Code*. Washington, DC: International Code Council.

IWBI. 2020. *WELL Building Standard Version 2*. New York: International WELL Building Institute. v2.wellcertified.com

Kira, A. 1966. *The Bathroom*. New York: Random House Inc.

Lowry, W. P. 1988. *Atmospheric Ecology for Designers and Planners*. New York, NY: Van Nostrand Reinhold.

Milne, M. 1976. *Residential Water Conservation. National Technical Information Service*. Washington, DC: U.S. Department of Commerce.

National Drinking Water Clearinghouse. 1996a. *Tech Brief: Filtration*. Washington, DC: U.S. Department of Agriculture.

National Drinking Water Clearinghouse. 1996b. *Tech Brief: Disinfection*. Washington, DC: U.S. Department of Agriculture.

National Drinking Water Clearinghouse. 1997a. *Tech Brief: Corrosion Control*. Washington, DC: U.S. Department of Agriculture.

National Drinking Water Clearinghouse. 1997b. *Tech Brief: Ion Exchange and Demineralization*. Washington, DC: U.S. Department of Agriculture.

National Institute of Environmental Health Sciences. 2025. *Perfluoroalkyl and Polyfluoroalkyl Substances (PFAS)*. Durham, NC. www.niehs.nih.gov/health/topics/agents/pfc

Plante, R. H. 1983. *Solar Domestic Hot Water: A Practical Guide*. Hoboken, NJ: John Wiley & Sons.

Reif, D. K. 1983. *Passive Solar Water Heaters*. Andover, MA: Brick House Publishing.

U.S. Access Board. 2004. *Americans with Disabilities Act Accessibility Guidelines for Buildings and Facilities (ADAAG)*. Washington, DC: U.S. Architectural and Transportation Barriers Compliance Board. federalregister.gov/documents/2004/07/23/04-16025/americans-with-disabilities-act-ada-accessibility-guidelines-for-buildings-and-facilities

USGS. 2024. *PFAS in US Tapwater Interactive Dashboard*. Reston, VA: United States Geological Survey. geonarrative.usgs.gov/pfasustapwater

CHAPTER 22

Liquid Waste

AMONG OTHER ISSUES, CHAPTERS 20 AND 21 CONSIDERED the waste of resources inherent in the use of potable water to flush toilets. This chapter examines some consequences of this conventional approach to bodily waste removal, along with alternatives that use no water at all, and indeed that offer to convert waste to a useful resource. For typical U.S. residences, the potential impact of such alternatives on water usage and treatment is shown in Figure 22.1

Almost every plumbing fixture within buildings is provided with both water supply and waste pipes. Because the toilet is usually the largest user of water (Figure 22.1), as well as one of the worst water polluters, this chapter begins with recycling/waterless alternatives to the flush toilets and urinals that were discussed in Chapter 21. Then, more conventional systems using water to move waste, and conventional treatment systems from individual to large-scale, are discussed. Discussions of graywater and stormwater treatment complete this chapter.

22.1 WATERLESS TOILETS AND URINALS

The array of waterless alternatives includes toilets in which chemicals or oil are substituted for water. These devices are commonly found in airplanes, vehicles, and boats, as well as in remote and environmentally sensitive areas. The chemicals must be frequently recharged and the waste products removed. Other waterless toilets temporarily treat the waste awaiting discharge to a sewer by freezing it, burning it, or otherwise packaging it so that it remains inoffensive. Such devices can become energy-intensive approaches to waterless waste disposal. Also, they are still dependent upon public sewer systems.

(a) Composting Toilets

Once considered too unconventional to ever gain mainstream acceptance, these ecologically attractive devices are now appearing not only in homes, but also in institutional buildings (Figure 22.2). Perhaps several generations needed to pass between regular use of the historic outhouse and use of these modern composting toilets before widespread acceptance could be observed. They offer a self-contained and low-energy strategy to provide waterless toilets. Wherever water supplies are unreliable, or waste discharge is difficult or prohibited, composting toilets offer an alternative. Silent and immune from potential water damage (freezing pipes or overflows), these systems rely upon aerobic digestion of waste (i.e., that which occurs in the presence of oxygen). By and large, aerobic systems are essentially odor-free, and the exhaust air they produce is rich in CO_2 (not necessarily good) and water vapor (harmless). In contrast, anaerobic decomposition (i.e., that which occurs without oxygen) is malodorous and produces

Mechanical and Electrical Equipment for Buildings, Fourteenth Edition. Walter Grondzik, Daniel Overbey, and Alison Kwok.

Companion website: www.wiley.com/go/meeb14e

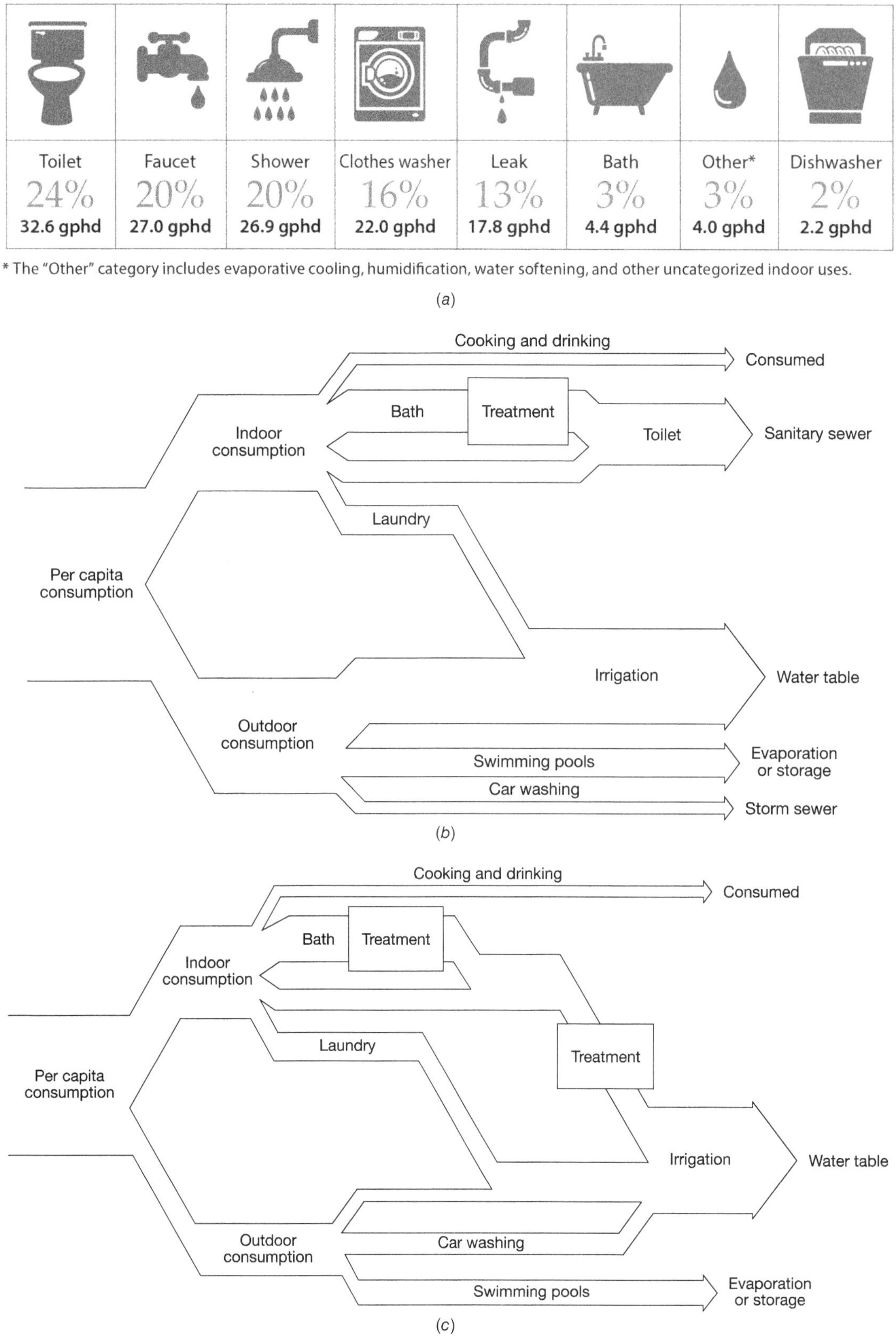

Figure 22.1 *Opportunities for water conservation for the typical U.S. family. (a) At present, the average U.S. residential usage is about 138 gphd (gallons per household per day) [522 Lphd; liters per household per day]—all of it potable—but representing an amazing reduction in usage over the past 40 years. (b) With attention to recycling and the matching of water quality to requirements, potable water usage could be cut by roughly 25%. (c) With further on-site treatment and recycling, potable water usage could drop by around 40%—and the need for public sewage treatment (for residences) would disappear. (a: DeOreo et al., 2016, © Water Research Foundation; used with permission; b, c: from Milne, 1976.)*

methane gas as a key by-product. See Sections 22.7 and 22.8 for more about methane and anaerobic decomposition.

Waste material deposited in composting toilets builds up in a mound, which retards the flow of air for decomposition. Wood shavings are commonly added with each use to encourage toppling of the mound and to add some carbon to the resulting compost. Some composting toilets include a means of regular mechanical stirring of the compost chamber, aiding aerobic decomposition.

Ventilation is important, both to reduce odors and to facilitate evaporation of excess moisture. A ventilation stack is an essential feature; it is often assisted by a very small (3–5 W) fan. Problems with insects hatching within the composting chamber can be lessened by keeping vegetable scraps out of the chamber, as they are frequent sources of insect eggs.

The C.K. Choi Building at the University of British Columbia eliminated a connection to the city sewer system by implementing a combination of composting toilets (Figure 22.2) and a graywater treatment system. The composting toilets in this three-story, 30,000-ft^2 [2787-m^2] facility for the Institute of Asian Research are estimated to save 1500 gal [5680 L] of water per day. At an estimated 30,000 uses per year, the environmental impact of this composting system is significant. The liquid end product will be added to the graywater treatment system at the rate of about 3–5 gal [11–19 L] per day. The composting chambers are located in a small basement below the toilet rooms at the center of the building.

The graywater includes waste from lavatories and sinks, as well as liquid from the composting toilets. This water drains into a subsurface "graywater trench" containing a variety of phragmite plant material. The roots support microbial life that is known for digesting and neutralizing bacteria. This graywater is then used for irrigation. Graywater systems are discussed further in Section 22.9.

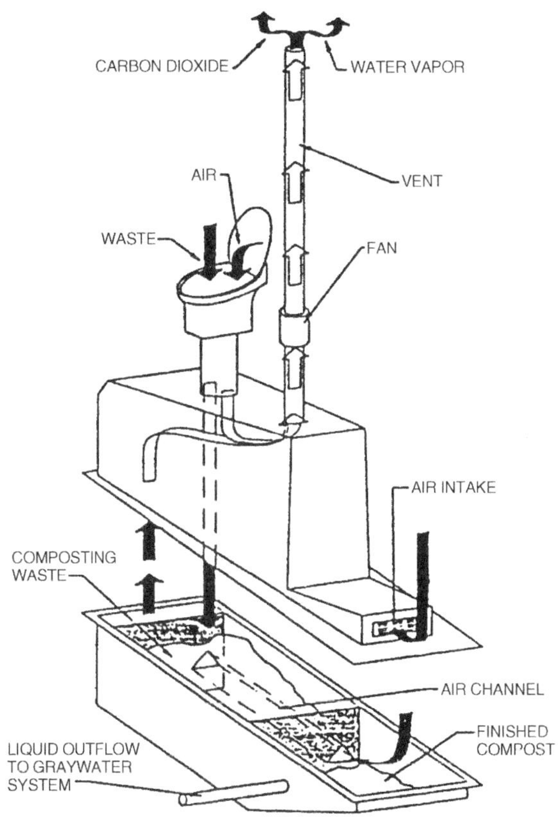

Figure 22.2 *Composting toilets save water and eliminate the need for a sewer connection. They are used at the C.K. Choi Building, University of British Columbia, Canada, where the toilet rooms are concentrated near the center stairs, with a small basement below containing composting chambers and a rainwater cistern for irrigation. The composting toilets produce both dry humus (nutrient-rich soil) and a liquid outflow to a graywater recycling system. (Courtesy of Matsuzaki Wright, architects.)*

(b) Vault-Type Composting Toilets

The Clivus Multrum system (Figure 22.3) has a large decomposition chamber that must be below the toilet (and perhaps also the kitchen), from which it readily accepts organic waste. It must also be accessible from the ground floor in order to remove the humus, varying between 3 and 10 gal [11 and 38 L] of soil per person per year.

The relatively large chamber (about the size of a Volkswagen "bug") and its low position/access requirement can have a significant impact on design. Also of significance is the device's appetite for fresh air. The more air it has, the speedier the process of aerobic decomposition and, therefore, the less odor. In winter, this air flow could cause increased heat losses in a building via infiltration. The natural ventilation due to the stack effect is at least 15 cfm [7 L/s] in the Clivus Multrum. A ventilating fan in the stack will increase this rate significantly. However, if outdoor air is brought directly to the chamber, it could be too cold for proper decomposition; a chamber temperature of 98°F [37°C] is optimum. Solar heating of input air offers one alternative. However, adding too much heat to compost piles encourages oxidation rather than aerobic decomposition; inferior compost results.

Another composting toilet application is shown on the shores of Oregon's Crater Lake National Park in Figure 22.4. Consider the challenge: This facility is at the bottom of a 1-mile trail down from the crater rim; public sewers are miles away and hundreds of feet higher in elevation; severely cold winters and deep snow close the trail for more than half of the year; and Crater Lake is one of the clearest in the world, so water

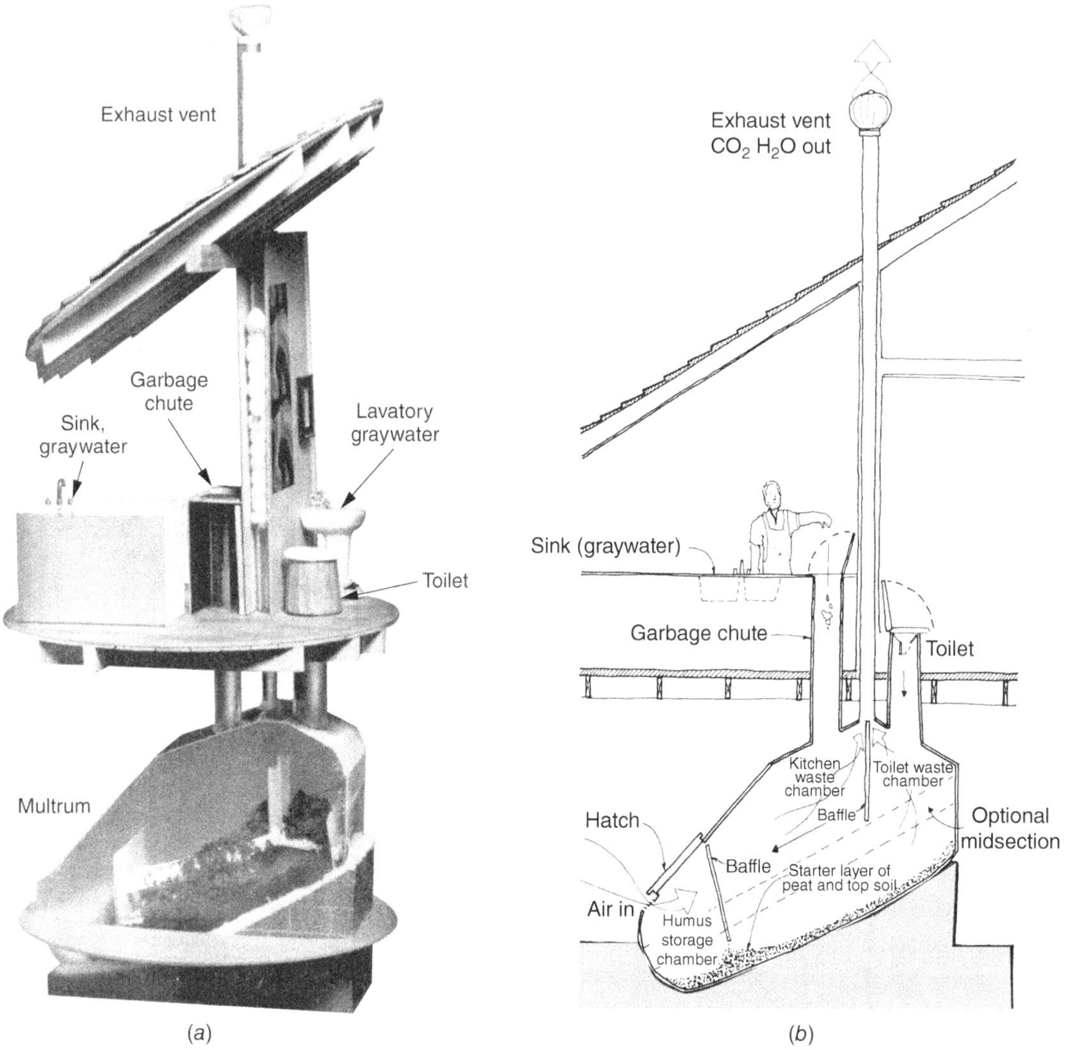

Figure 22.3 Clivus Multrum system, with composting of both kitchen and toilet organic wastes. (a) The lower composting chamber requires 4 ft × 8 ft × 7 ft [1.2 m × 2.4 m × 2.1 m] of floor space, a generous supply of air (warm air speeds decomposition), and access for periodic removal of the garden-ready humus. (b) It must be arranged so that its upper end receives toilet wastes; as the bottom slopes downward, kitchen wastes are deposited on top of the decomposing toilet wastes. The vent stack ensures continuous airflow, preventing odors from entering the bathroom or the kitchen. The toilet seat and cover must be kept closed when not in use to keep air flowing through the composting chamber and out of the stack. The toilet bowl (containing no water) swings open when the seat is occupied, exposing the composting chamber below. (Courtesy of Clivus Multrum USA, Inc.)

pollution is absolutely to be avoided. Three Phoenix composting toilet tanks sit within a solar-heated attached sunspace.

The Phoenix composting tanks (Figure 22.5) feature rotatable tines that speed aerobic decomposition. They also can be fitted with an evaporation system, typically using a small photovoltaic-driven pump that delivers liquid to the top of a drip tank filled with plastic balls. This establishes a very large surface area, and a small fan also aids evaporation.

(c) Heater-Type Composting Toilets

Another approach puts the composting chamber within the toilet itself. The more compact the composting chamber, the more likely that heat must be added to speed decomposition and to evaporate liquid. Although electricity provides an easy and compact way to heat, using such high-quality energy to treat waste reduces the compact toilet's ecological advantage.

In the most compact of Sun-Mar's products (Figure 22.6), a variable-diameter Bio-drum is turned by a fold-out hand crank at the front of a rather conventional-looking toilet. At this size, it is rated for continuous residential use by one adult or a family of two (or for weekend/vacation use, by three adults or a family of four). Sun-Mar uses a three-chamber approach, separating the quick composting of waste

Figure 22.4 On the shore of Crater Lake, Oregon, a building with three Phoenix composting toilets serves visitors. An attached sunspace contains and warms the composting chambers and contains an evaporator for liquid waste. The facility is closed all winter.

and toilet paper, the evaporation of liquid, and the safe storage of the finished compost.

(d) Waterless Urinals

Urine is often kept out of composting toilets for several reasons. First, excess liquid in the chamber creates a discharge issue, complicating an otherwise simple end result of occasional harvesting of garden-ready humus. Second, urine is nearly sterile and an excellent fertilizer, and its separate capture could be useful. If not captured, urine can be directed either to a graywater or to a wastewater system.

Waterless No-Flush™ urinals utilize a floating layer of BlueSeal® liquid that forms a barrier to sewer vapors but allows urine to readily pass through (Figure 22.7). This liquid does not dissolve, mix, or react chemically with urine. It is more than 95% biodegradable and does not evaporate at 100°F [38°C]. Because 1 quart [0.9 L] of this liquid lasts for 15,000–20,000 urinal uses, it replaces an estimated 15,000–60,000 gal [56,780–227,120 L] of water that would have been used by conventional flush urinals. The trap retains sediments and should be emptied occasionally, and the liquid is periodically replenished (3 oz [89 mL] per 1500 uses).

22.2 PRINCIPLES OF DRAINAGE

Early in the history of indoor plumbing, waste drainage was a simple matter; a pipe containing wastewater led to a sewer (Figure 22.8*a*). Before long, though, the noxious gases that were created by the anaerobic conditions in sewers became a threat to the health of those indoors. Thus, the trap was invented (Figure 22.8*b*) to block the pipe so that gases could not pass. However, as moving water fills the pipe downstream from a trap, the mass of water acts as a plunger, creating higher pressures in front and negative pressures behind. The positive pressures might force sewer gas through the water in other traps; worse, the negative pressure could suck (or siphon) the water from the trap, leaving it open to gas passage. The installation of vents provides a way to deal with these pressures so that the suction draws air down through the vents rather than water from the traps (Figure 22.8*c*). A typical arrangement of fixtures, traps, and vents is shown in Figure 22.9.

(a) Traps

The only separation between the unpleasant and dangerously unhealthy gases in sanitary drainage pipes and the air breathed by room occupants is the water caught in the fixture trap after each discharge from a fixture. Sufficient water must flow, especially in water closets, to keep this residual water clean. Traps are made of steel, cast iron, copper, plastic, or brass—except those in water closets and urinals, which are often made of vitreous china cast integrally with the fixture. The deeper the seal, the more resistance to siphonage, but the greater the fouling area; therefore, a minimum depth of 2 in. [50 mm] and a maximum depth of 4 in. [100 mm] are common standards. All traps should be self-cleaning, that is, capable of being completely flushed each time the trap operates so that no sediment will remain inside to decompose.

There are a few exceptions to the rule that each fixture should have its own trap. Common exceptions include two laundry trays and a kitchen sink connected to a single trap; not more than three laundry trays using one trap; and three lavatories on a single trap. In the case of the laundry trays and sink, the sink is equipped with the trap and set closest to the stack (see rightmost stack in Figure 22.9).

Traps are usually placed within 2 ft [610 mm] of the fixture and should be accessible for cleaning through a bottom opening that is otherwise closed by a plug. Overflow pipes from fixtures are connected to the inlet side of the trap. In long runs of horizontal pipe, so-called running traps are used only near drains

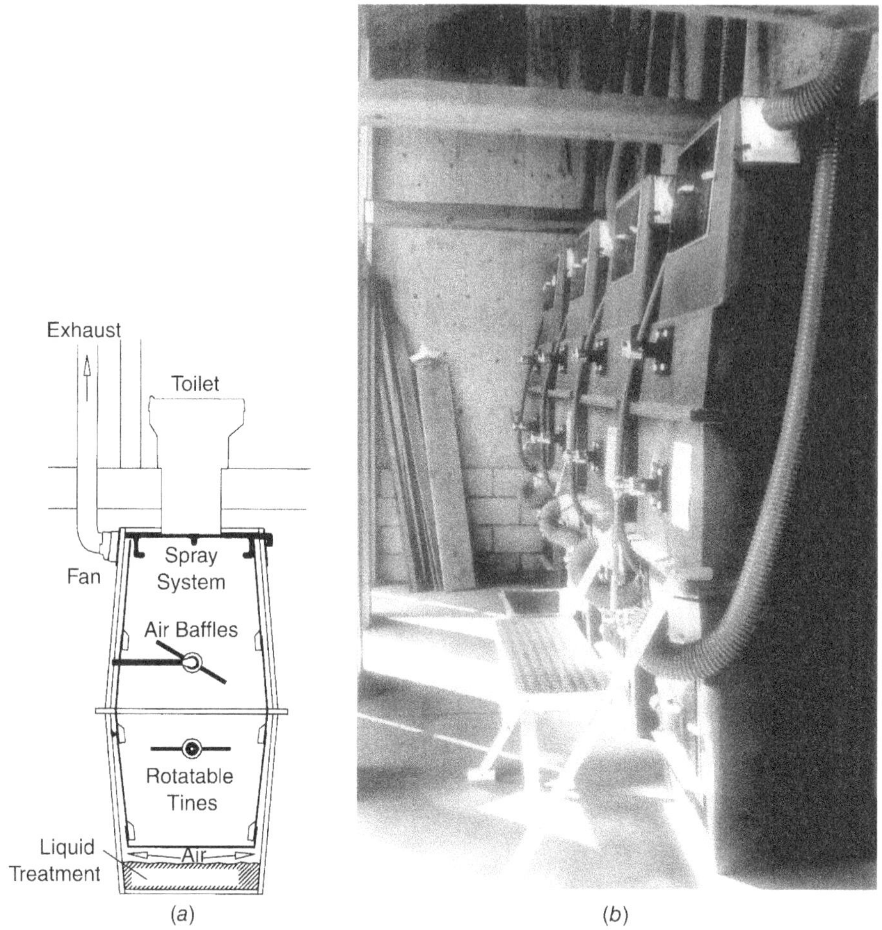

Figure 22.5 *The Phoenix composting toilet's tank serves either two toilets, or one toilet and one food waste inlet. (a) Schematic section: rotatable tines mix waste and control the movement of compost to the access area for eventual removal. The fan exhausts air from the Phoenix tank to control odors. Passing behind the air baffles, air makes frequent contact with the compost pile, ensuring aerobic decomposition. After passing through the compost pile and receiving secondary treatment in the bottom of the Phoenix, liquid is periodically sprayed on top of the compost pile to keep the pile moist, and to introduce decomposing organisms to the fresh waste. Each tank is 39 in. [1000 mm] wide, 61 in. [1550 mm] long, and requires another 60 in. [1525 mm] in front for maintenance access; they are available in three heights: 55 in. [1400 mm], 73 in. [1850 mm], and 92 in. [2350 mm]. Another 12 in. [300 mm] above the top of the tank is convenient for maintenance. (b) Four chambers with their rotatable tines serve the Vogelsang High Sierra Camp at Yosemite National Park, California. (Courtesy of Advanced Composting Systems.)*

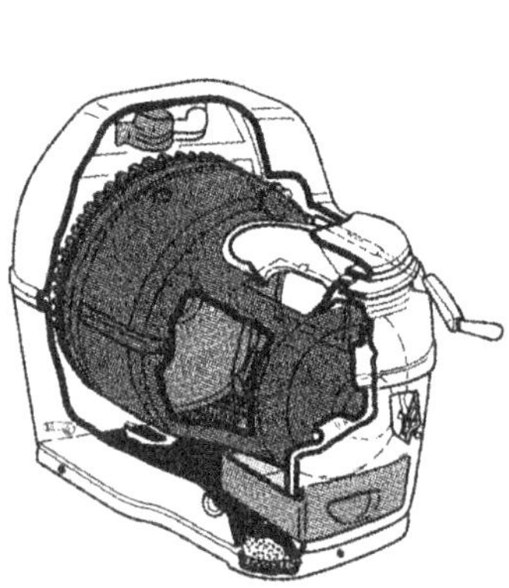

Figure 22.6 *Sun-Mar's "compact" composting toilet has a Bio-Drum™ that is turned by a hand crank. The crank folds back into the body of the toilet after use. It also uses a 200 W electric heater to speed decomposition, a 25 W fan, and a 2-in. [50-mm] vent that attaches at the rear. The self-contained unit is designed for light residential use. (Courtesy of Sun-Mar Corporation.)*

in floors, areaways, or yards and should be provided with handhole cleanouts. "Island" sinks pose a special problem when the vent line cannot lead up from such an exposed location. The sink's waste line can be taken to a distant sump, which is then itself trapped and vented.

When fixtures are used very infrequently, the water in traps can evaporate into the air, breaking the seal of the trap. In contemplating the possible frequency of use, this fact should be kept in mind by the designer. Unoccupied residences (such as weekend or vacation homes) may experience sewer gas penetration through traps emptied by evaporation. Otherwise, evaporation to a dangerous degree rarely occurs,

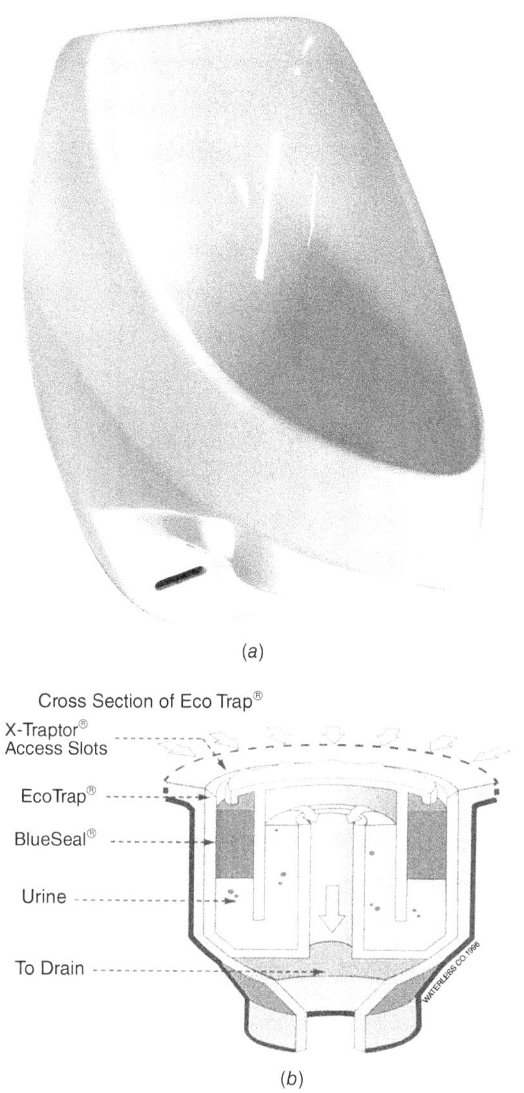

Figure 22.7 *Using no water, (a) the Waterless No-Flush™ urinal employs a special liquid to allow urine to pass through (b) while keeping sewer gas trapped in the waste pipe. (Courtesy of the Waterless Company LLC.)*

except in the case of floor drains. Used to carry away the water used in washing floors or drained from heating/cooling equipment, floor drains may often lose their water seal between their infrequent uses. Some authorities are reluctant to approve floor drains connected to a building sewer, requiring instead that they be separately connected to a dry well. In either case, the use of a special hose bibb, affording a source of water directly above the drain, is a wise precaution. It can easily be used to manually refill the trap of the drain. Other strategies include using a deep trap drain and/or providing a drip source of water to maintain the trap seal.

(b) Vents

To admit air and discharge gases, soil and waste stacks are extended through roofs, and a system of air vents, largely paralleling the drainage system, is provided. As in the case of drainage stacks, the ventilating stacks extend through the roof or vent through the drainage stack. The functions of venting are often misunderstood. It is true that one important purpose is to ventilate the system by allowing air from the fresh-air inlet (or from the sewer, if there is no house trap or fresh-air inlet) to rise through the system and carry away offensive gases. This provides some purification for the piping. However, several other purposes are served by the vent piping. The introduction of air near a fixture (and, in the case of circuit vents, at the branch soil line) breaks the possible siphonage of water out of a trap. Under other circumstances—namely, when drainage fluids descend to a fixture group through the soil stack—the foul gases, under pressure, could bubble through the trap seals of that group. The vent system provides a local escape for these gases. Circuit venting permits air and gases to pass in and out of the soil or waste branch, instead of at each fixture (as is the case of continuous venting in individual fixture venting); this can prevent the siphonage of trap seals or their penetration by gases (Figure 22.10). Some codes may prohibit circuit venting.

(c) Air Gaps and Vacuum Breakers

Nearly every plumbing fixture is supplied with pure water at one point, and most discharge contaminated fluids at another. The proximity of sewage to potable water at most fixtures is inescapable; sewage could accidentally be siphoned into a pipe carrying potable water. Consider an improperly placed faucet whose outlet is below the rim of a fixture. If the fixture overflow is plugged and the fixture bowl full, the faucet can easily project into the foul drainage water. If, in this circumstance, the water piping is drained while the faucet is open, contaminated water could be drawn by suction into the water piping.

In water closets served by flushometers (flush valves), the water supply unavoidably enters the bowl below the rim. A vacuum breaker placed in the flushometer closes with water pressure but opens to admit air if there is suction in the water pipe. This prevents siphonage in much the same way that a vent prevents trap siphonage (Figure 22.11). The use of vacuum breakers at dishwashers and clothes washers was diagrammed in Chapter 21. These are especially important locations because, in both of these appliances, pumps force wastewater into a drain line.

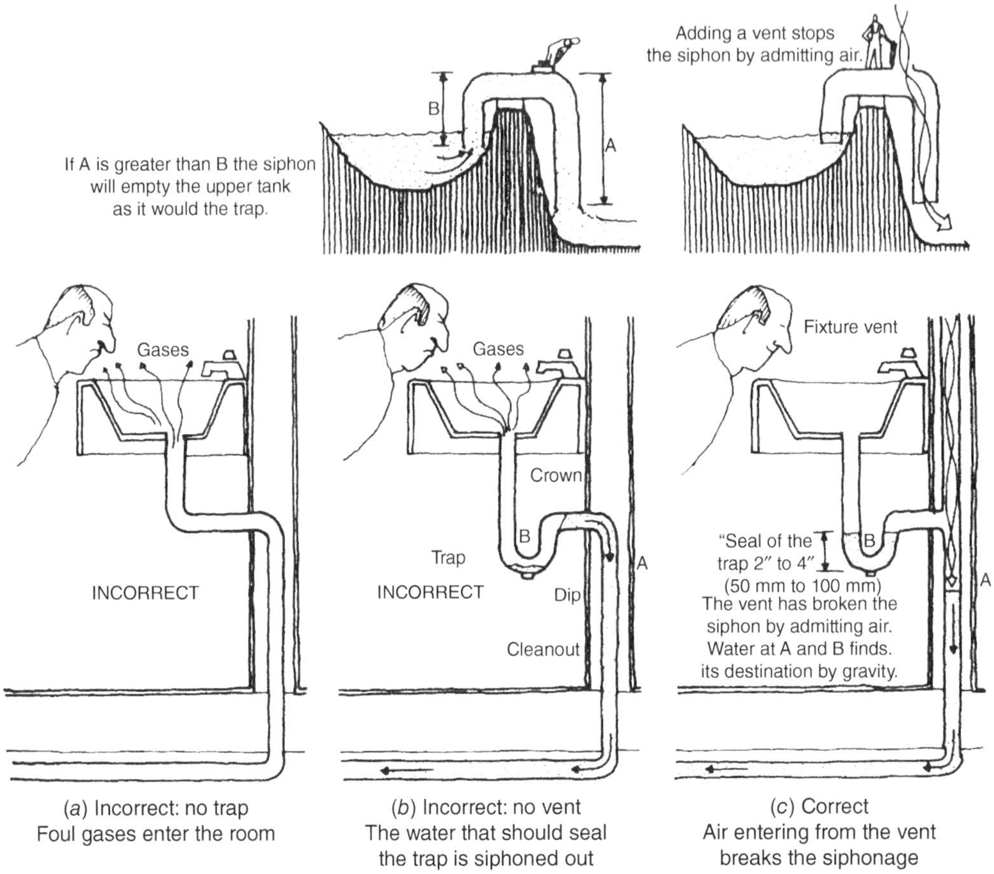

Figure 22.8 *Function of a trap and one of the several functions of a vent, preventing siphonage.*

22.3 PIPING, FITTINGS, AND ACCESSORIES

(a) Piping and Fittings

The materials used within buildings for soil and waste piping, and for venting, include cast iron, copper, acrylonitrile butadiene styrene (ABS) plastic, and polyvinyl chloride (PVC) type drainage, waste, and vent (DWV) pipe (Table 22.1, Figure 22.12). Galvanized steel is sometimes chosen for vents and for tall stacks in high-rise structures. Sometimes, different materials are used in the same system. Where dissimilar metals are connected, dielectric unions are used to prevent corrosion due to electrolysis. Materials used underground for sewage disposal (depending upon local codes) include vitrified clay tile, cast iron, copper, asbestos-cement, ABS plastic, PVC type DWV, and concrete pipe.

Cast Iron. Used first in Germany around 1562 and appearing in the United States about 1813, cast iron supplanted the tubing and culverts of earlier eras that employed clay, lead, bronze, and wood. Cast iron was thus the earliest of the modern materials used for piping. Its durability and resistance to corrosion make it appropriate for a wide range of uses, from small residential work to the stacks and branches of tall buildings.

Typical fittings for sanitary drainage appear in Figures 22.12 and 22.13. In sanitary flow systems that are composed of any material, changes in direction must be made with easy bends. To prevent clogging or fouling by the solid materials flowing in the piping, right-angle connections are not used. Thus, for sanitary drainage systems, the available options in Figure 22.13 would be the one-eighth bend connected to a 45° Y, or the one-quarter bend long sweep. The 90° T, in the position shown, would be used only in a vent system.

The three cast-iron soil pipe joints shown in Figure 22.14 represent semi-rigid, watertight, and

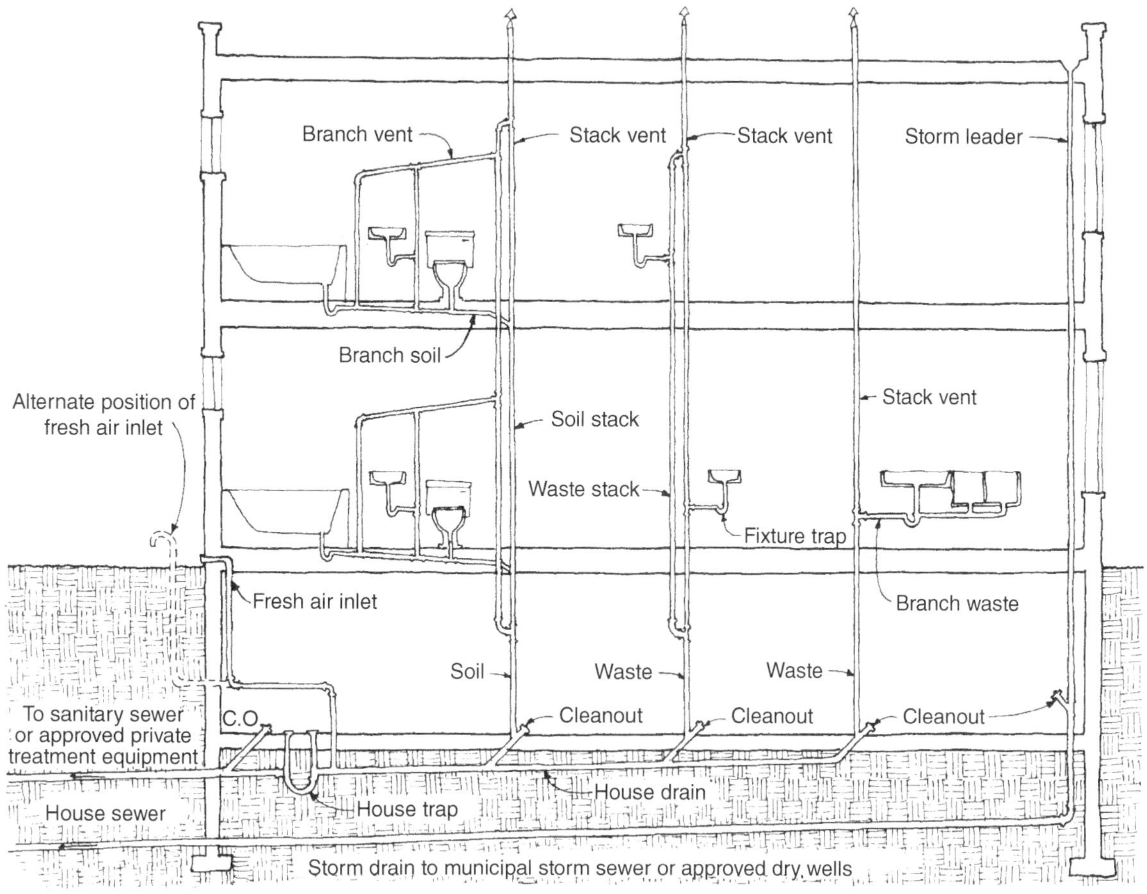

Figure 22.9 *Typical sanitary drainage system, which separates stormwater from sewage. This installation combines blackwater (from water closets) and graywater, although their separation would encourage graywater recycling. The house trap is optional under some plumbing codes and illegal under others because sanitary sewers could be vented at street level through the fresh air inlet.*

gastight connections of two or more pieces of pipe or fittings in a sanitary system. Types b and c provide a quieter plumbing system and slightly more flexible joints than type a. See Figure 22.15 for cast-iron soil pipe and fittings in bathroom groups.

Copper Tube and Fittings for DWV. There are several tube classifications for the copper products used in plumbing systems: types K, L, and M are for water supply systems, and type DWV is for drainage, waste, and vent installations. Connections between copper tubing and its couplings or fittings involve a sliding fit between components (see Figure 22.12*b*). Bridging the mating surfaces is a cylindrical capillary space filled with solder. The process of making the joint is a simple one, utilizing a flux, heat, and solder. Properly made, the joint will be airtight and capable of withstanding high pressure (although such pressure is not normal in waste lines). To undo the joint for repair or renovation, one simply reheats it until the solder melts. Like cast iron, copper has a history of use in ancient installations. Updated and highly developed in recent decades, it is now in widespread use.

Plastic Materials for DWV. Along with copper and cast iron, plastics are suitable for sanitary drainage systems. Table 22.1 lists the plastics suitable for DWV, as well as for building drains and sewers. Acrylonitrile butadiene styrene (ABS) is identified and further described by the labeling shown in Figure 22.16. One of several steps used in making a "solvent-weld" connection is seen in Figure 22.16, as is a method of support used in wood frame construction. Figure 22.17 shows assembled

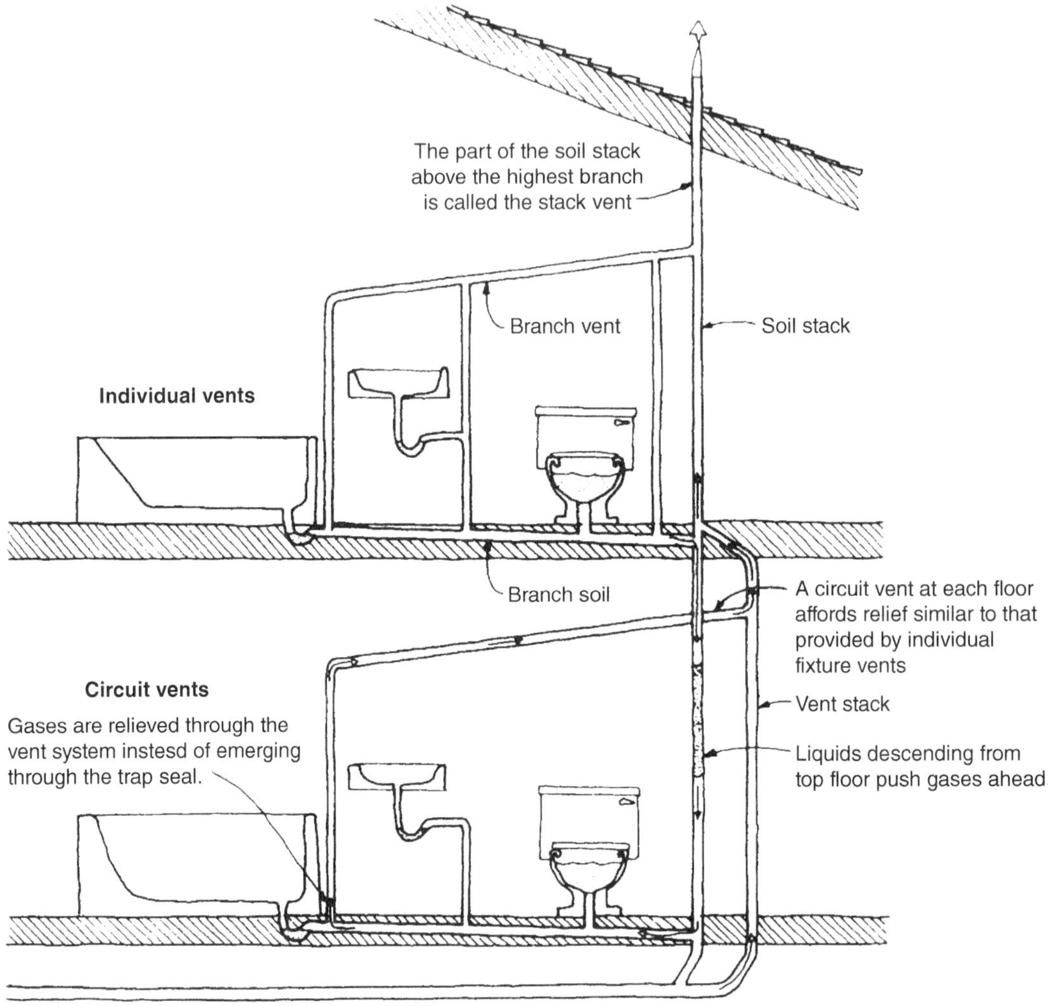

Figure 22.10 *Sewer gas relief through vents. Gases that are pressurized, either via hydraulic action or by expansion due to putrefaction, have an escape path through the vent system—keeping said gases from entering the rooms. The upper floor shows individual vents (a vent at each fixture); the lower floor shows circuit vents, permitted by some codes, where one vent serves the entire branch.*

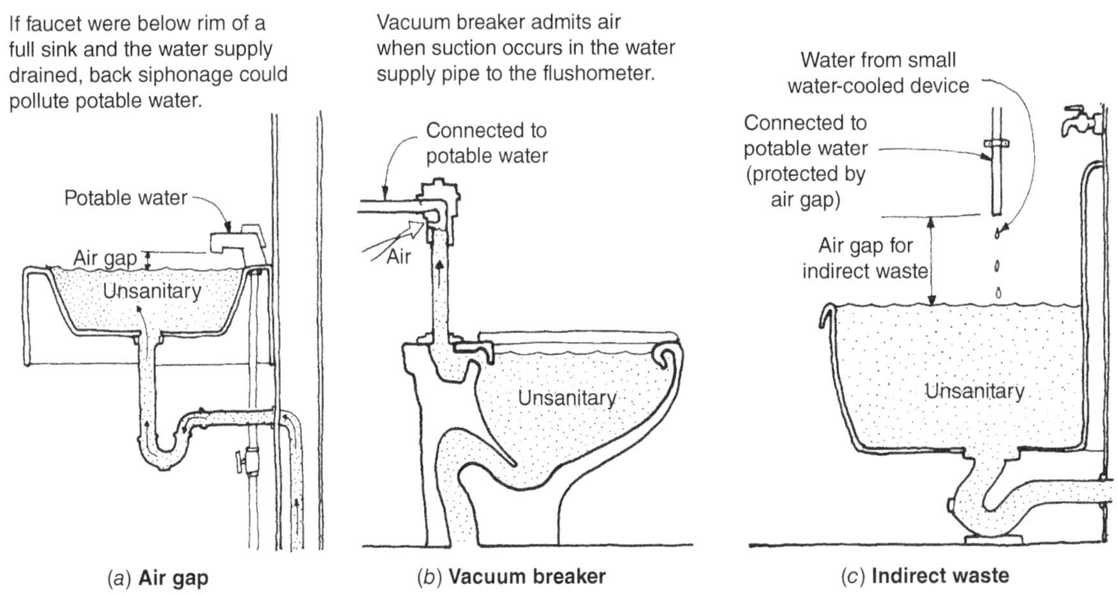

Figure 22.11 *Backflow preventers. Unsanitary fluid wastes cannot be siphoned into the potable water piping in these three examples.*

TABLE 22.1 Materials for Waste Piping

		Underground	
Material	**Above-ground DWV**	**Building Drainage and Vent**	**Building Sewer**
ABS (acrylonitrile butadiene styrene)	✓	✓	✓
Cast-iron	✓	✓	✓
Concrete			✓
Copper or copper-alloy (type)	✓ (K, L, M, or DWV)	✓ (K, L, M, or DWV)	✓ (K or L)
Galvanized steel	✓		
Glass	✓		
PE (polyethylene) (type)		✓(SDR-PR)	✓ (corrugated, profile wall, or SDR-PR)
Polyolefin	✓	✓	
PP (polypropylene)			✓
PVC (polyvinyl chloride)	✓	✓	✓
PVDF (polyvinylidene fluoride)	✓	✓	✓
Stainless steel drainage system (type)	✓ (304 or 316 L)	✓ (316 L)	✓ (304 or 316 L)
Vitrified clay			✓

Source: This table is a copyrighted excerpt from the 2024 International Plumbing Code.

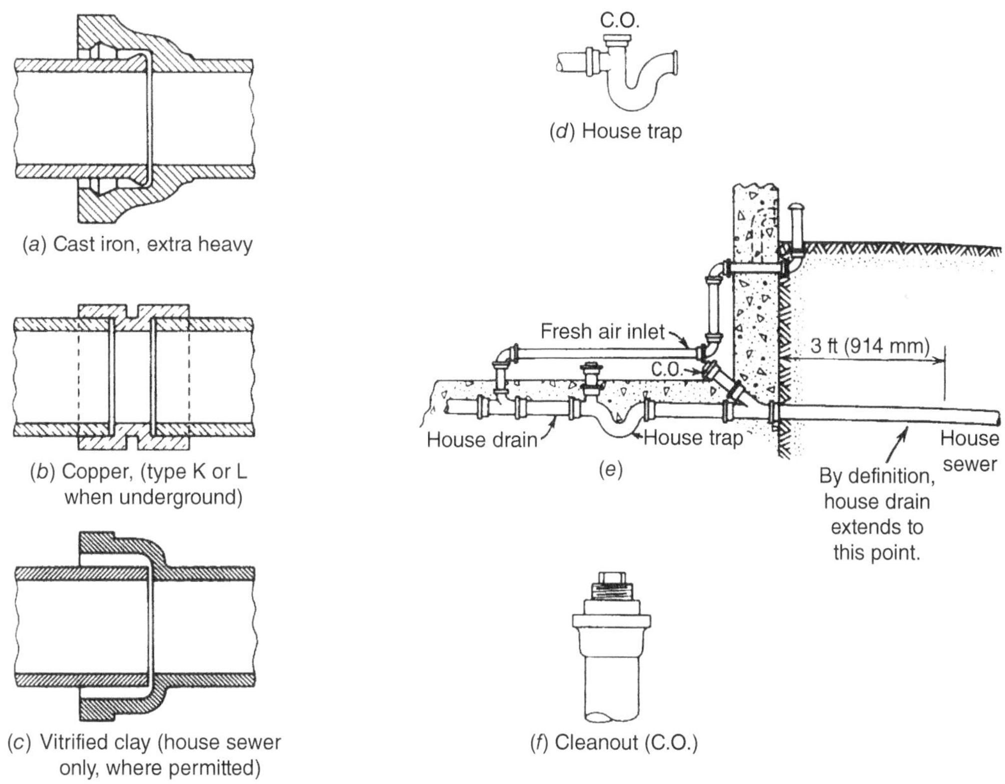

Figure 22.12 *Piping and fittings. (a) Connection of cast-iron piping. (b) Coupling to connect copper tubing. (c) Connection of vitrified clay piping. (d) Detail of house trap fitting. (e) House drain, house trap with cleanouts and vent (fresh air inlet), and house sewer. (f) Cleanout showing removable threaded plug. For large buildings, the terms building drain, building sewer, and so on, supplant the terms house drain, house sewer, etc. Local codes differ on inclusion or omission of the house trap. Note: plastic pipe connections are similar to that in (b).*

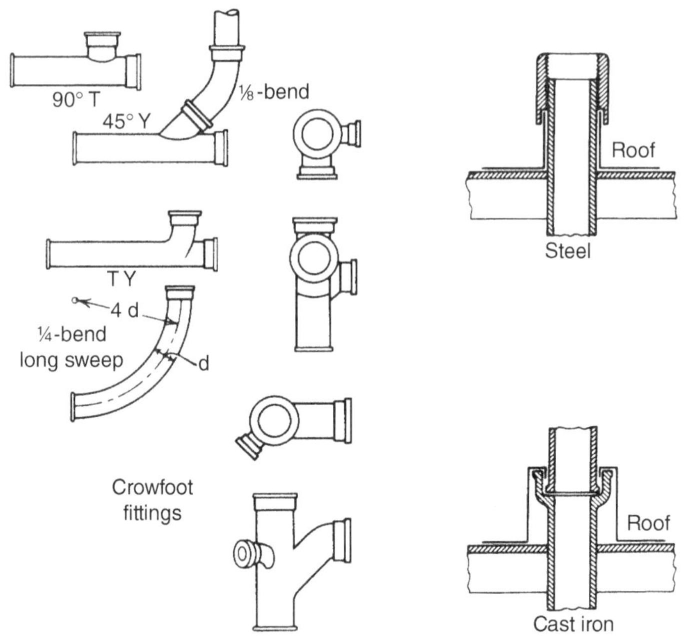

Figure 22.13 *Principal types of cast-iron fittings and method of flashing at roofs for steel and cast iron.*

bathroom piping in place. One advantage of plastic piping is its relatively light weight, allowing some preassembly in a location with conditions better than might be available at the final installation location—the assembly then being carried to its place of installation.

(b) Accessories

Among the many special devices that can form part of a plumbing system are floor drains, backwater valves, ejectors, and interceptors.

Floor Drains. When floors in buildings must be washed down after operations such as food preparation and cooking, floor drains are usually necessary. Figure 22.18 shows a typical floor drain. Because these drains are often connected to sanitary drainage systems (rather than dry wells) and, in long periods of disuse, might lose their trap seals by evaporation, special precautions are necessary to preserve the trap seal and avoid odors and unsanitary conditions in the room.

Backwater Valves. These devices (Figure 22.19) are sometimes used when plumbing fixtures are installed at low elevations, such as in basements, or in other locations that are near the level of the sewer. They cannot be used to protect the entire plumbing system and should be used only when necessary. They must be accessible for maintenance. An alternative to the use of backwater valves is the sewage sump.

Sewage Sumps and Ejectors. Whenever subsoil drainage, fixtures, or other equipment are situated below the level of public sewers, a sump pit or receptacle must be installed. The drainage from the low fixtures may flow into this pit by gravity, and from it the contents are then lifted (ejected) up into the building sewer. Sewer ejectors may be motor-driven centrifugal pumps (Figure 22.20), or they may be operated by compressed air. The latter have no revolving parts within the receptacle. An air compressor is started when a float within the sump reaches a certain level, and air at a pressure greater than 0.433 psi for each foot [10 kPa for each meter] of lift is delivered into the space above the liquid. The air pressure closes the inlet and opens the outlet check valves, expelling the contents of the sump and elevating it to the sewer.

Interceptors. Sanitary drainage installations ultimately discharge their waste matter into private or public sewage treatment plants that attempt to digest or cope with anything that may come through the pipes. Because it is impossible to control what will be discarded into the plumbing drains, trouble can occasionally be expected. The problem can be reduced somewhat by devices known as interceptors, which catch foreign matter before it travels too far into the system.

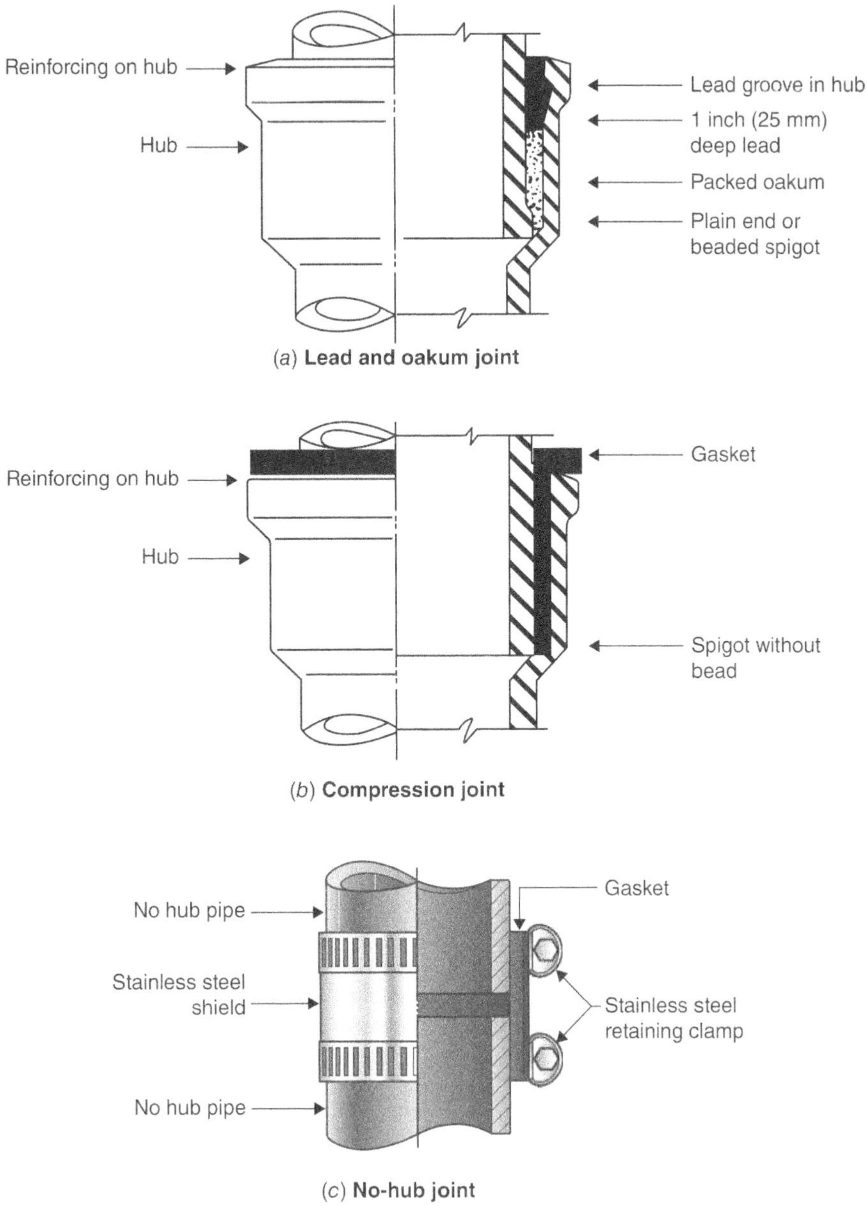

Figure 22.14 *Various joints used to connect cast-iron soil pipe and fittings. (Courtesy of the Cast Iron Soil Pipe Institute.)*

Interceptors require periodic servicing. Interceptors for as many as 25 different kinds of extraneous material are listed by manufacturers. They include devices to catch hair, grease, plaster, lubricating oil, glass grindings, or material from many industrial processes. One of the few interceptors that is sometimes used in homes, but more often in institutional kitchens, is the grease interceptor, or grease trap. As the waste is passed from a kitchen sink through the circuitous path within the grease interceptor, the grease floats to the top, where it is trapped between baffles, while the more fluid wastes pass through at a lower level (Figure 22.21). If not so intercepted, grease congeals within piping and thus physically retards the sewage collection process.

Cleanouts. Clogging of sanitary drainage piping systems is very possible—if not likely. From all plumbing fixtures to the end of the disposal process, all parts of a sanitary piping system should be accessible through cleanouts (Figure 22.12) and/or other points of access to relieve clogging that will often occur in the piping (as well as in the septic tank or public disposal plant).

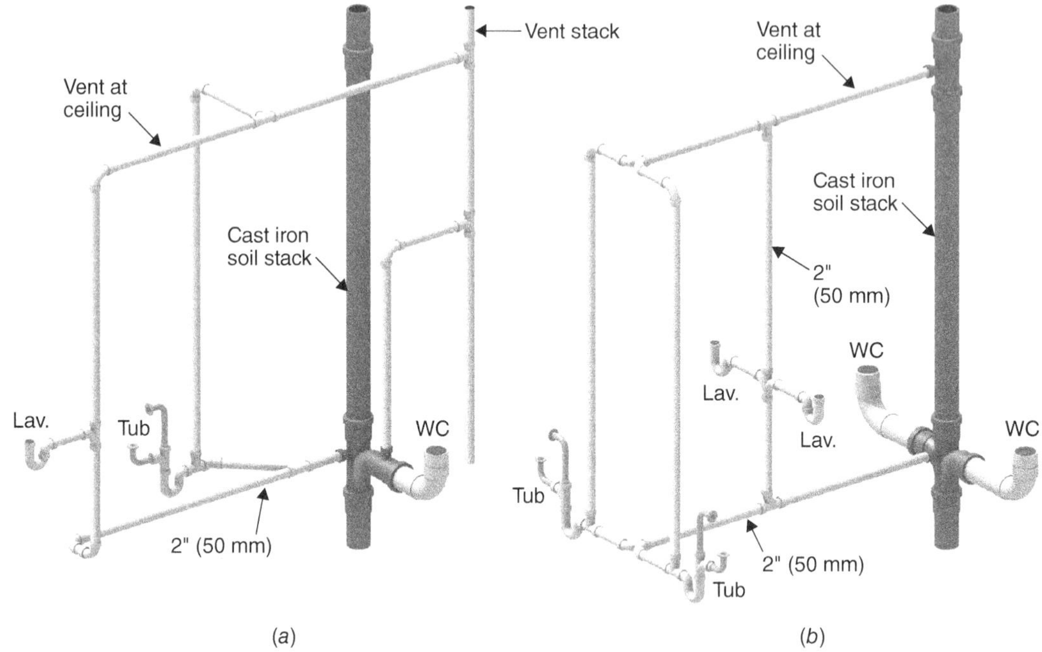

Figure 22.15 Two typical piping arrangements for a water closet, lavatory, and tub. (a) For a multistory installation with each fixture vented; (b) for a single-story, back-to-back installation. (Courtesy of the Cast Iron Soil Pipe Institute.)

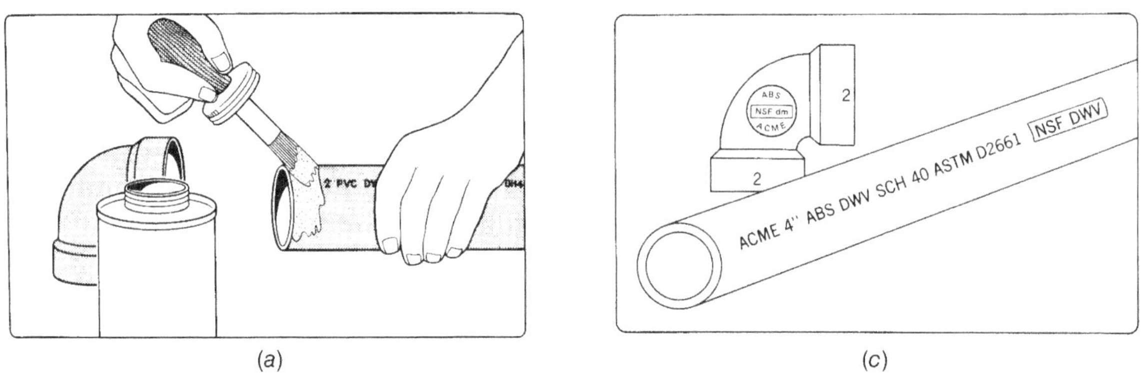

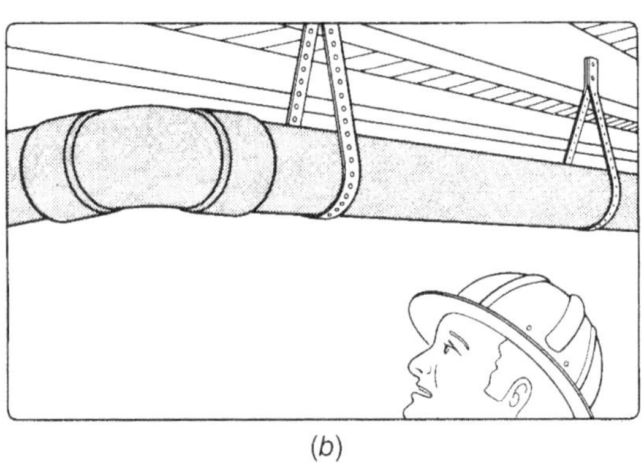

Figure 22.16 Details of the use of plastic pipe. (a) One of the steps in making a solvent weld of a plastic pipe to a plastic fitting. (b) In wood frame construction, plastic pipe assemblies can be supported by metal straps nailed to wood joists. The supports should be more closely spaced than those for metal piping because of plastic's increased flexibility. (c) Typical identification symbols on plastic pipe: ACME—the name of the manufacturer; 4-in. [102-mm]—the diameter of the pipe; ABS—acrylonitrile butadiene styrene, the material; DWV—suitable for drainage, waste, and vent; Scheme 40—Schedule 40, this identifies the wall thickness of the pipe; ASTM D2661—the applicable American Society for Testing and Materials standard; NSF DWV—tested by the National Sanitation Foundation Testing Laboratory, the pipe meets or exceeds the current standards for sanitary service. (Courtesy of the Plastic Pipe Institute.)

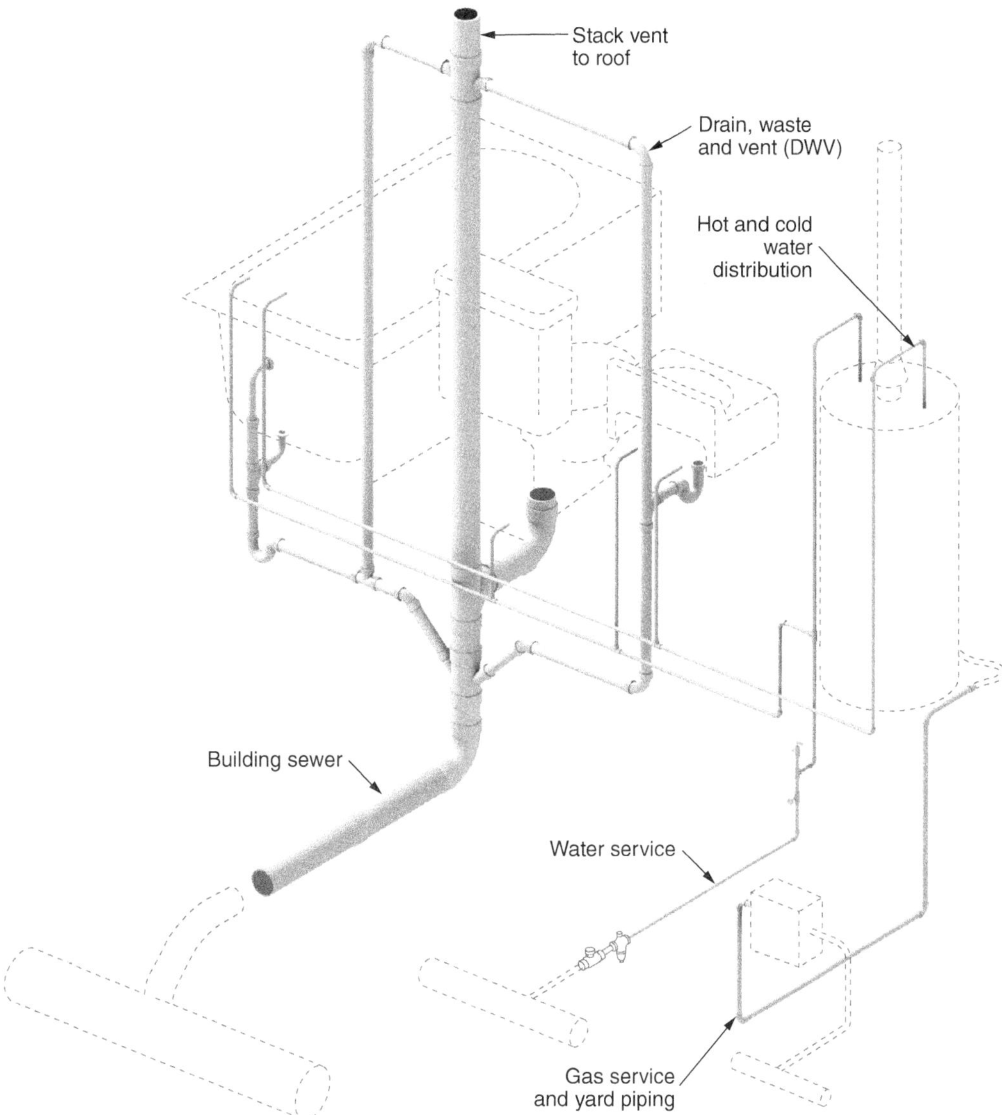

Figure 22.17 *Plastic piping for water service, gas service, hot and cold water supply, and for drainage, waste, and vent. Gas service below grade can be PE, PB, or PVC. (Courtesy of the Plastic Pipe Institute.)*

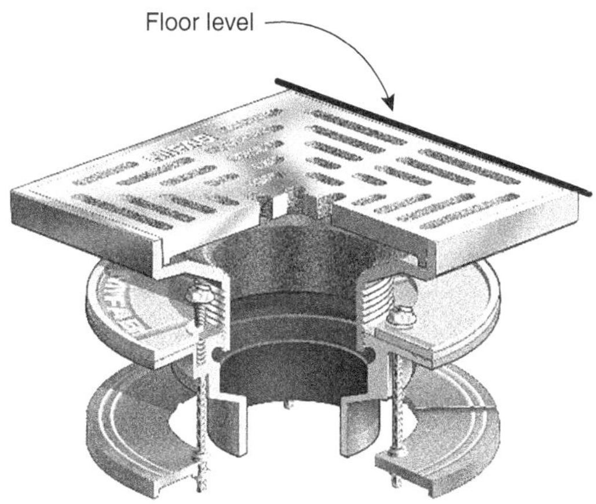

Figure 22.18 *Floor drain. (MIFAB, Inc.; used with permission.)*

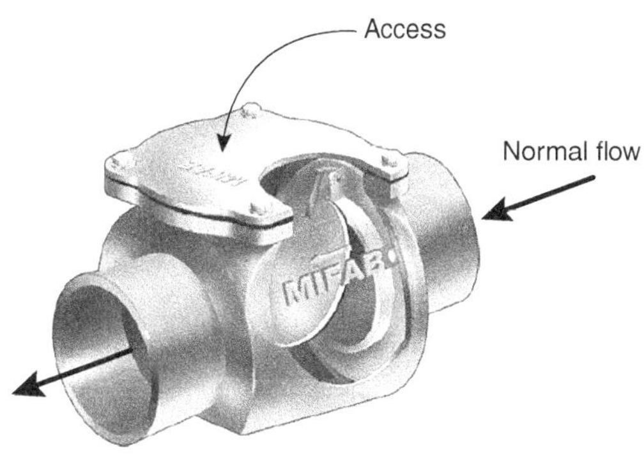

Figure 22.19 *Backwater valve, with access to facilitate unclogging and cleaning. (MIFAB, Inc.; used with permission.)*

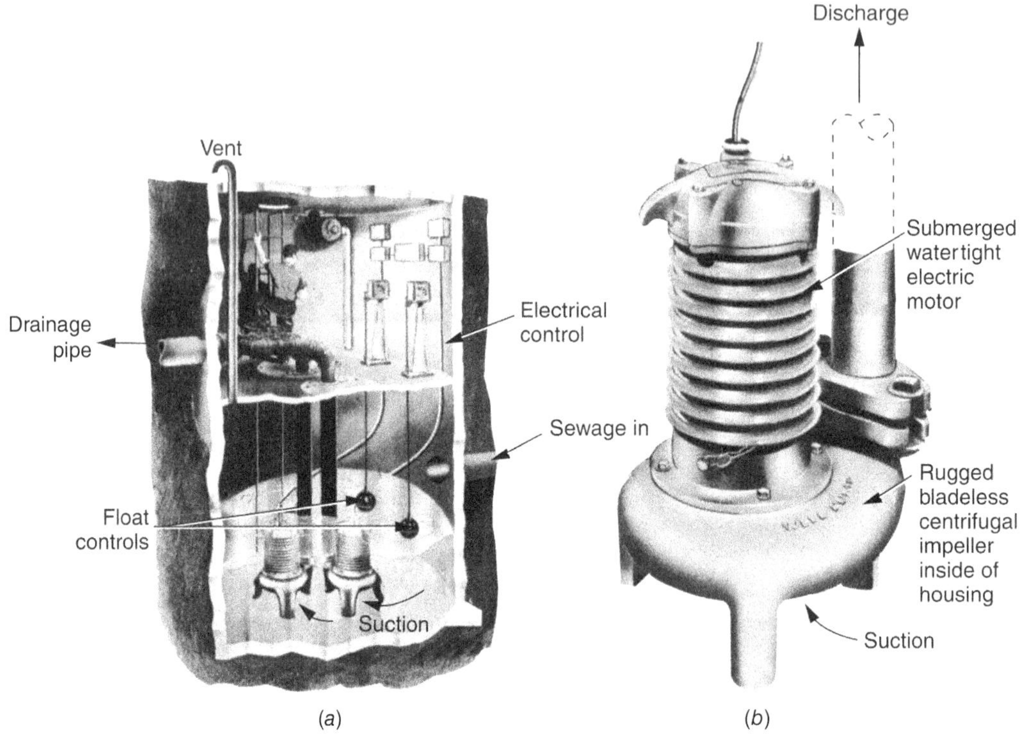

Figure 22.20 *Sump (a) and ejector pump (b). A submersible-type centrifugal pump for raising sewage to a higher level. Shown here for an outdoor subgrade sump installation, it may be used in basement applications within buildings. Venting must be carried to the roof. (Courtesy of the Weil Pump Company.)*

22.4 DESIGN OF RESIDENTIAL WASTE PIPING

In residential work, the piping assemblies may often look like a "flag," where the mast is the soil stack, the horizontal top of the flag is the branch vent, the bottom is the soil or waste branch, and the outer edge is the vertical pipe of the last fixture. In wood or steel frame construction, the flag usually fits into a 6-in. [150-mm] partition. Fixture branches project from the surface of the flag. There is considerable advantage in back-to-back layouts for baths and kitchens; this allows the piping assembly to pick up the drainage from fixtures on both sides of the piping. When all the fixtures are on nearly the same level, it is unnecessary to have a separate vent stack next to the soil stack, as is often the case in multistory construction. Generally, in one-story construction, the upper part of the soil stack is extended to form a vent called a stack vent, to which the branch vents connect. A separate major vertical vent is called a vent stack.

The task of fitting two generally parallel plumbing "distribution trees" (supply and waste) into available horizontal and vertical chases can be difficult. This is especially true of the waste system, which has larger pipes, because it is not under pressure and must carry solids as well as liquids. To compound the problem, because of gravity flow requirements, these larger pipes must slope continuously downward, from fixtures to the sewer. In residences, the best available "vertical chase" is sometimes nothing more than an extra-thick wall, increasing the difficulty of coordination.

In residential applications (and other relatively small buildings), certain fairly standard minimum sizes (such as a 4-in. [102-mm] soil stack and building drain) are usually adequate. Horizontal fixture branches from individual fixtures should be the same size as the fixture trap. Horizontal fixture branches from groups of fixtures are sized by the drainage fixture units of the group. For individual fixture vents, the vent size is usually the same as the size of the fixture's horizontal branch. Under many codes, vertical vents that penetrate the roof must increase to a 4-in. [102-mm] size to prevent blocking by icing in freezing weather.

Tables 22.2 through 22.5 list minimum pipe sizes to carry waste and serve for venting.

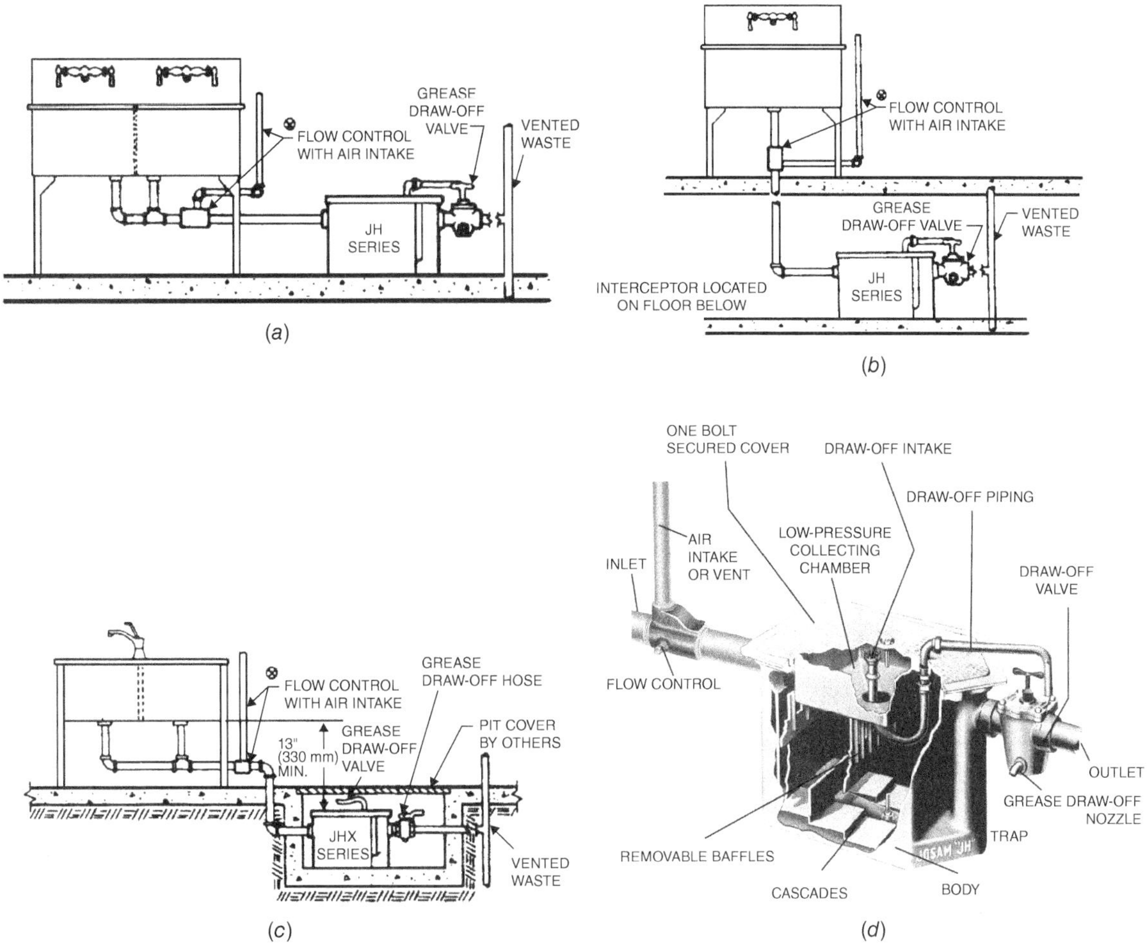

Figure 22.21 *One type of institutional interceptor, a grease trap. Choice of three locations—adjacent to the sink (a), on the floor below (b), or in a pit (c). Periodic cleaning is both necessary and unpleasant; easy access is helpful. (d) Cutaway view with identification of component parts. (Courtesy of the Josam Manufacturing Company.)*

EXAMPLE 22.1 Design, lay out, and size the piping for the sanitary drainage system for the house shown in Figure 22.22. Use I-P units.

SOLUTION

The first step is to identify the locations where hot and cold water is needed at fixtures, and where soil or waste drains must be provided. Figure 22.23*a* illustrates how this is done. A plan layout for the drains in both levels follows (Figure 22.23*b*).

Next comes the plumbing section diagram (Figure 22.24). The local administrative authority usually requires this to be submitted for approval. Note the given pipe sizes. Drainage fixture units (dfu) for this system are summarized in Table 22.6 from data given in Table 22.2.

Although Table 22.3 permits a 3-in. [76-mm] branch for up to 20 dfu, it is common practice to use a 4-in. [102-mm] branch for every water closet (or every group of water closets totaling 20 dfu or less). Because Table 22.3 allows 160 dfu for a 4-in. [102-mm] branch, this size is acceptable for any branch to which a water closet is connected. This size is also more than adequate for the stack in this example.

A water closet must have a 2-in. [51-mm] vent. Table 22.4 shows that a 2-in. [51-mm] vent, for developed vent lengths not exceeding 150 ft [45.7 m], will serve 20 dfu. This size is acceptable for all branch vents in this example. However, the branch vent serving the bar sink can be downsized to 1 ¼ in. [32 mm].

The house drain should not be less than 4-in. [102-mm]. From Table 22.5, at a ¼-in. fall per foot (2.1% slope), a 4-in. [102-mm] house drain will carry 216 dfu, which is more than adequate for the 26-dfu system. ■

There are prefabricated bathrooms in which a manufacturer assembles the piping for preselected fixtures. Going further, there are a few examples of entirely one-piece bathrooms, which incorporate the maintenance advantage of having no seams between fixtures, walls, and floors. This, of course, makes fixture replacement expensive and difficult, and tends to put the burden of access to plumbing on adjacent rooms.

TABLE 22.2 Drainage Fixture Units (dfu)

Part A. By Type of Fixture

Fixture(s)	Drainage Fixture Units (dfu)	Minimum Trap Size in.	Minimum Trap Size mm[a]
Automatic clothes washers: commercial[b,c]	3	2	51
Automatic clothes washers; residential[b]	2	2	51
Bathroom group: Water closet (1.6 gpf [6 Lpf]), lavatory, and bathtub or shower; with or without a bidet and emergency floor drain	5	—	—
Bathroom group: Water closet (>1.6 gpf [6 Lpf]), lavatory, and bathtub or shower; with or without a bidet and emergency floor drain[d]	6	—	—
Bathtub[e] (with or without overhead shower or whirlpool)	2	1 ½	38
Bidet	1	1 ¼	32
Combination sink and tray	2	1 ½	38
Dental lavatory	1	1 ¼	32
Dental unit or cuspidor	1	1 ¼	32
Dishwashing machine[f], domestic	2	1 ½	38
Drinking fountain	0.5	1 ¼	32
Emergency floor drain	0	2	51
Floor drains[g]	2[g]	2	51
Floor sinks	*Note h*	2	51
Kitchen sink, domestic	2	1 ½	38
Kitchen sink, domestic, with food waste disposer, dishwasher, or both	2	1 ½	38
Laundry tray (1 or 2 compartments)	2	1 ½	38
Lavatory	1	1 ¼	32
Shower (based on the total flow rate through showerheads and body sprays)			
Flow rate: 5.7 gpm or less	2	1 ½	38
Greater than 5.7gpm to 12.3 gpm	3	2	51
Greater than 12.3–25.8 gpm	5	3	76
Greater than 25.8–55.6 gpm	6	4	102
Service sink	2	1 ½	38
Sink	2	1 ½	38
Urinal	4	*Note d*	
Urinal, 1 gal [3.8 L] per flush or less	2[h]	*Note d*	
Urinal, nonwater supplied	0.5	*Note d*	
Wash sink (circular or multiple) each set of faucets	2	1 ½	38
Water closet, flushometer tank, public or private	4[h]	*Note d*	
Water closet, private (1.6 gpf [6 Lpf])	3[h]	*Note d*	
Water closet, private (flushing >1.6 gpf [6 Lpf])	4[h]	*Note d*	
Water closet, public (1.6 gpf [6 Lpf])	4[h]	*Note d*	
Water closet, public (flushing >1.6 gpf [6 Lpf])	6[h]	*Note d*	

Part B. By Size of Trap

Fixture Drain or Trap Size in.	Fixture Drain or Trap Size mm[a]	Drainage Fixture Unit (dfu) Value
1 ¼	32	1
1 ½	38	2
2	51	3
2 ½	64	4
3	76	5
4	102	6

Source: This table is a copyrighted excerpt from the 2024 International Plumbing Code. Copyright © 2023. International Code Council, Inc. All rights reserved. Reproduced with permission. www.ICCSAFE.org.

[a]SI conversions provided by the authors of this book (not part of original source).

[b]For traps larger than 3 in. [76 mm], see Part B of this table.

[c]See the current edition of the *International Plumbing Code* for sizing requirements for fixture drain, branch drain and drainage stack for an automatic clothes washer standpipe.

[d]For fixtures added to a bathroom group, add the dfu value of those additional fixtures to the bathroom group fixture count.

[e]A showerhead over a bathtub or whirlpool bathtub attachments do not increase the dfu value.

[f]For methods of computing unit value of fixtures not listed in this table or for rating of devices with intermittent flows, see the current edition of the *International Plumbing Code*.

[g]See the current edition of the *International Plumbing Code*.

[h]For the purpose of computer loads on building drains and sewers, water closets and urinals shall not be rated at a lower drainage fixtures unit unless the lower values are confirmed by testing.

[i]Trap size shall be consistent with fixture outlet size.

TABLE 22.3 Horizontal Fixture Branches and Stacks[a]

Diameter of Pipe		Maximum Number of Drainage Fixture Units (dfu)			
			Stacks[b]		
in.	mm[d]	Total for Horizontal Branch	Total Discharge into One Branch Interval	Total for Stack of Three Branch Intervals or Less	Total for Stack Greater Than Three Branch Intervals
1½	38	3	2	4	8
2	51	6	6	10	24
2½	64	12	9	20	42
3	76	20	20	48	72
4	102	160	90	240	500
5	127	360	200	540	1100
6	152	620	350	960	1900
8	203	1400	600	2200	3600
10	254	2500	000	3800	5600
12	305	3900	1500	6000	8400
15	381	7000	*Note c*	*Note c*	*Note c*

Source: This table is a copyrighted excerpt from the 2024 International Plumbing Code.

[a]Does not include branches of the building drain. Refer to Table 22.5.

[b]Stacks shall be sized based on the total accumulated connected load at each story or branch interval. As the total accumulated connected load decreases, stacks may be reduced in size. Stack diameters shall not be reduced to less than one-half the diameter of the largest stack size required.

[c]Sizing load based upon design criteria.

[d]SI conversions provided by the authors of this book (not part of original source).

For the more ordinary bathroom, the two maintenance questions of greatest concern are: (1) How easy is cleaning around and within the fixture?, and (2) How accessible are those parts of the fixture most likely to need repair or replacement? Ease of cleaning is often determined more by the space around the fixture than by the design of the fixture itself. This is particularly true of toilets, where a generous amount of open floor on either side makes maintenance easy and therefore likely to be more frequent. Access to fixture parts may be more difficult. An access panel in the wall of the room behind the fixtures is often provided, encouraging speedy repair and replacement for components at tubs, showers, and lavatories. Ideally, accessibility to all plumbing lines—whether via access panels in walls, trenches in concrete floors, exposed basement ceilings, or adequately deep crawl spaces—should be provided. Most bathrooms are likely to undergo thorough remodeling, including fixture replacement, as styles change and water/energy conservation becomes more important.

22.5 DESIGN OF LARGER-BUILDING WASTE PIPING

(a) Basic Planning

Multi-story construction, especially in office buildings, is often designed to be flexible and free of random partitions that would interfere with the periodic renovation and reorganization of interior spaces. Building "cores" contain elevators, stairs, and shafts for plumbing, mechanical, and electrical equipment. Cores are often placed in the central section of the building, freeing the surrounding areas for access to daylight. A hole in the floor for each pipe is often chosen in preference to a slot or shaft. This method usually interferes less with the floor construction (Figure 22.25).

Offices often need a single lavatory or a complete toilet room for executives at locations away from the central core of the building. The greater the horizontal distance from the core, the more vertical clearance that will be needed to allow the drain to slope. When such vertical clearance becomes difficult, "wet" columns with a full complement of plumbing pipes offer a solution. If the pipes are to accompany a column in a steel building, structural coordination must be sought early in the planning (Figure 22.26).

In some installations, the branch soil and waste piping perforates a floor and crosses below the slab to join the stack. Tubing has been developed, however, that sits above the structural slab, obviating the need for hung ceilings below (Figure 22.27). A lightweight concrete fill is cast to cover the tubing, raising the floor by 5 or 6 in. [125 or 150 mm], creating a raised floor in the toilet room—along with associated access problems. Given this, the higher floor level is usually carried throughout the entire story, forming a convenient space into which the electrical conduit can be placed at a time later than would have been required if it were to be placed in the structural slab. Such raised floors are becoming more common, along with the HVAC strategy of underfloor air distribution.

TABLE 22.4 Size and Developed Length of Stack Vents and Vent Stacks

Diameter of Soil or Waste Stack in. [mm][a]	Total Fixture Units Being Vented (dfu)	Maximum Developed Length[b] of Vent, ft [m][a] Diameter of Vent, in. [mm][a]									
		1¼ [32]	1½ [38]	2 [51]	2½ [64]	3 [76]	4 [102]	5 [127]	6 [152]	8 [203]	10 [254]
1¼ [32]	2	30 [9.1]									
1½ [38]	8	50 [15.2]	150 [45.7]								
1½ [38]	10	30 [9.1]	100 [30.5]								
2 [51]	12	30 [9.1]	75 [22.9]	200 [61.0]							
2 [51]	20	26 [7.9]	50 [15.2]	150 [45.7]							
21/2 [64]	42		30 [9.1]	100 [30.5]	300 [91.0]						
3 [76]	10		42 [12.8]	150 [45.7]	360 [109.7]	1040 [317]					
3 [76]	21		32 [9.8]	110 [33.5]	270 [82.3]	810 [246.9]					
3 [76]	53		27 [8.2]	94 [28.7]	230 [70.1]	680 [207.3]					
3 [76]	102		25 [7.6]	86 [26.6]	210 [64.0]	620 [189.0]					
4 [102]	43		25 [7.6]	35 [10.7]	85 [25.9]	250 [76.2]	980 [298.7]				
4 [102]	140		25 [7.6]	27 [8.2]	65 [19.8]	200 [61.0]	750 [228.6]				
4 [102]	320			23 [7.0]	55 [16.8]	170 [51.8]	640 [195.0]				
4 [102]	540			21 [6.4]	50 [15.2]	150 [45.7]	580 [176.8]				
5 [127]	190				28 [8.5]	82 [25.0]	320 [97.5]	990 [301.8]			
5 [127]	490				21 [6.4]	63 [19.2]	250 [76.2]	760 [231.6]			
5 [127]	940				18 [5.5]	53 [16.2]	210 [64.0]	670 [204.2]			
5 [127]	1400				16 [4.9]	49 [14.9]	190 [57.9]	590 [179.8]			
6 [152]	500					33 [10.1]	130 [39.6]	400 [121.9]	1000 [304.8]		
6 [152]	1100					26 [7.9]	100 [30.5]	310 [94.5]	780 [237.7]		
6 [152]	2000					22 [6.7]	84 [25.6]	260 [79.2]	660 [201.2]		
6 [152]	2900					20 [6.1]	77 [23.5]	240 [73.2]	600 [182.9]		
8 [203]	1800						31 [9.4]	95 [29.0]	240 [73.2]	940 [286.5]	
8 [203]	3400						24 [7.3]	73 [22.3]	190 [57.9]	729 [222.4]	
8 [203]	5600						20 [6.1]	62 [18.9]	160 [48.8]	610 [185.9]	
8 [203]	7600						18 [5.5]	56 [17.1]	140 [42.7]	560 [170.7]	
10 [254]	4000							31 [9.4]	78 [23.8]	310 [94.5]	960 [292.6]
10 [254]	7200							24 [7.3]	60 [18.3]	240 [73.2]	740 [225.6]
10 [254]	11,000							20 [6.1]	51 [15.5]	200 [61.0]	630 [192.0]
10 [254]	15,000							18 [5.5]	46 [14.0]	180 [54.9]	571 [174.2]

[a] SI conversions provided by the authors of this book (not part of original source).

[b] The developed length is measured from the vent connection to the open air.

TABLE 22.5 Building Drains and Sewers

Diameter of Pipe		Maximum Number of dfu Connected to Any Portion of the Building Drain or Building Sewer, Including Branches of the Building Drain[a]			
		Slope, in. per ft (% fall)			
in.	mm[b]	1/16 in. (0.5%)	1/8 in. (1.04%)	1/4 in. (2.1%)	1/2 in. (4.2%)
1 1/4	32			1	1
1 1/2	38			3	3
2	51			21	26
2 1/2	64			24	31
3	76		36	42	50
4	102		180	216	250
5	127		390	480	575
6	152		700	840	1000
8	203	1400	1600	1920	2300
10	254	2500	2900	3500	4200
12	305	3900	4600	5600	6700
15	381	7000	8300	10,000	12,000

Source: This table is a copyrighted excerpt from the 2024 International Plumbing Code. Copyright © 2023. International Code Council, Inc. All rights reserved. Reproduced with permission. www.ICCSAFE.org.

[a]The minimum size of any building drain serving a water closet shall be 3 in. [76 mm].

[b]SI conversions provided by the authors of this book (not part of original source).

(b) Roughing-In

This is the process of getting all pipes installed, capped, and pressure-tested before the fixtures are installed. An example of roughing-in for an office building is shown in Figure 22.27. The roughing-in of supply and waste piping for school lavatories is shown in Figure 22.28.

Institutions such as schools have extensive requirements for durability and ease of maintenance. The fixtures are made of such wear-resistant materials as stainless steel, chrome-plated cast brass, precast stone or terrazzo, or high-impact fiberglass. The fixture controls are designed to withstand heavy use—or misuse—and the fixtures are securely tied into the structure with concealed mounting hardware designed to resist extraordinary forces. (Some schools even move the lavatories into the hallway for better visual control of at least a part of the restroom facilities.)

In prisons, extreme measures are taken to prevent plumbing fixtures from becoming damaged, or being used as weapons. Heavy-gauge stainless steel fixtures with nonremovable fittings are provided, at very high cost, for both the fixture and its tamperproof installation.

EXAMPLE 22.2 Select drainage and vent piping sizes for the plumbing in an office building—for which the fixtures are shown in Figure 22.29.

SOLUTION

Selected dfu values from Table 22.2 are applied to each section of the piping and totaled for each branch and stack, as well as for the building drain and the building sewer. Individual fixture branches should not be less than the size indicated in Table 22.3 for the minimum size of trap for each fixture. An example of a fixture-unit summary, and sample sizes of individual branches that connect into a typical branch of the men's toilet group on any floor, are shown in Table 22.7.

Table 22.3 indicates that a 3-in. [76-mm] horizontal fixture branch is inadequate for the men's toilet group because this size pipe will handle only 20 dfu. Therefore, a 4-in. [102-mm] pipe is selected (also a wise minimum choice whenever water closets are served). Its capacity of 160 dfu will be more than enough for the 36 fixture units needed here. Table 22.3 also indicates that the soil stack diameter can be 4 in. [102 mm], and it is run thus for its entire height. Its capacity of 90 dfu per story (or interval) is sufficient for the 64 dfu that connect in at each T-Y connection. For three branch intervals, this 4-in. [102-mm] diameter soil stack will carry 240 dfu (this example requires $3 \times 64 = 192$ dfu); for its entire five-story height, the soil stack will carry 500 dfu (this example requires $5 \times 64 = 320$ dfu).

As noted in Figure 22.29, the building exhibits a 3-in. [76-mm] vent stack serving up to 338 dfu and the 4-in. [102-mm] soil stack must serve a maximum 70-ft [21.3-m] developed length. From Table 22.4, a 3-in. [76-mm] vent serving 540 dfu has a maximum developed length of 150 ft [45.7 m], well over the minimum requirement for this five-story building. All vent stack diameters will increase to 4 in. [102 mm] as they pass through the roof.

According to Table 22.5, the building drain and the building sewer, at a pitch of 1/4 in. per foot (2.1%), should carry at least a total of 350 dfu. A 5-in. [127-mm] building drain or building sewer has a capacity of 480 dfu.

Although opinions may vary about the relative merits of continuous or circuit venting, either system,

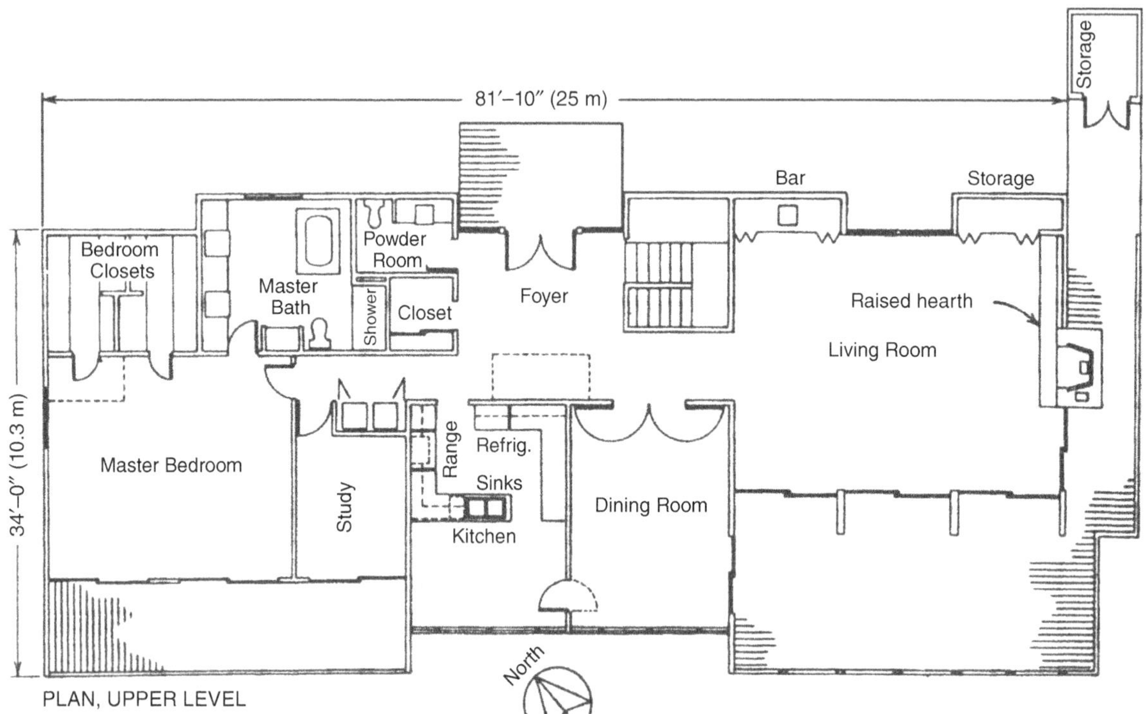

(a)

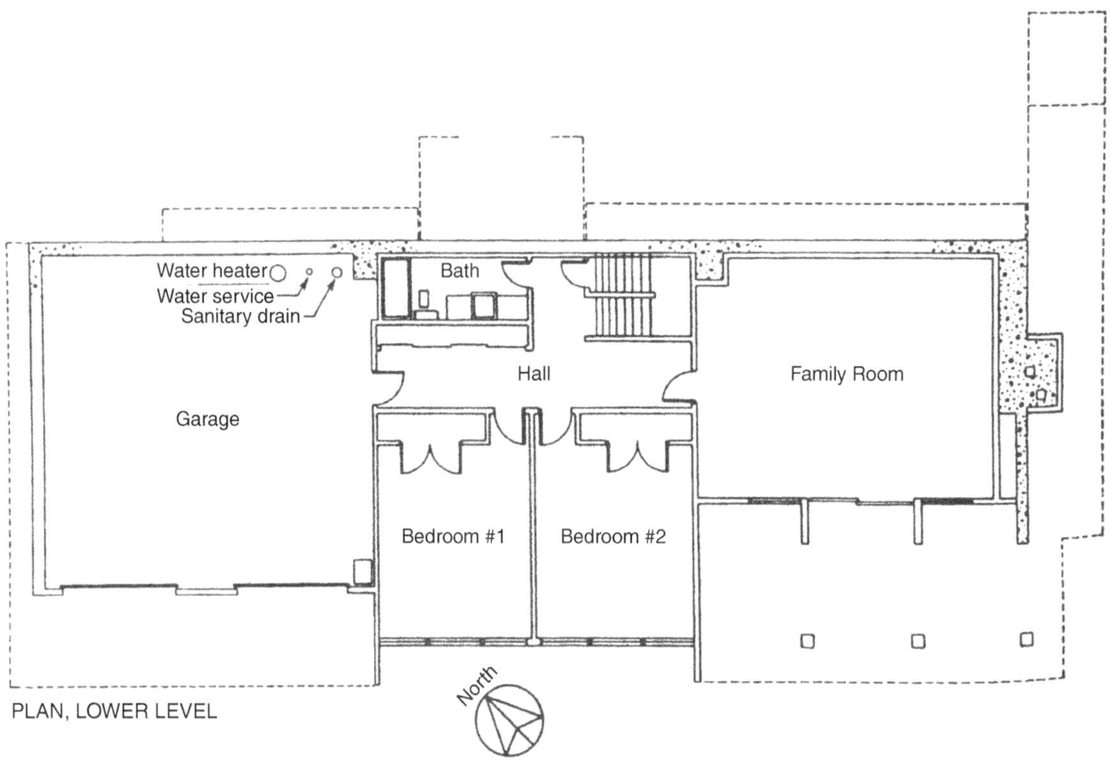

(b)

Figure 22.22 *Floor plans of a house on Long Island, New York—used in* Example 22.1. *(Courtesy of Budd Mogensen, architect and planner.)*

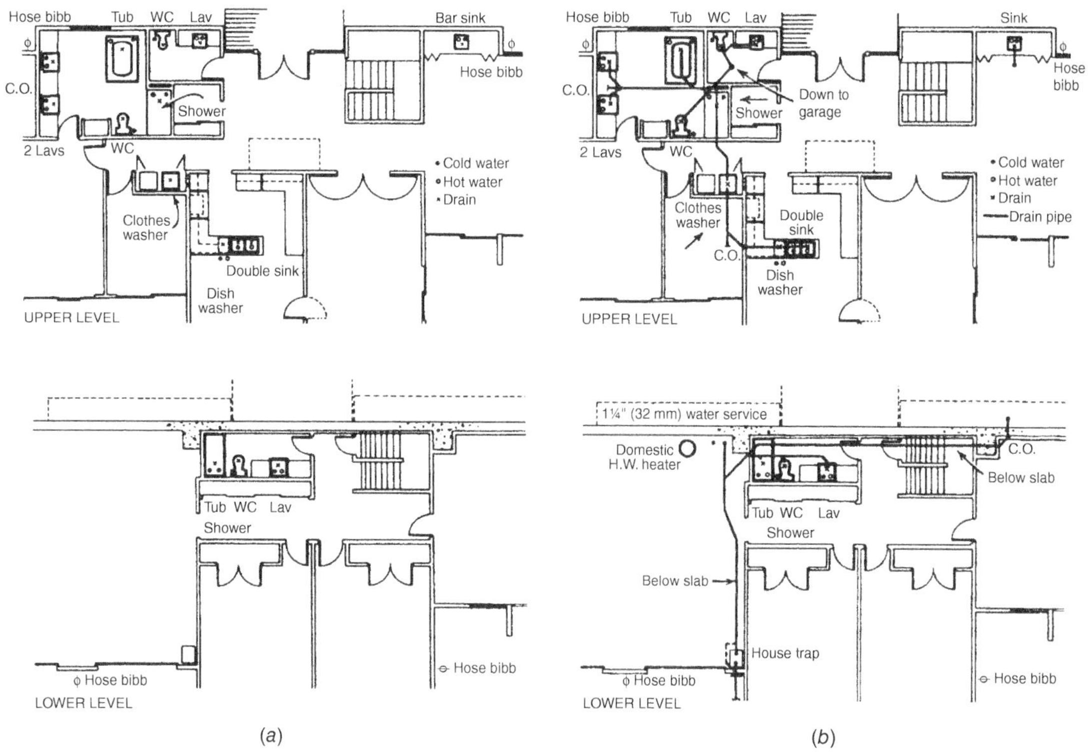

Figure 22.23 From Example 22.1. (a) Plumbing requirements, water supply, and partial sanitary drainage. (b) Sanitary drainage plan.

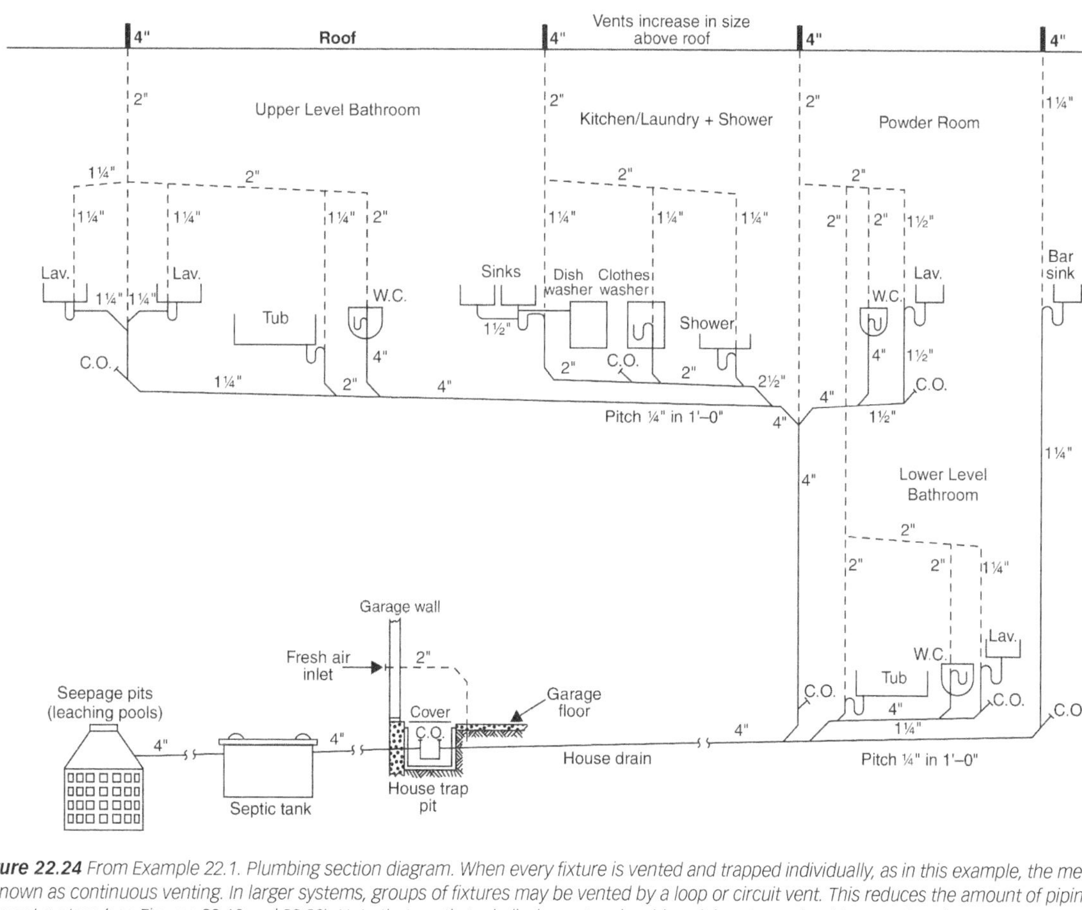

Figure 22.24 From Example 22.1. Plumbing section diagram. When every fixture is vented and trapped individually, as in this example, the method is known as continuous venting. In larger systems, groups of fixtures may be vented by a loop or circuit vent. This reduces the amount of piping in the vent system (see Figures 22.10 and 22.29). Note that septic tank discharge to a leaching pit is not permitted in many locations; see Section 22.6.

TABLE 22.6 Drainage Fixture Units (dfu) for Example 22.1

Residential Fixture	Drainage Fixture Units (dfu)
Bar sink	2
Kitchen sink with dishwasher	2
Lavatory	1
Water closet, private (flushing >1.6 gpf [6 Lpf])	4
Automatic clothes washer	2
Upper-level bathroom group: Water closet (>1.6 gpf [6 Lpf]), lavatory, and bathtub or shower; with or without a bidet and emergency floor drain	6
Extra lavatory	1
Shower	2
Lower-level bathroom group: Water closet (>1.6 gpf [6 Lpf]), lavatory, and bathtub or shower; with or without a bidet and emergency floor drain	6
Total	**26**

The dfu values are from Table 22.2. The hose bibb drains to the ground; roof drainage is taken to dry wells.

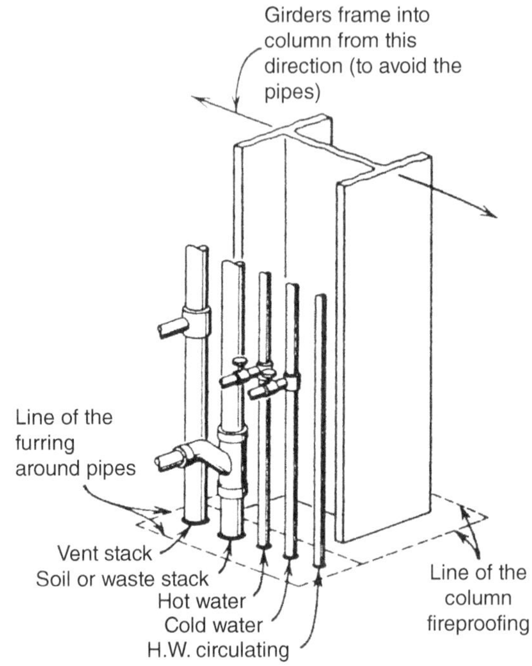

Figure 22.26 *Piping at a "wet" column. In large office buildings, there are usually several of these, remote from the core, in the general office area.*

Figure 22.25 *Plumbing risers in a fireproof multistory building. Pipes, tubes, conduits, and ducts adjoin toilet rooms and utility spaces. Ventilation ducts and a master 5-in. (127-mm) copper hot-water riser are just to the left of center. Soil and vent stacks with hot and cold water supplies—all copper—are to the right of this group. At the left, in a lighter tone, are the galvanized steel feeder conduit and the distribution circuit conduits for a local electrical control panel box. Note that some pipes and tubes are supported at this floor by bolted clamps. After testing and before pipes are enclosed, covering will be completed. (Courtesy of the Copper Development Association.)*

properly designed, will effectively prevent the siphoning of traps and relieve air pressures that could cause foul gases to bubble through the traps into the occupied space. Another system (Section 22.5c), especially suitable for high-rise buildings, eliminates the vent stack completely with equal effectiveness. ■

(c) The Sovent System

This essentially ventless system changes the nature of the effluent (the wastes discharged from the fixtures) instead of coping with the pressures and suctions that normal effluent would cause (see Figures 22.30 and 22.31).

The "plunger" effect of a descending "slug" of water/waste within pipes was described in Section 22.2. If the "effectiveness" of the plunger can be reduced, the negative and positive pressures created by it will also be reduced. If their values can be brought down below the holding power of the several inches of water in the trap, no vents will be necessary. In the single-stack Sovent system illustrated in Figure 22.31, this is done by dealing with the normal liquid effluent at each floor. Aeration there produces a foam that lacks the stack-filling tendency of unaerated liquid effluent. Thus, through the creation of a soft plunger, pressure variations in the single stack are minimized.

Tests have shown that the positive and negative pressures produced by normal liquid effluent during its descent (to be relieved by the vent piping) are often

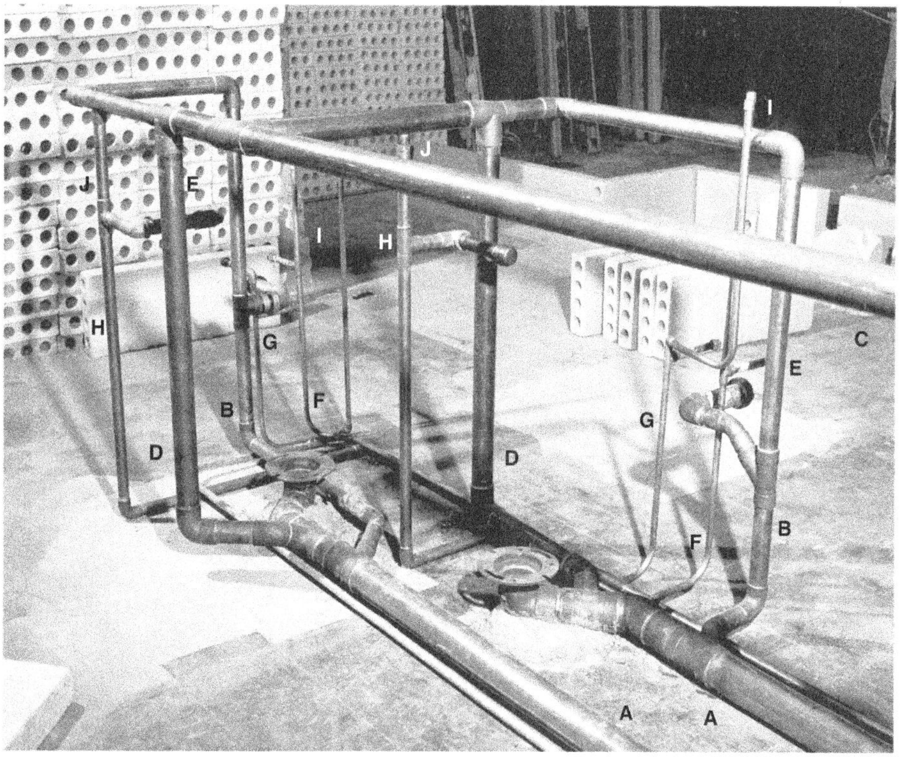

(a)

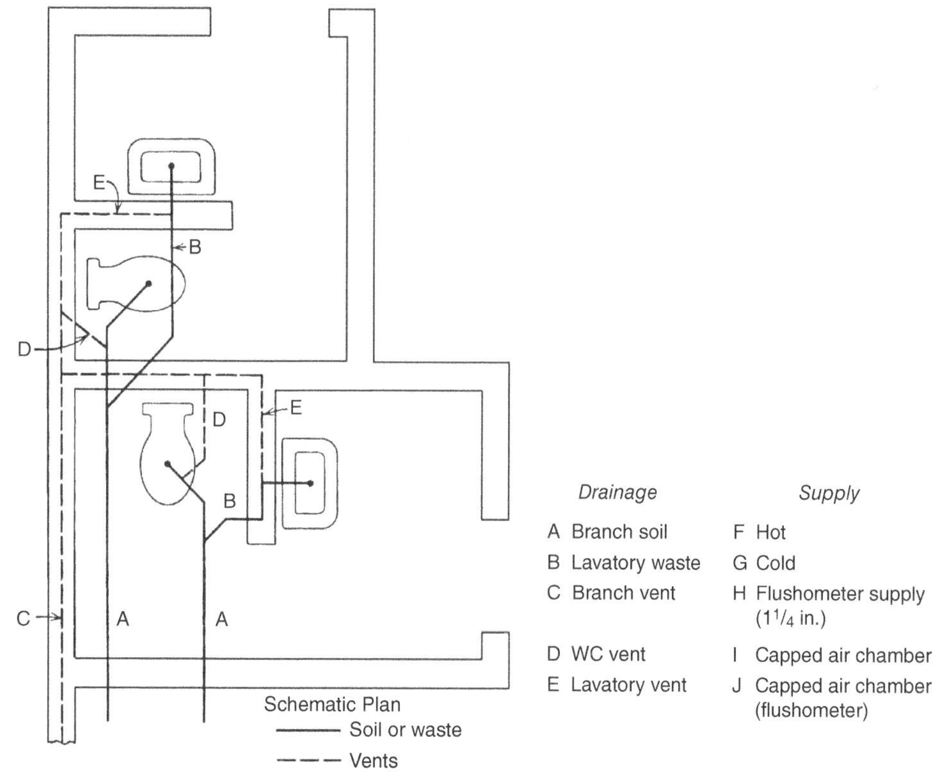

(b)

Figure 22.27 *(a) Horizontal waste piping above the structural slab, in an example of roughing-in plumbing for two lavatory rooms in a concrete office building. A lavatory and water closet (WC) in each room (see plan in b) are served by soil and waste branches below and vent branches above. Hot- and cold-water tubing with air chambers can be seen. Although the extensions of the water tubing above the two flushometer connections appear to connect into the horizontal vent branches, they actually do not; they are capped and merely touch the bottoms of the vent branches. Note that soil branches are above the structural slab. A fill of 5 or 6 in. [125–150 mm] is necessary to cover the tubing. All vertical tubing will be within the masonry block used to enclose the cubicles. (Courtesy of the Copper Development Association.)*

Figure 22.28 Roughing-in plumbing for four lavatories in the toilet room of a school. The waste branches have been capped and the system tested for leakage. The waste branches are cast iron, the vents galvanized steel, and the water supply lines are copper with soldered fittings. At the center of the supply array are vertical capped expansion and shock tubes, one for hot and one for cold supply water.

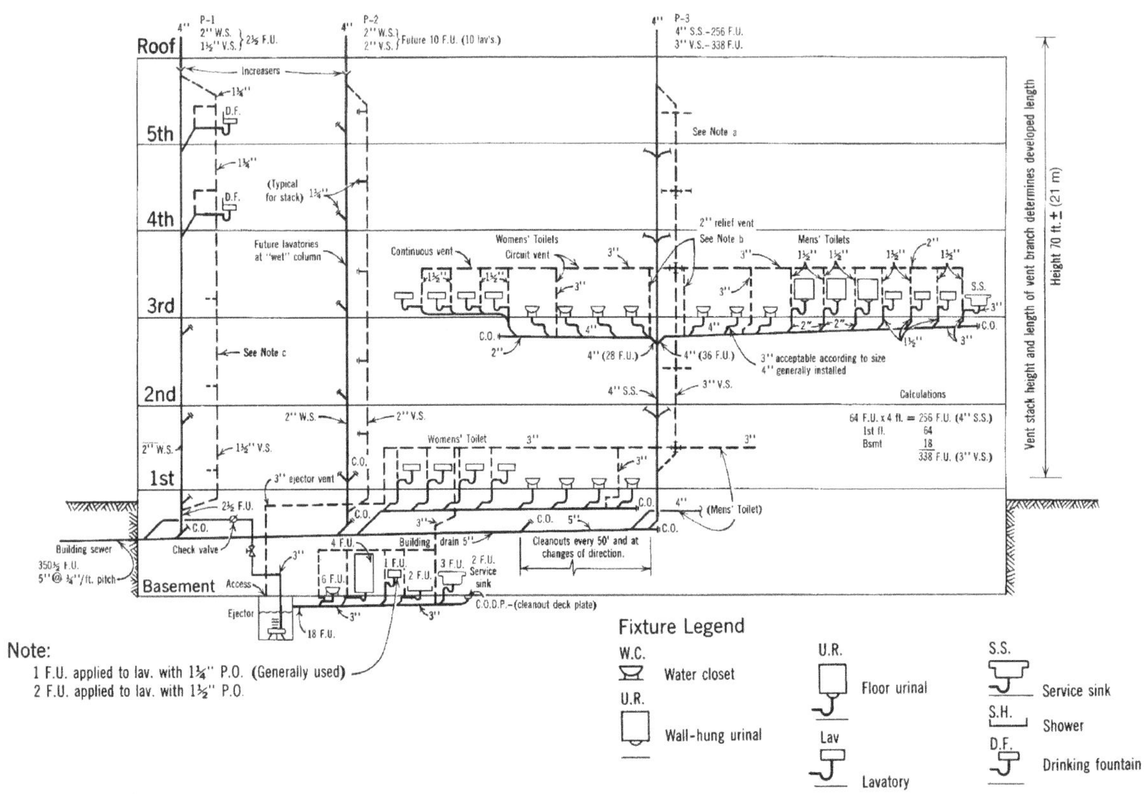

Figure 22.29 Plumbing section of an office building. Circuit vents serve branch soil lines, but some codes require continuous venting (individual fixture vents). The house trap and fresh-air inlet are omitted from the building drain. Note that a relief vent (1) is not required on the top floor. (2) Men's and women's toilets on the third floor are typical and would be repeated on the first, second, fourth, and fifth floors. (3) Drinking fountains on the fifth and fourth floors would be repeated on the first, second, and third floors.

TABLE 22.7 Total Fixture Units for a Men's Toilet Branch for Example 22.2

Features	Units per Fixture	Total Fixture Units	Diameter, Fixture Branch in.	mm
1 service sink (2-in. [51-mm] trap)	3	3	3	76
3 lavatories	1	3	3	76
3 urinals	4	12	2	51
3 water closet, public (flushing >1.6 gpf [6 Lpf])	6	18	4	102
Total fixture units, men's toilet branch		**36**		

about 5–12 in. [127–305 mm] water gauge. If vents were not provided, the 2–4 in. [51–102 mm] of water seal in the traps would be vulnerable to penetration by gases from pipes under greater positive pressure (or siphonage of water seals into pipes that may be under negative pressure).

Figures 22.31 and 22.32 illustrate the components and the action of the Sovent system. Effluent, already aerated and descending from upper stories, is diverted in the stack at each lower story. The aerator fitting there affords a passage for this diverted flow and also an air space into which the effluent from the local branch soil or waste can drop. Here it spatters, mixing with air to form a rarefied mixture of air and liquid. Tests show that this mixture does not produce pressures, positive or negative, of more than 1 in. [25 mm] on the water gauge. Thus, a trap seal of 2 in. [51 mm] or more is safe against siphonage or penetration.

At the foot of the single stack the aerated effluent is compacted—a process aided by a baffle in the path of the flow in the deaerator fitting (see Figure 22.32). If not relieved, air piling up at this point could cause pressures in the stack at the first floor. An air-discharge pipe provides relief of air from the deaerator fitting to the upper part of the building drain, above the liquid flow.

The Sovent system was invented by Fritz Sommer of Switzerland, who tested it in a 10-story drainage test tower. Since its introduction in 1962, it has been installed and used in hundreds of buildings in Europe and Africa. Canada used the Sovent method in the Habitat apartments at the 1967 Montreal Exposition. Sovent was first granted U.S. code acceptance in 1968 in Richmond, California. Following this success, its code acceptance grew rapidly during the early 1970s. However, not all jurisdictions allow Sovent installations.

22.6 ON-SITE INDIVIDUAL-BUILDING SEWAGE TREATMENT

The great majority of stand-alone building sewage treatment systems in North America use a septic tank (Figure 22.33) as a primary treatment device, where the settling of solids and anaerobic digestion take place. Subsequently, the effluent receives secondary treatment, which usually consists of a filtering process. Four common filtration systems are seepage pits, drain fields, mounds, and sand filters. Occasionally, a tertiary treatment (usually disinfection with chlorine) must be used, as when outflows from secondary treatment will flow directly into surface waterways.

(a) Primary Treatment: Septic Tanks

These septic tanks (Figure 22.34) are commonly constructed of precast concrete, although steel, fiberglass, and polyethylene tanks are also available. The sewage enters the first chamber, where solids sink to the bottom as sludge, and scum forms on the surface. Anaerobic decomposition proceeds, producing methane gas. The liquid moves through the submerged opening in the middle of the tank to the second chamber, where finer solids continue to sink, and less scum forms on the surface. Finally, the effluent, about 70% purified, leaves the septic tank for secondary treatment.

The longer the sewage stays in the septic tank, the less polluted the effluent. This is why water conservation measures are so welcome with septic tank systems; the less the flow, the longer the water remains in the tank. The anaerobic decomposition process, while malodorous, is so thorough that the sludge needs only occasional removal—once in several years is about average for residences. The methane and its odor are kept within the tank, and access hatches are sometimes covered with soil (as in Figure 22.34), even though this makes their occasional need to be located far more difficult.

Most systems with septic tanks will eventually experience failure, usually due to a breakdown in the secondary treatment rather than in the septic tank. With periodic removal of sludge, the septic tank itself is a reliable and simple primary treatment device. However, some types of domestic waste can disrupt the anaerobic process within the tank. The worst offenders are paints, varnishes, thinners, waste oil, photographic solutions, and pesticides. Also to be avoided are coffee grounds, dental floss, disposable diapers, kitty litter, sanitary napkins, tampons, cigarette butts, condoms, gauze bandages, fat and grease, and paper towels.

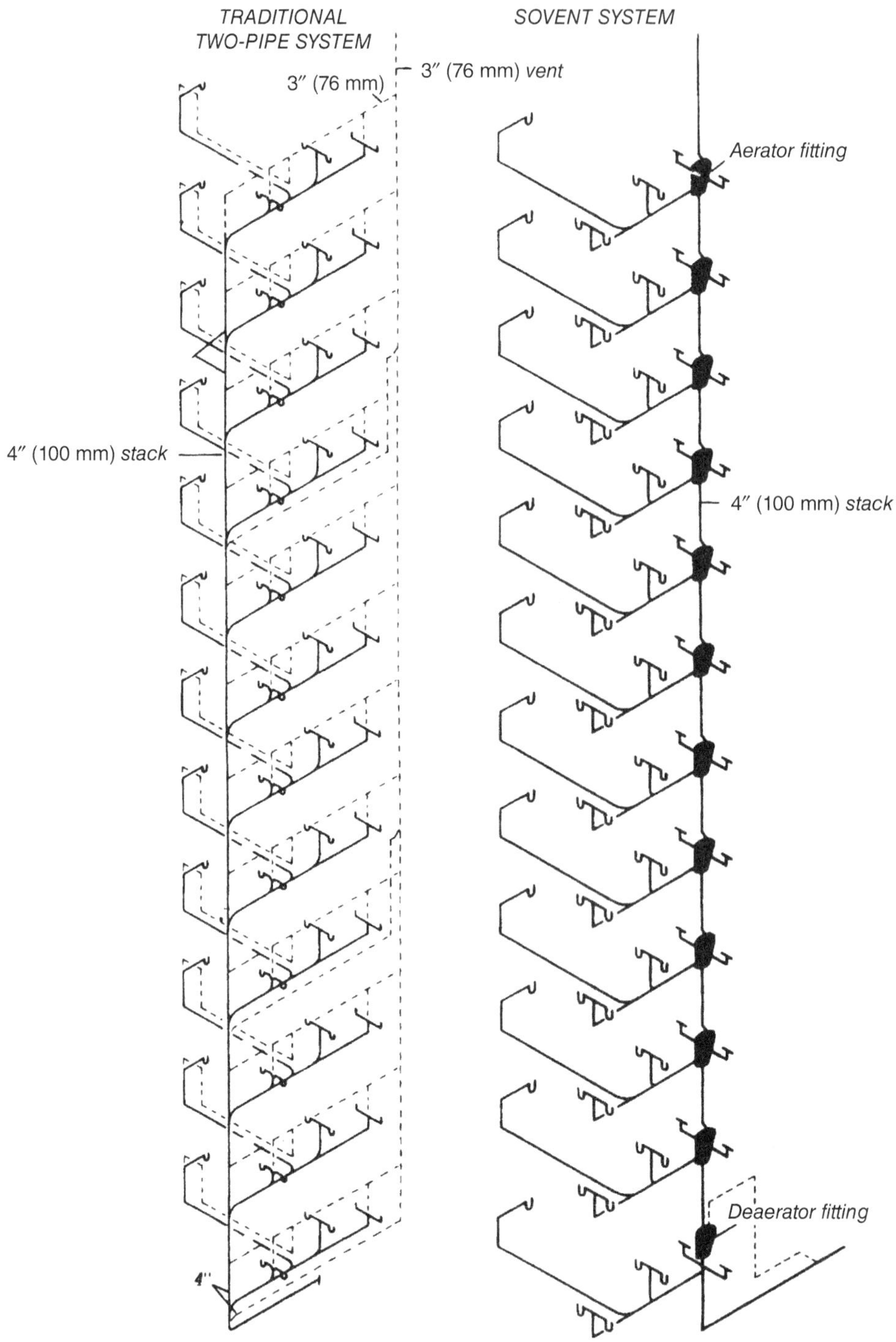

Figure 22.30 *Comparing a two-pipe system (for a 12-story stack serving an apartment grouping) to the Sovent system. (Courtesy of the Copper Development Association.)*

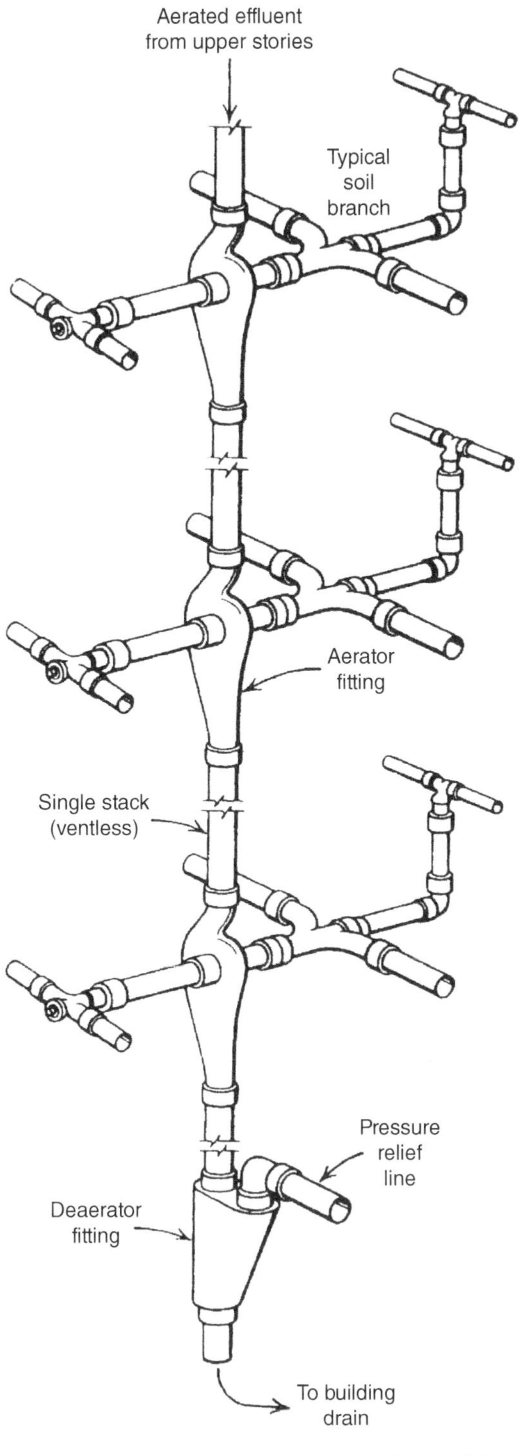

Figure 22.31 The Sovent drainage stack consists of aerator fittings that join the horizontal branches to the stack at each floor level and a deaerator fitting at the bottom of the stack. The stack is open to the atmosphere above the roof at the top. (Courtesy of the Copper Development Association.)

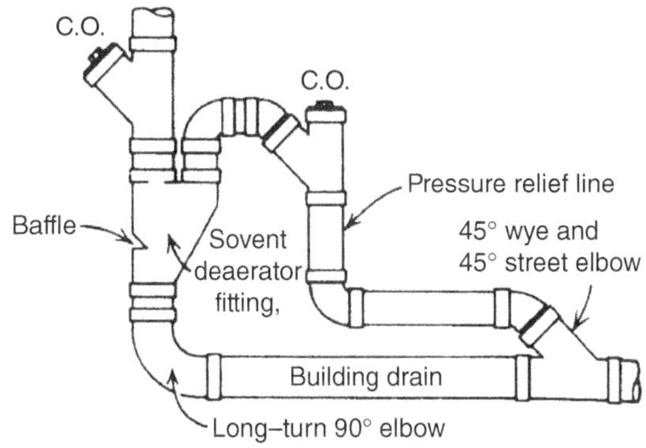

Figure 22.32 The deaerator consists of an air separation chamber with an internal nosepiece, a stack inlet, a pressure relief outlet at the top, and a stack outlet at the bottom. The deaerator fitting at the bottom of the stack functions in combination with the aerator fittings above to make the single stack self-venting. The deaerator is designed to overcome the tendency for the falling waste to build up excessive back pressure at the bottom of the stack when the flow decelerates at the bend into the horizontal drain. (Courtesy of the Copper Development Association.)

Septic tank sizes are commonly based on code requirements that consider the number of bedrooms in residences (Table 22.8), or the number of waste fixture units served (see Table 22.2). (As shown in Table 22.9, maximum size may also be related to the soil condition.) Sewage flow rates are also considered. Oversized septic tanks are more expensive to install, but they release cleaner effluent and prolong the life of the secondary treatment system.

The size of the secondary treatment system is usually based upon the size of the septic tank it serves or on the expected total flow over a 24-hour period. (Flow can be estimated from tabulated information in Chapter 22.) Where effluent must be raised for secondary treatment, or when the total length of the secondary treatment disposal line exceeds 500 ft [154 m], a dosing tank (also called a siphon or pumping chamber; see Figure 22.33*c*, *d*) is used to automatically discharge the septic tank's outflow chamber to the secondary treatment lines or sand filters.

(b) Primary Treatment: Aerobic Treatment Units

Active aerobic treatment units (ATUs, Figure 22.35) are an increasingly popular alternative to septic tanks for primary sewage treatment. They are frequently used to replace a passive septic tank in a troubled

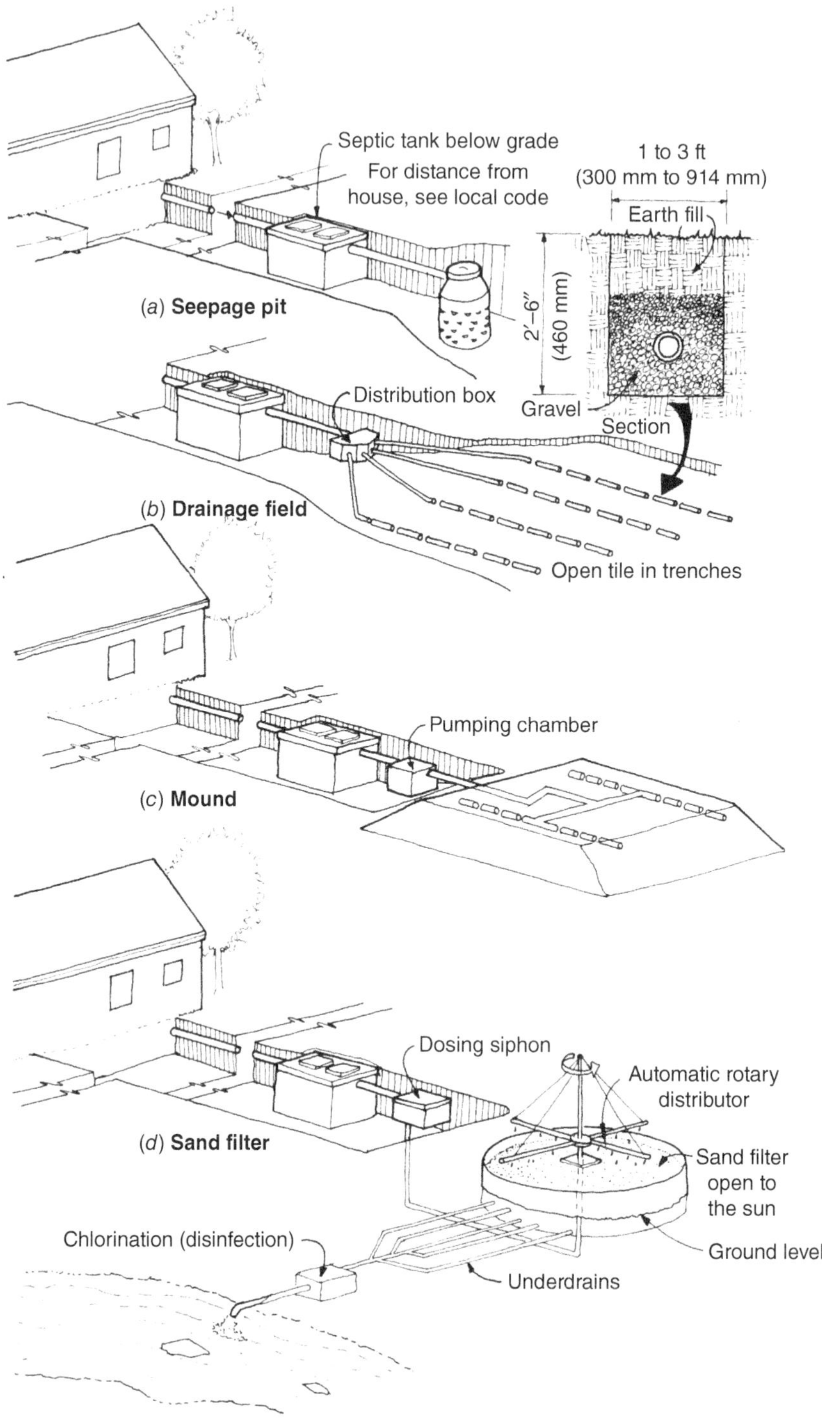

Figure 22.33 *For individual sewage treatment systems, septic tanks are commonly used for primary treatment. Four options for secondary treatment are shown here. (Tertiary treatment usually is required only for effluent discharge into waterways.) (a) Seepage pits are rarely allowed. (b) Drain fields are the most commonly used options. (c, d) Mounds and sand filters are more expensive to construct and are used where high water tables preclude the use of option (b).*

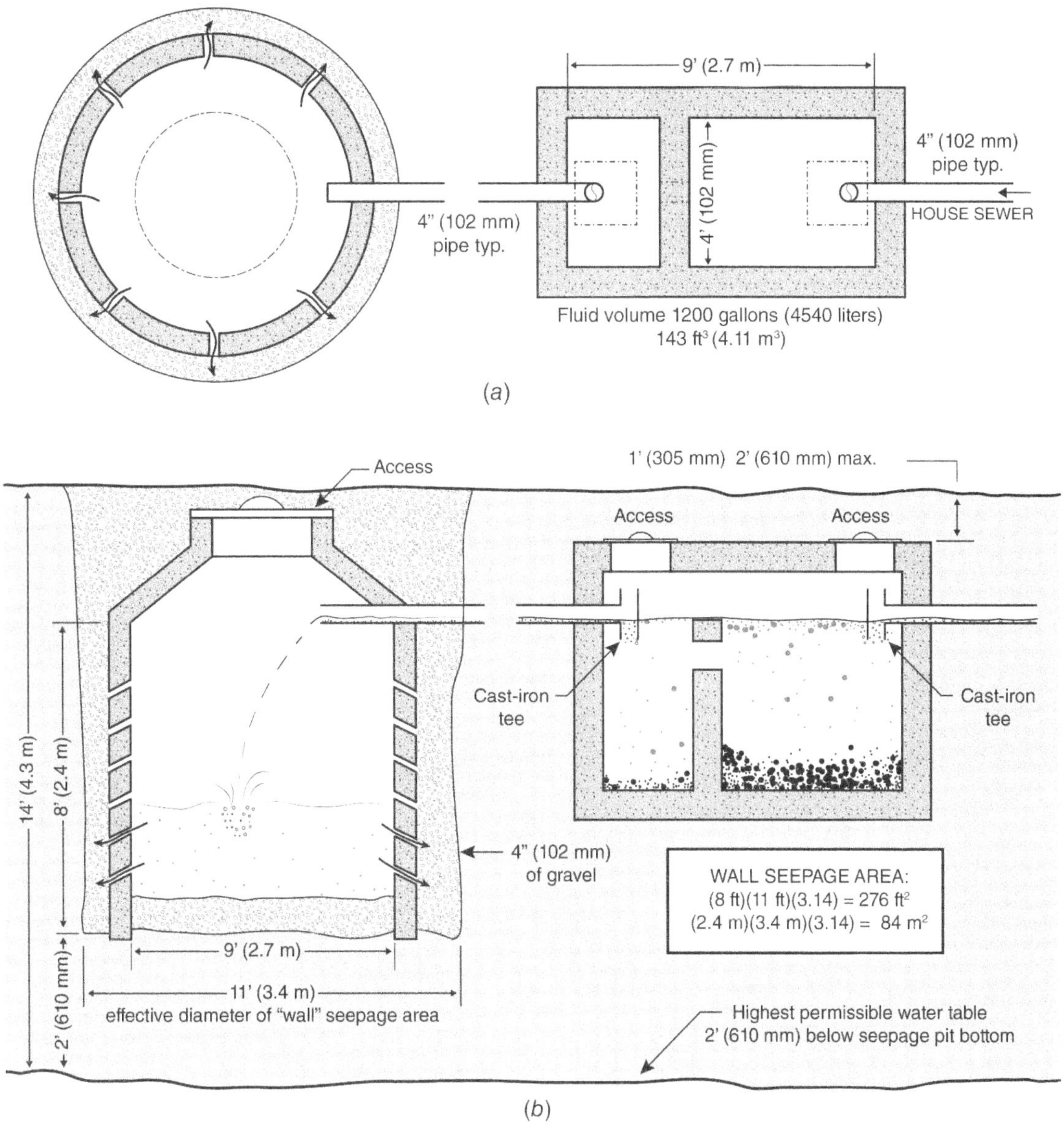

Figure 22.34 *Plan (a) and section (b) of a septic tank and seepage pit for a four-bedroom house in sandy loam soil. A pit is suitable only when the earth is absorbent and the water table low (below the pit bottom). (Redrawn by Nathan Leigh.)*

system; this "upgrade" can rejuvenate an existing drain field and extend its life. ATUs depend upon air bubbled through the sewage to achieve aerobic digestion, which is faster than anaerobic digestion; hence, they can be smaller than septic tanks. They are energy-intensive and require more maintenance than anaerobic tanks. A secondary treatment process is required as well. The effluent typically is less polluted than that of septic tanks; the biochemical oxygen demand (BOD) is reduced by 90% in ATUs compared to the 50% typical of septic tanks.

The sewage first enters an aeration chamber, where it is kept in turmoil so that air can continue to percolate through it. Either air is forced through the sewage by distribution lines fed by an air compressor or the sewage is stirred by a variety of devices, depending upon the manufacturer. After about a day's retention in the aeration chamber, the aerated wastewater enters

TABLE 22.8 Septic Tank Capacity[a]

Single-Family Dwellings—Number of Bedrooms	Multiple Dwelling Units or Apartments—One Bedroom Each	Other Uses: Maximum Fixture Units Served[b,c]	Minimum Septic Tank Capacity, Gal [L][c]
1 or 2		15	750 [2838]
3		20	1000 [3785]
4	2 units	25	1200 [4542]
5 or 6	3	33	1500 [5678]
	4	45	2000 [7570]
	5	55	2250 [8516]
	6	60	2500 [9463]
	7	70	2750 [10,409]
	8	80	3000 [11,355]
	9	90	3250 [12,301]
	10	100	3500 [13,248]
Extra bedroom: 150 gal [568 L] each			
Extra dwelling units over 10: 250 gal [946 L] each			
Extra fixture units over 100: 25 gal [95 L] per fixture unit			

Source: Adapted from (IAPMO, 2023). Reprinted with the permission of the International Association of Plumbing and Mechanical Officials. This copyright material and all points or statements in using this material have not been reviewed by IAPMO. The opinions expressed herein are not representations of fact from IAPMO.

The SI conversions were appended by the authors of this book.

[a]Septic tank sizes in this table include sludge storage capacity and the connection of domestic food waste disposal units without further volume increase.

[b]See Table 22.2.

[c]For larger or nonresidential installations in which sewage flow rate is known, size the septic tank as follows:

Flow up to 1500 gpd [5678 L/d]: flow × 1.5 = septic tank capacity

Flow over 1500 gpd [5678 L/d]: (flow × 0.75) + 1125 = septic tank capacity in gallons [(flow × 0.75) + 4258 = capacity in liters]

Secondary system shall be sized for total flow per 24 hours.

TABLE 22.9 Septic Tank and Leaching Area Design Criteria for Five Typical Soils

Type of Soil	Required ft² of Leaching Area per 100 gal [m²/L]	Maximum Absorption Capacity, gal/ft² of Leaching Area for a 24-h Period [L/m²]	Maximum Septic Tank Size Allowable	
			Gallons	Liters
1. Coarse sand or gravel	20 [0.005]	5 [203.7]	7500	28,387
2. Fine sand	25 [0.006]	4 [162.9]	7500	28,387
3. Sandy loam or sandy clay	40 [0.010]	2.5 [101.9]	5000	18,925
4. Clay with considerable sand or gravel	90 [0.022]	1.10 [44.8]	3500	13,247
5. Clay with small amount of sand or gravel	120 [0.029]	0.83 [33.8]	3000	11,355

Source: Adapted from (IAPMO, 2023). Reprinted with the permission of the International Association of Plumbing and Mechanical Officials. This copyright material and all points or statements in using this material have not been reviewed by IAPMO. The opinions expressed herein are not representations of fact from IAPMO.

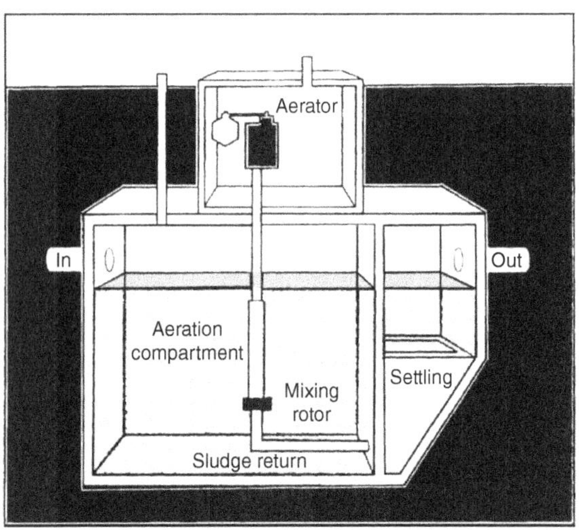

Figure 22.35 *Aerobic treatment unit (ATU), requiring more energy and maintenance, but with cleaner effluent than that of a passive anaerobic septic tank. (National Small Flows Clearinghouse)*

a settling chamber, allowing remaining solids to settle and to be filtered out before the effluent leaves, for further treatment in a secondary process.

(c) Secondary Treatment: Seepage Pits

Also called cesspools (Figure 22.36), these follow treatment by either septic tanks or ATUs. They are increasingly rare, and are appropriate only in very porous soil, where the water table is at least 2 ft [0.6 m] below the bottom of the pit. Because these pits are commonly 10–15 ft [3–4.6 m] below the earth's surface, a very low water table may be required. Another common usage of precast seepage pits is for dry wells that receive runoff from paved areas during rainstorms.

Seepage pits are sized by the square footage of the wall area exposed to the earth—the leaching area, as listed in Table 22.9. Placement of pits relative to buildings, water sources, waterways, and property

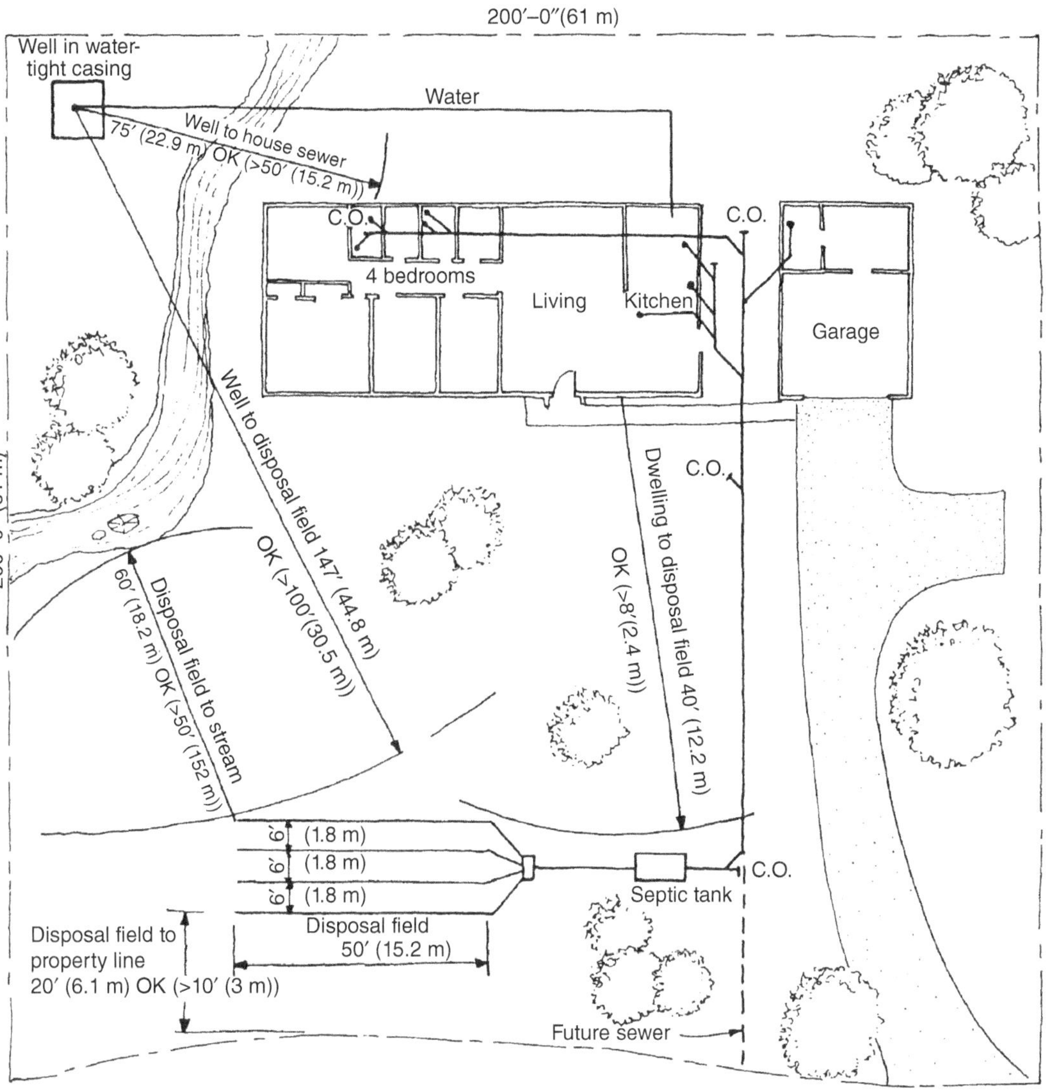

Figure 22.36 *Checking for required clearances on a suburban lot. Although the 40,000-ft² [3716-m²] lot is more than the minimum required, space should be left for a backup disposal field (of equal size) in case of failure of the original field.*

lines is strictly controlled—see Table 22.10 and Figure 22.36.

(d) Disposal Fields

These are very commonly used as a secondary treatment method (Figure 22.37) because they are relatively inexpensive to build and do not require a water table so deep—or soil so permeable—as do seepage pits. Drain lines consisting of either perforated pipes (of many approved materials) or square-edge agricultural tile that is 4 in. [102 mm] or more in diameter (with the ends of the tiles separated by ¼-in. [6-mm] openings) are typically used. These lines are placed in shallow trenches, on a bed of gravel, and covered with gravel. The effluent runs out of these lines and stands in the interstices of the gravel until it seeps into the earth. In effect, the gravel provides a space that acts as a dry well—receiving the fluids and accommodating them until they slowly sink into the ground.

Disposal fields are located according to regulations such as those given in Table 22.10 and sized in relation to total sewage flow and septic tank size (see Tables 22.9 and 22.11). Although Table 22.9 shows the maximum absorption capacities for soils based on gal/ft^2 [L/m^2] over a 24-hour period, a common alternative is to use percolation tests. For such tests, a trial pit is dug and water poured into it. The absorption in gal/ft^2 [L/m^2] over a 24-hour period is determined from this test. Some codes require design sizes based upon percolation test

TABLE 22.10 Location of On-Site Sewage Disposal System

Minimum Horizontal Distance Clear Required from:	Building Sewer ft	Building Sewer mm	Septic Tank ft	Septic Tank mm	Disposal Field ft	Disposal Field mm	Seepage Pit or Cesspool ft	Seepage Pit or Cesspool mm
Buildings or structures[a]	2	610	5	1542	8	2438	8	2438
Property line adjoining private property	Clear[b]		5	1542	5	1542	8	2438
Water supply wells	50[c]	15,240	50	15,240	100	30.5 m	150	45.7 m
Streams and other bodies of water	50	15,240	50	15,240	100[d]	30.5 m	150[d]	45.7 m
Trees	—	—	10	3048	—		10	3048
Seepage pits or cesspools[e]	—	—	5	1542	5	1542	12	3658
Disposal field[e]	—	—	5	1542	4[f]	1219	5	1542
On-site domestic water service line	1[g]	305	5	1542	5	1542	5	1542
Distribution box	—	—	—		5	1542	5	1542
Pressure public water main	10[h]	3048	10	3048	10	3048	10	3048

Source: Adapted from (IAPMO, 2023). Reprinted with the permission of the International Association of Plumbing and Mechanical Officials. This copyright material and all points or statements in using this material have not been reviewed by IAPMO. The opinions expressed herein are not representations of fact from IAPMO.

The SI conversions were appended by the authors of this book.

[a]Including porches and steps, whether covered or uncovered, breezeways, roofed porte cocheres, roofed patios, carports, covered walks, covered driveways, and similar structures or appurtenances.

[b]See Section 312.3 of the *2024 Uniform Plumbing Code* (IAPMO, 2023) for additional information.

[c]Drainage piping shall clear domestic water supply wells by not less than 50 ft [15.2 m]. This distance shall be permitted to be reduced to no less than 25 ft [7.6 m] when the drainage piping is constructed of materials approved for use within a building.

[d]These minimum clear horizontal distances shall also apply to disposal fields, seepage pits, and the mean high-tide line.

[e]When disposal fields, seepage pits, or both are installed in sloping ground, the minimum horizontal distance between any part of the leaching system and ground surface shall be 15 ft [4.6 m].

[f]Plus 2 ft [0.6 m] for each additional 1 ft [0.3 m] of depth in excess of 1 ft [0.3 m] below the bottom of the drain line.

[g]See Section 720.0 of the *2024 Uniform Plumbing Code* (IAPMO, 2023) for additional information.

[h]For parallel construction. For crossings, approval by the Health Department shall be required.

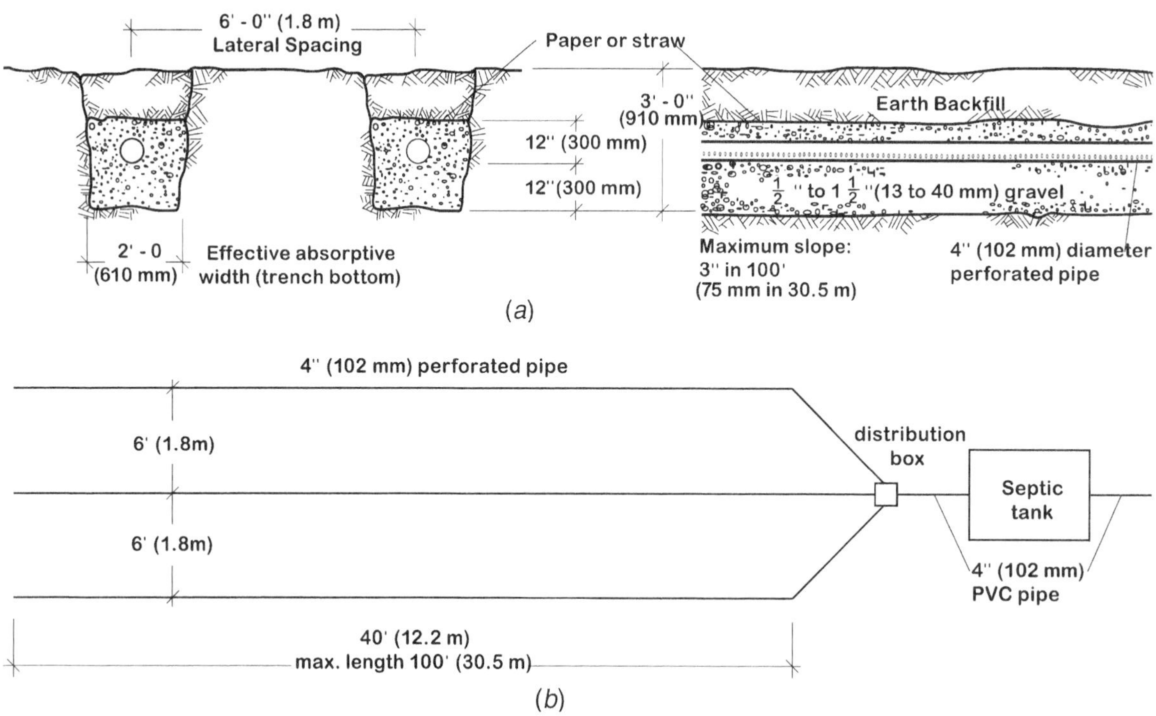

Figure 22.37 *Tile drain field for a four-bedroom house. Although the drawings are not to scale, the dimensions show a required area of about 20 ft × 50 ft [6 m × 15 m] on the lot. Because paving is not permitted over drain fields, sewage treatment on a small lot demands considerable creativity. (a) Transverse and longitudinal sections. (b) Schematic plan. (Redrawn with SI units by Nathan Majeski.)*

TABLE 22.11 Disposal Field Trenches

Part A. Dimensions		
	Minimum	**Maximum**
Number of drain lines per field	1	—
Length of each line	—	100 ft [30.5 m]
Bottom width of trench	18 in. [457 mm]	36 in. [0.9 m]
Spacing of lines, center-to-center[a]	6 ft [1.8 m]	—
Depth of earth cover of lines (preferred 18 in. [457 mm])	12 in. [305 mm]	—
Grade of lines	level	3 in./100 ft [25 mm/m]
Filter material under drain lines	12 in. [305 mm]	—
Filter material over drain lines	2 in. [51 mm]	—
Part B. Leaching Areas		
Trench bottom[b]: minimum 150 ft² [14 m²] per system		
Trench sidewall: minimum 12-in. [305-mm], maximum 36-in. [914-mm] depth below the leach line[c]		

Source: Adapted from (IAPMO, 2023). Reprinted with the permission of the International Association of Plumbing and Mechanical Officials. This copyright material and all points or statements in using this material have not been reviewed by IAPMO. The opinions expressed herein are not representations of fact from IAPMO.

The SI conversions were appended by the authors of this book.

[a]Minimum spacing of drain lines: 4 ft [1.2 m] plus 2 ft [0.6 m] for *each* additional foot [0.3 m] of depth more than 1 ft [0.3 m] below the bottom of the drain line.

[b]Exclusive of any hard pan, rock, clay, or other impervious formations.

[c]The UPC criteria for trench sidewalls is predicated on the notion that a deeper trench can contribute an effective absorption surface area, at least up to the 36-in. [914-mm] maximum.

data. Codes often require that an area equal to the prescribed drainage field size be set aside for use in the event of failure in the original field.

Design guidelines for sizing disposal fields were presented in Chapter 22, but a more detailed method is shown here. The trench width, depth, and spacing shown in Figure 22.37 are typical, but other combinations can be used, provided that they meet the requirements listed in Table 22.11.

EXAMPLE 22.3 See Figure 22.37. Design a septic tank and drain field system for a suburban residence under the following conditions:

Bedrooms	4
Soil	Sandy loam
Depth to water table	7 ft [2.1 m]

SOLUTION

Septic tank capacity (Table 22.8) = 1200 gal [4542 L]. Fluid volume of the tank (minimum required) is:

$$1200 \text{ gal} \div 7.48 \text{ gal/ft}^3 = 160 \text{ ft}^3$$

$$[4542 \text{ L} \div 1000 \text{ L/m}^3 = 4.5 \text{ m}^3]$$

Dimensions for a tank are chosen:

$$4.5 \text{ ft} \times 4.0 \text{ ft} \times 9.0 \text{ ft} = 162 \text{ ft}^3 \text{ volume}$$

$$[1.4 \text{ m} \times 1.2 \text{ m} \times 2.7 \text{ m} = 4.5 \text{ m}^3 \text{ volume}]$$

$$162 \text{ ft}^3 \text{ (proposed)} > 160 \text{ ft}^{3\,\text{(minimum)}}$$

[same conclusion with SI units]

Drain field size is established from Table 22.9: sandy loam (soil type 3) requires 40 ft² [3.7 m²] of leaching area per 100 gal [378 L]. With a septic tank capacity of 1200 gal [4.5 m³]:

$$40 \text{ ft}^2 \text{ area} \times 1200\text{-gal capacity}/100 \text{ gal} = 480 \text{ ft}^2 \text{ minimum absorption area}$$

$$[3.7 \text{ m}^2 \text{ area} \times 4542\text{-L capacity}/378 \text{ L} = 45 \text{ m}^2 \text{ minimum absorption area}]$$

The effective absorption area of the typical trench depth and spacing shown in Figure 22.37 is:

Trench width	2.0 ft²/ft [0.6 m²/m] of length
Trench sides	2.0 ft²/ft [0.6 m²/m] of length (12 in. [305 mm] on each side)
Total absorption	4.0 ft²/ft [1.2 m²/m]

The required absorption area was determined to be 480 ft² [45 m²]. Because the trench has 4.0 ft²/ft [1.2 m²/m] of length, the total trench length must be:

$$480 \text{ ft}^2/(4 \text{ ft}^2/\text{ft of length}) = 120 \text{ ft}$$

$$[45 \text{ m}^2/(1.2 \text{ m}^2/\text{m of length}) = 38 \text{ m}]$$

A three-line disposal field is selected, so 120 ft/(3 lines) = 40 ft [37 m/(3 lines) = 12.3 m] per line. Given the required clearances from the edges of the disposal field (Table 22.10) of 5 ft [1.5 m] to the property lines, the disposal field area can be calculated as:

width:

$$5 \text{ ft} + 12 \text{ ft (two spaces of 6 ft each)} + 5 \text{ ft} = 22 \text{ ft}$$

$$[1.5 \text{ m} + 3.7 \text{ m (two spaces at 1.85 m each)} + 1.5 \text{ m} = 6.7 \text{ m}]$$

length:

$$5 \text{ ft} + 40\text{-ft line} + 5 \text{ ft} = 50 \text{ ft}$$

$$[1.5 \text{ m} + 12.3 \text{ m line} + 1.5 \text{ m} = 15.3 \text{ m}]$$

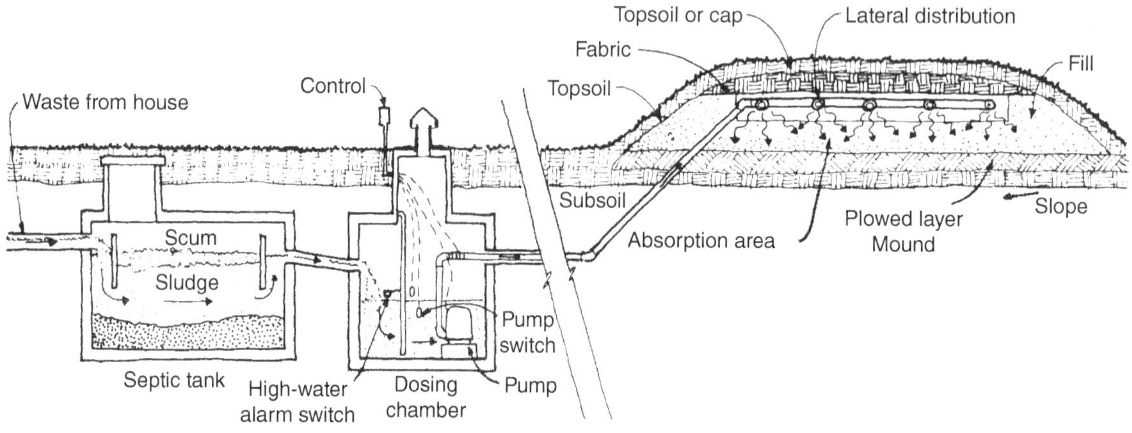

Figure 22.38 Mounds with leaching beds offer a disposal option when the water table is high. The system serves a two- or three-bedroom home. (Adapted from Converse et al., 1978.)

The site surface area, then, is 22 ft × 50 ft = 1100 ft² [6.7 × 15.3 m = 102 m²]. Note that double this area (for a second disposal field in the event of failure of the first field) is often required. ■

(e) Mounds with Leaching Beds

These mounds (Figure 22.38) represent a newer solution in the United States and thus may require special approval. The guidelines for leaching-bed sizing are essentially similar to those for drainage tile disposal fields. The absorption area for leaching beds, however, must be 50% greater than that required for trenches. The bottom of the leaching bed generally must be at least 5 ft [1.5 m] above the water table, although in water-scarce areas, officials may reduce this requirement.

(f) Buried Sand Filters

Slow sand filters for treating supply water were discussed in Chapter 21. Buried sand filters (Figure 22.39) work in a similar way, using primarily biological but also physical and chemical processes to clean wastewater. The medium is most commonly sand, although other locally available materials such as crushed glass, mineral tailings, bottom ash, and so on, have been used. The grain size ranges from 0.01 to 0.1 in. [0.3 to 3 mm] in diameter.

The layout is similar to that of the disposal fields described in Section 22.6(e), but these filters, themselves 24–36 in. [610–914 mm] deep, usually require an excavation of 4–5 ft [1.2–1.5 m]. The filter bed must be level and sited to avoid contact with groundwater

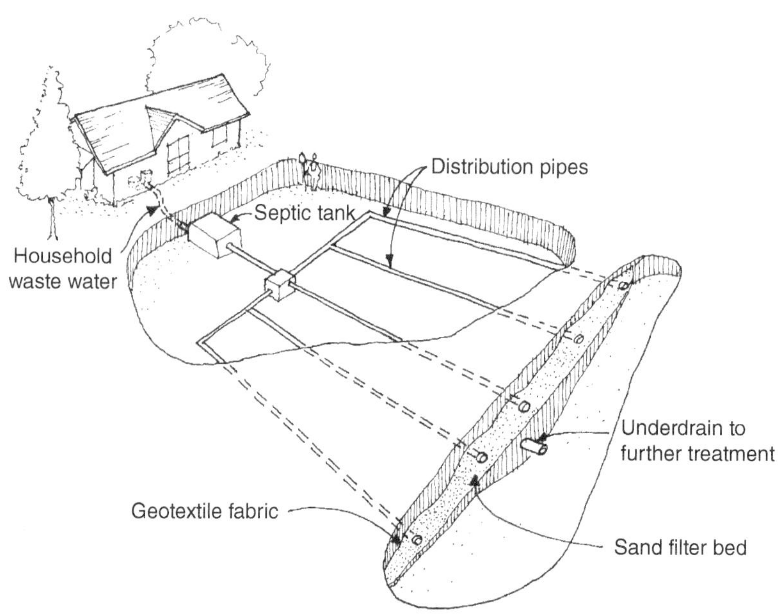

Figure 22.39 Buried sand filters appear to be like the common disposal fields, but are deeper and may require an impermeable membrane liner. (Adapted from National Small Flows Clearinghouse, n.d.)

TABLE 22.12 Sand Filter Design Values[a]

Design Factor	Buried Filters	Open Filters	Recirculating Filters
Pretreatment	All filters must be preceded by settling and removal of solids		
Media			
Materials	Washed, durable granular material		
Uniformity coefficient	Less than 4.0 between smallest and largest size particles		
Depth	24–36 in. [610–914 mm]		
Effective size	0.3–1 mm	0.3–1 mm	0.8–3 mm
Hydraulic loading	<1.5 gpd/ft^2 [<61 L/day m^2]	2–5 gpd/ft^2 [82–204 L/day m^2]	3–5 gpd/ft^2[b] [122–204 L/day m^2]
Media temperature	>41°F [>5°C]	>41°F [>5°C]	>41°F [>5°C]
Dosing frequency	<2 per day	<2 per day	5–10 min./30 min.[c]
Recirculation ratio[d]	(Not applicable)	(Not applicable)	3:1 to 5:1

Source: Adapted from National Small Flows Clearinghouse, 1997.

[a]These values show typical design criteria and do not represent all possibilities.

[b]Forward flow only, not including recirculated effluent.

[c]Dosage lasts for 5–10 minutes every ½ hour.

[d]Water recirculated: water discharged.

and excess surface water runoff. Some authorities require the sand filter to be contained in an impermeable membrane liner. Underdrain pipes and a graded layer of washed gravel or crushed rock are placed on the bottom of the filter bed, with the finer gravel on top to keep the medium from washing into the underdrains. Then a layer of fine gravel first, with coarser gravel above, is placed around and over the distribution pipes. A geotextile fabric covers the top of the entire filter bed, which in turn is covered by backfill material. A properly constructed buried sand filter that receives properly digested waste from a well-maintained septic tank should last for up to 20 years without maintenance.

Buried sand filters are one remedy for failed disposal fields. They are also used where high groundwater, shallow bedrock, or poor soil preclude use of the simpler disposal field. Although buried sand filters may require considerable site area, the ground surface can be used for lawns or other nonpaved surface activities. More detailed information on the sand filter types discussed here is available from the National Small Flows Clearinghouse (n.d.).

In general, sand filters are designed for normal dosing twice a day; they need time to allow the filter medium to drain between doses. The water that filters through the medium is collected in underdrains that are sloped toward an outlet and vented to the surface at the upstream end. This vent is a potential source of odors. The water is then taken either to a disinfection (usually chlorination) unit or to a disposal field for subsurface discharge.

Water from sand filters has been cleansed of many pathogens, such as harmful bacteria and viruses, and is therefore more likely to be approved for a down-line disposal field. Typical values for three types of sand filters are shown in Table 22.12. (Open and recirculating sand filters are discussed in Section 22.7(b).)

22.7 ON-SITE MULTIPLE-BUILDING SEWAGE TREATMENT

The first approach to a multiple-building treatment system involves providing septic tanks at each building for primary treatment. Then the outflows from the individual septic tanks are combined and enter a secondary treatment process. Either of two specific types of sand filters can be employed for this purpose. There are several advantages to this approach. The effluent lines from the septic tanks carry no solids; they can be much smaller in diameter and placed in shallower trenches than conventional sewer lines. The runs are far shorter, because distant centralized treatment plants are not involved. With sand filter secondary treatment, the septic tank wastewater is more likely to find a recycling use, such as irrigation.

(a) Open Sand Filters

Also called intermittent sand filters, these are very similar in construction to slow sand filters (Chapter 21). They are nearly identical in performance to buried sand filters (Figure 22.39)—the difference is that these are at least partially above ground. A detail of the filter medium is shown in Figure 22.40. Table 22.12 shows that they are able to accept a somewhat higher flow than a buried filter, thus requiring less area than the latter, but also leaving the surface no longer usable for any other purpose. (A removable cover is typically used for odor control, despite the "open" name.) They are

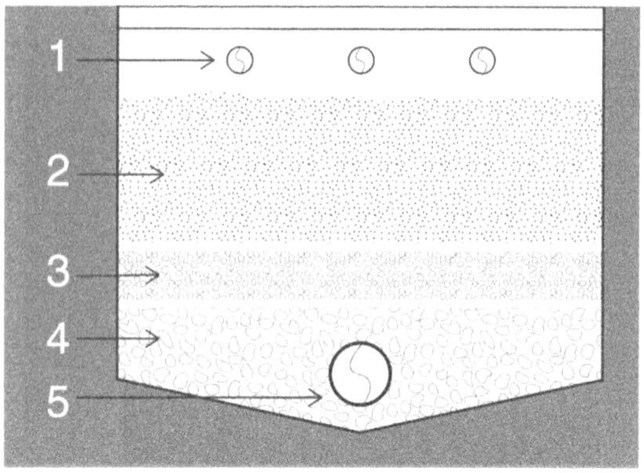

Figure 22.40 *Open sand filter receives wastewater from above through (1) distribution pipes (or a thin layer of flooded wastewater). A filter medium (2) such as sand is on top, then (3) fine gravel, then (4) coarse gravel, surrounding the (5) underdrain. Slow sand filters are very similar. (Redrawn by Nathan Leigh, after National Small Flows Clearinghouse.)*

used on systems with flows of up to 120,000 gpd [454,240 L/day] of wastewater. Frequently, two such open filters are built so that one can rest while the other is in use. Alternatively, two filters may be used in series, with an even cleaner resulting final effluent.

(b) Recirculating Sand Filters

These eliminate odors by ensuring an adequate supply of oxygen to the wastewater. The recirculating sand filter (Figure 22.41) first receives water (from the septic tank or other primary treatment) into a recirculation tank. This tank has a pump, a timing mechanism, and float valves. Either in timed doses or when triggered by the float valve, doses of water are pumped to the sand filter. The treated wastewater collects in the underdrain, where the majority (75% to 80%) is directed back to the recirculation tank. Thus, mixing with the septic tank effluent and being pumped to the sand filter, the result is weaker (cleaner) effluent, containing more oxygen, entering the sand filter. This eliminates odor and allows for a slightly larger filter grain size, making the system less prone to clogging. Table 22.12 compares sand filter types, showing the highest allowable flow rates per surface area for the recirculating sand filter.

At Stonehurst, a 47-lot subdivision in Contra Costa County, California, 1500-gal [5680-L] septic tanks serve each unit; each tank has a screened effluent vault, keeping solids out of the effluent lines. The effluent is taken through small-diameter, variable-grade sewers (located under the roadway) to a recirculating granular medium filter. (Homes lower than the roadway use effluent pumps at the septic tank outflows.) Outflow from the recirculating filter is then taken to a UV recirculation tank, where it is treated by UV radiation (see Chapter 21). From this tertiary treatment area, it is passed to a subsurface drip irrigation system serving a community park. In winter, a 2.5-acre [1-ha] community soil absorption field is utilized.

(c) Lagoons

These basins need sun, wind, and more land area than other methods, but are very simple to maintain and very low in energy use. They usually must be lined to prevent wastewater from polluting the groundwater. Sometimes the outflow must undergo further treatment before release. If lagoons are not square or round, their length should not exceed three times their width. Outflow and inflow should be at opposite ends. Detailed information is available from the National Small Flows Clearinghouse (1997).

Anaerobic lagoons are usually used as the first of at least two lagoons in series. Working much like a septic tank, they hold wastewater for 20–50 days and are relatively deep, 8–15 ft [2.4–4.6 m]. With a surface of floating scum that blocks oxygen from the wastewater, they are unsightly and malodorous.

Aerobic lagoons are shallower than other lagoons, so sunlight and oxygen from wind and air can better penetrate. Aerobic bacteria and algae do the work of cleaning the wastewater. These are best in warm climates where freezing is not a threat. They hold water for 3–50 days. The bottoms are either paved or lined to prevent weeds from growing within the lagoon.

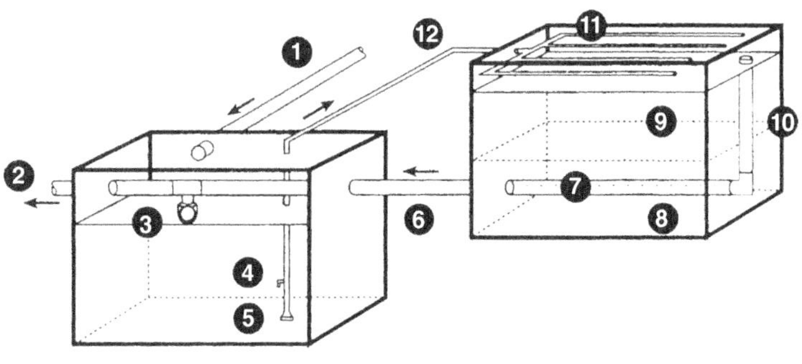

Figure 22.41 *Recirculating sand filters periodically distribute a cleaner, more oxygen-rich effluent to the sand filter. This eliminates odor and allows a larger grain size for the medium. (1) Effluent from pretreatment (such as a septic tank). (2) To disinfection/discharge. (3) Float valve. (4) Bleed line. (5) Submersible pump. (6) Recirculated effluent line. (7) Perforated underdrain piping. (8) Layered support gravel. (9) Filter media. (10) Riser pipe. (11) Distribution piping. (12) Recirculation pump discharge to recirculating filter. (National Small Flows Clearinghouse.)*

Aerated lagoons take the aerobic lagoon a step further, actively stirring and adding oxygen to the lagoon. Because they speed aerobic action, they require shorter retention time and less land area than a passive aerobic lagoon.

Facultative lagoons (also called stabilization, oxidation, photosynthetic, or aerobic–anaerobic ponds) are the type most commonly used by small communities and individual households. They work in most climates and require no machinery. They require about 1 acre (0.4 ha) for every 50 homes (or for every 200 people) served, are 3–8 ft [0.9–2.4 m] deep, and are designed to hold wastewater for 20–150 days (the longer period in cold weather).

Three layers form in these lagoons: the top is an aerobic zone, the middle is the facultative zone, and at the bottom is the anaerobic zone. The depth of the surface aerobic zone depends upon how much sunlight, wind, and rain can contribute oxygen; the deeper this zone is, the better it controls odors rising from the anaerobic zone. Therefore, these lagoons should be sited with full access to both sun and wind.

Down in the anaerobic zone, anaerobic bacteria convert sludge to gases, including hydrogen sulfide, ammonia, and methane. As these gases rise, they provide food for both the aerobic bacteria and the algae in the aerobic zone. Sludge accumulates more quickly in cold climates; sludge removal should be expected every 5–10 years.

(d) Advanced Integrated Wastewater Pond System (AIWPS™)

At California Polytechnic State University, San Luis Obispo, there was an innovative proposal to create an integrated infrastructure facility combining campus wastewater treatment and solid waste processing. In addition to the waste treatment and resources recovery for this campus of 18,000, the schematic design for the Energy Efficient Resource Recovery (E2R2) Facility includes an educational facility, a wildlife habitat, and park facilities. The wastewater and related wetlands proposals are shown in Figure 22.42; the solid waste facility appears in Chapter 23.

Three ponds are used in an AIWPS facility (Oswald, 1990; Oswald et al., 1963). The first is the advanced facultative pond (AFP; Figure 22.42*a*), which includes an in-pond digester (IPD) or anaerobic fermentation pit at the bottom, and an oxygen-rich aerobic surface water layer. Two AFPs used in parallel allow fail-safe operation. This design enables isolation and treatment of toxic substances in the event that such material is accidentally released into the community sewer system. Conventional sewage treatment plants typically cannot isolate or treat accidental toxic releases in wastewater due to their short processing time.

The raw sewage enters the system near the bottom of the IPD, where anaerobic bacteria slowly reduce biosolids to ash (Oswald et al., 1994). Biochemical oxygen demand (BOD) is reduced to 60–80%, and minimal accumulation of refractory solids (sludge) has been demonstrated in installations since the late 1960s.

Using earthwork construction for the ponds allows them to be quite large and deep, without the substantially higher costs of concrete and steel construction typical of conventional wastewater treatment facilities. A berm surrounding the IPD prevents mixing of aerobic water with the influent below, whether by wind or thermal inversion. The aerobic upper layer oxidizes any malodorous biogases rising from the IPD. The biogas generated in the IPD is scrubbed as it passes through the water column and the aerobic surface layer. With a campus population of 18,000, between 3000 and 5000 ft^3 [85 and 142 m^3] per day of methane (CH_4) were estimated to be available for collection from the IPD and use as fuel (Green and Bernstone et al., 1995).

The second pond is the high-rate pond (HRP) designed to grow algae (Figure 22.42*b*). The lined channels in this pond are approximately 3 ft [1 m] or less in water depth and include a paddle wheel (see Figure 22.49) to slowly mix the primary effluent. The HRP design allows maximum oxygen production by algal photosynthesis due to optimum exposure to sunlight and abundant nutrients. The pond's aerobic and facultative bacteria oxidize most of the soluble and readily biodegradable BOD.

The next pond in the series is an algal settling pond (ASP; Figure 22.42*c*). The quiet waters of this pond allow 50% to 80% of the algae to settle over several days' residence time. The settled algae are periodically harvested, and the algal biomass does not produce objectionable odors. When properly disinfected, the harvested algae, rich in nitrogen and phosphorus, can be used as fish or livestock feed, or as a nitrogen source in compost.

Near St. Helena, California, an AIWPS facility has functioned since the late 1960s. During this time period, no sludge removal has been necessary, attesting to the efficacy of the methane fermentation process. The facility is near several large wineries and is considered a good neighbor without odor problems (Oswald et al., 1970).

While a maturation pond usually follows the ASP in an AIWPS facility, various appropriate treatment components can subsequently be used to achieve advanced tertiary treatment—the highest wastewater reclamation and reuse standard required by the state of California (Green et al., 1996). In the proposed Cal Poly design, the ASP effluent flows into a constructed wetland.

(e) Constructed Wetlands

Wetland surrogates constructed specifically for waste treatment are shown in Figure 22.42*c*. In the schematic design, two parallel-constructed wetlands are proposed—allowing for a comparison between their outflows.

The free surface wetlands (FSW) consist of shallow open basins or channels that are lined to prevent seepage. Soil supports a succession of wetlands vegetation, nourished by a continuing flow of effluent from the initial wastewater treatment system(s) (such as a secondary AIWPS facility, septic tank, etc.). Although the water surface is exposed to beneficial air and sunlight, proper design is required to mitigate human contact with the treated effluent in FSWs. It is also necessary to address potential problems such as mosquitoes or other water-borne insects.

The subsurface flow wetlands utilize a lined treatment basin filled with a coarse medium such as large gravel or crushed rock. The large voids (if properly designed) allow the effluent to flow freely through rocks and plant root systems. The relationships among gravel size, slope of the bed, and rate of flow are complex and are described in detail in an EPA document (EPA, 1993). A layer of soil is used to cover the gravel bed.

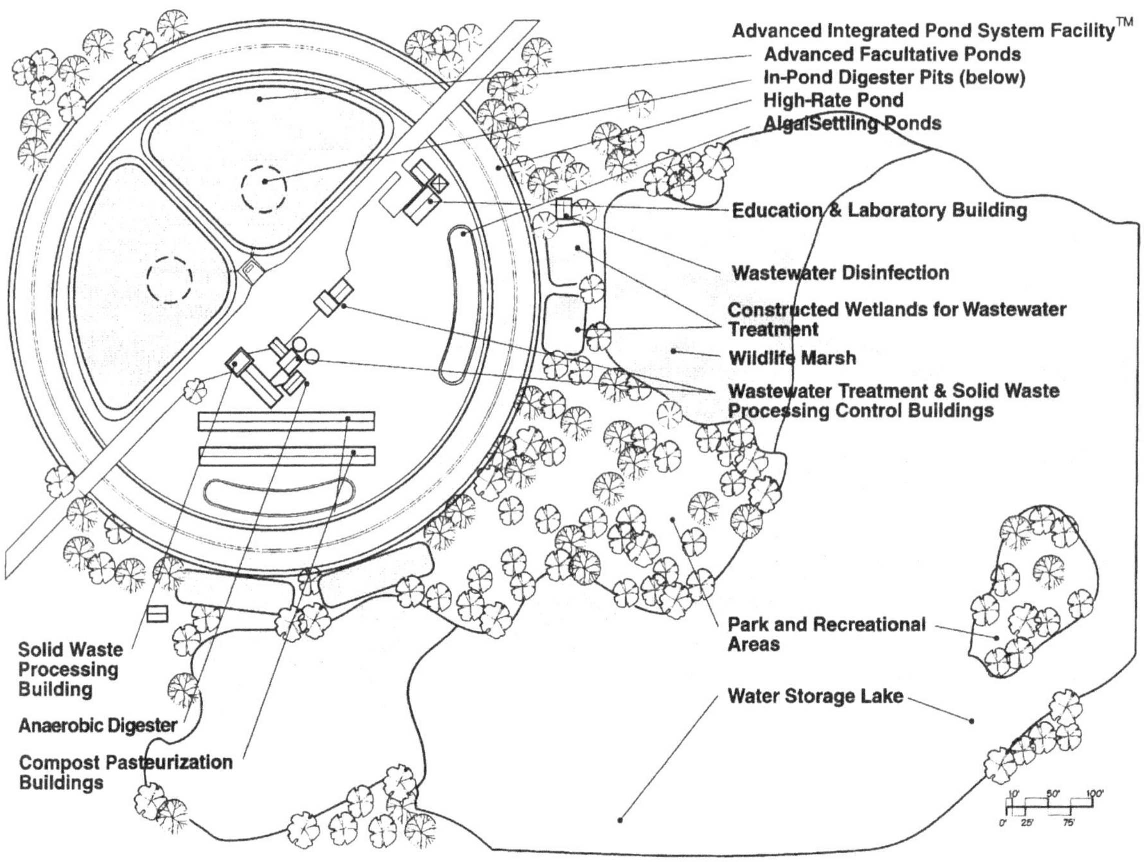

(*a*)

Figure 22.42 Schematic design proposal for wastewater treatment at the Energy Efficient Resource Recovery (E2R2) Facility at Cal Poly, San Luis Obispo. (a) Site plan. (b) Effluent enters the advanced facultative pond (AFP) through the in-pond digester, where solids settle, undergo methane fermentation, and are reduced to ash. Rising biogas is channeled by a proprietary submerged gas collector and passes through a highly aerobic surface layer where impurities are removed, yielding methane-rich biogas at the surface. (c) AFP effluent then goes to the high-rate pond, where it is slowly moved along the shallow channels by an energy-efficient paddle wheel providing optimum solar exposure and enabling the profusely growing algae to supply abundant oxygen for oxidation of wastewater organics and disinfection of pathogens. Subsequently, effluent moves into the quiescent Algal Settling Pond, where algae settles and is periodically removed for beneficial use. (d) Effluent then moves into constructed wetlands, where the plants in this system create an ideal environment for invertebrates and vertebrates (in free surface wetlands) and aerobic and anaerobic microbial populations to thrive and provide further treatment and effluent polishing. Finally, the effluent from the wetlands is disinfected by ozone or UV disinfection systems and stored in a lake that is part of a campus park and recreational facility. (Courtesy of Dr. William Oswald, Professor Emeritus of Civil and Environmental Engineering and Public Health at the University of California, Berkeley, and Principal of Oswald Engineering Associates, Inc.; Dr. Bailey Green, Research Engineer at the University of California, Berkeley, and Principal of Oswald Engineering Associates, Inc.; and Professor Daniel Panetta, Architecture Department, Cal Poly, San Luis Obispo.)

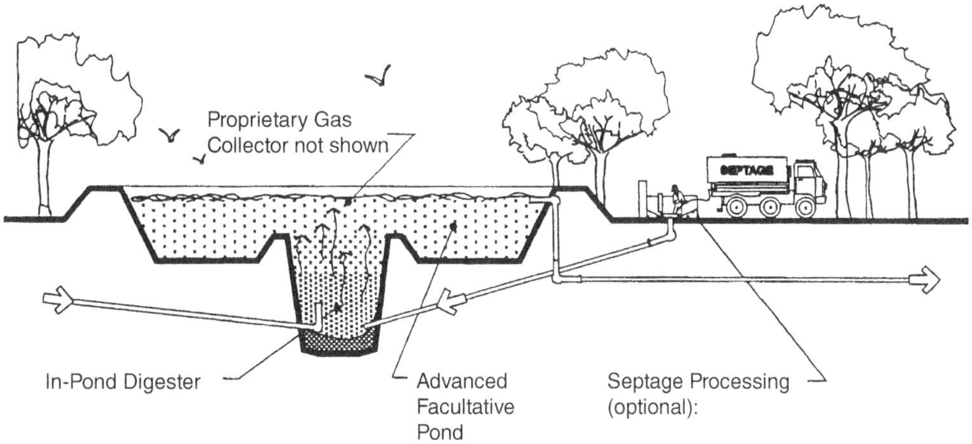

(*b*)

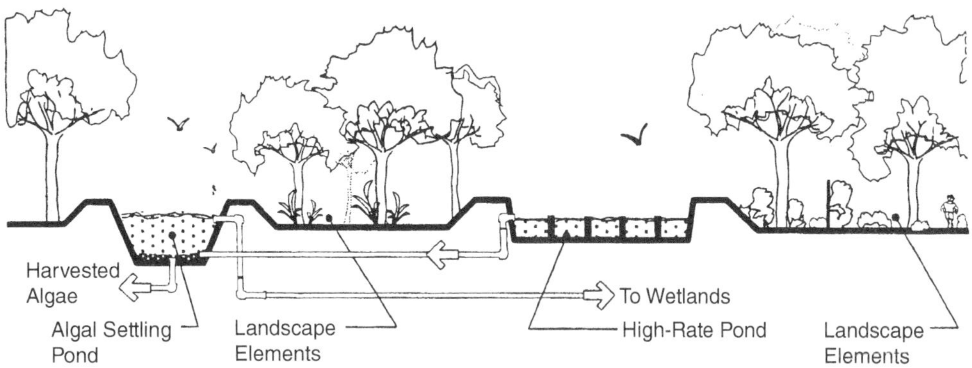

(*c*)

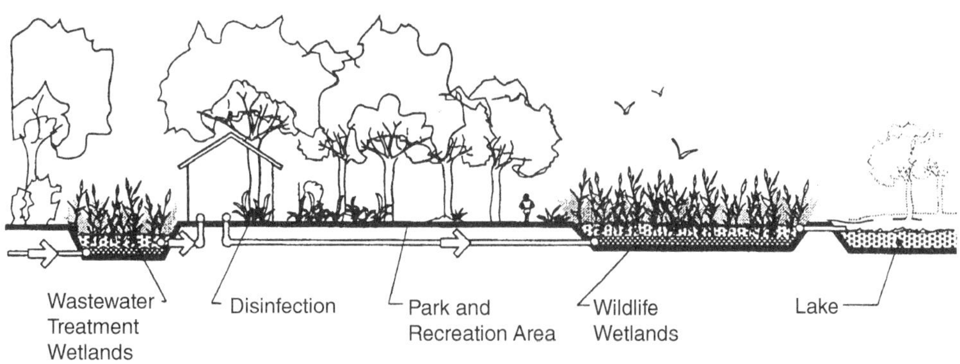

(*d*)

Figure 22.42 *(continued)*

so subsequent human contacts with the effluent and insect problems are precluded.

In either type of constructed wetland system, saturated and nutrient-rich conditions encourage the growth of both aerobic and anaerobic microbes and certain invertebrates. The most common plants include varieties of *scripus* (bulrushes), *phragmites* (reeds), *typha* (cattails), *canna*, and *iris*. These plants not only provide large amounts of surface area for microbial activities, but also create an aerobic zone underwater by transporting air through their stems to their roots. A thin aerobic environment results among the root hairs.

Constructed wetlands become a vivid, colorful demonstration of how, in nature, there is no such thing as waste. After passage through a slow sand filter (see again Chapter 21) and final disinfection, the reclaimed water from the E2R2 facility can provide campus landscape irrigation water, be released into the campus creek to assist with habitat restoration, or be used for aquifer recharge. Arcata, California, completed its wetlands treatment system in 1986. This system draws several hundred species of birds, as its abundant vegetation removes nutrients from, and allows habitat reuse of, the city's treated wastewater.

(f) Greenhouse Ecosystems

When constructed wetlands are moved indoors, greater control and less required area become possible. As pioneered by marine biologist John Todd at Ocean Arks International, Living Machines are a series of tanks, each with a particular ecosystem, that form a "stream." The stream is followed by an indoor marsh; together they achieve a high degree of tertiary wastewater treatment.

These enclosed systems typically cost less to construct and about the same to maintain as conventional sewage treatment plants, but use less energy because photosynthesis (solar energy) and gravity flow largely replace energy-intensive pumps and large-scale aerators. These systems do not harm the environment with a final dose of chlorine, and they produce about one-quarter of the sludge of conventional systems. Because they are pleasant to look at and smell much like commercial greenhouses, they can be located within the neighborhoods they serve. This can save the huge costs associated with large-diameter, deep-trench sewer lines and pumping stations necessary with distant, centralized treatment plants.

A diagram of one greenhouse ecosystem is shown in Figure 22.43. This Living Machine begins with an anaerobic bioreactor, in effect a septic tank where solids are reduced to sludge and methane gas is produced. From here the wastewater passes to aerated tanks, and then to ecological fluidized beds, each with a particular combination of bacteria, algae, snails, and fish. The further along the treatment process, the more advanced the ecosystem, with goldfish (carp) growing in the final clarifier. This process depends upon the following food chain: (1) aerobic bacteria consume suspended organic matter and convert ammonia into nitrites and then nitrates; (2) algae and duckweed feed on the products of the bacterial action; (3) snails and zooplankton eat the algae, while floating duckweed provides shade that discourages algal growth in the later stages of the process; (4) fish eat the zooplankton and the snails. Plants range from water hyacinth in the early stages to an increasing variety of specialized "accumulators" that remove troublesome phosphorus, heavy metals, and so on. Papyrus, canna lilies, bald cypress, willows, and eucalyptus (among other plants) all have a role. When trees outgrow the greenhouse, they are transplanted outdoors. It should be kept in mind that said trees are long-term storage devices for the heavy metals they absorb, not final solutions, because heavy metals will be returned to the earth when the trees die and decompose. Plants can be sold to nurseries or ground up as compost; the fish also enrich the compost. The smaller fish (shiners) can be sold as bait fish. These are small enough to dart back and forth in the piping between tanks, choosing their water quality. If a toxic slug of waste enters the system, they swim to later, clearer tanks until the danger is past.

The greenhouse solar aquatic system in Figure 22.44*a* cleans the wastewater from a mobile home park on Vancouver Island, British Columbia. In Figure 22.44*b*, a striking Living Machine design occupies the center of an urban block in Kolding, Denmark.

In milder climates and with specialized wastewater, these processes can be achieved without the greenhouse enclosure. The Sonoma Mountain Brewery's Living Machine (Figure 22.45) near Sonoma, California, occupies about 1 acre [0.4 ha] and treats a maximum flow of 7800 gpd [29,525 L/day] of waste from the brewery process. Two streams move in parallel, allowing a comparison of their outflows as experiments with detailed ecosystems are conducted. A part-time operator manages this treatment plant. After an underground anaerobic holding tank, flow moves in the following manner:

1. Closed aerobic reactors, each holding 2500 gal [9463 L]. Air is blown through the tanks from the bottom (an adjacent small building encloses the blower); wastewater spends about 15 hours in these tanks.
2. Next are open aerobic tanks, four per stream. Plants, snails, and organisms can migrate among

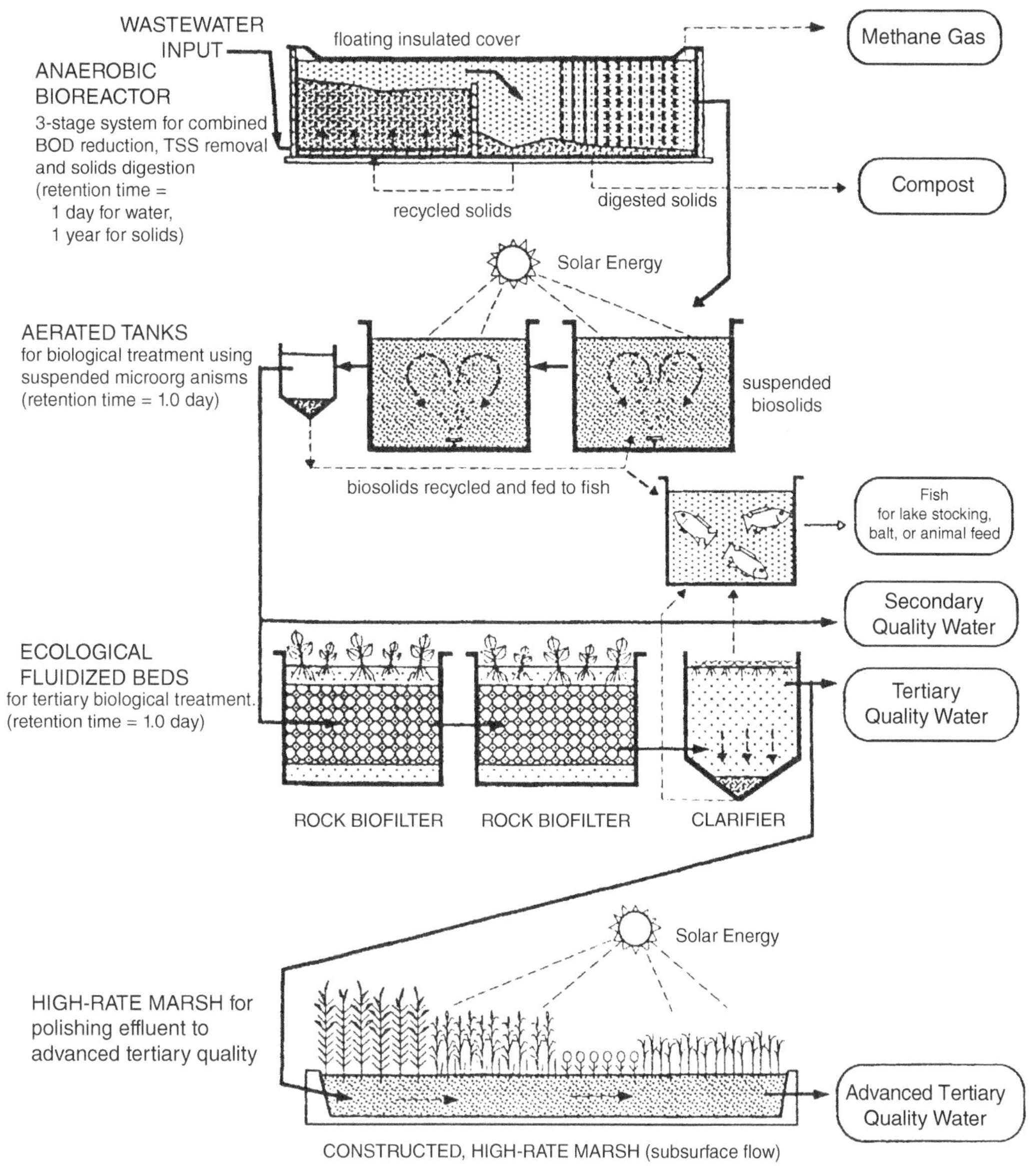

Figure 22.43 *A Living Machine enclosed within a greenhouse uses a food chain in a series of ecosystems to treat wastewater. (Living Technologies, Inc.)*

the agitated tanks, choosing their preferred level of nutrients ("pollution"). Water leaving the last of these tanks is much cleaner.

3. Water and sludge collected from the tanks are taken next to the clarifier. Shaded by water hyacinth, algae cannot grow in the water at this advanced stage. In this quiet-water environment, solids settle out.
4. The solids are then pumped to an adjacent open-air reed bed ("composter"), where they are dried and then composted. The compost will serve as nutrient for the hops used in the brewing process. Excess water percolates through an underlying sand bed and is returned to the beginning of the Living Machine streams.
5. Meanwhile the clarified water passes next to ecological fluidized beds (two per stream), whose purpose is to further polish the wastewater. Water circulates many times through this section, pumped from the plant-shaded perimeter into a rock bed at the center. In this bed, microorganisms continue to clean the water.
6. From the second ecological fluidized bed, water moves to an underground irrigation pond.
7. From there, it is filtered through a constructed wetland for further purification.

Finally, it is welcomed for irrigation of the hops and grapevines grown on site.

(*a*)

(*b*)

Figure 22.44 *(a) A greenhouse in a mobile home park in British Columbia encloses the 12 solar tanks and marsh of a Solar Aquatics System, designed by Ecological Engineering Associates and ECO-TEK, Inc. (b) In the middle of an urban block in Denmark, a Living Machine is a good neighbor. (a: Photo by Charles Rusch; b: Photo by Robert Peña.)*

The IslandWood environmental education campus on Bainbridge Island, Washington, prominently features a Living Machine (Figure 22.46) as a focal point near a major classroom. This location brings waste processing into the consciousness of those using the campus and reminds viewers that human wastes do not literally just disappear if piped off site. In this cool climate setting, the greenhouse provides a thermal environment to support the biological waste treatment processes that are the heart of a Living Machine.

(g) Pasveer Oxidation Stream

This method of sewage treatment (Figures 22.47 and 22.48) has been adopted at the New York Institute of Technology. Serving the 450-acre (182-ha) campus at Old Westbury, Long Island, New York, this system

(*a*)

(*b*)

(*c*)

Figure 22.45 *In the mild climate of California's Sonoma Valley, the waste from the Sonoma Mountain Brewery (a) is treated by an outdoor Living Machine (b). Flow is from left tanks to right tanks, then to a marsh. (c) Calla lilies and other plants lend beauty as well as function to this open aerobic tank. (Photo by Living Technologies, Burlington, VT.)*

(a)

(b)

Figure 22.46 *(a) Greenhouse enclosing a Living Machine on the IslandWood campus on Bainbridge Island, Washington. (b) Interior view of the IslandWood Living Machine, showing plant materials that are an integral aspect of this biological waste treatment approach. (Photos © Alison Kwok.)*

ACTIVE ENVIRON-MENTAL SYSTEMS

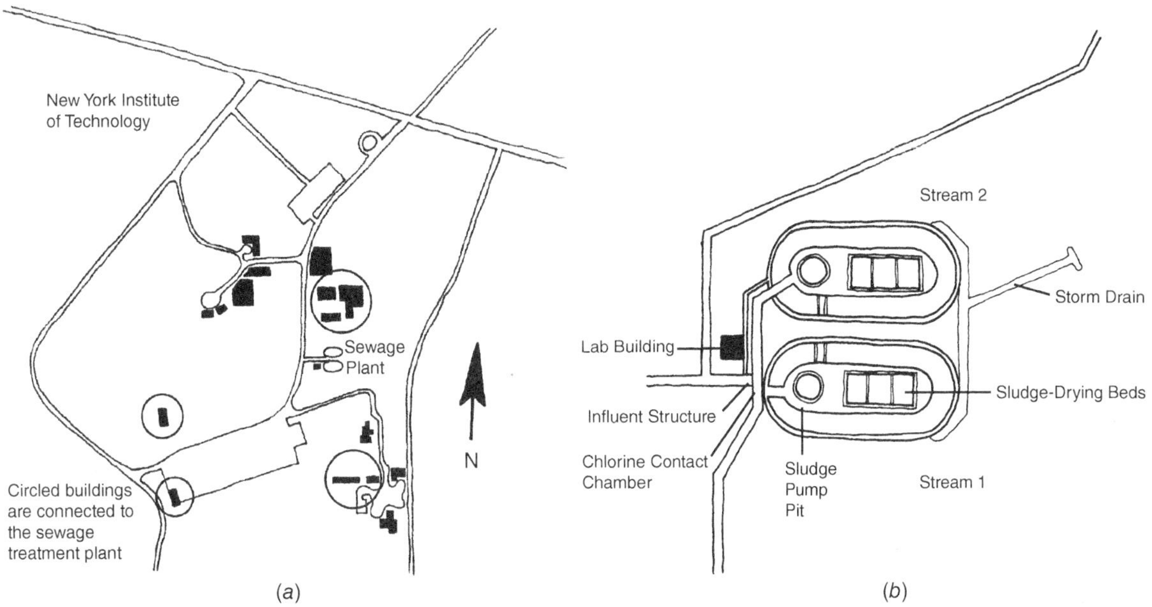

Figure 22.47 *Sewage treatment plant, New York Institute of Technology (NYIT). (a) The treatment plant's location is central to all of the buildings it serves—an efficient choice. (b) Layout of the treatment plant. Because of the plant's odorless operation, its proximity to campus buildings poses no problem. (Courtesy of Bogen Jenal, Engineers, P.C. Redrawn by Amanda Clegg.)*

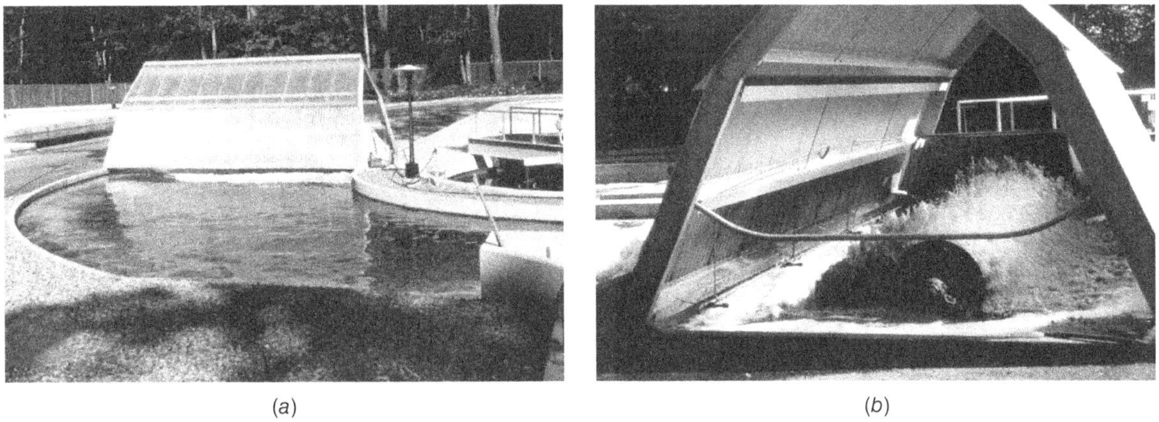

Figure 22.48 *Details of NYIT treatment plant. (a) Rotor with a plastic "greenhouse" cover. (b) The rotor induces oxidation in this oval circulating stream. A similar mechanism serves the high-rate pond in Figure 22.49.*

provides on-campus sewage treatment, which returns purified effluent to the ground through 48 leaching wells located under the athletic field. The groundwater, thus restored, provides a contributing source of water for 400-ft [122-m] deep wells, distantly located, that furnish part of the water supply for the campus buildings (see Figure 22.47).

A few of the original small buildings, later converted to administration offices and classrooms, are still served by septic tanks and leaching fields. There was no public sewer near the campus, and in the 1970s the health authorities ruled out the use of septic tanks for the numerous additional buildings that were being contemplated.

The oxidation stream process applied here was developed by the Netherlands Research Institute for Public Health Engineering, and is now in operation in many U.S. locations as well as in Europe. It is considered to be a modified form of the activated sludge process (see later discussion of the Oceanside Plant in Section 22.8b) composed of aerobic digestion and periodic sludge removal.

The mechanical aerator (Figure 22.48) that keeps the stream of sewage moving, while providing the oxidation necessary for aerobic digestion, is an important feature. In this design, sludge-drying beds are placed on an island, surrounded by the continuously moving oxidation stream. Another feature of the system is its low profile (Figure 22.49), readily screened by trees. The aerobic digestion process is not malodorous. Sound is produced by the splashing water-wheel action of the mechanical

(a)

(b)

Figure 22.49 *NYIT sewage treatment plant at completion. (a) The only projections above ground level are the office/laboratory and the plastic covers over the rotors. (b) The oxidation stream does not appear turbulent here, but it becomes so under the action of the mechanical aerator (rotor).*

aerators; no air compressor is required (reducing noise generation).

The plant has a full-time accredited operator; there are also two assistants and one relief operator. This design provides for an eventual population of 4330, with a 340,000-gpd [1,287,010-L/day] flow. Sludge was removed from the drying beds twice in the first five years of operation.

22.8 LARGER-SCALE SEWAGE TREATMENT SYSTEMS

Two examples of community-wide sewage treatment are presented here—one for a small, water-scarce community near San Diego, California, and one for an oceanside urban neighborhood in San Francisco.

(a) Padre Dam Municipal Water District

The Santee Water Reclamation Plant (Figure 22.50) incorporates innovations in water conservation and recycling developed during the 1960s and 1970s. The Padre Dam Municipal Water District is a region where rainfall is less than 15 in. [380 mm] per year. It has no local water supplies, so water is imported 300 miles [480 km] from the Colorado River. If conventionally treated, the wastewater resulting from water usage would be discharged into the Pacific Ocean. Alternative sewage treatment (shown here) processes this fluid waste for secondary uses such as irrigation and recharging of groundwater. Many municipal potable water supplies are taken from rivers that are often heavily polluted with sewage. After treatment, water from such sources is pumped into domestic water supply mains. Therefore, district officials here decided to install a system that would make use of purified wastewater for many secondary uses.

However, would such a concept be acceptable to the public? An effective public relations program concurrent with the incorporated technological advances has proved successful.

The project involved building a sewage treatment plant and utilizing seven pits left over from prior surface mining of sand and gravel. After partial purification of the sewage at the plant through the first two stages of a conventional activated sludge process, the effluent is discharged to form the lakes adjacent to the plant. This provides tertiary (oxidation) treatment, after which the cleaner effluent is pumped to a filter area at the north end of the complex, where it is further purified through sand and gravel. Chlorination is performed, and the water then flows through a series of seven lakes, Nos. 7–6–5–4–3–2–1, in that order.

The appearance of new lakes in this semi-arid region created initial interest that was rapidly augmented by a very clever program. At first, the lakes were fenced in. Then the fences were removed. Next, the seven lakes were made available for boating. They were stocked with fish, and careful studies were made that indicated that the fish were healthy and flourishing. Fishing was permitted, but for a while, all fish had to be returned to the water. Anglers were later permitted to keep and consume the fish. Finally, swimming was permitted. The overflow from the last lake, No. 1, discharges into Sycamore Canyon Creek and is used for irrigation of a golf course and the recharging of groundwater.

(b) Oceanside Water Pollution Control Plant

Imagine the challenge of locating a major new sewage treatment plant on the shore of the Pacific Ocean, within the city of San Francisco, immediately adjacent to both the city zoo and a well-established neighborhood. This plant treats all the sewage in the western San Francisco watershed, an average dry-weather flow of 21 million gal/day [79,491,720 L/day]. The treated effluent is discharged through an ocean-bottom pipeline, 4.5 miles [7.2 km] offshore, with effluent and ocean water quality monitored by an on-site laboratory. Sludge is shipped north to serve as the required daily cover for a landfill near Novato. The plant itself is concealed by large earth berms, through which entry and exit tunnels carry plant traffic. And the odor? It is completely contained within a huge underground complex of wastewater basins and channels, over which the zoo proposed to build a Mammal Conservation Center on 5 ft [1.5 m] of topsoil cover. Exhaust air from the underground plant is filtered through activated carbon and potassium permanganate before release.

The process is shown in simplified form in Figure 22.51*a*. Incoming sewage passes through a bar screen and grit chambers, which remove the largest objects as well as grit, sand, and gravel (these are trucked to a landfill). Then the water flows to the primary clarifier, in which floating scum is skimmed off, and sludge settling on the bottom is pumped to the egg-shaped anaerobic digesters (Figure 22.51*b*). The wastewater flows next to aeration tanks where bacteria grow fat, feeding on and breaking down the remaining dissolved and suspended solids. Next is the secondary clarifier, where the fattened bacteria (activated sludge) sink to the bottom. Part of this activated sludge is returned to the aeration tanks to bolster the bacterial culture there. The remainder is thickened and pumped to the digesters.

Thus, the anaerobic digesters receive both primary sludge and thickened, activated sludge. As this sludge

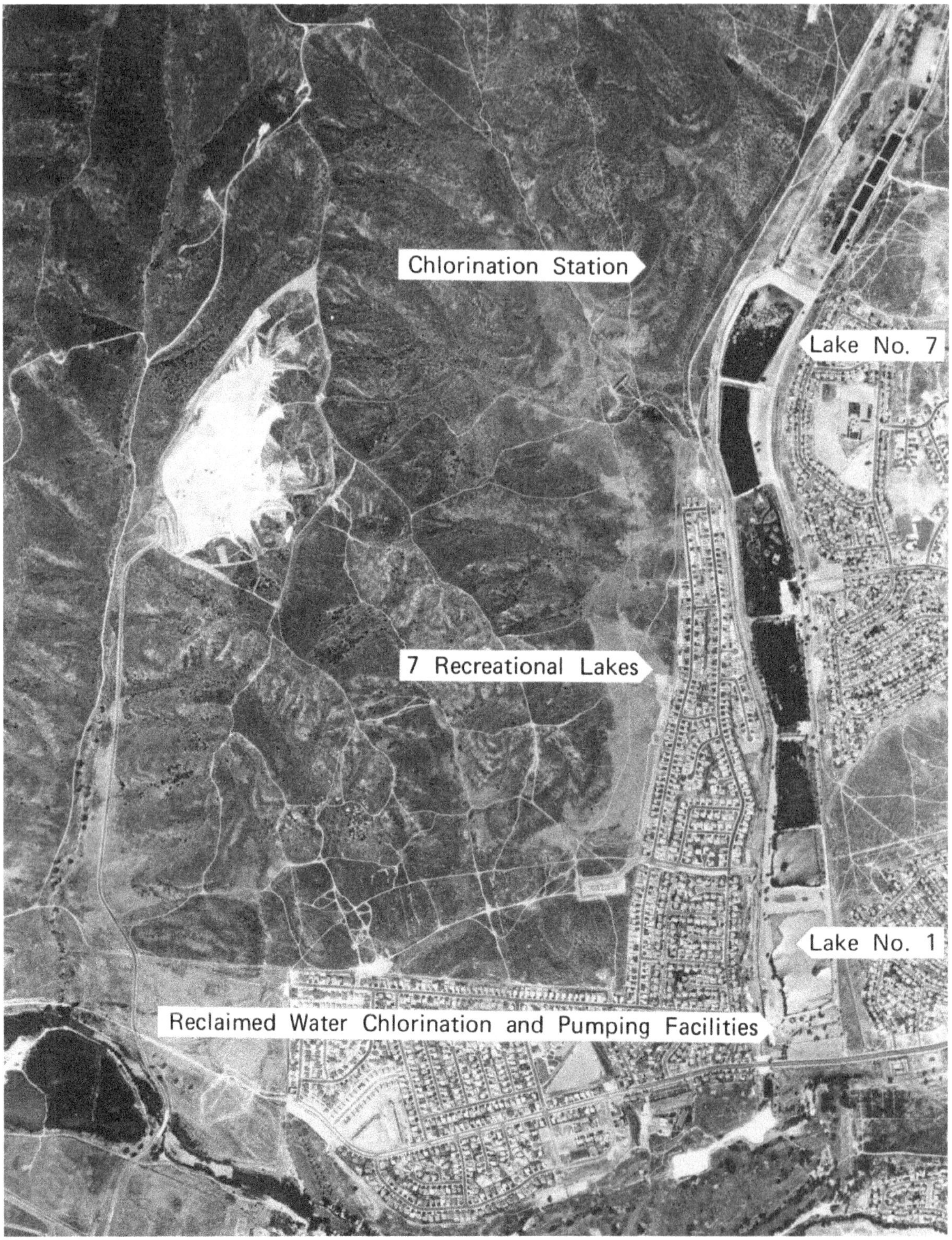

(a)

Figure 22.50 *Santee Water Reclamation Plant and Santee Park and Recreational Facilities, Padre Dam Municipal Water District, California. (a) Aerial photograph of the Santee Plant, including the seven recreational lakes. Note the location of the San Diego River at the bottom of the photograph. The water reclamation plant and the stabilization ponds do not appear in this aerial view. (b) Raw sewage from the community of Santee enters the treatment plant, which is located at the top of this diagram. The process then proceeds south to the point where reclaimed water is pumped to irrigation or recharges groundwater. Sludge is pumped to the San Diego Metro system. The plant is not burdened by stormwater, which is recharged to the ground locally in the community.*

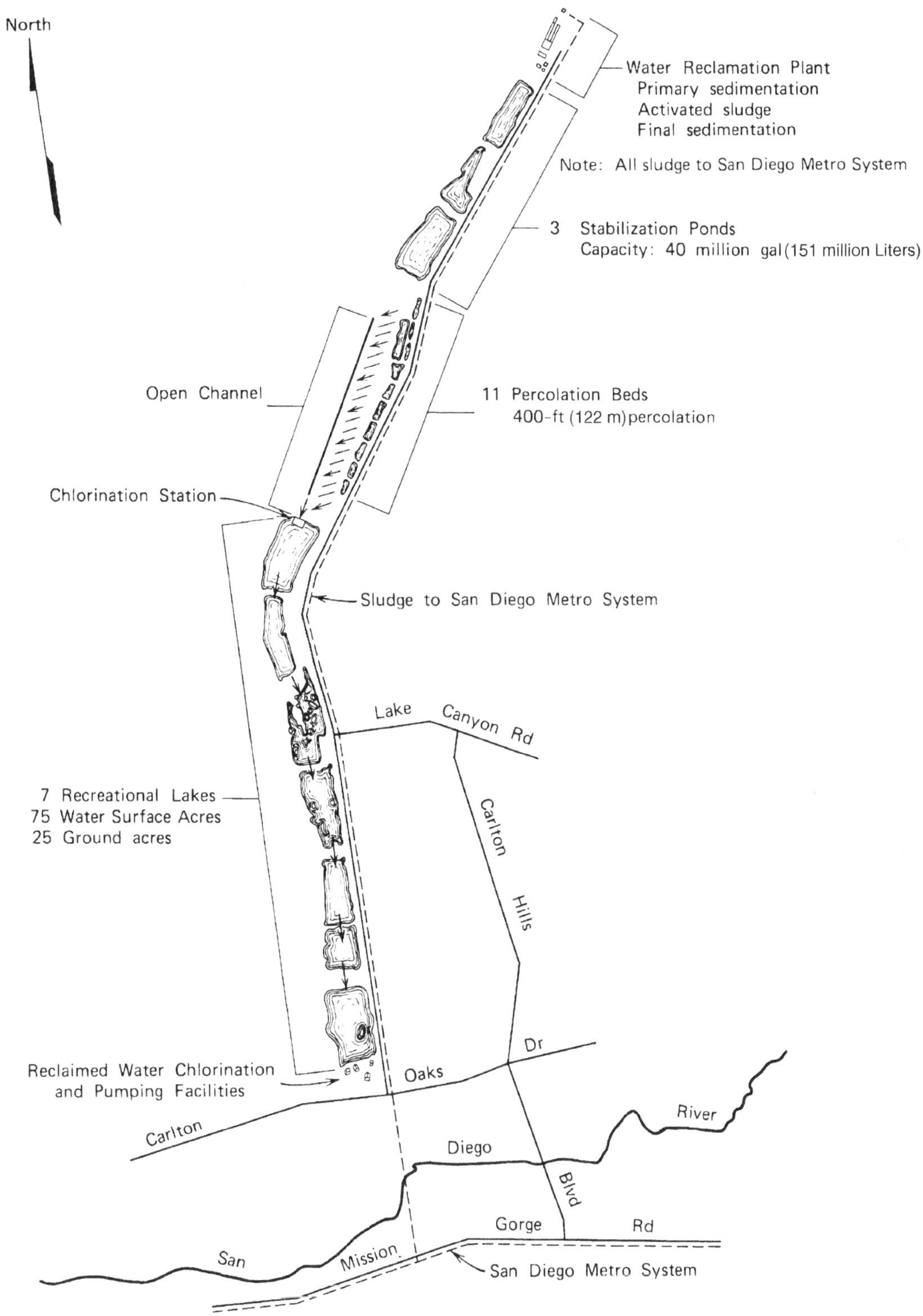

(*b*)

Figure 22.50 *(continued)*

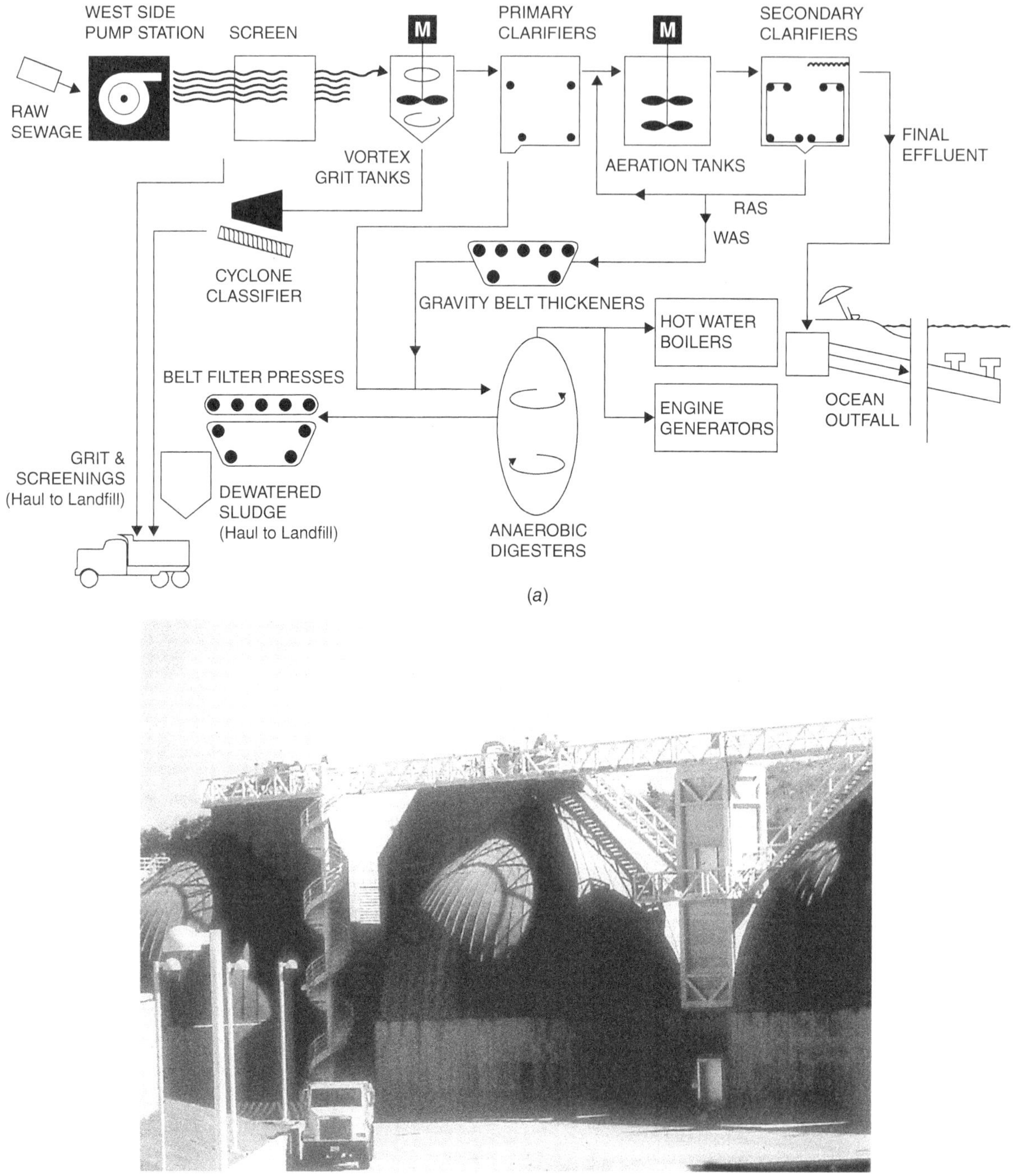

Figure 22.51 *The Oceanside Water Pollution Control Plant, San Francisco, is largely underground, as well as hidden behind high earth berms. (a) Flow diagram. (b) View of some of the egg-shaped anaerobic digesters from within the berm walls. Biogas produced within the four 750,000-gal [28,375-m^3] digesters supplies about 20% of the electricity used in the treatment plant. (Courtesy of CH2M Hill and the City/County of San Francisco.)*

is stabilized and reduced in volume, biogas is produced: enough to heat water for both process and facility heating, and to generate about one-fifth of the electricity used within the plant. Sludge leaving the digesters is dewatered in belt presses, before being hauled to the landfill. During summer, biosolids are spread on agricultural land for soil enrichment.

Questions remain about such use of biosolids. Although a 1998 EPA national survey of sewage sludge found toxic chemicals at such low levels that regulation was deemed unnecessary, concentrations of certain industries may produce local chemical or radioactive sludge pollutants at higher levels. Intercepting such wastes at their source, rather than allowing them into the sewage stream, seems a clear imperative.

22.9 WATER REUSE AND RECYCLING

These chapters on water and waste have illustrated four common "grades" of water in buildings:

- *potable water:* usually treated, suitable for drinking
- *rainwater:* usually collected on-site or nearby
- *graywater:* wastewater not from toilets or urinals
- *blackwater:* water containing toilet or urinal waste

Two more categories are:

- *dark graywater:* from washing machines with dirty diaper loads, kitchen sinks, and dishwashers—usually prohibited for reuse
- *clearwater:* backwash water usually from reverse osmosis water treatment and/or condensation from a cooling coil

After considering the many blackwater sewage treatment complications described in the previous section, it might be expected that graywater reuse would, by contrast, be a simple matter. However, there are lingering prejudices against wastewater of any kind that complicate, or directly prohibit, such water recycling. This section presents some future approaches that may be more actively implemented if water conservation is to be taken seriously, followed by some code-mandated approaches.

Rainwater, given an adequate supply, could meet the need for many uses of water in and around a building: bathing and laundry, toilet flushing, irrigation, and evaporative cooling, for example. Little or no treatment of rainwater is needed for such uses.

Blackwater requires much more extensive treatment, given its high concentration of human waste. However, this may be the easiest waste stream to eliminate, given the several waterless alternatives discussed in Section 22.1.

Graywater (also spelled greywater) reuse opportunities are more limited than those of rainwater, because graywater carries increased threats from pathogens. It must undergo treatment before reuse in toilet flushing. It is likely to contain soap, hair, and, occasionally, human waste (from soiled clothes in the laundry). If it is to be used in drip irrigation systems, filtering is essential. If kitchen wastes are included, grease and food solids add to the problems. Graywater may well be dirtier than rainwater, but its supply is likely far more predictable and constant. However, codes tend to limit the use of graywater to underground landscape irrigation for single-family houses.

The collection of these separate types of water poses a problem. In the early years of indoor plumbing, rainwater, graywater, and blackwater were mixed together, which severely overloaded sewage treatment facilities in rainy weather. Today, graywater and blackwater are typically mixed, with rainwater being kept separate. There are precedents for keeping all three separate, however (Figure 22.52).

(a) Graywater and Future Recycling

Notwithstanding present code restrictions, the potential for residential graywater reuse (given separate waste collection systems) was shown earlier (Figure 22.1*b*). The 21 g/cd (gallons per capita per day; [80 L/cd]) produced by bathing can meet much of the 32 g/cd [120 L/cd] water consumption of a conventional water closet and all of the water needs of a low-flush water closet. The 14 g/cd [53 L/cd] produced by doing laundry can help with irrigation; limiting contact between irrigation water and people seems prudent, hence the previous emphasis on underground distribution. Table 22.13 relates both the quantity and quality of residential graywater to its use in irrigation. Another potential benefit of separate graywater collection is that the heat content of the water might be partially reclaimed (Table 22.14). In England, the Aquasaver Company diverts and cleans water from lavatories, baths, and showers for reuse in toilet flushing, clothes washing, car washing, and irrigation. This low-pressure system, installed behind panels in a bathroom, uses a pump to push graywater through a series of filters to remove soap, detergents, and other impurities and then to a storage tank (in the attic or above points of use). Here it undergoes weir and trickle filtration and then treatment with nonhazardous cleaning agents. Finally, before reuse, it passes through a carbon filter network.

Storage raises a health issue. To avoid bacteria buildup in graywater, storage should be avoided and the filtered water taken directly to irrigation. However, storage allows for controlled dosage, allows solids to

(*a*)

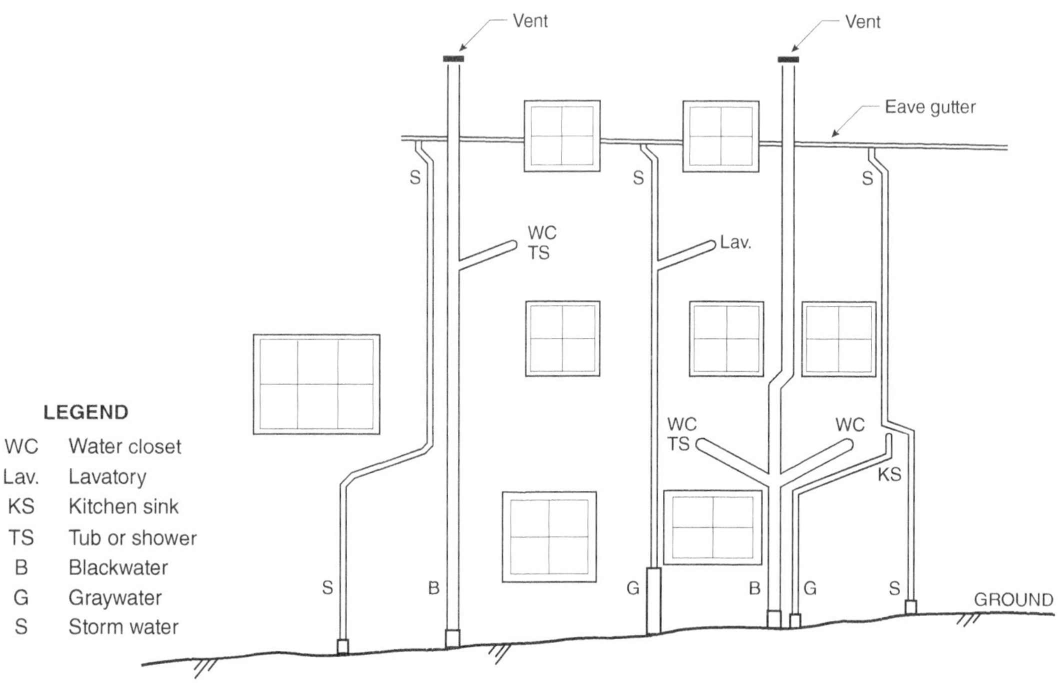

(*b*)

Figure 22.52 *Residence of the 27th Vicar of the Parish Church at Bibury (established 1086 CE) in the English Cotswolds. (a) Clearly, the plumbing was added after the construction of the residence—perhaps several hundred years later. In England, the drainage system for an older building often appears on the outside of the building. (b) It seemed inappropriate to ask the vicar's permission to inspect his indoor facilities, so probable uses have been assigned. Note that (1) offsets of the vertical piping are made with "easy bends" and are pitched down in the direction of the flow; (2) vents are located at high outdoor points; and (3) stacks carrying blackwater (from toilets) are of larger pipe size than those carrying graywater. In this sparsely settled region of the Cotswolds, the dispersal of stormwater to the ground and the private septic tank treatment of blackwater plus graywater can be satisfactory. (a: Photo by William McGuinness; b: Redrawn by Nathan Leigh.)*

TABLE 22.13 Residential Graywater Sources for Irrigation

Part A. Sources with Pumps		
Source	**Average Outflow Water Quantity**	**Outflow Water Relative Quality**
Clothes washing machine	30–50 gal [114–190 L] per load, top loader, 10 gal [38 L] per load, front loader. Average 1.5 loads per week per adult, 2.5 loads per week per child	Good: medium concentration of soaps, lint. With biocompatible cleaners, quality can improve to excellent; diapers degrade quality to poor.
Automatic dishwasher	5–10 gal [19–38 L] per load	Poor: low to high quantity of solids depending on degree of pre-rinsing; high salt and pH from conventional cleaning compounds
Part B. Gravity Flow Sources		
Source	**Outflow Water Quantity**	**Outflow Water Quality**
Shower	20 gal [76 L] per day per person, low-flow	Excellent: minimal concentration of soap and shampoo, but hair can clog distribution pumps and lines
	40 gal [151 L] per day per person, high-flow	
Tub	40 gal [151 L] per bath	Concentration in tub water likely less than shower water
Bathroom lavatory	1–5 gal [4–19 L] per day per person	Good: concentration of soap, shaving cream, and toothpaste can be high
Kitchen sink	5–15 gal [19–57 L] per day per person	Mixed: high in nutrients but also high in solids, grease, and soap
Reverse-osmosis water purifier	3–5 gal per gallon of drinking water [3–5 L per L]	Excellent: usually no suspended solids; contains 25% more concentration of the same dissolved solids as tap water
Water softener backwash	5% of indoor water use	Very bad: extremely high in salt
Softened water	(All graywater from softened water source)	Poor: contains salt (unless potassium chloride is used in softening process)

Source: Adapted from Ludwig, 1997.

TABLE 22.14 Heat Recovery from Graywater

Source	Volume/Use, gal [L]	Flow Rate, gal [L]	°F	°C	Quality	Coincident[a]
Kitchen sink	Up to 5 [up to 20]	2.5 [10]/min	85	30	Poor	No
Lavatory	Up to 1.5 [up to 5]	1.25 [5]/min	85	30	Fair	No
Bath	32 [120]	5 [20]/min	100	37	Good	No
Shower	15 [50]	2 [7]/min	100	37	Good	Yes
Washer	40 [150]	7 [25]/min	60	15	Moot	No

Source: Adapted from Nelson, 1981.

[a]Coincident flow of warm wastewater and input water to heater. If "no," a holding tank is needed for heat exchange. If "yes," a countercurrent heat exchanger can be built into the drain line (see Figure 22.53).

settle, allows for some potential heat recovery, provides for overflow to the sewer, and is often mandated by code.

Various stages of treatment are appropriate for different kinds of graywater (Figure 22.53). Kitchen sink waste contains grease, which should be trapped (and periodically removed) before it can clog the filters and heat exchangers in a graywater recycling system. Similarly, lavatory showers and laundry waste contain lint and hair that must be intercepted quickly. Devices that do so (called interceptors) were described in Figure 22.21.

Currently, many building codes tend to sharply limit graywater recycling, generally presume there will be no filtering, and rigidly constrict its use. It is hoped that such restrictions will loosen, with due regard for human health, as conservation of resources becomes more of an imperative than it already is. Adopters of the Living Building Challenge have worked to change (and challenge) graywater policies in cities in order to put in graywater systems.

An elegant and simple use of recovering energy from shower graywater (Figure 22.54) is used in a residence built to Passive House criteria in Salem, Oregon.

(b) Subsurface Irrigation

The *Uniform Building Code* notes that water only from bathroom lavatories, showers, and tubs, and clothes-washing machines and laundry tubs, and only in single-family residences, can only be used for subsurface "irrigation" on the same site as the residence. The associated use of mandated trench constructions, as a means of avoiding any surface graywater distribution, may deliver very little water to plants.

Site conditions must meet the requirements of Table 22.15 and Figure 22.55. The bottom of

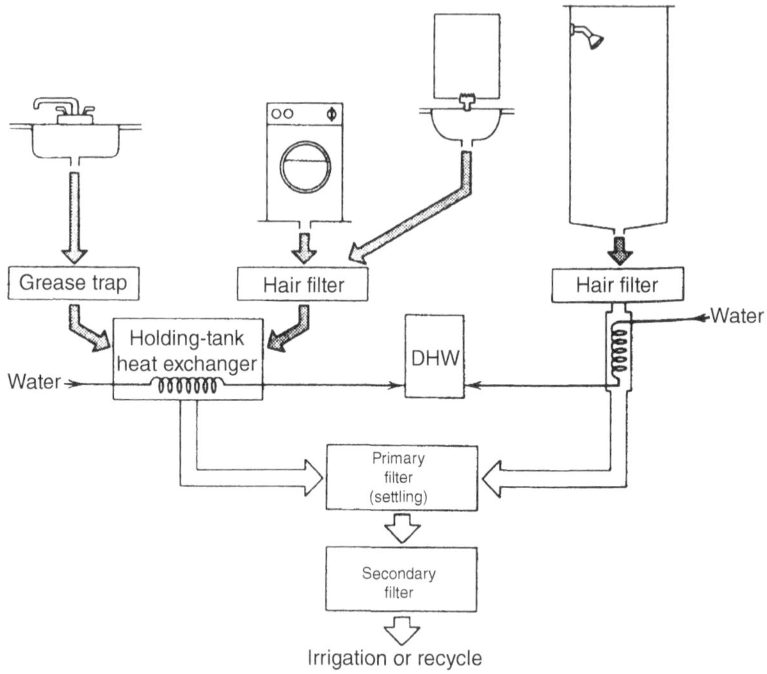

Figure 22.53 Sequence of water treatment and heat reclamation for domestic graywater, where such systems might be approved.

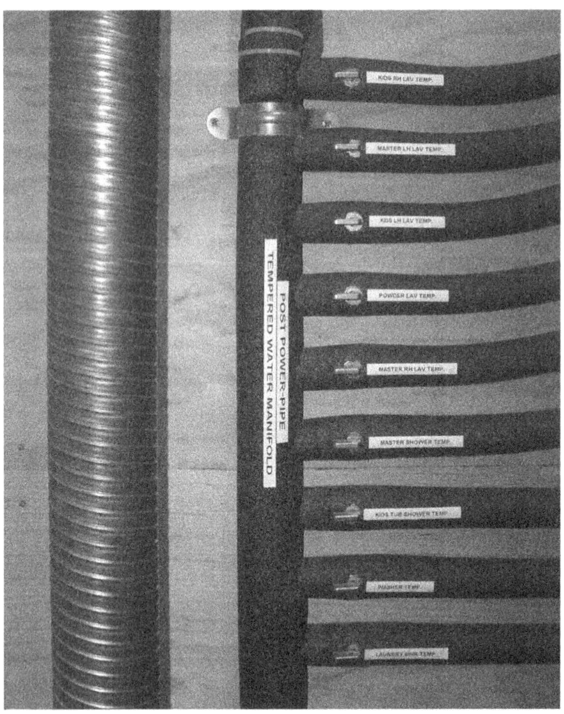

Figure 22.54 Coils of copper tubing wrap a drainage pipe (left) to form a heat exchanger to recover heat energy from the warm water in the shower above in a Passive House in Salem, Oregon. (© Bilyeu Homes; used with permission.)

the trenches must be at least 5 ft [1.5 m] above the highest-known seasonal groundwater. Flow estimates are as follows:

- number of occupants: two (2) for the first bedroom and one (1) for each additional bedroom.
- combined showers, bathtubs, and washbasins, flow per occupant: 25 gal/day [95 L/day]
- laundry, flow per occupant: 15 gal/day [57 L/day]

A holding tank is required with a minimum 50-gal (189-L) capacity. An unvalved overflow must connect to the building sewer (ahead of any septic tank).

The irrigation/disposal field must be divided into a minimum of three valved zones (allowing the occupants to better direct flow rates). The total area of this field is the aggregate length of the perforated pipe times the width of the proposed field. The required area is based on the estimated graywater flow (or size of the holding tank, whichever is larger), the soil types and acceptance rates, as well as the trench design guidelines, shown in Table 22.16. If percolation tests are performed, the 24-hour percolation rate results must be within these limits:

$$\text{Minimum, } 0.83\ \text{gal/ft}^2\ [33.8\ \text{L/m}^2]$$

$$\text{Maximum, } 5.12\ \text{gal/ft}^2\ [208.5\ \text{L/m}^2]$$

One of several graywater system types is shown in Figure 22.56.

EXAMPLE 22.4 A four-bedroom single-family residence is located on sandy clay soil more than 5 ft [1.5 m] above the highest seasonal groundwater level. If the bathroom and laundry facilities contribute to

TABLE 22.15 Location of Residential Graywater System[a]

	Minimum Horizontal Distance in Clear Required			
	To Surge Tank		To Subsurface and Subsoil Irrigation Field and Mulch Bed	
From	ft	m	ft	m
Building structures[b]	5[c,d]	1.5	2[e,f]	0.6
Property line adjoining private property	5	1.5	5[f]	1.5
Water supply wells[g]	50	15.2	100	30.5
Streams and lakes[g]	50	15.2	50[h]	15.2
Sewage pits or cesspools	5	1.5	5	1.5
Sewage disposal field[i]	5	1.5	4[j]	1.2
Septic tank	0	0	5	1.5
On-site domestic water service line	5	1.5	5	1.5
Pressurized public water main	10	3.1	10[a]	3.1

Source: Adapted from (IAPMO, 2023). Reprinted with the permission of the International Association of Plumbing and Mechanical Officials. This copyright material and all points or statements in using this material have not been reviewed by IAPMO. The opinions expressed herein are not representations of fact from IAPMO.

SI units converted from mm to m by the authors of this book.

[a]For parallel construction or crossings, approval by the Authority Having Jurisdiction shall be required.

[b]Including porches and steps, whether covered or uncovered, breezeways, roofed porte-cocheres, roofed patios, carports, covered walks, covered driveways, and similar structures or appurtenances.

[c]The distance may be reduced to 0 ft [m] for above ground tanks when first approved by the Administrative Authority.

[d]The distance shall be permitted to be reduced to 0 ft for surge tanks of 75 gal [284 L] or less.

[e]Assumes a 45 degree (0.79 rad) angle from the foundation.

[f]The distance shall be permitted to be reduced to 1 ½ feet [457 mm] for drip and mulch basin irrigation systems.

[g]Where special hazards are involved, the distance required shall be increased as may be directed by the Authority Having Jurisdiction.

[h]These minimum clear horizontal distances shall also apply between the irrigation/disposal field and the ocean mean higher tide line.

[i]Where irrigation or disposal fields are installed in the sloping ground, the minimum horizontal distance between a part of the distribution system and ground surface shall be 15 ft [4572 mm].

[j]Add 2 ft [610 mm] for each additional foot of depth more than 1 ft [305 mm] below the bottom of the drain line.

a graywater system, how large an irrigation/disposal field is required?

SOLUTION

The total number of occupants:

$$2 + 1 + 1 + 1 = 5 \text{ occupants}$$

Estimated graywater flow is:

$$(25 + 15) \text{ gal/day per occupant} \times 5 \text{ occupants} = 200 \text{ gal/day}$$

$$[(95 + 57) \text{ L/day per occupant} \times 5 \text{ occupants} = 760 \text{ L/day}]$$

Required field size (Table 22.16) of each valved zone is:

$$60 \text{ ft}^2/100 \text{ gal} \times 200 \text{ gal/day} = 120 \text{ ft}^2$$

$$[1.5 \text{ m}^2/100 \text{ L} \times 760 \text{ L/day} = 11.4 \text{ m}^2]$$

If each valved zone has two lines at the minimum spacing of 4 ft [1.2 m], the minimum width of the irrigation/disposal field is 4 ft [1.2 m]; the minimum length of the two perforated lines is therefore:

$$120 \text{ ft}^2/4 \text{ ft} = 30 \text{ ft}$$

$$[11.4 \text{ m}^2/1.2 \text{ m} = 9.5 \text{ m}]$$

Thus, three irrigation/disposal fields, each 4 ft by 30 ft [1.2 by 9.5 m], in trenches constructed as shown in Figure 22.55, will meet code minimums. ■

22.10 STORMWATER TREATMENT

Until relatively recently, stormwater runoff from buildings and paved surfaces was considered harmless and was conducted without treatment, and as rapidly as possible, to receiving streams, rivers, and lakes. However, almost three-quarters of the U.S. population lives on only about 7% of the land. And as our urban watersheds become more and more paved and/or roofed, two problems develop. First, a heavy rain greatly increases peaks in stream flows (because less rain is retained for later, slower release by the action of unpaved surfaces); this accentuated peak–valley pattern of flows makes life more difficult for aquatic flora and fauna. Second, pollutants (especially those accumulated on roadways) are washed into streams—oil, gasoline, antifreeze, fragments of brake linings and tires, and so on, are especially unwelcome deposits. Some design strategies to minimize these impacts, such as roof retention, porous pavement, and on-site groundwater recharging, were discussed in Chapter 20.

With minimal changes in street design, stormwater receptacles can be constructed not only to provide access for maintenance, but also to screen out sediment, capture oil and other floating contaminants, and provide a more even outflow (Figure 22.57). But much more can be done.

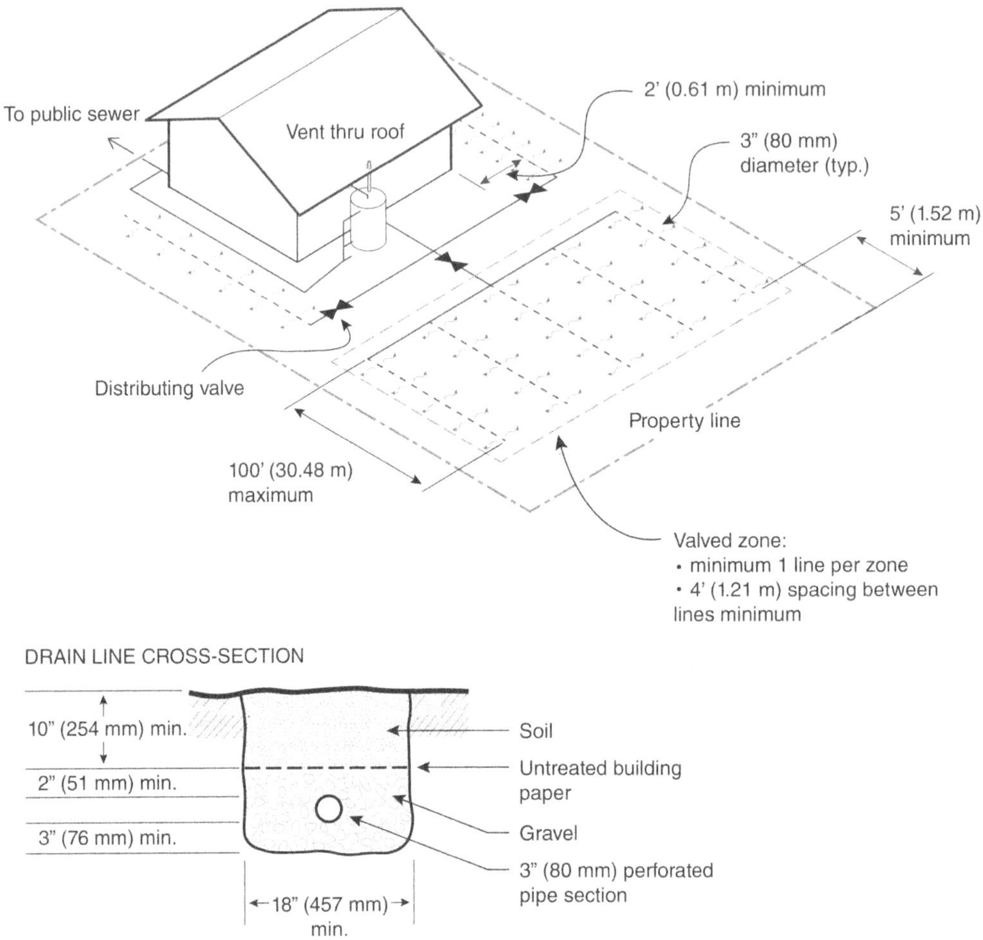

Figure 22.55 *Example of a residence whose daily graywater flow requires an irrigation/disposal field (using a valved zone as shown in the detail). In this example, five such valved zones are shown (a minimum of three such zones are required). (Redrawn by Nathan Leigh after © IAPMO, 2006. Reprinted with the permission of the International Association of Plumbing and Mechanical Officials. This copyright material and all points or statements in using this material have not been reviewed by IAPMO. The opinions expressed herein are not representations of fact from IAPMO.)*

TABLE 22.16 Graywater Design Criteria, Six Typical Soils

Part A. Area of Irrigation/Leaching Field				
	Minimum Area of Irrigation per Estimated Graywater Discharge per Day		Maximum Absorption Capacity of Irrigation/Leaching per 24-Hour Period	
Type of Soil	**ft²/100 gal**	**m²/L**	**gal/ft²**	**L/m²**
Coarse sand or gravel	20	0.005	5.0	203.7
Fine sand	25	0.006	4.0	162.9
Sandy loam	40	0.010	2.5	101.8
Sandy clay	60	0.015	1.7	69.2
Clay with considerable sand or gravel	90	0.022	1.1	44.8
Clay with small amounts of sand or gravel	120	0.030	0.8	32.6
Part B. Subsoil Irrigation Field Construction				
	Minimum		**Maximum**	
	I-P	**SI**	**I-P**	**SI**
Number of lines per valved zone	1	1		
Length of each perforated line			100 ft	30,840 mm
Bottom width of trench	12 in.	305 mm	18 in.	457 mm
Spacing of lines, center-to-center	4 ft	1219 mm		
Depth of earth cover of lines	10 in.	254 mm		
Depth of filter material cover of lines	2 in.	51 mm		
Depth of filter material beneath lines	3 in.	76 mm		
Grade of perforated lines level	level	2 mm/m	3 in./100 ft	2 mm/m

Source: Adapted from (IAPMO, 2023). Reprinted with the permission of the International Association of Plumbing and Mechanical Officials. This copyright material and all points or statements in using this material have not been reviewed by IAPMO. The opinions expressed herein are not representations of fact from IAPMO.

The SI conversions were appended by the authors of this book.

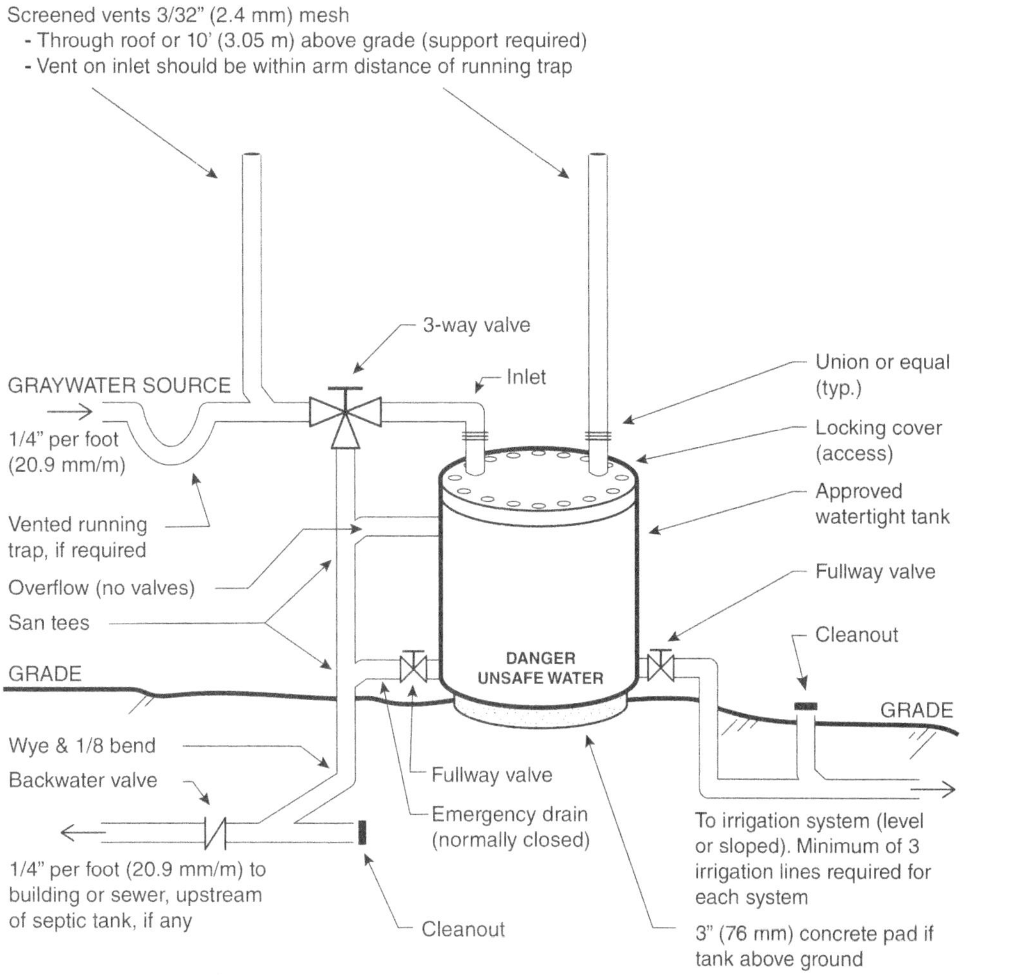

Figure 22.56 *Schematic diagram of a graywater system with an aboveground, gravity-emptied tank. (Redrawn by Nathan Leigh after © IAPMO, 2006. Reprinted with the permission of the International Association of Plumbing and Mechanical Officials. This copyright material and all points or statements in using this material have not been reviewed by IAPMO. The opinions expressed herein are not representations of fact from IAPMO.)*

Phytoremediation. Phytoremediation is the science of cleaning polluted soil and water with plants. (The constructed wetlands and algal ponds described in Sections 22.7 and 22.8 apply phytoremediation to sewage treatment.) Phytoremediation extracts, degrades, or contains contaminants. Plants that extract take up and accumulate contaminants in their shoots and leaves—when the vegetation is removed from the site, so are the contaminants. Plants that degrade break down contaminants (such as hydrocarbons and other organic compounds) until they are no longer toxic. Some plants degrade toxins in their root zone, whereas others use elements of organic toxins as food. Plants that contain toxins essentially immobilize toxins in long-term storage. However, when the plants die and decompose, long-lived toxins may be released.

For treating stormwater, plants are typically incorporated into drainage strips between paved parking areas. The mix of plants is similar to those cited earlier in wetland construction, with sedges and rushes especially effective at cleansing (see Figure 22.2 for a sample trench). Shade trees could add welcome summer thermal benefits, as well as intercepting and storing some pollutants. The capacity of the planting strip should be sufficient to intercept the "first flush" of rain, which carries the heaviest load of pollutants. Underground drains intercept excess runoff and lead to a capture chamber where floatable contaminants can be periodically removed. The screened, regulated flow from the capture basin will be much gentler on the receiving stream.

A new community in Bellevue, Washington, incorporated a stormwater treatment facility in a 20-acre [8.1-ha] public park. Beneath tennis courts, ballparks, picnic areas, and other playing fields, a 300,000-ft^3 [8495-m^3] sediment vault (surface area about 8 acres

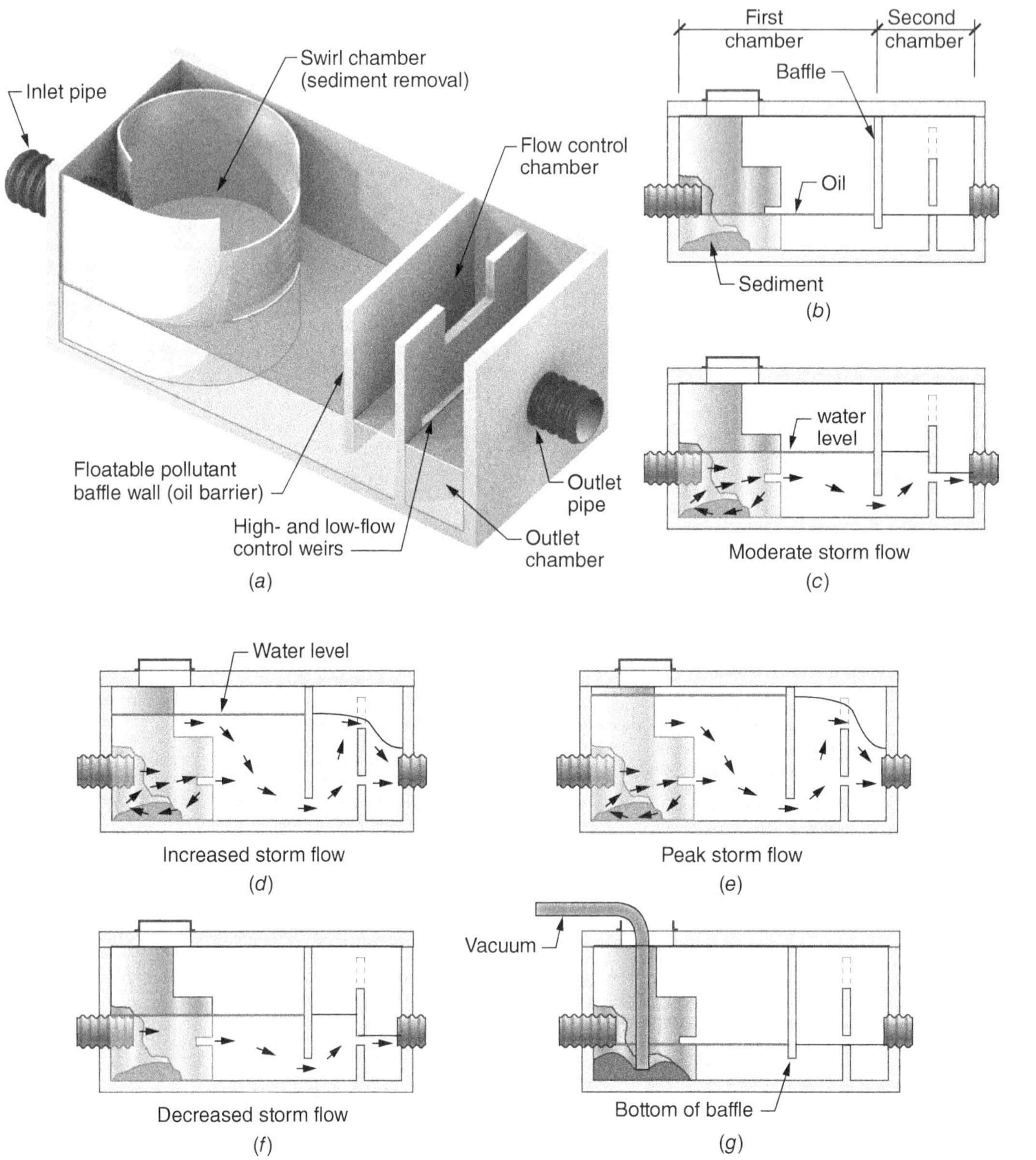

Figure 22.57 *This Vortechs™ Stormwater Treatment System (a) is designed to separate both floating oily contaminants and sediment from stormwater. (b) Rather like a septic tank, two chambers are separated by a baffle. Oil scum and sediment remain in the first chamber. (c) In a moderate storm, water rises above the top of the inlet pipe, greatly reducing velocity and turbulence within the first chamber, encouraging sediment to precipitate to the bottom. (d) As flow increases, rising above the low-flow control (second chamber), oily contaminants accumulated from previous storms float upward, and more sediment falls out. (e) At the height of stormwater flow, usually designed for a 10-year storm, the high-level outlet begins to discharge. Sometimes a peak flow bypass is used. Oily scum and sediment continue to be trapped in the manhole. (f) As the flow subsides, water outflow is controlled, lessening the strain on the storm drainage system. (g) Cleanout, at the lowest water levels, is best done by a vacuum truck. As the grit chamber empties, oily liquids and floating debris drain back toward the inlet and can be removed along with the sediment. Note that the bottom of the baffle between the chambers always remains submerged.*

[3.2 ha]) slows the flow of stormwater, encouraging sediment to drop out. The water then passes to a filter basin of vegetation, sand, and peat for the primary purpose of phosphorus removal. From there it is slowly released to the receiving stream, which then drains into a lake.

Building designers rarely encounter such community-wide pollution control strategies, but parking lots and on-site roadways are encountered on virtually every project. Design for local pollution control, even for seemingly benign stormwater, is part of the architect's responsibilities.

22.11 CASE STUDY—WATER CONSERVATION AND RESOURCE DESIGN

Hassalo on Eighth: NORM

PROJECT BASICS

- Location: 1088 NE Seventh Avenue, Portland, Oregon, USA
- Latitude: 45.53N; longitude: 122.66W; elevation: 135 ft [41 m]
- Heating degree days: 4522 base 65°F [2512 base 18.3°C]; cooling degree days: 2517 base 50°F [1398 base 10°C] at Portland airport; annual precipitation 36 in. [914 mm]
- Building type: New construction of housing, retail; renovation of office tower
- Building area: 92,600 gross square feet [55,100 sq m] of housing, 58,100 gross square feet [5400 sq m] of retail; renovation of 271,600 gross square feet [25,200 sq m] of office space
- Project Cost: $192 million
- Completed October 2015
- Client: American Assets Trust
- Design team—GBD: architecture; BioHabitats: NORM sewage treatment system; PLACE: landscape; Glumac: mechanical/electrical/plumbing/energy

(a)

(b)

Figure 22.58 *(a) Lloyd superblock before construction and (b) Hassalo on Eighth at completion. (© GBD Architects; used with permission.)*

Background and Context. Hassalo on Eighth is a private market-rate multi-use project in Portland's Lloyd district. The site, built in the 1960s and the 1970s, was a typical "tower in the parking lot" development, one of many, surrounding the large Lloyd Center shopping mall. The entire Lloyd district was a pioneer of what became a standard model for American suburbanism: shopping mall and parking surrounded by mid-rise office towers. In this instance, the towers were located on "superblocks" created from four of Portland's standard 200 ft x 200 ft [60 m x 60 m] street blocks (Figure 22.58a).

In the twenty-first century, it became evident that Portland needed a new urbanism more appealing to the new century and to the pressing environmental challenges. Inner-city neighborhoods were being redeveloped around alternative transportation opportunities and incorporating a blend of office, retail, and housing uses. Following several other successful inner-city redevelopment projects, Portland planners looked to the Lloyd district as an opportunity to raise the bar for environmentally sensitive intensification of an existing, underused neighborhood through creation of the Lloyd EcoDistrict. In this case, one of the critical environment issues they faced was pollution caused by the combined storm and sanitary sewer system and an undersized, aging piping and treatment infrastructure. In the event of a major rainstorm, the treatment plant was at times overwhelmed by stormwater volume, leading to release of raw sewage into the Willamette River. This problem was made worse by the very high percentage of pavement and other hard surfaces in the Lloyd district. These streets and parking areas were also responsible for large amounts of pollution caused by rainwater washing vehicular pollutants from parking and street surfaces into the river.

Hassalo on Eighth responded to these challenges by creating a private eco-district within Portland's larger Lloyd Ecodistrict, built on an existing super-block (one development site of four contiguous standard blocks from which the interior streets were removed), retaining and improving the existing Lloyd 700 office tower and infilling three new buildings on what had been a sea of parking (Figure 22.58*b*).

Design Intent. The project concept was to preserve and improve the existing Lloyd 700 office tower while creating an integrated eco-district through the construction of a neighborhood of new residential and retail buildings. These would supplement the existing building and replace the 1970s "tower in the parking lot" urbanism with a more contemporary, bike-/pedestrian-/transit-oriented, vibrant community. Parking would be reduced and moved underground, freeing up building sites and open space for an attractive urban streetscape that includes visible sewage treatment facilities.

Hassalo on Eighth is one of the first projects in the larger Lloyd EcoDistrict, and would establish goals largely through integrated design and would find synergies within and between major systems.

The project would set very high blackwater and graywater goals of preventing any increased flows into the municipal sewer system due to new building construction and substantially limit stormwater discharge. Although a market-rate private project, it would work cooperatively with the City of Portland to negotiate cost sharing of on-site water treatment strategies traded off against avoidance of major, expensive municipal sewer system improvements.

In addition, the project would examine various ways to integrate the energy systems of the existing and new buildings to find synergies between the various office, retail, and residential uses.

Design Goals

- Avoid increasing the sanitary (blackwater and graywater) load on Portland's combined sanitary/storm sewer system despite a major intensification of use on the site.
- Substantially decrease peak flow and decrease stormwater pollution discharged to combined sewer system.
- Reveal and celebrate advanced sanitary and stormwater systems in the new public spaces.
- Repair and improve the urban fabric by supplementing the existing office tower with infill of programmatically diverse buildings.
- Build transit-/bike-/pedestrian-friendly development.
- Attain LEED/Platinum for Homes—Mid-rise.
- Attain LEED/Platinum, Neighborhood Development.
- Market-rate: no external subsidies other than system development charge reductions that match savings to local government due to avoided costs.

Design Features

NORM Sewage Treatment and Recycling. Nicknamed Natural Organic Recycling Machine (NORM), the system diverts nearly all 45,000 gal/day [170,344 L/day] of sewage (blackwater and graywater) from the three new buildings to an integrated on-site sewage disposal system, showcasing the treatment in the pedestrian street and in the adjacent plaza. NORM's low-energy systems primarily use various bacteria, among other methods, to treat the wastewater. One of the more interesting of these are the "tidal basins," capped by plantings, as shown in Figure 22.59. These gradually fill to within 12–18 in. [305–457 mm] of grade, then gradually empty. To quote Pete Munoz of Biohabitats, "NORM is complex, but not complicated."

As diagrammed in Figure 22.59, wastewater is taken sequentially through:

- settling tanks (1) below grade with monitored emergency overflow bypass to the municipal sewer system
- a below-grade anoxic reactor (2)
- trickling filters that also serve as sculptural elements in the public pedestrian street (3)

Constructed treatment below plantings in the pedestrian street, consisting of:

- two tidal wetland tanks (4), which fill and empty in alternation about 12 times per day

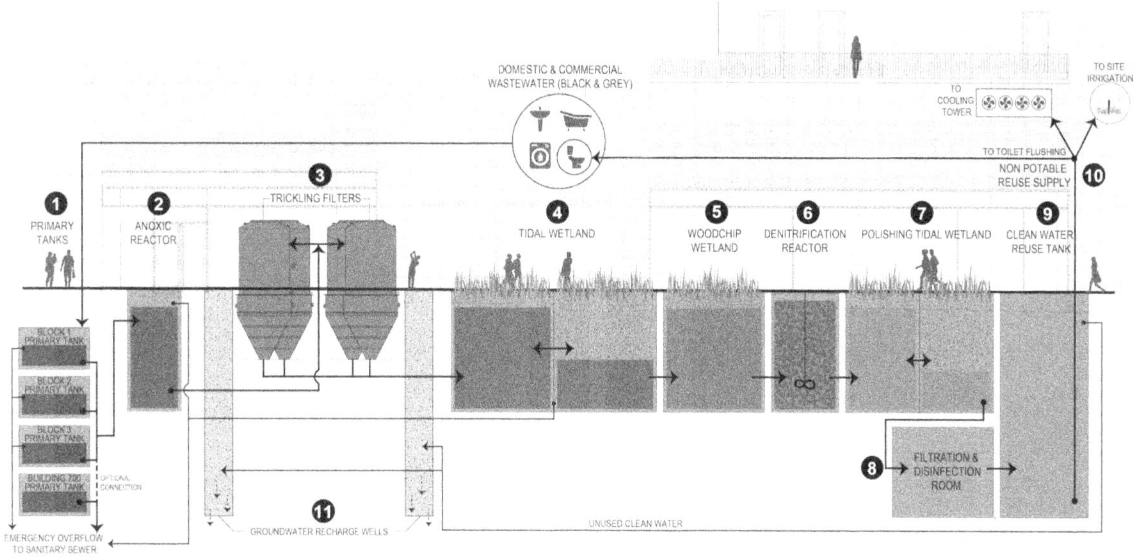

Figure 22.59 *NORM sewage treatment and reuse system, schematic section. (© GBD Architects; used with permission.)*

- a wood chip wetland (5) whose carbon further breaks down pollutants
- a fully enclosed denitrification reactor (6) which primarily uses bacteriological control
- two polishing tidal wetland tanks (7)
- final treatment and disinfection to Oregon Class A reuse standards (8)
- storage of effluent (9) for toilet flushing and irrigation for all four buildings (10), and potentially for cooling tower make-up water as well
- excess treated water is injected into groundwater via dry wells (11) to compensate for groundwater depletion due to urban hardscape

Solids are removed quarterly and further processed off-site for reuse as fertilizer.

One of the striking features of this system is that the vegetated treatment areas (Figures 22.60 and 22.61) are virtually indistinguishable from the other planted areas of the pedestrian urban fabric, other than small signs embedded in the edge of the paving identifying the plantings as part of a sewage treatment facility.

Stormwater Volume Control and Pollution Reduction

The pollution problems of Portland's combined sewer system are most acute during and after major rainstorms, as the run-off overwhelms the treatment facility's capacity and untreated sewage goes directly into the Willamette River. The previously existing parking on the site was a significant contributor to pollution in the stormwater (Figure 22.58*a*).

To combat this problem, the City of Portland asked the project design team to focus their stormwater efforts on reducing peak flow while also improving water quality. This was achieved through extensive vegetated roofs on all three new buildings (Figure 22.62), stormwater treatment planters at street level (Figures 22.63 and 22.64), a 65,000-gal [246,052 L] holding tank below grade for gradual release of stormwater into the combined municipal sewer as well as for the recirculating landscape water feature, and locating all parking underground to eliminate a major pollution source. The results ensure that all storm events—up to the design of a 25-year storm—are contained on site until municipal systems can treat them, and the water quality of the effluent is much improved.

Energy

A new, highly efficient central plant was provided for the existing Lloyd 700 office tower, and lighting at Lloyd 700 was replaced with LED lamps, as a first phase of energy improvements for the building. In addition, the project extended the condenser water loop from the office tower's central plant to retail and common spaces in the three new buildings, where heat pumps boost the temperature before use. This allows waste heat to be moved from the office tower to the new buildings when they need heat.

Figure 22.60 Red trickling filters are exposed in the pedestrian street next to a café/grocery store, with tidal wetlands in the foreground. (GBD Architects / photo © Christian Columbres; used with permission.)

Figure 22.61 Trickling filters and tidal wetlands adjacent to a grocery store and café. (GBD Architects / photo © Christian Columbres; used with permission.)

Figure 22.62 Vegetated roofs and plaza plantings retain and treat site stormwater. (GBD Architects / photo © Christian Columbres; used with permission.)

Figure 22.63 *Stormwater treatment and reuse in residential courtyard off pedestrian street. (GBD Architects / photo © Lincoln Barbour, lincolnbarbour.com; used with permission.)*

Figure 22.64 *Artificial stream at Hassalo on Eighth running along pedestrian street. Water source is stored stormwater. (GBD Architects / photo © Christian Columbres; used with permission.)*

Pedestrian Street and Plaza

The project creates new pedestrian streets in the alignment of the former streets of the superblock (Figure 22.64) and also builds a major plaza at the foot of the existing Lloyd 700 office tower (Figure 22.65). The urban character of the new pedestrian areas is tightly integrated with the treatment and celebration of water. The settling tanks contribute major vertical sculptural elements, and the nearby tidal basins are fully integrated into the larger scheme for the north–south pedestrian street which showcases and celebrates the water systems along its length, including through a small artificial stream (Figure 22.64) fed entirely by stored stormwater. This street is also the primary pedestrian connection to transit, to other nearby buildings, and to future areas slated for intensification and redevelopment.

Performance Data and Awards

- Sewage treatment: NORM treats up to 45,000 gal [170,344 L]/day, diverting nearly all sewage (blackwater and graywater) from the three new buildings, with potential to add preexisting Lloyd 700 if excess treatment capacity becomes available.
- Potable water savings from NORM sewage treatment and reuse: over 7 million gallons [26.5 L] per year.

Figure 22.65 Plaza south of Lloyd 700 office tower contains element of NORM as well as stormwater plantings. (GBD Architects / photo © Christian Columbres; used with permission.)

- Reclaimed water from NORM provides for nearly all of the toilet flushing and all irrigation needs for all four buildings, roughly 470,000 ft^2 [43,700 m^2].
- Roughly 50% of treated water is recycled for use and roughly 50% is returned to the aquifer via injection wells. Almost none enters the municipal sewer system.
- Energy conservation at existing office tower: Ongoing changes have reduced site EUI by about 25%.
- LEED Platinum/Homes—Mid-rise
- LEED Platinum/Neighborhood Development
- USGBC LEED Homes Award Project of the Year, 2016
- *Engineering News-Record* Best Residential Hospitality Project—Northwest Region

FOR FURTHER INFORMATION

Davis, K. and K. Andersen. 2016. "Rethinking Water at Hassalo on Eighth." *Consulting Specifying Engineer*. csemag.com/articles/rethinking-water-at-hassalo-on-eighth

Guo, A. 2013 "Re-defining the Lloyd District." *Center for Real Estate Quarterly Report 7* (No. 4, Fall 2013). Portland, OR: Portland State University.

ULI. n.d. "Hassalo on Eighth," in *ULI Case Studies*. Washington, DC: Urban Land Institute. uli.org/wp-content/uploads/ULI-Documents/Hassalo-on-Eighth.pdf

Winter, M. J. *Human Waste Treatment and Water Re-use at Hassalo on Eighth*. Olympia, WA: Northwest EcoBuilding Guild.

References and Resources

Composting Toilet World. compostingtoilet.org

Converse, J. et al. 1978. *Design and Construction Manual for Mounds. Small Scale Waste Management Project*. Madison, WI: University of Wisconsin Extension.

DeOreo, W. B., P. Mayer, B. Dziegelewski, and J. Kiefer. 2016. *Residential End Uses of Water, Version 2: Executive Report*. Report #4309A. Denver, CO: Water Research Foundation.

EPA. 1993. *Subsurface Flow Constructed Wetlands for Wastewater Treatment*. EPA 832-R-93-008. Washington, DC: U.S. Environmental Protection Agency.

Green, F. B., L. Bernstone, T. J. Lundquist, J. Muir, R. B. Tresan, and W. J. Oswald. 1995. "Methane Fermentation, Submerged Gas Collection, and the Fate of Carbon in Advanced Integrated Wastewater Pond Systems." *Water Science and Technology* 31(12): 55–65.

Green, F. B., L. Bernstone, T. J. Lundquist, and W. J. Oswald. 1996. "Advanced Integrated Wastewater Pond Systems for Nitrogen Removal." *Water Science and Technology* 33(7): 207–217.

Green, F. B., T. J. Lundquist, and W. J. Oswald. 1995. "Energetics of Advanced Integrated Wastewater Pond Systems." *Water Science and Technology* 31(12): 9–20.

IAPMO. 2006. *2006 Uniform Plumbing Code*. Walnut, CA: International Association of Plumbing and Mechanical Officials.

IAPMO. 2023. *2024 Uniform Plumbing Code*. Walnut, CA: International Association of Plumbing and Mechanical Officials.

ICC. 2023. *2024 International Plumbing Code*. Washington, DC: International Code Council.

Ludwig, A. 1997. *Create an Oasis with Greywater*. Santa Maria, CA: Poor Richard's Press.

Milne, M. 1976. *Residential Water Conservation*. National Technical Information Service. Washington, DC: U.S. Department of Commerce.

National Small Flows Clearinghouse. 1997. "Lagoon Systems Can Provide Low-Cost Wastewater Treatment." *Pipeline* 8(2): 1.

National Small Flows Clearinghouse. n.d. actat.wvu.edu/wastewater/onsite-wastewater-systems/alternative-technologies/aerobic-treatment-unit

Nelson, G. 1981. "Graywater Heat Recovery." *Solar Age* 1981: 50–53.

Oswald, W. J. 1990. "Advanced Integrated Waste-water Pond Systems: Supplying Water and Saving the Environment for Six Billion People," in *Proceedings of the 1990 American Society of Civil Engineers Convention*. San Francisco, CA: Environmental Engineering Division.

Oswald, W. J., C. G. Golueke, R. C. Cooper, H. K. Gee, and J. C. Bronson. 1963. "Water Reclamation, Algal Production and Methane Fermentation in Waste Ponds." *International Journal of Air and Water Pollution* 7: 627–648.

Oswald, W. J., F. B. Green, and T. J. Lundquist. 1994. "Performance of Methane Fermentation Pits in Advanced Integrated Wastewater Pond Systems." *Water Science and Technology* 30(12): 287–295.

Oswald, W. J., A. Meron, and M. D. Zabat. 1970. "Designing Waste Ponds to Meet Water Quality Criteria," in *Proceedings of the Second International Symposium on Waste Treatment Lagoons*. Kansas City, MO.

West Virginia University. n.d. "Filtration," in *Appalachian Community Technical Assistance and Training*. Morgantown, WV: West Virginia University. actat.wvu.edu/wastewater/onsite-wastewater-systems/alternative-technologies/filtration

CHAPTER 23

Solid Waste

FOR SOME DESIGNERS, the last building distribution system considered is the one involving the bulkiest items: the flow of supplies in and solid waste out of the building. Because this system usually is not seen as consuming building energy or requiring specialized equipment, it tends to become a low-priority system in the design process. Yet provisions for delivery of supplies, and especially for the collection and storage of solid wastes, can be more space-consuming than water/waste systems, can present a fire danger, and can create serious local environmental problems. The separation of solid waste for resource recovery involves significant energy and environmental consequences. Finally, mechanical equipment associated with solid waste is now commonly part of waste management systems.

23.1 WASTE AND RESOURCES

The concept of waste is very much a human-made construct. As architect William McDonough and chemist Michael Braungart framed it in their book *Cradle to Cradle* (2002), in nature, "waste" from one system becomes "food" for another system (i.e., "waste equals food"). Interconnected systems in nature connect outputs as inputs, constituting what may be termed a "biological metabolism." Humans have endeavored to realize this concept through industrial processes whereby the things we manufacture are designed for reconstitution within a "technological metabolism." The cradle-to-cradle ideology advances the notion that we should strive to eliminate the concept of waste. Pursuant to this ideology is the circular economy, which is a model of resource production and consumption that involves sharing, leasing, reusing, repairing, refurbishing, and recycling existing materials and products for as long as possible (EP, 2023).

In essence, "waste" occurs when the interconnected network of system outputs acting as the other system's inputs is broken—and this can occur in both nature and industry when systems become isolated. When we design and produce things without consideration for reconstitution at the end of their life-cycle, we create refuse and too often the most convenient solution is to quite literally throw it away into giant piles where we may or may not bury it (i.e., landfills). Municipal solid waste (MSW) constitutes everyday items that we throw away. Product packaging, grass clippings, furniture, clothing, bottles, food scraps, appliances, paint, batteries—we are talking about all of the items that we call "trash" or "garbage." According to the United Nations, global MSW generation is predicted to grow from 2.3 billion tons in 2023 to 4.2 billion tons by 2050 (UNEP, 2024). Data from the U.S. Environmental Protection Agency (EPA, 2023) indicate that the United States generated around 292.4 million tons of MSW in 2018, accounting for around 12% of the global total (despite being only 4% of the world population). On a per capita basis, MSW generation in the United States is around 4.9 pounds per person per day (compared to 2.4 in Sweden, 3.6 in Germany, or 2.6 in the United Kingdom). The EPA's 2018 data

Mechanical and Electrical Equipment for Buildings, Fourteenth Edition. Walter Grondzik, Daniel Overbey, and Alison Kwok.

Companion website: www.wiley.com/go/meeb14e

also suggest the top waste streams in the United States were:

Paper and cardboard	23.1%
Food	21.6%
Plastics	12.2%
Yard trimmings	12.1%
Metals	8.8%
Wood	6.2%

The trend for each waste stream varies, but on the basis of total tonnage, nearly all are either holding steady or increasing. Since the late 1940s, there has been a marked increase in the amount of packaging material used for consumer products. Where shoppers once refilled reusable containers for bulk supplies at the market, for example, they now buy food in bags or cans that are discarded after use. This trend has increased space demands in the store, where shelves of cans are required instead of a bin of bulk products—and the same is true for the home, where such packaging soon turns into waste and must be stored until garbage collection day. Energy is required to make the boxes, bags, cans, and other containers; transport them; and collect them as trash. Devices such as trash compactors add both space and energy requirements to the use/disposal process. Landfills for garbage disposal are filling rapidly as solid waste flows increase. Methane and leachate from these landfills are potential environmental problems.

There may not be much that building designers can do about these increased packaging trends, but it is important to anticipate and plan for the waste generation that occurs within buildings as they often contain important resources (Tables 23.1 and 23.2), and a designer can help society to recover those resources rather than burying them in landfills or dumping them in the ocean. According to the EPA, only about 24% of MSW is recovered and recycled (the vast majority of which is paper and cardboard; metals are a distant second). Furthermore, buildings themselves can be designed for construction materials recovery upon remodeling or demolition. With proper planning and contemporary best practices, the majority of construction and demolition (C&D) waste can be diverted from landfills. C&D waste is defined as the materials generated and unused during the construction, renovation, and/or demolition of buildings, roads, and other components of the built environment—such as concrete, masonry, wood, glass, metals, and plastics.

The resources within solid waste can be divided into the high-grade resources represented by recyclable materials and the low-grade resources of heat and byproducts obtainable from the burning or decomposing of combustible solid wastes.

(a) High-Grade Resources

These include metals such as aluminum and steel, paper and paperboard, wood, and some plastics. For building designers, these pose the problem of storage while awaiting collection. For many of these materials, reuse in their present form is desirable, which eliminates compaction as a storage strategy. Table 23.1 compares uncompacted and compacted volumes of various recyclable materials.

Glass is especially suitable for recycling when it can be reused after simple washing, as in the case of beer and soft-drink bottles. Returnable (for deposit) bottles and cans have measurably decreased roadside litter in states that require such policies. Newspaper is so readily stored and recycled that in many communities, charitable groups or service organizations recover newspapers for profit. Recycled paperboard can easily save 50% of the energy that would be required to produce pulp from virgin material.

Recovery and reprocessing of high-grade waste save resources and energy. Consider aluminum: energy conservation in aluminum production had developed to the point that only 82% as much energy was needed to make a pound of aluminum in 1982 as in 1972. More than a third of that reduction was attributable to aluminum-recycling programs; the remainder came from improvements in the production process. According to the EPA, aluminum can now be recycled using less than 5% of the energy otherwise required for production. As for steel, recycling can yield between 60% and 74% energy savings compared to the use of virgin material (EPA, 2016).

Wood has become more valuable as fiber than as fuel. Wood chips and scraps that once were burned or buried are now recycled to become oriented strand board (OSB), used in structural insulated panels (SIPs), window frames, and many other aspects of building and furniture construction.

Plastics are more difficult to recycle. Regulations and consumer preferences discourage the use of recycled plastics in food-related items. The plastics from items such as soft-drink containers, margarine tubs and lids, and milk jugs, however, can be reprocessed into plastic pellets, which are cheaper than virgin plastic pellets. These pellets are then made into nonfood items—toys, building products, sports products, and other things.

(b) Low-Grade Resources

These resources include materials for which recycling is impractical but that are combustible. Low-grade resources include gaseous wastes, liquid and semiliquid wastes, and solid wastes. Industrial and commercial

TABLE 23.1 Typical Recyclable Materials from Building Operations

Category	Product	Recycling Label	Material Description	Conversion, Volume to Weight	
				I-P Units	SI Units
Paper	Ledger paper, white letterhead	SWL	Sorted white ledger, high-grade white paper	Uncompacted: 1 yd^3 = 500 lb Compacted: 1 yd^3 = 750 lb	Uncompacted: 1 m^3 = 297 kg Compacted: 1 m^3 = 445 kg
	Computer paper	CPO	Computer printout	Uncompacted: 1 yd^3 = 500–600 lb Compacted: 1 yd^3 = 1000–1200 lb	Uncompacted: 1 m^3 = 297–356 kg Compacted: 1 m^3 = 593–712 kg
	Colored paper	SCL	Sorted color ledger	Uncompacted: 1 yd^3 = 500 lb Compacted: 1 yd^3 = 750 lb	Uncompacted: 1 m^3 = 297 kg Compacted: 1 m^3 = 445 kg
	Newspaper	Mix	Newsprint	Uncompacted: 1 yd^3 = 350–500 lb Compacted: 1 yd^3 = 750–1000 lb	Uncompacted: 1 m^3 = 208–297 kg Compacted: 1 m^3 = 445–593 kg
	Magazines	Mix	Clay-coated paper	Not available	Not available
	Telephone books	Mix	Mixed papers-adhesives	1 book = 1–3 lb	1 book = 0.5–1.4 kg
	Cereal boxes	Mix	Coated paperboard	Not available	Not available
	Shipping boxes	OCC	Old corrugated cardboard	Uncompacted: 1 yd^3 = 285 lb Compacted: 1 yd^3 = 500 lb	Uncompacted: 1 m^3 = 169 kg Compacted: 1 m^3 = 297 kg
Glass	Food jars, beverage bottles	1 2 3	Amber glass Green glass Clear glass	Loose, whole: 1 yd^3 = 600 lb Manually crushed: 1 yd^3 = 1000 lb Mechanically crushed: 1 yd^3 = 1800 lb	Loose, whole: 1 m^3 = 356 kg Manually crushed: 1 m^3 = 593 kg Mechanically crushed: 1 m^3 = 1068 kg
Plastic	Beverage containers	PET	Polyethylene terephthalate	Whole: 1 yd^3 = 30 lb	Whole: 1 m^3 = 18 kg
	Milk containers	HDPE	High-density polyethylene	Whole: 1 yd^3 = 25 lb Crushed: 1 yd^3 = 50 lb Compacted: 1 yd^3 = 600 lb	Whole: 1 m^3 = 15 kg Crushed: 1 m^3 = 30 kg Compacted: 1 m^3 = 356 kg
	"Clamshell" containers	—	Polystyrene plastic foam	Not available	Not available
	Film plastic	LDPE	Low-density polyethylene	Not available	Not available
Metals	Beverage cans		Aluminum/bimetal	Whole: 1 yd^3 = 50–70 lb Crushed: 1 yd^3 = 300–450 lb	Whole: 1 m^3 = 30–42 kg Crushed: 1 m^3 = 178–267 kg
	Food and beverage cans		Steel with tin finish	Whole: 1 yd^3 = 125–150 lb Crushed: 1 yd^3 = 500–850 lb	Whole: 1 m^3 = 74–89 kg Crushed: 1 m^3 = 297–504 kg
Miscellaneous	Pallets		Wood	Not available	Not available
	Food waste and liquids		Organic solids	55-gal drum = 415 lb	208 L drum = 188 kg
	Yard waste		Organic solids	Leaves, uncompacted: 1 yd^3 = 250 lb Leaves, compacted: 1 yd^3 = 450 lb Wood chips: 1 yd^3 = 500 lb Grass clippings: 1 yd^3 = 400 lb	Leaves, uncompacted: 1 m^3 = 148 kg Leaves, compacted: 1 m^3 = 267 kg Wood chips: 1 m^3 = 297 kg Grass clippings: 1 m^3 = 237 kg
	Used motor oil		Petroleum product	1 gal = 71 lb	1 L = 8.5 kg
	Tires		Rubber	1 passenger car = 20 lb 1 truck = 90 lb	1 passenger car = 9 kg 1 truck = 41 kg

Source: Reproduced with permission from Ramsey and Sleeper, 2000.

TABLE 23.2 Solid Waste Sources in Building Operation

Classification	Occupancy	Types of Waste Generated (percent, where available)	Quantities of Waste Generated: I-P Units	Quantities of Waste Generated: SI Units
Residential	Studio or 1-bedroom apartment 2- or 3-bedroom apt. or single-family house	Newspaper (38/43)[a] Plastic (18/7)[a] Miscellaneous (13/18)[a] Metals (14/9)[a] Compost (10/15)[a] Glass (2/8)[a]	1–1.5 yd^3/unit/mo (200–250 lb) or 2.5 lb/person/d 1.5–2 yd^3/unit/mo (250–400 lb) or 2.5 lb/person/d	0.8–1.1 m^3/unit/mo (91–113 kg) or 1.1 kg/person/d 1.1–1.5 m^3/unit/mo (113–181 kg) or 1.1 kg/person/d
Commercial	Office	Plastics, compost, used oil, metals, and glass (30)[b] High-grade paper (29)[b] Mixed papers (23)[b] Newspaper (10)[b] Corrugated cardboard (8)[b]	1.5 lb/employee/d or 1 yd^3/10,000 ft^2/d (includes 0.5 lb of high-grade paper/employee/d)	0.7 kg/employee/d, or 0.8 m^3/1000 m^2/d (includes 0.2 kg of high-grade paper/employee/d)
	Department store Wholesale/retail Shopping center	Corrugated cardboard, compost, wood pallets, high-grade paper, and plastic film	1 yd^3/2500 ft^2/d 70 lb/$1000 sales/d 2.5 lb/100 ft^2/d	0.8 m^3/250 m^2/d 32 kg/$1000 sales/d 1.1 kg/100 ft^2/d
	Supermarkets	Corrugated cardboard, compost, and wood pallets	1 yd^3/2500 ft^2/d	0.8 m^3/250 m^2/d
	Restaurants/ entertainment	Compost (38)[b] Corrugated cardboard (11)[b] Newsprint (5)[b] High-grade paper (4)[b]	Cafeteria, 1 lb/meal Fast food, 200 lb/$1000 sales Restaurant, 1.5 lb/meal	Cafeteria, 0.45 kg/meal Fast food, 91 kg/$1000 sales Restaurant, 0.7 kg/meal
	Drug stores	Corrugated cardboard and high-grade paper	1 yd^3/2000 ft^2/d	0.8 m^3/200 m^2/d
	Banks, insurance companies	High-grade paper, mixed paper, and corrugated cardboard	0.75 lb high-grade paper/person/d (survey required)	0.34 kg high-grade paper/person/d (survey required)
Hotels and motels (not including restaurants)	High occupancy Average occupancy	Glass, aluminum, plastic, high-grade paper, newspaper, and corrugated cardboard	0.5 yd^3/room/wk 3.2 lb/room/d 0.17 yd^3/room/wk 1.7 lb/room/d	0.38 m^3/room/wk 1.45 kg/room/d 0.13 m^3/room/wk 0.8 kg/room/d
Institutional	Hospitals, Nursing homes, Retirement homes	Compost, high-grade paper, biomedical waste, corrugated cardboard, glass, and plastics	1 yd^3/5 occupied beds/d 1 yd^3/15 persons/d 1 yd^3/20 persons/d	0.8 m^3/5 occupied beds/d 0.8 m^3/15 persons/d 0.8 m^3/20 persons/d
Educational	Grade school High school Universities	High-grade paper, mixed paper, newspaper, corrugated cardboard, compost, plastic, glass, and metals	1 yd^3/8 rooms/d 1 yd^3/10 rooms/d (Survey required)	0.8 m^3/8 rooms/d 0.8 m^3/10 rooms/d (Survey required)

Source: Reproduced with permission from Ramsey and Sleeper, 2000.

[a]Percent by volume/percent by weight.

[b]Percent by volume.

processes can generate wastes of all types, including some with very high heat content and some that are very toxic. Although recovering heat by burning such materials seems better than landfilling them, air pollution regulations have severely restricted simple trash burning as a means of solid waste disposal. Incinerators must meet increasingly strict regulations and, as a result, are very rarely installed in buildings. A better option is often to compost such materials in landfills. The methane generated in landfills can then be used as a high-grade fuel.

Where large quantities of mixed trash, garbage, and other refuse are collected, special resource-recovery plants can be built to recover materials, produce useful steam for electricity generation, and reduce the flow of waste to landfills. In this energy-intensive process, mixed garbage is shredded and blown through large "air classifiers" that separate the organic (burnable) wastes from metals and glass. The burnable wastes are then used as fuel, under controlled combustion, to generate electricity. The metals are further separated magnetically into ferrous and nonferrous classes; these and the glass are then recycled.

The Energy Efficient Resource Recovery (E2R2) integrated infrastructure facility that was proposed for California Polytechnic State University, San Luis Obispo, combining wastewater treatment and solid waste processing, was introduced in Chapter 22. The solid waste facility is shown in Figure 23.1.

The schematic design for solid waste processing (SWP) in the proposed E2R2 facility involves transporting unsorted solid waste to a central processing

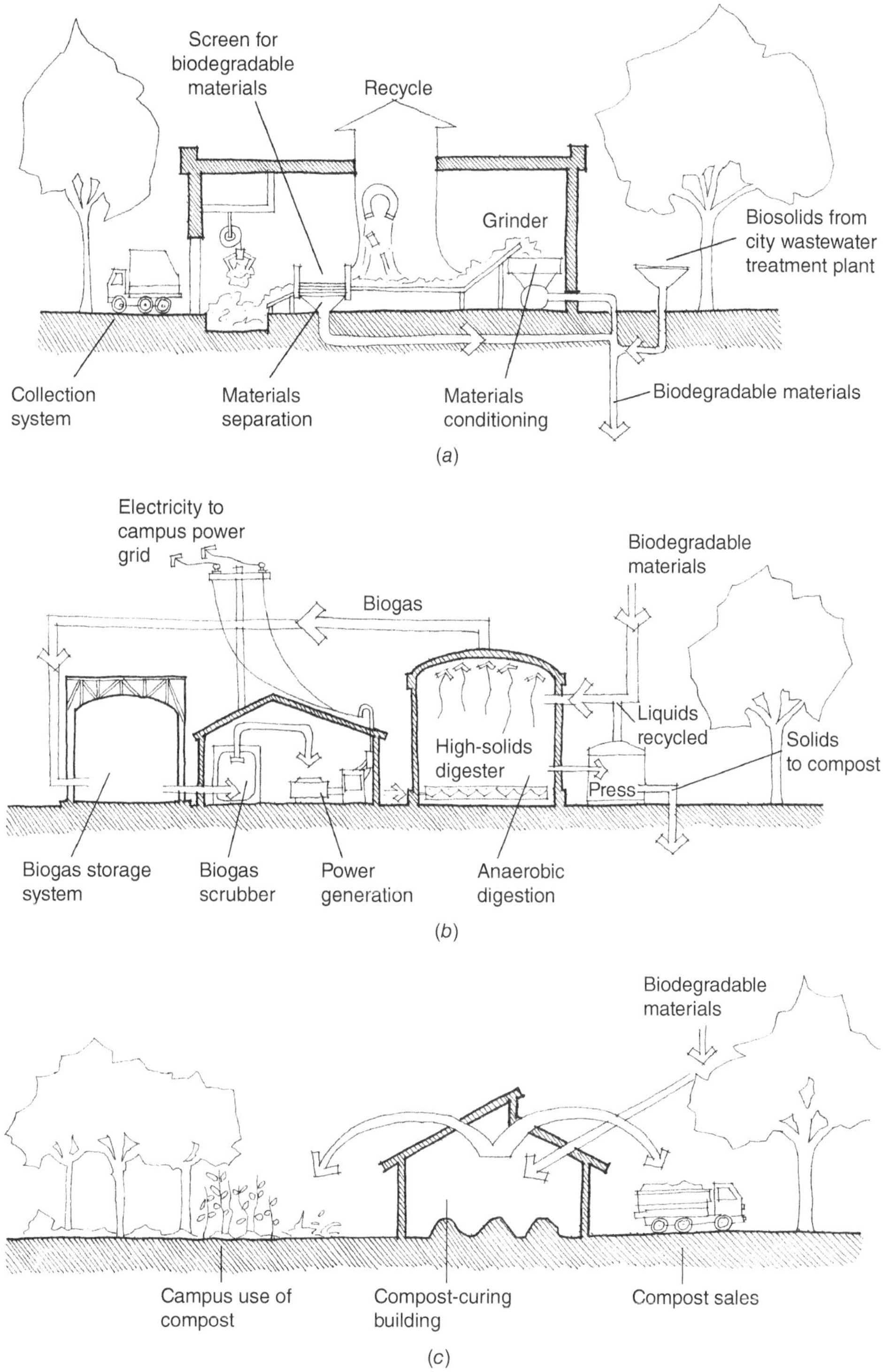

Figure 23.1 *Schematic design proposal for solid waste processing at the proposed E2R2 facility at California Polytechnic State University, San Luis Obispo, California. Unsorted solid wastes are separated into recyclable, organic, and inert materials (a). The organic materials are ground, combined with sludge from the city's sewage treatment plant, and loaded into the high-solids digester (b) to undergo anaerobic fermentation. The biogas generated by the fermentation process is scrubbed to remove impurities from the methane, making it suitable for use as a fuel to heat the digester and to power a micro fuel cell to generate electricity for the campus. Biosolids from the digester are pressed, and the liquid, rich in anaerobic bacteria, is then used to inoculate the next anaerobic fermentation cycle. Pressed biosolids are mixed with nutrient-rich algae from the adjacent wastewater facility and pasteurized in compost-curing buildings (c). Once cured, the compost can be used on landscaped areas or farmlands or sold to local gardeners. (Courtesy of Professor Douglas Williams, Agricultural Engineering Department, and Professor Daniel Panetta, Architecture Department, Cal Poly, San Luis Obispo. Redrawn by Dain Carlson.)*

ACTIVE ENVIRONMENTAL SYSTEMS

facility located on campus. Materials will then be sorted using mechanical equipment and processing. The composition of the solid waste stream is expected to be 70% organic materials, suitable for composting, and 815 tons [739 Mg] per year of recyclable materials (i.e., metals, glass, paper, and cardboard). Roughly 10% of the solid wastes are anticipated to be unsuitable for either recycling or composting. These "inert" materials will be disposed of at the local landfill.

Once the nonrecyclable, organic materials are separated, the design calls for them to be uniformly ground and combined with sludge (from the city of San Luis Obispo's wastewater treatment plant) or manure (from the university's farming operations) to achieve a proper carbon-to-nitrogen balance. Given the dynamic nature of the recyclable materials market, it is anticipated that, at times, some potentially recyclable paper will be included in the compost operations.

After mixing, the organic materials will be loaded into a high solids anaerobic digester to undergo methane fermentation. The biogas generated by the fermentation process will be harvested and stored on site. Prior to use, the biogas will be scrubbed (processed) to remove impurities, making it suitable for use as fuel. A portion of the methane will be used to heat the digester, and the excess will power a micro fuel cell to generate electricity for campus use. Other potential uses include using methane in the campus boiler, as an alternative to gasoline in the university's automobile and truck fleet, and even to power an ultra-high-temperature hazardous waste disposal system.

The anaerobic digestion process is projected to generate 180,000 ft^3 [5098 m^3] per day of biogas. The energy content of the methane in this biogas is estimated at 35×10^9 Btu [36.9×10^9 kJ] per year.

After being unloaded from the high solids digester, the remaining biosolids will be pressed to recover bacteria-rich liquids, which are useful in inoculating the next batch of materials for the digester. The dewatered biosolids will then be mixed with algae from the adjacent AIWPS (Advanced Integrated Wastewater Pond System) facility to create a nutrient-rich compost. This compost will be available for use on the campus's landscape and its farmlands or for sale to local gardeners. Yearly compost production is anticipated to exceed 5000 tons [4535 Mg] air-dried weight.

As an added touch, straw bale construction is proposed for the E2R2 facility buildings. Currently, plans call for an education and laboratory building, a shared control office for wastewater treatment and solid waste processing operations, a solid waste processing building, and on-site student housing. Straw bale construction is advantageous because of its affordability, as well as its exceptional insulation and noise mitigation characteristics. In addition, using rice straw for construction will help provide an alternative to burning waste straw from California's rice crop.

Conventional landfills accept unsorted garbage from a huge variety of sources. Over many years, anaerobic combustion in such enclosed landfills generates methane gas, a usable resource, as described previously.

23.2 RESOURCE RECOVERY: CENTRAL OR LOCAL?

In general, the more thoroughly mixed the different types of solid waste, the harder it is to recover their high- and low-grade resources. From an energy conservation viewpoint, solid wastes should be kept as separate as possible; glass bottles should be washed and reused rather than broken and recycled, and unrecyclable but burnable solid wastes should be kept clean and dry until they can be burned or composted. The earlier that different types of metals are separated, for example, the less energy spent later to separate aluminum from steel, and so forth. Organic food wastes could be composted for use on site rather than ground up and added to the load on the sewage treatment system.

There are two disadvantages, however, to local solid waste separation (at the point of their discard)—one cultural, one physical. Keeping solid wastes separate requires somewhat more effort and time on the part of the consumer. Rather than dumping everything in a garbage can, the resource-conscious consumer will take the time to understand the local recycling procedures and separate garbage and recyclables in accordance with such requirements. Composting in urban apartments presents a unique space-odor challenge.

This disadvantage of slightly more work and time for waste separation is compounded by the physical disadvantage of the floor or cabinet space taken up by multiple containers. In communities fortunate enough to have recycling-oriented garbage service (such a service, it should be noted, is expanding throughout the United States), these containers can be carried out and lined up to await garbage collection. The trucks used for this purpose are complex assemblages of various bins rather than one massive cavern with a compactor. In many communities, however, each pile of recyclable material at home must be dealt with separately, either collected by a specialized handler (nonprofit agencies, for example) or taken to a different collection point—a disadvantage in terms of both the time and energy used in transportation. Clearly, the designer's incentive to include recycling in building function will be stronger where recycling collection is well established.

TABLE 23.3 Space Planning for Solid Waste

Part A. I-P Units			
		Exterior Area Required	
Occupancy	Building Area, ft^2	For Trash, ft^2	For Recyclable Materials, ft^2
Nonresidential	0–5000	12	12
	5001–10,000	24	24
	10,001–25,000	48	48
	25,000+	Each additional 25,000 ft^2 require 48 ft^2 each, trash and recyclables	
Multifamily Residential	2–6 Units	12	12
	7–15 Units	24	24
	16–25 Units	48	48
	25+ Units	Each additional 25 dwelling units require 48 ft^2 each, trash and recyclables	
Part B. SI units			
		Exterior Area Required	
Occupancy	Building Area, m^2	For Trash, m^2	For Recyclable Materials, m^2
Nonresidential	0–465	1.1	1.1
	466–929	2.2	
	930–2323	4.5	
	2323+	Each additional 2323 m^2 require 4.5 m^2 each, trash and recyclables	
Multifamily Residential	2–6 Units	1.1	1.1
	7–15 Units	2.2	2.2
	16–25 Units	4.5	4.5
	25+ Units	Each additional 25 dwelling units require 4.5 m^2 each, trash and recyclables	

Source: Reproduced with permission from Ramsey and Sleeper, 2000.

The characteristics of local waste separation, then, are increased consumer time and building space requirements; see Tables 23.2 and 23.3. Central waste separation is characterized by energy-intensive (and noisy) industrial processes. Sections 23.3 and 23.4 offer a detailed look at the consequences of local waste separation for some common building types.

The massive quantities of solid waste generated in urban areas give rise to the terms urban mines and urban forests. To the extent that these solid wastes can be turned into resources, savings can be realized in various areas—transportation costs and energy consumption, land area set aside for landfills, and the replacement of virgin materials by these recycled resources.

23.3 SOLID WASTE IN SMALL BUILDINGS

Today, most occupants of small buildings have to choose between the separation of waste or the mixing of garbage. Where food preparation is involved, as in residences or restaurants, the variety of special containers can reach surprising complexity (Figure 23.2).

As the point of origin for so many types of solid waste, the kitchen is often the location of waste separation and storage. There is an inherent conflict, however, between the kitchen's frequently hot, humid environment and the need for a cool, dry, well-aired place for solid wastes. This suggests that a space—pantry, air-lock entry, cabinet, or closet—should be provided that opens to the kitchen on one side, and to the outside on the other (Figure 23.3). Through such an arrangement, both the daily depositing of solid waste and weekly waste removal are made easy, and the near-outdoor conditions are better for the waste in many climates. It is also important that the waste storage area be designed in a manner that makes it easy to clean.

(a) Garbage Disposer

This common appliance (Figure 23.4) is usually installed below the kitchen sink, often in conjunction with a dishwasher. It grinds up organic food scraps and sends them on to the sewer. This device is a boon to central garbage collection because it lightens the weight of the garbage can and adds less moisture to the garbage (which thus can be burned more efficiently). Finely chopped organic matter has a better chance to biodegrade at the wastewater treatment plant than in a tightly packed landfill. However, garbage disposer units require both water and energy. Water must be kept running during the grinding process in order to keep the blades free of debris, and cool the grinder's motor. In all, 2–4 gal [7.5–15 L] are required for

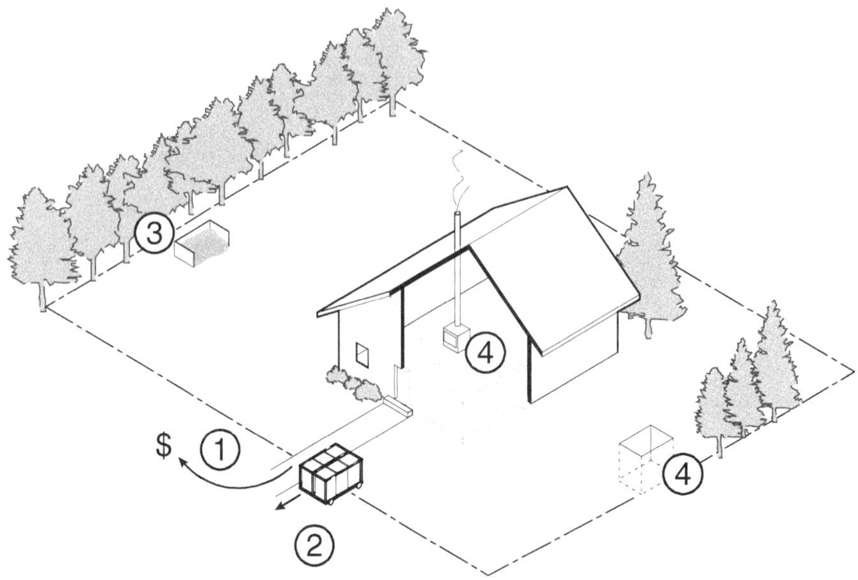

1. Returnable cans and bottles go to store for refund.
2. Recycling is sorted and delivered to central collection.
3. Compostable food scraps are sent to compost pile.
4. Garbage—non recyclable waste—is burned or buried (details vary by jursidiction).

Figure 23.2 *Local separation of solid wastes can require many containers. In communities that offer recycling garbage collection service, the consumer can recycle without transporting individual containers to many different collection points. Some recycling services ask for glass to be separated into green, brown, and clear types; some offer to recycle white paper separately from lower-grade paper. (Redrawn by Nathan Leigh.)*

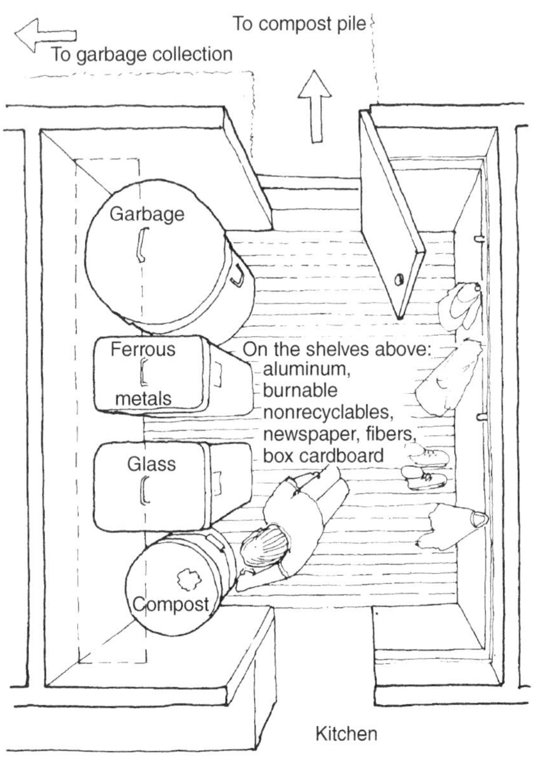

Figure 23.3 *An entry vestibule (or air lock) to a kitchen can serve as a depository for solid wastes, as well as for coats, boots, and other items. Heavy, dirtier items belong on the floor, and many recyclables can be readily and conveniently stored on shelves. An outdoor hose bibb is useful for washing out soiled compost containers and garbage cans.*

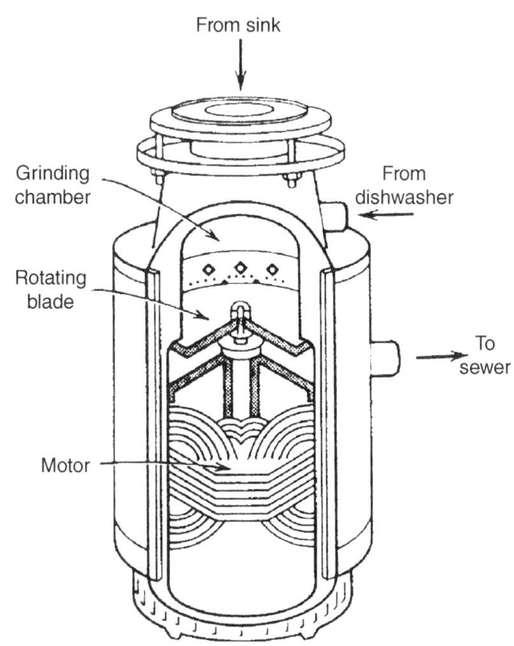

Figure 23.4 *The garbage disposer (or grinder) diverts solid waste (food scraps) from landfills to sewage treatment plants. Occupying very little space, it saves garbage storage space and some of the user's time, but it requires more energy and much more water than treatment of the food scraps as garbage. Composting is another alternative. (From Milne, 1976.)*

one minute of operation. Because more water and solid waste are deposited in the sewer system, moreover, the central sewage treatment plant requires more energy to operate.

(b) Garbage Compactor

In some small buildings, this device allows for a much less bulky storage arrangement. Used selectively, it can compact several of the stacks of items shown in Figure 23.2, such as aluminum for recycling, ferrous metals, and box cardboard. Used indiscriminately, it can make the central process of garbage separation more difficult by crushing dissimilar items together. For a recycling-conscious single-family household, it is questionable whether a compactor will save much more storage volume than it takes for itself. However, small stores and institutions with large quantities of a bulky waste (such as cardboard) could save considerable storage space.

(c) Compost Pile

An alternative to a resource-hungry garbage disposal is the compost pile, familiar to most home gardeners as a source of excellent soil conditioners. Urban opportunities for composting seem very limited; however, raised growing beds on a balcony or window flower boxes are logical recipients of a family's compost. The problem may not be so much what to do with the final product (humus) as where to locate the compost pile. Several self-contained composting containers are commercially available.

The outdoor compost pile has several characteristics that challenge the designer. At its best, it is a frequently turned, quite warm, damp, well-aired source of rich humus (and red worms) for gardens; odors are noticeable only while the pile is turned. At its worst, it is a source of unpleasant odors, a breeding place for vermin, and a fast-food attraction for rats, dogs, and raccoons. Where odors are not objectionable, the heat generated in a frequently fed and tended compost pile might be welcome against the exterior walls of a residence. Clearly, these walls must have nonorganic exterior materials.

(d) Storage Areas

As groups of residences are combined into large apartment complexes, the necessary solid waste systems can make more significant demands on the designer. Approximate exterior storage areas are listed in Table 23.3. Where central storage compounds are provided, garbage cans should be fenced to ward off dogs and other marauders. Different bins for recyclable materials might be provided. Where space is limited, a single bin could accept specific materials on designated days—newspapers on Mondays, metals on Wednesdays—if collection schedules permit. A central compost pile could provide humus for site landscaping. (An incinerator, used to recover heat from nonrecyclable burnable wastes, is almost never approved for installation today.) This combination of space and equipment has special environmental needs, including collection truck access, noise control, and location of both incinerator stacks and the compost pile with respect to prevailing winds.

23.4 SOLID WASTE IN LARGE BUILDINGS

In large buildings, it becomes more likely that solid wastes will be handled several times during the storage and collection process. This may inhibit the separated-waste approach because it is not only the employees who generate the waste, but also the custodians who collect and store it, who must understand what items go into which bins. Larger buildings, however, tend to generate large and concentrated kinds of wastes, which makes recycling more attractive, while also increasing the demand for storage space.

Consider the multistory office building. Office operations are likely to discard large quantities of white paper and smaller quantities of newspaper, box cardboard, and unrecyclable burnable trash (including floor sweepings). Much smaller quantities of food scraps (coffee grounds), metals, and glass are also generated. Given the high cost of floor space, the pressures are very high for a simple mixed-garbage can (rather than multiple separate bins). However, the cost of hauling trash to increasingly scarce landfills is now so high that many recycling programs can pay for themselves by avoiding landfill charges and by taking advantage of the value of recycled materials.

In larger buildings, the collection of solid waste is typically a three-stage process (Figure 23.5). The first stage is the generation of the waste itself by employees who might be provided with one wastebasket each. If each employee is expected to separate wastes, this suggests a redesigned receptacle. At the typical desk in an office building, white paper, recyclables, compostable materials, and garbage would be deposited in separate compartments. Waste separation at the point of origin thus may not require much more floor area, just more attention by the worker.

The second stage begins as custodians work with these individual baskets, dump them into separate bins on a collection cart, and reconnect the individual empty baskets for the next day's deposits. At various stations, special wastebaskets can be supplied: for white

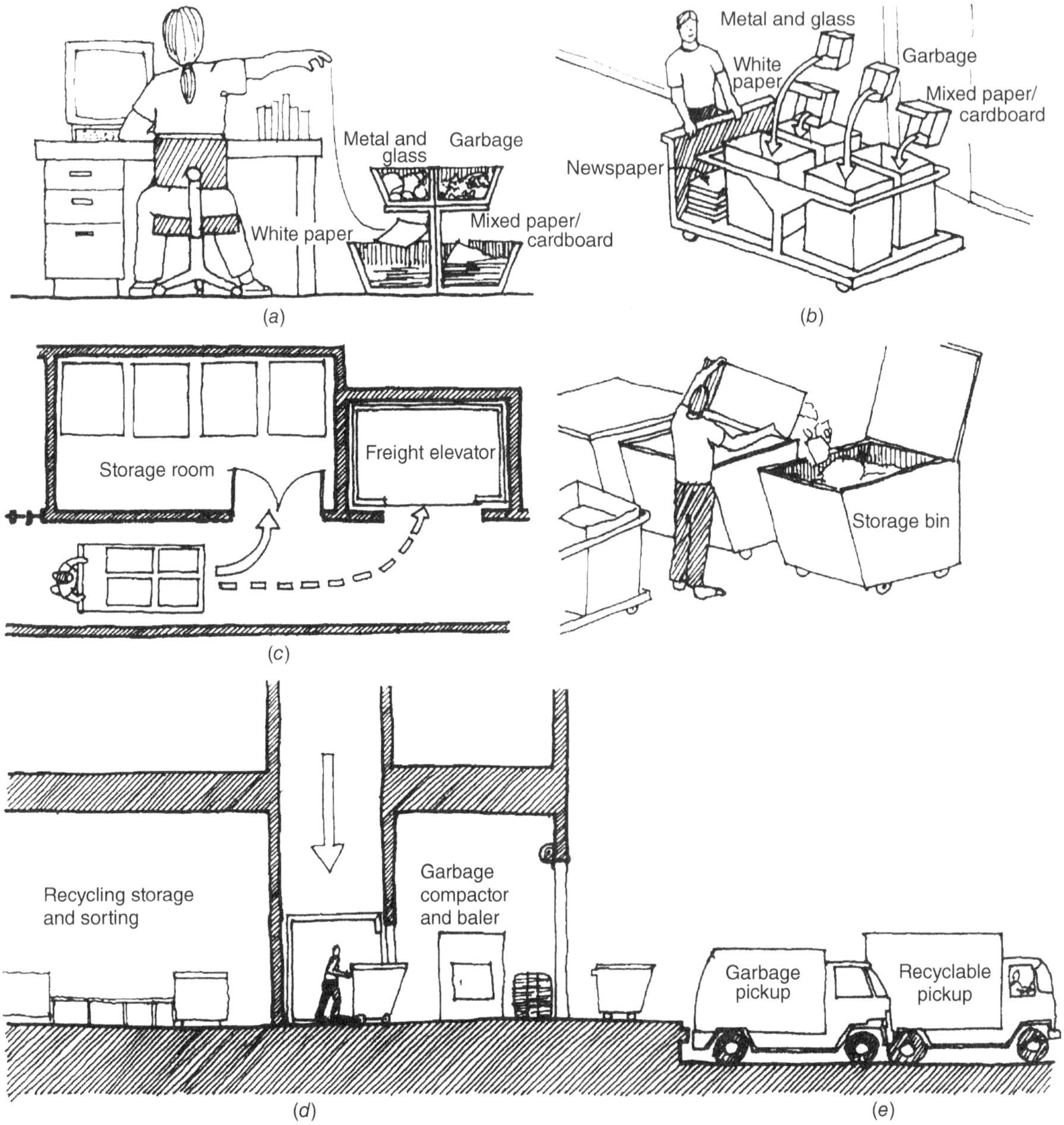

Figure 23.5 *Hypothetical three-stage solid waste collection process with four-category separation for office buildings. (a) At each workstation, a four-compartment waste receptacle is provided. White paper is the predominant waste product. (b) Custodians begin the second stage by collecting waste in separate bins. Floor sweepings are added to the garbage. (c) At the end of the second stage, the four categories of waste are deposited in separate bins. Service sinks, paper shredders, and other maintenance items can be incorporated in a service closet. (d) The third stage begins with compactors at the base of the service core. White paper is compacted, baled, and stored; recyclables are sorted and stored; garbage is compacted, baled, and stored. (e) At the end of the third stage, white paper, metal, plastic, and glass are collected for recycling, and garbage is collected for separation at a central plant.*

paper at the computer room and the copying machine, for compostable materials and garbage at the employee lounge. Floor sweepings are added to garbage. When the collection cart is full, it is wheeled to a service closet, which is probably located within the core of the building. Here there is a container for each category of waste, along with a service sink to wash out the garbage bin (and perhaps a paper shredder to be used selectively by employees).

The third stage begins at the ground floor of the service elevator, where white paper and other recyclables are perhaps shredded and stored until collection; compostable materials are stored or sent to a roof garden compost pile; and garbage is compacted and bagged. In the storage space, a cool, dry, fresh air supply is desirable. A sprinkler (fire protection) system is often required. The compactors and shredders are noisy and must be vibration-isolated from the floor. At the end of the third stage, a truck or van from a recycling center collects the recyclables, and a garbage truck collects the garbage bags.

23.5 EQUIPMENT FOR THE HANDLING OF SOLID WASTE

One approach for the handling of solid wastes is to integrate multiple steel chutes of moderate diameter in the building's vertical core. Another approach is to use only one vertical chute, with a rotating receiving facility at the base (Figure 23.6). With up to six receiving bins on a carousel and a control panel at each floor opening, considerable floor space is saved, compared to six individual chutes. (Also, custodial workers have fewer recyclables to separate and move.) The user first checks the control panel setting (it will be set at whatever bin the previous user had selected). After the user has made a selection, the carousel below turns to position the appropriate bin. The access panel then opens (interlocked so that only one floor can be opened at a time), and the recyclable material (or garbage) can be deposited. The receiving room at the base of the chute is about 12 ft [3.7 m] square, with a minimum height of 8 ft [2.4 m].

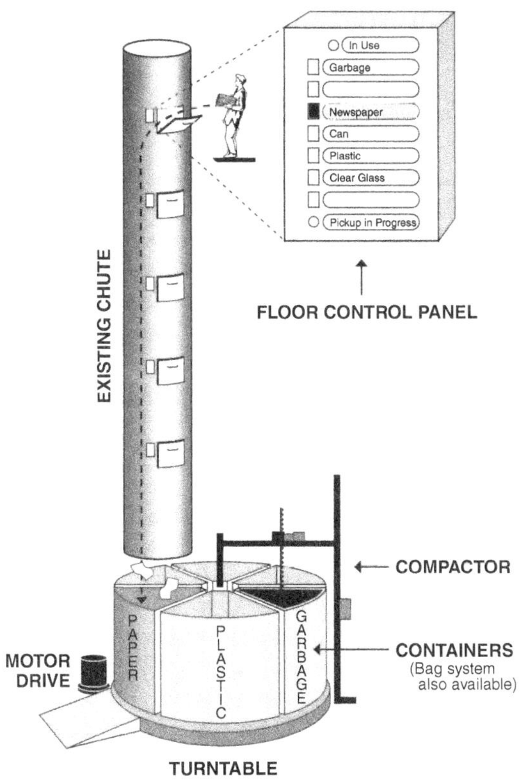

Figure 23.6 One chute serves many different recyclables as well as garbage. Interlocking access doors on each floor allow only one user at a time; the user selects the type of recyclable being deposited using the control panel at the opening. (Courtesy of Hi-Rise Recycling Systems, Inc.)

Where large amounts of solid waste are generated and little space is available in which to store it, various types of equipment can be used to change the volume or composition of the waste to facilitate its transportation and storage. The first step is to determine the probable daily flow of solid waste, as is done in Tables 23.1 and 23.2. When the extent of the problem is known, equipment can be selected.

(a) Compactors

Of the wide variety of compacting devices (Figure 23.7), most are able to reduce the volume of solid waste to as

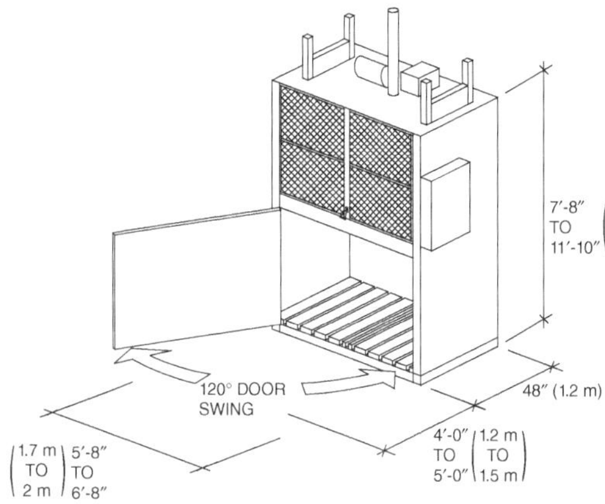

Figure 23.7 A refuse baler/compactor may save storage space where particular waste stream volumes are high. Cardboard, metal cans, and plastic bottles may be more attractive to recyclable haulers when their volume is thus reduced. (Reproduced with permission from Ramsey and Sleeper, 2000.)

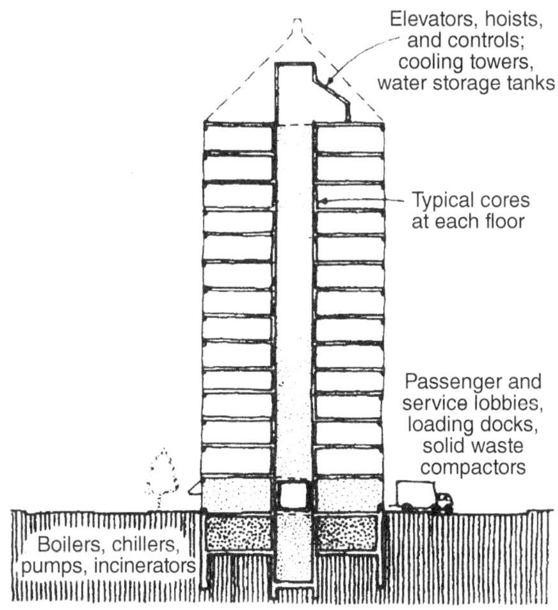

Figure 23.8 Schematic section through a core of a multistory building. At the top and bottom, floor space and ceiling height requirements increase. The consequences on the ground floor (or delivery/collection) are particularly great.

little as 10% of the original volume. Among the many choices to be made are: vertical versus horizontal compaction, automatic chute-fed versus manual free-standing, whether wastes are to be bagged or baled, and the final size of each unit of compacted waste. Compactors can be noisy; since heat is generated in the compaction process, they can also be prone to fires. Many compactors have built-in sprays for both fire control and disinfecting. Access to wash water and a floor drain are both highly desirable.

(b) Vacuum Systems

The primary advantage of a vacuum system is that it reduces the building volume consumed by waste chutes or storage cans on each floor. A grinder-plus-evacuated-tube system is similar to the vacuum sewage systems described in Chapter 22. Vacuum conveying systems are now used in several industrial applications, such as ash and wood-chip transporting. Vacuum systems that use air, not water, as the transport medium are frequently used for linens in hotels and for trash. Separate vacuum systems for trash and linens are desirable, both to lessen the soiling of the linens and to allow them to be delivered to a destination different from that for the trash.

The advantage of such pressurized systems is that lines can be small, and the contents can be moved horizontally or even up. This allows far greater

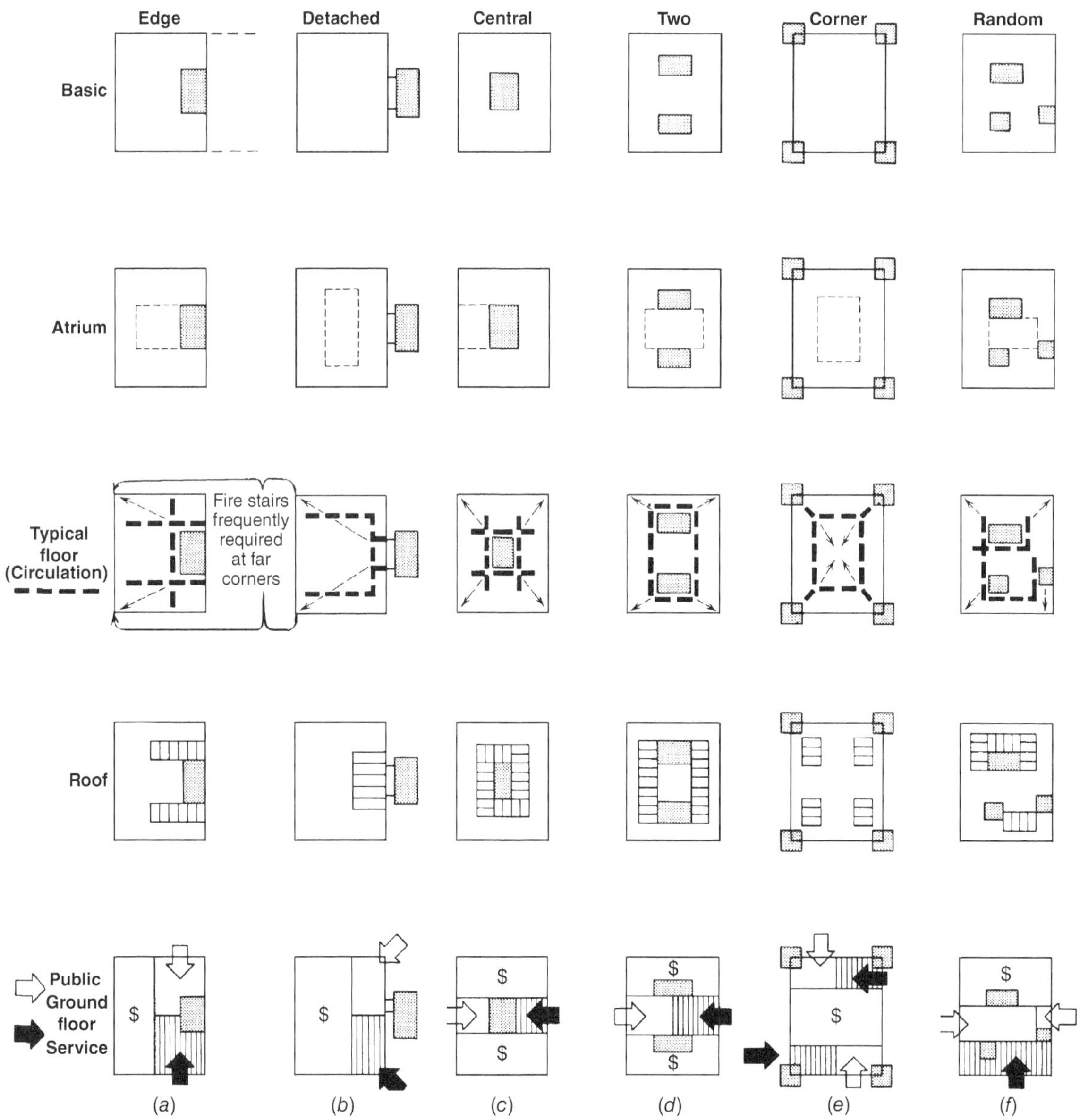

Figure 23.9 *Comparison of core arrangements and building plans. (a, b) The edge and the detached cores give great flexibility to the rental floor area, with light and view for core spaces. (c) The central core expands readily at the roof and the ground floor, and has clear circulation and fairly flexible rental space. (d) Two-core is a popular, workable arrangement. (e) The corner cores give great flexibility to rental floors but are difficult at the roof and ground floor. (f) Random cores generally occur in low-rise buildings, in which the benefits of repetitious plans are minimal.*

TABLE 23.4 Building Characteristics and Core Placement

Core designs

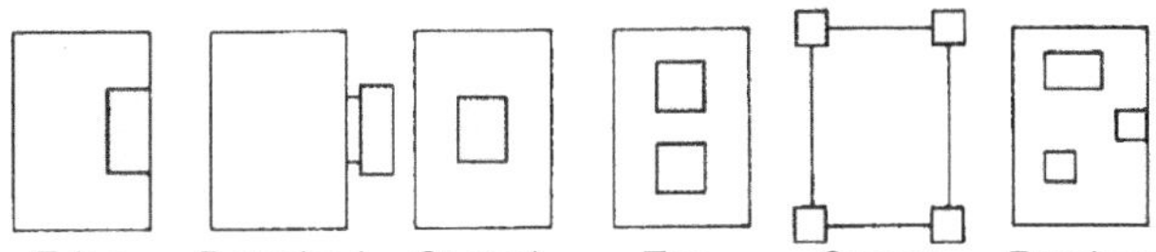

	Edge	Detached	Central	Two	Corner	Random
Flexibility of typical rental areas	Good	○	Average	Fair	Good	○
Perimeter for rental areas	Fair	Average	○	○	○	Good
Ground-floor high-rent area	Average	○	Average	Fair	Good	○
Distance of travel from core, typical	Fair	○	Good	○	Average	Average
Clarity of circulation, typical	Average	Fair	Good	○	Average	○
Daylight, view for core spaces	Good	○	○	○	○	Fair
Core expansion at roof	Average	○	○	Good	Fair	Average
Core expansion at ground floor	Average	Fair	Good	○	○	Average

Excellent **Good** **Average** **Fair** **Poor**

flexibility in design than is possible with gravity systems.

(c) Summary

There are many options to be considered within the present waste-handling sequence of grind, crush, burn, and bury. As Earth's nonrenewable resources are stripped away, the recovery of materials from solid waste becomes increasingly attractive. The situation is similar to energy-and-design issues: After decades of energy-intensive building trends, designers are now allocating space to daylighting atria and allowing surfaces to act as thermal mass. A modest increase in floor space and equipment to encourage resource recovery from solid waste seems an increasingly worthwhile design investment.

23.6 THE SERVICE CORE

Medium- and high-rise buildings usually concentrate many building services within a core, from which services can be distributed as needed throughout each floor. The core typically contains stairs; passenger and freight elevators; toilet rooms; service closets; mechanical, plumbing, and electrical chases; electrical/telephone closets with local switching capabilities; fire protection equipment; and supply closets. The size of such cores varies widely; for example, the taller the building, the more elevators.

Often, these service cores are identical in plan from one floor to the next. Alternatively, the vertical services can be identical (stairs, elevators, chases) and the arrangement of other elements (toilet rooms, supply closets) allowed to vary. Whether or not minor floor-plan variations occur, the cores usually depart radically from the typical plan—and ceiling height—at both the roof and the ground floor (Figure 23.8). (Interruptions also occur where intermediate mechanical floors are provided.) Depending upon the type of elevator machinery selected, the machine room's ceiling height above the top floor being served might be three times the ordinary floor height (see Chapter 33). On the delivery/loading floor level, large trucks must be accommodated. For chillers, boilers, incinerators, and fans, higher ceilings may be required to accommodate the flues, ducts, and pipes to which they must be connected; see Chapter 16 for the dimensions required. Considering the entrance lobby as an extension of the core, a building's service core can expand to nearly fill the ground floor. This core expansion typically happens on the roof and in the basement as well.

A service core can be related to the remaining service floor area in any of several ways (Figure 23.9). Common arrangements are: a single central core or one that is detached, two cores symmetrically placed, corner cores, or core services dispersed somewhat randomly.

Each core location has advantages and drawbacks. From the viewpoint of rentable space, important factors are the flexibility of the served (rental) floor area; the exposure of rental area to the perimeter, and thus to light, air, and view; and the extent to which ground-floor high-rent space gets commercial exposure. For fire safety and occupant convenience, the distance of travel from the farthest rental floor area to the service core, and the clarity of circulation within the rental area, are important. For the environment within the core—toilet rooms, stairs, elevator waiting areas—access to light, air, and view is desirable. For the mechanical services, ease of core-area expansion on the roof, at the loading dock, and in the basement is important, as is the length of travel for ducts and other such elements within each floor. These factors are compared for various types of core designs in Table 23.4.

References and Resources

EP. 2023. *Circular Economy: Definition, Importance and Benefits*. Strasbourg: European Parliament. europarl.europa.eu/topics/en/article/20151201STO05603/circular-economy-definition-importance-and-benefits

EPA. 2016. *Environmental Factoids*. Washington, DC: U.S. Environmental Protection Agency. archive.epa.gov/epawaste/conserve/smm/wastewise/web/html/factoid.html

EPA. 2023. *National Overview: Facts and Figures on Materials, Wastes and Recycling*. Washington, DC: U.S. Environmental Protection Agency. epa.gov/facts-and-figures-about-materials-waste-and-recycling/national-overview-facts-and-figures-materials

McDonough, M. and M. Braungart. 2002. *Cradle to Cradle: Remaking the Way We Make Things*. New York: North Point Press.

Milne, M. 1976. *Residential Water Conservation. National Technical Information Service*. Washington, DC: U.S. Department of Commerce.

Ramsey, C. G. and H. R. Sleeper. 2000. *Architectural Graphic Standards* (10th ed.). New York: American Institute of Architects/John Wiley & Sons.

UNEP. 2024. *Global Waste Management Outlook 2024*. Nairobi, Kenya: United Nations Environment Programme and International Solid Waste Association. doi.org/10.59117/20.500.11822/44939

PART VI

ACOUSTICS

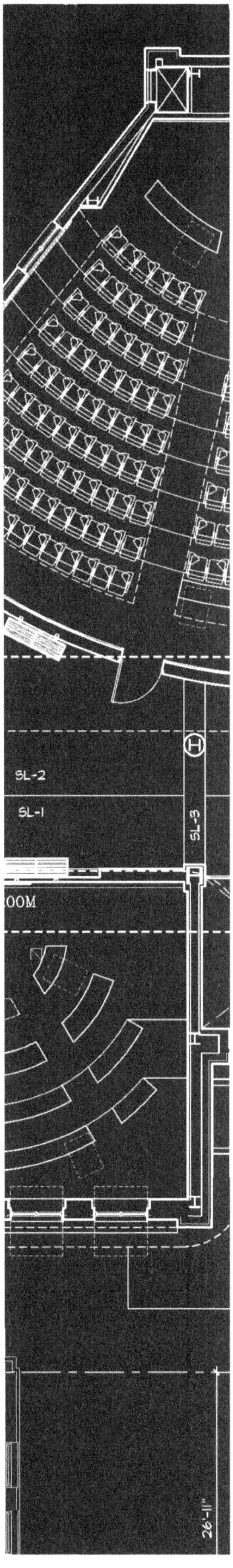

In Chapter 8, the distinction between the terms sound and noise was described as subjective—sound is often desirable, noise is never so. The function of architectural acoustics is to follow this logic: to enhance desired sounds and to attenuate noises.

Chapter 24 covers room acoustics. It provides an explanation of absorption and reverberation and presents acoustic design criteria for various indoor activities. The chapter concludes with a description of sound reinforcement systems.

Chapter 25 is devoted to noise control in and around buildings. It begins with a discussion of the use of absorptive materials for room noise reduction. It treats the problem of interspace noise transfer as two parts (airborne noise and structure-borne noise) because the solutions are different. Relevant criteria, including sound transmission class (STC) and impact insulation class (IIC), are introduced and explained. The problem of speech privacy for both enclosed spaces and open-plan areas (often offices) is considered in detail. The chapter also offers information on mechanical system noise control and ends with a discussion of exterior acoustics. Reference information for acoustical design is also provided.

CHAPTER 24

Sound in Enclosed Spaces

ARCHITECTURAL ACOUSTICS IS THE DESIGN OF SPACES, buildings, and mechanical/electrical systems to meet the hearing needs of occupants and users. With proper design efforts, wanted sounds can be heard properly and unwanted sounds (noise) can be attenuated or masked to the point where they do not cause annoyance. Achieving good acoustics, however, has become increasingly difficult for a variety of reasons. To lower construction costs, the weight of materials used in many of today's buildings has been reduced from those prevalent 50–100 years ago. Since light structures generally transmit sound more readily than heavy ones, lightweight buildings have the potential for major acoustical problems. Population density in office spaces has steadily increased, thus raising the amount of noise generated, and decreasing the distance between occupants. Worse yet—from the acoustics point of view—many offices today are designed as open areas with, at best, only thin, partial-height dividers (cubicles) separating workers. Forty percent or more of a building budget may be allocated for mechanical systems—most of which generate noise. Outside noise sources, such as cars, trucks, trains, and airplanes, can also present problems and require isolation of interior spaces from exterior sounds. This is particularly problematic where natural ventilation is used.

Building owners and tenants are aware that quality acoustic environments are required for high productivity and comfort in buildings, and hence for competitive rental or resale values. The architect is expected to provide such acoustic quality. A clear understanding of the principles explained in this and the following chapters will assist the architect in preparing straightforward designs, either alone or by knowledgeable collaboration with an acoustic consultant. Proper acoustic design responses early in the design process are critically important, since after-the-fact acoustic "repair" is often difficult (and, therefore, costly) and sometimes impossible without substantial structural alterations (which are very costly).

All acoustical situations have three common elements: a sound source, a sound transmission path or paths, and a receiver of the sound. Through design, a source can be made louder or quieter, and a path can be made to transmit more or less sound. Working with sources and paths throughout the various design phases (and into construction and often into the occupancy phases) is the bread and butter of architectural acoustics. The listener's perception of sound also may be influenced, though this is not normally an "architectural" solution.

Chapter 8 presented the fundamentals of sound and its measurement and expression, the characteristics of human hearing and speech, and a basic introduction to acoustical comfort. This chapter will discuss the interactions between an enclosed space and a sound generated within it; these interactions and their manipulation constitute room acoustics—which is basically the celebration of wanted sounds. Chapter 25 presents information on noise control—the mitigation of unwanted sounds.

Mechanical and Electrical Equipment for Buildings, Fourteenth Edition. Walter Grondzik, Daniel Overbey, and Alison Kwok.

Companion website: www.wiley.com/go/meeb14e

24.1 SOUND IN ENCLOSURES

When a continuous sound is generated in an enclosure, fields are set up as described in Section 24.7. When the sound is not a continuous tone or noise, but a series of discrete sounds following one another and containing information, as in speech or music, the room must be designed to maintain and enhance the information's intelligibility. That is what is meant by the design of room acoustics.

Generated sound radiates out from a source until it strikes a room boundary or another large surface. Before reaching this surface, the sound intensity is attenuated by distance (see the inverse square law, Chapter 8) and by absorption in the air. The latter is appreciable only in large rooms and at frequencies above 2000 Hz. When the sound reaches a wall, it is partially reflected and partially absorbed, and a small portion is transmitted into adjoining spaces. The energy transmitted has little effect on the space within which the sound originates, although, as discussed in Chapter 25, it may be very important in the surrounding spaces.

The ratio between the energy absorbed and the energy reflected by a surface will significantly affect what one hears within a space. Specifically, if little energy is absorbed and much is reflected, two effects will be noticeable. Intermittent sounds will be mixed together (which may make speech less intelligible or music more pleasant), and steady sounds will accumulate into a reverberant field, making the space noisy. Conversely, if much energy is absorbed and little reflected, the room will sound quiet for speech and "dead" for music. Quantification of these two primary characteristics of an enclosed space—absorption and reverberation (echo)—is the subject of the next six sections.

ABSORPTION

24.2 SOUND ABSORPTION

When sound energy impinges on a material, part is reflected and the remainder is absorbed (in the sense that it is not reflected). Some of the "absorbed" energy is transmitted, although that part is so small that for our discussion here it will be ignored. Materials are neither perfect reflectors nor perfect absorbers. The term used to define a material's sound absorption characteristic is its coefficient of absorption, which is usually represented by the lowercase Greek letter alpha (α). This sound absorption coefficient is defined as

$$\alpha = \frac{I_a}{I_i} \qquad \text{(Eq. 24.1)}$$

where

I_a = sound power density (intensity) absorbed by the material, W/cm^2
I_i = intensity impinging on the material, W/cm^2
α = absorption coefficient, with no units since it represents a ratio

This absorption coefficient α can also be thought of as a measure of absorption effectiveness; the larger the absorption coefficient, the more effective a sound absorber the material is. Thus, an absorption coefficient of 1.0 indicates 100% absorption and zero reflection of the impinging sound energy. We have already said that no material is a perfect absorber, but an absence of material, that is, an open space, transmits (absorbs) all the impinging energy and therefore can be considered a perfect absorbing "material."

Since open space has this characteristic, α has also been defined as the ratio between the absorption of a given material and that of an open window of the same area. By definition, then, for an opening (open window, open door, etc.)

$$\alpha = 1.0$$

The total absorption A of a given quantity of material is proportional to its area and its absorption coefficient, that is,

$$A = S\alpha \qquad \text{(Eq. 24.2)}$$

where

A = total absorption, sabins
S = surface area, square feet or square meters
α = absorption coefficient

Since α is a ratio and thus unitless, and S is a unit of area, $S\alpha$ should be in units of area as well. Instead, sound absorption units are called sabins in honor of W. C. Sabine, a pioneer in architectural acoustics. One sabin (m^2) is the sound absorption equivalent of an open window 1 m^2 in area. Similarly, 1 sabin (ft^2) is equivalent to 1 ft^2 of open window. As expected, 1 sabin (m^2) equals 10.76 sabins (ft^2). In this text, when not otherwise specified, a sabin (ft^2) is intended.

All rooms are constructed of several materials, each having a different absorption coefficient. In addition, most rooms contain furnishings, which have their own individual coefficients of absorption. Thus, to determine the total absorption of a room, it is necessary to sum the component absorptions, that is,

$$\sum S\alpha = S_1\alpha_1 + S_2\alpha_2 + \cdots + S_n\alpha_n$$

or

$$\sum A = A_1 + A_2 + \cdots + A_n \quad \text{(Eq. 24.3)}$$

where

$\sum S\alpha$ = total absorption in the room, sabins
S_1, S_2, etc. = surface area of each material
α_1, α_2, etc. = absorption coefficient of each material
A_1, A_2, etc. = total absorption of each material

If S is expressed in square feet, then A is in sabins (ft^2); if S is expressed in square meters, then A is sabins (m^2).

Absorption coefficients for some common materials, sound-absorbing materials, and auditorium furnishings are tabulated in Table 24.1. It is important to note that for most common materials, absorption (and therefore α) vary with frequency. Thus, accurate absorption calculations must be made individually for the frequencies being studied.

24.3 MECHANICS OF ABSORPTION

At this point, it is appropriate to examine absorption as an acoustic phenomenon so that we may understand the use of absorptive materials. In an untreated room of normal construction (Figure 24.1*a*), when sound waves strike the walls or ceiling, a small portion of the sound is transmitted, a small portion is absorbed, and most of it is reflected. The exact proportions depend upon the nature of the construction. When acoustical treatment is applied to the room surfaces, as in Figure 24.1*b*, some of the energy in the sound waves is dissipated before the sound reaches the wall. The transmitted portion is slightly reduced, but the reflected portion is greatly reduced.

It is important to understand, however, that the principal effect of absorptive material is on the reflected sound. The transmitted sound energy is essentially determined by the mass of the solid airtight barrier between the two spaces. This is graphically represented in Figure 24.1*c*. The effect of added acoustic absorption in a space is shown in Figure 24.2 and is calculated in Section 24.9.

24.4 ABSORPTIVE MATERIALS

We will now examine acoustic materials and the effect of varying type, quantity, thickness, and installation methods. There are three broad families of sound absorption: fibrous materials, panel resonators, and volume resonators. All three types absorb sound by changing sound energy into heat energy. Only fibrous materials and panel resonators are used commonly in buildings. Volume resonators—also known as Helmholtz resonators, after their originator, Hermann von Helmholtz—are used principally as devices for absorbing a narrow band of frequencies. The discussion in this section refers to fibrous absorbers; the other two types are discussed in Chapter 25.

Fibrous materials absorb acoustic energy by the frictional drag of air moving in the tiny spaces between the fibers. The absorption provided by a specific material depends upon its thickness, density, porosity, and resistance to airflow. Since the action depends upon absorbing energy by "pumping" air through the material, the air paths must extend from one side to the other. A fibrous material with sealed pores is almost useless as an acoustic absorbent. (Therefore, painting will generally ruin a porous absorber.) A simple test is to blow smoke through the material. If the smoke passes through freely and the material is porous, fibrous, and thick, it should be a good sound absorbent. Absorbency increases with increasing material porosity up to approximately 70% porosity; above that condition, absorbency remains essentially constant.

Table 24.1 gives absorption coefficients for fibrous absorbent materials and for some other building materials and furnishings. Several important conclusions can be drawn from examination of this table and Figure 24.3:

1. For absorbent materials, absorption is normally higher at high frequencies than at low ones.
2. Absorption is not always proportional to thickness but depends upon the type of material used and the method of installation (see Figure 24.3). It is clear from this figure that, beyond a nominal thickness, little is to be gained by additional thickness except at very low frequencies, or when material is installed discontinuously, as described in the next item.
3. It is possible to obtain effective α values greater than 1.0 by using very thick blocks. Because they are installed at a distance from each other, and their edge absorption is very large (particularly at high frequencies), they exhibit α greater than 1.0.
4. Installation methods have a pronounced effect, as discussed in the following section.
5. All other factors remaining constant, the thicker an absorbent material, the better its low-frequency absorption.

TABLE 24.1 Octave Band Average Sound Absorption Coefficients[a]

		Absorption Coefficients (α)						
General Building Materials and Furnishings[b]		**125 Hz**	**250 Hz**	**500 Hz**	**1000 Hz**	**2000 Hz**	**4000 Hz**	**NRC[c]**
Brick, unglazed		0.03	0.03	0.03	0.04	0.05	0.07	0.005
Brick, unglazed, painted		0.01	0.01	0.02	0.02	0.02	0.03	0.00
Carpet, heavy, on concrete		0.02	0.06	0.14	0.37	0.60	0.65	0.29
Carpet, heavy, on 40-oz [1.1 kg] hair felt or foam rubber		0.08	0.24	0.57	0.69	0.71	0.73	0.55
Concrete block, coarse		0.36	0.44	0.31	0.29	0.39	0.25	0.35
Concrete block, painted		0.10	0.05	0.06	0.07	0.09	0.08	0.05
Fabrics								
Light velour, 10 oz/yd^2 [237 g/m^2], hung straight, in contact with wall		0.03	0.04	0.11	0.17	0.24	0.35	0.15
Medium velour, 14 oz/yd^2 [332 g/m^2], draped to half-area		0.07	0.31	0.49	0.75	0.70	0.60	0.55
Heavy velour, 18 oz/yd^2 [427 g/m^2], draped to half-area		0.14	0.35	0.55	0.72	0.70	0.65	0.60
Floors								
Concrete or terrazzo		0.01	0.01	0.015	0.02	0.02	0.02	0.00
Linoleum, asphalt, rubber, or cork tile on concrete		0.02	0.03	0.03	0.03	0.03	0.02	0.05
Wood		0.15	0.11	0.10	0.07	0.06	0.07	0.10
Glass								
Large panes of heavy plate glass		0.18	0.06	0.04	0.03	0.02	0.02	0.05
Ordinary window glass		0.35	0.25	0.18	0.12	0.07	0.04	0.15
Gypsum board, ½ in. [13 mm], nailed to 2 x 4 s, 16 in. [50 x 100, 406 mm] o.c.		0.10	0.08	0.05	0.03	0.03	0.03	0.05
Marble or glazed tile		0.01	0.01	0.01	0.01	0.02	0.02	0.00
Openings								
Stage, depending on furnishings					0.25–0.75			
Deep balcony, upholstered seats					0.50–1.00			
Grilles, ventilating					0.15–0.50			
Plaster, gypsum or lime, smooth finish on tile or brick		0.013	0.015	0.02	0.03	0.04	0.05	0.05
Plaster, gypsum or lime, on lath		0.14	0.10	0.06	0.05	0.04	0.03	0.05
Plywood paneling, 3/8-in. [9-mm] thick		0.28	0.22	0.17	0.09	0.10	0.11	0.15
Rough wood, as tongue-and-groove cedar		0.24	0.19	0.14	0.08	0.13	0.10	0.14
Slightly vibrating surface (e.g., hollow core door)		0.02	0.02	0.03	0.03	0.04	0.05	0.03
Readily vibrating surface (e.g., thin wood paneling on 16-in. [406-mm] studs)		0.10	0.07	0.05	0.04	0.04	0.05	0.05
Water surface, as in a swimming pool		0.008	0.008	0.013	0.015	0.020	0.025	0.00
Absorption of Seats and Audience[d]		**125 Hz**	**250 Hz**	**500 Hz**	**1000 Hz**	**2000 Hz**	**4000 Hz**	**NRC[c]**
Audience, in upholstered seats, per ft^2 [0.09 m^2] of floor area		0.60	0.74	0.88	0.96	0.93	0.85	—
Unoccupied cloth-upholstered seats, per ft^2 [0.09 m^2] of floor area		0.49	0.66	0.80	0.88	0.82	0.70	—
Wooden pews, occupied, per ft^2 [0.09 m^2] of floor area		0.57	0.61	0.75	0.86	0.91	0.86	—
Students in tablet-arm chairs, per ft^2 [0.09 m^2] of floor area		0.30	0.42	0.50	0.85	0.85	0.84	—
Acoustic Absorptive Materials	**Mtg[e]**	**125 Hz**	**250 Hz**	**500 Hz**	**1000 Hz**	**2000 Hz**	**4000 Hz**	**NRC[c]**
High-performance vinyl-faced fiberglass ceiling panels								
1-in. [25 mm] thick	E405	0.73	0.88	0.71	0.98	0.96	0.77	0.90
1.5-in. [38 mm] thick	E405	0.79	0.98	0.83	1.03	0.98	0.80	0.95
Painted nubby glass cloth panels								
¾-in. [19 mm] thick	E405	0.81	0.94	0.65	0.87	1.00	0.96	0.85
1-in. [25 mm] thick	E405	0.78	0.92	0.79	1.00	1.03	1.10	0.95
Random fissured ¾-in. [19 mm]-thick panels	E405	0.52	0.58	0.60	0.80	0.92	0.80	0.70
Perforated metal panel with infill 1-in. [25 mm] thick	E405	0.70	0.86	0.74	0.88	0.95	0.86	0.85
Typical averages, mineral fiber tiles and panels								
¾-in. [19 mm] fissured	E405	0.47	0.50	0.52	0.76	0.86	0.81	0.65
¾-in. [19 mm] textured	E405	0.49	0.55	0.53	0.80	0.94	0.83	0.70
⅝-in. [16 mm] fissured	E405	0.28	0.33	0.66	0.73	0.74	0.75	0.60
⅝-in.[16 mm] textured	E405	0.29	0.35	0.66	0.63	0.44	0.34	0.50
⅝-in. [16 mm] perforated	E405	0.27	0.29	0.55	0.78	0.69	0.53	0.60
3-in. [76 mm] thick x 16-in. [406 mm] square on 24-in. [610 mm] centers	A	0.40	0.61	1.92	2.54	2.62	2.60	

[a]This table is useful in making preliminary calculations. Complete tables of absorption coefficients for the various materials that normally constitute the interior finish of rooms may be found in books on architectural acoustics.

[b]Selected data courtesy of Owens Corning Fiberglass.

[c]Noise reduction coefficient: the arithmetic average of the α values at 250 Hz, 500 Hz, 1000 Hz, and 2000 Hz.

[d]When the audience is randomly spaced, use an average of 5.0 sabins ft^2 [54 sabins m^2] per person.

[e]See Figure 24.4 for mounting methods.

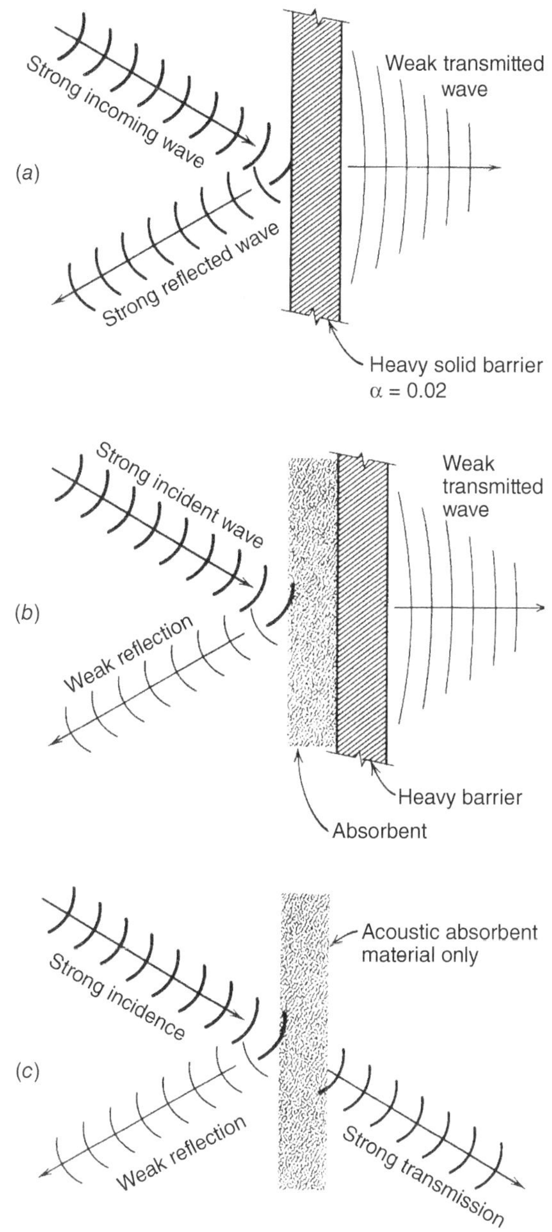

Figure 24.1 *(a) Action of an incoming sound wave striking a heavy barrier. Much of the energy is reflected, some is absorbed, and a little is transmitted. (b) When absorbent material is applied to a heavy wall, it "traps" sound, preventing reflection, while the wall mass acts to reduce transmission. (c) Action of acoustic absorbent material alone. Very little energy is reflected, some is absorbed, and most is transmitted.*

24.5 INSTALLATION OF ABSORPTIVE MATERIALS

Coefficient ratings for absorptive materials are always given with respect to mountings corresponding to ASTM (American Society for Testing and Materials) requirements (ASTM, 2023). The most common standardized mounting methods are shown in Figure 24.4. Installation of absorbent material directly on a wall or ceiling is the least effective means, since exposure to sound energy is minimal. When an air gap is left between the porous layer and the rigid surface, the combination acts almost as well in midfrequencies as an absorbent layer equivalent in thickness to the air plus the porous material (Figure 24.5). One problem with this technique, however, is that at the $\lambda/2$ node of a standing wave there is a severe drop in absorption, as can be seen in Figure 24.5*c*. At 1000 Hz, one-half wavelength is approximately 7 in. [178 mm]. At that distance, α drops severely but is a maximum at $\lambda/4$ or 3.5 in. [89 mm].

Recall that frequency of sound (f) can be calculated as the velocity of sound (c) divided by wavelength (λ). Thus, for a ceiling tile hung at 16 in. [405 mm] below the slab (Figure 24.4, type E 405 mounting method), and with consideration of the speed of sound at 1130 ft per second [344 m per second], the drop in absorption occurs at:

$$\lambda/2 = 16\text{ in. } [405\text{ mm}]$$

$$\lambda = 32\text{ in.} = 2.67\text{ ft } [810\text{ mm}]$$

$$f = \frac{1130}{2.67} = 423\text{ Hz}$$

which is midfrequency. This factor should be considered in applying absorptive material. Avoid a spacing corresponding to a drop in absorption at a sensitive frequency. To obtain good low-frequency absorption, it is essential that a deep air space be provided behind the absorbent material and that walls be treated in addition to the ceiling.

In increasing order of effectiveness, absorbent material can be applied:

1. directly to the room surface
2. hung below the ceiling and supported away from the walls
3. hung from the ceiling as louvers or baffles
4. made up into shapes such as cubes or tetrahedrons and suspended from the ceiling

The last two techniques are extremely effective because they expose a very large surface of porous material—much larger than could be obtained with wall or ceiling covering. Of course, these suspended objects become architectural elements and must be handled accordingly. In contrast, surface coverings are relatively architecturally neutral. In general, treatment should not be limited to one room surface such as the ceiling. All three principal surfaces in the direction of sound propagation—that is, the ceiling, floor, and back wall—should be treated approximately equally for best results. The common practice of treating the ceiling

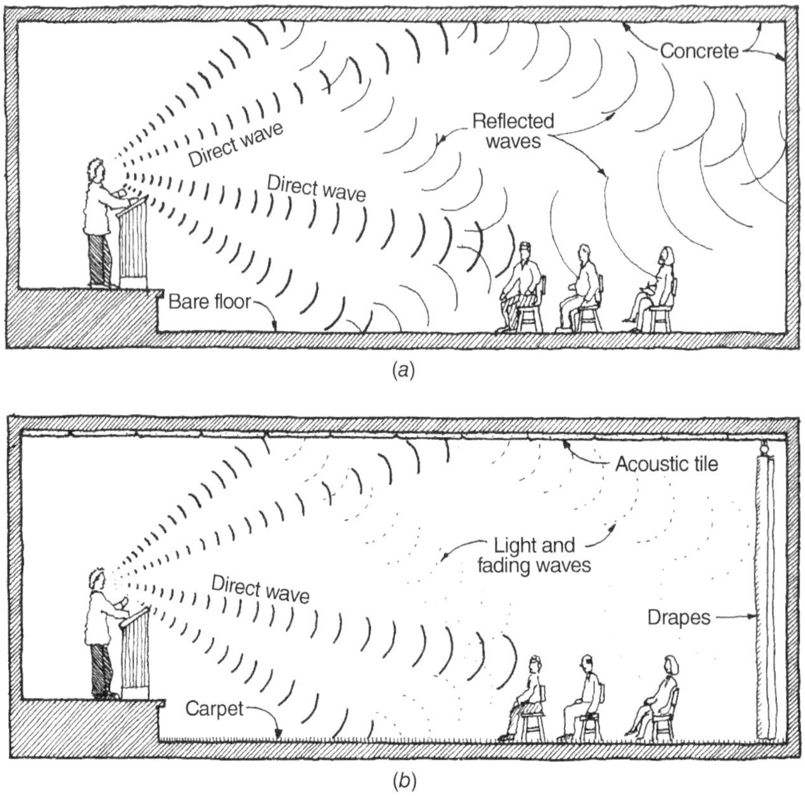

Figure 24.2 *In an untreated space (a) reverberant (reflected) sound constitutes a large portion of received sound in much of the room. These reflections are largely eliminated in (b) by wall and ceiling absorption. Note that direct wave sound is completely unaffected.*

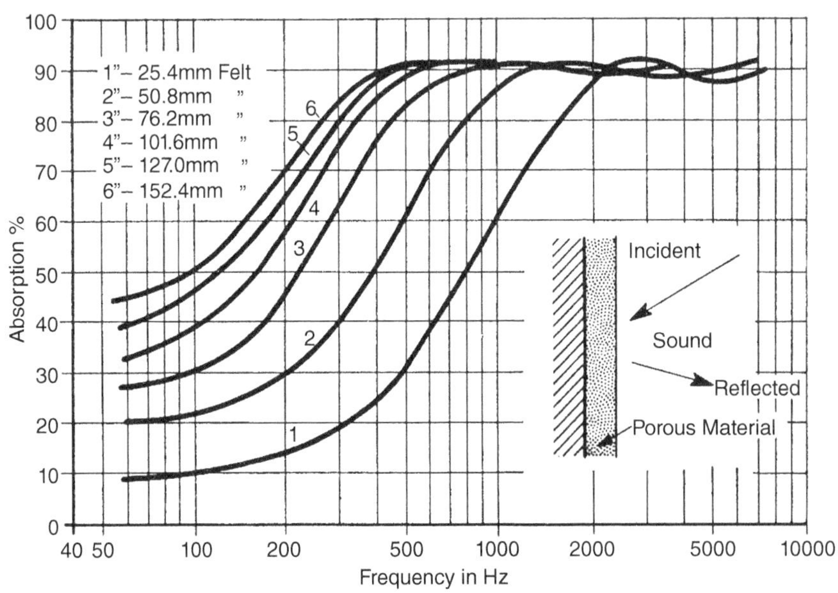

Figure 24.3 *Variation of the absorption coefficient with the thickness of felt absorbent. Note particularly that beyond 1 kHz, all thicknesses give the same α, whereas at low frequencies, absorption is proportional to thickness. Note also that a very heavy layer is required to give appreciable absorption at low frequency. (Courtesy of Brüel and Kjær.)*

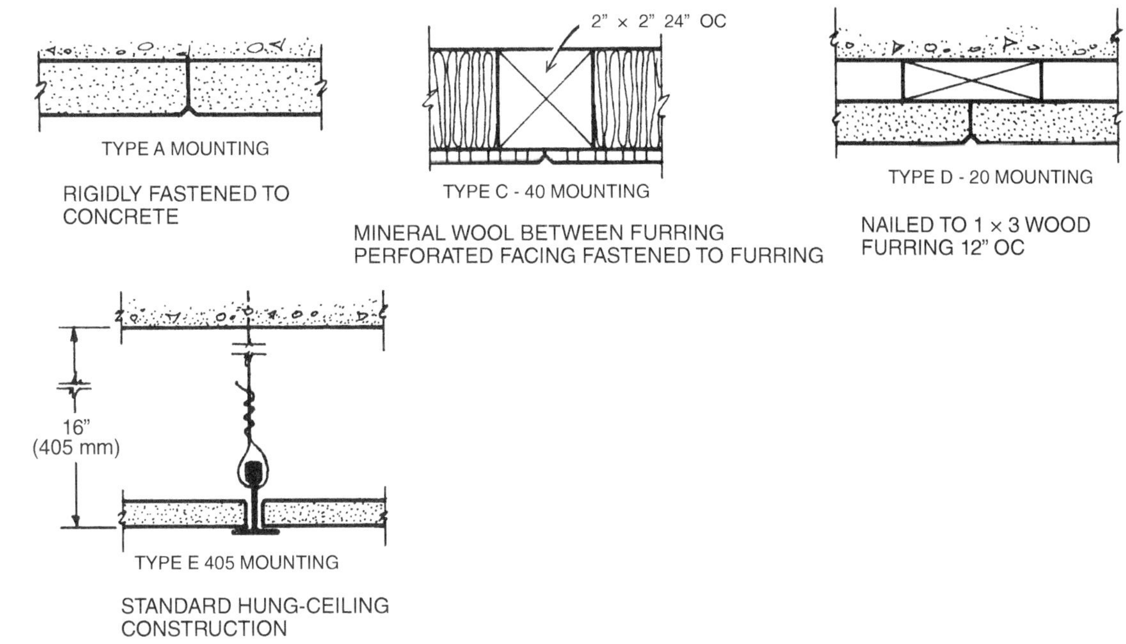

Figure 24.4 *Standard test-mounting methods for absorptive material, in accordance with which absorption coefficients are given by manufacturers. The number following the letter E represents the distance in millimeters from the mounting surface.*

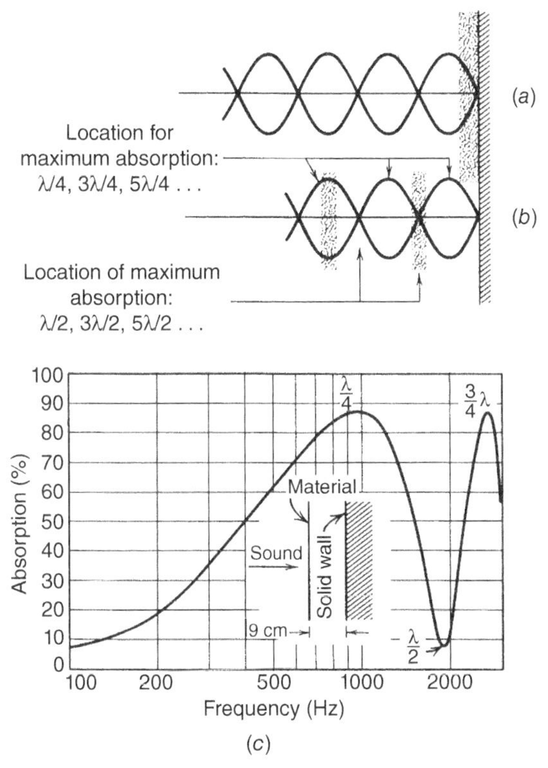

Figure 24.5 *Sound waves striking a surface of a large mass will create standing waves at certain frequencies (a), depending upon the room dimensions. Only insulation placed at the peaks (b) of these waves (λ/4, 3λ/4 ···) is effective, as seen in (c). (Reproduced with permission of Brüel and Kjær.)*

only is generally inadvisable, since high frequencies are highly directive and may not reach the ceiling until the third reflection.

In order to fully understand the effects of absorbent materials in an enclosed space, it is necessary to introduce two new concepts: reverberation and sound fields. This will be done in the next two sections, after which we will return to our discussion of the use and effects of acoustic absorptive material in enclosed spaces.

ROOM ACOUSTICS

24.6 REVERBERATION

Reverberation is the persistence of sound after the sound source has ceased. Such persistence is a result of repeated reflections in an enclosed space. Reverberation time (T_R) is defined as the time required for the sound level to decrease 60 dB after the source has stopped producing sound. For rooms of usual size and shape, the reverberation time at a specific frequency may be found by the formula

$$T_R = K \times \frac{V}{\sum A} \text{ seconds} \qquad \text{(Eq. 24.4)}$$

where

K = a constant, equal to 0.05 when measurements are in feet and 0.16 when in meters
V = room volume, ft^3 or m^3
$\sum A$ = total room absorption, sabins (ft^2 or m^2) at the frequency in question

(For spaces of unusual shapes, see Beranek, 1988.) Reverberation is one of the most pronounced hearing reactions in an enclosed space. It is the ear's reaction to echoes, giving a subjective impression of "liveness" or "deadness" to a space.

A space with highly reflective surfaces, and therefore a low average absorption coefficient ($\bar{\alpha} < 0.2$), sounds live. Conversely, a highly absorptive, nonreflective environment ($\bar{\alpha} > 0.4$) sounds dead. ($\bar{\alpha}$ is the average absorption coefficient of the entire space, as explained in detail in Section 24.8.) Since room absorption is related to total surface area, which in turn is related by room proportions to room volume, it is possible to relate all three factors in a single diagram. Figure 24.6 is drawn using room proportions of 2:1.5:1 for L:W:H, based on Figure 24.22 for preferred room proportions. These proportions (2:1.5:1) represent

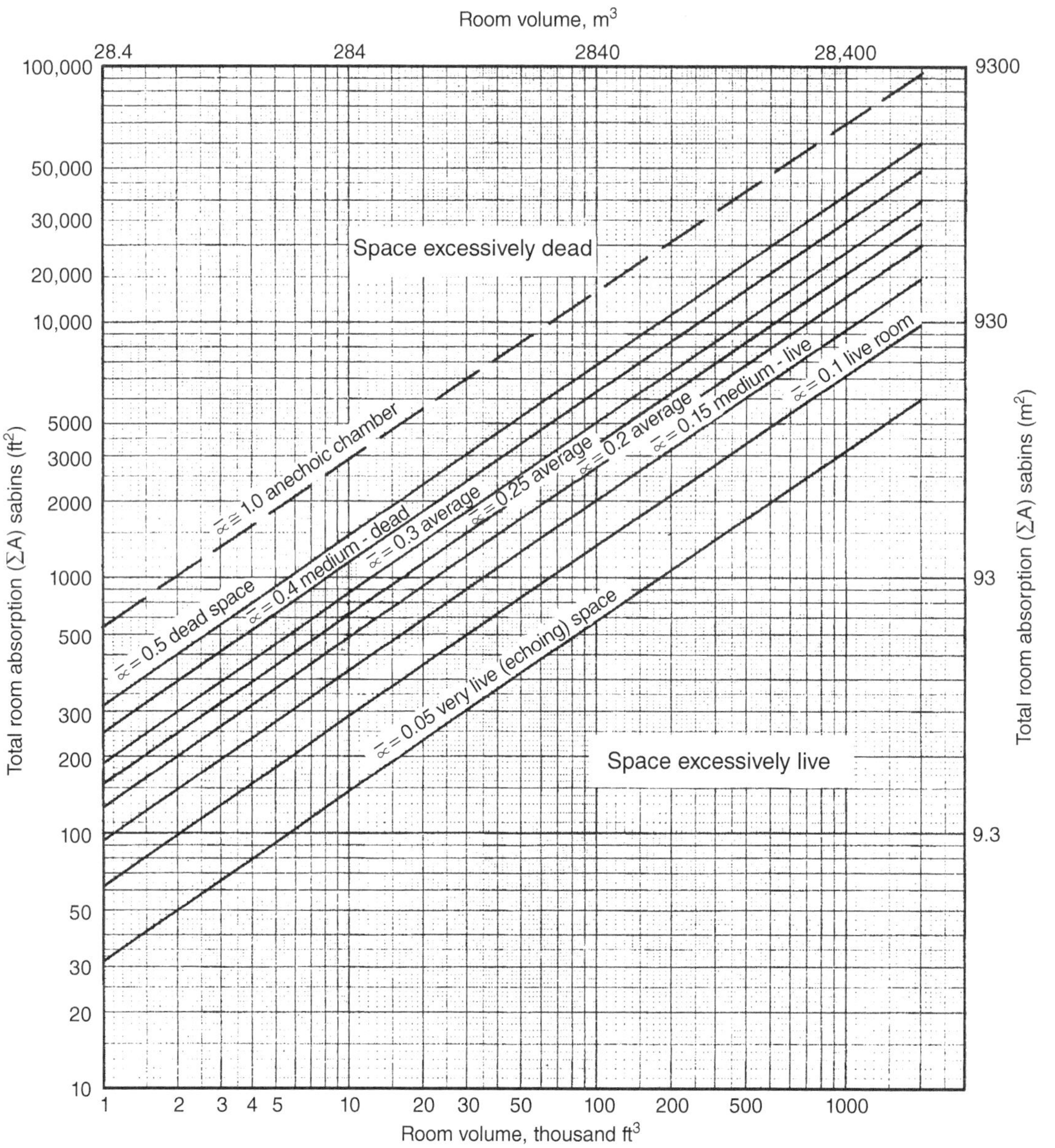

Figure 24.6 *Chart indicating a room's "liveness" as a function of its volume and sabins of total absorption. The L:W:H room proportions chosen—2H:1.5H:H—represent the average of the three extreme points of the recommended room proportion triangle in Figure 24.22. For these proportions,* $S = 6.25\,V^{2/3}$ *and therefore* $A = S\,\bar{\alpha} = 6.25\,\bar{\alpha}\,V^{2/3}\,\bar{\alpha}$.

the average of the extremes of Figure 24.22. The maximum differential introduced by using average proportions is 2.5%—well within engineering accuracy.

The ultimate dead space is one where there is no reflection at all—as is the case outdoors in a flat open area. Indoors, we receive an auditory cue in the form of reflected sound (feedback) that helps us regulate our sound power output (voice level). Outdoors, the absence of this cue, to which we are so accustomed by our largely indoor existence, automatically causes us to raise our output. For this reason, most people tend to speak excessively loudly outdoors. Conversely, a highly reflective indoor condition gives a large feedback signal, which usually results in lowered vocal output.

In common room acoustics studies, reverberation times are calculated at 125 Hz, 500 Hz, 1000 Hz, and 2000 Hz. The midfrequency (500 Hz to 1000 Hz) range is generally the reference used in specifying the reverberation time of a room when studying the speech characteristics of the space.

Reverberation can be considered as a mixture of previous and more recent sounds. The converse of reverberation, or reverberance, is articulation. An articulate environment keeps each sound event separate rather than running them together. Spaces for speech activities should be less reverberant—more articulate—than those designed for the performance of music. See Sections 24.11 and 24.12 for reverberation criteria for speech and music rooms, respectively.

24.7 SOUND FIELDS IN AN ENCLOSED SPACE

The inverse square law described in Chapter 8 holds true for the acoustic far field, which is a sound field sufficiently far from a source that intensity is proportional to power and inversely proportional to distance. This type of acoustic field is developed in open, obstruction-free space. Propagation in an enclosed space is quite different. There, when a sound reaches a wall or another large (with respect to wavelength) obstruction, part of the sound energy is reflected and part is absorbed. As a result, the sound at any point in the room is a combination of direct sound from the source plus reflected sound from walls and other obstructions. If the reflections are so dominant that the sound level becomes uniform throughout the room, the acoustic field within the room is termed a diffuse one (no shadows), and intensity measurements with respect to a specific source are meaningless. Of course, if it is our intention to measure sound pressure level (SPL) at a specific point, such as a seat in an auditorium, the type of acoustic field in the room is irrelevant.

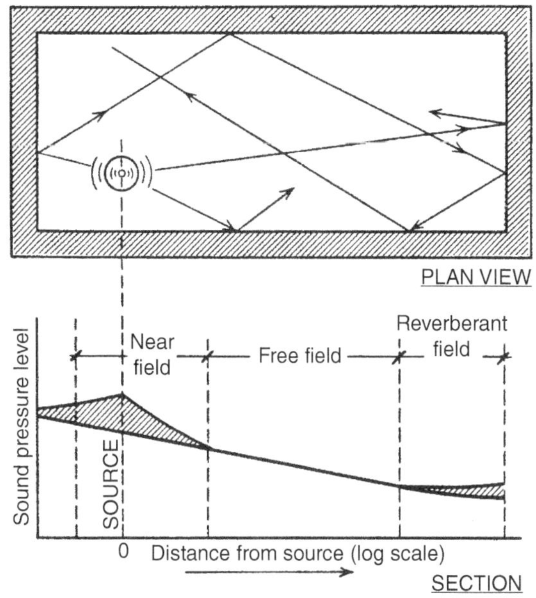

Figure 24.7 The type of sound field in an enclosed space depends largely on reflections (reverberation) and absorption. In a typical room there is a near field adjacent to the sound, a free field beyond that, and a reverberant field adjacent to the walls. In a large hall or auditorium, the reverberant field dominates, and the sound pressure level is approximately constant.

Most indoor spaces do not have such a high level of reflection that a diffuse field is created. Instead, there is a near field near the source, a free field beyond the near field, and a reverberant field near the walls (Figure 24.7). These can be recognized as follows:

1. The near field is generally within one wavelength of the lowest frequency of sound produced by the source. Within this distance, sound pressure level measurements vary widely and are not meaningful. (The maximum wavelength for the human male voice is about 11 ft [3.4 m].)
2. Close to large obstructions such as walls, the reverberant field is dominant and approaches a diffuse condition. In well-designed music auditoriums, the reverberant (diffuse) field predominates, and the sound pressure level remains relatively constant beyond the free field area.
3. The free (far) field exists between the near and reverberant fields, and there the intensity varies directly with pressure and inversely with distance squared. In this field, sound pressure level drops 6 dB with each doubling of distance from the source, and it is in this field that meaningful sound pressure level measurements can be made with respect to a specific small source.

24.8 SOUND POWER LEVEL AND SOUND PRESSURE LEVEL

Sound power level (PWL) is a measure of the amount of sound generated by a source, independent of its environment. Sound pressure level (SPL) can be thought of as the sound in an enclosed space resulting from a source in that space, as affected by the characteristics of the space and the position of the listener. These two acoustic conditions are further discussed in Chapter 8.

Since it is desirable to be able to predict the SPL in a space during design, before construction, it is useful to have equations and/or a graphic means of relating PWL to SPL. PWL data are supplied by manufacturers of equipment, referenced to either octave or one-third octave bands. In free space, SPL and PWL are related by the inverse square law, whereas in enclosed spaces the room characteristics come into play. Roughly speaking, an analogy to lighting can be drawn: SPL corresponds to illuminance (that is, footcandles or lux), and PWL corresponds to the lumen output of a light source.

The basic relationship between SPL and PWL for a single small source in a room large enough to have both free and reverberant fields is:

I-P units:

$$SPL = PWL + 10 \log\left(\frac{Q}{4\pi r^2} + \frac{4}{R}\right) + 10.5 \qquad \text{(Eq. 24.5)}$$

SI units:

$$SPL = PWL + 10 \log\left(\frac{Q}{4\pi r^2} + \frac{4}{R}\right) + 0.2 \qquad \text{(Eq. 24.6)}$$

where

Q = is a directivity constant
SPL = sound pressure level, dB
PWL = sound power level, dB
r = distance from source, ft [m]
R = room factor, ft² [m²]

The factor R can be calculated from

$$R = \frac{\sum S\bar{\alpha}}{1-\bar{\alpha}}$$

where

$\sum S$ = total room surface area, ft² [m²]
$\bar{\alpha}$ = average absorption coefficient of all materials in the room

that is,

$$\bar{\alpha} = \frac{\sum A \text{ (total room absorption)}}{\sum S \text{ (total room surface area)}} \qquad \text{(Eq. 24.7)}$$

or

$$\bar{\alpha} = \frac{S_1\alpha_1 + S_2\alpha_2 + \cdots + S_n\alpha_n}{S_1 + S_2 + \cdots + S_n} \qquad \text{(Eq. 24.8)}$$

The directivity constant Q is either inherent in the sound source, and as such will be part of the given data, or can be obtained from Figure 24.8 for a nondirectional source made directional by adjacent reflecting surfaces.

Thus, for a source suspended or supported at least ½ wavelength (at its lowest frequency) from a ceiling or floor, $Q = 1$; a source near a floor and distant from the walls relates to $Q = 2$; and so on. In a space containing more than one sound source, the SPLs can be combined as explained in Chapter 8. In the two extreme cases (i.e., rooms with only far [direct] fields and rooms with only reverberant [diffuse] fields, corresponding to dead and live rooms, respectively) for a nondirectional source located on the floor and away from walls ($Q = 2$), Equations 24.5 and 24.6 reduce to:

Dead room, direct field:

I-P units:

$$SPL = PWL - 20 \log r + (Q - 0.6)$$
$$= PWL - 20 \log r + 1.4 \qquad \text{(Eq. 24.9)}$$

SI units:

$$SPL = PWL - 20 \log r + (Q - 10.9)$$
$$= PWL - 20 \log r - 8.9 \qquad \text{(Eq. 24.10)}$$

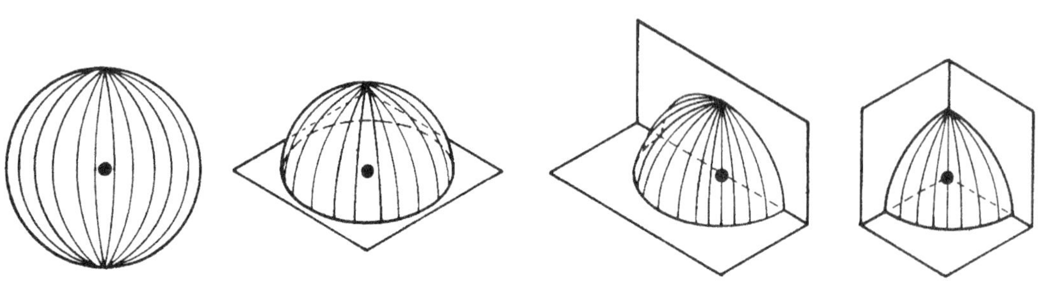

Figure 24.8 *Diagrams illustrating directivity factors for either inherently directive sources or nondirective sources placed adjacent to large reflecting surfaces. (Courtesy of Barry Blower Co.)*

ACOUSTICS

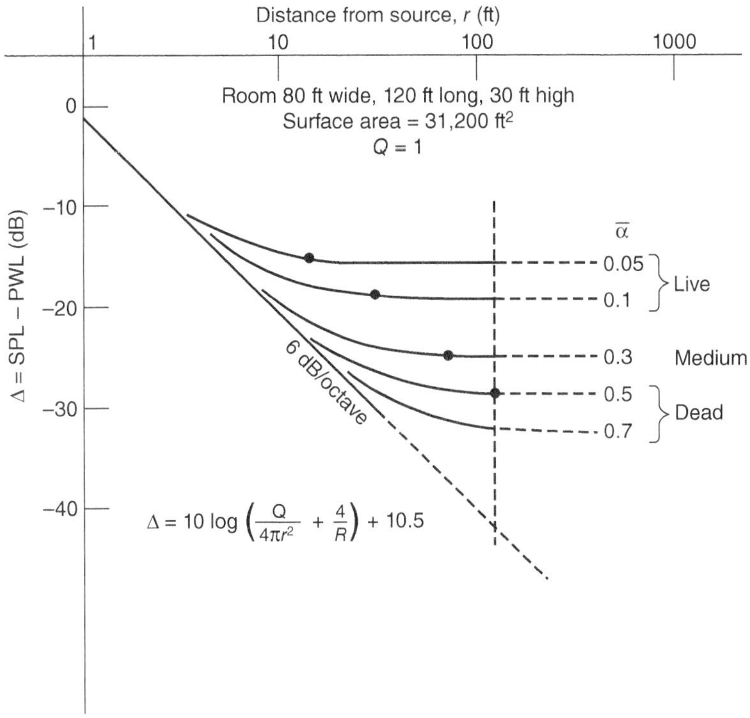

Figure 24.9 Calculated curves show that the development of a reverberant field (constant SPL, or flat curve) is dependent upon a room's absorption characteristics. On each curve, the point at which a reverberant field begins is indicated by a dot on the curve. In a live room ($\bar{\alpha} = 0.1$), the reverberant field begins 20 ft [6.1 m] from the source. In a dead room ($\bar{\alpha} = 0.5$), it begins 120 ft [36 m] from the source, that is, at the back wall. Effectively, then, there is no reverberant field in such a room (or in any room with $\bar{\alpha} = 0.5$). The slope of the asymptote is 6 dB per octave, that is, inverse square attenuation.

Live room, diffuse-reverberant field:

I-P units:

$$SPL = PWL - 10\ \log \sum A + 16.3 \qquad \text{(Eq. 24.11)}$$

SI units:

$$SPL = PWL - 10\ \log \sum A + 6.0 \qquad \text{(Eq. 24.12)}$$

To understand the effect of room absorption on the development of the reverberant field in a large space, refer to Figure 24.9. The field is plotted from Equation 24.5 for a room 80 ft × 120 ft × 30 ft high [24.4 m × 36.6 m × 9.1 m], using average absorptions ($\bar{\alpha}$) ranging from 0.05 to 0.7, that is, from very live to very dead. Note particularly that the addition of sufficient absorbent material can, by drastically reducing sound reflections from room surfaces, entirely prevent the development of a reverberant (diffuse) field. This is particularly important in noisy industrial interiors where the building materials commonly employed are generally nonabsorbent, and the noise sources are both numerous and loud. An example of the use of absorptive material to affect a considerable degree of in-room noise reduction is demonstrated in Section 24.9. Such an effect is almost exclusively in the room where the absorptive material is used, since the acoustic energy transmitted to adjacent spaces is only minimally affected. Sound transmission from one enclosed space to an adjoining space through a common barrier depends almost entirely upon the type of barrier. This is discussed in detail in Chapter 25.

24.9 NOISE REDUCTION BY ABSORPTION

Equation 24.6 relates SPL and PWL as a function of distance from the source and room absorption. Example 24.1 shows the application of this equation and the result of adding absorption to the space.

EXAMPLE 24.1 An open blower is installed on the floor, away from the walls, in a large, enclosed space that is 6 m [20 ft] long, 12 m [39 ft] wide, and 4 m [13 ft] high. The floor is concrete ($\alpha = 0.01$); the walls and ceiling are painted block ($\alpha = 0.07$). The PWL values supplied by the manufacturer are 90 dB at 500 Hz and 87 dB at 2000 Hz. Calculate the SPL at distances of 5 m and 10 m [16.4 and 32.8 ft] from the blower outlet: (a) in the original room, (b) with double the absorption, and (c) with quadruple the absorption. The surface areas are: floor (6 x 12) = 72 m²; walls 2 times (6 x 4) + (12 x 4) = 144 m²; ceiling (6 x 12) = 72 m² [floor = 775 ft²; walls = 1550 ft²; ceiling = 775 ft²].

SOLUTION

1. From Equation 24.6, we have the expression:

$$SPL = PWL + 10\ \log_{10}\left(\frac{Q}{4\pi r^2} + \frac{4}{R}\right) + 0.2$$

   In our example, $Q = 2$ (see Figure 24.8).

2. The first step is to calculate the room factor R for the three absorption situations (a, b, and c).

$$R = \frac{S\bar{\alpha}}{1 - \bar{\alpha}}$$

$$S = (72) + (72) + (144) = 288\ m^2\ [3100\ ft^2]$$

From Equation 24.7:

$$\bar{\alpha}\ \text{original}\ (\bar{\alpha}_a) = \frac{\sum A}{\sum S}$$

$$= \frac{(72\ m^2 \times 0.07) + (144\ m^2 \times 0.07) + (72\ m^2 \times 0.01)}{288\ m^2}$$

$$= \frac{(5.0\ m^2) + (10.1\ m^2) + (0.7\ m^2)}{288\ m^2}$$

$$= \frac{15.8\ m^2}{288\ m^2} = 0.05$$

Therefore, if $\bar{\alpha}_a = 0.05$, then:

$$R_a = \frac{288\ m^2 \times 0.05}{1 - 0.045} = \frac{15.8\ m^2}{0.95} = 16.6\ m^2$$

If $\bar{\alpha}_b = 0.05 \times 2 = 0.1$, then:

$$R_b = \frac{288\ m^2 \times 0.1}{1 - 0.1} = \frac{29\ m^2}{0.9} = 32\ m^2$$

If $\bar{\alpha}_b = 0.05 \times 4 = 0.2$, then:

$$R_b = \frac{288\ m^2 \times 0.2}{1 - 0.2} = \frac{57.6\ m^2}{0.8} = 72.0\ m^2$$

3. Calculating SPL values and tabulating, we obtain:

	500 Hz		2000 Hz	
SPL for:	**5 m**	**10 m**	**5 m**	**10 m**
(a) Original room	85.5	85.4	82.5	82.4
(b) Double $\bar{\alpha}$	82.4	82.3	79.4	79.3
(c) Quadruple $\bar{\alpha}$	79.2	78.9	76.2	75.9

The results indicate the very important fact that doubling the absorption decreases the noise (sound pressure level) by only 3 dB. Therefore, it requires a quadrupling of absorption to make the decrease noticeable. This is an expensive procedure with diminishing returns.

If the entire space is considered to be a reverberant field, then from Equation 24.11 the sound intensity level (*IL*) can be expressed as follows:

$$IL = PWL - 10 \log \sum A + 16.3\text{dB} \qquad \text{(Eq. 24.13)}$$

where

$\sum A$ = total absorption in room, sabins (ft^2)
IL = intensity level, dB
PWL = sound power level, dB

Although increasing absorption decreases the sound/noise level, the level cannot be reduced below the free field level for that distance from the source because the free field situation corresponds to outdoors, where $\bar{\alpha} = 1.0$.

The amount of noise reduction provided by additional absorption may be determined by noise reduction (*NR*):

$$NR = IL_1 - IL_2$$

$$= 10 \log \sum A_2 - 10 \log \sum A_1$$

Therefore,

$$NR = 10 \log \frac{\Sigma_{A2}}{\Sigma_{A1}} \qquad \text{(Eq. 24.14)}$$

where

NR = noise reduction, dB
$\sum A_2$ = total absorption, final condition
$\sum A_1$ = total absorption, initial condition

From Equation 24.14, it is seen that doubling the absorption results in a noise reduction of 3 dB, since $10 \log_{10} 2 = 3$ (expressed as dB). ■

It is of interest to work out a practical example using Equations 24.13 and 24.14 and compare the results with the more precise relation given in Equation 24.5.

EXAMPLE 24.2 Referring to Figure 24.10, calculate the original sound level and the subsequent noise reduction by three steps of sound absorption treatment, assuming a completely reverberant field in the space, that is, SPL independent of location.

Original condition: Painted concrete block chamber,

10 ft × 10 ft × 10 ft

[approximately 3 m × 3 m × 3 m]

Fan sound power level:

At 500 Hz = 88 dB

At 2000 Hz = 78 dB

SOLUTION

Frequency (Hz)	Area (ft^2)	α	Total Absorption (ΣA)
500	600	0.06	36 sabins (ft^2)
2000	600	0.09	54 sabins (ft^2)

a. Sound intensity level before treatment

At 500 Hz:

$$IL = \text{sound power level} - 10 \log \sum A + 16.3\ \text{dB}$$

$$= 88\ \text{dB} - 10 \log 36 + 16.3\ \text{dB}$$

$$= 88\ \text{dB} - 15.6\ \text{dB} + 16.3\ \text{dB}$$

$$= 88.7\ \text{dB}$$

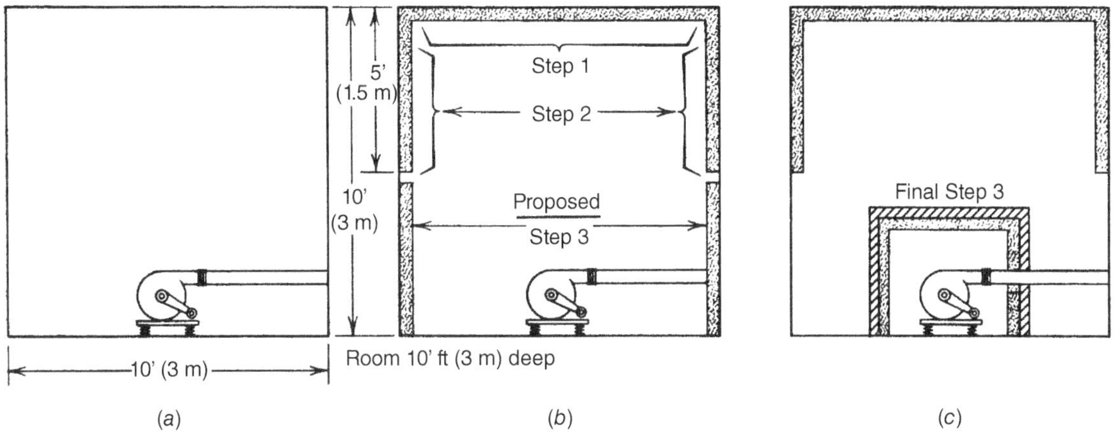

Figure 24.10 *Quieting a room (a) by addition of absorptive material is cost-effective only through Step 2 (b). Further quieting would be best accomplished locally (c), which might obviate the necessity for wall treatment (Step 2). Ceiling treatment should remain.*

At 2000 Hz:

$$IL = 78 \text{ dB} - 10 \log 54 + 16.3 \text{ dB}$$
$$= 78 \text{ dB} - 17.3 \text{ dB} + 16.3 \text{ dB}$$
$$= 77 \text{ dB}$$

b. Ceiling treatment only

At 500 Hz:

$$\alpha = 0.82$$
$$\text{added absorption} = 100\,(0.82 - 0.06)$$
$$= 76 \text{ sabins}$$
$$NR = 10 \log \frac{76 + 36}{36}$$
$$= 4.9 \text{ dB}$$

At 2000 Hz:

$$\alpha = 0.94$$
$$\text{added absorption} = 100\,(0.94 - 0.09) = 85$$
$$NR = 10 \log \frac{85 + 54}{54}$$
$$= 4.1 \text{ dB}$$

c. Ceiling and half-wall treatment

At 500 Hz:

$$\text{added absorption} = 300\,(0.82 - 0.06)$$
$$= 228 \text{ sabins}$$
$$NR = 10 \log \frac{228 + 36}{36}$$
$$= 8.7 \text{ dB}$$

At 2000 Hz:

$$\text{added absorption} = 300\,(0.94 - 0.09)$$
$$= 255 \text{ sabins}$$
$$NR = 10 \log \frac{255 + 54}{54}$$
$$= 7.5 \text{ dB}$$

d. Ceiling and full-wall treatment

At 500 Hz:

$$\text{added absorption} = 500\,(0.82 - 0.06)$$
$$= 380 \text{ sabins}$$
$$NR = 10 \log \frac{380 + 36}{36}$$
$$= 10.6 \text{ dB}$$

At 2000 Hz:

$$\text{added absorption} = 500\,(0.94 - 0.09)$$
$$= 425 \text{ sabins}$$
$$NR = 10 \log \frac{425 + 54}{54}$$
$$= 9.5 \text{ dB}$$

Summary

	IL (SPL)[a]	
	500 Hz	**2000 Hz**
Bare room	88.7	77
Ceiling treatment	−4.9 dB	−4.1 dB
	(−4.9)[a]	(−4.2)[a]
Ceiling and half-wall treatment	−8.7 dB	−7.5 dB
	(−8.7)[a]	(−7.7)[a]
Ceiling and full-wall treatment	−10.6 dB	−9.5 dB
	(−10.7)[a]	(−9.6)[a]

[a]The numbers in parentheses are the SPL differences as calculated from Equation 24.5 using Q = 2 and r = 5 ft [1.5 m].

Two important conclusions can be drawn from these results:

1. The third step of adding absorptive material is not worthwhile since negligible additional room quieting is accomplished. This clearly demonstrates the law of

diminishing returns as applied to room quieting with absorptive materials. Starting with a live room, the initial application is effective. Beyond that, additional quieting by absorption is not economical, and the same outlay would be better used in quieting the machine itself, probably with a machine enclosure, as indicated in Figure 24.10.

2. Comparing the values in parentheses that resulted from a full calculation (Equation 24.5) to those arrived at using the much simpler equations applicable only to a fully diffuse field (Equation 24.11), the differences are negligible. This means that the shortcut method can always be used in preliminary calculations and frequently for final analysis as well. The diffuse (reverberant) field calculation also lends itself nicely to a graphical solution.

Using the diffuse field Equation 24.11:

$$SPL = PWL - 10 \log \sum A + 16.3$$

and remembering that reverberation time is expressed in terms of total absorption and room volume, that is,

$$T_R = \frac{KV}{\sum A} \qquad \text{(Eq. 24.15)}$$

where K is 0.05 when room volume is in cubic feet and 0.16 when it is in cubic meters. Using these equations, we can then rewrite Equations 24.11 and 24.12 in terms of reverberation time and room volume:

For room volume in ft³:

$$SPL - PWL = -10 \log \frac{0.05V}{T_R} + 16.3 \qquad \text{(Eq. 24.16)}$$

For room volume in m³:

$$SPL - PWL = -10 \log \frac{0.16V}{T_R} + 6.0 \qquad \text{(Eq. 24.17)}$$

These two equations are plotted graphically in Figures 24.11 and 24.12. The charts are accurate for live-room, diffuse (reverberant) fields and provide a close approximation for other situations. An illustrative example will demonstrate the use of the charts. ■

EXAMPLE 24.3 A generator in a 14,000-ft³ [396-m³] room has a PWL of 95 dB (re: 10^{-12} W) at 400 Hz.

Using the chart in Figure 24.11, find the SPL at that frequency. Assume that the sound field in the room is reverberant. The room reverberation time is 1.0 s at 400 Hz.

Check the result using the chart in Figure 24.12.

SOLUTION

I-P units: Enter the chart in Figure 24.11 at 14,000 ft³ and draw a vertical line until it intersects the T_R = 1.0-second sloping line. Extend a line from that intersection point to the X-axis and read −12 dB.

Then, since

$$SPL - PWL = -12 \text{ dB}$$

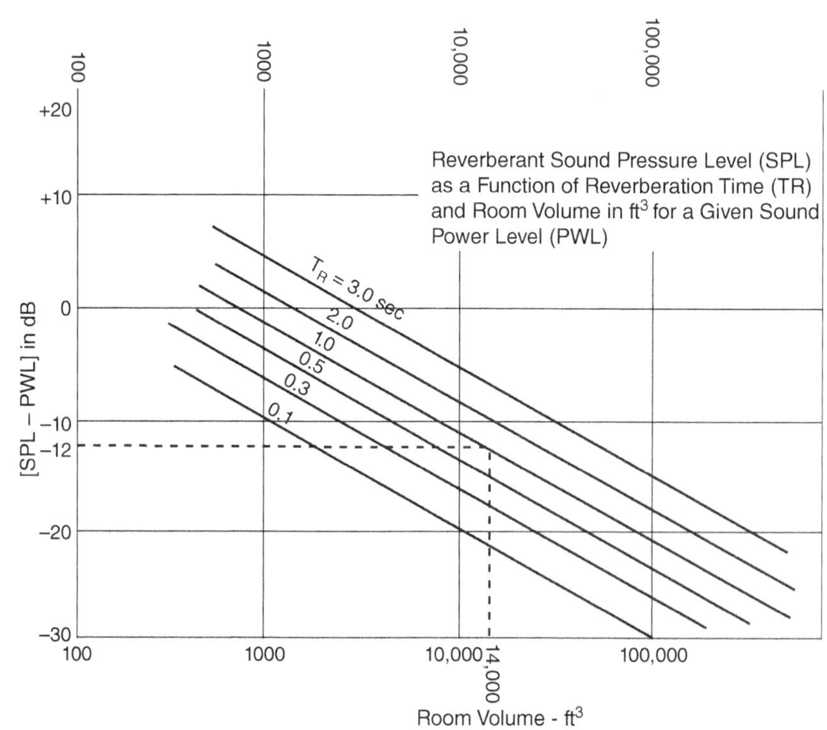

Figure 24.11 I-P chart for determining the reverberant sound pressure level (SPL) when the sound power level (PWL) of an item of equipment is known. Room volume in cubic feet and reverberation time in seconds are the variables.

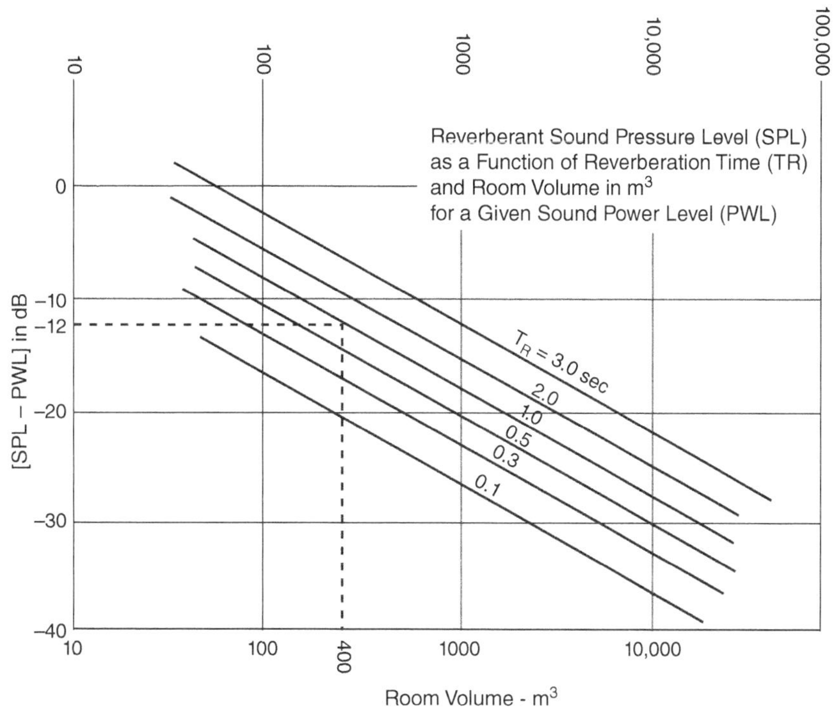

Figure 24.12 *SI chart for determining the reverberant sound pressure level (SPL) when the sound power level (PWL) of an item of equipment is known. Room volume in cubic meters and reverberation time in seconds are the variables.*

and PWL is given as 95 dB,

$$SPL = 95 - 12 = 83 \text{ dB}$$

SI units: The room volume is approximately 400 m^3. Enter the chart in Figure 24.12 and extend a line upward to T_R = 1.0-second and left to the X-scale. Read −12 dB. The two answers check, as they should. ■

24.10 NOISE REDUCTION COEFFICIENT

The last column in Table 24.1 is labeled NRC—noise reduction coefficient. This figure is the arithmetic average of the absorption coefficients at 250 Hz, 500 Hz, 1000 Hz, and 2000 Hz. The name is ill chosen because it cannot be used directly for noise control efforts, as the words seem to imply. It is simply useful as a single-number indicator of the midband effectiveness of a porous absorber. For critical acoustical situations, or if high- and low-end frequencies are of interest, NRC is nearly useless, and a detailed analysis of absorption over the entire frequency range must be made. Two materials with the same NRC can perform quite differently, since NRC is an average, and few materials have a flat absorption characteristic. Losing sight of such a difference by reliance upon NRC is not a good move. Even more importantly, sound absorption is not the same as noise reduction. An acoustical ceiling tile, which exhibits high sound absorption, is a poor noise control barrier.

ROOM DESIGN

24.11 REVERBERATION CRITERIA FOR SPEECH ROOMS

The overriding criterion for speech is intelligibility. Since speech consists of short, disconnected sounds 30–300 ms in length (see Section 24.4), among which are high-frequency, low-energy phonemes, the ideal room must ensure the ear's undistorted reception of these phonemes. This requires keeping reverberation to a minimum. We can obtain a good approximation of the subjective feeling of liveness of a room, for purposes of speech, from the relation:

$$T_R \text{ (speech)} = 0.3 \log \frac{V}{10} \qquad \text{(Eq. 24.18)}$$

where

T_R (speech) = optimum reverberation time in seconds, for speech
V = room volume, m^3

For instance, a typical classroom might have a volume of 150 m^3 [5300 ft^3]. Optimum reverberation time is:

$$T_R = 0.3 \log 15 = 0.35 \text{ s}$$

Reverberation times longer than this would sound live, shorter ones dead and flat. Indeed, an increase of

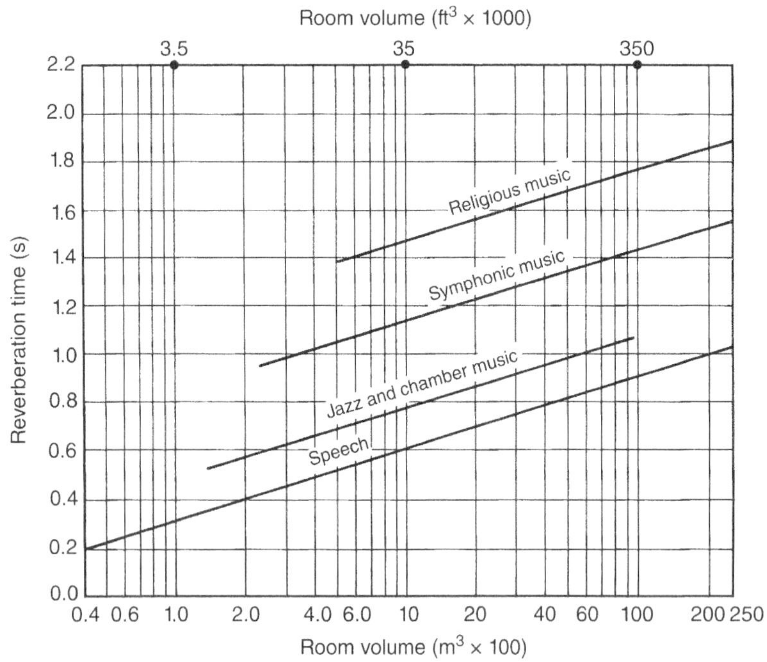

Figure 24.13 *Optimum reverberation times for the frequency range 500–1000 Hz as a function of room volume. (The cubic foot volume figures are based upon the approximate conversion factor of 35 ft³/m³). (From Magrab, 1975.)*

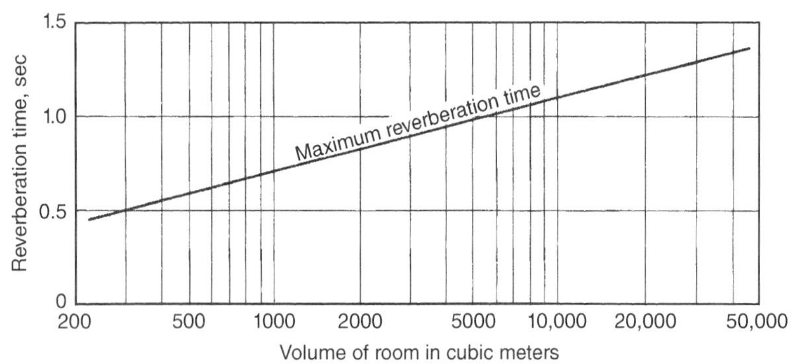

Figure 24.14 *Maximum recommended reverberation time for speech in auditoriums and lecture halls. (From Brüel, 1951.)*

20% in reverberation time would make the room excessively live and "boomy" and would negatively affect speech intelligibility. Figure 24.13 gives optimum midfrequency reverberation times as a function of room size and use. Figure 24.14 gives maximum reverberation times for speech in large rooms. For good speech intelligibility, reverberation time should remain essentially flat down to 100 Hz.

The reflected sounds associated with reverberation can have either a salutary or a deleterious effect. The ear cannot distinguish between sounds that arrive within a maximum of 50 ms (1/20 s) of each other (some authorities use 40 ms, i.e., 1/25 s). Sounds arriving within this time reinforce the direct-path signal and appear to come from the source. Sounds arriving after this time are apprehended as a fuzzy echo or elongation of the sound, reducing intelligibility and directivity.

Since the range of 40–50 ms corresponds at 1130 fps [344 m/s] to 45–56 ft [13.7–17.2 m], a speech room should be so arranged that the difference between the first reflection path and the direct path is no greater than 56 ft [around 17 m], and preferably 46 ft [14 m] or less at midfrequency (Figure 24.15). For more details concerning this and related factors on reflected paths, refer to Section 24.13.

Too low a reverberation time (very high absorption, minimum reflection) is also undesirable for the following reasons.

1. It limits the size of the room to that which can be covered by direct sound only.
2. It is disturbing to the speaker, since absence of reflection prevents him or her from gauging the proper voice level, which tends to cause excessive effort (shouting, as when outdoors).

Thus, proper design of a room for speech is a compromise between the need for some reflection and the desire to minimize reflection to preserve intelligibility.

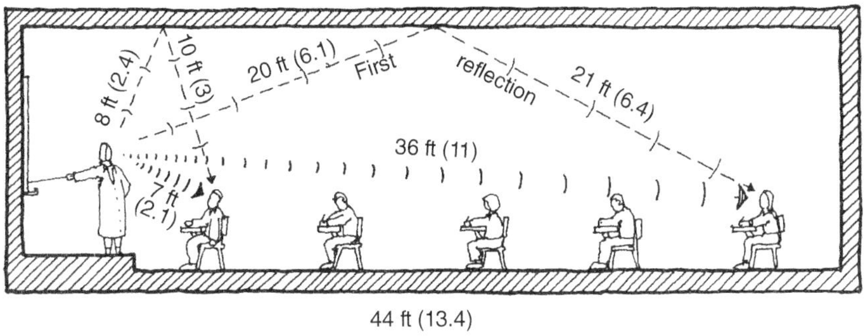

Figure 24.15 *Sound paths in a typical medium-sized lecture room. Note that for both extremes of listener position, the maximum path-length difference between direct and first reflection is 11 ft [3.4 m]. Thus, the signal is reinforced, and intelligibility should be excellent if room absorption is provided to limit reverberation time to about 0.5 second maximum (see Figure 24.14). Numbers in parentheses are dimensions in meters.*

24.12 CRITERIA FOR MUSIC PERFORMANCE

Adequate design for a music space requires recognition of the following:

1. Large-volume spaces require direct-path sound reinforcement by reflection.
2. Relatively long reverberation time is needed to enhance the music—the exact amount depending upon the type of music (Figures 24.13 and 24.16). Designers should keep in mind that reverberation time recommendations vary as much as 100% among respected sources.
3. It is generally agreed that reverberation time should vary inversely with frequency (i.e., T_R should be longer at lower frequencies [than the midfrequency recommendation] and shorter at higher frequencies). The longer T_R at low frequencies adds fullness

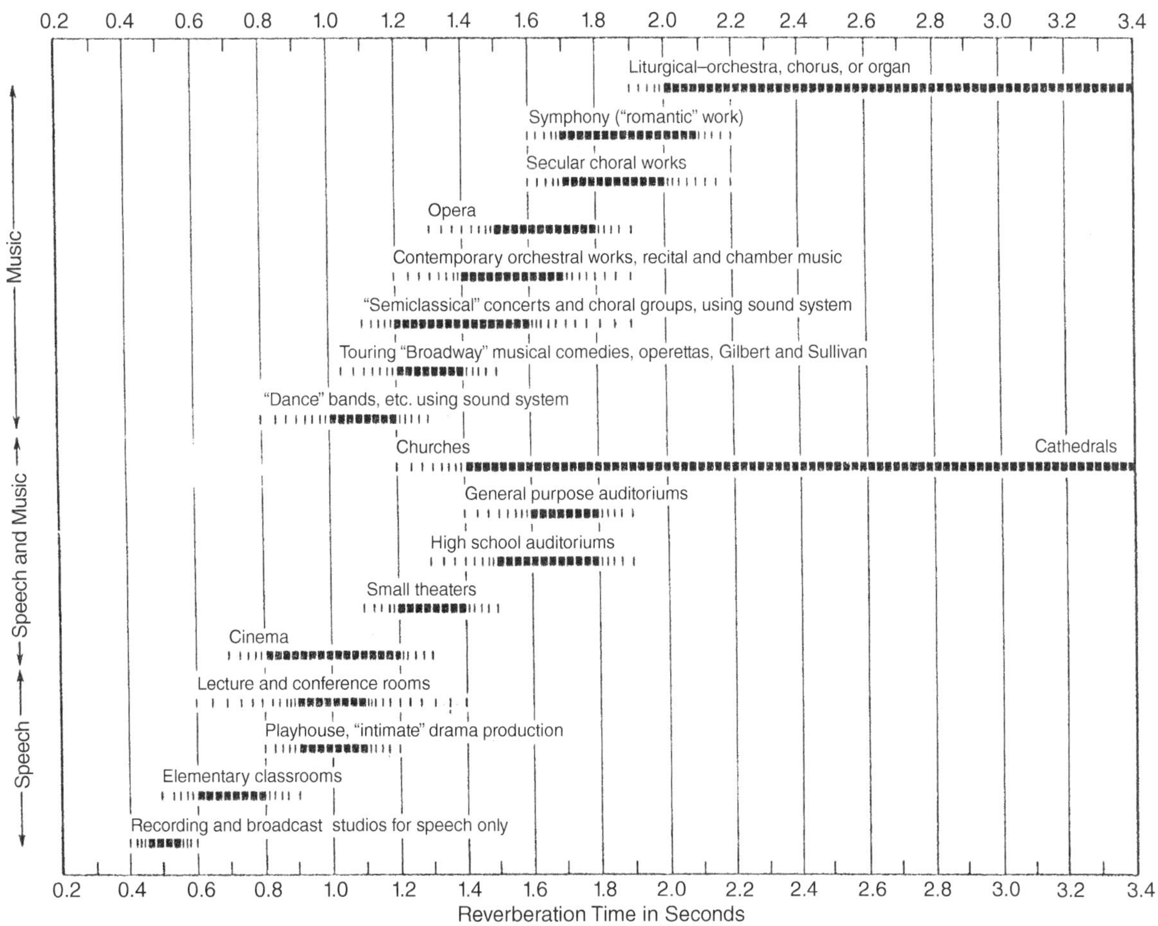

Figure 24.16 *Optimum reverberation times at midfrequencies (500–1000 Hz) for various types of facilities. The wide range shown reflects the effects of room proportions, shape, and volume on the optimum T_R, as well as differences of opinion among designers.*

to music and "body" to speech. Thus, T_R at 100 Hz should be, according to most researchers, 35% to 75% longer than T_R at the center frequencies.

4. Short T_R at upper frequencies adds directivity to the music. With large ensembles, directivity gives the sense of depth and instrument location necessary for proper appreciation. This is often referred to as clarity or definition in music. With a solo instrument, this problem is reduced.
5. Brilliance of tone is primarily a function of high-frequency content. Since these frequencies are most readily absorbed, a good direct path must exist between the sound source and the listener. Since our eyes and ears are close together, a good sound path exists when a good vision path exists. At the other end of the spectrum, lack of sufficient bass expresses itself as a loss of "fullness," which is often caused by resonant absorption.

The actual design of a music performance space is a very complex procedure involving extensive calculations of absorption, reverberation time, and ray diagramming, as well as juggling of materials, dimensions, and wall angles. Simulation techniques and acoustic models are also often employed.

Recent research and simulation studies of concert and recital halls have demonstrated that the sensation of fullness of music, or what is today referred to as sound envelopment, is enhanced by lateral reflections that reinforce the direct signal. It has also been found that the subjective judgment of reverberance is more strongly affected by early decay time (the time required for a 10-dB decrease in signal strength) than by the conventional 60-dB decay time. Finally, crispness or clarity of the music (particularly important in recital halls and for chamber music) depends upon reflections arriving within 40–70 ms. All these factors are considered both in the original design and in the often lengthy "tuning" process of a space intended for music performance. Most modern design solutions also use movable reflector panels and other active variables. After construction is completed, extensive tests are conducted and field adjustments are made. Application of the commissioning process to such spaces seems mandatory.

24.13 SOUND PATHS

Sound paths refer to the routes that sound waves travel from a source to a receiver. These "paths" may be direct or could involve reflections, refractions, and diffractions as the sound waves interacts with various surfaces in an environment. Ideally, every listener in a lecture hall, theater, or concert hall should hear the speaker or performer with the same degree of loudness and clarity. Since this is impossible by direct-path sound, the essential design task is to devise methods for reinforcing desirable reflections and minimizing and controlling undesirable ones. Normally only the first reflection is considered in ray diagramming (discussed in Section 24.14(a)) since it is the strongest. Second and subsequent reflections are usually attenuated to the point that they need not be considered except for the special situations of flutter, echoes, and standing waves, discussed in the following sections.

(a) Specular Reflection

Specular reflection occurs when sound reflects off a hard, polished surface. This characteristic can be used to good advantage to create an effective image source. In ancient Greek and Roman theaters, seats were arranged on a steep conical surface around the performers. The virtue of this arrangement (Figure 24.17*a*) is that the sound energy travels to each location with minimal attenuation. The same effect can be accomplished by placing the sound source above the seats. This is not practical physically, but it can be accomplished effectively using a reflecting panel (Figure 24.17*b*). The panel dimension must be at least one wavelength at the lowest frequency under consideration. Figure 24.18 is a chart for converting from frequency to wavelength in feet and meters.

(b) Echoes

Echoes occurs when sound paths are such that sound waves arrive to a listener's ear after a noticeable delay from the direct sound (Figure 24.19). As explained in Section 24.11, a clear echo is caused when reflected sound at sufficient intensity reaches a listener more than 50 ms after he or she has heard the direct sound. (Some authorities place this figure as high as 80 ms.) Echoes, even if not distinctly discernible, are undesirable. They make speech less intelligible and make music sound "mushy." The relative undesirability depends upon the time delay and loudness relative to the direct sound, which, in turn, are dependent upon the size, position, shape, and absorption of the reflecting surface.

Typical echo-producing surfaces in an auditorium are the back wall and the ceiling above the proscenium. Chapter 25 shows these problems and suggests remedies. Note that the energy that produced the echoes can be redirected to places where it becomes useful reinforcement. If echo control by absorption alone were used on the ceiling and back wall, that energy would be wasted. The rear wall, since its area cannot be reduced too far, may have to be made more sound-absorptive to reduce the loudness of the reflected sound.

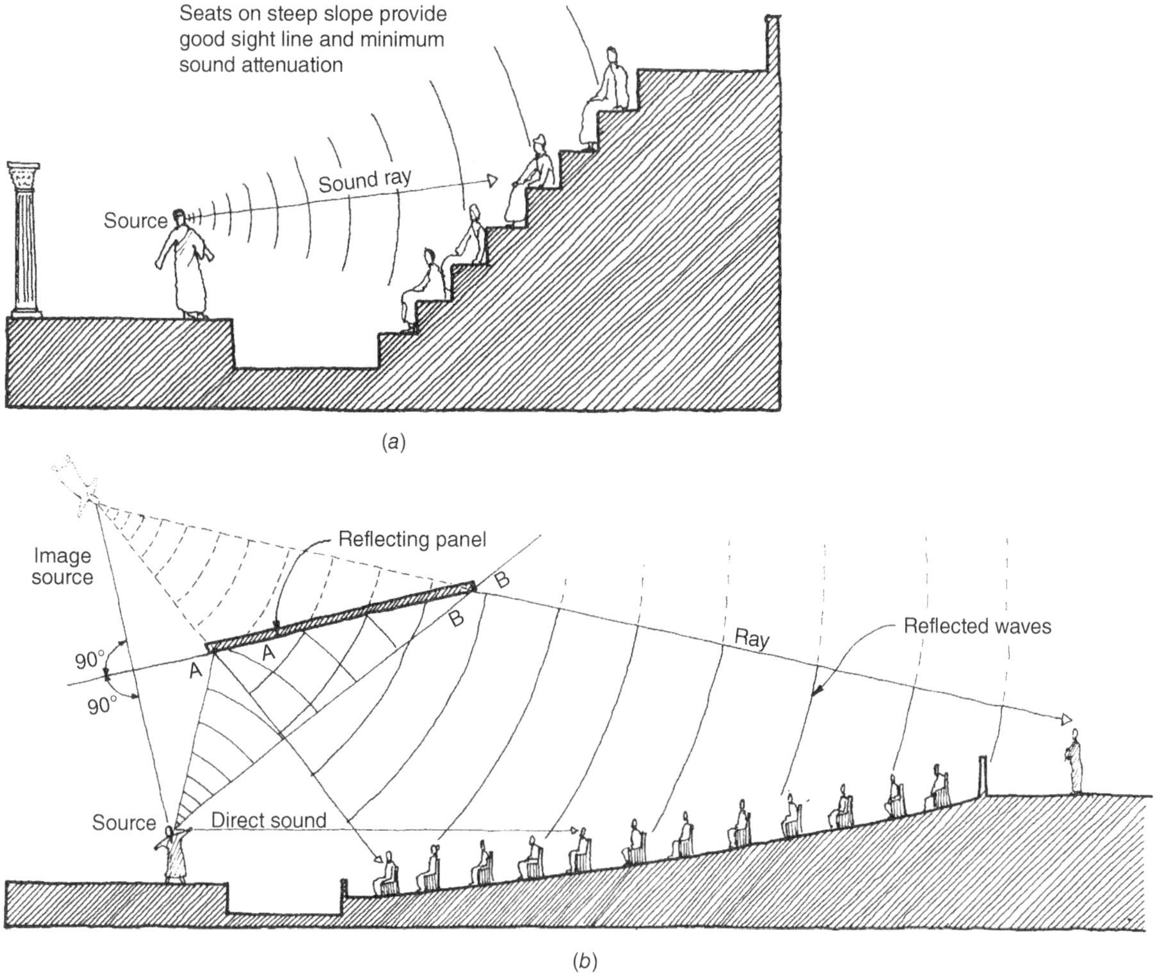

Figure 24.17 *Use of an angled reflector panel (b) creates an image source that stands in approximately the same relation to the audience as does the performer (a) in the classic Greek theater.*

(c) Flutter

A flutter, perceived as a buzzing or clicking sound, comprises repeated echoes traversing back and forth between two nonabsorbing parallel (flat or concave) surfaces. Flutters often occur between shallow domes and hard, flat floors. The remedy for a flutter is either to change the shape of the reflectors or their parallel relationship, or to add absorption. The solution chosen will depend upon reverberation requirements, cost, and aesthetics.

(d) Diffusion

Diffusion is the converse of focusing and occurs primarily when sound is reflected from convex surfaces. A degree of diffusion is also provided by flat horizontal and inclined reflectors (Figure 24.20). In a diffuse sound field, the sound level remains relatively constant throughout the space, an extremely desirable property for musical performances.

(e) Focusing

Concave domes, vaults, or walls will focus reflected sound into certain areas of rooms. This has several disadvantages. For example, it will deprive some listeners of useful sound reflections and cause hot spots at other audience positions (Figure 24.21*a*).

(f) Creep

Creep describes the reflection of sound along a curved surface from a source near the surface. Although the sound can be heard at points along the surface, it is inaudible away from the surface. Creep is illustrated in Figure 24.21*b*.

(g) Standing Waves

Standing waves and flutter are very similar in principle and cause but are heard quite differently. When an impulse (such as a hand clap) is the energy source, a

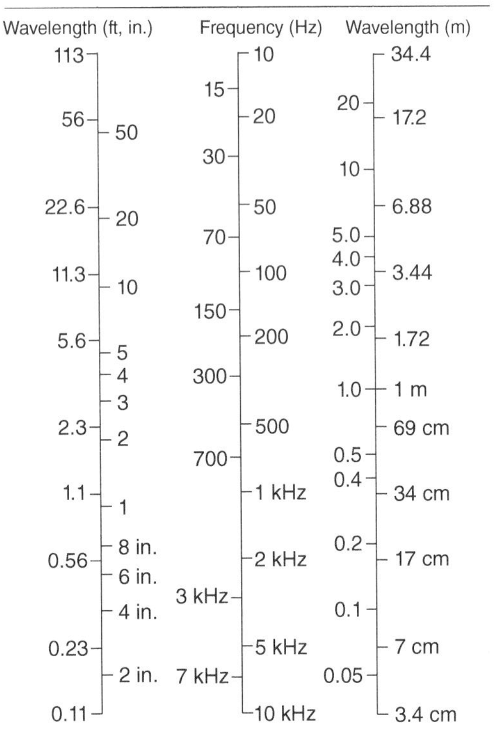

Figure 24.18 *Nomograph for determining wavelength in feet or meters, given frequency in hertz, or vice versa. Speed of sound is taken as 1130 ft/s [344 m/s]. To use it, hold a straightedge horizontally across the nomograph and read the values directly.*

flutter will occur between two highly reflective parallel walls. It is perceived as a slowly decaying buzz. When a steady, pure tone is the source, a standing wave will occur, but only when the parallel walls are spaced apart at some integral multiple of a half-wavelength.

When parallel walls are exactly one-half wavelength apart, the tone will sound very loud near the walls and very quiet halfway between them. This is because at the center, the reflected waves traveling in one direction are exactly one-half wavelength away from those traveling in the other direction, and are thus equal but opposite in pressure, which results in total cancelation. In other rooms, standing waves are noted as points of quiet and loudness in the room. Standing waves are important only in rooms that are small with respect to the wavelengths generated (smallest room dimension <30 ft [< 9 m] for music or < 15 ft [< 4.5 m] for speech).

Another effect of standing waves, called resonance, is the accentuation of a particular frequency that will cause a standing wave. Thus, if one speaks (or plays a musical instrument) while standing near a wall of a room about 8 ft × 8 ft [2.4 m × 2.4 m] in size, one will notice an abnormal and sometimes unpleasant loudness in the sound at about 280 Hz.

Similarly, when a musician plays a scale, one note may seem far louder than the adjacent ones,

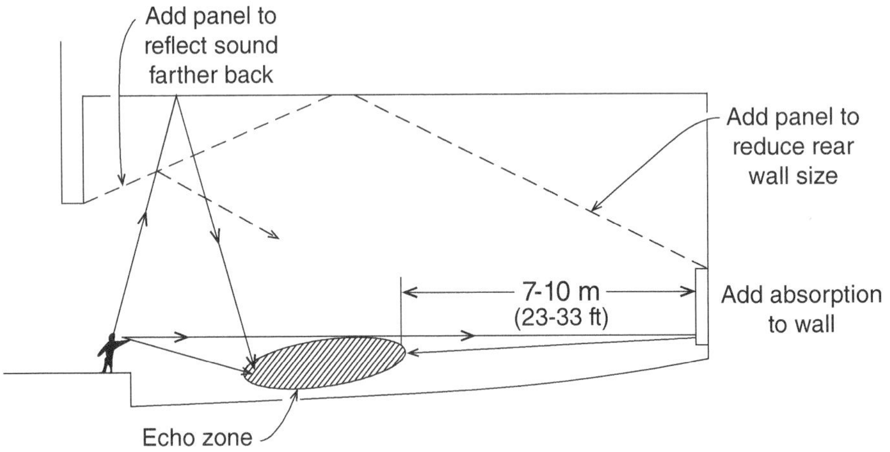

Figure 24.19 *Auditorium section showing the causes and remedies for two typical echoes.*

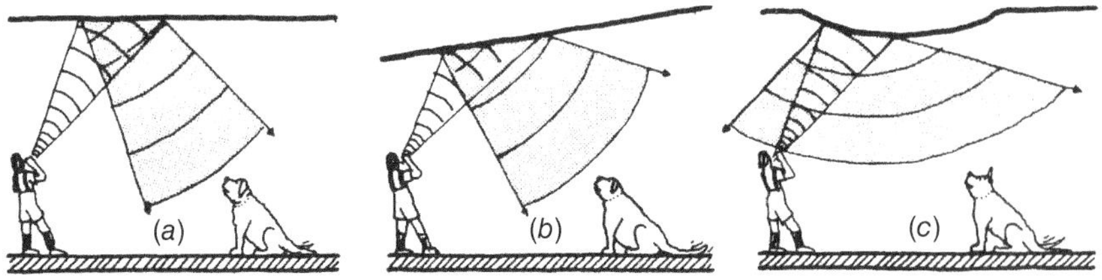

Figure 24.20 *Sound diffusion can be created with reflectors of different shapes, including horizontal flat (a), inclined flat (b), and convex (c) reflectors. Diffusion improves from (a) to (c).*

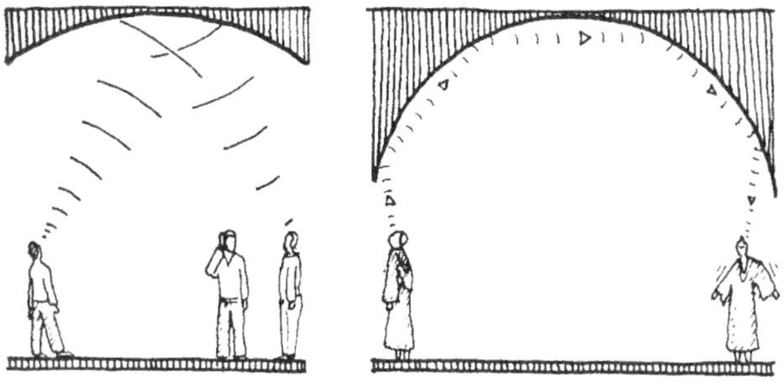

Figure 24.21 Two undesirable phenomena in room acoustics.

and listeners in one section of the room may hear a quality of sound different from that heard by those in other sections. This effect must be avoided for music performance but is merely an annoyance in rooms designed for speech use. This is one of the reasons that one finds music rehearsal rooms, broadcast studios, and the like with nonparallel walls and undulating ceilings. These irregularities prevent most of the undesirable effects described. However, since rooms with nonparallel walls and undulating ceilings are unacceptable for most applications, and since standing waves and other resonant phenomena are related to room geometry, it is possible to calculate room proportions with conventional geometries that will minimize these effects. Figure 24.22 shows such room proportions, which are applicable to the bass frequencies (i.e., below 100 Hz) that are the problematic frequencies. At higher frequencies (and in large rooms) standing waves may be common, but the overall effect is much less disturbing and frequently hardly noticeable.

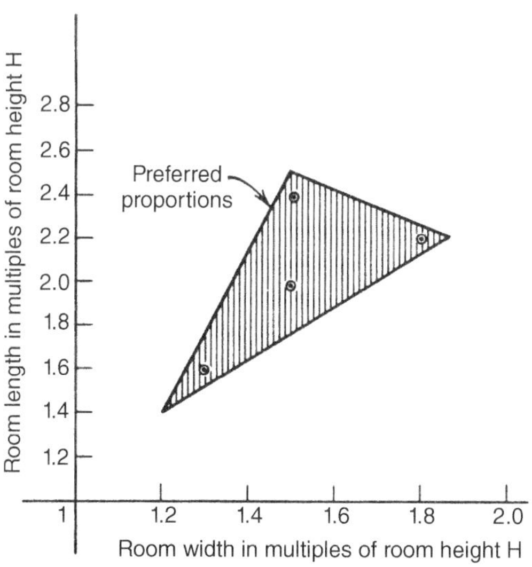

Figure 24.22 The triangular shape contains the room proportions recommended for avoidance of disturbing low-frequency acoustic phenomena such as flutter, standing waves, and resonances, particularly at frequencies below 100 Hz. The points shown at the three apexes were averaged to obtain the center point with a proportion of L = 2H; W = 1.5H; H = H.

24.14 RAY DIAGRAMS

Ray diagramming is a design procedure for analyzing the reflected sound distribution throughout a hall, using the first reflection only. Figure 24.23 shows a ray diagram. The rays are drawn normal (perpendicular) to the spherically propagating sound waves. Specular reflection is assumed—that is, at reflecting panels, the angles of the incident and reflected rays are always equal. Thus, in addition to direct sound, each listener is receiving reflected sound energy. It is as though there were additional sound sources, the real one and numerous image sources. Figure 24.23 shows the application of a ray diagram to the design of a lecture hall. In Figure 24.23*a*, the stage height and seating slope are arranged to provide good sight lines, and the ceiling height is established by reverberation requirements, aesthetics, cost, and so on. Less than half of the ceiling is providing useful reflection. Dividing the ceiling into two panels (Figure 24.23*b*) allows people in the rear of the room to perceive the direct source plus two image sources—increasing the useful reflecting area by 50%. In Figure 24.23*c*, the shape has been further refined to include a lighting slot and a loudspeaker grille.

As mentioned previously, the sensation of musical envelopment that is so satisfying to a listener, particularly when dealing with large groups playing symphonic music, is due in large measure to reflections received from the side (lateral reflections). This being so, ray diagramming, which was originally performed on room sections to determine ceiling shapes and the proper placement of ceiling-hung reflector panels, is now done in plan as well. This is particularly important in fan-shaped auditoriums, where a canted wall does not provide useful lateral reflections. The solution to this problem is to build a sawtooth wall or to

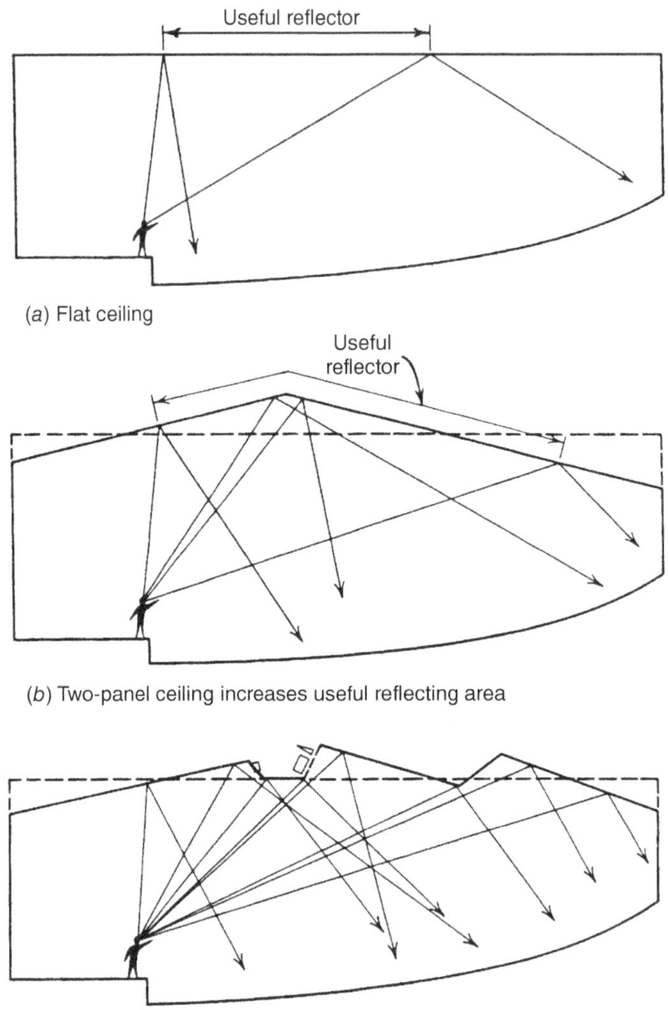

(*a*) Flat ceiling

(*b*) Two-panel ceiling increases useful reflecting area

(*c*) Multifaceted ceiling incorporates lighting fixtures and loudspeakers

Figure 24.23 *Section through a typical lecture room showing the use of ray diagrams.*

use reflector panels along the wall that will provide the desired lateral reflection. Nonhorizontal ceiling reflector panels can also provide a measure of lateral reflection, particularly in balconies.

Although they are a useful design tool, ray diagrams have certain restrictions. Design solutions always will require compromise between ray diagram results for various speaking positions on a stage. Thus, a paraboloid may be a perfect shape for one source position but a very poor shape for other positions.

24.15 AUDITORIUM DESIGN

Auditorium is a general term used to describe a space where people sit and listen to speech or music. Acoustical design of an auditorium includes room acoustics, noise control, and sound system design. Noise control is covered in Chapter 25, and sound systems are discussed in Sections 24.16 through 24.18. If the program for the auditorium includes activities that require different acoustical environments (as is frequently the case), then it must be decided early whether the acoustics will be a compromise between the program extremes or adjustable for various activities. Acoustical environments can be altered by changing the space volume, moving reflecting surfaces, and adding or subtracting sound-absorbing treatment. Figure 24.24 illustrates several examples of acoustical adjustability.

Factors that influence acoustical design include: audience size, range of performance activities, and sophistication of the potential audience and performers. A small school auditorium (Figure 24.25*a*) and a professional theater (Figure 24.25*b*) will have widely divergent demands from both audiences and performers. The audience size determines the basic floor area of an auditorium, assuming no balconies. Once this area has been fixed, the volume of the room is developed according to reverberation requirements of the space.

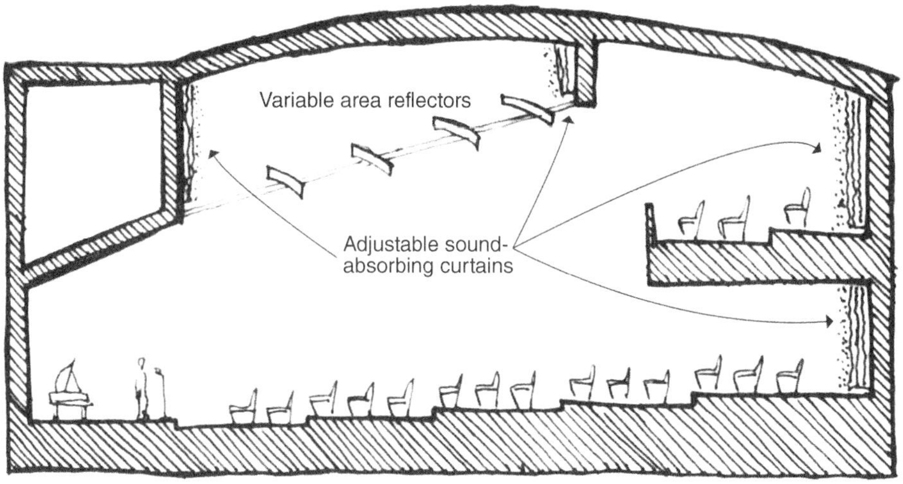

Figure 24.24 *Adjustable acoustic elements in an auditorium. (Redrawn by Zeta Fernando.)*

(*a*)

(*b*)

Figure 24.25 *(a) An institutional, shoebox-style auditorium. University of Oregon's Beall Hall, designed by Ellis F. Lawrence, is a simplified rendition of the Boston Symphony Hall. The combination of pyramidal sound diffusers, a coffered ceiling, and sound-absorptive wall curtains provides optimal sound conditions for performances within the School of Music. (b) Auditorium with a variety of sound-absorbing and sound-reflecting materials. The Hult Center's Silva Concert Hall in Eugene, Oregon, features convex wall and ceiling panels, diffusing panels, and sound-absorptive upholstery to reduce reverberation and increase sound clarity. (a: © Karen Tse; used with permission; b: © Tyler Mavichien; used with permission.)*

ACOUSTICS

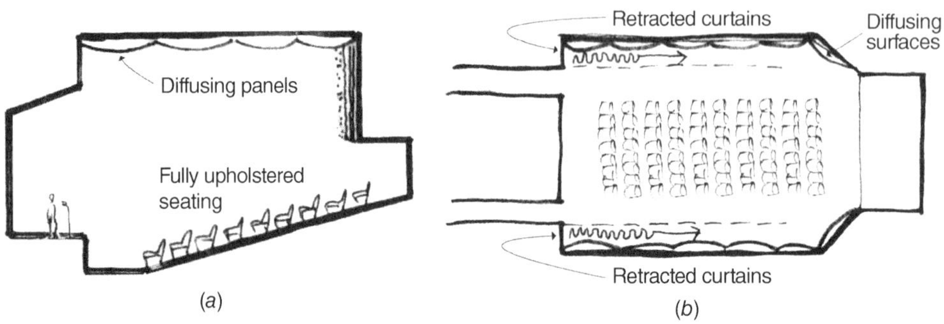

Figure 24.26 Auditorium analysis illustrating the use of surface treatments for control of reflections and reverberation.

Figure 24.26 shows a typical auditorium in plan and section. The shape of the wall and ceiling surfaces is developed to provide proper distribution of sound and eliminate focusing or echoes. Essential characteristics of the design include the following features:

1. Ceiling and side walls at the front of the auditorium distribute sound to the audience. These surfaces must be close enough to the performers to minimize time delays between direct sound and reflected sound.
2. Ceiling and side walls provide diffusion.

Acoustics must be considered in the selection of materials used in an auditorium. All auditoriums use both sound-reflecting and sound-absorbing materials. Since the largest area of sound-absorbing material in any auditorium is the audience, the difference in acoustical characteristics that occurs without an audience may be minimized by using fully upholstered seating.

Chairs with fully upholstered seats and backs, covered in an open-weave material, will have absorption characteristics approximating those of an audience. Using the auditorium in Figure 24.26 as an example, the reverberation characteristics of an auditorium with various materials may be examined—as shown in the following analysis in Table 24.2. In one configuration (A), the room use is assumed to be for

TABLE 24.2 Simplified Calculations of Midfrequency (500 and 1000 Hz) Average Reverberation Time

$$\text{Reverberation time } (T_R) \text{ in seconds} = \frac{0.05 \times \text{volume in ft}^3}{\text{total absorption in sabins}}$$

$$\left[\text{Reverberation time } (T_R) \text{ in seconds} = \frac{0.16 \times \text{volume in m}^3}{\text{total absorption in sabins}}\right]$$

Room volume = 155,500 ft³ [4404 m³]

	Configuration A: more reverberant condition; with curtains retracted				**Configuration B:** less reverberant condition; with curtains exposed		
	Area, ft² [m²]	**Alpha (α)**	**Absorption, ft² [m²]**		**Area, ft² [m²]**	**Alpha (α)**	**Absorption, ft² [m²]**
Seating and stage; with audience and performers	3323 [309]	0.92	3060 [284]	Seating and stage; with audience and performers	3323 [309]	0.92	3060 [284]
Wall area; concrete block	8000 [743]	0.2	1600 [149]	Wall area; concrete block, balance covered by curtains	3600 [334]	0.2	720 [67]
				Curtains	4400 [409]	0.45	1970 [184]
Lower rear wall; permanent sound-absorbing treatment	450 [42]	0.88	396 [37]	Lower rear wall; permanent sound-absorbing treatment	450 [42]	0.88	396 [37]
Total absorption			**5056 [470]**	**Total absorption**			**6146 [572]**
Reverberation time (T_R): more reverberant condition			$\frac{0.05 \times 155,500}{5056}$ = 1.5 sec	Reverberation time (T_R): less reverberant condition			$\frac{0.05 \times 155,500}{6146}$ = 1.2 sec
			$\left[\frac{0.16 \times 4404}{470}\right]$ = 1.5 sec				$\left[\frac{0.16 \times 4404}{572}\right]$ = 1.2 sec

music performances; the only sound absorption is that provided by the audience and seating. In the second configuration (B), absorptive curtains are installed along the rear wall and a portion of the side wall; this configuration might be used for lectures in a room that is adjustable between speech and music configurations. A third configuration might use permanent sound-absorbing treatment installed on the ceiling and rear and side walls; because of its low reverberation time, this configuration would be appropriate only for movies and lectures, not for music activities.

These simple examples indicate the effect of changes in the amount of absorption on the characteristics of a room. Adjustable treatments permit the characteristics of the room to be modified to any point between the extremes to meet the acoustic program requirements of a multipurpose hall.

Existing spaces may require remedial treatment to eliminate unwanted phenomena such as focusing and echoes, as shown in Figure 24.27. In the first example, the surface of the dome was covered with sound-absorbing material to eliminate focusing; in the second, sound-absorbing treatment was applied to a curved rear wall to eliminate an echo. Such treatment will also affect the reverberation characteristics.

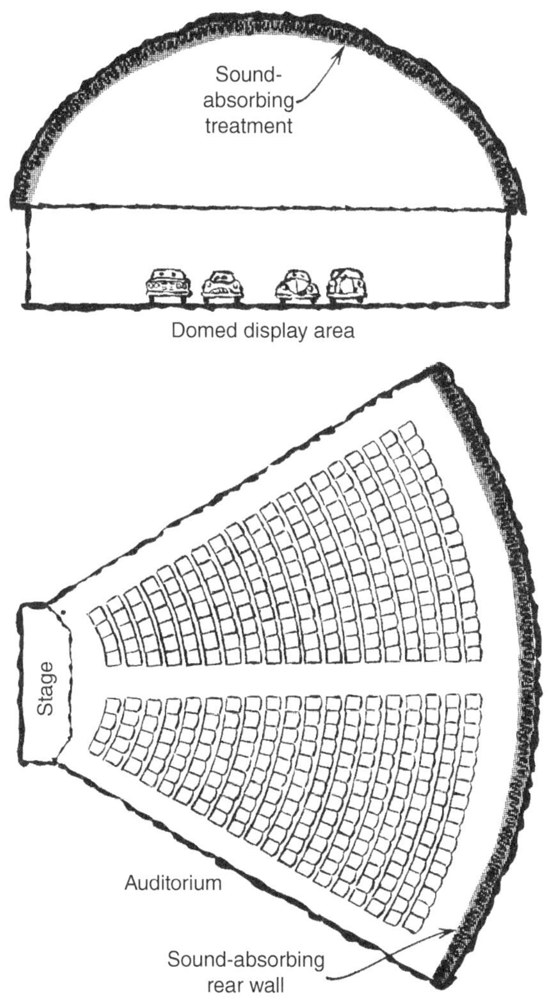

Figure 24.27 *Sound-absorbing treatment used to eliminate focusing from the dome and the curved auditorium wall.*

SOUND REINFORCEMENT SYSTEMS

24.16 OBJECTIVES AND CRITERIA

The purpose of a sound reinforcement system is just what the name indicates—to reinforce sound levels that would otherwise be inadequate. An ideal sound-reinforcing system will give the listener the same loudness, quality, directivity, and intelligibility as would result if the source of sound were immediately adjacent to the listener—roughly 2–3 ft [0.6–0.9 m] for speech, and farther for music (depending upon the type and number of instruments). In this ideal system, there are several factors that should remain constant—loudness, intelligibility, quality, and so on. Of these factors, loudness and intelligibility have previously been discussed. Quality means that the frequency response is linear so that reproduced sound bears the same relation to its frequency components as the original sound. Quality is then field-adjusted by voicing or equalization, as discussed in Section 24.17.

Directivity is the characteristic whereby the sound appears to be coming from the originating source, that is, the loudspeakers should be directionally "invisible," and the listener must have the impression of hearing the actual source. Note that sound systems cannot correct a poor acoustic design completely, although they can improve a bad situation.

Generally, sound systems will be required in spaces larger than 50,000 ft^3 [$\approx$1400 m^3]. In terms of occupancy, this volume translates into 550 persons in lecture rooms (15-ft [4.6-m] average ceiling height and 6 ft^2 [0.6 m^2] per person) and 325 persons in theaters (20-ft [6.1-m] average ceiling height and 7.5 ft^2 [0.7 m^2] per person). In such a room, a normal speaking voice can maintain a sound pressure level of only 55–60 dB, depending upon room design, voice strength, and frequency. With background noise at NC 30 (see Chapter 25), a speaker will be heard; at higher background noise levels, intelligibility will suffer.

24.17 COMPONENTS AND SPECIFICATIONS

All sound systems consist of three basic elements: input devices, amplifier(s), and loudspeaker systems.

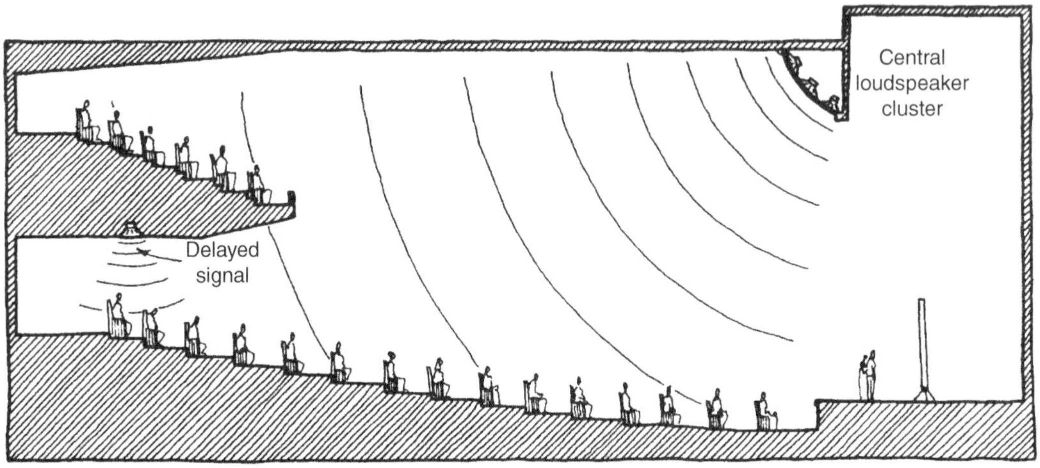

Figure 24.28 Loudspeaker system using a delayed signal to the under-balcony area.

(a) Input

Input usually means a microphone, a source of commercial broadcast material of various types, and a means of reproducing recorded material in all common commercial formats. Connections to local computers and computer networks are available in sophisticated systems.

(b) Amplifiers and Controls

Amplifiers must be rated to deliver sufficient power to produce intensity levels of 80 dB for speech, 95 dB for light music, and 105 dB for symphonic music. This assumes a maximum background noise level of 60 dBA. Thus, an 80-dB speech intensity will be 20 dB higher—or four times as loud as the noise level. If the noise level is known to be below 60 dBA maximum, amplifier and loudspeaker power ratings can be reduced accordingly. The amplifier should carry technical specifications for signal-to-noise ratio, linearity, and distortion. Exact values depend upon the application and are left to the acoustics specialist or sound engineer to supply.

In addition to the usual volume, tone mixing, and input/output selector controls, the amplifier must contain special equalization controls for signal shaping. These are highly critical filter networks that, by selective amplification and attenuation of portions of the overall audio frequency spectrum, voice or equalize a system after installation. Equalization is mandatory in a good sound system; without it, the system will howl, sound rough, give insufficient and poorly distributed gain and sound level, and generally produce bad sound. Essentially, voicing tailors the system to the acoustic properties of the space. A system not equipped for equalization is not a professional system, and the results will verify it. Furthermore, the specification must provide for the services of a competent sound engineer to perform the equalization after construction and system installation are complete.

Another control frequently required in theater systems is a delay mechanism or circuit that can introduce a time delay into a signal being fed to a loudspeaker. Figure 24.28 shows a sound system that covers a majority of an auditorium from a central loudspeaker cluster. The under-balcony seating areas are hidden from the central cluster and receive reinforced sound from distributed loudspeakers in the under-balcony soffit. To provide directional realism, the signal to the under-balcony loudspeakers must be delayed to allow the weaker signal from the central speakers to arrive first. Delay is necessary because electrical signals travel at the speed of light, whereas sound is much slower (one-millionth of light speed, approximately). With this arrangement, sound will seem to come from the source, and the directivity so necessary to realism is maintained.

(c) Loudspeakers

These are the heart of any sound system and must be of the same high quality as the remainder of the system. Indeed, scrimping on system costs will show up much more quickly in loudspeaker performance than in any other component. Selection of speakers is a complex technical task beyond the scope of our discussion. Nevertheless, a few general remarks are in order. The best systems with traditional components use central-speaker arrays consisting of high-quality, sectional (multicell), directional, high-frequency horns and large-cone woofers. These assemblies are frequently very large, and the architect should be aware of the dimensions that must be accommodated. Smaller units with folded horns can be used, at a sacrifice in low-frequency response. If only speech is to be reproduced, these units will perform adequately. Distributed systems use small (4- to 12-in. [100- to 305-mm]) diameter low-level speakers, ceiling-mounted and firing directly down.

Recent developments in loudspeaker design have produced units much smaller than those previously required for high-power, high-quality, low-frequency sound reproduction. Here again, the considerations are highly technical, and speakers should be supplied under a performance specification that guarantees user satisfaction subject to specified test procedures.

24.18 LOUDSPEAKER CONSIDERATIONS

Loudspeaker system design and placement must be coordinated with the architectural design. The two principal types of loudspeaker systems are central and distributed. The loudspeakers in a conventional central system are a carefully designed array of directional high-frequency units, combined with less directional low-frequency units placed above and slightly in front of the primary speaking position. In most theaters, this location is just above the proscenium on the centerline of the room. Located in this position, the system provides directional realism and is simple in its design.

A distributed loudspeaker system consists of a series of low-level loudspeakers located overhead throughout the space. Each loudspeaker covers a small area, in a manner similar to downlights. This type of system is used in low-ceiling areas where a central loudspeaker cluster cannot provide proper coverage. It also can be used for public address functions where directional realism is not essential, in spaces such as exhibition areas, airline terminals, and offices. In public areas such as transportation terminals, where the absence of absorptive material makes such spaces highly reverberant, care must be taken in speaker positioning and volume levels. Failure to do so will result in the unfortunately very common condition of extremely loud yet unintelligible speech.

Distributed loudspeaker systems provide flexibility for use in spaces where source and listener locations vary according to the use of the space, since loudspeakers can easily be switched to provide proper coverage. In general, a listening position should receive sound from only one loudspeaker. Systems that cover seating areas with signals from several scattered loudspeakers will increase the loudness of the sound but tend to produce garbled speech. This rule is the principal reason that the arrangements shown in Figure 24.29*e*–*h* will guarantee a bad job. The common practice of placing one loudspeaker on either side of a proscenium opening (Figure 24.29*e*), or rows of speakers on one or both sides of a room (Figure 24.29*f*), therefore, is particularly to be deplored.

Location and design of the sound system control position can create problems for the architect. The sound system operator must be within the coverage pattern of the loudspeakers. For proper operation, he or she should be able to hear the sound as it is heard by the audience. Some current auditorium designs locate sound system controls within the audience seating area. Other designs place a control room with a completely open wall or a large window at the rear of the auditorium. Monitor loudspeakers and earphones are inadequate substitutes for actual listening within the auditorium. In churches, the control equipment can usually be located at the rear of the congregation area.

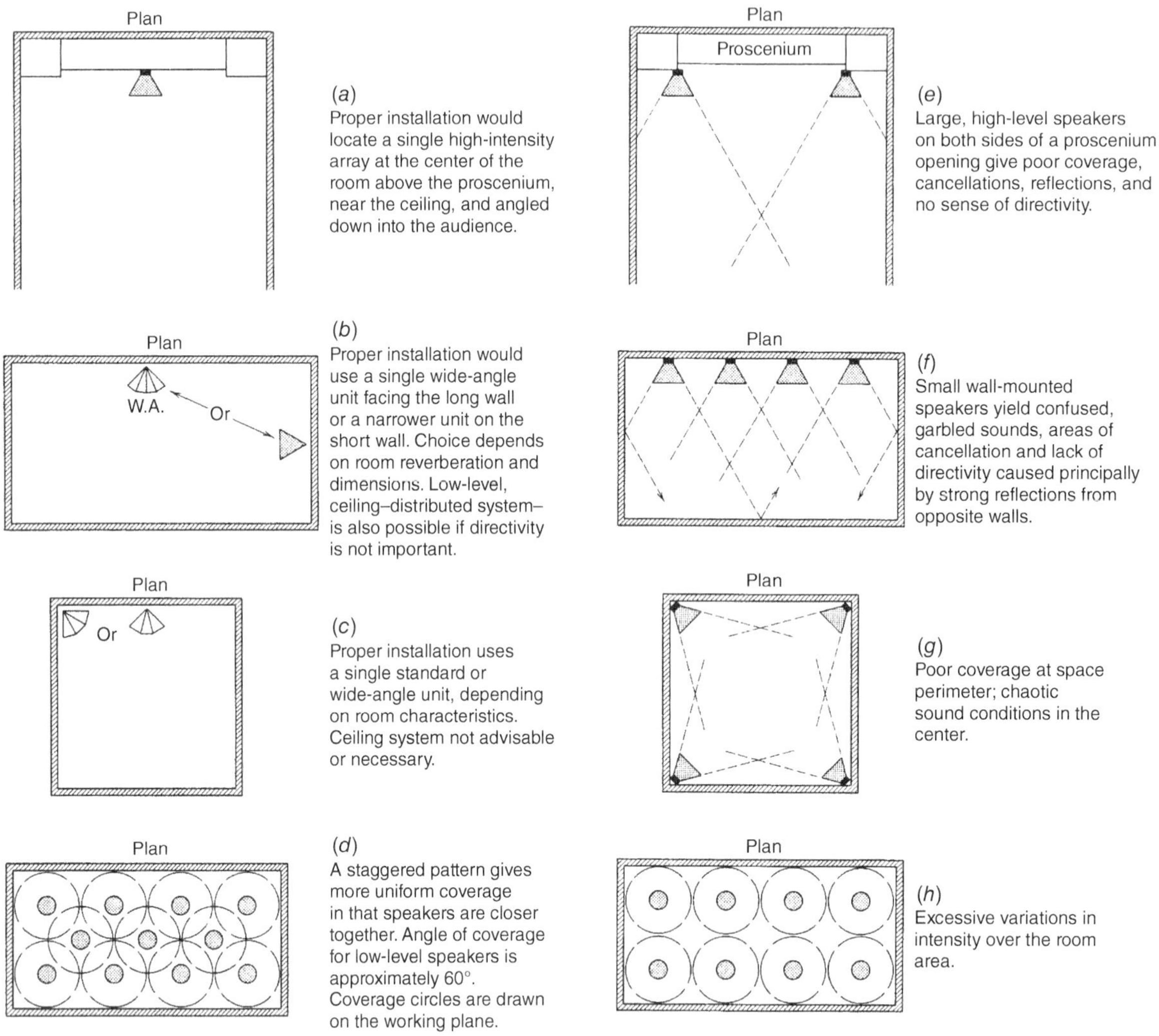

Figure 24.29 *Good speaker layouts (a–d) and poor layouts (e–h). Wide-angle speakers are available to fit most needs. Ceiling speakers give effective coverage on a 60° cone. The working plane is taken to be 4 ft [1.2 m] above the finished floor (AFF) for a seated audience and 6 ft [1.8 m] AFF for standing listeners.*

24.19 CASE STUDY—ARCHITECTURAL ACOUSTICS

Technical Building, Skywalker Ranch, Nicasio, California, USA

PROJECT BASICS

- Location: Nicasio, California, USA
- Latitude: 30.1°N; longitude: 122.7°W; elevation: 194 ft [59 m] above sea level
- Heating degree days: 2542 base 65°F [1412 base 18.3°C]; cooling degree days: 3549 base 50°F [1972 base 10°C]; annual precipitation: 36 in. [914 mm] (San Rafael, California data)
- Building type: sound post-production facility
- Size: 155,000 ft^2 [14,400 m^2]
- Completed 1987
- Client: Lucasfilm, Ltd.
- Design team: Bracken, Arrigoni & Ross, architects; Charles M. Salter Associates, acoustical consultants; Guttmann & MacRitchie, mechanical engineers; Hansen & Slaughter, electrical engineers; Ernie Schase, owner's representative (acoustics)

Background and Context. The Technical Building (Figure 24.30) is a post-production facility located on the Skywalker Ranch complex in San Marin County, California, north of San Francisco. Skywalker Ranch was developed as the corporate head-quarters for Lucasfilm, Ltd. (of *Star Wars* fame).

ACOUSTICS

Figure 24.30 *The Technical Building at Skywalker Ranch. (Mike McCune / Wikimedia Commons / CC BY 2.0.)*

The acoustical requirements for the Technical Building were exacting and challenging. All facets of architectural acoustics came into play in the design and construction of this project, including room acoustics, architectural noise control, mechanical/electrical equipment noise control, and vibration control.

The Technical Building includes a 300-seat screening room, a music scoring stage, ADR (automatic dialog replacement), Foley (synchronized sound effects), and pre- and final-sound-mix and editing suites. The recording stages established new thresholds for quiet background noise produced by ventilation systems. The room acoustics design for the critical listening and mixing studios was based on new criteria for speech intelligibility. Although the Technical Building has been around for some time now, its diverse and demanding acoustical requirements make it an excellent case study project.

Owner's Project Requirements (OPR). The OPR constitute the issues, intents, and criteria that, appropriately addressed, will collectively provide a successful project from the user perspective. Several overarching and fundamental concerns were identified for the Technical Building:

- Mechanical and electrical systems will handle the loads generated by the electronic systems that are the core of the facility's purpose.
- HVAC systems will respond to high summer temperature experienced in this location.
- HVAC and electrical systems will be aesthetically integrated.
- Long system life is desired.

Beyond these project-wide considerations, numerous specific acoustical constraints were imposed on the project design and construction. Excellent acoustics was a compelling objective across all aspects of the project. Three space types were identified based on their acoustic needs:

- Type-A: requiring background sound pressure levels from 5 to 20 NC (noise criteria curve values); such spaces included recording and dubbing rooms, Foley, ADR, premix, final mix, and screening spaces; this was a prevalent acoustic condition
- Type-B: requiring NC 20–30; spaces included sound transfer rooms and dubbing spaces
- Type-C: requiring NC 30; spaces included sound design and editing rooms

In addition to the stringent NC design targets noted above, several of the Technical Building spaces had reverberation time (RT) criteria that also had to be met (Figure 24.31). Efforts to address these

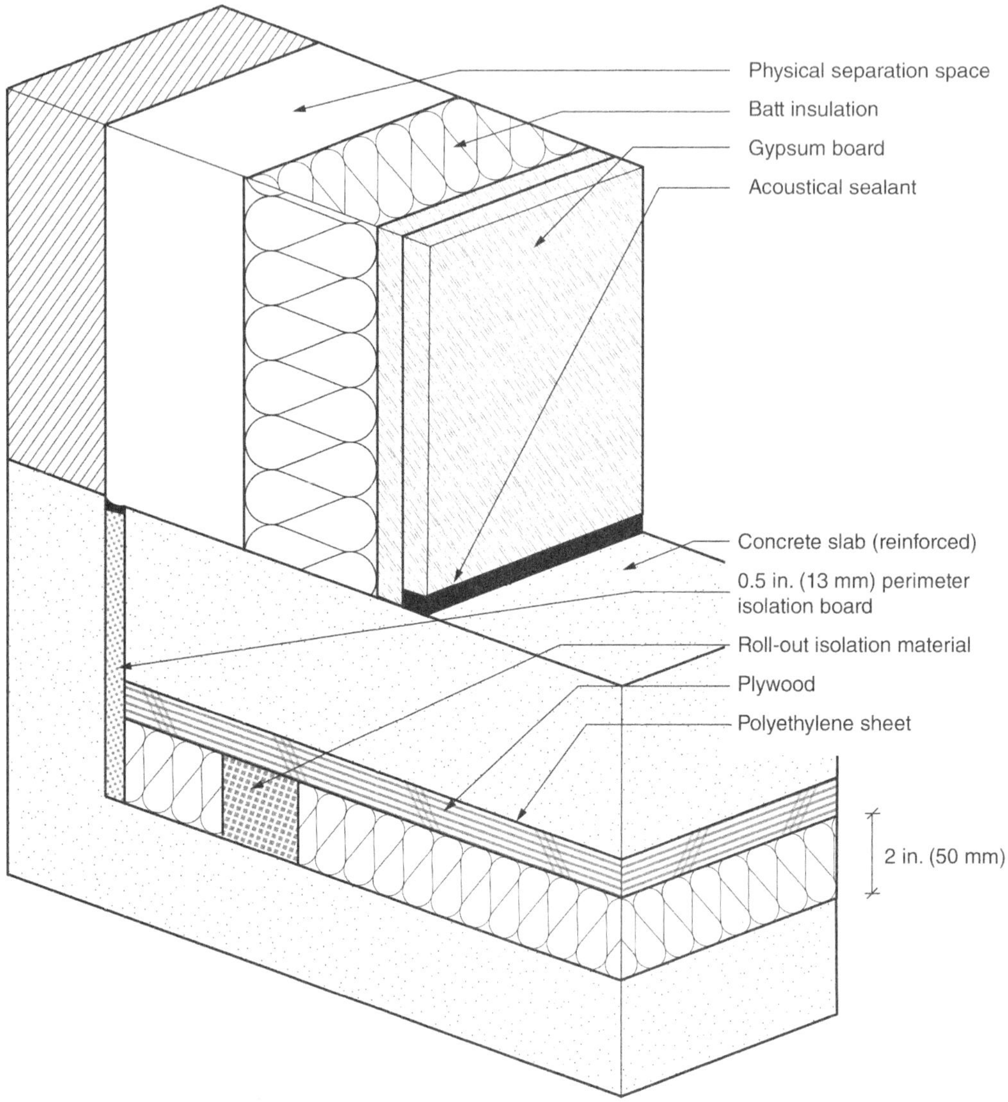

Figure 24.31 *Typical detail at intersection of floating floor and partition. (Drawn by Alexa Thornton; after information developed by Charles M. Salter Associates as presented in the Journal of the Society of Motion Picture and Television Engineers.)*

challenging acoustics conditions are summarized in the following sections of the case study.

Room Acoustics and Noise Control. Meeting the NC limits for spaces required design for sound isolation (including isolation from external noises, from internal sources, and from building equipment). The first design move related to isolation involved physical separation—putting distance between spaces with conflicting acoustic environments. Although a powerful (and often low-cost) solution, some undesirable adjacencies crop up on most projects; in the case of the Technical Building, this included the spatial relationship between the final mix spaces and the underground parking garage. The mix space floor construction was specifically designed as shown in Figure 24.32 (using a floating floor technique with resilient pads) to mitigate the sound pressure level produced by a noisy sports car.

Close attention to construction assembly STC (sound transmission class) was part of the design solution. In this building context, low-frequency transmission (not well represented by STC ratings) was a special concern. Heavyweight construction was required to provide the sound isolation demanded by the design criteria. Attention to wall-floor-ceiling interconnections was necessary to deliver the sound isolation anticipated. Doors and windows were also potential weak links in the isolation performance and were given increased attention. Vestibule entries were used in several situations instead of higher-STC doors (to keep door weight from becoming unmanageable for occupants).

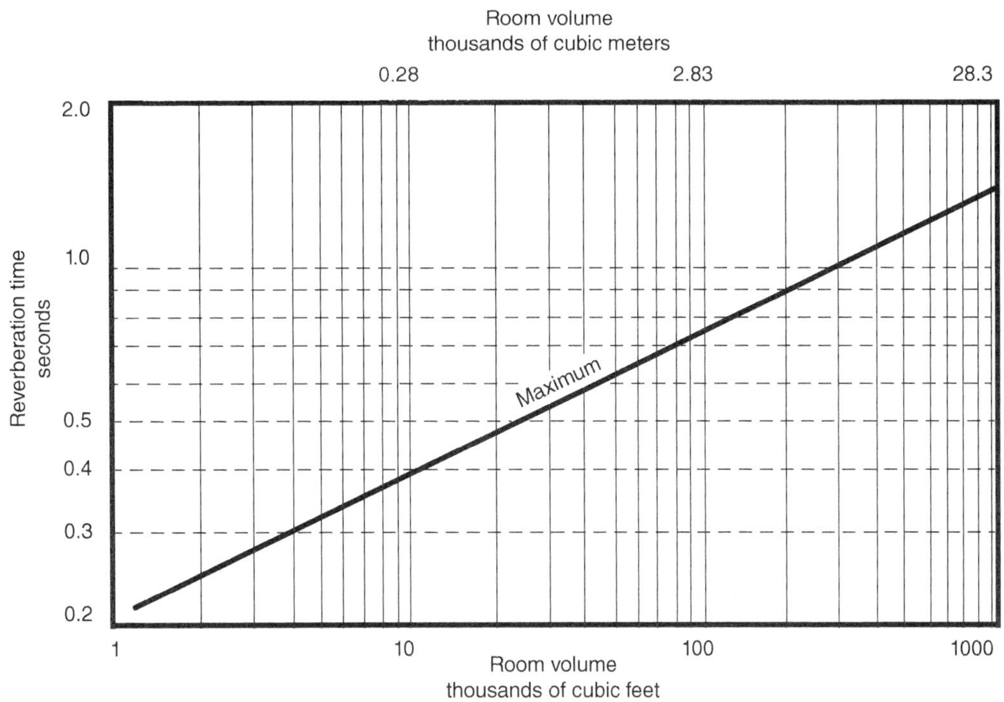

Figure 24.32 *Project criteria for reverberation time in the 500-Hz octave band. (Drawn by Alexa Thornton; after information developed by Charles M. Salter Associates as presented in the Journal of the Society of Motion Picture and Television Engineers.)*

The Scoring Stage had the most critical acoustical constraints in the Technical Building. The enclosure for the space was designed as a double envelope—with a massive inner structure and a lighter-weight external assembly resiliently supported from the inner assembly. To meet the design requirements for isolation from an airplane or helicopter flying 500 ft [150 m] above the Scoring Stage, the entire ceiling was resiliently suspended using spring vibration isolators.

The range of space types found within the Technical Building involved a similar range of preferred reverberation times. Interactions between the client and the design team led to a rethinking of conventional reverberation criteria. Project-specific criteria were then developed—which specified shorter reverberation times than those previously developed by Beranek. The reverberation time (RT) criteria for most spaces was based upon a desire for excellent speech intelligibility. There were two exceptions to this rule: The Scoring Stage (as a musical performance space) required fairly high reverberation, while the Foley space required low reverberation. Design decisions regarding finish materials considered these RT criteria (Figure 24.33).

Scale-model testing was employed as a design verification tool for sound diffusion performance of an important architectural detail. Well-developed drawings and specifications, regular site observation, and strategic testing of critical elements were important to the delivery of results as expected by the client and users. Attention to detail is the bottom line in the design, construction, and turnover of such a facility.

Mechanical and Electrical Noise Control Mechanical and electrical equipment can contribute to the noise that needs to be mitigated through architectural and engineering design efforts. This is true of most building typologies—but is of particular concern in an acoustically sensitive facility such as the Technical Building. The acoustical requirements of the Technical Building are substantial, and the mechanical–electrical systems are complex and robust due to the building's functions. Put another way, many spaces in the building must be very quiet while the systems are likely to be inherently noisy.

Close coordination between architectural, mechanical, and electrical design efforts began early in schematic design—with the objective of keeping noise-producing equipment and distribution elements away from noise-sensitive spaces; this resulted in equipment locations (including an underground vault for chiller equipment) that are not the norm in less acoustically critical buildings. Concern for operational flexibility in use of the 32 editing rooms led to the choice of fan-coil units for delivery and distribution of heating and cooling effect; this decision, however, also reduced the potential for crosstalk that would

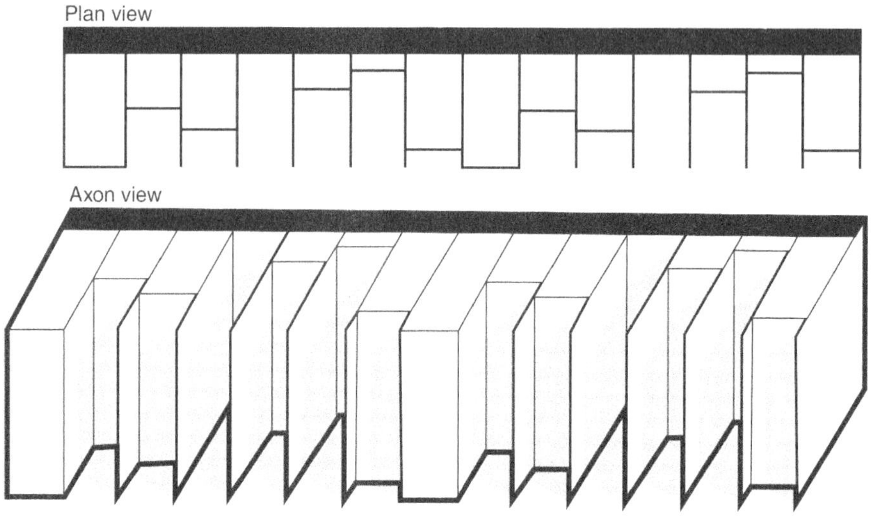

Figure 24.33 *Concept diagram showing arrangement of sound diffusing panels. (Drawn by Alexa Thornton; after information developed by Charles M. Salter Associates as presented in the Journal of the Society of Motion Picture and Television Engineers.)*

be present if an all-air-distribution system had been selected.

Acoustical concern and coordination continued beyond schematic/conceptual design and into design development and construction, as evidenced by the following examples:

- End-discharge and end-return air-handling units were specified (versus the bottom supply and return typical of rooftop installations) as a means of achieving better sound attenuation capabilities.
- Active mechanical equipment was not located above spaces with NC 30 or lower design targets.
- Only distribution runs serving the space were allowed to enter the area above sensitive rooms; noncritical plenum spaces were used for general distribution runs.
- Ductwork was sized to decrease noise generation at the entry and exit to low-NC spaces; ductwork was stiffened (by using bracing or thicker material) to reduce noise that can be generated as ducts expand and contract with changes in airflow; fiberglass ducts were used in especially sensitive spaces as a means of increasing attenuation and reducing resonance; and flexible ductwork (which can lead to noise generation in many circumstances) was avoided in all critical areas of the building.
- Air supply and return techniques were reconsidered, resulting in a novel trickle-supply delivery system employing the plenum space and having return air pass through an acoustically absorptive cavity prior to entering conventional return-air ducts (Figure 24.34).
- Large expansion plenums were incorporated into the air distribution system to achieve the necessary low-frequency sound absorption demanded of some production spaces.
- A standalone noise and vibration control specification was included in the contract documents as a means of emphasizing the importance of acoustics to the owner's project expectations.
- One-third octave-band criteria covering a wide range of frequencies were developed and enforced (versus the more typical octave-band criteria).

Vibration control (distinct from noise control) was an important design consideration for the Skywalker Ranch Technical Building. The project demands for a quiet environment, system requirements for rotating equipment, and the natural frequencies of building components all were considered in developing a comprehensive vibration control solution. Springs and neoprene pads were used as isolation elements as appropriate. In one sensitive location, an air-handling unit was installed such that airbag isolation could be easily added should it prove necessary following in situ performance testing. All piping and ductwork were isolated to break contact with the building structure. Seismic constraints were required in many situations due to California's active earthquake environment.

Construction and preoccupancy activities associated with the acoustics at Skywalker Ranch included:

- weekly site visits over a two-year period by a highly qualified owner's representative (who was a broadcast construction specialist)
- special attention during construction to vibration isolation elements, installations, and alignments

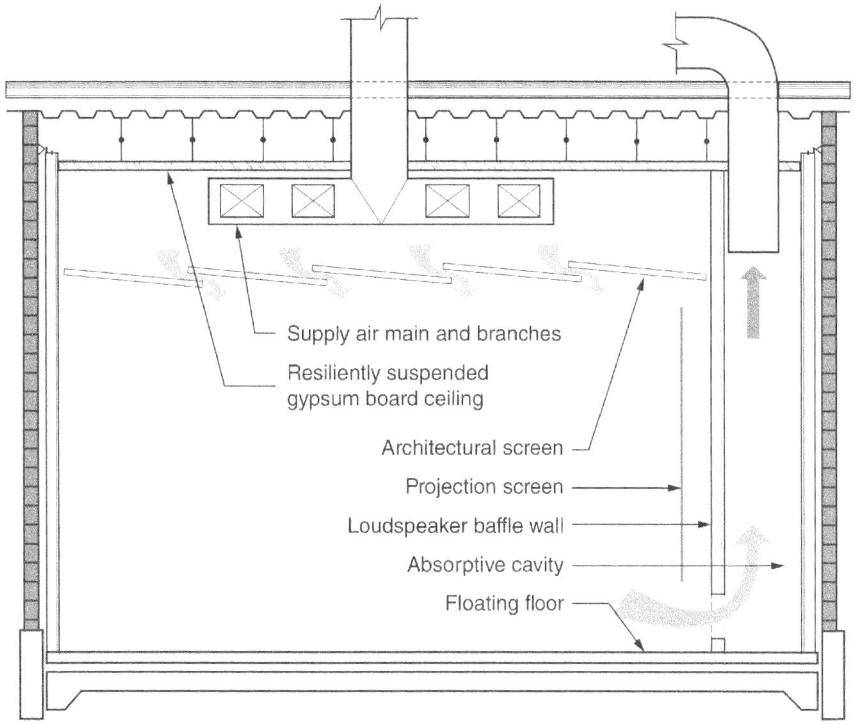

Figure 24.34 *Acoustically sensitive HVAC layout for post-production stage. (Drawn by Alexa Thornton; after information developed by Charles M. Salter Associates as presented in the Journal of the Society of Motion Picture and Television Engineers.)*

- installed condition testing of a typical fan-coil unit—which exposed a systemic noise problem associated with misoperation of a pressure relief fan
- post-construction measurement of background sound pressure levels in all noise-critical spaces in the Technical Building—which revealed a defective piece of process-cooling equipment that was identified through its definitive noise signature

FOR FURTHER INFORMATION

Charles M. *Salter Associates*. 2025. www.cmsalter.com

Schindler, T. 1989. "Acoustical Design for the Technical Building at Skywalker Ranch—Part 2: Mechanical and Electrical Acoustic Noise Control." *Journal of the Society of Motion Picture and Television Engineers* (Feb.): 106–112.

Schwind, D. 1989. "Acoustical Design for the Technical Building at Skywalker Ranch—Part 1: Sound Isolation and Room Acoustics." *Journal of the Society of Motion Picture and Television Engineers* (Feb.): 100–105.

Skywalker Ranch. 2025. skysound.com/ranch

References and Resources

ASTM. 2023. *ASTM E795-23: Standard Practices for Mounting Test Specimens During Sound Absorption Tests*. West Conshohocken, PA: ASTM International.

Beranek, L. L. 1988. *Noise and Vibration Control*, revised edn. Washington, DC: Institute of Noise Control Engineering.

Brüel, P. V. 1951. *Sound Insulation and Room Acoustics*. London: Chapman and Hall.

Magrab, E. B. 1975. *Environmental Noise Control*. New York: John Wiley & Sons.

CHAPTER 25

Building Noise Control

NOISE CONTROL IN BUILDINGS INVOLVES THREE KEY CONCEPTS:

1. Reduction of noise generation at the source by proper selection and installation of equipment.
2. Reduction of noise transmission from point to point (along the transmission path) by proper selection of construction materials and appropriate construction techniques.
3. Reduction of noise at the receiver through acoustical treatment of the relevant spaces to meet the noise criteria discussed in Chapters 8 and 24.

Speech privacy is achieved by manipulation of all of the foregoing, plus the use of masking noise where necessary.

Noise reduction is essentially the science of converting acoustical energy into another, less disturbing form of energy—heat. Since the amounts of energy involved are minute—130 dB corresponds to 1/1000 of a watt, or 0.003 Btu/h—the heat produced is completely negligible. This energy conversion is accomplished by absorption of sound energy by the room contents and surface coverings and also by the structure itself. The former controls noise levels within a space and the latter noise transmission between spaces. The reasons for this will become clear as our discussion proceeds.

ABSORPTION

25.1 THE ROLE OF ABSORPTION

Absorptive noise control treatment of a room will affect the reverberant noise level within that room but will have a minimal effect on the noise level in adjoining spaces. Refer to Figure 25.1 for a graphic presentation of this fundamental fact. The best that can be accomplished with acoustic room treatment is elimination of the reverberant field, that is, making the intensity at the room boundaries what it would have been in free space, as in Figure 25.1*d*. (Even this is extremely difficult; the actual field at the wall would be above 72 dB, except in a completely anechoic chamber.) Adding a further wall (or other) acoustic absorbent (as in Figure 25.1*e*) does nothing in the room itself and has a minimal effect on the overall transmission loss, since the transmission loss in the acoustic material itself is very low, as can be seen in Figure 25.1.

The subject of acoustic energy absorption and absorptive materials, and their effect on room acoustics (including noise reduction) is treated extensively. Porous absorptive materials are the primary focus. Two other types of absorptive material are in use, although much less commonly: panel resonators and cavity resonators. These are addressed first.

Mechanical and Electrical Equipment for Buildings, Fourteenth Edition. Walter Grondzik, Daniel Overbey, and Alison Kwok.

Companion website: www.wiley.com/go/meeb14e

(*a*) TV set in free space produces 75-dB sound level, which drops 6 dB for each doubling of distance. Attenuation by inverse square law.

(*b*) TV still produces 75 dB. In the free field, sound drops to 72 dB but builds up to 74 dB at the wall due to reverberant field reinforcement. Wall attenuation is 30 dB. Sound on other side of the wall is 74 – 30 = 44 dB.

(*c*) Acoustic tile ceiling acts to reduce room reverberant field. Free field is extended. Level at wall is 73 dB. Level in second space is 73 – 30 = 43 dB.

(*d*) The room is acoustically treated, effectively eliminating reverberant field. Room is "dead." Level on second side of wall is 72 dB less acoustic tile loss, less wall loss (that is, 72 – 2 – 30 = 40 dB).

(*e*) Add another 2¼ in. of acoustic wall treatment. Room is "dead." Level at wall 72 dB. Level in second space = 72 – 4 – 30 = 38 dB.

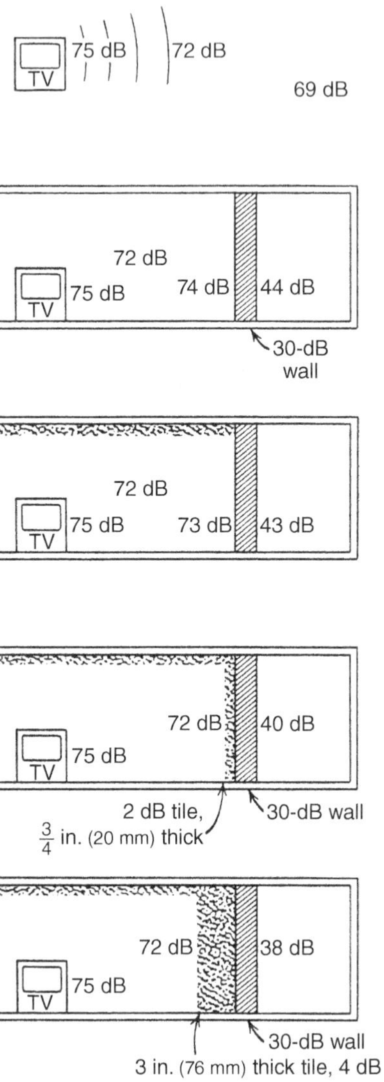

Figure 25.1 *Graphic description of the effect of porous absorptive material on sound fields and sound pressure level (SPL) in adjoining spaces.*

25.2 PANEL AND CAVITY RESONATORS

Panel resonators are built with a membrane such as thin plywood or linoleum in front of a sealed air space generally containing absorbent material. The panel is set in motion by the alternating pressure of the impinging sound wave. The sound energy is converted into heat through internal viscous damping. Panel resonators are used where efficient low-frequency absorption is required and middle- and high-frequency absorption is unwanted or provided by another treatment (Figure 25.2). Panel resonators are often used in recording studios.

A volume or cavity resonator (Helmholtz resonator) is an air cavity within a massive enclosure connected to the surroundings by a narrow neck opening. The impinging sound causes the air in the neck to vibrate, and the air mass behind causes the entire construction to resonate at a specific frequency. At that frequency, absorption approaches 100% but drops sharply above and below this frequency (see Figure 25.2). Adjusting the neck opening and cavity dimensions allows a unit to be tuned to resonate at different frequencies. This makes it extremely useful when a major single frequency is present, as with 120-Hz transformer hum. Concrete blocks can be used as resonators by tuning their cavities. Their absorption characteristic over the entire frequency band is improved by adding absorptive material in the cavities. Fibrous filler can be used in the block to increase high-frequency absorption. The blocks also serve as standard concrete construction blocks. See Figure 25.3 for typical block details and Figure 25.4 for a common application.

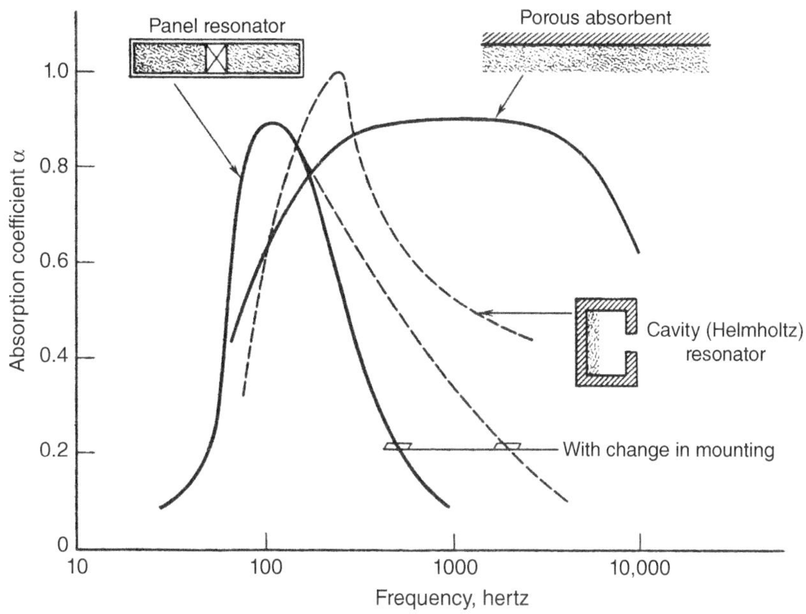

Figure 25.2 *Typical absorption curves for the three major types of sound absorbers. The absorption characteristic of each type can be changed by varying the design, as discussed in the text.*

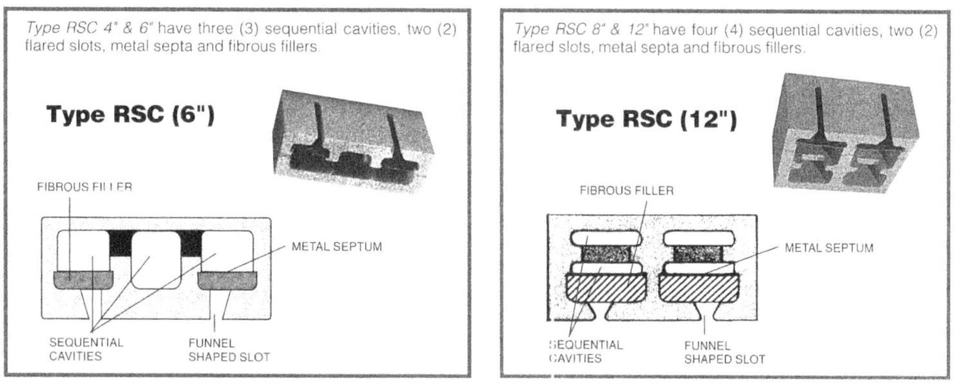

(*a*) (*b*)

Sound Absorption Coefficients — Type RSC

Size	Type	Surface	Exposed Slots/ Cavities	FREQUENCY – Hertz 125	160	200	250	315	400	500	630	800	1000	1250	1600	2000	2500	3150	4000	5000	NRC
6"	RSC	PAINTED	2/3	.48	.70	.93	1.14	1.05	.97	.91	.84	.75	.76	.77	.70	.67	.68	.56	.51	.59	**.85**
8"	RSC	PAINTED	2/4	.48	.85	1.17	.99	.90	.88	.98	.79	.62	.58	.60	.61	.70	.69	.70	.64	.51	**.80**
12"	RSC	PAINTED	2/4	.57	*	*	.76	*	*	1.09	*	*	.94	*	*	.54	*	*	.59	*	**.85**

(*c*)

Figure 25.3 *Details of a Helmholtz resonator design using concrete blocks, with tuned block cavities as the resonating chambers. (a) A 6-in. [150-mm] block with three sequential cavities tuned with a metal divider (septum). (b) A 12-in. [305-mm] block with four cavities and added absorptive filler. (c) Sound absorption coefficients. (Courtesy of The Proudfoot Co., Inc.)*

Figure 25.4 Application of concrete block resonators in a school gymnasium. (Courtesy of The Proudfoot Co., Inc.)

25.3 ACOUSTICALLY TRANSPARENT SURFACES

The soft, porous material of which acoustic absorbents are constructed may be covered with perforated metal or other materials to provide physical protection and act as stiffeners. These coverings are generally acoustically transparent except at higher frequencies. The frequency at which a noticeable reduction in absorption occurs for a perforated metal cover with circular holes can be estimated as:

$$f = \frac{40p}{d} \quad \text{(Eq. 25.1)}$$

where

f = frequency, Hz
p = percentage of open area
d = diameter of holes, in.

Thus, for 1/4-in. [6 mm] holes and 60% open area, which is a typical commercial material,

$$f = \frac{40(60)}{0.25} = 9600 \text{ Hz}$$

which is very high and generally not of major concern. It is always preferable, given a fixed percentage of open area, to use covers with small holes rather than large ones since, as seen from the formula, this raises the frequency at which absorption drops. It is also desirable to stagger the holes, because this improves absorption.

An open-weave fabric is almost completely transparent to sound and is often used as a decorative cover on absorbent wall coverings.

25.4 ABSORPTION RECOMMENDATIONS

To summarize the discussions in previous chapters, absorption techniques are useful and effective in the following situations:

1. To change room reverberation characteristics.
2. In spaces with distributed noise sources, such as offices, schools, restaurants, and machine shops.
3. In spaces with hard surfaces and little absorptive content.
4. Where listeners are in the reverberant field (no amount of absorptive material can reduce intensity levels in the free field).

Concentrated noise sources are better handled by individual equipment enclosures than by room treatment, since enclosures reduce the amount of sound emitted into a room, which room surface treatment cannot do. In addition to ceiling tiles and wall panels, acoustic absorbent material can be applied in the form of a sprayed-on finish (acoustic plaster), suspended unit absorbers, carpets, draperies, and the like.

25.5 CHARACTERISTICS OF ABSORPTIVE MATERIALS

1. *Acoustic tile:* This is available in size multiples of 12 in. [nominal 305 mm], from 12 in. × 12 in. [305 mm × 305 mm] up to 48 in. × 96 in. [1.2 m × 2.4 m], in a huge variety of patterns and finishes, including units with fire ratings. Installation methods include lay-in, nailing to furring strips, and gluing. Tile materials are generally mineral fiber or faced fiberglass, with noise reduction coefficient (NRC) ratings in the range of 0.45–0.75 for mineral fiber tiles and up to 0.95 for fiberglass. The latter, which are frequently used in open-office applications, have Articulation Class (AC) ratings of 170–210 (see Section 25.20(d)). Tiles for use in high-humidity areas should be certified by the manufacturer for that application (see Figure 25.5).
2. *Perforated metal-faced units:* These are usually installed in a lay-in suspension ceiling, although nailing to furring strips is possible. Units range in size from 12 in. × 24 in. [305 mm × 610 mm] to 24 in. × 96 in. [0.6 m × 2.4 m]—and larger on special order. The fill material is either wrapped mineral wool or fiberglass, with NRC ratings somewhat lower than those of acoustic tile of the same material. The metal finish is generally baked enamel in a range of colors. The units are applicable to all spaces and have the advantages of easy cleaning, high luminous reflectivity, and incombustibility. With the acoustic backing removed, a perforated unit can be used for air return (Figure 25.5*c*).
3. *Acoustic panels (boards):* These are made of treated wood fibers, bonded with an inorganic cement binder, in sizes ranging from 12 in. × 24 in. [305 mm × 610 mm] to 24 in. × 120 in. [610 mm × 3050 mm], with thicknesses from 1 in. to 3 in. [25 mm to 75 mm] and a smooth or "shredded" finish. Panels are used in ceiling suspension systems, or nailed or glued when applied to walls and structural ceilings. Their principal advantage is their high structural strength, which makes them applicable to installations requiring acoustic treatment combined with strength and abuse-resistance. A second advantage is their excellent flame-spread rating. Typical applications include full-span corridor ceilings, long-span direct-attached ceiling finish, wall panels in school gyms and corridors, and the like. NRC ratings range from 0.40 to 0.70. Acoustic panels are usually resistant to humidity, but usage in high-humidity spaces should be confirmed with the product manufacturer. This is particularly true for panels with "reveal" edges (Figure 25.6).
4. *Acoustic plaster:* This is a material comprising a plaster-type base into which is introduced fibrous

(*a*) (*b*) (*c*) (*d*)

Figure 25.5 *Four of the many available types of acoustic ceiling tiles. (a) and (b) are 24 in. × 48 in. × 1.5 in. [600 mm × 1200 mm × 36 mm] fiberglass ceiling tiles with an acoustically transparent plastic humidity shield facing and a foil back for good sound dispersion in the plenum of open-plan offices. (a) Tile has 1.0 NRC, 200 AC (Articulation Class), 0.89 light reflectance, and a thermal R value of 4.0. (b) Tile has similar acoustic and thermal characteristics, but due to its coarser surface, its light reflectance is 0.85. Both tiles are applicable to open-plan offices and are usable with indirect lighting systems. (c) Perforated metal cassette panels are used in lobbies, corridors, entries, outdoor canopies, and soffits, and areas with high humidity. They are usable for HVAC system air returns and, when supplied with an acoustic infill (mineral fiber or fiberglass), have good acoustic properties. (d) General-purpose, coarse-textured mineral fiber ceiling tile is rated 0.70 NRC, 25 CAC (Ceiling Attenuation Class), and 0.73 light reflectance. It can be used acoustically in offices, libraries, restaurants, and public spaces. (Courtesy of Armstrong World Industries.)*

Figure 25.6 Acoustic panel material made of pressed organic fiber with an inorganic binder can be used as ceiling tile or wall panels. The latter has cloth covering as an outer finish. (Courtesy of Tectum, Inc.)

Figure 25.7 Application of hanging acoustic baffles to reduce reverberation in the high pyramid-shaped volume of a hotel lobby in Hong Kong. (Photo by Alison Kwok.)

or light aggregate, and is useful for application to curved and other nonlinear surfaces, in thicknesses of up to 1.5 in. [38 mm]. Its advantages are ease of application and a high fire rating; its disadvantages are its inability to resist even mild abuse and its inapplicability to humid atmospheres. The noise absorption characteristics of acoustic plaster vary widely with composition, thickness, and application technique, and are generally below those of acoustic tile and panels.

5. *Sound blocks, baffles, and hanging panels:* These are simply masses of absorptive acoustic material that achieve absorption coefficients in excess of 1.0 by exposing more than a single absorptive surface to the impinging sound. For one product (an 8 ft^2 [0.7 m^2] baffle), absorption coefficients are reported as 0.66 @ 250 Hz, 1.31 @ 500 Hz, and 1.74 @ 1000 Hz. In all cases, because of their prominent appearance, necessitated by their shape, they obtrude into the space and frequently become a major architectural element. This is especially true of hanging baffles (Figure 25.7).
6. *Wall panels:* These consist of a wood or metal backing on which is mounted a mineral fiber or fiberglass substrate and a fabric covering. NRC coefficients vary from 0.5 for direct-mounted 1-in. [25 mm] mineral fiber substrate to as high as 0.85 for strip-mounted 1.5-in. [38 mm] fiberglass substrate panels. Wall panels are available in widths ranging from 18 in. to 48 in. [457 mm to 1220 mm] and lengths of up to 120 in. [3050 mm]. Fabric coverings generally carry a fire-spread rating. These panels are frequently used in offices, conference rooms, auditoriums, theaters, teleconferencing centers, and educational facilities (see Figure 25.8).
7. *Resonator sound absorbers:* These are available in a wide variety of sizes and shapes. Although some designs are available off the shelf, most are tailored to the specific acoustic needs of a project, using one of several standard designs. One fairly common type is shown in Figure 25.9. In general, resonators are large and must therefore be integrated into the architectural design of a space. Exterior shapes can be altered, and the units can be installed less obtrusively to fulfill this architectural requirement.
8. *Carpeting and drapery:* These can be used to cover large acoustically reflective surfaces in a space. Carpeting can be selected in almost any degree of density, looping, and depth, plus an additional depth of padding, to produce a high degree of absorption in middle and high frequencies (see Chapter 24). In general, absorption is proportional to pile height and density, and increases when the carpet is installed on a thick, fibrous pad. Where drapery is not feasible and wall panels are impractical, carpeting can be installed on walls. In such instances, installation on furring strips with an enclosed air space behind will increase

Figure 25.8 Acoustic wall panel consists of a rigid backing covered with either mineral fiber or fiberglass and finished with any of a wide selection of fabrics. NRC averages 0.6 for mineral fiber substrate and 0.8 for fiberglass with contact mounting (A mounting), and 0.7–0.9 when the panels are mounted on 20-mm [nominal 3/4-in.] furring strips (D-20 mounting). The latter mounting is particularly effective in increasing absorption below 500 Hz. (Courtesy of Armstrong World Industries.)

absorption over the entire acoustic spectrum and especially at low frequencies (where direct-contact installation exhibits poor absorption). Draperies are essentially acoustically transparent and provide appreciable absorption only in the middle and upper frequencies with heavy, dense, fuzzy fabrics, particularly when draped with a high degree of fold. As with carpeting, absorption increases over the entire spectrum when a heavy folded drapery forms an air space between itself and the wall. Approximate absorption coefficients are given in Chapter 24.

SOUND ISOLATION

25.6 AIRBORNE AND STRUCTURE-BORNE SOUND

In contrast to the preceding material, which was concerned with in-room sound reduction by absorption, the following sections will discuss the characteristics of sound transmission between enclosed spaces. A distinction is often made between airborne and structure-borne sound, although in reality they differ only in the origin of the sounds. Airborne sound originates in a space with any sound-producing source, and although it changes to structure-borne sound when the sound wave strikes the room boundaries, it is still referred to as airborne because it originated in the air. Structure-borne sound is generally understood as energy delivered by a vibrating or impacting source directly contacting the structure. Hence, a child crying in an adjoining apartment is contributing airborne sound; the same child bouncing a ball on the floor is creating structure-borne sound, in this case, by impact. Pumps that were installed without proper damping mounts create structure-borne sound by vibration.

In reality, all sound transmission is both airborne and structure-borne since, once having entered the structure, the sound travels along the structure and causes the structure to vibrate, in turn generating airborne sound. Figure 25.10 should assist in understanding this action. In Figure 25.10*a*, the sound is airborne, originating in the air on one side of the partition. The incidence of sound energy causes the partition to vibrate, generating sound on the other side. Sound does not "pass through" unless an air path exists. If the partition is airtight, then the sound energy causes the structure to become a secondary source. The partition vibrates primarily in the direction of the sound, that is, in the vertical plane. It also vibrates in other modes, causing some sound energy to pass into the floor and ceiling, depending upon the details of attachment. This energy becomes structure-borne sound.

In Figure 25.10*b*, the process is similar but reversed. Energy is introduced into the structure directly (and efficiently) by mechanical contact, that is, by vibration and impact. Sound travels along the structure, as shown, and, by causing the structure to vibrate, creates airborne sound. In a structure with rigid wall-to-floor connections, these sounds are clearly heard throughout the building. It is a common misconception that in a (relatively) massive concrete structure with masonry walls, such as a multistory residential building, light impacts such as footfalls will not be transmitted. On the contrary, the rigidity of the structure—particularly the rigid, airtight connections between partitions, floors, and ceilings—provides an excellent path for structure-borne sound. Only impact absorption by heavy carpeting will attenuate the sound of footsteps, and resilient floor-wall connections will attenuate structure-borne sound.

Airborne sound (originating in the air) is generally much less disturbing than structure-borne sound, since its initial energy is very small and it attenuates rapidly

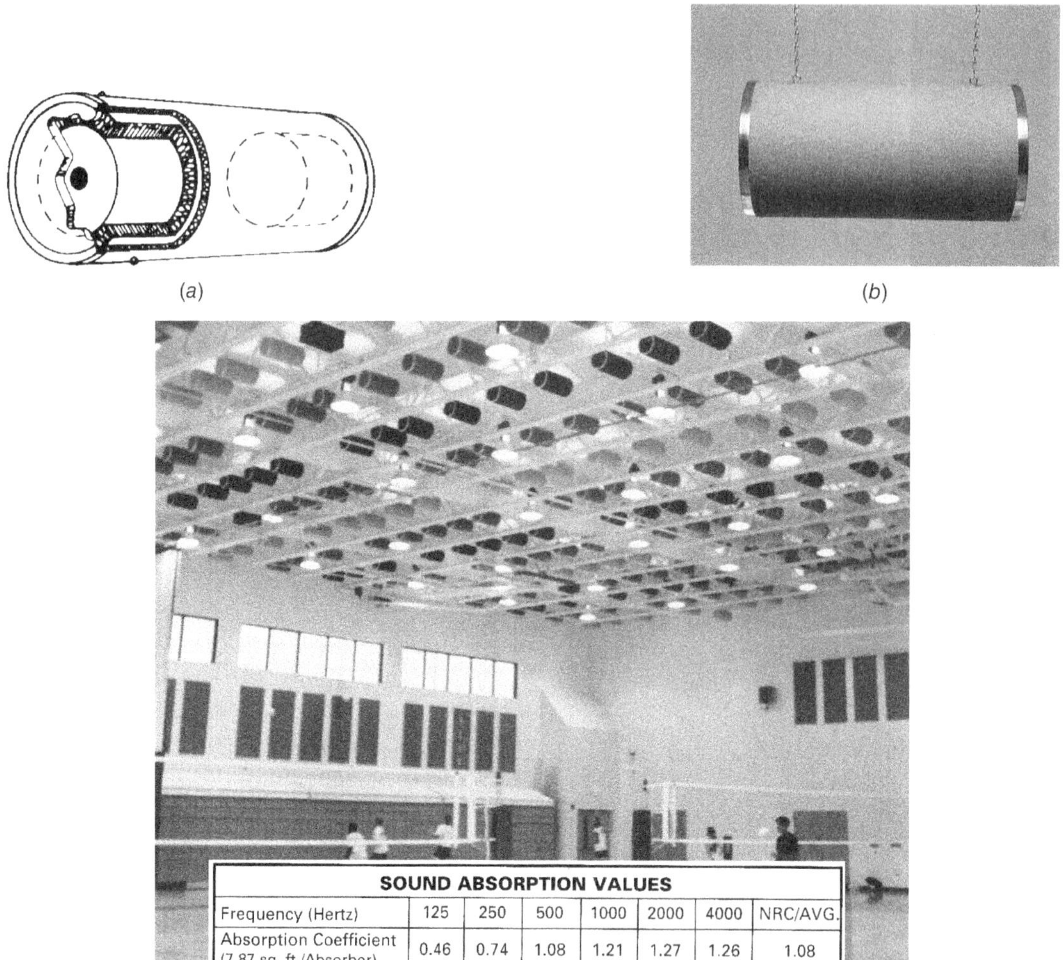

(a) (b)

SOUND ABSORPTION VALUES							
Frequency (Hertz)	125	250	500	1000	2000	4000	NRC/AVG.
Absorption Coefficient (7.87 sq. ft./Absorber)	0.46	0.74	1.08	1.21	1.27	1.26	1.08
Sabins/Absorber	3.65	5.84	8.48	9.51	9.98	9.90	8.45

(c)

Figure 25.9 *(a) Resonator/absorber unit consists of a molded fiberglass cylinder, of 12-in. [305-mm] diameter and 24-in. [610-mm] length, containing a metal resonator at each end for low-frequency absorption and a fill of acoustic fiberglass for full-spectrum absorption. (b) The unit, which weighs about 6 lb [2.7 kg], is suspended on thin wires or chains that are essentially invisible when the unit is mounted 10 ft [3 m] above the finished floor (AFF) or higher. (c) A typical application in a school gymnasium. Because of their large size, the units constitute a major architectural element. By color selection and hanging patterns, they can be effectively used as such. The absorption characteristics of these units are listed in the table as a function of frequency. (Courtesy of The Proudfoot Co., Inc.)*

at the boundaries. Structure-borne sound generally has a much higher initial energy level and attenuates slowly as it travels through a structure, thereby causing disturbance over large sections of a building. This disturbance is magnified by the "sounding board effect."

We are all familiar with the fact that a tuning fork must be held up to the ear to be heard directly, but if its handle is placed on a table the sound is amplified. This action is not really amplification but an increase in the efficiency of energy transfer. In general, the efficiency of a radiator is proportional to the ratio of its surface dimensions to the sound wavelength. A tuning fork vibrating at concert A (440 Hz) with a wavelength of 2.5 ft [0.75 m] cannot efficiently couple its energy into the air. It is simply too small. By placing the instrument on a table whose dimensions are approximately one wavelength, we permit it to transfer its energy efficiently, hence the amplification. The same effect can be extremely troublesome in structure-borne sound. A vibrating pump itself makes little sound. However, it transfers a large amount of energy into the structure, and that energy will appear as audible sound at each partition, floor, and wall that is rigidly coupled to the structure. Soft (damping) connections prevent energy transfer, thereby greatly attenuating the transmission of sound energy into connecting efficient radiating surfaces—hence the desirability of such flexible connections.

Airborne sound changes direction easily (diffracts), with low frequencies being most flexible in

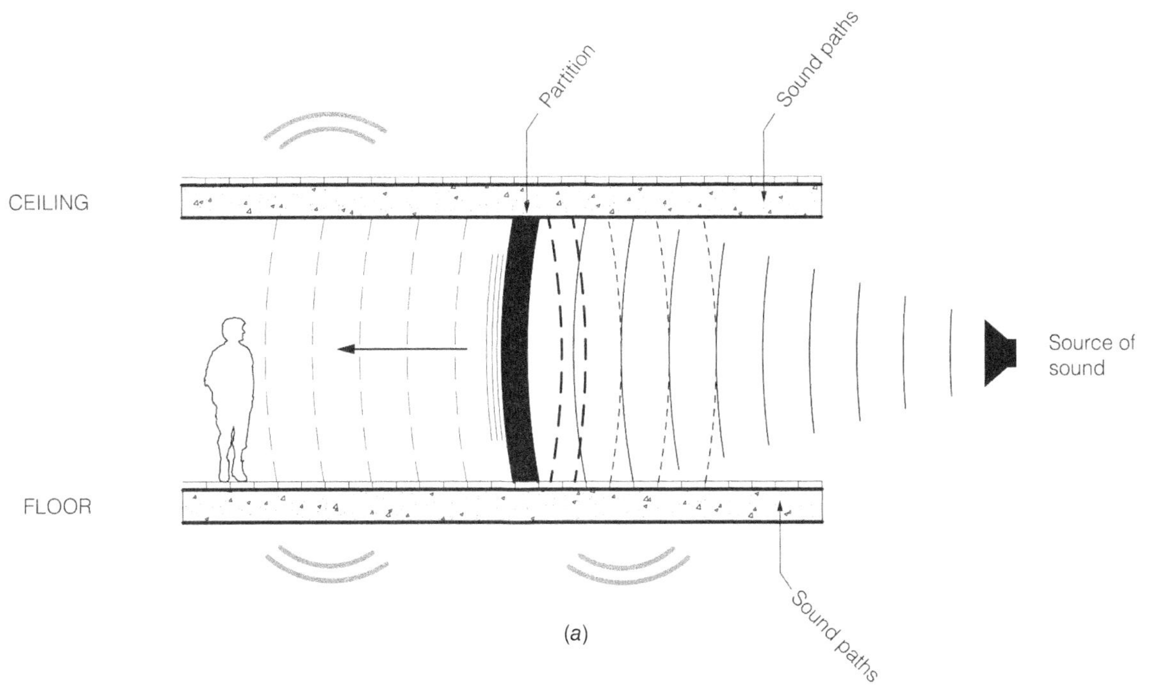

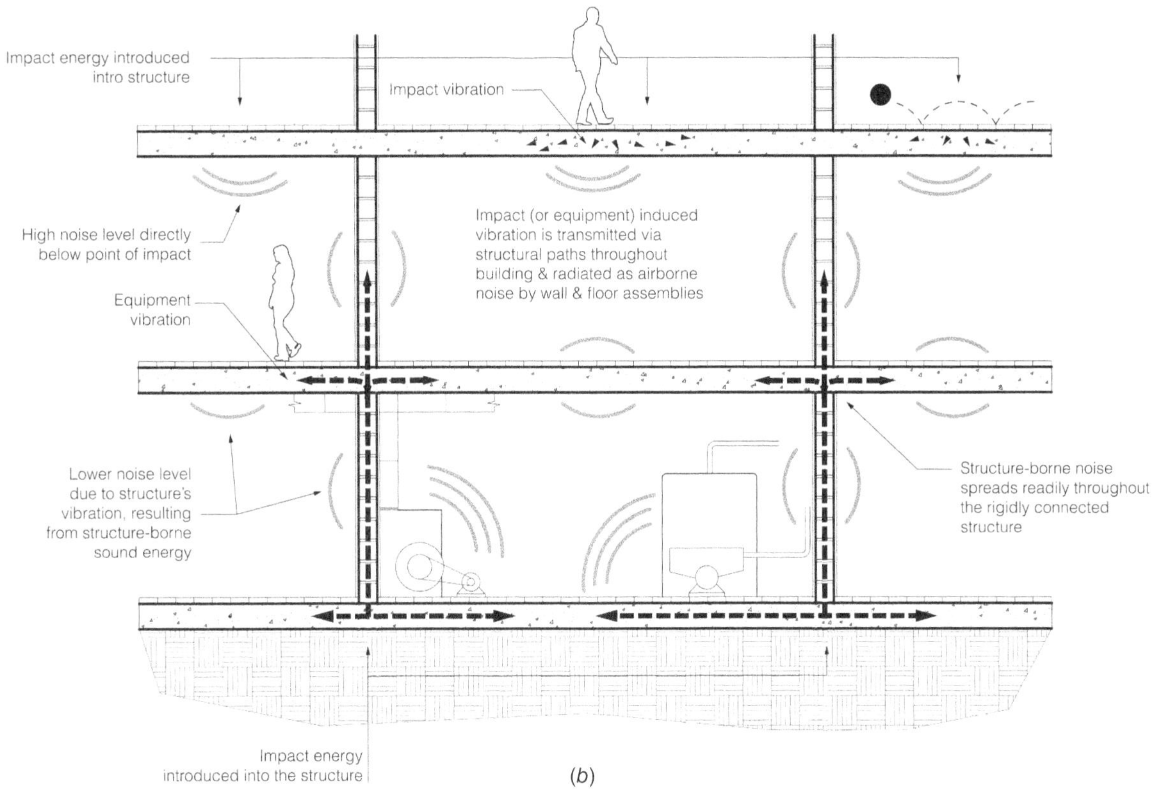

Figure 25.10 *(a) Airborne sound is so called because it originates in air; its energy level is low. Here the loudspeaker is the source of acoustic energy, which is converted to a vibration of the partition (shown greatly exaggerated), which in turn becomes a secondary source of airborne sound for the adjoining space. A small amount of energy is reradiated by the ceiling and floor of both spaces directly, and indirectly via the rigid partition-to-floor and partition-to-ceiling connections. (b) Structure-borne sound originates from mechanical contact between the structure and vibrating or impacting sources. As a result, its energy level is usually much higher than that of airborne sound. This energy is transmitted with little attenuation throughout the structure via the rigid partition-to-floor and partition-to-ceiling connections. The entire structure is then set into vibration as shown, converting the (large) structure-borne sound energy to noise throughout the building. (Redrawn by Wesley Thompson, from HUD, 1968.)*

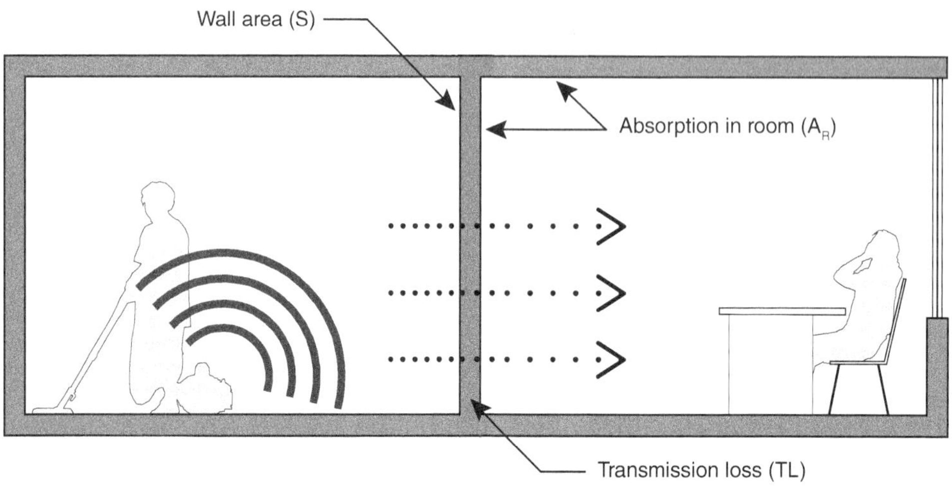

Figure 25.11 A simple case of airborne sound transmission between adjacent rooms through a common barrier. With a sound source in one room, the transmitted sound level is dependent not only on the transmission loss of the barrier but also on the area of the barrier and the receiving-room absorption. The background noise level in the receiving room determines whether the transmitted sound will be noticed.

this regard. Structure-borne sound travels much more rapidly than airborne sound (see Chapter 8) and with attenuation as low as 1 dB per kilometer (about 0.5 dB per mile). A sound traveling along a massive structure will radiate outward from the structure only minimally (although enough to be very annoying) because the large mass minimizes vibration in that direction. Thus, in Figure 25.10*b*, noise from impact on the floor above will be louder (for equal impact energy) than noise from machines below, because the former generates sound directly downward, while the latter introduces energy into the entire network of parallel paths.

The sections immediately following deal with airborne sound and the means for controlling it (Figure 25.11). Impact noise (a form of structure-borne sound) is covered in Sections 25.22 to 25.34.

AIRBORNE SOUND

25.7 TRANSMISSION LOSS AND NOISE REDUCTION

The transmission loss (*TL*) of a barrier is the ratio, expressed in decibels, of the acoustic energy reradiated by the barrier to the acoustic energy incident on it. This number is a figure of merit for the sound-isolating quality of the wall itself and is obtained from controlled laboratory tests. (In Europe, transmission loss is referred to as sound reduction index, *R*.) However, the number that is of greater end importance to the building designer is the actual noise reduction (*NR*) between two spaces separated by a barrier, that is, the action of the barrier in context. This noise reduction is defined as the difference between the sound intensity levels in the two rooms, that is,

$$NR = IL_{\text{room 1}} - IL_{\text{room 2}} \qquad \text{(Eq. 25.2)}$$

and is related to the *TL* of the barrier by the expression

$$NR = TL - 10 \log \frac{S}{A_R} \qquad \text{(Eq. 25.3)}$$

where

NR = noise reduction, dB
TL = barrier transmission loss, dB
S = area of the barrier, ft² [m²]
A_R = total absorption of the receiving room, sabins ft² [m²]

We see, therefore, that noise reduction and transmission loss are not equal but are related by the size of the dividing barrier, *S*, and the absorption characteristic of the receiving room, A_R. A moment's thought will confirm the logic of this relation. When sound energy impinges on the barrier, the barrier in turn becomes the sound source, radiating into the receiving room. Therefore, the amount of sound energy transferred is proportional to the (log of) area *S* of the common barrier between the two spaces.

The sound level in the receiving room is related to its own reverberance (absorption characteristic, A_R), as we have seen repeatedly. Thus, if the receiving room is a reverberant, live space, A_R is low and *NR* is less than *TL*. Conversely, if the receiving room is dead, A_R is large and *NR* can be greater than *TL*, depending upon the ratio of the barrier wall size to the room area (see Figure 25.11). In lieu of precise calculations, the following generalizations can be used:

1. For a live receiving room,

$$NR = TL - 1 \text{ dB}$$

2. For a medium receiving room,

$$NR = TL + 4 \text{ dB}$$

3. For a dead receiving room,

$$NR = TL + 7 \text{ dB}$$

The extreme case of "deadness" of a receiving room is one with no walls, that is, sound transmission from inside to the exterior. In such cases, *NR* exceeds *TL* by 10–15 dB, depending upon the size of the exterior opening and the point outside where *IL* is measured. To acquire a facility with sound-insulation techniques, the designer must become familiar with the relationship of transmission loss to the barrier's physical characteristics: its mass, rigidity, material of construction, and method of construction and attachment. These considerations are the subject of the following sections.

Note: Be careful when encountering the term noise reduction (NR), since a similar and completely unrelated term—noise reduction coefficient (NRC)—also exists. The latter is very poorly named. NR is the difference in sound pressure level (SPL) between two adjacent rooms caused by the TL of an intervening barrier. NRC is the average sound absorption coefficient measured at the four 1/3-octave bands centered on 250, 500, 1000, and 2000 Hz frequencies.

25.8 BARRIER MASS

Sound transmission between spaces requires that a barrier be set into vibration by the incident sound energy. Although this was stated in the preceding sections, we repeat it here to emphasize the fundamental importance of this simple statement. (We are assuming a barrier that is impervious to air—i.e., a solid barrier. Otherwise, the moving air molecules bearing the sound will simply pass through with minimal transmission loss.) The impinging sound energy acts as a force on the wall. Since force equals mass multiplied by acceleration ($F = MA$, Newton's Second Law of Motion), the larger the mass, the less it will vibrate. When other factors (particularly angle of incidence) are taken into account, the resultant acoustical relationship is known as the mass law. It states that for a nonporous, homogeneous structure of low stiffness, the sound transmission loss is proportional to the logarithm of the surface mass (the weight of the wall per unit of surface area) and to the frequency of vibration. Thus, doubling the mass (or frequency) will, theoretically, cause an increase of 6 dB in the transmission loss; stated otherwise, the slope of *TL* versus the frequency times mass (*fM*) curve is 6 dB. Figure 25.12 is a graphic representation of mass law operation. With sound incident at 90°, maximum energy is imparted

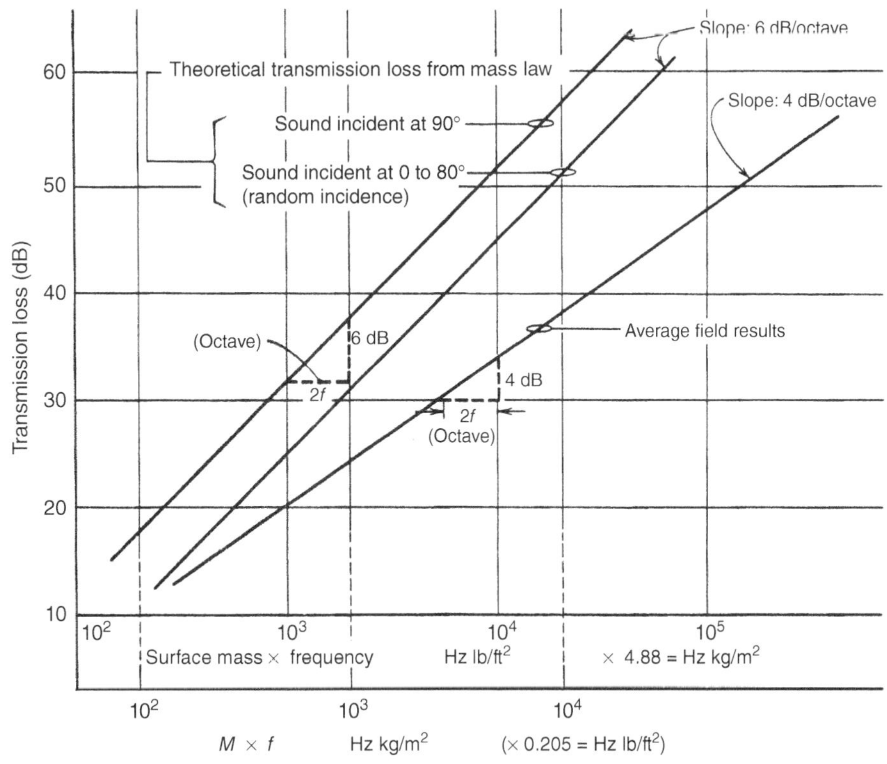

Figure 25.12 *Graphic representation of mass law action in attenuation of transmitted sound. Perpendicular (90°) sound incidence results in maximum transmission loss. Field or random incidence (0° to 80°) is approximately 6 dB lower but maintains the 6-dB-per-octave slope. Field results are lower due to flanking and stiffness effects, and the slope of the curve averages only 4 dB per octave (frequency doubling).*

ACOUSTICS

to the barrier, and the entire mass resists, resulting in maximum transmission loss. In practice, however, sound is incident from 0° to 80° (called field or random incidence), reducing the mass effect but keeping the slope at 6 dB per octave. Due to nonhomogeneity, porousness, and stiffness, actual field results indicate transmission losses closer to 4 dB per octave, as shown by the lower curve in Figure 25.12.

25.9 STIFFNESS AND RESONANCE

The stiffness of a barrier is a function of its material composition and the rigidity of its mounting. The former depends upon its internal cohesiveness—that is, its modulus of elasticity—and the latter depends upon its boundary restraints—whether the barrier is tightly or loosely held. A homogeneous material of high Young's modulus (such as steel) has great cohesiveness between its molecules. As soon as one molecule is set in motion by incident sound energy, the motion is passed to the next molecule, and so on, making the material an excellent sound conductor. Homogeneous materials with a low modulus of elasticity have high internal damping (the motion of molecules is not transmitted well), and they are good sound insulators. Composite materials such as concrete and organic materials such as wood do not conform to these general rules.

Rigidity of mounting can be likened to a drumhead—the tighter it is stretched, the better it resounds. Rigidity (stiffness) in a panel barrier resists damping and assists vibration, making it a good transmitter and, conversely, a poor noise transmission insulator. As a result, a material such as lead, which has a high mass and a low modulus of elasticity and resists rigid mounting, is an excellent sound attenuator.

The effects of stiffness and mass both vary with frequency, unfortunately in opposite directions. Stiffness acts to reduce transmission loss as frequency increases, while, as we have seen, mass acts to increase it; the combined effect is shown in Figure 25.13. Therefore, stiffness is most effective at low frequencies and mass at high frequencies. At very low frequencies the mass and stiffness effects negate each other, giving the resonance dips shown. Beyond approximately 200 Hz, most common wall construction enters the mass law range and continues with it until the critical frequency. Deviations from a smooth 4-to-6-dB-per-octave slope are due to the non-homogeneous nature of most wall constructions. At the critical frequency, the phase of incident sound waves corresponds to or coincides with the phase of vibration (shear wave) of the barrier in

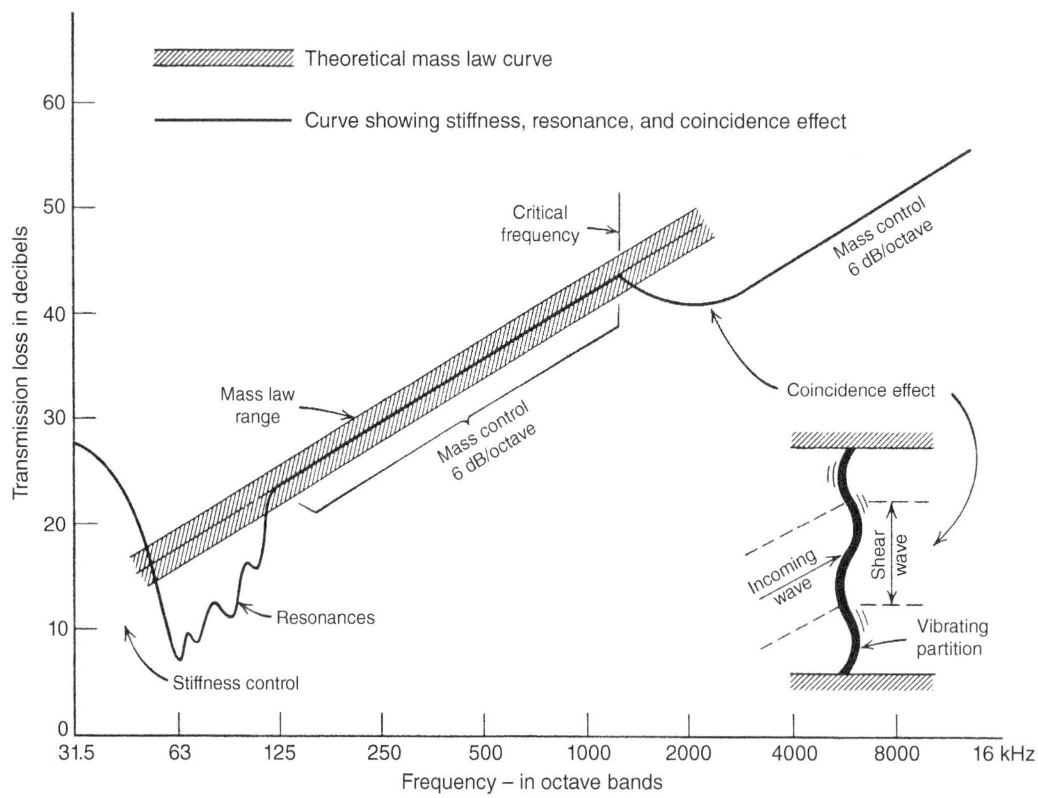

Figure 25.13 *Partition transmission loss as a function of frequency showing the effects of stiffness, resonance, and coincidence. At frequencies below resonance, control is almost purely a stiffness function; at frequencies above critical, control is almost purely a surface mass function.*

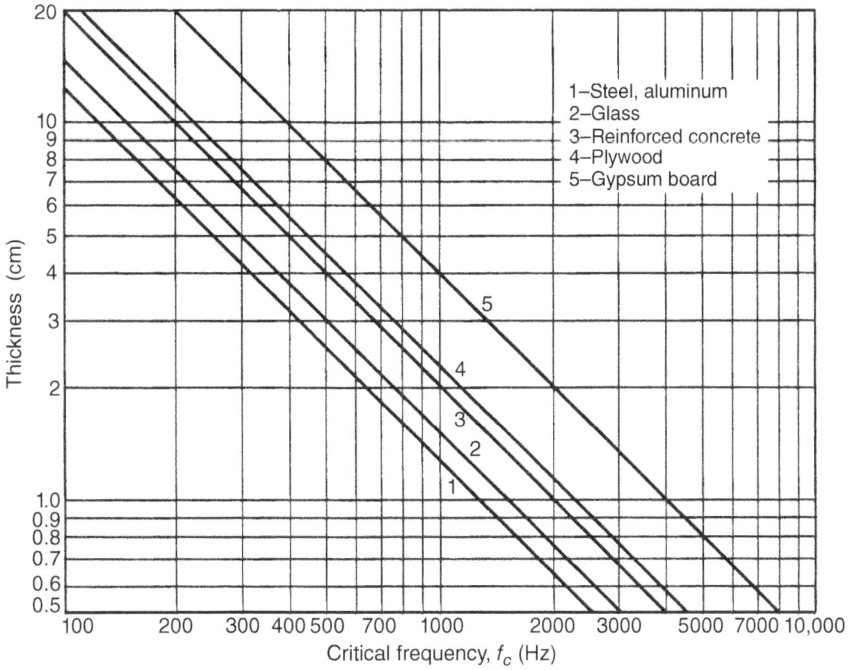

Figure 25.14 *Critical frequency as a function of thickness for several common materials. (From Magrab, 1975.)*

such a way as to pass a large portion of the incident energy. See the insert in Figure 25.13, which shows this effect as the coincidence dip. This effect is most pronounced in thin, homogeneous partitions and light, stiff ones.

Critical frequency, f_c, as a function of panel thickness for common materials, is plotted in Figure 25.14. To avoid a coincidence dip in the audible range, partitions can be either very heavy and/or very stiff (which greatly decreases the critical frequency) or heavy and limp (resilient—which greatly increases the critical frequency). In practical terms, cost effectiveness is heavily in favor of the latter alternative. Thus, for instance, the transmission loss of a wood partition can be improved by grooving it to increase its flexibility, thereby increasing the critical frequency. The dramatic improvement in transmission loss, resonances, and coincidence dip achieved using a resilient mounting with a simple masonry partition is shown in Figure 25.15. Both walls have the same weight—21 lb/ft^2 [102 kg/m^2]. The solid wall A has better attenuation below 200 Hz in the stiffness-controlled range. Above that frequency the resilient-mounted partition is 10 dB better, which means that the transmitted noise is only one-half as loud. The sound transmission class (STC)—which is a figure of merit for partition sound transmission, as explained in Section 25.11—of wall A is 40; that of wall B is 51. Both show a coincidence dip at approximately 250 Hz (see Figure 25.14), but that of B is shallow, whereas that of A is deep and wide. Furthermore, wall B is consistently better than mass law attenuation would predict, wall A consistently worse (due to stiffness).

25.10 COMPOUND BARRIERS (CAVITY WALLS)

Since the maximum theoretical increase in transmission loss with an increase of mass is 6 dB per doubling of mass, adding more mass as a means of transmission loss improvement rapidly reaches the limits of practicality. Indeed, as we have seen, the transmission loss of actual single homogeneous walls falls below the mass law curve. This is because mass increase brings with it stiffness increase, which acts to reduce transmission loss. If, however, a barrier is constructed of two separate layers without rigid interconnection, its performance exceeds the calculated transmission loss based on mass alone. Note that even the nonrigid wire ties of wall B in Figure 25.16 lower the STC by five points. At low frequencies, where stiffness controls the transmission loss (see Figure 25.13), the cavity in wall C (Figure 25.16) acts as a rigid connection between the layers, adding stiffness and increasing transmission loss. At higher frequencies, in the mass law range, the air in the cavity acts as a damping coupling to reduce stiffness. The net result is an improvement in performance throughout the frequency range.

Transmission loss for the entire cavity wall increases with the width of the air space at the rate of approximately 5 dB per doubling. Performance

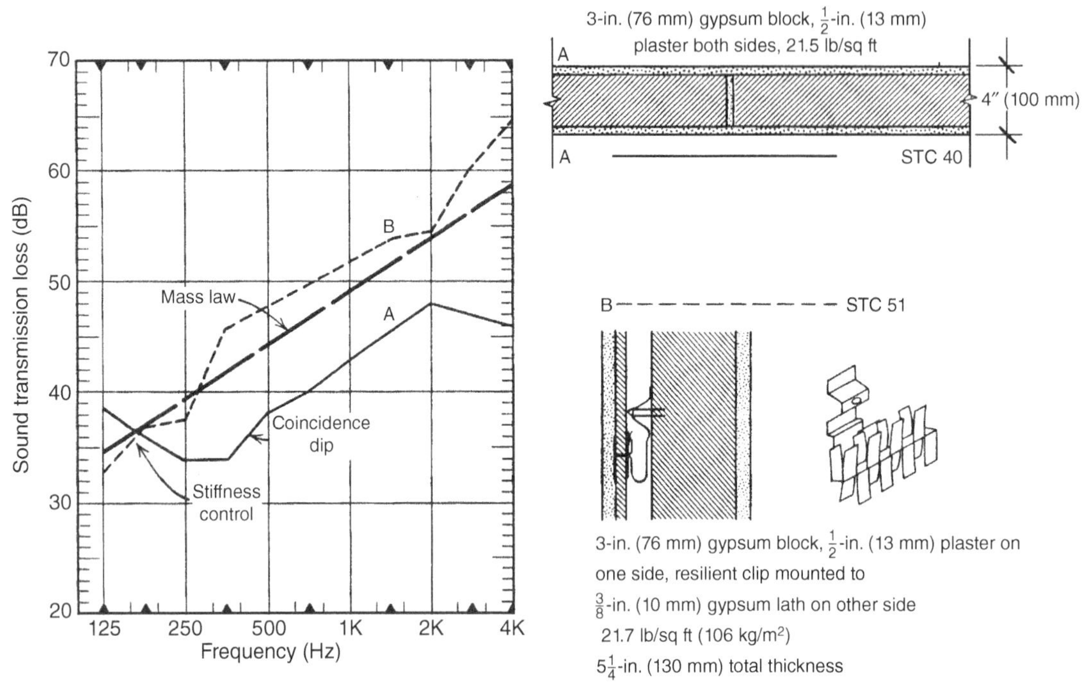

Figure 25.15 Transmission loss characteristics of two equal-weight partitions with similar boundary constraints. The solid partition A performs worse than the mass law, due to stiffness. The resilient-mounted wall B performs better than the mass law and much better than wall A, except at the lowest frequencies. (Adapted from HUD, 1968.)

can be improved still further by filling the void with porous, sound-absorbent material. This acts to further decrease the stiffness of the compound structure and to absorb the sound energy that reflects back and forth between the two inside surfaces. The performance of cavity walls is reduced by any rigid interconnections between leaves. Thus, a common stud wall with frequent rigid interconnections acts little better than a single homogeneous wall. However, a stud wall with staggered studs exhibits greatly improved performance over a single-material wall or a common stud wall. These effects are illustrated in Figures 25.16 and 25.17 (see also Appendix K). The effects of mass (Section 25.8), stiffness (Section 25.9), and compound barriers with and without filler are shown qualitatively and graphically in Figure 25.18.

25.11 SOUND TRANSMISSION CLASS

Various attempts at using a single-number average transmission loss to describe a barrier's characteristics have been made, with only limited success. Indeed, such averages can be misleading, since they ignore both deficiencies and proficiencies at particular frequencies. Their use, therefore, in all but rough work, is to be discouraged.

To avoid the shortcomings of averages and yet to benefit from the indisputable convenience of single-number ratings, a system of standard contours was developed in the United States called sound transmission class (STC) contours. (A similar system, conforming to ISO 717-1 : 2020, is used in Europe, involving a weighted sound reduction index R_W.) In practice, the STC number for a particular barrier construction is derived by comparing actual test results measured in a series of 16 ⅓-octave bands to the standard STC contours according to a fixed procedure. The technique is illustrated in Figure 25.19. Figure 25.20 shows two transmission loss curves and the STC ratings of each. Because STC fails to give credit for performance above the established requirements, octave band transmission loss data, rather than STC ratings, should be used in all critical areas such as music rooms or mechanical rooms where certain particular frequencies may be dominant.

Figure 25.21 gives three standard STC contours that are of interest because they are used by the Federal Housing Administration (FHA) to specify grades of construction. The criteria for their application are found in Section 25.33. An appreciation of the degree of speech sound isolation provided by walls with different STC ratings is given in Section 25.17 (see Table 25.5). Since the subjective reaction on the quiet side depends upon the background sound level, the table gives this reaction for two NC curve levels. To assist the designer, extensive sound transmission testing has been performed on most types of standard wall and

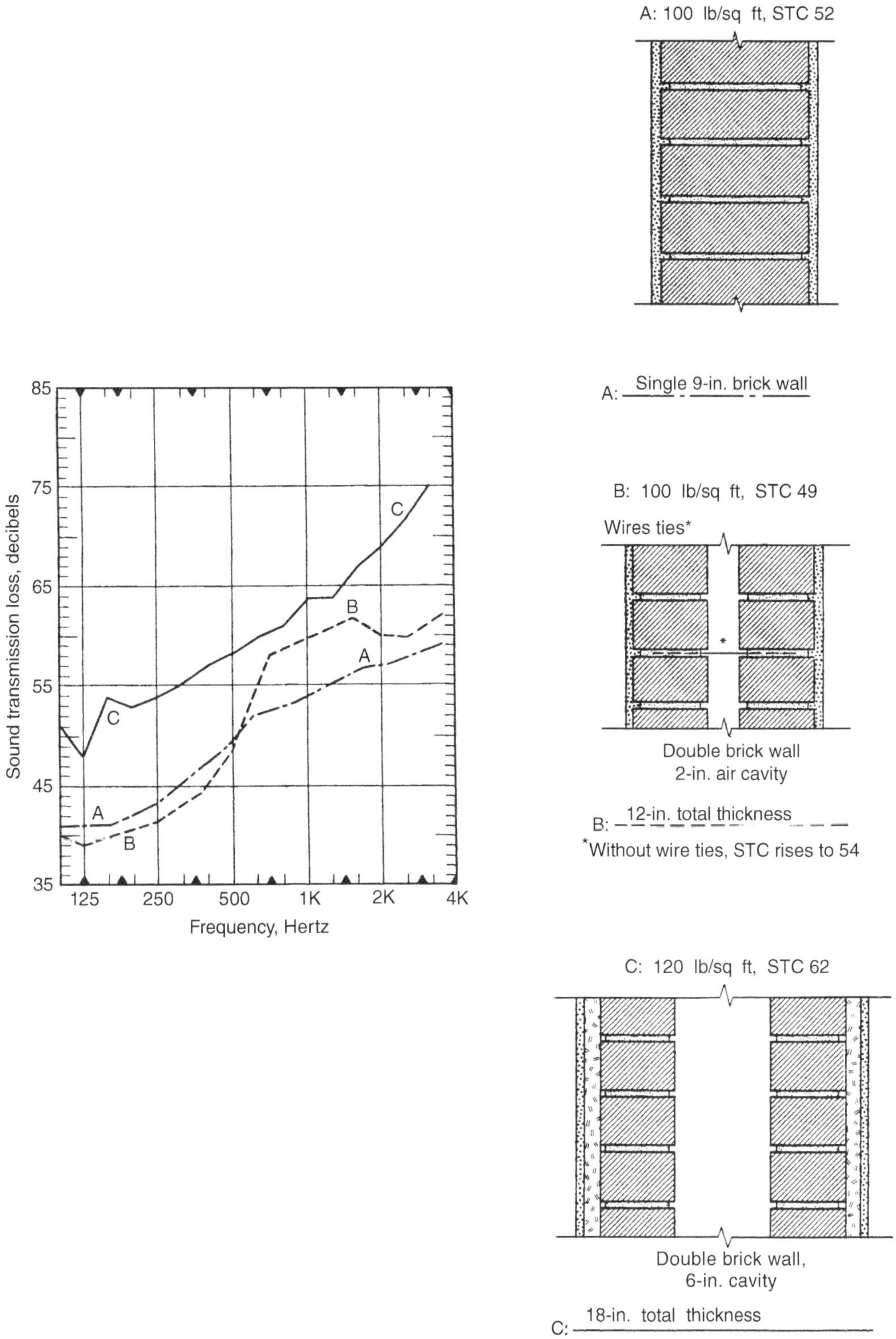

Figure 25.16 *Transmission loss curves (left) showing the effect of air space on heavy wall construction. All three walls (right) are approximately the same mass. The 2-in. [50-mm] air space of wall B is not significant until the higher frequencies, whereas the large 6-in. [200-mm] air space of wall C is effective throughout the frequency spectrum. (Adapted from HUD, 1968.)*

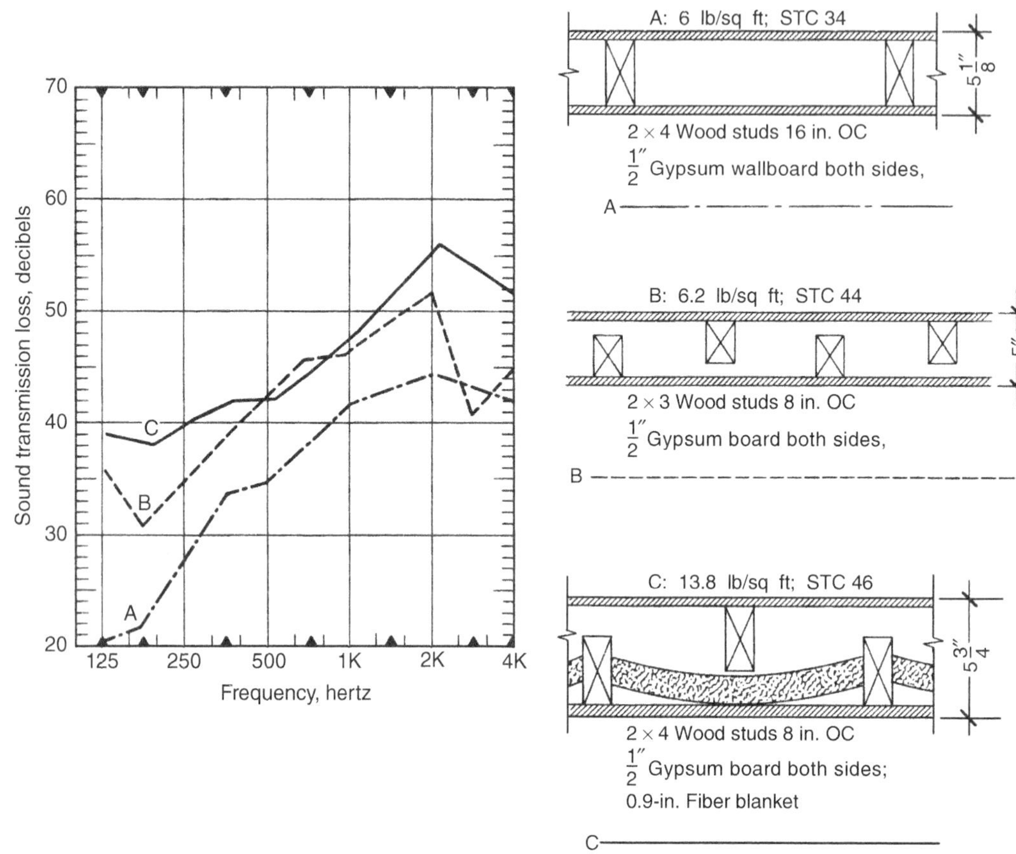

Figure 25.17 *Transmission loss curves illustrating the effect on lightweight walls of stiffness reduction and addition of absorptive material in the cavity. Curve A is a standard stud wall typically found in frame construction. Curve B shows the advantage of staggered studs over the entire frequency range. The dip at 3 kHz is a coincidence dip for a single leaf. Addition of absorptive material (curve C) improves the attenuation characteristic at both ends of the spectrum and is particularly useful for its low-frequency improvement. (Adapted from HUD, 1968.)*

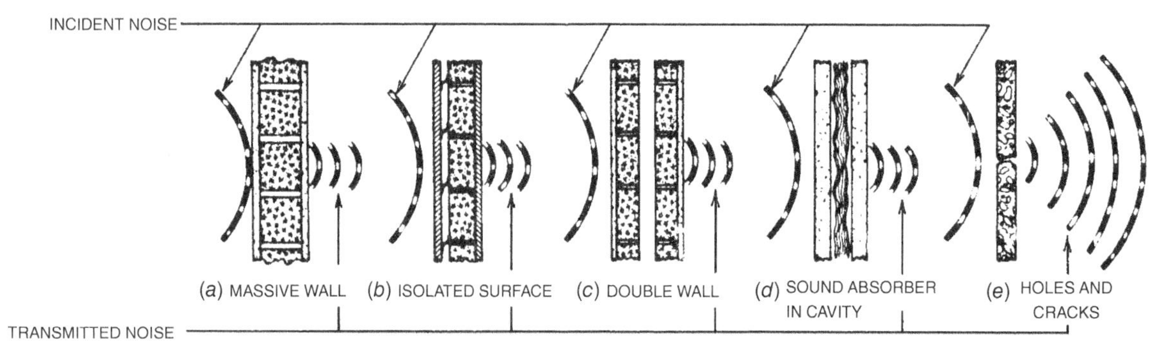

Figure 25.18 *(a–d) The various techniques used to increase the transmission loss of a partition. In contrast, even a very small air path through the partition (e) can effectively destroy its effectiveness as a sound barrier. (Adapted from HUD, 1968.)*

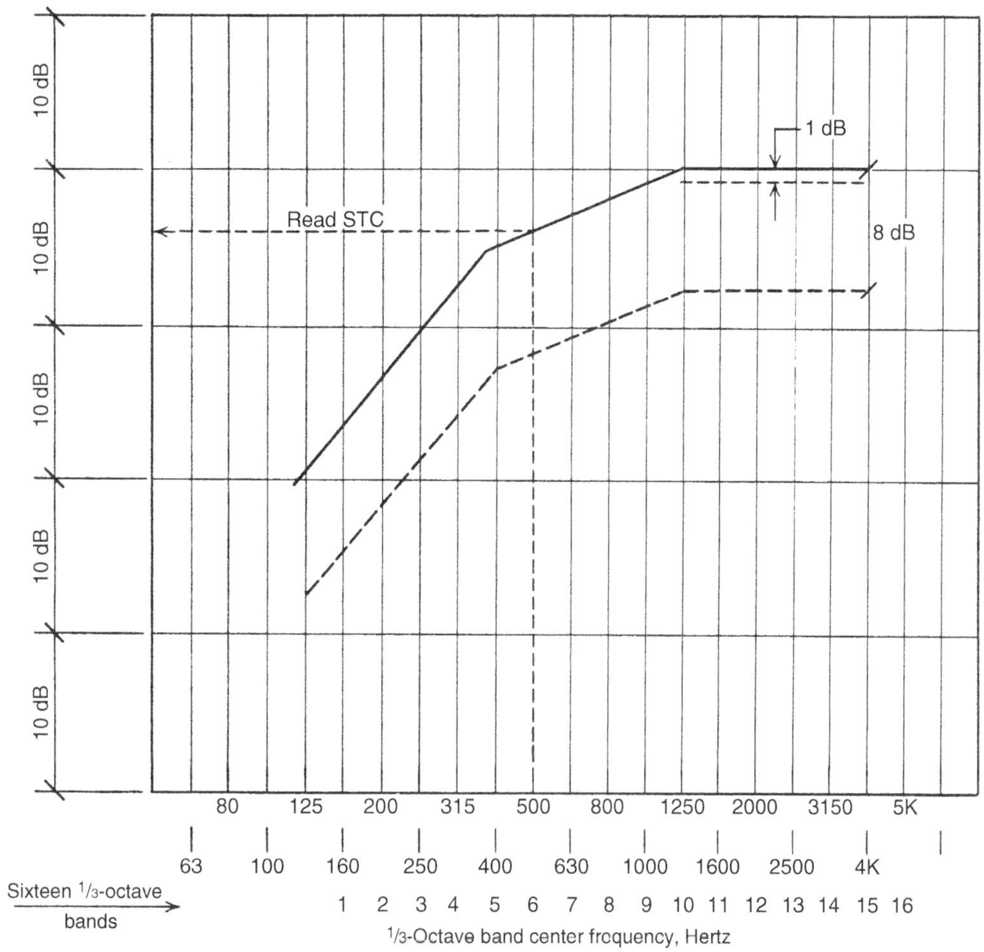

The STC is determined by comparison with a transparent overlay of this graph on which the STC contour is drawn. The STC contour is shifted vertically, relative to the test curve, until some of the measured *TL* values for the test specimen fall below those of the STC contour (the solid line) and the following conditions are fulfilled:

1. The sum of the deficiencies (i.e., the deviations below the contour) shall not be greater than 32 dB.
2. The maximum deficiency at a single test point shall not exceed 8 dB [the broken (dashed) line beneath the STC contour].

When the contour is adjusted to the highest value (in integral dB) that meets the above requirements, the sound transmission class for the specimen is the *TL* value corresponding to the intersection of the contour and the 500-Hz ordinate.

Figure 25.19 *Overlay from which sound transmission class (STC) is determined graphically.*

partition construction and the results published. Tables 25.1 and 25.2 and Appendix K give descriptions of constructions with typical details, transmission loss data, STC ratings, and other pertinent data.

25.12 COMPOSITE WALLS AND LEAKS

It is frequently necessary to determine the transmission loss of a composite wall—that is, a wall with a window, door, louver opening, and the like. It should be appreciated that the two elements are "in parallel," to borrow an electrical concept, and the acoustical behavior is similar to that situation. That is, the overall performance will be strongly affected by the poorer of the two, with some tempering of the degradation when the poorer barrier is much smaller in area than the other barrier element. Figure 25.22 enables us to analyze situations of this type.

Since an opening in a wall is effectively a second material of $TL = 0$, the curves in Figure 25.22 can be replotted for this situation as in Figure 25.23. Note that the curves very rapidly flatten out; thus, any wall with a 1% open area will have a maximum transmission loss of 20 dB, which makes it all but useless as a sound barrier. For this reason, it is imperative that

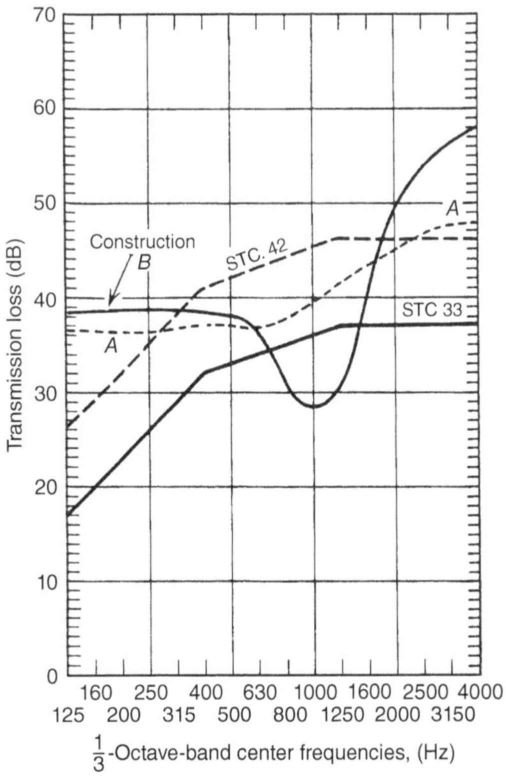

Figure 25.20 *Curves A and B are two different construction types with the same average transmission loss. However, application of the STC curve criteria (given in Figure 25.19) yields an STC of only 33 for construction B because of its deep center dip, compared to an STC of 42 for construction A. (From Magrab, 1975.)*

TABLE 25.1 Improvements in STC Rating of Stud[a] Partitions[b]

Description	STC[c]
Basic partition: single wood studs, 16 in. [406 mm] on center, ½-in. [13-mm] gypsum board on both sides, air cavity	35
Add to basic partition:	
Double gypsum board, one side	+2
Double gypsum board, both sides	+4
Single-thickness absorbent material in air cavity	+3
Double-thickness insulation	+6
Resilient channel supports for gypsum board	+5
Staggered studs	+9
Double studs	+13

[a]For application to metal stud partitions, use adders as in note b, but begin with STC = 40 for a 3 5/8-in. [92-mm] basic partition.

[b]When using two improvements, add an additional +2; for three improvements, add +3.

Example: Improvements to 35 STC basic partition:

Staggered wood studs	+9
Double gypsum board, one side	+2
Single-thickness insulation	+3
Adder (3 improvements)	+3
Total	+17
Total STC	35 + 17 = 52

[c]The STC values are conservative. Other sources list the same constructions with 1–5 points higher STC.

all openings be completely sealed, particularly those around doors and windows. A hairline crack degrades a wall 6 dB, a keyhole degrades a door 3 dB, and so on. Special considerations for doors and windows are discussed in the next section. Care must also be taken

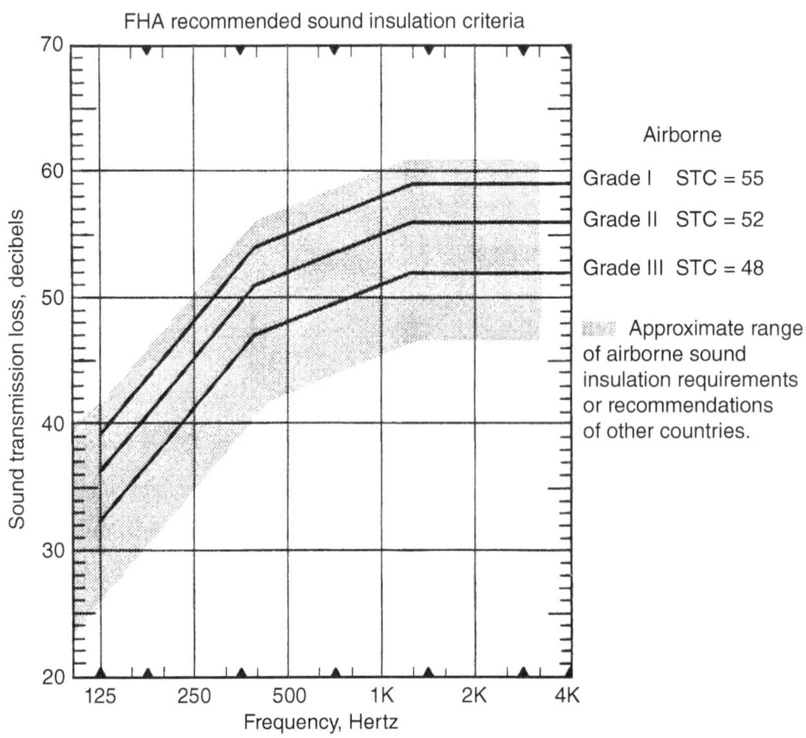

Figure 25.21 *Sound insulation criteria recommended by the FHA. (From HUD, 1968.)*

TABLE 25.2 STC Ratings of Masonry Walls

Description	STC[a]
4-in. [102-mm] lightweight[b] hollow block	36
4-in. [102-mm] dense hollow block	38
6-in. [152-mm] lightweight hollow block	41
6-in. [152-mm] dense hollow block	43
8-in. [203-mm] lightweight hollow block	46
8-in. [203-mm] dense hollow block	48
12-in. [305-mm] lightweight hollow block	51
12-in. [305-mm] dense hollow block	53
4-in. [102-mm] brick	41
6-in. [152-mm] brick	45
8-in. [203-mm] brick	49
12-in. [305-mm] brick	54
6-in. [152-mm] solid concrete	47
8-in. [203-mm] solid concrete	50
10-in. [254-mm] solid concrete	53
12-in. [305-mm] solid concrete	56

[a]See note c, Table 25.1.

[b]All ratings of lightweight block assume sealing with paint. Note that this reduces absorption.

Modifications	
Add sand to cores of hollow blocks	+3
Add plaster to one side	+2
Add plaster to both sides	+4
Add furring strips, lath and plaster	
One side	+6
Two sides	+10
Add plaster via resilient mounting	
One side	+10
Two sides	+15

with such common acoustic leaks as back-to-back electric outlets, pipes passing through walls, and medicine cabinets—in fact, any break in the integrity of a partition. All such openings must be caulked to make an airtight joint if any appreciable degree of sound insulation is to be maintained.

Examples 25.1 and 25.2 show how barriers of high transmission loss are degraded by standard openings. To maintain the integrity of a barrier, special care must be taken in the design of windows and doors, as explained in Section 25.13.

EXAMPLE 25.1 Given a 9-ft × 18-ft [2.7-m × 5.5-m] wall with a transmission loss of 52 dB at 1000 Hz, containing a 3-ft × 7-ft [0.9-m × 2.1-m], 6-in. [150-mm] hollow core door of 22 dB transmission loss at that frequency, find the overall transmission loss of the composite wall.

SOLUTION

Refer to Figure 25.22.

$$TL_1 - TL_2 = 52 - 22 = 30 \text{ dB}$$

$$\frac{S_2}{S_1} = \frac{3 \times 7.5}{9 \times 18} \times 100 = 13.9\%$$

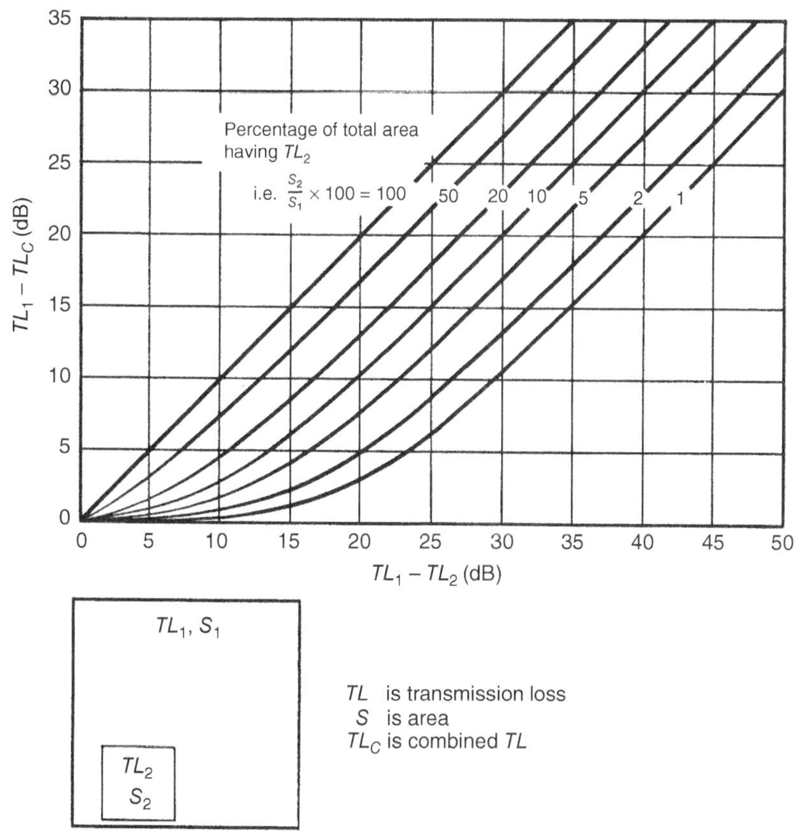

Figure 25.22 Transmission loss of a two-element composite barrier as a function of the relative transmission loss of the components. (From Magrab, 1975.)

ACOUSTICS

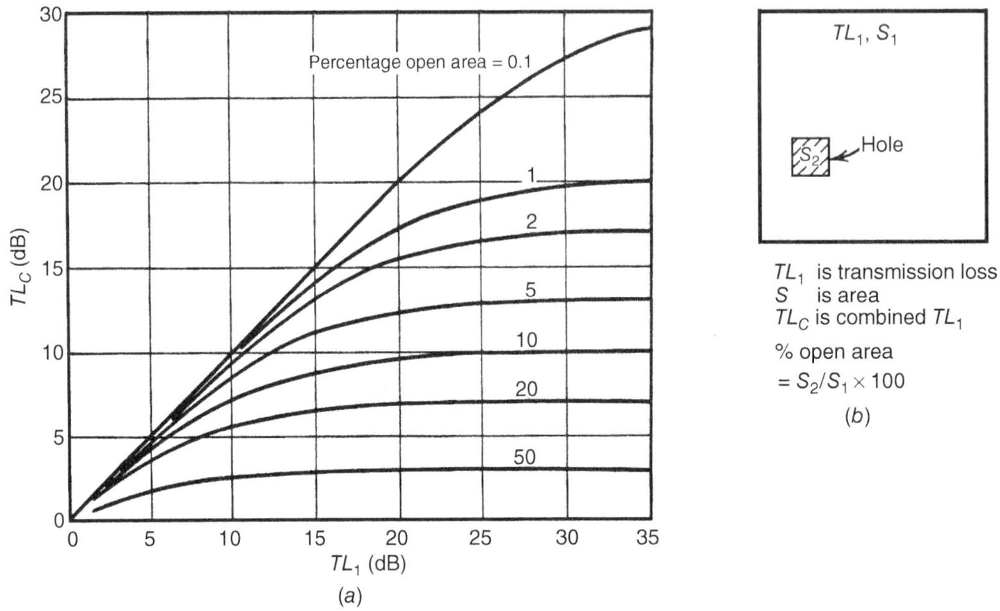

Figure 25.23 *Effect of a hole of a given size (as a percentage of the total area) on a partition with a given transmission loss. (From Magrab, 1975.)*

From the curves in Figure 25.22:

$$TL_1 - TL_C = 21.5 \text{ dB}$$

$$TL_C = 52 - 21.5 = 30.5 \text{ dB}$$

Thus, a door with an area of only 13.9% of the entire wall reduces the transmission loss of the structure from 52 to 30.5 dB—from excellent to very poor. ■

EXAMPLE 25.2 An exterior brick/frame wall having a transmission loss of 54 dB at 1000 Hz, measuring 8 ft × 16 ft [2.4 m × 4.9 m], is pierced by two wood-frame windows, each of area 3 ft × 4 ft [0.9 m × 1.2 m], with single ⅛-in. [3 mm] glass, with a transmission loss of 34 dB at 1000 Hz. Find the combined transmission loss.

SOLUTION

$$TL_1 - TL_2 = 54 - 34 = 20 \text{ dB}$$

$$\frac{S_2}{S_1} = \frac{2 \times 3 \times 4}{8 \times 16} \times 100 = 18.8\%$$

From Figure 25.22:

$$TL_1 - TL_C = 12.5 \text{ dB}$$

$$TL_C = 52 - 12.5 = 41.5 \text{ dB}$$

Again, the result is a reduction from an excellent wall to a poor one. ■

25.13 DOORS AND WINDOWS

As can be appreciated from the preceding section, doors and windows can in large measure determine the overall transmission loss of a wall. Since in almost every instance doors and windows have a lower acoustic transmission loss than the wall in which they are mounted, care must be taken not to degrade performance further with air leaks.

(a) Doors

Figure 25.24 gives nominal transmission loss data for the most common types of doors as a function of frequency. Average transmission loss values for doors—that is, the arithmetic average of the octave band transmission losses in the range of 150 Hz to 3000 Hz—are not useful, for two reasons:

1. The very important low-frequency attenuation data are absent.
2. Sharp peaks and valleys in the curves (see, for instance, the 6-dB peaks at 500 Hz in Figure 25.24*a*) are unrecognized. As a result, a particularly troublesome frequency may not be sufficiently attenuated. In the absence of a complete frequency analysis, the STC rating of a door is a better indication than an average transmission loss figure. Typical STC values are given in Table 25.3.

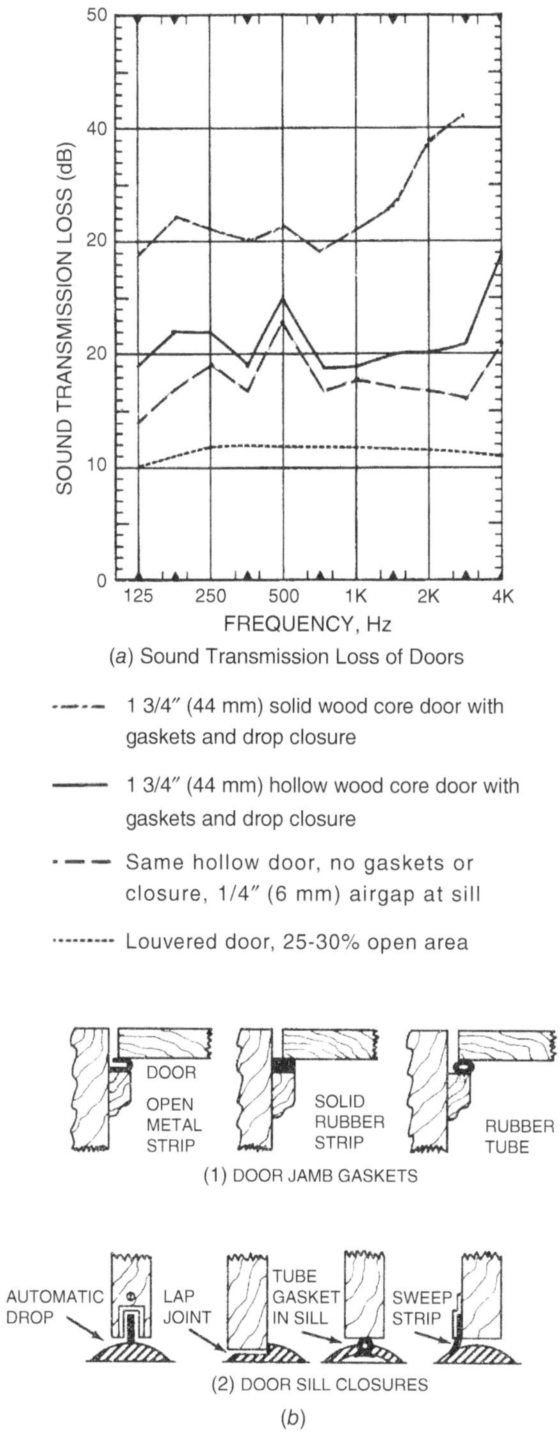

Figure 25.24 *(a) Chart of typical transmission loss values for representative door constructions as a function of frequency.*
(b) Method for gasketing a door edge enclosure (1) and sealing the gap between the bottom of the door and the door saddle (2). (From HUD, 1968.)

TABLE 25.3 Typical STC Values for Doors

Door Construction	STC
Louvered door	15
Any door, 2-in. [51-mm] undercut	17
1½-in. [38-mm] hollow core door, no gasketing	22
1½-in. [38-mm] hollow core door, gaskets and drop closure	25
1¾-in. [45-mm] solid wood door, no gasketing	30
1¾-in. [45-mm] solid wood door, gaskets and drop closure	35
Two hollow core doors, gasketed all around, with sound lock	45
Two solid core doors, gasketed all around, with sound lock	55
Special commercial construction, with lead lining and full sealing	45–65

Conclusions that can be drawn from inspection of Figure 25.24*a* are:

1. Louvered doors (and doors undercut to permit air movement) are useless as sound barriers.
2. The most important step in soundproofing doors is complete sealing around the opening. A door in the closed position should exert pressure on gaskets, making the joints airtight (see Figure 25.24*b*).

When a single door does not provide sufficient attenuation, and specially constructed high-attenuation commercial acoustic doors are not practical, a simple and very effective technique is the construction of a sound lock consisting of two doors, preferably with sufficient space between them to permit full door swing (see Figure 25.25). All surfaces in the sound lock should be covered completely by absorbent material and the floor carpeted. Such an arrangement will increase attenuation across the spectrum by at least 10 dB and by as much as 20 dB at some frequencies, depending upon the shape of the sound lock and the type, amount, and mounting of absorptive material in the sound lock. The two doors of the sound lock must be gasketed, as explained in the preceding discussion. In addition, each door should close quickly after user passage—but not slam shut.

Another important consideration with respect to sound intrusion via doors is the location of a door relative to sources of unwanted sound. This is particularly important in multiple-resident buildings of all types, including private homes, apartment houses, dormitories, hotels, and rooming houses. The same principle applies to commercial spaces where numerous private spaces such as offices open onto a common lobby or foyer. Figure 25.26 shows options for improvements in door placement.

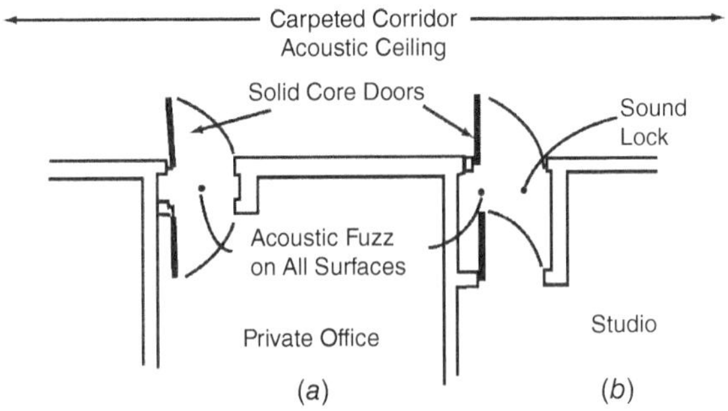

Figure 25.25 A two-door sound lock should increase the transmission loss of a door assembly by a minimum of 10 dB for the small lock (a) and 15 dB for the larger one (b), depending upon the type, thickness, and mounting technique of the absorptive material in the lock. The solid core doors must be sealed by one of the techniques shown in Figure 25.24(b).

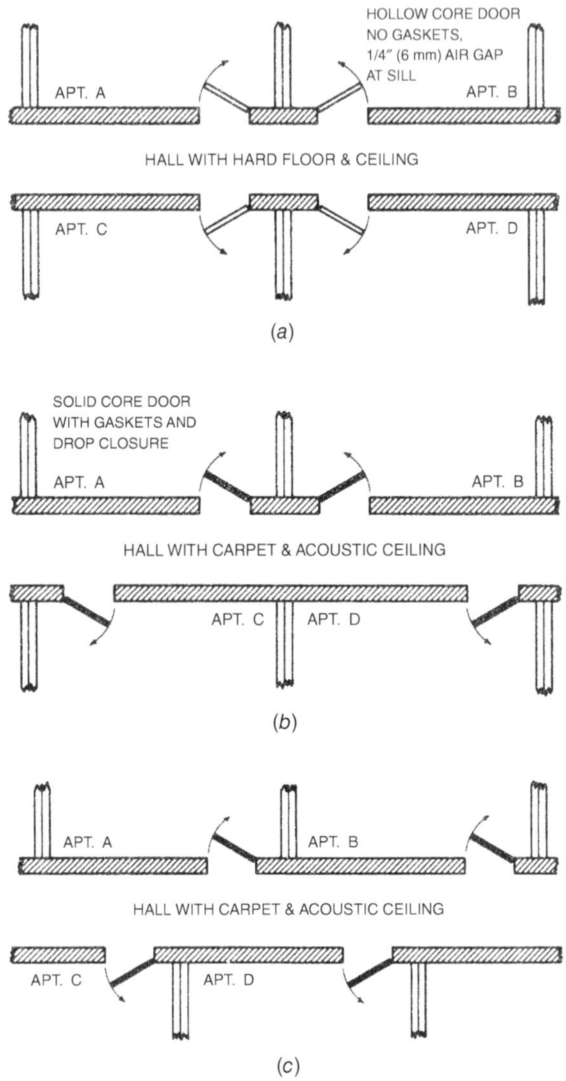

Figure 25.26 Proper arrangement of doors to rooms on a common corridor can diminish noise transfer in the area. (a) Poor arrangement because any noise emanating from one of the rooms or from the corridor has a very short and unattenuated path into the remaining rooms. (b) Better arrangement than plan (a) because noise from any source must travel a minimum of a room width along an absorbent corridor to reach any other room. Noise to the remaining rooms is further attenuated. The weak point of this plan is a noise short circuit via adjacent doors for Apt. A and Apt. B. Plan (c) is best because there are no short circuits for sound travel. Although the A–C and B–D paths are slightly shorter than in plan (b), the difference would not be noticeable. (From HUD, 1968.)

(b) Windows

Windows are critically important to blocking exterior noise, since exterior wall construction is generally of high STC, making the window the deciding factor in the composite exterior wall transmission loss. Sound leaks through cracks in closures of operable windows will normally establish a window's rating, regardless of the type of glazing. Fortunately, the attention now given to the sealing of windows for thermal purposes has had a positive effect on their acoustic properties. As with doors, the importance of proper gasketing and sealing cannot be overemphasized. Double glazing is effective only when the two panes are separated by a wide air gap (Figure 25.27). A narrow air gap acts as a stiff spring between the panes and transmits sound energy almost unattenuated. The result is approximately that of a single pane of double weight. Note that here too, as with absorptive material, the requirements of acoustic and thermal insulation are opposed. A small sealed air space between panes is desirable for thermal insulation, because a large space allows convection currents to transfer heat. For acoustical purposes, a small sealed space is not very useful, as explained previously, whereas a large space traps acoustic energy and is an effective noise barrier, as is clearly seen in Figure 25.28.

In addition to a window's sound transmission characteristic when closed, it is important to consider the transmission loss when open to accommodate ventilation and passive cooling requirements. The sound attenuation between the center of a room with a clear-through open window and a point some distance outside is 5–15 dB. This drops to about 5 dB as the receiver-observer approaches the open window. By making the path from inside to outside indirect, the open window attenuation can be increased to as much as 25 dB, but with considerable reduction of airflow and hence ventilation capacity. Several possible arrangements with approximate midfrequency transmission loss figures are given in Figure 25.29. This principle can be applied advantageously when exterior noise reduction is important but sealed windows are undesirable. Window-opening style and placement can also influence the amount of exterior noise admitted, as shown in Figures 25.29 and 25.30. Typical STC ratings of common window constructions are given in Table 25.4.

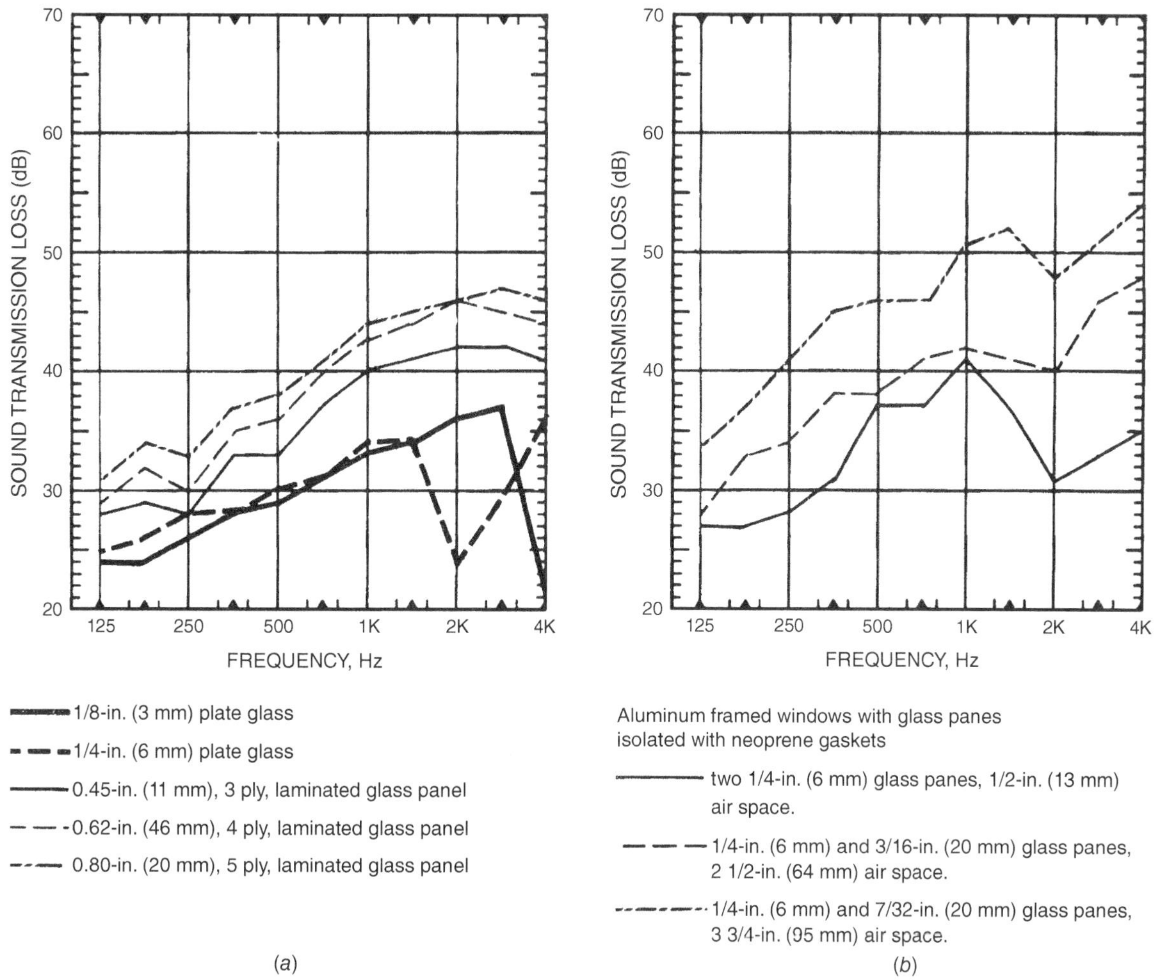

Figure 25.27 *(a) Sound transmission loss frequency spectrum shows the effect of thickening glass. The ¼-in. [6-mm] plate shows a very sharp (10-dB) coincidence dip at 2 kHz, making it less effective than a ⅛-in. [3-mm] plate between 1500 Hz and 2500 Hz. Further thickening with laminated glass eliminates the coincidence drop but reaches a practical limit at about ½-in. [13-mm] thickness. (b) Note that two ¼-in. [6-mm] panes with a ½-in. [13-mm] air space act as a stiff, thick pane and exhibit the sharp coincidence drop at 2 kHz. Larger sealed inter-pane air spaces markedly improve the acoustic insulation performance.*

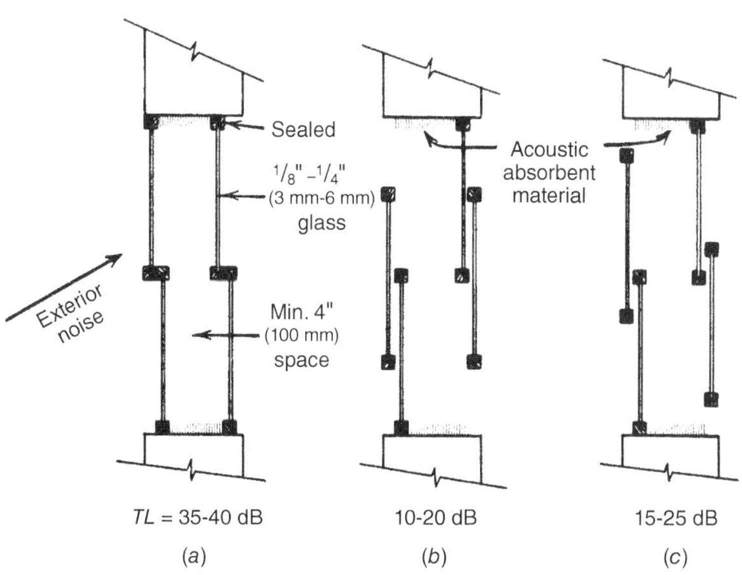

Figure 25.28 *The degree of attenuation of external noise can be regulated with acoustic sealant and absorbent materials when using pairs of double-hung (consider this a section view) or horizontally sliding (consider this a plan view) windows. Ventilation airflow varies inversely with transmission loss.*

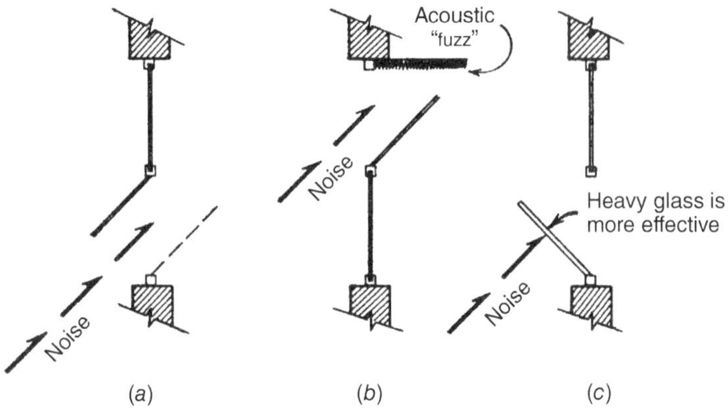

Figure 25.29 Alternative arrangements of the same basic "hopper" window design can yield results differing by as much as 10 dB. Design (a) is entirely open, and the noise path is unobstructed deep into the room. Design (b) is about 5 dB better than (a) at frequencies above 1 kHz because of higher absorption and less diffraction. Lower frequencies diffract readily around the window leaf and are less affected by absorptive material. Design (c) can be 10 dB better than (a), particularly at high frequencies, because it interposes a rigid barrier into the noise path. In this arrangement, the glass thickness is important.

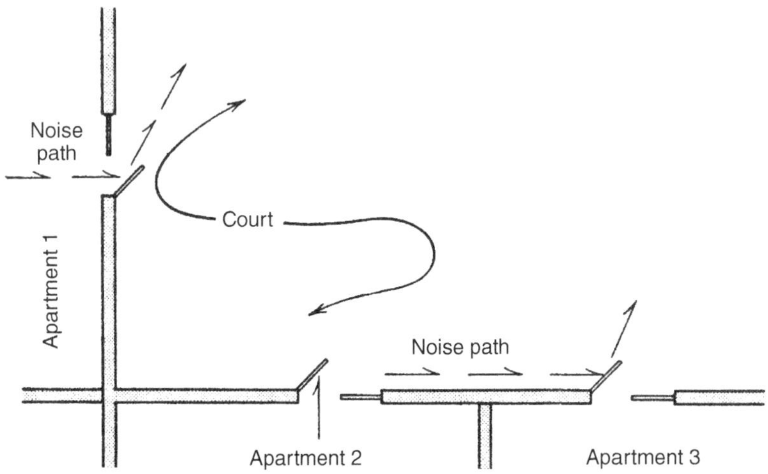

Figure 25.30 Noise transfer between contiguous corner spaces, as in Apartments 1 and 2, can be particularly severe if windows are improperly designed. Swinging windows, as shown, are preferable to double-hung or hopper windows because they reflect sound away from the adjacent space. Similarly, adjacent spaces on the same wall, such as Apartments 2 and 3, can benefit from this type of swinging arrangement, which is preferable to sliding or double-hung designs.

TABLE 25.4 Typical STC Values for Windows

Window Construction	STC
Operable wood sash, ⅛-in. [3.2-mm] glass, unsealed	23
Operable wood sash, ¼-in. [6.4-mm] glass, unsealed	25
Operable wood sash, ¼-in. [6.4-mm] glass, gasketed	30
Operable wood sash, laminated glass, unsealed	28
Operable wood sash, double-glazed, ⅛-in. [3.2-mm] panes, ⅜-in. [9.5-mm] air space, gasketed	29
Fixed sash, double ⅛-in. [3.2-mm] panes, 3-in. [76-mm] air space, gasketed	44
Fixed sash, double ⅛-in. [3.2-mm] panes, 4-in. [102-mm] air space, gasketed	48

25.14 DIFFRACTION: BARRIERS

The physical process by which sound passes around obstructions and through very small openings is called diffraction. Simply stated, diffraction is a process whereby any point on a sound wave establishes a new wave when passing an obstacle. Thus, although much of a sound wave is blocked by a small opening, the portion that does get through establishes a new wave front (see Figure 25.18*e*). The amplitude of the diffracted wave is determined by the relationship between the size of the opening and the wavelengths of the signal components. For a small hole, short wavelengths (high frequencies) are attenuated less than long wavelengths (low frequencies). See Figures 25.31 and 25.32.

When sound encounters a finite-length barrier, it diffracts around and over it, approximately as shown in Figure 25.32. The attenuation of the diffracted sound depends upon the frequency, type of source, and dimensions of the barrier. For a point source, with only a single practical path around the obstruction (barrier), the noise reduction in decibels can be calculated from Maekawa's empirical equation (in I-P units):

$$NR = 20\log\left[\frac{\sqrt{2\pi N}}{\tanh\sqrt{2\pi N}}\right] + 5\text{ dB} \qquad \text{(Eq. 25.4)}$$

where

N = (f/565) $(A + B - d)$
NR = noise reduction, dB
f = frequency, Hz
$A + B$ = shortest path length around the barrier, ft (over or around)
d = straight-line distance, source-to-receiver, ft
h = effective height of the barrier, ft

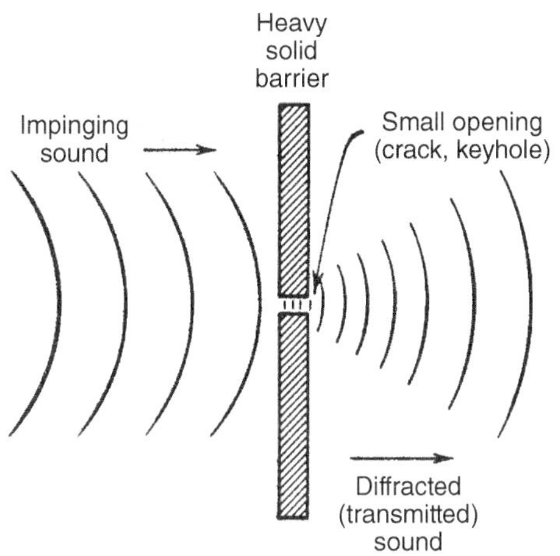

Figure 25.31 *Sound passes through small openings by diffraction. The intensity of the transmitted sound is proportional both to its frequency and to the size of the opening. It is always less than the intensity of the impinging sound.*

Note the following points about this equation:

1. It is applicable only to exterior barriers where sound passing over the barrier is partially diffracted and partially attenuated by distance. In an interior situation, sound passing over a partial-height barrier (see Figure 25.40) strikes the ceiling and is reflected down, increasing the received sound and effectively reducing barrier attenuation. Maximum exterior barrier attenuation is 24 dB, as compared to about 15 dB for a partial-height interior partition.
2. It assumes that the barrier is very long (or very high), so that only one sound path exists. In practice, a barrier whose length (height) is at least four times the distance between the source and the wall is sufficient if the barrier is close to the source. If the barrier is close to the receiver, it must be longer (higher) still.
3. It assumes a point source. Line sources (such as traffic) show 20–25% less attenuation for the same barrier.

The equation will, however, give reliable, usable results when the dimensions of the source are small with respect to the barrier, as is the case for: speech; individual motors, fans, engines, and other mechanical devices; and individual motor vehicles. The chart in Figure 25.23 relates barrier dimensions and position to traffic noise reduction. Note that frequency is not a variable on the chart, since it has been plotted for an average attenuation at 220 Hz, which is the center frequency for random car and truck traffic.

The best location for a barrier is either very close to the source or very close to the receiver. The worst position for attenuation is halfway between them. All

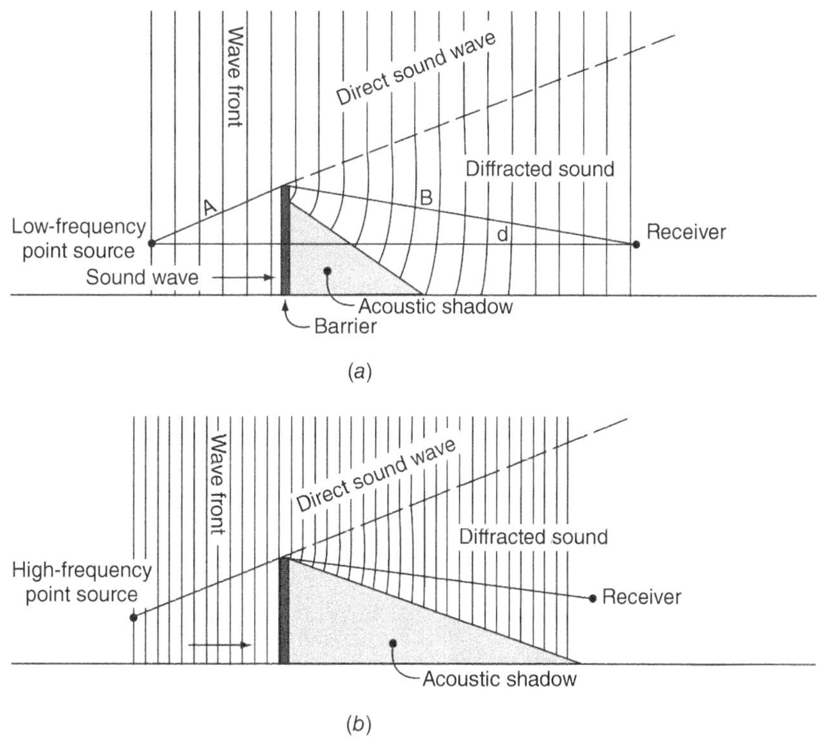

Figure 25.32 *Comparison of the effect of a barrier on sources of different frequencies. The low-frequency sound (a) diffracts more readily over the barrier than the high-frequency sound (b) because of its longer wavelength. Thus, the lower the frequency, the smaller the acoustic shadow and the lower the barrier attenuation. The shadow is not as sharply defined as shown; it represents increasing attenuation closer to the barrier.*

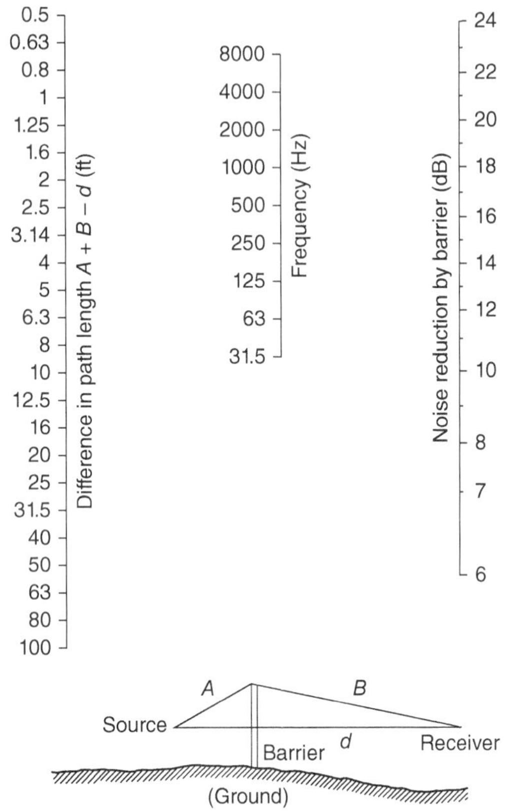

Figure 25.33 *This nomograph for estimating the noise reduction afforded by a barrier is based on Equation Eq. 25.4 and assumes a point (small) source and only a single path around the barrier. The dimensions A, B, and d are taken from the insert sketch. Dimensions A plus B represent the shortest path around the barrier—which may be over or around it. (From Fader, 1981.)*

effective barriers are assumed to be opaque and to have a minimum surface density of 5 lb/ft^2 [~25 kg/m^2]. The inherent transmission loss of the barrier need not be very high; a massively thick barrier has only marginally higher attenuation than one with the minimum surface weight. Absorptive material placed on the source side of a barrier will reduce the noise reflected toward the source but will not effectively increase the barrier's attenuation with respect to the receiver. Although the maximum theoretical noise reduction of an exterior barrier is about 24 dB, in practice, it rarely exceeds 20 dB. Figure 25.33 is a nomograph based upon Equation Eq. 25.4.

25.15 FLANKING

Just as sound will pass through the acoustically weakest part of a composite wall, it will also find parallel or flanking paths, that is, an acoustic short circuit. Proper design of window locations to avoid flanking paths has already been shown in Figure 25.30. The same situation pertains with respect to doors, as shown in Figure 25.26, and any other openings between spaces. Thus, in Figure 25.34, a high-STC wall between the two spaces is in large measure defeated by flanking paths F5, F6, and F7. In other spaces, the most common flanking path is via the plenum, as in Figure 25.34 (path F1) and in Figures 25.35*b*, *d*. Ductwork (with registers or grilles in various rooms) acts as an excellent intercom system unless it is completely lined with sound-absorptive material (see Section 25.27). Even then, low-frequency sound is only minimally attenuated, and special measures must be employed if good transmission loss is required. This subject is discussed further in Sections 25.25 through 25.27.

SPEECH PRIVACY

25.16 PRINCIPLES OF SPEECH PRIVACY BETWEEN ENCLOSED SPACES

The subject of speech privacy is of paramount importance in office design. Numerous studies have demonstrated that productivity and noise are related inversely when the noise carries information. When noise does not carry information, it can be annoying and therefore counterproductive, or it can be useful as a masking sound, depending upon its frequency content, intensity level, and constancy. Referring to Section 25.7, which discusses the noise reduction of an airtight barrier between two spaces, we saw that the sound intensity levels in the source room (1) and the receiving room (2) are related by the expression

$$IL_2 = IL_1 - NR$$

where NR is noise reduction, and IL_2 and IL_1 are sound intensity levels in the receiving and source rooms, respectively. If the receiving room is completely quiet and has no sound source other than the transmitted sound (essentially IL_2), then that sound will always be a potential source of annoyance to the occupant of the receiving room, as long as its intensity level is above the hearing threshold. If, however, there is a constant ambient sound level in the receiving room, then, depending upon its characteristics, it may mask the transmitted sound IL_2, even to the extent of making it completely inaudible. In most instances, however, it simply reduces or eliminates annoyance without completely masking the source. What we hear (and therefore what can potentially be a source of disturbance) depend upon our level of attention both to what we are doing and to the intrusive sound. (A remarkable exception is the ability of some, generally young, students to study in the presence of very loud, familiar—and therefore information-bearing—music. Indeed, some claim that they can only study that way.)

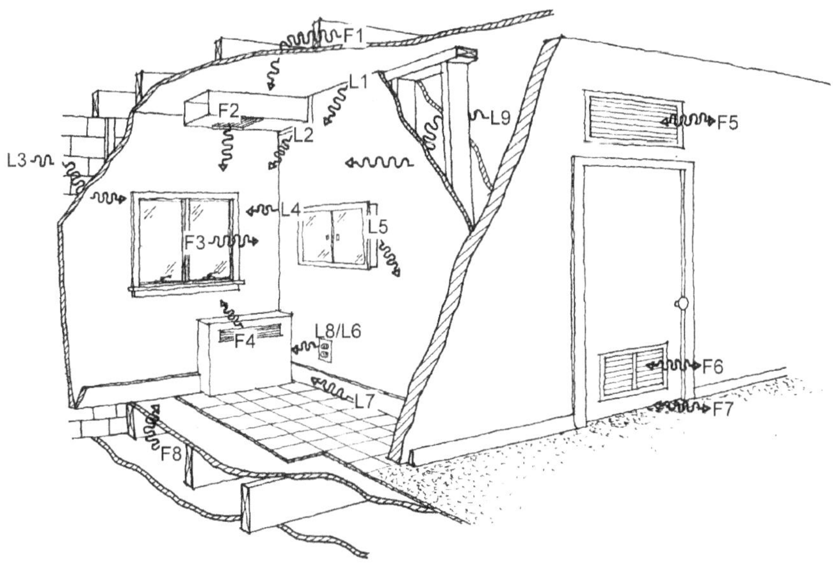

FLANKING NOISE PATHS

F1 OPEN PLENUMS OVER WALLS, FALSE CEILINGS
F2 UNBAFFLED DUCT RUNS
F3 OUTDOOR PATH, WINDOW TO WINDOW
F4 CONTINUOUS UNBAFFLED INDUCTOR UNITS
F5 HALL PATH, OPEN VENTS
F6 HALL PATH, LOUVERED DOORS
F7 HALL PATH, OPENINGS UNDER DOORS
F8 OPEN TROUGHS IN FLOOR-CEILING STRUCTURE

NOISE LEAKS

L1 POOR SEAL AT CEILING EDGES
L2 POOR SEAL AROUND DUCT PENETRATIONS
L3 POOR MORTAR JOINTS, POROUS MASONRY BLK
L4 POOR SEAL AT SIDEWALL, FILLER PANEL, ETC.
L5 BACK-TO-BACK CABINETS, POOR WORKMANSHIP
L6 HOLES, GAPS AT WALL PENETRATION
L7 POOR SEAL AT FLOOR EDGES
L8 BACK-TO-BACK ELECTRICAL OUTLETS

OTHER POINTS TO CONSIDER, RE: LEAKS ARE (A) BATTEN STRIP A/O POST CONNECTIONS OF PREFABRICATED WALLS, (B) UNDER-FLOOR PIPE OR SERVICE CHASES, (C) RECESSED, SPANNING LIGHT FIXTURES, (D) CEILING & FLOOR COVER PLATES OF MOVABLE WALLS, (E) UNSUPPORTED A/O UNBACKED WALL BOARD JOINTS, (F) EDGES & BACKING OF BUILT-IN CABINETS & APPLIANCES, (G) PREFABRICATED, HOLLOW METAL EXTERIOR CURTAIN WALLS.

Figure 25.34 *Flanking transmission of airborne noise. (From HUD, 1968. Redrawn by Jonathan Meendering.)*

Tests have shown that most adults will not consider an intruding noise level IL_2 to be annoying if the intensity of a properly designed background sound is either greater than or no more than 2 dB less than IL_2. Thus, a transmitted IL_2 of 40 dBA will not be considered annoying by most people if the level of the background sound is at least 38 dBA. The upper level of usable background masking sound is usually taken to be about 50 dBA. Any higher intensity level will itself become a source of annoyance. Figure 25.36 gives a graphic representation of the relation between transmitted and background sound levels in a receiving room. By way of summarizing the preceding discussion, we can simply state that the degree of speech privacy in a space is a function of two factors:

- the degree of sound isolation provided by the barriers between rooms
- the ambient sound level in the receiving room

25.17 SOUND ISOLATION DESCRIPTORS

Since the degree of intrusiveness of extraneous noise in a space varies, depending as it does upon both transmitted and ambient sound, having a descriptive scale of some sort is appropriate. If we restrict our discussion to speech sounds, since often we are interested in office design, where speech is the primary sound source, then a descriptive scale, as shown in Table 25.5, can be established. With these absolute descriptions in mind and remembering that the hearing condition in a receiving room can be altered by changing either the barrier characteristics or the background sound level, or both, we can express the effectiveness of a construction as a speech-sound barrier in terms of its STC for a given ambient sound level. Since the ambient sound (noise) level is frequency-dependent, it can be approximated by an NC value. This is particularly useful when the ambient sound level is generated by machinery or

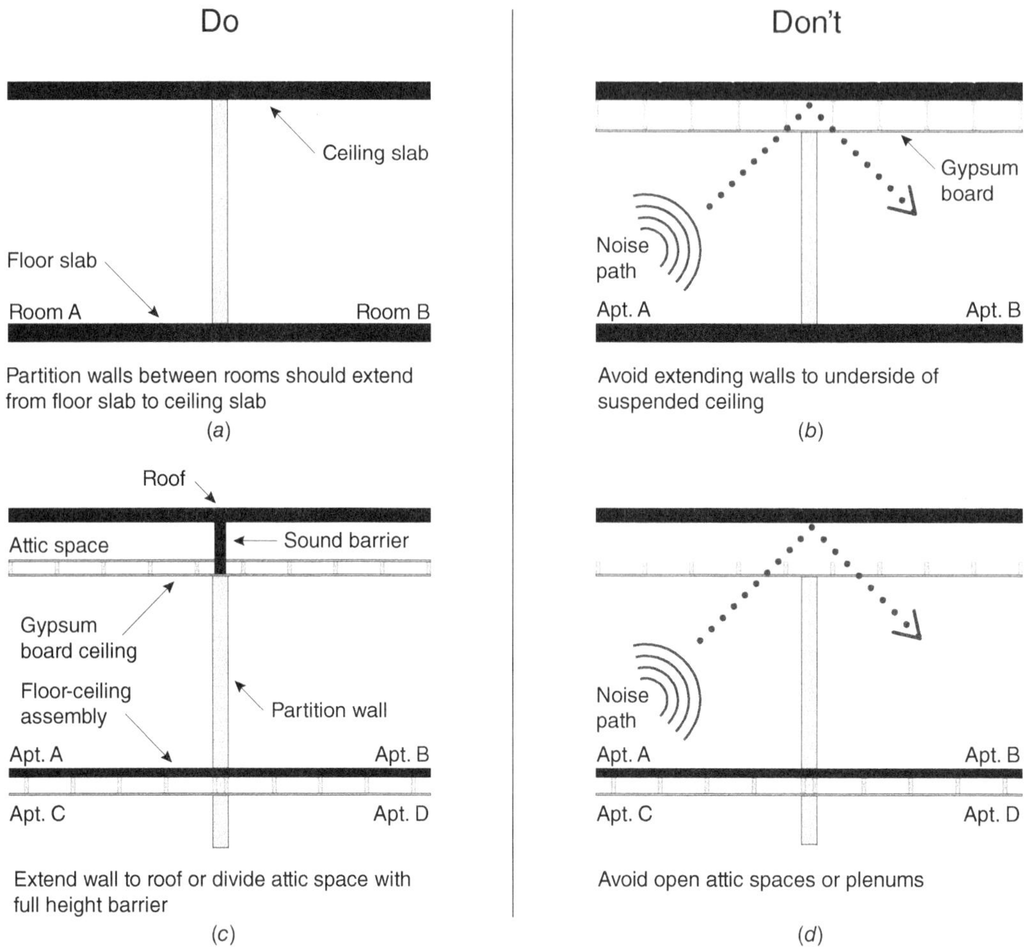

Figure 25.35 *Construction technique recommendations to avoid flanking paths. (Drawn by Nathan Leigh.)*

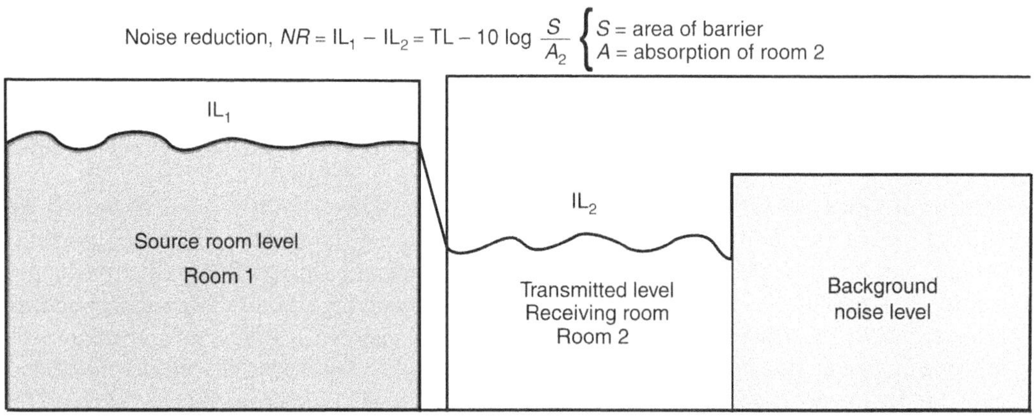

Figure 25.36 *The background noise level determines whether a transmitted sound will be heard. In this case, because the background noise level is considerably higher than the transmitted level from the adjacent room, the transmitted noise will not be noticeable (i.e., it will be masked).*

by the sound of an air-conditioning system rather than by a shaped signal from an electronic masking sound system.

Table 25.6 shows the hearing conditions in a receiving room with an NC-25 background noise as a function of the barrier STC rating. If the background noise level were raised to NC-30, then each descriptor would roughly increase one level in quality (i.e., poor would become fair, fair would become good, and so forth). Stated otherwise, the apparent isolation provided by a barrier may be increased by raising the background (masking) sound level in the receiving

TABLE 25.5 Relative Quality of Sound Isolation

Ranking	Descriptor	Hearing Condition[a]
6	Total privacy	Shouting barely audible
5	Excellent	Normal voice levels not audible. Raised voices barely audible but not intelligible.
4	Very good	Normal voice levels barely audible. Raised voices audible but largely unintelligible.
3	Good	Normal voice levels audible but generally unintelligible. Raised voices partially intelligible
2	Fair	Normal voice levels audible and intelligible some of the time. Raised voices generally intelligible.
1	Poor	Normal voice audible and intelligible most of the time
0	None	Normal voice levels always intelligible

[a]Hearing condition in the presence of ambient noise, if any.

TABLE 25.6 Relation between Barrier STC and Hearing Condition on the Receiving Side, Background Noise Level at NC-25

Barrier STC	Hearing Condition	Descriptor and Ranking[a]	Application
25	Normal speech can be understood quite easily and distinctly through the wall	Poor/1	Space divider
30	Loud speech can be understood fairly well. Normal speech can be heard but not easily understood.	Fair/2	Room divider where concentration is not essential
35	Loud speech can be heard but is not easily intelligible. Normal speech can be heard only faintly, if at all.	Very Good/4	Suitable for offices next to quiet spaces
42–45	Loud speech can be faintly heard but not understood. Normal speech is inaudible.	Excellent/5	For dividing noisy and quiet areas; party wall between apartments
46–50	Very loud sounds (such as loud singing, brass musical instruments, or a radio at full volume) can be heard only faintly or not at all	Total Privacy/6	Music room, practice room, sound studio, bedrooms adjacent to noisy areas

[a]See Table 25.5.

room. Figure 25.37 shows two conditions of adjacent spaces. Although the source room level is uniform and partitions on both sides of the source room are identical, the background sound in the two receiving rooms is different. In room A, the background is NC-35; in room B, it is NC-25. The occupant of room A is not disturbed by the little heard from the source room. The occupant of room B hears it clearly. Occupant A will probably praise the partition, whereas occupant B will complain. Although the levels of reradiated sound are identical in the two receiving rooms, the intruding signal is masked by the background sound in A, but it is clearly audible in B. Thus, the apparent noise reduction is substantially higher in A than in B. This clearly demonstrates the effectiveness of masking sound in providing apparent sound isolation and speech privacy.

Sound isolation can also be improved by careful planning. Storage and circulation areas can serve as buffers for noise-sensitive areas. Physical separation of noisy areas from quiet ones often eliminates the need for complicated and expensive compound barriers.

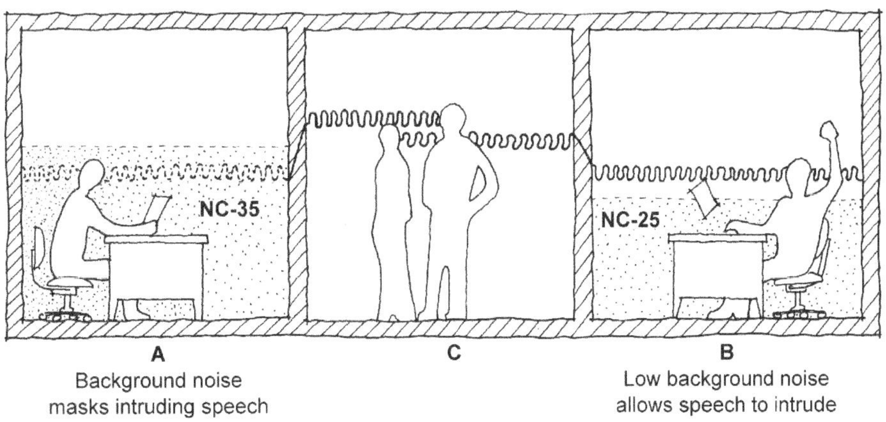

Figure 25.37 *The occupant in room A with background noise NC-35 (≈45 dBA) is unaware of the noise (loud speech from room C) that is so disturbing to occupant B, whose NC-25 (≈36 dBA) is insufficient to mask the transmitted noise. (Drawing by Jonathan Meendering.)*

ACOUSTICS

25.18 SPEECH PRIVACY DESIGN FOR ENCLOSED SPACES

The study of speech privacy has received considerable emphasis with the advent of open-plan offices (office landscaping), although the same issues prevail with both open and enclosed office designs. Essentially, the problem was to determine the factors affecting speech privacy and to quantify them with a degree of accuracy sufficient for design purposes. It rapidly became evident that although the physical principles are the same, the solutions to speech privacy design problems are radically different for closed spaces and for open-plan offices. In the former, the acoustic character of an airtight barrier between two spaces is the crucial element in the design, because all the interspace sound transmission is through this barrier. In contrast, sound transmission between adjacent "cubicles" in an open office is primarily the result of reflected and refracted sound, with the direct component passing through the barrier being of secondary importance. As a result, the acoustic transmission characteristics of the barrier also become secondary in importance.

Studies indicate that six factors are involved in enclosed-space speech privacy, which can be subsumed under two headings (Figure 25.38):

Heading 1: Speech rating—of the source room (Room No. 1):

1. *Speech effort:* A measure of the loudness of speech.
2. *Source room factor:* Gives the approximate effect of room absorption on the speech level in the source room. The scale in Figure 25.38 is drawn for average absorption. For live rooms, raise the factor by 2 points; for dead rooms, lower it by 2 points. Factors $a+b$ give the approximate source-room voice level.
3. *Privacy allowance:* What is the measure of privacy required? The privacy criteria definitions used in this step (Figure 25.38, Step 3) are:
   a. *Normal privacy:* Such that the receiving-room occupant can understand a small portion of normal voice conversation in the source room by listening intently. This corresponds to "Good" and a ranking of 3 in Table 25.5. It was found that most occupants can work normally with this level of speech intrusion.
   b. *Confidential privacy:* Assumes that only a few words will occasionally be intelligible. This privacy level corresponds to "Excellent" and a ranking of 5 in Table 25.5. The six speech-rating points between Normal Privacy (9 points) and Confidential (15 points) in Figure 25.38 correspond roughly to a 5-dB difference. If we remember that a sound intensity differential of 3 dB is barely perceptible, 6 dB is clearly noticeable, and 10 dB is a doubling of perceived sound (see Chapter 8), we can appreciate that the difference between normal and confidential privacy is small and calls for accurate design, plus a measure of field adjustment of masking sound.

Heading 2: Isolation rating—of the receiving room (Room No. 2):

4. *The STC rating of the barrier:* Table 25.7 gives some typical STC ratings for office partitions.
5. *The noise reduction factor:* A_2/S is an indication of receiving-room absorption, that is, the difference between *NR* and *TL*, where A_2 is the area of the receiving room, and S is the area of the barrier between the rooms. Absorption is assumed to be average. For live rooms, lower this factor 2 points; for dead rooms, raise it 2 points.

For the recommended background noise level in the receiving room, use Table 25.8.

An analysis sheet for enclosed spaces is provided in Figure 25.38 (see also Cavanaugh et al., 1962, and Young, 1965). The two examples of this analysis that follow should clarify its use. The reader should follow the analysis with Figure 25.38 in hand. The numbered steps in the examples correspond to the numbers in the figure.

EXAMPLE 25.3 Evaluate the effectiveness of a proposed office partition.

Source room:

General clerical office: 40 ft × 60 ft × 9 ft [12.2 m × 18.3 m × 2.7 m], higher-than-average $\bar{\alpha}$, 16-ft [4.9-m] long full-height partition, STC 40

Receiving room:

Conference room: 16 ft × 24 ft [4.9 m × 7.3 m], medium-dead room

Background sound level: 40 dBA (NC-30) (from Table 25.8)

Privacy analysis (using I-P units as per Figure 25.38):

(a)	Speech Rating:	
	Step 1. Speech effort: Raised	66
	Step 2. Source room floor area factor: $A_1 > 1000$ ft^2	0
	Step 3. Privacy allowance: Normal	9
	Speech rating (*a*) =	75
(b)	Isolation Rating:	
	Step 4. STC of common barrier (given)	40
	Step 5. Noise reduction factor (A_2/S) (16 × 24)/(16 × 9) = 2.6, corresponding to 2 on the scale; add 1 for higher than average $\bar{\alpha}$	3
	Step 6. Adjacent room background noise level	40
	Isolation rating (*b*) =	83
	Speech privacy rating, *a* − *b* =	−8

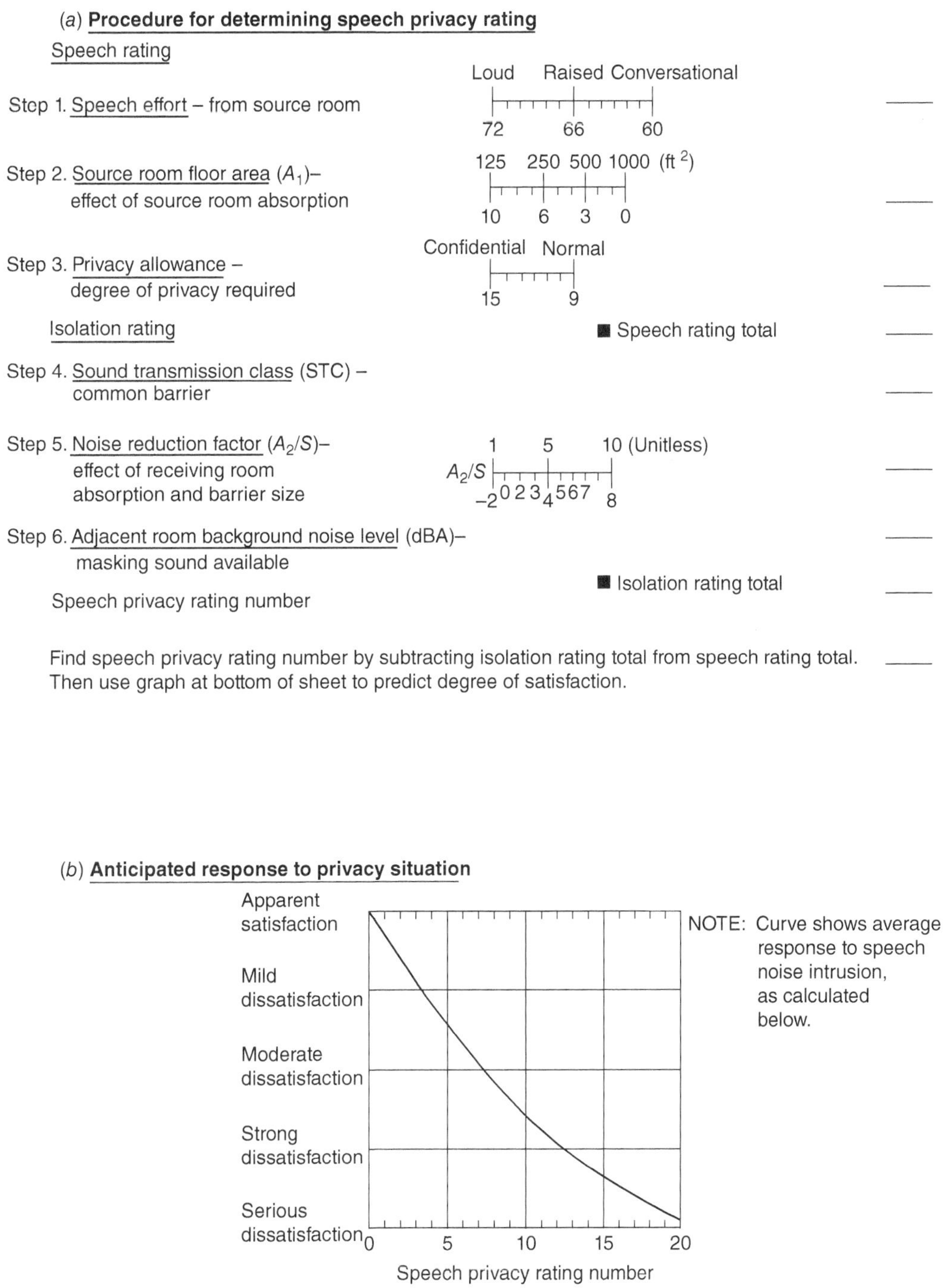

Figure 25.38 *Speech privacy analysis sheet. (Based on Cavanaugh et al., 1962.)*

TABLE 25.7 Typical STC Ratings of Interior Partitions

Type of Partition	STC
Demountable partition	20–30
Drywall partition up to acoustical ceiling	30
Drywall partition extending 12 in. [305 mm] above acoustical ceiling tile system into ceiling plenum	35
Drywall partition with cavity insulation, full height to the underside of slab above	40–45
Two-layer drywall partition with insulation, erected full height to underside of slab above	50

The partition performance is acceptable. In fact, the STC rating of the partition could be reduced to 32 without affecting speech privacy. ■

EXAMPLE 25.4 Evaluate speech privacy.

Source room:

Design studio: 20 ft × 30 ft [6.1 m × 9.1 m], medium-live

Common wall: 12 ft × 8 ft [3.7 m × 2.4 m] high, STC 26 (half-glass, with door)

Receiving room:

Supervisor's office: 12 ft × 14 ft × 8 ft [3.7 m × 4.3 m × 2.4 m], average absorption

Background sound level: 35 dBA

Privacy analysis (using I-P units as per Figure 25.38):

1. Speech effort: Conversation — 60
2. Source room floor area factor:
   For area: 600 ft^2 — +2
   For liveness — +1 — Total + 3
3. Privacy allowance: (almost) confidential — 13
   Speech rating $a = 76$
4. STC of common barrier (given) — 26
5. Noise reduction factor (A_2/S)
   $(12 \times 14)/(12 \times 8) = 1.8$, corresponding to a reduction factor of 1.6; gives 2.0 on the scale since the method uses whole numbers only; no adder required for average absorption
   Therefore, $2 + 0 = 2$
6. Adjacent room background noise level (given) — 35
   Isolation rating $b = 63$
   Speech privacy rating $a - b = 13$
   which indicates strong dissatisfaction

The suggested corrections are to increase the common wall STC to 36 by gasketing the door and to increase the background sound level in the receiving room to 40 dBA (NC-30). This would give a speech privacy rating of −2, as follows:

STC of barrier	36
A_2/S	2
Background noise	40
Isolation rating	78

Speech privacy rating $= a - b = 76 - 78 = -2$

This result should be satisfactory according to the chart in Figure 25.28. ■

TABLE 25.8 Suggested Noise Criteria Ranges for Steady Background Noise

Type of Space (and Acoustical Requirements)	NC Curve	Equivalent[a] dBA
Concert halls, opera houses, and recital halls (for listening to faint musical sounds)	10–20	20–30
Broadcast and recording studios (distant microphone pickup used)	15–20	25–30
Large auditoriums, large drama theaters, and houses of worship (for excellent listening conditions)	20–25	30–35
Broadcast, television, and recording studios (close microphone pickup only)	20–25	30–35
Small auditoriums, small theaters, small churches, music rehearsal rooms, large meeting and conference rooms (for good listening), or executive offices and conference rooms for 50 people (no amplification)	25–30	35–40
Bedrooms, sleeping quarters, hospitals, residences, apartments, hotels, motels, and so forth (for sleeping, resting, relaxing)	25–35	35–45
Private or semiprivate offices, small conference rooms, classrooms, libraries, and so forth (for good listening conditions)	30–35	40–45
Living rooms and similar spaces in dwellings (for conversing or listening to radio and TV)	35–45	45–55
Large offices, reception areas, retail shops and stores, cafeterias, restaurants, and so forth (for moderately good listening conditions)	35–50	45–60
Lobbies, laboratory workspaces, drafting and engineering rooms, general secretarial areas (for fair listening conditions)	40–45	50–55
Light maintenance shops, office and computer equipment rooms, kitchens, and laundries (for moderately fair listening conditions)	45–60	55–70
Shops, garages, power-plant control rooms, and so forth (for just acceptable speech and telephone communication). Levels above PNC-60 are not recommended for any office or communication situation.	—	—
For workspaces where speech or telephone communication is not required, but where there must be no risk of hearing damage	—	—

Source: From Magrab, 1975.

[a]For information only. These data are not part of the NC information and do not appear in the source.

25.19 PRINCIPLES OF SPEECH PRIVACY IN OPEN-AREA OFFICES

The huge increase in the service sectors of the world economy has brought with it a corresponding increase in desk jobs, each of which is often equipped with a computer console. This trend has also necessitated increased space density for office workers, made possible by the general elimination of paper storage (files) and the corresponding elimination of the necessity for employees to continually move about. The increased density problem has been largely solved by open-office plans with ever-smaller "cubicles," usually with single occupancy, but recently also with dual occupancy. This situation has aggravated the serious problem of annoyance due to the intrusion of speech sounds from neighboring workers, that is, a lack of speech privacy.

Since production is adversely affected by the inability of a worker to concentrate because of annoyance with speech intrusion—an annoyance that usually increases over time—the proper design of open office plans can have major economic benefits.

(a) Sound Paths in Open Offices

In contrast to the single, or at most dual, sound paths that exist with full-height enclosures (Figure 25.39), the sound paths in an open-plan arrangement (including first reflections) are direct, diffracted, ceiling-reflected, and laterally reflected (Figures 25.40 and 25.41). These illustrations disclose several important facts affecting speech privacy in open offices.

1. The angles of reflection of sound waves from the ceiling depend upon the location and height of the source and the ceiling height (Figure 25.40*a*). Measurements have shown that these angles vary from a minimum of 30° for a standing speaker in the center of a cubicle to a maximum of 60° for a speaker close to, and facing, a partition for ceiling heights of up to 9 ft [2.7 m]. Since much of the sound energy reaching an adjacent occupant does so after reflecting off the ceiling at these angles (30°–60°), a ceiling material with high absorption at these angles of incidence is required for speech privacy. This effectively negates the use of the noise reduction coefficient (NRC) as a useful factor to describe a ceiling material's absorption characteristic in an open-office design, since the NRC averages absorption at all angles. All major ceiling material manufacturers have tested, and will supply, accurate angular absorption data for their products. A single-number descriptor that relates to this characteristic, called the Articulation Class, is discussed in detail in Section 25.20.
2. Since absorption at the ceiling increases with the angle of incidence of a sound wave, it is always desirable for speech privacy to have maximum ceiling height. Most authorities recommend a minimum ceiling height of 9 ft [2.7 m].
3. Absorption efficiency of ceiling tiles varies inversely with the STC, which is to be expected, since sound transmission depends upon mass and absorption at air pockets. Light fiberglass ceiling tiles will typically have an absorption coefficient α of 0.95 at voice frequencies, with a Ceiling Attenuation Class (CAC) of 22–24 compared to more massive mineral fiber tiles with a maximum α of 0.8–0.85 but an STC of 34–36. Since we are interested in maximum absorption at the ceiling, the tile of choice is always that with the highest α at voice

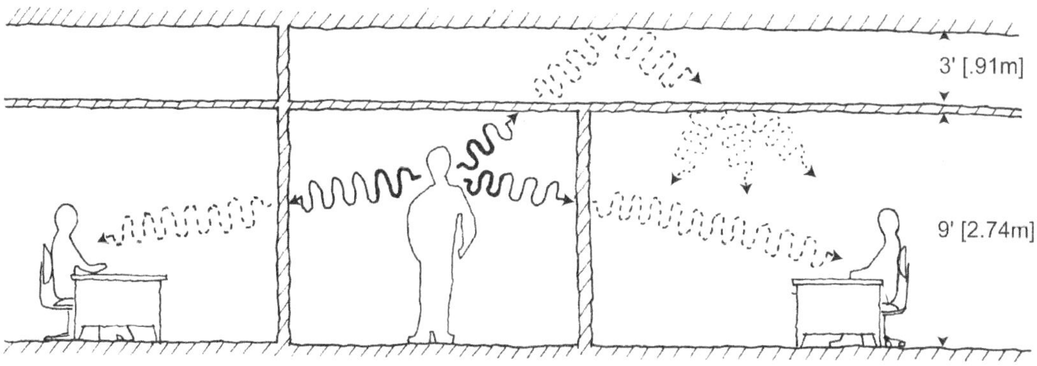

Figure 25.39 Sound paths between enclosed spaces are determined by the type of barrier separating them. In the case of a full-height barrier reaching to the underside of the ceiling slab, the only sound path is through the barrier, and its STC determines the sound pressure (noise) level in the receiving room. In the case of a ceiling-height partition with an overhead plenum, most of the sound energy will travel the upper, less-attenuating path. Factors affecting the level of received sound are the ceiling's CAC rating (Ceiling Attenuation Class, indicative of its transmission characteristic) and the acoustic characteristics of the plenum, including all of its contents. In all cases, sound within a reasonably absorptive space attenuates with increasing distance. (Drawn by Jonathan Meendering.)

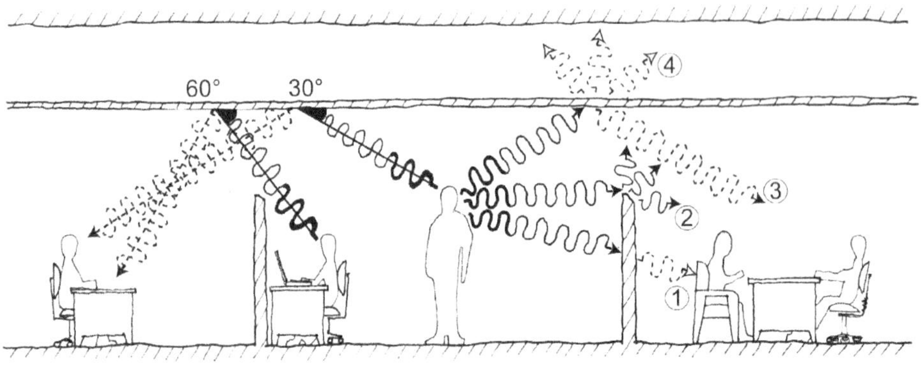

Principal Sound Paths

(1) Direct path through 5'-6"- [1.68 m]-high partition

(2) Diffracted path over partition

(3) Reflected path over partition

(4) Sound absorbed and diffused

(*a*)

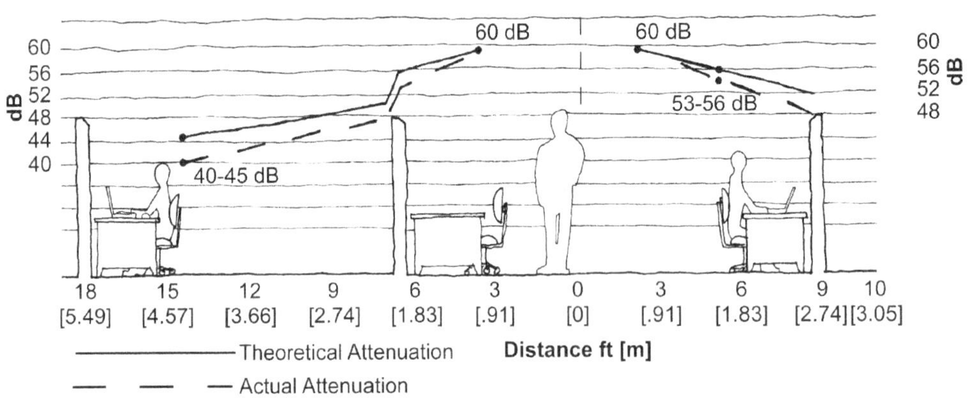

(*b*)

Figure 25.40 *(a) The principal sound paths between occupancies in an open office plan are (1) direct, (2) diffracted, (3) ceiling-reflected, and (4) reflected from the slab above the plenum. Only paths (2) and (3) are problematic from the speech privacy viewpoint, requiring masking sound. Note that the angle of incidence of the sound wave at the ceiling varies from 30° for a standing speaker to 60° for a seated speaker, requiring a ceiling material of high absorption between these angles. Large VDT screens (17–19 in. [432–482 mm]), in common use, create another path of strongly reflected sound. (b) Sound is attenuated by distance, dropping 4–6 dB for each doubling of the distance from the source (6 dB in the open; 4 dB in enclosed spaces due to interreflections). Average SPL of conversational speech is 60 dBA at 3 ft [0.9 m] from a speaker. Attenuation of a diffracted signal at a partial-height partition is 4–8 dB. Note that the diffracted sound (without a contribution from ceiling-reflected sound) drops to 40–45 dB in the adjacent cubicle. In a two-person open office, the received signal from a standing speaker is at least 53–56 dBA (i.e., perfectly intelligible, even in the presence of maximum [50 dB] masking noise). (Drawn by Jonathan Meendering.)*

frequencies. The fact that more energy will pass through a tile of lower CAC (STC) does not affect the acoustic result—most of the sound energy enters the plenum above the ceiling through openings in the far-from-airtight suspension system and is then largely absorbed and dissipated by the spray fireproofing, sound-absorbing material, ductwork, and structural members typically found in a plenum.

4. Since sound will always find the path of least resistance to travel, very little sound energy will pass through partitions between cubicles; the paths over the partition by diffraction and reflection are much less resistant. That being so, the STC of these partitions need not be high. Where the source is close to the partition, as is the case with a seated speaker facing a partition and delivering sound energy directly at the partition at a height of approximately 44–48 in. [1.1–1.2 m] from a distance of about 3 ft [0.9 m], the STC of the partition should be 25–26. For speakers at greater distances and heights, an STC of 20–22 is usually sufficient.

An exception to this general rule may occur when a large (17–19 in. [430–480 mm]) computer display is interposed between a seated speaker and an absorbent partition. If such digital screens are composed of a smooth, highly reflective glass surface, a strong sound path to the rear of the speaker is created, and the sound energy absorbed by the partition behind the digital screen is decreased. The variables, however, are so numerous that conservative design will use the higher STC rating. (Contrast these values with those required of a full-height, fixed-partition construction typical of enclosed spaces, as listed in Table 25.7.)

The absorption coefficient α of a partial-height partition at voice frequencies should be a minimum of 0.8 and preferably 0.85–0.95. Some manufacturers have assigned an Articulation Class rating to their partition products, although that descriptor is usually reserved for ceiling tiles, as explained in point 3. The recommended ratings for partitions range from 180 to 220.

5. Partitions must be tall enough to block direct line-of-sight voice transmission, since such a path is unattenuated except by distance. The median mouth height of a standing American male is 63 in. [1.6 m]—compared to 57.5 in. [1.5 m] for a female. The higher median mouth height is the basis of the widely accepted recommendation that partitions between adjoining cubicles should not be lower than 65 in. [1.7 m] and preferably 66–72 in. [1.7–1.8 m]. Since a 72-in. [1.8 m] high partition blocks vision for all but the tallest people, giving a subjective closed-in sensation to the occupant of a (small) cubicle, this height is normally used only between departments, with intradepartment partitions being 63–66 in. [1.6–1.7 m] high. Increasing the height of a partition from 65 to 72 in. [from 1.7 to 1.8 m] will increase path attenuation to an adjacent cubicle by 1–3 dB, depending upon ceiling height and speaker height and location.
6. Refer to Figure 25.40*b*. This figure indicates signal attenuation due to distance for two different paths. Speech intensity at a conversational level is approximately 60 dBA at 3 ft [0.9 m] from the speaker. Using the fact that sound in a free field attenuates 6 dB for every doubling of distance, and making the assumption that the sound field in a cubicle approaches that of a free field because of the large amount of highly absorptive material in the area, we obtain a sound intensity level of 54 dBA at 6 ft [1.8 m] from the speaker and 48 dBA at 12 ft [3.7 m]. In practice, the received sound level is several decibels higher because, despite the high α of the space, there are intraspace reflections that increase the sound level.

Referring again to Figure 25.40*b*, we see that the minimum sound level at a receiver within the cubicle would be 55–56 dB, a level that no practical amount of background sound can mask. Thus, two occupants of a single large cubicle will always hear each other quite clearly. The attenuation of a partition in the diffracted paths (over and around a partition) depends upon the location and height of the speaker and varies from 4 to 8 dB. The attenuation of the signal in the transmitted path (through the partition) will be at least 10 dB. Based on these figures, it is recommended that the minimum horizontal distance between occupants of adjoining cubicles when seated at their workstations be 10 ft [3 m] for minimum speech privacy. (Degrees of speech privacy are discussed in Section 25.20.) Speech levels in teamwork areas readily reach 66 dB. This necessitates either locating such areas away from normal working spaces or the use of full-height fixed or demountable partitions to completely enclose such areas. The spaces that require careful siting or complete enclosure because of raised voice levels (64–66 dB) include videoconferencing rooms, telecommunications spaces, and areas where workers use speakerphones or voice-activated computers. The latter two devices are usually forbidden in densely populated work areas where a reasonable degree of speech privacy is required for the conduct of regular business tasks.
7. Refer to Figure 25.41. As pointed out previously, sound will be received via paths of least resistance. These are often flanking paths that do not become evident except in plan view. In Figure 25.41, the flanking paths are particularly important because the first reflection occurs at a window. Glass has

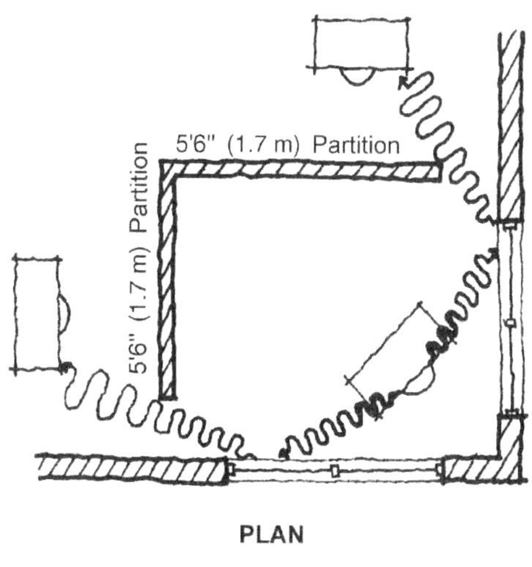

Figure 25.41 *Confidentiality in offices on exterior building walls is difficult to maintain because of the lack of attenuation of sound reflected from windows. The problem is compounded by the custom of placing management personnel in such offices.*

ACOUSTICS

negligible absorption and, because of its smoothness, exhibits specular reflection. As a result, the corner office occupant's voice will be heard clearly via the flanking paths shown, thus destroying the confidentiality of conversation in that office. Since offices on the building perimeter are usually reserved for middle and upper management, and since the large windows in such offices act to minimize the speech privacy so important to managerial personnel, the space designer has several options to ameliorate this condition:

a. Use full-height fixed partitions, with fixed glass vision panels if required, and doors rather than openings.
b. Use heavy drapes over "offending" glass windows, although this option defeats the visual and daylighting purpose of the windows.
c. Group spaces requiring confidentiality together, using unoccupied areas such as storage rooms as sound-buffers from open-office spaces.

8. Refer to Figure 25.42. The furniture arrangement establishes the source location of speech energy and consequently all of the sound paths that contribute to speech privacy—or, more accurately, to the lack of speech privacy. In layout (*a*), sound power, unattenuated except by distance, reflects off the back of the opposing aisle partition and travels to the occupant of the neighboring cubicle. Since the back of an acoustic partition is usually metallic and nonabsorbent, this arrangement would entail the additional expense of an absorbent rear surface on the corridor panels to maintain a degree of speech privacy between cubicles.

Changing the desk location in the same-shaped cubicle to that shown in Figure 25.42*b* improves speech privacy considerably by reducing the SPL of both the reflected and flanking paths, since the speakers face a highly absorptive surface. It may be unnecessary to use absorbent material on the rear of the corridor panels. A disadvantage of this arrangement lies in the short distance between speakers and neighbors if the same pattern of cubicles is continued longitudinally. A second possible disadvantage may be employee dissatisfaction with a working position facing a blank wall.

The arrangement in Figure 25.42*c* uses the same area per cubicle. By changing its shape and adding a 6-ft-high [1.8 m] acoustic partition whose top 18 in. [457 mm] is glass, the designer has reduced the sound energy levels in all paths while overcoming any employee resentment related to working in a blank, unexposed corner. Furthermore, the problematic flanking path in Figure 25.42*a* has been eliminated, as has probably any requirement for exterior acoustic absorption on the corridor partitions.

25.20 OPEN-OFFICE SPEECH PRIVACY LEVELS AND DESCRIPTORS

(a) Factors

The factors involved in determining the level of acoustic privacy that can be expected at a specific open-office location, due to neighboring speech sources, should be understood at this point. Restated, they are as follows:

1. The loudness of the source(s).
2. The acoustic characteristics of the source(s). These include location, height above the floor, directivity, frequency spectrum, and information content. The last factor refers to the ability of a listener to make sense of what is being said or heard. Therefore, most types of non-information-bearing sound would not be considered intrusive sources of noise. A completely foreign language should also fall into

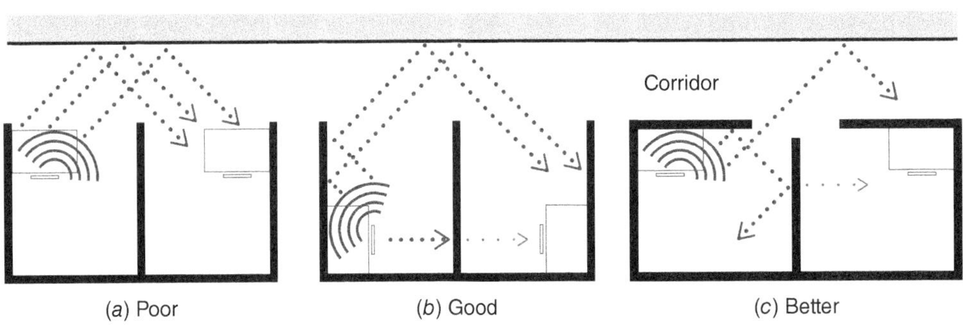

Figure 25.42 *Open-office shapes and furniture arrangements have a marked effect on speech privacy. (a) A strong, unattenuated signal reflects from the opposite corridor wall directly into the adjoining office. The side-to-side 15-ft [4.6-m] total spacing with an intervening partition is barely satisfactory. (b) The sound reflected from the corridor wall is much weaker than in (a) because of increased distance and decreased voice intensity from the side. Lateral sound is also weak due to the attenuation at the first reflection off the absorbent partition. (c) Reflection from the corridor is eliminated by the use of a front closure partition with glass at the upper 18 in. [457 mm] for visual purposes. Side-to-side sound transfer is weak due to wide spacing and the intervening partition. (Drawn by Nathan Leigh. © Walter Grondzik; all rights reserved.)*

ACOUSTICS

this category, although in practice that is so only when the words and syllables are muffled. Clearly heard words in a foreign language do constitute a source of speech annoyance, although to a far lesser extent than those in a comprehended language. Speech privacy calculation procedures ignore foreign languages and assume that all speech is a possible source of annoyance.

3. The signal attenuation along each path between the source and the receiver. This factor is different for every normal position of even a single source, as, for instance, for a seated or standing source person.
4. The level and frequency content of background sound (deliberate masking, HVAC noise, continuous machinery noise, and the like).
5. The degree of privacy required. This is discussed in detail in the next subsection. It is important to keep in mind, however, that the three classifications—confidential, normal, and transitional (minimal, marginal)—are assumed to remain constant if the physical factors involved do not change. This ignores the oft-demonstrated fact that intrusive noise causes psychological and physiological reactions in some people that tend toward aggravation and that increase in severity with the passage of time. A speech privacy situation originally classified as confidential may deteriorate to normal as the hearer's increasing sensitivity to noise causes him or her to strain to hear, and thereby to hear sound that ordinarily is masked. This is, in effect, another aspect of the cocktail party effect discussed in Chapter 8.

(b) Levels (Degrees) of Speech Privacy

Speech privacy is often achieved by masking intruding speech with background sound. By consensus among acousticians, the definitions of the three levels of speech privacy in an open office are (to reiterate):

1. *Confidential privacy:* Normal voice levels are audible but generally unintelligible. Raised voices are partially intelligible. Noise level is minimal. To achieve this level of privacy, the background sound level must be no more than 2 dB less than the intruding sound, and no more than about 3 dB more than the intruding sound, to satisfy the minimal noise level requirement. In this acoustic situation, approximately 95% of people will not sense any sound-intrusive disturbance and will be able to concentrate on most types of work.
2. *Normal privacy:* At this level, normal voice levels from adjacent spaces are heard but are not intelligible without concentration (i.e., by straining to hear and catch every syllable). Raised voices are generally intelligible. The overall noise level is low. This level of speech privacy is achieved when the background sound level is within 6 dB of (less than) the intruding speech level. This corresponds roughly to an intruding speech level of 50–54 dB, and a background sound level of 44–45 dB, that together give a range of 51–55 dB, levels that can be considered to meet the low-noise requirement.
3. *Transitional (minimal, marginal) privacy:* At this speech privacy level, speech at normal voice levels in adjacent open offices is readily understood most of the time, and the overall noise level is average. This noise level occurs when the intruding speech level is at least 10 dB more than the background sound level. Since background sound is limited to about 50 dB, this privacy level would mean an intruding speech intensity of 60 dB or more. Since 60 dB is approximately normal speech at 3 ft [0.9 m], this "privacy" level would occur with two occupants in a single office or a single occupant receiving intruding noise from at least three neighboring offices. This speech intrusion level would be considered intolerable by about 40% of people and would negatively affect the work efficiency of a higher percentage.

In summary, it is interesting to compare these three open-office speech privacy levels with the "absolute" grading given in Table 25.5.

Open-Office Class	Table 25.5 Rating
Confidential	Good
Normal	Fair
Marginal	Poor

By its very nature, an open office cannot achieve the top three grades of privacy listed in Table 25.5. If those levels of privacy are desired or required, fully enclosed spaces are necessary.

(c) Articulation Index (AI)

In order to quantify speech privacy for an open-office design, a single-number metric called the Articulation Index (AI) was developed in the 1970s by the acoustics consulting firm of Bolt, Beranek and Newman. This work was based in part upon studies of speech intelligibility by Bell Labs in the 1940s. Essentially, the AI relates speech intelligibility, speech intensity, and background sound level at the center of the five octave band frequencies that encompass the spectrum of the human voice: 250, 500, 1000, 2000, and 4000 Hz. AI is determined by measuring the percentage of individual words that can be understood under specific speech and background sound levels. An AI of 0%

TABLE 25.9 Articulation Index (AI) and Speech Privacy

AI	Persons Satisfied with Speech Privacy (%)	Open-Office Speech Privacy Descriptor
0–0.05	95–92	Confidential
0.06–0.2	90–80	Normal
0.21–0.3	79–65	Minimal
>0.3	<65	Unacceptable

indicates zero intelligibility and therefore ideal speech privacy. This, of course, does not mean silence; it means that with a specific combination of speech intensity level and background sound level, intelligibility is nil, and therefore speech privacy—defined as a lack of intelligible intruding speech—is ideal. At the other end of the scale, a high percentage of intelligible words yields a high AI and a correspondingly low speech privacy rating. In practice, the resultant AI figures are related to listener satisfaction and speech privacy descriptors, as shown in Table 25.9.

The calculation procedure for AI involves the use of weighting factors that are applied to intensity-level differences between speech and background sound levels at different frequencies, in order to reflect the connection between intelligibility and frequency. As pointed out in Chapter 24, most of the information in English words is carried by consonants, whose frequencies are generally above 2 kHz. Thus, the problem frequently encountered in telephone conversations, of distinguishing between *f* and *s*, *b* and *v*, *t* and *d*, and so on, is due to excessive attenuation of the high frequencies that distinguish these letters from each other. The AI calculation emphasizes the importance of high frequencies to intelligibility by using the following weighting factors:

Octave Band Center Frequency (Hz)	Relative Weighting Factor
250	1.0
500	2.5
1000	3.5
2000	5.0
4000	4.0

The calculation of AI values for an actual open-office design is complex and laborious, since it considers all sound paths between each source and each receiver, including the acoustic characteristics of all reflective and absorptive surfaces in the path. The results are a specific AI factor for every receiver location. Because of the very large number of calculations involved, the analysis is done by computer. Changes in materials, plan arrangements, and dimensions can be made if the calculated AI does not satisfy the space's speech privacy requirement.

These changes can be predicted fairly accurately, since it has been demonstrated that a 3-dB change in the relative level of an intruding speech signal with respect to the background sound level will result in a 0.1-level change in AI. Thus, an increase of 3 dB in the background sound level or a decrease of 3 dB in the intruding signal (by increased absorption or path length) will have the effect of increasing the AI between the involved workstations by 0.10, which is the difference between normal and poor speech privacy. Since 3 dB is a barely perceptible change in intensity (see Chapter 8), we can appreciate how sensitive speech privacy is to small changes in intensities.

(d) Articulation Class (AC)

Numerous measurements in actual open-office installations have indicated that the absorption characteristics of the ceiling are the most important factors in speech privacy design. As noted in the previous section, the angles of incidence of speech sound on the ceiling range between 30° and 60°, with most of the sound energy falling at the top of this range. A figure of merit for absorption, called the Articulation Class (AC), was established that indicates absorption effectiveness at angles of incidence between 45° and 55°. The usual range of AC is between 180 and 220 (no units), with higher numbers representing better absorption.

25.21 DESIGN RECOMMENDATIONS FOR SPEECH PRIVACY IN OPEN OFFICES

(a) General Factors

The architectural arrangement of spaces in an open-office design has a marked influence on speech privacy. Areas should be grouped according to their speech privacy requirements. Spaces rated as "confidential" should be placed on the perimeter of the open area to limit their exposure to speech intrusion—with the caveats relating to exterior windows and walls, discussed previously, being considered. The design emphasis for these areas should be not only on an AI between 0.0 and 0.05, but also on a low overall sound level, including background noise. Similarly, high-noise-producing areas should be grouped and placed on the perimeter at a maximum distance from confidential speech privacy areas. Use of demountable full-height partitions for such spaces should be considered.

Because of reflection from perimeter walls, open-area spaces should be as large as practical, with absorbent perimeter walls. Ceiling height should be no less than 9 ft [2.7 m] clear, with a 3-ft [0.9-m] plenum above. Extreme care must be taken with air-conditioning ductwork, which, if untreated, will act as an excellent speech and noise conduit via multiple ceiling outlets (diffusers). Furthermore, these outlets, which were once relied upon to produce an even level of background sound, generally no longer do so. Most HVAC systems today are variable air volume (VAV) designs, whose noise levels vary ±10 dB, making them useless as a reliable source of masking sound.

(b) Individual Office (Cubicle) Design

Offices should be designed for maximum closure and maximum partition length. Separation between occupants of adjacent offices should never drop below 10 ft [3 m], with a 12-ft [3.7-m] minimum as a design target for normal privacy and 16 ft [4.9 m] for confidential privacy. Minimum office area should be 80 ft^2 [7.4 m^2], with a design target of 100–120 ft^2 [9.3–11.1 m^2] for normal privacy and 200 ft^2 [18.6 m^2] for confidential privacy. Desk arrangements should be checked for optimum speech paths (for privacy), recognizing that office furniture arrangements need not be uniform in all offices (see Section 25.19).

(c) Ceilings

Because the ceiling is the most important design element in speech privacy, care must be taken to avoid unintentional strongly reflective speech paths, as from metal pan air diffusers, flat lighting fixture diffusers, and the like. If the use of such or similar items is unavoidable, highly absorptive vertical baffle strips may be placed on their perimeter to block sound paths. In general, ceiling tiles should have an AC rating of 220 minimum, and minimum absorption coefficients (α) at incidence angles of 30–60° as follows:

Frequency (Hz)	α
250	0.65
500	0.65–0.75
1000	0.85
2000	0.90
4000	0.90

(d) Partitions

As explained in subsection 25.19(a), the minimum height of partitions should be 65 in. [1.7 m], with 72-in. [1.8-m] high units separating offices from aisles and dividing departmental groups. The AC rating, if available, should range from 200 to 220. STC ratings, as explained in Section 25.11, depend to an extent upon speaker locations and vary from 20 to 26. Joints between partitions should be carefully sealed, because even small openings can seriously compromise a partition's already limited efficiency. All partitions should reach the floor, although the lower portion is not always absorptive in low-speech-privacy areas.

(e) Floors

Although carpeted floors do not seriously affect overall sound absorption, they do drastically reduce chair-movement and footfall sounds. For this reason, all floors in open office areas should be carpeted. The difference in effectiveness of shallow-pile carpet compared to deep-pile carpet is minimal, and the same differential can be achieved by using a polyurethane cushion backing in lieu of the more common jute pad. The principal purpose of carpeting is to cushion the footfall impact so that its energy is not introduced into the structure. This subject is covered in detail in the discussion of structure-borne sound that follows.

(f) Lighting Fixtures

Flat-bottom lighting fixtures must never be used. Fixtures should not be placed directly over partitions, to avoid an interoffice speech reflection path. Experience has shown that the best lighting fixture (from the speech privacy point of view) is one with deep parabolic reflector cells and overall dimensions of 1 ft × 4 ft [0.3 m × 1.2 m] or 2 ft × 4 ft [0.6 m × 1.2 m].

(g) Masking Sound

It is imperative that the level of masking sound be uniform throughout an open office area, and at as low a level as will yield the desired speech privacy. Nonuniformity will immediately be noticed as people move about, and the masking sound itself will become a source of auditory annoyance. For a similar reason, loudspeakers should not be visible. Visible units become themselves a source of interest initially and then a source of annoyance. Speakers should be placed in the plenum, preferably facing up to increase dispersion and improve uniformity. Speakers mounted in the ceiling and facing down should be avoided. Most ceiling tiles in open-office spaces have a low CAC so that sound will easily penetrate into the office area below the ceiling.

A masking sound system should comprise a signal (noise) generator, a sophisticated equalizer for shaping the signal, an amplifier with appropriate controls, and a distribution system to feed the speakers. Speakers are normally 12 in. [305 mm] and are installed in a grid on 12–16-in. [305–406 mm] centers. The amplifier should be arranged so that volume levels can be remotely controlled. This permits time control so that the background sound volume can be reduced automatically after working hours. This is necessary so that the few people working late are not annoyed by relatively loud background sound in the absence of intruding speech.

The noise produced by a masking sound system is variously described as white noise, noise of air rushing through an opening (whoosh sound), noise of water in piping, and the like. The actual sound can be tailored to the user's preference by adjusting the filters in the system's equalizer. Generally, masking sound emphasizes low frequencies, because higher frequencies are immediately noticed as an annoying hiss. As noted in Section 25.16, the background sound level should not exceed 48–50 dBA. In some installations, the masking sound system doubles as a public address system, although this practice is not recommended because the sound/noise stops during an announcement, and when it returns it is noticed. Background sound must be designed so as to blend into the ambience of the background, and anything that disturbs the hidden quality of masking sound is to be avoided.

(h) Design Procedure

Unfortunately, due to the large number of variables involved in open-office speech privacy design, a straightforward manual design method that will yield reliable results does not exist. However, computer programs are available that will calculate the AI for any location as a result of a specified speech source intrusion. Based on these calculations, changes can be made to the design and the program rerun to achieve improvements where the calculated AI is excessive. The most effective way to perform a complete design is then, on the basis of the preceding calculations, to construct a full-scale mock-up that can be field-tested and "tuned." Although this is expensive and time-consuming, it is frequently far cheaper than making the requisite changes after construction of an unacceptable solution.

One of the distinct advantages of this two-step design procedure is that it enables a designer to equalize the acoustic absorption "strength" of various paths so that the attenuations of major paths are equal. There is no economic or engineering sense in a system that is much more effective for one path than for another, since sound will always follow the path of least acoustic resistance. To accomplish this balancing, the designer has many variables to juggle. They include barrier height and material, ceiling material, baffle sizes and positions (if used), distances between the source and receiver, position and directions of sources, and level of background sound.

(i) Standards

The acoustic design and testing of open offices are covered by a group of ASTM International standards (see Section 25.38 for these and other resources) that should be in the hands of anyone engaged in open-office design. Additionally, two related international standards from ISO (International Organization for Standardization) will be useful to designers.

STRUCTURE-BORNE NOISE

25.22 STRUCTURE-BORNE IMPACT NOISE

The term "noise" will be used in lieu of "sound" in the following discussion of structure-borne, impact, and equipment noise control. Although use of the term "noise" assumes that a decision has been made that a given sound is unwanted, this is a very reasonable assumption for the situations to be discussed. Use of the term "noise" gets to the point.

Structure-borne noise is at least as serious a problem as airborne noise for the following reasons:

1. There is no air cushion between the source and the structure; thus, high-intensity energy is introduced into the structure, through which it travels with minimum attenuation and at great speed.
2. Sound, once introduced into the structure, is attenuated well only by discontinuities in the structure. Since the structure must have structural integrity to carry the loads, discontinuities of the type that will stop noise are complex and expensive.
3. The entire structure constitutes a network of parallel paths for sound. Therefore, partial solutions are useless, since sound will find flanking paths. The entire structure must be soundproofed to yield good results.
4. Unlike the case of airborne noise, additional mass does not usually block structure-borne noise, particularly in long spans where a floor can act as a diaphragm, thereby improving the structure-to-air noise transfer efficiency (like a drum).
5. The increasing use of exposed structural ceilings eliminates the attenuation that can be introduced by a plenum above a hung ceiling. This is particularly bad, since most structure-borne noise is carried by floor structures (rather than walls), which radiate sound up and down. The discussion that follows will be limited to impact noise. Refer to Section 25.26 for a brief treatment of vibration, which is felt rather than heard and is, in effect, a very low-frequency noise. Many of the practices and techniques that will minimize impact noise will also reduce vibration.

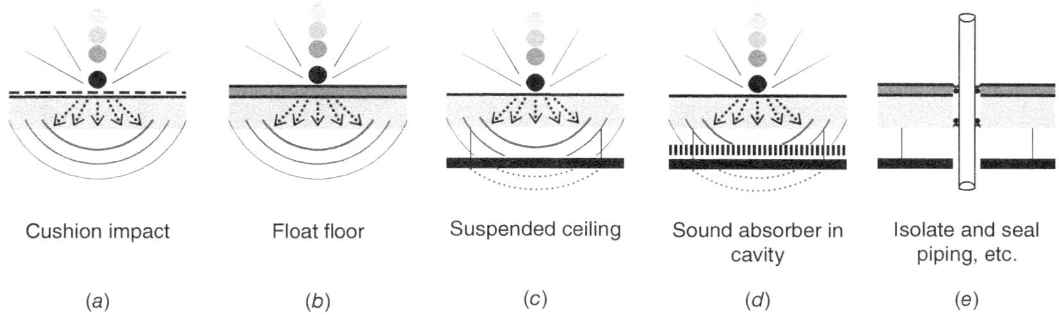

Figure 25.43 *Methods of controlling impact sound transmission through floors. (Redrawn by Nathan Leigh from NBS, 1986.)*

25.23 CONTROL OF IMPACT NOISE

Impact noise problems can be controlled in two ways—by preventing or minimizing the impact and by attenuating it once it has occurred. Prevention is discussed first; attenuation is covered in Section 25.24. Impact on floors is more serious than wall impact because the latter is partially attenuated at the wall/floor joint, whereas the former is introduced directly into the building framework. The following discussion addresses each of the solutions shown in Figure 25.43.

(a) Cushion the Impact

See Figure 25.43*a*. This solution will frequently eliminate all but severe problems. Resilient cushioning materials in common use are floor tile of rubber and cork, or carpeting on pads, in ascending order of impact insulation. See Section 25.24 and Appendix K for quantitative data on impact isolation.

(b) Float the Floor

See Figure 25.43*b*. Since the key to elimination of structure-borne sound is isolation, separating the impacted floor from the structural floor by a resilient element is extremely effective. This element can be rubber or mineral wool pads, or blankets, or special spring metal sleepers. The effectiveness depends upon the mass of the floating floor, compliance of the resilient support, and degree of isolation of the floating floor. The last element is extremely important, since flanking paths via end contacts with walls can short-circuit the floating element's sound impedance and defeat the system. With floating floors it is important that the following features are attended to:

1. The mass of the floating floor should be large enough to spread the loads properly. Otherwise, the pad will compress and deform sufficiently to transmit the impact.
2. Total construction should be airtight. Airtight is soundtight.
3. Care should be exercised where partitions rest on the floating floor (see Figure 25.44*a*).
4. Short circuits at walls or by penetrations should be avoided; see Figure 25.10*b*. Details of proper construction techniques are given in *A Guide to Airborne, Impact and Structure-Borne Noise Control in Multifamily Dwellings* (HUD, 1968).
5. Construction throughout should be consistent. Mixed construction types invite flanking noise paths (see Figure 25.44*b*).

(c) Suspend the Ceiling—and Use an Absorber in the Cavity

See Figure 25.43*c* and *d*. As stated, the most disturbing noise is that radiated down from the ceiling. A flexibly suspended ceiling with an acoustic absorbent layer suspended in it can be very effective if not flanked by paths leading into the walls and from there reradiating into the space below. It is imperative that the entire floor slab above be decoupled from the walls below by resilient separators.

(d) Isolate All Piping

See Figure 25.43*e*. All rigid structures such as piping must be isolated so as not to form a flanking path, and penetrations must be caulked with resilient sealant so as not to constitute an air–sound leakage path.

25.24 IMPACT INSULATION CLASS

The impact insulation class (IIC) is a single-number, impact isolation rating for floor construction, similar in intent and derivation to STC wall ratings.

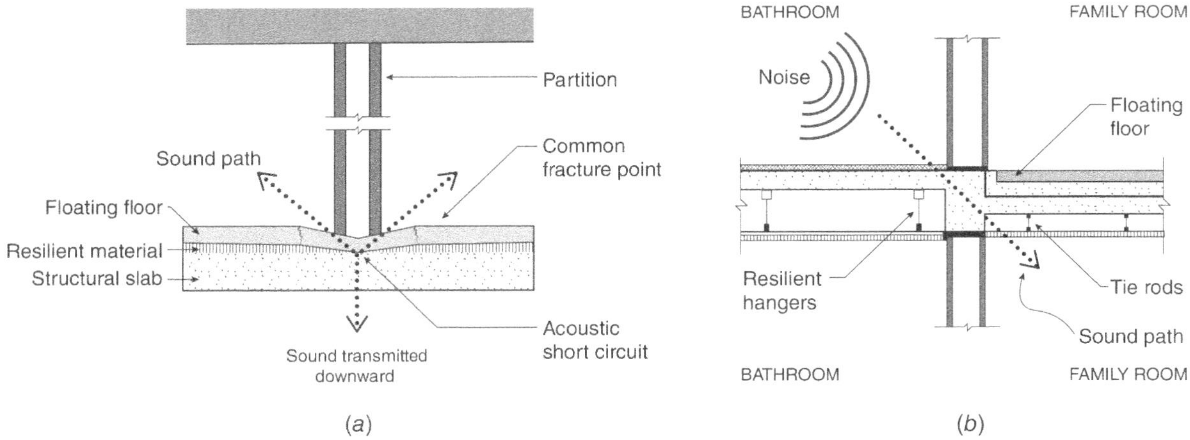

Figure 25.44 *(a) Caution must be exercised when supporting partitions on floating floors to prevent structural failures or short-circuiting of the floating element, as illustrated. (b) Flanking paths in mixed-construction type floors. The FHA does not recommend mixing construction types unless provisions have been made to prevent flanking (e.g., expansion joints or breaks in all structural paths between spaces). (Redrawn by Nathan Leigh from HUD, 1968.)*

The IIC rating is defined and utilized in several ASTM standards, namely:

ASTM E989-21: Standard Classification for Determination of Single-Number Metrics for Impact Noise: This standard defines the classification for determining an IIC rating, which is a single-number rating that assesses the impact sound insulation of floor-ceiling assemblies.

ASTM E492-09(2016)e1: Standard Test Method for Laboratory Measurement of Impact Sound Transmission Through Floor-Ceiling Assemblies Using the Tapping Machine: This standard defines the laboratory test method for measuring impact sound transmission through floor-ceiling assemblies using a standardized tapping machine. The data obtained are used to determine the IIC rating.

ASTM E1007-21: Standard Test Method for Field Measurement of Tapping Machine Impact Sound Transmission Through Floor-Ceiling Assemblies and Associated Support Structures: This standard defines the field measurement test method for measuring impact sound transmission through floor-ceiling assemblies and associated support structures. The data obtained are used to determine the field impact insulation class (FIIC) rating, which is analogous to IIC, but for field conditions.

Under standard laboratory conditions per *ASTM E492*, tests are made with a standard tapping machine and noise levels measured in $\frac{1}{3}$-octave bands. These are plotted and compared to a standard contour, approximately as with the sound transmission class.

The ICC rating for floor/ceiling assemblies typically falls somewhere between 25 and upwards of 85. A higher IIC rating signifies better noise reduction. IIC ratings may be generalized as follows:

IIC < 50	Poor performance: impact noise transmission (caused by footsteps, running, and furniture moving) will be very audible
IIC 50	Typical standard: impact noises will be somewhat audible
IIC 55–65	High-performance: impact noises are significantly reduced
IIC > 70	Superior performance: most impact noises are not audible

Details of typical floor constructions along with IIC ratings are given in Appendix K. Resilient floor finishes on any of the floor constructions not specifically provided with them will add to the IIC ratings approximately as follows:

$\frac{1}{16}$-in. [1.6 mm] vinyl tile	0
$\frac{1}{8}$-in. [3 mm] linoleum or rubber tile	4 ± 1
$\frac{1}{4}$-in. [6 mm] cork tile	10 ± 2
Low-pile carpet on fiber pad	12 ± 2
Low-pile carpet on foam rubber pad	18 ± 3
High-pile carpet on foam rubber pad	24 ± 3

MECHANICAL SYSTEM NOISE CONTROL

25.25 MECHANICAL NOISE SOURCES

Mechanical devices make noise. And generally, the more power they consume, the more noise they make. In many of today's buildings, 40% of the total construction budget is spent on mechanical systems located throughout a building.

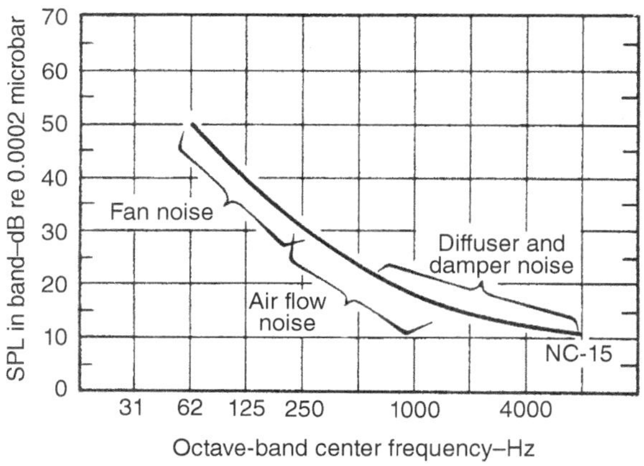

Figure 25.45 *Sound pressure level frequency spectrum of noise from HVAC system components.*

In most buildings, the primary sources of mechanical noise are the components of the air-conditioning and air-handling systems, such as fans, compressors, cooling towers, condensers, ductwork, dampers, mixing boxes, induction units, and diffusers. The curve of Figure 25.45 depicts typical air-handling system noise and indicates the portions of the spectrum produced by each group of components. Pumps are another source of mechanical noise, which (along with the noise of flowing liquid) is transmitted along pipes to locations throughout the building.

Elevators, escalators, and freight elevators also introduce mechanical noise into buildings. Escalators and freight elevators pose few problems, since they are localized in a specific area and have low operation speeds. Passenger elevator car operation, however, is rapid, and it affects large areas. In addition, the motors and controls are located on or above the prime upper floors of a building. Motor, shaftway, and other equipment noise must be properly controlled to prevent annoyance to building tenants located near the shaftways, elevator penthouses, or mechanical equipment rooms. Vibration isolation of these major components is a specialized problem beyond the scope of this book.

25.26 QUIETING OF MACHINES

Machines cause noise by vibration. This noise is imparted directly to the surrounding air and by vibrational contact to the surrounding structure. Therefore, there are three ways to reduce this noise:

1. Reduce the vibration itself.
2. Reduce the airborne noise by decoupling the vibration from efficient radiating sources.
3. Decouple the vibrating source from the structure.

Refer to Figure 25.46. Items 1, 3, and 4 reduce vibration; items 4, 5, 6, and 7 reduce and decouple the

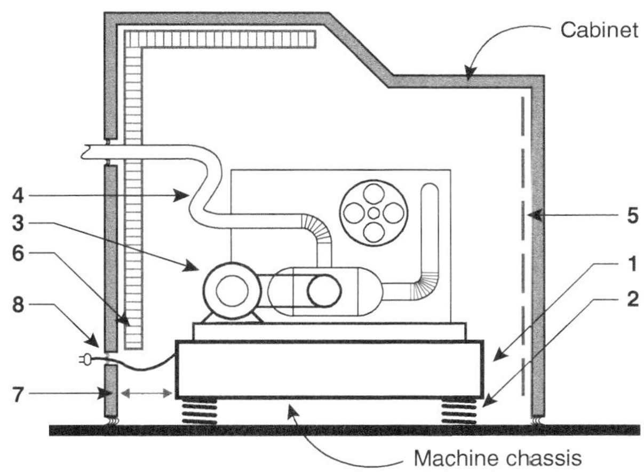

1. Install motors, pumps, fans, etc. on most massive part of the machine.
2. Install such components on resilient mounts or vibration isolators.
3. Use belt drive or roller drive systems in place of gear trains.
4. Use flexible hoses and wiring instead of rigid piping and stiff wiring.
5. Apply vibration damping materials to surfaces undergoing most vibration.
6. Install acoustical lining to reduce noise buildup inside machine.
7. Minimize mechanical contact between the cabinet and the machine chassis.
8. Seal openings at the base and other parts of the cabinet to prevent noise leakage.

Figure 25.46 *Techniques used to reduce the transmission of airborne and structure-borne noise from machines and appliances. (Redrawn by Nathan Leigh from HUD, 1968.)*

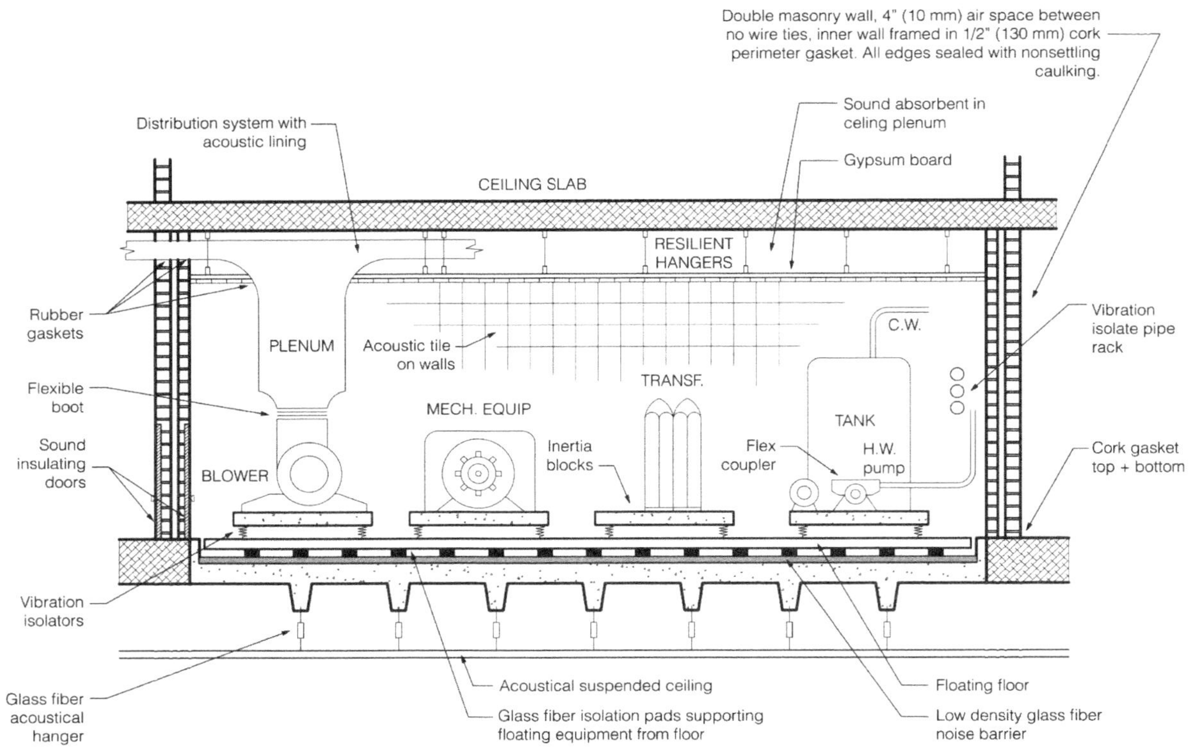

Figure 25.47 *Soundproofing a mechanical equipment room. Additional noise reduction to the space below can be achieved by inserting a layer of highly absorbent material in the space above the suspended acoustical ceiling. (Redrawn by Wesley Thompson from HUD, 1968.)*

vibration from the radiating cabinet; and items 2 and 8 decouple the vibrating source from the structure. Once a noise becomes airborne or structure-borne, the previously mentioned isolation techniques are employed.

Vibration reduction takes two forms: damping and isolation. One form of damping is accomplished by rigidly coupling the vibrating source to a large mass, frequently called an inertia block. Much of the energy is absorbed and dissipated as friction; the remainder results in lower-amplitude vibration (Figure 25.47). Isolation is accomplished by supporting the vibrating mass on resilient supports. These take many forms and can be used in combinations. Thus, machines are supported on fibrous, rubber, or spring steel vibration isolators, and the entire mass can be supported on a floating floor, which in turn rests on resilient vibration isolators, as in Figure 25.47. Large machines are supported on special commercial "sandwiches" of asbestos, lead, cork, and other strong, resilient materials. Piping is supported on cork pads and hung on resilient hangers.

Use of a diesel-driven electric generator for load peaking or cogeneration can cause very serious noise and vibration problems. The ideal solution to this situation is to completely isolate the unit in a separate outbuilding that is designed specifically to contain the very high noise level produced. If that is not practical and an inside location is necessary, a complete enclosure may be required to ameliorate the noise problem.

Vibration damping can be an even more serious problem, which can be solved satisfactorily with sufficient mass and proper vibration isolation. In the case of vibrating sheet metal, soft foam-type damping material glued directly to the metal is effective in damping. Flexible joints in all pipes and ducts connected to vibrating machines are mandatory. This includes flexible conduit connectors to all motors, transformers, and lighting fixtures using magnetic ballasts.

25.27 DUCT SYSTEM NOISE REDUCTION

Design of a quiet duct system entails more than specifying an absorptive duct lining. Air turbulence generates noise. Turbulence increases as the velocity of airflow increases and anywhere in the duct system where smooth laminar flow is disturbed, such as at sharp bends. The permissible "sharpness" of a bend depends in turn upon air velocity; the higher the air velocity, the more aerodynamic the duct system must be to prevent turbulence and, therefore, noise. Table 25.10 demonstrates this principle. The farther

TABLE 25.10 Maximum Air Speeds in Ducts to Yield NC-15 or NC-25 Background Levels[a]

Location	Supply		Return	
	NC-15	NC-25	NC-15	NC-25
Slot speed at min. ½-in. [13-mm] opening	250 fpm[b]	350 fpm	300 fpm	420 fpm
10 ft [3 m] of duct before opening	300	420	350	490
Next 20 ft [6 m]	400	560	450	630
Next 20 ft [6 m]	500	700	570	800
Next 20 ft [6 m]	640	900	700	980
Next 20 ft [6 m]	800	1120	900	1260
Next 20 ft [6 m]	000	1400	1100	1540
Next 20 ft [6 m]	1300	1820	1450	2030
Next 20 ft [6 m]	1600	2240	1800	2520

[a]Ducts with 1- to 2-in. [25- to 50-mm] thick inside duct lining, all duct sizes.
[b]Multiply fpm by 0.00508 to obtain m/s.

one proceeds upstream from a point of turbulence, the higher the air velocity may be from a noise perspective, since truly laminar flow is essentially noiseless. In principle, velocities should be as low as practical, since air turbulence noise increases exponentially with velocity.

Sound travels as easily against as with the airflow in ductwork. Therefore, both supply and return systems must be lined to control transmission of fan noise. Some fan noise reduction may occur at bends in the ductwork. But reducing fan noise at the source is a better strategy. Because 90° bends introduce turbulence that generates noise, introducing bends as fan noise attenuators can be counterproductive at air velocities above 600 fpm [3 m/s]. Another disadvantage of bends is added system friction and the additional energy and cost required to move air. This point will be discussed further in Section 25.28.

Other design approaches that create a quiet system include smooth transitions at changes of duct size and large-radius bends with turning vanes, the purpose of which is to reduce turbulence. Attenuation drops rapidly as duct size increases; therefore, ducts should not be deliberately oversized. Cross-talk between rooms and between ducts can be minimized by using lined ducts, separating adjacent ducts as much as possible, and gluing damping material on the outside and lining on the inside. Damping material is particularly effective in preventing the thin metal walls of ducts from resonating. Mufflers and silencers are effective in reducing the high-frequency components of fan noise, but much less so with low frequencies. The pressure drop these devices introduce, which can be considerable, must be compensated for in the fan selection.

The *ASHRAE Handbook—HVAC Applications* should be consulted for recommendations on noise control in air-handling units, plenums, housings, and ducts. Figure 25.48 shows some of the ways in which cross-talk and flanking noises can be reduced. Figure 25.49 shows some techniques employed for quieting duct noise. Active noise cancellation (see Section 25.28) is particularly useful in duct systems since it does not reduce airflow, as do liners, baffles, and other mechanical silencing devices, and it is effective at low frequencies, whereas these devices are not.

The increased use of variable air volume (VAV) systems has introduced some noise problems that should not be neglected. VAV system noise can be minimized by following a few basic design rules. Maintain minimum system static pressure, since fan noise increases exponentially with static pressure. Select the air volume modulating device at the fan with care, because it can be a noise source. Since outlet air volume control involves duct area restriction, with attendant velocity increase and resultant noise, such a design must include some sort of downstream silencing equipment. Ceiling diffuser acoustic characteristics must be coordinated with design air velocity and with any requirement for masking sound. Finally, avoid the use of throttling dampers on ceiling diffusers since a partially closed damper can generate very high noise levels.

25.28 ACTIVE NOISE CANCELLATION

Since noise consists of acoustic wave energy transmitted at certain frequencies, it is theoretically possible to eliminate (not mask) noise by simultaneous transmission of identical wave energy exactly out of phase with the noise. The sum of the two energy waves is zero—hence silence (Figure 25.50). The phase matching, however, must be exact; if the injected out-of-phase noise is not precisely a negative image of the original noise, then not only will the signals not cancel, but they may even increase the noise level. The technique for accomplishing noise cancellation is straightforward in theory but much less so in practice. A microphone samples a noise source and feeds that signal into an analysis/synthesis device. This, as the name implies, analyzes the frequency and amplitude content of the noise and synthesizes the anti-noise, which is then fed to a loudspeaker in the original noise

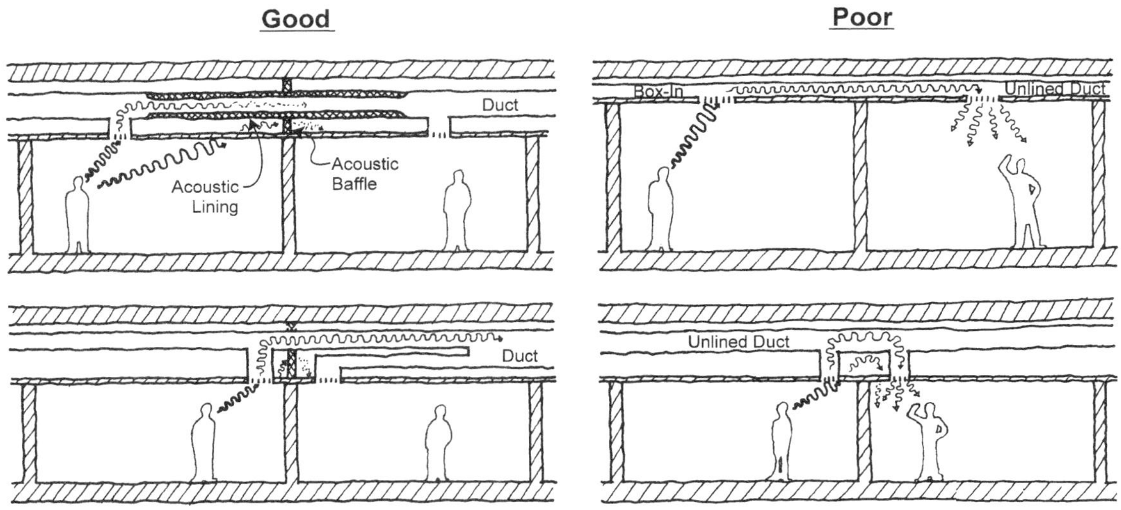

Figure 25.48 Since ducts are efficient sound transmission paths, precautions must be taken to avoid cross-talk, ventilation air noise, and equipment noise. Avoid running ducts as a common supply or return between rooms unless they are properly baffled and lined with sound-absorbing material. The common practice, in wood-frame structures, of using troughs between joists as a common return duct between rooms and between separate dwelling units results in serious noise transmission problems. Caulk or seal around ducts at all points of penetration through partitions. Use double-wall ducts, acoustical lining, flexible boots, and resilient hangers where required. Dwelling units should be serviced by separate supply and return ducts that branch off a main duct system. (Redrawn by Jonathan Meendering from HUD, 1968.)

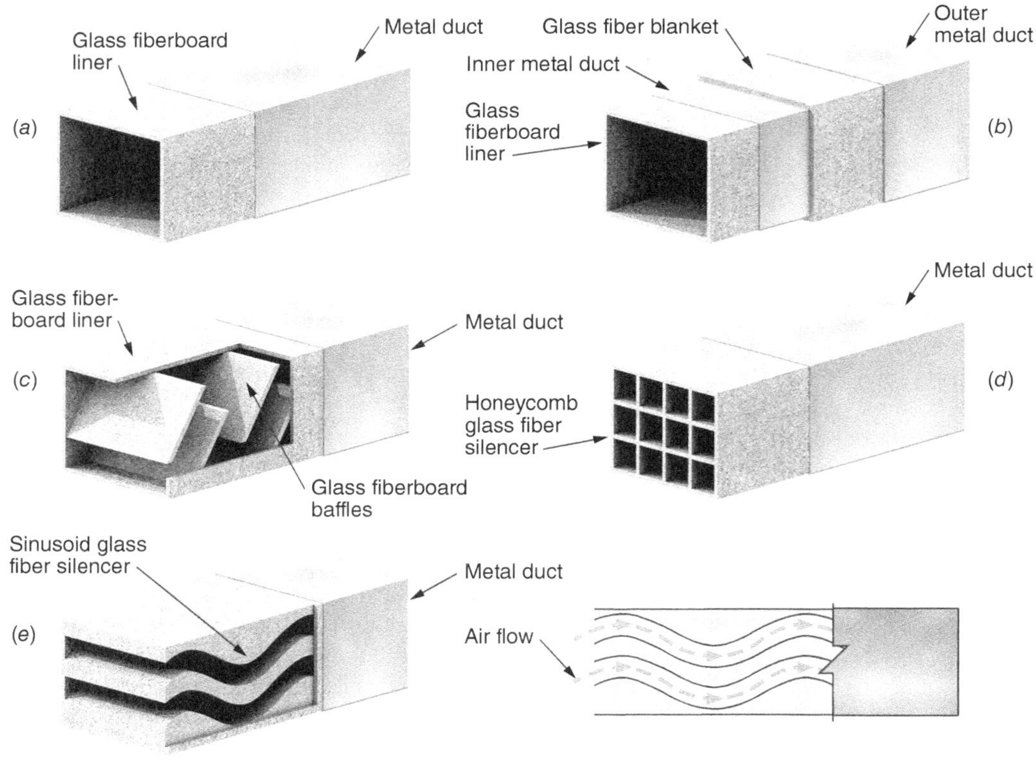

Figure 25.49 Unlined duct has negligible sound attenuation. Inside lining (a) gives 2–3 dB attenuation per foot in the range 1–2 kHz, dropping rapidly above and below those frequencies and giving negligible low-frequency attenuation. Double lining (b) gives higher attenuation and reduces cross-talk between ducts. Duct silencers and baffles (c–e) give high broadband attenuation: a maximum of 10–12 dB/ft [3.0–3.7 dB/m] in the range 1–2 kHz and lower above and below. They are useful to reduce fan noise in short runs but cause considerable pressure drop.

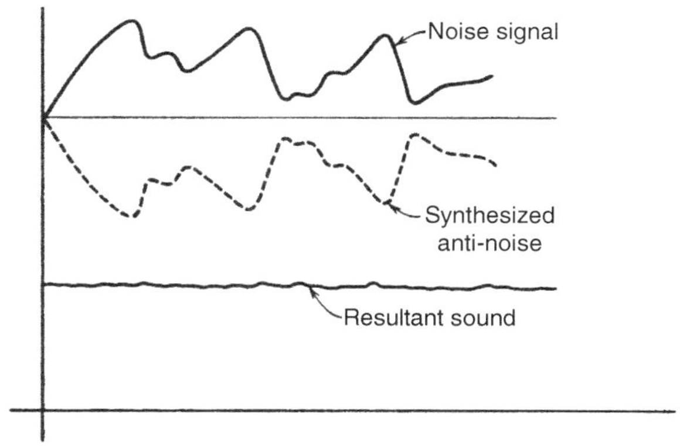

Figure 25.50 *Silencing of noise by introduction of a synthesized noise signal exactly out of phase with the original signal. The resultant sound is effectively zero—that is, silence.*

path. The resultant residual noise is detected by a downstream microphone and fed back into the controller as a feedback correction signal. This entire procedure is shown in block diagram form in Figure 25.51. The technical problems that must be overcome are formidable. Without delving deeply into the physics involved, we can describe them qualitatively.

1. It takes a finite amount of time to analyze the frequency content of a noise and to synthesize the anti-noise. If the noise is random, it will have changed by the time the anti-noise signal is injected, and will therefore not be cancelled. As a result, the only type of noise that can effectively be attenuated by anti-noise is one that is continuous and/or predictable. Random noise, such as that of a barking dog, cannot be actively attenuated with the present technology. Candidates for noise cancellation are sounds like those produced by operating machinery, that is, continuous and repetitive sounds. This includes very-low-frequency noise.
2. The analysis and synthesis process is achieved by a device called an adaptive digital filter. For technical reasons, the higher the frequencies (pitch) involved, the more complex and expensive are the required digital electronics, microphones, and loudspeakers. This further narrows the range of practical noise-cancellation candidates to those producing low frequencies, such as blowers, fans, rotors, internal combustion engines and their exhausts, transformers (hum), air movement in ducts, fluid movement in pipes, and the like.
3. The waveform of the noise, its modes, and its dispersion in space impact heavily on the type and amount of equipment required to effect attenuation economically (i.e., commercially). The simplest type of noise to treat is one that is confined by some sort of waveguide, at low frequency, and exhibits constant sound pressure and phase. Such a wave is known as a plane wave, and it can be "treated" with a single sampling microphone, a single loudspeaker, and a single processor.

Considering the three stated criteria, an ideal candidate for economic noise-cancellation treatment is duct noise, which is low frequency, continuous, and a plane wave. Figure 25.52 shows the construction of a typical commercial duct-noise-suppression unit, plus application photographs. Practical duct-noise-cancellation equipment in use today can reduce levels from NC-50 to NC-35 and is particularly effective at the "rumble" frequencies below 200 Hz.

Important auxiliary advantages of active duct-noise cancellation are energy conservation and economic benefits. Passive duct-noise reduction, particularly at low frequency, requires large noise absorbers, requiring 7–10 ft [2.1–3.0 m] of duct for installation. See Figure 25.49*c–e*. In addition to the high first cost of these devices and their installation, they introduce a static pressure loss of ½–1 ½ in. w.g. [125–375 Pa], depending upon airflow and speed. This, in turn, requires a higher horsepower (and noisier) fan. Economic analysis of such designs generally shows an advantage for active noise-cancellation equipment. Such an analysis should be performed for every design where duct silencers are desired or required.

Another important application of active noise cancellation, already in wide use, is for hearing protection for people exposed to high noise levels at work. Here the out-of-phase noise is introduced into miniature loudspeakers (earphones) in an acoustically transparent headset. This allows the wearer to hear random sounds (such as speech) clearly, while repetitive

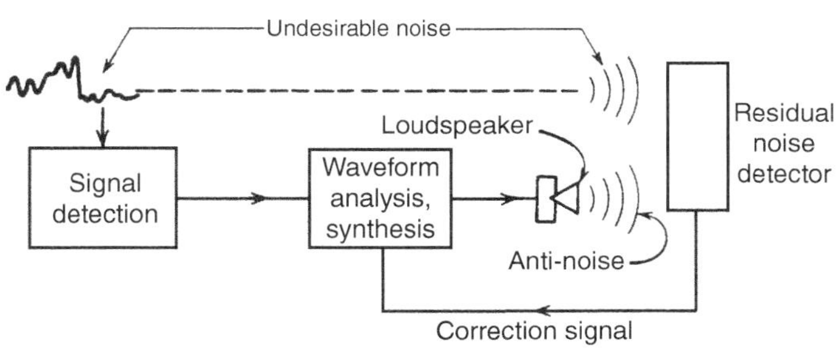

Figure 25.51 *Block diagram of an active noise-cancellation system. The noise signal is detected, its periodicity determined, its waveform analyzed, and an out-of-phase noise is synthesized and injected into the acoustic environment. A residual noise detector provides a feedback signal that acts to improve noise cancellation.*

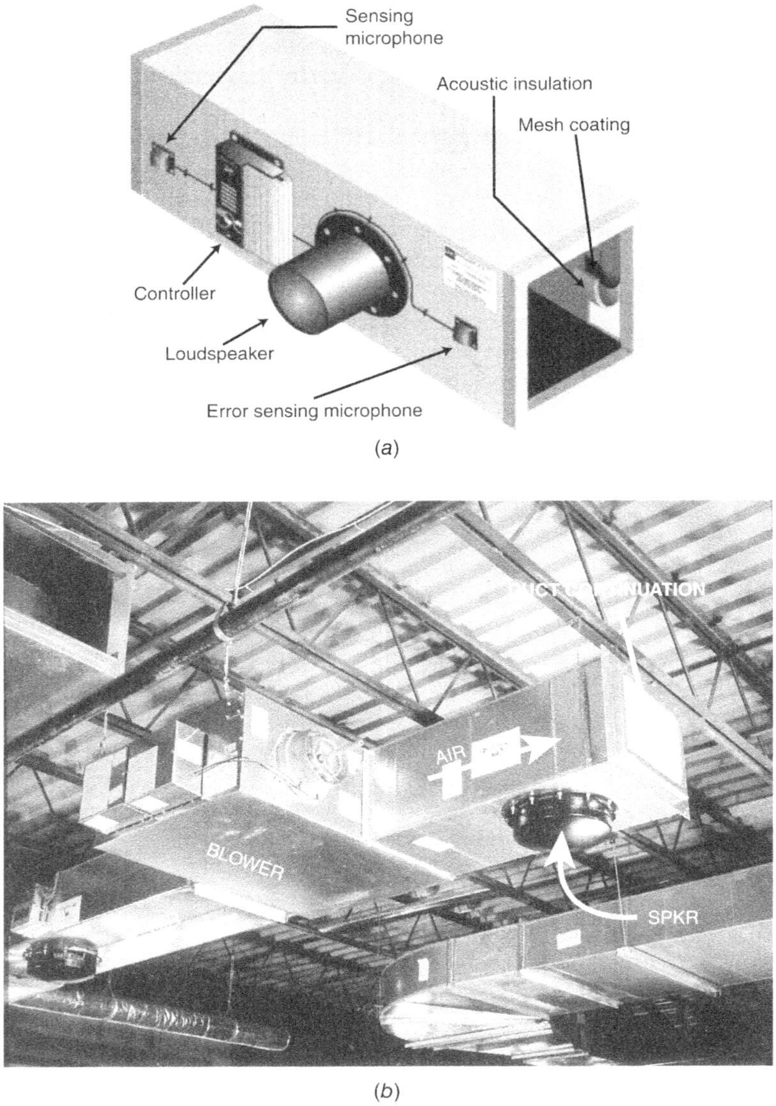

Figure 25.52 Active duct-noise-cancellation equipment. (a) The system components—consisting of the sensing and error-signal microphones, controller, and a loudspeaker—are built into a duct section of the required size, which is inserted into the duct system. (b) Photograph of a multichannel duct-noise-cancellation system in a new office building (designed with large, unlined HVAC ducts). (Courtesy of Digisonix.)

cyclic noise from engines and the like is attenuated (Figure 25.53).

Other areas where active noise cancellation is already in use include engine exhausts, heavy machine vibration, interiors of luxury automobiles, military and space vehicles, and selected shipboard spaces. Applications will undoubtedly increase with advances in digital signal processing technology, equipment miniaturization, and reduction in equipment costs.

25.29 PIPING SYSTEM NOISE REDUCTION

As with airflow, noise increases exponentially with liquid flow velocity. Piping is not a major noise source normally, since the radiating diameter is small, except for flow velocities much above 8 fps [2.4 m/s] where a pipe is in contact with the structure. This is, of course, most serious where a pipe passes through NC-15 to NC-25 areas (see Table 25.8). Domestic water system mains should be limited to 50 psi [345 kPa] in other than tall buildings, and pressure in branches limited to 35 psi [240 kPa]. In high-rise structures, pressure-reducing valves will be required in high-pressure mains to meet these recommendations. Piping must be designed to prevent water hammer, and noise sources must be located away from quiet areas.

Pumps, like all rotating equipment, are sources of vibration and noise and should be treated as described in Section 25.26. Figure 25.54 shows a typical pump installation with appropriate noise reduction measures. For at least a distance of 100 pipe diameters beyond the pump, resilient pipe hangers should be used. With centrifugal pumps, as with fans and blowers, machine

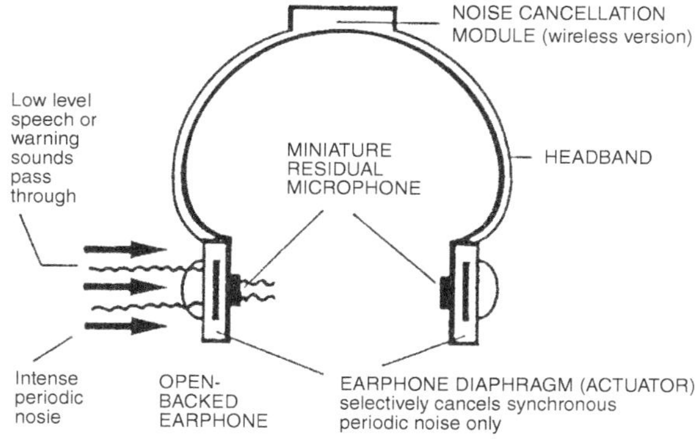

Figure 25.53 *Schematic diagram of a headset with active noise cancellation. Because of the physical proximity of all the elements, only a single microphone is used. (Courtesy of Noise Cancellation Technologies.)*

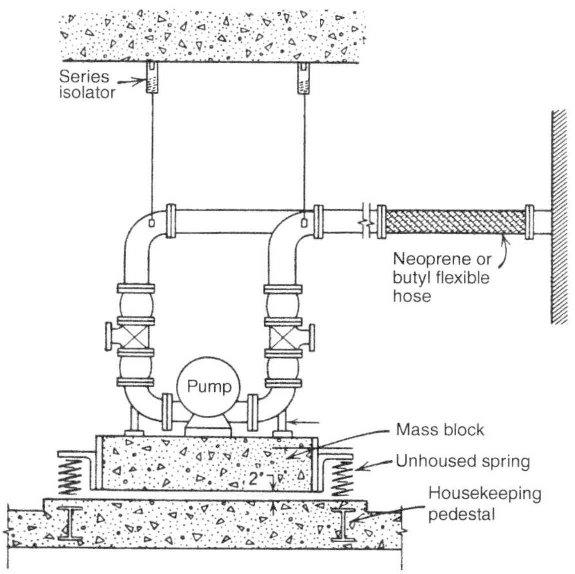

Figure 25.54 *Typical pump installation with appropriate vibration isolation and damping measures.*

sound concentrates in narrow bands and, if extremely disturbing, can be attenuated with resonant filters. Reciprocating pumps are more difficult to control, as the pulsations are more vibration than noise. Flexible connections and U-joints in the piping will absorb much of this vibration.

25.30 ELECTRICAL EQUIPMENT NOISE

Electrical equipment is generally overlooked as a noise source, and this is unwise. Most electrical noise is a 120-Hz hum. This can be very disturbing because the frequency is so low and, as we have noted repeatedly, low-frequency noise is difficult to attenuate passively. Transformer noise levels are dictated by the National Electrical Manufacturers Association (NEMA) and the American National Standards Institute (ANSI) standards. For a premium price, lower-noise units are obtainable. Table 25.11 lists the maximum average audible sound levels for dry-type units when tested in accordance with *NEMA ST 20-2021: Dry-Type Transformers for General Applications* (NEMA, 2022). Most manufacturers warranty noise below these levels. Oil- and silicone-filled units are normally quieter than dry-type transformers, as are units designed for lower temperature rise. Transformer noise can be minimized by these steps:

1. Mount the unit on vibration isolators.
2. If the transformer is wall-hung, use resilient hangers. If it is floor-mounted, place it on a massive slab if possible.
3. Locate the unit so that reflections do not amplify the sound. Sound-absorbent material on the walls behind the units is not useful at 120 Hz. Only cavity resonators will absorb appreciable amounts of sound at that frequency.
4. Use only flexible conduit connections.
5. Avoid locating transformers adjacent to, or immediately outside, quiet areas. A common error in this regard is placing a transformer pad immediately below the window of an NC 15–25 area.

The second major source of 120-Hz hum is conventional core-and-coil discharge lamp ballasts. These include magnetic ballasts for fluorescent and all high-intensity discharge (HID) sources. Fortunately, electronic ballasts, which are practically noiseless, are rapidly replacing core-and-coil ballasts in fluorescent fixtures, and to a lesser extent in HID units. The use of LED lamps also reduces the use of ballasts—but may introduce noise problems associated with LED drivers (especially when used for dimming). Table 25.12 lists recommended applications of nonelectronic fluorescent ballasts. Conventional coil-type HID ballasts can be very noisy, and care must be exercised in their placement. With all ballasts, the method of mounting has a marked effect on the radiated noise. As pointed out earlier, when a small vibrating source is coupled rigidly to a larger body, noise is amplified because of increased source-to-air coupling. Since core-and-coil fluorescent ballasts for linear fluorescent lamps are necessarily closely coupled to large metal fixtures for heat dissipation purposes, the sound radiation is greatly amplified. Many such fluorescent fixtures mounted in a plenum can create a serious noise problem. Solution of the problem lies either in ballast replacement or in the use of absorptive material in plenums, flexible conduit connection to fixtures, and resilient fixture hanging.

TABLE 25.11 Maximum Average Sound Levels: Self-Cooled Ventilated and Self-Cooled Sealed Transformers

	Average Sound Level, Decibels[a]			
	Self-Cooled Ventilated[b]			Self-Cooled Sealed
Equivalent Winding kVA Range	A K-Factor = 1, 4, and 9	B K-Factor = 13 and 20[c]	C Forced Air When Fans Running	D
3.00 and below	40	40	67	45
3.01–9.00	40	40	67	45
9.01–15.00	45	45	67	50
15.01–30.0	45	45	67	50
30.01–50.00	45	48	67	50
50.01–75.00	50	53	67	55
75.01–112.50	50	53	67	55
112.51–150.00	50	53	67	55
150.01–225.00	55	58	67	57
225.01–300.00	55	58	67	57
300.01–500.00	60	63	67	59
500.01–700.00	62	65	67	61
700.01–1000.00	64	67	67	63
Greater than 1000	Consult Factory			

Source: From NEMA, 2022 / National Electrical Manufacturers Association.

Notes:

[a]Sound levels are measured using the A-weighted scale, dB(A).

[b]Columns A and B apply when the fans are not running.

[c]Consult factory for non-linear requirements exceeding a K-factor rating of 20. K-factor is an indication of the anticipated extent of harmonic loading on a transformer.

TABLE 25.12 Acoustic Criteria for Selection of Conventional Core-and-Coil Fluorescent Lamp Ballasts

For an Installation in:	Use of Ballasts with This Rating Will Usually Be Satisfactory
TV or radio station, church, synagogue	A
Office, residence, library, reception or reading room, school study hall	B
Noisy office, doctor's or dentist's office, classroom	C
Industrial applications	D

In severe cases, ballasts can be remote-mounted. Coil-type HID ballasts are inherently noisier than fluorescent ones but, like ballasts for compact fluorescent lamps, are generally less troublesome, being coupled to small radiating bodies.

25.31 NOISE PROBLEMS DUE TO EQUIPMENT LOCATION

Roof-mounted HVAC units have proven to be very economical and very noisy. Vibration, short duct runs, and sound reflections are serious problems that can be solved with vibration isolators, sound mufflers, and careful location of equipment. Roof-mounted cooling towers are a problem when they are located adjacent to a taller building. This problem has led to lawsuits and noise control legislation in many cities. Attention should be paid to all exterior equipment during the design process.

In high-rise buildings, problems are caused by conflicts between the stringent noise requirements of the prime upper floor space and the near presence of elevator machine rooms, mechanical equipment rooms, and cooling towers. These problems are almost impossible to solve after construction and require the services of an acoustics expert during design.

25.32 SOUND ISOLATION ENCLOSURES, BARRIERS, AND DAMPING

In buildings with concentrated high-level noise sources, such as certain types of machinery, it is always more desirable to reduce the noise at its source than to attempt to treat the larger enclosing space. This is most effectively accomplished by enclosing the noise source with materials that provide a combination of reverberant noise reduction by absorption (as explained in Chapter 24) and blocking of airborne sound with high transmission loss (as detailed in Sections 25.7 to 25.12). These materials are available in the form of curtains, panels, and prefabricated partial and full enclosures tailored to the specific

Figure 25.55 *Flexible, lightweight noise absorber. The quilted noise absorber can achieve a noise reduction coefficient of up to 0.88 at a frequency of 500 Hz, dissipating sound energy via transmission from the outermost membrane to the inner fibers. The absorber is ASTM E84 Class I fire rated, and highly resistant to abrasion. (Courtesy of Industrial Noise Control, Inc.)*

characteristics of the noise source (Figure 25.55). Such enclosures are not normally the responsibility of the building designer. It is, however, important to know that they will be used, as well as their characteristics, so that appropriate isolation can be designed into the building for the residual sound that is radiated from the enclosures.

Where a noise is at least partially the result of vibrating (sheet) metal enclosures, as, for instance, on laundry machines, mixers, bins, chutes, polishing drums, and the like, a very effective noise control technique is to damp the vibration. This can be done by permanent attachment of a layer of foam to the vibrating metal, which, by minute flexing of the foam mass, converts the vibration energy to heat. The noise reduction can be further enhanced by adding a heavy, limp barrier material to the outside of the foam. Such combinations are called composite damping barrier materials, several of which are illustrated in Figure 25.56. The foam layer also acts to decouple the barrier and thereby increase the noise attenuation.

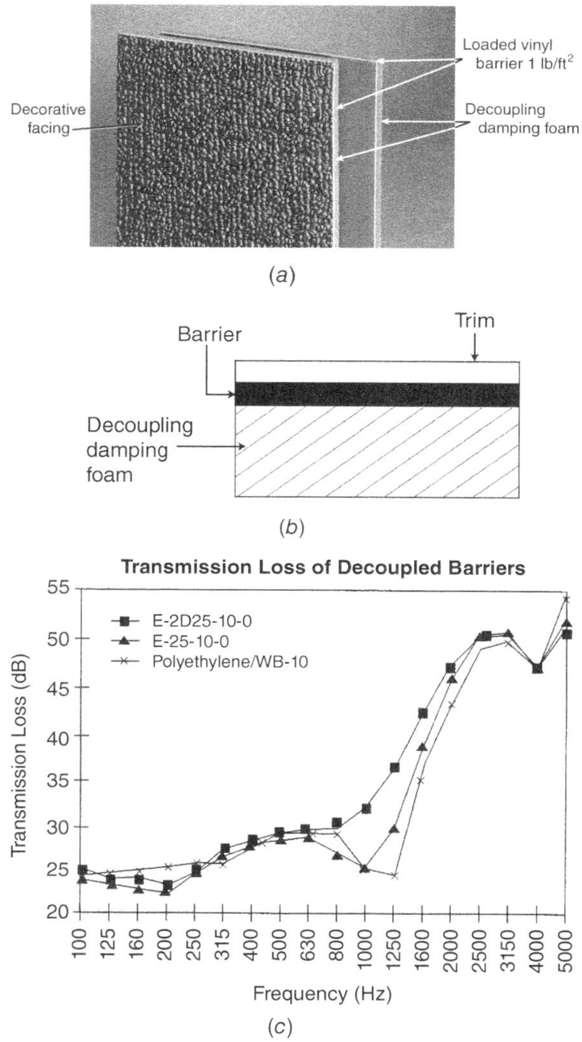

Figure 25.56 *Composite damped, decoupled barrier material comprising a layer of damping foam plus an integral layer of loaded vinyl barrier material. The exterior may be covered (a) with a decorative cover layer. The foam layer acts to damp the vibrating surface to which it adheres (b) and to decouple the barrier layer. Preliminary attenuation test data (c) for a composite consisting of ¼ in. [6.4 mm] foam and 1 lb/ft² [4.9 kg/m²] loaded vinyl. (Courtesy of E-A-R Specialty Composites, division of Aearo Co.)*

STC AND IIC RECOMMENDATIONS AND CRITERIA

Recommendations for background noise levels (NC criteria) are given in Table 25.8. Recommendations for STC ratings for partitions separating spaces with different occupancy types are offered in Table 25.13. Detailed criteria for partition isolation (STC) and impact isolation (IIC) are given in the following sections.

25.33 MULTIPLE-OCCUPANCY RESIDENTIAL STC/IIC CRITERIA

The most important acoustical design criteria for residential work in the United States are issued by the Department of Housing and Urban Development in conjunction with the Federal Housing Administration (HUD/FHA). The reader is referred to the longstanding 1968 issue of *A Guide to Airborne, Impact and Structure-Borne Noise Control* (HUD, 1968) and to subsequent HUD/FHA publications. Tables 25.14 and 25.15 give the essential data presented in the current edition of the *Guide*.

TABLE 25.13 Recommended STC Ratings for Partitions: Specific

Type of Occupancy	Wall, Partition, or Panel Between: Room Being Considered	and Adjacent Area	Sound Isolation Requirement: Background Level in Room Being Considered: Quiet	Normal
Normal school buildings without extraordinary or unusual activities or requirements	Classrooms	Adjacent classrooms	STC 42	STC 40
		Corridor or public areas	STC 40	STC 38
		Kitchen and dining areas	STC 50	STC 47
		Shops	STC 50	STC 47
		Recreation areas	STC 45	STC 42
		Music rooms	STC 55	STC 50
		Mechanical equipment rooms	STC 50	STC 45
		Toilet areas	STC 45	STC 42
	Music practice rooms	Adjacent practice rooms	STC 55	STC 50
		Corridor and public areas	STC 45	STC 42
Executive areas, doctors' suites; confidential privacy requirements	Office	Adjacent offices	STC 50	STC 45
		General office areas	STC 48	STC 45
		Corridor or lobby	STC 45	STC 42
		Washrooms and toilet areas	STC 50	STC 47
Normal office; normal privacy requirements; any occupancy using rooms for group meetings	Office	Adjacent offices	STC 40	STC 38
		Corridor, lobby, exterior	STC 40	STC 38
		Washrooms, kitchen, dining	STC 42	STC 40
	Conference rooms	Other conference rooms	STC 45	STC 42
		Adjacent offices	STC 45	STC 42
		Corridor or lobby	STC 42	STC 40
		Exterior of building	STC 40	STC 38
		Kitchen and dining areas	STC 45	STC 42
Large offices, drafting areas, banking floors, etc.	Large general office areas	Corridors, lobby, exterior data-processing area	STC 38	STC 35
			STC 40	STC 38
		Kitchen and dining areas	STC 40	STC 38
Motels and urban hotels, hospitals and dormitories	Bedrooms	Adjacent bedrooms[a]	STC 52	STC 50
		Bathroom[a]	STC 50	STC 45
		Living rooms[a]	STC 45	STC 42
		Dining areas	STC 45	STC 42
		Corridor, lobby, or public spaces	STC 45	STC 42

Source: Courtesy of U.S. Gypsum.

[a]Separate occupancy.

The recommendations are divided into Grades I, II, and III. Grade II is the most common category and is applicable primarily in residential urban and suburban areas considered to have an average noise environment. The nighttime exterior noise levels may be about 40–45 dBA, and the permissible interior noise environment characteristics should not exceed NC 25–30. Grade I is suburban, with a quiet noise environment characterized by a nighttime exterior noise level of about 35–40 dBA. Grade I STC/IIC criteria are 3 points higher than those of Grade II.

The fundamental criteria for airborne sound insulation between dwelling units are, for Grade II:

Wall partitions	STC > 52
Floor-ceiling assemblies	IIC > 52

These apply where similar function spaces are contiguous, such as bedroom to bedroom and living room to living room. Where this is not the case, the isolation must be increased to meet the higher sensitivity requirement.

Grade III recommendations are minimal and can be characterized as noisy environments, with an average nighttime exterior noise level of about 55 dBA or higher. Grade III STC/IIC recommendations are 4 points lower than those of Grade II.

25.34 SPECIFIC OCCUPANCIES

(a) Schools

School buildings house spaces of many kinds—classrooms, auditoriums, gymnasiums, cafeterias, shop areas, swimming pools, and music suites—that pose acoustical problems.

1. *Auditoriums:* All auditoriums require a sound system for some of the activities accommodated (see Chapter 24). The most difficult aspect, architecturally, is integration of a loudspeaker system into the design. To provide proper sound reinforcement, loudspeakers must be located properly without large

TABLE 25.14 Criteria for Airborne Sound Insulation of Partitions between Dwelling Units

Partition Function between Dwellings			
Apt. A		Apt. B	Grade II STC
Bedroom	to	Bedroom	52
Living room	to	Bedroom[a]	54
Kitchen[b]	to	Bedroom[a]	55
Bathroom	to	Bedroom[a]	56
Corridor	to	Bedroom[a,c]	52
Living room	to	Living room	52
Kitchen[b]	to	Living room[a]	52
Bathroom	to	Living room	54
Corridor	to	Living room[a,c,d]	52
Kitchen	to	Kitchen[e]	50
Bathroom	to	Kitchen	52
Corridor	to	Kitchen[a,c,d]	52
Bathroom	to	Bathroom	50
Corridor	to	Bathroom[a,c]	48

Source: Adapted from HUD, 1968.

Note: For Grade I, add 3 points; for Grade III, subtract 4 points.

[a]Whenever a partition wall may serve to separate several functional spaces, the highest criterion must prevail.

[b]Or dining or family or recreation room.

[c]It is assumed that there is no entrance door leading from the corridor to the living unit.

[d]Criterion applies to the partition. Doors in corridor partitions must have the rating of the partition, not vice versa.

[e]Double wall construction is recommended to minimize kitchen impact noises.

TABLE 25.15 Criteria for Airborne and Impact Sound Insulation of Floor-Ceiling Assemblies between Dwelling Units

Assembly Function between Dwellings			Grade II	
Apt. A		Apt. B	STC	IIC
Bedroom	above	Bedroom	52	52
Living room	above	Bedroom[a]	54	57
Kitchen[b]	above	Bedroom[a,c]	55	62
Family room	above	Bedroom[a,d]	56	62
Corridor	above	Bedroom[a]	52	62
Bedroom	above	Living room[e]	54	52
Living room	above	Living room	52	52
Kitchen	above	Living room[a,c]	52	57
Family room	above	Living room[a,d]	54	60
Corridor	above	Living room[a]	52	57
Bedroom	above	Kitchen[c,e]	55	50
Living room	above	Kitchen[c,e]	52	52
Kitchen	above	Kitchen[c]	50	52
Bathroom	above	Kitchen[a,c]	52	52
Family room	above	Kitchen[a,c,d]	52	58
Corridor	above	Kitchen[a,c]	48	52
Bedroom	above	Family room[e]	56	48
Living room	above	Family room[e]	54	50
Kitchen	above	Family room[e]	52	52
Bathroom	above	Bathroom[c]	50	50
Corridor	above	Corridor	48	48

Source: Adapted from HUD, 1968.

Note: For Grade I, add 3 points; for Grade III, subtract 4 points.

[a]This arrangement requires greater impact sound insulation than the converse, where a sensitive area is above a less sensitive area.

[b]Or dining or family or recreation room.

[c]It is assumed that plumbing fixtures, appliances, and piping are installed with proper vibration isolation.

[d]The airborne STC criteria in this table apply as well to vertical partitions between these two spaces.

[e]This arrangement requires greater impact sound insulation than the converse, where a sensitive area is above a less sensitive area.

obstructions. To accomplish this, the loudspeaker system should be incorporated into the earliest design stages.

In general, a school auditorium is a multipurpose facility. It should be designed to meet speech requirements and also should be suitable for the school's music activities. Often a modified gymnasium (gymnatorium) or cafeteria (cafetorium) functions as an auditorium. Acoustic compromises will need to occur in such facilities. Large areas of sound-absorbing treatment in either kind of space make them unsuitable as auditoriums and for most events that require speech amplification.

2. *Classrooms:* Typical classrooms are approximately 30 ft square [2.8 m^2] with 10-ft [3-m] ceilings. Adequate speech communication is easily achieved in a room of this size. Classroom acoustic design usually involves:

   a. locating sound-absorbing treatment to reduce classroom noise levels
   b. ensuring adequate privacy between adjacent spaces
   c. control of air-handling system noise. Acoustic tile ceilings provide adequate sound absorption for most classrooms. An NRC of 0.7 is recommended (see Section 24.10).

   Partition systems must produce enough isolation to prevent disturbance from activities in other classrooms and corridors. Such partitions should run full-height from floor to ceiling slab or roof construction. If return air transfer ducts are needed, their noise reduction characteristics must be as good as those of the walls or doors that they penetrate (for NC data, see Table 25.8). Unit ventilators commonly used for classrooms produce approximately the required level of background sound.

3. *Music suites:* School music programs usually range from individual instruction to band and choral concerts. The teaching spaces required for such a program include practice rooms, ensemble rooms, and large rehearsal spaces. Both room acoustics design and sound isolation are important in music suites. Privacy between adjacent spaces is critical, since simultaneous use is necessary.
4. *Dining areas:* The activity in cafeterias or lunchrooms usually generates a great deal of noise. The kitchen and serving areas should be separated from

the eating spaces. Ceilings and wall areas in the cafeteria should be treated with sound-absorbing material. Unless the ceiling is completely treated with a highly effective sound-absorbing material, the environment will be unsatisfactory due to its high noise level. The minimum NRC of this material should be 0.8.

5. *Gymnasiums:* Activities in gymnasiums create so much noise that even extensive treatment will not quiet these spaces. A quiet gymnasium probably would be unsatisfactory in any case, since spectators are conditioned to consider the noise as an enjoyable aspect of athletic events. However, to provide a proper environment for normal sports activities, the ceiling area should absorb sound. In addition, if a sound amplification system is to be used, sound-absorbing wall treatment may be required to eliminate echoes that would reduce the intelligibility of announcements. An NRC of 0.7 is suggested with sound-absorbent material to be ceiling-mounted.

   If a gymnasium will also serve as an auditorium, loudspeaker system placement requires special consideration. For example, the loudspeakers should be located above the source location for speeches and plays.
6. *Swimming pools:* The acoustic environment of swimming pools is often chaotic. Most sound-absorbing materials disintegrate in the high-humidity conditions prevalent in pool areas. Use special sound-absorbing units that have moisture-resistant properties.
7. *Shops:* Metal, woodworking, and scenery shops in schools contain many noise sources—saws, planers, drill presses, and manual tools. Each generates high airborne and structure-borne noise levels. Consolidating noisy areas and maximizing the distance between them and quiet spaces are essential. Ceiling and wall absorptive treatment with an NRC of at least 0.75 is recommended.

(b) Houses of Worship

The basic activities of these buildings usually combine speech and music. Thus, the worship environment must be acoustically hospitable to both. The architectural plan also must respond to religious requirements, including the relative positioning of pulpits, lecterns, the altar (if any), and the choir.

Successful acoustics can be achieved by designing the overall environment for music and providing special assistance for speech. Large congregational spaces frequently include a sound-reflecting canopy over the pulpit to direct the speaker's voice to the congregation. In some large buildings, a loudspeaker located above the canopy further reinforces speech from the pulpit. The choir and organ communicate with the entire volume of the building and, therefore, benefit from a reverberant environment.

(c) Offices

Although office buildings may contain public spaces, auditoriums, and restaurants, prime occupancy is in office areas. Most acoustics problems in office buildings relate to privacy—either between spaces within a single firm or between adjacent firms. Speech privacy is discussed at length in Sections 25.16 through 25.21, including consideration of open-plan offices. Mechanical and electrical equipment noise problems are discussed in Sections 25.25 through 25.32.

(d) Apartment Buildings

Large apartment buildings house hundreds and even thousands of residents. Privacy and freedom from annoyance are high on the list of tenant requirements. See the HUD/FHA criteria in Section 25.33.

The performance of partitions is compromised in many designs by careless planning of convenience outlets, medicine cabinets, and mechanical services. Direct-exhaust duct connections between apartments and back-to-back placement of medicine cabinets result in loss of privacy. Back-to-back convenience outlets must be avoided.

Installation of rugs or carpeting provides the best protection against footfall noise. Many leases now require that a tenant provide such impact-reducing floor covering over most of the floor area in an apartment. Good design also dictates that similar spaces in adjacent apartments ought to be grouped—bedrooms next to bedrooms, for example. Absorptive material in bedrooms should be ceiling mounted. A minimum NRC of 0.6 is recommended.

Apartment house site selection seldom includes consideration of acoustics. Nevertheless, truck routes, highways, and airports can be annoying "neighbors." Cooling towers serving adjacent buildings must be considered during the planning stages.

OUTDOOR ACOUSTIC CONSIDERATIONS

25.35 SOUND POWER AND PRESSURE LEVELS IN THE OUTDOORS

The equations related to sound power and sound pressure levels offered in Chapter 24 are not applicable to outdoor sound propagation, in which the large reflective component of the indoor condition is absent. Although the propagation of sound outdoors may not appear to be of immediate importance in architectural acoustics, outdoor noise sources such as traffic, cooling towers, and aircraft are frequently loud enough to disturb activities within or immediately adjacent to a building. Conversely, the noise made by building equipment such as cooling towers, air-cooled condensers, and even window air conditioners may be loud enough to disturb neighbors in a nearby building. For this reason, it is desirable to have some basic understanding of outdoor sound propagation.

For preliminary evaluation of an outdoor noise problem, assuming a small nondirectional source on the ground, the sound pressure level can be determined from equations offered in Chapter 24. For large sources such as cooling towers and traffic, which do not exhibit inverse square properties, sound level estimates are best made based upon experience and empirical data beyond the scope of this book (see Magrab, 1975; Schaudinischky, 1976). For small outdoor sources, the equipment power level can be estimated by measuring the sound pressure level at 5 ft [1.5 m] and adding 15 dB. Other factors (such as moisture in the air, the presence of trees, wind, and temperature gradients) will affect outdoor sound propagation to some extent, but they can be ignored except when great distances (i.e., over 1000 ft [305 m]) are involved. Barriers, which are the most effective outdoor attenuators, were discussed in Section 25.14.

25.36 BUILDING SITING

Building siting, relative to exterior noise sources, is as important as interior structural design. Since this subject is somewhat beyond our scope of concern, the discussion is brief. Buildings should be sited, with respect to noise sources:

1. to use natural terrain noise barriers (Figure 25.57*a*)
2. regarding trees as noise barriers, to rely only on very thickly wooded areas (Figure 25.57*b*)
3. to avoid naturally poor sites (Figure 25.57*c*)
4. to avoid sound reflection from other buildings (Figure 25.57*d*)

Point 4 is also important in a multi-wing building. Avoid U-shapes or other configurations where a central court can become an echo chamber.

Where avoidance of an exterior noise source is impossible, quiet zones can be buffered from noises by placing higher-noise areas on the noisy side of a building. Thus, in a school, classrooms and offices can be buffered by a cafeteria and gym; in a residence, bedrooms by living rooms and corridors; in an office building, private offices by noisier clerical offices; and so on.

REFERENCE MATERIAL

25.37 GLOSSARY

A-scale A sound meter filtering system with characteristics that roughly match the response characteristics of the human ear. Referred to as dBA.

Absorption coefficient (α) Refer to *sound absorption coefficient*.

Anechoic room A room that provides a free field acoustic testing environment like the outdoors. All the sound emanating from a source is essentially absorbed at the surfaces of the room.

ANSI American National Standards Institute, a nonprofit national technical association that publishes standards covering definitions, test methods, recommended practices, and specifications of materials.

Articulation class (AC) A figure of merit for acoustic absorption indicating the absorption efficiency of a material for angles of incidence of the sound wave between 45° and 55°. The range is 150–250 (no units), with higher numbers indicating better absorption.

Articulation index (AI) A numerical index in the range of 0.0–1.0 indicating the degree of speech intelligibility in an open-office design, measured with background noise, including masking sound, if any. Also, a measure of speech privacy. An index of 0.0 indicates no speech intelligibility, hence ideal speech privacy; an index of 1.0 indicates perfect speech intelligibility and therefore no speech privacy.

ASTM International (ASTM) Formerly the American Society for Testing and Materials, a nonprofit national technical society that publishes definitions, standards, test methods, recommended installation practices, and specifications for materials. ASTM is a consensus group of the building materials industry. It sets standards for products and establishes methods for testing.

Baffle or barrier, sound A shielding structure or partition used to increase the effective length of a sound transmission path between two locations.

Ceiling attenuation class (CAC) A single number or range of numbers for evaluating the effectiveness of an acoustical ceiling construction in isolating audible airborne sound transmission, tested at 16 1/3-octave frequencies. Higher numbers indicate more effectiveness in preventing noise between rooms.

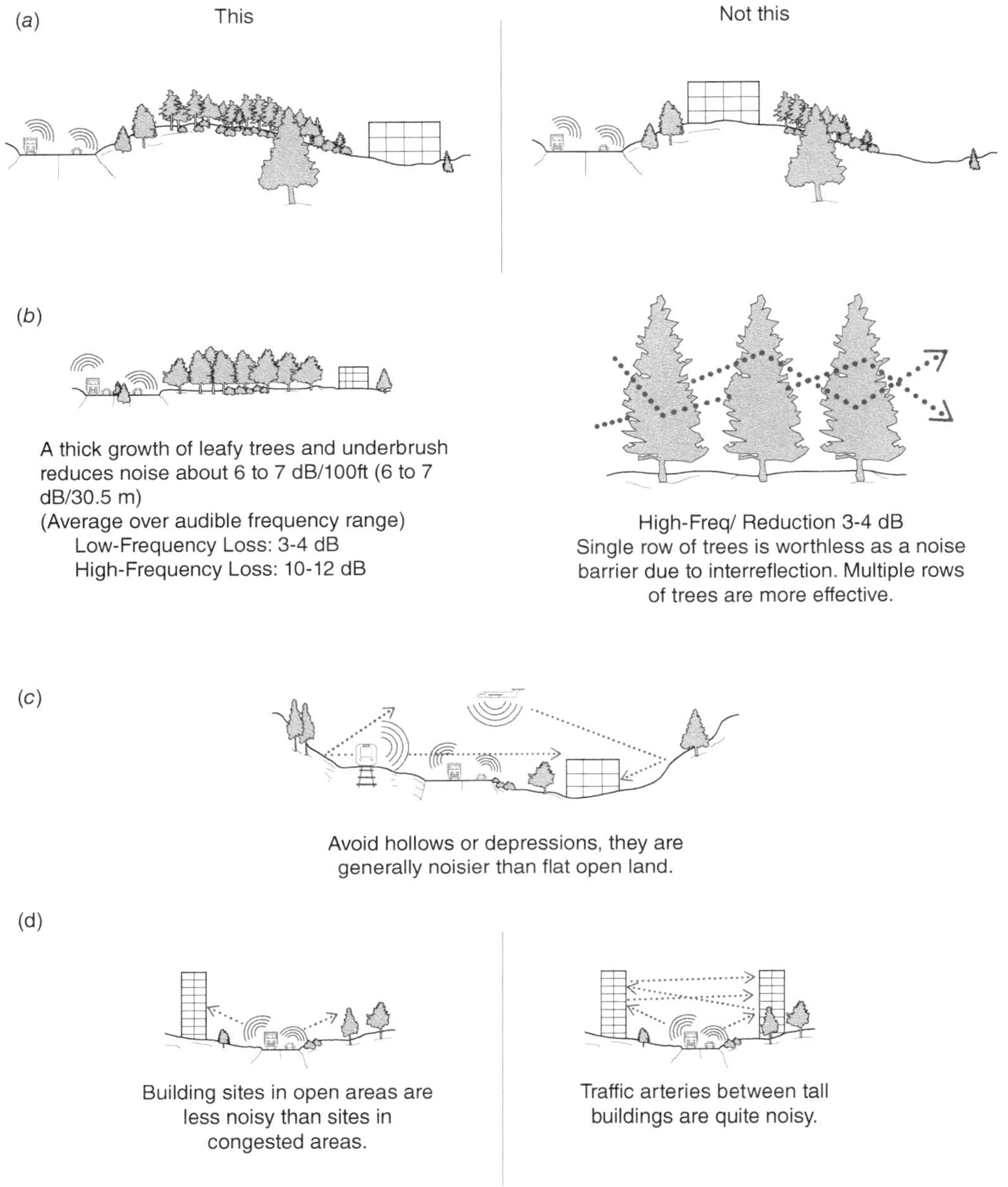

Figure 25.57 *(a) Use of natural noise barriers. (b) Effectiveness of wooded areas as noise barriers, showing noise reduction of trees. (c) An example of a poor building site. (d) Building site issues near traffic arteries and other buildings. (Redrawn by Nathan Leigh from HUD, 1968.)*

Tested in accordance with *ASTM E1414*. Previously described as the Ceiling Sound Transmission Class (CSTC) or Sound Transmission Class (STC).

Critical frequency The lowest frequency at which the wavelength of a bending wave, traveling in a structure, is the same as the wavelength in air at that frequency.

Damping Dissipation of structure-borne noise. This is usually accomplished by using a material with a high internal energy-absorbing capacity (i.e., high internal damping).

Decibel (dB) A descriptive notation adopted for convenience in representing vastly different sound quantities.

Diffraction The tendency of sound waves to flow readily around obstacles that are small in comparison to the wavelength of the sound.

Diffuse sound field A region where sound at any given point is made up of sound waves with all angles of incidence.

Direct sound field A region in which all or most of the sound arrives directly from the source without reflection.

Field impact insulation class (FIIC) A rating determined by an actual field test performed per *ASTM E1007*.

Field sound transmission class (FSTC) A rating determined by an actual field test performed per *ASTM E336*.

Flanking sound path The transmission of sound or noise from one room to another by indirect paths rather than directly through an intervening partition.

Flutter A multiple echo set up between parallel reflecting surfaces.

Free sound field (free field) A region in a homogeneous medium free from boundaries. In a free field, the sound pressure level decreases 6 dB for a doubling of the distance from a point source.

Impact insulation class (IIC) A single-figure rating that provides an estimate of the impact-sound-isolating performance of a floor-ceiling assembly.

Intensity The amount of sound energy per second that is carried across a unit area.

Intensity level (IL) A measure of the acoustic power passing through a unit area expressed in the decibel scale and referenced to some standard base (usually 10^{-12} W/cm^2).

Interference The destructive or reinforcing action of two or more waves arriving at the same position simultaneously.

Loudness A subjective human definition of the intensity of a sound.

Masking The presence of a background sound increases the level to which a sound signal must be raised in order to be heard or distinguished. If the level of the background sound is significantly higher than that of the sound signal, the signal cannot be heard. This effect is known as masking.

Mass law States that the transmission loss of walls (in part of the frequency range) is controlled entirely by the mass per unit area of the panel. It also states that the transmission loss increases 6 dB for each doubling of frequency or each doubling of the panel mass per unit area.

Noise Any undesired sounds, usually of different frequencies, resulting in an objectionable or irritating sensation.

Noise reduction (NR) (1) The reduction in sound pressure level caused by making an alteration to a sound source. (2) The difference in sound pressure level measured between two adjacent rooms caused by the transmission loss of an intervening barrier.

Noise reduction coefficient (NRC) The average sound absorption coefficient (to the nearest 0.05) measured at the four 1/3-octave bands centered on frequencies of 250, 500, 1000, and 2000 Hz.

Octave band A range of frequency where the highest frequency of the band is double the lowest frequency. The band is usually specified by its center frequency.

Phon Loudness level, at a given frequency, equal to the 1000-Hz decibel level of that equal-loudness contour.

Pink noise Wide-spectrum noise whose amplitude drops 3 dB per octave with increasing frequency (equal energy per octave). Useful for masking.

Random noise A noise whose magnitude and/or frequency cannot be predicted precisely at any given time. A rough approximation of random noise is the static heard on a radio between stations (see *Noise, Pink noise, White noise*).

Reverberation A persistence or echoing of previously generated sound caused by reflection of acoustic waves from the surfaces of enclosed spaces.

Reverberation time The time required for a sound to decay to a value one-millionth of its original intensity or to reduce 60 dB after the sound source has stopped.

Sabin The unit of acoustic absorption. One sabin ft^2 [m^2] is the absorption of 1 ft^2 [m^2] of perfect sound-absorbing material or open space with no reflecting surfaces.

Sound absorption coefficient (α) The ratio of the sound absorbed to the sound incident on a material or device. Stated differently, it is the fraction of the incident energy absorbed (not reflected) by a material when a sound wave strikes it, it is the sound absorption coefficient of that material. Usually represented by the Greek letter alpha (α).

Sound barrier A material installed to prevent the passage of sound from one area to another. Sound-deadening board and lead sheet or special insulation make good sound barriers.

Sound level meter An instrument for the direct measurement of sound pressure level. Sound level meters may also incorporate octave band filters for measuring sound directly in octave bands.

Sound power level (PWL) A measure of the total airborne acoustic power generated by a noise source, expressed in the decibel scale and referenced to some standard base (usually 10^{-12} W).

Sound pressure level (SPL) A measure of the air pressure change caused by a sound wave. Expressed in the decibel scale and referenced to some standard base (usually 0.0002 μbar).

Sound transmission class (STC) A single-number rating of a building element's efficacy in blocking the transmission of sound compared to a standard transmission attenuation/frequency curve. See also *Ceiling attenuation class.*

Transmission loss (TL) The reduction of airborne sound power caused by placing a wall or barrier between the reverberant sound field of a source and its receiver. Transmission loss is a property of the wall or barrier.

White noise Noise of a wide frequency range in which the amplitude of the noise is essentially the same in all frequency bands (equal energy per frequency band).

25.38 REFERENCE STANDARDS

Those standards with the greatest direct applicability to the design of building noise control solutions are summarized here.

ASTM E90-09(2016): Standard Test Method for Laboratory Measurement of Airborne Sound Transmission Loss of Building Partitions and Elements. This laboratory test method specifies the procedures that can be used to assess how effectively building components block sound transmission from one space to another. This test method produces the single-number sound transmission class (STC) rating.

ASTM E336-20: Standard Test Method for Measurement of Airborne Sound Attenuation between Rooms in Buildings. This field test method specifies the procedures that can be used to estimate lower limits for sound transmission through a partition element. This test method produces the following single-number ratings: Noise Isolation Class (NIC), Normalized Noise Isolation Class (NNIC), and Apparent Sound Transmission Class (ASTC).

ASTM E413-22: Classification for Rating Sound Insulation. This standard defines the rating procedures for single-number acoustical ratings for laboratory and field measurements of sound attenuation obtained in ⅓-octave bands. However, the name given to each rating is assigned by its respective test method.

ASTM E596-96(2016): Standard Test Method for Laboratory Measurement of Noise Reduction of Sound-Isolating Enclosures. This laboratory test method covers the reverberation room measurement of the noise reduction of sound-isolating enclosures. The measured noise reduction is used to determine the sound-isolating enclosure's Noise Isolation Class (NIC).

ASTM E795-23: Standard Practices for Mounting Test Specimens During Sound Absorption Tests. This document offers practices for test specimen mountings to be used during sound absorption testing performed in accordance with *ASTM C423, Standard Test Method for Sound Absorption and Sound Absorption Coefficients by the Reverberation Room Method.*

ASTM E1110-06(2019): Standard Classification for Determination of Articulation Class. This classification provides a single-figure rating that can be used for comparing building systems and subsystems for speech privacy purposes. Excluded from this classification are applications involving female speakers and children, languages other than English, and sound spectra other than speech.

ASTM E1111/E1111M-14(2022): Standard Test Method for Measuring the Interzone Attenuation of Open Office Components. This test method is intended to provide measurements of the speech privacy characteristics of ceiling systems (with partial-height partitions), furniture panels, and vertical panels.

ASTM E1130-16(2021): Standard Test Method for Objective Measurement of Speech Privacy in Open Plan Spaces Using Articulation Index. This method describes a field test for measuring speech privacy objectively between locations in open plan spaces. It relies upon acoustical measurement, published information on speech levels, and standard methods for assessing speech communication. This test method does not measure the performance of individual open-plan components that affect speech privacy; it measures the privacy that results from a particular configuration of components. This method relies upon the AI, which predicts the intelligibility of speech for a group of talkers and listeners.

ASTM E1179-13(2019): Standard Specification for Sound Sources Used for Testing Open Office Components and Systems. This specification states the requirements for sound sources used for measuring the speech privacy between open offices or for measuring the laboratory performance of acoustical components. The sound source is a loudspeaker located in an enclosure and driven with an appropriate test signal.

ASTM E1264-22: Standard Classification for Acoustical Ceiling Products. This classification covers ceiling products that provide acoustical performance and interior finish in buildings. It classifies acoustical ceilings by type, pattern, and certain ratings for acoustical performance, light reflectance, and fire safety.

ASTM E1374-18e1: Standard Guide for Open Office Acoustics and Applicable ASTM Standards. This guide discusses the acoustical principles and interactions that affect the acoustical environment and acoustical privacy in an open office. In this context, it describes the application and use of the series of ASTM standards that apply to open offices.

ASTM E1375-90(2002): Standard Test Method for Measuring the Interzone Attenuation of Furniture Panels Used as Acoustical Barriers (formally withdrawn in 2005; now part of *ASTM E1111*; retained here as a historical reference). This test method covered the measurement of the interzone attenuation of furniture panels used as acoustical barriers in open-plan spaces to provide speech privacy or sound isolation between working positions.

ASTM E1376-90(2002): Standard Test Method for Measuring the Interzone Attenuation of Sound Reflected by Wall Finishes and Furniture Panels (formally withdrawn in 2005; now part of *ASTM E1111*; retained here as a historical reference). This laboratory test method measured the degree to which reflected sound is attenuated by the most commonly found vertical surfaces in open-plan spaces. The vertical surfaces covered included wall finishes such as sound-absorbent panels and furniture panels or screens. It did not cover window finishes or furniture other than panels.

ASTM E1414/E1414M-21a: Standard Test Method for Airborne Sound Attenuation Between Rooms Sharing a Common Ceiling Plenum. This laboratory test method simulates a pair of horizontally adjacent small offices or rooms separated by a partition and sharing a common plenum space.

ASTM E1573-18: Standard Test Method for Evaluating Masking Sound in Open Offices Using A-Weighted and One-Third Octave Band Sound Pressure Levels. This test method specifies the procedures that can be used to evaluate the spatial and temporal uniformity of masking sound in open offices using A-weighted sound levels. It also specifies the procedure for evaluating the masking sound spectrum and level using 1/3-octave band sound pressure levels.

ISO 717-1:2020: Acoustics—Rating of Sound Insulation in Buildings and of Building Elements—Part 1: Airborne Sound Insulation. This standard outlines the basic measurement procedures for airborne and impact sound insulation of building elements using laboratory test facilities.

ISO 717-2:2020: Acoustics—Rating of Sound Insulation in Buildings and of Building Elements—Part 2: Impact Sound Insulation. This standard outlines the basic measurement procedures for impact sound insulation of building elements using laboratory test facilities.

ASTM E492-09(2016)e1: Standard Test Method for Laboratory Measurement of Impact Sound Transmission Through Floor-Ceiling Assemblies Using the Tapping Machine. This standard describes a test method for the laboratory measurement of the impact sound transmission of floor-ceiling assemblies using a tapping machine.

ASTM E1007-25: Standard Test Method for Field Measurement of Tapping Machine Impact Sound Transmission Through Floor-Ceiling Assemblies and Associated Support Structures. This standard describes a test method for the field measurement of the impact sound transmission generated by a standard tapping machine through floor-ceiling assemblies and associated supporting structures.

ASTM E989-21: Standard Classification for Determination of Single-Number Metrics for Impact Noise. This classification provides a method for determining ratings to compare the levels of impact noise transmitted through different floor-ceiling assemblies.

25.39 UNITS AND CONVERSIONS

The I-P and SI systems of units are used throughout this book. Acoustical measurements will occasionally be expressed in two historic systems of units (that are now generally obsolete). See Table 25.16 for MKS units (employing the meter, kilogram, and second as basic elements), CGS units (employing the centimeter, gram, and second), and common conversions.

TABLE 25.16 Acoustic Units and Conversions

Variable	MKS units		CGS units
Force	kilogram-meter/s^2 = newton		gram-cm/s^2 = dyne
Intensity	watts/meter^2		watts/cm^2
Pressure	newton meter^2 = pascals		dynes/cm^2 = microbars
	In Conversion:		
Quantity	**Multiply**	**By**	**To Obtain**
Force	newtons	10^5	dynes
	dynes	10^{-5}	newtons
Intensity	watts/cm^2	10^4	watts/m^2
	watts/m^2	10^{-4}	watts/cm^2
Pressure	pascals	10	microbars
	microbar	10^{-1}	pascals

Note: One atmosphere = 1 bar = 10^6 μbar.

ACOUSTICS

25.40 SYMBOLS

See Table 25.17 for abbreviations and symbols that are typically encountered in architectural acoustics.

TABLE 25.17 Symbols and Abbreviations Commonly Used in Acoustics

A	Total absorption, sabins; area in unit being used
A_R	Absorption in receiving room, sabins
$A_{1,2}$ ...	Total absorption of each material in a space, sabins
AC	Articulation class
c	Velocity of sound, feet per second
CAC	Ceiling attenuation class
d	Distance from source, meters or feet
dB	Decibel
f	Frequency of sound, hertz (Hz)
I	Intensity, W/cm^2
I_a	Absorbed energy, W/cm^2
I_i	Incident energy, W/cm^2
I_0	Reference intensity, 10^{-16} W/cm^2
IIC	Impact insulation class, no units
IL	Intensity level, decibels
NC	Noise criterion, no units
NR	Noise reduction, decibels
NRC	Noise reduction coefficient, no units
p	Pressure, pascals or microbars
p_0	Reference base pressure, 2×10^{-5} Pa
P	Acoustic power, watts
Pa	Pascal, unit of pressure (SI)
PWL	Sound power level, decibels
R	Room constant, square feet (square meters)
r	Distance from source, meters or feet
S	Surface area, in unit being used
SPL	Sound pressure level, decibels
STC	Sound transmission class, no units
T_R	Reverberation time, seconds
TL	Transmission loss, decibels
V	Volume (geometric)
W, P	Sound power, watts
W_0	Reference base sound power, 10^{-12} W
a, α	Absorption coefficient (no units)
$\bar{a}, \bar{\alpha}$	Average absorption coefficient (no units)
λ	Wavelength, feet or meters
Σ	Sum of, or total (no units)
$\Sigma S\alpha = \Sigma A$	Total absorption, sabins
Δ	Change in a quantity or difference between two quantities

Note: Where definitions are expressed in I-P units, SI units are also understood, with proper conversion factors, and vice versa.

References and Resources

ASHRAE. 2024. *ASHRAE Handbook—HVAC Applications*. Atlanta, GA: American Society of Heating, Refrigerating and Air-Conditioning Engineers.

Cavanaugh, W. J., W. R. Farrell, P. W. Hirtle, and B. G. Watters. 1962. "Speech Privacy in Buildings." *Journal of the Acoustical Society of America* 34(4): 475–492.

Fader, B. 1981. *Industrial Noise Control*. New York: John Wiley & Sons.

HUD. 1968. *A Guide to Airborne, Impact, and Structure-Borne Noise Control in Multifamily Dwellings*. Washington, DC: U.S. Department of Housing and Urban Development.

Magrab, E. B. 1975. *Environmental Noise Control*. New York: John Wiley & Sons.

NBS. 1986. *Quieting: A Practical Guide to Noise Control*. Gaithersburg, MD: National Bureau of Standards.

NEMA. 2022. *NEMA ST 20–2021: Dry-Type Transformers for General Applications*. Rosslyn, VA: National Electrical Manufactures Association.

Schaudinischky, L. H. 1976. *Sound, Man and Building*. London: Applied Science Publishers.

Young, R. W. 1965. "Revision of the Speech Privacy Calculation." *Journal of the Acoustical Society of America* 38(4): 524–530.

Note: Numerous standards from ASTM and ISO are described in this chapter but not listed in this section to avoid redundancy.

PART VII

FIRE PROTECTION

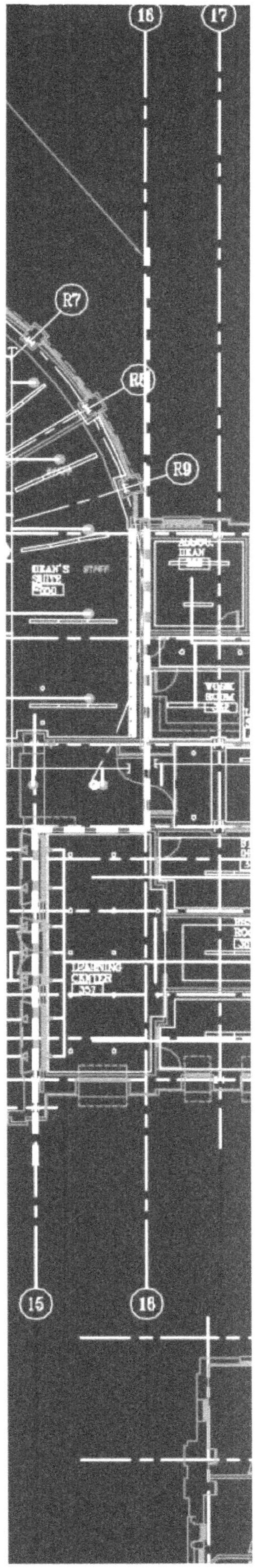

Part VII consists of Chapter 26 which broadly embraces passive and active fire protection strategies, detection and alarm systems, and fire and smoke control approaches.

Providing protection against the effects of a building fire is challenging. It is not pleasant to imagine a building burning and considering how its occupants, and then the building itself (along with its contents), can be saved.

To complicate this critically important concern, there are inherent conflicts between some fire protection features and useful characteristics of design strategies such as daylighting, passive cooling, or HVAC systems. There are even potential design conflicts between systems for the safe evacuation of people and systems for the suppression of fire—because both people and fire thrive on oxygen.

Many aspects of fire protection are rigidly governed by building codes that have evolved over the past century in response to disasters and milestone events. On the other hand, most large buildings include some feature (such as an atrium) that is unique and not clearly addressed by otherwise explicit fire codes. Designers must sometimes find an acceptable balance between respecting the codes and showing ingenuity to move beyond the codes when required.

CHAPTER 26

Fire Protection

JUST AS BUILDINGS CAN BE DESIGNED to minimize heating, cooling, and lighting equipment and thus to reduce energy consumption, they can also be designed to reduce the size of firefighting systems and retard the spread of smoke and fire. As the average size of fires within buildings continues to decline, the emphasis is shifting to minimizing water and smoke damage. Whole-building conflagrations, while spectacular, are relatively rare in large buildings in North America, now that both automatic fire detection and fire-extinguishing systems are widespread.

This discussion of fire protection begins with basic design considerations for fire resistance. Smoke management (for safe evacuation and for limited smoke damage) is considered next, followed by fire-suppression systems such as sprinklers and non-water-based approaches. Lightning protection is then discussed, along with the many fire detection and alarm systems that are keyed to the major stages of the typical building fire.

Throughout this chapter, the influence of the National Fire Protection Association (NFPA) is apparent. The NFPA is a U.S.-based international nonprofit organization dedicated to eliminating loss from fire, electrical, and related hazards. They maintain a range of model codes, standards, and other information resources. The NFPA *Fire Protection Handbook* (NFPA, 2008) and *NFPA 101: Life Safety Code* (NFPA, 2023c) are especially useful sources of information for design, from which several figures have been derived and illustrations have been reprinted in this text.

FIRE RESISTANCE, EGRESS, AND EXTINGUISHMENT

26.1 DESIGN FOR FIRE RESISTANCE

Fire is a special kind of oxidation known as combustion. Oxidation—discussed in previous chapters in terms of rust within water supply equipment and aerobic digestion in waste disposal systems—is a process in which molecules of a fuel are combined with molecules of oxygen, producing a mixture of gases and energy. When this occurs rapidly, as in a fire, energy is released as heat and light, and some gases become visible as smoke. Table 26.1 lists the stages of a fire and the factors that influence its growth. Many of these factors involve decisions made by building designers.

Fire has a triangle of needs: fuel, high temperature, and oxygen. If deprived of any of these needs, building fires will be extinguished. In general, this triangle's influence on building design is as follows. The fuel is the building's structure and contents; the designer controls the choice of structural and finish materials, but rarely the final contents. The temperatures achieved in fires are well beyond the ability of building cooling systems to control, so special water systems (in the form of sprinklers) are often installed to deprive fire of the high temperatures it needs. Oxygen may be denied to a fire partly by limitations on ventilation, but reduced oxygen can have serious safety consequences.

Mechanical and Electrical Equipment for Buildings, Fourteenth Edition. Walter Grondzik, Daniel Overbey, and Alison Kwok.

Companion website: www.wiley.com/go/meeb14e

TABLE 26.1 Major Factors Influencing Fire Growth

Realm	Approximate Ranges of Fire Sizes	Major Factors That Influence Growth
1 Preburning	Overheat to ignition	Amount and duration of heat flux, surface area receiving heat, material ignitability
2 Initial burning	Ignition to radiation point (10 in. [254 mm] high flame)	Fuel continuity, material ignitability, thickness, surface roughness, thermal inertia of the fuel
3 Fire growth	Radiation point to enclosure point (10 in. to 5 ft [250 mm to 1.5 m]) high flame	Interior finish, fuel continuity, heat feedback, material ignitability, thermal inertia of the fuel, proximity of flames to walls
4 Full item involvement	Enclosure point to ceiling point (5 ft [1.5 m]) high flame to flame touching ceiling	Interior finish, fuel arrangement, heat feedback, height of fuels, proximity of flames to walls, ceiling height, room insulation, size and location of openings, HVAC operation
5 Full room involvement	Ceiling point to full room involvement	Fuel arrangement, ceiling height, length/width ratio, room insulation, size and location of openings, HVAC operation

Source: Reproduced with permission of NFPA from the Fire Protection Handbook, 21st edition. Copyright © 2023, National Fire Protection Association. For a full copy of the Fire Protection Handbook, please go to www.nfpa.org.

Another design response is to install fire-suppression systems that either cover the fuel (foam, dry chemicals) or displace oxygen with another gas—for example, carbon dioxide or "clean agents" that inhibit the chemical action of the flame itself.

(a) Sources of Ignition

Buildings commonly contain three basic sources of ignition: chemical, electrical, and mechanical. In chemical combustion, most commonly known as spontaneous combustion, some chemicals reach ignition at ordinary temperatures within buildings. Chemical combustion depends upon the rate of heat generation (related to the degree of saturation of combustible products, determined by the chemicals involved), the air supply (enough to supply oxygen but not enough to lower the temperature), and the insulation provided by the immediate surroundings (the more insulation, the easier the attainment of combustion temperatures).

Electrical heat energy is most commonly supplied by resistance heating, a familiar process in many appliances and in space-heating equipment. Less common are electric ignition by induction, dialectic process, arcing, and static electricity. Lightning is an infrequent but enormously destructive electrical energy source.

Mechanical heat energy is produced by friction (including sparks), by overheating of machinery, and occasionally by the heat of compression.

(b) Products of Combustion

Everyone has experienced to some degree the dangers of the thermal products of combustion (Figure 26.1)—flame and heat. These visible and tactile elements of fire can cause burns, shock, dehydration, heat exhaustion, and fluid blockage of the respiratory tract, but they are responsible for only about one-quarter of the deaths resulting from building fires. Most fire deaths are caused by the nonthermal products—smoke and its wide range of constituent gases, liquids, and solids. Smoke can usually be seen and smelled. Made up of droplets of flammable tars and small particles of carbon suspended in gases, it irritates the eyes and nasal passages, sometimes blinding and/or choking a person. Gases are especially dangerous because, without visible smoke, they are so often difficult to detect. Some gases are directly toxic, but all are dangerous because they displace oxygen. Common gases released in building fires include carbon monoxide and carbon dioxide.

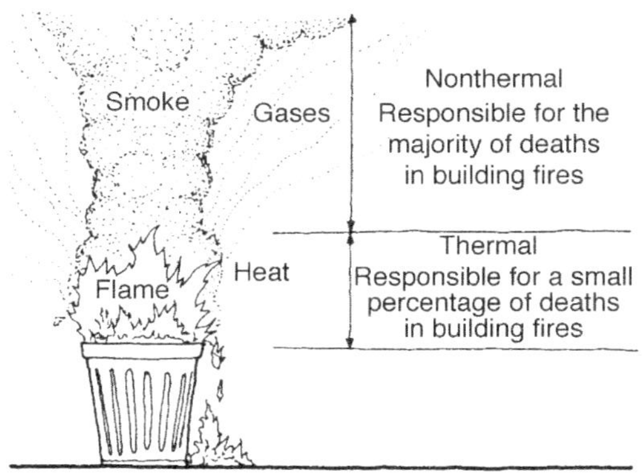

Figure 26.1 *Although the thermal products from a fire—flame and heat—make strong visible and tactile impressions, fire's nonthermal products—smoke and its constituent components—pose the greater threat to life.*

Carbon monoxide (CO) is a deadly product of combustion and is often the most abundant such product. It is produced when insufficient oxygen is available to completely oxidize the burning material. It is more readily attached to hemoglobin molecules in red blood cells than is oxygen, thus depriving the brain and muscles of needed oxygen. This leads to irrational behavior and loss of consciousness, then to death. Carbon dioxide (CO_2) is also likely to result from combustion but, unless intentionally introduced into a space as a means of fire suppression, should not be present in concentrations with serious health impacts (beyond potential headaches and dizziness).

Other dangerous and commonly encountered building-fire gases include hydrogen sulfide, sulfur dioxide, ammonia, oxides of nitrogen, cyanide, phosgene, and hydrogen chloride (from burning polyvinyl chlorides found in computers, electronic equipment, and cable jackets). These types of gases may cause eye, skin, nose, and throat irritation—and in sufficient concentrations can be psychotropic and/or toxic.

In indoor fires, oxygen commonly becomes insufficient because the fire consumes it so rapidly. The normal concentration of oxygen in air is about 21%. At less than 17%, muscular coordination and judgment are diminished; at 14% down to 10%, people remain conscious but become irrational, and fatigue is rapid; at 10% down to 6%, collapse occurs, but revival is possible when increased oxygen is supplied. The technique of starving the fire of oxygen can therefore pose a threat to human beings, both by increasing the chances of carbon monoxide production and by depriving people of oxygen.

(c) Objectives in Fire Safety

Many older buildings were designed at a time when building fires usually resulted in the loss of several adjacent structures; hence, the primary objective of firefighting was to limit the conflagration to as few city blocks as possible. With the increased use of fire-resistant construction and code control of site and building planning, fires typically were confined to one building at a time. As fire-suppression systems came into common usage within buildings, it then came to be expected that fires could be confined to one or two floors within a building. Now that automatic detection/suppression systems are technically advanced, fires can usually be confined to one room or to even smaller areas. In the United States, most fires in sprinklered buildings are now extinguished with one to five sprinklers operating.

Four common design intents related to building fire safety, in order of usual importance, are:

1. Protection of life
2. Protection of building
3. Protection of contents
4. Continuity of operation

Many of the elements of building fire safety are covered by building codes, but it is important to remember that codes specify the minimum acceptable protection. Designers can go much further than the codes require in order to enhance fire safety. It is also important to realize that codes cannot cover all aspects of fire safety, as there are too many variables in building design.

Codes typically prescribe design strategies that are passive means of limiting the spread of fire and protecting life: wall, floor, and ceiling constructions; maximum open floor areas; maximum distances to exits; and so on. Codes commonly allow some relaxing of such prescriptions when active fire-suppression systems (such as sprinklers) are designed into a building. Alternatively, a detailed computational analysis of fire spread and occupant evacuation within a proposed project may allow for even greater distances to exits, larger open floor areas, and alternative constructions. This is a performance-based approach to design (rather than a prescriptive-based approach). A performance-based approach requires close cooperation between designers and fire code enforcement authorities.

(d) Fire Safety and Other Environmental Control Systems

In some instances, the desirable design approaches for fire safety will track design approaches for lighting, thermal, acoustic, and water systems. Although these opportunities may be uncommon, it would be unfortunate (because of a discipline-centric design process) to miss such synergies because of the following reasons:

- *Thermal mass* is useful for passive heating and cooling systems, for acoustic isolation of airborne sound, and for fire barriers (most thermally massive materials will not burn easily).
- *High ceilings* are useful for daylight distribution and displacement ventilation, but also for collecting a large quantity of smoke before it reaches the occupants and for allowing smoke and/or flames from a fire to be seen from a greater indoor distance.
- *Windows*, useful for daylight, ventilation, and view, also allow access for firefighting and rescue, provide escape routes, relieve smoke accumulation with fresh air, and thus relieve some of the stress of trapped occupants.
- *Solid (noncombustible) overhangs over windows* not only provide sunshading, but also discourage the vertical spread of fire over the building face and can serve as emergency exterior places of refuge.
- *Elevated water storage tanks* provide both adequate water pressure for plumbing fixtures and water for firefighting in the first few minutes of a fire before firefighters arrive.

In many other instances, there are potential design conflicts between building performance under

TABLE 26.2 Summary of Life Safety Code Requirements for Interior Finish (Walls, Ceilings, and Floors)

Occupancy	Exits	Exit Access Corridors	Other Spaces
Assembly—New: >300 people	A	A or B	A or B
	I or II	I or II	NA
≤300 people	A	A or B	A, B, or C
	I or II	I or II	NA
Assembly—Existing: >300 people	A	A or B	A or B
≤300 people	A	A or B	A, B, or C
Educational—New	A	A or B	A or B; C on low partitions*
	I or II	I or II	NA
Educational—Existing	A	A or B	A, B, or C
Day Care Centers—New	A	A	A or B
	I or II	I or II	NA
Day Care Centers—Existing	A or B	A or B	A or B
Day Care Homes—New	A or B	A or B	A, B, or C
	or II		NA
Day Care Homes—Existing	A or B	A, B, or C	A, B, or C
Health Care—New	A	A	A
	NA	B on lower portion of corridor wall*	B in small individual rooms*
	I or II	I or II	NA
Health Care—Existing	A or B	A or B	A or B
Detention and Correctional— New (Sprinklers Mandatory)	A or B	A or B	A, B, or C
	I or II	I or II	NA
Detention and Correctional—Existing	A or B	A or B	A, B, or C
	I or II	I or II	NA
One- and Two-Family Dwellings and Lodging or Rooming Houses	A, B, or C	A, B, or C	A, B, or C
Hotels and Dormitories—New	A	A or B	A, B, or C
	I or II	I or II	NA
Hotels and Dormitories—Existing	A or B	A or B	A, B, or C
	I or II*	I or II*	NA
Apartment Buildings—New	A	A or B	A, B, or C
	I or II	I or II	NA
Apartment Buildings—Existing	A or B	A or B	A, B, or C
	I or II*	I or II*	NA
Residential Board and Care (See Chapters 32 and 33 or the 2024 NFPA 101)			
Mercantile—New	A or B	A or B	A or B
	I or II		NA
Mercantile—Existing			
Class A or Class B Stores	A or B	A or B	Ceilings—A or B; Walls—A, B, or C
Class C Stores	A, B, or C	A, B, or C	A, B, or C
Business and Ambulatory Health Care—New	A or B	A or B	A, B, or C
	I or II		NA
Business and Ambulatory Health Care—Existing	A or B	A or B	A, B, or C
Industrial	A or B	A, B, or C	A, B, or C
	I or II	I or II	NA
Storage	A or B	A, B, or C	A, B, or C
	I or II	I or II	NA

Source: Reproduced with permission of NFPA from the Fire Protection Handbook, 21st edition. Copyright © 2023, National Fire Protection Association. For a full copy of the Fire Protection Handbook, please go to www.nfpa.org.

*See corresponding chapters within 2024 NFPA 100 for details.

NA: Not applicable.

Notes: Class A Interior Wall and Ceiling Finish—flame spread 0–25, (new) smoke developed 0–450.

Class B Interior Wall and Ceiling Finish—flame spread 26–75, (new) smoke developed 0–450.

Class C Interior Wall and Ceiling Finish—flame spread 76–200, (new) smoke developed 0–450.

Class I Interior Floor Finish—critical radiant flux, minimum 0.45 W/cm^2.

Class II Interior Floor Finish—critical radiant flux, minimum 0.22 W/cm^2.

Automatic sprinklers — where a complete standard system of automatic sprinklers is installed, interior wall and ceiling finish with a flame spread rating not exceeding Class C is permitted to be used in any location where Class B is required, and Class B interior wall and ceiling finish is permitted to be used in any location where Class A is required; similarly, Class II interior floor finish is permitted to be used in any location where Class I is required, and no interior floor finish classification is required where Class II is required. These provisions do not apply to new detention and correctional occupancies.

Exposed portions of structural members complying with the requirements of heavy timber construction may be permitted.

ordinary conditions and building performance in the extraordinary event of a fire. Escalators invite shoppers to explore the upper levels of a department store or connect hotel lobbies to ballrooms; they are also efficient apertures for the vertical spread of fire and smoke. (Special fire protection strategies for escalators are found in Chapter 34.) Daylighting and natural ventilation are best served by high ceilings and low partitions, which encourage light and air from the perimeter to pervade building interiors. If unchecked by sprinklers, however, fire and smoke can easily spread through such open floor plans. (On the other hand, smoke and fire can build up much more rapidly in small, enclosed rooms that retain heat.) Forced-air systems that heat, ventilate, and cool are also potential pathways for smoke and fire; this hazard is especially serious when the systems penetrate floors, because vertically spreading fires are harder to fight than horizontally spreading ones. (Then again, carefully designed forced-air systems can aid in smoke management.) Windowless buildings that rely on electric lighting in place of daylight are especially dangerous in fires because firefighters cannot easily evacuate occupants or gain access to the building. Sunscreens that completely cover windows are disadvantageous for the same reason; nonoperable windows, although considered advantageous for tightly controlled air conditioning, must be broken for fire evacuation/access. The higher the window, the greater the danger from falling shards of glass. Many excellent insulating materials will burn readily, and some will give off toxic gases. Some of the loveliest interior finishes are both flammable and deadly sources of gases in a fire. Table 26.2 summarizes fire requirements for interior finishes.

The dichotomy between ensuring comfort throughout a building's normal life and safety at the moment it is threatened by fire is the principal reason that codes alone cannot ensure fire safety. The dilemma of ordinary versus extraordinary performance must be faced by the cooperative actions of the designer, owner, and occupants of a building.

(e) Protection of Life

The NFPA *Fire Protection Handbook* (NFPA, 2023d) discusses human behavior in fires in great detail. Although panic behavior drives many code requirements, such behavior has been found to be rare. Designers should consider how building occupants make decisions in a fire. In the first phase, cues are detected—the smell of smoke, sounds associated with a fire (breaking glass, sirens, alarm bells), and, more rarely, the sight of flames. Open plan offices (with longer visible indoor distances) are more amenable to exposing such clues to a wider population. In the second phase, the occupants define the situation: Just how serious is this fire? The more numerous the cues, the more rapid the definition phase. How other people are reacting is influential, and in the absence of strong cues can actually lead to a group refusal to evacuate in the early stages of a fire. In the third phase, coping behavior begins: fight or flight? Clear exit pathways and access to firefighting equipment are critical to rationally making this decision.

For most low-rise buildings, a reasonable goal is the evacuation of all occupants in the time interval between the detection of a fire and the arrival of the firefighters. Designers can provide clearly defined pathways to exits (exit access) that can be kept relatively clear of smoke (Figure 26.2). To accommodate a wheelchair, a minimum clear width of 32 in. [813 mm] is required. Exits can take a variety of forms. Vertical exits (Figures 26.3 and 26.4) include smokeproof towers, exterior and interior stairs and ramps, and escalators that meet specific requirements. Vertical exits do not include elevators; they are too easily stalled or, worse, opened at the floor of a fire by malfunctioning signal equipment. Exits in the horizontal plane include doors leading directly to the outside, 2-hour fire-rated enclosed hallways, and moving walks. Special horizontal exits are provided by internal firewalls penetrated by two fire doors—one swinging open in either direction. Exit discharge is the area outside an exit that leads to a public way and may still need protection in a fire.

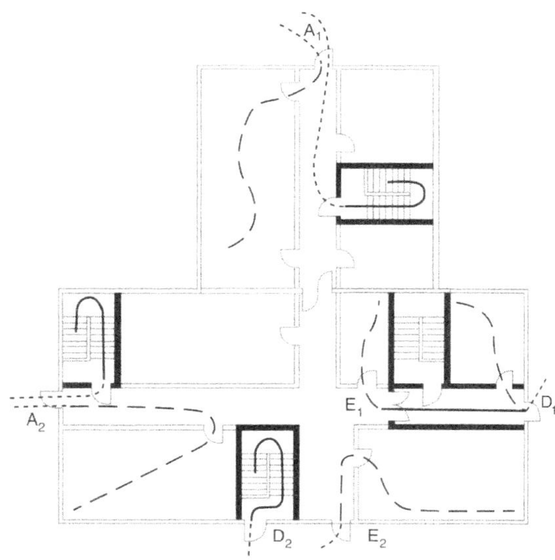

Figure 26.2 *Exit access, exit, and exit discharge on the first floor of a multistory building. Doors A_1, A_2, E_1, and E_2 are exits, and the path (dashed line) is the exit access. To the person emerging from the exit enclosure, doors A_1 and A_2 and the paths (dotted lines) are the exit discharge. Doors D_1 and D_2 are exit discharge doors. Solid-line paths are within the exit. (Reproduced with permission of NFPA from the Fire Protection Handbook, 20th edition. Copyright © 2008, National Fire Protection Association. For a full copy of the Fire Protection Handbook, please go to www.nfpa.org.)*

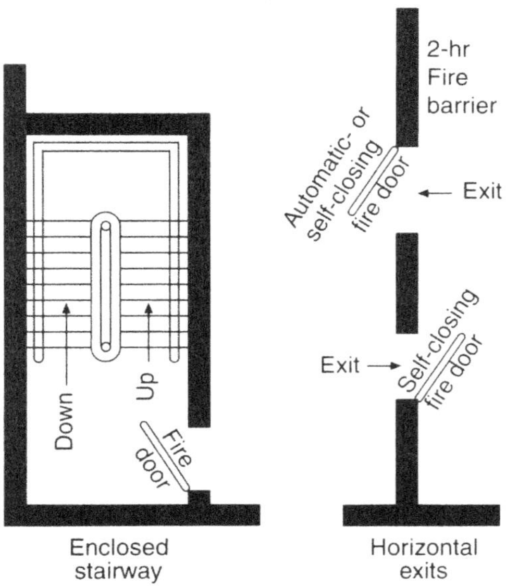

Figure 26.3 *Plan views of exits. Stair enclosure prevents a fire on any floor from trapping the persons above. A smokeproof tower is better, as it opens to the air at each floor, largely preventing the chance of smoke in the stairway. A horizontal exit provides a quick refuge and lessens the need for a hasty flight down the stairs. Fire-rated doors must be arranged to be self-closing or automatic-closing by smoke detection. (Reproduced with permission of NFPA from the Fire Protection Handbook, 20th edition. Copyright © 2008, National Fire Protection Association. For a full copy of the Fire Protection Handbook, please go to www.nfpa.org.)*

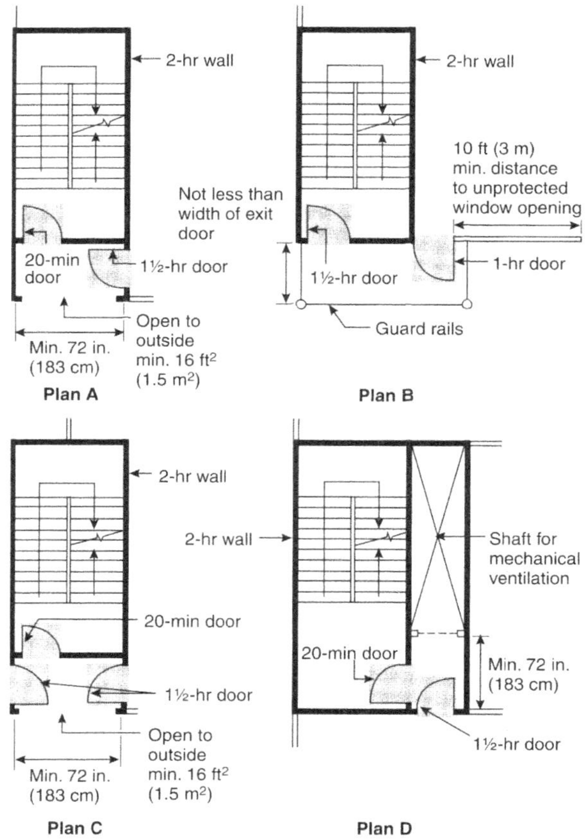

Figure 26.4 *Four variations of smokeproof towers. Plan A has an open-air vestibule opening from a corridor. Plan B shows an entrance by way of an outside balcony. Plan C could provide a stair tower entrance common to two areas. In plan D, smoke and gases entering the vestibule would be exhausted by a natural or induced draft in the open air shaft. In each case, a double entrance to the stair tower with at least one side open or vented is characteristic of this type of construction. Pressurization of the stair tower in the event of fire provides an attractive alternative for tall buildings and is a means of eliminating the entrance vestibule. (Reproduced with permission of NFPA from the Fire Protection Handbook, 21st edition. Copyright © 2023, National Fire Protection Association. For a full copy of the Fire Protection Handbook, please go to www.nfpa.org.)*

Maximum allowable distances to exits are specified in Table 26.3. These are influenced by past experience with people exiting through smoke, especially when it may involve briefly traveling toward the perceived fire location (as in the case of a dead-end corridor). When automatic fire suppression systems such as sprinklers are used, the allowable distances to exits are increased. Designers should remember, however, that at least 30% of building fire deaths result from fire cutting off the paths to exits.

Minimum egress widths per floor are found with the help of Table 26.4. First, calculate the floor area (net or gross, as specified in the table). Divide the floor area by the occupant load to determine the number of occupants for whom exits must be provided for that floor. Then calculate the exit capacity based on its clear width. Stairs are sized to meet the requirements of Table 26.5.

EXAMPLE 26.1 A multistory office building is 80 ft [24 m] wide by 300 ft [80 m] long. What exit capacity is required per floor?

SOLUTION (I-P UNITS)

The gross floor area:

$$80\ \text{ft} \times 300\ \text{ft} = 24{,}000\ \text{ft}^2$$

From Table 26.4, "Business Use" categories are based on one person per gross 150 ft^2.

The population per floor is, therefore:

$$(24{,}000\ \text{ft}^2)/(150\ \text{ft}^2/\text{person}) = 160\ \text{people}$$

Required total width for exit doors (to stairway):

$$160\ \text{people} \times 0.2\ \text{in./person} = 32\ \text{in. total}$$

Provided total width: One 34-in. clear door at each of two stairs:

$$2\ \text{doors} \times 34\ \text{in.} = 68\ \text{in. (more than the minimum)}$$

Required stairway width:

$$160\ \text{people} \times 0.3\ \text{in./person} = 48\ \text{in. total}$$

Provided stairway width:

$$2\ \text{stairs} \times 44\ \text{in. each}$$

$$= 88\ \text{in. (more than the minimum)}$$

■

TABLE 26.3 Common Path, Dead End, and Travel Distance

Type of Occupancy	Common Path Limit		Dead-End Limit		Travel Distance Limit	
	Unsprinklered, ft [m]	Sprinklered, ft [m]	Unsprinklered, ft [m]	Sprinklered, ft [m]	Unsprinklered, ft [m]	Sprinklered, ft [m]
Assembly						
New	20/75[a] [6.1/23]	20/75[a] [6.1/23]	20[c] [6.1]	20[c] [6.1]	200[d] [61]	250[d] [76]
Existing	20/75[a] [6.1/23]	20/75[a] [6.1/23]	20[b] [6.1]	20[b] [6.1]	200[c] [61]	250[c] [76]
Educational						
New	75 [23]	100 [30]	20 [6.1]	50 [15]	150 [45]	200 [61]
Existing	75 [23]	100 [30]	20 [6.1]	50 [15]	150 [45]	200 [61]
Day-Care						
New	75 [23]	100 [30]	20 [6.1]	50 [15]	150[d] [45]	200[d] [61]
Existing	75 [23]	100 [30]	20 [6.1]	50 [15]	150[d] [45]	200[d] [61]
Health Care						
New	100 [30]	100 [30]	30 [9.1]	30 [9.1]	200[d] [61]	200[d] [61]
Existing	NL	NL	NL	NL	150[d] [45]	200[d] [61]
Ambulatory						
New	75[e] [23]	100 [30]	20 [6.1]	50 [15]	150[d] [45]	200[d] [61]
Existing	75[e] [23]	100[e] [30]	50 [15]	50 [15]	150[d] [45]	200[d] [61]
Detention and Correctional						
New-Use Conditions II, III, IV	100 [30]	100 [30]	50 [15]	50 [15]	200[d] [61]	200[d] [61]
New-Use Condition V	100 [30]	100 [30]	20 [6.1]	20 [6.1]	200[d] [61]	200[d] [61]
Existing-Use Conditions II, III, IV, V	50[f] [15]	100[f] [30]	NL	NL	150[d] [45]	200[d] [61]
Residential						
One- and Two-Family Dwellings	NL	NL	NL	NL	NL	NL
Lodging or Rooming	NL	NL	NL	NL	NL	NL
Hotels and Dormitories						
New	35[g] [10.7]	50[g] [15]	35 [10.7]	50 [15]	175[d,h] [53]	325[d,h] [99]
Existing	35[g] [10.7]	50[g] [15]	50 [15]	50 [15]	175[d,h] [53]	325[d,h] [99]
Apartments						
New	35[g] [10.7]	50[g] [15]	35 [10.7]	50 [15]	175[d,h] [53]	325[d,h] [99]
Existing	35[g] [10.7]	50[g] [15]	50 [15]	50 [15]	175[d,h] [53]	325[d,h] [99]
Board and Care						
New-Small	NL	NL	NL	NL	NL	NL
New-Large	NP	75[i] [23]	NP	30 [9.1]	NP	250[d,h] [76]
Existing-Small	NL	NL	NL	NL	NL	NL
Existing-Large	110 [33]	160 [49]	50 [15]	50 [15]	175[d,h] [53]	325[d,h] [99]
Mercantile						
Class A, B, C						
New	75 [23]	100 [30]	20 [6.1]	50 [15]	150 [45]	250 [76]
Existing	75 [23]	100 [30]	50 [15]	50 [15]	150 [45]	250 [76]
Open-Air Covered Mall	NL	NL	0 [0]	0 [0]	NL	NL
New	75 [23]	100 [30]	20 [6.1]	50 [15]	150 [45]	400[j] [120]
Existing	75 [23]	100 [30]	50 [15]	50 [15]	150 [45]	400[j] [120]

TABLE 26.3 Common Path, Dead End, and Travel Distance (*Continued*)

Type of Occupancy	Common Path Limit		Dead-End Limit		Travel Distance Limit	
	Unsprinklered, ft [m]	Sprinklered, ft [m]	Unsprinklered, ft [m]	Sprinklered, ft [m]	Unsprinklered, ft [m]	Sprinklered, ft [m]
Business						
New	75[k] [23]	100[k] [30]	20 [6.1]	50 [15]	200 [61]	300 [91]
Existing	75[k] [23]	100[k] [30]	50 [15]	50 [15]	200 [61]	300 [91]
Industrial						
General	50 [15]	100 [30]	50 [15]	50 [15]	200 [61]	250[l] [76]
Special Purpose	50 [15]	100 [30]	50 [15]	50 [15]	300 [91]	400 [122]
High Hazard	NP	NP	NP	NP	NP	75 [23]
Aircraft-Servicing Hangars, Ground Floor[m]	50[n] [15]	100[n] [30]	50[n] [15]	50[n] [15]	*Note a*	*Note a*
Aircraft-Servicing Hangars, Mezzanine Floor[m]	50[n] [15]	75[n] [23]	50[n] [15]	50[n] [15]	75 [23]	75 [23]
Storage						
Low Hazard	NL	NL	NL	NL	NL	NL
Ordinary Hazard	50 [15]	100 [30]	50 [15]	100 [30]	200 [61]	400 [122]
High Hazard	NP	NP	NP	NP	75 [23]	100 [30]
Parking Garages, Open	75 [23]	75 [23]	50 [15]	50 [15]	300 [91]	400 [122]
Parking Garages, Enclosed	75 [23]	75 [23]	50 [15]	50 [15]	150 [45]	200 [60]
Aircraft Storage Hangars, Ground Floor[m]	50[n] [15]	100[n] [30]	50[n] [15]	50[n] [15]	Note *a*	Note *a*
Aircraft-Servicing Hangars, Mezzanine Floor[m]	50[n] [15]	75[n] [23]	50[n] [15]	50[n] [15]	75 [23]	75 [23]
Underground Spaces for Bulk Storage	50[n] [15]	100[n] [30]	50[n] [15]	100[n] [30]	200 [61]	400 [122]

Source: Reproduced with permission of NFPA from NFPA 101, Life Safety Code, 2024 edition. Copyright © 2023, National Fire Protection Association. For a full copy of NFPA 101, please go to www.nfpa.org.

NL = No limit; NP = Not permitted.

[a]For common path serving >50 persons, 20 ft [6.1 m]; for common path serving ≤50 persons, 75 ft [23 m].

[b]Dead-end corridors of 20 ft [6.1 m] permitted; dead-end aisles of 20 ft [6.1 m] permitted.

[c]See *NFPA 101* (NFPA, 2023c) for special considerations for smoke-protected assembly seating in arenas and stadia.

[d]This dimension is for the total travel distance, assuming incremental portions have fully utilized their allowable maximums. For travel distance within the room, and from the room exit access door to the exit, see the appropriate occupancy chapter of *NFPA 101* (NFPA, 2023c).

[e]See appropriate occupancy chapter of *NFPA 101* (NFPA, 2023c) for special considerations.

[f]See *NFPA 101* (NFPA, 2023c) for special considerations for existing common paths.

[g]This dimension is from the room/corridor or suite/corridor exit access door to the exit; thus it applies to corridor common path.

[h]See appropriate occupancy chapter of *NFPA 101* (NFPA, 2023c) for special travel distance considerations for exterior ways of exit access.

[i]See the appropriate occupancy chapter of *NFPA 101* (NFPA, 2023c) for second exit access based on room area.

[j]See 36.4.4 and 37.4.4 of *NFPA 101* (NFPA, 2023c) for special travel distance considerations in covered malls considered pedestrian ways.

[k]See *NFPA 101* (NFPA, 2023c) for special common path considerations for single-tenant spaces.

[l]See *NFPA 101* (NFPA, 2023c) for industrial occupancy special travel distance considerations.

[m]See *NFPA 101* (NFPA, 2023c) for special requirements on spacing of doors in aircraft hangars.

[n]See *NFPA 101* (NFPA, 2023c) for special requirements if high hazard.

TABLE 26.4 Summary of *NFPA 101, Life Safety Code®*, Provisions for Occupant Load and Exit Capacity

Occupancy	Occupant Load, ft² [m²] per Person	Stairways (width/person), in. [mm]	Level Components and Ramps (width/person), in. [mm]
Assembly			
Concentrated use without fixed seating	7 [0.65] Net	0.3 [7.6]	0.2 [5]
Less-concentrated use without fixed seating	15 [1.4] Net	0.3 [7.6]	0.2 [5]
Bench-type seating	1 person/18 linear in. [455 linear mm]	0.3 [7.6]	0.2 [5]
Fixed seating	Use number of fixed seats	0.3 [7.6]	0.2 [5]
Business use (general, nonconcentrated)	150 [14] Gross	0.3 [7.6]	0.2 [5]
Day-care use	35 [3.3] Net	0.3 [7.6]	0.2 [5]
Educational Use			
Classrooms	20 [1.9] Net	0.3 [7.6]	0.2 [5]
Shops, laboratories, vocational rooms	50 [4.6] Net	0.3 [7.6]	0.2 [5]
Health care use		AS NAS	AS NAS
Inpatient treatment departments	240 [22.3] Gross	0.3 [7.6] 0.6 [15]	0.2 [5] 0.5 [13]
Sleeping departments	120 [11.1] Gross	0.3 [7.6] 0.6 [15]	0.2 [5] 0.5 [13]
Mercantile			
Sales area on street floor	30 [2.8] Gross	0.3 [7.6]	0.2 [5]
Sales area on two or more street floors	40 [3.7] Gross	0.3 [7.6]	0.2 [5]
Sales area on floor below street floor	30 [2.8] Gross	0.3 [7.6]	0.2 [5]
Sales area on floors above street floor	60 [5.6] Gross	0.3 [7.6]	0.2 [5]
Residential Use			
Hotels and dormitories	200 [18.6] Gross	0.3 [7.6]	0.2 [5]
Apartment buildings	200 [18.6] Gross	0.3 [7.6]	0.2 [5]
Board and care, large	200 [18.6] Gross	0.4 (10)	0.2 [5]
Storage Use			
In mercantile occupancies	300 [27.9] Gross	0.3 [7.6]	0.2 [5]
Malls	See Code	0.3 [7.6]	0.2 [5]
Industrial Use			
General- and high-hazard industrial	100 [9.3] Gross		
Detention and correctional use	120 [11.1] Gross		

Source: Reproduced with permission of NFPA from NFPA 101, Life Safety Code, 2024 edition. Copyright © 2023, National Fire Protection Association. For a full copy of NFPA 101, please go to www.nfpa.org.

AS = sprinklered; NAS = nonsprinklered.

TABLE 26.5 Requirements for Exit Stairs

	New Stairs	Existing Stairs
Minimum stair width clear of all obstructions[a]:		
Total occupant load[b]		
< 50 persons	36 in. [915 mm]	36 in. [915 mm]
> 50 persons; < 2000 persons	44 in. [1120 mm]	36 in. [915 mm]
> 2000 persons	56 in. [1420 mm]	36 in. [915 mm]
Maximum height of risers	7 in. [178 mm]	8 in. [203 mm]
Minimum height of risers	4 in. [100 mm]	
Minimum tread depth	11 in. [280 mm]	9 in. [230 mm]
Minimum headroom	6 ft 8 in. [2030 mm]	6 ft 8 in. [2030 mm]
Maximum height between landings	12 ft [3660 mm]	12 ft [3660 mm]
Minimum dimension of landings in direction of travel	Note[c]	Note[d]
Doors opening immediately on stairs, without a landing at least the width of door	Note[e]	Note[e]

Source: Reproduced with permission of NFPA from NFPA 101, Life Safety Code, 2024 edition. Copyright © 2023, National Fire Protection Association. For a full copy of NFPA 101, please go to www.nfpa.org.

[a]Except projections not exceeding 4½ in. [114 mm] at and below handrail height on each side.

[b]Total occupant load includes all floors served by stairway.

[c]Every landing shall have a dimension, measured in the direction of travel, that is not less than the width of the stair. Such dimension need not be required to exceed 48 in. [1220 mm] in the direction of travel, provided that the stair has a straight run.

[d]Stairways and intermediate landings shall continue with no decrease in width along the direction of egress travel.

[e]In existing buildings, a door assembly at the top of a stair shall be permitted to open directly to the stair, provided that the door leaf does not swing over the stair and the door opening serves an area with an occupant load of fewer than 50 persons.

The building population as estimated for fire safety is usually much greater than the population for which HVAC, water, or elevator service is designed. The stairs must be designed to allow those already within the stairwell to continue down without interference from access doors on any floor. Stairs with direct access to outdoor air at each floor—so-called smokeproof towers—are the safest kind. A fire stair must allow firefighters to move up while occupants are moving down. Another phenomenon is reentry, in which occupants who have exited decide to reenter despite the danger. (Rescue of family members, pets, and valuables is a common reason.) This greatly complicates downward flow.

High-rise buildings present much more difficult problems because firefighting equipment can ordinarily reach no higher than seven floors (about 90 ft [27 m]) and because, typically, only two exit stairways are provided. This difficulty is recognized in building codes, which classify buildings with an occupiable floor more than 75 ft [23 m] above the lowest fire department access as high-rise with special design considerations. Downward flow rates in stairs were formerly assumed at about 45 persons/minute/22 in. [559 mm] of width, but more recently, peak flows of only 24 persons/minute/22 in. [559 mm] have been observed. With a typical exit stair, a 15-story building housing 60 persons per floor per stair can be evacuated in about 9 minutes. However, with the same stair in a 50-story building housing 240 persons per floor per stair, evacuation will take at least 2 hours, 11 minutes! When doors are held open at each floor to admit occupants fleeing the fire, smoke can readily enter the stairwell. Moreover, it is becoming increasingly common for people to refuse to evacuate a high-rise building, placing their faith in fire-extinguishing systems instead of facing a daunting descent through many floors of fire stairs.

Because of the impracticality of rapid evacuation, larger buildings are required to provide refuge areas where smoke penetration is less likely. For example, stairs can be designed to hold all the occupants, in which case, the design of discharge is based more on capacity than on flow. At maximum crowding, about 3 ft^2 [0.28 m^2] per person is recommended. Again, smokeproof towers will be safer in such a design, although stairs can be pressurized with outdoor air. Those details related to smoke control are presented in Section 26.2.

(f) Property Protection

One of the earliest design concerns in this category is that the site should permit access for firefighting equipment. Ideally, fire trucks should be able to pull alongside each exterior wall. If accessibility is limited by adjacent buildings, and roadways alongside buildings and other measures are impractical, more reliance must be placed on internal fire-suppression systems. Another factor is the amount of time it will ordinarily take for firefighters to reach a site. In congested urban areas or remote rural ones, the time that elapses between the alarm and the arrival of firefighters can permit a fire to grow to unstoppable proportions. In these cases, emphasis again shifts to internal fire-suppression systems.

Another design concern is adequate water to fight the fire. Reliance on city water mains is not always a good solution. Elevated tanks on buildings can help in the early moments of firefighting, but their capacity is soon exhausted. Some buildings therefore rely on lakes or enclosed reservoirs for a firefighting water supply (see the discussion in Section 26.3).

Exposure protection is becoming common in areas where highly flammable surroundings pose a serious threat of fires originating outside a building. Candidates for exposure protection include buildings surrounded by older wooden buildings, bordered by lumberyards or other commercial activities that utilize highly flammable materials, or even bordered by open fields of dry grass or brush. Exposure protection guards against heat transfer by radiation and convective currents and against direct fire transfer via flying embers. Exposure protection begins with the use of nonflammable materials for the building's exterior. Erecting firewalls between the building and a fire-threatening neighbor is a more drastic, but sometimes necessary, step. Exposure protection sometimes includes external water sprinkler systems, in which a spray head is placed over (or under) openings such as windows or doors. Sprinkler systems that soak the roof can play a cooling role on summer days, as well as a fire exposure protection role. Exterior doors can be chosen for their fire-delaying characteristics. Windows can also be protected by fire-rated shutters that are designed to close automatically at high temperatures (Figure 26.5). Sometimes exterior protection is necessary to protect the exit discharge from flames originating inside the building, as with narrow alleyways like that in Figure 26.5 in downtown Pittsburgh.

Compartmentation has become increasingly important as buildings have become lightweight structures incorporating decreased fire resistance and open floor areas that encourage the spread of fires. Building codes establish maximum floor area limits based upon construction type and occupancy. If a building's floor area exceeds such limits, it must be subdivided by

Figure 26.5 *Rolling metal shutters, housed in valences above each window, punctuate the exterior of this downtown Pittsburgh building opening onto a narrow street. The shutters can protect the building's contents from an exterior fire, but can also protect the narrow street from a fire inside the building. Note the exterior fire escapes across this narrow street. These are no longer permitted, partly because smoke plumes rising from windows can make such stairs unusable long before heat from a fire becomes a threat.*

firewalls into areas that fall within the code limitations. Automatic fire sprinklers allow for increased floor area limits, which can be a motivator for including sprinklers in a building context where they are not mandatory.

Openings in firewalls and similar protective constructions must be protected by fire-rated doors. Fire dampers must be installed in HVAC ductwork wherever ducts pass through a fire-rated construction (Figure 26.6). The intent is not to degrade fire resistance by the inclusion of weaker-link components. As important as compartmentation is in buildings with large floor areas, however, the vertical spread of fire (its natural path) poses a more serious problem. Fire-resistive barriers to limit vertical fire spread are also part of the compartmentation requirements found in codes. Thus, strict requirements to protect vertical openings and pathways are found in the fire codes.

Concealed spaces are found in many contemporary buildings, especially over suspended ceilings but also behind walls, within pipe chases, in attics, under raised floors, and in other places. All such spaces can offer paths for the spread of fire. The designer can utilize noncombustible materials, as far as possible, in such spaces. Another important step is to include automatic fire detection and suppression systems in these uninhabited spaces—and often the use of oxygen-deprivation approaches to fire suppression.

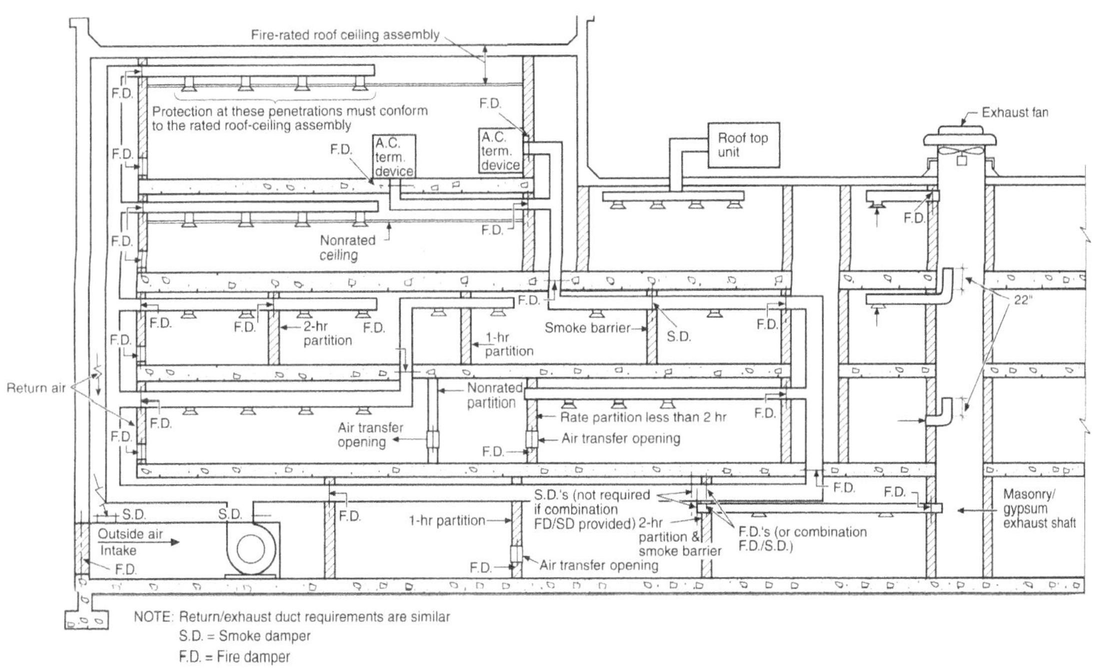

Figure 26.6 *Locations of fire dampers in a typical ducted HVAC system—at all penetrations through firewalls, fire-rated ceiling/floor assemblies, and vertical chase enclosures. (Reproduced with permission of NFPA from the Fire Protection Handbook, 20th edition. Copyright © 2008, National Fire Protection Association. For a full copy of the Fire Protection Handbook, please go to www.nfpa.org.)*

Another design response is compartmentation: using firestops (or firewalls) to break up otherwise continuous concealed spaces. Many codes, for example, require firestopping around each 1000 ft^2 [93 m^2] of suspended ceiling area and every 2000 ft^2 [186 m^2] of attic floor area.

Structural protection, another important requirement, allows a building to continue to stand during a fire and enables it to be salvaged rather than demolished after a fire. Codes require various protective layers for structural materials. In order of importance, it is usually most critical to protect columns, girders, beams, and, finally, the floor slab.

(g) Continuity of Operations

For most building functions, it is desirable to minimize the disruption of operation that a fire will cause. Design strategies to encourage continuity of operations include special fire alarm/suppression systems for especially critical operations areas (control rooms, for example), the design of HVAC systems to allow for 100% outdoor air (to aid in purging a building of smoke), and provision for the speedy removal of the water dumped on a fire by a sprinkler system. The floors in sprinkler-served buildings should be (yet rarely are) waterproof; waterproofing should also be carried up walls, columns, pipes, and other elements to a height of 4–6 in. [100–150 mm] above the floor. These provisions will help to minimize the water damage associated with fire-suppression efforts.

26.2 SMOKE CONTROL

As experience has made it increasingly evident that smoke kills more people in building fire events than either heat or structural collapse, it has also become the norm to design for smoke control as well as for resistance to fire and for fire suppression. The objectives of smoke control are the same as those for fire resistance: to reduce deaths and property damage due to smoke and provide for continuity of operations with minimal smoke interference. Airflow management systems have been used for several generations; laboratory buildings and commercial kitchens equipped with fume/exhaust hoods offer a familiar example. Several options for smoke control are now readily available for ordinary building design and are required by most building codes.

(a) Factors in Smoke Control

The heat of a fire produces air pressure and buoyancy that aid the spread of smoke well beyond the scene of the fire itself. In low buildings, the smoke is spread primarily by heat-induced convective air motion and by the differential air pressures caused by the expansion of gases as the temperature increases. In tall buildings, the stack effect complicates this pattern of smoke spread, encouraging the rapid rise of heated air within vertical shafts. Wind forces from outside are also more likely to be a factor in tall buildings. (Wind velocities are usually higher with increasing distance from the ground surface.) Forced-air systems, so common in high-rise buildings, can also contribute to the spread of smoke. The interactions among fire, the stack effect, wind, building geometry, and HVAC systems are complex. Several detailed calculation techniques are available, as explained in *NFPA 92: Standard for Smoke Control Systems* (NFPA, 2023b). These methods include building and testing a scale model, using a series of closed-form algebraic equations, and modeling a compartment fire with the help of a computer. In the last category, a more modest zone approach divides a fire into two zones (upper and lower), and the calculations can be done on a personal computer. The more ambitious field approach uses computational fluid dynamics (CFD), requiring greater computer capacity but allowing the testing of numerous variations in approaches to fire and smoke control; fire growth, smoke spread, structural behavior, and occupant evacuation can be examined together. Results can be surprising, leading at times to counterintuitive solutions.

(b) Confinement

A passive design response to smoke migration is to try to confine it to the fire area itself. Another technique is to exclude smoke from specially protected areas, called areas of refuge. Compartmentation is important in these approaches: Where firewalls cannot be used, partial smoke barriers (often called curtain boards) are suspended from the ceiling (Figure 26.7) in an effort to trap the initial layer of hot air and smoke, and therefore to set off the fire detection/suppression system more quickly. As useful as these partial barriers can be in a fire's early stages, they quickly lose effectiveness as the smoke layer thickens, or as air pressure forces the smoke below the boards. Even with firewalls with fire-rated doors, small gaps and cracks can provide smoke paths, aided by the air pressure differences that fires can quickly produce. The typical "fire" barrier is not very resistant to spreading smoke, and smoke control based solely on confinement must be closely linked to an effective early detection/suppression technique such as a sprinkler system.

Compartmentation may not always be the best strategy. In the Potomac Mills Mall in Prince William

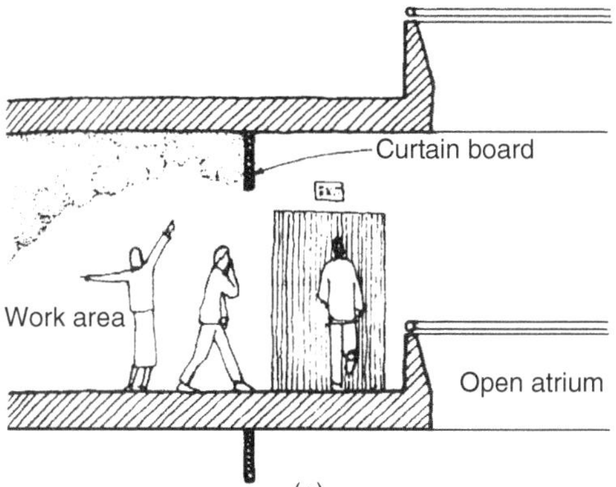

(a)

(b)

Figure 26.7 *(a) Smoke barriers (curtain boards) are useful in a fire's early stages. By confining the initial layer of heated air and smoke produced by a fire, they help to slow the spread of smoke while making the early detection and suppression of the fire more likely. (b) Glass used as curtain board in the atrium of the Lewis Integrative Science Building, University of Oregon. (© Ayush Vaidya; used with permission.)*

County, Virginia, a 156,000-ft^2 [14,493-m^2] addition to the shopping mall uses low walls and a high ceiling. The walls separating tenant spaces and public corridors are 12 ft [3.7 m] high, while the ceiling is at 17 ft [5.2 m], resulting in a large open volume of air. Smoke rising from a fire within a store can certainly spread well beyond that store, but will collect overhead in such a large space that evacuation of the mall can be completed well before the smoke reaches the occupied zone near the floor. A computer model indicated that 20 minutes into a fire (and assuming no sprinkler operation), the smoke layer would have dropped to 12.3 ft [3.75 m] above the floor. If the walls were extended to the ceiling, this same building would experience, in the same 20 minutes, within the origin store of the fire, smoke all the way down to the floor. CFD analysis may be required before approval for such a solution is granted by the code authority.

(c) Dilution

For a limited time in a fire's early stages, the dilution of smoke with 100% outdoor air (provided by the HVAC system) may make conditions bearable during occupant evacuation. Large quantities of fresh air are needed in so short a time, however, that dilution alone is rarely sufficient for smoke control, particularly when the smoke contains toxic components. When a dilution system is combined with both confinement and early detection/suppression systems, it becomes more feasible and attractive. Dilution systems that dump large quantities of outdoor air into refuge areas (including stairs) can create such high pressures within the refuge that smoke cannot enter through cracks around doors. When, however, doors to the area are opened (as when more people enter the refuge), dilution systems that rely on a conventional HVAC system can rarely provide sufficient pressure to exclude smoke.

Fan-driven dilution systems raise several issues. Is the fan located where fire will not affect its performance and where its air intake can be kept free of smoke? Given the multiple possibilities for building fire locations, this is a difficult promise to make. "Dedicated" systems, independent of the building's HVAC system, are often favored to enable such emergency equipment to be isolated and because such huge quantities of air for relatively small spaces are well beyond the capability of the typical HVAC system. Where should the fresh air supply outlets be located? Most certainly, below the level of smoke accumulation so as not to drive smoke downward. (This may suggest using an under-floor air supply distribution system for such a purpose.) In a stair, at what intervals should fresh air be supplied? Some fire engineers say every three floors, others say every 10, but supplying all the fresh air from either the top or the bottom is not recommended; it is too likely that open doors near a single source location would deplete fresh air for the rest of the stair. At what velocity should fresh air be supplied? The recommended maximum is 200 fpm [1 m/s] to avoid stirring the smoke layer and inviting its descent. The designer must consider that high rates of flow at low velocities will result in very large grilles being required within the space.

(d) Exhaust

Single-purpose smoke exhaust systems that function only in a fire are becoming more common. These systems employ both air velocity and air pressure to help control smoke movement. As with dilution systems, high velocities are avoided so that fresh air is not drawn up through the smoke layer. Such systems are particularly useful in large-volume atrium spaces, as in Figure 26.8. Smoke accumulating at the ceiling is removed, while fresh air is supplied lower in the space. If smoke can be removed at the same rate at which it is generated, the smoke layer can be kept

(a)

(b)

Figure 26.8 *The Rock and Roll Hall of Fame and Museum on Cleveland's lakefront (a) features a tall, pyramidal interior space. (b) At the apex, a large horizontal grille marks the smoke exhaust system intake. Smoke naturally accumulates here, and can be removed while dedicated sources of fresh air operate at ground level around the perimeter. (When hot days alternate with cool nights, such a system could be used for night ventilation of thermal mass.)*

from descending. Although they involve a greater initial expense, because they require special fans and special smoke-exhaust shafts, such systems have several advantages over simple confinement or dilution approaches:

1. They can help prevent toxic gases from entering refuge areas (particularly when there is a dilution or outdoor air supply to such areas).
2. They can help reduce concentrations of dangerous gases, potentially making conditions somewhat more tenable for those exposed to smoke-affected spaces.
3. By creating desirable patterns of air pressure, they can reduce smoke migration to areas of a building not directly involved with a fire event.
4. With them, the stack effect, complicated by buoyancy and wind, is less likely to overcome the smoke-control system in a tall building.
5. They can help to remove smoke that remains after the fire is extinguished.

Each of the two Petronas Towers in Kuala Lumpur, Malaysia, is occupied by as many as 18,000 workers during the day. These were the tallest buildings in the world at the turn of the twenty-first century, placing a majority of the workers well beyond the range of ordinary fire-rescue operations. Smoke exhaust, stair pressurization, and vestibule pressurization systems are installed, and each floor has isolation dampers, controlled by the fire alarm system, wherever air ducts serve multiple floors. Because each floor has its own air-handling system (using floor-by-floor fan rooms), however, vertical ductwork is minimized. There are both a main and a secondary fire command center for each of the towers and associated lower buildings.

(e) HVAC Systems, Sprinklers, and Smoke

Two systems within a building must be closely coordinated with smoke exhaust systems: the HVAC system and the fire detection/suppression (usually sprinkler) system. As the fire detection system activates the smoke exhaust fans, it must also override the conventional HVAC system operation, usually by switching to 100% outdoor supply air and simultaneously blocking the return air system. This action pressurizes each HVAC zone to form a barrier against smoke (Figure 26.9) and keeps smoke out of the return duct system. If the HVAC system is variable air volume (VAV), all supply control valves (dampers) must be moved to their full open position. If additional protection is needed for refuge areas, additional air supply to such areas can be provided, perhaps by a separate (smoke dilution) system, as described earlier.

An automatic sprinkler system can reduce the extent of smoke development by quickly extinguishing a fire; conversely, a sprinkler system can affect the functioning of smoke exhaust systems, both by creating a curtain of water that inhibits the movement of smoke and by cooling the smoke, thus reducing its buoyancy. As less buoyant smoke descends, visibility is

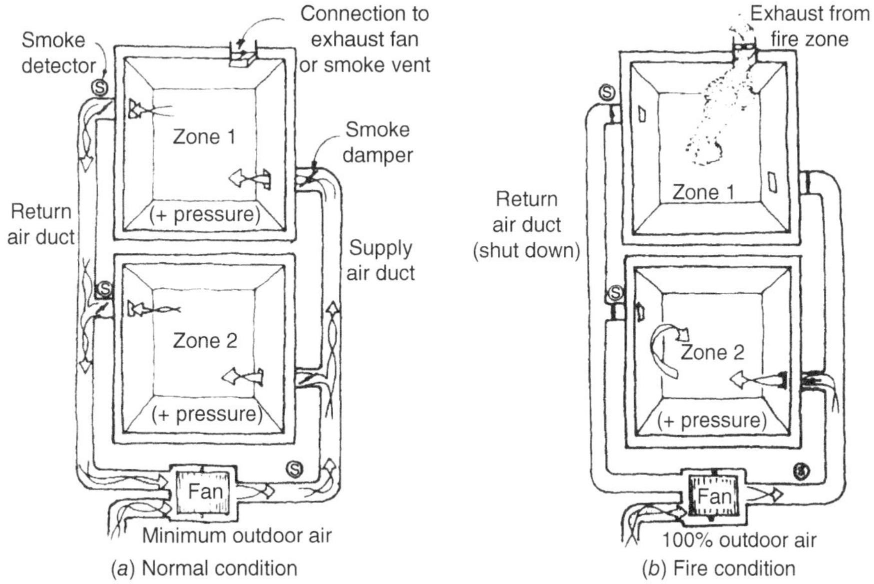

Figure 26.9 *Smoke control by a smoke exhaust system. (a) The conventional HVAC system is shown operating under normal minimum outdoor air supply conditions. (b) When smoke detectors trigger the smoke exhaust system, dedicated fans begin to exhaust smoke, through a special duct, from the affected area only. Simultaneously, the conventional system undergoes two changes: All supply air becomes 100% outdoor air (for smoke dilution), and all return openings are closed off (to pressurize all nonfire areas, to keep smoke out of them and out of the return duct system).*

reduced and the danger of smoke inhalation increases, particularly in spaces with ordinary ceiling heights. The buoyancy of smoke is a key factor in the success of a smoke exhaust shaft, whose intake is located at ceiling level within each zone. Although these two systems may appear to work at cross-purposes, suppression of a fire is as important a goal as smoke control. Sprinkler systems typically suppress fires so quickly that the size of the accompanying smoke exhaust systems can generally be reduced.

When the fire-suppression system relies on oxygen displacement (such as with carbon dioxide or "clean agents"), smoke exhaust systems clearly pose a threat—special attention is required if both systems are to be used. Luckily, oxygen displacement systems are often used in nonhabitable spaces, where smoke evacuation is less of an issue.

(f) Automatic Smoke and Heat Vents

Smoke and heat vents (which are mechanically controlled, but operate without fans) are required by the *International Building Code* and *International Fire Code* in selected occupancies—such as stages and certain warehouse/storage facilities (ICC, 2024a, 2024b). They may be installed (Figure 26.10) in a wide range of building types as part of an engineered smoke control system—often taking the form of hatches. The hatches open individually as heat or smoke from a fire triggers their control devices. As a result, indoor conditions near the fire are improved for firefighters, and firefighters on the roof are quickly alerted to the general location of the fire below.

In summary, designers can provide several components that aid in the exhaust of smoke and the provision of ventilation and improved visibility for firefighters and occupants in almost any building: dedicated smoke exhaust and dilution systems, emergency controls on HVAC systems, operable windows and skylights (to facilitate ventilation and occupant escape), and heat-and-smoke hatches in roofs.

26.3 WATER-BASED FIRE SUPPRESSION SYSTEMS

The most popular medium for building fire suppression is water, which is readily available and relatively low in cost. Water cools, smothers, emulsifies, and dilutes. As it turns to vapor, it removes 970 Btu/lb of water [2256 kJ/kg] at atmospheric pressure, and its volume increases 1700 times—a process that hinders access to the oxygen needed by the fire. Water has several disadvantages that sometimes preclude its use for fire suppression: It damages most building contents, including interior surfaces; as a stream, it conducts electricity readily (less readily as a spray); and many flammable oils will float on water while continuing to burn. Where these factors are major considerations, other suppression media can be considered (see Section 26.4).

Fire-suppressing systems are commonly combined with smoke control systems; Figure 26.11 shows a typical combination of provisions in a one-story building. Taller buildings, especially those with multistory interior spaces (atria), present a special challenge, as illustrated in Figure 26.12. A typical approach is to construct an approximately 6-ft [1.8-m]-deep curtain board (or fire spandrel) at the opening to the atrium on each floor. Smoke detectors, motorized dampers, and sprinklers can be combined in various ways. At the lobby level, the atrium floor level, and at any office floors open to the atrium, sprinklers at about 6 ft [1.8 m] o.c. (on center) will provide a water curtain at the edge of the atrium opening. Where glazing is used at the atrium's perimeter, the glazing frames

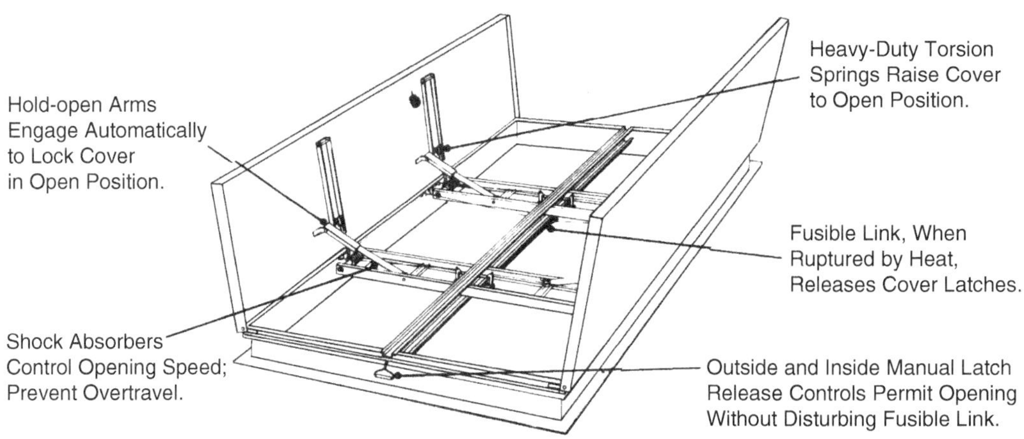

Figure 26.10 *Automatic smoke and heat roof hatch ventilators, primarily for one-story commercial and industrial buildings. (Courtesy of Inryco, Inc.)*

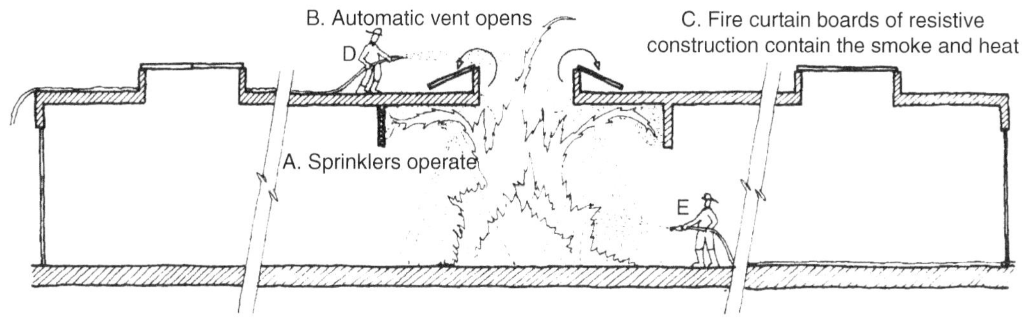

Figure 26.11 *Fire protection for large-area, one-story buildings showing the fire suppression (sprinkler) system (A) and the smoke control system (B, C). Locations D and E are relatively safe positions for firefighters.*

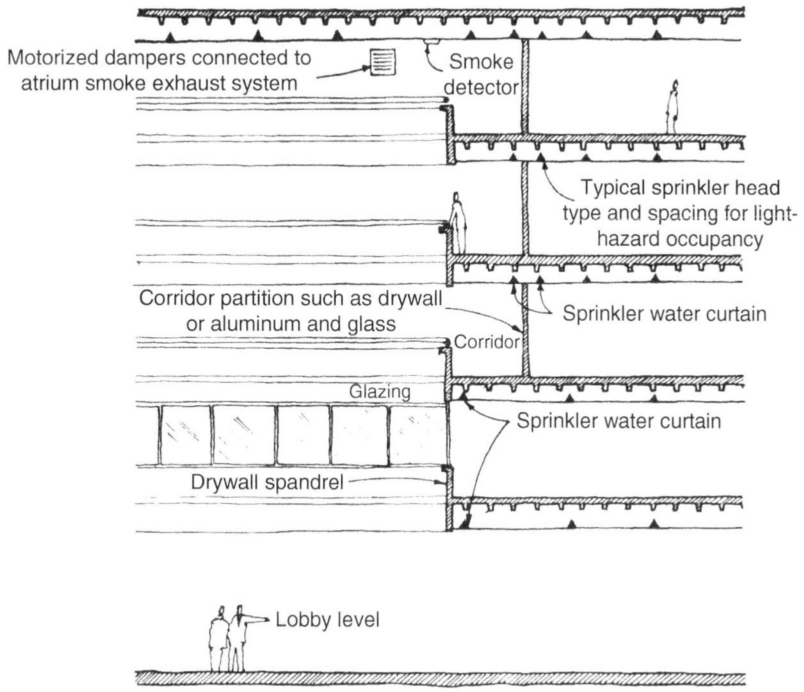

Figure 26.12 *Fire protection in an atrium-type office building: section showing the detection/suppression system and provisions for smoke control.*

sometimes can be designed for considerable thermal expansion, and sprinklers can provide a water curtain on the office side of the glazing. Fire-rated glazings and their frames are listed for both heat resistance and water pressure from a hose stream. Where balcony corridors adjoin the atrium, two sprinkler water curtains can be provided, one on either side of the partitions between the corridor and the office space. The smoke exhaust system often can provide six air changes per hour (ACH) for all spaces that open onto the atrium.

(a) Standpipes and Hoses

Standpipes and hoses (Figure 26.13) with a separate water reserve, upfeed pumping, and/or fire department connections are listed in three system classes and five system types. Defined in *NFPA 14: Standard for the Installation of Standpipe and Hose Systems* (NFPA, 2023a), the major differences are whether the system is for first-aid or full-scale firefighting and whether the system has an automatic water supply or a manual one.

The three standpipe system classes are as follows:

Class I systems: These are for full-scale firefighting, and are typically required in both sprinklered and unsprinklered buildings more than three stories high, as well as in malls. (This requirement is based on the substantial time required to lay out hoses from outdoor fire connections in such buildings.) This system is for use by trained firefighters using $2\frac{1}{2}$-in. [64-mm] hose connections at designated locations. Class I is becoming the system of choice as use of the next two system types declines.

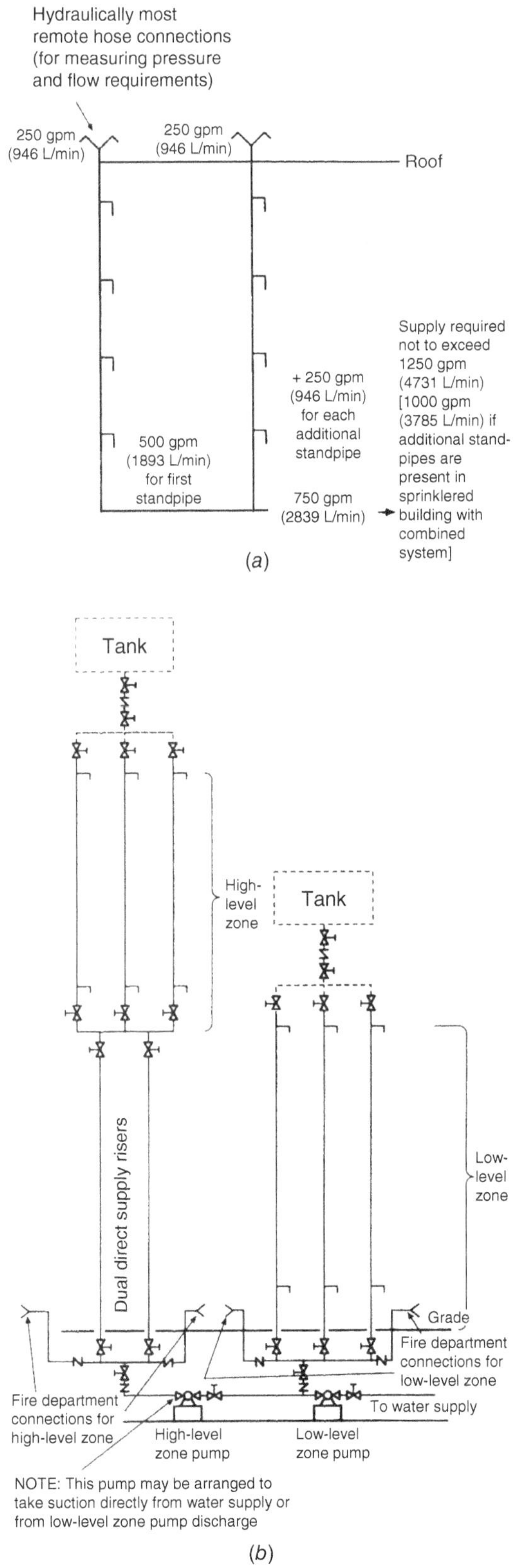

Figure 26.13 *Standpipe system variations. (a) Basic flow rate requirements for Class I and Class III systems. (b) Multiple-zone system. (Reproduced with permission of NFPA from the Fire Protection Handbook, 20th edition. Copyright © 2008, National Fire Protection Association. For a full copy of the Fire Protection Handbook, please go to www.nfpa.org.)*

Class II systems: These are for first-aid firefighting before the fire trucks arrive. These systems use 1 1/2-in. [38-mm] hose connections and typically provide access to a hose, nozzle, and hose rack (Figure 26.14) in each specified location. The difficulty experienced by untrained people trying to manage a 100-ft [30-m] hose containing a huge flow (100 gpm [378 L/min] or more) has led to a decline in the use of this system, although it can be found in large unsprinklered buildings and in special hazard areas such as stages. Many fire departments would rather the building occupants evacuate than try to manage such large hoses and water quantities. Significant water damage is a potential threat when hoses cannot be controlled.

Class III systems: These combine the characteristics of Classes I and II, thus serving both for first-aid and for full-scale firefighting. Hose connections of both sizes are provided (with adapters) at each specified location. The disadvantages of Class II systems also apply to Class III, and their use is declining for the same reasons.

"Combined" systems are either Class I or III standpipe systems that also supply water to a sprinkler system.The standpipe system types are as follows:

Automatic-wet systems: The pipes are filled with water and are connected to a water supply capable of automatically meeting the firefighting demands. Water flows immediately upon opening of a hose valve.

Automatic-dry systems: The pipes are filled with pressurized air, and are connected to a water supply capable of automatically meeting the firefighting demands. Through a device such as a dry-pipe valve, water replaces the air when a hose valve is opened.

Semiautomatic-dry systems: The pipes are filled with air and are connected to a water supply capable of automatically meeting the firefighting demands. Through a device such as a deluge valve, water replaces the air when both a remote sensing device at the hose station and a hose valve are opened.

Manual-dry systems: The pipes are filled with air, and there is no connection to a water supply system other than that provided by the fire department (as in Figure 26.14*c*).

Manual-wet systems: The pipes are filled with water, with a connection to a domestic water source that is used merely to fill and test the system. Water for firefighting is provided by the fire department.

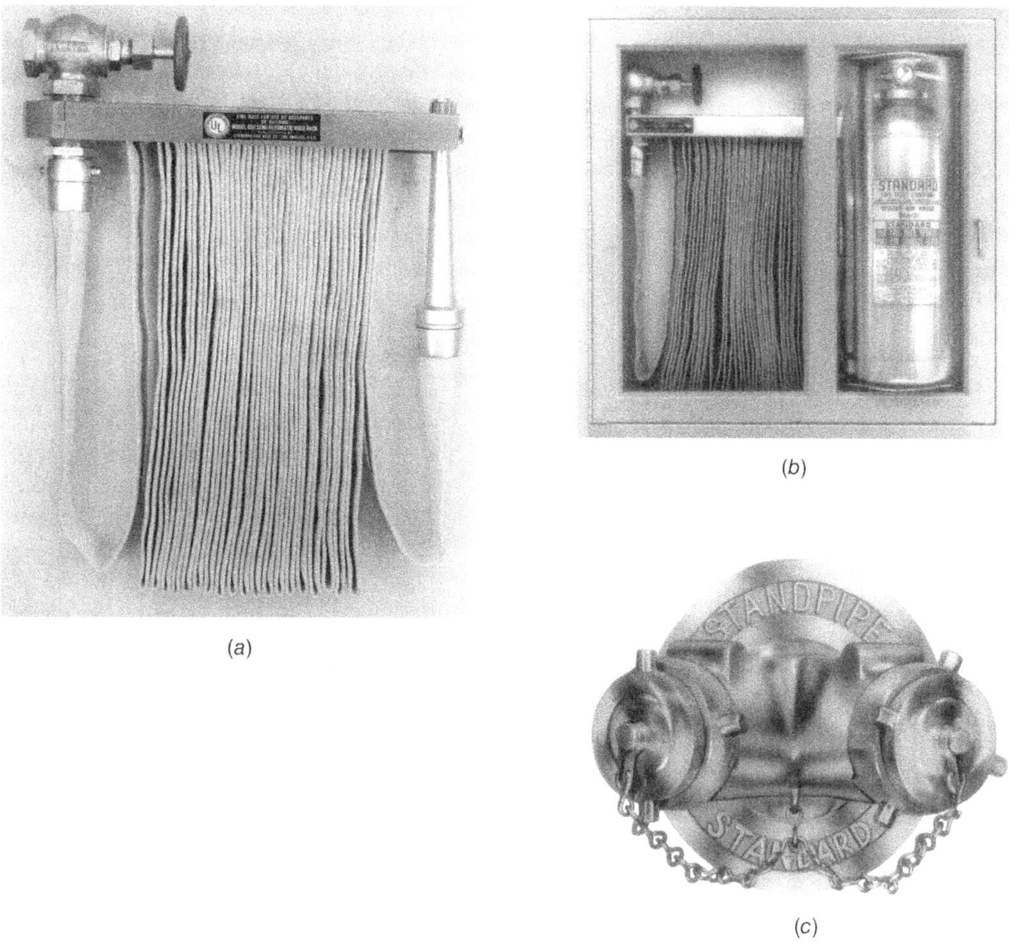

Figure 26.14 *Components of a standpipe and hose system. (a) Standard hose rack. (b) Hose rack and fire extinguisher in a cabinet with a glass door. (c) Siamese connection through which fire department pumping equipment supplies water to the standpipe system. The similar sprinkler siamese is marked "sprinkler." Color codes sometimes differentiate the two.*

Minimum flow rates and minimum and maximum pressure are all addressed during the design of standpipe systems. Two methods of sizing, the pipe schedule and the hydraulic method, are used. Neither is detailed here; see *NFPA 14* (NFPA, 2023a) for the procedures. The required supply flow rates are outlined in Figure 26.13*a*. As a preliminary design guideline (Figure 26.13*b*), Class I and Class III standpipes not exceeding 100 ft [30 m] in height must be a minimum of 4 in. [102 mm] nominal pipe size. Class I and III standpipes more than 100 ft [30 m] in height must be a minimum of 6 in. [152 mm] nominal pipe size (although the topmost 100 ft [30 m] may be a minimum of 4 in. [102 mm] nominal pipe size). For combined standpipe and sprinkler systems, regardless of height, a minimum of 6 in. [152 mm] nominal pipe size is required.

For Class I and Class III systems, a minimum hose pressure of 100 psi [690 kPa] is now required because of the widespread use of fog nozzles (rather than stream nozzles). The maximum hose pressure (also the maximum for sprinklers) is 175 psi [1207 kPa].

Water supply systems for firefighting may use roof tanks or fire pumps. It is not practical to store enough water on the roof for a protracted firefighting period, and it is usually assumed that a half-hour supply will be more than enough to provide for the time it takes the fire engines to arrive. When the system is used by the fire department, its pumps are attached to the street siamese (Figure 26.14*c*) to deliver water from street hydrants or the building's "secondary source." The check valve closest to the siamese in use opens, and the check valves at the tank close to prevent the water from rising uselessly in the tank. After the engines are disconnected from the siamese, the water between the siamese and the adjacent check valve drains out through a ball drip so that it does not freeze.

An overhead tank is considered a most dependable source, but the height required can be architecturally undesirable and a considerable structural disadvantage. In such a case, upfeed fire pumps (Figure 26.13*b*) operating automatically to deliver water to higher stories may be used. Another option in this case is a pneumatic tank that delivers water using the pressure of the air that is compressed in the upper portion of the tank.

A standpipe zone in a high-rise building is determined by hydraulic calculation so that maximum hose pressures are not exceeded. Fire standpipes and their hoses (for full-scale firefighting) are now recommended to be located on the landings of fire stairs, from which personnel or firefighters can approach a fire with charged hoses. This enables the hose connection to be made in a safe (relatively smoke-free) place, half a floor below the fire. Awkwardly, it means that the door to the stair will be held open by the hose, but the risk to occupants using the stairs is considered to be outweighed by the advantage to the firefighters.

(b) Sprinkler System Design Impacts

Unlike a fire hose, a sprinkler is likely to be already positioned above the point of a fire and is capable of deploying in seconds, not minutes. Sprinkler systems are widely relied on as proven automatic fire suppressers. (They have come a very long way from the early water-filled bucket suspended by a black powder fuse; when the fuse was lit by a fire, it blew up the bucket, dispersing the water and theoretically dousing the fire.)

However, designers must remember that the use of sprinklers does not give one a license to ignore fire code limitations, even though many codes provide more leeway for sprinklered buildings. In addition, provision must be made for an adequate water supply, adequate water pressure, and backup power for pumping (if pumps are required). It may be wise to provide for water drainage during and after a fire, particularly in multistory buildings. A sprinkler system may require large supply pipes and valves, fire pumps, and access for system monitoring and maintenance. These elements are generally considered unsightly and must be integrated into the architectural design. Sprinklers can afford opportunities for integration with energy-conserving HVAC systems (Section 26.3 k). Also, the cost of a sprinkler system installation can usually be recovered rather quickly through reductions in fire insurance premiums.

Automatic sprinkler systems usually consist of a horizontal pattern of pipes placed just below or within the ceilings of industrial buildings, warehouses, stores, theaters, offices, homes, and other structures at risk from fire. These pipes are provided with sprinkler heads constructed such that abnormally high temperatures will cause them to open automatically and emit a fine water spray. *NFPA 13: Standard for the Installation of Sprinkler Systems* (NFPA, 2024a) lists the detailed requirements for sprinkler systems (briefly referred to here), as well as numerous other regulations. In addition to pipes and sprinklers (discussed in the remainder of this section), several other system components are usually included.

Alarm Gong. An alarm gong mounted on the outside of the building warns of water flow through the alarm valve upon activation of a sprinkler head. This warning gives the building personnel an opportunity to make additional firefighting arrangements that can minimize loss and speed the termination of the fire; in this way, the sprinklers can be turned off in a timely manner to prevent excessive water damage to building contents after the fire is out. Sprinkler alarms commonly are also connected to private regional supervisory offices that communicate promptly with municipal fire departments upon receipt of a signal. All public buildings, and other buildings as required, should be provided with fire detection and alarm systems that indicate, in the custodian's office, the location of the fire.

Siamese Connections. Siamese connections permit fire engines to pump into the sprinkler system in a manner similar to that used for standpipe systems (Figure 26.14*c*).

Provisions for Drainage. Sprinkler heads can release a great deal of water, most of which will remain unvaporized and quickly collect at floor level. In addition to waterproofing the floors and lower walls, columns, and other elements, provision should be made, where possible, for gravity drainage of water. Scuppers in exterior walls are preferable to floor drains, which are more easily clogged by debris. Scuppers are provided with hoods that protect against infiltration, birds, and/or insects.

Water Supply. In tall buildings, sprinklers (and standpipes) can be supplied with water from elevated storage tanks used for domestic water. These tanks supply a constant pressure on the distribution lines, store sufficient water to balance supply and demand, prevent excessive starting and stopping of a fire pump, and provide a dependable fire reserve. The last factor has been critical in the calculation of fire insurance rates. When gravity tanks are used with sprinkler systems, they should provide enough water to operate 25% of the sprinkler heads for 20 minutes. As in the case of standpipe and hose systems, this provision gives the fire department a chance to arrive and take over.

The principal objections to the use of tanks have been their unsightliness, the problem of freezing, and—in the case of large buildings—the tanks' tremendous weight, with resulting added structural costs.

Skidmore, Owings & Merrill, in their design for General Mills' central office building in suburban Minneapolis, turned to automatically controlled pumping systems instead of elevated storage (Figure 26.15). For the General Mills building, the fire underwriters could have required at least 50,000 gal [189,250 L] of residual water in the tank for emergency purposes, which would have meant about 100,000 gal [378,500 L] of elevated storage. The alternative chosen was a reinforced-concrete structure placed underground to one side of the building and covered with 3–5 ft [0.9–1.5 m] of earth. Small vents rising from this reservoir blend in with the lawn and landscaped shrubbery above.

The saving in structural steel tended to make the overall cost comparable to that of an elevated tank of the same capacity. The savings were made possible chiefly by refined automatic-pump control. The underground reservoir eliminates the problems of an unsightly appearance and great weight, but the other usual advantages of an elevated tank—reliability in case of fire, minimum starting and stopping of motors, and the maintenance of pressure while balancing supply and demand—must be equaled in the automatic-pump control circuitry. This is not as simple as it might at first seem. Factors such as fluctuations in demand, friction within the pipes, elevations, starting surges from the pumps, and pressure-flow characteristics of the pumps themselves must be considered. Various combinations of these issues undoubtedly account for the continued use of elevated tanks. The trend, however, is toward the use of more sophisticated pump control.

A continuous flow from the General Mills deep-well pumps through both domestic and fire reservoirs prevents the water from becoming stale and rancid. The fire reservoir is given necessary priority over the domestic reservoir by means of a simple weir. Even if the domestic reservoir were completely empty, the fire reservoir would remain full. Pressure for the sprinklers is supplied by a small 20-gpm [1.26-L/s] jockey pump. Signals from the sprinkler system bring in a 750-gpm [47.3-L/s] main pump. If this fails, a diesel engine-driven pump of equal capacity automatically takes over. Normally, the system satisfies the heavy demands of air conditioning, fire control, and domestic water supply in this modern, rural, isolated office building.

Valves. Valves are required to allow the sprinkler system to be shut off for maintenance, system modification, or replacement of sprinkler heads that have operated after a fire. Indicating valves of various types are required so that it is always clear whether such crucial valves are open (as they nearly always should be) or

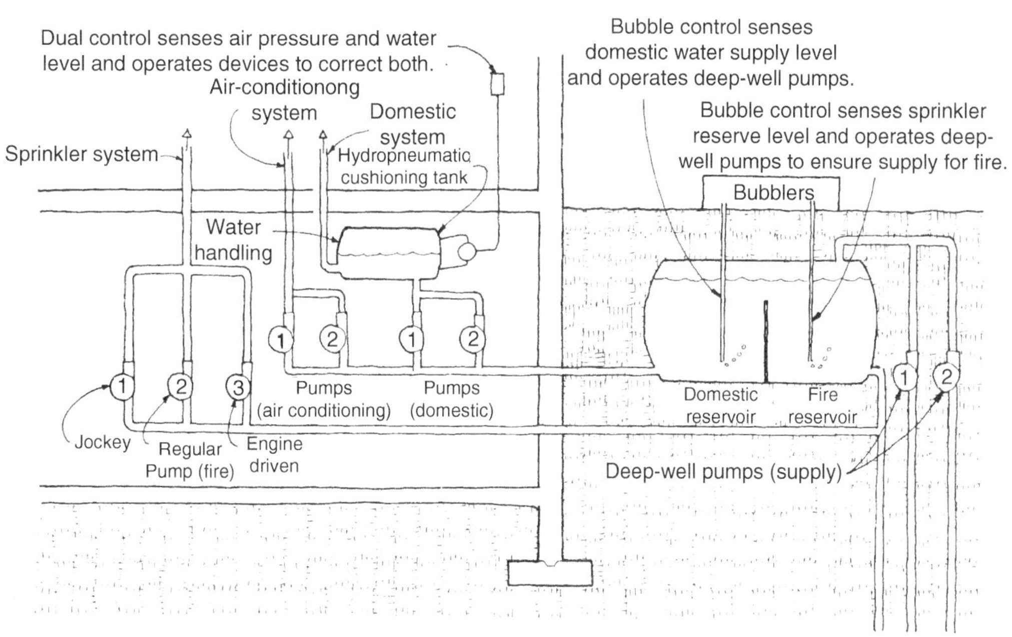

Figure 26.15 *Diagram of an upfeed pumping system to supply sprinklers and building demands for domestic water. In the General Mills Building, a large subsurface concrete water tank provides a secondary water source that reassures fire insurance carriers. Thus, the wells and their pumps constitute a primary water source. If the well yield fails, the concrete tank supply can be tapped. Backup engine-driven pumping operates in the event of a utility company power outage. Automatic controls and piping for the storage tank reserve system are not shown in this illustration.*

TABLE 26.6 Common Failure Modes for Automatic Sprinklers

Failure Mode	Potential Causes
Water supply valves are closed when sprinkler operates	Valve supervision is inadequate Maintenance policies are not effective
Water does not reach sprinkler	Dry pipe accelerator or exhauster malfunctions Preaction system malfunctions Maintenance and inspection are inadequate
Sprinkler fails to open when expected	Response time and temperature of thermal element are inappropriate for the area protected Sprinkler thermal element is protected from heat Sprinkler thermal element is painted, taped, bagged, or corroded
Water cannot contact fuel. (*Note*: The intent of this failure mode is to ensure that discharge is not interrupted in a manner that will prevent fire control by a sprinkler.)	Fuel is shielded New construction (walls, ductwork, ceilings) obstructs water spray
Water discharge density is not sufficient	Discharge needs are insufficient for the type of fire and the rate of heat release Change in combustible contents occurs Number of sprinklers open is too great for the water supply Water pressure is too low Water droplet size is inappropriate for the fire size
Enough water does not continue to flow	Water supply is inadequate because of original deficiencies, changes in water supply, or changes in the combustible contents Pumps are inadequate or unreliable Power supply malfunctions System is disrupted

Source: Reproduced with permission of NFPA from the Fire Protection Handbook, 21st edition. Copyright © 2023, National Fire Protection Association. For a full copy of the Fire Protection Handbook, please go to www.nfpa.org.

closed. Table 26.6 lists common failure modes in sprinkler systems; an improperly closed valve is the leading reason for system failure.

(c) Sprinkler Construction, Orientation, and Temperature Rating

The common sprinkler head blocks water flow by a plug or cap that is held tightly against the orifice (i.e., the small opening in a fire sprinkler head through which water is discharged when activated) by levers or other restraining devices. These, in turn, are held in place by the arms of the sprinkler body. In the past, the restraining device was usually a fusible metal link that melted at a predetermined temperature. Today, the restraining device is typically a glass bulb that contains both a colored liquid and an air bubble. As the liquid expands due to increasing temperature resulting from the heat of a fire, it compresses the air bubble until it is absorbed. Then continued expansion ruptures the bulb (again, at a predetermined temperature) and releases the water in a solid stream through the orifice.

A deflector converts this solid stream of water into a spray, whose pattern is determined by the design of the deflector. Most sprinklers direct the spray down and horizontally rather than upward toward the ceiling. This yields better water distribution close to the sprinkler head and more effective coverage of burning material below. The spray pattern from a typical upright or pendent sprinkler is shown in Figure 26.16.

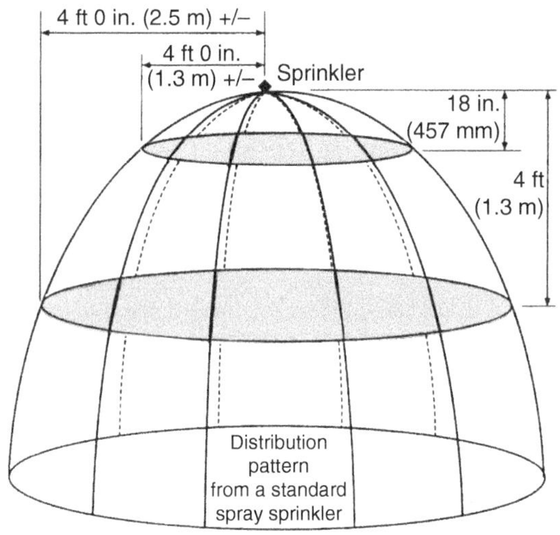

Figure 26.16 *Typical distribution pattern from a standard spray sprinkler. Obstructions (whether from structure or furnishings) are to be avoided within this volume. (Reproduced with permission of NFPA from NFPA 13, Automatic Sprinkler Systems Handbook, 2025 edition. Copyright © 2024, National Fire Protection Association. For a full copy of NFPA 13, please go to www.nfpa.org.)*

Sprinklers are designed to be installed in specific orientations. This allows a degree of design flexibility, but it is critical that sprinklers are installed as intended. As defined by *NFPA 14*, the common types of sprinkler heads are upright, pendent, and sidewall:

- *Standard spray upright (SFU):* Heads installed on top of piping, with the deflector positioned above the sprinkler (Figure 26.17*a*).
- *Standard spray pendent (SSP):* Heads hang downward from the piping, with the deflector located below

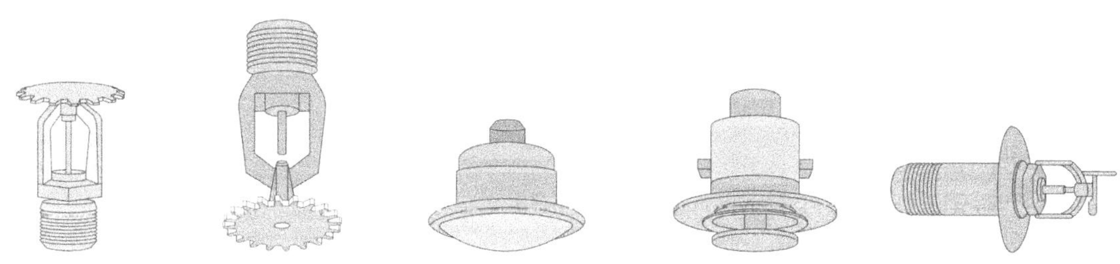

Figure 26.17 *Sprinkler heads. (a) Upright; it sits above exposed piping, just below the structural deck—where the hottest gases will first accumulate. (b) Pendent, projecting through a suspended ceiling in most installations. Both (a) and (b) use a quartzoid bulb, a transparent bulb with a colored liquid that ruptures at a preset temperature to release the water stream; heads can be manufactured as plated, polished, or colored—but never painted on site. (c) concealed and (d) flush pendent, styled for a more contemporary appearance. (e) Wall-mounted sidewall, with a deflector to ensure that the sprinkler discharges water toward the floor surface. (From Ramsey and Sleeper, 2016.)*

the sprinkler (Figure 26.17*b*). Pendent heads have a number of variations:

- *Recessed pendent* heads have part of the sprinkler body and the deflector below the ceiling.
- *Concealed pendent* heads are entirely above a ceiling cover plate that falls away in a fire, exposing the thermosensitive element (Figure 26.17*c*).
- *Flush pendent* heads have only the thermosensitive element projecting below the ceiling (Figure 26.17*d*).
- *Ornamental pendent* heads are manufacturer-coated (never field-coated!) to match a desired decor.

- *Standard sidewall spray:* Heads mounted horizontally on walls, with the deflector positioned to one side (Figure 26.17*e*). These heads are usually located adjacent to one wall of a smaller room—common in hotels, apartments, and so on—and throw a one-quarter-sphere spray of water entirely across such rooms. Typically, only one sidewall sprinkler head per small room is used.

The temperature at which a sprinkler is triggered is usually specified, as in Table 26.7. It should be at least 25°F [14°C] higher than the maximum ceiling temperature ordinarily expected. Ordinary sprinkler heads operate at between 135° and 170°F [57° and 77°C].

(d) Sprinkler Response Types

Buildings to be protected by sprinkler systems fall into several hazard groups, as described in Table 26.8. The response time of a sprinkler system should be appropriate for the hazard level of the space. Sprinkler response types are defined by their Response Time Index (RTI), which indicates how fast the sprinkler can absorb sufficient heat from its surroundings to cause activation. It is expressed as the square root of meters-seconds. In addition to their response time, sprinklers also can feature different thermal elements that are designed to activate at specific temperatures.

There are two basic sprinkler response types:

Standard response sprinklers: This type has a thermal element with an RTI of 80 or more—often substantially more. Conventional automatic sprinklers generally have an RTI of 250–300.

Quick-response (QR) sprinklers: Also referred to as fast-response sprinklers, this type is equipped with a thermal element with an RTI or 50 or less. As such, QR sprinklers are designed to activate quickly upon detecting heat. All light hazard occupancies are now required to have QR sprinklers. These include hotels, motels, offices, and other buildings where faster sprinkler operation could enhance life safety. The more thermally sensitive heads of QR sprinklers tend to open sooner than ordinary heads, and therefore often counteract fires with even fewer heads operating. The heads may even open with extraordinary heat that is not fire-associated. As such, QR sprinklers are superior for life protection compared to standard-response sprinklers.

Because of the thermal lag inherent in the glass bulb (or fusible link), the sprinkler body, and the water within sprinkler pipes, air temperature around a sprinkler may reach 1000°F [538°C] before a standard sprinkler, rated at 175°F [79°C], actually opens. The QR sprinkler's operating element has a smaller mass, enabling it to respond to the air temperature rise more quickly.

(e) Sprinkler Spacing

The spacing of sprinkler heads and the sizing of supply pipes are complex matters, and most sprinkler systems are designed by professionals working for sprinkler manufacturers. However, some guidelines for preliminary sprinkler location are included here. The first consideration is the degree of hazard faced by the occupants, as listed in Table 26.8. Once this is known, sprinklers and pipes can be approximately located in plan view with the aid of Table 26.9. Maximum floor

TABLE 26.7 Sprinkler Heads

Discharge Characteristics				
Nominal Orifice Size		K-Factor Range		
in.	mm	gpm/(psi)$^{1/2}$	L/min/(bar)$^{1/2}$	Percent of Nominal K-5.6 (80) Discharge[a]
1/4	6.4	1.3–1.5	19–22	25
5/16	8.0	1.8–2.0	26–29	33.3
3/8	10	2.6–2.9	38–42	50
7/16	11	4.0–4.4	57–63	75
1/2	13	5.3–5.8	76–84	100
17/32	13	7.4–8.2	107–118	140
5/8	16	11.0–11.5	159–166	200
3/4	20	13.5–14.5	195–209	250

Temperature Classifications						
Max. Ceiling Temp.		Temperature Rating				
°F	°C	°F	°C	Temperature Classification	Color Code	Glass Bulb Colors
100	38	135–170	57–77	Ordinary	Uncolored or black	Orange or red
150	66	175–225	79–107	Intermediate	White	Yellow or green
225	107	250–300	121–149	High	Blue	Blue
300	149	325–375	163–191	Extra high	Red	Purple
375	191	400–475	204–246	Very extra high	Green	Black
475	246	500–575	260–302	Ultra high	Orange	Black
625	329	650	343	Ultra high	Orange	Black

Source: Reproduced with permission of NFPA from NFPA 13. Automatic Sprinkler Systems Handbook, 2025 edition. Copyright © 2024, National Fire Protection Association. For a full copy of NFPA 13, please go to www.nfpa.org.

[a]This relates to the nominal discharge from a 1/2 orifice.

TABLE 26.8 Relative Fire Hazard for Various Occupancies, as Related to Sprinkler Installations

Classification of Occupancies	
Occupancy classifications for this standard shall relate to sprinkler design, installation, and water supply requirements only. They are not intended to be a general classification of occupancy hazards. Examples in the listings below are intended to represent the norm for those occupancy types. Unusual or abnormal fuel loadings or combustible characteristics and susceptibility for changes in these characteristics, for a particular occupancy, are considerations that should be weighed in the selection and classification.	
Light Hazard Occupancies	
Occupancies or portions of other occupancies where the quantity and/or combustibility of contents is low, and fires with relatively low rates of heat release are expected.	
The Light Hazard classification is intended to encompass residential occupancies; however, this is not intended to preclude the use of listed sprinklers in residential occupancies or residential portions of other occupancies.	
Light hazard occupancies include occupancies having uses and conditions similar to the following:	
Animal shelters	Libraries, except large stack rooms
Churches	Museums
Clubs	Nursing or convalescent homes
Eaves and overhangs, if combustible construction with no combustibles beneath	Office, including data processing
Educational	Residential
Hospitals, including animal hospitals and veterinary facilities	Restaurant seating areas
Institutional	Theaters and auditoriums, excluding stages and prosceniums
Kennels	Unused attics
Ordinary Hazard Occupancies	
Group 1 (OH1): Occupancies or portions of other occupancies where combustibility is low, quantity of combustibles is moderate, stockpiles of combustibles do not exceed 8 ft [2.4 m], and fires with moderate rates of heat release are expected.	
Ordinary hazard (Group 1) occupancies include occupancies having uses and conditions similar to the following:	
Automobile showrooms	Electronics plants
Bakeries	Glass and glass product manufacturing
Beverage manufacturing	Laundries
Canneries	Restaurant service areas
Dairy products manufacturing and processing	Porte cocheres
Electric fire pump room	Mechanical rooms

TABLE 26.8 (*Continued*)

Ordinary Hazard Occupancies	
Group 2 (OH2): Occupancies or portions of other occupancies where quantity and combustibility of contents are moderate to high, stockpiles do not exceed 12 ft [3.7 m], and stockpiles of contents with high rates of heat release that do not exceed 8 ft [2.4 m].	
Ordinary hazard (Group 2) occupancies include occupancies having uses and conditions similar to the following:	
Agricultural facilities	Paper and pulp mills
Automobile parking garages	Paper process plants
Barns and stables	Piers and wharves
Cereal mills	Plastics fabrication, including blow molding, extruding, and machining; excluding operations using combustible hydraulic fluids
Chemical plants—ordinary	Post offices
Confectionery products	Printing and publishing
Distilleries	Racetrack stable/kennel areas, including those stable/kennel areas, barns, and associated buildings at state, county, and local fairgrounds
Dry cleaners	Repair garages
Exterior loading docks	Resin application area
Feed mills	Stages
Horse stables	Textile manufacturing
Leather goods manufacturing	Tire manufacturing
Libraries—large stack room areas	Tobacco products manufacturing
Machine shops	Wood machining
Metal working	Wood product assembly
Mercantile	
Extra Hazard Occupancies	
Occupancies or portions of other occupancies where quantity and combustibility of contents are very high and flammable and combustible liquids, dust, lint, or other materials are present, introducing the possibility of rapidly developing fires with high rates of heat release. Extra hazard occupancies involve a wide range of variables that may produce severe fires.	
Extra Hazard Occupancies (Group 1 or EH1) include occupancies with very high quantity and combustibility of contents and where dust, lint, or other materials are present, introducing the probability of rapidly developing fires.	
Extra Hazard Occupancies (Group 2 or EH2) include occupancies with very high quantity and combustibility of contents; substantial amounts of combustible or flammable liquids; and where shielding of combustibles is extensive.	
Extra hazard (Group 1) occupancies include occupancies having uses and conditions similar to the following:	
Aircraft hangars (except as governed by *NFPA 409: Standard on Aircraft Hangars*)	Printing (using inks having flash points below 100°F [38°C])
Combustible hydraulic fluid use areas	Rubber reclaiming, compounding, drying, milling, vulcanizing
Die casting	Saw mills
Metal extruding	Textile picking, opening, blending, garnetting, or carding, combining of cotton, synthetics, wool shoddy, or burlap
Plywood and particleboard manufacturing	Upholstering with plastic foams
Extra hazard (Group 2) occupancies include occupancies having uses and conditions similar to the following:	
Asphalt saturating	Plastics manufacturing
Flammable liquids spraying	Solvent cleaning
Flow coating	Varnish and paint dipping
Manufactured home or modular building assemblies (where finished enclosure is present and has combustible interiors)	Car stackers and car lift systems with two cars stacked vertically and using ceiling-only protection
Open oil quenching	Diesel fire pump houses and rooms containing pump drivers and fuel tanks
Special Occupancy Hazards	
See *NFPA 13* (NFPA, 2024a) for a listing of standards by occupancy.	

Source: Reproduced with permission of NFPA from NFPA 13. Automatic Sprinkler Systems Handbook, 2025 edition. Copyright © 2024, National Fire Protection Association. For a full copy of NFPA 13, please go to www.nfpa.org.

[a]Note that exterior loading docks only used for loading and unloading of ordinary combustibles should be classified as OH2. For the handling of flammable and combustible liquids or hazardous materials, or where utilized for storage, exterior and interior loading docks should be protected based upon the actual occupancy and the materials handled on the dock, as if the materials were actually stored in that configuration.

TABLE 26.9 Protected Area, Maximum Spacing, and Distances below Ceiling for Sprinklers

Part A. Standard Spray Upright (SSU) and Standard Spray Pendent (SSP) Sprinklers

	Light Hazard		Ordinary Hazard		Extra Hazard		High-Piled Storage	
Construction Type	**Protection Area, ft² [m²]**	**Spacing (max.), ft [m]**	**Protection Area, ft² [m²]**	**Spacing (max.), ft [m]**	**Protection Area, ft² [m²]**	**Spacing (max.), ft [m]**	**Protection Area, ft² [m²]**	**Spacing (max.), ft [m]**
Noncombustible obstructed and unobstructed, and combustible unobstructed[a]	225 [20.9]	15 [4.6]	130 [12]	15 [4.6]	100[b] [9]	12[b,c] [3.7]	100[b] [9]	12[b,c] [3.7]
Combustible obstructed with exposed members 3 ft [910 mm] or more on center	168 [15.6]	15 [4.6]	130 [12]	15 [4.6]	100[b] [9]	12[b,c] [3.7]	100[b] [9]	12[b,c] [3.7]

Part B. Standard Sidewall Spray Sprinklers

	Light Hazard		Ordinary Hazard	
	Combustible Ceiling Finish	**Noncombustible or Limited-Combustible Ceiling Finish**	**Combustible Ceiling Finish**	**Noncombustible or Limited-Combustible Ceiling Finish**
Maximum distance along the wall (*S*), ft [m]	14 [4.3]	14 [4.3]	10 [3.0]	10 [3.0]
Maximum room width (*L*), ft [m]	12 [3.7]	14 [4.3]	10 [3.0]	10 [3.0]
Maximum protection area ft² [m²]	120 [11]	196 [18]	80 [7.4]	100 [9.3]

Part C. Extended Coverage (EC) Upright and Pendent Spray Sprinklers

	Light Hazard		Ordinary Hazard		Extra Hazard		High-Piled Storage	
Construction Type	**Protection Area, ft² [m²]**	**Spacing (max.), ft [m]**	**Protection Area, ft² [m²]**	**Spacing (max.), ft [m]**	**Protection Area, ft² [m²]**	**Spacing (max.), ft [m]**	**Protection Area, ft² [m²]**	**Spacing (max.), ft [m]**
Unobstructed	400 [37]	20 [6.1]	400 [37]	20 [6.1]	—	—	—	—
	324 [30]	18 [5.5]	324 [30]	18 [5.5]	—	—	—	—
	256 [24]	16 [4.9]	256 [24]	16 [4.9]	—	—	—	—
	—	—	196 [18]	14 [4.3]	196 [18]	14 [4.3]	196 [18]	14 [4.3]
	—	—	144 [13]	12 [3.7]	144 [13]	15 [4.6]	144 [13]	15 [4.6]
Obstructed noncombustible (when specifically listed for such use)	400 [37]	20 [6.1]	400 [37]	20 [6.1]	—	—	—	—
	324 [30]	18 [5.5]	324 [30]	18 [5.5]	—	—	—	—
	256 [24]	16 [4.9]	256 [24]	16 [4.9]	—	—	—	—
	—	—	196 [18]	14 [4.3]	196 [18]	14 [4.3]	196 [18]	14 [4.3]
	—	—	144 [13]	12 [3.7]	144 [13]	15 [4.6]	144 [13]	15 [4.6]
Obstructed combustible	N/A	N/A	N/A	N/A	N/A	N/A	N/A	N/A

TABLE 26.9 Protected Area, Maximum Spacing, and Distances below Ceiling for Sprinklers (*Continued*)

Part D. Extended Coverage (EC) Sidewall Sprinklers				
	Light Hazard		**Ordinary Hazard**	
Construction Type	**Protection Area, ft^2 [m^2]**	**Spacing (max.), ft [m]**	**Protection Area, ft^2 [m^2]**	**Spacing (max.), ft [m]**
Unobstructed, smooth, flat	400 [37]	28 [8.5]	400 [37]	24 [7.3]
Part E. Control Mode Specific Application (CMSA) Sprinklers				
Construction Type	**Protection Area, ft^2 [m^2]**		**Spacing, ft [m]**	
Noncombustible unobstructed	130 [12]		12 [3.7]	
Noncombustible obstructed	130 [12]		12 [3.7]	
Combustible unobstructed	130 [12]		12 [3.7]	
Rack storage combustible obstructed	100 [9]		12 [3.7]	
Rack storage unobstructed and noncombustible obstructed	100 [9]		12 [3.7]	
Part F. Early Suppression Fast-Response (ESFR) Sprinklers				
	Ceiling/Roof Heights Up to 30 ft [9.1 m]		**Ceiling/Roof Heights Over 30 ft [9.1 m]**	
Construction Type	**Protection Area, ft^2 [m^2]**	**Spacing, ft [m]**	**Protection Area, ft^2 [m^2]**	**Spacing, ft [m]**
Noncombustible unobstructed	100 [9]	12 [3.7]	100 [9]	
Noncombustible obstructed	100 [9]	12 [3.7]	100 [9]	10 [3.0]
Combustible unobstructed	100 [9]	12 [3.7]	100 [9]	10 [3.0]
Combustible obstructed	100 [9]	12 [3.7]	100 [9]	10 [3.0]
Part G. Maximum Allowable Deflector Distances Below the Ceiling for Various Sprinklers				
Sprinkler Type	**Maximum Distance, in. [mm]**			
SSU/SSP	12 [300]			
Standard sidewall	6 [150]			
EC upright and pendent	12 [300]			
EC sidewall	8 [200]			
CMSA	14 [350]			
ESFR	14 [350]			

[a]Hydraulically calculated system type.

[b]Hydraulically calculated with density ≥ 0.25 gpm/ft^2 [10.2 mm/min].

[c]In buildings where solid structural members create bays up to 25 ft [7.6 m] wide, maximum spacing between sprinklers is permitted up to 12 ft 6 in. [3.8 m].

TABLE 26.10 Maximum Floor Areas for Sprinkler Systems

The maximum floor area on any one floor to be protected by sprinklers supplied by any one sprinkler system riser (or combined system riser) is:

Hazard	Area, ft^2 [m^2]
Light hazard	52,000 [4830]
Light hazard protected with a wet pipe system and the system water flow and supervisory alarms are transmitted to a supervising station in accordance with *NFPA 72*	78,000 [7250]
Ordinary hazard	52,000 [4830]
Extra hazard, hydraulically calculated	40,000 [3720]
High-piled storage and storage covered by other NFPA standards	40,000 [3720]
In-rack storage	40,000 [3720]

Source: Reproduced with permission of NFPA from NFPA 13. Automatic Sprinkler Systems Handbook, 2025 edition. Copyright © 2024, National Fire Protection Association. For a full copy of NFPA 13, please go to www.nfpa.org.

Notes: Floor area occupied by mezzanines shall not be included in the above areas. Where single systems protect extra hazard or high-piled storage and ordinary or light-hazard areas, the extra hazard or storage area coverage shall not exceed the floor area specified for that hazard and the total area coverage shall not exceed 52,000 ft^2 [4831 m^2].

areas (per floor) to be protected by any single sprinkler system are shown, again by hazard, in Table 26.10. Within each space, sprinklers should be located so as to detect a fire readily and to discharge water over the greatest area (considering obstacles such as joists and beams, partial height partitions, etc.; see Figure 26.16).

Piping for the sprinkler supply can be hydraulically designed, using fundamental water flow principles. A complicating factor is the expectation that only a small percentage of the sprinklers will actually open: more than 50% of the fires studied over a 49-year period were extinguished by two or fewer sprinklers. A detailed sizing procedure would consider both the available pressure at the highest sprinkler and the expected flow rate, which can vary from 150 to 25,000 gpm [10 to 158 L/s]. The sprinklers' actual performance is determined by:

$$Q = K\sqrt{p}$$

where

Q = flow rate, gpm
K = K-factor (see Table 26.7)
p = pressure, psi

In SI units:

Q = flow rate, L/min
K = K factor (14.3 × I-P K factor)
p = pressure in bars

Sprinkler systems are usually designed for a maximum working pressure of 175 psi [1206 kPa]. As a preliminary design guideline:

Light hazard systems need a minimum residual pressure of 15 psi [104 kPa] and 500–750 gpm [32–47 L/s] at the base of the system riser for 30 to 60 minutes.

Ordinary hazard systems need a minimum residual pressure of 20 psi [138 kPa] and 850–1500 gpm [54–95 L/s] at the base of the system riser for 60 to 90 minutes.

Manufacturers publish a K-factor for every sprinkler head. This is a discharge coefficient that determines the amount of water a sprinkler can discharge. Essentially, the K-factor indicates the orifice size needed for a specific flow and pressure. A larger K-factor indicates a bigger orifice, more water flow, and requires a larger sprinkler system capacity. A smaller K-factor indicates a smaller orifice, less water flow, and a need for less systems capacity. The K-factor for sprinkler heads will range based on the type of sprinkler and its intended use. Generally, designers may anticipate the following:

- residential sprinklers: K-3.0 to K-5.8 [43 to 83]
- typical 1/2 in. orifice sprinklers: K-5.6 [80]
- CMSA and ESFR sprinklers: minimum K-11.2 [160]
- storage sprinklers: K-16.8 to K-25.2 [240 to 360]

Consult *NFPA 13* (NFPA, 2024a) for detailed K-factor requirements specific to project details.

In the past, the pipe schedule design method made system layout and sizing fairly simple. Now there are requirements for more complicated hydraulically designed systems, which are beyond the scope of this book.

(f) Special Sprinkler Models

Special sprinkler models are available for certain types of spaces. Such models include:

- *Extra-large orifice (ELO):* Used to deliver large water quantities where water pressures are relatively low.
- *Multilevel sprinklers:* Used where other sprinklers are at a higher plane within the same space. The lower sprinklers may otherwise have their action retarded by water from the higher-elevation heads.

(g) Residential Sprinklers

Now that some building codes require sprinklers in all residential occupancies, some new issues arise and another special sprinkler model has come to market. The residential sprinkler is a fast-response device having a thermal element with an RTI of 50 or less with a tested ability to enhance survivability in the room of the fire origin and is thus listed for protection of dwelling units. It is sensitive to both smoldering and rapidly developing fires, opening quickly to fight a fire with only one or two heads operating. This is important because residences normally do not have a water supply with sufficient capacity for standard sprinkler systems. Toxic gases and smoke quickly fill the small rooms typical of residences; a fast response is important for life safety.

Most codes that otherwise require residential sprinklers in all areas make an exception for bathrooms no larger than 55 ft^2 [5.1 m^2]; for closets with the least dimension not exceeding 3 ft [0.9 m]; for open porches, garages, and carports; for uninhabited attics and crawl spaces (if not used for storage); and for entrance foyers that are not the sole means of egress.

Residential sprinklers also have a special water distribution pattern capable of delivering water to the walls and high enough on the walls to prevent the fire from getting above the sprinklers. This water near the ceiling also tends to cool gases at the ceiling, reducing the likelihood of excessive sprinkler openings.

The added cost of residential sprinkler systems may be recovered in several ways. As with other buildings, reductions in fire insurance rates will be helpful, although in residences the payback time is rather long. Zoning could permit smaller parcels of land for sprinkler-protected residences, since separation between buildings is less important when sprinklers protect both buildings.

(h) Extended Coverage (EC) Sprinklers

EC sprinklers cover more area than standard sprinklers, therefore fewer sprinklers are typically required in a given area. See Table 26.9, Parts C and D. These sprinklers are limited to a type of unobstructed construction consisting of flat, smooth ceilings of a slope not exceeding 2 in. per foot [158 mm/m]. Note that a smooth ceiling means that luminaires and air grilles are flush or recessed, not suspended from the ceiling. Such sprinklers, however, can also be specifically listed for "noncombustible obstructed" construction, or as upright and pendent sprinklers within trusses or bar joists having web members not more than 1 in. [25 mm] thick, or where specifically listed for flat, smooth ceilings of a slope not exceeding 4 in. per foot [316 mm/m].

(i) Control Mode Specific Application (CMSA) Sprinklers

See Table 26.9, Part E. Control Mode Specific Application (CMSA) sprinklers are designed for applications such as high-challenge storage occupancies where the solution is to "control" the fire, not suppress it. CMSA sprinklers are a type of spray sprinkler that can produce larger water droplets and are listed for their capability to provide fire control within spaces with specific fire hazards. CMSA sprinklers have largely replaced "large-drop" sprinklers that depended on the weight of the water droplet to penetrate the fire plume. These sprinklers require a minimum pressure of 50 psi [3.4 bar] for K-11.2 [160] sprinklers and 22 psi [1.5 bar] for K-16.8 [240] sprinklers. For K-19.6 [2880] or larger, *NFPA 13* (NFPA, 2024a) exhibits varying requirements.

(j) Early Suppression Fast-Response (ESFR) Sprinklers

See Table 26.9, Part F. Early Suppression Fast-Response (ESFR) sprinklers feature rapid-response elements to detect and react to fires quickly, which is crucial for preventing rapid fire spread in storage settings. ESFR sprinkler heads are tested for their ability to suppress specific high-challenge fire hazards encountered in high-piled storage. They operate at a higher pressure and flow, and the water droplets produced depend upon momentum rather than gravity to penetrate to the bottom of high-velocity fire plumes. These sprinkler heads require a minimum water pressure of 50 psi [345 kPa] and a minimum flow of 100 gpm [6.3 L/s].

(k) Quick Response Extended Coverage (QREC) Sprinklers

Quick response extended coverage (QREC) sprinklers are engineered to provide rapid activation and cover larger areas per sprinkler head, reducing the total number of sprinklers needed. This approach uses the same principles as an ESFR sprinkler, but with a smaller orifice suitable to lighter-hazard occupancies. The QRES approach never really materialized, replaced instead by CMSA sprinklers.

(l) Future Sprinkler System Developments

Sprinkler technology continues to advance. In future systems, different types of sprays may be ejected from a single sprinkler head: one spray of larger droplets to penetrate the fire plume and thereby cool burning surfaces as well as adjacent surfaces, and another, finer spray to cool the ceiling itself.

Additional sprinkler types have recently come to market. Though, not as common as other types, they are worth noting:

- *Polymer sprinklers:* A residential sprinkler type, polymer sprinklers are lighter than conventional sprinklers.
- *Electrically activated sprinklers (EAS):* Also referred to as electronically operated sprinklers, this type is managed through an electrical signal provided by an electronic control system using detection and control algorithms. EAS sprinklers are emerging in high-challenge storage applications and may utilize a combination of heat and smoke detectors.

(m) Wet-Pipe Systems

There are two main categories of fire sprinkler systems: wet-pipe and dry-pipe systems. The basic difference between them is how the system is filled with water and how quickly it can deliver water. Wet-pipe systems are filled with pressurized water and deliver it almost immediately, while dry-pipe systems are filled with pressurized air (or nitrogen) and water delivery entails a short delay. As such, wet-pipe systems are standard for most residential and business settings, whereas dry-pipe systems are better-suited for many other applications.

Dry-pipe systems (see Section 26.3(o)) are the most common and the simplest systems, as shown in Figure 26.18. They are filled with water under pressure and are limited to spaces in which the air temperature does not fall below 40°F [4.4°C]. (Wet-pipe systems that contain antifreeze, and admit ordinary water when a sprinkler head opens, are included in this category; the type of antifreeze is limited when potable water supplies the sprinkler system.) In the wet-pipe system, sprinklers in the affected area are opened by heat-sensitive elements within the sprinkler heads themselves and immediately emit water.

A typical automatic wet-pipe sprinkler system is shown in Figure 26.19. The building, a printing and publishing plant, is in the category of "Ordinary Hazard, Group 2" (see Table 26.8). The sprinkler design results in a nozzle spacing such that one nozzle (sprinkler head) takes care of 130 ft^2 [12 m^2] of floor area.

The Transamerica Building in San Francisco (Figure 26.20) is an urban high-rise building with a wet-pipe system. Two fire pumps can deliver 750 gpm

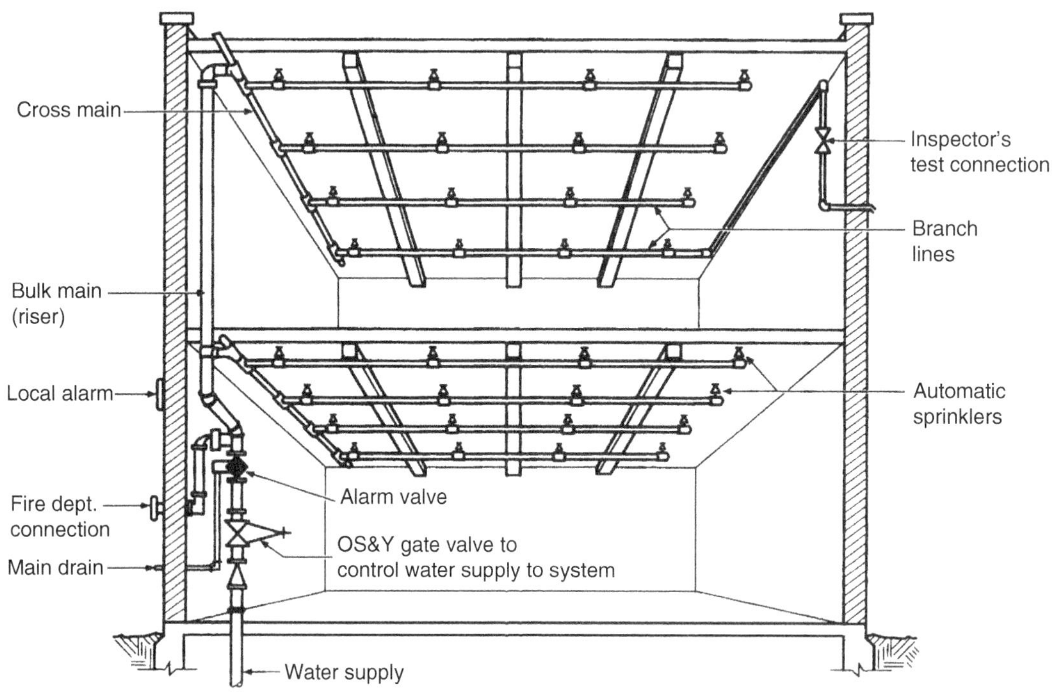

Figure 26.18 *Wet-pipe sprinkler system. This most typical system is under water pressure at all times so that water will be discharged immediately when an automatic sprinkler opens. The automatic alarm valve shown causes a warning signal to sound when water flows through the sprinkler piping. OS&Y means "outside stem and yoke," describing a valve that clearly indicates the degree to which it is open or closed. (Reproduced with permission of NFPA from the Fire Protection Handbook, 20th edition. Copyright © 2008, National Fire Protection Association. For a full copy of the Fire Protection Handbook, please go to www.nfpa.org.)*

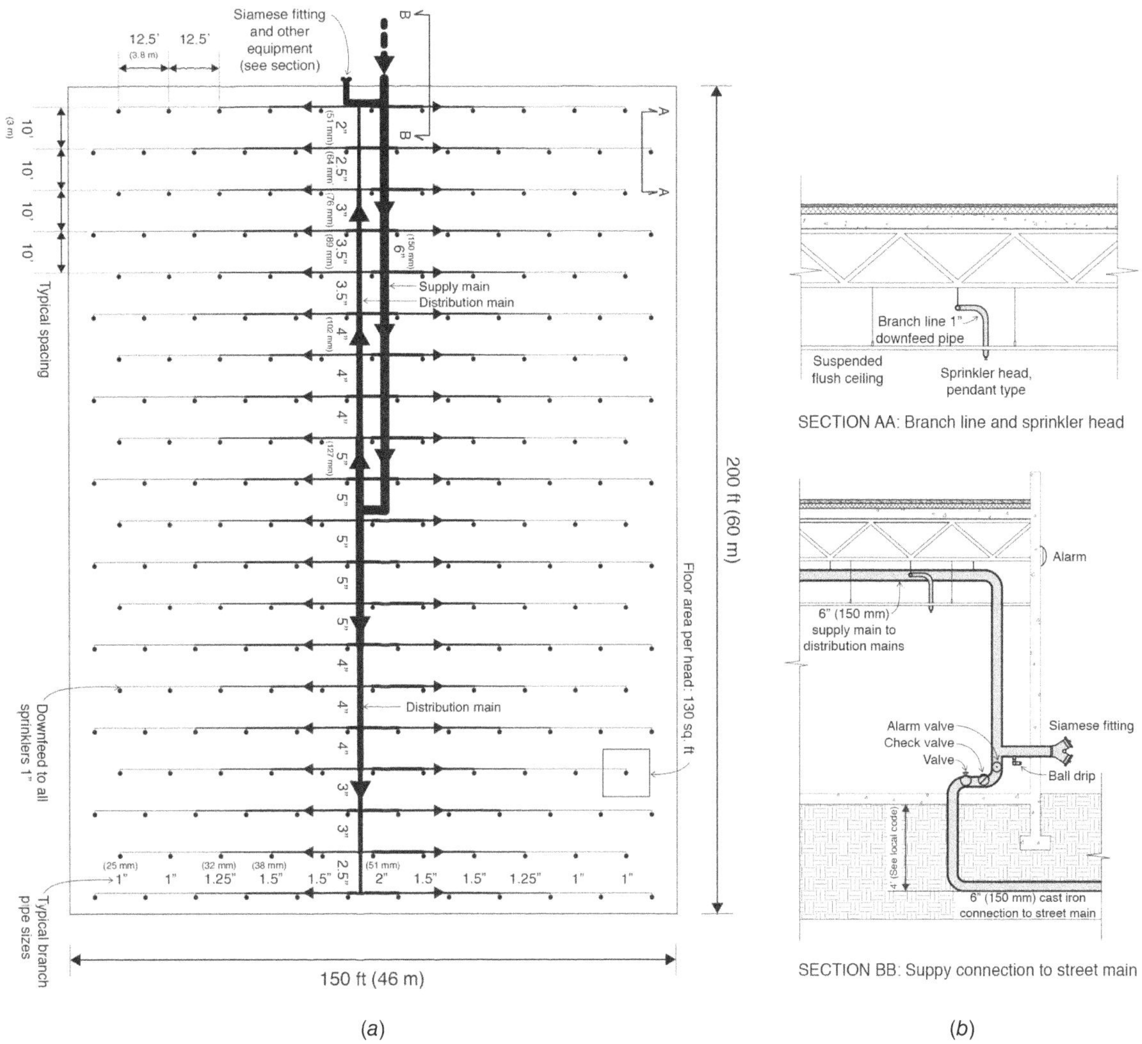

Figure 26.19 *(a) Plan of a sprinklered industrial building, ordinary hazard occupancy, whose 30,000-ft² [2790-m²] floor area is protected by 230 sprinkler heads (average 130 ft² [12 m²] per sprinkler head). Sprinklers are closer together than the maximum allowed. (b) Sprinklers (and standpipes) may use water from street mains when pressure is adequate. Either system may use pneumatic or gravity tanks. When the latter are used to supply both systems, an independent sprinkler reserve occupies the bottom, and the fire standpipe supply occupies the top. An auxiliary fire engine feed connection by siamese should be provided in all cases. Auxiliary sources and standby pressurization may be required if street main adequacy is questionable. (Drawn by Kelli Kimura.)*

at 275 psi [47.3 L/s at 1896 kPa] discharge pressure; they draw water from the city mains at 50 psi [345 kPa] or from a 5000-gal [18,925-L] closed tank in the basement. These pumps feed into two 6-in. [15-mm] diameter "express" risers, one in each stair tower, that run the full height of this 48-story office building. The risers serve both the sprinkler system and the fire department hose lines. In each of three 16-story zones, "local" 6-in. [150-mm] diameter risers branch off to feed a 2-in. [50-mm] looped main at each floor.

The sprinkler piping is carried above suspended ceilings; pendent sprinkler locations are coordinated with the modular grid that also locates partitions, air diffusers, and utility jacks. There are provisions for moving sprinkler heads as tenants change. At regular intervals, tees have been provided with one outlet stubbed and capped for future use. Typical office spaces are served by fully recessed, ½-in. orifice, 165°F [12.7-mm, 74°C] pendent sprinklers. Exposed pendent sprinklers are used in toilet rooms and service areas.

Although the typical sprinkler system is served by a single riser with a main line and branch lines, there are two variations that increase reliability by providing some redundancy in the supply lines. Figure 26.21*a* shows the gridded system, where each branch is served from either end, allowing each sprinkler head to receive water from either direction in the branch line. Figure 26.21*b* shows the loop system, where each branch line can receive water from either direction.

(n) Circulating Closed-Loop Systems

These wet-pipe systems use rather large sprinkler piping to circulate water for water-source heat pumps. Water is not normally removed from this system, merely circulated. Water temperature in these systems must not exceed 120°F [49°C] or fall below 40°F [4°C]. Such a system used for heating and cooling in a motel is shown in Figure 26.22.

(o) Dry-Pipe Systems

As shown in Figure 26.23, dry-pipe systems are filled with compressed air (or nitrogen) rather than with water. They are used in unheated areas, including loading docks and cold-storage areas. As soon as a sprinkler head opens, the compressed air rushes out, allowing water to enter the formerly dry-pipe network through a dry-pipe valve. The system then functions like a wet-pipe system. The dry-pipe valve must be within a heated enclosure, since water under pressure is on one side of the valve. Also, due to the delay in delivering water throughout a previously dry piping system, a maximum system capacity of 750 gal [2839 L] is recommended.

Dry-pipe systems require a device to maintain design air pressure within the pipes. Air pressure might be maintained by a compressor, by an air receiver tank, or by connection to an existing pressurized air system (as in manufacturing operations). Dry-pipe systems also require a heated main control valve housing and the pitching of all piping to allow thorough drainage after use.

(p) Preaction Systems

In these dry-pipe systems (Figure 26.24), the pipes are filled with air that may or may not be under pressure. In addition to the sprinkler heads, either a heat- or a smoke-detection system is installed. The detectors are more sensitive than the sprinkler head. Water is held back by the preaction valve. When the heat or smoke detectors are activated, they open the preaction valve, an alarm is sounded, and water fills the pipes. The system then functions like a wet-pipe system, with water flow into a space occurring only upon the opening of a sprinkler head.

A variation on this system, known as a combined dry pipe-preaction (or double-interlock preaction) system, has pipes filled with compressed air. Heat or smoke detectors release the preaction valve and sound the alarm, but air pressure keeps water out of the piping until a sprinkler head opens.

Preaction systems are popular where the building's contents are especially subject to water damage—computer rooms, retail stores, museums, and so on—because the early alarm provided by water filling the piping often permits the fire to be found and extinguished manually, before any sprinklers open.

Designers must remember, however, that when the onset of the water discharge is delayed, a fire can quickly grow in size.

(a)

Figure 26.20 *Transamerica Building, San Francisco, California; William Pereira & Associates, architects. (a) The pyramid form provides rentable floor areas ranging from 22,000 to 3000 ft^2 [2044 to 279 m^2]. (b) Sixth-floor sprinkler plan showing two risers, a looped feed main, and branch lines. (Courtesy of the Copper Development Association.)*

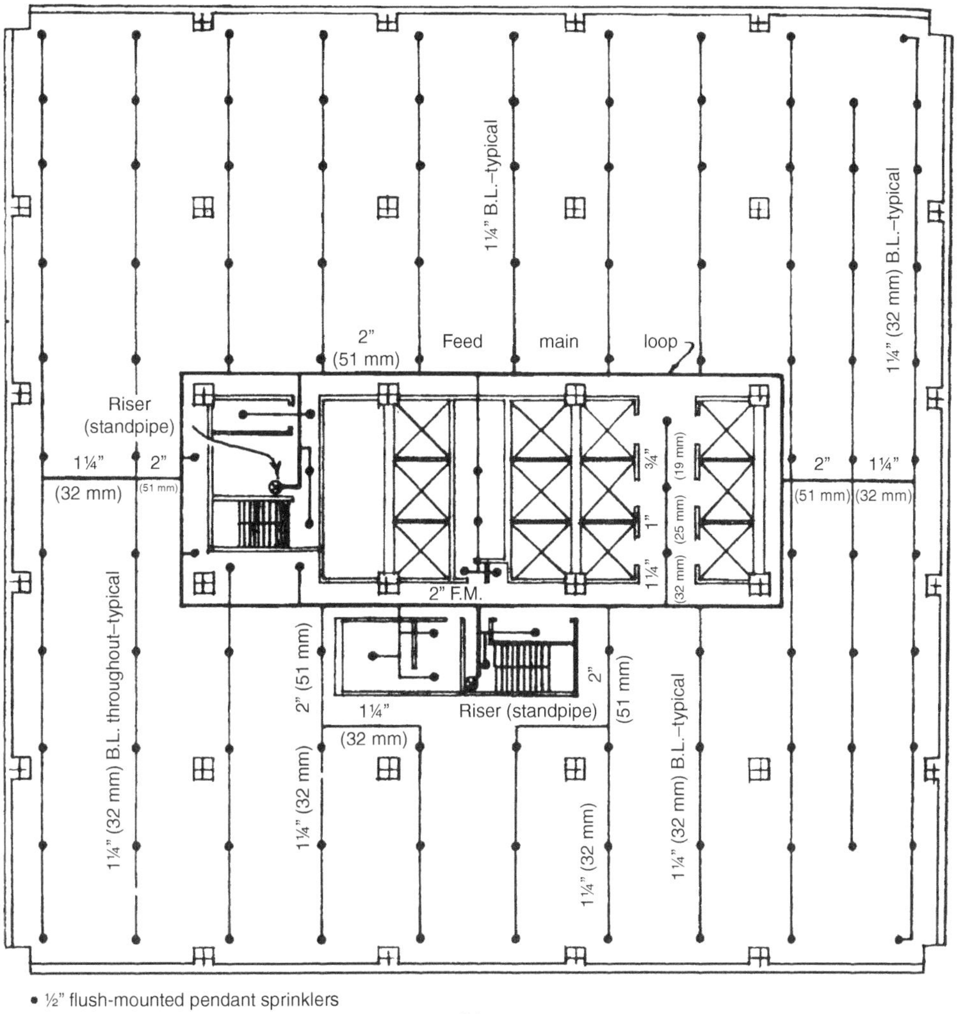

(*b*)

Figure 26.20 *(continued)*

(q) Deluge Systems

These dry-pipe systems (Figure 26.25) have (perpetually) open sprinkler heads. As with preaction systems, a separate heat- or smoke-detection system is installed. The detectors control a deluge valve, which, once opened, floods the system with water, and all heads emit water. Huge quantities of water are thus released. Deluge systems are used where extremely rapid fire spread is expected—in aircraft hangars, for example, or other places where flammable liquid fires may break out.

(r) Mist Systems

Also known as water mist fire suppression systems, mist systems use small water droplets to put out or control fires. Unlike other sprinkler systems, mist systems do use water in a conventional sense. Mist systems are discussed last because they introduce the next section, fire-suppression alternatives to water. With a history of success in shipboard fires, they are being considered for wider application. *NFPA 750: Standard on Water Mist Fire Protection Systems* (NFPA, 2022b) is the relevant standard.

A mist system offers fast initiation of an alarm as well as more rapid response to a fire. Smaller volumes of water mean less water damage, and the mist can move more easily around obstructions.

As an alternative to halon or other clean agent gases (Section 26.4), mist systems are more tolerant of small amounts of ventilation, reduce the radiant heat transfer from a fire, and eliminate the residues

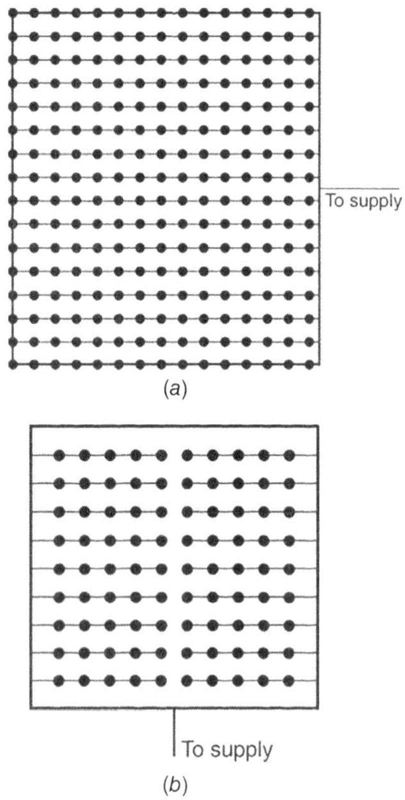

Figure 26.21 Alternative piping arrangements that provide more than one path for water. (a) Gridded systems supply water from either end of a branch line, offering the potential for water to come from either side of each sprinkler location. (b) Looped system, with the potential for each branch line to get water from either direction. (Reproduced with permission of NFPA from NFPA 13, Automatic Sprinkler Systems Handbook, 2025 edition. Copyright © 2024, National Fire Protection Association. For a full copy of NFPA 13, please go to www.nfpa.org.)

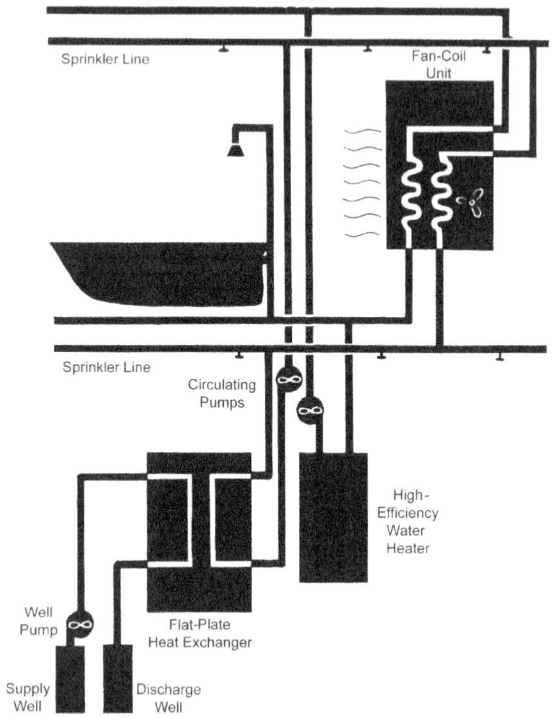

Figure 26.22 Circulating closed-loop sprinkler system, whose water is used by heating/cooling systems. This application is for a motel; a ground-source heat exchanger is supplemented by a water heater when required. (Redrawn by Amanda Clegg; Courtesy of Montana Power.)

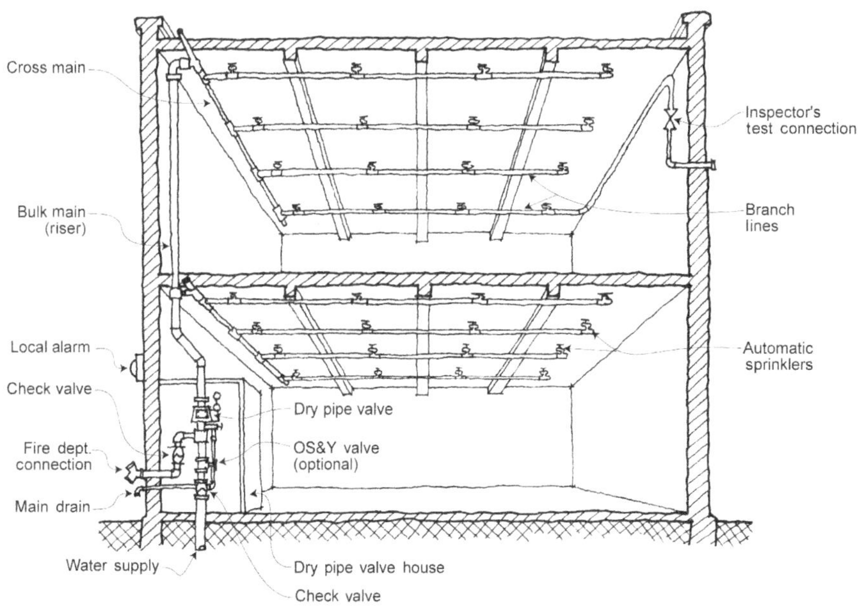

Figure 26.23 Dry-pipe sprinkler system for unheated occupancies. The system riser, cross-mains, and branch lines are maintained at above-atmospheric air pressure. The dry-pipe valve and water supply must be within a heated enclosure. (Reproduced with permission of NFPA from the Fire Protection Handbook, 20th edition. Copyright © 2008, National Fire Protection Association. For a full copy of the Fire Protection Handbook, please go to www.nfpa.org.)

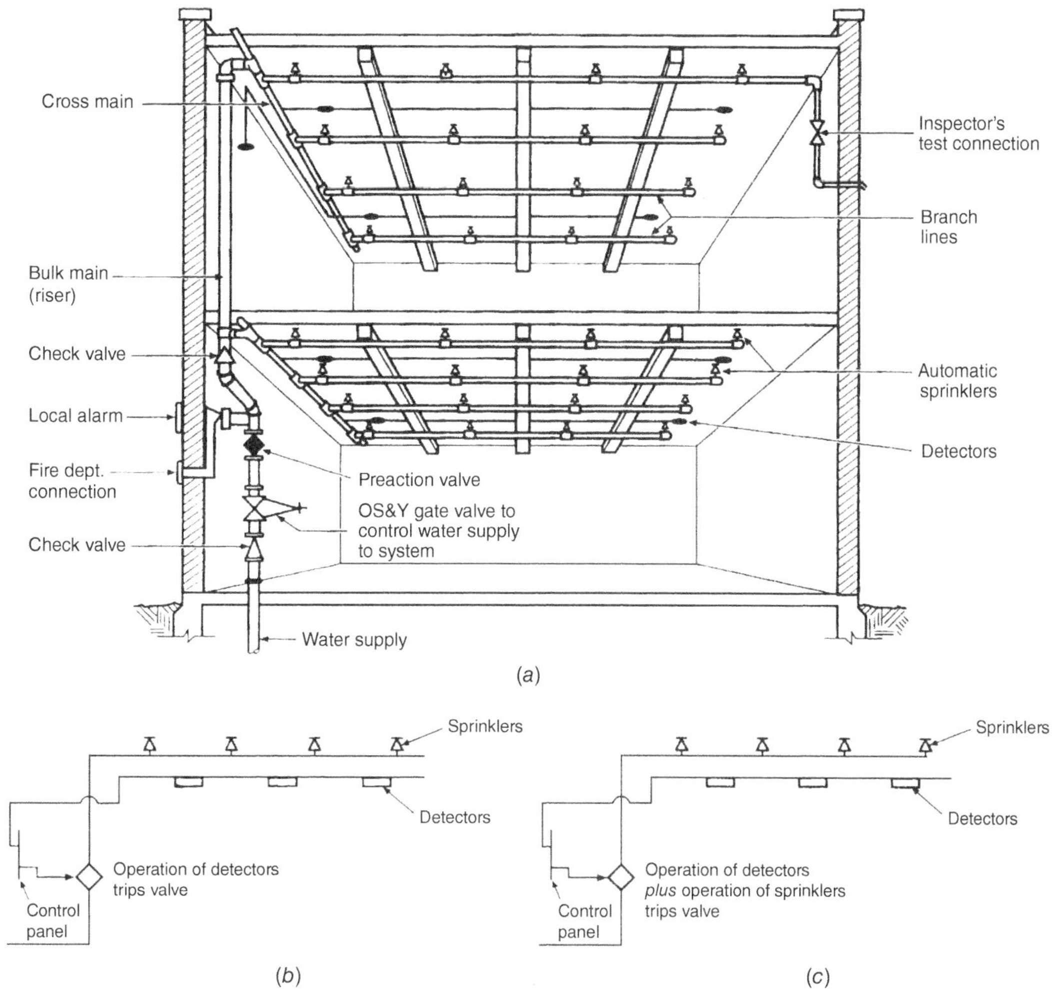

Figure 26.24 *(a) Preaction sprinkler system. Sprinkler piping contains air until the preaction valve allows water to enter the system piping. (b) In a preaction system, detectors open the valve. (c) In a double-interlock preaction system, both the detectors and then a sprinkler must operate before the valve allows water to enter the system piping. (Reproduced with permission of NFPA from the Fire Protection Handbook, 20th edition. Copyright © 2008, National Fire Protection Association. For a full copy of the Fire Protection Handbook, please go to www.nfpa.org.)*

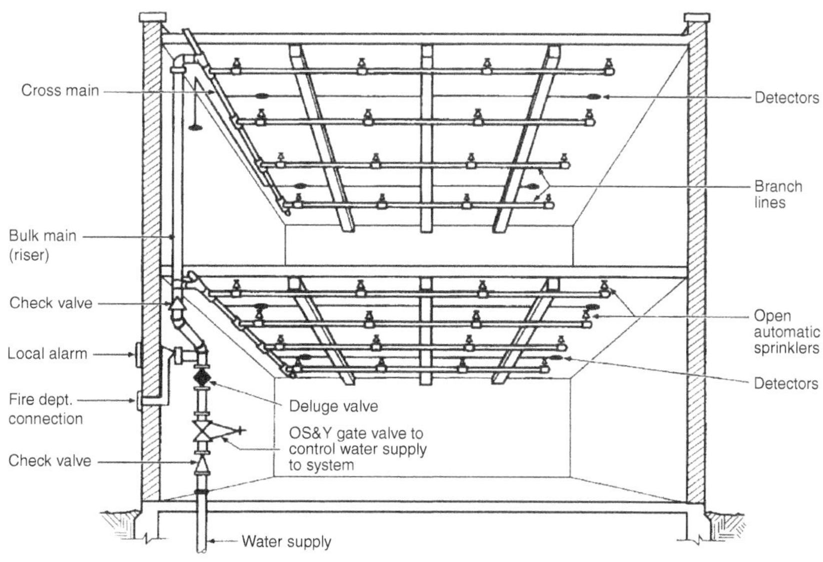

Figure 26.25 *Deluge system. All sprinklers are open; therefore, all operate when water is allowed to flow into the system piping. (Reproduced with permission of NFPA from the Fire Protection Handbook, 20th edition. Copyright © 2008, National Fire Protection Association. For a full copy of the Fire Protection Handbook, please go to www.nfpa.org.)*

associated with many clean agent gases. They also eliminate the cost of refilling a system with expensive clean agent gases and allow a faster return to service after discharge.

These systems produce a much smaller water droplet, thanks to inlet pressures ranging from 45 to 4100 psi [310 to 28,270 kPa], depending upon the design of the sprinkler head. The heads are typically spaced closer together and have more sensitive thermal elements.

The NFPA *Fire Protection Handbook* suggests three classes of mist systems:

Class I mists have a droplet size ≤ 200 μm, the finest mists. This is achieved at the expense of flow rate and spray velocity, and requires significant input of energy to produce useful quantities. These mists are most suitable where enclosure reduces the need for spray momentum and fuel wetting is not critical (examples are liquid fuel fires and spray fires in enclosed spaces).

Class II mists have a droplet size from 200 to 400 μm. With larger drops, it is easier to achieve higher mass flow rates, and considerable surface wetting is possible. These mists are likely to be effective on fires involving ordinary combustibles.

Class III mists have a droplet size from 400 to 1000 μm. They can be generated by small-orifice sprinklers and fire hose fog nozzles, and deliver the highest mass flow rates.

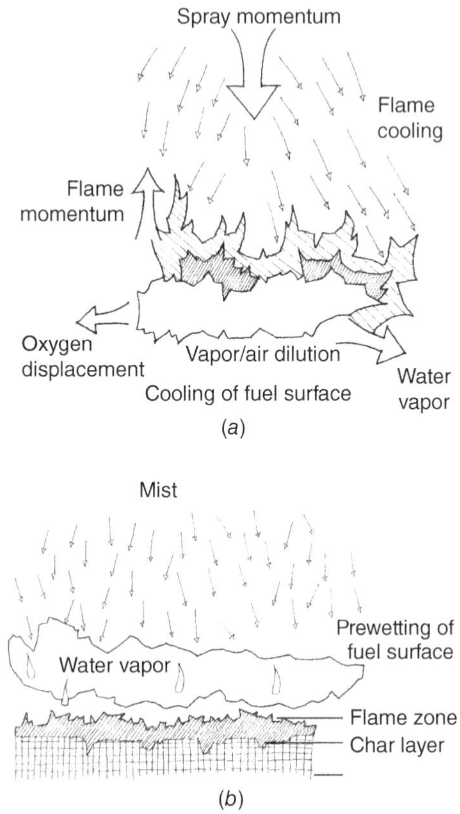

Figure 26.26 *Water mist fire-extinguishing mechanisms. (a) Interaction of mist with fire in a pool of flammable liquid. (b) Mist and flame interaction on a solid fuel with charring. (Reproduced with permission of NFPA from the Fire Protection Handbook, 20th edition. Copyright © 2008, National Fire Protection Association. For a full copy of the Fire Protection Handbook, please go to www.nfpa.org.)*

The fire-extinguishing mechanisms of mists are shown in Figure 26.26. Two primary mechanisms are heat extraction (through rapid evaporation of the finely divided water droplets) and oxygen displacement. Which mechanism dominates depends on whether the fire is poorly or well ventilated and the properties of the fuel. (Mist systems perform best in smaller enclosures with restricted ventilation, aided by heat entrapment and relatively easy oxygen displacement.) A third primary mechanism is the blocking of radiant heat. Secondary mechanisms in pool and spray fires are vapor/air dilution (mixing of water vapor and entrained air in the flammable vapor zone above the pool surface) and kinetic effects (the velocity of the flame front may be inhibited by small water droplets dispersed in a volume of flame).

Tests in 1998 by the Institute for Research in Construction, National Research Council of Canada, demonstrated that cycling, rather than continuous, discharge from a mist system extinguished most test fires within a shorter time and ended all test fires with less water. A Class II spray with a twin-fluid (water and compressed air distributed separately, mixed at delivery) nozzle was used. The cycling discharges were "long water off" (50 seconds on, 30 seconds off) and "short water off" (30 seconds on, 20 seconds off). Compared to continuous discharge, cycling discharge produced higher room air temperatures near the ceiling and more rapid oxygen depletion. Both are threats to the contents and the occupants, yet both also make the mist system more efficient at extinguishing the fire. This resulted in less water used and, in most cases, a shorter fire-extinction period.

26.4 OTHER FIRE-MITIGATING METHODS

When water damage poses almost as much of a threat to a structure or its contents as does fire, a variety of other fire-mitigating methods are available. The most passive of such measures are intumescent materials, which expand rapidly as they are touched by fire. This process creates air pockets that insulate a surface from the fire or swell a material until it blocks openings through which fire (or smoke) could have passed. Intumescent paints, caulks, and putties are available. Some intumescent materials come in 1/4-in. [6.35-mm]-thick sheets, with various facing materials.

(a) The Rise and Fall of Halon 1301

Halogenated hydrocarbons, known commonly as halons, are gases (stored as liquids) in which one or more hydrogen atoms have been replaced by halogen atoms. Although the original hydrocarbons are often highly flammable gases, the substitution of halogen not only makes the gas itself nonflammable, but also gives it a flame-extinguishing capability by interrupting the chemical chain reaction in a fire.

Until the mid-1990s, the most widely used of these compounds was *Halon 1301*; the name signifies that its molecule contains one carbon atom, three fluorine atoms, no chlorine atoms, and one bromine atom. (When a fifth number is listed for halons, it refers to the number of iodine atoms. Hydrogen atoms are not accounted for.)

The advantages of Halon 1301 (and other halons) are that it can be released with relative safety through a flooding system into areas such as computer server rooms, quickly extinguishing a fire with little harm to the contents, no oxygen displacement to threaten life support, little demonstrated harm to people, and no residue left on electronic components. It is also lightweight and space-saving relative to other fire suppressants, as seen in the following subsections. Until the late twentieth century, Halon 1301 was used on all commercial aircraft and in many special building applications, such as computer rooms, museums and libraries, telephone exchanges, and kitchens.

Halon 1301, however, was subsequently identified as a long-lived, significant threat to the Earth's protective stratospheric ozone layer. Its production was thus phased out as of 1994. The search for replacements for Halon 1301 (and 1211) has led to mists (discussed previously), foams, inerting gases, and clean agents (discussed later), as well as special-application approaches beyond the scope of this book.

Inerting gas and clean agent replacements for Halon 1301, unlike water, actually protect the building's contents rather than its structure. They leave no sticky residue, such as most dry chemicals produce, and are therefore a popular choice where a clean fire-suppressing agent is required, people are present, and there are objects or processes of high value. Occasionally, they are used where water availability is low or where space cannot be found for systems that use other, bulkier fire suppressants.

(b) Foams

Because foams (masses of gas-filled bubbles) are lighter than water and flammable liquids, they float on the surface of a burning liquid, smothering and cooling the fire and sealing in vapors. Foams can be designed to inundate a surface or to fill cavities; they can be thin and rapidly spread or thick, tough, and heat-resistant. Some foams can be spread by foam-water sprinkler systems. Many foaming agents suited for various purposes are available. The NFPA *Fire Protection Handbook* (NFPA, 2023d) presents a concise summary of foam extinguishing agents and systems.

Foams are defined by their expansion ratio (final foam volume to original foam solution volume before air is added). Low-expansion foam has an expansion ratio up to 20:1 and is used principally to extinguish burning flammable or combustible liquid spill or tank fires. Medium-expansion foam has an expansion ratio from 20:1 to 200:1; high-expansion foam has an expansion ratio from 200:1 to 1000:1. These last two types are used for indoor fires in confined spaces, to fill enclosures such as basement areas or holds of ships. They are also (at about 500:1) used to control liquefied natural gas (LNG)-spill fires and to help disperse the resulting vapor cloud.

Because foam breaks down and vaporizes its water content under attack by heat and flame, it must be applied at a sufficient volume and rate to compensate for water loss, leaving enough to produce a residual foam layer. Foam may easily be broken down by a water hose stream; turbulent air or rising combustion gases may divert foam from the burning area, especially if foam distribution occurs directly above a fire. Foam solutions are conductive, and therefore are not recommended for use on electrical fires. Firefighter entry to a foam-filled passage requires self-contained breathing apparatus and a life line (vision and hearing are reduced).

An example of high-expansion foam is shown in Figure 26.27, the former North Central Airlines Hangar Building at the Metropolitan Airport in Detroit (Albert Kahn Associated Architects & Engineers, Inc.). The foam-generating equipment in this installation was designed to fill the 38,400-ft^2 [3567-m^2] hangar with 1,400,000 ft^3 [39,644 m^3] of foam to a height of 36 ft [10.8 m] in less than 12 minutes. Automatic devices, upon sensing an abnormal temperature increase, operate the foam generators, open roof vents, start smoke control exhaust fans in the vents, and transmit a fire alarm signal to the Airport Fire Department. Foam discharge is delayed 30 seconds while evacuation sirens sound to permit occupants to leave the fire area. Manual "override" controls allow personnel to start the system in the event of failure of the automatic controls, or to stop it if the fire is small and controllable by other methods.

The high-expansion foam is created by wetting a nylon net with a mixture of water and a special detergent soap concentrate. A large blower directs an air

Figure 26.27 *Detergent foam (high-density) discharged from four units (two of which are seen in the illustration) after 3 minutes of operation. This installation at a hangar building at Detroit Metropolitan Airport smothers fire, prevents its spread, and will not harm airplanes or machinery. The system can handle combustible liquid fires. (Note also the supplementary grid of upright sprinklers directly below the roof surface.)*

current through the net, producing an avalanche of foam. Suds blanket the fire, attacking it in several ways. The water in the suds converts to steam, absorbing the heat of the fire. The expansion of the foam into steam reduces the oxygen content to about 7%, which is insufficient to support active combustion. A cooling effect is achieved by the wetting action of the breaking bubbles. The movement of air currents toward the fire to replace the rising hot gases draws the foam, supplied from the sides (rather than directly over the fire), to the center of the fire. There it blocks the airflow and cuts off the supply of oxygen. The fire, thus contained and diminished, can be approached by firefighters for further control.

Aircraft are not harmed by the foam. Delicate machinery that might be injured by high-velocity streams of water is undamaged and left clean when the foam is rinsed away. The structure—in this case, an open steel frame with a metal roof deck—is protected from high temperatures that might weaken it and cause it to collapse.

(c) Carbon Dioxide (CO_2)

Among other inerting gases, carbon dioxide (CO_2) has long been used to prevent ignition of potentially flammable mixtures and extinguish fires involving flammable liquids or gases. Although inerting gases certainly help to extinguish fire by displacing oxygen, they are even more effective by acting as a heat sink, absorbing combustion energy and reducing the temperature of the flame/vapor mixture below that necessary to sustain combustion.

Because of its oxygen displacement, CO_2 is often used in tightly confined spaces that are free of people or animals—for example, display cases, mechanical or electrical chases, and unventilated areas above suspended ceilings or below raised floors. It also finds applications in data centers, telecommunications equipment, and electrical equipment rooms—in short, wherever water damage is a major concern.

CO_2 is stored in cylinders as a liquid under great pressure, requiring roughly four times as many cylinders as would Halon 1301 for the same hazard. When it is released as a gas, it absorbs about 120 Btu/lb [280 kJ/kg] to provide cooling as well as smothering action. CO_2 is noncombustible and will not react with most substances. It does not conduct electricity and will not normally damage sensitive electronic equipment. There is no residual clean-up associated with its use as a fire-suppressing agent. It spreads upon discharge as a gas, but will stratify over time. When properly vented, the gas escapes to the atmosphere after a fire has been extinguished.

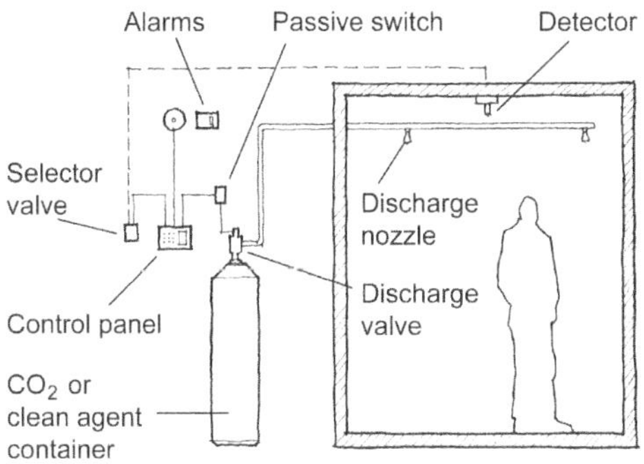

Figure 26.28 *Schematic of a CO_2 automatic fire detection/suppression system that provides total flooding of a space. Clean agent systems use similar components. (Drawing by Dain Carlson; © Walter Grondzik; all rights reserved.)*

The main problem with CO_2 is that it must be used at concentrations of 21% to 62% of that of the air, depending on the fire's fuel. At a CO_2 concentration of 9%, however, loss of consciousness will occur after a few minutes. This may allow enough time for an awake occupant to escape, but it poses problems for firefighters or trapped persons. Another potential problem is that after the initial smothering and subsequent dissipation of the CO_2, smoldering embers might reignite.

A typical CO_2 automatic detection/suppression system (Figure 26.28) has storage cylinders, a detection system, and a discharge valve that releases CO_2 into a piping system with discharge nozzles (rather similar to a sprinkler system).

(d) Clean Agent Gases

An array of replacement gases for Halon 1301 can be expected to meet demand for this type of fire suppressant. Initially, hydrochlorofluorocarbons (HCFCs) and hydrofluorocarbons (HFCs) are likely to be popular substitutes. One currently available alternative uses a potassium-based aerosol. Systems will typically resemble those used with CO_2 (Figure 26.28). Confined and vital spaces such as control rooms, computer and communication facilities, and emergency response centers are candidates for clean agent systems; consult the NFPA *Fire Protection Handbook* (NFPA, 2023d) for system descriptions.

One replacement to gain approval is FM-200® (heptafluoropropane, CF_3-CHF-CF_3, ASHRAE designation HFC-227ea). It works in a manner similar to that of halons but with zero ozone-depletion potential. With a much shorter atmospheric lifetime than halons, it is also less threatening as a greenhouse gas contributing to global warming. It will typically require about 50% more cylinders than Halon 1301 for the same hazard.

FM-200 is a clean, nontoxic, odorless, electrically-non-conductive suppression agent consisting of carbon, fluorine, and hydrogen. It is stored in pressurized steel containers, and, when released, vaporizes and interrupts the combustion process by mixing with the air. Upon actuation, it leaves no residue.

(e) Portable Fire Extinguishers

Most fires in buildings can be extinguished at an early stage with these commonly seen devices. Portable fire extinguishers are rated (or classified) based on the type of fire they are designed to fight. How many extinguishers are required and where they are located depend upon the hazard of the occupancy (as listed in Table 26.8). The following material is a short summary of *NFPA 10: Standard for Portable Fire Extinguishers* (NFPA, 2021).

Portable fire extinguishers are labeled as follows:

Class 1A to 40A: These extinguishers are used on "Class A" fires: involving ordinary combustibles such as wood, cloth, paper, rubber, and many plastics, and effectively suppressed by the heat-absorbing, cooling effects of water, the coating effects of dry chemicals, or the interruption of the combustion chain reaction provided by dry chemicals. The numerals refer to the relative extinguishing potential (a 40-capacity unit will extinguish 40 times as much as a 1). "A" refers to the contents: water, aqueous film-forming foam (AFFF), film-forming fluoroprotein foam (FFFP), and multipurpose dry chemical (ammonium-phosphate-base). (Halogenated agents are being phased out.)

Class 5B to 40B: These extinguishers are used on "Class B" fires: involving flammable or combustible liquids, flammable gases, greases, and similar materials that are best suppressed by excluding oxygen, by inhibiting the release of combustible vapors, or by interrupting the combustion chain reaction. The numerals refer to the approximate area (in square feet) of deep-layer liquid fire that an inexperienced operator can extinguish. "B" refers to the contents: smothering or flame-interrupting chemicals such as CO_2, dry chemicals, AFFF, or FFFP. (Halogenated agents are being phased out.)

Class C: "Class C" fires involve energized electrical equipment. The extinguisher contents are non-electrically-conducting, such as CO_2 or dry chemicals. (Halogenated agents are being phased out.)

Class A:B:C: Multipurpose dry chemical extinguishers filled primarily with ammonium phosphate and suitable for use with all three classes of fire. Unfortunately, ammonium phosphate has the disadvantage of leaving an especially hard residue unless promptly and thoroughly cleaned as soon as a fire is extinguished.

Class D: Class D fires involve combustible metals or metal alloys. The extinguisher contains dry powders, such as copper or graphite compounds or sodium chloride. These extinguishers are used on a variety of combustible metals; the specific combustible metal for which the extinguisher is intended is printed on the extinguisher's nameplate.

Class K: "Class K" fires involve cooking appliances that have vegetable oils and animal oils and fats—such as those found in commercial kitchens, restaurants, and cafeterias. These extinguishers contain a potassium acetate-based, low-PH agent that discharges as a fine mist to help prevent fires from grease splashes while the appliance cools.

A typical portable extinguisher is shown in Figure 26.14*b*. A common Class A extinguisher contains 2 1/2 gal [9.5 L] of water, weighs about 30 lb

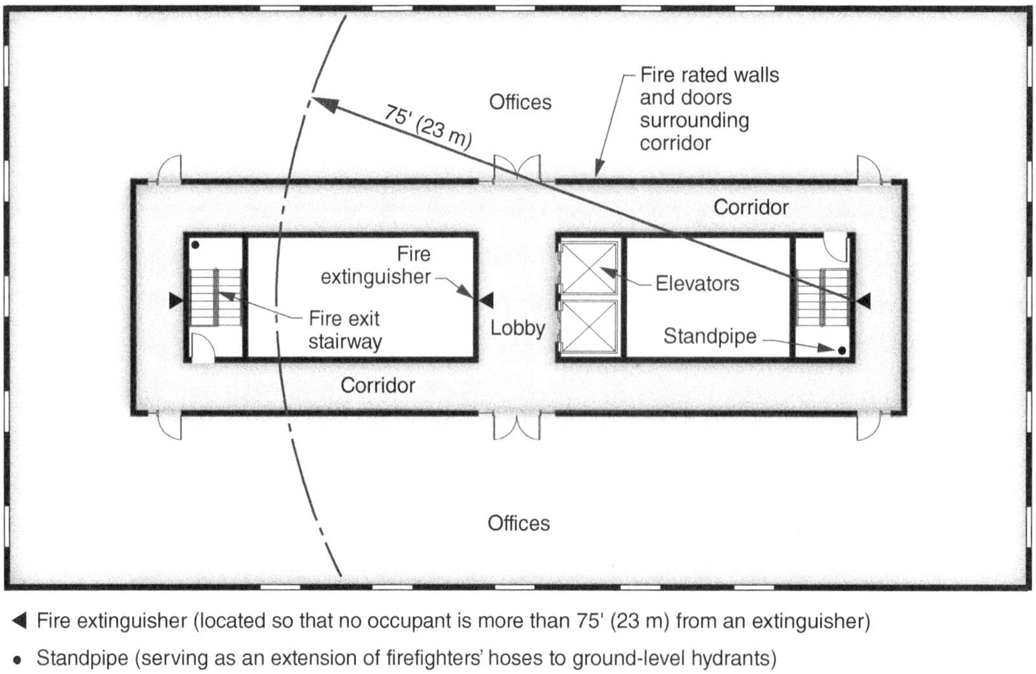

Figure 26.29 *Location of portable fire extinguishers for a typical multistory office building. Extinguishers should be easily reached and placed in conspicuous locations along normal paths of protected egress, away from potential fire hazards. Note that three locations are shown.*

TABLE 26.11 Class A Fire Extinguishers

Part A. Size and Placement for Class A Hazards			
	Occupancy		
	Light Hazard	Ordinary Hazard	Extra Hazard
Minimum rated single extinguisher	2-A[a]	2-A[a]	4-A[b]
Maximum floor area per unit of A, ft² [m²]	3000 [279]	1500 [139]	1000 [92.9]
Maximum floor area for extinguisher, ft² [m²]	11,250 [1045][c]	11,250 [1045][c]	11,250 [1045][c]
Maximum travel distance to extinguisher, ft [m]	75 [22.9]	75 [22.9]	75 [22.9]
Part B. Maximum Area to Be Protected per Extinguisher, ft² [m²]			
	Occupancy		
Class A Rating on Extinguisher	**Light Hazard**	**Ordinary Hazard**	**Extra Hazard**
1-A	—	—	—
2-A	6000 [557]	3000 [279]	—
3-A	9000 [836]	4500 [418]	—
4-A	11,250 [1045]	6000 [557]	4000 [372]
6-A	11,250 [1045]	9000 [836]	6000 [557]
10-A	11,250 [1045]	11,250 [1045]	10,000 [929]
20-A, 30-A, and 40-A	11,250 [1045]	11,250 [1045]	11,250 [1045]

Source: Reproduced with permission of NFPA from NFPA 10, Standard for Portable Fire Extinguishers, 2022 edition. Copyright © 2021, National Fire Protection Association. For a full copy of NFPA 10, please go to www.nfpa.org.

[a]Up to two water-type extinguishers, each with 1-A rating, shall be permitted to be used to fulfill the requirements of one 2-A-rated extinguisher.

[b]Two 2 ½-gal [9.5-L] water-type extinguishers can be used to fulfill the requirements of one 4-A rated extinguisher.

[c]This is the maximum area of a square that can be inscribed in a circle of 75-ft [22.9-m] radius where 75 ft [22.9 m] is the maximum distance to a fire extinguisher.

[13.6 kg], and emits a stream that can travel 30–40 ft [9–12 m]. Fire extinguishers are rarely considered aesthetically pleasing, yet they need to be located in conspicuous places along regular paths of egress (Figure 26.29). The requirements for maximum floor area served and maximum path to an extinguisher are shown in Table 26.11.

26.5 LIGHTNING PROTECTION

Lightning is nature's most destructive force: The average lightning discharge is estimated at 200 million V, 30,000 A, and courses through a grounded object in less than a thousandth of a second. "Cold" lightning bolts have ample current, voltage, and duration to shatter and kill, but not to ignite combustibles. "Hot" bolts will ignite combustibles as well.

Lightning protection can protect not just a building itself, but also serve to protect a building's electrical distribution system. The decision on whether to protect a structure against lightning depends upon an evaluation of these factors:

1. Frequency and severity of thunderstorms
2. Value and nature of building and contents
3. Hazard to building occupants
4. Building exposure; buildings in open and exposed areas are more susceptible to lightning than urban buildings, although very tall urban towers remain vulnerable
5. Indirect effects, for example, loss of a water tower will seriously affect fire prevention and other services

If a decision is reached to protect a building, it should be done completely and properly, with Underwriters Laboratories (UL) label equipment (Label A and B) and UL-approved installation (Label C). A partially protected building is in reality an improperly protected building, which might be worse than one with no protection at all.

The relevant Standards are *NFPA 780: Standard for the Installation of Lightning Protection Systems* (NFPA, 2022c), and *UL 96A: Installation Requirements for Lightning Protection Systems* (UL, 2023a). The subject of lightning protection is complex, and the design of an adequate system is best left to specialists. Design considerations and available materials are discussed briefly in the following material.

There is no known method of protection that will prevent the occurrence of a lightning stroke. The basic principle in lightning protection is, therefore, to provide a continuous metallic path to solid (low-resistance) ground for a lightning stroke. This approach will prevent the stroke from passing through the nonconductive portions of a building, accompanied by intense heat and very large mechanical forces that result from the high resistance of such a path.

Any lightning protection system includes one (usually more) air terminals (pointed copper or aluminum rods projecting above a structure); ground terminals embedded in highly conductive soil (or via a network of buried wires if the soil is nonconductive, such as bedrock or very dry soil); and down conductors that connect air and ground terminals. The conductors may be exposed on the building's exterior or concealed within a structure.

Three techniques for protecting structures from lightning are shown in Figure 26.30. All three offer

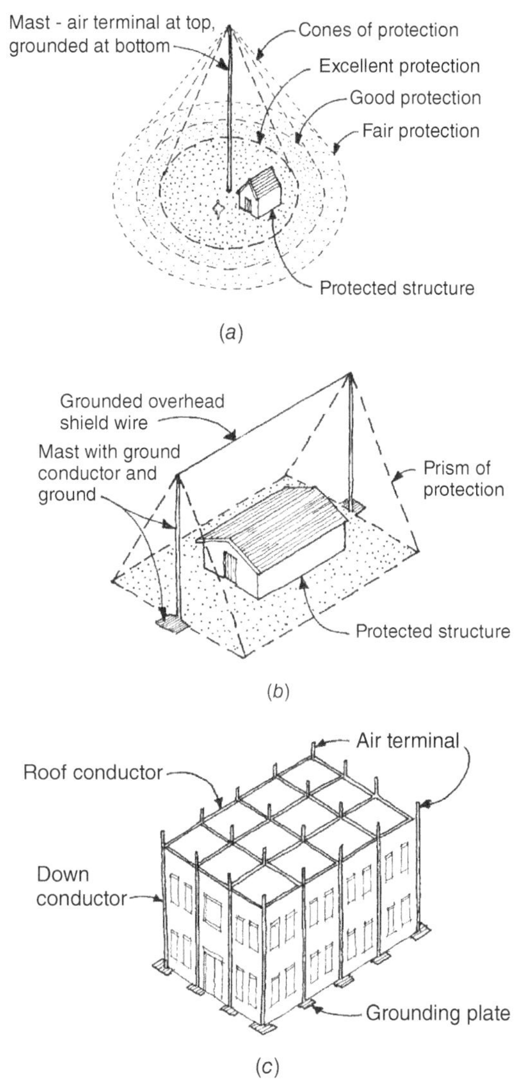

Figure 26.30 *Three approaches to lightning protection. (a) The Franklin cone is provided by a central mast that acts as an air terminal, conductor, and ground. (b) The overhead ground shield wire provides a triangular prism protection zone. (c) The Faraday cage wraps a protective mesh around a structure.*

the lightning stroke a good conductor above the level of the protected nonconductive structure. The lightning stroke is drawn to the conductor, then conducted harmlessly to the earth. (Trees can be similarly protected.)

(a) Franklin Cone

A Franklin cone (Figure 26.30*a*), named for Ben Franklin, is simply a mast with a conductor running straight to the ground. A "cone of protection" is formed that protects the objects within it from strikes by absorbing the lightning stroke at the mast and grounding it harmlessly. The closer an object to the ground and the mast, the better the protection: Buildings within an interior solid angle of 60° get excellent protection, those within an angle of 90° get good protection, and those within an angle of up to 126° get fair protection. (A more complex "geometric" version is described in the NFPA *Fire Protection Handbook*, NFPA, 2023d.)

(b) Overhead Ground Shield Wire

Forming a continuous air terminal, this shield (Figure 26.30*b*) is linear and horizontal, so the protected volume is a triangular prism, rather than a cone. This is most commonly used to protect overhead transmission lines.

It is recommended that both the single mast and the overhead ground shield project above any structure within the protected zone by at least 6 ft [1.8 m], and more height is better.

(c) Faraday Cage

Named for Michael Faraday, this approach (Figure 26.30*c*) depends upon an open interconnected mesh covering a large, nonconducting mass (a building) to draw a lightning stroke and conduct it to earth. This leaves large areas of the roof exposed, so the higher the air terminals and the denser the mesh, the better the protection. An office tower, with high air terminals at close intervals, is shown in Figure 26.31. Because this is the most common method used to protect buildings, some details follow.

All metallic objects on the roof should be bonded to a looped conductor that joins the air terminals. These air terminals commonly are placed both at roof edges and ridges at intervals of about 20 ft [6 m]. The conductors usually are made of copper and are best housed in a plastic pipe conduit. Ground conductors, often called counterpoise conductors, are also installed to connect the earth terminals.

Metal roof and siding must not be substituted for air terminals and down conductors, unless constructed of 3/16-in. [4.8-mm] minimum sheet metal that has been made electrically continuous by bonding or an approved interlocking contact.

Figure 26.31 *Lightning rods (air terminals) enliven the silhouette of the Bank of America Corporate Center in Charlotte, North Carolina. (Photo by Dale Brentrup.)*

In reinforced-concrete buildings, concrete column steel reinforcing can be used for lightning conduction if the reinforcing steel is welded rather than merely tied. Where tied, and in buildings constructed with precast concrete panels, reinforcing steel (and other metal) within a few feet of lightning conductors should be bonded to the conductors to avoid arcing. Such arcing, as noted previously, is accompanied by heat and mechanical forces that can produce major structural damage and fire. Precast concrete buildings present a challenge because the reinforcing steel within precast panels is not typically interconnected.

In steel structure buildings, the columns relatively easily become lightning conductors. Care must be taken to adequately bond both the tops and foundations of such columns to the air and ground terminals, respectively (Figure 26.32).

(d) Lightning Arresters

A lightning arrester is a special type of surge arrester generally connected at one end to an overhead electrical line and at the other end to the ground.

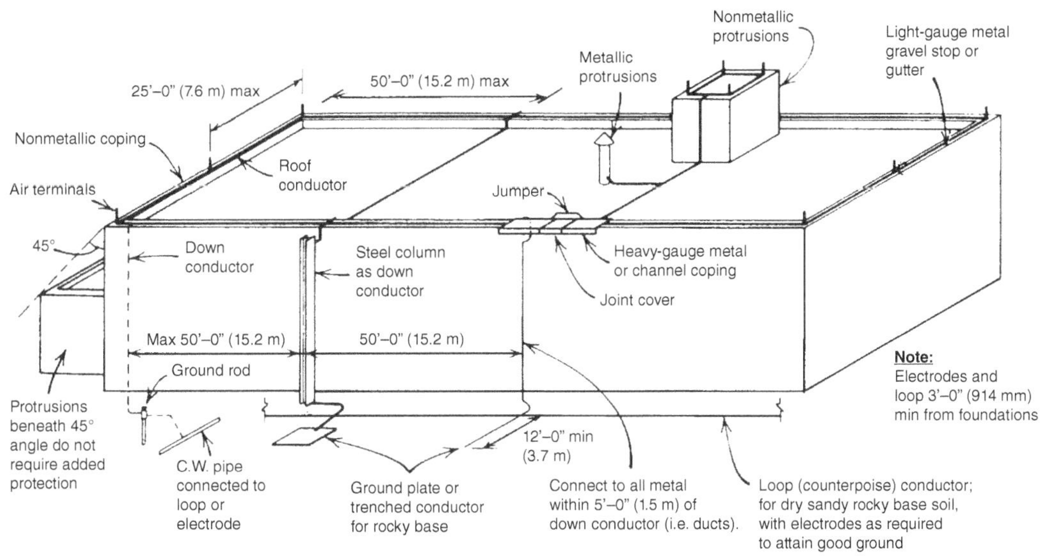

Figure 26.32 *Diagram of a typical lightning protection system.*

It thus suppresses any lightning voltage surges that appear in the electrical line so that equipment downstream on that line is not damaged. A lightning arrester operates only when it senses a voltage surge. At that point, it conducts the surge to the ground through its own low resistance and partially dissipates the energy of the surge in the body of the arrester. Once the surge has passed, the arrester returns to its high-impedance quiescent state, presenting an open-circuit connection to the normal voltage line to which it is connected.

(e) High-Rise Buildings

Lightning protection systems are particularly important for very tall buildings. Figure 26.33 shows the schematic diagram for the John Hancock Center in Chicago, a structural steel building of unusual height. Intermediate loops at the mechanical equipment floors help to overcome the problem posed by the poorer conductivity of metal piping systems, which can become relatively high-impedance paths for lightning, with resultant destructive effects. The intermediate loops are connected to each outside column, as well as to the main electrical, plumbing, air-conditioning, and fire protection risers. These loops are also connected to both the roof conductor and the ground counterpoise conductor.

FIRE ALARM SYSTEMS

26.6 GENERAL CONSIDERATIONS

A fire alarm system serves primarily to protect life and secondarily to prevent property loss. Because buildings vary in occupancy, flammability, type of construction, and value, a fire alarm system must be tailored to the needs of a specific facility. In schools, for instance, where the paramount consideration is rapid, orderly evacuation, the same general type of system will be used from project to project, although the means of automatic detection will vary with construction type, building height, specific area use, furnishings, and staffing. The fire alarm is part of the overall fire protection system of a building. In particular, it overlaps with the design of safe egress and smoke/fire control in matters such as fan control and smoke venting, smoke door closers, rolling shutters, elevator capture, and the like. These automatic functions are initiated by operation of the alarm system but are designed in accordance with the overall fire protection plan.

Like other alarm systems, a fire alarm system has three basic parts: signal initiation, signal processing, and alarm indication (see Figure 26.34). The signal initiation can be manual (pull stations or telephones) or automatic (fire and smoke detectors and/or water flow switches). The alarm signal is processed by some sort of control equipment, which in turn activates audible and visible alarms and, in some cases, alerts a central fire station or municipal authorities.

26.7 FIRE CODES, AUTHORITIES, AND STANDARDS

NFPA model codes (i.e., documents written in a manner conducive to legislative adoption) and standards are published by the National Fire Protection Association (NFPA). They bear directly on fire alarm system arrangements. There are probably more codes and

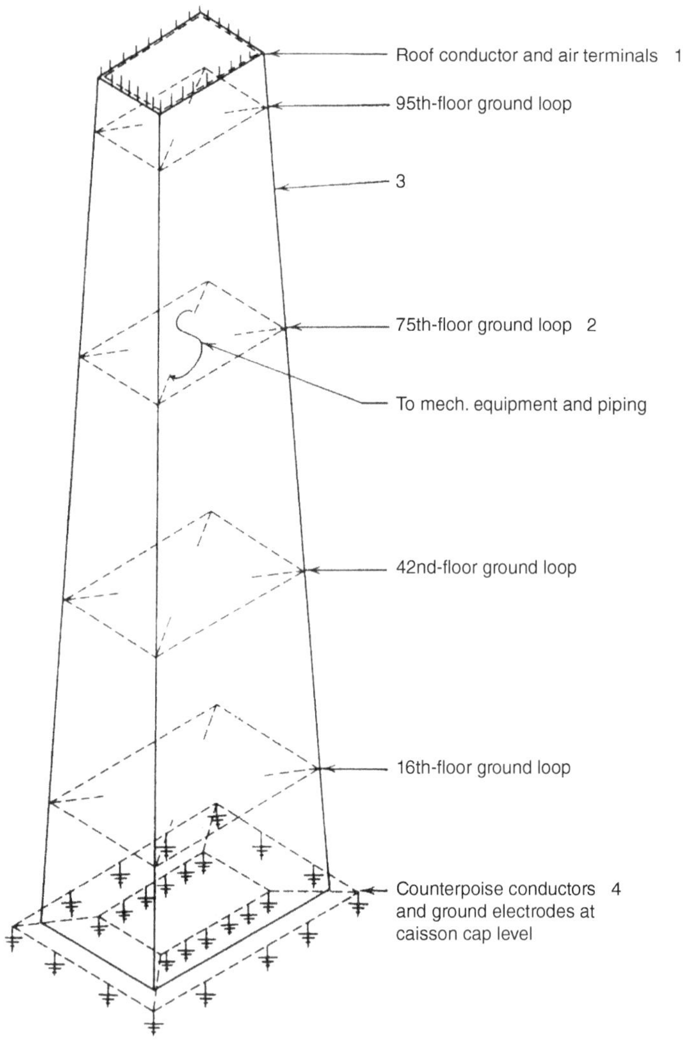

Figure 26.33 *Lightning protection for the John Hancock Center, Chicago. Air terminals with roof conductors (1) are bonded to building steel and looped to ensure that a single wiring break will not fault the system. Intermediate-floor ground loops (2) with connections to building mechanical equipment and piping protect against side-effect flashover to people or equipment. Building structural steel (3), properly bonded together, is used as a down conductor. Counterpoise conductors (4) cross-connect ground electrodes with ground conductors located below the minimum groundwater level to ensure adequate and uniform dissipation of the charge. (From Jensen, 1975.)*

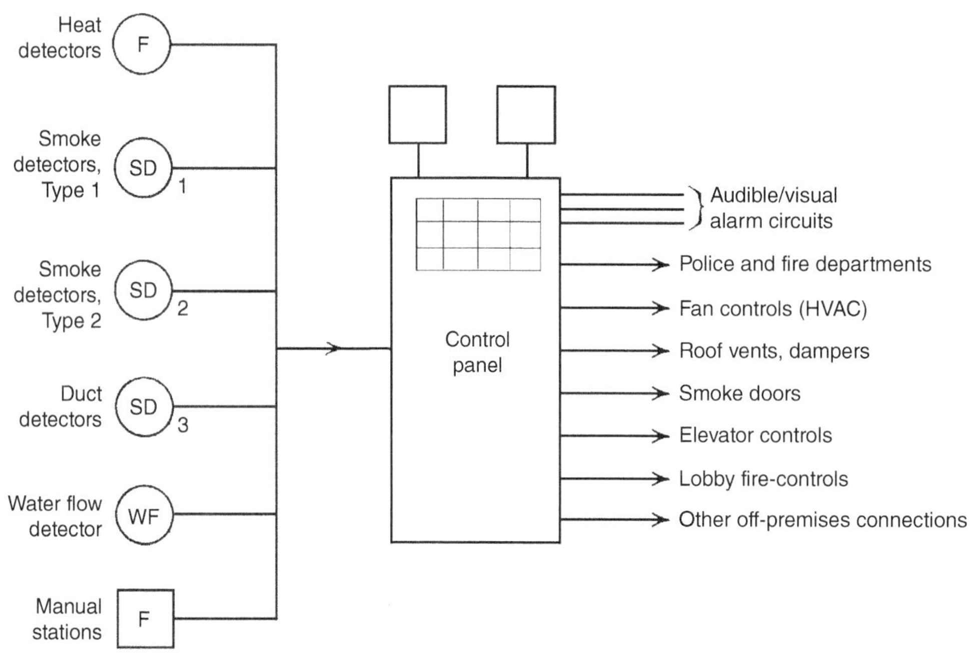

Figure 26.34 *Typical simplified arrangement of a building fire alarm system. Multiple circuits, zoning, and other accessories such as annunciators are not shown. In a protected-premises (local alarm) system, the off-premises connections to police, fire department, and other central stations do not exist.*

standards in the area of fire protection than in any other area, with the possible exception of structural design. Although the following documents are devoted primarily to fire protection and life safety, they also govern the type of fire alarm system to be installed, its components, and its installation. Key NFPA model codes and standards include:

NFPA 10: Standard for Portable Fire Extinguishers.

NFPA 13: Standard for the Installation of Sprinkler Systems.

NFPA 14: Standard for the Installation of Standpipe and Hose Systems.

NFPA 70: National Electrical Code (NEC).

NFPA 72: National Fire Alarm and Signaling Code.

NFPA 90A: Standard for the Installation of Air-Conditioning and Ventilating Systems.

NFPA 90B: Standard for the Installation of Warm Air Heating and Air-Conditioning Systems.

NFPA 92: Standard for Smoke Control Systems. This code provides a one-stop source for smoke control system design rules.

NFPA 99: Health Care Facilities Code.

NFPA 101: Life Safety Code. Specifies the means of egress, the type of fire alarm system, and the type and location of fire and smoke detection equipment for all types of occupancies.

NFPA 110: Standard for Emergency and Standby Power Systems.

NFPA 111: Standard on Stored Electrical Energy Emergency and Standby Power Systems.

NFPA 750: Standard on Water Mist Fire Protection Systems.

Additionally, organizations other than the NFPA have developed model codes and standards. Many of these are ANSI accredited. ANSI (the American National Standards Institute) does not develop standards; it accredits standards that are prepared by professional and technical societies and associations. Approval and listing by ANSI validate an organization's standards development process. The ICC (the International Code Council) publishes a series of building codes that have replaced legacy regional codes in the United States. HUD (the Department of Housing and Urban Development) promulgates regulations affecting a range of housing typologies.

Additional fire protection standards that may be relevant to building project teams thus include:

The model *International Fire Code (IFC)* contains regulations to safeguard life and property from fires and explosion hazards. The ICC states that "topics include general precautions, emergency planning and preparedness, fire department access and water supplies, automatic sprinkler systems, fire alarm systems, special hazards, and the storage and use of hazardous materials."

ICC/ANSI 117.1: Accessible and Usable Buildings and Facilities. This standard specifies requirements to make buildings accessible to and usable by people with physical disabilities.

HUD publishes requirements for residential buildings and care-type facilities. Federal Housing Administration (FHA) minimum property standards are included here.

Numerous performance standards are published by Underwriters Laboratories, Inc. (UL). UL laboratories test equipment for inherent safety and for performance according to the manufacturer's specification. A product is "listed" if the equipment passes UL safety and performance tests. Some UL standards related to fire protection include the following.

UL Detection Standards:

UL 38: Standard for Manual Signaling Boxes for Fire Alarm Systems.

UL 217: Standard for Smoke Alarms.

UL 268: Smoke Detectors for Fire Alarm Systems.

UL 268A: Standard for Smoke Detectors for Duct Application.

UL 346: Standard for Waterflow Indicators for Fire Protective Signaling Systems.

UL 521: Standard for Heat Detectors for Fire Protective Signaling Systems.

UL 539: Single and Multiple Station Heat Alarms.

UL 985: Standard for Household Fire Warning System Units.

UL 1730: Standard for Smoke Detector Monitors and Accessories for Individual Living Units of Multifamily Residences and Hotel/Motel Rooms.

UL Alarm Standards:

UL 228: Standard for Door Closers-Holders, With or Without Integral Smoke Detectors.

UL 464: Audible Signaling Devices for Fire Alarm and Signaling Systems, Including Accessories.

UL 827: Central-Station Alarm Services.

UL 1480: Speakers for Fire Alarm and Signaling Systems, Including Accessories.

UL 1638: Visible Signaling Devices for Fire Alarm and Signaling Systems, Including Accessories.

UL 1971: Standard for Signaling Devices for the Deaf and Hard of Hearing.

Property insurance groups (such as Factory Mutual) issue standards that may apply to a specific facility type—in addition to local codes, standards, and fire marshal regulations. As with other aspects of construction, the architect/designer must ascertain which regulations have jurisdiction before proceeding with the design. In addition, all codes and standards undergo a cyclical updating process with new versions appearing on a regular basis. The specific recommendations given in the following sections are representative of good practice. Actual design must be based upon and must meet the requirements of the current NFPA standards plus all codes having jurisdiction in the particular locale of a project.

26.8 FIRE ALARM DEFINITIONS AND TERMS

The following short list can serve as a primer for terminology that a designer may encounter in fire alarm work.

Addressable analog (smoke) detector Also called an intelligent detector. This addressable (see next definition) smoke detector continuously measures the concentration of smoke or other products of combustion in its test chamber. It continuously transmits an analog (voltage or current) signal to its (intelligent) control panel that is proportional to this measurement. It does not transmit an alarm signal (see Figure 26.46).

Addressable detector A smoke or fire detector that communicates an alarm condition along with an individual identification (address), thereby immediately establishing the alarm's location. It responds to polling (testing) when it senses its unique identification (address) signal, "replying" with a signal indicating OK, trouble, or alarm.

Air-sampling detector A system consisting of aspiration equipment and piping that draws air from a protected area to the detector, where analysis of the sampled air occurs.

Alarm signal An audible and/or visual signal indicating a fire emergency, generally requiring immediate building evacuation.

Alarm verification A technique for reducing false alarms. Common procedures include a requirement for alarming for a specific minimum time and a repeat alarm after initial reset.

Automatic system A system in which an alarm-initiating device operates automatically to transmit or sound an alarm signal.

Auxiliary fire alarm system See Section 26.9(c).

Break-glass A false-alarm deterrent installed in manual fire alarm stations; a glass rod is placed across the pull-lever that breaks easily when the lever is pulled.

Coded alarm signal An alarm signal that indicates the location of the fire alarm station that is operated by the pattern (code) of the audible signal.

Coded system One in which no fewer than three rounds of coded alarm signals are sounded, after which the system may either be silenced or set to sound a continuous alarm.

Control unit (fire alarm panel) The controls, microprocessors, relays, switches, and associated circuits necessary (1) to furnish power to a fire alarm system, (2) to receive and process signals from alarm-initiating devices and transmit them to indicating devices and accessory equipment, and (3) to electrically supervise the system's circuitry.

Drift compensation Adjustment function that compensates for detector sensitivity changes due to aging and environmental conditions. The adjustment may be automatic at the detector or manual/automatic at the control panel. Also referred to as automatic gain control in beam-type smoke detectors.

Dual-coded system See Section 26.12(d).

Intelligent control panel A control center that receives analog signals from detectors and, based on predetermined conditions (usually programmed into software or a microprocessor), determines whether an alarm situation exists. Sensitivity control, polling, and remote diagnostic testing of connected detectors are other functions usually found in these panels.

Listed or listing Inclusion of a device in a list published by a recognized testing organization, such as UL, indicating that the device meets the requirements of the referenced standard.

Local fire alarm systems See Section 26.9.

Manual system One in which the alarm-initiating device is operated by direct human action to transmit or sound an alarm signal.

Master-coded system See Section 26.12(b).

Multiplex system An arrangement whereby a single pair of wires can carry more than one message (signal) at a time in either or both directions.

Noncoded system See Section 26.12(a).

Presignal system See Section 26.12(f).

Private mode An alarm system installed in a location that has personnel specifically trained to act on receipt of an alarm signal.

Proprietary fire alarm system See Section 26.9(e).

Public mode An alarm system that alerts all of the protected spaces and occupants audibly and/or visually.

Rate-of-rise detector A temperature detector that alarms when the rate of temperature rise exceeds a specific design level, usually 15°F/minute [8.3°C/min].

Remote station fire alarm system See Section 26.9(d).

Selective-coded system See Section 26.12(e).

Station, fire alarm A manually operated alarm-initiating device. It may be equipped to generate a continuous signal (noncoded station) or a series of coded pulses (coded station).

Supervised system A system in which a break or ground in the wiring that prevents the transmission of an alarm signal actuates a trouble signal.

Trouble signal A signal indicating trouble of any nature, such as a circuit break or ground, occurring in a device or circuitry associated with a fire alarm system.

Zone-coded system See Section 26.12(c).

26.9 TYPES OF FIRE ALARM SYSTEMS

Fire alarm systems can be classified according to several different characteristics, including location, application, connections, coding, and the degree of automation of the detection system. The last factor is the one most often found in manufacturers' literature. To avoid confusion, the classification system found in the *NFPA 72: National Fire Alarm and Signaling Code* (NFPA, 2024b) is used herein. This code classifies fire alarm systems essentially by location and function:

- household fire warning systems
- protected-premises systems (local alarm)
- off-premises systems (connections between local alarms and off-premises equipment and systems). This category includes auxiliary, remote station, proprietary, central station, and municipal fire alarm systems.

(a) Household Fire Warning Systems

The term is self-explanatory. These systems are discussed at some length in Section 26.26.

(b) Protected-Premises Fire Alarm System

This arrangement (as per Figure 26.34), as the name indicates, is intended to sound an alarm only in the protected premises. Action in response to an alarm must be taken locally, either manually or automatically. Thus, notification to the fire department is manual, although fire-suppression systems can be set into operation automatically. This arrangement is applicable to privately owned facilities. When a building is unoccupied, notification to the fire department can come only incidentally, perhaps from a passerby. Local systems and their components are discussed in detail in Section 26.11.

(c) Auxiliary Fire Alarm System

This is simply a local system equipped with a direct connection to a municipal fire alarm box (Figure 26.35). The received alarm signal is identical to that resulting from a manual alarm at that city box. Since the fire department is aware of all city box connections, arriving firefighters would always check the protected premises. This type of system is usually applied to public buildings such as schools, government offices, museums, and the like.

(d) Remote Station Protective Signaling System

This system (see Figure 26.36) is similar to an auxiliary system, except that the alarm or trouble signal is transmitted via a leased telephone line to a remote location (a police station or a telephone-answering service) that is manned 24 hours a day. Notice of the alarm is then telephoned to the fire department. This arrangement is used in private buildings, such as stores and offices, which are unoccupied for considerable periods and for which reliance on a chance passerby to turn in a fire signal is unacceptable. An audible alarm circuit extended from a local system to a nearby building—as, for instance, from a store to a nearby residence—does not constitute a remote station system unless all the requirements of *NFPA 72* are met.

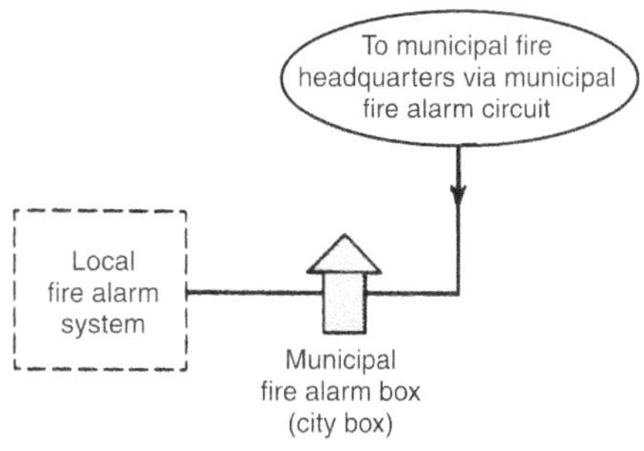

Figure 26.35 *An auxiliary fire alarm system is hard-wired to the municipal fire department, generally via a city box. This type of connection is usually restricted to public buildings.*

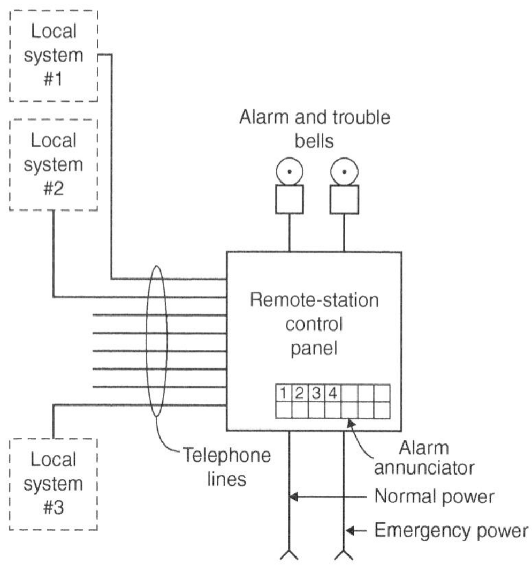

Figure 26.36 A remote station system transmits alarm, supervisory, and trouble signals to a remote location that is continuously attended. The required actions will be taken by those attending the remote stations.

(e) Proprietary Fire Alarm System

This system (see Figure 26.37), which is applicable to large multibuilding facilities such as universities, manufacturing plants, and the like, utilizes a dedicated central supervisory station to receive signals from all buildings. In a proprietary system, this station is on the site and is manned by persons associated with the facility. A common arrangement is for the station to be located in a guardhouse or similar supervisory location, from which point alarms are sent manually to a fire department and/or on-site fire brigades. Other actions that must be taken on receipt of alarm or trouble signals are performed by facility personnel at the central supervisory location.

Information on the exact locale or zone within each building at which an alarm occurs is transmitted to the central supervisory location. The central station has an audible alarm, some sort of visual display that indicates the alarm location, and a printer that makes a permanent record of each alarm. As mentioned, these central supervisory locations are frequently multipurpose, covering all aspects of facility security, plus, frequently, control functions such as energy management.

(f) Central Station Fire Alarm System

This arrangement is similar to that described for proprietary systems except that the system supervision and normally all of its equipment are owned and operated by a service company. In lieu of multiple buildings under one "ownership," a central station system supervises many individual unrelated local systems for a fee. Here too, operators of the central supervisory installation receive all signals from individual users and provide the required services, including alarm verification, fire reporting, and repairs when trouble signals are received. As with proprietary systems, central station consoles often supervise access control, intrusion alarms, and related systems.

26.10 CIRCUIT SUPERVISION

This common term refers to the circuit arrangements in fire alarm systems that indicate a malfunction in the wiring of alarm (and other) devices by sounding a trouble bell. The trouble signal is separate and distinct from an alarm signal. The extent of circuit supervision required varies with the type of facility and the specific code adopted by a jurisdiction. Minimally,

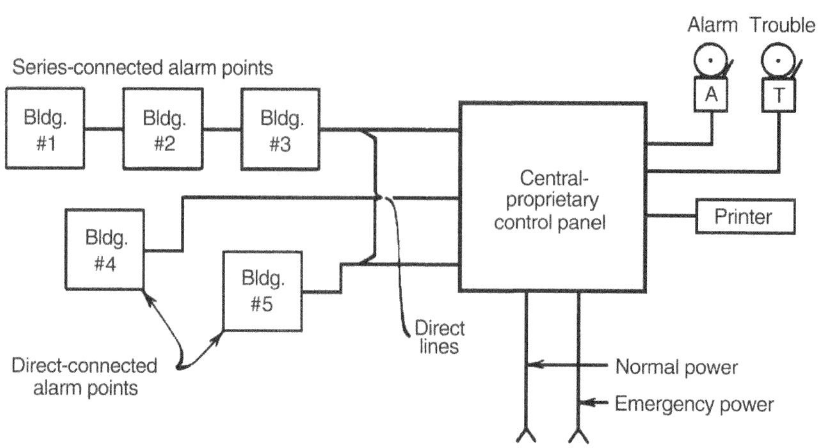

Figure 26.37 Block diagram of a proprietary fire alarm system. The entire system, including all local devices plus the central supervisory equipment, is the property of a single owner. The central supervisory installation is attended by facility personnel who are trained to respond appropriately to alarm, trouble, and supervisory signals.

an open circuit (break) or a ground in any part of the detector and manual station wiring will cause a trouble signal. In large, sophisticated systems, all of the wiring and circuitry is supervised, including the control panel(s), alarm device circuits, and even annunciator wiring.

In conventional fire alarm systems (see Section 26.11), the supervision does not indicate where a fault lies, but only that a fault exists. It is up to the system operator to troubleshoot the system in order to locate the fault. Because this may be a protracted procedure, all codes require that the system wiring be such that a single break or a single nonsimultaneous ground will not prevent an alarm signal from being sounded beyond the break (Class B circuits) or on the entire circuit (Class A circuits).

26.11 CONVENTIONAL SYSTEMS

Manufacturers classify fire alarm systems by the type of information transmitted to the control panel by the detectors and the alarm devices. There are two basic types:

Conventional (or traditional) systems: Devices such as smoke and heat detectors are connected to a control panel that monitors and manages the components of alarm system.

Addressable (intelligent) systems: Advanced technology is leveraged for improved reliability and greater user control by assigning every fire detection device with a unique address on an electronic network, allowing the control panel to identify and communicate with each device individually.

The simplest (and oldest) system is the conventional (or traditional) system (Figure 26.38). A conventional system uses detectors and manual stations that transmit an alarm signal only. When they are in standby or quiescent state, they do not transmit. Such detectors and manual stations are called conventional units; hence, the same name is used for the entire system. The signals transmitted by all detectors, and by manual stations as well, are identical and therefore indistinguishable from each other.

In theory, the number of alarm-initiating devices (such as fire and smoke detectors) that can be wired to a single detector circuit is unlimited. (In practice, there are electrical limitations.) As a result, when an alarm signal is received from such a multiple-device circuit, there is no way of knowing which of the devices alarmed and whether the alarm represents an actual fire condition or a false alarm caused by dirt, moisture, a puff of smoke, a short circuit in the wiring, or a malfunction of a detector.

Smoke detectors of all types (ionization, photoelectric) are subject to false alarms activated by particulate matter in the air. Detectors are, by nature, threshold devices; raising sensitivity increases false alarms, and decreasing sensitivity shortens the crucial early-warning period. False alarms are not merely nuisances; in facilities such as hospitals, and in places of public assembly such as theaters, public office buildings, and mass eating facilities, a false fire alarm can result in serious business disruption, property loss, personal injury, and even death if panic ensues. As a result, conventional detectors in such buildings require continual maintenance and field sensitivity checks to minimize false alarms, which is an expensive and time-consuming procedure.

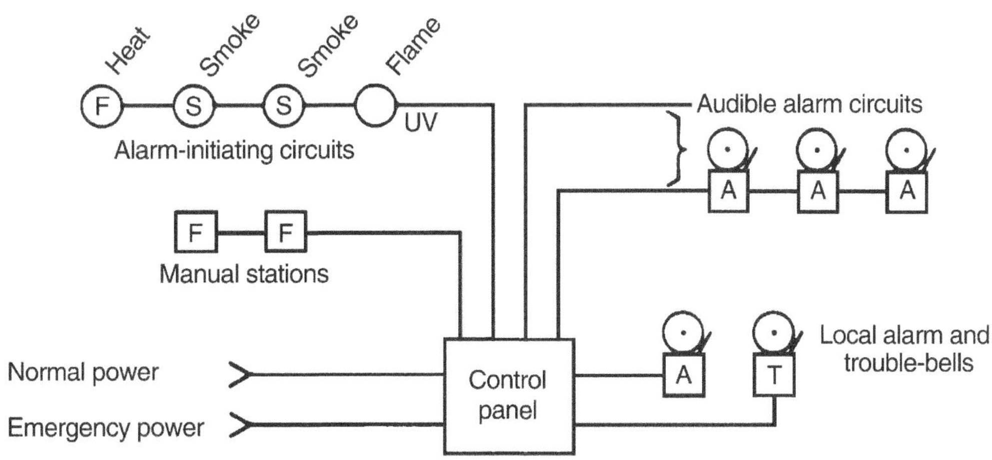

Figure 26.38 *Simplified schematic diagram of a conventional fire alarm system. Alarm-initiating devices are connected on detector circuits, as are manual stations and audible/visual alarm devices. The control panel receives identical signals from all detectors, manual stations, and alarm-initiating circuits.*

Because such facilities are so difficult and even dangerous to evacuate, most have fire and evacuation plans that call for some type of alarm verification before a general evacuation alarm is sounded. Because field experience has demonstrated that alarm verification can reduce false alarms appreciably without seriously degrading the early warning performance of combustion product detectors, recent editions of some fire codes now permit, or even require, alarm verification. This verification can be accomplished in conventional (hard-wired, zoned, nonaddressable) systems in a number of ways, including:

- requiring activation of at least two cross-zoned detectors in a single area
- requiring that a detector repeat its alarm after being reset (applicable only to systems with remote reset capabilities)
- requiring that a smoke detector continue to alarm for a minimum period of time, thus eliminating smoke puffs as the cause of an alarm
- physically inspecting the site protected by the detector to visually determine whether the alarm is actual or false

This last technique for alarm verification is probably the most foolproof, but it is also the most difficult because it requires knowledge of the exact location of the alarming detector. This is possible only if detectors are grouped into zones and each zone is annunciated individually (Figure 26.39). To assist in localizing areas in a building by annunciation, zones are wired to contain as few devices as possible, the ultimate arrangement being a separate zone for each detector. As can be seen in Figure 26.39, as the number of zones increases, so does the necessary wiring and, as a result, so do the installation costs. Thus, a balance must be struck between two requirements: the need for more zones to localize and thereby simplify alarm identification and verification, and limitation of the increased cost of zoning. The balance point depends upon the type of facility being protected. All modern conventional systems are solid-state, and all but the smallest use a wiring system called multiplexing.

Multiplexing uses a time-sharing electronic technique to transmit and receive multiple signals on a single two-wire communications circuit. This has the great advantage of reducing the primary cost item in large hard-wired systems—individually wired circuits. The alarm and audible signal devices in standard multiplexed systems are the same as those in conventional hard-wired systems. Only the system wiring and the panel arrangement change. Zones, and even coding, are substantially unaltered, although the reduction

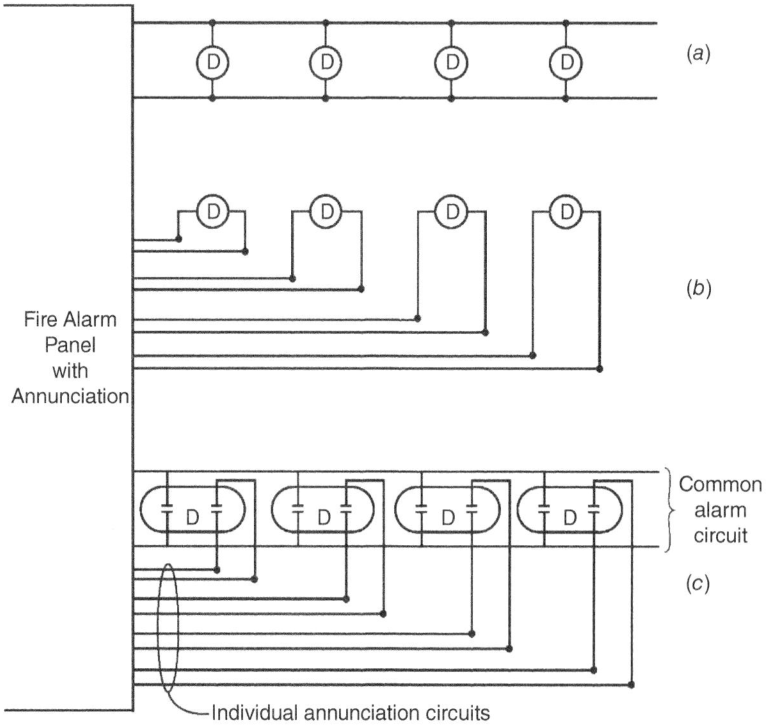

Figure 26.39 *Identification of detectors, D, can be accomplished in three ways. (a) Each circuit, with multiple detectors, is annunciated as a zone. This arrangement is applicable where detectors are close together. (b) Each detector is wired individually, comprising a separate zone. This is applicable for large or scattered detector spacing. (c) Multiple detectors are wired on a single alarm circuit but individually annunciated. This is applicable where the number of detector circuits in the panel is limited, but individual detector identification is required.*

in wiring costs permits the use of smaller zones. Some multiplexed systems are hybrid, using multiplexed detector circuits and hard-wired audible/visual alarm circuits. An additional advantage of multiplex panels is that they can be reprogrammed readily, which is emphatically not the case with hard-wired relay panels.

26.12 SYSTEM CODING

(a) Noncoded Systems

Noncoded systems are continuous ringing evacuation arrangements using manual and automatic alarm initiation. If desired, the devices can be zoned and, if the system is sufficiently large, annunciation can be provided. Audible devices are continuous ringing (vibrating) bells and horns. Visual devices are lights and strobes.

(b) Master-Coded Systems

This system, also called common-coded and fixed-coded, generates four rounds of code that are sounded and flashed on all of the building alarm devices when any signal-initiating device operates. It utilizes a single code transmitter at the panel. Normally, the system stops after four rounds of code, although it can readily be arranged to sound and flash continuously thereafter. When the code is set to ring bells at 108 strokes per minute, it is known as march time because of the rhythmic cadence. This beat aids in the rapid, panic-free evacuation of a building and therefore is frequently used in schools.

(c) Zone-Coded Systems

Identification of the alarmed zone in a system can be accomplished with zone lights, with an annunciator, or by coding. In the first two cases, it is necessary to go to the panel or annunciator to determine the location of the operated device, which entails a possibly critical delay. All coded systems avoid this drawback by sounding the code on all the gongs in a building, thus immediately identifying the station and permitting the building staff to quickly investigate the cause of the alarm and take appropriate measures.

If a coded system is desired, but by zone rather than by device (this is less expensive by far), noncoded manual stations are used, along with automatic detectors, grouped by circuit into zones or individually zoned (Figure 26.39). Each zone circuit trips a zone's transmitters in the panel, which in turn ring the zone's code on single-stroke gongs or chimes. As with all coded systems, four rounds of coded signal are sounded, after which the system is silenced. In all coded systems, it is advisable to include a device that records all alarms in plain language, including the time of receipt and the code sounded.

(d) Dual-Coded Systems

This arrangement represents a combination of noncoded and zone-coded systems. When an alarm device operates, it initiates two separate functions—an identifying coded alarm and a continuous ringing evacuation alarm. The alarms are sounded simultaneously—the coded alarm in the building's maintenance office, and the evacuation alarm on separate audible and visual devices throughout the building. A requisite to the application of this system is a continuously staffed office in which the coded identifying signal can be received and acted upon.

(e) Selective-Coded Systems

These are fully coded systems in which all manual devices are individually coded and all automatic devices are arranged to trip code transmitters at the panel. Each manual station can be immediately identified by its distinctive code. Automatic devices may be grouped in any fashion desired and annunciated if desired. The combinations and circuitry are entirely in the hands of the designer. In large conventional systems, which fully selective-coded systems usually are, sprinkler transmitters and smoke detectors operate as integral subsystems of the main fire alarm panel.

(f) Presignaling

When it is desired to alert only key personnel, a system called presignaling is used. Small bells or chimes are activated only at their work locations. Because these systems are always selectively coded, the personnel alerted can immediately investigate and, if necessary, manually turn in a general alarm by key operation of a station. Because of the delay involved, this type of system is used only in buildings where evacuation is difficult and sufficient staff is available to immediately investigate the cause of an alarm.

26.13 SIGNAL PROCESSING

Once a hazard has been detected, a signal is transmitted to a fire control center and appropriate action taken. In conventional systems, the alarm signal is transmitted over dedicated conductors ("hard wiring") to a control panel, consisting of either electromechanical relays (in older systems) or solid-state switching circuitry (in modern systems). The control panel, in turn, actuates audible and visible device circuits, illuminates annunciator panels, controls fans and door releases, and so on, all via dedicated wiring. This arrangement has the advantages of reliability and simplicity. As a system grows, however, the wiring becomes heavy, complex, and expensive; panels become large and bulky; and changes become difficult. Also, troubleshooting of system faults becomes time-consuming. Furthermore, minimizing false alarms is problematic because alarm verification in systems with large zones containing many alarm devices is very difficult within the extremely short time span permitted for this operation.

As a result of these problems, a system was developed in which every device can be individually identified and remotely checked, without separate wiring to each. It is called an addressable system because all the devices are individually addressable.

26.14 ADDRESSABLE SYSTEMS

An addressable fire alarm system is, by definition, one that uses "addressable" fire detection devices—both automatic and manual. These detectors are essentially identical to conventional detectors except that they are equipped with electronic circuitry, usually mounted in a special base, that effectively makes each detector a separate zone. Thus, in the event of an alarm, the control can require alarm confirmation from an adjacent detector or can require a repeat alarm from the same detector after a remote reset.

Addressable devices are continuously polled from the control panel, and each replies to the poll by reporting its condition as OK–standby, alarm, or trouble. Because each device is identified by its address, an alarm or trouble signal from a detector can be quickly and easily confirmed. Detector identification is accomplished either visually on an annunciator panel or by alphanumeric readout on an LCD readout panel. Wiring is simplified; up to 100 detectors can be wired on a single multiplex line. Hardware costs of addressable systems are higher than those of a conventional system because of the additional electronics in the detector bases and the control panel. Maintenance costs, however, are somewhat lower because the panel's polling operation checks some (but not all) of a detector's electronics. Wiring costs for large systems are considerably lower than those of a conventional system. The process of measuring and calibrating detector sensitivity is essentially the same as in a conventional system; the detector must be physically demounted and checked. This can be a major expense in a large system because periodic sensitivity checks are mandated by NFPA codes.

26.15 ADDRESSABLE ANALOG (INTELLIGENT) SYSTEMS

Addressable analog systems are known in the fire alarm industry as intelligent despite the fact that the word "intelligent" is being used to refer to an inanimate device or system. To further confuse the issue, there is no accord among manufacturers on the exact meaning or use of the word when so applied. As a result, "very intelligent" and "extremely intelligent" devices have already appeared and other superlatives are sure to follow. The more accurate technical term "addressable analog" is used herein.

As mentioned previously, conventional and addressable detectors are identical except for the electronic identification coding of the latter. Both types test the adjacent space for temperature characteristics and/or products of combustion, compare the result to a preset threshold (adjustable), and latch into alarm position when the threshold is exceeded. The decision on whether to alarm is made at the detector, and the control panel acts only to receive the alarm signal and translate it into an audiovisual building-wide alarm. Similar expensive periodic testing and adjustment of detectors in the field are necessary for both systems.

An analog detector gathers information from its sensor in analog form (voltage, current, or another variable characteristic proportional to the condition it senses) and transmits it to its control panel. That is, an analog detector acts only as a sensor; it does not make the threshold comparison leading to an alarm/no-alarm decision. That decision is made by a microprocessor at the control center. In modern systems, the data received from each detector are combined with other data—including reports from adjacent detectors, "historical" data on the condition of the detectors' time patterns at each detector that might falsely indicate an alarm, plus other relevant information. Analysis of all of this information leads to the decision on whether to sound a general alarm. This arrangement has the following advantages:

- False alarms are sharply reduced because of the alarm verification procedures at the panel.
- Threshold sensitivity is set at the panel, permitting adjustment without accessing the detector.
- A history of each detector's output is recorded, and sensitivity degradation can be compensated for automatically. This decreases maintenance costs by increasing the time period between cleanings.

- Any malfunction of a detector is noted immediately at the control panel, since all detectors are polled every few seconds.

Some manufacturers place an integral microprocessor in each detector that is capable of performing almost all of the functions described previously as panel functions. These detectors act as normal addressable analog detectors, transmitting analog signals to a control panel. In the event of a connection failure, however, the on-board microprocessor converts the unit to a stand-alone, self-contained detector/controller.

Figure 26.41 *Addressable analog (intelligent) ionization-type smoke detector with an integral microprocessor that operates the unit's detection, diagnostic, and error-checking procedures. The detector is self-adjusting to compensate for sensitivity changes due to environmental conditions. Operating sensitivity is remotely adjustable from the control panel. (Courtesy of Cerberus Pyrotronics.)*

26.16 AUTOMATIC FIRE DETECTION: INCIPIENT STAGE

Fire authorities agree that most fires pass through four stages, the last of which is the visible flaming fire. These stages are shown in Figure 26.40 as a function of duration and degree of hazard. Also shown is the type of automatic detector recommended for detection at each stage.

In the incipient stage, the combustion products comprise a significant quantity of microscopic particles (0.01–1.0 μm), which are best detected by ionization-type detectors (Figure 26.41). These detectors contain a small amount of radioactive material that serves to ionize the air between two charged surfaces, causing a current to flow. Combustion particles entering the detector chamber reduce air ion mobility, thus reducing current flow and increasing voltage. These changes are sensed, and the alarm is set off.

The response time of this type of detector depends on how rapidly the combustion particles can reach the detector—a factor that varies with room air currents and with the type of material "burning." Once the particles reach the detector, the response is essentially instantaneous. Therefore, ionization detectors are best

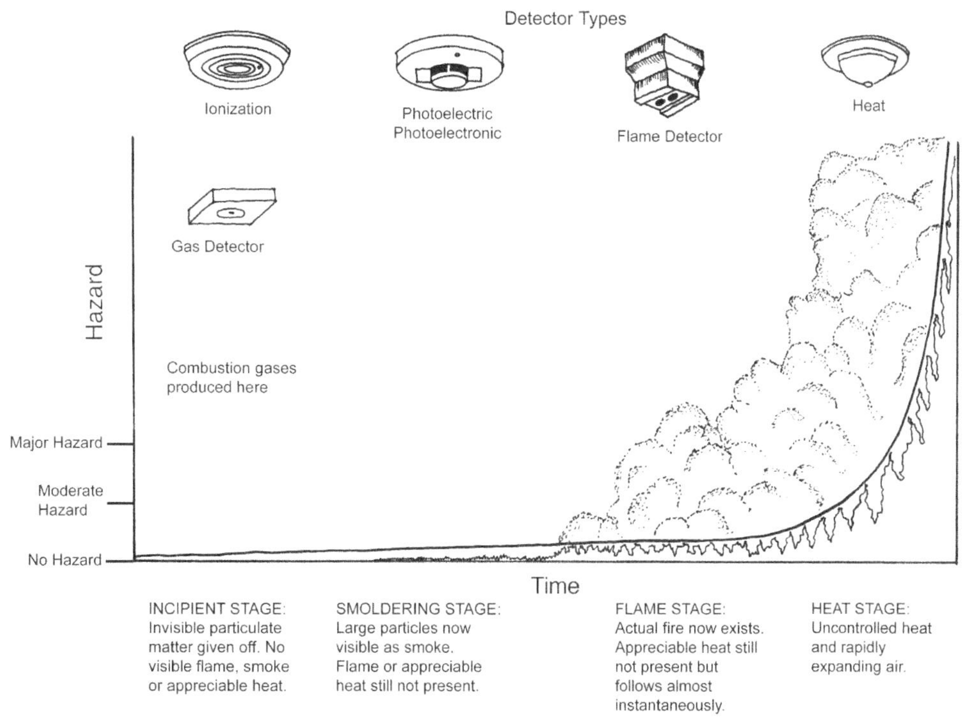

Figure 26.40 *Four stages of a fire. Early detection of each stage requires a different type of detector that responds to a developing fire's evolving characteristics (fire signature). (Drawn by Amanda Clegg.)*

applied indoors, in spaces with stagnant air or low air velocity (below 50 fpm [0.25 m/s]) and in which little visible smoke (large particles) is expected. (Some manufacturers make a special unit whose sensitivity increases with air velocity for installation in spaces with air velocities of up to 500 fpm [2.5 m/s].)

Ionization detectors should not be installed on warm or hot ceilings, or in any other location where hot air concentrates, because the hot air layer prevents the combustion particles from reaching the detectors. As a corollary, ionization detector sensitivity is higher at low ambient temperature.

Because ionization detectors react to minute particles of combustion, they should not be applied in, or close to, areas where normal activities produce such particles, as the result will be nuisance alarms. Thus, kitchens, bakeries, welding and brazing shops, workshops using open flames and burners, and areas with concentrated engine exhaust fumes are all inappropriate for the application of this type of device. Ionization detectors respond best to fast-burning flaming fires. Because dust settling in the ionization chambers reduces sensitivity, the units need periodic cleaning and recalibration.

As can be seen in Figure 26.40, the incipient stage of a fire also produces changes in the gas content of the environment. These combustion gases, which are not normally present in the air, are detectable by devices known as gas-sensing fire detectors. The principle of operation of these detectors depends upon the type. A semiconductor type reacts to a change in conductivity due to the presence of combustion gases, whereas a catalytic element type detects a change in the electrical resistance of the element caused by the presence of gases. Gas detectors are frequently used in conjunction with another type of detector so that both gases and particulate matter are detectable.

Coverage of incipient stage detectors varies from 150 to 900 ft^2 [14 to 83 m^2] per unit, depending upon the unit used, the type of combustible material in the space, and ambient conditions. Once a fire has passed the incipient stage and becomes smoky, an ionization detector loses sensitivity and photoelectric detectors are recommended. The same is true if, even in the incipient stage, the combustion products are the large particles typical of visible smoke. Ionization, gas, and photoelectric detectors are classified as early-warning types.

26.17 AUTOMATIC FIRE DETECTION: SMOLDERING STAGE

(a) Photoelectric Smoke Detection

The smoldering stage of a fire (see Figure 26.40) is characterized by particles up to 10 μm in size. Such particles, although small, are visible to the naked eye as smoke and are best detected by photometric means. The simplest type of photoelectric smoke detector operates on the principle of beam obscuration, as shown schematically in Figure 26.42. A beam of light is directed onto a photosensor, and a steady-state, no-smoke circuit condition is established. The presence of smoke in sufficient concentration partially obscures the beam, changing current flow in the photocell circuit and setting off an alarm response. Sensitivity is typically set at 0.5% to 4% obscuration per foot [0.3 m] for gray smoke and 0.5% to 10% for black smoke, to comply with *UL 268: Smoke Detectors for Fire Alarm Systems* (UL, 2023b).

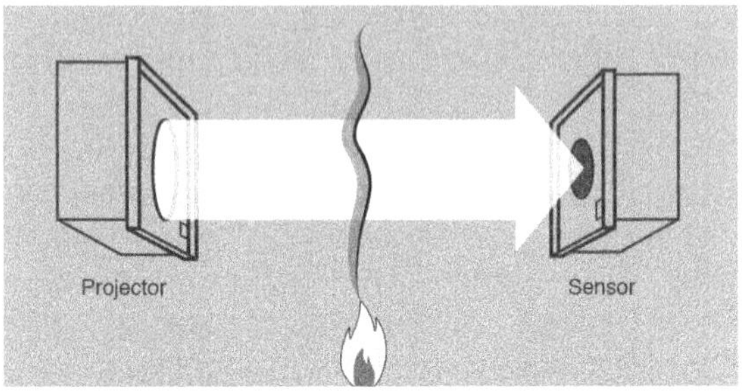

Figure 26.42 *Principle of smoke detection by beam obscuration. This design is used in two ways: as illustrated, with a projector in one location and a photocell at a remote location, or with the entire assembly in a single small housing, with the light source and receiver on opposite sides of a 2- to 3-in. [50- to 75-mm]-wide smoke chamber.*

In this design, accumulation of dust and dirt and the presence of heavy fumes from industrial processes cause gradual obscuration of both the photocell and the lamp. This, combined with lamp aging and consequent light reduction, results in increased sensitivity, causing false alarms that are not only a nuisance, but also can constitute a hazard. To correct this condition, continuous maintenance and periodic recalibration are necessary. This design is no longer recommended for spot-type smoke detection but is used in beam detectors, as discussed next.

(b) Projected Beam Photoelectric Smoke Detector

Beam-type photoelectric detectors (Figure 26.43) consist of two separate units: a beam transmitter and a beam receiver, normally wall-mounted on opposite sides of a space, somewhat below the ceiling. They operate on the simple obscuration principle. They are best applied in areas that do not lend themselves to the application of spot-type detectors. Among these are the following:

- High-ceiling areas such as atria, churches, malls, auditoriums, and the like. In these spaces,

Figure 26.43 *This projected beam smoke detector utilizes an infrared (IR) beam so that ambient light will not affect its operation. Two sensitivity levels are available that are selected to match the distance between transmitter and receiver. Because heat attenuates or deflects the IR beam, the receiver translates this beam strength reduction as obscuration, identical to smoke obscuration. This effect makes the IR projected beam detector sensitive to high-heat flaming fires as well as smoky smoldering ones. Transmitter (right) and receiver (left) are shown. (Photo courtesy of Notifier, a division of Pittway Corp.)*

ceiling-mounted spot-type detectors present serious maintenance problems. This is particularly true in spaces such as theaters and auditoriums.

- Spaces with medium- to high-velocity airflow at the ceiling level. This condition severely dilutes smoke entering the test chamber of a spot detector. Although beam detectors are also negatively affected, their long throw and wide "vision" angle make the dispersion and dilution problems much less serious.
- Closed areas with little airflow, resulting in a hot air layer at the ceiling that prevents smoke from reaching ceiling-mounted spot detectors (Figure 26.44). Exposed ceiling beams also tend to trap hot air and reduce the effectiveness of ceiling spot detectors.
- Environments that militate against spot detector use, such as those that are extremely dirty, corrosive, humid, very hot, or very cold. Beam-type detectors can be physically shielded against these conditions in a manner that interferes only minimally with their light-beam transmission and reception characteristics.

Beam-type detectors have an effective throw (from transmitter to receiver) ranging from 30 to 330 ft [9 to 100 m], depending upon ambient conditions, and can be spaced 30–60 ft [9–18 m] apart. This gives a coverage range of 900 to almost 20,000 ft^2 [84–1860 m^2], compared to spot detectors, whose maximum coverage is about 900 ft^2 [84 m^2]. Furthermore, beam detectors are wall-mounted, which reduces maintenance problems.

Finally, beam detectors are usually drift-compensated (thus further reducing maintenance expense) and are arranged not to alarm with full obscuration, a condition that frequently occurs during routine building maintenance and repair work.

Disadvantages of beam-type detectors, compared to spot units, include the following:

- Response is to the second (smoke) stage of a fire, and therefore not in the same early-warning category as ionization or gas detectors.
- High cost.
- Inapplicable in spaces that lack an unobstructed view, as where pendent lighting fixtures, exposed ductwork, and the like are installed.

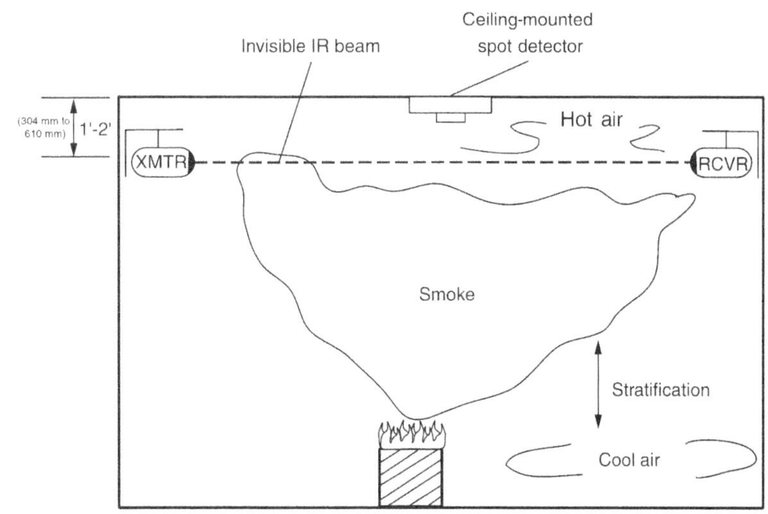

Figure 26.44 *Stratification of air in a closed space leads to a concentration of hot air at the ceiling that can prevent smoke from reaching a ceiling-mounted spot smoke detector. A projected beam detector mounted 1–2 ft [0.3–0.6 m] below the ceiling will alarm due to obscuration, as shown. XMTR is the transmitter; RCVR is the receiver.*

As with all types of detectors, the use of beam-type detectors in a particular application must be in accord with the requirements of all fire and administrative codes having jurisdiction in the particular locale.

(c) Scattered-Light Photoelectric Smoke Detector

A second design, usually referred to as a photoelectronic, scattered-light, or Tyndall-effect detector, is illustrated in Figures 26.45, 26.46, and 26.47. In this design, a pulsed light-emitting diode (LED) beam is directed at a supervisory photocell, which serves to provide a baseline reference signal. The alarm photocell is shielded and normally receives no light from this beam. When smoke enters the unit, light is reflected from (scattered by) smoke particles and strikes the alarm cell. This changes the cell's resistance, and the resultant signal is amplified electronically, causing an alarm. An alternative design depends upon refraction of light to set off the alarm. These designs are not sensitive to normal dust and dirt accumulation or to light source depreciation, and high sensitivity can be maintained without continual maintenance. Thus, they are usually found in commercial and high-quality residential applications. Sensitivity is usually set at 1% obscuration per foot [0.3 m].

(d) Laser Beam Photoelectric Detector

A very-high-sensitivity laser diode source, scattered-light-type photoelectric smoke detector has been developed that is classified by its manufacturer as an early warning device (Figure 26.48). The high sensitivity results from the ability of the very sharply focused laser beam and its associated control panel software to differentiate between a smoke particle and a dust particle. It is best applied in clean environments.

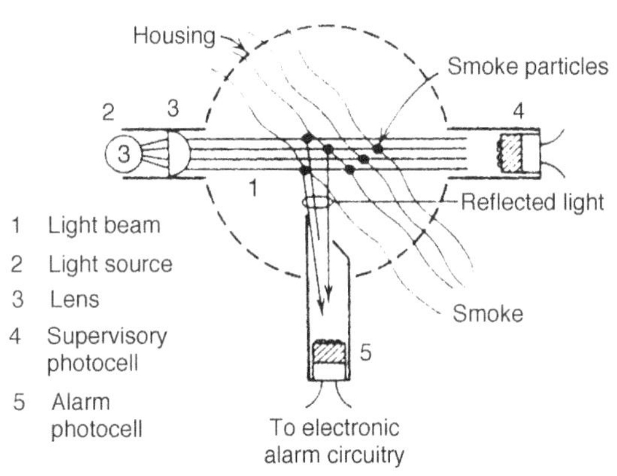

Figure 26.45 *Principle of operation of a scattered-light photoelectronic spot smoke detector. A pulsed beam of light (1) from a source (2) is focused by a lens (3) on a supervisory cell (4). Smoke entering the unit reflects the light onto the alarm cell (5), changing its electrical characteristics and setting off an alarm. Photoelectric units with sensitivities of up to 2% per foot [0.3 m] are usually classified as early-warning units.*

Figure 26.46 *Addressable analog (intelligent) scattered-light-type photoelectric smoke detector with an on-board microprocessor that negates false-alarm signals caused by radio frequency and electromagnetic interference (RFI/EMI). It also validates trouble signals. If required, this unit can be equipped with a resettable heat detection element. The unit is approximately 5½ in. [140 mm] deep and 3 in. [75 mm] high. (Courtesy of Cerberus Pyrotronics.)*

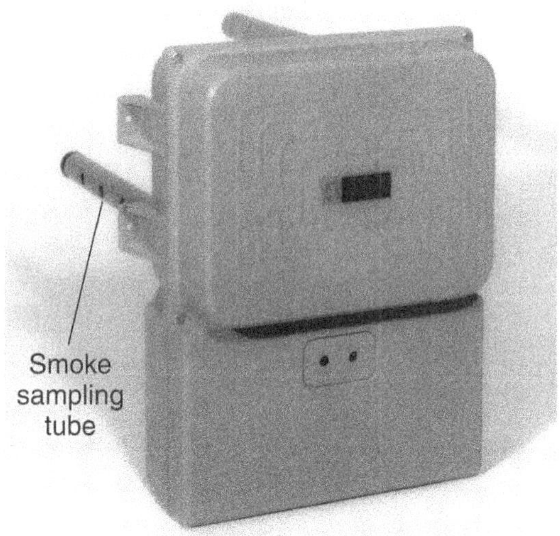

Figure 26.47 *This smoke detector, intended for duct mounting, can utilize either a photoelectronic or ionization-type head. The unit, which is usually mounted in return-air ducts rather than supply ducts to minimize smoke dilution effects, uses an air-sampling tube that extends across the full width of the duct. Duct smoke detectors are intended primarily to prevent smoke dispersal throughout a building by causing shutdown of the air-handling system and control of smoke damper operation. (Courtesy of Simplex.)*

(e) Air-Sampling Detection System

Another system that takes advantage of the high sensitivity of laser-beam-based photoelectric smoke

Figure 26.48 This unit is an addressable analog laser-beam type of scattered-light photoelectric smoke detector. The sharply focused laser beam permits discrimination between smoke and dust particles, resulting in a high-sensitivity unit without false alarming. The unit is shown mounted on a sounder base containing a horn assembly, making it applicable to small, enclosed rooms. (Courtesy of Notifier, a division of Pittway Corp.)

detection is the air-sampling system (Figure 26.49). In principle, the system is quite simple. Instead of waiting for air that may be carrying incipient smoke particles to reach the detector(s) by thermal currents, this system samples air throughout the protected space by aspiration and brings it to the detector for testing. The aspiration and air conduction system is simply a piping system with holes at sampling points, and is powered by a fan. The advantage of the system is that air throughout the space is sampled, thus, in effect, converting each sampling opening into a highly sensitive laser-beam detector.

Because the piping can be zoned into manageable areas that can readily be checked for alarm verification, the disadvantage of nonaddressability of the aspirating openings is avoided. Addition of a microprocessor to a multizone control unit, which adjusts the alarm sensitivity of each zone based on a history of environmental smoke levels, all but eliminates false/nuisance alarms.

(f) Application of Photoelectric Detectors

Photoelectric smoke detectors of all designs will detect particles from about 0.2 to about 1000 μm. Thus, they are useful not only for smoldering fires but also for the smoky fires that characterize the burning of certain plastics and chemicals. Also, particle agglomeration, which increases with the distance smoke travels, does not reduce their sensitivity, as it does with ionization types.

Maximum recommended spacing for photoelectric detectors, as for other types, is given by UL and Factory Mutual standards. Closer spacing is often mandated by the particular application or by the structure's characteristics. Manufacturers' recommendations should be obtained for all installations. In order to provide early-warning detection of a wider range of combustion products than is possible with either the photoelectronic or ionization-type detectors individually, several manufacturers produce a unit that combines a multichamber ionization detector with a photoelectronic detector.

26.18 AUTOMATIC FIRE DETECTION: FLAME STAGE

As noted in Figure 26.40, the appearance of flame is followed almost instantaneously by heat buildup and the rapid spread of flame, with a large increase in hazard. Detection of flame is not an "early warning," and the prime requirement for a detector at this stage is speed. The actions taken as a result of a flame detection alarm, such as fire suppression and/or

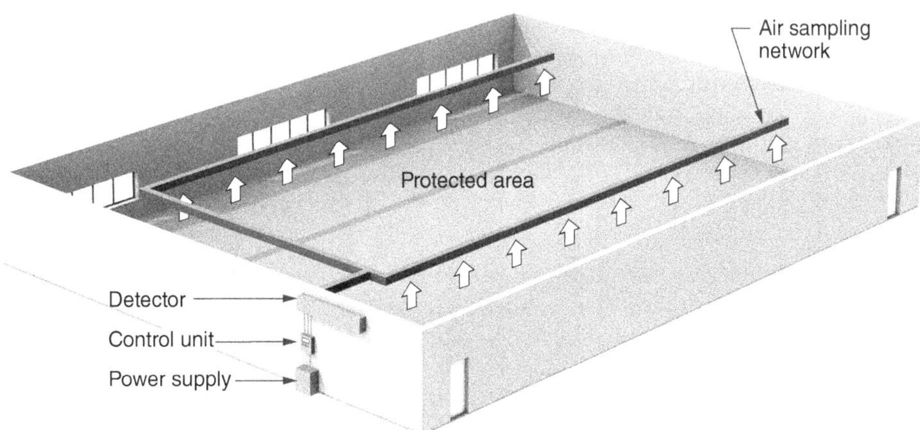

Figure 26.49 In an air-sampling system, air is aspirated at selected points in the protected area and piped to a highly sensitive laser beam photoelectric smoke detector for testing. Large areas are zone-piped so that trouble and alarm signals can be readily checked in the protected areas.

evacuation, must also be very rapid. This is not the case with early-warning equipment alarms, which normally allow several minutes or more to investigate the authenticity of the alarm, if desired, before taking appropriate action.

Flame detectors are of two types: those that detect ultraviolet radiation and those that detect infrared radiation. Both types of radiation are present at the beginning of the flame stage.

Ultraviolet (UV) radiation detectors operate by optically detecting the UV radiation produced by flames, which is typically in the 170- to 290-nm range. (The entire UV range is 40–390 nm.) Hydrocarbon (organic material) fires in particular produce strong radiation in this range. A quartz filter in front of the UV sensor tube limits the detector's reception range to between 190 and 260 nm, which is below the visible spectrum. This desensitizes the detector to sunlight and to incandescent and fluorescent sources. The detector need not "see" a flame directly; it can also detect UV radiation reflected from walls, ceilings, and other objects. The more direct the path, the stronger the radiation and the more rapidly the detector will respond. Because many devices—in particular, electric welders—produce UV radiation in sufficient quantity to activate such a sensor, flame detectors are usually programmed to respond only to flickering sources at the usual flame flicker rate of 5–30 times per second.

In sum, UV detectors are long-range and very sensitive; they react in milliseconds and respond to most types of fires. They are best applied in highly flammable or explosive storage and work areas, either indoor or outdoor. Modern units are equipped with microprocessors that perform self-test diagnostics. UV detectors are best applied to detect hydrocarbon-fueled fires such as burning alcohol, methane, propane, acetone, and all types of petroleum products. Their principal disadvantage is that they are "blinded" by thick black smoke.

Infrared (IR) radiation detectors are sensitive to radiation in the IR region, between 650 and 6500 nm. Most commercial units are filtered to be sensitive to radiation in the 3800- to 4300-nm range and to have maximum sensitivity at 4300 nm (4.3 μm), which corresponds to the radiation emitted by hot CO_2. They differ from ambient heat detectors in that they respond to radiant energy and are essentially optical detectors. IR units have about half the distance range of UV detectors, are sensitive (although not as sensitive as UV detectors), react in seconds, and must be programmed for flicker response to avoid false alarms, which can occur if they are exposed to other sources of IR radiation, such as sunlight. Because of interference problems that cause false alarms, IR units are normally applied to enclosed spaces such as sealed storage vaults and the like. IR flame detectors are best applied to fires that result in rapid flaming combustion and the production of CO_2. Typical materials in this risk category are petroleum products, wood and paper products, coal, and plastics.

Figure 26.50 *The combined ultraviolet-infrared (UV/IR) flame detector has peak sensitivity at 200 and 4300 nm, respectively, which corresponds to the maximum emission points of a typical hydrocarbon (e.g., petroleum products) fire. This unit is equipped with a microprocessor that performs diagnostic procedures and ensures proper calibration. The detector, which measures approximately 5 × 5⅔ × 6 in. [127 × 144 × 152 mm] deep is suitable for use in specific-class hazardous areas. (Courtesy of Meggitt Avionics.)*

Combination UV/IR detectors, such as the one illustrated in Figure 26.50, cover a wide range of risk applications, including aircraft hangars, fueling stations of all sorts, tank farms, and flammable storage areas. The illustrated unit contains a microprocessor that performs self-diagnostics, not only of the optical system but also of the electronic circuitry and calibration. The presence in a single unit of both UV and IR detectors permits the integral microprocessor to analyze the UV/IR ratio in any radiation detected and its time signature. This analysis is compared to known fire data in the microprocessor's memory in order to discriminate between actual fires and spurious radiation, thereby reducing false alarms.

26.19 AUTOMATIC FIRE DETECTION: HEAT STAGE

Again referring to Figure 26.40, the heat stage is the last and most hazardous stage because, by this time, a fire is burning openly and producing great heat, incandescent air, and smoke. Spread of the fire depends upon fuel, air currents, and the construction of the space in which the fire is burning. Detectors intended for use at this stage respond to heat and are referred to as heat-actuated, thermal, thermostatic, or simply temperature detectors. They act much like the fusible link in a sprinkler head. Effective application is restricted to locations where the subsequent alarm permits adequate countermeasures to be taken in time to prevent injury and minimize loss. Keep in mind that the heat stage follows the smoke stage and that smoke, not heat, is responsible for most casualties in fires.

Heat detectors have two designs: spot units, which are mounted in the center of the area that they protect, and linear units, which sense heat along their entire length. Both types respond to high ambient temperatures caused by hot air convection from a fire. The linear type can also sense the overheating of an object or surface with which it is in contact without the presence of fire.

Figure 26.51 Detectors in the illustrated design are made with fixed-temperature, nonresetting elements that melt out at 130°F [54°C] or 200°F [93°C] and rate-of-rise elements that will alarm when the temperature rise is faster than 15°F [8°C] / minute. The illustrated unit is a 200°F [93°C] fixed-temperature-only model. It is applicable to spaces with unusually violent temperature fluctuations and ceiling temperatures between 100° and 150°F [37° and 65°C]. (Courtesy of Cerberus Pyrotronics.)

(a) Spot-Type Heat Detectors

Spot heat detectors are of two types: fixed-temperature units and rate-of-rise units. In the former, a set of contacts operates when a preset (nonadjustable) temperature is reached—usually 135°F or 185°F [57°C or 85°C]. The rate-of-rise type operates when the rate of ambient temperature change exceeds a predetermined amount (usually 15°F/minute [8°C/min]), which is indicative of the heat stage of a fire. The rate-of-rise unit is normally combined with a fixed-temperature unit in a single housing.

The fixed-temperature unit is available in either a one-time nonrenewable design, which utilizes a low-melting-point alloy plug, or an automatically resetting unit that operates in the same fashion as a thermostat. For most applications, a resettable unit is preferred. Spot units are best applied in spaces that are separated from occupied areas and are subject to rapid-temperature-rise fires. Three different designs of heat detectors are shown in Figures 26.51, 26.52, and 26.53.

(b) Linear Heat Detectors

In these devices, the entire length of a cable-like element is heat-sensitive. This type of detector is applied to long, narrow elements, which are impractical to protect with spot detectors. Typical applications include cable trays, cable bundles of all sorts, conveyors, and large, long equipment. The heat-sensitive element is installed close to, or in direct continuous contact with, the protected item. As a result, any hot spot along the protected equipment will be detected.

The detectors themselves are available in several designs. One type uses a pair of steel wires under tension, held apart by thermoplastic insulation (Figure 26.54). When exposed to its rated alarm temperature, the insulation melts and the wires come into contact, changing the current flow in the circuit to which they are connected and setting off an alarm. This type is available in a range of alarm temperatures

(a)

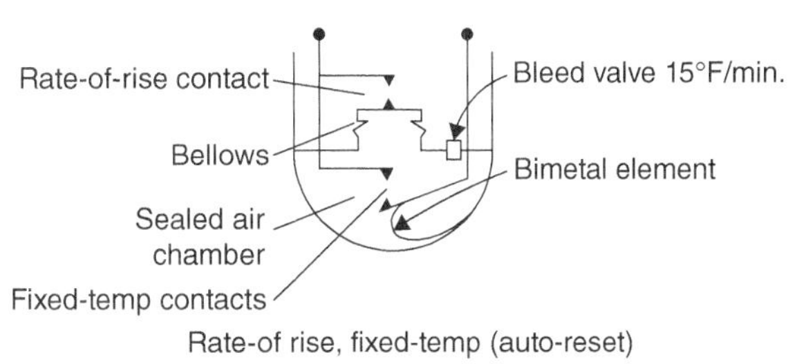

(b)

Figure 26.52 (a) A combination rate-of-rise and fixed-temperature detector. The detector and mounting plate fit on a standard 4-in. [100-mm]-square outlet box. (b) The principles of device operation are shown schematically. The rate-of-rise unit consists of an air chamber with a restricted bleed valve. Rapid temperature rise causes the bellows to expand before air is lost by bleeding, thereby setting off the alarm. The thermostatic element is a fixed-temperature, self-restoring bimetallic unit.

(a)

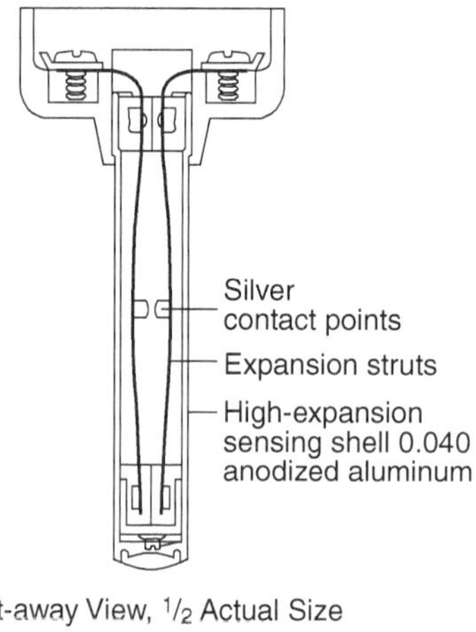

(b)

Figure 26.53 *(a) A rate-of-temperature rise and fixed-temperature-type (135° or 200°F [57° or 93°C]) detector. Unlike the rate-of-rise detector in Figure 26.52, this unit uses the principle of differential thermal expansion for both actions. (Courtesy of Cerberus Pyrotronics.) (b) Cutaway of a detector with a different base than that shown in (a). The detector consists of an aluminum tubular shell containing two curved expansion struts under compression fitted with a pair of normally open opposed contact points that are insulated from the shell. The tubular shell and the struts have different coefficients of expansion. When subjected to a rapid heat rise, the tubular shell expands and lengthens slightly. The interior struts lengthen but at a slower rate than the shell. The rapid lengthening of the shell pulls the struts together, closing the contact points and initiating the alarm. When subjected to a slow heat rise, the tubular shell and the interior struts lengthen at approximately the same rate. At the detector's set temperature points of 135°F or 200°F [57° or 93°C], the interior struts are fully extended, closing the contact points and initiating an alarm. (Drawing courtesy of Notifier, a division of Pittway Corp.)*

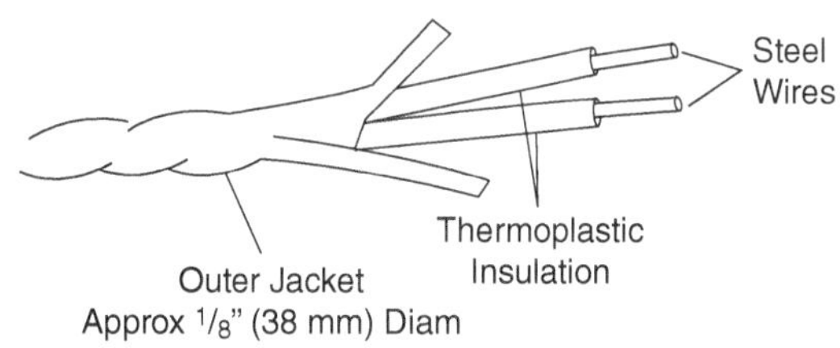

Figure 26.54 *In this linear heat detector, the thermoplastic insulation on the stiff, twisted steel conductors melts, allowing the two conductors to touch and thereby register an alarm.*

(155°F [68°C], 190°F [88°C], 280°F [138°C]) and can be used in open- or closed-circuit configurations in lengths of up to 2500 ft [760 m]. Meters are available that indicate the exact point of a fault, so that in long runs where overheating rather than fire is detected, countermeasures can be taken rapidly, without the delay caused by having to seek out the fault point. The major disadvantage of this type of detector is that it is nonrenewable; once it has faulted, the melted section must be cut out and replaced.

Another type of linear detector is basically a linear thermistor—that is, a device whose electrical resistance varies with temperature (Figure 26.55). The resistance is monitored at a control panel, and small changes are noted. Depending upon the thermistor material selected, the detection range can be set anywhere from 70°F [21°C] to 1200°F [650°C]. This type of linear detector is more sensitive, rapid, and expensive than the preceding type. When linear detectors are installed in direct contact with the protected equipment, they detect surface temperature rather than air temperature and therefore act as early-warning devices. They are used extensively in this fashion to continuously monitor the surface temperature of such devices as transformers, switchgear, generators, and all sorts of hazardous equipment.

A third design for linear heat detectors uses optical fiber with a thermoplastic cladding (coating). Optical fiber transmits light by successive internal reflection at the fiber/cladding interface throughout the entire length of the fiber. Excessive temperature anywhere

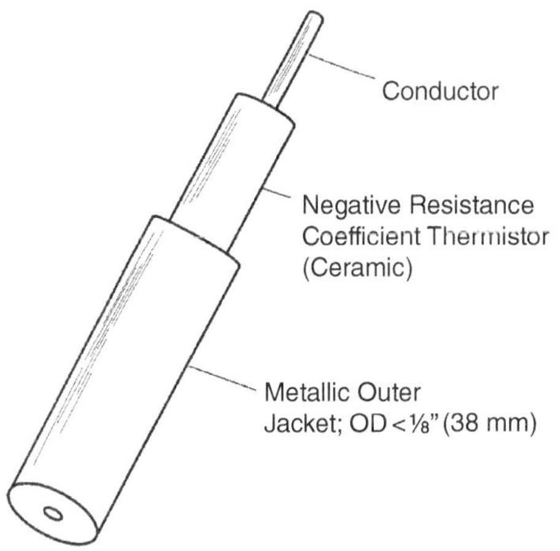

Figure 26.55 *This linear heat detector is connected to an electrical control that monitors circuit current and thereby resistance. Any appreciable temperature rise along the entire cable length causes a sharp drop in the ceramic material's electrical resistance, which is immediately noted and pinpointed at the control point.*

along the length of the fiber causes the cladding to melt and thereby appreciably changes the intensity of light received by the light sensor at the cable's end. This change is recorded and interpreted automatically as either a trouble signal or an overheat alarm signal. This design is available in lengths of up to 1 mi [1.6 km] or more. Other linear sensor designs include variable electrical resistance polymers, phase-change eutectic salt, and pneumatic tubing. Each type has its own advantages, drawbacks, and specific applicability.

Factors involved in the choice of a linear detector include:

- physical characteristics and value of the protected elements
- ease in locating the alarm area
- requirements for resetting the system after an alarm
- maintenance requirements
- costs (first and life-cycle)

26.20 SPECIAL TYPES OF FIRE DETECTORS

Several other sophisticated detectors are available for early-warning detection during the incipient and smoldering fire stages. Among the most sensitive is a unit based on the operating principle of a Wilson cloud chamber (see Figures 26.56 and 26.57). When microscopic particles, such as those produced by the early stages of fire, are introduced into a saturated atmosphere (cloud chamber), they act as nuclei around which water vapor condenses to form visible droplets. The detector operates by continuously sampling air from the protected space and setting off an alarm when the cloud chamber indicates the presence of particulate matter.

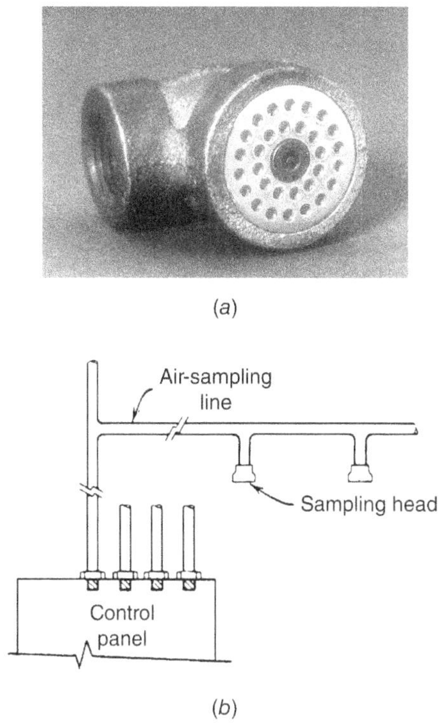

Figure 26.56 *This very early warning system detects microscopic combustion products in air aspirated from the protected area, using a cloud chamber effect. (a) Tamperproof head housing, applicable to high-vandalism areas and correctional institutions. (b) Sampling heads are placed in all spaces to be monitored, including inside cabinets. (Courtesy of Protec Fire Detection, plc.)*

The continuous sampling procedure and the unit calibration make the system insensitive to dust and other noncombustion particulate matter and generally free of nuisance false alarms. The system's disadvantages are the need for piping and its high cost in small installations. For large installations (30 or more detection points), the cost is competitive with those of comparable systems. This detector is particularly applicable to high-value installations, such as museums, data-processing spaces, libraries, clean rooms, and facility control rooms, where its extremely high sensitivity may provide a critically important advance warning of an incipient fire.

26.21 FALSE ALARM MITIGATION

Smoke detectors are subject to false alarms due to moisture and particulate matter in the air. Technological advancement has reduced this problem; technology alone, however, will not entirely eliminate false alarms; the fire alarm system designer and the system user can contribute appreciably to minimizing annoying and

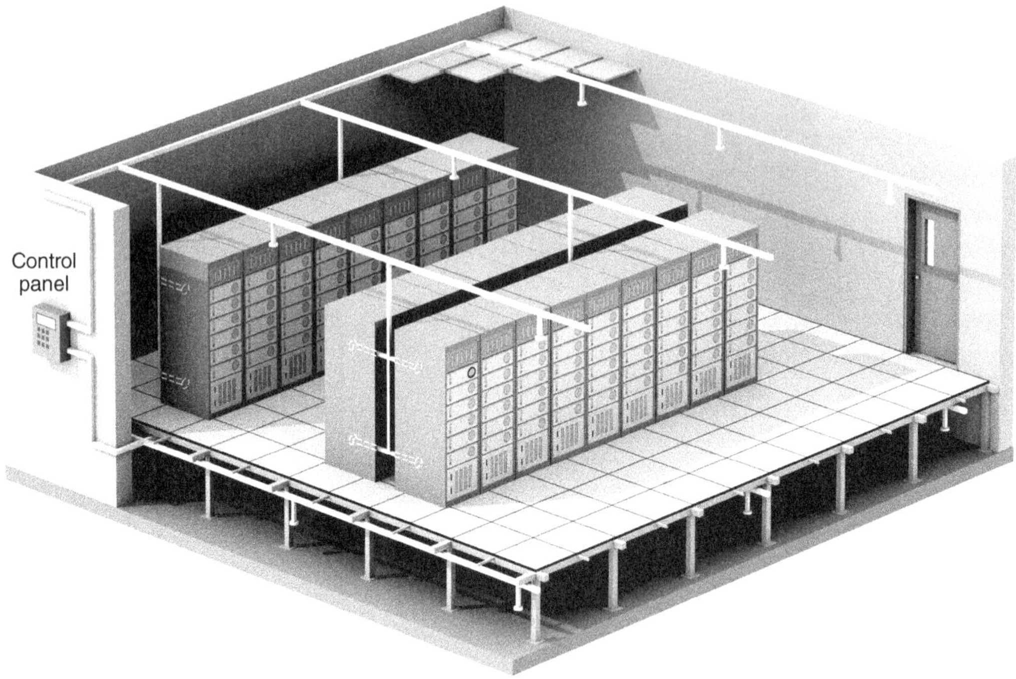

Figure 26.57 *Schematic diagram of a fire detection installation in a computer facility showing aspirating heads and piping. Note that air is sampled inside cabinets and under the raised floor in addition to sampling in the room itself.*

potentially dangerous false alarms by following some basic guidelines, including the following:

- Select appropriate smoke detectors. Choosing an ionization type where a photoelectric type is indicated, or vice versa, almost guarantees false alarms.
- Use detectors with drift compensation to permit accommodation of dirt/high temperature/humidity and air current ambient conditions, all of which increase false alarms.
- Where necessary, use radio-frequency interference (RFI) filters on detector circuits because line radio noise (spikes, surges) can cause alarms. This issue should be investigated in any installation with a heavy solid-state equipment load.
- Use control panels and detectors that have self-learning software and multispace-type algorithms that match sensitivity patterns to the spaces in which the detectors are installed.
- Use detector covers in areas where any type of construction work is taking place. This prevents fouling of detectors.
- When installing conventional systems, establish a regular maintenance and testing program that includes field testing and recalibration of all detectors.
- Avoid detector placement in areas where ambient conditions can cause problems. Where such placement is unavoidable, use compensating detectors, special maintenance routines, and appropriate verification procedures in the event of an alarm signal. These areas include:
  - kitchens, laundries, boiler rooms, shower rooms, and other spaces with high humidity and steam
  - repair shops, laboratories, and other areas where open flames are used in normal work
  - garages and engine test facilities where exhaust gases are present
  - smoking rooms and areas near spaces designated as smoking areas
  - areas in which heavy accumulation of dust and dirt can be expected
  - areas of high air movement, such as near loading docks and exit doors and near the discharge from diffusers or registers

26.22 MANUAL STATIONS

In contrast to automatic detectors, manual stations are operated by hand (Figure 26.58). Manual stations can serve to spread an alarm that has already been detected by other means, either human or automatic. In conventional systems, manual stations are either coded or noncoded. (See Section 26.12 for an explanation of coding.) If identification of the exact manual noncoded station operated is desirable, an annunciation panel can be added to the system; this is equivalent to using each station as a noncoded indicating zone. Because of wiring costs, annunciated systems become expensive; beyond 10 stations, coding should be considered (see Figure 26.39).

When the system design is such that immediate building-wide aural identification of the operated station is necessary, a coded station is used. The station code is received at the control panel, processed, and then transmitted audibly on the system gongs. At least three rounds of code, and normally four rounds, are transmitted. The code usually comprises three or four digits (e.g., 2-3-2) with a pause between the ringing groups and a longer pause between the rounds. The first number may identify the building floor, the second digit the wing, and the third digit the individual station. Establishment of codes is left to the user.

Manual stations are placed in the normal path of egress from a building so that an alarm may be turned in by a person as he or she exits. It is imperative, therefore, that stations be well marked and easily found. Architects who place fire alarm stations in nooks and corners and in camouflaged cabinets because they spoil the decor of the lobby are defeating the purpose of the system. Similarly, placement of bells inside hung ceilings because they are unattractive is not only foolish but dangerous and should never be done, regardless of the circumstances. Loss of property and even of life may result from such ill-conceived aesthetic considerations.

Addressable manual stations obviate the necessity for a code wheel for each manual station but do not eliminate coding of alarm bells if building-wide identification of the operated station is required. (Addressable manual stations are applicable only to addressable systems, not to conventional systems.) The purpose of gong coding is to permit rapid alarm verification by persons scattered about a building. An addressable manual station is identified only at the control panel, from which location a person must be sent to verify the alarm. In a small building, that is not a problem; in a large building, the time required to make such a verification is critical, because a manual alarm is already the result of a verified emergency, unless it is a deliberate false alarm. The decision, then, as to whether to code the gongs must be made on the basis of design decisions for a particular building. A typical addressable manual station is shown in Figure 26.58*c*.

(*a*)

(*b*)

(*c*)

Figure 26.58 *(a) Manual single-action fire alarm available with terminals or wire leads. The unit requires a key to reset; this avoids the necessity of replacing the break-glass rod used in older units. (b) This dual-action manual pull station latches in the down (operated) position when pulled and can only be reset by key. The manual switch can be polled from the control panel to verify that it has indeed been operated and that the alarm signal is not a circuit malfunction. Maintenance personnel can open the station for inspection without setting off the alarm. (c) A flush mount manual explosion and weatherproof pull station with polycarbonate case that is located outdoors. The protective case helps prevent false alarms.*

26.23 SPRINKLER ALARMS

Water flow switches are placed in sprinkler pipelines and operate when a sprinkler head goes off (Figure 26.59). In electrical terms, a water flow switch is a set of contacts similar to a temperature detector. It can be used to trip a coded transmitter, setting off a sprinkler code; to show up on a sprinkler annunciator

Figure 26.59 *Typical water flow indicator. The unit bolts onto a sprinkler pipe with the paddle inside the pipe. Any water motion deflects the paddle, causing a signal to be transmitted from the microswitch mounted in the box on top of the pipe. (Courtesy of Simplex/Potter.)*

board, called a sprinkler alarm panel; or to act as a zone in a noncoded system. Wiring of water flow switches, or of switches activated by other extinguishing systems, is the same as that for a manual station.

26.24 AUDIBLE AND VISIBLE ALARM DEVICES

The type and placement of audible and visible alarms in a (public) building must meet the requirements of *NFPA 72* and the ADA (Americans with Disabilities Act). There are some disagreements between these requirements, as of this writing. In such cases, the fire alarm system designer, in concert with the building architect, should settle these differences to the satisfaction of the authority having jurisdiction.

(a) Audible Signals

Where an evacuation signal will be sounded throughout a building (public mode) to alert all of its occupants, the requirements of *NFPA 72* are normally used because they are more stringent than those of the ADA. There are several requirements, the most important of which requires a minimum alarm sound pressure level of 15 dBA above the average ambient sound level or 5 dB above the maximum sound level (whichever is greater) for at least one minute. This means that the designer must actually measure or predict, with a high degree of accuracy, the anticipated sound levels in the various portions of a building. Average sound level estimates are available from many sources, including the NFPA code itself; maximum sound levels are not. As a result, most designers use a considerable factor of safety to ensure compliance. The upper limit permissible in any area is 120 dBA. See Chapter 24 for technical information on architectural acoustics and sound levels.

The minimum private mode requirement is the greater of 10 dBA above the average ambient level or 5 dBA above the maximum ambient sound pressure for at least one minute. Because this requirement generally applies to an enclosed space that houses the alarm panel or an adjacent space that is continuously occupied, the acoustic design is much simpler than for the public mode alarm.

The codes also specify the minimum required sound pressure levels for sleeping areas and mechanical equipment rooms, in addition to directives regarding the location of audible devices. The selection and placement of audible alarm devices are a technical task requiring acoustic analysis of spaces, occupancy, and the characteristics of available audible devices.

(b) Visible Signals

The requirements for these signals, the primary purpose of which is to alert hearing-impaired people, are covered in *NFPA 72*, *UL 1971*, *ANSI 117.1*, ADA, and various building code requirements. This is an area in which, at this writing, discrepancies exist between ADA requirements and those of *NFPA 72*, *UL 1971*, and *ANSI 117.1* (which are in general agreement). All of the standards require that the visible alarm be a strobe light (usually a xenon flashtube), flashing at less than 2 Hz to minimize problems for persons with photosensitive epilepsy. This does not, however, solve this particular problem, because a person viewing light from more than one strobe (as at a corridor junction) can be exposed to a higher flash rate unless all strobes are synchronized. Careful strobe placement can help to minimize this problem.

(c) Strobe Intensity

For nonsleeping areas, the ADA requires a strobe intensity of 75 candelas (cd) and 50 ft [15 m] maximum unit spacing, whereas the other standards (*NFPA 72*, *UL 1971*, and *ANSI 117.1*) require an intensity dependent upon the size of the room. For sleeping areas, the ADA requirement remains the same, whereas the other standards require a 110-cd source if wall-mounted and a 177-cd source if ceiling-mounted. For corridors, the ADA requirement remains 75 cd at 50 ft [15 m] strobe spacing, whereas the other standards have a requirement that varies with corridor length.

Location of the strobes according to *NFPA 72* must be such that the strobe light can be seen at any point in any space, regardless of the viewer's orientation, with the additional requirement that the maximum distance

Figure 26.60 *A combination audiovisual fire alarm signal is capable of sounding eight different warning tones that can be field selected. The 15-cd strobe flashes once every 1.5 seconds (2/3 Hz). The unit measures 4 in. [100 mm] square and can be mounted surface, semi-flush, or flush.*

between strobes will not exceed 100 ft [30 m]. A typical combination audiovisual fire alarm device is shown in Figure 26.60.

26.25 GENERAL FIRE ALARM RECOMMENDATIONS

The specific building system descriptions that follow are based upon good present practice. Specific design decisions must meet the requirements of the latest edition of NFPA standards plus other codes having jurisdiction. Special attention is directed to Article 760 of the *NFPA 70, National Electrical Code* (*NEC*) (NFPA, 2022a). It covers fire alarm system wiring in its broadest sense (i.e., "all circuits controlled and powered by the fire alarm system," including "fire detection and alarm notification, guard's tour, sprinkler ... systems ... safety functions, damper control and fan shutdown"). In office buildings the fire alarm system includes elevator capture and intercom, which, if powered and controlled by the fire alarm system, fall under *NEC* Article 760.

26.26 RESIDENTIAL FIRE ALARM BASICS

Refer to *NFPA 72* (NFPA, 2024b) and *NFPA 101* (NFPA, 2023c) for detailed requirements. Residential occupancies include single-family residences, multi-unit complexes, and many-unit apartments. The alarm requirements for such a range of building types vary. In general, the system should provide sufficient time for the evacuation of residents and, in larger buildings, for appropriate countermeasures to be initiated. The elements of the system are the various alarm-initiating devices, the wiring and control panel, and the audible alarm devices. These are self-contained in small systems and assembled on site for large systems.

A good basic system for a multi-unit installation requires the following:

- A listed smoke detector, outside and adjacent to each sleeping area, in each sleeping room, at the head of every stair, with at least one on every level, including the basement. Combined smoke/heat detectors should be installed in the boiler room, kitchen, garage, and attic. Listed heat detectors in the attic, kitchen, and boiler room are frequently set at 185°F [85°C] because of high ambient temperatures. Other units are set at 135°F [57°C].
- An alarm in any detector produces an alarm in all audible and visual units.
- Control unit (central panel) is annunciated to show the location of an alarmed device and arranged to shut off oil and gas lines and the attic fan (to prevent the spread of smoke). Also, it should turn on lighting both inside and out, operate an automatic dialer to ring a neighbor's phone or a commercial central station, and give a distinctive alarm sound when the phone is answered. An outside bell and some other device that transmits an alarm outside the residence are important.
- Backup power for the system—a supervised storage battery with a trickle charger.
- Wiring of all devices connected to a system control unit should be on supervised circuits that will sound a trouble alarm in the event of a fault. The trouble alarm should be distinct from the fire alarm.

Additional requirements for detectors, alarms, power supply, supervision signals, and other relevant items—including special requirements for fire alarm systems that are combined with other signal systems—are found in the current edition of the previously referenced standards.

26.27 MULTIPLE-DWELLING ALARM SYSTEMS

The exact requirements for the location and type of detectors and their action are specified in *NFPA 72* (NFPA, 2024b) and *NFPA 101* (NFPA, 2023c). Multiple dwellings include apartment houses, dormitories, hotels, motels, and rooming houses. The prime directive for these structures in the event of a fire emergency is early warning and orderly egress, with consideration of the fact that a fire emergency may occur when most of the building's occupants are asleep. Design guidelines are as follows:

- Audible/visual alarms must be positioned so that all sleeping persons, including those with hearing and/or sight impairment, will be wakened. Code requirements for audible levels are minimum and are based upon average spaces and furnishings. Unusually large or oddly shaped spaces should be designed individually.
- Consider the possibility of living rooms being used as sleeping areas on a regular basis.
- Provide smoke detection in corridors, service spaces, and utility and storage rooms.
- Battery-powered detectors may not be used because, unlike homeowners, apartment dwellers rely on the building's management for all building services, and periodic battery checks and replacement might not be carried out.
- Provide standby power to all fire alarm circuits.
- All alarms should be identifiable, either by addressing or by annunciation. Annunciators should be located at the system control panel in the building management office or at the lobby desk in the case of hotels and dormitories. In all buildings, a lobby annunciator for the use of firefighters is advantageous.
- An alarm light over the door of each apartment or suite to indicate the location of an alarm is desirable. This is particularly important if the central panel only provides zone annunciation.
- In high-rise buildings, an emergency voice/alarm communication system should be provided.

26.28 COMMERCIAL AND INSTITUTIONAL BUILDING ALARM SYSTEMS

The requirements for these buildings are so varied that no specific recommendations can be made. A few suggestions, however, are in order.

- Presignaling, where permitted, is recommended for buildings that do not readily tolerate an evacuation alarm.
- In schools, particularly for the elementary grades, rapid, orderly evacuation of the building is the primary requirement. Consideration must be given to the uniqueness of the sound of the fire alarm gongs to allow no possibility of confusion with program gongs where the latter are used. Also, because regular fire drills are mandatory in most schools, the system must be arranged to provide this capability.
- Public buildings should have an auxiliary alarm connection to the fire department.
- For medium-sized buildings, an accurate cost estimate frequently indicates an advantage of addressable analog systems over addressable systems because of the high detector maintenance costs of the latter. Similarly, addressable systems frequently have an economic advantage over hard-wired systems because of the high labor costs for wiring.

Figures 26.61, 26.62, and 26.63 show a few modern fire alarm control panels. Also known as the fire alarm control unit (FACU) or fire alarm panel,

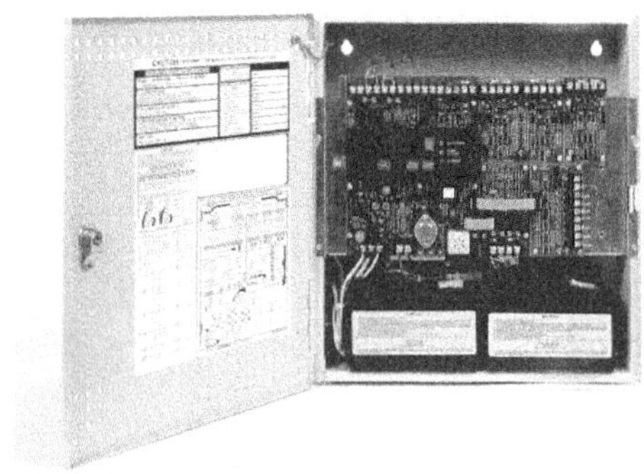

Figure 26.61 Conventional hard-wired eight-zone (maximum) solid-state fire alarm control panel for supervised Class A or Class B circuits. A single annunciator light indicates the alarmed zone. Alarm verification for each zone is provided. Both audible and visible alarms can be used on two alarm-signal circuits. The audible evacuation signal is a slow (20-BPM [beats per minute]) or fast (120-BPM) march time. Note the two standby batteries at the bottom of the cabinet. The unit measures approximately 16 in. [400 mm] wide and 17 in. [432 mm] high and is designed for surface mounting on 16-in. [400-mm] o.c. studs. (Courtesy of Simplex.)

Figure 26.62 Fire alarm control panel for a hard-wired conventional system that can accommodate up to 36 detection zones. The panel has a 2-line, 80-character alphanumeric display that indicates the zone alarm, trouble signals, and diagnostics information. The panel, which is 24 in. [610 mm] square, is microprocessor controlled and field programmable. A separate annunciator can be readily connected. (Courtesy of Simplex.)

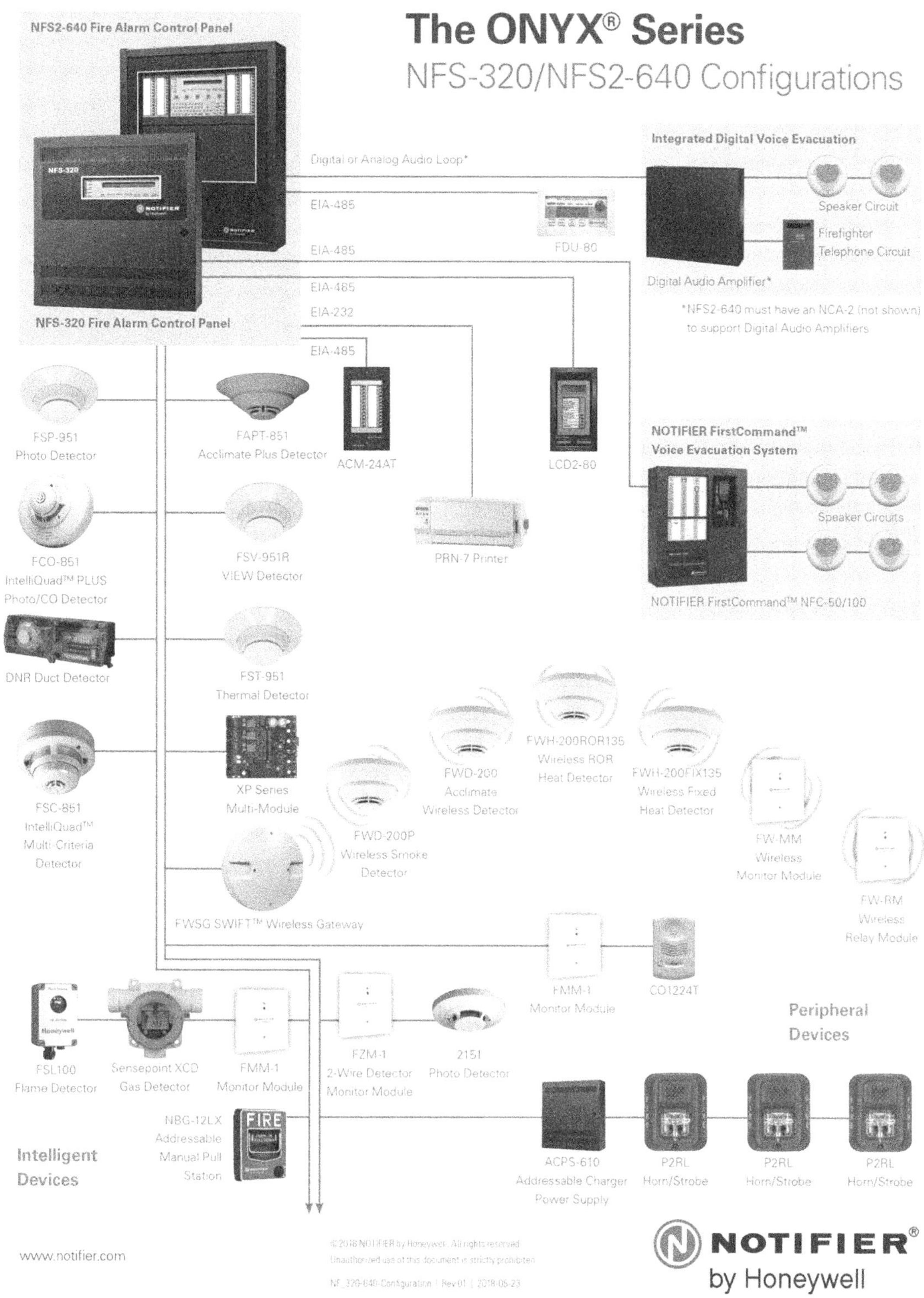

Figure 26.63 System block diagram for a compact addressable analog fire system for a medium-sized building. The panel, which measures approximately 15 in. [380 mm] wide by 16 in. [400 mm] high by 6 in. [150 mm] deep, has a capacity of 99 analog detectors and 99 monitor/control modules. Integral programming provides drift compensation and sensitivity adjustment for analog detectors, detector maintenance (dirt accumulation) alert, and automatic cyclic sensitivity adjustment, plus standard diagnostics. Custom graphics and the video display can be arranged to display floor plans, device locations, and the like. Relay control can be used for control of the building air-handling unit, elevator interlock, and security/access system interlock. Notification circuits connect to the central station, remote station, proprietary alarms, or municipal alarms. (© Notifier, Honeywell International.)

the fire alarm control panel serves as the brain of the fire alarm system by monitoring all the inputs and controlling all the outputs.

26.29 HIGH-RISE OFFICE BUILDING FIRE ALARM SYSTEMS

Sad experience has demonstrated that high-rise buildings, once thought to be fireproof, are emphatically not so. Indeed, due to their size, they have particularly severe fire protection problems, one of which is reliable communications during fire emergencies. As a result, fire codes now require that high-rise buildings be equipped with an emergency voice/alarm communication system. This system provides full control of transmission and building-wide distribution of all tones, alarm signals, and voice announcements on a selective or all-call basis. Specifically, it makes possible the following:

- Two-way active communications from the firefighters' control and command post (usually in the lobby) to at least one fire station per floor, all mechanical plant rooms, elevator machine rooms, and air-handling (fan) rooms
- Distribution to selected areas of alert tones, signals, and prerecorded messages on independent channels. (A complete system of loudspeakers covering all areas of the building is an integral part of the system.)
- Communication with the fire department or central station

In addition to communication functions, the (lobby) fire command post provides:

- Visual display (annunciation) of all fire alarm devices, including sprinkler valves and water flow indicators
- Fire pump status indication
- Controls for any automatic stair door unlocking system (security access system)
- Emergency generator status
- Elevator location indicators plus operation and capture controls
- Control of smoke control devices (doors, dampers, etc.)
- Means for testing all circuits and devices

The exact equipment supplied depends upon the building and the local fire code. The distinctive characteristics of this system are the communications system, the visual display of alarm locations, and the remote control of air-handling equipment.

26.30 INDUSTRIAL FACILITY ALARMS

In addition to manual stations at points of egress, these buildings use the following:

- Temperature and smoke detectors in storage areas and laboratories
- Smoke and flame detectors in record rooms and continuous process laboratories
- Water flow switches on all sprinklers

The annunciators, control panel, and alarm register are best placed in a guardroom. If none is available, an auxiliary or remote-station circuit should be added to allow remote monitoring.

Because of the high ambient noise level in many plants, horns are substituted in such areas for bells and gongs, which might be inaudible.

In summary, the specific occupancy recommendations given here are representative of good current practice. Codes, however, are constantly being changed in this particularly sensitive field, and they vary with project locale. In actual design situations, all codes and regulations having jurisdiction must be complied with.

References and Resources

DOJ. 2012. *2010 ADA Standards for Accessible Design*. Washington, DC: U.S. Department of Justice. ada.gov/law-and-regs/design-standards/2010-stds

Egan, M. D. 1978. *Concepts in Building Fire Safety*. Hoboken, NJ: John Wiley & Sons.

ICC. 2024a. *2024 International Building Code*. Washington, DC: International Code Council.

ICC. 2024b. *2024 International Fire Code*. Washington, DC: International Code Council.

Jensen, R. (ed.). 1975. *Fire Protection for the Design Professional*. New York: Van Nostrand Reinhold.

NFPA. 2008. *Fire Protection Handbook*, 20th edn. Quincy, MA: National Fire Protection Association.

NFPA. 2021. *2022 NFPA 10: Standard for Portable Fire Extinguishers*. Quincy, MA: National Fire Protection Association.

NFPA. 2022a. *2023 NFPA 70: National Electrical Code (NEC)*. Quincy, MA: National Fire Protection Association.

NFPA. 2022b. *2023 NFPA 750: Standard on Water Mist Fire Protection Systems*. Quincy, MA: National Fire Protection Association.

NFPA. 2022c. *2023 NFPA 780: Standard for the Installation of Lightning Protection Systems*. Quincy, MA: National Fire Protection Association.

NFPA. 2023a. *2024 NFPA 14: Standard for the Installation of Standpipe and Hose Systems*. Quincy, MA: National Fire Protection Association.

NFPA. 2023b. *2024 NFPA 92: Standard for Smoke Control Systems*. Quincy, MA: National Fire Protection Association.

NFPA. 2023c. *2024 NFPA 101: Life Safety Code*. Quincy, MA: National Fire Protection Association.

NFPA. 2023d. *Fire Protection Handbook*, 21st edn. Quincy, MA: National Fire Protection Association.

NFPA. 2024a. *2025 NFPA 13: Standard for the Installation of Sprinkler Systems*. Quincy, MA: National Fire Protection Association.

NFPA. 2024b. *2025 NFPA 72: National Fire Alarm and Signaling Code*. Quincy, MA: National Fire Protection Association.

NFPA. 2024c. *2025 NFPA 13: Automatic Sprinkler Systems Handbook*. Quincy, MA: National Fire Protection.

Ramsey, C. G. and H. R. Sleeper. 2016. *Architectural Graphic Standards* (12th edn. American Institute of Architects ed.). Hoboken, NJ: John Wiley & Sons.

UL. 2023a. *UL 96A: Installation Requirements for Lightning Protection Systems*. Northbrook, IL: Underwriters Laboratories, Inc.

UL. 2023b. *UL 268: Smoke Detectors for Fire Alarm Systems*. Northbrook, IL: Underwriters Laboratories, Inc.

U.S. Access Board. 2004. *Americans with Disabilities Act Accessibility Guidelines for Buildings and Facilities (ADAAG)*. Washington, DC: U.S. Architectural and Transportation Barriers Compliance Board. www.federalregister.gov/documents/2004/07/23/04-16025/americans-with-disabilities-act-ada-accessibility-guidelines-for-buildings-and-facilities

PART VIII

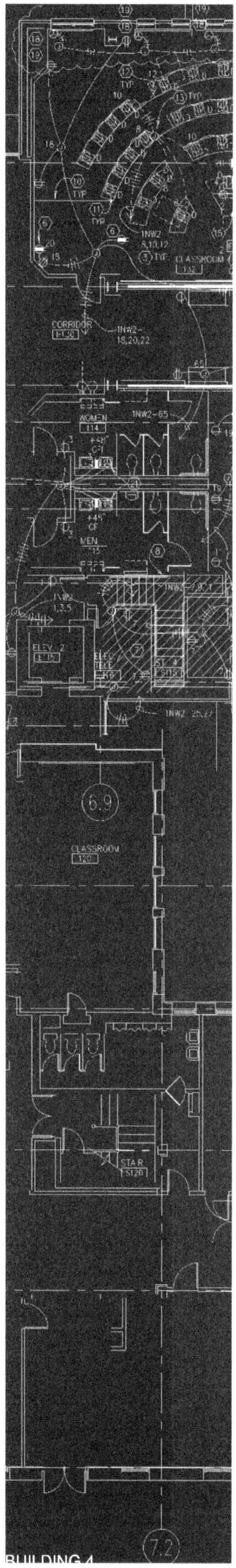

ELECTRICITY

Electricity is the most prevalent form of energy in a modern building. It not only powers convenience outlets and electric lighting, but also provides the motive force for HVAC equipment, moves elevators and escalators, and enables all modern signal and communications equipment. Given the complete dependence upon electric power for normal operation that is characteristic of most modern buildings, design team members must be familiar with the basic concepts and equipment of electrical systems.

An electric power failure can paralyze a facility. Such power failures often occur at a district or regional scale, and are beyond the control of a building designer. A properly designed facility can, however, quickly return to partial operation through the action of emergency equipment that can furnish part of the facility's electricity needs for a limited time. Such capabilities are important in the information age and in the face of concerns over extreme events and a desire for resiliency.

Chapter 27 reviews fundamental electrical relationships, with emphasis on electric circuits, power, energy, energy costs, and methods of energy management and electric load control—and closes with reflections on emerging changes to how buildings interact with the electrical grid. Chapter 28 describes electrical service, utilization, and emergency/standby power equipment. Also addressed are energy efficiency considerations and costs. Chapter 29 introduces electrical equipment ratings and capacity, and continues with a description of wiring systems and their components. Chapter 30 draws on information given in the three preceding chapters to demonstrate basic design methods for building electrical systems. Chapter 31 presents information on photovoltaic (PV) systems, which are being increasingly adopted as part of green, net-zero-energy, and carbon-neutral building design solutions.

Modern buildings, from the simplest residence to the most complex industrial facility, depend extensively upon electrical signal, alarm, control, and communication systems for their day-to-day functions. The systems discussed in Chapter 32 deal with security, music/sound, intercom, clock and program, paging, and building automation. (Fire and smoke detection and alarms are covered in Chapter 26.)

Building automation is essentially a complex control system with sophisticated hardware. Thus, the discussion of automation is limited to system architecture and application issues. A discussion of hardware, protocols, and software would be too detailed and would necessarily lag the times in a field that doubles its computing power, and therefore its control capability, roughly every 18 months.

CHAPTER 27

Principles of Electricity

HISTORICALLY, USABLE ENERGY WAS most often produced by burning a fossil fuel such as coal or oil. The resultant energy output was used directly as heat and/or light or converted by machines into motion. Only since the end of the nineteenth century has such heat, in turn, been used to create another very usable form of energy—electricity. Nuclear reactors, geothermal resources, and concentrating solar collectors may also be used to produce heat for electricity generation. Hydropower is an interesting example of non-heat-based generation. Fuel cells and photovoltaics produce electricity directly (with heat as a waste product). It is well to remember that, in terms of consumption of fuel resources, electricity is a resource-expensive form of energy because the efficiency of the overall heat-to-electricity conversion, on a commercial scale, rarely exceeds 40%.

27.1 ELECTRIC ENERGY

Electricity consists of the movement of electrons between two points when there is a potential difference between them. It is referred to as a secondary energy source because it is produced by converting primary energy sources (e.g., coal, natural gas, solar radiation, nuclear fission, etc.). Electricity is also referred to as an energy carrier because it can be converted to other forms of energy, such as mechanical energy or thermal energy (heat). Electricity occurs naturally, but only in unusable forms such as lightning and other static discharges or in natural galvanic cells (which cause corrosion). The primary problem in the utilization of electric energy is that, unlike fuels or even heat, it cannot be readily stored and therefore must be generated and utilized in the same instant. This requires an entirely different concept of utilization than, for example, a heating system with its fuel source, burner, piping, and associated equipment.

The bulk of electric energy produced and purchased today is in the form of alternating current (AC), output by AC generators commonly called alternators. The majority of building electrical loads are nominally AC (although these are often DC behind the scenes). Direct-current (DC) producers are utilized for applications requiring large quantities of DC. Local quantities of DC, furnished either by batteries, photovoltaic (PV) equipment, or rectifiers, are utilized for telephone and signal equipment, controls, and computers.

The building sector is a primary consumer of electricity in developed countries. According to the U.S. Department of Energy, buildings are responsible for approximately 40% of total energy use in the United States—including about 75% of all utility-scale electricity generation (Shoemaker, 2023). Interestingly, despite a prevalence of AC being produced and delivered at the infrastructure level, around a third of a building's electrical load may be DC power due to the increasing use of LED lamps and various electronic devices that utilize DC (e.g., smartphones, laptops, televisions, and the like). This proportion will likely increase with the market's uptake of electric vehicles and HVAC systems utilizing DC motors.

Mechanical and Electrical Equipment for Buildings, Fourteenth Edition. Walter Grondzik, Daniel Overbey, and Alison Kwok.

Companion website: www.wiley.com/go/meeb14e

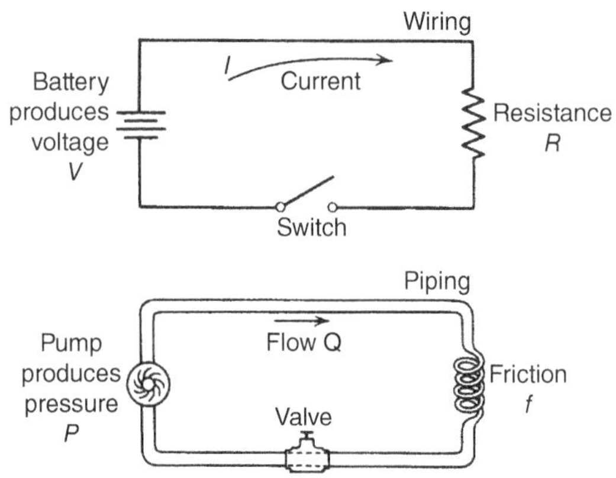

Figure 27.1 Electric-hydraulic analogy. The circuits show that voltage is analogous to pressure, current to flow, friction to resistance, wiring to piping, and switches to valves.

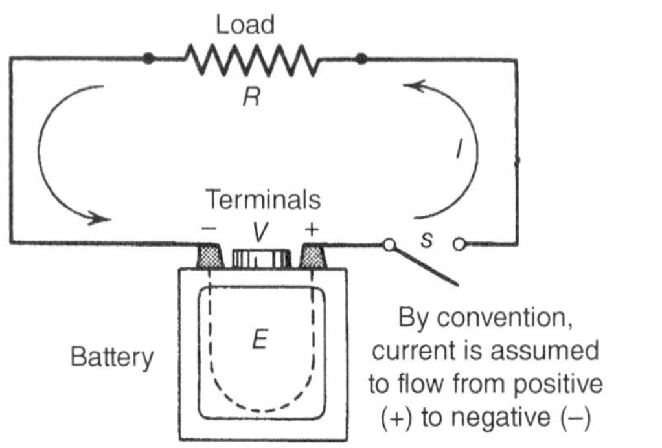

Figure 27.2 Current flows in the electric circuit as a result of the voltage (potential difference) V that exists between the terminals of the battery. By convention, current direction is from positive (+) to negative (−) in the circuit (and from − to + inside the battery).

27.2 UNIT OF ELECTRIC CURRENT: THE AMPERE

Electricity flowing in a conductor is called current, which is measured in amperes, abbreviated amp, amps, or simply *A*. When current is expressed in an equation, it is usually represented by the letter *I* or *i*.

It is convenient to establish an analogy between electrical systems and mechanical systems as an aid to understanding the characteristics of electricity. Current is a measure of flow and, as such, corresponds to water flow in a hydraulic system (Figure 27.1). The analogy is not complete, however, because in a hydraulic system, the velocity of water flow can vary, whereas in an electrical system the velocity of (electric) propagation is constant and may be considered instantaneous.

27.3 UNIT OF ELECTRIC POTENTIAL: THE VOLT

The movement of electrons (and their associated energy, which constitutes electricity) is caused by creating a higher positive electric charge at one point on a conductor than exists at another point on the same conductor. This difference in charge can be created in a number of ways. The oldest and simplest method is by electrochemical action, as in a battery.

In the ordinary dry cell, or in a storage battery, chemical action causes positive charges (+) to collect on the positive terminal and electrons—that is, negative charges (−)—to collect on the negative terminal. Assume for the moment that nothing is connected to the battery terminals. There is a tendency for flow between the electrified particles concentrated at the positive and negative terminals. Potential difference or voltage is the name given to this tendency, or force. It is analogous to pressure in a hydraulic or pneumatic system. Just as the pressure produced by a pump or blower causes water or air to flow in a connecting pipe, the potential (voltage) produced by a battery (or generator) causes current to flow in a conductor connecting the terminals between which a voltage exists (Figure 27.2). The higher the voltage (pressure), the higher the current (flow) for a given resistance (friction). Other means of producing voltage, both direct (DC) and alternating (AC), are discussed in Sections 27.8 and 27.9. The unit of voltage is the volt, abbreviated *V* or *v*.

27.4 UNIT OF ELECTRIC RESISTANCE: THE OHM

The flow of fluid in a hydraulic system is resisted by friction; the flow of current in an electric circuit is also resisted. In a DC circuit, this force is called resistance and is abbreviated *R*; in an AC circuit, it is called impedance and is abbreviated Z. The unit of measurement is the ohm. The scientific convention of naming units after persons whose work is closely related to the field is followed here. Thus, the units ampere, volt, and ohm are derived from André Ampère, Alessandro Volta, and Georg Ohm.

Materials display different resistances to the flow of electric current. Metals generally have the least resistance and are therefore called conductors. The best conductors are the precious metals—silver, gold, and platinum—with copper and aluminum (two conductors currently and historically used in electrical wiring, respectively) being only slightly inferior. Conversely, materials that resist the flow of current are called insulators. Glass, mica, rubber, oil, distilled water, porcelain, and certain synthetics exhibit nonconductive properties and are used to insulate electric conductors. Common examples are rubber and plastic wire coverings, porcelain lamp sockets, and oil-immersed switches.

27.5 OHM'S LAW

The current I that will flow in a DC circuit is directly proportional to the voltage V and inversely proportional to the resistance R of the circuit. In AC circuits, the same relation holds true except that instead of DC resistance, impedance is used. Expressed as an equation, this is the basic form of Ohm's law:

$$I = \frac{V}{R} \qquad \text{(Eq. 27.1)}$$

Ohm's law is frequently written in another form:

$$V = IR \qquad \text{(Eq. 27.2)}$$

which expresses the mathematical relationship that voltage is the product of current and resistance. This form has no intuitive or logical basis; therefore, remembering the form of Equation 27.1 will prove more useful. It clearly states the physical situation, that is, as a result of voltage V, a current I is produced that is proportional to the electric voltage (pressure) V and inversely proportional to the electric resistance (friction) R. If V increases, I will increase; if R increases, I will decrease.

Example 27.1 illustrates the application of Ohm's law. The example chosen is applicable to both AC and DC because the load device is purely resistive. When a load is purely resistive, resistance and impedance are equal. This is more fully explained in the subsequent discussion of alternating current.

EXAMPLE 27.1 An incandescent lamp having a hot resistance of 66 Ω is put into a socket that is connected to a 115-V supply. What current flows through the lamp (after it reaches operating temperature)?

SOLUTION

$$I = \frac{V}{R} = \frac{115 \text{ volts}}{66 \text{ ohms}} = 1.74 \text{ A}$$

(These values correspond to a legacy 200-W general service incandescent lamp; an obsolete reference, but intuitive.) ■

Hot resistance is mentioned in Example 27.1 because the electrical resistance of some materials changes with temperature. A typical example of this is a tungsten-filament lamp that, when first turned on (cold) accepts, for a fraction of a second, 10–15 times the steady-state hot-filament current.

27.6 CIRCUIT ARRANGEMENTS

An electric circuit may be defined as a closed-loop path through which charged electrons are intended to flow. The two basic electric circuit arrangements are series and parallel. These concepts are the same for both DC and AC. As previously, the discussion focuses upon purely resistive circuits so that calculations are applicable to both DC and AC. In other than purely resistive circuits, AC circuit calculations are different and much more complicated than those for their DC counterparts.

(a) Series Circuits

In a series arrangement the elements are connected one after another, that is, in series. Thus, resistances and voltages add. This is indicated graphically in Figure 27.3. An electric circuit may be defined as a complete conducting path that carries current from a source of electricity to and through some electrical device (or load) and back to the source. A current cannot flow unless there is a complete (closed) circuit. Due to the arrangement of components in a series circuit, the current is the same in all parts of the circuit. A somewhat more complicated circuit is shown in Figure 27.4 and is analyzed in Example 27.2.

It is customary to refer to connection points on wiring diagrams by letters, as a, b, c, d, and so on. The battery voltage may then be called V_{ab} = 12 V; the voltage across the load resistance, V_{cd} = 11.5 V; and the

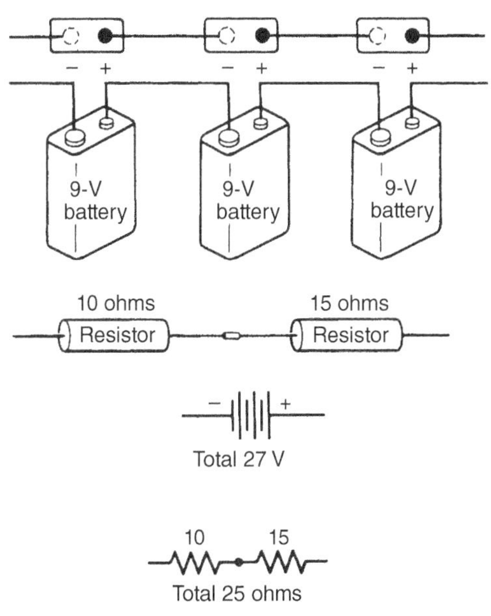

Figure 27.3 Physical and graphic representation of a series connection of batteries and resistors.

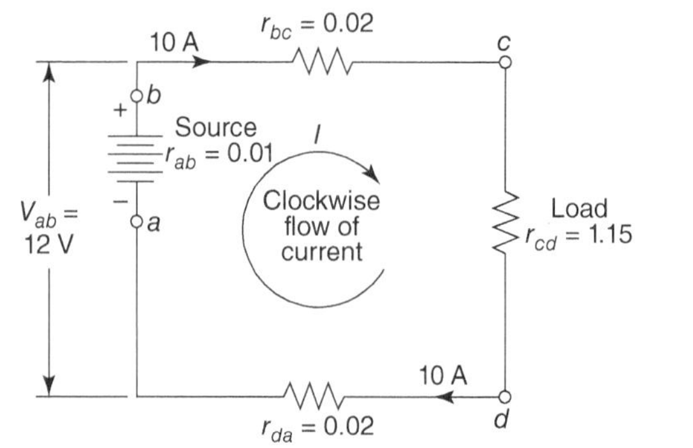

Figure 27.4 A series circuit always contains a voltage source (here, the battery) and a load (R). This circuit also shows the resistance of the connecting wires for accuracy. Such resistances are normally neglected because they usually are very small compared to the load resistance.

resistance of the two wires $r_{bc} + r_{da} = 0.04\ \Omega$. The positive and negative terminals of the battery are always shown.

EXAMPLE 27.2 The battery in Figure 27.4 is rated at 12 V; the total line resistance (the total for both wire segments) is 0.04 Ω; the battery internal resistance is 0.01 Ω; and the load resistance is 1.15 Ω. Determine (a) the current flowing in the circuit, and (b) the voltage across the load (V_{cd}).

SOLUTION

(a) The current flowing is:

$$I = \frac{V}{R} = \frac{V_{ab}}{r_{ab} + r_{bc} + r_{cd} + r_{da}}$$

$$= \frac{12}{0.01 + 0.02 + 1.15 + 0.02} = \frac{12}{1.2} = 10\ \text{A}$$

(b) The voltage drop across the load is:

$$V_{cd} = I \times R_{cd} = 10 \times 1.15 = 11.5\ \text{V}$$ ■

Series circuits find very limited application in building wiring because failure of any one load (such as a burned-out lamp) will open the circuit (shutting off power to all loads on the circuit).

(b) Parallel Circuits

When two or more branches with loads are connected between the same two points, they are said to be connected in parallel or multiple. Such a parallel arrangement and its hydraulic equivalent are shown in Figure 27.5. In the circuit of Figure 27.6, the voltage across each load is the same, but the current in each load (branch) depends upon the resistance of that load. Parallel loads, in effect, constitute separate circuits. In this arrangement, the total current in the circuit is the sum of the individual currents flowing in the branches—that is,

$$I_T = I_1 + I_2 + I_3$$

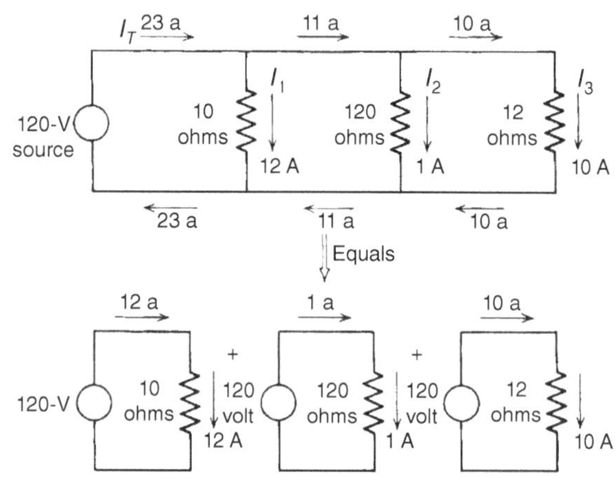

Figure 27.6 Note that loads connected in parallel are equivalent to separate circuits combined into a single circuit. Each load acts as an independent circuit unrelated to, and unaffected by, the other circuits.

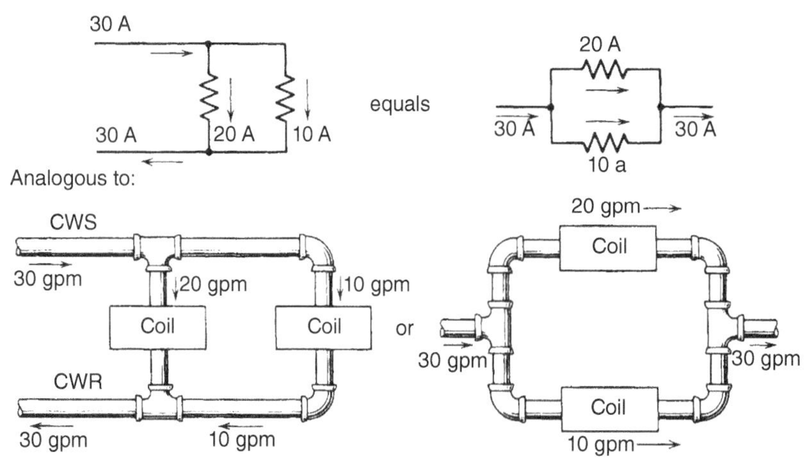

Figure 27.5 In a parallel connection, the flow divides between the branches, but the pressure is the same in each branch.

Note in Figure 27.6 that the total current flowing in the circuit is the sum of the currents in all the branches, but that the current in each branch is determined by a separate Ohm's law calculation. Thus, in the 10-Ω load, a 12-A current flows, and so forth.

The parallel circuit is the standard arrangement in all building wiring. A typical lighting and receptacle arrangement for a large room is shown in Figure 27.7. Here the lighting fixtures constitute one parallel (multiple) grouping, and the convenience wall outlets constitute a second parallel grouping. The fundamental principle to remember is that loads in parallel are additive for current and that each load has the same voltage across it.

One additional point is important to appreciate. Remember from Ohm's law that current is inversely proportional to resistance. Thus, as resistance drops, current rises. Now look at the circuit of Figure 27.7. Normally, that circuit carries the (vector) sum of the currents in the two branches. (Vectorial addition is discussed in Section 27.11, Example 27.5c.) If, by some mischance, such as deterioration of the wiring insulation, a connection forms between points a and b, the circuit is shortened so that there is no resistance in parallel branch ab. The current rises instantly to a very high value, and there is a short circuit. If the circuit is properly protected, a fuse or circuit breaker opens and the circuit is deenergized. If not, the heat generated by the very high current will probably start a fire.

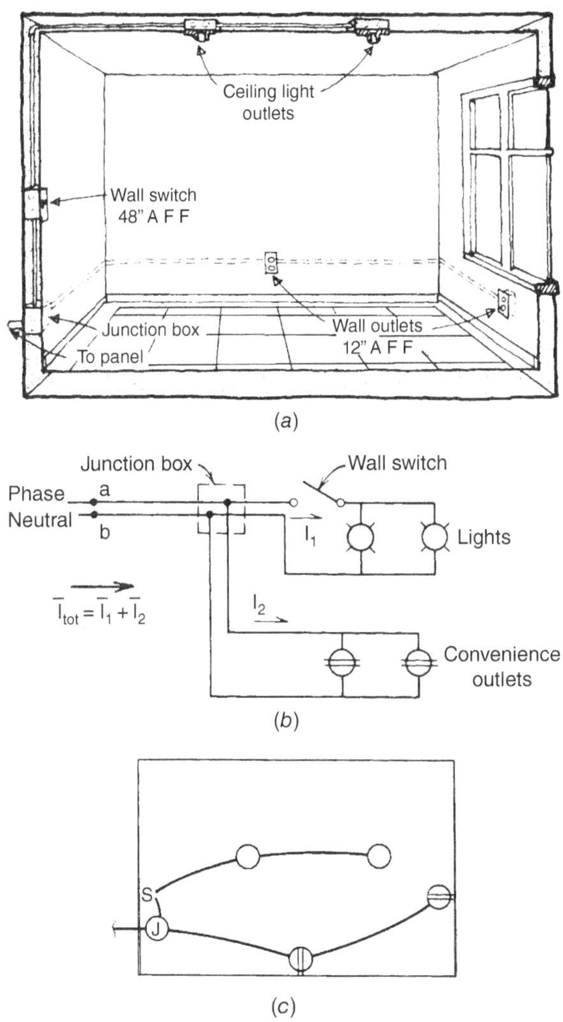

Figure 27.7 *Parallel groupings of ceiling lamp outlets and wall outlets are, in turn, connected in parallel to each other. The circuit is shown (a) pictorially, (b) schematically, and (c) as on an electrical floor plan. The horizontal bar over the currents I_1 and I_2 in (b) signifies the use of vectorial addition (see Example 27.5).*

27.7 DIRECT CURRENT AND ALTERNATING CURRENT

A flow of electric current that takes place at a constant time rate, practically unvarying and in the same direction around a circuit, is called a direct current (DC or dc). Figure 27.8 shows DC voltages of 1.5 V positive polarity and 1.0 V negative polarity.

When the flow of current is periodically varying in time and in direction, as indicated by the symmetrical positive and negative loops, or sine waves, in Figure 27.9, it is called an alternating current (AC or ac). The distance along the time axis spanned by a positive and a negative AC loop is called one cycle. The number of such cycles occurring in one second is known as the frequency of the AC current. Modern AC systems in the United States and Canada operate at a frequency of 60 cycles per second, or 60 hertz (after Heinrich Hertz). Thus, current at 60 hertz (abbreviated Hz) is delivered to the consumer. In Europe and much of Asia, 50-Hz distribution is standard.

An AC circuit differs from a DC circuit in a number of important respects and, because the normal current supply to a building is 60 Hz AC, it is important to

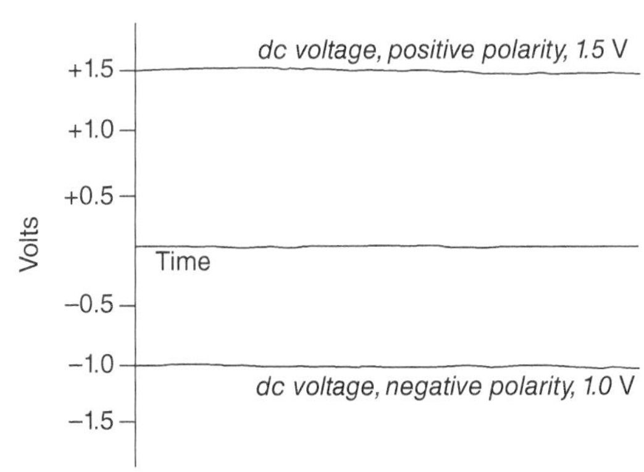

Figure 27.8 *Graphic representation of DC voltages with positive and negative polarity.*

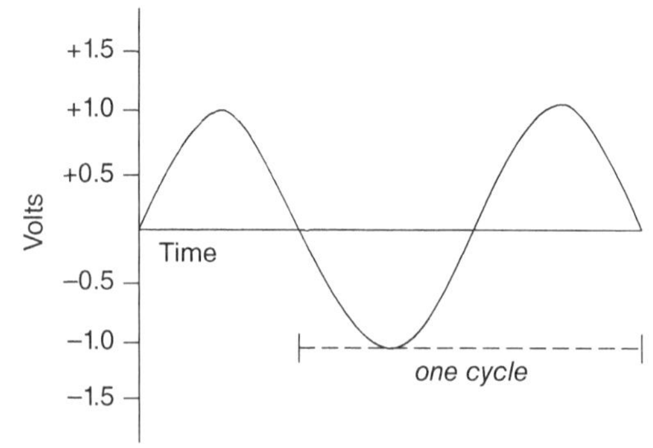

Figure 27.9 *Graphic representation of a pure single-frequency alternating current (AC). The figure shown is a sine wave. Note that a complete cycle includes both the positive and negative loops. The number of full cycles in one second is defined as the frequency of the current (or voltage).*

understand AC terminology and usage. Instead of resistance, the corresponding parameter in an AC circuit is impedance, which is (also) measured in ohms.

Depending upon the circuit load, impedance can be markedly different from the DC resistance. For an AC circuit, Ohm's law is:

$$I = \frac{V}{Z} \quad \text{(Eq. 27.3)}$$

where Z is the symbol for impedance. This book does not go into AC circuit calculations, primarily because such calculations are not especially useful to the general building designer. What is useful and important is an understanding of power and energy in both DC and AC circuits. These concepts are discussed in the following sections.

27.8 ELECTRIC POWER GENERATION: DC

With respect to the generation of large amounts of DC power, piezoelectric and thermoelectric effects can be ignored. Photovoltaic power generated from PV modules—as discussed in Chapter 31—is becoming increasingly important as a practical source of electricity for specialized uses and as part of green/carbon-neutral building design efforts. Thus, the battery, PV, and the DC generator are the most common sources of DC electricity. Because the DC generator is in reality an AC generator with a device (commutator) attached that rectifies the AC to DC, the battery and PV are the major direct sources of DC. (There are some special types of generators that produce DC directly, but their use to date has been extremely limited.) A discussion of the application of batteries for emergency and standby power supplies can be found in Chapter 30.

Another source of DC power is rectification of AC. This can be accomplished on any desired scale to provide as much DC power as there is available AC power. The principal application of DC in older buildings is for elevator motors. In many buildings DC is used for standby power. Small amounts of DC are also used for controls and telephones.

27.9 ELECTRIC POWER GENERATION: AC

Alternating current is produced commercially by an AC generator, generally called an alternator. Its prime mover may be any type of engine or turbine. The process by which electricity is produced is illustrated in Figure 27.10; it is based upon the fundamental discovery in 1831 by Michael Faraday of the principle of electromagnetic induction. Briefly, this principle states that when an electrical conductor is moved in a magnetic field, a voltage is induced in the conductor. The direction of movement determines the polarity of the induced voltage, as shown in Figure 27.10.

If a conductor is formed into a coil and rotated in a magnetic field, a voltage of alternating polarity is produced—that is, alternating current. It does not matter whether the conductor or the magnet moves; the motion of the conductor and the field with respect to each other produces the voltage (see Figure 27.11). It is only one step (that of commercial development) from this rudimentary AC generator to the large, powerful alternator that produces AC in a modern power plant. The frequency of the voltage generated is a function of the machine design and the speed at which it is driven. Normal generator frequency in the United States, as noted previously, is 60 Hz, whereas, in Europe and the Middle East, it is 50 Hz.

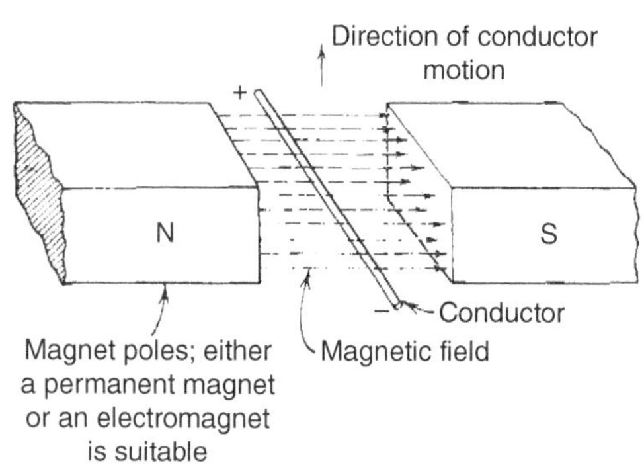

Figure 27.10 *The action fundamental to all generators. When a conductor of electricity moves through a magnetic field, a voltage is induced in the conductor, with polarity as shown.*

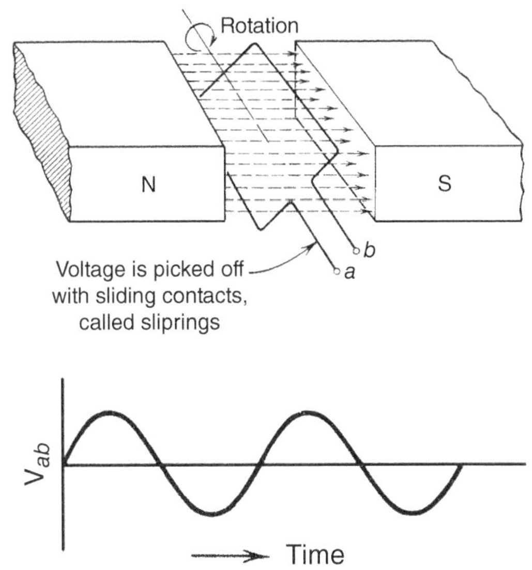

Figure 27.11 *Rotating a coil in a magnetic field produces an alternating sinusoidal voltage at terminals a and b because of the alternating polarity.*

27.10 POWER AND ENERGY

It is critically important that the distinction between power and energy be clearly understood. All too frequently the terms are used interchangeably, which is incorrect and can lead to bad design decisions. Energy is the technical term for the more common expression "work." Power is the rate at which energy is used. For example, it takes a fixed amount of energy to lift a weight a given distance either quickly (in five minutes) or slowly (over five hours), but the faster it is lifted, the more power is required. Energy (work) is the product of power and time, that is,

$$\text{energy} = \text{power} \times \text{time} \qquad \text{(Eq. 27.4)}$$

In practical terms, energy is synonymous with fuel consumption and, therefore, with cost (and resource depletion, for fossil fuels). Energy can be expressed as barrels (liters) of oil, cubic feet (cubic meters) of gas, tons (kilograms) of coal, kilowatt-hours of electricity usage, and/or dollars of fuel cost. The outcome of energy-efficiency efforts for buildings can be stated in terms of annual usage of oil, gas, and electricity or, alternatively, in terms of dollars spent for fuel or utilities. In technical terms, energy is expressed in British thermal units (Btu), calories (cal), joules (J), and kilowatt-hours (kWh). Conversion factors between the most common units of energy are found in Appendix A. In terms of fuel, 1 kilowatt-hour (kWh) of energy is roughly equivalent to 0.5 lb [0.23 kg] of coal, 0.07 gal [0.26 L] of oil, 7.7 ft^3 [0.22 m^3] of natural gas, and 8200 gal [31,000 L] of water in a hydroelectric dam.

Power is the rate at which energy is used or work is done (energy and work being synonymous). The term power implies continuity—that is, the use of energy at a particular rate, over a given, generally considerable, span of time. The concept of power necessarily involves the factor of time because it is, as stated previously, the rate at which work is done. Multiplying power by time yields energy consumption. Typical units of power in the I-P system are horsepower (hp), Btu per hour (Btu/h), watt (W), and kilowatt (kW). In the SI system, the corresponding units are joule per second (J/s), calorie per second (cal/s), watt, and kilowatt.

In physical terms, power is also the rate at which fuel (energy) is used. Thus, power can also be expressed as gallons (liters) of oil per hour, cubic feet (cubic meters) of gas per minute, and tons (kilograms) of coal per day. (When using fuel figures, it is important to ascertain the assumed fuel-to-useful-energy conversion efficiency. When not stated, "energy content" [i.e., energy availability at 100% conversion efficiency] is assumed.) Most U.S. building codes are based upon site energy use, rather than source energy—making conversion efficiency a hidden (but important) issue.

27.11 POWER IN ELECTRIC CIRCUITS

The unit of electric power is the watt (W); 1000 watts is the commonly encountered kilowatt (kW). Both units are used in building design situations depending upon the magnitude of the load—as in a 40-W fluorescent lamp, or a 5-kW resistance heating coil. The power input in watts to any electrical device having a resistance (R) and a current (I) is given by the equation

$$W = I^2R = (I)(IR) \qquad \text{(Eq. 27.5)}$$

where W is wattage. This is true for both DC and AC circuits. Because the resistance of a device or load is generally not known, whereas the circuit voltage and current are known, it would be preferable to be able to calculate power using these latter two quantities. This can be done, but differently for DC and AC.

In DC circuits, by Ohm's law, $V = IR$, and because $W = I(IR)$ from Equation 27.5, we obtain

$$W = VI \qquad \text{(Eq. 27.6)}$$

where

W = watts
R = ohms
I = amperes
V = volts

In AC circuits, impedance consists of a combination of DC resistance and AC resistance (called reactance). Reactance causes a phase difference between voltage and current. This phase difference is represented by an angle, the cosine of which is called the power factor (*pf*). The power factor is extremely important in that it enables us to calculate power in an AC circuit. The AC power equation is similar to that for DC (see Equation 27.6) with the addition of the AC power factor term; that is,

$$W = VI \times pf \qquad \text{(Eq. 27.7)}$$

If *pf* is not included in the equation, the product of voltage and current gives a quantity known as volt-amperes (VA). In a purely resistive circuit, such as one with only electric heating elements, impedance equals resistance, power factor equals 1.0, and wattage equals volt-amperage. In other circuits this is not the case. A few examples should help make the application of these equations clear.

EXAMPLE 27.3 Referring to Example 27.1, calculate the power drawn using Equations 27.5, 27.6, and 27.7.

SOLUTION

Because an incandescent lamp is purely resistive and therefore has unity (1.0) power factor, it does not matter whether the circuit is AC or DC.

From Example 27.1,

$$R = 66 \text{ ohms}$$

$$I = 1.74 \text{ amps}$$

$$V = 115 \text{ volts}$$

(a) In a DC circuit, we would use Equation 27.6:

$$W = VI = 115 \times 1.74 = 200 \text{ W}$$

(b) In an AC circuit, we would use Equation 27.7:

$$W = VI \times \text{pf} = 115 \times 1.74 \times 1.0 = 200 \text{ W}$$

(c) In either a DC or an AC circuit, we can use Equation 27.5:

$$W = I^2R = (1.74)^2 \times 66 = 200 \text{ W}$$

■

EXAMPLE 27.4 Using the data given in Example 27.2 and Figure 27.4, determine (a) the power loss in the wiring and (b) the power input to the load.

SOLUTION

(a) The total power loss in the wiring is

$$W = I^2R = I^2(r_{bc} + r_{da})$$
$$= (10)^2 \times 0.04 = 4 \text{ W}$$

(b) The power taken by the load is

$$W = I^2R = I^2R_{cd} = (10)^2 \times 1.15$$
$$= 115 \text{ W (or 0.115 kW)}$$

Alternatively, we can find this power by multiplying voltage and current. The voltage on the load is

$$IR = 10 \times 1.15 = 11.5 \text{ V}$$

and

$$W = VI = 11.5 \times 10 = 115 \text{ W (or 0.115 kW)}$$

■

EXAMPLE 27.5 Refer to Figure 27.7. Assume a legacy 150-W incandescent lamp at each ceiling outlet. Also assume the load connected to one convenience outlet to be a 10-A hair dryer with a power factor of 0.80. Calculate the current and power in the two branches of the circuit, and the total circuit current, assuming a 120-V AC source.

SOLUTION

(a) In the circuit branch feeding the lamps, the power consumption is for two 150-W lamps; that is

$$P = 2 \text{ lamps} \times 150 \text{ watts} = 300 \text{ watts}$$

To calculate the current, we would use Equation 27.7, which expresses power in an AC circuit:

$$W = VI \times pf$$

Because incandescent lamps are effectively resistive loads, their power factor is 1.0. Therefore:

$$\text{power} = VI$$

$$300 \text{ watts} = 120 \text{ volts} \times I$$

$$I = \frac{300 \text{ watts}}{120 \text{ volts}} = 2.5 \text{ amps}$$

(b) In the second branch, we have a 10-A, 0.8-*pf* load.

$$\text{power in watts} = \text{volts} \times \text{amperes} \times \text{power factor}$$
$$= 120 \times 10 \times 0.8 = 960 \text{ watts}$$

However, the branch volt-amperes are

$$V \times A = 120 \times 10 = 1200 \text{ VA}$$

This latter value is important for sizing electrical equipment.

(c) To calculate the total current flowing from the panel to both branches of the circuit, we must combine a purely resistive current (lamp circuit) with a reactive one (dryer circuit). The exact magnitude of the current is the vectorial sum of the two branch currents, which calculates to be 12.1 A. (Vectorial addition is a technique that considers the phase angle between two currents. Phase angle, or phase difference, was explained previously.) In normal circuit design practice, the arithmetic sum of 12.5 amps would be used. The error introduced is only 3.2%, and because it results in slightly oversized circuit components, it is on the safe side. Only where power factors are very low would a designer use vectorial addition to more accurately size a circuit. ■

One further example will demonstrate the importance of the power factor in normal situations.

EXAMPLE 27.6 The nameplate of a single-phase motor shows the following data: 3 hp, 240 V, AC, 17 A. Assume an efficiency of 90%. Calculate the motor (and therefore circuit) power factor.

SOLUTION

From Appendix A, the conversion 1 hp = 746 W is referenced. Therefore,

$$3\text{ hp} = 3 \times 746 = 2238\text{ W}$$

This represents the motor output. Because for any device:

$$\text{efficiency} = \frac{\text{output}}{\text{input}}$$

then

$$\text{power input} = \frac{2238\text{ W}}{0.9} = 2487\text{ W}$$

However, for AC,

$$\text{power} = \text{volts} \times \text{amperes} \times \text{power factor}$$

so

$$2487\text{ W} = 240\text{ V} \times 17\text{ A} \times \text{power factor}$$

therefore

$$\text{power factor} = \frac{2487\text{ W}}{240\text{ V} \times 17\text{ A}} = 0.61$$

Note the large difference between volt-amperes and watts:

$$V \times I = 240\ V \times 17\ A = 4080\text{ VA}$$

whereas, P, as before = 2487 W

with P designating power. This difference is important in circuit design, as will be seen in the discussion of methods for sizing circuit components. ■

27.12 ENERGY IN ELECTRIC CIRCUITS

Because energy = power × time, the amount of energy used is directly proportional to both the power of a system and the length of time it is in operation. Because power is expressed in watts or kilowatts and time in hours (seconds and minutes are too small for practical use), the units of energy used are watt-hours (Wh) or kilowatt-hours (kWh).

EXAMPLE 27.7 Consider daily energy consumption and cost.

(a) Find the daily energy consumption of the appliances listed if they are used for the length of time shown.

Toaster (1340 W)	15 min
Coffee maker (500 W)	2 h
Fryer (1560 W)	½ h
Iron (1400 W)	½ h

(b) Assuming that the average cost of energy is $0.12 per kilowatt-hour, find the daily operating cost.

SOLUTION

(a)	Toaster:	1340 W	= 1.34 kW × ¼ h
			= 0.335 kWh
	Coffee maker:	500 W	= 0.5 kW × 2 h
			= 1.00 kWh
	Fryer:	1560 W	= 1.56 kW × ½ h
			= 0.78 kWh
	Iron:	1400 W	= 1.4 kW × ½ h
			= 0.70 kWh
	Total		= 2.815 kWh per day

(b) The daily operating cost is 2.815 kWh × $0.12/kWh = $0.3378 (say, 34 cents). ■

The power used by a residential household varies with the time of day. A graph showing the power used by a typical American household during a normal weekday might look something like the one in Figure 27.12. The average power demand of the household is much lower than the maximum. The ratio between the two is called the overall load factor (LF) and runs between 20% and 30% for a typical household. The energy used by this household for the 24-hour period is represented by the area under the curve of Figure 27.12. This can be determined only by integration because it varies continuously. Such integration is exactly what a kilowatt-hour meter does, as explained in Section 27.15 (dealing with electrical measurements).

EXAMPLE 27.8 It has been estimated that the average power demand of an American household with an electric stove is 1.8 kW. Calculate the monthly electric bill of such a household, assuming a flat rate of $0.12 per kilowatt-hour.

ELECTRICITY

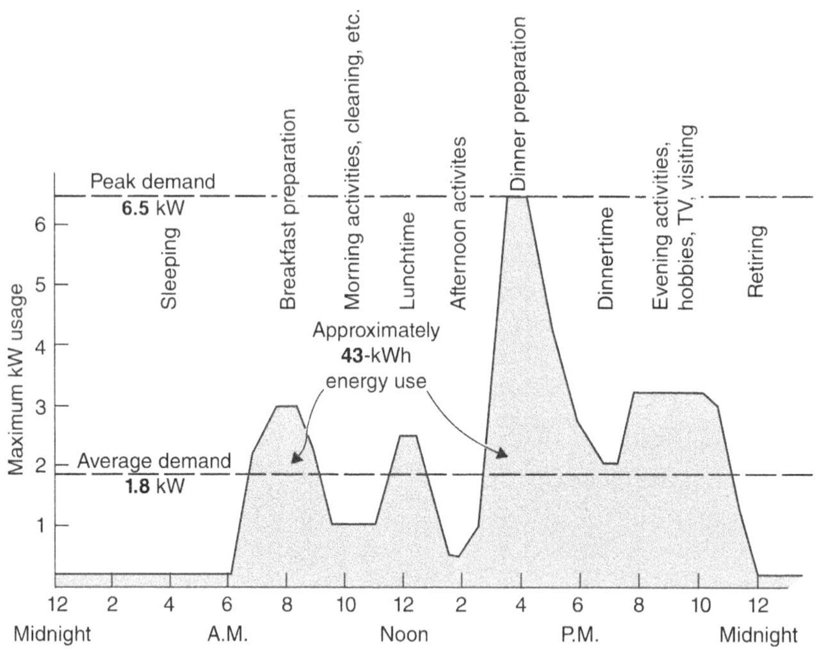

Figure 27.12 *Hypothetical graph of power usage for a typical U.S. household. Electric cooking is assumed. The area under the curve represents energy usage. Maximum kW demand (vertical axis) is based on a 15-minute integrated demand, thus eliminating spikes in demand, such as those caused by starting a refrigeration (air-conditioning) compressor. This curve has a 24-hour use of approximately 43 kWh, giving an average 24-hour demand of 1.8 kW. The ratio between the average demand and the peak demand is called the load factor, which here is 27.7%.*

SOLUTION

Monthly energy use:

$$1.8\ \text{kW} \times \frac{24\ \text{hours}}{\text{day}} \times \frac{30\ \text{days}}{\text{month}} = 1296\ \text{kWh per month}$$

Monthly electric bill:

$$1296\ \text{kWh} \times \$0.12/\text{kWh} = \$155.52$$

■

In Example 27.8, the bill was based on a flat rate of 12 cents per kWh. Flat rates are very common for residential users. In the contiguous United States, residential rates vary from lows of around 12 cents per kWh in many southern states to highs of over 26 cents per kWh in several northeastern states. Hawaii has the highest rates in the United States (around 40 cents). The difficulty in making blanket statements regarding the cost-effectiveness of any proposed energy-efficiency investment is illustrated by this wide range in electricity costs.

Prior to the 1990s, most energy utility companies in the United States were regulated by state governments. In other words, regulated utilities were allowed to operate as monopolies over their local markets in exchange for government regulation. At the time, this philosophy was grounded in the public interest. However, some energy markets began to deregulate in the 1990s, and these utility companies were broken up to allow new energy companies to operate such that market competition might better serve the public good. Today, there are both regulated and unregulated utilities across the United States.

Electric utilities (by franchise agreement) must provide for a customer's maximum power demand, whereas the monthly energy billing only compensates them for average demand, which is always lower. One technique used by electric utilities to account for this condition is to impose a demand charge for short-term power (kW) requirements in addition to the normal energy (kWh) billing for overall consumption. This technique has long been standard for industrial and commercial user rates, but it is still unusual for residential users. The demand charge is especially useful as a means of encouraging users to reduce their peak loads. In so doing, energy use is also somewhat reduced. As utilities continue to adjust to deregulation and rapidly fluctuating fuel costs, and incur difficulties meeting growing demands for service, expect to see more innovative (and more complex) billing schemes being implemented.

27.13 ELECTRIC DEMAND CHARGES

Electric utility companies normally impose a demand charge on all but individual residential and a few special-category customers. Varying with the utility company involved, this monthly charge runs between \$2 and \$15 per kW of maximum average demand

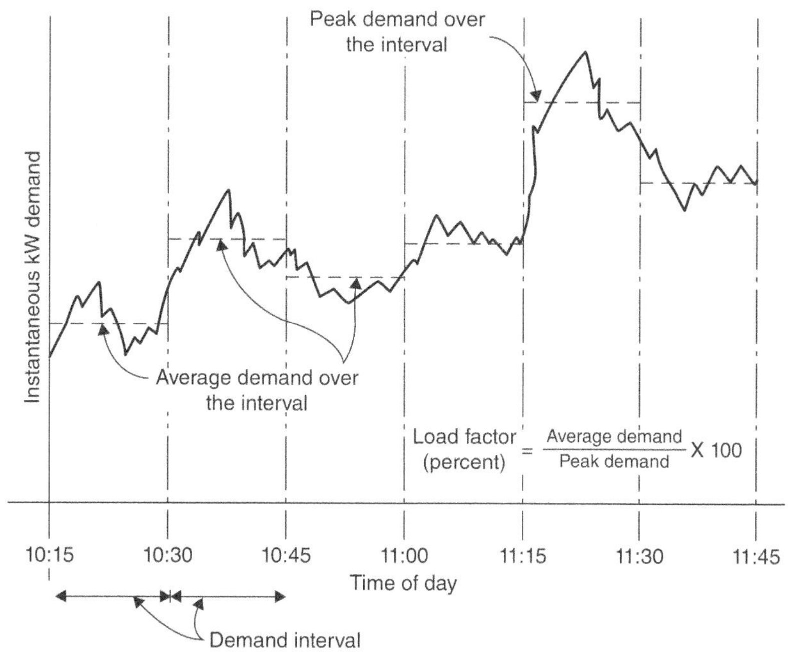

Figure 27.13 *Typical instantaneous load curve for a commercial facility. The utility demand meter records the average demand in each period (here, 15 minutes). The maximum interval demand—in this case, between 11:15 and 11:30—is used as the basis for monthly billing. A high load factor (utilization factor) indicates that little can be gained by demand control; a low load factor indicates the reverse.*

in any demand interval for a given month. Demand intervals vary, usually being either 15 or 30 minutes (see Figure 27.13).

Many utilities use a sliding-window interval-timing technique that starts a new interval every minute and updates the maximum interval demand accordingly. This enables them to find and bill for the maximum electric power demand in any 15- or 30-minute period in a month. Some companies also include a ratchet clause that imposes a demand charge for several months based on the maximum demand in any single month. This penalizes users with seasonal highs—that is, users with a low yearly load factor. The load factor (sometimes referred to as utilization factor) is a measure of uniformity of power demand; a low load factor indicates short-time demand peaks for which the user is heavily charged. The justification for the imposition of a demand charge and the significance of load factors can best be demonstrated by an example.

Assume that a pottery manufacturer, whose average 8-hour daily load is 20 kW for lighting and pottery wheels, operates two 50-kW electric kilns twice a month for a 4-hour period each time. Further assume an energy charge of \$0.10 per kWh. The total monthly energy bill for the operation of the two kilns would be:

$$\text{energy cost} = 2 \times 50\,\text{kW} \times 8\,\text{h} \times \frac{\$0.10}{\text{kWh}} = \$80$$

Thus, were it not for a demand charge, the utility company, which is required by law to supply the maximum customer demand, would have to provide and maintain 100 kW of generation, transmission, and distribution facilities in return for a payment that is the equivalent of 1.11 kW of average continuous load, that is

$$\text{equivalent continuous load} = \frac{\$80.00}{720\,\text{hr/month} \times \$0.10}$$
$$= 1.11\,\text{kW}$$

The user's load factor can be calculated readily. By definition,

$$\textit{load factor} = \frac{\text{average power demand}}{\text{maximum power demand}} \qquad \text{(Eq. 27.8)}$$

For a given time interval (e.g., day, month, or year), the average power demand is equal to the energy used in the period divided by the period length.

$$\text{average period power demand} = \frac{\text{kWh energy use}}{\text{hours of use}} \text{ for that time period} \qquad \text{(Eq. 27.9)}$$

Therefore, the average monthly power demand equals the average monthly energy use divided by 720 hours. Substituting the monthly version of Equation 27.9 as the general expression for the load factor in Equation 27.8 gives

$$\text{monthly load factor (LF)} = \frac{\text{monthly kWh energy use} \div 720\text{ hours}}{\text{maximum demand}} \qquad \text{(Eq. 27.10)}$$

For the case under consideration, the monthly load factor is

$$LF = \frac{((23 \text{ days} \times 8 \text{ h} \times 20 \text{ kW}) + (2 \text{ days} \times 4 \text{ h} \times 100 \text{ kW})) \div 720 \text{ h}}{120 \text{ kW}}$$

$$= \frac{(3680 + 800 \text{ kWh/month}) \div 720 \text{ h}}{120 \text{ kW}}$$

$$= 0.052 \text{ or } 5.2\%$$

This is a very poor load factor, which results in the imposition of a high demand charge. Assuming an $8.00 per kW demand tariff, this pottery manufacturer would be billed, monthly, an additional

$$\textit{demand charge} = 120 \text{ kW} \times \$8.00 = \$960$$

With an energy bill of only

$$\text{energy cost} = 4480 \text{ kWh} \times \$0.10 = \$448$$

This pottery producer is paying heavily for high peak power use.

Although the illustration selected is somewhat extreme in its pattern of electricity use, it is not uncommon to find demand charges of the same order of magnitude as energy charges. It is impossible to eliminate demand spikes entirely, but it is certainly possible (and frequently very simple) to reduce them. Such reduction is in the interest of the user, the utility, and the public at large: the user benefits for economic reasons; the utility is able to make more efficient use of its facilities; and the general public avoids unnecessary power plant construction and associated inefficient use of fuel during partial generator loading, and benefits from overall reduction in fuel use. The last item is a secondary benefit of demand control.

The next section discusses user electric demand control, the primary function of which is to reduce electric power demand. This reduces demand charges and, incidentally (secondarily), somewhat reduces energy consumption. Electric demand control is different from energy management, which is primarily concerned with reduction of all types of energy use, including electricity. Electric demand control is frequently included as one part of an overall energy-management system.

27.14 ELECTRIC DEMAND CONTROL

Electric demand control refers to internal strategies that utility customers use to manage and optimize their own electric power demand. Demand control methods vary greatly in complexity and in degree of automation, but all basically perform the same task—enabling efficient utilization of available energy to produce a high load factor, resulting in a lowering of demand charges. An ancillary but important benefit is the improved utilization of building electrical power equipment, which normally runs underloaded. When demand control is incorporated during the original design (instead of as a retrofit), the result is smaller equipment, a lower first cost, and less space allocated for equipment.

For several years, beginning in the 1980s, many utilities offered their customers rebates that covered up to 40% of the cost of equipment and renovations that would reduce maximum power demand and overall energy use. These programs acted as a clear financial incentive for the development of cost-effective demand control and energy conservation equipment and techniques that have since become widely adopted. The result has been a considerable reduction in nationwide per capita electric power and energy use. These rebate programs reduced the investment payback period to such an extent that many (perhaps most) large power users invested large sums in electric demand control and energy conservation and management equipment.

The advantage to the energy user is the cost: lower electric bills. The advantage to the utility is equally simple. It costs $3000–$7000 per kilowatt of generating capacity for new power plant construction (depending upon the location and required auxiliary construction), whereas rebates run from $150 to $1000 per kW conserved, with the larger amounts paid for by peak demand reduction. Because a utility must supply all the power demanded by customers, it is very much in the interest of any utility that is generating power output near the maximum capacity of its equipment to reduce loads in general and peak loads in particular.

The oldest and simplest utility-sponsored demand control scheme, still in use today, is the time-of-day-dependent, variable utility rate schedule. This schedule can be leveraged for time-variable pricing (TVP) in which utilities will vary prices at different times of the day (and often seasonally). TVP encourages off-peak use of electricity by offering a lower energy rate for consumption during off-peak hours. TVP types include time-of-use (TOU) rates, real-time pricing (RTP), day-ahead hourly pricing, or block-and-index pricing (or black-and-swing pricing).

Utilities offering TVP may install, at no charge to the user, a separate time-controlled circuit switch for use with time-deferrable loads such as water heaters, well pumps, battery chargers, and the like. The circuit is energized only during off-peak hours, which typically are mid-afternoon and after midnight. One such arrangement is shown in Figure 27.14. Note that the switch itself has no programming buttons. All programming is done by the utility with an auxiliary programmer (shown) that can be used to program many user switches.

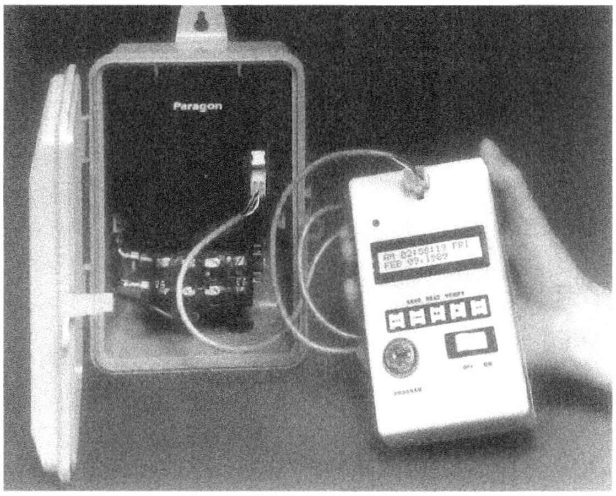

Figure 27.14 Programmable electronic time-control switch designed for time-of-use (off-peak) utility-rate schedules. The programming device is separate from the switch and can be used by a utility to program many customer switches. The illustrated unit is arranged for 365-day scheduling, which permits full coordination with utility schedules that vary with the seasons of the year. Typical controlled loads include water heaters, thermal storage units, water and air accumulators, and any other load that either inherently or by design can be delayed for several hours. (Courtesy of Paragon Electric Co., Inc.)

Advances in building automation technology and internet-enabled "smart" devices have expanded the concept of programmable electronic time-controlled switches that now entail sophisticated programs by which individual equipment may be enabled for demand-response (DR), which is a short-term and voluntary decrease in electricity consumption triggered by grid reliability and/or high wholesale market prices. A growing number of electric utilities now offer incentive-based demand response programs that reward commercial and industrial consumers who agree to change their usage patterns when the utility (or independent system operator [ISO] or curtailment services provider [CSP]) sends an alert—known as a demand response event (or curtailment event)—to the building automation system (BAS)—also called a building management system (BMS)—a programmable load control, or the building's operator (for manual response). DR programs set a maximum number of events that can be announced and specify the time frames in which they may occur. During a DR event, customer-side action typically entails either the ability to reduce consumption and/or use a behind-the-meter generator to serve a load that otherwise need be served by the utility. Such pre-programmed DR measures may involve programming DR capabilities into the BAS/BMS. Some customers may need to coordinate their information technology (IT) infrastructure with the local utility in order to allow DR capabilities.

The basic concept of DR programs is simple: electric loads are disconnected and reconnected in such a fashion that demand peaks are leveled off and the load factor is thereby improved. The extent to which a user's electric loads can tolerate this type of switching is an indication of the potential effectiveness of a demand controller. An installation with large uninterruptible loads, such as from computers or other productivity-related equipment, benefits minimally from demand control. Most industrial and commercial installations, however, contain a large percentage of interruptible loads (interruptions may be very short), and DR programs frequently accomplish a 15% to 20% reduction in electric bills with a resultant short payback period on equipment investment.

The proliferation of demand control equipment and demand response programs has also produced a proliferation of nomenclature, including load shedding control, automated load control, peak demand control, and programmable load control. Descriptions that include the term "energy management" refer to devices whose primary function is the control of energy consumption and that secondarily are equipped to provide electric demand control.

Demand control devices are intended to control power, which is timed energy use. Demand control produces energy savings as a secondary benefit. The expression electric demand control is used for simplicity in the following sections. Various demand control schemes are discussed in some detail to enable the reader to differentiate them by recognizing their specific characteristics. Much equipment on the market today provides additional functions not directly related to demand control. Thus, a good understanding of the essentials of demand control schemes will greatly assist the prospective user. As with other areas of design, manufacturers' literature usually comprises a lengthy list of equipment abilities, few (if any) shortcomings, and little indication of specific or comparative applicability.

(a) Level 1: Load Scheduling and Duty-Cycle Control

This level is the simplest and most direct approach and is applicable to all types of facilities. A facility's electric loads are analyzed and scheduled to restrict demand. Accordingly, large loads can be shifted to off-peak hours and controlled to avoid coincident operation. The scheduling can take advantage of special night and weekend utility rates for loads (such as battery charging and transfer pumping) that do not require a specific time of operation. The demand control device used is essentially a programmable time switch (see Chapter 26) applied to a number of circuits or loads. It is not, strictly speaking, a demand controller, in that no real-time measurement is made of the actual continuous electric load. Instead, the control operates on a preset timed duty cycle relying entirely on a prior analysis of the building loads. Typical applications of this device are control of HVAC loads, lighting loads, and process loads in small commercial, institutional, and industrial buildings.

(b) Level 2: Demand Metering Alarm

When, in conjunction with a duty-cycle controller, some type of continuous demand metering is installed that goes into alarm mode when a predetermined demand level is exceeded, a basic load (demand) control system is established. The load analysis discussed previously is structured and used to determine load priorities, so that when a preset maximum demand load is exceeded and the alarm sounds, loads can be shed (disconnected) manually in a preestablished order of priority—and subsequently reconnected, also in order of priority. This type of control is practical only for a moderate-size installation, inasmuch as the load-switching activity is manual.

(c) Level 3: Automatic Instantaneous Demand Control

This type of control (also called rate control) is, in effect, an automated version of the level 2 system described in the preceding section. The control unit accepts instantaneous kilowatt load information from the utility system, compares this information to a preset kilowatt limit, and acts automatically to disconnect and reconnect loads as required. These units do not recognize the utility's metering interval but act continuously on the basis of load comparison data. For this reason, these units are also referred to as load comparator controllers. Figure 27.15 will help explain the unit's operation.

Typical sheddable loads might include nonessential lighting, some cooling load, domestic hot water heating, snow melting, and the like. Nonsheddable loads (that do not tolerate even short interruption) might include essential lighting, elevators, communications equipment, computers, process control, emergency equipment, and the like. The nonsheddable loads are fed directly from the power line. The sheddable loads are fed through a panel of control relays that respond to on/off instructions from the demand controller. Although the resulting energy use with or without the controller is theoretically the same, in practice, energy savings of 15% and more are common.

The principal drawback of this system is that the load shedding is preprogrammed. This results in an inability to readily adapt to varying load patterns resulting from variable production schedules, time schedules, changes in weather, and so on. As a result of this limitation, this system is most useful in applications where operating modes do not change frequently and the facility is not very large. Thus, stores, supermarkets, warehouses, small industrial facilities, and commercial installations are well served with this type of system if they have at least 20% sheddable loads and their connected electric load is at least 150 kV-amperes (kVA).

This level of demand control—as well as levels 4 and 5, described in the following sections—is most often supplied as one of many functions of a larger building energy control system. Because of this bundling, it is necessary to ensure that the type of demand control to be supplied by the overall building control hardware and software is what the electrical system designer intended.

(d) Level 4: Ideal Curve Control

This control function operates by comparing the actual rate of energy usage to an ideal rate and controlling kilowatt demand by adjusting the total energy used within a metering interval. The utility company determines the demand over the demand interval by integrating the kilowatt-hour energy consumption over the interval and dividing by the interval time.

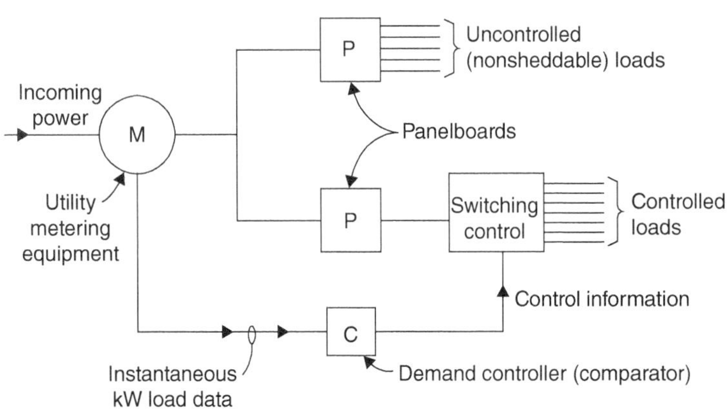

Figure 27.15 *Block diagram of a system for automatic electric power demand control. The demand controller receives instantaneous load data from the metering equipment, compares them to preset limits, and disconnects and reconnects controllable loads automatically to keep kW demand (load) within these limits.*

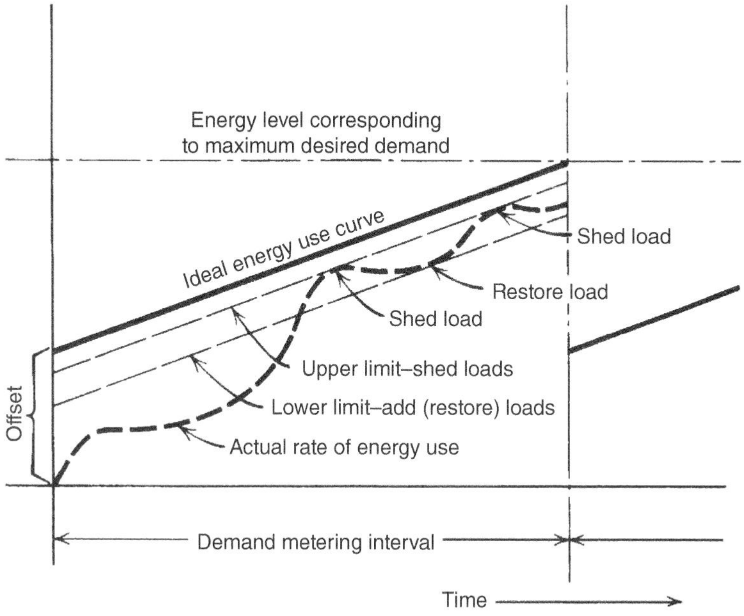

Figure 27.16 Graph of the action of an ideal rate controller. Note that toward the latter part of the period the pattern has been established, and the actual rate of energy use corresponds closely to the ideal. This type of control is often simply a part of the software in a large, centralized building control system.

Thus, the user is actually given a block of energy (kWh) that can be utilized at any desired rate, not necessarily at a constant rate. The desirable rate of energy use is shown as the "ideal curve" on the typical usage curve shown in Figure 27.16.

Shed points can be programmed independently for each load according to a predetermined priority, and priorities can be readily adjusted and rescheduled. Loads are normally shed only toward the end of an interval—when the permissible energy total for the interval is approached—and all loads are restored at the beginning of the next interval. Thus, during each interval, sheddable loads are off for only a few minutes at most. Controllers operating on the ideal curve principle are considerably more flexible than the kilowatt rate controller described in the level 3 system, and are applicable to facilities of widely divergent load size, but with at least a 300-kW connected load. As with other controllers, the principal savings will be in demand charges, but almost always with considerable savings in energy billings.

(e) Level 5: Forecasting Systems

These systems are by far the most sophisticated, the most expensive, and the most effective. They are best applied to large facilities where the number of loads, load patterns, and the complexity of operation preclude the effective use of the preceding systems. Because of the large amount of load data that needs to be processed, these systems are usually installed as part of a central control system (e.g., building automation system). Details of operation are too complex to be described here, but the basic operation can be outlined. These control units operate by continuously forecasting the amount of "unpenalized" energy remaining in the demand interval, based upon kilowatt-hour data received. They then examine the status and priority of each of the connected loads and decide on a course of action. Loads that in other systems are classified as nonsheddable are, in this system, controlled because of the accuracy and rapidity of the control function. A pneumatic compressor, for instance, that supplies process air might, in lower-level control systems, be classified as noncontrollable, despite the fact that it has long off periods. With computer control, such a load would be classified as "delayable" or "inhibited" because a 30-second or 1-minute delay in activation after the pressure switch closes its contact is normally acceptable.

The advantage of these systems is that, if programmed properly, they can make small, accurate load changes throughout an interval, resulting in minimum load cycling and maximum efficiency.

27.15 ELECTRICAL MEASUREMENTS

The preceding sections have explained the fundamental electric quantities of voltage and current and defined the units involved as volts and amperes, respectively. As is true of all physical quantities used in practical applications, a need exists for a simple means of measuring these quantities. This need was initially met by the development of the meter movement, illustrated in Figure 27.17.

Everyone at one time or another has felt the repulsion between like poles of two magnets held close

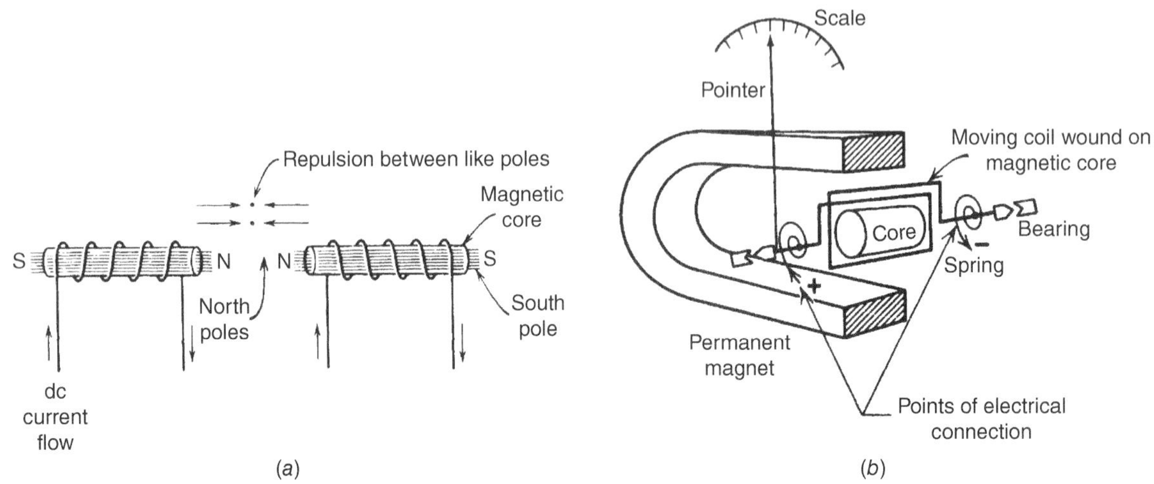

Figure 27.17 *(a) Diagram showing the basic electromagnetic principle and the interaction between electromagnets. Any iron core becomes an electromagnet when current flows in a coil wound around it, as shown. (b) The principle of the electromagnet is used in this basic meter movement. Current flowing in the movable coil forms an electromagnet whose field interacts (see a) with the permanent magnet's field, causing a deflection proportional to the current flow. (Courtesy of Wechsler, a division of Hughes Corp.)*

together and, conversely, the attraction between opposite poles. This principle was used in the first basic meter movement: causing a deflection of the pointer as a result of the repulsion between the field of a permanent magnet and an electromagnet. The electromagnet is formed when current flows in the coil, and its strength is proportional to the amount of current. Thus, a strong current causes a larger deflection of the needle and therefore a higher reading on the dial. A spring (see Figure 27.17*b*) provides restraining torque on the pointer. To make this very sensitive unit usable for large currents (it is intrinsically a micro-ammeter, sensitive to millionths of an ampere), the device simply diverts, or shunts away, most of the current, allowing only a few microamperes to actually flow in the meter coil.

To use the same device as a voltmeter, a large resistance called a multiplier is placed in series with the meter, thereby again limiting the current flow to a few microamperes. The meter scale is then calibrated in volts. All DC meters are made in this fashion. Most AC meters operate on basically the same principle, except that instead of a permanent magnet, an electromagnet is used. Thus, when the polarity reverses, the deflecting force retains the same direction. A DC meter connected to an AC circuit simply will not read because inertia prevents the needle from bouncing up and down 60 times a second.

The meters just described are conventional analog devices that read electrical values in proportion to mechanical forces exerted within the device—that is, by analogy. Modern electronics has produced a line of solid-state electrical meters (Figure 27.18) that display the measured electrical values in analog mode (needle on a dial), digital mode, or both. They operate in a number of ways, all of which are different from that described previously and are beyond the scope of this book.

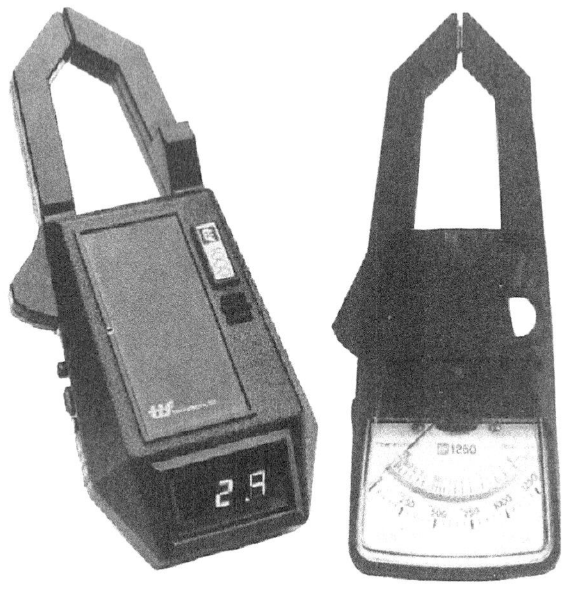

(*a*) (*b*)

Figure 27.18 *(a) Solid-state clamp-on type meter with digital readout and automatic ranging. The latter feature eliminates the necessity to preselect a meter range and is particularly useful where the magnitude of current or voltage is unknown. The clamp-on feature permits use without wiring into, or otherwise disturbing, the circuit being measured. The meter measures approximately 8 × 3 × 1.5 in. [200 × 75 × 40 mm], weighs less than 1 lb [0.5 kg], and is battery-powered. Scales are 0.1–1000 amp AC, 0.1–1000 V AC, and 0.1–1000 Ω resistance, with ±2% accuracy. (b) Solid-state auto-ranging clamp-on AC meter with analog-type readout. It is similar in design to the meter in (a), but with somewhat larger range scales and additional features such as peak current measurement. (Courtesy of TIF Instruments, Inc.)*

The fact that a meter reads digitally does not necessarily mean that it is highly accurate. Accuracy depends upon the quality of the internal circuitry. Digital meters are easier to read than analog meters because no visual interpretation or interpolation is involved. This advantage, plus the constantly declining cost of sophisticated electronics, will undoubtedly lead to solid-state digital meters replacing analog types, except for special applications.

In buildings, the measurement of current and voltage is generally not as important as the measurement of power and energy. To measure power, the fact that power is proportional to the product of the voltage and current in a circuit is employed. Although actual construction is complex, the theory of operation of a conventional coil-type wattmeter is simple. The meter has two coils: a current coil that is similar in connection to an ammeter and a voltage coil that is similar in connection to a voltmeter (Figure 27.19). By means of the physical coil arrangement, the meter deflection is proportional to the product of the two and therefore to the circuit power. The meter can be calibrated as desired, depending upon the size of the shunts and multipliers.

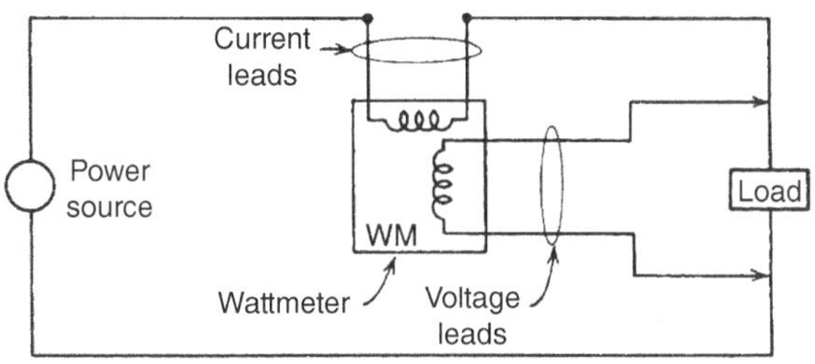

Figure 27.19 *Schematic arrangement of wattmeter connections. Note that the current coil is in series with the circuit load, whereas the voltage leads are in parallel.*

To measure energy, the factor of time must be introduced because:

$$\text{energy} = \text{power} \times \text{time}$$

Direct-current energy meters are available but are not currently of general interest because of the rarity of DC power. Alternating-current watt-hour meters are basically small motors, whose speed is proportional to the power being used. The number of rotations is counted on the dials, which are calibrated directly in kilowatt-hours. A diagram of the basic construction of an AC kilowatt-hour meter is shown in Figure 27.20. As can be seen, kilowatt-hour energy consumption and maximum interval kilowatt demand can be read directly from the dials. If the numbers involved are too large (because of calibration), a multiplying factor is required to arrive at the proper kilowatt-hour consumption. This number is written directly on the meter nameplate, and the meter reading is multiplied by it to get the actual kilowatt-hours. Several types of multifunction and energy meters are shown in Figures 27.21, 27.22, and 27.23.

Continuous monitoring of the electrical characteristics of an entire distribution system, or of a portion fed from a specific switchboard, can be readily accomplished by use of solid-state devices. Such programmable system analyzers can be arranged to measure power usage, perform power quality and harmonic analyses (and the like), and display the results locally and/or remotely.

ELECTRICITY

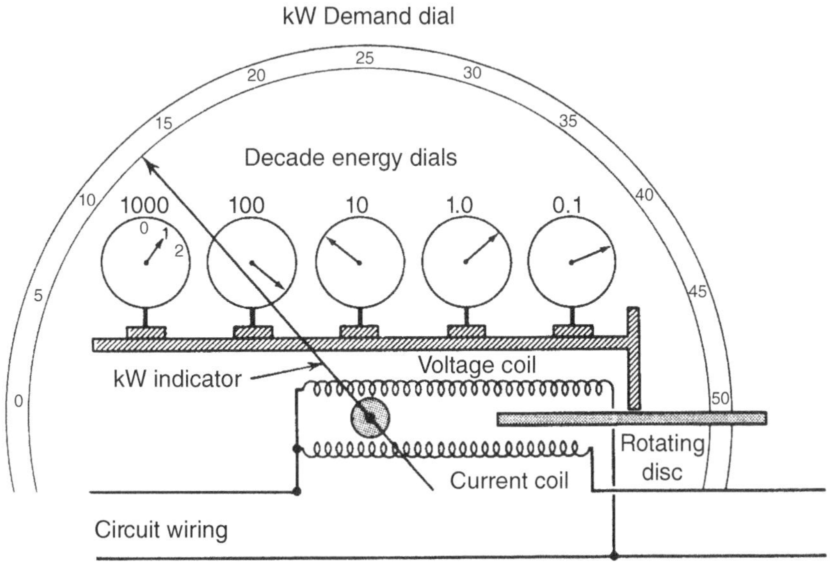

Figure 27.20 *Typical induction-type kilowatt-hour meter with kilowatt demand dial. Dials register total disc revolutions, which are proportional to energy. Disc rotational speed is proportional to power. Note that the current coil is in series with the load and that the voltage coil is in parallel.*

Figure 27.21 A multifunction solid-state kWh meter. This single-phase meter includes a mechanical register like that shown in Figure 27.20 plus a large liquid crystal display. It can be arranged to provide either time-of-use billing or demand/time-of-use billing in addition to its energy measurement function. It can also be configured to provide load control, demand threshold alert, end-of-interval alert, and load profile recording. (Courtesy of Landis and Gyr.)

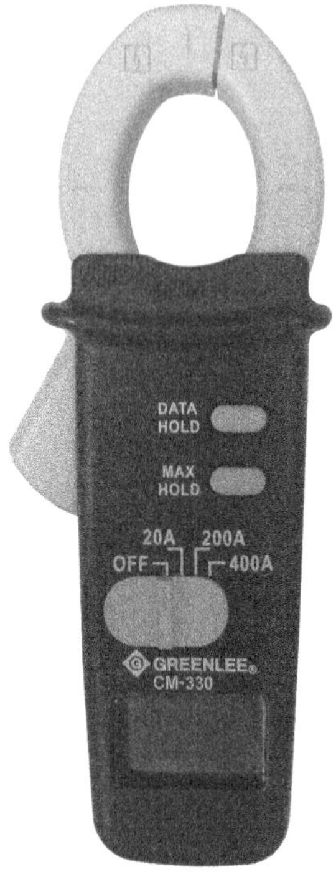

Figure 27.23 Handheld clamp-on ammeter. This ammeter has a maximum jaw width of 1.18 in. [30 mm] and can measure up to 400 amps of alternating current.

Figure 27.22 A digital kWh meter. Some digital meters allow homeowners to track their energy consumption in one-hour increments online. (Courtesy of Landis and Gyr.)

Several interesting, labor-saving kilowatt-hour meters have been developed to make electricity billing and control easier. One type is equipped with a programmable electro-optical automatic meter-reading system that can be activated locally or from a remote location. The meter data are transmitted electrically to a data-processing center, where they may be used by the utility to prepare subscribers' bills; prepare customer load profiles; and study, in combination with other data, area load patterns, equipment loadings, and so on. The customer can use instantaneous data to control loads, as explained in the preceding section.

Another meter type is equipped with a miniature radio transmitter that can be remotely activated to transmit the current kilowatt-hour reading. The meter reader moves along a street and remotely activates a meter by entering a customer code into a digital pad. A special receiver not only receives the transmitted kilowatt-hour data but encodes and records the data automatically. These and other "smart" kilowatt-hour meters are more costly than traditional meters, but they are gradually being introduced because of large reductions in labor costs and increases in the quantity and quality of data made available. In this regard, programmable meters are currently being deployed among residential customers in some areas to allow utility company control of demand (with the agreement of the consumer—usually resulting from a more favorable rate structure). Air-conditioning load-shedding controls are fairly common in many parts of the United States.

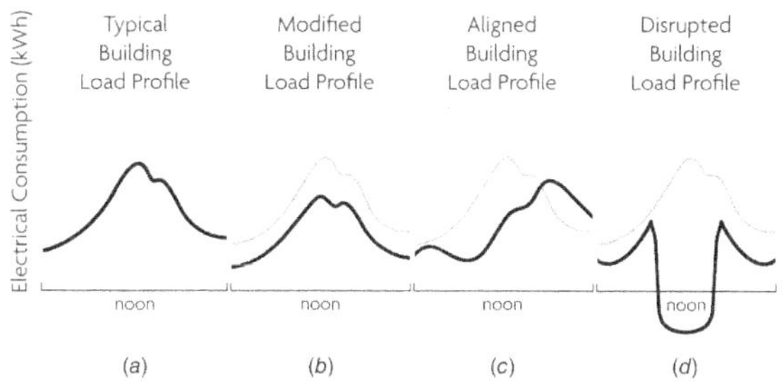

Figure 27.24 *Schematic representation of potential load profile changes brought about revised by building design and use. (Drawn by Evelyn Dinh after information from RMI, 2018, and Phius, 2024.)*

27.16 ELECTRIFICATION, SMART GRIDS, AND MICROGRIDS

Now, and surely looking forward, it is likely that load modifiers, load aligners, and load disruptors will play an increasingly important role in the design of building electrical systems. Electrification, smart grids, and microgrids are either drivers of change or responses to changes in how we consume energy. These trends are broadly discussed in this section. But first, the basic concept of a load profile must be engaged.

Load Profiles. A load profile is simply a graph of building load (in this case, electrical load) versus time. Figure 27.12 is an example of a typical residential load profile. Historically, electrical load profiles were of little interest to the design of residential-scale buildings because such buildings were not subject to electrical demand charges. Load profiles were of interest to many commercial and institutional building owners and designers because such buildings were subject to demand charges—which could be substantial and were determined by the peaks on the load profile.

Electric utilities were always interested in load profiles because they collectively established the need for base load and peak load generating and distribution capacity, which was expected to be met under the terms of government-granted operating licenses. Large-scale grid failures tend to be caused by loads that exceed generating capacity (often occurring under stressful weather conditions).

Traditional load profiles (see Figure 27.24*a*) are being acted upon by several factors that change their shapes—and thus, how buildings interact with the electrical grid. These transformative factors can be classified as modifiers, aligners, or disruptors:

Load modifiers: These are factors that tend to change the magnitude of a load profile without changing its basic shape (Figure 27.24*b*). The most common example of a load modifier is energy efficiency, which depresses the magnitude of a load profile more or less equally at all points. Load modifiers have an impact on electric utilities, but such impacts are generally positive (and have been supported by utilities through educational programs, grants, and rebates).

Load aligners: These move loads around on the time axis of a load profile. Such movement changes the fundamental shape of a load profile (Figure 27.24*c*) and may have a positive or a negative impact on the providing utility. Section 27.14 addresses controls that reduce building demand, which is a common load alignment strategy (with a focus on reducing costs for the building owner). Load adjustments may be accomplished by building controls that respond to signals from the grid. Thus, the related term "smart grid." Examples of load aligners include equipment or appliance schedulers (that might defer use of a dishwasher until after midnight) or storage (electric or thermal, that allows loads to be met with asynchronous operation of large equipment, such as a chiller or boiler).

Load disruptors: These radically change the shape and/or magnitude of a load profile (Figure 27.24*d*). Such changes to load profiles can have a substantial impact on both the utility and the building electrical system. Examples include electric vehicle charging, onsite renewable energy generation, and electrification.

Electrification. A push toward electrification is occurring in the United States with numerous proponents extolling the benefits of converting buildings to all-electric operation. Decarbonization is the driving logic behind this change in energy direction, with the assumption that the future electric grid will be low-carbon and therefore a preferable energy resource (versus fossil fuels). Electrification is clearly a load

ELECTRICITY

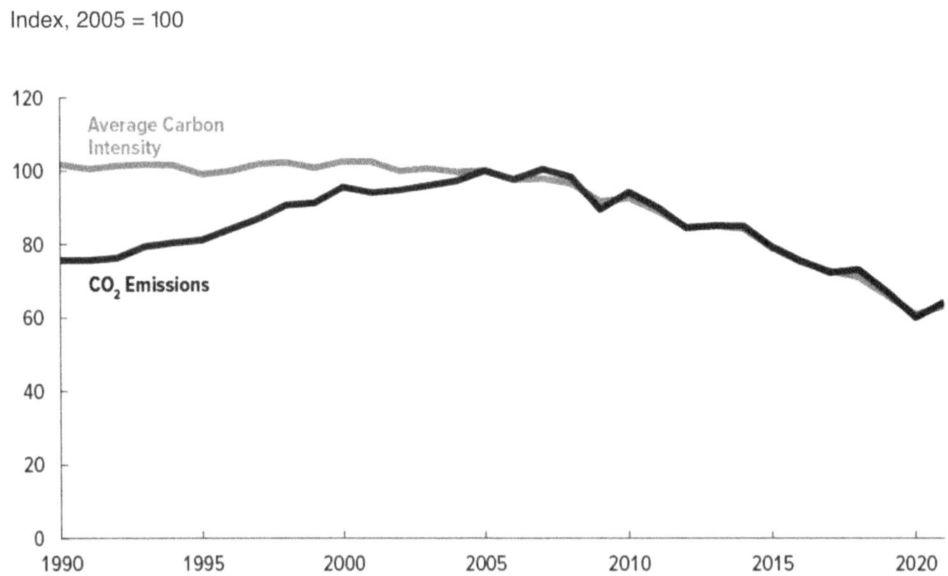

Figure 27.25 *Emissions of carbon dioxide and the carbon intensity of the electric power sector. Congressional Budget Office (CBO, 2022).*

disruptor, with major impacts on electrical generation and transmission capacity and upon building electrical system capacity and components. The effects of electrification on both sides of the meter will be profound.

Electrification is a rational idea in that the electrical grid has recently decarbonized and is likely to continue to do so. It is, however, not correct to state that grid electricity today is low-carbon energy. Such a claim lies sometime in the future. Figure 27.25 (CBO, 2022) shows a multiyear pattern of carbon intensity associated with electrical generation in the United States. Carbon intensity declined from about 0.6 metric tons (MT) of CO_2 per megawatt-hour of electricity produced in 2005 to about 0.4 MT of CO_2 per megawatt-hour in 2021. This is a change in the right direction and is expected to continue as dirtier generation is replaced by cleaner generation (such as from renewables).

Smart Grids. A smart grid is an advanced electricity distribution network that utilizes "intelligent" digital communication systems and controls to monitor and manage the generation and distribution of electricity. Smart grid technologies allow a two-way flow of electricity along with communication capabilities that facilitate power delivery improvements and increased infrastructure stability.

A smart grid can engage grid-interactive efficient buildings (GEBs) and/or distributed energy resources (DERs) to provide demand flexibility, reduce energy costs, and moderate utility loads via load shifts and load shedding—which can potentially reduce a utility's peak demand, mitigate its greenhouse gas emissions, and increase its reliability and resilience (Roth, 2019).

Microgrids. A microgrid is a set of interconnected and distributed energy resources coupled with loads that function as a single controllable entity with regard to the grid (the broader infrastructure for electricity generation and distribution; also referred to as the macrogrid). Figure 27.26 provides a schematic representation of a microgrid.

Microgrids are designed to connect and disconnect from the macrogrid as is most advantageous, thus operating in either a grid-connected or island mode. As such, one of the key advantages of microgrids is that they enable local power generation assets—often including small-scale renewable energy resources and distributed power storage technologies—to keep a defined collection of loads energized even when the larger grid experiences interruptions or failure. Thus, microgrids can potentially improve the reliability of electricity delivery when the broader infrastructure experience disturbances. Microgrids can also provide services (such as voltage support or frequency regulation) to a macrogrid. For more detailed information on microgrids, see the *Architect's Guide to Ultra-Low-Energy Buildings, Microgrids, & Direct Current* (Phius, 2024).

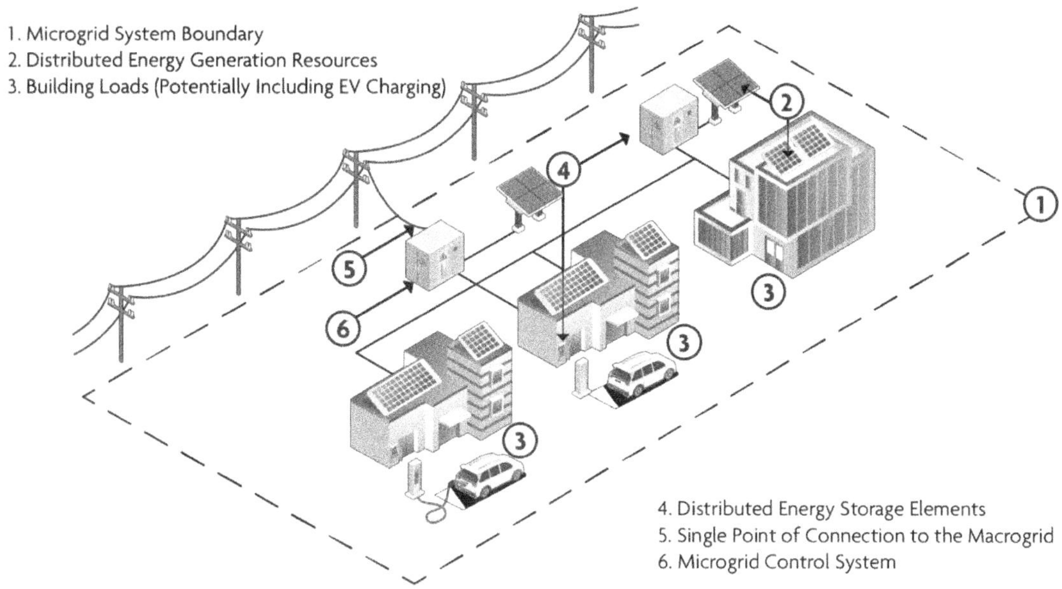

Figure 27.26 *Key components and their arrangement for a small-scale microgrid. (Drawn by Evelyn Dinh after Phius, 2024.)*

References and Resources

CBO. 2022. *Emissions of Carbon Dioxide in the Electric Power Sector*. Washington, DC: Congressional Budget Office. cbo.gov/publication/58860

NREL. *Microgrids*. Golden, CO: National Renewable Energy Laboratory nrel.gov/grid/microgrids.html.

Phius. 2024. *Architect's Guide to Ultra-Low-Energy Buildings, Microgrids, & Direct Current*. Chicago, IL: Phius. phius.org/architects-guide-ultra-low-energy-buildings-microgrids-direct-current

RMI. 2018. *Grid-Integrated Buildings: A Profitable Linchpin to Decarbonization*. Basalt, CO: Rocky Mountain Institute. rmi.org/grid-integrated-buildings/

Roth, A. 2019. *Grid-Interactive Efficient Buildings Technical Report Series: Whole-Building Controls, Sensors, Modeling, and Analytics*. Washington, DC: U.S. Department of Energy. nrel.gov/docs/fy20osti/75478.pdf

Shoemaker, S. 2023. *NREL Researchers Reveal How Buildings Across United States Do—and Could—Use Energy*. Golden, CO: National Renewable Energy Laboratory. nrel.gov/news/features/2023/nrel-researchers-reveal-how-buildings-across-the-united-states-do-and-could-use-energy.html

CHAPTER 28

Electrical Systems and Materials: Service and Utilization

THE FIRST STEP IN DEVELOPING AN UNDERSTANDING of building electrical systems is to examine how electric service is brought into a facility.

28.1 ELECTRIC SERVICE

The codes and standards that apply to electric service include:

1. *National Electrical Code © (NEC)*, National Fire Protection Association (NFPA) (NFPA, 2022).
2. *National Electrical Safety Code (NESC)*, published by the Institute of Electrical and Electronics Engineers (IEEE). This code is recognized by the American National Standards Institute (ANSI). It deals with clearances for overhead lines, grounding methods, underground construction, and related matters.
3. Standards of the utility supplying the electric service.

Public utility franchises in the United States require only that service be made available at the private property line. Thus, service is normally connected to the utility lines at a mutually agreeable point at or beyond the property line. The service tap may be a connection on a pole with an overhead service drop or an underground service lateral to the building, or a connection to an underground utility line with a service lateral to the building. Electrical construction work on private property is usually at the owner's expense.

Under certain conditions, the owner can influence the type of service connection used by the electric utility company in conveying electric service to a site. This is often the case in large tract developments and in places where owners are willing to share some of the cost of better-grade construction. Also, in many areas, the utilities themselves have instituted "beautification" programs in an effort to decrease the objectionable appearance of much of their equipment.

Service from a utility line to a building may be run overhead or underground, depending upon the following factors:

- length of the service run
- type of terrain
- customer participation in the cost-of-service installation
- service voltage
- size and nature of the electric load
- importance of appearance
- local practices and ordinances
- maintenance and service reliability
- weather conditions
- type of interbuilding distribution, if applicable

Mechanical and Electrical Equipment for Buildings, Fourteenth Edition. Walter Grondzik, Daniel Overbey, and Alison Kwok.

Companion website: www.wiley.com/go/meeb14e

28.2 OVERHEAD SERVICE

The principal advantage (to the utility) of overhead electric lines is low cost. Depending upon terrain and other factors, the cost saving of overhead compared to underground installation has historically ranged from 10% to 50% (the latter when compared to direct burial cable installation). This accounts for the majority of installations being overhead. In recent years, special techniques in underground installation have lowered that cost, making it a reasonable economic alternative when its advantages are considered.

Where the length of the service run is several hundred feet [meters] or more, voltages higher than the facility's utilization level may be involved. This heavily favors installation of overhead lines, particularly with voltages exceeding 5000 V. Similarly, when terrain is rocky and the electrical load is heavy, the cost of underground installation is prohibitive. Because overhead lines are easily maintained and repaired and faults easily located, service continuity with overhead lines is generally acceptable. In areas with severe weather conditions, called heavy loading areas, where combinations of snow, wind, and ice increase the possibility of outages on overhead lines, underground service is preferable. This is particularly true when even short service outages cause hardship or financial loss. Reliability of overhead service can be improved markedly by taking service from two separate, and preferably separated, overhead lines. A final decision on the type and location of electrical service will be made after meetings among the building's architect, the electrical engineer, and the electric utility's technical personnel.

Overhead cables are of several types: bare, weatherproof, or preassembled aerial cable. Bare copper cables supported on porcelain or glass insulators on crossarms are normally used for high-voltage lines (2.4 kV and higher). Low-voltage circuits (600 V and below) are generally run on porcelain spool secondary racks using single-conductor (1/c) weatherproof cable. Preassembled aerial cable consists of three or four insulated cables wrapped together with a metallic tape and suspended by hooks from poles. This type of construction may be used for voltages of up to 15 kV (Figure 28.1) and often proves to be more economical than crossarm or rack installation and more resistant to damage from severe weather conditions.

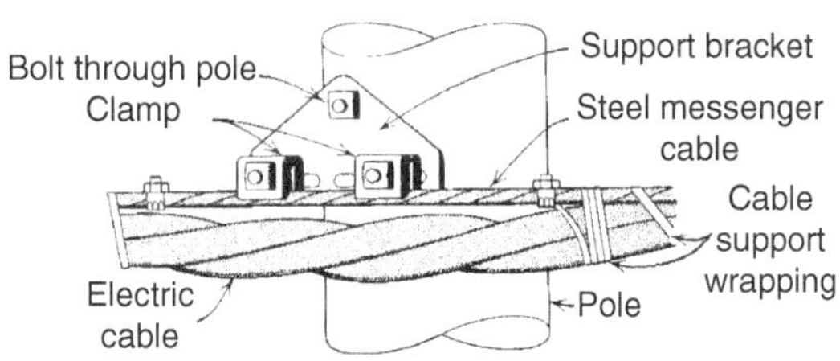

Figure 28.1 *Preassembled aerial messenger cables are carried by steel cables clamped to poles.*

A typical detail of an overhead electric service entrance to a multifamily residential building is shown in Figure 28.2.

28.3 UNDERGROUND SERVICE

The advantages of underground electric service are attractiveness (lack of physical and visual clutter overhead), service reliability, and long life. The principal disadvantage is high cost. To overcome this, utilities frequently use direct burial techniques that, by eliminating a raceway, reduce costs considerably. Because direct buried cable cannot be pulled out if it faults, as is the case with raceway-installed cable, restoration of service after a cable fault is time-consuming. It is recommended that the decision on which technique will be used be based upon consideration of these factors:

- the cost premium for underground raceway installation, including handholes if required (see Section 28.4)
- the history of outages for direct burial installation by this installer, in the immediate area
- cost and availability of repair service (utilities frequently will repair customer-owned underground service laterals for a fee)
- impact of electric service outage in terms of time delays, inconvenience, necessity of digging up lawns and paved areas, and cost impact in the case of a commercial facility

28.4 UNDERGROUND WIRING

The methods available for underground wiring are:

- direct burial (Figure 28.3)
- installation in Type I, concrete-encased duct (Figure 28.4*a*)
- installation in Type II, direct burial duct (Figure 28.4*b*).

The first alternative offers low cost and ease of installation, with the disadvantage regarding repairs stated previously. The second offers high strength and permanence, but at the highest price of the three. The last offers median cost but little strength. It is applicable only for installations on undisturbed earth and/or under light paving.

Nonmetallic duct (conduit) intended for underground electrical use is commercially available in two wall thicknesses. NEMA (National Electrical

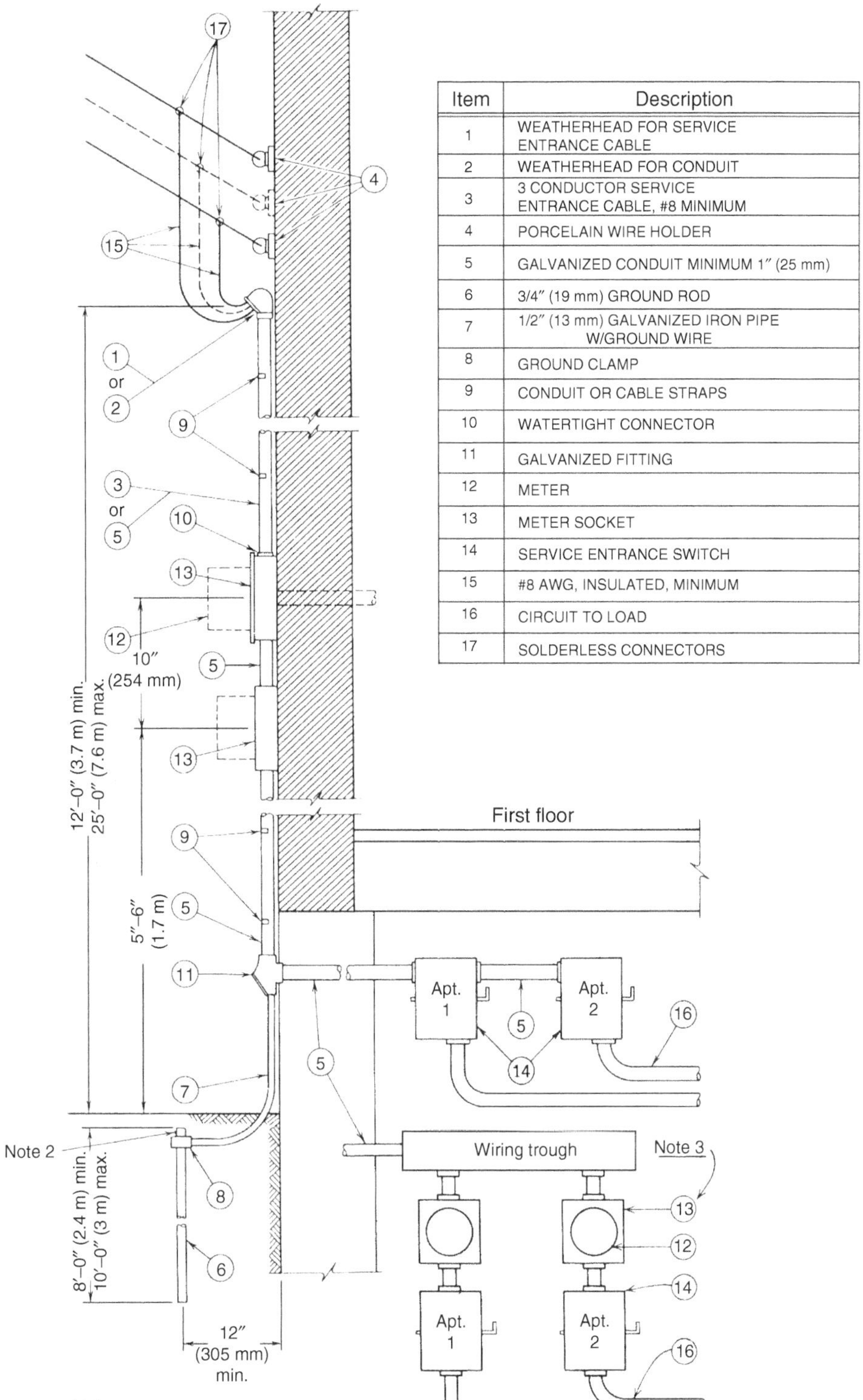

Item	Description
1	WEATHERHEAD FOR SERVICE ENTRANCE CABLE
2	WEATHERHEAD FOR CONDUIT
3	3 CONDUCTOR SERVICE ENTRANCE CABLE, #8 MINIMUM
4	PORCELAIN WIRE HOLDER
5	GALVANIZED CONDUIT MINIMUM 1″ (25 mm)
6	3/4″ (19 mm) GROUND ROD
7	1/2″ (13 mm) GALVANIZED IRON PIPE W/GROUND WIRE
8	GROUND CLAMP
9	CONDUIT OR CABLE STRAPS
10	WATERTIGHT CONNECTOR
11	GALVANIZED FITTING
12	METER
13	METER SOCKET
14	SERVICE ENTRANCE SWITCH
15	#8 AWG, INSULATED, MINIMUM
16	CIRCUIT TO LOAD
17	SOLDERLESS CONNECTORS

Note:

1. Omit item #10 if conduit is used.
2. Cold water pipe ground may be used in lieu of ground rod.
3. Meters may alternatively be placed inside the building.

Figure 28.2 *Detail of typical overhead electric service to a four-family residence. Note that meters in small residential buildings are frequently placed on the exterior of the building. Alternatively, they can be installed inside, provided that ready access is available or some type of remote meter-reading system is installed.*

ELECTRICITY

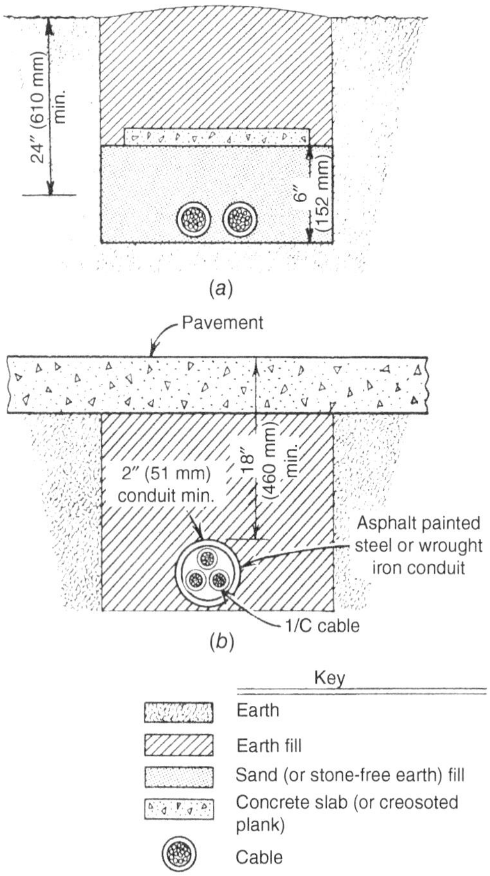

Figure 28.3 *(a) Technique for installation of direct-burial cable. (b) Under highways, streets, and other high-load areas, cable should be installed in metal conduit.*

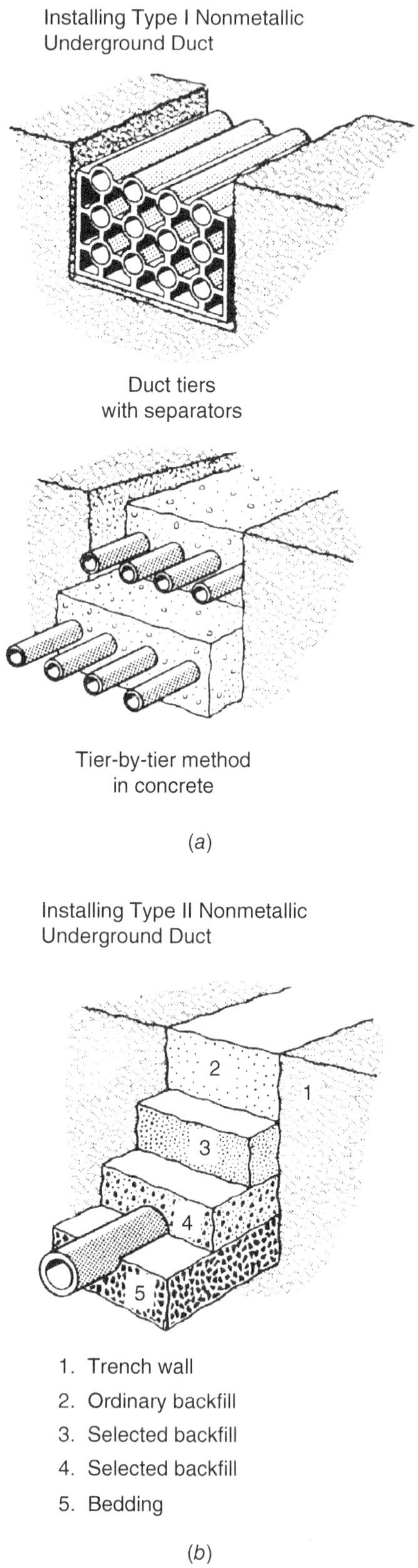

Figure 28.4 *Underground duct installation. (a) The concrete duct bank should have at least 6 in. [150 mm] of earth cover. (b) Heavy-wall duct should be buried at least 24 in. [600 mm] in ordinary traffic areas and 36 in. [900 mm] in areas subject to heavy traffic. Each layer is about 8 in. [200 mm] thick.*

Manufacturers Association) Type II with a heavy wall provides the physical protection required for direct burial installation with no concrete encasement. Type I is manufactured with a thinner wall and is intended for encasement in a minimum of 2 in. [50 mm] of concrete. Such ducts may be manufactured from a variety of materials and are sold under several trade names. Plastic conduit is typically comprised of polyvinyl chloride and is, thus, commonly referenced by the abbreviation PVC.

Nonmetallic conduit is most frequently used without concrete encasement for low-voltage and signal wiring and with encasement for high-voltage wiring. It offers several advantages over steel conduit for underground work, such as lower cost and freedom from corrosion.

When underground electric wiring is duct-installed and the run extends over several hundred feet [meters]—the exact distance depending upon the pulling tension—a pulling handhole or manhole is necessary. Handholes are used for low-voltage power and signal cables and for runs with a small number of cables. Manholes are used for high-voltage cables and where large duct banks must be accommodated. Precast handholes and manholes are readily available in many standard sizes and are usually cheaper than field-formed and poured units.

Cable used in underground wiring must be specially manufactured and approved for that purpose. Type SE is the basic service entrance cable, constructed with a moisture- and flame-resistant covering. When it is provided with waterproofing for underground use, the designation is SE type U, or simply USE. Underground cable for other than service runs is classified as type UF (underground feeder).

28.5 SERVICE EQUIPMENT

Referring to Chapter 29, Figure 29.1, note that a block labeled "transformer" is located between the high-voltage incoming utility lines and the secondary service conductors. This equipment item is required whenever the building utilization voltage is different from the service voltage. A transformer may be pole- or pad-mounted outside the building or installed in a room or vault inside the building, as discussed later.

28.6 TRANSFORMERS

A transformer is a device that changes or transforms alternating current (AC) of one voltage to alternating current of another voltage. Transformers used in building work consist essentially of an iron core on which are wound at least two coils: a primary (coil) winding and a secondary (coil) winding. A voltage impressed on the primary winding induces (through the iron core) a voltage in the secondary winding in proportion to the ratio of turns in the two coils. Thus, a step-down transformer has a larger number of turns in its primary winding than in its secondary winding, and a step-up transformer has the reverse. In theory, transformers are reversible, although in practice they are rarely used that way. Transformers cannot be used on DC.

A transformer typically is used to step down an incoming 4160-V service to 480-V for distribution within a building. Another transformer is then used to step down the 480 to 120 V for use with receptacle circuits. Ordinarily, 120, 208, 240, 277, and 480 V are called low or secondary voltages, and 2400, 4160, 7200, 12,470, and 13,200 V are high or primary voltages.

Transformers are available in single-phase or three-phase construction. Transformer power capacity is rated in kilovolt-amperes (kVA). For a single-phase unit, this figure is the product of the full load current and the voltage. Because the voltages on the primary and secondary are different, so are the primary and secondary currents—because the kVA remains constant.

For example, a single-phase 100-kVA, 2400/120-V transformer will carry at full load:

$$\text{primary current} = \frac{100{,}000\ \text{VA}}{2400\ \text{V}} = 41.6\ \text{A}$$

$$\text{secondary current} = \frac{100{,}000\ \text{VA}}{120\ \text{V}} = 832\ \text{A}$$

That is, the ratio of primary to secondary current is exactly the reverse of the ratio of primary to secondary voltage, because their product, which is the transformer's kVA rating, remains constant. Expressed in an equation:

$$\text{primary current} \times \text{primary voltage} = \text{secondary current} \times \text{secondary voltage}$$

or

$$V_p I_p = V_s I_s \qquad \text{(Eq. 28.1)}$$

where

V_p is the primary voltage
I_p is the primary current
V_s is the secondary voltage
I_s is the secondary current

Heat is generated by the passage of current through the transformer coils due to the winding cable resistance. This heat is transferred to the unit's cooling medium, where it is radiated or otherwise disposed of. The unit's cooling medium is a property of major importance. Transformers are either dry (air-cooled) or liquid-filled. The choice depends upon the required electrical characteristics, the proposed physical location of the transformer, and costs. Although detailed considerations are beyond the scope of this book, some selection criteria are presented in the following sections. In general, units rated above 5 kV are liquid-filled, and units in the 600-V class are dry. Units installed indoors, except in vaults, are normally of the dry type, and are intended for general-purpose lighting and power circuits (Table 28.1). Load center transformers are installed in unit substations, both indoor and outdoor. Distribution transformers are mounted on a pole or on a concrete pad outdoors. Substation transformers are large and are always mounted on a concrete pad.

The insulation class of a transformer affects its permissible temperature rise and operating temperature, and, as a direct result, its physical size, electrical power losses, overload capacity, and life.

The physical size of a transformer of a given kVA rating and voltage depends upon the type of insulation

TABLE 28.1 Typical Transformer Data

Transformer Type (Application)	Maximum Capacity[a] (kVA)	Insulating (Cooling) Medium	Voltage Range	
			Primary	Secondary
General-purpose, dry type	1000	Air	120–600 V	120–600 V
Load center pad-mounted unit substation	3000	Air	2500–15,000 V	120–600 V
Load center pad-mounted unit substation	5000	Oil Silicone fluid[b] Epoxy/resin[c]	2500–34,500 V	120–4800 V
Distribution system	10,000	Oil Silicone	2500–67,000 kV	120–4800 V

[a]Three-phase bank.

[b]High-fire-point paraffinic hydrocarbon fluid, manufactured expressly as a transformer dielectric-coolant.

[c]Solid dielectrics used for special installation, such as in high-hazard areas.

TABLE 28.2 Air-Cooled Transformer Electrical Insulation Temperature Ratings (Based on 40°C Ambient)

Insulation Class System[a]	Insulation Type	Average Conductor Temperature Rise	Ambient Temperature	Hot-Spot Temperature Gradient	Total Maximum Temperature
105°C	Organic (A)	55°C	40°C	10°C	105°C
150°C	Mica, glass, resins (B)	80°C	40°C	30°C	150°C
185°C	Asbestos (F)	115°C	40°C	30°C	185°C
220°C	Silicones (H)	150°C	40°C	30°C	220°C

Note: Ratings are expressed in °C. Equivalent I-P values are: (a) for temperatures—40°C [104°F], 105 [221], 150 [302], 185 [365], 220 [428]; (b) for temperature increments—10°C [18 °F], 30 [54], 55 [99], 80 [144], 115 [207], 150 [270].

[a]Modern terminology for insulation class uses system in lieu of class.

used. In order of decreasing physical size and increasing operating temperature, we have, for dry transformers, 105, 150, 185, and 220°C [221, 302, 365, and 428°F] systems that represent organic, inorganic (two types), and silicone insulating materials, respectively (Table 28.2).

Dimensional data for transformers vary considerably from one manufacturer to another. Therefore, Table 28.3 is useful only to provide a sense of bulk volume and weight.

Although 220°C [428°F] system insulation transformers can withstand a 150°C [270°F] rise (see Table 28.2), some users specify 220°C [428°F] system insulation with a 115°C [207°F] rise or even an 80°C [144°F] rise; that is, a better grade of insulation is used with an underrated transformer. There are four advantages to this design decision:

- longer life
- higher overload capacity
- lower operating cost
- increased capability to withstand the heating effects of harmonics commonly found in commercial and industrial electrical systems (see Section 28.18)

A 220°C [428°F] system transformer operated at full load continuously (an unusual situation) has a short life—estimated on the basis of accelerated aging tests to be between 3 and 10 years. The same class insulation transformer (220°C [428°F] system) rated at 80°C [144°F] has a life expectancy of more than 100 years. With respect to operating cost, the same issues apply here as with the selection of high-temperature insulation discussed in Chapter 29 and summarized in Table 28.2. The adage that you get what you pay for can be expressed here as follows: In return for a smaller, lighter, cheaper, and hotter transformer (220°C [428°F] system), there will be higher losses, and therefore a higher operating cost and a shorter life.

It is good practice in buildings with a total transformer capacity in excess of 300 kVA to complete a calculation comparing the operating costs of various types of transformers in order to balance energy costs with necessary operating flexibility, reliability, and safety.

Such a calculation (using generic cost data) was performed assuming a daily/weekly operating cycle and loading that are representative of commercial use. The two transformers studied were of the dry type, rated 750 kVA, one designed for an 80°C [144°F] rise and the second for a 150°C [270°F] temperature rise. At an energy cost of 7 cents per kWh (average at the time of the study), the lower energy waste of the more expensive 80°C [144°F] unit repays the first-cost differential in about four years. A higher electric energy cost reduces the payback time, and vice versa. Beyond the payback period, the advantage is entirely on the

TABLE 28.3 Typical Dry-Type Transformer: Dimensions and Weights[a]

kVA Output Continuous	Temperature Rise, °F [°C]	Approximate Dimensions, in. [mm] H × W × D	Approximate Weight, lb [kg]
		PRIMARY: 480 V SECONDARY: 120/240 V	
		Single-phase	
3	207 [115]	13 × 8 × 8 [330 × 203 × 203]	60 [27]
5	207 [115]	14 × 12 × 12 [356 × 305 × 305]	120 [54]
10	207 [115]	15 × 12 × 12 [381 × 305 × 305]	150 [68]
15	270 [150]	24 × 16 × 16 [610 × 406 × 406]	200 [91]
25	270 [150]	24 × 16 × 16 [610 × 406 × 406]	230 [104]
50	270 [150]	30 × 20 × 20 [762 × 508 × 508]	430 [195]
100	270 [150]	40 × 24 × 24 [1016 × 610 × 610]	650 [295]
		PRIMARY: 480 V SECONDARY: 208Y/120 V	
		Three-phase	
45	270 [150]	26 × 26 × 15 [660 × 660 × 381]	380 [172]
75	270 [150]	30 × 30 × 20 [762 × 762 × 508]	600 [272]
150	270 [150]	40 × 36 × 26 [1016 × 914 × 660]	1100 [499]
225	270 [150]	44 × 36 × 26 [1118 × 914 × 660]	1400 [635]

[a]For general reference. Specific dimensions and weights vary among manufacturers. All values increase with higher primary voltage and with lower temperature rise.

side of the 80°C [144°F] unit. Using a 30-year life, 8% fixed capital cost, and 3% annual cost escalation, the life-cycle cost of the 80°C [144°F] unit is $10,000 less than that of the 150°C [270°F] unit. (At the assumed loading, even the 150°C [270°F] unit will probably last 30 years, the 80°C [144°F] unit much longer.)

Furthermore, with a total permissible hot-spot temperature of 220°C [428°F], the location of the 150°C-rise [270°F-rise] unit must be very carefully chosen because it can create a serious heat generation and radiation problem.

To summarize, a transformer is specified by type, phase, voltages, kVA rating, sound power level, and insulation class. Thus, a 112.5-kVA, three-phase, 480/120–208-V, air-cooled, indoor, dry-type transformer with a 220°C [428°F] insulation system and 115°C [207°F] rise, and a 45-dBA (decibel, A-scale) maximum sound power level, represents an adequate transformer description. Sound ratings of transformers, as well as installation techniques and acoustical treatment, are discussed in Part V.

A service transformer bank is necessary when the facility utilization voltage is different from the utility voltage (Figure 28.5*a*). A building design may call for a step-up, step-down arrangement when the service run is so long that the conductor cost, when run at a low voltage, would be excessive because of large cables. In such instances, the cost of the double transformer installation must be more than offset by the savings in feeder cost, to be economically justifiable (Figure 28.5*b*). More commonly, step-down transformers are the norm.

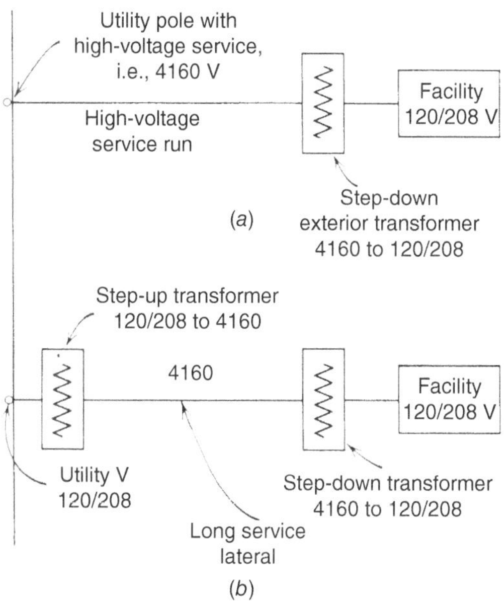

Figure 28.5 *Service transformer arrangements. (a) High-voltage service with a step-down service transformer at the facility and (b) low-voltage service, with transformers at both ends of a long service run.*

28.7 TRANSFORMERS OUTDOORS

The advantages of an outdoor transformer installation are:

- no enclosed building space is required
- reduced probability of a noise problem within the building
- lower first cost

- ease of maintenance and replacement
- no interior heat production
- opportunity to use low-cost, long-life, oil-filled units

Although an exterior installation comes with pluses, it is frequently easier to find space indoors (preferably in a basement) than to find a holistically suitable exterior location. Noise may be more disturbing in the context of some available exterior spaces, such as courtyards, than with a basement location. Costs can be high if long secondary voltage runs are required; heat can often be handled by louvers or areaways adjoining a basement, or even used beneficially if the transformer load is fairly constant. Because exposure to direct solar radiation decreases a transformer's rating by increasing its temperature, a shaded exterior location is preferable, but may be difficult to find. Furthermore, an exterior transformer, other than pole-mounted, is a questionable choice in an area with a high incidence of vandalism, regardless of the sturdiness of construction.

Finally, the appearance of an exterior unit may be objectionable. This point has received much attention from manufacturers, and numerous designs have been developed that minimize aesthetic concerns (Figure 28.6). The most popular type of exterior transformer installation is the pad mount. It has all the advantages listed previously in addition to extreme simplicity of installation—it is set on a simple concrete pad. Consult manufacturers' catalogs for dimensional data.

(a)

(b)

Figure 28.6 (a) Pad-mounted exterior transformers are neat, compact, and, if sited properly, unobtrusive. Large units can be partially screened by shrubbery. (b) When the size is such that visual screening becomes a problem, consideration should be given to a structural screen such as a decorative brick wall. (Courtesy of General Electric, USA.)

28.8 TRANSFORMERS INDOORS: HEAT LOSS

When an indoor transformer installation is indicated, special consideration must be given to the transformer's heat-generating properties. Between 1% and 1.5% of a transformer's load rating, depending upon the type, is converted to heat at full load. For a 750-kVA, dry-type, 150°C-rise [270°F-rise] unit, 12 kW, or 41,000 Btu/h, of heat loss is generated at full load! Losses are lower for 80°C-rise [144°F-rise] units. Liquid-filled units have approximately the same losses as 80°C-rise [144°F-rise] dry units. Unless the heat can be used, sufficient ventilation—either natural or forced—must usually be provided to keep the transformer room's ambient temperature from exceeding 40°C [104°F].

Ventilation by natural convection is most desirable, requiring the transformer room to be located on an exterior wall (adjacent to an areaway if the room is located below grade). The required size (the free area) for a ventilation opening is 3 in.^{2}/kVA [1935 mm^2/kVA] of capacity, plus an additional 1 in.^{2}/kVA [645 mm^2/kVA] for switchgear heat losses, if any. If a louver is used to protect the ventilation opening, the size of the opening must usually be doubled (total: 8 in.^{2}/kVA [5160 mm^2/kVA]) because most louvers have a 50% free area. For good convection, it is desirable to divide the opening areas in half, placing one-half near the ceiling and the remaining half near the floor. To provide for future equipment removal, if an areaway in a basement installation is large enough, it may be useful to increase the overall opening area and make the louver removable for access. A bird screen is also desirable.

Because outdoor air temperature varies, it is advisable to use a temperature-controlled, adjustable louver. In extremely cold climates, heat loss from the electrical equipment may not be sufficient to warm the room in winter. In such instances, a unit heater should be installed in the room with a thermostat set at 55°F [13°C].

28.9 TRANSFORMERS INDOORS: SELECTION

When transformers are installed indoors, they are subject to stringent *National Electrical Code* (*NEC*) regulations that are designed to make the installations intrinsically safe. These regulations are detailed in *NEC* Article 450. The essential considerations are presented herein.

(a) Oil-Insulated Transformers

These present a fire hazard when installed indoors because flammable oil can spread from a tank leak or rupture. To prevent this, most oil-filled transformers must be installed in a fire-resistant vault, the construction of which involves substantial cost. (See *NEC* Article 450 for exceptions.) Advantages offsetting this cost are an oil-filled transformer's small size, low first cost, low losses, long life, excellent electrical characteristics, low noise level, and high overload capacity. Despite these, the vault requirement typically restricts oil-filled units to industrial facilities and other structures where electrical considerations favor their use.

(b) "Less-Flammable" Liquid-Insulated Transformers

Transformers rated 35 kV or less that are insulated with a liquid ("less-flammable" liquid) whose fire point is not less than 300°C [572°F] may be installed indoors without a vault. A liquid confinement area is required, plus either a fire-extinguishing system or restrictions on the combustibility of the building's contents. The insulating liquids used are special types of silicones and hydrocarbons, and the transformer installation must meet all requirements for the specific liquid involved. Selection of these units achieves the electrical and physical advantages of oil-insulated units without the cost of vault construction and installation. The first cost of the units depends upon the fluid type and the specific electrical characteristics required. In general, they are somewhat less expensive than nonflammable fluid-insulated units with similar electrical ratings.

(c) Nonflammable Fluid-Filled Transformers

Certain nonflammable liquid coolants, called askarels, generally containing polychlorinated biphenyl (PCB), were once very widely used. Use of PCB coolants in new equipment is prohibited by federal regulation due to their negative environmental impact, and existing PCB transformers are being phased out. Newer nonflammable fluid coolants have increased the price of these transformers considerably (Table 28.4). Such units, nevertheless, have most of the advantages of oil-filled units and do not require a vault unless the voltage is very high. They do, however, require a sump or catch basin of sufficient capacity for all of the contained liquid. This and the high first cost are the negative aspects of this design.

(d) Dry-Type Transformers

These are the transformers of choice in the majority of indoor installations, despite having shorter life, higher losses, higher noise level, greater weight, and larger size than liquid-filled units. The principal advantages are simplicity of installation and almost unrestricted choice of location. As explained previously, using an underrated transformer (220°C [428°F] system, 80°C [144°F] rise) can reduce losses (heat) and extend life. Also, for a price premium, the sound power level can be reduced.

TABLE 28.4 Relative Installation Costs for a 300- to 1000-kVA Transformer

Transformer Type	Temperature Rise[a], °F [°C]	Relative Transformer Cost	Construction Cost	Total Relative First Cost
Oil-filled	117 [65]	1.00	50–100[b]	1.50–2.00
Less-flammable and nonflammable fluids	117 [65]	1.15–1.30	20–40[c]	1.35–1.70
Dry, ventilated	144 [80]	1.65	—	1.65
	144 [80]	1.50	—	1.50
	270 [150]	1.35	—	1.35
	270 [150]	1.20	—	1.20

[a]Transformers of equal temperature rise have approximately equal life and equal losses.

[b]Cost of the vault depends on local labor costs and size of the transformer. The relative cost decreases with increasing transformer size.

[c]Cost of catch basin. As in the case of a vault, relative cost depends on labor rates and transformer size.

Table 28.4 gives a comparison of the installed costs for the four classes of transformers discussed in this section.

28.10 TRANSFORMER VAULTS

A transformer vault is basically a fire-rated enclosure provided because of the possibility of fire caused by rupture of an oil-filled transformer case. This should not, however, be construed as implying that transformers are hazardous or delicate devices prone to faults. On the contrary, transformers are extremely tough, sturdy, long-lived, and capable of sustaining large and prolonged overloads. They are among the most reliable elements of an electrical system. However, faults do occur, and an oil-filled transformer is a potential fire hazard.

Transformer vaults should be located, to the extent possible, where they can be ventilated using outdoor air without flues or ducts. The combined net area of all ventilating openings (gross area less screens, louvers, etc.) should be as noted in the preceding section: not less than 3 in.^{2}/kVA [1935 mm^2/kVA], and in no case less than 1 ft^2 [0.09 m^2]. Further ventilation recommendations plus details of enclosure construction materials, fire rating, door and sill details, and other relevant design information are provided by *NEC* Article 450.

28.11 SERVICE EQUIPMENT ARRANGEMENTS AND METERING

Metering must be provided ahead (electrically) of the building service entrance switch(es). The metering is accomplished either at the utility or the facility voltage, and either at the service point or inside the building. The choice is generally left to the owner, with the understanding that interior meter equipment must be readily accessible to utility personnel. Although in increasingly large numbers of facilities regular meter reading is accomplished remotely, the meter equipment must still be available for utility company inspection and service.

If high-voltage service is purchased, then transformers and all other equipment beyond the service connection will be furnished by the owner. Conversely, if low-voltage service is purchased, all the equipment necessary to provide such low voltage is furnished by the utility. As might be expected, the billing rates for low-voltage service are higher than those for high-voltage service, in order to compensate the utility for the cost of providing and maintaining the needed step-down transformer and associated equipment. Therefore, it is often advisable for the owner of a large facility to investigate the economics of purchasing power at high voltage. Because many owners are not equipped to maintain high-voltage equipment, arrangements can sometimes be made to pay the utility to provide and maintain transformers while taking the cost advantage of high-voltage service.

Figure 28.7 *Three-section commercial submetering switchboard comprising a center service-entrance section rated 4000-A, 3-phase, 480-V maximum, equipped (usually) with a main switch or circuit breaker. Each of the end cubicles will accept six tenant meters, each on a 320-A 3-phase circuit, maximum, protected by an integral circuit breaker or fused switch. Tenant wiring enters the top or bottom of the board. Switchboards of this type can be installed on every floor of a large multitenant commercial office building. (Courtesy of Cutler-Hammer.)*

For a single-occupant building or a building in which electric energy is included in the rental charge, only a single meter is necessary. Provision for such metering may be provided in the main switchboard, or the meter may be independently mounted. In both cases, the meter is furnished and installed by the utility company. Where submetering is required, such as in apartment houses, banks of meter sockets are installed to accommodate the multiple meters. Federal regulations forbid master metering in new multiple-dwelling constructions because it encourages energy waste (Figure 28.7). Submetering should be considered in any facility where the energy consumption of specific building systems is of interest to conservation and/or ongoing commissioning efforts.

A low-voltage underground service plan as it would appear on a set of contract drawings, including relevant details, is given in Figure 28.8. Note that here the service switch and meters are separately mounted.

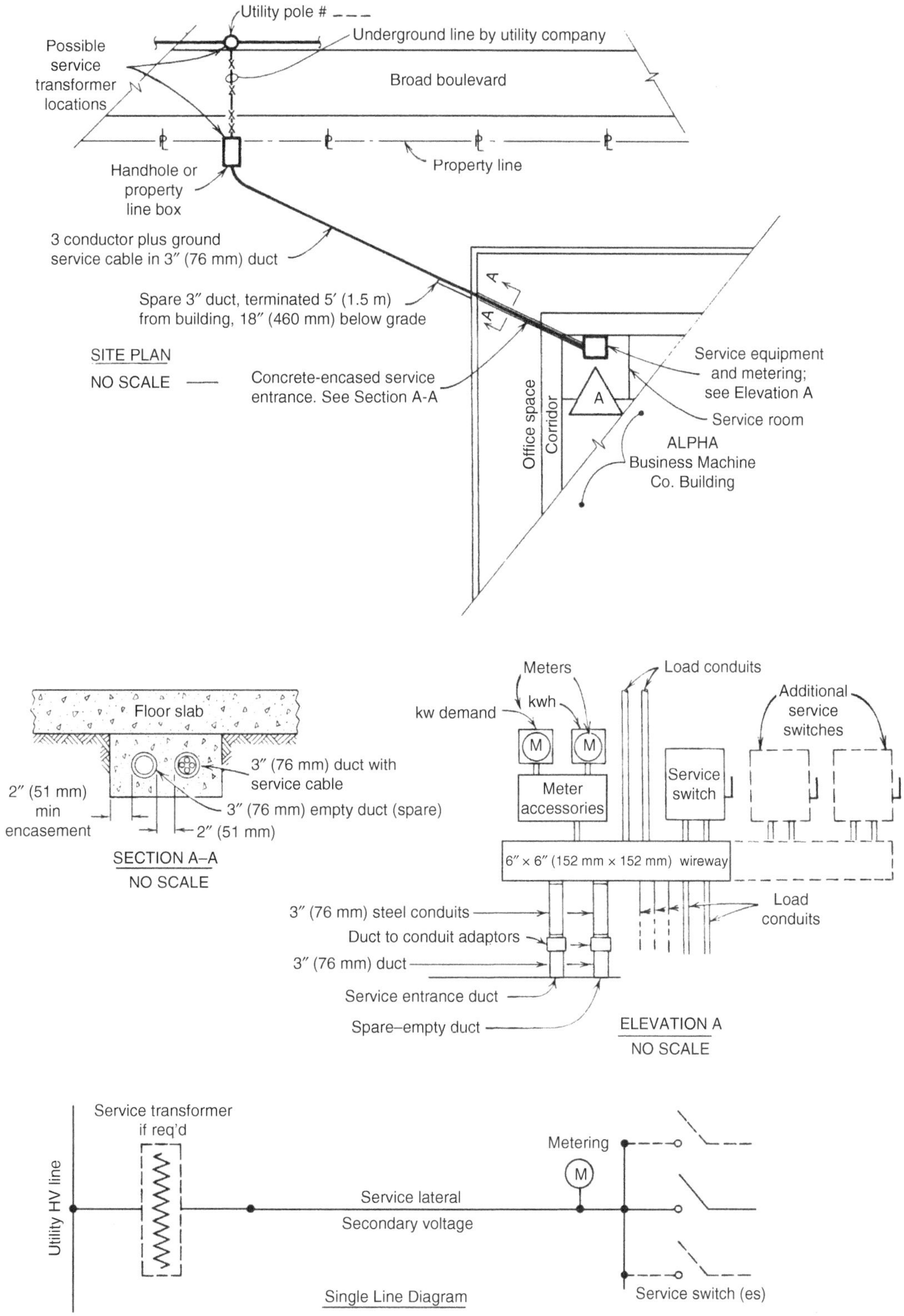

Figure 28.8 *Typical electrical plan and details of an underground low-voltage service (600-V class) to an industrial facility. Note that the portion of underground service beneath the building must be encased in concrete. A service transformer (if required) could be mounted on the utility pole or on a pad at the property line.*

28.12 SERVICE SWITCH(ES)

The purpose of electric service switch(es) is to disconnect the normal service to the building. (See Chapter 30 for a discussion of emergency electric service.) In the event of a fire, for example, the switch can be used to ensure that no electrical hazard will face firefighters. This disconnecting apparatus must be located at a readily accessible spot near the point where the service conductors enter a building. If such a location is not feasible, service conductors may be run in concrete encasement under a building—and are considered "outside the building" up to the point at which they emerge from the floor in the building (see Figure 28.8). At that point, the service disconnect switch must be installed. The service switch or, more accurately, the service-disconnecting means, may comprise between one and six properly rated switches. These are frequently assembled into a switchboard. Before discussing switchboards, however, a description of switches, circuit breakers, and fuses (the components of which switchboards are constructed) is in order.

28.13 SWITCHES

Traditional electrical switching devices are discussed in this and the following two sections. Mechanical switches close and open an electric circuit by physically moving two electrical conductors into contact with each other to close the circuit and physically separating them to open the circuit. The motive force can be supplied by hand, an electrical coil, a spring, or a motor; the device name may change accordingly, but the action is the same. Solid-state switches perform the same electrical switching function but by a completely different process, without moving parts. They are discussed in Section 28.16.

An electrical switch is rated by current and voltage, duty, poles and throw, fusibility, and enclosure. The current rating is the amount of current that the switch can carry continuously and interrupt safely. Switches intended for motor control are also rated in horsepower (kW). The voltage rating of a switch is noted, as for other electrical equipment, by voltage class. Thus, a switch may be rated 250 V, 600 V, or 5 kV, as required. Switches intended for normal (occasional) use in lighting and power circuits are called general-duty safety switches. Switches intended for frequent interrupting, high-fault currents, and ease of maintenance are rated HD for heavy duty.

Figure 28.9*a* should clarify what is meant by the number of poles and throws of a switch. Unless otherwise noted, a switch is assumed to be single-throw. Because the *NEC* states generally that the grounded neutral conductor of a circuit should not be broken, most switches carry the neutral through unbroken by means of a solid link within the switch. This gives rise to the term solid-neutral (SN) switch. Switches are available in 1-, 2-, 3-, 4-, and 5-pole construction. Poles are indicated by a "P"; thus, 3-pole is written "3P," and so on.

A switch may be constructed with or without provision for fusing. If provided, the switch is fusible; if not, the switch is nonfusible. All individual enclosed switches are mounted in an appropriate cabinet. NEMA standardized the nomenclature and application of enclosures for all electrical control equipment, of which switches are only one item. Summarizing, the following adequately describes a switch: switch, HD, 3P & SN, 200A/150AF, 600 V, in NEMA 12 enclosure. This "translates" as a heavy-duty switch, 3 poles and solid neutral, 200-amp rating with 150-amp fuses, 600-V rating, in an industrial use enclosure (Table 28.5). Such a switch is illustrated in Figure 28.10.

28.14 CONTACTORS

A contactor is a switch. Instead of a handle-operated, movable blade, a contactor uses contact blocks of silver-coated copper, which are forced together to make (close) or are separated to break (open) a circuit. The common wall-mounted light switch is a small, mechanically operated contactor. A relay is an electrically operated contactor. Most contactors are operated by means of an electromagnet that causes the contacts to close. They open either by spring action or by gravity.

Contactor terminology is somewhat different from that of switches. Its condition when deenergized is its normal state. Thus, a contactor whose contacts are open when the coil is not energized is referred to as normally open (NO), whereas a contactor whose contacts are closed when deenergized is referred to as normally closed (NC). Units intended for motor connection are called motor starters and are discussed in Section 28.26. Current, voltage, and number of poles have the same nomenclature for contactors and relays as for switches.

The great advantage of contactors over switches is their facility for remote control. Switches must be manually thrown—or, in very special cases, thrown by a motor. The magnetic contactor, by contrast, is inherently a remotely controlled device, making it ideal for myriad control functions. It can be controlled by a manual or remote pushbutton or by automatic devices such as timers, float switches, thermostats, pressure switches, and so on. Because control can be both remote and automatic, contactors are universally used in the control of lighting, heating, air conditioning,

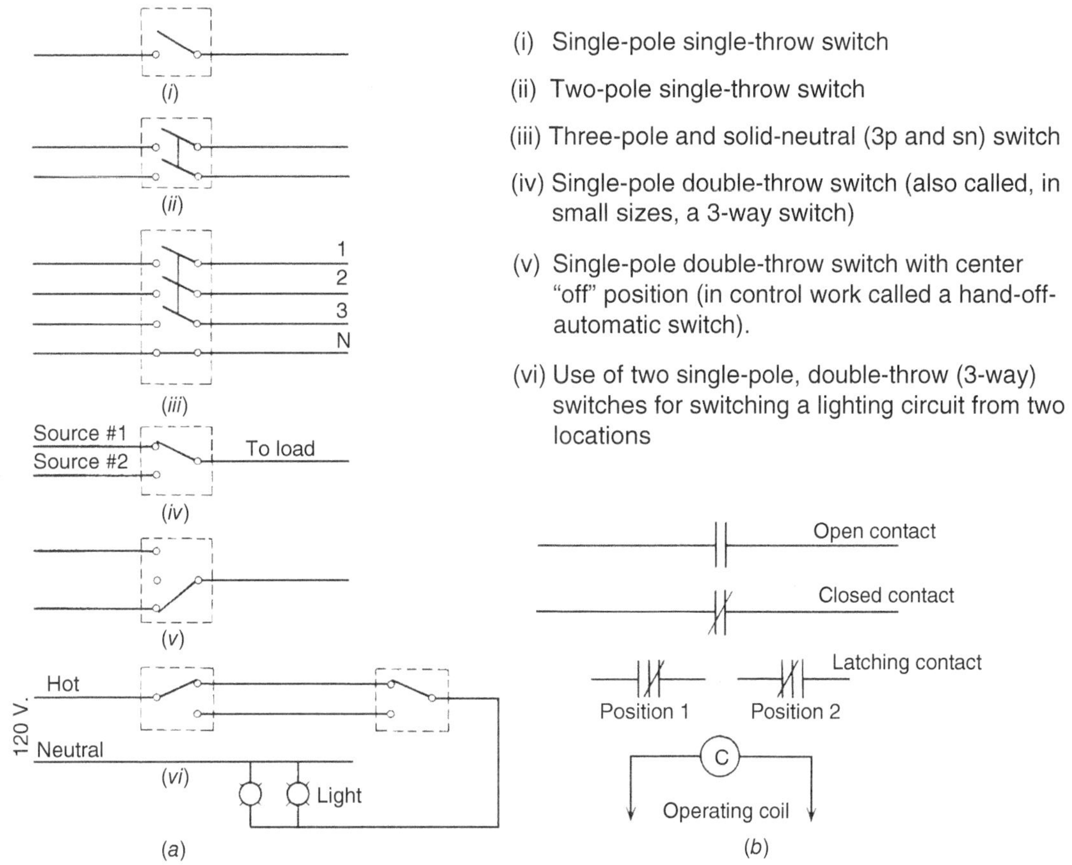

Figure 28.9 *(a) Typical switch configurations. Note that switches are always shown in the open position. (b) Graphic representation of switch (contact) positions in a contactor or relay. Note that a latching contactor shows the switch in one of two positions to indicate the latching nature of the contactor.*

TABLE 28.5 Control Equipment Enclosures

NEMA Designation: Type	Description	Application
1	General-purpose	Dry, indoor use
2	Drip-proof	Indoor, subject to dripping
3	Dust-tight, rain-tight, and sleet-resistant	Indoor/outdoor, where subject to windblown dust and water
3R	Rainproof and sleet-resistant	Outdoor, subject to falling rain, snow, and sleet
3S	Dust-tight, rain-tight, and sleet-proof	Outdoor, subject to windblown water, dust, and sleet; most severe exterior duty
4	Watertight and dust-tight	Indoor/outdoor, subject to water from all directions; not sleet-proof
4X	Watertight, dust-tight, corrosion-resistant	Same as type 4 with added corrosion protection
7–9	Hazardous	Differing in application by class and group of hazardous use; see *NEC*
12	Industrial use, dust-tight, and drip-tight	Indoor only, general use, industrial and other "dirty" environments

motors, and the like. A graphic representation of contactors is shown in Figure 28.9*b*.

28.15 SPECIAL SWITCHES

Many special types of switches are available. Most of these types are beyond the scope of this book, except for the following, which are used extensively in building design work.

(a) Remote-Control (RC) Switches

A contactor that latches mechanically after being operated is known as a remote-control (RC) switch. It differs from a relay in that the latching operation performs the dual function of latching the contacts and disconnecting the electric circuit. To unlatch the contacts, the electric circuit must be reenergized. Such a device is therefore also known as an electrically operated mechanically held contactor. This characteristic

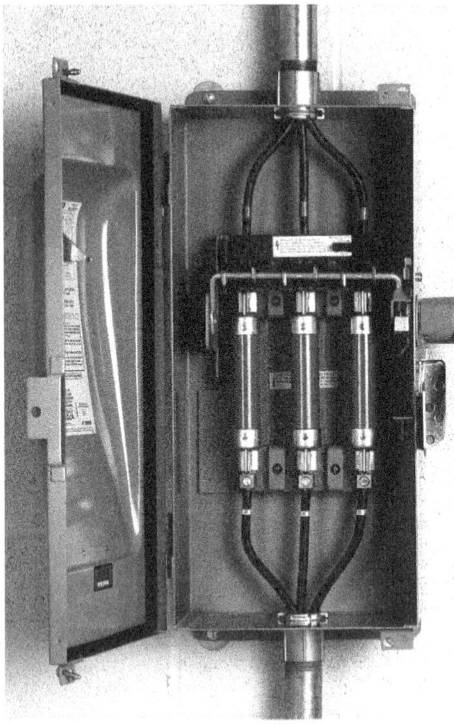

Figure 28.10 *The internal construction of a heavy-duty, fused, 3-pole, 3-wire switch in an industrial NEMA 12 enclosure. Note the gasketing on the inside of the cover, which seals the sheet-steel box and makes it usable in environments containing lint, dust, dirt, sawdust, and the like. The external handle is a double-lockable indicating type for safety. (Courtesy of Siemens Energy & Automation, Inc.)*

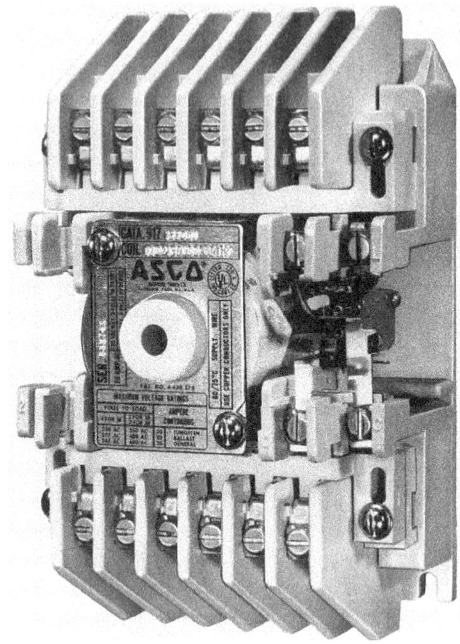

Figure 28.11 *This remote-control switch (mechanically held, electrically operated contactor) is a 12-pole, 20-A, 600-V unit measuring approximately 7 in. H × 6 in. W × 4 in. D [178 × 150 × 100 mm], making it suitable for stud-wall mounting. Compact solid-state control modules are available that mount directly onto the unit and convert it to a flush, wall-mounted multicircuit programmable controller (see Section 28.16). (Courtesy of Automatic Switch Co.)*

is advantageous where the position of the contacts will be maintained for a long period (as in a lighting control circuit) and it is undesirable to hold the contacts by the continuous energizing of a coil, as would be required with an ordinary relay. A typical unit is shown in Figure 28.11, and a typical application is shown in Chapter 30. A common application of these switches is for controlling a "block" load, such as exterior lighting, whole-floor or whole-building lighting, and the like.

(b) Automatic Transfer Switch

This device, which is an essential part of all emergency and standby power arrangements, is basically a double-throw switch—generally 3-pole—so arranged that on failure of normal service it automatically transfers to the emergency service. When normal service is restored, it automatically retransfers to it. (Retransfer can also be arranged to have a preset minimum delay, or it can be entirely manual.) The switch control is accomplished by voltage sensors that sense the condition of the service and operate the switch accordingly. Auxiliary devices can be added to the basic switch, the most common being emergency generator starting equipment. Figure 28.12 illustrates a typical unit. Transfer switches used in uninterrupted power supplies are normally solid-state devices because these involve no switching time lapse.

(c) Time-Controlled Switches

This category includes all switches whose operation is time-based. The timer can be either the familiar electromechanical device consisting of a low-speed miniature drive motor, to which some type of contact-making device is physically connected, or it can be a solid-state electronic timer, which in turn controls either a relay or a solid-state switch. The latter arrangement is the basis of all modern programmable time controls (Section 28.16), which find wide application in lighting control (Chapter 17), energy management and automated building control (Section 28.13 and Chapter 32), and clock and program systems (Chapter 32).

Motor-operated timers (other than hand-wound types) depend for their accuracy on power-line frequency, have moving parts that wear, must be reset after power outages (although some units have spring-wound reserve power motors), and become cumbersome and expensive with an increase in the number of controlled events. Electronic units are small, cool, quiet, independent of line frequency for accuracy (in many designs), able to function during power

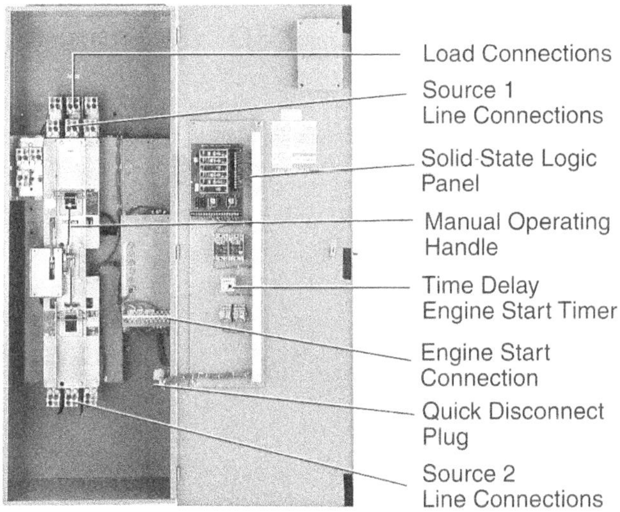

Figure 28.12 *Automatic transfer switch with solid-state logic. The unit can be operated manually if required. It is equipped with engine-starting contacts, auxiliary contacts for monitoring, electrical safety interlocks, and automatic differential voltage sensing on the normal source. An 800-A, 3-phase unit of this design, in a NEMA indoor sheet-metal enclosure, measures 73 in. H × 26 in. W × 21 in. D [1.9 × 0.7 × 0.5 m] and weighs approximately 600 lb [272 kg]. (Courtesy of Cutler-Hammer.)*

Figure 28.13 *Adjustable time switch. This unit allows users to specify a singular, automatic "on" and "off" time.*

outages when equipped with a standby battery, and virtually unlimited in their event-control capacity. They have largely replaced mechanical timers except for the simplest applications (Figure 28.13).

28.16 CONTROLLERS

A solid-state switch is an electronic device with a conducting state and a nonconducting state, corresponding to a conventional switch in its closed and open positions. The change between the two states is accomplished by the application of a control signal (voltage). The change of state is instantaneous and noiseless, and occurs without arcing. The current ratings of solid-state switches are approximately the same as those of their mechanical counterparts, ranging from fractions of an ampere to hundreds of amperes. Voltage ratings are usually limited to secondary systems values (0–600 V) due to the nature of the semiconductors comprising the active elements of these switches, although higher-voltage equipment is available for special applications.

If an electronic timing device is added to a solid-state switch, a time-controlled electronic switch is created that functions exactly like the electromechanical time-based switches described in Section 28.15. These solid-state switches are independent of the utility line frequency for time control, and there are no moving parts to wear out. If a small, programmable memory circuit, in the form of an erasable programmable read-only memory (EPROM) chip is added, the device becomes a programmable time switch. The simplest of these are single-circuit 15-A devices intended to replace common wall switches (see Figure 28.46).

A preprogrammed device, until reprogrammed, gives fixed, invariable instructions to a (multichannel) switch, as is the case with a programmable time switch. If a microprocessor (i.e., a logic/memory device programmed to respond according to a particular algorithm [control plan] to given input signals) is installed, a readily reprogrammable controller results (Figures 28.14–28.16). By definition, a programmable controller is an electronic device that uses a programmable memory chip for internal storage of instructions that in turn implement specific functions such as logic, timing, counting, and so on.

The internal control/memory is frequently referred to as a microprocessor; with respect to the overall programmable controller, it acts as a central processing unit (CPU). The programmable controller is therefore a special type of computer. It differs from a conventional computer in that its program is relatively short and exists in hardware (a microprocessor),

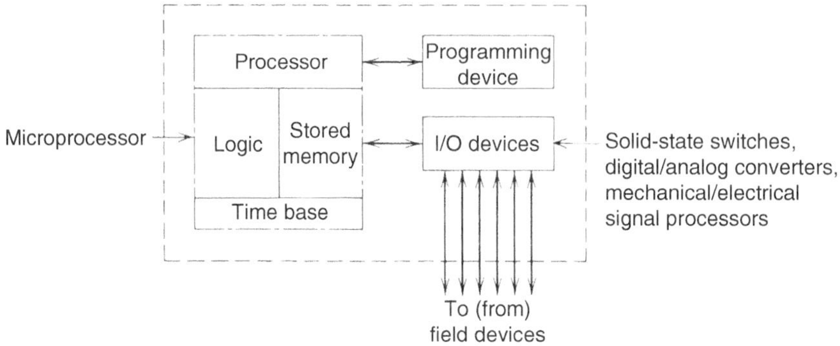

Figure 28.14 Block diagram of a programmable controller. These devices are designed to supervise field conditions continuously and are therefore used in industrial applications as well as for lighting and environmental system control.

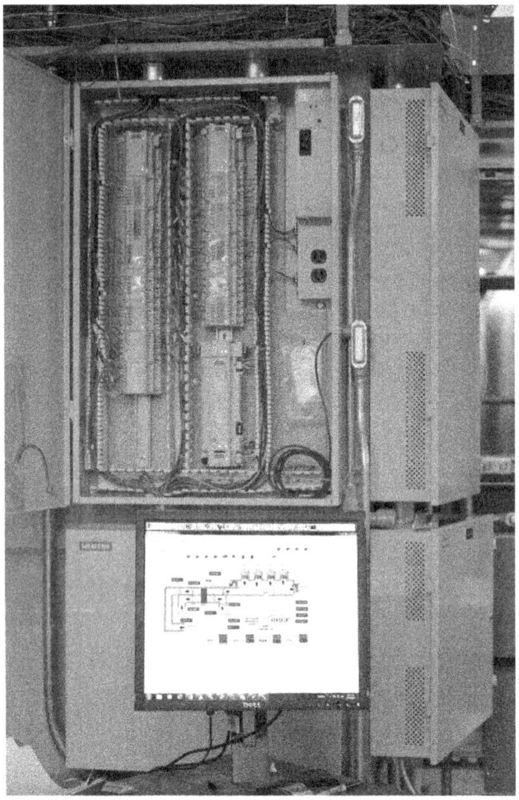

Figure 28.15 Programmable building monitoring system (BMS) from Siemens Corporation. Utilizing building system control devices, BMS software, and numerous sensors, the system allows for close monitoring of electric lighting, temperature, electrical demand, time signals, central control signals, and other energy-consuming inputs. (© Fred Tepfer; used with permission.)

whereas a computer reads its program, which can be very lengthy, into its memory from software. Furthermore, the programmable controller is designed for a specific function, whereas a conventional computer is a general-use calculating device that processes almost any type of software.

Programmable controllers have all but replaced the ubiquitous hard-wired relay panel in applications such as industrial control, elevator control, process control, and the like, because of the ease with which such a controller can be reprogrammed (in lieu of rewiring a relay panel), the speed and accuracy of control, and the vastly increased complexity possible.

28.17 EQUIPMENT ENCLOSURES

Proper NEMA nomenclature, descriptions, and applications for the more common enclosures are found in Table 28.5. It is important to note that there is no enclosure described as WP or weatherproof. Equipment intended for outdoor use should be specified for installation:

1. in a type 3R enclosure to protect against rain (Figure 28.17)
2. in a type 3S enclosure to protect against wind-driven rain and sleet
3. in a type 4 enclosure to protect against rain, wind-driven rain and sleet, plus splashing and condensation

Note also that the type 12 industrial enclosure is similar to type 1 except that it is gasketed for dust and drip resistance and therefore is well applied in all "dirty" indoor environments, including those in commercial and institutional spaces (see Figure 28.10).

28.18 CIRCUIT-PROTECTIVE DEVICES

To protect insulation, wiring, switches, and other apparatus from the destructive effects of overload

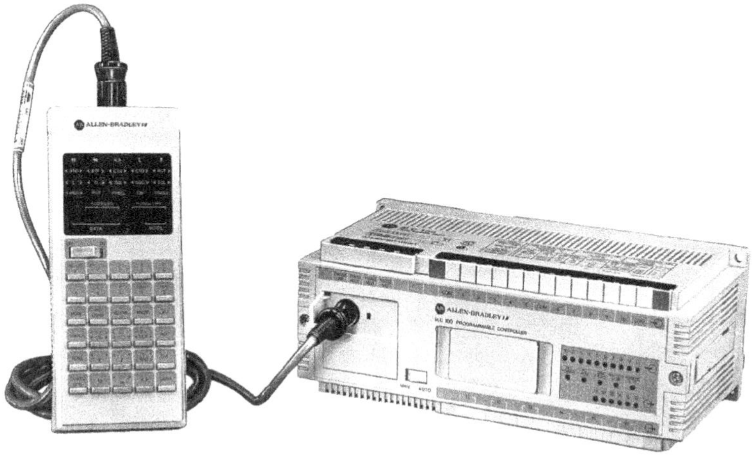

Figure 28.16 General-purpose programmable controller. The unit contains the central processing unit (CPU), random access memory (RAM), and power supply. Peripherals include the illustrated handheld programming unit and EEPROM (electrically erasable programmable read-only memory) module, and various computer/software interfaces (not shown). The unit illustrated, which measures approximately 10 × 9 × 3 in. [255 × 230 × 75 mm], has 16 input/output (I/O) circuits and an 885-word memory. It is readily adaptable to a variety of automated control processes in manufacturing, building control, and material handling. (Courtesy of Allen-Bradley Co.)

Figure 28.17 Rain-tight NEMA type 3R outdoor circuit breaker enclosure. This is the type usually intended when "weatherproof" is specified. (Courtesy of General Electric, USA.)

and short-circuit currents, an automatic means for opening the circuit is required. The two most common devices employed to fulfill this function are the fuse and the circuit breaker, the latter frequently abbreviated *c/b*.

(a) Fuses

The fuse is a simple device consisting of a fusible link or wire of low melting temperature that, when enclosed in an insulating fiber tube, is called a cartridge fuse and, when enclosed in a porcelain cup, is known as a plug fuse. Figure 28.18 shows common types of fuses. Plug fuses, such as those in residential use, are rated 5–30 A, 150 V to ground, maximum. Cartridge fuses of various designs are available up to 6000 A and 600 V.

(b) Circuit Breakers

A circuit breaker is an electromechanical device that performs the same protective function as a fuse and, in addition, acts as a switch. Thus, it can be used in lieu of a switch-and-fuse combination to both protect and disconnect a circuit. Most circuit breakers are equipped with both thermal and magnetic trips. The thermal trip acts on overload, whereas the magnetic trip acts on short circuit. These thermal and magnetic actions have inverse-time characteristics: That is, the heavier the overload, the faster the trip action. Modern circuit breakers in commercial and industrial applications are frequently equipped with solid-state electronic tripping control units, which provide fully adjustable overload, short-circuit, and ground-fault protection.

Air-circuit breakers are available in two types: the molded-case breaker and the large air-circuit breaker. Molded-case breakers consist of a complete mechanism encased in a molded phenolic case. A light-duty molded-case 50-A frame, plug-in circuit breaker is illustrated in Figure 28.19*a*. The molded-case breaker shown in Figure 28.19*b* has a 400-A frame and can be equipped with an adjustable electronic tripping unit, ground fault trip, plus other features formerly

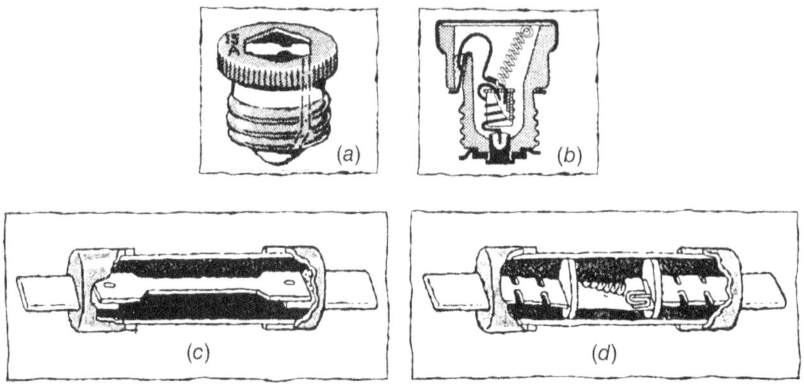

Figure 28.18 *Plug-type fuses are made in two physical types. (a) The nonrenewable type with a standard (Edison) base screws directly into a standard socket. (b) Nonrenewable NEC type S fuses have a smaller base than type (a) and therefore require an adapter to fit into a standard socket. The adapter is current-rated and nonremovable. This prevents deliberate or accidental use of a type S fuse of incorrect rating. Type S is required in new construction where plug fuses are used. Both fuses shown are nonrenewable and must be replaced after blowing. The type S fuse shown in (b) is a dual-element time-delay fuse. Cartridge fuses are available in a variety of designs. Illustrated are (c) the nonrenewable, single-element type and (d) the nonrenewable, dual-element, time-delay type. Because fuses are inherently very-fast-acting devices, time delay must be built into a fuse to prevent blowing on short-time overloads such as those caused by motor starting. A dual-element fuse as shown in (b) allows the heat generated by temporary overloads to be dissipated in the larger center metal element, preventing the fuse from blowing. If the overload reaches dangerous proportions, the metal will melt, releasing the spring and opening the circuit. The notched metal portions of the fuse element, at both ends of the dual center element, provide short-circuit protection. The time required to clear (blow) a fuse is generally inversely proportional to the amount of current. In renewable-cartridge fuses, the spent (blown) element is replaceable, allowing reuse of the original outer cartridge and blade connectors.*

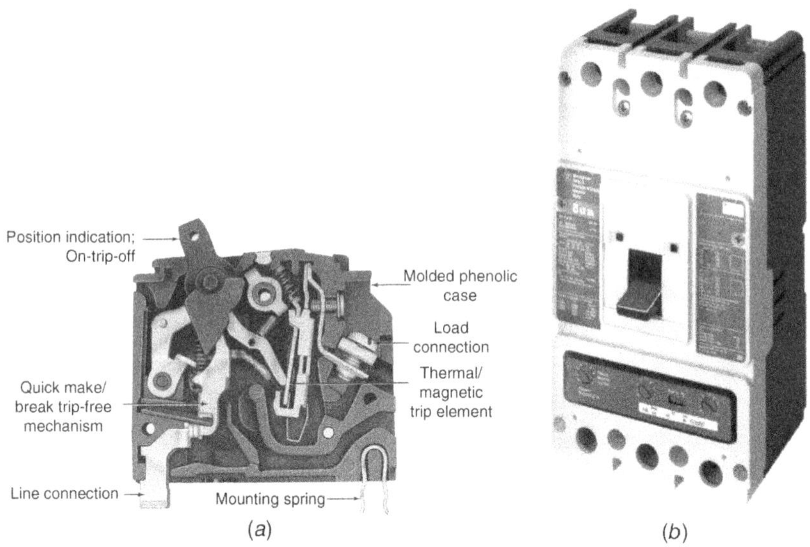

Figure 28.19 *(a) Essential elements of a plug-in-type molded-case circuit breaker are shown in cutaway. Details vary with breakers designed for specific applications. The unit shown is typical of breakers in the 50- to 100-A size. (b) Heavy-duty molded-case circuit breaker, 400-A frame, 2- to 4-pole, 240–600 V. This circuit breaker is available with either a standard thermal-magnetic trip or an adjustable electronic trip with ground-fault protection. Among the accessories available are auxiliary contacts, shunt trip, undervoltage release, and an electrical solenoid operator that permits remote control of the breaker. The unit is suitable for individual or panelboard mounting. A typical 3-pole unit measures 5½ in. W × 10 in. H × 4 in. D [140 × 255 × 100 mm] and weighs about 12 lb [5.5 kg]. (a: courtesy of The Square D Company; b: courtesy of Cutler-Hammer.)*

available only on large air-circuit breakers. The large air-circuit breaker is a more complicated and highly adjustable device that can be used in applications that preclude the use of molded-case breakers. A modern solid-state adjustable trip controller for a large air-circuit breaker is shown in Figure 28.20, along with a circuit breaker with a similar solid-state trip unit.

All breakers can be equipped with remote trip and auxiliary contacts, and all good breakers have trip-indicating handles and are trip-free (i.e., they will trip out harmlessly if closed on a short-circuited line). Low-voltage (600-V class) circuit breakers are available in frame sizes ranging from 50 to 4000 A and 1 to 3 poles. Their characteristics vary widely and are beyond the scope of this book.

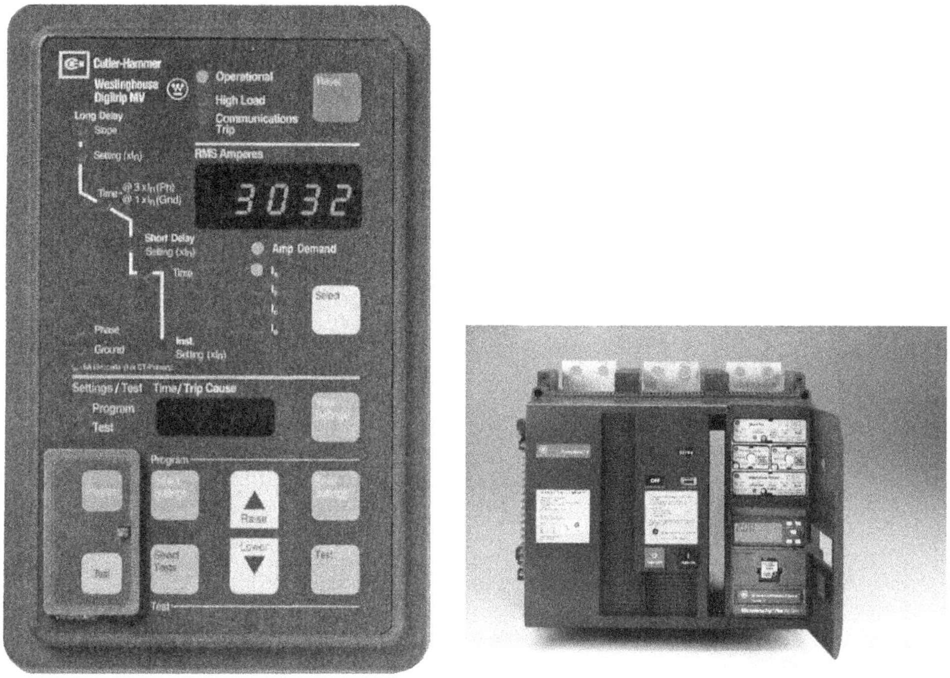

(a) (b)

Figure 28.20 (a) A multifunction microprocessor-based trip control unit for a large circuit breaker. It provides true root-mean-square (rms) sensing of phase and ground currents, selectable tripping characteristics, and trip alarm contacts. In addition, the unit can be used to display and transmit information on phase and ground currents. (b) Electrically operated large air-circuit breaker with electronic microprocessor-controlled trip unit and additional remote-control accessories. The illustrated unit, rated 2000 A, is shown before a stationary (nondrawout) mounting in a switchboard. (a: courtesy of Cutler-Hammer; b: Courtesy of General Electric, USA.)

(c) Characteristics of Fuses and Circuit Breakers

Although both fuses and circuit breakers are circuit-protective devices, their characteristics differ markedly, as seen in Table 28.6.

These characteristics demonstrate that there is no single answer to the question "Which are preferable—fuses or breakers?" The answer depends upon the specific application involved and is often based on highly technical factors requiring detailed analysis. Circuit breakers are generally used in lighting and appliance panelboards.

28.19 SWITCHBOARDS AND SWITCHGEAR

A switchboard is a large, freestanding assembly of switches and fuses (and/or circuit breakers), which normally provides switching and overcurrent protection to a number of circuits connected to a single electric source. Metering and other instrumentation are also often included. A switchboard may be represented in a single-line diagram, as shown in Figure 28.21. This equipment serves to distribute, with adequate protection, bulk power into smaller "packages." Thus, in a hydraulic analogy, the main buswork of the switchboard is equivalent to a main water supply header, the switches to on/off valves, the fuses to flow-limiting devices, and the feeders to subpiping connected to the main header.

Modern switchboards (Figures 28.22–28.25) are all dead front—that is, they have all circuit breakers, switches, fuses, and live parts completely enclosed in a metal structure. The operator controls all devices by means of pushbuttons and insulated handles on the front panel. Circuit breakers equipped with bayonet-type contacts, each mounted in a movable drawer (like the drawers of a standard letter file) in a switchboard, are described as the drawout type. This drawout arrangement facilitates emergency replacements, inspection, and repairs and is illustrated in Figure 28.24.

No clear distinction is made between the terms switchboard and switchgear, although the terms suggest a sense of scale. Generally, low-voltage switchboards with large circuit breakers and all high-voltage equipment (above 600 V) are referred to as switchgear. When molded-case circuit breakers are utilized in a switchboard, it is often referred to as a building-type switchboard.

TABLE 28.6 Comparison of Fuses Versus Circuit Breakers

Fuses—Switch-and-Fuse Combination
Advantages
Simple and foolproof Constant characteristics (no aging) Initial economy Very high interrupting capacity (IC) No maintenance Instantaneous; energy-limiting
Disadvantages
Fuses are single-pole only Necessity for storage of replacement fuses Nonrenewable (one-time operation) Nonadjustable Nonindicating[a] No electric or remote control[a] Not trip-free
Circuit Breakers
Advantages
Usable as switches Multipole No storage of replacements required Resettable Indicates trip Trip-free Remote-control potential Adjustable
Disadvantages
Low to medium IC, except for special units Periodic maintenance required High initial cost Complex construction changes with age

[a]Can be accomplished with accessories.

Recommended minimum space requirements for various types of switchgear are shown in Figure 28.26. Working space around all types of electrical equipment must meet the requirements of *NEC* Article 110. Main metal-clad switchgear for commercial, industrial, and public buildings is almost invariably located in a basement and housed in a separate well-ventilated electrical switchgear room. Smaller subdistribution switchboards require no special room. A wire screen enclosure to prevent tampering or vandalism plus a large "DANGER—HIGH VOLTAGE" sign are usually adequate. The architect must provide adequate exits, hallways, and/or hatches for the installation and removal of all equipment. Specifications for switchgear should state the maximum overall dimensions of sections that will be transported and installed in one piece.

When switchgear is to be installed outdoors, one of three methods is employed: build a small structure to enclose normal indoor gear, utilize weatherproof outdoor gear, or utilize switchgear that is built into its own exterior enclosing structure, as seen in Figure 28.25. These housings are equipped with heating and lighting and often prove to be the most economical choice.

28.20 UNIT SUBSTATIONS (TRANSFORMER LOAD CENTERS)

An assembly, comprising a primary voltage switch-and-fuse or circuit breaker, a step-down transformer, meters, controls, buswork, and secondary (low-voltage) switchgear, is known as a unit substation or a load-center substation. Its function is to accept an incoming high-voltage power supply, transform the high voltage to a voltage that can be utilized in the facility, and distribute the low-voltage power through associated low-voltage (secondary) switchgear. A dimensional physical sketch of a typical

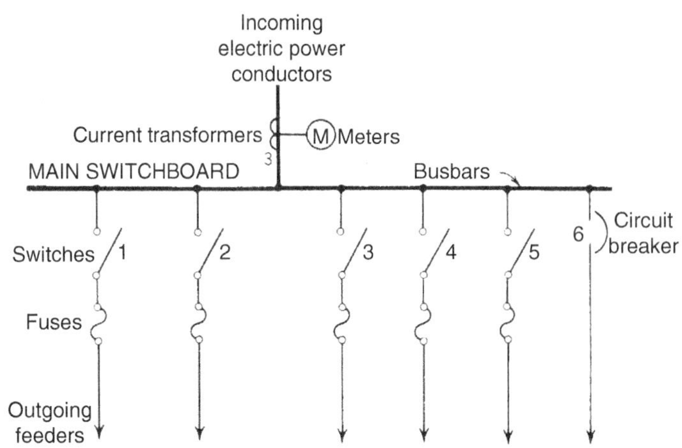

Figure 28.21 One-line diagram of a typical switchboard. Switches are normally shown in the open position. Switches must be on the line (supply) side of fuses. Each line in a single-line diagram represents a 3-phase circuit. If circuit breakers were used, the entire board would be composed of units as depicted in circuit 6.

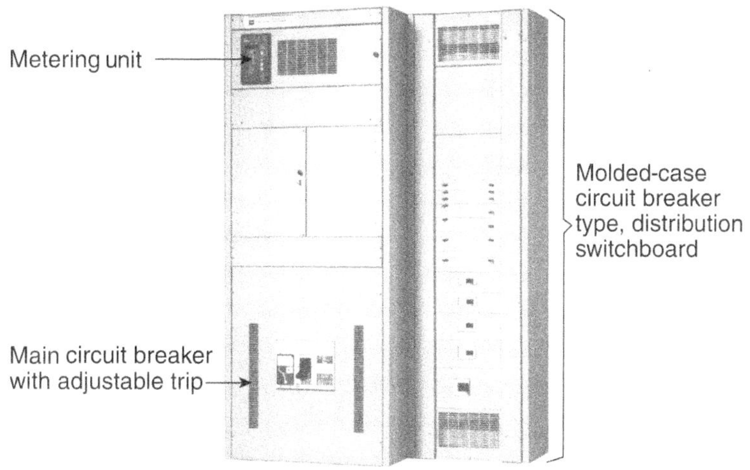

Figure 28.22 *Freestanding, indoor-service, low-voltage (600-V) switchboard with group-mounted, fixed molded-case circuit breakers. The deeper section contains the service entrance main circuit breaker (maximum 3000 A) plus instrumentation. Section width varies from 30 to 45 in. [0.75 to 1.1 m] and depth from 18 to 36 in. [0.5 to 0.9 m], depending upon the main circuit breaker rating. The smaller section contains branch circuit breakers. All units are 90 in. [2.3 m] high. (Courtesy of Cutler-Hammer.)*

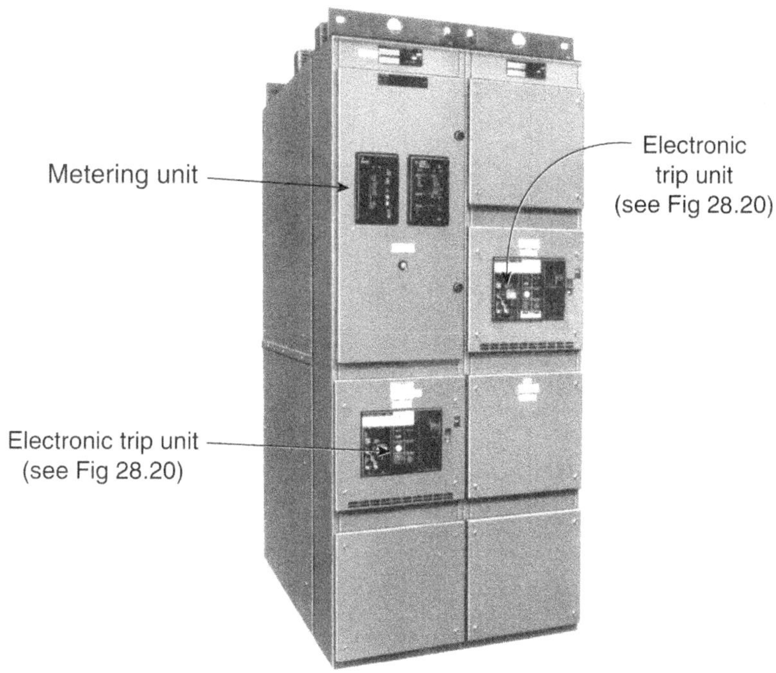

Figure 28.23 *Metal-enclosed, indoor, low-voltage power switchboard with fixed or drawout large circuit breakers. Maximum bus capacity is 5000 A. Section dimensions vary with the size and type of contained equipment. All units are 90 in. [2.3 m] high. Each circuit breaker is equipped with a microprocessor-controlled trip unit. It can be seen clearly on the front of each unit. In addition, another programmable solid-state metering/analysis unit can be seen at the left of the top compartment of the entry (left) cubicle of both this switchgear and the switchboard in Figure 28.22. (Courtesy of Cutler-Hammer.)*

unit substation, along with its electrical single-line diagram, is shown in Figure 28.27. Equipment is available for indoor or outdoor installation.

If installed indoors, the location of a unit substation is governed by the type of transformer utilized, as explained in the discussion on indoor transformer installations. For this reason, almost all indoor unit substations utilize dry-type (air-filled) transformers. A basement location is most often selected, with ventilation requirements as detailed previously. Access should be restricted to authorized persons. Unit substations are selected to obtain the economies inherent in prefabricated construction using coordinated components.

Figure 28.24 Three sections of medium-voltage metal-clad drawout switchgear, suitable for indoor use as shown and for outdoor use with cabinet modification. Voltage ratings are 5–27 kV; current ratings are 1200–3000 A. Each indoor section measures 36 in. W × 95 in. H × 96 in. D [0.9 × 2.4 × 2.4 m] and contains two compartments for circuit breakers and/or instrumentation. Due to the drawout characteristic (shown), a 72-in. [1.8-m] front aisle is required. Similar outdoor aisle-less switchgear measures 36 in. W × 115 in. H × 101 in. D [0.9 × 2.9 × 2.6 m]. (Courtesy of Cutler-Hammer.)

28.21 PANELBOARDS

An electrical panel, or panelboard, serves essentially the same function as a switchboard except on a smaller scale. It accepts a relatively large block of power at some downstream point in a system and distributes it in smaller blocks. Like a switchboard, it comprises main buses to which are connected circuit-protective devices (breakers or fuses) that feed smaller circuits. The panelboard level of a system usually represents the final distribution point, thence feeding out to the branch circuits that contain the electrical utilization apparatus and devices, such as lighting, motors, and so on. Small panels, particularly in residential work, are frequently referred to as load centers.

The panelboard components—the buses, breakers, and so on—are mounted inside an open metal cabinet called a backbox. The backbox is prefabricated with knockouts at the top, bottom, and sides to permit connection of conduits carrying circuit conductors (Figure 28.28). The main feeders that supply power to the panel's busbars enter through the large knockouts

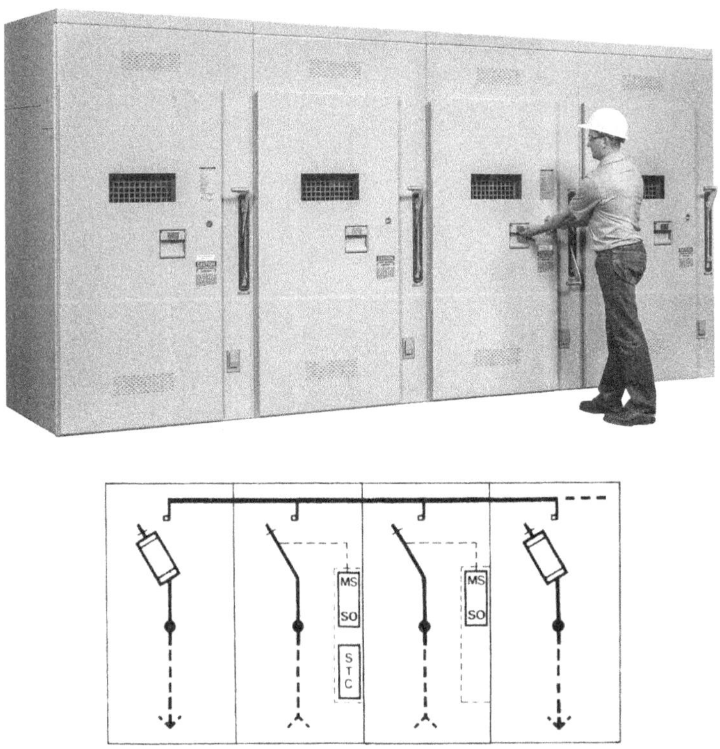

One-line diagram

Figure 28.25 Four-bay (section), power-operated, high-voltage service entrance switchgear rated 13.8 kV, consisting of (see the single-line diagram inset, and reading from left to right) a fused feeder section, two bays with parallel incoming lines and automatic transfer in the event of power failure, and a second switch and fuse feeder section. Fuses are electronic, with circuitry designed to provide a trip signal when excessive current is sensed, rather than simply relying on the thermal characteristic of a fusible element. This gear, which measures 93 in. H × 172 in. W × 44 in. D [2.4 × 4.4 × 1.1 m] and weighs 8000 lb [3630 kg], is suitable for interior or exterior installation.

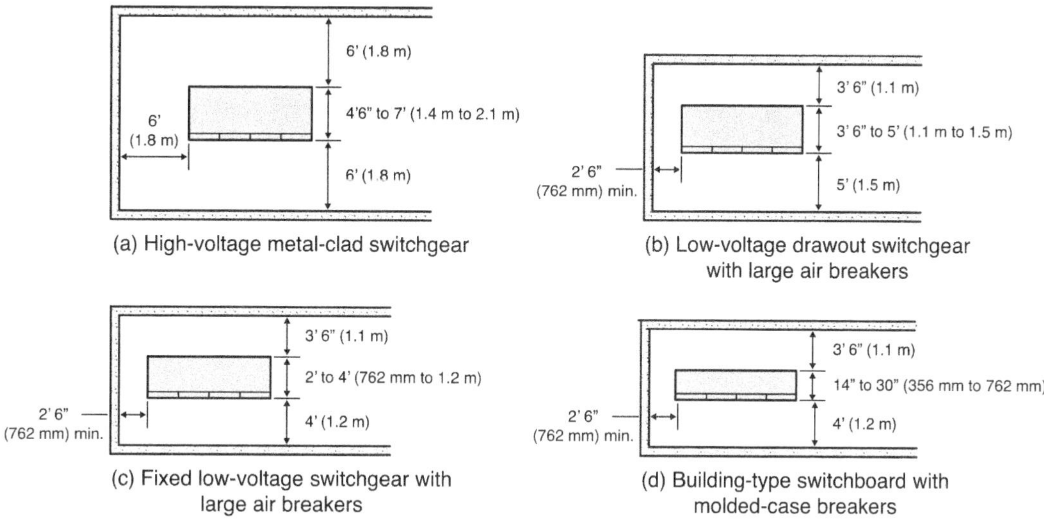

Figure 28.26 *Typical nominal switchgear space requirements. Each room should have two doors when switchgear is connected to high-capacity systems or operates at high voltage. Switchgear is shown in plan view. For detailed space requirements, see NEC Article 110.*

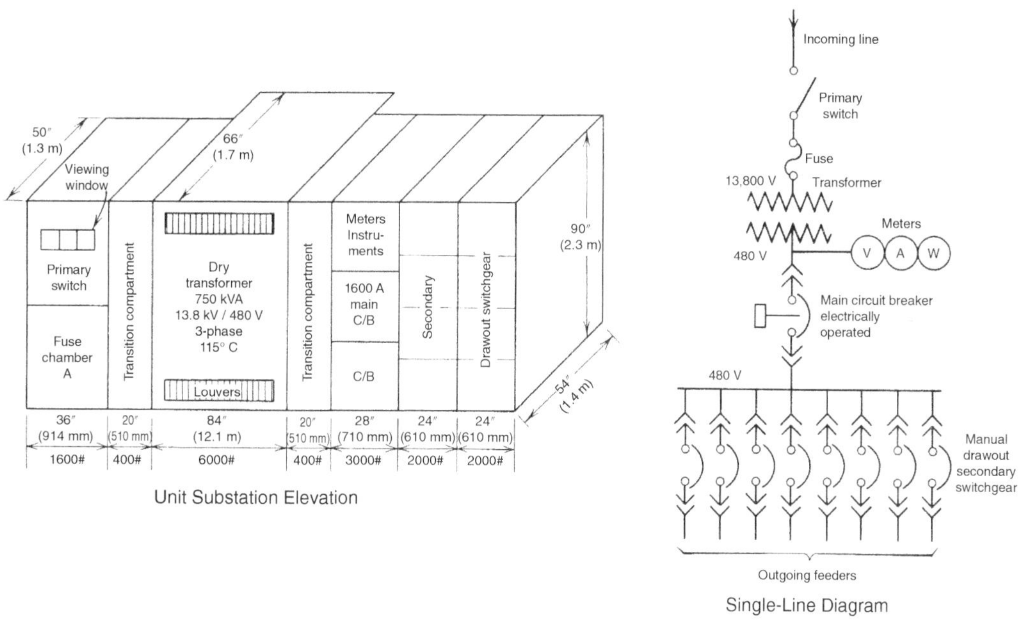

Figure 28.27 *Approximate sizes and weights of a typical large single-ended unit substation with related single-line diagram. Such a unit would supply a building with a maximum demand of 750 kVA. The incoming 13,800-V cables enter cubicle A and connect to the primary switch and fuses. The load side of the fuses connects to the transformer, which in turn connects to the secondary switchgear. The main secondary switchgear then feeds various switchboards and panelboards distributed throughout the building.*

in the metal backbox, as in Figure 28.28. Details of panelboard construction are presented in Figure 28.29. A panel may be equipped with a main circuit breaker whose function is to disconnect the entire panel in the event of a major fault. Figures 28.28 and 28.29 show panelboards with only branch circuit devices and no main breaker or switch. These panels are described as "lugs in mains only," which means that the panel has only connectors (lugs) on the main busbars for connection of the main feeder cables and no main protective device. The backbox is closed with a protective front panel with an access door. Three basic panelboard types are shown in Figure 28.30.

The line terminal of each circuit-protective device (breaker or fused switch) is connected to the busbars of the panelboard. The load terminal of the device then feeds the outgoing branch circuit. This is shown schematically in Figure 28.31, which is a line-drawing

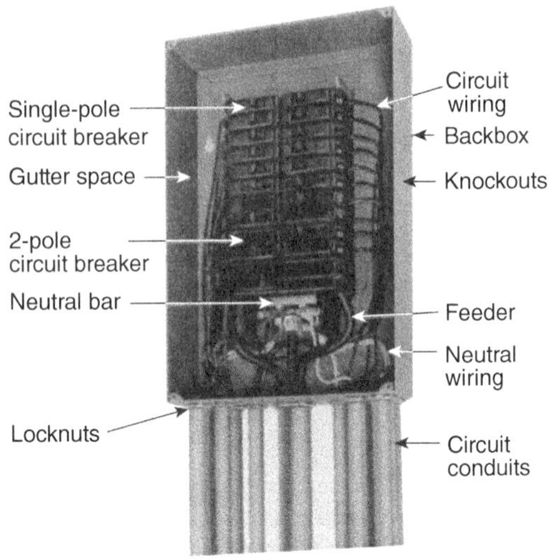

Figure 28.28 *A wired panelboard is shown with the cover removed. This panel is single-phase, 3-wire, meaning that it has two phase bars and a neutral (note the three heavy feeder cables). Lighting and appliance panels such as this are 4½–6 in. [115–150 mm] deep, 16–20 in. [400–500 mm] wide, and of sufficient height to accommodate the devices. The top device may be no higher than 78 in. [2 m] above the finished floor (AFF) and the bottom one no lower than 18 in. [450 mm] AFF to meet NEC requirements.*

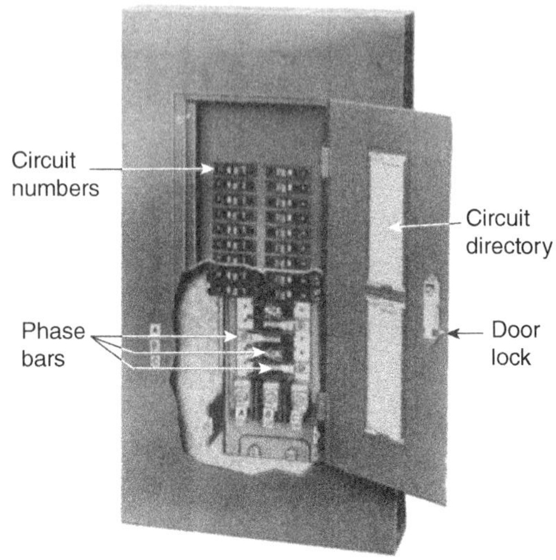

Figure 28.29 *Cutaway of a typical lighting and appliance panelboard. Branch circuit-protective devices are normally arranged in two vertical rows, even in 3-phase panels. The circuit directory on the inside of the panelboard door gives a brief description of the loads connected to each branch circuit breaker.*

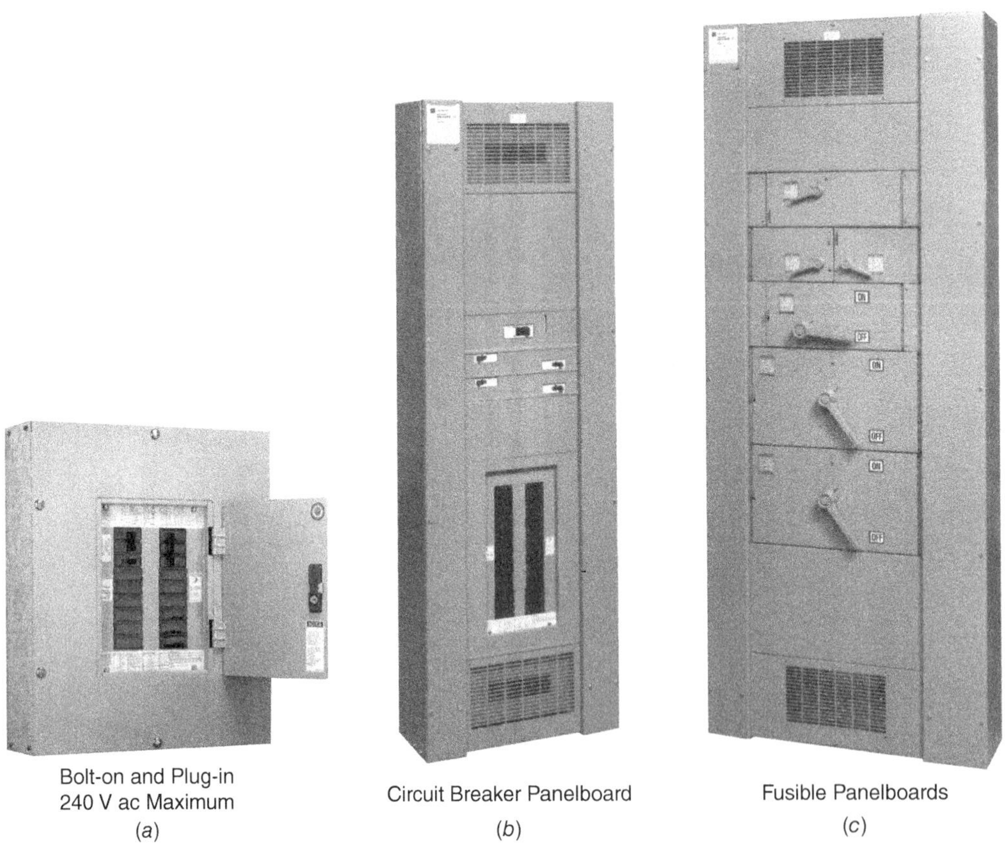

Figure 28.30 *Three different designs of panelboards. (a) Circuit breaker–type panelboard using bolt-on and plug-in molded-case circuit breakers 1- to 3-pole, 15–100 A. A main device of up to 400 A can be installed. The panel is rated for 240 V AC. The illustrated panel, which is arranged for surface mounting, measures 30 in. H × 20 in. W × 5¾ in. D [760 × 510 × 145 mm]. (b) Circuit breaker distribution-type panelboard with maximum bus ratings of 1200 A and 600 V AC. Branch circuit breakers can be 15–100 A, single-pole and 15–1200 A, 2- or 3-pole. These panels are freestanding but can be wall-mounted, provided that the highest device does not exceed 6 ft, 6 in. [2 m] AFF. Similar units are available with a front door. Depending upon the contents, cabinets can be 24–36 in. [610–915 mm] W × 60–90 in. [1.5–2.3 m] H × 6–8 in. [150–200 mm] D. (c) Switch- and fuse-type distribution panelboard, freestanding, with ratings similar to those of the circuit breaker type in (b). Cabinets are 36–44 in. [915–1115 mm] W, 74–90 in. [1.8–2.3 m] H, and 10½ in. [265 mm] D. (Courtesy of Cutler-Hammer.)*

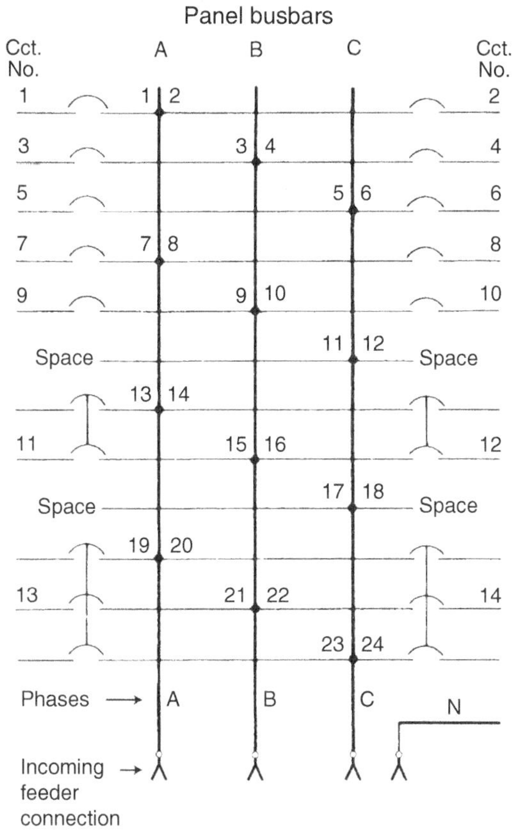

Figure 28.31 *Typical schematic diagram for a panel. Note that single-pole, 2-pole, and 3-pole circuit breakers are connected to 1-, 2-, and 3-phase buses, respectively. They supply 120-V 2-wire, 120/208-V 3-wire, and 120/208-V 3-phase 4-wire, respectively. This panelboard would be described as a 3-phase 4-wire lighting and appliance panel, with 10 SP, 2–2P, and 2–3P circuit breakers, lugs in mains only. The voltage of the panel and the ampere ratings of buses and all circuit breakers would then be specified.*

representation of an electrical panelboard. Notice that the circuit breakers are arranged in two vertical rows, corresponding to actual construction practice. In the illustrated line drawing, a 3-phase panel is shown (i.e., 3-phase busbars and a neutral bar). The busbars of the panelboard are energized by a feeder from a switchboard or large power distribution panel.

Panelboards are described and specified by type, bus arrangement, branch breakers, main breaker, voltage, and mounting—although not necessarily in that order. A typical description might be: lighting and appliance panel, 3-phase, 4-wire; 200A mains; main c/b, 225A frame, 150A trip. Branch breakers—all 100A frame; 8 ea. SP-20A, 4 ea. 2P-20A, 4 ea. 3P-20A; flush with hinged, locked door.

28.22 PRINCIPLES OF ELECTRIC LOAD CONTROL

The key to energy conservation and electric demand limitation is load control. In the past, the extent to which both of these desirable outcomes were implemented was essentially an economic decision: Would the savings in electricity costs justify the expense of installing the required control? Today, and increasingly in the future, building codes (and a desire to exceed the codes) will mandate energy consumption limitations. *ASHRAE/IES Standard 90.1* requires lighting controls in nonresidential buildings, and the trend toward additional and more stringent requirements is clear. The rationale for load control to limit demand charges was discussed previously.

Lighting control strategies are discussed in Chapter 17. The essential control principle that applies to all loads, not only to lighting, is straightforward: The load is switched or modulated in response to a control signal (Figure 28.32*a*). In terms of hardware, the process is somewhat more complex. Figure 28.32*b* shows in principle the equipment and wiring required in a common application: time control of exterior lighting, with photocell override. Figure 28.33 shows a more complex arrangement for area lighting control with local override, as might be installed in conjunction

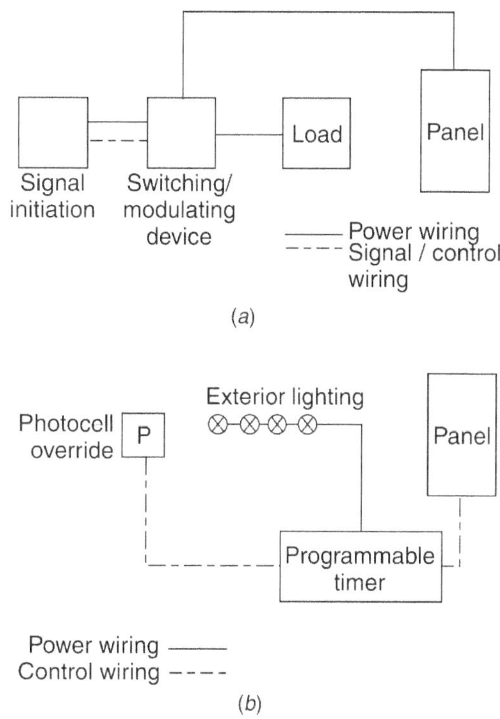

Figure 28.32 *(a) A low-voltage lighting control system always includes the elements shown. Signal initiation can come from a sensor such as a photocell or from a control device such as a timer, an override signal, a central computer signal, and the like. The signal activates the power-switching device, which in turn switches or modulates the load. The power source is always the associated lighting panelboard.*
(b) Schematic arrangement of a control system for exterior lighting. If the system is compact, a switching timer would be used. For a spread system, timer-controlled switching relays at each fixture would be used with attendant additional control wiring. Grouping of the fixtures into different control sequences (road, parking lot, building entrance, etc.) further complicates the wiring.

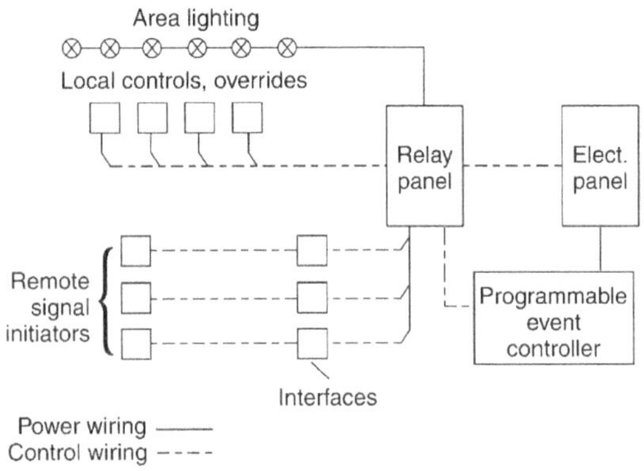

Figure 28.33 *Schematic diagram showing equipment and connections that might be used for control of a block of lighting in a large commercial building. The wiring can be even more complex and extensive if control of smaller lighting groups is desired or if control is desired from more than one location.*

with a commercial office building. In each case, there are signal-initiation devices, one or more relays that perform the actual electrical switching, and control and power wiring—in addition to the main power wiring from the panelboard.

Many past users found that the payback from reduced electric billing was insufficient to cover the control equipment cost, particularly where local and/or remote override was required, and as a result relied on inefficient and generally ineffective manual control. That option, however, is limited today and will become even more so in the future. As a result, a more economical way to accomplish the required control functions was developed, thanks in large measure to modern miniaturized programmable control elements. These can be installed directly in the lighting panelboard. Such a panel is known in the industry by the generic term intelligent panelboard and by various trade names that identify specific manufacturers.

28.23 INTELLIGENT PANELBOARDS

The idea behind an intelligent panelboard is straightforward. If many of the necessary electric load control and switching functions are incorporated into the panelboard by use of a compact, centralized, programmable microprocessor, the corresponding external devices and associated wiring are eliminated. The switching function is made possible by the use of motor- or solenoid-operated panel circuit breakers, thus taking advantage of their switching capability and eliminating the need for relays, contactors, and switches. The internal central controller permits local programmable control of each panel circuit breaker individually.

This central panel-mounted controller can also accept signal data from individual remote or network sources and can provide status reports, alarm signals, operational logs, and local bypass and override functions. Control cards in an adjacent cabinet, connected to the controller, can provide local switch override, daylight dimming and on/off control, telephone override energy control, electric demand control, and networking. With careful planning, the high initial cost of an intelligent panelboard can be offset by elimination of remote relays, relay panels, programmable time switches, remote-control switches, and all of their associated wiring. Simplified and improved facility operation, plus reduced maintenance costs and electric bills, are additional benefits that can be achieved with proper application and careful design. Figure 28.34 shows the essential elements of an intelligent panelboard and a typical network arrangement using the remote-control capabilities of such panels.

28.24 ELECTRIC MOTORS

Motors are frequently supplied along with the equipment they drive (such as fans, blowers, and so on) as part of a complete package. The actual choice of motor is left to the driven-equipment supplier, because the supplier is presumed best qualified to select a motor that will optimally match the driven-equipment requirements, for whose proper operation the supplier is responsible. In practice, however, a supplier is frequently guided primarily by price—low first cost being a recurring theme in building design and construction, very often to the detriment of low life-cycle cost. The designer, therefore, should be sufficiently knowledgeable so that the best motor for the application can be specified. The following sections are written with that purpose in mind, and are primarily concerned with application.

(a) Direct-Current Motors

These motors are not normally used in building work. The fine speed control (for which they were primarily used) is now available, more economically, with AC motors, as explained in Section 28.26.

(b) Alternating-Current Motors

These motors fall into three general classifications: polyphase induction motors, polyphase synchronous motors, and single-phase motors. Within these categories there are further subdivisions. Of these many types, most motors used in building equipment are squirrel-cage induction machines; therefore, this type is studied in some detail.

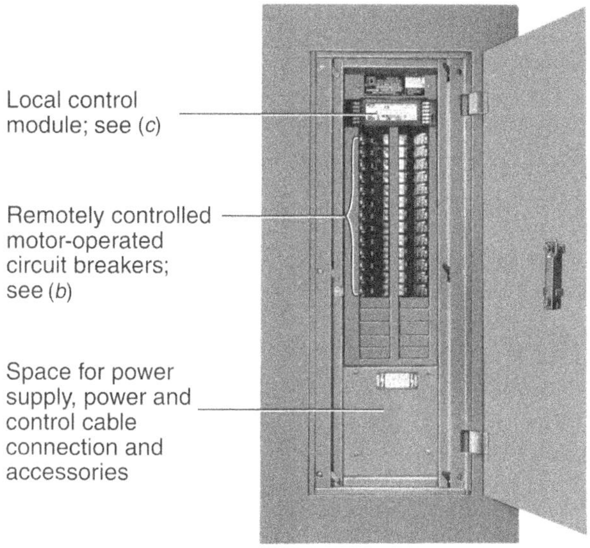

(a) Intelligent panelboard

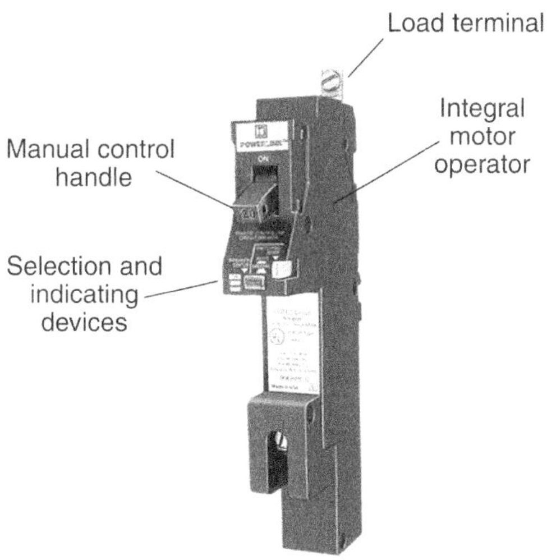

(b) Single-pole motor-operated circuit breaker

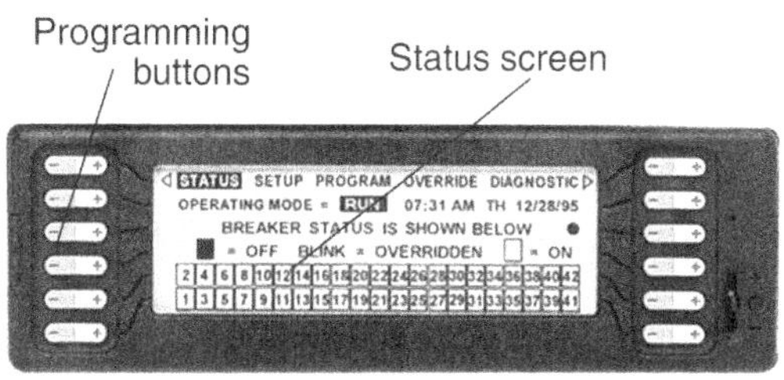

(c) Control module

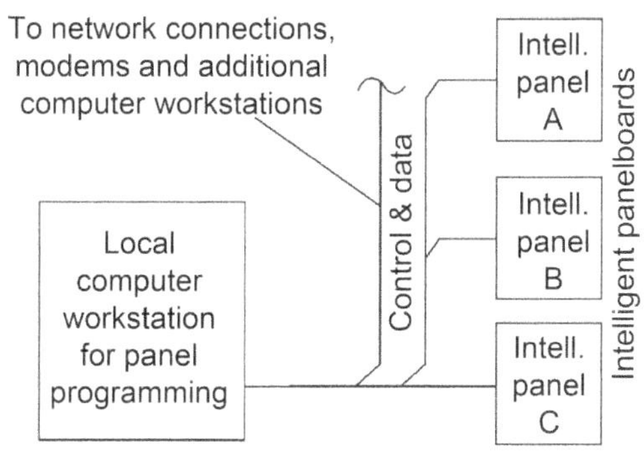

(d) Block diagram

Figure 28.34 *(a) This intelligent panelboard contains (from top to bottom) a microprocessor control module, 26 single-pole, remote-controllable (motor-operated) circuit breakers, space for 10 additional poles, and a compartment containing a DC power supply (for circuit breaker operation) and accessories. (b) This motorized single-pole circuit breaker can be manually or remotely controlled. Electric operation latches the circuit breaker so that in the event of power failure, all breakers remain in their pre-outage condition. The illustrated unit is a 100-A frame, 20-A trip unit. A status window on each unit displays its condition, while electronic sensors in each unit sense voltages and indicate on the control panel the status of each unit. (c) The control module, mounted at the top of the panelboard, is equipped with programming buttons that allow control of each circuit breaker individually. The illustrated basic control module programs and controls essential functions, including configuration, circuit breaker status, resetting, zoning, settings for hardware characteristics, override modes, screen selection, and self-diagnostics. Additional control modules (not shown) supply time-based programming, logging functions, and timed functions. The illustrated breaker status screen is one of five displays. The others are setup, program, override, and diagnostic screens. (d) Intelligent panels can be incorporated into a large control network by use of interface connections of various types (not shown). The illustration is a block diagram showing one of the networking possibilities. (Courtesy of The Square D Company.)*

ELECTRICITY

(c) Squirrel-Cage Induction Motors

This type of induction motor owes its interesting name to an early design in which the rotor consisted of a group of bars welded together into a cylindrical cage-type shape. The design, invented by Nikola Tesla, is basically unchanged today except for refinements. Squirrel-cage motors are manufactured in four different NEMA designs to meet different application requirements. Of these, the most common are:

Type B: Standard design, high efficiency and high power factor, normal torque; applicable to fans, blowers, and pumps

Type C: High starting torque, fair efficiency and power factor; applicable to compressors, conveyors, and other devices that start under load

A motor nameplate gives important information on a motor that is not self-evident:

1. *Type:* This is the manufacturer's designation and indicates primarily the enclosure. Common enclosures are open drip-proof, totally enclosed, fan-cooled, and weather-protected.
2. *Duty (time rating):* Continuous or intermittent.
3. *Service factor:* Permissible overload; generally 15%.
4. *kVA code:* Indicates by a letter the maximum starting current per horsepower. This is useful in selecting motor-protective devices. More recently, actual locked rotor amperes have been given.
5. *Frame:* A NEMA standard number that indicates the motor's physical dimensions.
6. *Motor voltage:* The standard motor voltages are 208, 230/460, and 575 V. Induction motors generally operate satisfactorily at +/− 10% voltage. Only 208-V motors should be used on 208-V systems because the actual line voltage may be as low as 200 V. Using a 230-V motor on a 200-V circuit will result in sharply reduced torque, increased temperature rise, and poor overload capacity.
7. *Motor nominal full-load efficiency:* As stated, a requirement of *ASHRAE/IES Standard 90.1* and EPACT.
8. *Design letter:* Indicates inherent motor design characteristics.

(d) Electric Motor Energy Considerations

The federal Energy Policy Act of 1992 (EPACT; subsequently updated/amended/morphed) required that the most common designs of induction motors, accounting for 70% to 80% of all polyphase induction motors sold in the United States, meet higher efficiency and power-factor standards. Specifically, these efficiency standards apply to single-speed, polyphase, squirrel-cage induction motors, designs A and B, continuous rating, and operating at 230/460 V, 60 Hz. These motors are known as premium efficiency or PE motors.

The decision on whether to use a PE motor in applications not mandated by law is essentially an economic and environmental one. Many motor manufacturers make available straightforward software that enables a designer to rapidly make a detailed life-cycle cost comparison among motor types, with operating time, loading, first cost, energy and maintenance costs, and so on, as the variables. As a guideline, motor applications involving low starting torque and frequent motor operation at partial load are likely candidates for short payback periods on the price premium of PE motors. Higher efficiency also results in reduced consumption of the typically nonrenewable energy resources used to generate electricity.

28.25 MOTOR CONTROL STANDARDS

American designers and equipment manufacturers historically utilized motor control equipment manufactured to NEMA standards exclusively. It was also common practice in the construction industry to split the motor and motor-control package by having equipment manufacturers supply the motorized equipment while the electrical construction contractor supplied the motor control equipment. Few coordination problems between controller and motor drive were encountered because control equipment sized and built to NEMA standards is large, heavy-duty, and applicable to a wide range of motor sizes and characteristics.

The International Electrotechnical Commission (IEC) is a technical organization with 62 full (including the United States) and 27 associate member nations. It develops recommended standards for electrical equipment, including motor controls. The IEC design concept regarding a motor and its controller involves close coordination. Such application sensitivity—coupled with flexible ratings for a single piece of equipment, depending upon its application—makes equipment built to IEC standards smaller, lighter, and cheaper than similarly rated equipment built to NEMA standards. As a result, many US manufacturers utilize IEC motor controls in conjunction with their motive equipment, particularly on international jobs. Because US architects, engineers, and contractors are involved in a large number of construction projects outside of the territorial United States, those responsible for designing, specifying, and purchasing electrical equipment in general, and motor control equipment in particular, should be aware of the comparative characteristics of NEMA-based and IEC-based equipment.

28.26 MOTOR CONTROL

(a) Fundamentals

A conventional AC motor controller is basically a contactor (see Section 28.14) designed to handle the heavy inrush currents involved in starting an AC motor. Its function is twofold: to start and stop the motor and to protect it from overload. These two separate and distinct functions are accomplished by combining a set of contacts for on/off control with a set of thermal overload elements for overload protection in a single unit. When the contacts are operated by hand, the controller is called a manual starter; when the contacts are operated by a magnetic coil controlled by pushbuttons, thermostats, or other devices, the unit is known as a magnetic controller or, simply and more commonly, a magnetic starter.

Motors of 1 hp [0.75 kW] or less are generally controlled by a manual switch that contains an overload protection device. It is advisable to utilize such a device for all fractional-horsepower (low-kW) motors.

Most starters are of the full-voltage across-the-line (ATL) type; that is, the contacts place the motor directly onto the line, and the motor starts up immediately. When such a procedure is undesirable because of voltage dip and flicker caused by the large inrush current or because of utility company limitations, a reduced-voltage starter, sometimes called a compensator, is used. These units initially apply reduced voltage to the motor, thus reducing the starting inrush current and line voltage drop. This in turn reduces dimming of lights, flicker, and other undesirable effects.

Older reduced-voltage starters use a stepping arrangement whereby voltage is increased incrementally, thus limiting inrush current. Unfortunately, starting torque is also limited, thus restricting the application of these starters. Solid-state starters that have been available since the late 1980s provide continuous (stepless), controlled motor starting. These units not only limit inrush current, but also, by adjusting the acceleration time, allow the required starting torque of the motor to be maintained. Such starters (Figure 28.35) provide soft starts (i.e., starts that minimize the mechanical stresses caused by rapid application of accelerating torque). Additional advantages of these units are reduced size and weight, long life, more sophisticated motor protection, and additional operating functions such as jogging and reversal. Disadvantages are higher cost and possible radio frequency noise problems.

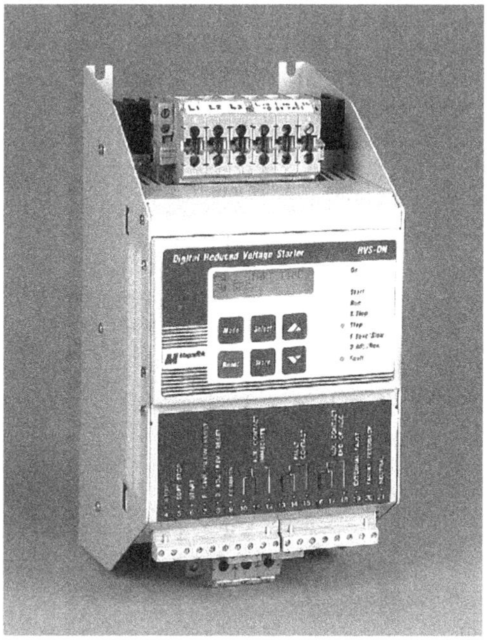

Figure 28.35 *Soft starter type for 3-phase induction motors of up to 1400 hp [1045 kW], shown without an enclosure. This microprocessor-based unit provides two selectable starting characteristics with stepless acceleration, plus additional control and protection functions including reversing control, over- and undervoltage and current protection, and two normal motor operating speeds. (Courtesy of Magnetek.)*

(b) Motor Speed Control

Many applications in heating, ventilation, and air-conditioning (HVAC), fluid flow, and industrial systems require speed control of electric motors. Until recently, this constituted a problem whose solution was expensive, often space-consuming, and rarely energy-efficient. The reason for this is that the common squirrel-cage induction motor is essentially a constant-speed device, where speed is determined by power line frequency and motor design. Speed drops slightly (slips) as the load increases. Until the advent of cheap electronic power equipment, speed control was usually accomplished by using a wound-rotor motor in lieu of a squirrel-cage unit. This drastically increased both motor and controller costs while providing only limited speed control.

Advances in power electronics have made practical a variable-voltage variable-frequency (VVVF) controller that gives smooth, continuous speed control over a range exceeding 30:1 while maintaining motor torque. These highly reliable controllers, also known as variable frequency drives (VFD), provide considerable energy economies that usually result in rapid payback of the relatively high first cost of equipment. The essentials of the VVVF control scheme are shown in Figure 28.36. Voltage control, in addition to frequency

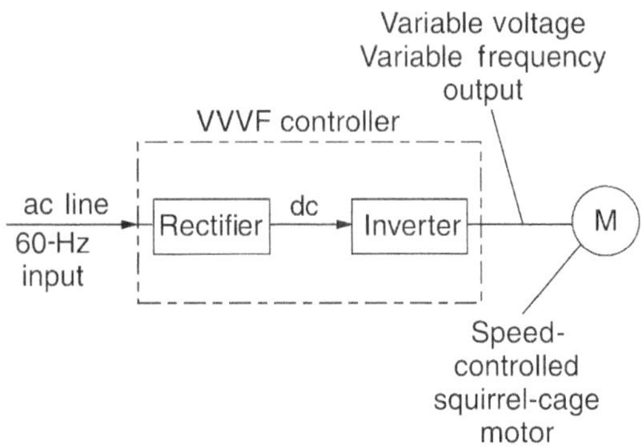

Figure 28.36 Schematic representation of the action of a variable-voltage, variable-frequency (VVVF) motor speed controller. Line voltage at 60 Hz is rectified to DC and the inverted voltage to AC. Frequency and voltage control are built into the inverters, designed to maintain a constant voltage-to-frequency ratio. This constancy is necessary to prevent malfunction of the motor as its speed is varied.

control, is necessary in order to maintain a constant voltage/frequency ratio, which is necessary for efficient, safe operation. An increase in this ratio results in overheating of the motor, whereas a decrease results in insufficient torque. Additional advances in power electronics have produced motor speed controllers with controlled speed ratios approaching 1000:1.

One disadvantage of these speed controllers has been the production of line harmonics and radio noise of sufficient strength to constitute a serious engineering problem. At this writing, these problems have been mitigated, and will undoubtedly be pursued to the point that, except for highly sensitive areas, the interference will be negligible. At this point, however, the designer must investigate these effects for the particular application. Typical adjustable-frequency controllers are shown in Figure 28.37.

28.27 MOTOR CONTROL EQUIPMENT

All the starters discussed previously are available in a wide range of sizes, voltages, and enclosure types. Every motor controller is required by the *NEC* to have a disconnecting means for safety reasons. Where convenient, this disconnect switch may be combined with the starter into a single unit known as a combination starter. A circuit breaker or fused switch is often used in such an arrangement, which then constitutes the branch circuit protection and disconnecting means (Figure 28.38).

A typical, brief description of a conventional motor controller would be similar to the following: combination circuit-breaker type, across-the-line (ATL) motor controller, NEMA size 2, three O.L. (overload) elements, 208 V, in a NEMA 1 enclosure; starter shall contain integral on/off pushbuttons.

Table 28.7 gives the approximate dimensions of one particular type of combination starter. As with other electrical equipment, a current manufacturer's catalog must be consulted for reliable data on a particular manufacturer's products. Dimensions vary somewhat among manufacturers (while still meeting NEMA specifications). As noted previously, physical characteristics for control equipment of similar ratings can vary greatly among items made to meet IEC specifications and those made to NEMA specifications.

Where several motors are installed in close proximity to one another, it is often convenient from the control perspective, as well as economically, to combine the motor starters, disconnect switches, motor controls, and indicating devices into a single large assembly. Such an assembly is called a motor

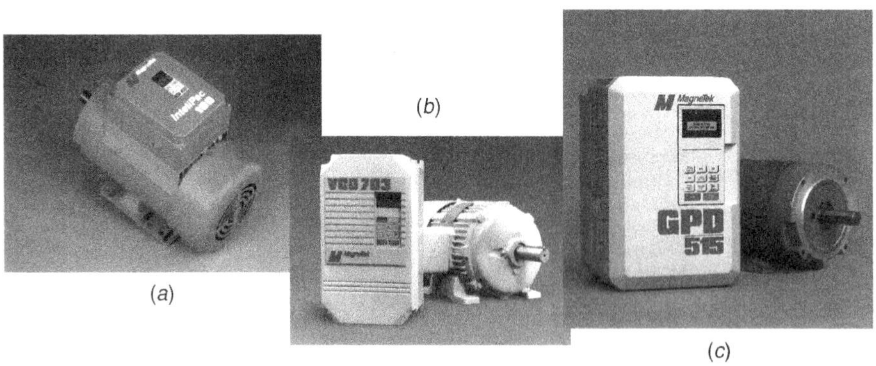

Figure 28.37 Three programmable, microprocessor-controlled motor controllers. (a) Integrated motor and controller for motors of up to 1 hp [0.75 kW]. The controller provides a soft start, a speed control range of 40:1, high efficiency, multiple speed settings, and full motor protection. (b) Controller for motors of up to 3-phase, 400 hp [300 kW]; this type of unit provides programmable functions including speed control from standstill to double-rated speed, reversing, adjustable acceleration, and full motor protection. (c) Controller that can be programmed for three different control modes. Speed control range, regulation, and torque curves vary in each mode to suit the specific controlled load. The controller has low harmonic distortion plus fault diagnostic software, in addition to the normal overload and protective characteristics. It also has terminals for remote monitoring and control plus network control capabilities. A 20-hp [15-kW], 230-V controller in a NEMA 1 (indoor) enclosure measures approximately 16 in. × 10 in. × 9 in. [400 × 250 × 230 mm] and weighs 24 lb [11 kg]. (Courtesy of Magnetek.)

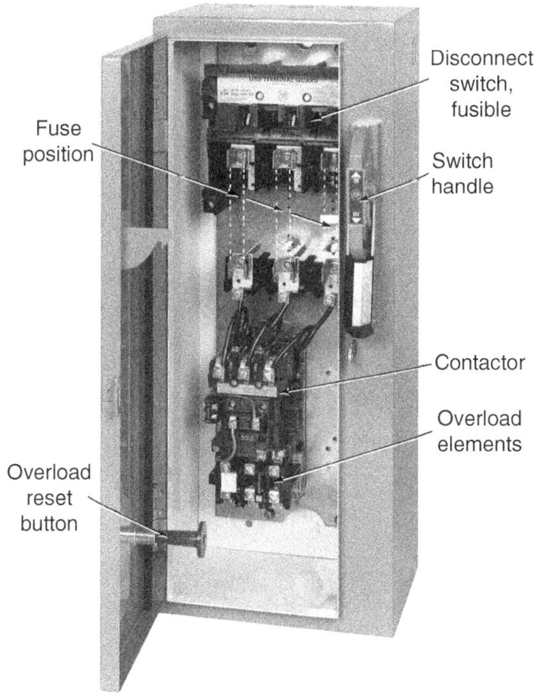

Figure 28.38 Interior of a conventional (non-solid-state) combination fused switch-type, across-the-line motor controller. Note that the unit is essentially a switch and a starter wired together and installed in a single cabinet. Approximate dimensions of the illustrated unit are 20 in. W × 48 in. H × 9 in. D [510 × 1140 × 230 mm], with a 75-hp [56-kW] capacity. (Courtesy of Allen-Bradley Co.)

TABLE 28.7 Rating and Approximate Dimensions[a] of AC Full-Voltage Conventional Single-Speed Motor Controllers, 3-Phase Combination Circuit Breaker Type[b]

NEMA Size[c] Designation	Maximum Horsepower, hp [kW][c]	Width	Height	Depth
			in. [mm]	
0	3 [2.2]	10 [254]	24 [610]	7 [178]
1	7½ [5.6]	10 [254]	24 [610]	7 [178]
2	15 [11.2]	10 [254]	24 [610]	7 [178]
3	30 [22]	20 [508]	24 [610]	9 [229]
4	50 [37]	20 [508]	48 [1219]	9 [229]
5	100 [75]	20 [508]	56 [1422]	11 [279]

[a]Dimensions vary among manufacturers.

[b]All starters are housed in a NEMA 1 indoor ventilated enclosure.

[c]Maximum hp [kW] that can be controlled at 208–230 V. When operating at 460 V, a starter one size smaller can generally be used.

control center (MCC). Typical MCC types are shown in Figure 28.39.

28.28 WIRING DEVICES: GENERAL DESCRIPTION

The general term wiring devices includes all devices that are normally installed in wall outlet boxes, including receptacles, switches, dimmers, fan controls, and so on. Attachment plugs, also called caps, and wall plates are also included in wiring devices.

There are three grades of device quality, using NEMA and UL grading nomenclature: hospital grade, federal specification grade, and UL general-purpose grade—in descending order of price and quality. Inasmuch as these grades correspond to published specification requirements, they are the only reliable gauge of construction quality. Hospital-grade devices are identified by a green dot on the device face. These devices are built to withstand severe abuse while maintaining reliable operation and must meet UL requirements for this grade.

Manufacturers usually grade their devices as hospital grade, premium or industrial specification grade, commercial specification grade, and residential grade, in descending order of quality and price. Hospital grade is approximately the same among manufacturers because industry standards must be met. Industrial and commercial specification grades correspond roughly to federal specification grade, and residential grade to UL general-purpose grade. Manufacturers' grades are qualitative, although reputable makers list the NEMA and UL specifications that each of their grades meets. It is only on this standardized basis that a technical quality comparison can be made.

In application, industrial specification grade equipment is usually used in industrial and high-grade commercial construction; commercial specification grade in most educational and good residential buildings; and standard or residential grade in low-cost construction of all types. The grade of wiring devices should, as with all electrical equipment, be consistent with the quality of construction in the facility as a whole.

Although historically the term wiring device referred to devices that operate at line voltage (120, 208, 240, and 277 V) in a branch circuit, the term today also includes low-voltage lighting control devices, fan controls, and the like. The accepted criteria seem to be that only devices rated 30 amps or less, that can be mounted in a small wall box, are considered wiring devices. Communication attachments are not usually considered wiring devices, even when they are supplied by the electrical contractor. They are part of premise wiring, which is the term commonly used to describe data and communication system wiring, raceways, and devices (see Chapter 32).

28.29 WIRING DEVICES: RECEPTACLES

A receptacle is, by *NEC* definition, "a contact device installed at the outlet for the connection of a single

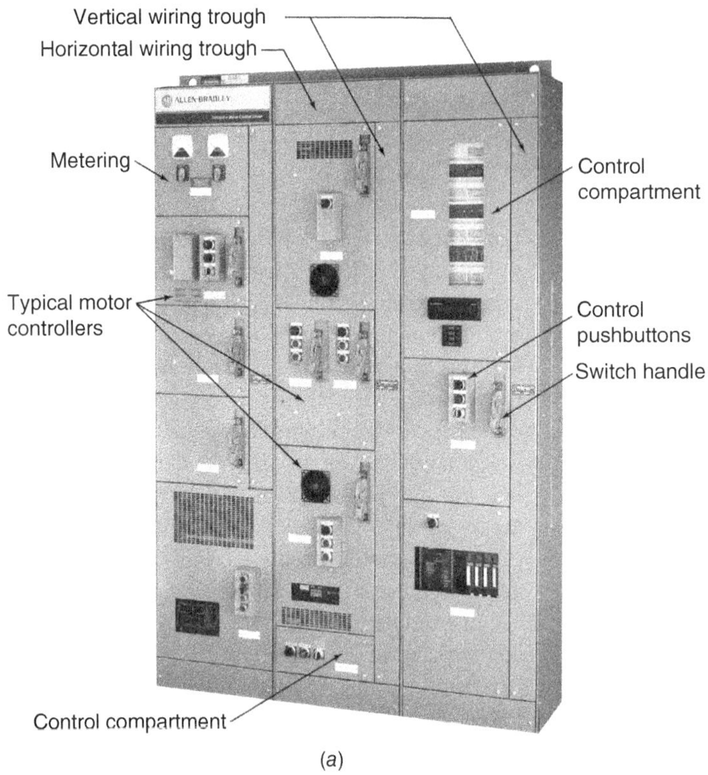

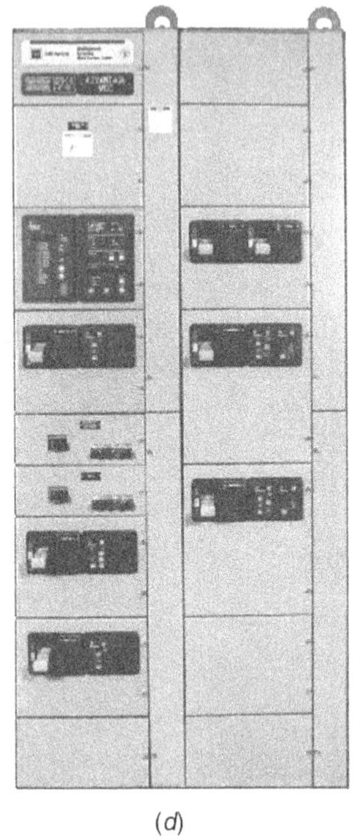

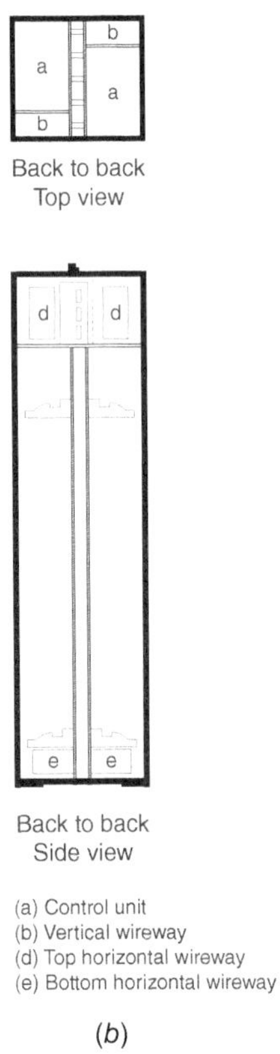

Figure 28.39 *Motor control centers (MCC) of various constructions. (a) Conventional MCC construction with magnetic motor starters, manual local control, analog metering, and conventional overload and protective equipment. (b) Back-to-back construction of conventional MCC adds only 5 in. [125 mm] to the basic 15-in. [380-mm] depth. (c) Units are 90 in. [2.3 m] high and 20 in. [500 mm] wide per section. Solid-state controllers can be installed in this unit, as well as in the MCC shown in a, along with conventional magnetic contactors. (d) MCC that uses solid-state programmable controllers with heaterless overload protection. The unit also provides protection against phase loss, single phasing, and ground faults. Single side units in this construction are 90 in. [2.3 m] H × 20 in. [500 mm] W × 16–20 in. [400–500 mm] D (depending upon starter size). Back-to-back construction is 21 in. [534 mm] D. (a: courtesy of Allen-Bradley Co.; b, c: Courtesy of Siemens-Furnas; d: Courtesy of Cutler-Hammer.)*

attachment plug." This usually takes the form of the common wall outlet or, as illustrated later, larger and more complex devices. A comment about terminology is required. The common wall outlet is properly called a convenience receptacle outlet, a receptacle outlet, or a convenience outlet. The term wall plug, which is heard so often, is incorrect. A plug is another name for the attachment device (cap) on the wire that carries electricity to the appliance. The device at the end of the line cord is plugged into the wall, hence plug.

Because by *NEC* definition a receptacle is a contact device for the connection of a single attachment plug, and because the normal wall convenience receptacle will accept two attachment plugs, it is properly called a duplex convenience receptacle or duplex convenience outlet. Most people shorten this term to duplex receptacle or duplex outlet.

Receptacles are identified by the number of poles, the number of wires, and whether the device is designed for connection of a separate equipment ground. (This ground may connect to a separate equipment-grounding wire or to the metallic conduit system, according to system design.) The number of receptacle poles equals the number of current-carrying contacts, including the neutral. The equipment ground connection (pole) is not counted because it does not (normally) carry current. The number of wires includes the equipment ground connection because it is wired. The equipment ground connection (grounding pole) should not be confused with the system ground (usually the neutral pole), nor may the wiring for the two be interchanged (see Section 29.4).

In a typical application, a receptacle for an electric dryer with a 4800-W, 208-V heating element and a 1/6-hp [0.12 kW], 115-V motor would be NEMA 14–50R (Figure 28.40). The motor would connect across W and X, the heater across X and Y, and the appliance case to G. Receptacles installed on

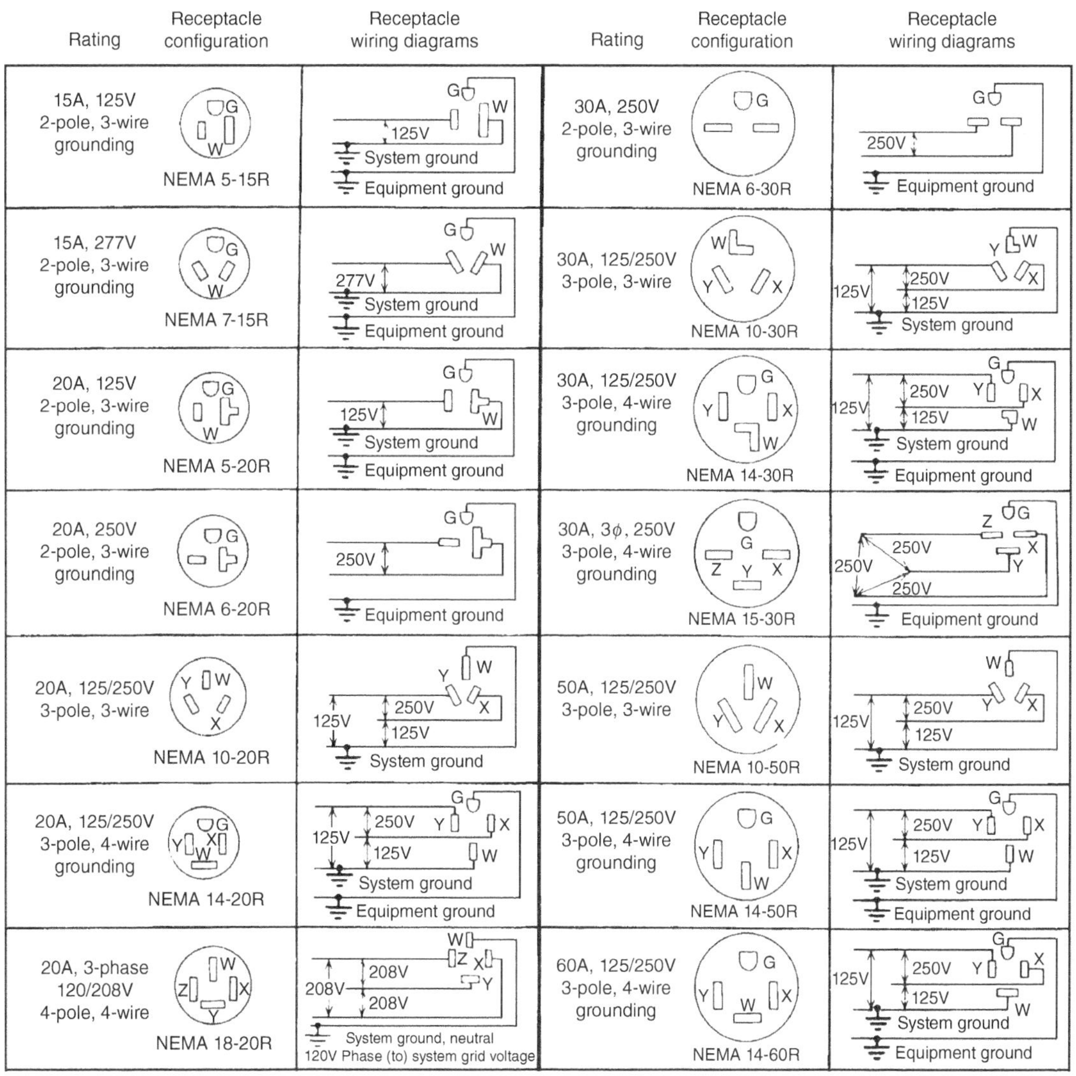

Note: "W" denotes white system ground or neutral
"G" denotes green equipment ground.

Figure 28.40 *Receptacle configuration chart with selected common general-purpose, nonlocking devices with associated NEMA designations.*

standard 15- or 20-A branch circuits must be of the grounding type. Receptacles connected to different voltages, frequencies, or current type (AC or DC) on the same premises must be polarized so that attachment plugs are not interchangeable. Figure 28.40 shows some of the standard receptacle configurations and their ratings. Figure 28.41 shows typical receptacle construction plus enclosures for use in damp and wet locations.

Receptacles are readily available from 10 to 400 A, 2 to 4 poles, and 125 to 600 V. In addition, special types such as locking, explosion-proof, tamper-proof, and decorative design units are made. Also, specific-usage devices, such as electric range receptacles, are available. All receptacles other than the normal 15/20-A, 3-wire, parallel-slot type should be specified to be furnished with at least two matching caps (plugs). A typical receptacle specification would be: receptacle, duplex, 2-pole, 3-wire, grounding type, 20 A, 250 V, federal specification grade, for indoor use. Receptacles are normally mounted vertically between 12 and 18 in. [305–457 mm] above the finished floor (AFF), except that in shops, labs, and other areas where tables are placed against the walls, 42 in. [1.07 m] is the usual mounting height.

In addition to these types, several manufacturers produce a receptacle with built-in ground-fault circuit protection (Figure 28.42). For a discussion of these ground-fault circuit interrupter (GFCI) receptacles, see Chapter 30.

The sensitivity of certain modern electronic equipment to voltage surges and electrical noise (random, spurious electrical voltages) has led to the development of two special receptacles. Receptacles with built-in surge suppression protect connected equipment from overvoltage spikes. Receptacles with an insulated equipment-grounding terminal separate the device ground terminal from the system (raceway) ground

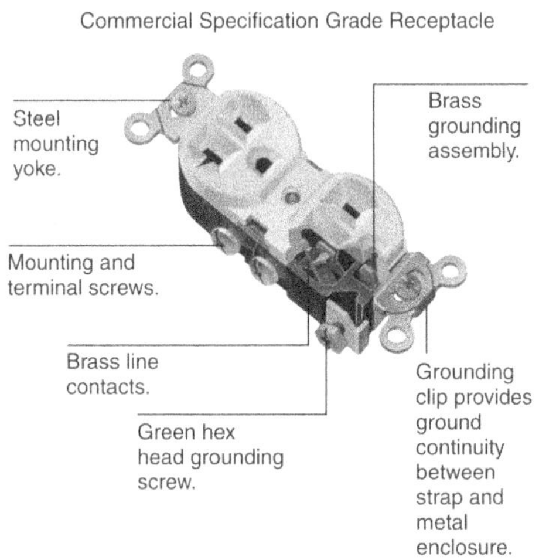

(a)

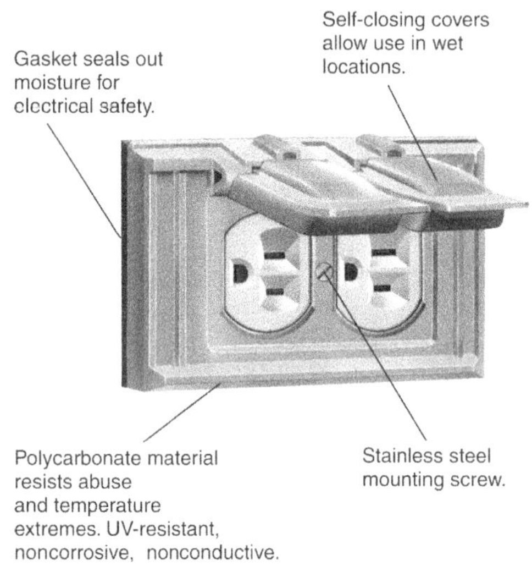

(b)

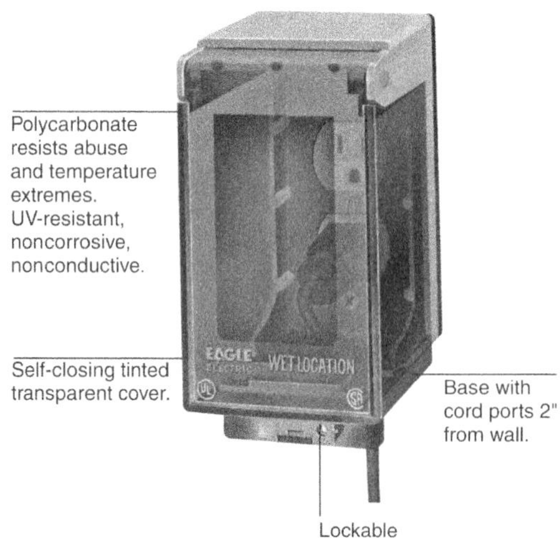

(c)

Figure 28.41 *(a) Cutaway of a commercial, specification-grade duplex convenience receptacle showing its construction. (b) Receptacles installed in damp locations, outdoors but not exposed to the rain, or in wet locations but used to energize portable equipment and continuously attended, should have a cover that is weatherproof when the attachment plug cap is removed, of which one type is illustrated. (c) Receptacles in wet locations (that may be in use unattended) must have a weatherproof enclosure, the integrity of which is not affected by the attachment of a plug. One such enclosure is illustrated. (Courtesy of Eagle Electric Manufacturing Co. Inc.)*

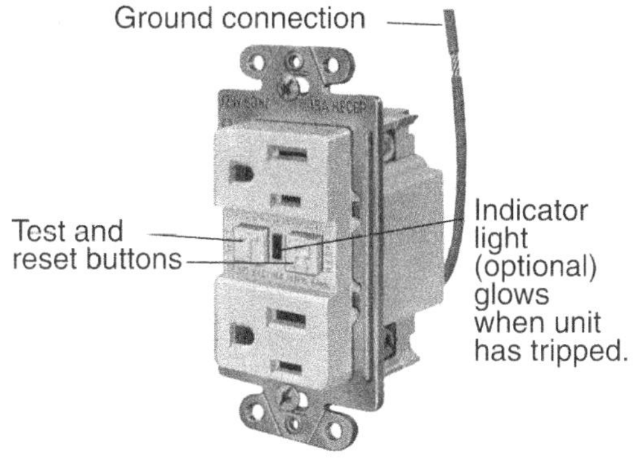

Figure 28.42 *Where it is desired to provide ground-fault protection at a single outlet only, and thereby localize power interruption (rather than using a GFCI circuit breaker on the entire circuit), a receptacle containing a ground-fault interrupter, as illustrated, may be used. A test button is provided to permit periodic testing, and a reset button reenergizes the receptacle after it has tripped on ground fault and the fault has been cleared. (Courtesy of Eagle Electric Manufacturing Co., Inc.)*

Figure 28.43 *Hospital-grade, insulated (isolated) ground receptacle used where heavy duty is expected, and both reliability and freedom from electronic noise are required. Hospital grade is denoted by a green dot on the upper left of the receptacle face, and the insulated ground by an orange triangle on the upper right. Some manufacturers make the entire receptacle orange. (Courtesy of Leviton Manufacturing Co.)*

because it has been determined that much unwanted electrical noise can be eliminated by such disconnection. The receptacle's insulated ground terminal is connected to an isolated green ground conductor that carries through the entire system but is connected to the system ground at the service entrance only; hence the name isolated ground. Isolated ground receptacles are identified by an orange triangle on their face (see Figure 28.43).

28.30 WIRING DEVICES: SWITCHES

Switches of up to 30 A that can be outlet-box mounted fall into this category. Generally, because of their better construction, AC-only switches are preferable to the AC/DC type. The usual AC switch rating is 15, 20, or 30 A at 120 or 120/277 V. Usual constructions are single-pole, 2-pole, 3-way, 4-way, momentary-contact, 2-circuit, and maintained-contact SPDT (single-pole, double-throw) and DPDT (double-pole, double-throw). Operating handles are toggle, key, push, touch, rocker, rotary, and tap-plate types. Mercury and AC quiet types are relatively noiseless; toggle, tumbler, and AC/DC types are generally not. A typical switch specification might be: switch, single-pole, AC, quiet type, federal specification grade, 15 A, 120 V, with press handles lighted when off, for side wiring only.

A switch incorporating a solid-state rectifier is readily available that gives high/off/low control for incandescent lamps and costs very little more than an ordinary switch. Typical applications are in areas where a lower illumination level is often acceptable and always desirable as an energy-conserving measure. In high-security areas where an easily defeated normal key-operated switch is inadequate, a tumbler lock-controlled unit can be used. Loads that can be timed out, such as bathroom heaters and ventilating fans, can be controlled by a spring-wound timer switch, as illustrated in Figure 28.44 (which also illustrates

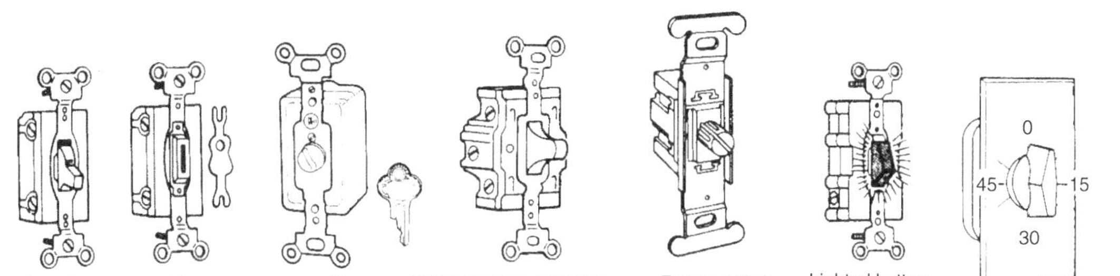

Figure 28.44 *A few of the most common wiring device switches, generally installed in a small wall box and used for control of lighting circuits.*

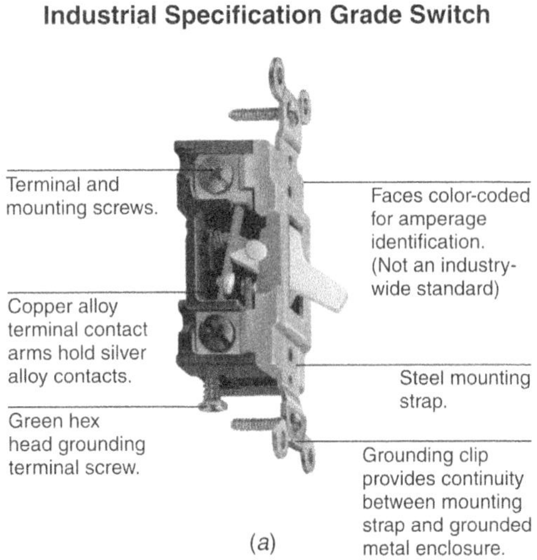

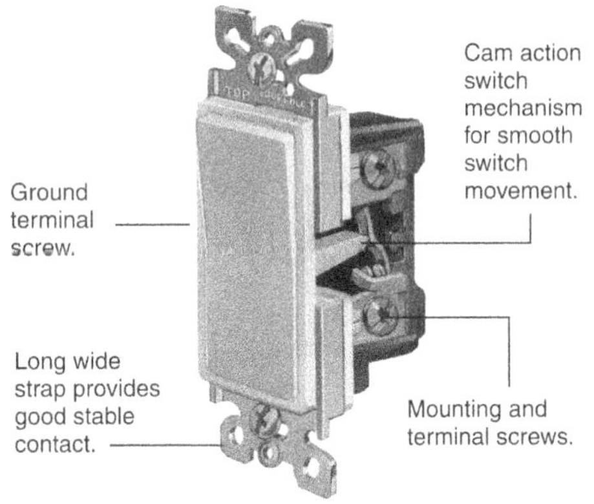

Figure 28.45 *Cutaway drawing of two common wall-switch designs showing operating mechanisms and construction details. (a) Conventional toggle design. (b) Rocker plate design. (Drawings courtesy of Eagle Electric Manufacturing Co.)*

a few common switch types). Figure 28.45 shows two common switch designs in cutaway, revealing their construction and operating mechanisms. A solid-state switch with adjustable time delay is shown in Figure 28.46.

Thanks to miniature electronics, a programmable switch is available that fits into a wall-outlet box in lieu of an ordinary switch. The unit acts as a solid-state 15-A switch and can be readily programmed to switch the controlled circuit or device at preset times (see Figure 28.46).

28.31 WIRING DEVICES: SPECIALTIES

The most common device in this group, which also includes pilot lights, fan controls, and other small motor controls, is the lighting dimmer. The original lighting dimmers were variable autotransformers.

Figure 28.46 *Solid-state adjustable electronic switch timer that mounts in a common wall-switch box. The unit switches off, with delay adjustable from zero to 15 minutes in five discrete steps. Units such as these can control incandescent and fluorescent lamps and small motors. They are applicable where either direct manual control or occupancy sensors are inapplicable. (Courtesy of Leviton Manufacturing Co.)*

They were large, heavy, and expensive, and therefore found application mainly in theater lighting, displays, and the like. The advent of small electronic units (silicon-controlled rectifiers, SCRs) made the modern wall-box dimmer possible. The simplest units are rotary, with limited capacity and applicable only to incandescent lamp loads. One of their disadvantages, still found in inexpensive units, is the production of annoying radio frequency interference (RFI) and line harmonics. Newer models are available that provide preset control, combine dimming and switching, are usable for smooth fan speed control, can be remotely controlled (in addition to local control), provide automatic fadeout, and can dim fluorescent lamps (provided with dimming ballasts). A few types are shown in Figures 28.47 and 28.48.

28.32 LOW-VOLTAGE SWITCHING

The switches illustrated and discussed in Section 28.30 are all full-voltage types; that is, they are wired directly into the load circuit and operate at line voltage and full current. A control scheme that uses light-duty, low-voltage (24-V) switches to control line voltage relays, which in turn do the actual circuit

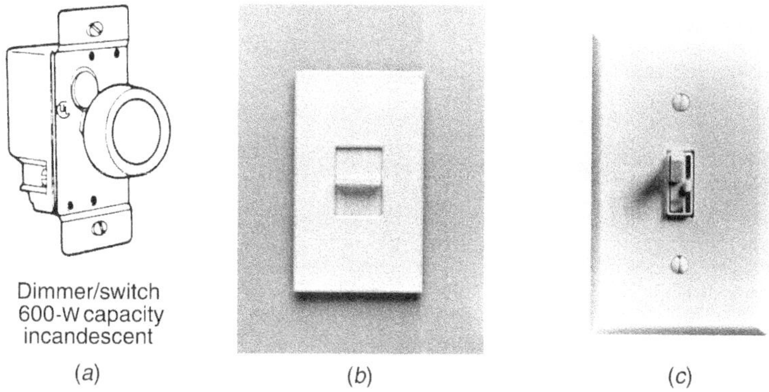

Figure 28.47 (a) Simple solid-state dimmer, suitable for a maximum 600-W incandescent load only. The switch permits level preset in a pushbutton design. Inexpensive units may cause considerable radio frequency interference (RFI). (b) Slide dimmer for a 600-W incandescent load, including on/off and level preset. This dimmer is available with a wireless receiver that responds to a handheld infrared remote control. (c) Combined on/off switch and slide dimmer provides preset level control. Units similar to (b) and (c) are available for loads of up to 1000 W, fluorescent lamp control (with appropriate ballast), and fan speed control. (b, c: courtesy of Lutron Electronics Co., Inc.)

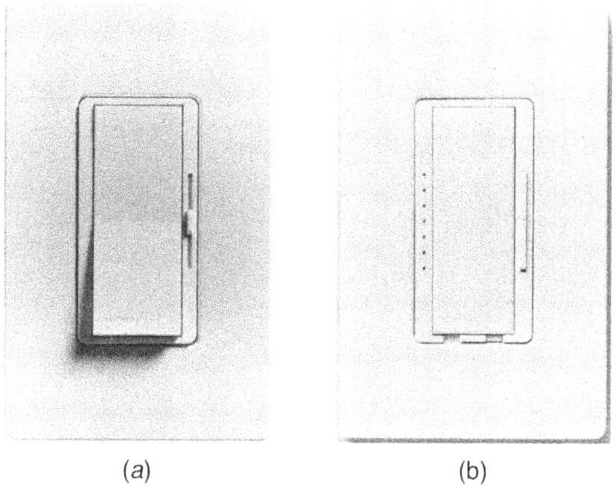

Figure 28.48 (a) Combined tap switch and slide dimmer provides preset level control. A built-in night-light suggests a residential application. (b) Multifunction tap switch in the center permits switching to full brightness, gradual brightening to preset, 10-second fade-out, and off. Rocker switch at the right control levels. LEDs behind pinholes at the left provide an indication of the preset level. Both illustrated unit designs are available for incandescent and fluorescent (with special ballast) lamp control and fan speed control. Both are designed with RFI suppression. (Courtesy of Lutron Electronics Co., Inc.)

switching, is illustrated in Figures 28.49 and 28.50. The system is variously referred to as low-voltage switching, remote-control switching, and low-voltage control.

Because the loads in this system are relay-controlled, any number of switches and control devices can be wired in parallel to operate a relay and thereby effect the switching. Thus, the control can be local, remote, and master, with individual override by local control devices such as occupancy sensors, and central control or override at a central controller by timers, daylight controllers, and so on. The advantages of this system of control over full-voltage switching are:

- flexibility of control location
- individual load control override by local control devices (occupancy sensors, photocells)
- group load override by central control devices such as timers, daylight controllers, and energy management systems
- low cost of low-voltage, low-current wiring compared to full-voltage wire and conduit
- inherent system flexibility and simplification of alterations
- monitoring of the status of individual loads at a centralized control panel by the use of relays with auxiliary contacts (see Figure 28.49)

The second and third points are particularly important from an energy conservation perspective, as it has been amply demonstrated that reliance on manual controls to limit energy usage yields disappointing results.

The basic components of the entire system are the control relay and the switches, shown in Figure 28.51. Small systems use hard wiring (physically wired connections), with the hard-wired relays installed either dispersed at each controlled fixture (or lighting group) or grouped in a relay panel. When the number of relays exceeds about 20, these panels become so densely wired that control changes become difficult and tedious. As a result, prewired computer- or microprocessor-controlled panels are used in large systems, where connections can be made and changed easily. The interior of one such panel is shown in Figure 28.52. The logic of a medium-sized computer-controlled system is shown in Figure 28.53. Manufacturers have made software available that permits rapid, accurate design of complex, large systems. A typical computer screen in such a system is shown in Figure 28.54.

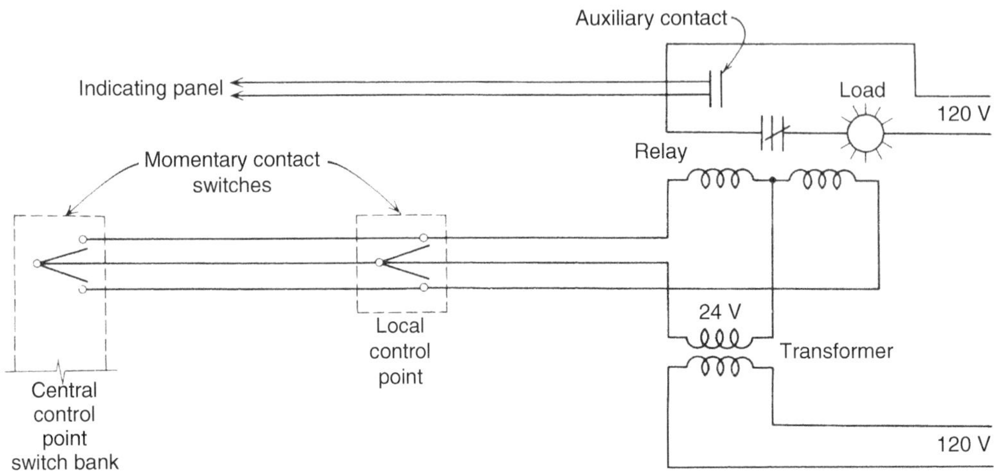

Figure 28.49 *Low-voltage switching control. Multipoint control and central control are illustrated. The diagram shows the relays located at the load. Central relay cabinets are used in dense load areas.*

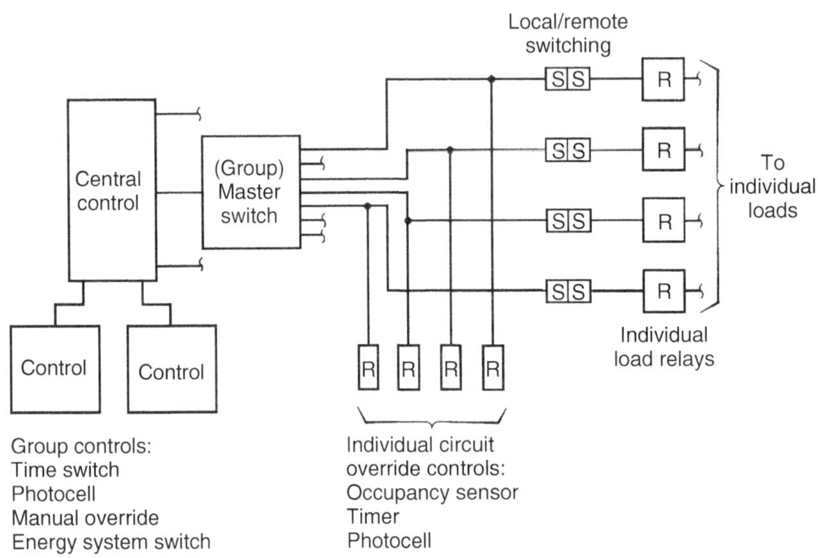

Figure 28.50 *Schematic arrangement of a large low-voltage switching system. Individual loads can be switch-controlled from multiple locations, including a group (master) control point. In addition, local control devices such as photocells and occupancy sensors can override the local switches to maximize energy conservation. Central (group) control devices can perform the same override function at the group level.*

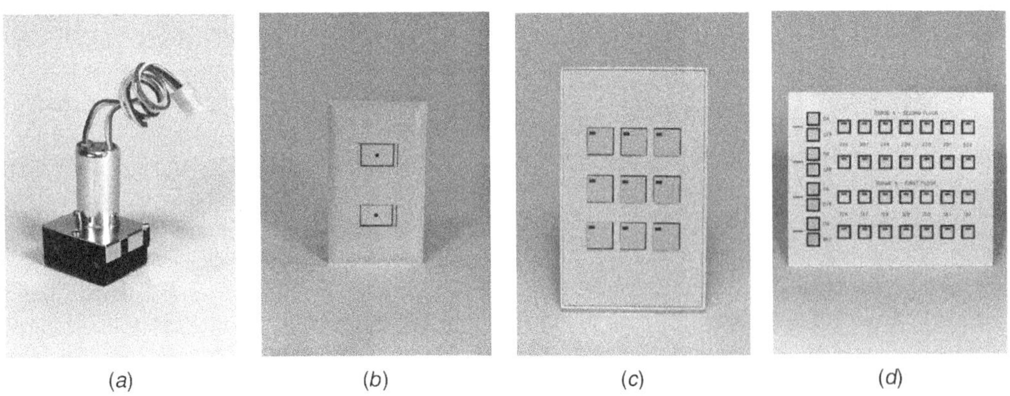

Figure 28.51 *Components of low-voltage switching and control systems for lighting and other electric loads. (a) The basic switching device is a single-pole, double-throw latching relay with a 24-V coil and 20-A, 120–277-V contacts. It is shown with a plug-in connector on its leads, making it suitable for rapid mounting in a relay panel (see Figure 28.52). The same relay without the connector is suitable for hard-wiring at a controlled device. (b) Low-voltage (24-V) wall switches can be mounted singly or ganged in groups of two to eight switches. Switches are available in button, rocker, key-operated, and lighted designs (as in b–d). (c) Switching station with pushbutton low-voltage switches, each with an LED to indicate switch position (on/off). (d) Switch bank with illuminated pushbutton switches identified with room number and floor. The on/off switches at the left of the bank are override switches for the corresponding horizontal line. (Courtesy of Touchplate Technologies, Inc.)*

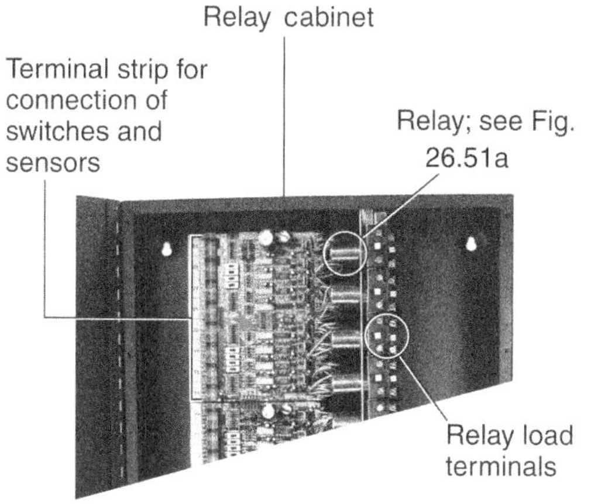

Figure 28.52 Prewired relay cabinet (partial). Field wiring consists of control wiring to switches and sensors and power wiring to the controlled loads. (Courtesy of Touchplate Technologies, Inc.)

28.33 WIRELESS SWITCHING AND CONTROL

Full-voltage local switching was described in Section 28.30, and low-voltage switching, both local and remote, was described in Section 28.32. The next logical step is remote wireless control and switching. These systems have undergone extensive development. They impinge on the design considerations of architects and building engineers to the extent that facilities to accommodate such systems must be provided if such controls are to be used on a project.

Aspects of wireless control and switching, as well as system functioning and application to lighting control, are discussed in detail in Chapter 17.

28.34 POWER LINE CARRIER SYSTEMS

Although the advantages inherent in flexible, programmable load-switching systems such as those described in the preceding sections are undeniable, particularly in connection with energy management systems, their high first cost frequently makes their use uneconomical. This is particularly true when retrofitting large, complex facilities with energy management controls, where the cost of wiring, even when using low-voltage control wires, can

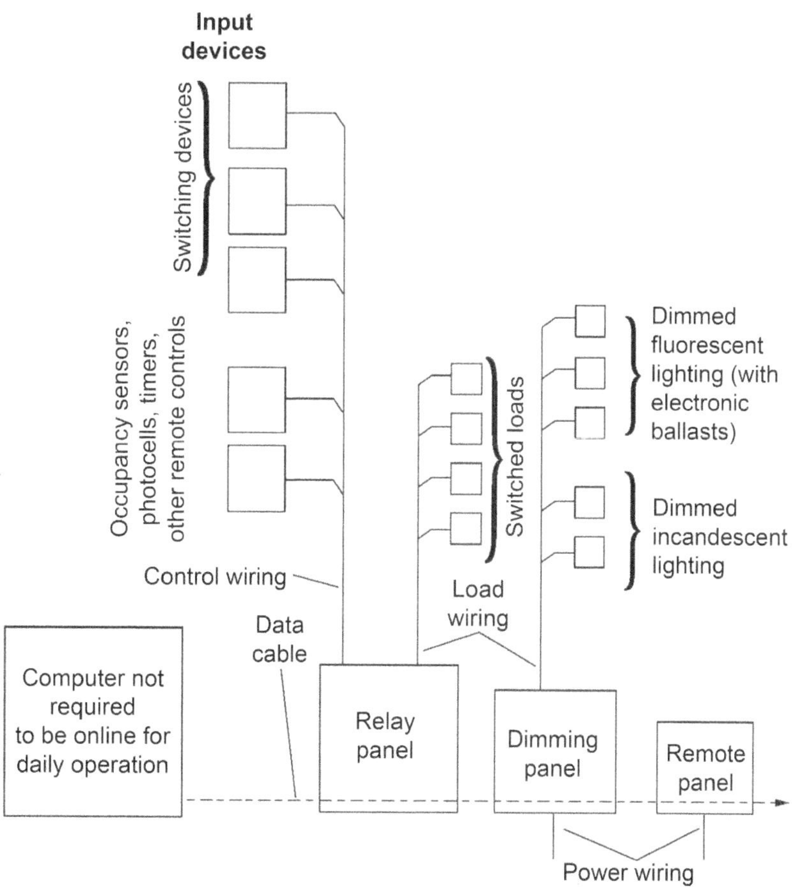

Figure 28.53 Block diagram of a large, computer-controlled, low-voltage control system. The relay panel contains the incoming and outgoing power and control wiring, the switching relays, and the computer interface card. The system arrangement is controlled by computer software. (Courtesy of Touchplate Technologies, Inc.)

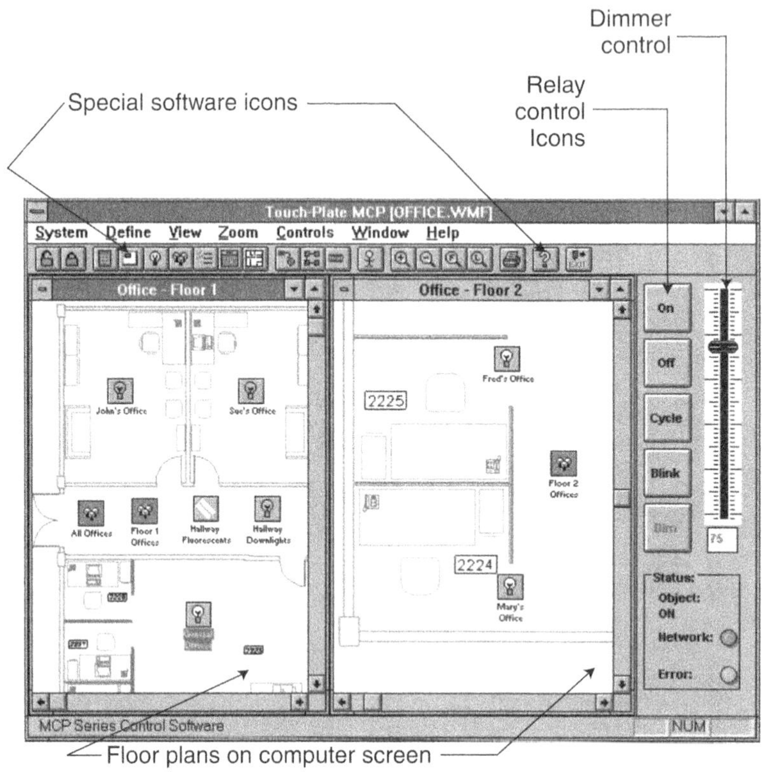

Figure 28.54 *Typical computer screen of low-voltage control software showing facility floor plans (scanned in) and controlled loads. The icons at the top of the screen are customized and are used to control relays, make connections, check status, print, and log events by time and date. Other screens enable group and event definition, labeling, establishment of priorities, and other global sequential and individual control functions. (Courtesy of Touchplate Technologies, Inc.)*

amount to more than half of the total cost of the project.

To minimize control wiring costs, a system of control signal transmission that uses the electric power wiring (either new or existing) to conduct signals was developed, thus removing the need for and expense of dedicated control wiring. This system is called a power line carrier (PLC) system because the power line carries the control signal. An ancillary benefit of such an arrangement is that not only can it be added to an existing system, it can also be easily removed, thus making possible and economical the leasing of energy management control systems.

The system operates by injecting into the power wiring a series of low-voltage, high-frequency, binary-coded control signals, which then disperse over the entire power network. Only a receiver that is "tuned" to receive a particular code reacts, generally by operating a local device, which in turn connects, disconnects, dims, or brightens a particular load. In practice, the control signal generator can be a small, manually programmed controller, as in a residence, or a computer-operated energy management or lighting controller in a commercial facility. The receivers are normally of four types, most of which are designed to fit into an ordinary wiring device box:

1. a wall-switch module, which acts as a normal manual wall light switch but also contains a coded receiver and a fully rated relay (20 A, 125 V or 277 V)
2. a wall receptacle module, which contains a 20-A receptacle along with a receiver and relay that serve to switch power on and off
3. a switching module, which can be connected to activate contactors, small motors, and the like
4. a dimming module, which provides both local and remote control

Some of these components are shown in Figure 28.55. Larger, multiple-circuit switching modules are also available for more extensive systems. Each miniature receiver can be manually tuned in the field (i.e., its code can be set and reset readily), giving the entire system extreme flexibility.

The system operates in the same manner as a low-voltage switching system or an energy

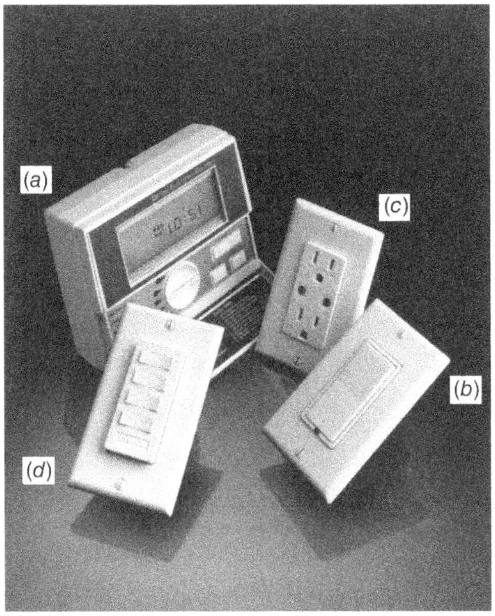

Figure 28.55 *Components of a residential-type PLC system. (a) Control unit, at which both manual and automatic (programmed) control of a maximum of 256 address codes is possible. (b) Wall-switch module has local dimming and switching functions in addition to being remotely controlled from the control center (a). (c) Duplex wall receptacle, either split-wired with local programming of one-half of the outlet or complete remote control. (d) Outlet-box wall-mounted control switch that provides local group control (switching and/or dimming) for one to four devices (address codes). (Courtesy of Leviton Manufacturing Co.)*

management or lighting control system, with the essential difference that no dedicated control wiring is required. In retrofit jobs, existing wall switches and receptacle outlets are replaced with receiver-controlled units; in new installations, these devices are installed initially.

Because the control signals travel over the power wiring and are therefore attenuated by high-resistance connections, poor grounds, and faulty wire insulation, PLC installations require a high-quality power wiring installation to operate properly. In retrofit work, preliminary tests occasionally indicate the impracticability of a PLC installation because of the condition of the power wiring system. Another problem encountered in PLC systems is interference with the PLC control signals from radio noise generated by faulty power equipment and by improperly shielded or grounded electronic equipment. These problems can frequently be overcome by careful testing, installation of filters and jumpers, and repair of improperly operating equipment. In addition to energy management and normal lighting control, PLC control can be applied to group switching of other loads, such as office machines, in addition to simplifying control of temporary light and power for special events.

28.35 POWER CONDITIONING

(a) General Information

Power conditioning is a term that describes the process of converting utility-supplied electrical power, which is frequently characterized by transient surges, spikes, radio frequency noise, and voltage fluctuation, to a pure (noiseless), accurately voltage-regulated sinusoidal waveform, normally referred to as computer-grade power.

Some form of power conditioning is required by all large-scale data-processing and telecommunication equipment and by many other types of electronic equipment. This is due to their extreme sensitivity to fluctuations in the power supply voltage and particularly to random high-frequency voltages superimposed on the power supply voltage, known as radio frequency interference (RFI) or simply radio noise. In the case of data-processing equipment (computers and peripherals), radio noise can cause data errors; slow voltage fluctuations can cause overheating, data loss, and premature equipment failures; and large, rapid voltage fluctuations (spikes, transients) can cause equipment burnout and system collapse. The economic ramifications of such effects on large commercial data-handling systems are so severe that power conditioning has become almost universal. (Power conditioners must not be confused with uninterruptible power supplies [UPS], which function to maintain power during utility failure, although in many commercial installations, both functions are served by a single equipment package.)

(b) Source of Disturbance

Utility power systems are designed to maintain voltage and frequency within certain limits on an average, long-term basis. For valid technical reasons, utility equipment cannot react to the short-time disturbances that are so inimical to electronic equipment. Furthermore, much of the noise in utility lines is introduced by users and is therefore beyond the utility's point of control. Disturbances on the electrical power line can generally be classified into three types: voltage variations, electrical noise, and transients.

1. *Voltage variations:* These can be caused by short-time current drain, such as during motor starts, deliberate voltage reductions (brownouts), load switching of emergency medical services (EMS) systems, excessive line voltage drop due to heavy loading or current leakage, and similar problems. These variations are relatively slow and long-lasting.

2. *Electrical noise:* This is a low-amplitude, higher-frequency voltage that, when superimposed on the power line, results in a distorted waveform. Sources of such extraneous voltages are electronic equipment power supplies, lighting dimmers, solid-state motor controls, PLC systems, arc welding, switching transients caused by branch circuit switches and relays, and voltages induced by local magnetic fields.
3. *Transients:* These are large, short-duration voltage variations, also known as surges or spikes. They are the principal culprits in major equipment failures. They are caused by lightning strikes to overhead lines (possibly many miles from an affected installation), by electrical system faults, and by utility switching of high-voltage lines, transformers, capacitor banks, generators, and circuit reclosers.

28.36 POWER-CONDITIONING EQUIPMENT

Each of the three types of problems described in the preceding section requires its own type of correction. Voltage variations are corrected with voltage regulators; noise problems are corrected by electrical isolation, filtering, and noise suppression; and transients are treated with surge suppressors. Each corrective device is a separate entity that can be applied individually to suit the particular problem encountered or in combination with another conditioning device. The choice in a specific installation is a highly technical decision and should be made by the facility's electrical engineer after careful analysis of the quality of the utility service. A few general recommendations can be made.

1. All computer installations, including the smallest, should be protected from line transients with an appropriate surge suppressor. Multitap plug-in strips (low-cost) with integral surge suppressors are unsatisfactory unless they have surge current, clamping voltage, and surge-energy specifications, and these meet the installation requirements.
2. Major data-processing installations normally require all three types of line treatment. The most economical way to provide them is with a single integrated power-conditioning unit.
3. Considerable improvement in the quality of electrical power can frequently be achieved by the simple expedient of running separate electrical feeders for sensitive loads.
4. Physical isolation of sensitive equipment areas is frequently helpful, particularly when disturbances are being induced by proximity to switching, arcing, and rectifying equipment. Discharge-type lighting, including fluorescent, mercury, sodium-vapor, and metal-halide, can cause interference, especially when used with electronic ballasts.
5. It was found some years ago that much of the noise in electrical power systems was introduced through the (equipment) grounding pole of an electrical receptacle. Because this connection is required for electrical safety, it cannot be eliminated. It can, however, be separated from the wiring system ground and still maintain its function. Receptacles so constructed are color-coded either entirely orange or have an orange triangle on the face (see Section 28.29). This "fix" is particularly effective where electronic dimmers, ballasts, and switching devices are present. Because the insulated ground receptacle (sometimes referred to as an isolated ground receptacle) is a special item, it must be clearly indicated in the contract documents (drawings and specifications). Furthermore, because these receptacles require special isolated wiring, the entire matter must be studied carefully during the electrical design stage and before issuance of construction contracts.
6. A problem related to equipment that produces radio noise is that this same equipment also produces harmonics that appear on the system neutral and can cause overloading. This subject is covered in Chapter 30.

The issue of power quality is so important in modern commercial installations that the following steps should be taken essentially as a matter of course during design:

- A detailed report should be obtained from the electric utility that will be supplying electric power, showing power outages, voltage constancy, frequency control, and, if possible, oscillograms of waveforms at random intervals over the past year. Particularly important are periods of peak loading and periods of weather extremes.
- A power-quality study should be made in the area in which the facility will be (or is) located. Sophisticated instrumentation for this purpose as well as specialists in this field are readily available.
- An in-depth study of all of the electrical equipment to be used in the building should be made for the purpose of:
  - Reducing the sources of electrical disturbances.
  - Determining the tolerance or sensitivity of the data-processing equipment to be used. In the absence of reliable data for the second purpose, as is the case in speculative construction, a high degree of sensitivity should be assumed. In any case, for equipment that is known to produce

TABLE 28.8 Power-Conditioning Equipment

Equipment Functions				Dimensions, in. [mm]		
Voltage Regulation	**Noise Suppression**	**Surge Suppression**	**kVA[a]**	**H**	**W**	**D**
×			5	15 [381]	20 [508]	20 [508]
			10	35 [889]	40 [1016]	20 [508]
			20	40 [1016]	40 [1016]	30 [762]
×	×		0.25	6 [152]	6 [152]	10 [254]
			0.5	7 [178]	7 [178]	12 [305]
			1.0	8 [203]	10 [254]	15 [381]
			3.0	10 [254]	12 [305]	20 [508]
×	×		5	10 [254]	10 [254]	15 [381]
			10	20 [508]	30 [762]	15 [381]
			20	30 [762]	30 [762]	20 [508]
×	×	×	5	40 [1016]	30 [762]	30 [762]
			10	50 [1270]	35 [889]	30 [762]
			20	70 [1778]	40 [1016]	35 [889]
		×	NA[a]			
			15 A	3 [76]	3 [76]	16 [406]
			100 A	10 [254]	12 [305]	12 [305]
			500 A	15 [381]	15 [381]	12 [305]

[a]Surge suppressors are rated by transient energy-handling capability, and therefore a kVA rating is not applicable. However, because transient energy increases with the kVA size of protected equipment, a service ampere rating is given.

radio noise, such as electronic lighting ballasts and dimmers, the specifications should include limitations, plus a requirement for submission of laboratory tests on random samples.

Power companies may offer "computer-grade power" that meets stated limitations to voltage fluctuation and the like. This type of service is desirable for all new office space or full building renovations. For individual office spaces, separate power-conditioning equipment is required.

Table 28.8 and Figure 28.56 give some idea of the physical characteristics of power-conditioning equipment. Dimensions are typical and are indicative only of equipment bulk. Designs vary widely among manufacturers, as do, therefore, individual dimensions. Voltage-regulating equipment containing regulating transformers is very heavy, and this must be considered during design. Power-conditioning equipment is frequently combined with UPS equipment (see Section 28.38).

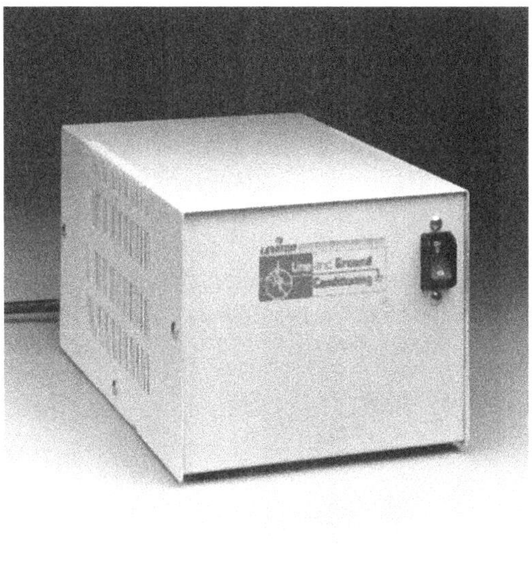

Figure 28.56 *This single-phase power conditioner supplies noise filtering and surge suppression by use of an isolation transformer, ground noise filters, and category A and B surge suppression (see Section 28.37). It is rated 5 A, 120 V, measures 6 in. H × 7 in. W × 11 in. D [150 × 180 × 280 mm] and weighs 18 lb [8 kg]. A similar unit rated 20 A, 120 V measures 8 in. H × 8½ in. W × 14 in. D [200 × 215 × 355 mm]. (Courtesy of Leviton Manufacturing Co.)*

28.37 SURGE PROTECTIVE DEVICES

A wide range of power anomalies can wreak havoc on sensitive electronic equipment. One ongoing and common threat to electronic equipment is dangerous voltage surges, which can be caused by factors ranging from lightning strikes to routine loads switching across the utility grid. Safeguarding equipment with a common type of power-conditioning equipment known as surge protectors or surge protective devices (SPDs)—also known as surge suppressors or surge suppression equipment (SSE)—is a simple and effective way to better protect electrical equipment. A common technical term for this aspect of power conditioning is transient voltage surge suppression (TVSS). Because transient voltages, also called surges or spikes, can cause major physical damage to computer systems and other types of electronic equipment, it is this aspect of system protection that has generally received the most attention.

SPDs are available in a wide variety of designs and ratings, ranging from the ubiquitous cord-connected, multiple-outlet strip to large, 3-phase, service-entrance units. They all have a single purpose, however: to suppress (limit) a voltage surge to a level that the protected equipment can withstand without damage. This is done in two ways:

1. By placing one or more devices that present a high impedance in series with the incoming voltage transient, thereby limiting the let-through current.
2. By placing one or more devices across the incoming power line, in parallel with the protected load, that present a low impedance to the high transient voltage and thereby bypass (shunt away) the incoming disturbance's current. Many SPDs use both techniques in a single "hybrid" unit.

There are two basic types of devices: clamp-type devices and crowbar-type devices. Clamp-type devices have faster response times but are limited in their currently handling ability. Crowbar-type devices have slower response times but can handle much higher current.

Please note that surge protective devices are not the same as surge arrestors. SPDs are used for low-voltage applications, such as those encountered in residential and commercial structures. Surge arrestors are primarily used in industrial and utility applications to protect equipment from high-voltage surges—this includes lighting in the case of lighting arrestors.

(a) Terminology

To be able to select SPDs and compare their ratings, a basic technical vocabulary is necessary.

Clamping voltage. The voltage at which a shunting device (such as an avalanche diode) begins to conduct (and thereby clamp) an incoming voltage surge. The lower a device's clamping voltage, the better its suppression action.

EMI/RFI filtering. A measure of the attenuation, by an SPD, of electromagnetic interference (EMI) and radio frequency interference (RFI). An SPD, due to its internal filters, always provides some radio noise attenuation. This characteristic is not of primary significance to selection of a device used primarily for surge suppression.

Isolation transformer. A transformer with two separate windings and no conductive electrical connection between them. By virtue of the inductive coupling of the windings, considerable attenuation of sharp voltage spikes is achieved. It can be used in high-capacity SPDs in lieu of multiple solid-state components in parallel. It is also useful in attenuating high-frequency noise at any voltage level. An isolation transformer offers physical and electrical separation between two circuits (hence, "isolation") and allows electricity to transfer via a magnetic coupling. This serves the ultimate purpose of mitigating voltage spikes in the supply line.

Joule rating. A joule, equal to 1 watt-second, is the SI unit of energy. A device's joule rating is a measure of its capacity to absorb and dissipate heat generated by its action. Although the joule rating is important where large amounts of energy are to be absorbed, as in (hard-wired) units at a service entrance or at a power distribution point, published ratings can be misleading because there is, at this writing, no industry-wide standard calculation procedure for joule rating. As a result, the industry standard specification for SPDs, *NEMA LS 1-1992*, does not include this rating. (Note that NEMA rescinded the Standard in 2000 and there have been no superseding NEMA documents issued therefore, the standard is maintained as a historical reference.)

Maximum surge current. A measure of a device's ability to divert and dissipate surge current without failing. It is a useful metric for SPDs used at service entrances and, for reasons explained previously, is more reliable than published joule ratings.

Response time. A measure of the rapidity with which an SPD begins to clamp a voltage surge, usually measured in microseconds or nanoseconds. It is not of great significance because both SADs and MOVs (found in high-grade SPD designs) are faster than the voltage buildup time of an incoming surge. Response time is not included in the NEMA Standard specification, apparently because it is not considered a very important characteristic.

Sine wave tracking. A marketing term of uncertain meaning found frequently in the manufacturers' literature.

(b) SPD Component Types

There are several different types of SPDs, each designed for certain applications and surge capacities, which require specific types of core components. Accordingly, these devices are defined by the capabilities of that component. The most common device components are SAD, MOV, and GDT. Clamp-type devices use SADs and MOVs. Crowbar-type devices use GDTs.

Gas Discharge Tube (GDT). Generally stronger than a SAD or MOV, a GDT is a high-energy shunt device, typically found in high-capacity, good-quality SPDs that are subject to large surges (such as those caused by lightning). It consists of a gas-filled tube with two electrodes. The tube conducts when a high-voltage surge causes its electrodes to arc over. Because of low arc impedance, the current drain and energy absorption are very high, thus clamping the incoming surge. Its principal drawback is its relatively slow response time, which necessitates its use in conjunction with other, faster devices.

Metal-Oxide Varistors (MOV). This component type is commonly used in surge protective stripes intended to protect consumer devices plugged into outlets. An MOV is a shunt device somewhat slower than a SAD but with much higher energy absorption capability and higher clamping voltage. MOVs are very widely used in SPDs. Like SADs, MOVs suffer degradation with use, thus limiting the useful life of the entire SPD.

Silicon Avalanche Diode (SAD). Also known as an avalanche breakdown diode (ABD), this component type is commonly used in surge protective technology for data and network devices and other sensitive electronic equipment. A solid-state device placed in parallel with a protected load, a SAD responds to overvoltage in nanoseconds by conducting (i.e., placing) its very low impedance across the incoming voltage. This results in a large ("avalanche") bypass current flow that serves to reduce the incoming voltage. It has a low clamping voltage (see Section 28.37(d)) and low energy absorption capacity. This latter characteristic causes "wear" and eventual failure.

Hybrid Solutions. Hybrid surge protectors come in various configurations, but the most common pairing is MOV with a GDT—which can combine the fast response time of a MOV with the higher capacity of a GDT.

(c) SPD Operation

Figure 28.57 is intended to be self-explanatory. Each element acts to reduce the magnitude of the incoming voltage surge by either bypassing (shunting away) current or presenting a high impedance to the voltage surge. Both actions reduce the voltage. A commercial SPD may contain any or all of the illustrated components, singly or in multiple, depending upon the current and voltage it is expected to handle.

(d) SPD Standards

Standards applicable to SPDs are offered below. The Institute of Electrical and Electronic Engineers (IEEE) maintains a set of ANSI-approved standards commonly referred to as the "trilogy" with regard to addressing surge equipment:

- *IEEE C62.41.1-2002: Guide on the Surge Environment in Low-Voltage (1000 V and less) AC Power Circuits* (IEEE, 2003a)
- *IEEE C62.41.2-2002: Recommended Practice on Characterization of Surges in Low-Voltage (1000 V and less) AC Power Circuits* (IEEE, 2003b)

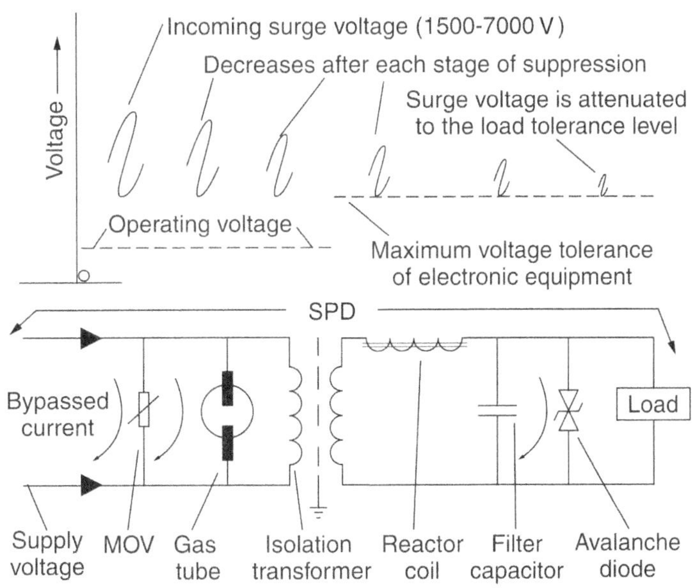

Figure 28.57 *All of the elements usually found in a surge protective device (SDP) are shown. Each acts to reduce voltage surge—parallel elements by shunting away (bypassing) surge current, and series elements by increasing the electrical impedance to the voltage surge.*

- *IEEE C62.45-2002: Recommended Practice on Surge Testing for Equipment Connected to Low-Voltage (1000 V and less) AC Power Circuits (IEEE, 2003c)*

Other notable standards include:

- *UL 1449: Surge Protective Devices, last published in 2021 (UL, 2021)*
- *NEMA LS 1-1992 (R2000): Low-Voltage Surge-Protection (LVSP) Devices, last published in 1992 and rescinded in 2000, but maintained as a historical reference (NEMA, 1992)*

Surge-suppression equipment can be subjected to widely varied voltage stresses, depending upon where in an electrical system a unit is installed. *IEEE C62.41.1* categorizes installation points according to the severity of service expected, that is, the (electrical) environment in which the surge-suppression equipment operates.

Category C (outdoors) represents the most severe environment, corresponding to outdoor locations on the line side of the facility service switches. This environment is subject to the full force of lightning and switching surges and requires robust equipment. As a voltage surge travels through a building's electrical system, it is attenuated by cables, transformers, and other electrical equipment. Therefore, the further along one travels in the building's electrical distribution system, the less severe the voltage surges become. All SPDs used in category C environments are hard-wired (Figure 28.58).

Category B (indoors) covers environments on the load side of the service equipment, including heavy feeder distribution panels and short branch circuits.

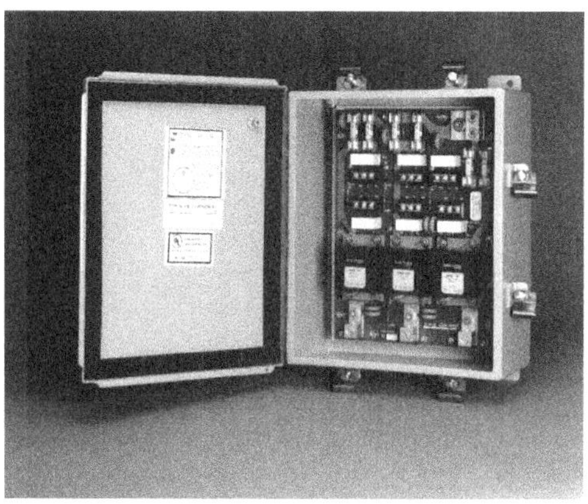

Figure 28.58 *Surge suppressor designed for use on the incoming service of a facility. It is rated for category C3 application: shunt type, 120/208 V, 3-phase, unlimited current. It is provided with RFI/EMI filtering and replaceable surge-suppressor modules, and arranged to indicate the power and protection status of each phase. (Courtesy of Surge Control Ltd.)*

Figure 28.59 *SPD (formerly known as a TVSS unit) designed to be used at service and branch panelboards in residences and small commercial installations. The illustrated unit is rated single-phase, 120/240 V, 3-wire service, category B3. The unit measures 7 in. W × 7 in. H × 4 in. D [180 × 180 × 100 mm] and is equipped with indicating lights that note power and suppression status. (Courtesy of Leviton Manufacturing Co.)*

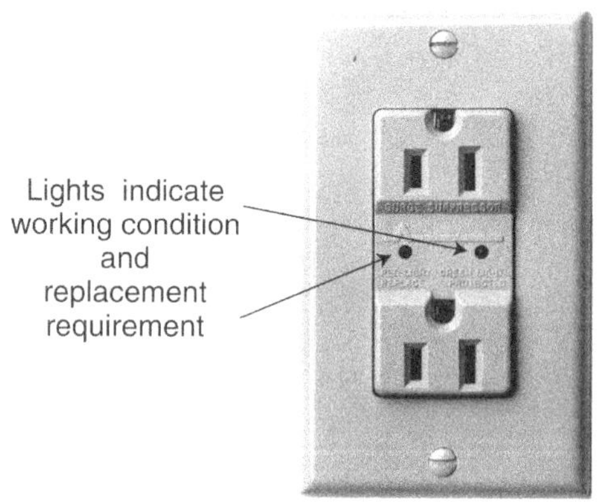

Figure 28.60 *Typical category A utilization surge protective receptacle equipped with indicating lights that show the condition of the surge-suppressor circuitry. (Courtesy of Eagle Electric Manufacturing Co., Inc.)*

Most SPDs in category B areas are hard-wired (Figure 28.59).

Category A (indoors), which represents the least-severe service, includes utilization outlets and long branch circuits (because the long cables attenuate a voltage surge). Most category A SPDs are arranged for loads to be plugged in directly. Devices used in category A environments include the common cord-connected multioutlet strip and surge-suppression-equipped receptacles (Figure 28.60).

Each of these three major categories is in turn subdivided into three minor categories: C3, C2, C1; B3, B2, B1; and A3, A2, and A1—where 3 is the most severe and 1 the least. The differentiation depends upon the electrical capacity of the facility's service and distribution system. Large systems have higher surges and less attenuation in the system, thus requiring larger and heavier SPD equipment than smaller installations. The three major categories are shown graphically with annotations in Figure 28.61.

IEEE C62.45 established the test procedures and test voltage levels to be used by manufacturers in testing and classifying their surge protective devices in order to simulate the voltage stresses found in A, B, and C environments.

UL 1449 describes standards for SPD safety and performance. An important aspect of this standard is that it establishes clamping voltage ratings. The ones of interest to building designers are the lowest four, which are 330, 400, 500, and 600 V. *UL 1449* requires testing for actual clamping levels, which are then rounded upward to one of the standard ratings. Thus, an SPD that clamps at 240 V and one that clamps at 320 V would both be rated at 330 V, despite the fact that the 240-V unit gives better protection. This distinction is particularly important when protecting computers and other electronic equipment equipped with switching power supplies, and requires a comparison of the actual clamping voltage of potential devices rather than simple reliance on the voltage classification.

NEMA LS 1 standardizes the specification format for suppression units, and in so doing establishes a basis for comparison among various manufacturers' products. Important characteristics that must be specified include line voltage, maximum operating current and voltage, maximum surge current, and clamping voltage

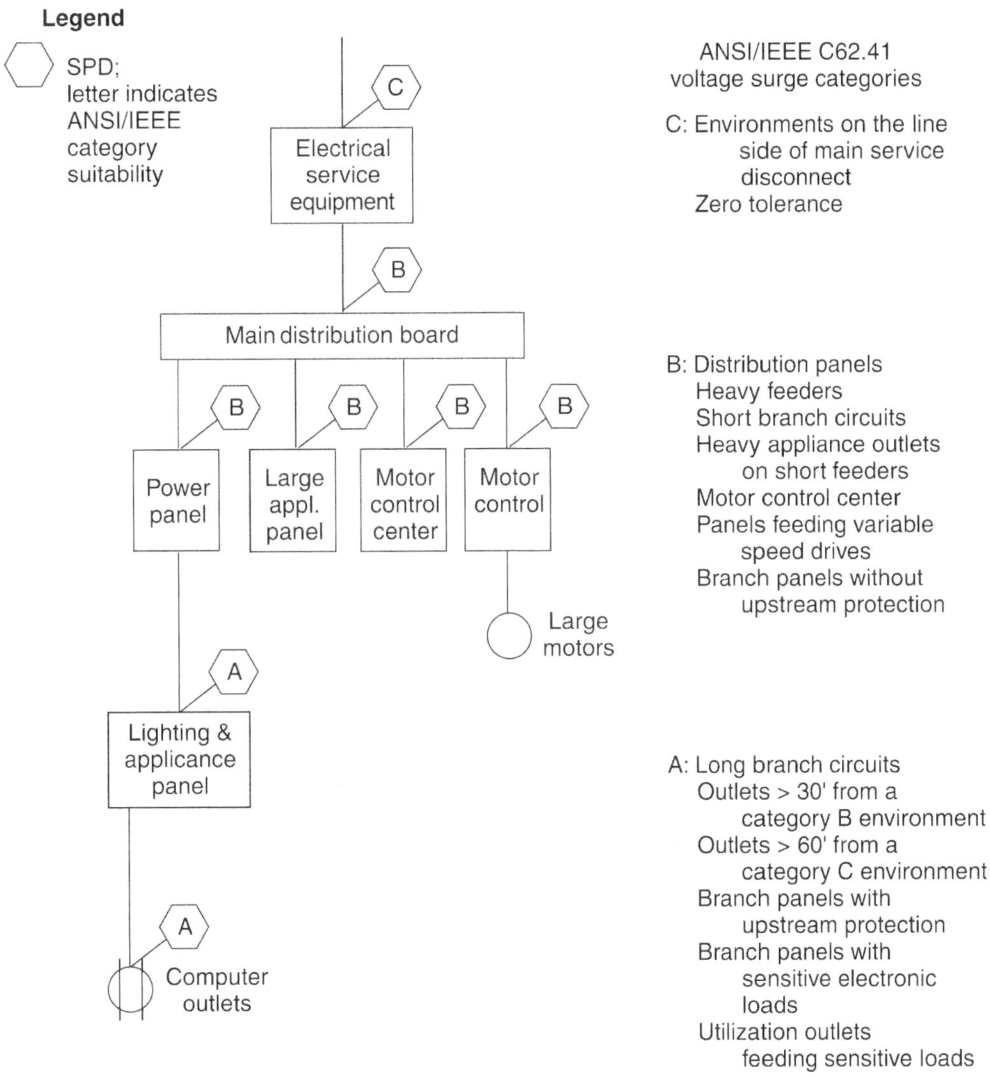

Figure 28.61 *Illustration based on* IEEE C62.41.1 *showing application of transient voltage-suppression devices in a typical electrical facility, based on the three categories of electrical environments (C, B, A) defined by the Standard.*

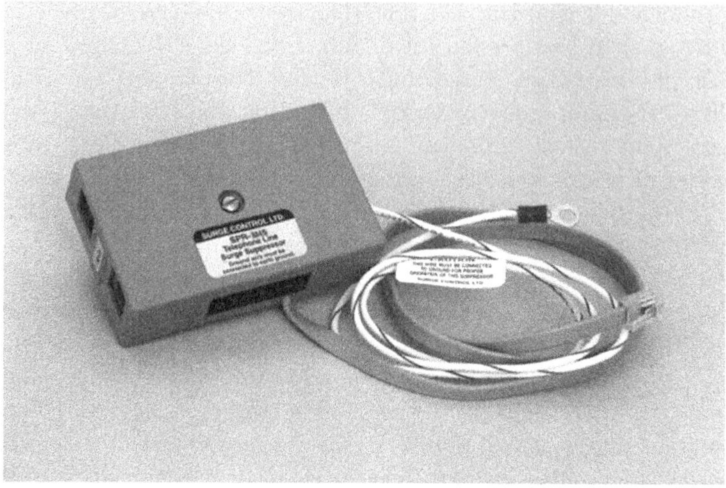

Figure 28.62 Surge suppressor designed for use with standard snap-in low-voltage telephone and industry jacks. The illustrated unit is designed to protect telephone/telecommunication lines, both analog and digital. The units are rated 185 V, with a maximum surge current of 5000 A and a test clamping voltage of 240 V. The units can be used with telephones, modems, fax machines, and the like. They measure approximately 4 in. W × 2½ in. H × 1 in. D [100 × 64 × 25 mm]. (Courtesy of Surge Control Ltd.)

rating. (Actual clamping voltage is not supplied unless specifically requested, as it involves specific testing.)

(e) Application of SPDs

Selection and application of surge-suppression devices for even the simplest application require considerable technical background, for which the foregoing material is intended as a primer. Important suggestions to assist in this area are:

- Every power level should have its own SPD. Small systems can function with IEEE C62.41.1 categories B and A only. Large systems may require protection for two or three B levels (B3, B2, B1) and more than one A level.
- In the absence of accurate data on the surge tolerance of electronic equipment, use protection for the most sensitive level. At the utilization level, this is category A3 and US Government classification Grade A, Class 1, Mode 1.
- Mode 1 (also called Mode A) indicates protection line-to-line or line-to-neutral. This mode should be specified for stand-alone items. It is preferable to Mode 2 (or B), which is connected line-to-ground, because the latter can introduce random signals (contaminants) into the equipment-grounding system.
- The primary criteria for selection and comparison are the actual clamping voltage and maximum surge current capacity. Other important characteristics are those listed in NEMA LS 1.
- SPDs should be installed as close as possible to the loads being protected.
- Telephone and data lines must also be protected, particularly if the service is overhead. Most telephone companies install an appropriate SPD at the main distribution frame near the service entrance. This may not be sufficient to protect electronic equipment from a disturbance that can cause distortion of data. A specialist should be consulted to determine the required protection levels and select appropriate equipment (Figures 28.62 and 28.63).

Figure 28.63 SPD designed to protect video lines from voltage transients. The unit plugs into a standard outlet to establish a ground connection for current bypass. This unit is equipped with BNC (Bayonet Neill-Concelman) connections; others have telephone, PL (coaxial), and RS (serial device) connectors for use on telephone, communication, and data lines. The illustrated unit has a clamping voltage of 15 V and a maximum surge current capacity of 5000 A. It measures approximately 2 in. W × 4 in. H × 1 in. D [50 × 100 × 25 mm]. (Courtesy of Leviton Manufacturing Co.)

28.38 UNINTERRUPTIBLE POWER SUPPLY

As explained previously, power-conditioning equipment can supply clean utility power; it cannot, however, supply any power during a utility outage. That eventuality is addressed by providing an alternate supply of power. Facilities with desktop computers, servers, and other data-processing equipment cannot tolerate power outages in excess of about 8 to 50 milliseconds (ms) without serious risk of data loss, and all the negative ramifications of such a loss.

An uninterruptible power supply (UPS) is an arrangement of normal and backup power supplies that transfers a facility's critical load from the normal to the backup supply in so short a time that no computer malfunction results. This transfer time varies somewhat among different schemes and manufacturers but is always less than 8.3 ms, which is the minimum period of power outage that computers must tolerate without disturbance to meet the computer industry's manufacturing guidelines. This time period is double the maximum transfer time required by *IEEE 446-1995: Recommended Practice for Emergency and Standby Power Systems for Industrial and Commercial Applications*, of $\frac{1}{4}$ cycle at 60 Hz, or 4.16 ms. Thus, all computer systems fed from a UPS that meets this standard have a safety factor of at least 2 with respect to transfer time.

The period of time after transfer that the equipment will run on the standby source depends upon system design. In most cases it is 5–10 minutes, which is usually enough time to permit an orderly manual or automatic shutdown. Where shutdown is not a viable alternative, as with computer-controlled manufacturing processes or critical server equipment, the standby power system can readily be designed to supply power indefinitely.

The selection of a UPS system for a facility is a complex process beyond the scope of this book. The material that follows is intended to provide sufficient familiarity with these systems to permit preliminary selection and planning.

(a) Alternate Power Source

For most applications, battery backup is sufficient because only an orderly shutdown is required. Where this is not the case, a two-stage transfer can be used—the first transfer being to a battery backup that can carry the load for up to 1 $\frac{1}{2}$ hours (or more), depending upon the load magnitude, followed by a second transfer to a long-term standby generator set. Thereafter, the availability of generated standby power is limited only by the fuel supply for the generator. Alternatively, the generator can be maintained online, and the load picked up directly upon the first transfer.

In large industrial installations, the standby source may also be a second utility line. In such cases, only a single transfer is required. Questions of service reliability and equipment redundancy are both technical and economic, and must be studied carefully for each individual installation.

(b) Equipment Arrangement: Classic Standby and Online Topologies

Most UPS systems today are described by their manufacturers as online or simply as UPS rather than standby, primarily because there is no generally accepted industry-wide definition of "online" and also because many systems are hybrid and do not fall easily into either category. Figure 28.64 shows the usual static (nonrotary) UPS equipment arrangement that applies, in principle, to both standby and online modes.

In the standby mode (Path A), utility power is normally passed through some power-conditioning equipment to shield against surges and remove random noise, and then, via position A in the static transfer switch, to the load. The AC line also provides a small current to the small battery charger that keeps the battery fully charged. The DC to AC inverter is open-circuited at the transfer switch and delivers no power. In the event of a utility power failure, the transfer switch moves to position B, and power flows from the battery through the inverter and transfer switch to the load (Path B). With proper equipment design, the transfer is usually accomplished smoothly, although an instantaneous change from no-load to full load on the inverter can sometimes cause a transfer voltage loss. For this reason, online schemes are preferred.

In the straightforward online arrangement shown in Figure 28.65, power Path A (normal) is through the rectifier and inverter to the load via position B in the transfer switch. The battery floats on the DC line. Failure of utility power causes the power path to change to Path B (i.e., battery to inverter to load). The transfer switch remains in the B position. As a result, voltage to the load is undisturbed. If one of the Path A components fails, the transfer switch moves to position A, and power is supplied directly from the utility lines via Path C, which is referred to as a utility bypass. The advantage of the online arrangement is that it eliminates reliance upon the quality of utility power. Power conditioning and surge suppression are provided by the solid-state equipment.

The disadvantages of this classic online topology are high cost due to the full load capacity (size) of the charger/rectifier and an overall efficiency in Path A of

ELECTRICITY

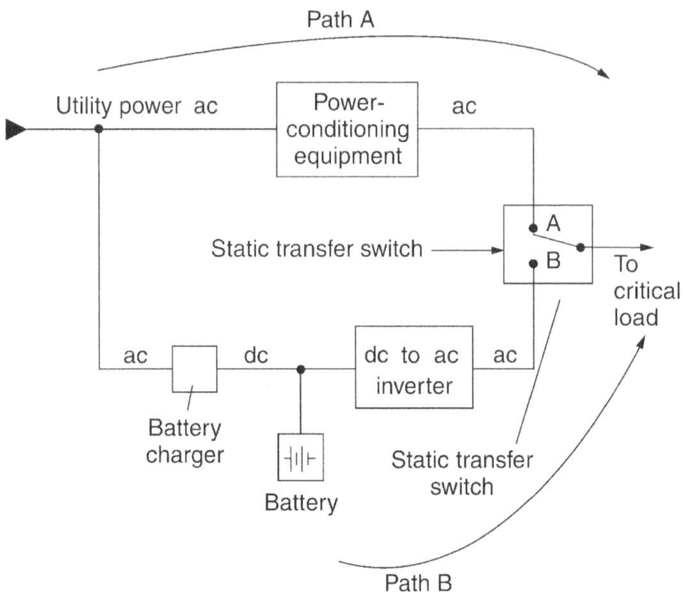

Figure 28.64 *In this UPS standby arrangement, power normally flows from the utility to the critical (computer) loads through position A of the transfer switch (Path A). The battery floats until a failure of utility power puts it online through the inverter and position B in the solid-state, effectively instantaneous, transfer switch (Path B). The inverter is normally inactive. The battery charger is small because it carries only charging current.*

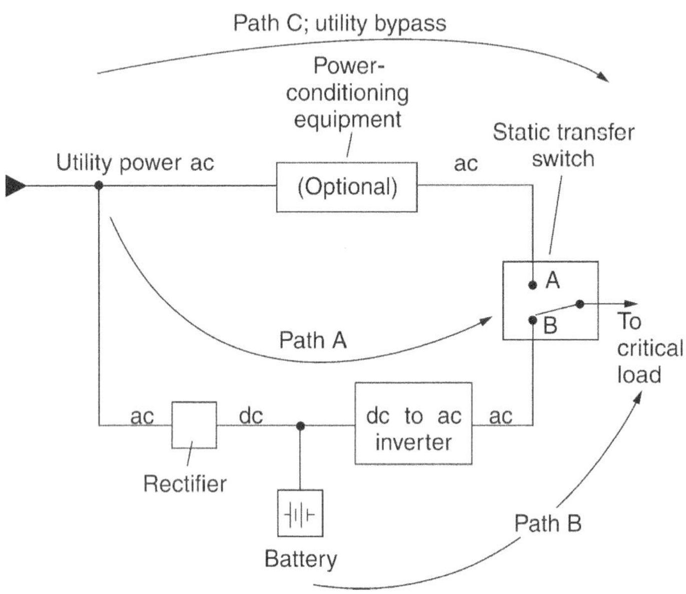

Figure 28.65 *In this UPS online mode, the normal power Path A is from the utility through a full-size rectifier to the load via an inverter. The battery floats, taking a small charging current from the rectifier. Utility failure brings the battery line through the inverter (Path B). In both instances, the transfer switch remains in position B, and the load senses no change. Only in the event of UPS equipment failure is power taken directly from the utility (Path C).*

about 75% to 80%. The resultant heat production can be problematic and is certainly expensive in terms of power consumption.

(c) Additional UPS Topologies

1. A money-saving arrangement called online without bypass is simply the bottom portion of Figure 28.65. The first cost is reduced by elimination of the transfer switch and considerable cabling. Because failure of any of the components disables the system completely, as there is no backup power path, this system is not recommended for critical loads.
2. A system variously known as line-interactive, standby line-interactive, and hybrid interactive is shown in Figure 28.66. In its normal mode (Path A), power flows from the utility lines through power-conditioning equipment and position A of the transfer switch to the load. (This is identical to the normal mode of the standby system shown in Figure 28.64.) In addition, however, utility AC is tapped at the power-conditioning equipment to feed the inverter that, operating in reverse as a rectifier (AC to DC), serves as a battery charger.

In the event of a utility power failure, power is drawn from the battery; passing through the inverter, the power is converted to AC. It then connects to the load via the B connection in the transfer switch. The power path is indicated on Figure 28.66 as Path B. The advantage of this topology is the elimination of a large rectifier by dual use of the inverter and

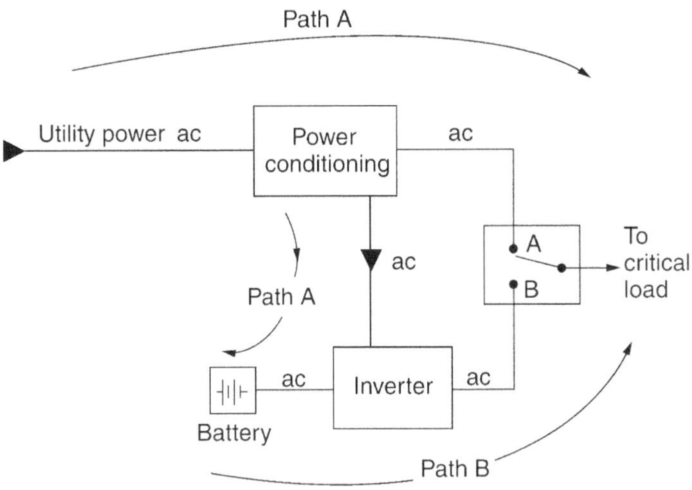

Figure 28.66 *In this UPS hybrid (line-interactive) mode, the normal power Path A takes power directly from the utility but keeps the inverter online, working in reverse to furnish DC charging current to the battery. Utility failure causes a change to power Path B (i.e., from battery through the inverter now operating normally to invert DC to AC). The static switch changes to position B.*

elimination of transients caused by switching on the inverter at full load. The system is efficient, produces little heat, and is highly reliable.

(d) System Selection and Comparison

The preceding system descriptions are intended to provide a topical familiarity with UPS equipment arrangements. Actual selection depends upon the types and magnitudes of the critical loads, detailed outage histories, space considerations, cost factors, equipment and service redundancy, and overall system efficiency. The last characteristic can be a major decision factor where large loads are being supplied, because overall system efficiencies vary from a high of about 95+% for classic standby systems (Figure 28.64) to as low as 65% for some arrangements. Because all of the power lost to inefficiency turns to heat, a low-efficiency 100-kVA/80-kW system would produce about 27 kW or about 92,000 Btu/h. Unless this heat can be used effectively, power costs and ventilation requirements should be factored into the selection process.

(e) Ancillary Characteristics of UPS Systems

Programmable microprocessors and effective sensors and transducers have been added to basic UPS systems to produce what is known as intelligent UPS equipment. These additional devices now make readily available such operating and diagnostic data as battery status, equipment operating temperatures, load data, waveform displays, online power quality analyses, operational and event logs, and the like. These data help an informed operator optimize system functioning, anticipate problems, and keep an accurate printed operational record. Table 28.9 gives some typical dimensional and run-time data for UPS units of different topologies.

28.39 EMERGENCY/STANDBY POWER EQUIPMENT

The *NEC* makes a clear distinction between emergency systems and standby systems, covering the former in Article 700 and the latter in Articles 701 and 702. The equipment, circuitry, and arrangement of both are similar; the purpose is somewhat different. The reader is also referred to *NFPA Standard 110: Standard for Emergency and Standby Power Systems*, which covers equipment requirements for these systems.

Emergency systems are intended to supply electric power to equipment essential for human safety upon

TABLE 28.9 Typical Dimensional Data for UPS Equipment

Topology	Capacity, kVA	Dimensions H × W × D, in. [mm]	Weight[a], lb [kg]	Full Load Run Time, min.
Standby	0.3	6 × 6 × 12 [152 × 152 × 305]	15 [6.8]	10
Single phase	0.6	7 × 6 × 14 [178 × 152 × 356]	25 [11.3]	5
Online	3.0	30 × 12 × 24 [762 × 305 × 610]	250 [114]	20
Single phase	7.5	36 × 12 × 36 [914 × 305 × 915]	500 [227]	10
Three-phase	10.0	60 × 24 × 36 [1524 × 610 × 915]	1200 [544]	15
Interactive	0.4	6 × 6 × 12 [152 × 152 × 305]	20 [9.1]	5
Single phase	1.0	8 × 8 × 15 [203 × 203 × 381]	40 [18]	5
	2.0	16 × 8 × 20 [406 × 203 × 508]	105 [48]	20
	3.5	18 × 10 × 24 [457 × 254 × 610]	140 [64]	10

[a]With batteries.

interruption of the normal power supply. Included in this classification are illumination in areas of assembly (to permit safe exiting and prevent panic) and such other vital functions as fire detection and alarm systems, elevators, fire pumps, public address and communication systems, and orderly shutdown or maintenance of hazardous processes.

Standby systems are divided into two categories: those legally required (Article 701) and optional systems (Article 702). The former are intended to power processes and systems (other than those classified as emergency systems) whose stoppage might create hazards or hamper firefighting operations. This classification goes beyond the emergency systems and could include HVAC systems, water supply equipment, and industrial processes whose interruption could cause a safety or health hazard. It is intended primarily as a safety measure.

Optional standby systems can cover any or all loads in a facility at the discretion of the owner and are normally intended to protect property and prevent financial loss in the event of a normal service interruption. Examples might include a critical industrial process or equipment for an ongoing research project.

Health care facilities are covered by a separate set of regulations: *NEC* Article 517 and *NFPA Standard 99: Health Care Facilities Code*, both of which are referenced as legally binding in the vast majority of jurisdictional codes. It is important to note that for both emergency and legally required standby systems, the provision of the system must be mandated by the authority having jurisdiction over construction, whether it is a local, state, or federal agency or a combination of these. The *NEC* dictates how the system is to be designed and constructed; its existence depends upon another authority.

A case in point is the fundamental issue of exit lighting. This is mandated by the *NFPA Life Safety Code* and Title 29, Subpart E of the Occupational Safety and Health Administration's (OSHA) Code of Federal Regulations (CFR). It is not required by the *NEC*, and reference to *NEC* will not ensure its provision. The code(s) having authority and jurisdiction must specifically require an emergency and/or standby system—by that nomenclature—in order for *NEC* provisions to apply. The precise items of equipment to be powered are selected by the designer, keeping in mind the specific and general requirements of the codes having jurisdiction. The majority of codes make the provisions of the *NEC* and the relevant NFPA standards legally binding. Most codes require emergency systems; far fewer require standby systems, and then only for essential water and water-treatment systems and a few other essential uses. The designer must thoroughly investigate the matter of jurisdictional codes before considering emergency electric power systems and equipment.

System design arrangements are discussed in Chapter 30. Emergency lighting equipment and system design is covered in Chapter 19, which falls into two principal categories—that is, generator and battery installations—and is discussed later. Optional standby systems normally use a fueled prime mover rather than batteries.

(a) Engine–Generator Sets

An engine–generator set installation comprises basically three components: a fuel system (including storage, if necessary); the set itself, plus exhaust facilities; and the space housing the equipment (Figure 28.67). The principal advantages of an engine–generator set are unlimited kVA capacity, duration of power limited only by the size of the fuel tank, use for peak-load shaving, and, if properly maintained, indefinite life. The disadvantages are noise, vibration, the nuisance of exhaust piping and exhaust, the need for constant maintenance and regular testing, and difficulties with fuel storage. Gasoline can be stored for only a year at most, and subsequent disposal is difficult. Diesel fuel keeps somewhat longer, but disposal is also difficult. Use of natural gas for an engine obviates the fuel storage problem but poses the alternate problem of availability of gas service during emergencies. In some large cities, steam is commercially available as an energy source. Here too, service reliability, particularly in the event of a widespread electric service failure, must be carefully investigated.

(b) Battery Equipment

A battery energy storage system (BESS), or battery storage power station, is a type of energy storage solution that uses a group of batteries to store electrical energy. BESS solutions range in scale and technology. Systems can be designed to serve the utility grid, a microgrid, or individual buildings.

At the building scale, a BESS is often used to supply limited amounts of emergency power for critical services, essential lighting, and they may serve UPS systems, as detailed previously. Increasingly, these systems are being used in concert with on-site renewable energy generation systems (e.g., photovoltaic solar systems) and leveraged to reduce peak loads.

In a building-scale BESS, batteries are mounted in individual cabinets or in racks for large installations and are always provided with automatic charging equipment. For residential and small commercial

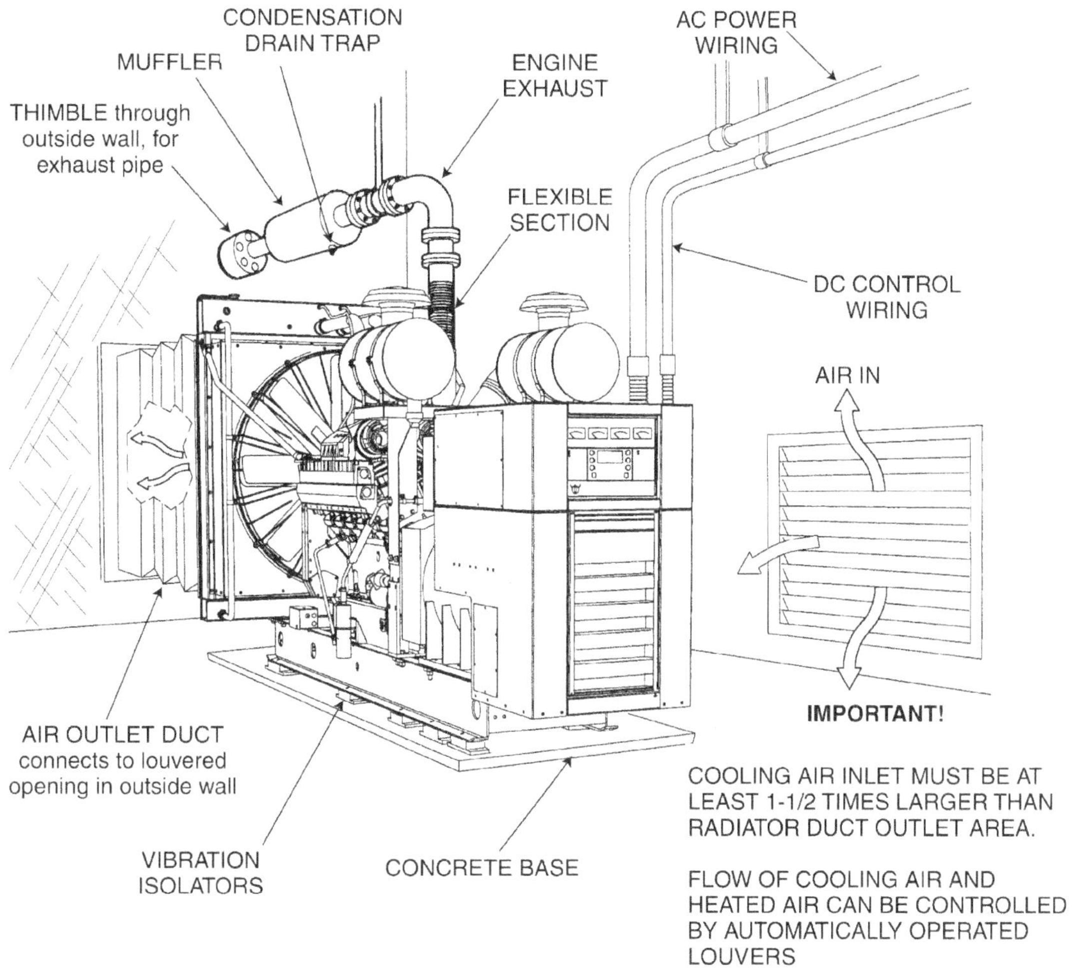

Figure 28.67 *Typical installation diagram of a radiator-cooled diesel engine–generator set. The set should be oriented so that the hot radiator cooling air vented through an outside wall flows in the same direction as prevailing winds. A solid masonry wind and noise barrier not less than 6 ft [1.8 m] from the radiator air outlet is recommended, as well as an elevated exhaust pipe outlet. (Courtesy of Onan Corp.)*

buildings, battery systems that are essentially consumer products are available for use in conjunction with on-site power generation and electric vehicles.

Battery types are undergoing intensive development. At this writing, the types principally in use in buildings are lithium-ion (Li-ion), lead-acid, "flow" batteries (which utilize a water-based solution of zinc-bromide), sodium-nickel chloride, lead–calcium cells, and several alkaline types. Once prevalent, nickel-cadmium (Ni-Cd) batteries are no longer favored due to environmental concerns. Lead-antimony may be encountered, but these are also less commonly used today. The choice of battery type depends upon the application. Installation requirements such as ventilation, gas detection, isolation, and the like depend entirely upon the type of battery, size of the installation, and battery voltage. Thus, no general guidelines for battery equipment can be given except to follow *NEC* Article 480 requirements and consult a battery specialist for other data.

Batteries have the distinct advantage that they can be installed either in a central system with distribution of the battery power throughout the facility or in small package units around a building. Central systems are 24–125 V, DC or AC, and normally feed emergency lighting only. Individual packs are used often to supply AC power via built-in inverters. The great disadvantage of battery systems is their limited duration of power and their environmental effects upon disposal. The *NEC* requires that batteries maintain loads for a 1 1/2-hour minimum, but larger capacity is frequently installed.

28.40 SYSTEM INSPECTION

Each electric wiring system is inspected at least twice by the local inspection authorities: once after raceways (roughing) have been installed and before the wiring and closing-in of walls, and once after the entire job is complete. The purpose of these inspections is to determine whether the design, material, and installation techniques meet the national and local code requirements. Quality of installation is the responsibility of the contractor. The designer, however, must be familiar with installation work and the equipment's physical characteristics in order to properly design an electrical system that will not present the contractor with unwarranted difficulties. The designer must understand and be aware of equipment substitutions by a contractor, who, having submitted a bid on the basis of plans and specifications, should be required to supply the specified equipment. Commissioning of electrical systems to the owner's project requirements (OPR) is recommended.

References and Resources

IEEE. 2003a. *IEEE C62.41.1-2002: Guide on the Surge Environment in Low-Voltage (1000 V and less) AC Power Circuits*. Piscataway, NJ: Institute of Electrical and Electronics Engineers.

IEEE. 2003b. *IEEE C62.41.2-2002: Recommended Practice on Characterization of Surges in Low-Voltage (1000 V and less) AC Power Circuits*. Piscataway, NJ: Institute of Electrical and Electronics Engineers.

IEEE. 2003c. *IEEE C62.45-2002: Recommended Practice on Surge Testing for Equipment Connected to Low-Voltage (1000 V and less) AC Power Circuits*. Piscataway, NJ: Institute of Electrical and Electronics Engineers.

IEEE. 2022. *NESC® 2023: National Electrical Safety Code*. Piscataway, NJ: Institute of Electrical and Electronics Engineers.

NEMA. 1992. *NEMA LS 1-1992 (R2000): Low-Voltage Surge-Protection (LVSP) Devices*. Rosslyn, VA: National Electrical Manufacturers Association.

NFPA. 2022. *2023 NFPA 70: National Electrical Code (NEC)*. Quincy, MA: National Fire Protection Association.

UL. 2021. *UL 1449: Surge Protective Devices*. Northbrook, IL: Underwriters Laboratories, Inc.

CHAPTER 29

Electrical Systems and Materials: Wiring and Raceways

ELECTRICITY

THE MAJOR COMPONENTS OF A BUILDING ELECTRICAL system can be grouped into three categories: wiring and raceways, power-handling equipment, and utilization equipment. The first category includes conductors and raceways of all types; the second includes transformers, switchboards, panelboards, large switches, and circuit breakers; and the last includes utilization equipment such as lighting, motors, controls, and wiring devices. After an introduction applicable to all electrical materials, this chapter discusses in detail the items in the first of these three categories—that is, the wiring and raceway system. Chapter 32 covers most of the items in the other two categories, with the exception of lighting equipment, which is discussed in Part V. Signal systems, building control, and automation are discussed in Chapter 32.

29.1 SYSTEM COMPONENTS

Figure 29.1 illustrates how power distribution equipment proceeds from the service point to the utilization points in a progression of decreasing circuit capacity. This is analogous to the arrangement of water supply systems and HVAC systems; the distribution equipment is largest at the supply point and decreases in size on its way to the farthest utilization points. The reverse is true in collection systems such as those for drainage and venting; in collection systems, piping is smallest at the initial collection points, growing larger (in steps) as the quantities of fluid increase.

A typical single-line diagram does not differentiate by line weight between heavy and light (large and small) conductors (the heavier the conductor, the greater the amount of power being carried), but the single-line diagram in Figure 29.1 does differentiate in order to show relative power levels throughout the system. This "size" differentiation is more clearly shown in Figure 29.2, which is a pictorial representation of a system similar to that of Figure 29.1—but in somewhat greater detail and omitting items beyond the panelboard. Only standard items are shown; special items, however common, such as emergency and uninterruptible power sources, energy controls, service transformers, and the like, are omitted for the sake of clarity.

29.2 NATIONAL ELECTRICAL CODE

The *National Electrical Code* (*NEC*) of the National Fire Protection Association (NFPA) defines the fundamental safety measures that must be followed in the selection, construction, and installation of electrical equipment and systems. This code is used by all inspectors, electrical designers, engineers, contractors, and operating personnel in the United States. Having been incorporated into the Occupational Safety and Health Act

Mechanical and Electrical Equipment for Buildings, Fourteenth Edition. Walter Grondzik, Daniel Overbey, and Alison Kwok.

Companion website: www.wiley.com/go/meeb14e

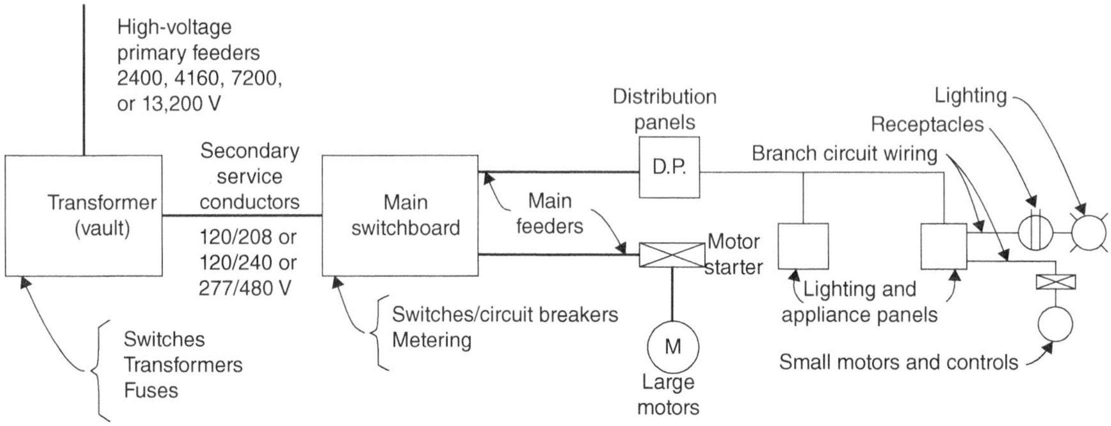

Figure 29.1 *Single-line diagram of a typical building electrical distribution system, from the incoming service to the utilization items at the end of the system. This is also referred to as a block diagram because the major components are shown as rectangles, or blocks. When this same type of information is presented showing the vertical spatial relationship between components, it is called a riser diagram; when electrical symbols are used in lieu of the blocks, it is called a one-line or single-line diagram. The connecting conductors between the major system components are drawn here with varying thickness to reflect size, and thereby power-handling capacity. In normal practice, all connecting lines are shown the same weight. Also, branch circuits are not usually shown; the typical diagram ends at the lighting and appliance panels.*

of 1970 (OSHA—which is also the acronym for the Occupational Safety and Health Administration, which administers the Act) it has, in effect, the force of law throughout the United States. Frequent references are made to the *NEC* throughout this chapter. In addition to the *NEC*, many jurisdictions (such as New York, Massachusetts, and Washington, D.C.) have their own electrical codes that, although similar to the *NEC*, contain numerous special requirements.

In order to ensure a minimum standard of intrinsic electrical safety for electrical equipment, a single organization was needed to establish standards and to test and inspect electrical equipment. Underwriters Laboratories (UL) in the United States is such an organization, and it publishes extensive lists of inspected and approved electrical equipment. These listings are universally accepted, and many local codes state that only electrical materials bearing the UL label (of approval) are acceptable.

29.3 ECONOMIC AND ENVIRONMENTAL CONSIDERATIONS

The selection of electrical materials involves not only choosing a material or assembly that is functionally adequate and, where necessary, visually acceptable, but also the consideration of costs. Usually, many types of equipment (from a range of manufacturers) will fulfill the technical requirements of a project. In most cases, the final selection of a device or approach hinges upon cost. Ideally, "cost" would be life-cycle cost; more realistically, it is often first cost.

The selection decision is relatively simple when various options differ only slightly from each other and a straightforward first-cost comparison is all that is required. Often, however, the choice is not so simple, because materials and equipment may vary considerably in characteristics other than functional suitability, demanding a more detailed cost study. Such economic analyses are frequently performed to make comparisons among competing HVAC systems. This happens less often when dealing with electrical systems because, from an energy point of view, electrical systems are generally less critical (relative to consumption) than HVAC systems, and it is energy costs that are often the decisive factor in economic analyses. Life-cycle system/equipment costs (over the life of the structure) are expressed in present-value dollars or annual owning and operating costs, including equipment amortization costs. The type of analysis used will depend upon the situation. Such comparisons are useful, however, only when both the initial cost and the operating costs will be borne by the same individual—that is, an owner-operator. In the case of a speculative building venture, first cost is usually the deciding factor for a developer because others will bear the operating costs.

It is often difficult to perform a life-cycle cost analysis because accurate data regarding service life and maintenance costs for electrical equipment may not be readily available. Still, even if a precise analysis cannot be done, the principle involved in life-cycle analysis should be considered in the selection of all electrical materials and equipment, bearing in mind the relative importance to an environmentally responsive building,

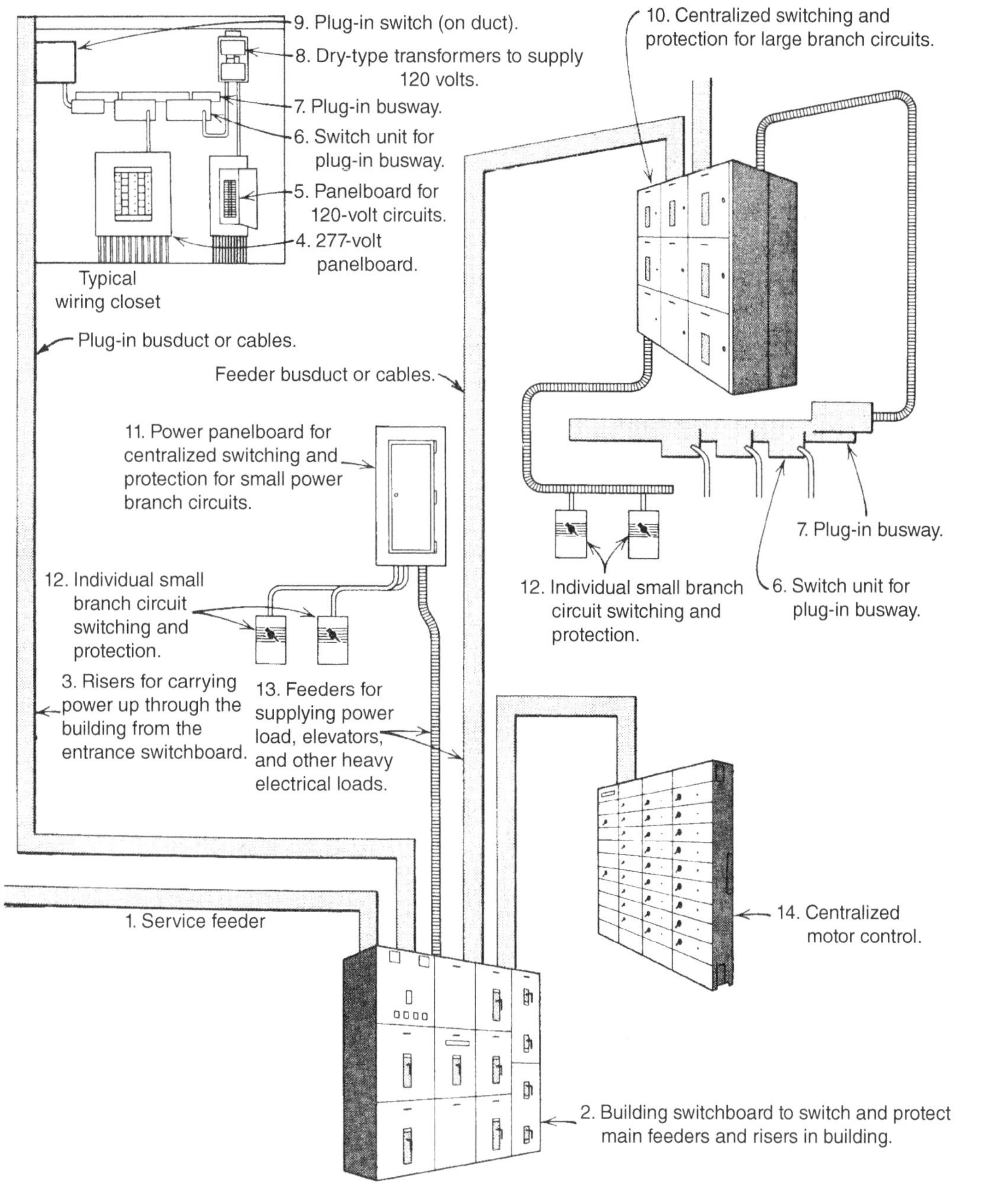

Figure 29.2 *Diagrammatic (pictorial) representation of a typical building electrical power system with relative power capacity indicated by the size of the conductors. This diagram does not extend beyond the local panelboard and includes only commonly used items. Note also that the entire wiring system is, in effect, jacketed in steel. (Courtesy of General Electric.)*

of energy-utilizing and energy-conveying equipment. See Appendix J for an overview of cost analysis techniques. Economic issues for particular items are found in the following technical discussions.

Energy costs are a major factor in economic analysis. Energy considerations, however, are at least as important in and of themselves because of ever-increasing concerns about limited energy and natural resources. This concern is addressed in the context of building energy budgets and their components, including lighting, elevators, and electric motors that are a part of many building systems. Building codes in the United States only lightly deal with electrical energy system efficiency—primarily through the requirements presented in *ASHRAE Standard 90.1*. No explicit recommendation regarding electrical system design is provided in any of the current green building guidelines.

29.4 ELECTRICAL EQUIPMENT RATINGS

All electrical equipment must be rated for the normal service it is intended to perform. These ratings may refer to voltage, current, duty, horsepower, kilowatt (kW), kilo-volt-amperes (kVA), temperature, enclosure, and so on. Ratings related to specific equipment are discussed in the sections that follow. The ratings that are specifically and characteristically electrical are those of voltage and current.

(a) Voltage

The voltage rating of an item of electrical equipment is the maximum voltage that can safely be applied to the unit continuously. It frequently, but not always, corresponds to the voltage applied in normal use. An ordinary wall electrical receptacle, for example, is normally rated at 250 V maximum, although in normal use only 120 V are applied to it. The rating is determined by the type and quantity of insulation used and the physical spacing between electrically energized parts.

(b) Current

The current rating of a device is determined by the maximum temperature at which its components can operate at full load. That in turn depends upon the type of insulation used. As a case in point, consider an electric motor. The current flowing in the motor windings causes a power loss (I^2R), which generates heat. If the windings are insulated with varnished cotton braid, with a maximum safe operating temperature of 65°C [150°F], the maximum permissible current (to which the horsepower [kW] rating of the motor is directly related) is the current that produces this operating temperature. If these same windings are insulated with a silicone or glass compound with a maximum operating temperature of 150°C [300°F], more current can be carried safely, and the horsepower (kW) rating is consequently larger. Thus, although a motor is rated in horsepower (or kW), a transformer is rated in kVA, and a cable (discussed later) is rated in amperes, the actual criterion on which all these ratings are based is the maximum permissible operating temperature of the device's insulation (and other components).

29.5 INTERIOR WIRING SYSTEMS

At this point, it is helpful to survey the different types of interior wiring systems before commencing a discussion of components. When the primary purpose of a system is to distribute electrical energy, it is referred to as an electrical power system; when the purpose is to transmit information, it is referred to as an electrical signal or a communication system. This chapter deals with electrical power systems, except that the discussion of raceways covers equipment also used by communication systems.

Due to the nature of electricity, its distribution within a structure for power use poses a single basic problem: how to construct a distribution system that safely provides the energy required at the desired locations. The safety consideration is all-important because all parts of an interior distribution system are connected to the utility's powerful network, and the very real potential for physical damage, injury, and fire is always present. The solution to this problem is to isolate all electrically conducting elements from the building structure, except at those specific points, such as wall receptacles, where contact is desired. This isolation is generally accomplished by insulating the conductors and placing them in protective raceways.

The principal types of interior wiring systems in use today are exposed insulated cables, insulated cables in open raceways, and insulated conductors in closed raceways.

(a) Exposed Insulated Cables

This category includes, using the *NEC* nomenclature, cable types NM ("Romex") and AC ("BX"). (See Sections 29.10 and 29.11.) Also included are other types where the cable construction itself provides the necessary electrical insulation and mechanical protection.

(b) Insulated Cables in Open Raceways (Trays)

This system is specifically intended for industrial applications, and it relies on both the cable and the tray for safety.

(c) Insulated Conductors in Closed Raceways

This system is the most general type and is applicable to all types of facilities. In general, the raceway is installed first and the wiring is pulled in or laid in later. The raceways themselves may be as follows.

1. Buried in the structure—for example, conduit in the floor slab or underfloor duct (see Sections 29.17, 29.18, 29.24, and 29.25).
2. Attached to the structure—for example, all types of surface raceways, including conduit and wireways suspended above hung ceilings (see Sections 29.22 and 29.29).

3. Part of the structure—for example, cellular concrete and cellular metal floors (see Sections 29.26 and 29.27).

(d) Combined Conductor and Enclosure

This category is intended to cover all types of factory-prepared and factory-constructed integral assemblies of conductor and enclosure. Included here are all types of busway, busduct, and cablebus (Section 29.14); flat cable assemblies and lighting track (Section 29.15); flat cable intended for under-carpet installation (Section 29.29); and manufactured wiring systems (Section 29.30).

29.6 CONDUCTORS

Electrical conductors ("wires") are the means by which current is conducted through the electrical system, corresponding to the piping of a hydraulic system. The standard of the U.S. wire and cable industry for round cross-section conductors is the American Wire Gauge (AWG). By convention, a single insulated conductor No. 6 AWG or larger, or several conductors of any size assembled into a single unit, are referred to as cable. Single conductors No. 8 AWG and smaller are called wire.

All wire sizes up to No. 0000 (also written as No. 4/0) are expressed in AWG. The AWG numbers run in reverse order to the size of the wire—that is, the smaller the AWG number, the larger the size. Thus, No. 10 is a heavier wire than No. 12 wire and is lighter (thinner) than No. 8 wire. The No. 4/0 size is the largest AWG designation, beyond which a different designation called kcmil (thousand circular mil) is used. In this designation, wire diameter increases with number; thus, 500 kcmil is a heavier wire (double the area) than 250 kcmil (the former designation for this unit was "thousands of circular mils" or MCM, a designation that is still used by some resources).

Outside of the United States, where SI units are in general use, conductor sizes are given simply by their diameter in millimeters. Table 29.1 gives dimensional and stranding data for common wire sizes and includes the millimeter equivalent of each size. This will prove useful in relating American gauges to SI sizes.

29.7 CONDUCTOR AMPACITY

As noted, conductor current-carrying capacity, or ampacity, is determined by the maximum safe operating temperature of the insulation used on the conductor. Heat generated as a result of current flow is dissipated into the environment. Thus, for a given installation context (open-air, buried in earth, or enclosed), ampacity increases with increasing conductor size and with maximum permissible insulation temperature. This is shown in Table 29.2. If more than three conductors are placed in a conduit, the resultant increase in temperature requires that the conductors be derated by the amount shown in Table 29.3 to maintain safe operating conditions.

Because heat dissipation from a conductor in free air is much greater than that from the same conductor enclosed in conduit or directly buried, its corresponding allowable ampacity is also greater. Conversely, if the

TABLE 29.1 Physical Properties of Bare Copper Conductors

Size AWG kcmil (MCM)	Area Circular Mils	Diameter (in.)		Diameter (mm)		DC Resistance at 75°C [167°F] (Uncoated) ohms/1000 ft [305 m]
		Overall	Stranding	Overall	Stranding	
18 AWG	1620	0.040	—	1.02	—	7.77
16	2580	0.051	—	1.29	—	4.89
14	4110	0.064	—	1.63	—	3.07
12	6530	0.081	—	2.05	—	1.93
10	10,380	0.102	—	2.59	—	1.21
8	16,510	0.128	—	3.26	—	0.764
6	26,240	0.184	0.061	4.67	1.56	0.491
4	41,740	0.232	0.077	5.89	1.96	0.308
2	66,360	0.292	0.097	7.42	2.47	0.194
1	83,690	0.332	0.066	8.43	1.69	0.154
0 (1/0)	105,600	0.372	0.074	9.45	1.89	0.122
00 (2/0)	133,100	0.418	0.084	10.62	2.13	0.097
000 (3/0)	167,800	0.470	0.094	11.94	3.39	0.077
0000 (4/0)	211,600	0.528	0.106	13.41	2.68	0.061
250 kcmil	250,000	0.575	0.082	14.61	2.09	0.052
300 kcmil	300,000	0.630	0.090	16.00	2.29	0.043
400 kcmil	400,000	0.728	0.104	18.49	2.64	0.032
500 kcmil	500,000	0.813	0.116	20.65	2.95	0.026

Source: Reprinted with permission from 2023 NFPA-70: National Electrical Code®, Copyright © 2022, National Fire Protection Agency, Quincy, MA, please go to www.nfpa.org.

ELECTRICITY

TABLE 29.2 Allowable Ampacities of Insulated Conductors Rated 0 Through 2000 V, 60° to 90°C [140° to 194°F], Not More Than Three Current-Carrying Conductors in Raceway or Cable or Earth (Directly Buried), Based on Ambient Temperature of 30°C [86°F]

PART A

	Temperature Rating of Conductor[a,b]		
	60°C [140°F] Type TW, UF	75°C [167°F] Types RHW, THHW, THW, THWN, XHHW, XHWN, USE, ZW	90°C [194°F] Types TBS, SA, SIS, FEP, FEPB, MI, PFA, RHH, RHW-2, THHN, THHW, THW-2, THWN-2, USE-2, XHH, XHHW, XHHW-2, XHWN, XHWN-2, XHHN, Z, ZW-2
Size AWG or kcmil		**COPPER**	
14 AWG[c]	15	20	25
12[c]	20	25	30
10[c]	30	35	40
8	40	50	55
6	55	65	75
4	70	85	95
2	95	115	130
1	110	130	145
0	125	150	170
00	145	175	195
000	165	200	225
0000	195	230	260
250 kcmil	215	255	290
300	240	285	320
350	260	310	350
400	280	335	380
500	320	380	430

PART B

Ambient Temperature		Correction Factors for Ambient Temperatures Other Than 30°C [86°F], Multiply the Allowable Ampacities Shown Above by the Appropriate Correction Factor Shown Below		
°C	°F			
21–25	69–77	1.08	1.05	1.04
26–30	78–86	1.00	1.00	1.00
31–35	87–95	0.91	0.94	0.96
36–40	96–104	0.82	0.88	0.91
41–45	105–113	0.71	0.82	0.87

Source: Reprinted with permission from 2023 NFPA-70: National Electrical Code®, Copyright © 2022, National Fire Protection Agency, Quincy, MA, please go to www.nfpa.org.

[a]See Table 29.5 for conductor insulation and application information.

[b]Consult the *NEC* for specifications regarding application provisions.

[c]Unless otherwise permitted by the *NEC*, the overcurrent protection for these conductors shall not exceed 15 A for 14 AWG, 20 A for 12 AWG, and 30 A for 10 AWG, after application of correction factors for ambient temperature and number of conductors in raceway.

TABLE 29.3 Adjustment Factors for More Than Three Current-Carrying Conductors[a]

Number of Current-Carrying Conductors in Raceway[b]	Adjustment Factor[c]
4–6	0.80
7–9	0.70
10–20	0.50
21–30	0.45
31–40	0.40
41 and above	0.35

Source: Reprinted with permission from 2023 NFPA-70: National Electrical Code®, Copyright © 2022, National Fire Protection Agency, Quincy, MA. For a full copy of NFPA 70, please go to www.nfpa.org.

[a]These factors are to be applied after application of the factors for ambient temperatures other than 30°C [86°F]. See Table 29.2.

[b]Number of conductors is the total number of conductors in the raceway or cable, including spare conductors. The count shall be adjusted in accordance with 2023 *NEC* Article 310.15(E) and (F). The count shall not include conductors that are connected to electrical components that cannot be simultaneously energized.

[c]When a load diversity factor is applicable, factors as given in 2023 *NEC* Annex B Table B.2(11) shall be used.

TABLE 29.4 Typical Ambient Temperatures

Location	Temperature
Well-ventilated, normally heated buildings	30°C [86°F]
Buildings with major heat sources such as power stations or industrial processes	40°C [104°F]
Poorly ventilated spaces such as attics	45°C [113°F]
Furnace and boiler rooms (min.)	40°C [104°F]
Furnace and boiler rooms (max.)	60°C [140°F]
Outdoors in shade in air	40°C [104°F]
In thermal insulation	45°C [113°F]
Direct solar exposure	55°C [131°F]

ambient temperature around a conductor is higher than 30°C [86°F], the temperature upon which all standard ampacity tables are based, the permissible ampacity must be reduced.

Ampacity tables for conductors in free air, for cable types not shown in Table 29.2, and derating factors for high ambient temperatures are found in the *NEC*. Typical ambient temperatures are given in Table 29.4.

29.8 CONDUCTOR INSULATION AND JACKETS

Most conductors are covered with some type of insulation that provides both electrical isolation and a degree of physical protection. Additional physical shielding, where necessary, is provided by a jacket placed over the insulation. Insulation is rated by voltage. Ordinary building wiring is rated for either 300 or 600 V. Common types of building wire insulation are listed in Table 29.5, with their associated trade names, code letters, maximum permitted operating temperatures, and special provisions.

29.9 COPPER AND ALUMINUM CONDUCTORS

Aluminum has an inherent weight advantage over copper, which brings with it lower installation costs. Economy usually lies with copper in small and medium-sized cable, because weight is not a problem, and the smaller conduit required for the smaller copper conductors generally makes the combined installation cheaper. In larger cable sizes, the aluminum weight advantage offsets the economy of smaller copper size and conduit, and aluminum generally proves less expensive, particularly in urban areas with high labor costs.

Aluminum and copper both exhibit the low electrical resistivity necessary for a good electrical conductor. There are, however, difficulties inherent in splicing and terminating aluminum. These difficulties—which can be overcome with the use of proper equipment, techniques, and workmanship—stem from aluminum's cold-flow characteristic when under pressure (causing joints to loosen) and aluminum's oxide. This oxide, which forms within minutes on any exposed aluminum surface, is an adhesive, poorly conductive film that must be removed and prevented from reforming if a successful, long-life joint or termination is to be made. If this is not done, the oxide causes a high-resistance joint with excessive heat generation and possible incendiary effects. The oxide problem can be largely overcome by the use of copper-clad aluminum wire, but the cold-flow problem remains. Furthermore, when used in residential branch circuits, aluminum can create problems (even if properly installed initially) when wiring devices are replaced by unskilled homeowners.

As a result of a number of unfortunate incidents, some jurisdictions in the United States have banned the use of aluminum wire in branch circuitry. Heavy feeders are normally installed by experienced and skilled contractors, and the risk of a poor joint is minimized. We recommend that the use of aluminum conductors be restricted to sizes no smaller than No. 4 AWG and that installation be permitted only by contractors who certify expertise in the specialized techniques involved. Also, local codes and electrical inspectors should be consulted. All references in this text, including all tables and illustrations, are to copper conductors. The following sections provide a brief description of the principal building wire types.

29.10 ARMORED CABLE

Among the most common types of exposed wiring is *NEC* type AC armored cable, commonly known in the smaller sizes by the trade name "BX." It is an assembly of insulated wires, bound together and enclosed in a protective armor made of a spiral-wound interlocking strip of steel tape (Figure 29.3). The cable is installed with simple U-clamps or staples holding it against beams, walls, and so on. This type of installation is frequently used in residences and in the rewiring of existing buildings. Use of type AC cable is generally restricted to dry locations. For application and installation details and restrictions, see *NEC* Article 320.

A similar construction with much broader application (covered in *NEC* Article 330) is metal-clad (MC) cable. This cable (Figure 29.4) may be used exposed or concealed and in cable trays, and, when covered with a moisture-impervious jacket, in wet and outdoor locations as well.

29.11 NONMETALLIC-SHEATHED CABLE

NEC types NM and NMC (Figure 29.5), also known by the trade name "Romex," are restricted to small building applications—that is, residential and other structures not exceeding three floors above grade.

TABLE 29.5 Conductor Insulation and Application[a]

Trade Name	Type Letter	Maximum Operating Temperature	Application Provisions
Fluorinated ethylene propylene	FEP or FEPB	90°C [194°F]	Dry and damp locations
		200°C [392°F]	Dry locations—special applications[b]
Mineral insulation (metal sheathed)	MI	90°C [194°F]	Dry and wet locations
		250°C [482°F]	For special applications[b]
Moisture-, heat-, and oil-resistant thermoplastic	MTW	60°C [140°F]	Machine tool wiring in wet locations
		90°C [194°F]	Machine tool wiring in dry locations. Informational Note: See NFPA 79-2021, *Electrical Standard for Industrial Machinery*
Paper	—	85°C [185°F]	For underground service conductors, or by special permission
Perfluoro-alkoxy	PFA	90°C [194°F]	Dry and damp locations
		200°C [392°F]	Dry locations—special applications[b]
Perfluoro-alkoxy	PFAH	250°C [482°F]	Dry locations only. Only for leads within apparatus or within raceways connected to apparatus (nickel or nickel-coated copper only)
Thermoset	RHH	90°C [194°F]	Dry and damp locations
Moisture-resistant thermoset	RHW	75°C [167°F]	Dry and wet locations
	RHW-2	90°C [194°F]	Dry and wet locations
Silicone	SA	90°C [194°F]	Dry and damp locations
		200°C [392°F]	For special application[b]
Thermoset	SIS	90°C [194°F]	Switchboard and switchgear wiring only
Thermoplastic and fibrous outer braid	TBS	90°C [194°F]	Switchboard and switchgear wiring only
Extended polytetra-fluoro-ethylene	TFE	250°C [482°F]	Dry locations only. Only for leads within apparatus or within raceways connected to apparatus, or as open wiring (nickel or nickel-coated copper only).
Heat-resistant thermoplastic	THHN	90°C [194°F]	Dry and damp locations
Moisture- and heat-resistant thermoplastic	THHW	75°C [167°F]	Wet locations
		90°C [194°F]	Dry locations
Moisture- and heat-resistant thermoplastic	THW	75°C [167°F]	Dry and wet locations
		90°C [194°F]	Special applications within electric discharge lighting equipment. Limited to 1000 open-circuit volts or less. (Size 14-8 only as permitted in *NEC* Article 410.68.)
	THW-2	90°C [194°F]	Dry and wet locations
Moisture- and heat-resistant thermoplastic	THWN	75°C [167°F]	Dry and wet locations
	THWN-2	90°C [194°F]	
Moisture-resistant thermoplastic	TW	60°C [140°F]	Dry and wet locations
Underground feeder and branch-circuit cable—single conductor (for Type UF cable employing more than one conductor, see Part II of *NEC* Article 340).	UF	60°C [140°F]	See Part II of *NEC* Article 340
		75°C [167°F][c]	
Underground service-entrance cable—single conductor (for Type USE cable employing more than one conductor, see Part II of *NEC* Article 338)	USE USE-2	75°C [167°F][c]	See Part II of *NEC* Article 338
		90°C [194°F]	Dry and wet locations
Thermoset	XHH	90°C [194°F]	Dry and damp locations
	XHHN	90°C [194°F]	Dry and damp locations
Moisture-resistant thermoset (no outer covering)	XHHW	90°C [194°F]	Dry and damp locations
		75°C [167°F]	Wet locations
	XHHW-2	90°C [194°F]	Dry and wet locations
Moisture-resistant thermoset (with nylon jacket or equivalent outer covering)	XHWN	75°C [167°F]	Dry and wet locations
	XHWN-2	90°C [194°F]	Dry and wet locations
Modified ethylene tetrafluoro-ethylene	Z	90°C [194°F]	Dry and damp locations
		150°C [302°F]	Dry locations—special applications[c]
Modified ethylene tetrafluoro-ethylene	ZW	75°C [167°F]	Wet locations
		90°C [194°F]	Dry and damp locations
		150°C [302°F]	Dry locations—special applications[c]
	ZW-2	90°C [194°F]	Dry and wet locations

Source: Reprinted with permission from 2023 NFPA-70: National Electrical Code®, Copyright © 2022, National Fire Protection Agency, Quincy, MA, please go to www.nfpa.org.

Note: Conductors in this table shall be permitted to be rated up to 1000 V if listed and marked.

[a]Consult *NEC* Table 310.4(1) for additional information, including information regarding insulation, thickness of insulation, and out covering requirements (where applicable).

[b]Higher temperature rated constructions shall be permitted where design conditions require maximum conductor operating temperatures above 90°C [194°F].

[c]The ampacity of Type UF cable shall be limited in accordance with *NEC* Article 340.80.

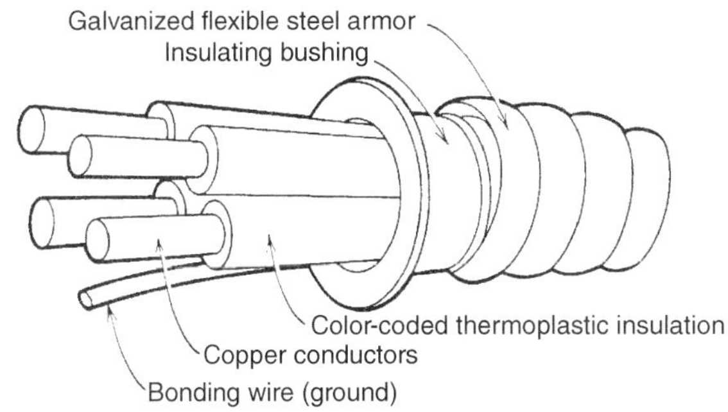

Figure 29.3 *Flexible armored cable (NEC type AC, trade name "BX"). Note the insulating bushing that is always installed on the end of the armor to protect the wires from damage from the sharp edges of the cut steel armor. (Courtesy of AFC Cable Systems.)*

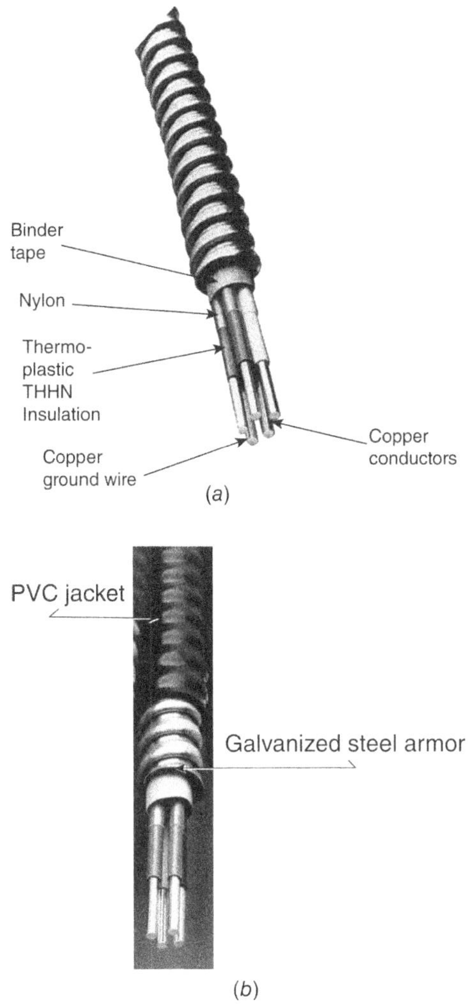

Figure 29.4 *(a) Metal-clad cable (NEC type MC, Article 330) with aluminum armor in lieu of the more common galvanized steel armor. Use is similar to that of the steel-armored cable, with the weight advantage of aluminum. Conductors are factory-installed, color-coded, and covered with type THHN insulation and nylon jacket. Cables of similar construction, using steel armor, are available for almost all power and control applications. (Courtesy of AFC Cable Systems.) (b) Jacketed-type MC cable has a wide variety of applications, including in wet locations, because of its water-impervious outer PVC jacket. Where specifically indicated by the manufacturer, this cable may be used for direct burial in earth or installed in concrete. (Courtesy of AFC Cable Systems.)*

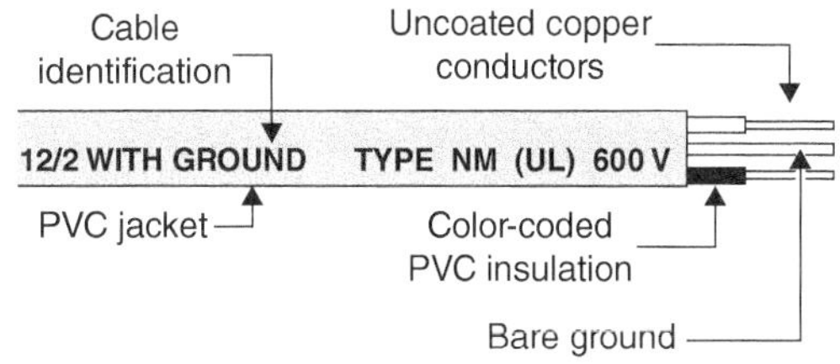

Figure 29.5 *Construction of typical NEC type NM cable. The illustrated cable is a two-conductor, No. 12 AWG with ground, insulated for 600 V. Normally shown are the manufacturer, cable trade name, and the letters (UL), which indicate listing of this product by Underwriters Laboratories, Inc. The ground wire is bare or covered, and the entire cable may be obtained flat (illustrated), oval, or round.*

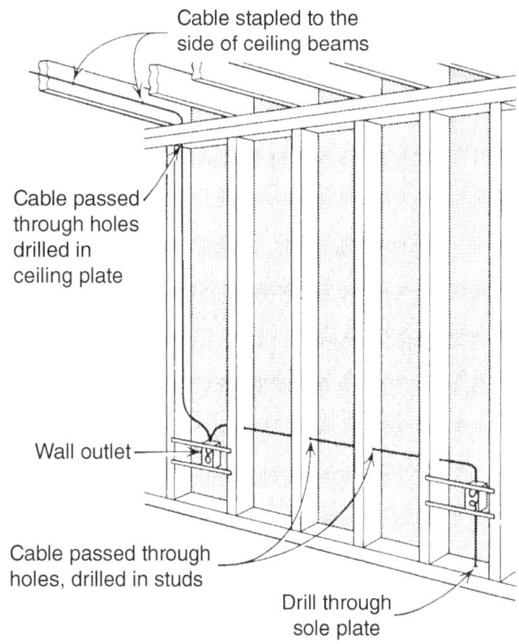

Figure 29.6 *Typical wiring technique using types NM (Romex) or AC (BX) in wood stud construction. With metal stud construction, AC cables are passed through precut openings in lieu of field-drilled holes. Where cables are exposed to damage from nails, screws, and other hazards, protective metal cover plates are required. For details, see the installation limitations in the NEC.*

The plastic outer jacket, unlike the armor on type AC, makes type NM easier to handle but more vulnerable to physical damage. For application details and restrictions, see *NEC* Article 334. The typical installation technique is shown in Figure 29.6.

29.12 CONDUCTORS FOR GENERAL WIRING

Under Article 310, the *NEC* lists the wire types that are generally installed in raceways and are referred to by the term wire. The most common types are listed in Table 29.5. These wires consist of a copper conductor

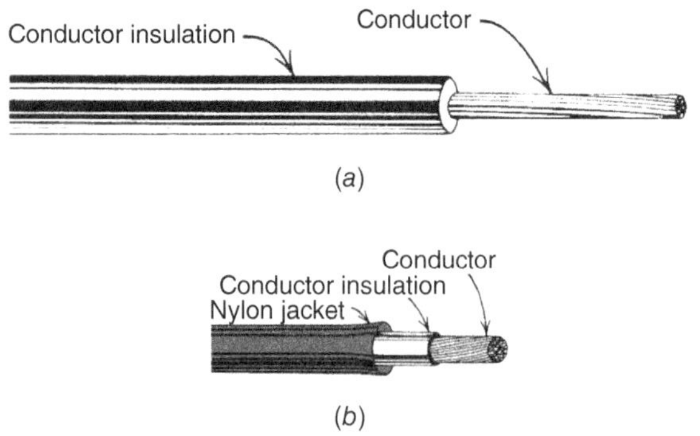

Figure 29.7 *(a) Typical construction of unjacketed building wire such as type THW (see Table 29.5). (b) The illustrated construction is typical for any nylon-jacketed cable such as THWN or THHN. The first three letters indicate the type of insulation, and the final N indicates the nylon jacket.*

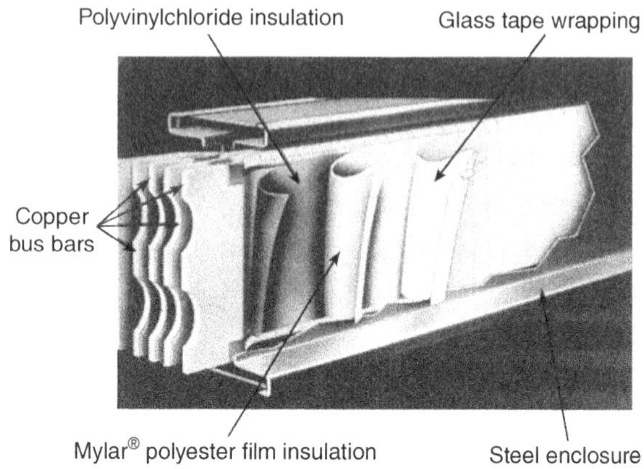

Figure 29.8 *Cutaway view showing construction of a typical feeder busduct. This design is highly compact and rigid, which gives desirable electrical characteristics as well as the advantage of small size. (Courtesy of Siemens Energy and Automation, Inc.)*

covered with insulation and, in some instances, with a jacket (Figure 29.7).

29.13 SPECIAL CABLE TYPES

Although most building wiring is accomplished with plastic-insulated 300- and 600-V conductors of the types described in the preceding sections, some applications require the use of special cables. These include high-voltage cables, armored cables, corrosion-resistant jacketed cables, underground cables, and so on. The reader is referred to manufacturers' catalogs and the *NEC* for construction and application details. Service-entrance cables and their installation are discussed in Chapter 28.

29.14 BUSWAY/CABLEBUS

Defined in *NEC* Article 368, a busway (also referred to as a busduct) is a prefabricated assembly of copper or aluminum bars in a rigid metallic housing (Figure 29.8). Its use is almost always preferable, from an economic viewpoint, in two instances: when it is necessary to carry large amounts of current (power), and when it is necessary to tap onto an electrical power conductor at frequent intervals along its length.

In the case of a heavy current requirement, the alternatives to using busways are to use paralleled sets of round conductors or a single large conductor. Paralleled sets of conductors are almost always more expensive than a busway of similar current capacity because of the high installation cost of multiple conduits. Alternatively, using a single, large-diameter conductor becomes increasingly inefficient as cable size increases because large round conductors require more cross-section per ampere of current-carrying capacity than do the flat conductors (busbars) used in busduct (Figure 29.9).

Where many power taps are required along an electric feeder run consisting of cable in conduit, costs become very high because of the large amount of expensive hand labor involved, since a connection must be made to each conductor in the run. The preferable alternative is to use a "plug-in" busway to which connections can be made easily and rapidly with a

Figure 29.9 *Sectional view of a busduct (on top) shows the tight assembly of insulated conductors within a metal housing. This design, unlike the ventilated type, can be mounted in any position because heat dissipation is by conduction from the busbars to the housing. The eight sets of cable (shown on the bottom) have the same current-carrying capacity as the busduct. (Courtesy of Square D Company.)*

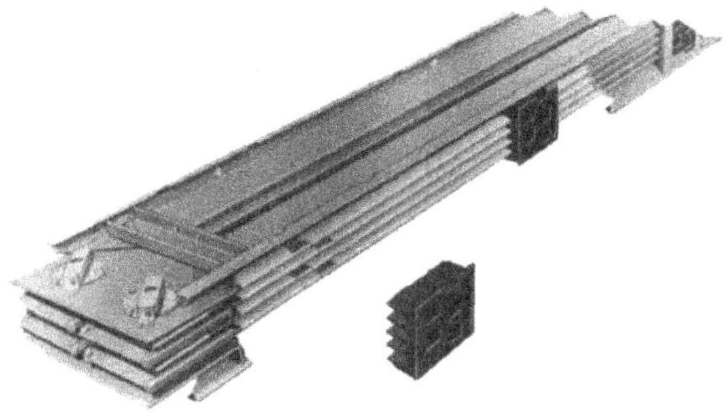

Figure 29.10 Construction of one type of plug-in busduct. Plug-ins are evenly spaced on alternate sides to facilitate connection of plug-in breakers, switches, transformers, or cable taps. Housing is of sheet steel with openings for ventilation. The cover plate is not shown. (Courtesy of Square D Company.)

plug-in device, similar to the insertion of a common plug into a wall receptacle. This has the additional advantage of convenience; connection/disconnection is simply a matter of inserting or withdrawing the plug-in device, whereas cable taps are permanent connections (Figure 29.10).

A typical application of a heavy-duty busduct might be a vertical feeder in a high-rise building connecting the basement switchboard to the penthouse machine room. The same building might also use a heavy-duty plug-in busduct as vertical riser(s) with taps feeding individual floors (Figure 29.2). Typical applications for a light-duty plug-in busduct (70–100 A) could be any machine shop or workshop. The electrical supply to individual machine tools is made very simply and flexibly with a tap-on device (see Figure 29.10).

Busduct is specified by type, material, number of buses, current capacity, and voltage (e.g., aluminum feeder busduct, 4-wire, 1000 A, 600 V, or copper plug-in busway, 100 A, 3-wire, 600 V). Feeder busduct (no plug-in capability) is available in ratings from 400 to 4000 A. Plug-in busway is available from 30 A for lighting or light-machinery circuits (see Section 29.15) to 3000 A. A wide variety of fittings and joints are available for all busways to permit easy installation (Figure 29.11). Devices are available for indoor and outdoor application.

Defined in *NEC* Article 370, a cablebus (or cable bus) is similar to a ventilated busduct, except that it uses insulated cables instead of busbars. The cables are rigidly mounted in an open space frame. The advantage of this construction is that it carries the ampacity rating of its cables in free air, which is much higher than for the same cables in conduit, thus giving a high amperes-per-dollar first-cost figure. Its principal disadvantages are bulkiness and difficulty in making taps.

An example of the type of economic analysis that should be made when considering an item as fundamental as a heavy-current electrical feeder, is

Figure 29.11 Typical installation of compact-design busduct. Note that the individual busducts are supported by channels hung from the ceiling and that the same hangers support more than one level of bus. Right-angle turns are easily made in the same plane (horizontal or vertical) and between vertical and horizontal planes. (Reproduced by permission of Square D Company.)

summarized in Table 29.6, which shows the results of such a study in terms of relative costs. Note that when considering first cost alone, the advantage lies with a cable tray (with interlocked armor cables) and cablebus. Adding energy-loss considerations shifts the advantage to a cable tray and wire in conduit. No general conclusion should be drawn from Table 29.6 regarding costs. A change in feeder length, number of taps, hours of operation, energy costs, or any of the other factors can shift the advantage to a different system. The point of the study is to demonstrate that life-cycle costs and first costs often lead to entirely different conclusions and that this type of study is truly required before a rational engineering decision can be made. (Life-cycle cost in this example was taken as the present value of all costs over the installation's life-cycle—in this case, 20 years.)

Two additional items are worthy of note:

1. The very factors that yield a lower first cost operate to yield a higher operating cost. The smaller copper sizes in busduct and cablebus, permitted by high-temperature insulation and good ventilation, cause increased power loss because of their higher resistivity.
2. If the heat loss from the busduct or cablebus can be used to advantage, the related energy cost can be credited instead of being considered a total loss, and life-cycle costs can be changed considerably. Conversely, the heat generated can negatively affect the building cooling load.

TABLE 29.6 Life-Cycle Relative Cost Comparison of 2000-A, 208-V Feeder Installation

Feeder System Description[a]	Material Cost	Labor Cost[b]	Total First Cost	Power Loss per 100 ft[c] (kW)	Annual Energy Loss[d] (kWh)	Annual Energy Cost[e]	Life-Cycle Energy Cost[f]	Total Life-Cycle Cost[g]
Cable tray	0.63	0.37	1.00	2.90	12,536	0.086	1.081	2.08 100%
Wire and conduit	0.68	0.61	1.29	2.90	12,536	0.086	1.081	2.37 114%
Busduct	1.14	0.24	1.38	5.60	24,207	0.166	2.087	3.47 167%
Cablebus	0.69	0.45	1.14	5.91	25,547	0.174	2.203	3.34 161%

[a]Equipment rating is 600 V, cable is copper, conduit is rigid steel, cable tray is aluminum.

[b]Labor cost includes overhead.

[c]Based on published resistivity data for cable and bus, assuming 80% demand (1600 A) and all conductors in the system equally loaded. (Loss is equivalent to kW per 30.5 m.)

[d]Based on 80% demand, 12 hours per day, 360 days per year.

[e]Using $0.10 per kWh as the combined net rate, including demand charges.

[f]Using a 20-year life-cycle, 8% fixed capital cost, and 3% annual escalation in energy cost.

[g]Sum of the fourth and eighth columns.

29.15 LIGHT-DUTY BUSWAY, FLAT CABLE ASSEMBLIES, AND LIGHTING TRACK

Special prefabricated assemblies that act as light-duty (branch circuit) plug-in electrical feeders are widely used because of their simplicity of installation and, more importantly, because of their plug-in mode of connection.

(a) Light-Duty Plug-In Busway

This construction, which may be used either for feeder or branch circuit applications, is covered by the *NEC* Article 368 on busways, with restrictions when applied as branch circuit wiring. Light-duty busways are rated from 20 to 60 A at 300 V, in 2- and 3-wire construction. A somewhat heavier design rated 60–100 A at 600 V is available in 3- and 4-wire construction. Their application is principally for direct connection (with overcurrent protection) of light machinery and industrial lighting (Figures 29.12 and 29.13).

(b) Flat Cable Assemblies

Type FC, defined in *NEC* Article 332, is a specially designed cable consisting of two, three, or four

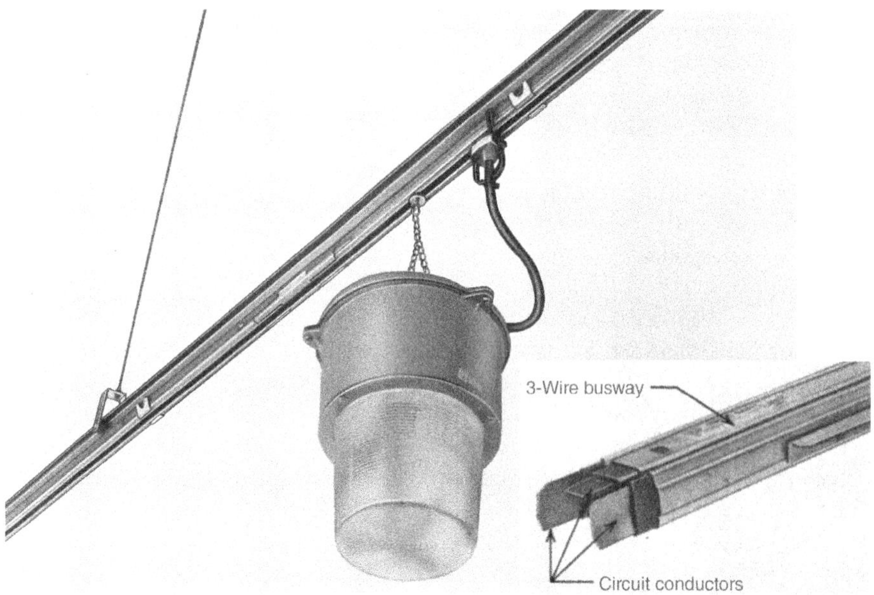

Figure 29.12 *Light-duty busway is rated 20–60 A, 300 V, and either 2- or 3-wire. Power takeoff devices twist into the bus to make contact with the circuit conductors. Within NEC restrictions of overcurrent protection, this busway may be used for standard and heavy-duty lighting fixtures and for other electrical devices such as electrically powered tools. (Courtesy of Siemens Energy and Automation, Inc.)*

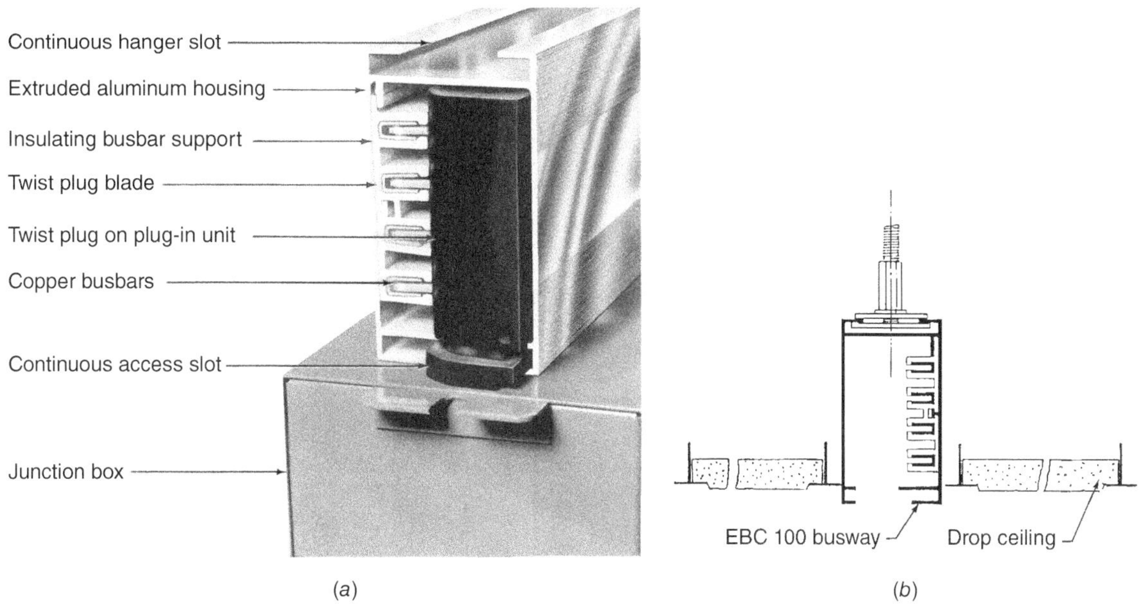

Figure 29.13 (a) Plug-in busway rated 100 A, 3-phase, 4-wire, 600 V measures approximately 2½ in. W × 4½ in. H [64 × 115 mm]. The twist-in plug, which is integrally attached to a connection means (junction box in the illustration), is rated 30–100 A, single- or 3-phase, as required. The attached junction box (or receptacle, circuit breaker, or fuse box) then feeds the utilization device (e.g., heavy-duty lighting or machinery). (b) This bus can also be installed in a hung ceiling. (Courtesy of Universal Trolley.)

No. 10 AWG conductors, and is field-installed in a rigidly mounted standard 1⅝-in. [41-mm] square structural channel. Power-tap devices, installed where required, puncture the insulation of one of the phase conductors and the neutral. Electrical connection is then made to the pigtail wires that extend from the tap devices. This connection can extend directly to the device or to an outlet box with a receptacle, which then acts as a disconnecting means for the electric device being served. In this fashion, lighting fixtures, small motors, unit heaters, and other single-phase, light-duty devices can be served without the necessity of "hard" (conduit and cable) wiring. Figure 29.14 illustrates similar equipment that is specifically intended to feed industrial lighting fixtures.

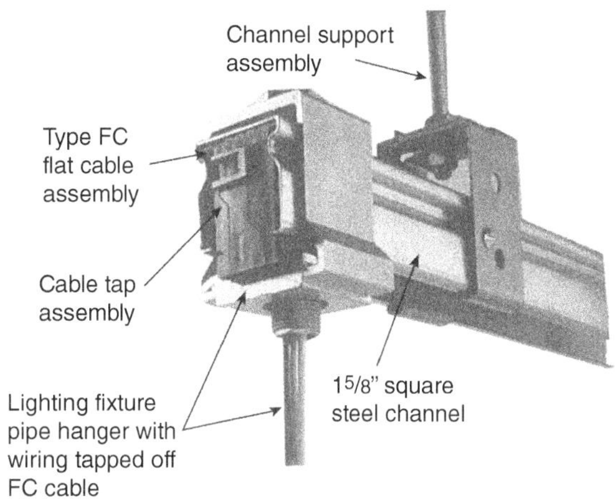

Figure 29.14 A 3-phase, 4-wire flat cable assembly installed in a steel channel is shown. Taps into the conductors are made by tightening the tap device. Taps can be made phase-to-ground to give 120 V and phase-to-phase to give 208 V. (If the cable is connected to a 277/480-V, 3-phase system, then the phase-to-ground and phase-to-phase voltages will be 277 and 480 V, respectively.) After a tap device is removed, the puncture made by the tap "heals" itself. (Courtesy of Chan-L-Wire/Wiremold Company.)

(c) Lighting Track

This is defined in *NEC* Article 410, Part XIV, as a factory-assembled channel with conductors for one to four circuits permanently installed in the track. Power is taken from the track by special tap-off devices that contact the track's electrified conductors and carry the power to the attached lighting fixture, which can be positioned anywhere along the track. The tracks are generally rated at 20 A and, unlike FC cable assemblies, may feed only lighting fixtures. Taps to feed convenience receptacles are not permitted. A typical design is shown in Figure 29.15.

29.16 CABLE TRAY

This system, defined in *NEC* Article 392, is a unit or assembly of units and associated fittings forming a rigid structural system that functions to securely fasten or support approved cables and raceways. When used as a general wiring system, the cables

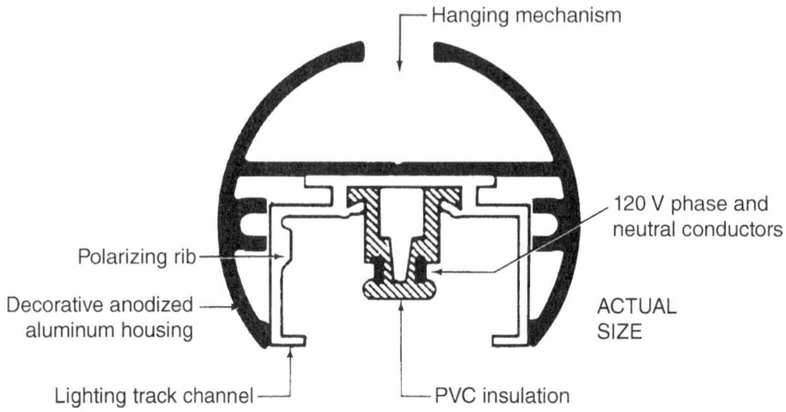

Figure 29.15 Decorative lighting track. The circular housing can be obtained in a variety of finishes and hanger arrangements. The actual lighting track, shown roughly full size, is readily available without the housing for direct surface mounting. Track is also available in a 4-conductor configuration (three 120-V conductors plus neutral). (Courtesy of Swivelier.)

must be self-protected. The advantages of this system are free-air-rated cable ampacities, easy installation and maintenance, and relatively low cost. The disadvantages are bulkiness and accessibility requirements. Cable trays are used primarily in industrial applications.

29.17 DESIGN CONSIDERATIONS FOR RACEWAY SYSTEMS

The following sections deal with closed wiring raceways, which completes the discussion of raceways. The details of construction and application are not discussed here because such information is readily available from manufacturers and applicable *NEC* articles. However, enough material is provided for the reader to become familiar with the types of raceways, their common applications and limitations, and, where applicable, comparative characteristics.

Although this chapter covers equipment used primarily in electrical power systems, empty raceways intended for signal, data, and communications wiring (to be provided by others) are normally specified under the electrical section of a construction contract. The function of a raceway in such systems is largely the same as it is for power wiring: protection and isolation of the wiring.

Prior to the widespread use of computers in buildings of all types, raceway space requirements for communication and signal wiring were easily established because such wiring consisted of small telephone cables plus miscellaneous signal and alarm wires. These requirements were readily satisfied with empty conduits or cells in floor raceway systems. Today, when virtually every commercial/institutional building uses some type of data-processing equipment, and communication networking is commonplace even in small facilities, raceways for communication cabling have become a major design consideration. They often far exceed in cross-sectional area the space required for power cabling. Their space requirements are sometimes so great that they, like ductwork, have substantial architectural impact and must therefore be considered early in the design process.

Sizing of raceways for power wiring is an exact process (see Chapter 30), based as it is upon maximum permissible temperatures for specific materials in a given environment. This is not so for communication cabling, even when a system's present requirements are known, because of the extremely rapid growth of demand for networking and data interchange. The problem is all the more difficult when designing commercial space for rental to an unknown client. The advisable approach in such cases is either to provide a reasonable amount of floor-level raceway space for main cabling (see Sections 29.24 and 29.25) and to rely on add-on systems such as under-carpet wiring (see Section 29.29) and surface or ceiling raceways (Section 29.30) for additional raceway area, or, alternatively, to use a structural system that provides virtually unlimited wiring space (Figures 29.26 through 29.32). Because the latter is a major structural/architectural decision, it must be made during the preliminary stages of design. A clear understanding of the owner's project requirements will assist in decision making.

Design considerations for power system raceways are discussed in the following sections and in Chapter 30. Design of raceways for data and communication system wiring includes the following considerations:

1. Number, type, and location of data-processing terminals.
2. Networking requirements:

a. The type of local area network largely determines the communication media (i.e., coaxial cable, shielded and unshielded wire, and fiber-optic cables), which affects the raceway space requirements. The cable type also determines the type of connectors needed (and their space requirements) and the type of floor outlets used for machine connection.
b. Cable topology (i.e., interconnection arrangements): This item is frequently not within the domain of the architectural designer, although the raceway space availability seriously affects the cabling arrangement and vice versa.
c. Requirement for interconnection of networks and connection to remote networks.

3. Number, location, and characteristics of major peripheral devices, such as mass storage, printing, and plotting.
4. Location and type of major subsystems, such as computer-aided design/manufacturing spaces.
5. Location of presentation spaces that require interconnection to computer networks.

In view of these highly technical and rapidly changing requirements, engaging the services of a consultant with specialization in this area is suggested.

29.18 STEEL CONDUIT

The purpose of conduit is to do the following:

1. Protect the enclosed wiring from mechanical injury and damage from the surrounding atmosphere.
2. Provide a grounded metal enclosure for the wiring in order to avoid a shock hazard.
3. Provide a system ground path.
4. Protect surroundings against a fire hazard as a result of overheating or arcing of the enclosed conductors.
5. Support the conductors.

For these reasons, the *NEC* generally requires that all power wiring be enclosed in a rigid metallic corrosion-resistant conduit. To this latter end, steel conduit is manufactured in several types, among which are hot-dip galvanized, sherardized (coated with zinc dust), enameled, and plastic-covered.

There are three types of steel conduit that differ basically only in wall thickness. They are, in order of decreasing weight:

1. Rigid metal conduit (RMC), a heavy-wall conduit of galvanized steel, stainless steel, or aluminum, covered by *NEC* Article 344.
2. Intermediate metal conduit (IMC), comprised of steel with a protective coating or stainless steel, covered by *NEC* Article 342.
3. Electrical metallic tubing (EMT), commonly known as thin-wall conduit and comprised of galvanized steel, stainless steel or aluminum, covered by *NEC* Article 358.

The differences are shown in Table 29.7. Several types of conduit (including rigid steel) are shown in Figure 29.16. Standard rigid steel conduit dimensions and wall thicknesses are provided in Table 29.8.

Rigid conduit and IMC use the same threaded fittings. As a result of its thin wall, EMT is not threaded: instead, it uses set screw (or setscrew) and pressure fittings. The thinner walls of EMT and IMC yield a larger inside diameter (ID) and, therefore, easier wire pulling. The combination of lower weight and easier wire pulling gives EMT and IMC a distinct labor-cost advantage over rigid conduit, which is further enhanced in jobs with a great deal of field bending and handling of conduit. Both, however, have application restrictions, which are detailed in the *NEC*.

Generally, no conduit smaller than ½-in. [13-mm] nominal trade diameter is used. Ordinary steel pipe may not be used as electric conduit, and all electric steel conduit is distinctively marked as such.

When steel conduit is installed in direct contact with the earth, it is advisable to use the hot-dip galvanized type and to coat the joints with bitumen (asphaltum). If the earth is very wet, the entire conduit system should be coated with an appropriate waterproofing compound. Alternatively, a plastic-jacketed conduit can be used.

Conduit is fastened to the building structure in much the same way as piping: with pipe straps and clamps. The vertical load at floor openings is taken with special support clamps. Trapeze mounting is common for conduit banks hung from the ceiling, as in Figure 29.17.

Conduit size depends not only on the maximum permissible temperature of the contained conductors, but also on the number and diameter of the wires that may be drawn into the conduit without injuring the wire. The number and radius of bends in the conduit, as well as its total length, affect the degree of abrasion to the wiring insulation during installation. No wires should be installed until a conduit system has been inspected and approved.

For structural reasons, conduits in concrete slabs are run close to the bottom surface (in the portion of the slab in tension) or near the center. If a large number of conduits must be embedded, it may be necessary to increase the slab thickness. In many instances, the structural slab is covered with a concrete topping, in

TABLE 29.7 Comparative Dimensions and Weights of Metallic Conduit[a]

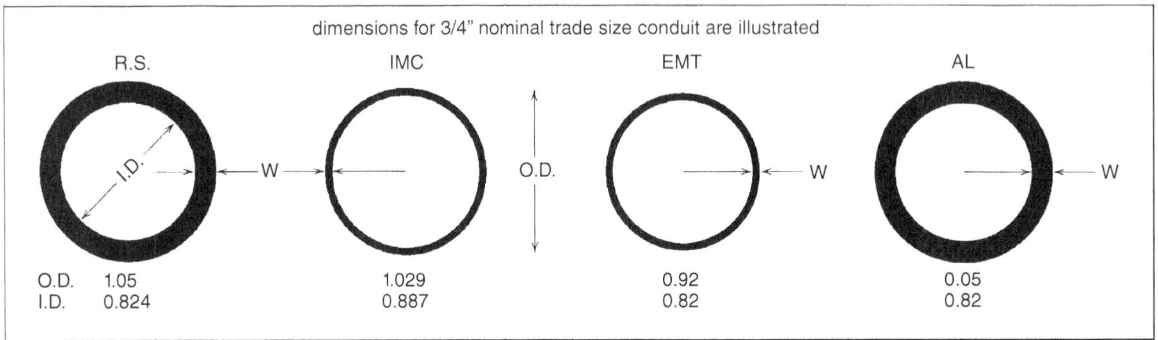

Nominal or Trade Size, in. [mm]	Outside Diameter, in. [mm]				Inside Diameter, in. [mm]				Weight per 10-ft [3.05-m] Length, lb [kg][b]			
	RS[c]	IMC[d]	EMT[e]	AL[f]	RS	IMC	EMT	AL	RS	IMC	EMT	AL
½	0.84	0.82	0.71	0.84	0.62	0.69	0.62	0.62	7.9	5.3	2.9	2.7
[16]	[21.3]	[20.8]	[18.0]	[21.3]	[15.8]	[17.5]	[15.8]	[15.8]	[3.6]	[2.4]	[1.3]	[1.2]
¾	1.05	1.03	0.92	1.05	0.82	0.89	0.82	0.82	10.9	7.2	4.4	3.6
[21]	[26.7]	[26.2]	[23.4]	[26.7]	[20.8]	[22.6]	[20.8]	[20.8]	[4.9]	[3.3]	[2.0]	[1.6]
1	1.32	1.29	1.16	1.32	1.05	1.13	1.05	1.05	16.5	10.6	6.4	5.3
[27]	[33.5]	[32.8]	[29.5]	[33.5]	[26.7]	[28.7]	[26.7]	[26.7]	[7.5]	[4.8]	[2.9]	[2.4]
1¼	1.66	1.64	1.51	1.66	1.38	1.47	1.38	1.38	21.5	14.4	9.5	7.0
[35]	[42.2]	[41.7]	[38.4]	[42.2]	[35.1]	[37.3]	[35.1]	[35.1]	[9.8]	[6.5]	[4.3]	[3.2]
1½	1.90	1.88	1.74	1.90	1.61	1.70	1.61	1.61	25.8	17.7	11.0	8.6
[41]	[48.3]	[47.8]	[44.2]	[48.3]	[40.9]	[43.2]	[40.9]	[40.9]	[11.7]	[8.0]	[5.0]	[3.9]
2	2.38	2.36	2.20	2.38	2.07	2.17	2.07	2.07	35.2	23.6	14.0	11.6
[53]	[60.5]	[59.9]	[55.9]	[60.5]	[52.6]	[55.1]	[52.6]	[52.6]	[16.0]	[10.7]	[6.4]	[5.3]
2½	2.88	2.86	2.88	2.88	2.47	2.61	2.73	2.47	56.7	38.2	20.5	18.3
[63]	[73.2]	[72.6]	[73.2]	[73.2]	[62.7]	[66.3]	[69.3]	[62.7]	[25.7]	[17.3]	[9.3]	[8.3]
3	3.50	3.48	3.50	3.50	3.07	3.23	3.36	3.07	71.4	46.9	25.0	23.9
[78]	[88.9]	[88.4]	[88.9]	[88.9]	[78.0]	[82.0]	[85.3]	[78.0]	[32.4]	[21.3]	[11.3]	[10.8]
3½	4.00	3.97	4.00	4.00	3.55	3.72	3.83	3.55	86.0	54.7	32.5	28.8
[91]	[101.6]	[100.8]	[101.6]	[101.6]	[90.2]	[94.5]	[97.3]	[90.2]	[39.0]	[24.8]	[14.7]	[13.1]
4	4.50	4.47	4.50	—	4.03	4.21	4.03	—	97.2	62.5	37.0	—
[103]	[114.3]	[113.5]	[114.3]	—	[102.4]	[106.9]	[102.4]	—	[44.1]	[28.4]	[16.8]	—

[a] Data vary slightly among manufacturers. O.D. is outside diameter; I.D. is inside diameter; W. is wall thickness.

[b] Standard length including one coupling.

[c] RS: rigid steel type of RMC.

[d] IMC: intermediate metal conduit.

[e] EMT: electrical metallic tubing.

[f] AL: aluminum conduit.

which conduit may be installed without affecting slab integrity. In all cases, local building codes should be consulted for limitations on embedded conduits. In any event, the top of any conduit shall be at least ¾ in. [19 mm] below the finished floor surface to prevent cracking. When heavy truck traffic is expected, this allowance should be increased to 1½ in. [38 mm] minimum.

In general, the following rules should be observed and included in all specifications for conduit work in concrete slabs:

1. Conduits shall have an outside diameter (OD) no greater than one-third the slab thickness, as measured at its thinnest point.
2. Conduits running parallel to each other shall be spaced not less than three times the OD of the largest conduit center-to-center.
3. Conduit crossings shall be as close to a right angle as possible.
4. Minimum cover over conduits shall be ¾ in. [19 mm].

29.19 ALUMINUM CONDUIT

The use of aluminum conduit has increased greatly in recent years because of the weight advantage that aluminum has over steel, being even lighter than EMT. The savings in labor costs usually more than offset

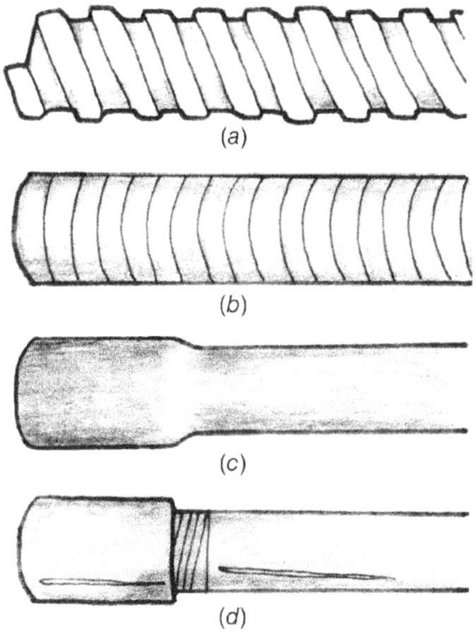

Figure 29.16 *Conduits: (a) flexible aluminum conduit, suitable for power and lighting circuits, (b) flexible steel, liquid-tight conduit, resistant to moisture, oil, and sunlight, (c) polyvinyl chloride (PVC) conduit, resistant to heat and corrosion, (d) galvanized steel conduit, protects against magnetic fields and damage to electrical cables. (Drawing by Karen Tse; © Alison Kwok, all rights reserved.)*

TABLE 29.8 Rigid Steel Conduit Sizes

	Dimensions, in. [mm]		
Conduit Size, in. [mm]	**Outside Diameter (OD)**	**Inside Diameter (ID)**	**Wall Thickness (W)**
½ [16]	0.84 [21.3]	0.63 [16.1]	0.10 [2.6]
¾ [21]	1.05 [26.7]	0.84 [21.3]	0.11 [2.7]
1 [27]	1.31 [33.4]	1.06 [27.0]	0.13 [3.2]
1 ¼ [35]	1.66 [42.2]	1.39 [35.4]	0.13 [3.4]
1 ½ [41]	1.90 [48.3]	1.63 [41.3]	0.14 [3.5]
2 [53]	2.37 [60.3]	2.08 [52.9]	0.15 [3.7]
2 ½ [63]	2.87 [73.0]	2.49 [63.2]	0.19 [4.9]
3 [78]	3.50 [88.9]	3.09 [78.5]	0.20 [5.2]
3 ½ [91]	4.00 [101.6]	3.57 [90.6]	0.22 [5.5]
4 [103]	4.50 [114.3]	4.05 [102.9]	0.22 [5.7]

Source: Steel Tube Institute, steelconduit.org.

the additional cost of the material itself. In addition, aluminum has better corrosion resistance in most atmospheres; it is nonmagnetic, giving a lower voltage drop; it is nonsparking; and, generally, it does not require painting.

Its major drawback is its deleterious effect on many types of concrete, causing spalling and cracking when embedded. Although manufacturers can demonstrate cases of embedding in concrete without harmful effect, this procedure should be avoided unless the concrete additives are rigidly controlled and the conduit is coated to prevent contact with the concrete. It is also inadvisable to bury aluminum in earth, with or without asphalt or another coating, because of the rapid corrosion often encountered. Other difficulties frequently encountered are mechanical freezing of threaded joints (because of thread deformation) and difficulty in obtaining electrical contact with grounding straps. With the exceptions noted, aluminum conduit may be used in all locations where steel conduit is used.

29.20 FLEXIBLE METAL CONDUIT

Reduced wall flexible metal conduit, defined in *NEC* Article 348 as type FMC, consists of an empty, spirally wound, interlocked armor steel or aluminum raceway. Commonly known by the trade name Greenfield, FMC is used principally for motor connections and other locations where vibration is present, where movement is encountered, or where physical obstructions make its use necessary. The vibration isolation provided by flexible conduit is one of its most important applications. It should always be used in connections to motors, transformers, ballasts, and the like. Another common application is for wiring inside metal partitions. When covered with a liquid-tight plastic jacket, it is suitable for use in wet locations (Figure 29.18). In this configuration, it is most often known by the trade name Sealtite.

29.21 NONMETALLIC CONDUIT

A separate classification of rigid conduit (addressed by several *NEC* articles) covers raceways that are formed from such materials as fiber, asbestos-cement (not as serious an environmental concern as it might sound), soapstone, rigid polyvinyl chloride (PVC), and high-density polyethylene.

For use aboveground, such conduits must be flame-retardant, tough, and resistant to heat distortion, sunlight, and low-temperature effects. For use underground, the last two requirements are waived. Generally, nonmetallic conduit may be used without restriction in nonhazardous areas within the physical limitations of the material involved. Thus, plastic conduit has a temperature limitation, asbestos-cement has considerable physical strength limitations, and so on. As a result of these limitations, PVC conduit is the material of choice for indoor exposed use, and asbestos-cement, fiber, and PVC plastic for outdoor and underground use. A separate ground wire must

Figure 29.17 Typical overhead conduit bank installation. Note that due to field conditions the insert (a) for hanger rods was inadequate and an additional rod (b) was provided. This conduit bank uses EMT, which has a pipe-wall thickness approximately one-third that of heavy-wall rigid conduit. The resulting weight difference in a large bank such as this is very pronounced. EMT joints are made with set-screw fittings (c). Note how individual conduits are fixed by clamps to the trapeze channel (d). (Courtesy of Republic Steel Corp.)

Figure 29.18 This is a good application for liquid-tight, flexible conduit because it provides weatherproofing and mechanical isolation of the vibration-producing equipment. (Courtesy of Electri-Flex Company.)

be provided because the ground provided by a metallic conduit is absent.

29.22 SURFACE RACEWAYS (METALLIC AND NONMETALLIC)

These raceways are covered in *NEC* Articles 386 and 388. Surface raceways and multioutlet assemblies may be utilized only in dry, nonhazardous, noncorrosive locations and may generally contain only wiring operating below 300 V. Such raceways are normally installed exposed, in places not subject to physical injury.

The principal applications of surface raceways are:

1. where economy in construction weighs very heavily in favor of surface raceways and where expansion is anticipated (Figure 29.19)
2. where outlets are required at frequent intervals and where rewiring is required or anticipated (Figure 29.20)
3. where access to equipment in the raceways is required and/or where necessary due to the nature of the wiring (Figures 29.19, 29.21, 29.22, and 29.23)

Figure 29.19 Multisection nonmetallic surface-type raceway measures 4 in. H and 1.5 in. D [100 × 38 mm]. The sections are intended for power, telephone, and data cabling. The outlet box protrudes from the raceway surface so as not to lessen the raceway wiring space. Vertical dividers in the raceway keep the three distinctly different types of wiring completely separated. (Courtesy of Hubbell, Inc.)

Figure 29.20 Large-capacity surface metal raceways are particularly useful for wiring in full-access floors (see Section 29.28) because of the heavy wiring (see wall box) and frequent rewiring. The perforated floor tile in the foreground is used to supply laminar airflow in this clean room at an integrated circuit manufacturer's facility.

4. where the extensive and expensive cutting and patching required to "bury" a raceway during rewiring are to be avoided (Figure 29.24)

29.23 OUTLET AND DEVICE BOXES

These boxes are generally of galvanized stamped sheet metal. The most common sizes are the 4-in. [100-mm] square and 4-in. [100-mm] octagonal boxes used for fixtures, junctions, and devices and the $4 \times 2\,^1\!/_8$ in. [100 × 54 mm] box used for single devices where no splicing is required. Box depths vary from $1\,^1\!/_2$ to 3 in. [38 to 76 mm]. Nonmetallic boxes may be used with NM and NMC cable and with nonmetallic conduit installations. In wet locations and for outdoor work, cast-iron or cast-aluminum boxes are recommended.

An *NEC* Article 300.21 requirement that electrical penetrations in fire-rated floors be designed to maintain fire ratings has spurred electrical manufacturers to produce a line of poke-through fittings to meet this need. (This requirement applies also to rated walls, ceilings, and partitions.) One such design is shown in Figure 29.25. These electrical penetrations have become increasingly prevalent in existing commercial spaces where the expanded need for desktop power and data wiring can be met most economically and rapidly by through-the-floor feeds from accessible wiring in the suspended ceiling plenum below. In addition, these fittings facilitate the electrical wiring relocations so common in rental office occupancies.

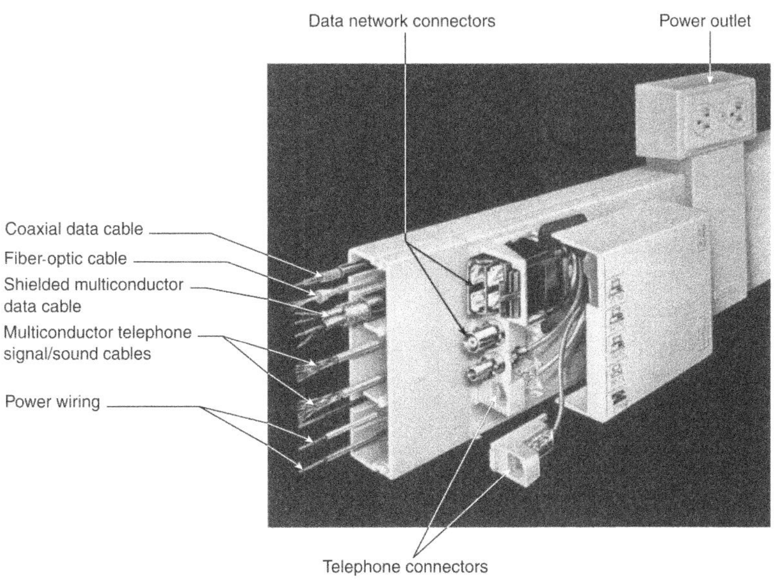

Figure 29.21 Multichannel nonmetallic surface raceway with snap-in connector modules for data network, signal, and power wiring systems. (Similar metallic raceways are also available.) The raceway itself measures $4\,^1\!/_2$ in. H × 1 in. D [115 × 25 mm]. The internal dividers are movable or entirely removable, which permits varying the number and size of the wiring channels. The principal application of this type of wireway is in commercial occupancies using extensive desktop data-processing and communication equipment. (Courtesy of Panduit.)

Figure 29.22 The frequent wiring changes required for theater and exhibition lighting are easily made when the wiring is run in a suspended surface raceway.

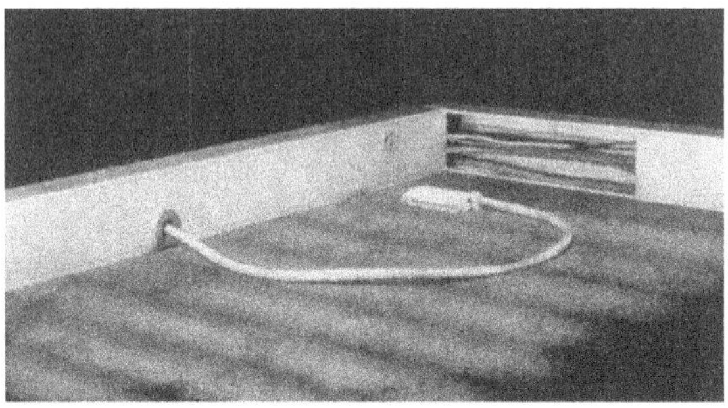

Figure 29.23 The basic raceway illustrated is $1\frac{7}{8}$ in. D × $3\frac{3}{16}$ in. W [48 × 81 mm]. It is shown with a divider installed, permitting use of the top section for power wiring and the bottom section for low-voltage wiring. Because data and communication cables are frequently supplied with factory-installed terminations (as in the photo), a raceway where the cable can be laid in rather than pulled in is required. Also, terminal strips and other equipment can be installed in the low-voltage section of these large raceways, making the use of separate terminal cabinets unnecessary.

29.24 FLOOR RACEWAYS

In commercial spaces with large open floor areas, it is common practice to place desks and other workstations throughout the space, at considerable distances from permanent walls containing electrical services. Because each workstation in a modern office requires electric power for a computer, desk lamp, and other common equipment, plus a telephone line, a computer network connection, and possibly a data outlet, the problem of bringing these services to the workstations with a minimum of exposed wiring is challenging. The required outlets can be installed on the floor adjacent to or under the workstations or, if partial-height partitions are used, within these partitions.

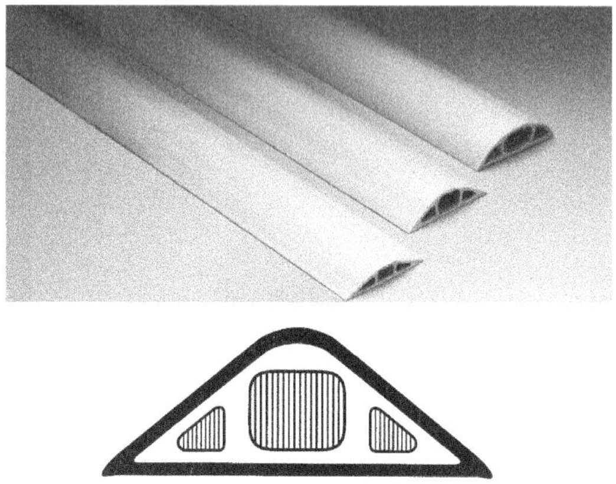

FloorTrak	**2**	**3**	**4**
Width	2.68"	2.985"	2.975"
Height	.465"	.610"	.885"
Ctr. Hole	.50x.312"	.75x.500"	1.0x.750"

Figure 29.24 Where low-voltage wire and cable cannot be concealed (buried), shallow, properly shaped floor track can be used. Such installations should avoid foot-traffic areas, where possible, to minimize trip hazards. (Courtesy of Hubbell Premise Wiring.)

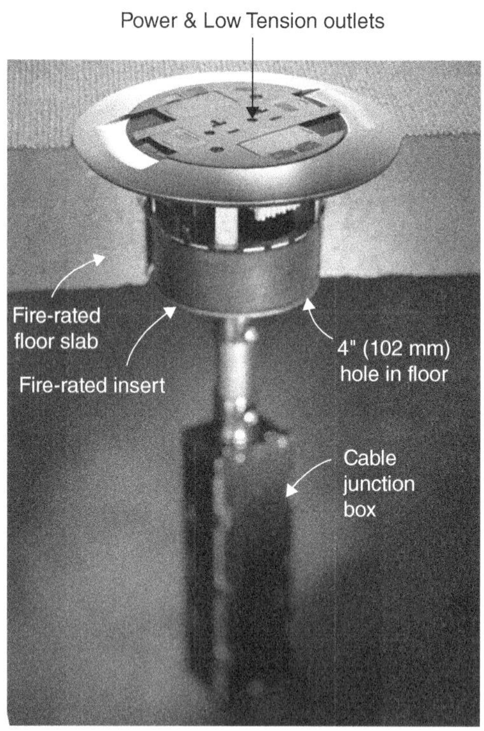

Figure 29.25 Typical poke-through electrical fitting mounts in a 4-in. [100-mm] hole. It is wired from underneath with the required power, telephone, signal, and data cables. Power and low-voltage cables are separated as required by code. Units are available prewired or suitable for field wiring, and adaptable for varying floor thicknesses. The floor fitting is provided with power, telephone, and data cable outlets as required for the specific installation.

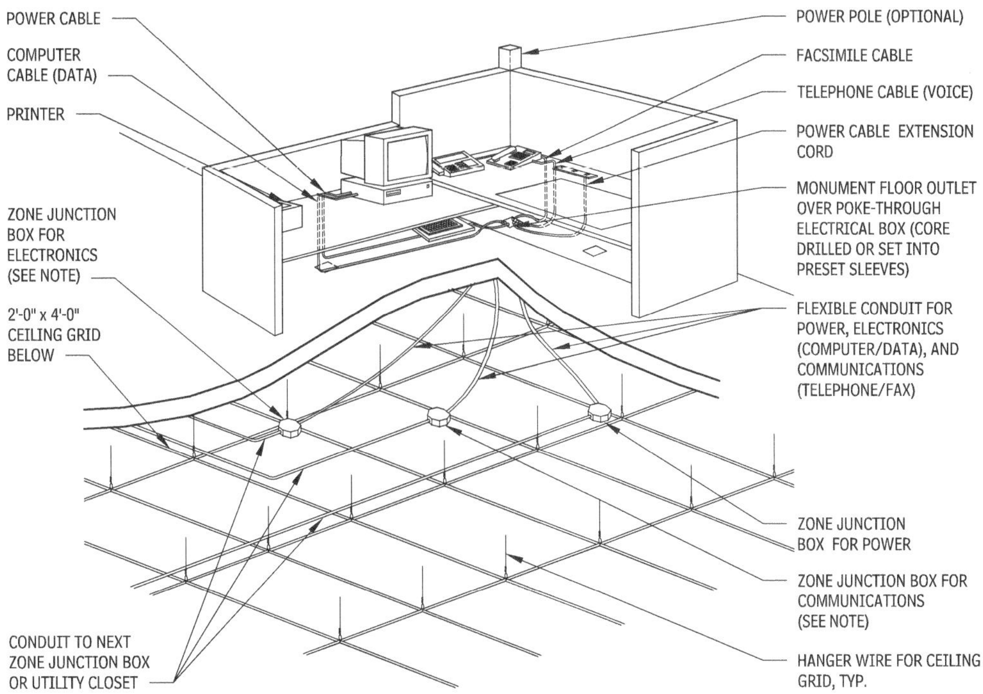

DESK EQUIPMENT LAYOUT

Computer and telephone cabling is often combined as an integrated voice/data cabling system, eliminating the need for three raceways, except when extra capacity is needed.

Figure 29.26 *Typical application of a through-the-floor system to provide power, telephone, and data service to a modern workstation. This drawing shows the electrical services being tapped at junction boxes in a hung ceiling conduit system on the floor below. The ceiling wiring system can also be a raceway network in lieu of the hard-wiring shown here. (From Ramsey and Sleeper, 2016.)*

To bring the various electrical and communication services to the user, in the absence of a well-designed overall floor raceway system, the installing contractor has one of four choices:

1. Channel the floor and install a conduit in the channel, connecting it to the nearest wall outlet. Patch the chased portion of the floor.
2. Install a surface floor raceway. The usefulness of this technique is very limited because it presents a tripping hazard and problems with routine floor cleaning.
3. Drill through the floor twice and connect the new outlet to a nearby existing floor or wall outlet via a conduit on the underside of the floor slab. Floor penetrations must be fireproof.
4. Drill through the floor and run a conduit in or on the ceiling below. For using this technique, special poke-through fittings are available that restore the fire rating of the slab (see Figure 29.25). These fittings are designed to carry all the electric services normally required at a workstation. They can then be connected to a single-location multiservice floor outlet group, as in Figure 29.25, or used to wire the partitions in a workstation, as in Figure 29.26.

All four of these methods have serious disadvantages; method 1 is labor-intensive, method 2 is unsightly and presents a safety hazard, methods 3 and 4 disturb the occupants below, and all four methods are generally inflexible and therefore unsuitable for spaces where reasonable changes in wiring and workstation location are anticipated. For these reasons, overall-access in-floor and underfloor raceway systems were developed and are widely used in high-grade commercial and institutional spaces.

The *NEC* recognizes three types of in-floor raceways:

1. underfloor raceways—Article 390
2. cellular metal floor raceways—Article 374
3. cellular concrete floor raceways—Article 372

All three types are applicable to a wide range of structures, and none may be used in corrosive or hazardous areas. The fundamental difference between them is that underfloor raceways are added onto the structure, whereas cellular floor raceways are part of the structure itself—and therefore have a pronounced effect on architectural coordination. (Underfloor duct systems antedate poke-through fittings, which are a relatively recent development.)

29.25 UNDERFLOOR DUCT

These raceways may be installed beneath or flush with the floor. They find their widest application in office spaces because their use permits placement of power, data, and signal outlets close to desks and other furniture, regardless of spatial layout. Underfloor duct systems were widely employed until the introduction of what may be called over-the-ceiling ducts (in contrast to under-the-floor ducts) and flat cable under-carpet wiring. These systems are discussed in Sections 29.29 and 29.30. The reason that alternative systems were developed is simply economic: underfloor duct systems are expensive and, because they are inflexible, being literally cast in concrete, they are frequently underutilized in one area while being inadequate in another. Before discussing the relative merits of systems, however, an understanding of what underfloor duct systems are and how they are assembled and utilized is necessary.

An underfloor duct system is simply an arrangement of parallel rectangular metal or heavy plastic raceways laid on the structural slab and covered with concrete fill. Access to the wiring in these distribution ducts is via inserts that connect to openings in the ducts and terminate in floor fittings for both power and signal/data wiring. Cable feeds to the distribution ducts are supplied by a second set of rectangular raceways called feeder ducts, usually laid at right angles to the distribution ducts.

In a single-level underfloor duct system, the distribution and feeder ducts are on the same level, and the interwiring between them is accomplished in junction boxes. The advantage of a single-level system is shallow concrete fill, normally 2½ to 3 in. [64–76 mm]. The limiting constraint of a single-level system is the junction box, which becomes more complex and multisectioned with an increasing number of ducts and wires. Newer systems utilize a one-piece triple-cell duct for both distribution and feeder ducts, with factory set inserts every 24 in. [610 mm] that straddle all three cells at once (Figure 29.27). By placing distribution ducts on 5 ft [1.5 m] centers with adequate crosswise feeder ducts and utilizing large flat junction boxes, a cost-effective installation adequate for all but the heaviest wiring demands can be assembled. For demanding areas, distribution ducts can be arranged to feed under-carpet cables (see Section 29.29).

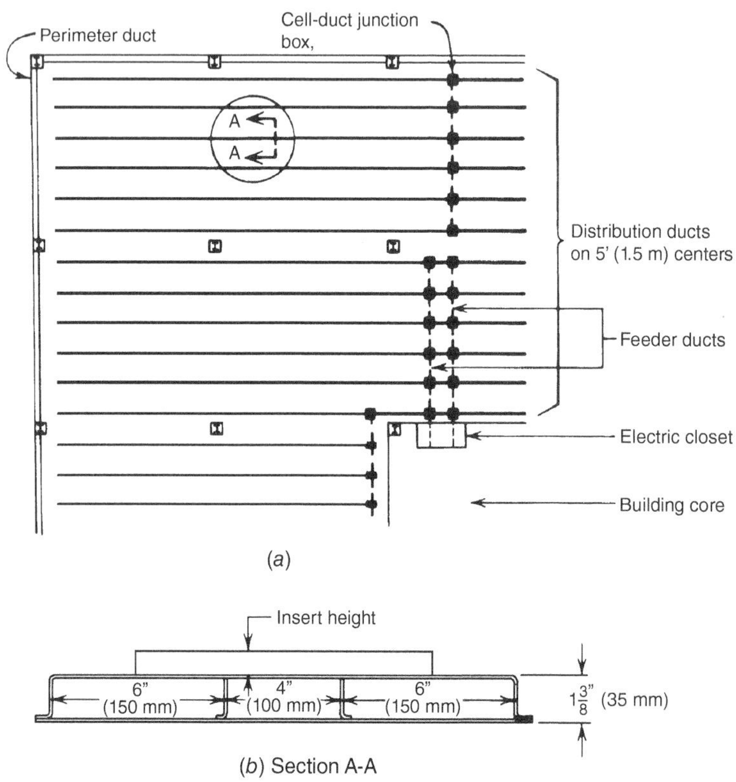

Figure 29.27 *Details of a single-level underfloor duct system utilizing three-section cell duct for both distribution and feeder ducts. (a) Portion of a typical large open floor space in a commercial facility. Distribution ducts may be placed as close as 5 ft [1.5 m] on-center to satisfy dense desk spacing. (b) Three-cell distribution duct utilizes a 4-in. [100-mm]-wide center cell (4.9 in.^{2} [3160 mm^2]) for power and either 3-in. [75-mm]-wide (3.7 in.^{2} [2386 mm^2]) or 6-in. [150-mm]-wide (7.4 in.^{2} [4773 mm^2]) outer cells for signal and data cabling. Minimum concrete fill depth is 2½ in. [64 mm], resulting in a minimum 1-in. [25-mm] cover over the distribution ducts. Service fittings are flush with the floor. (Courtesy of Square D Company.)*

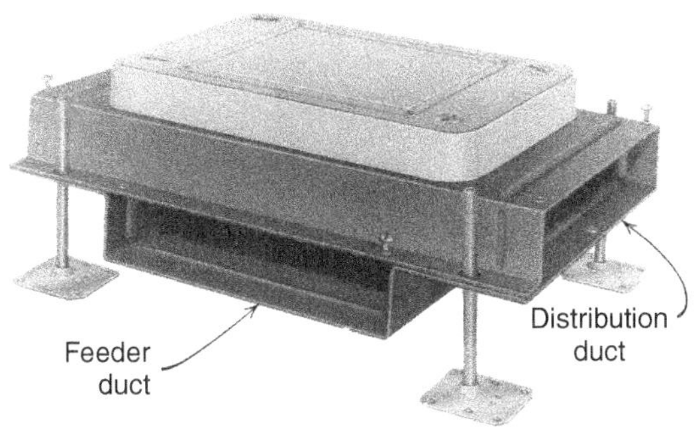

Figure 29.28 *Typical two-level junction box demonstrates the simplicity of the two-level system. (Courtesy of Square D Company.)*

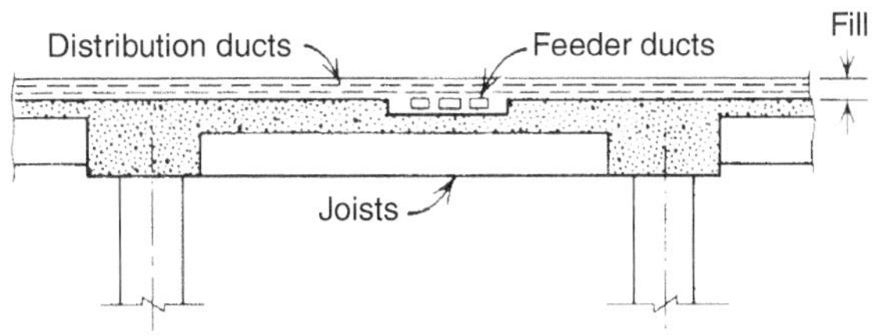

Figure 29.29 *Setting a two-level underfloor duct system. To avoid thickening the fill, a depression in the slab can accept feeder ducts. Ducts would be run near the bay center to avoid the negative steel required of joists near columns.*

Because the initial cost of a full underfloor system is high, an alternative arrangement utilizes only feeder ducts on approximately 25-ft [7.6-m] centers, with flat (under-carpet) cable box connectors spaced approximately every 20 ft [6.1 m] along the feeder ducts. The low-tension (voltage) portion of this system relies completely on flat telephone and data transmission cables, including optical fiber—also commonly referred to as fiber optic (FO)—cables. Because these cables are generally precut and factory terminated, the system requirements must be carefully analyzed (see Section 29.29) before committing to a complete under-carpet wiring system.

A two-level underfloor duct system is essentially the same as a single-level system except that the distribution ducts and feeder ducts are on different levels (Figure 29.28). This arrangement has the advantages of simplifying junction boxes and of giving the system unlimited feeder capacity, but the distinct disadvantage of requiring a minimum of 3 ⅝ in. [92 mm] of concrete fill. This additional slab thickness can frequently be avoided by depressing part of the slab to accommodate feeder ducts run under the distribution ducts, as shown in Figure 29.29.

A typical two-level system is illustrated in Figure 29.30. Here, the feeder ducts run above the distribution ducts and intersect at a specially

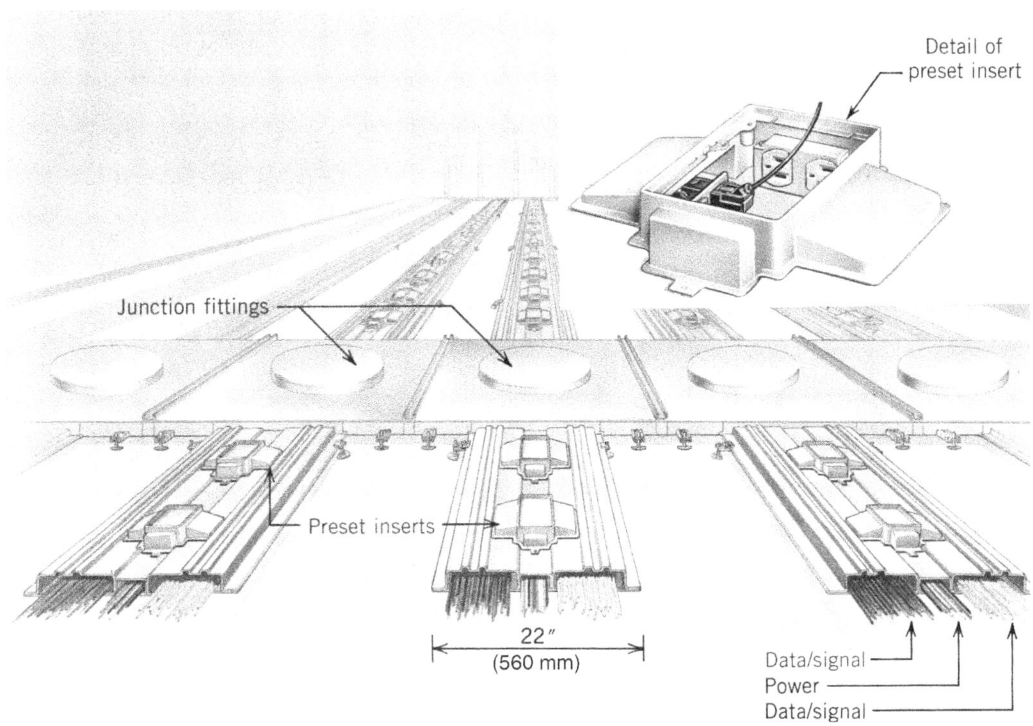

Figure 29.30 *Modified two-level underfloor duct system requires 3 in. [75 mm] of concrete fill for 2-in. [50-mm]-high ducts and 4 in. [100 mm] of fill for 3-in. [75-mm]-high ducts. Preset inserts (see detail) straddle all three cells and provide two duplex power receptacles, plus connections to both low-voltage cells. Distribution ducts recess into feeder-header ducts at junction boxes to reduce the overall height. (Courtesy of Walker Systems, a division of The Wiremold Company.)*

constructed junction fitting, into which the distribution ducts partially recess in order to reduce overall system height. The required concrete fill is either 3 or 4 in. [75 or 100 mm], depending upon the depth of the distribution cells (2 or 3 in. [50 or 75 mm]).

In all underfloor duct systems, the principal cable capacity bottleneck is usually the supply point to the feeder ducts. One solution to this problem uses a special feed arrangement at the panels (Figure 29.31). Another possibility is to subdivide large floor areas, supplying each via a system of multiple feed points arranged in closets or at wall panels. In such systems, care must be taken to ensure sufficient interconnection capacity between feed points because data networks are not only floor-wide but frequently building-wide.

Underfloor ducts may be cast into the structural slab in lieu of being installed in fill or topping, but the slab must be designed to accommodate them. The use of a fill or topping on the structural slab for an underfloor duct installation has these advantages:

1. Ducts can be run in any direction, without conflict with structural elements.
2. Finishing is simplified.
3. Coordination is simplified.
4. Formwork and construction sequence are simplified.

The disadvantages are:

1. Additional concrete increases costs directly by increasing the weight of the structure. This is particularly expensive in seismic designs.
2. The building height may be increased.

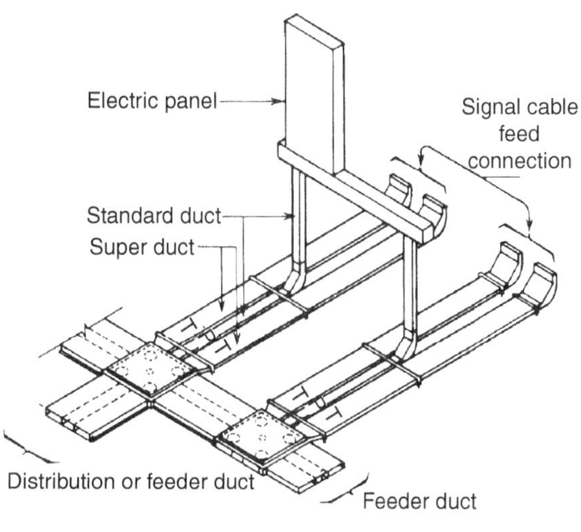

Figure 29.31 *Due to the large capacity of both distribution ducts and feeder ducts, central cable feed points—at electric closets, for instance—can cause bottlenecks. Illustrated is one possible solution, consisting of a double-duct feed arrangement. Signal cable would feed in from cable boxes (not shown). P is a power duct, T a signal duct. (Courtesy of Square D Company.)*

In retrofit jobs where underfloor duct is selected rather than one of the other floor or ceiling raceway systems, the ducts will be placed in a new (added) floor fill.

In conclusion, some general comments on the application of underfloor duct systems are in order. Underfloor duct systems are expensive. They can add 50% to the building's electric system cost, without consideration of the other construction costs involved. To justify their use, therefore, a building should meet these criteria:

1. There are open floor areas, with a requirement for outlets at locations removed from walls and partitions.
2. An under-carpet wiring system is inapplicable.
3. Outlets from ceiling systems are unacceptable.
4. Frequent rearrangement of furniture and other items requiring electrical and signal service is anticipated.

The facilities that may meet these criteria include many office buildings, museums, galleries (and other display-case spaces), high-cost merchandising areas, and selected areas in industrial facilities. Bear in mind that even in high-cost office construction, underfloor duct systems are difficult to justify economically unless the spatial/furniture layout will be likely to change. In doubtful cases, alternate arrangements can be planned and an intelligent choice made after costs and the impact on the building structure are studied.

29.26 CELLULAR METAL FLOOR RACEWAY

The underfloor duct system described previously is best applied to rectilinear arrangements. More free-flowing arrangements, such as those found in office landscaping layouts, require a fully accessible floor—if the floor is to be used for electrification. This may be provided by a cellular (metal) floor that is an integrated structural/electrical system. The floor can be partially or completely electrified. One of the many available structural element approaches is shown in Figure 29.32.

The cellular floor is part of the structural system and is designed accordingly. Electrical wiring is fed into the cells from header ducts and/or trenches that run perpendicular to the floor cells and constitute a system of underfloor ducts in themselves. The header ducts, in turn, are fed from electric panels and signal, data-transmission, and telephone cabinets in much the same manner as underfloor ducts are fed.

Three types of wiring systems generally run in separate floor cells and header ducts—electric power,

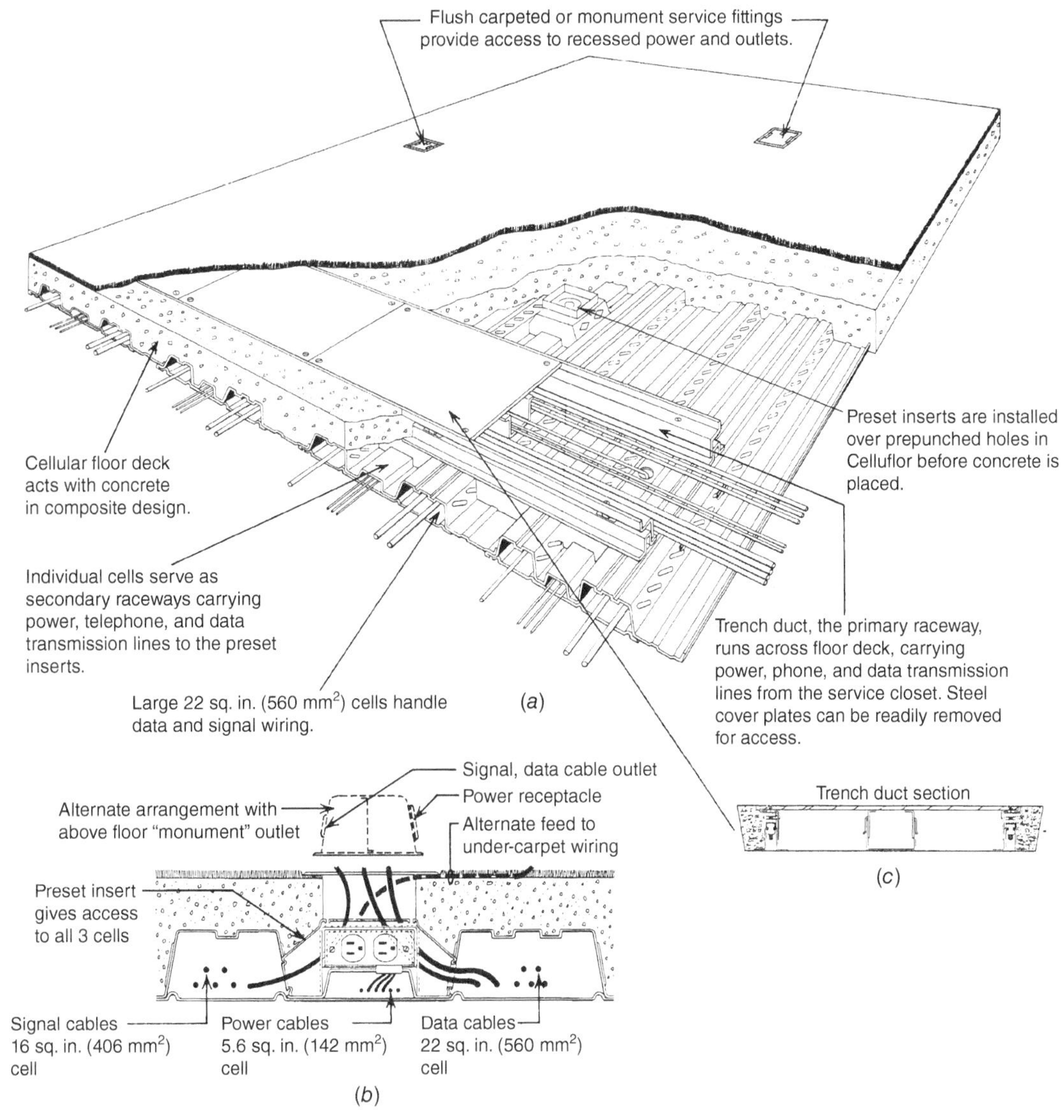

Figure 29.32 *(a) One of several options for electrified cellular floors. The floor cells are available in many designs, depending primarily upon the structural requirements. The trench (illustrated in c) that straddles the cells provides the electrical feeds through precut holes in the cells. The trench itself is completely accessible from the top and, when opened, exposes all the wiring and the cells below. (b) Activated preset insert. Note that the insert straddles the center (power) cell and provides access to the two adjoining low-voltage wiring cells. Power and signal wiring are completely separated at all times by metal barriers. If desired, a standard surface "monument" fitting can be mounted on the floor, or a connection can be made to under-carpet cables in lieu of the flush plate shown. When an insert is to be deactivated, the flush cover plate is simply replaced with a blank plate. (c) Section through a trench duct, which acts as a feeder for distribution ducts. The trench is available with or without bottom, in any required height, in widths from 9 to 36 in. [230 to 915 mm], and one, two, or three compartments, depending upon floor cell design and cabling requirements. (Courtesy of Walker Systems.)*

data-transmission wiring, and telephone and signal systems. The last two may be combined in a single cell only if the signal system voltage and power level are low and the local telephone company approves. A complete range of outlets and fittings is available.

29.27 PRECAST CELLULAR CONCRETE FLOOR RACEWAYS

This structural concrete system is similar to a cellular metal floor in application and has the same advantages: large capacity, versatility in that each cell is a potential raceway, and flexibility in outlet placement and movement. Here too, as with metal cell constructions, the first cost is higher than that of standard underfloor duct installation, although the life-cycle cost is frequently lower, depending upon space use and reconfigurations.

A raceway cell is defined in the *NEC* as a "single enclosed tubular space in a cellular metal or concrete floor member, the axis of the cell being parallel to the axis of the floor member." A feed for these cells is provided, as with metal cellular floor construction, by header ducts. Although header ducts are normally installed in concrete fill above the hollow-core structural slab, a header arrangement with feed from the ceiling below is also entirely practical. As with a metallic cellular floor, the cells can be used for air distribution and even for piping, although these elements are generally installed in a hung ceiling.

29.28 RAISED ACCESS FLOOR

Raised access (or full-access) floor systems are applicable to spaces with very heavy cabling requirements, particularly if frequent recabling and reconnection are required. The system provides rapid and complete access to an underfloor plenum. It was originally developed for data-processing areas that require large, fully accessible cable spaces and large quantities of conditioned air. The solution to both of these requirements is an infinite-access floor, usually constructed of panels—which may come in a variety of materials including lightweight die-cast aluminum, concrete, concrete with steel core, hollow steel, or wood—supported by a network of adjustable steel or aluminum pedestals. Panels are commonly available in modular sizes ranging from 18 × 18 in. to 36 × 36 in. [457 × 457 mm to 915 × 915 mm], and the floor level may be installed at a variety of heights, typically ranging from 6 to 48 in. [152 to 1219 mm], although taller pedestals are available. The subfloor space thus created can be used for cabling and also to carry conditioned air either in ducts or by using the entire space as an air plenum. (In the latter case, the wiring system must be suitable for air plenum use; as in Figure 29.22.) The construction is usually fireproof. Sufficient floor-to-floor height is necessary to accommodate the raised floor. This approach to electrical distribution may coordinate quite well with an underfloor air distribution (UFAD) system—see Chapter 16.

Where air requirements are limited or nonexistent and the floor is intended primarily for cabling, pedestals as short as 6 in. [152 mm] can be used, thus reducing ceiling height problems (Figure 29.33). In such access floor spaces, use of multiservice distribution modules and modular wiring avoids cable tangles and reduces labor costs significantly (Figures 29.33–29.36).

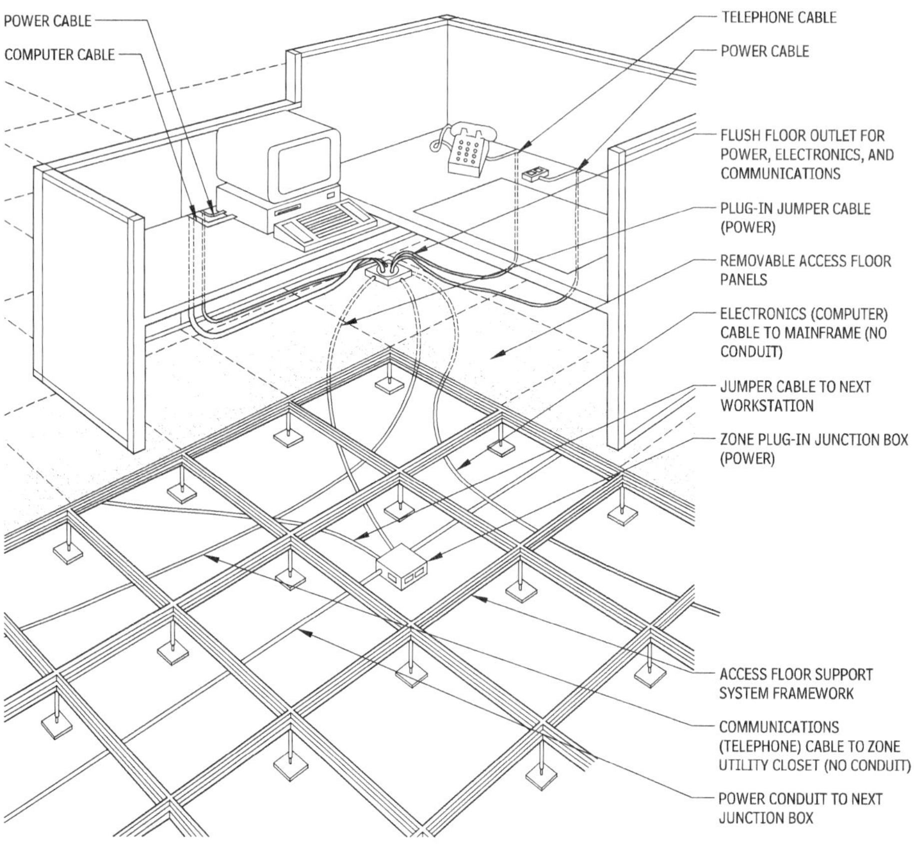

Figure 29.33 *Pictorial representation of a raised access floor designed to provide complete electrical, data, and signal services to a modern workstation layout. The infinite access and unlimited space are ideal for heavily wired, rapidly changing workstation arrangements. (From Ramsey and Sleeper, 2016.)*

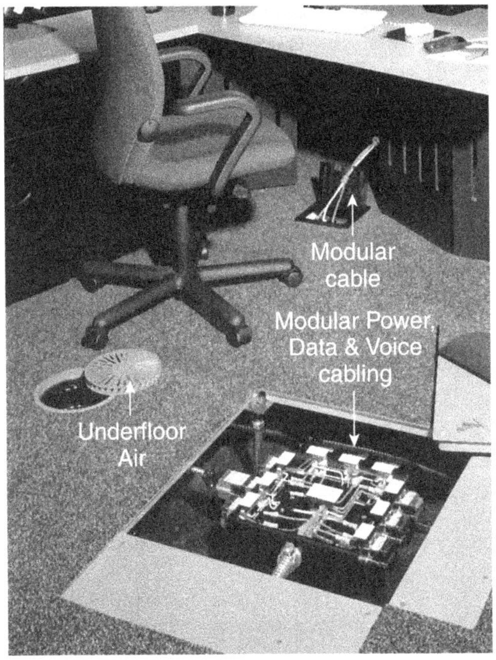

Figure 29.34 *Raised access floors provide infinite access to the myriad cables used in a modern commercial office. The modular wiring, using snap-on connectors and preassembled junction boxes (see Figure 29.36a), drastically reduces the labor cost of installation and frequent cabling changes. (Courtesy of Tate Access Floors, Inc.)*

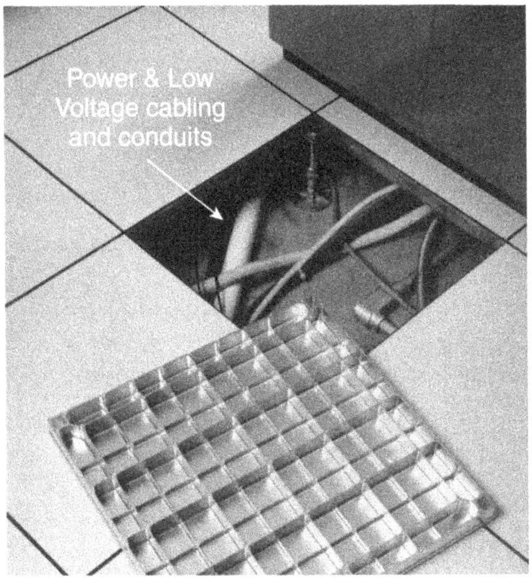

Figure 29.35 *Reinforced nonferrous (aluminum) floors are used frequently in high-tech applications requiring the large wiring capacity and convenience of full-access floors. Floor construction can be stringerless, as illustrated, or with stringers, as in Figure 29.36b. (Courtesy of Tate Access Floors, Inc.)*

29.29 UNDER-CARPET WIRING SYSTEM

An under-carpet (or undercarpet) wiring system, which is covered in *NEC* Article 324, was originally developed as both an inexpensive alternative to an underfloor or cellular floor system and as a means for providing a flexible floor-level branch circuit wiring system. Essentially, the system consists of a factory-assembled flat conductor cable (*NEC* type FCC), approved for floor installation only under carpet squares, plus the accessories necessary for connection to 120-V power outlets. The cable itself consists of three or more flat copper conductors, placed edge to edge and enclosed in an insulating

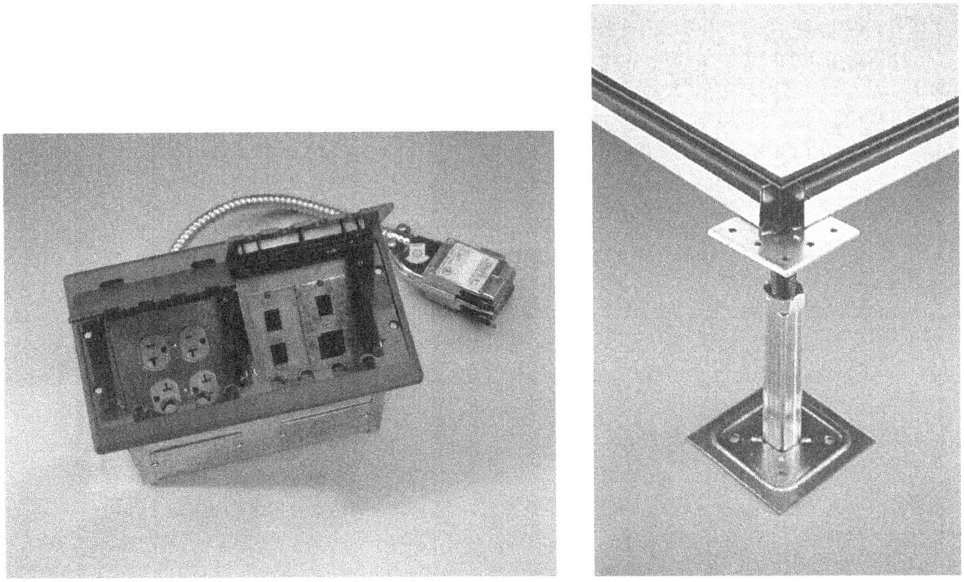

(a) (b)

Figure 29.36 *(a) Typical power/low-voltage floor box for use with a full-access floor. (b) Steel-bolted stringer support for access floor panels. Many stringer and stringerless designs are available (see Figure 29.35). (Courtesy of Tate Access Floors, Inc.)*

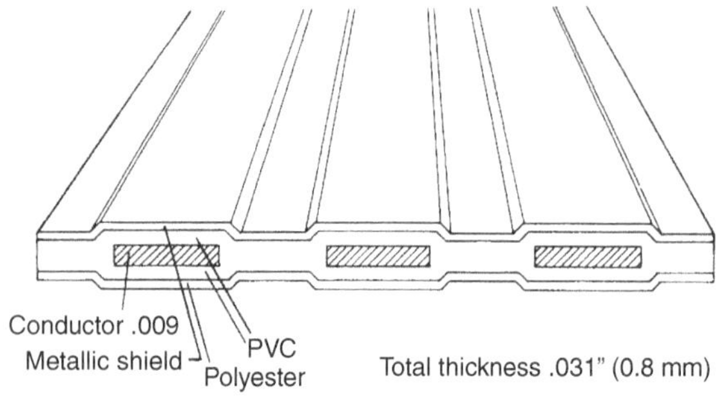

Figure 29.37 Schematic section through one design of NEC type FCC under-carpet cable. The copper conductors illustrated are the equivalent of No. 12 AWG. The PVC acts as insulation, and the polyester as both insulation and physical protection. All designs require a metallic top shield and a metallic or nonmetallic bottom shield for physical protection.

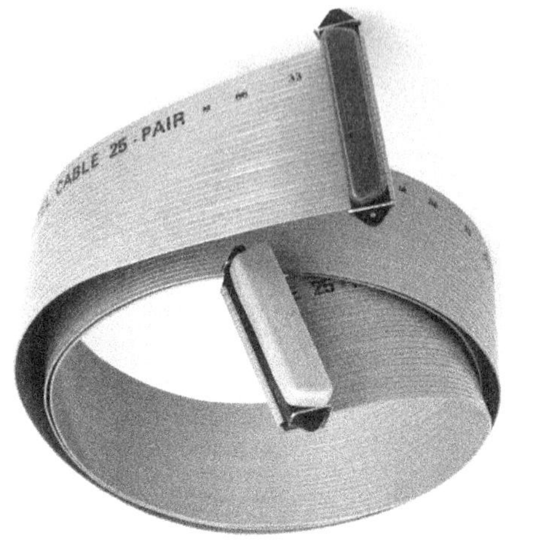

Figure 29.38 Preterminated 25-pair under-carpet telephone cable. These cables are commercially available in lengths from 5 to 50 ft [1.5 to 15 m] in 5-ft [1.5-m] increments. (With permission of AMP.)

material (Figure 29.37). The entire assembly is covered with a grounded metal shield, which, like a metal conduit, provides both physical protection and a continuous electrical ground path. In addition, a bottom shield is required, which is usually heavy PVC or metal.

The cable, when properly installed on a hard, flat surface, is approximately 0.03 in. [0.8 mm] high, and thus essentially undetectable when covered with carpet. Because carpet squares are designed to be readily removable, the entire system can be repositioned to meet changing furniture layout requirements with a minimum of disruption and no structural work. The cable is designed to carry normal physical loads such as office traffic and furniture placement without affecting its electrical performance.

The attractiveness and simplicity of the system led to the development of similarly designed flat, low-tension (voltage) cables for signal and communication wiring (Figure 29.38) and, more recently, both electrical and fiber optic cables and accessories for data transmission. Manufacturers also offer a complete line of junction fittings, connectors, adapters, and receptacles (Figure 29.39).

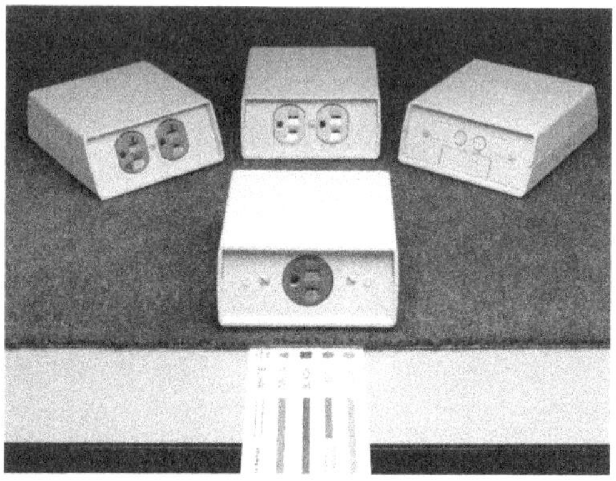

Figure 29.39 Typical components of an under-carpet wiring system. The under-carpet FCC power cable is shown without the metallic top shield required in actual installation. It is a color-coded, five-conductor cable (neutral, equipment ground, and three circuit conductors or two circuit conductors and an isolated ground conductor). The floor outlets shown are front, single-power outlet; rear (left to right), duplex power outlet (one of which has isolated ground); standard duplex power outlet; and combination data cable, communications, and telephone outlet. (Courtesy of Hubbell, Inc.)

The problems inherent in this type of on-the-floor wiring system, such as cable crossings, splicing, interfacing with round cable systems, interconnections at floor boxes and fittings, and feed connections from cabinets, underfloor ducts, floor cells, and through-the-floor fittings, have all been solved by a full line of manufactured devices designed for specific situations.

Because under-carpet wiring systems are separate and distinct from wire and conduit systems, they, like underfloor duct systems, are usually shown on a separate electrical plan. A small plan typical of this type is shown in Figure 29.40. Figures 29.41 and 29.42 are photographs of essential portions of such an installation. Note that a complex system such as that shown in Figure 29.42 requires recessing a floor box into the slab, which to an extent contradicts the essential simplicity and flexibility of an under-carpet system. Although these systems, at least in their simplest form, are particularly applicable to retrofit work, their low cost, combined with the inherent advantages of a flexible floor-level wiring system—particularly in open office areas—has made them a widely used first choice in new construction as well.

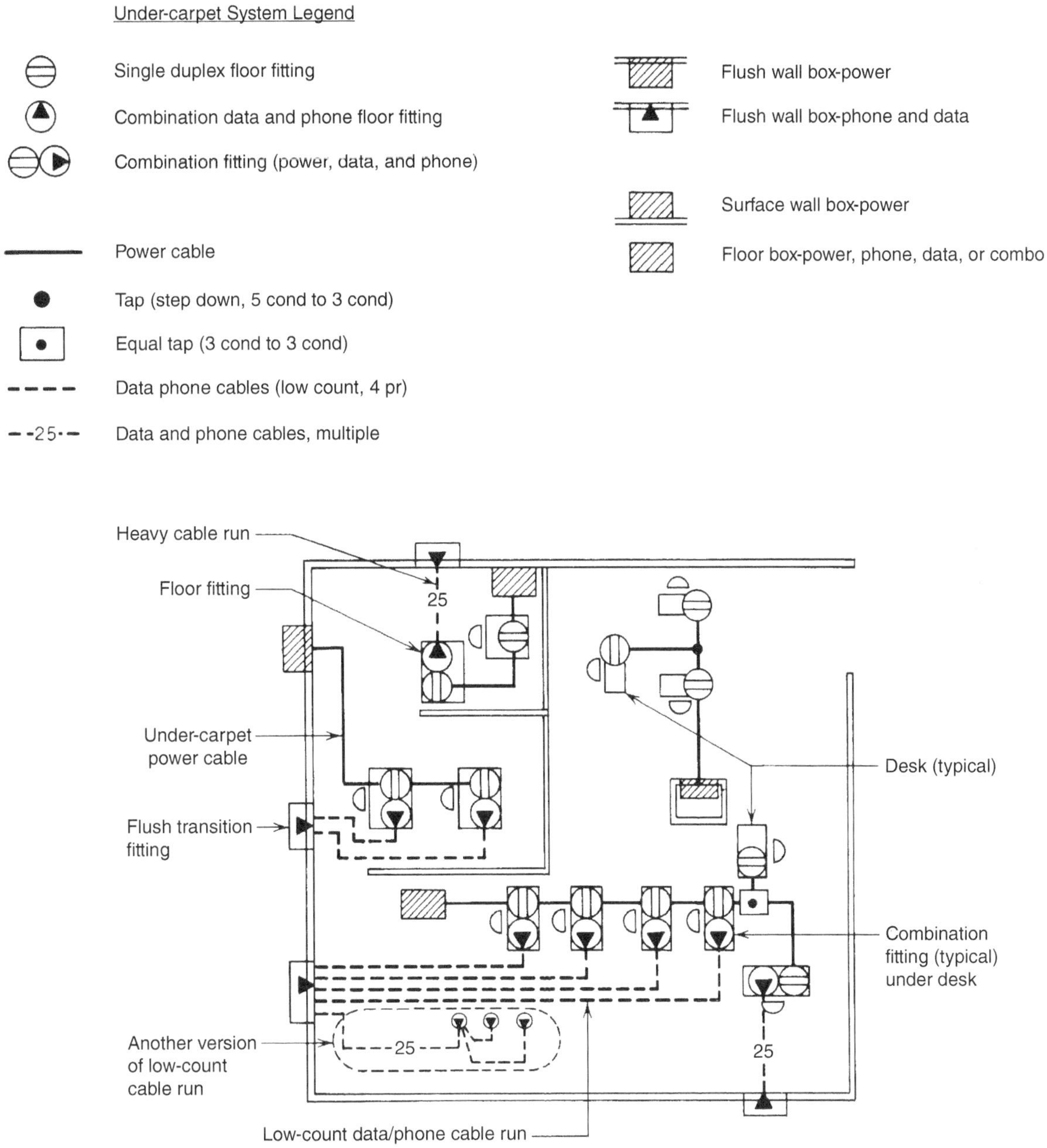

Figure 29.40 *Typical layout of under-carpet power and low-voltage/data cabling for a small office. A power, phone, and data cable connection on the floor is provided under or immediately adjacent to each desk (see Figures 29.41 and 29.42 for typical details). (Reproduced with permission of AMP.)*

29.30 CEILING RACEWAYS AND MANUFACTURED WIRING SYSTEMS

The need for flexibility in a facility's electrical system coupled with the high cost of underfloor electrical raceway systems encouraged the development of equivalent over-the-ceiling systems. These systems are actually more flexible than their underfloor counterparts because they energize lighting, provide power and telephone facilities, and even supply outlets for the floor above, in addition to permitting very rapid layout changes at low cost. This last characteristic is particularly desirable in stores where frequent display changes necessitate corresponding electrical facility changes. Beyond the extreme flexibility made possible by the ceiling raceway system, it has the additional advantage that the system itself, not being cast in concrete like its underfloor counterpart, can be altered

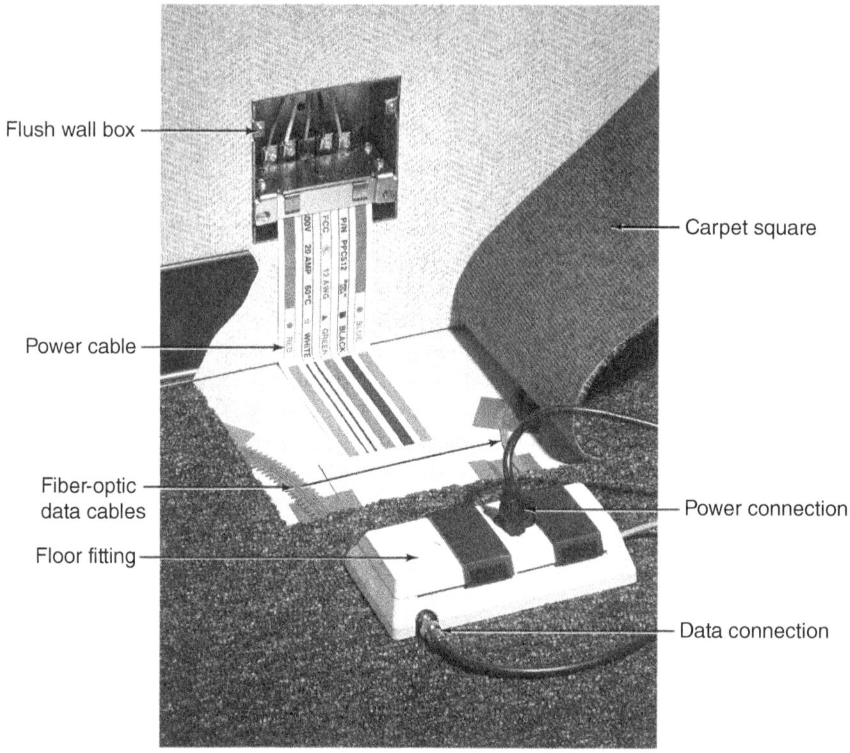

Figure 29.41 *Under-carpet flat power cables connect to round supply cables in a flush wall box and then extend to a combination power/low-voltage/data floor fitting. Data cables are connected to the combination floor fitting from their system boxes, either individually or via other floor fittings. (Courtesy of Walker Systems, a division of The Wiremold Company.)*

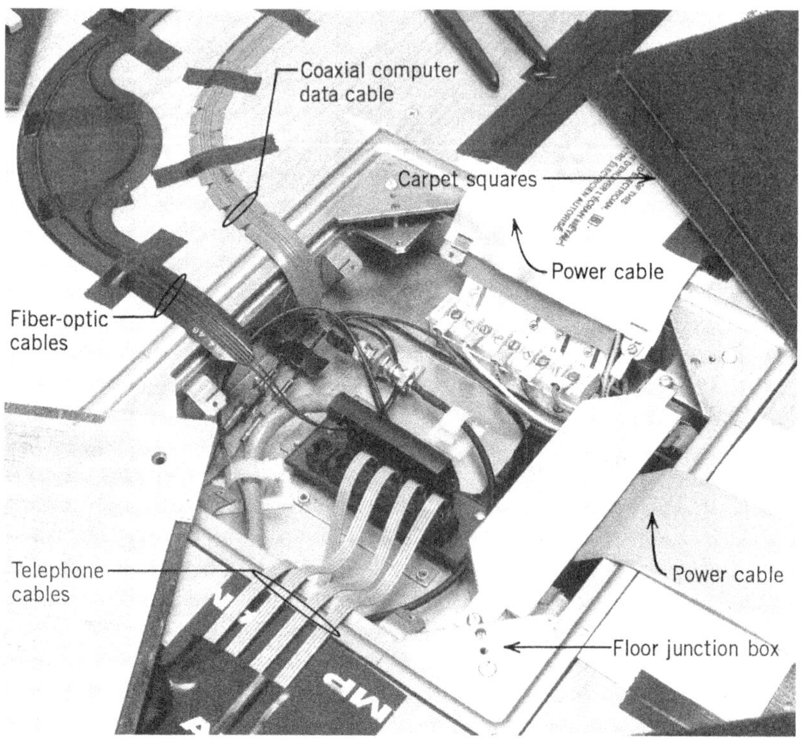

Figure 29.42 *Four service floor-junction boxes handle under-carpet power, telephone, low-voltage electric, and FO cables. The floor box measures approximately 14 in. [356 mm] square and 2 in. [50 mm] deep. It can accommodate up to four 3-phase power circuits, eight 25-pair telephone/data cables, 24 coaxial cables, or, as illustrated, a mixture of cable types. (Reprinted with permission of AM.)*

at will. Thus, not only layout changes (as mentioned previously) but also changes in space function can readily be accommodated. This is a particularly important characteristic in merchandising and educational facilities, where space function can be repeatedly changed during the course of a building's life.

Details vary among manufacturers, but the systems are essentially the same and, in principle, resemble underfloor systems. A typical system is constructed of metallic or nonmetallic surface-type raceways arranged in a tree formation (i.e., large trunk [header] raceways feed multiple smaller branch [distribution] raceways, and so on). The raceways are hung in the ceiling plenum from the concrete slab above. The hung ceiling must consist of lift-out panels because this type of wiring system is not permitted in spaces rendered inaccessible by the building structure. Header ducts (large area raceways) are fed from electrical panels and from signal, data, and telephone cabinets in the electrical and low-voltage wiring service closets, respectively. Data headers are normally larger than the power header and can carry other low-voltage, low-power signal wiring as well. Distribution ducts (laterals) tap onto the headers. These laterals act as subdistribution raceways, feeding lighting fixtures and data, signal, telephone, and power outlets on the same floor and, via poke-through fittings, outlets on the floor above.

The standard method for extending wiring from ceiling plenum raceways to floor-level or desk-level signal and power outlets uses vertical multisection raceways fed from the top (see Figure 29.43). These service poles are available in a large variety of designs, finishes, and cross-sectional raceway areas and are easily installed in almost any location. They may be

Figure 29.43 Typical floor-to-ceiling electrical/communication raceway poles. Units are available in a wide variety of shapes, sizes, and cross-sectional configurations in aluminum and steel. (Courtesy of Hubbell Premise Wiring.)

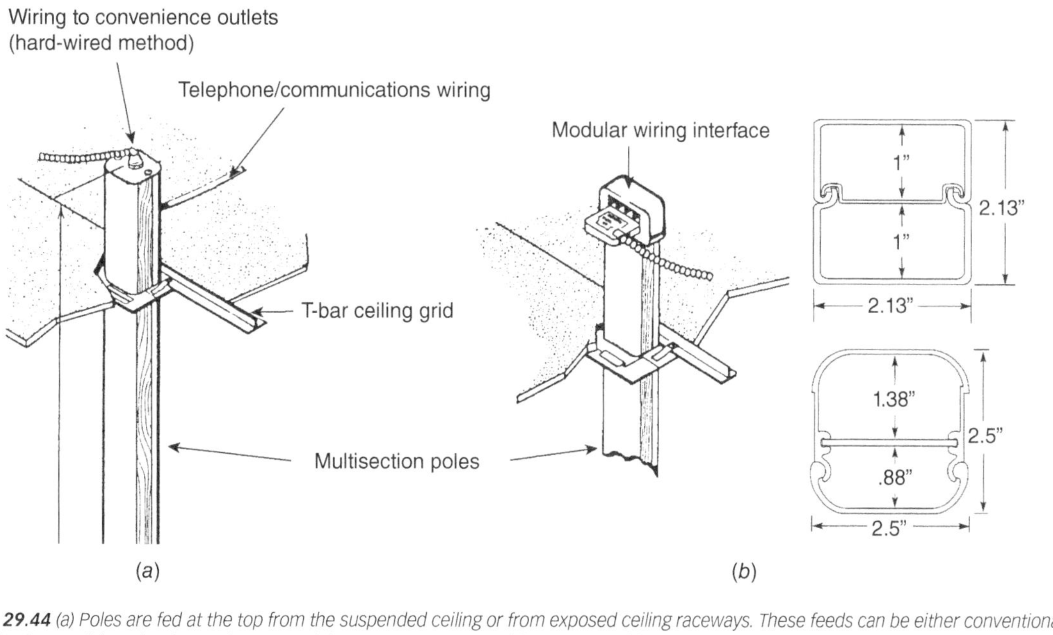

Figure 29.44 (a) Poles are fed at the top from the suspended ceiling or from exposed ceiling raceways. These feeds can be either conventional hard-wired or modular (plug-in), as shown. Modular connectors are used for power and low-voltage (telephone, communication, data) wiring. (b) Two of the many cross-sectional configurations available are illustrated. Other designs divide the pole into three sections to suit the specific application. (Courtesy of Hubbell Premise Wiring.)

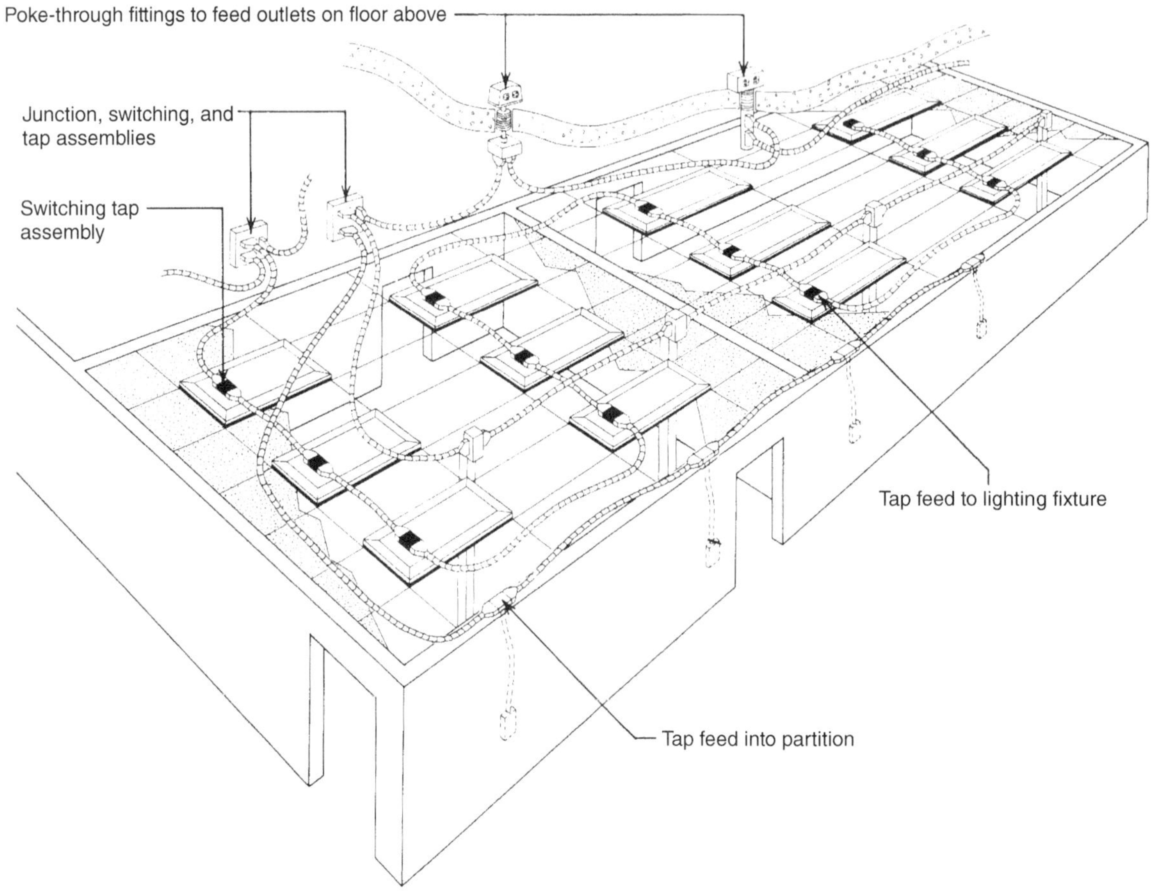

Figure 29.45 *Manufactured modular wiring assemblies are used for tap connections to feed ceiling lighting fixtures or any other ceiling connection, as well as a complete range of junction, switching, tap, and poke-through units. (Courtesy of AFC Cable Systems.)*

prewired and usually contain several power outlets, a telephone connection, and possibly data cable outlets. The result in a hung-ceiling office area or an exposed ceiling slab area (Figure 29.44) is certainly less elegant than that of a floor-level wiring system, but for most users it is satisfactory, and its low cost compared to any type of floor-duct system is a prime redeeming feature.

When electrical connections to poles, lighting fixtures, receptacles, and communication/data outlets are made with hard wiring, considerable field labor is required, with a corresponding high cost. Furthermore, the relative permanence of such wiring lessens the inherent flexibility of the raceway system. To solve both problems, a number of manufacturers have developed a line of modular branch-circuit wiring elements. These, covered in *NEC* Article 604 under the very logical name of manufactured wiring systems, consist of metal-clad or armored cable sets terminating in polarized plugs. The polarization prevents accidental interconnection of low-voltage, 120- and 277-V systems. Ceiling raceways can be equipped with matching receptacles, and connection to fixtures, poles, and other devices becomes a simple matter of plug insertion (Figures 29.45 and 29.46).

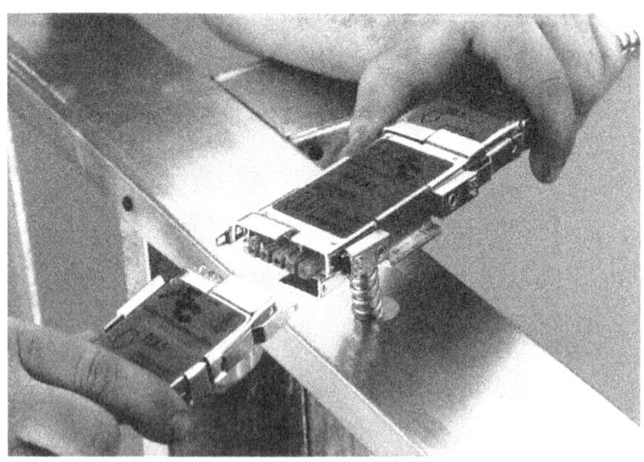

Figure 29.46 *Manufactured modular wiring connectors simplify installation and reduce labor costs initially and during any subsequent rearrangements. (Courtesy of AFC Cable Systems.)*

The result is a wiring system of extreme flexibility in which even extensive changes can be made very rapidly with minimal disruption and virtually no mess.

Manufactured wiring systems are only permitted in accessible areas, for logical reasons. They are also applicable to access floor spaces, as seen in Figures 29.33 to 29.36. The additional cost of manufactured wiring elements is frequently offset by the labor savings, even upon initial installation and certainly after one or two field changes. Cable sets are available for power (120 and 277 V), telephone, and all types of low-voltage signal equipment. The cables must be approved for use in conditioned-air plenums and suspended ceilings.

To take full advantage of the potential labor-cost savings inherent in the system, field labor must be minimized. This is accomplished by the factory equipping all utilization equipment, including lighting fixtures, with appropriate plug-in connectors.

References and Resources

NFPA. 2022. *2023 NFPA 70: National Electrical Code (NEC)*. Quincy, MA: National Fire Protection Association.

Ramsey, C. G., and H. R. Sleeper. 2016. *Architectural Graphic Standards*, 12th edn. New York: John Wiley & Sons.

CHAPTER 30

Electric Wiring Design

THERE ARE NUMEROUS POSSIBLE SOLUTIONS TO EACH WIRING DESIGN PROBLEM—some good, some fair, and some poor. Experience guides designers to a solution that best suits the job, within the context of establishing the most cost-effective design, while considering the project design intent and criteria. This chapter opens with a discussion of typical design criteria and continues with wiring design. An electrical design solution intended for construction must meet all the requirements of local and national building codes. In the United States, building codes generally refer to *NFPA 70:* the *National Electrical Code* (*NEC*) published by the National Fire Protection Association (NFPA), as the minimum standard for all electrical design; thus, its requirements are made legally binding on the design team. The electrical design data and procedures presented in this chapter, and in other sections of this book, are based upon *NEC* requirements and are, to the extent possible, current. However, they are intended only as informative materials. Code requirements can change fairly rapidly; the *NEC*, for example, is revised every three years. The designer is strongly advised to obtain and study the latest edition of all codes having jurisdiction, including the *NEC*, to ensure that a proposed electrical design meets the requirements of the current and applicable codes.

30.1 GENERAL CONSIDERATIONS

(a) Flexibility

Every wiring system should incorporate sufficient flexibility in branch circuitry, feeders, and panels to accommodate all reasonably probable patterns, arrangements, and locations of electric loads. The degree of flexibility to be incorporated depends largely upon the type of facility. Thus, laboratories, research facilities, and small educational buildings typically require far more flexibility than do residential, office, and fixed-purpose industrial installations. Part of designing for flexibility includes providing for building and electric service expansion, as experience has demonstrated that most facilities will grow, both physically and in electrical demand. Overdesign, however, is as bad as underdesign, being wasteful of money and resources, both initially and in operation. An understanding of the client's needs, as defined in an owner's project requirements (OPR) document, can greatly assist in reaching a reasonable compromise between current and future needs.

(b) Reliability

The reliability of electrical power within a facility is determined by two factors: the incoming utility service and the building's electrical system. The service record of the utility should be studied along with the economic impacts of a power outage to determine whether, and to what extent, standby power equipment is justified (see Chapter 28 and Section 30.21). Emergency equipment requirements for the safety of building occupants are established by local, state, and national building codes.

Beyond the electrical service point, the reliability of power is entirely dependent upon the wiring and distribution system. Here, too, economic studies must be made to determine the quality of equipment and

Mechanical and Electrical Equipment for Buildings, Fourteenth Edition. Walter Grondzik, Daniel Overbey, and Alison Kwok.

Companion website: www.wiley.com/go/meeb14e

the extent of redundancy (duplicate equipment) to be installed. The subject of reliability is complex, and only a few general principles are noted here.

1. The reliability of an electric system is only as good as that of its weakest element. Therefore, it may be necessary to provide redundancy at anticipated weak points in the system.
2. The electrical service and the building distribution system act together. An extremely reliable (and expensive) service is of little use if the power cannot reach the desired points of usage.
3. Critical loads within the facility should be defined to determine how best to serve them reliably. This can be done by establishing reliable power paths or by furnishing individual standby power packages for such loads. The latter course is often chosen for health-care facilities and critical loads in other buildings.

(c) Safety

Although rigid adherence to the requirements of the *NEC* and other applicable NFPA codes ensures an initially safe electrical installation, the designer must be alert to the potential for electric hazards caused by misuse or abuse of equipment or by equipment failure. Also, attention to the physical size of equipment during design can eliminate the often-encountered hazards caused by obstruction of access spaces, passages, closets, and surfaces with electric equipment. Finally, lightning protection should also be considered under the heading of safety; this topic is discussed in Chapter 26.

(d) Economic Considerations

This area of concern can readily be divided into two frequently interrelated items: first cost and life-cycle cost. All other factors being equal, first cost depends largely upon whether an owner is interested in the minimum first cost or the minimum overall ownership cost. These two costs frequently stand in inverse relationship to one another (exceptions are mentioned later). Low first-cost equipment generally results in higher energy costs, higher maintenance costs, and shorter life—with a resulting higher life-cycle cost.

The electrical energy cost related to building operations is directly related to energy consumption, with one exception. That exception is the utility company's demand charge—which is discussed in Chapter 27 along with means for minimizing this cost. Here, there is a coincidence, whereby a reduction in both first cost and operating costs can be achieved. Load-leveling equipment permits the electrical distribution system to be sized without consideration of coincident load peaks, thus resulting in smaller equipment that operates more efficiently—near its full-load capacity. All other reductions in electric energy cost flow directly from the corresponding reduction in energy consumption.

(e) Energy Considerations

These concerns are complex, involving consideration of energy codes and budgets, energy conservation and energy efficiency techniques, and energy control approaches.

Buildings designed under a number of scenarios may be effectively subject to energy budget limitations (expressed in Btu/ft^2/yr [kWh/m^2 per year]). Although the lion's share of this budget is taken by heating/cooling and lighting systems, the electrical distribution system is also subject to limits in some codes/standards. Most important among these is *ASHRAE Standard 90.1* (ASHRAE, 2022). Detailed energy conservation considerations are discussed in Section 30.5.

(f) Space Allocations

The general impression that electrical equipment is small and easily concealed is accurate only for wire and conduit. Panels, motor control centers, busduct, distribution centers, switchboards, transformers, and so on can be large, bulky, noisy, and highly sensitive to tampering and vandalism. Thus, space allocations must be concerned with ease of maintenance, ventilation potential, expandability, centrality (to limit the length of distribution runs), limiting access to authorized personnel, and noise control, in addition to the fundamental consideration of space adequacy.

(g) Special Considerations

These depend upon the nature of the facility and may include items such as security, central and/or remote controls, interconnection with other facilities, and the like.

30.2 LOAD ESTIMATING

When initiating the electrical system design for a facility, it is necessary to estimate the total building load in order to plan for such spaces as transformer rooms, conduit chases, and electrical closets. An estimated load is also required by the local power company well in advance of the start of construction. An exact load tabulation can be made only after the design is completed, but because this is often several months later, a good preliminary estimate is

TABLE 30.1 Electric Load Estimating[a]

I	II	III	IV	V	VI
			Volt-Amperes per ft^2 [per m^2][b] Air Conditioning[f]		Ten-Year Percent Load Growth
Type of Occupancy	Lighting[c,d]	Misc. Power[e]	Electric	Nonelectric	
Auditorium					
General	1.0–2.0 [10.8–21.5]	0	12–20 [129–215]	5–8 [54–86]	20–40
Stage	20–40 [215–430]	0.5 [5.4]			
Art gallery	2.0–4.0 [21.5–43.1]	0.5 [5.4]	5–7 [54–75]	2.0–3.2 [21.5–34.5]	20–40
Bank	1.5–2.5 [16.2–26.9]		5–7 [54–75]	2.0–3.2 [21.5–34.5]	30–50
Cafeteria	1.0–1.6 [10.8–17.2]	0.5 [5.4]	6–10 [65–108]	2.5–4.5 [26.9–48.4]	20–40
Church and synagogue	1.0–3.0 [10.8–32.3]	0.5 [5.4]	5–7 [54–75]	2.0–3.2 [21.5–34.5]	10–30
Computer area	1.2–2.1 [12.9–22.6]	2.5 [26.9][g]	12–20 [129–215]	5–8 [54–86]	50–200
Department store					
Basement	3–5 [32.3–53.8]	1.5 [16.2]			
Main floor	2.0–3.5 [21.5–37.7]	1.5 [16.2]	5–7 [54–75]	2.0–3.2 [21.5–34.5]	50–100
Upper floor	2.0–3.5 [21.5–37.7]	1.0 [10.8]			
Dwelling (not hotel)					
0–3000 ft^2 [0–280 m^2]	3.0 [32.3]	5.0 [53.8]	—	—	50–100
3000–120,000 ft^2 [280–11,150 m]	0.4 [4.3]	0.15 [1.6]	—	—	—
Above 120,000 ft^2 [>11,150 m]	1.5–2.5 [16.2–26.9]	2.0 [21.5]	—	—	—
Garage (commercial)					10–30
Hospital	1.0–3.5 [10.8–37.7]	1.5 [16.2]	5–7 [54–75]	2.0–3.2 [21.5–34.5]	40–80
Hotel	1.0–2.0 [10.8–21.5]	0.5 [5.4]			
Lobby	1.0–1.5 [10.8–16.2]	1.0 [10.8]	5–8 [54–86]	2.0–3.5 [21.5–37.7]	
Rooms (no cooking)	2.0–3.0 [21.5–32.3]	5–20 [54–215]	3–5 [32–54]	1.5–2.5 [16.2–26.9]	30–60
Industrial loft building	1.2–2.2 [12.9–23.7]	0.5 [5.4]	—	—	50–100
Laboratories	1.5–3.0 [16.2–32.3]	1.5 [16.2]	6–10 [65–108]	2.5–4.5 [26.9–48.4]	100–300
Library	1.0–2.0 [10.8–21.5]	0.5 [5.4]	5–7 [54–75]	2.2–3.2 [23.7–34.5]	30–40
Medical center	1.5–3.5 [16.2–37.7]	2.5 [26.9]	4–7 [43–75]	1.5–3.2 [16.2–34.5]	50–80
Motel	1.2–2.5 [12.9–26.9]	0.5 [5.4]	—	—	30–60
Office building	1.5–2.8 [16.2–30.1]	2.5 [26.9]	4–7 [43–75]	1.5–3.2 [16.2–34.5]	50–80
Restaurant			6–10 [65–108]	2.5–4.5 [26.9–48.4]	20–40
School	2.0–2.5 [21.5–26.9]	2.0 [21.5]	3.5–5.0 [38–54]	1.5–2.2 [16.2–23.7]	50–80
Shops	2.0–3.5 [21.5–37.7]	0.5 [5.4]			
Barber and beauty	2.0–3.0 [21.5–32.3]	0.5 [5.4]			
Dress	2.0–3.0 [21.5–32.3]	0.5 [5.4]	5–9 [54–97]	2–4 [22–43]	40–80
Drug	2.0–3.5 [21.5–37.7]	0.25 [2.7]			
Miscellaneous goods	0.25–1.0 [2.7–10.8]	—	4–7 [43–75]	1.5 to 3.2 [16.2–34.5]	
Hat, shoe, specialty	0.3 [3.2]	—			
Warehouse (storage)	0.25 [2.7]	—	—	—	10–30
In the buildings listed except single dwellings:					
Halls, closets, corridors	0.5 [5.4]	—	—	—	
Storage spaces	0.25 [2.7]	—	—	—	

[a]Values assume energy-conservation techniques are applied.

[b]These values do not include allowance for future loads.

[c]The values given in *NEC* Article 220 are minimum values for calculation of electric feeder sizes.

[d]See Chapter 18 for a discussion of power limits.

[e]These values are based on experience and must be applied judiciously.

[f]Includes the loads of air-handling equipment and pumps.

[g]This value does not include the power used by the computer (which can vary widely depending upon the type being used).

required. Such an estimate can be made from the figures given in Table 30.1. These values represent averages. When it appears that a building will have heavier or lighter loads because of lighting levels, a design (such as a green building) that exceeds energy codes, load-management equipment, the aggressive use of passive systems, or other factors, the values should be modified accordingly. The figures given in the individual categories should be added without application of demand or diversity factors in order to obtain the maximum load for which the building service equipment must be sized—in the absence of electric load-control (load-leveling) equipment. (Noncoincident loads, such as heating and cooling in the same space, are not combined.)

At this point in the design process, an analysis can be made to determine the feasibility of incorporating electric load-control equipment into the facility. Inputs to such a study include the utility's complete rate schedule, including all penalty clauses, a detailed analysis of the building's equipment load patterns, and any external constraints such as maximum loads imposed by power and energy budgets. Equipment load patterns must be carefully analyzed because they determine load "sheddability." Thus, for kitchen equipment, load interruption may be unacceptable, but shifting of cooking time by half an hour may be quite feasible. For heating, ventilation, and air-conditioning (HVAC) equipment, building thermal inertia and extending maximum and minimum temperature and humidity limits may permit considerable latitude without adverse effects. As explained in Chapter 27, the degree and duration of load shedding are a function of the type of control equipment utilized. To repeat what was previously stated—load control affects maximum demand, with only minor effect on total energy consumption. The external constraints referred to are the energy budgets recommended or required by codes, legislation, and funding bodies.

After a load-control analysis is complete, or simultaneously with such an analysis, a building energy consumption analysis must be performed. This may be done manually, but numerous computer programs are readily available that not only increase accuracy but also permit consideration of more factors. The results of this analysis will indicate whether the target electrical energy budget is being met. If not, loads must be modified by reconsideration of proposed systems (and perhaps system design criteria), by incorporating energy conservation devices and techniques into the electrical system design, and by preparing energy-use guidelines that will be applied when the building is occupied. Because this last item depends for success upon the day-to-day voluntary actions of the building's operators and occupants, experience has shown that it should not be considered a major conservation factor. Ongoing building commissioning can be a powerful assist in changing this situation. Conservation measures are covered in Section 30.5.

The electrical loads in any facility can be categorized as follows:

1. Lighting
2. Miscellaneous power, which includes data-processing equipment, convenience outlets, and small motors
3. HVAC equipment
4. Plumbing and other piping-based systems
5. Mechanized transportation equipment and fixed material-handling equipment
6. Kitchen equipment
7. Special equipment

Category 1 is covered by column II of Table 30.1. Note, however, that lighting loads are carefully prescribed in *ASHRAE Standard 90.1* (ASHRAE, 2022) and that these prescribed loads should be used where this standard has jurisdiction (which is most of the United States). Contradictions between requirements in applicable codes/standards must be adjudicated by local authorities before a design is established. Consult the current version of *ASHRAE Standard 90.1* for details of prescribed lighting loads.

Category 2, column III, includes, in addition to receptacles and small motors, such items as desktop computers and data-processing terminals, including all peripheral equipment, plus plug-in heaters, water fountains, and so on.

Category 3, column IV, includes all loads imposed by the HVAC equipment. This includes fuel pumps, boiler motors, air-handling units, exhaust fans, and so forth. Also included in column IV, for air-conditioning loads, are refrigeration compressors. This item is omitted in column V because such air conditioning utilizes absorption machines that do not use electricity for primary power. When cooling is not anticipated, the HVAC load is still appreciable because of heating and ventilation requirements. A rough estimate for the load in this situation would be two-thirds of the loads in column V.

Category 4 includes all loads associated with water and sanitary systems, including house water pumps, air compressors and vacuum pumps, sump pumps and ejectors, well pumps and fire pumps, water heaters, and pneumatic tubes, plus special items such as display fountain pumps. Also included in this category are electric loads associated with fixed piping-based systems such as cooking gas, medical gas piping, distilled water systems, and so on.

Because these loads vary widely with local conditions and with specific facility design as much as with the type of facility, an estimate cannot be made on a volt-amperes per unit area basis by type of building. Thus, no value is included in Table 30.1. If actual data cannot be used, it is helpful to remember that plumbing/piping electrical loads are relatively small, rarely exceeding 20% of the HVAC system electric load (although, for the most part, they are unrelated to the HVAC components).

Category 5, vertical transportation, is related to floor area in some types of buildings but not in others. A close estimate of power and energy requirements can be made by the project's elevator consultant. Lacking this, a fair estimate can be made from the data given in Chapters 33 and 34. In addition to elevators, escalators, moving walkways, and ramps, these loads should include dumbwaiters, horizontal and vertical conveyors, trash and linen transport systems, automated container delivery, and fixed conveyors. These loads are readily available from a

TABLE 30.2 Nominal Service Size in Amperes

Nominal service sizes are 100 A, 150 A, 200 A, 400 A, 600 A

	Area in ft² [m²]				
Facility	**1000 [93]**	**2000 [186]**	**5000 [465]**	**10,000 [929]**	**Remarks**
Single-phase, 120/240 V, 3-wire					
Residence	100 A	100 A	150 A	—	Minimum 100 A
Store[a]	100 A	150 A	—	—	
School	100 A	100 A	150 A	—	
Church[a]	100 A	150 A	—	—	
3-phase, 120/208 V, 4-wire					
Apartment House	—	—	100 A	150 A	
Hospital[a]	—	—	200 A	400 A	
Office[a]	—	100 A	400 A	600 A	
Store[a]	—	100 A	400 A	600 A	
School	—	100 A	100 A	200 A	

[a]Fully air-conditioned using electric-driven compressors. Based upon values in Table 30.1.

material transport consultant, who is frequently also the elevator consultant.

Category 6, kitchen equipment, is also not included in Table 30.1, although present in all restaurants, most hospitals, and some office and religious-use buildings. The reason for this omission is that the primary power for the major load, cooking equipment, may be either gas or electric. Other large-energy-use equipment such as dishwashers can be electric, gas, or steam-fed. Furthermore, no correlation can be made between the facility type, floor area, and load, even if electrically powered, because population and schedule are also major factors. When kitchens are planned, a kitchen consultant or other experienced planner can supply an estimate of the electric power requirements.

Category 7, special equipment, is so varied that no values can be listed. This category includes such items as laboratory equipment, shop loads, display area loads, floodlighting, canopy heaters, display window lighting, industrial loads, and so on. Data for these loads must be gathered for individual items of equipment and added to the foregoing totals.

Table 30.2 gives a tabulation of service entrance size in amperes, based on single- and 3-phase service for typical occupancies. These figures are intended for rough estimating purposes and should be adjusted after the design is completed.

30.3 SYSTEM VOLTAGE

Several voltage systems (arrangements) are commonly available in the United States and Canada. (See *NEC* Article 230 for requirements.)

(a) 120-V, Single-Phase, 2-Wire

This arrangement (Figure 30.1) is used for the smallest facilities, such as outbuildings, and isolated small loads of up to 6 kVA. The load is calculated by multiplying current and voltage. For 60-A service, which is the normal limit for this type of service, no more than 50 A are usually drawn. Thus:

$$VA = 120 \times 50 = 6000\,\text{VA} = 6\,\text{kVA}$$

The nominal system voltage is 120 V, although it is also referred to as 110 V and 115 V.

(b) 120/240-V, Single-Phase, 3-Wire

This system (Figure 30.2) is for somewhat heavier loads. The code requires that all one-family residences with six or more 2-wire circuits or a net computed load (by code calculation) of 10 kVA or more have a minimum of 100-A, 3-wire service. Service disconnect for 100-A service would be a 100-A, 2-pole, solid neutral switch, fused at no more than 80% of the

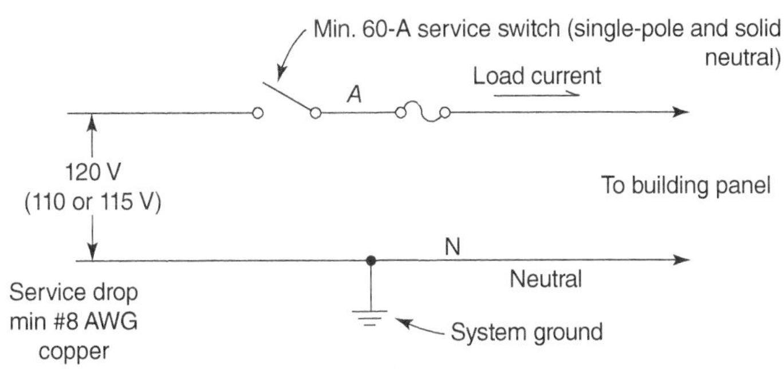

Figure 30.1 *Shown is a 120-V, single-phase, 2-wire service. This is also the arrangement of a typical branch circuit.*

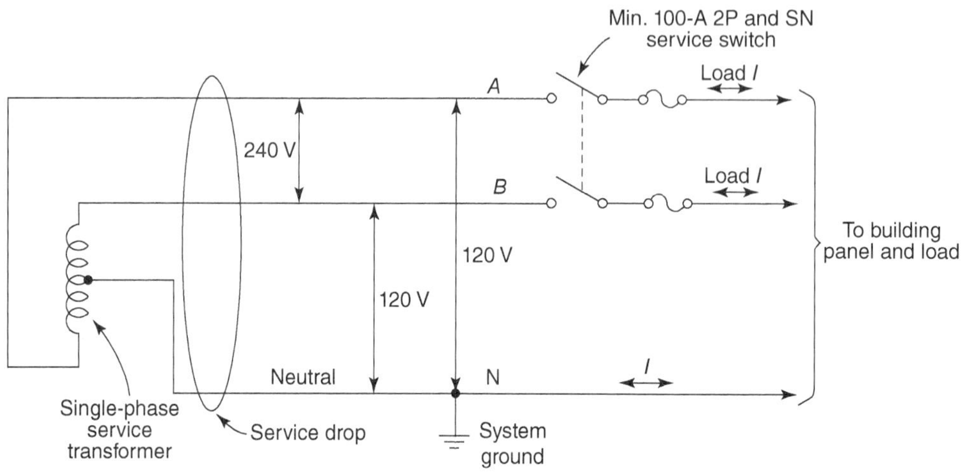

Figure 30.2 *Shown is a 120/240-V, single-phase, 3-wire service. The single-phase transformer, which is normally located on the power company's pole along a street or road, is center-tapped to establish a neutral. The neutral connection is always grounded.*

rating, or 80 A. This is usually written 100A, 2P & SN, 80AF. This service is used principally for residences, small stores, and other occupancies where the load does not exceed 80 A or 19.2 kVA. The load is calculated thus:

$$\text{kVA} = \frac{V \times I}{1000} = \frac{240 \times 80}{1000} = 19.2$$

Although it may appear otherwise, the neutral carries no more than full-load current. Note that each "hot leg" of the 3-wire system carries line current. Thus, total load can also be calculated:

$$\text{load kVA} = \text{twice load on each line}$$

Assuming a balanced 80-A load:

$$\begin{aligned}\text{total kVA} &= 2 \times 80\,A \times 120\,V \\ &= 2 \times 9600\text{ VA} = 19{,}200\text{ VA} \\ &= 19.2\text{ kVA}\end{aligned}$$

If the loads were unbalanced with, say, 30 A in one line and 50 A in the other, the total load would be:

$$\begin{aligned}120 \times 30 + 120 \times 50 &= 3600 + 6000 = 9600\text{ VA} \\ &= 9.6\text{ kVA}\end{aligned}$$

Actual system voltages can be 120/240, 115/230, or 110/220, although 120/240 is the accepted industry standard. Normal receptacle and lighting loads at 120 V are connected between line and neutral. Heavier loads, such as for a clothes dryer or an electric stove, are connected between phase lines at 240 V.

For example, to find the line currents caused by the three loads shown in Figure 30.3:

1. 120-V, 1200-W iron, line A to neutral
2. 120-V, 1440-W hair dryer, line B to neutral
3. 240-V, 4800-W clothes dryer, line A to line B

Calculate:

$$I_1 = \frac{1200}{120} = 10\text{ A}$$

$$I_2 = \frac{1400}{120} = 12\text{ A}$$

$$I_3 = \frac{4800}{240} = 20\text{ A}$$

Note that the neutral only carries the unbalance of 2 A (see Figure 30.3).

$$\begin{aligned}&\text{Total current in line A} = 20 + 10 = 30\text{ A} \\ &\text{Total current in line B} = 20 + 12 = 32\text{ A} \\ &\text{Total current in neutral N} = 2\text{ A} \\ &\text{Total load} = 120(30) + 120(32) \\ &\qquad = 3600 + 3840 = 7440\ VA\end{aligned}$$

or

$$\text{Loads are } 1200 + 1440 + 4800 = 7440\text{ VA}$$

Because the loads are almost entirely resistive heating loads (small hair dryer motor and approximately a 1/6-hp [0.12 kW] dryer motor), the entire load

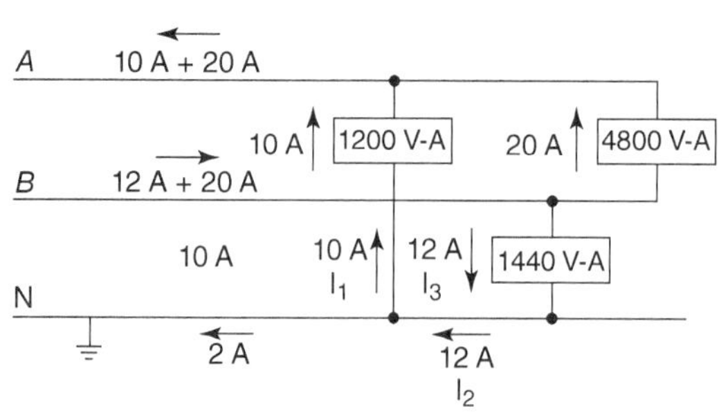

Figure 30.3 *An example 120/240 V, 3-wire distribution. Note that the neutral carries the difference in current between the A and B legs and therefore a maximum that is equal to the current in one of the legs (when the other is zero).*

has a power factor of almost 1.0. This, however, affects only energy calculations because equipment is sized, in general, by kVA capacity. The 120/240-V, single-phase system is derived from a center-tapped, single-phase transformer.

(c) 120/208-V, Single-Phase, 3-Wire

Although this system (Figure 30.4) appears similar to the one described in Section 30.3*b*, it is really part of a 3-phase system. It is most often found within a building that takes 3-phase service rather than constituting a service voltage arrangement. It is used to serve a load that does not require 3-phase, 4-wire but does require a voltage higher than 120 V. Calculation of loads and line currents is considerably more complex than in Section 30.3*b* because of the 120° phase displacement between phases A and B. Here, as before, the neutral carries no more than line current, regardless of whether the system is balanced.

(d) 120/208-V, 3-Phase, 4-Wire

This system (Figure 30.5) is a widely used 3-phase arrangement applicable to all facilities except very large ones. In the latter, lengths of feeders and sizes of loads become so great that a higher system voltage is required. In this system, 120-V loads such as lighting, computers and accessories, receptacles, and so on, are fed at 120 V by connection between a phase leg

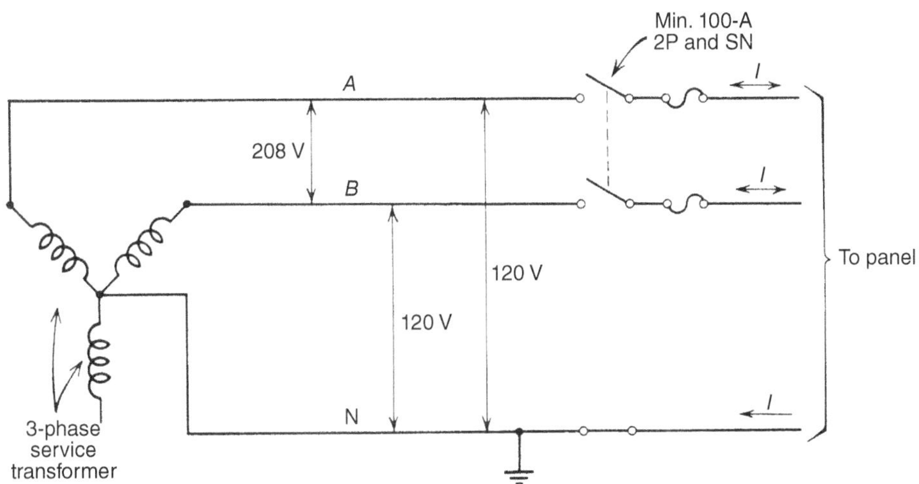

Figure 30.4 *Shown is a 120/208-V, single-phase, 3-wire service. This arrangement comprises two-thirds of the full 120/208-V, 3-phase, 4-wire connection shown in Figure 30.5.*

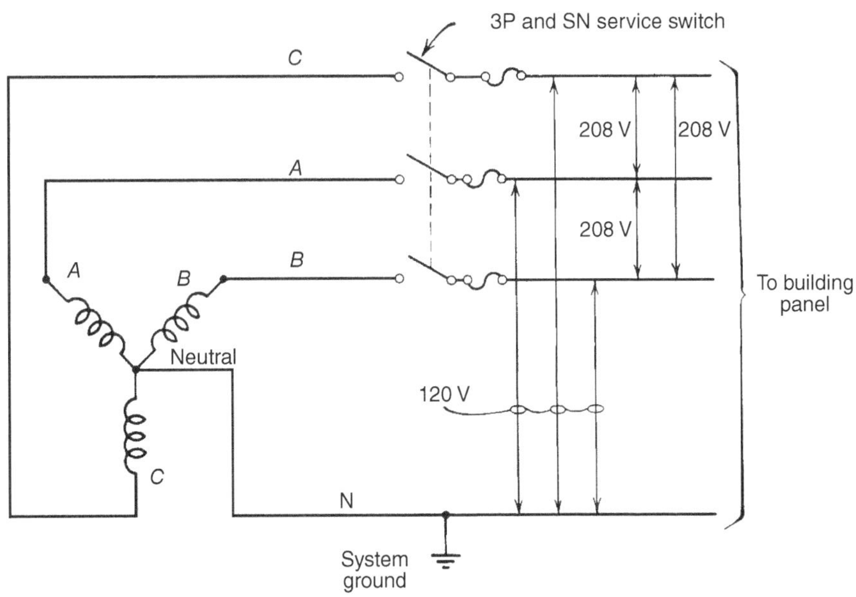

Figure 30.5 *Shown is a 120/208-V, 3-phase, 4-wire system. The neutral connection is connected to the system ground and is not broken by the service switch.*

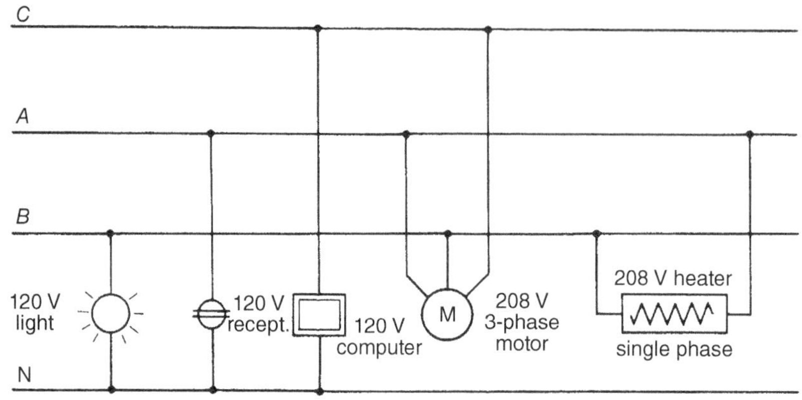

Figure 30.6 The flexibility of the 120/208-V system is illustrated; this accounts for its wide usage. Loads are shown schematically. In practice, the loads are fed via protective devices in the panel (omitted here for clarity). A, B, C, and N represent the panel buses.

(Figure 30.6) and neutral. Motors larger than ½ hp [0.37 kW] and all 3-phase loads are fed at 208 V by connection to the 3-phase legs. Single-phase, 208-V loads such as heaters are accommodated as described in Section 30.3*c* by connection between two phase legs. Such loads are often referred to as 2-pole loads, alluding to the 2-pole current breakers used to feed them.

(e) 277/480-V, 3-Phase, 4-Wire

This system (Figures 30.7 and 30.8) is applicable to large buildings (either horizontally or vertically) where lighting is principally fluorescent and/or high-intensity discharge (HID), and the 120-V load does not exceed one-third of the total load. It provides 277 V for fluorescent and HID lighting and 480 V for machinery. Small (3- to 25-kVA) dry-type transformers are used to step down from 480 V for 120-V loads. This system is ideally suited to multistory office buildings and large single-level or multilevel industrial buildings. Cost savings are generated by the smaller feeder and conduit sizes and smaller switchgear, which more than offset the additional cost of step-down transformers for the 120-V loads.

As an example of the economies possible with this system, consider the wiring required for a 15-kW heater. At 3-phase, 208 V:

$$I = \frac{15{,}000\ W}{\sqrt{3} \times 208\ V} = 42\ \text{A}$$

requiring No. 8 AWG wire (45-A capacity). At 480 V:

$$I = \frac{15{,}000\ W}{\sqrt{3} \times 480\ V} = 18\ \text{A}$$

requiring only No. 12 AWG wire (20-A capacity). This voltage system is also referred to as 265/460 V and 255/440 V.

Voltages above 150 V to ground are generally avoided in residential branch circuits but may be used in commercial and industrial facilities within the guidelines established by the *NEC*.

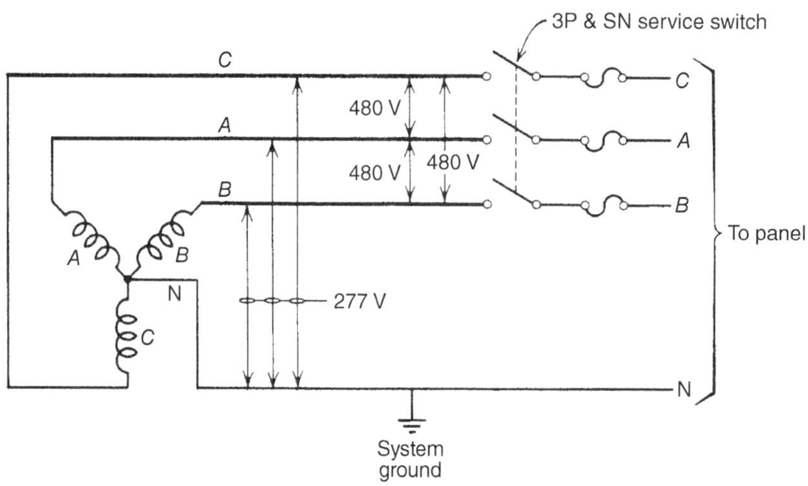

Figure 30.7 Shown is a 277/480-V, 3-phase, 4-wire service system. The system is identical to the 120/208-V system shown in Figure 30.5 except for the voltages.

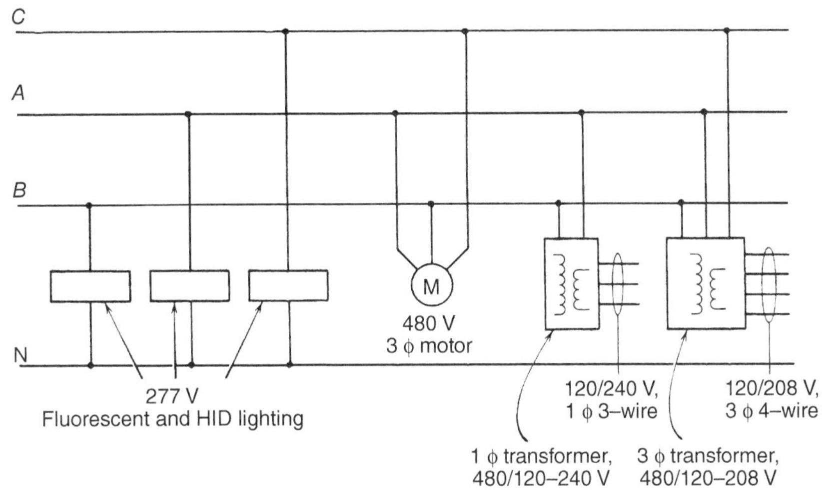

Figure 30.8 *Normal load arrangement for a 277/480-V system is illustrated. The lighting can be fluorescent or HID. Transformers, either single-phase or 3-phase, supply 120 V for receptacles and 208 V for loads requiring that voltage.*

(f) 2400/4160-V, 3-Phase, 4-Wire

This system is used only in very large commercial or industrial buildings with machinery requiring these voltages. The cost of running these voltages within a building is high because of *NEC* requirements and the inherently higher cost of 5-kV equipment. A detailed cost and engineering analysis by an electrical engineer is required for each case. Voltages above this level are widely used in large industrial plants but are beyond the scope of this book.

Reference was made earlier to the varied "voltages" assigned to the same voltage system. Thus, at the lowest level, there is 110, 115, and 120 V; at the next level, 200, 208, 220, 230, and 240 V; at the next, 255, 265, and 277 V; and finally, 440, 465, and 480 V. These voltage terminology differences arise because of the historic difference between transformer voltage standards, which establish the system voltage, and motor voltage standards, which govern utilization voltage (Table 30.3). Present motor voltage standards are in agreement with the system voltage standards. Note the close correspondence between motor standard voltage and system voltage with a (normal) 4% feeder voltage drop. Thus, on 240- and 480-V systems, respectively, 230- and 460-V motors are suitable.

Difficulties arise in the application of 230- and 240-V motors to a 208-V system. Although motors will operate at plus or minus 10% voltage, 230- and 240-V motors should not be used on 208-V systems. Instead, motors specially wound for 200 V should be specified. A brief summary of the effects of undervoltage is given in Table 30.4.

30.4 GROUNDING AND GROUND-FAULT PROTECTION

The vast majority of secondary wiring systems are solidly grounded. The reasons for this arrangement

TABLE 30.3 Standard System and Utilization Voltages[a]

System Voltage (Transformers)		Utilization Voltage[b] (Motors)	
Nominal	**With 4% Drop[b]**	**Current Standard[c]**	**Obsolete Standard**
120	115.2	115	110
208[d]	199.7	200	208
240[d]	230.4	230	220
480	460.8	460	440
600	576.0	575	550

Source: From ANSI C84.1-2020.

[a]When specifying transformers, use system voltages; for motors, use utilization voltage.

[b]Note that utilization voltage corresponds to a 4% drop from the system voltage, well within the normal motor tolerance.

[c]Motors for 208-V systems are rated 200 V. Motors for 240-V systems are rated 230 V. They cannot be used interchangeably without seriously affecting motor performance.

TABLE 30.4 Effects of Undervoltage on Utilization Equipment[a]

Load	10% Undervoltage
Lighting	
Incandescent	Output reduced 30%
Fluorescent	Output reduced, poor start
Mercury vapor	Low output, poor start
LED	Little effect
Motors	20% lower torque, hotter operation, reduced life, overloading
Heaters	20% reduction in output
Small tools	Stalling, low power

[a]Computers and peripherals are generally supplied with voltage-regulating power supplies that make them tolerant of ±10% variation in steady-state supply voltage. Conversely, they are highly intolerant of rapid supply voltage fluctuation.

are several and varied. Among them are the following reasons.

1. To prevent sustained contact between a low-voltage secondary system and a high-voltage primary system in the event of an insulation failure. Such contact could cause a breakdown of the secondary system insulation and severely endanger system users.
2. To prevent single grounds from going unnoticed until a second ground occurs, which would extensively disable the secondary system.
3. To permit locating ground faults with ease.
4. To protect against voltage surges.
5. To establish a neutral at zero potential for safety and for reference.

Points 2 and 4 are highly technical, and a full explanation is beyond the scope of this book. Point 5 requires that the neutral in a single secondary system should have the following conditions applied.

1. It should never be interrupted by switches or other devices.
2. It should be connected to ground only at one point—the service entrance.
3. It should be color-coded (in the United States) white, natural gray, or by three continuous white stripes on any insulation color other than green, along the entire conductor length, for easy recognition.

A typical service-grounding diagram is given in Figure 30.9. Universal acceptance and use of grounded secondary 120-V systems introduce another shock

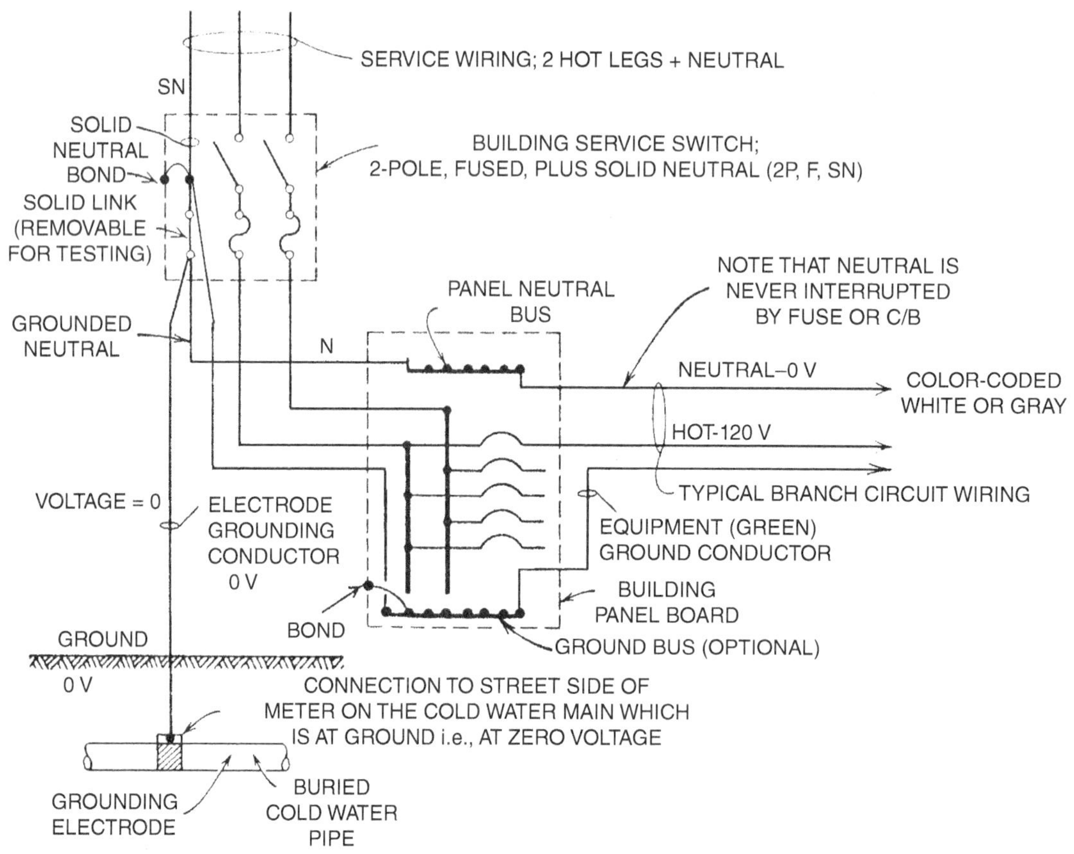

Figure 30.9 *Typical service-grounding arrangement. Note that the grounded neutral is unbroken throughout. The ground bus, if present, is separate and distinct from the neutral bus, and both are grounded at the service entrance point. For the sake of clarity, not all bonding connections, which are required to ground all non-current-carrying metal parts of the electrical system (boxes, enclosures, conduits, etc.), are shown.*

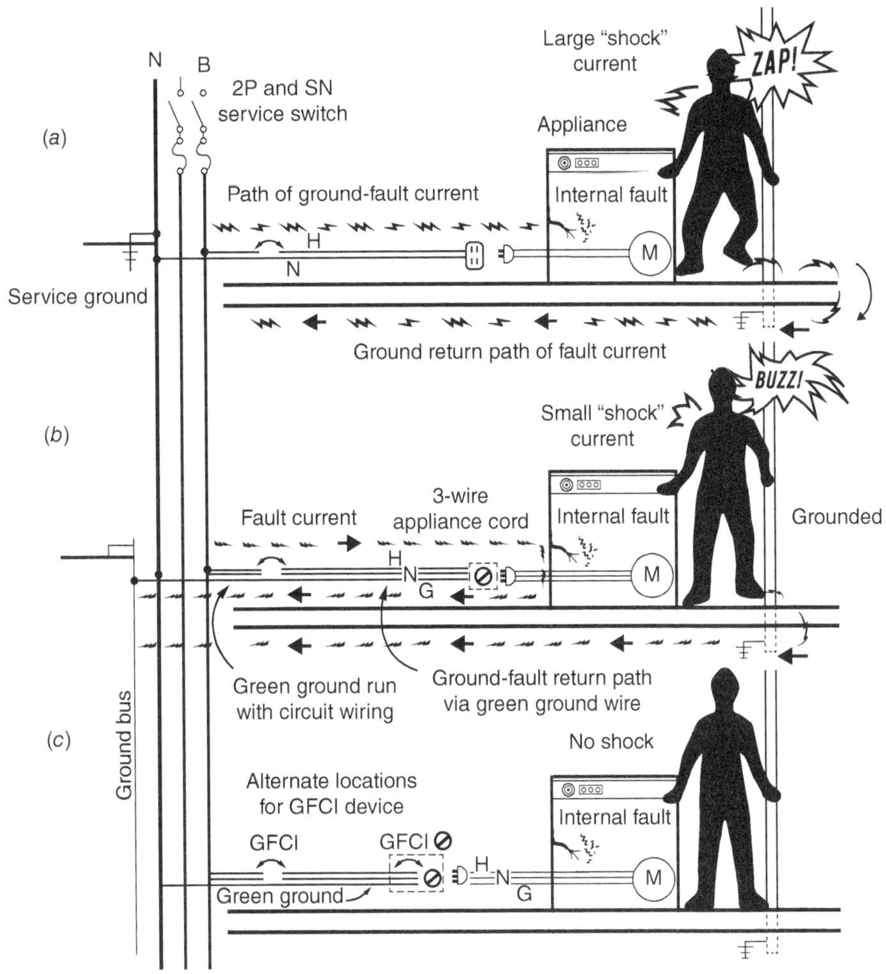

Figure 30.10 Three types of circuit arrangements. (a) A 2-wire, grounded-neutral circuit with no means of preventing shocks from ground faults. (b) A similar arrangement that includes a green ground wire, which considerably lessens the danger of ground-fault shocks. Note, however, that if the ground-fault current is below the rating of the branch circuit protective device (i.e., below 15–20 A), it will continue to flow within the appliance, causing overheating, arcing, and eventual destruction of the appliance. (c) GFCIs are mandatory for bathrooms, outdoor locations, swimming pool circuits, and other areas where ground faults are common and particularly hazardous. (Drawn by Martin Lee.)

hazard while eliminating the dangers described previously. This is shown in Figure 30.10*a*. An accidental fault within an appliance could connect the metal case of the appliance to the line. This can occur with such common devices as an electric saw, a clothes washer or dryer, or a food mixer. A person contacting the appliance housing and simultaneously a ground, such as a water pipe, would receive a nasty 120-V shock. If the contact were made with wet hands, the shock could be fatal. Unfortunately, however, until such an incident occurred, the internal fault would remain an unnoticed source of danger.

To eliminate this hazard, appliance manufacturers have always recommended that appliance housings be grounded to a cold-water pipe. In addition, the appliances are supplied with 3-wire plugs: two wires connected to the appliance and the third wire to the housing. To accommodate such plugs and to provide a ground path, the *NEC* requires all receptacles to be of the grounding type and all wiring systems to provide a ground path, separate and distinct from the neutral conductor (see Chapter 28). The result of such wiring is shown in Figure 30.10*b*, where the ground current passes harmlessly through the internal fault, along the ground-wire path, and back to the panel. A person contacting the appliance housing establishes a parallel ground path. However, because this path is usually of much higher resistance than the ground-conductor path, only a very small current will flow. Wet hands materially reduce contact resistance, and shock current can increase to a dangerous level. If the ground current is sufficiently high, the branch circuit breaker or fuse opens, disconnecting the circuit.

When wiring systems are installed in metallic conduit, the conduit itself or the conduit plus a separate conductor within the conduit may be used as the grounding path. The latter method, with the additional ground wire, is much preferable, as explained in

Figure 30.10. When nonmetallic or flexible metallic wiring (NM "Romex" or AC "BX") is used, a separate grounding conductor run with the regular circuit conductors must be used.

All insulated ground conductors must have their covering colored green for identification as a grounding conductor. Many industrial installations use complete "green-ground" systems in an attempt to eliminate shock hazard and reduce insulation failures. This has not been entirely successful for the reason noted previously—in order to clear the ground fault, its current must be high enough to trip the branch circuit protective device. Otherwise, the ground fault continues to "leak," unnoticed by the system's protective devices. Unfortunately, ground faults are by nature low-current, leak-type faults, because they result from weak spots in insulation, dirt accumulation, and so on. Therefore, although the shock hazard is greatly reduced by the green-ground path, the fault continues to leak and arc until it becomes hot enough to cause a major breakdown, frequently accompanied by fire.

To eliminate this potentially dangerous situation, which occurs any time there is a leak of current to ground in an electric circuit, the ground-fault circuit interrupter (GFCI) was developed (Figure 30.10*c*). This device compares, with extreme precision, the current flowing in the hot and neutral legs of a circuit; if there is a difference, it indicates a ground fault and the device trips out. The rapidity of this action—approximately one-half second—eliminates the possibility of a potentially dangerous shock hazard, which exists even in a properly grounded circuit, as in Figure 30.10*b*, and all the more so for the circuit arrangement in Figure 30.10*a*. The separate ground path required by the *NEC* (bonded metallic conduit system, green-ground, or both) serves to minimize the shock current experienced by a person before the GFCI operates. The device can be applied at the panel to replace a normal circuit breaker (see Chapter 28) or at an individual outlet to replace a normal receptacle device. See the *NEC* for locations where GFCI use is mandatory, such as outdoors and in kitchens and bathrooms. It is advisable to use GFCI devices on all appliance circuits. Application to lighting circuits is generally not required because fixtures are usually out of reach and are switch-controlled. In mixed lighting and receptacle circuits, a GFCI is best applied at the outlet that is to be protected.

30.5 ENERGY CONSERVATION CONSIDERATIONS

Before proceeding with a detailed description of design procedures, the many energy conservation ideas and techniques applicable to electrical distribution systems should be surveyed. This is done to provide focus on this important aspect of design, as individual techniques appear throughout the lengthy design procedure discussion and may be overlooked. The following recommendations are intended to assist with meeting a design intent to conserve energy.

1. Establish an energy budget based upon projected loads and normal operations; then, set an energy reduction figure of 10% to 20% and meet this goal by using the specific techniques that follow. Annual energy consumption estimates are best made with the aid of one of the many computer programs available for this purpose.
2. Learn to recognize the energy-use characteristics of all equipment and systems specified (see Table 30.7). In general, select high-efficiency equipment (motors, transformers, etc.). If these are not identified as such, use materials and equipment with the lowest temperature rise, because these have the lowest losses. This generally indicates the selection of choice. When comparisons are close, a detailed analysis may be necessary. Economic justification can be established with a life-cycle cost analysis. To avoid making a detailed cost analysis on every item, use one of the many available shortcut calculations for payback time (see Appendix J).
3. Provide electric load control equipment (demand control) either as part of an overall building control system or separately.
4. In any multitenant residential building, provide individual user metering. Metering of heating and cooling energy should be part of the HVAC contract. All tenants should be made financially responsible for the energy they use. Exceptions to this rule would be made in the case of hotels, dormitories, and other transient facilities.
5. Where a choice of service voltages is available, consider the highest available. Similarly, consider the highest voltage in each class for interior distribution systems. This means 480 V in the 600-V class, 4 kV in the 5-kV class, and 13 kV in the 15-kV class. The result will be lower line losses, smaller panelboards at the branch circuit level, and generally a lower electrical contract cost (see Section 30.3).
6. In any building, the maximum total voltage drop shall not exceed 3% in branch circuits or feeders, for a total of 5% to the farthest outlet, based on steady-state design load conditions.
7. Avoid the use of electric heating elements if alternatives are available. Electric heat is an inefficient

use of natural resources because of the low overall efficiency of fuel-to-electricity conversion.

8. Provide metering points (for fixed or plug-in meters) throughout the system to permit accurate analysis of power and energy use. Meters, both instantaneous reading and recording types, provide essential data regarding equipment loading, load patterns, load coincidence, power factor, load voltage, power demand, and energy consumption. Analysis of these data can indicate how to program and shift loads for maximum operational efficiency and lowest energy cost. A flexible design that permits such load shifting is assumed (see item 10).
9. Include provisions for power-factor correction in the system, both at devices and at the feeder level. Then, if metering (see item 8) indicates the necessity, add capacitors as required. High power factor reduces line losses, permits maximum utilization of equipment capacity, and avoids utility penalty charges. Utilization equipment rated greater than 1000 W and lighting equipment greater than 15 W, with an inductive reactance load component, should have a power factor of not less than 85% under rated load conditions. Utilization equipment with an inherent power factor of less than 85% should be corrected to at least 90% under rated load conditions.
10. Size equipment as closely as possible to match the load. This normally results in improved efficiency and a higher power factor. Where load varies considerably—for example, between day and night or on weekends—consider splitting the loads so that part of the equipment can be switched off when the load is low, and the remaining load can be fed from a "night/weekend" feeder. Such a design allows the shutdown of whole systems rather than operating at a very low load with the associated high losses and low power factor. The design must be sufficiently flexible to permit shifting of loads between feeders if measurements on the facility in operation indicate that this is desirable (see item 8). The aim is to operate equipment as closely to the rated load as possible and to deenergize lightly loaded sections by shifting load to other, partly loaded equipment.
11. Use the most efficient type of control. This means solid-state controls for motors, variable-voltage, variable-frequency (VVVF) controls for variable-speed equipment (which should be considered for all variable building loads), remote switch controls for blocks of lighting, electronic control systems for elevators, and so on.
12. Arrange automatic time controls for equipment that is only needed for part of the day (such as ventilation fans, water coolers, vending machines, and computers) that might otherwise be active for 24 hours.
13. Seal all electric riser shafts to avoid heat loss by stack effect.
14. Generally, select the coolest possible locations for electric equipment. Low ambient temperature (below 40°C [104°F]) permits the use of smaller equipment for the same load, with resultant lower cost and losses. If below-grade space is available, it is well suited for this purpose.
15. Provision for future expansion should be made by means of a design that accommodates additional equipment in lieu of oversizing equipment initially. Here again, the higher cost of two pieces of equipment can normally be justified by a detailed owning-and-operating (life-cycle) cost analysis using realistic cost escalation and capital cost figures.
16. Energy-conservation techniques for lighting and lighting control are found in Chapters 17 and 18.
17. Energy-conservation techniques for vertical transportation are found in Chapter 33.

30.6 ELECTRICAL WIRING DESIGN PROCEDURE

The steps involved in the electrical wiring design of any facility are outlined in this section. These steps may in some instances be performed in a different order, or two or more steps may be combined, but the procedure normally used is the following:

1. Make an electrical load estimate based upon the areas involved, available building data, and any other pertinent information (see Section 30.2).
2. In cooperation with the local electric utility, decide upon the point of service entrance, type of service run, service voltage, metering location, and building utilization voltage (see Chapter 28 and Section 30.5, item 5).
3. Determine, with the active input of the client, the proposed usage of all areas and information about all client-furnished equipment (including specific electric ratings and service connection requirements).
4. Determine from other consultants (such as those dealing with HVAC, plumbing, elevators, kitchens, and the like) the exact electrical ratings for all the equipment within their realm of design. This determination is often made after conferences during which the electrical consultant makes valuable recommendations to these other specialists about the comparative characteristics and

costs of equipment (see Section 30.5, items 2, 7, 9, and 10). Such timely sharing of expertise is the foundation of the integrated building design process.

5. Determine the location and estimate the size of all required electric equipment spaces including switchboard rooms, emergency equipment spaces (or areas), electric closets, and so forth. Panelboards are normally located in closets but may be located in corridor walls or elsewhere. These decisions are necessary early in the design process to enable the architect to reserve the appropriate spaces (in appropriate locations) for electrical equipment. Once the design is developed in detail, the estimated space requirements can be checked and necessary adjustments made.
6. Design the lighting for the facility. This step is complex and involves continued interaction between the architect and the lighting designer. Coordination between daylighting and electric lighting systems is especially critical.
7. Depending upon the type of facility, it may be necessary to separate the lighting plans from the plan(s) for other devices, such as signaling, low-voltage systems, and receptacles. When an underfloor wiring system is used, it is customary to show it on a separate plan to avoid clutter and confusion. A decision about drawing conventions is made at the preliminary design stage by the project engineer. Once the decision has been made as to how this is to be handled, the lighting fixture layout can be made.
8. On the plan(s), locate all electrical apparatus including receptacles, switches, motors, and other power-consuming equipment. Underfloor, under-carpet, and over-ceiling wiring and raceway systems would be shown at this stage, generally on a separate plan.
9. On the plan(s), locate data-processing and signal apparatus such as telecommunication outlets, network connections, phone outlets, speakers, microphones, TV outlets, fire and smoke detectors, and so on. At this stage, provision is also made for load control wiring, building automatic control wiring, computer control, and the like. The decision is also made as to whether intelligent panelboards will be used (see Chapter 28). Because some of these systems may be covered by separate contracts, the division of responsibility must be clearly defined. Often, only an empty conduit system and power outlets or sources are required in the electrical work contract. The information in this chapter deals only with power wiring design; further discussion of signal equipment is reserved for Chapter 31.
10. Circuit all lighting devices and power equipment to the appropriate panels and prepare the panel schedules. Include in this step the circuitry for emergency equipment.
11. Compute panel loads.
12. Prepare a riser diagram. This includes the design of distribution panels, switchboards, and service equipment.
13. Compute feeder sizes and all protective equipment ratings.
14. Check the preceding work.
15. Coordinate the electrical work with the other disciplines and with the architectural plans. This is not really a separate step, but rather a continuing process starting at step 9 and covering all subsequent stages of the work. Including the building electrical system in the scope of the commissioning process is advisable; and highly recommended for complex and mission-critical buildings.

The background for items 1, 2, 3, 4, 6, 7, and 9 is covered elsewhere in this book. The remaining steps, that is, 5, 8, 10, 11, 12, 13, and 15, are discussed in order in the following sections.

30.7 ELECTRICAL EQUIPMENT SPACES

The spaces required for electrical equipment in a facility vary greatly, depending upon the design and the nature of the building. The working spaces required around major pieces of electrical switchgear and transformers were discussed previously (see Chapter 28). *NEC* Article 110 further specifies the minimum working space required in front of electrical equipment.

(a) Residences

In private residences, the service equipment and the building panelboard are generally incorporated into a single unit. The main disconnect is usually installed as the main switch/breaker of the panel. A number of typical residential service-panel arrangements are shown in Figure 30.11. The panel is normally placed in the garage, utility room, or basement. To minimize voltage drop, the panel should be placed as close to the major electrical loads as practical, without sacrificing valuable space or making the panel inaccessible. Frequently, a smaller panel can be subfed from the main panel to feed the kitchen and laundry loads. In apartments, panels are normally placed in the kitchen or the corridor immediately adjoining the kitchen. This location is chosen so that the panel circuit breaker can act as the required disconnecting means for most fixed appliances (see *NEC* Article 422, Part III).

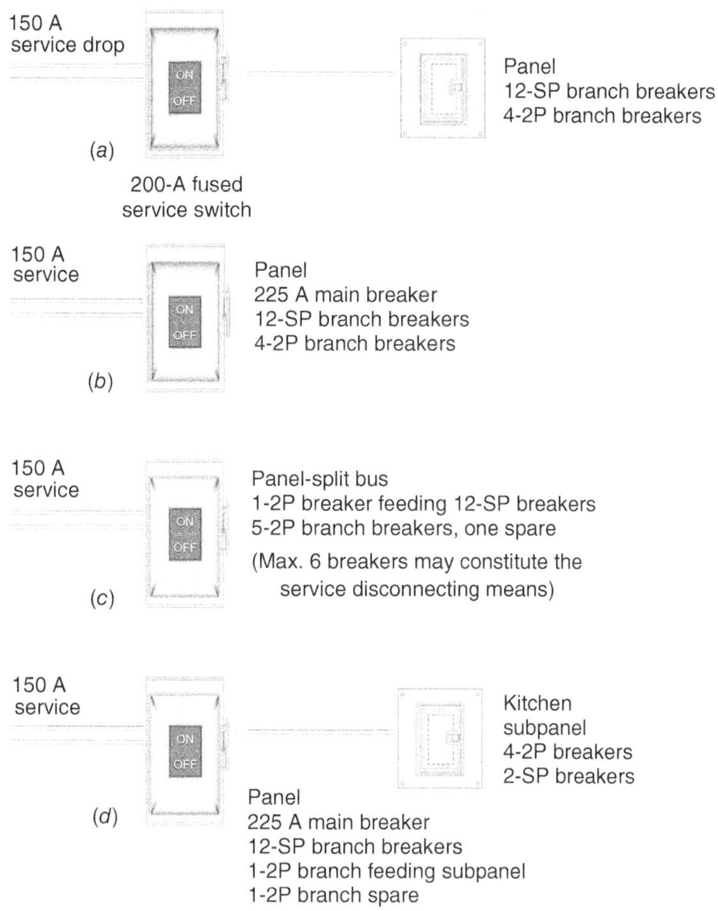

Figure 30.11 *Typical 150-A service arrangements, applicable to residences and small commercial buildings. The service switch can be a separate unit (a) or combined with the branch circuit panelboard (b–d). The panel may be a single unit (a, b), two units in a single enclosure, fed by separate service switches (c), or a central panel and a subfed single-use panel (d)—in this case, a kitchen subpanel. (Drawn by Martin Lee.)*

(b) Commercial Spaces

The location of the required panelboards depends upon their type and number and upon availability of space. In the example research building (for which Figure 30.12 is a partial plan), lighting panels are recessed into the corridor wall because the building is only two stories high, and the panels can be vertically stacked and fed by a single conduit. If this building were six or more stories high, an electric closet (see Section 30.8) would be advisable to accommodate the panel and riser conduits. When panels are installed in finished areas, such as corridors, flush mounting is required.

To limit the voltage drop on a branch circuit in accordance with code requirements, panelboards should be located so that no circuit exceeds 100 ft [30 m] in length. If 15-A or 20-A branch circuits longer than this are unavoidable, No. 10 AWG wire should be used for runs of 100–150 ft [30–46 m] and No. 8 AWG for longer circuits. These circuits are normally wired with No. 12 AWG wire.

The laboratory between the two offices of Figure 30.12*a* is intended to function as a self-contained unit and is therefore equipped with its own panel. Multi-outlet assemblies, all wiring within the room, and the panel itself are surface-mounted to allow ready access to all components in order to accommodate the frequent rewiring encountered in laboratories. A main circuit breaker should be provided in such a panel to act as a main disconnect, whether required by code or not. Where panels are convenient to the load controlled, the panel circuit breakers may be used for switching.

Panels supplying large blocks of load simultaneously switched, such as auditorium house lighting, lobby lighting, large single-use office areas, store lighting, and the like, can be constructed with built-in electrically operated, mechanically held contactors to switch the entire load, with control at any desired remote location. These remote-control (RC) switches are discussed in Chapter 28. If only parts of the panel circuits are so arranged (i.e., for remotely controlled block switching), a split bus panel can be used, with the RC switch controlling only that part of the panel load (Figure 30.13).

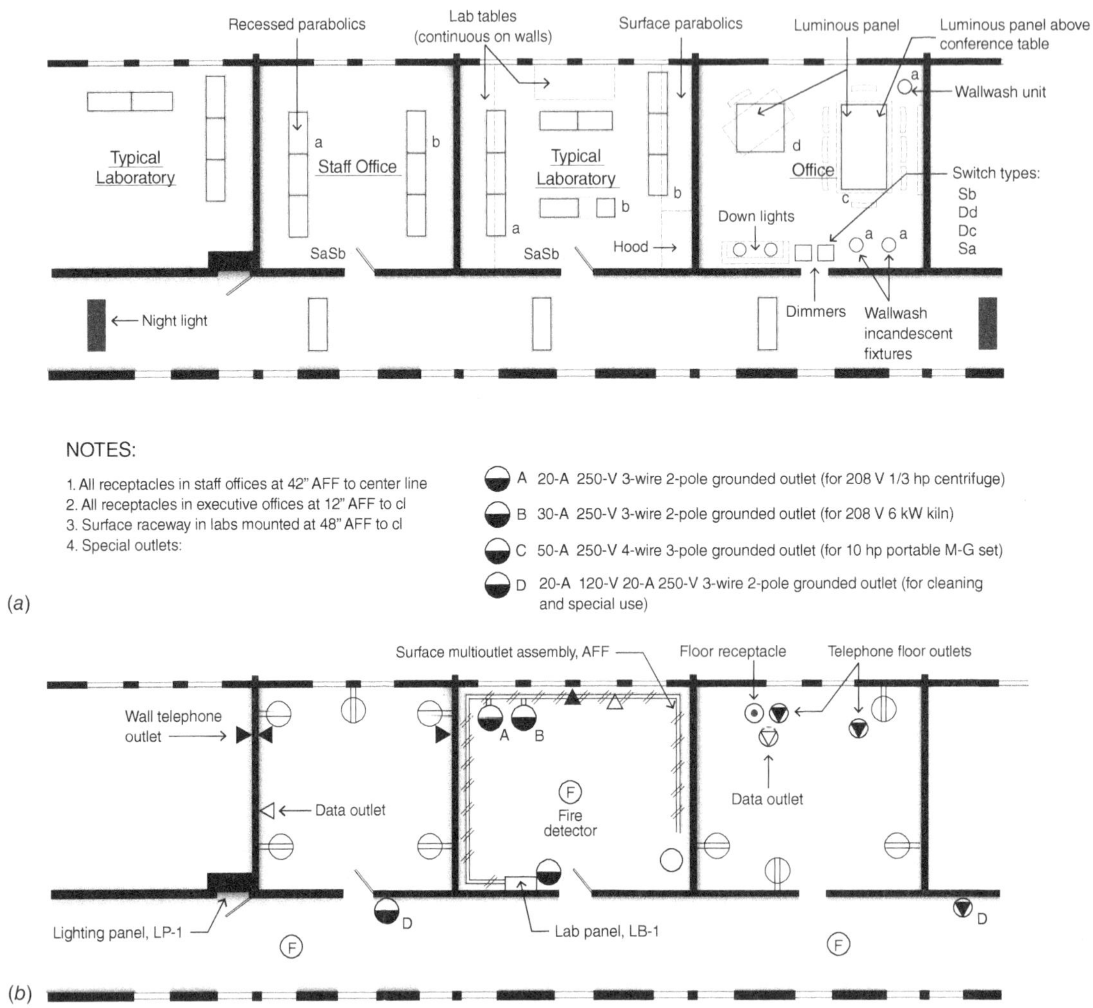

Figure 30.12 *Typical floor plans for lighting and power for a section of an office–laboratory building. Separate lighting (a) and power (b) plans are drawn for the sake of clarity. For circuited plans, see Figures 30.21 and 30.22. Data outlets handle data-processing and telecommunications cabling.*

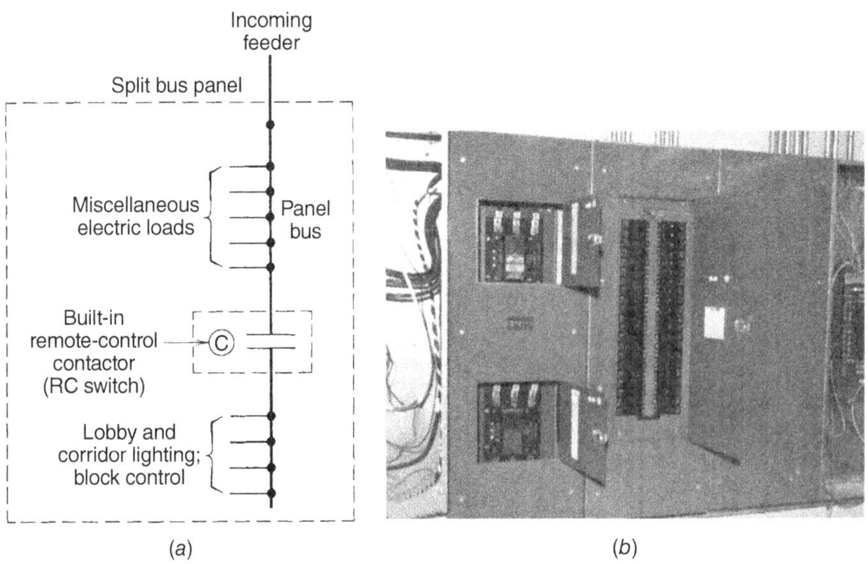

Figure 30.13 *(a) Single-line diagram of a split bus panel. This design is used when a block of load is to be controlled as a unit. The control point is either remote, at the panel, or both. The contactor (RC switch) is electrically operated and mechanically latched. (b) RC switches (mechanically held contactors) are shown to the left of a bank of panels. The contactors were installed to provide photocell and timer control of department store lighting. (Courtesy of Automatic Switch Co.)*

Small offices, stores, and similar buildings frequently have lighting panels mounted in a convenient finished area and utilize the panels' circuit breakers for load switching. In large buildings, strategically located electric closets are provided to house all electrical supply equipment. Power panels and distribution panels are located as required by the loads fed through them.

In general, branch circuit panels, distribution panels, and switchboards are best located near the electrical load center. This minimizes feeder length and reduces voltage drop, making it the most economical arrangement.

Every completely enclosed switchgear room, emergency generator room, or transformer vault should be equipped with an emergency light source. In generator rooms, these should be battery-operated to give illumination for generator repairs in the event of generator failure during a power outage.

30.8 ELECTRICAL CLOSETS

When designing a building electric system, particularly for multistory construction, it is often advantageous and convenient to group the electrical equipment in a small room called an electrical closet (Figure 30.14). The shape of this space can be varied to fit the architectural requirements, and it should provide the following features.

1. It should have one or more locking doors.
2. It should have vertical stacking, above and below other electric closets, and located so as not to block conduits entering or leaving horizontally. Thus, locations on outside walls and adjoining shafts, columns, and stairs are poor choices.
3. A space should be left free of other utilities, such as piping or ducts passing through the closet, either horizontally or vertically.
4. There should be sufficient wall space to mount all requisite and future panels, switches, transformers, telephone cabinets, and communication equipment. Wall cabinet space must be coordinated with raceway connections to underfloor ducts and over-the-ceiling raceway systems (see Chapter 29).
5. There should be floor slots or sleeves of sufficient size for all present and future conduit or bus risers.

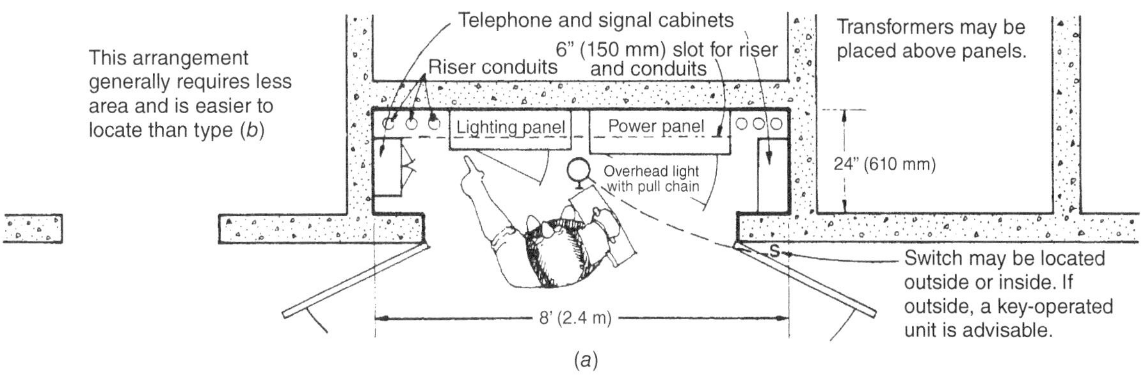

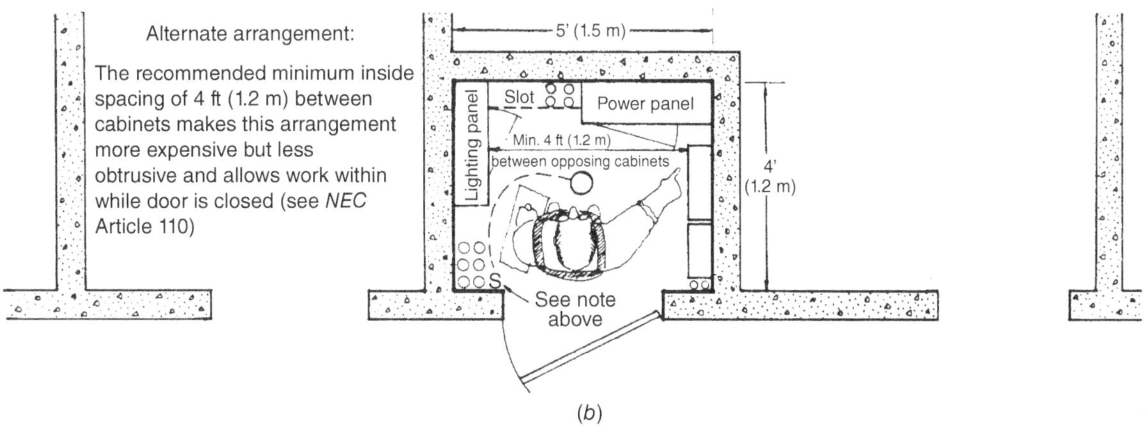

Figure 30.14 Typical electric closets (a, b) with some of the usual equipment. If warranted due to the amount of equipment, separate closets may be used for signal and telephone conduits and cabinets. Smaller spaces may be used, depending upon the types of equipment cabinets and their arrangement (see NEC *Article 110). In all cases, enough space must be provided to allow full opening of all cabinet doors. Empty wall space for future cabinets and equipment should also be considered.*

6. There should be sufficient floor space so that an electrician can work comfortably and safely on initial installation and repair.
7. There should be adequate illumination and ventilation.

30.9 EQUIPMENT LAYOUT

Wiring devices, principally receptacles and switches, are located on electrical plans as required by the equipment and by the anticipated use of an area.

Switches for control of lighting or receptacles are normally placed on the strike side of a door. Other devices, such as plug-in strips on walls and special purpose receptacles, are shown and identified. Signal outlet locations are often noted but generally remain uncircuited on floor plans, a riser or floor raceway plan being utilized to show interconnections. Signal devices include fire alarm equipment, telephone and intercom equipment, data and communications, radio and television outlets, and so on. These devices may be identified by a special symbol or note where a standard symbol is not available.

As mentioned previously, lighting fixture outlets are normally placed on the same drawing as wiring devices unless the large number of the latter precludes showing the lighting without unduly complicating the drawings. In such an event, the lighting is shown on one drawing and receptacles on another, with signals shown on the one less cluttered. A ceiling or underfloor wiring system would necessitate such separation. Motors, heaters, and other fixed and permanently wired equipment are shown and identified on the receptacle drawings (also called power drawings, in contrast to lighting drawings). Equipment furnished with a cord and plug is not normally shown. However, the receptacle intended for supplying such a device is shown and identified. An abbreviated symbol list is given in Figure 30.15.

30.10 APPLICATION OF OVERCURRENT EQUIPMENT

Before beginning an explanation of circuiting, it is necessary to explain the principles underlying overcurrent protection. As outlined in Chapter 28, the function of an overcurrent device is to open (interrupt) a circuit when the current rating of the equipment being protected is exceeded. All equipment must be protected in accordance with its current-carrying capacity. Where ratings do not correspond exactly, the next larger standard size for protective devices up to 800 A may be used; above this, use the next lower rating (see *NEC* Article 240 for further restrictions). The following general rules govern the application of overcurrent protection devices:

1. Overcurrent devices must be placed on the line or supply side of the equipment being protected (Figure 30.16).
2. Overcurrent devices must be placed on all ungrounded conductors of the protected circuit.
3. In general, conductor sizes shall not be reduced in a circuit or tap unless the smallest-size wire is protected by the circuit overcurrent protection devices (Figure 30.17).
4. Overcurrent devices shall be located so as to be readily accessible.

30.11 BRANCH CIRCUIT DESIGN

A branch circuit (or branch-circuit), by *NEC* definition, refers only to the circuit conductors between the final overcurrent device protecting the circuit and the outlet(s), although for our purposes, and in trade parlance, it includes the protective device and the outlets served. Such circuits may be the multioutlet general-purpose type (Figure 30.18*a*), the multioutlet appliance type (Figure 30.18*b*), or the single-outlet type intended for a specific piece of equipment (Figure 30.18*c*). The multioutlet types are limited to 50 A in capacity, whereas the single-outlet type is governed in size only by the requirements of the item being served and may be 200 or 300 A in size.

In its simplest form, a branch circuit comprises only two circuit wires. However, multiwire branch circuits carrying 2- to 3-phase wires plus a neutral are also widely used. Generally, each branch circuit should be sized for the load connected to it plus the load expansion that is expected. These general rules of good practice should be followed:

1. In all but the smallest installations, connect lighting, convenience receptacles, and appliances on separate (groups of) circuits, although this is not an *NEC* requirement.
2. General-purpose branch circuits should be 20 A and wired with No. 12 AWG wire. Switch legs may be No. 14 AWG if the lighting load permits and the wiring meets the tap requirements of *NEC* Article 240.
3. Limit the circuit load on 15-A and 20-A circuits to the values shown in Table 30.5. This provides the required building load expansion capability in the branch circuitry; that is, bringing the loads on the branch circuits up to maximum allows additional building loads to be absorbed. Because it is not always economical or feasible to expand existing

26 05 00-003	ID	electrical, duct, underfloor junction box 260500-003-ID ELEC DUCT UNDERFL J-BOX	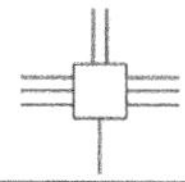
26 05 00-004	ID	electrical, earth ground 260500-004-ID ELEC EARTH G	
26 05 00-005	ID	electrical, junction box 260500-005-ID ELEC J-BOX	
26 05 00-006	ID	electrical, ladder cable tray 260500-006-ID ELEC LAD CABLE TRAY	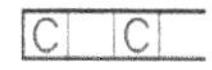
26 05 00-007	ID	electrical, panelboard, home run to (arrowheads indicate the number of circuits) 260500-007-ID ELEC PNLBD HOME RUN TO (ARROWHEADS INDICATE THE NO OF CKTS)	
26 05 00-008	ID	electrical, pressure switch- close on increase 260500-008-ID ELEC PRESS SW- CLOSE ON INCREASE	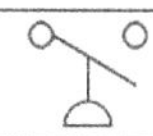
26 05 00-009	ID	electrical, pressure switch- open on increase 260500-009-ID ELEC PRESS SW- OPEN ON INCREASE	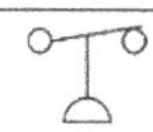
26 05 00-010	ID	electrical, pull box 260500-010-ID ELEC PB	
26 05 00-011	ID	electrical, switch, multiposition 260500-011-ID ELEC SW MULTIPOS	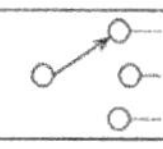
26 05 00-012	ID	electrical, switch, normally closed float 260500-012-ID ELEC SW NC FLOAT	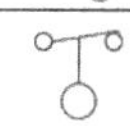
26 05 00-013	ID	electrical, switch, normally closed foot operated 260500-013-ID ELEC SW NC FOOT OPERATED	
26 05 00-014	ID	electrical, switch, normally closed limit 260500-014-ID ELEC SW NC LIMIT	
26 11 00-001	ID	electrical, interconnection with substation, aboveground 261100-001-ID ELEC INTERCONNECTION WITH SUBSTA ABOVEG	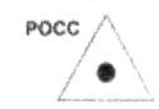
26 11 01-002	ID	electrical, interconnection with substation, underground 261100-002-ID ELEC INTERCONNECTION WITH SUBSTA UNDERG	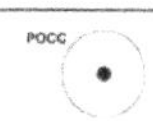
26 20 00-001	ID	electrical, busway 262000-001-ID ELEC BUSWAY	
26 20 00-003	ID	electrical, fuse with rating 262000-003-ID ELEC FUSE WITH RATING	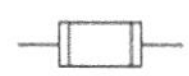
26 20 00-004	ID	electrical, normally closed relay contact 262000-004-ID ELEC NC RELAY CONTACT	
26 20 00-005	ID	electrical, normally open relay contact 262000-005-ID ELEC NO RELAY CONTACT	
26 20 00-007	ID	electrical, push button 262000-007-ID ELEC PB	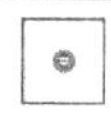
26 24 00-001	ID	electrical, distribution panel 262400-001-ID ELEC DISTR PNL	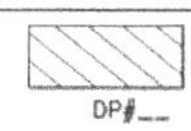
26 24 00-002	ID	electrical, lighting panel 262400-002-ID ELEC LTING PNL	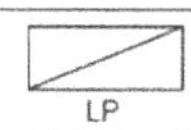
26 24 00-003	ID	electrical, panelboard cabinet, flush mounted 262400-003-ID ELEC PNLBD CAB FLUSH MTD	
26 24 00-004	ID	electrical, panelboard cabinet, surface mounted 262400-004-ID ELEC PNLBD CAB SURF MTD	
26 24 00-005	ID	electrical, power panel 262400-005-ID ELEC PWR PNL	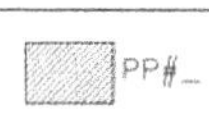

Figure 30.15 *Selected electrical symbols from the United States National CAD Standard® version 6. (Reproduced with permission of the National Institute of Building Sciences 2014.)*

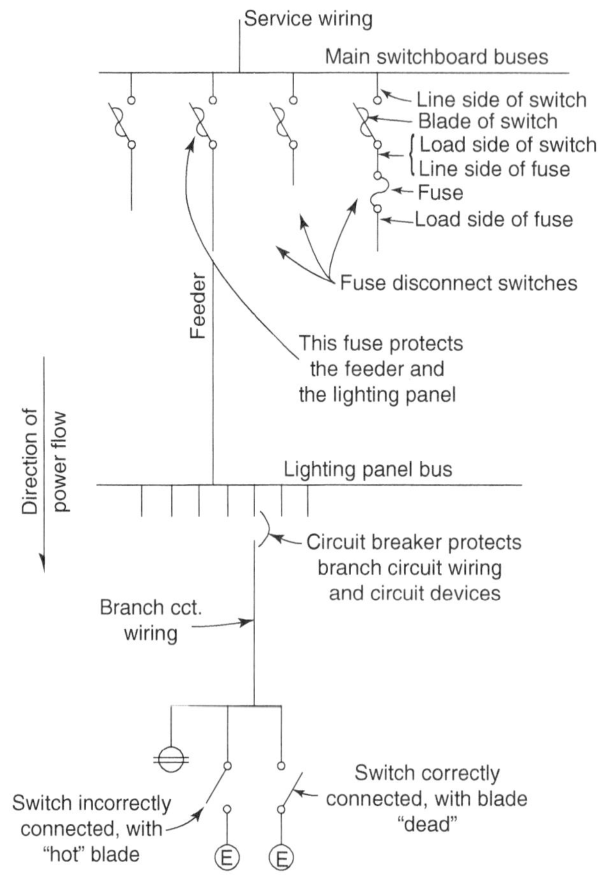

Figure 30.16 *Location of overcurrent protective devices. Protective devices should always be located at the point where the conductor receives its source of supply so that when the protective device operates, the current supply is cut off.*

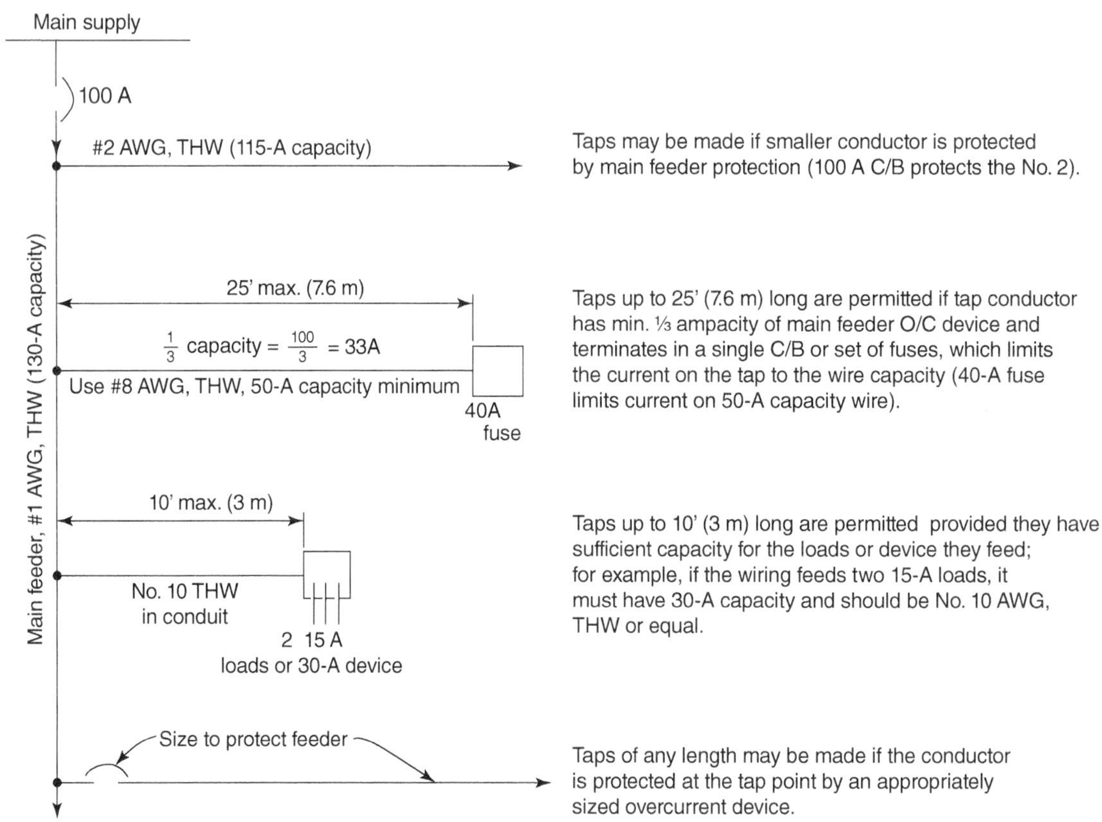

Figure 30.17 *Some permissible tap arrangements. See* NEC *Article 240 for restrictions on taps in addition to those given in the figure.*

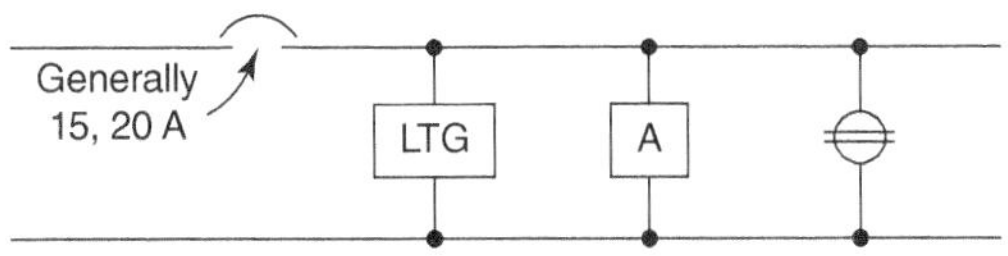

(a) General-purpose branch circuit. Supplies outlets for lighting and appliances, including convenience receptacles.

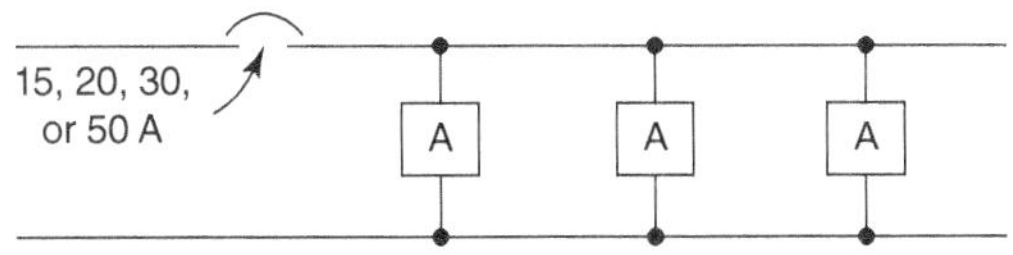

(b) Appliance branch circuit. Supplies outlets intended for feeding appliances. Fixed lighting not supplied.

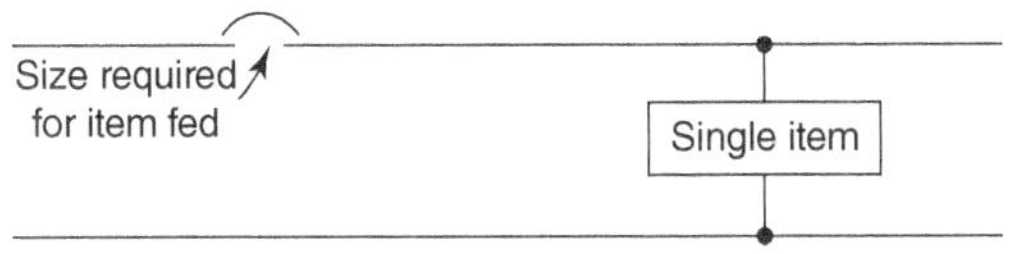

(c) Individual branch circuit, designed to supply a single specific item.

Figure 30.18 *Branch circuit types (a, b, c). See* NEC *Article 210 for detailed descriptions.*

circuits, however, panels are always equipped with spare breakers. These can also be utilized to pick up building load expansion, or the user may employ a combination of the two techniques, expanding some existing circuits and adding new ones. (See Section 30.15 for a discussion of this subject.)

Because lighting and specific devices are circuited according to their nameplate ratings, the only circuitry item left to the judgment of the designer is the number of convenience receptacles per circuit. The *NEC* specifies that plug outlets (convenience receptacles) be counted, in totaling loads, at 1.5 A each unless included in the load for general lighting. This point requires some clarification. The receptacle load for general illumination in dwellings is included in the 3 V-A/ft^2 [32.3 V-A/m^2] specified by *NEC* Table 220-12 and found in Table 30.1. This is because receptacle outlets on general illumination circuits are assumed to supply illumination (i.e., lamps), and are therefore necessarily included in the lighting load. Application of the 1.5-A (180-V-A) per outlet criterion to such outlets would, in the opinion of the *NEC*, unnecessarily limit the number of outlets permitted on such circuits.

The same is true for appliance circuits. On both of these types, according to the *NEC*, the number of receptacle outlets is not limited. On 15- and 20-A receptacle circuits, which are neither general illumination nor small appliance branch circuits, the number of outlets is limited to 10 (1800 V-A) on a 15-A circuit and 13 (2340 V-A) on a 20-A circuit by applying the 1.5-A per outlet criterion.

It appears that the distinction between circuit types is not completely clear, and that good practice dictates the limitation of convenience receptacles per circuit to a considerably smaller number. Thus, following the guidelines of Table 30.5 of 9- and 12-A loading on 15- and 20-A circuits, respectively, this approach would yield:

$$\text{15-A circuit: } \frac{9\text{ A}}{1.5\text{ A per outlet}} = 6 \text{ outlets per circuit}$$

$$\text{20-A circuit: } \frac{12\text{ A}}{1.5\text{ A per outlet}} = 8 \text{ outlets per circuit}$$

These values must be used judiciously. If the devices to be energized are small, these quantities may be used. Such would be the case in a business office, where only an electric pencil sharpener or a desk lamp (100 V-A) would be plugged in. However, for

TABLE 30.5 Recommended Branch Circuit Loads[a]

Size Circuit	Circuit[d] Amperes	Volt-Amperes[d] at 120 V	Volt-Amperes at 277 V
0% Expansion[b,c]			
15 A	12	1440	—
20 A	16	1920	4440
25% Expansion[e]			
15 A	9.6	1150	—
20 A	12.8	1520	3600
50% Expansion[f]			
15 A	8	960	—
20 A	11	1300	3000

[a]The loading shown will provide the specified expansion in the branch circuits. Where branch circuit expansion is not practical, the required expansion can be obtained from panel spares.

[b]See Table 30.1 for anticipated load growth.

[c]For 0% expansion, initial load = 80% of the circuit rating, which is the maximum permissible for continuous loads. See Table 30.6, note 6.

[d]For branch circuits feeding utilization equipment such as air conditioners in addition to convenience outlets, see *NEC* Article 210.23.

[e]To accomplish 25% expansion, the circuit is derated to 80%, that is, 80% of 12 A = 9.6 A, etc.

[f]To accomplish 50% expansion, the circuit is derated by 1/3, that is, two-thirds of 12 A = 8 A, etc. (Then 50% expansion on 8 A yields 12 A.)

TABLE 30.6 Branch Circuit Requirements

	Branch Circuit Size				
	15 A	**20 A**	**30 A**	**40 A**	**50 A**
Minimum-size copper conductors	No. 14	12	10	8	6
Minimum-size taps	No. 14	14	14	12	12
Overcurrent device rating	15 A	20	30	40	50
Lampholders permitted	Any type	Any type	Heavy duty	Heavy duty	Heavy duty
Receptacle rating permitted (see note 7)	15 A	15 or 20	30	40 or 50	50
Maximum load (see notes 6 & 8)	15	20	30	40	50

Notes:

1. Wiring shall be types RHW, RHH, T, THW, TW, THWN, THHN, XHHW in raceway or cable.
2. On a 15-A circuit, the maximum single appliance shall draw 12 A. On a 20-A circuit, the maximum single appliance shall draw 16 A. If combined with lighting or portable appliances, any fixed appliance shall not draw more than 7.5 A on a 15-A circuit and 10 A on a 20-A circuit.
3. On a 30-A circuit, maximum single appliance draw shall be 24 A.
4. Heavy-duty lampholders are units rated not less than 750 W.
5. 30-, 40-, and 50-A circuits shall not be used for fixed lighting in residences.
6. When loads are connected for long periods (3 hours or more), the actual load shall not exceed 80% of the branch circuit rating. Conversely, continuous-type loads shall be figured at 125% of the actual load in all load calculations.
7. A single receptacle on an individual branch circuit shall have a rating not less than the circuit—for example, 15 A on a 15-A circuit. Also, 15-A receptacles on a 20-A circuit shall not supply a load greater than 12 A for appliances; 20-A receptacles on a 20-A circuit shall be limited to a 16-A load.
8. Rating of a single piece of cord-connected utilization equipment shall not exceed 24 A when connected to a 30-A circuit.
9. For additional restrictions and data on branch circuit use and application, see NEC Table 210-24.

computer-equipped offices, service and repair facilities, and the like, plan for no more than four receptacles on a 20-A circuit—and preferably two or three receptacles maximum on a 15-A circuit. Of course, diversity of use is an all-important factor, and the more accurately it can be estimated, the better the design result.

A further note of caution is in order: receptacles should be arranged, if at all possible, so that the loss of a single circuit does not deprive an entire area of power. That is, for the sake of reliability, circuitry should be alternated to give each space access to parts of different circuits.

NEC Article 210.24 specifies certain requirements for conductors, devices, and loads permissible on general-purpose branch circuits. These are excerpted and summarized in Table 30.6.

30.12 BRANCH CIRCUIT DESIGN GUIDELINES: RESIDENTIAL

1. The *NEC* requires for residences sufficient circuitry to supply a load of 3 V-A/ft2 [32.3 V-A/m^2] in the building, excluding unfinished spaces such as porches, garages, and basements. Using Table 30.5, which already includes a 20% derating of circuits to provide for continuous loading, this works out to 480 ft^2 [45 m^2] on a 15-A circuit (1440 V-A) and 640 ft^2 [60 m^2] on a 20-A circuit (1920 V-A). Allowing for some expansion, such as a finished basement or enclosing a porch to expand the living area, results in a good-practice recommendation of 400–480 ft^2 [37–45 m^2] per 15-A circuit and 530–640 ft^2 [49–60 m^2] per 20-A circuit. These good-practice figures normally provide enough circuits for all but the most heavily loaded residences. If actual design analysis, considering the owner's project requirements, indicates the need for additional circuits, however, they clearly must be provided.
2. *NEC* Article 210.11 requires a minimum of two 20-A appliance branch circuits (Figure 30.18b) to feed all the receptacle outlets in the kitchen, pantry, breakfast, and/or dining room and similar areas, and only these outlets. This is because any receptacle in these areas is a potential appliance outlet and must be fed and circuited as such. Permanently installed appliances such as a food disposal, dishwasher, fan hood, and the like may not be connected to these appliance circuits. An exception is made to permit clock outlets to be fed from these circuits. Furthermore, the *NEC* requires that all kitchen outlets intended to serve countertop areas must be fed from at least two of these appliance circuits (these circuits may also feed appliance outlets in the other spaces specified previously). Thus, not all countertop workspace will be deenergized by the failure of a single circuit.

Although the code does not limit the number of appliance outlets to be wired into each circuit, good practice dictates that these receptacles be circuited with no more than four such outlets per 20-A circuit. This in turn usually requires more than the code minimum of two appliance circuits. Among other requirements, the *NEC* states that for kitchen and dining areas:

1. No point on a wall behind a countertop shall be more than 24 in. [610 mm] from an outlet.

2. All countertop convenience receptacles shall be of the GFCI type.
3. Locations utilized primarily or frequently for workshop-type activities, such as garages, utility rooms, and basements, should be provided with receptacles wired in appliance-type circuits (i.e., 20-A receptacles on 20-A circuits), with no more than four such receptacles to a circuit. Receptacles in garages, sheds, crawlspaces, below-grade finished or unfinished basements, and outdoors must be of the GFCI type. (For exceptions to this GFCI rule, see *NEC* Article 210.8.)
4. Additional circuits similar to appliance circuits (no fixed lighting outlets) should be furnished to supply one outlet in each bedroom of a house that is not centrally air-conditioned. Such outlets are intended for window air conditioners. (Good design practice may provide a window arrangement, attic ventilation, insulation, sunscreening, and the like, that can reduce the demand for these noisy energy users.)
5. The *NEC* requires that at least one 20-A appliance circuit supply the laundry outlets only. This requirement is good practice. If an electric clothes dryer is anticipated (and it should be, unless it is definitely known that a gas dryer will be used), an individual branch circuit (distinct from the laundry outlets circuit[s] and rated for the load) should be supplied to serve this load via a heavy-duty receptacle. (Facilities for hanging clothes should be provided for those who prefer not to consume electrical energy on this simple task.)
6. Lay out convenience receptacles so that no point on a wall is more than 6 ft [1.8 m] from an outlet. Use 20-A, grounding-type receptacles only. Do not combine receptacles and switches into a single outlet except where convenience of use dictates high mounting of receptacles.
7. Circuit the lighting and receptacles so that each room has parts of at least two circuits. This includes basements and garages.
8. Avoid placing all the lighting in a building on a single circuit.
9. Supply at least one 20-A wall-mounted receptacle adjacent to each bathroom lavatory (sink) location. Such receptacles must have GFCI protection. They should be fed from a 20-A circuit that energizes only such bathroom receptacles. Bathroom lighting, exhaust fan, heaters, or other outlets should not be connected to the bathroom receptacle circuit.
10. Provide at least two GFCI-protected and weatherproof receptacles on the outside of the house, one in front and one at the rear. Switch control of the outside receptacles from inside the house is a convenience.
11. In rooms without overhead lighting, provide switch control for one-half of a strategically located receptacle that is intended to supply a lamp (see Figure 30.19 for the wiring arrangement in such a case).
12. Provide switch control for closet lighting. Pull chains are a nuisance (but are considerably cheaper).
13. In bedrooms, supply two duplex outlets at either side of the likely bed location to accommodate electric blankets, clocks, radios, lamps, and other such appliances.
14. A disconnecting means, readily accessible and within sight of the controlled item, must be provided for electric ranges, cooktops, and ovens. It is good practice to utilize a small kitchen panel recessed into a kitchen wall to control the large kitchen appliances and provide the required disconnecting means.
15. Perimeter lighting, sensor control, and manual override can do much to lessen vandalism and discourage prowlers.
16. To accommodate the very rapidly increasing number of persons who work at home (either full- or part-time) every study and work room or, in their absence, every large master bedroom should be electrically equipped to double as a home office. This room or area should be supplied minimally with these electrical services:

    a. Six duplex 15-A or 20-A receptacles connected to at least two different circuits, one of which should serve no other outlets. All of these receptacles should be equipped with appropriate surge suppression (see Chapter 28).
    b. An additional separate insulated and isolated ground wire, connected only at the service point, should be run to the box(es) containing two of the receptacles required in

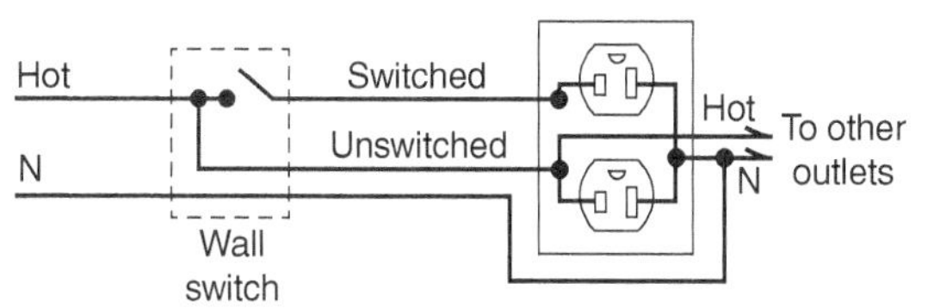

Figure 30.19 *Split wiring of a duplex receptacle. The upper half is switch-controlled; the lower half is "hot" all the time. This allows wall switch control of a lamp or another device while keeping part of the receptacle live for independent use. Notice that the receptacle is mounted with the grounding pole at the top. This is the safest way to install receptacles because a metallic item such as a paper clip falling on an inserted cap will contact the ground pole only.*

the preceding point, and there terminated, clearly marked, and labeled. These grounds are intended for possible use with isolated ground (IG) receptacles, if it is found that the normal receptacles are too electrically noisy.

c. Two telephone jacks in recessed boxes should be provided in the area. In addition, an empty ¾-in. [19-mm] conduit from the telephone entry service point to an empty 4-in. [100-mm] square box in the area should be provided. An appropriate surge suppressor must be provided for the incoming telephone service lines.

d. For lighting design of this area, refer to the discussion on lighting for spaces with numerous computers in Chapter 18.

A tabulation of residential electrical equipment, including recommended circuit and receptacle descriptions, is shown in Table 30.7. A complete residential wiring plan for a small house is shown in Figure 30.20. Although residential plans are frequently left uncircuited, a completely circuited layout is shown here for informational purposes. The plan does not show the electrical equipment recommended for a home office that is described in item 16 for two reasons:

1. The plan is so compact that there is no single preferable location for the numerous outlets required.
2. Adding the required outlets to any selected area would clutter the plan to the point of illegibility.

TABLE 30.7 Load, Circuit, and Receptacle Chart for Residential Electrical Equipment

Appliance	Typical Connected Volt-Amperes[a]	Volts	Wires[b]	NEMA Circuit Breaker or Fuse, amp	Outlets on Circuits	Device[d] and Configuration (see Figure 28.40)
		Kitchen				
Range[c,e,i]	12,000	115/230	3 #6	60	1	14–60R
Oven (built-in)[c,i]	4500	115/230	3 #10	30	1	14–30R
Range tops[c,i]	6000	115/230	3 #10	30	1	14–30R
Dishwasher[c]	1200	115	2 #12	20	1	5–20R
Waste disposer[c,i]	300	115	2 #12	20	1	5–20R
Microwave oven	1000	115	2 #12	20	1 or more	5–20R
Refrigerator[f]	300	115	2 #12	20	1 or more	5–20R
Freezer[f]	350	115	2 #12	20	1 or more	5–20R
		Laundry				
Washing machine	1200	115	2 #12	20	1	5–20R
Dryer[c,i]	5000	115/230	3 #10	30	1	14–30R
Hand iron[e]	1650	115	2 #12	20	1	5–20R
		Living Areas				
Workshops[e,j]	1500	115	2 #12	20	1 or more	5–20R
Portable heater[e]	1600	115	2 #12	20	1	5–20R
Television	300	115	2 #12	20	1 or more	5–20R
Audio center[g]	350	115	2 #12	20	1 or more	5–20R
DVD or VCR[g]	150	115	2 #12	20	1 or more	5–20R
Personal computer and peripherals[h]	1000	115	2 #12	20	1 or more	5–20R
		Fixed Utilities				
Fixed lighting	1200	115	2 #12	20	1 or more	—
Air conditioner, ¾ hp [0.56 kW][i,j]	1200	115	2 #12	20 or 30	1	5–20R 14–30R
Central air conditioner[c,i,j]	5000	115/230	3 #10	40	1	—
Sump pump[j]	300	115	2 #12	20	1 or more	—
Heating plant (i.e., forced-air furnace)[j,k]	600	115	2 #12	20	1	—
Attic fan[j]	300	115	2 #12	20	1 or more	5–20R

[a]Wherever possible, use the actual equipment rating.

[b]Number of wires does not include equipment grounding wires. Ground wire is No. 12 AWG for a 20-amp circuit and No. 10 AWG for 30- and 50-amp circuits.

[c]For a discussion of disconnect requirements, see *NEC* Article 422.

[d]Equipment ground is provided in each receptacle.

[e]Heavy-duty appliances regularly used at one location should have a separate circuit. Only one such unit should be attached to a single circuit at the same time.

[f]A separate circuit serving only one other outlet is recommended.

[g]Surge protection recommended.

[h]Isolated ground may be required.

[i]A separate circuit is recommended.

[j]It is recommended that all motor-driven devices be protected by a local motor-protection element unless motor protection is built into the device.

[k]Connect through the disconnect switch equipped with a motor-protection element.

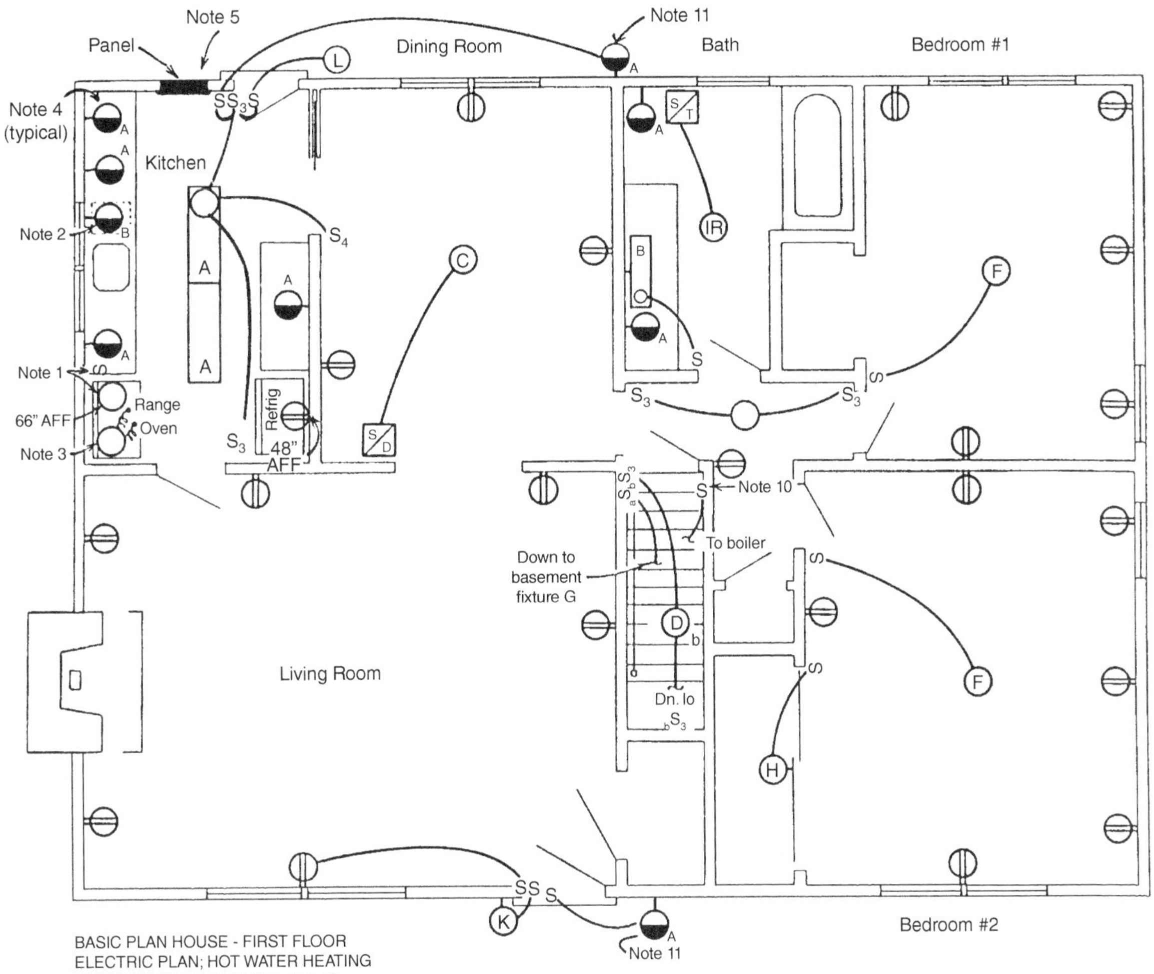

BASIC PLAN HOUSE - FIRST FLOOR
ELECTRIC PLAN; HOT WATER HEATING
EQUIPMENT AND DEVICE LAYOUT

Fixture Schedule: see Figure 30.20 *c*
Symbol List: see Figure 30.20 *d*
Circuited Plan: see Figures 30.20 *e, f*
Plan, Schedule: see Figure 30.30 *g*

Notes:

1. Switch and outlet for exhaust fan. Switch wall mtd. above counter-blacksplash. Outlet with blank cover mounted adjacent to fan wall opening. Separate switch may be omitted if fan is supplied with integral switch.
2. Dishwasher receptacle wall mtd. behind unit, 6" AFF.
3. Range and oven outlet boxes wall mtd., 36" AFF. Flexible connection to units.
4. Receptacles at countertop locations to be wall mounted 2" above backsplash.
5. Max. ht. of top c/b to be 78" AFF.
6. Wiring shown as run exposed indicates absence of finished ceiling in basement level. All BX to be run through framing members. Attachment below ceiling joists not permitted.
7. Connect to two type G fixtures ceiling mounted at ⅓ point of crawl space.
8. Connect to one type G fixture at center of crawl space.
9. Connect to shutdown switch at top of stairs. Boiler control wiring by others. See Note 10.
10. Boiler wiring safety disconnect. Provide RED wall plate, clearly marked "BOILER ON–OFF."
11. Equipped with self-closing gasket WP cover.

(*a*)

Figure 30.20 *Electrical plan of a small house. (a) Layout of lighting and electrical devices for the main floor. (b) Layout of lighting and electrical devices for the basement. (c) Lighting fixture schedule. (d) Symbols and abbreviations. (e) Circuited plan, main floor of house. (f) Circuited plan, basement. (g) Panel schedule for the house.*

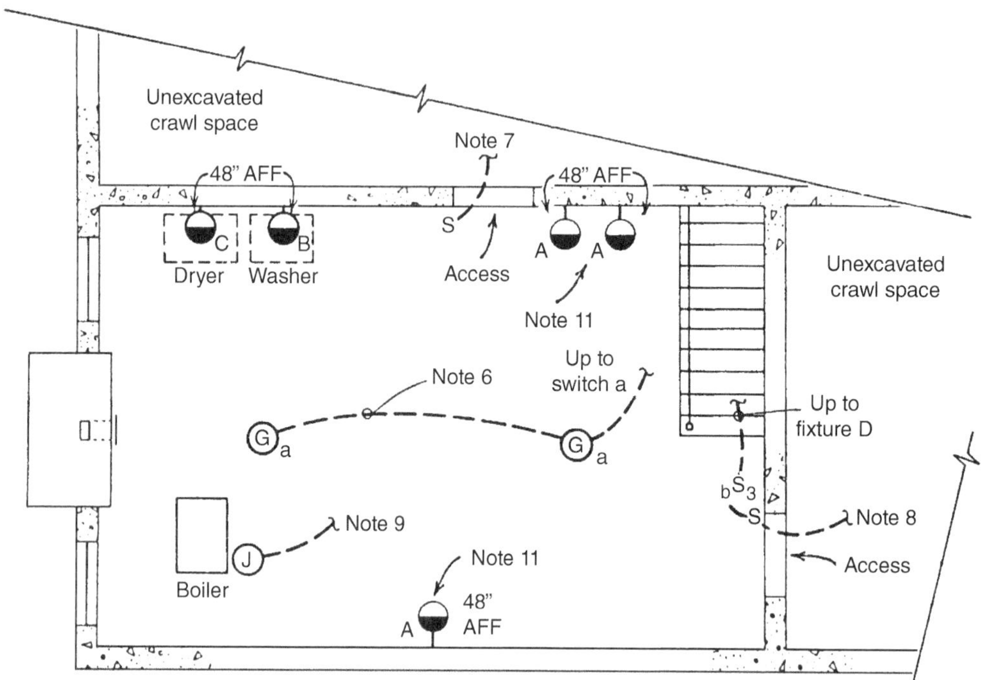

BASIC HOUSE PLAN – BASEMENT
ELECTRIC PLAN; H.W. HEATING
EQUIPMENT AND DEVICE LAYOUT

(*b*)

LIGHTING FIXTURE SCHEDULE			
TYPE	DESCRIPTION	MANUFACTURER	REMARKS
A	48" L X 12" W X 4" DEEP NOMINAL, 2 LAMP/FLUORES-CENT, WRAPAROUND ACRYLIC LENS, 40 W (NOM.) T8 LAMPS. SURFACE MTD.	BRITE–LITE CO. CAT. #2740/KFF OR EQUAL	4" DEPTH MAXIMUM
B	24" L, 1 LAMP 20W FLUOR. FIXTURE, WRAPAROUND WHITE DIFFUSER, MOUNT ABOVE MEDICINE CABINET.	BRITE–LITE CO. CAT. #1/20/BFF OR EQUAL	MAX. MTG. HT. 78" TO ℄.
C	ADJUSTABLE HEIGHT PENDANT INCANDESCENT, 3–75W MAX., BUILT–IN 3–POSITION SWITCH.	HOMELAMP CO. CAT. #3/75/DRP OR EQUAL	———
D	10" D. DRUM–TYPE FIXTURE, WHITE GLASS DIFFUSER, CENTER LOCK–UP, 2–60W INCAND. MAX., SURF. MTD.	BRITELITE CO. CAT. #2/60/HF OR EQUAL	6" MAX. DEPTH.
F	12" D. DRUM FIXTURE, CONCEALED HINGE ON OPAL GLASS DIFFUSER FOR RELAMPING WITHOUT GLASS REMOVAL, 2–75W INCAND. MAX. SURFACE MTD.	DENMARK LIGHTING SPECIAL UNIT #374821	NO SUBSTITUTION WILL BE ACCEPTED.
G	PORCELAIN LAMPHOLDER, PULL CHAIN WITH WIRE GUARD, 100W. INCAND. SURF. MTD.	———	———
H	SAME AS TYPE G, EXCEPT W/O GUARD.	———	———
K	DECORATIVE OUTDOOR LANTERN, MAX. 150W INCAND., WALL MTD. 84" AFF TO ℄.	TO BE CHOSEN BY OWNER	———
L	UTILITY OUTDOOR LIGHT, ANODIZED ALUMINUM BODY AND CYLINDRICAL OPAL GLASS DIFFUSER. 1–100W INCAND. MAX. 84" AFF TO ℄.	UTIL–LITE CO. CAT. #1/100/BP OR EQUAL	IF VANDALISM IS OF CONCERN, SUBST. PLASTIC DIFFUSER.
IR	RECESSED FIXTURE WITH 150 W INFRARED HEAT LAMP	HEAT-LIGHT CO.	———

(*c*)

Figure 30.20 *(continued)*

SYMBOLS AND ABBREVIATIONS

- BX CABLES RUN CONCEALED; TICS INDICATE NUMBER OF CONDUCTORS EXCLUDING GROUND WIRES. 2 #12 + BARE GROUND, UON
- SAME AS ABOVE EXCEPT RUN EXPOSED
- WIRING RUN TURNING DOWN; WIRING TURNING UP
- 1 3 — HOME RUN TO PANEL; ARROWS AND NUMERALS IDENTIFY CIRCUITS; TICS INDICATE WIRING – AS NOTED ABOVE
- OUTLET BOX AND FINAL CONNECTION TO EQUIPMENT WITH FLEXIBLE CONDUIT (OR BX.)
- D a — CLG. OUTLET WITH INCANDESCENT LTG. FIXTURE D, SWITCH CONTROL – 'a'
- H — WALL OUTLET W/INCAND. FIXT. 'H', M.HT. SHOWN
- A — CLG. OUTLET W/FLUOR. FIXT. 'A'
- B — WALL OUTLET W/FLUOR. FIXT. 'B', M.HT. SHOWN
- J — JUNCTION BOX
- DUPLEX CONVENIENCE RECEPTACLE, 20 A, 2P, 3W, 125V, GROUNDING, WALL MTD., VERTICAL, ℄ 12" AFF NEMA 5–20 R
- A — DUPLEX CONVENIENCE RECEPTACLES, 15A, 2P, 3W, 125V, GROUNDING. W/INTEGRAL GFCI
- B — SINGLE RECEPTACLE, 20 A, 2 P, 3 W, GND'G., NEMA 5–20 R
- C — SINGLE RECEPTACLE, 30 A, 125/250 V, 3 POLE–4 WIRE GROUNDING NEMA 14–30 R; SUPPLIED WITH MATCHING CAP
- S_a — SINGLE POLE SWITCH, 15 A, 125 V, ℄ 50" AFF, UON, CONTROLLING OUTLET (S) 'a'
- $_aS_3$ — SWITCH 3 WAY, 15 A, 125 V, ℄ 50" AFF, UON, CONTROLLING OUTLETS 'a'
- S/T — MANUAL TIMER SWITCH, 1 SET 15 AMP N.O. CONTACTS
- S/D — OUTLET BOX MTD. SWITCH AND DIMMER, INCAND. LOAD ONLY, 600 W MAXIMUM, ℄ 50" AFF
- FLUSH MTD. PANELBOARD

AFF	ABOVE FINISHED FLOOR
MH	MOUNTING HEIGHT
T	THERMOSTAT
UON	UNLESS OTHERWISE NOTED
GFCI	GROUND-FAULT CIRCUIT INTERRUPTER
WP	WEATHERPROOF
NO	NORMALLY OPEN

(*d*)

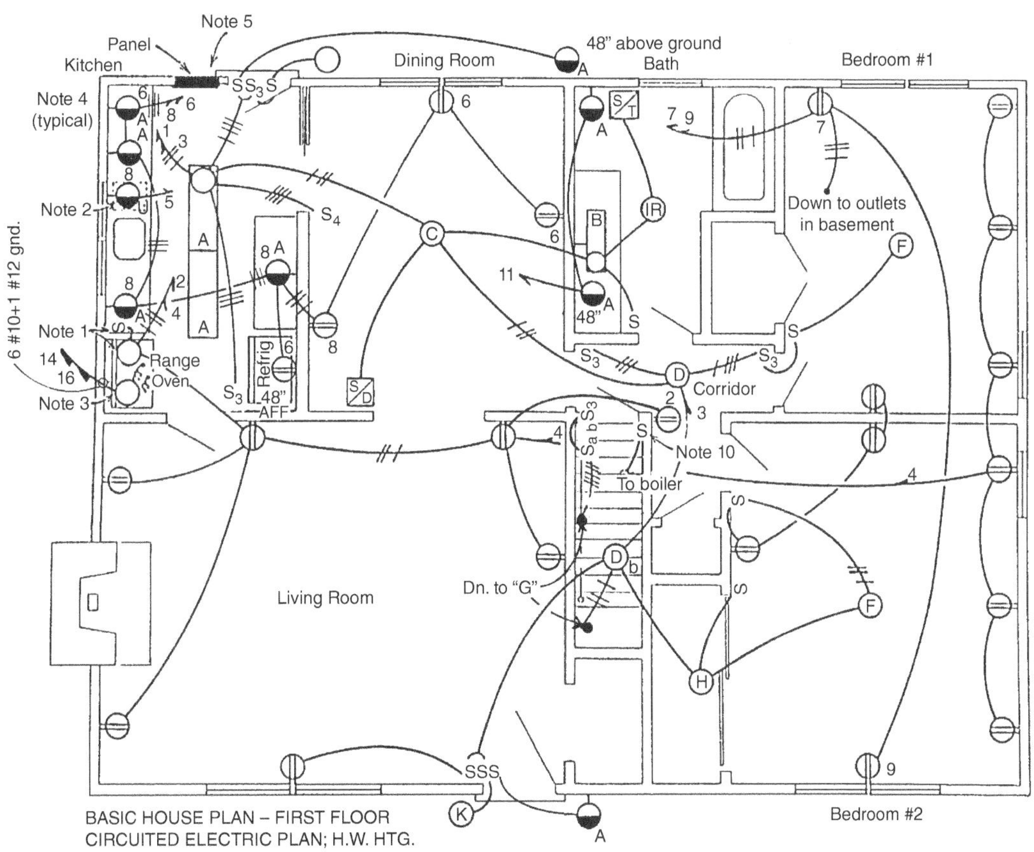

BASIC HOUSE PLAN – FIRST FLOOR CIRCUITED ELECTRIC PLAN; H.W. HTG.

(*e*)

Figure 30.20 *(continued)*

ELECTRICITY

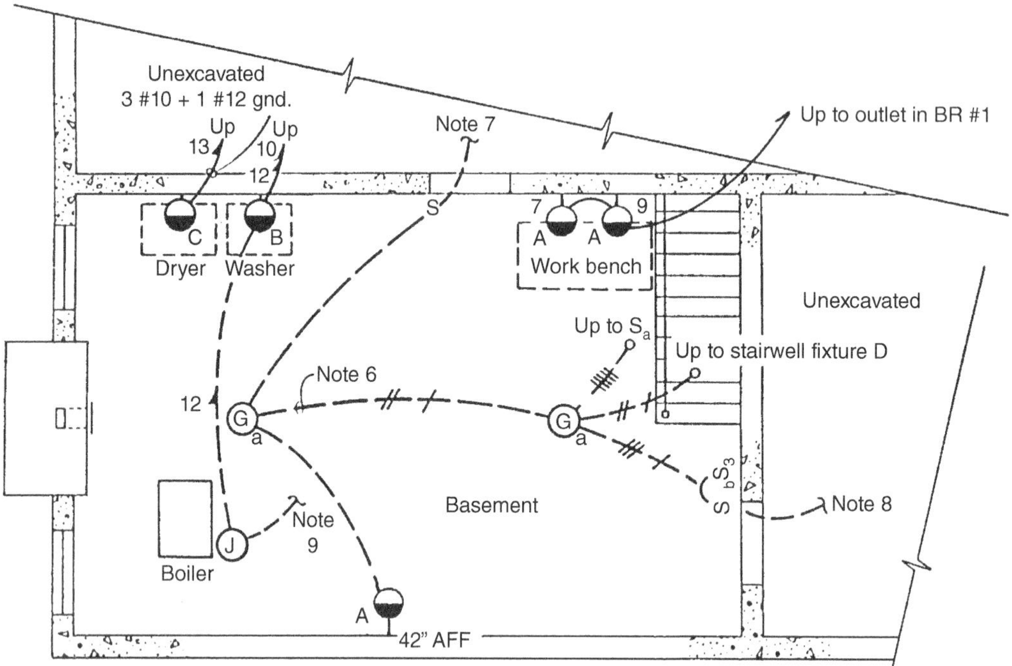

BASIC HOUSE PLAN – BASEMENT
CIRCUITED ELECTRIC PLAN,
H.W. HTG.

(*f*)

Panel schedule

Circ. no.	Description	LOAD V-A.	Circuit breakers	Load V-A.	Description	Circ. no.
1	LTG – {Kit. Dr., Br. #1, Hall, Outside, bath + Φ}	1045 1 R	20 1 2 20	30 6 R	Outlets – LR. & corridor + Exh. fan	2
3	LTG – {Outside Lr., Stair, Br. #2, Bsmt. + 6 Φ}	935 6 R	20 A 3 4 20	6 R	Outlets – BR. 1 & 2	4
5	Dishwasher	1500	20 5 6 20	1500	Outlets – Kit., Dr.	6
7	Outlets – BR 1, Bsmt	2 R	20 7 8 20	1500	Outlets – Kit., Dr.	8
9	Outlets – Bsmt., BR 2	2 R	20 9 10 20	1500	Laundry outlet – Bsmt.	10
11	Bathroom outlets	2 R	20 11 12 20	1300	Boiler	12
13	Spare	—	20 13 14 30 A 2 P	6Kw	Range	14
15	Spare	—	20 15 16			
17	Clothes dryer	5Kw	30 A 2 P 17 18 19 20 30 A 2 P	4800w	Oven	16
	Space for 2 – 1P or 1 – 2 pole		21 22 23 24		Space for 1 – 2P or 2 – 1 pole	
Load total – phase A					Load total, phase B	

<u>Panel Data</u>
Mains, GND. BUS: 150 A MNS., 60 A GND. BUS
Mains C/B 150/100
Branch C/B INT. CAP. AMP. 10,000
Mounting – recess
Remarks: Front suitable for painting

Voltage 120/208
1 PH, 3 wire

(*g*)

Figure 30.20 *(continued)*

30.13 BRANCH CIRCUIT DESIGN GUIDELINES: NONRESIDENTIAL

(a) Schools

Because schools comprise an assembly of varied-use spaces, including instruction, lab, shop, assembly, office, and gymnasium, plus special areas such as swimming pools, photographic labs, and so on, it is not possible to generalize on branch circuit design considerations except for the following:

1. To accommodate the audio-visual equipment frequently used in classrooms, 20-A outlets wired two to a circuit should be placed at the front and back of each such room. A similar receptacle, wired six or eight to a circuit, should be placed on each remaining wall.
2. To accommodate fixed computers, a detailed layout of the space showing the proposed computer locations is required. A two-section surface-mounted or recessed raceway on the wall behind the computers should be provided. The power section should have two duplex 20-A receptacles at each computer station, wired on alternate circuits. The second section of the raceway is intended for low-voltage wiring, such as network cabling, and wiring to computer peripherals. Providing for good cable management can substantially improve the appearance and maintainability of this critical area.
3. Switching for lighting should provide the following features:

    a. High–low levels for energy conservation and to permit low-level lighting for screen viewing. With fluorescent lighting, this can be accomplished by alternate ballast wiring and switching, thus avoiding the high cost of dimming equipment. The multiple modes of classroom use must be considered when establishing switching scenarios.
    b. Separate switching of the lighting fixtures on the window side of the room, which is often well lit by daylight. Control should be initiated automatically by a photocell.

4. Provide appropriate outlets for all special equipment in labs, shops, rooms with cooking activities, and the like.
5. Use heavy-duty devices and key-operated switches for public area lighting (corridors, etc.), plastic instead of glass in fixtures, and vandal-proof equipment wherever possible. All panels must be locked and should be located in locked closets.
6. The *NEC* requires sufficient branch circuitry to provide a minimum of 3 V-A/ft^2 [32.3 V-A/m^2] for general lighting in schools (refer to *NEC* Article 220). Unlike the value for residential occupancy, this requirement does not include receptacles. Receptacles are calculated separately at 180 V-A each for ordinary convenience outlets.

(b) Office Space

1. In small office spaces (less than 400 ft^2 [37 m^2]), provide either one convenience outlet for every 40 ft^2 [3.7 m^2] or one outlet for every 10 linear ft [3 m] of wall space, whichever is greater. In larger office spaces, provide one outlet for every 100–125 ft^2 [9.3–11.6 m^2] beyond the initial 400 ft^2 [37 m^2]. These outlets are intended for miscellaneous electrical devices, in addition to anticipating the constantly increasing number of data system accessories. In addition, provide at every desk (on an adjacent wall, in a "power pole," or in the floor) at least one duplex 20-A receptacle for a computer. These receptacles should be circuited at no more than six to a 20-A branch circuit, and fewer if the specific equipment to be powered so dictates. Figure 30.21 shows one possible circuiting arrangement for the room layouts shown in Figure 30.12. Although other arrangements are possible, the net result is the same (see also Figure 30.22).
2. Corridors should have a 20-A, 120-V outlet every 50 ft [15 m] to provide power for cleaning and waxing machines.
3. As with all nonresidential buildings, convenience receptacles are figured at 180 V-A each.
4. Only specification grade equipment should be used.

(c) Industrial Spaces

These areas are so specialized that no meaningful general guidelines can be given.

(d) Stores

In stores, good practice requires at least one convenience outlet receptacle for every 300 ft^2 [28 m^2] in addition to outlets required for loads such as lamps, show windows, and demonstration appliances.

30.14 LOAD TABULATION

While the designer is circuiting loads, he or she draws up a panel schedule that lists the circuit numbers, load descriptions, wattage (actually, volt-amperes), and the current rating and number of poles of the circuit-protective device feeding each circuit. Spare circuits are included to the extent that the designer

ELECTRICITY

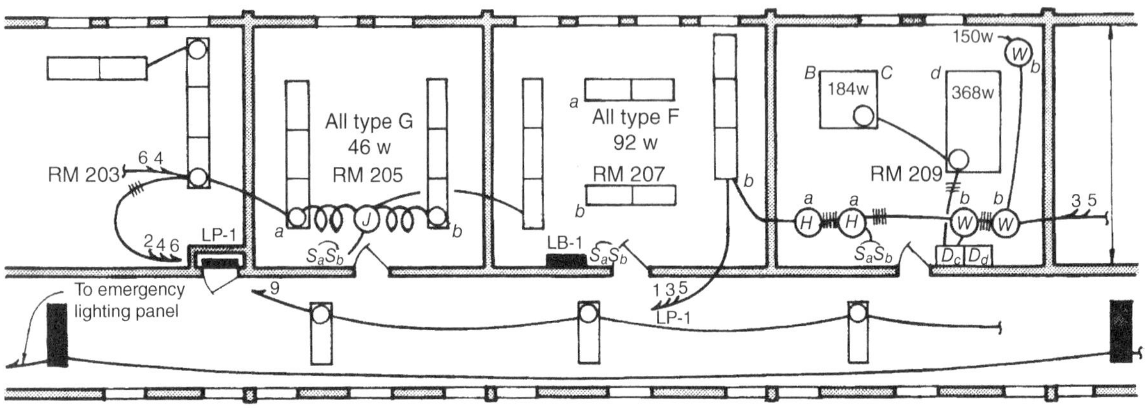

Figure 30.21 *Alternative methods of circuiting are shown in different rooms. Room 205 shows the actual junction box location, with flexible connections to the box at each fixture. Room 207 shows circuit numbers and switch designations only; the placement of junction boxes is understood, and conduit runs are omitted for clarity and because they usually are not representative of the actual field installation. Room 209 shows an outlet box at each fixture, with schematic conduit connections. All of these methods are in common use.*

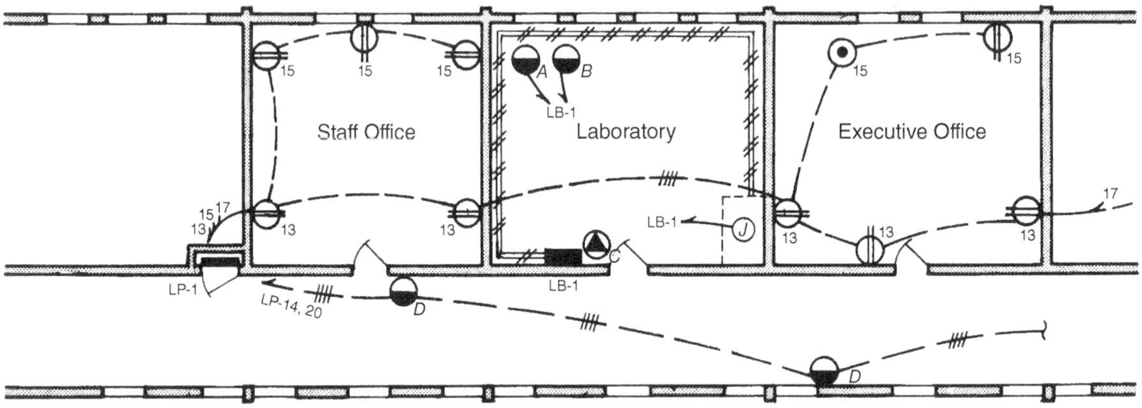

Figure 30.22 *Typical receptacle circuiting of several rooms in an office–laboratory building. Lighting (see Figure 30.21) and power (receptacles) are shown on separate plans to avoid crowding. See Figure 30.15 for symbols and Figure 30.12 for notes. Lighting in offices is recessed; lighting in labs is surface-mounted for flexibility. Note the double circuiting of the type-D receptacles.*

considers them necessary and consistent with design intent (including budget), but normally no less than 20% of the number of active circuits. Finally, spaces are left for future circuit breakers in approximately the same quantity as the number of spare circuits, but always so as to round off the total number of circuits.

Panelboards are normally manufactured with an even number of poles. Thus, if a panel had, with spares, 21 poles, the designer would ask for (at least) three spaces, to give a 24-circuit box. A typical panel schedule is shown in Figure 30.23, which serves the laboratory of Figures 30.21 and 30.22.

In calculating panel loads, the following rules apply:

1. Each specific appliance, device, lighting fixture, or other load is taken at its nameplate rating, except certain kitchen and laundry appliances for which the *NEC* allows application of a demand factor (see *NEC* Article 220).
2. Each convenience outlet, in other than residential spaces, is counted as 1.5 A (180 V-A).
3. Loads for special areas and devices such as show window lighting, heavy-duty lampholders, and multioutlet assemblies are taken at the figures given in *NEC* Article 220.
4. Spare circuits are figured at approximately the same load as the average active circuits (1200 to 1500 V-A).
5. Spaces are not counted in the load.
6. Continuous loads such as lighting are calculated at 125% of their actual value. See *NEC* Article 210.19, which refers to load calculations for conductor sizing and feeder overcurrent protection, respectively, for combinations of continuous and noncontinuous loads. These calculations apply to panel load calculations as well and are best accomplished by adding 25% to the panelboard load, as shown in Figure 30.23.

No.	ELEC. PANEL-LP-1 120/208 V 3∅ 4W LOAD	LOAD IN WATTS ∅A	∅B	∅C
1	Lighting	1200		
2	Lighting	1200		
3	Lighting		1450	
4	Lighting		1050	
5	Lighting			1100
6	Lighting			1200
7	Lighting	800		
8	Lighting	1100		
9	Lighting		700	
10	Lighting		1050	
11	Lighting			1000
12	Lighting			1200
13	Receptacle [2]	900		
14	Receptacle [3]	900		
15	Receptacle		900	
16	Receptacle		900	
17	Spare			1200
18	Spare			1200
19	Spare	1200		
20	Receptacle [4]		1000	
				1000
21	Spare	1200		
			1200	
22-26	Spaces only			
	25% addition for			
	continuos loads [5]	1025	1063	1125
	Phase totals	9525	9313	9025
	Panel total		27863	V-A
	Max ∅ current		77 amp	
	25% spare capacity	~	20 amp	(Future)
	Max. Phase I		97 amps	

Main breaker 225A 3pole (see text example 17.1, section 17.18)
Trip 100 A
Feeder size 4 #2 THW in 1¼"C.

NOTES
1. All C/B 1P; 50 A frame, 20 A trip except ccts 20 and 21, which are 2P, 50 AF/20 AT.
2. 5 receptacles @ 1.5 A each.
3. Corridor receptacle; 120 V section.
4. Corridor receptacle; 208 V section.

Figure 30.23 *Schedule for lighting panel LP-1 (see Figure 30.22).*

In calculating total panel loads, as shown in Figure 30.23, no demand factor may be applied except as specifically stated in the NEC (see NEC Article 220). This is true despite the knowledge that most often the usage will be such that the average load will be lower than the maximum demand (see Section 30.17 for multipanel feeders). If it is known that certain loads will not or cannot be used simultaneously, the load total should reflect only the larger of the two. Thus, heating and cooling loads are generally not concurrent. Nor is building night floodlighting concurrent with the business office load, but it may be with general interior lighting. Note in Figure 30.20g that 2-pole loads (208 single-phase) appear in two columns. Similarly, 3-phase loads would appear in three columns. Also, note that the phase loads are not equal. It is the responsibility of the designer (or contractor) to circuit the loads so that the phases are as closely balanced in load as possible. If this is not done, one phase will carry considerably more current than the others. Because the panel feeder must be sized for the maximum phase current, this may lead to an oversized feeder and therefore a waste of money.

Once the loads have been tabulated, balanced, and totaled by phase, the maximum current is calculated. A portion of the spare capacity available in the branch circuits is added to the total as the basis for the calculation of the feeder load. This spare capacity, shown in Table 30.5, is between 25% and 50%. The exact amount to be initially added in feeder sizing is developed in the ensuing discussion.

30.15 SPARE CAPACITY

Load calculations and examples for residential occupancies and other spaces are detailed in *NEC* Article 220, and examples are given in Chapter 2 of the *NEC*. Because these calculations are specialized (but routine) and are covered there in detail, they are not repeated here. Code requirements, it must be remembered, are minimum; a design solution to meet a client's needs often exceeds these basic requirements.

Once the panel load totals have been determined as detailed previously, the next step is to size the conductors feeding the panel. To do this, an examination of the spare capacity of the panel and of the feeders is necessary in order that the system design be consistent, giving equal capacity for future growth in all of its components. Considering the panel circuitry first, let us examine the effect of load expansion, including spares and spaces (Table 30.8).

TABLE 30.8 Panel Initial and Expanded Loads[a]

	Panels in Facilities Expecting Limited Expansion; Circuits Initially Loaded to Give 25% Expandability (See Table 30.5) (%)	Panels in Facilities Expecting Extensive Expansion; Circuits Initially Loaded to Give 50% Expandability (See Table 30.5) (%)
Initial load	100	100
Initial plus spares	120	120
Load after all circuits including spares are loaded to maximum allowable	150	180
Load after utilizing 20% spaces also	175	210

[a]For development of these values, see McGuinness et al. (1980), pp. 690–691.

ELECTRICITY

As noted in Section 30.11, spare capacity is built into the branch circuitry and into the panels. Most often, expansion is accomplished by additional loading on some circuits and also by adding new circuits via spare circuit breakers in the panel. Table 30.8 gives the ultimate capacity of the panel—that is, fully loaded circuits and fully utilized spares and spaces. Because this ultimate capacity is rarely achieved, panel feeders need only be sized for initial loads as detailed in Section 30.14, and provision made for rewiring to meet anticipated expansion by one of the techniques listed in Section 30.16.

These results can be summarized as follows: For panels in buildings expecting limited expansion, for which branch circuits are loaded to 80% of capacity (25% branch circuit expansion; see Table 30.5, note *e*), the ultimate panel load without new conduit work (i.e., merely by filling out circuits) is 1.5 L, or 50% beyond the initial load. By adding breakers in the spaces, this load can be expanded to 75% beyond the initial load. The corresponding figures for panels that are lightly loaded (66% capacity, i.e., 50% branch circuit expansion, as in Table 30.5, note *f*), in anticipation of considerable load growth, are 80% to 110%, respectively. These results are summarized in Table 30.8.

30.16 FEEDER CAPACITY

To achieve economy, the panel feeder must accommodate the initial load plus some portion of the future load. Spare capacity in feeders (to accommodate a considerable portion of the panel spare capacity, as shown in Table 30.8) is provided by one or more of the following procedures:

1. Provide feeder (and conduit) capacity initially to handle the entire eventual load. This method is the most expensive, requiring an initial outlay for (potentially) no return, and.is rarely used.
2. Provide feeder capacity for the initial load plus spare capacity, with properly sized conduit. Conduit is sized for type THW or RHW without covering. This method yields limited spare capacity.
3. Provide feeder capacity for the initial load plus spare capacity, with conduit oversized by one size. Size conduit for type THW wire, which is very widely used because of its attractive price and excellent electrical properties. Some additional costs are entailed here.
4. If the initial wiring is done with type TW wire, the effect is approximately the same as that of oversizing the conduit by one size and wiring initially with THW wire. This is the result of the lower ampacity rating of TW compared to THW, resulting in a larger conduit size for the same initial ampacity. Exact figures for each can be worked out with the help of information in Chapter 29 along with Table 30.9—or by using the tables in *NEC* Article 310.
5. Provide feeder capacity for the initial load plus spare capacity and oversize conduit by two sizes. This yields most of the capacity necessary in facilities anticipating large expansion.
6. Provide for the initial load plus spare capacity and install an empty conduit for future use. This method is expensive because of the high conduit cost and is advisable only infrequently.

In approaches 2, 3, and 4, the future capacity beyond that initially supplied is handled by the use of larger-gauge wire in the existing conduit. To examine exactly what these alternatives provide in spare capacity, Table 30.10 tabulates the maximum ampacity of various sizes of conduits and the future capacity obtainable. Table 30.9 is taken directly from the *NEC*. Note from Table 30.10 that simply by rewiring, up to 30% additional capacity can be obtained, whereas if the conduit had been oversized, the additional capacity would be 35% to 146%.

Answering the question of how large to make the feeder for a given panel load requires balancing the future panel load, the initial cost of feeder, and the future capacity of existing conduit. It is best to avoid the installation of empty conduits because this is expensive. Rewiring, however, is relatively inexpensive, and oversizing the conduit is the method of choice if rapid expansion is anticipated.

Referring to Table 30.10, note that normal design uses THW cable. Design with T or TW is, in effect, a first step in oversizing the conduit and is generally not economical. The second step is a deliberate oversizing of the conduit that results in much increased conduit ampacity, as reflected by the figures in the table. Using these values in practice requires that the designer juggle cable and conduit costs against anticipated load growth to arrive at the most economical long-term solution. Applying these numbers to concepts previously developed, suggests that buildings designed for 25% branch circuit expansion (see Table 30.5) have a panel capacity of 1.75 times the load (Table 30.8). If it is desired to design the panel feeder to carry the full expansion possible, then calculate the feeder on the basis of panel load plus 20% spares and oversize the feeder by 20%. This gives a feeder capacity of:

$$1.20 \times 1.20 = 1.44 \times \text{initial load}$$

TABLE 30.9 Maximum Number of Conductors in Trade Sizes of Rigid Metal Conduit

		Conduit Trade Size (in.)									
Type Letters	**Conductor Size AWG (kcmil)**	**½**	**¾**	**1**	**1¼**	**1½**	**2**	**2½**	**3**	**3½**	**4**
TW, THW, RHW, RHH (without outer covering), and THHW	6	1	3	5	8	11	18				
	4	1	1	3	6	8	14	20			
	2	1	1	2	4	6	10	14			
	0		1	1	2	3	6	8	13		
	00		1	1	2	3	5	7	11		
	000		1	1	1	2	4	6	9	13	
	0000			1	1	1	3	5	8	10	
	250			1	1	1	3	4	6	8	11
	300			1	1	1	2	3	5	7	9
	350				1	1	1	3	5	6	8
	400				1		1	2	3	5	6
	500				1	1	1	2	3	5	6
	600				1	1	1	1	3	4	5
THHN	6	2	4	7	12	16	27				
	4	1	2	4	7	10	16	24			
	2	1	1	3	5	7	12	17	26		
	1	1	1	1	4	5	9	12	19	26	
XHHW (4–500 kcmil)	0	1	1	1	3	4	7	10	16		
	00		1	1	2	3	6	9	13		
	000		1	1	1	3	5	7	11	15	
	0000		1	1	1	2	4	6	9	12	16
	250			1	1	1	3	5	7	10	13
	300			1	1	1	3	4	6	9	11
	350			1	1	1	2	3	6	7	10
	400			1	1	1	2	3	5	7	9
	500				1	1	1	2	4	5	7
	600				1	1	1	1	3	4	

Source: From NFPA, 2023 / The National Electrical Code.

TABLE 30.10 Maximum Wire and Ampacity of a Conduit, and Ampacity Gain on Rewiring

Initial Installation with THW Cable			Rewiring with XHHW				
			Using Original Conduit			Capacity Increase Having Oversized Conduit	
Initial Conduit Size (in.)	**Maximum[a] Wire AWG/kcmil**	**Maximum Amperes**	**Maximum Wire**	**Maximum Amperes**	**Capacity Increase (%)**	**One Size (%)**	**Two Sizes (%)**
1½	1	130	1/0	170	30	100	146
2	3/0	200	4/0	260	30	60	115
2½	250	255	300	320	25	69	100
3	400	335	500	430	28	35	81
3½	600	420	600	475	13	64	—

[a]Assuming four single conductors in conduit: 3-phase conductors plus neutral.

Table 30.10 indicates that rewiring adds another 20% on average. This gives:

$$1.20 \times 1.44 = 1.73 \times \text{initial load}$$

which corresponds very closely to the 1.75 desired.

For a building with 50% branch circuit expansion, utilizing the full panel space capacity creates an ultimate panel capacity of 2.1 times the initial load (see Table 30.8). This is accomplished as follows: Oversize the feeder by 15% and oversize the conduit by one size. The latter step gives an average expansion of 65%. Therefore,

$$\text{feeder capacity} = 1.15 \times 1.20 \times 1.65 = 2.3$$

which is approximately the desired value.

If, as in the case of laboratories, more than 100% expansion is anticipated (see Table 30.1), conduit should be oversized by two sizes and initial wiring oversized by approximately 25%. Feeders thus arranged will handle the new panels required to meet the anticipated expansion.

Two factors should be carefully noted here. First, the smaller conduits offer the largest expandability, although, in dollars per amperes, they are more expensive. Second, in order to take advantage of spaces in a panel, conduit stubs should be taken from the panel and extended into suspended ceilings (or another procedure used to make the panel circuitry easily accessible in the future).

ELECTRICITY

30.17 PANEL FEEDER LOAD CALCULATION

EXAMPLE 30.1 Refer to Figure 30.23. The panel is for a laboratory/office area. Because substantial expansion is anticipated, circuitry follows the bottom section of Table 30.5. The ultimate panel load would be:

26 circuits at 1920 V-A = 50 kVA = 138 A

Thus, the initial feeder is sized for 115 A (indicated in the schedule as 4 #2 THW, 1 ¼ in. C), but rewiring with XHHW will allow as much as 150 A in the same conduit. A 225-A frame circuit breaker is chosen initially because eventually the trip will be raised to 150 A.

The *NEC* in Article 220 specifies minimum volt-amperes per unit area (V-A/ft^2 [V-A/m^2]) values for various occupancies, for lighting, and for miscellaneous power loads. Proper design procedure therefore requires that after detailed design of an area, the actual loading should be compared to these minima, and the larger of the figures, concerning the number of circuits and feeder load, should be used. An example will help to make this clear. ■

EXAMPLE 30.2 Assume a single floor of an office building of 100 × 200 ft [30.5 × 61 m]. Assume also that 15% of the area is corridor and storage, equally divided between the two. Calculate the load and feeder size. Assume a good-grade speculative construction venture.

SOLUTION

Office space	= 85% of 20,000 ft^2 [1860 m^2]
	= 17,000 ft^2 [1581 m^2]
Corridor and storage	= 15% of 20,000 ft^2 [1860 m^2]
	= 3000 ft^2 [279 m^2]

The *NEC* specifies a minimum of 3 ½ V-A/ft^2 [37.7 V-A/m^2] for lighting and 1 V-A/ft^2 [10.8 V-A/m^2] for miscellaneous receptacles in office space. It further specifies ½ V-A/ft^2 [5.4 V-A/m^2] minimum for corridors and ¼ V-A/ft^2 [2.7 V-A/m^2] for storage. Therefore:

Lighting	
Office load:	17,000 × 3 ½ = 59.5 kVA
	[1581 × 37.7 = 59.5 kVA]
Corridor:	1500 × ½ = 0.74 kVA
	[139 × 5.4 = 0.74 kVA]
Storage:	1500 × ¼ = 0.38 kVA
	[139 × 2.7 = 0.38 kVA]
Minimum lighting load	= 60.6 kVA
RECEPTACLES	

17,000 × 1 V-A/ft^2 = 17 kVA
[1581 × 10.8 = 17 kVA]

These values would then be compared to the actual design loads. Receptacles are counted at 180 V-A each, as noted in Section 30.14, item 2.

If the code minima exceed the design load (as they well may if lighting is carefully designed), panels must be equipped with sufficient additional circuits to make up the difference. The number of such circuits is up to the designer, because circuit loading is not specified.

For instance, suppose that the lighting design was accomplished at 2 V-A/ft^2 [21.5 V-A/m^2]. Panel circuits would have to be provided for the additional 1 ½ V-A/ft^2 [16.2 V-A/m^2] as follows:

$$17{,}000\ \text{ft}^2 \times 1.5\ \text{V-A/ft}^2 = 25.5\ \text{kVA}$$

$$[1581\ \text{m}^2 \times 16.2\ \text{V-A/m}^2 = 25.5\ \text{kVA}]$$

Assuming a 30% to 50% future load expansion, we would circuit the lighting loads at approximately 1300 V-A per circuit (see Table 30.5). Therefore, we would provide:

$$\frac{25{,}500\ \text{V-A}}{1300\ \text{V-A}} = 20\ \text{additional circuits}$$

With respect to the minimum feeder load, the *NEC* specifies that it be increased by 25% if loads are continuous (3 or more hours). This requirement allows for breakers to heat up in panels while carrying a continuous load and is waived for circuit breakers that are ambient compensated—that is, are rated to carry 100% load. Because we have established 80% of the breaker rating as the maximum load (see Table 30.5), we have already accounted for this factor in circuitry but must utilize it in the feeder calculation. Assuming that the code minima are the design loads, then, for feeder calculation:

Lighting load = 60.6 kVA × 125% = 75.75 kVA
Receptacle load, as given = 17.0 kVA
Minimum feeder load = 92.75 kVA
25% future load = 23.2 kVA
Design feeder load = 116 kVA

Because this load would be divided among several panels, the building electrical design might be such that the panels are not all fed by one feeder (see Figure 30.24). Assuming they are, however, the feeder would be calculated in terms of 3-phase current thus:

$$I = \frac{kVA}{0.360} = \frac{116}{0.360} = 322\ \text{A}$$

■

Using THW cable, a minimum of 400 kcmil would be required. Conduit would be a minimum of 3 in. [76 mm] but might be increased to 3 ½ or 4 in. [89 or 102 mm], according to the considerations of spare capacity discussed in the previous section.

Table 30.11 lists the relevant relations that will assist in computing currents. A note of caution is

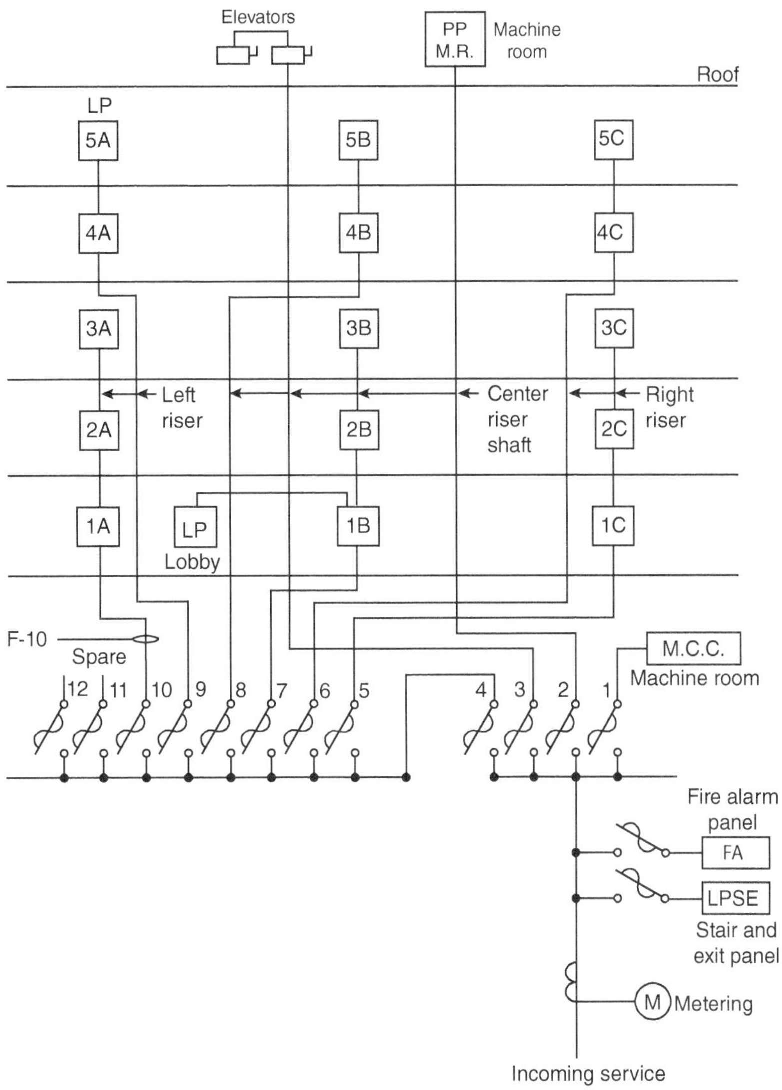

Figure 30.24 *Typical power riser diagram. Ordinarily, the main switchboard would be shown as a large rectangle with the feeders emanating from it, and a switchboard schedule would detail the contents. Here, because of the unusual bus arrangement, the main switchboard appears as it would on a single-line diagram.*

TABLE 30.11 Current and Volt-Ampere Relationships

120-V Single-Phase	120/240-V 3-Wire	120/208-V Single-Phase, 3-Wire	120/208-V 3-Phase	277/480-V 3-Phase	277-V Single-Phase
$I = (V\text{-}A)/120$	$I = (V\text{-}A)/240$	$I = (V\text{-}A)/208$	$I = (V\text{-}A)/360$	$I = (V\text{-}A)/830$	$I = (V\text{-}A)/277$

required regarding the use of computer programs to perform the previous calculations. It should be very clear that considerations of economy and anticipated expansion must be considered when designing a feeder system. Most software will not make such necessary comparisons and calculations unless specifically directed, and this frequently requires a multistage calculation. Most programs are arranged to select feeders on the basis of current loads, plus a specified spare capacity, without regard to expansion or economic considerations. It is important to understand the methods used by any particular computer program in order to apply its output intelligently.

30.18 HARMONIC CURRENTS

Large harmonic currents have been the cause of considerable difficulty in modern electrical installations. Without going into detail on this highly technical subject, a brief description of the problem and its causes is given here.

Conventional electrical loads such as lighting, resistive devices (heaters), motors, and the like are linear (i.e., the load impedance remains essentially constant, regardless of instantaneous voltage). This is not the case with most electronic equipment. Computers, modems, printers, electronic lighting ballasts, variable-speed motor drives, and solid-state equipment of all types are essentially nonlinear loads. As such, they produce harmonic currents, of which the odd-order ones are additive in the power system neutral conductor. The most troublesome of these are the third harmonic and its odd multiples (9th, 15th, 21st, ⋯). These currents can become so large in a modern computerized office (especially with electronic ballasts) that instead of the neutral conductors carrying the unbalanced current in a 3-phase system (zero in a balanced system), they actually carry more current than the phase wires.

Other serious negative effects of harmonic currents are:

- deterioration of electronic equipment performance; continuous or sporadic computer malfunctions
- overheating of the neutral—possibly causing neutral burnout and resulting in equipment being subjected to severe voltage variations
- overheating and premature failure of transformers—even when the transformer nameplate rating seems adequate
- overheating of motors because of operation with a distorted voltage waveform
- nuisance tripping of circuit breakers and adjustable-speed drives
- telephone interference
- capacitor fuse blowing

The problem of destructive harmonic currents becomes progressively more severe as the amount of electronic equipment in use increases (as it does continuously).

Today, at least half of the electric load in a modern office-type facility is composed of nonlinear, harmonic-producing equipment. It follows that all such facilities, both existing and under design, must take necessary corrective measures. In the past, these measures consisted of oversizing equipment to avoid overload burnout; adding passive harmonic filters (which act to reduce harmonic content) in the electric distribution system; using isolation transformers at sensitive loads; selecting power sources with low output impedance to minimize voltage distortion; using controls that are relatively insensitive to harmonic distortion; adding meters throughout the system that measure true-RMS (root mean square) voltage and current rather than the average values shown by conventional meters; and other expensive (and essentially passive) power conditioning equipment (see Chapter 28). In view of the increasing severity of the problem, computer-controlled variable power-conditioning equipment (called active line conditioning) has become available. Such power-conditioning equipment operates in a fashion similar to that described for active noise cancellation in Chapter 25. The conditioner instantaneously and continuously analyzes the harmonic content of the line voltage and injects an equal but exactly out-of-phase voltage to cancel the harmonics and produce a pure sinusoidal voltage supply. The harmonic currents that are required by nonlinear loads are supplied by a digital signal generator. Other techniques are also used; their description is beyond the scope of this book.

In any retrofit work, the electrical designer must obtain a detailed electrical system analysis for the existing system, performed by engineers experienced in the field of power quality. Many existing systems carry as much as 70% to 80% harmonic current and constitute a major system failure waiting to happen. A proper power quality study, performed with such instruments as true-RMS meters, harmonic analyzers, frequency selective voltmeters, and spectrum analyzers, will yield a true picture of an existing system and permit the electrical rehab work to be engineered with harmonic limitations as one of the important design parameters.

30.19 RISER DIAGRAMS

When all devices are circuited and panels are located and scheduled, a riser diagram can be prepared. A typical diagram, shown in Chapter 28, represents a block version of a single-line diagram except that, as the name implies, vertical relationships are shown. All panels, feeders, switches, switchboards, and major components are shown, up to, but not including, branch circuiting. This diagram is essentially a vertical section taken through a building electrical system.

EXAMPLE 30.3 Feeder F10 of Figure 30.24 serves lighting panels 1A, 2A, and 3A. Calculate the required feeder size, considering loads, future expansion, and voltage drop.

SOLUTION

The connected loads on these panels have been computed in accordance with the previous considerations and are:

LP-1A: 110A
LP-2A: 125 A
LP-3A: 100 A
335 A

These values include connected load, spares, and a 25% future capacity factor (the *NEC* requires that feeders be sized minimally for loads calculated by using the given *NEC* unit area loads plus any demand factors listed; see *NEC* Article 220). If actual panel circuit loads are larger than the *NEC* minima, then the actual loads are used in feeder calculations, and any reasonable demand may be used, provided that at no time do the totals drop below the minima specified in the *NEC*. Therefore, diversity factors may be applied between panel loads in a judicious manner.

In office building work, typical diversity factors are as follows:

Lighting Panels Fed from a Single Feeder	Diversity Factor
1 or 2	1.00
3 or 4	1.09
5, 6, or 7	1.18
8, 9, or 10	1.33

Thus, the load on feeder F10, using 100% demand per panel and 1.09 diversity between panels, would be (335 A × 1.0)/1.09 = 307 A. ■

Methods for handling future expansion were discussed previously. In this case, the feeder, before voltage drop considerations, would be (from Chapter 29 and Table 30.9) 4–350 kcmil THW in 3-in. C. In this case, the values in Table 30.10 can be misleading if not read carefully. A 3-in. conduit will take a maximum of 400 kcmil THW. However, because the initial design is using only 350 kcmil, which also requires a 3-in. conduit, the expansion possible by rewiring with XHHW is:

$$\frac{\text{ampacity of 500 kcmil, XHHW in 3-in.C}}{\text{ampacity of 350 kcmil, THW in 3-in.C}}$$
$$= \frac{430\text{ A}}{310\text{ A}}$$
$$= 1.39 \text{ (i.e., 39\% expansion)}$$

This should be sufficient.

The final consideration in sizing a feeder is voltage drop. The *NEC* suggests (but does not require) that sizing branch circuit conductors for a maximum of 3% voltage drop to the farthest outlet, and sizing feeders so that the maximum combined feeder and branch circuit voltage drop to the farthest outlet does not exceed 5%, will give reasonably efficient operation. These figures apply to all types of loads.

It is suggested that it makes economic sense to restrict branch circuit voltage drop to 2% to the farthest outlet while maintaining the overall 5% limitation. This allows a minimum 3% drop in feeders, which are frequently long and, because of size and ampacity inefficiency, relatively more expensive than small branch circuit wiring.

Many tables and curves are published by manufacturers from which voltage drop can be obtained. Such a set of curves is shown in Figure 30.25, which shows the maximum length of run for a 1% drop. Applying these curves to Example 30.3 and assuming an 80-ft [24-m] run gives the following.

From the curves, 307 A on 350 kcmil cable gives a 1% drop in 50 ft [15 m]. Therefore, the drop in 80 ft [24 m] will be 1.6%, which is well within our criteria, assuming that branch circuits were limited to a 2% voltage drop.

In summary, then, feeders are sized in accordance with load (actual or based upon unit area, whichever is larger) and voltage drop. Conduit may be oversized to address the potential for substantial future load expansion.

30.20 SERVICE EQUIPMENT AND SWITCHBOARD DESIGN

The main switchboard shown in Figure 30.24 constitutes a combination of service equipment and feeder switchboard. The service equipment portion of the board comprises the metering and four main switches feeding the risers, motor control center (MCC), rooftop machine room, and elevators. The feeder board comprises switches 5 through 12. Such an arrangement is permissible inasmuch as the *NEC* allows up to six fused switches or circuit breakers to serve as the service disconnect means. This arrangement was chosen in order to separate to the greatest extent possible the motor loads (elevators, air-conditioning equipment, basement power, etc.) from the lighting. Such a procedure minimizes lighting fluctuations resulting from motor starting and yields simpler maintenance. Also, the size of the main switch is reduced. This switchboard would be of the metal-clad dead front type with switches or circuit breakers, as desired.

Other recommendations affecting service equipment are as follows:

- All equipment used for service, including cable, switches, meters, and so on, should be approved for that purpose.

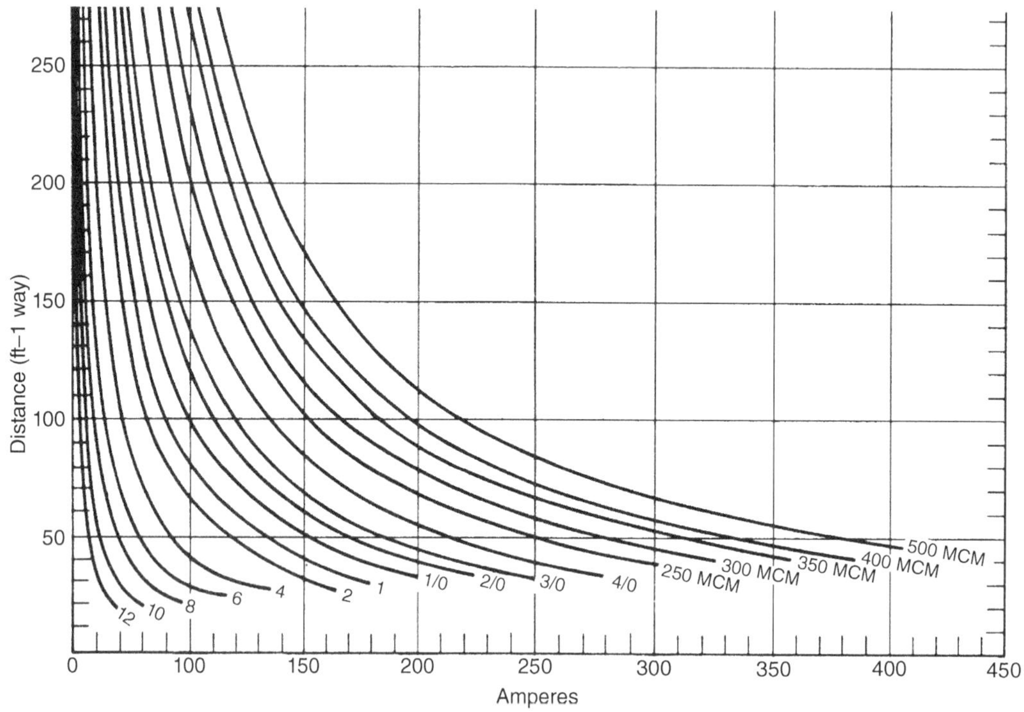

Figure 30.25 *Curves for determining voltage drop in copper cables. Curves show the maximum one-way circuit length for a 1% voltage drop.*

- A minimum of 100-A, 3-wire, 120/240-V service be provided for all individual residences.
- No service switch smaller than 60 A, or circuit breaker frame smaller than 100 A, should be used.
- In multiple-occupancy buildings, tenants must have access to their own disconnect means.
- All building equipment should be connected on the load side of the service equipment, except that service fuses, metering, fire alarm and signal equipment, and equipment serving emergency systems may be connected ahead of the main disconnect (see Figure 30.24).
- For additional information on electric service, see *NEC* Article 230.

30.21 EMERGENCY SYSTEMS

(a) General Information

Some considerations relevant to power reliability are discussed in Section 30.1*b*, and a brief review of the equipment available to supply emergency power is presented in Chapter 28. Emergency lighting equipment is covered in Chapter 19. This section discusses possible emergency power supply arrangements. The choice of arrangement and the size and type of equipment depend largely upon the requirements of local codes, which determine the loads that must be supplied from the emergency system. Although this section uses the term "emergency," the concepts involved are equally applicable to standby systems. Where differences occur, they are pointed out.

The three classes of electrical power supply systems that are included under the general category of emergency electric supply systems are:

- Emergency systems: Covered by *NEC* Article 700
- Legally required standby systems: Covered by *NEC* Article 701
- Optional standby systems: Covered by *NEC* Article 702

The essential differences among the three classifications involve the purpose, application, and duration of the specific non-normal power supply. All three types are activated when normal power fails, but only the first two types are legally required when so specified by a governmental agency having jurisdiction. Optional standby systems are, as their name states, optional; they are entirely dependent upon the project requirements established by the facility's owner or operator.

Emergency systems are those essential to human safety. As such, and depending upon the type of facility involved, they may supply power for fire detection, lighting, alarm and communication systems, elevators, fire pumps, and such loads as ventilation, refrigeration, and industrial processes where power interruption would imperil human safety. Generally, emergency systems must pick up the loads automatically within 10 seconds of power interruption. Where the emergency power supply consists of batteries, they must have a full-load capacity of 90 minutes. Where the power supply is an engine–generator set, an on-site fuel supply that will suffice for a minimum of two hours of full load is required. The entire wiring system for an emergency power system must be separate and distinct from the normal power system, except where both systems are connected to the same item of equipment.

Legally required standby systems are those required to provide power for essentially the same types of loads as emergency systems, except that their absence does not involve immediate danger to human life. Systems for firefighting, control of health hazards, long-term rescue operations, and industrial hazard prevention would be among the electrical loads served. For these systems, loads must be picked up automatically within 60 seconds of failure of normal power and be maintained for the same minimum periods of time specified for emergency power systems. Wiring may be run in the same enclosures as those used for the normal power system.

Optional standby power systems are intended essentially to minimize economic loss and, therefore, can supply any load identified by the building owner. Connection of the selected standby source can be manual or automatic, and the duration of service is entirely at the owner's discretion. No restrictions are placed on the wiring system beyond those applied to the normal electrical wiring system.

(b) Relevant NFPA Codes

The *NEC* does not determine whether an emergency system is required; that decision is made by the applicable jurisdictional authorities and is generally a response to the requirements of *NFPA 101: Life Safety Code*®. Once a determination to provide an emergency system has been made, however, the equipment and installation must comply with the requirements of *NFPA 70: National Electrical Code*®. Other relevant NFPA standards (each of which is updated on a regular basis) include:

NFPA 70. National Electrical Code®

NFPA 99. Health Care Facilities Code Online Training Series

NFPA 101. Life Safety Code®

NFPA 110. Standard for Emergency and Standby Power Systems

NFPA 111. Standard on Stored Electrical Energy Emergency and Standby Power Systems

(c) Technical Considerations

In general, when emergency power is discussed, it is assumed to be replacing normal power; that is, the assumption underlying most codes and standards is that power must be supplied to selected loads within the building because of a utility power outage. In good-practice design, cognizance must also be taken of situations in which normal power has not failed, and an outage is localized because of an in-building equipment failure. This aspect of design—reliability—is left to the designer and owner to decide in the context of a given building situation. Some of the arrangements discussed here differentiate between the nature of outages—that is, a utility or general outage versus an equipment or local outage. An exception to this generalization occurs with health-care facilities, where *NEC* Article 517 specifies an internal electrical design that largely covers both types of outages down to the distribution level.

An emergency system includes all devices, wiring, raceways, and other electrical equipment, including the emergency source that is intended to supply electric power to the selected load(s). An important aspect of code requirements permits the use of a single power source for (1) emergency, (2) legally required standby, and (3) optional standby systems, provided that it is equipped with automatic selective load pickup and load-shedding equipment that will ensure adequate power to the three types of systems in the order of priority stated (1, 2, 3).

1. Where emergency loads are not substantial, a storage battery arranged to be connected automatically on power outage may be used. Where all emergency loads can be operated on DC, the arrangement of Figure 30.26a is used. If AC power is required, the arrangement of Figure 30.26b can be utilized.
2. Where emergency loads are larger than can be practically supplied by a battery installation, and where a start-up and power transfer time of up to 10 seconds is tolerable, a generator set is employed. A combination of emergency sources may be used in a single building. For instance, a generator can supply bulk power loads and a battery installation selected lighting loads.

   The system can be arranged with a single transfer switch that senses normal power loss, as in Figure 30.27a, or it can use multiple switches, each one of which will sense power loss at its downstream location, as in Figure 30.27b. The latter system provides greater power reliability, provided that the design is such that the emergency power uses an independent power path to the transfer switches. Otherwise, a fault in a junction box or an

ELECTRICITY

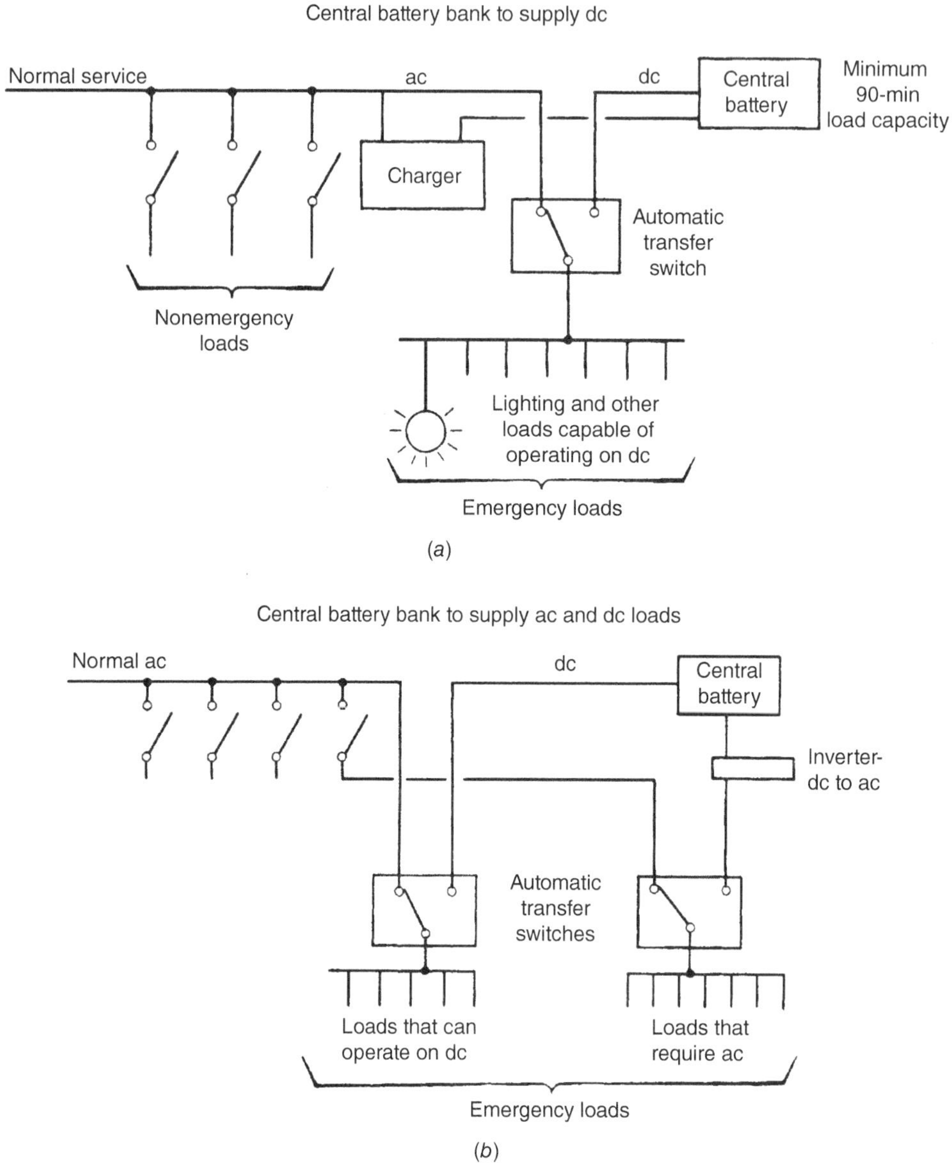

Figure 30.26 *Use of a central battery for emergency power supply. When all loads can be energized with DC, the arrangement of (a) is satisfactory. When AC as well as DC must be distributed, a central inverter is added, as in (b), and the AC and DC emergency loads are fed separately.*

item of equipment that interrupts normal power downstream will also prevent emergency power from reaching that point.

3. Some codes permit the use of two separate electric services in lieu of a normal service plus an emergency source, provided that the two sources are truly independent—that is, come from different utility transformers or feeders; enter the building at different points and, preferably, from different directions; and use separate service drops or laterals. The point is that the type of reliability desired can be obtained only by minimizing the possibility of a single event interrupting both services. The usual arrangement is for one such utility service to be normal and the other standby.
4. The least reliable arrangement is one in which the emergency loads are connected ahead of the main disconnects and are so arranged that a downstream fault within the building will not affect these items. This situation is illustrated in the riser diagram of Figure 30.24, where the stair and exit panel (which supplies egress lighting) and the fire alarm panel are connected ahead of the building main disconnect and protected with their own fuses.

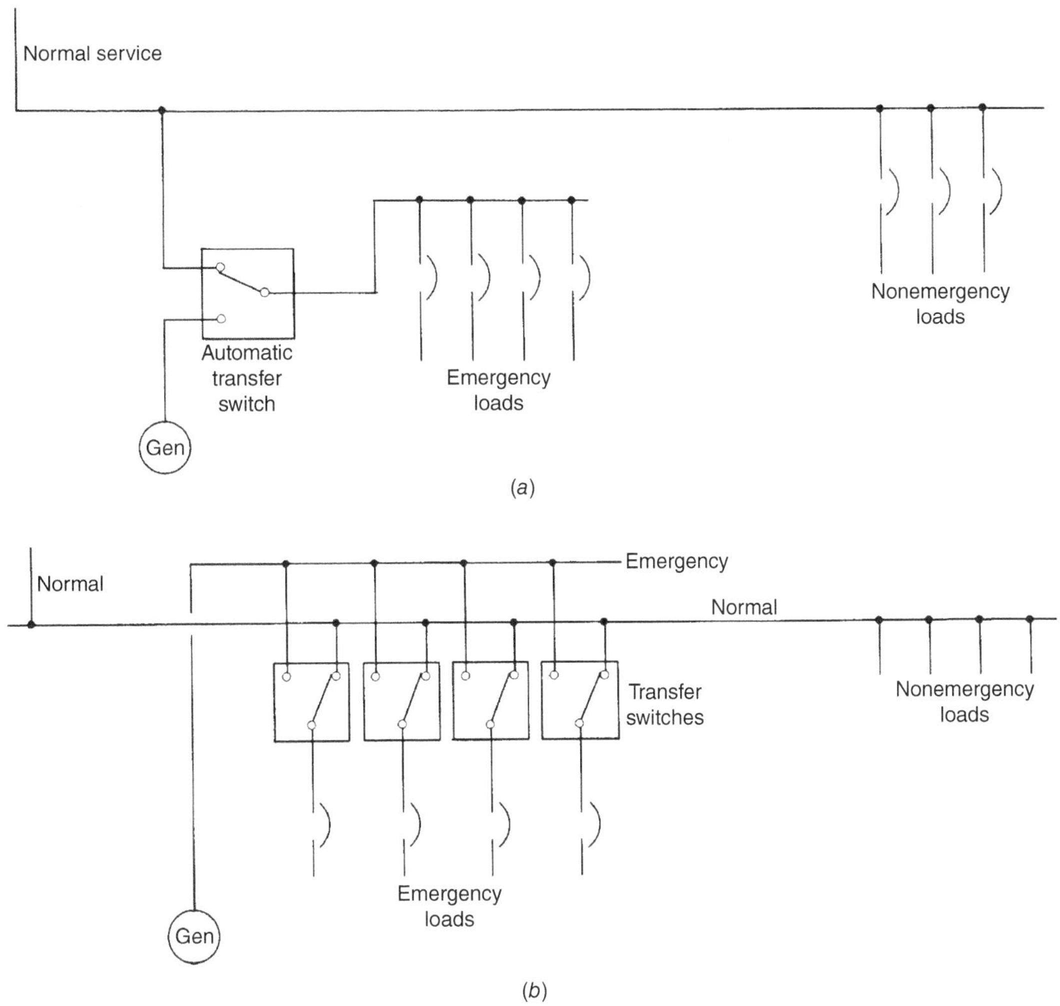

Figure 30.27 *Alternative arrangements of emergency/normal power feed. In (a) a single transfer switch serves the normal power and transfers to the generator upon power failure. In (b) there are multiple smaller transfer switches, thus reducing the chance of a single equipment failure faulting out the entire emergency power system.*

This arrangement is permitted only for Legally Required Standby Systems (as defined in *NEC* Article 701) and not for Emergency Systems (as defined in *NEC* Article 700). The items that may be connected ahead of the main service equipment are generally limited to emergency lighting, fire alarm, fire pumps, standby power equipment, and other alarm and protective equipment, each of which must be provided with a separate disconnect means and overcurrent protection (see *NEC* Article 230.82). This power arrangement can do nothing in the event of a power outage. It was once very common, but is now falling into disuse as a result of more stringent codes and its serious shortcomings.

5. The *NEC* recognizes a category of equipment for emergency illumination called unit equipment. These types of devices, discussed and illustrated in Chapter 19, consist of individual self-contained packages with a battery, charger, and light source permanently mounted and wired at required locations. The panel device feeding these units should be clearly identified.

References and Resources

McGuinness, W. J., B. Stein, and J. S. Reynolds. 1980. *Mechanical and Electrical Equipment for Buildings*, 6th edn. New York: John Wiley & Sons.

ANSI. 2020. *ANSI C84.1-2020: American National Standard for Electric Power Systems and Equipment—Voltage Ratings (60 Hertz)*. Washington, DC: American National Standards Institute.

ASHRAE. 2022. *ASHRAE Standard 90.1-2022: Energy Standard for Sites and Buildings Except Low-Rise Residential Buildings*. Atlanta, GA: American Society of Heating, Refrigerating and Air-Conditioning Engineers.

NFPA. 2022. *2023 NFPA 70: National Electrical Code (NEC)*. Quincy, MA: National Fire Protection Association.

National Institute of Building Sciences. 2014. *United States National CAD Standard, Version 6*. Washington, DC: National Institute of Building Sciences. nationalcadstandard.org

Note: Numerous additional standards from NFPA dealing with emergency power systems and other aspects of electrical system design feature minor references in this chapter.

CHAPTER 31

Photovoltaic Systems

ELECTRICITY

31.1 A CONTEXT FOR PHOTOVOLTAICS

A PHOTOVOLTAIC (PV) SYSTEM PRODUCES ELECTRICITY through direct conversion of incident solar radiation—primarily, but not solely, radiation within the visible portion of the solar radiation spectrum. The amount of electricity produced from PV systems depends on the characteristics of the radiation arriving to the system (such as intensity and wavelengths) and multiple performance attributes of the system itself. This is an incredibly useful transformation process. Virtually every building of any size or function requires and consumes electricity. Solar radiation is a renewable resource, which will generally impinge upon a building in useful quantities in most climate zones (whether or not the building uses it for beneficial purposes). PV then seems an ideal design solution—providing a needed energy form from an otherwise often unused resource. In an ideal world, every building would have a PV system (Figure 31.1). Unfortunately, the economics of PV power generation dampen this idyllic picture. PV power production generally costs an owner more than utility-provided power, even considering reasonable life-cycle costing scenarios. In fact, the economics are currently such that PV systems for buildings are generally the exception rather than the rule. There has long been hope that the economic picture would change through the evolving availability of much cheaper PV components. This has not really happened to the extent anticipated. In fact, recent volatility in the source of PV module supplies has led to some questions regarding long-term product quality. Any future shift in PV system economics will likely occur mainly as the result of higher cost for competing electricity sources (utilities) and/or rebates or incentives from governments or utilities for distributed, greener power sources.

PV electric power systems, which until the middle to late 1980s were little more than a curiosity in the United States because of their very limited applicability and discouraging economics, have recently begun to make a noticeable impact on the commercial electric power scene. This has come about for several reasons, the most important of which are federal government-financed research and development, and state and federal government-sponsored legislation in the electric utility field. The PV Manufacturing Technology project (PVMaT) involving the U.S. Department of Energy (DOE) and 20 private companies (launched in 1990), plus the 1992 Building Opportunities in the United States for Photovoltaics (PV:BONUS) program (also sponsored by the DOE), resulted in the reduction of PV module costs from upward of \$5.00 to about \$2.00–\$3.00 per installed watt, depending on specific circumstances. In addition, these initiatives have spurred the development of new PV module materials, construction techniques, and product forms; increases in PV module efficiency; and continued overall decreases in the cost of power produced.

The legislative impetus has taken the form of utility deregulation and specific legislation requiring utilities to purchase power produced by small, private installations. Several recent government efforts to foster the development of PV systems are underway. In the United States, the Energy Policy Act of 2005 spurred market update of PV by leveraging the Investment Tax

Mechanical and Electrical Equipment for Buildings, Fourteenth Edition. Walter Grondzik, Daniel Overbey, and Alison Kwok.

Companion website: www.wiley.com/go/meeb14e

Figure 31.1 Photovoltaic system installed atop the roof of the West Perry Branch of the Indianapolis Public Library in Indianapolis, Indiana. (Photo by Telamon Energy Solutions.)

Credit (ITC)—a federal policy tool first introduced in 1962—to offer a 30% tax credit for residential and commercial installations. The Inflation Reduction Act (IRA) of 2022 extended the ITC until 2032.

The 1978 federal Public Utilities Regulatory Policy Act (PURPA) required electric utilities to buy electric power at a price equal to the costs they avoid by not having to produce that power. The utilities generally interpret that "avoided cost" as fuel cost only, without considering other avoided costs such as additional plant construction. As a result, energy is often bought under this law at about \$0.03/kWh by the same utilities that sell energy at \$0.08–0.15/kWh. This spread does not encourage the construction of grid-connected PV installations (see Section 31.8). Subsequently, many states adopted net-metering laws that reduced this price imbalance.

In a net-metering arrangement, the utility buys power during peak PV generation periods (see Figure 31.2) at the same rate at which it sells power.

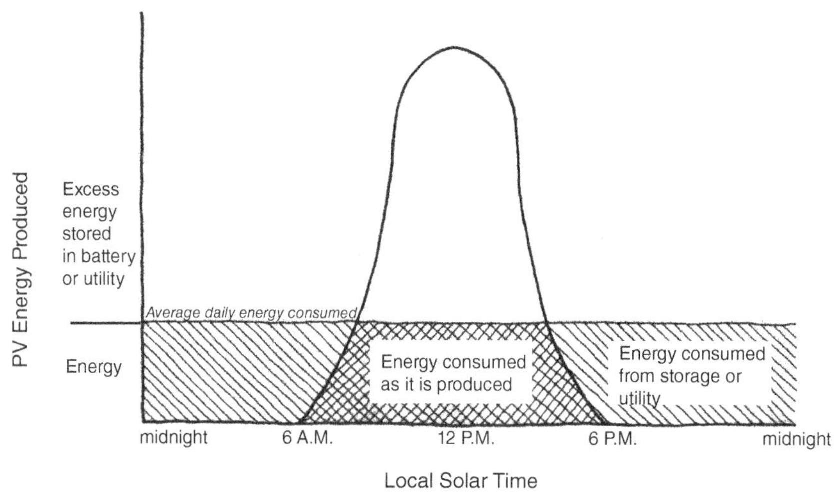

Figure 31.2 Graph representing electricity produced and used by a residence equipped with a PV array. Daily energy use is shown as a uniform average over a 24-hour period to indicate overall quantity but not an actual usage pattern. The bell-shaped PV energy production curve is centered on solar noon and extends from sunrise (solar 6:00 a.m.) to sunset (solar 6:00 p.m.). The area of the bell curve above the average consumption line indicates electricity that would be purchased by a utility in a grid-connected system or stored in batteries in a stand-alone system. (Adapted from Harris et al., 1985. Redrawn by Amanda Clegg.)

At the end of a billing period (usually monthly), any difference between utility-purchased and customer-purchased power is paid for at the given rates (i.e., if the customer used more power than he/she generated, that power is paid for at the conventional utility rate). Conversely (and less commonly), if the customer generated more power than was used, the utility would buy it at the avoided-cost rate. In some jurisdictions, a feed-in tariff (FIT) program ensures that renewable energy producers receive a set price for their electricity, which may be higher than the retail price of electricity. Jurisdictions with regulated utilities often not only lack a FIT program but may implement policies to purchase customer-generated renewable energy at a rate significantly less than the retail cost of purchasing electricity from the utility—severely disincentivizing on-site PV generation systems.

In many instances, a net-metering arrangement is attractive economically to both utility and customer. A utility buys power that offsets its peak afternoon load, which is particularly severe in warm weather, and sells power in its off-peak morning and evening periods. A residential customer sells power at the PV installation's peak output, which is also usually a period of low user consumption (Figure 31.2), and buys it back at the same price in the evening, when PV output is low but power demand is high. The arrangement removes the necessity for an expensive battery installation that, in a stand-alone system, provides power during periods of low PV energy production. This advantage, combined with other advantages such as having a nonpolluting power source, has led to a large increase in grid-connected installations. Indeed, the attractiveness of having a green, environmentally beneficial electric power supply is so great that a number of electric utilities have instituted a (primarily residential) PV installation program in which the utility installs and maintains a PV system on the customer's premises (usually on the roof), for which the customer pays a nominal surcharge on the monthly electric power bill. The advantages of net billing, however, are available only to customers who initiate and maintain a PV system themselves.

31.2 PV CELLS

The production of an electric charge when solar radiation (primarily light) strikes some metallic surfaces has been a recognized physical phenomenon since the mid-nineteenth century. Around 1905, Einstein established a mathematical basis for what has come to be known as the photoelectric effect. The explanation of this effect, in simplified and qualitative terms, is approximately as follows. Light exhibits both wave characteristics and characteristics of a stream of energetic particles called photons. When a photon strikes a photoelectric metal surface, it dislodges a single electron from its normal orbit, causing a charge to appear. When pure silicon is exposed to an intense stream of photons, as from sunlight, many electrons are dislodged from their orbits and proceed to wander about the crystal lattice structure of the silicon crystals. By a process called doping, impurities are deliberately added to pure silicon to create a P–N (positive–negative) junction, across which the electrons flow to create an electric current and give the doped silicon the properties of a semiconductor. If the semiconducting doped silicon is exposed to light and the negative and positive sides of the junction are connected through a load, an electron flow (a current) commences from the negative side of the junction to the positive side, doing work on the load. Work, of course, is energy. The energy comes from the fast-moving photons (light) and is imparted to the dislodged electrons by impact. The overall effect is to create a current flow proportional to the intensity of photon bombardment—that is, proportional to light intensity. Figure 31.3 is a schematic representation of this process.

The photons in solar radiation are not uniformly energetic. It has been found experimentally that a photon must have an energy of at least 1.08 electron-volts (corresponding to a wavelength of 1150 nm) to dislodge an electron. Photons at longer wavelengths do not have sufficient energy to create the photoelectric effect. Because shorter-wavelength photons are more energetic, the entire visible spectrum (light) is useful in PV action, the magnitude of current produced depending upon the intensity of light. According to Einstein's PV theory, a single photon can only dislodge a single electron; thus, any photon energy in excess of the minimum (dislodgement) level is dissipated as heat. For this reason, PV cells generate heat as well as electricity, and some means of heat collection or dissipation is required in practical PV array construction. Commercial PV cells operate at an overall insolation-to-electric energy conversion efficiency of around 15% to 20%. By layering different strata of semiconductor material, each of which is sensitive to a different limited-frequency bandwidth (called the bandgap), efficiencies as high as 35% to 40% can be achieved. Economic considerations, however, have limited today's commercial PV modules to a maximum of two layers with a maximum conversion efficiency of about 15%.

The PV cell (or solar cell) is the smallest electricity-producing unit in a PV system. Cells are not sold individually as commercial products but are arranged by the manufacturer into modules. A PV module is the smallest commercially available electricity-producing

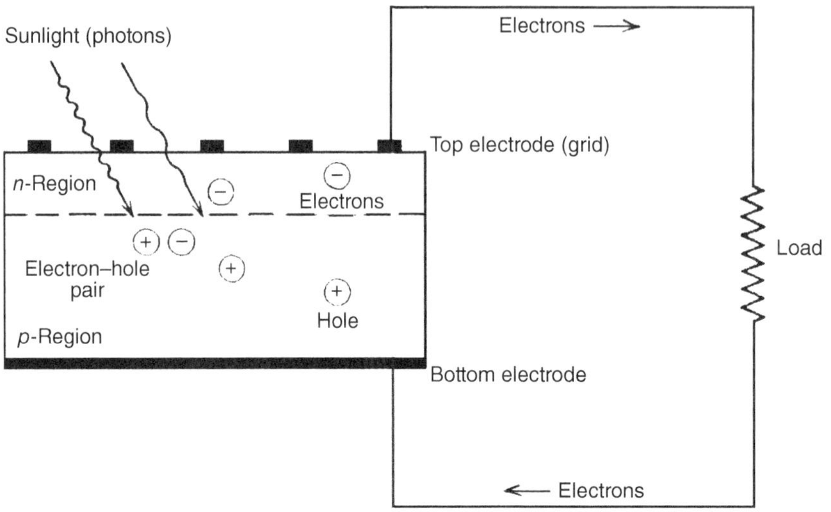

Figure 31.3 *Schematic drawing showing the conversion of solar energy to electrical energy by a PV cell. Photons from solar radiation dislodge electrons in the silicon semiconductor surface, creating a positively charged "hole." The P–N junction creates an electric field that "sweeps" electrons across the junction. When the top and bottom electrodes are electrically connected through a load, an electric current flows. This current is capable of doing work (supplying energy). (Meyers, 1983.)*

increment. The size of a module and the number of modules required to produce a desired electrical output depend upon the type of PV cell used, the manufacturer's product line, and the intent of the PV system designer. If more than one module is required, as is often the case, multiple modules are assembled on site into arrays. A module is a self-contained product; a PV array is a field-assembled group of modules.

An individual PV cell is typically small and produces between one and two watts. Most PV cells are made from crystalline silicon. There are two common types of silicon-based solar cells:

- *Monocrystalline cells:* Produced from a single silicon crystal (hence "mono"); whereas polycrystalline cells consist of many crystals, melted and blended. Monocrystalline cells have a more seamless, homogeneous structure and therefore exhibit a higher production efficiency (typically, 15%–22%).
- *Polycrystalline cells:* Less expensive, have a less efficient interior structure, and exhibit a lower production efficiency (15%–17%).

As mentioned previously, intensive research and development work on solar cell construction has yielded a generation of PV materials that are referred to as thin-film (TF) solar cells, using various copper and cadmium compounds and amorphous rather than large crystal silicon modules. These materials offer varying efficiencies (typical efficiency between 6% and 12%) and costs (sometimes a fraction of the costs of crystalline cells), can be manufactured into large elements, and, because they do not require a glass cover, can be integrated into standard building material formats (as building-integrated PV elements). Typical data for two commercial TF products using this technology are:

Amorphous silicon (a-Si) module

Dimensions: 2 ft × 4 ft × 2 in.
[610 mm × 1220 mm × 50 mm]

Weight: 33 lb [15 kg]

Electrical: 50 W, 72 V, 0.7 A

Thin-film (TF) polycrystalline module

Dimensions: 2 ft × 4 ft × 2 in.
[610 mm × 1220 mm × 50 mm]

Weight: 22 lb [10 kg]

Electrical: 75 W, 17 V, 4.5 A

These next-generation PV materials have high efficiency and a flexible physical format, distinct advantages over the older, large-crystal silicon wafer elements. However, the latter are cheaper, extremely hardy, reliable, and long-lived, as has been demonstrated in many satellite and space vehicle applications. The newer units are initially somewhat unstable and have a shorter life than the large crystal silicon units. It is also important to acknowledge that PV system electrical output changes as weather varies. Extreme heat can lower the electricity production of PV cells by around 20%.

31.3 PV ARRAYS

A PV array is a complete and connected set of modules mounted and ready to deliver electricity. Two basic

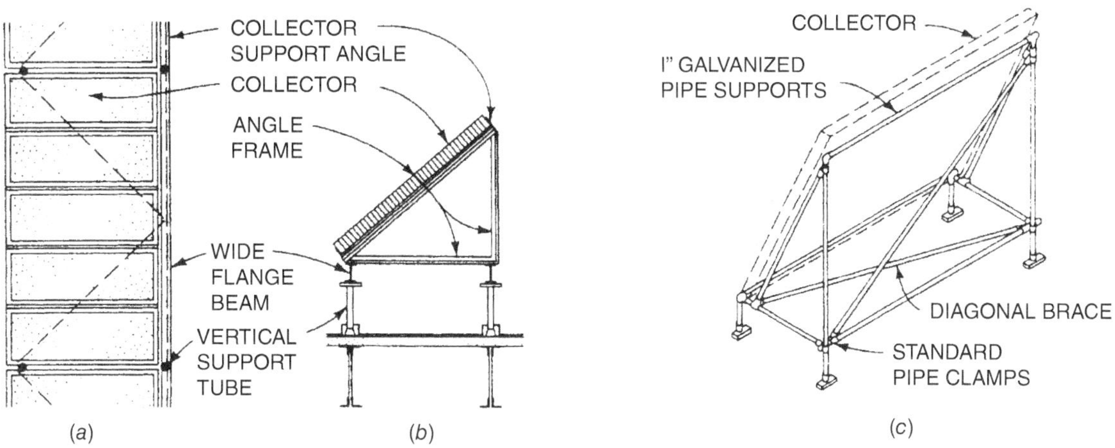

Figure 31.4 *Typical mounting techniques for PV arrays. (a) Plan of the arrangement for mounting large arrays on a horizontal surface. (b) Section through (a) showing mounting and arrays. (c) Simple pipe rack mounting for a small array. For all installations, ensure that wind and seismic loads, snow collection, and array cooling have been considered. (Adapted from Ramsey and Sleeper, 2000.)*

array arrangements exist: stationary and tracking. Building-mounted arrays are typically stationary and often have the advantage of not requiring a substantial support structure, although even a simple roof mounting on a pitched roof can be expensive due to the labor cost involved. The materials usually used for ground-mounted array structures are concrete, galvanized steel, and aluminum. Wood and painted iron are not recommended because of their high maintenance costs and relatively short life. The design of an array mounting will depend upon the array size, weight, mounting angle, and wind loads and is therefore unique to each installation. Mounting on a wooden pole or steel pipe with a single support point is limited to relatively small arrays in low-wind areas because of the stress on the support arrangement. Roof-mounted PV arrays are often ballasted, meaning they employ a mounting method that does not penetrate the surface. Several typical fixed mounting arrangements are shown in Figures 31.4 through 31.6. Some roofing manufacturers offer innovative PV racking solutions that are integrated with roofing membrane systems in order to mitigate water intrusion concerns (Figure 31.7).

The angle, measured from the horizontal, at which a flat-plate collector is mounted is known as the tilt angle. For a stationary (nontracking) array, total annual insolation is maximum for a tilt angle equal to the latitude of the site. To maximize winter insolation, when the sun is lowest, the tilt angle can be increased to latitude plus 15°. Conversely, to maximize summer insolation, when the sun is highest, the tilt angle can be reduced to latitude minus 15°. The tilt angle chosen for a stationary array depends upon PV output usage schedules, the amount of insolation for the months of the year, and economic considerations.

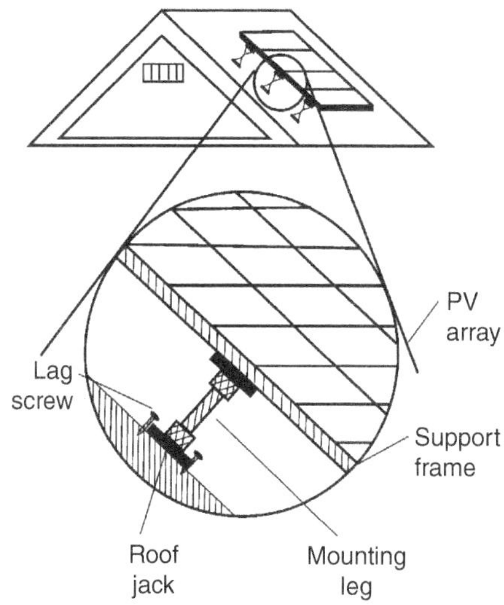

Figure 31.5 *Typical roof mount for a PV array. Array tilt and azimuth can be adjusted through the mount design. Exercise caution in snow areas so that accumulating snow does not block the array and does not slide off onto pedestrians. (From Sandia Laboratories, 1995.)*

Tracking arrays follow the motion of the sun and thereby substantially increase the insolation per unit area of solar cells. Single-axis tracking follows the sun's motion from east to west while maintaining a constant array tilt angle. This type of tracking can increase insolation by 35% to 50% throughout the year and is most effective at low latitudes. Completely passive, sealed compressed gas, single-axis tracking drives are available for small to medium-sized pole-mounted arrays in low-wind areas. Where wind velocity is high, a motorized drive using PV-generated power may be necessary. Dual-axis tracking (following solar altitude and azimuth) for flat-plate collectors and altitude tracking for various designs of concentrating collectors are beyond the scope of this book.

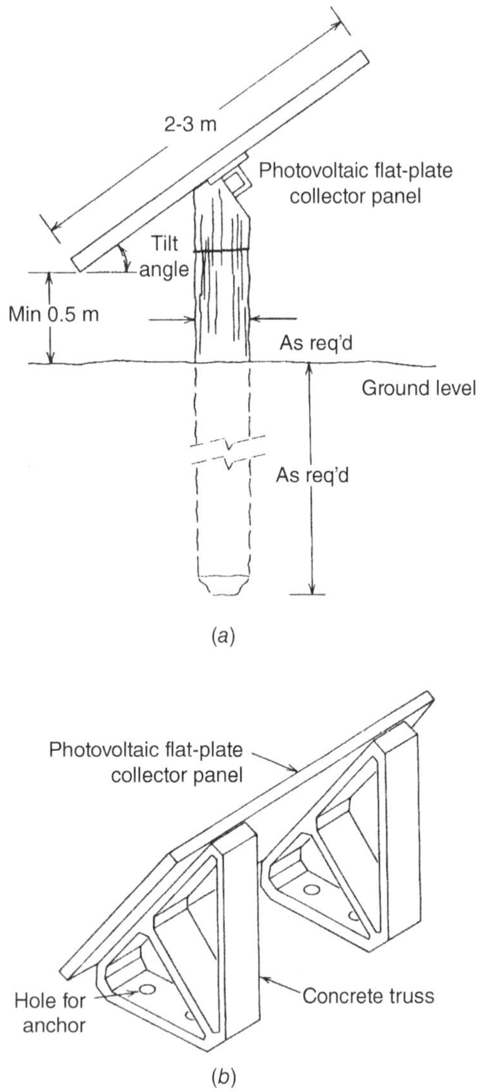

Figure 31.6 *Flat-plate PV collector support structures. (a) Wooden-pole torsion tube support. (b) Concrete truss support. (Meyers, 1983.)*

Figure 31.7 *The Sika SolaRoof® System is a PV racking solution that requires no roof penetrations. This is accomplished through PVC injection-molded "clicks" that are made from the same formulation as the roof membrane to eliminate compatibility issues. The clicks are attached to the sides of the PV system's steel support structure and then hot-air welded to fuse the clicks to the roofing membrane. (Photo by Sika Corporation.)*

31.4 PV SYSTEM TYPES AND APPLICATIONS

There are two basic types of PV systems: stand-alone and grid-connected. These names well describe their essential characteristic, which is their relationship to an external source of electricity such as a utility connection.

(a) Stand-Alone Systems

A stand-alone system is just that; it is isolated from any outside electrical connections and is designed to do a carefully defined specific job. Stand-alone systems are the oldest PV installations because they were essentially the only solution to the problem of electrifying a remote and/or unattended load. Stand-alone installations typically supply electricity for sign lighting, railroad crossing lights, unattended pumps, lighthouses, unattended navigational aids, microwave repeaters, motor homes, sailboats and yachts, isolated small residences, and the like. The common characteristic of all these installations is the impossibility or impracticality of feeding them from a utility grid. When loads become larger than can be supplied by a practical solar array (from a physical or economic viewpoint), and particularly when the peak load is periodic rather than continuous, a fuel-powered generator is usually added to the system. Such a system is known as a hybrid stand-alone system and is used frequently for remote residences where the connection charge levied by the local utility is so high that a PV system is an economical and environmentally friendly solution to the problem of providing electricity. Schematic diagrams of stand-alone system types are shown in Figures 31.8 and 31.9.

The difference between the systems shown in Figures 31.8 and 31.9 is the use of an energy storage medium—in most cases, a battery. The

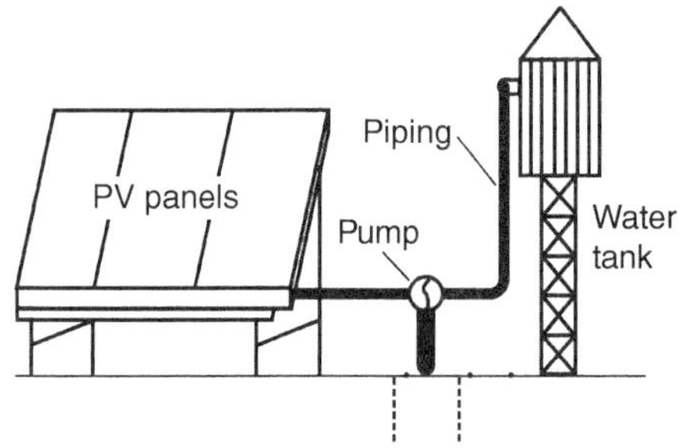

Figure 31.8 *Typical direct-connected PV installation feeds electric power to a positive-displacement pump that pumps water into an elevated tank.*

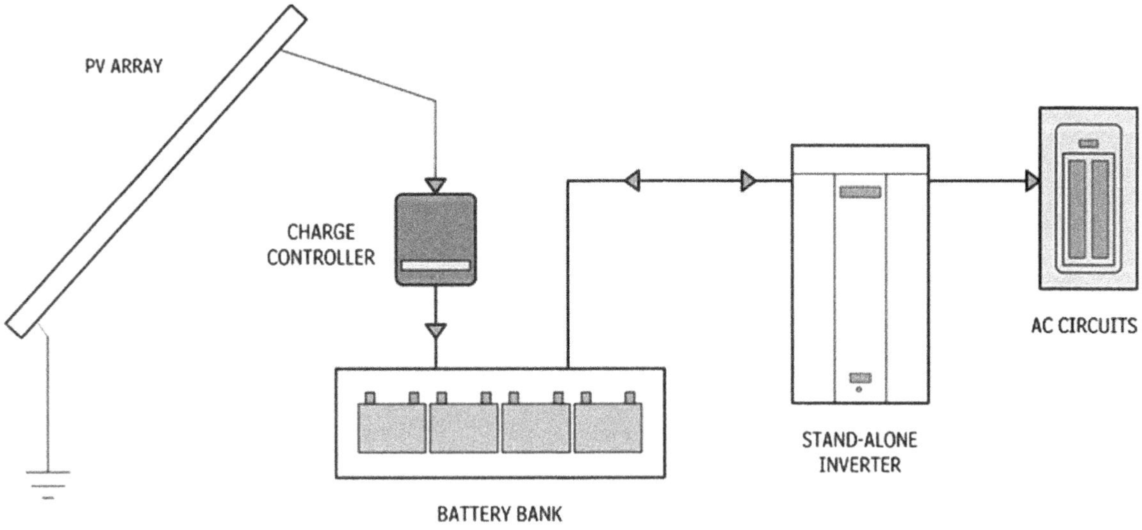

Figure 31.9 *Schematic diagram of the most common type of AC stand-alone photovoltaic system. (Ramsey and Sleeper, 2016.)*

direct-connected system in Figure 31.8 is applicable only with loads that tolerate variable power levels ranging from zero at night to maximum at solar noon. The most common of these is water pumping, using a DC motor to drive a positive displacement (piston) pump. Such installations are frequently used to fill elevated water tanks because a slow, interruptible fill rate does not adversely affect the water system's usability.

The vast majority of stand-alone systems are assembled as shown in Figure 31.9, using a storage battery to store excess electricity produced during peak insolation hours for use during periods of reduced solar resource (cloudy days) and full darkness (nighttime). With proper component selection, as is discussed in detail in subsequent sections, stand-alone PV systems can be used to adequately supply remote year-round residences, small mercantile establishments, and other off-grid loads.

A stand-alone system may contain a DC-to-AC inverter if any of the loads require AC that cannot be supplied by an equipment-specific inverter. The choice of whether to use a single central inverter or distributed smaller inverters is largely an economic decision. Fluorescent lamps, a reasonably efficient source of light, are available with integral inverter ballasts. Some kitchen appliances and power tools, however, are not readily available in a DC or battery-powered format, although that situation is rapidly changing and warrants careful investigation during system design.

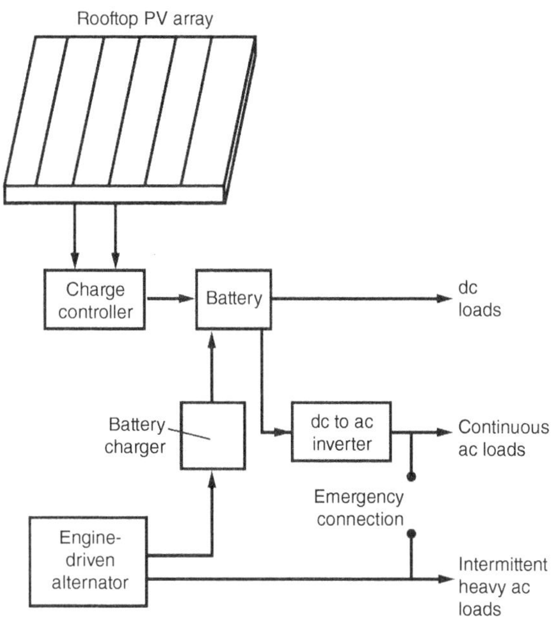

Figure 31.10 *One arrangement for a hybrid stand-alone PV system (control and safety devices are omitted for clarity). A diesel-, gasoline-, or propane-driven AC generator (alternator) supplies large, intermittent AC-only loads, whereas the PV array and battery supply DC loads and continuous AC loads through an inverter. During long, cloudy periods the alternator can charge the battery and, via an emergency connection, can also supply the regular AC loads. Other arrangements are possible, such as using a larger inverter to supply an all-AC system, with the alternator supplying intermittent heavy AC loads and all loads during low-battery periods.*

The hybrid stand-alone system shown in Figure 31.10 is similar to the stand-alone system in Figure 31.9, with the addition of a small AC generator. This type of system, as noted previously, is used where operation of a heavy AC load, such as a clothes washer or another relatively large AC motor-driven appliance, is necessary. Where these loads are fairly continuous, as with refrigeration and air-conditioning equipment, a larger PV system should be considered. A continuously running engine-generator set is a source of noise and air pollution, and the ongoing consumption of a fossil fuel resource in such an installation makes it contradictory in spirit to a completely silent, pollution-free, renewable-resource-driven, solar-powered PV installation.

(b) Grid-Connected Systems

Prior to the advent of net-metering, the number of stand-alone PV systems in areas served by electric utility grids was small because of the overwhelming economic advantage of purchased power. Even today, with net-metering, the payback period for an initial PV system investment seldom justifies an installation on purely economic grounds, although the continuous development work in PV materials may change that situation.

In a stand-alone arrangement, battery costs over the life of the system can exceed those of the PV array. Lead-acid batteries are relatively cheap but have a short life; nickel-cadmium batteries have a longer life but are much more expensive. By replacing the batteries that furnish energy storage and carryover in a stand-alone system with a utility connection that effectively serves the same purposes, and by taking advantage of the economies available from building-integrated PV elements (see Section 31.4*c*) in new construction, an economical PV system with a reasonable payback can be constructed. A schematic diagram of one such system arrangement is shown in Figure 31.11.

The inverter shown in Figure 31.11 creates utility-grade AC from the PV array DC output and feeds it to the utility through an electric (kWh) meter. The inverter shown is equipped with a safety cutoff that instantly disconnects the PV system from the grid if a utility outage is sensed. This is a requirement of the electric utilities to ensure that their maintenance personnel are not endangered by PV electric power being fed into an apparently de-energized line.

The power company supplies house power through a second meter, as shown. In net-metering, the difference between the readings of the two meters is either billed to the consumer at the normal utility rate or credited to the consumer at the avoided-cost rate, depending on which meter reads higher. A consumer will notice no difference whatever in electric service, except in the lower, or possibly even negative, electric utility bills.

More recently, to encourage wider use of nonpolluting PV power, some utilities have permitted the installation of small, individual PV modules in existing buildings. These units plug into convenience outlets in the building, supplying power to the building itself. Any excess electrical energy not required by the structure is fed back into the utility via their reversible energy meter. This arrangement is shown schematically in Figure 31.12. Each PV module (panel) is equipped with a built-in, high-quality inverter that produces utility-grade power and is equipped with the safety devices described previously. A common commercially

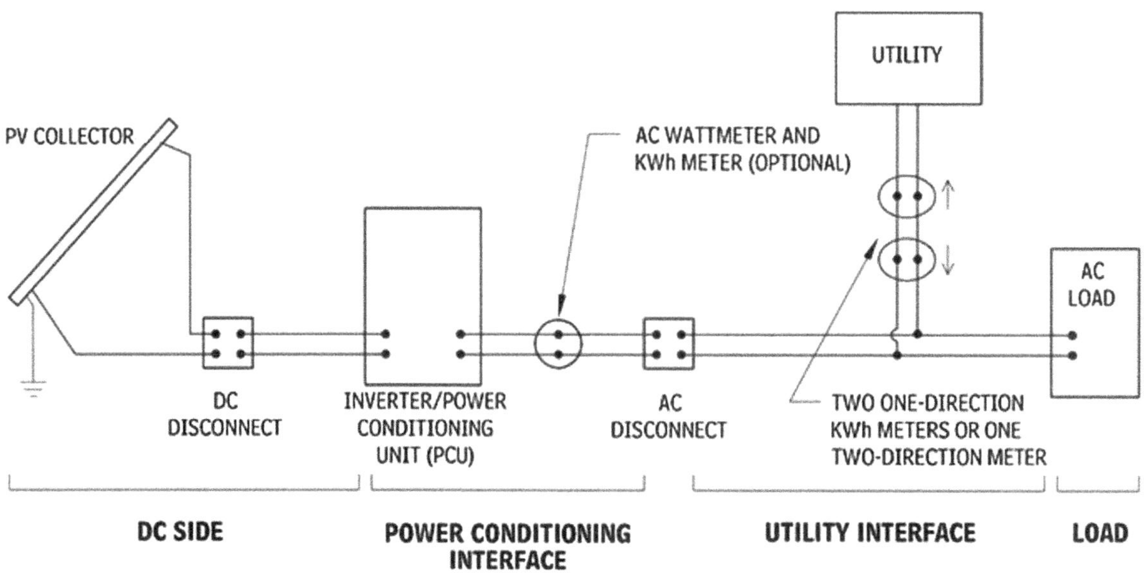

Figure 31.11 *Schematic drawing of a typical grid-connected photovoltaic system arranged for net-metering. Power from the array is converted from DC to utility-grade AC power (constant voltage and frequency; low harmonic content) by the inverter and fed to the utility connection through a kWh meter that measures the electricity sold to the utility. A second (utility company) two-directional meter (or two one-directional meters) measures the total power used by the structure. The difference in meter readings represents the positive or negative (credit) utility billing. (Ramsey and Sleeper, 2016.)*

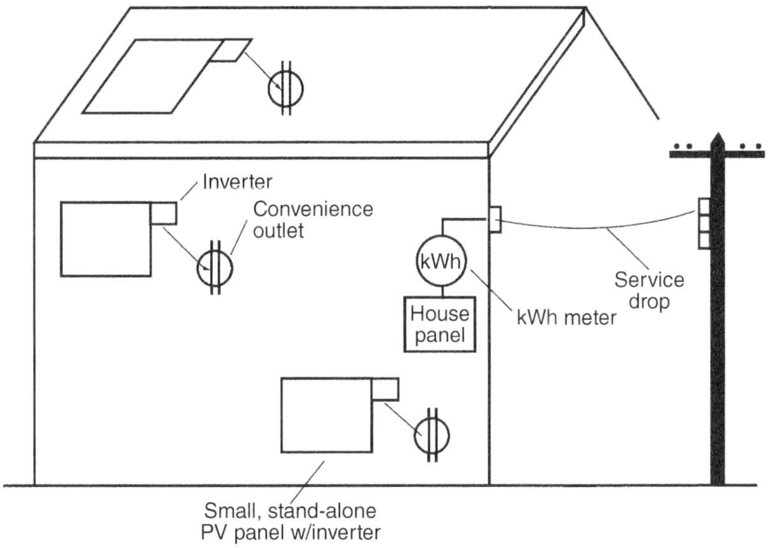

Figure 31.12 *Schematic drawing of a grid-connected PV system using small, distributed PV arrays connected to the house distribution system via convenience outlets. In this arrangement, the utility meter is unratcheted (reversible) so that excess PV-produced electricity reverses the power company's meter. Any net monthly credit is calculated at the avoided-cost rate.*

available unit of this type is rated at 100 W peak, 85 W nominal, 120 V, 60 Hz; measures approximately 2 ft × 4 ft × 4 in. [610 mm × 1220 mm × 50 mm]; and weighs 25 lb [11 kg]. The advantage gained with these units is that a user can begin with one or two modules and then expand as desired, without a relatively high first cost investment and without a centralized installation requiring professional construction and electrical personnel. Some power companies require separate disconnect switches for each PV unit rather than relying only on physical disconnection of the unit plug from the wall outlet.

The National Renewable Energy Laboratory (NREL) developed and maintains a design tool for analysis of grid-connected PV systems. The tool, called PVWatts, is freely available online (NREL, n.d.). Performance and cost analyses can be quickly run for a fairly wide range of locations, array orientations, and array tilts. As with most easy-to-use design tools, there are several assumptions built into the program—but all are noted, and most can be modified (either explicitly or implicitly) by the user.

(c) Economic Considerations

Economic issues can be complex, and they are often the deciding factor in the choice between full PV, a PV-grid combination, and a conventional (grid) electric service. Among the cost factors are the following:

- cost of a grid connection
- cost of power from the grid
- life-cycle cost of each component of the system over the life of the entire installation. This is necessary because the life span of individual components varies sharply; PV modules last about 20 years, batteries about 5 years, electrical components 10–20 years (or more), PV module mountings 15–25 years, tracking mechanisms (if used) 5–10 years, and so on.
- life-cycle maintenance costs (parts and labor) for each item
- financing costs
- cost credits for PV systems, including such items as use of a system battery as an uninterruptable power supply (UPS) source for computers and peripherals, savings from avoided grid-power service interruptions, and construction cost credits when PV modules are used in lieu of building elements

The last factor has taken on considerable importance in recent years with the development of building-integrated photovoltaic (BIPV) elements. A traditional PV installation consists of PV modules mounted near or on a building or facility. Often, the modules are roof-mounted if a roof surface of appropriate orientation (generally south-facing) and sufficient size is available. The development of thin-film amorphous silicon modules permits incorporation of PV cells into assemblies that can replace traditional building construction elements (typically in the form of roof panels, roofing tiles, wall panels, or skylights). These assemblies fulfill the dual function of producing electricity and weatherproofing the building enclosure. In the jargon of design intent, they are transformer elements, replacing historically used barrier or filter elements. In such cases, only the difference in cost between standard construction and BIPV construction should be taken as an additional cost of the PV system.

31.5 PV SYSTEM BATTERIES

Batteries are required in a stand-alone system to store excess PV-produced electricity (from high-production periods) for use during peak electricity consumption periods (the pattern illustrated in Figure 31.2). For residential installations, there is usually almost no coincidence between these two time periods because PV output peaks at solar noon, whereas residential usage peaks in the evening and is low at noon. In comparison, commercial stand-alone PV installations such as off-grid stores, shops, cottage industries, and the like have a much higher coincidence. This means that during normal, everyday use, the drain on the battery is shallow, and a cheaper battery will frequently suffice. These considerations are expanded in the system design discussions in Sections 31.7 and 31.8. Several manufacturers are currently marketing battery systems as a consumer product—for use with PV systems, electric cars, and DC power systems.

A PV system battery is usually expected to be capable of supplying most or all of an installation's electrical requirements for a given period—usually three days of cloudy weather if no other specific-use data are available. This means that the battery selected:

- must be capable of repeated discharge to 20% capacity or less without damage
- must be designed to accept slow, lengthy recharge

  In addition, the battery should fulfill the following conditions:
- have long life
- be suitable for the physical conditions in which it will be used (i.e., space, ambient temperature, ventilation, and maintenance availability)

Pure lead automobile batteries (also known as lead-acid batteries) are not suitable for PV systems because they are designed for rapid, shallow discharge (engine starting) and very rapid recharge. If such a battery is repeatedly discharged more than about 20% (shallow discharge), or if it is not completely recharged after discharge, its life will be extremely short. Batteries made for use in electric vehicles such as golf carts and forklifts are constructed with heavy plates and are designed for the deep-cycle discharge needed in PV systems. Most of these are flooded batteries, meaning that the plates must be completely covered with electrolyte to prevent damage. This requires frequent maintenance, making flooded batteries unsuitable for unattended sites.

Lead-calcium batteries are almost maintenance-free, rarely requiring addition of water or electrolyte. As a result, they are frequently used in unattended systems. They do not cycle as deeply as heavy-plate, deep-cycle units, and this must be considered during design.

Sealed batteries are completely maintenance-free. They are frequently constructed with a gel-type electrolyte that does not "boil" off on recharge and therefore needs no replacement. This makes them ideal for unattended sites, provided that they are specifically designed for deep-cycle service (80% discharge) and are so labeled. Most maintenance-free batteries are shallow-cycle units and therefore are unsuitable. An important characteristic of sealed units is that they require very accurate charge and discharge control to prevent overcharge or overdischarge, or over-rapid charge or discharge. Any of these conditions severely shortens the battery life. As a result, gel-cell batteries require the use of a high-quality charge controller.

Lead-antimony batteries are very sturdy and tolerate deep discharge and slow or rapid charge while maintaining their rated life. They require occasional water replenishment and adequate ventilation to disperse gases created during charging cycles.

Nickel-cadmium batteries are highly desirable in PV systems because they are maintenance-free, can be discharged to essentially 100% of capacity without injury or shortening of life, are much less temperature-sensitive than lead-acid batteries, and have a longer life. Their principal drawback is their high price. Nickel-cadmium batteries should not be shallow-discharged because they tend to develop a "memory" of the discharge cycle and after many shallow cycles may not supply sufficient current for a deep discharge. Therefore, in essentially clear-sky climates, where a 3-day period of cloudiness is rare, nickel-cadmium batteries may not be applicable because storage sized for a 3-day system supply would normally be operating on shallow discharge.

Additional considerations in battery selection include:

Temperature: Lead-acid battery capacity declines with temperature. Nominal battery capacity ratings are referenced to a temperature of 77°F [25°C]. Batteries operating at lower temperatures should be derated (in the absence of specific manufacturer's recommendations) as follows:

Ambient Temperature	Derating Factor
70°F [21°C]	0.96
60°F [16°C]	0.90
50°F [10°C]	0.84
40°F [4°C]	0.77
30°F [−1°C]	0.71
20°F [−7°C]	0.63

It is important to note that a discharged lead-acid battery contains essentially water and therefore freezes at about 30°F [−1°C]. A fully charged battery will not freeze until the temperature dips below 0°F [−18°C]. Nickel-cadmium batteries are much less affected by temperature.

Voltage: Battery terminal voltage varies with its state of charge, ranging for a 12-V battery at 77°F [25°C] from approximately 15 V at full charge to just over 12 V immediately after a deep discharge. If the battery is not charged immediately, voltage drops to a low of slightly over 11 V.

Life: A properly maintained deep-cycle lead-acid battery should last for a minimum of five years. A nickel-cadmium battery should last 1.5–2 times as long.

Rating: The manufacturer's stated ampere-hour (A-H) rating of a battery refers to a single continuous low-rate discharge under laboratory conditions. The actual field-use A-H capacity of a battery depends heavily upon the discharge rate, and it is this value, as calculated for the particular installation, that should be used to compare batteries of the same type. All reputable manufacturers have readily available discharge curves that show voltage and duration at different discharge rates.

The basic storage battery is rated at 12 V, although 6-V units are available. Figure 31.13 shows typical battery connection arrangements for the most common system voltages, which are 12, 24, 48, and 72 V. Table 31.1 shows some typical physical data for batteries.

31.6 BALANCE OF SYSTEM

The balance of system (BOS) consists of a charge controller, inverter, if used, and accessories such as blocking diodes, a lightning arrester, an electrical grounding system, and other electrical specialty items. Refer to Figures 31.9 through 31.11 for a sense of how these components fit into a PV system. Although they are important to overall system performance and safety, a full description of these items is beyond the scope of this section. Refer to the references at the end of this

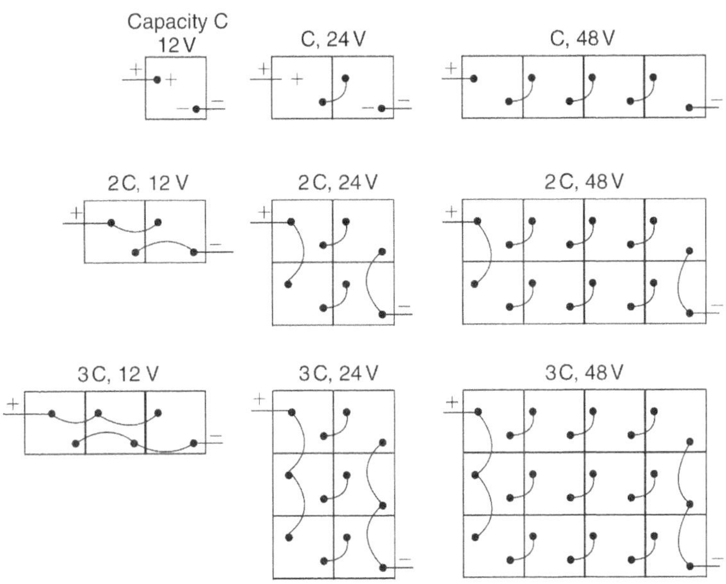

Figure 31.13 *Wiring arrangements for 12-V battery packs to obtain increased capacity, increased voltage, or both.*

TABLE 31.1 Physical Characteristics of Typical 12-V[a] Batteries

Battery Type	Capacity in Amp-Hours	Dimensions of a 12-V Unit L × W × H, in. [mm]	Weight, lb [kg]
	(6/20/100-hr discharge rates)		
Deep-cycle lead-acid	500/625/775	33 × 6.5 × 24 [838 × 165 × 609]	500 [227]
	675/840/1025	40 × 6.5 × 24 [1016 × 165 × 609]	700 [318]
	850/1050/1275	50 × 6.5 × 24 [1270 × 165 × 609]	800 [363]
	(20-h discharge rate)		
Sealed gel-cells	850	40 × 8 × 24 [1016 × 203 × 609]	650 [295]
	1050	40 × 9 × 24 [1016 × 229 × 609]	770 [349]
	1275	40 × 10 × 24 [1016 × 254 × 609]	930 [422]
	1500	40 × 12 × 24 [1016 × 305 × 609]	1100 [499]

[a]Use two units for a 24-V battery and four units for a 48-V battery.

chapter for a few of the many excellent resources that address these components. When pricing photovoltaic installations, it is important to note whether the BOS components are incorporated as they may constitute between a third and half of the total system cost.

31.7 DESIGN OF A STAND-ALONE PV SYSTEM

There is a fundamental difference in the procedures for array sizing and system arrangement between a stand-alone system and a grid-connected PV system. A stand-alone system must supply the entire electric load; there is no utility backup. Design decisions are primarily technical. A grid-connected system may supply part, or all, of the electric load. Design decisions are influenced by economics and owner intent, within the context of technical feasibility. These two procedures are addressed separately in this and the following section.

Figure 31.9 shows the essentials of a stand-alone PV system. Because building electricity demands rarely exactly match the output of a PV system, the system operates successfully by using a PV array to charge a battery, which then supplies the building load. Any occasional coincidence between load and PV production is incidental to the design process. The three factors that establish the system requirements are:

- the daily electricity usage
- the period of time for which the battery must supply the electric load without recharge
- the available insolation

The first factor is a function of building design. The second factor is a function of local weather conditions and would most often be chosen as the number of successive solid overcast or rainy days usually encountered. Designing for highly unusual weather results in a system that is oversized for normal (say, 95th percentile) weather patterns and, therefore, is essentially uneconomical. That scenario weighs heavily in favor of a hybrid system (see Figure 31.10), which defeats, at least in part, the environmental benefits of the PV system.

The PV system design procedure is generally straightforward and is as follows:

STEP 1. Determine the average daily energy usage in watt-hours.

STEP 2. Determine the maximum period for which the system battery must supply the entire electrical load.

STEP 3. Determine the required system battery capacity, which is equal to the daily load (Step 1) multiplied by the number of days that the battery must supply the load (Step 2). System losses are considered in this calculation. (In practice, this capacity can be reduced by switching off nonessential loads during unusually long sunless periods.)

STEP 4. Determine the preliminary size of the PV array. The array must supply all of the required electricity, that is, the facility's daily requirements (Step 1) plus all system losses. The preliminary array size is this capacity in terms of watt-hours. To determine the physical size of the array, the designer must determine the number of watt-hours produced per unit area of PV module. This involves knowledge of the site's solar radiation, the type and efficiency of the PV module used, the mounting approach (stationary or tracking), and the tilt angle (module angle measured from the horizontal). With these data in hand, the area of PV array required is simply the total load in watt-hours divided by the PV unit's energy production (perhaps obtained from PVWatts) in watt-hours per unit area.

STEP 5. Check that the array selected is sufficiently large to charge the battery capacity established in Step 3. Assuming, for instance, a 3-day period during which the battery must supply the entire load, about 3 ½ consecutive sunny days at the same insolation will be needed for full recharge (considering losses of approximately 20%). If the facility is to be used year-round and the PV array is sized on the basis of summer insolation, it is necessary to do a calculation for a winter month as well. Insolation data are usually available for both seasons, and average monthly insolation accounts for the appreciable energy available from cloudy skies as well as sunny skies. The result generally indicates a larger PV array requirement for winter months. In such cases, on purely economic grounds, the decision is often to use a small engine generator for battery-charging for winter use only. This is not the same as a hybrid system (Figure 31.10), in which the generator is used year-round to pick up heavy AC loads. Some designers base the PV array size on minimum yearly (winter) insolation, reasoning that in other months any wasted overproduction from an array is the price paid for year-round operation without fossil fuels. Others use a median insolation figure, reasoning that power outages will be few and of short duration. Using summer insolation values will minimize the PV array size but probably will require inclusion of a small generator.

An illustrative example using the outlined design procedure should clarify the issues and calculations involved.

EXAMPLE 31.1 Design a stand-alone PV system for a 650-ft^2 [60-m^2] occupied-year-round, off-line cottage located near Prescott, Arizona. In addition to the usual appliances, an evaporative cooler is used during hot summer days. A PV array can be mounted on the south-facing sloping roof at any required angle. Determine the required battery and PV array sizes. State all assumptions.

SOLUTION

STEP 1. Determine the average electric energy consumption for the cottage. Although refrigerators that run on bottled gas are readily available, assume use of an electric refrigerator. Water heating is solar thermal, with a small gas booster heater. Cooking is with bottled gas.

24-V DC Appliances	Power Draw (W)	Daily Use (h)	Daily Energy Need (Wh)
Mid-sized refrigerator/ freezer	35	24	840
Small color TV	60	4	240
Evaporative cooler[a]	50	8	400
Stereo system	40	2	80
Ceiling fans[a] (3)	3 at 25	4	300
Radios	5	10	50
Incandescent lighting	50	8	400
		Daily use DC	2310 Wh
Summer only[a]			
120-V AC Appliances	**Power Draw (W)**	**Daily Use (h)**	**Daily Energy Need (Wh)**
120-V AC Appliances			
Fluorescent lighting	80	10	800
Clothes washer	500	0.5	250
Vacuum cleaner	600	0.25	150
Kitchen appliances	400	1	400
Iron	1000	0.25	250
Miscellaneous	100	3	300
		Daily use[a] AC	2150 Wh
[a]Maximum coincident ac load.			
Total daily energy usage:			
DC =			2310
AC = 2150 Wh × 1.15 (inverter loss)			2473
Total			4783 Wh
Ampere-hours = (4783 Wh/24 V) = 199, say, 200 A-H			

STEP 2. Assume that investigation of local climate conditions indicates occasional periods throughout the year, but primarily in the winter, of 2–3 consecutive days of heavy overcast, clouds, and rain.

STEP 3. Sizing the system battery. Assume use of a deep-cycle flooded-cell lead battery with an appropriate charge controller. These batteries can supply about 80% of their rated ampere-hour capacity. Since the battery will be located indoors, but in an unheated room, it should be derated 15% for ambient temperature for a conservative design. Required battery size in ampere-hours equals:

$$\frac{(3\text{ days})(200\text{ A-H daily load})(1.2\text{ discharge factor})}{0.85\text{, temperature derating}}$$

$$= 850\text{ A-H}$$

One type of 850-A-H, 24-V deep-cycle lead-acid battery (using two at 12-V units in series) measures approximately 40 in. long × 15 in. wide × 24 in. high [1020 mm × 380 mm × 610 mm] and weighs about 1500 lb [680 kg].

STEP 4. Sizing the PV array. Because the cottage will be occupied all year, and because the critical performance period is winter, when the sun is low, select a tilt of (latitude +15°) to maximize winter PV production. As the array is roof-mounted and will be relatively large, use a stationary rather than a tracking array. The array must supply the daily energy requirement of 4783 Wh or 4.8 kWh.

Referring to Figure 31.14, which gives solar radiation data for the Prescott location, obtain the following data: average solar radiation for a PV array facing south at a tilt angle of latitude +15°:

January: 5.5 kWh/m^2/day; average monthly temperature 2.3°C [0.51 kWh/ft^2/day; 36°F]

June: 6.0 kWh/m^2/day; average monthly temperature 19.6°C [0.56 kWh/ft^2/day; 67°F]

Figure 31.15 was drawn using the computer program PVFORM, developed at Sandia National Laboratories. It gives daily average PV array production in Wh/ft^2 as a function of monthly average temperature and monthly average solar radiation in kWh/m^2/day. Using the January and June radiation figures stated previously, obtain from Figure 31.15:

January: 52 Wh/ft^2 of a PV array [560 Wh/m^2]

June: 53 Wh/ft^2 of a PV array [570 Wh/m^2]

The PV array output is essentially constant throughout the year (with radiation availability and ambient temperature effects offsetting), and no difficulty will be encountered in an "off" season. If the PV modules to be used have an efficiency greater than 10%, then the array output would be increased proportionately.

The required area of PV array, assuming 10% conversion efficiency, is:

PV area

$$= (4783\text{ Wh/day required})/(52\text{ Wh/day/ft}^2\text{ output})$$

$$= 91\text{ ft}^2\ [8.5\text{ m}^2]$$

STEP 5. Because the solar radiation in Prescott, Arizona, is essentially identical for summer and winter, PV production in 3 to 3½ clear days will fully recharge the battery, as explained previously. However, in sizing PV systems, the need to provide charge to discharged batteries while simultaneously providing power to building loads should be considered. A system sized just large enough to meet all loads under design conditions will generally not have enough output to recharge depleted batteries.

Summary. The suggested PV system consists of approximately 90 ft^2 [8.4 m^2] of PV collector, an 850-A-H, 24-V

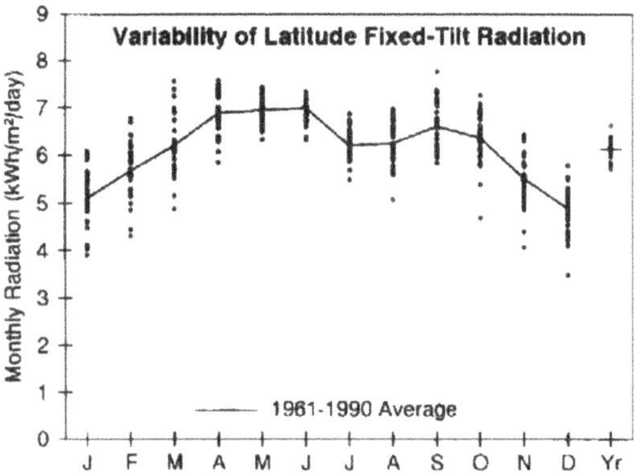

Prescott, AZ

WBAN NO. 23184

LATITUDE: 34.65° N
LONGITUDE: 112.43° W
ELEVATION: 1531 meters
MEAN PRESSURE: 847 millibars

STATION TYPE: Secondary

Solar Radiation for Flat-Plate Collectors Facing South at a Fixed Tilt (kWh/m²/day), Uncertainty ±9%

Tilt (°)		Jan	Feb	Mar	Apr	May	June	July	Aug	Sept	Oct	Nov	Dec	Year
0	**Average**	**3.1**	**3.9**	**5.1**	**6.6**	**7.5**	**8.0**	**6.9**	**6.3**	**5.7**	**4.6**	**3.4**	**2.8**	**5.3**
	Min/Max	**2.6/3.5**	**3.2/4.5**	**4.2/6.1**	**5.7/7.1**	**6.8/8.0**	**7.2/8.5**	**6.1/7.7**	**5.1/7.0**	**5.1/6.7**	**3.6/5.2**	**2.8/3.9**	**2.2/3.2**	**5.1/5.7**
Latitude -15	Average	4.4	5.1	5.9	7.0	7.5	7.7	6.7	6.5	6.5	5.8	4.8	4.1	6.0
	Min/Max	3.5/5.1	4.0/6.0	4.7/7.2	6.0/7.7	6.8/8.0	6.9/8.1	5.9/7.5	5.3/7.2	5.7/7.6	4.4/6.6	3.7/5.5	3.0/4.8	5.6/6.5
Latitude	**Average**	**5.1**	**5.7**	**6.2**	**6.9**	**7.0**	**7.0**	**6.2**	**6.3**	**6.6**	**6.4**	**5.5**	**4.9**	**6.1**
	Min/Max	**3.9/6.1**	**4.3/6.8**	**4.9/7.6**	**5.8/7.6**	**6.3/7.4**	**6.3/7.3**	**5.5/6.9**	**5.1/7.0**	**5.8/7.8**	**4.7/7.3**	**4.1/6.4**	**3.5/5.8**	**5.7/6.6**
Latitude +15	Average	5.5	5.9	6.1	6.4	6.1	6.0	5.4	5.7	6.4	6.5	5.9	5.4	5.9
	Min/Max	4.2/6.6	4.4/7.1	4.8/7.5	5.4/7.1	5.6/6.5	5.4/6.3	4.8/5.9	4.6/6.3	5.6/7.5	4.7/7.5	4.3/6.9	3.7/6.4	5.5/6.4
90	**Average**	**5.0**	**4.9**	**4.3**	**3.5**	**2.6**	**2.3**	**2.3**	**2.9**	**4.0**	**5.0**	**5.2**	**5.0**	**3.9**
	Min/Max	**3.8/6.1**	**3.6/5.9**	**3.4/5.2**	**3.0/3.8**	**2.5/2.8**	**2.2/2.3**	**2.1/2.4**	**2.4/3.1**	**3.6/4.7**	**3.6/5.8**	**3.6/6.2**	**3.4/6.0**	**3.5/4.2**

Solar Radiation for 1-Axis Tracking Flat-Plate Collectors with a North-South Axis (kWh/m²/day), Uncertainty ±9%

Axis Tilt (°)		Jan	Feb	Mar	Apr	May	June	July	Aug	Sept	Oct	Nov	Dec	Year
0	**Average**	**4.8**	**5.9**	**7.4**	**9.4**	**10.6**	**11.3**	**9.3**	**8.8**	**8.4**	**7.1**	**5.4**	**4.4**	**7.7**
	Min/Max	**3.7/5.7**	**4.4/7.2**	**5.7/9.5**	**7.9/10.6**	**9.4/11.5**	**9.9/12.0**	**7.9/10.8**	**7.0/10.1**	**7.1/10.2**	**5.0/8.2**	**4.0/6.3**	**3.2/5.2**	**7.1/8.4**
Latitude -15	Average	5.7	6.8	8.1	9.8	10.6	11.2	9.3	9.0	9.0	8.0	6.3	5.4	8.3
	Min/Max	4.3/6.9	5.0/8.3	6.2/10.3	8.2/11.0	9.4/11.6	9.8/11.9	7.8/10.7	7.1/10.4	7.6/10.9	5.6/9.3	4.6/7.5	3.8/6.4	7.5/9.0
Latitude	**Average**	**6.3**	**7.2**	**8.3**	**9.8**	**10.3**	**10.7**	**8.9**	**8.8**	**9.1**	**8.4**	**6.9**	**6.0**	**8.4**
	Min/Max	**4.7/7.6**	**5.3/8.8**	**6.3/10.6**	**8.1/11.0**	**9.2/11.3**	**9.4/11.4**	**7.6/10.4**	**7.0/10.2**	**7.7/11.1**	**5.8/9.8**	**4.9/8.2**	**4.1/7.2**	**7.6/9.2**
Latitude +15	Average	6.6	7.4	8.2	9.4	9.7	10.1	8.4	8.4	8.9	8.5	7.2	6.3	8.3
	Min/Max	4.9/8.1	5.3/9.1	6.2/10.6	7.7/10.6	8.6/10.6	8.8/10.7	7.1/9.7	6.7/9.8	7.5/10.9	5.8/9.9	5.1/8.6	4.3/7.6	7.5/9.0

Average Climatic Conditions

Element	Jan	Feb	Mar	Apr	May	June	July	Aug	Sept	Oct	Nov	Dec	Year
Temperature (°C)	2.3	3.9	5.9	9.6	14.2	19.6	22.8	21.3	18.1	12.6	6.7	2.6	11.7
Daily Minimum Temp	-5.6	-4.4	-2.2	0.8	5.1	9.9	14.4	13.2	9.3	3.4	-1.9	-5.4	3.1
Daily Maximum Temp	10.2	12.2	14.1	18.3	23.3	29.2	31.2	29.4	26.7	21.8	15.3	10.6	20.2
Record Minimum Temp	-18.3	-18.3	-13.9	-6.1	-3.3	2.2	8.3	4.4	1.1	-7.2	-13.3	-22.8	-22.8
Record Maximum Temp	23.3	25.0	27.2	30.6	35.0	39.4	38.9	37.2	35.0	32.2	26.1	21.1	39.4
HDD, Base 18.3°C	496	403	384	262	138	32	0	0	41	182	350	489	2775
CDD, Base 18.3°C	0	0	0	0	11	68	139	96	32	4	0	0	351
Relative Humidity (%)	59	53	49	38	33	27	44	51	47	46	51	59	47
Wind Speed (m/s)	3.1	3.5	4.1	4.2	4.3	4.1	3.5	3.1	3.4	3.3	3.3	3.0	3.6

Figure 31.14 *Solar radiation data for Prescott, Arizona. (From National Renewable Energy Laboratory, 1994.)*

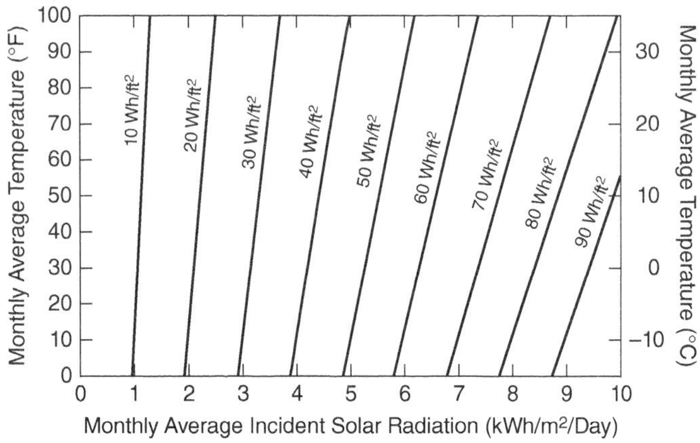

Figure 31.15 *Daily average PV array production, Wh/ft², as a function of monthly average temperature and incident solar radiation. Production was calculated using PVFORM for a 10% efficient solar cell array at 77°F [25°C] ambient temperature. (From Association of Collegiate Schools of Architecture, 1993.)*

battery, and a 1500-W inverter to handle the initial load plus the anticipated addition of small AC appliances. The summer-only DC cooling loads will likely be offset in winter by additional use of lighting and television and other in-cottage activities.

Additional Considerations. Using accurate radiation and weather data for a site is necessary for the design of an efficient and workable system. Unfortunately, accurate radiation data, particularly for tilted surfaces, are not always available, and tilted-surface data are not easily derived from horizontal-surface or normal-to-the-sun data. PV installations are frequently planned for sites outside of large cities, and it is usually for these areas that a paucity of data exists. In such instances, data for nearby areas can be used but must be applied judiciously, as insolation conditions can vary sharply within a relatively small geographic area. ■

31.8 DESIGN OF A GRID-CONNECTED PV SYSTEM

The design procedure for an interactive system (with feedback into the utility grid; see Figure 31.11) differs from the typical stand-alone system design procedure described in Section 31.7. The principal differences are as follows:

- The loads in an interactive system are all AC rather than a mix of AC and DC because the building is fed from the utility line or from the PV array DC-to-AC inverter; DC is not available to the loads.
- The existence of a PV system has no bearing on the choice of fuel for appliances. Therefore, for instance, whether to use gas or electric cooking is a matter of fuel availability, costs, and owner preferences.
- Loads are generally heavier in interactive systems because one cannot normally run out of electric power, as is possible with a stand-alone system. As a result, an all-electric building or a heavily automated building can be a candidate for a PV system as much as any other building.
- Load data can be obtained for a specific design or from general utility-supplied data similar to those shown in Figures 31.16 and 31.17.
- Because no battery and associated control equipment are used, space allocation in the structure for PV-related equipment is small, and system maintenance is minimal. Also, system life is equivalent to PV array life.
- Design note: Some areas in North America have recently been hit with uncharacteristic ice storms that have caused unexpected and relatively lengthy power outages. Some owners with grid-connected PV systems in these areas have added a small battery to their systems to meet their essential needs (lights, heating, computer use) through such outages.

Assuming a net-metering arrangement, there is no advantage to coincidence of building electrical loads and peak PV production. This is emphatically not the case where PV production buyback is based upon avoided costs. In such situations, careful scrutiny of the utility's rate schedule may yield cost-saving opportunities in matching loads to lower-rate time periods. For nonresidential buildings, careful consideration of future electric loads should be reflected in architectural planning for future (larger) PV arrays. In this connection, the possibility of using distributed arrays, as in Figure 31.12, should be studied.

The general design procedure for a grid-connected PV system is as follows:

STEP 1. Establish with the power company the metering arrangement to be used, both physical and financial. The utility can also generally supply reliable data on solar radiation, load patterns, PV array sizing and tracking feasibility, and other relevant design data.

STEP 2. Using a detailed cost analysis, determine the optimum size of the PV array in today's market. Although this analysis will involve technical data, the decision on an optimum size is essentially economic—trading utility cost savings against additional first costs for the PV system within a feasibility window established by the owner. This economic feasibility window will likely be influenced by qualitative factors such as a desire to reduce dependence on the "grid," a desire to be more environmentally friendly, or a desire to use a PV system to obtain green building certification points. Sensitivity analysis can be incorporated into the economic analysis to suggest ranges of reasonable system sizes given estimated changes in economic variables, such as utility rates, tax credits, or future PV array costs.

STEP 3. Using specific data or generalized information similar to that given in Figures 31.16 and 31.17, determine the required performance characteristics and physical size of the PV array established in step 2.

31.9 CODES AND STANDARDS

PV installations are electrical systems and therefore fall under the purview of *NFPA 70*, the *National Electrical Code* (*NEC*), published by the National Fire Protection Association. Article 690 of the *NEC*—"Solar Photovoltaic Systems"—is devoted specifically to such systems. Its requirements, plus those of applicable articles in the remainder of the *NEC*, must be observed in all areas where the *NEC* is mandated by local jurisdiction. In addition, for grid-connected systems,

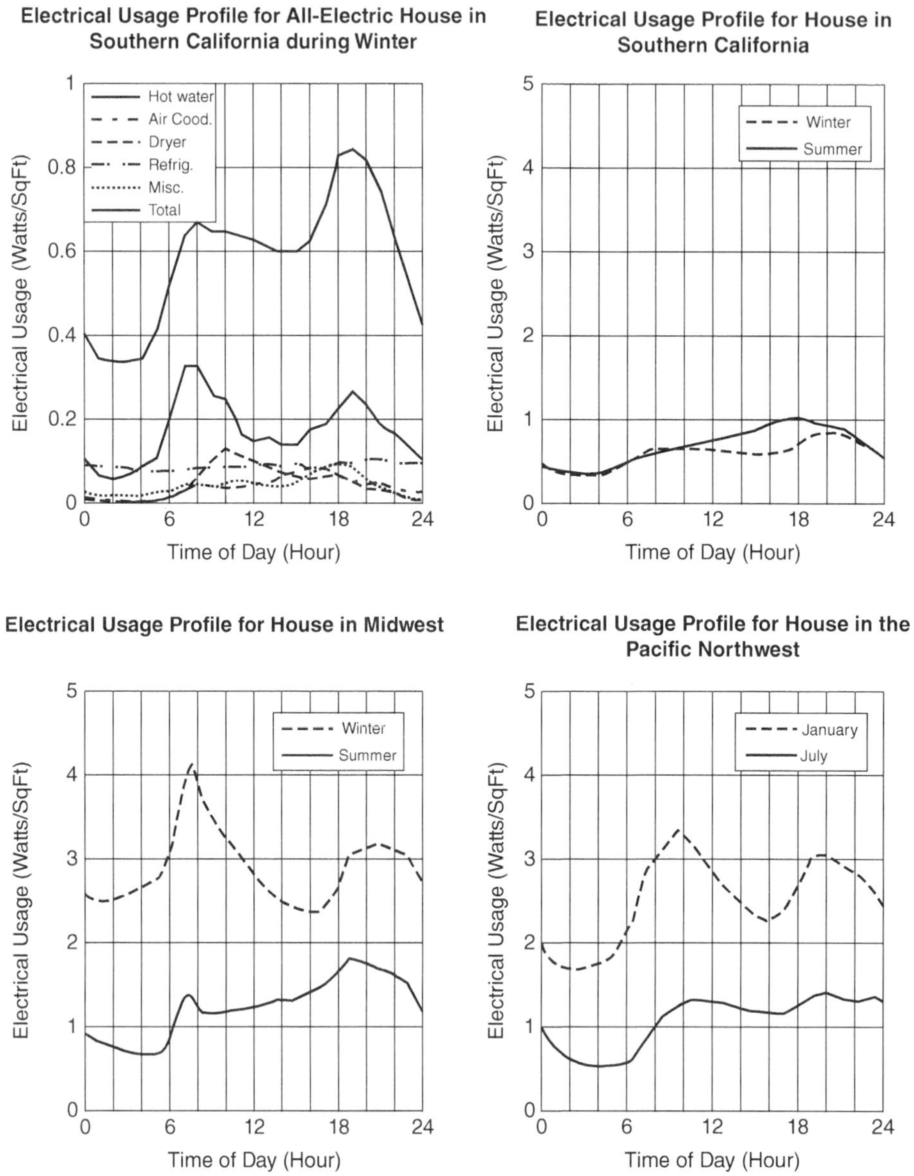

Figure 31.16 *Typical residential electricity usage profiles based on a 1500-ft² [140-m²] house. (From Association of Collegiate Schools of Architecture, 1993.)*

the specific requirements of the local utility must be followed, all equipment must be approved by them, and the final installation subject to their inspection.

Other codes and standards that contain important information and may be applicable if required by local authorities include the following by the IEC = International Electrotechnical Commission; the IEEE = Institute of Electrical and Electronics Engineers; and the UL = Underwriters Laboratories:

IEC TS 61836:2016: Solar Photovoltaic Energy Systems—Terms, Definitions and Symbols

IEC 61727:2004: Photovoltaic (PV) Systems—Characteristics of the Utility Interface

IEC 61215:2021: Terrestrial Photovoltaic (PV) Modules—Design Qualification and Type Approval (various parts)

IEC 60364-7-712:2002: Electrical Installations of Buildings—Part 7-712: Requirements for Special Installations or Locations—Solar Photovoltaic (PV) Power Supply Systems

IEC 61277:1995: Terrestrial Photovoltaic (PV) Power Generating Systems—General and Guide (withdrawn, but available for purchase)

IEC 61702:1995: Rating of Direct Coupled Photovoltaic (PV) Pumping Systems (withdrawn, but available for purchase)

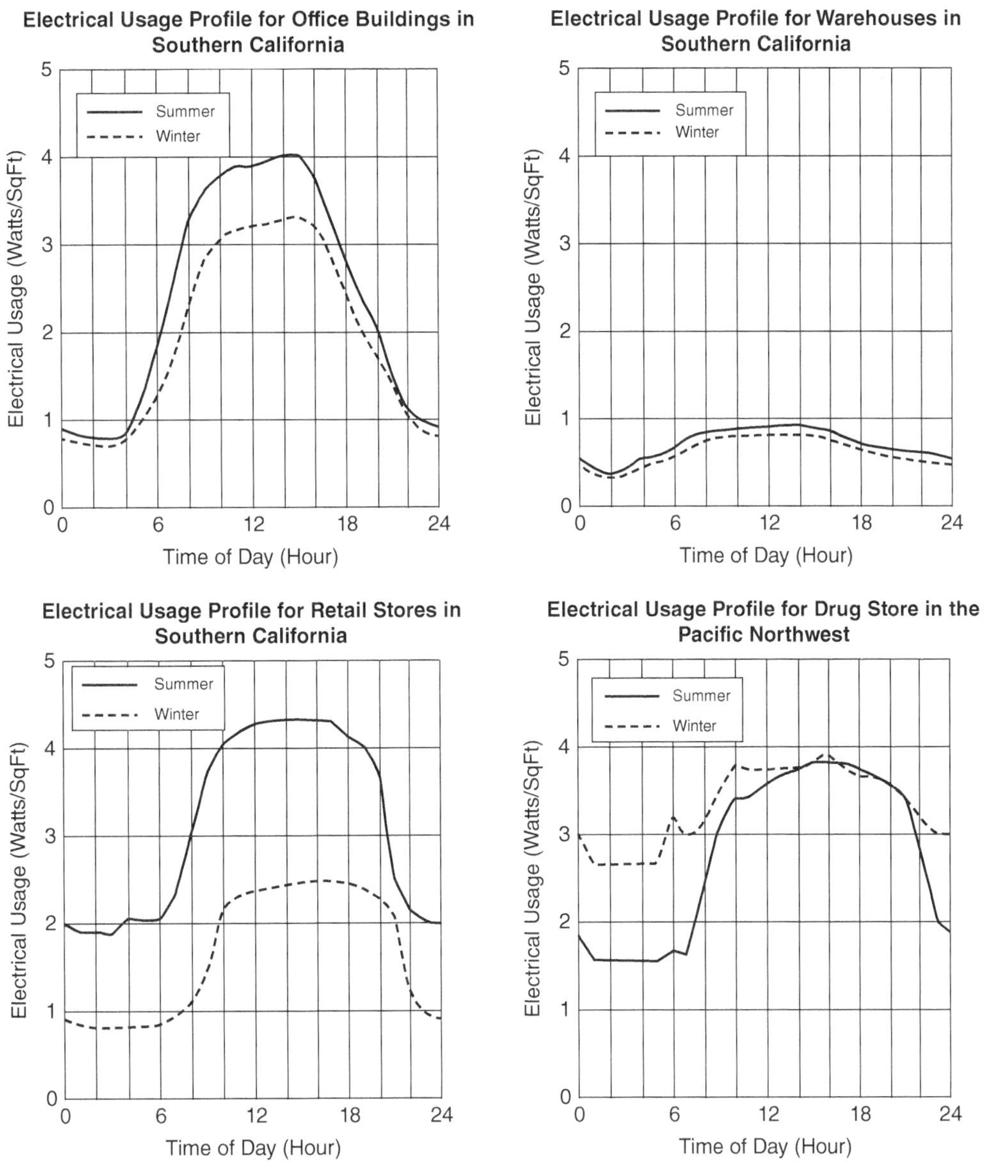

Figure 31.17 *Typical electric load profiles for nonresidential U.S. buildings. (From Association of Collegiate Schools of Architecture, 1993.)*

IEEE 1013-2019: Recommended Practice for Sizing Lead-Acid Batteries for Stand-Alone Photovoltaic (PV) Systems

IEEE 937-2019: Recommended Practice for Installation and Maintenance of Lead-Acid Batteries for Photovoltaic (PV) Systems

UL 1703-2024: Standard for Flat-Plate Photovoltaic Modules and Panels

UL 1741-2024: Standard for Inverters, Converters, Controllers and Interconnection System Equipment for Use with Distributed Energy Resources

As with most standards, these are regularly updated; verification of the most current version is advised.

31.10 CURRENT PV INSTALLATIONS

PV power appears to be one of the dominant trends of the future. PV and buildings are too logical a connection to not develop into the norm. Even today, with arguably unfavorable economics, numerous notable examples of PV installations exist. Several interesting examples are presented here.

Example 1: PV integrated with solar hot-water collectors (Figure 31.18). This is an interesting application that matches resource with need and obviates the need for storage of the PV system's energy production. The PV output drives a pump that circulates solar-heated water. The pump operates only when there is enough solar radiation to heat the water,

Figure 31.18 Synergistic PV application—pumping hot water produced by solar collectors for a residence in the southwestern United States. (© Walter Grondzik; all rights reserved.)

which means there is also enough radiation to drive the pump via the PV module. No batteries or grid backup are required.

Example 2: Many parking lots feature canopies of photovoltaic arrays that also serve as a shading source. The Springs Preserve parking lot (Figure 31.19) in Las Vegas, Nevada, uses small, two-sided, single-axis tracking modules to provide 409 kW of peak capacity.

Example 3: Photovoltaics at the Woods Hole Research Center (Figure 31.20), Falmouth, Massachusetts. This green building features a substantial PV installation. The system is a net-metering arrangement (as described earlier).

Example 4: PV on a solar racer (Figure 31.21), Australia, 1983. An early example of TIPV (transportation-integrated photovoltaics)? The lessons presented by this example actually apply well to buildings—reduce the load and optimize the collectors.

Figure 31.19 Six, single-axis photovoltaic arrays at the Springs Preserve in Las Vegas, Nevada, are positioned over highly reflective fabric panels. (© Alison Kwok; all rights reserved.)

Figure 31.21 The PV car, a solar racer, after crossing Australia (not a minor feat). (© Walter Grondzik; all rights reserved.)

(a) (b)

Figure 31.20 PV at the Woods Hole Research Center, Falmouth, Massachusetts. (a) Low-slope PV array on the roof of the building. (b) An inverter (one of several) and a disconnect switch are also located on the roof. (© Walter Grondzik; all rights reserved.)

31.11 FUTURE PV INSTALLATIONS

The title "future" PV installations is not fully correct—as some of what is discussed herein is being done today. The options presented here, however, are not yet the norm and will require some catalyst to enter the mainstream of design thinking.

DC Power Grids. Photovoltaic cells produce direct current (DC) power. Although DC is sometimes used directly from a PV array in building applications, by far the more common approach is to invert the DC to AC in order to access a broader and more consumer-friendly range of appliances and electrical equipment. The conversion of DC to AC involves efficiency losses. These losses are relatively low as inverter efficiencies have improved, but they are not zero. In addition, a fair amount of the AC power that drives electronic equipment is converted back to DC—introducing more efficiency losses. Add in a desire for battery storage and electrical vehicle charging and a rational argument can be made for going to DC power distribution in buildings.

The founding argument of AC versus DC power distribution (Edison/GE versus Tesla/Westinghouse) was decided over 100 years ago in favor of AC. This is not likely to change in the foreseeable future due to the huge infrastructure invested in AC. But DC is emerging as a growing consideration at the building scale. This leads to some interesting opportunities, such as microgrids.

Microgrids and Nanogrids. A microgrid is a small electrical grid; a nanogrid is an even smaller grid. There is no unanimity at this time on formal definitions for these terms; and they do seem relative. Recent Solar Decathlon exhibitions have involved a competition microgrid designed to handle the nominal 20 projects installed at the competition site. The term microgrid has been used to describe the interconnection of, say, 100 homes in a neighborhood; it has also been used to describe the interconnection of six condo units in a single building. The term nanogrid has been used to describe the electrical distribution arrangement for each condo unit within a building.

The benefit of a microgrid or a nanogrid is that it can be operated to best benefit the needs of its member units. These needs come into focus when the local situation differs substantively from the conditions that exist at a larger scale (such as on a regional power grid). On-site power production, a desire to act as an "energy commons," interest in on-site energy storage—these might all drive consideration of use of a microgrid.

On-Site Energy Storage. Batteries have long been a staple of off-the-grid building installations. The storage capacity provided by batteries bridges the gaps between the power production profile of a local PV or wind electric generation system and the building's energy consumption profile. As a minimum, battery storage allows loads to be met when there is no solar for PV production or wind for a wind machine. What has changed recently is the ready availability of battery packages.

The Tesla Powerwall and the SonnenBatterie eco are just two examples of an expanding field of electric power storage devices that are essentially consumer appliances. As appliances, these systems typically have numerous control settings that can be accessed via dashboards and modified by smart apps. It appears that with today's products, battery system sizing may be dependent upon the ability of the system to meet peak connected load and its ability to provide power over time.

Electric Vehicles. Electric vehicles (EVs), as with the technologies noted above, are not yet the norm; neither are they an exotic and untested product. It is reasonable to forecast a growing role for electric vehicles in addressing carbon emissions from the transportation sector. More intriguingly, EVs are also potentially a significant part of an integrated building energy system. Electric vehicles are energy-dense; a typical EV can power a normal single-family residence for two to four days—with the higher end of the range applying to low-energy buildings (net-zero, passive house).

The potential for vehicle-building interactions includes powering an EV from site-produced renewable energy; using an EV as a battery for the storage of excess electricity production (when other options, such as feeding to the utility grid, are precluded or make little economic sense); and/or powering a house from an EV in the face of on-site production challenges (such as periods of exceptionally cloudy weather, persistent snow, etc.).

Islanding. Turning a noun into a verb is common (so what the heck, we'll go with the flow). Islanding refers to the practice or capability of isolating some part of a larger system when doing so is beneficial for the individual unit or for the whole. Islanding can be part of designing for resilience or actively assisted passive survivability (see Chapter 1). In the event of a system-wide power outage or failure, it might be beneficial for a single home with PV to isolate itself from the grid so that the power produced on site can be beneficially used to operate the building. With a micro-grid, several PV-equipped homes (or buildings) might be able to "carry" several other buildings on a small and isolated grid through a local or regional power outage.

31.12 TERMINOLOGY AND DEFINITIONS

A generally accepted terminology has developed in the PV field. The list that follows contains common terms and their accepted definitions. Attention should be paid to two terms that are often incorrectly used interchangeably: insolation and irradiance. Irradiance is the amount of solar power impinging on a specific area, usually measured in units of watts per square meter—and represents an instantaneous value. Insolation is the amount of solar energy received by a given area, measured in Wh/m^2, kWh/m^2, Btu/ft^2, or MJ/m^2 (with $1\,kWh/m^2 = 317.2\,Btu/ft^2$)—and represents a time-averaged value.

Avoided Cost The minimum amount an electric utility is required to pay an independent power producer under the Public Utilities Regulatory Policy Act of 1978.

Balance of System (BOS) All components other than the PV modules themselves, which include the system electronics, support structure, and storage.

BIPV Building-integrated photovoltaic (system); a PV system with cells/modules incorporated as an integral part of the building enclosure (rather than simply tacked on to the building shell).

Blocking Diode A diode used to block reverse flow of current into a PV source circuit.

Interactive System A PV system that operates in parallel with the electric utility lines and may be designed to deliver power to the utility.

Inverter Commonly known as a power conversion system (PCS), an inverter is a device that changes direct current (DC) to alternating current (AC). An inverter is not the same as a power-conditioning unit (PCU).

Microinverter An inverter that serves a single solar panel.

Panel A collection of modules mechanically fastened together, wired, and designed to provide a field-installable unit.

Peak Sun Hours The number of hours per day at an irradiance of $1\,kW/m^2$ that is equivalent to the total daily insolation energy.

PV Array A mechanically integrated assembly of modules or panels with a support structure and other components, as required, including tracking apparatus where used, forming a power-producing unit.

PV Cell The basic PV device that generates electricity when exposed to solar radiation.

PV Module The smallest complete, environmentally protected assembly of PV cells and other components normally sold by a manufacturer; comprising several (or many) PV cells.

PV System All the components and subsystems that, in combination, convert solar energy into electrical energy suitable for connection to a load.

Stand-Alone System A PV system that supplies electrical energy without interconnection to any other power source.

Thick-Crystal Photovoltaics The original commercial type of PV material.

Thin-Film Photovoltaics PV devices made of a semiconductor material, such as copper indium diselenide or amorphous silicon, a few micrometers thick, deposited on substrates of glass, ceramic, or another compatible material.

31.13 CASE STUDY—PHOTOVOLTAICS

Lillis Business Complex, the University of Oregon

Project Basics

- Location: Eugene, Oregon, USA
- Latitude: 44.1°N; longitude: 123.2°W; elevation: 425 ft [130 m] above sea level
- Heating degree days: 4546 base 65°F [2528 base 18.3°C]; cooling degree days: 300 base 65°F [167 base 18.3°C]; annual precipitation: 43 in. [1092 mm]; cloud cover: 75 clear days a year, 82 partly cloudy, and 209 cloudy
- Building type: Institutional—classrooms, offices, common areas
- Building area: 196,500 ft^2 [18,255 m^2]; four occupied stories
- Completed October 2003
- Client: University of Oregon

Figure 31.22 *South façade of the Lillis Business Complex, Eugene, Oregon. The atrium glazing incorporates PV cells in this façade example of building-integrated PVs. (© Karen Tse; used with permission.)*

- Design team: SRG Partnership with Solar Design Associates (PV) and consultants

Background. The Lillis Business Complex is an academic facility situated adjacent to a major circulation axis on the University of Oregon campus in Eugene, Oregon. The building (Figure 31.22) has the size, character, and location to serve as a teaching tool for thousands of students who walk by on any given day. North American universities as a rule (with notable exceptions) have not been in the forefront of the renewable energy movement. This project is one of the exceptions.

Context. The Lillis Business Complex is an example of what can happen when various positive forces come into alignment on a project. In this case, an aware client seeking innovation partnered with a local utility promoting the use of alternative energy sources on a campus with an active Energy Studies in Buildings Laboratory (ESBL). The resulting synergy enabled this project to evolve into its final form without too much compromising of objectives. Utility support for the PV system was critical to its inclusion in the project; the utility will buy PV power output at $0.25 per kWh under a 10-year agreement.

Design Intent. The Lillis Business Complex was intended to be an example of innovation, particularly with respect to energy and the environment. An aggressive energy goal of 40% beyond the minimum efficiency required by code set a challenging benchmark for many design decisions. Other green design features flowed from the intent to achieve Leadership in Energy and Environmental Design (LEED) certification for the project.

Design Criteria and Validation. A desire for LEED certification provided the general criteria for the green features of the building. The specific "40% better" energy benchmark provided the criterion for energy performance. Extensive modeling and analysis of concepts and systems were conducted by various consultants (including ESBL) during the course of the design process as a means of verifying design approaches and implementation methods. The building was commissioned.

Key Design Features

- The third-largest PV array in Oregon, with a peak capacity of about 45 kW of electricity.
- A green roof, the first application of this design approach at the university.
- Extensive use of daylighting, involving lightshelves and solar shading devices.
- Efficient electric lighting using dimmable fluorescent lamps with electronic ballasts and daylight sensor controls.
- "Smart plugs" that are wired to occupancy sensors to turn off devices in unoccupied offices.
- An integrated natural ventilation system to accommodate outdoor air needs (for acceptable indoor air quality [IAQ]), to cool the space and people, and to cool the thermal mass; ceiling fans augment the ventilation system.
- Thermal mass to provide storage capacity for ventilation cooling and help the building heat up more slowly in summer and cool more slowly in winter.

Figure 31.24 *Close-up view of an embedded polycrystalline PV cell at the Lillis Business Complex with interconnecting conductor strips. (© Alison Kwok; all rights reserved.)*

Post-Occupancy Validation Methods and Performance Data. The total PV system output is approximately 45 kW peak (DC). There are four distinct PV elements in the building. These include the PV cells integrated into the south-facing glass curtain wall at the entry (Figures 31.23 and 31.24), which produce about 6 kW (DC), and 2.7 kW (DC) of PV cells incorporated into skylights above the atrium. These two PV applications are good examples of BIPV. The PV cells are integrated into the glazing units (Figure 31.25), such that these

Figure 31.23 *Glazing panel on the Lillis Business Complex showing embedded PV cells. The glazing essentially acts as a power plant while still providing for views and daylighting. (© Alison Kwok; all rights reserved.)*

Figure 31.25 *View from the atrium of the Lillis Business Complex; the integrated PV cells provide partial shading and a lively lighting pattern. (© Alison Kwok; all rights reserved.)*

Figure 31.26 Fish-eye view of the Lillis Business Complex atrium. The south-facing panels of the rectangular skylights also contain embedded PV cells. (© Alison Kwok; all rights reserved.)

units are also the PV modules. In addition, another 6 kW (DC) of PV units are installed on sloping penthouse roof panels (Figure 31.26), and another 30 kW (DC) installed horizontally. Each element of the PV system is connected to separate inverter systems to maximize system output.

FOR FURTHER INFORMATION

Brown, G. Z. et al. 2004. "A Lesson in Green." *Solar Today* 18(2): 28–31.

References and Resources

ACSA. 1993. *Design for PV*. Washington, DC: Association of Collegiate Schools of Architecture.

Harris, N. C., C. E. Miller, and I. E. Thomas. 1985. *Solar Energy Systems Design*. Hoboken, NJ: John Wiley & Sons.

Meyers, R. 1983. *Handbook of Energy Technology and Economics*. Hoboken, NJ: John Wiley & Sons.

NIBS. 2024. "Building Integrated Photovoltaics (BIPV)," in *Whole Building Design Guide*. Washington, DC: National Institute of Building Sciences. wbdg.org/resources/building-integrated-photovoltaics-bipv

NREL. n.d. *PVWatts Calculator*. Golden, CO: National Renewable Energy Laboratory. pvwatts.nrel.gov

NREL. 1994. *Solar Radiation Data Manual*. Golden, CO: National Renewable Energy Laboratory.

Ramsey, C. G. and H. R. Sleeper. 2000. *Architectural Graphic Standards*. American Institute of Architects. (10th ed.). New York: John Wiley & Sons.

Ramsey, C. G. and H. R. Sleeper. 2016. *Architectural Graphic Standards*. American Institute of Architects. (12th ed.). Hoboken, NJ: John Wiley & Sons.

Sandia National Laboratories. 1995. *Stand-Alone Photovoltaic Systems: A Handbook of Recommended Design Practices*. Albuquerque, NM: Sandia National Laboratories.

CHAPTER 32

Signal and Communication Systems

32.1 INTRODUCTION

NO AREA OF EQUIPMENT DESIGN AND APPLICATION in buildings has seen such rapid and sustained change as that of signal equipment. Signal equipment encompasses all communication and control equipment, the function of which is to assist in ensuring proper building operation. Included are surveillance equipment, such as that for fire and access control; audio and visual communication equipment, such as telephone, intercom, and television (both public and closed-circuit); and timing devices such as clock and program equipment and all types of time-based controls. Specifically excepted is the area of occupant computing, which is not within the province of the building designer, except for equipment space allocations and indoor environmental quality.

Clock and program equipment, which once was limited to schools and some industrial facilities, is now incorporated into building mechanical equipment control systems. Closed-circuit television (CCTV), formerly limited to classroom and college use, is now commonplace in mercantile area surveillance systems. The hundreds of signals generated throughout a large facility are logged, channeled, and controlled by means of specially programmed computers and microprocessors. Signal systems that once were separate and distinct are now frequently combined and serve multiple purposes.

A detailed study of the design and application of such diverse equipment is beyond the scope of this book. This chapter, however, discusses the basic operation of the various systems, some of the equipment available, its application to different types of facilities, and the impact of these systems on building spaces. The types of facilities considered are single and multiple residences, schools, stores, office buildings, and industrial facilities. Hospitals and laboratories are combinations of these types, but they are too highly specialized to be discussed here.

32.2 SENSOR TECHNOLOGY

To understand the design of intrusion detection systems, it is necessary first to understand the characteristics of the commonly available intrusion detectors (sensors) on which these systems are based. Once an intrusion event has been detected by a sensor, the signal must be processed and appropriate measures taken. This may include sounding loud alarms, turning on lights, and/or sending signals to central proprietary or private surveillance services or police.

(a) Sensors with Normally Open (NO) Contacts

These devices are no longer used in any reputable system because the circuits are unsupervised. As a

Mechanical and Electrical Equipment for Buildings, Fourteenth Edition. Walter Grondzik, Daniel Overbey, and Alison Kwok.

Companion website: www.wiley.com/go/meeb14e

result, a defective circuit does not indicate trouble, and the system is thereby rendered ineffective.

(b) Simple Normally Closed (NC) Contact Sensors

These come in a variety of designs, the most common of which are magnetic contacts for doors and windows, spring-loaded plunger contacts for doors and windows, window foil, and pressure/tension devices. They operate to transmit an alarm signal, as noted previously. Used in closed, supervised circuits, they provide a trouble signal when the sensors or circuits are damaged.

(c) Mechanical Motion Detectors

Where window foil or fixed contacts are impractical, a mechanical motion detector can be used. This type of device is basically a spring-mounted contact suspended inside a second contact surface. Any appreciable motion of the surface on which the device is placed causes the contacts to momentarily "make," turning on an alarm. These devices are very sensitive and can be activated by sonic booms, wind, and even a heavy truck passing by. For this reason, most units are provided with sensitivity adjustment.

(d) Photoelectric Devices

These devices operate on the simple principle of beam interruption. When the beam is received, a contact in the receiver is closed. Interruption of the beam causes the contact to open, setting off the alarm. Older devices of this design use a visible beam (light) and rely for concealment on the fact that light is seen only when reflected from an intervening object. This is quite effective indoors, but outdoors, dust, insects, birds, and so on can show the location of the beam, permitting it to be circumvented. Birds and small animals can set it off, too. Dispersion of light also limits the throw of the devices when used outdoors. Modern units use lasers or infrared (IR) beams, which are less easily detected and can be arranged to distinguish between intruders and other disturbances. These latter devices have a longer effective range for exterior use. When a laser beam is used, the signal can be picked up, amplified, and retransmitted in a different direction, thus establishing a perimeter security "fence" from a single source.

(e) Passive Infrared (PIR) "Presence" Detector

This device (Figure 32.1*a*) acts on the principle that all objects emit IR radiation, (or heat). The amount radiated depends primarily on the object's temperature and secondarily on its material, color, and texture. The PIR sensor uses a lens or mirror that focuses on a small area and concentrates the IR radiation collected from the conic volume of space that falls between the device and the target area of the IR sensor, forming, in effect, a conical beam of coverage called a zone. IR radiation in an area that is undisturbed (not necessarily unoccupied) changes very slowly (because object temperatures change very slowly). As a result, any rapid change in the IR reading of a zone indicates an object entering (or leaving) the space, and this triggers

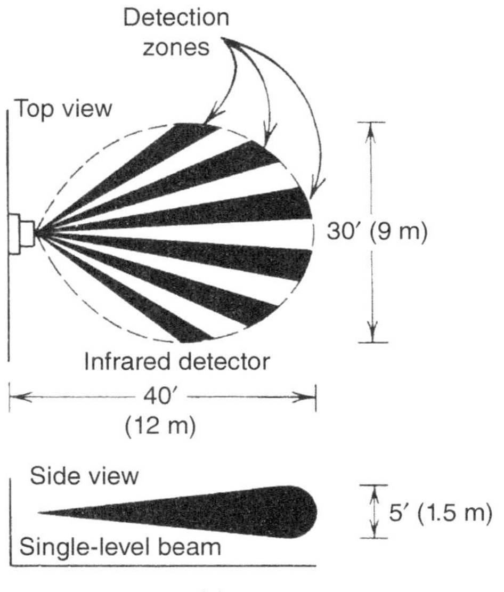

(*a*)

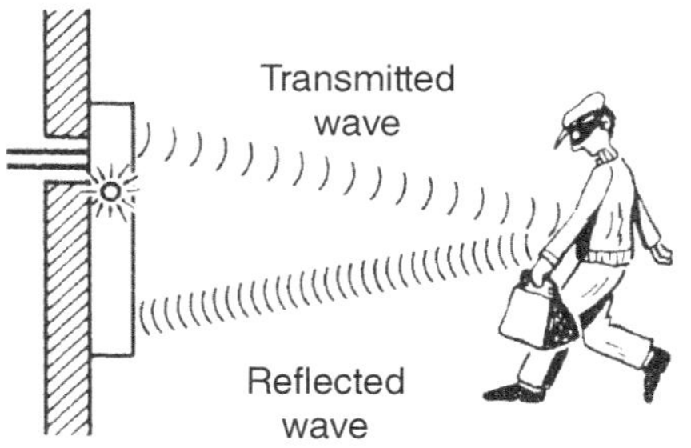

(*b*)

Figure 32.1 *(a) Passive infrared detectors generally give a 30 × 40 ft [9 m by 12.2 m] oval protective zone, starting as a narrow beam and widening with distance. Focusability permits exact coverage of any area in a space. Units are also readily available with multiple vertical sensitivity stacked beams that give the same type of high-level interbeam vertical sensitivity as is available from multiple horizontal beams (zones). (b) Motion detectors depend upon the Doppler effect. They detect changes in the frequency of a signal reflected from a moving object. Sensitivity is highest when relative motion is greatest, that is, when the intruder is moving directly toward (or away from) the detector.*

an alarm. These detectors can be used as occupancy sensors to turn off lights when occupants leave a space.

Covering larger spatial volumes requires detectors with multiple lenses or mirrors to create more than one focus zone. Also, since the IR detector is not intrinsically sensitive to motion, but only to heat, it is usable where motion in the monitored area is unavoidable. The principal disadvantage of PIR detectors is that rapid temperature changes caused by direct solar radiation, a cold breeze, a heater turning on, and the like can cause false alarms. PIR detectors are applied as sensitive motion detectors by using a multibeam (zone) unit. Motion is detected as changes in the IR radiation of adjacent zones or in the radiation of a zone (target) with respect to the background, both of which characterize motion.

(f) Motion Detectors

These devices (Figure 32.1*b*), which operate at either microwave or ultrasonic frequencies, detect motion in the protected area by the Doppler effect. (Named after the Austrian physicist Christian Doppler, who described the phenomenon in 1842, this is the effect that changes the perceived sound of a car horn or train whistle as the vehicle passes.) Any moving body changes the received frequency of the signal it reflects, and an alarm sounds. However, because the Doppler effect depends on relative motion between the source and the moving body, an intruder moving laterally may go undetected if sensitivity has been reduced to avoid false alarms. Therefore, units should be located so that the path of an intruder is as nearly as possible directly toward or away from the detector. Ultrasonic units are more expensive than microwave units because they are more complex to install and program but are more sensitive (and can thus be disturbed by strong air turbulence and very loud noises). Microwave units can cover larger areas and are undisturbed by air or noise, but because they (like ordinary TV signals) penetrate solids, they can be affected by motion outside the protected area, and they tend to have higher false alarm rates than ultrasonic sensors.

(g) Acoustic Detectors

These units alarm when the sound pressure level exceeds a preset maximum. Alternatively, they can be arranged to respond to a particular range of frequencies corresponding to the sound of breaking glass, forced entry, or whatever is desired. Although applied principally in security systems, they can also be used as occupancy sensors for control of lighting.

A selection of the devices described in this section is shown in Figure 32.2.

As with fire detectors, a balance must be struck between the sensitivity of detectors and the nuisance of false alarms, as, unfortunately, increasing the former increases the latter as well. One very effective method of reducing nuisance alarms is to use multiple detectors with different technologies that, in effect, verify each other. Thus, as with occupancy detectors, units are available that combine PIR and ultrasonic detectors in a single housing. Such a unit does not transmit an alarm until both detectors indicate intrusion. This dual technique is applicable to area and perimeter protection as well as portal surveillance.

The previous discussion and that immediately following refer only to intrusion detection (i.e., indication of a human presence where it should not be). Access control, which is another area of security design, deals with controlled entry to an area. It is generally more complex than intrusion control and uses different technologies and a different philosophy of design (see Section 32.25).

32.3 WIRELESS COMMUNICATION TECHNOLOGIES

The field of information technology (IT) is rapidly expanding. Moreover, communication technology, in particular, is increasingly ubiquitous and wireless. Technologies that once were exclusive to facilities serving commercial and industrial enterprises are now as common as personal cellular phones. The emergence of consumer-based wireless communication networks and internet-connected devices has ushered in the age of "smart" solutions—everything from "smartphones," "smart homes," to "smart cars."

Wireless communication technologies continue to evolve and have enveloped signal systems in buildings. The following is a basic primer of some of the most commonly encountered technologies, which are referenced throughout this chapter.

(a) Infrared (IR)

IR communication is a wireless technology that uses electromagnetic radiation from the infrared range of the spectrum to transmit data. IR communication is a common, inexpensive, and easy-to-use technology. IR LEDs, which produce nonvisible radiation with a wavelength around 950 nm, are commonly used by remote controls and turn on/off tens of thousands of times per second to send information to IR photoelectric sensors in a variety of electronic devices such as televisions, stereos, and projectors. Encoding IR signals may involve binary signals (e.g., "1" and "0") in terms of marks and spaces; other methods

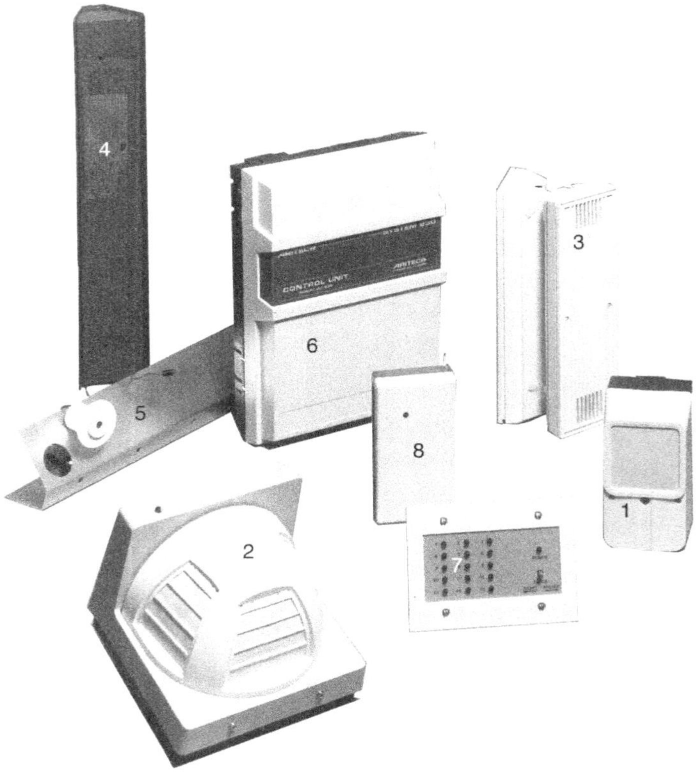

Figure 32.2 *Typical intrusion detection equipment: (1) Passive infrared (PIR) detector. Maximum sensitivity is a 4F° [2C°] differential between target and background and a target motion of $2\frac{1}{2}$ in./s [64 mm/s]. The unit is approximately 6 in. [150 mm] deep. These detectors are available with different coverage patterns and a maximum coverage of 2000 ft^2 [185 m^2]. (2) High-sensitivity, multizone, wide-angle passive infrared detector gives coverage of 17 zones horizontally and vertically. (3) Ultrasonic detectors arranged to give broad coverage of approximately 200 × 50 ft [61 by 15 m]. Each unit is $11 \times 5\frac{1}{2} \times 3\frac{1}{2}$ in. [280 × 140 × 90 mm]. (4) High-sensitivity, balanced signal type of ultrasonic detector differentiates between an intrusion signal and random environmental signals, thus reducing false alarms. (5) Unobtrusive, button-type ultrasonic sensor intended for ceiling mounting. (6–8) System control panel, annunciator, and mechanical alarm interface units. (Courtesy of Sentrol, an SLC Technologies Company.)*

include length encoding or a serial data signaling system called Manchester encoding (named after the University of Manchester, where the system was developed).

(b) Radio Frequency (RF)

Radio communication is not new. Physicist Heinrich Hertz established the existence of radio waves in the 1880s. Radio waves are a type of electromagnetic radiation with the lowest frequencies (i.e., 3000 hertz to 3000 gigahertz—one gigahertz is a billion hertz) and the longest wavelengths in the electromagnetic spectrum—longer than microwaves, which are longer than infrared wavelengths.

Unlike some shorter-wave radiation, such as gamma rays, X-rays, and certain ultraviolet (UV) portions of the electromagnetic spectrum, radio waves are non ionizing, meaning they do not have enough energy to remove electrons from atoms—thus, radio waves cannot cause significant damage to biological tissue at typical exposure levels. Radio waves are practical for modern electronic devices because radio frequencies are a limited, safe, and shared resource.

Many types of modern electronics—including radios, wireless headsets, and televisions—utilize radio waves for communication. Radio frequency (RF) communication requires that a transmitter generate a radio wave, which is then detected by a receiver.

RF communication is a huge component of IT infrastructure in buildings today. Some of the most common RF communication systems encountered in buildings are defined below.

Cellular Networks. A very common and contemporary example of RF communication is the cellular phone (or cell phone). A cell phone transmits signals using radio waves to communicate with the closest cell tower. The cell tower serves as a single base station, which covers a certain geographic area—called a cell. The network of cells from various stations is what constitutes a cellular network. Once a cell tower receives a signal from a cell phone, it transmits the signal to a switching center and the cellular connection from the phone will switch between base stations seamlessly as better connections are found within the device's range.

Various building systems can communicate with cellular networks including security systems, asset trackers, smart meters, and heating, ventilation, and air conditioning (HVAC) systems with remote control capabilities.

Global Positioning System (GPS). The Global Positioning System (GPS) is a public satellite-based radio navigation system, owned by the United States Space Force, that provides geolocation and time information to GPS-enabled receivers anywhere on (or near) the Earth with an unobstructed line of sight to four or more GPS-networked satellites. Similar systems, under different ownership, exist globally.

Unlike cellular devices, which communicate with a terrestrial cellular network, GPS radio waves are transmitted from satellites—providing a much wider coverage area and generally higher accuracy for pinpointing locations. Compared to cellular signals, which communicate voice and data transmissions, GPS signals operate at higher frequencies and are utilized specifically for precise two-dimensional (i.e., latitude and longitude) positioning. As such, few building systems have a practical use for GPS communication. However, some personal electronic devices (such as smartwatches), automobiles, and public transportation assets may utilize GPS for positioning, navigation, and timing (PNT) services.

Radio Frequency Identification (RFID). Unlike cellular wireless communication, which enables long-range communication and data transfer, radio frequency identification (RFID) is used primarily for short-range identification and tracking of objects. RFID communication entails radio waves transmitting data between a reader and a tag. The reader emits radio waves and the tag, which contains a microchip and an antenna, receives and transmits data. The integrated tag is often adhered to a substrate. RFID tags allow for a one-way exchange within a relatively short range—typically up to several feet or meters for "passive" (i.e., no battery power source) RFID tags; although "active" (i.e., battery-powered) RFID tags can range hundreds of feet or meters—making them ideal for asset tracking and identification purposes. RFID readers can process multiple tags simultaneously. RFID facilitates relatively quick, low security, and highly customizable active and passive communication solutions.

Near-Field Communication (NFC). Near-field communication (NFC) is a set of RF communication protocols that enables data exchange between two electronic devices over very short distances up to about 4 in. [100 mm]. Evolved from the one-way exchange RFID technology, NFC can allow a two-way exchange. As with RFID technology, NFC tags consist of microchips that store data and are often embedded in stickers, labels, or magnets. Once an NFC tag is activated, small amounts of data can be exchanged with an NFC-enabled device in a two-way manner—meaning that the device (such as a smartphone) can both read data from and write data to an NFC tag, facilitating more interactive and dynamic applications such as peer-to-peer (P2P) communication between devices—enabling functions such as data exchange, payments, or access control.

Wi-Fi. Wi-Fi (sometimes abbreviated "WiFi") is a wireless networking technology that allows devices to connect to the internet via radio frequencies. Contrary to common misconception, the term "Wi-Fi" is not short for "wireless fidelity." Rather, Wi-Fi is a trademarked phrase referenced in the family of 802.11 standards from the Institute of Electrical and Electronics Engineers (IEEE).

Wi-Fi dates back to Hawaii in 1971, where a wireless ultrahigh-frequency (UHF) packet network known as ALOHAnet was used to connect the islands. Later protocols developed in 1991 by NCR and AT&T were released under the brand name WaveLAN.

In 1997, IEEE completed *Standard 802.11* for wireless local area networks. A wireless local area network (WLAN) is a variation of a local area network (LAN), which is a collection of computers and other devices connected in one physical location, allowing for group sharing of data and devices, including access to the internet. WaveLAN-II was released in 1998 and offered compatibility with the IEEE *Standard 802.11-1997*.

The Wi-Fi Alliance, an organization of leading wireless equipment and software providers with the mission of certifying all 802.11-based products for interoperability, was formed in 1999 and currently owns the Wi-Fi registered trademark. The organization defines Wi-Fi as any WLAN products based on the IEEE 802.11 Standards. IEEE continues to advance the 802.11 Standards for Wi-Fi, along the way establishing a naming convention through the Wi-Fi Alliance to help the general public distinguish between various IEEE 802.11 implementations. For instance, IEEE 802.11b-1999 goes by the naming convention "Wi-Fi 1", while IEEE 802.11be-2024 goes by "Wi-Fi 7" (IEEE SA, 2023).

Wi-Fi, in essence, is a type of WLAN that uses RF communication to connect devices within a small, localized area without cables. A cornerstone of any WLAN is an access point (AP), which broadcasts a wireless signal that computers and other devices with wireless network adapters can detect and use to establish connections to the network.

Access to the internet requires that a LAN/WLAN connect to a wide area network (WAN) through a router, where the router's WAN port is connected to an internet service provider (ISP) via a service cable. In a sense, the router is a device that serves as a bridge between the local network and the internet.

If an ISP provides "broadband" internet access, that means high-speed. As of 2024, the term broadband in the sense of a fixed (i.e., wired; not mobile) connection to the internet is defined by the Federal Communication Commission (FCC)—an independent agency of the US government that regulates interstate and international communications through cable, radio, television, satellite, and wire—as internet access with a minimum download speed of 100 megabits per second (Mbps) and an upload speed of 20 Mbps. Broadband can be provided over different platforms, including digital subscriber line (DSL) via existing telephone lines, cable modem, fiber, wireless, and satellite.

Bluetooth. In 1996, Intel, Ericsson, and Nokia collaborated to pursue a standardization of short-range radio technology. "Bluetooth" was a temporary code name (honoring the tenth-century King Harald "Bluetooth" Gormsson, who famously united Denmark and Norway in 958 CE). This name would eventually be officially retained. The nonprofit Bluetooth Special Interest Group (SIG) was founded in 1998 and owns the Bluetooth trademark. The organization manages the development of Bluetooth Standards. The Bluetooth brand stands today as a term synonymous with short-wave wireless technology. It allows electronic devices to communicate with each other over short distances using radio waves. First introduced in audio devices, the technology has expanded into other applications that utilize internet connectivity.

Products that use Bluetooth technology must be qualified and pass interoperability testing by the Bluetooth SIG. Bluetooth originally adhered to the IEEE 802.15.1 Standard, which defined a short-range wireless communication protocol for cordless devices such as keyboards, mice, and headsets (IEEE SA, 2005). However, the Standard has been effectively superseded by newer versions of the Bluetooth Standard that incorporate additional features and improvements. Thus, manufacturers and developers primarily use the updated Bluetooth specifications rather than the IEEE Standard.

A more recent technological advance, referred to as Bluetooth Low Energy (Bluetooth LE or BLE), was introduced in 2011 and can achieve a much lower energy consumption rate than standard Bluetooth technology. BLE devices basically remain in sleep mode until a connection is initiated—at which time a connection is established for a few milliseconds as data are exchanged. BLE is not ideal for music or talking on the phone but can provide a low-energy option for applications that only periodically exchange data.

32.4 INFORMATION TECHNOLOGY (IT) NETWORKS

The Internet of Things (IoT) encompasses a widening variety of electronic devices in both residential and nonresidential buildings that send signals containing sensor data, such as temperature, pressure, humidity, or motion detection, transmitted through wireless protocols—including Bluetooth, Wi-Fi, or cellular networks. IoT infrastructure uses two basic types of communication:

Device-to-Device (D2D): Allowing direct communication between IoT devices without the need for a central hub or network infrastructure.

Device-to-Cloud (D2C): Allowing IoT devices to send data to cloud-based applications for storage, analysis, and further action.

In either case, the IoT devices must be connected to a network in order to interconnect with each other and/or the internet. Information technology (IT) networking is a rapidly evolving and expanding field. Building design and construction professionals should maintain a basic understanding of how IT networks are structured.

(a) Open System Interconnection (OSI) Model of Networking

The Open Systems Interconnection (OSI) model is a conceptual framework used to describe how a network functions. A network in this context refers to a system of interconnected computational devices (e.g., computers, servers, and printers) that share data through wired or wireless connections. There are many network types. The OSI model is not a set of rules; rather, it serves as a tool for explaining how networks function. A grossly generalized listing of common network types by relative scale would be:

WAN = wide area network (largest; the internet is a type of WAN)

VPN = virtual private network

RAN = radio area network

MAN = metropolitan area network
CAN = campus area network
LAN = local area network
WLAN = wireless local area network (Wi-Fi is a type of WLAN)
HAN = home area network
PAN = personal area network
WPAN = wireless personal area network
LR-WPAN = low-rate wireless personal area network
BAN = body area network
NFC = near-field communication

At a basic level, a node is a physical electronic device connected to a network, for example, computer, router, etc. Nodes can be set up adjacent to one other (i.e., node A connects directly to node B) or there may be an intermediate node (i.e., a switch or a router between node A and node B). Links connect nodes on a network. Links can be wired (such as with an Ethernet, which uses cables to connect devices to LANs), or wireless (such as Wi-Fi). A protocol is a mutually agreed upon set of rules that govern data transmission between nodes on a network. While anyone can create a protocol, the most widely adopted protocols are based on published Standards. A topology describes how nodes and links fit together in a network configuration. Topology types include point to point, bus, ring, star, tree, mesh, and hybrid.

The OSI model consists of seven layers of networking:

1. Physical layer
2. Data link layer
3. Network layer
4. Transport layer
5. Session layer
6. Presentation layer
7. Application layer

A comprehensive definition of the seven layers is beyond the scope of this book. However, the first two layers of the OSI model are of concern to building mechanical and electrical systems and should be noted.

- The first layer—the physical layer (PHY)—encompasses a lot of technology, including IoT-enabled devices, signal transmission methods, cabling, and more.
- The second layer—the data link layer—defines how data are formatted for transmission, including how much data can flow between nodes (i.e., "flow control") as well as error detection and correction (i.e., "error control"). There are two distinct sublayers within this second layer:
  - *Media Access Control (MAC):* This is a network protocol that governs how devices on a network can access and share a communication medium.
  - *Logic Link Control (LLC):* This is a data link protocol that provides error control, flow control, and multiple application communication on a local network.

IEEE 802.15.4 is a technical standard that defines the operation of LR-WPANs, which are wireless networks that use short-range radio frequency transmissions to provide low-cost, low-power, and low-data-rate connectivity between devices. The standard specifies the PHY and MAC for such networks. There are several wireless communication protocols that use the IEEE 802.15.4 Standard, including ZigBee, Thread, and 6LoWPAN. These protocols are used by everything from IoT devices such as lamps, sockets, plugs, motion sensors, and other home automation devices to smart energy meters, electric vehicle (EV) charging stations, and industrial building controls that can communicate with each other over area networks with small, low-power digital radios (IEEE SA, 2024).

Wireless communication protocols such as ZigBee and Bluetooth are typically used in IoT system D2D communication due to their low-power consumption and ability to create a mesh network, a collective communication network that significantly boosts the reach and resilience of IoT systems. A mesh network can be considered a variation of a WPAN, often utilizing protocols that comply with the IEEE *802.15.5 Standard*. Essentially, a mesh network can function as a WPAN if the devices involved are within the close proximity required for a personal area network.

(b) Remote Monitoring and Predictive Maintenance

By collecting and analyzing data from sensors and other field devices, remote monitoring can equip facility managers and commissioning providers (CxP) with insights into the performance of building systems, processes, and people. Sensor data can be accessed in real time from any location. Moreover, the data can be logged for long-term analytics. Such remote monitoring solutions are giving way to advanced data analytics and enhanced commissioning services leveraging continuous monitoring.

Wireless sensor networks are essential for remote monitoring capabilities. Historically, such networks were difficult to install and complicated to maintain over time. However, advances over the past several years have brought to market monitoring solutions that offer a reliable wireless communication solution integrated into a single, inexpensive unit via wireless remote I/O (i.e., input/output) devices. Collectively, a remote I/O system serves as an interface between the field devices (e.g., building equipment) and the building automation system (BAS).

Building systems with integrated software applications and D2C communication offer opportunities for machine learning as well as digital twins—a virtual replica of a physical object, process, or system, including entire buildings or a campus/district, that uses real-time data to simulate its behavior by leveraging artificial intelligence (AI). Such tools are quickly changing how building systems are operated and maintained by allowing better-informed, data-driven decisions and enabling predictive maintenance.

For some enterprises, D2C communication by building systems may pose a significant IT security vulnerability. An improperly secured network could allow unauthorized access to a building's IT infrastructure through weak points in the network—such as an unsecured Wi-Fi-enabled "smart" variable air volume terminal

Although network security is a paramount concern for IT managers, for a variety of reasons, there may be a desire for remote monitoring of a building system. In such instances, an edge controller may be desirable. Such a device is situated at the "edge" of a network, near the source of data generation, and collects and processes real-time information from sensors and other devices, before sending a summarized or filtered version to a central monitoring system in the cloud. This allows for faster response times, reduced bandwidth usage, and localized decision-making capabilities within a specific area or machine. Moreover, by processing data closer to the source (i.e., on the "edge"), edge controllers can facilitate IT security by allowing faster threat detection and response, local security measures such as encryption, and reduce the amount of sensitive data transmitted across the network.

Regarding the physical location of sensors and network components, flexibility and accessibility are essential to ensure long-term adaptability and, thus, utility. Remote sensors and network system components need to be physically accessed periodically and, over time, system sensors and components must be replaced.

PRIVATE RESIDENTIAL SYSTEMS

32.5 GENERAL INFORMATION

Modern private residences utilize a variety of signal devices that greatly enhance their functional value. Indeed, automation in "smart home" residences, which includes control of and by a host of signal systems, is today a key factor in perceived property value. Figure 32.3 shows a conventional, nonautomated residential plan, providing what would be considered minimally adequate sound and signal equipment.

In general, all signal systems require a signal source; equipment to process the signal, including transmitting it; and finally, a means of indicating the signal, whether audibly, visually, or on a permanent hard copy. A complex system still falls into this threefold category except that the individual items of equipment and their functions become more sophisticated. Most designers are familiar with the sophisticated automated systems available in the residential market that handle security, fire alarm, time functions, lighting, and so on. They function on dedicated wiring, a control bus, or power line carrier (PLC) signals (high-frequency signals impressed on electric power wiring). To focus on the basics, the following descriptions consider such systems separately, despite the clear trend toward consolidated control.

Residential fire alarm systems are covered in Chapter 26. The components and design of such systems, including detectors, circuit arrangement, and the like, are also discussed there.

Table 32.1 lists the systems and equipment found in the residence shown in Figure 32.3, organized by the functions of source, processing, and indication. Note that the fire alarm, smoke detection, and intrusion alarm systems have been combined into a single system. This simplifies operation and avoids unnecessary equipment duplication. As the more complex systems are discussed, it will be seen that the basic functions remain unchanged.

As shown in Figure 32.3, a single control panel can serve multiple residential systems. An annunciator, either integral with the panel or in a separate adjacent enclosure, displays the nature and location of the alarm device that has "tripped." A riser diagram for this residence is shown in Figure 32.3*d*. The alarm devices themselves are not shown on the riser because they appear on the plans.

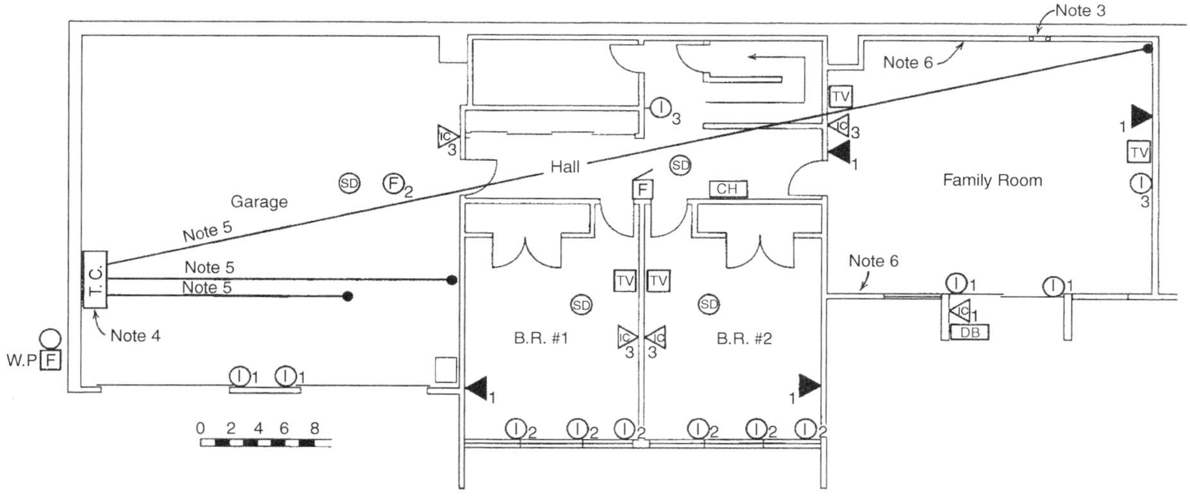

(*a*)

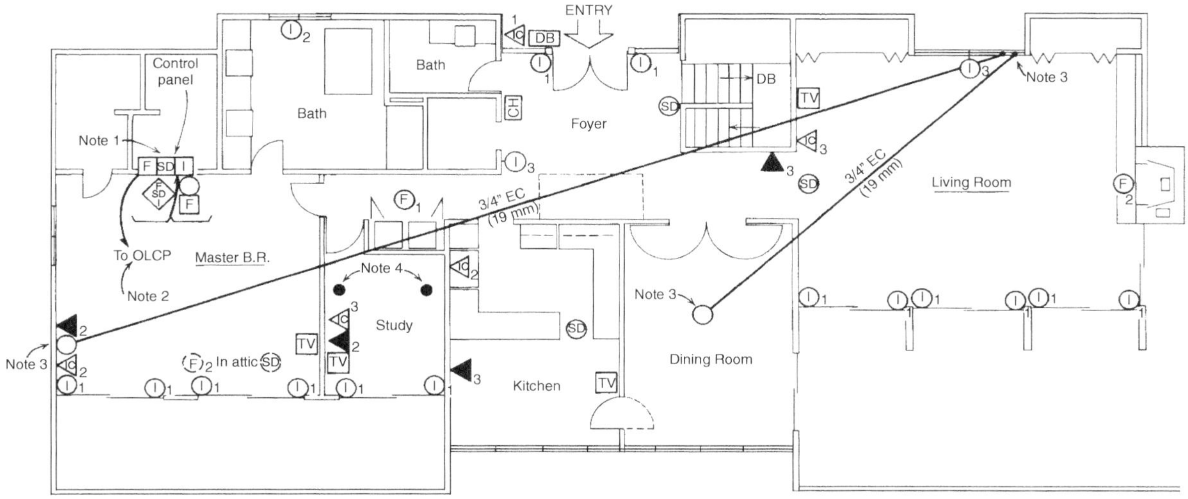

(*b*)

Notes:

1. The fire detection, smoke detection, and intrusion alarm devices all operate from a single control panel; see Figure 32.3*d*. The alarm bell is common. The annunciator indicates the device that operated and its location.
2. The connection between the signal control panel and the OLCP (outside lighting control panel) activates all outside light fixtures when a signal device trips. Selected light fixtures inside the house can also be connected to turn on. See riser diagram (Figure 32.3*d*).
3. Two ¾-in. [20-mm] empty plastic conduits, extending from 4-in. [100-mm] boxes in living room wall down to family room and terminating in 4-in. [100-mm] flush boxes. Boxes to be 18 in. [460 mm] above the finished floor (AFF) and fitted with blank covers. Also, extend a ¾-in. [20-mm] empty plastic conduit from one 4-in. [100-mm] box in living room to 12-in. [300-mm] speaker backbox recessed in dining room ceiling. Locate in the field. From the second 4-in. [100-mm] box in living room extend a ¾-in. [20-mm] empty plastic conduit to an empty 4-in. [100-mm] box in the master bedroom, 18 in. [460 mm] AFF. Finish with blank cover.
4. Coordinate the location and size of the telephone entrance service cabinet (or box) with the phone company. Extend a ¾-in. [20-mm] empty plastic conduit from the telephone entrance cabinet to each of the signal raceways in the study. Extend one 1¼-in. [32-mm] or two ¾-in. [20-mm] empty plastic conduits from the cabinet to the empty signal raceway around the perimeter of the family room.
5. Provide a ¾-in. [20-mm] electrical conduit through the wall and capped at both ends, for entry of cables from an exterior satellite dish. Coordinate location with TV/cable TV/CATV/satellite dish contractor.
6. Mechanical room to contain smoke detector, fixed 190°F [88°C] heat detector, and remote station intercom outlet.
7. Wiring or electrical conduit penetrations through the exterior envelope to be air-sealed.

Figure 32.3 *Signal systems in a residence. (a) Electrical signal equipment plan—lower level. (b) Electrical signal equipment plan—upper level. (c) Symbol list for signal equipment. (d) Signal equipment riser diagram.*

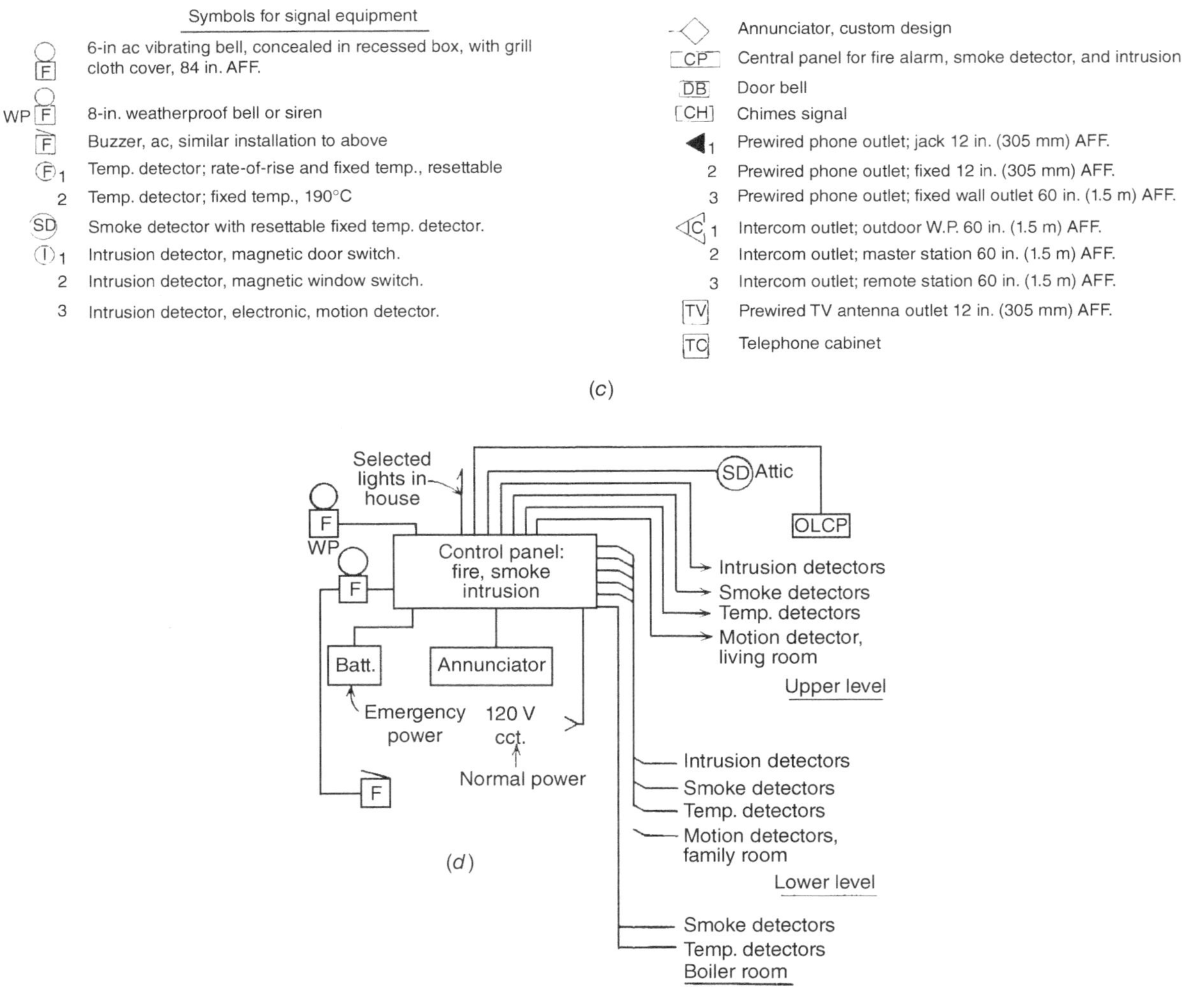

Figure 32.3 *(continued)*

TABLE 32.1 Elements of Residential Signal Systems

System Type	Signal Generator	Signal Processor[a]	Signal Transducer
Fire alarm	Temperature and smoke detectors	Control cabinet(s)	Bells, siren, annunciator, buzzer
Intrusion alarm	Door and window contact switches, motion detectors	Control cabinet	Bells, buzzer, annunciator, siren
Emergency call system[b]	Pull cord, pushbutton	Control cabinet	Bells, annunciator, corridor lights
Intercom	Microphone, speaker–microphone	Amplifier	Speakers in various stations

[a]Proper wiring and switching are included under this heading in all cases.
[b]More common in senior living and multifamily housing.

32.6 RESIDENTIAL INTRUSION ALARM SYSTEMS

Although any (or all) of the devices described in Section 32.2 may be used, residences normally utilize door and window magnetic contact switches, as well as PIR and/or motion detectors, as shown in Figure 32.3. A switch at the end of a long cord is also often provided so that a resident may set off the alarm manually if an intruder is heard. If the system employs the same audible signal devices as the fire system, the resulting sound should be distinctive so that the nature of the alarm can be discerned aurally. Intrusion alarm systems can be continuously supervised by connection with central stations of security companies that monitor the system and either respond directly to an alarm call or notify homeowners and/or local police of any illegal entry.

32.7 RESIDENTIAL INTERCOM SYSTEMS

Interest in home intercom systems seems to come and go. Although available with various features, a basic intercom system comprises one or more masters and several remote stations, one of which monitors the front door, allowing it to be answered from various points within the home. Where desired, a closed-circuit television (CCTV) system can be added so that visual identification at entrances, in addition to voice communication, is available. In general, master stations allow selective calling, whereas remote stations operating through the masters are nonselective. The systems are particularly useful when left in the open (monitor) position for remote "babysitting." The applicability of such systems to residences with outbuildings should be apparent. As the wiring is low voltage and low power, multiconductor color-coded intercom cable is generally used, run concealed within walls, attics, and basements. In addition, battery-operated, Wi-Fi-enabled cameras for door monitoring/video and plug-in Wi-Fi-enabled virtual assistant devices for intercom purposes eliminate the need for hardwiring in homes.

Systems are also available that impose voice signals on the house power wiring. This has the advantage of eliminating separate wiring and making remote stations portable—they are connected simply by plugging into a power outlet.

32.8 RESIDENTIAL TELECOMMUNICATION AND DATA SYSTEMS

With the widespread adoption of mobile phones in recent decades, hard-wired telephones (landlines) are on the decline. If a landline is necessary, the telephone company normally follows the route of the electric service, entering the building overhead or underground as desired. In both cases, a separate service entrance means must be provided: if aerial, a sleeve through the wall; if underground, a separate entrance conduit. Unless a residence has many service lines, no source of power is required for the telephone equipment.

Hard-wired broadband connections for cable TV and internet are still commonly used in many homes. However, smart devices within homes such as smart TVs are increasingly being used to stream content using a Wi-Fi connection to a centrally located, cable-connected internet hub. This smart technology has eliminated the complexity and cost associated with running data cable to spaces within a home and has eliminated unsightly exposed or surface-mounted cable lines on the exterior and interior of homes. At this writing, cellular service providers are beginning to offer home internet capabilities that rival traditional cable internet, which suggests that hard-wired broadband connections to homes may decrease in the future and be replaced by small, compact, wireless internet hubs.

The availability of wireless internet access within homes has made it possible to work from home and is resulting in an exponential increase in work-at-home situations.

32.9 PREMISE WIRING

As noted in Section 32.8, premise wiring refers generally to communication system wiring, including raceways and outlets and, paradoxically, frequently not including wiring. The "wiring" is installed by the various communication contractors and includes fiber-optic (FO) cabling under the category of wiring, despite the fact that wiring is traditionally understood to be metallic. FO cables are now used for most data cabling because of proven advantages in bandwidth, freedom from interference, and high-level security. However, the prevalence of smart electronics and Wi-Fi-enabled devices in commercial and residential buildings means that dedicated cabling using raceways and wall-mounted ethernet port connections is on the decline.

Most premise wiring raceways are surface-mounted because access is frequently required and because data cables are often preterminated, making it difficult to pull them into recessed raceways. Finally, premise wiring raceways are usually large and as such are much easier and more economical to install as surface-mounted units. Figures 32.4 and 32.5 show a few of the hundreds of devices and raceways used in premise wiring, also referred to as wire management. See Chapter 29 for a cutaway view of a typical premise wiring raceway.

MULTIPLE-DWELLING SYSTEMS

See Chapter 26 for a description of multiple-occupancy residential fire detection and alarm systems.

32.10 MULTIPLE-DWELLING ENTRY AND SECURITY SYSTEMS

Apartment houses and other large residential buildings often combine the functions of access control with the familiar lobby-to-apartment communication system. The most basic system is a series of pushbuttons in the lobby and an intercom speaker or telephone with

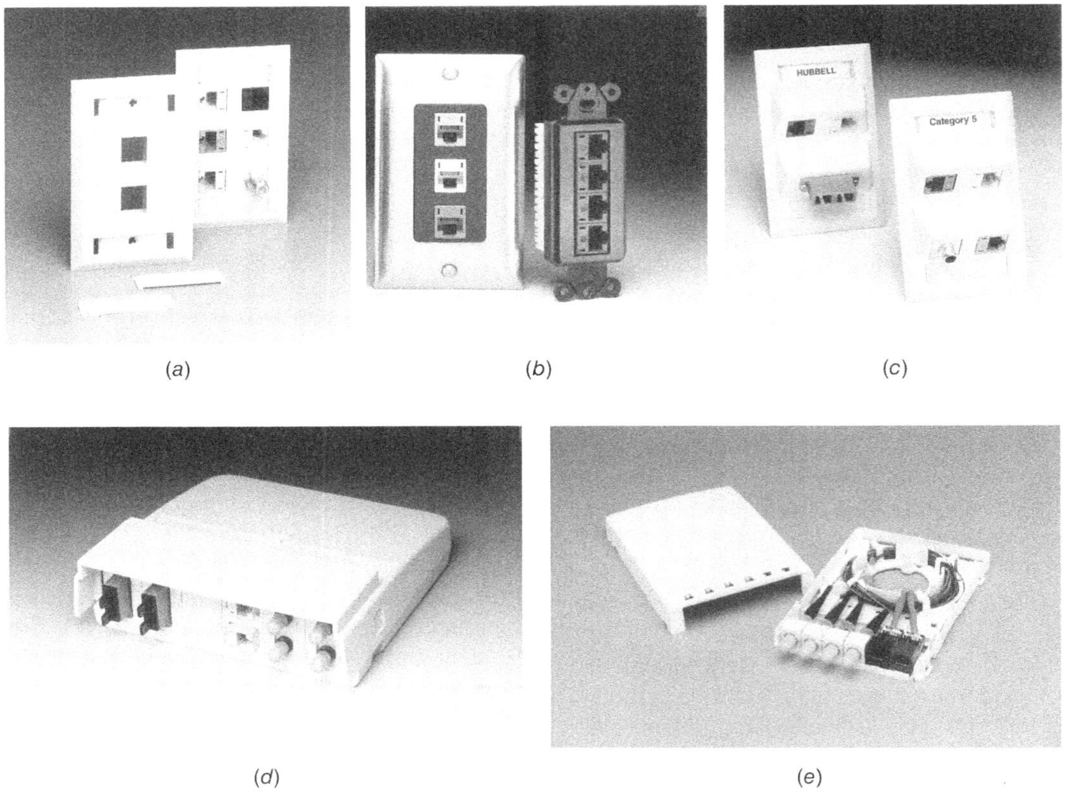

Figure 32.4 *Typical boxes, devices, and plates used in data and telecommunication wiring systems. (a) Single- and double-gang flush wall plates with various jacks and connectors. (b) Communication outlets styled to match standard rectangular electrical wiring device plates. (c) Angled faceplates with a variety of jacks and connectors. (d) Multimedia outlet that accepts both copper and fiber-optic (FO) cable connections. (e) Multimedia FO outlet with a built-in storage spool for storage of spare FO cable. (a–d: courtesy of Hubbell Premise Wiring; e: Courtesy of Panduit Corp.)*

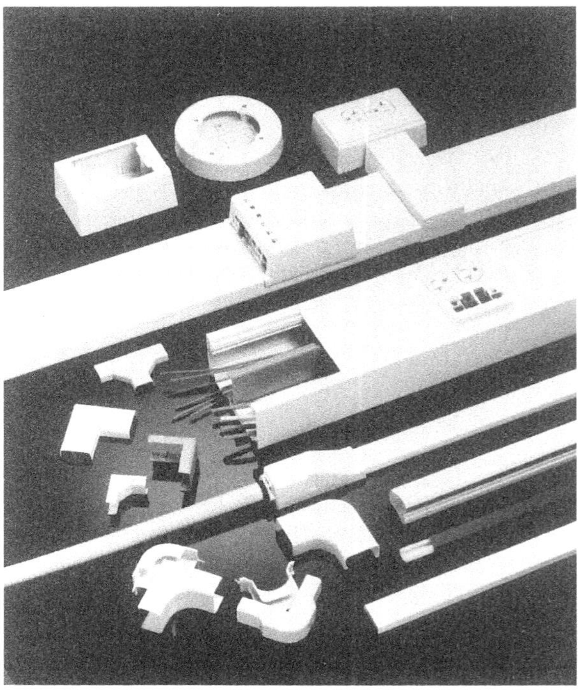

Figure 32.5 *A range of raceway sizes and shapes is available to meet all signal cabling requirements. (Courtesy of Panduit Corp.)*

which to communicate with residents. At the other end, the tenant has a speaker microphone plus a lobby door-opener button. This system can also be arranged to utilize the tenants' regular telephones or mobile phones. When the number of occupants is large, an alphabetical roster is added to the apartment-button panel to avoid the nuisance of scanning all the apartment names when the sought party's apartment number is not known. When the number is larger yet, a simple pushbutton per apartment arrangement becomes cumbersome and is usually replaced by an alphabetical tenant register plus a dial or button phone. CCTV is frequently added to the lobby system, enabling the occupant not only to converse with, but also see, the caller. Such a system increases the electrical contract cost for an average apartment house by 5% to 7%.

In addition to the security provisions provided by apartment-to-lobby audio and video connections, additional security and alarm devices are often used, such as emergency call buttons within the apartments. These perform alarm functions required to deal with an intruder who manages to bypass the lobby security system. In geriatric housing units, these buttons typically unlock the apartment door to allow helpers to enter if summoned by lights and alarms. In luxury apartment buildings, apartment doors can be monitored from a central security desk and any unscheduled door movement will be subject to immediate investigation. These systems are custom-designed to meet the needs and requirements of the owner.

A security problem applicable to all facilities, including residential ones, involves limiting entry to unsupervised areas to authorized persons (i.e., access control). The problems with keys and locks are well known, despite advances in that field. More sophisticated means include magnetic cards and electronic combination locks, which, because of the ease of code changes, are particularly useful in residential facilities that cater to transients. Remote locking technology can now also be controlled via mobile devices such as smartphone and tablet apps to secure or provide access at doors even if the resident is not home.

Another aspect of security that is particularly appropriate to housing for the elderly and handicapped, although applicable to all housing installations, is the emergency call system. The purpose of this system is to alert outsiders to an emergency situation inside a closed apartment. This alarm system is essentially a way to call for help in time of illness or other distress. Many construction and housing codes include descriptions of the required equipment. Most often they prescribe a call initiation button in each bedroom and bathroom that will register an audible (alarm) and visible (annunciated) signal at a location that is monitored, locally or remotely, 24 hours a day. Additional signals are required in the floor corridor and at the apartment, the purpose of which is to alert immediate neighbors to a distress call.

32.11 MULTIPLE-DWELLING TELEVISION SYSTEMS

All modern multiple-dwelling residences supply each room with one or more cable/internet outlets. The "cable" jack may supply house signals from a rooftop satellite dish or cable TV provider. As these systems are always subcontracted, the design requirements in new construction include a system of empty conduits connecting to cable pulling points in cabinets. Raceways should be sized liberally because of the constant expansion of options in the electronic entertainment field. As pointed out previously, the ideal raceway for signal cabling is 100% accessible (i.e., a surface-mounted raceway with a removable cover). However, this solution is rarely used in new work because of its unsightliness, plus the fact that, for most cabling, concealed raceways are adequate. In low-budget construction, only floor and wall sleeves are supplied, and coaxial cables are run exposed in residential spaces. The cable contractor arranges for the electric power needed for any local amplification. Normally, a 15-A dedicated circuit is sufficient. Rooftop satellite dish or cable TV/internet services can now be replaced with cellular internet services, which are a type of mobile broadband that connects a Wi-Fi-enabled internet hub to smart devices in the dwelling without the need for cable runs.

32.12 MULTIPLE-DWELLING TELEPHONE SYSTEMS

Though their utility in new construction is diminishing due to the ubiquitous uptake of personal cellular phones, hard-wired "landline" telephone service may still be encountered—especially in existing buildings. Telephone service cable normally follows the same entrance path and method of entrance as the electric power service. This is common in single- and multiple-dwelling buildings alike. For the sake of economy in underground construction, the two services often share the same trench, although in different raceways, and utilize twin manholes where these are required. Typical entrance arrangements for any large building, residential or other, are shown in Figure 32.6.

The service entrance space requirements vary with the size of the building and the telephone capacity. For a small apartment house of the garden or three-story type, a clear wall space of 4–6 ft [1.2–1.8 m] is sufficient. A terminal (equipment) room is required only in large buildings (see Section 32.22). The telephone system is one part of the overall telecommunications system in most modern residences, preparations for which fall into the premise wiring area discussed previously. Therefore, generalizations are not useful, and planning must consider the type of occupancy and future requirements.

Rental apartment buildings and dormitories differ from commercial structures in that the layouts of all floors are similar, so that the arrangement of risers is relatively simple. It is common practice to utilize cable only in risers that extend through vertically aligned closets in apartments. To accommodate these cables, sleeves through the floor between closets are necessary. If a riser is located in a shaft other than a closet, conduit is normally utilized to allow for easy installation, protection, and repair. If the location is accessible, as in an alcove, only a sleeve is provided. When the riser is located outside the apartment, each dwelling unit is connected to the riser by a conduit with a junction box at either end. Beyond the apartment service point, the individual rooms can normally be prewired entirely without conduit or with only a few short sleeves. In condominium and coop structures, large raceways are run from the service entrance point to cabinets on each

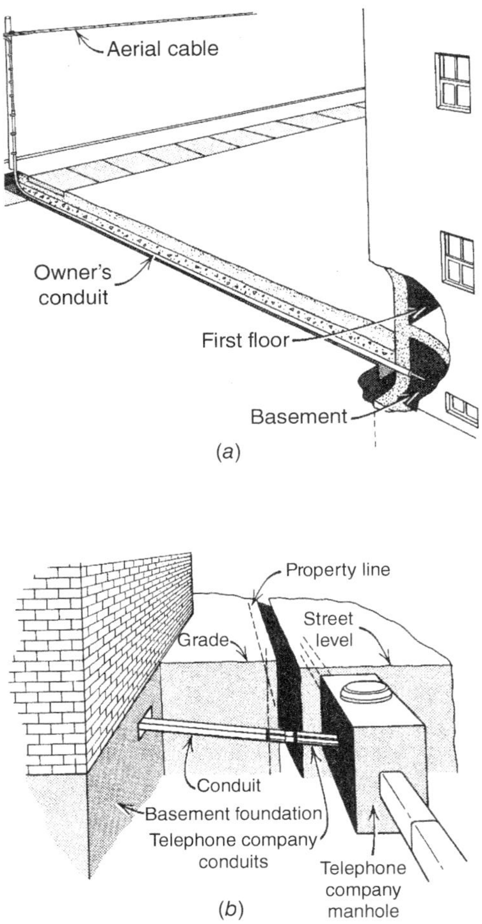

Figure 32.6 *Telephone service cable may enter a building underground after originating on overhead lines (a) or via manholes (b).*

floor of the building. From terminal devices in these cabinets, the required service can be extended into each apartment. Prewiring of the apartment avoids the unsightliness of exposed cabling or of cabling within surface raceways.

32.13 HOTELS AND MOTELS

(a) Security

Because of the transient population of these facilities and the need to provide maximum service and security to guests during their stay, the principal problem in addition to room access security is hotel equipment security (i.e., prevention of what is euphemistically known as shrinkage).

Room Access Security. The ineffectiveness of key locks in preventing undesired entry (even in private homes) is well known; in hotels, a key is little more than a psychological barrier. As a result, most modern hotels have installed electronic room locks whose opening device code is changed with every new guest. These lock systems typically entail one of the following systems:

1. *magnetic stripe cards*, by which a guest inserts a card into a slot on the door lock, which reads a magnetized pattern and verifies a match from a central lock–security console
2. *RFID cards*, by which a guest taps a card against a reader on the door—bringing it close enough for the lock to detect and read the data wirelessly
3. *smart cards*, by which a guest may insert a card into a reader or hold it near a contactless reader on the door (Figure 32.7).

The card lock system may be supplemented by smart phone applications, which may offer "digital keys." This method typically uses BLE or NFC technology between the phone and the hotel door lock. Although less common for guest room access security, hotels may also utilize programmable electronic lock and coded-key types of device (Figure 32.8). All of these systems have advantages and disadvantages, depending upon the number of rooms, whether the installation is new construction or retrofit, and the average length of the guest stay.

Figure 32.7 *Electronically coded card keys are often used in hotels. The coding is changed after each room use. (Courtesy of Bruce Haglund.)*

Figure 32.8 *This electronic door-locking system uses a programmable electronic lock (top) with a non-duplicatable fixed-code metal key (bottom). The lock, which can be reprogrammed using a portable programmer, reads the key code with an internal infrared reader. The keys, which can be programmed with up to 1.5 million codes, are discarded after use. Each lock accepts up to 3000 different codes and can be programmed in two modes: passage mode (to admit a user during a specific time slot) and time mode (to accept specific keys during specific time slots). In addition, the locks can be audited to reveal access records, invalid attempts, activation/deactivation, and mechanical key override.*

Equipment Security. Expectations for amenities require that guest rooms be equipped with a television where movies are typically handled on-demand through cable companies and meeting rooms with TVs or projectors that can connect to personal computers via Wi-Fi, Bluetooth, or high-definition multimedia interface (HDMI) cable. Theft control is a specialty on its own and is beyond the scope of this book. One such system (equally applicable to equipment in hotels, schools, office buildings, and industrial facilities) senses the disconnection of equipment from its power connection (wall plug) and transmits an alarm over the power lines to an annunciator at a selected control location. This arrangement has the advantage of alarming immediately on equipment removal, thus normally permitting appropriate retrieval action to be taken in time.

(b) Telecommunication: Data

The telephone and data communication system in large modern hotels is important and complex. Hotels catering to businesspersons must provide access to the internet that can be used at the very least in rooms dedicated to that purpose and, almost universally today, in guest rooms as well. Therefore, the building designer must provide adequate raceway (and cabling) facilities, as discussed in Chapter 29. The current practice of holding business meetings and technical conferences in hotels also requires that conference and meeting rooms be arranged for very heavy concentrations of electronic equipment that can be installed and rearranged in short order. This may call for some sort of access floor and modular cabling, as described in Chapter 29. Meeting attendees are increasingly expecting wireless internet access throughout a hotel.

SCHOOL SYSTEMS

32.14 GENERAL INFORMATION

The proper operation of a modern school requires that flexible and efficient signal and communications equipment be available to the administrative and teaching staff. Such equipment, engineered to meet the needs of the individual institution, can improve utilization of staff and student time. School fire detection and alarm systems are discussed in Chapter 26.

32.15 SCHOOL SECURITY SYSTEMS

Although intrusion alarms and security systems are not historical school requirements, this situation has unfortunately changed. Sensing devices on doors and windows can be arranged both to trip local alarm devices and, via auxiliary circuits, to notify police headquarters. Often, vandals can be frightened off by having the alarm system actuate a protective lighting system that illuminates the building exterior and any desired interior areas, such as records rooms. A perimeter alarm detection system of the types described in Section 32.2 can be installed in particularly vandalism-prone areas to assist in preventing entry to school premises after hours. Although expensive, they are very frequently cost-effective.

An exit-control alarm is a type of exterior door security device applicable to schools (and other facilities) with doors that are locked from the outside but that must be openable from the inside in the event of an emergency. This device (Figure 32.9) alarms as desired (i.e., audibly or visually, locally or remotely) when a

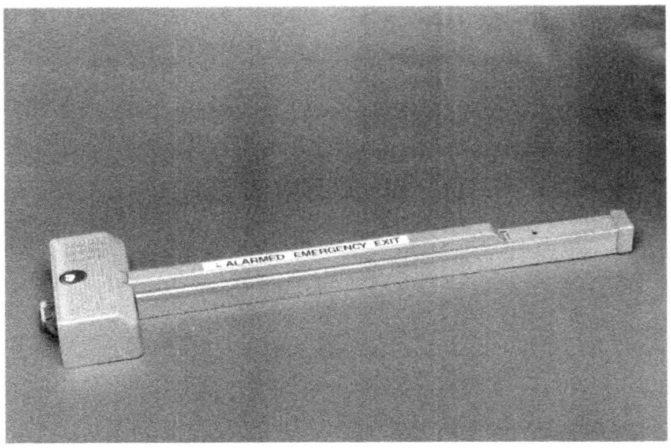

Figure 32.9 *Delayed-egress exit alarm. When the bar is pushed, an alarm sounds immediately, but egress is delayed 15 seconds to permit security personnel to investigate before the door release operates. In the event of fire or another emergency, the door releases instantly. The alarm can be bypassed and reset by key only. (Courtesy of NAPCO.)*

monitored door is opened. It is normally equipped with a timed bypass that allows authorized personnel to key-operate the bypass but prevents the door from being held open without alarming. One model of door-opener alarm requires a continuous pressure on the door-opening bar for 10–20 seconds before the door opens, with the alarm being activated immediately on application of pressure. This arrangement (usable only where fire codes permit) allows sufficient time for the facility's staff to investigate an attempted exit before the door opens and is applicable only where exit control is the overriding consideration. As mentioned in Chapter 26, stair and exit-door exit locks can be arranged to be centrally released from a fire control center.

32.16 SCHOOL CLOCK AND PROGRAM SYSTEMS

The clock system and the program system were once separate and distinct, sharing only the time-keeping facilities provided by a master clock. Now that electromechanical programming devices are effectively obsolete, the two systems are actually one, but the traditional two-system name remains. As shown in Figure 32.10, the heart of the system is the time base (electronic clock). This is the same device that provides timing for all programmable switches and controllers. In a clock and program device, it controls clock signals, audible devices, and, if desired, other switching functions.

The clock system may use analog clocks (with hands), digital units, or both. The former are usually locally powered, and a correction signal is periodically transmitted via dedicated wiring from the master clock at the controller. This same controller continuously transmits a binary-coded signal to digital "satellite" clocks, which can be either of the self-illuminated light-emitting diode (LED) type or of the liquid crystal display (LCD) type. LCD units are easily visible only in high ambient illumination and when viewed directly (not at an acute angle). LED units are best applied in areas of low ambient illuminance. Large-face analog clocks are easily visible in all ambient light situations.

The programming function of the controller serves to delineate audibly the various time periods into which the school day and week are divided. A single-circuit unit is utilized in an institution that operates entirely on one schedule, such as an elementary school on

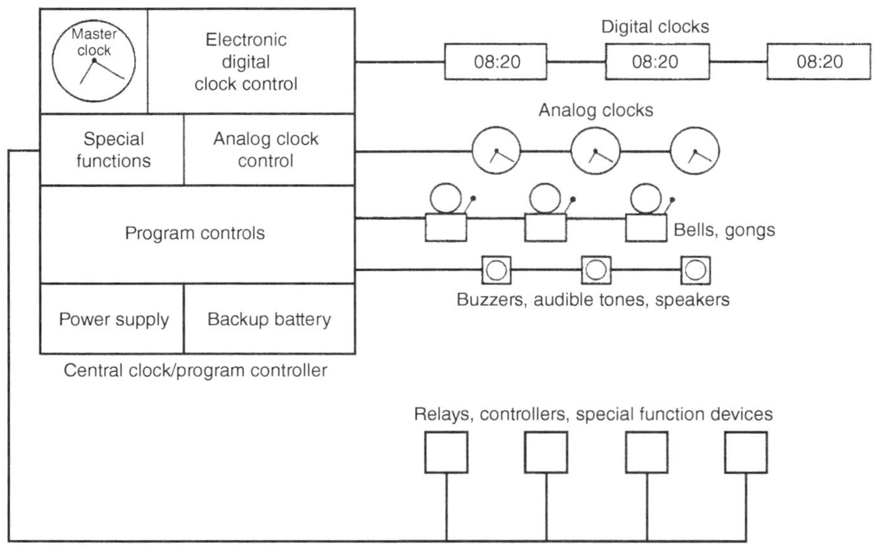

Figure 32.10 *A central clock and program device uses its electronic time base to control clocks of all types, time-based audible signals, and event controllers.*

a morning period–lunch–afternoon period regimen. For a school employing different schedules for its various parts, the controller can provide multiple program schedules on different circuits, depending upon its design. Controllers are user-programmable and are normally provided with a crystal-controlled master clock to ensure accurate timekeeping, regardless of line frequency variation; a backup power source to maintain user programming and master-clock local display; and various conveniences such as daylight saving time correction, security arrangements, event timers, and so on. If desired, the clock and program controllers can also be used for mechanical system control by the simple expedient of adding relays and switching devices, as shown in Figure 32.10. Typical clock and program controllers are shown in Figures 32.11 and 32.12.

The audible devices in a program system may be bells, gongs, buzzers, horns, or a tone reproduced on a classroom loudspeaker. The last system has the following advantages:

1. Clear audibility in each classroom, with an adjustable volume to accommodate both quiet and noisy areas.
2. No possibility of confusion between the program tone and other signals such as fire alarm gongs.
3. Multiple use of the speaker unit for classroom sounds as well as program tones.
4. Complete flexibility of programming that is not possible with hall gongs. This is particularly desirable in schools with special programs for particular groups of students.

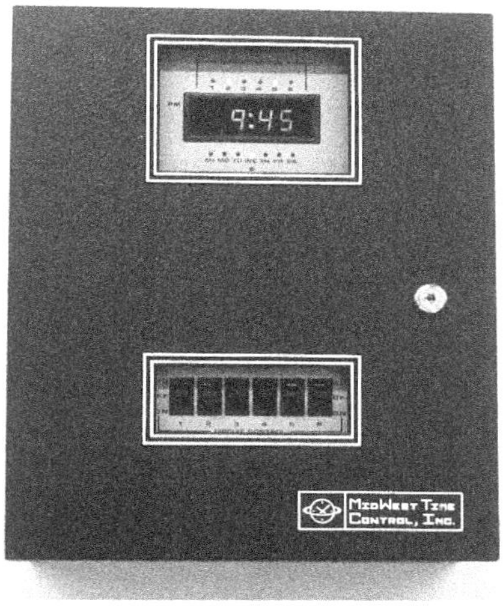

Figure 32.11 Typical microprocessor-based master time clock and program device. This unit, which measures 14 × 12 × 5 in. deep [355 × 305 × 127 mm], provides time correction signals to synchronous wired clocks of the digital or analog type and can be adapted to control minute impulse clocks. Programming of output circuits can be daily or weekly and can accommodate daylight saving, holiday, and leap-year updates with a total program capacity in excess of 600 events. Programming is done locally or via a remote computer. This type of device is applicable to any facility requiring a clock and/or program system. (Courtesy of Midwest Time Control Inc.)

32.17 SCHOOL INTERCOM SYSTEMS

Various types of intercom systems are available, depending on the needs of the school building involved. In small schools, a simple wired intercom system connecting the various offices is usually sufficient. This is supplemented by outside telephones in the administrative offices and a functional paging system that is normally part of the school's sound system. With larger buildings and correspondingly larger numbers of extensions and multiple-function demands, more sophisticated equipment is required. The unit illustrated in Figure 32.13 is typical of modern school intercom equipment and is actually a private telephone system with considerable flexibility. Such a system is generally interfaced with the school sound system and may provide these functions:

1. Intercom between staff members and offices
2. Direct communication with classrooms, including selective and all-call capability
3. Paging zone, group call, and conference call functions
4. Interconnection with the outside phone system
5. Class-change signals via interface with the master clock system
6. School intercomputer communication via a network

A combination program/intercom controller is shown in Figure 32.14. These systems use direct

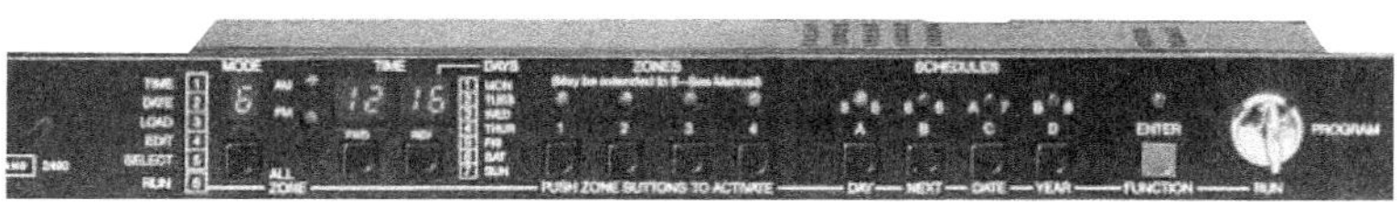

Figure 32.12 User-programmable master clock and program controller. This unit has nonvolatile memory, eight program zones, automatic daylight saving time and holiday adjustment, and a maximum of four schedules; it is usable with most types of secondary clocks. (Courtesy of Rauland-Borg.)

Figure 32.13 Intercom controller. The illustrated microcomputer design interfaces with the existing telephone system. In addition, it provides paging, tone distribution, and program distribution, with control from a telephone. The unit, which measures 19 in. × 9.5 in. × 3.4 in. high [483 mm × 241 mm × 86 mm] and weighs 9 lb [4 kg], also has a built-in master clock. (Courtesy of Rauland-Borg.)

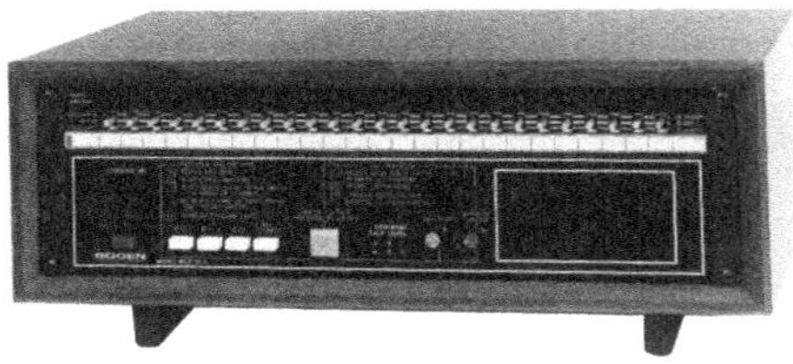

Figure 32.14 Compact program/intercom control center. This unit, which measures 20.5 in. × 11 in. × 8.5 in. high [521 mm × 280 mm × 216 mm], has a separate power amplifier for intercom and program material. It distributes program material or provides intercom with 25 loudspeaker locations and has additional capabilities, such as emergency/all-call paging, tone calling, time signal, night bell, and telephone paging. Program material is accepted from separate external sources. (Courtesy of Bogen Communications Inc.)

pushbutton dialing and programming. All stations are coded with three-digit codes, and all switching is solid-state, minimizing maintenance problems. Such systems are adequate for all but the largest institutions.

32.18 SCHOOL SOUND SYSTEMS

An integrated sound-paging-radio system designed for school use offers several modes of operation and considerable flexibility. Its function is to provide a means for distributing signals from recordings, broadcasts, or live sound to preselected areas of the school. Thus, a simple system might provide a single microphone input and a single channel to all the speakers in the school, whereas a complex system can be arranged to operate with three simultaneous input signals distributed to six different areas of the school (see Figure 32.14).

A conventional system consists of a control console containing most of the input units, amplifiers, switching devices, and connections to the remote loudspeakers. The input units vary and include microphones. One microphone is normally located at the console, with others in the principal's office, auditorium, school office, or other selected locations. If desired, microphone outlets can be spotted around the school, and a spare microphone and stand supplied that can be plugged in at any of these points (Figure 32.15).

Loudspeakers, located in classrooms, the gymnasium, the auditorium, the cafeteria, and outdoors, receive the amplified signal through the switching mechanisms located in the console. The function of

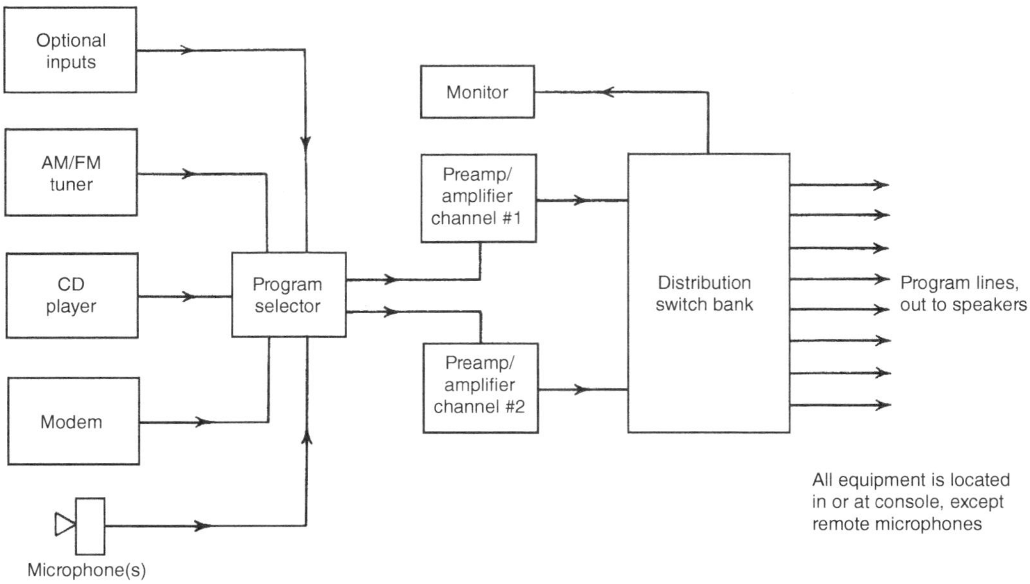

Figure 32.15 Block diagram of a two-channel sound system. Options that can be added to the console are a tape deck, private telephone communications, modem connection, and equipment for rebroadcast of signals between areas. An intercom and a master clock and program can also be incorporated or interconnected.

these switches is to deliver the program material to the various loudspeaker circuits, called program lines. Thus, using a system with multiple amplifiers, music can be piped to the cafeteria, an educational broadcast program to selected classrooms, and instructions to an outdoor gym class or team during practice. An all-call feature also allows announcements to reach all speakers in the system simultaneously. The intercom system discussed previously can be interconnected with the sound system to allow conversation between classrooms and the console or other points. A small system can be contained in a compact desktop console (see Figure 32.14). A large system frequently requires a separate console. A console is usually built in a desk arrangement, and it is advisable to provide sufficient space for it and for the person who operates it. Often an alcove of 30–50 ft^2 [2.8–4.7 m^2] is reserved for it and a library of recordings.

Loudspeakers may be placed flush or in surface baffles at the discretion of the designer. Gymnasium, cafeteria, and auditorium units are normally flush-mounted in the ceiling. For such large areas, it is well to provide a volume control, enclosed in a recessed wall box with a locking cover. A common variation of this system uses separate subsystems for the cafeteria, auditorium, ball field, and other areas that utilize sound equipment frequently. These smaller systems have their own input, amplification, and control devices but utilize speakers in common with the central console. Normally, the console has an override feature that allows it to override local systems.

For a discussion of high-quality sound systems required for recital halls in music schools and the like, refer to Chapter 24.

OFFICE BUILDING SYSTEMS

32.19 GENERAL INFORMATION

This section addresses systems found in office, professional, and sales-type buildings. Such buildings house tenants with varying schedules and requirements. This diversity must be considered in the design of the signal and communications systems for such buildings.

Although in many medium-sized and large buildings the various control, alarm, and security functions are combined in multiuse apparatus and consoles (discussed later in the chapter), the basic systems are essentially separate despite having shared-use equipment. They are discussed individually to demonstrate function and equipment and then as combined to show economies and modern practice.

32.20 OFFICE BUILDING SECURITY SYSTEMS

Although automatic surveillance systems are applicable to office and mercantile occupancies, they are more frequently found in industrial facilities and are discussed under that heading. Office buildings normally utilize some type of manual watchman's tour system so that surveillance of unoccupied areas is conducted on a regular basis.

The simplest and oldest type is nonelectric and comprises a number of small cabinets, each containing a key, placed at intervals around the interior and exterior of the building. The watchman uses these keys to operate a special clock that he carries about, thus recording the exact time at which he "clocked in" at any specific location. Alternatively, the clock is wall-mounted, and the guard carries only a key (Figure 32.16). Modern electronic guard patrol systems are also available. In place of the watchman's clock, a device called a patrol reader/wand is used to scan electronic devices located around the building, which record the time in the wand. At the end of a shift, the watchman transfers the data from the wand to a computer.

Guard tour systems are also available that permit constant supervision and are particularly effective where more than one person is on duty. Such systems show on a panel the location and progress of the watchman by means of lights that glow when the device at each location is operated. Because part of the effectiveness of these systems lies in the timing of the tour, a system can be arranged to sound an alarm if a particular station is not operated within a

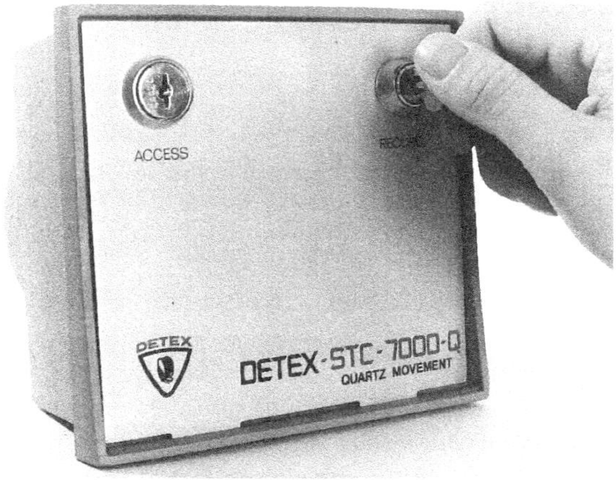

Figure 32.16 *Watchman's tour station. Here the clock and recording tape are inside. The station is operated by simply inserting a key, as shown, and giving it a quarter-turn. An alternative version of this system, commonly used, consists of a portable clock and tape operated by a key in a wall box at each station. (Courtesy of Detex Corp.)*

ELECTRICITY

specific time period. For protection of areas housing extremely valuable items or documents, an intrusion alarm system may be employed.

32.21 OFFICE BUILDING COMMUNICATIONS SYSTEMS

Planning for this type of system requires considering four functions, which are frequently melded into a single network:

1. Intraoffice voice communication, or intercom
2. Interoffice and intraoffice data communication using telephone and communication cabling
3. Outside-the-building communication via telephone company (or data) lines
4. Paging function

Today, in most locations, the user has the choice of purchasing or leasing as much of the communication system as is desired. In other words, the communication system, including cables, instruments, switching equipment, and so on, can be all privately owned, all supplied by the local communications company, or acquired in almost any other arrangement desired. Except for a small interoffice intercom, however, duplication is eliminated—that is, the same instruments and switching equipment are used for both intercom and outside connection. What to buy, rent, or lease constitutes an economic decision because the required functions are satisfied by either private or externally provided equipment.

32.22 OFFICE BUILDING COMMUNICATIONS PLANNING

Planning for the telephone and other communications equipment in an office building is of prime importance because of the large amounts and critical locations of required space. Therefore, it must be done simultaneously with other space planning. Exact requirements for office space are generally unknown at the time of design, and even if they were known, planning would have to account for changes in space usage as well as increased communications and data transmission services. For this reason, all planning is based on usable office area. Planning is essentially focused upon spaces, from incoming service space to final instrument locations, because cabling and equipment are furnished and installed either by a private equipment installer or a leased services provider.

The planning information that follows is applicable to office buildings with average telephone and data transmission loads. Buildings whose tenants include brokerage houses, insurance companies, headquarters of multibranch operations, and other heavy telephone and data transmission users may need more space. Advances in technology, however, including the use of fiber-optic (FO) cables, act to reduce equipment size and space requirements. As a result, the best course of action is to design for current requirements with a reasonable estimate for future needs and technologies based upon advice from experts in the field.

Space is required for the following:

1. A service entrance room that houses incoming cable (network cable), terminated empty conduits for expansion and data cables, a network cable splice box, connection (network interface) cabinets, and equipment required to interface the building's equipment, including telephones, modems, Wi-Fi hubs, and the like, to the main connection cabinets. This space is usually referred to as the equipment room.
2. Riser spaces (shafts) and riser closets. These are stacked vertically and carry main cables.
3. Satellite closets where required.
4. Horizontal distribution between closets and devices, including conduit, boxes and cabinets, underfloor raceways, and over-the-ceiling systems.

A riser diagram of a typical system is shown in Figure 32.17.

(a) Service Entrance

At least two 4- or 5-in. [100- or 125-mm] conduits extend from the exterior service connection to the basement or ground-floor service equipment room. One conduit will contain the service (network) cable; the other is a spare. Spare conduits should be terminated 12 in. [300 mm] above the finished floor (AFF) of the equipment room, in a threaded fitting, and sealed against entrance of dirt, moisture, or rodents. The service entrance cable terminates in a splice box or cabinet, from which, in large buildings, additional cables extend to one or more wall cabinets, each measuring about 3-ft wide, 6-ft high, and 2-ft deep [$0.9 \times 1.8 \times 0.6$ m]. Each cabinet can serve a maximum of 30,000–100,000 ft^2 [2790–9290 m^2] of tenant area, depending upon the density of service required. To accommodate these cabinets from which risers extend, a clear wall space of 6 ft [1.8 m] for every cabinet should be provided. This wall area should

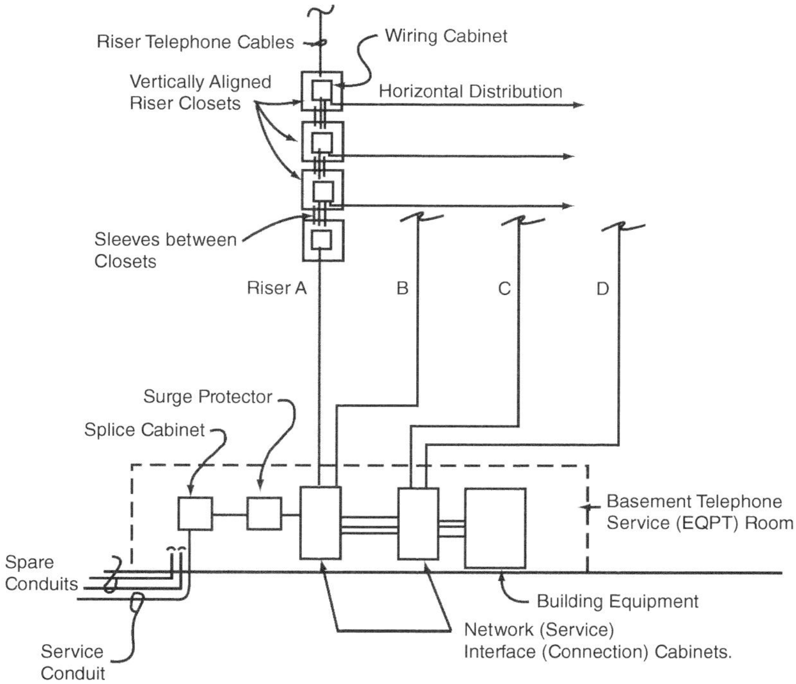

Figure 32.17 *Riser diagram of the telecommunications system raceway and space arrangements for a typical commercial office building.*

be covered with ¾-in. [20-mm] marine plywood for cabinet mounting. A 1-in. [25-mm] PVC conduit should extend between cabinets from a suitable electrical system ground point, with a ground wire sized according to communication company requirements. The room should be dry, ventilated, well lit for close work in wiring and color recognition, and supplied with at least two 20-A duplex convenience outlets on a separate circuit.

If additional connection, switching, or communication equipment is required, the project electrical engineer should obtain the equipment size and electrical requirements, provide the necessary space requirements to the architect, and design the electrical elements. In large equipment rooms, circulation and egress must be considered, as well as emergency light and power.

(b) Riser Shafts

These accept cables extending beyond the equipment room and carry them vertically through the building. Connection from the wall cabinets in the equipment room up to the riser shaft should preferably be in conduit.

The riser shafts provide the means for the cables to extend vertically and terminate at each floor. Ideally, the riser spaces comprise a series of vertically aligned closets connected by ½- or 4-in. [12- or 100-mm] sleeves set in the floors and extending 1 in. [25 mm] above the floor. Sleeves are preferable to slots because they can easily be sealed and fireproofed; slots cannot. A minimum of four 4-in. [100-mm] conduit sleeves should be installed between closets vertically, plus an additional 4-in. [100-mm] sleeve for every additional 50,000 ft^2 [4645 m^2] of office space to be served. It is preferable to separate communications closets from electrical power closets. Where multiple risers are used, shafts should be interconnected by several 2-in. [50-mm] conduits to allow for interconnection of systems. Riser cables are also known as backbone cables.

(c) Riser Closets

Here, cables from the riser system are interconnected to switching and power equipment, as well as to the cables that radiate from the closet to station locations throughout the floor area. These closets may also be called zone closets or apparatus closets, particularly if they function with an underfloor raceway system. The walls of the closet should be lined with plywood at least ¾ in. [20 mm] thick to support the weight of switching and connection panels, power equipment, terminals, connecting blocks, and other hardware. Each riser (apparatus) closet must be provided with a switched ceiling light and a separate 20-A, 120-V circuit with two duplex receptacles. A source of emergency power is desirable to avoid curtailment of communication service during power outages.

Riser closets should preferably have a minimum net area of 20 ft^2 [2 m^2] and a minimum clear wall width of 5 ft [1.5 m] for cabinet mounting. Each closet can supply telecommunication service for 8000–10,000 ft^2 [740–930 m^2] of floor area. Telephone equipment being utilized should not be more than 250 ft [75 m] from a closet.

(d) Satellite Closets

Unlike riser (apparatus) closets, satellite closets do not contain switching and power equipment. Their primary use is to provide cable-connecting and cable-terminating facilities in large, complex buildings, where riser closet space is insufficient.

(e) Auxiliary Equipment Rooms

Where extensive cross-connection is required or tenants utilize private switchboard (PBX) systems, the required equipment is placed in a relatively small auxiliary equipment room. These spaces, which are actually small closets or alcoves, should contain a 20- to 30-A, 2P, 120/208-V circuit, a 20-A, 120-V outlet, a grounding point, good lighting, and sufficient equipment space. Space requirements vary with each installation and can be obtained from the equipment supplier. Ventilation is essential, as is absorptive acoustic material on the ceiling and at least one wall. Connections between these spaces and other communication equipment closets should be via floor or ceiling ducts or multiple ½-in. [12-mm] conduits.

(f) Horizontal Distribution

Cabling from riser, satellite, apparatus, and equipment closets to individual outlets and instruments can be run underfloor, in ceilings and plenum spaces, under carpet, or in surface raceways. Because of the large raceway volumes required, conduit is infrequently used. When underfloor raceways are used, a header capacity of 2 in.^{2} [1290 mm^2] per workstation is reasonable, based on one multiline telephone and one video display unit per station.

(g) FO Cables

Fiber-optic (FO) cables are used in lieu of copper cabling in installations with very heavy data transmission loads, in those using video systems, and/or in applications for which the high-security, low-noise, and broad-bandwidth characteristics of FO are necessary or desirable. Information on space requirements and connection accessories can be obtained from manufacturers. Because cables are often supplied with factory-applied terminations, raceways must be of the large-capacity, full-access type.

32.23 OFFICE BUILDING CONTROL AND AUTOMATION SYSTEMS

As mechanical and electrical systems in modern office buildings increased in complexity, the need arose for a central point of supervision, control, and data collection. This provides a single locale from which to survey and control an entire building's functioning, plus increased opportunities for automation. From a supervisory location, water, air-conditioning, heating, ventilation, electrical, and other systems can be controlled with accuracy and convenience. Data on temperatures, pressures, flow, current, voltage, and the many other parameters of mechanical and electrical systems can be available instantly so that operational decisions can be made and automated. All systems can be monitored centrally and all alarms instantly and automatically acted upon.

Such supervisory control centers are now installed as a matter of course in office buildings. They are equipped with computers that process the data received to make operational decisions intended to optimize system performance. As a result, they can provide considerable savings in operating and maintenance costs. These systems in their most generalized form are referred to as building automation systems (BAS) and are discussed at length in Sections 32.27–32.33.

Architecturally, the spaces housing BAS centers require good lighting, good ventilation, and extensive raceway space but little area. Because systems are custom-tailored to a specific building, no guidelines can be stated for space requirements.

INDUSTRIAL BUILDING SYSTEMS

32.24 GENERAL INFORMATION

All industrial facilities, ranging from a loft housing a small hand-assembly plant to an immense steel manufacturing plant, require a variety of signal and communication equipment. Fire alarm systems for industrial buildings are discussed in Chapter 26. Audible alarms for building security in any industrial facility must be selected with the likelihood of high ambient noise levels in mind. See Figures 32.18 and 32.19 for recommendations.

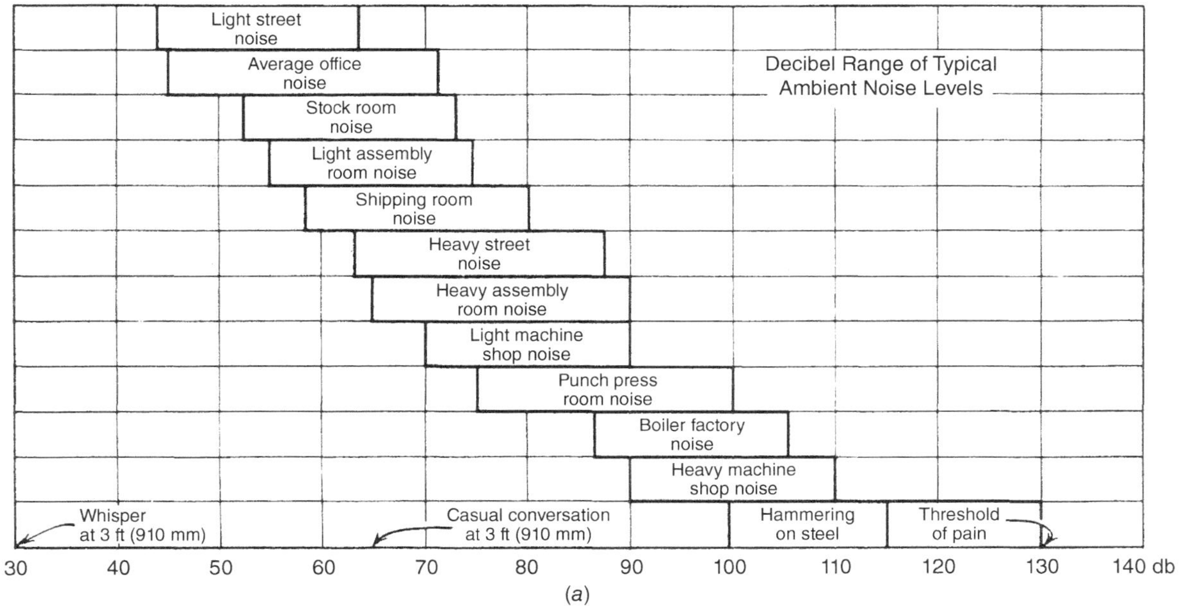

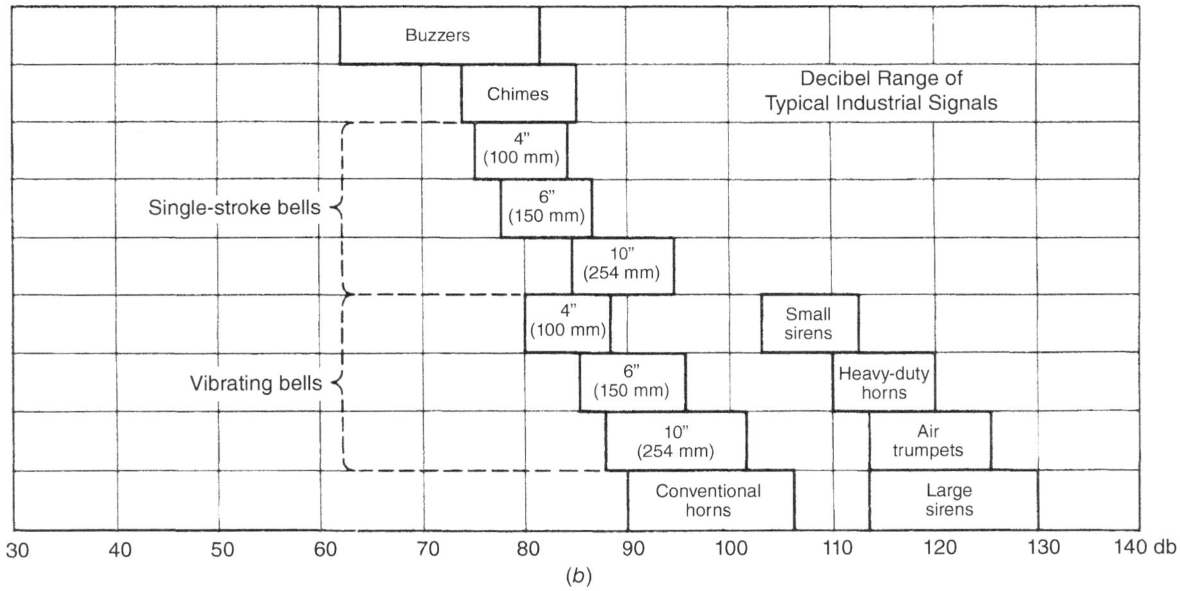

Figure 32.18 *(a) Decibel range of typical industrial noise situations. (b) The decibel rating of typical industrial signals may be compared with the anticipated ambient noise level (a) to facilitate selection.*

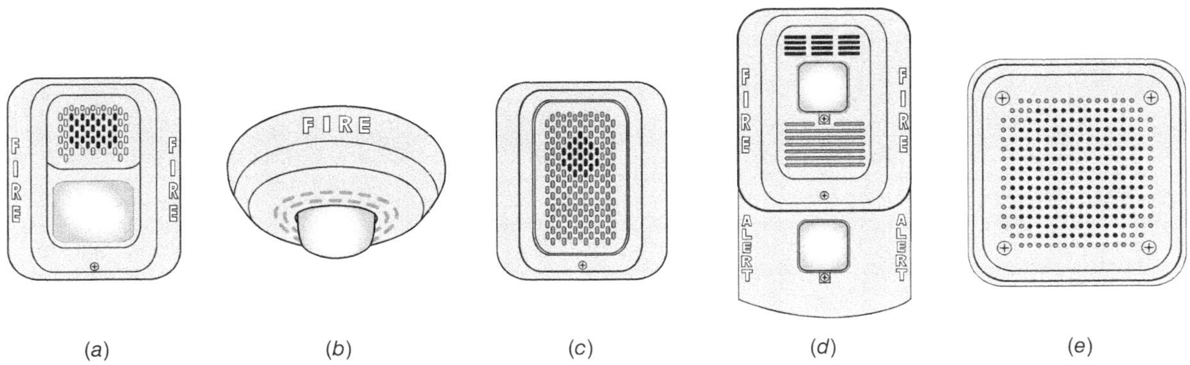

Figure 32.19 *Selected array of typical fire signals found in industrial facilities: (a) horns and strobes, (b) speakers/strobes, (c) chimes, (d) emergency communication systems, and (e) exit directional speakers. (Drawn by Kelli Kimura.)*

32.25 INDUSTRIAL BUILDING PERSONNEL ACCESS CONTROL

The design and engineering of personnel access control systems are a specialty that has burgeoned since the late 1980s and is now an independent profession. The increasingly diverse variety of identification technologies, plus the control and supervisory capabilities of computers, has greatly enhanced the access control capabilities of even a relatively small system. As a result, this section does not attempt to discuss system design but focuses instead upon the operation and application of the subsystems that compose the whole.

The traditional access control question was always "Who is permitted to pass through a specific portal?" The means by which identification is made to answer this question varies with the importance of access control at that portal. Thus, entrance to a large, multipurpose space used by many people must be rapid and must avoid the delay engendered by a physical barrier. One means commonly used is one or more guards who visually, and in relatively cursory fashion, inspect a badge with an ID photo. The next level of supervised, barrier-free access control may be a "turnstile" that requires presentation of an access card to an electronic proximity reader. One such unit is illustrated in Figure 32.20. This type of barrier is limited to a maximum pass-through rate of about one person per second. Because these two control methods are generally barrier-free, the action taken on alarm is to activate physical barriers and sound the appropriate alarms.

When a physical access barrier is involved, passage is slow, depending only partly on the identification process. Even with the most rapid electronic identification, a door or other portal closure must be physically released and operated, which can be very time-consuming.

Unattended physical barriers can be released by a multitude of differing identification technologies, depending upon the required level of scrutiny. At the

Figure 32.20 *Optical turnstiles in an office lobby. Turnstiles with barriers, also called "speedgates," are preferred due to the deterrent effect and the obstruction that slows or blocks (with a locking barrier) passage. Such turnstiles require the presentation of access credentials, often a proximity card, but possibly a biometric or smartphone, to be granted passage through the lane into the building. Unauthorized access initiates local audible and visual alarms and other measures as designed to secure the building from intruders. Frequently, such optical turnstiles are installed with a guard desk nearby. (Courtesy of Smarter Security, Inc.)*

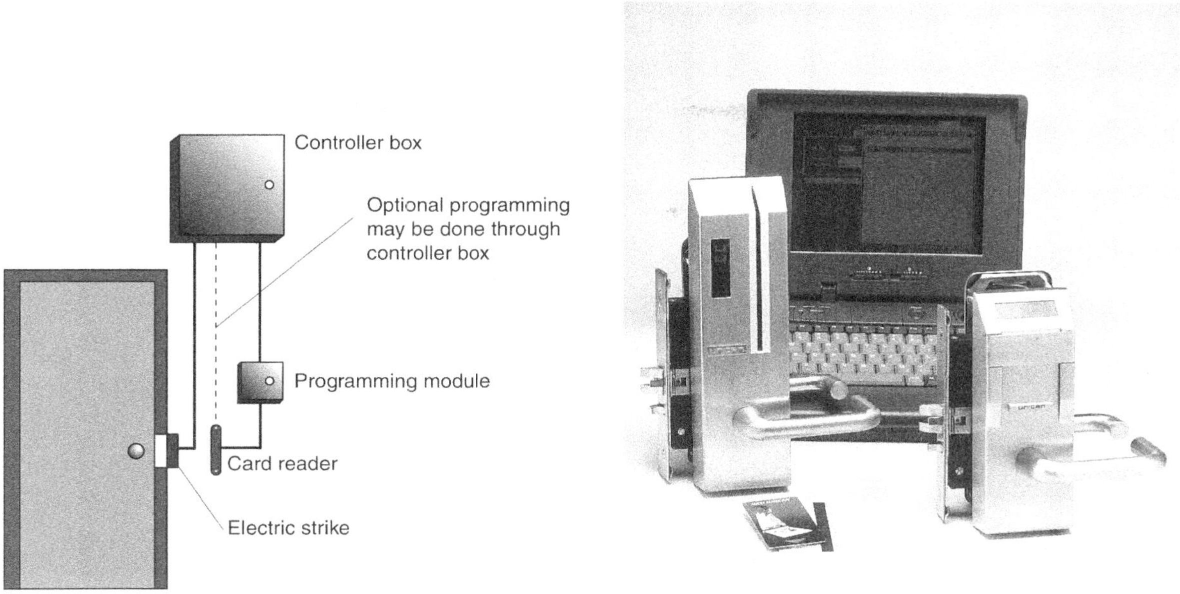

(a) (b)

Figure 32.21 *Stand-alone electronic locking system. (a) The arrangement and interconnection of equipment at each portal are shown. The card reader (swipe or insert type) is built into the door mechanism, which also contains the electric strike mechanism. (b) Each lock (swipe style on the left and insert type on the right) is individually programmed using the laptop computer seen in the background. Each lock can be programmed for up to 1500 users, each on a time-schedule basis, covering not only working hours, holidays, weekends, and the like but also individual time schedules. All programming is performed locally, and the passage record of each lock can be downloaded easily. (Courtesy of Ilco-Unican Corp., Electronic Access Control Div.)*

lower end of the security scale are barriers that require some embedded knowledge and/or inspect an object presented by the user, such as a magnetic, bar-coded, or proximity-reader card. Because locks can be controlled electronically and remotely, they can be programmed to give access at certain times only or to specific groups of cards or individual cards only, or to deny access to specific cards (e.g., cards reported lost), or for any other selection criterion. Time-controlled access is flexible; it integrates extremely well with intrusion security systems because portals can be easily coordinated with changing work schedules, both general and specific. Thus, a person can be barred from an area to which he or she should not have access at some particular time of the day (Figure 32.21).

The most sophisticated identification methods in use today are biometric: examining a specific physical characteristic that is unique to each person. This is considered a higher level of entry control because it identifies a person, not an object. Several of these methods are illustrated in Figures 32.22 through 32.24.

In view of the plethora of available access-control technologies, access is being redefined beyond its original physical-admittance meaning. Access control today is also used to limit access to copy machines, fax machines, phone lines, and other office facilities frequently used by employees for other than purely business purposes.

Figure 32.22 *This foolproof identification procedure relies on the fact that no two persons have identical retinal vascular patterns. The illustrated device, usually placed at an unmanned controlled-access portal, scans the entrant's retina, compares it to a previously recorded pattern, and then either permits or denies entry. (Courtesy of Eyedentify Inc.)*

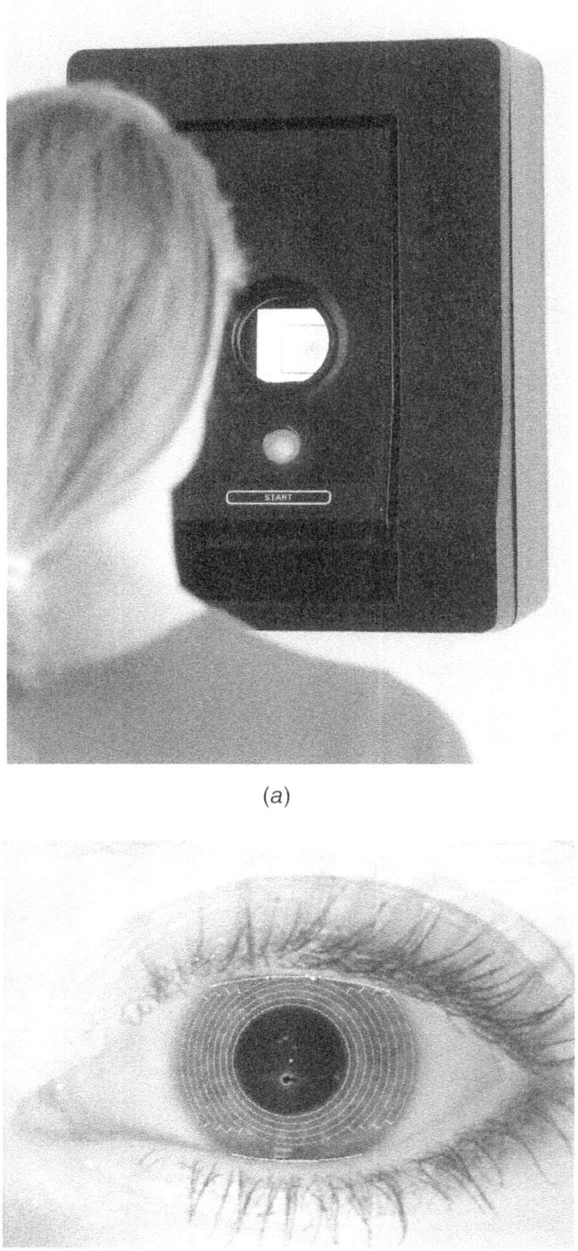

(a)

(b)

Figure 32.23 Iris-scanning system. (a) When the system is used as an access control device at a portal, the entrant stands about 10–12 in. [250–300 mm] away from the scanner as the iris is scanned. The technology relies on the fact that each person's iris has unique and randomly formed features. (b) After scanning, the image is digitized (as shown) and compared to a computer record. The entire procedure is very fast; entry access or denial is automatic. (Courtesy of IriScan Inc.)

32.26 INDUSTRIAL BUILDING SOUND AND PAGING SYSTEMS

Cellular phones are frequently ineffective in industrial facilities because of high ambient noise levels.

Figure 32.24 This biometric identification device takes more than 90 hand measurements that are then converted into a unique identification "template" that is stored in the system's computer. Once identification is made, it can be used for access control. Access points can be programmed to permit entry to authorized personnel on any time/date schedule desired.

In such areas, voice paging at high audio levels (see Figures 32.18 and 32.19), radio-frequency paging calls on earphones or speakers built into sound-reducing ear protection (see Chapter 25), and paging calls using paging lights are used. Similar methods are employed in noisy areas for announcements and special voice messages. In less noisy areas, such as offices adjoining manufacturing spaces, sound-reinforced conventional office-type systems are usually employed.

AUTOMATION

32.27 GENERAL INFORMATION

It is important that the distinction between remote control and automation be clearly understood. Remote control is simply a technique by which an action that can be performed manually at the device being controlled is performed from a remote location by some intermediate means. The means may be low-voltage wiring (see Chapter 28), a high-frequency wired signal (see Chapter 28), a wireless signal such as from a household remote control, a similar IR control unit, or RF wireless control. In each case, there is a single-stage action that is manually initiated. When the signal initiation is nonmanual—that is, automatic, from a timing device, a sensor such as a daylight or occupancy sensor, or a programmable device such as a microprocessor or computer—the control comes from an automated system. The system can be elementary, as just described, in which an automated signal controls a single action. It can also be very complex, with many levels, yet treat only a single function, such as an automated lighting system with sensors activating or overriding scene-presets that activate dimmers and switches. Such a system is referred to as a stand-alone (automated) system. When several of these stand-alone systems are interconnected and supervised by a higher-level controller, each is referred to as a subsystem, and the overall system is referred to as an integrated control system. Such an integrated system, when applied to the individual systems in a building, is referred to as a building automation system (BAS).

32.28 STAND-ALONE LIGHTING CONTROL SYSTEMS

A straightforward stand-alone lighting control system is shown in Figure 32.25. The initial programming is entered from a laptop computer into the control panel using system (the manufacturer's) software. The control panel contains the required microprocessors, timers, and output control devices to establish lighting levels and settings in the entire structure via dimming and switching panels. In addition, through interfaces (connection devices), the control panel reacts to signals from daylight and occupancy sensors in a preprogrammed fashion to switch and dim certain luminaires. Window-shade control is two-way: shading devices may close automatically when direct solar radiation is detected in a specific area (while also causing illuminance to increase) or they may open when direct sunlight is absent (but useful daylight is abundant). In addition, a conference room arranged for audiovisual display may provide control of lighting and shades via a local lectern-mounted, scene-preset controller, and so on. When another stand-alone system such as a fire alarm system (which itself may be part of an overall security/safety system) sends an emergency signal, the control panel reacts as programmed. It may turn on all lighting to full output, or it may brighten exit paths and dim other lighting, and so forth. This interconnection between two stand-alone systems is characteristic of a BAS.

32.29 BUILDING AUTOMATION SYSTEMS (BAS)

The clear trend in buildings, except for small or simple structures, is to use integrated system design plus centralized monitoring and control of building systems. The subsystems almost always included in a BAS are HVAC, energy management, and lighting control. Inclusion of security (cameras), life safety (fire alarm, fire control and suppression, electrical systems, plus emergency aspects of vertical transportation), material handling, special ventilation (fume collection), and some aspects of communications depends upon the specific needs of the building. This trend toward building automation, which previously was economically justifiable only in large owner-user

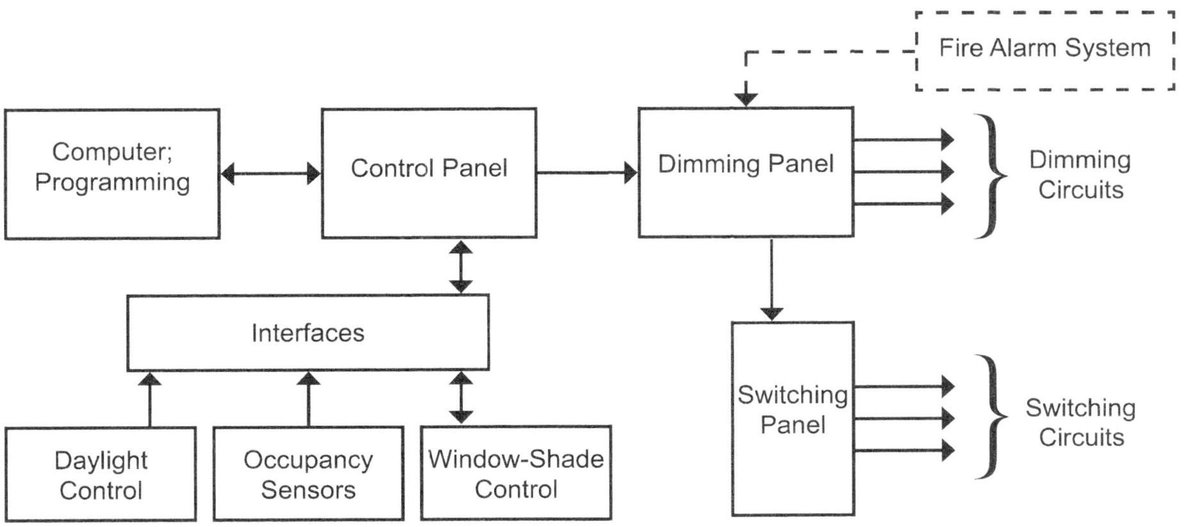

Figure 32.25 Block diagram of a typical stand-alone automated lighting control system. A connection to another stand-alone system (fire alarm), shown dotted, would exist at the overall building control level.

ELECTRICITY

facilities, is today not only economically feasible but also very nearly an economic necessity because of high labor costs and the relatively low cost of computer and microprocessor controls. What the rapid advances in microelectronics and computer technology have done is to make possible and practical detailed multipoint monitoring and control in real time, the result of which (assuming proper programming and BAS operation) can be a highly efficient, environmentally conscious, and money-saving facility. Indeed, the advantages of such an arrangement are so great that retrofitting of existing buildings, which is more costly than new construction, has become a major industry.

Because the technology of building automation has grown so rapidly in the past few years, the terminology surrounding it and its equipment suffers from a lack of standardization and uniformity. Thus, terms such as building automation system (BAS), building management system (BMS), building control system (BCS), smart/intelligent buildings, computerized building control, integrated computer control, supervisory data systems, integrated building control, facilities management systems, and so on are used almost interchangeably. The expression building automation system is used in the discussion that follows. Every BAS is by definition programmed; thus, any discussion necessarily involves basic control and computer terminology, in addition to that which specifically applies to a BAS. The short glossary that follows in Section 32.35 should prove helpful in this regard.

32.30 BAS ARRANGEMENT

A block diagram of a typical BAS is given in Figure 32.26. Notice that information is transferred vertically between levels within a stand-alone system, whereas horizontal intersystem information transfer occurs only at the highest level (Level 1), because that is the level at which systems are interconnected. (In practice, some interconnection also occurs at Level 2.) This means that a smoke alarm from Level 5 of the fire alarm system travels to Level 1, where the central controller tells the other stand-alone systems to take the required action. These systems in turn then signal downward, essentially to Level 4, where appropriate commands are given to controls for supply fans, elevators, security, and so on. The stand-alone systems are essentially independent and in practice cannot be interconnected as easily as might be thought because of the absence of standardization in wiring and computer protocols. The ostensibly simple problem of connector standardization has yet to be fully solved, as anyone who has done hands-on work in the communication field can attest, although the situation is much improved over the near chaos that once reigned. Buildings with fully or even partially integrated systems are not as common as might be suspected. Such buildings may reasonably be called "intelligent buildings" and are discussed in Section 32.31.

Returning to the block diagram (Figure 32.26), any specific BAS may contain more or fewer monitors, sensors, dependent and stand-alone local controllers, and interconnected systems, but the overall

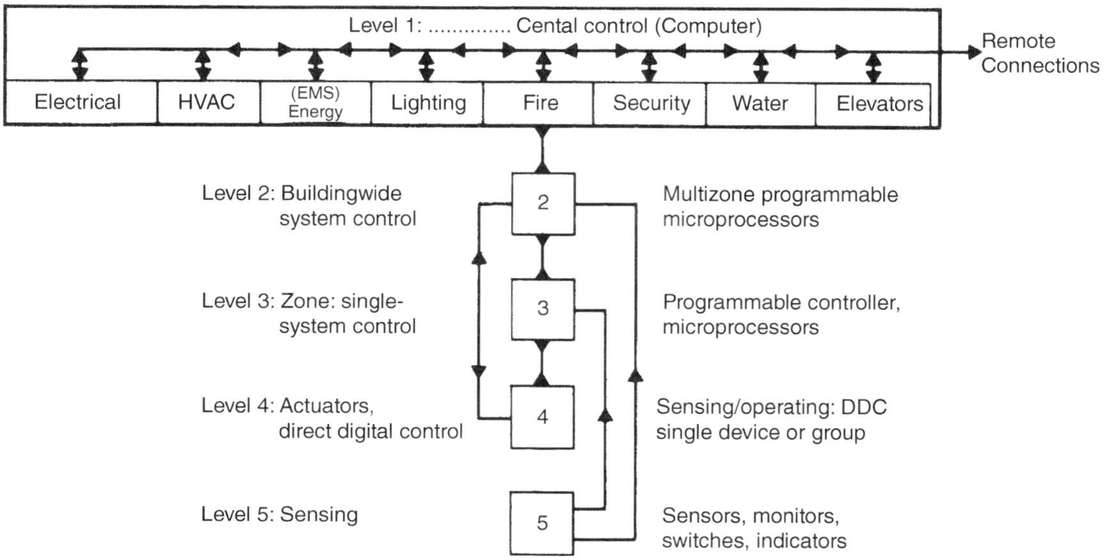

Figure 32.26 *Overall block diagram of a building automation system (BAS). In an actual system, the various levels and devices can be combined and telescoped by use of equipment that combines several level functions in a single housing.*

arrangement remains generally the same. Depending upon the equipment manufacturer and the system complexity, some of the functions and levels can be telescoped or grouped. This discussion deals with the general case. To solidify understanding, fire control and evacuation system terminology is used as a relatively simple and straightforward example of overall system design, function, and interconnections.

Level 5 in the diagram is the lowest system level. It contains automatically operated passive sensors such as flame, smoke, ionization, temperature, and water-flow detectors; physical condition indicators such as smoke damper and fire-door position sensors; and manual fire alarm pull stations. All yield a digital signal (on–off), which is transmitted to Level 3. As Level 5 is purely a sensor/monitor level, it does not receive any control or operating signals.

Level 4 contains sensor/operators or actuators (i.e., devices that both transmit a condition signal and receive an operation signal). In this example, Level 4 could contain smoke damper and fire-door actuators, smoke vent controls, sprinkler system valves, and fan control (override) switches. The last item can either reside at this level or can be controlled through the HVAC system via Levels 1, 2, or 3. Control at this level would typically involve hard-wiring of a single device or block.

Level 3 could represent a floor, zone, or area controller, with either relay circuitry or microprocessor control (programmable controller), depending upon system complexity. This level receives information from Levels 4 and 5, processes it, returns control information to Level 4, operates specific local audible and visible alarms directly, and sends alarm and condition information to the Level 2 building controller and to the master controller (computer) at Level 1. It also receives and processes signals from Levels 1 and 2, containing such information as alarm reset, alarm device operation for building emergency evacuation, and any reprogram or reset function.

Level 2 contains the building's central fire and evacuation system (life safety) controls. This level can contain some of the Level 1 programming logic relating to firefighter control of elevator operations and HVAC fan functioning, or it can simply act as an I/O device to Level 1 for these functions, again depending upon system design and complexity. In some facilities, these life-safety system functions may be controlled from a separate console that is interconnected with the BAS. In a multibuilding facility, this level contains the individual building fire system central control and interconnections to other building fire system controllers, as well as external connections for city alarm and external supervisory services.

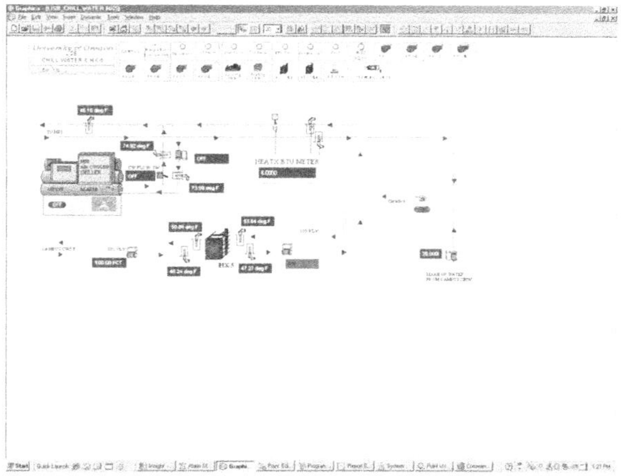

Figure 32.27 *Screenshot of a portion of a building automation system (BAS) and the status of all components. Here, are displayed the chiller status (on–off, COP, tons, run time), water temperature and flow in supply and return lines, and condenser water temperature. (Courtesy of Fred Tepfer.)*

Level 1 is the central computer console. This level receives status reports on the individual devices from Level 4 or Level 5; on areas, zones, and subsystems from Level 3; and on the entire fire/evacuation system from Level 2. Hard-copy printouts of all alarms and periodic status reports are generally made at this level, although they can also be made at Level 2. Computer monitors at Level 1 can be used to view the status of any area and all the devices in the fire/evacuation system, or for any other system (Figures 32.27 and 32.28), either graphically or as a table. Pictorial graphics for nontechnical end users are also possible.

At Level 1, information is transferred to HVAC, elevator, security, and lighting systems in preprogrammed alarm modes, with manual override possible. A fire emergency scenario might interconnect with the HVAC, elevator, security, water, electrical, and lighting systems to activate exhaust and pressurizing fans and deactivate supply fans, place elevators in fire-evacuation mode, override access barriers to permit unhindered facility evacuation, connect fire pumps to emergency power feeders, disconnect high-voltage feeders, and activate emergency lighting, all via preprogrammed intersystem functions. The system operator at Level 1 could then view any part of any system and reactivate or override preprogrammed functions as desired. With this system design, this is the level at which human supervisory intervention is intended. At other levels, such intervention occurs only if trouble arises, periodic maintenance is scheduled, or an alteration is required such as a sensor adjustment or replacement.

Returning to Figure 32.26, we can consider a much more complex nonemergency system: energy.

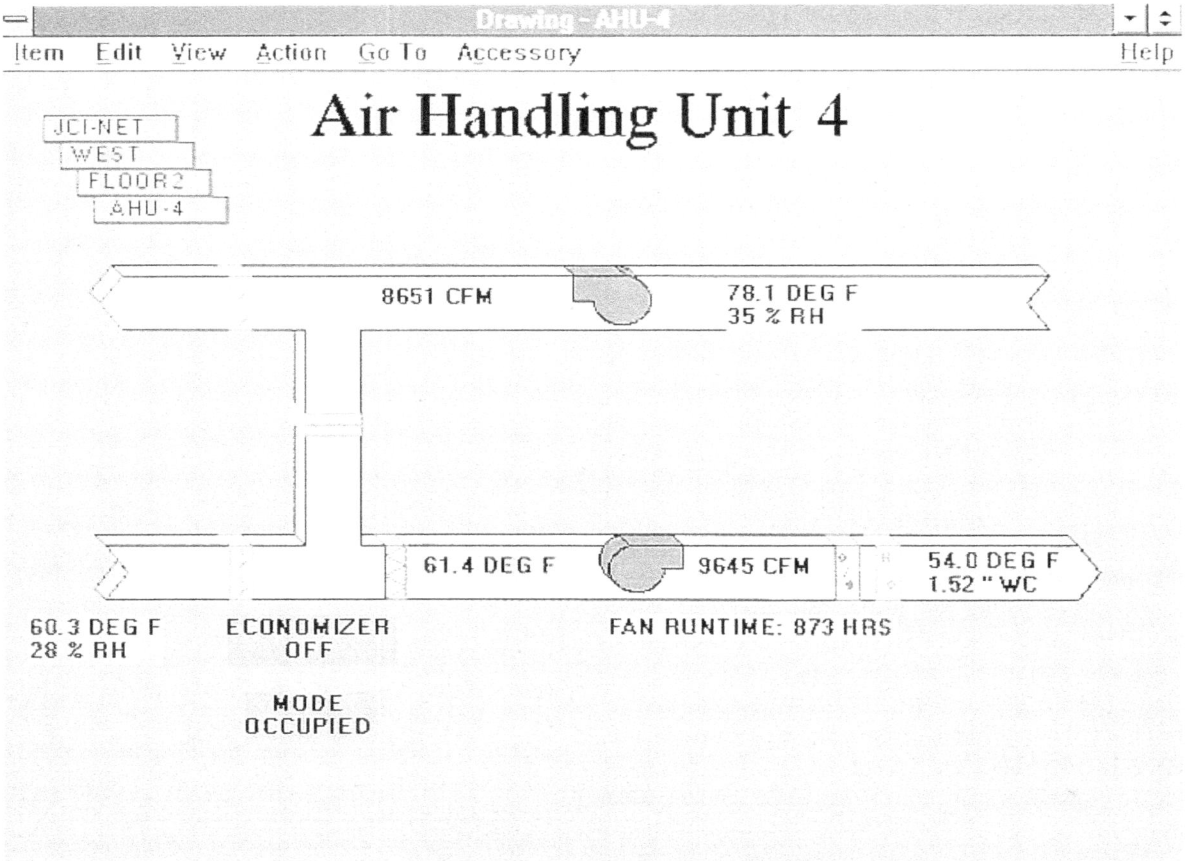

Figure 32.28 *A workstation screen can display any portion of any system in the BAS. Here, an air-handling unit (AHU) is displayed with data on airflow, air temperatures, and relative humidity in each duct section. In addition, cumulative fan run time, economizer status, and area occupancy status are displayed. (Redrawn by Tyler Mavichien from source: Johnson Controls.)*

Here, Level 5 senses and continuously monitors indoor and outdoor temperature and humidity, duct air temperatures, space pressurization, airflows, and the operating and load status of all fans, compressors, and refrigeration devices. Level 4 contains monitors and direct digital controls for valves, boilers, chillers, fans, humidifiers, dampers, cooling towers, and compressors. Levels 2 and 3 contain electric load managers for duty cycling and demand limiting, HVAC controls for economizers, run-time optimizers, supply air and supply water reset, chilled and condenser water reset, enthalpy switching, load sequencing, etc. These functions are performed either by a stand-alone energy management system (EMS) controller (see Chapter 28) or by individual demand, HVAC, and chiller load managers controlled by Level 1 logic. Finally, Level 1 contains the programming for all of the foregoing functions, any of which can be displayed at a Level 1 workstation terminal at any time. The system operator can then reprogram any function after analyzing system performance at the Level 1 computer and obtain immediate visual and hard-copy readout of the changes.

32.31 INTELLIGENT BUILDINGS

Although the term intelligent building (IB) seems to attribute a decidedly human capability to an inanimate object, the term's meaning (as defined by the Intelligent Buildings Institute (IBI) in 1987) is: "a building which provides a productive and cost-effective environment through optimization of its four basic elements—structure, systems, services, and management—and the interrelationships between them ··· Successful building intelligence is the matching of solutions to occupant need." In the framework of such a definition, the intelligence of a building depends upon the ability of the design team to understand the occupants' present needs (via a current owner's project requirements document) and to make an educated forecast of the client's and occupants' future needs. Because a building's useful life depends largely upon the accuracy of this prediction, building systems are being designed with open control architecture that lends itself easily and economically to change, both minor and major. The term "open architecture" is generally understood to mean, in terms of a BAS, a system design that uses non proprietary standards and protocols to make it easier to add, update, and replace components and that, at the very least, can do the following:

- Utilize equipment from any manufacturer whose equipment meets industry-wide specifications
- Readily reprogram equipment to serve a variety of functions
- Supply required system information at any I/O port
- Permit access to databases at any workstation
- Utilize networks for information transfer at one or more levels

These requirements demand open protocols, standardized transmission media and connectors, and virtually unfettered cabling space. (The last requirement is easily fulfilled by using any of the underfloor or cellular floor raceway systems described in Chapter 29.) Decades of proprietary system interests and the prevalence of stand-alone system designs stifled the move to open architecture in the United States. There are, at this writing, signs within the industry–such as evolving standards, industry partnerships, and new emerging technologies–that are pointing toward more flexible, non proprietary open architecture systems. One noteworthy movement toward standardization was the development and application of protocols such as BACnet (ASHRAE, 2020).

Referring to Figure 32.26, which shows a prevalent BAS architecture, an open system would eliminate at least two of the levels, reducing the system to a three-level structure that might be organized as follows:

Level 1, the uppermost level, contains workstations connected to a main data bus. This level, which might be called the management or information level, has direct access to databases and information networks. It serves to establish operational programming and procedures. All points are accessible from any workstation.

Level 2, the intermediate level, might be called the supervisory or network level. Network controllers serve as a network platform—both supervising stand-alone controllers from various systems at the third level and transferring information to other network controllers supervising other areas of the building via the LAN. In addition, the network controllers exchange information with one or more workstations. A network controller can also act as a process controller. A typical unit might appear as in Figure 32.29.

Figure 32.29 *Network control unit acts as a LAN platform, performing supervisory functions for a portion of a building. It connects to operator workstation(s) at the upper management level, to other network controllers at its level via the LAN, and to specific stand-alone controllers at the lower, control level. This modular server rack has radiant cooling.*

Level 3, the control level, contains dedicated controllers in stand-alone systems. The units are highly flexible and can be locally reprogrammed to control the same function in a wide variety of different systems. Two such controllers are shown in Figures 32.30 and 32.31.

32.32 HOME AUTOMATION

The operational principles of remote control and automation discussed previously in this chapter have found ready application in home automation either through an ad hoc mesh network of IoT devices with D2D communication and/or an internet-connected WLAN, such as Wi-Fi, and IoT devices with D2C communication with a cloud-based platform. Residences so equipped and variously referred to as

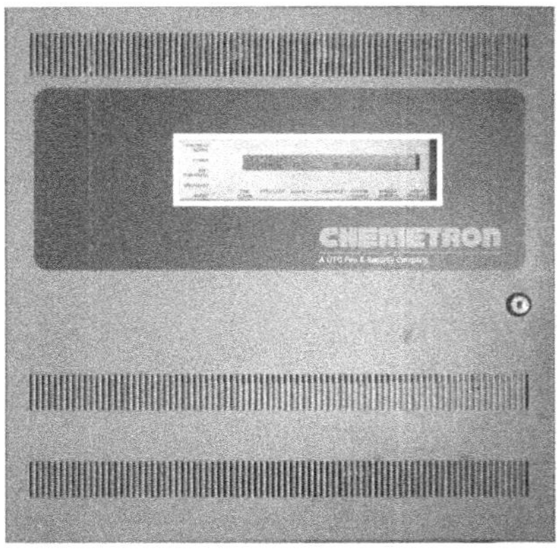

Figure 32.30 *Fire control unit. This device controls intelligent addressable analog devices in an intelligent fire control system. (Courtesy of Johnson Controls.)*

Figure 32.31 *A unitary controller is designed to control single-zone HVAC subsystems such as heat pumps, fan coils, and the like. It is preset from the network level but can also be locally monitored and adjusted to satisfy local requirements. These adjustments are made with a portable plug-in control unit. (Courtesy of Johnson Controls.)*

intelligent, smart, or automated appear increasingly in the media—especially press and advertisements from the tech sector touting virtual assistants (just say, "Hey, Siri ··· ").

The degree of automation in homes varies considerably, the simplest being no more than a low-voltage control system, with time-based programming using a relatively simple microprocessor. More complex systems use touch-screen computers for control or a portable plug-in programmable microprocessor to set various portions of the system. The best (and most expensive) systems achieve a level of user–system integration not yet available in a complex BAS, in that the user's smartphone can control any portion of the system directly. This is possible because only one level or, at most two levels, of control are involved—that is, the commands are on/off/set, where the last refers to a particular electronic contact. This contact in turn establishes a new preprogrammed condition such as night-setback, lighting preset for extended absence, reset of second-floor temperatures to daytime levels, and so forth. These systems also allow relatively straightforward reprogramming to meet the occupant's needs.

All new residential construction includes some degree of automation that will increase in the future. Traditionally, residences, both single and multiple, were provided with the cabling and/or raceway system that was the backbone of an automation system. However, there is now a prevalence of wireless options for residential automation that take advantage of Wi-Fi, Bluetooth, etc. Wireless connections to the smart home network are generally easier to install, more flexible over time, and less costly than hard-wired options. With changes to the ways spaces in our homes are used, and with many spaces now performing multiple functions, space-specific automation capabilities appear less attractive (e.g., a dedicated home office). However, some smart home features that should be considered include: learning thermostats, lighting, appliances, plugs/electrical outlets, audio speakers, security cameras, locks, and virtual assistants. These smart features can be preprogrammed and accessed remotely via smartphone/tablet apps and voice command.

32.33 ARTIFICIAL INTELLIGENCE

The realm of artificial intelligence (AI)—a branch of computer science concerning software platforms that "learn" based on data and modify their actions accordingly—is manifesting in a profound way across all aspects of the building sectors, not the least of which being building systems automation. A burgeoning marketplace of both proprietary and open-source "AI-powered" building automation platforms is ushering in an era of data analytics and machine learning regarding building operations and maintenance. AI algorithms can process vast amounts of data from various network-connected systems—including HVAC system components, lighting, sensors, energy and water meters, and various IoT devices—to adjust system operations in real time and/or provide facility managers improved feedback loops. AI-integrated building systems may potentially facilitate:

- Energy performance optimization
- Predictive maintenance
- Enhanced occupant comfort
- Automated fault detection and diagnostics
- Data-driven decision-making
- Automated demand response programming
- Monitoring-based commissioning

By leveraging AI, building automation systems can become more intelligent and responsive, leading to significant improvements in energy performance, maintenance, occupant comfort, and overall operational effectiveness.

32.34 BUILDING PHYSICAL SECURITY

Although not directly related to "signal systems," the issue of physical security for buildings is an escalating concern for a wide range of building types and clients. It is beyond the scope of this book to address this problem in any detail; however, it should be mentioned that the design of mechanical and electrical systems for a "secure" building will involve considerations beyond those seen in a typical design situation. For example, physical security is now a primary design consideration for specific building types including airports, schools, government buildings/embassies, etc. Resources to assist system designers in these circumstances may be found through a range of sources, including the US Department of Homeland Security (DHS) and the Federal Emergency Management Agency (FEMA).

32.35 TERMINOLOGY AND DEFINITIONS

Address A number used to identify an I/O (input/output) channel or a module.

Algorithm A detailed description, often in flow-diagram or mathematical form, of a method for solving a problem. The outline used for computer or microprocessor programming.

Alphanumeric Characters including letters of the alphabet, symbols, and numbers.

Analog data Numerical information about physical variables given in a physically representative, continuously variable form (as with dial meters, temperature gauges).

Artificial intelligence (AI) A branch of computer science concerning computer programs that "learn" based on data and then modify actions accordingly.

ASCII A 7-bit standard alphanumeric code established by the American National Standards Institute (ANSI). The acronym ASCII stands for American Standard Code for Information Interchange.

Baud Rate at which information is transmitted (in bits per second).

Binary A numeric system using only 1 and 0, normally used to represent the on and off circuit conditions.

Bit The number 1 or 0; a contraction of binary digit.

Bluetooth A short-range wireless technology standard used for exchanging data between fixed and mobile devices over short distances and over personal area networks.

Building automation system (BAS) A computer-based system that monitors and controls selected aspects of building operations. A BAS is understood to include all of the equipment involved and its interconnections.

Bus An electrical data channel.

Byte A string of bits; 8 bits unless otherwise specified.

Cellular network A radio frequency communication system that divides a geographic area into cells, each served by a base station (e.g., cell tower), enabling cellular devices to connect and maintain communication across regions by seamlessly transitioning between cells.

Central processing unit (CPU) That portion of a computer, microprocessor, or programmable controller that processes data and executes programs; the "brains" of a computer.

Cloud A metaphor for the distributed network of off-site servers that host software, data collection, and IT infrastructure for personal and/or enterprise utility.

Community antenna television (CATV) A cable-based television system that receives broadcasts by an antenna and relays them by cable to subscribers in areas where direct antenna broadcast reception is either poor or not possible.

Converter General term for a device that converts data or energy from one form to another. Data converters are analog to digital (ADC) or digital to analog (DAC). Energy-form converters (e.g., temperature or illuminance to voltage) are properly called *transducers*.

Digital data Information that is stored and transmitted as a series of symbols, usually binary (ones and zeros), that can be understood by a computer or other device.

Direct digital control (DDC) Control arrangement that processes digital input data according to a programmed algorithm and yields a control function (e.g., the action of a local microprocessor controller or a computer control function).

Duplex, duplex transmission Simultaneous bidirectional data transmission.

EIA Electronic Industries Alliance (formerly Electronic Industries Association). Establishes data communication standards (e.g., RS-232).

Energy management system (EMS) Computer-based system for monitoring and controlling facility energy use. Frequently part of a BAS.

Global positioning system (GPS) A public satellite-based radio navigation system that provides geolocation and time information to compatible receivers anywhere on (or near) the Earth with an unobstructed line of sight to four or more GPS-networked satellites.

Hardware All physical equipment associated with a computer or control system, as opposed to software.

IB Acronym for intelligent building. A building that uses technology to create a comfortable, safe, and productive environment for occupants and improves efficiency for owners and maintenance staff.

IEEE Institute of Electrical and Electronic Engineers. Establishes best practices for a wide range of technologies through industry standards.

I/O Input/output. A fundamental concept of computing involving the process of sending and receiving data from devices, networks, and databases.

IR Acronym for infrared.

IT Acronym for information technology.

IT infrastructure Defined broadly as a set of software and network components that serve as the foundation of an IT service.

LCD Acronym for liquid crystal display.

LED Acronym for light-emitting diode.

Load control, load management Arrangement whereby energy-using devices are monitored and controlled. Part of an EMS.

Local area network (LAN) Signal network providing intercomputer communications.

Logic Factors involved in the design of a CPU (or control scheme).

Microprocessor A CPU, usually in the form of a single integrated circuit (IC), preprogrammed to perform specific functions in response to specific input.

Modem Device that connects a data-originating device to a communication line; from modulator–demodulator.

Multiplex(er) (MX, MUX) Use of a single communications line for simultaneous transmission of multiple signals.

Near-field communication (NFC) A short-range wireless technology that allows devices to communicate when they are close together.

Network Two or more devices that can communicate through interconnection using wired or wireless means.

Nonvolatile memory Memory that remains intact when electric power is shut off.

Operating system Software that controls input to, output from, and operation of the CPU.

Parity Method of verifying the accuracy of recorded data.

Point General term used to describe either a sensor/monitor location or a specific control signal in a BAS.

Port An I/O connection point.

Programmable controller Electronic device containing a microprocessor, a programming means or device, and I/O interfaces.

Protocol Conventions governing the format and timing of signaling between communication devices (e.g., between computers in area networks).

Radio frequency (RF) Refers to the range of electromagnetic radiation frequencies from 3 kilohertz (kHz) to 300 gigahertz (GHz), encompassing the frequencies used for communications signals such as radio and television broadcasting, cellular phones, satellite transmissions, and radar signals.

Radio frequency identification (RFID) A radio frequency communication system that uses radio waves to identify and track nearby objects through communication between a reader and tags.

RAM, ROM, PROM, EPROM Random access memory, read-only memory, programmable ROM, erasable PROM. Computers process information stored in their memory. Two types of semiconductor memory are RAM (temporary storage) and ROM (semi-permanent storage).

Real-time Arrangement in which a computer receives and processes data without introduction of any deliberate time delay (effectively providing an immediate response).

Router A device that forwards data packets between networks.

RS-232 Standard for interconnecting digital communications devices, such as computers, printers, and monitors.

Smart home A home equipped with lighting, HVAC, electronic devices (appliances), and other systems that can be controlled remotely by a smartphone, tablet, computer, or voice command.

Software Computer program; as opposed to hardware.

Stand-alone Descriptive term for a control device or system that can perform independently and usually can also interconnect as a subsystem (controller) in a larger system.

UPS Uninterruptible power supply.

Volatile memory Memory that loses its information on loss of power.

Wi-Fi A wireless networking technology that allows devices to connect to the internet via radio frequencies and is used in residential and commercial buildings to allow LANs to operate without cables.

Acknowledgments

Thomas Collins, Ph.D., Ball State University, is acknowledged as a significant contributor to the contents of this chapter.

References and Resources

ANSI/BICSI. 2024. *ANSI/BICSI 007, Standards for Intelligent Buildings*. Tampa, FL: Building Industry Consulting Service (BICSI).

ASHRAE. 2020. *Standard 135-2020: BACnet - A Data Communication Protocol for Building Automation and Control Networks*. Atlanta, GA: American Society of Heating, Refrigerating and Air-Conditioning Engineers. automatedbuildings.com

DHS. 2013. *Building and Infrastructure Protection Series: Designing Buildings to Withstand Almost Anything*. Washington, DC: U.S. Department of Homeland Security. dhs.gov/building-and-infrastructure-protection-series-designing-buildings-withstand-almost-anything

IEEE SA. 2005. *IEEE Standard for Information Technology – Local and Metropolitan Area Networks - Specific Requirements – Part 15.1a: Wireless Medium Access Control (MAC) and Physical Layer (PHY) Specifications for Wireless Personal Area Networks (WPAN)*. Piscataway, NJ: Institute of Electrical and Electronics Engineers Standards Association. standards.ieee.org/ieee/802.15.1/3513

IEEE SA. 2023. *The Evolution of Wi-Fi Technology and Standards*. Piscataway, NJ: Institute of Electrical and Electronics Engineers Standards Association. standards.ieee.org/beyond-standards/the-evolution-of-wi-fi-technology-and-standards

IEEE SA. 2024. *IEEE Standard for Low-Rate Wireless Networks*. Piscataway, NJ: Institute of Electrical and Electronics Engineers Standards Association. standards.ieee.org/ieee/802.15.4/11041

PART IX

TRANSPORTATION

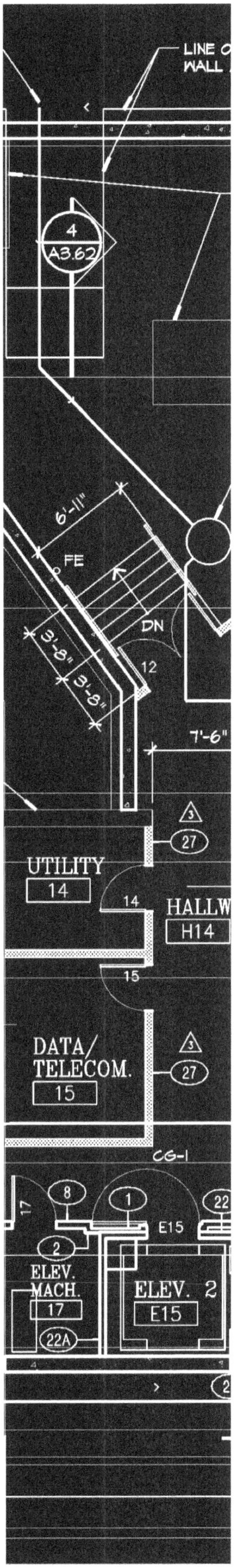

No systems in a multistory building are taken for granted more consistently than elevators and escalators. They are relied upon to move people quickly and safely under all conditions, including emergencies. This movement should be rapid, trouble-free, agreeable, and economical. Because vertical transportation accounts for 10% to 15% of the construction budget in tall buildings and a somewhat lesser percentage of building floor area—and is a determining factor in building shape, core layout, and lobby design—elevator and escalator selection and design integration are major tasks for the building designer.

Chapter 33 introduces elevators with a description of components, including traction and hydraulic elevator equipment, cars, safety devices, and systems of control and supervision. Fundamental topics, such as codes, standards, and accessibility requirements, are also covered. Criteria for elevator performance are presented for use in conjunction with the owner's project requirements to preliminarily select the size, number, and speed of elevators for a project. Principles of elevator zoning are presented. The chapter continues with a discussion of spatial requirements for elevators and addresses special-case elevators; concludes with building integration considerations, such as energy use, safety, security, and noise. The chapter also covers alternative design options, including double-deck and observation cars.

Chapter 34 explains escalator system components, capacities, sizes, speeds, and methods of selection. This is followed by topics including lighting, fire protection, power/energy requirements, and budget estimating. Information on mechanized walkways and ramps is also presented. The chapter briefly describes material-handling devices: dumbwaiters, horizontal and vertical conveyors, pneumatic systems, container delivery systems, and automated self-propelled vehicles. This information responds to the need to preplan for material handling in commercial and institutional buildings.

CHAPTER 33

Elevators

TRANSPORTATION

OF THE MANY DECISIONS THAT MUST BE MADE BY THE DESIGNER OF A MULTISTORY BUILDING, few are more important than the selection of the vertical transportation equipment, which includes passenger and freight elevators. These items represent a major building expense, perhaps as much as 10% of the construction cost in a 25-story office building, and the quality of elevator service is an important factor in a tenant's choice of space in competing buildings.

Although the final decision as to the type of equipment rests with the architect, the factors affecting it are so numerous that the building designer should consult with an elevator expert. Such services are available from consultants and from the major elevator manufacturers. The purpose of this chapter is to familiarize readers with the nature and application of vertical transportation equipment and to enable wise preliminary design decisions and effective interactions with consultants.

PASSENGER ELEVATORS

33.1 INTRODUCTION

This chapter is concerned primarily with general-purpose electric traction and hydraulic passenger elevators. Information on freight elevators and special elevator designs is also provided. Beginning in ancient Greece, designers have experimented with ways of moving people or goods between floors of buildings. Early elevators were human-, animal-, or water-powered using winches, counterweights/pullies, screw mechanisms, etc. Two technological innovations in the nineteenth century made elevators more popular and useful in building design. First, steam and electric power allowed elevators to be pushed upward using hydraulics or pulled upward using roped electric traction motors. Second, Elisha Graves Otis pioneered the first elevator safety brake in the 1850s, which greatly increased public confidence in elevator safety.

An ideal elevator system will provide minimum waiting time for a car at any floor level; comfortable acceleration; a rapid ride; smooth braking; accurate automatic leveling at landings; and quick loading and unloading at all stops. The elevator system must also provide quick, quiet operation of doors; good floor status and travel direction indication (both in the cars and at landings); easily operated car and landing call buttons (or other devices); smooth, quiet, and safe operation of mechanical equipment under all conditions of loading; comfortable lighting; reliable emergency and security equipment; and a generally pleasant car atmosphere.

In addition to passenger-satisfaction considerations, elevators have architectural design impacts. Cars and shaftway doors must be treated in a manner consistent with the architectural unity of the building. More important, however, elevator shaftways are major space elements whose integration into a building is a prime factor in composition, as is the design of the elevator lobby.

Mechanical and Electrical Equipment for Buildings, Fourteenth Edition. Walter Grondzik, Daniel Overbey, and Alison Kwok.

Companion website: www.wiley.com/go/meeb14e

33.2 CODES AND STANDARDS

Perhaps more than any other element of construction, elevators are governed by strict installation codes. The definitive resource in the United States is the American Society of Mechanical Engineers' *ASME A17.1: Safety Code for Elevators and Escalators* (it is also incorporated by the Canadian Standards Association as *CSA B44*). A certain edition of this model document has been adopted as code in many jurisdictions of the United States and Canada. Two related standards should also be noted: *ASME A17.3: Safety Code for Existing Elevators and Escalators* covers existing elevators and escalators and *ASME A17.4-2015: Guide for Emergency Personnel* covers emergency evacuation of passengers from elevators. Some states and municipalities have their own elevator codes that are generally based upon these ASME documents but may be significantly amended. The primary European standards regarding elevators are *EN 81-20* and *EN 81-50*. Developed by the European Committee for Standardization (CEN), the scope of *EN 81-20* covers the safety requirements for the construction and installation of elevators, whereas *EN 81-50* defines the examination, testing, and certification requirements for elevators' components. Readers should note that these and other European standards commonly use the term "lift" rather than "elevator" in accordance with British English conventions.

In addition to the elevator code, other construction and installation requirements influence elevator work. The *NFPA 101: Life Safety Code* (from the National Fire Protection Association) details fire safety requirements; the *NFPA 70: National Electrical Code* (NEC) governs some of the electrical aspects of elevator construction; and state and local laws can add a multitude of requirements and restrictions bearing on fire safety, emergency power, security concerns, and accommodations for persons with disabilities. Provisions for the disabled are specifically covered by *ICC A117.1: Accessible and Usable Buildings and Facilities* (from the International Code Council; ICC), by the requirements of the Americans with Disabilities Act (ADA), and, in most locations, by local law.

Like most large industries, the elevator industry is self-regulating and standardized. The National Elevator Industry, Inc. (NEII) publishes standard elevator layouts for traction and hydraulic installations. Elevator consultants and elevator company representatives are normally knowledgeable about the codes and standards in force, but this does not relieve the architect-engineer of legal responsibility for system design. Thus, in the preliminary planning stage all regulations concerning vertical transportation should be accessed, reviewed, and understood.

TRACTION ELEVATORS

33.3 PRINCIPAL COMPONENTS

The car, cables, elevator machine, control equipment, counterweights, hoistway, rails, penthouse, and pit are the principal parts of a traction elevator installation. An idea of the functioning and orientation of these items of equipment can be obtained from Figure 33.1.

The car is the elevator component with which the average building user is most familiar. Building prestige can depend upon appropriate design of the elevator car. Essentially, a traction car is a cage of some fire-resistant material supported on a structural frame; lifting cables are fastened to the top of the frame. The car is guided in its vertical travel in the shaft by means of guide shoes on the side of the frame. The car is provided with safety doors, operating-control equipment, floor-level indicators, lighting, emergency exits, and ventilation. It is designed for long life, quiet operation, and low maintenance.

The cables (ropes) that are connected to the crosshead (top beam of the elevator) and carry the weight of the car and its live load are made of groups of steel wires specially designed for this application. Four to eight cables, depending upon car speed and capacity, are placed in parallel. Although multiple ropes are used primarily to increase the traction area on the drive sheaves, they also increase the elevator safety factor—as each rope is normally capable of supporting the entire load. The minimum factor of safety varies from 7.6 to 12.0 for passenger elevators (from 6.6 to 11.0 for freight elevators). At the traction machine the cables from the top of the car pass over a motor-driven cylindrical sheave (grooved for the cables) and then go downward to the counterweight. Coated cables can reduce energy consumption with no reduction in safety.

The counterweight is comprised of steel plates stacked in a frame attached to the opposite ends of the cables to which the car is fastened. It is guided in its travel up and down the shaft by two guide rails typically installed on the back wall of the shaft. Its weight equals that of the empty car plus 40% of the rated live load. The counterweight serves several purposes: to provide adequate traction at the sheave for car lifting, to reduce the size of the traction machine (drive motor),

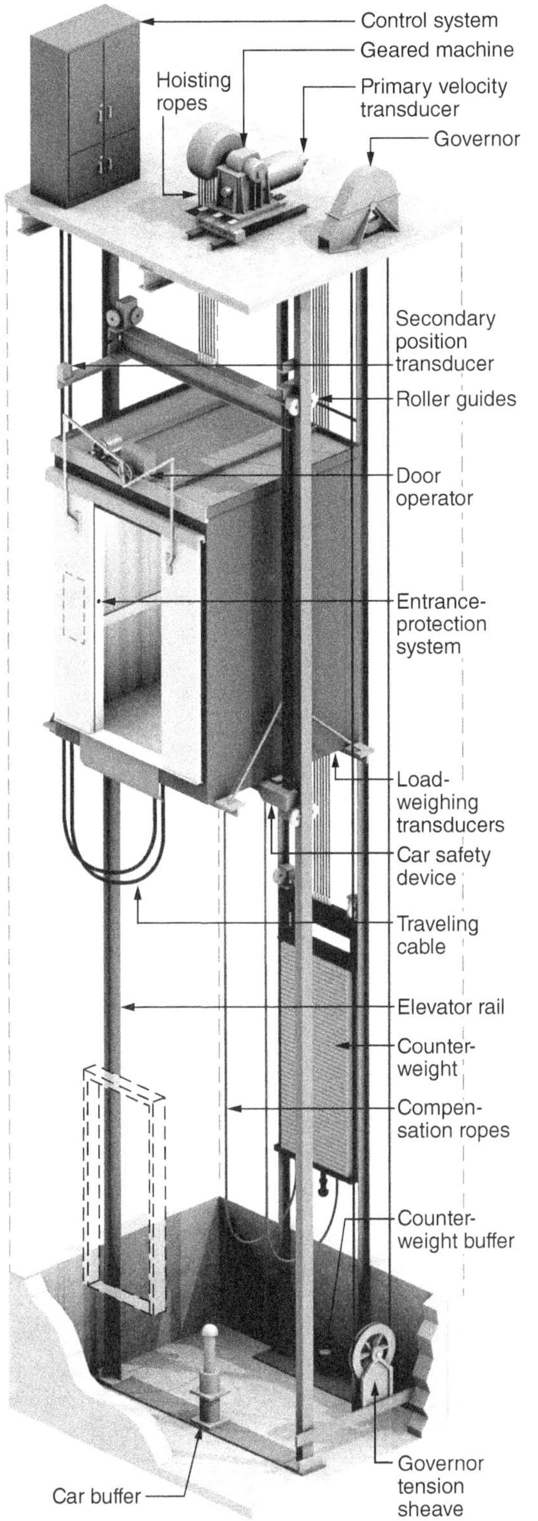

Figure 33.1 *Components of a geared traction elevator installation with solid-state control and motor drive. (Courtesy of Otis Elevator Co.)*

and to reduce power demand and energy use. These advantages come at the cost of having to strengthen the overhead machine room floor, which must carry the additional structural load of the counterweight.

Approximately 75% of the energy expended in lifting a car is returned to the system by regeneration when the car is lowered. Regeneration is the process by which the traction motor becomes a generator when the car is lowered and feeds power back into the electrical system. The unrecovered energy appears as heat, primarily in the machine room. See Section 33.37 for a discussion of traction elevator system power and energy requirements.

To compensate for the hoist rope weight, which becomes an important factor in high-rise elevator installations, cables are attached to the bottom of the car and the counterweight, thus equalizing loads, regardless of the car position. These cables can be seen in Figure 33.1.

The elevator machine turns the sheave and lifts or lowers the car. The machine consists of a heavy structural frame on which are mounted the sheave and driving motor, the gears (if any), the brakes, the magnetic safety brake, and other auxiliaries. In some existing installations, the elevator driving (traction) motor receives its energy from a separate motor-generator (m-g) set, which operates when the elevator is available to handle traffic. This m-g set is properly considered a part of the elevator machine, although it may be located some distance from it. In newer installations, solid-state power and control equipment replace the m-g set. A governor, which limits the car to safe speeds, is mounted on or near the elevator machine.

The control equipment is usually divided into three functional groups:

1. *Drive (motion) control:* This is concerned with the velocity, acceleration, position determination, and leveling of the car.
2. *Operating control:* This covers car door operation and functioning of car signals, including floor call buttons and indicating devices.
3. *Supervisory control:* This is concerned with group operation of multiple-car installations.

The devices in these control systems were historically electromechanical (analog), but solid-state (digital) components are used in modern installations. The indicating and control devices that are seen by the elevator user, including car and hallway buttons,

TABLE 33.1 Elevator Capacity and Speed, Approximate SI and I-P Product Equivalents

Weight		Speed	
SI Standard (kg)	Approximate U.S. I-P (lb)	SI Standard (m/s)	Approximate U.S. I-P (fpm)
1000	2000	1.0	200
1250	2500–3000	1.6	300
1600	3500	2.0	400
2000	4000	2.5	500
		3.15	600
		4.0	800
		5.0	1000

lanterns (location indicators), and audible devices, are coordinated into an overall operational control scheme, to provide rapid, safe, and comfortable vertical transportation.

The shaft, also known as a shaftway or hoistway, is the vertical passageway for the car and counterweights. On its sidewalls are the car guide rails and mechanical and electrical auxiliaries of the control apparatus. At the bottom of the shaft are the car and counterweight buffers. At the top is the structural platform on which the elevator machine rests. The elevator machine room (which may occupy one or two levels) is usually directly above the hoistway. It contains the traction machine and the solid-state control (or motor generator set) that supplies energy to the elevator machine and control equipment. Machinery and control equipment are designed for quiet, vibration-free operation.

33.4 GEARLESS TRACTION MACHINES

A gearless traction machine consists of a DC or AC motor, the shaft of which is directly connected to a brake wheel and driving sheave. The elevator hoist ropes are placed around this sheave. The absence of gears means that the motor must run at the same relatively low speed as the driving sheave. As it is not economically practical to build motors for operation at very low speeds, a gearless machine is utilized for medium and high-speed elevators—that is, 500 fpm [2.5 m/s] and above. The motors range from 20 to 400 hp [15 to 300 kW].

Gearless machines are generally utilized for passenger service, with car capacities of 2000–4000 lb [907–1814 kg], although special cars of up to 10,000 lb [4536 kg] have been built. Below 500 fpm [2.5 m/s], geared machines are used. At this writing, maximum car speed is limited to around 4000 fpm [20 m/s]. Faster drives have been developed and will likely play a role in the development of the next generation of very-high-rise buildings.

A gearless traction machine is considered superior to a geared machine because it is more efficient and quieter in operation, requires less maintenance, and has longer life. Also, a gearless motor is more compact because it does not require a gearbox, which can avoid the need for a designated machine room. The decision as to whether these advantages are worth the additional cost involved can be made only after a careful analysis. Generally, a gearless machine is chosen where the rise is more than 250 ft [76 m], and very smooth, high-speed operation is desired. In the intermediate range of rise and speeds (i.e., 150–250 ft [46–76 m] height and 400–500 fpm [2–2.5 m/s]), excellent equipment, both geared and gearless, is available.

Outside of the United States, elevator size and speed are specified in kilograms and meters per second, making it useful to be aware of the SI equivalents for key design variables. Because 1 kg is 2.2 lb and 1 m/s is 196.86 fpm, approximate conversion factors of 2 and 200, respectively, can be used (Table 33.1). SI units shown herein are generally more precise conversions of I-P units.

33.5 GEARED TRACTION MACHINES

A geared traction machine has a worm and gear interposed between the driving motor and the hoisting sheave (Figure 33.2). The driving motor can

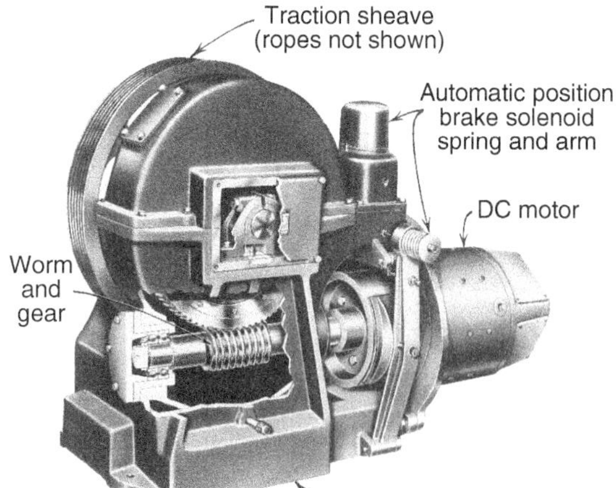

Figure 33.2 *Cutaway of a typical DC geared traction elevator machine. Note the grooves for multiple ropes in the traction sheave. (Courtesy of Schindler Elevator Corp.)*

therefore be a smaller, cheaper, high-speed unit, rather than a large, low-speed unit required by a gearless installation.

The motor used in a geared installation, as in a gearless one, depends upon the type of drive system and may be either DC or AC. Characteristics of the various drive systems and their application are discussed in detail in Sections 33.15 to 33.18. Geared machines are used for car speeds of up to 500 fpm [2.5 m/s] and a maximum rise of about 350 ft [107 m]. With an appropriate drive and control system, a geared traction machine can give almost the same high-quality, accurate, smooth ride as is available from a gearless installation.

33.6 ARRANGEMENT OF ELEVATOR MACHINES, SHEAVES, AND ROPES

The simplest method of arranging vertical travel of a car is to pass a rope over a sheave and counterbalance the weight of the car by a counterweight. Then, rotating the sheave makes the car move up or down and requires very little energy to do so. This is essentially the scheme that is used on a majority of high-speed passenger elevators, as illustrated in Figure 33.3*a*.

When the four or more supporting ropes merely pass over the sheave (*T*) and connect directly to the counterweights, the lifting power is exerted by the sheave through the traction of the ropes in the parallel grooves on the sheave. This system is referred to as the single-wrap traction elevator machine. The function of the sheave (*S*) is merely that of a guide pulley; it is called the deflector sheave.

The arrangement shown in Figure 33.3*b* is called double-wrap 1:1 roping. It provides greater traction than the single-wrap arrangement and is used in many automatic high-speed installations.

A 1:1 roping arrangement (Figure 33.3*a*, *b*, *d*) gives no mechanical advantage. The 2:1 roping (Figure 33.3*c*) has a mechanical advantage of 2, which permits use of a high-speed, low-power (and, therefore, lower-cost) traction machine. This arrangement is used for a wide variety of installations varying from medium-speed (500–700 fpm [2.5–3.6 m/s]) gearless passenger elevators to low-speed, heavy-duty freight elevators.

In types *a*, *b*, and *c* in Figure 33.3, the elevator machines are located in a machine room penthouse at the top of a hoistway. If, for architectural or other reasons, it is desired to eliminate the penthouse, a basement machine room can be used with the roping shown in Figure 33.3*d*. This arrangement uses geared traction equipment with speeds of up to 400 fpm [2.0 m/s]. All the illustrated ropings are applicable to the full range of car capacities.

33.7 SAFETY DEVICES

The main brake of an elevator is mounted directly on the shaft of the elevator machine (see Figure 33.2). The elevator is first slowed by dynamic braking of the motor, and the brake then operates to clamp the brake drum, thus holding the car still at the floor.

A dual safety system, designed to stop an elevator car automatically before its speed becomes excessive, is normally used. The device that acts first is a centrifugal governor or an electronic speed sensor that cuts off the power to the traction motor and sets the brake in the event of a limited overspeed. This usually stops the car. Should the speed still increase, the governor

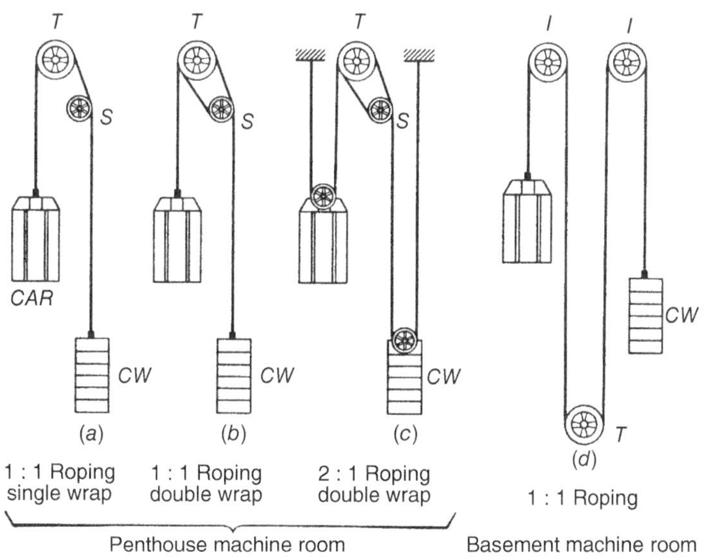

Figure 33.3 Elevator roping and sheave arrangements. (a) Basic single-wrap rope arrangement. In (b) and (c), the rope passes over traction sheave T and sheave S, doubles back over T, and then extends past S to the counterweight CW. This double-wrap arrangement provides additional traction at the drive sheave. (d) Roping arrangement for a basement machine room.

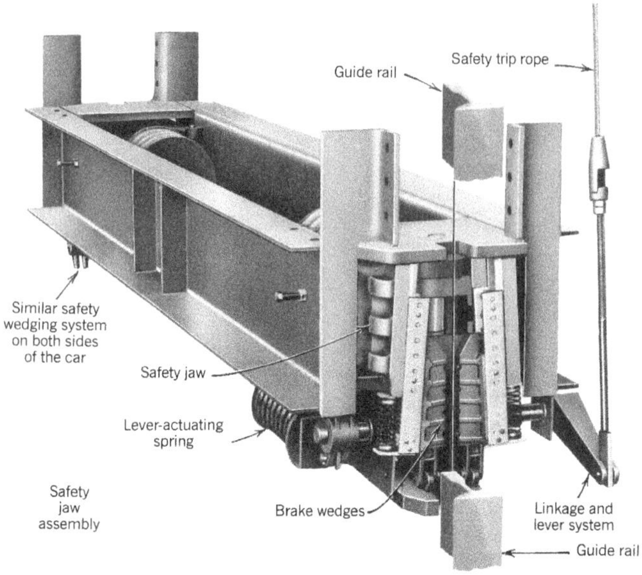

Figure 33.4 *Elevator safety devices. The governor or velocity transducer senses overspeed, clamping the safety trip rope and releasing the safety jaws, which exert a constant retarding force on the car rails, thus bringing the car to a gradual and safe stop.*

actuates two safety rail clamps, which are mounted at the bottom of the car, one on either side. They clamp the guide rails by wedging action, bringing the car to a smooth stop (Figure 33.4).

Oil or spring buffers are usually placed in the elevator pit. Their purpose is not to stop a falling car but to bring it to a somewhat cushioned stop if it overtravels the lower terminal. If a car overtravels (down or up), travel sensors deenergize the traction motor and set the main brake. Safety arrangements under emergency conditions (fire or power failure) are discussed in Section 33.40.

HYDRAULIC ELEVATORS

33.8 CONVENTIONAL PLUNGER-TYPE HYDRAULIC ELEVATORS

The elevators discussed in the preceding sections are traction types, that is, they are raised and lowered as a result of friction that transmits force from the drive mechanism through cables attached to or passing under the car. These cables in turn are raised and lowered by a motor-driven traction sheave. In contrast, the conventional hydraulic (or plunger) elevator car is raised and lowered by means of a movable rod (plunger) rigidly fixed to the bottom of the car. The absence of cables, drums, traction motors, elaborate controllers, safety devices, and penthouse equipment makes this system inherently less expensive and often the indicated choice for low-speed (up to 200 fpm [1 m/s]), low-rise (up to 65 ft [20 m]) applications, where construction of the plunger pit does not present difficulties, and/or the absence of a penthouse is desirable.

The first hydraulic elevators used water as the system fluid, supplied at sufficient pressure from roof-mounted water tanks. The tanks were kept full by a building's water pumps. All hydraulic elevators today use oil and obtain their motive power from a sealed oil-piping circuit powered by an oil pump.

The components of a typical hydraulic unit are shown in Figure 33.5. This system operates the same way as a hydraulic automobile jack. Oil from a reservoir is pumped under the plunger, thereby raising it and the car. The pump is stopped during downward motion, the car being lowered by gravity and controlled by the action of bypass valves, which also control the positioning of the car during upward motion. Control systems that are normally used are similar to those for traction elevators—for example, collective and selective collective approaches. Similarly, door arrangements are the same as in traction elevators—that is, single-slide, center-opening, and two-speed arrangements. Automatic leveling is readily available and is standard on all units with automatic controls.

From a design and construction perspective, the major advantage of hydraulic units is the absence of an overhead machine room, a penthouse, and traction equipment. As shown in Figure 33.5, only the guide rails project above the car and, if these are camouflaged, the impression of a freestanding elevator car is given. This effect can be used to good advantage inside large, open spaces such as in shopping malls and stores; when combined with glass-enclosed, observation-type cars, the effect can be striking (see Figure 33.44).

Additional advantages of hydraulic elevators over traction units include:

- The elevator load is carried by the ground and not by the structure. In contrast, traction units place a large

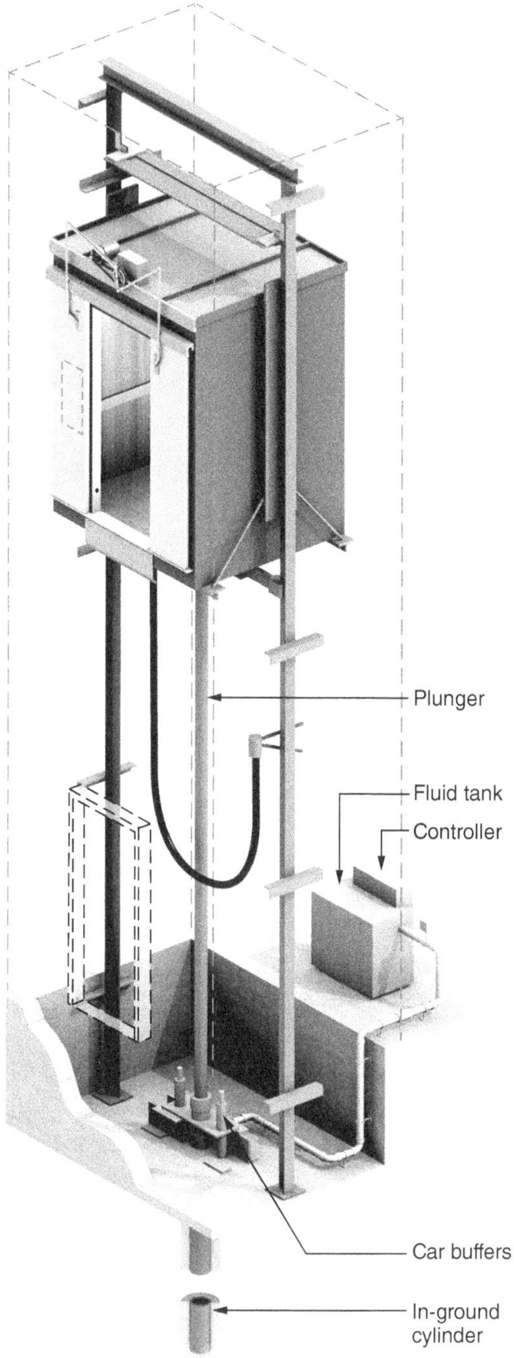

Figure 33.5 *Phantom view of a conventional "holed" hydraulic elevator, so called because the elevator car rests on a hydraulically activated steel plunger that descends into a hole in the ground. The hole is actually a buried hollow steel cylinder into which oil is pumped, under pressure, to raise the car. The oil pump is inside the fluid tank.*

structural load on the penthouse and machine room floors and on overhead steel as well.

- The hoistway is smaller due to the absence of a counterweight and its guide rails.

TABLE 33.2 Comparative Energy Use of Various Elevator Arrangements

Elevator/Control Arrangement	Relative Energy Use
Hydraulic elevator	10
UMV motion control; geared DC traction motor	7–7.5
UMV motion control; gearless DC traction motor	6–6.5
Thyristor control; geared DC traction motor	5–5.5
Thyristor control; gearless DC traction motor	4–4.5
Variable-voltage, variable-frequency control, AC traction motor	2.5–3

- Cars can be lowered manually by the operation of oil valves. This is particularly useful and important in the event of control equipment failure or power outage.
- There is essentially no limit to the load that can be lifted.

The major disadvantage of a standard hydraulic elevator is its operating expense. Because it is not counterweighted, it requires a relatively large motor to drive the oil pump, and all the energy is lost in heat (see Table 33.2). As an example of this operating cost disadvantage, consider a 3500-lb [1588-kg], 125-fpm [0.6-m/s] hydraulic elevator in a department store. Such a unit requires a 40-hp [30-kW] motor. Assuming the unit to be in operation 10 hours a day, 6 days a week, and assuming a normal 60% time-in-operation value, analysis shows (remembering that the motor operates only in the up direction):

$$\text{energy used/day} = \frac{40\text{ hp}}{0.82\text{ eff}} \times \frac{0.746\text{ kW}}{\text{hp}} \times 0.6 \times 10\text{ h} \times \frac{1}{2} = 109\text{ kWh}$$

At an electricity cost of $0.08/kWh,

$$\text{monthly energy cost} = \frac{109 \times 6\text{ days}}{\text{week}} \times \frac{4.33\text{ weeks}}{\text{month}} \times \$0.08 = \$227/\text{month}$$

Compare this to the calculated (in Section 33.37) $72 monthly energy cost for a 3500-lb [1588-kg], 600-fpm [3-m/s] traction car using solid-state equipment and $126 with an m-g set power supply, to get an appreciation of the value of a counterweight.

Other disadvantages of hydraulic units as compared to traction units include:

- They are limited to low-rise, low-speed applications.
- Ride quality is inferior to that of a good traction unit, although it is entirely acceptable for residential, mercantile, and industrial applications.

TRANSPORTATION

- Because oil viscosity changes with temperature, the ambient temperature of the space containing the pump and the oil storage tank must be controlled to maintain ride quality and performance.
- The high inrush current required by the pump each time it starts (which is every time the elevator travels upward) necessitates a "stiff" power supply to avoid problems of flickering lamps and other undesirable line-voltage fluctuations.
- Noise from the pump and motor plus piping noise can be disturbing. This problem can be ameliorated by skillfully locating the pump mechanism (which can be placed up to 50 ft [15 m] from the elevator shaft).

Plunger-type hydraulic elevators have experienced some disfavor because of problems caused by oil seepage into the ground. The plunger travels in a buried steel casing that, despite butyl or other liquid-tight coatings, can corrode after an extended period and leak oil. In most locations, this will violate groundwater pollution regulations. Because repair of such an oil leak is expensive and entails extended elevator outage, other arrangements (as detailed in the next two sections) are increasingly used.

Hydraulic elevators are best applied to low-speed, low-rise applications such as office buildings and residential buildings up to four stories in height, low-rise department stores, malls, basement and parking garage shuttles, theater elevators, and stage lifts and freight applications of all sorts—in particular, those intended for very heavy loads. Hydraulic elevators are commonly used in low-rise buildings to provide ADA-mandated vertical accessibility. A typical layout and dimensional data for standard plunger units are given in Figure 33.6, along with capacities and application recommendations.

33.9 HOLELESS HYDRAULIC ELEVATORS

A holeless hydraulic elevator may be suitable where drilling a plunger hole presents difficulties, yet a hydraulic installation using a telescoping plunger or a roping arrangement is feasible. The telescoping jack design, in a single-jack arrangement, is shown schematically in Figure 33.7. As this causes a lateral stress in the building due to the cantilevered car, a dual-jack arrangement is used more frequently. When supported on both sides, as shown in Figure 33.8, the entire vertical elevator load is transferred directly to the ground. The telescoping jack, in which all sections move simultaneously, is more complex than a simple plunger unit and requires more maintenance, although maintenance is simplified by the fact that the entire length of the jack is readily accessible. The ride with the telescoping jack arrangement is not as smooth as that on a straight plunger elevator because the simultaneous movement of the jack's telescoping sections causes a degree of jerk.

33.10 ROPED HYDRAULIC ELEVATORS

The roped hydraulic arrangement is simpler than the telescoping plunger unit because it uses only a single moving jack section, compared to two or even three in a telescoping unit for the same rise. It accomplishes this by using 2:1 roping, which means that the car travels twice as far as the piston. This is accomplished by passing the rope over a pulley in the piston crosshead. One end of the rope is attached to a fixed point in the pit below the car, and the other end is attached to the base of the car (Figure 33.9). The piston (plunger, jack) lifts the crosshead, which in turn lifts the car twice as far.

The arrangement shown in Figure 33.9 uses a single jack and a cantilevered car. Other arrangements use two jacks to eliminate the lateral building load (Figure 33.10). A 2:1 roping is standard in the United States. The simplicity and reliability of the single- or double-jack roped arrangement have made it by far the most common choice for low-rise, light- to medium-duty hydraulic elevators. Because it is a roped unit, it is equipped with a slack-rope safety in addition to the other safeties used on direct-connected hydraulic systems.

OTHER ELEVATOR DRIVES

33.11 RACK-AND-PINION ELEVATORS

Rack-and-pinion lifts are used both indoors and outdoors, primarily in industrial environments, for vertical transport of both passengers and materials. These elevators operate on the straightforward principle of a rotating cogged-wheel pinion attached to a vertical rack. Rotation of the pinion forces the attached car up and down along the rack (see Figure 33.11). The advantages of this system are its inherent simplicity and safety, unlimited rise, and low maintenance and operating costs, plus minimum space requirements. The last characteristic was primarily responsible for the selection of a rack-and-pinion design for the 210-ft-rise [64-m] emergency elevator installed as part of the renovation and rehabilitation of the Statue of Liberty in 1986.

The Statue of Liberty elevator is used primarily to evacuate victims of heart attacks. Carrying them down the 171-step spiral staircase that connects the main

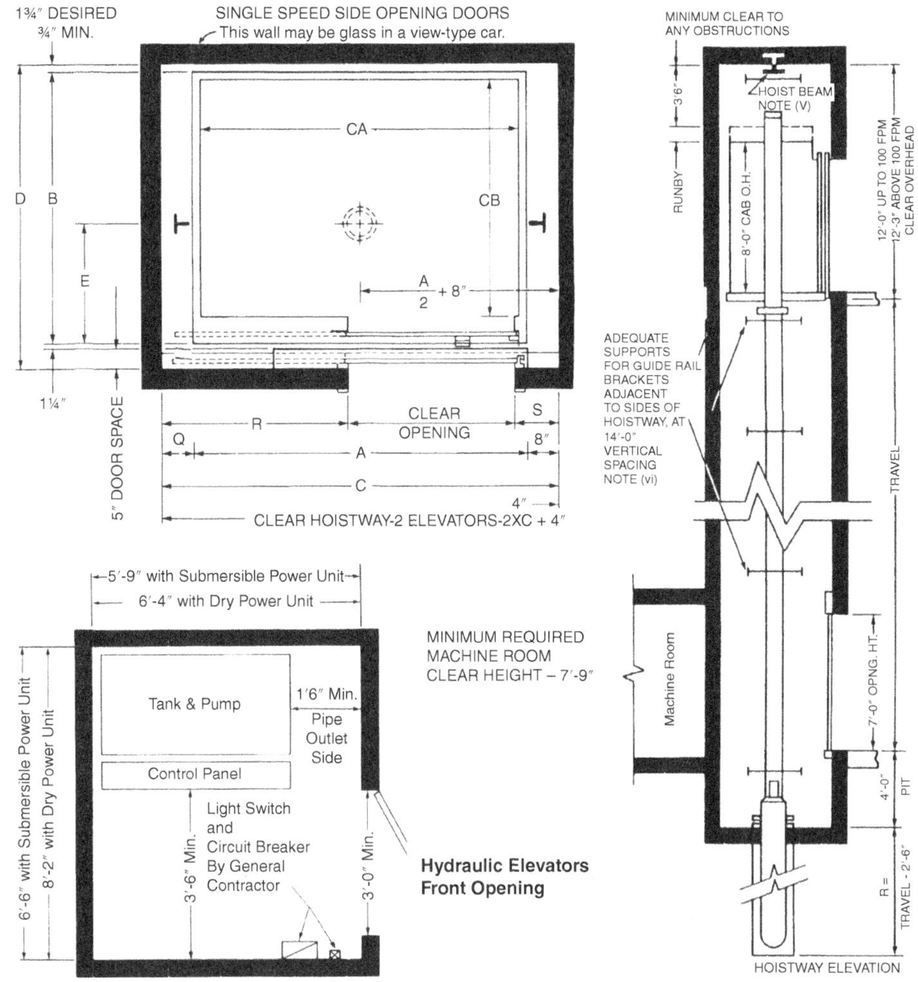

(a)

CAPACITY (LBS)	STD. SPEED (FPM)	MAXIMUM STANDARD OPENINGS	MAXIMUM STANDARD TRAVEL (i)	CLEAR OPENING (ii)	PLATFORM SIZE A WIDTH	PLATFORM SIZE B DEPTH	MINIMUM CLEAR CAB INSIDE CA WIDTH	MINIMUM CLEAR CAB INSIDE CB DEPTH	HOISTWAY C CLEAR WIDTH	HOISTWAY D(iv) WALL TO WALL	R	S	E	Q
2000	100	8	70′-0″	3′-0″	6′0″	5′-1″	5′-8″	4′-3″	7′-4″	5′-9″	3′-5″	11″	2′-1″	8″
	125		77′-0″											
2500	100	8	70′-0″	3′-6″	7′0″	5′-1″	6′-8″	4′-3″	8′-4″	5′-9″	3′-11″	11″	2′-1½″	8″
	125		76′-0″											
	150		76′-0″											
3000	100	8	70′-0″	3′-6″	7′0″	5′-7″	6′-8″	4′-9″	8′-4″	6′-3″	3′-11″	11″	2′-4⅛″	8″
	125		72′-0″											
	150		73′-0″											
3500	100	8	69′-0″	3′-6″	7′0″	6′-3″	6′-8″	5′-5″	8′-4″	6′-11″	3′-11″	11″	2′-8⅛″	8″
	125		69′-0″											
	150		69′-0″											
4000	100	8	66′-0″	4′-0″	8′0″	6′-3″	7′-8″	5′-5″	9′-4″	6′-11″	—	—	2′-8½″	8″
	125		66′-0″											
	150		66′-0″											

(b)

Figure 33.6 *(a) Typical dimensional, capacity, and (b) layout data for a conventional plunger-type hydraulic elevator. Door systems are single-slide (SS) or center-opening (CO). The car can be used as a viewing-type unit by utilizing a glass wall at the back of the car. (From published data of Schindler Elevator Corp.)*

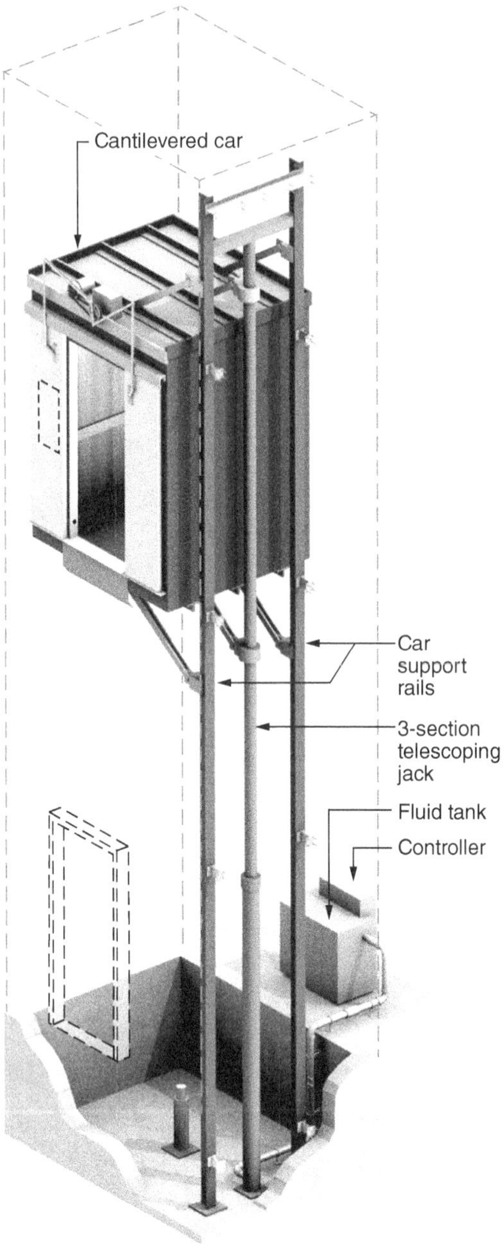

Figure 33.7 Holeless hydraulic elevator driven by a single telescoping jack. The cantilevered car exerts a lateral structural load on the building. This arrangement is used less frequently than the two-jack mechanism shown in Figure 33.8.

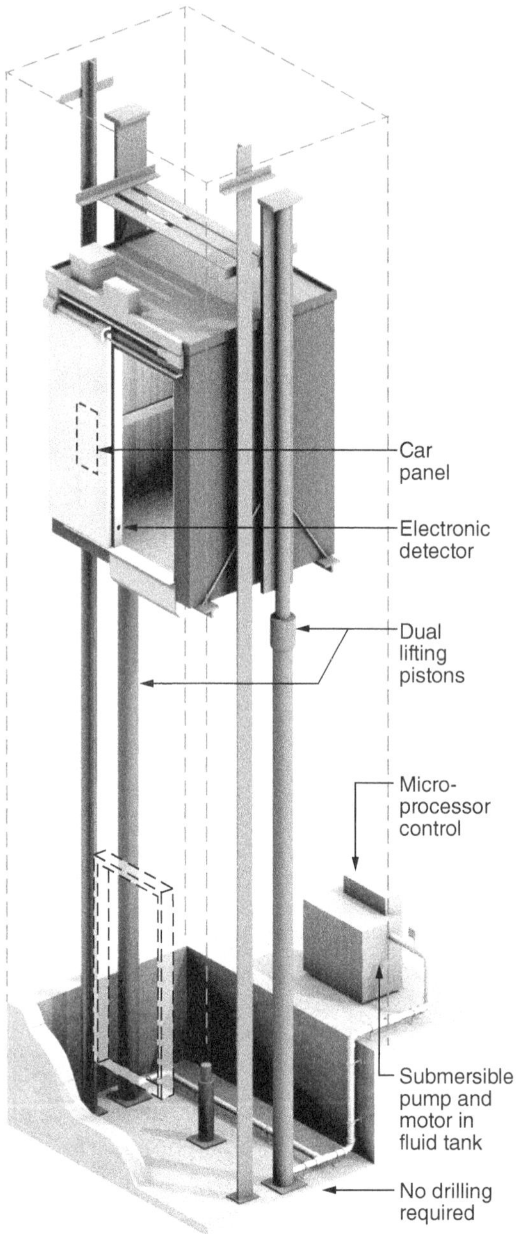

Figure 33.8 Dual-jack holeless hydraulic elevator. The balanced vertical load is borne by the ground only, with no building component. The unit illustrated is suitable only for a low-rise (two-stop) installation, with a car weighing no more than 2500 lb [1134 kg].

upper landing to the observation platform in the crown is not practical. The tight headroom available—less than 9 ft [2.7 m]—ruled out a traction unit, as did the extremely tight shaftway, which measures only 2 ft, ½ in. × 4 ft, 10 in. [0.62 by 1.47 m]. The outside dimensions of the car itself (see Figure 33.12) are only 1 ft, 11 in. × 3 ft, 10 in [0.58 by 1.17 m]. Space was so tight that the drive was divided between two motors, both of which are mounted under the passenger compartment.

33.12 INNOVATIVE MOTOR DRIVES

An elevator drive design developed by Otis in the late 1980s utilizes a linear induction motor built into the counterweight frame. Motive power is supplied at the counterweight, thus eliminating the overhead traction machine and machine room. The inherent architectural, construction, energy use, maintenance, and cost advantages of this design make it an attractive option for building designers. A schematic diagram of the

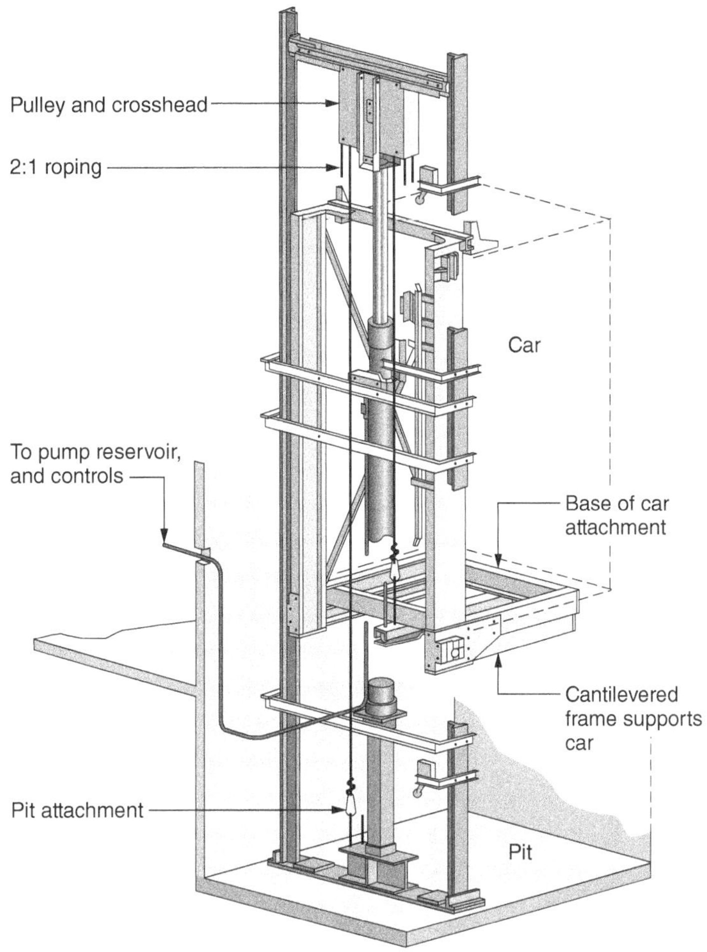

Figure 33.9 *Roped hydraulic low-rise residential-type elevator. The power unit, which includes the oil tank, pumps, and control, is usually mounted at the lower level. Control is automatic, including automatic leveling. Depending upon the specific design, the car is a 700–750-lb [318–340-kg], 30–36-fpm [0.15–0.18-m/s] unit, normally with a single-story rise. The shaftway is 16–26 ft² [1.5–2.4 m²], with the larger size used for a car intended to accommodate a wheelchair.*

design is shown in Figure 33.13. Other proprietary innovations have led to the common use of traction elevator systems that require no penthouse machine room, with the drive motor being incorporated into the hoistway space.

PASSENGER INTERACTION CONSIDERATIONS

33.13 ELEVATOR DOORS

The choice of a door for a car and hoistway affects the speed and quality of elevator service considerably. Doors for passenger elevators are power-operated and are synchronized with the leveling controls so that the doors are fully opened by the time a car comes to a complete stop at a landing. The closing time varies with the type of door and the size of the opening. For safety reasons, the kinetic energy of an automatic door is limited to 7 ft-lb [9.5 N m] and its closing pressure to 30 lb [135 N]. To provide the fastest closing within this energy limitation, a center-opening door is used. Also, to reduce passenger transfer time and avoid discomfort, a clear opening of 42 in. [1.07 m] is used in most commercial installations, which permits simultaneous loading and unloading without undue passenger contact. Some consultants feel that simultaneous passenger transfer is practical only with a 48-in. [1.2-m] clear opening (Figure 33.14). When an opening narrower than 36 in. [915 mm] is used, loading is delayed until unloading is complete, and the speed and quality of service are thereby markedly reduced. Such small doors are applicable only in residential or small, light-traffic buildings. Available door types are shown in Figure 33.15.

A two-speed door design is used where space conditions dictate or where a wide opening is required.

TRANSPORTATION

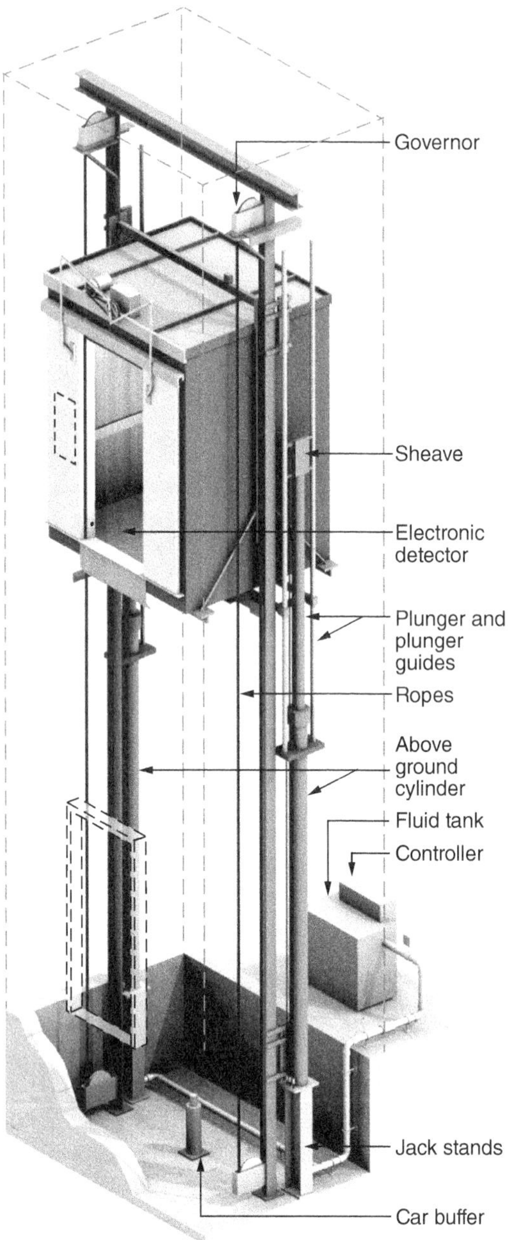

Figure 33.10 *For heavier loads than can be readily handled by the cantilevered car in a single-jack design (Figure 33.9), a balanced dual jack (cylinder) is used, as seen in this phantom view. This design, which can accomplish somewhat higher rise than a single-jack design, in addition to carrying much heavier loads, also requires a larger shaftway to accommodate the second cylinder and rails. A second pulley and crosshead are omitted for clarity in the drawing.*

Figure 33.11 *The basis of the rack-and-pinion drive is simply a driven pinion on a stationary rack; rotation of the pinion forces the structure attached to the pinion to move along the rack. (Courtesy of Alimak Elevator Co.)*

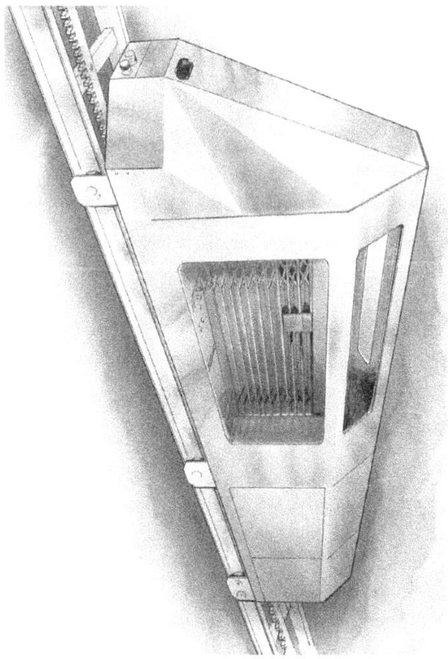

Figure 33.12 *The rack-and-pinion principle is used in the very small Statue of Liberty emergency elevator. The rack extends the entire length of the shaft, and the car moves along the rack, driven by two motorized pinions below the passenger compartment. (Courtesy of Alimak Elevator Co.)*

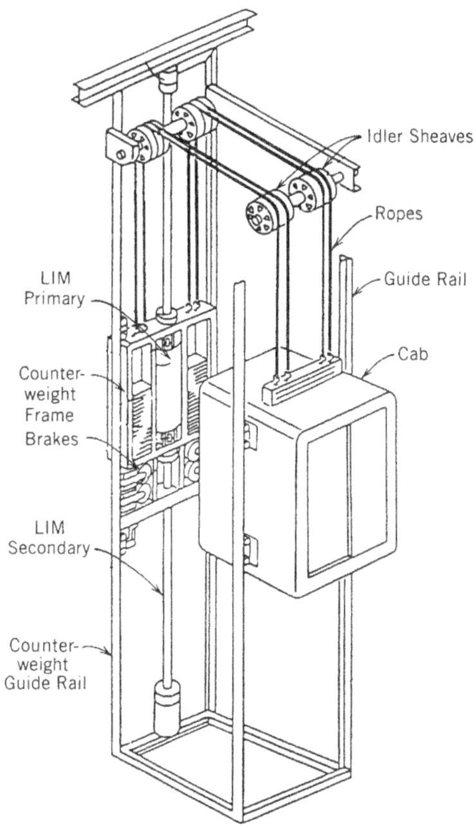

Figure 33.13 *Schematic drawing of Otis Japan's linear induction motor (LIM) elevator design. Because the motor is built into the counterweight and supplies motive power linearly (vertically), the conventional overhead rotary traction machine is entirely eliminated, as is the machine room. The system operates more efficiently, quietly, and smoothly than a conventional drive. (Courtesy of Otis Elevator Co.)*

The term two-speed reflects the fact that the two halves of the door must travel at different speeds to complete their travel simultaneously (see Figure 33.15*c*).

Installations can be equipped with an electronic sensing device that detects passengers in a wide area on the landing in front of the car door rather than only directly in the door's path. Such detection, often accompanied by an audible signal, causes the car door to remain open for a predetermined length of time, or causes a closing door to reverse. These devices are particularly useful in installations where passengers cannot approach the entrance or cannot enter the car quickly—for example, riders with baggage or holding children, people in wheelchairs, and employees moving bulky objects such as beds or carts in hospitals or wheeled objects in office and industrial facilities.

All automatic elevators, whether or not equipped with detection beams, are required by ANSI to have a safety edge device on the car doors that causes the car and hoistway doors, which operate in synchrony, to reopen when the safety edge meets any obstruction. Some cars feature doors that are designed to "nudge" when almost closed and/or after a specific time period. ADA requirements for elevators are discussed in Section 33.15.

33.14 CARS AND SIGNALS

An area where the architect has a generally free hand in choosing equipment relates to the decor of the cars

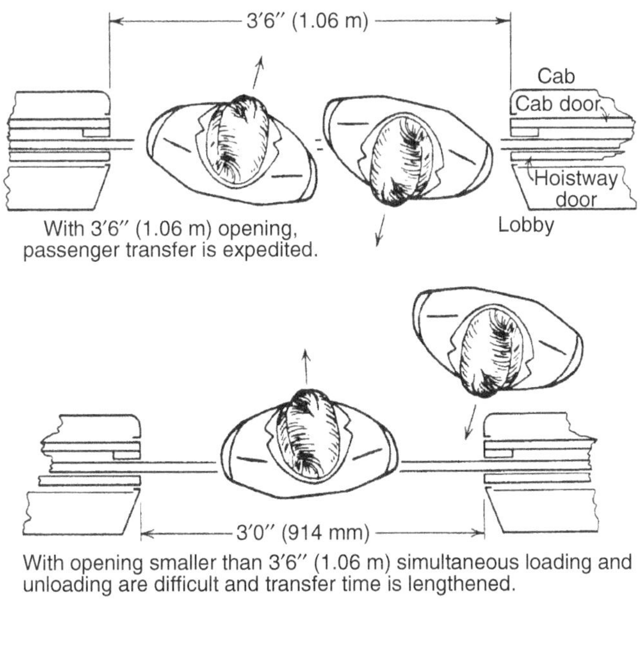

Figure 33.14 *Transfer of passengers with door openings of different widths. With openings smaller than 3 ft, 6 in. [1.07 m], simultaneous loading and unloading are difficult and transfer time is lengthened. With a 42-in. [1.07-m] opening, large people or people with bulky outerwear may brush against each other in passing. For complete isolation of passengers, a 48-in. [1.22-m] opening may be necessary.*

TRANSPORTATION

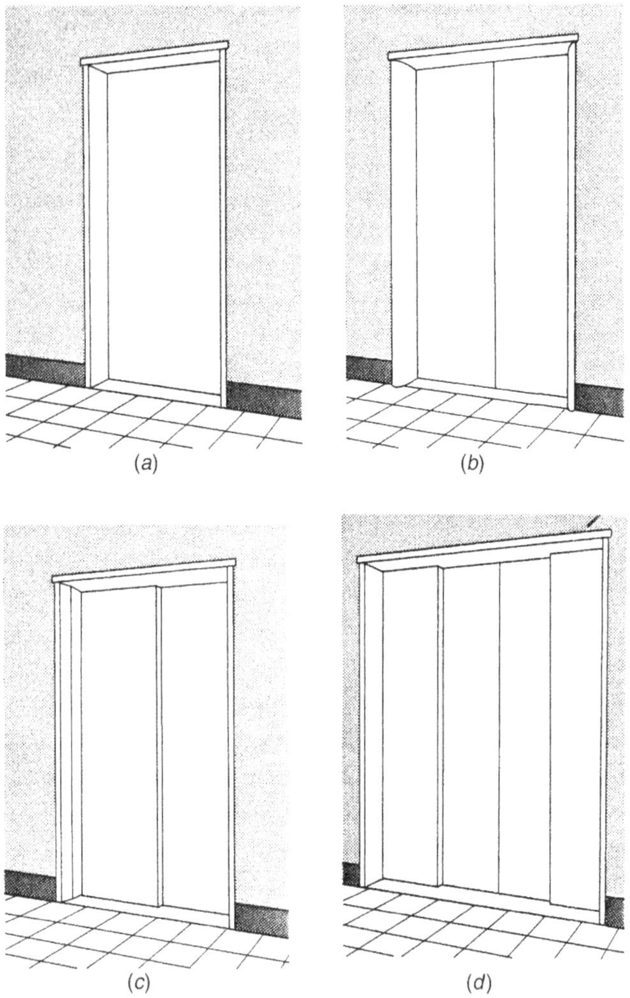

Figure 33.15 Typical hoistway doors and applications. (a) Single-slide door, 24–36 in. [610–914 mm] wide, for small commercial building or residential use. (b) Standard commercial 42-in. [1.07-m] center-opening door for office building use or 48–60-in. [1.22–1.52-m] center-opening door for hospital or service car. (c) Two-speed, 42-in. [1.07-m] general commercial use door. (d) Two-speed, center-opening, 60-in. [1.52-m] department store door for freight, passenger, and nonautomatic service.

and the selection of hallway and car signals. A typical elevator specification is functional and describes the intended operation of the equipment, while usually including an amount to cover decor of the cars. The type and functions of signal equipment are also specified, but finish and styling are optional. Car interiors may be finished in wood paneling, plastic (Micarta or Formica), stainless steel, or almost any material desired. Floors may be tile, wood, or carpeting, as selected. Illumination may be from ceiling fixtures, coves, or a luminous ceiling of standard or special design. For each bank of elevators, it is wise to furnish at least one set of wall mats to protect wall finishes when cars are being used to move furniture. This is especially important where no separate service car has been provided.

Car and hallway signals and lanterns should be designed to fulfill their basic functions, consider the needs of the disabled, and coordinate with the decor of the cars and corridors. (For a discussion of the car operating panel, see Section 33.27.) The call buttons should indicate the desired direction of travel, and by visual means confirm that a call has been placed. The hall lantern located at each car entrance must visually indicate the direction of travel of an arriving elevator and preferably its present location (Figure 33.16). An audible signal should announce a car's imminent arrival. This feature allows waiting passengers to move toward the arriving car, which speeds service when there are multiple elevators in a bank. Hall stations can be equipped with special switches for fire, priority, and limited-access service, as required. Destination group systems allow passengers waiting in lobbies to specify their destination before getting into the elevator cab, which helps the system to allocate the appropriate elevator for groups of passengers traveling to similar areas of the building, reduces the numbers of stops, decreases round-trip times, and enhances system handling capacities.

Within the car, travel direction and present location can be indicated either with separate fixtures or by indicators built into the car panel. Most manufacturers can provide voice synthesizers built into the car panel that announce the floor, direction of

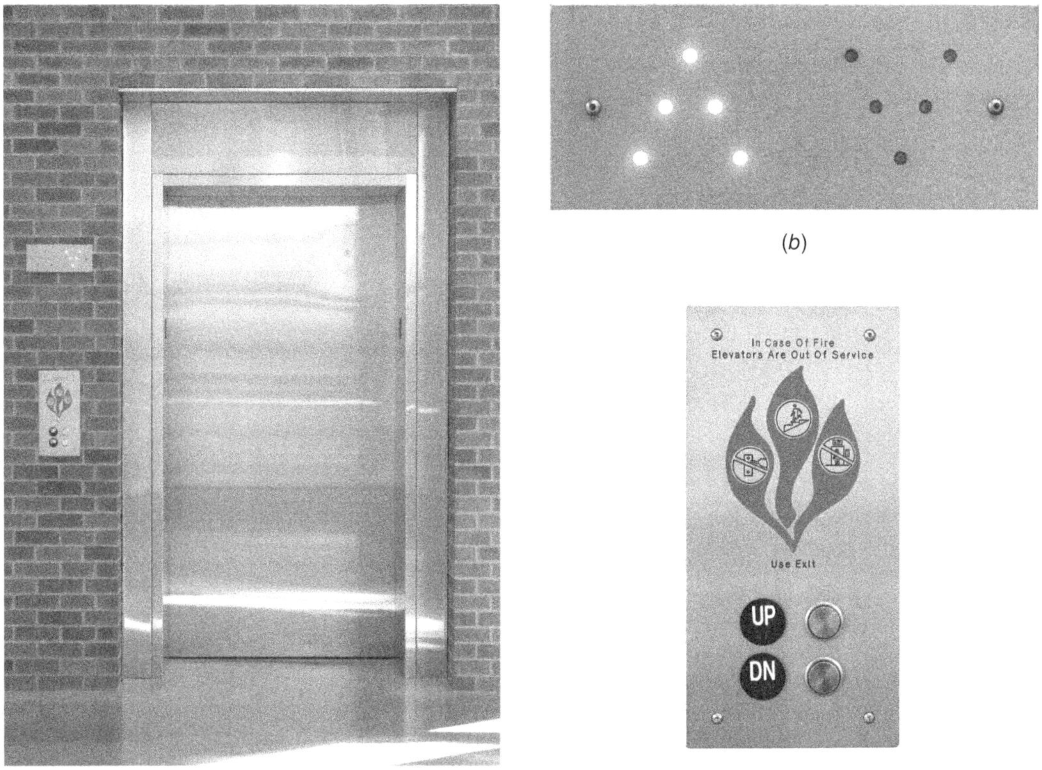

(a) (b) (c)

Figure 33.16 *(a) Typical single-slide-door elevator with call buttons and travel direction indicators. (b) LED arrowheads to indicate the direction of elevator car travel. (c) Elevator pushbuttons to call the elevator. (Courtesy of Karen Tse.)*

travel, and any other desired message such as safety or emergency messages.

33.15 ACCESSIBILITY REQUIREMENTS

Elevator manufacturers follow the latest guidelines set by the Americans with Disabilities Act (ADA) Accessibility Standards as a minimum, and may add additional conveniences for the disabled, to meet specific owner's project requirements or local codes. No planning should begin until the project architect has assembled and reviewed all applicable codes and standards. Information regarding elevator accessibility requirements (which incorporates *ASME A17.1: Safety Code for Elevators and Escalators* by reference) may be found on the United States Access Board's website (www.access-board.gov).

The primary physical limitations addressed by the ADA are those of ambulation, sight, and hearing. Thus, to ease access for passengers in wheelchairs (or with walking aids), the ADA requires excellent car leveling, 36-in. [915-mm] minimum clear door opening (42 in. [1.1 m] recommended—and required on center door cabs), delayed door closing, detection beams that reopen the door without contact on sensing a passenger in the door opening, inside car dimensions that permit turning a wheelchair, buttons and emergency controls within easy reach, and appropriate car furnishings.

For those with sight impairment, the ADA requires audible signals in addition to easily seen and recognized visual ones, both in the car and at landings, to indicate call registration, car approach, car landing, direction of travel, floor, car position, and so on. In this connection, a voice synthesizer is of assistance. In addition, Braille plates adjacent to car floor buttons, and large, easily recognizable symbols adjacent to passenger-enabled emergency controls are to be used. Hearing-impaired passengers are assisted by the large visual signals required for sight-impaired passengers. In addition, call buttons must visually indicate that a call has been placed and likewise indicate when the call has been answered. Many of these requirements make life easier for all passengers. A note of caution: Delayed door closing may increase overall travel time appreciably. In buildings with substantial traffic peaks, this factor needs to be considered when designing the elevator system.

Requirements of the ADA that apply to elevators are presented in Figure 33.17, with accompanying illustrations (figure and chapter numbers within the illustration are ADA references).

407 Elevators

407.1 General. Elevators shall comply with 407 and with ASME A17.1 (incorporated by reference, see "Referenced Standards" in Chapter 1). They shall be passenger elevators as classified by ASME A17.1. Elevator operation shall be automatic.

> **Advisory 407.1 General.** The ADA and other Federal civil rights laws require that accessible features be maintained in working order so that they are accessible to and usable by those people they are intended to benefit. Building owners should note that the ASME Safety Code for Elevators and Escalators requires routine maintenance and inspections. Isolated or temporary interruptions in service due to maintenance or repairs may be unavoidable; however, failure to take prompt action to effect repairs could constitute a violation of Federal laws and these requirements.

407.2 Elevator Landing Requirements. Elevator landings shall comply with 407.2.

407.2.1 Call Controls. Where elevator call buttons or keypads are provided, they shall comply with 407.2.1 and 309.4. Call buttons shall be raised or flush.

EXCEPTION: Existing elevators shall be permitted to have recessed call buttons.

407.2.1.1 Height. Call buttons and keypads shall be located within one of the reach ranges specified in 308, measured to the centerline of the highest *operable part*.

EXCEPTION: Existing call buttons and existing keypads shall be permitted to be located at 54 inches (1370 mm) maximum above the finish floor, measured to the centerline of the highest *operable part*.

407.2.1.2 Size. Call buttons shall be ¾ inch (19 mm) minimum in the smallest dimension.

EXCEPTION: Existing elevator call buttons shall not be required to comply with 407.2.1.2.

407.2.1.3 Clear Floor or Ground Space. A clear floor or ground *space* complying with 305 shall be provided at call controls.

> **Advisory 407.2.1.3 Clear Floor or Ground Space.** The clear floor or ground space required at elevator call buttons must remain free of obstructions including ashtrays, plants, and other decorative elements that prevent wheelchair users and others from reaching the call buttons. The height of the clear floor or ground space is considered to be a volume from the floor to 80 inches (2030 mm) above the floor. Recessed ashtrays should not be placed near elevator call buttons so that persons who are blind or visually impaired do not inadvertently contact them or their contents as they reach for the call buttons.

407.2.1.4 Location. The call button that designates the up direction shall be located above the call button that designates the down direction.

EXCEPTION: Destination-oriented elevators shall not be required to comply with 407.2.1.4.

> **Advisory 407.2.1.4 Location Exception.** A destination-oriented elevator system provides lobby controls enabling passengers to select floor stops, lobby indicators designating which elevator to use, and a car indicator designating the floors at which the car will stop. Responding cars are programmed for maximum efficiency by reducing the number of stops any passenger experiences.

407.2.1.5 Signals. Call buttons shall have visible signals to indicate when each call is registered and when each call is answered.

EXCEPTIONS: 1. Destination-oriented elevators shall not be required to comply with 407.2.1.5 provided that visible and audible signals complying with 407.2.2 indicating which elevator car to enter are provided.

2. Existing elevators shall not be required to comply with 407.2.1.5.

407.2.1.6 Keypads. Where keypads are provided, keypads shall be in a standard telephone keypad arrangement and shall comply with 407.4.7.2.

Figure 33.17 *ADA requirements for elevators. (From 2010 ADA Standards, U.S. Department of Justice.)*

407.2.2 Hall Signals. Hall signals, including in-car signals, shall comply with 407.2.2.

407.2.2.1 Visible and Audible Signals. A visible and audible signal shall be provided at each hoistway entrance to indicate which car is answering a call and the car's direction of travel. Where in-car signals are provided, they shall be visible from the floor area adjacent to the hall call buttons.

EXCEPTIONS: 1. Visible and audible signals shall not be required at each destination-oriented elevator where a visible and audible signal complying with 407.2.2 is provided indicating the elevator car designation information.

2. In existing elevators, a signal indicating the direction of car travel shall not be required.

407.2.2.2 Visible Signals. Visible signal fixtures shall be centered at 72 inches (1830 mm) minimum above the finish floor or ground. The visible signal *elements* shall be 2-½ inches (64 mm) minimum measured along the vertical centerline of the *element*. Signals shall be visible from the floor area adjacent to the hall call button.

EXCEPTIONS: 1. Destination-oriented elevators shall be permitted to have signals visible from the floor area adjacent to the hoistway entrance.

2. Existing elevators shall not be required to comply with 407.2.2.2.

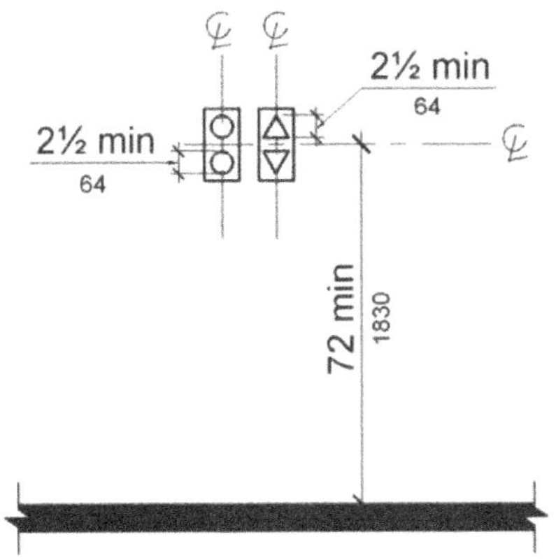

Figure 407.2.2.2
Visible Hall Signals

407.2.2.3 Audible Signals. Audible signals shall sound once for the up direction and twice for the down direction, or shall have verbal annunciators that indicate the direction of elevator car travel. Audible signals shall have a frequency of 1500 Hz maximum. Verbal annunciators shall have a frequency of 300 Hz minimum and 3000 Hz maximum. The audible signal and verbal annunciator shall be 10 dB minimum above ambient, but shall not exceed 80 dB, measured at the hall call button.

EXCEPTIONS: 1. Destination-oriented elevators shall not be required to comply with 407.2.2.3 provided that the audible tone and verbal announcement is the same as those given at the call button or call button keypad.

2. Existing elevators shall not be required to comply with the requirements for frequency and dB range of audible signals.

Figure 33.17 *(continued)*

407.2.2.4 Differentiation. Each destination-oriented elevator in a bank of elevators shall have audible and visible means for differentiation.

407.2.3 Hoistway Signs. Signs at elevator hoistways shall comply with 407.2.3.

407.2.3.1 Floor Designation. Floor designations complying with 703.2 and 703.4.1 shall be provided on both jambs of elevator hoistway entrances. Floor designations shall be provided in both *tactile characters* and braille. *Tactile characters* shall be 2 inches (51 mm) high minimum. A *tactile* star shall be provided on both jambs at the main entry level.

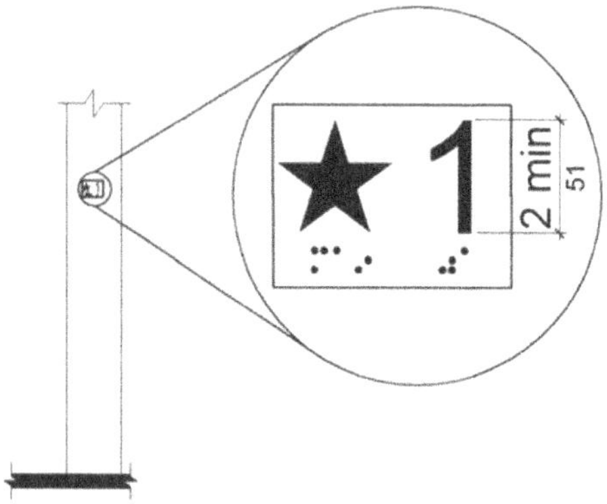

Figure 407.2.3.1
Floor Designations on Jambs of Elevator Hoistway Entrances

407.2.3.2 Car Designations. Destination-oriented elevators shall provide *tactile* car identification complying with 703.2 on both jambs of the hoistway immediately below the floor designation. Car designations shall be provided in both *tactile characters* and braille. *Tactile characters* shall be 2 inches (51 mm) high minimum.

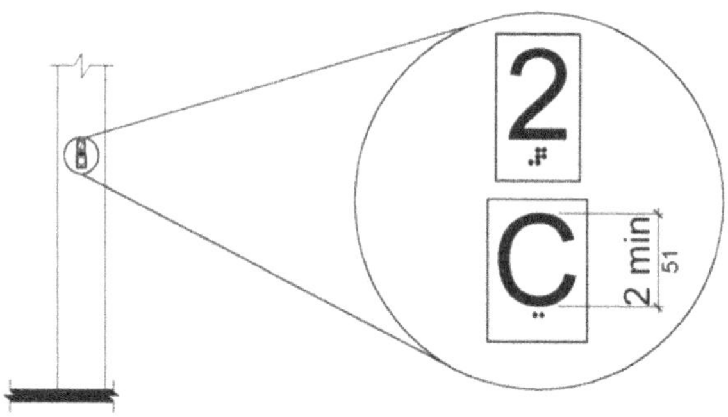

Figure 407.2.3.2
Car Designations on Jambs of Destination-Oriented Elevator Hoistway Entrances

Figure 33.17 *(continued)*

407.3 Elevator Door Requirements. Hoistway and car doors shall comply with 407.3.

407.3.1 Type. Elevator doors shall be the horizontal sliding type. Car gates shall be prohibited.

407.3.2 Operation. Elevator hoistway and car doors shall open and close automatically.
EXCEPTION: Existing manually operated hoistway swing doors shall be permitted provided that they comply with 404.2.3 and 404.2.9. Car door closing shall not be initiated until the hoistway door is closed.

407.3.3 Reopening Device. Elevator doors shall be provided with a reopening device complying with 407.3.3 that shall stop and reopen a car door and hoistway door automatically if the door becomes obstructed by an object or person.
EXCEPTION: Existing elevators with manually operated doors shall not be required to comply with 407.3.3.

407.3.3.1 Height. The device shall be activated by sensing an obstruction passing through the opening at 5 inches (125 mm) nominal and 29 inches (735 mm) nominal above the finish floor.

407.3.3.2 Contact. The device shall not require physical contact to be activated, although contact is permitted to occur before the door reverses.

407.3.3.3 Duration. Door reopening devices shall remain effective for 20 seconds minimum.

407.3.4 Door and Signal Timing. The minimum acceptable time from notification that a car is answering a call or notification of the car assigned at the means for the entry of destination information until the doors of that car start to close shall be calculated from the following equation:

$T = D/(1.5\ \text{ft/s})$ or $T = D/(455\ \text{mm/s}) = 5$ seconds minimum where T equals the total time in seconds and D equals the distance (in feet or millimeters) from the point in the lobby or corridor 60 inches (1525 mm) directly in front of the farthest call button controlling that car to the centerline of its hoistway door.
EXCEPTIONS: 1. For cars with in-car lanterns, T shall be permitted to begin when the signal is visible from the point 60 inches (1525 mm) directly in front of the farthest hall call button and the audible signal is sounded.
2. Destination-oriented el evators shall not be required to comply with 407.3.4.

407.3.5 Door Delay. Elevator doors shall remain fully open in response to a car call for 3 seconds minimum.

407.3.6 Width. The width of elevator doors shall comply with Table 407.4.1.
EXCEPTION: In existing elevators, a power-operated car door complying with 404.2.3 shall be permitted.

407.4 Elevator Car Requirements. Elevator cars shall comply with 407.4.

407.4.1 Car Dimensions. Inside dimensions of elevator cars and clear width of elevator doors shall comply with Table 407.4.1.

Figure 33.17 *(continued)*

TRANSPORTATION

EXCEPTION: Existing elevator car configurations that provide a clear floor area of 16 square feet (1.5 m^2) minimum and also provide an inside clear depth 54 inches (1370 mm) minimum and a clear width 36 inches (915 mm) minimum shall be permitted.

Table 407.4.1 Elevator Car Dimensions

	Minimum Dimensions			
Door Location	**Door Clear Width**	**Inside Car, Side to Side**	**Inside Car, Back Wall to Front Return**	**Inside Car, Back Wall to Inside Face of Door**
Centered	42 inches (1065 mm)	80 inches (2030 mm)	51 inches (1295 mm)	54 inches (1370 mm)
Side (off-centered)	36 inches (915 mm)[1]	68 inches (1725 mm)	51 inches (1295 mm)	54 inches (1370 mm)
Any	36 inches (915 mm)[1]	54 inches (1370 mm)	80 inches (2030 mm)	80 inches (2030 mm)
Any	36 inches (915 mm)[1]	60 inches (1525 mm)[2]	60 inches (1525 mm)[2]	60 inches (1525 mm)[2]

1. A tolerance of minus 5/8 inch (16 mm) is permitted.
2. Other car configurations that provide a turning *space* complying with 304 with the door closed shall be permited.

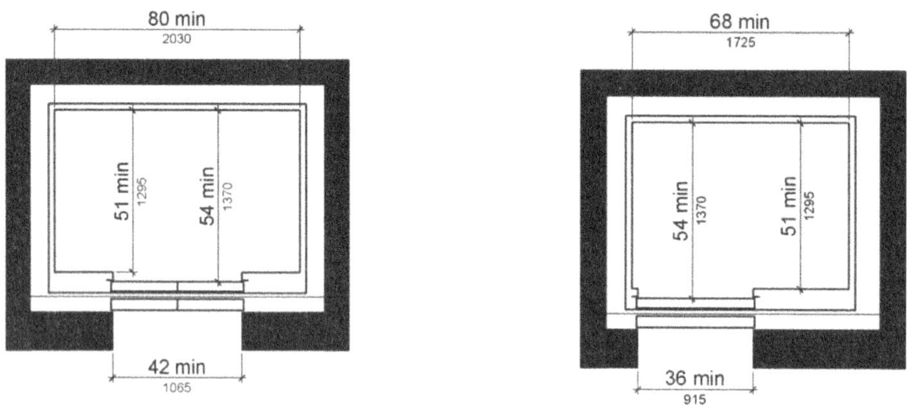

(a)
centered door

(b)
side (off-centered) door

Figure 407.4.1
Elevator Car Dimensions

Figure 33.17 *(continued)*

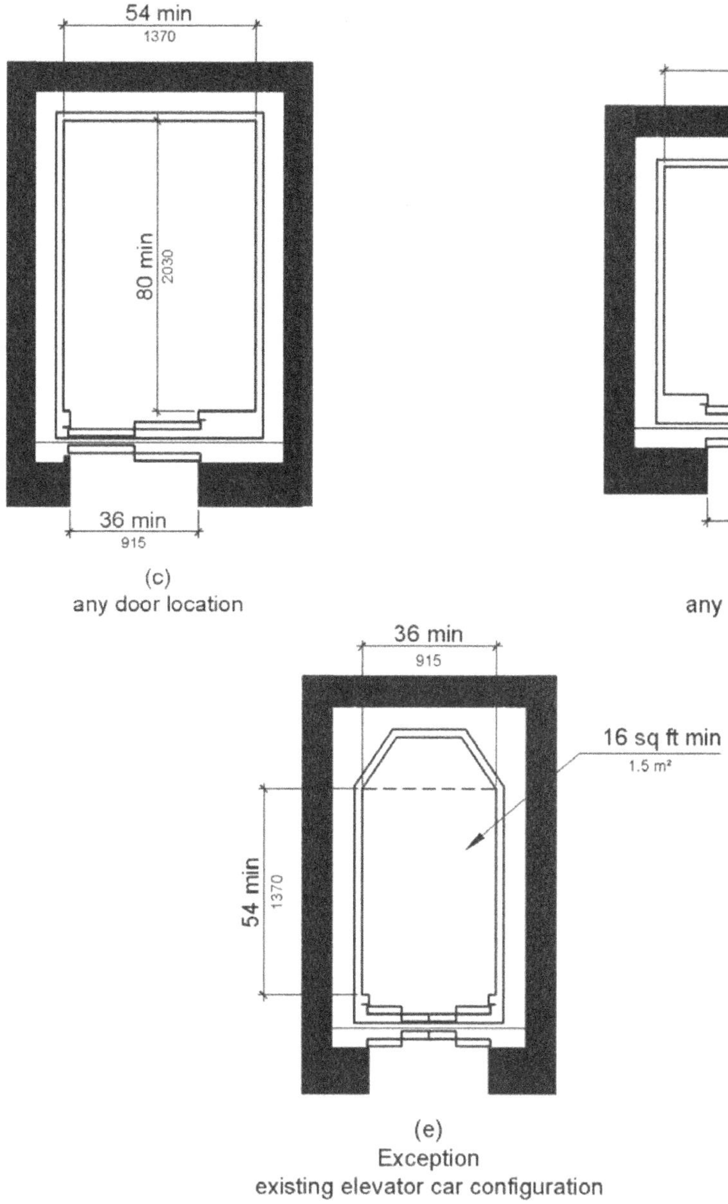

Figure 407.4.1
Elevator Car Dimensions

Figure 33.17 *(continued)*

407.4.2 Floor Surfaces. Floor surfaces in elevator cars shall comply with 302 and 303.

407.4.3 Platform to Hoistway Clearance. The clearance between the car platform sill and the edge of any hoistway landing shall be 1¼ inch (32 mm) maximum.

407.4.4 Leveling. Each car shall be equipped with a self-leveling feature that will automatically bring and maintain the car at floor landings within a tolerance of ½ inch (13 mm) under rated loading to zero loading conditions.

407.4.5 Illumination. The level of illumination at the car controls, platform, car threshold and car landing sill shall be 5 foot candles (54 lux) minimum.

407.4.6 Elevator Car Controls. Where provided, elevator car controls shall comply with 407.4.6 and 309.4.

EXCEPTION: In existing elevators, where a new car operating panel complying with 407.4.6 is provided, existing car operating panels shall not be required to comply with 407.4.6.

407.4.6.1 Location. Controls shall be located within one of the reach ranges specified in 308.

EXCEPTIONS: 1. Where the elevator panel serves more than 16 openings and a parallel approach is provided, buttons with floor designations shall be permitted to be 54 inches (1370 mm) maximum above the finish floor.

2. In existing elevators, car control buttons with floor designations shall be permitted to be located 54 inches (1370 mm) maximum above the finish floor where a parallel approach is provided.

407.4.6.2 Buttons. Car control buttons with floor designations shall comply with 407.4.6.2 and shall be raised or flush.

EXCEPTION: In existing elevators, buttons shall be permitted to be recessed.

407.4.6.2.1 Size. Buttons shall be 3/4 inch (19 mm) minimum in their smallest dimension.

407.4.6.2.2 Arrangement. Buttons shall be arranged with numbers in ascending order. When two or more columns of buttons are provided they shall read from left to right.

407.4.6.3 Keypads. Car control keypads shall be in a standard telephone keypad arrangement and shall comply with 407.4.7.2.

407.4.6.4 Emergency Controls. Emergency controls shall comply with 407.4.6.4.

407.4.6.4.1 Height. Emergency control buttons shall have their centerlines 35 inches (890 mm) minimum above the finish floor.

407.4.6.4.2 Location. Emergency controls, including the emergency alarm, shall be grouped at the bottom of the panel.

407.4.7 Designations and Indicators of Car Controls. Designations and indicators of car controls shall comply with 407.4.7.

Figure 33.17 *(continued)*

EXCEPTION: In existing elevators, where a new car operating panel complying with 407.4.7 is provided, existing car operating panels shall not be required to comply with 407.4.7.

407.4.7.1 Buttons. Car control buttons shall comply with 407.4.7.1.

407.4.7.1.1 Type. Control buttons shall be identified by *tactile characters* complying with 703.2.

407.4.7.1.2 Location. Raised *character* and braille designations shall be placed immediately to the left of the control button to which the designations apply.
EXCEPTION: Where *space* on an existing car operating panel precludes *tactile* markings to the left of the controls, markings shall be placed as near to the control as possible.

407.4.7.1.3 Symbols. The control button for the emergency stop, alarm, door open, door close, main entry floor, and phone, shall be identified with *tactile* symbols as shown in Table 407.4.7.1.3.

Table 407.4.7.1.3 Elevator Control Button Identification

Control Button	Tactile Symbol	Braille Message
Emergency Stop		"ST"OP Three cells
Alarm		AL"AR"M Four cells
Door Open		OP"EN" Three cells
Door Close		CLOSE Five cells
Main Entry Floor		MA"IN" Three cells
Phone		PH"ONE" Four cells

407.4.7.1.4 Visible Indicators. Buttons with floor designations shall be provided with visible indicators to show that a call has been registered. The visible indication shall extinguish when the car arrives at the designated floor.

Figure 33.17 *(continued)*

407.4.7.2 Keypads. Keypads shall be identified by *characters* complying with 703.5 and shall be centered on the corresponding keypad button. The number five key shall have a single raised dot. The dot shall be 0.118 inch (3 mm) to 0.120 inch (3.05 mm) base diameter and in other aspects comply with Table 703.3.1.

407.4.8 Car Position Indicators. Audible and visible car position indicators shall be provided in elevator cars.

407.4.8.1 Visible Indicators. Visible indicators shall comply with 407.4.8.1.

407.4.8.1.1 Size. *Characters* shall be ½ inch (13 mm) high minimum.

407.4.8.1.2 Location. Indicators shall be located above the car control panel or above the door.

407.4.8.1.3 Floor Arrival. As the car passes a floor and when a car stops at a floor served by the elevator, the corresponding *character* shall illuminate.

EXCEPTION: Destination-oriented elevators shall not be required to comply with 407.4.8.1.3 provided that the visible indicators extinguish when the call has been answered.

407.4.8.1.4 Destination Indicator. In destination-oriented elevators, a display shall be provided in the car with visible indicators to show car destinations.

407.4.8.2 Audible Indicators. Audible indicators shall comply with 407.4.8.2.

407.4.8.2.1 Signal Type. The signal shall be an automatic verbal annunciator which announces the floor at which the car is about to stop.

EXCEPTION: For elevators other than destination-oriented elevators that have a rated speed of 200 feet per minute (1 m/s) or less, a non-verbal audible signal with a frequency of 1500 Hz maximum which sounds as the car passes or is about to stop at a floor served by the elevator shall be permitted.

407.4.8.2.2 Signal Level. The verbal annunciator shall be 10 dB minimum above ambient, but shall not exceed 80 dB, measured at the annunciator.

407.4.8.2.3 Frequency. The verbal annunciator shall have a frequency of 300 Hz minimum to 3000 Hz maximum.

407.4.9 Emergency Communication. Emergency two-way communication systems shall comply with 308. *Tactile* symbols and *characters* shall be provided adjacent to the device and shall comply with 703.2.

Figure 33.17 *(continued)*

ELEVATOR CONTROLS

The movement of an elevator car and all its parts is controlled by three different systems that combine and interact to provide a unified control system. The supervisory system controls a bank of elevators as a group and dictates which car answers which call. The operational control system determines when and where physical motion of a car and its doors should occur. This system deals with the operation of the car doors and the integration of car buttons, lanterns, and passenger-operated devices into the overall control and indicating system. The operating control system passes information about car and door control to the motion control system (also known as drive control). Motion control determines the car's acceleration, velocity, braking, leveling, and regenerative braking, plus all aspects of door motion.

33.16 MOTION CONTROL

Elevator car acceleration and deceleration are accomplished by controlling the speed of the motor that drives the elevator traction machine. Speed control can be accomplished in several ways:

- thyristor control of an asynchronous (squirrel cage) AC traction motor
- thyristor control of a DC traction motor
- motor-generator (m-g) set control of a DC traction motor (Ward–Leonard system)
- variable-voltage, variable-frequency (VVVF) control of an asynchronous AC traction motor

Each system is described briefly in the following sections, and applicability, advantages, and disadvantages are noted. A thyristor is a semiconductor device that can be used in elevator control systems to manage the flow of current and voltage to drive the elevator car. A rheostat is a type of variable resistor, which also can be used to manage current and has historically been used in some types of elevators; however, rheostatic control of single and multispeed AC motors is essentially obsolete and therefore only mentioned referentially in this chapter.

(a) Thyristor Control, AC and DC Motors

Alternating Current (AC). High-power transistors make accurate speed control of inexpensive AC squirrel-cage motors practical by the simple expedient of supplying carefully regulated variable voltage to the motor. The resultant speed control is smooth and stepless, making it applicable to passenger elevator installations. However, the high motor slip caused by the constant frequency of the variable-voltage AC supply causes large thermal losses with resultant low operating efficiency. In addition, the "chopper," which is used to provide the necessary voltage control, introduces harmonics into the power system in considerable quantity. These harmonics cause undesirable radio noise and can cause system component overheating. Finally, the system's low power factor increases line losses and necessitates increased feeder sizes. As a result, this option, which is applicable to low- and mid-rise passenger service (maximum rise 250 ft [76 m] and maximum car speed of 350 fpm [1.8 m/s]), using geared machines, has been largely replaced by VVVF control (see Section 33.18).

Direct Current (DC). Excellent ride quality can be had by utilizing the DC version of thyristor control (see Figure 33.18*a*) to supply variable voltage to a DC traction motor. This arrangement provides the high-quality ride and leveling characteristic of DC drives and removes the rise and speed limitations of its AC counterpart, but not the system's inherent low power factor, heavy line harmonics, and high machine room heat gains. Line harmonics can be sharply reduced, however, by the use of power line filters. DC thyristor control is widely used today, with both geared and gearless traction machines as applicable, for installations of all rises and car speeds. Its extremely smooth ride makes it one of the three systems of choice in high-quality construction. The other two systems are the traditional Ward–Leonard variable-voltage DC system and the VVVF AC system.

(b) Variable-Voltage DC Motor Control

Before the development of electronic motor control, the only practical way of obtaining the precise motor speed control necessary for smooth, stepless elevator operation was to provide a variable DC voltage to a DC traction motor. This variable DC voltage was obtained from an auxiliary m-g set comprising an AC motor and a DC generator. This arrangement (see Figure 33.18*b*) is known as a Ward–Leonard system or as a unit multivoltage (UMV) drive. Although this arrangement has disadvantages, it is a classic high-quality elevator drive system and is found in the vast majority of the quality geared and gearless installations built before 1990. The disadvantages of this system—low overall efficiency, expensive machines, high thermal losses in machine rooms, high maintenance costs for three

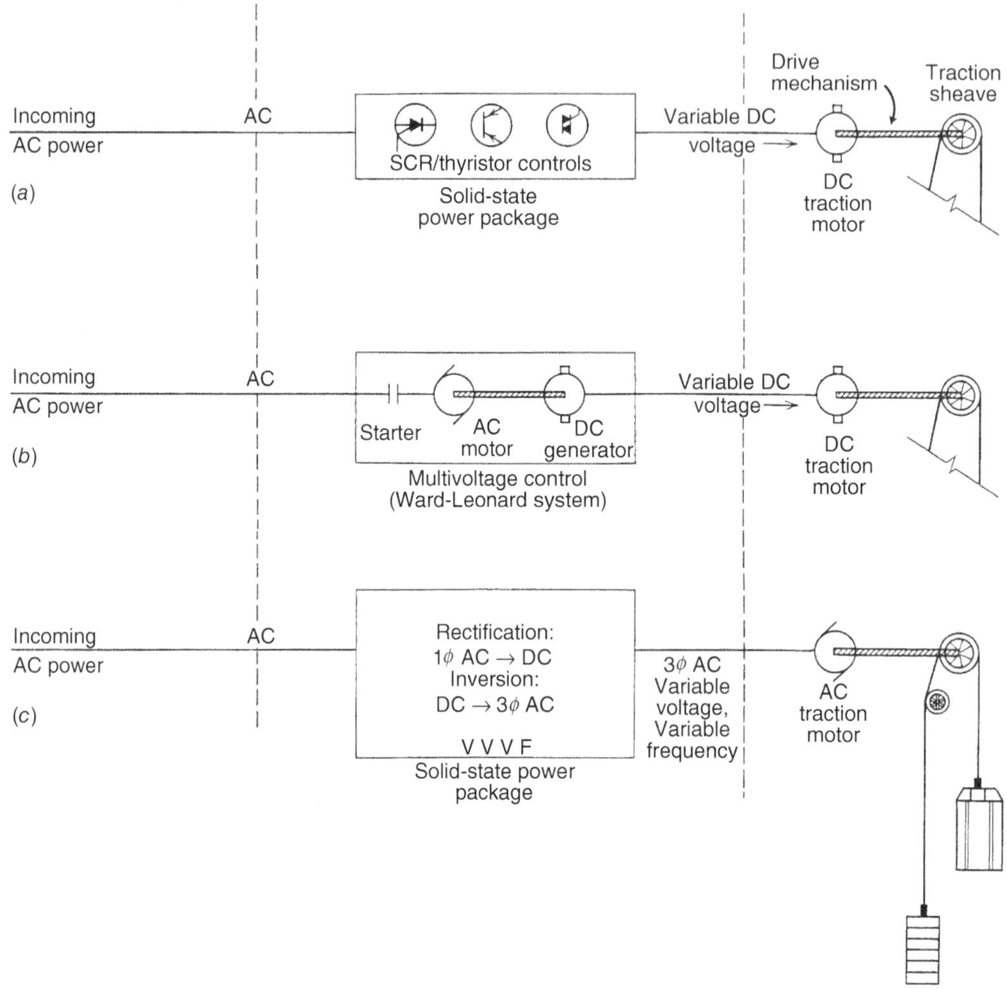

Figure 33.18 *Arrangement of electric speed-control equipment for elevator drives. (a) Solid-state, silicon-controlled rectifier (SCR) thyristor control produces variable-voltage DC and finds application in good-quality new and retrofit medium- and high-speed installations. (b) The traditional Ward–Leonard (m-g) system, which produces finely controlled, variable-voltage DC. This is the system in use today in the vast majority of existing high-quality installations. (c) Variable-voltage, variable-frequency (VVVF) AC is energy-efficient, highly accurate, and applicable to all types of installations.*

rotating machines, and high noise levels—were, until the development of a practical VVVF system, accepted as the price of a system that operated faultlessly, with great accuracy at all speeds, and for long periods of time, given proper maintenance. As noted previously, it is today still one of the three principal drive control systems being used in new installations. Its primary advantage over VVVF control is that it is "forgiving" of a less than ideal building power supply.

In modernization work, buildings with existing Ward–Leonard systems are frequently converted to solid-state DC thyristor control. This change maintains the excellent ride characteristics of variable-voltage DC traction machine installations while improving system performance by replacing the auxiliary m-g set with a low-loss, low-maintenance, quiet, solid-state thyristor drive control.

(c) Variable-Voltage, Variable-Frequency AC Motor Control

This system (see Figure 33.18*c*), which is entirely solid-state, is (in the opinion of many elevator professionals) the drive system of choice for all high-quality new installations of any speed or rise. The system consists of a rectifier, which changes the incoming AC to DC, and an inverter, which creates variable-voltage, infinitely variable-frequency, three-phase AC from the (rectified) DC. This AC is then applied to a standard squirrel-cage AC motor, which operates essentially at the speed corresponding to the frequency of the input. By maintaining a constant voltage-to-frequency ratio input to the traction motor, it is possible to provide continuously variable, highly accurate speed control at extremely high efficiencies throughout the full speed

TABLE 33.3 Comparative Characteristics of Elevator Drive Systems

Type	Rise, ft [m]	Speed, fpm [m/s]	Control	Initial Cost	Operating Cost	Performance
Geared AC	250 [76]	150–350 [0.8–1.8]	Thyristor	Medium	Medium	Fair
	300 [91]	150–450 [0.8–2.3]	VVVF	High	Low	Excellent
Geared DC	175 [53]	50–450 [0.3–2.3]	UMV	High	High	Excellent
	250 [76]	50–450 [0.3–2.3]	Thyristor	Medium	Low	Very good
Gearless DC	Unlimited	400–1200 [2.0–6.1]	UMV	High	High	Excellent
			Thyristor	Medium	Low	Very good
Gearless AC	Unlimited	400–2000 [2.0–10.2]	VVVF	Medium	Low	Excellent

Note: Life of equipment is generally indefinite except for the worm and gear of a geared unit, which have a 30- to 40-year life.

range of the motor. This is precisely what is required for high-quality geared and gearless installations, without the energy losses associated with the auxiliary m-g set found in UMV control. The VVVF speed control system eliminates most of the disadvantages of the solid-state thyristor control system and has the following characteristics:

- overall system efficiency is high at all motor speeds
- traction motors are economical single-speed squirrel-cage AC
- motors are 10% to 15% smaller than those of other drive systems
- system power factor is close to unity
- line harmonics are lower than with thyristor controls
- speed control and leveling are equal to DC traction motor control
- equipment is 98% solid-state, thus requiring minimum maintenance
- electric energy use and peak loads are reduced, thus reducing elevator electric billings by up to 35%
- the system is applicable to all rises and speeds
- machine rooms are smaller, cooler, and quieter than those using Ward–Leonard UMV controls

Table 33.3 summarizes the preceding discussion with comparisons of the various drive systems in use today. The exact extent of improvement possible when changing to solid-state equipment in a retrofit job depends on the quality of both the old and new systems. It is greatest when a microcomputer group control system replaces a relay-logic terminal dispatch system.

33.17 OPERATING CONTROL

Assuming an elevator system is energized and at rest, registration of a call from a lobby station or an upper-floor corridor activates the system. The particular elevator car that answers the call is selected by the supervisory system. In a UMV system, the car's m-g set is started, whereas with solid-state motion control power is available immediately. Switching devices in the car control panel release the brake, energize the elevator motor, and accelerate the car to its rated speed. Reverse operations are initiated when decelerating and stopping (landing) the car. When the car stops, the brake holds the sheave and elevator stationary.

The motion of a single car is determined by the action of three principal items of equipment: the car controller, the motion controls, and the system supervisory equipment (Sections 33.20 to 33.24 discuss supervisory systems). The function of the car controller is to provide information on the car's exact location, car panel calls, and hall calls. This information is fed into the supervisory system and the motion control equipment, which in turn act to initiate all the procedures necessary to answer all calls via the individual car controller panels. The car controller panel also supplies the necessary signals to car and hall lanterns that indicate the car position and direction of travel.

33.18 SYSTEM CONTROL

An operating system provides for the automatic response of a group of cars to calls for service. An effective system must process information regarding hall calls and car calls, car travel directions, and car position (in relation to each other and in relation to the call requirements), plus the trends of traffic. The last is required to allow the system to anticipate demand rather than just react to it. Because traffic and calls are never static, a control system that can satisfy all these demands in a large elevator system is an extremely sophisticated one. On small systems, the operating control is much simpler, as described in later sections. Throughout the discussion that follows, solid-state systems are assumed.

(a) Single Automatic Pushbutton Control

This system is the simplest of the passenger-operated automatic control schemes. It handles only one call at a time, providing an uninterrupted trip for each call. A single corridor button at each level can register a call only when the car is not in motion. This system is used only in private residences and for light-use freight elevators.

(b) Collective Control

Cars stop at each floor registering a call irrespective of direction, hence the term collective. This leads to slow and annoying service. As a result, this system is no longer used in new installations in the United States, although it is common in other countries.

(c) Selective Collective Operation

This type of collective operation is "selective" in that it is arranged to collect all waiting "up" calls on the trip up and all "down" calls on the trip down. The control system stores all calls until they are answered and automatically reverses the direction of travel at the highest and lowest calls. When all calls have been cleared, the car will remain at the floor of its last stop awaiting the next call. Any hall button call will set the car in operation.

Selective collective control is standard in locations where service requirements are moderate, such as in apartment houses, small offices, and professional buildings. Because these locations often require more than one car, a group control scheme for up to three cars automatically assigns each hall call to the car best situated to answer it, prevents more than one car from answering a call, allows one car to be detached for freight duty, and automatically parks cars at the ground floor when they are not required.

Although selective collective control is common in residential and other buildings with light- to moderate-service requirements, its inherent tendency toward bunching of cars can result in long waiting times. This characteristic is particularly annoying with groups of three cars. Frequently, a passenger arrives at a landing to find that all three cars have just passed, going in the same direction. The result is that service is only slightly better than would be rendered by a single car, except that the load (handling) capacity is greater. For this reason, operation of more than two cars with this system is not recommended, and operation of more than three cars is not feasible.

(d) Computerized System Control

Prior to the advent of computer supervisory control, large banks of elevators were controlled by lobby-based human "starters" who attempted, with limited success, to recognize and anticipate traffic patterns and thus speed service. Due to the huge amount of information that had to be processed, service was frequently less than satisfactory—particularly during heavy traffic. This situation changed with the advent of computerized, microprocessor-controlled operating systems.

A satisfactory control system must continuously monitor demand and control each car's motion in response to demand only; that is, it must analyze all the possibilities and answer each call in optimum fashion. The definition of optimum depends, of course, on the system design strategy, and this varies among manufacturers. Such a system is possible only with the aid of a central computer combined with programmable microprocessor-controlled peripherals, because the amount of data that must be collected and instantaneously processed is enormous.

All manufacturers attempt to optimize the parameters by which system quality is measured—specifically, to minimize interval, hall waiting time, and average trip time. How this is accomplished depends on the relative weight assigned to each item of input data in an extremely complex computer algorithm. One manufacturer uses an algorithm that calculates a figure of merit for each car to answer each waiting hall call. The number arrived at represents the weighted sum of the projected passenger hall waiting time (interval and hall wait time) and traveling passenger delay (average trip time). The car with the best figure of merit (minimum time) is assigned to the landing call. Another manufacturer uses a somewhat simpler dynamic call allocation algorithm that relies on the high-speed calculation capabilities of a central computer to make a last-moment decision as to which car answers which call. Another algorithm, in addition to analyzing each car's capability to answer a call, calculates the effect of any decision on the overall elevator service quality for the building and uses this factor as well in the final decision.

Still another algorithm uses a car dispatch program based on preprogrammed and learned traffic patterns, modified by the history of the previous few minutes of operation. In the basic program, each car computes its own response time to a waiting hall call—considering its own position, velocity, and car calls—and compares it with that of all other cars in the group. The car that will provide the optimum hall-call service time (waiting time plus trip time) answers the hall call. To facilitate this type of car "bidding" for calls, some systems are arranged so that each car controller can act as a master group controller. At any one time, one car in the group acts as the master, but if it is taken out of service, another car controller becomes the master.

In addition, some programs use an artificial intelligence module to learn traffic data and history and continuously change the car-dispatching mode. These changes consist in part of sectoring or zoning the building so that cars are grouped to provide optimal service. This type of system is particularly useful in buildings with single-floor or multifloor tenant groups with repetitive traffic patterns. Most algorithms permit overriding the program mode in response to crowd sensors or to particularly heavy, concentrated service demand, such as might occur if the occupants of an entire floor left the building at an unusual hour. Internet protocol (IP) addressable "smart elevators" are coming to market with Internet of Things (IoT) network integration capabilities. This will allow large amounts of elevator and passenger data to be analyzed to track and optimize performance.

The actual program logic for even a small group supervisory traffic control system is beyond the scope of this book, but its guidelines are not. An adequate system should do the following:

- Be programmed initially for the anticipated service needs. These needs can be analyzed in existing buildings with computerized traffic analyzers (Figure 33.19).
- Be reprogrammable to meet changes in building needs at nominal cost and with minimum shutdown time. It must be possible to detach cars from the system during testing, reprogramming, and routine maintenance, and to provide minimal service even during off-hours.
- Provide for priority calls (based on landing waiting times), statistical analysis of traffic in order to anticipate patterns, adaptive (zoned) car parking to meet specific needs, adjustable door timing based on the type of call (lobby, landing, car), backup dispatch means (in case of dispatch system failure), and automatic call cutout for constant (stuck) signals.
- Provide means for viewing elevator traffic information from selected locations (lobby or management office) and for obtaining a hard copy of this information—both as stored and real-time data. The information-handling equipment must provide a fault mode that stores, displays, and diagnoses system operating faults. The owner/manager unit may have on-site control and reprogramming capabilities if so required.
- Provide additional functions such as emergency power elevator selection and control (see Section 33.45), priority service, selective hall/car call cutout, swing (separate) car operation, and hoistway access controls for maintenance.
- Provide, where specifically required by the building management, a riot-control feature (allowing access limits at entrance levels), crossover floor operation in a zoned system, convention service (intense short-time usage at selected levels), and controlled access at given floors and for specific occupants.
- Be fully coordinated with the fire protection system in accordance with the local fire regulations (see Section 33.46).
- Coordinate with elevator security equipment so that operation of security/alarm devices initiates automatic elevator motion control procedures. This too must be coordinated with the local security authorities, and the automatic procedures must be subject to manual override (see Section 33.47).

Proper operation of the system should also result in the following:

1. All floors getting equal service, including the basement, if required.

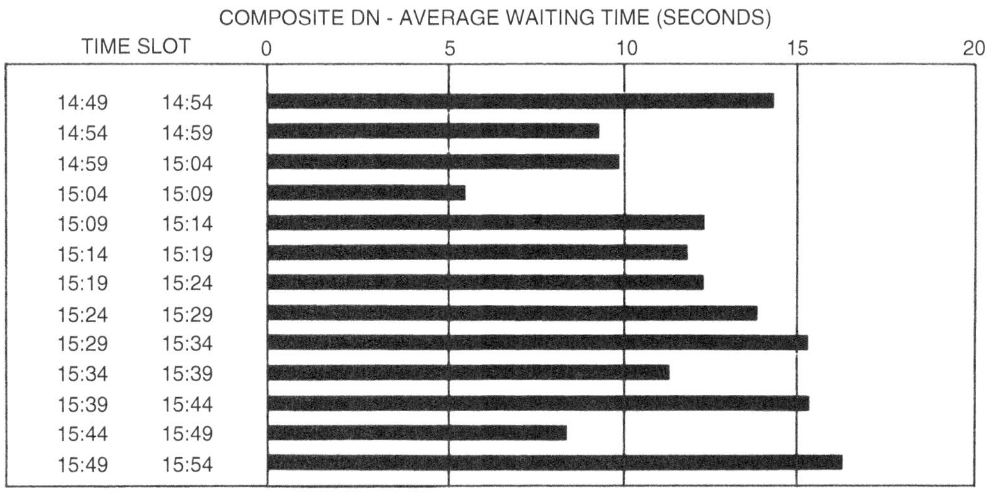

Figure 33.19 *Building elevator traffic studies can be performed quickly and accurately using a software-based traffic analyzer with multiple inputs. Accumulated data on call registration, response time, waiting time, maximum interval, cars in service, and so on, can be presented graphically. This report describes down (DN) traffic. (Courtesy of UNITEC Parts Co.)*

2. Proper handling of multi-floor tenants in office buildings to permit efficient interfloor traffic.
3. Appropriate action in emergencies, such as general or local power failure or any type of abnormal car-to-signal operation.

33.19 LOBBY ELEVATOR PANEL

The traditional lobby elevator control and information panel for each elevator bank, which was usually wall-mounted adjacent to the related elevators, has become one or more computer screens positioned at a lobby desk and/or in the building maintenance office. In addition, an information-only screen is frequently wall-mounted adjacent to the related elevators for the edification of waiting passengers.

The information displayed on the screen includes car locations, movement direction, waiting corridor calls, and any special-status data. The control functions available at the computer terminal permit intervention to establish special types of operation including:

- car movement without operating the usual audible and visual signals (inconspicuous riser)
- one or more cars removed from supervisory control and operated manually (attendant or independent service)
- cars selected for night or weekend service while the other cars are shut down
- car(s) assigned to a particular floor on a fixed- or priority-basis (convention feature or priority)

Among other control functions are those concerned with emergency service, including the "fireman's return" feature required by ANSI and many local fire codes (Section 33.46) and controls related to switching of power between cars in the event that operation on emergency generator power is necessary (Section 33.39). In addition, means of two-way communication with each car and other selected locations are provided at the control center.

33.20 CAR OPERATING PANEL

A typical car operating panel is illustrated in Figure 33.20. Every car panel (station) is equipped with full-access buttons for call registry, door-open, alarm, emergency stop, and firefighter control. An intercom device that permits communication with the building control office is also always provided. A door-close button is sometimes provided if extensive hand operation of the car is anticipated. It is activated only when the car is under manual control.

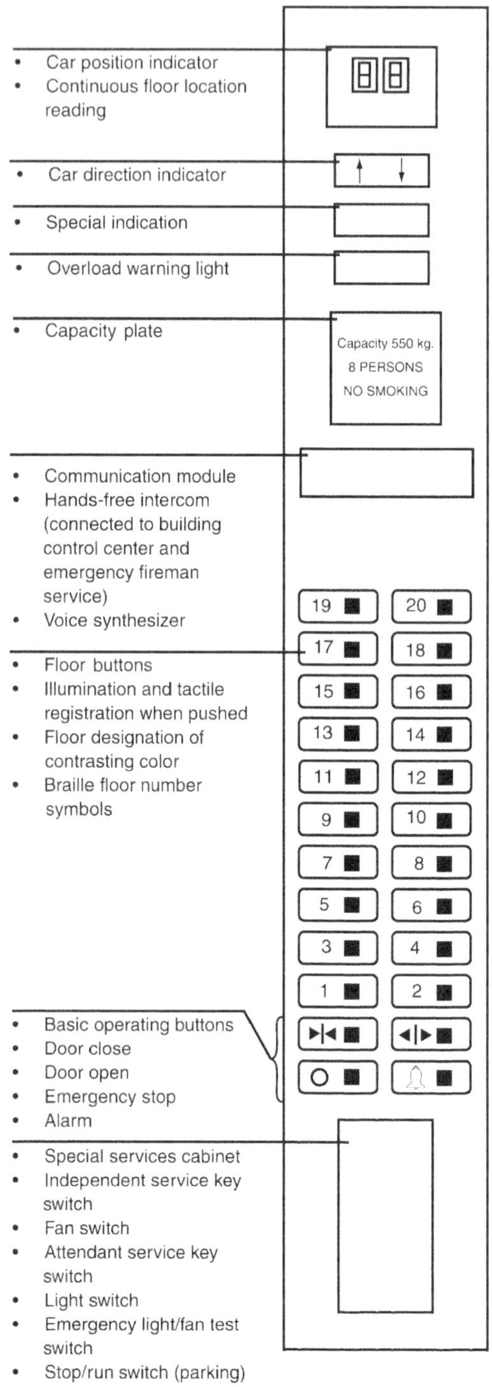

Figure 33.20 *Typical car operating panel. Designs of these panels vary widely, but the essential components are as shown.*

Controls that do not concern the normal passenger are grouped in a locked compartment in the car panel. These include a hand-operation switch; light, fan, and power switches; and any special controls, such as security and emergency devices. Finally, a compartment accessible only to technicians contains the devices controlling door motion, car signals, door and car position transducers, load-weighing control, door and platform detection beam equipment, speech synthesizer (if used), and visual displays.

PASSENGER ELEVATOR SELECTION

33.21 GENERAL CONSIDERATIONS

The selection of elevators for any but the simplest of buildings requires the simultaneous consideration of several factors: adequate elevator service for the intended building usage, economics, and the architectural integration of spaces assigned to elevators, including lobbies, shafts, and machine rooms. These factors should be captured in the owner's project requirements. System and equipment performance options are also numerous. In large buildings, many viable system options are possible because of the interdependence and interaction of building and system design variables. The selection of an optimum system for such buildings is only practical with the aid of a computer or simulator, and their use has become standard practice in the industry. Hand computation, following certain guidelines, can yield good results for small, straightforward buildings and reliable preliminary data for almost all buildings.

The key elevator-specific design criteria usually used to express elevator service quality are:

- interval
- average waiting time
- handling capacity
- travel time

The elevator system selection process involves matching these criteria with estimated performance indicators. Project-specific design intent (for example, acceptable service or excellent service) will inform the choice of numeric values for the criteria.

33.22 DEFINITIONS

Definitions of important terms, including variant usages, follow.

Average lobby time or average lobby waiting time The average time spent by a passenger between arriving in the lobby and leaving the lobby in a car. This is a key selection criterion.

Handling capacity (HC) The maximum number of passengers that a system can handle in a given time period—usually 5 minutes, thus the term 5-minute handling capacity. When expressed as a percentage of the building population, it is called percent handling capacity (PHC). This is a key selection criterion.

Interval (I) or lobby dispatch time The average time between departures of cars from the lobby.

Registration time Waiting time at an upper floor after a call is registered.

Round-trip time (RT) The average time required for a car to make a round trip—starting from the lower terminal and returning to it. The time includes a statistically determined number of upper-floor stops in one direction and, when calculating elevator requirements based on up-peak traffic, an express return trip.

Travel time or average trip time (AVTRP) The average time spent by passengers from the moment they arrive in the lobby to the moment they leave the car at an upper floor. This is a key selection criterion.

Zone A group of floors in a building that is considered as a unit with respect to elevator service. It may consist of a physical entity—a group of upper floors above and below which are blind shafts—or it may be a product of the elevator group control system, changing with system needs.

33.23 INTERVAL AND AVERAGE LOBBY WAITING TIME

In an ideal installation, at least from the user's point of view, a car would be waiting at the lower terminal on the rider's arrival or would be available after a short wait. Because cars leave the lobby separated in time by the interval (I) and passengers arrive at the lobby in random fashion, the average waiting time in the lobby should be half (50%) the interval. Field measurements show, however, that it is longer than this. The figure most often used in the industry is 60%—that is,

$$\text{average lobby waiting time} = 0.6 \times I$$

Table 33.4 lists intervals and suggested values for office buildings and the related average waiting time based on the foregoing relationship. Because some control systems zone the building in such a way that some cars do not return to the lobby, the interval as a figure of merit may be somewhat misleading in such buildings. The table also lists recommended intervals for other types of buildings.

With intervals in the recommended range, riders are not conscious of any irksome delay in elevator service. Consciousness of delay is considered a major drawback in rental desirability and should be avoided for all traffic conditions except morning and evening peaks, when a certain delay is expected and therefore

TRANSPORTATION

TABLE 33.4 Recommended Elevator Intervals and Related[a] Lobby Waiting Time

Facility Type	Interval (seconds)	Waiting Time[a] (seconds)
Office Buildings		
Excellent service	15–24	9–14
Good service	25–29	15–17
Fair service	30–39	18–23
Poor service	40–49	24–29
Unacceptable service	50+	30+
Residential		
Prestige apartments	50–70	30–42
Middle-income apartments	60–80	36–48
Low-income apartments	80–120	48–72
Dormitories	60–80	36–48
Hotels—first quality	30–50	18–30
Hotels—second quality	50–70	30–42

[a]Based on the relationship: waiting time = 0.6 × interval.

tolerated, however grudgingly. Even in peak periods, a modern group supervisory system will recognize a timed-out call—that is, a call with a registration time exceeding 50 seconds—as a priority call. Priority calls are answered by the first available car, usually within 15 seconds. If a considerable amount of interfloor traffic is expected during peak periods, as may be the case when a large company occupies several floors of a building, elevator capacity should be increased by 20% to 40% over the capacity otherwise calculated, to maintain proper intervals.

33.24 HANDLING CAPACITY

The frequency, or interval, with which a car appears at the main building lobby is one of the two factors that determine the passenger capacity of an elevator system. The other is the size of the elevator car. The system's handling capacity (HC) is completely determined by these two factors—car size and interval—and is independent of the number of cars. This can be best understood by visualizing an elevator system as a single set of doors that opens periodically (the interval) to remove a given number of passengers (the car capacity) from a patiently waiting group of would-be passengers. Whether the set of doors represents a single car or many cars that take turns is immaterial. The only factors that determine the handling capacity are passenger load (car capacity) and frequency of loading (interval) (Table 33.5).

During peak traffic periods, cars are not loaded to maximum capacity but typically only to about 80%—a value determined by actual count in many existing installations.

As a convenient measure of capacity, the handling capacity of a system for 5 minutes is taken as a standard. This is because a 5-minute rush period is historically used as a measure of a system's ability to handle traffic. This may be expressed as:

$$\text{handling capacity } (HC) = \frac{\text{passengers}}{\text{car}} \times \frac{\text{cars}}{\text{sec}} \times 5\text{ min} \times \frac{60\text{ sec}}{\text{min}}$$

Because the number of cars per second is the reciprocal of the interval (e.g., 30 seconds between cars is the same as 1/30th of a car per second), this equation reduces to:

$$HC = \frac{\text{passengers/car}}{\text{interval} \times 300}$$

or

$$HC = \frac{300p}{I}$$

where p is car loading (number of passengers/car). When the interval is 30 seconds, the system's handling capacity is $10p$, a convenient value to remember.

To establish criteria for building service, the system HC must be related to building size. This is normally done by establishing the minimum percentage of the building population that the system must handle in 5 minutes, called the PHC. A good system for a diversified office building will handle no less than 12% of the building population. Similar values are shown in Table 33.6 for various types of facilities.

TABLE 33.5 Car Passenger Capacity (p)

Elevator Capacity, lb [kg]	Maximum Passenger Capacity	Normal Passenger[a] Load per Trip
2000 [907]	12	10
2500 [1134]	17	13
3000 [1361]	20	16
3500 [1588]	23	19
4000 [1814]	28	22

[a]The number of passengers carried on a trip during peak conditions is approximately 80% of the car capacity.

TABLE 33.6 Minimum Percent Handling Capacity (PHC)

Facility	Percent of Population to Be Carried in 5 Minutes
Office Buildings	
City center	12–14
Investment	11.5–13
Single-purpose	14–16
Residential	
Prestige	5–7
Other	6–8[a]
Dormitories	10–11
Hotels—first quality	12–15
Hotels—second quality	10–12

[a]Due to more urgent traffic demands, particularly at the school and work exodus.

TABLE 33.7 Population of Typical Buildings for Estimating Elevator and Escalator Requirements

Building Type	Net Area
Office Buildings	**ft² per person [m²]**
Diversified (multiple tenancy)	
Normal	110–130 [10–12][a]
Prestige	150–250 [14–23]
Single tenancy	
Normal	90–110 [8–10]
Prestige	130–200 [12–19]
Hotels	Persons per sleeping room
Normal use	1.3
Conventions	1.9
Hospitals	Visitors and staff per bed[b]
General private	3
General public (large wards)	3–4
Apartment houses	Persons per bedroom
High-rental housing	1.5
Moderate-rental housing	2.0
Low-cost housing	2.5–3.0

[a]Density may vary for different floors. Clerical and stenographic areas may have a population density as high as 70 ft² [6.5 m²] per person.

[b]If visiting hours are restricted, the visitor population will determine elevator requirements. If visiting is not restricted to certain hours, staff requirements may determine elevator design. Where traffic is heavy, a combination of passenger cars and larger "hospital" cars should be used to provide optimum service.

TABLE 33.8 Office Building Efficiency

Building Height	Net Usable Area as Percentage of Gross Area	
0–10 floors	Approximately 80%	
0–20 floors	Floors	1–10 approximately 75%
		11–20 approximately 80%
0–30 floors	Floors	1–10 approximately 70%
		11–20 approximately 75%
		21–30 approximately 80%
0–40 floors	Floors	1–10 approximately 70%
		11–20 approximately 75%
		21–30 approximately 80%
		31–40 approximately 85%

Source: From Strakosch, 1983.

Note: Applicable to buildings with 15,000–20,000 gross ft² [1400–1850 m²] per floor.

Building population must be estimated to establish elevator requirements. This is particularly difficult in speculative, diversified-use buildings. A fair estimate can be made, however, based on rental cost, area, and building type. Population estimates for office buildings are based upon net area—that is, actual available area for tenancy. Table 33.7 gives suggested density figures, and Table 33.8 gives average office building efficiency values for use in calculating net area.

33.25 TRAVEL TIME OR AVERAGE TRIP TIME

The average trip time (or time to destination) is the sum of the lobby waiting time plus travel time to a median floor stop. Car round-trip time is also used as a performance criterion, but it is not as meaningful as trip time. In a commercial building, a trip of less than 1 minute is highly desirable, a 75-second trip is acceptable, a 90-second trip is annoying, and a 120-second trip is the limit of toleration. In the more relaxed atmosphere of a residence, where interval alone can account for a minute or more of trip time, these maxima can be revised upward. Figures 33.21 and 33.22 show typical average trip times for a range of car sizes and speeds—assuming a 9.5 ft [2.9 m] and 12 ft [3.7 m] floor height, respectively.

An important reservation on the foregoing statements must be noted. The curves presented in Figures 33.21 and 33.22 are based upon statistical calculations, empirical data, and field observations, as discussed in the next section. Thus, the estimates that these curves provide should be considered ±15% accurate, and borderline cases can be shifted either way. Designs that show high travel time on paper frequently work out well in the field, because lobby loading is often less than 80%, upper-floor stops take less than the statistically predicted time due to groups of people going to the same floor, and staggered working hours relieve traffic peaks. Also, a feature called high-call reversal takes account of the fact that cars do not travel to the top of the shaft on each trip, but reverse at the topmost call. This can reduce the average trip time by 5% to 10%. Finally, sophisticated solid-state traffic controls allow for rapid acceleration and deceleration without discomfort, variable door-closing time, and very efficient selection of landing call responses, all of which can further reduce the trip time by another 5%.

33.26 ROUND-TRIP TIME

Round-trip time during up-peak traffic conditions, used for calculating elevator requirements, is the sum of four factors: (1) time to accelerate and decelerate, (2) time to open and close doors at all stops, (3) time to load and unload, and (4) running time. Physically, round-trip time is the time from door opening at the lower terminal to door opening at the same terminal at the end of a round trip. Because the actual number of stops to be made by a car is unknown, a statistical probability value is used, based upon the passenger capacity of the car and the number of local floors above the lower terminal. In calculating round-trip time (*RT*), it is assumed that a car will depart the lower terminal when loaded. No intentional delay is included at either the lower or upper terminal. The *RT* thus calculated is a median figure, with any single actual round trip taking more or less time. In detail, *RT* consists of the time expended in

1. loading at the lobby
2. door closing at the lobby

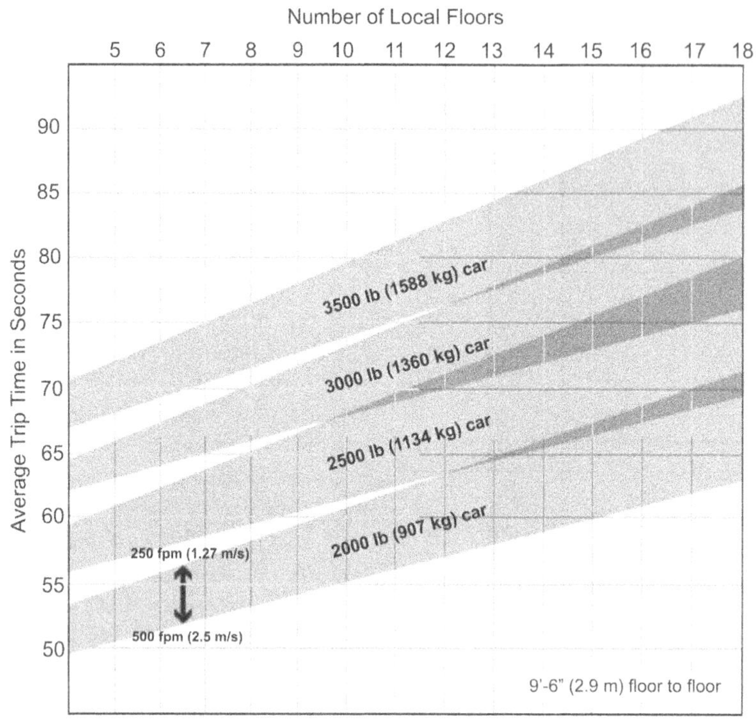

Figure 33.21 *Estimates of average trip time (AVTRP) for various car speeds and capacities with a 9-ft, 6-in. [2.9-m] floor height and a 30-second interval. (Tom Collins from information in Grondzik and Kwok, 2014.)*

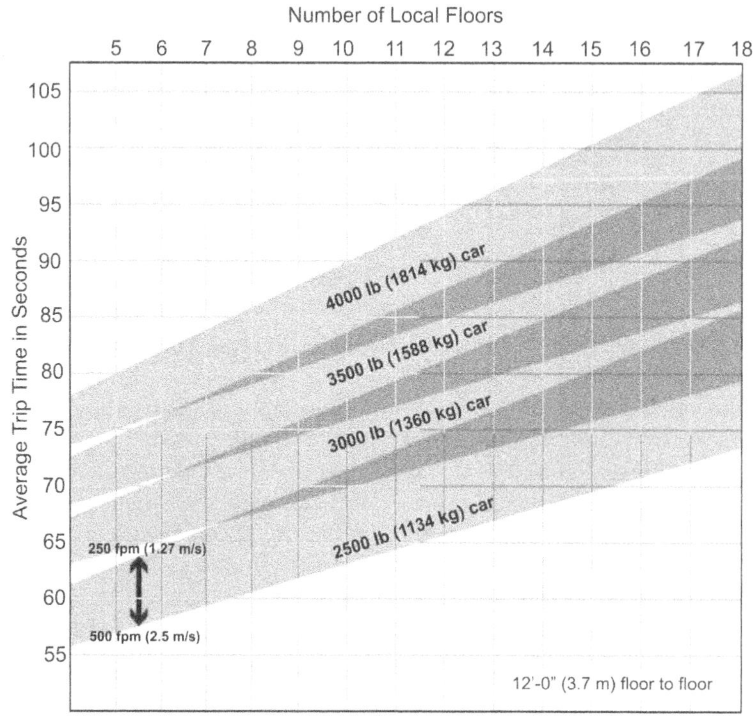

Figure 33.22 *Estimates of average trip time (AVTRP) for various car speeds and capacities for a 12-ft [3.7-m] floor height and a 30-second interval. (Tom Collins from information in Grondzik and Kwok, 2014.)*

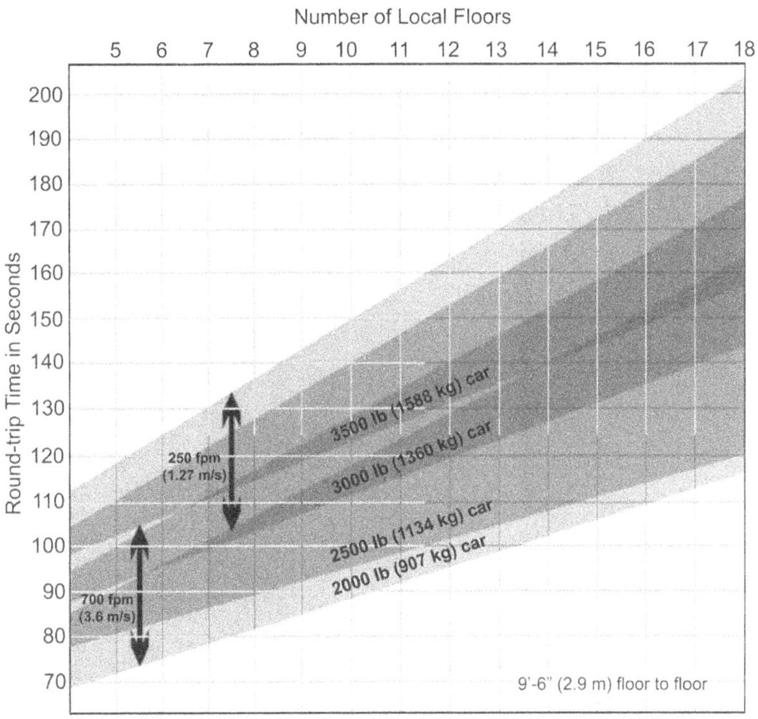

Figure 33.23 *Plots of round-trip time (RT) for various car speeds and capacities with a 9-ft, 6-in. [2.9-m] floor height and a 30-second interval. (Tom Collins from information in Grondzik and Kwok, 2014.)*

3. accelerating from the lobby and from each stop
4. decelerating at each stop
5. passenger transfer at each stop
6. door operations at each stop
7. running time at rated speed between stops
8. return express run from the last stop to the lobby

Figures 33.23 and 33.24 show anticipated round-trip times for a range of car sizes and speeds—assuming a 9.5 ft [2.9 m] and 12 ft [3.7 m] floor height, respectively. Figure 33.25 shows one-way express time estimates for high- and low-speed car options.

These values are obtained as follows:

1. *Field observations:* Items 1 and 5 are based upon a 3-ft, 6-in. [1.07-m] door opening. A smaller door opening increases passenger transfer time.
2. *Calculations:* Items 2, 3, 4, 6, 7, and 8. Door-closing time is based on a 3-ft, 6-in. [1.07-m] center-opening door with adjustable speed.

Acceleration and deceleration times are calculated using a maximum value of 4 ft/s/s [1.2 m/s/s] because anything beyond that results in physical discomfort for the passengers. Running time at rated speed takes place after the car has accelerated and before it begins to decelerate. Considering that it takes between 20 and 30 ft [6–9 m] to accelerate to 700 fpm [3.6 m/s], depending upon the rate of acceleration, in local runs a car never gets to the rated speed. It simply accelerates and decelerates. Higher-speed equipment with a larger motor accelerates more quickly and gives some time advantage on the return express run, but it has no great time advantage overall. In determining *RT* for cars serving upper zones, it is necessary to know the time required to traverse the express floors. This involves a one-way express run.

33.27 SYSTEM RELATIONSHIPS

The symbols used to describe elevator performance are:

p	Individual car capacity, equal to 80% of the maximum during peak hours
h	5-minute capacity of a single car
N	Number of cars in a system
HC	System 5-minute handling capacity, expressed in number of persons
RT	Round-trip time, in seconds
AVTRP	Average trip time, in seconds
I	Interval, in seconds
D	Population density, in ft^2 [m^2] per person
PHC	Percent of the population to be moved in 5 minutes, and expressed as a percentage

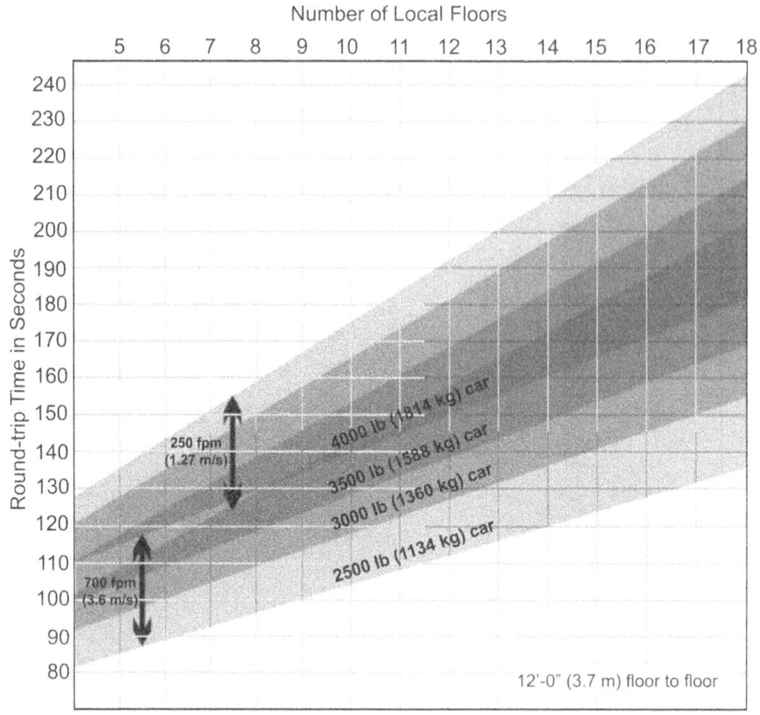

Figure 33.24 *Plots of round-trip time (RT) for various car speeds and capacities with a 12-ft floor [3.7-m] height and a 30-second interval. (Tom Collins from information in Grondzik and Kwok, 2014.)*

Considering the definitions of interval, handling capacity, average trip time, and round-trip time presented above, the interrelationships among these quantities can be explored, along with other factors governing the performance of elevator systems.

HC is determined by car capacity p and interval I:

$$HC = \frac{300\,p}{I} \quad \text{(Eq. 33.1)}$$

In a system consisting of a single car, the interval (I) is equal to the round-trip time (RT). In a system with more than one car, the interval is reduced in proportion to the number of cars. Thus,

$$I = \frac{RT}{N} \quad \text{(Eq. 33.2)}$$

The 5-minute handling capacity of a single car (h) is then:

$$h = \frac{300\,p}{RT} \quad \text{(Eq. 33.3)}$$

remembering that for a single car, its interval is its round-trip time. It follows that if the handling capacity of a single car is h, then the handling capacity (HC) of N cars is N times as much. Thus,

$$HC = N \times h$$

or

$$N = \frac{HC}{h} \quad \text{(Eq. 33.4)}$$

33.28 CAR SPEED

The selection of an appropriate car speed is a matter of trial and error, with the final selection being the speed that will give an RT that in turn provides an acceptable interval. To establish a starting point, it has been found that a minimum car speed corresponding to a given building height—or, in elevator terms, rise—can be established. Similarly, although car size can be selected at any value, it has been shown that for certain facility types, specific car sizes are indicated. These recommendations are given in Table 33.9.

Bear in mind that elevator equipment falls into distinct speed categories. Thus, most manufacturers use geared equipment through 400 fpm [2 m/s] and gearless equipment thereafter. The next category is 500 fpm [2.5 m/s] gearless, followed by 600 fpm [3.0 m/s] to 700 fpm [3.6 m/s], and so on. It is wise to avoid moving into the next higher—and more expensive—equipment category if possible. This may mean exceeding the recommended interval or dropping slightly below desired handling capacity. It will be found, however, that this can usually be done without injury to the elevator system performance, provided that a high-quality group supervisory control system is employed. The record for elevator speed is currently 4000 fpm [20 m/s] with manufacturers

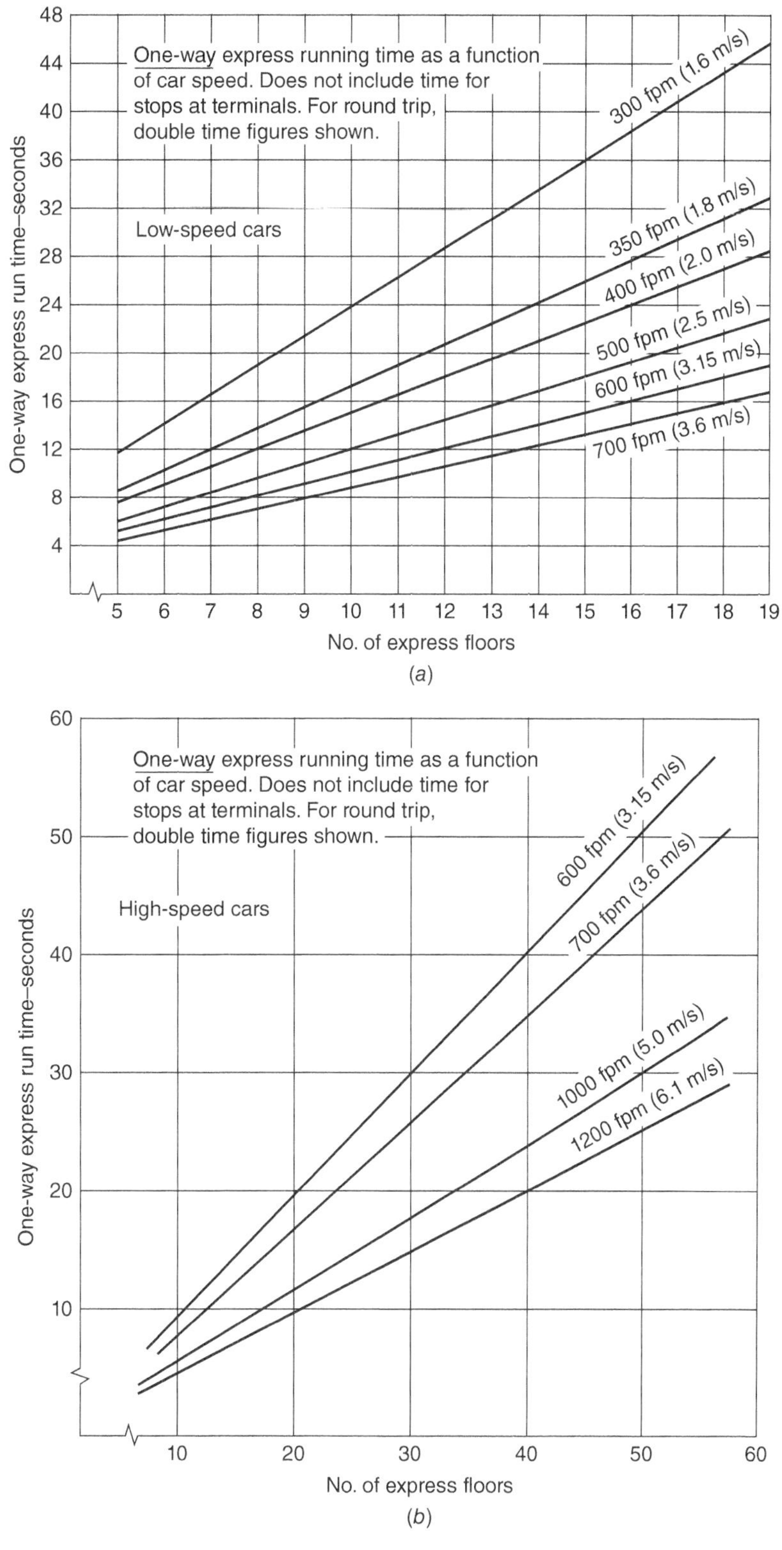

Figure 33.25 *One-way express running time, not including terminal time; (a) low-speed cars, (b) high-speed cars.*

TABLE 33.9 Elevator Equipment Recommendations

	Car Capacity[a]		Rise		Minimum[a] Car Speed	
Building Type	**lb**	**kg**	**ft**	**m**	**fpm**	**m/s**
Office building	2500	1134	0–125	0–40	350–400	2.0
	3000	1361	126–225	41–70	500–600	2.5
	3500	1588	226–275	71–85	700	3.15
			276–375	86–115	800	4.0
			>375	>115	1000	5.0
Hotel	2500	1134	As above		As above	
	3000	1361				
Hospital	3500	1588	0–60	0–20	150	0.63
	4000	1814	61–100	21–30	200–250	1.0
			101–125	31–40	250–300	1.6
			126–175	41–55	350–400	2.0
			176–250	56–75	500–600	3.15
			>250	>75	700	4.0
Apartment	2000	907	0–75	0–25	100	0.63
	2500	1134	76–125	26–40	200	1.0
			126–200	41–60	250–300	1.6
			>200	>60	350–400	2.0
Store	3500	1588	0–100	0–30	200	1.0
	4000	1814	101–150	31–45	250–300	1.6
	5000	2268	151–200	46–60	350–400	2.0
			>200	>60	500	2.5

[a]Car capacity is determined by building size; car speed is determined by rise.

exploring higher speed systems for the world's very tallest buildings. Physically achievable elevator speeds must, however, be balanced with passenger comfort due to increased pressure on the human body.

33.29 SINGLE-ZONE SYSTEMS

Having established the relationships that govern the design and performance of an elevator system comprising a single zone, an illustrative example is useful.

EXAMPLE 33.1 Office building, downtown, diversified use, 14 rentable floors above the lobby, each 12,000 ft^2 [1115 m^2] net. Floor-to-floor height—12 ft [3.7 m]. Determine a workable elevator system arrangement.

SOLUTION

From Table 33.6, consider a minimum PHC of 13%.
From Table 33.4, consider a borderline good/excellent interval of 25 seconds.
From Table 33.7, consider an average (mid-normal range) population density of 120 ft^2 [11 m^2] per person.

Trial 1
Building population:

14 floors at 12,000 ft^2)/120 ft^2 per person = 1400 persons (I-P)

14 floors at 1115 m^2)/11 m^2 per person ≈ 1400 persons (SI)

Suggested minimum handling capacity:

$$PHC = 13\%$$

$$HC = 0.13 \times 1400 = 182 \text{ persons}$$

$$\text{rise} = 14 \text{ floors at } 12 \text{ ft } [3.7 \text{ m}] = 168 \text{ ft } [51 \text{ m}]$$

From Table 33.9, select a car size of 3500 lb [1588 kg] at a minimum car speed of 500 fpm [2.5 m/s].

Then, from Figures 33.24 and 33.22, respectively, one may ascertain:

$$RT = 155 \text{ seconds}$$

$$AVTRP = 82 \text{ seconds}$$

Single-car capacity: $h = 300p/RT$
From Table 33.5, p is determined as 19.

$$h = \frac{300(19)}{155} = 36.8 \text{ persons}$$

$$N = \frac{HC}{h} = \frac{182}{36.8} = 4.9, \text{ say, } 5 \text{ cars}$$

$$I = \frac{RT}{5} = \frac{155}{5} = 31 \text{ seconds}$$

$$\text{actual } PHC = \frac{5(13\%)}{4.9} = 13\%$$

These results are acceptable, but faster cars might reduce the interval. So, try 700 fpm [3.6 m/s].

Trial 2
Consider the same 3500-lb [1588-kg] car size but at a minimum car speed of 700 fpm [3.6 m/s].

$$RT = 151 \text{ seconds}$$

$$AVTRP = 81 \text{ seconds (by extrapolation)}$$

$$h = \frac{300(19)}{151} = 37.7 \text{ persons}$$

$$N = \frac{182}{37.7} = 4.8, \text{ say, 5 cars}$$

$$I = \frac{RT}{N} = \frac{151}{5} = 30 \text{ seconds}$$

$$\text{actual } PHC = \frac{5(13\%)}{4.8} = 13.5\%$$

This solution is only marginally better than the previous 500 fpm [2.5 m/s] solution, and the increased cost would not be justified. A trial using smaller cars with shorter *RT* is called for.

Trial 3
Consider a 3000-lb [1361-kg] car size at a minimum car speed of 500 fpm [2.5 m/s].

$$RT = 143 \text{ seconds}$$

$$AVTRP = 76 \text{ seconds}$$

$$h = \frac{300(16)}{143} = 33.6 \text{ persons}$$

$$N = \frac{HC}{h} = \frac{182}{33.6} = 5.4 \text{ cars}$$

Using five cars:

$$I = \frac{RT}{N} = \frac{143}{5} = 28.4 \text{ seconds}$$

$$\text{actual } PHC = \frac{5}{5.4} \times 0.13 = 12\%$$

Trial 4
Using six 3000-lb [1361-kg] cars with 500-fpm [2.5-m/s] minimum speeds:

$$I = \frac{RT}{N} = \frac{143}{6} = 23.8 \text{ seconds}$$

and

$$PHC = \frac{6}{5.4} \times 0.13 = 14.4\%$$

Both solutions are acceptable.
Tabulating the calculation results, we have:

Trial	Cars, lb [kg]	Speed, fpm [m/s]	*RT* (sec)	*AVTRP* (sec)	*I* (sec)	*PHC* (%)
1	5 @ 3500 [1588]	500 [2.5]	155	82	31	13
2	5 @ 3500 [1588]	700 [3.6]	151	81	30	13.5
3	5 @ 3000 [1361]	500 [2.5]	143	76	28.4	12
4	6 @ 3000 [1361]	500 [2.5]	143	76	23.8	14.4

Solutions from trials 1, 3, and 4 are acceptable. Solution 2 was discounted due to the high cost. Interestingly, solution 3, using smaller cars than solution 1, and therefore being more economical, gives better results except for HC. Although the best solution is number 4, which gives excellent interval and HC, the additional cost of a sixth car plus the revenue loss from the rentable area occupied by the sixth elevator shaft and the cost of additional maintenance weigh heavily against this option. A trial with five 3000-lb [1361-kg] cars at 700 fpm [3.6 m/s] is in order, with the knowledge that a considerable cost increase would result because 700-fpm [3.6-m/s] cars require gearless equipment, whereas 500-fpm [2.5-m/s] cars are available in either geared or gearless format, both giving excellent service.

Trial 5
Using five 3000-lb [1361-kg] cars with 700-fpm [3.6-m/s] minimum speeds:

$$RT = 140 \text{ seconds}$$

$$AVTRP = 73 \text{ seconds}$$

$$h = \frac{300(16)}{140} = 34.3 \text{ persons}$$

$$N = \frac{182}{34.3} = 5.3 \text{ cars, say, 5 cars}$$

$$I = \frac{RT}{N} = \frac{140}{5} = 28 \text{ seconds}$$

$$PHC = \frac{5}{5.3} \times 0.13 = 12.3\%$$

As expected, improvement over the performance at 500 fpm [2.5 m/s] is very slight: an interval of 28 seconds rather than 28.4 seconds (trial 3), and a handling capacity of 12.3% versus 12%. The large increase in first cost for gearless equipment would not be justified. ■

At this point, the final selection would be made based on cost. When considering cost, note that first cost is the governing factor only in a speculative venture. With an owner-operator building, the cost comparison should be on a life-cycle basis. Cost figures must reflect the impact of elevator space requirements on net rentable area in the building. Comparative cost estimates are given in Table 33.10.

As mentioned earlier, and shown in Figure 33.19, planners of new buildings today generally take advantage of elevator selection software provided by consultants, by manufacturers, or via in-house capabilities. The results from one such program are shown in Figure 33.26 (this particular analysis was prepared by Otis Elevator Co.). Note that projected system performance under a variety of operational and functional scenarios can be evaluated and compared.

33.30 MULTIZONE SYSTEMS

In general, buildings with fewer than 15 stories are elevatored with a single zone (i.e., all cars serve all

TABLE 33.10 Relative First Cost[a] Figures for Passenger Elevators of Various Speeds and Drive Systems

	Hydraulic, fpm [m/s]	Geared Traction, fpm [m/s]			Gearless Traction, fpm [m/s]			
Car Size. lb [kg]	100 [0.63]	200 [1.0]	350 [2.0]	500 [2.5]	500 [2.5]	700 [4.0]	1000 [5.0]	1200 [6.0]
2000 [907]	40	80	100	130	165	170	220	235
2500 [1134]	43	85	115	145	175	180	235	250
3000 [1361]	50	90	120	150	180	185	250	265
3500 [1588]	58	95	125	155	190	195	265	275
4000 [1814]	60	100	135	165	200	205	280	300
4500 [2041][b]	70	120	150	185	225	230	300	325
5000 [2268][b]	75	130	160	200	240	250	330	350

Note: See Table 33.9 for speed/rise recommendations.

[a]Costs are +/− 10%; based on standard fixtures, cabs, and entrances, and average rise for the speed indicated.

[b]Service elevator or hospital elevator.

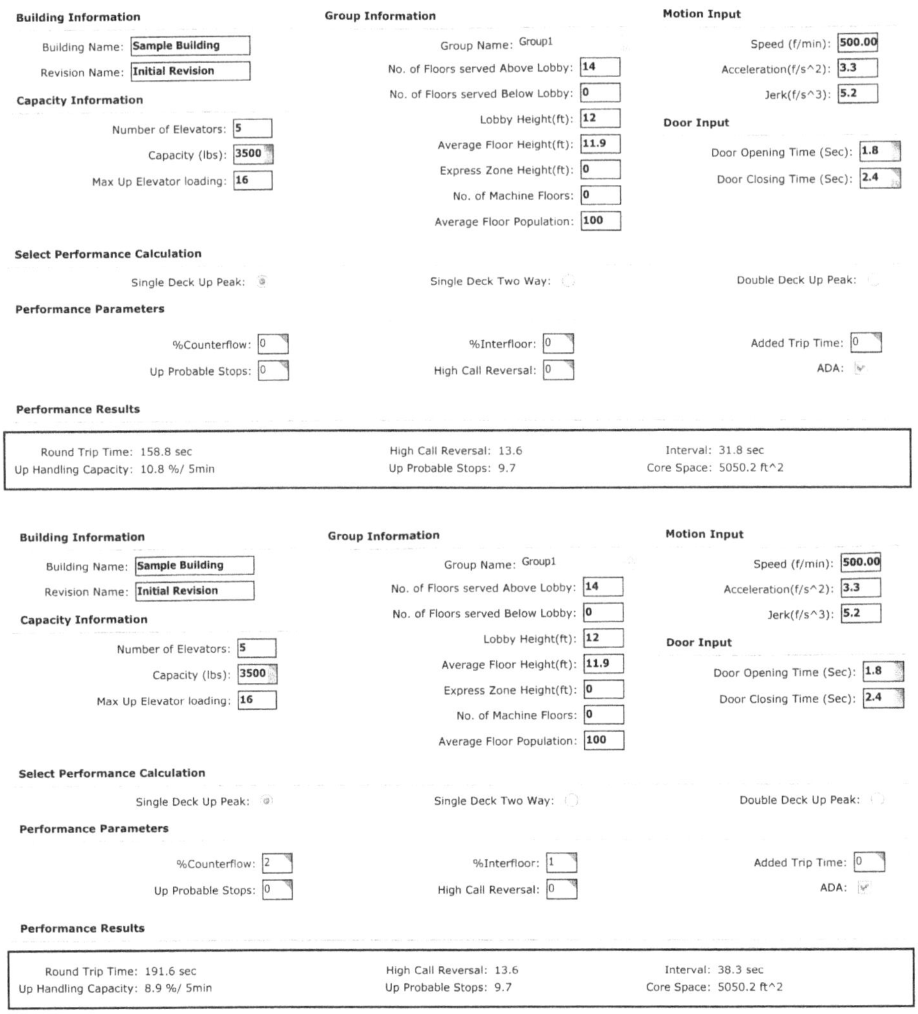

Figure 33.26 *Printouts from a computerized elevator selection program. This analysis shows that the addition of even light counterflow and interfloor traffic can seriously affect system performance, as can be seen from the round-trip, interval, and handling capacity figures. (Courtesy of Otis Elevator Co.)*

floors), and buildings with more than 20 stories are split into two or more zones. Buildings in between these limits—16–19 stories—can go either way, depending upon the population density and the required interval. A modern group supervisory system can automatically zone a building when traffic requires it. Such an arrangement, although efficient, is expensive in terms of both equipment and construction because it does not take advantage of the considerable savings provided by blind lower shaftways for upper-zone elevators.

Analysis of multizone systems is complex and is rarely done by hand. See Stein et al. (1986) for a detailed explanation of the technique involved. Most designers and consultants use one of many available computerized simulation and selection programs. These can, in addition to using basic criteria and building parameters, also consider the effect of variations in traffic control. Furthermore, a good program can evaluate the engineering and economic impacts of such factors as varying rental rates for different floors, rental space, machine room and hoistway space, core layout, and the structural ramifications of the elevator system.

33.31 ELEVATOR SELECTION FOR SPECIFIC OCCUPANCIES

(a) Office Buildings

Project-appropriate design criteria can be selected from Table 33.7. Supervisory group control is normally microprocessor-based. Approximately 1 service car per 10 passenger cars should be provided or, alternatively, 1 service car for every 300,000 ft^2 [27,870 m^2] of net area. Service cars should be 5000 lb [2268 kg] or larger without a dropped ceiling and, if also used for passenger service, equipped with wall pads. An oversized door (e.g., 4 ft, 0 in. [1.2 m] or 4 ft, 6 in. [1.4 m]) is particularly useful in handling furniture. Service elevators should have a shaftway door at every level plus easy access to the truck dock (or other freight entrance) as well as the lobby. These cars operate as service cars normally but can serve as passenger cars in peak periods to reduce congestion and delay. This flexibility is particularly useful in marginally acceptable systems. See Table 33.11 for approximate building costs.

TABLE 33.11 Office Buildings: Cost of Elevator and Electrical Work (%)

	Number of Stories		
Item	**20**	**35**	**60**
Elevator work	10.9	11.9	12.2
Electrical work	13.3	12.6	12.2

(b) Apartment Buildings

Studies indicate that apartment building traffic depends not only upon the population but also on the location and type of tenant. Buildings with many children experience a school-hour peak; buildings in midtown with predominantly adult tenancy exhibit evening peaks due to homecoming workers and outgoing dinner traffic. Where two cars are required, the second car should function both as a service car and as a passenger car. The cars may be banked or separated, as desired. If a single car is used, it should be of service-elevator size.

Self-service collective control is the general choice, with provision for attendant control in prestigious buildings. With small cars and a short rise, a swing-type manual corridor door is acceptable; in larger installations, both the car and the corridor door should be the power-operated sliding type.

Service elevators must be large enough to handle bulky furniture and should be at least 4000 lb [1814 kg], with a 48-in. [1.2-m] door and a high ceiling. Hoistways must be isolated from sleeping rooms by lobbies or other acoustically non-sensitive spaces. Similarly, machine rooms must be isolated because the starting and stopping of motors and other machine room noises are a detriment to sleep. Security arrangements are discussed in Section 33.41.

(c) Hospitals

As mentioned in Table 33.7, the governing factor in the determination of elevator requirements may be either normal hospital traffic or visitor traffic, depending upon the arrangement of visiting times. Due to the large volume of vehicular traffic—such as stretcher carts, wheelchairs, beds, linen carts, and laundry trucks—hospital elevator cars are much deeper than a normal passenger car. A hospital car, when used for passenger service, holds more than 20 persons and therefore gives slow service. For this reason, it is occasionally advisable to utilize some normal passenger cars in addition to hospital-size cars, particularly in large hospitals.

Food service carts can impose a considerable load upon the elevator system before, during, and after meals, and passenger service is seriously disrupted. To reduce this congestion and delay, many architects and hospital administrators prefer the use of dumbwaiter cars or another of the many types of materials-handling systems. These systems can also be used for transporting pharmaceuticals and other items and are discussed in Chapter 34.

Elevators should be grouped centrally, although separated by type of use. Car control is normally self-service collective. The population of a hospital may be estimated from Table 33.7. Experience has shown that a carrying capacity of 45 passengers in a 5-minute period is adequate (estimating each vehicle as equivalent to 9 passengers). Intervals should not exceed 1 minute. All recommendations regarding service for the disabled should be adopted (see Section 33.14).

(d) Retail Stores

Retail stores present a unique problem in vertical transportation because the objective is partially to transport users to specific levels and partially to expose the passengers (customers) to displayed merchandise. For this reason, stores rely heavily on escalators, with one or two elevators provided for use by staff, shoppers with strollers, and those with disabilities. When, for some reason, it is desired to equip a store exclusively with elevators, use the recommendations shown in Table 33.9, calculated for a load of 10% to 20% of the store's population. Control should be automatic, selective collective. Cars are arranged in a straight line to facilitate loading and waiting.

33.32 REHABILITATION WORK

The primary reason for rehabilitating an existing elevator system is to improve its operating performance. Traction machinery has extremely long life, particularly in the gearless configuration. Thus, the usual rehabilitation project consists of replacing the m-g set, car control panel, and group controller with solid-state, microprocessor-controlled programmable equipment, while retaining the original traction equipment. The car door operator, which controls door action, is usually also replaced because door opening and closing characteristics are an important factor in overall trip time and therefore in system performance.

Major manufacturers have developed computer-based elevator system simulators. Such programs enable architects and owners to input design (or as-is) building data and receive graphic and text output on the performance of proposed systems. Because such programs are interactive, the user can change the input data and the characteristics of the proposed equipment until the desired performance is reached. These programs are particularly useful in modernization work because the owner sees in advance the operation of a proposed system in his/her building and can make appropriate decisions based upon good information.

SPATIAL REQUIREMENTS FOR ELEVATORS

33.33 SHAFTS AND LOBBIES

Elevator lobbies and shafts are a major architectural space issue. The elevator lobby on each floor is the focal point from which corridors radiate for access to all rooms, stairways, service spaces, and so forth. These lobbies must be located above each other. The ground-floor elevator lobby (also called the lower terminal) must be conveniently located with respect to the main building entrances. Equipment within or adjacent to this area should include a building directory, elevator indicators, and possibly a control desk.

Lobbies should provide adequate area for the peak-load gathering of passengers to ensure rapid and comfortable service to all. The number of people contributing to the period of peak load (15- to 20-minute peak) determines the required lobby area.

Not less than 5 ft^2 [0.5 m^2] of floor space per person should be provided at peak periods for waiting passengers at a given elevator or bank of elevators. The hallways approaching and adjacent to the lobby should also provide at least 5 ft^2 [0.5 m^2] per person. Under self-adjusting relaxed conditions, density is about 7 ft^2 [0.65 m^2] per person. Crowding occurs during peak periods, however, reducing this to 3–4 ft^2 [0.3–0.4 m^2] per person. An acceptable compromise is 5 ft^2 [0.5 m^2] per person.

The main lower terminal of elevator banks is generally on the street-floor level, although it may be on a mezzanine level when the elevations of the street entrances vary so that one side of the building is at mezzanine level, whereas another entrance is lower. Such a situation is ideal for the use of escalators, which can economically and rapidly carry large numbers of people between levels, thus making practical and efficient a single main lower elevator terminal. The upper terminal is usually the top floor of the building. Typical dimensional data for hoistways and lobby arrangements are shown in Figures 33.27, 33.28, and 33.29.

33.34 DIMENSIONS AND WEIGHTS

Most manufacturers and elevator consultants will, upon request, supply standard layouts for elevators—including dimensions, weights, and structural loads. Furthermore, to assist in preliminary design, major manufacturers have agreed upon and publish a set of Standard Elevator Layouts via their trade organization,

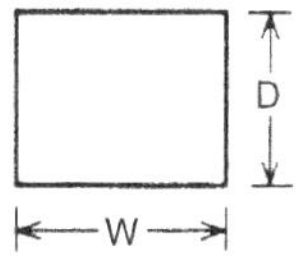

Car size	D	W
2500 lb	70"	8' 6"
3000 lb	7' 6"	8' 6"
3500 lb	8' 2"	8' 6"
4000 lb	8' 2"	9' 6"

(*a*)

Car size kg	speed m/s	D mm	W mm
1000	2.5	2200	2400
1250	2.5	2200	2600
1600	2.5	2500	2600
2000	2.5	2500	2600
1000	4.0	2200	2400
1250	4.0	2200	2600
1600	4.0	2500	2600
1250	6.0	2250	2500

(*b*)

Figure 33.27 *Rough hoistway dimensional data for use in schematic design. (a) I-P elevator sizes and dimensions. (b) SI elevator sizes and dimensions.*

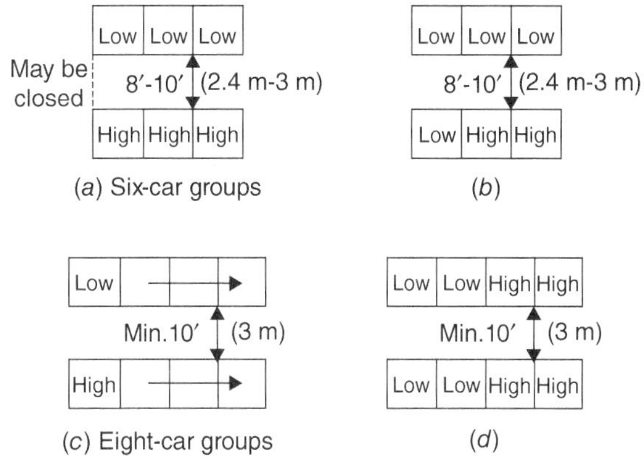

Figure 33.29 *Lobby groupings for multiple zone systems. Arrangement (a) is preferable to (b), and (c) to (d). Groups with more than four cars in a row are not used because end-to-end walking time would excessively lengthen landing stops and hence total travel time.*

the National Elevator Industry, Inc. (NEII). One such standard is reproduced in Figure 33.30 for 500- to 700-fpm [2.5–3.6-m/s] gearless units in the full range of car capacities. These standards are available from the NEII.

As may be seen from Figure 33.30*b*, providing for an elevator installation requires consideration of such factors as the depth of the pit, the dimensions of the hoistway, the clearance from the top of the hoistway to the floor of the penthouse, the size of the penthouse, and the loads that must be carried by the supporting beams.

The penthouse floor (and the secondary-level floor, where required) are located above the shaft of each elevator and need approximately 1 ½ stories of additional height above the top of the support beam of a given elevator when it is at its top-floor location. The actual floor area required by the elevator traction machine and its controls is roughly two times the area of the elevator shaft itself. The machine room contains the bulk of the elevator machinery. Because some of this equipment must be moved for maintenance, it is advisable to furnish an overhead trolley beam that can be used during installation as well. The maximum beam load is supplied by the elevator manufacturer.

Some typical machine room dimensional data are listed in Table 33.12, taken from actual installations. Because of multiple drive options and flexibility in equipment arrangements, no general conclusions can

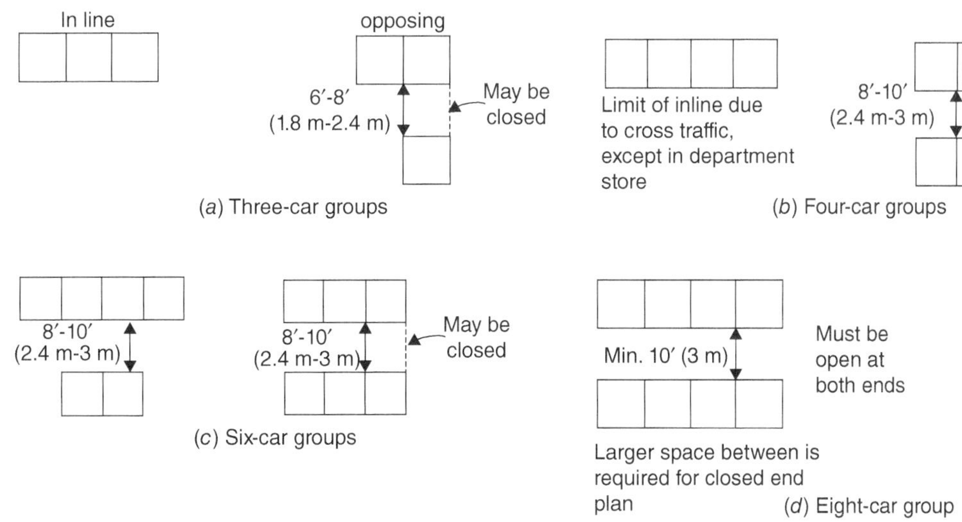

Figure 33.28 *Lobby groupings for single-zone systems: (a) three-, (b) four-, (c) six-, and (d) eight-car groups.*

TRANSPORTATION

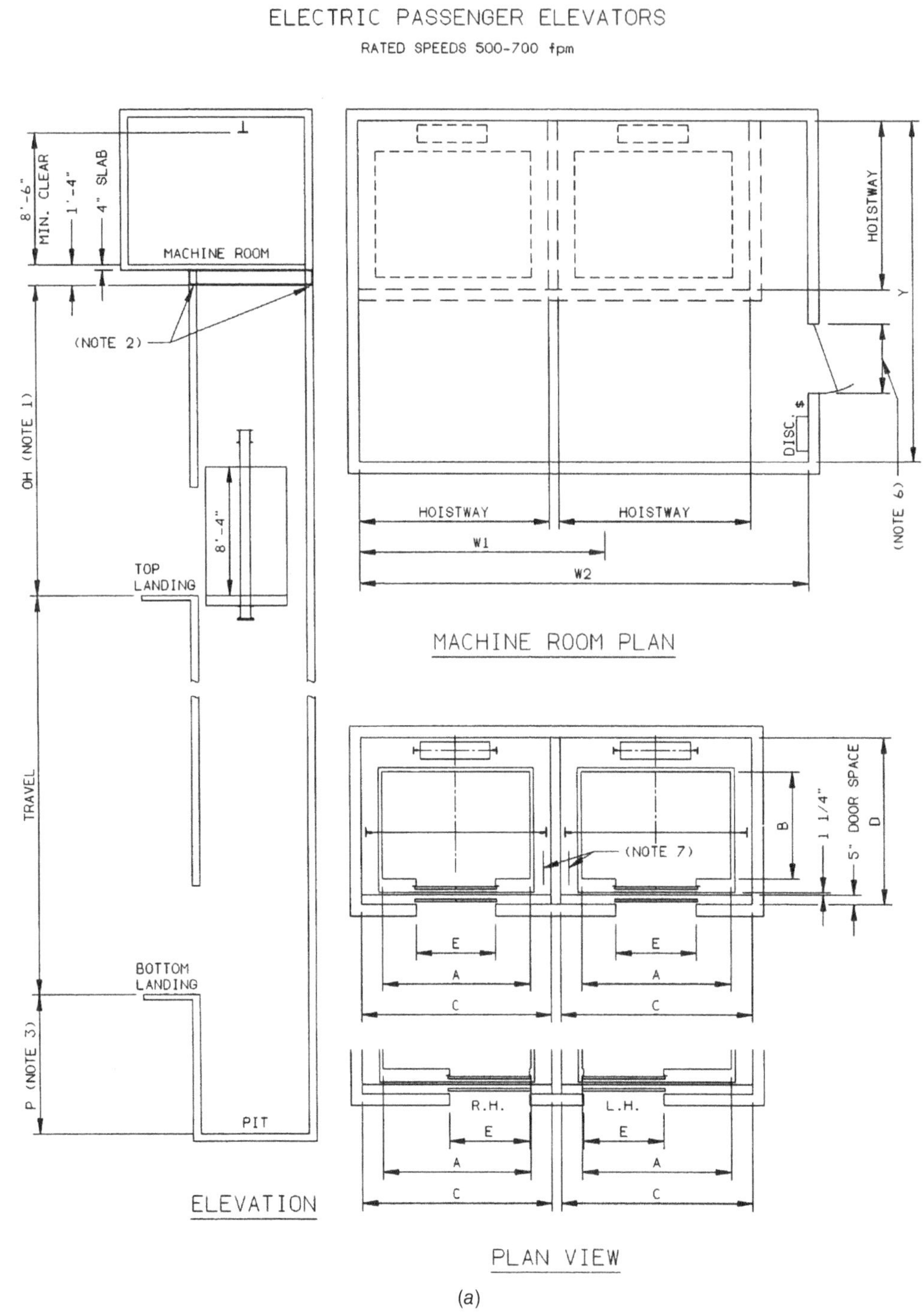

(a)

Figure 33.30 *a, b. Typical elevator installation dimensional data. (with permission of National Elevator Industry, Inc., 1992.)*

"OH" BOTTOM OF BEAM (Note 1) ft-in.					
SPEED (fpm)	RATED LOAD (lb)				
	2000	2500	3000	3500	4000
500	17-6	17-6	18-3	18-3	17-6
600	18-6	18-6	18-6	18-6	18-6
700	20-6	20-6	20-6	20-6	20-6

CAR & HOISTWAY								
RATED LOAD (lb)	AREA ■	ft-in. (Note 11)					ENTRANCE (Note 10)	
		A	B	C	D	E	TYPE	
2000	24.2	5-8	4-3	7-4	6-11	3-0	SSSO ▲	SSCO
2500	29.1	6-8	4-3	8-4	6-11	3-6	SSSO *	SSCO ▲
3000	33.7	6-8	4-7	8-4	7-5	3-6	SSSO *	SSCO ▲
3500	38.0	6-8	5-3	8-4	8-1	3-6	SSSO *	SSCO ▲
4000	42.2	7-8	5-3	9-6	8-1	4-0	SSCO *	

"P" PIT DEPTH (Note 3) ft-in.					
SPEED (fpm)	RATED LOAD (lb)				
	2000	2500	3000	3500	4000
500	10-1	10-1	10-1	10-1	10-1
600	11-5	11-5	11-5	11-5	11-5
700	11-5	11-5	11-5	11-5	11-5

MACHINE ROOM ft-in.					
RATED LOAD (lb)	W1 WIDTH	W2 WIDTH	"Y" DEPTH FOR RATED SPEED (fpm)		
			500	600	700
2000	7-4	15-0	18-6	18-6	18-6
2500	8-4	17-0	18-6	18-6	18-6
3000	8-4	17-1	18-6	18-6	18-6
3500	8-4	17-1	18-6	18-6	18-6
4000	9-6	19-5	18-6	18-6	18-6

Notes

1. "OH" Dimensions are based on a 8'-4" overall car height.
2. Supports for elevator machine beams not by elevator supplier.
3. For travel greater than 400'-0" increased pit depth may be required. Consult elevator supplier.
4. 3'-6" x 7'-0" Recommended.
5. Pit ladder not by elevator supplier.
6. Dividing beams, not by elevator supplier, to be designed to sustain rail forces. Consult elevator supplier.
7. When compliance with seismic risk zone 2 or greater requirements is anticipated, provide additional hoistway space.

Notes

- ■ Maximum allowable inside car area in ft^2 per ASME A17.1, Rule 207.1.
- * These car dimensions and entrance types provide wheelchair accessibility and accommodate an ambulance type stretcher (76 in. x 24 in.) in the horizontal position.
- ▲ These car dimensions and entrance types provide wheelchair accessibility.

(*b*)

Figure 33.30 *(continued)*

TABLE 33.12 Typical Elevator Machine Room Dimensions

Bank Description Number, lb [kg]	fpm [m/s]	Approximate Machine Room Dimensions, ft [m]
2–2500 [1134]	125 [0.6]	16 × 18 [4.9 × 5.5]
3–2500 [1134]	350 [1.8]	19 × 26 [5.8 × 7.9]
3–2500 [1134]	700 [3.6]	21 × 26 [6.4 × 7.9]
6–3500 [1588]	500 [2.5]	26 × 29 [7.9 × 8.8]
6–3500 [1588]	700 [3.6]	27 × 27 [8.2 × 8.2]
6–3500 [1588]	1000 [5.1]	27 × 27 [8.2 × 8.2]
6–5000 [2268]	450 [2.3]	31 × 33 [9.5 × 10.1]

be drawn from these figures; they are listed simply to give a general picture of requirements. A manufacturer's layout giving dimensional data for the hoistway and machine room is shown in Figure 33.31.

When penthouse space is not available and a hydraulic unit is not desired, a basement traction unit, also referred to as an underslung arrangement, can be used. These units are always low-speed (100–350 fpm [0.5–1.8 m/s]) and are therefore applicable only where rise is limited and traffic is light to medium. Figure 33.32 shows a typical shaft section for this design with car and dimensional data.

33.35 STRUCTURAL STRESSES

For structural design, it is necessary to know the overhead load that must be supported by the foundations, by structural columns extending up to the penthouse, and by the main beams that support the penthouse floor and subfloor. These loads (reactions) are supplied by manufacturers and usually include the actual dead weights of equipment when the elevator is not in motion, plus the added weight caused by the momentum of all moving

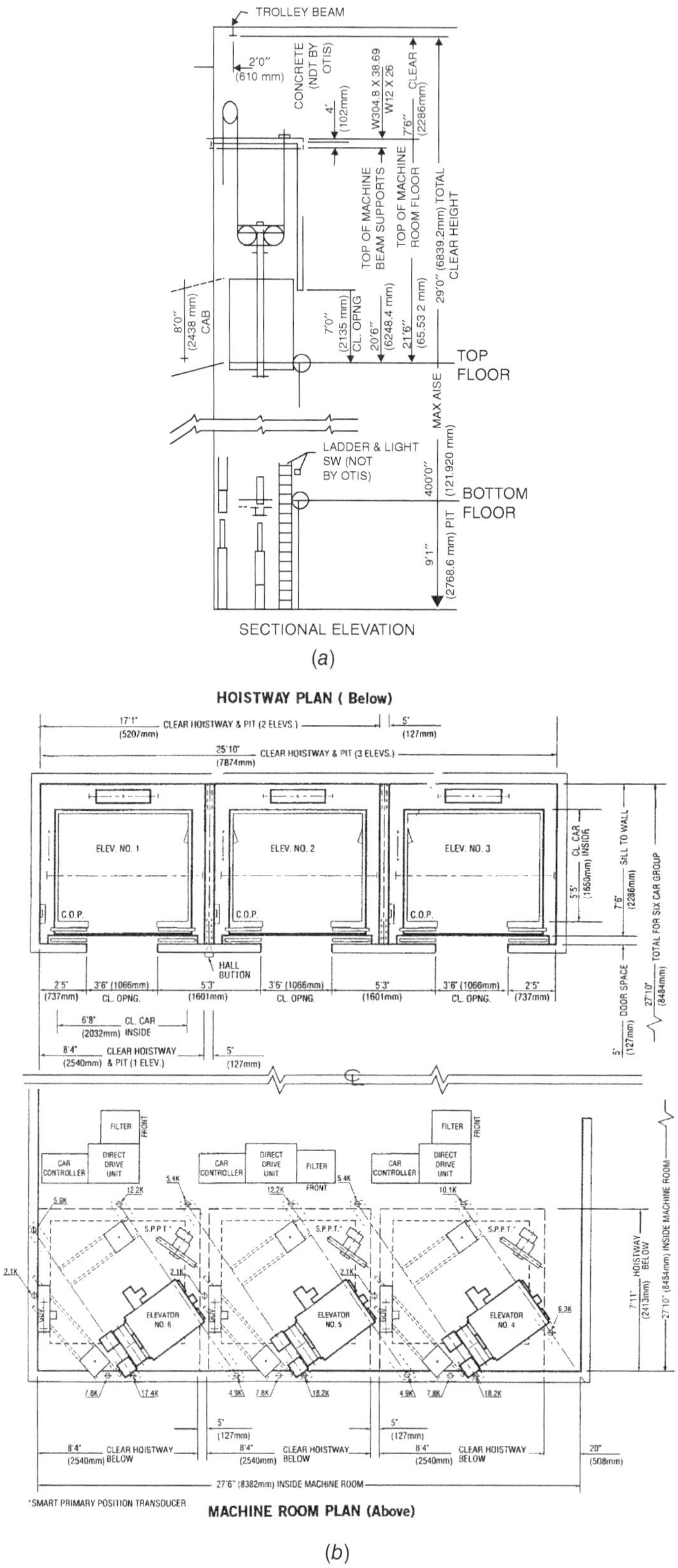

Figure 33.31 *Manufacturer's layout data for a bank of six 3500-lb [1588-kg], 700-fpm [3.45-m/s] gearless passenger elevators. Equipment shown in the machine room is the thyristor control for DC traction machines. Because each controller provides group supervisory control in this design (Otis Elevonic 411), no separate group supervisory equipment is shown. No additional space would be required if a UMV drive with m-g sets were selected rather than thyristor control. (Courtesy of Otis Elevator Co.)*

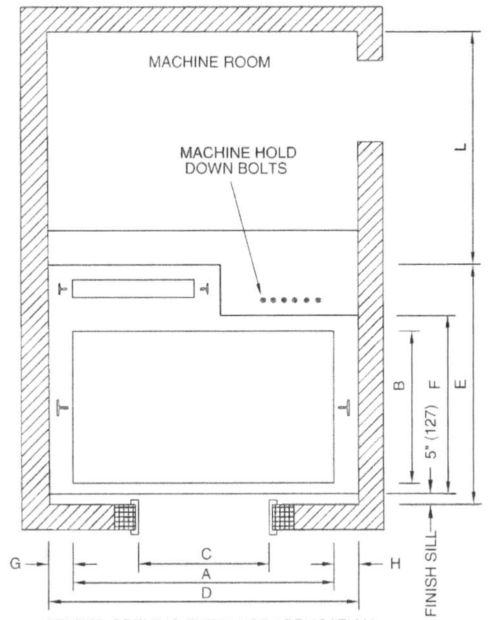

CENTER OPENING ENTRANCE APPLICATION
(N/A FOR 2000 LB. CAP.)
(SINGLE SLIDE OPTIONAL)

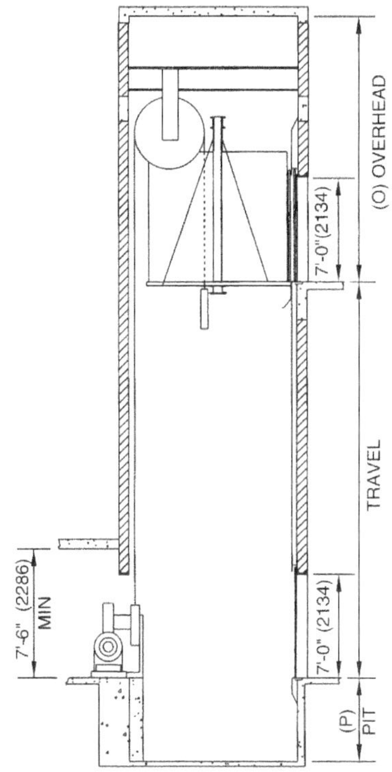

2:1 ROPING

2:1 ROPING ARRANGEMENT

Overhead dimension is reduced to a minimum. Greater pit depth is required due to the sheave arrangement as compared to a 1:1 installation.

SIZES & CAPACITIES

CAPACITY	2000 lbs. (907 kg)	2500 lbs. (1134 kg)	3000 lbs. (1361 kg)	3500 lbs. (1588 kg)
A	6'-0" (1829)	7'-0" (2134)	7'-0" (2134)	7'-0" (2134)
B	5'-0" (1524)	5'-0" (1524)	5'-6" (1676)	6'-2" (1880)
C	3'-0" (914)	3'-6" (1067)	3'-6" (1067)	3'-6" (1067)
D	8'-0" (2438)	9'-0" (2743)	9'-0" (2743)	9'-0" (2743)
E	6'-10" (2083)	6'-10" (2083)	7'-4" (2235)	8'-0" (2438)
F	5'-4" (1626)	5'-4" (1626)	5'-10" (1778)	6'-6" (1981)
G	1'-0" (305)	1'-0" (305)	1'-0" (305)	1'-0" (305)
H	1'-0" (305)	1'-0" (305)	1'-0" (305)	1'-0" (305)

MINIMUM MACHINE ROOM OVERHEAD & PIT DIMENSIONS

SPEED	100 FPM (0.51 m/s)	200 FPM (1.02 m/s)	250 FPM (1.27 m/s)
L	10'-6" (3200)	10'-6" (3200)	10'-6" (3200)
O	13'-0" (3962)	13'-2" (4013)	13'-7" (4140)
P	5'-6" (1676)	6'-6" (1981)	6'-11" (2108)

NOTES:

1. If a counterweight safety is applied, consult Montgomery KONE
2. Layouts and dimensions shown are for machine located at rear of hoistway.
3. Add 4" (102) to dimension D in seismic zones.
4. Other capacities, speeds, and arrangements are available.

Figure 33.32 *Typical data for a basement traction machine (underslung) arrangement, used where a penthouse is unavailable or undesirable. (Courtesy of Montgomery-KONE.)*

parts and passengers when the elevator is at top speed and is suddenly stopped by the safety devices.

POWER AND ENERGY

33.36 POWER REQUIREMENTS

The power required by an elevator drive is that which is needed to provide the necessary traction and to overcome friction. Because power is equal to the rate at which work is done, elevator motor size is directly proportional to the speed of the system. In other words, it requires proportionately more power to lift a 3000-lb [1361-kg] car at 700 fpm [3.6 m/s] than at 200 fpm [1.0 m/s]. This relationship is demonstrated in Figure 33.33, which shows the minimum size of a DC elevator traction motor as a function of speed for cars of different capacity. (For power data on hydraulic elevators, see Section 33.8.) As friction is higher in a geared machine than in a gearless unit, the geared machine traction motor must be larger for the same car speed. The size of the traction machine shown in Figure 33.33 is independent of the power supply design (m-g set, VVVF, thyristor control) because it is determined purely by traction system requirements. In practice, however, traction motors with VVVF control are frequently smaller because they operate more efficiently.

An elevator moves only about 50% of the time, the remainder being spent standing at various landings. As the number of cars in a bank increases, the probability of all the cars being in operation simultaneously decreases, resulting in a system demand factor of less than 1.0. The factor for different group sizes is shown in Figure 33.33. Conceptually, demand factor is similar to diversity in a plumbing system.

As an example of the use of the power curves, consider a bank of five 3500-lb [1588-kg], 600-fpm [3.0-m/s] units. From Figure 33.33, each car requires 48 hp [36 kW]:

$$\text{group demand factor} = 0.67$$

$$\text{total instantaneous power required}$$

$$= 5 \times 48 \times 0.67 = 160 \text{ hp (I-P)}$$

$$= 5 \times 36 \times 0.67 = 120 \text{ kW (SI)}$$

Note that this is the traction motor power requirement. If an m-g set with an overall efficiency of 80% is used to drive the traction motor, the elevator system power requirement is

$$\text{system power} = (160 \text{ hp}/0.80) = 200 \text{ hp (I-P)}$$

$$= (120 \text{ kW}/0.80) = 150 \text{ kW (SI)}$$

which must be provided by the building electrical system. If a solid-state power supply system with a

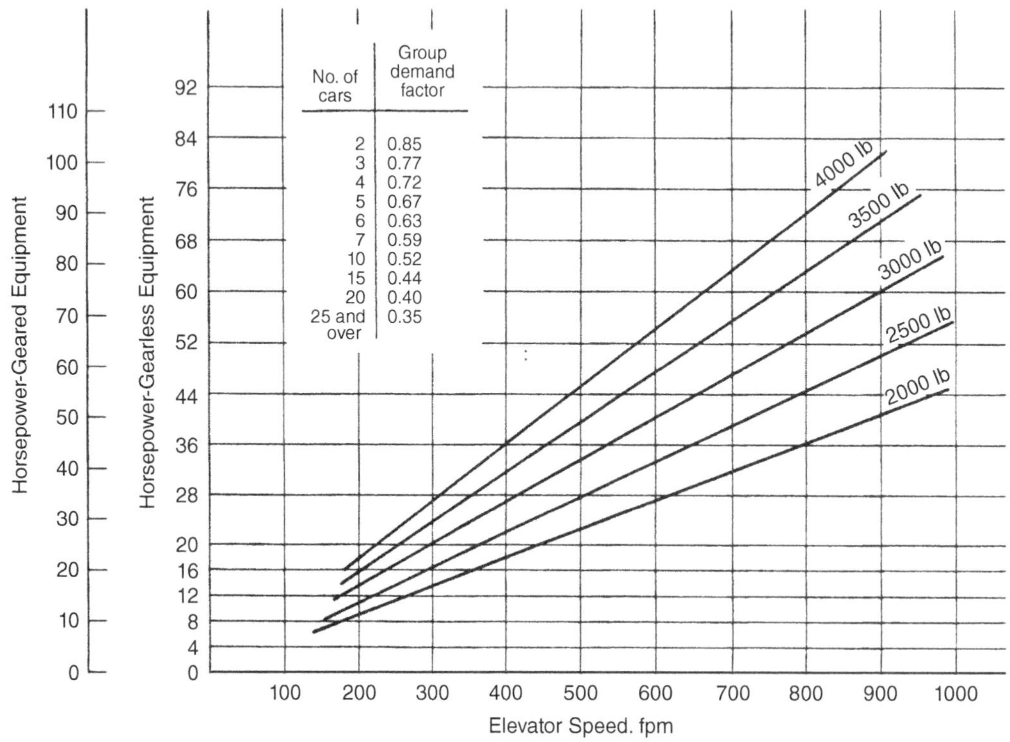

Figure 33.33 *Elevator traction motor power requirements per car. An m-g set drive (if used) is approximately 20% larger than a traction machine.*

typical efficiency of 92% is used, the system power requirement will be only

$$\text{system power} = (160\text{ hp}/0.92) = 174\text{ hp (I-P)}$$
$$= (120\text{ kW}/0.92) = 130\text{ kW (SI)}$$

which is a 13% reduction from the previously calculated 200-hp [150-kW] requirement.

33.37 ENERGY REQUIREMENTS

As a rule-of-thumb, anticipate elevators to constitute 2–5% of the total energy consumption of a typical building. The energy used by an elevator is essentially the system friction, including the heat generated by the brakes plus the electrical losses in the traction motor and power supply equipment (rotary or solid-state). The energy expended in raising a car and its passengers is simply stored as potential energy. It is returned to the power system when the car and passengers descend via the system of regenerative braking used in almost all elevator systems. Refer to Figure 33.34, which shows the approximate efficiencies of the components of a typical system. With these data, it is possible to calculate a system's energy consumption.

EXAMPLE 33.2 Given a system of five 3500-lb [1588-kg], 600-fpm [3.0-m/s] gearless cars, calculate:

1. The heat generated in the machine room during peak periods; assume solid-state control.
2. The approximate monthly energy cost; using a combined demand/energy rate of $0.13/kWh.

SOLUTION

During peak periods, the traction motor operates approximately 50% of the time and is at standstill the other half. Assume that, while operating, it draws 90% of the full load (with a VVVF power supply, this figure is reduced considerably).

Therefore, for one car, from Figure 33.33:

$$\text{traction motor} = 48\text{ hp [36 kW]}$$

Total loss per machine:
In controls:

$$48\text{ hp} \times 90\%\text{ load} \times 50\%\text{ operation} \times 10\%\text{ loss}$$
$$= 2.16\text{ hp [1.6 kW]}$$

In traction motor:

$$48\text{ hp} \times 90\%\text{ load} \times 50\%\text{ operation} \times 20\%\text{ loss}$$
$$= 4.32\text{ hp [3.2 kW]}$$

Therefore, in total:

$$= 6.48\text{ hp [4.8 kW], equivalent to 16,488 Btu/h}$$

Because five elevators are operating, the total heat generated is:

$$5\text{ elevators} \times 16{,}488\text{ Btu/h}$$
$$= 82{,}440\text{ Btu/h [24.2 kW]}$$

This is roughly the heating capacity of a home furnace. As a result, machine room temperatures in warm climates frequently reach 120°F [49°C]. No diversity is taken because all the machines are operating and the heating is additive; diversity is applicable only in calculating instantaneous load.

As solid-state elevator equipment is much less tolerant of high ambient temperatures than the electromechanical switches and relays previously used, an elevator machine room should be held to a maximum dry-bulb temperature of 90°F [32°C]. Temperatures above 90°F [32°C] can result in unreliable elevator system performance. This limit can sometimes be accomplished by thermostatically controlled forced ventilation, particularly if spill-over air from an air-conditioned space is available. However, because machine rooms are frequently on the building roof and exposed on all surfaces to ambient temperatures and solar radiation, air conditioning may be necessary. It is also important to

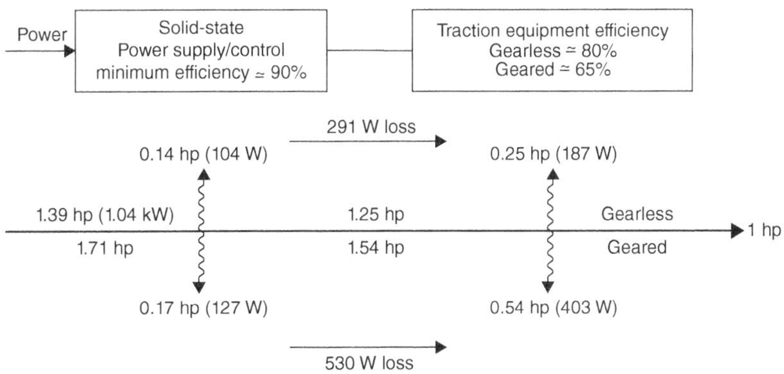

Figure 33.34 *Block diagram showing losses in the system per horsepower [kW] delivered to the elevator car and the equivalent wattages. Note that the losses in a geared system are almost double those of a gearless one. Values shown are for solid-state thyristor controls.*

prevent machine room temperature from dropping below 55°F [13°C]. This can usually be done with one or more unit heaters, which will normally operate only during the winter, and then only on nights and weekends.

In actual design situations, accurate heat loss values, which are available from manufacturers, would be used, along with accurate heat gain and heat loss calculations for the specific machine room being designed. Frequently, use of thermal insulation, thermostatically controlled louvers, and sunshading can ease the thermal load and result in appreciable savings in money and energy. Because of the high initial and operating costs of air conditioning, some elevator manufacturers use control components that are tolerant of high temperatures. This point should be carefully examined with proposed elevator equipment manufacturers and the conclusions reflected in the elevator system specifications.

To calculate a monthly energy cost, an estimate must be made of the total usage of the system. Assuming the system to be in an office building, a reasonable breakdown of operation during a 24-hour day would be:

2 hours at peak use
2 hours at 70% of peak
6 hours at 50% of peak
14 hours at 10% of peak

This gives a weighted average of 30% of peak load for the elevator bank. Therefore, per car:

$$
\begin{aligned}
\text{energy} &= 30\% \times \text{total losses} \times 24 \text{ hours} \\
&= 0.3 \times 6.72 \text{ hp [or 5 kW]} \times 24 \text{ hours} \\
&= 48 \text{ hp-h [or 36 kW-h]} \\
&= 36 \text{ kWh per day per car}
\end{aligned}
$$

Monthly cost (assuming 25 operating days) would be:

$$
\begin{aligned}
&36 \text{ kWh/day} \times 25 \text{ days} \times \$0.13/\text{kWh} \\
&= \$117/\text{month per car} \\
&= \$585/\text{month for the elevator bank}
\end{aligned}
$$

The cost would be lower with a VVVF power supply and higher for a Ward–Leonard (m-g set) arrangement. ■

33.38 ENERGY CONSERVATION

A reduction in energy consumption can be accomplished by implementing the following recommendations:

For existing elevators

1. Increase the interval during nonpeak hours.
2. Replace m-g sets with a solid-state DC power supply or AC traction motors with a VVVF power supply. This conserves energy not only due to the higher efficiency of the power supply, but also because energy consumption of idling machines is eliminated.
3. Reclaim machine room waste heat.
4. Shut down some units completely during off hours.

For a building in the planning stage

1. Base the design on the maximum recommended trip time.
2. Use the lowest speeds possible within a drive type (geared or gearless).
3. Use gearless equipment whenever possible.
4. After construction, implement the energy conservation recommendations for existing elevators.

Because elevator shafts can induce a powerful stack effect, measures should be taken to counteract the potential loss of heat from the building that may occur during the heating season. In addition, the placement of thoughtfully located and visually pleasant stairs between floors can reduce short elevator trips, save energy, and promote social interaction among building occupants.

33.39 EMERGENCY POWER

Major power failures and local brownouts have demonstrated forcefully the need for a standby or emergency power source of adequate size to operate an affected building's elevators. Few experiences are so harrowing as being trapped in the crowded confines of a small box suspended in a long vertical shaft, with little or no light, and complete strangers for companions.

A common misconception about elevators is that on failure of power, the cars will automatically descend to the nearest landing, where an exit is then possible. In reality, the car brake is set immediately upon power outage and the car remains stationary. Hydraulic cars can be lowered by operation of a manual valve; small traction cars can be cranked to a landing by hand, but large cars are fixed in position. This is particularly bad for cars in blind shafts—that is, express shafts with no shaftway doors. In such cases, escape from the cars via a hatchway is not practical; when emergency power is not available, the undesirable option of breaking through the shaftway walls is the only recourse.

In addition to simple inconvenience, loss of elevator service in facilities such as hospitals and penal institutions constitutes a danger to life. For this reason, most codes require that emergency power be available in specific building types to operate at least one elevator at a time, and for elevator lighting and communications. Many installations separate the emergency power functions, providing a generator for elevator traction power and separate individual elevator battery packs for communications, lighting, and, preferably, the car fan. The last two items can be provided as car options by elevator manufacturers.

The generator is normally sized to supply one elevator motor at a time, with manual or automatic switching arranged between unit controllers. Thus, each car in turn can be brought to a landing, and thereafter a single car can be retained in service. If it is desired to operate more than one car, a larger generator can be installed. This might be the case in a multiwing building with critical service requirements, such as a hospital.

The amount of power required, the size of the emergency generator, and the equipment size necessary to absorb regenerative power are all data that can be furnished by a consulting engineer and the elevator manufacturer.

OTHER DESIGN CONSIDERATIONS

33.40 FIRE SAFETY

Most fire codes specify the procedures that standard elevator control equipment must implement when a fire emergency is initiated. Details vary somewhat among jurisdictions, but in general the actions are as follows:

1. All cars close their doors and return nonstop to the lobby or another designated floor, where they park with the doors open. Thereafter, they are operable in manual mode only, by use of the firefighter's key in the car panel.
2. All car and hall calls are canceled, and call-registered lights and directional arrows deactivated.
3. The fire emergency light or message panel in each car is activated to inform passengers of the nature of the alert and that cars are returning to a designated terminal.
4. Door sensors and in-car emergency stop switches are deactivated.
5. Traveling cars stop at the next landing without opening their doors and then proceed to the designated terminal.

In the event of a false alarm, the emergency procedure can be overridden at the main control point, and the system returned to normal while the source of the alarm is located. This is a particularly important feature in large buildings with automatic fire alarm systems containing hundreds of fire, smoke, and water-flow detectors.

There are two special uses for elevators during a fire emergency: fire service access elevators (FSAEs) and occupant evacuation elevators. FSAEs can be used by trained personnel to transport firefighters and equipment and for evacuation. In buildings over 120 ft [36 m] in height from the street access level, codes typically require two FSAEs or that all elevators have 3500-lb [1588-kg] capacity. In very tall buildings, egress from upper floors can be problematic for occupants. Codes now allow for occupant evacuation elevators to be used in addition to egress stairs. Such elevators have special requirements for two-way communication systems, elevator monitoring, instructional signage, view windows, etc. FSAEs and occupant evacuation elevators have special fire protection system requirements including fire-rated hoistway doors and no sprinklers in the hoistways or machine rooms. In addition, cabs are often sized to accommodate a stretcher.

33.41 ELEVATOR SECURITY

Elevator security has two key aspects: physical security of riders and consideration of the elevator as an access portal in a building security system.

(a) Rider Security

This design problem is difficult because a traveling elevator is an enclosed space that can be rendered inaccessible simply by pressing the emergency stop button. Thereafter, an attacker can escape at a floor of his/her choice. To reduce this danger (to some extent), elevators are equipped with alarm buttons that alert residents and security personnel (if any). Every elevator, by code, must be equipped with communication equipment. A two-way communication system with "no-hands" operation in the car is particularly effective for security. When a closed-circuit TV monitor is added, utilizing a wide-angle camera in each car (Figure 33.35), the security problem will have been addressed to a considerable extent. Using a communication and TV system assumes continuous monitoring of the building security desk so that an incident will be detected—requiring coordination between design decisions and building operations.

(b) Access Control

This is often a matter of restricting access to (and from) a floor or car. This can be accomplished by pushbutton combination locks or coded cards, the proper use of which will permit access. However, if an unauthorized

Figure 33.35 Wide-angle camera coverage intended for elevator car surveillance. A prominent printed warning in the car is an integral part of the system's effectiveness.

person accompanies (intentionally or unintentionally) an authorized person, the effectiveness of this type of access control is seriously compromised. The most effective security system is a combination of automatic monitoring and access devices coupled with continuous supervision by persons who know the appropriate actions to take in an emergency.

33.42 ELEVATOR NOISE

Elevator operation, with its rotating, sliding, and vibrating masses, can be a cause of serious noise disturbance to quiet areas such as sleeping rooms, libraries, and certain types of office space. Noise and vibration can be reduced by the appropriate application of noise control strategies and vibration isolators (e.g., between guide rails and the structure), but primarily by placing noise-sensitive areas away from elevator shafts and machine rooms. The clatter and whirring sound associated with older machine rooms (caused by relays, step switches, m-g sets, and sliding contacts) can be entirely eliminated by the use of solid-state equipment.

33.43 ELEVATOR SPECIFICATIONS

Two basic types of specifications for elevator systems are utilized—as is true for other building systems. These are the prescriptive (equipment-based) and performance (outcome-based) approaches. Performance specifications describe job conditions and invite contractors to submit detailed proposals that will meet explicit design criteria. In a prescriptive approach, an architect or engineer designs the system and requires that installed components meet the code requirements. In a performance approach, the owner lays out expected outcomes and receives a bid from a contractor for a system that will provide the intended performance (while still meeting prescriptive specs for safety, ADA, etc.).

In recent years, the use of performance specifications has increased because of the advent of preengineered, premanufactured systems. These are supplied by the major elevator manufacturers and have the following advantages:

1. approximately 10% lower cost than a custom-designed system
2. a completely engineered and tested system whose performance and cost are known exactly
3. rapid delivery
4. minimum supervision required by the owner and architect

If an owner decides to use a custom-designed system, the design team must prepare detailed drawings and specifications. The design must address:

- elevator type, rated load, and speed
- maximum travel
- number of landings and openings
- type of control and supervisory system
- details of car and shaft doors
- signal equipment
- characteristics of the power supply
- finishes (often left as an allowance to provide owner flexibility; going above the allowance would be a cost add-on)

Because the selection of, and technical specifications for, elevators are specialized and complex, the services of an elevator consultant are usually required. In addition to the technical portions of the specifications, it is imperative that the following contractual items be covered in detail.

(a) General Contractor's Responsibility

The General Contractor (under contract with the owner) normally provides the following:

1. the hoistway, including a properly designed, lighted, drained, waterproofed, and ventilated machine room and pit

2. access doors, ladders, and required guards
3. guide rail bracket supports, and support for machine and sheave beams
4. electric feeder terminating in a switch in the machine room
5. hoistway outlets for lighting, power, and telephone
6. temporary lighting and power during construction
7. concrete machine foundations
8. vents, holes, and other work to satisfy fire codes
9. all cutting, patching, and fabricating of walls, beams, masonry, and so on
10. coordination of all work
11. any special work, as negotiated and specified

(b) Elevator Contractor's Responsibility

The elevator contractor shall provide a complete, working, tested, and approved system in accordance with specifications, plus any special work such as painting, special tests, work scheduling, and temporary elevator service. The system is "inserted" into the building framework described directly above.

(c) Special Job Conditions

These include work restrictions, scheduling, penalties or bonuses, test reports, and the like. If the elevator system is to be part of a building commissioning effort, this must be made known to all the parties involved in design and construction.

In all elevator contract work, bids should be solicited from parties named on qualified bidder lists. A complete elevator contract includes a warranty and provisions for maintenance of the installation for a specific period after completion. In alteration and modernization work, coordination problems are complex, and an elevator contractor well-experienced in this type of work should be selected.

33.44 INNOVATIVE EQUIPMENT

(a) Rope-Free and Two-Dimensional Elevators

The elevator industry is constantly developing new equipment to improve system operation and safety. In addition, novel designs that differ from standard traction arrangements are being developed to increase the efficiency of space use, to decrease the high cost of standard traction machinery, and to increase handling capacities. Among the most interesting designs being developed permits an elevator car to travel vertically and horizontally—or two-dimensionally—rather than the normal vertical-only motion. Another innovation involves the use of two separate elevator cars traveling in the same shaft. Both developments are possible because of the removal of hoist ropes from elevator shafts combined with the use of new linear motors (as opposed to a rotating unit) to supply electric traction power. TK Elevator's MULTI system is an example of such advanced elevator development—which may be referred to as rope-free or cable-free elevators.

(b) Machine Room-Less (MRL) Elevators

A recently developed interesting variation of the conventional traction design that effects a considerable space reduction is shown in Figures 33.36 through 33.39. Its principal applications are in low-speed, low-rise installations typically serviced by hydraulic elevators, but with higher speeds and rises under development. The novelty of the design lies in the use of a flat (disc-shaped), synchronous AC gearless hoisting motor, which, due to its flat disc shape, can be mounted directly on the main car guide rail at one side of the shaft (see Figure 33.38). This essentially removes the need for a penthouse and a large machine room above the hoistway. As such, this solution is commonly called a machine room-less (MRL) elevator. Due to the traction motor's position at the side of the hoistway, the car is roped in an underslung arrangement, as shown in Figure 33.37. Additional space economy is achieved by the use of a small drive controller built into an alcove at the top landing (see Figure 33.38). The pictorial hoistway representation in Figure 33.39 shows the equipment arrangement, demonstrating the absence of a penthouse and the limited machine room space requirement. An additional advantage of this arrangement is that the elevator loads and reactions are borne by the (stiffened) guide rail and transferred

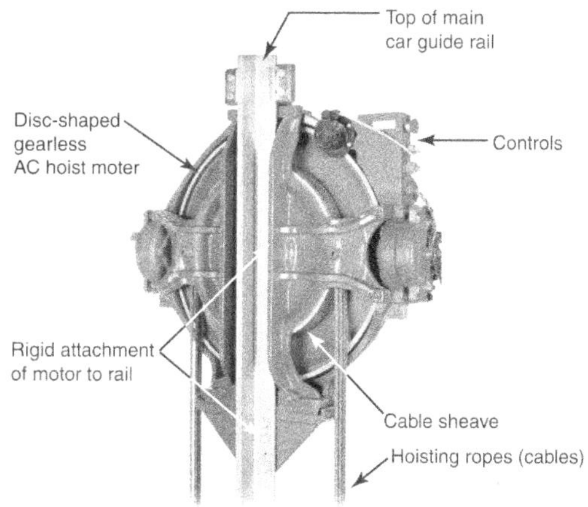

Figure 33.36 *Disc-shaped hoisting motor rigidly mounted on the elevator guide rail. The AC synchronous motor is connected directly to the hoisting cable drive sheave with no intervening gears. Brakes and controls are built into the assembly.*

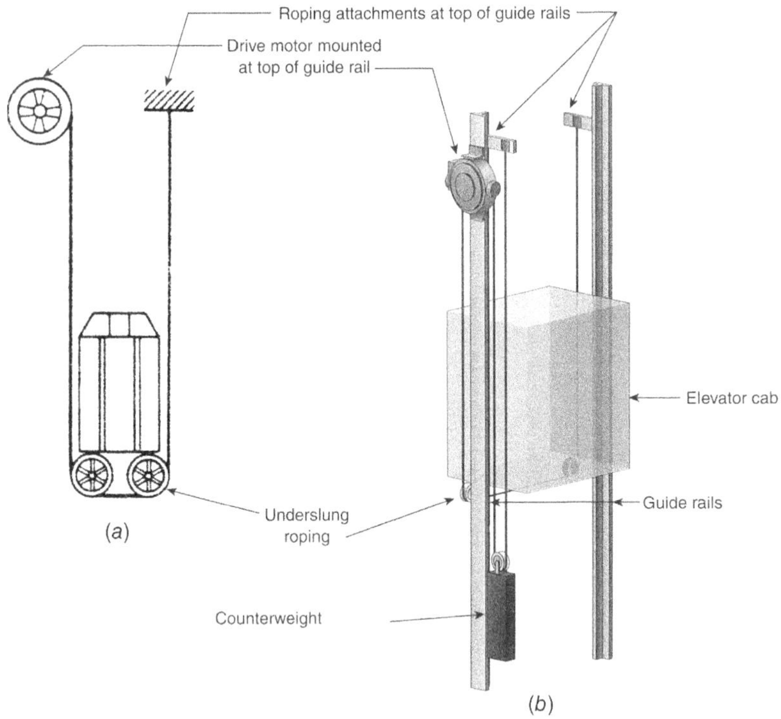

Figure 33.37 *Schematic (a) and pictorial (b) representations of the disc-motor-driven elevator arrangement.*

directly to the concrete pit below the bottom landing. This reduces the reactions borne by the machine room level in a conventional traction design and results in reduced structural loads. Due to their efficient motors and regenerative drives, gearless MRLs can potentially save more than a third of the energy than what would be required for a traditional elevator solution.

SPECIAL SHAFT ARRANGEMENTS

The fact that each elevator in a building typically has a full-height basement-to-penthouse shaft that occupies valuable building space—and that the elevator car only uses part of the space and only for part of the time—has always disturbed finance-conscious building operators. Development of high-speed drives and very sophisticated control systems has been spurred, in part, by the desire to increase the efficiency with which elevators use building space. An alternate approach to this problem lies in innovative use of shaft space so that more than one car can use a single shaft. A detailed analysis of these solutions is beyond the scope of this chapter; however, a short review is possible.

33.45 SKY LOBBY ELEVATOR SYSTEM

For skyscrapers and high-rise multiple-use buildings (such as the John Hancock Center in Chicago)—which are, in effect, stacked multiple buildings—the elevator solution may involve transporting large groups of people from the street lobby to an upper lobby called a sky lobby or plaza. At this point, the passengers transfer to another elevator to continue their upward journey.

The traditional approach in tall, single-purpose buildings is to vertically zone the structure and to provide banks of elevators that serve groups of floors. The difficulty with this solution is that the upper-zone cars travel through a long, expensive blind shaft to reach the floors being served. The sky plaza approach stacks (figuratively) two (or more) shafts vertically, with a lobby in between. The result is to have two cars operating in the equivalent of a single full-height shaft. To mitigate the annoyance felt by passengers taking a lengthened (interrupted) trip, the sky lobby is attractively appointed and may serve as an observation floor. Also, an elevator trip broken up by a lobby stop is thought to be less annoying than a somewhat shorter but still long, uninterrupted one. Finally, this plan is used principally where a clear differentiation in building use occurs, usually consisting of office

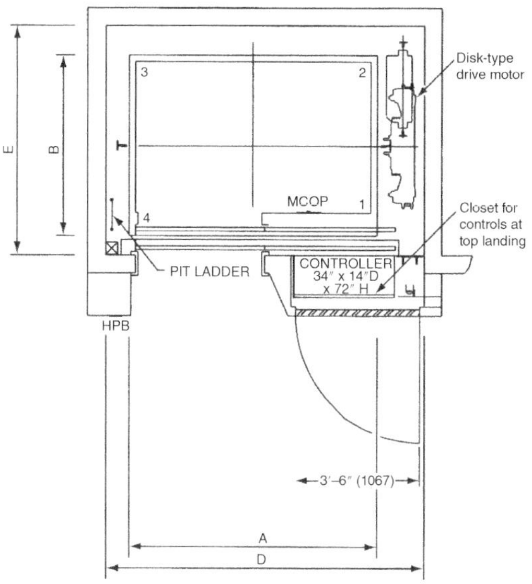

(a)

MonoSpace Elevator System	
Speed	200 FPM (1.0 m/s)
Capacity	2500 lb. (1134 kg)
A	7'-0" (2134)
B	5'-0" (1524)
D	9'-0" (2744)
E	6'-4" (1931)
Entrance Types	Single Slide
Overhead (O)	12'-9" (3887) min
Minimum Travel	8'-4" (2540)
Maximum Travel	80'-0" (24384)
Landings Served	2 - 10
Pit (P)	5'-6" (1677) min
Machine Room (W x D x H)	3'-9" x 16" x 8'-0" (1143 x 407 x 2439)

(b)

Figure 33.38 *(a) Section through the top of a hoistway showing dimensional data for a single 2500-lb [1134-kg], 200-fpm [1-m/s] installation with a rise of up to 80 ft [24 m]. Note that the drive motor occupies less than 2 ft [0.6 m] in the width of the hoistway and that the elevator motion and operating controls (Section 33.4) are installed in a closet 42 in. [1.07 m] wide and approximately 20 in. [508 mm] deep at the top landing. (b) Elevator system basic data for the simplex (single) unit shown in (a). (Courtesy of Montgomery-KONE.)*

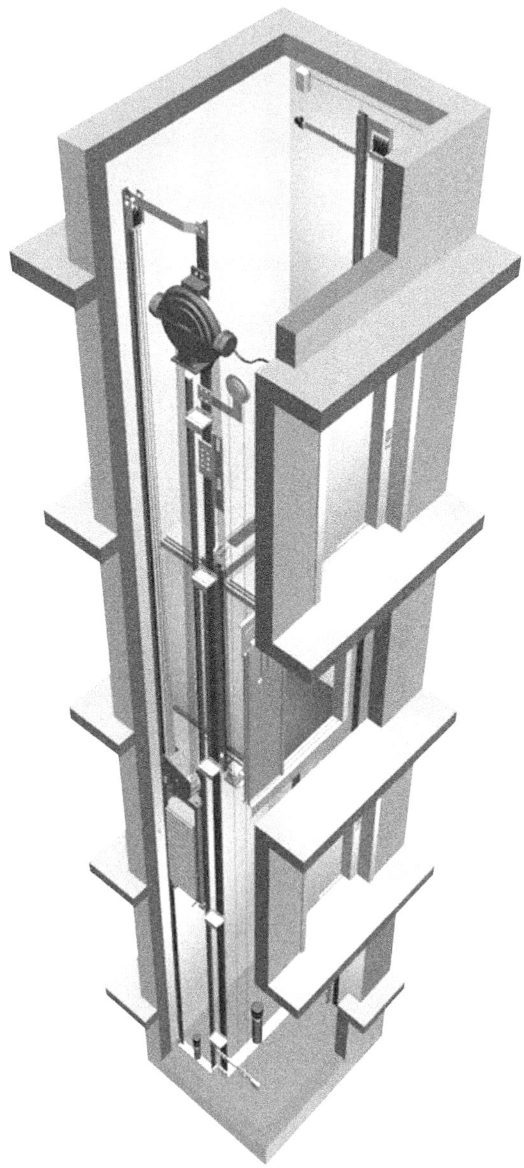

Figure 33.39 *Pictorial representation of a disc-type traction motor hoistway showing the system's essentials.*

areas below and residential units above, as in the John Hancock building. Thus, most lower-section users never encounter the sky lobby. Passengers headed for upper zones can use a sky lobby "shuttle" that travels from the entry level to the sky lobby and then continue the trip in an upper-section car.

Some designs also provide for a single-shaft upper-zone elevator group that serves upper-zone passengers who prefer not to use the sky lobby. A schematic drawing of the elevator arrangement in the John Hancock Building suggests the spatial advantages of the system (Figure 33.40).

33.46 DOUBLE-DECK ELEVATORS

The double-deck elevator is a recently revived and revised technique to answer the needs of tall buildings such as the Willis Tower (formerly the Sears Tower) in Chicago and Citicorp Tower in New York City. Its principal purpose is to limit the otherwise prohibitively large amount of space occupied by elevator shafts (Figure 33.41). The double-deck car increases shaft capacity, decreases the number of local stops, and

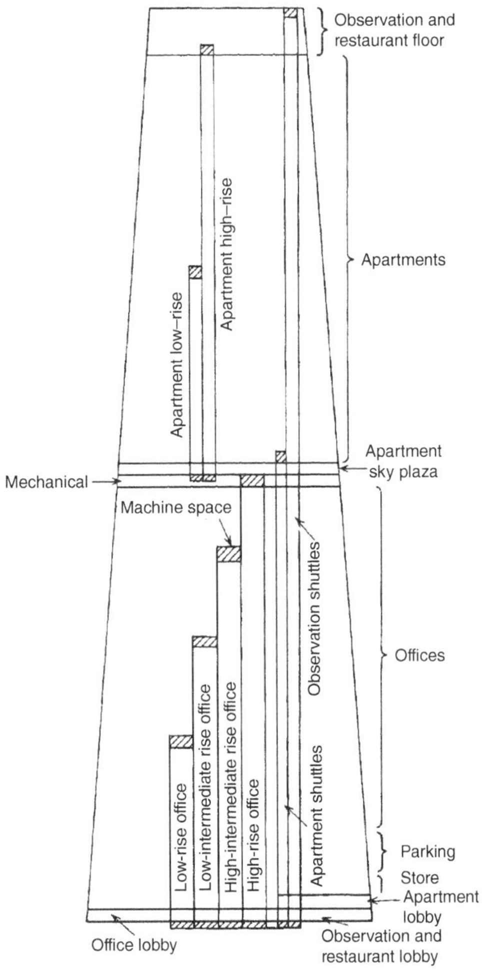

Figure 33.40 The use of a sky plaza in this dual-purpose building (the John Hancock Center, Chicago) eliminates the blind shafts below the upper-section (residential) elevators. The lengthened trip time is well tolerated by riders destined for the upper residential section, whereas the lower office section remains unaffected and retains its short interval and trip time. (From Strakosch, 1983.)

increases the available rental area. This technique can also be combined with sky lobbies for further space economy, as was done in the Willis Tower.

33.47 OBSERVATION CARS

By placing the traction lifting mechanism behind the car (Figure 33.42) attaching the car at the back, and using a glass-enclosed, observation-style car, a spectacular unit can be constructed that becomes an attraction in itself. Basic construction and examples are seen in Figures 33.42, 33.43 and 33.44. If the back screen is treated properly, the car gives the impression of movement with no apparent motive force or machinery.

The same effect can be accomplished by using a single-jack hydraulic lift mechanism and a cantilevered car, as suggested in Figure 33.44. Observation cars, which are placed on the outside of a wall, do not require shaft space. Thus, in addition to an interesting and attractive appearance, they increase the amount of usable interior space, thereby effecting a considerable cost economy.

33.48 INCLINED ELEVATORS

Although elevators normally travel vertically, this is not necessarily mandatory. Slanted or inclined elevators have been constructed in numerous locations where a building is built onto an inclined surface, usually a hillside or mountainside. One such application is shown in Figure 33.45. The motive mechanism used varies with the angle of inclination. In most instances, the car rides on inclined rails and is pulled up by a traction cable. It is counterweighted either by a weight riding on another set of rails in the case of a single car, or by the weight of another car in a two-car installation.

FREIGHT ELEVATORS

The objective of a freight elevator is similar to that for passenger elevators: to transport a given load (in this case, freight) effectively, economically, and quickly. A "service" car in a building can be used as a freight elevator, but, if utilized for passenger duty at all, it must meet passenger service requirements. If passenger duty is not required or if substantial freight is to be handled, a car designed specifically for freight should be used.

33.49 GENERAL INFORMATION

Factors to be considered in freight elevator selection, in addition to tonnage movement per hour, are size of load, method of loading, travel, type of load, type of doors, and speed and capacity of cars. Due to the interrelation of these factors, the selection process involves making assumptions based on recommendations and then arriving at a solution, as is done for passenger elevators.

A detailed discussion of the selection of materials-handling elevators is beyond the scope of this book

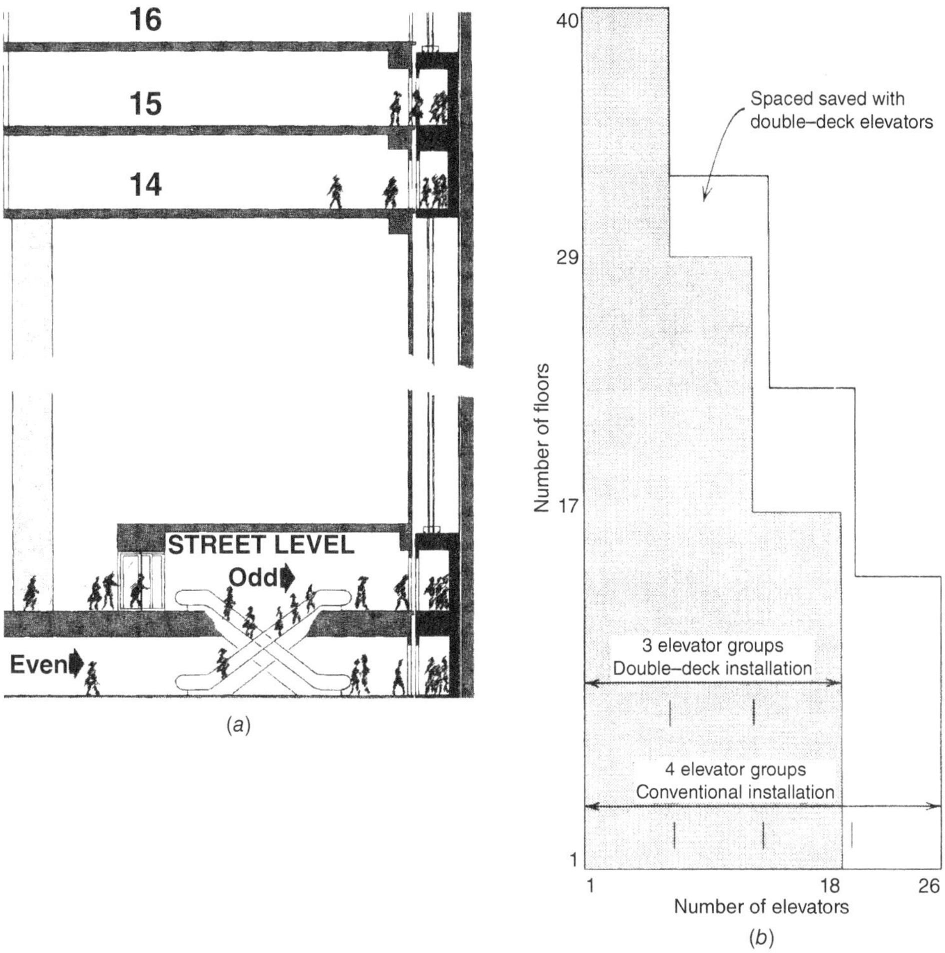

Figure 33.41 (a) A double-deck car increases car capacity and decreases shaft space demands. Coincidence of calls in the upper and lower cars reduces the number of local stops made by the double-deck unit. (b) Graphical representation of the space saved in a 40-story building by the use of double-deck elevators. (Courtesy of Otis Elevator Co.)

because of the large number of considerations involved. Therefore, the following sections are restricted to descriptive material and recommendations. Since freight elevators form such an important link in industrial processes, a careful and detailed material-flow study should be made before freight elevators are selected. Elevator manufacturers' representatives and materials-handling consultants can be very helpful in this regard.

33.50 FREIGHT CAR CAPACITY

Figure 33.46 is a section through a typical traction-type freight car shaft. Capacities corresponding to a specific platform size result from the varying unit-area loads that are permissible. This is considered by *ASME A17.1*, which has established five load classifications for freight elevators:

Class A: General Freight Loading by hand truck. Single items may not exceed 25% of the car-rated load. The rated load is based on 50 pounds per square foot (psf) [244 kg/m^2] of net inside platform area.

Class B: Motor Vehicle Loading. The elevator car will carry automobiles or automobile trucks. The rating is based on a load of 30 psf (146 kg/m^2) of net inside platform area.

Class C1: Industrial Truck Loading; truck carried.

Class C2: Industrial Truck Loading; truck not carried.

Class C3: Concentrated Loading; no truck used; increments greater than 25% of rated capacity.

(*a*)

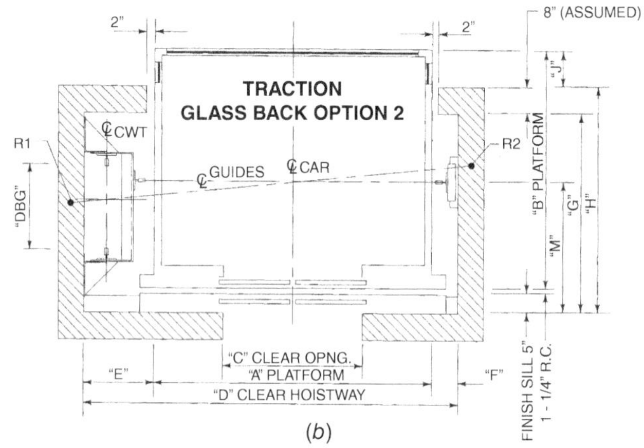

(*b*)

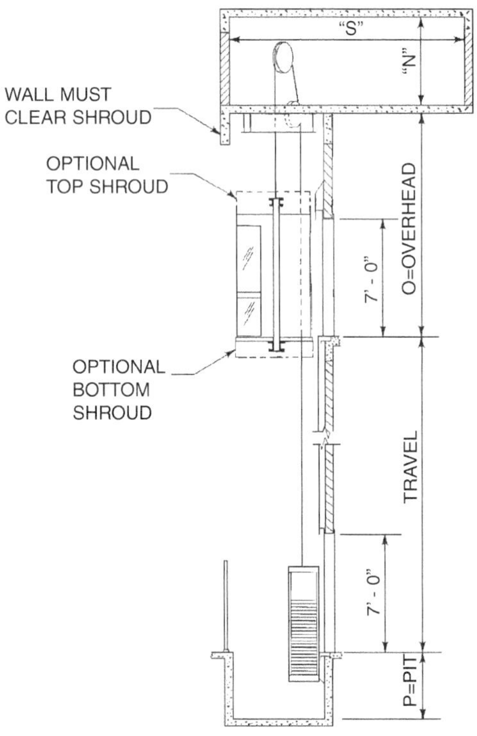

(*c*)

Notes (Hydraulic and Traction)

State and local code requirements may vary. Data is for general application.

All layout details are based upon the use of center opening type entrances.

Elevator machine room temperature should be maintained between 65°F and 100°F.

Please consult Montgomery KONE for exact information for working drawings and for designs which may vary from those depicted.

(*d*)

Figure 33.42 *Details of a geared, traction-type observation elevator. (a) The glass-enclosed observation portion of this elevator faces the rear. The shaftway doors are seen in the wall behind the elevator. The traction ropes are also clearly seen above each elevator. (b) Plan drawing of the unit. Note that the counterweight is mounted at the side of the car to be essentially invisible and not detract from the relatively uncluttered view of the car in motion. (c) Vertical section of the installation, including the overhead machine room. (d) Notes applicable to drawings (b) and (c). (Courtesy of Montgomery-KONE.)*

BASIC DIMENSIONS: TRACTION OBSERVATION ELEVATORS						
DIMENSION	CAPACITY					
	3000 lbs.		3500 lbs.		4500 lbs.	
	Option 2		Option 2		Option 2	
"DBG"	25"		31"		37"	
A	7' - 0"		7' - 0"		7' - 0"	
B	5' - 8"		6' - 6"		7' - 6"	
C	3' - 6"		3' - 6"		3' - 6"	
D	9' - 5"		9' - 5"		9' - 5"	
E	1' - 8"		1' - 8"		1' - 8"	
F	9"		9"		9"	
G	4' - 7 1/2"		5' - 1 1/2"		5' - 7 1/2"	
H	5' - 3 1/2"		5' - 9 1/2		6' - 3 1/2"	
J	10 3/4"		1' - 2 3/4"		1' - 8 3/4"	
K	NA		NA		NA	
L	NA		NA		NA	
M	3' - 4 1/2"		3' - 10 1/2"		4' - 4 1/2"	
N	7' - 6"		7' - 6"		7' - 6"	
S	13' - 8"		13' - 8"		13' - 8"	
SPEED	200 fpm	350 fpm	200 fpm	350 fpm	200 fpm	350 fpm
*O(1)	14' - 8"	15' - 4"	14' - 8"	15' - 4"	15' - 2"	15' - 6"
**O(1)	15' - 2"	15' - 10"	15' - 2"	15' - 10"	15' - 2"	15' - 6"
*P	5' - 0"	5' - 6"	5' - 0"	5' - 6"	5' - 4"	5' - 10"
**P	5' - 4"	5' - 10"	5' - 4"	5' - 10"	5' - 4"	5' - 10"

NOTE: * Dimension when no shroud is used.
** Dimension when 2' - 0" shroud is used.
(1) Based on standard height cab (8' - 0" to underside of canopy).
→ The relationships between dimensions G, H, J and K are based on an assumed thickness of 8" for the rear wall of the hoistway.
→ The hoistway dimensions shown are based on a counterweight not requiring safeties.

(*a*)

VERTICAL REACTIONS	TRACTION ELEVATOR CAPACITY			
	2500 lbs.	3000 lbs.	3500 lbs.	4500 lbs.
R1	17500 lbs.	18900 lbs	20900 lbs.	24400 lbs.
R2	13900 lbs.	15000 lbs.	16300 lbs.	19100 lbs.

NOTE: Reactions include allowance for impact but DO NOT include weight of concrete slab. Reactions are for preliminary use only. Exact reactions will be provided when exact conditions are known.

(*b*)

Figure 33.43 *(a) Dimensions applicable to drawings (b) and (c) in Figure 33.42. (b) Vertical reactions required for structural design of the elevator in Figure 33.42. (Courtesy of Montgomery-KONE.)*

For classes C1, C2, and C3, the rated load is based on 50 psf [244 kg/m^2]. Elevator cars have automatic leveling.

33.51 FREIGHT ELEVATOR DESCRIPTION

As speeds are generally between 50 and 500 fpm [0.25 to 2.5 m/s], a geared-type traction machine or a hydraulic unit is used. The preferred system of control is collective, with a variable-voltage DC supply, either UMV or VVVF. A two-speed AC rheostatic control may be considered if the car is used infrequently (fewer than five trips a day), economy is very important, accurate leveling is not essential, and a rougher ride is tolerable.

For low-rise installations, a hydraulic unit is most often employed. These, like variable-voltage DC traction units, provide accurate control, smooth operation, and accurate automatic leveling. Hydraulic units rarely exceed 65 ft [20 m] in height and operate at speeds of up to 150 fpm [0.76 m/s]. Accessories such as governors, safeties, and brakes are similar to those for passenger elevators.

General-purpose freight elevators, whether traction (Figure 33.47) or hydraulic (Figure 33.48), in load ranges of up to 20,000 lb [9072 kg], are standard design items applicable to all types of commercial and industrial buildings. Heavier units are individually engineered. Units of 20,000 lb [9072 kg] and more require special safeties. As with passenger elevators,

(a)

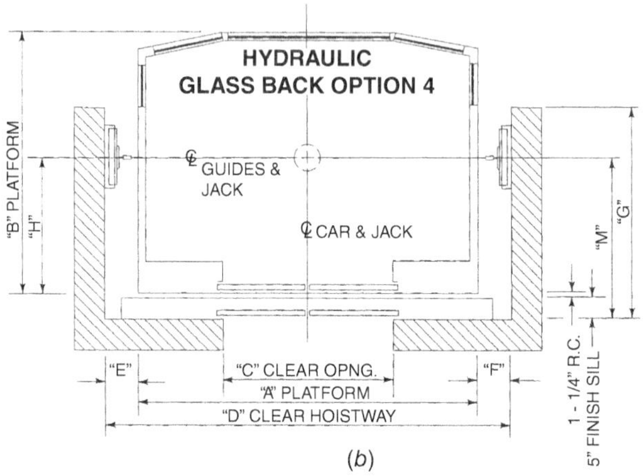

(b)

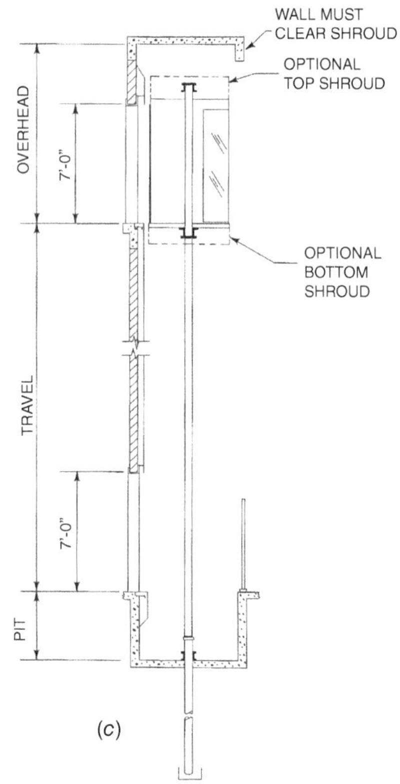

(c)

BASIC DIMENSIONS: HYDRAULIC OBSERVATION ELEVATORS				
	CAPACITY			
	2500 lbs.	3000 lbs.	3500 lbs.	
Dimension	OPTION 1,2,3,4	OPTION 1,2,3,4	OPTION 1,2,3,4	OPTION 5
A	7' - 0"	7' - 0"	7' - 0"	7' - 0"
B	5' - 3"	5' - 11"	6' - 6"	7' - 2"
C	3' - 6"	3' - 6"	3' - 6"	3' - 6"
D	8' - 4"	8' - 4"	8' - 4"	8' - 4"
E	8"	8"	8"	8"
F	8"	8"	8"	8"
G	4' - 3"	4' - 7"	4' - 10 1/2"	4' - 10 1/2"
H	2' - 8 3/4"	3' - 0 3/4"	3' - 4 1/4"	3' - 4 1/4"
M	3' - 3"	3' - 7"	3' - 10 1/2"	3' - 10 1/2"

Basic dimensions outlined above are based upon conventional inground jack application.

(d)

HYDRAULIC MACHINE ROOM SIZES (Typically recommended)		
SIZE	CAPACITY	
	2500 lbs. & 3000 lbs.	3500lbs.
WIDTH	9' - 0"	9' - 6"
LENGTH	6' - 8"	6' - 9"
DOOR	3' - 6" × 7' - 0"	

NOTE: Hydraulic machine room location should be at the lowest landing adjacent to the hoistway. Consult your Montgomery KONE Professional for alternative locations and optimum sizes. Material above is typical for 150 FPM speed.

(e)

Figure 33.44 *Hydraulic observation elevator. (a) The hydraulic plunger is clearly visible below each car against the background of the shaftway doors. (b) Dimensioned sectional drawing of the car and its enclosure. (c) Vertical section of the elevator travel. (d) Dimensions applicable to drawings (b) and (c). (e) Dimensional data for the required machine room. Note that in drawing (c), the machine room is not shown because it can be remote from the elevators, connected only by piping. (Courtesy of Montgomery-KONE.)*

Figure 33.45 The St. Louis Gateway Arch has a 10-passenger inclined elevator in each leg. Placement of doors, arrangement of the counterweight, and size of the shaft all depend upon the angle of incline. The car moves 82 ft [25 m] horizontally during its 386 ft [118 m] of total travel at an incline of approximately 12°. (Courtesy of Bethlehem Steel Corporation, which supplied the elevator rope for this installation.)

structural reactions for traction units are supplied by the manufacturer to the architect, who is responsible for providing adequate structural support. Structure is of great importance in larger car installations, because traction unit rails must be supported every few feet, and additional steel must be provided to accomplish this.

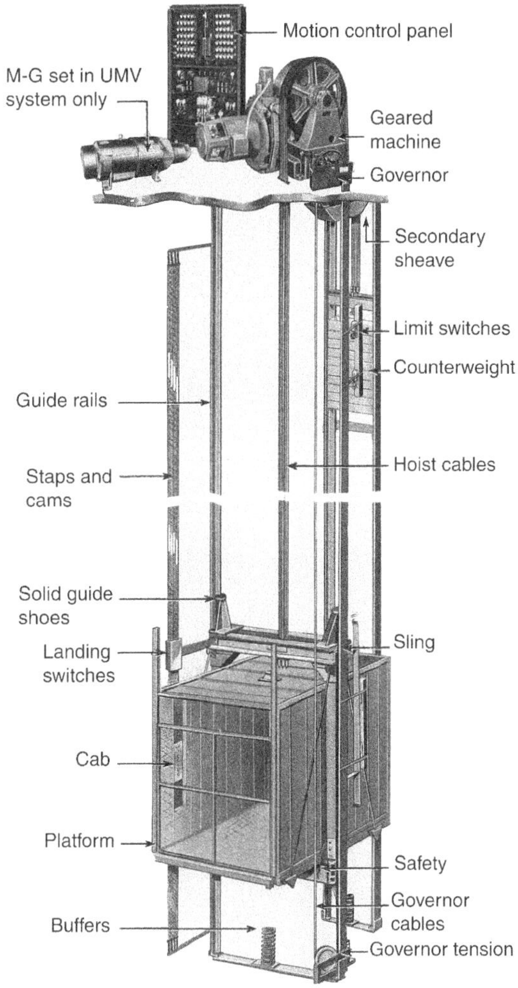

Figure 33.46 Components of a typical freight elevator installation utilizing a variable-voltage-controlled, geared traction machine. The sling that lifts the car is frequently arranged with double sheaves over which the hoisting ropes pass. This roping arrangement increases the mechanical advantage of the lifting ropes.

33.52 FREIGHT ELEVATOR CARS, GATES, AND DOORS

Cars for freight service are normally built of heavy-gauge steel with a multilayer wooden floor, the entire unit being designed for hard service. Guarded ceiling light fixtures are required. Car gates slide vertically and are a minimum of 6 ft [1.8 m] high. Hoistway doors are normally vertical lift, center-opening, manual or power-operated. Both car gates and hoistway doors are counterweighted and open fully to give complete floor and head clearance.

33.53 FREIGHT ELEVATOR COST DATA

The cost of a freight elevator installation, as with passenger elevator installations, is dependent upon many factors, principally capacity, type of control, use, and type of door operation.

Exact pricing is outside the scope of this discussion. A reputable manufacturer or elevator consultant should be sought for such information. Some general remarks on pricing, however, follow.

1. Variable-voltage-controlled equipment, depending on the type, costs 20% to 50% more than rheostatically controlled equipment.
2. Above a basic two-floor rise, the cost increases linearly with rise.
3. Electric door operation can increase the cost of a car installation 10% to 25%.

As an example of comparative pricing, using a nominal 100 monetary units for an 8000-lb [3630-kg], 75-fpm [0.38-m/s], four-floor, manual-door car with AC rheostatic control and automatic leveling, the same car with variable-voltage control, 150 fpm [0.76 m/s], and electrically operated doors would cost approximately 180.

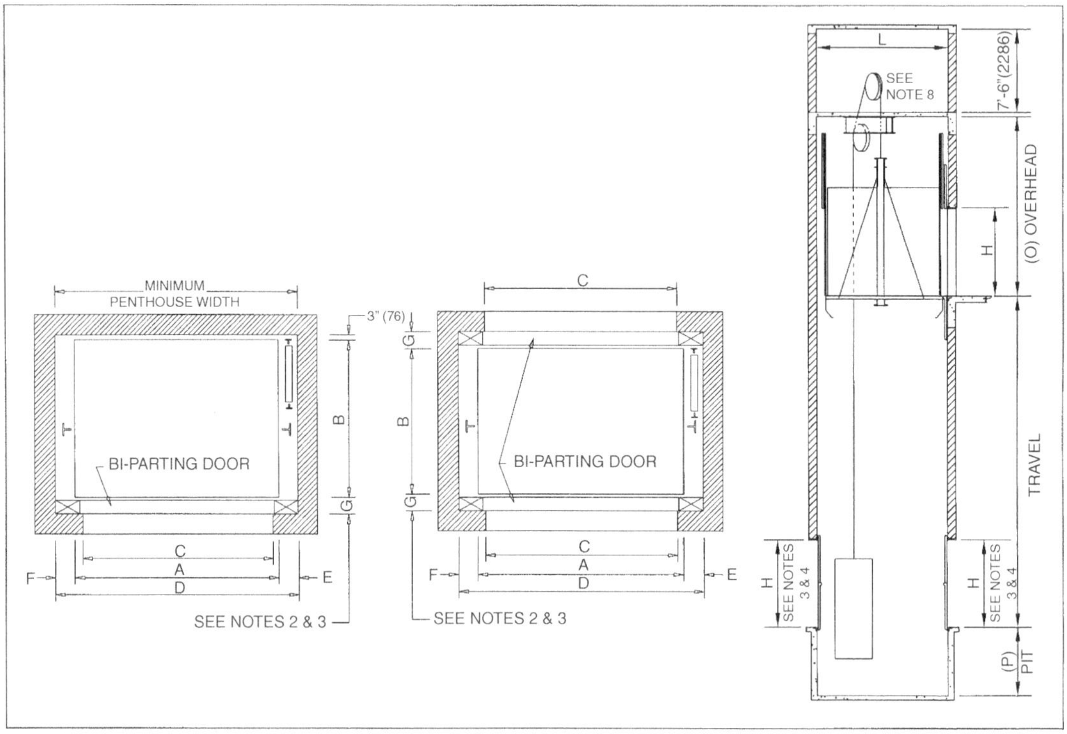

(a)

LIGHT AND MEDIUM DUTY FREIGHT ELEVATORS					
CAPACITY	2500 lbs. (1134 kg)	3000 lbs. (1361 kg)	4000 lbs. (1814 kg)	6000 lbs. (2722 kg)	8000 lbs. (3629 kg)
A	5'-4" (1626)	6'-4" (1930)	6'-4" (1930)	8'-4" (2540)	8'-4" (2540)
B	7'-0" (2134)	8'-0" (2438)	8'-0" (2438)	10'-0" (3048)	10'-0" (3048)
C	5'-0" (1524)	6'-0" (1829)	6'-0" (1829)	8'-0" (2438)	8'-0" (2438)
D	7'-10"(2388)	8'-10"(2692)	8'-10"(2692)	10'-10"(3302)	10'-10"(3302)
E	1'-7" (483)	1'-7" (483)	1'-7" (483)	1'-7" (483)	1'-7" (483)
F	0'-11" (279)	0'-11" (279)	0'-11" (279)	0'-11" (279)	0'-11" (279)
L	13'-0" (3962)	14'-0" (4267)	14'-0" (4267)	14'-0" (4267)	14'-0" (4267)

HEAVY DUTY POWER TRUCK LOADING FREIGHT ELEVATORS					
CAPACITY	10,000 lbs. (4536 kg)	12,000 lbs. (5443 kg)	16,000 lbs. (7258 kg)	18,000 lbs. (8165 kg)	20,000 lbs. (9072 kg)
A	8'-4" (2540)	10'-4" (3150)	10'-4" (3150)	10'-4" (3150)	12'-4" (3759)
B	12'-0" (3658)	14'-0" (4267)	14'-0" (4267)	16'-0" (4877)	20'-4" (6198)
C	8'-0" (2438)	10'-0" (3048)	10'-0" (3048)	10'-0" (3048)	12'-0" (3658)
*D	11'-4" (3454)	13'-6" (4115)	14'-0" (4267)	14'-2" (4318)	16'-6" (5029)
E	1'-7" (483)	1'-7" (483)	1'-7" (483)	1'-7" (483)	1'-7" (483)
F	0'-11" (279)	0'-11" (279)	0'-11" (279)	0'-11" (279)	0'-11" (279)
L	14'-0" (4267)	15'-0" (4572)	15'-0" (4572)	17'-0" (5182)	21'-0" (6400)

MINIMUM OVERHEAD & PIT DIMENSIONS FOR LIGHT & MEDIUM DUTY FREIGHT ELEVATORS				
SPEED	50 FPM (0.25 m/s)	75 FPM (0.38 m/s)	100 FPM (0.51 m/s)	200 FPM (1.02 m/s)
O	16'-0"(4877)	16'-0" (4877)	16'-0"(4877)	16'-0"(4877)
P	5'-6"(1676)	5'-6" (1676)	5'-6"(1676)	6'-0"(1829)

NOTES:

2. Dimension G = 5" (127) for regular type counterbalanced hoistway doors and 6 3/4" (172) for pass type counterbalanced hoistway doors.
3. Pass type hoistway doors are required when floor heights are less than 11'-0" (3353) for 7'-0" (2134) openings and less than 12'-6" (3810) for 8'-0" (2438) openings.
4. Dimension H = 7'-0" (2134) on light and medium duty and 8'-0" (2438) (or as required) for heavy-duty doors. Doors higher than 8'-0" (2438) require additional overhead.
8. 2:1 Roping recommended for heavy duty power truck loading freight elevators

* D dimension includes space required when vertical columns are added (inside hoistway) for rail support. If no columns required, D=A+E+F.

(b)

Figure 33.47 *Typical data for traction-type freight elevators: (a) drawings and (b) dimensions. (Courtesy of Montgomery-KONE.)*

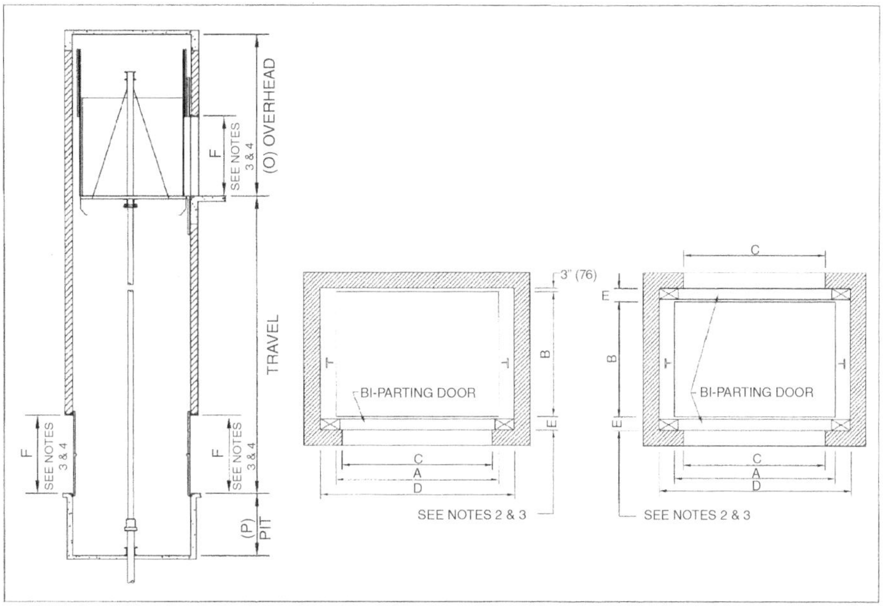

(a)

LIGHT AND MEDIUM DUTY FREIGHT ELEVATORS						
CAPACITY	2000 lbs. (907 kg)	3000 lbs. (1361 kg)	4000 lbs. (1814 kg)	5000 lbs. (2268 kg)	6000 lbs. (2722kg)	8000 lbs. (3629 kg)
A	5'-0" (1524)	5'-6" (1676)	6'-6" (1981)	8'-6" (2591)	8'-6" (2591)	8'-6" (2591)
B	6'-0" (1829)	7'-0" (2134)	8'-0" (2438)	10'-0" (3048)	12'-0" (3658)	12'-0" (3658)
C	4'-8" (1422)	5'-2" (1575)	6'-2" (1880)	8'-2" (2490)	8'-2" (2490)	8'-2" (2490)
D: **manual doors**	6'-4" (1930)	6'-10" (2083)	7'-10" (2388)	9'-10" (2997)	10'-0" (3048)	10'-6" (3200)
D: **power doors**	6'-10" (2083)	7'-4" (2235)	8'-4" (2540)	10'-4" (3150)	10'-6" (3200)	10'-6" (3200)
O: 7'-0" door ht. (2134)	13'-2" (4013)	13'-2" (4013)	13'-2" (4013)	13'-2" (4013)	13'-2" (4013)	13'-2" (4013)
P: 7'-0" door ht. (2134)	4'-0" (1219)	4'-0" (1219)	4'-0" (1219)	4'-6" (1372)	4'-6" (1372)	5'-0" (1524)

HEAVY DUTY POWER TRUCK LOADING FREIGHT ELEVATORS					
CAPACITY	10,000 lbs. (4536 kg)	12,000 lbs. (5443 kg)	16.000 lbs. (7258 kg)	18,000 lbs. (8165 kg)	20,000 lbs (9072 kg)
A	10'-6" (3200)	10'-6" (3200)	10'-6' (3200)	10'-6" (3200)	12'-6" (3810)
B	14'-0" (4267)	14'-0" (4267)	16'-0" (4877)	16'-0" (4877)	20'-0" (6096)
C	10'-2" (3098)	10'-2" (3098)	10'-2" (3098)	10'-2" (3098)	12'-2" (3708)
D: **manual doors**	12'-6" (3810)	12'-6" (3810)	12'-6" (3810)	12'-6" (3810)	14'-6" (4420)
D: **power doors**	12'-6" (3810)	12'-6" (3810)	12'-6" (3810)	12'-6" (3810)	14'-6" (4420)
O: 7'-0" door ht. (2134)	13'-2" (4013)	13'-2" (4013)	13'-2" (4013)	13'-2" (4013)	13'-2" (4013)
O: 8'-0" door ht. (2438)	14'-2" (4318)	14'-2" (4318)	14'-2" (4318)	14'-2" (4318)	14'-2" (4318)
P: 7'-0" door ht. (2134)	5'-0" (1524)	6'-0" (1829)	6'-0" (1829)	6'-0" (1829)	6'-0" (1829)
P: 8'-0" door ht. (2438)	5'-0" (1524)	6'-0" (1829)	6'-0" (1829)	6'-0" (1829)	6'-0" (1829)

NOTES:

2. Dimension E = 5" (127) for regular type counterbalanced hoistway doors and 6¾" (172) for pass type counterbalanced hoistway doors.
3. Pass type hoistway doors are required when floor heights are less than 11'-0" (3353) for 7'-0" (2134) openings and less than 12'-6" (3810) for 8'-0" (2438) openings.
4. Layout and dimensions shown for freight elevators based on bi-parting counterbalanced type hoistway doors – dimension F.
5. Dimensions O and P are minimums based on car speeds up to 150 FPM (0.76 m/s).

(b)

Figure 33.48 *Typical data for hydraulic freight elevators: (a) drawings and (b) dimensions. (Courtesy of Montgomery-KONE.)*

TRANSPORTATION

33.54 CASE STUDY—VERTICAL TRANSPORTATION

Hong Kong International Commerce Center

Project Basics

- Location: West Kowloon, Hong Kong, People's Republic of China
- Latitude: 22.3°N; longitude: 114.1°W; elevation: 60 ft [18.3 m] above sea level
- Heating degree days: 425 base 65°F [236 base 18.3°C]; cooling degree days: 8284 base 50°F [4602 base 10°C]; annual precipitation: 88 in. [2235 mm]
- Building type: New construction; mixed-use composed of offices, a hotel, recreation areas, retail space, and an observation deck
- Size: 2,800,000 ft^2 [260,000 m^2]
- Completed 2010
- Client: Sun Hung Kai Properties
- Design team: Kohn Pedersen Fox Associates, Wong & Ouyang Architects Ltd., and J. Roger Preston (Building Services Engineer) are key collaborators on the design development of the vertical transportation for the project.

Background. The 118-story International Commerce Center (ICC), as shown in Figures 33.49 and 33.50 designed by Kohn Pedersen Fox Associates (KPF), was the winning entry in an international design competition. A mixed-use building, composed of offices, a hotel, recreation areas, retail space, and an observation deck, the ICC as a whole serves an important role in the larger Union Square reclamation project. As of 2018, it is the tallest building in Hong Kong. The ICC is embedded with intelligent elevator, heating/cooling, and water-reclamation systems, while also being integrated into the municipal transit network. Design principles, client values, and a collaborative as well as interdisciplinary design process were important to the success of the project.

Context. The ICC was designed as an iconic structure that would support high-end finance, tourism, shopping, and hospitality. Kowloon Station, with which the ICC is closely integrated, aside from supporting 11 million passenger-journeys per day (Malott, 2011) is linked, via high-speed rail, subway, bus, and ferry terminals, to both mainland China and Hong Kong International Airport. In dealing with this programmatic juxtaposition, KPF built upon precedents such as the Met Life building over Grand Central Station and the JR Central Towers in Nagoya, Japan. Anticipating 30,000 visitors per day, the ICC pursued a multi-deck elevator system, with intelligent access controls (Figures 33.51–33.53, and 33.54), in order to increase the efficiency of the interior vertical transportation network, both in terms of energy expended, and time spent in the elevator.

Design Intent. The design was meant to integrate high-density development, high-technology efficiency, and transit-network integration.

The leading principles behind the design were as follows:

- set an iconic precedent for dense, integrated, and sustainable development for Hong Kong
- provide for the functional needs of the mixed uses that the ICC supports
- implement an intelligent elevator system that satisfies both the expected passenger loads and the multi-use nature of the building
- use intelligent HVAC systems
- use water reclamation/reuse systems

The architects came up with an effective vertical transportation strategy that provides good quality lift service, and at the same time, developed floor plans that optimize space utilization for all zones of the building (Figures 33.55 and 33.56).

The office portion of the tower is divided into five office zones. Double-deck elevator cabs minimize the requirement for lift shafts and are employed to serve the first four zones. Zones 1 and 2 are served from the two entry levels. Zones 3 and 4 are reached via the sky-lobby. Zone 5 has an independent drop-off with direct access via single-deck elevators.

The Ritz-Carlton Hotel has its own drop-off and a distinctive hotel entrance with a magnificent view of the Victoria Harbor. Guests take the shuttle lifts to the reception area at the 103rd floor, before using local lifts to access the top-zone guestrooms. Sandwiched between the Ritz-Carlton and the office portion is Sky100 with two observation decks. The general

Figure 33.49 *The ICC tower and its immediate context. (© SHK Properties; used with permission.)*

Figure 33.50 *Lower lobby level showing entry to the lower half of the double-decker elevator cars. (© Julia Lau; used with permission.)*

(a) (b)

Figure 33.51 *(a) Card key entry system for security. (b) Card is placed under the monitor for identification. (© Julia Lau; used with permission.)*

public can take the shuttle lifts from the ticketing office in the podium below.

Design Criteria and Validation. The project was intended to serve as a novel prototype by which skyscrapers could be designed and understood. Aside from the sheer scale of the project, the nature of its mixed-use functions required KPF to approach the project in a collaborative manner. Kohn Pedersen Fox Associates, Wong & Ouyang Architects Ltd., and J. Roger Preston [building services engineer], along with other specialist consultants, collaborated to achieve this end, and to help synchronize, more fully, the original design intent with the outcome.

Performance Data. Information available to date suggests success in a range of fields, but particularly in the areas of urban transportation integration, and vertical passenger movement. In terms of the latter, the ICC utilizes a wide variety of elevators, including double-deck passenger, service, emergency service, VIP, and freight elevators. This helps to move an estimated 30,000 people/day through the building in an efficient and less-energy-consuming manner. In order to optimize system performance, an intelligent identification system also helps to group passengers according to their destination requirements.

The building contains 80 Schindler elevators, 40 of which are double-deck elevator cars. They travel at speeds anywhere from 11.5 fps [3.5 m/s] to 29.5 fps [8.9 m/s], and cover distances from 236 ft [71.9 m] to 1555 ft [474 m] (Schindler Group, 2009). Such double-deck systems, optimized via intelligent transit management systems, are crucial for multi-use and

(a)

(b)

Figure 33.52 *(a) Main "double" lobby showing escalators and elevator entry. (b) Entry to elevators to specific floors. (© Julia Lau; used with permission.)*

Figure 33.53 *(a) Card reader to interior elevator lobby. (b) Multiple entries for high traffic flow. (© Julia Lau; used with permission.)*

Figure 33.54 *View from upper lobby to lower lobby at street level. (© SHK Properties; used with permission.)*

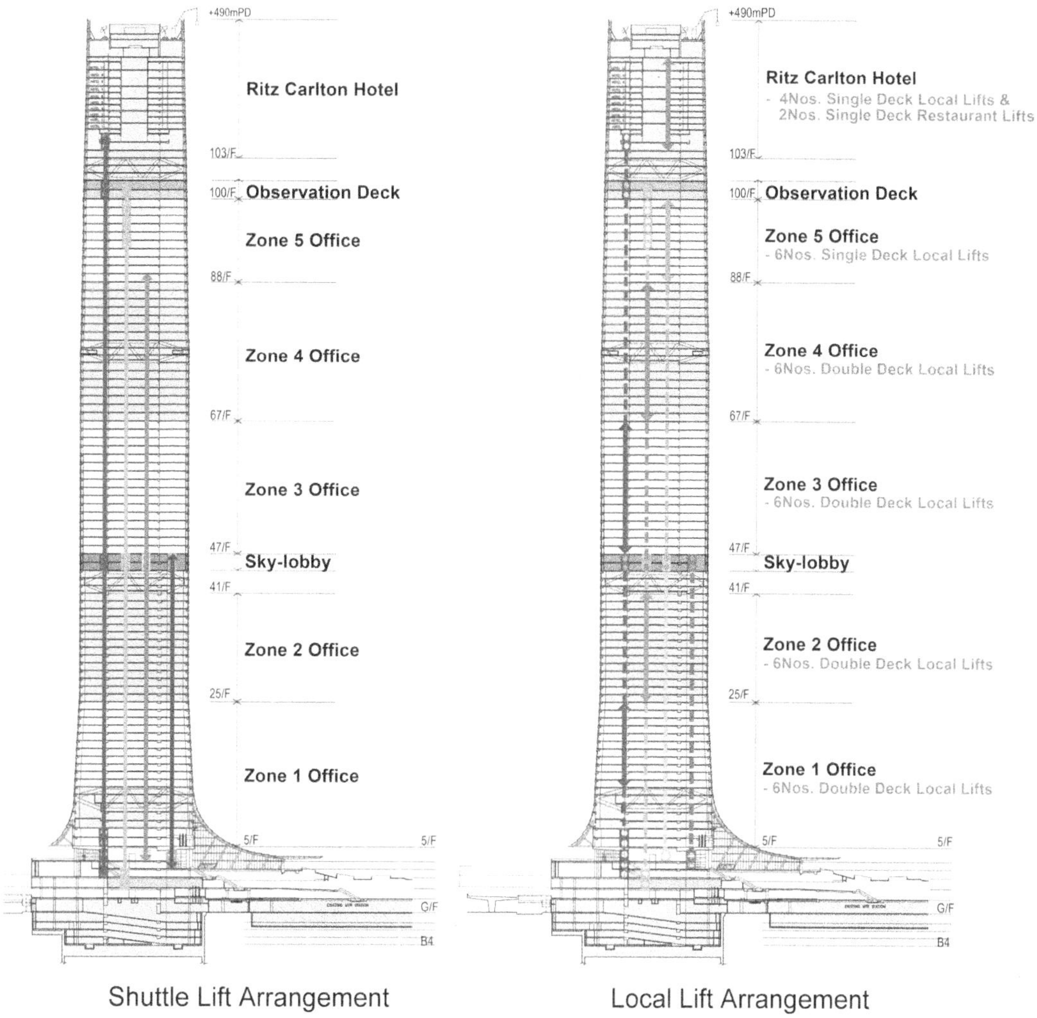

Figure 33.55 *Shuttle lift arrangement (left) shows fast travel to specific destinations, and local lift arrangement (right) shows intermittent stops between and within zones. (© Wong & Ouyang [HK] Ltd.; used with permission.)*

super-high structures such as the ICC to satisfy the large daily inflow of visitors, while not sacrificing rentable floor space. In dealing with buildings even taller than the ICC, consultants have proposed the use of triple-deck elevator systems (Knight, 2005).

FOR FURTHER INFORMATION

Knight, W. 2005. "Architects Plan Kilometre-High Skyscraper." *New Scientist*. newscientist.com/article/dn8445-architects-plan-kilometrehigh-skyscraper.html#.UfYUZKzgeS0

Kohn Pedersen Fox Associates. kpf.com

Malott, D. 2011. "International Commerce Center." *Elevator World*, May.

Schindler Group. 2009. East Meets West. *Next Floor*, August: 12–15. schindler.com/content/dam/website/us/docs/media/next-floor-magazine/next-floor-08-2009.pdf/_jcr_content/renditions/original./next-floor-08-2009.pdf

Schindler Elevators. schindler.com

Sun Hung Kai Properties. shkp.com

Wong & Ouyang Architects Ltd. www.wongouyang.com

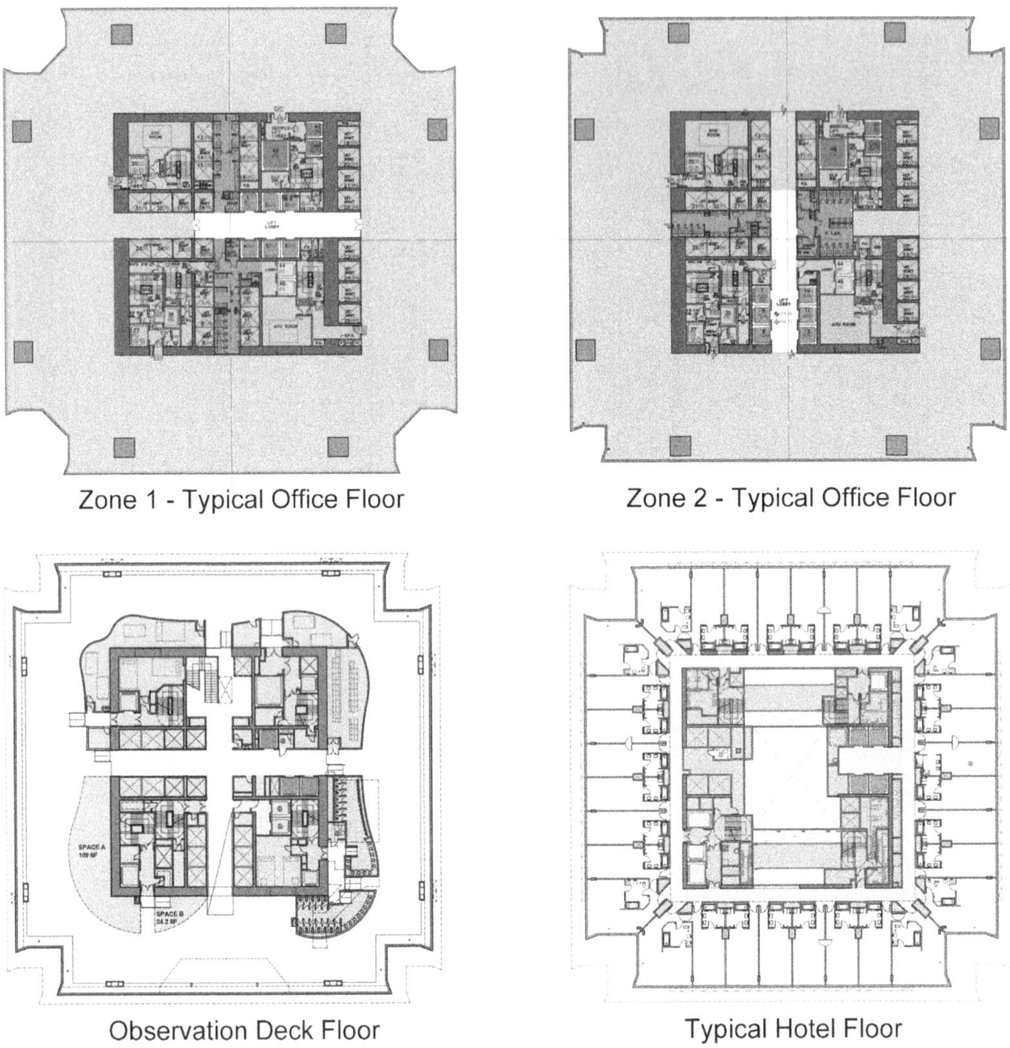

Figure 33.56 *Typical floor arrangements for office zones, observation deck, and the hotel. (© Wong & Ouyang [HK] Ltd.; used with permission.)*

References and Resources

ASME. 2015. *ASME A17.4-2015: Guide for Emergency Personnel*. New York: American Society of Mechanical Engineers.

ASME. 2022. *ASME A17.1/CSA B44-2022: Safety Code for Elevators and Escalators*. New York: American Society of Mechanical Engineers.

ASME. 2023. *ASME A17.3-2023: Safety Code for Existing Elevators and Escalators*. New York: American Society of Mechanical Engineers.

Cardinale, E. et al. 2004. *Emergency Evacuation Guidebook*. Chicago, IL: Council on Tall Buildings and Urban Habitat.

CEN. 2020a. *EN 81-20:2020: Safety Rules for the Construction and Installation of Lifts – Lifts for the Transport of Persons and Goods – Part 20: Passenger and Goods Passenger Lifts*. Brussels: European Committee for Standardization.

CEN. 2020b. *EN 81-50:2020: Safety Rules for the Construction and Installation of Lifts – Examinations and Tests – Part 50: Design Rules, Calculations, Examinations and Tests of Lift Components*. Brussels: European Committee for Standardization.

DOJ. 2012. *2010 ADA Standards for Accessible Design*. Washington, DC: U.S. Department of Justice. ada.gov/law-and-regs/design-standards/2010-stds

Elevator History. elevatorhistory.net/elevator-history

Goetz, A. 2003. *Up, Down, Across*. London: Merrell Publishers.

Grondzik, W. and A. Kwok. 2014. *Mechanical and Electrical Equipment for Buildings*, 12th edn. Hoboken, NJ: John Wiley & Sons.

ICC. 2017. *ICC A117.1-2017: Standard for Accessible and Usable Buildings and Facilities*. Washington, DC: International Code Council.

National Elevator Industry, Inc. 1992. *Vertical Transportation Standards*, 7th edn. New York: National Elevator Industry, Inc.

NFPA. 2022. *2023 NFPA 70: National Electrical Code (NEC)*. Quincy, MA: National Fire Protection Association.

NFPA. 2023. *2024 NFPA 101: Life Safety Code*. Quincy, MA: National Fire Protection Association.

Stein, B., J. Reynolds, and W. McGuinness. 1986. *Mechanical and Electrical Equipment for Buildings*, 7th edn. New York: John Wiley & Sons.

Strakosch, G. R. 1983. *Vertical Transportation, Elevators and Escalators*, 2nd edn. New York: John Wiley & Sons.

Strakosch, G. R. (ed.). 1998. *The Vertical Transportation Handbook*. New York: John Wiley & Sons.

U.S. Access Board. 2004. *Americans with Disabilities Act Accessibility Guidelines for Buildings and Facilities (ADAAG)*. Washington, DC: U.S. Architectural and Transportation Barriers Compliance Board. federalregister.gov/documents/2004/07/23/04-16025/americans-with-disabilities-act-ada-accessibility-guidelines-for-buildings-and-facilities

CHAPTER 34

Escalators and Other Systems

CHAPTER 33 DEALT WITH THE CONSTRUCTION, CONTROL, AND SELECTION OF PASSENGER AND FREIGHT ELEVATORS. This chapter discusses escalators; moving walks and ramps; and materials handling systems.

ESCALATORS

34.1 INTRODUCTION

The escalator, also referred to as a moving stairway, existed in a primitive, human/animal-powered form during the construction of the great Egyptian pyramids, with stone blocks moving up an incline covered in rolling tree trunks. Modern escalators emerged with steam and electric power in the mid-nineteenth century with early pioneers such as Jesse Reno and Charles Seeberger, both of whom sold their patents to the Otis Elevator Company. One of the first escalators used by the public was at the Paris Exposition in 1900.

The word escalator was derived from the Latin term for "stair" (*scala*) and the word "elevator" and was a proprietary term until the 1950s. Its modern successors deliver passengers comfortably, rapidly, safely, and continuously at constant speed and usually with no delay at the boarding level. The annoyance of waiting for elevators is eliminated. Also, no time is lost by acceleration, retardation, leveling, and door operation or by passenger interference while getting in or out of the cars. Instead of requiring formal lobbies and hallways leading to a bank of elevators on each floor and a ride in a small, enclosed box, the electric stairway is always in motion, inviting passengers to ride on an open, airy, conveyance that can never trap them due to equipment or power failure. In contrast to the generally utilitarian function of an elevator, an escalator also has a decorative/design function, and its open nature is frequently used to expose the rider to intentional visual panoramas.

34.2 COMPONENTS

The major components of a standard escalator installation are shown in Figure 34.1. Safety devices are discussed in Section 34.6.

The truss is a welded steel frame that supports the entire apparatus (see Figure 34.4). The tracks are steel angles attached to the truss on which the step rollers are guided, thus controlling the motion of the steps. The sprocket assemblies, chains, and machine provide the motive power for the unit, somewhat like the chain drive of a bicycle.

The handrail is driven by sheaves powered from the top sprocket assembly. It is synchronized with the tread motion and carefully calibrated to provide stability to riding passengers and support for entering and leaving passengers. Handrails disappear at inaccessible points at the newels.

Transparent balustrades are made of glass and are frequently referred to as crystal balustrades. In these units, the handrail is pinch-driven within the truss. In addition to metal and glass balustrade materials, fiberglass, wood, and various plastic materials are used.

Mechanical and Electrical Equipment for Buildings, Fourteenth Edition. Walter Grondzik, Daniel Overbey, and Alison Kwok.

Companion website: www.wiley.com/go/meeb14e

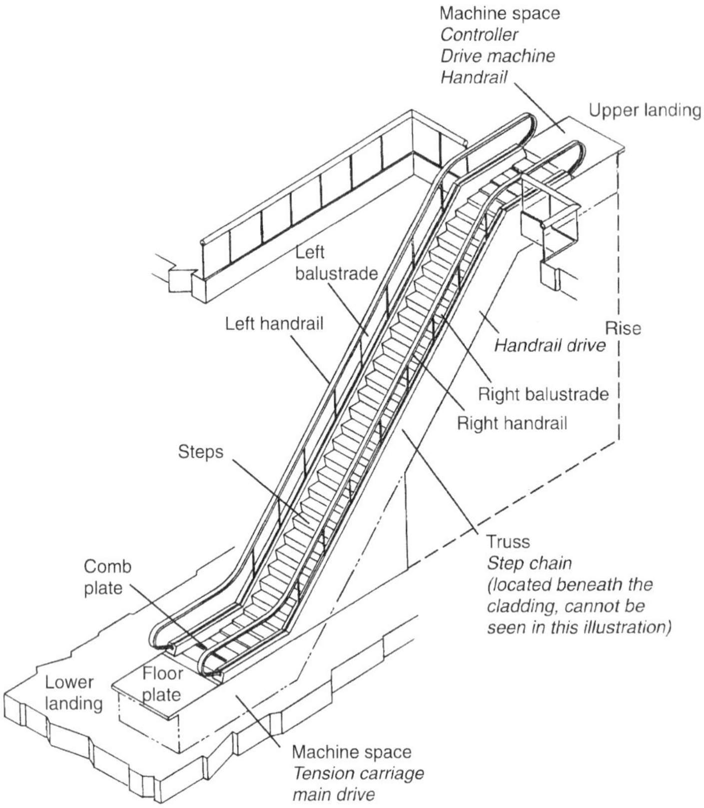

Figure 34.1 *View of a standard escalator showing the principal parts. (Courtesy of Otis Elevator Co.)*

The control cabinet, which is normally located near the drive machine, contains malfunction indicators in addition to the drive controls. The cabinet may also contain a microprocessor malfunction analyzer and communication means for transmitting escalator operating conditions to a central control point. Key-operated control switches at the top and bottom newels start, stop, and reverse the stairway.

34.3 PARALLEL AND CRISSCROSS ARRANGEMENTS

Moving stairs can be installed in three ways—in a crisscross arrangement or one of two parallel arrangements. These descriptive terms become evident in the diagrams and photos that follow.

The essential difference between the two basic arrangements lies in how a user transfers from one escalator to the next. In the crisscross arrangement, paired escalators are separated by the horizontal length of an escalator. In either parallel arrangement, paired escalators face in the same direction.

(a) The Crisscross Arrangement

This layout is simpler to visualize and more common (refer to Figures 34.2*a* and 34.2*b*). Notice that the escalator construction in both options (spiral and walk-around) is identical; the difference occurs in the direction of operation of the second level of escalator. In the spiral crisscross arrangement of Figure 34.2*a*, the rider begins an upward trip on escalator L1A and, by means of a 180° turn, continues the trip uninterrupted on escalator L2A—traveling, effectively, in an upward spiral. The downward-traveling rider performs the same spiral trip on escalators L2B and L1B. This arrangement is rapid, pleasant, and very economical of space because the escalators nest into each other. It can be used with as many as five floors without excessive annoyance to the rider.

Reversing the direction of the second-level stairs L2A and L2B (Figure 34.2*b*) forces the upward-traveling passenger to leave L1A at the first upper landing and walk around the entire length of the stair to continue the trip on L2B. Similarly, the downward-traveling passenger begins the trip on L2A but, at the first landing, must traverse the escalator length to reach the next down escalator L1B; hence the descriptive name walk-around crisscross. This arrangement requires floor space around the escalators, which is used in stores to display merchandise. Indeed, the potential for sales is the reason that stores force passengers to endure the potentially annoying walk-around.

A similar, but generally much shorter and therefore less objectionable, walk-around characterizes the spiral crisscross plan, where the escalators are

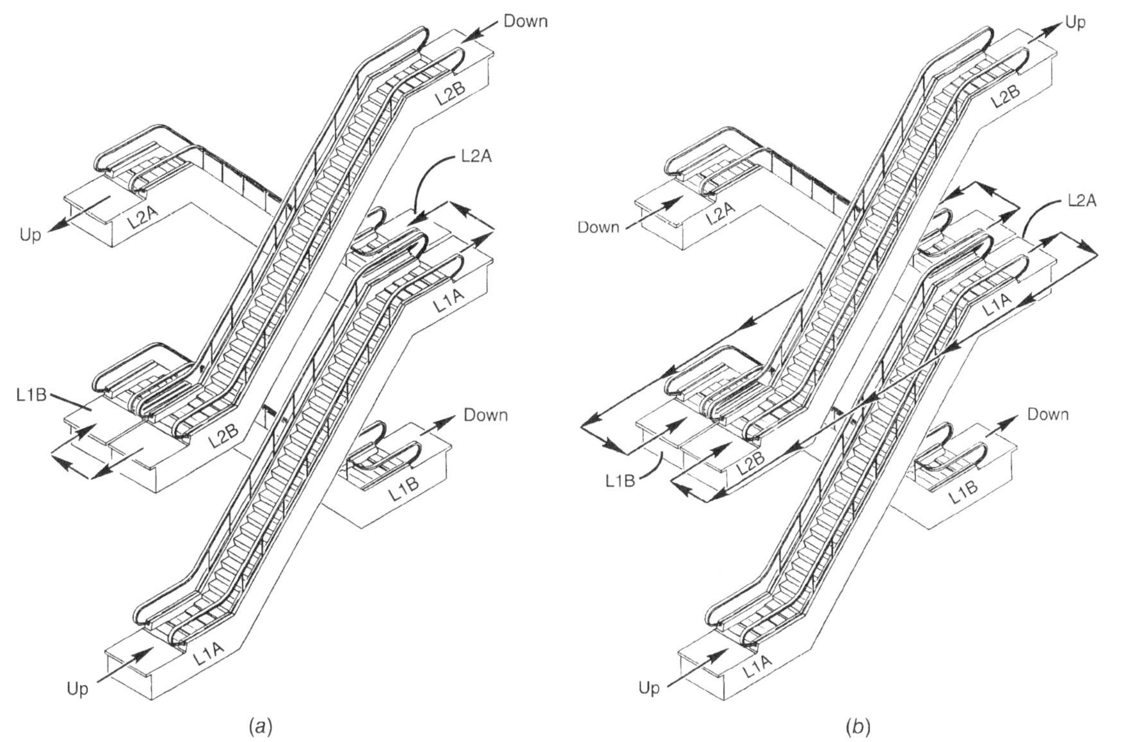

Figure 34.2 *Escalators constructed in a crisscross arrangement. (a) When escalators L1A and L2A are operated in the up direction, a passenger travels upward in a spiral, with only a turnaround at each level. The same is true for a down passenger. This system is called a spiral operation, and the entire installation is referred to as a crisscross spiral operation. (b) When the direction of upper-level escalators L2A and L2B is reversed, passengers at each intermediate landing (only one is shown) must traverse the entire horizontal escalator length to reach the next escalator traveling in the desired direction (see arrows on the drawing). This arrangement is known as the crisscross walk-around operation. (Courtesy of Otis Elevator Co.)*

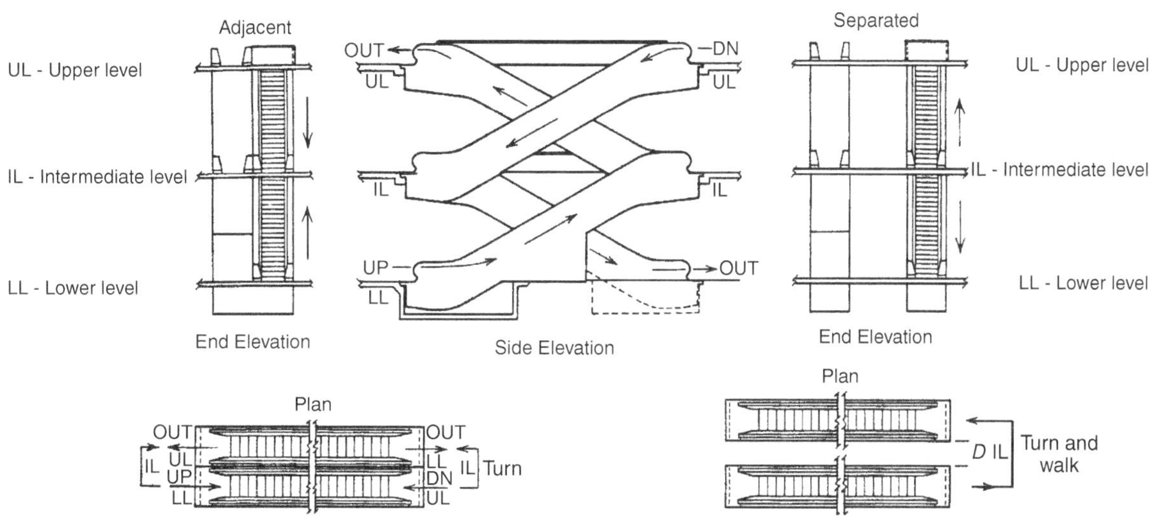

Figure 34.3 *Side and end elevations and plan views of the crisscross escalator arrangements. In the spiral operation mode (see text), separating the escalators forces the rider to walk distance D to continue a trip. In the walk-around operation mode, escalators are frequently separated to provide a walk-around path.*

separated by distance *D* (Figure 34.3). This distance is usually limited to about 10 ft [3 m] because any greater distance places the trip-continuation escalator out of sight, causing confusion and annoyance. Separation of the escalators is frequently an architectural consideration and has the advantage of easier mixing of riders entering at the various levels with riders making a continuous trip. Designers must be aware of

Traffic guide at turn-around

Figure 34.4 *Bank of glass balustrade escalators with mirrored sides in a spiral operation, crisscross plan. Note the dividers at the turnaround at the intermediate level, whose function is to guide traffic smoothly, either away from the escalator or to the turn point. This effectively eliminates undesirable bunching and crowding. (Courtesy of O&K Rolltreppen.)*

the possibility of a negative reaction to the separation of escalators, which can be reinforced when:

1. insufficient floor space is provided for the transit between escalators, causing crowding, pushing, and delay;
2. insufficient elevator service is provided for passengers wishing to travel at least three floors. This situation forces people to make a multistory escalator trip, which can be wearying, particularly when carrying parcels. If such a trip is further lengthened by an enforced walk-around at each floor, it can become a source of severe irritation.

Figure 34.3 shows the plan and elevation of the crisscross arrangement for both spiral and walk-around operations (layouts are identical for both). Figures 34.4 and 34.5 illustrate the crisscross plan. Figures 34.6 and 34.7 show shopping mall applications of the spiral and walk-around configurations.

(b) Parallel Escalators

This configuration requires more floor space than the crisscross arrangement and is therefore used less often. It is installed in either of two designs: the parallel spiral arrangement, shown in Figure 34.7*a*, or the stacked parallel arrangement, with forced walk-around, shown in Figure 34.7*b*. The principal advantage of the parallel arrangement is its impressive appearance, as can be seen in Figures 34.8 through 34.14. The stacked arrangement must be used with caution due to the inconvenience to the rider of an enforced long walk-around to continue the trip. This arrangement is found most often in mass-purchase facilities and in malls—where this arrangement is much less objectionable, as many people are there to browse and window-shop rather than to purchase and leave, as is the case in a single-purpose store. In a multistory store, the inconvenience of this walk—which is frequently compounded by the crowds of people who normally congregate at special sale merchandise displays and counters—rapidly breeds annoyance. Thus, the stacked arrangement is seldom used above two floors (requiring one walk-around), and when it is used above two floors, riders can be expected to gravitate to elevators.

Escalators between two contiguous levels do not present the continued trip problem and therefore are frequently used in the parallel arrangement. This configuration is particularly common in public buildings, transportation terminals, and other heavy traffic areas, where the advantage of a single location for the entrance to the bank of escalators eliminates confusion and the safety hazard engendered by hesitant, confused riders in heavy traffic.

Figure 34.5 *Crisscross design applicable to either spiral or walk-around operation. The horizontal area in the right foreground can be used to display impulse-buying merchandise to walk-around riders during holiday seasons, after which operation can be returned to the spiral mode. (Courtesy of O&K Rolltreppen.)*

(*a*) (*b*)

Figure 34.6 *(a, b) Crisscross design featuring seven stories of spiral escalators at the Westfield San Francisco Centre, an urban shopping mall located in San Francisco, California. Entries to stores surround the atrium. (a: © Robert Marcial, used with permission; b: © Alison Kwok, used with permission.)*

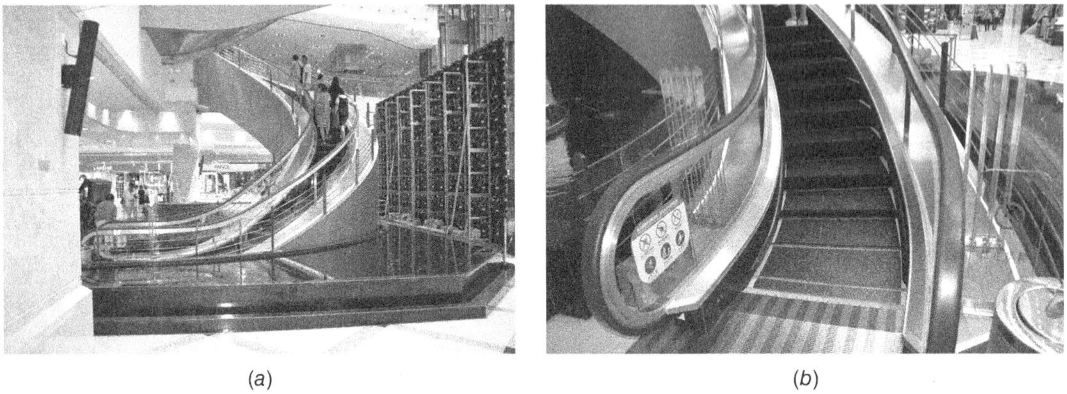

Figure 34.7 (a) A spiral escalator at the New World Dawan Department Store in Shanghai, China, and (b) the curved treads with yellow demarcation stripes added to show the edge of the tread. (© Moses Ling; used with permission.)

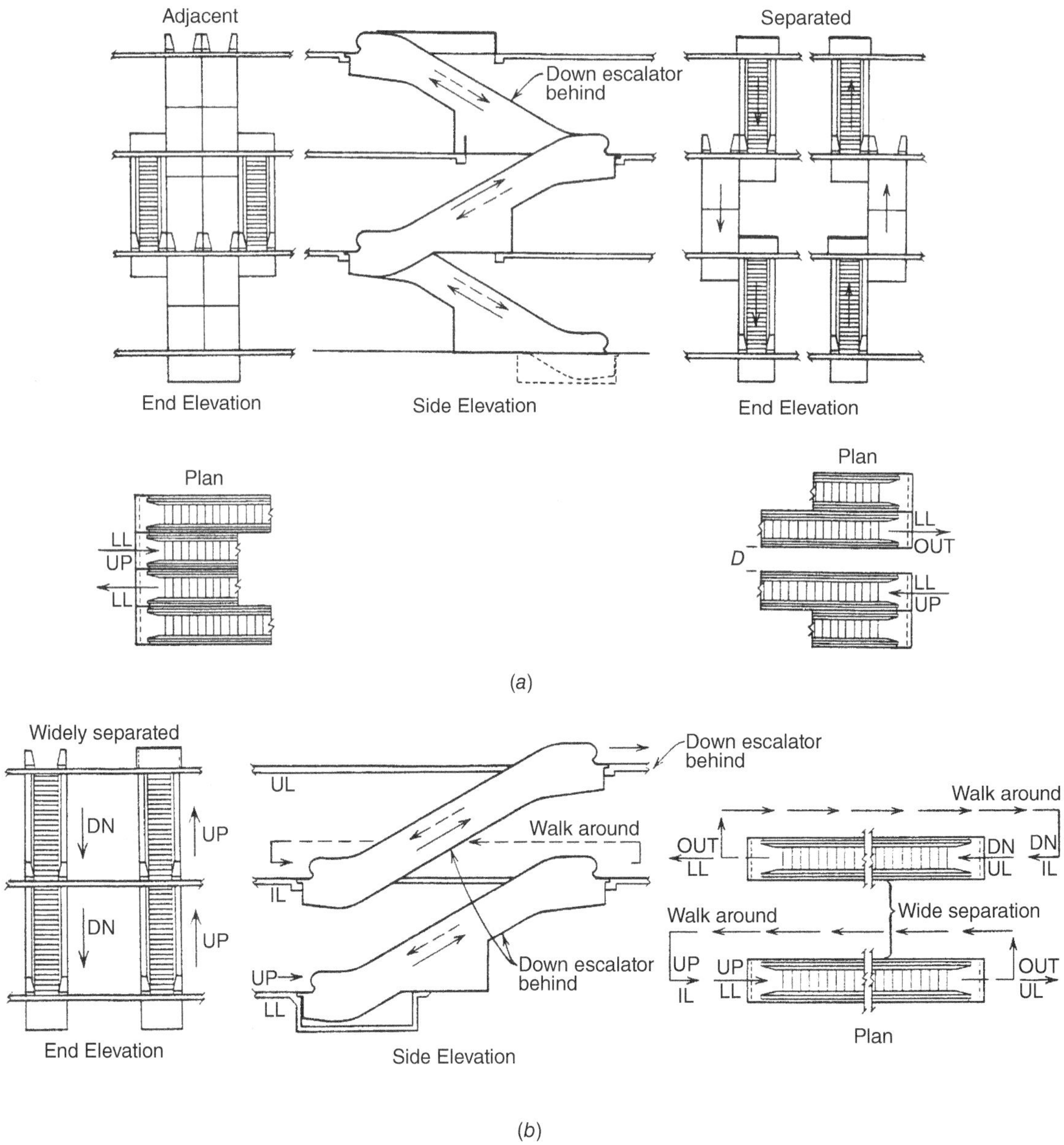

Figure 34.8 Side and end elevations and plan views of escalators in (a) spiral parallel and (b) stacked parallel arrangements. In the spiral parallel arrangement (a), the separation has no effect on the rider, who simply makes a turn to continue in the same up or down direction. In the stacked arrangement (b), the rider must traverse the entire length of the escalator at each level to continue.

Figure 34.9 *An unusual design of stacked parallel escalators with a forced walk-around in a continuous trip. The exposed truss and drive elements, combined with a transparent balustrade, add considerable visual interest to the installation. (Courtesy of O&K Rolltreppen.)*

Figure 34.11 *Escalators in transportation terminals are subject to periods of very heavy use. To avoid crowding at the entrance to the up (left) escalator, the balustrade is effectively extended with a glass-sided divider. Passengers on the down (right) escalator are guided by large overhead signs and a standing sign at the base, as well as balustrade extension to avoid hesitation and bunching. Note also the large open area in front of both units. (Courtesy of O&K Rolltreppen.)*

Figure 34.10 *An installation of parallel escalators adapted to a long four-story building. In the more common instance of a building that is taller than it is long, the escalators are stacked as in Figure 34.8. Here, the design is adapted to the large central court/atrium, with offset stepped landings and entrances to the buildings. (Courtesy of O&K Rolltreppen.)*

Planning for the division of rider traffic between elevators and escalators in a store is important. The general philosophy of store owners is to make escalators the primary means of vertical transportation to increase merchandise exposure. As a result, elevators are frequently placed at one or both ends of a store, whereas the escalator banks are central. However, care must be taken to provide sufficient elevatoring in stores exceeding three floors, particularly in stores using the stacked or walk-around crisscross plans, because of the inconvenience of multifloor escalator trips.

34.4 LOCATION

Because escalators are constantly moving and are generally part of a horizontal and vertical trip, they must be placed directly in the main line of traffic. This contrasts with an elevator bank, which can be set off as an element on its own for people to approach and utilize. Escalators must therefore be placed in the area served (often with a dominating presence) and provide a clear sense of destination. This allows potential riders to immediately:

1. locate the escalators
2. recognize an individual escalator's destination
3. move easily and comfortably toward the escalator

Figure 34.12 *Among the longest escalators in Europe are those installed in the Stockholm subway system. This parallel bank of three escalators is 230 ft [70 m] long with a 108-ft [33-m] rise. Unlike American standards of 30° incline, 100-fpm [0.51-m/s] speed, and 32- or 48-in. [0.81- or 1.22-m] width, these units are at 27.3° incline, 147.6-fpm [0.75-m/s] speed, and 40-in. [1.02-m] width. (Courtesy of O&K Rolltreppen.)*

Figure 34.13 *At 2625 ft [800 m], the Central–Mid-Levels escalator in Hong Kong is the longest outdoor, covered escalator system in the world. The goal was to give office workers from higher mid-level residential areas an easy, congestion-free passage to and from Central, the business district. There is no cost for people to take the escalator. (© Alison Kwok, used with permission.)*

Figure 34.14 *The Central–Mid-Levels escalator in Hong Kong offers stairways adjacent to 20 escalators and 3 inclined, moving walkways. The system operates 6:00 a.m. to 10:00 a.m. downhill and in the uphill direction 10:00 a.m. to midnight. (© Alison Kwok, used with permission.)*

Sufficient lobby space must be provided for queuing where anticipated, particularly at discharge points. A restricted or poorly marked discharge area causes passenger hesitation and traffic backup. Because the escalator discharges continuously, backup of traffic is dangerous and intolerable. This is particularly crucial in theaters and stadiums, where even momentary hesitation during peak traffic periods can be disastrous. To avoid this, four design steps, in descending order of importance, are taken:

1. provide well-marked escalators with sufficient traffic-carrying capacity
2. provide breathing room at intermediate landings so that passenger pressure can be relieved
3. provide a physical divider at intermediate landing turnaround points that guides riders away from the discharge point, and provide adequate space (and

time) for riders either to leave at that level or to follow the guide around and continue the trip (see Figures 34.3 and 34.4)
4. provide a slight setback for the next escalator so that the necessary 180° turn can readily be negotiated

At the exit terminus, an escalator should discharge into an open area with no turns or choice of direction. When such clarity is absolutely unachievable, large, unambiguous signs should make hesitation unnecessary (see Figure 34.11). The landing space beyond the escalator newels should be a minimum of 5 ft 6 in. [1.7 m] for 32-in. [0.81 m] units and 8 ft 2 in. [2.5 m] for 48-in. [1.22 m] units for a standard 100-fpm [0.5-m/s] speed. For escalators that will be reversed to accommodate a change in traffic direction, such landing space must be provided at the top and bottom.

The crisscross arrangement has the advantages of lower cost, minimum floor space being occupied, and the least structural requirements. The parallel arrangement, being less efficient and more expensive, has as a compensating virtue a very impressive appearance that strongly draws people to it. For this reason, it is frequently employed, particularly in banks of three or four units, in transportation terminals (see Figures. 34.11 and 34.12). In such large installations, flexibility is maintained by operating all but one escalator in the direction of heaviest traffic and reversing the direction of the escalators when needed.

34.5 SIZE, SPEED, CAPACITY, AND RISE

ASME A17.1/CSA B44-2022: Safety Code for Elevators and Escalators sets certain safety criteria in North America. Escalators are also built according to manufacturer and industry standards and are, therefore, available in standard designs. However, all major manufacturers also produce special designs for particular applications. The data that follow refer to standard designs.

All escalators in the United States are installed at an angle of 30° from the horizontal, with a minimum vertical clearance of 7 ft [2.1 m] for passengers. The 30° inclination means that the rise is equal to 57% of the unit's projected floor area for its inclined portion. The length of the horizontal portions of the stairway depends on the specific design. To meet universal design and the Americans with Disabilities Act (ADA) requirements, elongated newels with at least two horizontal treads before the landing plate are needed. Although the maximum linear speed permitted by *ASME A17.1* is 125 fpm [0.64 m/s], the industry has standardized on a single speed of 100 fpm [0.5 m/s].

ASME 17.1 now defines the width of an escalator as the width of the stair tread (in inches). The previous width designation referred to the distance between balustrades, a value difficult to define because of design variations. That older measurement is now called size. The standard sizes and widths are as shown.

Escalator Size, in. [m]	Tread Width, in. [m]
32 [0.81]	24 [0.61]
48 [1.22]	40 [1.02]

Table 34.1 lists theoretical, nominal (design), and observed average escalator passenger capacities. Maximum loads assume approximately 1 ¾ persons per tread for a 32-in. [0.81-m] unit and almost 2 persons per tread for a 48-in. [1.22-m] unit. In reality, maximum capacity is approached only during peak-load periods in transportation terminals and stadiums. At other times, a full (heavy) load is represented by the nominal capacity value and an average (extended period) load by the observed capacity value. Although a 40-in. [1.02-m] tread can indeed carry two persons, psychological factors, plus physical ones such as bulky clothing, packages, purses, and briefcases, militate against such loading. As a result, on a 40-in.-wide [1.02-m] tread, one person uses each step in a diagonal pattern, and on a 24-in. [0.61-m] escalator, one person occupies every other tread.

Some major manufacturers make two standard models (not styles) of escalators: (1) a standard-duty unit intended for general indoor use, which provides low to medium rise, and (2) a heavy-duty, sturdier unit intended for all-weather heavy-traffic use such as at transport terminals. The latter, because of the heavier construction, can provide a greater maximum rise than standard units. Maximum rise for off-the-shelf design

TABLE 34.1 Escalator Passenger Capacity

Size, in. [m]	Tread Width, in. [m]	Speed, fpm [m/s]	Passengers per Hour		
			Maximum[a]	Nominal[b]	Observed[c]
32 [0.81]	24 [0.61]	100 [0.51]	5200	4000	2300
48 [1.22]	40 [1.02]	100 [0.51]	9000	6750	4500

[a]Theoretical maximum (see text).
[b]Heavy loading (see text).
[c]Average long-term loading.

TRANSPORTATION

TABLE 34.2 Approximate Maximum Escalator Rise

Unit Size, in. [m]	Type	Supports	Maximum Rise, ft [m]
32 [0.81]	Standard	Ends	24 [7.3]
48 [1.22]			16 [4.9]
32 [0.81]	Standard	Ends and center	30 [9.1]
48 [1.22]			20 [6.1]
32 [0.81]	Heavy	Ends	24 [7.3]
48 [1.22]			18 [5.5]
32 [0.81]	Heavy	Ends and center	40 [12.2]
48 [1.22]			20 [6.1]

units varies among manufacturers. Approximate maximum values are given in Table 34.2. Specially designed units are available with rises of up to about 60 ft [18 m].

The motive power is delivered to an escalator at one point; the drive motor drives the main chain, which drives the top sprocket, which drives the step chain, which pulls up the steps, causing the entire assembly to move. This arrangement is suitable for moderate rises up to approximately 25 ft [7.6 m]; beyond that, the design becomes increasingly inefficient. As the rise increases, the loads on all the drive components, including chains and sprockets, increase sharply. Furthermore, to accommodate the heavier equipment, truss width increases, as do wellway size and balustrade decks. For rises above 25–35 ft [7.6–10.7 m]—depending on unit width—the drive motor is too large to fit inside the truss and requires a separate machine room below the truss, with associated cost. These factors combine to limit escalators to a maximum rise of about 60 ft [18 m], which varies slightly among manufacturers.

34.6 SAFETY FEATURES

Protection of passengers during normal escalator operation is ensured by several safety features:

- Handrails and steps travel at exactly the same speed to provide stability for users; the handrails extend beyond the point where the steps appear and disappear to ensure steadiness and balance for passengers stepping on or off the combplates (a safety device at the bottom and top of the escalator); handrails are pinch-resistant and the extension at the top and bottom avoids hands being caught at the point where the handrail disappears.
- The steps are large and steady and are designed to prevent slipping.
- Step design and step leveling with the combplates at each landing prevent passengers from tripping upon entering or leaving the escalator; this is accomplished with two or three (depending upon the manufacturer) horizontal steps at either end of the escalator to allow passengers to adjust.
- Balustrade and understep lighting are available to increase visibility of moving parts and to guide passengers on and off.
- The balustrade is designed to prevent catching of passengers' clothing or packages; close clearances provide safety near the combplates and step treads; glass balustrades are tempered glass.
- Adequate illumination is provided at all landings, at the combplates, and completely down the escalator; some escalator designs provide built-in lighting, as discussed in Section 34.8.
- A strip of nylon bristles guides passengers' feet and hands away from the gap between the floor plate/skirt panel and the first/last tread.
- Soft-start gently accelerates to full speed from a stopped position.
- An emergency stop button wired to the controller and placed near or on the escalator housing at both ends, stops the drive machine and applies the brake.
- An automatic service brake will bring the stairway to a smooth stop if

    - the drive chain or the step chain is broken or abnormally stretched
    - a foreign object is jammed into the handrail inlet, between the skirt guard and step or between steps, causing them to separate
    - a power failure occurs
    - the emergency stop button is operated (one is located at either end of the escalator)
    - any of the fire safety system devices operates (see Section 34.7)
    - a tread sags, rises, or breaks
    - a drive motor malfunction occurs

- In locations with seismic concerns, codes often require deeper pits and additional structural bracing.

In case of overspeed or underspeed, an automatic governor shuts down the escalator, prevents reversal of direction (up or down), and operates the service brake. If an escalator is stopped by operation of a safety device, passengers can walk the steps as they would on a stationary stairway.

34.7 FIRE PROTECTION

Four methods of providing fire protection near escalators are available: the rolling shutter, the smoke guard,

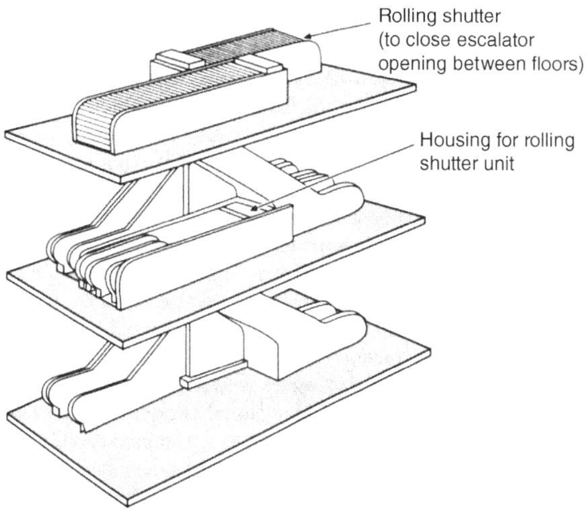

Figure 34.15 *Rolling-shutter method of wellway fire protection. (Reproduced with permission of NFPA from the Fire Protection Handbook, 20th edition. Copyright © 2008, National Fire Protection Association. For a full copy of the Fire Protection Handbook, please go to www.nfpa.org.)*

the spray-nozzle curtain, and the sprinkler-vent. One of these methods is required by code when more than two floors are pierced by an escalator opening (well). Figure 34.13 illustrates how the wellway at a given floor level may be entirely closed off by a fire shutter, thus preventing draft and the spread of fire upward through escalator wells. Temperature and smoke detectors automatically actuate the motor-driven shutters. The shutter in Figure 34.15 is shown at the third-floor level, but other shutters may be installed at any floor. This approach appears to be little implemented in North America, but is fairly common in the United Kingdom and Europe.

Figure 34.16 illustrates the smoke-guard method of protection. It consists of fireproof baffles surrounding the wellway and extending downward about 20 in. [510 mm] below the ceiling level. These guards can be designed with unobtrusive transparent glass panels with sealed joints. Smoke and flames rising upward to the escalator floor opening meet a curtain of water automatically released from conventional sprinkler heads, shown at the ceiling level. The baffle is a smoke and flame deflector. The vertical shields between adjacent sprinklers ensure that the spray from one will not prevent the opening of adjacent sprinklers.

The spray-nozzle curtain of water (not shown) is similar to the smoke-guard protection. Here, closely spaced, high-velocity water nozzles form a compact water curtain to prevent smoke and flames from rising through the wellways. Independent heat or smoke sensors open all nozzles simultaneously.

The sprinkler-vent fire control system is shown in Figure 34.17. The fresh air intake on the roof contains a blower to drive air downward through escalator floor openings, while the exhaust fan on the roof creates a strong draft upward through an exhaust duct; this duct in turn draws air from separate ducts just under the ceiling of each moving stairway floor opening. Three such separate wellway ducts are shown. Each duct has several smoke-pickup relays that automatically start the fans. The spray nozzles on the ceiling near the stairways aid in quenching the fire.

34.8 LIGHTING

The three primary goals for escalator lighting are: to enhance passenger safety; to provide clear directional guidance to passengers; and to attract passengers to the escalator as a preferred means of vertical circulation.

Adequate illumination of an escalator, particularly at the landings, is important to aesthetics as well as safety. Where general-area lighting does not provide sufficient illumination for an escalator, appropriate lighting fixtures are installed on the ceiling above the escalator, with special emphasis on lighting the combplate. In Figure 34.12, banks of luminaires are placed across the escalators at frequent intervals along the rise. Note the additional lighting provided at the combplate. Two different lighting treatments of similar installations are shown in Figures 34.18 and 34.19 (see also the balustrade lighting in Figure 34.20).

New approaches to escalator lighting include continuous light-emitting diode (LED) skirt bands that provide uniform illuminance at the base of the handrail along the treads. LEDs can provide light and color effects to enhance the aesthetic appeal of the escalator. LEDs can also be used as an alternative to fluorescent lamps under handrails. The structural trusses can be exposed and lit from within for aesthetic appeal or to highlight the escalator mechanicals. Soffit lighting underneath the escalator can be effective with parallel and crisscross escalators servicing multiple levels. Traffic lighting in the form of stop and go lights can be used at the entrance to escalators to direct or inform passengers.

34.9 ESCALATOR APPLICATIONS

1. Escalators on the main floor of a building should be placed in the direct flow of traffic to ensure maximum use.
2. Vertical arrangements should be designed to accomplish specific intents, such as exposure of merchandise, maximum passenger capacity, and/or maximum accessibility to various areas.
3. Reversible escalators should be considered in applications where major traffic flow is unidirectional. Light traffic in the reverse direction can be handled

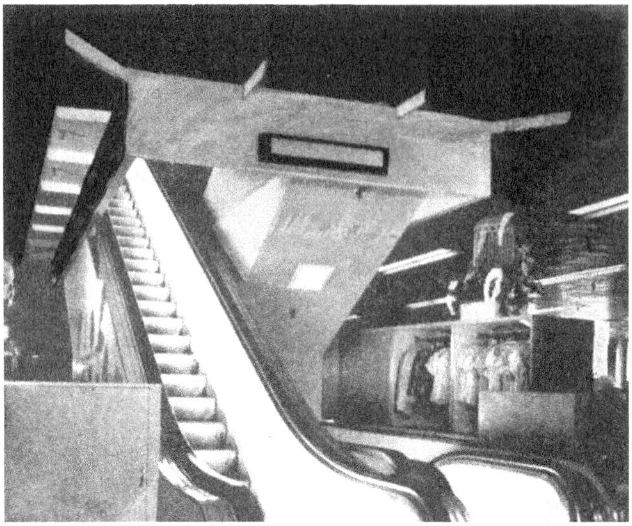

(a)

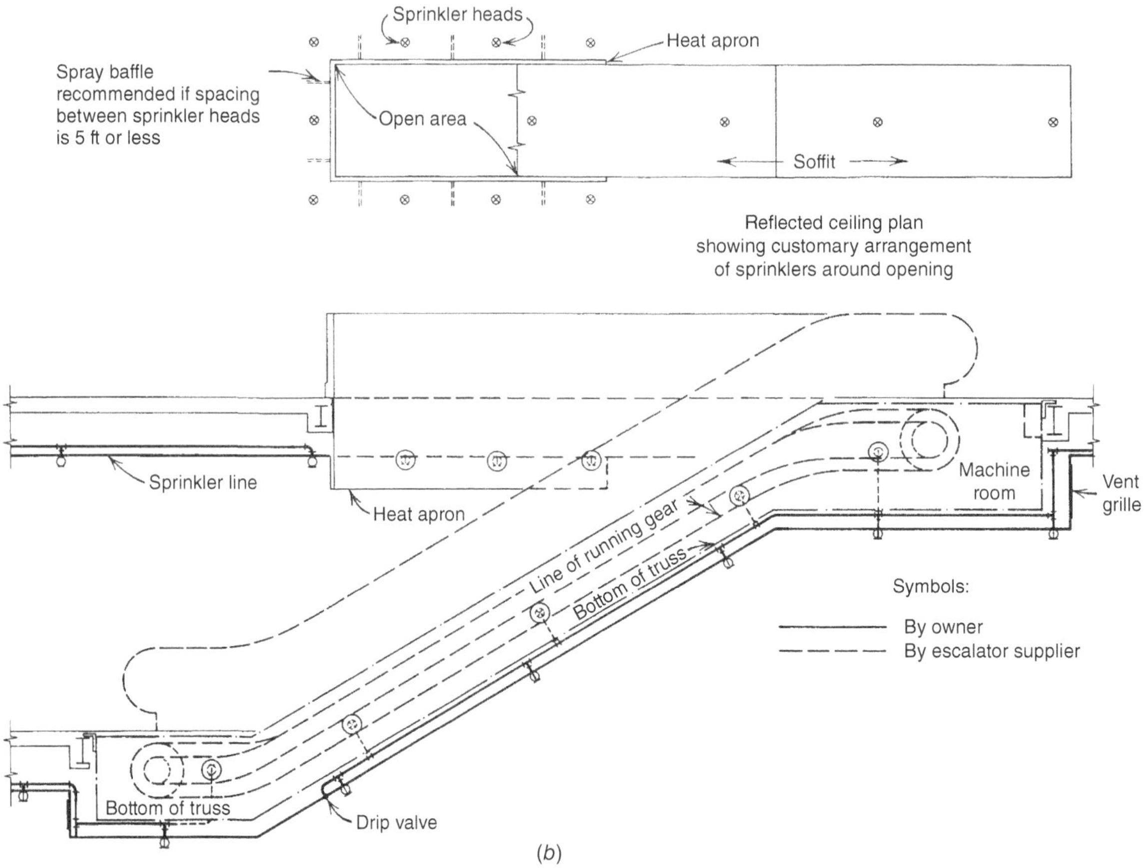

(b)

Figure 34.16 *(a) Smoke-guard method of fire protection for a 32-in. [0.81-m] escalator, crisscross type. The floor opening (per floor) is approximately 4 ft, 4 in. × 14 ft, 8 in. [1.3 by 4.5 m]. (b) Reflected ceiling plan and section showing baffle and sprinkler layout. (Courtesy of Otis Elevator Co.)*

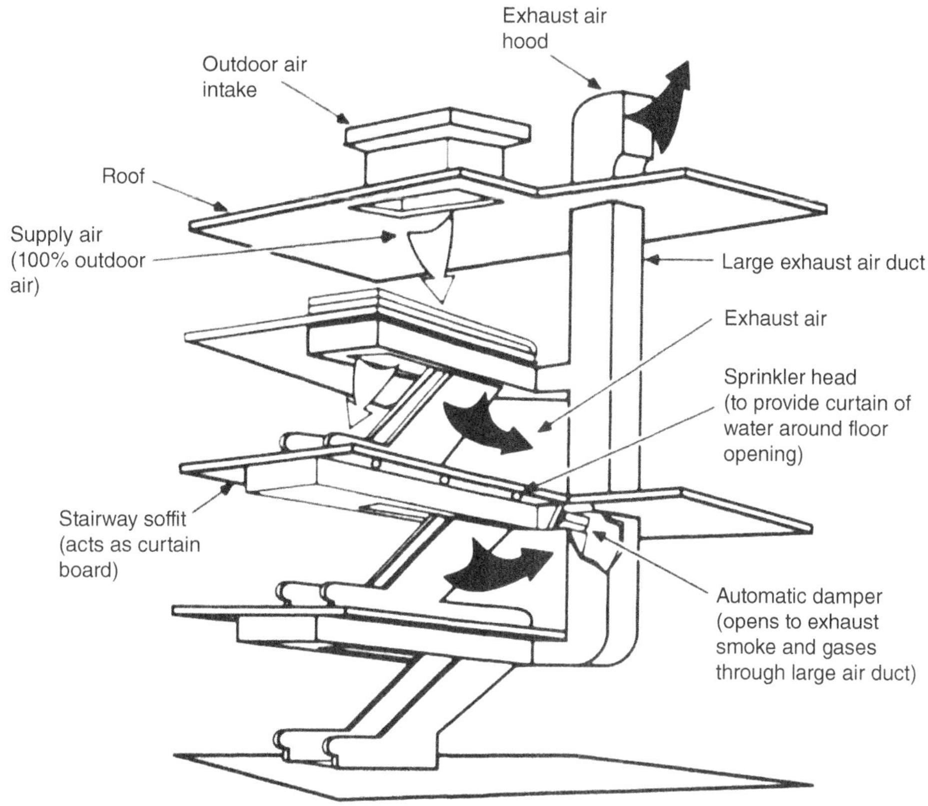

Figure 34.17 *Sprinkler-vent fire protection for escalator openings—an exception (with control) to the rule against perforations in floors. (Reproduced with permission of NFPA from the Fire Protection Handbook, 20th edition. Copyright © 2008, National Fire Protection Association. For a full copy of the Fire Protection Handbook, please go to www.nfpa.org.)*

Figure 34.18 *A continuous fluorescent source is placed under the handrail of the crystal balustrade in this covered mall in West Germany. Note the additional light at the base for illuminating the combplate. (Courtesy of O&K Rolltreppen.)*

Figure 34.19 *Lighting for escalator treads and combplates is supplied by a continuous fluorescent strip at the base of the balustrade in the Seattle Public Library. (© Alison Kwok, used with permission.)*

by a normal fixed stair adjacent to the escalator (see Figure 34.20). Similarly, a bank of two escalators can operate either both up, both down, or one up and one down, to handle variable traffic conditions in areas such as office buildings and transportation terminals.

4. Exterior escalators can provide an interesting and economical way of transporting people to elevated entry points in a building (see Figure 34.20); proper drainage should be included and attention paid to durability and maintenance of materials exposed to the elements.

Figure 34.20 An all-weather exterior escalator is often the best solution to an elevated building entrance or exit. Note the elongated left balustrade, which prevents bunching at the escalator entrance, and the luminaire built into the balustrade, which provides nighttime tread lighting. (Courtesy of O&K Rolltreppen.)

34.10 ELEVATORS AND ESCALATORS

Elevators and escalators should be considered together—as an integrated vertical transportation solution customized for the particular facility being designed (see Figure 34.21). Elevators allow people and goods to be moved over substantial vertical distances but at moderate effective speed. Escalators move people (but not goods) quickly but with vertical travel limits. Paired, these two systems can work in synergy in many building types (see Figure 34.22). Escalators can act as feeders for elevators in situations where having multiple "main" elevator lobbies would seriously degrade elevator system performance.

Figure 34.21 An escalator, an elevator, and a communicating stair together at Philadelphia International Airport offering passengers three different means of vertical transportation. (Courtesy of Tom Collins.)

In certain facilities, there are times when demand for vertical transportation is so great that elevators alone are not a feasible solution. A prime example is a school building. During class change, virtually the entire building population moves, with as many as 80% moving between floors. Since class change time is at most 10 minutes, the only reasonable solution is the combined use of fixed stairs and escalators. In other buildings, such as multifloor stores, the escalator provides for short trips of one or two floors, and the elevator generally transports passengers traveling three or more stories.

A comparison of travel time between escalators and elevators is of interest. For an escalator, using a normal speed of 100 fpm [0.51 m/s], a 12-ft [3.7-m] floor height requires 14.4 seconds for travel plus about 5 to 6 seconds to transfer to the next section, for a

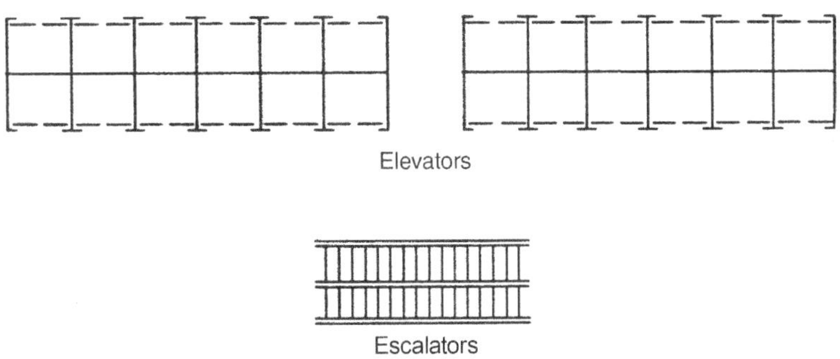

Figure 34.22 Comparative space requirements for equivalent passenger-handling capacity. Note the substantial space savings offered by escalators.

total of approximately 20 seconds. Thus, a four-story trip would take approximately 75 seconds. A similar elevator trip would take perhaps 60 seconds but involve lobby waiting time. The additional escalator time is not noticed amid the activity of boarding, turning, and riding. A trip of more than four stories, however, becomes tiresome, all the more so when there is an enforced walk-around at each floor.

34.11 ELECTRIC POWER REQUIREMENTS

Standard North American escalators are driven by three-phase, 60-Hz, AC induction motors at standard voltages (208, 230, and 460 V). Approximate horsepower data for drive motors are shown in Table 34.3.

It is recommended that no more than four escalators be served by a single electric feeder and further, that not all the escalators in an installation, whatever the number, be served from the same feeder.

Since one cannot be trapped on an escalator, emergency power is rarely required. Ventilation for the drive machinery should be supplied to handle the approximately 40% of the power expected to be dissipated as heat in a confined space. Thus, a 10-hp [7.5-kW] motor would require the dissipation of 0.40 × 10 × 2500 (Btu/h per hp), or approximately 10,000 Btu/h [2.9 kW].

34.12 ENERGY CONSERVATION

A reduction in energy consumption can be accomplished using several approaches. Escalators in buildings traditionally were run continuously throughout the day even when no passengers required service. Escalators, however, can be controlled to stop when not in use and to start when they detect passengers. Detection can be in form of infrared sensors, photocell sensors, or pressure mats and is typically coupled with a soft-start feature that slowly increases speed for the ease and safety of passengers. However, stopped escalators can be mistaken for broken escalators by passengers. One alternative is an escalator that slows considerably when not in use through variable voltage motors and controls. As with elevators, regenerative braking can be used on down escalators.

TABLE 34.3 Typical Standard-Duty Escalator Motor Sizes

Escalator Maximum Size, in. [m]	Rise, ft [m]	Size of Motor, hp [kW]
32 [0.81]	14 [4.3]	5 [3.7]
	22 [6.7]	7½ [5.6]
	30 [9.1]	10 [7.5]
48 [1.22]	10 [3.1]	5 [3.7]
	15 [4.6]	10 [7.5]
	20 [6.1]	15 [11.2]

34.13 SPECIAL DESIGN ESCALATORS

Like elevators, escalators of nonstandard design are available on special order. These include units with slopes other than 30°, speeds other than 100 fpm [0.51 m/s], and rises beyond normal. The most unusual of these special designs is the curved escalator, which made its debut overseas and is now available in the United States. This design, which solves the very common problem of directing passengers to make a 90° turn as they leave the escalator, is an engineering *tour de force*. The entire drive mechanism had to be designed anew because the tread speed along the inside curve must be lower than that on the outside curve (similar to an airport baggage carousel).

34.14 PRELIMINARY DESIGN DATA AND INSTALLATION DRAWINGS

At the preliminary design stage, an architect requires rough dimensional and structural data for a proposed escalator installation. At this stage of design, a specific escalator manufacturer has not typically been selected from whom data can be obtained. In addition, data vary from one supplier to another. Thus, the information in Figure 34.23 and Table 34.4 is given to provide a range that covers most of the major manufacturers. Similar data are available from the National Elevator Industry, Inc. Once construction contracts have been awarded and an escalator supplier selected, precise space requirements and structural reactions are available from the vendor. Two "working points" (see Figure 34.23) are usually identified, from which all other measurements are made—that is, locating the centerline of the truss sections, placing the lower and upper landing truss support beams, and so on.

34.15 BUDGET ESTIMATING FOR ESCALATORS

The first cost of an escalator includes the cost of the associated mechanical and electrical equipment, plus delivery and installation charges. The manufacturer provides engineering services and a field erector who supervises the installation.

TRANSPORTATION

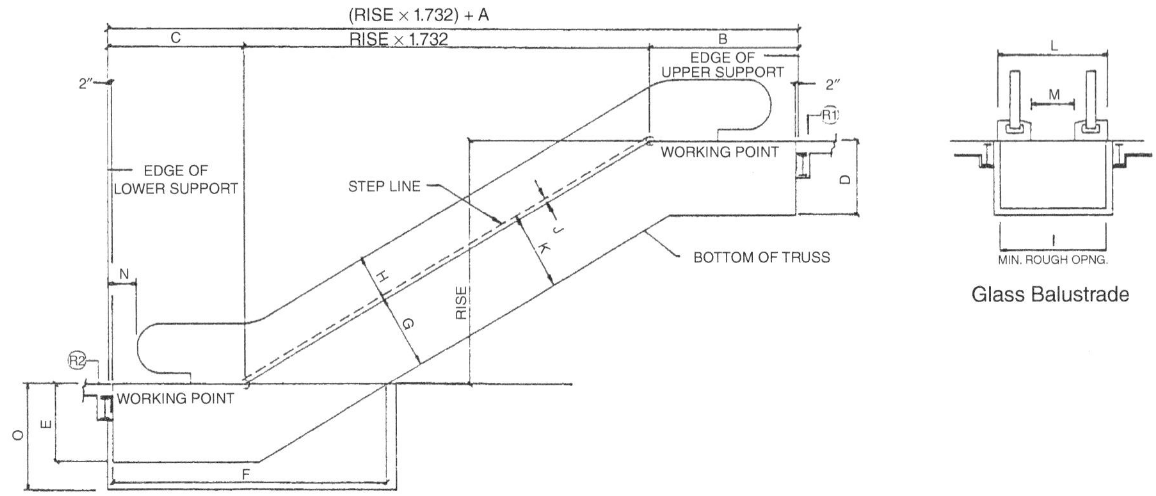

Figure 34.23 *Generic escalator dimensional data. See Table 34.4 for I-P and SI values. This information should be used for preliminary planning only.*

TABLE 34.4 Typical Escalator Dimensions

Dimension	Range (I-P)	Range (SI)
A	13′-2″ to 16′-10″	4.0 to 5.1 m
B	7′-5″ to 9′-4″	2.3 to 2.9 m
C	5′-9″ to 7′-6″	1.8 to 2.3 m
D	3′-3″ to 4′-2″	1.0 to 1.3 m
E	3′-3″ to 3′-9″	1.0 to 1.1 m
F	12′-5″ to 14′-10″	3.8 to 4.5 m
G	2′-11″ to 3′-1″	889 to 940 mm
H	2′-7″ to 2′-9″	787 to 838 mm
I: 32″ [812 mm]	3′-11″ to 4′-6″	1.2 to 1.4 m
I: 48″ [1220 mm]	5′-3″ to 5′-8″	1.6 to 1.7 m
J	0	0
K	2′-11″ to 3′-2″	889 to 965 mm
L: 32″ [812 mm]	4′-0″ to 4′-4″	1.2 to 1.3 m
L: 48″ [1220 mm]	5′-3″ to 5′-10″	1.6 to 1.8 m
M: 32″ [812 mm]	2′-0″	610 mm
M: 48″ [1220 mm]	3′-4″	1.0 m
N	1′-9″ to 2′-2″	533 to 660 mm
O	3′-6″ to 4′-0″	1.1 to 1.2 m

Note: See Figure 34.23 for locations of dimensions A through O.

TABLE 34.5 Relative Escalator Costs

	Size	
Rise, ft [m]	**32 in. [0.81 m]**	**48 in. [1.22 m]**
14 [4.3]	100[a]	118
16 [4.9]	103	121
18 [5.5]	105	124
20 [6.1]	108	127
22 [6.7]	111	130
24 [7.3]	113	133
26 [7.9]	115	135
28 [8.5]	118	138
30 [9.1]	120	141

[a]Base price value is 100 monetary units for the shortest, narrowest, slowest unit (i.e., 14-ft [4.3-m] rise, 32-in. [0.81-m] wide). All units are 100 fpm [0.51 m/s]. Add 5% for a glass balustrade for any unit.

Relative escalator prices, on a base of 100 monetary units, are given in Table 34.5. Prices for units with a rise above 35 ft [10.7 m] increase very rapidly and depend upon the type of unit used. Only actual manufacturer's quotes should be used for project estimating. To these base values must be added the cost of the contractor's work, wellway protection, lighting, wellway balustrades, fire protection devices, and associated walls/soffits.

MOVING WALKS AND RAMPS

34.16 GENERAL INFORMATION

Moving walks and ramps are different from escalators in application, function, construction, and capacity. The first moving walks were used at the 1893 Columbian World's Exhibition in Chicago and the 1900 Exposition Universelle in Paris to transport people around the fairgrounds. The Paris device was innovative in its use of three stepped tiers of parallel walks moving at different speeds, which allowed passengers to easily step from standing position to a slowly moving, then to a slightly faster moving platform, and then to a platform moving at 6 mph [10 km/h]. Airports are a common building type where moving walks are found; Dallas was the first installation in 1958. When transportation of wheeled vehicles or large parcels is required, the use of an escalator is awkward, if not unsafe and impossible. A moving walk or ramp serves such situations very well.

Herein, a moving walk is defined as having an incline not exceeding 5°, where the principal function is horizontal motion, and any slight inclined motion is incidental to the horizontal. A moving ramp is a device with an incline limited by code to 15°, where vertical motion is generally more important than the horizontal component. A moving walk and a moving ramp are physically the same concept, but differently applied.

34.17 APPLICATION OF MOVING WALKS

The principal uses of moving walks, also known as autowalks, are to do the following:

1. eliminate and/or accelerate burdensome walking
2. eliminate congestion
3. force movement
4. easily transport large, bulky objects.

Anyone who has walked the seemingly endless distances in a major airport, carrying or dragging a heavy suitcase, can appreciate the benefits of a moving walkway. For this reason, air transport terminals (see Figure 34.24) have become major users. Other transportation facilities, such as rail and ship terminals, can often find excellent applications for a moving walk, since much heavy and bulky luggage is moved in these areas.

The apparent travel compression that moving walks can provide permits the placement of parking areas in more distant locations. Thus, a store can extend its parking area with no annoyance to patrons who must make a long trip to their cars with bulky packages or shopping carts. These advantages are even more appreciated by persons with a walking impediment.

A second application of moving walks, as noted previously, is the routing of traffic to avoid congestion, milling about, and lost time and motion. This is particularly applicable in transportation terminals, where people are always traveling in opposite directions through the same—and often restricted—area, such as in "fingers" leading from airplanes to the main air terminal.

Figure 34.24 *Two pairs of moving walks. At Chicago O'Hare International Airport, horizontal moving walks ease circulation congestion and the hassle of traveling with heavy carry-on items. The walkway connecting the concourses is complemented with vibrant colors that change in pattern, designed to be an enlivening experience for travelers. (© Alison Kwok, used with permission.)*

Figure 34.25 Unusual angular design of multiple moving ramps in a shopping mall in Brisbane, Australia. The use of ramps rather than escalators permits shopping carts and bulky packages to be moved with ease between the building levels. (Courtesy of O&K Rolltreppen.)

Moving walks are also useful to move people past a display window or some other point where congestion caused by stopping is undesirable. This "movement of objects" application suggests that the moving walk is simply a large conveyor belt, regardless of its construction. To make users aware of the transition from moving surface to stationary surface, most moving walk installations provide an announcement as you near the end.

34.18 APPLICATION OF MOVING RAMPS

A moving ramp, which combines horizontal and vertical movement, is principally applicable:

1. to move persons and wheeled vehicles vertically
2. to move persons who lack the agility required to use an escalator
3. to move large, bulky objects vertically.

Ramps have found an important field of application in multilevel stores (Figure 34.25), where escalators are not feasible for shopping-cart users. Such stores may also utilize rooftop parking that is made accessible to cart users via a moving ramp. Since luggage carriers are not easily used on escalators, transportation terminals, which are almost always multilevel, also find extensive application for moving ramps (Figures 34.26 and 34.27).

34.19 SIZE, CAPACITY, AND SPEED

The speed, physical dimensions, and passenger capacity of moving walks and ramps are not as extensively standardized as with escalators. Manufacturers utilize several different tread widths, combined with various speeds and ramp angles of incline. The combinations are designed to suit the application. Furthermore, since the maximum permissible walk or ramp speed varies with the angle of slope, and with the design of the point of entry, passenger capacity ratings vary with

Figure 34.26 Multiple inclined moving ramp installation in a European department store (three up, two down). Of interest in this installation are the level entries, the lighted balustrades, the ease with which shopping carts are handled on the wide pallet, and the use of the space between the transparent balustrades for merchandising. (Courtesy of O&K Rolltreppen.)

Figure 34.27 *Inclined moving ramp installation at a European airport allowing passengers to roll luggage with ease on the ramp, approximately 5 ft [1.5 m] versus carrying luggage up the stairs shown on the left. (© Alison Kwok, used with permission.)*

each design. By code, higher speeds are allowed for a level entrance than for a sloping entrance, since a level entrance is easier to board. Tables 34.6, 34.7 and Figure 34.28 give data for typical units.

TABLE 34.6 Moving Walks/Ramps: Commercial Ratings

Angle	Pallet Width, in. [m]	Device Speed, fpm [m/s]
0–3° Walk	32 [0.81]	100 [0.5], 130 [0.65], 150 [0.75]
	40 [1.02]	100 [0.5], 130 [0.65], 150 [0.75]
	62 [1.58]	90 [0.45], 100 [0.5], 130 [0.65]
10°, 11°, 12° Ramp	32 [0.81]	100 [0.5], 130 [0.65]
	40 [1.02]	90 [0.45], 100 [0.5], 130 [0.65]

TABLE 34.7 Maximum Permissible Operating Speeds of Moving Ramps

	Maximum Speed, fpm [m/s]	
Angle of Incline	Level Entrance	Sloping Entrance
0–3°	180 [0.9]	180 [0.9]
3–5°	180 [0.9]	160 [0.8]
5–8°	180 [0.9]	140 [0.7]
8–12°	140 [0.7]	130 [0.65]
12–15°	140 [0.7]	125 [0.625]

Since capacity varies with width, speed, and type of entrance, exact capacity values must be obtained for each specific design. The maximum practical length at present is approximately 1000 ft [305 m], with longer units under consideration. Wide units (55 in. [1.4 m]) are now extensively used in transportation terminals, since they permit passengers with luggage trolleys who walk on the moving walkway to easily pass other passengers with luggage carts who prefer to stand still and move at the relatively slow rate of 100–130 fpm [0.51–0.66 m/s]—about half of normal walking speed. Installations of such wide autowalks are found at Heathrow, Gatwick, and Munich airports, among others.

34.20 COMPONENTS

Moving walks and ramps are manufactured in only one design: a derivative of the escalator, which uses a flattened pallet in place of a step. In all other respects—the drive mechanism, safeties, brake, handrails, and so on—they are similar to an escalator. Preliminary layout data and detailed plans are available

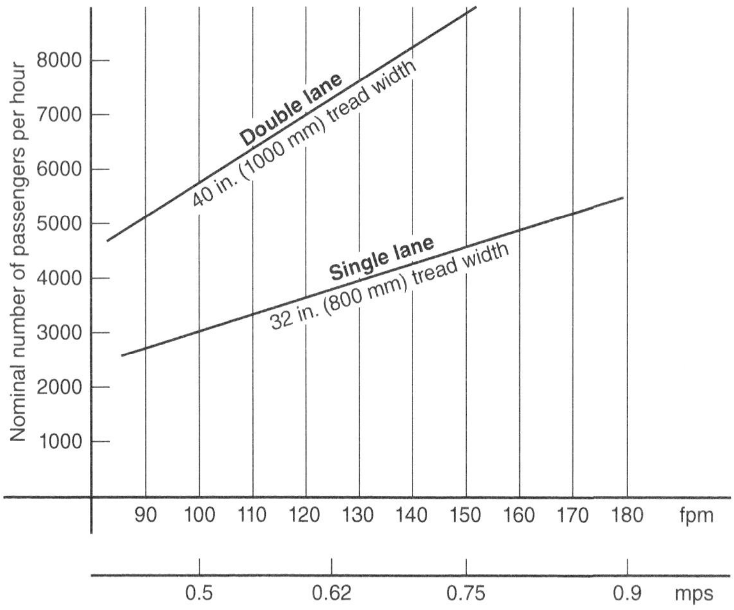

Figure 34.28 *The capacity of moving walks varies with speed, angle of incline, and tread width. The capacity shown is for the maximum incline permitted at that walk speed. Because handrail support is required, tread widths greater than double lane are not often utilized.*

from manufacturers, who frequently are the same companies that supply escalators.

SPECIAL PASSENGER CONVEYANCE SYSTEMS

34.21 AERIAL TRAMS

An aerial tram is a type of lift or cable car, typically with one or two fixed cables in an arrangement whereby one loop of the cable (haulage rope) grips the wheel that rolls on the two fixed cables. Passenger cabins move as an electric motor drives the haulage rope to propel the aerial tramway.

The Portland Aerial Tram (Figure 34.29) uses cables to hoist a passenger car that carries commuters uphill 500 ft [150 m] to the Oregon Health & Science University, from Portland's South Waterfront District 3300 ft [1000 m] away. Two cars, each carrying up to 79 people, pass each other along the cable span. The system is powered by an electric motor at the bottom, which moves the cable pulling one car down and the other car up.

Figure 34.29 *The Portland Aerial Tram ride takes about 3 minutes each way, connecting employees, parking, transportation, housing, and health care on the waterfront to the university hospital on the hill. (© Alison Kwok, used with permission.)*

34.22 RESIDENTIAL ELEVATORS AND CHAIR LIFTS

Although the needs of the disabled have been officially recognized in legislated requirements only since the late 1970s, the elevator industry has been providing for the mobility-impaired for years. Small private-residence elevators can double as wheelchair lifts and are available in a wide range of designs, including winding-drum (Figure 34.30), roped hydraulic, and worm and screw units. Most residential elevators are low-speed, limited-load units. Standard traction elevators requiring overhead space and standard hydraulic

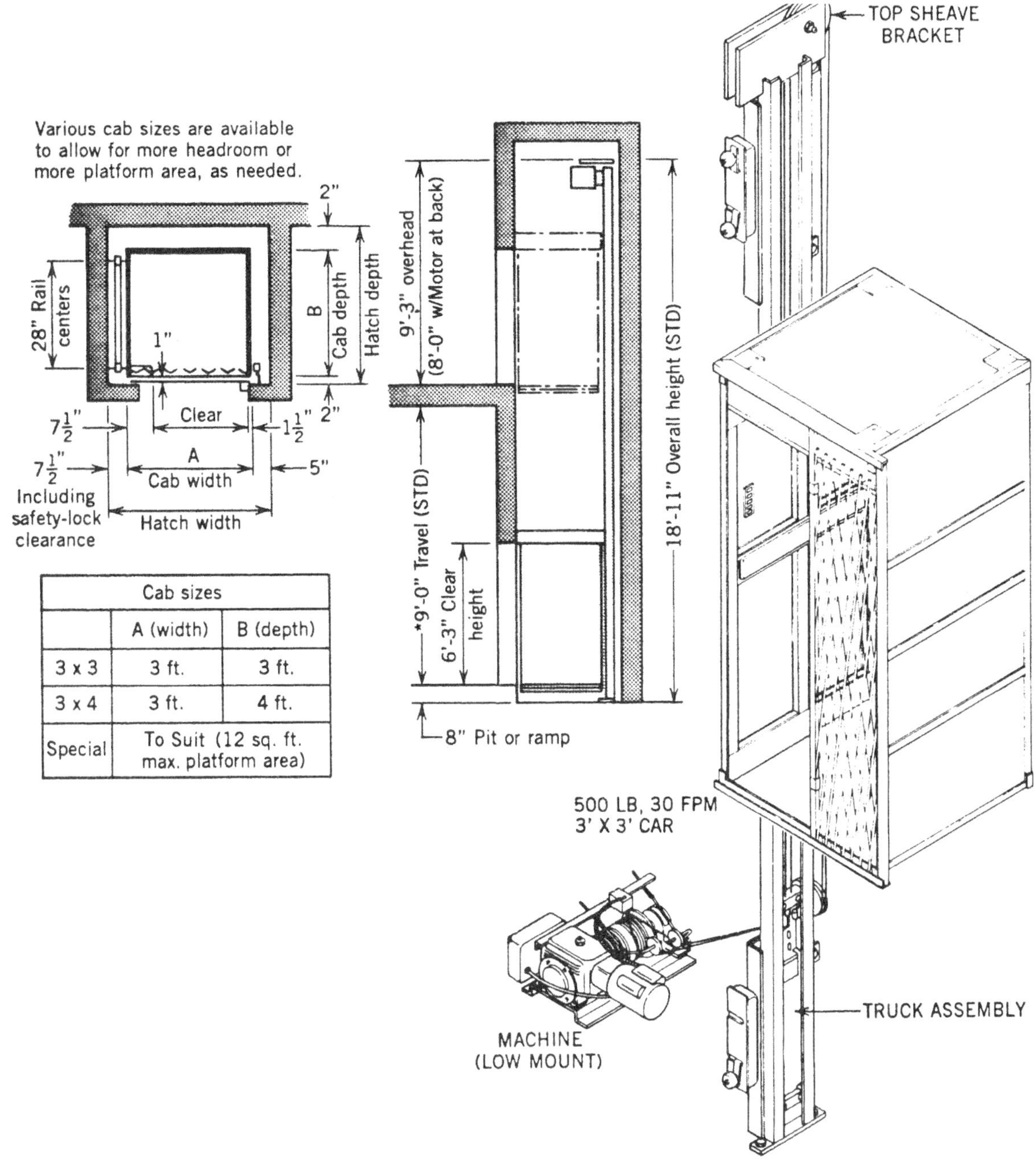

Cab sizes		
	A (width)	B (depth)
3 x 3	3 ft.	3 ft.
3 x 4	3 ft.	4 ft.
Special	To Suit (12 sq. ft. max. platform area)	

Figure 34.30 *Isometric drawing of a typical residential elevator of the winding-drum design. Units of this type are usually limited to 500-lb [227-kg] car capacity, 40-ft [12.2-m] rise, and 30-fpm [0.15-m/s] speed. The car is rigidly attached in cantilever fashion to a rolling truck, which is raised and lowered by cables attached to the winding drum. The driving motor and winding drum—shown here at the base of the assembly—can be installed at any floor stop along the hoistway or overhead. Control is normally pushbutton automatic with limit-switch leveling. A 6-in. [150-mm] pit is required at the bottom of the hoistway. Door interlock, cable failure, and overrun safeties are standard items in such installations, which must meet elevator code requirements. (Courtesy of Waupaca Elevator Co., Inc.)*

elevators requiring a plunger borehole are rarely used in residential applications.

In recognition of the different cost, load, speed, and space considerations in residences, ASME developed a section of the *A17.1* safety code to cover such units. The section applies to Limited Use/Limited Application (LU/LA) elevators, defined as "a power passenger elevator where the use and application is limited by size, capacity, speed and rise, intended primarily to provide vertical transportation for people with physical disabilities." The code section limits LU/LA elevators to a maximum size of 18 ft^2 [1.7 m^2], a load of 1400 lb [635 kg], a rise of 25 ft [7.6 m], and a speed of 30 fpm [0.15 m/s]. Due to these limitations, safety requirements are different from those of standard traction elevators, drives may be different (e.g., winding-drum cable lifts), space requirements are smaller, and overall costs are considerably reduced. Elevators that exceed the defined limitations must conform to the requirements of the *ASME A17.1* safety code.

Chair lifts and wheelchair platform lifts also come in a variety of designs. The chair lift shown in Figure 34.31 uses a rack and pinion drive that permits it to negotiate a turn, as illustrated. For straight chair lifts, a winding-drum design similar to that used for residential elevators is often used. Other drives in use include worm gear, chain and screw worm, and cog. The drive selection depends upon the load, rise, duty, and, of course, cost. The wheelchair platform lift illustrated in Figure 34.32 uses a ball screw drive. Other designs use some of the drives listed previously, including a roped-hydraulic drive for units requiring a large rise. The hydraulic scissor-jack type of wheelchair platform lift is most frequently found in public and institutional buildings because it is particularly applicable to short-rise, heavy-load, frequent-duty applications and requires minimal maintenance.

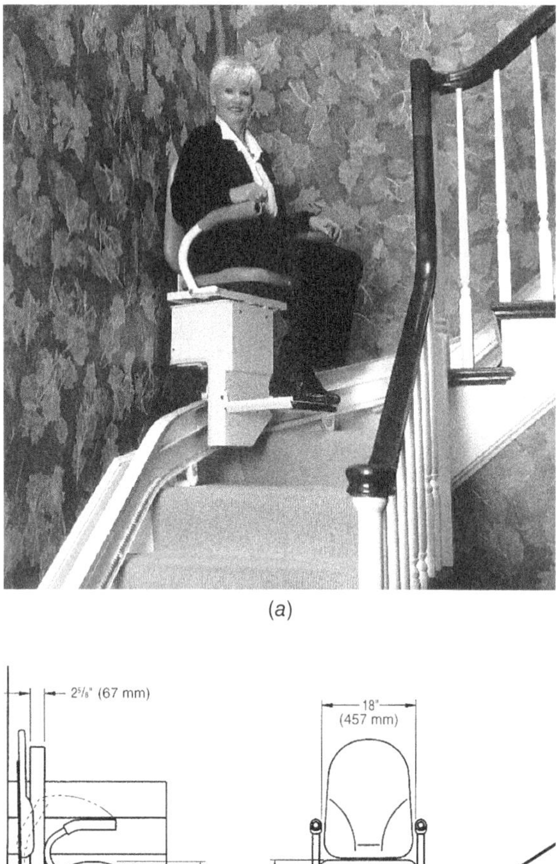

(*a*)

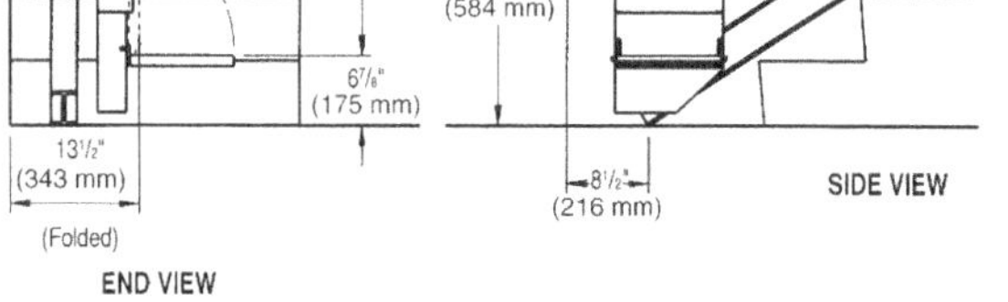

(*b*)

Figure 34.31 *(a) Chair lift with a rack and pinion drive permits the lift to negotiate a turn in the stairs. This unit is limited to a live load of 300 lb [135 kg], runs at 25 fpm [0.13 m/s], climbs at any incline up to 51°, and is driven by a ½-hp [0.4-kW] motor. Maximum rise is 60 ft [18 m]. (b) Dimensional data for the chair lift shown. (Courtesy of Access Industries, Inc.)*

All of these units must be installed in accordance with code requirements, including safeties and controls. Residential units are normally arranged to operate on 120-V AC, although heavy-duty units may require 240-V service.

MATERIALS HANDLING

34.23 GENERAL INFORMATION

The materials-handling equipment briefly discussed herein finds application in commercial and institutional buildings. Industrial materials handling is a separate subject beyond the scope of this section.

The need to transport things within a building has always existed and, until about the late 1970s, was done largely manually with mechanical assistance. Thus, offices used messengers, and hospitals used dumbwaiters, service elevators, conveyors, and chutes. An exception to this situation was the extensive use of pneumatic tube systems in large stores. Today's transport systems accomplish the same end—the transfer of materials—but do so automatically and, in general, much more rapidly. The first cost of these systems is frequently high, but the reduction in labor and the increase in speed generally yield a short payback period.

Modern commercial materials-handling systems can be grouped into four broad categories:

1. *Elevator-type systems:* These are vertical-lift, car-type systems including the common dumbwaiter and ejection lifts (which are basically automated dumbwaiters).
2. *Conveyor-type systems:* These include horizontal and vertical conveyors.
3. *Pneumatic systems:* These include sophisticated pneumatic tube systems and pneumatic trash and linen systems that use air under pressure.
4. *Other systems:* Systems that do not fit easily into any of the previously mentioned categories, including automated messenger carts and automatic track-type container delivery systems.

(a)

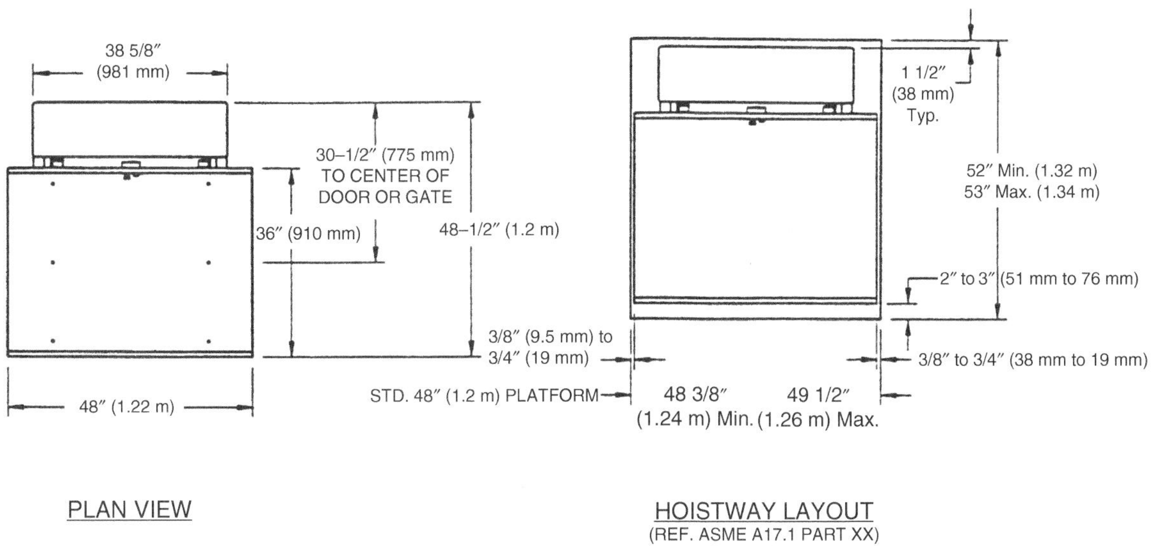

(b)

Figure 34.32 *The wheelchair platform lift shown in (a) is designed for a maximum load of 750 lb [340 kg] and a maximum lifting height of 171 in. [4.34 m]. The drive can be either a motor-driven recirculating ball screw unit or a 1:2 roped-hydraulic unit. Controls are mounted at the top of the right post of the gate frame. The unit is a straight-through design with gates at both ends. Dimensions are shown in (b). (Courtesy of Access Industries, Inc.)*

34.24 MANUAL LOAD/UNLOAD DUMBWAITERS

Dumbwaiters often provide the most convenient and economical means of transporting relatively small articles between levels. In department stores, dumbwaiters transport merchandise from stock areas to sales or pickup counters; in hospitals, dumbwaiters often transport food, drugs, linens, and other small items. In multilevel restaurants, dumbwaiters are almost always used for delivery of food from the kitchen and for return of soiled dishes.

Dumbwaiter cars are limited to a platform area of 9 ft^2 [0.84 m^2] and a maximum height of 4 ft [1.2 m]. The car may be, and frequently is, fitted with shelves. Normal speeds are 50–150 fpm [0.25–0.75 m/s], with a capacity of up to 500 lb [227 kg]. Cars may be of the traction (counterweighted) or drum (direct pickup) type. Control is normally "call and send" between two floors, although multibutton selector switch or central dispatching arrangements are available for applications with more than two floors. Loading may be at floor, counter, or any other specified height (see Figure 34.33 for typical layouts).

34.25 AUTOMATED DUMBWAITERS

These units are also known as ejection lifts because of the method of delivery (Figure 34.34). They find their best application in institutions and other facilities that require rapid vertical movement of relatively large items. Thus, this device is ideally suited for delivery of

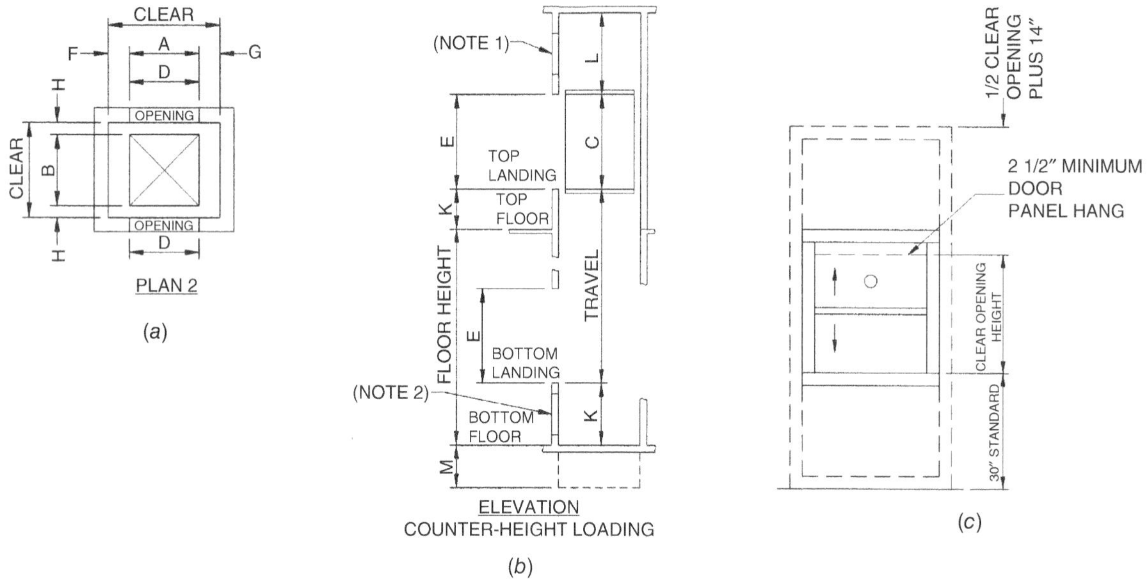

RATED LOAD (lb)	VOL. ■ ft^3	COUNTER-HEIGHT LOADING DIMENSIONS in.											
		CAR			DOORS ✱		HOISTWAY						
		A WIDTH	B DEPTH	C HEIGHT	D WIDTH	E HEIGHT	F	G	H	J	K ▲	L	M ✱✱
PLAN 1 OR 2 (MACHINE OVERHEAD OR BELOW) RATED SPEED 50–150 fpm													
APPLICATION – BULK SUPPLY, FOOD SERVICE (TRAY) & MATERIAL HANDLING													
100	8.33	20	24	30	20	30	6 1/2	7	5	3	30 ▲	48	6
200	12	24	24	36	36	24	6 1/2	7	5	3	30 ▲	48	12
300	17.5	30	24	42	30	42	6 1/2	7	5	3	30 ▲	48	18
400	30	36	30	48	48	48	6 1/2	7	5	3	30 ▲	48	24
500	36	36	36	48	48	48	6 1/2	7	5	3	30 ▲	48	24

✱ ONE OPENING MUST BE LARGER THAN THE CAR SIZE SO THE ASSEMBLED CAR CAN BE PLACED INTO, OR REMOVED FROM THE DUMBWAITER HOISTWAY.

✱✱ DEPTH OF PIT REQUIRED IF SLIDE DOWN TYPE DOORS ARE USED.

● 48 in. IF BI-PARTING HOISTWAY DOORS ARE USED.

▲ IF BASEMENT MACHINE (DRUM OR TRACTION) 42 in.

■ CAR SIZES ARE OPTIONAL UP TO 9 ft^2 OF FLOOR AREA AND UP TO 48 in. HIGH.

NOTES
1. 2'-0" X 2'-0" ACCESS DOOR IF MACHINE ABOVE.
2. 2'-0" X 2'-0" ACCESS DOOR IF MACHINE BELOW.

(d)

Figure 34.33 *Typical layout of a counter-height dumbwaiter with manual doors. The shaft openings (a, b) may be single (one side) or dual access. Doors can be biparting (c), slide up, or slide down. In addition to the sizes shown in the table (d), light-load cars rated 25 and 50 lb [11 and 23 kg] at 50 fpm [0.25 m/s] are available in standard designs. (Reproduced with permission of National Elevator Industry, Inc., 1992.)*

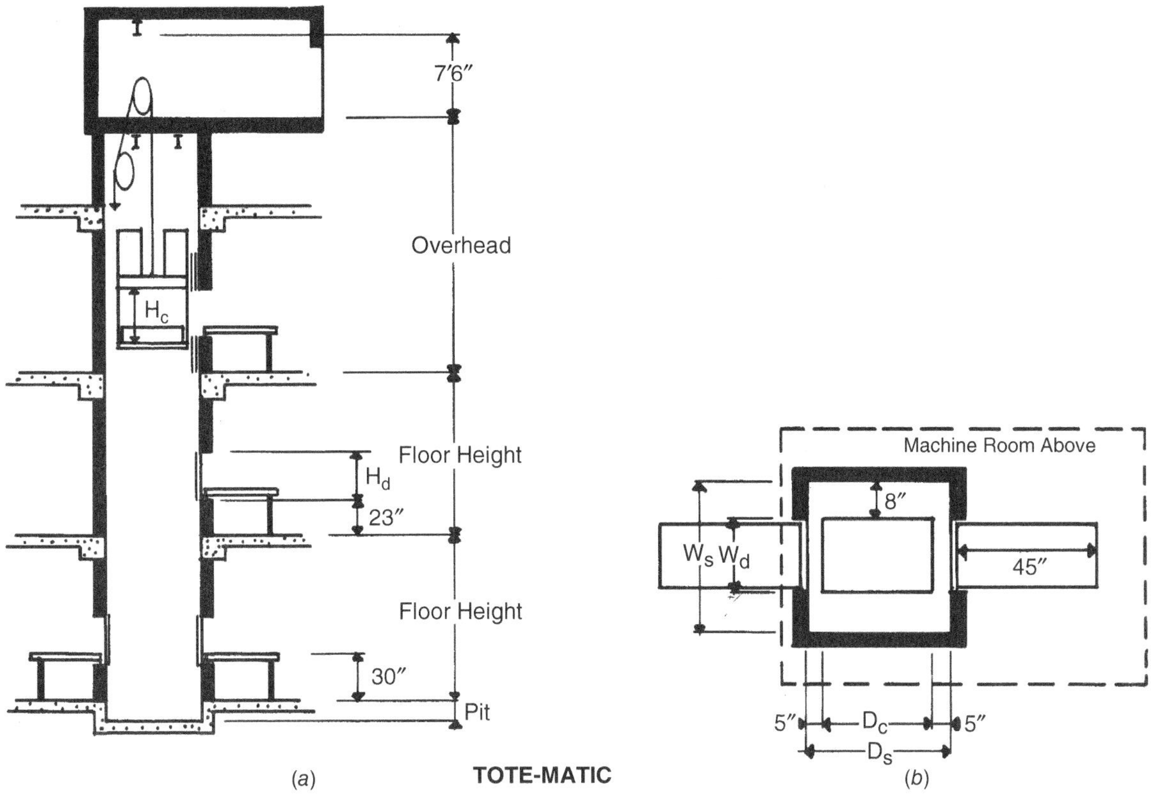

Lift Model	Capacity	Inside Car Dimensions			Inside Saft Dimensions		Hoistway Door Frame Dimensions		Pit (a)	Overhead (a)	Suggested Vertical Transportation Speed	
	Lbs.	WC	Dc	Hc	Ws	Ds	Wd	Hd			Landings	Feet per Minute
Tote-Matic CC-2	100	20"	31"	37"	40"	41"	20"	34"	12"	13'4"	7-10	150
											4-6	100
											2-3	50

(a) These figures must meet the requirements of ANSI/ASME 17.1 and local codes.

(c)

Figure 34.34 *This automated dumbwaiter or vertical ejection lift is designed to eject (unload) a container automatically at a preselected station. (Note the dispatching station to the right of the doors.) The tote box is carried on an ejection conveyor, which electrically senses arrival at its destination and effects ejection. Station doors are automatically electrically operated. The dumbwaiter car drive mechanism is a counterweighted traction drive. Car capacity is 100 lb [45 kg], although containers are normally limited to 50 lb [23 kg] for ease of handling. Car speeds (see part c) are normally 50–150 fpm [0.25–0.75 m/s]. A typical hoistway section (a) and plan (b) are shown. Vertical travel is not limited. Dimensions are provided in (c). (Courtesy of Courion Industries, Inc.)*

food carts, linens, dishes, bulk-liquid containers, and so on. The load can be a cart (Figure 34.35) containing the items being transported. At the delivery terminal, the item must be picked up and transferred horizontally to its final destination (if remote from the delivery point). Loading can be manual or automatic. Sophisticated ejection lifts use programmable controllers for automated loading, dispatch, and ejection; electronic sensors to determine whether space is available for a load; and automated return of the unloaded cart.

Payload capacity for cart systems is available up to 1000 lb [454 kg] and car speeds up to 350 fpm [1.8 m/s]. Maximum cart size is approximately 32 in. W × 68 in. L × 70 in. H [0.8 by 1.7 by 1.8 m]. The round-trip time for a 200-fpm [1-m/s] unit with five loading stations is approximately 2 minutes. Major design considerations for these units are their relatively high cost and the large shaft area required.

34.26 HORIZONTAL CONVEYORS

Although horizontal conveyors (sometimes referred to as horizontal belt conveyers) find their best application in industrial facilities, they are also usable in commercial buildings such as mail-order houses, which require a continuous flow of material. Restrictions in application stem from inflexible right-of-way

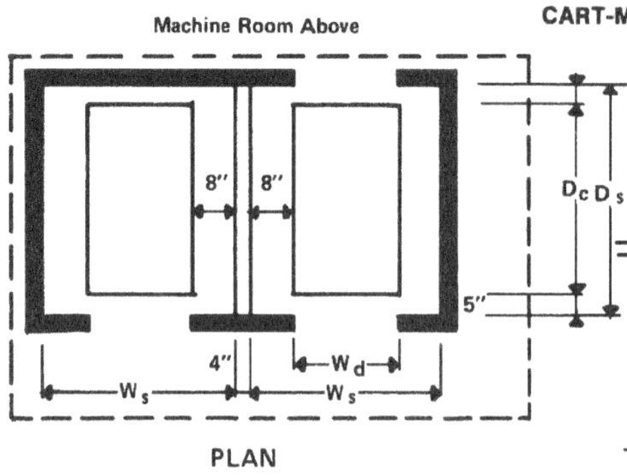

TABLE OF STANDARDS

Capacity	Inside Car Dimensions			Inside Shaft Dimensions		Hoistway Door Frame Dimensions		Pit **	Overhead **
Lbs.	Wc	Dc	Hc	Ws	Ds	Wd	Hd		
500	28"	46"	48"	48"	56"	28"	50"	46"	14'4"
1000	30"	55"	67"	56"	65"	30"	69"	60"	16'8"
1000	34"	72"	72"	60"	82"	34"	74"	60"	17'4"

** May vary with local codes

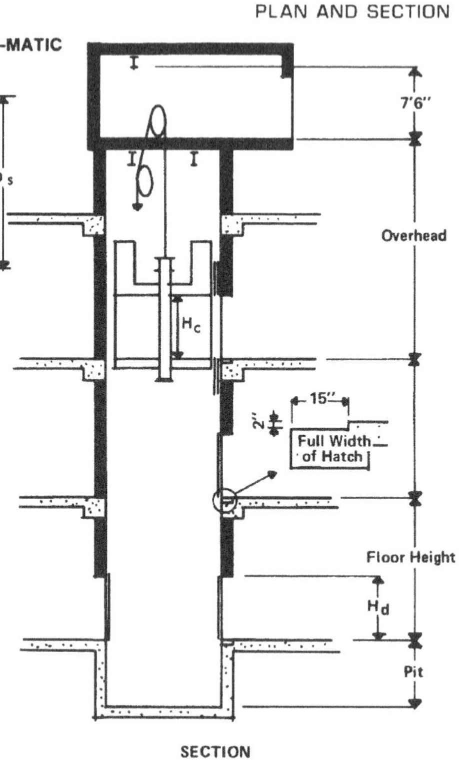

Figure 34.35 *Plan, section, and dimensions for a vertical ejection lift system (automated dumbwaiter). (Courtesy of Courion Industries, Inc.)*

requirements, noise generation, and a degree of danger if conveyors are left unprotected or exposed to unauthorized persons. The cost is relatively low, and the capacity is virtually unlimited.

34.27 VERTICAL RECIPROCATING CONVEYORS

The action of this system is like that of an automated dumbwaiter in that the system transfers materials vertically and automatically loads and unloads, but the similarity ends there. Vertical reciprocating conveyors (VRCs) are constructed with a moving continuous-loop chain with attached carriages that pick up and deliver tote boxes (called trays). At sending and receiving stations, the operator places the items to be moved (up to 60 lb [27 kg]) in the tote box, "addresses" the box in one of several ways depending on the system, and places it at a pickup point. The first empty carriage on the chain picks up the box and delivers it to its address. Modern vertical conveyor operations are monitored by a microprocessor, which tracks system performance and furnishes maintenance data and operational diagnostics. Drawbacks of this system are the large shaft required, noise, and cumbersome arrangements when interfacing with horizontal conveyors. The cost is moderate. VRCs are not considered elevators and should never be used to transport people. In the United States, VRCs are specifically exempt from *ASME A17.1* because they have their own national model code, *ASME B20.1: Safety Standard for Conveyors and Related Equipment.*

34.28 PNEUMATIC TUBES

Where information must be moved, electronic data transfer has largely replaced the flow of paper between two points. This well-tried system, however, will undoubtedly continue to be used where physical transfer of an item is required. Pneumatic tube systems are available with a 2¼-to-7-in. [57-to-178-mm] range of tube diameters (special shapes are also used) and with single or multiple loops, but 4-in. and 6-in. [102-mm and 152-mm] diameter tubes are the most common. These systems are still widely used in hospitals and drive-in banks.

Whereas older systems were generally pressurized using a single large, noisy compressor, newer systems are computer-controlled, utilize a small blower in each of the zones, operate basically on vacuum but also on pressure, are relatively quiet, and are capable of being constructed in unlimited system length. Color can be used to code the carriers for identification of departments/senders or uses. Carriers typically travel at 25 fps [7.6 m/s], but some systems allow for variable speeds to deal with delicate or sensitive loads. The computerized control center provides information and control of all system components, including status, traffic data, station assignments, scheduling, and the like. Some systems allow for user controls where a security code must be entered or a name badge barcode scanned before a carrier is dropped at the receiving end. These controls are used where limited access and tracking are a priority, such as for hospitals transporting confidential information or medications. Overall, pneumatic tube systems perform their task reliably, rapidly, and efficiently, at relatively low cost if installed during initial building construction. Typical system components are shown in Figure 34.36.

(a)

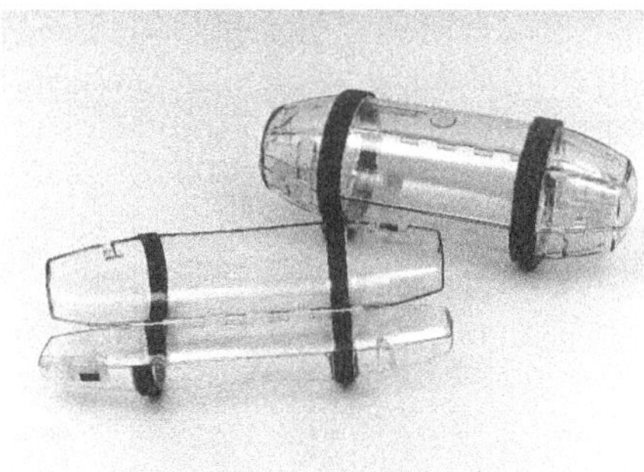

(b)

Figure 34.36 *(a) The recessed tube station illustrated is a full-facility combined sending and receiving point. Dispatching and receiving share the same tube but operate independently and automatically. Arriving carriers (b) are decelerated by an air cushion and drop into the front bin on arrival. (b) Carriers are typically $3\frac{3}{4}$ in. D × $15\frac{1}{2}$ in. L or $5\frac{3}{4}$ in. D × 16 in. L [95 × 394 mm and 146 × 406 mm] and are made of impact-resistant transparent polycarbonate plastic. Liners are used with fragile items. The bands around the carriers that maintain the air seal in the transport tubes are replaceable. (Courtesy of Swisslog Healthcare Solutions.)*

34.29 PNEUMATIC TRASH AND LINEN SYSTEMS

The purpose of this type of system is to rapidly move bagged or packaged trash/recycling and/or linens from numerous outlying stations to a central collecting point. Health codes require separate tubes for trash and linens. Linen systems are found generally in hospitals; trash systems are found in various facilities, frequently in conjunction with compactors. Trash and recycling systems are found in buildings but are also increasingly being used in cities as low-carbon alternatives to moving material to disposal or sorting centers. Examples of such municipal trash systems are Barking Riverside in London and the Seocho Gu district in Seoul. The basic system is a network of large pipes, negatively pressurized, with numerous loading stations throughout the building. Pipe sizes are 16, 18, or 20 in. [406, 457, 508 mm], operating at high static pressure. A system normally can handle only one load at a time but moves it so quickly (20–30 fps [6.1–9.1 m/s]) that system capacity is large and delays are not encountered. Material placed into a loading station is picked up as soon as the previous load clears. Compressors are large and very noisy, requiring considerable space and sound isolation/attenuation. In addition to the main vacuum system, a high-pressure air line is required to operate the doors, and sprinkler heads must be installed every few floors. Some systems incorporate features such as sensors that can detect clogged lines or time released bags to prevent back-ups. Overall costs are low to moderate. For the specific task performed, a cheaper and more efficient transfer technique is difficult to find.

34.30 AUTOMATED CONTAINER DELIVERY SYSTEMS

This system employs captive and secure containers locked onto a motorized carriage that, in turn, is locked onto a track system. Power for the motor in the carriage is picked off a third rail at 24 V DC. The entire assembly moves horizontally or vertically with equal ease. Containers move at a constant 120 fpm [0.6 m/s] horizontally but more slowly on rises—depending upon the container load and the steepness of the rise. Containers are available in several shapes and volumes to suit the particular installation requirements. Two standard sizes are approximately 18 in. × 6 in. × 13 in. [457 mm × 152 mm × 330 mm] narrow profile and 18 in. × 12 in. × 8 in. [457 mm × 305 mm × 203 mm] low profile. The normal payload is 20 lb [9 kg]; although with only horizontal runs, considerably heavier loads can be carried.

In simple single- or dual-track point-to-point or loop systems, right-of-way conflicts cannot occur, and routing is simply a matter of addressing the car. For complex systems involving loops and branches, routing is decided by a central computerized controller that finds the shortest route for each car (up to 250 cars per system), side-tracks cars, and parks cars; in effect, operating a miniature railroad-type system. The system is easily added to a facility as a retrofit because of track flexibility and small size. Its major drawback is its high cost.

34.31 AUTOMATED GUIDED VEHICLES AND AUTONOMOUS MOBILE ROBOTS

Automated guided vehicles (AGVs) are driverless devices used in a variety of applications but most commonly as material handlers or load carriers. Programming of routes, stops, and vertical interfaces, changing of the cycle, and other functions are accomplished with an external programming device and an internal programmable controller. AGVs use advanced technology to consistently and reliably operate with minimal human intervention—a solution referred to as automation.

Automated guided vehicles are often conflated with autonomous mobile robots (AMRs), but these technologies are distinct from one another. AGV devices are relatively simpler and use a "fixed" navigation system to follow a predetermined route, which is often determined by passive magnetic floor tape or a wired navigation path embedded into a facility's floor. AMRs are far more technologically advanced, exhibiting "intelligent" navigation capabilities such as sensors and/or cameras to detect objects in an environment (including moving ones) and operate accordingly. AMRs may navigate an environment based on a range of technologies including laser target navigation, inertial (gyroscopic) navigation that respond to transponders embedded into the floor, camera-based vision guidance, or sensor-based geoguidance. Some highly advanced AMRs may combine camera and sensor technologies in a method called LiDAR (Light Detection And Ranging) by which millions of distance point measurements are aggregated to piece together a real-time three-dimensional map of the AMR's surroundings such that it can navigate a dynamic environment with moving objects. Advanced ("AI-powered") algorithms allow AMRs to gather data (i.e., "learn") from past activities on its environment to more efficiently and effectively operate in the future—a process referred to as machine learning.

References and Resources

ASME. 2022. *ASME A17.1/CSA B44-2022: Safety Code for Elevators and Escalators*. New York: American Society of Mechanical Engineers.

ASME. 2024. *ASME B20.1-2024: Safety Standard for Conveyors and Related Equipment*. New York: American Society of Mechanical Engineers.

DOJ. 2012. *2010 ADA Standards for Accessible Design*. Washington, DC: U.S. Department of Justice. ada.gov/law-and-regs/design-standards/2010-stds

Elevator History. elevatorhistory.net/elevator-history

Goetz, A. 2003. *Up, Down, Across*. London: Merrell Publishers.

National Elevator Industry, Inc. 1992. *Vertical Transportation Standards* (7th ed.). New York: National Elevator Industry, Inc.

NFPA. 2008. *Fire Protection Handbook* (20th ed.). Quincy, MA: NFPA.

Strakosch, G. R. 1983. *Vertical Transportation, Elevators and Escalators* (2nd ed.). New York: John Wiley & Sons.

Strakosch, G. R. (ed.). 1998. *The Vertical Transportation Handbook*. New York: John Wiley & Sons.

U.S. Access Board. 2004. *Americans with Disabilities Act Accessibility Guidelines for Buildings and Facilities (ADAAG)*. Washington, DC: U.S. Architectural and Transportation Barriers Compliance Board. federalregister.gov/documents/2004/07/23/04-16025/americans-with-disabilities-act-ada-accessibility-guidelines-for-buildings-and-facilities

PART X

APPENDICES

TING / TEMPERING / P

HEATING CFM	SIZE (INCHES) HxL(No. Coils)
9300	43x79(1)
9000	43x72(1)
4620	27x51(1)
4390	32x43(1)
2910	23x47(1)
940	12x20(1)
5830	35x40(1)
3640	45x22(1)
3740	45x22(1)
970	6x501)
1640	12x411)
950	9x34(1)
1010	12x30(1)
1040	9x37(1)
1030	9x37(1)
620	6x38(1)
890	6x44(1)
790	6x40(1)
780	6x40(1)
860	6x44(1)
4570	32x46(1)
2690	32x30(1)
2460	41x45(1)

LING COIL SCHEDUL

COOLING CFM	SIZE (INCHES) HxL(No. Coils)
9000	27x48(2)
1040	6x50(1)
2180	12x50(1)
1060	9x34(1)
1140	12x30(1)
16100	33x79(2)
13000	25x75(2)
1150	9x37(1)
1140	9x37(1)
780	6x38(1)
6200	35x49(1)
990	6x44(1)
1260	9x45(1)
1310	9x45(1)
1200	9x38(1)
6200	33x54(1)
4760	30x46(1)
940	12x31(1)
3640	45x26(1)
3740	45x26(1)

The purpose of Part X is to present important reference and ancillary data (which is often extensive) in a separate, easily accessed area that does not detract from the flow of the text in the chapters. The appendices are:

A. Metrication, SI Units, and Conversions
B. Climatic Conditions for the United States, Canada, and Mexico
C. Solar and Daylighting Design Data
D. Solar Geometry
E. Thermal Properties of Materials and Assemblies
F. Ventilation and Infiltration
G. Heating and Cooling Design Guidelines and Information
H. Standards/Guidelines for High-Performance Building Design
I. Heating Degree-Day and Solar Radiation Data
J. Economic Analysis
K. Sound Transmission Properties
L. Building Performance Simulation Tools and Resources

APPENDIX A

Metrication, SI Units, and Conversions

A.1 GENERAL COMMENTS ON SI UNITS

The building industry in the United States has been slow (glacially so) in adopting the metric (more accurately, Le Système International d'Unités, or SI) system of units for many reasons. The rest of the world, however, is generally fully SI (or metric). Many professional societies use a mixture of I-P (inch-pound, previously "English" units) and SI units as a means of addressing their international audiences. Many older units are also so entrenched that changing them is a seemingly impossible task. This book attempts to provide dual units where feasible, but that is not always possible. For this reason, tables of conversions and approximations are presented in the following material to enable the user to work with both systems. This appendix also presents some facts that should make use of the SI system a bit easier.

Table A.1 Common SI Unit Prefixes

Multiples and Submultiples	Prefixes	Symbols
1,000,000,000 = 10^9	giga	G
1,000,000 = 10^6	mega	M[a]
1000 = 10^3	kilo	K[a]
100 = 10^2	hecto[b]	H
10 = 10	deka	Da
0.1 = 10^{-1}	deci	D
0.01 = 10^{-2}	centi	C[a]
0.001 = 10^{-3}	milli	m[a]
0.000,001 = 10^{-6}	micro	μ[a]
0.000,000,001 = 10^{-9}	nano	N
0.000,000,000,001 = 10^{-12}	pico	P

[a]Most commonly used.

[b]A hectare is a square hectometer (i.e., 10^4 m^2).

A.2 SI NOMENCLATURE AND SYMBOLS

For a full discussion of the SI system, see *AIA Metric Building and Construction Guide*, edited by S. Braybrooke (New York: John Wiley & Sons, 1980) and the American Society of Heating, Refrigerating and Air-Conditioning (ASHRAE) publication "SI for HVAC&R" (May 1999), available from the ASHRAE website (ashrae.org). The units in common use include the basic SI staples—the meter, kilogram, and second (MKS)—plus a host of derived, supplementary, and non-SI units such as the pascal (pressure), radian (solid angle), and kilowatt-hour (energy). Also, multiple and submultiple units such as the liter, metric ton, and millibar are so common that they stand as separate units instead of being expressed as 10^{-3} m^3, 10^3 kg, and 10^{-2} Pa.

Table A.1 lists SI unit prefixes with their accepted symbols. Symbols do not change in the plural. That is, 6 millimeters is written as 6 mm and 20 kilograms is written as 20 kg. All units and prefixes except Fahrenheit and Celsius are in lower-case when written out (as in megaton or meter).

A.3 COMMON USAGE UNITS

1. *Length:* meter (m), kilometer (km), millimeter (mm), micrometer (μm), nanometer (nm). The micrometer (μm) is also called a micron in the U.S.
2. *Area:* square meter (m^2), hectare (ha).

APPENDICES

Mechanical and Electrical Equipment for Buildings, Fourteenth Edition. Walter Grondzik, Daniel Overbey, and Alison Kwok.

Companion website: www.wiley.com/go/meeb14e

3. *Volume:* cubic meter (m^3), liter (L).
4. *Flow:* cubic meters per second (m^3/s).
5. *Velocity—airflow:* meters per second (m/s).
6. *Weight:* kilogram (kg), gram (g).

The SI system clearly differentiates between mass (kg) and force (kg • m/s^2), the latter being given a separate name and symbol, the newton (N). Weight is not used, because it is a force that depends upon acceleration and is therefore variable. However, the construction industry largely continues to use the terms *mass*, *weight*, and *force* interchangeably.

7. *Force:* newton (N), kilonewton (kN). A newton is the force required to accelerate 1 kg at 1.0 m/s^2.
8. *Pressure:* pascal (Pa), kilopascal (kPa). A pascal is a newton per square meter (N/m^2).
9. *Energy, work, quantity of heat:* joule (J), kilojoule (kJ), megajoule (MJ). A joule is a watt-second (W • s).
10. *Temperature:* degree Celsius (°C), degree kelvin (K).

SI temperature in degrees kelvin is not as commonly used as °C. (*Celsius* is the accepted term; centigrade is obsolete.) The Celsius and kelvin scales are subdivided equally but start at different points—that is, 0 K is −273.15°C. Therefore, to determine kelvin from Celsius, simply add 273.15. Increments are equal because of equal subdivisions; that is, a change of 10 K is the same as a 10°C change. Because of its common usage in design work, a Fahrenheit/Celsius conversion chart is given in Figure A.1.

11. *Illumination:* see Table 7.1.
12. *Acoustics:* see Table 25.16.
13. *CGS/MKS conversions:* see Table 25.16.
14. *Abbreviations:* see Table A.2.
15. *Approximations:* see Table A.3.

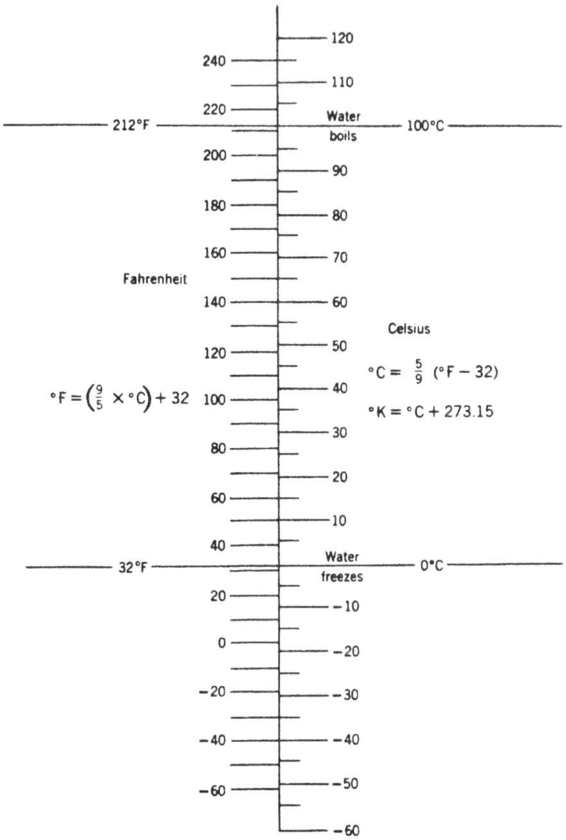

Figure A.1 Conversion nomograph: Fahrenheit temperature/Celsius temperature.

A.4 CONVERSION FACTORS

Table A.4 is an alphabetized list of useful conversion factors. Decimal notation has been used in this table (versus scientific notation), that is, 0.00378 versus 3.78×10^{-3} or 3.78e3.

Table A.2 Typical Abbreviations: All Systems of Units

atmospheres	atm	gallons	gal
British thermal units	Btu	gallons per hour	gph, gal/h
British thermal units per hour	Btu/h	gallons per minute	gpm, gal/min
calorie	cal	grams	g
cubic feet	cf, ft^3	hectares	ha
cubic feet per minute	cfm, ft^3/min	horsepower	hp
cubic feet per second	cfs, ft^3/s	inches	in.
cubic meters	m^3	inches of mercury	in. Hg
feet	ft	joules	J

Table A.2 (*Continued*)

feet per second	fps, ft/s	kilocalories	kcal
kilograms	kg	miles per hour	mph
kilograms per second	kg/s	millimeters	mm
kilojoules	kJ	millimeters of mercury	mm Hg
kilometers	km	newtons	N
kilometers per hour	kph, km/h	ounces	oz
kilonewtons	kN	pounds	lb
kilopascals	kPa	pounds of force	lbf
kilowatts	kW	pounds per cubic foot	lb/ft^3
kilowatt-hours	kWh	second	sec, s
liters	L	square feet	ft^2
liters per second	L/s	square inches	$in.^2$
megajoules	MJ	square meters	m^2
meganewtons	MN	watts	W
megapascals	MPa	watts per square meter	W/m^2
meters	m	yards	yd
meters per second	m/s		

Table A.3 Common Approximations

Approximate Common Equivalents					
1 inch	=	25 millimeters	1 millimeter	=	0.04 inch
1 foot	=	0.3 meter	1 liter	=	61 cubic inches
1 yard	=	0.9 meter	1 meter	=	3.3 feet
1 mile	=	1.6 kilometers	1 meter	=	1.1 yards
1 square inch	=	6.5 square centimeters	1 kilometer	=	0.6 mile
1 square foot	=	0.09 square meter	1 square centimeter	=	0.16 square inch
1 square yard	=	0.8 square meter	1 square meter	=	11 square feet
1 acre	=	0.4 hectare	1 square meter	=	1.2 square yards
1 cubic inch	=	16 cubic centimeters	1 hectare	=	2.5 acres
1 cubic foot	=	0.03 cubic meter	1 cubic centimeter	=	0.06 cubic inch
1 cubic yard	=	0.8 cubic meter	1 cubic meter	=	35 cubic feet
1 quart	=	1 liter	1 cubic meter	=	1.3 cubic yards
1 gallon	=	0.004 cubic meter	1 liter	=	1 quart
1 ounce	=	28 grams	1 cubic meter	=	250 gallons
1 pound	=	0.45 kilogram	1 kilogram	=	2.2 pounds
1 horsepower	=	0.75 kilowatt	1 kilowatt	=	1.3 horsepower

Table A.4 Useful Conversion Factors: Alphabetized

Multiply	By	To Get
acres	4047	square meters
atmospheres	33.93	feet of water
atmospheres	29.92	inches of mercury
atmospheres	760.0	millimeters of mercury
Btu (energy)	0.252	kilocalories
Btu (energy)	1.055	kilojoules
Btu/h (power)	0.2931	watts
Btu/h ft^2 (energy transfer)	3.155	watts per square meter
Btu/°F (heat capacity)	1.897	kilojoules per kelvin[a]
Btu/lb °F (specific heat)	4.184	kilojoules per kilogram per kelvin[a]
Btu/h °F ft (thermal conductivity[b])	1.729	watts per kelvin[a] per meter
Btu/h °F ft^2 (thermal conductance[c])	5.673	watts per kelvin[a] per square meter
Btu/°F day (building load coefficient, BLC)	0.022	watts per kelvin[a]
Btu/°F day ft^2 (load–collector ratio, LCR)	0.236	watts per kelvin[a] per square meter
cubic feet	0.02832	cubic meters
cubic feet	7.481	gallons
cubic feet	28.32	liters
cubic feet per minute	0.4719	liters per second
cubic feet per second	28.32	liters per second
cubic inches	16.39	cubic centimeters
cubic meters	35.32	cubic feet
cubic meters	1.308	cubic yards
cubic meters	264.2	gallons
cubic yards	0.765	cubic meters
feet	0.3048	meters
feet	304.8	millimeters
feet per second	0.3048	meters per second
footcandle	10.764	lux
foot-pounds of force per second	1.356	watts
gallons	3.785	liters
gallons per hour	1.052	liters per second
gallons per minute	0.0022	cubic feet per second
gallons per minute	0.06308	liters per second
grams	0.035	ounces (avoirdupois)
hectares	2.471	acres
horsepower	0.746	kilowatts
horsepower	746	watts
inches	25.4	millimeters
inches of mercury	0.033	atmospheres
inches of mercury	1.133	feet of water
inches of mercury (60°F [15.6°C])	3377	newtons per square meter
inches of mercury	0.491	pounds per square inch
inches of water	0.002458	atmospheres
inches of water	0.036	pounds per square inch

Table A.4 (*Continued*)

Multiply	By	To Get
inches of water (60°F) [15.6°C]	248.8	newtons per square meter
kilocalories	3.968	British thermal units
kilocalories	4190	joules
kilograms	2.205	pounds
kilograms per cubic meter	1.686	pounds per cubic yard
kilograms per square meter	0.0033	feet of water
kilograms per square meter	0.0029	inches of mercury
kilograms per square meter	0.205	pounds per square foot
kilograms per square meter	0.001422	pounds per square inch
kilojoules	0.948	British thermal units
kilojoules per kilogram	0.430	British thermal units per pound
kilometers	0.621	miles
kilometers per hour	0.621	miles per hour
kilonewtons	0.1004	tons of force
kilonewtons	224.8	pounds of force
kilopascals	20.89	pounds of force per square foot
kilowatts	1.341	horsepower
kilowatt-hours	3.6	megajoules
liters	0.03532	cubic feet
liters	61.02	cubic inches
liters	0.2642	gallons
liters	1.057	quarts
liters per second	2.119	cubic feet per minute
liters per second	951.0	gallons per hour
liters per second	15.85	gallons per minute
megajoules	0.278	kilowatt-hours
meganewtons	100.36	tons of force
megapascals	145.04	pounds of force per square inch
megapascals	9.324	tons of force per square foot
meters	3.281	feet
meters	1094	yards
meters per second	196.86	feet per minute
meters per second	2.237	miles per hour
miles	1.609	kilometers
miles per hour	1.609	kilometers per hour
miles per hour	0.447	meters per second
milliliters	0.061	cubic inches
milliliters	0.035	fluid ounces
millimeters	0.039	inches
millimeters of mercury	133.3	newtons per square meter
million gallons per day	18.94	cubic meters per hour
newtons	0.225	pounds of force
ounces (avoirdupois)	28.35	grams
ounces (fluid)	29.6	milliliters
pounds	0.4536	kilograms

Table A.4 (*Continued*)

Multiply	By	To Get
pounds of force	4.448	newtons
pounds of force per square foot	47.88	pascals
pounds of force per square inch	6.895	kilopascals
pounds per cubic foot	16.02	kilograms per cubic meter
pounds per cubic yard	0.593	kilograms per cubic meter
pounds per square foot	4.882	kilograms per square meter
square feet	0.0929	square meters
square inches	645.2	square millimeters
square kilometers	0.386	square miles
square meters	10.76	square feet
square meters	1.196	square yards
square miles	2.590	square kilometers
square yards	0.836	square meters
tons of force	9.964	kilonewtons
tons of force per square foot	107.25	kilopascals
tons of force per square inch	15.44	megapascals
torr (millimeters of mercury at 0°C [2°F])	133.3	newtons per square meter
watts	3.412	British thermal units per hour
watts	0.738	foot-pounds of force per second
watts per square meter	0.317	British thermal units per square foot
yards	0.914	meters
Add your own conversion factors here.	By	To Get

[a]K or °C.

[b]Thermal conductivity (k).

[c]Thermal conductance (C) or transmittance (U).

APPENDIX B

Climatic Conditions for the United States, Canada, and Mexico

This appendix contains the following information:

- Table B.1 Outside Design Conditions: United States and Canada (I-P Units)
- Table B.2 Outdoor Design Conditions: Mexico and Puerto Rico (I-P Units)
- Table B.3 Outside Design Conditions: United States and Canada (SI Units)
- Table B.4 Outdoor Design Conditions: Mexico and Puerto Rico (SI Units)
- Figure B.1 Climate Zones by United States Counties
- Table B.5 Climate Zones by United States Counties

The climate data presented in Tables B.1 through B.4 are taken from *ANSI/ASHRAE Standard 169-2021: Climatic Data for Building Design Standards* (ASHRAE, 2020). The information is reprinted with permission of the American Society of Heating, Refrigerating and Air-Conditioning Engineers, Inc. (www.ashrae.org).

Station Designations (where identified):

AAF = Army Airfield
AP = Airport
AFB = Air Force Base
CO = Urban areas, influenced by surroundings
No designation = semirural, similar to airport locations

Figure B.1 and Table B.5 are also taken from *ASHRAE Standard 169-2021: Climatic Data for Building Design Standards* (www.ashrae.org). Figure B.1 is a graphic depiction of every county in the United States and their corresponding climate zones. Table B.5 offers the information in a tabulated manner. Climate zone definitions are provided in *Standard 169-2021*. Please note that county climate zone designations can shift over time as climate conditions change. As such, users should always note the year/edition when referencing climate zone designations.

B.1 WINTER DESIGN CONDITIONS

- *HDD65:* Annual heating degree-days at base 65°F (18.3°C), obtained from *ANSI/ASHRAE Standard 169-2021*. (HDD65 values are also given in Appendix C; those values may vary slightly from these values because they are based upon data from a different set of years.)
- *Design dry-bulb (99%):* Outdoor temperatures that will, in a normal year, be the lowest experienced during 99% of the hours in a year. This means that approximately 1% of the hours (88 hours) will experience temperatures lower than this value. These data were obtained from *ASHRAE Standard 169-2021*. Note that ASHRAE implicitly specifies use of 99.6% values for compliant designs—and the necessary climate data reflecting these values may be obtained from that standard. The 99.6% values are more stringent than the 99% values—they represent

Mechanical and Electrical Equipment for Buildings, Fourteenth Edition. Walter Grondzik, Daniel Overbey, and Alison Kwok.

Companion website: www.wiley.com/go/meeb14e

somewhat more extreme conditions, which may be good for conservative design purposes, but will also result in larger systems and (theoretically) more energy use. The 99% values given here are believed to represent a more "green" approach to selecting design climate data.

B.2 SUMMER DESIGN CONDITIONS

- *Design dry-bulb and mean coincident wet-bulb (1%):* Outdoor temperatures that will, in a normal year, be the highest experienced during all but 1% of the hours in a year. This means that approximately 88 hours will experience temperatures higher than these values. The word "coincident" means that the given wet-bulb temperature is likely to occur in concert with the paired dry-bulb temperature. These data were obtained from *ASHRAE Standard 169-2021* (ASHRAE, 2020). Note that *ASHRAE/IES Standard 90.1-2022: Energy Standard for Sites and Buildings Except Low-Rise Residential Buildings* (ASHRAE, 2022) explicitly specifies use of the 1% dry-bulb values as cooling design temperatures. The 1% values are more stringent than the 2% values—they represent somewhat more extreme conditions, which may be good for conservative design purposes, but may also result in larger systems and (theoretically) more energy use. The 1% values are particularly advisable for passive system component sizing as the lifecycle of such components constitutes the full life of a building. Unless specified otherwise, the examples in this book utilize the 1% values as cooling design temperatures.
- *Design dry-bulb and mean coincident wet-bulb (2%):* Outdoor temperatures that will, in a normal year, be the highest experienced during all but 2% of the hours in a year. This means that approximately 176 hours will experience temperatures higher than these values. The word "coincident" means that the given wet-bulb temperature is likely to occur in concert with the paired dry-bulb temperature. There is a philosophical approach that considers using the 2% values to represent a more "green" approach to mechanical system components and "right sizing," as such equipment has a life-cycle limited to around 15–20 years.
- *Mean daily range:* The average difference between daily maximum and minimum dry-bulb temperatures—in this case, for the month of July (often used as the summer design condition for solar radiation). These data were obtained from *ASHRAE Standard 169-2021*.
- *Design wet-bulb (2%):* The outdoor temperature that will, in a normal year, be the highest experienced during all but 2% of the hours in a year. This value would be used for the selection and sizing of evaporative cooling components. These data were obtained from *ASHRAE Standard 169-2021*.
- *CDD50:* Annual cooling degree-days at base 50°F (10°C), obtained from *ASHRAE Standard 169-2021*. (CDD50 values are also given in Appendix G; they may vary slightly from these values because they are based upon data from a different set of years.)
- *Other stations:* Many more stations in each state or province, and for other countries, are found in the current *ASHRAE Handbook—Fundamentals*, *ASHRAE Standard 169-2021*, and through various other online resources.

B.3 INTERPRETATIONS BETWEEN STATIONS

Generally, weather data from listed stations can be adjusted for nearby locations in these ways:

- *Adjustment for elevation:* Elevations for many locations are given in Appendix C, Table C.19. For lower elevations, which tend to be warmer, increase the design temperatures:
  - dry-bulb temperature: 1°F per 200 feet of elevation (0.9°C per 100 meters of elevation)
  - wet-bulb temperature: 1°F per 500 feet of elevation (0.7°C per 200 meters of elevation)

- Adjustment for air mass (at coastlines): Along the West Coast, both dry-bulb and wet-bulb design temperatures increase with distance from the ocean. Along the Gulf Coast, dry-bulb temperatures increase for the first 200–300 miles [320–480 km] inland, while wet-bulb temperatures decrease slightly. (Beyond this 200–300-mile [320–480-km] belt, both dry-bulb and wet-bulb values decrease.)

Table B.1 Outdoor Design Conditions: United States and Canada (I-P Units)

		Winter	Summer						
State/Province and Station[a]	HDD65°F	Design Dry-Bulb (°F) (99%)	Design Dry-Bulb (°F) (1%)	Mean Coincident Wet-Bulb (°F) (1%)	Design Dry-Bulb (°F) (2%)	Mean Coincident Wet-Bulb (°F) (2%)	Mean Daily Range (°F)	Design Wet-Bulb (°F) (2%)	CDD50°F
ALABAMA									
Birmingham AP	2540	24.9	93.2	74.5	91.1	74.3	17.7	76.7	5740
Huntsville AP	2969	22.7	93.2	74.6	91.1	74.1	18.3	76.8	5335
Mobile AP	1600	30.9	92.3	76.8	90.7	76.4	16.5	78.5	6739
Montgomery AP	2039	27.9	94.7	76.0	92.9	75.7	19.3	77.8	6322
Tuscaloosa AP	2396	26.5	94.7	75.7	92.6	75.4	19.6	77.9	5941
ALASKA									
Anchorage AP	9859	−2.8	69.2	57.9	66.7	56.8	11.8	58.1	812
Fairbanks AP	13,366	−37.6	77.7	59.9	74.5	58.4	17.7	60.1	1094
Juneau AP	8295	9.8	70.5	58.4	67.0	56.7	12.3	58.2	668
Nome AP	13,436	−22.5	64.2	55.2	61.1	54.2	10.1	55.0	238
ARIZONA									
Bellemont (Flagstaff)	5232	13.5	92.8	57.8	90.6	57.6	28.0	61.8	3365
Phoenix AP	874	41.9	108.5	69.0	106.7	68.9	20.6	74.0	9328
Prescott AP	4032	21.2	92.4	60.0	90.3	59.6	25.1	64.5	3705
Tucson CO	1328	34.7	103.6	65.7	101.5	65.4	22.8	71.2	7635
Winslow AP	4369	15.6	96.3	60.4	94.0	59.9	29.6	64.2	3939
Yuma AP	637	44.8	108.9	72.2	107.6	71.6	23.1	77.2	9604
ARKANSAS									
Bentonville AP	3953	15.9	91.7	75.2	90.2	74.5	19.1	75.6	4479
Fort Smith AP	3063	22.3	96.9	76.3	94.3	76.0	19.6	77.9	5540
Little Rock AP	2881	24.0	95.6	77.1	93.3	76.7	18.4	78.3	5675
Texarkana AP	2432	26.9	96.6	75.8	94.1	75.6	19.1	77.7	6054
CALIFORNIA									
Bakersfield AP	1957	36.0	100.9	69.0	98.6	68.0	25.7	70.1	6122
Burbank AP	1345	41.3	94.3	66.8	91.2	66.1	23.2	69.3	5662
Crescent City AP	4674	36.7	65.5	58.8	63.7	57.8	9.4	59.1	1213
Fresno AP	2138	34.8	101.3	68.6	98.8	67.7	29.1	69.9	5807
Long Beach AP	1139	43.9	88.2	66.0	85.1	65.4	16.9	69.4	5504
Los Angeles AP	1256	47.0	81.3	64.2	78.6	64.3	10.3	68.2	4894
Oakland AP	2602	39.2	79.5	63.1	75.4	62.1	16.6	63.8	3176
Palm Springs AP	723	43.9	109.7	70.2	107.8	69.8	25.8	75.3	9275
Sacramento AP	2475	33.6	97.5	69.1	94.6	67.8	32.6	69.4	4657
San Diego AP	1101	47.2	81.6	65.6	79.0	65.8	8.6	69.3	5157
San Francisco AP	2606	42.2	78.1	62.0	74.5	61.5	16.0	63.0	3128
San Luis Obispo AP	2155	36.5	85.3	63.3	81.8	62.9	23.6	64.6	3737
Santa Barbara AP	2205	37.3	80.1	63.4	77.2	63.0	15.6	66.1	3555
Santa Maria AP	2554	35.9	80.9	61.8	77.4	61.3	20.3	63.8	3170
Stockton AP	2377	33.1	98.3	68.8	95.3	68.0	32.8	69.1	4802
COLORADO									
Alamosa AP	8215	−7.8	83.7	54.6	81.4	54.4	33.6	57.5	1527
Colorado Springs AP	6013	7.4	88.7	58.4	86.1	58.3	26.0	61.6	2626
Denver AP	5874	5.8	92.3	59.7	89.6	59.5	27.5	62.5	3009
Fort Collins	6096	4.8	87.2	60.4	84.4	60.1	24.5	62.5	2550
Grand Junction AP	5416	10.9	95.5	60.3	93.1	59.8	27.9	63.1	3672
Pueblo AP	5378	7.2	96.2	62.1	93.5	61.7	31.7	64.9	3404

Table B.1 (*Continued*)

State/Province and Station[a]	Winter HDD65°F	Winter Design Dry-Bulb (°F) (99%)	Summer Design Dry-Bulb (°F) (1%)	Mean Coincident Wet-Bulb (°F) (1%)	Design Dry-Bulb (°F) (2%)	Mean Coincident Wet-Bulb (°F) (2%)	Mean Daily Range (°F)	Design Wet-Bulb (°F) (2%)	CDD50°F
CONNECTICUT									
Bridgeport AP	5193	15.3	85.4	72.0	83.0	70.8	14.0	73.8	3272
Hartford-Bradley AP	5828	9.1	88.7	71.9	85.8	70.4	20.2	73.4	3089
DELAWARE									
New Castle, Wilmington AP	4677	16.9	89.5	74.1	87.0	73.2	17.1	75.7	3815
DISTRICT OF COLUMBIA									
Washington DC, Reagan AP (Arlington)	3856	20.7	92.0	74.7	89.6	73.7	15.7	76.3	4604
FLORIDA									
Daytona Beach AP	709	39.6	90.9	77.2	89.5	77.1	14.6	78.8	7913
Fort Myers, SW Florida AP	295	45.6	92.2	76.6	91.0	76.5	15.5	78.7	9024
Gainesville AP	1108	33.3	92.4	75.8	90.9	75.6	17.9	78.0	7264
Jacksonville AP	1268	32.8	92.6	76.8	90.9	76.4	17.8	78.4	7040
Key West AP	63	57.8	89.9	79.3	89.1	79.1	8.3	80.2	10328
Miami NHC	112	52.7	90.9	77.7	90.0	77.6	11.6	79.3	10022
Orlando AP	512	42.4	92.4	76.3	91.1	76.0	15.9	78.3	8468
Palm Beach AP	207	48.5	90.5	77.8	89.4	77.7	12.4	79.0	9488
Panama City	1242	35.7	91.1	76.9	90.1	76.7	13.7	79.3	7227
Pensacola AP	1367	33.8	92.0	77.4	90.3	77.1	14.2	79.4	7103
Tallahassee	1441	30.0	94.3	75.8	92.5	75.5	18.3	78.3	7014
Tampa AP	481	43.7	91.3	77.0	90.3	76.9	12.9	79.1	8752
GEORGIA									
Athens AP	2698	26.3	93.2	74.2	90.9	73.9	19.7	76.1	5315
Atlanta AP	2578	26.4	91.6	73.6	89.7	73.3	16.5	75.3	5514
Augusta AP	2325	26.1	94.9	75.6	92.8	75.3	21.2	77.5	5797
Columbus AP	1991	29.4	94.1	74.3	92.3	74.1	18.0	76.5	6295
Lawson AAF	2253	25.6	94.8	75.9	92.7	75.8	21.4	78.4	5864
Robins AFB	2084	28.0	94.8	75.7	92.7	75.4	19.9	77.7	6072
Rome AP	2967	23.4	93.3	74.0	91.2	73.9	20.5	76.3	5166
Savannah AP	1688	30.7	93.3	76.8	91.4	76.4	17.8	78.4	6588
Valdosta AP	1456	30.6	94.4	76.3	92.6	75.9	19.7	78.4	6895
HAWAII									
Hilo AP	0	62.9	85.2	74.0	84.2	73.6	12.8	75.3	8817
Honolulu AP	0	65.1	88.7	73.7	87.8	73.3	11.8	75.6	10196
IDAHO									
Boise AP	5311	16.4	95.9	62.9	93.1	61.9	29.4	63.4	3297
Coeur D'Alene AP	6734	11.5	89.7	62.9	85.5	61.6	28.7	62.8	2007
Lewiston AP	5024	19.4	95.4	64.4	91.8	63.3	28.3	64.7	3123
Pocatello AP	6866	4.8	92.0	60.8	89.1	59.9	34.8	62.0	2249
ILLINOIS									
Champaign, Willard AP (Savoy)	5655	3.7	89.6	74.7	87.1	73.4	19.3	75.7	3518
Chicago, Midway AP	5814	5.0	89.5	73.2	86.8	72.0	15.7	74.4	3497
Moline, Quad Cities AP	6040	1.2	89.8	74.9	87.2	73.2	19.4	75.6	3442
Peoria AP	5669	3.7	89.7	75.2	87.3	73.7	18.4	76.0	3620
Quincy AP	5448	4.9	90.4	75.4	88.0	74.2	18.7	75.9	3734
Springfield AP	5284	6.2	90.6	75.9	88.4	74.4	19.4	76.7	3815

Table B.1 (*Continued*)

State/Province and Station[a]	Winter		Summer						
	HDD65°F	Design Dry-Bulb (°F) (99%)	Design Dry-Bulb (°F) (1%)	Mean Coincident Wet-Bulb (°F) (1%)	Design Dry-Bulb (°F) (2%)	Mean Coincident Wet-Bulb (°F) (2%)	Mean Daily Range (°F)	Design Wet-Bulb (°F) (2%)	CDD50°F
INDIANA									
Evansville AP	4353	13.8	91.4	75.4	89.4	74.6	18.5	76.8	4352
Fort Wayne AP	5948	4.5	88.1	73.0	85.5	71.7	19.7	74.3	3202
Indianapolis AP	5224	7.5	88.9	73.9	86.6	72.7	17.6	75.1	3758
Muncie AP	5502	6.7	88.2	72.9	85.9	71.6	18.3	74.3	3494
South Bend AP	6187	5.0	87.5	72.4	84.9	71.2	18.8	73.7	3068
Terre Haute AP	5165	7.7	90.0	75.4	87.7	74.0	19.7	76.0	3711
West Lafayette AP	5561	4.9	89.2	74.3	86.6	72.8	19.4	75.3	3496
IOWA									
Burlington, SE Iowa AP	5715	2.4	90.0	76.0	87.6	74.4	18.3	76.5	3654
Des Moines AP	6065	0.4	90.1	75.0	87.4	73.6	17.8	75.6	3576
Dubuque AP	7005	−3.2	85.6	73.4	83.2	71.6	18.1	73.9	2789
Mason City AP	7525	−6.5	87.3	73.2	84.5	71.5	20.8	74.0	2706
Sioux City AP	6689	−2.4	90.1	74.4	87.5	73.2	20.8	75.5	3208
Waterloo AP	6977	−4.7	88.1	73.8	85.5	72.3	19.9	74.8	2986
KANSAS									
Dodge City AP	4834	9.8	97.6	69.6	94.6	69.6	25.3	72.1	4185
Goodland AP	5734	6.2	94.8	66.8	92.0	66.5	27.4	69.1	3303
Manhattan AP	5053	8.0	96.8	75.5	93.1	74.8	22.4	76.6	4175
Topeka AP	4823	8.9	94.9	76.1	92.0	75.2	19.9	76.9	4294
Wichita AP	4370	12.4	97.3	74.2	94.1	73.9	21.0	75.7	4653
KENTUCKY									
Bowling Green	4001	15.3	88.9	74.6	87.0	73.5	17.1	75.4	4407
Lexington, Blue Grass AP	4483	13.3	89.4	73.4	87.3	72.6	18.0	74.9	4057
Louisville AP	4010	15.7	91.7	75.8	89.6	74.0	16.8	76.3	4678
LOUISIANA									
Alexandria AP	1837	29.8	95.0	76.8	93.0	76.6	19.9	78.7	6670
Baton Rouge AP	1537	31.5	93.3	77.3	91.9	76.9	16.9	79.0	6981
Lafayette AP	1388	33.5	93.2	77.6	91.6	77.3	16.3	79.4	7247
Lake Charles	1387	33.7	93.2	77.8	91.7	77.6	15.1	79.9	7231
Monroe AP	2146	28.1	95.8	77.4	93.7	77.1	19.0	79.2	6388
New Orleans AP	1193	36.6	92.8	77.7	92.1	77.5	14.0	79.6	7574
Shreveport AP	2043	29.0	96.9	75.9	94.6	75.9	19.8	78.2	6572
MAINE									
Bangor AP	7601	−1.6	84.0	69.2	81.1	67.2	20.0	69.6	2149
Caribou AP	9092	−8.9	81.4	66.9	78.4	65.4	19.4	68.1	1677
Portland AP	6891	5.0	83.3	69.9	80.4	68.2	17.1	70.6	2240
MARYLAND									
Baltimore CO	3858	20.7	92.9	74.3	90.5	73.4	15.0	75.9	4747
Salisbury AP	4225	18.5	90.6	75.4	88.3	74.4	19.3	76.8	4038
MASSACHUSETTS									
Boston AP	5498	12.8	87.7	71.6	84.5	70.0	14.7	72.9	3072
Worcester AP	6561	6.8	83.4	69.7	81.0	68.2	15.9	71.1	2533
MICHIGAN									
Alpena AP	7883	−0.1	84.6	69.5	81.5	67.9	22.1	70.2	2018
Detroit AP	5988	8.8	87.6	71.9	85.3	70.5	17.2	73.0	3186

Table B.1 (*Continued*)

	Winter		Summer						
State/Province and Station[a]	HDD65°F	Design Dry-Bulb (°F) (99%)	Design Dry-Bulb (°F) (1%)	Mean Coincident Wet-Bulb (°F) (1%)	Design Dry-Bulb (°F) (2%)	Mean Coincident Wet-Bulb (°F) (2%)	Mean Daily Range (°F)	Design Wet-Bulb (°F) (2%)	CDD50°F
Escanaba AP	8436	−3.7	80.6	68.7	77.3	67.2	19.8	69.1	1696
Flint AP	6636	3.9	87.1	71.9	84.4	70.2	21.1	72.6	2749
Grand Rapids AP	6486	6.8	86.7	71.7	84.2	70.2	19.3	72.8	2839
Muskegon AP	6495	9.0	84.0	70.9	81.8	69.7	17.5	72.3	2633
Sault Sainte Marie AP	8532	−4.9	80.7	67.8	77.9	66.1	19.3	68.4	1755
Traverse City AP	7162	5.0	85.8	70.1	82.8	68.6	20.3	71.0	2408
MINNESOTA									
Bemidji AP	9494	−17.5	82.4	67.4	80.7	66.0	20.6	68.4	1856
Duluth AP	9173	−12.1	81.4	67.5	78.6	65.7	19.0	68.1	1754
International Falls AP	9984	−20.8	82.2	67.7	79.5	66.0	23.3	68.5	1611
Minneapolis-St. Paul AP	7396	−6.0	87.9	72.0	85.2	70.4	17.4	73.0	2977
Rochester AP	7779	−8.1	84.7	71.9	82.3	0.5	18.1	72.9	2548
MISSISSIPPI									
Columbus AFB	2599	24.6	94.2	76.7	92.4	76.4	19.6	78.7	5725
Jackson AP	2210	26.8	94.3	75.9	92.3	75.7	18.9	77.8	6164
Meridian AP	2307	25.9	94.1	76.1	92.2	75.9	20.1	78.1	5936
MISSOURI									
Columbia AP	4843	8.9	91.9	75.9	89.2	75.0	19.2	76.6	4062
Joplin (Asbury)	4398	12.2	92.5	75.2	89.9	75.3	20.4	76.6	4265
Kansas City AP	4977	7.2	92.4	76.3	89.6	75.5	18.5	77.1	4112
St. Louis AP	4379	12.2	93.5	76.2	91.2	75.1	16.8	77.0	4639
MONTANA									
Billings AP	6746	−2.7	91.7	61.9	88.4	61.3	26.6	63.2	2628
Bozeman AP	8160	−6.5	88.6	60.3	85.3	59.4	34.5	61.1	1736
Cut Bank AP	8543	−12.4	85.8	58.8	82.1	57.9	28.9	59.4	1491
Glasgow AP	8460	−14.6	91.2	64.1	87.5	63.0	27.0	65.3	2317
Great Falls AP	7593	−8.9	89.5	60.1	86.1	59.3	30.7	60.9	1942
Helena AP	7466	−5.3	90.1	60.7	86.8	59.8	29.7	61.4	2158
Kalispell, Glacier Park AP	7921	1.9	87.2	61.5	83.8	60.5	32.8	61.8	1520
Lewistown	8889	−10.6	83.2	56.0	80.2	55.6	30.3	57.4	1104
Miles City AP	7430	−9.1	94.5	64.6	91.0	64.0	27.4	66.3	2747
Missoula AP	7331	4.3	90.0	61.1	86.5	60.2	32.7	61.6	1947
NEBRASKA									
Grand Island, Central NE AP	6051	2.0	92.4	73.4	89.4	72.3	21.3	74.6	3443
Lincoln AP	5913	2.2	93.2	74.8	90.4	73.7	21.3	75.5	3675
North Platte AP	6463	1.3	93.0	70.7	89.8	70.1	25.8	72.2	2994
Omaha AP	5701	4.5	91.3	74.7	89.1	73.6	18.3	75.7	3744
Scottsbluff, Western NE AP	6433	1.2	93.6	65.2	90.5	64.8	29.4	67.6	2838
NEVADA									
Elko AP	6835	4.5	93.0	58.3	90.4	57.4	37.8	59.4	2240
Ely AP	7282	1.4	89.4	55.7	87.3	55.0	37.2	57.7	1875
Las Vegas AP	1841	35.4	106.7	66.4	104.6	65.7	21.5	69.9	7546
Lovelock, Derby Field AP	5407	11.6	98.1	61.1	95.5	60.2	37.5	62.0	3165
Reno AP	4817	19.5	94.4	60.4	91.9	59.3	31.7	61.0	3293
Tonopah AP	5278	15.4	93.4	58.4	91.1	57.4	31.8	60.2	3126

Table B.1 (*Continued*)

State/Province and Station[a]	HDD65°F	Winter: Design Dry-Bulb (°F) (99%)	Summer: Design Dry-Bulb (°F) (1%)	Mean Coincident Wet-Bulb (°F) (1%)	Design Dry-Bulb (°F) (2%)	Mean Coincident Wet-Bulb (°F) (2%)	Mean Daily Range (°F)	Design Wet-Bulb (°F) (2%)	CDD50°F
Winnemucca AP	6136	9.3	95.8	60.5	93.3	59.5	39.3	60.9	2505
NEW HAMPSHIRE									
Concord AP	7010	1.9	87.2	69.9	84.3	68.6	22.6	71.4	2427
Portsmouth AP	6390	7.3	86.2	71.3	83.2	69.8	18.1	72.2	2592
NEW JERSEY									
Atlantic City AP	4734	16.0	89.5	74.0	86.8	73.0	17.6	75.8	3664
Newark AP	4646	16.2	91.1	72.8	88.3	71.9	16.1	74.9	3926
NEW MEXICO									
Albuquerque AP	3873	22.4	93.4	59.6	91.2	59.4	22.9	63.7	4340
Farmington, Four Corners AP	5189	13.3	93.5	58.9	91.3	58.7	29.6	62.5	3389
Gallup AP	6056	6.6	89.9	56.9	87.7	56.3	32.0	60.5	2436
Roswell AP	3051	21.6	99.2	64.7	96.8	64.7	25.8	68.9	5336
Santa Fe AP	5275	14.7	90.3	57.8	88.1	57.6	28.0	61.4	3015
Taos AP	6833	3.2	86.3	55.6	83.8	55.1	32.0	58.2	1998
Tucumcari AP	3724	16.8	98.7	64.2	96.3	64.2	27.3	67.9	4789
NEW YORK									
Albany AP	6418	4.3	86.3	71.1	83.7	69.8	18.9	72.4	2830
Binghamton AP (Johnson City)	7034	3.9	82.3	68.6	79.8	67.4	17.0	69.8	2339
Buffalo AP	6449	6.8	83.9	70.0	81.7	69.1	16.0	71.6	2722
Ithaca	6898	4.6	85.5	71.0	82.7	69.5	21.3	71.6	2465
Massena AP	7807	−7.6	84.6	70.9	82.1	69.5	20.4	71.6	2307
New York City, Kennedy AP	4761	17.5	86.7	72.1	84.2	71.2	13.4	74.7	3573
Rochester AP	6404	6.6	86.0	71.4	83.2	69.8	18.7	72.2	2729
Syracuse AP	6504	4.1	86.4	71.4	83.8	69.9	18.8	72.2	2765
NORTH CAROLINA									
Asheville AP	4000	19.4	85.9	70.2	83.9	69.6	18.4	71.7	3702
Charlotte AP	3030	24.8	91.9	74.0	89.8	73.5	18.6	75.4	5031
Greensboro AP	3017	25.1	92.5	75.8	90.4	75.3	19.6	77.7	5141
Raleigh-Durham AP	3188	23.6	92.6	75.1	90.3	74.5	19.1	76.4	4970
Wilmington AP	2355	27.5	91.2	76.9	89.1	76.3	15.6	78.1	5730
NORTH DAKOTA									
Bismarck AP	8414	−11.8	89.4	69.3	86.1	68.0	24.9	70.4	2401
Fargo AP	8685	−13.9	87.0	70.4	84.3	68.8	20.9	71.5	2482
Minot AP	8733	−13.0	87.7	68.5	84.2	66.5	22.3	69.1	2273
OHIO									
Akron-Canton AP	5923	7.3	86.3	71.6	84.0	70.2	18.5	72.5	3078
Cincinnati AP	4776	12.6	89.8	74.4	87.6	73.5	19.9	75.7	3791
Cleveland AP	5737	9.0	87.2	72.4	84.6	71.1	16.8	73.2	3207
Columbus AP	5161	9.8	88.9	72.5	86.6	71.3	18.0	73.8	3683
Dayton AP	5442	7.5	87.9	72.8	85.5	71.4	17.9	73.8	3504
Mansfield AP	6056	5.9	85.9	71.8	83.6	70.5	18.7	72.8	2999
Toledo AP	5885	6.6	88.8	72.7	86.1	71.4	18.4	73.9	3298
Youngstown-Warren AP	6113	7.1	85.6	71.2	83.3	69.7	20.2	72.0	2829

APPENDICES

Table B.1 (*Continued*)

State/Province and Station[a]	HDD65°F	Winter: Design Dry-Bulb (°F) (99%)	Summer: Design Dry-Bulb (°F) (1%)	Mean Coincident Wet-Bulb (°F) (1%)	Design Dry-Bulb (°F) (2%)	Mean Coincident Wet-Bulb (°F) (2%)	Mean Daily Range (°F)	Design Wet-Bulb (°F) (2%)	CDD50°F
OKLAHOMA									
Oklahoma City AP	3398	19.0	97.6	74.2	94.7	74.3	21.6	76.1	5250
Tulsa AP	3411	18.2	97.2	76.2	94.6	76.0	19.0	77.6	5388
OREGON									
Astoria AP	4870	31.1	72.4	62.1	69.3	61.1	13.7	62.1	1530
Eugene AP	4629	27.2	88.3	65.7	84.6	64.4	30.2	65.5	2287
Medford AP	4195	26.7	95.5	65.6	92.3	64.4	31.3	65.8	3333
Pendleton AP	5233	15.7	93.8	63.7	90.3	62.6	30.4	63.8	2799
Portland AP	4179	29.4	87.5	66.3	83.8	64.9	21.6	66.2	2764
Salem AP	4436	27.8	88.5	65.9	84.6	64.5	26.9	65.5	2482
PENNSYLVANIA									
Allentown AP	5435	12.5	88.5	72.7	86.0	71.4	19.9	73.9	3282
Erie AP	6021	9.7	84.4	71.8	82.1	70.6	14.6	72.5	2925
Harrisburg AP	5026	14.8	89.5	74.0	86.8	72.9	17.5	75.1	3700
Philadelphia AP	4410	17.8	90.8	73.9	88.4	72.5	16.1	75.5	4136
Pittsburgh AP	5518	9.2	86.4	70.8	84.3	69.6	18.0	72.3	3242
Williamsport AP	5749	10.1	87.7	71.2	85.0	69.9	20.8	72.7	3100
RHODE ISLAND									
Providence AP	5477	12.8	86.8	72.1	84.0	70.6	16.8	73.6	3072
SOUTH CAROLINA									
Anderson AP	2858	25.8	92.8	73.6	90.6	73.3	19.1	75.5	5149
Charleston AP	1821	30.6	92.3	77.4	90.3	76.9	16.1	78.7	6404
Columbia AP	2374	27.0	94.8	74.8	92.8	74.6	19.1	77.0	5941
Greenville AP	2762	26.8	92.0	72.9	90.1	72.4	17.2	74.8	5322
SOUTH DAKOTA									
Huron AP	7648	−8.2	90.0	72.8	87.0	71.3	22.7	73.8	2847
Pierre 24 S	7315	−4.5	91.3	69.4	87.7	68.4	24.3	71.0	2818
Rapid City AP	7113	−3.0	92.4	65.9	88.8	65.3	26.8	68.0	2575
Sioux Falls AP	7442	−6.3	88.3	73.1	85.6	71.7	20.3	74.0	2851
TENNESSEE									
Blountville, Tri-Cities AP	4131	17.4	88.4	71.4	86.5	70.9	19.8	73.1	3915
Chattanooga AP	3037	23.4	92.6	73.8	90.6	73.4	18.2	75.7	5179
Knoxville AP (Alcoa)	3486	21.0	90.4	73.3	88.4	72.7	17.7	75.0	4709
Memphis AP	2856	23.2	94.3	76.5	92.4	76.1	16.4	78.1	5801
Nashville AP	3430	19.5	92.3	74.5	90.4	73.9	18.0	76.1	4994
TEXAS									
Abilene AP	2413	24.4	98.4	70.9	96.2	71.0	21.2	74.0	6253
Amarillo	3952	16.0	96.1	65.9	93.6	66.1	24.9	69.2	4439
Austin AP	1648	29.9	98.4	74.6	96.5	74.7	22.4	77.5	7154
Brownsville AP	499	42.4	94.9	78.3	93.7	78.2	16.7	80.3	9194
Corpus Christi AP	821	38.1	95.4	77.9	93.9	77.9	18.3	80.1	8435
Dallas/Fort Worth AP	2113	27.4	99.1	74.6	96.9	74.7	19.1	77.1	6814
Del Rio AP	1269	34.5	100.5	72.5	98.8	72.6	20.8	76.4	7926
El Paso AP	2203	28.8	99.1	63.4	96.9	63.3	24.9	68.2	6326
Galveston AP	957	39.6	90.9	79.5	90.1	79.4	9.6	81.0	8054
Houston, Hobby AP	1115	36.6	94.4	77.3	92.8	77.2	15.8	79.3	7802
Laredo AP	768	39.0	102.9	74.7	101.2	74.7	21.8	78.2	9320

APPENDICES

Table B.1 (*Continued*)

State/Province and Station[a]	HDD65°F	Winter: Design Dry-Bulb (°F) (99%)	Summer: Design Dry-Bulb (°F) (1%)	Mean Coincident Wet-Bulb (°F) (1%)	Design Dry-Bulb (°F) (2%)	Mean Coincident Wet-Bulb (°F) (2%)	Mean Daily Range (°F)	Design Wet-Bulb (°F) (2%)	CDD50°F
Lubbock AP	3217	20.1	97.3	67.0	94.9	67.0	23.1	70.6	5197
Midland AP	2475	24.6	99.2	67.4	97.0	67.5	22.7	71.3	6156
Nacogdoches AP	2093	27.6	96.8	76.0	93.3	75.9	21.9	78.0	6332
Port Arthur AP	1286	34.9	93.4	78.0	91.9	78.0	16.5	80.0	7388
San Angelo AP	2138	25.8	100.0	70.2	97.9	70.2	23.6	73.5	6555
San Antonio AP	1352	33.1	97.7	73.6	95.9	73.8	19.7	76.9	7590
Waco AP	2003	27.9	99.3	74.7	97.3	74.8	21.3	77.2	6855
Wichita Falls AP	2565	24.0	101.3	71.8	99.1	71.3	22.1	75.2	6387
UTAH									
Cedar City AP	5874	7.9	91.6	58.6	89.4	57.8	30.2	61.2	2835
Salt Lake City AP	5329	15.6	95.9	61.9	93.4	61.3	24.9	64.0	3685
VERMONT									
Burlington AP	7145	−1.7	85.7	69.9	83.0	68.6	18.8	71.1	2580
Rutland AP	7344	−0.3	82.3	68.7	80.9	67.5	19.1	69.6	2177
VIRGINIA									
Norfolk AP	3139	25.8	91.2	76.0	89.0	75.3	14.2	77.3	5017
Richmond AP	3635	21.2	92.4	75.0	90.1	74.2	18.6	76.4	4654
Roanoke AP	3927	19.8	89.7	71.9	87.5	71.2	18.2	73.3	4197
WASHINGTON									
Bellingham AP	5265	24.7	76.8	63.9	73.6	62.2	16.3	63.2	1661
Olympia AP	5319	24.6	83.8	64.6	80.1	63.3	25.5	64.2	1747
Seattle AP	4299	29.8	82.2	63.9	79.2	62.7	18.0	63.9	2410
Seattle–Tacoma AP	4621	30.0	82.2	63.9	78.7	62.7	19.0	63.6	2147
Spokane AP	6539	11.8	90.1	61.6	86.6	60.5	26.2	61.9	2276
Walla Walla AP	4762	18.2	94.8	64.7	91.1	63.6	26.7	64.9	3327
Yakima AP	5730	14.4	94.1	65.3	90.7	63.9	32.8	65.1	2653
WEST VIRGINIA									
Charleston AP	4385	14.8	88.8	72.4	86.7	71.7	18.9	74.2	3897
Huntington AP	4383	14.7	89.1	73.3	86.9	72.4	18.5	74.7	3977
Martinsburg AP	4830	15.4	89.9	73.0	87.4	72.0	20.6	74.2	3622
Morgantown AP	4939	11.7	87.0	71.0	84.8	70.1	18.6	72.7	3525
WISCONSIN									
Eau Claire AP	7810	−8.5	86.6	71.0	83.9	69.3	21.0	72.1	2573
Green Bay AP	7519	−3.2	85.0	71.9	82.5	70.3	20.3	72.5	2415
La Crosse AP	6929	−4.5	88.6	73.0	85.9	71.5	19.4	73.9	3089
Madison AP	7053	−1.9	86.4	72.4	83.9	71.1	19.3	73.2	2731
Milwaukee AP	6647	2.8	86.4	72.3	83.5	70.6	15.4	73.0	2784
WYOMING									
Casper AP	7336	−1.0	91.4	58.9	88.6	58.3	33.7	60.5	2217
Cheyenne AP	6972	3.1	87.3	57.6	84.5	57.3	26.5	60.5	2142
Lander	7376	−0.2	88.8	55.8	86.1	55.6	27.4	57.8	2301
Laramie AP	8530	−3.3	82.9	54.5	80.7	54.0	30.2	56.9	1411
Rock Springs AP	7977	1.1	85.9	54.4	83.4	53.7	27.9	55.9	1949
Sheridan AP	7470	−4.0	92.0	62.9	88.7	62.2	32.0	64.2	2189
ALBERTA									
Calgary AP	9098	−12.0	80.3	59.9	77.0	58.8	22.2	60.3	1241
Edmonton AP	10,465	−20.1	79.0	63.2	76.1	61.5	22.9	63.5	1089
Medicine Hat AP	8416	−14.8	88.0	62.4	84.5	61.5	26.4	63.0	1943

Table B.1 (*Continued*)

State/Province and Station[a]	Winter: HDD65°F	Winter: Design Dry-Bulb (°F) (99%)	Summer: Design Dry-Bulb (°F) (1%)	Summer: Mean Coincident Wet-Bulb (°F) (1%)	Summer: Design Dry-Bulb (°F) (2%)	Summer: Mean Coincident Wet-Bulb (°F) (2%)	Summer: Mean Daily Range (°F)	Summer: Design Wet-Bulb (°F) (2%)	CDD50°F
BRITISH COLUMBIA									
Prince George AP	9221	−12.2	78.5	59.8	75.0	58.1	22.7	59.5	998
Prince Rupert AP	6851	20.7	64.5	57.9	62.5	57.1	8.9	58.2	736
Vancouver AP	5171	26.6	74.7	64.3	72.5	63.2	13.8	63.9	1796
Victoria AP	5324	28.4	76.6	62.4	73.6	61.1	18.0	61.8	1575
MANITOBA									
The Pas AP	11,646	−24.7	79.9	64.4	76.9	63.0	18.3	65.0	1408
Winnipeg AP	10,254	−21.2	83.6	68.5	80.9	66.8	21.1	69.0	1870
NEW BRUNSWICK									
Fredericton AP	8174	−4.5	82.6	68.1	79.7	66.3	20.4	68.8	1871
Saint John AP	8371	−2.7	76.5	64.3	73.8	62.9	18.2	65.3	1376
NEWFOUNDLAND									
Gander AP	8998	4.5	76.8	64.4	73.9	63.2	17.5	65.3	1158
St. John's AP	8497	10.3	74.3	65.3	71.8	64.0	13.8	65.7	1087
NORTHWEST TERRITORIES									
Fort Smith AP	12,974	−31.8	79.7	61.2	76.7	59.7	20.6	61.8	1147
Inuvik AP	16,604	−35.3	74.9	59.4	71.3	57.9	17.4	59.0	603
Yellowknife AP	14,491	−35.8	75.0	59.6	72.2	58.4	14.5	60.1	972
NOVA SCOTIA									
Halifax AP	7548	4.6	79.2	67.0	76.6	65.8	16.5	68.3	1787
Sydney AP	7978	6.5	78.8	67.3	75.9	65.8	16.0	67.8	1516
Yarmouth AP	7046	11.6	72.1	65.8	70.2	64.8	13.0	66.3	1470
ONTARIO									
Ottawa AP	8012	−6.5	84.5	69.7	81.6	68.4	18.8	70.7	2325
Sudbury AP	9328	−12.6	81.1	66.2	78.2	64.6	19.0	67.2	1767
Thunder Bay AP	9926	−16.1	80.9	66.6	77.9	65.0	22.3	66.9	1412
Toronto, Pearson AP	6803	3.8	85.5	70.8	82.6	69.4	17.1	71.4	2628
PRINCE EDWARD ISLAND									
Charlottetown AP	8119	1.3	78.0	67.7	75.6	66.3	14.7	68.2	1697
QUEBEC									
Montréal, Trudeau AP	7662	−4.3	83.7	70.0	81.3	68.6	15.8	70.9	2496
Québec City, Jean Lesage AP	8933	−9.4	81.1	68.3	78.4	66.7	18.0	69.0	1826
Sept-Îles AP	10,980	−13.3	70.6	60.3	68.2	59.4	14.9	61.4	798
Trois-Rivières	8208	−6.1	79.2	69.4	77.2	68.3	13.5	70.3	2138
Val-d'Or AP	10,620	−20.6	80.9	65.4	77.9	64.0	20.0	66.7	1443
SASKATCHEWAN									
Prince Albert AP	11,085	−25.7	80.6	64.7	77.8	62.8	20.9	64.8	1388
Regina AP	10,296	−21.9	83.9	65.1	80.7	63.7	23.6	65.8	1611
Saskatoon AP	10,535	−23.3	82.9	64.5	79.7	63.3	22.6	65.2	1529
YUKON TERRITORY									
Whitehorse	11,982	−28.3	74.4	56.0	70.8	54.6	20.0	55.7	652

[a]U.S. solar data are available from the National Climatic Data Center, Federal Building, Asheville, NC 28801. Canadian solar data are available from the Canadian Climate Center, Atmospheric Environment Service, 4905 Dufferin St., Downsview, Ontario M3H 5T4.

Table B.2 Outdoor Design Conditions: Mexico and Puerto Rico (I-P Units)

	Winter		Summer						
Country/Territory and Station	HDD65°F	Design Dry-Bulb (°F) (99%)	Design Dry-Bulb (°F) (1%)	Mean Coincident Wet-Bulb (°F) (1%)	Design Dry-Bulb (°F) (2%)	Mean Coincident Wet-Bulb (°F) (2%)	Mean Daily Range (°F)	Design Wet-Bulb (°F) (2%)	CDD50°F
MEXICO									
Guadalajara AP	627	37.7	89.9	58.8	88.1	58.3	31.7	66.5	6171
Mérida AP	2	60.2	99.4	76.1	97.6	76.2	23.4	81.4	11444
Mexico City AP	1023	40.6	82.5	53.8	80.6	53.7	24.2	59.3	4844
Monterrey AP	679	40.8	100.0	72.8	98.3	72.8	22.0	77.6	8689
PUERTO RICO									
San Juan AP	0	71.0	89.8	77.8	89.0	77.9	10.1	79.8	11230

Table B.3 Outdoor Design Conditions: United States and Canada (SI Units)

	Winter		Summer						
State/Province and Station[a]	HDD18.3°C	Design Dry-Bulb (°C) (99%)	Design Dry-Bulb (°C) (1%)	Mean Coincident Wet-Bulb (°C) (1%)	Design Dry-Bulb (°C) (2%)	Mean Coincident Wet-Bulb (°C) (2%)	Mean Daily Range (°C)	Design Wet-Bulb (°C) (2%)	CDD10.0°C
ALABAMA									
Birmingham AP	1411	−3.9	34.0	23.6	32.9	23.5	9.8	24.8	3189
Huntsville AP	1649	−5.2	34.0	23.7	32.9	23.4	10.2	24.9	2964
Mobile AP	889	−0.6	33.5	24.9	32.6	24.7	9.2	25.8	3744
Montgomery AP	1133	−2.3	34.8	24.4	33.8	24.3	10.7	25.5	3512
Tuscaloosa AP	1331	−3.0	34.8	24.3	33.7	24.1	10.9	25.5	3301
ALASKA									
Anchorage AP	5477	−19.4	20.7	14.4	19.3	13.8	6.6	14.5	451
Fairbanks AP	7426	−38.6	25.4	15.5	23.6	14.7	9.8	15.6	608
Juneau AP	4608	−12.3	21.4	14.7	19.4	13.7	6.8	14.5	371
Nome AP	7464	−30.3	17.9	12.9	16.1	12.3	5.6	12.8	132
ARIZONA									
Bellemont (Flagstaff)	2907	−10.3	33.8	14.3	32.6	14.2	15.5	16.5	1869
Phoenix AP	486	5.5	42.5	20.6	41.5	20.5	11.4	23.3	5182
Prescott AP	2240	−6.0	33.6	15.5	32.4	15.4	13.9	18.0	2058
Tucson CO	738	1.5	39.8	18.7	38.6	18.6	12.7	21.8	4242
Winslow AP	2427	−9.1	35.7	15.8	34.4	15.5	16.4	17.9	2188
Yuma AP	354	7.1	42.7	22.3	42.0	22.0	12.8	25.1	5336
ARKANSAS									
Bentonville AP	2196	−8.9	33.2	24.0	32.4	23.6	10.6	24.2	2488
Fort Smith AP	1702	−5.4	36.1	24.6	34.6	24.5	10.9	25.5	3078
Little Rock AP	1601	−4.5	35.3	25.0	34.1	24.8	10.2	25.7	3153
Texarkana AP	1351	−2.8	35.9	24.3	34.5	24.2	10.6	25.4	3363
CALIFORNIA									
Bakersfield AP	1087	2.2	38.3	20.5	37.0	20.0	14.3	21.1	3401
Burbank AP	747	5.2	34.6	19.3	32.9	19.0	12.9	20.7	3146

Table B.3 (*Continued*)

State/Province and Station[a]	HDD18.3°C	Winter: Design Dry-Bulb (°C) (99%)	Summer: Design Dry-Bulb (°C) (1%)	Mean Coincident Wet-Bulb (°C) (1%)	Design Dry-Bulb (°C) (2%)	Mean Coincident Wet-Bulb (°C) (2%)	Mean Daily Range (°C)	Design Wet-Bulb (°C) (2%)	CDD10.0°C
Crescent City AP	2597	2.6	18.6	14.9	17.6	14.3	5.2	15.0	674
Fresno AP	1188	1.6	38.5	20.3	37.1	19.9	16.2	21.1	3226
Long Beach AP	633	6.6	31.2	18.9	29.5	18.6	9.4	20.8	3058
Los Angeles AP	698	8.3	27.4	17.9	25.9	18.0	5.7	20.1	2719
Oakland AP	1446	4.0	26.4	17.3	24.1	16.7	9.2	17.7	1764
Palm Springs AP	402	6.6	43.2	21.2	42.1	21.0	14.4	24.1	5153
Sacramento AP	1375	0.9	36.4	20.6	34.8	19.9	18.1	20.8	2587
San Diego AP	612	8.5	27.5	18.7	26.1	18.8	4.8	20.7	2865
San Francisco AP	1448	5.7	25.6	16.7	23.6	16.4	8.9	17.2	1738
San Luis Obispo AP	1197	2.5	29.6	17.4	27.7	17.2	13.1	18.1	2076
Santa Barbara AP	1225	3.0	26.7	17.5	25.1	17.2	8.7	19.0	1975
Santa Maria AP	1419	2.1	27.2	16.6	25.2	16.3	11.3	17.6	1761
Stockton AP	1321	0.6	36.8	20.5	35.1	20.0	18.2	20.6	2668
COLORADO									
Alamosa AP	4564	−22.1	28.7	12.5	27.4	12.4	18.7	14.1	848
Colorado Springs AP	3341	−13.7	31.5	14.7	30.1	14.6	14.4	16.4	1459
Denver AP	3263	−14.6	33.5	15.4	32.0	15.3	15.3	16.9	1672
Fort Collins	3387	−15.1	30.7	15.8	29.1	15.6	13.6	16.9	1417
Grand Junction AP	3009	−11.7	35.3	15.7	34.0	15.4	15.5	17.3	2040
Pueblo AP	2988	−13.8	35.7	16.7	34.2	16.5	17.6	18.3	1891
CONNECTICUT									
Bridgeport AP	2885	−9.3	29.7	22.2	28.3	21.6	7.8	23.2	1818
Hartford-Bradley AP	3238	−12.7	31.5	22.2	29.9	21.3	11.2	23.0	1716
DELAWARE									
New Castle, Wilmington AP	2598	−8.4	31.9	23.4	30.6	22.9	9.5	24.3	2119
DISTRICT OF COLUMBIA									
Washington DC, Reagan AP (Arlington)	2142	−6.3	33.3	23.7	32.0	23.1	8.7	24.6	2558
FLORIDA									
Daytona Beach AP	394	4.2	32.7	25.1	31.9	25.0	8.1	26.0	4396
Fort Myers, SW Florida AP	164	7.5	33.4	24.8	32.8	24.7	8.6	25.9	5013
Gainesville AP	616	0.7	33.6	24.4	32.7	24.2	10.0	25.5	4036
Jacksonville AP	704	0.5	33.7	24.9	32.7	24.7	9.9	25.8	3911
Key West AP	35	14.4	32.2	26.3	31.7	26.2	4.6	26.8	5738
Miami NHC	62	11.5	32.7	25.4	32.2	25.3	6.4	26.3	5568
Orlando AP	284	5.8	33.5	24.6	32.8	24.4	8.8	25.7	4704
Palm Beach AP	115	9.2	32.5	25.4	31.9	25.4	6.9	26.1	5271
Panama City	690	2.1	32.8	24.9	32.3	24.9	7.6	26.3	4015
Pensacola AP	759	1.0	33.3	25.2	32.4	25.1	7.9	26.3	3946
Tallahassee	801	−1.1	34.6	24.3	33.6	24.2	10.2	25.7	3897
Tampa AP	267	6.5	33.0	25.0	32.4	25.0	7.1	26.2	4862

Table B.3 (*Continued*)

State/Province and Station[a]	HDD18.3°C	Winter: Design Dry-Bulb (°C) (99%)	Summer: Design Dry-Bulb (°C) (1%)	Mean Coincident Wet-Bulb (°C) (1%)	Design Dry-Bulb (°C) (2%)	Mean Coincident Wet-Bulb (°C) (2%)	Mean Daily Range (°C)	Design Wet-Bulb (°C) (2%)	CDD10.0°C
GEORGIA									
Athens AP	1499	−3.2	34.0	23.5	32.7	23.3	10.9	24.5	2953
Atlanta AP	1432	−3.1	33.1	23.1	32.0	22.9	9.2	24.1	3063
Augusta AP	1292	−3.3	34.9	24.2	33.8	24.0	11.8	25.3	3221
Columbus AP	1106	−1.4	34.5	23.5	33.5	23.4	10.0	24.7	3497
Lawson AAF	1252	−3.5	34.9	24.4	33.7	24.3	11.9	25.8	3258
Robins AFB	1158	−2.2	34.9	24.3	33.7	24.1	11.1	25.4	3373
Rome AP	1648	−4.8	34.1	23.3	32.9	23.3	11.4	24.6	2870
Savannah AP	938	−0.7	34.1	24.9	33.0	24.7	9.9	25.8	3660
Valdosta AP	809	−0.8	34.7	24.6	33.7	24.4	10.9	25.8	3831
HAWAII									
Hilo AP	0	17.2	29.6	23.4	29.0	23.1	7.1	24.1	4898
Honolulu AP	0	18.4	31.5	23.2	31.0	22.9	6.5	24.2	5664
IDAHO									
Boise AP	2951	−8.7	35.5	17.2	33.9	16.6	16.3	17.4	1832
Coeur D'Alene AP	3741	−11.4	32.1	17.1	29.7	16.4	15.9	17.1	1115
Lewiston AP	2791	−7.0	35.2	18.0	33.2	17.4	15.7	18.1	1735
Pocatello AP	3814	−15.1	33.3	16.0	31.7	15.5	19.3	16.6	1249
ILLINOIS									
Champaign, Willard AP (Savoy)	3142	−15.7	32.0	23.7	30.6	23.0	10.7	24.3	1954
Chicago, Midway AP	3230	−15.0	32.0	22.9	30.5	22.2	8.7	23.6	1943
Moline, Quad Cities AP	3356	−17.1	32.1	23.8	30.7	22.9	10.8	24.2	1912
Peoria AP	3149	−15.7	32.1	24.0	30.7	23.2	10.2	24.4	2011
Quincy AP	3027	−15.0	32.5	24.1	31.1	23.4	10.4	24.4	2074
Springfield AP	2936	−14.4	32.6	24.4	31.3	23.6	10.8	24.8	2119
INDIANA									
Evansville AP	2418	−10.1	33.0	24.1	31.9	23.7	10.3	24.9	2418
Fort Wayne AP	3304	−15.3	31.2	22.8	29.7	22.1	10.9	23.5	1779
Indianapolis AP	2902	−13.6	31.6	23.3	30.3	22.6	9.8	23.9	2088
Muncie AP	3057	−14.1	31.2	22.7	29.9	22.0	10.2	23.5	1941
South Bend AP	3437	−15.0	30.8	22.4	29.4	21.8	10.5	23.2	1704
Terre Haute AP	2869	−13.5	32.2	24.1	31.0	23.3	11.0	24.4	2062
West Lafayette AP	3089	−15.1	31.8	23.5	30.3	22.6	10.8	24.1	1942
IOWA									
Burlington, SE Iowa AP	3175	−16.5	32.2	24.4	30.9	23.6	10.1	24.7	2030
Des Moines AP	3369	−17.5	32.3	23.9	30.8	23.1	9.9	24.2	1987
Dubuque AP	3892	−19.6	29.8	23.0	28.4	22.0	10.1	23.3	1549
Mason City AP	4181	−21.4	30.7	22.9	29.0	22.0	11.5	23.4	1503
Sioux City AP	3716	−19.1	32.3	23.6	30.8	22.9	11.5	24.2	1782
Waterloo AP	3876	−20.4	31.2	23.2	29.7	22.4	11.1	23.8	1659

Table B.3 (*Continued*)

State/Province and Station[a]	HDD18.3°C	Winter: Design Dry-Bulb (°C) (99%)	Summer: Design Dry-Bulb (°C) (1%)	Mean Coincident Wet-Bulb (°C) (1%)	Design Dry-Bulb (°C) (2%)	Mean Coincident Wet-Bulb (°C) (2%)	Mean Daily Range (°C)	Design Wet-Bulb (°C) (2%)	CDD10.0°C
KANSAS									
Dodge City AP	2686	−12.3	36.4	20.9	34.8	20.9	14.1	22.3	2325
Goodland AP	3186	−14.3	34.9	19.3	33.3	19.2	15.2	20.6	1835
Manhattan AP	2807	−13.3	36.0	24.2	33.9	23.8	12.5	24.8	2319
Topeka AP	2679	−12.9	35.0	24.5	33.3	24.0	11.1	25.0	2386
Wichita AP	2428	−10.9	36.3	23.4	34.5	23.3	11.7	24.3	2585
KENTUCKY									
Bowling Green	2223	−9.3	31.6	23.7	30.6	23.1	9.5	24.1	2448
Lexington, Blue Grass AP	2491	−10.4	31.9	23.0	30.7	22.6	10.0	23.8	2254
Louisville AP	2228	−9.1	33.2	23.8	32.0	23.3	9.3	24.6	2599
LOUISIANA									
Alexandria AP	1021	−1.2	35.0	24.9	33.9	24.8	11.1	26.0	3706
Baton Rouge AP	854	−0.3	34.1	25.2	33.3	25.0	9.4	26.1	3878
Lafayette AP	771	0.9	34.0	25.3	33.1	25.2	9.1	26.3	4026
Lake Charles	771	0.9	34.0	25.4	33.1	25.3	8.4	26.6	4017
Monroe AP	1192	−2.2	35.5	25.2	34.3	25.1	10.6	26.2	3549
New Orleans AP	663	2.6	33.8	25.4	32.9	25.3	7.8	26.4	4208
Shreveport AP	1135	−1.7	36.1	24.4	34.8	24.4	11.0	25.7	3651
MAINE									
Bangor AP	4223	−18.7	28.9	20.7	27.3	19.5	11.1	20.9	1194
Caribou AP	5051	−22.7	27.4	19.4	25.8	18.5	10.8	20.0	932
Portland AP	3828	−15.0	28.5	21.0	26.9	20.1	9.5	21.4	1244
MARYLAND									
Baltimore CO	2143	−6.3	33.8	23.5	32.5	23.0	8.3	24.4	2637
Salisbury AP	2347	−7.5	32.6	24.1	31.3	23.6	10.7	24.9	2243
MASSACHUSETTS									
Boston AP	3054	−10.6	30.9	22.0	29.1	21.1	8.1	22.7	1707
Worcester AP	3645	−14.0	28.6	20.9	27.2	20.1	8.8	21.7	1407
MICHIGAN									
Alpena AP	4379	−17.8	29.2	20.8	27.5	19.9	12.3	21.2	1121
Detroit AP	3327	−12.9	31.0	22.2	29.6	21.4	9.6	22.8	1770
Escanaba AP	4687	−19.8	27.0	20.4	25.1	19.5	11.0	20.6	942
Flint AP	3687	−15.6	30.6	22.2	29.1	21.2	11.7	22.5	1527
Grand Rapids AP	3603	−14.0	30.4	22.0	29.0	21.2	10.7	22.7	1577
Muskegon AP	3608	−12.8	28.9	21.6	27.7	20.9	9.7	22.4	1457
Sault Sainte Marie AP	4740	−20.5	27.1	19.9	25.5	18.9	10.7	20.2	975
Traverse City AP	3979	−15.0	29.9	21.2	28.2	20.3	11.3	21.7	1338
MINNESOTA									
Bemidji AP	5274	−27.5	28.0	19.7	27.1	18.9	11.4	20.2	1031
Duluth AP	5096	−24.5	27.5	19.7	25.9	18.7	10.5	20.0	974

Table B.3 (*Continued*)

State/Province and Station[a]	Winter		Summer							
	HDD18.3°C	Design Dry-Bulb (°C) (99%)	Design Dry-Bulb (°C) (1%)	Mean Coincident Wet-Bulb (°C) (1%)	Design Dry-Bulb (°C) (2%)	Mean Coincident Wet-Bulb (°C) (2%)	Mean Daily Range (°C)	Design Wet-Bulb (°C) (2%)	CDD10.0°C
International Falls AP	5547	−29.3	27.9	19.8	26.4	18.9	13.0	20.3	895
Minneapolis-St. Paul AP	4109	−21.1	31.1	22.2	29.5	21.3	9.6	22.8	1654
Rochester AP	4322	−22.3	29.3	22.1	27.9	21.4	10.1	22.7	1416
MISSISSIPPI									
Columbus AFB	1444	−4.1	34.6	24.8	33.5	24.7	10.9	25.9	3181
Jackson AP	1228	−2.9	34.6	24.4	33.5	24.3	10.5	25.4	3424
Meridian AP	1282	−3.4	34.5	24.5	33.4	24.4	11.2	25.6	3298
MISSOURI									
Columbia AP	2691	−12.8	33.3	24.4	31.8	23.9	10.6	24.8	2257
Joplin (Asbury)	2443	−11.0	33.6	24.0	32.2	24.1	11.3	24.8	2369
Kansas City AP	2765	−13.8	33.6	24.6	32.0	24.2	10.3	25.0	2284
St. Louis AP	2433	−11.0	34.2	24.5	32.9	23.9	9.4	25.0	2577
MONTANA									
Billings AP	3748	−19.3	33.1	16.6	31.3	16.3	14.8	17.4	1460
Bozeman AP	4533	−21.4	31.5	15.7	29.6	15.2	19.2	16.2	964
Cut Bank AP	4746	−24.6	29.9	14.9	27.8	14.4	16.1	15.2	828
Glasgow AP	4700	−25.9	32.9	17.8	30.8	17.2	15.0	18.5	1287
Great Falls AP	4218	−22.7	31.9	15.6	30.0	15.2	17.1	16.1	1079
Helena AP	4148	−20.7	32.3	15.9	30.4	15.5	16.5	16.3	1199
Kalispell, Glacier Park AP	4401	−16.7	30.7	16.4	28.8	15.8	18.2	16.6	844
Lewistown	4938	−23.7	28.4	13.4	26.8	13.1	16.8	14.1	613
Miles City AP	4128	−22.8	34.7	18.1	32.8	17.8	15.2	19.1	1526
Missoula AP	4073	−15.4	32.2	16.2	30.3	15.7	18.2	16.4	1082
NEBRASKA									
Grand Island, Central NE AP	3362	−16.7	33.6	23.0	31.9	22.4	11.8	23.6	1913
Lincoln AP	3285	−16.6	34.0	23.8	32.4	23.2	11.8	24.2	2042
North Platte AP	3591	−17.1	33.9	21.5	32.1	21.1	14.3	22.3	1663
Omaha AP	3167	−15.3	33.0	23.7	31.7	23.1	10.2	24.3	2080
Scottsbluff, Western NE AP	3574	−17.1	34.2	18.4	32.5	18.2	16.3	19.8	1577
NEVADA									
Elko AP	3797	−15.3	33.9	14.6	32.5	14.1	21.0	15.2	1244
Ely AP	4046	−17.0	31.9	13.2	30.7	12.8	20.7	14.3	1042
Las Vegas AP	1023	1.9	41.5	19.1	40.4	18.7	12.0	21.1	4192
Lovelock, Derby Field AP	3004	−11.3	36.7	16.1	35.3	15.7	20.8	16.7	1758
Reno AP	2676	−7.0	34.7	15.8	33.3	15.1	17.6	16.1	1829
Tonopah AP	2932	−9.2	34.1	14.6	32.9	14.1	17.6	15.7	1737
Winnemucca AP	3409	−12.6	35.5	15.8	34.1	15.3	21.8	16.1	1392

Table B.3 (*Continued*)

State/Province and Station[a]	HDD18.3°C	Winter: Design Dry-Bulb (°C) (99%)	Summer: Design Dry-Bulb (°C) (1%)	Mean Coincident Wet-Bulb (°C) (1%)	Design Dry-Bulb (°C) (2%)	Mean Coincident Wet-Bulb (°C) (2%)	Mean Daily Range (°C)	Design Wet-Bulb (°C) (2%)	CDD10.0°C
NEW HAMPSHIRE									
Concord AP	3894	−16.7	30.6	21.1	29.1	20.3	12.6	21.9	1348
Portsmouth AP	3550	−13.7	30.1	21.8	28.4	21.0	10.0	22.3	1440
NEW JERSEY									
Atlantic City AP	2630	−8.9	31.9	23.3	30.5	22.8	9.8	24.3	2036
Newark AP	2581	−8.8	32.8	22.7	31.3	22.2	8.9	23.8	2181
NEW MEXICO									
Albuquerque AP	2152	−5.3	34.1	15.3	32.9	15.2	12.7	17.6	2411
Farmington, Four Corners AP	2883	−10.4	34.2	15.0	33.0	14.8	16.4	16.9	1883
Gallup AP	3364	−14.1	32.2	13.8	30.9	13.5	17.8	15.8	1353
Roswell AP	1695	−5.8	37.3	18.1	36.0	18.2	14.3	20.5	2964
Santa Fe AP	2931	−9.6	32.4	14.3	31.2	14.2	15.6	16.3	1675
Taos AP	3796	−16.0	30.2	13.1	28.8	12.8	17.8	14.6	1110
Tucumcari AP	2069	−8.4	37.1	17.9	35.7	17.9	15.2	19.9	2661
NEW YORK									
Albany AP	3566	−15.4	30.1	21.7	28.7	21.0	10.5	22.4	1572
Binghamton AP (Johnson City)	3908	−15.6	28.0	20.3	26.6	19.7	9.4	21.0	1299
Buffalo AP	3583	−14.0	28.8	21.1	27.6	20.6	8.9	22.0	1512
Ithaca	3832	−15.2	29.7	21.6	28.2	20.8	11.8	22.0	1369
Massena AP	4337	−22.0	29.2	21.6	27.9	20.8	11.3	22.0	1282
New York City, Kennedy AP	2645	−8.0	30.4	22.3	29.0	21.8	7.4	23.7	1985
Rochester AP	3558	−14.1	30.0	21.9	28.5	21.0	10.4	22.3	1516
Syracuse AP	3613	−15.5	30.2	21.9	28.8	21.1	10.5	22.4	1536
NORTH CAROLINA									
Asheville AP	2222	−7.0	29.9	21.2	28.8	20.9	10.2	22.1	2057
Charlotte AP	1683	−4.0	33.3	23.4	32.1	23.1	10.3	24.1	2795
Greensboro AP	1676	−3.8	33.6	24.3	32.5	24.0	10.9	25.4	2856
Raleigh-Durham AP	1771	−4.6	33.6	23.9	32.4	23.6	10.6	24.7	2761
Wilmington AP	1308	−2.5	32.9	25.0	31.7	24.6	8.7	25.6	3183
NORTH DAKOTA									
Bismarck AP	4674	−24.4	31.9	20.7	30.1	20.0	13.8	21.3	1334
Fargo AP	4825	−25.5	30.6	21.3	29.1	20.5	11.6	21.9	1379
Minot AP	4852	−25.0	30.9	20.3	29.0	19.2	12.4	20.6	1263
OHIO									
Akron-Canton AP	3291	−13.7	30.2	22.0	28.9	21.2	10.3	22.5	1710
Cincinnati AP	2653	−10.8	32.1	23.5	30.9	23.0	11.1	24.3	2106
Cleveland AP	3187	−12.8	30.7	22.4	29.2	21.7	9.3	22.9	1782
Columbus AP	2867	−12.3	31.6	22.5	30.3	21.9	10.0	23.2	2046

Table B.3 (*Continued*)

State/Province and Station[a]	HDD18.3°C	Winter: Design Dry-Bulb (°C) (99%)	Summer: Design Dry-Bulb (°C) (1%)	Mean Coincident Wet-Bulb (°C) (1%)	Design Dry-Bulb (°C) (2%)	Mean Coincident Wet-Bulb (°C) (2%)	Mean Daily Range (°C)	Design Wet-Bulb (°C) (2%)	CDD10.0°C
Dayton AP	3023	−13.6	31.0	22.7	29.7	21.9	9.9	23.2	1947
Mansfield AP	3364	−14.5	30.0	22.1	28.7	21.4	10.4	22.7	1666
Toledo AP	3269	−14.1	31.6	22.6	30.0	21.9	10.2	23.3	1832
Youngstown-Warren AP	3396	−13.8	29.8	21.8	28.5	20.9	11.2	22.2	1572
OKLAHOMA									
Oklahoma City AP	1888	−7.2	36.5	23.4	34.9	23.5	12.0	24.5	2917
Tulsa AP	1895	−7.6	36.2	24.5	34.8	24.4	10.6	25.3	2993
OREGON									
Astoria AP	2706	−0.5	22.4	16.7	20.7	16.2	7.6	16.7	850
Eugene AP	2572	−2.6	31.3	18.7	29.2	18.0	16.8	18.6	1271
Medford AP	2331	−2.9	35.3	18.6	33.5	18.0	17.4	18.8	1852
Pendleton AP	2907	−9.1	34.3	17.6	32.4	17.0	16.9	17.7	1555
Portland AP	2322	−1.4	30.8	19.1	28.8	18.3	12.0	19.0	1536
Salem AP	2464	−2.3	31.4	18.8	29.2	18.0	14.9	18.6	1379
PENNSYLVANIA									
Allentown AP	3019	−10.9	31.4	22.6	30.0	21.9	11.0	23.3	1823
Erie AP	3345	−12.4	29.1	22.1	27.9	21.4	8.1	22.5	1625
Harrisburg AP	2792	−9.6	32.0	23.3	30.4	22.7	9.7	24.0	2056
Philadelphia AP	2450	−7.9	32.7	23.3	31.3	22.5	9.0	24.2	2298
Pittsburgh AP	3066	−12.7	30.2	21.6	29.0	20.9	10.0	22.4	1801
Williamsport AP	3194	−12.2	31.0	21.8	29.4	21.1	11.5	22.6	1722
RHODE ISLAND									
Providence AP	3043	−10.7	30.5	22.3	28.9	21.4	9.3	23.1	1707
SOUTH CAROLINA									
Anderson AP	1588	−3.4	33.8	23.1	32.5	23.0	10.6	24.1	2861
Charleston AP	1012	−0.8	33.5	25.2	32.4	24.9	8.9	26.0	3558
Columbia AP	1319	−2.8	34.9	23.8	33.8	23.7	10.6	25.0	3301
Greenville AP	1534	−2.9	33.3	22.7	32.3	22.5	9.6	23.8	2957
SOUTH DAKOTA									
Huron AP	4249	−22.3	32.2	22.7	30.6	21.8	12.6	23.2	1582
Pierre 24 S	4064	−20.3	33.0	20.8	31.0	20.2	13.5	21.7	1566
Rapid City AP	3952	−19.4	33.6	18.8	31.6	18.5	14.9	20.0	1431
Sioux Falls AP	4134	−21.3	31.3	22.8	29.8	22.0	11.3	23.3	1584
TENNESSEE									
Blountville, Tri-Cities AP	2295	−8.1	31.3	21.9	30.3	21.6	11.0	22.8	2175
Chattanooga AP	1687	−4.8	33.7	23.2	32.5	23.0	10.1	24.3	2877
Knoxville AP (Alcoa)	1937	−6.1	32.5	22.9	31.4	22.6	9.8	23.9	2616
Memphis AP	1587	−4.9	34.6	24.7	33.6	24.5	9.1	25.6	3223
Nashville AP	1906	−7.0	33.5	23.6	32.4	23.3	10.0	24.5	2774

APPENDICES

Table B.3 (*Continued*)

State/Province and Station[a]	Winter		Summer						
	HDD18.3°C	Design Dry-Bulb (°C) (99%)	Design Dry-Bulb (°C) (1%)	Mean Coincident Wet-Bulb (°C) (1%)	Design Dry-Bulb (°C) (2%)	Mean Coincident Wet-Bulb (°C) (2%)	Mean Daily Range (°C)	Design Wet-Bulb (°C) (2%)	CDD10.0°C
TEXAS									
Abilene AP	1341	−4.2	36.9	21.6	35.7	21.7	11.8	23.4	3474
Amarillo	2196	−8.9	35.6	18.9	34.2	18.9	13.8	20.7	2466
Austin AP	916	−1.2	36.9	23.6	35.8	23.7	12.5	25.3	3974
Brownsville AP	277	5.8	34.9	25.7	34.3	25.7	9.3	26.8	5108
Corpus Christi AP	456	3.4	35.2	25.5	34.4	25.5	10.2	26.7	4686
Dallas/Fort Worth AP	1174	−2.5	37.3	23.6	36.1	23.7	10.6	25.0	3786
Del Rio AP	705	1.4	38.1	22.5	37.1	22.6	11.5	24.7	4403
El Paso AP	1224	−1.8	37.3	17.4	36.1	17.4	13.8	20.1	3514
Galveston AP	532	4.2	32.7	26.4	32.3	26.4	5.3	27.2	4474
Houston, Hobby AP	619	2.6	34.7	25.2	33.8	25.1	8.8	26.3	4334
Laredo AP	427	3.9	39.4	23.7	38.4	23.7	12.1	25.6	5178
Lubbock AP	1787	−6.6	36.3	19.4	35.0	19.4	12.9	21.4	2887
Midland AP	1375	−4.1	37.3	19.7	36.1	19.7	12.6	21.9	3420
Nacogdoches AP	1163	−2.5	36.0	24.5	34.0	24.4	12.2	25.5	3518
Port Arthur AP	714	1.6	34.1	25.6	33.3	25.5	9.2	26.7	4104
San Angelo AP	1188	−3.5	37.8	21.2	36.6	21.2	13.1	23.1	3642
San Antonio AP	751	0.6	36.5	23.1	35.5	23.2	10.9	24.9	4217
Waco AP	1113	−2.3	37.4	23.7	36.3	23.8	11.9	25.1	3808
Wichita Falls AP	1425	−4.5	38.5	22.1	37.3	21.8	12.3	24.0	3548
UTAH									
Cedar City AP	3263	−13.4	33.1	14.8	31.9	14.4	16.8	16.2	1575
Salt Lake City AP	2961	−9.1	35.5	16.6	34.1	16.3	13.8	17.8	2047
VERMONT									
Burlington AP	3969	−18.7	29.9	21.0	28.3	20.3	10.5	21.7	1433
Rutland AP	4080	−18.0	27.9	20.4	27.1	19.7	10.6	20.9	1209
VIRGINIA									
Norfolk AP	1744	−3.5	32.9	24.5	31.7	24.0	7.9	25.2	2787
Richmond AP	2019	−6.0	33.6	23.9	32.3	23.4	10.3	24.6	2586
Roanoke AP	2182	−6.8	32.1	22.1	30.8	21.8	10.1	23.0	2332
WASHINGTON									
Bellingham AP	2925	−4.0	24.9	17.7	23.1	16.8	9.0	17.3	923
Olympia AP	2955	−4.1	28.8	18.1	26.7	17.4	14.1	17.6	971
Seattle AP	2388	−1.2	27.9	17.7	26.2	17.1	10.0	17.7	1339
Seattle–Tacoma AP	2567	−1.1	27.9	17.7	26.0	17.0	10.6	17.6	1193
Spokane AP	3633	−11.2	32.3	16.4	30.3	15.8	14.5	16.6	1264
Walla Walla AP	2646	−7.7	34.9	18.2	32.8	17.6	14.8	18.3	1848
Yakima AP	3183	−9.8	34.5	18.5	32.6	17.7	18.2	18.4	1474
WEST VIRGINIA									
Charleston AP	2436	−9.6	31.6	22.4	30.4	22.0	10.5	23.4	2165
Huntington AP	2435	−9.6	31.7	23.0	30.5	22.4	10.3	23.7	2209

APPENDICES

Table B.3 (*Continued*)

State/Province and Station[a]	HDD18.3°C	Winter: Design Dry-Bulb (°C) (99%)	Summer: Design Dry-Bulb (°C) (1%)	Mean Coincident Wet-Bulb (°C) (1%)	Design Dry-Bulb (°C) (2%)	Mean Coincident Wet-Bulb (°C) (2%)	Mean Daily Range (°C)	Design Wet-Bulb (°C) (2%)	CDD10.0°C
Martinsburg AP	2683	−9.2	32.1	22.8	30.8	22.2	11.4	23.4	2012
Morgantown AP	2744	−11.3	30.6	21.7	29.4	21.1	10.3	22.6	1958
WISCONSIN									
Eau Claire AP	4339	−22.5	30.4	21.7	28.8	20.7	11.7	22.3	1429
Green Bay AP	4177	−19.5	29.5	22.2	28.1	21.3	11.3	22.5	1342
La Crosse AP	3849	−20.3	31.4	22.8	30.0	21.9	10.8	23.3	1716
Madison AP	3918	−18.8	30.2	22.5	28.8	21.7	10.7	22.9	1517
Milwaukee AP	3693	−16.2	30.2	22.4	28.6	21.5	8.5	22.8	1547
WYOMING									
Casper AP	4076	−18.3	33.0	14.9	31.5	14.6	18.7	15.9	1232
Cheyenne AP	3873	−16.1	30.7	14.2	29.2	14.1	14.7	15.8	1190
Lander	4098	−17.9	31.5	13.2	30.1	13.1	15.2	14.3	1278
Laramie AP	4739	−19.6	28.3	12.5	27.0	12.2	16.8	13.8	784
Rock Springs AP	4432	−17.2	29.9	12.4	28.5	12.1	15.5	13.3	1083
Sheridan AP	4150	−20.0	33.3	17.1	31.5	16.8	17.8	17.9	1216
ALBERTA									
Calgary AP	5054	−24.5	26.8	15.5	25.0	14.9	12.3	15.7	689
Edmonton AP	5814	−29.0	26.1	17.3	24.5	16.4	12.7	17.5	605
Medicine Hat AP	4676	−26.0	31.1	16.9	29.2	16.4	14.7	17.2	1079
BRITISH COLUMBIA									
Prince George AP	5123	−24.5	25.8	15.4	23.9	14.5	12.6	15.3	554
Prince Rupert AP	3806	−6.3	18.0	14.4	16.9	14.0	5.0	14.5	409
Vancouver AP	2873	−3.0	23.7	17.9	22.5	17.3	7.7	17.7	998
Victoria AP	2958	−2.0	24.8	16.9	23.1	16.2	10.0	16.5	875
MANITOBA									
The Pas AP	6470	−31.5	26.6	18.0	25.0	17.2	10.1	18.3	782
Winnipeg AP	5697	−29.6	28.7	20.3	27.1	19.3	11.7	20.5	1039
NEW BRUNSWICK									
Fredericton AP	4541	−20.3	28.1	20.0	26.5	19.1	11.3	20.4	1039
Saint John AP	4651	−19.3	24.7	18.0	23.2	17.2	10.1	18.5	764
NEWFOUNDLAND									
Gander AP	4999	−15.3	24.9	18.0	23.3	17.3	9.7	18.5	643
St. John's AP	4721	−12.0	23.5	18.5	22.1	17.8	7.7	18.7	604
NORTHWEST TERRITORIES									
Fort Smith AP	7280	−35.5	26.5	16.2	24.8	15.4	11.5	16.6	637
Inuvik AP	9224	−37.4	23.8	15.2	21.8	14.4	9.7	15.0	335
Yellowknife AP	8051	−37.7	23.9	15.3	22.3	14.6	8.0	15.6	540
NOVA SCOTIA									
Halifax AP	4193	−15.2	26.2	19.5	24.8	18.8	9.2	20.2	993
Sydney AP	4432	−14.2	26.0	19.6	24.4	18.8	8.9	19.9	842
Yarmouth AP	3914	−11.3	22.3	18.8	21.2	18.2	7.2	19.0	817

Table B.3 (*Continued*)

State/Province and Station[a]	HDD18.3°C	Winter: Design Dry-Bulb (°C) (99%)	Summer: Design Dry-Bulb (°C) (1%)	Mean Coincident Wet-Bulb (°C) (1%)	Design Dry-Bulb (°C) (2%)	Mean Coincident Wet-Bulb (°C) (2%)	Mean Daily Range (°C)	Design Wet-Bulb (°C) (2%)	CDD10.0°C
ONTARIO									
Ottawa AP	4451	−21.4	29.2	20.9	27.6	20.2	10.4	21.5	1292
Sudbury AP	5182	−24.8	27.3	19.0	25.7	18.1	10.5	19.5	982
Thunder Bay AP	5514	−26.7	27.2	19.2	25.5	18.3	12.4	19.4	784
Toronto, Pearson AP	3779	−15.7	29.7	21.5	28.1	20.8	9.5	21.9	1460
PRINCE EDWARD ISLAND									
Charlottetown AP	4511	−17.1	25.6	19.8	24.2	19.0	8.1	20.1	943
QUEBEC									
Montréal, Trudeau AP	4257	−20.2	28.7	21.1	27.4	20.4	8.8	21.6	1387
Québec City, Jean Lesage AP	4963	−23.0	27.3	20.2	25.8	19.3	10.0	20.6	1014
Sept-Îles AP	6100	−25.1	21.4	15.7	20.1	15.2	8.3	16.4	443
Trois-Rivières	4560	−21.2	26.2	20.8	25.1	20.2	7.5	21.3	1188
Val-d'Or AP	5900	−29.2	27.2	18.6	25.5	17.8	11.1	19.3	802
SASKATCHEWAN									
Prince Albert AP	6158	−32.1	27.0	18.2	25.5	17.1	11.6	18.2	771
Regina AP	5720	−29.9	28.8	18.4	27.1	17.6	13.1	18.8	895
Saskatoon AP	5853	−30.7	28.3	18.0	26.5	17.4	12.5	18.4	849
YUKON TERRITORY									
Whitehorse	6657	−33.5	23.5	13.3	21.6	12.6	11.1	13.2	362

[a]U.S. solar data are available from the National Climatic Data Center, Federal Building, Asheville, NC 28801. Canadian solar data are available from the Canadian Climate Center, Atmospheric Environment Service, 4905 Dufferin St., Downsview, Ontario M3H 5T4.

Table B.4 Outdoor Design Conditions: Mexico and Puerto Rico (SI Units)

Country/Territory and Station	HDD18.3°C	Winter: Design Dry-Bulb (°C) (99%)	Summer: Design Dry-Bulb (°C) (1%)	Mean Coincident Wet-Bulb (°C) (1%)	Design Dry-Bulb (°C) (2%)	Mean Coincident Wet-Bulb (°C) (2%)	Mean Daily Range (°C)	Design Wet-Bulb (°C) (2%)	CDD10.0°C
MEXICO									
Guadalajara AP	348	3.1	32.2	14.9	31.2	14.6	17.6	19.1	3428
Mérida AP	1	15.7	37.4	24.5	36.4	24.6	13.0	27.5	6358
Mexico City AP	568	4.9	28.0	12.1	27.0	12.0	13.4	15.2	2691
Monterrey AP	377	4.9	37.8	22.7	36.8	22.6	12.2	25.3	4827
PUERTO RICO									
San Juan AP	0	21.7	32.1	25.5	31.6	25.5	5.6	26.6	6239

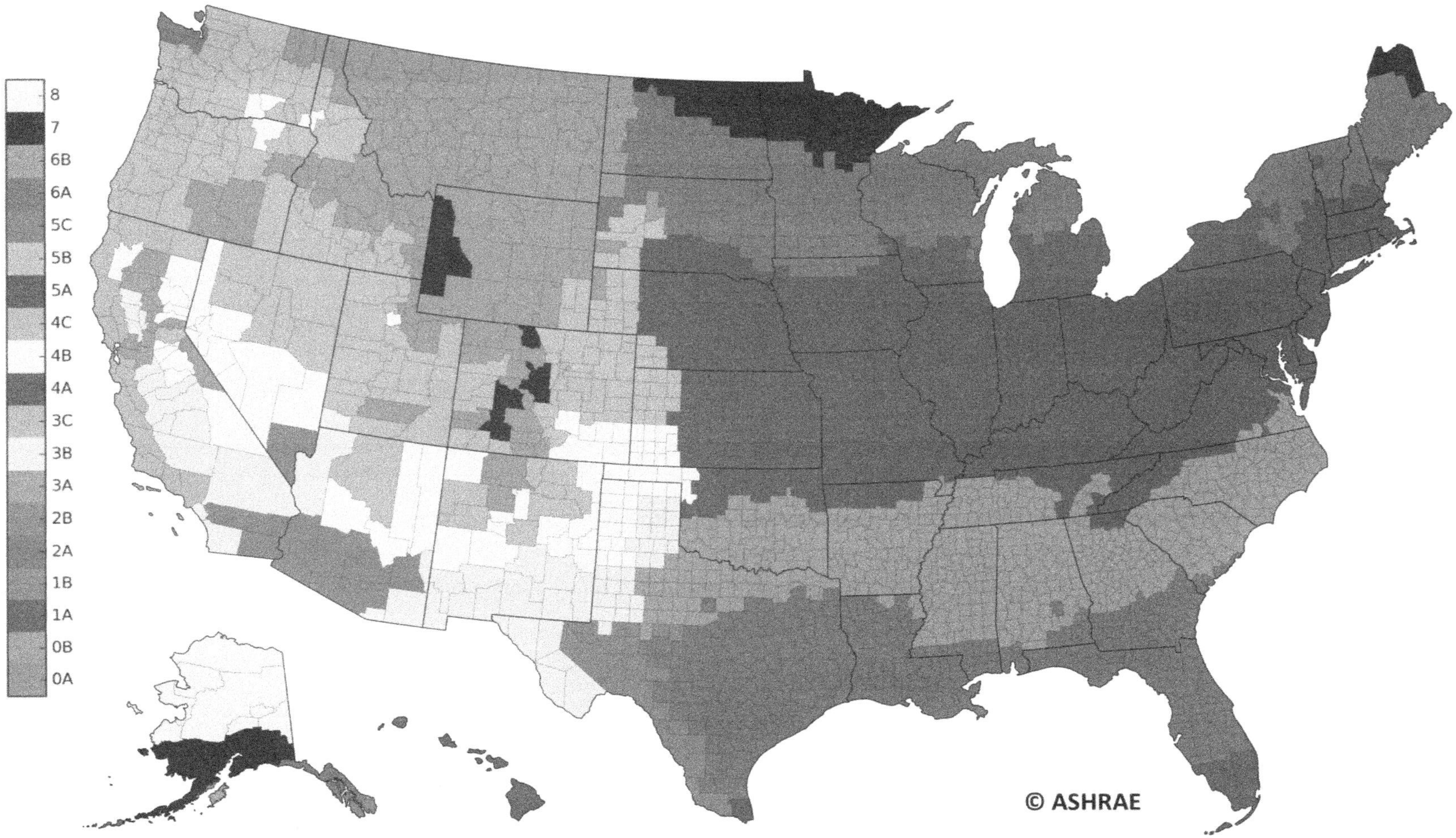

Figure B.1 *Climate Zones by United States Counties. (© ASHRAE, www.ashrae.org. 2020. ASHRAE Standard 169-2021.)*

Table B.5 Climate Zones by United States Counties

State	County	Climate Zone
Alabama (AL)		
	Autauga	3A
	Baldwin	2A
	Barbour	2A
	Bibb	3A
	Blount	3A
	Bullock	3A
	Butler	3A
	Calhoun	3A
	Chambers	3A
	Cherokee	3A
	Chilton	3A
	Choctaw	3A
	Clarke	3A
	Clay	3A
	Cleburne	3A
	Coffee	2A
	Colbert	3A
	Conecuh	3A
	Coosa	3A
	Covington	2A
	Crenshaw	3A
	Cullman	3A
	Dale	2A
	Dallas	3A
	DeKalb	3A
	Elmore	3A
	Escambia	3A
	Etowah	3A
	Fayette	3A
	Franklin	3A
	Geneva	2A
	Greene	3A
	Hale	3A
	Henry	2A
	Houston	2A
	Jackson	3A
	Jefferson	3A
	Lamar	3A
	Lauderdale	3A
	Lawrence	3A
	Lee	3A
	Limestone	3A
	Lowndes	3A
	Macon	3A
	Madison	3A
	Marengo	3A
	Marion	3A
	Marshall	3A
	Mobile	2A
	Monroe	3A
	Montgomery	2A
	Morgan	3A
	Perry	3A
	Pickens	3A
	Pike	3A
	Randolph	3A
	Russell	3A
	Shelby	3A
	St. Clair	3A
	Sumter	3A
	Talladega	3A
	Tallapoosa	3A
	Tuscaloosa	3A
	Walker	3A
	Washington	3A
	Wilcox	3A
	Winston	3A

Table B.5 (*Continued*)

State	County	Climate Zone
Alaska (AK)		
	Aleutians East	7
	Aleutians West	6A
	Anchorage	7
	Bethel	7
	Bristol Bay	7
	Denali	8
	Dillingham	7
	Fairbanks North Star	8
	Haines	6A
	Hoonah-Angoon	6A
	Juneau	6A
	Kenai Peninsula	7
	Ketchikan Gateway	5C
	Kodiak Island	6B
	Kusilvak	8
	Lake and Peninsula	7
	Matanuska-Susitna	7
	Nome	8
	North Slope	8
	Northwest Arctic	8
	Petersburg	6A
	Prince of Wales-Hyder	5C
	Sitka	6A
	Skagway	6A
	Southeast Fairbanks	8
	Valdez-Cordova	7
	Wrangell	6A
	Yakutat	6A
	Yukon-Koyukuk	8
Arizona (AZ)		
	Apache	4B
	Cochise	3B
	Coconino	5B
	Gila	4B
	Graham	2B
	Greenlee	3B
	La Paz	1B
	Maricopa	1B
	Mohave	3B
	Navajo	4B
	Pima	2B
	Pinal	2B
	Santa Cruz	3B
	Yavapai	4B
	Yuma	1B
Arkansas (AR)		
	Arkansas	3A
	Ashley	3A
	Baxter	4A
	Benton	4A
	Boone	4A
	Bradley	3A
	Calhoun	3A
	Carroll	4A
	Chicot	3A
	Clark	3A
	Clay	3A
	Cleburne	3A
	Cleveland	3A
	Columbia	3A
	Conway	3A
	Craighead	3A
	Crawford	3A
	Crittenden	3A
	Cross	3A
	Dallas	3A

Table B.5 (*Continued*)

State	County	Climate Zone
	Desha	3A
	Drew	3A
	Faulkner	3A
	Franklin	3A
	Fulton	4A
	Garland	3A
	Grant	3A
	Greene	3A
	Hempstead	3A
	Hot Spring	3A
	Howard	3A
	Independence	3A
	Izard	4A
	Jackson	3A
	Jefferson	3A
	Johnson	3A
	Lafayette	3A
	Lawrence	4A
	Lee	3A
	Lincoln	3A
	Little River	3A
	Logan	3A
	Lonoke	3A
	Madison	4A
	Marion	4A
	Miller	3A
	Mississippi	3A
	Monroe	3A
	Montgomery	3A
	Nevada	3A
	Newton	4A
	Ouachita	3A
	Perry	3A
	Phillips	3A
	Pike	3A
	Poinsett	3A
	Polk	3A
	Pope	3A
	Prairie	3A
	Pulaski	3A
	Randolph	4A
	Saline	3A
	Scott	3A
	Searcy	4A
	Sebastian	3A
	Sevier	3A
	Sharp	4A
	St. Francis	3A
	Stone	4A
	Union	3A
	Van Buren	3A
	Washington	4A
	White	3A
	Woodruff	3A
	Yell	3A
California (CA)		
	Alameda	3C
	Alpine	6B
	Amador	4B
	Butte	3A
	Calaveras	4B
	Colusa	3B
	Contra Costa	3A
	Del Norte	4C
	El Dorado	5B
	Fresno	3B
	Glenn	3B
	Humboldt	4C
	Imperial	2B
	Inyo	4B
	Kern	3B

Table B.5 (*Continued*)

State	County	Climate Zone
	Kings	3B
	Lake	3C
	Lassen	4B
	Los Angeles	3C
	Madera	3B
	Marin	3C
	Mariposa	4B
	Mendocino	3C
	Merced	3B
	Modoc	5B
	Mono	6B
	Monterey	3C
	Napa	3C
	Nevada	4B
	Orange	3B
	Placer	3A
	Plumas	4B
	Riverside	2B
	Sacramento	3A
	San Benito	3C
	San Bernardino	3B
	San Diego	3B
	San Francisco	4C
	San Joaquin	3A
	San Luis Obispo	3C
	San Mateo	3C
	Santa Barbara	3C
	Santa Clara	3C
	Santa Cruz	3C
	Shasta	3A
	Sierra	4B
	Siskiyou	4C
	Solano	3A
	Sonoma	3C
	Stanislaus	3B
	Sutter	3B
	Tehama	3A
	Trinity	4B
	Tulare	3B
	Tuolumne	4B
	Ventura	3C
	Yolo	3B
	Yuba	3A
Colorado (CO)		
	Adams	5B
	Alamosa	6B
	Arapahoe	5B
	Archuleta	5B
	Baca	4B
	Bent	4B
	Boulder	5B
	Broomfield	5B
	Chaffee	6B
	Cheyenne	5B
	Clear Creek	7
	Conejos	6B
	Costilla	6B
	Crowley	5B
	Custer	5B
	Delta	5B
	Denver	5B
	Dolores	6B
	Douglas	5B
	Eagle	6B
	El Paso	5B
	Elbert	5B
	Fremont	5B
	Garfield	5B
	Gilpin	7
	Grand	6B
	Gunnison	7

Table B.5 (*Continued*)

State	County	Climate Zone
	Hinsdale	7
	Huerfano	5B
	Jackson	7
	Jefferson	5B
	Kiowa	5B
	Kit Carson	5B
	La Plata	5B
	Lake	7
	Larimer	5B
	Las Animas	4B
	Lincoln	5B
	Logan	5B
	Mesa	5B
	Mineral	7
	Moffat	6B
	Montezuma	5B
	Montrose	5B
	Morgan	5B
	Otero	4B
	Ouray	6B
	Park	7
	Phillips	5B
	Pitkin	6B
	Prowers	4B
	Pueblo	4B
	Rio Blanco	6B
	Rio Grande	6B
	Routt	6B
	Saguache	6B
	San Juan	7
	San Miguel	6B
	Sedgwick	5B
	Summit	7
	Teller	5B
	Washington	5B
	Weld	5B
	Yuma	5B
Connecticut (CT)		
	Fairfield	4A
	Hartford	5A
	Litchfield	5A
	Middlesex	5A
	New Haven	5A
	New London	5A
	Tolland	5A
	Windham	5A
Delaware (DE)		
	Kent	4A
	New Castle	4A
	Sussex	4A
District of Columbia (DC)	District of Columbia	4A
Florida (FL)		
	Alachua	2A
	Baker	2A
	Bay	2A
	Bradford	2A
	Brevard	2A
	Broward	1A
	Calhoun	2A
	Charlotte	2A
	Citrus	2A
	Clay	2A
	Collier	1A
	Columbia	2A
	DeSoto	2A
	Dixie	2A

Table B.5 (*Continued*)

State	County	Climate Zone
	Duval	2A
	Escambia	2A
	Flagler	2A
	Franklin	2A
	Gadsden	2A
	Gilchrist	2A
	Glades	2A
	Gulf	2A
	Hamilton	2A
	Hardee	2A
	Hendry	1A
	Hernando	2A
	Highlands	2A
	Hillsborough	2A
	Holmes	2A
	Indian River	2A
	Jackson	2A
	Jefferson	2A
	Lafayette	2A
	Lake	2A
	Lee	1A
	Leon	2A
	Levy	2A
	Liberty	2A
	Madison	2A
	Manatee	2A
	Marion	2A
	Martin	2A
	Miami-Dade	1A
	Monroe	1A
	Nassau	2A
	Okaloosa	2A
	Okeechobee	2A
	Orange	2A
	Osceola	2A
	Palm Beach	1A
	Pasco	2A
	Pinellas	2A
	Polk	2A
	Putnam	2A
	Santa Rosa	2A
	Sarasota	2A
	Seminole	2A
	St. Johns	2A
	St. Lucie	2A
	Sumter	2A
	Suwannee	2A
	Taylor	2A
	Union	2A
	Volusia	2A
	Wakulla	2A
	Walton	2A
	Washington	2A
Georgia (GA)		
	Appling	2A
	Atkinson	2A
	Bacon	2A
	Baker	2A
	Baldwin	3A
	Banks	3A
	Barrow	3A
	Bartow	3A
	Ben Hill	2A
	Berrien	2A
	Bibb	3A
	Bleckley	3A
	Brantley	2A
	Brooks	2A
	Bryan	2A

Table B.5 (*Continued*)

State	County	Climate Zone
	Bulloch	2A
	Burke	3A
	Butts	3A
	Calhoun	2A
	Camden	2A
	Candler	2A
	Carroll	3A
	Catoosa	3A
	Charlton	2A
	Chatham	2A
	Chattahoochee	3A
	Chattooga	3A
	Cherokee	3A
	Clarke	3A
	Clay	2A
	Clayton	3A
	Clinch	2A
	Cobb	3A
	Coffee	2A
	Colquitt	2A
	Columbia	3A
	Cook	2A
	Coweta	3A
	Crawford	3A
	Crisp	2A
	Dade	3A
	Dawson	3A
	Decatur	2A
	DeKalb	3A
	Dodge	2A
	Dooly	3A
	Dougherty	2A
	Douglas	3A
	Early	2A
	Echols	2A
	Effingham	2A
	Elbert	3A
	Emanuel	3A
	Evans	2A
	Fannin	4A
	Fayette	3A
	Floyd	3A
	Forsyth	3A
	Franklin	3A
	Fulton	3A
	Gilmer	4A
	Glascock	3A
	Glynn	2A
	Gordon	3A
	Grady	2A
	Greene	3A
	Gwinnett	3A
	Habersham	4A
	Hall	3A
	Hancock	3A
	Haralson	3A
	Harris	3A
	Hart	3A
	Heard	3A
	Henry	3A
	Houston	3A
	Irwin	2A
	Jackson	3A
	Jasper	3A
	Jeff Davis	2A
	Jefferson	3A
	Jenkins	3A
	Johnson	3A
	Jones	3A
	Lamar	3A
	Lanier	2A
	Laurens	3A

Table B.5 (*Continued*)

State	County	Climate Zone
	Lee	2A
	Liberty	2A
	Lincoln	3A
	Long	2A
	Lowndes	2A
	Lumpkin	4A
	Macon	3A
	Madison	3A
	Marion	3A
	McDuffie	3A
	McIntosh	2A
	Meriwether	3A
	Miller	2A
	Mitchell	2A
	Monroe	3A
	Montgomery	2A
	Morgan	3A
	Murray	3A
	Muscogee	3A
	Newton	3A
	Oconee	3A
	Oglethorpe	3A
	Paulding	3A
	Peach	3A
	Pickens	3A
	Pierce	2A
	Pike	3A
	Polk	3A
	Pulaski	3A
	Putnam	3A
	Quitman	2A
	Rabun	4A
	Randolph	2A
	Richmond	3A
	Rockdale	3A
	Schley	3A
	Screven	2A
	Seminole	2A
	Spalding	3A
	Stephens	3A
	Stewart	3A
	Sumter	3A
	Talbot	3A
	Taliaferro	3A
	Tattnall	2A
	Taylor	3A
	Telfair	2A
	Terrell	2A
	Thomas	2A
	Tift	2A
	Toombs	2A
	Towns	4A
	Treutlen	3A
	Troup	3A
	Turner	2A
	Twiggs	3A
	Union	4A
	Upson	3A
	Walker	3A
	Walton	3A
	Ware	2A
	Warren	3A
	Washington	3A
	Wayne	2A
	Webster	3A
	Wheeler	2A
	White	4A
	Whitfield	3A
	Wilcox	2A
	Wilkes	3A
	Wilkinson	3A
	Worth	2A

Table B.5 (*Continued*)

State	County	Climate Zone
Hawaii (HI)		
	Hawaii	1A
	Honolulu	1A
	Kalawao	1A
	Kauai	1A
	Maui	1A
Idaho (ID)		
	Ada	5B
	Adams	6B
	Bannock	5B
	Bear Lake	6B
	Benewah	5B
	Bingham	5B
	Blaine	6B
	Boise	6B
	Bonner	6B
	Bonneville	6B
	Boundary	6B
	Butte	6B
	Camas	6B
	Canyon	5B
	Caribou	6B
	Cassia	5B
	Clark	6B
	Clearwater	5B
	Custer	6B
	Elmore	5B
	Franklin	5B
	Fremont	6B
	Gem	5B
	Gooding	5B
	Idaho	5B
	Jefferson	6B
	Jerome	5B
	Kootenai	5B
	Latah	5B
	Lemhi	6B
	Lewis	5B
	Lincoln	5B
	Madison	6B
	Minidoka	5B
	Nez Perce	4B
	Oneida	5B
	Owyhee	5B
	Payette	5B
	Power	5B
	Shoshone	6B
	Teton	6B
	Twin Falls	5B
	Valley	6B
	Washington	6B
Illinois (IL)		
	Adams	5A
	Alexander	4A
	Bond	4A
	Boone	5A
	Brown	5A
	Bureau	5A
	Calhoun	4A
	Carroll	5A
	Cass	5A
	Champaign	5A
	Christian	4A
	Clark	4A
	Clay	4A
	Clinton	4A
	Coles	4A
	Cook	5A
	Crawford	4A
	Cumberland	4A
	De Witt	5A
	DeKalb	5A
	Douglas	5A
	DuPage	5A
	Edgar	4A
	Edwards	4A
	Effingham	4A
	Fayette	4A
	Ford	5A
	Franklin	4A
	Fulton	5A
	Gallatin	4A
	Greene	4A
	Grundy	5A
	Hamilton	4A
	Hancock	5A
	Hardin	4A
	Henderson	5A
	Henry	5A
	Iroquois	5A
	Jackson	4A
	Jasper	4A
	Jefferson	4A
	Jersey	4A
	Jo Daviess	5A
	Johnson	4A
	Kane	5A
	Kankakee	5A
	Kendall	5A
	Knox	5A
	Lake	5A
	LaSalle	5A
	Lawrence	4A
	Lee	5A
	Livingston	5A
	Logan	5A
	Macon	4A
	Macoupin	4A
	Madison	4A
	Marion	4A
	Marshall	5A
	Mason	5A
	Massac	4A
	McDonough	5A
	McHenry	5A
	McLean	5A
	Menard	5A
	Mercer	5A
	Monroe	4A
	Montgomery	4A
	Morgan	4A
	Moultrie	4A
	Ogle	5A
	Peoria	5A
	Perry	4A
	Piatt	5A
	Pike	4A
	Pope	4A
	Pulaski	4A
	Putnam	5A
	Randolph	4A
	Richland	4A
	Rock Island	5A
	Saline	4A
	Sangamon	4A
	Schuyler	5A
	Scott	4A
	Shelby	4A
	St. Clair	4A
	Stark	5A

Table B.5 (*Continued*)

State	County	Climate Zone
	Stephenson	5A
	Tazewell	5A
	Union	4A
	Vermilion	5A
	Wabash	4A
	Warren	5A
	Washington	4A
	Wayne	4A
	White	4A
	Whiteside	5A
	Will	5A
	Williamson	4A
	Winnebago	5A
	Woodford	5A
Indiana (IN)		
	Adams	5A
	Allen	5A
	Bartholomew	4A
	Benton	5A
	Blackford	5A
	Boone	5A
	Brown	4A
	Carroll	5A
	Cass	5A
	Clark	4A
	Clay	4A
	Clinton	5A
	Crawford	4A
	Daviess	4A
	Dearborn	4A
	Decatur	4A
	DeKalb	5A
	Delaware	5A
	Dubois	4A
	Elkhart	5A
	Fayette	4A
	Floyd	4A
	Fountain	5A
	Franklin	4A
	Fulton	5A
	Gibson	4A
	Grant	5A
	Greene	4A
	Hamilton	5A
	Hancock	5A
	Harrison	4A
	Hendricks	4A
	Henry	5A
	Howard	5A
	Huntington	5A
	Jackson	4A
	Jasper	5A
	Jay	5A
	Jefferson	4A
	Jennings	4A
	Johnson	4A
	Knox	4A
	Kosciusko	5A
	LaGrange	5A
	Lake	5A
	LaPorte	5A
	Lawrence	4A
	Madison	5A
	Marion	4A
	Marshall	5A
	Martin	4A
	Miami	5A
	Monroe	4A
	Montgomery	5A
	Morgan	4A
	Newton	5A

Table B.5 (*Continued*)

State	County	Climate Zone
	Noble	5A
	Ohio	4A
	Orange	4A
	Owen	4A
	Parke	5A
	Perry	4A
	Pike	4A
	Porter	5A
	Posey	4A
	Pulaski	5A
	Putnam	4A
	Randolph	5A
	Ripley	4A
	Rush	4A
	Scott	4A
	Shelby	4A
	Spencer	4A
	St. Joseph	5A
	Starke	5A
	Steuben	5A
	Sullivan	4A
	Switzerland	4A
	Tippecanoe	5A
	Tipton	5A
	Union	4A
	Vanderburgh	4A
	Vermillion	5A
	Vigo	4A
	Wabash	5A
	Warren	5A
	Warrick	4A
	Washington	4A
	Wayne	5A
	Wells	5A
	White	5A
	Whitley	5A
Iowa (IA)		
	Adair	5A
	Adams	5A
	Allamakee	5A
	Appanoose	5A
	Audubon	5A
	Benton	5A
	Black Hawk	5A
	Boone	5A
	Bremer	5A
	Buchanan	5A
	Buena Vista	5A
	Butler	5A
	Calhoun	5A
	Carroll	5A
	Cass	5A
	Cedar	5A
	Cerro Gordo	6A
	Cherokee	5A
	Chickasaw	5A
	Clarke	5A
	Clay	6A
	Clayton	5A
	Clinton	5A
	Crawford	5A
	Dallas	5A
	Davis	5A
	Decatur	5A
	Delaware	5A
	Des Moines	5A
	Dickinson	6A
	Dubuque	5A
	Emmet	6A
	Fayette	5A
	Floyd	5A

APPENDICES

Table B.5 (*Continued*)

State	County	Climate Zone
	Franklin	5A
	Fremont	5A
	Greene	5A
	Grundy	5A
	Guthrie	5A
	Hamilton	5A
	Hancock	6A
	Hardin	5A
	Harrison	5A
	Henry	5A
	Howard	6A
	Humboldt	5A
	Ida	5A
	Iowa	5A
	Jackson	5A
	Jasper	5A
	Jefferson	5A
	Johnson	5A
	Jones	5A
	Keokuk	5A
	Kossuth	6A
	Lee	5A
	Linn	5A
	Louisa	5A
	Lucas	5A
	Lyon	6A
	Madison	5A
	Mahaska	5A
	Marion	5A
	Marshall	5A
	Mills	5A
	Mitchell	6A
	Monona	5A
	Monroe	5A
	Montgomery	5A
	Muscatine	5A
	O'Brien	6A
	Osceola	6A
	Page	5A
	Palo Alto	6A
	Plymouth	5A
	Pocahontas	5A
	Polk	5A
	Pottawattamie	5A
	Poweshiek	5A
	Ringgold	5A
	Sac	5A
	Scott	5A
	Shelby	5A
	Sioux	5A
	Story	5A
	Tama	5A
	Taylor	5A
	Union	5A
	Van Buren	5A
	Wapello	5A
	Warren	5A
	Washington	5A
	Wayne	5A
	Webster	5A
	Winnebago	6A
	Winneshiek	5A
	Woodbury	5A
	Worth	6A
	Wright	5A
Kansas (KS)		
	Allen	4A
	Anderson	4A
	Atchison	4A
	Barber	4A
	Barton	4A
	Bourbon	4A

Table B.5 (*Continued*)

State	County	Climate Zone
	Brown	4A
	Butler	4A
	Chase	4A
	Chautauqua	4A
	Cherokee	4A
	Cheyenne	5B
	Clark	4A
	Clay	4A
	Cloud	4A
	Coffey	4A
	Comanche	4A
	Cowley	4A
	Crawford	4A
	Decatur	5B
	Dickinson	4A
	Doniphan	4A
	Douglas	4A
	Edwards	4A
	Elk	4A
	Ellis	4A
	Ellsworth	4A
	Finney	4B
	Ford	4A
	Franklin	4A
	Geary	4A
	Gove	5B
	Graham	4A
	Grant	4B
	Gray	4B
	Greeley	5B
	Greenwood	4A
	Hamilton	4B
	Harper	4A
	Harvey	4A
	Haskell	4B
	Hodgeman	4A
	Jackson	4A
	Jefferson	4A
	Jewell	5A
	Johnson	4A
	Kearny	4B
	Kingman	4A
	Kiowa	4A
	Labette	4A
	Lane	4A
	Leavenworth	4A
	Lincoln	4A
	Linn	4A
	Logan	5B
	Lyon	4A
	Marion	4A
	Marshall	4A
	McPherson	4A
	Meade	4B
	Miami	4A
	Mitchell	4A
	Montgomery	4A
	Morris	4A
	Morton	4B
	Nemaha	4A
	Neosho	4A
	Ness	4A
	Norton	5A
	Osage	4A
	Osborne	4A
	Ottawa	4A
	Pawnee	4A
	Phillips	5A
	Pottawatomie	4A
	Pratt	4A
	Rawlins	5B

APPENDICES

Table B.5 (*Continued*)

State	County	Climate Zone
	Reno	4A
	Republic	5A
	Rice	4A
	Riley	4A
	Rooks	4A
	Rush	4A
	Russell	4A
	Saline	4A
	Scott	5B
	Sedgwick	4A
	Seward	4B
	Shawnee	4A
	Sheridan	5B
	Sherman	5B
	Smith	5A
	Stafford	4A
	Stanton	4B
	Stevens	4B
	Sumner	4A
	Thomas	5B
	Trego	4A
	Wabaunsee	4A
	Wallace	5B
	Washington	4A
	Wichita	5B
	Wilson	4A
	Woodson	4A
	Wyandotte	4A
Kentucky (KY)		
	Adair	4A
	Allen	4A
	Anderson	4A
	Ballard	4A
	Barren	4A
	Bath	4A
	Bell	4A
	Boone	4A
	Bourbon	4A
	Boyd	4A
	Boyle	4A
	Bracken	4A
	Breathitt	4A
	Breckinridge	4A
	Bullitt	4A
	Butler	4A
	Caldwell	4A
	Calloway	4A
	Campbell	4A
	Carlisle	4A
	Carroll	4A
	Carter	4A
	Casey	4A
	Christian	4A
	Clark	4A
	Clay	4A
	Clinton	4A
	Crittenden	4A
	Cumberland	4A
	Daviess	4A
	Edmonson	4A
	Elliott	4A
	Estill	4A
	Fayette	4A
	Fleming	4A
	Floyd	4A
	Franklin	4A
	Fulton	4A
	Gallatin	4A
	Garrard	4A
	Grant	4A

Table B.5 (*Continued*)

State	County	Climate Zone
	Graves	4A
	Grayson	4A
	Green	4A
	Greenup	4A
	Hancock	4A
	Hardin	4A
	Harlan	4A
	Harrison	4A
	Hart	4A
	Henderson	4A
	Henry	4A
	Hickman	4A
	Hopkins	4A
	Jackson	4A
	Jefferson	4A
	Jessamine	4A
	Johnson	4A
	Kenton	4A
	Knott	4A
	Knox	4A
	Larue	4A
	Laurel	4A
	Lawrence	4A
	Lee	4A
	Leslie	4A
	Letcher	4A
	Lewis	4A
	Lincoln	4A
	Livingston	4A
	Logan	4A
	Lyon	4A
	Madison	4A
	Magoffin	4A
	Marion	4A
	Marshall	4A
	Martin	4A
	Mason	4A
	McCracken	4A
	McCreary	4A
	McLean	4A
	Meade	4A
	Menifee	4A
	Mercer	4A
	Metcalfe	4A
	Monroe	4A
	Montgomery	4A
	Morgan	4A
	Muhlenberg	4A
	Nelson	4A
	Nicholas	4A
	Ohio	4A
	Oldham	4A
	Owen	4A
	Owsley	4A
	Pendleton	4A
	Perry	4A
	Pike	4A
	Powell	4A
	Pulaski	4A
	Robertson	4A
	Rockcastle	4A
	Rowan	4A
	Russell	4A
	Scott	4A
	Shelby	4A
	Simpson	4A
	Spencer	4A
	Taylor	4A
	Todd	4A
	Trigg	4A
	Trimble	4A

Table B.5 (*Continued*)

State	County	Climate Zone
	Union	4A
	Warren	4A
	Washington	4A
	Wayne	4A
	Webster	4A
	Whitley	4A
	Wolfe	4A
	Woodford	4A
Louisiana (LA)		
	Acadia	2A
	Allen	2A
	Ascension	2A
	Assumption	2A
	Avoyelles	2A
	Beauregard	2A
	Bienville	2A
	Bossier	2A
	Caddo	2A
	Calcasieu	2A
	Caldwell	2A
	Cameron	2A
	Catahoula	2A
	Claiborne	2A
	Concordia	2A
	De Soto	2A
	East Baton Rouge	2A
	East Carroll	3A
	East Feliciana	2A
	Evangeline	2A
	Franklin	2A
	Grant	2A
	Iberia	2A
	Iberville	2A
	Jackson	2A
	Jefferson	2A
	Jefferson Davis	2A
	Lafayette	2A
	Lafourche	2A
	LaSalle	2A
	Lincoln	2A
	Livingston	2A
	Madison	3A
	Morehouse	2A
	Natchitoches	2A
	Orleans	2A
	Ouachita	2A
	Plaquemines	2A
	Pointe Coupee	2A
	Rapides	2A
	Red River	2A
	Richland	2A
	Sabine	2A
	St. Bernard	2A
	St. Charles	2A
	St. Helena	2A
	St. James	2A
	St. John the Baptist	2A
	St. Landry	2A
	St. Martin	2A
	St. Mary	2A
	St. Tammany	2A
	Tangipahoa	2A
	Tensas	2A
	Terrebonne	2A
	Union	3A
	Vermilion	2A
	Vernon	2A
	Washington	2A
	Webster	2A
	West Baton Rouge	2A
	West Carroll	3A

Table B.5 (*Continued*)

State	County	Climate Zone
	West Feliciana	2A
	Winn	2A
Maine (ME)		
	Androscoggin	6A
	Aroostook	7
	Cumberland	5A
	Franklin	6A
	Hancock	6A
	Kennebec	6A
	Knox	6A
	Lincoln	6A
	Oxford	6A
	Penobscot	6A
	Piscataquis	6A
	Sagadahoc	6A
	Somerset	6A
	Waldo	6A
	Washington	6A
	York	6A
Maryland (MD)		
	Allegany	4A
	Anne Arundel	4A
	Baltimore	4A
	Baltimore	4A
	Calvert	4A
	Caroline	4A
	Carroll	4A
	Cecil	4A
	Charles	4A
	Dorchester	4A
	Frederick	4A
	Garrett	5A
	Harford	4A
	Howard	4A
	Kent	4A
	Montgomery	4A
	Prince George's	4A
	Queen Anne's	4A
	Somerset	4A
	St. Mary's	4A
	Talbot	4A
	Washington	4A
	Wicomico	4A
	Worcester	4A
Massachusetts (MA)		
	Barnstable	5A
	Berkshire	5A
	Bristol	5A
	Dukes	5A
	Essex	5A
	Franklin	5A
	Hampden	5A
	Hampshire	5A
	Middlesex	5A
	Nantucket	5A
	Norfolk	5A
	Plymouth	5A
	Suffolk	5A
	Worcester	5A
Michigan (MI)		
	Alcona	6A
	Alger	6A
	Allegan	5A
	Alpena	6A
	Antrim	6A
	Arenac	6A
	Baraga	6A
	Barry	5A
	Bay	5A

Table B.5 (*Continued*)

State	County	Climate Zone
	Benzie	5A
	Berrien	5A
	Branch	5A
	Calhoun	5A
	Cass	5A
	Charlevoix	6A
	Cheboygan	6A
	Chippewa	6A
	Clare	6A
	Clinton	5A
	Crawford	6A
	Delta	6A
	Dickinson	6A
	Eaton	5A
	Emmet	6A
	Genesee	5A
	Gladwin	6A
	Gogebic	6A
	Grand Traverse	6A
	Gratiot	5A
	Hillsdale	5A
	Houghton	6A
	Huron	5A
	Ingham	5A
	Ionia	5A
	Iosco	6A
	Iron	6A
	Isabella	5A
	Jackson	5A
	Kalamazoo	5A
	Kalkaska	6A
	Kent	5A
	Keweenaw	6A
	Lake	6A
	Lapeer	5A
	Leelanau	6A
	Lenawee	5A
	Livingston	5A
	Luce	6A
	Mackinac	6A
	Macomb	5A
	Manistee	6A
	Marquette	6A
	Mason	5A
	Mecosta	6A
	Menominee	6A
	Midland	5A
	Missaukee	6A
	Monroe	5A
	Montcalm	5A
	Montmorency	6A
	Muskegon	5A
	Newaygo	5A
	Oakland	5A
	Oceana	5A
	Ogemaw	6A
	Ontonagon	6A
	Osceola	6A
	Oscoda	6A
	Otsego	6A
	Ottawa	5A
	Presque Isle	6A
	Roscommon	6A
	Saginaw	5A
	Sanilac	5A
	Schoolcraft	6A
	Shiawassee	5A
	St. Clair	5A
	St. Joseph	5A
	Tuscola	5A
	Van Buren	5A
	Washtenaw	5A

Table B.5 (*Continued*)

State	County	Climate Zone
	Wayne	5A
	Wexford	6A
Minnesota (MN)		
	Aitkin	7
	Anoka	6A
	Becker	6A
	Beltrami	7
	Benton	6A
	Big Stone	6A
	Blue Earth	6A
	Brown	6A
	Carlton	7
	Carver	6A
	Cass	7
	Chippewa	6A
	Chisago	6A
	Clay	6A
	Clearwater	7
	Cook	7
	Cottonwood	6A
	Crow Wing	6A
	Dakota	6A
	Dodge	6A
	Douglas	6A
	Faribault	6A
	Fillmore	6A
	Freeborn	6A
	Goodhue	6A
	Grant	6A
	Hennepin	6A
	Houston	5A
	Hubbard	7
	Isanti	6A
	Itasca	7
	Jackson	6A
	Kanabec	6A
	Kandiyohi	6A
	Kittson	7
	Koochiching	7
	Lac qui Parle	6A
	Lake	7
	Lake of the Woods	7
	Le Sueur	6A
	Lincoln	6A
	Lyon	6A
	Mahnomen	7
	Marshall	7
	Martin	6A
	McLeod	6A
	Meeker	6A
	Mille Lacs	6A
	Morrison	6A
	Mower	6A
	Murray	6A
	Nicollet	6A
	Nobles	6A
	Norman	7
	Olmsted	6A
	Otter Tail	6A
	Pennington	7
	Pine	6A
	Pipestone	6A
	Polk	7
	Pope	6A
	Ramsey	6A
	Red Lake	7
	Redwood	6A
	Renville	6A
	Rice	6A
	Rock	6A
	Roseau	7
	Scott	6A

Table B.5 (*Continued*)

State	County	Climate Zone
	Sherburne	6A
	Sibley	6A
	St. Louis	7
	Stearns	6A
	Steele	6A
	Stevens	6A
	Swift	6A
	Todd	6A
	Traverse	6A
	Wabasha	6A
	Wadena	6A
	Waseca	6A
	Washington	6A
	Watonwan	6A
	Wilkin	6A
	Winona	5A
	Wright	6A
	Yellow Medicine	6A
Mississippi (MS)		
	Adams	3A
	Alcorn	3A
	Amite	2A
	Attala	3A
	Benton	3A
	Bolivar	3A
	Calhoun	3A
	Carroll	3A
	Chickasaw	3A
	Choctaw	3A
	Claiborne	3A
	Clarke	3A
	Clay	3A
	Coahoma	3A
	Copiah	3A
	Covington	3A
	DeSoto	3A
	Forrest	2A
	Franklin	3A
	George	2A
	Greene	2A
	Grenada	3A
	Hancock	2A
	Harrison	2A
	Hinds	3A
	Holmes	3A
	Humphreys	3A
	Issaquena	3A
	Itawamba	3A
	Jackson	2A
	Jasper	3A
	Jefferson	3A
	Jefferson Davis	3A
	Jones	3A
	Kemper	3A
	Lafayette	3A
	Lamar	2A
	Lauderdale	3A
	Lawrence	3A
	Leake	3A
	Lee	3A
	Leflore	3A
	Lincoln	3A
	Lowndes	3A
	Madison	3A
	Marion	2A
	Marshall	3A
	Monroe	3A
	Montgomery	3A
	Neshoba	3A
	Newton	3A

Table B.5 (*Continued*)

State	County	Climate Zone
	Noxubee	3A
	Oktibbeha	3A
	Panola	3A
	Pearl River	2A
	Perry	2A
	Pike	2A
	Pontotoc	3A
	Prentiss	3A
	Quitman	3A
	Rankin	3A
	Scott	3A
	Sharkey	3A
	Simpson	3A
	Smith	3A
	Stone	2A
	Sunflower	3A
	Tallahatchie	3A
	Tate	3A
	Tippah	3A
	Tishomingo	3A
	Tunica	3A
	Union	3A
	Walthall	2A
	Warren	3A
	Washington	3A
	Wayne	3A
	Webster	3A
	Wilkinson	2A
	Winston	3A
	Yalobusha	3A
	Yazoo	3A
Missouri (MO)		
	Adair	5A
	Andrew	4A
	Atchison	5A
	Audrain	4A
	Barry	4A
	Barton	4A
	Bates	4A
	Benton	4A
	Bollinger	4A
	Boone	4A
	Buchanan	4A
	Butler	4A
	Caldwell	4A
	Callaway	4A
	Camden	4A
	Cape Girardeau	4A
	Carroll	4A
	Carter	4A
	Cass	4A
	Cedar	4A
	Chariton	4A
	Christian	4A
	Clark	5A
	Clay	4A
	Clinton	4A
	Cole	4A
	Cooper	4A
	Crawford	4A
	Dade	4A
	Dallas	4A
	Daviess	4A
	DeKalb	4A
	Dent	4A
	Douglas	4A
	Dunklin	3A
	Franklin	4A
	Gasconade	4A
	Gentry	5A

Table B.5 (*Continued*)

State	County	Climate Zone
	Greene	4A
	Grundy	5A
	Harrison	5A
	Henry	4A
	Hickory	4A
	Holt	5A
	Howard	4A
	Howell	4A
	Iron	4A
	Jackson	4A
	Jasper	4A
	Jefferson	4A
	Johnson	4A
	Knox	5A
	Laclede	4A
	Lafayette	4A
	Lawrence	4A
	Lewis	5A
	Lincoln	4A
	Linn	5A
	Livingston	4A
	Macon	5A
	Madison	4A
	Maries	4A
	Marion	5A
	McDonald	4A
	Mercer	5A
	Miller	4A
	Mississippi	4A
	Moniteau	4A
	Monroe	4A
	Montgomery	4A
	Morgan	4A
	New Madrid	4A
	Newton	4A
	Nodaway	5A
	Oregon	4A
	Osage	4A
	Ozark	4A
	Pemiscot	3A
	Perry	4A
	Pettis	4A
	Phelps	4A
	Pike	4A
	Platte	4A
	Polk	4A
	Pulaski	4A
	Putnam	5A
	Ralls	4A
	Randolph	4A
	Ray	4A
	Reynolds	4A
	Ripley	4A
	Saline	4A
	Schuyler	5A
	Scotland	5A
	Scott	4A
	Shannon	4A
	Shelby	5A
	St. Charles	4A
	St. Clair	4A
	St. Francois	4A
	St. Louis	4A
	St. Louis	4A
	Ste. Genevieve	4A
	Stoddard	4A
	Stone	4A
	Sullivan	5A
	Taney	4A
	Texas	4A
	Vernon	4A
	Warren	4A

Table B.5 (*Continued*)

State	County	Climate Zone
	Washington	4A
	Wayne	4A
	Webster	4A
	Worth	5A
	Wright	4A
Montana (MT)		
	Beaverhead	6B
	Big Horn	6B
	Blaine	6B
	Broadwater	6B
	Carbon	6B
	Carter	6B
	Cascade	6B
	Chouteau	6B
	Custer	6B
	Daniels	6B
	Dawson	6B
	Deer Lodge	6B
	Fallon	6B
	Fergus	6B
	Flathead	6B
	Gallatin	6B
	Garfield	6B
	Glacier	6B
	Golden Valley	6B
	Granite	6B
	Hill	6B
	Jefferson	6B
	Judith Basin	6B
	Lake	6B
	Lewis and Clark	6B
	Liberty	6B
	Lincoln	6B
	Madison	6B
	McCone	6B
	Meagher	6B
	Mineral	6B
	Missoula	6B
	Musselshell	6B
	Park	6B
	Petroleum	6B
	Phillips	6B
	Pondera	6B
	Powder River	6B
	Powell	6B
	Prairie	6B
	Ravalli	6B
	Richland	6B
	Roosevelt	6B
	Rosebud	6B
	Sanders	6B
	Sheridan	6B
	Silver Bow	6B
	Stillwater	6B
	Sweet Grass	6B
	Teton	6B
	Toole	6B
	Treasure	6B
	Valley	6B
	Wheatland	6B
	Wibaux	6B
	Yellowstone	6B
Nebraska (NE)		
	Adams	5A
	Antelope	5A
	Arthur	5A
	Banner	5B
	Blaine	5A
	Boone	5A
	Box Butte	5B

Table B.5 (*Continued*)

State	County	Climate Zone
	Boyd	5A
	Brown	5A
	Buffalo	5A
	Burt	5A
	Butler	5A
	Cass	5A
	Cedar	5A
	Chase	5B
	Cherry	5A
	Cheyenne	5B
	Clay	5A
	Colfax	5A
	Cuming	5A
	Custer	5A
	Dakota	5A
	Dawes	5B
	Dawson	5A
	Deuel	5B
	Dixon	5A
	Dodge	5A
	Douglas	5A
	Dundy	5B
	Fillmore	5A
	Franklin	5A
	Frontier	5A
	Furnas	5A
	Gage	5A
	Garden	5B
	Garfield	5A
	Gosper	5A
	Grant	5A
	Greeley	5A
	Hall	5A
	Hamilton	5A
	Harlan	5A
	Hayes	5B
	Hitchcock	5B
	Holt	5A
	Hooker	5A
	Howard	5A
	Jefferson	5A
	Johnson	5A
	Kearney	5A
	Keith	5A
	Keya Paha	5A
	Kimball	5B
	Knox	5A
	Lancaster	5A
	Lincoln	5A
	Logan	5A
	Loup	5A
	Madison	5A
	McPherson	5A
	Merrick	5A
	Morrill	5B
	Nance	5A
	Nemaha	5A
	Nuckolls	5A
	Otoe	5A
	Pawnee	5A
	Perkins	5B
	Phelps	5A
	Pierce	5A
	Platte	5A
	Polk	5A
	Red Willow	5A
	Richardson	5A
	Rock	5A
	Saline	5A
	Sarpy	5A
	Saunders	5A

Table B.5 (*Continued*)

State	County	Climate Zone
	Scotts Bluff	5B
	Seward	5A
	Sheridan	5B
	Sherman	5A
	Sioux	6B
	Stanton	5A
	Thayer	5A
	Thomas	5A
	Thurston	5A
	Valley	5A
	Washington	5A
	Wayne	5A
	Webster	5A
	Wheeler	5A
	York	5A
Nevada (NV)		
	Carson City	4B
	Churchill	4B
	Clark	2B
	Douglas	4B
	Elko	5B
	Esmeralda	4B
	Eureka	5B
	Humboldt	5B
	Lander	5B
	Lincoln	4B
	Lyon	4B
	Mineral	4B
	Nye	4B
	Pershing	5B
	Storey	4B
	Washoe	4B
	White Pine	5B
New Hampshire (NH)		
	Belknap	6A
	Carroll	6A
	Cheshire	5A
	Coos	6A
	Grafton	6A
	Hillsborough	5A
	Merrimack	5A
	Rockingham	5A
	Strafford	5A
	Sullivan	6A
New Jersey (NJ)		
	Atlantic	4A
	Bergen	4A
	Burlington	4A
	Camden	4A
	Cape May	4A
	Cumberland	4A
	Essex	4A
	Gloucester	4A
	Hudson	4A
	Hunterdon	4A
	Mercer	4A
	Middlesex	4A
	Monmouth	4A
	Morris	4A
	Ocean	4A
	Passaic	4A
	Salem	4A
	Somerset	4A
	Sussex	5A
	Union	4A
	Warren	5A
New Mexico (NM)		
	Bernalillo	4B
	Catron	4B

Table B.5 (*Continued*)

State	County	Climate Zone
	Chaves	3B
	Cibola	5B
	Colfax	5B
	Curry	4B
	De Baca	4B
	Doña Ana	3B
	Eddy	3B
	Grant	3B
	Guadalupe	4B
	Harding	5B
	Hidalgo	3B
	Lea	3B
	Lincoln	3B
	Los Alamos	4B
	Luna	3B
	McKinley	5B
	Mora	5B
	Otero	3B
	Quay	4B
	Rio Arriba	6B
	Roosevelt	4B
	San Juan	4B
	San Miguel	5B
	Sandoval	6B
	Santa Fe	4B
	Sierra	3B
	Socorro	4B
	Taos	5B
	Torrance	5B
	Union	4B
	Valencia	4B
New York (NY)		
	Albany	5A
	Allegany	5A
	Bronx	4A
	Broome	5A
	Cattaraugus	5A
	Cayuga	5A
	Chautauqua	5A
	Chemung	5A
	Chenango	6A
	Clinton	6A
	Columbia	5A
	Cortland	6A
	Delaware	6A
	Dutchess	5A
	Erie	5A
	Essex	6A
	Franklin	6A
	Fulton	6A
	Genesee	5A
	Greene	5A
	Hamilton	6A
	Herkimer	6A
	Jefferson	6A
	Kings	4A
	Lewis	6A
	Livingston	5A
	Madison	6A
	Monroe	5A
	Montgomery	6A
	Nassau	4A
	New York	4A
	Niagara	5A
	Oneida	5A
	Onondaga	5A
	Ontario	5A
	Orange	5A
	Orleans	5A
	Oswego	5A
	Otsego	6A

Table B.5 (*Continued*)

State	County	Climate Zone
	Putnam	5A
	Queens	4A
	Rensselaer	5A
	Richmond	4A
	Rockland	4A
	Saratoga	5A
	Schenectady	5A
	Schoharie	5A
	Schuyler	5A
	Seneca	5A
	St. Lawrence	6A
	Steuben	5A
	Suffolk	4A
	Sullivan	5A
	Tioga	5A
	Tompkins	5A
	Ulster	5A
	Warren	6A
	Washington	5A
	Wayne	5A
	Westchester	4A
	Wyoming	5A
	Yates	5A
North Carolina (NC)		
	Alamance	3A
	Alexander	4A
	Alleghany	4A
	Anson	3A
	Ashe	4A
	Avery	5A
	Beaufort	3A
	Bertie	3A
	Bladen	3A
	Brunswick	3A
	Buncombe	4A
	Burke	4A
	Cabarrus	3A
	Caldwell	4A
	Camden	3A
	Carteret	3A
	Caswell	4A
	Catawba	3A
	Chatham	3A
	Cherokee	4A
	Chowan	3A
	Clay	4A
	Cleveland	3A
	Columbus	3A
	Craven	3A
	Cumberland	3A
	Currituck	3A
	Dare	3A
	Davidson	3A
	Davie	3A
	Duplin	3A
	Durham	3A
	Edgecombe	3A
	Forsyth	3A
	Franklin	3A
	Gaston	3A
	Gates	3A
	Graham	4A
	Granville	3A
	Greene	3A
	Guilford	3A
	Halifax	3A
	Harnett	3A
	Haywood	4A
	Henderson	4A
	Hertford	3A

Table B.5 (*Continued*)

State	County	Climate Zone
	Hoke	3A
	Hyde	3A
	Iredell	3A
	Jackson	4A
	Johnston	3A
	Jones	3A
	Lee	3A
	Lenoir	3A
	Lincoln	3A
	Macon	4A
	Madison	4A
	Martin	3A
	McDowell	4A
	Mecklenburg	3A
	Mitchell	5A
	Montgomery	3A
	Moore	3A
	Nash	3A
	New Hanover	3A
	Northampton	3A
	Onslow	3A
	Orange	3A
	Pamlico	3A
	Pasquotank	3A
	Pender	3A
	Perquimans	3A
	Person	4A
	Pitt	3A
	Polk	3A
	Randolph	3A
	Richmond	3A
	Robeson	3A
	Rockingham	4A
	Rowan	3A
	Rutherford	3A
	Sampson	3A
	Scotland	3A
	Stanly	3A
	Stokes	4A
	Surry	4A
	Swain	4A
	Transylvania	4A
	Tyrrell	3A
	Union	3A
	Vance	3A
	Wake	3A
	Warren	3A
	Washington	3A
	Watauga	4A
	Wayne	3A
	Wilkes	4A
	Wilson	3A
	Yadkin	3A
	Yancey	5A
North Dakota (ND)		
	Adams	6A
	Barnes	6A
	Benson	7
	Billings	6B
	Bottineau	7
	Bowman	6B
	Burke	7
	Burleigh	6A
	Cass	6A
	Cavalier	7
	Dickey	6A
	Divide	6B
	Dunn	6A
	Eddy	7
	Emmons	6A
	Foster	6A

Table B.5 (*Continued*)

State	County	Climate Zone
	Golden Valley	6B
	Grand Forks	7
	Grant	6A
	Griggs	7
	Hettinger	6A
	Kidder	6A
	LaMoure	6A
	Logan	6A
	McHenry	7
	McIntosh	6A
	McKenzie	6B
	McLean	6A
	Mercer	6A
	Morton	6A
	Mountrail	6A
	Nelson	7
	Oliver	6A
	Pembina	7
	Pierce	7
	Ramsey	7
	Ransom	6A
	Renville	7
	Richland	6A
	Rolette	7
	Sargent	6A
	Sheridan	6A
	Sioux	6A
	Slope	6B
	Stark	6A
	Steele	7
	Stutsman	6A
	Towner	7
	Traill	7
	Walsh	7
	Ward	6A
	Wells	6A
	Williams	6B
Ohio (OH)		
	Adams	4A
	Allen	5A
	Ashland	5A
	Ashtabula	5A
	Athens	4A
	Auglaize	5A
	Belmont	5A
	Brown	4A
	Butler	4A
	Carroll	5A
	Champaign	5A
	Clark	5A
	Clermont	4A
	Clinton	4A
	Columbiana	5A
	Coshocton	5A
	Crawford	5A
	Cuyahoga	5A
	Darke	5A
	Defiance	5A
	Delaware	5A
	Erie	5A
	Fairfield	4A
	Fayette	4A
	Franklin	4A
	Fulton	5A
	Gallia	4A
	Geauga	5A
	Greene	4A
	Guernsey	4A
	Hamilton	4A
	Hancock	5A
	Hardin	5A

APPENDICES

Table B.5 (*Continued*)

State	County	Climate Zone
	Harrison	5A
	Henry	5A
	Highland	4A
	Hocking	4A
	Holmes	5A
	Huron	5A
	Jackson	4A
	Jefferson	5A
	Knox	5A
	Lake	5A
	Lawrence	4A
	Licking	5A
	Logan	5A
	Lorain	5A
	Lucas	5A
	Madison	4A
	Mahoning	5A
	Marion	5A
	Medina	5A
	Meigs	4A
	Mercer	5A
	Miami	5A
	Monroe	4A
	Montgomery	4A
	Morgan	4A
	Morrow	5A
	Muskingum	4A
	Noble	4A
	Ottawa	5A
	Paulding	5A
	Perry	4A
	Pickaway	4A
	Pike	4A
	Portage	5A
	Preble	5A
	Putnam	5A
	Richland	5A
	Ross	4A
	Sandusky	5A
	Scioto	4A
	Seneca	5A
	Shelby	5A
	Stark	5A
	Summit	5A
	Trumbull	5A
	Tuscarawas	5A
	Union	5A
	Van Wert	5A
	Vinton	4A
	Warren	4A
	Washington	4A
	Wayne	5A
	Williams	5A
	Wood	5A
	Wyandot	5A
Oklahoma (OK)		
	Adair	4A
	Alfalfa	4A
	Atoka	3A
	Beaver	4B
	Beckham	3A
	Blaine	4A
	Bryan	3A
	Caddo	3A
	Canadian	3A
	Carter	3A
	Cherokee	4A
	Choctaw	3A
	Cimarron	4B
	Cleveland	3A
	Coal	3A

Table B.5 (*Continued*)

State	County	Climate Zone
	Comanche	3A
	Cotton	3A
	Craig	4A
	Creek	3A
	Custer	4A
	Delaware	4A
	Dewey	4A
	Ellis	4B
	Garfield	4A
	Garvin	3A
	Grady	3A
	Grant	4A
	Greer	3A
	Harmon	3A
	Harper	4B
	Haskell	3A
	Hughes	3A
	Jackson	3A
	Jefferson	3A
	Johnston	3A
	Kay	4A
	Kingfisher	3A
	Kiowa	3A
	Latimer	3A
	Le Flore	3A
	Lincoln	3A
	Logan	3A
	Love	3A
	Major	4A
	Marshall	3A
	Mayes	4A
	McClain	3A
	McCurtain	3A
	McIntosh	3A
	Murray	3A
	Muskogee	3A
	Noble	4A
	Nowata	4A
	Okfuskee	3A
	Oklahoma	3A
	Okmulgee	3A
	Osage	4A
	Ottawa	4A
	Pawnee	4A
	Payne	4A
	Pittsburg	3A
	Pontotoc	3A
	Pottawatomie	3A
	Pushmataha	3A
	Roger Mills	4A
	Rogers	4A
	Seminole	3A
	Sequoyah	3A
	Stephens	3A
	Texas	4B
	Tillman	3A
	Tulsa	3A
	Wagoner	3A
	Washington	4A
	Washita	3A
	Woods	4A
	Woodward	4A
Oregon (OR)		
	Baker	6B
	Benton	4C
	Clackamas	4C
	Clatsop	4C
	Columbia	4C
	Coos	4C
	Crook	5B
	Curry	4C

APPENDICES

Table B.5 (*Continued*)

State	County	Climate Zone
	Deschutes	5B
	Douglas	4C
	Gilliam	5B
	Grant	5B
	Harney	6B
	Hood River	5B
	Jackson	4C
	Jefferson	5B
	Josephine	4C
	Klamath	5B
	Lake	6B
	Lane	4C
	Lincoln	4C
	Linn	4C
	Malheur	5B
	Marion	4C
	Morrow	5B
	Multnomah	4C
	Polk	4C
	Sherman	5B
	Tillamook	4C
	Umatilla	4B
	Union	5B
	Wallowa	5B
	Wasco	5B
	Washington	4C
	Wheeler	5B
	Yamhill	4C
Pennsylvania (PA)		
	Adams	4A
	Allegheny	4A
	Armstrong	5A
	Beaver	5A
	Bedford	5A
	Berks	4A
	Blair	5A
	Bradford	5A
	Bucks	4A
	Butler	5A
	Cambria	5A
	Cameron	5A
	Carbon	5A
	Centre	5A
	Chester	4A
	Clarion	5A
	Clearfield	5A
	Clinton	5A
	Columbia	5A
	Crawford	5A
	Cumberland	4A
	Dauphin	4A
	Delaware	4A
	Elk	5A
	Erie	5A
	Fayette	5A
	Forest	5A
	Franklin	4A
	Fulton	5A
	Greene	5A
	Huntingdon	5A
	Indiana	5A
	Jefferson	5A
	Juniata	5A
	Lackawanna	5A
	Lancaster	4A
	Lawrence	5A
	Lebanon	4A
	Lehigh	5A
	Luzerne	5A
	Lycoming	5A
	McKean	5A

Table B.5 (*Continued*)

State	County	Climate Zone
	Mercer	5A
	Mifflin	5A
	Monroe	5A
	Montgomery	4A
	Montour	5A
	Northampton	5A
	Northumberland	5A
	Perry	4A
	Philadelphia	4A
	Pike	5A
	Potter	5A
	Schuylkill	5A
	Snyder	5A
	Somerset	5A
	Sullivan	5A
	Susquehanna	5A
	Tioga	5A
	Union	5A
	Venango	5A
	Warren	5A
	Washington	5A
	Wayne	5A
	Westmoreland	5A
	Wyoming	5A
	York	4A
Rhode Island (RI)		
	Bristol	4A
	Kent	5A
	Newport	4A
	Providence	4A
	Washington	4A
South Carolina (SC)		
	Abbeville	3A
	Aiken	3A
	Allendale	3A
	Anderson	3A
	Bamberg	3A
	Barnwell	3A
	Beaufort	2A
	Berkeley	3A
	Calhoun	3A
	Charleston	2A
	Cherokee	3A
	Chester	3A
	Chesterfield	3A
	Clarendon	3A
	Colleton	3A
	Darlington	3A
	Dillon	3A
	Dorchester	3A
	Edgefield	3A
	Fairfield	3A
	Florence	3A
	Georgetown	3A
	Greenville	3A
	Greenwood	3A
	Hampton	3A
	Horry	3A
	Jasper	2A
	Kershaw	3A
	Lancaster	3A
	Laurens	3A
	Lee	3A
	Lexington	3A
	Marion	3A
	Marlboro	3A
	McCormick	3A
	Newberry	3A
	Oconee	3A
	Orangeburg	3A

Table B.5 (*Continued*)

State	County	Climate Zone
	Pickens	3A
	Richland	3A
	Saluda	3A
	Spartanburg	3A
	Sumter	3A
	Union	3A
	Williamsburg	3A
	York	3A
South Dakota (SD)		
	Aurora	5A
	Beadle	6A
	Bennett	5A
	Bon Homme	5A
	Brookings	6A
	Brown	6A
	Brule	5A
	Buffalo	6A
	Butte	6A
	Campbell	6A
	Charles Mix	5A
	Clark	6A
	Clay	5A
	Codington	6A
	Corson	6A
	Custer	6B
	Davison	5A
	Day	6A
	Deuel	6A
	Dewey	6A
	Douglas	5A
	Edmunds	6A
	Fall River	5B
	Faulk	6A
	Grant	6A
	Gregory	5A
	Haakon	5B
	Hamlin	6A
	Hand	6A
	Hanson	6A
	Harding	6B
	Hughes	6A
	Hutchinson	5A
	Hyde	6A
	Jackson	5A
	Jerauld	6A
	Jones	5A
	Kingsbury	6A
	Lake	6A
	Lawrence	6A
	Lincoln	6A
	Lyman	5A
	Marshall	6A
	McCook	6A
	McPherson	6A
	Meade	5B
	Mellette	5A
	Miner	6A
	Minnehaha	6A
	Moody	6A
	Oglala Lakota	5B
	Pennington	5B
	Perkins	6A
	Potter	6A
	Roberts	6A
	Sanborn	6A
	Spink	6A
	Stanley	6A
	Sully	6A
	Todd	5A
	Tripp	5A
	Turner	6A

Table B.5 (*Continued*)

State	County	Climate Zone
	Union	5A
	Walworth	6A
	Yankton	5A
	Ziebach	6B
Tennessee (TN)		
	Anderson	3A
	Bedford	3A
	Benton	3A
	Bledsoe	4A
	Blount	3A
	Bradley	3A
	Campbell	4A
	Cannon	3A
	Carroll	3A
	Carter	4A
	Cheatham	3A
	Chester	3A
	Claiborne	4A
	Clay	4A
	Cocke	4A
	Coffee	3A
	Crockett	3A
	Cumberland	4A
	Davidson	3A
	Decatur	3A
	DeKalb	4A
	Dickson	3A
	Dyer	3A
	Fayette	3A
	Fentress	4A
	Franklin	3A
	Gibson	3A
	Giles	3A
	Grainger	4A
	Greene	4A
	Grundy	4A
	Hamblen	4A
	Hamilton	3A
	Hancock	4A
	Hardeman	3A
	Hardin	3A
	Hawkins	4A
	Haywood	3A
	Henderson	3A
	Henry	3A
	Hickman	3A
	Houston	3A
	Humphreys	3A
	Jackson	4A
	Jefferson	4A
	Johnson	4A
	Knox	3A
	Lake	3A
	Lauderdale	3A
	Lawrence	3A
	Lewis	3A
	Lincoln	3A
	Loudon	3A
	Macon	4A
	Madison	3A
	Marion	4A
	Marshall	3A
	Maury	3A
	McMinn	3A
	McNairy	3A
	Meigs	3A
	Monroe	4A
	Montgomery	4A
	Moore	3A
	Morgan	4A
	Obion	3A

APPENDICES

Table B.5 (*Continued*)

State	County	Climate Zone
	Overton	4A
	Perry	3A
	Pickett	4A
	Polk	4A
	Putnam	4A
	Rhea	3A
	Roane	3A
	Robertson	4A
	Rutherford	3A
	Scott	4A
	Sequatchie	4A
	Sevier	4A
	Shelby	3A
	Smith	4A
	Stewart	4A
	Sullivan	4A
	Sumner	4A
	Tipton	3A
	Trousdale	4A
	Unicoi	4A
	Union	4A
	Van Buren	4A
	Warren	3A
	Washington	4A
	Wayne	3A
	Weakley	3A
	White	4A
	Williamson	3A
	Wilson	4A
Texas (TX)		
	Anderson	2A
	Andrews	3B
	Angelina	2A
	Aransas	2A
	Archer	3A
	Armstrong	3B
	Atascosa	2A
	Austin	2A
	Bailey	4B
	Bandera	2A
	Bastrop	2A
	Baylor	3A
	Bee	2A
	Bell	2A
	Bexar	2A
	Blanco	2A
	Borden	3B
	Bosque	2A
	Bowie	3A
	Brazoria	2A
	Brazos	2A
	Brewster	3B
	Briscoe	3B
	Brooks	2A
	Brown	2A
	Burleson	2A
	Burnet	2A
	Caldwell	2A
	Calhoun	2A
	Callahan	3A
	Cameron	1A
	Camp	3A
	Carson	3B
	Cass	2A
	Castro	4B
	Chambers	2A
	Cherokee	2A
	Childress	3B
	Clay	3A
	Cochran	3B
	Coke	2B

Table B.5 (*Continued*)

State	County	Climate Zone
	Coleman	2A
	Collin	3A
	Collingsworth	3B
	Colorado	2A
	Comal	2A
	Comanche	2A
	Concho	2A
	Cooke	3A
	Coryell	2A
	Cottle	3A
	Crane	2B
	Crockett	2B
	Crosby	3A
	Culberson	3B
	Dallam	4B
	Dallas	2A
	Dawson	3B
	Deaf Smith	4B
	Delta	3A
	Denton	3A
	DeWitt	2A
	Dickens	3A
	Dimmit	2B
	Donley	3B
	Duval	2A
	Eastland	3A
	Ector	3B
	Edwards	2A
	El Paso	3B
	Ellis	2A
	Erath	2A
	Falls	2A
	Fannin	3A
	Fayette	2A
	Fisher	3A
	Floyd	3B
	Foard	3A
	Fort Bend	2A
	Franklin	3A
	Freestone	2A
	Frio	2A
	Gaines	3B
	Galveston	2A
	Garza	3A
	Gillespie	2A
	Glasscock	2B
	Goliad	2A
	Gonzales	2A
	Gray	3B
	Grayson	3A
	Gregg	2A
	Grimes	2A
	Guadalupe	2A
	Hale	4B
	Hall	3B
	Hamilton	2A
	Hansford	4B
	Hardeman	3A
	Hardin	2A
	Harris	2A
	Harrison	2A
	Hartley	4B
	Haskell	3A
	Hays	2A
	Hemphill	3B
	Henderson	2A
	Hidalgo	1B
	Hill	2A
	Hockley	3B
	Hood	2A
	Hopkins	2A
	Houston	2A

Table B.5 (*Continued*)

State	County	Climate Zone
	Howard	3B
	Hudspeth	3B
	Hunt	3A
	Hutchinson	3B
	Irion	2A
	Jack	3A
	Jackson	2A
	Jasper	2A
	Jeff Davis	3B
	Jefferson	2A
	Jim Hogg	2A
	Jim Wells	2A
	Johnson	2A
	Jones	3A
	Karnes	2A
	Kaufman	2A
	Kendall	2A
	Kenedy	2A
	Kent	3A
	Kerr	2A
	Kimble	2A
	King	3A
	Kinney	2B
	Kleberg	2A
	Knox	3A
	La Salle	2B
	Lamar	3A
	Lamb	4B
	Lampasas	2A
	Lavaca	2A
	Lee	2A
	Leon	2A
	Liberty	2A
	Limestone	2A
	Lipscomb	4B
	Live Oak	2A
	Llano	2A
	Loving	2B
	Lubbock	3B
	Lynn	3B
	Madison	2A
	Marion	2A
	Martin	3B
	Mason	2A
	Matagorda	2A
	Maverick	2B
	McCulloch	2A
	McLennan	2A
	McMullen	2A
	Medina	2A
	Menard	2A
	Midland	2B
	Milam	2A
	Mills	2A
	Mitchell	3A
	Montague	3A
	Montgomery	2A
	Moore	4B
	Morris	3A
	Motley	3B
	Nacogdoches	2A
	Navarro	2A
	Newton	2A
	Nolan	3A
	Nueces	2A
	Ochiltree	4B
	Oldham	4B
	Orange	2A
	Palo Pinto	3A
	Panola	2A
	Parker	3A
	Parmer	4B
	Pecos	2B
	Polk	2A
	Potter	4B
	Presidio	3B
	Rains	2A
	Randall	4B
	Reagan	2B
	Real	2A
	Red River	3A
	Reeves	2B
	Refugio	2A
	Roberts	3B
	Robertson	2A
	Rockwall	3A
	Runnels	2B
	Rusk	2A
	Sabine	2A
	San Augustine	2A
	San Jacinto	2A
	San Patricio	2A
	San Saba	2A
	Schleicher	2A
	Scurry	3A
	Shackelford	3A
	Shelby	2A
	Sherman	4B
	Smith	2A
	Somervell	2A
	Starr	1B
	Stephens	2A
	Sterling	3A
	Stonewall	3A
	Sutton	2B
	Swisher	3B
	Tarrant	2A
	Taylor	3A
	Terrell	2B
	Terry	3B
	Throckmorton	3A
	Titus	3A
	Tom Green	2A
	Travis	2A
	Trinity	2A
	Tyler	2A
	Upshur	2A
	Upton	2B
	Uvalde	2A
	Val Verde	2B
	Van Zandt	2A
	Victoria	2A
	Walker	2A
	Waller	2A
	Ward	2B
	Washington	2A
	Webb	1B
	Wharton	2A
	Wheeler	3B
	Wichita	3A
	Wilbarger	3A
	Willacy	1A
	Williamson	2A
	Wilson	2A
	Winkler	2B
	Wise	3A
	Wood	2A
	Yoakum	3B
	Young	3A
	Zapata	1B
	Zavala	2B

APPENDICES

Table B.5 (*Continued*)

State	County	Climate Zone
Utah (UT)		
	Beaver	5B
	Box Elder	5B
	Cache	5B
	Carbon	5B
	Daggett	6B
	Davis	5B
	Duchesne	6B
	Emery	5B
	Garfield	6B
	Grand	5B
	Iron	5B
	Juab	5B
	Kane	5B
	Millard	5B
	Morgan	6B
	Piute	5B
	Rich	6B
	Salt Lake	4B
	San Juan	5B
	Sanpete	5B
	Sevier	5B
	Summit	6B
	Tooele	5B
	Uintah	5B
	Utah	5B
	Wasatch	6B
	Washington	3B
	Wayne	5B
	Weber	5B
Vermont (VT)		
	Addison	6A
	Bennington	5A
	Caledonia	6A
	Chittenden	5A
	Essex	6A
	Franklin	6A
	Grand Isle	6A
	Lamoille	6A
	Orange	6A
	Orleans	6A
	Rutland	6A
	Washington	6A
	Windham	5A
	Windsor	6A
Virginia (VA)		
	Accomack	4A
	Albemarle	4A
	Alexandria	4A
	Alleghany	4A
	Amelia	4A
	Amherst	4A
	Appomattox	4A
	Arlington	4A
	Augusta	4A
	Bath	5A
	Bedford	4A
	Bland	4A
	Botetourt	4A
	Bristol	4A
	Brunswick	3A
	Buchanan	4A
	Buckingham	4A
	Buena Vista	4A
	Campbell	4A
	Caroline	4A
	Carroll	4A
	Charles City	4A
	Charlotte	4A
	Charlottesville	4A

Table B.5 (*Continued*)

State	County	Climate Zone
	Chesapeake	3A
	Chesterfield	4A
	Clarke	4A
	Colonial Heights	4A
	Covington	4A
	Craig	4A
	Culpeper	4A
	Cumberland	4A
	Danville	4A
	Dickenson	4A
	Dinwiddie	4A
	Emporia	4A
	Essex	4A
	Fairfax	4A
	Fairfax	4A
	Falls Church	4A
	Fauquier	4A
	Floyd	4A
	Fluvanna	4A
	Franklin	3A
	Franklin	4A
	Frederick	4A
	Fredericksburg	4A
	Galax	4A
	Giles	4A
	Gloucester	3A
	Goochland	4A
	Grayson	4A
	Greene	4A
	Greensville	4A
	Halifax	4A
	Hampton	3A
	Hanover	4A
	Harrisonburg	4A
	Henrico	4A
	Henry	4A
	Highland	5A
	Hopewell	4A
	Isle of Wight	3A
	James City	3A
	King and Queen	3A
	King George	4A
	King William	4A
	Lancaster	4A
	Lee	4A
	Lexington	4A
	Loudoun	4A
	Louisa	4A
	Lunenburg	4A
	Lynchburg	4A
	Madison	4A
	Manassas	4A
	Manassas Park	4A
	Martinsville	4A
	Mathews	3A
	Mecklenburg	3A
	Middlesex	3A
	Montgomery	4A
	Nelson	4A
	New Kent	4A
	Newport News	3A
	Norfolk	3A
	Northampton	4A
	Northumberland	4A
	Norton	4A
	Nottoway	4A
	Orange	4A
	Page	4A
	Patrick	4A
	Petersburg	4A
	Pittsylvania	4A
	Poquoson	3A

APPENDICES

Table B.5 (*Continued*)

State	County	Climate Zone
	Portsmouth	3A
	Powhatan	4A
	Prince Edward	4A
	Prince George	4A
	Prince William	4A
	Pulaski	4A
	Radford	4A
	Rappahannock	4A
	Richmond	4A
	Richmond	4A
	Roanoke	4A
	Roanoke	4A
	Rockbridge	4A
	Rockingham	4A
	Russell	4A
	Salem	4A
	Scott	4A
	Shenandoah	4A
	Smyth	4A
	Southampton	3A
	Spotsylvania	4A
	Stafford	4A
	Staunton	4A
	Suffolk	3A
	Surry	3A
	Sussex	3A
	Tazewell	4A
	Virginia Beach	3A
	Warren	4A
	Washington	4A
	Waynesboro	4A
	Westmoreland	4A
	Williamsburg	3A
	Winchester	4A
	Wise	4A
	Wythe	4A
	York	3A
Washington (WA)		
	Adams	5B
	Asotin	4B
	Benton	4B
	Chelan	5B
	Clallam	5C
	Clark	4C
	Columbia	5B
	Cowlitz	4C
	Douglas	5B
	Ferry	6B
	Franklin	4B
	Garfield	5B
	Grant	5B
	Grays Harbor	4C
	Island	5C
	Jefferson	5C
	King	4C
	Kitsap	5C
	Kittitas	5B
	Klickitat	5B
	Lewis	4C
	Lincoln	5B
	Mason	4C
	Okanogan	5B
	Pacific	4C
	Pend Oreille	6B
	Pierce	4C
	San Juan	5C
	Skagit	4C
	Skamania	5B
	Snohomish	4C
	Spokane	5B
	Stevens	6B

Table B.5 (*Continued*)

State	County	Climate Zone
	Thurston	4C
	Wahkiakum	4C
	Walla Walla	5B
	Whatcom	4C
	Whitman	5B
	Yakima	5B
West Virginia (WV)		
	Barbour	4A
	Berkeley	4A
	Boone	4A
	Braxton	4A
	Brooke	5A
	Cabell	4A
	Calhoun	4A
	Clay	4A
	Doddridge	4A
	Fayette	4A
	Gilmer	4A
	Grant	4A
	Greenbrier	5A
	Hampshire	4A
	Hancock	5A
	Hardy	4A
	Harrison	4A
	Jackson	4A
	Jefferson	4A
	Kanawha	4A
	Lewis	4A
	Lincoln	4A
	Logan	4A
	Marion	4A
	Marshall	5A
	Mason	4A
	McDowell	4A
	Mercer	4A
	Mineral	4A
	Mingo	4A
	Monongalia	4A
	Monroe	4A
	Morgan	4A
	Nicholas	4A
	Ohio	5A
	Pendleton	5A
	Pleasants	4A
	Pocahontas	5A
	Preston	4A
	Putnam	4A
	Raleigh	4A
	Randolph	5A
	Ritchie	4A
	Roane	4A
	Summers	4A
	Taylor	4A
	Tucker	5A
	Tyler	4A
	Upshur	4A
	Wayne	4A
	Webster	4A
	Wetzel	4A
	Wirt	4A
	Wood	4A
	Wyoming	4A
Wisconsin (WI)		
	Adams	6A
	Ashland	6A
	Barron	6A
	Bayfield	6A
	Brown	6A
	Buffalo	6A
	Burnett	6A

APPENDICES

Table B.5 (*Continued*)

State	County	Climate Zone
	Calumet	6A
	Chippewa	6A
	Clark	6A
	Columbia	5A
	Crawford	5A
	Dane	5A
	Dodge	5A
	Door	6A
	Douglas	6A
	Dunn	6A
	Eau Claire	6A
	Florence	6A
	Fond du Lac	5A
	Forest	6A
	Grant	5A
	Green	5A
	Green Lake	5A
	Iowa	5A
	Iron	6A
	Jackson	6A
	Jefferson	5A
	Juneau	6A
	Kenosha	5A
	Kewaunee	6A
	La Crosse	5A
	Lafayette	5A
	Langlade	6A
	Lincoln	6A
	Manitowoc	6A
	Marathon	6A
	Marinette	6A
	Marquette	5A
	Menominee	6A
	Milwaukee	5A
	Monroe	6A
	Oconto	6A
	Oneida	6A
	Outagamie	6A
	Ozaukee	5A
	Pepin	6A
	Pierce	6A
	Polk	6A
	Portage	6A
	Price	6A
	Racine	5A
	Richland	5A
	Rock	5A
	Rusk	6A
	Sauk	5A
	Sawyer	6A
	Shawano	6A
	Sheboygan	6A
	St. Croix	6A
	Taylor	6A
	Trempealeau	6A

Table B.5 (*Continued*)

State	County	Climate Zone
	Vernon	5A
	Vilas	6A
	Walworth	5A
	Washburn	6A
	Washington	6A
	Waukesha	5A
	Waupaca	6A
	Waushara	6A
	Winnebago	6A
	Wood	6A
Wyoming (WY)		
	Albany	6B
	Big Horn	6B
	Campbell	6B
	Carbon	6B
	Converse	6B
	Crook	6B
	Fremont	6B
	Goshen	5B
	Hot Springs	6B
	Johnson	6B
	Laramie	5B
	Lincoln	7
	Natrona	6B
	Niobrara	6B
	Park	6B
	Platte	5B
	Sheridan	6B
	Sublette	7
	Sweetwater	6B
	Teton	7
	Uinta	6B
	Washakie	6B
	Weston	6B

Source: © ASHRAE, www.ashrae.org. 2020. ASHRAE Standard 169-2021.

References

ASHRAE. 2020. *ASHRAE Standard 169-2021: Climatic Data for Building Design Standards*. Atlanta, GA: American Society of Heating, Refrigerating and Air-Conditioning Engineers, Inc.

ASHRAE. 2022. *ASHRAE/IES Standard 90.1-2022: Energy Standard for Sites and Buildings Except Low-Rise Residential Buildings*. Atlanta, GA: American Society of Heating, Refrigerating and Air-Conditioning Engineers, Inc.

APPENDIX C

Solar and Daylighting Design Data

This appendix contains the following information:

- Tables C.1–C.10 Solar Intensity and Solar Heat Gain Factors.

  These values are listed for five latitudes (16, 32, 40, 48, and 64°N) in both I-P and SI units (the tables are so labeled). Solar heat gain factors represent the solar radiation gain through one layer of double-strength sheet (DSA) glass. The data are based on an "average cloudless" day at the respective latitudes. Tables C.1–C.10 are copyrighted by the American Society of Heating, Refrigerating and Air-Conditioning Engineers, Inc., Atlanta, GA, and are reprinted by permission from the 1997 *ASHRAE Handbook—Fundamentals* (ASHRAE, 1997).
- Tables C.11–C.18 Solar Position and Clear-Day Global Irradiance (Insolation).

  These values are listed for four latitudes (32, 40, 48, and 64°N). The data are presented in I-P and SI units for the following tilt angles (degrees above the horizontal): latitude −10°, latitude, latitude +10°, latitude +20°, and vertical.
- Tables C.19 and C.20 Elevation, Latitude, Average Horizontal Insolation, Average Vertical Insolation, Average Temperature, and Heating Degree Days.

  These values are listed in I-P and SI units for January and July, along with yearly totals, for a number of locations (see Figure C.1).
- Tables C.21–C.26 Daylighting Coefficient of Utilization Values for Various Window and Sky Configurations.
- Table C.27 Reflectances of Materials Used to Construct Daylighting Models.

APPENDICES

Mechanical and Electrical Equipment for Buildings, Fourteenth Edition. Walter Grondzik, Daniel Overbey, and Alison Kwok.

Companion website: www.wiley.com/go/meeb14e

Table C.1 Solar Intensity and Solar Heat Gain Factors[a] for 16°N Latitude (I-P Units)[b]

			Solar Heat Gain Factors, Btu/h ft^2																	
Date	Solar Time, a.m.	Direct Normal, Btu/h ft^2	N	NNE	NE	ENE	E	ESE	SE	SSE	S	SSW	SW	WSW	W	WNW	NW	NNW	HOR	Solar Time, p.m.
Jan 21	7	141	5	6	44	92	124	134	126	96	49	6	5	5	5	5	5	5	14	5
	8	262	14	15	55	147	210	240	233	189	114	25	14	14	14	14	14	14	79	4
	9	300	21	21	32	122	200	244	251	219	152	58	22	21	21	21	21	21	150	3
	10	317	26	26	27	66	150	209	233	223	178	102	31	26	26	26	26	26	203	2
	11	325	29	29	29	31	77	148	195	210	194	146	75	31	29	29	29	29	236	1
	12	327	30	30	30	30	32	73	139	184	199	184	138	72	32	30	30	30	248	12
HALF-DAY TOTALS			110	112	196	461	760	1000	1096	1020	781	426	211	127	111	110	110	110	805	
Feb 21	7	182	8	17	84	138	169	172	150	103	36	8	8	8	8	8	8	8	25	5
	8	273	17	19	96	180	231	247	224	166	77	18	17	17	17	17	17	17	101	4
	9	305	23	24	64	153	214	242	233	188	110	30	23	23	23	23	23	23	174	3
	10	319	28	29	33	92	161	202	211	188	134	61	30	28	28	28	28	28	229	2
	11	326	32	32	32	37	83	136	167	172	149	102	49	33	32	32	32	32	263	1
	12	328	33	33	33	33	34	60	107	142	154	142	106	60	34	33	33	33	275	12
HALF-DAY TOTALS			124	137	321	609	865	1023	1034	885	582	287	174	132	124	124	124	124	930	
Mar 21	7	201	11	53	124	172	192	183	145	82	15	10	10	10	10	10	10	10	40	5
	8	272	20	50	140	205	239	235	195	123	35	19	19	19	19	19	19	19	120	4
	9	299	26	35	109	179	218	225	197	138	57	27	26	26	26	26	26	26	192	3
	10	312	31	33	61	120	165	182	172	134	76	34	32	31	31	31	31	31	247	2
	11	318	34	35	36	53	87	114	125	116	89	55	36	35	34	34	34	34	280	1
	12	320	35	35	36	36	37	47	69	87	93	86	68	47	37	38	36	35	291	12
HALF-DAY TOTALS			141	226	494	755	928	975	879	643	319	187	153	142	139	139	139	139	1025	
Apr 21	6	14	2	8	12	14	14	12	8	2	1	1	1	1	1	1	1	1	1	6
	7	197	24	94	153	187	191	167	117	45	14	13	13	13	13	13	13	13	53	5
	8	256	27	99	172	216	227	204	150	69	24	22	22	22	22	22	22	22	131	4
	9	280	31	79	149	193	208	193	147	77	31	29	29	29	29	29	29	29	197	3
	10	293	35	54	102	141	158	151	120	73	37	34	33	33	33	33	33	33	249	2
	11	299	38	40	54	72	86	88	78	60	43	38	38	36	36	36	36	37	279	1
	12	301	39	39	39	40	40	41	43	45	45	45	43	41	40	39	39	39	289	12
HALF-DAY TOTALS			179	403	674	859	922	851	653	352	174	159	157	156	155	155	155	156	1057	
May 21	6	44	14	30	41	45	43	34	19	4	3	3	3	3	3	3	3	3	5	6
	7	193	50	120	168	191	185	150	92	24	16	16	16	16	16	16	16	17	62	5
	8	244	52	132	189	218	215	179	115	38	25	24	24	24	24	24	24	25	135	4
	9	268	49	116	171	198	197	167	109	45	32	30	30	30	30	30	30	32	197	3
	10	280	47	89	130	151	150	126	84	44	37	35	35	35	35	35	35	37	245	2
	11	286	47	63	79	87	83	70	52	41	40	39	38	38	38	39	39	41	273	1
	12	288	46	46	44	43	42	41	41	41	41	41	41	41	42	43	44	46	282	12
HALF-DAY TOTALS			283	575	804	916	897	748	493	217	172	167	167	167	167	168	169	176	1058	
Jun 21	6	53	20	39	52	55	51	39	20	4	4	4	4	4	4	4	4	4	7	6
	7	188	62	128	172	190	179	141	80	20	16	16	16	16	16	16	16	18	64	5
	8	238	66	142	194	217	207	167	99	31	25	25	25	25	25	25	25	27	135	4
	9	261	63	130	178	198	190	154	93	37	31	31	31	31	31	31	31	33	194	3
	10	273	59	104	140	154	145	115	70	39	37	36	36	36	36	36	36	38	241	2
	11	279	57	76	90	92	82	63	46	41	40	39	39	39	39	40	41	43	268	1
	12	281	57	55	50	45	43	42	41	41	41	41	41	42	42	45	50	55	277	12
HALF-DAY TOTALS			356	648	850	929	876	700	430	194	174	171	171	171	172	173	176	190	1049	
Jul 21	6	41	14	29	39	42	40	31	18	4	3	3	3	3	3	3	3	3	6	6
	7	184	51	118	164	185	179	145	88	23	16	16	16	16	16	16	16	17	62	5
	8	236	55	132	187	214	210	174	111	37	25	25	25	25	25	25	25	26	133	4
	9	259	52	117	170	196	193	163	106	44	32	31	31	31	31	31	31	33	194	3
	10	272	50	92	131	151	148	123	81	44	38	36	36	36	36	36	36	38	241	2
	11	278	49	66	81	88	83	69	52	42	41	40	39	39	39	40	40	42	289	1
	12	279	49	48	46	44	43	42	42	42	42	42	42	42	43	44	46	48	277	12
HALF-DAY TOTALS			296	580	799	903	878	729	478	215	176	172	171	171	171	172	173	182	1043	
Aug 21	6	11	2	7	10	12	12	10	6	2	1	1	1	1	1	1	1	1	1	6
	7	180	26	92	145	176	180	156	109	42	15	14	14	14	14	14	14	14	53	5
	8	240	30	100	168	209	219	196	143	65	25	23	23	23	23	23	23	23	128	4
	9	266	33	82	148	190	203	187	142	74	33	30	30	30	30	30	30	30	193	3
	10	279	37	58	104	140	155	147	117	71	39	36	35	35	35	35	35	35	243	2
	11	285	40	43	57	75	86	87	76	59	44	40	39	38	38	38	38	39	273	1
	12	287	41	41	41	42	42	43	44	45	46	45	44	43	42	41	41	41	282	12
HALF-DAY TOTALS			191	410	666	837	891	817	624	339	180	167	165	164	163	163	163	164	1033	
Sep 21	7	179	12	50	114	158	176	168	133	76	15	11	11	11	11	11	11	11	39	5
	8	253	21	49	134	196	227	224	186	119	36	20	20	20	20	20	20	20	116	4
	9	281	28	36	106	173	211	217	191	134	57	28	27	27	27	27	27	27	185	3
	10	295	32	34	61	118	161	178	168	132	76	35	33	32	32	32	32	32	236	2
	11	302	35	36	37	54	86	113	123	114	88	56	38	36	35	35	35	35	271	1
	12	304	36	36	37	38	39	49	69	86	93	86	69	48	39	38	37	36	282	12
HALF-DAY TOTALS			146	226	475	722	885	931	842	622	319	192	159	148	145	144	144	144	991	
Oct 21	7	166	8	18	79	128	156	159	139	95	33	9	8	8	8	8	8	8	25	5
	8	259	17	20	95	174	223	237	215	159	74	19	17	17	17	17	17	17	99	4
	9	292	24	25	65	150	209	235	225	187	106	31	24	24	24	24	24	24	170	3
	10	307	29	30	34	92	158	197	205	183	130	60	31	29	29	29	29	29	224	2
	11	314	32	32	33	39	83	133	163	167	145	100	49	34	32	32	32	32	258	1
	12	316	33	33	33	34	35	60	105	139	150	138	104	60	35	34	33	33	270	12
HALF-DAY TOTALS			127	141	318	592	836	986	996	852	563	283	175	136	128	127	127	127	911	
Nov 21	7	134	5	6	43	89	119	129	120	92	47	6	5	5	5	5	5	5	14	5
	8	255	15	15	55	145	206	235	228	185	111	25	15	15	15	15	15	15	78	4
	9	295	21	21	33	121	197	241	247	215	150	57	22	21	21	21	21	21	149	3
	10	312	26	26	28	67	147	206	230	220	176	100	31	26	26	26	26	26	201	2
	11	320	29	29	29	31	77	146	192	207	191	144	74	31	29	29	29	29	234	1
	12	322	30	30	30	30	32	72	137	181	196	181	137	72	32	30	30	30	246	12
HALF-DAY TOTALS			112	113	197	456	749	983	1077	1001	767	420	210	128	112	112	112	112	799	
Dec 21	7	118	4	5	30	72	101	112	107	85	48	7	4	4	4	4	4	4	10	5
	8	255	13	14	41	132	198	233	231	193	124	33	13	13	13	13	13	13	69	4
	9	297	20	20	25	108	191	241	254	227	165	72	21	20	20	20	20	20	138	3
	10	315	25	25	26	56	144	208	239	233	192	117	35	25	25	25	25	25	191	2
	11	323	28	28	28	29	73	150	202	221	207	161	86	30	28	28	28	28	223	1
	12	325	29	29	29	29	30	77	149	197	212	196	149	76	30	29	29	29	234	12
HALF-DAY TOTALS			104	105	159	402	710	975	1099	1050	836	484	228	125	105	104	104	104	748	
			N	**NNW**	**NW**	**WNW**	**W**	**WSW**	**SW**	**SSW**	**S**	**SSE**	**SE**	**ESE**	**E**	**ENE**	**NE**	**NNE**	**HOR**	**PM**

[a]Total solar heat gains for DSA glass (based on a ground reflectance of 0.20).

[b]Half-day totals computed by Simpson's rule, time interval = 10 minutes.

APPENDICES

Table C.2 Solar Intensity and Solar Heat Gain Factors[a] for 32°N Latitude (I-P Units)[b]

Date	Solar Time, a.m.	Direct Normal, Btu/h ft²	Solar Heat Gain Factors, Btu/h ft²																	Solar Time, p.m.
			N	NNE	NE	ENE	E	ESE	SE	SSE	S	SSW	SW	WSW	W	WNW	NW	NNW	HOR	
Jan 21	7	141	5	6	44	92	124	134	126	96	49	6	5	5	5	5	5	5	14	5
	8	203	9	9	29	105	160	189	189	159	103	28	9	9	9	9	9	9	32	4
	9	269	15	15	17	91	175	229	246	225	169	82	17	15	15	15	15	15	88	3
	10	295	20	20	20	41	135	209	249	250	212	141	46	20	20	20	20	20	136	2
	11	306	23	23	23	24	68	159	221	249	238	191	110	29	23	23	23	23	166	1
	12	310	24	24	24	24	25	88	174	228	246	228	174	88	25	24	24	24	176	12
HALF-DAY TOTALS			79	79	107	284	570	856	1015	1014	853	553	264	112	80	79	79	79	512	
Feb 21	7	112	4	7	47	82	102	106	95	67	26	4	4	4	4	4	4	4	9	5
	8	245	13	14	65	149	205	228	216	170	95	17	13	13	13	13	13	13	64	4
	9	287	19	19	32	122	199	242	248	216	149	55	20	19	19	19	19	19	127	3
	10	305	24	24	25	62	151	213	241	232	189	112	31	24	24	24	24	24	176	2
	11	314	26	26	26	28	76	156	208	227	212	165	87	28	26	26	26	26	207	1
	12	316	27	27	27	27	29	79	155	204	221	204	155	79	29	27	27	27	217	12
HALF-DAY TOTALS			100	103	201	445	735	978	1080	1010	780	452	228	122	100	100	100	100	691	
Mar 21	7	185	10	37	105	153	176	173	142	88	20	9	9	9	9	9	9	9	32	5
	8	260	17	25	107	183	227	237	209	150	62	18	17	17	17	17	17	17	100	4
	9	290	23	25	64	151	210	237	227	183	107	30	23	23	23	23	23	23	164	3
	10	304	28	28	30	87	158	202	215	195	144	70	29	28	28	28	28	28	211	2
	11	311	31	31	31	34	82	142	179	188	168	120	59	32	31	31	31	31	242	1
	12	313	32	32	32	32	33	66	122	162	176	162	122	66	33	32	32	32	252	12
HALF-DAY TOTALS			124	162	359	629	875	1033	1041	888	589	326	193	136	125	124	124	124	874	
Apr 21	6	66	9	35	54	65	66	56	38	12	4	3	3	3	3	3	3	3	7	6
	7	206	17	80	146	188	200	182	136	65	16	14	14	14	14	14	14	14	61	5
	8	255	23	61	144	200	227	219	177	107	30	22	22	22	22	22	22	22	129	4
	9	278	28	36	103	168	206	212	187	133	58	29	28	28	28	28	28	28	188	3
	10	290	32	34	52	108	155	177	172	141	87	39	33	32	32	32	32	32	233	2
	11	295	35	35	36	47	83	118	135	132	108	70	40	36	35	35	35	35	262	1
	12	297	36	36	36	37	38	53	82	106	115	106	82	53	38	37	36	36	271	12
HALF-DAY TOTALS			161	296	550	792	952	992	889	645	360	228	177	157	153	152	152	152	1015	
May 21	6	119	33	77	108	121	116	94	56	13	8	8	8	8	8	8	8	9	21	6
	7	211	36	111	170	202	204	174	118	42	19	18	18	18	18	18	18	19	81	5
	8	250	29	94	165	208	220	199	149	73	27	25	25	25	25	25	25	25	146	4
	9	269	33	61	128	177	198	190	155	93	37	32	31	31	31	31	31	31	201	3
	10	280	36	40	76	121	150	156	138	99	54	37	35	35	35	35	35	35	243	2
	11	285	38	39	42	59	83	99	102	90	68	47	40	39	37	37	37	37	269	1
	12	286	38	39	40	40	41	47	59	70	74	70	59	47	41	40	40	39	277	12
HALF-DAY TOTALS			222	438	702	900	985	933	747	447	250	199	183	177	175	174	174	175	1098	
Jun 21	6	131	44	92	123	135	127	99	55	12	10	10	10	10	10	10	10	11	28	6
	7	210	47	122	176	204	201	168	108	35	20	20	20	20	20	20	20	21	88	5
	8	245	36	106	171	208	214	189	135	60	28	27	27	27	27	27	27	27	151	4
	9	264	35	74	137	178	193	180	139	77	35	32	32	32	32	32	32	32	204	3
	10	274	38	47	86	125	146	145	123	83	45	38	36	36	36	36	36	36	244	2
	11	279	40	41	47	64	82	91	89	75	56	43	41	40	39	39	39	39	269	1
	12	280	41	41	41	42	42	46	52	58	60	58	52	46	42	42	41	41	276	12
HALF-DAY TOTALS			261	504	762	935	985	897	678	372	225	197	189	185	184	184	183	186	1122	
Jul 21	6	113	34	76	105	117	113	90	53	12	9	9	9	9	9	9	9	9	22	6
	7	203	38	111	167	198	198	169	114	41	20	19	19	19	19	19	19	19	81	5
	8	241	31	95	163	204	215	194	145	70	28	26	26	26	26	26	26	26	145	4
	9	261	34	64	129	175	195	186	150	90	37	32	32	32	32	32	32	32	198	3
	10	271	37	42	78	121	148	153	134	96	53	38	36	36	36	36	36	36	240	2
	11	277	39	40	43	60	83	98	99	88	66	47	41	40	38	38	38	38	265	1
	12	279	40	40	41	41	42	48	58	68	72	68	58	48	42	41	41	40	273	12
HALF-DAY TOTALS			231	444	701	890	967	912	726	433	248	202	187	182	180	179	179	180	1088	
Aug 21	6	59	10	33	50	60	60	51	34	11	4	4	4	4	4	4	4	4	8	6
	7	190	19	79	141	179	190	172	128	61	17	15	15	15	15	15	15	15	61	5
	8	240	25	63	141	195	219	210	170	102	31	23	23	23	23	23	23	23	128	4
	9	263	30	39	104	166	200	206	181	127	57	31	29	29	29	29	29	29	185	3
	10	276	34	36	55	109	153	173	167	136	84	40	35	34	34	34	34	34	229	2
	11	282	36	37	39	50	84	116	131	127	104	69	41	38	36	36	36	36	256	1
	12	284	37	37	37	39	40	54	81	103	111	103	81	54	40	39	37	37	265	12
HALF-DAY TOTALS			171	303	546	774	922	955	854	618	352	231	184	166	162	161	160	160	999	
Sep 21	7	163	10	35	96	139	159	156	128	80	20	10	10	10	10	10	10	10	31	5
	8	240	18	26	103	173	215	224	198	143	60	19	18	18	18	18	18	18	96	4
	9	272	24	26	64	146	202	227	218	177	105	31	24	24	24	24	24	24	158	3
	10	287	29	29	32	86	154	196	208	189	141	70	31	29	29	29	29	29	204	2
	11	294	32	32	32	36	81	139	174	182	163	118	59	34	32	32	32	32	234	1
	12	296	33	33	33	33	35	66	120	158	171	158	120	66	35	33	33	33	244	12
HALF-DAY TOTALS			130	164	345	598	831	982	993	852	574	325	197	142	130	129	129	129	845	
Oct 21	7	99	4	7	43	74	92	96	85	60	24	5	4	4	4	4	4	4	10	5
	8	229	13	15	63	143	195	217	206	162	90	17	13	13	13	13	13	13	63	4
	9	273	20	20	33	120	193	234	239	208	144	54	21	20	20	20	20	20	125	3
	10	293	24	24	26	62	147	207	234	225	183	109	32	24	24	24	24	24	173	2
	11	302	27	27	27	29	76	152	203	221	207	160	85	29	27	27	27	27	203	1
	12	304	28	28	28	28	30	78	151	199	215	199	151	78	30	28	28	28	213	12
HALF-DAY TOTALS			103	106	200	433	708	941	1038	972	753	441	226	125	104	103	103	103	679	
Nov 21	7	2	0	0	0	1	1	1	1	1	1	0	0	0	0	0	0	0	0	5
	8	196	9	9	29	103	156	184	184	155	100	27	9	9	9	9	9	9	32	4
	9	263	16	16	17	90	173	225	241	221	166	80	17	16	16	16	16	16	88	3
	10	289	20	20	21	41	134	206	245	246	209	138	45	21	20	20	20	20	136	2
	11	301	23	23	23	24	67	157	218	245	234	188	109	29	23	23	23	23	165	1
	12	304	24	24	24	24	25	87	171	224	243	224	171	87	25	24	24	24	175	12
HALF-DAY TOTALS			80	81	108	282	561	841	996	995	838	544	261	113	81	80	80	80	509	
Dec 21	8	176	7	7	19	84	135	163	166	143	97	31	7	7	7	7	7	7	22	4
	9	257	14	14	15	77	162	218	238	222	171	89	15	14	14	14	14	14	72	3
	10	288	18	18	18	34	127	204	246	251	216	148	52	19	18	18	18	18	119	2
	11	301	21	21	21	22	63	157	222	252	243	197	116	29	21	21	21	21	148	1
	12	304	22	22	22	22	23	89	177	232	252	232	177	89	23	22	22	22	158	12
HALF-DAY TOTALS			71	71	84	227	500	792	965	986	852	578	275	107	71	71	71	71	440	
			N	**NNW**	**NW**	**WNW**	**W**	**WSW**	**SW**	**SSW**	**S**	**SSE**	**SE**	**ESE**	**E**	**ENE**	**NE**	**NNE**	**HOR**	**p.m.**

[a] Total solar heat gains for DSA glass (based on a ground reflectance of 0.20).

[b] Half-day totals computed by Simpson's rule, time interval = 10 minutes.

APPENDICES

Table C.3 Solar Intensity and Solar Heat Gain Factors[a] for 40°N Latitude (I-P Units)[b]

Date	Solar Time, a.m.	Direct Normal, Btu/h ft²	Solar Heat Gain Factors, Btu/h ft²																	Solar Time, p.m.
			N	NNE	NE	ENE	E	ESE	SE	SSE	S	SSW	SW	WSW	W	WNW	NW	NNW	HOR	
Jan 21	7	141	5	6	44	92	124	134	126	96	49	6	5	5	5	5	5	5	14	5
	9	239	12	12	13	74	154	205	224	209	160	82	13	12	12	12	12	12	55	3
	10	274	16	16	16	31	124	199	241	246	213	146	51	17	16	16	16	16	96	2
	11	289	19	19	19	20	61	156	222	252	244	198	118	28	19	19	19	19	124	1
	12	294	20	20	20	20	21	90	179	234	254	234	179	90	21	20	20	20	133	12
HALF-DAY TOTALS			61	61	73	199	452	734	904	932	813	561	273	101	62	61	61	61	354	
Feb 21	7	55	2	3	23	40	51	53	47	34	14	2	2	2	2	2	2	2	4	5
	8	219	10	11	50	129	183	206	199	160	94	18	10	10	10	10	10	10	43	4
	9	271	16	16	22	107	186	234	245	218	157	66	17	16	16	16	16	16	98	3
	10	294	21	21	21	49	143	211	246	243	203	129	38	21	21	21	21	21	143	2
	11	304	23	23	23	24	71	160	219	244	231	184	103	27	23	23	23	23	171	1
	12	307	24	24	24	24	25	86	170	222	241	222	170	86	25	24	24	24	180	12
HALF-DAY TOTALS			84	86	152	361	648	916	1049	1015	821	508	250	114	85	84	84	84	548	
Mar 21	7	171	9	29	93	140	163	161	135	86	22	8	8	8	8	8	8	8	26	5
	8	250	16	18	91	169	218	232	211	157	74	17	16	16	16	16	16	16	85	4
	9	282	21	22	47	136	203	238	236	198	128	40	22	21	21	21	21	21	143	3
	10	297	25	25	27	72	153	207	229	216	171	95	29	25	25	25	25	25	186	2
	11	305	28	28	28	30	78	151	198	213	197	150	77	30	28	28	28	28	213	1
	12	307	29	29	29	29	31	75	145	191	206	191	145	75	31	29	29	29	223	12
HALF-DAY TOTALS			114	139	302	563	832	1035	1087	968	694	403	220	132	114	113	113	113	764	
Apr 21	6	89	11	46	72	87	88	76	52	18	5	5	5	5	5	5	5	5	11	6
	7	206	16	71	140	185	201	186	143	75	16	14	14	14	14	14	14	14	61	5
	8	252	22	44	128	190	224	223	188	124	41	22	21	21	21	21	21	21	123	4
	9	274	27	29	80	155	202	219	203	156	83	29	27	27	27	27	27	27	177	3
	10	286	31	31	37	92	152	187	193	170	121	56	32	31	31	31	31	41	217	2
	11	292	33	33	34	39	81	130	160	166	146	102	52	35	33	33	33	33	243	1
	12	293	34	34	34	34	36	62	108	142	154	142	108	62	36	34	34	34	252	12
HALF-DAY TOTALS			154	265	501	758	957	1051	994	782	488	296	199	157	148	147	147	147	957	
May 21	5	1	0	1	1	1	1	1	0	0	0	0	0	0	0	0	0	0	0	7
	6	144	36	90	128	145	141	115	71	18	10	10	10	10	10	10	10	11	31	6
	7	216	28	102	165	202	209	184	131	54	20	19	19	19	19	19	19	19	87	5
	8	250	27	73	149	199	220	208	164	93	29	25	25	25	25	25	25	25	146	4
	9	267	31	42	105	164	197	200	175	121	53	32	30	30	30	30	30	30	195	3
	10	277	34	36	54	105	148	168	163	133	83	40	35	34	34	34	34	34	234	2
	11	283	36	36	38	48	81	113	130	127	105	70	42	38	36	36	36	36	257	1
	12	284	37	37	37	38	40	54	82	114	113	104	82	54	40	38	37	37	265	12
HALF-DAY TOTALS			215	404	666	893	1024	1025	881	601	358	247	200	180	176	175	174	175	1083	
Jun 21	5	22	10	17	21	22	20	14	6	2	1	1	1	1	1	1	1	2	3	7
	6	155	48	104	143	159	151	121	70	17	13	13	13	13	13	13	13	14	40	6
	7	216	37	113	172	205	207	178	122	46	22	21	21	21	21	21	21	21	97	5
	8	246	30	85	156	201	216	199	152	80	29	27	27	27	27	27	27	27	153	4
	9	263	33	51	114	166	192	190	161	105	45	33	32	32	32	32	32	32	201	3
	10	272	35	38	63	109	145	158	148	116	69	39	36	35	35	35	35	35	238	2
	11	277	38	39	40	52	81	105	116	110	88	60	41	39	38	38	38	38	260	1
	12	279	38	38	38	40	41	52	72	89	95	89	72	52	41	40	38	38	267	12
HALF-DAY TOTALS			253	470	734	941	1038	999	818	523	315	236	204	191	188	187	186	188	1126	
Jul 21	5	2	1	2	2	2	2	1	1	0	0	0	0	0	0	0	0	0	0	7
	6	138	37	89	125	142	137	112	68	18	11	11	11	11	11	11	11	12	32	6
	7	208	30	102	163	198	204	179	127	53	21	20	20	20	20	20	20	20	88	5
	8	241	28	75	148	196	216	203	160	90	30	26	26	26	26	26	26	26	145	4
	9	259	32	44	106	163	193	196	170	118	52	33	31	31	31	31	31	31	194	3
	10	269	35	37	56	106	146	165	159	129	81	41	36	35	35	35	35	35	231	2
	11	275	37	38	40	50	81	111	127	123	102	69	43	39	37	37	37	37	254	1
	12	276	38	38	38	40	41	55	80	101	109	101	80	55	41	40	38	38	262	12
HALF-DAY TOTALS			223	411	666	885	1008	1003	858	584	352	248	204	186	181	180	180	181	1076	
Aug 21	6	81	12	44	68	81	82	71	48	17	6	5	5	5	5	5	5	5	12	6
	7	191	17	71	135	177	191	177	135	70	17	16	16	16	16	16	16	16	62	5
	8	237	24	47	126	185	216	214	180	118	41	23	23	23	23	23	23	23	122	4
	9	260	28	31	82	153	197	212	196	151	80	31	28	28	28	28	28	28	174	3
	10	272	32	33	40	93	150	182	187	165	116	56	34	32	32	32	32	32	214	2
	11	278	35	35	36	41	81	128	156	160	141	99	52	37	35	35	35	35	239	1
	12	280	35	35	35	36	38	63	106	138	149	138	106	63	38	36	35	35	247	12
HALF-DAY TOTALS			164	273	498	741	928	1013	956	751	474	296	205	166	157	156	156	156	946	
Sep 21	7	149	9	27	84	125	146	144	121	77	21	9	9	9	9	9	9	9	25	5
	8	230	17	19	87	160	205	218	199	148	71	18	17	17	17	17	17	17	82	4
	9	263	22	23	47	131	194	227	226	190	124	41	23	22	22	22	22	22	138	3
	10	280	27	27	28	71	148	200	221	209	165	93	30	27	27	27	27	27	180	2
	11	287	29	29	29	31	78	147	192	207	191	146	77	31	29	29	29	29	206	1
	12	290	30	30	30	30	32	75	142	185	200	185	142	75	32	30	30	30	215	12
HALF-DAY TOTALS			119	142	291	534	787	980	1033	925	672	396	222	137	119	118	118	118	738	
Oct 21	7	48	2	3	20	36	45	47	42	30	12	2	2	2	2	2	2	2	4	5
	8	204	11	12	49	123	173	195	188	151	89	18	11	11	11	11	11	11	43	4
	9	257	17	17	23	104	180	225	235	209	151	64	18	17	17	17	17	17	97	3
	10	280	21	21	22	50	139	205	238	235	196	125	38	22	21	21	21	21	140	2
	11	291	24	24	24	25	71	156	212	236	224	178	101	28	24	24	24	24	168	1
	12	294	25	25	25	25	27	85	165	216	234	216	165	85	27	25	25	25	177	12
HALF-DAY TOTALS			88	89	152	351	623	878	1006	974	791	493	247	117	89	88	88	88	540	
Nov 21	8	136	5	5	18	69	108	128	129	110	72	21	6	5	5	5	5	5	14	4
	9	232	12	12	13	73	151	201	219	204	156	80	13	12	12	12	12	12	55	3
	10	268	16	16	16	31	122	196	237	242	209	143	50	17	16	16	16	18	96	2
	11	283	19	19	19	20	61	154	218	248	240	194	116	28	19	19	19	19	123	1
	12	288	20	20	20	20	21	89	176	231	250	231	176	89	21	20	20	20	132	12
HALF-DAY TOTALS			63	63	75	198	445	721	887	914	798	551	269	101	63	63	63	63	354	
Dec 21	8	89	3	3	8	41	67	82	84	73	50	17	3	3	3	3	3	3	6	4
	9	217	10	10	11	60	135	185	205	194	151	83	13	10	10	10	10	10	39	3
	10	261	14	14	14	25	113	188	232	239	210	146	55	15	14	14	14	14	77	2
	11	280	17	17	17	17	56	151	217	249	242	198	120	28	17	17	17	17	104	1
	12	285	18	18	18	18	19	89	178	233	253	233	178	89	19	18	18	18	113	12
HALF-DAY TOTALS			52	52	56	146	374	649	822	867	775	557	276	94	53	52	52	52	282	
			N	NNW	NW	WNW	W	WSW	SW	SSW	S	SSE	SE	ESE	E	ENE	NE	NNE	HOR	p.m.

[a]Total solar heat gains for DSA glass (based on a ground reflectance of 0.20).

[b]Half-day totals computed by Simpson's rule, time interval = 10 minutes.

Table C.4 Solar Intensity and Solar Heat Gain Factors[a] for 48°N Latitude (I-P Units)[b]

Date	Solar Time, a.m.	Direct Normal, Btu/h ft²	Solar Heat Gain Factors, Btu/h ft²																	Solar Time, p.m.
			N	NNE	NE	ENE	E	ESE	SE	SSE	S	SSW	SW	WSW	W	WNW	NW	NNW	HOR	
Jan 21	7	141	5	6	44	92	124	134	126	96	49	6	5	5	5	5	5	5	14	5
	9	185	8	8	8	53	118	160	176	166	129	69	10	8	8	8	8	8	25	3
	10	239	12	12	12	22	106	175	216	223	195	136	50	12	12	12	12	12	55	2
	11	261	14	14	14	15	53	144	208	239	233	190	116	26	14	14	14	14	77	1
	12	267	15	15	15	15	16	86	171	226	245	226	171	86	16	15	15	15	85	12
HALF-DAY TOTALS			43	43	46	117	316	567	729	776	701	512	259	85	43	43	43	43	203	
Feb 21	7	4	0	0	1	3	3	3	3	2	1	0	0	0	0	0	0	0	0	5
	8	180	8	8	36	103	149	170	166	136	82	17	8	8	8	8	8	8	25	4
	9	247	13	13	16	90	168	216	230	209	155	71	14	13	13	13	13	13	66	3
	10	275	17	17	17	38	131	203	242	244	207	138	44	18	17	17	17	17	105	2
	11	288	19	19	19	20	65	158	221	249	239	192	113	27	19	19	19	19	130	1
	12	292	20	20	20	20	22	89	176	231	250	231	176	89	22	20	20	20	138	12
HALF-DAY TOTALS			68	68	107	274	541	816	968	967	813	531	261	104	68	68	68	68	395	
Mar 21	7	153	7	22	80	123	145	145	123	80	23	7	7	7	7	7	7	7	20	5
	8	236	14	15	76	154	204	222	206	158	82	15	14	14	14	14	14	14	68	4
	9	270	19	19	3	121	193	234	239	207	142	52	20	19	19	19	19	19	118	3
	10	287	23	23	24	58	146	208	237	231	189	115	33	23	23	23	23	23	156	2
	11	295	25	25	25	26	74	156	210	232	218	172	94	28	25	25	25	25	180	1
	12	298	26	26	26	26	27	83	161	211	228	211	161	83	27	26	26	26	188	12
HALF-DAY TOTALS			100	118	250	494	775	1012	1100	1014	767	465	244	126	101	100	100	100	636	
Apr 21	6	108	12	53	86	105	107	93	64	23	6	6	6	6	6	6	6	6	15	6
	7	205	15	61	132	180	199	189	148	84	18	14	14	14	14	14	14	14	60	5
	8	247	20	32	111	179	219	225	196	138	55	21	20	20	20	20	20	20	114	4
	9	268	25	26	60	141	197	223	215	176	106	33	25	25	25	25	25	25	161	3
	10	280	28	28	31	77	148	193	209	194	150	80	31	28	28	28	28	28	196	2
	11	286	31	31	31	33	78	140	181	193	177	133	69	33	31	31	31	31	218	1
	12	288	31	31	31	31	34	71	131	172	186	172	131	71	34	31	31	31	226	12
HALF-DAY TOTALS			147	242	461	724	957	1098	1081	895	605	370	226	156	141	140	140	140	875	
May 21	5	41	17	31	40	42	39	29	14	3	3	3	3	3	3	3	3	3	5	7
	6	162	35	97	141	162	160	133	85	24	12	12	12	12	12	12	12	13	40	6
	7	219	23	90	158	200	212	191	142	68	21	19	19	19	19	19	19	19	91	5
	8	248	26	54	132	190	218	214	178	113	38	25	25	25	25	25	25	25	142	4
	9	264	29	32	82	151	194	208	192	147	77	32	29	29	29	29	29	29	185	3
	10	274	33	34	39	90	145	178	184	163	116	57	35	33	33	33	33	33	219	2
	11	279	35	35	36	40	79	126	155	160	142	101	54	37	35	35	35	35	240	1
	12	280	35	35	35	36	38	63	107	139	150	139	107	63	38	36	35	35	247	12
HALF-DAY TOTALS			215	388	645	893	1065	1114	1007	749	483	316	225	184	174	173	173	174	1045	
Jun 21	5	77	35	61	76	80	72	53	24	6	5	5	5	5	5	5	5	8	12	7
	6	172	46	110	155	175	169	138	84	22	14	14	14	14	14	14	14	16	51	6
	7	220	29	101	165	204	211	187	135	60	23	21	21	21	21	21	21	21	103	5
	8	246	29	64	139	191	215	206	168	101	34	27	27	27	27	27	27	27	152	4
	9	261	31	36	91	153	190	199	180	133	66	33	31	31	31	31	31	31	193	3
	10	269	34	36	45	94	143	169	171	148	101	50	36	34	34	34	34	34	225	2
	11	274	36	36	38	44	79	118	142	145	126	88	49	38	36	36	36	36	246	1
	12	275	37	37	37	38	40	60	96	124	134	124	96	60	40	38	37	37	252	12
HALF-DAY TOTALS			257	459	722	955	1095	1102	955	678	436	299	228	197	189	188	188	191	1108	
Jul 21	5	43	18	33	42	45	41	30	15	3	3	3	3	3	3	3	3	4	6	7
	6	156	37	96	138	159	156	129	82	24	13	13	13	13	13	13	13	14	41	6
	7	211	25	90	156	196	207	186	138	66	22	20	20	20	20	20	20	20	92	5
	8	240	27	56	132	187	214	209	174	110	38	26	26	26	26	26	26	26	142	4
	9	256	30	34	83	149	191	204	187	143	75	33	30	30	30	30	30	30	184	3
	10	266	34	35	41	90	143	174	180	158	113	56	36	34	34	34	34	34	217	2
	11	271	36	36	37	42	79	124	151	156	138	99	54	38	36	36	36	36	237	1
	12	272	36	36	36	37	39	63	104	136	146	136	104	63	39	37	36	36	244	12
HALF-DAY TOTALS			223	395	646	886	1050	1092	983	730	474	315	229	190	181	179	179	180	1042	
Aug 21	6	99	13	51	81	98	100	87	60	22	7	7	7	7	7	7	7	7	16	6
	7	190	17	61	128	172	190	179	141	79	19	15	15	15	15	15	15	15	61	5
	8	232	22	34	110	174	211	216	188	132	53	23	22	22	22	22	22	22	114	4
	9	154	27	28	63	139	192	216	108	169	102	34	27	27	27	27	27	27	159	3
	10	266	30	30	33	78	145	188	203	188	144	78	33	30	30	30	30	30	193	2
	11	272	32	32	32	36	78	137	175	187	171	129	68	35	32	32	32	32	215	1
	12	274	33	33	33	33	36	71	128	167	189	167	128	71	36	33	33	33	223	12
HALF-DAY TOTALS			157	251	459	709	929	1060	1040	862	587	366	231	165	151	149	149	149	869	
Sep 21	7	131	8	21	71	108	128	128	108	71	21	8	7	7	7	7	7	7	20	5
	8	215	15	16	72	144	191	207	193	148	77	16	15	15	15	15	15	15	65	4
	9	251	20	20	34	116	184	223	227	197	136	52	21	20	20	20	20	20	114	3
	10	269	24	24	25	58	141	200	228	221	182	112	34	24	24	24	24	24	151	2
	11	278	26	26	26	28	73	151	203	223	210	166	92	29	26	26	26	26	174	1
	12	280	27	27	27	27	29	82	156	204	220	204	156	82	29	27	27	27	182	12
HALF-DAY TOTALS			105	121	240	465	729	953	1040	963	737	453	243	131	106	105	105	105	614	
Oct 21	7	4	0	0	2	3	4	4	3	2	1	0	0	0	0	0	0	0	0	5
	8	165	8	9	35	96	139	159	155	126	77	16	8	8	8	8	8	8	25	4
	9	233	14	14	16	88	161	207	220	199	148	68	15	14	14	14	14	14	66	3
	10	262	18	18	18	39	128	196	233	234	199	133	43	18	18	18	18	18	104	2
	11	274	20	20	20	21	64	153	213	241	231	186	109	27	20	20	20	20	128	1
	12	278	21	21	21	21	23	87	171	223	242	223	171	87	23	21	21	21	136	12
HALF-DAY TOTALS			71	71	108	266	519	780	925	925	779	513	256	106	72	71	71	71	391	
Nov 21	8	36	1	1	4	18	29	34	35	30	20	6	1	1	1	1	1	1	2	4
	9	179	8	8	9	52	115	156	171	161	125	67	10	8	8	8	8	8	26	3
	10	233	12	12	12	22	104	172	212	218	191	133	49	13	12	12	12	12	55	2
	11	255	15	15	15	15	52	142	204	234	228	186	114	26	15	15	15	15	77	1
	12	261	15	15	15	15	17	85	168	222	240	222	168	85	17	15	15	15	85	12
HALF-DAY TOTALS			44	44	47	117	310	555	713	760	686	502	255	85	44	44	44	44	204	
Dec 21	9	140	5	5	6	36	86	120	133	127	100	56	8	5	5	5	5	5	13	3
	10	214	10	10	10	16	91	156	194	201	179	126	49	10	10	10	10	10	38	2
	11	242	12	12	12	13	46	134	195	225	220	180	111	25	12	12	12	12	57	1
	12	250	13	13	13	13	14	81	163	215	233	215	168	81	14	13	13	13	65	12
HALF-DAY TOTALS			33	33	34	73	233	458	610	665	616	468	247	76	34	33	33	33	141	
			N	NNW	NW	WNW	W	WSW	SW	SSW	S	SSE	SE	ESE	E	ENE	NE	NNE	HOR	p.m.

[a]Total solar heat gains for DSA glass (based on a ground reflectance of 0.20).

[b]Half-day totals computed by Simpson's rule, time interval = 10 minutes.

Table C.5 Solar Intensity and Solar Heat Gain Factors[a] for 64°N Latitude (I-P Units)[b]

Date	Solar Time, a.m.	Direct Normal, Btu/h ft²	Solar Heat Gain Factors, Btu/h ft²																	Solar Time, p.m.
			N	NNE	NE	ENE	E	ESE	SE	SSE	S	SSW	SW	WSW	W	WNW	NW	NNW	HOR	
Jan 21	7	141	5	6	44	92	124	134	126	96	49	6	5	5	5	5	5	5	14	5
	11	81	3	3	3	3	15	45	67	77	75	62	38	8	3	3	3	3	6	1
	12	100	3	3	3	3	4	33	67	89	96	89	67	33	4	3	3	3	8	12
HALF-DAY TOTALS			5	5	5	6	25	79	121	142	141	119	75	23	5	5	5	5	11	
Feb 21	8	18	1	1	3	10	15	17	17	14	9	2	1	1	1	1	1	1	1	4
	9	134	5	5	6	43	89	118	128	119	90	45	6	5	5	5	5	5	13	3
	10	190	8	8	8	18	87	144	176	180	157	108	38	9	8	8	8	8	28	2
	11	215	10	10	10	11	44	122	177	202	197	160	97	20	10	10	10	10	41	1
	12	222	11	11	11	11	12	73	147	194	210	194	147	73	12	11	11	11	45	12
HALF-DAY TOTALS			29	30	33	89	244	446	578	617	560	411	212	66	30	29	29	29	106	
Mar 21	7	95	4	11	47	74	90	91	79	53	17	4	4	4	4	4	4	4	9	5
	8	185	9	10	46	113	158	177	170	135	78	14	9	9	9	9	9	9	32	4
	9	227	13	13	16	88	159	203	215	194	143	64	14	13	13	13	13	13	59	3
	10	249	16	16	16	35	122	190	226	228	194	130	42	16	16	16	16	16	84	2
	11	260	17	17	17	18	60	148	209	236	228	184	109	25	17	17	17	17	99	1
	12	263	18	18	18	18	19	85	168	221	239	221	168	85	19	18	18	18	105	12
HALF-DAY TOTALS			68	74	150	334	596	854	984	958	779	504	257	104	68	68	68	68	335	
Apr 21	5	27	8	18	24	27	26	20	12	2	1	1	1	1	1	1	1	1	2	7
	6	133	12	59	102	127	132	118	84	35	8	8	8	8	8	8	8	8	21	6
	7	194	14	41	113	163	189	185	153	96	25	13	13	13	13	13	13	13	51	5
	8	228	17	19	79	153	201	217	201	153	79	19	17	17	17	17	17	17	85	4
	9	248	21	21	32	111	180	219	225	197	138	55	22	21	21	21	21	21	116	3
	10	260	23	23	24	51	134	194	225	221	185	118	38	24	23	23	23	23	140	2
	11	266	24	24	24	26	68	148	202	225	214	171	99	29	24	24	24	24	155	1
	12	268	25	25	25	25	27	83	159	208	224	208	159	83	27	25	25	25	160	12
HALF-DAY TOTALS			131	218	410	671	943	1150	1186	1036	763	487	273	149	121	120	120	120	651	
May 21	4	51	30	44	51	51	43	28	8	3	3	3	3	3	3	3	3	10	6	8
	5	132	48	95	125	135	125	96	50	11	9	9	9	9	9	9	9	11	26	7
	6	185	28	97	150	181	183	158	109	40	15	15	15	15	15	15	15	15	55	6
	7	218	21	63	138	189	211	201	161	94	24	19	19	19	19	19	19	19	90	5
	8	239	23	28	97	167	209	220	198	146	68	25	23	23	23	23	23	23	124	4
	9	252	26	27	45	122	183	215	215	184	123	46	27	26	26	26	26	26	152	3
	10	261	28	28	30	61	135	188	212	205	167	102	36	28	28	28	28	28	174	2
	11	265	30	30	30	32	72	141	188	207	195	154	87	33	30	30	30	30	188	1
	12	267	30	30	30	30	33	78	146	189	204	189	146	78	33	30	30	30	192	12
HALF-DAY TOTALS			247	425	680	950	1177	1291	1218	985	708	465	288	191	169	168	168	176	911	
Jun 21	4	93	53	83	96	94	78	50	14	7	7	7	7	7	7	7	7	21	16	8
	5	154	62	114	148	158	145	110	55	14	12	12	12	12	12	12	12	14	39	7
	6	194	36	107	162	191	192	163	110	39	18	17	17	17	17	17	17	18	71	6
	7	221	24	71	145	193	213	200	158	89	25	22	22	22	22	22	22	22	105	5
	8	239	25	33	104	170	208	216	192	139	62	27	25	25	25	25	25	25	137	4
	9	251	28	29	51	124	181	210	208	175	115	43	29	28	28	28	28	28	165	3
	10	258	30	30	32	65	134	183	204	195	157	94	36	30	30	30	30	30	186	2
	11	262	32	32	32	34	72	137	180	196	184	144	82	35	32	32	32	32	199	1
	12	263	32	32	32	32	35	76	138	179	193	179	138	76	35	32	32	32	203	12
HALF-DAY TOTALS			322	533	801	1061	1253	1317	1195	946	679	455	296	211	192	190	191	213	1021	
Jul 21	4	53	32	47	55	54	46	29	9	4	4	4	4	4	4	4	4	11	8	8
	5	128	49	94	123	133	124	95	50	11	10	10	10	10	10	10	10	11	28	7
	6	179	30	96	148	177	180	155	106	39	16	15	15	15	15	15	15	15	57	6
	7	211	22	64	137	186	207	197	157	92	25	20	20	20	20	20	20	20	92	5
	8	231	24	30	97	165	205	215	193	142	67	26	24	24	24	24	24	24	124	4
	9	245	27	28	47	121	180	211	211	179	120	46	28	27	27	27	27	27	152	3
	10	253	29	29	31	62	134	185	208	200	164	100	37	29	29	29	29	29	174	2
	11	257	31	31	31	33	72	139	185	202	191	151	86	34	31	31	31	31	187	1
	12	259	31	31	31	31	34	78	143	185	200	185	143	78	34	31	31	31	192	12
HALF-DAY TOTALS			258	434	684	946	1163	1269	1193	965	697	462	292	198	177	175	175	185	918	
Aug 21	5	29	9	20	27	30	28	22	13	2	2	2	2	2	2	2	2	2	3	7
	6	123	13	58	97	121	125	111	80	34	9	9	9	9	9	9	9	9	23	6
	7	181	15	42	109	157	180	176	145	92	26	14	14	14	14	14	14	14	53	5
	8	214	19	21	78	148	193	208	192	147	76	21	19	19	19	19	19	19	87	4
	9	234	22	22	34	109	174	211	217	189	133	55	23	22	22	22	22	22	117	3
	10	246	25	25	26	52	131	188	217	214	178	114	39	25	25	25	25	25	140	2
	11	252	26	26	26	28	69	144	196	217	207	166	97	31	26	26	26	26	154	1
	12	254	27	27	27	27	29	82	155	201	217	201	155	82	29	27	27	27	159	12
HALF-DAY TOTALS			142	226	410	657	914	1109	1141	997	740	478	275	158	131	130	130	130	656	
Sep 21	7	77	4	10	39	62	74	75	65	44	15	4	4	4	4	4	4	4	8	5
	8	163	10	10	43	103	143	160	154	123	71	14	10	10	10	10	10	10	31	4
	9	206	14	14	17	83	148	189	200	181	133	61	15	14	14	14	14	14	57	3
	10	229	16	16	17	35	116	179	213	214	183	123	41	17	16	16	16	16	81	2
	11	240	18	18	18	19	59	141	198	224	216	174	104	26	18	18	18	18	96	1
	12	244	19	19	19	19	21	82	160	209	227	209	160	82	21	19	19	19	101	12
HALF-DAY TOTALS			71	77	142	307	547	787	910	891	731	480	249	106	72	71	71	71	324	
Oct 21	8	17	1	1	3	10	14	16	16	13	8	2	1	1	1	1	1	1	1	4
	9	122	5	5	6	40	82	109	118	110	83	42	6	5	5	5	5	5	13	3
	10	176	9	9	9	18	83	135	165	169	147	102	36	9	9	9	9	9	29	2
	11	201	11	11	11	11	43	116	167	191	186	152	92	20	11	11	11	11	41	1
	12	208	11	11	11	11	13	70	140	184	199	184	140	70	13	11	11	11	46	12
HALF-DAY TOTALS			31	31	34	86	231	420	542	580	527	388	202	66	32	31	31	31	108	
Nov 21	10	23	1	1	1	1	10	17	21	22	20	14	5	1	1	1	1	1	1	2
	11	79	3	3	3	3	15	44	65	75	74	61	37	8	3	3	3	3	6	1
	12	97	4	4	4	4	4	32	66	87	93	87	66	32	4	4	4	4	8	12
HALF-DAY TOTALS			5	5	5	6	26	79	120	141	140	117	74	23	6	5	5	5	11	
Dec 21	11	4	0	0	0	0	1	2	3	4	4	3	2	0	0	0	0	0	0	1
	12	16	0	0	0	0	1	5	11	14	15	14	11	5	1	0	0	0	1	12
HALF-DAY TOTALS			0	0	0	0	1	5	9	11	11	10	7	3	0	0	0	0	1	
			N	NNW	NW	WNW	W	WSW	SW	SSW	S	SSE	SE	ESE	E	ENE	NE	NNE	HOR	p.m.

[a]Total solar heat gains for DSA glass (based on a ground reflectance of 0.20).

[b]Half-day totals computed by Simpson's rule, time interval = 10 minutes.

Table C.6 Solar Intensity and Solar Heat Gain Factors[a] for 16°N Latitude (SI Units)[b]

Date	Solar Time, a.m.	Direct Normal, W/m^2	Solar Heat Gain Factors, W/m^2 N	NNE	NE	ENE	E	ESE	SE	SSE	S	SSW	SW	WSW	W	WNW	NW	NNW	HOR	Solar Time, p.m.
Jan 21	7	445	17	19	138	291	390	424	397	303	155	19	17	17	17	17	17	17	43	5
	8	827	45	48	174	463	662	757	734	596	359	79	45	45	45	45	45	45	249	4
	9	948	67	67	102	384	630	770	791	690	481	183	69	67	67	67	67	67	472	3
	10	1001	82	82	86	209	474	658	737	704	563	321	97	82	82	82	82	82	640	2
	11	1025	92	92	92	96	242	467	614	663	612	462	236	96	92	92	92	92	745	1
	12	1032	95	95	95	95	100	228	438	580	628	580	438	228	100	95	95	95	782	12
HALF-DAY TOTALS			348	352	618	1453	2398	3153	3458	3217	2465	1344	666	401	350	348	348	348	2539	
Feb 21	7	575	24	55	265	435	532	544	474	326	113	26	24	24	24	24	24	24	80	5
	8	862	53	60	304	567	729	778	706	525	244	56	53	53	53	53	53	53	319	4
	9	961	74	77	202	482	676	763	733	592	347	96	74	74	74	74	74	74	549	3
	10	1006	90	91	104	292	508	636	665	593	423	193	94	90	90	90	90	90	722	2
	11	1027	99	99	102	118	262	428	527	542	471	323	154	103	99	99	99	99	831	1
	12	1034	103	103	103	105	108	189	336	448	487	448	336	189	108	105	103	103	868	12
HALF-DAY TOTALS			390	431	1013	1922	2730	3228	3263	2792	1836	906	547	417	393	390	390	390	2933	
Mar 21	7	634	36	167	393	544	606	578	458	260	47	33	33	33	33	33	33	33	126	5
	8	857	63	157	442	648	752	741	615	390	111	61	61	61	61	61	61	61	380	4
	9	943	84	110	343	565	689	709	622	435	180	86	82	82	82	82	82	82	606	3
	10	983	98	103	191	379	519	575	543	424	240	107	100	98	98	98	98	98	778	2
	11	1003	108	110	113	166	273	361	395	366	281	173	114	110	108	108	108	108	885	1
	12	1008	111	111	112	114	117	149	216	273	295	273	216	149	117	114	112	111	919	12
HALF-DAY TOTALS			443	712	1558	2383	2926	3076	2773	2028	1006	588	483	448	440	483	437	437	3233	443
Apr 21	6	44	7	24	37	43	43	37	24	7	2	2	2	2	2	2	12	2	4	6
	7	622	75	298	482	589	604	528	369	141	45	42	42	42	42	42	42	42	169	5
	8	807	85	312	543	682	718	644	473	217	74	70	70	70	70	70	70	70	413	4
	9	885	97	248	469	610	657	608	465	244	97	90	90	90	90	90	90	90	623	3
	10	924	112	171	321	444	499	476	380	231	118	109	166	106	106	106	106	106	784	2
	11	942	120	127	169	228	270	276	245	189	136	121	118	115	115	115	115	118	882	1
	12	948	123	123	124	125	126	129	135	141	143	141	135	129	126	125	124	123	911	12
HALF-DAY TOTALS			565	1272	2127	2711	2909	2684	2059	1111	547	503	494	491	490	489	490	492	3333	
May 21	6	138	43	94	128	141	134	106	59	11	9	9	9	9	9	9	9	10	195	6
	7	608	157	378	531	603	583	474	290	76	49	49	49	49	49	49	49	55	17	5
	8	771	165	415	598	689	677	564	362	121	78	76	76	76	76	76	76	80	425	4
	9	845	156	366	539	626	622	526	344	141	100	96	96	96	96	96	96	100	621	3
	10	883	149	281	410	478	474	398	264	139	116	111	111	111	111	111	111	116	772	2
	11	902	147	198	248	273	262	220	166	130	126	123	120	120	120	122	124	128	862	1
	12	907	146	144	140	134	132	131	130	129	129	129	130	131	132	134	140	144	890	12
HALF-DAY TOTALS			893	1813	2537	2891	2829	2360	1555	685	541	528	527	526	527	529	533	557	3338	
Jun 21	6	168	64	124	163	175	162	123	64	13	12	12	12	12	12	12	12	13	24	6
	7	593	195	404	543	598	565	445	252	63	52	52	52	52	52	52	52	57	203	5
	8	750	209	449	612	684	653	526	313	98	79	79	79	79	79	79	79	84	425	4
	9	823	199	409	560	626	601	487	294	118	98	98	98	98	98	98	98	105	613	3
	10	861	187	329	441	486	459	363	222	125	116	113	113	113	113	113	113	121	759	2
	11	879	181	241	283	290	258	200	146	130	126	122	122	122	122	125	128	136	847	1
	12	885	179	173	158	142	134	132	130	129	129	129	130	132	134	142	158	173	873	12
HALF-DAY TOTALS			1122	2043	2683	2930	2763	2207	1357	612	548	540	540	540	542	545	555	599	3308	
Jul 21	6	128	43	90	122	134	127	99	55	11	9	9	9	9	9	9	9	10	17	6
	7	579	161	373	518	585	564	456	277	74	51	51	51	51	51	51	51	55	194	5
	8	743	173	415	590	676	661	549	349	110	78	78	78	78	78	78	78	83	419	4
	9	818	165	371	537	618	610	513	334	138	102	99	99	99	99	99	99	104	611	3
	10	857	157	289	413	475	467	389	257	138	118	113	113	113	113	113	113	120	759	2
	11	876	154	207	255	277	262	218	164	133	128	125	122	122	122	125	128	131	848	1
	12	882	154	151	145	138	135	134	133	132	132	132	133	134	135	138	145	151	875	12
HALF-DAY TOTALS			933	1829	2521	2848	2770	2298	1507	680	554	541	539	539	540	542	547	575	3289	
Aug 21	6	136	7	21	31	37	36	31	20	6	2	2	2	2	2	2	2	2	4	6
	7	569	81	289	458	555	567	493	343	131	48	45	45	45	45	45	45	45	167	5
	8	757	94	315	531	660	691	617	451	206	79	74	74	74	74	74	74	74	404	4
	9	838	104	258	467	598	640	589	448	235	103	95	95	95	95	95	95	95	608	3
	10	879	118	183	328	443	490	464	368	224	122	114	110	110	110	110	110	110	766	2
	11	899	127	136	180	235	271	273	240	185	138	127	124	120	120	120	120	124	860	1
	12	905	129	130	130	131	132	134	139	143	145	143	139	134	132	131	130	130	889	12
HALF-DAY TOTALS			603	1294	2099	2640	2810	2578	1969	1069	568	528	519	516	515	514	515	518	3258	
Sep 21	7	565	38	157	360	497	554	529	419	240	48	35	35	35	35	35	35	35	122	5
	8	797	67	156	424	618	716	705	587	374	113	64	64	64	64	64	64	64	367	4
	9	887	87	114	335	547	665	684	602	424	181	90	86	86	86	86	86	86	585	3
	10	931	101	107	193	372	507	560	529	415	240	111	103	101	101	101	101	101	752	2
	11	952	111	114	118	169	272	356	389	361	279	176	118	114	111	111	111	111	856	1
	12	958	114	114	116	118	121	153	217	272	293	272	217	153	121	118	116	114	889	12
HALF-DAY TOTALS			461	712	1500	2276	2791	2937	2658	1963	1007	605	501	466	457	455	454	454	3126	
Oct 21	7	524	25	56	249	404	492	502	437	300	105	27	25	25	25	25	25	25	79	5
	8	816	55	64	299	548	702	747	677	502	234	59	55	55	55	55	55	55	313	4
	9	920	76	80	204	473	659	741	711	573	336	97	76	76	76	76	76	76	537	3
	10	968	92	94	108	291	499	621	647	577	412	189	97	92	92	92	92	92	707	2
	11	991	102	105	122	261	420	515	528	459	315	154	106	102	102	102	102	102	814	1
	12	997	105	105	105	107	111	189	330	437	474	437	330	189	111	107	105	105	850	12
HALF-DAY TOTALS			402	444	1002	1869	2637	3110	3141	2689	1776	891	553	428	404	402	402	402	2872	
Nov 21	7	423	17	19	134	280	375	406	379	289	147	19	17	17	17	17	17	17	43	5
	8	806	46	49	174	456	651	742	719	583	350	78	46	46	46	46	46	46	247	4
	9	929	67	67	105	382	623	779	779	679	472	180	70	67	67	67	67	67	468	3
	10	984	83	83	87	210	470	651	727	694	554	316	98	83	83	83	83	83	635	2
	11	1009	92	92	92	98	242	462	607	654	603	455	234	98	92	92	92	92	740	1
	12	1016	96	96	96	96	101	227	432	572	619	572	432	227	101	96	96	96	775	12
HALF-DAY TOTALS			352	357	620	1438	2363	3101	3396	3159	2421	1324	664	404	355	352	352	352	2520	
Dec 21	7	372	13	14	93	228	318	354	339	268	150	21	13	13	13	13	13	13	31	5
	8	803	42	43	129	416	625	734	728	609	390	105	42	42	42	42	42	42	219	4
	9	936	63	63	78	341	604	761	800	716	520	226	66	63	63	63	63	63	436	3
	10	993	78	78	81	178	445	657	753	735	605	369	112	79	78	78	78	78	602	2
	11	1019	87	87	87	92	231	474	638	698	654	506	270	94	87	87	87	87	704	1
	12	1026	91	91	91	91	95	241	469	620	670	620	469	241	95	91	91	91	739	12
HALF-DAY TOTALS			328	330	501	1269	2241	3076	3468	3312	2638	1527	721	393	331	328	328	328	2361	
			N	NNW	NW	WNW	W	WSW	SW	SSW	S	SSE	SE	ESE	E	ENE	NE	NNE	HOR	p.m.

[a]Total solar heat gains for DSA glass (based on a ground reflectance of 0.20).

[b]Half-day totals computed by Simpson's rule, time interval = 10 minutes.

APPENDICES

Table C.7 Solar Intensity and Solar Heat Gain Factors[a] for 32°N Latitude (SI Units)[b]

Date	Solar Time, a.m.	Direct Normal, W/m²	Solar Heat Gain Factors, W/m²																	Solar Time, p.m.
			N	NNE	NE	ENE	E	ESE	SE	SSE	S	SSW	SW	WSW	W	WNW	NW	NNW	HOR	
Jan 21	7	4	0	0	1	3	4	4	4	3	2	0	0	0	0	0	0	0	0	5
	8	640	28	29	93	330	505	597	596	502	326	88	29	28	28	28	28	28	102	4
	9	849	48	48	53	286	553	721	775	711	534	258	52	48	48	48	48	48	278	3
	10	931	63	63	64	129	427	659	784	788	670	444	144	64	63	63	63	63	430	2
	11	967	71	71	71	75	213	502	698	784	750	602	347	90	71	71	71	71	524	1
	12	977	74	74	74	74	79	277	548	718	777	718	548	277	79	74	74	74	556	12
HALF-DAY TOTALS			249	250	338	897	1797	2700	3201	3198	2692	1746	831	353	252	249	249	249	1614	
Feb 21	7	352	13	23	148	258	323	336	299	212	83	14	13	13	13	13	13	13	30	5
	8	771	41	43	204	469	646	719	683	538	300	53	41	41	41	41	41	41	200	4
	9	905	60	60	101	386	627	764	783	680	471	174	63	60	60	60	60	60	402	3
	10	964	75	75	78	195	475	670	760	733	595	352	99	75	75	75	75	75	556	2
	11	990	83	83	83	88	240	491	656	717	670	519	275	89	83	83	83	83	652	1
	12	998	86	86	86	86	91	250	489	644	696	643	489	250	91	86	86	86	684	12
HALF-DAY TOTALS			314	324	635	1405	2318	3086	3408	3188	2461	1426	718	384	316	314	314	314	2181	
Mar 21	7	583	30	116	332	483	556	545	447	276	63	29	29	29	29	29	29	29	100	5
	8	821	54	78	339	576	717	746	661	473	195	57	54	54	54	54	54	54	315	4
	9	914	74	77	203	475	662	746	716	578	339	95	74	74	74	74	74	74	516	3
	10	959	88	88	96	273	499	637	677	615	456	221	93	88	88	88	88	88	667	2
	11	980	96	96	98	108	258	447	564	592	529	380	185	101	96	96	96	96	762	1
	12	987	99	99	99	99	105	209	386	511	554	511	386	209	105	99	99	99	795	12
HALF-DAY TOTALS			393	512	1131	1985	2761	3258	3284	2801	1858	1030	609	429	394	391	391	391	2757	
Apr 21	6	210	29	110	172	205	207	177	119	38	11	11	11	11	11	11	11	11	23	6
	7	649	53	253	462	593	631	575	428	204	49	45	45	45	45	45	45	45	191	5
	8	804	73	192	453	632	715	689	559	337	95	69	69	69	69	69	69	69	408	4
	9	876	89	113	324	532	649	669	591	418	183	92	87	87	87	87	87	87	593	3
	10	913	101	106	165	342	490	599	543	445	275	123	104	101	101	101	101	101	736	2
	11	932	109	109	115	149	263	371	426	415	340	222	126	113	109	109	109	109	825	1
	12	937	112	112	112	116	120	167	260	335	363	335	260	167	120	116	112	112	854	12
HALF-DAY TOTALS			508	932	1734	2500	3003	3129	2806	2033	1135	721	559	496	482	479	479	479	3203	
May 21	6	374	104	244	340	381	367	295	175	39	26	26	26	26	26	26	26	28	67	6
	7	666	112	350	535	638	643	550	374	134	60	57	57	57	57	57	57	59	256	5
	8	787	93	295	519	655	694	629	470	229	86	80	80	80	80	80	80	80	461	4
	9	849	104	194	404	558	625	601	488	294	117	100	97	97	97	97	97	97	633	3
	10	882	114	127	240	383	473	491	435	313	170	117	110	110	110	110	110	110	766	2
	11	898	121	124	132	186	260	312	322	285	215	147	126	122	118	118	118	118	847	1
	12	904	121	123	125	127	130	148	186	220	223	220	186	148	130	127	125	123	873	12
HALF-DAY TOTALS			700	1382	2214	2840	3106	2943	2356	1409	788	627	577	558	551	550	548	551	3464	
Jun 21	6	412	140	289	388	425	401	314	174	39	32	32	32	32	32	32	32	35	89	6
	7	662	148	384	556	644	634	529	342	109	62	62	62	62	62	62	62	65	279	5
	8	773	115	335	540	656	677	597	427	188	89	84	84	84	84	84	84	84	476	4
	9	831	110	234	431	563	609	567	440	244	111	101	101	101	101	101	101	101	642	3
	10	863	120	150	272	395	461	458	387	261	143	120	114	114	114	114	114	114	770	2
	11	880	126	130	148	202	258	288	280	237	177	135	128	125	122	122	122	122	847	1
	12	885	128	129	130	132	134	144	164	183	191	183	164	144	134	132	130	129	871	12
HALF-DAY TOTALS			824	1589	2403	2950	3108	2829	2137	1174	710	621	595	585	582	579	579	586	3538	
Jul 21	6	358	106	240	332	370	355	285	168	39	27	27	27	27	27	27	27	30	70	6
	7	640	118	349	526	623	626	534	361	129	62	59	59	59	59	59	59	61	257	5
	8	761	98	300	515	645	680	613	456	221	89	82	82	82	82	82	82	82	457	4
	9	823	107	202	405	553	615	588	475	284	117	100	100	100	100	100	100	100	626	3
	10	856	118	133	247	383	467	481	424	303	167	120	113	113	113	113	113	113	757	2
	11	873	124	128	137	191	261	308	314	277	210	147	129	125	121	121	121	121	836	1
	12	879	126	127	128	130	133	150	184	214	226	214	184	150	133	130	128	127	861	12
HALF-DAY TOTALS			728	1402	2210	2809	3052	2877	2290	1365	783	637	591	574	568	566	565	568	3431	
Aug 21	6	187	30	103	159	188	189	162	108	35	12	12	12	12	12	12	12	12	25	6
	7	599	59	250	444	565	598	542	402	192	53	49	49	49	49	49	49	49	192	5
	8	756	79	200	446	614	690	662	536	321	96	74	74	74	74	74	74	74	403	4
	9	830	95	124	328	523	632	648	570	402	179	98	93	93	93	93	93	93	583	3
	10	869	106	113	175	344	482	544	526	429	266	126	110	106	106	106	106	106	722	2
	11	889	115	117	122	157	264	365	414	401	328	217	131	119	115	115	115	115	808	1
	12	894	117	117	117	122	126	170	255	324	350	324	255	170	126	122	117	117	837	12
HALF-DAY TOTALS			539	957	1721	2443	2910	3014	2694	1950	1109	729	581	524	510	507	506	506	3152	
Sep 21	7	514	32	109	302	437	502	492	405	252	62	30	30	30	30	30	30	30	97	5
	8	758	57	81	324	546	679	706	626	450	190	61	57	57	57	57	57	57	304	4
	9	857	77	82	201	459	637	717	688	557	330	99	77	77	77	77	77	77	498	3
	10	905	91	91	101	270	485	617	655	596	444	220	97	91	91	91	91	91	645	2
	11	928	100	100	102	113	257	437	549	576	516	373	187	106	100	100	100	100	737	1
	12	935	103	103	103	103	110	210	379	499	540	499	379	210	110	103	103	103	769	12
HALF-DAY TOTALS			409	518	1089	1888	2621	3097	3132	2689	1812	1025	620	446	411	408	408	408	2664	
Oct 21	7	312	13	24	136	233	291	301	268	190	75	14	13	13	13	13	13	13	30	5
	8	724	42	46	200	450	616	684	649	511	285	54	42	42	42	42	42	42	198	4
	9	862	63	63	104	378	608	738	755	655	454	170	65	63	63	63	63	63	395	3
	10	923	77	77	81	197	465	652	737	711	577	342	101	77	77	77	77	77	546	2
	11	952	86	86	86	91	239	481	639	697	651	505	269	93	86	86	86	86	639	1
	12	960	89	89	89	89	95	247	478	626	677	626	278	247	95	89	89	89	671	12
HALF-DAY TOTALS			325	335	632	1365	2234	2968	3275	3067	2376	1390	714	394	328	325	325	325	2143	
Nov 21	7	5	0	0	1	3	4	5	4	3	2	0	0	0	0	0	0	0	0	5
	8	619	29	29	93	323	493	581	579	488	316	86	29	29	29	29	29	29	102	4
	9	829	49	49	55	284	545	709	761	697	523	253	49	49	49	49	49	49	277	3
	10	912	64	64	65	131	423	650	772	775	659	437	142	65	64	64	64	64	428	2
	11	949	72	72	72	76	213	496	688	773	739	593	342	91	72	72	72	72	521	1
	12	959	75	75	75	75	80	274	541	708	766	708	541	274	80	75	75	75	553	12
HALF-DAY TOTALS			253	254	342	890	1769	2653	3143	3138	2643	1717	824	355	256	253	253	253	607	
Dec 21	8	556	22	22	59	265	426	515	523	452	305	98	23	22	22	22	22	22	69	4
	9	812	43	43	46	244	512	686	751	701	539	282	48	43	43	43	43	43	228	3
	10	908	57	57	57	106	402	642	777	792	683	466	163	59	57	57	57	57	375	2
	11	949	66	66	66	69	200	497	700	795	766	620	367	92	66	66	66	66	468	1
	12	960	69	69	69	69	73	281	559	733	794	733	559	281	73	69	69	69	500	12
HALF-DAY TOTALS			223	223	265	717	1577	2499	3043	3111	2687	1823	868	339	225	223	223	223	1389	
			N	NNW	NW	WNW	W	WSW	SW	SSW	S	SSE	SE	ESE	E	ENE	NE	NNE	HOR	p.m.

[a]Total solar heat gains for DSA glass (based on a ground reflectance of 0.20).

[b]Half-day totals computed by Simpson's rule, time interval = 10 minutes.

Table C.8 Solar Intensity and Solar Heat Gain Factors[a] for 40°N Latitude (SI Units)[b]

Date	Solar Time, a.m.	Direct Normal, W/m^2	Solar Heat Gain Factors, W/m^2																	Solar Time, p.m.
			N	NNE	NE	ENE	E	ESE	SE	SSE	S	SSW	SW	WSW	W	WNW	NW	NNW	HOR	
Jan 21	8	446	17	17	55	223	350	417	420	358	236	60	17	17	17	17	17	17	44	4
	9	753	37	37	41	233	485	648	706	658	504	260	42	37	37	37	37	37	173	3
	10	865	51	51	51	97	390	627	761	776	671	460	161	53	51	51	51	51	303	2
	11	913	59	59	59	62	193	493	699	796	769	623	372	89	59	59	59	59	390	1
	12	926	62	62	62	62	66	293	563	740	802	740	563	283	62	62	62	62	419	12
HALF-DAY TOTALS			194	194	231	628	1426	2316	2852	2941	2566	1770	860	318	196	194	194	194	1117	
Feb 21	7	175	6	10	71	127	160	167	150	107	43	6	6	6	6	6	6	6	11	5
	8	692	33	35	158	407	576	651	628	505	296	56	33	33	33	33	33	33	136	4
	9	854	52	52	70	337	587	738	773	689	496	209	55	52	52	52	52	52	309	3
	10	926	65	65	67	155	450	666	777	768	641	408	120	66	65	65	65	65	451	2
	11	958	73	73	73	77	224	504	690	769	730	579	325	86	73	73	73	73	538	1
	12	967	76	76	76	76	80	271	536	702	760	702	536	271	80	76	76	76	568	12
HALF-DAY TOTALS			267	271	478	1140	2043	2888	3308	3202	2591	1602	790	361	269	267	267	267	1730	
Mar 21	7	540	27	92	295	441	514	509	425	271	69	26	26	26	26	26	26	26	83	5
	8	789	50	57	288	534	686	732	665	494	232	54	50	50	50	50	50	50	268	4
	9	889	67	70	147	429	640	749	744	625	404	127	69	67	67	67	67	67	450	3
	10	938	80	80	85	226	482	653	722	682	539	299	91	80	80	80	80	80	587	2
	11	961	88	88	88	94	247	476	623	673	622	473	244	94	88	88	88	85	673	1
	12	968	91	91	91	91	97	238	458	602	650	602	458	238	97	91	91	91	702	12
HALF-DAY TOTALS			358	440	954	1777	2626	3265	3429	3055	2191	1270	694	417	360	357	357	357	2411	
Apr 21	6	282	36	144	228	275	279	241	164	56	16	15	15	15	15	15	15	15	34	6
	7	651	50	223	442	584	633	588	451	255	51	45	45	45	45	45	45	45	193	5
	8	795	69	140	402	601	706	703	594	391	130	69	67	67	67	67	67	67	389	4
	9	865	84	91	254	488	638	691	640	494	260	91	84	84	84	84	84	84	557	3
	10	901	96	99	117	291	480	589	608	538	381	177	101	96	96	96	96	96	685	2
	11	920	104	107	107	122	255	411	506	522	459	323	164	109	104	104	104	104	766	1
	12	926	106	106	106	109	114	196	341	448	486	448	341	196	114	109	106	106	794	12
HALF-DAY TOTALS			487	835	1580	2390	3020	3314	3135	2466	1539	935	628	495	467	464	464	464	3020	
May 21	5	3	1	2	3	3	3	2	1	0	0	0	0	0	0	0	0	0	0	7
	6	453	113	284	403	458	446	364	223	56	33	33	33	33	33	33	33	35	96	6
	7	681	90	320	520	638	659	580	412	172	63	59	59	59	59	59	59	59	276	5
	8	787	86	230	471	629	694	655	519	295	92	80	80	80	80	80	80	80	461	4
	9	844	99	131	330	518	620	632	551	382	168	101	96	96	96	96	96	96	616	3
	10	875	107	114	171	332	467	529	513	419	262	127	111	107	107	107	107	107	737	2
	11	891	115	115	121	152	256	357	409	400	331	222	133	120	115	115	115	115	812	1
	12	896	117	117	117	121	126	171	258	329	355	329	258	171	126	121	117	117	836	12
HALF-DAY TOTALS			679	1275	2102	2818	3231	3232	2778	1897	1129	780	630	568	554	550	550	552	3418	
Jun 21	5	68	32	54	68	70	63	46	20	5	4	4	4	4	4	4	4	7	8	7
	6	488	150	329	450	501	478	380	222	53	39	39	39	39	39	39	39	43	126	6
	7	681	118	355	543	648	654	562	385	145	68	65	65	65	65	65	65	67	306	5
	8	776	94	268	492	633	680	626	480	252	93	85	85	85	85	85	85	85	484	4
	9	829	105	160	359	524	607	601	507	332	142	104	100	100	100	100	100	100	633	3
	10	859	112	121	197	345	457	500	468	366	218	122	115	112	112	112	112	112	750	2
	11	874	119	123	128	166	254	332	367	347	279	188	130	124	119	119	119	119	821	1
	12	879	121	121	121	127	131	163	227	281	301	281	227	163	131	127	121	121	844	12
HALF-DAY TOTALS			799	1483	2314	2968	3275	3151	2580	1649	995	743	642	602	592	588	587	594	3551	
Jul 21	5	7	3	5	7	7	6	5	2	0	0	0	0	0	0	0	0	1	1	7
	6	435	116	281	395	447	433	352	216	55	34	34	34	34	34	34	34	37	100	6
	7	656	95	321	513	625	643	564	400	166	66	62	62	62	62	62	62	62	278	5
	8	762	90	236	468	620	680	639	505	285	94	83	83	83	83	83	83	83	459	4
	9	818	102	138	333	513	610	618	537	371	165	104	99	99	99	99	99	99	611	3
	10	850	110	118	177	333	462	519	501	407	255	129	114	110	110	110	110	110	729	2
	11	866	117	120	125	157	256	352	400	389	321	217	135	123	117	117	117	117	802	1
	12	871	120	120	120	125	130	172	253	320	345	320	253	172	130	125	120	120	826	12
HALF-DAY TOTALS			705	1296	2102	2792	3180	3164	2707	1842	1110	783	643	586	572	568	567	570	3395	
Aug 21	6	255	38	137	214	256	259	223	151	52	18	17	17	17	17	17	17	17	38	6
	7	603	55	223	426	557	602	557	426	222	55	49	49	49	49	49	49	49	196	5
	8	747	75	149	397	584	681	676	569	374	128	74	72	72	72	72	72	72	386	4
	9	819	89	97	259	481	621	669	618	475	251	97	89	89	89	89	89	89	549	3
	10	857	102	105	126	294	472	574	590	519	567	173	107	102	102	102	102	102	674	2
	11	876	109	109	113	130	257	403	492	505	443	313	165	116	109	109	109	109	753	1
	12	882	112	112	112	115	120	197	333	434	470	434	333	197	120	115	112	112	780	12
HALF-DAY TOTALS			518	861	1571	2338	2929	3196	3015	2370	1496	932	647	524	496	492	492	492	2983	
Sep 21	7	472	28	87	265	395	460	456	381	244	66	27	27	27	27	27	27	27	80	5
	8	725	52	61	275	504	646	689	626	467	224	57	52	52	52	52	52	52	258	4
	9	830	71	73	148	413	613	717	712	599	391	129	73	71	71	71	71	71	434	3
	10	882	84	84	89	224	468	631	697	659	522	293	96	84	84	84	84	84	567	2
	11	906	92	92	92	99	245	463	604	652	603	461	242	99	92	92	92	92	651	1
	12	914	95	95	95	95	101	237	446	584	631	584	446	237	101	95	95	95	679	12
HALF-DAY TOTALS			374	447	917	1683	2484	3092	3257	2918	2119	1250	699	432	376	373	373	373	2329	
Oct 21	7	153	6	10	64	113	142	148	132	94	38	7	6	6	6	6	6	38	12	5
	8	644	35	37	155	387	545	615	592	476	280	56	35	35	35	35	35	35	136	4
	9	811	54	54	73	329	567	710	743	661	476	202	57	54	54	54	54	54	305	3
	10	884	67	67	70	157	439	646	752	742	619	395	120	69	67	67	67	67	443	2
	11	917	76	76	76	80	223	491	670	745	707	562	317	89	76	76	76	76	529	1
	12	927	78	78	78	78	84	267	521	681	737	681	521	267	84	78	78	78	558	12
HALF-DAY TOTALS			277	282	479	1106	1965	2771	3173	3074	2494	555	780	369	280	277	277	277	1704	
Nov 21	8	430	17	17	55	217	339	404	406	346	228	66	18	17	17	17	17	17	44	4
	9	733	38	38	42	231	476	634	691	643	492	254	43	38	38	38	38	38	173	3
	10	846	52	52	52	99	385	617	748	763	659	145	159	54	52	52	52	52	303	2
	11	894	60	60	60	63	192	486	688	783	757	613	367	89	60	60	60	60	388	1
	12	908	63	63	63	63	67	280	555	728	789	728	555	280	67	63	63	63	418	12
HALF-DAY TOTALS			197	198	235	623	1403	2273	2797	2884	2516	1738	850	319	199	197	197	197	1115	
Dec 21	8	279	9	9	25	129	212	259	264	230	157	52	10	9	9	9	9	9	20	4
	9	685	31	31	33	188	427	584	646	611	477	260	40	31	31	31	31	31	124	3
	10	825	45	45	45	78	358	594	732	755	661	462	173	47	45	45	45	45	244	2
	11	882	53	53	53	55	177	477	685	786	765	624	379	90	53	53	53	53	327	1
	12	898	56	56	56	56	60	281	560	736	798	738	560	281	60	56	56	56	356	12
HALF-DAY TOTALS			165	165	178	461	1180	2046	2594	2736	2446	1757	870	298	167	165	165	165	891	
			N	NNW	NW	WNW	W	WSW	SW	SSW	S	SSE	SE	ESE	E	ENE	NE	NNE	HOR	p.m.

[a]Total solar heat gains for DSA glass (based on a ground reflectance of 0.20).

[b]Half-day totals computed by Simpson's rule, time interval = 10 minutes.

APPENDICES

Table C.9 Solar Intensity and Solar Heat Gain Factors[a] for 48°N Latitude (SI Units)[b]

Date	Solar Time, a.m.	Direct Normal, W/m²	Solar Heat Gain Factors, W/m²																	Solar Time, p.m.
			N	NNE	NE	ENE	E	ESE	SE	SSE	S	SSW	SW	WSW	W	WNW	NW	NNW	HOR	
Jan 21	8	116	4	4	13	57	90	109	110	94	63	18	4	4	4	4	4	4	7	4
	9	584	24	24	26	168	371	505	555	523	406	217	30	24	24	24	24	24	79	3
	10	754	37	37	37	69	333	554	682	702	615	429	159	39	37	37	37	37	174	2
	11	823	45	45	45	47	166	455	656	753	734	598	365	83	45	45	45	45	244	1
	12	842	48	48	48	48	51	271	541	713	772	713	541	271	51	48	48	48	269	12
HALF-DAY TOTALS			134	134	144	368	996	1788	2298	2448	2212	1616	817	267	136	134	134	134	639	
Feb 21	7	11	0	1	5	8	10	11	10	7	0	0	0	0	0	0	0	0	1	5
	8	568	24	25	114	324	470	537	524	428	259	52	24	24	24	24	24	24	78	4
	9	780	42	42	49	284	530	683	727	660	488	225	45	42	42	42	42	42	210	3
	10	869	54	54	55	120	415	641	764	768	653	434	137	55	54	54	54	54	331	2
	11	908	61	61	61	64	205	498	696	787	755	607	355	84	61	61	61	61	409	1
	12	920	63	63	63	63	68	280	555	728	790	728	555	280	68	63	63	63	435	12
HALF-DAY TOTALS			214	216	337	864	1708	2575	3054	3051	2565	1677	825	328	216	214	214	214	1246	
Mar 21	7	482	23	71	253	387	458	458	388	253	71	23	22	22	22	22	22	22	64	5
	8	744	44	48	239	486	644	701	651	498	257	49	44	44	44	44	44	44	214	4
	9	853	60	60	104	381	609	738	753	652	449	164	62	60	60	60	60	60	371	3
	10	906	72	72	75	183	460	656	749	728	596	363	104	72	72	72	72	72	493	2
	11	932	79	79	79	83	232	492	663	731	689	541	297	88	79	79	79	79	568	1
	12	939	81	81	81	81	86	261	509	666	720	666	509	261	86	81	81	81	594	12
HALF-DAY TOTALS			317	372	790	1558	2446	3193	3471	3199	2420	1466	769	399	319	316	316	316	2006	
Apr 21	6	340	39	167	271	330	337	294	203	74	20	19	19	19	19	19	19	19	46	6
	7	646	49	191	417	567	628	595	468	264	56	45	45	45	45	45	45	45	189	5
	8	779	64	99	350	566	690	709	619	435	173	67	64	64	64	64	64	64	359	4
	9	847	79	83	191	444	621	703	679	554	335	105	79	79	79	79	79	79	507	3
	10	884	90	90	97	244	466	610	660	613	472	251	97	90	90	90	90	90	618	2
	11	902	97	97	97	105	245	443	570	609	558	419	217	103	97	97	97	97	689	1
	12	908	99	99	99	99	106	224	414	543	587	543	414	224	106	99	99	99	713	12
HALF-DAY TOTALS			463	765	1454	2285	3019	3465	3411	2825	1908	1169	713	492	445	442	442	442	2761	
May 21	5	129	52	97	125	133	122	91	44	9	8	8	8	8	8	8	8	10	16	7
	6	511	112	305	443	512	504	418	267	75	38	38	38	38	38	38	38	40	125	6
	7	690	73	283	498	631	668	604	448	214	66	61	61	61	61	61	61	61	287	5
	8	782	83	170	418	599	689	675	562	358	120	79	79	79	79	79	79	79	449	4
	9	834	93	101	259	475	611	656	605	463	243	100	93	93	93	93	93	93	585	3
	10	864	103	106	124	283	458	561	579	513	366	179	109	103	103	103	103	103	690	2
	11	879	109	109	113	127	249	396	488	505	447	320	170	116	109	109	109	109	756	1
	12	884	111	111	111	114	120	198	336	439	474	439	336	198	120	114	111	111	778	12
HALF-DAY TOTALS			679	1224	2035	2817	3360	3515	3176	2363	1525	997	709	579	550	546	546	549	3297	
Jun 21	5	243	111	192	241	252	228	166	75	19	17	17	17	17	17	17	17	24	38	7
	6	544	146	348	488	552	534	434	266	70	46	46	46	46	46	46	46	49	162	6
	7	693	93	317	521	642	467	591	427	188	72	67	67	67	67	67	67	67	324	5
	8	775	90	203	440	604	678	651	529	320	107	84	84	84	84	84	84	84	479	4
	9	822	98	114	286	482	600	629	567	418	208	105	98	98	98	98	98	98	610	3
	10	849	108	113	142	296	450	534	540	465	319	157	114	108	108	108	108	108	711	2
	11	864	114	114	120	138	248	373	449	456	396	279	156	121	114	114	114	114	775	1
	12	869	116	116	116	120	126	189	303	391	423	391	303	189	126	120	116	116	796	12
HALF-DAY TOTALS			811	1446	2279	3013	3454	3477	3013	2140	1376	942	718	620	597	593	592	602	3495	
Jul 21	5	135	57	104	133	141	129	96	46	11	9	9	9	9	9	9	9	12	18	7
	6	492	116	302	436	501	492	407	259	75	40	40	40	40	40	40	40	43	130	6
	7	666	78	285	492	619	653	588	436	207	69	63	63	63	63	63	63	63	290	5
	8	757	86	176	416	590	675	660	547	348	119	81	81	81	81	81	81	81	448	4
	9	809	96	106	263	471	601	643	591	450	237	104	96	96	96	96	96	96	582	3
	10	839	106	109	130	285	453	550	566	500	356	177	112	106	106	106	106	106	684	2
	11	855	112	112	117	132	249	390	477	492	435	312	169	119	112	112	112	112	749	1
	12	859	114	114	114	117	124	198	329	428	462	428	329	198	124	117	114	114	771	12
HALF-DAY TOTALS			705	1247	2039	2795	3311	3446	3101	2303	1496	993	721	598	570	565	565	569	3287	
Aug 21	6	311	42	161	256	310	316	273	190	70	22	21	21	21	21	21	21	21	51	6
	7	599	53	193	403	543	598	565	444	250	59	49	49	49	49	49	49	49	193	5
	8	732	69	108	347	549	665	681	593	417	168	73	69	69	69	69	69	69	358	4
	9	801	84	90	198	437	605	681	655	534	323	108	84	84	84	84	84	84	502	3
	10	839	95	95	104	247	453	593	639	592	456	245	104	95	95	95	95	95	610	2
	11	858	102	102	102	112	247	433	553	589	539	406	215	110	102	102	102	102	679	1
	12	864	104	104	104	104	113	224	404	526	568	526	404	224	113	104	104	104	702	12
HALF-DAY TOTALS			496	790	1449	2237	2929	3343	3282	2720	1852	1156	728	521	475	471	471	471	2741	
Sep 21	7	414	24	66	224	342	403	403	342	224	66	24	23	23	23	23	23	23	62	5
	8	678	46	51	228	455	602	654	608	467	244	52	46	46	46	46	46	46	206	4
	9	792	63	63	107	366	581	702	716	621	430	163	66	63	63	63	63	63	358	3
	10	848	75	75	79	182	444	630	719	699	573	353	107	75	75	75	75	75	476	2
	11	876	82	82	82	88	230	476	639	704	664	524	292	93	82	82	82	82	549	1
	12	884	84	84	84	84	91	257	494	643	695	643	494	257	91	84	84	84	574	12
HALF-DAY TOTALS			332	381	758	1467	2300	3007	3280	3039	2324	1429	766	334	331	331	331	1937		
Oct 21	7	12	0	1	5	9	11	12	11	8	3	0	0	0	0	0	0	0	1	5
	8	522	25	27	111	304	439	501	489	398	242	51	25	25	25	25	25	25	79	4
	9	734	44	44	52	276	508	652	693	629	466	216	47	44	44	44	44	44	208	3
	10	825	56	56	58	122	403	618	736	739	628	418	136	58	56	56	56	56	327	2
	11	866	64	64	64	67	203	483	673	760	729	587	345	86	64	64	64	64	403	1
	12	878	66	66	66	66	71	274	538	704	763	704	538	274	71	66	66	66	429	12
HALF-DAY TOTALS			223	225	340	838	1637	2461	2918	2918	2459	1619	808	334	226	223	223	223	1233	
Nov 21	8	115	4	4	13	57	90	108	109	93	62	18	4	4	4	4	4	4	7	4
	9	565	25	25	27	165	363	492	540	509	394	211	31	25	25	25	25	25	81	3
	10	735	38	38	38	70	328	543	668	688	602	420	156	40	38	38	38	38	175	2
	11	804	46	46	46	48	165	448	645	739	720	587	358	83	46	46	46	46	244	1
	12	823	48	48	48	48	53	267	531	700	758	700	531	267	53	48	48	48	269	12
HALF-DAY TOTALS			138	138	147	368	979	1752	2250	2396	2165	1583	804	268	140	138	138	138	642	
Dec 21	9	442	16	16	18	113	272	378	420	401	316	176	25	16	16	16	16	16	42	3
	10	676	30	30	30	51	286	491	613	636	563	398	156	32	30	30	30	30	119	2
	11	765	38	38	38	40	146	421	615	708	694	569	351	80	38	38	38	38	181	1
	12	789	41	41	41	41	44	256	514	679	734	679	514	256	44	41	41	41	204	12
HALF-DAY TOTALS			105	105	107	231	735	1444	1924	2098	1944	1477	778	239	107	105	105	105	444	
			N	NNW	NW	WNW	W	WSW	SW	SSW	S	SSE	SE	ESE	E	ENE	NE	NNE	HOR	p.m.

[a] Total solar heat gains for DSA glass (based on a ground reflectance of 0.20).

[b] Half-day totals computed by Simpson's rule, time interval = 10 minutes.

Table C.10 Solar Intensity and Solar Heat Gain Factors[a] for 64°N Latitude (SI Units)[b]

Date	Solar Time, a.m.	Direct Normal, W/m^2	Solar Heat Gain Factors, W/m^2																	Solar Time, p.m.
			N	NNE	NE	ENE	E	ESE	SE	SSE	S	SSW	SW	WSW	W	WNW	NW	NNW	HOR	
Jan 21	10	69	2	2	2	4	29	51	63	66	58	41	16	2	2	2	2	2	4	2
	11	256	9	9	9	9	46	143	210	242	238	196	120	24	9	9	9	9	18	1
	12	316	11	11	11	11	12	104	211	280	302	280	211	104	12	11	11	11	24	12
HALF-DAY TOTALS			16	16	16	18	80	250	382	449	446	374	237	72	17	16	16	16	33	
Feb 21	8	56	2	2	9	31	46	53	53	44	27	6	2	2	2	2	2	2	3	4
	9	422	16	16	18	136	280	373	403	375	285	143	18	16	16	16	16	16	41	3
	10	601	26	26	26	57	276	453	554	567	495	341	119	28	26	26	26	26	90	2
	11	679	32	32	32	34	139	386	558	638	622	506	307	64	32	32	32	32	128	1
	12	701	34	34	34	34	37	231	464	613	662	613	464	231	37	34	34	34	143	12
HALF-DAY TOTALS			93	93	103	279	770	1408	1822	1947	1767	1298	668	210	95	93	93	93	334	
Mar 21	7	300	12	34	147	235	284	288	249	168	54	13	12	12	12	12	12	12	27	5
	8	582	29	31	147	357	499	559	536	427	246	44	29	29	29	29	29	29	100	4
	9	717	41	41	51	277	502	642	679	613	450	203	44	41	41	41	41	41	187	3
	10	786	49	49	50	111	386	598	714	719	613	410	132	51	49	49	49	49	264	2
	11	821	54	54	54	57	190	468	658	746	718	579	344	80	54	54	54	54	314	1
	12	831	56	56	56	56	61	269	530	696	754	696	530	269	61	56	56	56	331	12
HALF-DAY TOTALS			213	234	473	1052	1879	2695	3103	3021	2457	1591	811	330	215	213	213	213	1057	
Apr 21	5	85	24	56	77	85	81	64	36	6	4	4	4	4	4	4	4	5	8	7
	6	419	38	187	320	400	416	371	266	111	26	24	24	24	24	24	24	24	67	6
	7	613	43	129	355	516	595	584	482	303	80	41	41	41	41	41	41	41	162	5
	8	720	54	60	249	483	633	685	634	484	250	60	54	54	54	54	54	54	269	4
	9	783	65	65	100	351	567	691	710	620	436	174	68	65	65	65	65	65	367	3
	10	820	72	72	76	161	421	613	709	698	582	372	119	74	72	72	72	72	441	2
	11	839	77	77	77	82	215	468	637	710	676	540	312	93	77	77	77	77	448	1
	12	846	79	79	79	79	85	262	503	655	708	655	503	262	85	79	79	79	504	12
HALF-DAY TOTALS			413	687	1294	2116	2974	3628	3741	3268	2408	1537	861	471	381	378	378	378	2053	
May 21	4	160	94	139	162	161	135	87	25	10	10	10	10	10	10	10	11	31	20	8
	5	416	152	298	393	425	395	302	158	34	29	29	29	29	29	29	29	33	81	7
	6	584	90	304	474	570	579	499	343	126	49	46	46	46	46	46	46	46	175	6
	7	688	65	199	436	596	665	634	507	297	78	60	60	60	60	60	60	60	284	5
	8	754	72	89	307	527	660	694	623	459	215	78	72	72	72	72	72	72	390	4
	9	796	82	85	143	384	577	678	679	579	389	147	85	82	82	82	82	82	481	3
	10	822	89	89	94	193	427	593	668	645	528	322	114	89	89	89	89	89	549	2
	11	836	93	93	93	100	226	446	594	652	614	485	275	105	93	93	93	93	592	1
	12	841	95	95	95	95	103	248	460	597	644	597	460	248	103	95	95	95	606	12
HALF-DAY TOTALS			780	1341	2145	2998	3712	4073	3841	3108	2234	1467	909	603	533	529	530	554	2875	
Jun 21	4	294	181	263	302	297	247	156	43	21	21	21	21	21	21	21	23	67	50	8
	5	485	195	360	466	498	458	346	175	44	39	39	39	39	39	39	39	45	124	7
	6	614	113	338	510	603	605	516	347	122	57	55	55	55	55	55	55	57	223	6
	7	698	74	225	457	610	672	632	498	282	80	68	68	68	68	68	68	68	331	5
	8	754	79	105	327	535	657	682	605	438	197	85	79	79	79	79	79	79	433	4
	9	791	89	93	161	393	572	862	655	552	362	137	92	89	89	89	89	89	520	3
	10	814	96	96	102	205	423	577	642	615	497	297	114	96	96	96	96	96	585	2
	11	827	100	100	100	108	228	431	568	619	581	456	258	110	100	100	100	100	627	1
	12	831	101	101	101	101	110	240	436	566	610	566	436	240	110	101	101	101	641	12
HALF-DAY TOTALS			1016	1080	2526	3347	3953	4154	3769	2985	2142	1436	935	667	605	601	603	673	3219	
Jul 21	4	168	101	149	173	172	144	93	27	12	12	12	12	12	12	12	12	35	24	8
	5	405	154	297	389	420	390	298	156	36	32	32	32	32	32	32	32	36	88	7
	6	564	94	303	467	560	567	488	335	124	51	49	49	49	49	49	49	49	181	6
	7	665	69	201	431	585	652	620	495	290	80	63	63	63	63	63	63	63	289	5
	8	730	75	94	307	520	648	680	609	449	211	81	75	75	75	75	75	75	393	4
	9	771	85	88	148	382	567	665	664	566	380	146	88	85	85	85	85	85	481	3
	10	797	92	92	98	196	422	583	655	631	516	315	116	92	92	92	92	92	548	2
	11	812	96	96	96	104	226	439	582	638	601	475	271	108	96	96	96	96	590	1
	12	816	98	98	98	98	107	246	452	585	630	585	452	246	107	98	98	98	604	12
HALF-DAY TOTALS			814	1370	2157	2985	3669	4004	3763	3046	2198	1457	920	626	557	553	554	582	2985	
Aug 21	5	92	28	62	85	94	89	71	40	8	6	6	6	6	6	6	6	6	11	7
	6	388	42	182	306	380	395	352	252	107	30	27	27	27	27	27	27	27	73	6
	7	570	48	132	344	494	567	555	458	289	81	45	45	45	45	45	45	45	168	5
	8	675	59	66	247	468	609	657	607	464	241	66	59	59	59	59	59	59	273	4
	9	737	70	70	107	345	549	666	683	597	420	172	74	70	70	70	70	70	368	3
	10	775	78	78	82	165	412	594	685	674	562	361	122	80	78	78	78	78	440	2
	11	794	82	82	82	89	216	456	617	686	653	523	305	99	82	82	82	82	485	1
	12	801	84	84	84	84	92	260	489	635	684	633	489	260	92	84	84	84	501	12
HALF-DAY TOTALS			447	714	1293	2073	2884	3498	3600	3147	2333	1509	869	500	413	409	409	410	2069	
Sep 21	7	242	12	30	122	194	234	238	206	139	47	13	12	12	12	12	12	12	26	5
	8	513	30	33	136	324	451	505	484	387	225	45	30	30	30	30	30	30	97	4
	9	651	43	43	54	261	468	596	631	570	420	193	47	43	43	43	43	43	181	3
	10	722	52	52	53	111	366	563	672	677	577	388	131	54	52	52	52	52	255	2
	11	758	57	57	57	61	185	445	623	706	680	549	329	83	57	57	57	57	303	1
	12	769	59	59	59	59	65	260	504	660	715	660	504	260	65	59	59	59	320	12
HALF-DAY TOTALS			224	241	448	968	1727	2484	2871	2810	2305	1513	787	336	227	224	224	224	1021	
Oct 21	8	54	2	2	10	30	45	52	51	42	26	6	2	2	2	2	2	2	4	4
	9	383	17	17	19	127	259	345	372	346	263	133	19	17	17	17	17	17	42	3
	10	556	28	28	28	58	261	426	520	532	465	321	115	29	28	28	28	28	91	2
	11	633	34	34	34	36	135	367	528	603	588	479	292	64	34	34	34	34	130	1
	12	655	36	36	36	36	40	222	441	581	628	581	441	222	40	36	36	36	144	12
HALF-DAY TOTALS			98	98	109	273	728	1324	1711	1829	1661	1225	638	209	100	98	98	98	339	
Nov 21	10	72	2	2	2	5	31	53	67	69	62	43	17	3	2	2	2	2	4	2
	11	250	9	9	9	10	46	140	207	238	234	192	118	24	9	9	9	9	19	1
	12	307	11	11	11	11	13	102	207	274	295	274	207	102	13	11	11	11	25	12
HALF-DAY TOTALS			17	17	17	19	81	248	378	444	440	369	234	71	18	17	17	17	35	
Dec 21	11	12	0	0	0	0	2	7	10	12	11	9	6	1	0	0	0	0	1	1
	12	51	1	1	1	1	2	16	34	45	48	45	34	16	2	1	1	1	3	12
HALF-DAY TOTALS			1	1	1	1	3	16	28	35	36	32	22	8	1	1	1	1	2	
			N	NNW	NW	WNW	W	WSW	SW	SSW	S	SSE	SE	ESE	E	ENE	NE	NNE	HOR	p.m.

[a]Total solar heat gains for DSA glass (based on a ground reflectance of 0.20).

[b]Half-day totals computed by Simpson's rule, time interval = 10 minutes.

APPENDICES

Table C.11 Solar Position and Clear-Day Insolation, 32°N Latitude (I-P Units)

	Solar Time		Solar Position				Global Irradiance (Btu/h ft²)				
					Direct		South-Facing Elevation Angle				
Date	a.m.	p.m.	Alt	Azm	Normal	HOR	22	32	42	52	90
Jan 21	7	5	1.4	65.2	1	0	0	0	0	1	1
	8	4	12.5	56.5	203	56	93	106	116	123	115
	9	3	22.5	46.0	269	118	175	193	206	212	181
	10	2	30.6	33.1	295	167	235	256	269	274	221
	11	1	36.1	17.5	306	198	273	295	308	312	245
	12		38.0	0.0	310	209	285	308	321	324	253
			Surface Daily Totals		2458	1288	1839	2008	2118	2166	1779
Feb 21	7	5	7.1	73.5	121	22	34	37	40	42	38
	8	4	19.0	64.4	247	95	127	136	140	141	108
	9	3	29.9	53.4	288	161	206	217	222	220	158
	10	2	39.1	39.4	306	212	266	278	283	279	193
	11	1	45.6	21.4	315	244	304	317	321	315	214
	12		48.0	0.0	317	255	316	330	334	328	222
			Surface Daily Totals		2872	1724	2188	2300	2345	2322	1644
Mar 21	7	5	12.7	81.9	185	54	60	60	59	56	32
	8	4	25.1	73.0	260	129	146	147	144	137	78
	9	3	36.8	62.1	290	194	222	224	220	209	119
	10	2	47.3	47.5	304	245	280	283	278	265	150
	11	1	55.0	26.8	311	277	317	321	315	300	170
	12		58.0	0.0	313	287	329	333	327	312	177
			Surface Daily Totals		3012	2084	2378	2403	2358	2246	1276
Apr 21	6	6	6.1	99.9	66	14	9	6	6	5	3
	7	5	18.8	92.2	206	86	78	71	62	51	10
	8	4	31.5	84.0	255	158	156	148	136	120	35
	9	3	43.9	74.2	278	220	225	217	203	183	68
	10	2	55.7	60.3	290	267	279	272	256	234	95
	11	1	65.4	37.5	295	297	313	306	290	265	112
	12		69.6	0.0	297	307	325	318	301	276	118
			Surface Daily Totals		3076	2390	2444	2356	2206	1994	764
May 21	6	6	10.4	107.2	119	36	21	13	13	12	7
	7	5	22.8	100.1	211	107	88	75	60	44	13
	8	4	35.4	92.9	250	175	159	145	127	105	15
	9	3	48.1	84.7	269	233	223	209	188	163	33
	10	2	60.6	73.3	280	277	273	259	237	208	56
	11	1	72.0	51.9	285	305	305	290	268	237	72
	12		78.0	0.0	286	315	315	301	278	247	77
			Surface Daily Totals		3112	2582	2454	2284	2064	1788	469
Jun 21	6	6	12.2	110.2	131	45	26	16	15	14	9
	7	5	24.3	103.4	210	115	91	76	59	41	14
	8	4	36.9	96.8	245	180	159	143	122	99	16
	9	3	49.6	89.4	264	236	221	204	181	153	19
	10	2	62.2	79.7	274	279	268	251	227	197	41
	11	1	74.2	60.9	279	306	299	282	257	224	56
	12		81.5	0.0	280	315	309	292	267	234	60
			Surface Daily Totals		3084	2634	2436	2234	1990	1690	370

Table C.11 Solar Position and Clear-Day Insolation, 32°N Latitude (I-P Units) (*Continued*)

	Solar Time		Solar Position			Global Irradiance (Btu/h ft²)					
							South-Facing Elevation Angle				
Date	a.m.	p.m.	Alt	Azm	Direct Normal	HOR	22	32	42	52	90
Jul 21	6	6	10.7	107.7	113	37	22	14	13	12	8
	7	5	23.1	100.6	203	107	87	75	60	44	14
	8	4	35.7	93.6	241	174	158	143	125	104	16
	9	3	48.4	85.5	261	230	220	205	185	159	31
	10	2	60.9	74.3	271	274	269	254	232	204	54
	11	1	72.4	53.3	277	302	300	285	262	232	69
	12		78.6	0.0	279	311	310	296	273	242	74
			Surface Daily Totals		3012	2558	2422	2250	2030	1754	458
Aug 21	6	6	6.5	100.5	59	14	9	7	6	6	4
	7	5	19.1	92.8	190	85	77	69	60	50	12
	8	4	31.8	84.7	240	156	152	144	132	116	33
	9	3	44.3	75.0	263	216	220	212	197	178	65
	10	2	56.1	61.3	276	262	272	264	249	226	91
	11	1	66.0	38.4	282	292	305	298	281	257	107
	12		70.3	0.0	284	302	317	309	292	268	113
			Surface Daily Totals		2902	2352	2388	2296	2144	1934	736
Sep 21	7	5	12.7	81.9	163	51	56	56	55	52	30
	8	4	25.1	73.0	240	124	140	141	138	131	75
	9	3	36.8	62.1	272	188	213	215	211	201	114
	10	2	47.3	47.5	287	237	270	273	268	255	145
	11	1	55.0	26.8	294	268	306	309	303	289	164
	12		58.0	0.0	296	278	318	321	315	300	171
			Surface Daily Totals		2808	2014	2288	2308	2264	2154	1226
Oct 21	7	5	6.8	73.1	99	19	29	32	34	36	32
	8	4	18.7	64.0	229	90	120	128	133	134	104
	9	3	29.5	53.0	273	155	198	208	213	212	153
	10	2	38.7	39.1	293	204	257	269	273	270	188
	11	1	45.1	21.1	302	236	294	307	311	306	209
	12		47.5	0.0	304	247	306	320	324	318	217
			Surface Daily Totals		2696	1654	2100	2208	2252	2232	1588
Nov 21	7	5	1.5	65.4	2	0	0	0	1	1	1
	8	4	12.7	56.6	196	55	91	104	113	119	111
	9	3	22.6	46.1	263	118	173	190	202	208	176
	10	2	30.8	33.2	289	166	233	252	265	270	217
	11	1	36.2	17.6	301	197	270	291	303	307	241
	12		38.2	0.0	304	207	282	304	316	320	249
			Surface Daily Totals		2406	1280	1816	1980	2084	2130	1742
Dec 21	8	4	10.3	53.8	176	41	77	90	101	108	107
	9	3	19.8	43.6	257	102	161	180	195	204	183
	10	2	27.6	31.2	288	150	221	244	259	267	226
	11	1	32.7	16.4	301	180	258	282	298	305	251
	12		34.6	0.0	304	190	271	295	311	318	259
			Surface Daily Totals		2348	1136	1704	1888	2016	2086	1794

Source: Adapted from Lunde, 1980.

Table C.12 Solar Position and Clear-Day Insolation, 40°N Latitude (I-P Units)

	Solar Time		Solar Position			Global Irradiance (Btu/h ft²)					
							South-Facing Elevation Angle				
Date	**a.m.**	**p.m.**	**Alt**	**Azm**	**Direct Normal**	**HOR**	**22**	**32**	**42**	**52**	**90**
Jan 21	8	4	8.1	55.3	142	28	65	74	81	85	84
	9	3	16.8	44.0	239	83	155	171	182	187	171
	10	2	23.8	30.9	274	127	218	237	249	254	223
	11	1	28.4	16.0	289	154	257	277	290	293	253
	12		30.0	0.0	294	164	270	291	303	306	263
			Surface Daily Totals		2182	948	1660	1810	1906	1944	1726
Feb 21	7	5	4.8	72.7	69	10	19	21	23	24	22
	8	4	15.4	62.2	224	73	114	122	126	127	107
	9	3	25.0	50.2	274	132	195	205	209	208	167
	10	2	32.8	35.9	295	178	256	267	271	267	210
	11	1	38.1	18.9	305	206	293	306	310	304	236
	12		40.0	0.0	308	216	306	319	323	317	245
			Surface Daily Totals		2640	1414	2060	2162	2202	2176	1730
Mar 21	7	5	11.4	80.2	171	46	55	55	54	51	35
	8	4	22.5	69.6	250	114	140	141	138	131	89
	9	3	32.8	57.3	282	173	215	217	213	202	138
	10	2	41.6	41.9	297	218	273	276	271	258	176
	11	1	47.7	22.6	305	247	310	313	307	293	200
	12		50.0	0.0	307	257	322	326	320	305	208
			Surface Daily Totals		2916	1852	2308	2330	2284	2174	1484
Apr 21	6	6	7.4	98.9	89	20	11	8	7	7	4
	7	5	18.9	89.5	206	87	77	70	61	50	12
	8	4	30.3	79.3	252	152	153	145	133	117	53
	9	3	41.3	67.2	274	207	221	213	199	179	93
	10	2	51.2	51.4	286	250	275	267	252	229	126
	11	1	58.7	29.2	292	277	308	301	285	260	147
	12		61.6	0.0	293	287	320	313	296	271	154
			Surface Daily Totals		3092	2274	2412	2320	2168	1956	1022
May 21	5	7	1.9	114.7	1	0	0	0	0	0	0
	6	6	12.7	105.6	144	49	25	15	14	13	9
	7	5	24.0	96.6	216	114	89	76	60	44	13
	8	4	35.4	87.2	250	175	158	144	125	104	25
	9	3	46.8	76.0	267	227	221	206	186	160	60
	10	2	57.5	60.9	277	267	270	255	233	205	89
	11	1	66.2	37.1	283	293	301	287	264	234	108
	12		70.0	0.0	284	301	312	297	274	243	114
			Surface Daily Totals		3160	2552	2442	2264	2040	1760	724
Jun 21	5	7	4.2	117.3	22	4	3	3	2	2	1
	6	6	14.8	108.4	155	60	30	18	17	16	10
	7	5	26.0	99.7	216	123	92	77	59	40	14
	8	4	37.4	90.7	246	182	159	142	121	97	16
	9	3	48.8	80.2	263	233	219	202	179	151	47
	10	2	59.8	65.8	272	272	266	248	224	193	74
	11	1	69.2	41.9	277	296	296	278	253	221	92
	12		73.5	0.0	279	304	306	289	263	230	98
			Surface Daily Totals		3180	2648	2434	2224	1974	1670	610

Table C.12 Solar Position and Clear-Day Insolation, 40°N Latitude (I-P Units) (*Continued*)

Date	Solar Time a.m.	Solar Time p.m.	Solar Position Alt	Solar Position Azm	Direct Normal	Global Irradiance (Btu/h ft²) HOR	South-Facing Elevation Angle 22	32	42	52	90
Jul 21	5	7	2.3	115.2	2	0	0	0	0	0	0
	6	6	13.1	106.1	138	50	26	17	15	14	9
	7	5	24.3	97.2	208	114	89	75	60	44	14
	8	4	35.8	87.8	241	174	157	142	124	102	24
	9	3	47.2	76.7	259	225	218	203	182	157	58
	10	2	57.9	61.7	269	265	266	251	229	200	86
	11	1	66.7	37.9	275	290	296	281	258	228	104
	12		70.6	0.0	276	298	307	292	269	238	111
			Surface Daily Totals		3062	2534	2409	2230	2006	1728	702
Aug 21	6	6	7.9	99.5	81	21	12	9	8	7	5
	7	5	19.3	90.0	191	87	76	69	60	49	12
	8	4	30.7	79.9	237	150	150	141	129	113	50
	9	3	41.8	67.9	260	205	216	207	193	173	89
	10	2	51.7	52.1	272	246	267	259	244	221	120
	11	1	59.3	29.7	278	273	300	292	276	252	140
	12		62.3	0.0	280	282	311	303	287	262	147
			Surface Daily Totals		2916	2244	2354	2258	2104	1894	978
Sep 21	7	5	11.4	80.2	149	43	51	51	49	47	32
	8	4	22.5	69.6	230	109	133	134	131	124	84
	9	3	32.8	57.3	263	167	206	208	203	193	132
	10	2	41.6	41.9	280	211	262	265	260	247	168
	11	1	47.7	22.6	287	239	298	301	295	281	192
	12		50.0	0.0	290	249	310	313	307	292	200
			Surface Daily Totals		2708	1788	2210	2228	2182	2074	1416
Oct 21	7	5	4.5	72.3	48	7	14	15	17	17	16
	8	4	15.0	61.9	204	68	106	113	117	118	100
	9	3	24.5	49.8	257	126	185	195	200	198	160
	10	2	32.4	35.6	280	170	245	257	261	257	203
	11	1	37.6	18.7	291	199	283	295	299	294	229
	12		39.5	0.0	294	208	295	308	312	306	238
			Surface Daily Totals		2454	1348	1962	2060	2098	2074	1654
Nov 21	8	4	8.2	55.4	136	28	63	72	78	82	81
	9	3	17.0	44.1	232	82	152	167	178	183	167
	10	2	24.0	31.0	268	126	215	233	245	249	219
	11	1	28.6	16.1	283	153	254	273	285	288	248
	12		30.2	0.0	288	163	267	287	298	301	258
			Surface Daily Totals		2128	942	1636	1778	1870	1908	1686
Dec 21	8	4	5.5	53.0	89	14	39	45	50	54	56
	9	3	14.0	41.9	217	65	135	152	164	171	163
	10	2	20.7	29.4	261	107	200	221	235	242	221
	11	1	25.0	15.2	280	134	239	262	276	283	252
	12		26.6	0.0	285	143	253	275	290	296	263
			Surface Daily Totals		1978	782	1480	1634	1740	1796	1646

Source: Adapted from Lunde, 1980.

Table C.13 Solar Position and Clear-Day Insolation, 48°N Latitude (I-P Units)

Date	Solar Time		Solar Position		Direct Normal	Global Irradiance (Btu/h ft^2)					
							South-Facing Elevation Angle				
	a.m.	p.m.	Alt	Azm		HOR	22	32	42	52	90
Jan 21	8	4	3.5	54.6	37	4	17	19	21	22	22
	9	3	11.0	42.6	185	46	120	132	140	145	139
	10	2	16.9	29.4	239	83	190	206	216	220	206
	11	1	20.7	15.1	261	107	231	249	260	263	243
	12		22.0	0.0	267	115	245	264	275	278	255
			Surface Daily Totals		1710	596	1360	1478	1550	1578	1478
Feb 21	7	5	2.4	72.2	12	1	3	4	4	4	4
	8	4	11.6	60.5	188	49	95	102	105	106	96
	9	3	19.7	47.7	251	100	178	187	191	190	167
	10	2	26.2	33.3	278	139	240	251	255	251	217
	11	1	30.5	17.2	290	165	278	290	294	288	247
	12		32.0	0.0	293	173	291	304	307	301	258
			Surface Daily Totals		2330	1080	1880	1972	2024	1978	1720
Mar 21	7	5	10.0	78.7	153	37	49	49	47	45	35
	8	4	19.5	66.8	236	96	131	132	129	122	96
	9	3	28.2	53.4	270	147	205	207	203	193	152
	10	2	35.4	37.8	287	187	263	266	261	248	195
	11	1	40.3	19.8	295	212	300	303	297	283	223
	12		42.0	0.0	298	220	312	315	309	294	232
			Surface Daily Totals		2780	1578	2208	2228	2182	2074	1632
Apr 21	6	6	8.6	97.8	108	27	13	9	8	7	5
	7	5	18.6	86.7	205	85	76	68	59	48	21
	8	4	28.5	74.9	247	142	149	141	129	113	69
	9	3	37.8	61.2	268	191	216	208	194	174	115
	10	2	45.8	44.6	280	228	268	260	245	223	152
	11	1	51.5	24.0	286	252	301	294	278	254	177
	12		53.6	0.0	288	260	313	305	289	264	185
			Surface Daily Totals		3076	2106	2358	2266	2114	1902	1262
May 21	5	7	5.2	114.3	41	9	4	4	4	3	2
	6	6	14.7	103.7	162	61	27	16	15	13	10
	7	5	24.6	93.0	219	118	89	75	60	43	13
	8	4	34.7	81.6	248	171	156	142	123	101	45
	9	3	44.3	68.3	264	217	217	202	182	156	86
	10	2	53.0	51.3	274	252	265	251	229	200	120
	11	1	59.5	28.6	279	274	296	281	258	228	141
	12		62.0	0.0	280	281	306	292	269	238	149
			Surface Daily Totals		3254	2482	2418	2234	2010	1728	982
Jun 21	5	7	7.9	116.5	77	21	9	9	8	7	5
	6	6	17.2	106.2	172	74	33	19	18	16	12
	7	5	27.0	95.8	220	129	93	77	59	39	15
	8	4	37.1	84.6	246	181	157	140	119	95	35
	9	3	46.9	71.6	261	225	216	198	175	147	74
	10	2	55.8	54.8	269	259	262	244	220	189	105
	11	1	62.7	31.2	274	280	291	273	248	216	126
	12		65.5	0.0	275	287	301	283	258	225	133
			Surface Daily Totals		3312	2626	2420	2204	1950	1644	874

Table C.13 Solar Position and Clear-Day Insolation, 48°N Latitude (I-P Units) (*Continued*)

	Solar Time		Solar Position				Global Irradiance (Btu/h ft²)				
							South-Facing Elevation Angle				
Date	**a.m.**	**p.m.**	**Alt**	**Azm**	**Direct Normal**	**HOR**	**22**	**32**	**42**	**52**	**90**
Jul 21	5	7	5.7	114.7	43	10	5	5	4	4	3
	6	6	15.2	104.1	156	62	28	18	16	15	11
	7	5	25.1	93.5	211	118	89	75	59	42	14
	8	4	35.1	82.1	240	171	154	140	121	99	43
	9	3	44.8	68.8	256	215	214	199	178	153	83
	10	2	53.5	51.9	266	250	261	246	224	195	116
	11	1	60.1	29.0	271	272	291	276	253	223	137
	12		62.6	0.0	272	279	301	286	263	232	144
			Surface Daily Totals		3158	2474	2386	2200	1974	1694	956
Aug 21	6	6	9.1	98.3	99	28	14	10	9	8	6
	7	5	19.1	87.2	190	85	75	67	58	47	20
	8	4	29.0	75.4	232	141	145	137	125	109	65
	9	3	38.4	61.8	254	189	210	201	187	168	110
	10	2	46.4	45.1	266	225	260	252	237	214	146
	11	1	52.2	24.3	272	248	293	285	268	244	169
	12		54.3	0.0	274	256	304	296	279	255	177
			Surface Daily Totals		2898	2086	2300	2200	2046	1836	1208
Sep 21	7	5	10.0	78.7	131	35	44	44	43	40	31
	8	4	19.5	66.8	215	92	124	124	121	115	90
	9	3	28.2	53.4	251	142	196	197	193	183	143
	10	2	35.4	37.8	269	181	251	254	248	236	185
	11	1	40.3	19.8	278	205	287	289	284	269	212
	12		42.0	0.0	280	213	299	302	296	281	221
			Surface Daily Totals		2568	1522	2102	2118	2070	1966	1546
Oct 21	7	5	2.0	71.9	4	0	1	1	1	1	1
	8	4	11.2	60.2	165	44	86	91	95	95	87
	9	3	19.3	47.4	233	94	167	176	180	178	157
	10	2	25.7	33.1	262	133	228	239	242	239	207
	11	1	30.0	17.1	274	157	266	277	281	276	237
	12		31.5	0.0	278	166	279	291	294	288	247
			Surface Daily Totals		2154	1022	1774	1860	1890	1866	1626
Nov 21	8	4	3.6	54.7	36	5	17	19	21	22	22
	9	3	11.2	42.7	179	46	117	129	137	141	135
	10	2	17.1	29.5	233	83	186	202	212	215	201
	11	1	20.9	15.1	255	107	227	245	255	258	238
	12		22.2	0.0	261	115	241	259	270	272	250
			Surface Daily Totals		1668	596	1336	1448	1518	1544	1442
Dec 21	9	3	8.0	40.9	140	27	87	98	105	110	109
	10	2	13.6	28.2	214	63	164	180	192	197	190
	11	1	17.3	14.4	242	86	207	226	239	244	231
	12		18.6	0.0	250	94	222	241	254	260	244
			Surface Daily Totals		1444	446	1136	1250	1326	1364	1304

Source: Adapted from Lunde, 1980.

Table C.14 Solar Position and Clear-Day Insolation, 64°N Latitude (I-P Units)

	Solar Time		Solar Position				Global Irradiance (Btu/h ft^2) South-Facing Elevation Angle				
Date	a.m.	p.m.	Alt	Azm	Direct Normal	HOR	22	32	42	52	90
Jan 21	10	2	2.8	28.1	22	2	17	19	20	20	20
	11	1	5.2	14.1	81	12	72	77	80	81	81
	12		6.0	0.0	100	16	91	98	102	103	103
			Surface Daily Totals		306	45	268	290	302	306	304
Feb 21	8	4	3.4	58.7	35	4	17	19	19	19	19
	9	3	8.6	44.8	147	31	103	108	111	110	107
	10	2	12.6	30.3	199	55	170	178	181	178	173
	11	1	15.1	15.3	222	71	212	220	223	219	213
	12		16.0	0.0	228	77	225	235	237	232	226
			Surface Daily Totals		1432	400	1230	1286	1302	1282	1252
Mar 21	7	5	6.5	76.5	95	18	30	29	29	27	25
	8	4	12.7	62.6	185	54	101	102	99	94	89
	9	3	18.1	48.1	227	87	171	172	169	160	153
	10	2	22.3	32.7	249	112	227	229	224	213	203
	11	1	25.1	16.6	260	129	262	265	259	246	235
	12		26.0	0.0	263	134	274	277	271	258	246
			Surface Daily Totals		2296	932	1856	1870	1830	1736	1656
Apr 21	5	7	4.0	108.5	27	5	2	2	2	1	1
	6	6	10.4	95.1	133	37	15	9	8	7	6
	7	5	17.0	81.6	194	76	70	63	54	43	37
	8	4	23.3	67.5	228	112	136	128	116	102	91
	9	3	29.0	52.3	248	144	197	189	176	158	145
	10	2	33.5	36.0	260	169	246	239	224	203	188
	11	1	36.5	18.4	266	184	278	270	255	233	216
	12		37.6	0.0	268	190	289	281	266	243	225
			Surface Daily Totals		2982	1644	2176	2082	1936	1736	1594
May 21	4	8	5.8	125.1	51	11	5	4	4	3	3
	5	7	11.6	112.1	132	42	13	11	10	9	8
	6	6	17.9	99.1	185	79	29	16	14	12	11
	7	5	24.5	85.7	218	117	86	72	56	39	28
	8	4	30.9	71.5	239	152	148	133	115	94	80
	9	3	36.8	56.1	252	182	204	190	170	145	128
	10	2	41.6	38.9	261	205	249	235	213	186	167
	11	1	44.9	20.1	265	219	278	264	242	213	193
	12		46.0	0.0	267	224	288	274	251	222	201
			Surface Daily Totals		3470	2236	2312	2124	1898	1624	1436
Jun 21	3	9	4.2	139.4	21	4	2	2	2	2	1
	4	8	9.0	126.4	93	27	10	9	8	7	6
	5	7	14.7	113.6	154	60	16	15	13	11	10
	6	6	21.0	100.8	194	96	34	19	17	14	13
	7	5	27.5	87.5	221	132	91	74	55	36	23
	8	4	34.0	73.3	239	166	150	133	112	88	73
	9	3	39.9	57.8	251	195	204	187	164	137	119
	10	2	44.9	40.4	258	217	247	230	206	177	157
	11	1	48.3	20.9	262	231	275	258	233	202	181
	12		49.5	0.0	263	235	284	267	242	211	189
			Surface Daily Totals		3650	2488	2342	2118	1862	1558	1356

Table C.14 Solar Position and Clear-Day Insolation, 64°N Latitude (I-P Units) (*Continued*)

	Solar Time		Solar Position				Global Irradiance (Btu/h ft²) South-Facing Elevation Angle				
Date	a.m.	p.m.	Alt	Azm	Direct Normal	HOR	22	32	42	52	90
Jul 21	4	8	6.4	125.3	53	13	6	5	5	4	4
	5	7	12.1	112.4	128	44	14	13	11	10	9
	6	6	18.4	99.4	179	81	30	17	16	13	12
	7	5	25.0	86.0	211	118	86	72	56	38	28
	8	4	31.4	71.8	231	152	146	131	113	91	77
	9	3	37.3	56.3	245	182	201	186	166	141	124
	10	2	42.2	39.2	253	204	245	230	208	181	162
	11	1	45.4	20.2	257	218	273	258	236	207	187
	12		46.6	0.0	259	223	282	267	245	216	195
			Surface Daily Totals		3372	2248	2280	2090	1864	1588	1400
Aug 21	5	7	4.6	108.8	29	6	3	3	2	2	2
	6	6	11.0	95.5	123	39	16	11	10	8	7
	7	5	17.6	81.9	181	77	69	61	52	42	35
	8	4	23.9	67.8	214	113	131	123	112	97	87
	9	3	29.6	52.6	234	144	190	182	169	150	138
	10	2	34.2	36.2	246	168	237	229	215	194	179
	11	1	37.2	18.5	252	183	268	260	244	222	205
	12		38.3	0.0	254	188	278	270	255	232	215
			Surface Daily Totals		2808	1646	2108	2008	1860	1662	1522
Sep 21	7	5	6.5	76.5	77	16	25	24	24	23	21
	8	4	12.7	62.6	163	51	92	92	90	85	81
	9	3	18.1	48.1	206	83	159	159	156	147	141
	10	2	22.3	32.7	229	108	212	213	209	198	189
	11	1	25.1	16.6	240	124	246	248	243	230	220
	12		26.0	0.0	244	129	258	260	254	241	230
			Surface Daily Totals		2074	892	1726	1736	1696	1608	1532
Oct 21	8	4	3.0	58.5	17	2	9	9	10	10	10
	9	3	8.1	44.6	122	26	86	91	93	92	90
	10	2	12.1	30.2	176	50	152	159	161	159	155
	11	1	14.6	15.2	201	65	193	201	203	200	195
	12		15.5	0.0	208	71	207	215	217	213	208
			Surface Daily Totals		1238	358	1088	1136	1152	1134	1106
Nov 21	10	2	3.0	28.1	23	3	18	20	21	21	21
	11	1	5.4	14.2	79	12	70	76	79	80	79
	12		6.2	0.0	97	17	89	96	100	101	100
			Surface Daily Totals		302	46	266	286	298	302	300
Dec 21	11	1	1.8	13.7	4	0	3	4	4	4	4
	12		2.6	0.0	16	2	14	15	16	17	17
			Surface Daily Totals		24	2	20	22	24	24	24

Source: Adapted from Lunde, 1980.

Table C.15 Solar Position and Clear-Day Insolation for 32°N Latitude (SI Units)

	Solar Time		Solar Position			Global Irradiance (W/m^2)					
					Direct		South-Facing Elevation Angle				
Date	a.m.	p.m.	Alt	Azm	Normal	HOR	22	32	42	52	90
Jan 21	7	5	1.4	65.2	3	0	0	0	0	3	3
	8	4	12.5	56.5	641	177	293	334	366	388	363
	9	3	22.5	46.0	849	372	552	609	650	669	571
	10	2	30.6	33.1	931	527	741	808	849	865	697
	11	1	36.1	17.5	965	625	861	931	972	984	773
	12		38.0	0.0	978	659	899	972	1013	1022	798
			Surface Daily Totals		7755	4064	5802	6335	6682	6834	5613
Feb 21	7	5	7.1	73.5	382	69	107	117	126	133	120
	8	4	19.0	64.4	779	300	401	429	442	445	341
	9	3	29.9	53.4	909	508	650	685	700	694	499
	10	2	39.1	39.4	965	669	839	877	893	880	609
	11	1	45.6	21.4	994	770	959	1000	1013	994	675
	12		48.0	0.0	1000	805	997	1041	1054	1035	700
			Surface Daily Totals		9061	5439	6903	7257	7399	7326	5187
Mar 21	7	5	12.7	81.9	584	170	189	189	186	177	101
	8	4	25.1	73.0	820	407	461	464	454	432	246
	9	3	36.8	62.1	915	612	700	707	694	659	376
	10	2	47.3	47.5	959	773	883	893	877	836	473
	11	1	55.0	26.8	981	874	1000	1013	994	947	536
	12		58.0	0.0	987	906	1038	1051	1032	985	558
			Surface Daily Totals		9503	6575	7503	7582	7440	7086	4026
Apr 21	6	6	6.1	99.9	208	44	28	19	19	16	10
	7	5	18.8	92.2	650	271	246	224	196	161	32
	8	4	31.5	84.0	805	499	492	467	429	397	110
	9	3	43.9	74.2	877	694	710	685	641	577	215
	10	2	55.7	60.3	915	842	880	858	808	738	300
	11	1	65.4	37.5	931	937	988	965	915	836	353
	12		69.6	0.0	937	968	1025	1003	950	871	372
			Surface Daily Totals		9705	7541	7711	7433	6960	6291	2410
May 21	6	6	10.4	107.2	375	114	66	41	41	38	22
	7	5	22.8	100.1	666	338	278	237	189	139	41
	8	4	35.4	92.9	789	552	502	458	401	331	47
	9	3	48.1	84.7	849	735	704	659	593	514	104
	10	2	60.6	73.3	883	874	861	817	748	656	177
	11	1	72.0	51.9	899	962	962	915	846	748	227
	12		78.0	0.0	902	994	994	950	877	779	243
			Surface Daily Totals		9818	8146	7742	7206	6512	5641	1480
Jun 21	6	6	12.2	110.2	413	142	82	51	47	44	28
	7	5	24.3	103.4	663	363	287	240	186	129	44
	8	4	36.9	96.8	773	568	502	451	385	312	51
	9	3	49.6	89.4	833	745	697	644	571	483	60
	10	2	62.2	79.7	865	880	846	792	716	622	129
	11	1	74.2	60.9	880	965	943	890	811	707	177
	12		81.5	0.0	883	994	975	921	842	738	189
			Surface Daily Totals		9730	8310	7686	7048	6279	5332	1167

Table C.15 Solar Position and Clear-Day Insolation for 32°N Latitude (SI Units) (*Continued*)

	Solar Time		Solar Position			Global Irradiance (W/m^2)					
					Direct		South-Facing Elevation Angle				
Date	a.m.	p.m.	Alt	Azm	Normal	HOR	22	32	42	52	90
Jul 21	6	6	10.7	107.7	357	117	69	44	41	38	25
	7	5	23.1	100.6	640	338	274	237	189	139	44
	8	4	35.7	93.6	760	549	498	451	394	328	50
	9	3	48.4	85.5	823	726	694	647	584	502	98
	10	2	60.9	74.3	855	864	849	801	732	644	170
	11	1	72.4	53.3	874	953	947	899	827	732	218
	12		78.6	0.0	880	981	978	934	861	764	233
			Surface Daily Totals		9503	8070	7641	7099	6405	5534	1445
Aug 21	6	6	6.5	100.5	186	44	28	22	19	19	13
	7	5	19.1	92.8	599	268	243	218	189	158	38
	8	4	31.8	84.7	757	492	480	454	416	366	104
	9	3	44.3	75.0	830	681	694	669	622	562	205
	10	2	56.1	61.3	871	827	858	833	786	713	287
	11	1	66.0	38.4	890	921	962	940	887	811	338
	12		70.3	0.0	896	953	1000	975	921	846	357
			Surface Daily Totals		9156	7421	7534	7244	6764	6102	2322
Sep 21	7	5	12.7	81.9	514	161	177	177	174	164	95
	8	4	25.1	73.0	757	391	442	445	435	413	237
	9	3	36.8	62.1	858	593	672	678	666	634	360
	10	2	47.3	47.5	905	748	852	861	846	805	457
	11	1	55.0	26.8	928	846	965	975	956	912	517
	12		58.0	0.0	934	877	1003	1013	994	947	540
			Surface Daily Totals		8859	6354	7219	7282	7143	6796	3868
Oct 21	7	5	6.8	73.1	312	60	91	101	107	114	101
	8	4	18.7	64.0	722	284	379	404	420	423	328
	9	3	29.5	53.0	861	489	625	565	672	669	483
	10	2	38.7	39.1	924	644	811	849	861	852	593
	11	1	45.1	21.1	953	745	928	969	981	965	659
	12		47.5	0.0	959	779	965	1010	1022	1003	685
			Surface Daily Totals		8506	5218	6626	6966	7105	7042	5010
Nov 21	7	5	1.5	65.4	6	0	0	0	3	3	3
	8	4	12.7	56.6	618	174	287	328	357	375	350
	9	3	22.6	46.1	830	372	546	599	637	656	555
	10	2	30.8	33.2	912	524	735	795	836	852	685
	11	1	36.2	17.6	950	622	852	918	956	969	760
	12		38.2	0.0	959	653	890	959	997	1010	786
			Surface Daily Totals		7591	4038	5729	6247	6575	6720	5496
Dec 21	8	4	10.3	53.8	555	129	243	284	319	341	338
	9	3	19.8	43.6	811	322	508	568	615	644	577
	10	2	27.6	31.2	909	473	697	770	817	842	713
	11	1	32.7	16.4	950	568	814	890	940	962	792
	12		34.6	0.0	959	599	855	931	981	1003	817
			Surface Daily Totals		7408	3584	5376	5957	6360	6581	5660

Source: Adapted from Lunde, 1980.

Table C.16 Solar Position and Clear-Day Insolation for 40°N Latitude (SI Units)

	Solar Time		Solar Position			Global Irradiance (W/m²)					
					Direct		South-Facing Elevation Angle				
Date	a.m.	p.m.	Alt	Azm	Normal	HOR	22	32	42	52	90
Jan 21	8	4	8.1	55.3	448	88	205	233	256	268	265
	9	3	16.8	44.0	754	262	489	540	574	590	540
	10	2	23.8	30.9	864	401	688	748	786	801	704
	11	1	28.4	16.0	912	486	811	874	915	924	798
	12		30.0	0.0	928	517	852	918	956	965	830
			Surface Daily Totals		6884	2991	5237	5711	6013	6133	5446
Feb 21	7	5	4.8	72.7	218	32	60	66	73	76	69
	8	4	15.4	62.2	707	230	360	385	398	401	338
	9	3	25.0	50.2	864	416	615	647	659	656	527
	10	2	32.8	35.9	931	562	808	842	855	842	663
	11	1	38.1	18.9	962	650	924	965	978	959	745
	12		40.0	0.0	972	681	965	1006	1019	1000	773
			Surface Daily Totals		8329	4461	6499	6821	6947	6865	5458
Mar 21	7	5	11.4	80.2	540	145	174	174	170	161	110
	8	4	22.5	69.6	789	360	442	445	435	413	281
	9	3	32.8	57.3	890	546	678	685	672	637	435
	10	2	41.6	41.9	937	688	861	871	855	814	555
	11	1	47.7	22.6	962	779	978	988	969	924	631
	12		50.0	0.0	969	811	1016	1029	1010	962	656
			Surface Daily Totals		9200	5843	7282	7351	7206	6859	4682
Apr 21	6	6	7.4	98.9	281	63	35	25	22	22	13
	7	5	18.9	89.5	650	274	243	221	192	158	38
	8	4	30.3	79.3	795	480	483	457	420	369	167
	9	3	41.3	67.2	864	653	697	672	628	565	293
	10	2	51.2	51.4	902	789	868	842	795	722	398
	11	1	58.7	29.2	921	874	972	950	899	820	464
	12		61.6	0.0	924	905	1010	988	934	855	486
			Surface Daily Totals		9755	7174	7610	7320	6840	6171	3224
May 21	5	7	1.9	114.7	3	0	0	0	0	0	0
	6	6	12.7	105.6	454	155	79	47	44	41	28
	7	5	24.0	96.6	681	360	281	240	189	139	41
	8	4	35.4	87.2	789	552	498	454	394	328	79
	9	3	46.8	76.0	842	716	697	650	587	505	189
	10	2	57.5	60.9	874	842	852	805	735	647	281
	11	1	66.2	37.1	893	924	950	905	833	738	341
	12		70.0	0.0	896	950	984	937	864	767	360
			Surface Daily Totals		9970	8052	7705	7143	6436	5553	2284
Jun 21	5	7	4.2	117.3	69	13	9	9	6	6	3
	6	6	14.8	108.4	489	189	95	57	54	50	32
	7	5	26.0	99.7	681	388	290	243	186	126	44
	8	4	37.4	90.7	776	574	502	448	382	306	50
	9	3	48.8	80.2	830	735	691	637	565	467	148
	10	2	59.8	65.8	858	858	839	782	707	609	233
	11	1	69.2	41.9	874	934	934	877	798	697	290
	12		73.5	0.0	880	959	965	912	830	726	309
			Surface Daily Totals		10033	8354	7679	7017	6228	5269	1925

Table C.16 Solar Position and Clear-Day Insolation for 40°N Latitude (SI Units) (*Continued*)

	Solar Time		Solar Position			Global Irradiance (W/m²)					
							South-Facing Elevation Angle				
Date	a.m.	p.m.	Alt	Azm	Direct Normal	HOR	22	32	42	52	90
Jul 21	5	7	2.3	115.2	6	0	0	0	0	0	0
	6	6	13.1	106.1	435	158	82	54	47	44	28
	7	5	24.3	97.2	656	360	281	237	189	139	44
	8	4	35.8	87.8	760	549	495	448	391	322	76
	9	3	47.2	76.7	817	710	688	640	574	495	183
	10	2	57.9	61.7	849	836	839	792	722	631	271
	11	1	66.7	37.9	868	915	934	887	814	719	328
	12		70.6	0.0	871	940	969	921	849	751	350
			Surface Daily Totals		9661	7995	7600	7036	6329	5452	2215
Aug 21	6	6	7.9	99.5	256	66	38	28	25	22	16
	7	5	19.3	90.0	603	274	240	218	189	155	38
	8	4	30.7	79.9	748	473	473	445	407	357	158
	9	3	41.8	67.9	820	647	681	653	609	546	281
	10	2	51.7	52.1	858	776	842	817	770	697	379
	11	1	59.3	29.7	877	861	947	921	871	795	442
	12		62.3	0.0	883	890	981	956	905	827	464
			Surface Daily Totals		9200	7080	7427	7124	6638	5976	3086
Sep 21	7	5	11.4	80.2	470	136	161	161	155	148	101
	8	4	22.5	69.6	726	344	420	423	413	391	265
	9	3	32.8	57.3	830	527	650	656	640	609	416
	10	2	41.6	41.9	883	666	827	836	820	779	530
	11	1	47.7	22.6	905	754	940	950	931	887	606
	12		50.0	0.0	915	786	978	988	969	921	631
			Surface Daily Totals		8544	5641	6973	7029	6884	6543	4467
Oct 21	7	5	4.5	72.3	151	22	44	47	54	54	50
	8	4	15.0	61.9	644	215	334	357	369	372	316
	9	3	24.5	49.8	811	398	548	615	631	625	505
	10	2	32.4	35.6	883	536	773	811	823	811	640
	11	1	37.6	18.7	918	628	893	931	943	928	722
	12		39.5	0.0	928	656	931	972	984	965	751
			Surface Daily Totals		7742	4253	6190	6499	6619	6543	5218
Nov 21	8	4	8.2	55.4	429	88	199	227	246	259	256
	9	3	17.0	44.1	732	259	480	527	562	577	527
	10	2	24.0	31.0	846	398	678	735	773	786	691
	11	1	28.6	16.1	893	483	801	861	899	909	782
	12		30.2	0.0	909	514	842	905	940	950	814
			Surface Daily Totals		6714	2972	5162	5610	5900	6020	5319
Dec 21	8	4	5.5	53.0	281	44	123	142	158	170	177
	9	3	14.0	41.9	685	205	426	480	517	540	514
	10	2	20.7	29.4	823	338	631	697	741	764	697
	11	1	25.0	15.2	883	423	754	827	871	893	795
	12		26.6	0.0	899	451	798	868	915	934	830
			Surface Daily Totals		6241	2467	4669	5155	5490	5666	5193

Source: Adapted from Lunde, 1980.

Table C.17 Solar Position and Clear-Day Insolation for 48°N Latitude (SI Units)

	Solar Time		Solar Position			Global Irradiance (W/m²)					
					Direct		South-Facing Elevation Angle				
Date	a.m.	p.m.	Alt	Azm	Normal	HOR	22	32	42	52	90
Jan 21	8	4	3.5	54.6	117	13	54	60	66	69	69
	9	3	11.0	42.6	584	145	379	416	442	457	439
	10	2	16.9	29.4	754	262	599	650	681	694	650
	11	1	20.7	15.1	823	338	729	786	820	830	767
	12		22.0	0.0	842	363	773	833	868	877	805
			Surface Daily Totals		5395	1880	4291	4663	4890	4979	4663
Feb 21	7	5	2.4	72.2	38	3	9	13	13	13	13
	8	4	11.6	60.5	593	155	300	322	331	334	303
	9	3	19.7	47.7	792	316	562	590	603	599	527
	10	2	26.2	33.3	877	439	757	792	805	792	685
	11	1	30.5	17.2	915	521	877	915	928	909	779
	12		32.0	0.0	924	546	918	959	969	950	814
			Surface Daily Totals		7351	3407	5931	6222	6386	6241	5427
Mar 21	7	5	10.0	78.7	483	117	155	155	148	142	110
	8	4	19.5	66.8	745	303	413	416	407	385	303
	9	3	28.2	53.4	852	464	647	653	640	609	480
	10	2	35.4	37.8	905	590	830	839	823	782	615
	11	1	40.3	19.8	931	669	947	956	937	893	704
	12		42.0	0.0	940	694	984	994	975	928	732
			Surface Daily Totals		8771	4979	6966	7029	6884	6543	5149
Apr 21	6	6	8.6	97.8	341	85	41	28	25	22	16
	7	5	18.6	86.7	647	268	240	215	186	151	66
	8	4	28.5	74.9	779	448	470	445	407	357	218
	9	3	37.8	61.2	846	603	681	656	612	549	363
	10	2	45.8	44.6	883	719	846	820	773	704	480
	11	1	51.5	24.0	902	795	950	928	877	801	558
	12		53.6	0.0	909	820	988	962	912	833	584
			Surface Daily Totals		9705	6644	7439	7149	6670	6001	3982
May 21	5	7	5.2	114.3	129	28	13	13	13	9	6
	6	6	14.7	103.7	511	192	85	50	47	41	32
	7	5	24.6	93.0	691	372	281	237	189	136	41
	8	4	34.7	81.6	782	540	492	448	388	319	142
	9	3	44.3	68.3	833	685	685	637	574	492	271
	10	2	53.0	51.3	864	795	836	792	722	631	379
	11	1	59.5	28.6	880	864	934	887	814	719	445
	12		62.0	0.0	883	887	965	921	849	751	470
			Surface Daily Totals		10266	7831	7629	7048	6342	5452	3098
Jun 21	5	7	7.9	116.5	243	66	28	28	25	22	16
	6	6	17.2	106.2	543	233	104	60	57	50	38
	7	5	27.0	95.8	694	407	293	243	186	123	47
	8	4	37.1	84.6	776	571	495	442	375	300	110
	9	3	46.9	71.6	823	710	681	625	552	464	233
	10	2	55.8	54.8	849	817	827	770	694	596	331
	11	1	62.7	31.2	864	883	918	861	782	681	398
	12		65.5	0.0	868	905	950	893	814	710	420
			Surface Daily Totals		10449	8285	7635	6954	6152	5187	2757

Table C.17 Solar Position and Clear-Day Insolation for 48°N Latitude (SI Units) (*Continued*)

	Solar Time		Solar Position				Global Irradiance (W/m²) South-Facing Elevation Angle				
Date	a.m.	p.m.	Alt	Azm	Direct Normal	HOR	22	32	42	52	90
Jul 21	5	7	5.7	114.7	136	32	16	16	13	13	9
	6	6	15.2	104.1	492	196	88	57	50	47	35
	7	5	25.1	93.5	666	372	281	237	186	133	44
	8	4	35.1	82.1	757	540	486	442	382	312	136
	9	3	44.8	68.8	808	678	675	628	562	483	262
	10	2	53.5	51.9	839	789	823	776	707	615	366
	11	1	60.1	29.0	855	858	918	871	798	704	432
	12		62.6	0.0	858	880	950	902	830	732	454
			Surface Daily Totals		9963	7805	7528	6941	6228	5345	3016
Aug 21	6	6	9.1	98.3	312	88	44	32	28	25	19
	7	5	19.1	87.2	599	268	237	211	183	148	63
	8	4	29.0	75.4	732	445	457	432	394	344	205
	9	3	38.4	61.8	801	596	663	634	590	530	347
	10	2	46.4	45.1	839	710	820	795	748	675	461
	11	1	52.2	24.3	858	782	924	899	846	770	533
	12		54.3	0.0	864	808	959	934	880	805	558
			Surface Daily Totals		9143	6581	7257	6941	6455	5793	3811
Sep 21	7	5	10.0	78.7	413	110	139	139	136	126	98
	8	4	19.5	66.8	678	290	391	391	382	363	284
	9	3	28.2	53.4	792	448	618	622	609	577	451
	10	2	35.4	37.8	849	571	792	801	782	745	584
	11	1	40.3	19.8	877	647	905	912	896	849	669
	12		42.0	0.0	883	672	943	953	934	887	697
			Surface Daily Totals		8102	4802	6632	6682	6531	6203	4878
Oct 21	7	5	2.0	71.9	13	0	3	3	3	3	3
	8	4	11.2	60.2	521	139	271	287	300	300	274
	9	3	19.3	47.4	735	297	527	555	568	562	495
	10	2	25.7	33.1	827	420	719	754	764	754	653
	11	1	30.0	17.1	864	495	839	874	887	871	748
	12		31.5	0.0	877	524	880	918	928	909	779
			Surface Daily Totals		6796	3224	5597	5868	5963	5887	5130
Nov 21	8	4	3.6	54.7	114	16	54	60	66	69	69
	9	3	11.2	42.7	565	145	369	407	432	445	426
	10	2	17.1	29.5	735	262	587	637	669	678	634
	11	1	20.9	15.1	805	338	716	773	805	814	751
	12		22.2	0.0	823	363	760	817	852	858	789
			Surface Daily Totals		5263	1880	4215	4568	4789	4871	4550
Dec 21	9	3	8.0	40.9	442	85	274	309	331	347	344
	10	2	13.6	28.2	675	199	517	568	606	622	599
	11	1	17.3	14.4	764	271	653	713	754	770	729
	12		18.6	0.0	789	297	700	760	801	820	770
			Surface Daily Totals		4556	1407	3584	3944	4184	4303	4114

Source: Adapted from Lunde, 1980.

Table C.18 Solar Position and Clear-Day Insolation for 64°N Latitude (SI Units)

	Solar Time		Solar Position				Global Irradiance (W/m²) South-Facing Elevation Angle				
Date	a.m.	p.m.	Alt	Azm	Direct Normal	HOR	22	32	42	52	90
Jan 21	10	2	2.8	28.1	69	6	54	60	63	63	63
	11	1	5.2	14.1	256	38	227	243	252	256	256
	12		6.0	0.0	316	50	287	309	322	325	325
			Surface Daily Totals		965	142	846	915	953	965	959
Feb 21	8	4	3.4	58.7	110	13	54	60	60	60	60
	9	3	8.6	44.8	464	98	325	341	350	347	338
	10	2	12.6	30.3	628	174	536	562	571	562	546
	11	1	15.1	15.3	700	224	669	694	704	691	672
	12		16.0	0.0	719	243	710	741	748	732	713
			Surface Daily Totals		4518	1262	3881	4057	4108	4045	3950
Mar 21	7	5	6.5	76.5	300	57	95	91	91	85	79
	8	4	12.7	62.6	584	170	319	322	312	297	281
	9	3	18.1	48.1	716	274	540	543	533	505	483
	10	2	22.3	32.7	786	353	716	722	707	672	640
	11	1	25.1	16.6	820	407	827	836	817	776	741
	12		26.0	0.0	830	423	864	874	855	814	776
			Surface Daily Totals		7244	2940	5856	5900	5774	5477	5225
Apr 21	5	7	4.0	108.5	85	16	6	6	6	3	3
	6	6	10.4	95.1	420	117	47	28	25	22	19
	7	5	17.0	81.6	612	240	221	199	170	136	117
	8	4	23.3	67.5	719	353	429	404	366	322	287
	9	3	29.0	52.3	782	454	622	596	555	498	457
	10	2	33.5	36.0	820	533	776	754	707	640	593
	11	1	36.5	18.4	839	581	877	852	805	735	681
	12		37.6	0.0	846	599	912	887	839	767	710
			Surface Daily Totals		9408	5187	6865	6569	6108	5477	5029
May 21	4	8	5.8	125.1	161	35	16	13	13	9	9
	5	7	11.6	112.1	416	133	41	35	32	38	25
	6	6	17.9	99.1	584	249	91	50	44	38	35
	7	5	24.5	85.7	688	369	271	227	177	123	88
	8	4	30.9	71.5	754	480	467	420	363	297	252
	9	3	36.8	56.1	795	574	644	600	536	458	404
	10	2	41.6	38.9	823	647	786	741	672	587	527
	11	1	44.9	20.1	836	691	877	833	764	672	609
	12		46.0	0.0	842	707	909	864	792	700	634
			Surface Daily Totals		10948	7055	7294	6701	5988	5124	4531
Jun 21	3	9	4.2	139.4	66	13	6	6	6	6	3
	4	8	9.0	126.4	293	85	32	28	25	22	19
	5	7	14.7	113.6	486	189	50	47	41	35	32
	6	6	21.0	100.8	612	303	107	60	54	44	41
	7	5	27.5	87.5	697	416	287	233	174	114	73
	8	4	34.0	73.3	754	524	473	420	353	278	230
	9	3	39.9	57.8	792	615	644	590	517	432	375
	10	2	44.9	40.4	814	685	779	726	650	558	495
	11	1	48.3	20.9	827	729	868	814	735	637	571
	12		49.5	0.0	830	741	896	842	764	666	596
			Surface Daily Totals		11516	7850	7389	6682	5875	4915	4278

Table C.18 Solar Position and Clear-Day Insolation for 64°N Latitude (SI Units) (*Continued*)

	Solar Time		Solar Position			Global Irradiance (W/m²)						
							South-Facing Elevation Angle					
Date	a.m.	p.m.	Alt	Azm	Direct Normal	HOR	22	32	42	52	90
Jul 21	4	8	6.4	125.3	167	41	19	16	16	13	13
	5	7	12.1	112.4	404	139	44	41	35	32	28
	6	6	18.4	99.4	565	256	95	54	50	41	38
	7	5	25.0	86.0	666	372	271	227	177	120	88
	8	4	31.4	71.8	729	480	461	413	357	287	243
	9	3	37.3	56.3	773	574	634	587	524	445	391
	10	2	42.2	39.2	798	644	773	626	656	571	511
	11	1	45.4	20.2	811	688	861	814	745	653	590
	12		46.6	0.0	817	704	890	842	773	681	615
			Surface Daily Totals		10639	7092	7193	6594	5881	5010	4417
Aug 21	5	7	4.6	108.8	91	19	9	9	6	6	6
	6	6	11.0	95.5	388	123	50	35	32	25	22
	7	5	17.6	81.9	571	243	218	192	164	133	110
	8	4	23.9	67.8	675	357	413	388	353	306	274
	9	3	29.6	52.6	738	454	599	574	533	473	435
	10	2	34.2	36.2	776	530	748	722	678	612	565
	11	1	37.2	18.5	795	577	846	820	770	700	647
	12		38.3	0.0	801	593	877	852	805	732	678
			Surface Daily Totals		8859	5193	6651	6335	5868	5244	4802
Sep 21	7	5	6.5	76.5	243	50	79	76	76	73	66
	8	4	12.7	62.6	514	161	290	290	284	268	256
	9	3	18.1	48.1	650	262	502	502	492	464	445
	10	2	22.3	32.7	722	341	669	672	659	625	596
	11	1	25.1	16.6	757	391	776	782	767	726	694
	12		26.0	0.0	770	407	814	820	801	760	726
			Surface Daily Totals		6543	2814	5446	5477	5351	5073	4833
Oct 21	8	4	3.0	58.5	54	6	28	28	32	32	32
	9	3	8.1	44.6	385	82	271	287	293	290	284
	10	2	12.1	30.2	555	158	480	502	508	502	489
	11	1	14.6	15.2	634	205	609	634	640	631	615
			15.5	0.0	656	224	653	678	685	672	656
			Surface Daily Totals		3906	1129	3433	3584	3635	3578	3489
Nov 21	10	2	3.0	28.1	73	9	57	63	66	66	66
	11	1	5.4	14.2	249	38	221	240	249	252	249
	12		6.2	0.0	306	54	281	303	316	319	316
			Surface Daily Totals		953	145	839	902	940	953	947
Dec 21	11	1	1.8	13.7	13	0	9	13	13	13	13
	12		2.6	0.0	50	6	44	47	50	54	54
			Surface Daily Totals		76	6	62	73	76	80	80

Source: Adapted from Lunde, 1980.

Figure C.1 *U.S. and Canadian locations for which solar data are given in Tables C.19 and C.20. Basic temperature data for many more locations are found in Appendix B. (Adapted from Balcomb et al., 1983.)*

Table C.19 Average Insolation, Temperature, and DD Data (I-P Units)

Elevation in feet; latitude in degrees north; HS (horizontal surface) and VS (vertical surface) insolation in Btu/day ft^2; TA in degrees F; D50, D55, D60, and D65 in degree days F (for these various "base" temperatures).

United States

MONTGOMERY, ALABAMA				Elev 203	Lat 32.3
	HS	VS	TA	D50	D65
Jan	752	896	47.6	170	542
Jul	1841	820	82.2	0	0
Yr	1390	946	66.1	428	2039

PHOENIX, ARIZONA				Elev 1112	Lat 33.4
	HS	VS	TA	D50	D65
Jan	1021	1462	56.9	10	255
Jul	2486	964	95.3	0	0
Yr	1371	1326	75.5	27	874

TUCSON, ARIZONA				Elev 2556	Lat 32.1
	HS	VS	TA	D50	D65
Jan	1099	1539	53.7	37	353
Jul	2341	922	87.9	0	0
Yr	1874	1307	70.6	116	1328

FORT SMITH, ARKANSAS				Elev 463	Lat 35.3
	HS	VS	TA	D50	D65
Jan	744	996	40.6	321	757
Jul	2065	908	82.9	0	0
Yr	1406	1013	62.6	936	3063

FRESNO, CALIFORNIA				Elev 328	Lat 36.8
	HS	VS	TA	D50	D65
Jan	657	886	48.6	88	510
Jul	2685	1076	83.8	0	0
Yr	1714	1210	65.2	246	2138

LOS ANGELES, CALIFORNIA				Elev 105	Lat 33.9
	HS	VS	TA	D50	D65
Jan	926	1293	57.9	2	227
Jul	2307	942	69.3	0	2
Yr	1596	1157	63.4	4	1256

SACRAMENTO, CALIFORNIA				Elev 26	Lat 38.5
	HS	VS	TA	D50	D65
Jan	597	826	47.5	105	541
Jul	2688	1131	76.8	0	0
Yr	1646	1198	61.9	309	2475

SAN DIEGO, CALIFORNIA				Elev 30	Lat 32.7
	HS	VS	TA	D50	D65
Jan	976	1325	58.0	1	221
Jul	2186	902	70.1	0	1
Yr	1600	1151	64.1	1	1101

SAN FRANCISCO, CALIFORNIA				Elev 16	Lat 37.6
	HS	VS	TA	D50	D65
Jan	708	1023	51.3	29	424
Jul	2392	1034	63.3	0	74
Yr	1556	1156	58.3	84	2606

SANTA MARIA, CALIFORNIA				Elev 236	Lat 34.9
	HS	VS	TA	D50	D65
Jan	854	1198	52.9	27	374
Jul	2341	965	63.5	0	67
Yr	1610	1172	58.4	102	2554

DENVER, COLORADO				Elev 5531	Lat 39.7
	HS	VS	TA	D50	D65
Jan	840	1465	31.9	565	1027
Jul	2273	1053	75.0	0	4
Yr	1570	1334	51.2	2579	5874

GRAND JUNCTION, COLORADO				Elev 4839	Lat 39.1
	HS	VS	TA	D50	D65
Jan	791	1296	28.6	662	1127
Jul	2465	1094	79.7	0	0
Yr	1661	1346	53.6	2347	5416

HARTFORD, CONNECTICUT				Elev 180	Lat 41.9
	HS	VS	TA	D50	D65
Jan	477	694	27.0	713	1177
Jul	1649	861	74.4	0	2
Yr	1060	835	51.3	2617	5828

WASHINGTON, DC				Elev 289	Lat 38.9
	HS	VS	TA	D50	D65
Jan	572	793	36.7	426	877
Jul	1817	883	80.2	0	0
Yr	1210	912	59.0	1327	3856

MIAMI, FLORIDA				Elev 7	Lat 25.8
	HS	VS	TA	D50	D65
Jan	1057	1121	68.4	1	49
Jul	1763	787	84.0	0	0
Yr	1474	941	77.5	1	112

ORLANDO, FLORIDA				Elev 118	Lat 28.5
	HS	VS	TA	D50	D65
Jan	999	1151	60.5	15	184
Jul	1801	795	82.5	0	0
Yr	1488	984	73.1	27	512

TALLAHASSEE, FLORIDA				Elev 69	Lat 30.4
	HS	VS	TA	D50	D65
Jan	877	1033	52.1	90	408
Jul	1748	786	82.4	0	0
Yr	1434	969	68.6	208	1441

ATLANTA, GEORGIA				Elev 1033	Lat 33.6
	HS	VS	TA	D50	D65
Jan	718	884	44.2	237	644
Jul	1812	821	80.2	0	0
Yr	1347	941	63.3	648	2578

BOISE, IDAHO				Elev 2867	Lat 43.6
	HS	VS	TA	D50	D65
Jan	485	770	32.4	547	1012
Jul	2613	1309	77.9	0	3
Yr	1499	1255	53.4	2071	5311

CHICAGO, ILLINOIS				Elev 623	Lat 41.8
	HS	VS	TA	D50	D65
Jan	507	756	25.5	760	1224
Jul	1944	984	76.1	0	1
Yr	1217	960	52.1	2738	5814

MOLINE, ILLINOIS				Elev 594	Lat 41.4
	HS	VS	TA	D50	D65
Jan	535	803	23.1	835	1300
Jul	1939	974	75.4	0	2
Yr	1226	973	51.3	2980	6040

SPRINGFIELD, ILLINOIS				Elev 614	Lat 39.8
	HS	VS	TA	D50	D65
Jan	585	852	27.3	706	1168
Jul	2058	984	76.1	0	1
Yr	1304	1003	53.8	2414	5284

Table C.19 Average Insolation, Temperature, and DD Data (I-P Units) (*Continued*)

Elevation in feet; latitude in degrees north; HS (horizontal surface) and VS (vertical surface) insolation in Btu/day ft^2; TA in degrees F; D50, D55, D60, and D65 in degree days F (for these various "base" temperatures).

United States

INDIANAPOLIS, INDIANA				Elev 807	Lat 39.7
	HS	VS	TA	D50	D65
Jan	496	668	28.2	681	1140
Jul	1806	891	75.7	0	1
Yr	1167	873	53.9	2346	5224

SOUTH BEND, INDIANA				Elev 774	Lat 41.7
	HS	VS	TA	D50	D65
Jan	416	566	24.5	790	1254
Jul	1852	944	73.0	0	6
Yr	1140	864	50.3	2975	6187

DES MOINES, IOWA				Elev 965	Lat 41.5
	HS	VS	TA	D50	D65
Jan	581	912	22.6	849	1314
Jul	2097	1037	76.6	0	1
Yr	1314	1065	51.5	3040	6065

DODGE CITY, KANSAS				Elev 2582	Lat 37.8
	HS	VS	TA	D50	D65
Jan	827	1303	32.9	535	996
Jul	2295	1013	79.8	0	1
Yr	1562	1232	55.9	2023	4834

TOPEKA, KANSAS				Elev 886	Lat 39.1
	HS	VS	TA	D50	D65
Jan	681	1033	30.3	613	1075
Jul	2128	993	79.9	0	0
Yr	1387	1036	56.1	2081	4823

LEXINGTON, KENTUCKY				Elev 988	Lat 38
	HS	VS	TA	D50	D65
Jan	546	714	33.1	541	989
Jul	1850	881	76.2	0	0
Yr	1221	892	56.2	1809	4483

BATON ROUGE, LOUISIANA				Elev 75	Lat 30.5
	HS	VS	TA	D50	D65
Jan	785	889	51.9	103	421
Jul	1746	786	82.5	0	0
Yr	1380	913	68.4	251	1537

NEW ORLEANS, LOUISIANA				Elev 10	Lat 30.0
	HS	VS	TA	D50	D65
Jan	835	950	54.0	70	359
Jul	1813	801	83.4	0	0
Yr	1438	943	70.3	149	1193

SHREVEPORT, LOUISIANA				Elev 259	Lat 32.5
	HS	VS	TA	D50	D65
Jan	762	920	48.1	161	530
Jul	2014	864	83.8	0	0
Yr	1428	973	66.8	446	2043

PORTLAND, MAINE				Elev 62	Lat 43.6
	HS	VS	TA	D50	D65
Jan	450	689	23.5	823	1287
Jul	1659	894	69.8	0	12
Yr	1052	857	47.2	3254	6891

DETROIT, MICHIGAN				Elev 627	Lat 42.4
	HS	VS	TA	D50	D65
Jan	417	585	26.2	739	1203
Jul	1835	951	74.5	0	2
Yr	1122	869	51.1	2799	5988

SAULT STE. MARIE, MICHIGAN				Elev 725	Lat 46.5
	HS	VS	TA	D50	D65
Jan	325	492	15.6	1067	1532
Jul	1835	1045	65.8	0	61
Yr	1044	861	42.2	4593	8532

MINNEAPOLIS, MINNESOTA				Elev 837	Lat 44.9
	HS	VS	TA	D50	D65
Jan	464	768	15.9	1058	1523
Jul	1970	1071	74.5	0	4
Yr	1172	996	47.0	4064	7396

MERIDIAN, MISSISSIPPI				Elev 308	Lat 32.3
	HS	VS	TA	D50	D65
Jan	744	883	46.4	197	582
Jul	1823	815	81.1	0	0
Yr	1371	933	64.8	539	2307

SAINT LOUIS, MISSOURI				Elev 564	Lat 38.7
	HS	VS	TA	D50	D65
Jan	627	898	32.3	559	1014
Jul	2049	959	80.3	0	0
Yr	1329	1006	57.8	1804	4379

SPRINGFIELD, MISSOURI				Elev 1270	Lat 37.2
	HS	VS	TA	D50	D65
Jan	684	956	33.7	518	970
Jul	2063	936	78.5	0	0
Yr	1364	1011	57.0	1733	4369

HELENA, MONTANA				Elev 3898	Lat 46.6
	HS	VS	TA	D50	D65
Jan	419	719	23.4	826	1290
Jul	2334	1312	71.3	0	15
Yr	1266	1134	45.8	3700	7466

NORTH OMAHA, NEBRASKA				Elev 1325	Lat 41.4
	HS	VS	TA	D50	D65
Jan	634	1034	27.3	704	1169
Jul	2106	1038	75.7	0	3
Yr	1323	1078	51.6	2920	5981

ELY, NEVADA				Elev 6253	Lat 39.3
	HS	VS	TA	D50	D65
Jan	819	1380	26.5	730	1195
Jul	2447	1094	68.9	0	12
Yr	1675	1391	45.8	3421	7282

LAS VEGAS, NEVADA				Elev 2178	Lat 36.1
	HS	VS	TA	D50	D65
Jan	978	1553	49.6	74	479
Jul	2588	1039	93.2	0	0
Yr	1866	1431	70.0	232	1841

RENO, NEVADA				Elev 4400	Lat 39.5
	HS	VS	TA	D50	D65
Jan	800	1345	36.7	414	877
Jul	2692	1167	76.9	0	1
Yr	1764	1439	54.5	1663	4817

ALBUQUERQUE, NEW MEXICO				Elev 5312	Lat 35.0
	HS	VS	TA	D50	D65
Jan	1016	1562	38.0	372	836
Jul	2489	995	79.2	0	0
Yr	1830	1379	58.5	1253	3873

Table C.19 Average Insolation, Temperature, and DD Data (I-P Units) (*Continued*)

Elevation in feet; latitude in degrees north; HS (horizontal surface) and VS (vertical surface) insolation in Btu/day ft^2; TA in degrees F; D50, D55, D60, and D65 in degree days F (for these various "base" temperatures).

United States

LOS ALAMOS, NEW MEXICO				Elev 7380	Lat 35.9
	HS	VS	TA	D50	D65
Jan	893	1340	32.6	541	1006
Jul	2051	913	71.7	0	4
Yr	1579	1244	52.7	1989	5131

ALBANY, NEW YORK				Elev 292	Lat 42.7
	HS	VS	TA	D50	D65
Jan	456	674	23.8	812	1276
Jul	1725	908	72.8	0	4
Yr	1068	843	49.3	3085	6418

BUFFALO, NEW YORK				Elev 187	Lat 42.9
	HS	VS	TA	D50	D65
Jan	349	465	25.2	771	1233
Jul	1776	935	71.6	0	7
Yr	1037	780	49.0	3092	6449

NEW YORK, NEW YORK				Elev 187	Lat 40.8
	HS	VS	TA	D50	D65
Jan	500	708	33.0	530	994
Jul	1688	861	76.4	0	0
Yr	1101	849	54.9	1801	4761

SYRACUSE, NEW YORK				Elev 407	Lat 43.1
	HS	VS	TA	D50	D65
Jan	385	538	24.1	806	1269
Jul	1758	931	72.1	0	5
Yr	1037	791	48.9	3149	6504

CHARLOTTE, NORTH CAROLINA				Elev 768	Lat 35.2
	HS	VS	TA	D50	D65
Jan	719	944	42.0	287	714
Jul	1831	841	79.5	0	0
Yr	1346	981	61.5	843	3030

RALEIGH-DURHAM, NORTH CAROLINA				Elev 440	Lat 35.9
	HS	VS	TA	D50	D65
Jan	694	924	41.4	308	733
Jul	1776	832	79.9	0	0
Yr	1297	955	61.0	940	3188

BISMARCK, NORTH DAKOTA				Elev 1647	Lat 46.8
	HS	VS	TA	D50	D65
Jan	467	847	13.0	1148	1613
Jul	2184	1241	71.9	0	11
Yr	1251	1145	43.5	4784	8414

FARGO, NORTH DAKOTA				Elev 899	Lat 46.9
	HS	VS	TA	D50	D65
Jan	415	720	9.3	1261	1726
Jul	2120	1210	71.4	0	14
Yr	1206	1075	42.8	5119	8685

CINCINNATI, OHIO				Elev 889	Lat 39.1
	HS	VS	TA	D50	D65
Jan	500	659	31.9	572	1025
Jul	1771	869	75.8	0	0
Yr	1160	858	55.0	1959	4776

CLEVELAND, OHIO				Elev 804	Lat 41.4
	HS	VS	TA	D50	D65
Jan	388	507	27.7	697	1157
Jul	1828	929	73.5	0	4
Yr	1093	808	51.6	2616	5737

COLUMBUS, OHIO				Elev 833	Lat 40.0
	HS	VS	TA	D50	D65
Jan	459	606	29.4	648	1104
Jul	1755	876	75.4	0	0
Yr	1128	834	53.9	2268	5161

OKLAHOMA CITY, OKLAHOMA				Elev 1302	Lat 35.4
	HS	VS	TA	D50	D65
Jan	801	1114	39.3	355	799
Jul	2128	925	82.5	0	0
Yr	1463	1073	61.3	1133	3398

MEDFORD, OREGON				Elev 1299	Lat 42.4
	HS	VS	TA	D50	D65
Jan	407	565	40.5	297	759
Jul	2475	1207	75.7	0	2
Yr	1356	1033	56.0	1133	4195

SALEM, OREGON				Elev 200	Lat 44.9
	HS	VS	TA	D50	D65
Jan	332	471	41.9	258	716
Jul	2142	1154	69.0	0	18
Yr	1130	897	53.9	1065	4436

HARRISBURG, PENNSYLVANIA				Elev 348	Lat 40.2
	HS	VS	TA	D50	D65
Jan	536	763	30.4	611	1073
Jul	1764	883	77.0	0	0
Yr	1152	887	54.4	2091	5026

PHILADELPHIA, PENNSYLVANIA				Elev 30	Lat 39.9
	HS	VS	TA	D50	D65
Jan	555	792	33.8	511	968
Jul	1758	876	78.9	0	0
Yr	1170	905	56.8	1672	4410

PITTSBURGH, PENNSYLVANIA				Elev 1224	Lat 40.5
	HS	VS	TA	D50	D18.3
Jan	424	553	28.6	671	1128
Jul	1689	857	73.1	0	3
Yr	1071	793	52.1	2467	5518

PROVIDENCE, RHODE ISLAND				Elev 62	Lat 41.7
	HS	VS	TA	D50	D65
Jan	506	750	30.0	622	1085
Jul	1695	878	74.3	0	2
Yr	1114	884	52.2	2276	5477

CHARLESTON, SOUTH CAROLINA				Elev 39	Lat 32.9
	HS	VS	TA	D50	D65
Jan	744	904	49.5	133	486
Jul	1799	813	82.2	0	0
Yr	1346	940	66.7	322	1821

RAPID CITY, SOUTH DAKOTA				Elev 3169	Lat 44.0
	HS	VS	TA	D50	D65
Jan	542	928	24.9	778	1243
Jul	2223	1161	73.3	0	8
Yr	1344	1177	47.3	3553	7113

KNOXVILLE, TENNESSEE				Elev 981	Lat 35.8
	HS	VS	TA	D50	D65
Jan	621	785	38.8	375	812
Jul	1804	839	78.2	0	0
Yr	1275	909	59.8	1134	3486

Table C.19 Average Insolation, Temperature, and DD Data (I-P Units) (*Continued*)

Elevation in feet; latitude in degrees north; HS (horizontal surface) and VS (vertical surface) insolation in Btu/day ft^2; TA in degrees F; D50, D55, D60, and D65 in degree days F (for these various "base" temperatures).

United States

MEMPHIS, TENNESSEE				Elev 285	Lat 35.0
	HS	VS	TA	D50	D65
Jan	683	870	42.0	300	714
Jul	1972	879	82.4	0	0
Yr	1368	965	63.5	863	2856

NASHVILLE, TENNESSEE				Elev 591	Lat 36.1
	HS	VS	TA	D50	D65
Jan	580	721	38.9	379	810
Jul	1891	869	79.9	0	0
Yr	1272	891	60.5	1149	3430

AUSTIN, TEXAS				Elev 620	Lat 30.3
	HS	VS	TA	D50	D65
Jan	864	1008	50.9	108	447
Jul	2105	865	84.4	0	0
Yr	1478	974	68.8	297	1648

BROWNSVILLE, TEXAS				Elev 20	Lat 25.9
	HS	VS	TA	D50	D65
Jan	913	923	61.9	17	164
Jul	2212	867	85.1	0	0
Yr	1550	917	75.1	38	499

DALLAS, TEXAS				Elev 489	Lat 32.8
	HS	VS	TA	D50	D65
Jan	821	1035	47.1	175	559
Jul	2122	890	86.2	0	0
Yr	1470	1014	67.3	499	2113

EL PASO, TEXAS				Elev 3917	Lat 31.8
	HS	VS	TA	D50	D65
Jan	1125	1572	46.5	145	573
Jul	2450	934	84.0	0	0
Yr	1901	1327	66.2	422	2203

HOUSTON, TEXAS				Elev 108	Lat 30.0
	HS	VS	TA	D50	D65
Jan	772	852	55.0	57	331
Jul	1828	805	84.4	0	0
Yr	1353	884	71.0	138	1115

LUBBOCK, TEXAS				Elev 3241	Lat 33.6
	HS	VS	TA	D50	D65
Jan	1031	1497	41.4	289	730
Jul	2412	956	81.1	0	0
Yr	1768	1279	61.6	980	3217

SAN ANTONIO, TEXAS				Elev 794	Lat 29.5
	HS	VS	TA	D50	D65
Jan	895	1026	52.9	73	384
Jul	2121	863	85.0	0	0
Yr	1501	973	70.3	196	1352

BRYCE CANYON, UTAH				Elev 7588	Lat 37.7
	HS	VS	TA	D50	D65
Jan	914	1511	21.0	900	1365
Jul	2424	1044	64.1	0	61
Yr	1742	1404	41.3	4378	8673

SALT LAKE CITY, UTAH				Elev 4226	Lat 40.8
	HS	VS	TA	D50	D65
Jan	639	1017	31.1	586	1051
Jul	2590	1186	80.9	0	0
Yr	1606	1301	54.1	2189	5329

BURLINGTON, VERMONT				Elev 341	Lat 44.5
	HS	VS	TA	D50	D65
Jan	385	572	20.0	930	1394
Jul	1721	941	71.5	0	8
Yr	1023	815	47.0	3676	7145

NORFOLK, VIRGINIA				Elev 30	Lat 36.9
	HS	VS	TA	D50	D65
Jan	678	932	41.5	304	729
Jul	1853	868	80.1	0	0
Yr	1327	990	61.3	900	3139

RICHMOND, VIRGINIA				Elev 164	Lat 37.5
	HS	VS	TA	D50	D65
Jan	632	863	38.5	384	822
Jul	1774	849	79.3	0	0
Yr	1250	936	59.5	1202	3635

ROANOKE, VIRGINIA				Elev 1175	Lat 37.3
	HS	VS	TA	D50	D65
Jan	660	911	37.2	416	863
Jul	1796	854	76.9	0	1
Yr	1271	958	57.8	1347	3927

SEATTLE-TACOMA, WASHINGTON				Elev 400	Lat 47.4
	HS	VS	TA	D50	D65
Jan	262	378	42.3	243	704
Jul	2248	1299	66.5	0	40
Yr	1056	857	53.0	1067	4621

SPOKANE, WASHINGTON				Elev 2365	Lat 47.6
	HS	VS	TA	D50	D65
Jan	315	496	29.5	636	1101
Jul	2357	1368	71.2	0	22
Yr	1227	1068	48.5	2837	6539

CHARLESTON, WEST VIRGINIA				Elev 951	Lat 38.4
	HS	VS	TA	D50	D65
Jan	498	638	34.5	502	946
Jul	1682	827	75.1	0	0
Yr	1125	822	56.0	1699	4385

MADISON, WISCONSIN				Elev 860	Lat 43.1
	HS	VS	TA	D50	D65
Jan	515	822	19.4	949	1414
Jul	1934	1009	72.1	0	8
Yr	1193	978	47.5	3653	7053

MILWAUKEE, WISCONSIN				Elev 692	Lat 42.9
	HS	VS	TA	D50	D65
Jan	479	731	23.0	836	1301
Jul	1962	1017	72.6	0	9
Yr	1194	957	48.8	3241	6647

SHERIDAN, WYOMING				Elev 3966	Lat 44.8
	HS	VS	TA	D50	D65
Jan	517	900	24.4	795	1260
Jul	2329	1237	71.5	0	12
Yr	1333	1170	45.8	3716	7470

Table C.19 Average Insolation, Temperature, and DD Data (I-P Units) (*Continued*)

Elevation in feet; latitude in degrees north; HS (horizontal surface) and VS (vertical surface) insolation in Btu/day ft^2; TA in degrees F; D50, D55, D60, and D65 in degree days F (for these various "base" temperatures).

Canada

EDMONTON, ALBERTA				Elev 2220	Lat 53.6	HALIFAX, NOVA SCOTIA				Elev 136	Lat 44.6
	HS	VS	TA	D50	D65		HS	VS	TA	D50	D65
Jan	324	746	10.3	1232	1697	Jan	456	737	22.9	839	1304
Jul	1977	1378	61.4	1	135	Jul	1694	929	66.8	0	34
Yr	1114	1205	36.4	6037	10465	Yr	1076	907	44.9	3643	7548
SUFFIELD, ALBERTA				Elev 2549	Lat 50.3	OTTAWA, ONTARIO				Elev 377	Lat 45.5
	HS	VS	TA	D50	D65		HS	VS	TA	D50	D65
Jan	433	937	16.8	1029	1494	Jan	510	914	14.4	1103	1568
Jul	2173	1377	68.6	0	29	Jul	1875	1040	70.1	0	12
Yr	1239	1269	42.8	4557	8416	Yr	1158	1015	44.3	4406	8012
VANCOUVER, BRITISH COLUMBIA				Elev 310	Lat 37.5	TORONTO, ONTARIO				Elev 443	Lat 43.7
	HS	VS	TA	D50	D65		HS	VS	TA	D50	D65
Jan	254	395	39.9	315	778	Jan	487	777	23.1	836	1300
Jul	2021	1239	64.8	0	45	Jul	1958	1035	71.9	0	7
Yr	1060	916	51.1	1404	5171	Yr	1171	948	48.0	3347	6803
WINNIPEG, MANITOBA				Elev 820	Lat 50.0	NORMANDIN, QUEBEC				Elev 450	Lat 48.8
	HS	VS	TA	D50	D65		HS	VS	TA	D50	D65
Jan	461	1011	2.7	1466	1931	Jan	454	921	1.8	1493	1958
Jul	2025	1264	67.5	0	43	Jul	1707	1031	63.9	0	84
Yr	1190	1199	37.7	6345	10254	Yr	1092	1053	35.5	6600	10870

Sources: Balcomb et al., 1980 / American Solar Energy Society, Inc.

Except for TA, D50, and D65 data reprinted by permission from ASHRAE, 2021.

Table C.20 Average Insolation, Temperature, and DD Data (SI Units)

Elevation in meters; latitude in degrees north; HS (horizontal surface) and VS (vertical surface) insolation in W/day m^2; TA in degrees C; D10.0, D12.8, D15.6, and D18.3 in degree days C (for these various "base" temperatures).

United States

MONTGOMERY, ALABAMA — Elev 62 — Lat 32.3

	HS	VS	TA	D10.0	D18.3
Jan	2373	2827	8.7	94	301
Jul	5808	2587	27.9	0	0
Yr	4385	2985	19.0	238	1133

PHOENIX, ARIZONA — Elev 339 — Lat 33.4

	HS	VS	TA	D10.0	D18.3
Jan	3221	4613	13.8	6	142
Jul	7843	3041	35.1	0	0
Yr	4326	4184	24.2	15	486

TUCSON, ARIZONA — Elev 779 — Lat 32.1

	HS	VS	TA	D10.0	D18.3
Jan	3467	4856	12.0	21	196
Jul	7386	2909	31.1	0	0
Yr	5912	4124	21.4	64	738

FORT SMITH, ARKANSAS — Elev 141 — Lat 35.3

	HS	VS	TA	D10.0	D18.3
Jan	2347	3142	4.8	178	421
Jul	6515	2865	28.3	0	0
Yr	4436	3196	17.0	520	1702

FRESNO, CALIFORNIA — Elev 100 — Lat 36.8

	HS	VS	TA	D10.0	D18.3
Jan	2073	2795	9.2	49	283
Jul	8471	3395	28.8	0	0
Yr	5408	3818	18.5	137	1188

LOS ANGELES, CALIFORNIA — Elev 32 — Lat 33.9

	HS	VS	TA	D10.0	D18.3
Jan	2922	4079	14.4	1	126
Jul	7279	2972	20.7	0	1
Yr	5035	3650	17.4	2	698

SACRAMENTO, CALIFORNIA — Elev 8 — Lat 38.5

	HS	VS	TA	D10.0	D18.3
Jan	1884	2615	8.6	58	301
Jul	8481	3568	24.9	0	0
Yr	5193	3780	16.6	172	1375

SAN DIEGO, CALIFORNIA — Elev 9 — Lat 32.7

	HS	VS	TA	D10.0	D18.3
Jan	3079	4180	14.4	1	123
Jul	6897	2846	21.1	0	1
Yr	5048	3631	17.8	1	612

SAN FRANCISCO, CALIFORNIA — Elev 5 — Lat 37.6

	HS	VS	TA	D10.0	D18.3
Jan	2234	3228	10.7	16	236
Jul	7547	3262	17.4	0	41
Yr	4909	3647	14.6	47	1448

SANTA MARIA, CALIFORNIA — Elev 72 — Lat 34.9

	HS	VS	TA	D10.0	D18.3
Jan	2694	3780	11.6	15	208
Jul	7386	3045	17.5	0	37
Yr	5080	3698	14.7	57	1419

DENVER, COLORADO — Elev 1625 — Lat 39.7

	HS	VS	TA	D10.0	D18.3
Jan	2650	4622	−0.1	314	571
Jul	7171	3322	23.9	0	2
Yr	4953	4209	10.7	1433	3263

GRAND JUNCTION, COLORADO — Elev 1475 — Lat 39.1

	HS	VS	TA	D10.0	D18.3
Jan	2496	4089	−1.9	368	626
Jul	7777	3452	26.5	0	0
Yr	5240	4247	12.0	1304	3009

HARTFORD, CONNECTICUT — Elev 55 — Lat 41.9

	HS	VS	TA	D10.0	D18.3
Jan	1505	2190	−2.8	396	654
Jul	5203	2716	23.5	0	1
Yr	3344	2634	10.7	1454	3238

WASHINGTON, DC — Elev 88 — Lat 38.9

	HS	VS	TA	D10.0	D18.3
Jan	1805	2502	2.6	237	487
Jul	5733	2786	26.8	0	0
Yr	3818	2877	15.0	737	2142

MIAMI, FLORIDA — Elev 2 — Lat 25.8

	HS	VS	TA	D10.0	D18.3
Jan	3335	3537	20.2	1	27
Jul	5562	2483	28.9	0	0
Yr	4650	2969	25.3	1	62

ORLANDO, FLORIDA — Elev 36 — Lat 28.5

	HS	VS	TA	D10.0	D18.3
Jan	3152	3631	15.8	8	102
Jul	5682	2508	28.1	0	0
Yr	4695	3105	22.8	15	284

TALLAHASSEE, FLORIDA — Elev 21 — Lat 30.4

	HS	VS	TA	D10.0	D18.3
Jan	2767	3259	11.2	50	227
Jul	5515	2480	28.0	0	0
Yr	4524	3057	20.4	116	801

ATLANTA, GEORGIA — Elev 315 — Lat 33.6

	HS	VS	TA	D10.0	D18.3
Jan	2265	2789	6.8	132	358
Jul	5717	2590	26.8	0	0
Yr	4250	2969	17.4	360	1432

BOISE, IDAHO — Elev 874 — Lat 43.6

	HS	VS	TA	D10.0	D18.3
Jan	1530	2429	0.2	304	562
Jul	8244	4130	25.5	0	2
Yr	4729	3960	11.9	1151	2951

CHICAGO, ILLINOIS — Elev 190 — Lat 41.8

	HS	VS	TA	D10.0	D18.3
Jan	1600	2385	−3.6	422	680
Jul	6133	3105	24.5	0	1
Yr	3840	3029	11.2	1521	3230

MOLINE, ILLINOIS — Elev 181 — Lat 41.4

	HS	VS	TA	D10.0	D18.3
Jan	1688	2533	−5.0	464	722
Jul	6118	3037	24.1	0	1
Yr	3868	3070	10.7	1656	3356

SPRINGFIELD, ILLINOIS — Elev 187 — Lat 39.8

	HS	VS	TA	D10.0	D18.3
Jan	1846	2688	−2.6	392	649
Jul	6493	3105	24.5	0	1
Yr	4114	3164	12.1	1341	2936

APPENDICES

Table C.20 Average Insolation, Temperature, and DD Data (SI Units) (*Continued*)

Elevation in meters; latitude in degrees north; HS (horizontal surface) and VS (vertical surface) insolation in W/day m^2; TA in degrees C; D10.0, D12.8, D15.6, and D18.3 in degree days C (for these various "base" temperatures).

United States

INDIANAPOLIS, INDIANA				Elev 246	Lat 39.7
	HS	VS	TA	D10.0	D18.3
Jan	1565	2108	−2.1	378	633
Jul	5698	2811	24.3	0	1
Yr	3682	2754	12.1	1303	2902

SOUTH BEND, INDIANA				Elev 236	Lat 41.7
	HS	VS	TA	D10.0	D18.3
Jan	1312	1786	−4.1	439	697
Jul	5843	2978	22.8	0	3
Yr	3597	2726	10.1	1653	3437

DES MOINES, IOWA				Elev 294	Lat 41.5
	HS	VS	TA	D10.0	D18.3
Jan	1833	2877	−5.2	472	730
Jul	6616	3272	24.8	0	1
Yr	4146	3360	10.8	1686	3369

DODGE CITY, KANSAS				Elev 787	Lat 37.8
	HS	VS	TA	D10.0	D18.3
Jan	2609	4111	0.5	297	553
Jul	7241	3196	26.6	0	1
Yr	4928	3887	13.3	1124	2686

TOPEKA, KANSAS				Elev 270	Lat 39.1
	HS	VS	TA	D10.0	D18.3
Jan	2149	3259	−0.9	341	597
Jul	6714	3133	26.6	0	0
Yr	4376	3269	13.4	1156	2679

LEXINGTON, KENTUCKY				Elev 301	Lat 38.0
	HS	VS	TA	D10.0	D18.3
Jan	1723	2253	0.6	301	549
Jul	5837	2780	24.5	0	0
Yr	3852	2814	13.4	1005	2491

BATON ROUGE, LOUISIANA				Elev 23	Lat 30.5
	HS	VS	TA	D10.0	D18.3
Jan	2477	2805	11.0	57	234
Jul	5509	2480	28.1	0	0
Yr	4354	2881	20.2	139	854

NEW ORLEANS, LOUISIANA				Elev 3	Lat 30
	HS	VS	TA	D10.0	D18.3
Jan	2634	2997	12.2	39	199
Jul	5720	2527	28.6	0	0
Yr	4537	2975	21.3	83	663

SHREVEPORT, LOUISIANA				Elev 79	Lat 32.5
	HS	VS	TA	D10.0	D18.3
Jan	2404	2903	9.0	89	294
Jul	6354	2726	28.8	0	0
Yr	4550	3070	19.3	248	1135

PORTLAND, MAINE				Elev 19	Lat 43.6
	HS	VS	TA	D10.0	D18.3
Jan	1420	2174	−4.7	457	715
Jul	5234	2821	21.0	0	7
Yr	3319	2704	8.5	1808	3828

DETROIT, MICHIGAN				Elev 191	Lat 42.4
	HS	VS	TA	D10.0	D18.3
Jan	1316	1846	−3.2	411	668
Jul	5789	3000	23.6	0	1
Yr	3540	2742	10.6	1555	3327

SAULT STE. MARIE, MICHIGAN				Elev 221	Lat 46.5
	HS	VS	TA	D10.0	D18.3
Jan	1025	1552	−9.1	593	851
Jul	5789	3297	18.8	0	34
Yr	3294	2716	5.7	2552	4740

MINNEAPOLIS, MINNESOTA				Elev 255	Lat 44.9
	HS	VS	TA	D10.0	D18.3
Jan	1464	2423	−9.0	588	846
Jul	6215	3379	23.6	0	2
Yr	3698	3142	8.3	2258	4109

MERIDIAN, MISSISSIPPI				Elev 94	Lat 32.3
	HS	VS	TA	D10.0	D18.3
Jan	2347	2786	8.0	109	323
Jul	5752	2571	27.3	0	0
Yr	4326	2944	18.2	299	1282

SAINT LOUIS, MISSOURI				Elev 172	Lat 38.7
	HS	VS	TA	D10.0	D18.3
Jan	1978	2833	0.2	311	563
Jul	6465	3026	26.9	0	0
Yr	4193	3174	14.3	1002	2433

SPRINGFIELD, MISSOURI				Elev 387	Lat 37.2
	HS	VS	TA	D10.0	D18.3
Jan	2158	3016	1.0	288	539
Jul	6509	2953	25.8	0	0
Yr	4303	3190	13.9	963	2427

HELENA, MONTANA				Elev 1188	Lat 46.6
	HS	VS	TA	D10.0	D18.3
Jan	1322	2268	−4.8	459	717
Jul	7364	4139	21.8	0	8
Yr	3994	3578	7.7	2056	4148

NORTH OMAHA, NEBRASKA				Elev 404	Lat 41.4
	HS	VS	TA	D10.0	D18.3
Jan	2000	3262	−2.6	391	649
Jul	6644	3275	24.3	0	2
Yr	4174	3401	10.9	1622	3323

ELY, NEVADA				Elev 1906	Lat 39.3
	HS	VS	TA	D10.0	D18.3
Jan	2584	4354	−3.1	406	664
Jul	7720	3452	20.5	0	7
Yr	5285	4389	7.6	1901	4046

LAS VEGAS, NEVADA				Elev 664	Lat 36.1
	HS	VS	TA	D10.0	D18.3
Jan	3086	4900	9.8	41	266
Jul	8165	3278	34.0	0	0
Yr	5887	4515	21.1	129	1023

RENO, NEVADA				Elev 1341	Lat 39.5
	HS	VS	TA	D10.0	D18.3
Jan	2524	4243	2.6	230	487
Jul	8493	3682	24.9	0	1
Yr	5565	4540	12.5	924	2676

ALBUQUERQUE, NEW MEXICO				Elev 1619	Lat 35.0
	HS	VS	TA	D10.0	D18.3
Jan	3205	4928	3.4	207	464
Jul	7853	3139	26.2	0	0
Yr	5774	4351	14.7	696	2152

Table C.20 Average Insolation, Temperature, and DD Data (SI Units) (*Continued*)

Elevation in meters; latitude in degrees north; HS (horizontal surface) and VS (vertical surface) insolation in W/day m^2; TA in degrees C; D10.0, D12.8, D15.6, and D18.3 in degree days C (for these various "base" temperatures).

United States

LOS ALAMOS, NEW MEXICO				Elev 2249	Lat 35.9
	HS	VS	TA	D10.0	D18.3
Jan	2817	4228	0.3	301	559
Jul	6471	2881	22.0	0	2
Yr	4982	3925	11.5	1105	2851

ALBANY, NEW YORK				Elev 89	Lat 42.7
	HS	VS	TA	D10.0	D18.3
Jan	1439	2126	−4.5	451	709
Jul	5442	2865	22.7	0	2
Yr	3370	2660	9.6	1714	3566

BUFFALO, NEW YORK				Elev 215	Lat 42.9
	HS	VS	TA	D10.0	D18.3
Jan	1101	1467	−3.8	428	685
Jul	5603	2950	22.0	0	4
Yr	3272	2461	9.4	1718	3583

NEW YORK, NEW YORK				Elev 57	Lat 40.8
	HS	VS	TA	D10.0	D18.3
Jan	1578	2234	0.5	294	552
Jul	5326	2716	24.7	0	0
Yr	3474	2679	12.7	1001	2645

SYRACUSE, NEW YORK				Elev 124	Lat 43.1
	HS	VS	TA	D10.0	D18.3
Jan	1215	1697	−4.4	448	705
Jul	5546	2937	22.3	0	3
Yr	3272	2496	9.4	1749	3613

CHARLOTTE, NORTH CAROLINA				Elev 234	Lat 35.2
	HS	VS	TA	D10.0	D18.3
Jan	2268	2978	5.5	159	397
Jul	5777	2653	26.4	0	0
Yr	4347	3095	16.4	468	1683

RALEIGH-DURHAM, NORTH CAROLINA				Elev 134	Lat 35.9
	HS	VS	TA	D10.0	D18.3
Jan	2190	2915	5.2	171	407
Jul	5603	2625	26.6	0	0
Yr	4092	3013	16.1	522	1771

BISMARCK, NORTH DAKOTA				Elev 502	Lat 46.8
	HS	VS	TA	D10.0	D18.3
Jan	1473	2672	−10.6	638	896
Jul	6891	3915	22.2	0	6
Yr	3947	3612	6.4	2658	4674

FARGO, NORTH DAKOTA				Elev 274	Lat 46.9
	HS	VS	TA	D10.0	D18.3
Jan	1309	2272	−12.6	701	959
Jul	6689	3818	21.9	0	8
Yr	3805	3392	6.0	2844	4825

CINCINNATI, OHIO				Elev 271	Lat 39.1
	HS	VS	TA	D10.0	D18.3
Jan	1578	2079	0.0	318	569
Jul	5588	2742	24.3	0	0
Yr	3660	2707	12.8	1088	2653

CLEVELAND, OHIO				Elev 245	Lat 41.4
	HS	VS	TA	D10.0	D18.3
Jan	1224	1600	−2.4	287	643
Jul	5767	2931	23.0	0	2
Yr	3448	2549	10.9	1453	3187

COLUMBUS, OHIO				Elev 254	Lat 40.0
	HS	VS	TA	D10.0	D18.3
Jan	1448	1912	−1.5	360	613
Jul	5537	2764	24.1	0	0
Yr	3559	2631	12.2	1260	2867

OKLAHOMA CITY, OKLAHOMA				Elev 397	35.4
	HS	VS	TA	D10.0	D18.3
Jan	2527	3515	4.0	197	444
Jul	6714	2918	28.0	0	0
Yr	4616	3385	16.3	629	1888

MEDFORD, OREGON				Elev 396	Lat 42.4
	HS	VS	TA	D10.0	D18.3
Jan	1284	1783	4.7	165	422
Jul	7809	3808	24.3	0	1
Yr	4278	3259	13.3	629	2331

SALEM, OREGON				Elev 61	Lat 44.9
	HS	VS	TA	D10.0	D18.3
Jan	1047	1486	5.5	143	398
Jul	6758	3641	20.6	0	10
Yr	3565	2830	12.2	592	2464

HARRISBURG, PENNSYLVANIA				Elev 106	Lat 40.2
	HS	VS	TA	D10.0	D18.3
Jan	1691	2407	−0.9	339	596
Jul	5565	2786	25.0	0	0
Yr	3635	2798	12.4	1162	2792

PHILADELPHIA, PENNSYLVANIA				Elev 9	Lat 39.9
	HS	VS	TA	D10.0	D18.3
Jan	1751	2499	1.0	284	538
Jul	5546	2764	26.1	0	0
Yr	3691	2855	13.8	929	2450

PITTSBURGH, PENNSYLVANIA				Elev 373	Lat 40.5
	HS	VS	TA	D10.0	D18.3
Jan	1338	1745	−1.9	373	627
Jul	5329	2704	22.8	0	2
Yr	3379	2502	11.2	1371	3066

PROVIDENCE, RHODE ISLAND				Elev 19	Lat 41.7
	HS	VS	TA	D10.0	D18.3
Jan	1596	2366	−1.1	346	603
Jul	5348	2770	23.5	0	1
Yr	3515	2789	11.2	1264	3043

CHARLESTON, SOUTH CAROLINA				Elev 12	Lat 32.9
	HS	VS	TA	D10.0	D18.3
Jan	2347	2852	9.7	74	270
Jul	5676	2565	27.9	0	0
Yr	4247	2966	19.3	179	1012

RAPID CITY, SOUTH DAKOTA				Elev 966	Lat 44.0
	HS	VS	TA	D10.0	D18.3
Jan	1710	2928	−3.9	432	691
Jul	7014	3663	23.0	0	4
Yr	4240	3713	8.5	1974	3952

KNOXVILLE, TENNESSEE				Elev 299	Lat 35.8
	HS	VS	TA	D10.0	D18.3
Jan	1959	2477	3.8	208	451
Jul	5692	2647	25.7	0	0
Yr	4023	2868	15.4	630	1937

Table C.20 Average Insolation, Temperature, and DD Data (SI Units) (*Continued*)

Elevation in meters; latitude in degrees north; HS (horizontal surface) and VS (vertical surface) insolation in W/day m^2; TA in degrees C; D10.0, D12.8, D15.6, and D18.3 in degree days C (for these various "base" temperatures).

United States

MEMPHIS, TENNESSEE				Elev 87	Lat 35.0
	HS	VS	TA	D10.0	D18.3
Jan	2155	2745	5.6	167	397
Jul	6222	2773	28.0	0	0
Yr	4316	3045	17.5	479	1587

NASHVILLE, TENNESSEE				Elev 180	Lat 36.1
	HS	VS	TA	D10.0	D18.3
Jan	1830	2275	3.8	211	450
Jul	5966	2742	26.6	0	0
Yr	4013	2811	15.9	638	1906

AUSTIN, TEXAS				Elev 189	Lat 30.3
	HS	VS	TA	D10.0	D18.3
Jan	2726	3180	10.5	60	248
Jul	6641	2729	29.1	0	0
Yr	4663	3073	20.4	165	916

BROWNSVILLE, TEXAS				Elev 6	Lat 25.9
	HS	VS	TA	D10.0	D18.3
Jan	2881	2912	16.6	9	91
Jul	6979	2735	29.5	0	0
Yr	4890	2893	23.9	21	277

DALLAS, TEXAS				Elev 149	Lat 32.8
	HS	VS	TA	D10.0	D18.3
Jan	2590	3265	8.4	97	311
Jul	6695	2808	30.1	0	0
Yr	4638	3199	19.6	277	1174

EL PASO, TEXAS				Elev 1194	Lat 31.8
	HS	VS	TA	D10.0	D18.3
Jan	3549	4960	8.1	81	318
Jul	7730	2947	28.9	0	0
Yr	5998	4187	19.0	234	1224

HOUSTON, TEXAS				Elev 33	Lat 30.0
	HS	VS	TA	D10.0	D18.3
Jan	2436	2688	12.8	32	184
Jul	5767	2540	29.1	0	0
Yr	4269	2789	21.7	77	619

LUBBOCK, TEXAS				Elev 988	Lat 33.6
	HS	VS	TA	D10.0	D18.3
Jan	3253	4723	5.2	161	406
Jul	7610	3016	27.3	0	0
Yr	5578	4035	16.4	544	1787

SAN ANTONIO, TEXAS				Elev 242	Lat 29.5
	HS	VS	TA	D10.0	D18.3
Jan	2824	3237	11.6	41	213
Jul	6692	2723	29.5	0	0
Yr	4736	3070	21.3	109	751

BRYCE CANYON, UTAH				Elev 2313	Lat 37.7
	HS	VS	TA	D10.0	D18.3
Jan	2884	4767	−6.1	500	758
Jul	7648	3294	17.8	0	34
Yr	5496	4430	5.2	2432	4818

SALT LAKE CITY, UTAH				Elev 1288	Lat 40.8
	HS	VS	TA	D10.0	D18.3
Jan	2016	3209	−0.5	326	584
Jul	8171	3742	27.2	0	0
Yr	5067	4105	12.3	1216	2961

BURLINGTON, VERMONT				Elev 104	Lat 44.5
	HS	VS	TA	D10.0	D18.3
Jan	1215	1805	−6.6	517	774
Jul	5430	2969	21.9	0	4
Yr	3228	2571	8.3	2042	3969

NORFOLK, VIRGINIA				Elev 9	Lat 36.9
	HS	VS	TA	D10.0	D18.3
Jan	2139	2940	5.3	169	405
Jul	5846	2739	26.7	0	0
Yr	4187	3123	16.3	500	1744

RICHMOND, VIRGINIA				Elev 50	Lat 37.5
	HS	VS	TA	D10.0	D18.3
Jan	1994	2723	3.6	213	457
Jul	5597	2679	26.3	0	0
Yr	3944	2953	15.3	668	2019

ROANOKE, VIRGINIA				Elev 358	Lat 37.3
	HS	VS	TA	D10.0	D18.3
Jan	2083	2874	2.9	231	479
Jul	5666	2694	25.0	0	1
Yr	4010	3022	14.3	748	2182

SEATTLE-TACOMA, WASHINGTON				Elev 122	Lat 47.4
	HS	VS	TA	D10.0	D18.3
Jan	827	1193	5.7	135	391
Jul	7092	4098	19.2	0	22
Yr	3332	2704	11.6	593	2567

SPOKANE, WASHINGTON				Elev 721	Lat 47.6
	HS	VS	TA	D10.0	D18.3
Jan	994	1565	−1.4	353	612
Jul	7436	4316	21.8	0	12
Yr	3871	3370	9.1	1576	3633

CHARLESTON, WEST VIRGINIA				Elev 290	Lat 38.4
	HS	VS	TA	D10.0	D18.3
Jan	1571	2013	1.4	279	526
Jul	5307	2609	23.9	0	0
Yr	3549	2593	13.3	944	2436

MADISON, WISCONSIN				Elev 262	Lat 43.1
	HS	VS	TA	D10.0	D18.3
Jan	1625	2593	−7.0	527	786
Jul	6102	3183	22.3	0	4
Yr	3764	3086	8.6	2029	3918

MILWAUKEE, WISCONSIN				Elev 211	Lat 42.9
	HS	VS	TA	D10.0	D18.3
Jan	1511	2306	−5.0	464	723
Jul	6190	3209	22.5	0	5
Yr	3767	3019	9.3	1801	3693

SHERIDAN, WYOMING				Elev 1209	Lat 44.8
	HS	VS	TA	D10.0	D18.3
Jan	1631	2840	−4.2	442	700
Jul	7348	3903	21.9	0	7
Yr	4206	3691	7.7	2064	4150

APPENDICES

Table C.20 Average Insolation, Temperature, and DD Data (SI Units) (*Continued*)

Elevation in meters; latitude in degrees north; HS (horizontal surface) and VS (vertical surface) insolation in W/day m^2; TA in degrees C; D10.0, D12.8, D15.6, and D18.3 in degree days C (for these various "base" temperatures).

Canada

EDMONTON, ALBERTA				Elev 677	Lat 53.6
	HS	VS	TA	D10.0	D18.3
Jan	1022	2354	−12.1	684	943
Jul	6237	4348	16.3	1	75
Yr	3515	3802	2.5	3354	5814

HALIFAX, NOVA SCOTIA				Elev 41	Lat 44.6
	HS	VS	TA	D10.0	D18.3
Jan	1439	2325	−5.0	466	724
Jul	5345	2931	19.3	0	19
Yr	3395	2862	7.2	2024	4193

SUFFIELD, ALBERTA				Elev 777	Lat 50.3
	HS	VS	TA	D10.0	D18.3
Jan	1366	2956	−8.4	572	830
Jul	6856	4344	20.4	0	16
Yr	3909	4004	6.0	2532	4676

OTTAWA, ONTARIO				Elev 115	Lat 45.5
	HS	VS	TA	D10.0	D18.3
Jan	1609	2884	−9.8	613	871
Jul	5916	3281	21.2	0	7
Yr	3653	3202	6.8	2448	4451

VANCOUVER, BRITISH COLUMBIA				Elev 94	Lat 37.5
	HS	VS	TA	D10.0	D18.3
Jan	801	1246	4.4	175	432
Jul	6376	3909	18.2	0	25
Yr	3344	2890	10.6	780	2873

TORONTO, ONTARIO				Elev 135	Lat 43.7
	HS	VS	TA	D10.0	D18.3
Jan	1536	2451	−5.0	464	722
Jul	6177	3265	22.1	0	4
Yr	3695	2991	8.9	1859	3779

WINNIPEG, MANITOBA				Elev 250	Lat 50.0
	HS	VS	TA	D10.0	D18.3
Jan	1454	3190	−16.3	814	1073
Jul	6389	3988	19.7	0	24
Yr	3754	3783	3.2	3525	5697

NORMANDIN, QUEBEC				Elev 137	Lat 48.8
	HS	VS	TA	D10.0	D18.3
Jan	1432	2906	−16.8	829	1088
Jul	5386	3253	17.7	0	47
Yr	3445	3322	2.0	3667	6039

Sources except as noted below: Balcomb et al., 1980 / American Solar Energy Society, Inc. The original source provided only I-P units; the conversions to SI units were made by the authors of this book.

The TA, D50, and D65 data are reprinted by permission from ASHRAE, 2021.

Table C.21 Coefficient of Utilization from Window Without Blinds; Sky Component $E_{xvk}/E_{xhk}=0.75$

Room Depth/ Window Height	Depth	Window Width/Window Height 0.5	1	2	3	4	6	8	Infinite
	10	0.824	0.864	0.870	0.873	0.875	0.879	0.880	0.883
	30	0.547	0.711	0.777	0.789	0.793	0.798	0.799	0.801
1	50	0.355	0.526	0.635	0.659	0.666	0.669	0.670	0.672
	70	0.243	0.386	0.505	0.538	0.548	0.544	0.545	0.547
	90	0.185	0.304	0.418	0.451	0.464	0.444	0.446	0.447
	10	0.667	0.781	0.809	0.812	0.813	0.815	0.816	0.824
	30	0.269	0.416	0.519	0.544	0.551	0.556	0.557	0.563
2	50	0.122	0.204	0.287	0.319	0.331	0.339	0.341	0.345
	70	0.068	0.116	0.173	0.201	0.214	0.223	0.226	0.229
	90	0.050	0.084	0.127	0.151	0.164	0.167	0.171	0.172
	10	0.522	0.681	0.739	0.746	0.747	0.749	0.747	0.766
	30	0.139	0.232	0.320	0.350	0.360	0.366	0.364	0.373
3	50	0.053	0.092	0.139	0.163	0.174	0.183	0.182	0.187
	70	0.031	0.053	0.081	0.097	0.106	0.116	0.116	0.119
	90	0.025	0.041	0.061	0.074	0.082	0.089	0.090	0.092
	10	0.405	0.576	0.658	0.670	0.673	0.675	0.674	0.707
	30	0.075	0.134	0.197	0.224	0.235	0.243	0.243	0.255
4	50	0.028	0.050	0.078	0.094	0.104	0.112	0.114	0.119
	70	0.018	0.031	0.048	0.059	0.065	0.073	0.074	0.078
	90	0.016	0.026	0.040	0.048	0.053	0.059	0.061	0.064
	10	0.242	0.392	0.494	0.516	0.521	0.524	0.523	0.588
	30	0.027	0.054	0.086	0.102	0.111	0.119	0.120	0.135
6	50	0.011	0.023	0.036	0.044	0.049	0.055	0.056	0.063
	70	0.009	0.018	0.027	0.032	0.035	0.040	0.041	0.046
	90	0.008	0.016	0.023	0.028	0.031	0.034	0.035	0.040
	10	0.147	0.257	0.352	0.380	0.387	0.391	0.392	0.482
	30	0.012	0.026	0.043	0.054	0.060	0.067	0.070	0.086
8	50	0.006	0.013	0.021	0.026	0.029	0.033	0.035	0.043
	70	0.005	0.011	0.017	0.021	0.023	0.026	0.027	0.034
	90	0.004	0.010	0.015	0.019	0.021	0.023	0.025	0.030
	10	0.092	0.168	0.248	0.275	0.284	0.290	0.291	0.395
	30	0.006	0.014	0.026	0.032	0.036	0.041	0.044	0.059
10	50	0.003	0.008	0.014	0.017	0.019	0.022	0.024	0.032
	70	0.003	0.007	0.012	0.014	0.016	0.018	0.019	0.026
	90	0.003	0.006	0.011	0.013	0.015	0.016	0.017	0.024

Source: IES RP-23-1989; reprinted with permission of Illuminating Engineering Society.

Table C.22 Coefficient of Utilization from Window Without Blinds; Sky Component $E_{xvk}/E_{xhk} = 1.00$

Room Depth/ Window Height	Depth	Window Width/Window Height 0.5	1	2	3	4	6	8	Infinite
1	10	0.671	0.704	0.711	0.715	0.717	0.726	0.726	0.728
	30	0.458	0.595	0.654	0.668	0.672	0.682	0.683	0.685
	50	0.313	0.462	0.563	0.589	0.598	0.607	0.608	0.610
	70	0.227	0.362	0.478	0.515	0.527	0.530	0.532	0.534
	90	0.186	0.306	0.424	0.465	0.481	0.468	0.471	0.472
2	10	0.545	0.636	0.658	0.660	0.661	0.665	0.666	0.672
	30	0.239	0.367	0.459	0.484	0.491	0.499	0.501	0.506
	50	0.121	0.203	0.286	0.320	0.335	0.348	0.351	0.355
	70	0.074	0.128	0.192	0.226	0.243	0.259	0.264	0.267
	90	0.058	0.101	0.156	0.188	0.207	0.215	0.221	0.223
3	10	0.431	0.561	0.607	0.613	0.614	0.616	0.615	0.631
	30	0.133	0.223	0.306	0.337	0.348	0.357	0.357	0.366
	50	0.058	0.103	0.155	0.183	0.197	0.211	0.213	0.218
	70	0.037	0.064	0.098	0.119	0.132	0.147	0.150	0.154
	90	0.030	0.051	0.079	0.098	0.110	0.122	0.126	0.129
4	10	0.339	0.482	0.549	0.560	0.563	0.566	0.565	0.593
	30	0.078	0.139	0.204	0.234	0.247	0.258	0.260	0.272
	50	0.033	0.060	0.094	0.114	0.126	0.139	0.143	0.150
	70	0.022	0.039	0.061	0.074	0.083	0.095	0.099	0.104
	90	0.019	0.032	0.050	0.061	0.070	0.080	0.084	0.089
6	10	0.211	0.343	0.433	0.453	0.458	0.461	0.461	0.518
	30	0.033	0.065	0.103	0.123	0.135	0.145	0.148	0.167
	50	0.015	0.029	0.047	0.057	0.064	0.073	0.077	0.086
	70	0.011	0.021	0.033	0.040	0.045	0.051	0.054	0.060
	90	0.010	0.019	0.028	0.034	0.038	0.044	0.046	0.052
8	10	0.135	0.238	0.326	0.353	0.362	0.366	0.367	0.452
	30	0.016	0.034	0.058	0.072	0.080	0.090	0.094	0.116
	50	0.008	0.017	0.027	0.034	0.039	0.045	0.048	0.059
	70	0.006	0.013	0.021	0.026	0.028	0.032	0.035	0.043
	90	0.005	0.012	0.019	0.023	0.025	0.029	0.031	0.038
10	10	0.090	0.165	0.244	0.272	0.283	0.290	0.291	0.395
	30	0.009	0.020	0.036	0.045	0.052	0.060	0.064	0.087
	50	0.005	0.010	0.019	0.023	0.026	0.030	0.033	0.044
	70	0.004	0.009	0.015	0.018	0.020	0.023	0.025	0.033
	90	0.003	0.008	0.014	0.016	0.018	0.020	0.022	0.030

Source: IES RP-23-1989; reprinted with permission of Illuminating Engineering Society.

Table C.23 Coefficient of Utilization from Window Without Blinds; Sky Component $E_{xvk}/E_{xhk} = 1.25$

Room Depth/ Window Height	Depth	Window Width/Window Height							
		0.5	1	2	3	4	6	8	Infinite
1	10	0.578	0.607	0.614	0.619	0.621	0.633	0.634	0.635
	30	0.405	0.525	0.580	0.594	0.599	0.612	0.614	0.615
	50	0.287	0.423	0.519	0.547	0.556	0.569	0.571	0.573
	70	0.218	0.347	0.461	0.501	0.515	0.522	0.525	0.526
	90	0.186	0.307	0.428	0.473	0.491	0.483	0.486	0.487
2	10	0.472	0.549	0.566	0.569	0.570	0.574	0.575	0.581
	30	0.221	0.337	0.422	0.447	0.456	0.465	0.467	0.472
	50	0.120	0.202	0.285	0.321	0.337	0.353	0.357	0.361
	70	0.078	0.136	0.204	0.242	0.261	0.281	0.287	0.290
	90	0.064	0.112	0.174	0.211	0.233	0.244	0.251	0.253
3	10	0.377	0.488	0.527	0.533	0.534	0.536	0.536	0.549
	30	0.130	0.217	0.298	0.329	0.341	0.352	0.353	0.362
	50	0.062	0.110	0.165	0.195	0.211	0.228	0.231	0.237
	70	0.040	0.070	0.109	0.132	0.147	0.166	0.171	0.175
	90	0.033	0.057	0.090	0.112	0.127	0.142	0.148	0.152
4	10	0.300	0.424	0.484	0.494	0.497	0.499	0.499	0.524
	30	0.080	0.143	0.209	0.240	0.255	0.267	0.269	0.283
	50	0.036	0.066	0.104	0.126	0.140	0.156	0.160	0.168
	70	0.024	0.043	0.068	0.083	0.094	0.109	0.115	0.120
	90	0.021	0.036	0.056	0.070	0.080	0.092	0.099	0.103
6	10	0.193	0.314	0.395	0.415	0.420	0.423	0.423	0.476
	30	0.036	0.071	0.113	0.136	0.149	0.161	0.165	0.186
	50	0.017	0.033	0.053	0.065	0.074	0.084	0.089	0.100
	70	0.012	0.024	0.037	0.045	0.050	0.058	0.061	0.069
	90	0.011	0.021	0.031	0.038	0.043	0.049	0.053	0.060
8	10	0.128	0.226	0.310	0.337	0.346	0.351	0.352	0.433
	30	0.019	0.039	0.066	0.082	0.092	0.104	0.109	0.134
	50	0.009	0.019	0.031	0.040	0.045	0.052	0.056	0.069
	70	0.007	0.015	0.023	0.029	0.032	0.037	0.040	0.049
	90	0.006	0.013	0.021	0.025	0.028	0.032	0.035	0.043
10	10	0.088	0.164	0.241	0.270	0.282	0.290	0.291	0.396
	30	0.011	0.024	0.043	0.054	0.062	0.071	0.076	0.103
	50	0.005	0.012	0.022	0.026	0.030	0.035	0.038	0.052
	70	0.004	0.010	0.017	0.020	0.023	0.026	0.028	0.038
	90	0.004	0.009	0.016	0.018	0.020	0.023	0.025	0.034

Source: IES RP-23-1989; reprinted with permission of Illuminating Engineering Society.

Table C.24 Coefficient of Utilization from Window Without Blinds; Sky Component $E_{xvk}/E_{xhk} = 1.50$

Room Depth/ Window Height	Depth	Window Width/Window Height 0.5	1	2	3	4	6	8	Infinite
	10	0.503	0.528	0.536	0.541	0.544	0.557	0.558	0.559
	30	0.359	0.464	0.514	0.528	0.534	0.549	0.550	0.552
1	50	0.261	0.384	0.471	0.499	0.508	0.524	0.526	0.527
	70	0.204	0.325	0.432	0.470	0.485	0.497	0.499	0.500
	90	0.179	0.295	0.412	0.456	0.475	0.474	0.477	0.478
	10	0.412	0.477	0.490	0.492	0.493	0.498	0.499	0.505
	30	0.201	0.304	0.379	0.402	0.410	0.422	0.424	0.429
2	50	0.115	0.192	0.269	0.304	0.320	0.339	0.343	0.347
	70	0.078	0.136	0.204	0.241	0.261	0.286	0.292	0.295
	90	0.066	0.117	0.183	0.221	0.246	0.262	0.271	0.273
	10	0.331	0.426	0.458	0.461	0.462	0.465	0.465	0.477
	30	0.121	0.202	0.275	0.304	0.316	0.327	0.329	0.337
3	50	0.062	0.109	0.164	0.193	0.209	0.228	0.232	0.238
	70	0.041	0.073	0.114	0.138	0.154	0.176	0.183	0.188
	90	0.035	0.062	0.099	0.123	0.141	0.159	0.169	0.173
	10	0.265	0.372	0.422	0.430	0.433	0.435	0.435	0.456
	30	0.077	0.137	0.199	0.229	0.243	0.256	0.259	0.272
4	50	0.037	0.069	0.107	0.130	0.144	0.161	0.167	0.175
	70	0.026	0.046	0.073	0.089	0.101	0.119	0.126	0.132
	90	0.022	0.039	0.063	0.078	0.090	0.106	0.114	0.120
	10	0.173	0.281	0.351	0.368	0.373	0.375	0.375	0.422
	30	0.037	0.073	0.115	0.137	0.151	0.164	0.168	0.189
6	50	0.018	0.036	0.058	0.071	0.080	0.092	0.098	0.110
	70	0.013	0.026	0.040	0.049	0.056	0.064	0.069	0.078
	90	0.012	0.023	0.035	0.043	0.048	0.057	0.062	0.070
	10	0.117	0.207	0.282	0.305	0.314	0.319	0.320	0.393
	30	0.020	0.042	0.071	0.087	0.098	0.111	0.116	0.143
8	50	0.010	0.021	0.035	0.044	0.050	0.058	0.063	0.078
	70	0.007	0.016	0.026	0.032	0.036	0.041	0.045	0.055
	90	0.076	0.014	0.023	0.028	0.031	0.036	0.040	0.049
	10	0.082	0.153	0.224	0.250	0.262	0.269	0.271	0.368
	30	0.012	0.026	0.047	0.059	0.068	0.078	0.084	0.114
10	50	0.006	0.014	0.024	0.030	0.034	0.040	0.044	0.060
	70	0.005	0.011	0.019	0.022	0.025	0.029	0.032	0.043
	90	0.004	0.010	0.017	0.020	0.023	0.026	0.028	0.038

Source: IES RP-23-1989; reprinted with permission of Illuminating Engineering Society.

Table C.25 Coefficient of Utilization from Window Without Blinds; Sky Component $E_{xvk}/E_{xhk} = 1.75$

Room Depth/ Window Height	Depth	Window Width/Window Height 0.5	1	2	3	4	6	8	Infinite
1	10	0.435	0.457	0.465	0.471	0.474	0.486	0.488	0.489
	30	0.317	0.407	0.452	0.466	0.471	0.486	0.488	0.489
	50	0.234	0.343	0.422	0.447	0.456	0.472	0.475	0.476
	70	0.187	0.297	0.395	0.430	0.445	0.458	0.461	0.462
	90	0.168	0.276	0.384	0.426	0.444	0.447	0.450	0.451
2	10	0.357	0.412	0.422	0.424	0.424	0.430	0.431	0.436
	30	0.180	0.271	0.335	0.356	0.363	0.375	0.378	0.381
	50	0.106	0.177	0.246	0.278	0.293	0.313	0.318	0.321
	70	0.074	0.130	0.194	0.229	0.249	0.274	0.282	0.284
	90	0.065	0.116	0.181	0.219	0.244	0.264	0.273	0.276
3	10	0.288	0.369	0.394	0.397	0.397	0.400	0.401	0.411
	30	0.110	0.183	0.247	0.272	0.282	0.294	0.296	0.304
	50	0.058	0.104	0.154	0.181	0.196	0.215	0.221	0.226
	70	0.040	0.072	0.112	0.136	0.152	0.176	0.184	0.188
	90	0.035	0.063	0.101	0.126	0.144	0.166	0.177	0.182
4	10	0.232	0.324	0.365	0.371	0.373	0.375	0.375	0.394
	30	0.071	0.127	0.183	0.209	0.222	0.235	0.238	0.250
	50	0.036	0.067	0.104	0.125	0.139	0.157	0.163	0.171
	70	0.025	0.046	0.072	0.089	0.101	0.119	0.127	0.134
	90	0.022	0.041	0.065	0.082	0.095	0.114	0.124	0.130
6	10	0.153	0.247	0.307	0.320	0.324	0.326	0.327	0.367
	30	0.035	0.070	0.109	0.130	0.143	0.155	0.160	0.180
	50	0.018	0.036	0.058	0.071	0.080	0.091	0.098	0.110
	70	0.013	0.026	0.041	0.051	0.058	0.067	0.073	0.082
	90	0.012	0.023	0.037	0.046	0.052	0.062	0.069	0.078
8	10	0.104	0.184	0.249	0.269	0.276	0.281	0.282	0.346
	30	0.020	0.042	0.070	0.086	0.096	0.109	0.115	0.141
	50	0.010	0.022	0.036	0.046	0.052	0.060	0.066	0.081
	70	0.008	0.017	0.027	0.033	0.038	0.044	0.048	0.059
	90	0.007	0.015	0.024	0.030	0.034	0.040	0.044	0.054
10	10	0.074	0.138	0.201	0.223	0.233	0.240	0.242	0.328
	30	0.012	0.027	0.048	0.059	0.067	0.078	0.084	0.114
	50	0.006	0.014	0.026	0.032	0.036	0.043	0.047	0.064
	70	0.005	0.011	0.020	0.024	0.027	0.031	0.034	0.046
	90	0.004	0.010	0.018	0.022	0.024	0.028	0.031	0.042

Source: IES RP-23-1989; reprinted with permission of Illuminating Engineering Society.

Table C.26 Coefficient of Utilization from Window Without Blinds; Ground Component

Room Depth/ Window Height	Depth	Window Width/Window Height 0.5	1	2	3	4	6	8	Infinite
1	10	0.105	0.137	0.177	0.197	0.207	0.208	0.210	0.211
	30	0.116	0.157	0.203	0.225	0.235	0.241	0.243	0.244
	50	0.110	0.165	0.217	0.241	0.252	0.267	0.269	0.270
	70	0.101	0.162	0.217	0.243	0.253	0.283	0.285	0.286
	90	0.091	0.146	0.199	0.230	0.239	0.290	0.292	0.293
2	10	0.095	0.124	0.160	0.178	0.186	0.186	0.189	0.191
	30	0.082	0.132	0.179	0.201	0.212	0.219	0.222	0.225
	50	0.062	0.113	0.165	0.189	0.202	0.214	0.218	0.220
	70	0.051	0.093	0.141	0.165	0.179	0.194	0.198	0.200
	90	0.045	0.079	0.118	0.140	0.153	0.179	0.183	0.185
3	10	0.088	0.120	0.157	0.175	0.183	0.185	0.163	0.167
	30	0.059	0.107	0.154	0.176	0.187	0.198	0.193	0.198
	50	0.039	0.074	0.114	0.134	0.146	0.157	0.163	0.170
	70	0.031	0.055	0.085	0.101	0.111	0.122	0.127	0.130
	90	0.028	0.047	0.070	0.083	0.092	0.107	0.113	0.115
4	10	0.073	0.113	0.154	0.174	0.183	0.187	0.176	0.184
	30	0.040	0.082	0.127	0.148	0.159	0.170	0.177	0.185
	50	0.025	0.049	0.078	0.094	0.103	0.113	0.117	0.123
	70	0.020	0.036	0.054	0.065	0.071	0.079	0.083	0.087
	90	0.019	0.032	0.046	0.054	0.060	0.069	0.073	0.076
6	10	0.056	0.106	0.143	0.164	0.175	0.184	0.173	0.194
	30	0.021	0.050	0.081	0.098	0.107	0.117	0.123	0.138
	50	0.013	0.027	0.041	0.049	0.054	0.060	0.064	0.072
	70	0.011	0.021	0.029	0.033	0.035	0.039	0.041	0.046
	90	0.011	0.020	0.026	0.030	0.032	0.035	0.037	0.042
8	10	0.036	0.082	0.122	0.143	0.156	0.166	0.170	0.208
	30	0.011	0.029	0.050	0.062	0.070	0.078	0.082	0.101
	50	0.007	0.016	0.024	0.028	0.031	0.035	0.038	0.046
	70	0.006	0.013	0.018	0.020	0.021	0.023	0.025	0.030
	90	0.006	0.013	0.017	0.019	0.020	0.022	0.023	0.028
10	10	0.024	0.061	0.109	0.120	0.131	0.144	0.147	0.200
	30	0.006	0.017	0.034	0.040	0.046	0.053	0.056	0.076
	50	0.004	0.010	0.016	0.018	0.020	0.023	0.024	0.033
	70	0.004	0.009	0.013	0.014	0.015	0.016	0.016	0.022
	90	0.004	0.009	0.013	0.013	0.014	0.015	0.016	0.021

Source: IES RP-23-1989; reprinted with permission of Illuminating Engineering Society.

Table C.27 Reflectances of Daylighting Modeling Materials

Materials	Number	Reflectance (%)
Crescent Board Color		
Raven Black	989	6.7
Newport Blue	977	9.1
Wine	907A	12.6
Williamsburg Green	988	12.7
Volcano Blue	1081	15.5
Russet	996	15.7
Las Cruces Purple	1076	16.1
Marine Blue	1082	16.5
Madeira Red	1075	19.4
Avocado	1084	21.6
Persimmon	1087	26.4
Gibraltar Grey	1074	28.5
Burnt Orange	1077	33.5
Moss Point Green	1001	35.0
Bar Harbor Grey	976	38.5
Bimini Blue	1080	43.3
Mist Grey	1002	43.5
Suntan	1062	45.8
Stone Grey	975	48.5
French Blue	972	54.2
Cameo Rose	973	56.0
Biscay Blue	1073	56.0
Sauterne	1089	59.4
Lime	910A	59.7
Sandstone	1061	61.0
Madagascar Pink	1078	73.7
Mist	1088	76.3
Daffodil	971	88.8
Yellow	902A	93.1
Mat Board		
Arctic White	3297	91.4
Linen Board		
Cream Linen	2961A	91.0
French Grey	2962A	76.0
Museum Board		
2-ply White	1150	96.5
2-ply Antique	1157	91.5
2-ply Cream	1152	94.2

Source: Technical Reference Sheet, The Lighting Design Lab, Seattle, WA.

References and Resources

ASHRAE. 1997. *ASHRAE Handbook—Fundamentals*. Atlanta, GA: American Society of Heating, Refrigerating and Air-Conditioning Engineers.

ASHRAE. 2021. *ANSI/ASHRAE Standard 169-2021: Climate Data for Building Design Standards*. Atlanta, GA: American Society of Heating, Refrigerating and Air-Conditioning Engineers.

Balcomb, J.D., D. Barley, R. McFarland, J. Perry, W. Wray, and S. Noll. 1980. *Passive Solar Design Handbook*, Vol. 2: *Passive Solar Design Analysis*. Washington, DC: U.S. Department of Energy.

Balcomb, J.D., et al. 1983. *Passive Solar Design Handbook*, Vol. 3. Boulder, CO: American Solar Energy Society, Inc. IES. 1989. *IES RP-23-1989: Recommended Practice for the Lumen Method of Daylight Calculations*. New York: Illuminating Engineering Society of North America (withdrawn 2013).

Lunde, P.J. 1980. *Solar Thermal Engineering*. New York: John Wiley & Sons.

APPENDIX D

Solar Geometry

This appendix provides information on various aspects of solar geometry, including three sets of sunpath diagrams (sunpeg charts, horizontal projection charts, and vertical projection charts):

- Table D.1 Solar altitude and azimuth data for 30, 34, 38, 42, 44, and 48°N latitudes
- Figure D.1 Sunpeg charts for 28, 32, 36, 40, 44, 48, and 52°N latitudes
- Figure D.2 Horizontal projection (equidistant) sunpath charts for 24, 28, 32, 36, 40, 44, 48, and 52°N latitudes
- Figure D.3 Vertical projection sunpath charts for 28, 32, 36, 40, 44, 48, 52, and 56°N latitudes

APPENDICES

Mechanical and Electrical Equipment for Buildings, Fourteenth Edition. Walter Grondzik, Daniel Overbey, and Alison Kwok.

Companion website: www.wiley.com/go/meeb14e

D.1 SOLAR ALTITUDE AND AZIMUTH DATA FOR 30, 34, 38, 42, 44, AND 48°N LATITUDES

Table D.1 Typical Solar Altitude and Azimuth Data as a Function of Date and Time of Day

Latitude (°N)		Date	Solar Time[a] a.m.: 6 / p.m.: 6	7 / 5	8 / 4	9 / 3	10 / 2	11 / 1	Noon
30	Altitude	June 21	12	24	37	50	63	75	83
		Mar–Sept 21	—	13	26	38	49	57	60
		Dec 21	—	—	12	21	29	35	37
	Azimuth	June 21	111	104	99	92	84	67	0
		Mar–Sept 21	90	83	74	64	49	28	0
		Dec 21	—	60	54	44	32	17	0
34	Altitude	June 21	13	25	37	50	62	74	79
		Mar–Sept 21	—	12	25	36	46	53	56
		Dec 21	—	—	9	18	26	31	33
	Azimuth	June 21	110	103	95	90	78	58	0
		Mar–Sept 21	90	82	72	61	46	26	0
		Dec 21	—	—	54	43	30	16	0
38	Altitude	June 21	14	26	37	49	61	71	75
		Mar–Sept 21	—	12	23	34	43	50	52
		Dec 21	—	—	7	16	23	27	28
	Azimuth	June 21	109	101	90	83	70	46	0
		Mar–Sept 21	90	81	71	58	43	24	0
		Dec 21	—	—	54	43	30	16	0
42	Altitude	June 21	16	26	38	49	60	68	71
		Mar–Sept 21	—	11	22	32	40	46	48
		Dec 21	—	—	4	13	19	23	25
	Azimuth	June 21	108	99	89	78	63	39	0
		Mar–Sept 21	90	80	69	56	41	22	0
		Dec 21	—	—	53	42	29	15	0
46	Altitude	June 21	17	27	37	48	57	65	67
		Mar–Sept 21	—	10	20	30	37	42	44
		Dec 21	—	—	2	10	15	20	21
	Azimuth	June 21	107	97	88	74	58	34	0
		Mar–Sept 21	90	79	67	54	39	21	0
		Dec 21	—	—	52	41	28	14	0
48	Altitude	June 21	17	27	37	47	56	63	65
		Mar–Sept 21	—	10	20	29	36	40	42
		Dec 21	—	—	1	8	14	17	19
	Azimuth	June 21	106	95	85	72	55	31	0
		Mar–Sept 21	90	79	67	53	38	20	0
		Dec 21	—	—	52	41	28	14	0

[a]Solar time and clock time do not usually coincide. They are related by the equation of time (see Chapter 11).

D.2 SUNPEG CHARTS FOR 28, 32, 36, 40, 44, 48, AND 52°N LATITUDES

The sunpeg charts in Figure D.1 indicate the exact positions of solar radiation and shadow on a model of any scale, on any date, at any time of day between shortly after sunrise and shortly before sunset. Latitudes are indicated in the upper-left corner of each chart. See Chapter 11 for instructions on the use of these charts.

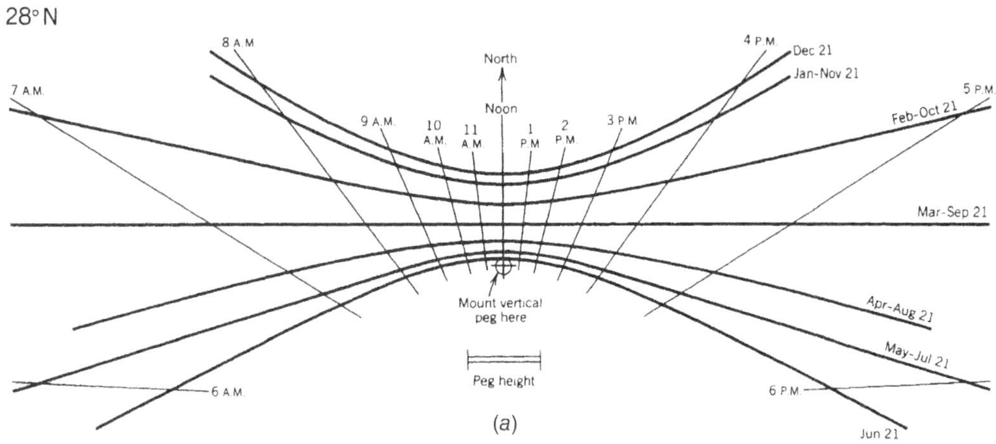

(a)

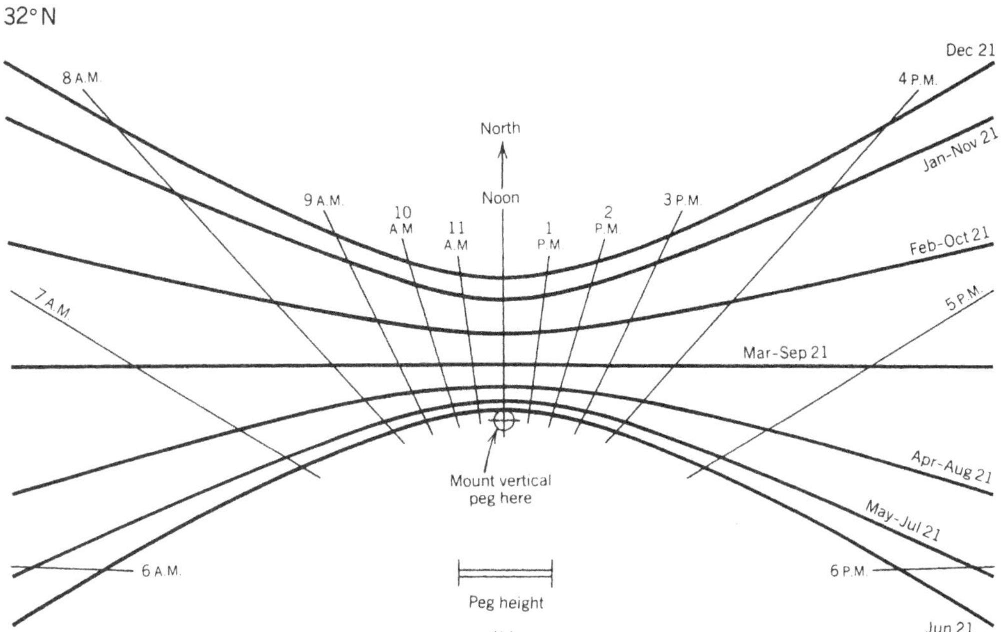

(b)

Figure D.1 *(a–g) Sunpeg shadow charts.*

APPENDICES

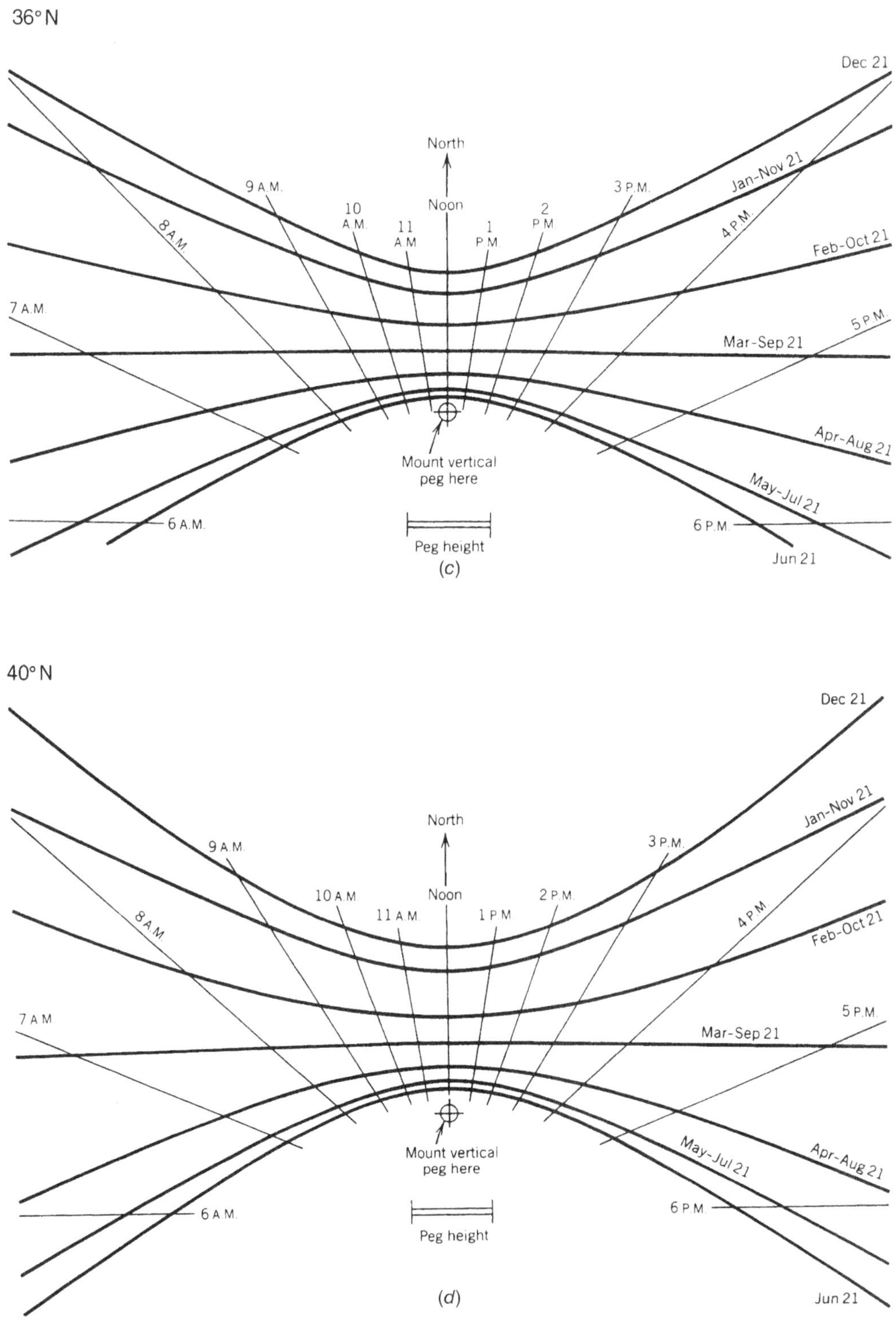

Figure D.1 *(continued)*

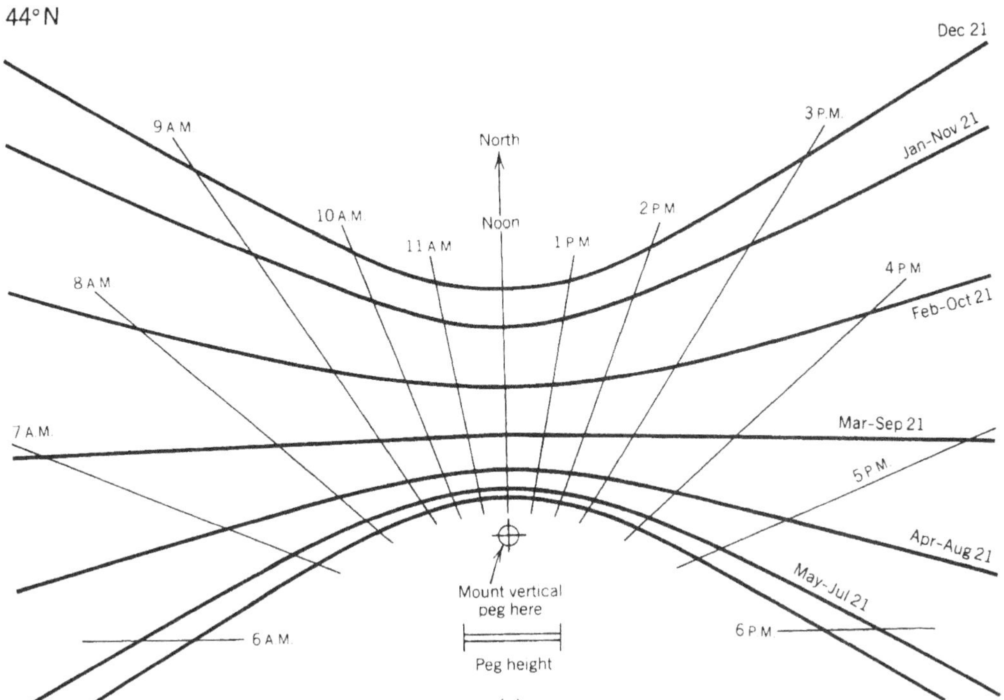

(e)

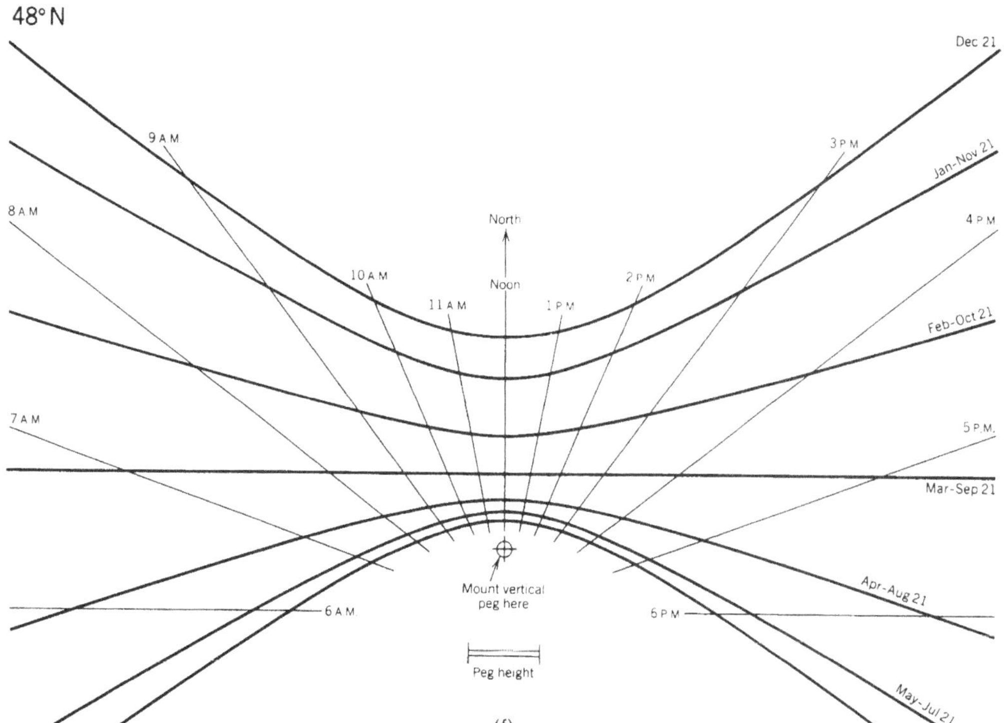

(f)

Figure D.1 *(continued)*

APPENDICES

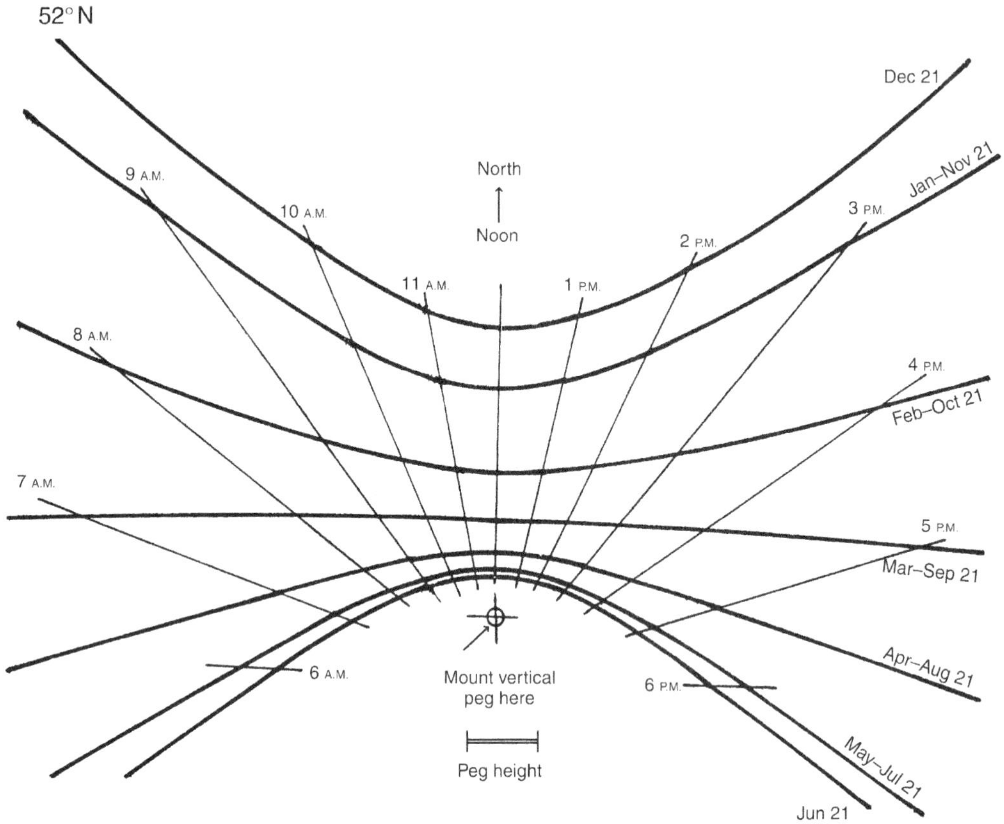

(g)

Figure D.1 *(continued)*

D.3 HORIZONTAL PROJECTION (EQUIDISTANT) SUNPATH CHARTS FOR 24, 28, 32, 36, 40, 44, 48, AND 52°N LATITUDES

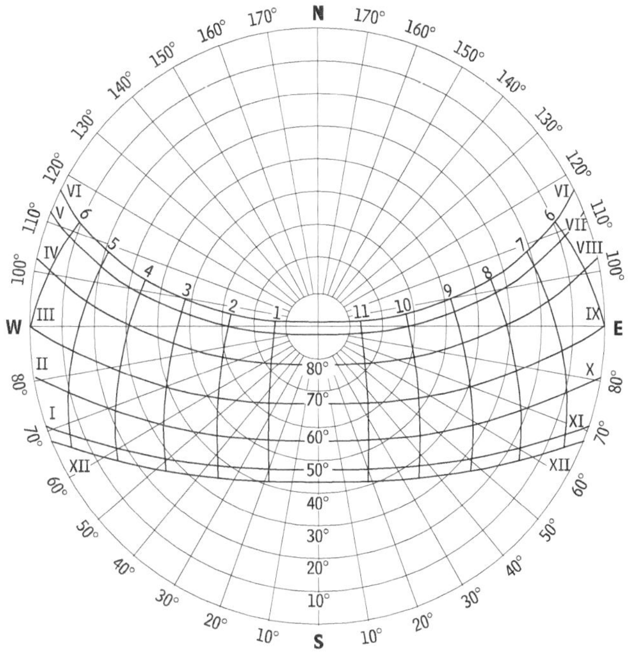

(*a*) **24°N LATITUDE**

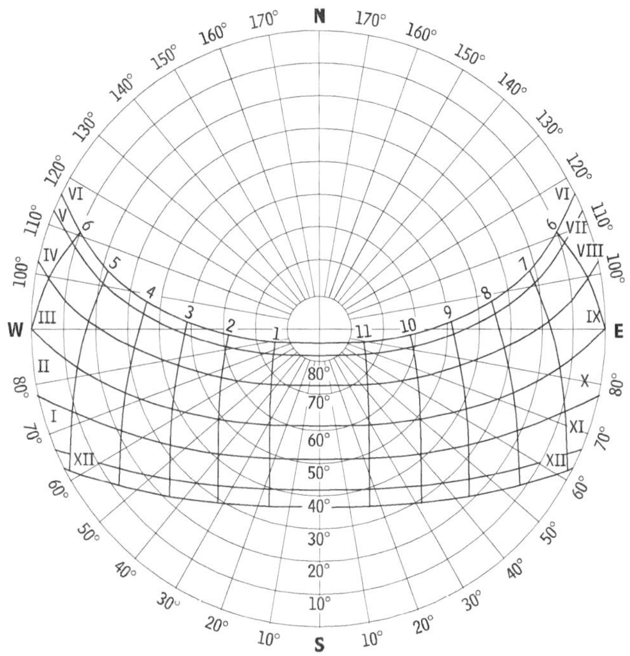

(*b*) **28°N LATITUDE**

Figure D.2 *(a–h) Set of horizontal projection (equidistant) sunpath charts at 4° latitude intervals. Latitudes are indicated below each chart. (Adapted from Ramsey and Sleeper, 2016.)*

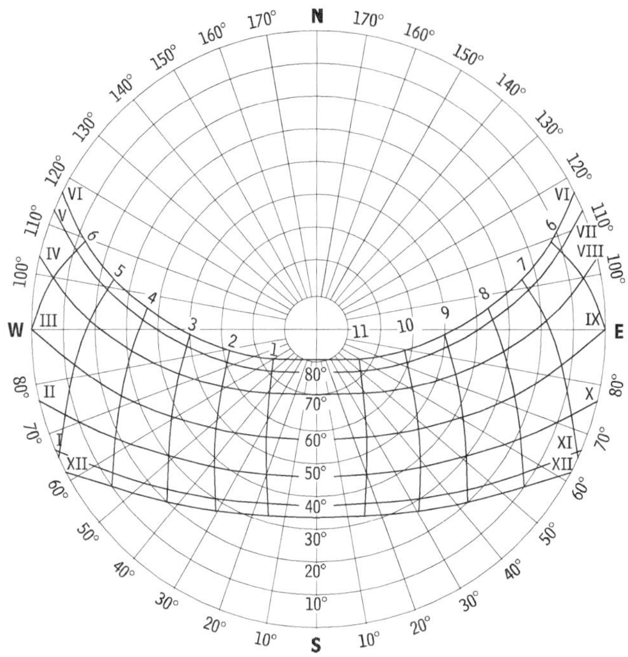

(c) **32°N LATITUDE**

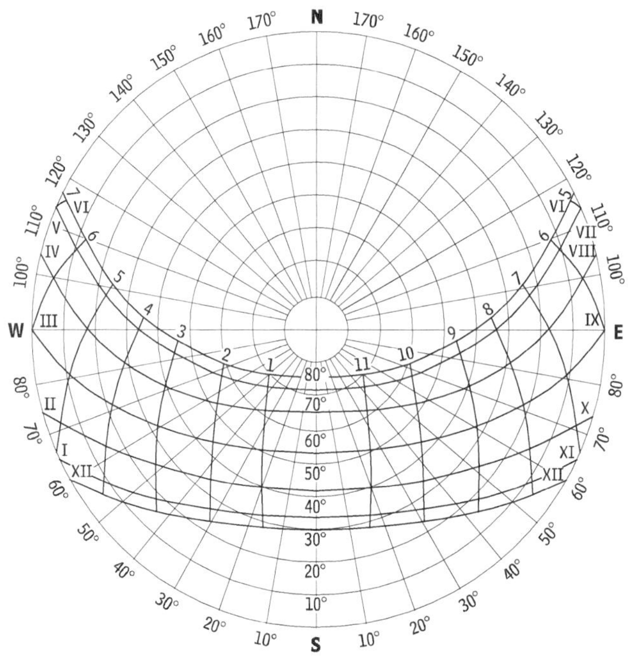

(d) **36°N LATITUDE**

Figure D.2 *(continued)*

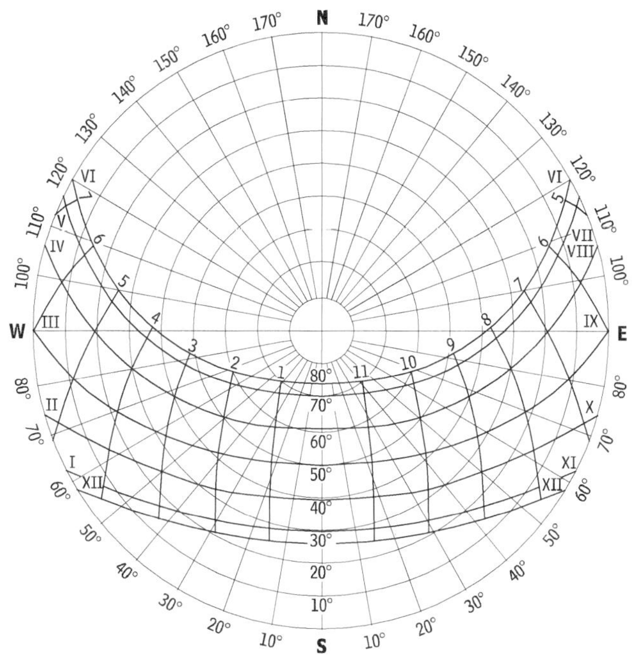

(*e*) **40°N LATITUDE**

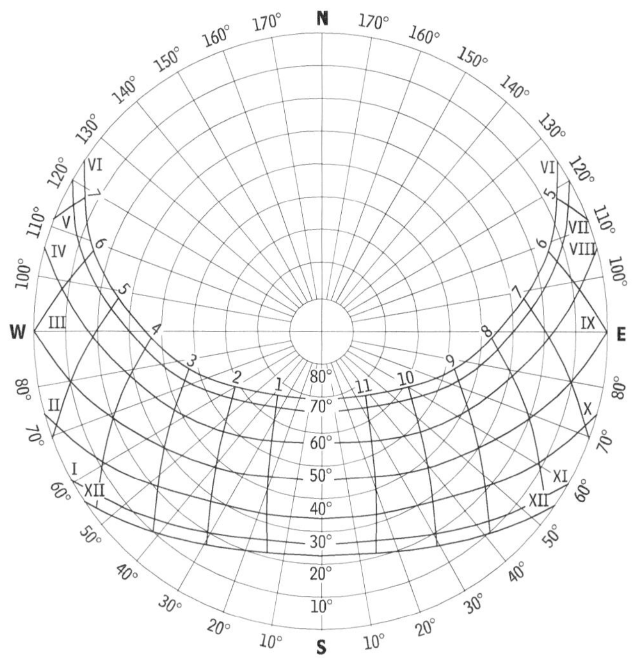

(*f*) **44°N LATITUDE**

Figure D.2 *(continued)*

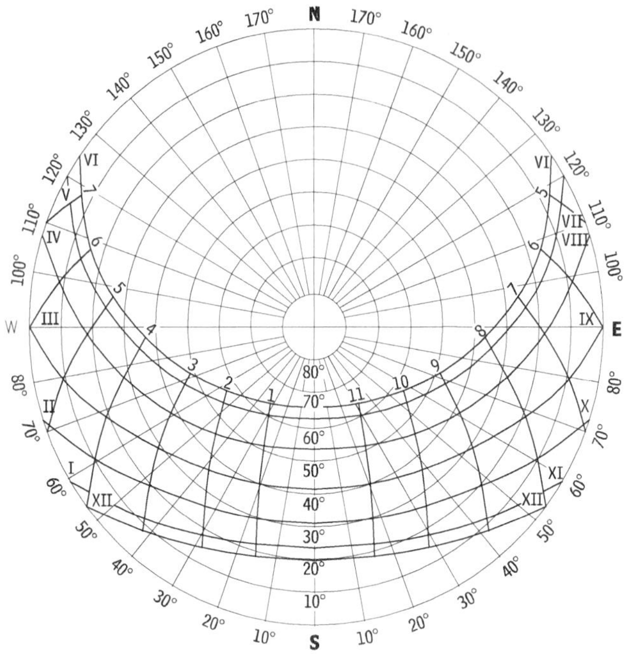

(g) **48°N LATITUDE**

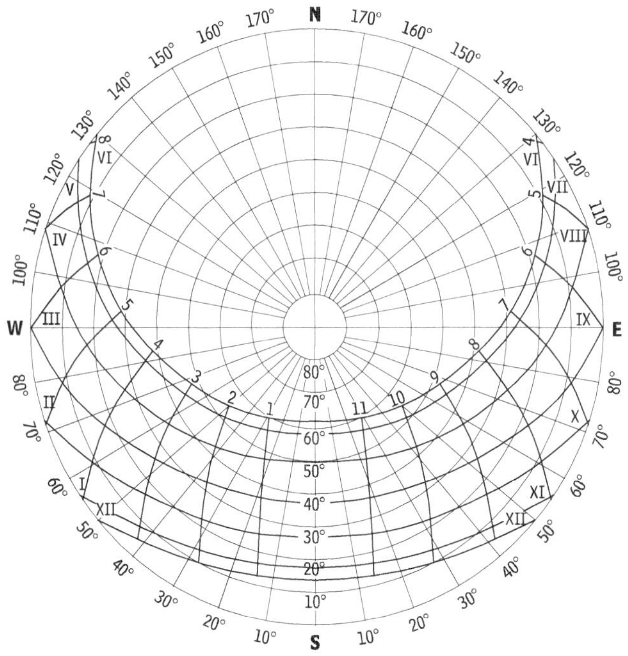

(h) **52°N LATITUDE**

Figure D.2 *(continued)*

D.4 VERTICAL PROJECTION SUNPATH CHARTS FOR 28, 32, 36, 40, 44, 48, 52, AND 56°N LATITUDES

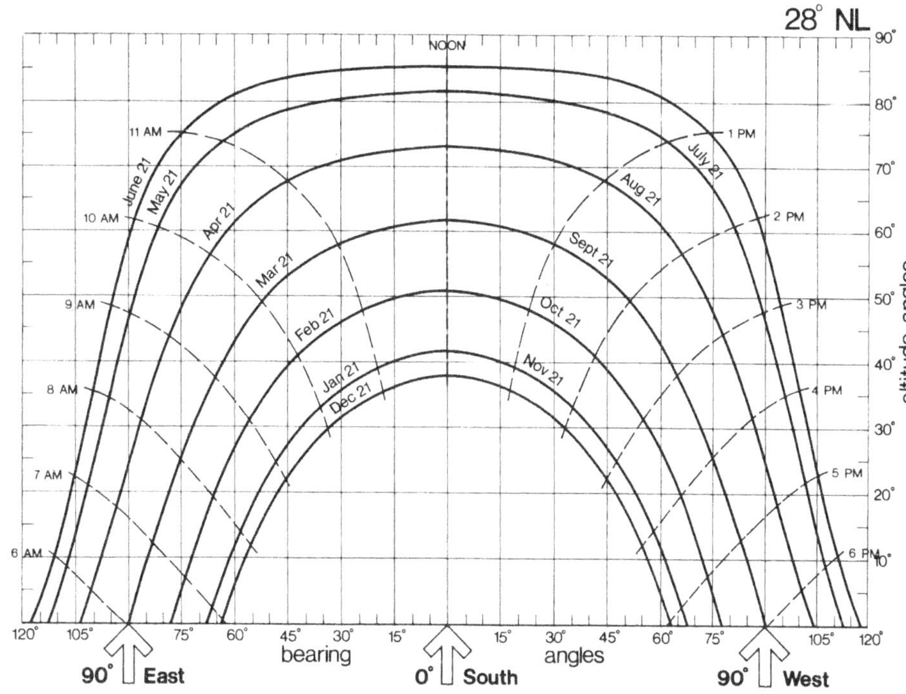

(a)

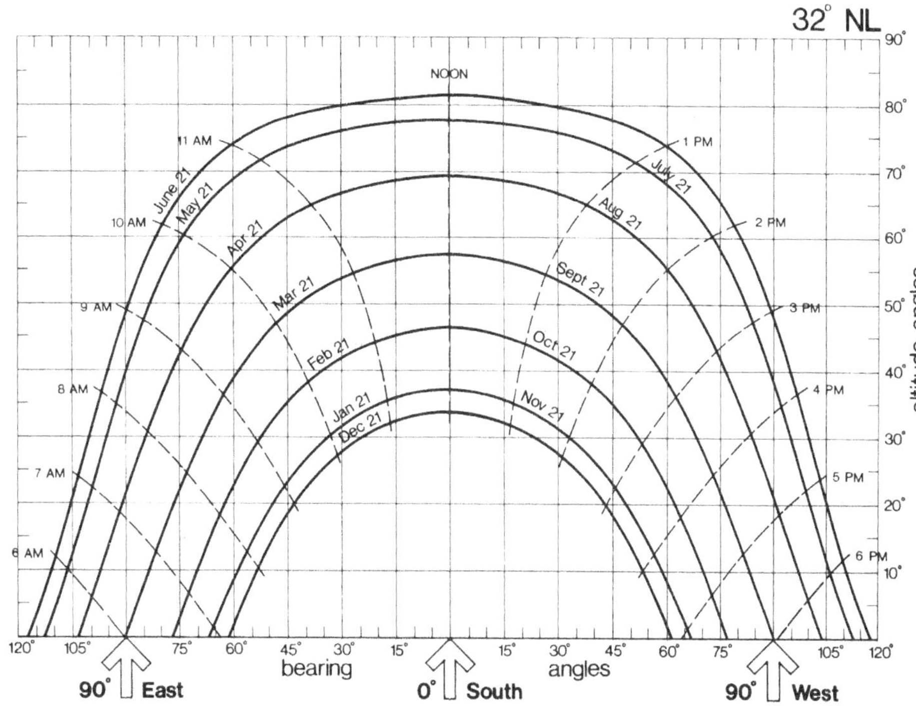

(b)

Figure D.3 *(a–h) Set of vertical projection sunpath charts at 4° latitude intervals. Latitudes are indicated in the upper-right corner of each chart. (From Mazria, 1979.)*

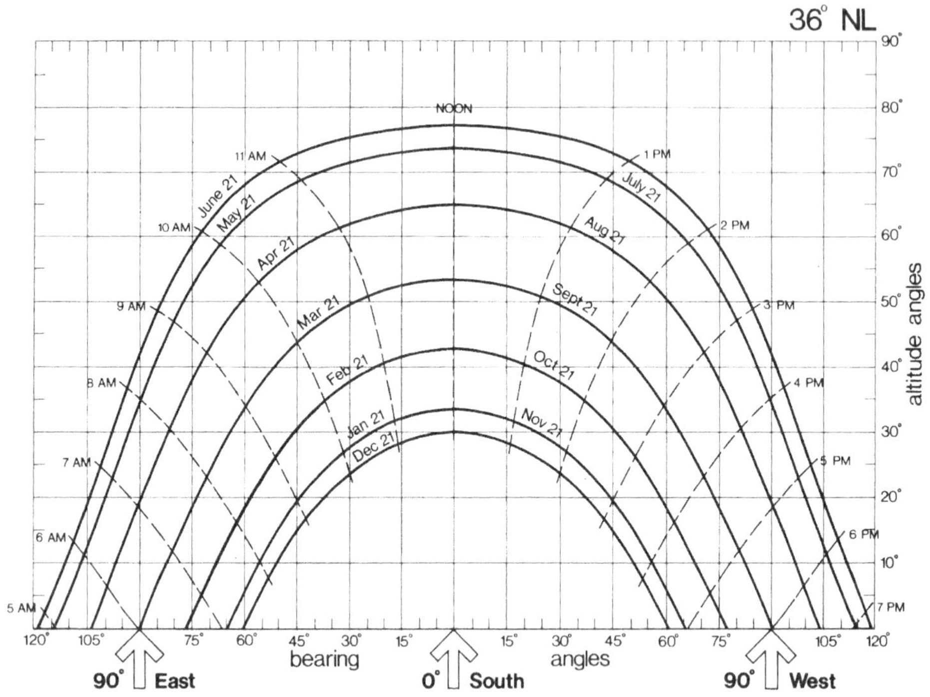

(c)

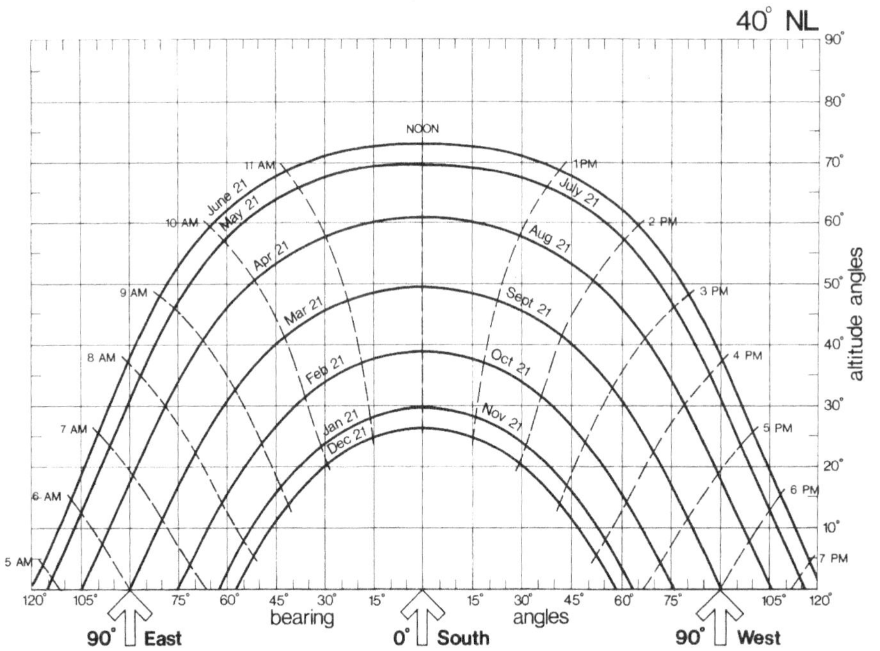

(d)

Figure D.3 *(continued)*

APPENDICES

283

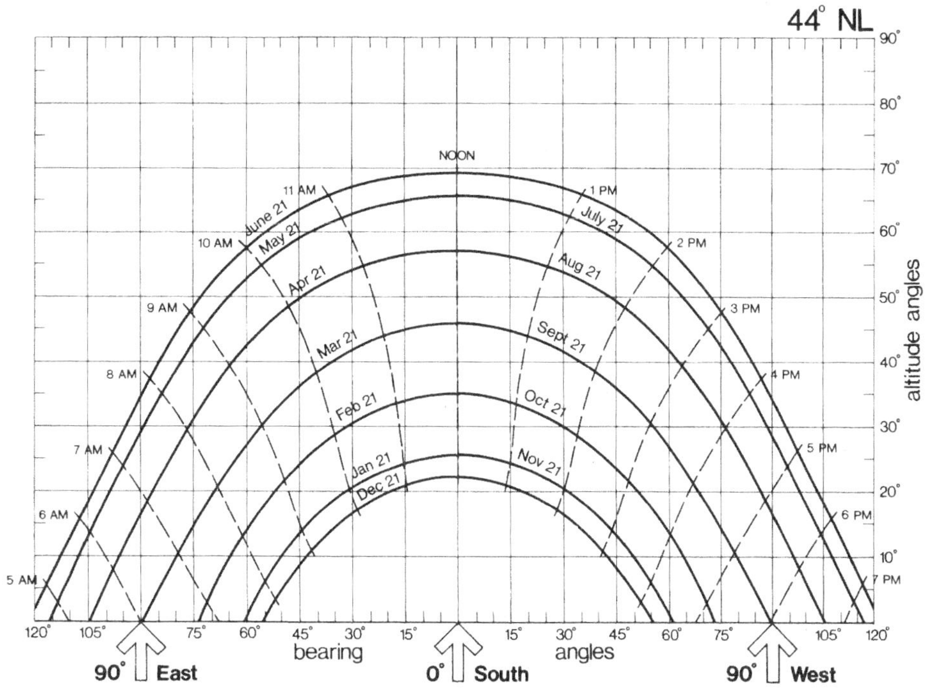

(e)

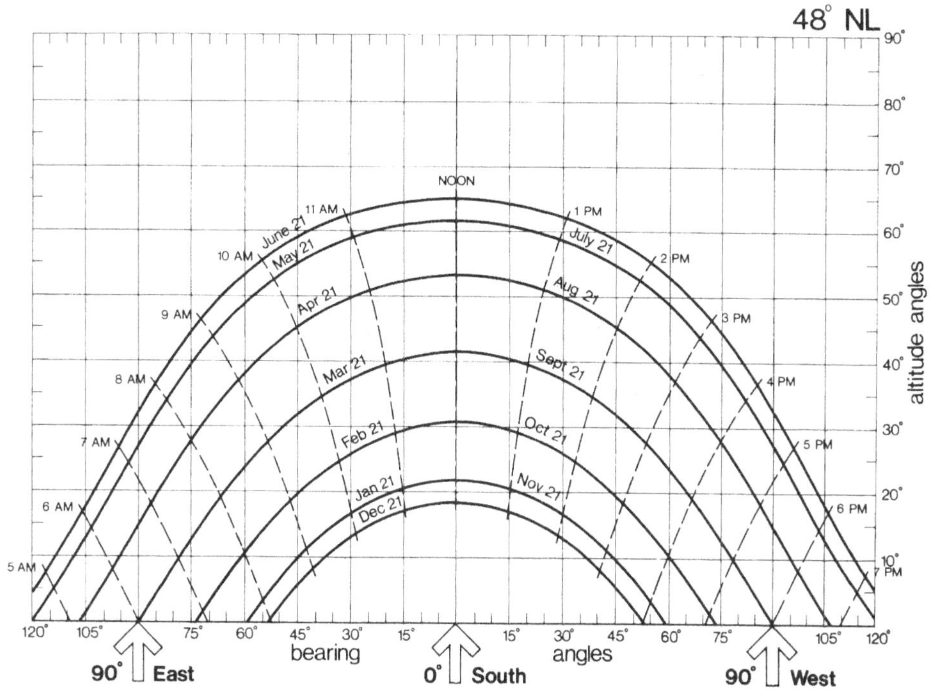

(f)

Figure D.3 *(continued)*

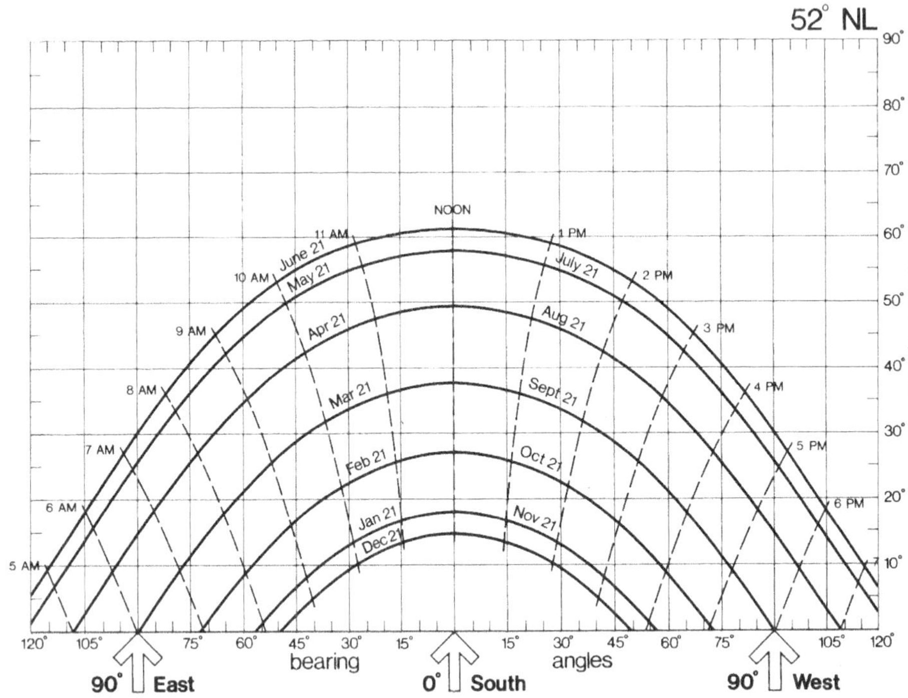

(g)

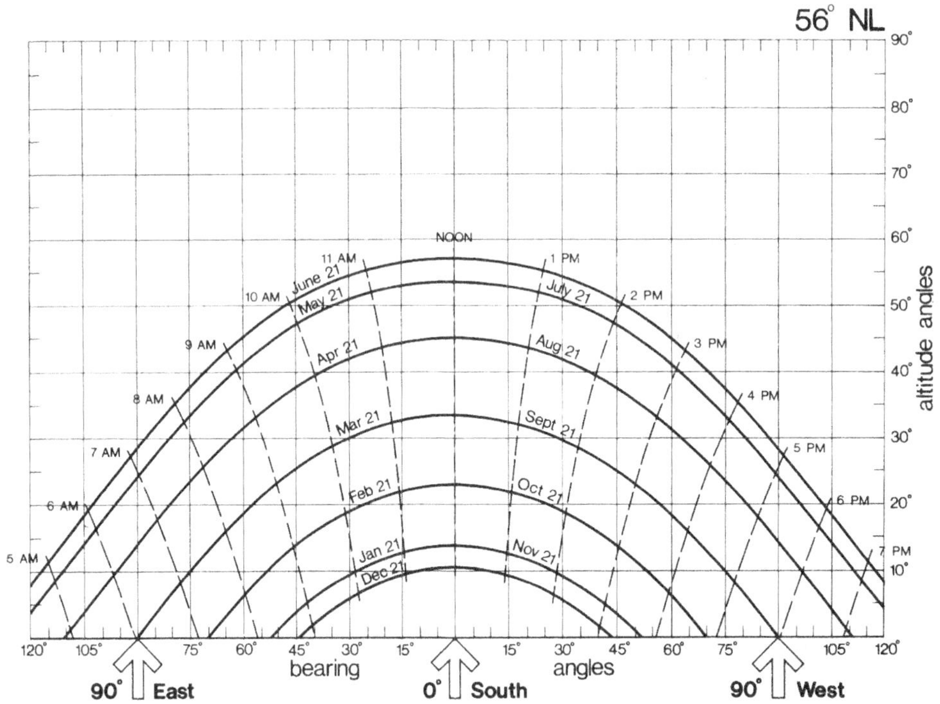

(h)

Figure D.3 *(continued)*

References

Mazria, E. 1979. *The Passive Solar Energy Book*. Emmaus, PA: Rodale Press.

Ramsey, C.G. and H.R. Sleeper. 2016. *Architectural Graphic Standards*, 12th edn. New York: John Wiley & Sons.

APPENDIX E

Thermal Properties of Materials and Assemblies

The tables in this appendix provide information on thermal (and occasionally visual) properties for a range of generic building materials and assemblies:

The information provided herein would typically be used early in the design process before specific products have been selected and/or to determine the properties of assemblies that are fabricated from generic materials (such as gypsum wallboard, bricks, and concrete masonry units). When specific products (such as insulations, windows, or skylights) have been selected for use, data for those specific materials and assemblies should be obtained from appropriate manufacturers' resources (catalogs or websites) rather than from these tables.

APPENDICES

Mechanical and Electrical Equipment for Buildings, Fourteenth Edition. Walter Grondzik, Daniel Overbey, and Alison Kwok.

Companion website: www.wiley.com/go/meeb14e

Table E.1 Building and Insulating Materials: Design Values[a]

Part A: I-P Units				
Description	**Density lb/ft³**	**Conductivity[b] (k) Btu-in/h ft² °F**	**Resistance (R) h ft² °F/Btu**	**Specific Heat Btu/lb °F**
Insulating Materials				
Blanket and Batt[c,d]				
Glass-fiber batts				0.20
	0.47–0.51	0.32–0.33	—	—
	0.61–0.75	0.28–0.30	—	—
	0.79–0.85	0.26–0.27	—	—
	1.4	0.23	—	—
Rock and Slag Wool Batts	—	—	—	0.20
	2–2.3	0.25–0.26	—	—
	2.8	0.23–0.24	—	—
Mineral Wool, Felted	1–3	0.28	—	—
	1–8	0.24	—	—
Board and Slabs				
Cellular Glass	7.5	0.29		0.20
Cement Fiber Slabs, Shredded Wood with Portland Cement Binder	25–27	0.50–0.53	—	—
With Magnesia Oxysulfide Binder	22	0.57	—	0.31
Glass Fiber Board	—	—	—	0.20
	1.5–6.0	0.23–0.24	—	—
Expanded Rubber (rigid)	4	0.2	—	0.40
Extruded Polystyrene, Smooth Skin	—	—	—	0.35
Aged per *CAN/ULC Standard S770-2003*	1.4–3.6	0.18–0.20		
Aged 180 days	1.4–3.6	0.20		
European Product	1.9	0.21		
Aged 5 years at 75°F	2–2.2	0.21	—	—
Blown with Low Global Warming Potential (GWP) (<5) Blowing Agent		0.24–0.25	—	—
Expanded Polystyrene, Molded Beads	—	—	—	0.35
	1.0–1.5	0.24–0.26	—	—
	1.8	0.23	—	—
Mineral Fiberboard, Wet Felted	10	0.26	—	0.20
Rock Wool Board	—	—	—	0.20
Floors and Walls	4.0–8.0	0.23–0.25	—	—
Roofing	10–11	0.27–0.29	—	0.20
Acoustical Tile[g]	21–23	0.36–0.37	—	0.14–0.19
Perlite Board	9	0.36	—	—
Polyisocyanurate	—	—	—	0.35
Unfaced, Aged Per *CAN/ULC Standard S770-2003*	1.6–2.3	0.16–0.17	—	—
With Foil Facers, Aged 180 days[e]	—	0.15–0.16	—	—
Phenolic Foam Board with Facers, Aged[f]	—	0.14–0.16	—	—
Loose Fill				
Cellulose Fiber, Loose Fill	—	—	—	0.33
Attic Application up to 4 in.	1.0–1.2	0.31–0.32	—	—
Attic Application > 4 in.	1.2–1.6	0.27–0.28	—	—
Wall Application, Dense Packed	3.5	0.27–0.28	—	—
Perlite, Expanded	2–4	0.27–0.31	—	0.26
	4–7.5	0.31–0.36	—	—
	7.5–11	0.36–0.42	—	—
Glass Fiber[d]				
Attics, ~4–12 in.	0.4–0.5	0.36–0.38	—	—
Attics, ~12–22 in.	0.5–0.6	0.34–0.36	—	—
Closed Attic or Wall Cavities	1.8–2.3	0.24–0.25	—	—

Table E.1 (*Continued*)

Part A: I-P Units					
Description		**Density lb/ft³**	**Conductivity[b] (k) Btu-in/h ft² °F**	**Resistance (R) h ft² °F/Btu**	**Specific Heat Btu/lb °F**
Rock and Slag Wool[d]					
Attics, ~3.5–4.5 in.		1.5–1.6	0.34	—	—
Attics, ~5–17 in.		1.5–1.8	0.32–0.33	—	—
Closed Attic or Wall Cavities		4.0	0.27–0.29	—	—
Vermiculite, Exfoliated		7.0–8.2	0.47	—	0.32
		4.0–6.0	0.44	—	—
Spray-Applied					
Cellulose, Sprayed into Open Wall Cavities		1.6–2.6	0.27–0.28	—	—
Glass Fiber, Sprayed into Open Wall or Attic Cavities		1.0	0.27–0.29	—	—
		1.8–2.3	0.23–0.26	—	—
Polyurethane Foam		—	—	—	0.35
Low Density, Open Cell		0.45–0.65	0.26–0.29	—	—
Medium Density, Closed Cell, Aged 180 days		1.9–3.2	0.14–0.20	—	—
Building Board and Siding Board					
Board					
Asbestos/Cement Board		120	4.0	—	0.24
Cement Board		71	1.7	—	0.20
Fiber/Cement Board		88	1.7	—	0.20
		61	1.3	—	0.20
		26	0.5	—	0.45
		20	0.4	—	0.45
Gypsum or Plaster Board		40	1.1	—	0.21
Oriented Strand Board (OSB)	7/16 in.	41	—	0.62	0.45
	1/2 in.	41	—	0.68	0.45
Plywood (Douglas Fir)	1/2 in.	29	—	0.79	0.45
	5/8 in.	34	—	0.85	0.45
Plywood/Wood Panels	3/4 in.	28	—	1.08	0.45
Vegetable Fiber Board					
Sheathing, Regular Density	1/2 in.	18	—	1.32	0.31
Intermediate Density	1/2 in.	22	—	1.09	0.31
Nail-based Sheathing	1/2 in.	25	—	1.06	0.31
Shingle Backer	3/8 in.	18	—	0.94	0.30
Sound Deadening Board	1/2 in.	15	—	1.35	0.30
Tile and Lay-in Panels, Plain or Acoustic		18	0.4	—	0.14
Laminated Paperboard		30	0.5	—	0.33
Homogeneous Board from Repulped Paper		30	0.5	—	0.28
Hardboard					
Medium Density		50	0.73	—	0.31
High Density, Service-Tempered Grade and Service Grade		55	0.82	—	0.32
High Density, Standard-Tempered Grade		63	1.0	—	0.32
Particleboard					
Low Density		37	0.71	—	0.31
Medium Density		50	0.94	—	0.31
High Density		62	1.18	0.85	—
Underlayment	5/8 in.	44	0.73	0.82	0.29
Waferboard		37	0.63	0.21	0.45

Table E.1 (*Continued*)

Part A: I-P Units					
Description		**Density lb/ft³**	**Conductivity[b] (k) Btu-in/h ft² °F**	**Resistance (R) h ft² °F/Btu**	**Specific Heat Btu/lb °F**
Shingles					
Asbestos/Cement		120	—	0.21	—
Wood, 16 in., 7 1/2 in. Exposure		—	—	0.87	0.31
Wood, Double, 16 in., 12 in. Exposure		—	—	1.19	0.28
Wood, Plus ins. Backer Board	5/16 in.	—	—	1.40	0.31
Siding					
Asbestos/Cement, Lapped	1/4 in.	—	—	0.21	0.24
Asphalt Roll Siding		—	—	0.15	0.35
Asphalt Insulating Siding (1/2 in. bed)		—	—	0.21	0.24
Hardboard Siding	7/16 in.	—	—	0.15	0.35
Wood, Drop, 8 in.	1 in.	—	—	0.79	0.28
Wood, Bevel		—			
8 in., Lapped	1/2 in.	—	—	0.81	0.28
10 in., Lapped	3/4 in.	—	—	1.05	0.28
Wood, Plywood, 3/8 in., Lapped		—	—	0.59	0.29
Aluminum, Steel, or Vinyl,[j,k] Over Sheathing					—
Hollow-backed		—	—	0.62	0.29[i]
Insulating-board-backed	3/8 in.	—	—	1.82	0.32
Foil-backed	3/8 in.	—	—	2.96	—
Insulated Vinyl Siding	3/4–1 1/4 in.			2.0–2.7[m]	
Architectural (Soda-lime Float) Glass		158	6.9	—	0.21
Building Membrane					
Vapor-permeable Felt		—	—	0.06	—
Vapor: Seal, 2 Layers of Mopped 15 lb Felt		—	—	0.12	—
Vapor: Seal, Plastic Film		—	—	Negligible	—
Finish Flooring Materials					
Carpet and Rebounded Urethane Pad	3/4 in.	7	—	2.38	—
Carpet and Rubber Pad (One-piece)	3/8 in.	20	—	0.68	—
Pile Carpet with Rubber Pad	3/8–1/2 in.	18	—	1.59	—
Linoleum/Cork Tile	1/4 in.	29	—	0.51	—
PVC/Rubber Floor Covering		—	2.8	—	—
Rubber Tile	1.0 in.	119	—	0.34	—
Terrazzo	1.0 in.	—	—	0.08	0.19
Metals (see ASHRAE, 2021, Chapter 33, Table 3)					
Roofing					
Asbestos/Cement Shingles		120	—	0.21	0.24
Asphalt (Bitumen with Inert Fill)		100	2.98	—	—
		119	4.0	—	—
		144	7.97	—	—
Asphalt Roll Roofing		70	—	0.15	0.36
Asphalt Shingles		70	—	0.44	0.30
Built-up Roofing	3/8 in.	70	—	0.33	0.35
Mastic Asphalt (Heavy, 20% Grit)		59	1.32	—	—
Reed Thatch		17	0.62	—	—
Roofing Felt		141	8.32	—	—
Slate	1/2 in.	—	—	0.05	0.30
Straw Thatch		15	0.49	—	—
Wood Shingles, Plain and Plastic-Film-Faced		—	—	0.94	0.31

APPENDICES

Table E.1 (*Continued*)

Part A: I-P Units					
Description		**Density lb/ft³**	**Conductivity[b] (k) Btu-in/h ft² °F**	**Resistance (R) h ft² °F/Btu**	**Specific Heat Btu/lb °F**
Plastering Materials					
Cement Plaster, Sand Aggregate		116	5.0	—	0.20
Sand Aggregate	3/8 in.	—	—	0.08	0.20
	3/4 in.	—	—	0.15	0.20
Gypsum Plaster		70	2.63	—	—
		80	3.19	—	—
Lightweight Aggregate	1/2 in.	45	—	0.32	—
	5/8 in.	45	—	0.39	—
On Metal Lath	3/4 in.	—	—	0.47	—
Perlite Aggregate		45	1.5	—	0.32
Sand Aggregate		105	5.6	—	0.20
On Metal Lath	3/4 in.	—	—	0.13	—
Vermiculite Aggregate		30	1.0	—	—
		40	1.39	—	—
		45	1.7	—	—
		50	1.8	—	—
		60	2.08	—	—
Perlite Plaster		25	0.55	—	—
		38	1.32	—	—
Pulpboard or Paper Plaster		38	0.48	—	—
Sand/Cement Plaster, Conditioned		98	4.4	—	—
Sand/Cement/Lime Plaster, Conditioned		90	3.33	—	—
Sand/Gypsum (3:1) Plaster, Conditioned		97	4.5	—	—
Masonry Materials					
Masonry Units					
Brick, Fired Clay		150	8.4–10.2	—	—
		140	7.4–9.0	—	—
		130	6.4–7.8	—	—
		120	5.6–6.8	—	0.19
		110	4.9–5.9	—	—
		100	4.2–5.1	—	—
		90	3.6–4.3	—	—
		80	3.0–3.7	—	—
		70	2.5–3.1	—	—
Clay Tile, Hollow					
1 Cell Deep	3 in.	—	—	0.80	0.21
	4 in.	—	—	1.11	—
2 Cells Deep	6 in.	—	—	1.52	—
	8 in.	—	—	1.85	—
	10 in.	—	—	2.22	—
3 Cells Deep	12 in.	—	—	2.50	—
Lightweight Brick		50	1.39	—	—
		48	1.51	—	—
Concrete Blocks[h i]					
Limestone Aggregate					
8 in., 36 lb, 138 lb/ft³ Concrete, 2 Cores		—	—	—	—
With Perlite-Filled Cores		—	—	2.1	—
12 in., 55 lb, 138 lb/ft³ Concrete, 2 Cores		—	—	—	—
With Perlite-Filled Cores		—	—	3.7	—
Normal-Weight Aggregate (Sand and Gravel)					
8 in., 33–36 lb, 126–136 lb/ft³ Concrete, 2 or 3 Cores		—	—	1.11–0.97	0.22

APPENDICES

Table E.1 (*Continued*)

Part A: I-P Units				
Description	Density lb/ft³	Conductivity[b] (k) Btu-in/h ft² °F	Resistance (R) h ft² °F/Btu	Specific Heat Btu/lb °F
With Perlite-filled Cores	—	—	2.0	—
With Vermiculite-Filled Cores	—	—	1.92–1.37	—
12 in., 50 lb, 125 lb/ft³ Concrete, 2 Cores	—	—	1.23	0.22
Medium-Weight Aggregate (Combinations of Normal and Lightweight Aggregate)				
8 in., 26–29 lb, 97–112 lb/ft³ Concrete, 2 or 3 Cores	—	—	1.71–1.28	—
With Perlite-Filled Cores	—	—	3.7–2.3	—
With Vermiculite-Filled Cores	—	—	3.3	—
With Molded-EPS-Filled (Beads) Cores	—	—	3.2	—
With Molded EPS Inserts in Cores	—	—	2.7	—
Lightweight (low-mass) Aggregate (Expanded Shale, Clay, Slate or Slag, Pumice)				
6 in., 16–17 lb, 85–87 lb/ft³ Concrete, 2 or 3 Cores	—	—	1.93–1.65	—
With Perlite-Filled Cores	—	—	4.2	—
With Vermiculite-Filled Cores	—	—	3.0	—
8 in., 19–22 lb, 72–86 lb/ft³ Concrete	—	—	3.2–1.9	0.21
With Perlite-Filled Cores	—	—	6.8–4.4	—
With Vermiculite-Filled Cores	—	—	5.3–3.9	—
With Molded-EPS-Filled (Beads) Cores	—	—	4.8	—
With UF Foam-Filled Cores	—	—	4.5	—
With Molded EPS Inserts in Cores	—	—	3.5	—
12 in., 32–36 lb, 80–90 lb/ft³ Concrete, 2 or 3 Cores	—	—	2.6–2.3	—
With Perlite-Filled Cores	—	—	9.2–6.3	—
With Vermiculite-Filled Cores	—	—	5.8	—
Stone, Lime, or Sand	180	72	—	—
Quartzitic and Sandstone	160	43	—	—
	140	24	—	—
	120	13	—	0.19
Calcitic, Dolomitic, Limestone, Marble, and Granite	180	30	—	—
	160	22	—	—
	140	16	—	—
	120	11	—	0.19
	100	8	—	—
Gypsum Partition Tile				
3 by 12 by 30 in., Solid	—	—	1.26	0.19
4 Cells	—	—	1.35	—
4 by 12 by 30 in., 3 Cells	—	—	1.67	—
Limestone	150	3.95	—	0.20
	163	6.45	—	0.20
Concretes				
Sand and Gravel or Stone Aggregate Concretes	150	10.0–20.0	—	—
(Concretes with >50% Quartz or Quartzite Sand Have Conductivities in Higher End of Range)	140	9.0–18.0	—	0.19–0.24
	130	7.0–13.0	—	—
Lightweight (low-mass) Aggregate or Limestone Concretes	120	6.4–9.1	—	—
Expanded Shale, Clay, or Slate; Expanded Slags; Cinders	100	4.7–6.2	—	0.20

APPENDICES

Table E.1 (*Continued*)

Part A: I-P Units				
Description	**Density lb/ft³**	**Conductivity[b] (k) Btu-in/h ft² °F**	**Resistance (R) h ft² °F/Btu**	**Specific Heat Btu/lb °F**
Pumice (with Density up to 100 lb/ft³); Scoria (Sanded Concretes Have Conductivities in Higher End of Range)	80	3.3–4.1	—	0.20
	60	2.1–2.5	—	—
	40	1.3	—	—
Gypsum/Fiber Concrete (87.5% Gypsum, 12.5% Wood Chips)	51	1.66	—	0.20
Cement/Lime, Mortar, and Stucco	120	9.70	—	—
	100	6.7	—	—
	80	4.5	—	—
Perlite, Vermiculite, and Polystyrene Beads	50	1.8–1.9	—	—
	40	1.4–1.5	—	0.15–0.23
	30	1.1	—	—
	20	0.8	—	—
Foam Concretes	120	5.4	—	—
	100	4.1	—	—
	80	3.0	—	—
	70	2.5	—	—
Foam Concretes and Cellular Concretes	60	2.1	—	—
	40	1.4	—	—
	20	0.8	—	—
Aerated Concrete (Oven-Dried)	27–50	1.4	—	0.20
Polystyrene Concrete (Oven-Dried)	16–50	2.54	—	0.20
Polymer Concrete	122	11.4	—	—
	138	7.14	—	—
Polymer Cement	117	5.39	—	—
Slag Concrete	60	1.5	—	—
	80	2.25	—	—
	100	3.0	—	—
	125	8.53	—	—
Woods (12% Moisture Content)[l]				
Hardwoods	—	—	—	0.39[n]
Oak	41–47	1.12–1.25	—	—
Birch	43–45	1.16–1.22	—	—
Maple	40–44	1.09–1.19	—	—
Ash	38–42	1.06–1.14	—	—
Softwoods	—	—	—	0.39[n]
Southern Pine	36–41	1.00–1.12	—	—
Southern Yellow Pine	31	1.06–1.16	—	—
Eastern White Pine	25	0.85–0.94	—	—
Douglas Fir/Larch	34–36	0.95–1.01	—	—
Southern Cypress	31–32	0.90–0.92	—	—
Hem/Fir, Spruce/Pine/Fir	24–31	0.74–0.90	—	—
Spruce	25	0.74–0.85	—	—
Western Red Cedar	22	0.83–0.86	—	—
West Coast Woods, Cedars	22–31	0.68–0.90	—	—
Eastern White Cedar	23	0.82–0.89	—	—
California Redwood	24–28	0.74–0.82	—	—
Pine (Oven-Dried)	23	0.64	—	0.45
Spruce (Oven-Dried)	25	0.69	—	0.45

APPENDICES

Table E.1 (*Continued*)

Part B: SI Units				
Description	**Density kg/m³**	**Conductivity[b] (k) W/(m K)**	**Resistance (R) (m² K)/W**	**Specific Heat kJ/(kg K)**
Insulating Materials				
Blanket and Batt[c d]				
Glass-Fiber Batts				0.80
	7.5–8.2	0.046–0.048	—	—
	9.8–12	0.040–0.043	—	—
	13–14	0.037–0.039	—	—
	22	0.033	—	—
Rock and Slag Wool Batts	—	—	—	0.80
	32–37	0.036–0.037	—	—
	45	0.033–0.035	—	—
Mineral Wool, Felted	16–48	0.040	—	—
	16–130	0.035	—	—
Board and Slabs				
Cellular Glass	120	0.042		0.80
Cement Fiber Slabs, Shredded Wood with Portland Cement Binder	400–430	0.072–0.076	—	—
With Magnesia Oxysulfide Binder	350	0.082	—	1.30
Glass Fiber Board	—	—	—	0.80
	24–96	0.033–0.035	—	—
Expanded Rubber (Rigid)	64	0.029	—	1.70
Extruded Polystyrene, Smooth Skin	—	—	—	1.50
Aged per *CAN/ULC Standard S770-2003*	22–58	0.026–0.029		
Aged 180 days	22–58	0.029		
European Product	30	0.030		
Aged 5 Years at 24°C	32–35	0.030	—	—
Blown with Low Global Warming Potential (GWP) (<5) Blowing Agent		0.035–0.036	—	—
Expanded Polystyrene, Molded Beads	—	—	—	1.50
	16–24	0.035–0.037	—	—
	29	0.033	—	—
Mineral Fiberboard, Wet Felted	160	0.037	—	0.80
Rock Wool Board	—	—	—	0.80
Floors and Walls	64–130	0.033–0.036	—	—
Roofing	160–180	0.039–0.042	—	0.80
Acoustical Tile[g]	340–370	0.052–0.053	—	0.6–0.8
Perlite Board	140	0.052	—	—
Polyisocyanurate	—	—	—	1.50
Unfaced, Aged per *CAN/ULC Standard S770-2003*	26–37	0.023–0.025	—	—
With Foil Facers, Aged 180 Days[e]	—	0.020–0.023	—	—
Phenolic Foam Board with Facers, Aged[f]	—	0.020–0.023	—	—
Loose Fill				
Cellulose Fiber, Loose Fill	—	—	—	1.40
Attic Application up to 100 mm	16–19	0.045–0.046	—	—
Attic Application > 100 mm	19–26	0.039–0.040	—	—
Wall Application, Dense Packed	56	0.039–0.040	—	—
Perlite, Expanded	32–64	0.039–0.045	—	1.10
	64–120	0.045–0.052	—	—
	120–180	0.052–0.061	—	—

APPENDICES

Table E.1 (*Continued*)

Part B: SI Units					
Description		**Density kg/m³**	**Conductivity[b] (k) W/(m K)**	**Resistance (R) (m² K)/W**	**Specific Heat kJ/(kg K)**
Glass Fiber[d]					
Attics, ~100–600 mm		6.4–8.0	0.052–0.055	—	—
Attics, ~600–1100 mm		8–9.6	0.049–0.052	—	—
Closed Attic or Wall Cavities		29–37	0.035–0.036	—	—
Rock and Slag Wool[d]					
Attics, ~90–115 mm		24–26	0.049	—	—
Attics, ~125–430 mm		24–29	0.046–0.048	—	—
Closed Attic or Wall Cavities		64	0.039–0.042	—	—
Vermiculite, Exfoliated		112–131	0.068	—	1.30
		64–96	0.063	—	—
Spray-Applied					
Cellulose, Sprayed into Open Wall Cavities		26–42	0.039–0.040	—	—
Glass Fiber, Sprayed into Open Wall or Attic Cavities		16	0.039–0.042	—	—
		29–37	0.033–0.037	—	—
Polyurethane Foam		—	—	—	1.50
Low Density, Open Cell		7.2–10	0.037–0.042	—	—
Medium Density, Closed Cell, Aged 180 Days		30–51	0.020–0.029	—	—
Building Board and Siding Board					
Board					
Asbestos/Cement Board		1900	0.57	—	1.00
Cement Board		1150	0.25	—	0.84
Fiber/Cement Board		1400	0.25	—	0.84
		1000	0.19	—	0.84
		400	0.07	—	1.88
		300	0.06	—	1.88
Gypsum or Plaster Board		640	0.16	—	1.15
Oriented Strand Board (OSB)	9–11 mm	650	—	0.11	1.88
	12.7 mm	650	—	0.12	1.88
Plywood (Douglas Fir)	12.7 mm	460	—	0.14	1.88
	15.9 mm	540	—	0.15	1.88
Plywood/Wood Panels	19.0 mm	450	—	0.19	1.88
Vegetable Fiber Board		650	—	0.11	1.88
Sheathing, Regular Density	12.7 mm	290	—	0.23	1.30
Intermediate Density	12.7 mm	350	—	0.19	1.30
Nail-Based Sheathing	12.7 mm	400	—	0.19	1.30
Shingle Backer	9.5 mm	290	—	0.17	1.30
Sound Deadening Board	12.7 mm	240	—	0.24	1.26
Tile and Lay-In Panels, Plain or Acoustic		290	0.058	—	0.59
Laminated Paperboard		480	0.072	—	1.38
Homogeneous Board from Repulped Paper		480	0.072	—	1.17
Hardboard					
Medium Density		800	0.105	—	1.30
High Density, Service-Tempered Grade and Service Grade		880	0.12	—	1.34
High Density, Standard-Tempered Grade		1010		—	1.34
Particleboard					
Low Density		590	0.102	—	1.30

APPENDICES

Table E.1 (*Continued*)

Part B: SI Units					
Description		**Density kg/m³**	**Conductivity[b] (k) W/(m K)**	**Resistance (R) (m² K)/W**	**Specific Heat kJ/(kg K)**
Medium Density		800	0.135	—	1.30
High Density		1000	1.18	—	—
Underlayment	15.9 mm	640	—	1.22	1.21
Waferboard		700	0.072	—	1.88
Shingles					
Asbestos/Cement		1900	—	0.037	—
Wood, 400 mm, 190 mm Exposure		—	—	0.15	1.30
Wood, Double, 400 mm, 300 mm Exposure		—	—	0.21	1.17
Wood, Plus ins. Backer Board	8 mm	—	—	0.25	1.30
Siding					
Asbestos/Cement, Lapped	6.4 mm	—	—	0.037	1.01
Asphalt Roll Siding		—	—	0.026	1.47
Asphalt Insulating Siding (12.7 mm bed)		—	—	0.26	1.47
Hardboard Siding	11 mm	—	—	0.12	1.17
Wood, Drop, 200 mm	25 mm	—	—	0.14	1.17
Wood, Bevel		—			
200 mm, Lapped	13 mm	—	—	0.14	1.17
250 mm, Lapped	19 mm	—	—	0.18	1.17
Wood, Plywood, 9.5 mm, Lapped		—	—	0.10	1.22
Aluminum, Steel, or Vinyl,[j,k] Over Sheathing				—	
Hollow-backed		—	—	0.11	1.22[i]
Insulating-Board-Backed	9.5 mm	—	—	0.32	1.34
Foil-Backed	9.5 mm	—	—	0.52	—
insulated Vinyl Siding	19–32 mm			0.35–0.48[m]	
Architectural (Soda-lime Float) Glass		2500	1.0	—	0.84
Building Membrane					
Vapor-permeable Felt		—	—	0.011	—
Vapor: Seal, 2 Layers of Mopped 0.73 kg/m² Felt		—	—	0.21	—
Vapor: Seal, Plastic Film		—	—	Negligible	—
Finish Flooring Materials					
Carpet and Rebounded Urethane Pad	19 mm	110	—	0.42	—
Carpet and Rubber Pad (One-piece)	9.5 mm	320	—	0.12	—
Pile Carpet with Rubber Pad	9.5–12.7 mm	290	—	0.28	—
Linoleum/Cork Tile	6.4 mm	465	—	0.09	—
PVC/Rubber Floor Covering		—	0.40	—	—
Rubber Tile	25 mm	1900	—	0.06	—
Terrazzo	25 mm	—	—	0.014	0.80
Metals (see ASHRAE, 2021, Chapter 33, Table 3)					
Roofing					
Asbestos/Cement Shingles		1920	—	0.037	1.00
Asphalt (Bitumen with Inert Fill)		1600	0.43	—	—
		1900	0.58	—	—
		2300	1.15	—	—
Asphalt Roll Roofing		920	—	0.027	1.51
Asphalt Shingles		920	—	0.078	1.26
Built-Up Roofing	10 mm	920	—	0.059	1.47

Table E.1 (*Continued*)

Part B: SI Units					
Description		**Density kg/m³**	**Conductivity[b] (k) W/(m K)**	**Resistance (R) (m² K)/W**	**Specific Heat kJ/(kg K)**
Mastic Asphalt (Heavy, 20% Grit)		950	0.19	—	—
Reed Thatch		270	0.09	—	—
Roofing Felt		2250	1.20	—	—
Slate	13 mm	—	—	0.009	1.26
Straw Thatch		240	0.07	—	—
Wood Shingles, Plain and Plastic-film-faced		—	—	0.166	1.30
Plastering Materials					
Cement Plaster, Sand Aggregate		1860	0.72	—	0.84
Sand Aggregate	10 mm	—	—	0.013	0.84
	20 mm	—	—	0.026	0.84
Gypsum Plaster		1120	0.38	—	—
		1280	0.46	—	—
Lightweight Aggregate	13 mm	720	—	0.056	—
	16 mm	720	—	0.066	—
On Metal Lath	19 mm	—	—	0.083	—
Perlite Aggregate		720	0.22	—	1.34
Sand Aggregate		1680	0.81	—	0.84
On Metal Lath	19 mm	—	—	0.023	—
Vermiculite Aggregate		480	0.14	—	—
		600	0.20	—	—
		720	0.25	—	—
		840	0.26	—	—
		960	0.30	—	—
Perlite Plaster		400	0.08	—	—
		600	0.19	—	—
Pulpboard or Paper Plaster		600	0.07	—	—
Sand/Cement Plaster, Conditioned		1560	0.63	—	—
Sand/Cement/Lime Plaster, Conditioned		1440	0.48	—	—
Sand/Gypsum (3:1) Plaster, Conditioned		1550	0.65	—	—
Masonry Materials					
Masonry Units					
Brick, Fired Clay		2400	1.21–1.47	—	—
		2240	1.07–1.30	—	—
		2080	0.92–1.12	—	—
		1920	0.81–0.98	—	0.80
		1760	0.71–0.85	—	—
		1600	0.61–0.74	—	—
		1440	0.52–0.62	—	—
		1280	0.43–0.53	—	—
		1120	0.36–0.45	—	—
Clay Tile, Hollow					
1 Cell Deep	75 mm	—	—	0.14	0.88
	100 mm	—	—	0.20	—
2 Cells Deep	150 mm	—	—	0.27	—
	200 mm	—	—	0.33	—
	250 mm	—	—	0.39	—
3 Cells Deep	300 mm	—	—	0.44	—
Lightweight Brick		800	0.20	—	—
		770	0.22	—	—

Table E.1 (*Continued*)

Part B: SI Units				
Description	**Density kg/m³**	**Conductivity[b] (k) W/(m K)**	**Resistance (R) (m² K)/W**	**Specific Heat kJ/(kg K)**
Concrete Blocks[h,i]				
Limestone Aggregate				
~200 mm, 16.3 kg, 2200 kg/m³ Concrete, 2 Cores	—	—	—	—
With Perlite-Filled Cores	—	—	0.37	—
~300 mm, 25 kg, 2200 kg/m³ Concrete, 2 Cores	—	—	—	—
With Perlite-Filled Cores	—	—	0.65	—
Normal-Weight Aggregate (Sand and Gravel)				
~200 mm, 16 kg, 2100 kg/m³ Concrete, 2 or 3 Cores	—	—	0.20–0.17	0.92
With Perlite-Filled Cores	—	—	0.35	—
With Vermiculite-Filled Cores	—	—	0.34–0.24	—
~300 mm, 22.7 kg, 2000 kg/m³ Concrete, 2 Cores	—	—	0.217	0.92
Medium-weight Aggregate (Combinations of Normal and Lightweight Aggregate)				
~200 mm, 13 kg, 1550–1800 kg/m³ Concrete, 2 or 3 Cores	—	—	0.30–0.22	—
With Perlite-Filled Cores	—	—	0.65–0.41	—
With Vermiculite-Filled Cores	—	—	0.58	—
With Molded-EPS-Filled (Beads) Cores	—	—	0.56	—
With Molded EPS Inserts in Cores	—	—	0.47	—
Lightweight (Low-Mass) Aggregate (Expanded Shale, Clay, Slate or Slag, Pumice)				
~150 mm, 7 1/2 kg, 1400 kg/m³ Concrete, 2 or 3 Cores	—	—	0.34–0.29	—
With Perlite-Filled Cores	—	—	0.74	—
With Vermiculite-Filled Cores	—	—	0.53	—
200 mm, 8–10 kg, 1150–1380 kg/m³ Concrete	—	—	0.56–0.33	0.88
With Perlite-Filled Cores	—	—	1.20–0.70	—
With Vermiculite-Filled Cores	—	—	0.93–0.69	—
With Molded-EPS-Filled (Beads) Cores	—	—	0.85	—
With UF Foam-Filled Cores	—	—	0.79	—
With Molded EPS Inserts in Cores	—	—	0.62	—
300 mm, 16 kg, 1400 kg/m³ concrete, 2 or 3 Cores	—	—	0.46–0.40	—
With Perlite-Filled Cores	—	—	1.6–1.1	—
With Vermiculite-Filled Cores	—	—	1.0	—
Stone, Lime, or Sand	2880	10.4	—	—
Quartzitic and Sandstone	2560	6.2	—	—
	2240	3.46	—	—
	1920	1.88	—	0.88
Calcitic, Dolomitic, Limestone, Marble, and Granite	2880	4.33	—	—
	2560	3.17	—	—
	2240	2.31	—	—
	1920	1.59	—	0.88
	1600	1.15	—	—
Gypsum Partition Tile				
75 by 300 by 760 mm, Solid	—	—	0.222	0.79

APPENDICES

Table E.1 (*Continued*)

Part B: SI Units				
Description	**Density kg/m³**	**Conductivity[b] (k) W/(m K)**	**Resistance (R) (m² K)/W**	**Specific Heat kJ/(kg K)**
4 Cells	—	—	0.238	—
100 by 300 by 760 mm, 3 Cells	—	—	0.294	—
Limestone	2400	0.57	—	0.84
	2600	0.93	—	0.84
Concretes				
Sand and Gravel or Stone Aggregate Concretes	2400	1.4–2.9	—	—
(Concretes with >50% Quartz or Quartzite Sand Have	2240	1.3–2.6	—	0.80–1.00
Conductivities in Higher End of Range)	2080	1.0–1.9	—	—
Lightweight (Low-Mass) Aggregate or Limestone Concretes	1920	0.9–1.3	—	—
Expanded Shale, Clay, or Slate; Expanded Slags; Cinders, Pumice (with	1600	0.68–0.89	—	0.84
Density up to 1600 kg/m³); Scoria (Sanded Concretes Have Conductivities	1280	0.48–0.59	—	0.84
in Higher End of Range)	960	0.30–0.36	—	—
	640	0.18	—	—
Gypsum/Fiber Concrete (87.5% Gypsum, 12.5% Wood Chips)	800	0.24	—	0.84
Cement/Lime, Mortar, and Stucco	1920	1.40	—	—
	1600	0.97	—	—
	1280	0.65	—	—
Perlite, Vermiculite, and Polystyrene Beads	800	0.26–0.27	—	—
	640	0.20–0.22	—	0.63–0.96
	480	0.16	—	—
	320	0.12	—	—
Foam Concretes	1920	0.75	—	—
	1600	0.60	—	—
	1280	0.44	—	—
	1120	0.36	—	—
Foam Concretes and Cellular Concretes	960	0.30	—	—
	640	0.20	—	—
	320	0.12	—	—
Aerated Concrete (Oven-Dried)	430–800	0.20	—	0.84
Polystyrene Concrete (Oven-Dried)	255–800	0,37	—	0.84
Polymer Concrete	1950	1.64	—	—
	2200	1.03	—	—
Polymer Cement	1870	0,78	—	—
Slag Concrete	960	0.22	—	—
	1280	0.32	—	—
	1600	0.43	—	—
	2000	1.23	—	—
Woods (12% Moisture Content)[l]				
Hardwoods	—	—	—	1.63[n]
Oak	660–750	0.16–0.18	—	—
Birch	680–725	0.17–0.18	—	—
Maple	635–700	0.16–0.17	—	—
Ash	615–670	0.15–0.16	—	—
Softwoods	—	—	—	1.63[n]
Southern Pine	570–660	0.14–0.16	—	—

APPENDICES

Table E.1 (*Continued*)

Part B: SI Units				
Description	**Density kg/m³**	**Conductivity[b] (k) W/(m K)**	**Resistance (R) (m² K)/W**	**Specific Heat kJ/(kg K)**
Southern Yellow Pine	500	0.13	—	—
Eastern White Pine	400	0.1	—	—
Douglas Fir/Larch	535–580	0.14–0.15	—	—
Southern Cypress	500–515	0.13	—	—
Hem/Fir, Spruce/Pine/Fir	390–500	0.11–0.13	—	—
Spruce	400	0.09	—	—
Western Red Cedar	350	0.09	—	—
West Coast Woods, Cedars	350–500	0.10–0.13	—	—
Eastern White Cedar	360	0.1	—	—
California Redwood	390–450	0.11–0.12	—	—
Pine (Oven-Dried)	370	0.092	—	1.88
Spruce (Oven-Dried)	395	0.1	—	1.88

Source: Adapted with permission; © ASHRAE (www.ashrae.org); from the 2021 ASHRAE Handbook—Fundamentals, Chapter 26, Table 1.

[a]Values are for a mean temperature of 75°F [24°C]. Representative values for dry materials are intended as design (not specification) values for materials in normal use. Thermal values of insulating materials may differ from design values depending on their in situ properties (e.g., density, moisture content, orientation) and variability experienced during manufacture. For properties of a particular product, use the value supplied by the manufacturer or by unbiased tests.

[b]Symbol λ also used to represent thermal conductivity.

[c]Does not include paper backing and facing, if any. Where insulation forms a boundary (reflective or otherwise) of an air space, see Tables E.3 and E.4/E.5 for the insulating value of an air space with the appropriate effective emittance and temperature conditions of the space.

[d]Conductivity varies with fiber diameter. (See ASHRAE, 2021, Chapter 25, "Heat, Air, and Moisture Control in Building Assemblies – Fundamentals.") Batt, blanket, and loose-fill mineral fiber insulations are manufactured to achieve specified R-values, the most common of which are listed in the table. Due to differences in manufacturing processes and materials, the product thicknesses, densities, and thermal conductivities vary over considerable ranges for a specified R-value.

[e]Values are for aged products with gas-impermeable facers on the two major surfaces. An aluminum foil facer of 0.001 in. [25 μm] thickness or greater is generally considered impermeable to gases. For change in conductivity with age of expanded polyisocyanurate, see Society of the Plastics Industry, SPI Bulletin U108.

[f]Cellular phenolic insulation may no longer be manufactured.

[g]Insulating values of acoustical tile vary, depending on the density of the board and on the type, size, and depth of perforations.

[h]Values for fully grouted block may be approximated using values for concrete with a similar unit weight.

[i]Values for concrete block and concrete are at moisture contents representative of normal use.

[j]Values for metal or vinyl siding applied over flat surfaces vary widely, depending on amount of ventilation of air space beneath the siding; whether air space is reflective or nonreflective; and on thickness, type, and application of insulating backing used. Values are averages for use as design guides, and were obtained from several guarded hot box tests (*ASTM Standard* C 236) on hollow-backed types and made using backing of wood fiber, foamed plastic, and glass fiber. Departures of +/– 50% or more from these values may occur.

[k]Vinyl specific heat = 0.25 Btu/lb °F [1.0 kJ/kg K].

[l]See Adams (1971). "Supporting Cryogenic Equipment with Wood." *Chemical Engineering* (May): pp. 156–158); MacLean (1941). "Thermal Conductivity of Wood." *ASHVE Transactions* 47:323; and Wilkes (1979). "Thermophysical Properties Data Base Activities at Owens-Corning Fiberglas." *Proceedings of the ASHRAE/DOE-ORNL Conference, Thermal Performance of the Exterior Envelopes of Buildings*, ASHRAE SP 28:662–677. Conductivity values listed are for heat transfer across the grain. The thermal conductivity of wood varies linearly with the density, and the density ranges listed are those normally found for the wood species given. If the density of the wood species is not known, use the mean conductivity value. For extrapolation to other moisture contents, refer to the equation presented in footnote "l" of ASHRAE, 2021, Chapter 26, Table 1.

[m]Dimension referenced is taken at the maximum siding profile thickness. The range of R values and associated thicknesses represent values for products tested to ASTM *Standard* D7793, which required applying 15 mph [6.7 m/s] airstream perpendicular to surface of siding during testing.

[n]For an empirical equation for the specific heat of moist wood at 75°F [24°C], refer to the equation presented in footnote "n" of ASHRAE, 2021, Chapter 26, Table 1.

Table E.2 Thermal Properties of Alternative[a] Building and Insulating Materials

Part A: I-P Units

Description	Density lb/ft³	Conductivity (k) Btu-in/h ft² °F	Resistance Per Inch Thickness (1/k) h ft² °F/Btu-in.	Resistance for Thickness Listed h ft² °F/Btu
Building Materials				
Adobe[b] Typically 10–24 in.	106	3.3	0.3	—
Cob[c] Typically 18–36 in.	—	4	0.25	—
Rammed Earth[d] 18 in. Typical, Uninsulated	106	4	0.25	—
Straw Bale[e] Typically 16–24 in.	5–10	0.67–0.33	1.5–3.0	—
Structural Insulated Panels (SIPs), EPS[f], Various Thicknesses	—	0.25	4	—
Structural Insulated Panels (SIPs), XPS[g], Various Thicknesses	—	0.20	5	—
Structural Insulated Panels (Sips), Polyisocyanurate[h], Various Thicknesses	—	0.17–0.14	6–7	—
Insulating Materials				
AirKrete Insulation[i]		2.2	—	3.9
Insulating Concrete Forms (ICF)[j]		—	0.06–0.03	17–40

Part B: SI Units

Description	Density kg/m²	Conductivity (k) W/m K	Resistance Per Meter Thickness (1/k) K m/W	Resistance for Thickness Listed K m²/W
Building Materials				
Adobe[b] Typically 250–610 mm	1600	0.476	2.1	—
Cob[c] Typically 460–915 mm	—	0.577	1.7	—
Rammed Earth[d] 460 mm Typical	1600	0.577	1.7	—
Straw Bale[e] Typically 400–610 mm	80–160	0.097–0.048	10.4–20.8	—
Structural Insulated Panels (SIPs), (EPS)[f], Various Thicknesses	—	0.036	27.7	—
Structural Insulated Panels (SIPs), XPS[g], Various Thicknesses	—	0.029	34.7	—
Structural Insulated Panels (SIPs), Polyisocyanurate[h], Various Thicknesses	—	0.025–0.02	41.6–48.5	—
Insulating Materials				
AirKrete Insulation[i]	35	0.037	27.0	
Insulating Concrete Forms (ICF)[j]	—	—		3.0–7.0

Sources: Various (see references below).

[a]It is generally difficult to obtain reliable and consistent information on the thermal properties of alternative building materials that are not sold as a proprietary product. The data in this table have been obtained from a variety of sources (as noted); note the wide range of values and lack of data for some properties (such as specific heat).

[b]Adobe: U.S. Department of Energy, Office of Building Technology, State and Community Programs, 1995, *House of Straw — Straw Bale Construction Comes of Age* (out of print).

[c]Cob: Historical data from toolbase.org. The information resource is no longer available.

[d]Rammed earth: Historical data from toolbase.org. The information resource is no longer available.

[e]Straw bale: www.buildinggreen.com/news-analysis/r-value-straw-bales-lower-previously-reported; recent tests suggest lower R-values than previously reported, with a typical R-value 1.5 per inch and around 26–28 per bale.

[f]Structural insulated panel with expanded polystyrene insulation: www.energy.gov/energysaver/energy-saver

[g]Structural insulated panel with extruded polystyrene insulation: www.eere.energy.gov/consumerinfo/factsheets/bd1.html

[h]Structural insulated panel with polyurethane/isocyanurate insulation: www.energy.gov/energysaver/energy-saver; R-values are for aged panels (as opposed to new panel R-values of 7–9 as typically reported).

[i]AirKrete: www.airkrete.com

[j]Insulating concrete forms come in a variety of configurations; the listed R-17 value is for a 9-inch wall section (4 in. polystyrene and 5 in. poured concrete); www.energy.gov/energysaver/energy-saver. Another manufacturer forms that provide R-values of 22, 32, and 40.

APPENDICES

Table E.3 Thermal Properties of Surface Air Films and Air Spaces

Part A. Surface Conductances[a] (h) and Resistances[b] (R) for Surface Air Films

		Surface Emittance, ε											
		I-P Units[c]						SI Units[d]					
		Nonreflective		Reflective				Nonreflective		Reflective			
Position of Surface	**Direction of Heat Flow**	ε = 0.90 h_i	ε = 0.90 R	ε = 0.20 h_i	ε = 0.20 R	ε = 0.05 h_i	ε = 0.05 R	ε = 0.90 h_i	ε = 0.90 R	ε = 0.20 h_i	ε = 0.20 R	ε = 0.05 h_i	ε = 0.05 R
Still Air													
Horizontal	Upward	1.63	0.61	0.91	1.10	0.76	1.32	9.26	0.11	5.17	0.19	4.32	0.23
Sloping—45°	Upward	1.60	0.62	0.88	1.14	0.73	1.37	9.09	0.11	5.00	0.20	4.15	0.24
Vertical	Horizontal	1.46	0.68	0.74	1.35	0.59	1.70	8.29	0.12	4.20	0.24	3.35	0.30
Sloping—45°	Downward	1.32	0.76	0.60	1.67	0.45	2.22	7.50	0.13	3.41	0.29	2.56	0.39
Horizontal	Downward	1.08	0.92	0.37	2.70	0.22	4.55	6.13	0.16	2.10	0.48	1.25	0.80
Moving Air (any position)		h_o	R					h_o	R				
Winter Wind													
15 mph (6.7 m/s)	Any	6.00	0.17					34.0	0.030				
Summer Wind													
7.5 mph (3.4 m/s)	Any	4.00	0.25					22.7	0.044				

Part B. Emittance Values of Various Surfaces and Effective Emittances of Air Spaces

		Effective Emittance, ε_{eff}, of Air Space	
	Average Emittance ε	**One Surface Emittance ε; Other Surface 0.9**	**Both Surfaces Emittance ε**
Aluminum Foil, Bright	0.05	0.05	0.03
Aluminum Foil with Condensate Just Visible	0.30	0.29	—
Aluminum Foil with Condensate Clearly Visible	0.70	0.65	—
Aluminum Sheet	0.12	0.12	0.06
Aluminum-coated Paper, Polished	0.20	0.20	0.11
Steel, Galvanized, Bright	0.25	0.24	0.15
Aluminum Paint	0.50	0.47	0.35
Building Materials: Wood, Paper, Masonry, Nonmetallic Paints	0.90	0.82	0.82
Regular Glass	0.84	0.77	0.72

Source: Reprinted with permission; © ASHRAE (www.ashrae.org); from the 2021 ASHRAE Handbook—Fundamentals, Chapter 26, Tables 10 and 2.

[a]Conductances are for surfaces of the stated emittance facing virtual blackbody surroundings at the same temperature as the ambient air. Values are based on a surface–air temperature difference of 10°F [5.5°C] and for surface temperatures of 70°F [21°C].

[b]No surface has both an air space resistance value and a surface resistance value.

[c]I-P units: surface conductance h_i and h_o measured in Btu/h ft^2 °F; resistance R measured in h ft^2 °F/Btu.

[d]SI units: surface conductance h_i and h_o measured in W/m^2 K; resistance R measured in m^2 K/W.

Table E.4 Thermal Resistances of Plane[a] Air Spaces (I-P Units)

All resistance are values expressed in ft^2 °F h/Btu. Values apply only to air spaces of uniform thickness bounded by plane, smooth, parallel surfaces with no leakage of air to or from the space. These conditions are not normally present in standard building construction. Thermal resistance values for multiple air spaces must be based on careful estimates of mean temperature differences for each air space.

Position of Air Space	Direction of Heat Flow		Air Space Mean Temp,[b] °F	Air Space Temp Diff,[b,e] °F	0.5-in. Air Space[d] Value of E[b,c] 0.03	0.05	0.2	0.5	0.82	0.75-in. Air Space[d] Value of E[b,c] 0.03	0.05	0.2	0.5	0.82
Horizontal	Up	↑	90	10	2.13	2.03	1.51	0.99	0.73	2.34	2.22	1.61	1.04	0.75
			50	30	1.62	1.57	1.29	0.96	0.75	1.71	1.66	1.35	0.99	0.77
			50	10	2.13	2.05	1.60	1.11	0.84	2.30	2.21	1.70	1.16	0.87
			0	20	1.73	1.70	1.45	1.12	0.91	1.83	1.79	1.52	1.16	0.93
			0	10	2.10	2.04	1.70	1.27	1.00	2.23	2.16	1.78	1.31	1.02
			−50	20	1.69	1.66	1.49	1.23	1.04	1.77	1.74	1.55	1.27	1.07
			−50	10	2.04	2.00	1.75	1.40	1.16	2.16	2.11	1.84	1.46	1.20
45° Slope	Up	↗	90	10	2.44	2.31	1.65	1.06	0.76	2.96	2.78	1.88	1.15	0.81
			50	30	2.06	1.98	1.56	1.10	0.83	1.99	1.92	1.52	1.08	0.82
			50	10	2.55	2.44	1.83	1.22	0.90	2.90	2.75	2.00	1.29	0.94
			0	20	2.20	2.14	1.76	1.30	1.02	2.13	2.07	1.72	1.28	1.00
			0	10	2.63	2.54	2.03	1.44	1.10	2.72	2.62	2.08	1.47	1.12
			−50	20	2.08	2.04	1.78	1.42	1.17	2.05	2.01	1.76	1.41	1.16
			−50	10	2.62	2.56	2.17	1.66	1.33	2.53	2.47	2.10	1.62	1.30
Vertical	Horizontal	→	90	10	2.47	2.34	1.67	1.06	0.77	3.50	3.24	2.08	1.22	0.84
			50	30	2.57	2.46	1.84	1.23	0.90	2.91	2.77	2.01	1.30	0.94
			50	10	2.66	2.54	1.88	1.24	0.91	3.70	3.46	2.35	1.43	1.01
			0	20	2.82	2.72	2.14	1.50	1.13	3.14	3.02	2.32	1.58	1.18
			0	10	2.93	2.82	2.20	1.53	1.15	3.77	3.59	2.64	1.73	1.26
			−50	20	2.90	2.82	2.35	1.76	1.39	2.90	2.83	2.36	1.77	1.39
			−50	10	3.20	3.10	2.54	1.87	1.46	3.72	3.60	2.87	2.04	1.56
45° Slope	Down	↘	90	10	2.48	2.34	1.67	1.06	0.77	3.53	3.27	2.10	1.22	0.84
			50	30	2.64	2.52	1.87	1.24	0.91	3.43	3.23	2.24	1.39	0.99
			50	10	2.67	2.55	1.89	1.25	0.92	3.81	3.57	2.40	1.45	1.02
			0	20	2.91	2.80	2.19	1.52	1.15	3.75	3.57	2.63	1.72	1.26
			0	10	2.94	2.83	2.21	1.53	1.15	4.12	3.91	2.81	1.80	1.30
			−50	20	3.16	3.07	2.52	1.86	1.45	3.78	3.65	2.90	2.05	1.57
			−50	10	3.26	3.16	2.58	1.89	1.47	4.35	4.18	3.22	2.21	1.66

Table E.4 Thermal Resistances of Plane[a] Air Spaces (I-P Units) (*Continued*)

Position of Air Space	Direction of Heat Flow		Air Space: Mean Temp,[b] °F	Air Space: Temp Diff,[b,e] °F	0.5-in. Air Space[d] Value of E[b,c]: 0.03	0.05	0.2	0.5	0.82	0.75-in. Air Space[d] Value of E[b,c]: 0.03	0.05	0.2	0.5	0.82
Horizontal	Down	↓	90	10	2.48	2.34	1.67	1.06	0.77	3.55	3.29	2.10	1.22	0.85
			50	30	2.66	2.54	1.88	1.24	0.91	3.77	3.52	2.38	1.44	1.02
			50	10	2.67	2.55	1.89	1.25	0.92	3.84	3.59	2.41	1.45	1.02
			0	20	2.94	2.83	2.20	1.53	1.15	4.18	3.96	2.83	1.81	1.30
			0	10	2.96	2.85	2.22	1.53	1.16	4.25	4.02	2.87	1.82	1.31
			−50	20	3.25	3.15	2.58	1.89	1.47	4.60	4.41	3.36	2.28	1.69
			−50	10	3.28	3.18	2.60	1.90	1.47	4.71	4.51	3.42	2.30	1.71

Position of Air Space	Direction of Heat Flow		Air Space: Mean Temp,[b] °F	Air Space: Temp Diff,[b,e] °F	1.5-in. Air Space[d] Value of E[b,c]: 0.03	0.05	0.2	0.5	0.82	3.5-in. Air Space[d] Value of E[b,c]: 0.03	0.05	0.2	0.5	0.82
Horizontal	Up	↑	90	10	2.55	2.41	1.71	1.08	0.77	2.84	2.66	1.83	1.13	0.80
			50	30	1.87	1.81	1.45	1.04	0.80	2.09	2.01	1.58	1.10	0.84
			50	10	2.50	2.40	1.81	1.21	0.89	2.80	2.66	1.95	1.28	0.93
			0	20	2.01	1.95	1.63	1.23	0.97	2.25	2.18	1.79	1.32	1.03
			0	10	2.43	2.35	1.90	1.38	1.06	2.71	2.62	2.07	1.47	1.12
			−50	20	1.94	1.91	1.68	1.36	1.13	2.19	2.14	1.86	1.47	1.20
			−50	10	2.37	2.31	1.99	1.55	1.26	2.65	2.58	2.18	1.67	1.33
45° Slope	Up	↗	90	10	2.92	2.73	1.86	1.14	0.80	3.18	2.96	1.97	1.18	0.82
			50	30	2.14	2.06	1.61	1.12	0.84	2.26	2.17	1.67	1.15	0.86
			50	10	2.88	2.74	1.99	1.29	0.94	3.12	2.95	2.10	1.34	0.96
			0	20	2.30	2.23	1.82	1.34	1.04	2.42	2.35	1.90	1.38	1.06
			0	10	2.79	2.69	2.12	1.49	1.13	2.98	2.87	2.23	1.54	1.16
			−50	20	2.22	2.17	1.88	1.49	1.21	2.34	2.29	1.97	1.54	1.25
			−50	10	2.71	2.64	2.23	1.69	1.35	2.87	2.79	2.33	1.75	1.39
Vertical	Horizontal	→	90	10	3.99	3.66	2.25	1.27	0.87	3.69	3.40	2.15	1.24	0.85
			50	30	2.58	2.46	1.84	1.23	0.90	2.67	2.55	1.89	1.25	0.91
			50	10	3.79	3.55	2.39	1.45	1.02	3.63	3.40	2.32	1.42	1.01
			0	20	2.76	2.66	2.10	1.48	1.12	2.88	2.78	2.17	1.51	1.14
			0	10	3.51	3.35	2.51	1.67	1.23	3.49	3.33	2.50	1.67	1.23
			−50	20	2.64	2.58	2.18	1.66	1.33	2.82	2.75	2.30	1.73	1.37
			−50	10	3.31	3.21	2.62	1.91	1.48	3.40	3.30	2.67	1.94	1.50

45° Slope	Down	↘	90	10	5.07	4.55	2.56	1.36	0.91	4.81	4.33	2.49	1.34	0.90
			50	30	3.58	3.36	2.31	1.42	1.00	3.51	3.30	2.28	1.40	1.00
			50	10	5.10	4.66	2.85	1.60	1.09	4.74	4.36	2.73	1.57	1.08
			0	20	3.85	3.66	2.68	1.74	1.27	3.81	3.63	2.66	1.74	1.27
			0	10	4.92	4.62	3.16	1.94	1.37	4.59	4.32	3.02	1.88	1.34
			−50	20	3.62	3.50	2.80	2.01	1.54	3.77	3.64	2.90	2.05	1.57
			−50	10	4.67	4.47	3.40	2.29	1.70	4.50	4.32	3.31	2.25	1.68
Horizontal	Down	↓	90	10	6.09	5.35	2.79	1.43	0.94	10.07	8.19	3.41	1.57	1.00
			50	30	6.27	5.63	3.18	1.70	1.14	9.60	8.17	3.86	1.88	1.22
			50	10	6.61	5.90	3.27	1.73	1.15	11.15	9.27	4.09	1.93	1.24
			0	20	7.03	6.43	3.91	2.19	1.49	10.90	9.52	4.87	2.47	1.62
			0	10	7.31	6.66	4.00	2.22	1.51	11.97	10.32	5.08	2.52	1.64
			−50	20	7.73	7.20	4.77	2.85	1.99	11.64	10.49	6.02	3.25	2.18
			−50	10	8.09	7.52	4.91	2.89	2.01	12.98	11.56	6.36	3.34	2.22

Source: Reprinted with permission; © ASHRAE (www.ashrae.org); from the 2021 ASHRAE Handbook—Fundamentals, Chapter 26, Table 3.

[a]Thermal resistance values were determined from the relation R = 1/C, where $C = h_c + \varepsilon_{eff}\, h_r$; h_c is the conduction–convection coefficient; $\varepsilon_{eff}\, h_r$ is the radiation coefficient $= 0.0068\varepsilon_{eff}\,[(t_m + 460)/100]^3$; and t_m is the mean temperature of the air space. For extrapolation from this table to air spaces less than 0.5 in. (as in insulating window glass), assume that $h_c = 0.159(1 + 0.0016t_m)/l$, where l is the air space thickness in inches and h_c is heat transfer through the air space only.

[b]Interpolation is permissible for other values of mean temperature, temperature difference, and effective emittance ε_{eff}. Interpolation and moderate extrapolation for air spaces greater than 3.5 in. are also permissible.

[c]Effective emittance, ε_{eff}, of the air space is given by $1/\varepsilon_{eff} = 1/\varepsilon_1 + 1/\varepsilon_2 - 1$, where ε_1 and ε_2 are emittances of the surfaces of the air space (from Table E.3).

[d]A single resistance value cannot account for multiple air spaces. Each air space requires a separate resistance calculation that applies only for the established boundary conditions.

[e]Resistances of horizontal spaces with heat flow downward are substantially independent of temperature difference.

Table E.5 Thermal Resistances of Plane[a] Air Spaces (SI Units)

All resistance values are expressed in K m²/W. Values apply only to air spaces of uniform thickness bounded by plane, smooth, parallel surfaces with no leakage of air to or from the space. These conditions are not normally present in standard building construction. Thermal resistance values for multiple air spaces must be based on careful estimates of mean temperature differences for each air space.

			Air Space		13-mm Air Space[d] Value of E[b c]					20-mm Air Space[d] Value of E[b c]				
Position of Air Space	Direction of Heat Flow		Mean Temp,[b] °C	Temp Diff,[b e] °C	0.03	0.05	0.2	0.5	0.82	0.03	0.05	0.2	0.5	0.82
Horizontal	Up	↑	32.2	5.6	0.37	0.36	0.27	0.17	0.13	0.41	0.39	0.28	0.18	0.13
			10.0	16.7	0.29	0.28	0.23	0.17	0.13	0.30	0.29	0.24	0.17	0.14
			10.0	5.6	0.37	0.36	0.28	0.30	0.15	0.40	0.39	0.30	0.20	0.15
			−17.8	11.1	0.30	0.30	0.26	0.20	0.16	0.32	0.32	0.27	0.20	0.16
			−17.8	5.6	0.37	0.36	0.30	0.22	0.18	0.39	0.38	0.31	0.23	0.18
			−45.6	11.1	0.30	0.29	0.26	0.22	0.18	0.31	0.31	0.27	0.22	0.19
			−45.6	5.6	0.36	0.35	0.31	0.25	0.20	0.38	0.37	0.32	0.26	0.21
45° Slope	Up	↗	32.2	5.6	0.43	0.41	0.29	0.19	0.13	0.52	0.49	0.33	0.20	0.14
			10.0	16.7	0.36	0.35	0.27	0.19	0.15	0.35	0.34	0.27	0.19	0.14
			10.0	5.6	0.45	0.43	0.32	0.21	0.16	0.51	0.48	0.35	0.23	0.17
			−17.8	11.1	0.39	0.38	0.31	0.23	0.18	0.37	0.36	0.30	0.23	0.18
			−17.8	5.6	0.46	0.45	0.36	0.25	0.19	0.48	0.46	0.37	0.26	0.20
			−45.6	11.1	0.37	0.36	0.31	0.25	0.21	0.36	0.35	0.31	0.25	0.20
			−45.6	5.6	0.46	0.45	0.38	0.29	0.23	0.45	0.43	0.37	0.29	0.23
Vertical	Horizontal	→	32.2	5.6	0.43	0.41	0.29	0.19	0.14	0.62	0.57	0.37	0.21	0.15
			10	16.7	0.45	0.43	0.32	0.22	0.16	0.51	0.49	0.35	0.23	0.17
			10	5.6	0.47	0.45	0.33	0.22	0.16	0.65	0.61	0.41	0.25	0.18
			−17.8	11.1	0.50	0.48	0.38	0.26	0.20	0.55	0.53	0.41	0.28	0.21
			−17.8	5.6	0.52	0.50	0.39	0.27	0.20	0.66	0.63	0.46	0.30	0.22
			−45.6	11.1	0.51	0.50	0.41	0.31	0.24	0.51	0.50	0.42	0.31	0.24
			−45.6	5.6	0.56	0.55	0.45	0.33	0.26	0.65	0.63	0.51	0.36	0.27
45° Slope	Down	↘	32.2	5.6	0.44	0.41	0.29	0.19	0.14	0.62	0.58	0.37	0.21	0.15
			10.0	16.7	0.46	0.44	0.33	0.22	0.16	0.60	0.57	0.39	0.24	0.17
			10.0	5.6	0.47	0.45	0.33	0.22	0.16	0.67	0.63	0.42	0.26	0.18
			−17.8	11.1	0.51	0.49	0.39	0.27	0.20	0.66	0.63	0.46	0.30	0.22
			−17.8	5.6	0.52	0.50	0.39	0.27	0.20	0.73	0.69	0.49	0.32	0.23
			−45.6	11.1	0.56	0.54	0.44	0.33	0.25	0.67	0.64	0.51	0.36	0.28
			−45.6	5.6	0.57	0.56	0.45	0.33	0.26	0.77	0.74	0.57	0.39	0.29

Horizontal	Down	↓	32.2	5.6	0.44	0.41	0.29	0.19	0.14	0.62	0.58	0.37	0.21	0.15
			10.0	16.7	0.47	0.45	0.33	0.22	0.16	0.66	0.62	0.42	0.25	0.18
			10.0	5.6	0.47	0.45	0.33	0.22	0.16	0.68	0.63	0.42	0.26	0.18
			−17.8	11.1	0.52	0.50	0.39	0.27	0.20	0.74	0.70	0.50	0.32	0.23
			−17.8	5.6	0.52	0.50	0.39	0.27	0.20	0.75	0.71	0.51	0.32	0.23
			−45.6	11.1	0.57	0.55	0.45	0.33	0.26	0.81	0.78	0.59	0.40	0.30
			−45.6	5.6	0.58	0.56	0.46	0.33	0.26	0.83	0.79	0.60	0.40	0.30

			Air Space		40-mm Air Space[d] Value of E[b,c]					90-mm Air Space[d] Value of E[b,c]				
Position of Air Space	Direction of Heat Flow		Mean Temp,[b] °C	Temp Diff,[b,e] °C	0.03	0.05	0.2	0.5	0.82	0.03	0.05	0.2	0.5	0.82
Horizontal	Up	↑	32.2	5.6	0.45	0.42	0.30	0.19	0.14	0.50	0.47	0.32	0.20	0.14
			10.0	16.7	0.33	0.32	0.26	0.18	0.14	0.27	0.35	0.28	0.19	0.15
			10.0	5.6	0.44	0.42	0.32	0.21	0.16	0.49	0.47	0.34	0.23	0.16
			−17.8	11.1	0.35	0.34	0.29	0.22	0.17	0.40	0.38	0.32	0.23	0.18
			−17.8	5.6	0.43	0.41	0.33	0.24	0.19	0.48	0.46	0.36	0.26	0.20
			−45.6	11.1	0.34	0.34	0.30	0.24	0.20	0.39	0.38	0.33	0.26	0.21
			−45.6	5.6	0.42	0.41	0.35	0.27	0.22	0.47	0.45	0.38	0.29	0.23
45° Slope	Up	↗	32.2	5.6	0.51	0.48	0.33	0.20	0.14	0.56	0.52	0.35	0.21	0.14
			10.0	16.7	0.38	0.36	0.28	0.20	0.15	0.40	0.38	0.29	0.20	0.15
			10.0	5.6	0.51	0.48	0.35	0.23	0.17	0.55	0.52	0.37	0.24	0.17
			−17.8	11.1	0.40	0.39	0.32	0.24	0.18	0.43	0.41	0.33	0.24	0.19
			−17.8	5.6	0.49	0.47	0.37	0.26	0.20	0.52	0.51	0.39	0.27	0.20
			−45.6	11.1	0.39	0.38	0.33	0.26	0.21	0.41	0.40	0.35	0.27	0.22
			−45.6	5.6	0.48	0.46	0.39	0.30	0.24	0.51	0.49	0.41	0.31	0.24
Vertical	Horizontal	→	32.2	5.6	0.70	0.64	0.40	0.22	0.15	0.65	0.60	0.38	0.22	0.15
			10.0	16.7	0.45	0.43	0.32	0.22	0.16	0.47	0.45	0.33	0.22	0.16
			10.0	5.6	0.67	0.62	0.42	0.26	0.18	0.64	0.60	0.41	0.25	0.18
			−17.8	11.1	0.49	0.47	0.37	0.26	0.20	0.51	0.49	0.38	0.27	0.20
			−17.8	5.6	0.62	0.59	0.44	0.29	0.22	0.61	0.59	0.44	0.29	0.22
			−45.6	11.1	0.46	0.45	0.38	0.29	0.23	0.50	0.48	0.40	0.30	0.24

Table E.5 Thermal Resistances of Plane[a] Air Spaces (SI Units) (*Continued*)

Position of Air Space	Direction of Heat Flow	Air Space: Mean Temp,[b] °C	Air Space: Temp Diff,[b,e] °C	40-mm Air Space[d] Value of E[b,c]: 0.03	0.05	0.2	0.5	0.82	90-mm Air Space[d] Value of E[b,c]: 0.03	0.05	0.2	0.5	0.82
		–45.6	5.6	0.58	0.56	0.46	0.34	0.26	0.60	0.58	0.47	0.34	0.26
45° Slope	Down ↘	32.2	5.6	0.89	0.80	0.45	0.24	0.16	0.85	0.76	0.44	0.24	0.16
		10.0	16.7	0.63	0.59	0.41	0.25	0.18	0.62	0.58	0.40	0.25	0.18
		10.0	5.6	0.90	0.82	0.50	0.28	0.19	0.83	0.77	0.48	0.28	0.19
		−17.8	11.1	0.68	0.64	0.47	0.31	0.22	0.67	0.64	0.47	0.31	0.22
		−17.8	5.6	0.87	0.81	0.56	0.34	0.24	0.81	0.76	0.53	0.33	0.24
		–45.6	1.1	0.64	0.62	0.49	0.35	0.27	0.66	0.64	0.51	0.36	0.28
		–45.6	5.6	0.82	0.79	0.60	0.40	0.30	0.79	0.76	0.58	0.40	0.30
Horizontal	Down ↓	32.2	5.6	1.07	0.94	0.49	0.25	0.17	1.77	1.44	0.60	0.28	0.18
		10.0	16.7	1.10	0.99	0.56	0.30	0.20	1.69	1.44	0.68	0.33	0.21
		10.0	5.6	1.16	1.04	0.58	0.30	0.20	1.96	1.63	0.72	0.34	0.22
		−17.8	11.1	1.24	1.13	0.69	0.39	0.26	1.92	1.68	0.86	0.4	0.29
		−17.8	5.6	1.29	1.17	0.70	0.39	0.27	2.11	1.82	0.89	0.44	0.29
		–45.6	11.1	1.36	1.27	0.84	0.50	0.35	2.05	1.85	1.06	0.57	0.38
		–45.6	5.6	1.42	1.32	0.86	0.51	0.35	2.28	2.03	1.12	0.59	0.39

Source: Reprinted with permission; © ASHRAE (www.ashrae.org); from the 2021 ASHRAE Handbook—Fundamentals, Chapter 26, Table 3.

[a]Thermal resistance values were determined from the relation R = 1/C, where $C = h_c + \varepsilon_{eff} h_r$; h_c is the conduction–convection coefficient; $\varepsilon_{eff} h_r$ is the radiation coefficient $= 0.227\varepsilon_{eff}[(t_m + 273)/100]^3$; and t_m is the mean temperature of the air space. For extrapolation from this table to air spaces less than 12.5 mm (as in insulating window glass), assume that $h_c = 21.8\,(1 + 0.00274 t_m)/l$, where l is the air space thickness in inches, and h_c is the heat transfer in W/m² K through the air space only.

[b]Interpolation is permissible for other values of mean temperature, temperature difference, and effective emittance e_{eff}. Interpolation and moderate extrapolation for air spaces greater than 90 mm are also permissible.

[c]Effective emittance, ε_{eff}, of the air space is given by $1/f/\varepsilon_{eff} = 1/\varepsilon_1 + 1/\varepsilon_2 - 1$, where ε_1 and ε_2 are emittances of the surfaces of the air space (from Table E.3).

[d]A single resistance value cannot account for multiple air spaces. Each air space requires a separate resistance calculation that applies only for the established boundary conditions.

[e]Resistances of horizontal spaces with heat flow downward are substantially independent of temperature difference.

Table E.6 U-Factors for Common Wall, Roof, and Floor Assemblies

Assembly	Basic Construction	Other Thermal Components	I-P Insulation R-value °F ft^2 h/Btu	SI Insulation R-value K m^2/W	I-P Assembly U-factor Btu/°F ft^2 h	SI Assembly U-factor W/K m^2
Wall	Wood studs, nominal 2 in. × 4 in., 16 in. o.c. [50 mm × 100 mm, 400 mm o.c.]	Exterior air film, stucco, exterior gypsum board, interior gypsum board, interior air film	11	1.9	0.096	0.55
	As above	Above plus R-4 [SI: R-0.71] continuous insulation	15	2.6	0.060	0.39
	Wood studs, nominal 2 in. × 6 in., 24 in. o.c. [50 mm × 150 mm, 400 mm o.c.]		19	3.3	0.065	0.37
	Steel studs, nominal 2 in. × 4 in., 16 in. o.c. [50 mm × 100 mm, 400 mm o.c.]		11	1.9	0.132	0.75
	As above	Above plus R-4 [SI: R-0.71] continuous insulation	15	2.6	0.080	0.46
	6-in. concrete masonry unit at 115 lb/ft^3 [150 mm, at 1840 kg/m^3]	Partly grouted, cells insulated	N/A	N/A	0.41	2.33
	As above	Above with R-5 [SI: R-0.88] continuous insulation with 1-in. [25-mm] metal clips at 24 in. [600 mm] o.c. and 16 in. [400 mm] vertically	8	0.88	0.17	0.98
Roof	Standard wood joists	Semiexterior air film, gypsum board, interior air film	30	5.28	0.034	0.19
	Steel joists	Exterior air film, metal deck, interior air film	30	5.28	0.041	0.23
	System with insulation entirely above roof deck	Exterior air film, metal deck, interior air film	30	5.28	0.032	0.18
Floor	Nominal 6-in. wood joists [150 mm]	Semiexterior air film, wood subfloor, carpet and pad, interior film air	11	1.94	0.074	0.42
	Steel floor joists	Semiexterior air film, metal deck, concrete slab, carpet and pad, interior air film	11	1.94	0.078	0.44

Source: Extracted and used with permission; © ASHRAE (www.ashrae.org); from ANSI/ASHRAE/IES Standard 90.1-2022: Energy Standard for Buildings Except Low-Rise Residential Buildings.

Note: The SI units for resistance and U-factor were appended to the ASHRAE I-P data by the authors of this book.

APPENDICES

Table E.7 Time-Averaged U-Factors for Walls with Movable Insulation

	Whole-Assembly U-Factor, Winter					
	No Night Insulation		With Night Insulation[a] of Specified R-Value			
Component	I-P Btu/°F ft² h	SI W/K m²	I-P R-4 Btu/°F ft² h	SI R-0.7 W/K m²	I-P R-9 Btu/°F ft² h	SI R-1.58 W/K m²
Direct Gain[b]	0.55	3.1	0.30	1.7	0.24	1.4
Trombe Wall, 18-in. [460 mm] Thick[b]	0.22	1.2	0.15	0.9	0.12	0.7
Water Wall[b]	0.33	1.9	0.20	1.1	0.17	1.0

[a]These are daily average U-factors rather than the instantaneous U-factors generally presented in these tables. Insulation is assumed to be in place from 5:00 p.m. to 8:00 a.m., solar time, for Trombe and water walls and 5:00 p.m. to 7:00 a.m. for direct gain.

[b]From Balcomb et al., 1980. SI values were appended to the I-P data by the authors of this book.

Table E.8 Effective R-Values for Wall, Roof, and Floor Systems with Steel Framing

		Cavity Insulation R-value			Effective Framing/ Insulation R-value	
Construction	Spacing of Framing	I-P °F ft² h / Btu	SI K m² / W	Correction Factor[a b]	I-P °F ft² h / Btu	SI K m² / W
Walls						
Nominal Cavity Depth (Stud Size)						
4 in. [100 mm]	16 in. [400 mm] o.c.	11	1.94	0.50	5.5	0.96
		13	2.29	0.46	6.0	1.06
		15	2.64	0.43	6.4	1.13
4 in. [100 mm]	24 in. [600 mm] o.c.	11	1.94	0.60	6.6	1.16
		13	2.29	0.55	7.2	1.27
		15	2.64	0.52	7.8	1.37
6 in. [150 mm]	16 in. [400 mm] o.c.	19	3.35	0.37	7.1	1.25
		21	3.70	0.35	7.4	1.30
6 in. [150 mm]	24 in. [600 mm] o.c.	19	3.35	0.45	8.6	1.52
		21	3.70	0.43	9.0	1.59
8 in. [200 mm]	16 in. [400 mm] o.c.	25	4.41	0.31	7.8	1.37
8 in. [200 mm]	24 in. [600 mm] o.c.	25	4.41	0.38	9.6	1.69
Roofs/Floors						
Insulation R	4 ft [1.2 m] o.c.					
11		11	1.94	0.91	10.0	1.76
13		13	2.29	0.90	11.7	2.06
15		15	2.64	0.88	13.2	2.32
19		19	3.35	0.86	16.3	2.87
21		21	3.70	0.84	17.6	3.10
25		25	4.41	0.81	20.2	3.56
30		30	5.28	0.79	23.7	4.17
35		35	6.16	0.76	26.6	4.68
40		40	7.04	0.73	29.2	5.14
45		45	7.92	0.71	32.0	5.63
50		50	8.80	0.69	34.5	6.07
55		55	9.68	0.67	36.9	6.49

Source: Extracted and used with permission; © ASHRAE (www.ashrae.org); from ANSI/ASHRAE/IES Standard 90.1-2022: Energy Standard for Buildings Except Low-Rise Residential Buildings.

Note: The SI units for resistance were appended to the ASHRAE I-P data by the authors of this book.

[a] Correction factors for walls are not exhibited explicitly in *ANSI/ASHRAE/IES Standard 90.1*; however, the values are derived mathematically from the information exhibited in the standard.

[b] Roof/floor correction factors are from *Standard 90.1* and are based upon 4-ft [1.2-m] o.c. metal trusses that penetrate the insulation and 0.66-in. [16.8-mm] crossbars every 1 ft [0.3 m].

APPENDICES

Table E.9 Comparison of U-Factors for Framed Walls and Structural Insulated Panels (SIPs)

Component	Cavity R-Value[a]		Center-of-Insulated Whole-Wall R-Value[b]		Wall U-Factor[c]	
	I-P °F ft² h/Btu	SI K m²/W	I-P °F ft² h/Btu	SI K m²/W	I-P Btu/°F ft² h	SI W/k m²
2-in. × 4-in. insulated stud wall at 16-in. o.c. [50 × 100 mm at 400 mm]	13.6	2.40	9.6	1.69	0.104	0.59
2-in. × 6-in. insulated stud wall at 24-in. o.c. [50 × 150 mm at 600 mm]	19.6	3.46	13.7	2.41	0.073	0.41
6½-in. [165 mm] EPS-insulated SIP system	25.1	4.35	21.6	3.80	0.046	0.26

Source: Adapted from *Environmental Building News* 7(5) May 1998.

[a]Center-of-insulated-cavity R-value is the maximum total R-value through a cross-section of the wall system, including insulation, sheathing, drywall, etc.

[b]Whole-wall R-value is an average value that includes adjustments for the framing as well as some openings in a standardized wall system.

[c]U-factors calculated by the authors of this book and appended to Oak Ridge R-value data; SI dimensions also appended by the authors of this book.

Table E.10 U-Factors for Wood and Steel Doors

Nominal Door Thickness in.	mm		Wood Storm Door[a] I-P Btu/h ft² °F	Wood Storm Door[a] SI W/m² K	Metal Storm Door[b] I-P Btu/h ft² °F	Metal Storm Door[b] SI W/m² K	No Storm Door I-P Btu/h ft² °F	No Storm Door SI W/m² K
Unglazed Wood Doors[c, d]								
1 3/8	35	Panel door with 7/16-in. [11-mm] panels[e]	0.33	1.87	0.37	2.10	0.57	3.24
1 3/8	35	Hollow core flush door	0.30	1.70	0.32	1.82	0.47	2.67
1 3/8	35	Solid core flush door	0.26	1.48	0.28	1.59	0.39	2.21
1 3/4	45	Panel door with 7/16-in. [11-mm] panels	0.32	1.82	0.36	2.04	0.54	3.07
1 3/4	45	Hollow core flush door	0.29	1.65	0.32	1.82	0.46	2.61
1 3/4	45	Panel door with 1 1/8-in. [29-mm] panels	0.26	1.48	0.28	1.59	0.39	2.21
1 3/4	45	Solid core flush door	—	—	0.26	1.48	0.40	2.27
2 1/4	57	Solid core flush door	0.20	1.14	0.21	1.19	0.27	1.53
Unglazed Steel Doors[d]								
1 3/4	45	Fiberglass or mineral wool core with steel stiffeners, no thermal break[f]			—	—	0.60	3.41
1 3/4	45	Paper honeycomb core without thermal break[f]			—	—	0.56	3.18
1 3/4	45	Solid urethane foam core without thermal break[c]			—	—	0.40	2.27
1 3/4	45	Solid fire-rated mineral fiberboard core without thermal break[f]			—	—	0.38	2.16
1 3/4	45	Polystyrene core without thermal break[f] (18-gage [1.31-mm] commercial steel)			—	—	0.35	1.99
1 3/4	45	Polyurethane core without thermal break[f] (24-gage [0.70-mm] residential steel)			—	—	0.29	1.65
1 3/4	45	Polyurethane core with thermal break and wood perimeter[f] (24-gage [0.70-mm] residential steel)			—	—	0.20	1.14
1 3/4	45	Solid urethane foam core with thermal break[c]			0.16	0.91	0.20	1.14

Source: Previously adapted, with the permission of ASHRAE (www.ashrae.org), from the 2001 ASHRAE Handbook—Fundamentals (Chapter 25, Table 6). This citation to an older version of the Handbook is intentional and provides access to historic reference information of ongoing interest.

Note: U-factors listed are for exterior doors with no glazing, except for the storm doors, which are in addition to the main exterior door. Any glazing area in exterior doors should be included with the appropriate glass type and analyzed as a window (see Table E.14). Interpolation and moderate extrapolation are permitted for door thicknesses other than those specified

[a]Values for wood storm door are approximately 50% glass.

[b]Values for metal storm door are for any percentage of glass area.

[c]Values are based on a nominal 32-in. x 80-in. [810-mm x 2030-mm] door size with no glazing.

[d]Outside air conditions: 15 mph [24 km/h] wind speed, 0°F [−18°C] air temperature; inside air conditions: natural convection, 70°F [21°C] air temperature.

[e]A 55% panel area.

[f]*ASTM C 236* hot box data on a nominal 3-ft x 7-ft [910-mm x 2130-mm] door size with no glazing.

Table E.11 Heat Loss Coefficients (F_2) for Slab-on-Grade Floors

		I-P Btu/h °F ft Perimeter Degree Days (65°F base)			SI W/K m Perimeter Degree Days (18°C base)		
Construction[a]	**Insulation**	**2950**[c]	**5350**	**7433**[c]	**1640**[c]	**2970**	**4130**[c]
(a) Block wall, 8 in. [200 mm], brick facing	Uninsulated	0.62	0.68	0.72	1.07	1.17	1.24
	Insulated from edge to footer: R-5.4 h ft^2 °F/Btu [SI: R-0.95 m^2 K/W]	0.48	0.50	0.56	0.83	0.86	0.97
(b) Block wall, 4 in. [100 mm], brick facing	Uninsulated	0.80	0.84	0.93	1.38	1.45	1.61
	Insulated from edge to footer: R-5.4 h ft^2 °F/Btu [SI: R-0.95 m^2 K/W]	0.47	0.49	0.54	0.81	0.85	0.93
(c) Metal stud wall, stucco	Uninsulated	1.15	1.20	1.34	1.99	2.07	2.32
	Insulated from edge to footer: R-5.4 h ft^2 °F/Btu [SI: R-0.95 m^2 K/W]	0.51	0.53	0.58	0.88	0.92	1.00
(d) Poured concrete wall with duct near perimeter[b]	Uninsulated	1.84	2.12	2.73	3.18	3.67	4.72
	Insulated from edge to footer, and 3 ft [910 mm] under floor slab R-5.4 h ft^2 °F/Btu [SI: R-0.95 m^2 K/W]	0.64	0.72	0.90	1.11	1.24	1.56

Source: Reprinted with permission; © ASHRAE (www.ashrae.org); from the 2021 ASHRAE Handbook—Fundamentals, Chapter 18, Table 24. The SI units for F_2 shown in the last three columns were appended to the ASHRAE I-P data by the authors of this book.

[a]See Figure E.1 for illustrations of the listed constructions.

[b]Weighted average temperature of heating duct was assumed to be 110°F [43°C] during the heating season (outdoor air temperature less than 65°F [18°C]).

[c]Information in this column has been adapted, with the permission of ASHRAE (www.ashrae.org) from ASHRAE, 2017. This citation to an older version of the *Handbook* is intentional and provides access to historic reference information of ongoing interest.

Note: To use this table:

$$q = F_2 P \Delta t$$

where

q = heat loss through perimeter (Btu/h [W])
F_2 = heat loss coefficient from above
P = perimeter of exposed slab edge (ft [m])
Δt = temperature difference between indoor and outdoor air (°F [°C])

Do not assume additional losses from the slab to the earth below. Heat gains are assumed to be nonexistent.

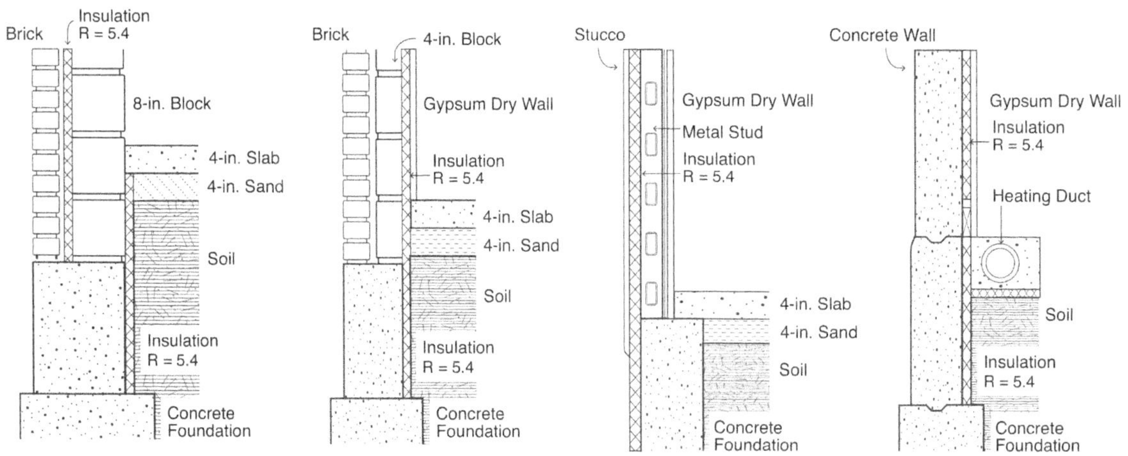

Figure E.1 *Insulation and construction configurations for Table E.11 data (heat loss coefficients for slab-on-grade floors). (Redrawn by Martin Lee; adapted from ASHRAE, 2001. This citation to an older version of the Handbook is intentional and provides access to historic reference information of ongoing interest.)*

Table E.12 Heat Loss Coefficients (F_2) for Slab-on-Grade Floors with Various Insulation Strategies

R-Value, Position, and Width (or Depth) of Insulation	F_2 I-P Btu/h ft °F	F_2 SI W/m K
Uninsulated slab	0.73	1.26
R-5 [SI: R-0.88] Horizontal insulation, 2 ft [0.6 m], no thermal break	0.70	1.21
R-10 [SI: R-1.76] Horizontal insulation, 2 ft [0.6 m], no thermal break	0.70	1.21
R-15 [SI: R-2.64] Horizontal insulation, 2 ft [0.6 m], no thermal break	0.69	1.19
R-5 [SI: R-0.88] Horizontal insulation, 4 ft [1.2 m], no thermal break	0.67	1.16
R-10 [SI: R-1.76] Horizontal insulation, 4 ft [1.2 m], no thermal break	0.64	1.11
R-15 [SI: R-2.64] Horizontal insulation, 4 ft [1.2 m], no thermal break	0.63	1.09
R-5 [SI: R-0.88] Vertical insulation, 2 ft [0.6 m]	0.58	1.00
R-10 [SI: R-1.76] Vertical insulation, 2 ft [0.6 m]	0.54	0.93
R-15 [SI: R-2.64] Vertical insulation, 2 ft [0.6 m]	0.52	0.90
R-5 [SI: R-0.88] Vertical insulation, 4 ft [1.2 m]	0.54	0.93
R-10 [SI: R-1.76] Vertical insulation, 4 ft [1.2 m]	0.48	0.83
R-15 [SI: R-2.65] Vertical insulation, 4 ft [1.2 m]	0.45	0.78
R-10 [SI: R-1.76] Fully insulated slab (insulated under entire slab as well as around edge)	0.36	0.62

Source: Extracted and used with permission; © ASHRAE (www.ashrae.org); from ANSI/ASHRAE/IES Standard 90.1-2022: Energy Standard for Buildings Except Low-Rise Residential Buildings.

Notes:

1. Insulation is extruded polystyrene, R = 5.0 h ft^2 °F/Btu-in. [SI: R = 34.7 m K/W].
2. Soil conductivity is 0.75 Btu/h ft °F [1.30 W/K m].
3. No thermal break at edge of slab, where so indicated. If a thermal break is provided with horizontal insulation, use the corresponding value for vertical insulation.
4. Values assume an unheated slab; F_2 values increase substantially when slab is heated.

Table E.13 Heat Flow through Below-Grade Walls and Floors

IP Units				
Part A: Average U-Factor for Basement Walls with Uniform Insulation				
	$U_{avg,bw}$ from Grade to Depth, Btu/h ft² °F			
Depth, ft	**Uninsulated**	**R-5**	**R-10**	**R-15**
1	0.432	0.135	0.080	0.057
2.6	0.331	0.121	0.075	0.054
3	0.273	0.110	0.070	0.052
4	0.235	0.101	0.066	0.050
5	0.208	0.094	0.063	0.048
6	0.187	0.088	0.060	0.046
7	0.170	0.083	0.057	0.044
8	0.157	0.078	0.055	0.043
Part B: Average U-Factor for Basement Floors				
	$U_{avg,bf}$ Btu/h ft² °F			
Depth of Floor Below Grade, ft	w_b (Shortest Width of Basement), ft **20**	**24**	**28**	**32**
1	0.064	0.057	0.052	0.047
2	0.054	0.048	0.044	0.040
3	0.047	0.042	0.039	0.036
4	0.042	0.038	0.035	0.033
5	0.038	0.035	0.032	0.030
6	0.035	0.032	0.030	0.028
7	0.032	0.030	0.028	0.026

SI Units				
Part A: Average U-Factor for Basement Walls with Uniform Insulation				
	$U_{avg,bw}$ from Grade to Depth, W/m² K			
Depth, m	**Uninsulated**	**R-0.88**	**R-1.76**	**R-2.64**
0.3	2.468	0.769	0.458	0.326
0.8	1.898	0.689	0.427	0.310
0.9	1.571	0.628	0.401	0.296
1.2	1.353	0.579	0.379	0.283
1.5	1.195	0.539	0.360	0.272
1.8	1.075	0.505	0.343	0.262
2.1	0.980	0.476	0.328	0.252
2.4	0.902	0.450	0.315	0.244
Part B: Average U-Factor for Basement Floors				
	$U_{avg,bf}$ W/m² K			
Depth of Floor Below Grade, m	w_b (Shortest Width of Basement), m **6**	**7**	**8**	**9**
0.3	0.370	0.335	0.307	0.283
0.8	0.310	0.283	0.261	0.242
0.9	0.271	0.249	0.230	0.215
1.2	0.242	0.224	0.208	0.195
1.5	0.220	0.204	0.190	0.179
1.8	0.202	0.188	0.176	0.166
2.1	0.187	0.175	0.164	0.155

Source: Reprinted with permission; © ASHRAE (www.ashrae.org); from the 2021 ASHRAE Handbook—Fundamentals, Chapter 18, Tables 22 and 23.

Notes:

1. Only heat losses are assumed because the interior temperature is normally higher than the ground temperature.
2. Soil conductivity assumed to be 0.8 Btu /h ft °F [1.4 W/m K].

Table E.14 U-Factors of Representative Window Assemblies

Glazing System Description	Aluminum without Thermal Break I-P: Btu/h ft^2 °F [SI: W/m^2 K]	Aluminum with Thermal Break I-P: Btu/h ft^2 °F [SI: W/m^2 K]	Insulated Wood/Vinyl I-P: Btu/h ft^2 °F [SI: W/m^2 K]	Fiberglass/Vinyl I-P: Btu/h ft^2 °F [SI: W/m^2 K]
Single glazing with uncoated ⅛ in. [3.2 mm] clear pane	1.23 [7.01]	1.07 [6.08]	0.91 [5.20]	0.85 [4.83]
Single glazing with uncoated ¼ in. [6.4 mm] acrylic/polycarbonate pane	1.10 [6.23]	0.94 [5.35]	0.80 [4.52]	0.74 [4.18]
Double glazing with ⅛ in. [3.2 mm] panes: uncoated clear | clear with ¼ in. [6.4 mm] air space	0.81 [4.62]	0.64 [3.61]	0.55 [3.14]	0.50 [2.84]
Double glazing with ⅛ in. [3.2 mm] panes: uncoated clear | clear with ½ in. [13 mm] air space	0.76 [4.30]	0.58 [3.31]	0.50 [2.86]	0.45 [2.58]
Double glazing with ⅛ in. [3.2 mm] panes: uncoated clear | low-ε (0.2) on surface 3 with ½ in. [13 mm] air space	0.65 [3.70]	0.48 [2.75]	0.41 [2.34]	0.37 [2.07]
Triple glazing with ⅛ in. [3.2 mm] panes: uncoated clear | clear with ½ in. [13 mm] air spaces	0.61 [3.46]	0.44 [2.47]	0.38 [2.14]	0.34 [1.90]
Triple glazing with ⅛ in. [3.2 mm] panes: uncoated clear | low-ε (0.2) on surfaces 3 and 5 with ½ in. [13 mm] air spaces	0.52 [2.95]	0.35 [1.99]	0.30 [1.69]	0.26 [1.48]
Quadruple glazing with ⅛ in. [3.2 mm] panes: uncoated clear | low-ε (0.1) on surfaces 3 and 5 with ½ in. [13 mm] air spaces	0.48 [2.71]	0.31 [1.77]	0.26[1.43]	

Source: Reprinted with permission; © ASHRAE (www.ashrae.org); from the 2021 ASHRAE Handbook—Fundamentals, Chapter 15, Table 4.

Notes:

1. Based upon an operable 3-ft × 5-ft [0.9-m × 1.5-m] aluminum-framed window.
2. Glazing surfaces are numbered starting with the surface closest to the sun; thus, surface 2 would be the inner surface of an exterior pane of glass.

Table E.15 Representative Window Characteristics[a]

Glazing Description and Reference Number	Layers of Glazing and Spaces (outside to inside)	Frame (and Spacer)	Total Window U-Factor		Total Window					Air Leakage			
			Btu/h ft² °F	W/m²K	SHGC[b]	VT[c]	LSG[d]	FHR[e]	FCR[e]	cfm/ lin ft	L/s m	cfm/ ft²	L/s m²
1. Single-glazed clear	⅛ in. [3 mm] clear	Aluminum, no thermal break	1.30	7.38	0.79	0.69	0.87	0	0	0.65	1.01	0.98	4.98
2. Single-glazed bronze	⅛ in. [3 mm] bronze	Aluminum, no thermal break	1.30	7.38	0.69	0.52	0.75	−2	8	0.65	1.01	0.98	4.98
3. Double-glazed clear	⅛ in. [3 mm] clear	Aluminum, thermal break (aluminum)	0.64	3.63	0.65	0.62	0.95	19	12	0.37	0.57	0.56	2.85
	½ in. [13 mm] air												
	⅛ in. [3 mm] clear												
4. Double-glazed bronze	⅛ in. [3 mm] bronze	Aluminum, thermal break (aluminum)	0.64	3.63	0.55	0.47	0.85	17	20	0.37	0.57	0.56	2.85
	½ in. [13 mm] air												
	⅛ in. [3 mm] clear												
5. Double-glazed clear	⅛ in. [3 mm] clear	Wood or vinyl (aluminum)	0.49	2.78	0.58	0.57	0.98	24	18	0.37	0.57	0.56	2.85
	½ in. [13 mm] air												
	⅛ in. [3 mm] clear												
6. Double-glazed bronze	⅛ in. [3 mm] bronze	Wood or vinyl (aluminum)	0.49	2.78	0.48	0.43	0.90	22	25	0.37	0.57	0.56	2.85
	½ in. [13 mm] air												
	⅛ in. [3 mm] clear												
7. Double-glazed low-ε[f]	⅛ in. [3 mm] clear	Wood or vinyl (stainless)	0.33	1.87	0.55	0.52	0.95	32	19	0.10	0.16	0.15	0.76
	½ in. [13 mm] argon												
	⅛ in. [3 mm] low-ε 0.20												
8. Double-glazed low-ε[f]	⅛ in. [3 mm] low-ε 0.08	Wood or vinyl (stainless)	0.30	1.70	0.44	0.56	1.27	32	27	0.10	0.16	0.15	0.76
	½ in. [13 mm] argon												
	⅛ in. [3 mm] clear												
9. Double-glazed spectrally selective[f]	⅛ in. [3 mm] low-ε 0.04	Wood or vinyl (stainless)	0.29	1.65	0.31	0.51	1.65	30	36	0.10	0.16	0.15	0.76
	½ in. [13 mm] argon												
	⅛ in. [3 mm] clear												

10. Double-glazed spectrally selective[f]	⅛ in. [3 mm] low-ε 0.10	Wood or vinyl (stainless)	0.31	1.76	0.26	0.31	1.19	27	40	0.10	0.16	0.15	0.76
	½ in. [13 mm] argon												
	⅛ in. [3 mm] clear												
11. Triple-glazed low-ε[f] superwindow	⅛ in. [3 mm] low-ε 0.08	Insulated vinyl (insulated)	0.15	0.85	0.37	0.48	1.30	38	33	0.05	0.08	0.08	0.41
	½ in. [13 mm] krypton												
	⅛ in. [3 mm] clear												
	½ in. [13 mm] krypton												
	⅛ in. [3 mm] low-ε 0.08												
12. Triple-glazed clear	⅛ in. [3 mm] clear	Wood or vinyl (stainless)	0.34	1.93	0.52	0.53	1.02	32	22	0.10	0.16	0.15	0.76
	½ in. [13 mm] air												
	⅛ in. [3 mm] clear												
	½ in. [13 mm] air												
	⅛ in. [3 mm] clear												

Source: Carmody et al., 1996. SI units appended by the authors of this book.

[a]Based upon a casement window, 2 ft × 4 ft [610 mm × 1220 mm].

[b]Solar heat gain coefficient (higher numbers mean more solar heat flow).

[c]Visible transmittance (higher numbers mean more light transmitted).

[d]Light-to-solar gain ratio, LSG = VT/SHGC. For typical center-of-glass values for VT and SHGC, see Table E.18.

[e]FHR (fenestration heating rating) and FCR (fenestration cooling rating) are heating season and cooling season (respectively) estimates of the percentage of energy saved in a typical residential application compared to using window 1.

[f]Low-ε ratings indicate what percentage of the long-wavelength radiant energy is admitted; 0.20 therefore admits 20% and reflects 80%.

Table E.16 Representative Skylight U-Factors[a]

Part A. I-P Units (Btu/h ft^2 °F)

Glazing	Skylight with Curb: Aluminum, Thermal Break, No	Skylight with Curb: Aluminum, Thermal Break, Yes	Skylight with Curb: Reinforced Vinyl/ Aluminum-Clad Wood	Skylight with Curb: Wood/ Vinyl	Skylight without Curb: Aluminum, Thermal Break, No	Skylight without Curb: Aluminum, Thermal Break, Yes	Skylight without Curb: Structural Glazing
Single- glazed	1.98	1.89	1.75	1.47	1.36	1.25	1.25
Double-glazed, ½-in. gap	1.30	1.10	1.04	0.84	0.81	0.69	0.65
Double-glazed, ½-in. gap, argon, low-ε 0.20 on surface[b] 2 or 3	1.15	0.95	0.89	0.68	0.66	0.55	0.51
Double-glazed, ½-in. gap, argon, low-ε 0.10 on surface[b] 2 or 3	1.13	0.93	0.87	0.67	0.65	0.53	0.49
Double-glazed, ½-in. gap, argon, low-ε 0.05 on surface[b] 2 or 3	1.11	0.91	0.85	0.65	0.63	0.52	0.47
Triple-glazed, ½-in. gaps	1.10	0.87	0.81	0.61	0.62	0.51	0.45
Triple-glazed, ½-in. gaps, argon, low-ε 0.10 on surfaces[b] 2 or 3 and 4 or 5	0.95	0.72	0.67	0.47	0.48	0.37	0.31

Part B. SI Units (W/m^2 K)

Glazing	Skylight with Curb: Aluminum, Thermal Break, No	Skylight with Curb: Aluminum, Thermal Break, Yes	Skylight with Curb: Reinforced Vinyl/ Aluminum-Clad Wood	Skylight with Curb: Wood/ Vinyl	Skylight without Curb: Aluminum, Thermal Break, No	Skylight without Curb: Aluminum, Thermal Break, Yes	Skylight without Curb: Structural Glazing
Single-glazed, 3 mm glass	11.24	10.73	9.94	8.34	7.73	7.09	7.09
Double-glazed, 12.7-mm gap	7.39	6.27	5.90	4.74	4.59	3.95	3.70
Double-glazed, 12.7-mm gap, argon, low-ε 0.20 on surface[b] 2 or 3	6.52	5.39	5.04	3.89	3.77	3.12	2.87
Double-glazed, 12.7-mm gap, argon, low-ε 0.10 on surface[b] 2 or 3	6.41	5.29	4.94	3.79	3.67	3.03	2.78
Double-glazed, 12.7-mm gap, argon, low-ε 0.05 on surface[b] 2 or 3	6.31	5.18	4.84	3.69	3.57	2.93	2.68
Triple-glazed, 12.7-mm gaps	6.22	4.92	4.62	3.48	3.51	2.88	2.56
Triple-glazed, 12.7-mm gaps, argon, low-ε 0.10 on surfaces[b] 2 or 3 and 4 or 5	5.38	4.07	3.79	2.66	2.72	2.09	1.78

Source: Excerpted with permission; © ASHRAE (www.ashrae.org); from ANSI/ASHRAE/IES Standard 90.1-2022: Energy Standard for Buildings Except Low-Rise Residential Buildings. The SI values were appended to ASHRAE's I-P data by the authors of this book. These values are applicable to skylights that have not been tested or labeled by a certification organization (such as NFRC).

[a] All glazings are ⅛-in. [3-mm] thickness. Winter conditions, 15-mph [24-km/h] wind with 0°F [–18°C] outside, 70°F [21°C] inside, no solar radiation. Glazing is sloped at 20° from the horizontal.

[b] Glazing layer surfaces are numbered from outside to inside.

APPENDICES

Table E.17 Solar Heat Gain Coefficients (SHGC) for Plastic Domed Horizontal Skylights

		Curb (See Sketch Below)		
Dome	**Light Diffuser (Translucent)**	**Height, in. [mm]**	**Width-to-Height Ratio**	**Solar Heat Gain Coefficient**
Clear ($\tau = 0.86^a$)	Yes ($\tau = 0.58$)	0 [0]	∞	0.53
		9 [230]	5	0.50
		18 [460]	2.5	0.44
Clear ($\tau = 0.86$)	None	0 [0]	∞	0.86
		9 [230]	5	0.77
		18 [460]	2.5	0.70
Translucent ($\tau = 0.52$)	None	0 [0]	∞	0.50
		18 [460]	2.5	0.40
Translucent ($\tau = 0.27$)	None	0 [0]	∞	0.30
		9 [230]	5	0.26
		18 [460]	2.5	0.24

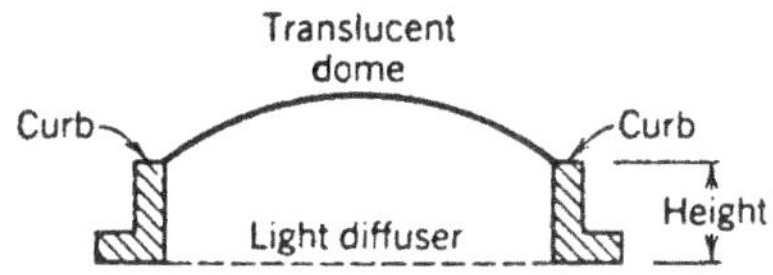

Source: Excerpted with permission; © ASHRAE (www.ashrae.org); from the 2021 ASHRAE Handbook—Fundamentals, Chapter 15, Table 11.

[a] τ refers to the transmittance of the dome or diffusing element, as noted.

Table E.18 Solar Optical Properties of Representative Glazings and Window Assemblies

Glazing System Description	Center of Glazing VT	Center of Glazing SHGC[a]	Window Assembly VT[a b]	Window Assembly SHGC[a b]
Single Glazings—with 1/4 in. [6.4 mm] Panes Except as Noted				
Uncoated 1/8 in. [3.2 mm] clear	0.90	0.86	0.80	0.78
Uncoated clear	0.88	0.81	0.78	0.74
Uncoated bronze	0.54	0.62	0.48	0.57
Uncoated green	0.76	0.60	0.68	0.55
Uncoated gray	0.46	0.59	0.41	0.54
Reflective stainless steel on clear (8%)	0.08	0.19	0.07	0.18
Reflective stainless steel on clear (20%)	0.20	0.31	0.18	0.29
Double Glazings—All with 1/4 in. [6.4 mm] Panes				
Uncoated clear | clear	0.78	0.70	0.69	0.64
Uncoated bronze | clear	0.47	0.49	0.42	0.45
Uncoated high-performance green | clear	0.59	0.39	0.53	0.36
Reflective stainless steel on clear (8%) | clear	0.07	0.13	0.06	0.13
Reflective titanium on clear (20%) | clear	0.18	0.21	0.16	0.21
Clear | low-ε (0.2) on surface[c] 3	0.73	0.65	0.65	0.59
Low-ε (0.2) on surface 2 | clear	0.73	0.60	0.64	0.55
Low-ε (0.1) on surface 2 | clear	0.72	0.60	0.64	0.55
Low-ε (0.05) on surface 2 | clear	0.70	0.37	0.62	0.34
Triple Glazings—All with 1/4 in. [6.4 mm] Panes				
Clear | clear | clear	0.70	0.61	0.62	0.56
High-performance green | clear | clear	0.53	0.32	0.47	0.30
Low-ε (0.2) on surface 2 | clear | clear	0.64	0.53	0.57	0.51
Low-ε (0.1) on surface 2 | clear | low-ε (0.1) on surface 5	0.59	0.36	0.53	0.34
Low-ε (0.05) on surface 2 | low-ε (0.05) on surface 4 | clear	0.55	0.26	0.49	0.25

Source: Excerpted with permission; © ASHRAE (www.ashrae.org); from the 2021 ASHRAE Handbook—Fundamentals, Chapter 15, Table 10.

[a]Values are for a normal angle of solar incidence; see ASHRAE, 2021 for SHGC values at other incidence angles.

[b]Based upon an operable 3-ft × 5-ft [0.9-m × 1.5-m] aluminum-framed window.

[c]Glazing surfaces are numbered starting with the surface closest to the sun; thus, surface 2 would be the inner surface of an exterior pane of glass.

Table E.19 Solar Optical Properties of Transparent Plastics

	Transmittance		
Type of Plastic	**Visible**	**Solar**	**SC[a]**
Acrylic			
Clear	0.92	0.85	0.98
Gray tint	0.16	0.27	0.52
Gray tint	0.33	0.41	0.63
Gray tint	0.45	0.55	0.74
Gray tint	0.59	0.62	0.80
Gray tint	0.76	0.74	0.89
Bronze tint	0.27	0.35	0.58
Bronze tint	0.49	0.56	0.75
Bronze tint	0.61	0.62	0.80
Bronze tint	0.75	0.75	0.90
Reflective[b]	0.14	0.12	0.21
Polycarbonate			
Clear, ⅛ in. [3 mm]	0.88	0.82	0.98
Gray or bronze, ⅛ in. [3 mm]	0.50	0.57	0.74

Source: Previously adapted, with the permission of ASHRAE (www.ashrae.org), from the 1997 ASHRAE Handbook—Fundamentals. This citation to an older version of the *Handbook* is intentional and provides access to historic reference information of ongoing interest.

[a]Shading coefficient; multiply SC by 0.87 for an approximate estimation of SHGC (solar heat gain coefficient).

[b]Aluminum metalized polyester film on plastic.

Table E.20 Approximate Shading Coefficients (SC) of External Shading Devices

Awnings	
Of venetian blind type, 2/3 drawn	0.43
Of venetian blind type, fully drawn	0.15
Dark or medium canvas	0.25
Shading screens	0.28–0.23
Louvers, movable	0.15–0.10
Overhang: continuous, completely shading window	0.25
Dense tree casting heavy shade	0.25

Source: Adapted from Olgyay, 1963.

Table E.21 Shading Coefficients (SC) for Louvered Sunscreens

Profile Angle, (deg)	Group 1[a]		Group 2[b]		Group 3[c]		Group 4[d]		Group 5[e]		Group 6[f]	
	Transmittance	SC	Transmittance	SC	Transmittance	SC	Transmittance	SC	Transmittance	SC	Transmittance	SC
10	0.23	0.35	0.25	0.33	0.40	0.51	0.48	0.59	0.15	0.27	0.26	0.45
20	0.06	0.17	0.14	0.23	0.32	0.42	0.39	0.50	0.04	0.11	0.20	0.35
30	0.04	0.15	0.12	0.21	0.21	0.31	0.28	0.38	0.03	0.10	0.13	0.26
≥40	0.04	0.15	0.11	0.20	0.07	0.18	0.20	0.30	0.03	0.10	0.04	0.13

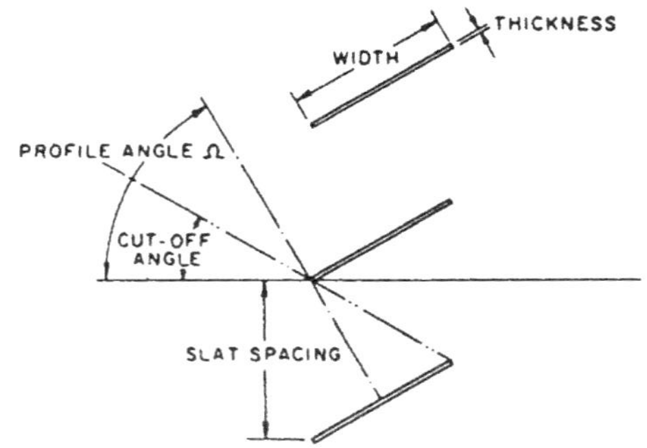

Source: Previously adapted, with the permission of ASHRAE (www.ashrae.org), from the 1997 ASHRAE Handbook—Fundamentals. This citation to an older version of the *Handbook* is intentional and provides access to historic reference information of ongoing interest.

Note: The *2021 ASHRAE Handbook—Fundamentals* (ASHRAE, 2021) presents a different and more complex methodology for determining solar gain through externally shaded glazing systems. The methodology and associated data are not presented here, as they are less intuitive and less suited to hand calculations than the older SC data presented in this table. Refer to the current *Handbook* for this new information.

[a] Group 1: Black, width over spacing ratio 1.15/1; 23 louvers/in. [1.1 mm between louvers], with ¼-in. [6.4-mm] single clear glass.

[b] Group 2: Light color; high reflectance, otherwise same as group 1.

[c] Group 3: Black or dark color; width over spacing ratio 0.85/1; 17 louvers/in. (1.5 mm between louvers), with ¼-in. [6.4-mm] single clear glass.

[d] Group 4: Light color or unpainted aluminum; high reflectance; otherwise, same as group 3.

[e] Group 5: Same as group 1, except two lights of ¼-in. [6.4-mm] clear glass with ½-in. [12.7-mm] air space.

[f] Group 6: Same as group 3, except two lights of ¼-in. [6.4-mm] clear glass with ½-in. [12.7-mm] air space.

Table E.22 Performance Characteristics of Adjustable Sunshading

Shading Type	Typical Orientations	Solar Heat Rejection	Daylight Distribution	Ventilation	View	Winter Night Insulation
Roll-down Shades Solid 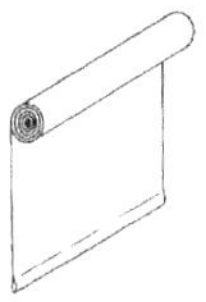	East and west, where it is rolled down daily during hours of direct sun exposure.	Blocks direct, diffuse, and reflected radiation. Heat builds up between shade and glass unless vented.	Translucent shades diffuse light evenly but may become a source of glare due to brightness. If shades are colored, incoming light will be colored.	Blocks ventilation airflow.	Blocks view.	Can be effective if airflow around the shade is blocked (providing dead air space).
Roll-down Shades Slatted 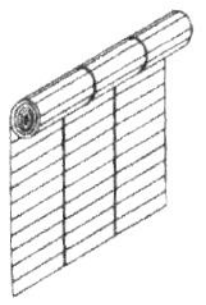	East and west, where it is rolled down daily during hours of direct sun exposure.	Blocks most direct, diffuse, and reflected radiation. Heat can convect through slats into space.	A thinly striped pattern of direct sun is admitted. Opaque slats will heavily filter daylight.	Greatly reduces ventilation airflow.	Substantially obstructs and filters view	No value unless slats can be fitted together to close openings and trap air between shade and glass.
Fold-out Retractable Awning 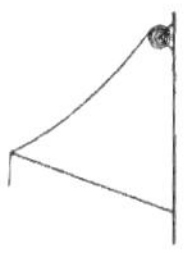	Any orientation. On south, awnings can be deep and be left in position for weeks at a time.	Blocks direct sun; admits most diffuse and reflected radiation. Heat collected on awning easily dissipated outside of building.	Translucent awnings evenly diffuse sunlight. (Colored awnings will color light.) Awnings can redirect ground-reflected light into building for distribution by ceiling.	Blocks view of sky; permits view of ground and some surroundings (depending upon positioning).	Reduces ventilation airflow somewhat, tends to direct wind upward into space.	If awning folds back against window, it can provide some insulation value—but likely limited.
Pivoting Louvers Horizontal 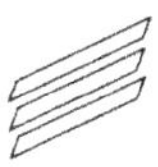	South; also used on other elevations (although less effective).	Blocks direct sun; admits ground-reflected and some diffuse radiation.	Skylight can be reflected to ceiling plane for distribution into building. Ground-reflected light can be reflected into space.	Reduces ventilation airflow somewhat (based upon louver spacing); directs wind upward into space.	Blocks sky view; filters ground and surrounding views (depending upon angle of louvers).	Closed louvers can provide some insulation value (depending upon tightness of closure).
Pivoting Louvers Vertical 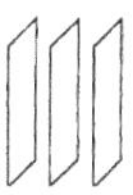	East and west; also used on north. Not as effective on south as horizontal louvers.	Blocks direct sun; admits most ground-reflected and diffuse radiation.	Blocks direct sunlight; admits most ground-reflected and diffuse light.	Reduces ventilation airflow somewhat (to extent of solidity); directs wind sideways into space.	Filters view, especially to sides.	Closed louvers can provide some insulation value (depending upon tightness of closure).

Table E.23 Shading Coefficients (SC) for Glazing with Integral or Interior Shading

Part A. SC for Insulating Glass[a] with Indoor Shading by Venetian Blinds or Roller Shades

Type of Glass	Nominal Thickness, Each Light	Solar Transmittance[b]		Type of Shading				
				Venetian Blinds[c]		Roller Shade		
						Opaque		Translucent
		Outer Pane	Inner Pane	Medium	Light	Dark	White	Light
Clear out	⅛ in. [3.2 mm]	0.87	0.87	0.62	0.58 (45° open)	0.71	0.35	0.40
Clear in				0.53	0.38 (closed)[d]			
Heat-absorbing[e] out, Clear in	¼ in. [6.4 mm]	0.46	0.80	0.39	0.36	0.40	0.22	0.30
Reflective coated glass								
$SC^f = 0.20$				0.19	0.18			
0.30				0.27	0.26			
0.40				0.34	0.33			

Part B. Shading Coefficients for Double Glazing with Between-Glass Shading

Type of Glass	Nominal Thickness, Each Pane	Solar Transmittance[b]		Description of Air Space	Type of Shading		
					Venetian Blinds		
		Outer Pane	Inner Pane		Light	Medium	Louvered Sunscreen
Clear out Clear in	⅛ in. [3.2 mm]	0.87	0.87	Shade in contact with glass or shade separated from glass by airspace	0.33	0.36	0.43
Clear out Clear in	¼ in. [6.4 mm]	0.80	0.80	Shade in contact with glass; voids filled with plastic	—	—	0.49
Heat-absorbing[e] out, Clear in	¼ in. [6 mm]	0.46	0.80	Shade in contact with glass or shade separated from glass by airspace	0.28	0.30	0.37
				Shade in contact with glass; voids filled with plastic	—	—	0.41

Source: Previously adapted, with the permission of ASHRAE (www.ashrae.org), from the 1997 ASHRAE Handbook—Fundamentals. This citation to an older version of the *Handbook* is intentional and provides access to historic reference information of ongoing interest.

[a]Factory-manufactured units with 3/16-in. [4.8-mm], ¼-in. [6.4-mm], or ½-in. [12.7-mm] air space.

[b]Refer to manufacturers' literature for exact values.

[c]For vertical blinds with opaque, white, or beige louvers, tightly closed, SC is approximately the same as for opaque white roller shades.

[d]Use these values only when operation is automated for solar gain reduction (as opposed to daylight use).

[e]Refers to bronze or green-tinted, heat-absorbing glass.

[f]SC for glass with no shading device.

Table E.24 Shading Coefficients (SC) for Single and Insulating Glass with Draperies

			Range of Shading Coefficients	
			Drapery Fabrics[b]	
Glazing	Glass Trans.	Glass SC[a]	High Transmittance, Low Reflectance[c]	Low Transmittance, High Reflectance[c]
Single glass				
1/8 in. [3.2 mm] clear	0.86	1.00	0.87	0.37
1/4 in. [6.4 mm] clear	0.80	0.95	0.80	0.35
1/2 in. [12.7 mm] clear	0.71	0.88	0.74	0.35
1/4 in. [6.4 mm] heat absorbing	0.46	0.67	0.57	0.33
1/2 in. [12.7 mm] heat absorbing	0.24	0.50	0.43	0.30
Reflective coated (see manufacturers' literature for exact values)	—	0.60	0.57	0.33
	—	0.50	0.46	0.31
	—	0.40	0.36	0.26
	—	0.30	0.25	0.20
Insulating glass, 1/4 in. [12.7 mm] air space				
Clear out and clear in	0.64	0.83	0.66	0.35
Heat absorbing out and clear in	0.37	0.55	0.49	0.32
Reflective coated (see manufacturers' literature for exact values)	—	0.40	0.38	0.28
	—	0.30	0.29	0.24
	—	0.20	0.19	0.15

Source: Previously adapted, with the permission of ASHRAE (www.ashrae.org), from the 1997 ASHRAE Handbook—Fundamentals. This citation to an older version of the *Handbook* is intentional and provides access to historic reference information of ongoing interest.

Note: The *2021 ASHRAE Handbook—Fundamentals* (ASHRAE, 2021) presents a different and more complex methodology for determining solar gain through internally shaded glazing systems—involving the variable IAC (interior attenuation coefficient). The methodology and associated data are not presented here, as they are less intuitive and less suited to hand calculations than the older SC data presented in this table. Refer to the *2021 Handbook* for this new information.

[a]For glass alone, with no drapery.

[b]Draperies of 100% fullness, loose hanging.

[c]See ASHRAE, 1997, Chapter 27, Table 29, for more detailed listings.

Table E.25 Properties of Thermal Storage Materials

Part A. IP Units								
Category	Material	Thermal Conductivity (Btu/h ft °F)	Specific Heat Capacity (Btu-in/h ft2 °F)	Heat of Fusion (Btu/lb)	Density (lb/ft^3)	Emissivity[a]	Solar Absorptance[b]	Melting Temperature (°F)
Concrete[a]	Heavyweight poured	1.2–2.0	0.22–0.24		137–150	0.85–0.9	0.4–0.5	
Earth	Adobe[c]	0.3–0.7	0.20–0.24		75–125	0.9	0.5–0.7	
	Rammed earth[c,d]	0.38–0.72	0.21–0.30		96–119	0.9	0.5–0.7	
	Brick[a]	0.4–0.6	0.19		120–123	0.75–0.93	0.5–0.7	
Stone[e–g]	Basalt	0.6–1.2	0.17–0.25		169–207	0.72	0.7–0.9	
	Sandstone	0.6–1.6	0.17–0.22		124–153	0.9	0.6–0.7	
	Granite	1.4–2.6	0.16–0.19		162–170	0.9	0.55	
Phase-change materials[h]	$CaCl_2 \cdot 6H_2O$	0.31 (liquid)	0.35	82	98 (liquid)	–	Untinted, in clear containers: 0.25–0.4	84
	$Na_2SO_4 \cdot 10H_2O$	0.32 (solid)	0.46	104	93 (solid)	–	Tinted or in dark containers: up to 0.9	90
	$Zn(NO_3)_2 \cdot 6H_2O$	0.27 (liquid)	0.32	63	114 (liquid)	–		97
	Paraffin C_{13}–C_{24}	0.12 (liquid)	–	81	47 (liquid) 56 (solid)	–		72–75
	Paraffin C_{18}	0.09 (liquid)	–	105	48 (liquid) 51 (solid)	–		82

Part B. SI Units

Category	Material	Thermal Conductivity (W/m K)	Specific Heat Capacity (J/kg K)	Heat of Fusion (kJ/kg)	Density (kg/m³)	Emissivity[a]	Solar Absorptance[b]	Melting Temperature (°C)
Concrete[a]	Heavyweight poured	2.0–3.5	900–1000		2200–2400	0.85–0.9	0.4–0.5	
Earth	Adobe[c]	0.5–1.2	840–1000		1200–2000	0.9	0.5–0.7	
	Rammed earth[c,d]	0.65–1.25	870–1260		1540–1900	0.9	0.5–0.7	
	Brick[a]	0.7–1.0	790–800		1920–1970	0.75–0.93	0.5–0.7	
Stone[e–g]	Basalt	1.0–2.0	720–1040		2700–3310	0.72	0.7–0.9	
	Sandstone	1.0–2.7	730–930		1990–2450	0.9	0.6–0.7	
	Granite	2.4–4.5	650–800		2600–2720	0.9	0.55	
Phase-change materials[h]	$CaCl_2 \cdot 6H_2O$	0.54 (liquid)	1450	191	1,562 (liquid)	–	Untinted, in clear containers: 0.25–0.4	29
	$Na_2SO_4 \cdot 10H_2O$	0.55 (solid)	1930	241	1485 (solid)	–		32
	$Zn(NO_3)_2 \cdot 6H_2O$	0.46 (liquid)	1340	147	1828 (liquid)	–		36
	Paraffin C_{13}–C_{24}	0.21 (liquid)	–	189	760 (liquid) 900 (solid)	–	Tinted or in dark containers: up to 0.9	22–24
	Paraffin C_{18}	0.15 (liquid)	–	244	774 (liquid) 814 (solid)	–		28

Sources: Various (see references below).

[a]ASHRAE. 2005. This citation to an older version of the *Handbook* is intentional and provides access to historic reference information of ongoing interest.

[b]The Engineering ToolBox, 2009.

[c]Minke, 2009.

[d]Allinson and Hall. 2010.

[e]Clauser and Huenges. 1995.

[f]Robertson, 1988.

[g]Olhoeft and Johnson. 1989. "Densities of Rocks and Minerals," in *CRC Practical Handbook of Physical Properties of Rocks and Minerals*. R.S. Carmichael (ed.). Boca Raton, FL: CRC Press.

[h]Cabeza, et al., 2011.

References and Resources

Allinson, D. and M. Hall. 2010. "Hygrothermal Analysis of a Stabilised Rammed Earth Test Building in the UK." *Energy and Buildings* 42: 845–852.

ASHRAE. 1997. *ASHRAE Handbook—Fundamentals*. Atlanta, GA: American Society of Heating, Refrigerating, and Air-Conditioning Engineers.

ASHRAE. 2001. *ASHRAE Handbook—Fundamentals*. Atlanta, GA: American Society of Heating, Refrigerating, and Air-Conditioning Engineers.

ASHRAE. 2005. *ASHRAE Handbook—Fundamentals*. Atlanta, GA: American Society of Heating, Refrigerating Engineers.

ASHRAE. 2017. *ASHRAE Handbook—Fundamentals*. Atlanta, GA: American Society of Heating, Refrigerating Engineers.

ASHRAE. 2021. *ASHRAE Handbook—Fundamentals*. Atlanta, GA: American Society of Heating, Refrigerating, and Air-Conditioning Engineers.

ASHRAE. 2022. *ANSI/ASHRAE/IES Standard 90.1-2022: Energy Standard for Buildings Except Low-Rise Residential Buildings*. Atlanta, GA: American Society of Heating, Refrigerating, and Air-Conditioning Engineers.

Balcomb, J. D., et al. 1980. *Passive Solar Design Handbook* (Vol. 2: *Passive Solar Design Analysis*). Washington, DC: U.S. Department of Energy.

Cabeza, L. F., A. Castell, C. D. Barreneche, A. De Gracia, and A. I. Fernández. 2011. "Materials Used as PCM in Thermal Energy Storage in Buildings: A Review." *Renewable and Sustainable Energy Reviews* 15(3): 1675–1695.

Carmody, J., S. Selkowitz, and L. Heschong. 1996. *Residential Windows: A Guide to New Technologies and Energy Performance*. New York: W. W. Norton & Company.

Clauser, C. and E. Huenges. 1995. "Thermal Conductivity of Rocks and Minerals," in *Rock Physics and Phase Relations: A Handbook of Physical Constants* (pp. 105–126), T. J. Ahrens (ed.). Washington, DC: American Geophysical Union.

Olgyay, V. 1963. *Design with Climate*. Princeton, NJ: Princeton University Press.

Minke, G. 2009. *Building with Earth: Design and Technology of a Sustainable Architecture*, 2nd edn. Boston, MA: Birkhauser.

Olhoeft, G. R. and G. R. Johnson. 1989. "Densities of Rocks and Minerals," in *CRC Practical Handbook of Physical Properties of Rocks and Minerals*, R. S. Carmichael (ed.). Boca Raton, FL: CRC Press.

Robertson, E. C. 1988. *Thermal Properties of Rocks. Open File Report 88–441 of USGS*. Reston, VA: U.S. Geological Survey.

The Engineering ToolBox. 2009. *Absorbed Solar Radiation*. Online: www.engineeringtoolbox.com/solar-radiation-absorbed-materials-d_1568.html

APPENDIX F

Ventilation and Infiltration

This appendix provides data to support the design and analysis process presented in Chapters 10, 12, 13, and 16:

- Table F.1 Minimum Ventilation Rates in Breathing Zone—For Buildings Except Low-Rise Residential
- Table F.2 Recommended Ventilation and Exhaust Air Requirements—Low-Rise Residential
- Table F.3 Estimated Overall Infiltration Rates for Small Buildings
- Table F.4 Approximate Infiltration through Doors and Windows of Small Buildings

Table F.1 Minimum Ventilation Rates in Breathing Zone—For Buildings Except Low-Rise Residential

	People Outdoor Air Rate R_p		Area Outdoor Air Rate R_A		Default Values Occupancy Density		
Occupancy Category	cfm/person	L/s person	cfm/ft²	L/s m²	#/1000 ft² or #/100 m²	Air Class (see notes)	OS (see notes)
Animal Facilities							
Animal exam room (veterinary office)	10	5	0.12	0.6	20	2	
Animal imaging (MRI/CT/PET)	10	5	0.18	0.9	20	3	
Animal operating rooms	10	5	0.18	0.9	20	3	
Animal postoperative recovery room	10	5	0.18	0.9	20	3	
Animal preparation rooms	10	5	0.18	0.9	20	3	
Animal procedure room	10	5	0.18	0.9	20	3	
Animal surgery scrub	10	5	0.18	0.9	20	3	
Large-animal holding room	10	5	0.18	0.9	20	3	
Necropsy	10	5	0.18	0.9	20	3	
Small-animal-cage room (static cages)	10	5	0.18	0.9	20	3	
Small-animal-cage room (ventilated cages)	10	5	0.18	0.9	20	3	
Correctional Facilities							
Booking/waiting 7.5 3.8	7.5	3.8	0.06	0.3	50	2	
Cell 5 2.5	5	2.5	0.12	0.6	25	2	
Dayroom 5 2.5	5	2.5	0.06	0.3	30	1	
Guard stations 5 2.5	5	2.5	0.06	0.3	15	1	

Mechanical and Electrical Equipment for Buildings, Fourteenth Edition. Walter Grondzik, Daniel Overbey, and Alison Kwok.

Companion website: www.wiley.com/go/meeb14e

Table F.1 (*Continued*)

Occupancy Category	People Outdoor Air Rate R_p		Area Outdoor Air Rate R_A		Default Values Occupancy Density	Air Class (see notes)	OS (see notes)
	cfm/ person	L/s person	cfm/ ft^2	L/s m^2	#/1000 ft^2 or #/100 m^2		
Educational Facilities							
Art classroom	10	5	0.18	0.9	20	2	
Classrooms (ages 5 to 8)	10	5	0.12	0.6	25	1	
Classrooms (age 9 plus)	10	5	0.12	0.6	35	1	
Computer lab	10	5	0.12	0.6	25	1	
Daycare sickroom	10	5	0.18	0.9	25	3	
Daycare (through age 4)	10	5	0.18	0.9	25	2	
Lecture classroom	7.5	3.8	0.06	0.3	65	1	✓
Lecture hall (fixed seats)	7.5	3.8	0.06	0.3	150	1	✓
Libraries	5	2.5	0.12	0.6	10		
Media center	10	5	0.12	0.6	25	1	
Multiuse assembly	7.5	3.8	0.06	0.3	100	1	✓
Music/theater/dance	10	5	0.06	0.3	35	1	✓
Science laboratories	10	5	0.18	0.9	25	2	
University/college laboratories	10	5	0.18	0.9	25	2	
Wood/metal shop	10	5	0.18	0.9	20	2	
Food and Beverage Service							
Bars, cocktail lounges	7.5	3.8	0.18	0.9	100	2	
Cafeteria/fast-food dining	7.5	3.8	0.18	0.9	100	2	
Kitchen (cooking)	7.5	3.8	0.12	0.6	20	2	
Restaurant dining rooms	7.5	3.8	0.18	0.9	70	2	
General							
Break rooms	5	2.5	0.06	0.3	25	1	✓
Coffee stations	5	2.5	0.06	0.3	20	1	✓
Conference/meeting	5	2.5	0.06	0.3	50	1	✓
Corridors	—	—	0.06	0.3	—	1	✓
Occupiable storage rooms for liquids or gels	5	2.5	0.12	0.6	2	2	
Hotels, Motels, Resorts, Dormitories							
Barracks sleeping areas	5	2.5	0.06	0.3	20	1	✓
Bedroom/living room	5	2.5	0.06	0.3	10	1	✓
Laundry rooms, central	5	2.5	0.12	0.6	10	2	
Laundry rooms within dwelling units	5	2.5	0.12	0.6	10	1	
Lobbies/prefunction	7.5	3.8	0.06	0.3	30	1	✓
Multipurpose assembly	5	2.5	0.06	0.3	120	1	✓
Miscellaneous Spaces							
Banks or bank lobbies	7.5	3.8	0.06	0.3	15	1	✓
Bank vaults/safe deposit	5	2.5	0.06	0.3	5	2	✓
Computer (not printing)	5	2.5	0.06	0.3	4	1	✓

Table F.1 (*Continued*)

Occupancy Category	People Outdoor Air Rate R_P cfm/ person	People Outdoor Air Rate R_P L/s person	Area Outdoor Air Rate R_A cfm/ ft²	Area Outdoor Air Rate R_A L/s m²	Default Values Occupancy Density #/1000 ft² or #/100 m²	Air Class (see notes)	OS (see notes)
Freezer and refrigerated spaces (<50°F [10°C])	10	5	0	0	0	2	
Manufacturing where hazardous materials are not used	10	5	0.18	0.9	7	2	
Manufacturing where hazardous materials are used (excludes heavy industrial and chemical processes)	10	5	0.18	0.9	7	3	
Pharmacy (prep. area)	5	2.5	0.18	0.9	10	2	
Photo studios	5	2.5	0.12	0.6	10	1	
Shipping/receiving	10	5	0.12	0.6	2	2	
Sorting, packing, light assembly	7.5	3.8	0.12	0.6	7	2	
Telephone closets	—	—	0	0	—	1	
Transportation waiting	7.5	3.8	0.06	0.3	100	1	✓
Warehouses	10	5	0.06	0.3	—	2	
Office Buildings							
Breakrooms	5	2.5	0.12	0.6	50	1	
Main entry lobbies	5	2.5	0.06	0.3	10	1	✓
Occupiable storage rooms for dry materials	5	2.5	0.06	0.3	2	1	
Office space	5	2.5	0.06	0.3	5	1	✓
Reception areas	5	2.5	0.06	0.3	30	1	✓
Telephone/data entry	5	2.5	0.06	0.3	60	1	✓
Public Assembly Spaces							
Auditorium seating area	5	2.5	0.06	0.3	150	1	✓
Courtrooms	5	2.5	0.06	0.3	70	1	✓
Legislative chambers	5	2.5	0.06	0.3	50	1	✓
Libraries	5	2.5	0.12	0.6	10	1	
Lobbies	5	2.5	0.06	0.3	150	1	✓
Museums (children's)	7.5	3.8	0.12	0.6	40	1	
Museums/galleries	7.5	3.8	0.06	0.3	40	1	✓
Places of religious worship	5	2.5	0.06	0.3	120	1	✓
Residential							
Common corridors	—	—	0.06	0.3		1	✓
Retail							
Sales (except as below)	7.5	3.8	0.12	0.6	15	2	
Barbershop	7.5	3.8	0.06	0.3	25	2	×
Beauty and nail salons	20	1.1	0.12	0.6	25	2	
Coin-operated laundries	7.5	3.8	0.12	0.6	20	2	
Mall common areas	7.5	3.8	0.06	0.3	40	1	×
Pet shops (animal areas)	7.5	3.8	0.18	0.9	10	2	
Supermarket	7.5	3.8	0.06	0.3	8	1	✓

Table F.1 (*Continued*)

Occupancy Category	People Outdoor Air Rate R_p cfm/person	People Outdoor Air Rate R_p L/s person	Area Outdoor Air Rate R_A cfm/ft²	Area Outdoor Air Rate R_A L/s m²	Default Values Occupancy Density #/1000 ft² or #/100 m²	Air Class (see notes)	OS (see notes)
Sports and Entertainment							
Bowling alley (seating)	10	5	0.12	0.6	40	1	
Disco/dance floors	20	1.1	0.06	0.3	100	2	✓
Gambling casinos	7.5	3.8	0.18	0.9	120	1	
Game arcades	7.5	3.8	0.18	0.9	20	1	
Gym, sports arena (play area)	20	1.1	0.18	0.9	7	2	
Health club/aerobics room	20	1.1	0.06	0.3	40	2	
Health club/weight rooms	20	1.1	0.06	0.3	10	2	
Spectator areas	7.5	3.8	0.06	0.3	150	1	✓
Stages, studios	10	5	0.06	0.3	70	1	✓
Swimming (pool and deck)	—	—	0.48	2.4	—	2	

Source: Reprinted with permission; © ASHRAE (www.ashrae.org); from ANSI/ASHRAE Standard 62.1-2022: Ventilation for Acceptable Indoor Air Quality.

General Notes:

1. Related Requirement. The rates in this table are based on all other applicable requirements of this standard being met.
2. Environmental Tobacco Smoke (ETS). This table applies to ETS-free areas. Refer to Section 5.13 of *Standard 62.1-2022* (ASHRAE, 2022) for requirements for buildings containing ETS areas and ETS-free areas.
3. Air Density. Volumetric airflow rates are based on dry air density of 0.075 lb_{da}/ft^3 [1.2 kg_{da}/m^3] at a barometric pressure of 1 atm [101.3 kPa] and an air temperature of 70°F [21°C]. Rates shall be permitted to be adjusted for actual density.
4. Default Occupant Density. The default occupant density shall be used where the actual occupant density is not known.
5. Unlisted Occupancies. Where the occupancy category for a proposed space or zone is not listed, the requirements for the listed occupancy category that is most similar in terms of occupant density, activities, and building construction shall be used.
6. Residential Facilities. Rates shall be determined in accordance with ASHRAE (2022).
7. Health Care Facilities. Refer to ASHRAE (2017).

Air Class:

Air (return, transfer, or exhaust air) leaving each space or location shall be designated at an expected air-quality classification not less than that shown in the table (or as approved by the authority having jurisdiction).

Class 1: Air with low contaminant concentration, low sensory-irritation intensity, and inoffensive odor.

Class 2: Air with moderate contaminant concentration, mild sensory-irritation intensity, or mildly offensive odors. (Class 2 air also includes air that is not necessarily harmful or objectionable but that is inappropriate for transfer or recirculation to spaces used for different purposes.)

OS:

For DCV zones in the occupied standby (OS) mode, breathing zone outdoor airflow (V_{bz}) shall be permitted to be reduced to zero for the occupancy categories indicated "OS" in Table F.1, provided that airflow is restored to V_{bz} whenever occupancy is detected.

DCV = *demand control ventilation*. DCV is defined as any means by which the breathing zone outdoor airflow (V_{bz}) can be varied to the occupied space or spaces based on the actual or estimated number of occupants, ventilation requirements of the occupied zone, or both.

OS = *occupied standby* mode permitted for DCV zone. OS mode is defined as when a zone is scheduled to be occupied and an occupant sensor indicates zero population within the zone.

V_{bz} = *breathing zone outdoor airflow*. Calculated using Equation 5.1 in Chapter 5,and the design value of the zone population (P_z), cfm (L/s).

Table F.2 Recommended Ventilation and Exhaust Air Requirements—Low-Rise Residential

Part A. Ventilation Air cfm [L/s]					
Floor Area ft² [m²]	**1 Bedroom**	**2 Bedrooms**	**3 Bedrooms**	**4 Bedrooms**	**5 Bedrooms**
<500 [<47]	30 [14]	38 [18]	45 [21]	53 [25]	60 [28]
501–1000 [47–93]	45 [21]	53 [24]	60 [28]	68 [31]	75 [35]
1001–1500 [93–139]	60 [28]	68 [31]	75 [35]	83 [38]	90 [42]
1501–2000 [140–186]	75 [35]	83 [38]	90 [42]	98 [45]	105 [49]
2001–2500 [186–232]	90 [42]	98 [45]	105 [49]	113 [52]	120 [56]
2501–3000 [232–279]	105 [50]	113 [52]	120 [56]	128 [59]	135 [63]
3001–3500 [279–325]	120 [56]	128 [59]	135 [63]	143 [66]	150 [70]
3501–4000 [325–372]	135 [63]	143 [66]	150 [70]	158 [73]	165 [77]
4001–4500 [372–418]	150 [70]	158 [73]	165 [77]	173 [80]	180 [84]
4501–5000 [418–465]	165 [77]	173 [80]	180 [84]	188 [87]	195 [91]

Part B. Exhaust Air

Continuous local exhaust airflow rates:

Enclosed Kitchen: Vented range hood (including appliance-range hood combinations): 100 cfm [50 L/s]; other kitchen exhaust fans, including downdraft: 300 cfm [150 L/s] or a capacity of 5 ACH

Bathroom: 20 cfm [10 L/s]

Demand-controlled local airflow rates:

Kitchen: 100 cfm [50 L/s] (vented range hood required if exhaust fan flow rate is less than 5 kitchen air changes per hour)

Bathroom: 50 cfm [25 L/s]

Source: Reprinted with permission; © ASHRAE (www.ashrae.org); from ANSI/ASHRAE Standard 62.2-2022: Ventilation for Acceptable Indoor Air Quality in Low-Rise Residential Buildings. See *Standard 62.2* for definitions, assumptions, implementation, and exceptions.

APPENDICES

Table F.3 Estimated Overall Infiltration Rates for Small Buildings

Part A. Construction Types

Construction Type	Description
Tight	Good multifamily residential construction with close-fitting doors, windows, and framing is considered tight. New houses with full vapor retarder, no fireplace, well-fitted windows, weather-stripped doors, one-story, and less than 1500 ft^2 [140 m^2] floor area fall into this category.
Medium	Medium structures include new two-story frame houses or one-story houses more than 10 years old with average maintenance, a floor area greater than 1500 ft^2 [140 m^2], average-fit windows and doors, and a fireplace with damper and glass closure. Below-average multifamily construction falls in this category.
Loose	Loose structures are poorly constructed single and multifamily residences with poorly fitted windows and doors. Examples include houses more than 20 years old, of average maintenance, having a fireplace without damper or glass closure, or having more than an average number of vented appliances. Average manufactured homes are in this category.

Part B. Design Infiltration Rate (ACH) for Winter: Indoors 68°F [20°C]; Wind Speed = 15 mph [6.7 m/s]

Construction Type	Winter Outdoor Design Temperature °F: 50 / °C: 10	40 / 4	30 / −1	20 / −7	10 / −12	0 / −18	−10 / −23	−20 / −29	−30 / −34	−40 / −40
Tight	0.41	0.43	0.45	0.47	0.49	0.51	0.53	0.55	0.57	0.59
Medium	0.69	0.73	0.77	0.81	0.85	0.89	0.93	0.97	1	1.05
Loose	1.11	1.15	1.2	1.23	1.27	1.3	1.35	1.4	1.43	1.47

Part C. Design Infiltration Rate (ACH) for Summer: Indoors 75°F [24°C]; Wind Speed = 7.5 mph [3.4 m/s]

Construction Type	Summer Outdoor Design Temperature °F: 85 / °C: 29	90 / 32	95 / 35	100 / 38	105 / 41	110 / 43
Tight	0.33	0.34	0.35	0.36	0.37	0.38
Medium	0.46	0.48	0.50	0.52	0.54	0.56
Loose	0.68	0.70	0.72	0.74	0.76	0.78

Part D. Infiltration Rates per Unit Floor Area

Ceiling Height	Air Flow	Air Changes per Hour (ACH) 0.3	0.4	0.5	0.6	0.7	0.8	0.9	1.0	1.1	1.2	1.3	1.4	1.5	1.6	1.7	1.8	1.9	2.0
7.5 ft	cfm/ft^2	0.04	0.05	0.06	0.08	0.09	0.10	0.11	0.13	0.14	0.15	0.16	0.18	0.19	0.20	0.21	0.23	0.24	0.25
2.3 m	L/s m^2	0.20	0.25	0.31	0.41	0.46	0.51	0.56	0.66	0.71	0.76	0.81	0.91	0.97	1.02	1.07	1.17	1.22	1.27
8.0 ft	cfm/ft^2	0.04	0.05	0.07	0.08	0.09	0.11	0.12	0.13	0.15	0.16	0.17	0.19	0.20	0.21	0.23	0.24	0.26	0.27
2.4 m	L/s m^2	0.20	0.25	0.36	0.41	0.46	0.56	0.61	0.66	0.76	0.81	0.86	0.97	1.02	1.07	1.17	1.22	1.32	1.37
8.5 ft	cfm/ft^2	0.04	0.06	0.07	0.09	0.10	0.11	0.13	0.14	0.16	0.17	0.18	0.20	0.21	0.23	0.24	0.26	0.27	0.28
2.6 m	L/s m^2	0.20	0.31	0.36	0.46	0.51	0.56	0.66	0.71	0.81	0.86	0.91	1.02	1.07	1.17	1.22	1.32	1.37	1.42
9.0 ft	cfm/ft^2	0.05	0.06	0.08	0.09	0.11	0.12	0.14	0.15	0.17	0.18	0.20	0.21	0.23	0.24	0.26	0.27	0.29	0.30
2.7 m	L/s m^2	0.25	0.31	0.41	0.46	0.56	0.61	0.71	0.76	0.86	0.91	1.02	1.07	1.17	1.22	1.32	1.37	1.47	1.52

Part E. Infiltration Heat Flow Rates per Unit Floor Area

Ceiling Height	Heat Flow	Air Changes per Hour (ACH) 0.3	0.4	0.5	0.6	0.7	0.8	0.9	1.0	1.1	1.2	1.3	1.4	1.5	1.6	1.7	1.8	1.9	2.0
7.5 ft	Btu/h ft^2 °F	0.04	0.05	0.07	0.08	0.09	0.11	0.12	0.14	0.15	0.16	0.18	0.20	0.20	0.22	0.23	0.24	0.26	0.27
2.3 m	W/m^2 K	0.23	0.28	0.40	0.45	0.51	0.63	0.68	0.80	0.85	0.91	1.02	1.14	1.14	1.25	1.31	1.36	1.48	1.53
8.0 ft	Btu/h ft^2 °F	0.04	0.06	0.07	0.09	0.10	0.12	0.13	0.14	0.16	0.17	0.19	0.22	0.22	0.23	0.24	0.26	0.27	0.29
2.4 m	W/m^2 K	0.23	0.34	0.40	0.51	0.57	0.68	0.74	0.80	0.91	0.97	1.08	1.25	1.25	1.31	1.36	1.48	1.53	1.65
8.5 ft	Btu/h ft^2 °F	0.05	0.06	0.08	0.09	0.11	0.12	0.14	0.15	0.17	0.18	0.20	0.23	0.23	0.24	0.26	0.28	0.29	0.30
2.4 m	W/m2 K	0.28	0.34	0.45	0.51	0.63	0.68	0.80	0.85	0.97	1.02	1.14	1.31	1.31	1.36	1.48	1.59	1.65	1.70
9.0 ft	Btu/h ft^2 °F	0.05	0.06	0.08	0.10	0.11	0.13	0.15	0.16	0.18	0.19	0.21	0.24	0.24	0.26	0.28	0.29	0.31	0.32
2.4 m	W/m^2 K	0.28	0.34	0.45	0.57	0.63	0.74	0.85	0.91	1.02	1.08	1.19	1.36	1.36	1.48	1.59	1.65	1.76	1.82

Sources: Parts A through C reprinted with permission of ASHRAE (www.ashrae.org), from the 2001 ASHRAE Handbook—Fundamentals; Parts D and E from the ASHRAE Cooling and Heating Load Calculation Manual, 1979.

Table F.4 Approximate Infiltration Through Doors and Windows of Small Buildings

Part A. Converting Wind Speed to Velocity Head Factor

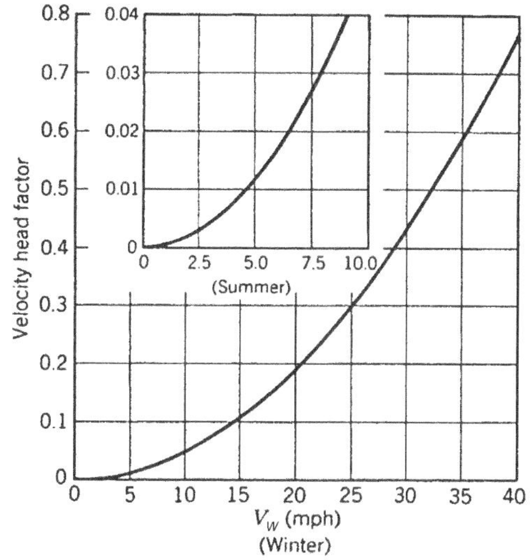

Note: Design conditions are often assumed as:

Winter wind V_w = 15 mph [6.7 m/s] corresponding to a velocity head factor of 0.105

Summer wind V_w = 7.5 mph [3.4 m/s] corresponding to a velocity head factor of 0.028

Part B. Infiltration Rates for Velocity Head Factors

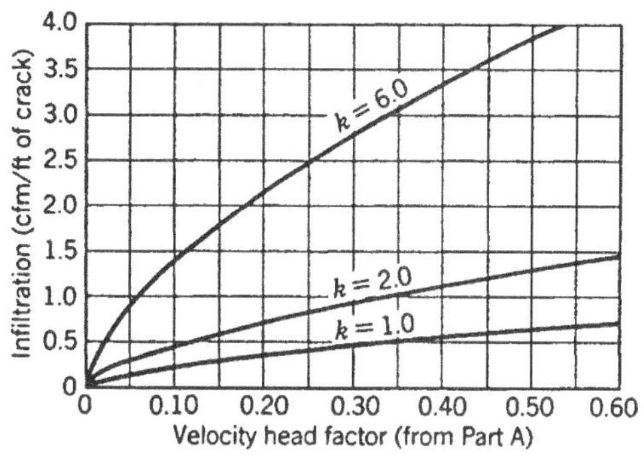

Note: Enter this graph with velocity head factor (from Part A) to find infiltration rate in cfm/ft of crack (using values of k found in Part C or D). L/s per m = (cfm/ft) (0.646).

Table F.4 (*Continued*)

Part C. Classifications of Windows for Infiltration		
Window Fit	**Wood Double-Hung (Locked)**	**Other Types**
Tight, $k = 1.0$	Weather stripped; average gap (1/64 in. [0.4 mm] crack)	Wood casement and awning windows; weather stripped Metal casement windows; weather stripped
Average, $k = 2.0$	Non-weather stripped; average gap (1/64 in. [0.4 mm] crack) or weather stripped; large gap (3/32 in. [2.4 mm] crack)	All types of vertical and horizontal sliding windows; weather stripped. If average gap (1/64 in. [0.4 mm] crack), this could be a tight-fitting window Metal casement windows; non-weather stripped If large gap (3/32 in. [2.4 mm] crack), this could be a loose-fitting window
Loose, $k = 6.0$	Non-weather stripped; large gap (3/32 in. [2.4 mm] crack)	Vertical and horizontal sliding windows; non-weather stripped

Part D. Classification of Residential-Type Doors for Infiltration	
Door Fit	**Comments**
Tight, $k = 1.0$	Very small perimeter gap and perfect fit weather stripping—often characteristic of new doors
Average, $k = 2.0$	Small perimeter gap having stop trim fitting properly around door; weather stripped
Loose, $k = 6.0$	Large perimeter gap having poor fitting stop trim; weather stripped or Small perimeter gap; no weather stripping

Source: Reprinted with permission of ASHRAE (www.ashrae.org), from Cooling and Heating Load Calculation Manual, 2nd edition, 1992. SI units are appended by the authors of this book.

APPENDICES

References and Resources

ASHRAE. 1979. *ASHRAE Cooling and Heating Load Calculation Manual*. 1st edn. Atlanta, GA: American Society of Heating, Refrigerating, and Air-Conditioning Engineers.

ASHRAE. 1992. *Cooling and Heating Load Calculation Manual*, 2nd edn. Atlanta, GA: American Society of Heating, Refrigerating, and Air-Conditioning Engineers.

ASHRAE. 2017. *ANSI/ASHRAE/ASHE Standard 170-2017: Ventilation for Health Care Facilities*. Atlanta, GA: American Society of Heating, Refrigerating, and Air-Conditioning Engineers.

ASHRAE. 2021. *2021 ASHRAE Handbook—Fundamentals*. Atlanta, GA: American Society of Heating, Refrigerating, and Air-Conditioning Engineers.

ASHRAE. 2022. *ANSI/ASHRAE Standard 62.1-2022: Ventilation for Acceptable Indoor Air Quality*. Atlanta, GA: American Society of Heating, Refrigerating, and Air-Conditioning Engineers.

APPENDIX G

Heating and Cooling Design Guidelines and Information

This appendix provides data to support the design and analysis processes presented in Chapters 13–15.

- Section G.1 Estimating Summer Heat Gains
- Section G.2 Design Temperature Differences for Opaque Enclosure Assemblies
- Section G.3 Heat Gains (Cooling Loads) Through Glass
- Section G.4 Heat Gains (Cooling Loads) Due to Infiltration/Ventilation
- Section G.5 Heat Gains from Building Occupants
- Section G.6 Heat Gains from Office Equipment
- Section G.7 Heat Gains from Appliances
- Section G.8 Design Data for Earth Tubes
- Section G.9 Psychrometric Charts

APPENDICES

Mechanical and Electrical Equipment for Buildings, Fourteenth Edition. Walter Grondzik, Daniel Overbey, and Alison Kwok.

Companion website: www.wiley.com/go/meeb14e

G.1 ESTIMATING SUMMER HEAT GAINS

Table G.1 Approximate Summer Heat Gains from Occupants, Equipment, Lighting, and Enclosure

Part A. Internal Heat Sources—People and Equipment

	Area per Person[a]		I-P Sensible Heat Gain (Btu/h ft² of Floor Area)			SI Sensible Heat Gain (W/m² of Floor Area)		
Function	ft²	m²	People[b]	Equipment[c]	Total	People[b]	Equipment[c]	Total
Office, U.S.[c,d]	180–100	16.7–9.3	1.3–2.3	0.4–1.1	1.7–3.4	4.1–7.3	1.2–3.4	5.3–10.7
Office, Europe[e]			1–1.6	2.2–4.2	3.2–5.8	3–5	7–13.1	10–18.1
School: elementary, U.S.	100–20	9.3–1.9	2.3–11.5	0–0.6	2.3–12.1	7.3–36.3	0–2.0	7.3–38.3
Schools, Europe[e]			3.8–8.0	0–0.6	3.8–8.6	12–25.2	0–2.0	12.0–27.2
School: secondary, college	150–100	13.9–9.3	1.7–2.6	0–0.6	1.7–3.2	5.4–8.2	0–2.0	5.4–10.2
Health care								
Sleeping (hospital)	240	22.3	0.9	0.6[e]	1.5	2.8	2.0[e]	4.8
In-patient (clinic)	120	11.1	1.9	Varies	1.9+	6	Varies	6.0+
Assembly: fixed seats	15	1.4	14	—	14	44.2	—	44.2
Standing space, concentrated use	15–7	1.4–0.7	21.0–45.0	0–0.5	21.0–45.5	66.3–142.0	0–1.6	66.3–143.6
Restaurant:[f]								
Fast food: dining area	15	1.4	17	3.4	20.4	53.6	10.7	64.3
Kitchen, refrigeration				17.1	17.1		54	54
Sit-down: dining area	25	2.3	10.2	5.1	15.3	32.2	16.1	48.3
Kitchen, refrigeration				7.2	7.2		22.7	22.7
Mercantile: street floor	50–30	4.7–2.8	6.3–10.5	3.4	9.7–13.9	19.9–33.1	10.7	30.6–43.8
Other sales floors	60–50	5.6–4.7	5.3–6.3	3.4	8.7–9.7	16.7–19.9	10.7	27.4–30.6
Shopping center, Europe[e]			3.2	0.3–1.3	3.5–4.5	10	1.0–4.0	11.0–14.0
Warehouse	1000–300	92.9–27.9	0.4–1.2	—	0.4–1.2	1.3–3.8	—	1.3–3.8
Hotels, nursing homes	300–200	27.9–18.6	0.8–1.2	3.4	4.2–4.6	2.5–3.8	10.7	13.2–14.5
Apartments[g]	300–200	27.9–18.6	0.8–1.2	See note[g]	See note *g*	2.5–3.8	See note[g]	See note[g]

Part B. Internal Heat Sources—Electric Lighting

	I-P Sensible Heat Gain[h] (Btu/h ft² of Floor Area)			SI Sensible Heat Gain[h] (W/m² of Floor Area)		
Function	DF (daylight factor) < 1	1 < DF < 4[h]	DF > 4[h]	DF < 1	1 < DF < 4[h]	DF > 4[h]
Office	5.1	2.0	0.5	16.1	6.3	1.6
School: elementary	6.3–6.8	2.5–2.7	0.6–0.7	19.9–21.5	7.9–8.5	1.9–2.2
School: secondary, college	6.3–6.8	2.5–2.7	0.6–0.7	19.9–21.5	7.9–8.5	1.9–2.2
Health care						
Sleeping (hospital)	6.8	2.7	0.7	21.5	8.5	2.2
In-patient (clinic)	6.8	2.7	0.7	21.5	8.5	2.2
Assembly	3.8	1.5	0.4	12.0	4.7	1.3
Restaurants[i]	6.3	2.5	0.6	19.9	7.9[i]	1.9
Mercantile	5.1–6.8	2.0–2.7	0.5–0.7	16.1–21.5	6.3–8.5	1.6–2.2
Warehouse	2.4	1.0	0.2	7.6	3.2	0.6
Hotels, nursing homes	6.8	2.7	0.7	21.5	8.5	2.2
Apartments[g]	Up to 6.8	Up to 2.7	Up to 0.7	Up to 21.5	Up to 8.5	Up to 2.2

Table G.1 (*Continued*)

Part C. Heat Gain Through Enclosure[j]

Function	Outdoor Design Temperature (Btu/h ft^2 of Floor Area) 90°F	100°F	Outdoor Design Temperature (W/m^2 of Floor Area) 32°C	38°F
I. Gains through externally shaded windows[k]:				
Find ratio, $\frac{\text{total window area}}{\text{total floor area}}$, then multiply by	16	21	50	66
II. Gains through opaque walls:				
Find ratio, $\frac{\text{total opaque wall area}}{\text{total floor area}} \times (U_{wall})$, then multiply by	15	25	8	14
III. Gains through roofs:				
Find ratio, $\frac{\text{total opaque roof area}}{\text{total floor area}} \times (U_{roof})$, then multiply by	35	45	19	25

Part D. Summary Gains

I. Passive cooling systems for (thermally) "open" buildings using (for example):

Cross ventilation

Stack ventilation

Nighttime or "open" hours of thermal mass/night ventilation

Total gains: Add Parts A, B, and C to obtain cooling load, in Btu/h ft^2 [W/m^2], of floor area.

II. Passive cooling systems for (thermally) "closed" buildings using (for example):

Roofponds

Evaporative cooling

Daytime or "closed" hours of thermal mass/night ventilation

Total gains: Add Parts A, B, C, and E to obtain cooling load, in Btu/h ft^2 [W/m^2], of floor area.

Part E. Gains from Infiltration/Ventilation of "Closed" Buildings

Function	Outdoor Design Temperature 90°F	100°F	Outdoor Design Temperature 32°C	38°F
Infiltration:				
Find ratio, $\frac{\text{total window + opaque wall area}}{\text{total floor area}}$, then multiply by	1.0	1.9	3.2	6.0
Ventilation:				
Find known, $\frac{\text{total cfm of outdoor air}}{\text{total floor area, ft}^2}$, then multiply by	16	27	—	—
Find known, $\frac{\text{total L/s of outdoor air}}{\text{total floor area, m}^2}$, then multiply by	—	—	9.9	16.8

[a]Lower density than elevator population estimates; higher density than fire population estimates. Lower density for whole-building estimates; higher density for single-space estimates.

[b]Sensible gains per adult from Table G.7 for activities as stated.

[c]The typical "miscellaneous" load of 1 W/ft^2 [10.8 W/m^2] produces 3.4 Btu/h ft^2.

[d]Komor, 1997. This is considerably lower than the typical 2–4 W/ft^2 [21.5–43.1 W/m^2] for offices used by many designers.

[e]Adapted from Hastings, 1994.

[f]Based on total area of restaurant + kitchen. From ACSA, 1994. Data from Electric Power Research Institute.

[g]Residential internal gains are often assumed at 230 Btu/h per occupant plus 1200–1600 Btu/h total from appliances [68 W per occupant, plus 350–470 W total from appliances].

[h]Adapted from Northwest Power Planning Council, 1983. Values shown for DF ≥ 1 assume automatic dimming in the presence of daylight.

[i]Lighting is often much lower in sit-down than in fast-food restaurants.

[j]Averaged from the more specific data found in Table G.2.

[k]If windows are not externally shaded, see Table G.3 for multipliers, which will vary by orientation.

G.2 DESIGN TEMPERATURE DIFFERENCES FOR OPAQUE ENCLOSURE ASSEMBLIES

Table G.2 Design Equivalent Temperature Differences (DETD)

Part A. Mass Walls, Roofs, and Floors																								
SI Units												Outdoor Design Temperature	I-P Units											
29.4°C		32.2				35.0		37.7		40.5	43.3	Daily Temperature	85°F		90				95		100		105	110
L	M	L	M	H	L	M	H	M	H	H	H	Range[a]	L	M	L	M	H	L	M	H	M	H	H	H
												Walls												
5.7	3.5	8.5	6.3	3.5	11.3	9.1	6.3	11.8	9.1	11.8	14.6	1. Masonry walls, 200-mm [8-in.] block or brick	10.3	6.3	15.3	11.3	6.3	20.3	16.3	11.3	21.3	16.3	21.3	26.3
5.0	2.7	7.7	5.5	2.7	10.5	8.3	5.5	11.1	8.3	11.1	13.8	2. Partitions, frame masonry	9.0	5.0	14.0	10.0	5.0	19.0	15.0	10.0	20.0	15.0	20.0	25.0
1.4	0.0	4.2	1.9	0.0	6.9	4.7	1.9	7.5	4.7	7.5	10.3		2.5	0	7.5	3.5	0	12.5	8.5	3.5	13.5	8.5	13.5	18.5
												Ceilings and Roofs												
												1. Ceilings under naturally vented attic or vented flat roof												
21.1	18.8	23.8	21.6	18.8	26.6	24.4	21.6	27.2	24.4	27.2	30.0	—dark	38.0	34.0	43.0	39.0	34.0	48.0	44.0	39.0	49.0	44.0	49.0	54.0
16.6	14.9	19.4	17.2	14.4	22.2	20.0	17.2	22.7	20.0	22.7	25.5	—light	30.0	26.0	35.0	31.0	26.0	40.0	36.0	31.0	41.0	36.0	41.0	46.0
												2. Built-up roof, no ceiling												
21.1	18.8	23.3	21.6	18.8	26.6	24.4	21.6	27.2	24.4	27.2	30.0	—dark	38.0	34.0	43.0	39.0	34.0	48.0	44.0	39.0	49.0	44.0	49.0	54.0
16.6	14.9	19.4	17.2	14.4	22.2	20.2	17.2	22.7	20.0	22.7	25.5	—light	30.0	26.0	35.0	31.0	26.0	40.0	36.0	31.0	41.0	36.0	41.0	46.0
5.0	2.7	7.7	5.5	2.7	10.5	8.3	5.5	11.1	8.3	11.1	13.8	3. Ceilings under unconditioned rooms	9.0	5.0	14.0	10.0	5.0	19.0	15.0	10.0	20.0	15.0	20.0	25.0
												Floors												
5.0	2.7	7.7	5.5	2.7	10.5	8.3	5.5	11.1	8.3	11.1	13.8	1. Over unconditioned rooms	9.0	5.0	14.0	10.0	5.0	19.0	15.0	10.0	20.0	15.0	20.0	25.0
0.0	0.0	0.0	0.0	0.0	0.0	0.0	0.0	0.0	0.0	0.0	0.0	2. Over basement, enclosed crawl space, or concrete slab-on-ground	0	0	0	0	0	0	0	0	0	0	0	0
5.0	2.7	7.7	5.5	2.7	10.5	8.3	5.5	11.1	8.3	11.1	13.8	3. Over open crawl space	9.0	5.0	14.0	10.0	5.0	19.0	15.0	10.0	20.0	15.0	20.0	25.0

Table G.2 Design Equivalent Temperature Differences (DETD) (Continued)

Part B. Frame Walls and Doors

SI Units												Outdoor Design Temperature Daily Temperature Range[a]	I-P Units											
29.4°C		32			35			38		41	43		85°F		90			95			100		105	110
L	M	L	M	H	L	M	H	M	H	H	H	Orientation	L	M	L	M	H	L	M	H	M	H	H	H
4	2	7	4	2	10	7	4	10	7	10	13	North	8	3	13	8	3	18	13	8	18	13	18	23
8	5	11	8	5	13	11	8	13	11	13	16	NE and NW	14	9	19	14	9	24	19	14	24	19	24	29
10	7	13	10	7	16	13	10	16	13	16	18	East and West	18	13	23	18	13	28	23	18	28	23	28	33
9	6	12	9	6	14	12	9	14	12	14	17	SE and SW	16	11	21	16	11	26	21	16	26	21	26	31
6	3	9	6	3	12	9	6	12	9	12	14	South	11	6	16	11	6	21	16	11	21	16	21	26

Source: Part A, Previously adapted, with the permission of ASHRAE (www.ashrae.org), from the 1981 ASHRAE Handbook—Fundamentals; Part B, Previously adapted, with the permission of ASHRAE (www.ashrae.org), from the 1997 ASHRAE Handbook—Fundamentals. This citation to an older version of the *Handbook* is intentional and provides access to historic reference information of ongoing interest.

[a]Daily temperature range: L (low), M (medium), H (high).

From Appendix B, mean daily range column: L is less than 16°F [9°C]; M is 16 to 25°F [9 to 14°C]; H is greater than 25°F [14°C].

G.3 HEAT GAINS (COOLING LOADS) THROUGH GLASS

Table G.3 Design Cooling Load Factors (DCLF) Through Glass

Part A. I-P Units: Btu/h ft²																					
	Regular Single Glass						Regular Double Glass						Heat-Absorbing Double Glass						Clear Triple Glass		
Outdoor Design Temp.[a]	**85**	**90**	**95**	**100**	**105**	**110**	**85**	**90**	**95**	**100**	**105**	**110**	**85**	**90**	**95**	**100**	**105**	**110**	**85**	**90**	**95**
No Awnings or Inside Shading																					
North	23	27	31	35	39	44	19	21	24	26	28	30	12	14	17	19	21	23	17	19	20
NE and NW	56	60	64	68	72	77	46	48	51	53	55	57	27	29	32	34	36	38	42	43	44
East and west	81	85	89	93	97	102	68	70	73	75	77	79	42	44	47	49	51	53	62	63	64
SE and SW	70	74	78	82	86	91	59	61	64	66	68	70	35	37	40	42	44	46	53	55	56
South	40	44	48	52	56	61	33	35	38	40	42	44	19	21	24	26	28	30	30	31	33
Horiz. skylight	160	164	168	172	176	181	139	141	144	146	148	150	89	91	94	96	98	100	126	127	129
Draperies or Venetian Blinds																					
North	15	19	23	27	31	36	12	14	17	19	21	23	9	11	14	16	18	20	11	12	14
NE and NW	32	36	40	44	48	53	27	29	32	34	36	38	20	22	25	27	29	31	24	26	27
East and west	48	52	56	60	64	69	42	44	47	49	51	53	30	32	35	37	39	41	38	39	41
SE and SW	40	44	48	52	56	61	35	37	40	42	44	46	24	26	29	31	33	35	32	33	34
South	23	27	31	35	39	44	20	22	25	27	29	31	15	17	20	22	24	26	18	19	21
Roller Shades Half-Drawn																					
North	18	22	26	30	34	39	15	17	20	22	24	26	10	12	15	17	19	21	13	14	15
NE and NW	40	44	48	52	56	61	38	40	43	45	47	49	24	26	29	31	33	35	34	35	35
East and west	61	65	69	73	77	82	54	56	59	61	63	65	35	37	40	42	44	46	49	49	50
SE and SW	52	56	60	64	68	73	46	48	51	53	55	57	30	32	35	37	39	41	41	42	43
South	29	33	37	41	45	50	27	29	32	34	36	38	18	20	23	25	27	29	25	26	26
Awnings[b]																					
North	20	24	28	32	36	41	13	15	18	20	22	24	10	12	15	17	19	21	11	12	13
NE and NW	21	25	29	33	37	42	14	16	19	21	23	25	11	13	16	18	20	22	12	13	14
East and west	22	26	30	34	38	43	14	16	19	21	23	25	12	14	17	19	21	23	12	13	14
SE and SW	21	25	29	33	37	42	14	16	19	21	23	25	11	13	16	18	20	22	12	13	14
South	21	24	28	32	36	41	13	15	18	20	22	24	11	13	16	18	20	22	11	12	13

Part B. SI Units: W/m²																					
	Regular Single Glass						Regular Double Glass						Heat-Absorbing Double Glass						Clear Triple Glass		
Outdoor Design Temp.[a]	29.4	32.2	35.0	37.7	40.5	43.3	29.4	32.2	35.0	37.7	40.5	43.3	29.4	32.2	35.0	37.7	40.5	43.3	29.4	32.2	35.0
No Awnings or Inside Shading																					
North	72.6	85.2	97.8	110.4	123.0	138.8	59.9	66.2	75.8	82.0	88.3	94.7	37.9	44.1	53.7	60.0	66.2	72.6	53.7	60.0	63.1
NE and NW	176.6	189.3	202.0	214.6	227.1	243.0	145.1	151.4	161.0	167.2	173.6	179.9	85.1	91.4	101.0	107.2	113.6	119.9	132.6	135.7	138.9
East and west	255.5	268.1	280.8	293.4	306.0	321.9	214.6	220.9	230.3	236.7	243.0	249.2	132.6	138.9	148.2	154.6	160.9	167.2	195.7	198.8	202.6
SE and SW	220.9	233.4	246.0	258.8	271.3	287.1	186.1	192.4	202.0	208.2	214.6	220.9	110.4	116.8	126.2	132.6	138.9	145.1	167.2	173.6	176.7
South	126.2	138.9	151.4	164.0	176.7	192.4	104.1	110.4	119.9	126.2	132.6	138.9	66.0	75.8	75.8	82.0	88.3	94.7	94.7	97.9	104.1
Horiz. skylight	504.8	517.4	530.0	542.7	555.2	571.0	438.6	444.9	454.3	460.7	467.0	473.2	280.8	296.6	296.6	302.9	309.1	315.6	397.6	400.7	407.0
Draperies or Venetian Blinds																					
North	47.3	60.0	72.6	85.1	97.9	113.6	37.9	44.1	53.7	60.0	66.2	72.6	28.3	34.7	44.1	50.4	56.8	63.1	34.7	37.9	44.2
NE and NW	101.0	113.6	126.2	138.9	151.4	167.2	85.1	91.4	101.0	107.2	113.6	119.9	63.1	69.4	78.9	85.1	91.4	97.9	75.8	82.0	85.1
East and west	151.4	164.0	176.7	189.3	202.0	217.7	132.6	138.9	148.2	154.6	161.0	167.2	94.7	101.0	110.4	116.8	123.0	129.3	119.9	123.0	129.3
SE and SW	126.2	138.9	151.4	164.0	176.7	192.4	110.4	116.8	126.2	132.6	138.9	145.1	75.8	82.0	91.4	97.9	104.1	110.4	101.0	104.1	107.2
South	72.5	85.1	97.9	110.4	123.0	138.9	63.1	69.4	78.9	85.1	91.4	97.9	47.3	53.7	63.1	75.8	75.8	82.0	56.8	60.0	66.2
Roller Shades Half-Drawn																					
North	56.8	69.4	82.0	92.7	102.2	123.0	47.3	53.7	63.1	69.4	75.8	82.0	31.6	37.9	47.3	53.7	60.0	66.2	41.0	44.2	47.3
NE and NW	126.2	138.9	151.4	164.0	176.7	192.4	119.9	126.2	135.7	142.0	148.2	154.6	75.8	82.0	91.4	97.9	104.1	110.4	107.2	110.4	110.4
East and west	192.5	205.0	217.7	230.3	243.0	258.8	170.3	176.7	186.1	192.4	198.8	205.0	110.4	116.8	126.2	132.5	138.9	145.1	154.6	154.6	157.8
SE and SW	164.0	176.7	189.3	202.0	214.6	230.3	145.1	151.4	161.0	167.2	173.6	179.9	94.7	101.0	110.4	123.0	123.0	129.3	129.3	132.6	135.7
South	91.4	104.1	116.8	129.3	142.0	157.8	85.1	91.4	101.0	107.2	113.6	119.9	56.8	63.1	72.6	78.9	85.1	91.4	78.9	82.0	82.0
Awnings[b]																					
North	63.1	75.8	88.3	101.0	113.6	129.3	41.0	47.3	56.8	63.1	69.4	75.8	31.6	37.9	47.3	53.7	60.0	66.2	34.7	37.9	41.0
NE and NW	66.2	78.9	91.4	104.1	116.8	132.6	44.2	50.4	60.0	66.2	72.6	78.9	34.7	41.0	50.4	56.8	63.1	69.4	37.9	41.0	44.2
East and west	69.4	82.0	94.7	107.2	119.9	135.7	44.2	50.4	60.0	66.2	72.6	78.9	37.9	44.2	53.6	60.0	66.2	72.6	37.9	41.0	44.2
SE and SW	66.2	78.9	91.4	104.1	116.8	132.6	44.2	50.4	60.0	66.2	72.6	78.9	34.7	41.0	50.4	56.8	63.1	69.4	37.9	41.0	44.2
South	66.2	75.8	88.3	101.0	113.6	129.3	41.0	47.3	56.8	63.1	69.4	75.8	34.7	41.0	50.4	56.8	63.1	69.4	34.7	37.9	41.0

Source: Previously adapted, with the permission of ASHRAE (www.ashrae.org), from the 1981 ASHRAE Handbook—Fundamentals. This citation to an older version of the *Handbook* is intentional and provides access to historic reference information of ongoing interest.

[a]Based on indoor design temperature of 75°F [23.8°C] and outdoor design temperatures as indicated. Interpolate to obtain factors for outdoor design temperatures other than those given.

[b]For other external shading devices that completely shade the glass at any orientation, use the values for "Awnings, north."

G.4 HEAT GAINS (COOLING LOADS) DUE TO INFILTRATION/VENTILATION

Table G.4 Sensible Cooling Load Factors Due to Infiltration and Ventilation

Design Temperature °C:									Design Temperature °F:					
29.4	32.2	35.0	37.7	41.5	43.3	Units	Condition	Units	85	90	95	100	105	110
2.2	3.5	4.7	6.0	6.9	8.2	W/m^2	Infiltration, per gross exposed wall area	Btu/h ft^2	0.7	1.1	1.5	1.9	2.2	2.6
6.8	9.9	13.6	16.7	19.8	23.6	W per L/s	Mechanical ventilation	Btu/h per cfm	11.0	16.0	22.0	27.0	32.0	38.0

Source: Previously adapted, with the permission of ASHRAE (www.ashrae.org), from the 1981. ASHRAE Handbook—Fundamentals. This citation to an older version of the *Handbook* is intentional and provides access to historic reference information of ongoing interest.

G.5 HEAT GAINS FROM BUILDING OCCUPANTS

Table G.5 Rates of Heat Gain from Occupants of Conditioned Spaces

		Heat Gain Btu/h				Heat Gain W			
Activity	Location	Adult Male	Adjusted, M/F[a]	Sensible Heat	Latent Heat	Adult Male	Adjusted, M/F[a]	Sensible Heat	Latent Heat
Seated at theater	Theater	390	350	245	105	114	103	72	31
Seated, very light work	Offices, hotels, apartments	450	400	245	155	132	117	72	45
Moderately active office work	Offices, hotels, apartments	475	450	250	200	139	132	73	59
Standing, light work; walking	Department store; retail store	550	450	250	200	161	132	73	59
Walking, standing	Drug store, bank	550	500	250	250	161	147	73	73
Sedentary work	Restaurant[b]	490	550	275	275	144	161	81	81
Light bench work	Factory	800	750	275	475	234	220	81	139
Moderate dancing	Dance hall	900	850	305	545	264	249	89	160
Walking 4.8 km/h [3 mph], light machine work	Factory	1000	1000	375	625	293	293	110	183
Bowling[c]	Bowling alley	1500	1450	580	870	440	425	170	255
Heavy work	Factory	1500	1450	580	870	440	425	170	255
Heavy machine work, lifting	Factory	1600	1600	635	965	469	469	186	283
Athletics	Gymnasium	2000	1800	710	1090	586	528	208	319

Source: Adapted with permission; © ASHRAE (www.ashrae.org); from the 2021 ASHRAE Handbook—Fundamentals, Chapter 18, Table 1.

Notes:

1. Tabulated values are based on 75°F [24°C] room dry-bulb temperature. For 80°F [27°C] room dry-bulb, total heat remains the same, but sensible heat values should be decreased by approximately 20%, and latent heat values increased accordingly.
2. Also see ASHRAE (2021), Table 4, Chapter 9, for additional rates of metabolic heat generation.
3. All values are rounded to nearest 5 Btu/h [1.5 W].
4. SI conversions added by the authors of this book.

[a]Adjusted heat gain is based on the normal percentage of men, women, and children for the application listed and assumes that gain from an adult female is 85% of that for an adult male, and gain from a child is 75% of that for an adult male.

[b]Adjusted heat gain includes 60 Btu/h [18 W] for food per individual (30 Btu/h [9 W] sensible and 30 Btu/h [9 W] latent).

[c]Figure one person per alley actually bowling, and all others sitting (400 Btu/h [117 W]) or standing or walking slowly (550 Btu/h [161 W]).

G.6 HEAT GAINS FROM OFFICE EQUIPMENT

Table G.6 Heat Gains from Office Equipment Recommended for Use in Design Calculations

Equipment Description	Nameplate Power,[a] W	Peak Heat Gain,[b] W
Desktop computer		
Manufacturer 1, 3.5 GHz processor, 8 GB RAM, *n* = *2*	NA	42
Manufacturer 1, 3.6 GHz processor, 16 GB RAM, *n* = *2*	NA	66
Manufacturer 1, 3.3 GHz processor, 16 GB RAM, *n* = *2*	NA	52
Manufacturer 1, 4.0 GHz processor, 16 GB RAM, *n* = *1*	NA	83
Manufacturer 2, 3.6 GHz processor, 32 GB RAM, *n* = *8*	NA	80
Manufacturer 2, 3.6 GHz processor, 16 GB RAM, *n* = *1*	NA	78
Manufacturer 2, 3.4 GHz processor, 32 GB RAM, *n* = *1*	NA	72
Manufacturer 2, 3.4 GHz processor, 24 GB RAM, *n* = *1*	NA	86
Laptop computer		
Manufacturer 1, 2.6 GHz processor, 8 GB RAM, *n* = *1*	NA	46
Manufacturer 2, 2.4 GHz processor, 4 MB RAM, *n* = *1*	NA	59
Average 15-min peak power consumption (range)	**53 (46–59)**	
Flat-panel monitor[c]		
Manufacturer 1, 27 in. [686 mm] LED flat screen, *n* = *2*	40	26
Manufacturer 1, 22 in. [546 mm] LED flat screen, *n* = *2*	29	25
Manufacturer 3, 23 in. [584 mm] LED flat screen, *n* = *3*	50	17
Manufacturer 3, 23 in. [584 mm] LED flat screen, *n* = *1*	38	21
Manufacturer 4, 24 in. [610 mm] LED flat screen, *n* = *1*	42	25
Average 15-min peak power consumption (range)	**21 (14–26)**	
Multifunction printer (copy, print, scan), multiuser, office type		
Max. printing speed up to 40 pages per minute	1010	540 (idle 29W)
Max. printing speed up to 30 pages per minute	1300	303 (idle 116W)
Max. printing speed up to 28 pages per minute	1500	433 (idle 28W)
Average 15-min peak power consumption (range)	**425 (303–540)**	
Vending machine		
Drinks, 280–400 items	NA	940
Snacks	NA	54
Food (e.g., sandwiches)	NA	465
Microwave oven, 7–9 gal	1000–1500	713–822
Coffee maker, 12 cups	950	780
Refrigerator, 18–28 ft^3 [0.5–0.80 m^3]	NA	387–430

Source: Excerpted and reprinted with permission; © ASHRAE (www.ashrae.org); from the 2021 ASHRAE Handbook—Fundamentals, Chapter 18, Tables 8, 9, and 10.

n = number of tested equipment of same configuration.

[a]Nameplate power for desktop computers is present on its power supply, which is mounted inside desktop, hence not accessible for most computers, where NA = not available. Screens with atypical size and shape are excluded for calculating average 15-minute peak power consumption. Voltage and amperage information for laptop computer is available on power supply nameplates.

[b]For equipment peak heat gain value, highest 15-minute interval of recorded data is listed in tables.

[c]Approximately 60% convective heat gain and 40% radiative heat gain.

G.7 HEAT GAINS FROM APPLIANCES

Table G.7 Recommended Rates of Radiant and Convective Heat Gain from Unhooded (Unvented) Electric Appliances During Idle (Ready-to-Cook)

Part A. I-P Units: Btu/h						
	Energy Rate		Rate of Heat Gain			
	Rated	Standby	Sensible Radiant	Sensible Convective	Latent	Total
Cabinet: hot serving (large), insulated	6800	1200	400	800	0	1200
Hot serving (large), uninsulated	6800	3500	700	2800	0	3500
Proofing (large)	17,400	1400	1200	0	200	1400
Proofing (small 15-shelf)	14,300	3900	0	900	3000	3900
Cheesemelter	8200	3300	1500	1800	0	3300
Coffee brewing urn	13,000	1200	200	300	700	1200
Drawer warmers, 2-drawer (moist holding)	4100	500	0	0	200	200
Egg cooker	8100	850	200	650	0	850
Espresso machine	8200	1200	400	800	0	1200
Food warmer: steam table (2-well-type)	5100	3500	300	600	2600	3500
Freezer (small)	2700	1100	500	600	0	1100
Fryer, countertop, open deep fat	15,700	1500	700	800	0	1500
Griddle, countertop	27,300	6100	2900	3200	0	6100
Hot dog roller	5500	4200	900	3300	0	4200
Hot plate: single element, high speed	3800	3400	1100	2300	0	3400
Hot-food case (dry holding)	31,100	2500	900	1600	0	2500
Hot-food case (moist holding)	31,100	3300	900	1800	600	3300
Induction hob, countertop	17,100	0	0	0	0	0
Microwave oven: commercial	5800	0	0	0	0	0
Oven: countertop conveyorized bake/finishing	17,100	13,500	2500	11,000	0	13,500
Panini	6100	2300	700	1600	0	2300
Popcorn popper	2900	400	100	300	0	400
Rapid-cook oven (quartz-halogen)	41,000	0	0	0	0	0
Rapid-cook oven (microwave/convection)	19,400	3900	300	3600	0	3900
Reach-in refrigerator	4800	1200	300	900	0	1200
Refrigerated prep table	2000	900	600	300	0	900
Rice cooker	5300	300	50	250	0	300
Soup warmer	2700	1300	0	200	1100	1300
Steamer (bun)	5100	700	100	600	0	700
Steamer, countertop	28,300	1200	0	800	400	1200
Toaster: 4-slice pop-up (large): cooking	6100	3000	200	1400	1000	2600
Contact (vertical)	8900	2600	600	2000	0	2600
Conveyor (large)	32,800	10,300	3000	7300	0	10,300
Small conveyor	6000	5800	1200	4600	0	5800
Tortilla grill	7500	3600	900	2700	0	3600
Waffle iron	9200	900	200	700	0	900

Table G.7 (*Continued*)

Part B. SI Units: W	Energy Rate		Rate of Heat Gain			
	Rated	Standby	Sensible Radiant	Sensible Convective	Latent	Total
Cabinet: hot serving (large), insulated	1993	352	117	234	0	352
Hot serving (large), uninsulated	1993	1026	205	821	0	1026
Proofing (large)	5099	410	352	0	59	410
Proofing (small 15-shelf)	4191	1143	0	264	879	1143
Cheesemelter	2400	976	443	533	0	976
Coffee brewing urn	3810	352	59	88	205	352
Drawer warmers, 2-drawer (moist holding)	1202	147	0	0	59	59
Egg cooker	2380	249	65	184	0	249
Espresso machine	2403	352	117	234	0	352
Food warmer: steam table (2-well-type)	1495	1026	88	176	762	1026
Freezer (small)	791	322	147	176	0	322
Fryer, countertop, open deep fat	4600	431	202	229	0	431
Griddle, countertop	8000	1771	848	923	0	1771
Hot dog roller	1600	1240	267	973	0	1240
Hot plate: single element, high speed	1100	982	314	668	0	982
Hot-food case (dry holding)	9115	733	264	469	0	733
Hot-food case (moist holding)	9115	967	264	528	176	967
Induction hob, countertop	5000	0	0	0	0	0
Microwave oven: commercial	1700	0	0	0	0	0
Oven: countertop conveyorized bake/finishing	5000	3932	718	3214	0	3932
Panini	1800	673	195	478	0	673
Popcorn popper	850	115	28	87	0	115
Rapid-cook oven (quartz-halogen)	12,016	0	0	0	0	0
Rapid-cook oven (microwave/convection)	5700	1141	96	1045	0	1141
Reach-in refrigerator	1407	352	88	264	0	352
Refrigerated prep table	586	264	176	88	0	264
Rice cooker	1550	82	14	68	0	82
Soup warmer	800	390	0	53	337	390
Steamer (bun)	1500	200	32	168	0	200
Steamer, countertop	8300	344	0	248	96	344
Toaster: 4-slice pop up (large): cooking	1788	879	59	410	293	762
Contact (vertical)	2600	759	180	579	0	759
Conveyor (large)	9613	3019	879	2139	0	3019
Small conveyor	1745	1702	358	1344	0	1702
Tortilla grill	2200	1034	254	780	0	1034
Waffle iron	2700	267	60	207	0	267

Source: Reprinted with permission; © ASHRAE (www.ashrae.org); from the 2021 ASHRAE Handbook—Fundamentals, Chapter 18, Table 5.

For recommended rates of radiant heat gain from hooded electric appliances during idle (ready-to-cook) conditions; recommended rates of radiant heat gain from hooded gas appliances during idle (ready-to-cook) conditions; and recommended rates of radiant and convective heat gain from warewashing equipment during idle (standby) or washing conditions—refer to ASHRAE (2021).

G.8 DESIGN DATA FOR EARTH TUBES

Table G.8 Earth Tube Applications[a]

Part A. I-P Units

Soil, No. of Tubes, Arrangement	Tube: Diameter (in.)	Tube: Length (ft)	Tube: Depth[b] (ft)	Temperatures (°F): Soil	Temperatures (°F): Intake Air[c]	Air Velocity (ft/min)	Total Flow Rate (ft^3/h)	Flow Per Tube (ft^3/h)	Peak Sensible Cooling (Btu/h f)
Sandy soil; 8 tubes, radial	10	100	8	73°	67–105°	220	57,600	7200	20
Silt loam, wet; 9 tubes, radial (3 trenches)	5	150	8–11.5	73°	62–95°	500	60,000	4000	11
Silt loam; 22 tubes, parallel at 4 ft	10	180	8–9	72°	70–94°	500	359,700	16,350	30
Silt loam; 5 tubes, parallel	12	260	10	63°	60–92°	530	123,600	24,720	40

Part B. SI Units

Soil, No. of Tubes, Arrangement	Tube: Diameter (cm)	Tube: Length (m)	Tube: Depth[b] (m)	Temperatures (°C): Soil	Temperatures (°C): Intake Air[c]	Air Velocity (m/s)	Total Flow Rate (m^3/h)	Flow Per Tube (m^3/h)	Peak Sensible Cooling (W/m)
Sandy soil; 8 tubes, radial	25	30.5	2.4	23°	19.4–40.6°	1.1	1632	204	19.7
Silt loam, wet; 9 tubes, radial (3 trenches)	13	46	2.4–3.5	23°	16.7–35°	2.5	1080	120	10.6
Silt loam; 22 tubes, parallel at 1.2 m	25	55	2.4–2.7	22°	21.1–34.4°	2.5	10,200	463	29.5
Silt loam; 5 tubes, parallel	30	79	3	17°	15.6–33.3°	2.7	3500	696	38.8

Source: Givoni, 1994. I-P conversions added by the authors of this book.

[a]Central Illinois farm applications, summertime.

[b]Tubes often slope to drain the condensation that occurs as air reaches the dew point within the tube.

[c]Intake air temperature range during the period of study.

G.9 PSYCHROMETRIC CHARTS

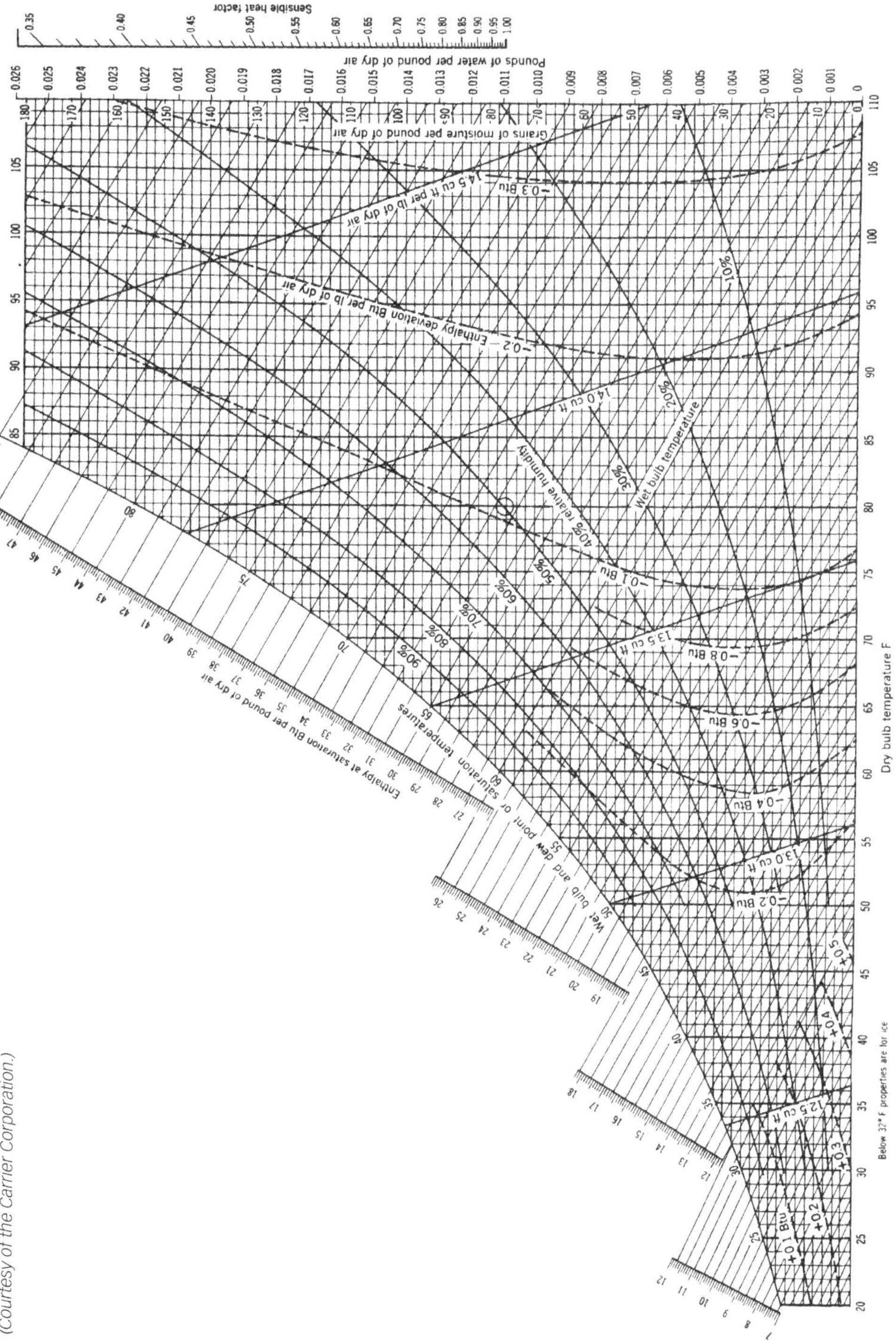

Figure G.1 Psychrometric chart, I-P units. Conditions at sea level, standard atmospheric pressure (29.921 in. Hg). (Courtesy of the Carrier Corporation.)

Figure G.2 *Psychrometric chart, SI units. Conditions at sea level, standard atmospheric pressure (101.325 kPa). (Adapted from Szokolay, 1980.)*

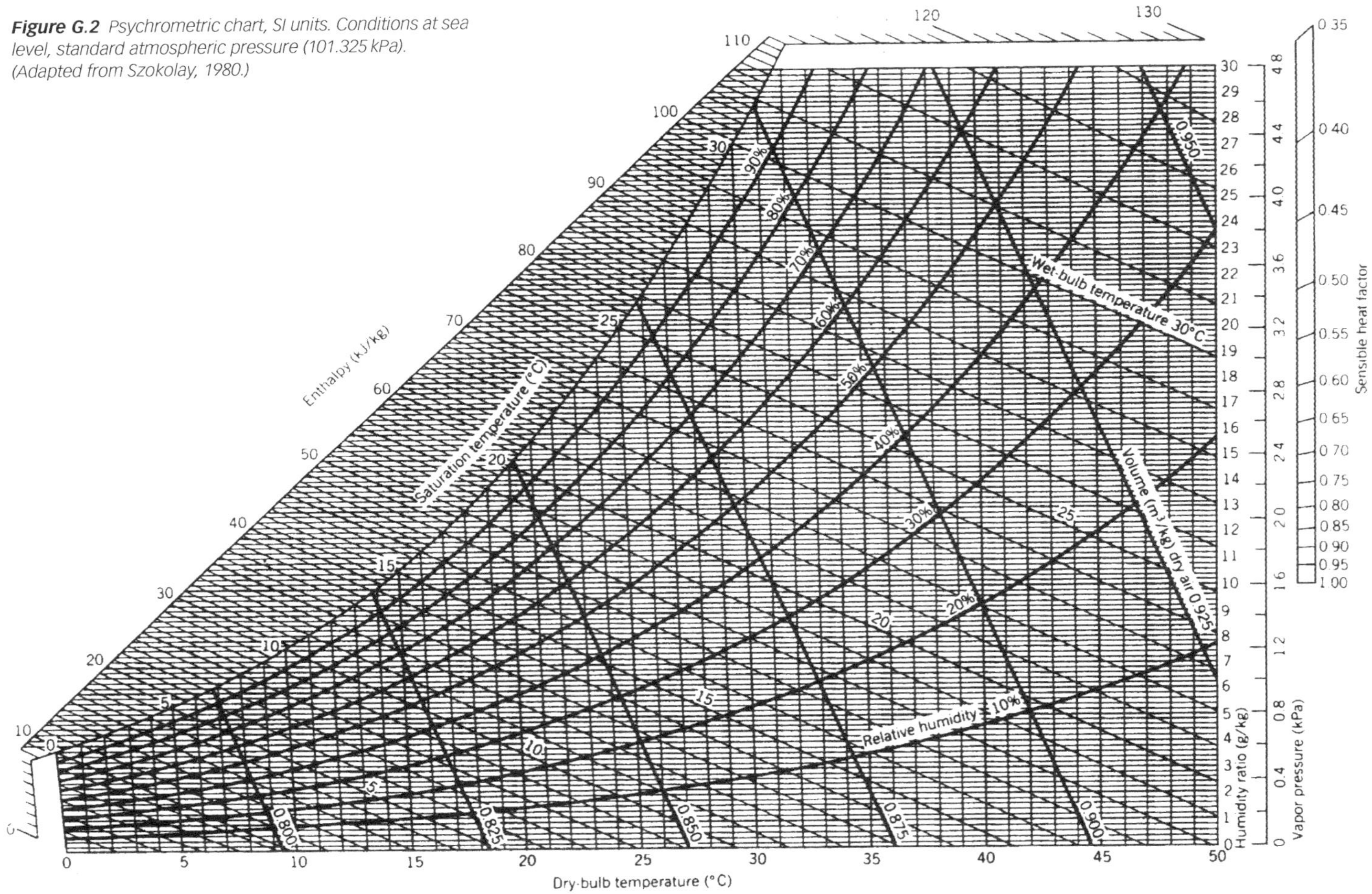

References and Resources

ACSA. 1994. *Design with PV*. Washington, DC: Association of Collegiate Schools of Architecture.

ASHRAE. 1981. *ASHRAE Handbook—Fundamentals*. Atlanta, GA: American Society of Heating, Refrigerating, and Air-Conditioning Engineers.

ASHRAE. 1997. *ASHRAE Handbook—Fundamentals*. Atlanta, GA: American Society of Heating, Refrigerating, and Air-Conditioning Engineers.

ASHRAE. 2021. *ASHRAE Handbook—Fundamentals*. Atlanta, GA: American Society of Heating, Refrigerating, and Air-Conditioning Engineers.

Givoni, B. 1994. *Passive and Low Energy Cooling of Buildings*. New York: Van Nostrand Reinhold.

Hastings, S. R. 1994. *Passive Solar Commercial and Institutional Buildings*, International Energy Agency. Chichester: John Wiley & Sons.

Komor, P. 1997. *Space Cooling Demand from Office Plug Loads*. December: *ASHRAE Journal*.

Northwest Power Planning Council. 1983. *Maximum Lighting Standards*. Portland, OR: Northwest Power Planning Council.

Szokolay, S. V. 1980. *Environmental Science Handbook*. New York: Longman Group Ltd.

APPENDIX H

Standards/Guidelines for High-Performance Building Design

This appendix provides extracts from *ASHRAE/IES Standard 90.1-2022: Energy Standard for Buildings Except Low-Rise Residential Buildings* (ASHRAE, 2022) and ASHRAE's *Advanced Energy Design Guide for Small to Medium Office Buildings* (ASHRAE, 2011) (Tables H1–H3). The purpose of these extracts is to show the general nature of these guidance documents and their formats. Obtain complete information regarding the most current versions of each of these documents from the American Society of Heating, Refrigerating and Air-Conditioning Engineers, Inc. (ASHRAE). Note that ASHRAE uses the term "envelope" to describe what in this book is termed "enclosure."

Mechanical and Electrical Equipment for Buildings, Fourteenth Edition. Walter Grondzik, Daniel Overbey, and Alison Kwok.

Companion website: www.wiley.com/go/meeb14e

H.1 SAMPLE OF PRESCRIPTIVE BUILDING ENVELOPE REQUIREMENTS EXTRACTED FROM *ASHRAE/IES STANDARD 90.1-2022: ENERGY STANDARD FOR BUILDINGS EXCEPT LOW-RISE RESIDENTIAL BUILDINGS* (ASHRAE, 2022)

Table H.1 *ASHRAE/IES Standard 90.1:* **Building Envelope Requirements for Climate Zone 4 (I-P Units)**

	Nonresidential		Residential		Semiheated	
Opaque Elements	**Assembly Maximum**	**Insulation Min. R-Value**	**Assembly Maximum**	**Insulation Min. R-Value**	**Assembly Maximum**	**Insulation Min. R-Value**
Roofs						
Insulation entirely above deck	U-0.032	R-30.0 *c.i.*	U-0.032	R-30 *c.i.*	U-0.093	R-10 *c.i.*
Metal building[a]	U-0.037	R-19 + R-11 *Ls* or R-25 + R-8 *Ls*	U-0.037	R-19 + R-11 *Ls* or R-25 + R-8 *Ls*	U-0.082	R-19
Attic and other	U-0.021	R-49	U-0.021	R-49	U-0.034	R-30
Walls, above Grade						
Mass	U-0.104	R-9.5 *c.i.*	U-0.090	R-11.4 *c.i.*	U-0.580	NR
Metal building	U-0.060	R-0 + R-15.8 *c.i.*	U-0.050	R-0 + R-19 *c.i.*	U-0.162	R-13
Steel-framed	U-0.064	R-13 + R-7.5 *c.i.*	U-0.064	R-13 + R-7.5 *c.i.*	U-0.124	R-13
Wood-framed and other	U-0.064	R-13 + R-3.8 *c.i.* or R-20	U-0.064	R-13 + R-3.8 *c.i.* or R-20	U-0.089	R-13
Wall, below Grade						
Below-grade wall	C-0.119	R-7.5 *c.i.*	C-0.092	R-10 *c.i.*	C-1.140	NR
Floors						
Mass	U-0.057	R-14.6 *c.i.*	U-0.051	R-16.7 *c.i.*	U-0.107	R-6.3 *c.i.*
Steel-Joist	U-0.038	R-30	U-0.038	R-30	U-0.052	R-19
Wood-framed and other	U-0.033	R-30	U-0.033	R-30	U-0.051	R-19
Slab-on-Grade Floors						
Unheated	F-0.520	R-15 for 24 in.	F-0.520	R-15 for 24 in.	F-0.730	NR
Heated	F-0.843	R-20 for 24 in.	F-0.688	R-20 for 48 in.	F-0.900	R-10 for 24 in.
Opaque Doors						
Swinging	U-0.370		U-0.370		U-0.370	
Nonswinging	U-0.310		U-0.310		U-0.360	

	Nonresidential			Residential			Semiheated		
Fenestration	**Assembly Max. U**	**Assembly Max. SHGC**	**Assembly Min. VT/SHGC**	**Assembly Max. U**	**Assembly Max. SHGC**	**Assembly Min. VT/SHGC**	**Assembly Max. U**	**Assembly Max. SHGC**	**Assembly Min. VT/SHGC**
Vertical Fenestration, 0% to 40% of Wall									
Fixed	0.36	0.36	1.10 (for all types)	0.36	0.36	1.10 (for all types)	0.50	NR (for all types)	NR (for all types)
Operable	0.45	0.33		0.45	0.33		0.65		
Entrance door	0.63	0.33		0.63	0.33		0.77		
Skylight, 0% to 3% of Roof									
All Types	0.50	0.40	NR	0.50	0.40	NR	0.75	NR	NR

Source: © ASHRAE, www.ashrae.org. ANSI/ASHRAE/IES Standard 90.1-2022.

Note: The following definitions apply:

c.i. = continuous insulation

Ls = liner system

NR = no (insulation) requirement

[a]When using the R-value compliance method for metal building roofs, a thermal spacer block is required (see ASHRAE, 2022).

Table H.2 ***ASHRAE/IES Standard 90.1:*** **Building Envelope Requirements for Climate Zone 4 (SI Units)**

	Nonresidential		Residential		Semiheated	
Opaque Elements	**Assembly Maximum**	**Insulation Min. R-Value**	**Assembly Maximum**	**Insulation Min. R-Value**	**Assembly Maximum**	**Insulation Min. R-Value**
Roofs						
Insulation entirely above deck	U-0.184	R-5.3 *c.i.*	U-0.184	R-5.3 *c.i.*	U-0.527	R-1.8 *c.i.*
Metal building[a]	U-0.210	R-3.3 + R-1.9 *Ls* or R-4.4 + R-1.4 *Ls*	U-0.210	R-3.3 + R-19 *Ls* or R-4.4 + R-1.4 *Ls*	U-0.466	R-3.3
Attic and other	U-0.119	R-8.6	U-0.119	R-8.6	U-0.192	R-5.3
Walls, above Grade						
Mass	U-0.592	R-1.7 *c.i.*	U-0.513	R-2.0 *c.i.*	U-3.293	NR
Metal building	U-0.341	R-0 + R-2.8 *c.i.*	U-0.286	R-0 + R-3.3 *c.i.*	U-0.920	R-2.3
Steel-framed	U-0.365	R-2.3 + R-1.3 *c.i.*	U-0.365	R-2.3 + R-1.3 *c.i.*	U-0.705	R-2.3
Wood-framed and other	U-0.365	R-2.3 + R-0.7 *c.i.* or R-3.5	U-0.365	R-2.3 + R-0.7 *c.i.* or R-3.5	U-0.504	R-2.3
Wall, below Grade						
Below-grade wall	C-0.678	R-1.3 *c.i.*	C-0.522	R-1.8 *c.i.*	C-6.473	NR
Floors						
Mass	U-0.321	R-2.6 *c.i.*	U-0.287	R-2.9 *c.i.*	U-0.606	R-1.1 *c.i.*
Steel-Joist	U-0.214	R-5.3	U-0.214	R-5.3	U-0.296	R-3.3
Wood-framed and other	U-0.188	R-5.3	U-0.188	R-5.3	U-0.288	R-3.3
Slab-on-Grade Floors						
Unheated	F-0.900	R-2.6 for 600 mm	F-0.900	R-2.6 for 600 mm	F-1.264	NR
Heated	F-1.459	R-3.5 for 600 mm	F-1.191	R-3.5 for 1200 mm	F-1.558	R-1.8 for 600 mm
Opaque Doors						
Swinging	U-2.101		U-2.101		U-2.101	
Nonswinging	U-1.760		U-1.760		U-2.044	

	Nonresidential			Residential			Semiheated		
Fenestration	**Assembly Max. U**	**Assembly Max. SHGC**	**Assembly Min. VT/SHGC**	**Assembly Max. U**	**Assembly Max. SHGC**	**Assembly Min. VT/SHGC**	**Assembly Max. U**	**Assembly Max. SHGC**	**Assembly Min. VT/SHGC**
Vertical Fenestration, 0% to 40% of Wall									
Fixed	2.04	0.36	1.10 (for all types)	2.04	0.36	1.10 (for all types)	2.84	NR (for all types)	NR (for all types)
Operable	2.56	0.33		2.56	0.33		3.69		
Entrance door	3.58	0.33		3.58	0.33		4.37		
Skylight, 0% to 3% of Roof									
All Types	2.84	0.40	NR	2.84	0.40	NR	4.26	NR	NR

Source: © ASHRAE, www.ashrae.org. ANSI/ASHRAE/IES Standard 90.1-2022.

Note: The following definitions apply:

c.i. = continuous insulation

Ls = liner system

NR = no (insulation) requirement

[a]When using the R-value compliance method for metal building roofs, a thermal spacer block is required (see ASHRAE, 2022).

H.2 SAMPLE OF RECOMMENDED BUILDING ENVELOPE REQUIREMENTS EXTRACTED FROM *ADVANCED ENERGY DESIGN GUIDE FOR SMALL TO MEDIUM OFFICE BUILDINGS* (ASHRAE, 2011)

Table H.3 ***Advanced Energy Design Guide for Small to Medium Office Buildings*; Climate Zone 4 Envelope Recommendations (50% Savings version)**

Item	Component	Recommendation
Roofs	Insulation entirely above deck	R-30.0 *c.i.*
	Attic and other	R-49.0
	Metal building	R-19.0 + R-11.0 *Ls*
	SRI	No recommendation
	Mass (HC > 7 Btu/ft^2)	R-13.3 *c.i.*
Walls	Steel framed	R-13.0 + R-7.5 *c.i.*
	Wood framed and other	R-13.0 + R-7.5 *c.i.*
	Metal building	R-0.0 + R-15.8 *c.i.*
	Below-grade walls	R-7.5 *c.i.*
	Mass	R-14.6 *c.i.*
Floors	Steel joist	R-38.0
	Wood framed and other	R-38.0
Slabs	Unheated	R-15.0 for 24 in.
	Heated	R-20.0 for 24 in.
Doors	Swinging	U-0.50
	Nonswinging	U-0.50
Vestibules	At building entrance	Yes
Continuous Air Barriers	Continuous air barrier	Entire building envelope
Vertical Fenestration	WWR	20% to 40%
	Window orientation	Area of W and E windows each less than area of S windows (N in southern hemisphere)
	Exterior sun control (S, E, and W only)	PF-0.5
	Thermal transmittance	Nonmetal framing windows = U-0.38 Metal framing windows = U-0.39
	SHGC	Nonmetal framing windows = 0.26 Metal framing windows = 0.38

Source: © ASHRAE, www.ashrae.org. Advanced Energy Design Guide for Small to Medium Office Buildings: Achieving 50% Energy Savings Towards a Net Zero Energy Building, 2011.

Note: The following definitions apply:

c.i. = continuous insulation

Ls = liner system

PF = projection factor

WWR = *window-to-wall ratio*

References and Resources

ASHRAE. 2022. *ANSI/ASHRAE/IES Standard 90.1-2022: Energy Standard for Sites and Buildings Except Low-Rise Residential Buildings*. Atlanta, GA: American Society of Heating, Refrigerating and Air-Conditioning Engineers.

ASHRAE. 2011. *Advanced Energy Design Guide for Small to Medium Office Buildings: Achieving 50% Energy Savings Towards a Net Zero Energy Building*. Atlanta, GA: American Society of Heating, Refrigerating, and Air-Conditioning Engineers.

APPENDICES

APPENDIX I

Heating Degree-Day and Solar Radiation Data

This appendix provides heating degree-day and incident solar radiation data to support examples and exercises, including those in Chapters 13 and 15.

Table I.1 HDD Data and Incident Solar Radiation on Tilted Surfaces

Heating degree-days at a base temperature of 65°F are designated by HDD65; those at a base temperature of 18.3°C are designated by HDD18.3. ©ASHRAE, www.ashrae.org. 2021 ASHRAE Standard 169.
Solar radiation values shown below represent total monthly solar radiation calculated from solar radiation data in Typical Meteorological Year 3 (TMY3) weather files using the algorithm of Perez et al. (1990).

Part A: I-P Units

ANCHORAGE, AK		**Solar radiation (kBtu/ft² mo) on a surface of tilt:**									
Mo	**HDD65**	**90°**	**80°**	**70°**	**60°**	**50°**	**40°**	**30°**	**20°**	**10°**	**0°**
Jan	1466	8	8	8	7	7	6	6	5	4	3
Feb	1203	23	24	24	24	23	21	18	16	12	9
Mar	1202	31	33	34	35	35	34	32	29	26	22
Apr	817	34	38	40	42	43	43	42	40	37	33
May	519	32	37	41	44	46	48	48	48	46	44
Jun	273	34	40	45	49	53	55	56	56	54	52
Jul	173	32	36	41	44	47	49	49	49	48	46
Aug	231	29	33	36	39	40	41	41	40	39	36
Sep	458	25	27	29	30	30	29	28	26	24	21
Oct	874	15	16	17	17	17	16	15	14	12	10
Nov	1234	13	13	13	13	12	11	9	8	6	4
Dec	1409	4	4	4	4	4	4	3	3	2	2
FAIRBANKS, AK		**Solar radiation (kBtu/ft² mo) on a surface of tilt:**									
Mo	**HDD65**	**90°**	**80°**	**70°**	**60°**	**50°**	**40°**	**30°**	**20°**	**10°**	**0°**
Jan	2267	6	6	6	6	5	5	4	3	2	1
Feb	1791	19	19	19	19	18	16	14	12	9	6
Mar	1680	37	39	40	41	40	38	35	31	27	22
Apr	934	46	51	54	56	56	56	53	50	45	40
May	454	41	47	52	55	58	59	59	57	55	51
Jun	148	37	42	47	51	54	56	57	57	55	53
Jul	100	36	41	46	50	52	54	54	53	52	49
Aug	258	36	40	44	46	47	47	47	45	42	38
Sep	561	27	29	31	31	31	31	29	27	24	21
Oct	1190	20	21	21	21	20	18	17	15	12	9
Nov	1828	10	10	10	9	9	8	7	5	4	2
Dec	2155	2	2	2	2	2	1	1	1	1	0

Mechanical and Electrical Equipment for Buildings, Fourteenth Edition. Walter Grondzik, Daniel Overbey, and Alison Kwok.

Companion website: www.wiley.com/go/meeb14e

Table I.1 (*Continued*)

BIRMINGHAM, AL		Solar radiation (kBtu/ft² mo) on a surface of tilt:									
Mo	**HDD65**	**90°**	**80°**	**70°**	**60°**	**50°**	**40°**	**30°**	**20°**	**10°**	**0°**
Jan	641	33	35	37	38	38	37	35	32	29	25
Feb	462	33	36	39	40	41	41	39	37	34	29
Mar	305	33	39	43	47	49	50	50	49	47	43
Apr	112	29	36	42	47	51	54	56	56	55	53
May	22	22	29	36	42	47	52	55	57	58	57
Jun	0	20	26	33	40	45	50	54	57	59	59
Jul	0	22	28	35	42	47	52	55	58	59	59
Aug	0	26	33	39	45	50	54	56	57	57	56
Sep	5	31	37	41	45	48	49	50	49	47	44
Oct	105	39	44	47	50	51	51	50	47	44	39
Nov	338	35	38	40	41	41	40	38	35	31	27
Dec	550	37	40	41	42	42	40	37	34	30	24

MOBILE, AL		Solar radiation (kBtu/ft² mo) on a surface of tilt:									
Mo	**HDD65**	**90°**	**80°**	**70°**	**60°**	**50°**	**40°**	**30°**	**20°**	**10°**	**0°**
Jan	443	37	40	42	43	43	42	40	37	32	28
Feb	297	26	29	31	32	33	33	32	31	28	25
Mar	176	31	36	41	45	47	49	49	48	46	43
Apr	50	28	36	42	48	53	56	59	59	59	57
May	5	20	27	33	39	44	49	52	54	56	55
Jun	0	18	25	33	40	47	53	58	61	63	64
Jul	0	20	26	33	40	46	51	56	59	60	60
Aug	0	25	33	40	46	51	55	58	60	60	59
Sep	1	30	36	41	46	49	51	52	52	50	47
Oct	47	44	49	54	57	59	59	58	55	51	45
Nov	207	46	50	53	55	55	53	51	47	41	35
Dec	374	37	40	42	42	42	41	39	35	31	26

FORT SMITH, AR		Solar radiation (kBtu/ft² mo) on a surface of tilt:									
Mo	**HDD65**	**90°**	**80°**	**70°**	**60°**	**50°**	**40°**	**30°**	**20°**	**10°**	**0°**
Jan	757	34	37	39	39	39	37	35	32	28	23
Feb	563	33	36	38	40	41	40	39	36	33	29
Mar	372	36	42	46	50	52	53	53	51	48	44
Apr	135	28	34	40	45	49	52	53	54	53	51
May	28	24	31	38	44	50	54	58	60	60	60
Jun	0	21	29	36	43	49	55	59	62	63	63
Jul	0	23	31	39	46	53	58	62	65	66	66
Aug	0	29	37	44	50	55	59	61	62	61	59
Sep	9	33	39	44	48	51	53	53	52	50	46
Oct	122	40	45	48	51	52	52	50	47	43	38
Nov	391	33	36	38	39	39	38	36	34	30	26
Dec	686	38	40	42	42	42	40	37	33	29	24

LITTLE ROCK, AR		Solar radiation (kBtu/ft² mo) on a surface of tilt:									
Mo	**HDD65**	**90°**	**80°**	**70°**	**60°**	**50°**	**40°**	**30°**	**20°**	**10°**	**0°**
Jan	708	34	37	38	39	39	38	35	32	29	24
Feb	536	29	32	34	36	36	35	34	32	29	25
Mar	356	31	36	40	43	45	47	47	45	43	40
Apr	123	29	35	41	46	49	52	53	53	52	50
May	22	23	30	36	42	48	52	55	57	57	57
Jun	0	21	28	35	42	48	53	57	60	62	62
Jul	0	23	31	39	46	52	57	61	64	65	64
Aug	0	29	37	44	51	56	60	62	63	63	61
Sep	5	32	38	43	47	50	51	52	51	49	46
Oct	115	40	45	49	51	53	52	51	48	44	39
Nov	375	34	36	38	39	40	39	37	34	30	26
Dec	641	33	36	37	38	37	36	34	30	26	22

FLAGSTAFF, AZ		Solar radiation (kBtu/ft² mo) on a surface of tilt:									
Mo	**HDD65**	**90°**	**80°**	**70°**	**60°**	**50°**	**40°**	**30°**	**20°**	**10°**	**0°**
Jan	1064	48	52	54	55	54	52	48	43	37	30
Feb	911	44	49	52	53	54	53	50	47	42	36
Mar	827	37	43	48	51	54	55	55	53	50	46
Apr	650	32	40	46	52	57	60	62	62	61	58
May	426	26	35	44	52	60	66	70	73	74	73
Jun	137	21	29	38	46	54	61	66	70	72	72

Table I.1 (*Continued*)

Jul	24	23	30	38	45	52	57	61	64	65	65
Aug	48	29	37	45	52	58	62	64	66	65	63
Sep	211	38	45	51	56	59	61	61	59	56	52
Oct	553	46	51	56	58	59	59	57	53	48	42
Nov	816	52	56	58	59	58	56	52	47	41	33
Dec	1077	49	53	54	55	54	51	47	42	35	28
PHOENIX, AZ		**Solar radiation (kBtu/ft² mo) on a surface of tilt:**									
Mo	**HDD65**	**90°**	**80°**	**70°**	**60°**	**50°**	**40°**	**30°**	**20°**	**10°**	**0°**
Jan	255	50	53	56	57	56	54	50	45	39	32
Feb	165	43	48	51	53	53	53	50	47	42	36
Mar	64	42	49	54	59	62	63	62	60	57	52
Apr	19	36	45	53	60	65	69	71	71	70	67
May	1	26	35	45	54	61	68	73	76	77	77
Jun	0	22	31	41	50	58	66	72	76	78	79
Jul	0	24	33	42	51	58	65	70	73	75	74
Aug	0	31	40	49	56	63	67	71	72	72	70
Sep	0	41	49	56	62	66	68	69	67	64	60
Oct	6	49	55	60	63	64	64	62	59	53	47
Nov	79	52	56	59	60	60	58	54	49	43	36
Dec	285	51	55	57	57	56	54	50	44	38	30
FRESNO, CA		**Solar radiation (kBtu/ft² mo) on a surface of tilt:**									
Mo	**HDD65**	**90°**	**80°**	**70°**	**60°**	**50°**	**40°**	**30°**	**20°**	**10°**	**0°**
Jan	510	22	24	25	26	26	26	25	23	21	18
Feb	358	34	37	39	40	40	40	38	35	32	28
Mar	235	39	44	49	53	55	56	55	53	50	46
Apr	131	37	45	52	58	63	66	67	67	65	61
May	30	29	38	46	54	61	66	70	73	73	72
Jun	2	25	34	43	52	60	67	72	76	78	77
Jul	0	27	36	45	54	62	68	73	76	77	77
Aug	0	34	44	53	61	67	72	75	75	75	72
Sep	2	41	48	54	59	62	64	64	62	59	54
Oct	53	47	52	56	59	60	60	58	54	49	43
Nov	288	36	39	41	42	41	40	38	35	31	26
Dec	529	29	31	32	32	32	31	29	26	23	19
LOS ANGELES, CA		**Solar radiation (kBtu/ft² mo) on a surface of tilt:**									
Mo	**HDD65**	**90°**	**80°**	**70°**	**60°**	**50°**	**40°**	**30°**	**20°**	**10°**	**0°**
Jan	227	38	41	43	43	43	42	39	36	32	27
Feb	215	33	36	39	41	42	41	40	37	34	30
Mar	191	36	42	47	51	54	55	55	53	51	47
Apr	147	31	38	45	50	54	57	59	59	58	55
May	77	25	32	39	46	52	57	60	62	63	62
Jun	15	22	30	38	46	53	59	63	66	68	68
Jul	2	25	34	42	51	58	64	68	71	72	72
Aug	1	31	39	47	54	60	64	67	68	68	66
Sep	3	35	42	48	52	56	58	58	57	55	51
Oct	29	41	46	50	52	54	54	52	50	46	41
Nov	116	43	46	49	50	50	48	45	42	37	31
Dec	233	42	45	47	48	47	45	42	37	32	26
SACRAMENTO, CA		**Solar radiation (kBtu/ft² mo) on a surface of tilt:**									
Mo	**HDD65**	**90°**	**80°**	**70°**	**60°**	**50°**	**40°**	**30°**	**20°**	**10°**	**0°**
Jan	541	24	26	27	28	28	27	25	23	20	17
Feb	398	28	30	32	33	33	33	31	29	27	23
Mar	304	41	47	52	55	57	58	57	55	51	46
Apr	187	38	46	52	58	62	65	66	66	63	60
May	56	30	39	48	56	63	68	72	73	74	72
Jun	6	27	36	45	54	61	68	73	76	77	77
Jul	0	29	39	48	57	64	70	75	77	78	77
Aug	0	36	45	54	61	67	71	73	74	73	69
Sep	6	42	48	54	58	61	62	62	60	56	52
Oct	85	40	45	48	50	51	51	49	46	42	36
Nov	342	35	38	39	40	40	38	36	33	28	24
Dec	550	30	32	33	33	32	31	29	26	22	18

Table I.1 (*Continued*)

SAN FRANCISCO, CA		**Solar radiation (kBtu/ft² mo) on a surface of tilt:**									
Mo	**HDD65**	**90°**	**80°**	**70°**	**60°**	**50°**	**40°**	**30°**	**20°**	**10°**	**0°**
Jan	424	31	33	34	35	35	33	31	29	25	21
Feb	336	35	38	40	41	42	41	39	36	32	28
Mar	305	38	43	48	51	53	53	53	51	47	43
Apr	251	32	39	45	50	53	56	57	57	56	53
May	191	27	34	42	49	54	59	62	64	64	63
Jun	112	24	32	40	48	54	60	64	67	68	68
Jul	74	27	35	44	51	58	63	67	70	71	70
Aug	56	32	40	47	54	59	63	65	66	65	63
Sep	57	40	47	53	57	60	62	61	60	57	52
Oct	113	40	44	48	50	51	50	49	46	42	37
Nov	266	34	37	39	39	39	38	36	32	29	24
Dec	421	36	39	40	40	39	37	34	30	26	21

DENVER, CO		**Solar radiation (kBtu/ft² mo) on a surface of tilt:**									
Mo	**HDD65**	**90°**	**80°**	**70°**	**60°**	**50°**	**40°**	**30°**	**20°**	**10°**	**0°**
Jan	1027	46	49	51	51	49	47	43	37	31	25
Feb	904	39	42	44	45	45	44	41	38	33	28
Mar	734	43	49	54	57	59	59	58	55	51	46
Apr	525	29	35	40	44	47	49	49	49	48	45
May	279	28	36	43	50	55	59	62	64	64	62
Jun	53	26	34	43	50	57	62	66	69	70	69
Jul	4	27	35	43	50	56	61	64	66	67	66
Aug	8	32	39	46	52	57	60	62	63	62	59
Sep	107	39	46	51	55	58	59	59	57	54	49
Oct	437	44	48	52	54	54	53	51	47	42	36
Nov	743	39	42	43	43	43	41	38	34	29	23
Dec	1053	41	43	44	44	43	40	37	32	26	20

GRAND JUNCTION, CO		**Solar radiation (kBtu/ft² mo) on a surface of tilt:**									
Mo	**HDD65**	**90°**	**80°**	**70°**	**60°**	**50°**	**40°**	**30°**	**20°**	**10°**	**0°**
Jan	1127	42	44	46	46	45	43	40	35	30	24
Feb	824	38	41	43	44	44	43	40	37	33	28
Mar	615	41	47	51	54	56	57	56	53	50	45
Apr	392	36	44	50	55	59	61	62	62	59	56
May	154	30	38	46	54	60	65	68	70	70	69
Jun	16	26	35	43	52	59	65	69	72	74	73
Jul	0	29	38	46	54	61	67	71	73	74	73
Aug	0	34	42	50	56	61	65	68	68	67	65
Sep	56	42	49	54	59	62	63	62	60	57	52
Oct	369	49	54	58	60	61	60	58	54	48	42
Nov	745	40	43	45	46	45	43	40	36	31	26
Dec	1118	42	45	46	46	44	42	38	33	28	22

HARTFORD, CT		**Solar radiation (kBtu/ft² mo) on a surface of tilt:**									
Mo	**HDD65**	**90°**	**80°**	**70°**	**60°**	**50°**	**40°**	**30°**	**20°**	**10°**	**0°**
Jan	1177	30	31	32	33	32	30	28	25	22	18
Feb	989	29	32	33	34	34	33	32	29	26	22
Mar	841	32	37	40	43	44	45	44	43	40	37
Apr	465	28	33	38	41	44	46	47	46	45	43
May	197	26	31	37	42	46	50	52	53	53	52
Jun	34	24	31	37	42	47	51	54	55	56	55
Jul	2	26	32	38	44	49	52	55	57	57	56
Aug	6	29	35	40	45	48	51	52	52	52	49
Sep	82	32	36	40	43	45	46	46	45	43	39
Oct	369	32	35	37	38	39	38	37	35	31	28
Nov	674	29	31	33	33	32	31	29	26	23	19
Dec	992	26	27	28	28	27	26	24	21	18	15

ORLANDO, FL		**Solar radiation (kBtu/ft² mo) on a surface of tilt:**									
Mo	**HDD65**	**90°**	**80°**	**70°**	**60°**	**50°**	**40°**	**30°**	**20°**	**10°**	**0°**
Jan	184	44	48	51	53	53	51	49	45	40	33
Feb	103	26	29	31	33	34	34	33	32	29	26
Mar	54	33	39	44	48	51	53	54	53	50	47
Apr	8	28	36	44	50	56	60	62	63	62	60
May	0	20	28	36	44	51	56	61	64	66	66

Table I.1 (*Continued*)

Jun	0	16	22	28	34	40	45	50	53	55	55
Jul	0	18	25	33	40	46	52	56	59	61	61
Aug	0	24	31	39	45	51	56	59	61	61	60
Sep	0	35	43	50	56	60	63	64	64	62	58
Oct	5	32	37	41	43	45	46	45	44	41	38
Nov	39	48	53	56	58	58	57	54	50	44	38
Dec	119	51	55	58	59	58	56	53	48	41	34

TALLAHASSEE, FL		**Solar radiation (kBtu/ft² mo) on a surface of tilt:**									
Mo	**HDD65**	**90°**	**80°**	**70°**	**60°**	**50°**	**40°**	**30°**	**20°**	**10°**	**0°**
Jan	408	36	39	42	43	43	42	40	37	33	28
Feb	271	35	38	41	43	44	43	42	39	36	31
Mar	161	32	38	43	47	50	51	52	51	48	45
Apr	47	29	36	43	48	53	57	59	59	59	56
May	3	20	26	33	40	46	51	55	57	58	58
Jun	0	17	22	29	35	40	45	49	51	53	53
Jul	0	19	25	32	38	43	48	52	54	56	56
Aug	0	22	29	35	41	46	50	53	54	55	54
Sep	0	29	35	40	44	47	49	50	50	48	46
Oct	36	36	41	46	49	51	51	50	49	45	41
Nov	183	36	39	42	44	44	43	42	39	35	31
Dec	332	36	39	41	42	42	40	38	35	31	26

ATLANTA, GA		**Solar radiation (kBtu/ft² mo) on a surface of tilt:**									
Mo	**HDD65**	**90°**	**80°**	**70°**	**60°**	**50°**	**40°**	**30°**	**20°**	**10°**	**0°**
Jan	644	37	40	42	43	43	42	40	36	32	27
Feb	470	33	37	39	41	41	41	39	37	33	29
Mar	318	36	42	46	50	52	54	54	52	49	46
Apr	118	30	36	43	48	52	55	57	57	56	54
May	22	24	32	39	46	51	56	60	62	63	62
Jun	1	21	29	36	43	49	55	59	62	64	64
Jul	0	22	29	36	43	49	54	58	61	62	62
Aug	0	27	34	40	46	51	55	57	59	59	57
Sep	5	28	33	37	41	44	45	46	46	44	41
Oct	103	37	41	45	48	49	49	48	46	42	38
Nov	342	36	39	41	43	43	42	40	36	33	28
Dec	555	37	39	41	42	41	40	37	34	30	25

DES MOINES, IOWA		**Solar radiation (kBtu/ft² mo) on a surface of tilt:**									
Mo	**HDD65**	**90°**	**80°**	**70°**	**60°**	**50°**	**40°**	**30°**	**20°**	**10°**	**0°**
Jan	1314	34	36	37	37	36	34	32	28	24	19
Feb	1064	34	37	39	40	40	38	36	33	30	25
Mar	791	35	39	43	45	46	47	46	44	41	37
Apr	406	32	38	43	47	50	52	53	52	50	47
May	148	27	34	41	47	51	55	58	59	59	58
Jun	14	26	34	41	48	53	58	62	64	64	63
Jul	1	28	35	42	49	55	59	62	64	64	63
Aug	3	32	39	45	51	55	58	60	60	58	56
Sep	71	38	44	48	52	54	55	54	52	49	45
Oct	361	39	43	46	47	48	47	45	42	37	32
Nov	744	32	34	36	36	35	34	31	28	24	20
Dec	1148	30	32	33	32	32	30	28	24	21	17

BOISE, ID		**Solar radiation (kBtu/ft² mo) on a surface of tilt:**									
Mo	**HDD65**	**90°**	**80°**	**70°**	**60°**	**50°**	**40°**	**30°**	**20°**	**10°**	**0°**
Jan	1012	26	28	28	29	28	27	25	23	20	17
Feb	772	30	33	34	35	35	33	32	29	26	22
Mar	617	35	39	42	45	46	46	45	43	40	36
Apr	432	36	43	48	52	56	57	58	57	55	51
May	210	31	38	45	52	57	61	63	64	64	62
Jun	57	30	38	46	54	60	66	69	72	72	71
Jul	3	33	42	51	59	66	71	75	76	76	74
Aug	5	38	47	54	61	66	69	71	70	69	65
Sep	79	44	50	55	58	60	61	60	57	53	47
Oct	381	43	47	50	51	52	51	48	45	40	34
Nov	724	29	31	32	32	31	30	28	25	21	17
Dec	1019	27	28	29	29	28	26	24	21	18	14

Table I.1 (*Continued*)

COEUR D'ALENE, ID		Solar radiation (kBtu/ft^2 mo) on a surface of tilt:									
Mo	**HDD65**	**90°**	**80°**	**70°**	**60°**	**50°**	**40°**	**30°**	**20°**	**10°**	**0°**
Jan	1093	26	27	27	27	26	24	22	19	15	12
Feb	930	34	36	37	37	36	35	32	28	24	20
Mar	818	43	47	50	53	53	52	50	47	43	37
Apr	581	35	41	46	50	52	54	54	53	51	47
May	335	29	36	41	46	50	53	55	56	55	53
Jun	171	29	36	43	49	54	58	60	62	62	61
Jul	32	35	43	51	58	64	68	71	72	71	69
Aug	36	41	48	55	60	64	67	67	66	64	60
Sep	193	34	38	41	43	45	45	44	42	39	35
Oct	570	36	39	41	42	42	41	39	35	31	26
Nov	856	17	18	19	19	19	18	16	15	13	11
Dec	1119	16	17	17	17	16	15	14	12	10	8
CHICAGO, IL		**Solar radiation (kBtu/ft^2 mo) on a surface of tilt:**									
Mo	**HDD65**	**90°**	**80°**	**70°**	**60°**	**50°**	**40°**	**30°**	**20°**	**10°**	**0°**
Jan	1224	28	30	31	31	30	29	27	24	21	17
Feb	1002	28	31	32	33	33	32	31	28	25	22
Mar	799	31	35	38	40	41	42	41	39	37	33
Apr	444	27	32	36	40	43	45	45	45	44	41
May	188	28	35	42	48	53	56	59	60	60	59
Jun	26	25	32	39	45	50	55	58	60	60	59
Jul	1	28	35	42	48	53	57	60	61	62	60
Aug	3	29	35	40	45	49	52	53	53	53	50
Sep	58	32	37	41	44	46	47	47	45	43	39
Oct	329	34	37	39	41	41	40	39	36	33	29
Nov	689	24	26	27	27	27	26	25	23	20	17
Dec	1051	26	27	28	28	27	26	24	21	18	15
INDIANAPOLIS, IN		**Solar radiation (kBtu/ft^2 mo) on a surface of tilt:**									
Mo	**HDD65**	**90°**	**80°**	**70°**	**60°**	**50°**	**40°**	**30°**	**20°**	**10°**	**0°**
Jan	1140	27	29	30	31	31	29	28	25	22	19
Feb	910	30	32	34	35	35	35	34	31	28	25
Mar	703	32	36	40	43	45	46	45	44	42	39
Apr	348	29	35	40	44	47	49	50	50	49	47
May	128	27	34	40	46	51	54	57	59	59	57
Jun	12	24	31	38	44	49	54	57	60	60	60
Jul	1	26	33	41	47	53	57	61	62	63	62
Aug	2	29	36	41	46	51	53	55	56	55	53
Sep	50	34	40	44	48	50	51	51	50	47	44
Oct	306	36	40	43	45	45	45	43	41	37	33
Nov	644	28	30	31	32	31	30	29	26	23	20
Dec	980	25	26	27	27	27	26	24	22	19	16
WICHITA, KS		**Solar radiation (kBtu/ft^2 mo) on a surface of tilt:**									
Mo	**HDD65**	**90°**	**80°**	**70°**	**60°**	**50°**	**40°**	**30°**	**20°**	**10°**	**0°**
Jan	980	39	42	43	44	43	41	38	34	29	23
Feb	771	38	41	43	44	44	43	41	37	33	28
Mar	551	37	42	46	50	52	52	52	50	47	42
Apr	273	31	37	43	48	51	54	55	55	54	51
May	75	26	33	40	46	51	56	59	60	61	60
Jun	3	23	31	38	45	51	57	61	63	65	64
Jul	0	26	34	43	50	57	62	66	69	69	69
Aug	0	30	37	44	50	55	58	60	61	60	58
Sep	25	37	43	49	53	55	57	57	55	52	48
Oct	221	42	47	50	52	53	53	51	48	43	38
Nov	555	39	42	44	45	44	43	40	36	31	26
Dec	916	41	43	44	45	43	41	38	33	28	22
LOUISVILLE, KY		**Solar radiation (kBtu/ft^2 mo) on a surface of tilt:**									
Mo	**HDD65**	**90°**	**80°**	**70°**	**60°**	**50°**	**40°**	**30°**	**20°**	**10°**	**0°**
Jan	931	29	31	33	33	33	31	30	27	24	20
Feb	726	30	33	35	36	36	35	34	32	29	25
Mar	538	32	37	40	43	45	46	45	44	41	38
Apr	224	29	35	40	45	48	50	51	51	50	48
May	63	25	31	38	43	48	52	55	57	57	56
Jun	3	23	31	38	45	51	56	59	62	63	63

Table I.1 (*Continued*)

Jul	0	25	33	40	46	52	57	60	62	62	62
Aug	0	29	35	42	47	51	54	56	57	56	54
Sep	21	33	38	42	46	48	49	49	48	46	42
Oct	203	36	39	42	44	45	45	43	41	37	33
Nov	507	30	33	34	35	35	34	32	29	26	22
Dec	794	28	29	31	31	30	29	27	25	22	18
NEW ORLEANS, LA		**Solar radiation (kBtu/ft² mo) on a surface of tilt:**									
Mo	**HDD65**	**90°**	**80°**	**70°**	**60°**	**50°**	**40°**	**30°**	**20°**	**10°**	**0°**
Jan	359	36	39	41	42	42	41	39	36	32	28
Feb	234	22	24	26	28	29	29	29	28	26	24
Mar	123	27	32	36	40	42	44	45	44	42	40
Apr	26	20	24	27	31	34	36	37	38	38	37
May	1	19	25	31	37	42	46	49	52	53	53
Jun	0	17	21	26	30	34	38	40	43	44	44
Jul	0	18	25	31	38	43	48	51	54	55	55
Aug	0	22	28	33	38	43	46	49	50	51	50
Sep	0	28	33	38	42	45	47	48	48	47	45
Oct	19	31	35	38	40	42	42	42	40	38	34
Nov	136	31	34	36	38	38	37	36	34	31	27
Dec	295	25	27	29	30	30	29	28	26	24	21
SHREVEPORT, LA		**Solar radiation (kBtu/ft² mo) on a surface of tilt:**									
Mo	**HDD65**	**90°**	**80°**	**70°**	**60°**	**50°**	**40°**	**30°**	**20°**	**10°**	**0°**
Jan	530	34	37	39	40	40	39	36	33	29	25
Feb	381	31	35	37	39	40	40	39	36	34	30
Mar	229	32	37	41	45	47	49	49	48	46	42
Apr	74	26	33	38	43	47	50	52	53	52	50
May	8	22	29	36	42	47	52	55	57	58	58
Jun	0	19	26	33	40	47	52	56	59	61	61
Jul	0	21	28	36	43	49	54	58	61	63	63
Aug	0	26	34	41	47	53	57	60	61	61	59
Sep	2	32	38	43	47	50	53	53	53	51	48
Oct	65	40	46	50	53	54	54	53	50	46	41
Nov	265	35	38	41	42	42	41	39	37	33	28
Dec	489	37	40	42	42	42	41	38	35	30	25
BOSTON, MA		**Solar radiation (kBtu/ft² mo) on a surface of tilt:**									
Mo	**HDD65**	**90°**	**80°**	**70°**	**60°**	**50°**	**40°**	**30°**	**20°**	**10°**	**0°**
Jan	1088	30	32	33	33	33	31	29	25	22	17
Feb	925	33	36	37	38	38	37	35	32	29	24
Mar	824	33	37	40	42	44	44	43	42	39	35
Apr	490	28	33	37	41	44	46	46	46	45	43
May	238	27	33	39	45	49	53	55	56	56	55
Jun	53	24	30	36	42	47	51	53	55	55	55
Jul	3	28	35	41	47	52	56	59	61	61	59
Aug	4	31	37	43	48	52	55	56	56	55	52
Sep	62	34	39	43	46	48	49	48	47	44	40
Oct	314	35	38	40	42	42	41	40	37	33	29
Nov	602	29	31	33	33	32	31	29	26	23	19
Dec	895	29	31	31	31	30	29	26	23	19	15
BALTIMORE, MD		**Solar radiation (kBtu/ft² mo) on a surface of tilt:**									
Mo	**HDD65**	**90°**	**80°**	**70°**	**60°**	**50°**	**40°**	**30°**	**20°**	**10°**	**0°**
Jan	961	32	35	36	36	35	34	31	28	24	20
Feb	797	30	33	35	36	36	35	34	31	28	24
Mar	642	35	40	43	46	47	48	47	45	42	38
Apr	311	30	36	41	46	49	51	52	52	51	48
May	111	25	32	38	43	48	52	55	56	56	55
Jun	9	24	31	38	45	50	55	59	61	62	61
Jul	0	24	31	38	44	49	53	57	59	59	59
Aug	0	28	34	40	45	49	52	53	54	53	52
Sep	31	30	35	39	43	45	46	47	46	44	41
Oct	255	38	42	45	47	48	47	45	43	39	34
Nov	544	32	34	35	36	36	34	32	29	25	21
Dec	814	30	32	33	33	32	31	29	26	22	18

Table I.1 (*Continued*)

BANGOR, ME		Solar radiation (kBtu/ft^2 mo) on a surface of tilt:									
Mo	**HDD65**	**90°**	**80°**	**70°**	**60°**	**50°**	**40°**	**30°**	**20°**	**10°**	**0°**
Jan	1438	26	28	28	28	27	26	24	21	18	14
Feb	1223	31	33	35	35	35	33	31	28	24	20
Mar	1067	25	28	30	31	32	32	32	30	28	26
Apr	665	27	31	35	38	40	42	42	42	40	38
May	336	24	29	34	38	42	45	47	47	47	46
Jun	102	26	32	38	44	49	53	56	57	57	56
Jul	15	26	33	39	44	48	52	54	55	55	53
Aug	24	29	35	40	44	48	50	51	51	50	48
Sep	173	26	30	33	35	36	37	37	36	34	31
Oct	512	35	38	40	41	41	40	38	35	31	26
Nov	836	29	30	31	31	30	29	26	23	20	16
Dec	1210	25	26	26	26	25	24	21	19	15	12

PORTLAND, ME		Solar radiation (kBtu/ft^2 mo) on a surface of tilt:									
Mo	**HDD65**	**90°**	**80°**	**70°**	**60°**	**50°**	**40°**	**30°**	**20°**	**10°**	**0°**
Jan	1287	34	36	37	37	36	34	31	28	23	18
Feb	1095	36	38	40	41	41	39	37	34	30	25
Mar	971	37	42	45	48	49	50	49	46	43	39
Apr	619	31	36	41	44	47	49	49	48	47	44
May	342	28	34	40	45	50	53	55	56	56	54
Jun	102	24	30	35	41	45	49	51	53	53	53
Jul	12	28	36	42	48	54	57	60	62	61	60
Aug	19	32	38	44	49	53	55	56	56	55	52
Sep	145	36	41	45	48	50	51	50	48	45	41
Oct	461	35	38	41	42	42	41	39	36	33	28
Nov	759	28	29	30	30	30	29	27	24	21	17
Dec	1079	30	31	32	32	31	29	26	23	19	15

DETROIT, MI		Solar radiation (kBtu/ft^2 mo) on a surface of tilt:									
Mo	**HDD65**	**90°**	**80°**	**70°**	**60°**	**50°**	**40°**	**30°**	**20°**	**10°**	**0°**
Jan	1203	25	26	27	27	27	26	24	22	19	16
Feb	1023	16	18	19	19	20	20	19	18	17	15
Mar	871	31	35	38	40	42	42	41	39	37	33
Apr	494	30	35	40	44	47	49	50	49	48	45
May	205	28	34	41	46	51	54	57	58	58	56
Jun	27	25	32	38	44	49	53	56	58	58	58
Jul	2	29	36	43	50	55	60	63	64	64	63
Aug	5	29	34	40	44	48	50	52	52	51	49
Sep	76	35	40	45	48	50	51	50	49	46	42
Oct	359	28	31	33	34	35	34	33	31	29	26
Nov	697	17	19	20	20	20	19	19	17	15	13
Dec	1026	21	22	23	23	22	21	20	18	15	12

SAULT STE MARIE, MI		Solar radiation (kBtu/ft^2 mo) on a surface of tilt:									
Mo	**HDD65**	**90°**	**80°**	**70°**	**60°**	**50°**	**40°**	**30°**	**20°**	**10°**	**0°**
Jan	1532	25	26	27	27	27	26	24	21	19	15
Feb	1327	35	38	39	39	39	37	35	32	28	23
Mar	1194	39	44	47	50	51	51	50	47	44	39
Apr	784	33	38	43	46	49	50	50	49	47	44
May	421	30	37	43	48	52	56	58	58	58	56
Jun	155	27	34	41	47	51	55	58	59	60	58
Jul	61	29	36	43	49	54	57	60	61	61	59
Aug	63	30	36	41	45	48	50	51	51	50	47
Sep	217	29	32	36	38	39	40	40	38	36	33
Oct	579	24	26	28	29	29	29	28	26	24	21
Nov	914	17	18	19	19	19	19	18	16	15	13
Dec	1285	20	21	21	21	21	20	19	17	15	13

DULUTH, MN		Solar radiation (kBtu/ft^2 mo) on a surface of tilt:									
Mo	**HDD65**	**90°**	**80°**	**70°**	**60°**	**50°**	**40°**	**30°**	**20°**	**10°**	**0°**
Jan	1679	30	32	33	33	32	30	27	24	20	16
Feb	1404	33	35	36	37	36	35	33	30	26	22
Mar	1188	36	41	44	46	47	47	46	44	40	36
Apr	774	34	40	45	49	51	53	53	52	50	46
May	429	31	38	44	50	54	57	59	60	59	57

Table I.1 (*Continued*)

Jun	157	28	34	40	46	51	54	57	58	59	58
Jul	50	29	36	42	47	52	56	58	59	58	57
Aug	66	32	38	43	48	51	53	54	54	52	49
Sep	247	31	35	38	41	42	43	42	41	39	35
Oct	639	26	29	30	31	32	31	30	28	25	22
Nov	1048	24	25	26	26	26	24	23	20	18	14
Dec	1492	25	26	26	26	25	23	21	18	15	12

MINNEAPOLIS, MN		**Solar radiation (kBtu/ft² mo) on a surface of tilt:**									
Mo	**HDD65**	**90°**	**80°**	**70°**	**60°**	**50°**	**40°**	**30°**	**20°**	**10°**	**0°**
Jan	1523	34	36	36	36	35	33	30	26	22	17
Feb	1254	34	37	38	39	38	37	34	31	27	22
Mar	990	36	41	44	46	48	48	46	44	41	37
Apr	541	31	37	42	45	48	50	50	50	48	45
May	219	30	37	44	49	54	58	60	60	60	58
Jun	33	28	35	41	47	53	57	60	61	62	61
Jul	4	28	35	42	48	53	56	59	60	60	59
Aug	8	31	37	43	47	51	53	54	54	53	50
Sep	116	33	38	42	45	46	47	46	45	42	38
Oct	476	35	38	40	41	41	40	39	36	32	28
Nov	893	27	28	29	29	28	27	25	22	19	15
Dec	1339	25	26	27	27	26	24	22	19	16	12

KANSAS CITY, MO		**Solar radiation (kBtu/ft² mo) on a surface of tilt:**									
Mo	**HDD65**	**90°**	**80°**	**70°**	**60°**	**50°**	**40°**	**30°**	**20°**	**10°**	**0°**
Jan	1108	38	41	42	42	41	39	36	33	28	22
Feb	876	32	34	36	37	37	36	34	32	28	24
Mar	635	36	41	46	49	51	51	51	49	46	41
Apr	316	31	37	43	47	51	53	54	53	52	49
May	102	27	34	41	47	52	56	59	60	61	59
Jun	6	24	31	38	45	51	56	59	62	63	62
Jul	0	26	33	41	47	53	58	61	63	64	63
Aug	1	30	38	44	50	55	58	60	61	60	58
Sep	48	36	42	47	50	53	54	54	52	49	45
Oct	281	40	44	47	49	50	49	47	44	40	35
Nov	619	34	37	38	39	38	37	35	31	27	23
Dec	985	35	37	38	38	37	35	32	28	24	19

ST LOUIS, MO		**Solar radiation (kBtu/ft² mo) on a surface of tilt:**									
Mo	**HDD65**	**90°**	**80°**	**70°**	**60°**	**50°**	**40°**	**30°**	**20°**	**10°**	**0°**
Jan	1014	33	35	37	37	36	35	32	29	25	21
Feb	792	30	33	35	36	36	35	33	31	28	24
Mar	571	34	39	43	45	47	48	47	45	43	39
Apr	253	28	33	38	42	45	47	48	48	47	45
May	71	24	31	37	43	47	51	54	55	56	55
Jun	5	24	31	38	45	51	56	60	62	63	63
Jul	0	26	34	41	48	54	59	62	64	65	64
Aug	0	28	35	41	46	50	53	55	56	55	53
Sep	27	35	41	46	50	52	53	53	52	49	45
Oct	222	40	44	47	49	50	49	47	44	40	35
Nov	545	30	33	34	35	35	34	32	29	26	22
Dec	879	32	34	35	35	34	33	30	27	23	19

JACKSON, MS		**Solar radiation (kBtu/ft² mo) on a surface of tilt:**									
Mo	**HDD65**	**90°**	**80°**	**70°**	**60°**	**50°**	**40°**	**30°**	**20°**	**10°**	**0°**
Jan	565	34	37	39	40	40	39	37	34	30	26
Feb	405	31	35	37	39	40	40	38	36	33	30
Mar	255	31	36	41	44	47	48	48	47	45	42
Apr	89	28	36	42	47	52	55	57	57	56	54
May	12	21	28	34	40	45	50	53	55	56	56
Jun	0	19	26	33	40	46	51	55	58	59	60
Jul	0	21	29	36	43	49	54	58	60	61	61
Aug	0	26	33	40	47	52	56	59	60	60	59
Sep	3	31	37	42	47	50	52	52	52	50	47
Oct	83	40	45	49	52	54	54	53	50	47	42
Nov	299	36	39	41	42	43	42	40	37	33	28
Dec	499	37	40	42	43	42	41	39	35	31	26

Table I.1 (*Continued*)

BILLINGS, MT		Solar radiation (kBtu/ft² mo) on a surface of tilt:									
Mo	**HDD65**	**90°**	**80°**	**70°**	**60°**	**50°**	**40°**	**30°**	**20°**	**10°**	**0°**
Jan	1173	31	32	33	33	32	30	27	24	20	16
Feb	1010	34	36	38	38	38	36	34	31	27	23
Mar	845	39	44	47	49	50	50	48	46	42	37
Apr	575	33	39	43	47	50	51	52	51	48	45
May	321	31	38	45	50	55	58	61	61	61	59
Jun	91	29	36	43	49	55	59	62	64	64	63
Jul	6	33	42	49	56	62	67	70	71	70	68
Aug	18	37	44	51	56	61	63	65	64	62	59
Sep	164	43	48	53	56	57	57	56	53	49	44
Oct	523	37	41	43	44	44	43	41	37	33	28
Nov	852	30	32	33	33	32	30	28	25	21	17
Dec	1168	31	33	33	33	31	29	26	23	19	14

MISSOULA, MT		Solar radiation (kBtu/ft² mo) on a surface of tilt:									
Mo	**HDD65**	**90°**	**80°**	**70°**	**60°**	**50°**	**40°**	**30°**	**20°**	**10°**	**0°**
Jan	1220	21	23	23	23	23	22	20	18	16	13
Feb	1006	28	29	31	31	31	29	28	25	22	18
Mar	845	31	34	37	38	39	39	38	36	34	30
Apr	610	32	38	42	46	48	50	50	49	47	44
May	367	29	36	42	47	51	54	56	57	56	55
Jun	172	29	36	42	48	53	57	60	62	62	61
Jul	25	32	40	48	54	59	63	66	67	67	65
Aug	37	35	42	48	53	57	60	61	61	59	56
Sep	224	39	44	48	51	53	53	52	49	46	41
Oct	631	38	41	43	44	44	43	40	37	32	27
Nov	956	22	23	24	24	24	23	21	19	17	14
Dec	1238	18	19	20	20	19	18	17	15	13	11

ASHEVILLE, NC		Solar radiation (kBtu/ft² mo) on a surface of tilt:									
Mo	**HDD65**	**90°**	**80°**	**70°**	**60°**	**50°**	**40°**	**30°**	**20°**	**10°**	**0°**
Jan	852	37	40	41	42	42	40	37	34	29	24
Feb	667	32	35	37	39	39	38	37	34	31	27
Mar	538	34	39	43	47	49	49	49	48	45	41
Apr	267	30	36	42	47	51	54	55	55	54	51
May	92	23	30	36	42	47	51	54	56	57	56
Jun	8	21	28	34	41	46	51	55	57	59	58
Jul	0	22	29	36	43	48	53	56	59	60	59
Aug	1	26	33	39	44	48	52	54	55	54	53
Sep	37	29	34	39	42	45	46	46	46	44	41
Oct	256	40	44	48	50	52	51	50	47	43	38
Nov	534	36	39	41	41	41	40	38	35	30	26
Dec	748	39	41	43	43	42	40	37	33	29	23

CHARLOTTE, NC		Solar radiation (kBtu/ft² mo) on a surface of tilt:									
Mo	**HDD65**	**90°**	**80°**	**70°**	**60°**	**50°**	**40°**	**30°**	**20°**	**10°**	**0°**
Jan	714	36	39	40	41	41	39	36	33	29	24
Feb	541	31	34	37	38	38	38	36	34	30	26
Mar	392	35	41	45	49	51	52	51	50	47	43
Apr	156	31	37	43	48	52	55	56	56	55	52
May	36	23	30	37	43	49	53	57	59	59	59
Jun	2	21	29	36	43	49	54	58	60	62	61
Jul	0	23	30	37	44	50	55	58	61	62	61
Aug	0	27	34	40	45	50	53	55	56	56	54
Sep	11	32	37	42	45	48	49	50	49	47	44
Oct	148	41	46	50	52	54	53	52	49	45	40
Nov	412	36	40	42	43	42	41	39	36	32	27
Dec	618	37	40	41	42	41	39	37	33	28	23

OMAHA, NE		Solar radiation (kBtu/ft² mo) on a surface of tilt:									
Mo	**HDD65**	**90°**	**80°**	**70°**	**60°**	**50°**	**40°**	**30°**	**20°**	**10°**	**0°**
Jan	1216	37	39	40	40	39	38	35	31	27	22
Feb	1027	31	33	35	36	36	35	34	31	28	24
Mar	708	35	39	43	46	47	48	47	45	42	38
Apr	380	30	35	40	44	47	49	49	49	47	44
May	134	28	35	42	48	53	57	60	61	61	59

Table I.1 (*Continued*)

Jun	6	26	33	40	47	52	57	61	63	63	62
Jul	1	28	35	42	49	54	58	62	63	63	62
Aug	4	31	38	44	49	53	56	58	58	57	55
Sep	57	37	42	47	50	53	54	53	51	48	44
Oct	358	42	46	49	50	51	50	47	44	39	34
Nov	706	35	38	39	39	39	37	34	31	26	21
Dec	1104	33	35	36	36	35	33	30	26	22	17
CONCORD, NH		**Solar radiation (kBtu/ft^2 mo) on a surface of tilt:**									
Mo	**HDD65**	**90°**	**80°**	**70°**	**60°**	**50°**	**40°**	**30°**	**20°**	**10°**	**0°**
Jan	1334	28	30	31	31	30	29	26	23	20	16
Feb	1125	32	34	36	37	37	36	34	31	28	24
Mar	986	33	37	41	43	44	44	44	42	39	35
Apr	592	31	36	41	45	48	50	50	50	48	45
May	288	26	32	38	43	47	50	52	53	53	52
Jun	77	25	31	37	43	48	52	55	56	57	56
Jul	11	28	35	41	47	52	56	59	60	60	59
Aug	23	30	36	42	47	50	53	54	55	54	51
Sep	154	33	38	41	44	46	47	46	45	42	39
Oct	488	33	36	38	39	39	39	37	35	31	27
Nov	795	27	28	29	30	29	28	26	23	20	17
Dec	1137	27	29	29	29	28	27	25	22	19	15
ALBUQUERQUE, NM		**Solar radiation (kBtu/ft^2 mo) on a surface of tilt:**									
Mo	**HDD65**	**90°**	**80°**	**70°**	**60°**	**50°**	**40°**	**30°**	**20°**	**10°**	**0°**
Jan	836	49	53	55	56	55	53	49	44	38	31
Feb	629	46	50	54	55	56	54	52	48	43	36
Mar	464	41	47	53	57	59	60	60	58	55	50
Apr	249	36	44	52	58	63	67	69	69	67	64
May	65	26	35	44	52	58	64	68	70	71	70
Jun	1	21	30	38	47	54	61	66	69	71	72
Jul	0	25	33	42	51	58	64	68	71	73	72
Aug	0	32	41	49	56	62	67	70	71	70	68
Sep	16	40	47	53	58	61	63	63	62	59	55
Oct	213	49	55	59	62	63	63	60	57	51	45
Nov	552	50	54	56	57	56	54	51	46	40	32
Dec	848	53	56	58	58	57	54	49	44	37	29
LAS CRUCES, NM		**Solar radiation (kBtu/ft^2 mo) on a surface of tilt:**									
Mo	**HDD65**	**90°**	**80°**	**70°**	**60°**	**50°**	**40°**	**30°**	**20°**	**10°**	**0°**
Jan	644	38	41	43	44	44	43	40	37	33	28
Feb	468	45	50	54	56	56	56	53	49	44	38
Mar	297	43	50	57	61	64	66	65	63	60	54
Apr	113	32	41	49	55	60	64	67	67	66	63
May	23	25	35	45	54	63	69	75	78	80	79
Jun	0	20	29	39	48	56	64	70	74	76	77
Jul	0	21	30	39	47	55	61	66	70	72	72
Aug	0	28	37	45	52	58	63	66	67	67	65
Sep	2	39	46	53	59	63	65	66	65	62	58
Oct	100	46	52	57	60	61	61	59	56	51	45
Nov	390	57	61	65	66	66	64	60	54	48	39
Dec	655	56	60	63	63	62	59	55	49	42	34
LAS VEGAS, NV		**Solar radiation (kBtu/ft^2 mo) on a surface of tilt:**									
Mo	**HDD65**	**90°**	**80°**	**70°**	**60°**	**50°**	**40°**	**30°**	**20°**	**10°**	**0°**
Jan	479	48	51	53	53	52	50	46	41	35	29
Feb	333	42	47	49	51	51	50	48	44	40	34
Mar	162	41	48	53	57	59	60	60	58	55	50
Apr	63	37	46	54	60	65	69	70	70	69	65
May	8	28	37	46	55	62	68	72	75	76	75
Jun	1	23	33	42	51	59	65	71	74	76	76
Jul	0	25	35	44	53	61	67	72	76	77	77
Aug	0	34	43	52	59	65	70	73	74	73	71
Sep	0	44	52	58	63	67	69	69	67	64	59
Oct	34	52	58	62	65	66	65	62	58	52	45
Nov	242	51	55	58	58	58	56	52	47	40	33
Dec	519	50	53	55	55	53	51	47	41	35	27

Table I.1 (*Continued*)

RENO, NV		Solar radiation (kBtu/ft^2 mo) on a surface of tilt:									
Mo	**HDD65**	**90°**	**80°**	**70°**	**60°**	**50°**	**40°**	**30°**	**20°**	**10°**	**0°**
Jan	877	36	39	40	41	40	38	35	32	27	22
Feb	707	39	43	45	46	46	45	42	38	34	29
Mar	589	39	44	48	51	52	53	52	49	46	42
Apr	428	34	41	47	53	56	59	60	60	58	55
May	205	30	38	46	53	60	64	68	70	70	68
Jun	48	26	35	44	53	60	66	70	73	75	74
Jul	1	29	38	47	56	63	69	73	76	77	76
Aug	3	37	46	55	62	68	72	75	75	74	71
Sep	65	44	51	57	62	64	66	65	63	59	54
Oct	338	49	54	57	59	60	59	56	52	46	40
Nov	648	44	47	49	50	49	46	43	38	33	26
Dec	908	41	43	44	44	43	40	37	32	27	21

ALBANY, NY		Solar radiation (kBtu/ft^2 mo) on a surface of tilt:									
Mo	**HDD65**	**90°**	**80°**	**70°**	**60°**	**50°**	**40°**	**30°**	**20°**	**10°**	**0°**
Jan	1276	25	26	27	27	27	26	24	22	19	16
Feb	1073	31	33	35	36	36	35	33	31	28	24
Mar	919	33	37	41	43	45	45	45	43	41	37
Apr	523	29	34	39	43	45	47	48	47	45	43
May	218	28	35	41	46	51	55	57	58	58	56
Jun	41	25	32	38	44	49	53	56	58	58	58
Jul	4	28	35	41	47	52	56	59	61	61	60
Aug	10	30	36	42	46	50	53	55	55	54	52
Sep	110	32	37	41	44	46	47	47	45	43	39
Oct	428	33	36	38	39	39	39	37	35	32	28
Nov	741	25	27	28	28	28	27	25	23	20	17
Dec	1075	24	25	26	26	25	24	23	20	18	14

BUFFALO, NY		Solar radiation (kBtu/ft^2 mo) on a surface of tilt:									
Mo	**HDD65**	**90°**	**80°**	**70°**	**60°**	**50°**	**40°**	**30°**	**20°**	**10°**	**0°**
Jan	1233	22	23	24	24	24	23	22	20	17	15
Feb	1079	23	25	27	27	28	27	26	25	23	20
Mar	960	32	36	39	41	43	43	42	40	38	34
Apr	583	29	34	38	42	44	46	47	46	45	43
May	255	28	34	41	46	51	54	56	57	57	55
Jun	53	27	34	41	47	53	57	60	62	63	62
Jul	7	28	35	41	47	52	56	59	60	61	59
Aug	11	29	35	41	46	50	52	54	54	53	51
Sep	104	31	36	39	42	44	45	44	43	41	37
Oct	407	28	31	33	34	35	34	33	31	28	25
Nov	717	19	20	21	22	22	21	20	18	17	14
Dec	1040	18	19	20	20	20	19	18	17	15	13

NEW YORK, NY		Solar radiation (kBtu/ft^2 mo) on a surface of tilt:									
Mo	**HDD65**	**90°**	**80°**	**70°**	**60°**	**50°**	**40°**	**30°**	**20°**	**10°**	**0°**
Jan	994	31	33	34	34	34	32	30	26	23	18
Feb	845	31	34	35	36	36	35	33	30	27	23
Mar	728	31	35	38	41	42	42	42	40	38	34
Apr	408	30	35	40	44	47	49	50	50	48	45
May	165	27	34	40	46	51	55	57	59	59	57
Jun	19	24	30	36	42	47	51	54	55	56	55
Jul	0	25	31	38	43	48	52	55	56	57	56
Aug	1	31	38	44	49	54	57	58	59	58	55
Sep	25	32	37	41	44	46	48	48	46	44	41
Oct	237	41	46	48	50	51	50	48	44	40	35
Nov	529	28	30	31	31	31	30	28	25	22	18
Dec	810	31	33	34	33	33	31	28	25	21	17

SYRACUSE, NY		Solar radiation (kBtu/ft^2 mo) on a surface of tilt:									
Mo	**HDD65**	**90°**	**80°**	**70°**	**60°**	**50°**	**40°**	**30°**	**20°**	**10°**	**0°**
Jan	1269	23	24	25	25	25	24	23	21	18	16
Feb	1093	27	29	31	32	32	31	30	28	25	22
Mar	961	33	37	40	42	43	43	42	41	38	34
Apr	558	29	34	38	42	45	47	47	47	46	43
May	238	29	36	43	48	53	57	59	60	60	58
Jun	52	25	31	38	43	48	52	55	57	58	57

Table I.1 (*Continued*)

Jul	5	27	34	41	47	52	56	58	60	60	59
Aug	12	31	37	42	47	51	53	54	54	53	51
Sep	113	31	36	39	42	44	44	44	43	40	37
Oct	413	29	31	33	35	35	34	33	31	28	25
Nov	728	19	21	22	22	22	22	21	19	17	15
Dec	1062	20	21	21	22	21	21	20	18	16	14
CLEVELAND, OH		**Solar radiation (kBtu/ft² mo) on a surface of tilt:**									
Mo	**HDD65**	**90°**	**80°**	**70°**	**60°**	**50°**	**40°**	**30°**	**20°**	**10°**	**0°**
Jan	1157	19	20	21	21	21	21	20	19	17	15
Feb	976	23	25	26	27	28	28	27	25	24	21
Mar	842	30	34	37	39	41	41	41	39	37	34
Apr	468	29	34	39	42	45	47	48	47	46	43
May	201	28	35	41	47	52	55	58	59	59	58
Jun	33	24	31	37	43	49	53	56	58	58	58
Jul	4	27	35	41	48	53	57	60	62	62	60
Aug	6	29	35	40	45	48	51	52	53	52	50
Sep	75	31	35	39	42	44	45	45	44	41	38
Oct	341	30	34	36	37	38	37	36	34	31	27
Nov	655	22	24	25	25	25	24	23	21	19	16
Dec	979	16	17	18	18	18	18	17	16	15	13
COLUMBUS, OH		**Solar radiation (kBtu/ft² mo) on a surface of tilt:**									
Mo	**HDD65**	**90°**	**80°**	**70°**	**60°**	**50°**	**40°**	**30°**	**20°**	**10°**	**0°**
Jan	1104	27	29	30	30	30	28	27	24	21	17
Feb	906	23	25	27	28	28	28	27	26	24	21
Mar	722	31	36	39	41	43	43	43	41	39	36
Apr	353	28	33	38	42	44	46	47	47	46	44
May	132	25	31	37	42	46	50	52	54	54	53
Jun	13	24	30	36	42	47	52	55	57	58	57
Jul	0	25	32	38	44	49	53	56	57	58	57
Aug	2	29	35	41	46	49	52	54	54	53	51
Sep	50	32	37	41	44	46	47	47	46	44	40
Oct	309	35	39	41	43	43	43	41	38	35	30
Nov	633	24	25	26	27	27	26	25	23	21	18
Dec	937	21	22	23	23	23	22	21	19	17	15
OKLAHOMA CITY, OK		**Solar radiation (kBtu/ft² mo) on a surface of tilt:**									
Mo	**HDD65**	**90°**	**80°**	**70°**	**60°**	**50°**	**40°**	**30°**	**20°**	**10°**	**0°**
Jan	799	42	45	47	48	47	45	42	38	33	27
Feb	612	33	37	39	41	42	41	40	38	34	30
Mar	415	36	42	46	50	52	53	53	52	49	45
Apr	181	29	36	42	47	51	54	55	55	54	52
May	42	24	31	38	44	49	54	57	59	60	59
Jun	1	21	29	36	44	50	56	60	63	64	64
Jul	0	24	33	41	49	56	62	66	69	70	70
Aug	0	29	37	44	51	56	60	63	64	63	61
Sep	13	37	43	49	53	56	58	58	57	54	50
Oct	152	43	48	51	54	55	55	53	50	45	40
Nov	434	39	42	44	45	45	43	41	37	33	27
Dec	749	41	44	45	46	45	43	40	35	30	24
EUGENE, OR		**Solar radiation (kBtu/ft² mo) on a surface of tilt:**									
Mo	**HDD65**	**90°**	**80°**	**70°**	**60°**	**50°**	**40°**	**30°**	**20°**	**10°**	**0°**
Jan	730	19	20	21	21	21	20	19	17	15	12
Feb	611	22	23	25	25	25	25	24	22	20	17
Mar	564	28	31	33	35	36	36	36	34	32	29
Apr	440	27	32	36	39	42	43	44	44	42	40
May	289	27	33	38	43	47	50	52	53	53	52
Jun	146	27	34	41	47	53	57	61	63	63	62
Jul	23	31	39	47	54	60	65	68	69	69	67
Aug	22	36	43	50	56	60	63	65	64	63	59
Sep	101	36	41	45	48	49	50	49	47	43	39
Oct	366	32	35	37	38	39	38	37	34	31	27
Nov	581	17	18	19	19	19	18	17	16	14	12
Dec	756	15	16	16	16	16	15	14	13	11	9

Table I.1 (*Continued*)

PORTLAND, OR		Solar radiation (kBtu/ft² mo) on a surface of tilt:									
Mo	**HDD65**	**90°**	**80°**	**70°**	**60°**	**50°**	**40°**	**30°**	**20°**	**10°**	**0°**
Jan	724	19	20	21	21	20	20	18	16	14	12
Feb	597	22	24	25	26	26	25	24	22	20	17
Mar	528	29	33	35	37	38	38	37	36	33	30
Apr	379	27	31	35	38	40	41	42	41	40	37
May	206	28	34	40	45	49	52	54	54	54	52
Jun	88	27	33	40	45	50	53	56	58	58	57
Jul	12	31	38	45	51	56	59	62	63	63	61
Aug	7	34	41	47	52	56	58	59	59	57	54
Sep	64	34	39	42	45	47	47	47	45	42	38
Oct	302	28	31	32	34	34	33	32	30	27	24
Nov	536	17	18	19	19	19	19	18	16	15	13
Dec	736	17	18	19	19	18	17	16	15	13	10

REDMOND, OR		Solar radiation (kBtu/ft² mo) on a surface of tilt:									
Mo	**HDD65**	**90°**	**80°**	**70°**	**60°**	**50°**	**40°**	**30°**	**20°**	**10°**	**0°**
Jan	960	28	30	31	31	30	28	26	23	20	16
Feb	826	29	31	33	33	33	32	30	28	24	21
Mar	764	36	40	44	46	47	47	46	44	41	37
Apr	616	35	41	46	50	54	55	56	55	53	50
May	398	32	40	47	53	58	61	64	65	64	62
Jun	206	30	39	47	54	60	65	69	71	71	70
Jul	48	33	42	50	58	64	69	73	74	74	72
Aug	59	39	47	54	61	65	68	70	69	67	64
Sep	225	43	49	54	57	59	60	59	56	52	47
Oct	548	43	47	50	51	51	50	47	44	39	33
Nov	788	29	30	31	31	31	29	27	24	21	17
Dec	1018	28	30	30	30	29	27	24	21	17	13

PHILADELPHIA, PA		Solar radiation (kBtu/ft² mo) on a surface of tilt:									
Mo	**HDD65**	**90°**	**80°**	**70°**	**60°**	**50°**	**40°**	**30°**	**20°**	**10°**	**0°**
Jan	968	33	35	36	37	36	34	32	28	24	20
Feb	806	31	34	36	37	37	36	34	31	28	24
Mar	653	34	38	42	44	46	46	46	44	41	37
Apr	313	29	34	39	43	46	48	49	49	47	45
May	100	26	33	39	45	49	53	56	57	57	55
Jun	8	24	31	38	44	49	54	57	59	60	59
Jul	0	25	31	37	43	48	52	55	56	57	56
Aug	0	31	38	44	49	53	56	58	58	57	55
Sep	21	32	38	42	45	48	49	49	47	45	41
Oct	222	37	40	43	45	46	45	43	40	37	32
Nov	517	31	34	35	36	35	34	32	29	25	21
Dec	802	31	32	33	33	33	31	28	25	21	17

PITTSBURGH, PA		Solar radiation (kBtu/ft² mo) on a surface of tilt:									
Mo	**HDD65**	**90°**	**80°**	**70°**	**60°**	**50°**	**40°**	**30°**	**20°**	**10°**	**0°**
Jan	1128	17	18	19	19	19	19	18	16	15	13
Feb	939	43	47	49	50	50	48	45	41	36	30
Mar	781	33	38	41	44	46	46	45	44	41	37
Apr	401	23	27	31	34	36	38	38	38	38	36
May	169	21	25	30	34	37	40	42	43	44	43
Jun	27	21	26	31	36	41	44	47	48	49	49
Jul	3	22	28	33	38	43	46	49	50	51	50
Aug	6	23	28	32	36	39	42	43	44	43	42
Sep	77	23	26	29	32	33	34	34	34	32	31
Oct	360	37	41	44	46	46	46	44	41	37	32
Nov	666	23	24	26	26	26	25	24	22	20	17
Dec	961	10	11	12	12	12	12	12	11	10	9

RAPID CITY, SD		Solar radiation (kBtu/ft² mo) on a surface of tilt:									
Mo	**HDD65**	**90°**	**80°**	**70°**	**60°**	**50°**	**40°**	**30°**	**20°**	**10°**	**0°**
Jan	1243	38	40	41	40	39	37	33	29	24	19
Feb	1082	39	41	43	44	43	41	39	35	30	25
Mar	915	40	45	49	51	52	52	51	48	44	40
Apr	619	33	39	44	48	51	53	53	53	51	47
May	342	31	38	45	51	55	59	61	62	62	60

Table I.1 (*Continued*)

Jun	87	28	35	43	49	55	60	63	65	65	64
Jul	8	31	39	47	54	60	64	67	69	69	67
Aug	17	36	44	51	56	61	63	65	64	63	59
Sep	160	39	45	50	53	55	56	55	53	49	44
Oct	534	44	47	50	51	51	50	47	43	38	32
Nov	888	35	37	39	39	38	36	33	30	26	21
Dec	1218	35	37	37	37	36	33	30	26	21	16
SIOUX FALLS, SD		**Solar radiation (kBtu/ft² mo) on a surface of tilt:**									
Mo	**HDD65**	**90°**	**80°**	**70°**	**60°**	**50°**	**40°**	**30°**	**20°**	**10°**	**0°**
Jan	1491	34	36	37	37	36	34	31	27	23	18
Feb	1226	37	39	41	42	41	40	37	34	30	25
Mar	972	34	38	41	43	45	45	44	42	38	35
Apr	560	33	39	44	48	51	52	53	52	50	47
May	247	28	35	41	47	52	55	57	58	58	56
Jun	39	26	34	41	47	52	57	60	62	62	61
Jul	7	30	38	45	52	58	62	65	66	66	65
Aug	13	32	39	45	50	54	57	58	58	57	54
Sep	132	36	41	45	48	50	51	50	49	45	41
Oct	495	38	42	45	46	46	45	43	40	36	31
Nov	914	30	32	33	33	33	31	29	26	22	18
Dec	1346	27	28	29	29	28	27	24	22	18	15
KNOXVILLE, TN		**Solar radiation (kBtu/ft² mo) on a surface of tilt:**									
Mo	**HDD65**	**90°**	**80°**	**70°**	**60°**	**50°**	**40°**	**30°**	**20°**	**10°**	**0°**
Jan	812	31	33	35	35	35	34	32	30	26	22
Feb	614	29	32	34	35	36	36	35	33	30	27
Mar	447	33	38	42	45	47	48	47	46	44	40
Apr	189	29	36	42	46	50	53	54	54	53	51
May	49	23	29	36	42	47	51	54	56	56	56
Jun	2	21	28	35	41	47	52	55	58	59	59
Jul	0	23	30	37	44	49	54	57	60	60	60
Aug	0	27	34	40	46	51	54	56	57	57	55
Sep	13	31	37	41	45	48	49	50	49	47	44
Oct	180	36	40	43	45	46	46	44	42	38	34
Nov	474	34	36	38	39	39	38	36	33	29	25
Dec	706	31	34	35	35	35	33	31	28	24	20
MEMPHIS, TN		**Solar radiation (kBtu/ft² mo) on a surface of tilt:**									
Mo	**HDD65**	**90°**	**80°**	**70°**	**60°**	**50°**	**40°**	**30°**	**20°**	**10°**	**0°**
Jan	714	33	35	37	38	37	36	34	31	27	22
Feb	533	28	31	33	34	35	34	33	31	28	25
Mar	353	33	38	42	46	48	49	49	48	45	42
Apr	123	29	35	41	46	50	52	54	54	53	50
May	21	23	30	37	43	49	53	56	58	59	59
Jun	0	22	29	36	43	50	55	59	62	63	63
Jul	0	24	32	39	46	53	58	62	64	65	65
Aug	0	29	37	44	51	56	60	62	63	63	61
Sep	5	33	38	43	47	50	52	52	51	49	46
Oct	112	39	43	47	49	50	50	49	46	43	38
Nov	370	33	36	38	39	39	38	36	33	30	25
Dec	625	34	36	37	38	37	36	33	30	26	22
NASHVILLE, TN		**Solar radiation (kBtu/ft² mo) on a surface of tilt:**									
Mo	**HDD65**	**90°**	**80°**	**70°**	**60°**	**50°**	**40°**	**30°**	**20°**	**10°**	**0°**
Jan	810	27	30	31	32	32	31	29	27	24	21
Feb	617	31	34	36	38	38	38	37	35	32	28
Mar	445	33	38	42	45	47	48	48	46	44	40
Apr	183	29	35	41	46	49	52	53	53	52	50
May	46	24	30	37	42	47	51	54	56	57	56
Jun	1	23	30	37	44	50	55	59	61	63	63
Jul	0	24	31	38	44	50	54	58	60	61	61
Aug	0	28	35	42	47	52	56	58	59	58	56
Sep	13	31	36	40	44	46	48	48	47	45	42
Oct	166	40	45	48	51	52	51	50	47	43	37
Nov	451	33	36	37	38	38	37	35	32	28	24
Dec	698	32	34	35	35	35	33	30	27	23	19

Table I.1 (*Continued*)

DALLAS, TX		Solar radiation (kBtu/ft^2 mo) on a surface of tilt:									
Mo	**HDD65**	**90°**	**80°**	**70°**	**60°**	**50°**	**40°**	**30°**	**20°**	**10°**	**0°**
Jan	559	37	40	43	44	43	42	40	37	32	27
Feb	402	34	38	41	42	43	43	41	39	36	32
Mar	240	34	39	44	48	50	51	51	50	48	44
Apr	72	29	36	43	48	52	55	57	58	57	54
May	10	22	29	36	43	48	53	57	59	60	60
Jun	0	21	28	36	43	50	56	61	64	66	66
Jul	0	22	31	39	47	54	60	64	67	69	69
Aug	0	27	35	42	49	54	59	62	63	63	62
Sep	2	35	42	47	52	55	57	57	56	54	50
Oct	58	44	50	54	57	59	59	57	54	50	45
Nov	255	35	38	40	41	42	41	39	36	33	28
Dec	515	38	41	43	43	43	41	39	35	31	26

EL PASO, TX		Solar radiation (kBtu/ft^2 mo) on a surface of tilt:									
Mo	**HDD65**	**90°**	**80°**	**70°**	**60°**	**50°**	**40°**	**30°**	**20°**	**10°**	**0°**
Jan	573	49	53	56	57	57	55	52	47	41	34
Feb	379	44	49	53	55	55	55	52	49	44	38
Mar	214	43	51	57	62	65	66	66	65	61	56
Apr	69	34	43	51	58	63	67	70	70	69	66
May	6	25	34	43	52	60	66	71	74	75	75
Jun	0	20	29	38	47	56	63	69	73	75	76
Jul	0	21	29	38	46	53	59	64	67	68	68
Aug	0	27	35	43	51	57	61	64	66	66	64
Sep	2	38	46	53	59	63	65	66	65	62	58
Oct	64	47	53	58	61	62	62	61	57	53	47
Nov	314	51	56	59	60	60	58	55	50	44	37
Dec	582	50	54	56	57	56	54	50	45	39	32

LUBBOCK, TX		Solar radiation (kBtu/ft^2 mo) on a surface of tilt:									
Mo	**HDD65**	**90°**	**80°**	**70°**	**60°**	**50°**	**40°**	**30°**	**20°**	**10°**	**0°**
Jan	730	44	47	50	50	50	48	45	41	36	30
Feb	555	39	42	45	47	48	47	45	42	38	33
Mar	384	40	46	51	55	58	59	59	57	53	49
Apr	170	32	40	47	53	58	61	63	63	61	59
May	44	24	32	40	47	53	58	62	65	66	65
Jun	1	20	28	37	44	52	58	63	66	68	68
Jul	0	21	29	37	45	51	57	62	65	66	66
Aug	0	28	37	44	51	57	61	64	65	65	63
Sep	16	34	40	46	50	53	55	55	54	52	48
Oct	158	45	50	54	57	59	58	57	53	49	43
Nov	435	44	47	50	51	50	49	46	42	37	30
Dec	724	47	50	52	52	51	49	46	41	35	28

SAN ANTONIO, TX		Solar radiation (kBtu/ft^2 mo) on a surface of tilt:									
Mo	**HDD65**	**90°**	**80°**	**70°**	**60°**	**50°**	**40°**	**30°**	**20°**	**10°**	**0°**
Jan	384	36	40	42	43	44	43	41	38	35	30
Feb	252	31	35	38	40	42	42	41	39	36	33
Mar	137	33	39	44	48	51	52	53	52	50	47
Apr	30	24	31	36	42	46	49	52	53	53	51
May	2	20	27	33	40	45	50	53	56	57	57
Jun	0	18	25	32	40	46	52	57	60	62	63
Jul	0	19	27	34	42	49	54	59	62	64	64
Aug	0	26	35	43	50	56	61	64	66	66	65
Sep	1	34	41	47	52	56	59	60	59	57	54
Oct	30	40	46	50	53	55	56	55	53	49	45
Nov	162	35	38	41	42	43	42	41	38	35	31
Dec	354	37	40	42	43	43	42	39	36	32	27

SALT LAKE CITY, UT		Solar radiation (kBtu/ft^2 mo) on a surface of tilt:									
Mo	**HDD65**	**90°**	**80°**	**70°**	**60°**	**50°**	**40°**	**30°**	**20°**	**10°**	**0°**
Jan	1051	28	30	31	31	31	30	28	25	22	19
Feb	815	33	35	37	38	38	38	36	33	30	26
Mar	626	35	40	44	47	48	49	48	46	43	39
Apr	430	34	40	46	50	54	56	56	56	54	50

Table I.1 (*Continued*)

May	195	28	36	43	49	54	59	62	63	63	62
Jun	40	27	36	45	52	59	65	69	72	73	72
Jul	0	29	38	47	54	61	66	70	72	72	71
Aug	1	35	43	50	57	62	65	67	67	66	63
Sep	64	42	49	54	58	61	62	61	59	55	50
Oct	363	44	48	51	53	54	53	50	47	42	36
Nov	707	31	33	34	35	35	33	32	29	26	22
Dec	1037	27	29	29	29	29	27	25	22	19	15
CHARLOTTESVILLE, VA		**Solar radiation (kBtu/ft² mo) on a surface of tilt:**									
Mo	**HDD65**	**90°**	**80°**	**70°**	**60°**	**50°**	**40°**	**30°**	**20°**	**10°**	**0°**
Jan	882	37	39	40	41	40	38	35	32	27	22
Feb	711	27	30	32	33	33	32	31	29	26	23
Mar	560	32	37	41	44	46	46	46	44	42	38
Apr	263	27	32	37	41	44	47	48	48	47	44
May	92	24	30	36	42	46	50	53	54	54	54
Jun	7	21	26	32	37	42	46	49	51	52	51
Jul	1	22	28	33	38	43	46	49	51	52	51
Aug	1	24	30	35	39	43	45	47	48	48	46
Sep	37	27	32	35	38	40	41	42	41	39	37
Oct	246	45	49	53	55	56	55	53	50	45	39
Nov	511	33	36	37	38	37	36	34	31	27	22
Dec	768	39	41	42	43	41	39	36	32	27	21
NORFOLK, VA		**Solar radiation (kBtu/ft² mo) on a surface of tilt:**									
Mo	**HDD65**	**90°**	**80°**	**70°**	**60°**	**50°**	**40°**	**30°**	**20°**	**10°**	**0°**
Jan	729	32	34	35	36	35	34	32	29	26	21
Feb	593	32	36	38	39	39	39	37	34	31	27
Mar	469	34	39	43	46	48	49	48	47	44	40
Apr	204	30	37	42	47	51	53	55	55	53	51
May	56	24	31	37	43	48	52	55	56	57	56
Jun	3	22	29	36	42	48	52	56	59	60	59
Jul	0	23	30	37	43	49	53	56	59	59	59
Aug	0	27	33	39	45	49	52	54	55	55	53
Sep	3	32	37	42	45	48	49	49	48	46	43
Oct	119	37	41	45	47	48	47	46	43	40	35
Nov	374	32	35	37	37	37	36	34	31	27	23
Dec	589	34	36	37	37	37	35	32	29	25	20
RICHMOND, VA		**Solar radiation (kBtu/ft² mo) on a surface of tilt:**									
Mo	**HDD65**	**90°**	**80°**	**70°**	**60°**	**50°**	**40°**	**30°**	**20°**	**10°**	**0°**
Jan	822	35	37	39	39	39	37	35	32	28	23
Feb	661	34	37	39	41	41	40	38	35	32	27
Mar	509	36	41	45	48	50	50	50	48	45	41
Apr	218	30	37	42	47	50	53	54	54	53	50
May	68	25	31	38	44	49	53	56	57	58	57
Jun	4	22	29	35	42	47	52	55	57	59	58
Jul	0	24	31	38	44	50	54	57	59	60	59
Aug	0	29	36	42	48	52	56	58	58	58	56
Sep	14	32	37	41	45	47	48	48	47	45	42
Oct	184	36	40	43	45	46	46	44	42	38	34
Nov	456	34	37	39	39	39	38	35	32	28	23
Dec	699	35	37	38	38	38	36	33	29	25	20
BURLINGTON, VT		**Solar radiation (kBtu/ft² mo) on a surface of tilt:**									
Mo	**HDD65**	**90°**	**80°**	**70°**	**60°**	**50°**	**40°**	**30°**	**20°**	**10°**	**0°**
Jan	1394	25	27	27	28	27	26	25	22	20	17
Feb	1189	28	30	32	32	32	31	30	28	25	21
Mar	1039	31	34	37	40	41	41	41	39	36	33
Apr	612	31	37	42	45	48	50	50	50	48	45
May	257	27	33	39	44	48	52	54	55	54	53
Jun	58	26	32	39	44	49	53	56	57	58	57
Jul	8	28	35	41	47	52	55	58	59	59	58
Aug	19	32	38	43	48	52	54	56	56	54	52
Sep	140	30	34	37	40	42	43	42	41	39	36
Oct	472	30	32	34	35	36	35	34	32	29	25
Nov	795	22	24	24	25	25	24	22	20	18	15
Dec	1162	20	21	21	21	21	19	18	16	14	11

Table I.1 (*Continued*)

SEATTLE, WA		Solar radiation (kBtu/ft² mo) on a surface of tilt:									
Mo	**HDD65**	**90°**	**80°**	**70°**	**60°**	**50°**	**40°**	**30°**	**20°**	**10°**	**0°**
Jan	681	16	17	17	17	17	16	15	14	12	10
Feb	590	22	24	25	25	25	24	23	21	18	16
Mar	546	27	30	33	34	35	35	34	33	30	27
Apr	402	30	35	39	43	45	46	46	46	44	41
May	233	31	37	42	47	51	54	56	56	56	54
Jun	108	28	35	41	46	51	54	57	58	58	56
Jul	23	32	39	46	52	57	60	62	63	63	61
Aug	18	32	38	43	48	51	53	54	54	52	50
Sep	94	32	36	39	42	43	44	43	41	39	35
Oct	345	28	30	32	33	33	32	31	29	26	23
Nov	550	16	17	17	17	17	17	16	15	13	11
Dec	709	16	17	18	17	17	16	15	13	11	9

EAU CLAIRE, WI		Solar radiation (kBtu/ft² mo) on a surface of tilt:									
Mo	**HDD65**	**90°**	**80°**	**70°**	**60°**	**50°**	**40°**	**30°**	**20°**	**10°**	**0°**
Jan	1557	31	32	33	33	32	30	28	25	21	17
Feb	1292	35	38	39	40	40	39	36	33	29	25
Mar	1041	36	40	43	45	46	46	45	43	40	36
Apr	593	32	38	42	46	49	50	51	50	48	45
May	264	29	35	41	46	51	54	56	57	57	55
Jun	52	27	34	40	46	51	55	58	60	60	59
Jul	11	28	35	41	46	51	55	57	58	58	57
Aug	22	31	37	42	47	51	53	54	54	53	50
Sep	160	32	37	40	43	45	45	45	43	41	37
Oct	524	31	34	36	37	37	37	35	33	30	26
Nov	927	25	26	27	27	27	25	24	21	18	15
Dec	1367	28	30	30	30	29	28	25	22	19	15

MADISON, WI		Solar radiation (kBtu/ft² mo) on a surface of tilt:									
Mo	**HDD65**	**90°**	**80°**	**70°**	**60°**	**50°**	**40°**	**30°**	**20°**	**10°**	**0°**
Jan	1414	31	32	33	34	33	32	29	26	23	19
Feb	1172	30	33	35	35	35	35	33	30	27	24
Mar	945	31	35	38	41	42	43	42	41	39	36
Apr	549	29	35	39	43	46	48	48	48	47	44
May	247	29	36	42	47	52	56	58	59	59	57
Jun	43	27	34	41	47	52	57	60	62	62	61
Jul	8	29	36	42	48	53	57	60	61	62	60
Aug	17	31	38	43	48	52	55	56	56	55	53
Sep	137	33	38	41	44	46	47	46	45	42	38
Oct	461	32	35	37	38	39	38	36	34	30	26
Nov	834	26	27	28	28	28	27	25	23	20	16
Dec	1226	26	27	28	28	27	26	24	22	19	15

MORGANTOWN, WV		Solar radiation (kBtu/ft² mo) on a surface of tilt:									
Mo	**HDD65**	**90°**	**80°**	**70°**	**60°**	**50°**	**40°**	**30°**	**20°**	**10°**	**0°**
Jan	1031	13	14	15	15	16	15	15	14	13	12
Feb	847	20	22	23	24	24	24	23	22	20	17
Mar	699	32	36	39	42	43	44	43	42	39	36
Apr	344	23	28	31	34	37	39	40	40	39	37
May	144	25	31	37	42	47	50	53	54	54	53
Jun	20	22	28	34	40	44	48	51	53	54	54
Jul	2	23	28	34	39	43	47	50	51	52	51
Aug	3	24	29	34	38	42	44	45	46	45	44
Sep	61	24	27	30	33	34	35	36	35	34	32
Oct	313	27	30	32	34	35	35	34	32	30	27
Nov	601	19	21	22	22	22	22	21	19	17	15
Dec	874	23	24	25	25	25	24	22	20	17	14

JACKSON HOLE, WY		Solar radiation (kBtu/ft² mo) on a surface of tilt:									
Mo	**HDD65**	**90°**	**80°**	**70°**	**60°**	**50°**	**40°**	**30°**	**20°**	**10°**	**0°**
Jan	1515	30	31	32	32	31	29	27	23	20	16
Feb	1338	32	34	35	36	36	34	32	29	26	21
Mar	1192	39	43	47	49	51	51	49	47	43	39
Apr	850	36	42	48	52	56	57	58	57	55	51
May	597	31	38	45	51	56	60	63	64	64	62

Table I.1 (*Continued*)

Jun	365	30	38	46	53	60	65	69	71	71	70
Jul	135	31	40	48	55	61	66	70	72	72	70
Aug	173	36	43	50	56	61	64	66	66	64	61
Sep	422	38	43	48	51	53	54	53	51	48	44
Oct	808	40	44	47	48	48	47	45	41	37	31
Nov	1142	34	36	37	37	36	35	32	28	24	19
Dec	1507	23	24	25	24	24	22	20	18	15	12
LARAMIE, WY		**Solar radiation (kBtu/ft² mo) on a surface of tilt:**									
Mo	**HDD65**	**90°**	**80°**	**70°**	**60°**	**50°**	**40°**	**30°**	**20°**	**10°**	**0°**
Jan	1317	40	43	44	44	42	40	36	32	27	21
Feb	1137	43	47	49	50	49	47	44	40	35	28
Mar	1019	52	59	64	67	69	68	66	63	57	50
Apr	805	33	40	45	49	53	55	55	55	53	50
May	546	29	37	44	50	56	60	63	64	64	63
Jun	212	27	35	43	51	57	62	66	69	70	69
Jul	45	28	36	43	50	56	60	64	66	66	65
Aug	85	32	39	46	52	56	59	61	61	60	57
Sep	311	32	37	41	44	46	47	46	45	43	39
Oct	703	41	45	48	50	50	49	47	44	39	34
Nov	1013	39	42	43	43	42	40	37	33	28	23
Dec	1337	39	41	42	42	40	38	34	30	24	19
Part B: SI Units											
ANCHORAGE, AK		**Solar radiation (kWh/m² mo) on a surface of tilt:**									
Mo	**HDD18.3**	**90°**	**80°**	**70°**	**60°**	**50°**	**40°**	**30°**	**20°**	**10°**	**0°**
Jan	814	24	24	24	23	22	20	18	15	12	9
Feb	668	74	76	77	75	71	65	58	49	39	29
Mar	668	96	104	108	110	109	106	100	91	81	69
Apr	454	107	119	128	133	136	135	132	125	115	103
May	288	101	115	128	139	146	151	152	151	146	138
Jun	152	108	126	142	156	166	173	176	175	172	164
Jul	96	100	115	129	140	148	154	156	155	151	144
Aug	128	92	104	114	122	128	130	130	128	122	114
Sep	254	79	86	91	93	94	92	88	82	75	65
Oct	486	48	51	53	53	52	51	48	43	38	33
Nov	686	40	41	41	40	37	34	30	25	19	13
Dec	783	14	14	14	13	13	12	10	9	7	5
FAIRBANKS, AK		**Solar radiation (kWh/m² mo) on a surface of tilt:**									
Mo	**HDD18.3**	**90°**	**80°**	**70°**	**60°**	**50°**	**40°**	**30°**	**20°**	**10°**	**0°**
Jan	1259	20	20	19	18	17	15	13	10	7	4
Feb	995	60	61	61	59	56	51	44	37	29	20
Mar	933	117	124	128	128	125	119	110	99	85	69
Apr	519	146	160	170	176	178	175	168	157	143	125
May	252	130	148	163	174	182	186	185	181	173	162
Jun	82	116	134	149	162	171	177	180	179	174	167
Jul	56	114	131	145	156	165	169	171	169	164	156
Aug	143	114	127	138	145	149	150	147	141	132	120
Sep	312	86	92	97	99	99	97	92	85	76	66
Oct	661	62	65	66	65	62	58	53	46	38	30
Nov	1016	32	32	31	30	28	25	21	17	12	8
Dec	1197	6	6	6	5	5	4	4	3	2	1
BIRMINGHAM, AL		**Solar radiation (kWh/m² mo) on a surface of tilt:**									
Mo	**HDD18.3**	**90°**	**80°**	**70°**	**60°**	**50°**	**40°**	**30°**	**20°**	**10°**	**0°**
Jan	356	103	111	117	120	119	116	111	102	91	78
Feb	257	103	114	122	127	129	128	123	116	106	93
Mar	169	105	122	136	147	155	159	159	155	148	137
Apr	62	92	113	133	149	162	171	176	177	174	166
May	12	71	92	113	132	149	163	173	180	182	181
Jun	0	62	83	105	125	144	159	172	181	185	186
Jul	0	68	90	111	131	149	164	175	182	185	185
Aug	0	82	104	124	143	158	169	177	181	181	176
Sep	3	98	116	131	143	151	156	157	155	148	138
Oct	58	122	137	149	157	161	161	158	150	138	123

Table I.1 (*Continued*)

Nov	188	110	120	126	130	130	126	120	110	98	84
Dec	306	117	125	131	133	131	126	118	107	93	77

MOBILE, AL Solar radiation (kWh/m² mo) on a surface of tilt:

Mo	HDD18.3	90°	80°	70°	60°	50°	40°	30°	20°	10°	0°
Jan	246	116	126	133	136	136	132	125	115	102	87
Feb	165	81	90	97	102	105	104	102	97	89	80
Mar	98	97	114	129	141	149	154	155	152	145	135
Apr	28	89	112	133	152	167	178	185	188	185	179
May	3	64	84	104	123	140	154	165	172	175	175
Jun	0	58	80	104	127	148	167	182	193	199	201
Jul	0	62	83	105	127	146	162	176	185	190	190
Aug	0	79	103	125	145	161	175	184	189	189	185
Sep	1	95	114	131	145	155	162	165	163	158	150
Oct	26	138	156	170	180	186	186	182	174	160	143
Nov	115	144	158	167	172	173	169	160	147	131	111
Dec	208	116	125	131	134	133	129	121	111	98	82

FORT SMITH, AR Solar radiation (kWh/m² mo) on a surface of tilt:

Mo	HDD18.3	90°	80°	70°	60°	50°	40°	30°	20°	10°	0°
Jan	421	108	116	122	123	122	118	110	100	87	72
Feb	313	103	114	121	126	128	126	122	115	104	92
Mar	207	114	132	146	157	164	167	166	161	152	140
Apr	75	89	108	126	142	154	163	168	169	166	160
May	16	74	97	119	139	157	171	182	189	191	189
Jun	0	67	90	114	136	156	173	186	195	200	200
Jul	0	73	98	123	146	166	183	196	205	209	208
Aug	0	91	115	138	157	173	185	193	196	194	187
Sep	5	105	123	139	152	161	166	167	164	157	146
Oct	68	126	141	152	160	163	163	158	149	137	121
Nov	217	105	114	120	123	124	121	115	106	95	81
Dec	381	119	127	132	133	131	126	117	106	91	75

LITTLE ROCK, AR Solar radiation (kWh/m² mo) on a surface of tilt:

Mo	HDD18.3	90°	80°	70°	60°	50°	40°	30°	20°	10°	0°
Jan	393	107	115	121	123	122	119	112	102	90	76
Feb	298	92	101	108	112	113	112	108	101	91	80
Mar	198	99	114	126	136	143	147	147	143	137	127
Apr	68	91	111	129	144	156	164	168	169	165	157
May	12	73	94	115	134	150	163	173	179	181	179
Jun	0	67	89	111	133	152	168	181	190	194	194
Jul	0	73	98	121	144	164	180	192	200	204	202
Aug	0	92	117	140	160	176	189	196	200	198	191
Sep	3	101	119	135	148	157	162	164	161	155	145
Oct	64	127	142	154	162	166	165	161	152	139	123
Nov	208	106	115	121	125	125	122	116	107	95	82
Dec	356	105	112	117	119	117	113	106	96	84	69

FLAGSTAFF, AZ Solar radiation (kWh/m² mo) on a surface of tilt:

Mo	HDD18.3	90°	80°	70°	60°	50°	40°	30°	20°	10°	0°
Jan	591	152	164	170	173	170	163	151	135	116	94
Feb	506	139	153	163	169	170	167	159	147	132	113
Mar	459	116	135	150	162	169	173	172	167	158	145
Apr	361	101	125	146	165	179	190	195	196	192	183
May	237	82	111	139	165	188	207	221	229	232	229
Jun	76	67	93	120	146	170	191	208	220	227	229
Jul	13	71	95	120	142	163	180	193	202	206	205
Aug	27	92	118	143	164	181	195	203	207	205	199
Sep	117	121	143	161	176	186	191	191	186	177	163
Oct	307	145	162	175	183	187	185	179	168	152	133
Nov	453	162	175	183	186	184	177	165	148	128	104
Dec	598	156	166	172	173	169	161	148	132	111	89

PHOENIX, AZ Solar radiation (kWh/m² mo) on a surface of tilt:

Mo	HDD18.3	90°	80°	70°	60°	50°	40°	30°	20°	10°	0°
Jan	142	156	168	176	179	177	170	159	143	124	102
Feb	92	136	150	161	167	169	166	159	148	133	115
Mar	36	133	154	172	185	194	198	197	191	180	164

Table I.1 (*Continued*)

Apr	11	113	141	167	189	206	218	224	225	220	210
May	1	82	112	142	169	194	214	230	240	244	242
Jun	0	69	98	128	158	184	207	226	239	246	248
Jul	0	75	104	133	160	184	204	220	231	235	235
Aug	0	97	126	153	177	197	213	223	228	227	220
Sep	0	131	156	178	195	208	215	217	212	203	188
Oct	3	155	174	189	199	203	202	196	185	169	148
Nov	44	164	178	187	191	190	183	172	156	136	112
Dec	158	162	173	180	181	178	170	157	140	119	95
FRESNO, CA		**Solar radiation (kWh/m^2 mo) on a surface of tilt:**									
Mo	**HDD18.3**	**90°**	**80°**	**70°**	**60°**	**50°**	**40°**	**30°**	**20°**	**10°**	**0°**
Jan	283	70	76	80	82	82	81	77	72	66	57
Feb	199	106	115	122	126	127	125	120	111	101	88
Mar	131	122	140	155	166	173	176	174	169	159	145
Apr	73	117	142	165	184	198	208	212	211	205	193
May	17	90	119	146	171	193	210	222	229	231	227
Jun	1	78	107	137	165	190	212	228	239	245	244
Jul	0	84	114	144	171	195	216	231	240	244	242
Aug	0	109	139	167	191	211	226	235	238	235	226
Sep	1	129	152	171	186	197	202	202	197	186	171
Oct	29	147	164	177	186	189	188	182	171	155	136
Nov	160	114	123	129	132	131	127	120	109	97	81
Dec	294	91	97	101	102	101	97	90	82	71	59
LOS ANGELES, CA		**Solar radiation (kWh/m^2 mo) on a surface of tilt:**									
Mo	**HDD18.3**	**90°**	**80°**	**70°**	**60°**	**50°**	**40°**	**30°**	**20°**	**10°**	**0°**
Jan	126	119	128	134	137	136	132	124	114	100	84
Feb	119	104	115	123	129	131	130	126	118	108	95
Mar	106	115	133	149	161	169	173	173	168	160	147
Apr	82	98	121	141	158	172	181	186	187	183	175
May	43	78	101	124	145	164	179	190	196	199	197
Jun	8	70	95	121	145	167	185	199	208	213	213
Jul	1	78	106	134	160	183	202	216	225	228	226
Aug	1	97	123	148	170	188	202	211	215	214	208
Sep	2	111	132	150	165	175	182	183	180	173	161
Oct	16	128	144	156	165	169	169	165	157	145	129
Nov	64	135	146	154	158	157	152	144	131	115	97
Dec	129	133	143	148	150	148	141	132	118	102	83
SACRAMENTO, CA		**Solar radiation (kWh/m^2 mo) on a surface of tilt:**									
Mo	**HDD18.3**	**90°**	**80°**	**70°**	**60°**	**50°**	**40°**	**30°**	**20°**	**10°**	**0°**
Jan	301	77	83	86	88	87	84	79	72	64	54
Feb	221	88	96	101	105	105	104	99	92	84	73
Mar	169	130	148	163	174	180	182	180	172	161	145
Apr	104	119	144	166	183	197	205	209	207	200	188
May	31	95	124	152	176	197	214	226	232	232	227
Jun	3	84	113	142	170	194	214	229	239	243	242
Jul	0	92	122	152	179	203	222	236	244	247	243
Aug	0	113	143	169	192	210	224	231	233	229	219
Sep	3	131	152	170	184	193	196	195	189	178	163
Oct	47	127	141	152	158	161	159	154	144	131	115
Nov	190	110	119	124	126	125	120	113	103	90	75
Dec	306	94	100	103	104	102	98	91	82	71	58
SAN FRANCISCO, CA		**Solar radiation (kWh/m^2 mo) on a surface of tilt:**									
Mo	**HDD18.3**	**90°**	**80°**	**70°**	**60°**	**50°**	**40°**	**30°**	**20°**	**10°**	**0°**
Jan	236	97	104	108	110	109	105	99	90	79	66
Feb	187	109	120	127	131	131	129	123	114	102	87
Mar	169	119	136	150	160	166	168	166	160	149	135
Apr	139	101	122	141	157	169	177	181	181	176	168
May	106	84	109	132	154	172	186	196	202	203	199
Jun	62	76	101	127	151	172	189	203	211	215	213
Jul	41	84	111	138	162	183	200	213	220	223	219
Aug	31	100	126	149	170	186	199	206	208	205	198
Sep	32	126	148	166	180	189	194	194	188	178	164
Oct	63	126	140	150	157	160	159	154	145	133	117

Table I.1 (*Continued*)

Nov	148	109	117	122	124	123	119	112	102	90	76
Dec	234	115	122	125	126	123	117	108	96	82	66
DENVER, CO		**Solar radiation (kWh/m² mo) on a surface of tilt:**									
Mo	**HDD18.3**	**90°**	**80°**	**70°**	**60°**	**50°**	**40°**	**30°**	**20°**	**10°**	**0°**
Jan	571	146	155	159	159	155	147	134	118	99	77
Feb	502	123	133	140	143	142	138	130	118	104	87
Mar	408	136	154	169	179	185	186	182	174	161	144
Apr	292	93	110	125	138	147	153	156	155	150	142
May	155	89	113	136	156	173	187	196	201	200	196
Jun	29	82	109	135	158	179	196	209	217	219	217
Jul	2	85	110	135	157	176	191	203	209	211	208
Aug	4	101	124	146	165	180	190	197	198	194	186
Sep	59	124	144	161	174	182	186	185	180	169	155
Oct	243	139	153	163	169	170	167	160	148	133	114
Nov	413	123	131	136	137	134	128	119	106	91	74
Dec	585	130	137	140	139	135	127	115	101	84	64
GRAND JUNCTION, CO		**Solar radiation (kWh/m² mo) on a surface of tilt:**									
Mo	**HDD18.3**	**90°**	**80°**	**70°**	**60°**	**50°**	**40°**	**30°**	**20°**	**10°**	**0°**
Jan	626	132	140	145	145	142	135	125	111	95	76
Feb	458	119	129	136	139	139	135	127	117	103	87
Mar	342	130	147	162	172	177	179	176	168	157	141
Apr	218	115	138	157	174	186	193	196	194	187	176
May	86	93	120	146	169	189	205	216	222	222	218
Jun	9	82	110	137	163	185	204	219	229	233	232
Jul	0	90	118	146	171	193	210	223	231	233	229
Aug	0	106	132	156	177	194	206	213	215	212	204
Sep	31	132	154	172	185	194	198	196	190	179	163
Oct	205	154	171	183	190	192	189	181	169	152	131
Nov	414	127	136	142	144	142	136	127	115	99	81
Dec	621	134	141	145	144	140	132	120	105	88	68
HARTFORD, CT		**Solar radiation (kWh/m² mo) on a surface of tilt:**									
Mo	**HDD18.3**	**90°**	**80°**	**70°**	**60°**	**50°**	**40°**	**30°**	**20°**	**10°**	**0°**
Jan	654	93	99	102	103	101	96	89	80	68	56
Feb	549	92	100	105	108	108	105	100	92	82	70
Mar	467	102	116	127	135	140	141	139	135	127	116
Apr	258	89	105	119	130	139	145	147	146	142	135
May	109	81	99	117	133	146	156	163	167	167	163
Jun	19	76	96	116	133	149	161	170	175	176	174
Jul	1	82	102	121	138	153	165	174	178	179	176
Aug	3	90	109	126	141	152	160	165	166	163	156
Sep	46	100	115	127	137	143	146	145	142	134	124
Oct	205	100	110	117	121	122	121	116	109	99	87
Nov	374	93	99	103	104	102	98	92	83	72	60
Dec	551	82	86	88	88	86	82	76	67	57	46
ORLANDO, FL		**Solar radiation (kWh/m² mo) on a surface of tilt:**									
Mo	**HDD18.3**	**90°**	**80°**	**70°**	**60°**	**50°**	**40°**	**30°**	**20°**	**10°**	**0°**
Jan	102	140	153	162	166	166	162	153	141	125	106
Feb	57	81	91	98	104	106	107	104	99	92	83
Mar	30	104	123	139	153	162	168	169	166	159	148
Apr	4	90	115	138	159	175	188	196	199	197	190
May	0	64	88	114	138	160	178	192	202	207	207
Jun	0	51	69	89	109	127	143	156	166	172	175
Jul	0	58	79	103	125	145	162	176	186	192	193
Aug	0	75	99	122	143	161	176	186	192	193	189
Sep	0	112	137	159	177	190	199	203	201	195	183
Oct	3	102	116	128	137	142	144	143	138	130	119
Nov	22	151	166	177	183	184	180	171	158	140	119
Dec	66	161	174	182	186	184	178	166	151	131	108
TALLAHASSEE, FL		**Solar radiation (kWh/m² mo) on a surface of tilt:**									
Mo	**HDD18.3**	**90°**	**80°**	**70°**	**60°**	**50°**	**40°**	**30°**	**20°**	**10°**	**0°**
Jan	227	114	124	131	135	135	132	126	116	103	88
Feb	151	109	121	130	136	138	137	132	124	113	99
Mar	89	102	120	136	148	157	162	163	160	153	142

APPENDICES

Table I.1 (*Continued*)

Apr	26	90	113	134	152	167	178	185	187	185	178
May	2	62	83	105	126	145	160	172	180	184	184
Jun	0	53	71	90	109	126	141	153	162	167	168
Jul	0	60	79	100	119	137	152	163	172	176	176
Aug	0	70	91	111	130	145	158	167	172	173	169
Sep	0	92	110	126	139	149	155	158	157	152	144
Oct	20	115	131	144	153	159	161	159	153	144	130
Nov	102	113	124	133	138	139	137	132	123	111	96
Dec	184	115	124	130	133	132	128	120	110	96	81

ATLANTA, GA		Solar radiation (kWh/m² mo) on a surface of tilt:									
Mo	**HDD18.3**	**90°**	**80°**	**70°**	**60°**	**50°**	**40°**	**30°**	**20°**	**10°**	**0°**
Jan	358	117	126	133	136	136	132	125	114	101	86
Feb	261	105	116	124	128	130	128	124	116	105	92
Mar	177	113	131	146	158	166	169	169	164	156	144
Apr	66	93	115	134	151	164	174	180	181	178	171
May	12	77	100	123	144	162	177	188	195	197	195
Jun	1	67	90	113	136	156	173	186	196	200	201
Jul	0	70	92	114	136	155	170	183	191	195	194
Aug	0	84	106	127	145	161	173	181	185	185	180
Sep	3	88	104	118	130	138	143	145	144	139	130
Oct	57	115	130	142	150	154	155	151	144	133	119
Nov	190	114	124	131	134	134	131	125	115	103	88
Dec	308	115	124	129	131	130	125	117	107	93	78

DES MOINES, IOWA		Solar radiation (kWh/m² mo) on a surface of tilt:									
Mo	**HDD18.3**	**90°**	**80°**	**70°**	**60°**	**50°**	**40°**	**30°**	**20°**	**10°**	**0°**
Jan	730	106	112	116	116	113	108	99	89	76	61
Feb	591	108	117	123	126	125	121	115	105	93	79
Mar	439	109	123	134	142	147	148	145	139	130	117
Apr	226	101	119	135	148	158	164	166	164	158	148
May	82	87	108	129	147	162	174	183	187	187	182
Jun	8	82	106	129	150	169	183	194	201	203	200
Jul	1	87	111	134	155	172	186	196	202	202	198
Aug	2	101	123	143	160	173	183	188	188	184	176
Sep	39	120	138	152	163	170	173	171	165	155	141
Oct	201	124	136	144	149	150	148	141	131	118	101
Nov	413	102	108	112	113	111	107	99	89	77	63
Dec	638	95	100	103	102	100	95	87	77	65	52

BOISE, ID		Solar radiation (kWh/m² mo) on a surface of tilt:									
Mo	**HDD18.3**	**90°**	**80°**	**70°**	**60°**	**50°**	**40°**	**30°**	**20°**	**10°**	**0°**
Jan	562	82	87	90	91	89	85	80	72	63	52
Feb	429	96	103	108	110	109	106	100	91	80	68
Mar	343	110	123	134	141	145	145	142	135	125	112
Apr	240	114	134	151	165	175	181	182	179	172	161
May	117	97	121	143	163	179	191	199	202	201	195
Jun	32	93	120	146	170	190	207	219	226	228	225
Jul	2	105	134	162	187	208	224	235	241	241	235
Aug	3	120	147	171	191	207	218	223	222	216	205
Sep	44	138	157	173	184	191	192	188	180	167	150
Oct	212	135	148	157	162	163	160	152	140	125	107
Nov	402	91	97	100	101	99	94	87	79	68	55
Dec	566	85	89	91	91	88	82	75	66	55	43

COEUR D'ALENE, ID		Solar radiation (kWh/m² mo) on a surface of tilt:									
Mo	**HDD18.3**	**90°**	**80°**	**70°**	**60°**	**50°**	**40°**	**30°**	**20°**	**10°**	**0°**
Jan	607	81	85	86	85	82	76	69	59	49	37
Feb	517	106	113	117	117	115	109	101	90	77	62
Mar	454	135	149	159	166	168	165	159	148	134	117
Apr	323	111	129	144	156	165	170	171	167	160	149
May	186	93	112	130	146	158	168	174	176	174	168
Jun	95	93	114	135	154	169	182	190	195	195	192
Jul	18	110	137	161	183	201	214	223	226	225	218
Aug	20	128	152	173	191	203	210	213	210	201	188
Sep	107	106	119	130	137	141	141	138	132	122	110
Oct	317	114	124	130	133	133	129	122	112	99	83

Table I.1 (*Continued*)

Nov	476	55	58	60	60	59	56	52	47	40	33
Dec	622	52	54	55	54	52	48	44	38	32	24
CHICAGO, IL		**Solar radiation (kWh/m² mo) on a surface of tilt:**									
Mo	**HDD18.3**	**90°**	**80°**	**70°**	**60°**	**50°**	**40°**	**30°**	**20°**	**10°**	**0°**
Jan	680	88	94	97	97	96	91	85	76	66	54
Feb	557	90	97	102	104	104	101	96	89	80	69
Mar	444	99	111	120	127	131	132	129	124	116	105
Apr	247	85	101	115	126	135	140	143	142	138	131
May	104	89	111	132	151	166	178	186	190	189	185
Jun	14	80	101	122	142	159	172	182	188	190	188
Jul	1	88	111	132	151	167	180	189	194	194	190
Aug	2	91	110	127	142	154	163	167	168	166	159
Sep	32	102	117	130	139	145	148	147	143	135	124
Oct	183	106	116	124	128	129	127	122	114	103	90
Nov	383	75	81	84	85	85	82	78	71	63	54
Dec	584	82	86	89	89	86	82	75	67	57	46
INDIANAPOLIS, IN		**Solar radiation (kWh/m² mo) on a surface of tilt:**									
Mo	**HDD18.3**	**90°**	**80°**	**70°**	**60°**	**50°**	**40°**	**30°**	**20°**	**10°**	**0°**
Jan	633	86	92	96	97	96	93	87	80	70	59
Feb	506	93	102	108	111	112	110	106	99	90	79
Mar	391	100	115	126	135	141	144	143	139	132	122
Apr	193	92	110	125	138	149	155	159	158	155	147
May	71	84	106	126	144	159	172	180	185	185	181
Jun	7	76	98	119	139	156	170	181	188	191	189
Jul	1	82	105	128	148	166	181	191	197	199	196
Aug	1	92	113	131	147	159	169	174	176	173	167
Sep	28	108	125	139	150	158	161	161	157	149	137
Oct	170	115	127	136	141	143	142	137	128	117	103
Nov	358	88	94	98	100	99	96	91	83	73	62
Dec	544	78	83	86	86	85	82	76	69	60	50
WICHITA, KS		**Solar radiation (kWh/m² mo) on a surface of tilt:**									
Mo	**HDD18.3**	**90°**	**80°**	**70°**	**60°**	**50°**	**40°**	**30°**	**20°**	**10°**	**0°**
Jan	544	124	132	137	138	135	129	120	107	91	74
Feb	428	118	129	136	140	140	136	129	118	105	88
Mar	306	117	133	147	156	163	165	163	157	147	134
Apr	152	97	117	135	150	162	170	174	174	170	162
May	42	81	104	125	145	162	175	185	191	192	189
Jun	2	72	96	120	143	162	179	191	200	204	202
Jul	0	82	109	135	158	179	196	209	216	219	216
Aug	0	95	118	139	158	173	184	190	193	190	183
Sep	14	117	137	153	166	175	179	179	174	165	152
Oct	123	133	147	158	165	168	166	160	150	136	119
Nov	308	123	132	138	141	139	134	126	114	99	82
Dec	509	128	136	140	140	137	130	119	105	89	70
LOUISVILLE, KY		**Solar radiation (kWh/m² mo) on a surface of tilt:**									
Mo	**HDD18.3**	**90°**	**80°**	**70°**	**60°**	**50°**	**40°**	**30°**	**20°**	**10°**	**0°**
Jan	517	92	99	103	104	103	99	93	85	74	62
Feb	403	95	103	110	113	114	112	107	100	90	78
Mar	299	102	116	127	136	142	144	143	138	130	120
Apr	124	92	110	127	141	151	159	162	162	158	151
May	35	77	98	118	136	152	165	173	179	180	177
Jun	2	74	97	120	141	160	175	187	195	199	197
Jul	0	80	103	125	146	164	178	189	195	197	194
Aug	0	91	112	131	148	162	172	178	179	177	170
Sep	12	103	120	134	145	152	156	156	152	145	134
Oct	113	112	124	134	139	142	141	136	128	117	103
Nov	282	96	103	108	110	109	106	100	92	81	68
Dec	441	87	93	96	97	96	92	86	78	68	57
NEW ORLEANS, LA		**Solar radiation (kWh/m² mo) on a surface of tilt:**									
Mo	**HDD18.3**	**90°**	**80°**	**70°**	**60°**	**50°**	**40°**	**30°**	**20°**	**10°**	**0°**
Jan	199	113	123	130	134	134	131	124	115	102	87
Feb	130	69	77	83	88	91	92	91	88	83	76
Mar	68	85	101	115	126	134	139	141	139	134	126

Table I.1 (*Continued*)

Apr	14	62	75	87	97	106	113	118	120	120	118
May	1	60	79	98	116	132	145	155	163	166	166
Jun	0	53	66	81	95	108	119	128	134	138	140
Jul	0	58	78	99	119	136	151	162	170	174	174
Aug	0	68	87	105	121	135	146	154	159	160	158
Sep	0	87	104	120	132	142	149	152	152	148	141
Oct	11	97	110	120	128	132	133	132	127	119	108
Nov	76	98	107	114	119	120	118	114	106	96	84
Dec	164	79	86	91	94	94	93	89	83	75	66

SHREVEPORT, LA		**Solar radiation (kWh/m² mo) on a surface of tilt:**									
Mo	**HDD18.3**	**90°**	**80°**	**70°**	**60°**	**50°**	**40°**	**30°**	**20°**	**10°**	**0°**
Jan	394	109	118	123	126	125	122	115	105	92	77
Feb	212	99	110	118	123	126	125	122	115	106	94
Mar	127	101	117	131	142	149	153	153	150	144	134
Apr	41	83	103	121	137	150	159	165	167	165	159
May	4	70	91	112	132	149	163	174	181	184	183
Jun	0	61	83	105	127	147	164	177	186	191	192
Jul	0	66	89	112	135	154	171	184	193	198	197
Aug	0	83	107	129	149	166	179	188	192	192	187
Sep	1	99	119	135	149	159	166	168	166	160	151
Oct	36	127	144	157	166	171	171	167	159	147	131
Nov	147	110	121	128	132	133	130	124	116	104	90
Dec	272	117	126	132	134	133	128	120	109	96	80

BOSTON, MA		**Solar radiation (kWh/m² mo) on a surface of tilt:**									
Mo	**HDD18.3**	**90°**	**80°**	**70°**	**60°**	**50°**	**40°**	**30°**	**20°**	**10°**	**0°**
Jan	604	96	102	105	105	103	98	90	80	68	55
Feb	514	104	112	118	120	120	117	110	101	90	77
Mar	458	103	116	126	134	138	139	137	132	123	112
Apr	272	87	103	117	129	138	144	146	146	142	134
May	132	85	105	125	142	156	167	174	178	177	173
Jun	29	75	95	115	132	147	160	169	174	175	172
Jul	2	87	109	130	149	165	178	186	191	191	187
Aug	2	98	118	136	152	164	172	177	177	173	166
Sep	34	107	123	135	145	151	153	152	147	139	127
Oct	174	110	120	128	132	133	131	125	117	106	92
Nov	334	93	99	103	104	102	98	92	83	72	60
Dec	497	92	97	99	99	96	90	82	72	60	47

BALTIMORE, MD		**Solar radiation (kWh/m² mo) on a surface of tilt:**									
Mo	**HDD18.3**	**90°**	**80°**	**70°**	**60°**	**50°**	**40°**	**30°**	**20°**	**10°**	**0°**
Jan	534	103	109	113	114	112	107	99	89	76	62
Feb	443	96	105	111	114	114	112	107	99	88	76
Mar	357	110	125	136	145	150	151	148	142	132	120
Apr	173	96	114	131	145	155	162	165	164	159	151
May	62	79	100	119	137	152	164	172	177	177	174
Jun	5	75	98	120	140	159	174	185	192	195	193
Jul	0	77	98	119	138	155	169	179	185	187	185
Aug	0	87	107	125	141	153	163	169	171	169	163
Sep	17	96	111	125	135	142	147	147	144	138	128
Oct	142	120	133	142	148	151	149	144	134	122	107
Nov	302	100	107	112	113	112	108	101	92	80	66
Dec	452	95	101	104	104	102	97	90	81	69	56

BANGOR, ME		**Solar radiation (kWh/m² mo) on a surface of tilt:**									
Mo	**HDD18.3**	**90°**	**80°**	**70°**	**60°**	**50°**	**40°**	**30°**	**20°**	**10°**	**0°**
Jan	799	82	87	89	89	86	82	75	66	56	44
Feb	679	98	105	110	111	110	105	98	89	77	63
Mar	593	78	87	94	99	102	102	100	96	89	81
Apr	369	84	98	110	120	128	132	133	132	127	119
May	187	75	92	108	121	133	141	147	150	149	145
Jun	57	81	102	121	139	154	167	175	180	180	177
Jul	8	83	103	122	139	153	163	170	174	173	169
Aug	13	91	110	126	140	151	158	162	162	158	151
Sep	96	83	94	103	110	115	116	116	112	106	98
Oct	284	109	119	125	128	128	125	119	109	97	83

Table I.1 (*Continued*)

Nov	464	90	95	98	98	96	91	83	74	63	50
Dec	672	78	82	83	82	79	74	67	59	49	37
PORTLAND, ME		**Solar radiation (kWh/m² mo) on a surface of tilt:**									
Mo	**HDD18.3**	**90°**	**80°**	**70°**	**60°**	**50°**	**40°**	**30°**	**20°**	**10°**	**0°**
Jan	715	107	113	116	116	113	107	98	87	73	58
Feb	608	113	121	127	129	128	124	116	106	93	78
Mar	539	117	132	143	151	155	156	153	147	137	124
Apr	344	97	114	128	140	148	153	155	153	147	138
May	190	87	107	126	143	156	167	174	177	176	171
Jun	57	74	93	111	128	142	154	162	167	168	166
Jul	7	89	112	134	153	169	181	190	194	194	189
Aug	11	100	121	139	154	166	174	178	177	173	164
Sep	81	113	129	142	152	157	159	157	152	142	129
Oct	256	111	121	128	132	133	130	124	115	103	88
Nov	422	87	92	95	96	94	90	84	75	65	53
Dec	599	94	98	100	100	97	91	83	73	61	48
DETROIT, MI		**Solar radiation (kWh/m² mo) on a surface of tilt:**									
Mo	**HDD18.3**	**90°**	**80°**	**70°**	**60°**	**50°**	**40°**	**30°**	**20°**	**10°**	**0°**
Jan	668	78	83	86	87	85	82	77	69	61	50
Feb	568	51	55	59	61	62	62	60	57	53	48
Mar	484	99	111	120	127	131	132	130	124	116	105
Apr	274	94	111	127	139	148	154	157	155	150	142
May	114	88	109	128	146	160	172	179	183	182	177
Jun	15	79	100	120	139	155	168	177	183	184	182
Jul	1	90	114	137	157	175	188	198	203	203	199
Aug	3	90	108	125	139	151	159	163	164	161	154
Sep	42	110	127	140	151	157	160	159	154	145	132
Oct	199	88	97	103	108	109	108	105	99	91	81
Nov	387	55	59	62	63	63	61	58	54	49	42
Dec	570	66	70	72	72	71	67	62	56	48	39
SAULT STE MARIE, MI		**Solar radiation (kWh/m² mo) on a surface of tilt:**									
Mo	**HDD18.3**	**90°**	**80°**	**70°**	**60°**	**50°**	**40°**	**30°**	**20°**	**10°**	**0°**
Jan	851	79	83	85	86	84	80	75	68	59	49
Feb	737	111	118	123	124	123	118	110	100	87	73
Mar	663	125	139	149	157	161	160	157	149	138	124
Apr	436	104	121	135	146	154	158	158	155	148	138
May	234	95	116	135	152	165	175	182	184	182	176
Jun	86	87	108	128	147	162	175	183	188	188	184
Jul	34	93	115	135	154	169	181	188	192	191	186
Aug	35	95	113	129	142	152	159	162	161	157	149
Sep	121	90	102	112	120	124	126	125	121	114	104
Oct	322	76	83	88	91	92	90	87	82	75	66
Nov	508	54	58	60	61	61	59	56	52	46	40
Dec	714	62	65	67	67	66	64	59	54	47	40
DULUTH, MN		**Solar radiation (kWh/m² mo) on a surface of tilt:**									
Mo	**HDD18.3**	**90°**	**80°**	**70°**	**60°**	**50°**	**40°**	**30°**	**20°**	**10°**	**0°**
Jan	933	96	101	103	103	100	94	86	76	64	51
Feb	780	104	111	115	116	115	110	103	94	82	68
Mar	660	115	128	138	145	149	149	145	138	127	113
Apr	430	109	126	142	153	162	166	167	164	157	146
May	238	99	120	140	157	171	181	187	190	188	181
Jun	87	87	108	127	145	160	172	180	184	185	182
Jul	28	91	112	132	150	164	175	182	185	184	179
Aug	37	101	120	136	150	161	167	170	169	164	155
Sep	137	97	110	120	128	133	135	134	129	122	111
Oct	355	83	91	96	99	99	98	94	87	79	69
Nov	582	76	80	83	83	81	77	72	64	55	45
Dec	829	78	81	82	81	78	73	66	58	48	37
MINNEAPOLIS, MN		**Solar radiation (kWh/m² mo) on a surface of tilt:**									
Mo	**HDD18.3**	**90°**	**80°**	**70°**	**60°**	**50°**	**40°**	**30°**	**20°**	**10°**	**0°**
Jan	846	107	113	115	114	111	104	95	83	69	53
Feb	697	109	116	121	122	121	116	109	98	86	71
Mar	550	114	128	139	146	150	150	146	139	129	115

Table I.1 (*Continued*)

Apr	301	99	116	131	143	152	157	159	156	150	141
May	122	96	118	138	156	171	181	188	191	189	183
Jun	18	87	109	130	150	166	179	189	194	195	191
Jul	2	90	112	132	151	166	178	186	190	190	185
Aug	4	98	117	134	149	160	167	171	171	167	159
Sep	64	105	120	132	141	146	148	147	141	133	121
Oct	264	109	119	126	130	130	127	122	113	101	87
Nov	496	84	89	92	92	90	85	79	70	60	48
Dec	744	80	83	85	84	81	76	69	60	49	38

KANSAS CITY, MO **Solar radiation (kWh/m² mo) on a surface of tilt:**

Mo	HDD18.3	90°	80°	70°	60°	50°	40°	30°	20°	10°	0°
Jan	616	121	128	133	133	131	124	115	103	88	70
Feb	487	100	109	115	118	118	115	109	100	89	76
Mar	353	115	131	144	154	159	161	159	153	144	131
Apr	176	99	118	135	149	160	167	169	168	163	154
May	57	84	107	129	148	164	177	186	191	191	187
Jun	3	74	98	120	142	160	176	188	195	198	196
Jul	0	81	105	128	149	168	182	193	200	202	199
Aug	1	96	119	140	158	173	183	190	192	189	181
Sep	27	114	132	148	159	167	171	170	165	156	143
Oct	156	125	138	148	154	157	155	149	139	126	110
Nov	344	109	116	121	123	121	117	109	99	86	71
Dec	547	110	117	120	120	117	111	101	90	76	60

ST LOUIS, MO **Solar radiation (kWh/m² mo) on a surface of tilt:**

Mo	HDD18.3	90°	80°	70°	60°	50°	40°	30°	20°	10°	0°
Jan	563	104	111	115	117	115	110	102	92	80	66
Feb	440	94	103	109	112	113	110	105	97	87	75
Mar	317	108	123	135	143	149	150	148	143	134	122
Apr	141	87	104	120	133	143	149	153	152	148	141
May	39	76	97	116	134	149	161	170	175	175	172
Jun	3	76	99	121	143	161	177	189	196	199	198
Jul	0	82	107	130	151	170	185	196	202	204	201
Aug	0	89	110	129	145	159	168	174	176	174	168
Sep	15	112	130	145	156	164	168	168	163	155	143
Oct	123	125	138	148	154	156	154	149	139	126	110
Nov	303	95	103	108	110	109	106	100	91	81	68
Dec	488	101	107	110	111	109	104	96	86	73	60

JACKSON, MS **Solar radiation (kWh/m² mo) on a surface of tilt:**

Mo	HDD18.3	90°	80°	70°	60°	50°	40°	30°	20°	10°	0°
Jan	314	107	116	122	125	125	122	116	107	96	82
Feb	225	98	109	117	123	125	125	121	115	106	94
Mar	142	99	115	129	140	147	151	152	149	142	133
Apr	49	90	112	132	149	163	173	179	181	178	171
May	7	67	88	108	126	143	157	167	174	177	176
Jun	0	60	81	103	125	144	160	173	183	188	188
Jul	0	68	90	113	134	153	169	182	190	194	193
Aug	0	82	106	128	147	164	177	186	190	190	186
Sep	2	98	117	134	147	157	163	165	163	157	148
Oct	46	127	143	156	165	170	171	167	159	147	131
Nov	166	112	122	130	134	134	131	125	116	104	89
Dec	277	118	127	133	135	134	129	121	111	97	81

BILLINGS, MT **Solar radiation (kWh/m² mo) on a surface of tilt:**

Mo	HDD18.3	90°	80°	70°	60°	50°	40°	30°	20°	10°	0°
Jan	652	96	101	103	103	99	94	86	75	63	50
Feb	561	106	114	119	120	119	115	108	98	86	72
Mar	469	124	137	148	155	158	157	153	144	133	118
Apr	319	104	122	137	149	157	162	163	159	153	142
May	178	99	121	141	159	174	184	191	193	191	185
Jun	51	90	113	136	156	173	186	196	201	202	198
Jul	3	104	131	156	178	196	210	220	224	222	216
Aug	10	116	139	161	178	191	200	204	203	197	186
Sep	91	134	152	166	175	181	181	177	168	155	139
Oct	291	118	128	135	138	138	135	128	118	105	89

Table I.1 (*Continued*)

Nov	473	95	100	103	103	101	95	88	78	67	54
Dec	649	99	103	104	103	99	92	83	72	58	44
MISSOULA, MT		**Solar radiation (kWh/m² mo) on a surface of tilt:**									
Mo	**HDD18.3**	**90°**	**80°**	**70°**	**60°**	**50°**	**40°**	**30°**	**20°**	**10°**	**0°**
Jan	678	68	71	73	73	72	69	64	57	50	41
Feb	559	87	93	97	98	96	93	87	79	69	58
Mar	469	97	108	116	121	124	124	121	115	106	96
Apr	339	102	119	133	145	153	157	158	155	149	139
May	204	93	113	131	148	161	171	177	179	178	172
Jun	96	90	112	133	152	168	181	190	195	195	191
Jul	14	102	127	150	170	187	200	208	212	211	205
Aug	21	111	133	152	168	181	189	192	191	186	176
Sep	124	124	140	152	161	166	167	163	156	144	129
Oct	351	119	129	136	139	138	134	127	116	102	86
Nov	531	69	73	76	76	75	72	67	60	52	43
Dec	688	57	60	62	62	60	57	53	47	41	33
ASHEVILLE, NC		**Solar radiation (kWh/m² mo) on a surface of tilt:**									
Mo	**HDD18.3**	**90°**	**80°**	**70°**	**60°**	**50°**	**40°**	**30°**	**20°**	**10°**	**0°**
Jan	473	117	125	131	133	131	126	118	107	92	76
Feb	371	100	110	117	122	123	121	116	108	97	84
Mar	299	107	123	137	147	153	156	155	150	142	130
Apr	148	93	114	133	148	161	169	174	174	170	162
May	51	72	93	114	132	148	162	171	177	179	176
Jun	4	65	87	108	129	147	162	173	181	185	184
Jul	0	70	92	114	134	152	167	178	185	188	186
Aug	1	82	103	122	139	152	163	170	172	171	166
Sep	21	93	109	122	133	141	145	147	144	139	130
Oct	142	125	140	152	159	163	162	157	149	136	120
Nov	297	113	122	128	131	130	126	119	109	96	81
Dec	416	122	130	134	136	133	127	118	105	90	73
CHARLOTTE, NC		**Solar radiation (kWh/m² mo) on a surface of tilt:**									
Mo	**HDD18.3**	**90°**	**80°**	**70°**	**60°**	**50°**	**40°**	**30°**	**20°**	**10°**	**0°**
Jan	397	114	122	128	129	128	123	115	104	91	75
Feb	301	98	108	115	120	121	119	114	106	96	84
Mar	218	112	129	143	153	160	163	162	158	149	137
Apr	87	97	118	137	153	165	174	178	178	174	165
May	20	74	96	117	137	154	168	178	185	187	185
Jun	1	68	90	113	134	153	169	182	190	194	194
Jul	0	72	95	118	139	157	172	184	191	194	193
Aug	0	85	106	126	143	157	168	175	177	176	170
Sep	6	100	117	132	143	151	156	157	154	147	137
Oct	82	129	144	157	165	169	169	164	155	142	126
Nov	229	115	125	131	134	134	130	123	113	100	84
Dec	343	117	125	130	131	129	124	115	104	90	73
OMAHA, NE		**Solar radiation (kWh/m² mo) on a surface of tilt:**									
Mo	**HDD18.3**	**90°**	**80°**	**70°**	**60°**	**50°**	**40°**	**30°**	**20°**	**10°**	**0°**
Jan	676	115	122	126	127	124	118	110	98	84	68
Feb	571	97	105	110	113	114	111	106	98	88	76
Mar	393	110	124	136	144	149	150	148	142	133	121
Apr	211	94	111	126	139	148	154	156	154	149	140
May	74	89	112	133	152	168	180	188	192	192	187
Jun	3	81	104	127	147	166	180	191	197	199	197
Jul	1	88	111	134	154	171	184	194	199	200	196
Aug	2	98	119	139	155	168	177	182	183	180	172
Sep	32	116	133	148	159	166	169	168	162	153	140
Oct	199	132	145	154	159	160	157	150	139	124	107
Nov	392	111	119	123	124	122	116	108	97	83	68
Dec	613	105	110	113	112	109	103	94	83	69	54
CONCORD, NH		**Solar radiation (kWh/m² mo) on a surface of tilt:**									
Mo	**HDD18.3**	**90°**	**80°**	**70°**	**60°**	**50°**	**40°**	**30°**	**20°**	**10°**	**0°**
Jan	741	89	95	97	97	95	90	83	74	62	50
Feb	625	101	109	114	117	116	113	107	99	88	75
Mar	548	105	118	128	135	139	140	138	132	123	112

Table I.1 (*Continued*)

Apr	329	97	115	130	142	151	157	159	157	152	143
May	160	82	101	119	135	148	158	165	168	167	163
Jun	43	78	98	117	135	150	163	172	178	179	177
Jul	6	87	109	130	148	164	177	185	189	190	185
Aug	13	94	114	132	147	159	167	172	173	169	162
Sep	86	104	118	130	140	145	147	146	141	133	122
Oct	271	103	113	120	124	125	122	117	109	99	86
Nov	442	84	89	92	93	92	88	82	74	64	53
Dec	632	86	90	92	92	89	85	78	69	58	47
ALBUQUERQUE, NM		**Solar radiation (kWh/m^2 mo) on a surface of tilt:**									
Mo	**HDD18.3**	**90°**	**80°**	**70°**	**60°**	**50°**	**40°**	**30°**	**20°**	**10°**	**0°**
Jan	464	154	166	173	175	173	166	154	139	120	98
Feb	349	145	159	169	175	176	172	163	151	134	114
Mar	258	129	150	166	179	187	191	189	183	173	158
Apr	138	114	140	164	184	200	211	217	217	212	201
May	36	83	111	138	163	184	202	215	222	225	221
Jun	1	67	94	121	147	171	191	207	219	225	227
Jul	0	77	106	134	160	183	202	216	225	229	227
Aug	0	101	129	155	178	197	211	220	224	222	215
Sep	9	125	148	167	183	194	199	200	196	186	172
Oct	118	155	173	187	196	200	198	191	179	162	141
Nov	307	157	169	177	180	178	171	160	144	125	102
Dec	471	166	177	183	183	179	170	156	137	115	90
LAS CRUCES, NM		**Solar radiation (kWh/m^2 mo) on a surface of tilt:**									
Mo	**HDD18.3**	**90°**	**80°**	**70°**	**60°**	**50°**	**40°**	**30°**	**20°**	**10°**	**0°**
Jan	358	119	129	135	139	138	134	127	116	103	87
Feb	260	143	158	170	176	178	175	168	156	140	120
Mar	165	136	159	178	193	203	207	206	200	188	172
Apr	63	102	129	153	174	191	203	210	212	209	200
May	13	80	111	142	171	197	219	236	247	252	250
Jun	0	64	93	123	152	178	201	220	233	241	243
Jul	0	68	95	123	149	173	194	210	221	226	226
Aug	0	89	116	142	164	183	198	208	213	212	206
Sep	1	122	147	168	185	197	205	207	204	195	182
Oct	56	146	165	179	189	193	193	187	177	162	143
Nov	217	178	194	204	209	208	201	189	172	150	124
Dec	364	177	190	198	200	196	188	174	155	132	106
LAS VEGAS, NV		**Solar radiation (kWh/m^2 mo) on a surface of tilt:**									
Mo	**HDD18.3**	**90°**	**80°**	**70°**	**60°**	**50°**	**40°**	**30°**	**20°**	**10°**	**0°**
Jan	266	150	161	167	168	165	158	146	131	112	90
Feb	185	134	147	156	161	162	159	151	140	125	107
Mar	90	130	150	166	179	187	191	189	183	173	158
Apr	35	118	145	170	190	206	217	222	222	216	205
May	4	88	117	146	173	196	214	228	237	239	236
Jun	1	74	103	132	160	185	206	223	235	241	241
Jul	0	80	110	140	167	192	213	228	239	243	242
Aug	0	106	136	163	187	206	221	230	233	231	223
Sep	0	138	163	184	200	211	217	217	212	201	185
Oct	19	163	181	195	204	207	205	197	183	165	143
Nov	134	161	174	182	184	182	175	163	147	127	104
Dec	288	157	167	172	173	169	160	147	130	109	86
RENO, NV		**Solar radiation (kWh/m^2 mo) on a surface of tilt:**									
Mo	**HDD18.3**	**90°**	**80°**	**70°**	**60°**	**50°**	**40°**	**30°**	**20°**	**10°**	**0°**
Jan	487	115	122	127	128	126	120	111	100	86	70
Feb	393	124	135	142	145	145	141	133	121	107	90
Mar	327	122	138	151	160	165	166	163	156	145	131
Apr	238	108	130	150	166	178	186	189	188	182	172
May	114	93	120	145	168	188	203	214	220	220	216
Jun	27	83	112	140	166	189	208	222	232	235	233
Jul	1	91	121	150	176	199	218	232	240	242	239
Aug	2	116	146	173	196	215	228	236	238	233	223
Sep	36	138	161	180	194	203	207	206	199	187	170
Oct	188	153	169	181	187	189	185	177	164	147	126

Table I.1 ***(Continued)***

Nov	360	140	149	155	156	153	146	135	121	103	83
Dec	504	129	136	139	138	134	127	115	101	85	66

ALBANY, NY — **Solar radiation (kWh/m² mo) on a surface of tilt:**

Mo	**HDD18.3**	**90°**	**80°**	**70°**	**60°**	**50°**	**40°**	**30°**	**20°**	**10°**	**0°**
Jan	709	78	82	85	86	85	82	76	69	61	51
Feb	596	97	105	110	113	113	110	105	97	87	75
Mar	511	104	118	128	136	141	143	141	136	128	117
Apr	291	92	109	123	134	143	148	150	149	143	135
May	121	88	109	129	146	161	172	179	183	182	178
Jun	23	80	100	120	139	155	168	177	183	184	182
Jul	2	89	110	131	149	165	178	186	191	192	188
Aug	6	94	113	131	147	159	168	173	174	171	165
Sep	61	102	117	130	139	145	148	147	142	135	124
Oct	238	103	112	119	123	124	122	117	110	99	87
Nov	412	78	84	87	88	87	84	79	71	63	52
Dec	597	75	79	82	82	80	77	71	64	55	46

BUFFALO, NY — **Solar radiation (kWh/m² mo) on a surface of tilt:**

Mo	**HDD18.3**	**90°**	**80°**	**70°**	**60°**	**50°**	**40°**	**30°**	**20°**	**10°**	**0°**
Jan	685	68	72	75	76	75	73	68	62	55	47
Feb	599	73	79	84	86	87	86	83	79	72	64
Mar	533	101	114	124	131	135	135	133	127	119	108
Apr	324	90	106	120	132	140	145	148	146	142	134
May	142	87	108	128	145	159	170	177	180	179	174
Jun	29	84	107	129	149	166	180	190	196	197	194
Jul	4	87	109	130	149	165	178	186	191	191	187
Aug	6	93	112	129	144	156	165	170	171	168	161
Sep	58	98	112	124	133	139	141	140	136	128	118
Oct	226	89	98	104	108	109	108	104	98	89	79
Nov	398	60	64	67	68	68	66	63	58	52	45
Dec	578	57	61	63	64	63	61	58	53	48	41

NEW YORK, NY — **Solar radiation (kWh/m² mo) on a surface of tilt:**

Mo	**HDD18.3**	**90°**	**80°**	**70°**	**60°**	**50°**	**40°**	**30°**	**20°**	**10°**	**0°**
Jan	552	98	104	108	108	106	101	93	83	71	58
Feb	469	98	106	111	114	114	110	104	96	85	72
Mar	404	97	110	121	128	133	134	132	127	119	108
Apr	227	94	112	127	140	150	156	158	157	152	143
May	92	85	107	127	146	161	173	181	185	185	180
Jun	11	74	95	114	132	147	160	169	175	176	174
Jul	0	78	99	118	136	152	164	173	178	179	175
Aug	1	98	119	139	155	169	178	184	185	182	174
Sep	14	100	116	129	140	147	150	150	146	139	128
Oct	132	130	144	153	159	160	157	151	140	126	109
Nov	294	88	94	97	99	97	94	88	79	69	57
Dec	450	98	103	106	106	103	97	89	79	66	52

SYRACUSE, NY — **Solar radiation (kWh/m² mo) on a surface of tilt:**

Mo	**HDD18.3**	**90°**	**80°**	**70°**	**60°**	**50°**	**40°**	**30°**	**20°**	**10°**	**0°**
Jan	705	72	77	79	80	79	77	72	66	58	49
Feb	607	85	92	97	100	100	99	94	88	80	70
Mar	534	103	116	125	132	136	137	134	128	119	108
Apr	310	91	107	121	133	141	147	149	148	144	136
May	132	91	114	134	153	168	179	187	190	189	183
Jun	29	79	99	118	136	152	165	174	180	182	180
Jul	3	85	107	128	147	163	175	184	189	189	185
Aug	7	97	116	134	148	160	168	172	172	168	160
Sep	63	99	113	124	133	138	140	139	134	126	116
Oct	229	91	99	105	109	110	109	105	98	89	79
Nov	404	61	66	69	70	70	68	65	60	54	47
Dec	590	62	65	68	68	68	65	62	56	50	43

CLEVELAND, OH — **Solar radiation (kWh/m² mo) on a surface of tilt:**

Mo	**HDD18.3**	**90°**	**80°**	**70°**	**60°**	**50°**	**40°**	**30°**	**20°**	**10°**	**0°**
Jan	643	58	63	66	67	67	66	63	59	54	47
Feb	542	72	78	83	86	88	87	85	80	74	67
Mar	468	95	107	117	124	128	130	128	124	117	107

Table I.1 (*Continued*)

Apr	260	92	108	122	134	143	148	150	149	144	136
May	112	88	109	130	148	163	175	183	187	187	182
Jun	18	77	98	118	137	153	167	176	182	184	182
Jul	2	86	109	131	150	167	180	189	194	195	191
Aug	3	90	109	126	141	152	161	166	167	164	157
Sep	42	97	112	124	133	140	142	142	138	131	121
Oct	189	96	106	113	118	119	118	114	107	98	87
Nov	364	69	74	78	79	79	76	73	67	60	51
Dec	544	50	54	56	57	57	56	54	51	46	41
COLUMBUS, OH		**Solar radiation (kWh/m^2 mo) on a surface of tilt:**									
Mo	**HDD18.3**	**90°**	**80°**	**70°**	**60°**	**50°**	**40°**	**30°**	**20°**	**10°**	**0°**
Jan	613	85	91	94	95	94	90	84	76	66	55
Feb	503	73	80	85	89	90	89	86	81	75	66
Mar	401	99	112	123	130	135	137	135	131	123	113
Apr	196	88	104	119	131	140	146	149	148	145	137
May	73	78	97	115	132	146	157	165	169	170	167
Jun	7	74	95	115	133	149	163	173	179	182	180
Jul	0	79	100	120	138	154	167	176	181	183	180
Aug	1	91	111	128	144	156	164	169	170	167	161
Sep	28	100	116	129	139	146	150	149	145	138	127
Oct	172	111	122	130	134	136	134	129	121	110	96
Nov	352	74	80	83	85	85	83	79	73	65	56
Dec	521	65	70	72	73	73	70	66	61	54	47
OKLAHOMA CITY, OK		**Solar radiation (kWh/m^2 mo) on a surface of tilt:**									
Mo	**HDD18.3**	**90°**	**80°**	**70°**	**60°**	**50°**	**40°**	**30°**	**20°**	**10°**	**0°**
Jan	444	134	143	149	151	149	143	134	121	104	86
Feb	340	105	116	124	129	131	130	126	118	108	95
Mar	231	114	132	146	158	165	168	168	163	154	142
Apr	101	93	114	132	148	160	169	174	174	170	163
May	23	75	97	119	139	156	170	181	187	189	187
Jun	1	68	91	115	138	158	175	189	198	203	203
Jul	0	77	103	130	155	177	195	209	218	221	220
Aug	0	93	117	140	160	177	189	197	201	199	192
Sep	7	116	137	154	168	177	182	183	179	170	157
Oct	84	135	150	162	170	173	172	166	157	143	125
Nov	241	122	132	139	142	141	137	129	117	103	87
Dec	416	129	138	143	144	141	135	125	112	96	77
EUGENE, OR		**Solar radiation (kWh/m^2 mo) on a surface of tilt:**									
Mo	**HDD18.3**	**90°**	**80°**	**70°**	**60°**	**50°**	**40°**	**30°**	**20°**	**10°**	**0°**
Jan	406	60	64	66	67	66	63	59	53	46	39
Feb	339	68	74	78	80	80	78	74	69	62	54
Mar	313	87	97	105	111	114	115	113	109	102	93
Apr	244	86	100	113	123	131	136	138	137	133	127
May	161	84	103	121	136	149	159	166	168	167	163
Jun	81	85	108	130	150	167	181	191	197	199	196
Jul	13	98	124	149	171	190	204	214	218	218	213
Aug	12	112	136	158	176	190	199	204	203	198	188
Sep	56	113	128	141	150	155	157	154	147	137	123
Oct	203	100	109	117	121	122	120	116	108	98	85
Nov	323	53	57	59	60	59	58	54	50	44	38
Dec	420	47	49	51	51	50	48	44	40	35	29
PORTLAND, OR		**Solar radiation (kWh/m^2 mo) on a surface of tilt:**									
Mo	**HDD18.3**	**90°**	**80°**	**70°**	**60°**	**50°**	**40°**	**30°**	**20°**	**10°**	**0°**
Jan	402	60	63	65	66	64	62	57	52	45	37
Feb	332	70	76	79	81	81	78	74	69	62	53
Mar	293	92	103	111	117	120	120	118	113	105	95
Apr	211	86	99	111	120	127	131	132	130	125	118
May	114	89	108	126	141	154	163	169	172	171	166
Jun	49	85	105	125	142	157	169	177	182	182	179
Jul	7	97	120	141	160	176	188	195	199	198	193
Aug	4	109	130	149	164	176	184	187	186	180	170
Sep	36	107	122	134	143	148	150	148	142	133	120
Oct	168	88	96	102	106	107	105	101	95	86	75

Table I.1 (*Continued*)

Nov	298	53	57	59	60	60	58	56	52	46	40
Dec	409	55	58	59	59	58	55	51	46	40	33
REDMOND, OR		**Solar radiation (kWh/m² mo) on a surface of tilt:**									
Mo	**HDD18.3**	**90°**	**80°**	**70°**	**60°**	**50°**	**40°**	**30°**	**20°**	**10°**	**0°**
Jan	533	90	94	97	97	94	89	82	73	62	50
Feb	459	91	98	103	105	104	101	95	87	77	65
Mar	424	113	127	138	145	149	150	147	140	130	117
Apr	342	110	129	146	159	169	175	177	174	168	157
May	221	102	125	147	166	182	194	201	204	202	196
Jun	114	96	123	148	171	191	207	218	224	225	222
Jul	27	105	133	159	183	203	219	229	235	234	229
Aug	33	123	149	172	191	206	215	220	219	213	201
Sep	125	136	155	170	180	187	188	185	177	164	147
Oct	304	136	148	157	161	162	158	150	138	122	104
Nov	438	90	96	99	99	97	92	85	76	65	53
Dec	566	90	94	95	94	91	85	77	67	55	42
PHILADELPHIA, PA		**Solar radiation (kWh/m² mo) on a surface of tilt:**									
Mo	**HDD18.3**	**90°**	**80°**	**70°**	**60°**	**50°**	**40°**	**30°**	**20°**	**10°**	**0°**
Jan	538	105	111	115	116	113	108	100	89	77	62
Feb	448	98	106	112	115	115	113	107	99	88	75
Mar	363	106	120	131	140	144	146	144	138	129	117
Apr	174	91	109	124	137	146	153	155	154	149	141
May	56	83	104	123	141	156	167	175	179	179	175
Jun	4	76	97	119	138	155	169	180	186	189	187
Jul	0	78	98	118	135	151	163	172	177	179	176
Aug	0	97	118	138	155	168	177	183	184	181	173
Sep	12	102	119	132	143	150	153	153	149	142	131
Oct	123	116	128	136	142	144	142	136	128	116	101
Nov	287	99	106	111	112	111	107	100	91	79	66
Dec	446	97	102	105	105	103	97	90	80	68	54
PITTSBURGH, PA		**Solar radiation (kWh/m² mo) on a surface of tilt:**									
Mo	**HDD18.3**	**90°**	**80°**	**70°**	**60°**	**50°**	**40°**	**30°**	**20°**	**10°**	**0°**
Jan	627	52	56	59	60	60	58	56	52	47	41
Feb	522	136	147	155	158	157	151	142	129	113	94
Mar	434	105	119	131	139	144	145	143	137	129	117
Apr	223	72	85	97	106	114	119	121	121	119	114
May	94	65	80	94	107	118	127	133	137	138	136
Jun	15	65	82	99	114	128	139	147	153	155	154
Jul	2	71	88	105	121	134	145	153	158	160	158
Aug	3	74	89	102	114	124	131	136	137	136	132
Sep	43	73	83	92	99	105	107	108	106	102	96
Oct	200	118	130	139	144	146	144	138	129	117	102
Nov	370	72	77	81	83	82	80	76	70	63	54
Dec	534	33	35	37	38	39	39	37	36	33	30
RAPID CITY, SD		**Solar radiation (kWh/m² mo) on a surface of tilt:**									
Mo	**HDD18.3**	**90°**	**80°**	**70°**	**60°**	**50°**	**40°**	**30°**	**20°**	**10°**	**0°**
Jan	691	119	125	128	127	123	116	105	92	77	59
Feb	601	122	130	136	138	136	130	122	110	95	78
Mar	508	127	142	154	162	165	165	161	152	140	125
Apr	344	105	123	139	152	161	166	168	166	160	150
May	190	97	120	141	159	175	186	194	197	195	189
Jun	48	87	111	134	155	174	188	199	205	206	203
Jul	4	98	124	148	170	188	202	212	217	216	211
Aug	9	114	138	160	177	191	200	204	203	197	187
Sep	89	124	142	157	167	174	176	173	166	155	139
Oct	297	138	150	158	162	162	157	149	136	120	102
Nov	493	111	118	122	123	120	114	106	94	80	65
Dec	677	111	116	118	117	112	105	95	82	68	51
SIOUX FALLS, SD		**Solar radiation (kWh/m² mo) on a surface of tilt:**									
Mo	**HDD18.3**	**90°**	**80°**	**70°**	**60°**	**50°**	**40°**	**30°**	**20°**	**10°**	**0°**
Jan	828	108	114	117	116	113	107	98	86	72	57
Feb	681	115	124	129	131	130	125	118	107	93	77

Table I.1 (*Continued*)

Mar	540	107	120	130	137	141	141	138	131	121	109
Apr	311	104	122	138	150	159	165	166	164	157	147
May	137	90	111	131	148	163	174	181	184	183	178
Jun	22	83	106	128	148	165	179	190	195	197	194
Jul	4	95	120	143	164	182	195	205	209	209	204
Aug	7	101	123	141	157	170	178	183	183	178	170
Sep	73	113	130	143	153	159	161	159	153	144	130
Oct	275	121	132	140	145	146	143	137	127	114	98
Nov	508	94	100	104	104	103	98	91	82	71	58
Dec	748	85	89	91	91	89	84	77	69	58	47
KNOXVILLE, TN		**Solar radiation (kWh/m² mo) on a surface of tilt:**									
Mo	**HDD18.3**	**90°**	**80°**	**70°**	**60°**	**50°**	**40°**	**30°**	**20°**	**10°**	**0°**
Jan	451	97	104	109	111	111	107	102	93	83	70
Feb	341	91	101	107	112	114	113	109	103	95	84
Mar	248	104	119	132	141	148	150	149	145	137	126
Apr	105	93	113	131	146	158	167	171	171	167	159
May	27	73	93	113	131	147	160	170	176	178	176
Jun	1	67	88	110	130	148	163	175	183	187	186
Jul	0	73	95	117	137	155	170	181	188	191	189
Aug	0	85	107	127	145	159	170	178	181	180	175
Sep	7	98	115	130	142	150	155	156	154	147	138
Oct	100	112	125	135	142	145	144	140	132	121	107
Nov	263	106	115	120	123	123	120	113	104	93	79
Dec	392	99	106	110	111	109	105	98	88	76	63
MEMPHIS, TN		**Solar radiation (kWh/m² mo) on a surface of tilt:**									
Mo	**HDD18.3**	**90°**	**80°**	**70°**	**60°**	**50°**	**40°**	**30°**	**20°**	**10°**	**0°**
Jan	397	104	112	117	119	118	114	107	97	85	71
Feb	296	88	97	103	107	109	108	104	97	88	78
Mar	196	104	120	134	144	152	155	155	151	143	132
Apr	68	92	112	129	145	157	165	169	169	166	158
May	12	74	96	117	136	153	167	178	184	186	185
Jun	0	68	91	115	137	156	173	186	195	199	198
Jul	0	75	100	124	146	166	183	195	203	206	205
Aug	0	93	117	140	160	176	189	196	200	198	191
Sep	3	103	121	137	149	158	164	165	162	155	145
Oct	62	122	136	147	155	159	158	154	146	134	119
Nov	206	104	113	119	123	123	120	114	105	93	80
Dec	347	106	113	118	119	118	113	105	95	83	68
NASHVILLE, TN		**Solar radiation (kWh/m² mo) on a surface of tilt:**									
Mo	**HDD18.3**	**90°**	**80°**	**70°**	**60°**	**50°**	**40°**	**30°**	**20°**	**10°**	**0°**
Jan	450	86	93	98	100	100	97	93	86	77	67
Feb	343	98	108	115	119	121	120	116	109	100	89
Mar	247	104	119	132	141	148	151	150	146	138	128
Apr	102	92	111	129	144	155	163	168	168	164	157
May	26	75	96	116	134	149	162	171	177	179	176
Jun	1	71	94	117	138	157	173	185	194	198	197
Jul	0	75	97	119	139	156	171	182	189	192	191
Aug	0	88	111	132	150	165	176	183	186	184	178
Sep	7	96	113	127	138	147	151	152	149	143	134
Oct	92	127	142	153	160	163	162	156	147	134	118
Nov	251	104	113	118	121	120	117	110	101	89	75
Dec	388	100	107	110	111	109	104	96	86	73	59
DALLAS, TX		**Solar radiation (kWh/m² mo) on a surface of tilt:**									
Mo	**HDD18.3**	**90°**	**80°**	**70°**	**60°**	**50°**	**40°**	**30°**	**20°**	**10°**	**0°**
Jan	311	118	128	134	137	137	133	126	115	102	87
Feb	223	107	119	128	134	136	135	131	123	113	100
Mar	133	106	124	138	150	158	162	162	158	150	139
Apr	40	92	114	134	151	165	175	180	182	179	172
May	6	71	93	114	135	152	167	178	186	189	188
Jun	0	65	89	113	137	158	177	191	202	208	209
Jul	0	71	97	123	148	170	189	203	213	217	217
Aug	0	85	110	133	154	172	186	195	200	200	195
Sep	1	110	131	149	163	173	179	181	177	170	158

Table I.1 (*Continued*)

Oct	32	139	157	171	181	186	186	181	172	158	140
Nov	142	109	119	127	131	131	129	123	114	103	89
Dec	286	120	129	134	136	135	130	122	111	98	81

EL PASO, TX — **Solar radiation (kWh/m² mo) on a surface of tilt:**

Mo	HDD18.3	90°	80°	70°	60°	50°	40°	30°	20°	10°	0°
Jan	318	156	169	177	181	180	173	163	148	129	107
Feb	211	140	155	166	173	175	172	165	153	138	119
Mar	119	137	160	179	194	205	210	210	204	193	178
Apr	38	106	134	160	182	200	213	220	222	219	210
May	3	78	107	136	164	188	208	224	234	238	237
Jun	0	63	91	121	149	176	199	217	230	238	240
Jul	0	66	92	119	144	166	185	200	211	216	216
Aug	0	85	112	137	159	178	193	203	208	208	203
Sep	1	121	146	168	185	198	205	207	204	196	183
Oct	36	148	167	181	191	196	196	191	181	166	147
Nov	174	161	175	185	190	189	184	173	158	139	116
Dec	323	159	171	178	181	178	171	159	143	123	101

LUBBOCK, TX — **Solar radiation (kWh/m² mo) on a surface of tilt:**

Mo	HDD18.3	90°	80°	70°	60°	50°	40°	30°	20°	10°	0°
Jan	406	139	149	156	159	158	152	143	129	113	94
Feb	308	122	134	143	148	150	148	141	132	119	103
Mar	213	125	145	162	175	183	186	185	179	169	154
Apr	94	101	126	148	167	182	192	198	198	194	185
May	24	75	100	125	148	168	184	197	204	207	205
Jun	1	64	89	115	140	163	182	198	209	215	216
Jul	0	68	92	117	141	162	181	195	204	209	209
Aug	0	90	116	140	161	179	193	202	206	205	199
Sep	9	108	127	144	158	167	173	174	171	163	152
Oct	88	140	158	171	180	185	184	179	169	154	136
Nov	242	137	149	156	160	159	154	145	132	115	96
Dec	402	148	158	164	165	162	155	144	129	110	89

SAN ANTONIO, TX — **Solar radiation (kWh/m² mo) on a surface of tilt:**

Mo	HDD18.3	90°	80°	70°	60°	50°	40°	30°	20°	10°	0°
Jan	213	114	125	132	137	138	135	130	121	109	94
Feb	140	99	111	121	128	131	132	129	124	115	104
Mar	76	105	123	138	151	160	165	166	164	158	148
Apr	17	77	96	115	131	145	156	163	166	166	162
May	1	63	84	105	125	142	157	169	177	181	181
Jun	0	57	78	102	125	146	164	179	190	196	198
Jul	0	61	84	109	132	153	171	186	196	202	202
Aug	0	83	109	135	157	177	192	203	209	209	205
Sep	1	108	130	149	165	178	185	189	187	181	171
Oct	17	126	144	158	168	175	176	174	167	156	141
Nov	90	109	120	129	134	136	134	129	121	110	97
Dec	197	116	126	133	136	135	132	125	114	101	86

SALT LAKE CITY, UT — **Solar radiation (kWh/m² mo) on a surface of tilt:**

Mo	HDD18.3	90°	80°	70°	60°	50°	40°	30°	20°	10°	0°
Jan	584	87	93	97	98	97	93	87	79	70	58
Feb	453	103	112	118	121	121	118	113	104	93	81
Mar	348	112	127	139	148	153	154	152	146	136	124
Apr	239	107	127	145	159	169	175	177	175	169	158
May	108	89	113	135	155	172	185	194	199	199	195
Jun	22	87	114	141	165	187	204	218	226	229	226
Jul	0	93	120	147	171	192	209	221	227	229	224
Aug	1	110	136	159	179	194	205	211	212	207	198
Sep	36	133	154	171	183	191	195	193	186	174	158
Oct	202	139	153	162	168	170	166	159	148	132	114
Nov	393	97	103	108	110	109	105	99	91	81	69
Dec	576	85	90	93	93	90	86	79	70	60	48

CHARLOTTESVILLE, VA — **Solar radiation (kWh/m² mo) on a surface of tilt:**

Mo	HDD18.3	90°	80°	70°	60°	50°	40°	30°	20°	10°	0°
Jan	490	116	123	128	129	126	121	112	100	86	69

Table I.1 (*Continued*)

Feb	395	87	95	100	104	104	102	98	91	82	71
Mar	311	102	117	129	138	144	146	145	140	131	120
Apr	146	85	102	117	130	140	147	150	150	147	140
May	51	75	95	114	131	146	158	166	171	172	169
Jun	4	65	83	101	118	132	145	154	160	163	162
Jul	1	70	88	105	121	135	147	155	161	163	161
Aug	1	76	93	109	123	134	143	149	151	150	146
Sep	21	86	100	111	121	127	131	131	129	124	115
Oct	137	140	156	167	174	176	174	168	157	142	124
Nov	284	105	112	117	119	118	114	107	97	85	70
Dec	427	123	130	134	134	131	123	113	100	84	66
NORFOLK, VA		**Solar radiation (kWh/m² mo) on a surface of tilt:**									
Mo	**HDD18.3**	**90°**	**80°**	**70°**	**60°**	**50°**	**40°**	**30°**	**20°**	**10°**	**0°**
Jan	405	99	107	111	113	112	108	101	92	81	67
Feb	329	102	112	119	123	124	122	117	109	98	85
Mar	261	108	123	136	145	151	154	152	147	138	126
Apr	113	95	115	133	149	160	168	172	172	168	160
May	31	76	97	117	136	151	164	173	178	179	176
Jun	2	69	91	112	133	151	166	177	185	188	187
Jul	0	74	95	117	137	154	168	178	185	187	185
Aug	0	85	106	125	141	155	165	171	174	172	166
Sep	2	100	117	131	143	151	155	156	153	146	136
Oct	66	117	130	141	147	150	150	145	137	125	111
Nov	208	102	110	115	118	117	113	107	98	86	73
Dec	327	106	113	117	118	116	110	102	91	78	63
RICHMOND, VA		**Solar radiation (kWh/m² mo) on a surface of tilt:**									
Mo	**HDD18.3**	**90°**	**80°**	**70°**	**60°**	**50°**	**40°**	**30°**	**20°**	**10°**	**0°**
Jan	457	110	118	122	124	123	118	110	100	87	72
Feb	367	107	117	124	128	129	126	120	112	100	86
Mar	283	113	128	141	151	157	159	157	151	142	129
Apr	121	96	115	133	148	159	167	171	170	166	158
May	38	78	99	120	138	154	166	176	181	182	179
Jun	2	70	91	112	131	148	163	174	181	185	184
Jul	0	76	98	120	140	157	171	181	187	189	186
Aug	0	92	113	133	151	165	176	182	184	182	176
Sep	8	100	116	130	141	148	152	152	149	141	131
Oct	102	113	126	136	142	145	144	140	132	121	107
Nov	253	108	117	122	124	123	119	111	101	89	74
Dec	388	110	117	120	121	118	113	104	93	79	63
BURLINGTON, VT		**Solar radiation (kWh/m² mo) on a surface of tilt:**									
Mo	**HDD18.3**	**90°**	**80°**	**70°**	**60°**	**50°**	**40°**	**30°**	**20°**	**10°**	**0°**
Jan	774	79	84	87	87	86	83	78	71	62	52
Feb	661	88	95	100	102	102	99	94	87	78	67
Mar	577	97	109	118	125	129	130	128	123	115	104
Apr	340	99	117	131	143	152	158	159	157	152	142
May	143	86	105	124	140	153	163	169	172	171	166
Jun	32	82	102	122	139	155	167	176	181	183	180
Jul	4	89	110	130	148	163	174	182	186	186	182
Aug	11	100	119	137	152	163	171	175	175	171	163
Sep	78	94	107	118	126	132	134	134	130	123	114
Oct	262	93	102	108	112	113	111	107	100	91	80
Nov	442	70	74	77	78	77	75	70	64	56	47
Dec	646	62	65	67	67	65	61	56	50	43	34
SEATTLE, WA		**Solar radiation (kWh/m² mo) on a surface of tilt:**									
Mo	**HDD18.3**	**90°**	**80°**	**70°**	**60°**	**50°**	**40°**	**30°**	**20°**	**10°**	**0°**
Jan	378	50	52	54	54	53	51	47	43	37	31
Feb	328	71	76	79	80	79	76	72	66	58	49
Mar	303	86	96	103	108	111	111	108	103	96	87
Apr	223	96	111	124	134	141	146	146	144	138	129
May	129	96	116	134	149	162	171	176	178	175	169
Jun	60	88	109	129	146	160	171	179	182	182	177
Jul	13	100	124	145	164	179	190	197	200	197	191

Table I.1 (*Continued*)

Aug	10	101	120	137	151	161	168	171	170	165	156
Sep	52	101	114	125	132	137	138	136	130	122	111
Oct	192	87	95	100	103	104	102	97	91	82	71
Nov	306	49	52	54	55	55	53	50	46	41	35
Dec	394	52	54	55	55	53	51	46	41	35	28
EAU CLAIRE, WI		**Solar radiation (kWh/m² mo) on a surface of tilt:**									
Mo	**HDD18.3**	**90°**	**80°**	**70°**	**60°**	**50°**	**40°**	**30°**	**20°**	**10°**	**0°**
Jan	865	97	102	104	104	101	96	88	78	66	53
Feb	718	111	119	125	127	126	122	114	105	92	78
Mar	578	113	126	136	143	147	147	143	136	126	113
Apr	329	102	119	133	145	153	158	159	157	151	142
May	147	90	111	130	146	160	171	178	181	179	174
Jun	29	85	106	127	145	161	174	183	188	189	186
Jul	6	88	109	129	146	161	172	180	184	183	179
Aug	12	98	117	134	148	160	167	171	171	167	159
Sep	89	102	116	127	136	141	143	141	136	128	117
Oct	291	98	107	114	117	118	116	111	104	94	82
Nov	515	79	83	86	86	84	80	74	66	57	46
Dec	759	89	94	95	95	92	87	80	70	59	47
MADISON, WI		**Solar radiation (kWh/m² mo) on a surface of tilt:**									
Mo	**HDD18.3**	**90°**	**80°**	**70°**	**60°**	**50°**	**40°**	**30°**	**20°**	**10°**	**0°**
Jan	786	96	102	105	106	104	99	93	83	72	60
Feb	651	96	104	109	112	112	109	104	96	86	74
Mar	525	98	111	121	128	133	135	134	129	122	112
Apr	305	93	109	123	135	144	150	152	151	147	139
May	137	92	113	132	150	164	175	183	186	185	180
Jun	24	84	106	128	148	165	178	188	194	196	193
Jul	4	90	112	133	152	168	181	190	194	194	190
Aug	9	99	119	137	152	164	172	177	178	175	168
Sep	76	104	119	131	140	146	148	146	141	132	121
Oct	256	102	111	117	121	122	119	114	106	96	83
Nov	463	81	86	89	90	88	85	79	71	62	51
Dec	681	81	86	88	88	86	82	76	68	59	48
MORGANTOWN, WV		**Solar radiation (kWh/m² mo) on a surface of tilt:**									
Mo	**HDD18.3**	**90°**	**80°**	**70°**	**60°**	**50°**	**40°**	**30°**	**20°**	**10°**	**0°**
Jan	573	42	45	47	49	49	48	47	44	41	36
Feb	471	63	68	72	75	76	75	72	68	62	55
Mar	388	100	113	124	132	137	139	137	132	124	113
Apr	191	73	87	99	109	117	122	125	125	122	117
May	80	79	98	116	133	147	158	166	170	170	167
Jun	11	70	89	108	125	140	153	162	168	171	170
Jul	1	71	89	107	123	137	148	156	161	163	161
Aug	2	77	93	108	121	131	139	143	145	143	138
Sep	34	75	86	95	103	108	111	112	111	107	101
Oct	174	85	95	102	107	110	110	107	102	94	85
Nov	334	60	65	68	70	70	68	65	61	55	47
Dec	486	72	77	79	80	79	75	70	63	55	45
JACKSON HOLE, WY		**Solar radiation (kWh/m² mo) on a surface of tilt:**									
Mo	**HDD18.3**	**90°**	**80°**	**70°**	**60°**	**50°**	**40°**	**30°**	**20°**	**10°**	**0°**
Jan	842	94	98	101	100	97	92	84	74	62	49
Feb	743	100	107	112	113	112	108	101	92	81	67
Mar	662	122	137	148	156	160	160	156	148	137	123
Apr	472	114	134	151	165	175	181	183	180	173	161
May	332	97	121	142	161	178	190	198	202	201	195
Jun	203	94	120	145	168	188	204	216	223	225	222
Jul	75	99	126	151	174	194	209	220	226	226	221
Aug	96	113	137	159	178	192	202	207	207	202	193
Sep	234	119	137	151	161	167	170	168	162	152	138
Oct	449	127	139	147	151	152	148	141	130	115	98
Nov	634	107	114	117	118	115	109	101	89	76	61
Dec	837	73	76	78	77	75	70	64	56	47	37

Table I.1 (*Continued*)

LARAMIE, WY		Solar radiation (kWh/m^2 mo) on a surface of tilt:									
Mo	**HDD18.3**	**90°**	**80°**	**70°**	**60°**	**50°**	**40°**	**30°**	**20°**	**10°**	**0°**
Jan	732	127	134	138	137	133	126	115	101	84	66
Feb	632	137	147	154	157	155	149	139	126	109	90
Mar	566	164	185	201	211	216	216	209	197	180	159
Apr	447	105	125	142	156	166	172	175	173	167	156
May	303	91	115	138	159	176	189	198	203	203	199
Jun	118	86	112	137	160	180	197	209	217	220	218
Jul	25	88	113	136	158	176	191	201	207	209	206
Aug	47	102	124	145	163	176	186	192	193	189	180
Sep	173	101	116	129	138	144	147	146	142	134	123
Oct	391	129	142	152	157	159	156	149	138	124	107
Nov	563	123	131	136	137	134	127	118	105	89	72
Dec	743	124	130	133	132	127	119	108	94	77	58

References and Resources

ASHRAE. 2021. *ANSI/ASHRAE Standard 169-2021, Climate Data for Building Design Standards*. Atlanta, GA: American Society of Heating, Refrigerating, and Air-Conditioning Engineers.

Perez, R., P. Ineichen, R. Seals, J. Michalsky, and R. Stewart. 1990. "Modeling Daylight Availability and Irradiance Components from Direct and Global Irradiance." *Solar Energy* 44: 271–289.

APPENDIX J

Economic Analysis

J.1 ECONOMIC DECISION MAKING

As has been emphasized in this book, economic factors are a major consideration in almost every decision in the building design process. The discussion of economic analysis that follows is necessarily brief but covers the essential principles. For in-depth information on economic analyses of the design and construction process, see the references at the end of this appendix. Publications of the National Institute of Science and Technology (NIST; formerly the National Bureau of Standards) by Marshall and Ruegg (1980a, 1980b) and Ruegg et al. (1978) are key sources for the following information.

The most common economic analysis questions are:

1. How can the cost of different systems that produce the same result be compared? This may apply to the purchase of a single item such as a motor, to the lighting of a room, or to the choice of the type of HVAC installation for an entire building. The principle involved is the same.
2. If improving a system, existing or proposed, will reduce operating costs and if it is known that many different types of improvement are possible, what preliminary economic guidelines can be established to determine whether any proposed investment appears cost-effective? Stated differently, how can the initial (simple) rate of return (ROR) of a proposed investment be determined so that it can be compared to the minimum expected rate of return on the investment?
3. Having determined, in answering the previous question, that several different proposals—each with their own investment and payback (savings) figures—are apparently cost-effective, how can these be compared to determine which one may be the most cost-effective? Put another way, which proposal has the maximum net benefit or savings? This is the type of decision that must be made when comparing energy conservation projects such as lighting control systems.
4. Having established which of the proposals may be the most cost-effective, how can the actual rate of return of the investment be determined? This figure, often referred to as the internal rate of return (IRR), can then be compared to an expected (or required) rate of return to determine whether an investment is economically desirable.
5. What is the (complete) payback period of the investment? At what point does the investment "pay for itself?"

These questions are considered individually in the discussion that follows.

J.2 LIFE-CYCLE COST

To address the first of the five economic analysis questions, let us compare first cost to life-cycle cost (LCC). Cost comparisons made on the basis of first cost are legitimate only when operating/replacement costs are nominal or hold no concern for an owner. Proposals for a concrete slab might represent the former situation, and speculative construction for immediate sale the latter. In such cases, cost comparison is simple—and based upon first cost. However, when cost comparison involves operating/maintenance/replacement costs that vary from system to system, the only reasonable way to determine and compare overall (true) system costs is by life-cycle costing. The basic theory and calculation techniques are well documented (see Fuller and Peterson, 1995; AIA, 1977; Marshall and Ruegg,

Mechanical and Electrical Equipment for Buildings, Fourteenth Edition. Walter Grondzik, Daniel Overbey, and Alison Kwok.

Companion website: www.wiley.com/go/meeb14e

1980a, 1980b; Ruegg et al., 1978), so they will be reviewed only briefly here. For discipline-specific information, see ASHRAE (2023) and IES (2020).

Stated simply, life-cycle cost represents the total cost of an item or system over its entire life-cycle—it is the sum of the first cost and all future costs, less salvage. When the life-cycle of a system being studied does not correspond to that of the building in which it is installed (as is usually the case), the designer must decide which life-cycle to use. For instance, a lighting system may have an estimated life of 15–20 years, whereas the building life is usually at least double that figure. It is often wiser to conduct the analysis over the shorter time period, since (almost certainly) changes in technology will present an entirely different picture when the time comes to replace the system. This is true of most mechanical/electrical systems being designed today.

Life-cycle cost, in equation form, is

$$LCC = IC + MC + AC + OC \qquad \text{(Eq. J.1)}$$

where

LCC = life-cycle cost

IC = investment cost (i.e., first cost minus salvage)

MC = maintenance and repair cost

AC = amortization (replacement) cost (not always included in life-cycle calculations)

OC = operating costs, including labor and energy

All costs are typically expressed in terms of present value, which is explained in this section. Costs can also be annualized. The present value approach tends to give a clearer answer to the economic decision questions posed in Section J.1. When calculating costs, the designer must include related ancillary costs. For instance, in comparing lighting systems, the costs must reflect the impact of lighting on the HVAC system (capacity and energy use) and the wiring system. For lighting system cost analysis information, see *IES RP-31-20: Recommended Practice: Economic Analysis of Lighting* (IES, 2020).

When the life-cycle analysis period is short—say, two years maximum—the last three components of Equation J.1 can be assumed to be constant without introducing an unacceptable error. For longer periods, however, this is not a valid assumption because of escalating labor and energy costs. Since these two factors weigh heavily in all economic comparisons, and since they generally apply in different proportions in the systems being considered, escalation of such costs will change the economic balance over the equipment life span. It is therefore necessary to conduct an analysis that reflects this escalation. Let us consider the present value and net present value of the decision being made.

Present value (*PV*) is a way to express the current (present) value of future (projected) cash flow after factoring in a discount rate (which factors in the "time value" of money, risk, and opportunity cost—i.e., we know a dollar today is not worth the same as a dollar in the future).

The current value of future payments of 1.00 unit, made at the end of each of a series of *n* periods, is

$$PV = \frac{(1+i)^n - 1}{i(1+i)^n} \qquad \text{(Eq. J.2)}$$

where

PV = present value

i = interest (discount) rate expressed as a decimal (e.g., 8% = 0.08)

n = number of periods

The factor *i* is sometimes called the "interest rate," but when calculating *PV*, it is more accurately called the "discount rate" being that it also reflects inflation (either actual or anticipated).

If, however, the future payments escalate, the *PV* expression becomes

$$PV = \frac{1 - \left(\dfrac{K}{1+i}\right)^n}{\dfrac{1+i}{K} - 1} \qquad \text{(Eq. J.3)}$$

where *K* is the escalation rate per period and the remaining items are as in Equation J.2. Thus, for a 5% annual escalation rate, $K = 1.05$, and so on.

The present value (*PV*) of various numbers (*n*) of future payments beginning at the end of the first payment period is provided for fixed interest (discount) rates without escalation in Table J.1 and for escalation rates (*K*) per period of 3%, 5%, 8%, and 10% in Tables J.2, J.3, and J.4 for interest (discount) rates (*i*) of 8%, 10%, and 12%, respectively. Other values can be calculated from Equation J.3.

Net present value (*NPV*) is more complex than *PV* as it constitutes the sum of the *PV* of all cash inflows and outflows associated with an investment such that

$$NPV = \sum \left[\frac{R_t}{(1+i)^t}\right] - C_0 \qquad \text{(Eq. J.4)}$$

where

NPV = net present value

t = time period

R_t = net cash inflow at time t

i = interest (discount) rate expressed as a decimal (e.g., 8% = 0.08)

C_0 = cost of initial investment

APPENDICES

Table J.1 Present Value of *n* Future Payments Beginning at the End of the First Period

n	Interest (Discount) Rate per Period						
	6%	8%	10%	12%	15%	20%	25%
1	0.94	0.93	0.91	0.89	0.87	0.83	0.80
2	1.83	1.78	1.74	1.69	1.63	1.53	1.44
3	2.67	2.58	2.49	2.40	2.28	2.11	1.95
4	3.47	3.31	3.17	3.04	2.86	2.59	2.36
5	4.21	3.99	3.79	3.61	3.35	2.99	2.69
6	4.92	4.62	4.36	4.11	3.78	3.33	2.95
7	5.58	5.21	4.87	4.56	4.16	3.61	3.16
8	6.21	5.75	5.34	4.97	4.49	3.84	3.33
9	6.80	6.25	5.76	5.33	4.77	4.03	3.46
10	7.36	6.71	6.15	5.65	5.02	4.19	3.57
11	7.89	7.14	6.50	5.94	5.23	4.33	3.66
12	8.83	7.54	6.81	6.19	5.42	4.44	3.73
13	8.85	7.90	7.10	6.42	5.58	4.53	3.78
14	9.30	8.24	7.37	6.63	5.72	4.61	3.82
15	9.71	8.56	7.61	6.81	5.85	4.68	3.86
16	10.11	8.85	7.82	6.97	5.95	4.73	3.89
17	10.48	9.12	8.02	7.12	6.05	4.78	3.91
18	10.83	9.37	8.20	7.25	6.13	4.81	3.93
19	11.16	9.60	8.37	7.37	6.20	4.84	3.94
20	11.47	9.82	8.51	7.47	6.26	4.87	3.95
21	11.76	10.02	8.65	7.56	6.31	4.89	3.96
22	12.04	10.20	8.77	7.65	6.36	4.91	3.97
23	12.30	10.37	8.88	7.72	6.40	4.93	3.98
24	12.55	10.53	8.99	7.78	6.43	4.94	3.98
25	12.78	10.68	9.08	7.84	6.46	4.95	3.99

As an example of *NPV*, let us consider an initial investment cost of \$100,000. The first year of cash inflow (R_1) is \$40,000. The same is true with the second year (R_2) and the third year (R_3). The discount rate (i) is 10% (0.10):

$$PV \text{ of } R_1 \text{ inflow} = \$40{,}000/(1+0.10)^1 = \$40{,}000/1.10 = \$36{,}364$$

$$PV \text{ of } R_2 \text{ inflow} = \$40{,}000/(1+0.10)^2 = \$40{,}000/1.21 = \$30{,}058$$

$$PV \text{ of } R_3 \text{ inflow} = \$40{,}000/(1+0.10)^3 = \$40{,}000/1.331 = \$30{,}053$$

Sum of the present values of cash inflows:

$$\$36{,}364 + \$30{,}058 + \$30{,}053 = \$96{,}475$$

Finally, subtract the initial investment to determine the *NPV*:

$$NPV = \$96{,}475 - \$100{,}000 = -\$3525$$

Since the *NPV* is a negative value (−\$3525), this \$100,000 investment would not be considered profitable. How is this possible? Because a \$40,000 inflow in three years is not worth as much presently. In order for the R_3 inflow to be worth \$40,000 in "today's dollars," one would have to factor in the 10% discount rate compounded by three years. If the *PV* of R_3 inflow is \$30,053, then it would require \$9947 additional dollars to maintain a *PV* of \$40,000.

It is apparent that the result of a life-cycle cost analysis depends heavily on accurate forecasting when using escalations, and that comparative results can readily shift not only with escalation estimates of the

Table J.2 Present Value of *n* Future Payments Beginning at the End of the First Period and Escalating at *K* per Period—8% Discount

	Interest (Discount) Rate per Period 8% Annual (Periodic) Escalation Rate *K*			
n	1.03	1.05	1.08	1.10
1	0.95	0.97	1.00	1.02
2	1.86	1.92	2.00	2.06
3	2.73	2.84	3.00	3.11
4	3.56	3.73	4.00	4.19
5	4.35	4.60	5.00	5.28
6	5.10	5.44	6.00	6.40
7	5.82	6.26	7.00	7.54
8	6.50	7.06	8.00	8.70
9	7.15	7.84	9.00	9.88
10	7.78	8.59	10.00	11.08
11	8.37	9.33	11.00	12.30
12	8.94	10.04	12.00	13.55
13	9.48	10.73	13.00	14.82
14	9.99	11.41	14.00	16.11
15	10.48	12.06	15.00	17.43
16	10.95	12.70	16.00	18.77
17	11.40	13.32	17.00	20.13
18	11.82	13.92	18.00	21.53
19	12.23	14.51	19.00	22.94
20	12.62	15.08	20.00	24.39
21	12.99	15.63	21.00	25.86
22	13.34	16.17	22.00	27.35
23	13.68	16.69	23.00	28.88
24	14.00	17.20	24.00	30.43
25	14.30	17.69	25.00	32.01

Table J.3 Present Value of *n* Future Payments Beginning at the End of the First Period and Escalating at *K* per Period—10% Discount

	Interest (Discount) Rate per Period 10% Annual (Periodic) Escalation Rate *K*			
n	1.03	1.05	1.08	1.10
1	0.94	0.96	0.98	1.00
2	1.81	1.87	1.95	2.00
3	2.63	2.74	2.89	3.00
4	3.40	3.57	3.82	4.00
5	4.12	4.36	4.73	5.00
6	4.80	5.12	5.63	6.00
7	5.43	5.84	6.51	7.00
8	6.02	6.53	7.37	8.00
9	6.57	7.18	8.22	9.00
10	7.09	7.81	9.05	10.00
11	7.58	8.41	9.87	11.00
12	8.03	8.98	10.67	12.00
13	8.46	9.53	11.46	13.00
14	8.85	10.05	12.23	14.00
15	9.23	10.55	12.99	15.00
16	9.58	11.27	13.74	16.00
17	9.90	11.48	14.47	17.00
18	10.21	11.91	15.19	18.00
19	10.50	12.32	15.90	19.00
20	10.76	12.72	16.59	20.00
21	11.02	13.09	17.27	21.00
22	11.25	13.45	17.94	22.00
23	11.47	13.80	18.59	23.00
24	11.68	14.12	19.24	24.00
25	11.87	14.44	19.87	25.00

Table J.4 Present Value of *n* Future Payments Beginning at the End of the First Period and Escalating at *K* per Period—12% Discount

	Interest (Discount) Rate per Period 12% Annual (Periodic) Escalation Rate *K*			
n	1.03	1.05	1.08	1.10
1	0.92	0.94	0.96	0.98
2	1.77	1.82	1.89	1.95
3	2.54	2.64	2.79	2.89
4	3.26	3.41	3.66	3.82
5	3.92	4.14	4.49	4.74
6	4.52	4.82	5.29	5.64
7	5.08	5.45	6.07	6.52
8	5.59	6.05	6.82	7.38
9	6.06	6.61	7.54	8.23
10	6.49	7.13	8.23	9.07
11	6.89	7.63	8.90	9.89
12	7.26	8.09	9.55	10.69
13	7.59	8.52	10.17	11.49
14	7.90	8.92	10.77	12.26
15	8.19	9.30	11.35	13.03
16	8.45	9.66	11.91	13.78
17	8.69	9.99	12.45	14.51
18	8.91	10.31	12.97	15.23
19	9.11	10.60	13.47	15.95
20	9.30	10.87	13.95	16.64
21	9.47	11.13	14.42	17.34
22	9.63	11.37	14.87	18.00
23	9.78	11.60	15.30	18.66
24	9.91	11.81	15.72	19.31
25	10.04	12.01	16.12	19.95

various components but also with the length of the life-cycle being analyzed.

J.3 INITIAL (SIMPLE) RATE OF RETURN AND SIMPLE RETURN ON INVESTMENT

This analysis addresses the second question in Section J.1: Is a proposed cost-saving investment worth examining in detail? That is, is the initial (simple) rate of return (ROR) sufficient? The difference between initial and actual rates of return is that the former is simply

$$\text{initial ROR} = (\text{annual returns}/\text{cost of investment}) \times 100 \quad \text{(Eq. J.5)}$$

where "annual returns" is a broad financial term that could entail annual savings or profit. Please note the difference between ROR (rate of return) and ROI (return on investment). While often conflated, ROR measures the performance of an investment over a specific period (e.g. annually). ROI looks at the entire investment and measures the overall profitability of the investment relative to its cost (e.g., the investment is projected to return X over a specific period).

As an example of ROR, let us consider two HVAC system options for a building. Option A is an energy-code-compliant system for $200,000. Option B costs 30% more but will reduce the projected $85,000 annual energy costs by 20%. What is the initial (simple) ROR and simple ROI for choosing system Option B and taking on its cost premium over Option A?

$$\begin{aligned}\text{initial ROR} &= (\$85{,}000 \times 0.20)/(\$200{,}000 \times 1.30)\\ &= (\$17{,}000)/(\$260{,}000)\\ &= 0.0654 \times 100\\ &= 6.54\%\end{aligned}$$

$$\begin{aligned}\text{simple ROI} &= \$17{,}000 \text{ annual savings}/\$60{,}000 \text{ premium}\\ &= 3.5 \text{ years}\end{aligned}$$

Contrast initial (simple) ROR with the actual IRR. IRR depends upon the total life-cycle of a project and the total savings accrued. Unlike initial (simple) ROR, the discount (interest) rate is factored into the actual IRR. The calculation method for the actual IRR is given in Section J.5.

Similarly, a simple ROI calculation is not the same calculation as the actual (discounted) payback period. The calculation method for the actual (discounted) payback period is explained in Section J.6. (Also, note that in the example above, the 3.5-year simple ROI is only for the "investment" of the HVAC Option B cost premium and not the complete system itself. The simple ROI for the entire system would be reciprocal of the initial ROR and equates to 15.3 years. Both figures may be valuable in their own right.)

Initial ROR is a useful criterion in that it can be very easily calculated and then compared to a minimum desired rate of return on investment. The minimum rate is emphasized, since the actual rate of return is always lower than the simple initial rate of return, except where savings escalate during the project life-cycle. If a preliminary comparison indicates that the simple rate of return exceeds the minimum by a margin of a few percentage points, it is likely that a project is economically feasible, and a detailed analysis can be undertaken.

J.4 COST-EFFECTIVENESS COMPARISON

This analysis is undertaken in answer to the third question in Section J.1. That is, several proposals are being considered, all of which meet the stated criteria. Which one is most desirable from a cost-effectiveness viewpoint? (If only one proposal is under consideration, the analysis should proceed directly to the IRR calculation described in the next section.)

The difference between life-cycle project cost and life-cycle project savings (i.e., net lifetime savings), is a measure of cost-effectiveness. Therefore, to compare the cost-effectiveness of two proposals, it is necessary simply to calculate this differential. This can be expressed as an equation:

$$\text{net savings} = \text{life-cycle savings} - \text{life-cycle cost} \quad \text{(Eq. J.6)}$$

Since this is true for any investment, a comparative differential can be calculated by using the differential in each term of the life-cycle cost equation, that is

$$\Delta_{\text{net savings}} = \Delta_{\text{savings}} - \Delta_{\text{cost}} \quad \text{(Eq. J.7)}$$

Taking a very simple example, assume the following conditions.

	Project A	Project B
Investment	$10,000	$15,000
Life-cycle (years)	10	8
Annual savings (US$)	2000	3200
Initial rate of return (%)	20	21.3
Discount rate (%)	12	12

Since Project B has an apparently higher initial ROR, we would formulate the differential as B minus A. Using Table J.1, the present value (*PV*) of lifetime savings for Project B (without escalation) is $3200 × 4.97 or $15,904; and that of Project A is

APPENDICES

$2000 × 5.65 or $11,120. Therefore, the differential cost-benefit (CB) of B minus A is

$$\Delta_{CB} = \Delta_{\text{lifetime savings}} - \Delta_{\text{project costs}} \qquad \text{(Eq. J.8)}$$

or

$$\begin{aligned}\Delta_{CB} &= (\$15{,}904 - \$11{,}120) - (\$15{,}000 - \$10{,}000)\\ &= \$4784 - \$5000\\ &= -\$216\end{aligned}$$

indicating that Project A is slightly more cost-effective despite the higher rate of return of Project B.

If we include escalation in cost-effectiveness comparison, the lifetime savings of the longer-life Project A will shrink more than the savings of shorter-life Project B. Use Table J.4 and assume 8% annual escalation (i.e., $K = 1.08$). The present value (*PV*) of lifetime savings for Project B would be $3200 × 6.82 or $21,824; and that of Project A would be $2000 × 8.23 or $16,460.

$$\begin{aligned}\Delta_{CB} &= (\$21{,}824 - \$16{,}460) - (\$15{,}000 - \$10{,}000)\\ &= \$5364 - \$5000\\ &= \$364\end{aligned}$$

With higher rates of escalation, Project B would be more cost-effective.

J.5 INTERNAL RATE OF RETURN

The fourth question in Section J.1 asks, having established which of the proposal may be the most cost-effective, how can the actual rate of return of the investment be determined?

The internal rate of return (IRR) represents the rate of return at which lifetime savings are exactly equal to lifetime costs. IRR cannot be calculated directly, but only by trial and error. Assume a rate of return, calculate the present value (*PV*) of savings, and compare the *PV* to the lifetime costs. Because of this difficulty, IRR is frequently neglected, and the initial (simple) ROR is used instead; although, as discussed in Section J.4, this can readily give misleading results. To illustrate the calculation of IRR, return to the very simple example in Section J.4 and calculate IRR for Projects A and B.

For Project A, a present value factor of (initial investment/annual savings) $10,000/$2000 = 5.0 is needed within its 10-year life-cycle. Referring to Table J.1, note that at a discount (interest) rate of 15%, the present value factor is 5.02, indicating that IRR is slightly above this required level of 5.0. For Project B, an 8-year present value factor of $15,000/$3200, or 4.69, is needed. Referring again to Table J.1, note that this required factor of 4.69 falls between the interest (discount) rates of 12% and 15%. A series of trials with Equation J.2 will eventually arrive at an interest (discount) rate of 13.7% required for Project B, which is 1.4% less than the 15% required for Project A, confirming the conclusion reached in Section J.4—that Project A is slightly more cost-effective despite the higher rate of return of Project B.

J.6 ACTUAL (DISCOUNTED) PAYBACK PERIOD

The fifth and final question in Section J.1 seeks to ascertain the (complete) payback period of the investment.

The payback period—or return on investment (ROI)—referred to in most proposals is the reciprocal of the initial rate of return (ROR) and is usually referred to as the simple payback period or simple ROI (e.g., using the example in Section J.3, the initial ROR of 6.54% for Option B would equate to a simple payback period of 15.3 years). However, just as the actual rate of return differs from the simple rate (and is usually lower), the actual or discount payback period is different from, and usually longer than, the simple one. This period is, logically, the period of time required for the accumulated net savings to equal the initial investment, with all figures expressed in present value dollars. To illustrate, return to the example given in Section J.4.

1. *Project A.* As in the IRR calculation, a present value factor of $10,000/$2000, or 5.0, is needed at a 12% discount rate. From Table J.1, find this to be about 8.1 years. This compares to a simple payback period of 5 years. Note that this is not the reciprocal of IRR.
2. *Project B.* Here a present worth factor of $15,000/$3200, or 4.7, is needed. At a 12% discount rate, from Table J.1, this is about 7.3 years. This compares to a simple payback period of 4.7 years.

Note from these results the very important fact that a shorter simple payback period does not necessarily indicate a more cost-effective investment. In this example, Project A has a better internal rate of return precisely because it yields good savings for a longer period. Summing up the results of the simple study, we have:

	Project A	Project B
Investment	$10,000	$15,000
Life-cycle (years)	10	8
Annual savings (US$)	2000	3200
Initial (simple) rate of return (ROR) (%)	20	21.3
Actual internal rate of return (IRR) (%)	15.1	13.7
Simple payback period (years)	5	4.7
Actual (discounted) payback period (years)	8.1	7.3

This shows very clearly that although initial rate of return and simple payback period are indications of a proposal's value, they are useless in comparative studies. Such studies should lean on the internal rate of return and the actual (discounted) payback period.

References and Resources

AIA. 1977. *Life-cycle Cost Analysis—A Guide for Architects*. Washington, DC: American Institute of Architects.

ASHRAE. 2023. *ASHRAE Handbook—HVAC Applications, Chapter 39*. Atlanta, GA: American Society of Heating, Refrigerating and Air-Conditioning Engineers, Inc.

Fuller, S. K. and S. R. Peterson. 1995. *Life-Cycle Costing Manual for the Federal Energy Management Program*. Gaithersburg, MD: National Institute of Standards and Technology. fire.nist.gov/bfrlpubs

IES. 2020. *ANSI/IES RP-31-20: Recommended Practice: Economic Analysis of Lighting*. New York: Illuminating Engineering Society.

Marshall, H. E. and R. T. Ruegg. 1980a. *Energy Conservation in Buildings: An Economics Guidebook for Investment Decision*. National Bureau of Standards Handbook No, 132. Washington, DC: NBS.

Marshall, H. E. and R. T. Ruegg. 1980b. *Simplified Energy Design Economics*. National Bureau of Standards Special Publication No. 544. Washington, DC: NBS.

Ruegg, R. T., et al. 1978. *Life-Cycle Costing*. National Bureau of Standards Publication No. 113. Washington, DC: National Institute of Standards and Technology.

APPENDIX K

Sound Transmission Properties

K.1 SOUND TRANSMISSION DATA FOR WALLS

To use the data presented in this appendix:

1. Find the desired wall construction Type Code in Table K.1.
2. Find the desired sound transmission class (STC) corresponding to the selected wall construction Type Code.
3. Refer to Table K.3 for details of wall construction to confirm that the choice is reasonable.
4. Refer to Table K.2 for wall thickness, weight, sound transmission class, fire rating, and the transmission losses at standard octave midpoints.

EXAMPLE K.1 An interior masonry partition with an STC between 50 and 55 is desired. Of interest is the TL at 1000 Hz.

1. From Table K.1, find Type Code "c."
2. From Table K.2, note that constructions W6, W12b, W15b, and W22b are acoustically suitable for the proposed use.
3. From consideration of the information in Table K.3, decide that construction W15b is most suitable for the proposed use.
4. From Table K.2, find that construction W15b has a transmission loss of 55 dB at 1000 Hz.

■

Mechanical and Electrical Equipment for Buildings, Fourteenth Edition. Walter Grondzik, Daniel Overbey, and Alison Kwok.

Companion website: www.wiley.com/go/meeb14e

Table K.1 Sound Transmission Class: Walls

Type Code		
a. Wooden stud	e. Staggered stud	i. Absorbent blankets or fill
b. Metal stud	f. Plaster	j. Fiber board
c. Masonry	g. Gypsum wallboard	k. Lead
d. Concrete	h. With resilient element	l. Gypsum core board

STC	Type	Item No.	STC	Type	Item No.
63	d,f	W4	46	a,f	W30a
62[a]	c,f,j,m	W23	46	a,f	W31a
56	c	W7	46	a,f,k	W31c
56[a]	c,f	W8	46	a,e,g,i	W37
55[a]	b,g,i	W63	45	c,d	W10b
54[a]	c,f,m	W22b	45	a,f,i	W38
54	b,f,h	W52	45	b,f,h	W50a
53[a]	d,f	W2	45	g,l,m	W85a
53	c,f,h	W15b	44	c	W12a
52[a]	c,f	W6	44	a,e,g	W34a
51	b,f	W44a	44	a,e,g	W35a
51	b,g,h,i	W67	44	a,f,h	W40a
50	c,k	W12b	43	c,d	W10a
50	b,g,i	W60	43[a]	d,j,m	W21
50	g,i,l,m	W85b	43	a,e,f	W35a
49[a]	c,f,m	W22a	43	a,e,f	W36a
48	c,d	W9	43	b,f,k	W43b
48	a,e,f,i	W36b	42[a]	c,f	W5
48	b,f,k	W43c	42	a,f	W14a
47	d	W1	41	b,f	W43a
47	a,g,k	W28b	41	b,g	W55a
47	a,f,k	W31b	40	c,f	W13
47	a,g,h	W39a	40	a,g	W32a
46	c,f	W11a	39	a,g	W28a
46	c,f,h	W15a	36	g,l	W80

Source: Adapted from HUD/FHA/NBS (1971).
[a]Field measurement.

APPENDICES

Reference

HUD/FHA/NBS. 1971. *Report*. Washington, DC: U.S. Department of Housing and Urban Development.

Table K.2 Sound Transmission Loss: Walls

Designation	Thickness, in. [mm]	Weight, lb/ft² [kg/m²]	Transmission Loss (dB) at frequency 125	250	500	1K	2K	4K	STC	Fire Rating (hr)
W1	3 [76]	39 [190]	35	40	44	52	58	64	47	0.5
W2	7 [178]	80 [390]	39	42	50	58	64	—	53	3
W4	≈ 16 [406]	184 [898]	50	54	59	65	71	68	63	4+
W5	5½ [140]	55 [268]	34	34	41	50	66	—	42	2.5
W6	10 [254]	100 [488]	41	43	49	55	57	—	52	4+
W7	12 [305]	121 [590]	45	45	53	58	60	61	56	4+
W8	25 [635]	280 [1366]	50	53	52	58	61	—	56	4+
W9	12 [305]	79 [386]	46.5	44	46	52	54	56	48	4
W10a	6 [152]	34 [166]	32	33	40	47	51	48	43	1
W10b	6 [152]	34 [166]	37	36	42	49	55	58	45	1
W11a	5¼ [133]	35.8 [175]	36	37	44	51	55	62	46	2
W12a	3¾ [133]	26.1 [127]	40	40	40	48	55	56	44	1.5
W12b	5 [127]	31 [151]	41	46	46	56	63	67	50	1.5
W13	4 [102]	21.5 [105]	39	34	38	43	48	46	40	3
W14a	5 [127]	23.4 [114]	37	42	39	44	49	49	42	4
W15a	5 [127]	27 [132]	38	37	44	51	56	59	46	3
W15b	6 [152]	31 [151]	45	44	50	55	56	59	53	4
W20	7 [178]	22.9 [112]	32	46	49	53	58	66	52	3
W21	10¼ [260]	37 [181]	41	42	46	51	52	—	43	Not available
W22a	12 [305]	100 [488]	37	41	48	60	60.5	—	49	4+
W22b	12 [305]	100 [488]	40	44	55	67.5	70	—	54	4+
W23	18 [457]	120 [586]	48	54	58	64	69	—	62	4+
W28a	5 [127]	6 [29]	21	28	35	42	45	41	39	0.5
W28b	≈ 5⅛ [130]	12 [58]	27	37	43	52	56	—	47	0.5
W30a	5¾ [146]	13.4 to 15.7 [65 to 77]	32	37	42	47	47	63	46	0.75
W31a	5¾ [146]	13.4 to 15.7 [65 to 77]	32	37	42	48	48	63	46	0.75
W31b	≈ 5⅞ [149]	—	33	41	45	52	55	65	47	0.75
W31c	≈ 5⅞ [149]	17-19 [83-93]	32.5	40	43	47	50	62	46	0.75
W32a	5½ [140]	8.2 [40]	27	31	39	45	52.5	48	40	1
W34a	5 [127]	6.2 [30]	36	36	40	47	52	45	44	0.5
W35a	6½ [165]	13.4 [65]	41	41	46	49	41	54	44	1.5
W35b	5¾ [146]	15.6 [76]	48	46	48	48	48	59	43	1
W36a	6¼ [159]	11.1 [54]	36	33	42	42	41	51	43	0.75
W36b	6¼ [159]	12.8 [62]	37	37	49	50	52	66	48	1
W37	5¾ [146]	13.8 [67]	39	40	42	47.5	55	51.5	46	0.5
W38	⅜; [137]	14.2 [69]	39	45	48	50	44	54	45	1
W39a	6¼ [159]	6.7 [33]	30	40	46	50	49	49	47	1
W40a	≈ 6½ [165]	14.4 [70]	43	41	48	50	42	56	44	1
W43a	3⅜; [86]	12.3 [60]	27	37	43	46	39	47	41	0.75
W43b	3½ [89]	15.2 [74]	35	43	45	47	48	58	43	0.75
W43c	≈ 3½ [89]	18.2 [89]	36	45	47	50	53	61	48	0.75
W44a	5 [127]	15.7 [77]	34	38	47	50	52	58	51	1
W50a	5¼ [133]	13 [63]	35	46	48	51	48	43	45	0.75
W52	5 [127]	19 [93]	50	52	55	56	52	60	54	1
W55a	4⅞ [124]	6 [29]	29	36	40	46	40	46	41	1
W60	3½ [89]	5.4 [26]	34	40	47	50	53	54	50	1
W63	6⅛ [156]	11.5 [56]	36	47	51	57	57	62	55	2
W67	6½ [165]	11.3 [55]	41	46	49	51	50	60	51	2
W80	2¼ [57]	10.2 [50]	34	34	37	38	39	45	36	1
W85a	5⅛ [130]	14.6 [71]	36	35	45	51	53	57	45	3
W85b	6 [152]	12.8 [62]	37	37	54	56	56	62	50	3

Source: Adapted from HUD/FHA/NBS (1971). SI units were added by authors of this book (not part of original source).

Table K.3 Construction Characteristics of Walls (for other data, see Table K.2)

Designation	Description	Section Sketch
Solid Concrete		
W1	3-in.-thick [76-mm] solid concrete wall poured in-situ in test opening. All surface cavities were sealed with thin mortar mix.	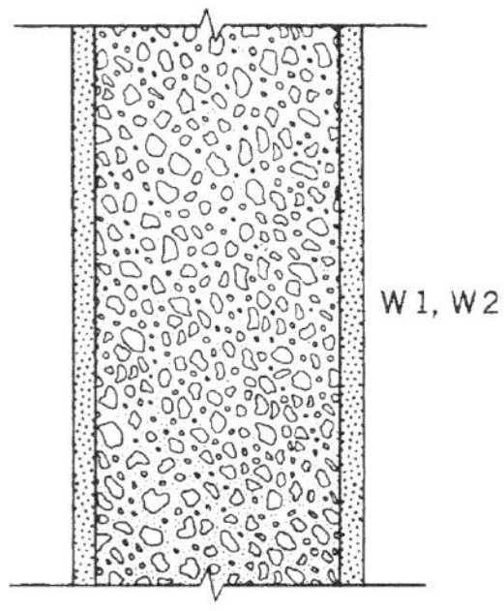
W2	6-in.-thick [152-mm] concrete wall with ½-in.-thick [13-mm] layer of plaster on both sides.	
W4	Wall of 4, 6, and 8 × 8 × 16 in. [102, 152, 204 × 204 × 406 mm] sand and gravel aggregate solid concrete blocks; on each side, ¼- to ½-in.-thick [6- to 13-mm] layer of cement, gypsum plaster, and sand.	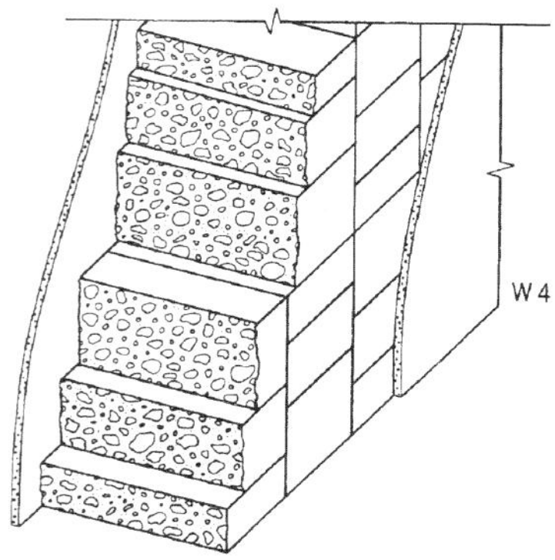
W5	4½-in.-thick [114-mm] brick wall with ½-in.-thick [13-mm] layer of plaster on each side.	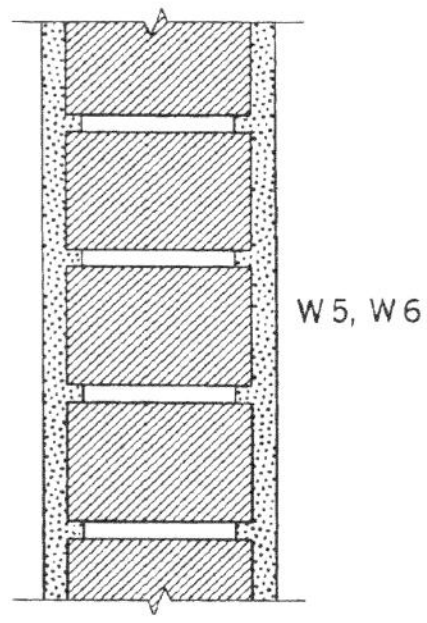
W6	9-in.-thick [229-mm] brick wall with ½-in.-thick [13-mm] layer of plaster on each side.	
W7	12-in.-thick [305-mm] brick wall.	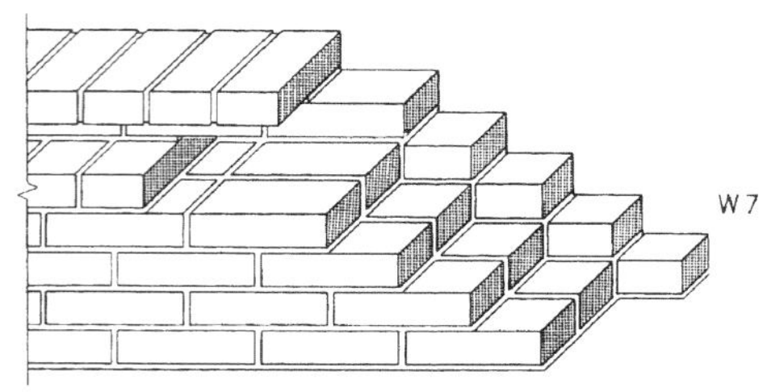
W8	24-in.-thick [610-mm] stone wall with ½-in.-thick [13-mm] layer of plaster on both sides.	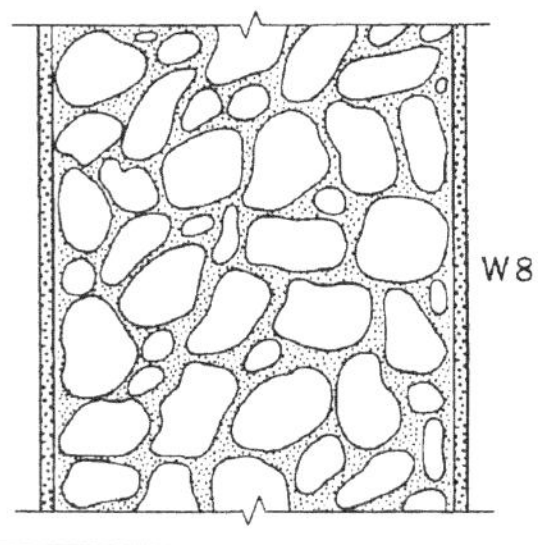

Table K.3 ***(Continued)***

Designation	Description	Section Sketch
Hollow Concrete Block		
W9	12-in. [305-mm] wall made of hollow 8 × 8 × 12 in. [204 × 204 × 305 mm] and 8 × 4 × 16 in. [204 × 102 × 406 mm] concrete blocks.	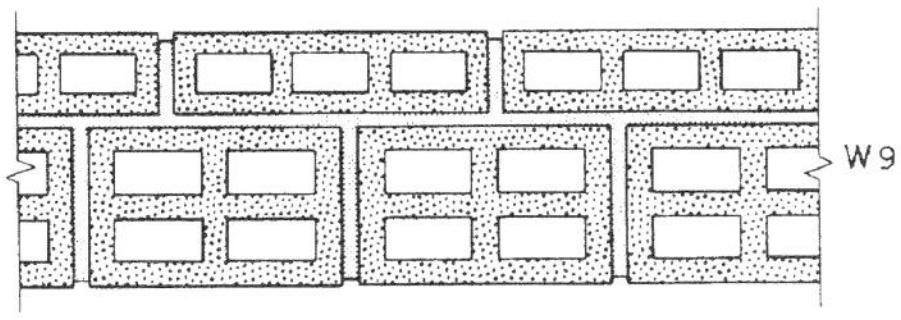
W10a	6-in. [152-mm] hollow concrete blocks constructed with vertical mortar joints staggered.	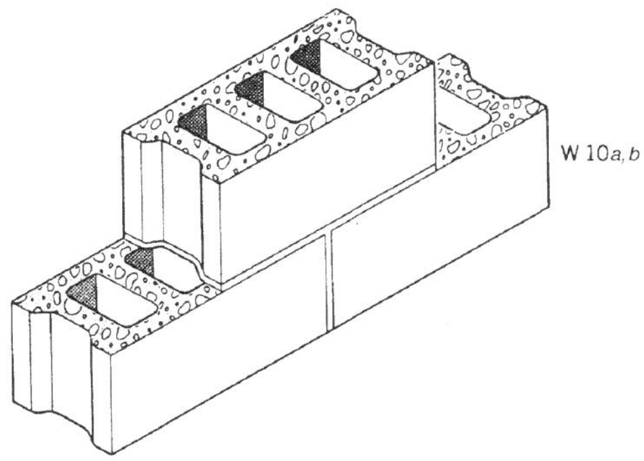
W10b	Similar to W10a except wall painted.	
Cinder Block		
W11a	4 × 8 × 16-in. [102 × 204 × 406-mm] hollow cinder blocks; on each side, $\frac{5}{8}$ in. [16 mm] of sanded gypsum plaster.	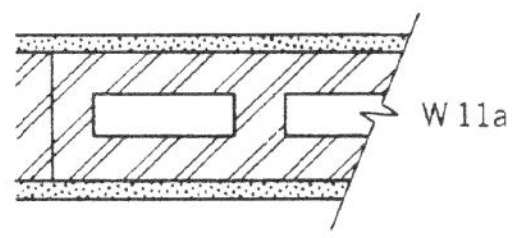
Cement Block		
W12a	$3\frac{5}{8} \times 7\frac{3}{4} \times 13\frac{1}{2}$-in. [92 × 197 × 343-mm] lightweight-aggregate cement blocks with $\frac{1}{2}$-in. [13-mm] mortar joints; three coats of masonry paint applied to each side of partition.	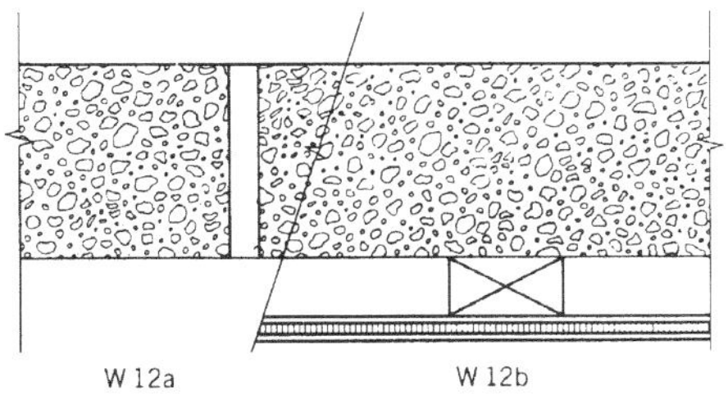
W12b	Same as W12a, except that 1 × 2-in. [25 × 50-mm] furring strips were nailed vertically to partition on one side; 1/16-in. [1.5-mm] layer of lead, 3.94 lb/ft^2 [19.2 kg/m^2], nailed to furring strips, $\frac{1}{4}$-in. [6-mm] plywood-covered lead with joints caulked.	
Hollow Gypsum Block		
W13	3-in. [76-mm] hollow gypsum blocks cemented together with $\frac{3}{8}$-in. [10-mm] mortar joints; on each side, $\frac{1}{2}$-in. [13-mm] sanded gypsum plaster.	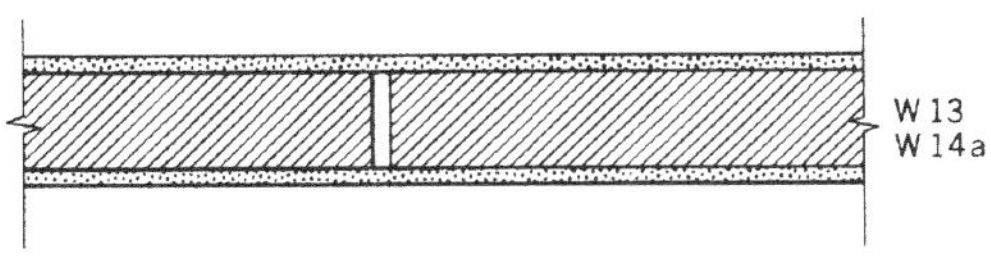
W14a	4-in. [102-mm] hollow gypsum blocks cemented together with $\frac{3}{8}$-in. [10-mm] mortar joints; on each side, $\frac{1}{2}$-in. [13-mm] sanded gypsum plaster.	
Hollow Gypsum Block, Resilient on One Side, Plaster on Both Sides		
W15a	3 × 12 × 30-in. [76 × 305 × 762-mm] hollow gypsum blocks with $\frac{1}{2}$-in. [13-mm] mortar joints. On one side, $\frac{7}{16}$-in. [11-mm] sanded gypsum plaster; on the other side, resilient clips, spaced 18 in. [457 mm] apart on centers vertically and 16 in. [406 mm] apart on centers horizontally, held to $\frac{3}{4}$-in. [19-mm] metal channels 16 in. [406 mm] on centers, to which expanded metal lath was wire-tied; $\frac{1}{16}$-in. [27-mm] sanded gypsum plaster; $\frac{1}{16}$-in. [1.5-mm] white-coat finish applied to both sides.	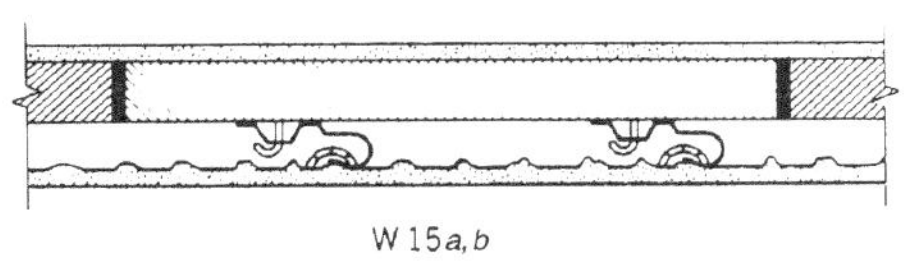

Table K.3 (*Continued*)

Designation	Description	Section Sketch
W15b	Similar to W15a except that 4 × 12 × 30-in. [102 × 305 × 762 mm] gypsum blocks were used.	
Hollow Concrete		
W21	Precast concrete hollow wall panels with in-situ concrete posts and beams. Panels have 1 ½-in.-thick [38-mm] concrete shells with 6 ¼-in. [159-mm] air space between them. Layer of fiberboard ½ in. [13 mm] thick is adhered to the exposed surfaces of the panel.	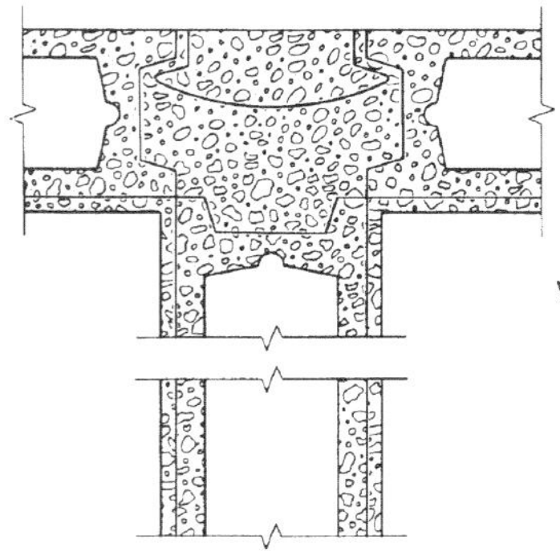 W 21
Double Walls		
W22a	Double wall with 4 ½-in.-thick [114-mm] brick leaves separated by a 2-in. [50-mm] cavity (wire ties between leaves); ½-in. [13-mm] plaster on exposed sides.	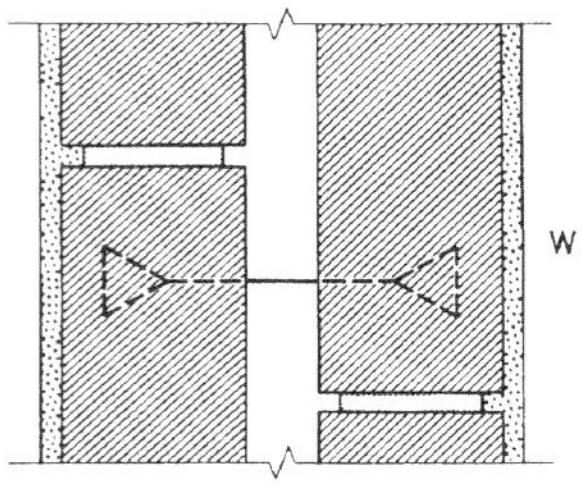 W 22
W22b	Similar to W22a but no wire ties between the leaves.	
W23	Double wall with 4 ½-in.-thick [114-mm] brick leaves, 6-in. [152-mm] cavity (no ties); on exposed sides, ½-in. [13-mm] plaster on 1-in.-thick [25-mm] wood-wool slabs mortared to the brick walls.	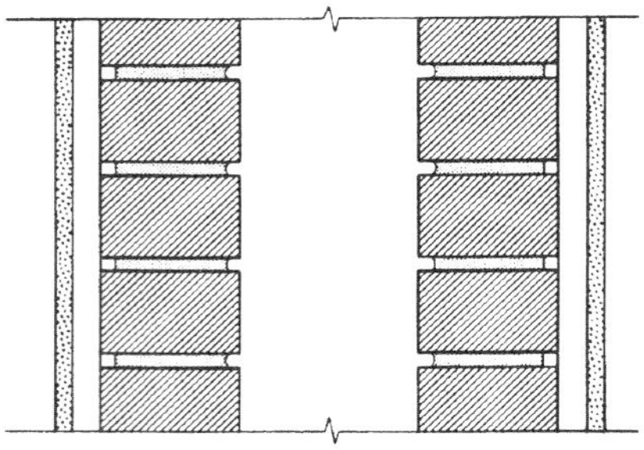 W 23
Wood Stud Walls		
W28a	2 × 4-in. [50 × 100-mm] wooden studs, 16 in. [406 mm] on centers, ½-in. [13-mm] gypsum wallboard nailed to each side. All joints taped and finished.	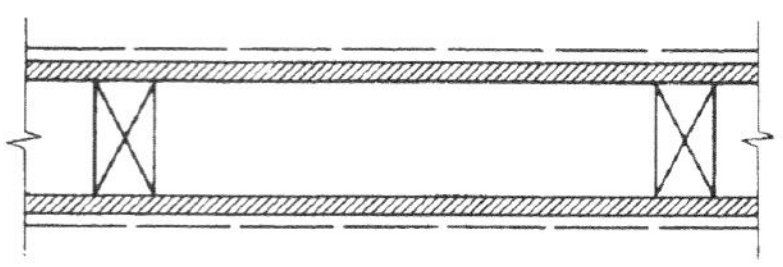 W 28*a, b*
W28b	Similar to W28a except that a layer of lead, 2.95 lb/ft^2 [14.4 kg/m^2], was laminated to each side of the panel.	
W30a	2 × 4-in. [50 × 100-mm] wooden studs, 16 in. [406 mm] on centers, attached to 2 × 4-in. [50 × 100-mm] wooden floor and ceiling plates, ⅜-in. [10-mm] gypsum lath nailed to studs on both sides, ½-in. [13-mm] sanded plaster with white-coat finish.	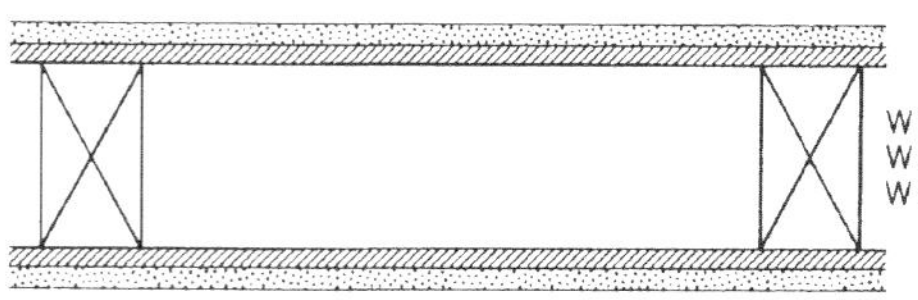 W 30*a* W 31*a, b, c* W 32*a*
W31a	2 × 4-in. [50 × 100-mm] wooden studs, 16 in. [406 mm] on centers, ⅜-in. [10-mm] gypsum lath nailed to studs on both sides, ½-in. [13-mm] sanded plaster with white-coat finish.	
W31b	Similar to W31a except that a 0.065-in.-thick [1.7-mm] layer of lead weighing 3.85 lb/ft^2 [18.8 kg/m^2] was laminated to each side of the panel.	

Table K.3 (*Continued*)

Designation	Description	Section Sketch
W31c	Similar to W31a except that a 0.13-in.-thick [3.3-mm] layer of lead weighing 7.9 lb/ft^2 [38.5 kg/m^2] was laminated to one side of the panel.	
W32a	2 × 4-in. [50 × 100-mm] wooden studs, 16 in. [406 mm] on centers; on each side, two layers of ⅜-in. [10-mm] gypsum wallboard were cemented together; joints in exposed surfaces taped and finished.	
Staggered Wood Stud Walls		
W34a	2 × 3-in. [50 × 75-mm] wooden studs, 16 in. [406 mm] on centers, staggered 8 in. [203 mm] on centers, attached to 2 × 4-in. [50 × 100-mm] wooden plates at ceiling and floor; ½-in. [13-mm] gypsum wallboard nailed 7 in. [178 mm] on centers on both sides to studs. All joints taped and finished.	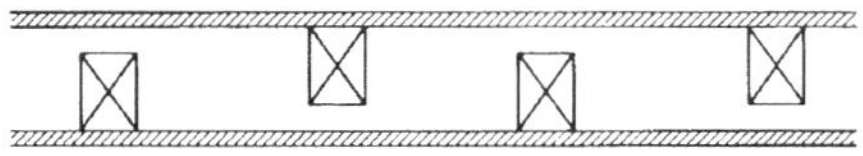 W 34a
W35a	2 × 3-in. [50 × 75-mm] wooden studs, 16 in. [406 mm] on centers, staggered 8 in. [203 mm] on centers [attached to 2 × 4-in. [50 × 100-mm] wooden plates at floor and ceiling]; two layers of ⅝-in. [16-mm] tapered-edge gypsum wallboard, first layer nailed 7 in. [178 mm] on centers, second layer nailed 16 in. [406 mm] on centers. All exposed joints taped and finished.	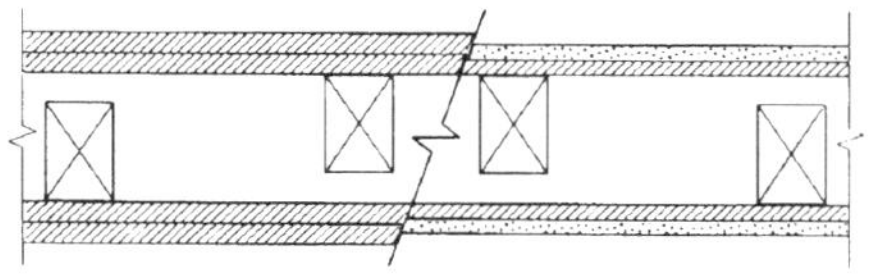 W 35a W 35b
W35b	Similar to W35a except that the wall was constructed with ⅜-in. [10-mm] perforated gypsum lath and ½-in. [13-mm] sanded gypsum plaster with white-coat finish.	
W36a	2 × 4-in. [50 × 100-mm] wooden studs, 16 in. [406 mm] on center, staggered 8 in. [203 mm] on center and offset ½ in. [13 mm]. On each side, ⅜-in. [10-mm] gypsum lath nailed to studs, ½-in. [13-mm] gypsum vermiculite plaster, machine-applied, and a hand-applied white-coat finish.	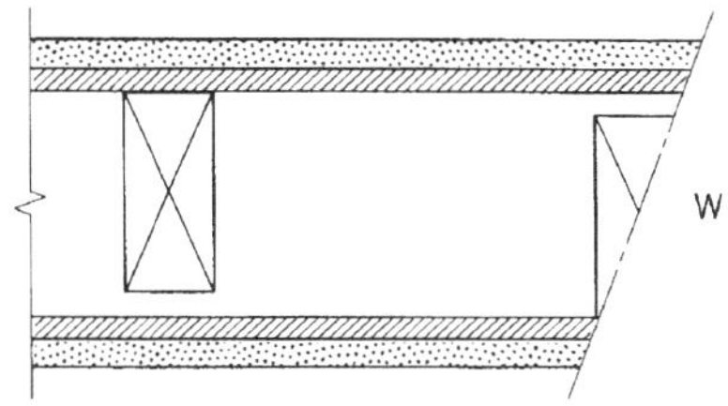 W 36a
W36b	Same as W36a except that the space between the studs contained vermiculite fill with a density of 6.3 lb/ft^3 [101 kg/m^3].	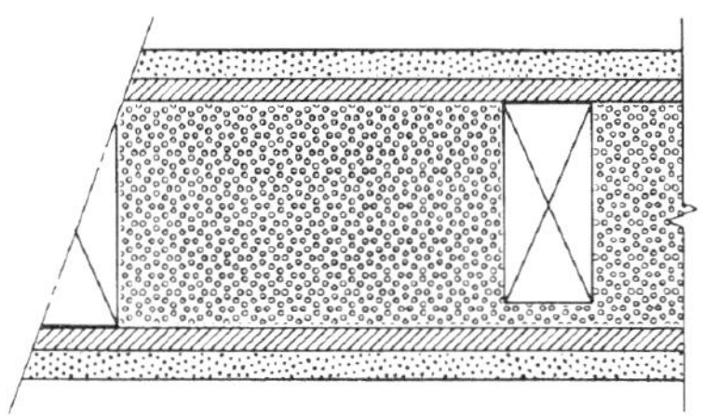 W 36b
W37	2 × 4-in. [50 × 100-mm] wooden studs, 16 in. [406 mm] on center, staggered 8 in. [203 mm] on center, attached to a 2 × 4¾-in. [50 × 120-mm] wooden floor and ceiling plates; ½-in. [13-mm] gypsum wallboard nailed on both sides to studs, 0.9-in. [23-mm] wood-fiber wool blanket stapled on the inside of one side of the wall. All joints taped and finished.	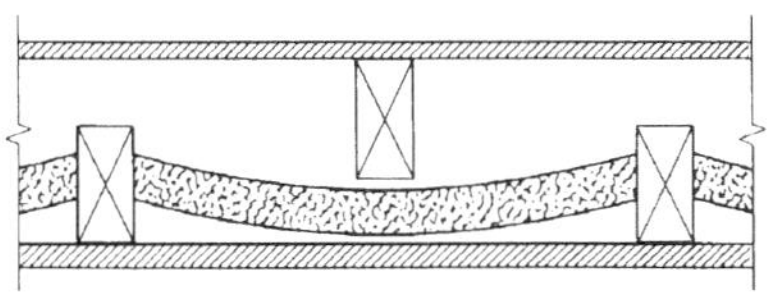 W 37

APPENDICES

Table K.3 (*Continued*)

Designation	Description	Section Sketch
Slotted Wood Studs		
W38	2 × 4-in. [50 × 100-mm] slotted wooden studs, 16 in. [406 mm] on centers, attached to 2 × 4-in. [50 × 100-mm] wooden floor and ceiling plates; ⅜-in. [10-mm] gypsum lath nailed 7 in. [178 mm] on center to studs, ½-in. [13-mm] gypsum plaster with white-coat finish applied to both sides. 3-in. [75-mm] mineral fiber batts stapled between studs.	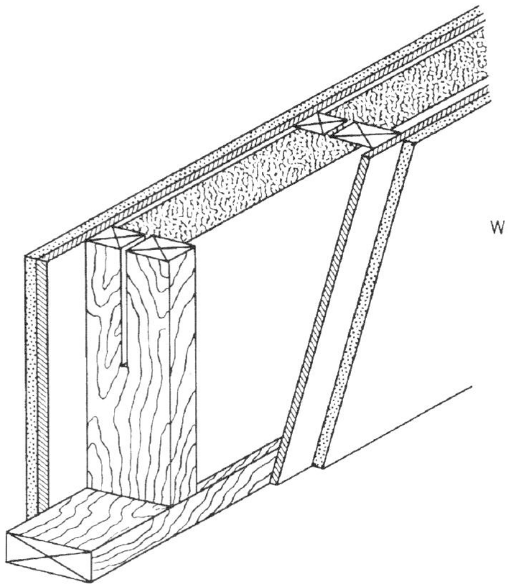 W 38
Wood Studs; Resilient Mounting		
W39a	2 × 4-in. [50 × 100-mm] wooden studs, 16 in. [406 mm] on centers, attached to 2 × 4-in. [50 × 100-mm] wooden floor and ceiling plates; resilient channels nailed horizontally to both sides of studs 24 in. [610 mm] on center, ⅝-in. [16-mm] gypsum wallboard screwed 12 in. [305 mm] on center to channels. All joints taped and finished.	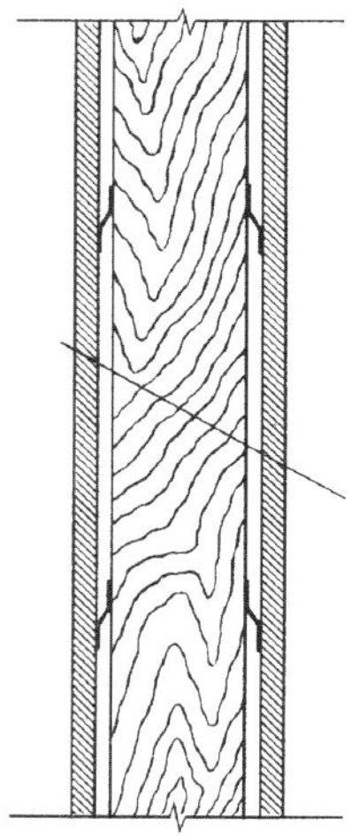 W 39a
W40a	2 × 4-in. [50 × 100-mm] wooden studs, 16 in. [406 mm] on center; resilient clips, nailed to studs on both sides, held ⅜-in. [10-mm] gypsum lath, ½-in. [13-mm] sanded gypsum plaster with white-coat finish.	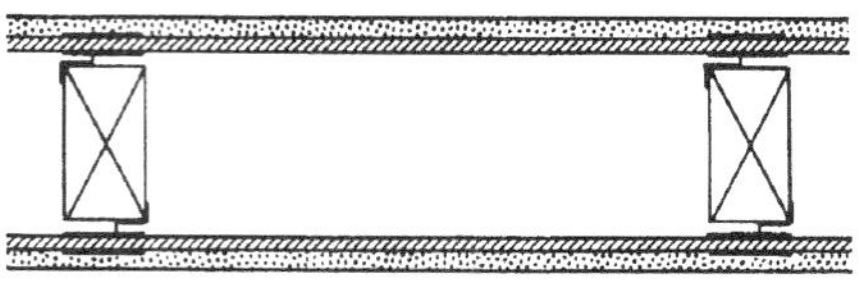 W 40a
Steel Truss Stud Wall		
W43a	1⅝-in. [41-mm] steel truss studs; ⅜-in. [10-mm] gypsum lath, ½-in. [13-mm] plaster on both sides.	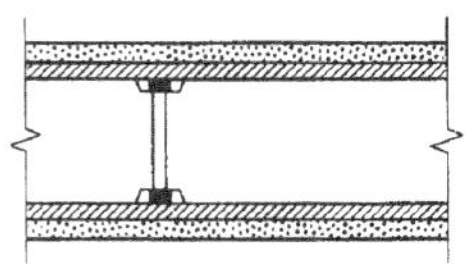 W 43*a, b, c* W 44*a*
W43b	Similar to W43a except that a layer of lead, 2.95 lb/ft^2 [14.4 kg/m^2], was laminated to one side of the partition.	
W43c	Similar to W43a except that a layer of lead, 2.95 lb/ft^2 [14.4 kg/m^2], was laminated to each side of the partition.	
W44a	3¼-in. [83-mm] steel truss studs, 24 in. [610 mm] on center, attached to metal floor and ceiling tracks; on both sides, ⅜-in. [10-mm] perforated gypsum lath attached with wire clips wire-tied to studs, ½-in. [13-mm] sanded gypsum plaster.	

Table K.3 (*Continued*)

Designation	Description	Section Sketch
Steel Truss Studs; Resilient Mounting		
W50a	2 ½-in. [64-mm] steel truss studs 16 in. [406 mm] on center, ⅜-in. [10-mm] gypsum lath attached with resilient clips to studs, ½-in. [13-mm] plaster applied to both sides.	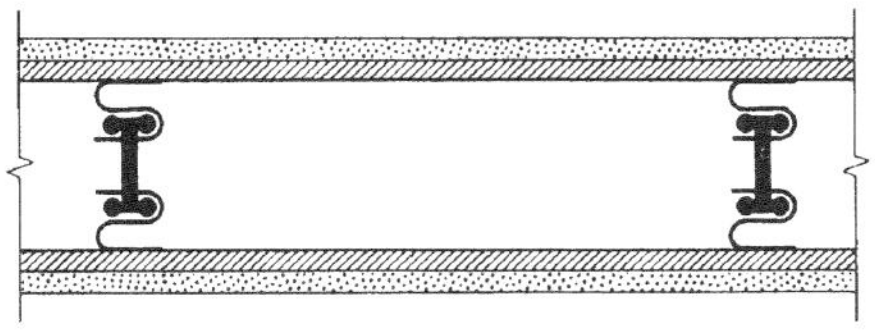 W 50a
W52	3 ¼-in. [83-mm] steel truss studs, 16 in. [406 mm] on center; on each side, resilient clips fastened 16 in. [406 mm] on center to studs, ¼-in. [6-mm] metal rod wire-tied to clips, diamond mesh metal lath wire-tied to metal rods, ¾-in. [19-mm] sanded gypsum plaster.	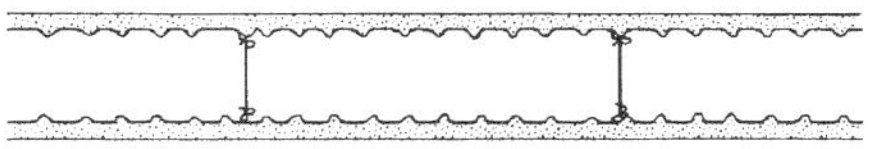 W 52
Metal Channel Stud Wall		
W55a	3 ⅝-in. [92-mm] metal channel studs, 24 in. [610 mm] on center, set into 3 ⅝-in. [92-mm] metal floor and ceiling runners; ⅝-in. [16-mm] gypsum wallboard screwed to studs on both sides. All joints taped and finished.	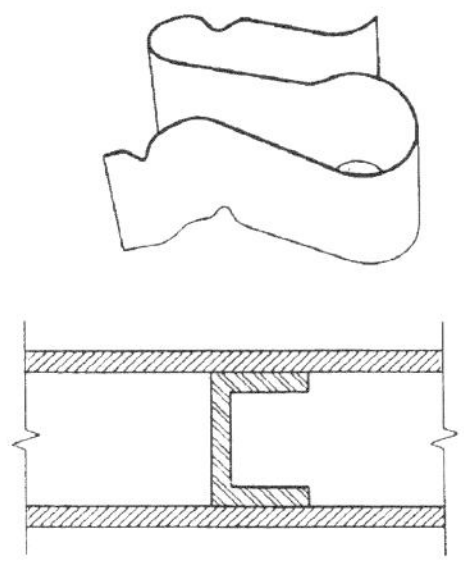 W 55a
W60	2 ½-in. [64-mm] metal channel studs, 24 in. [610 mm] on center, set in 2 ½-in. [64-mm] metal floor and ceiling runners; ½-in. [13-mm] vinyl-coated gypsum wallboard adhesively attached and screwed to studs on both sides. All joints sealed with caulking compound. Aluminum batten strips screwed 12 in. [305 mm] on center to gypsum board at joints; top and bottom finished with aluminum ceiling and base trim. 2-in. [50-mm] mineral fiber blankets hung between studs.	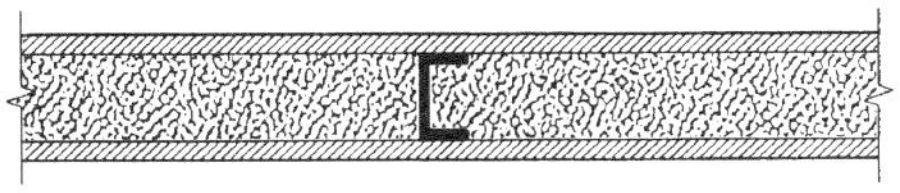 W 60
W63	3 ⅝-in. [92-mm] metal channel studs, 24 in. [610 mm] on center, set into 3 ⅝-in. [92-mm] metal runners, which were attached through continuous beads of nonsetting resilient caulking compound to floor and ceiling, respectively. Two layers of ⅝-in. [16-mm] gypsum wallboard attached to both sides of studs. First layer screwed 8 in. [203 mm] on center at joints and 12 in. [305 mm] on center in field; second layer laminated and screwed 24 in. [610 mm] on center to first layer, with joints staggered 24 in. [610 mm]. 1 ½-in.-thick [38-mm] mineral fiber felt, 3 lb/ft^3 [48 kg/m^3], stapled between studs. All exposed joints taped and finished. The ¼-in. [6-mm] clearance around the perimeter closed with a nonsetting resilient caulking compound.	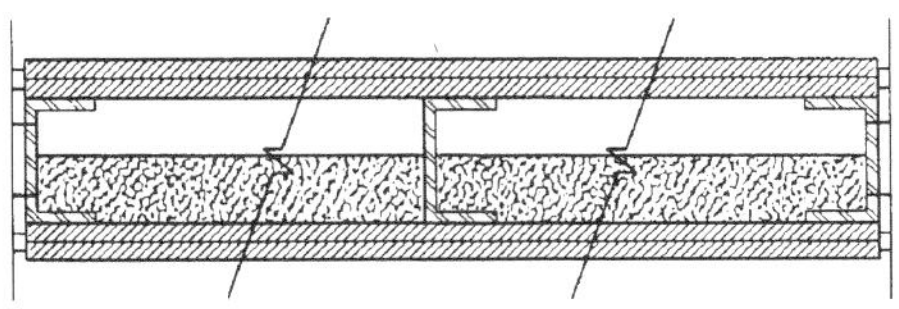 W63
W67	3 ⅝-in. [92-mm] metal channel studs, 24 in. [610 mm] on center, set in 3 ⅝-in. [92-mm] metal floor and ceiling runners; ⅝-in. [16-mm] gypsum wallboard screwed to studs on both sides. On one side, resilient channels screwed horizontally, 24 in. [610 mm] on center to inner layer; ⅝-in. [16-mm] gypsum wallboard screwed to channels. On the other side, ⅝ -in. [16-mm] gypsum wallboard laminated directly to inner layer. 3-in. [75-mm] mineral fiber blankets hung between studs. All exposed joints taped and finished.	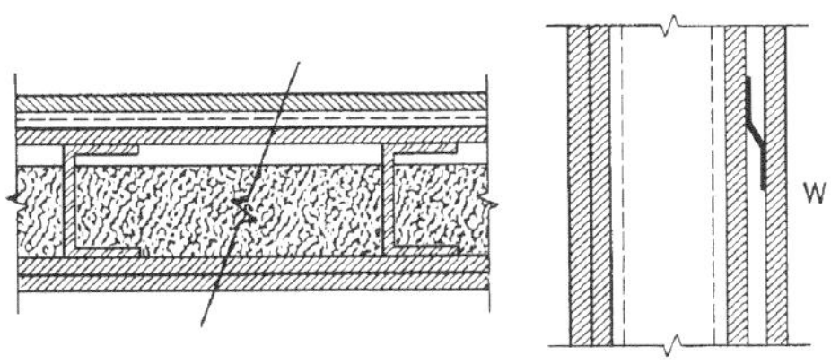 W 67
Gypsum Partitions		
W80	24-in.-wide [610-mm] panels constructed of 1 × 24-in. [25 × 610-mm] gypsum core board offset 1 ½ in. [38 mm] at edges to form tongue-and-groove edge; ⅝-in. [16-mm] vinyl-faced gypsum wallboard laminated to both sides of core board. Panels inserted into two-piece metal floor and ceiling tracks. Gypsum-to-gypsum screws at ¼ and ½ points along vertical edges of face boards.	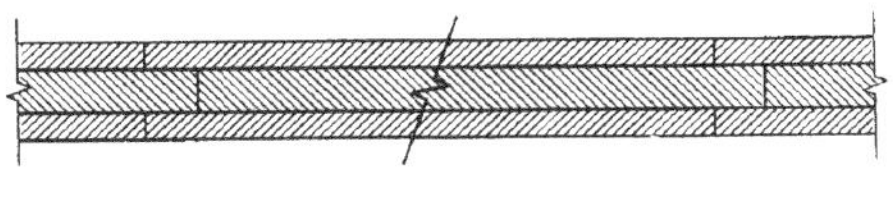 W 80

Table K.3 (*Continued*)

Designation	Description	Section Sketch
W85a	Double wall with $1\frac{3}{8}$-in. [35-mm] air space. Each leaf consisted of 24-in.-wide [610-mm] panels of $\frac{5}{8}$-in. [16-mm] gypsum core board strips, $7\frac{1}{2}$ and $4\frac{3}{8}$ in. [191 and 111 mm] wide, offset $1\frac{1}{2}$ in. [38 mm] at edges to form tongue and groove; $\frac{5}{8}$-in. [16-mm], vinyl-faced, gypsum wallboard laminated to both sides of core board strips. Panels screwed 12 in. [305 mm] on center to $1\frac{1}{4} \times 1$-in. [32 × 25-mm] angle floor and ceiling runners.	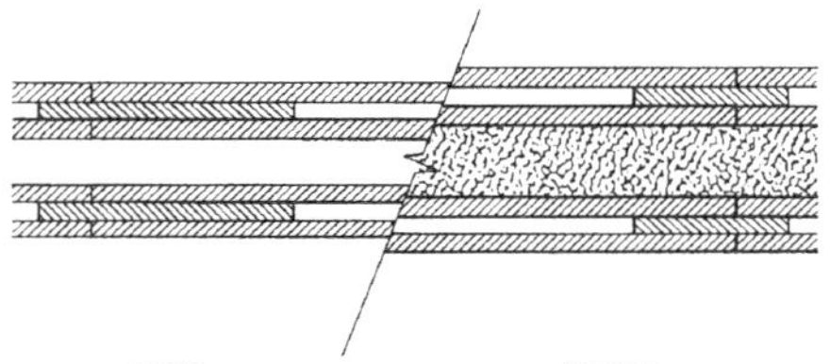
W85b	Similar to W85a except that space between leaves was $2\frac{1}{8}$ in. [54 mm] and contained 2-in. [50-mm] mineral fiber blankets stapled to one leaf. $\frac{1}{4}$-in. [6-mm] perimeter clearance closed with a nonsetting resilient caulking compound. Vertical face layer joints sealed with joint compound.	

Source: Adapted from HUD/FHA/NBS (1971). SI units were added by authors of this book (not part of original source).

K.2 SOUND TRANSMISSION AND IMPACT INSULATION DATA FOR FLOOR/CEILING CONSTRUCTIONS

To use these data:

1. Find the desired construction Type Code in Table K.4 or K.5. Table K.4 lists constructions by sound transmission class (STC). Table K.5 lists constructions by impact insulation class (IIC).
2. Find desired STC/IIC ratings corresponding to the selected Type Code.
3. Refer to Table K.7 for details of construction to confirm appropriateness of selection.
4. Refer to Table K.6 for thickness, weight, STC, IIC, fire rating, and transmission loss at standard octave midpoints.

EXAMPLE K.2 A standard wooden joist floor-ceiling construction is desired. The assembly must have a minimum IIC of 55; any STC above 35 is acceptable.

1. Since IIC is the determining factor, refer to Table K.5. The basic Type Code is "a."
2. From Table K.5, note that none of the listed "a" constructions gives an IIC of 55. However, from Section 23.24 of the text, note that the addition of carpeting will add 10 to 27 points to the impact insulation class (IIC), making items F34, F38a, F39a, and F30 all suitable. This approach avoids special and resilient constructions in order to comply with the requirement for standard (i.e., simple) construction.
3. From Table K.7, select construction F30 as being most appropriate for the application of carpeting.
4. From Table K.6, find (without carpet): thickness $9\frac{1}{2}$ in. [240 mm], STC 34, IIC 32 (+ carpet), fire rating $\frac{1}{4}$ hour. With high-pile carpeting on a foam rubber pad, this construction will have an IIC > 55, STC > 35.

■

References and Resources

HUD/FHA/NBS. 1971. *Report*. Washington, DC: U.S. Department of Housing and Urban Development.

Table K.4 Sound Transmission Class: Floor/Ceiling Constructions

Type Code	
a. Wooden joist	f. With resilient ceiling element
b. Metal joist	g. With resilient floor element
c. Concrete or masonry	h. With carpeting
d. Plaster ceiling	i. With absorbent blankets
e. Gypsum board ceiling	j. With separate ceiling joists

STC	IIC	Type	Item No.	STC	IIC	Type	Item No.
55[a]	57[a]	c,d,f	F14	46[a]	42[a]	c,d	F23
54[a]	64[a]	c,d,f	F17b	45	44	a,e,f	F44
52	80	a,e,i,j	F48	44	42	c,f	F3-2d
51[a]	53[a]	c,d,f	F10	44	41	c	F3-1a
51[a]	48[a]	c,d	F7a	44	80	c,h	F2-1a
50[a]	53[a]	c,e,f	F25	44	29	c	F1-c
50[a]	51[a]	c,e	F27	44	25	c	F1a
50[a]	48[a]	c,d	F16	43[a]	43[a]	a,d	F32b
49[a]	48[a]	c,d	F9	42[a]	32[a]	c	F22
48[a]	47[a]	c,d	F12	40	32	a,e,i	F40a
48	33	b,c,d	F60a	39[a]	37[a]	a,e	F34
47	62	b,c,e	F58	37	33	a,e	F38a
47[a]	42[a]	c,d	F24	37	32	a,e	F39a
47	59	b,c,d,h	F57b	34[a]	32[a]	a,e	F30
47	37	b,c,d	F57a	29[a]	32[a]	a,e	F35a
46	74	b,c,d	F60c	29	56	a,e,h	F35b

[a]Field measurement.

Table K.5 Impact Insulation Class: Floor/Ceiling Constructions

Type Code	
a. Wooden joist	f. With resilient ceiling element
b. Metal joist	g. With resilient floor element
c. Concrete or masonry	h. With carpeting
d. Plaster ceiling	i. With absorbent blankets
e. Gypsum board ceiling	j. With separate ceiling joists

IIC	STC	Type	Item No.	IIC	STC	Type	Item No.
80	52	a,e,h,i,j	F48	44	45	a,e,f	F44
80	44[b]	c,h	F2-1a	43[a]	43[a]	a,d,g	F32b
74	46	b,c,d,h	F60c	42[a]	47[a]	c,d	F24
64[a]	54[b]	c,d,g	F17b	42[a]	46[a]	c,d	F23
62	47	b,c,e,h	F58	42	44[b]	c,g	F3-2(d)
59	47[b]	b,c,d,h	F57b	41	44[b]	c	F3-1(a)
57[a]	55[a]	c,d,g	F14	37	47	b,c,d	F57a
56[a]	29[b]	a,e,h	F35b	37[a]	39[a]	a,e	F34
53[a]	51[a]	c,d,g	F10	33	48	b,c,d	F60a
53[a]	50[a]	c,e,g	F25	33	37	a,e	F38a
51[a]	50[a]	c,e	F27	32[a]	42[a]	c	F22
48[a]	51[a]	c,d	F7a	32	40	a,e,i	F40a
48[a]	50[a]	c,d	F16	32	37	a,e	F39a
48[a]	49[a]	c,d,g	F9	32[a]	34[a]	a,e	F30
47[a]	48[a]	c,d	F12	32[a]	29[a]	a,e	F35a
				29	44[b]	c	F1c
				25	44	c	F1a

[a]Field measurement.

[b]Estimated on the basis of similar structures.

Table K.6 Floor/Ceiling Sound Transmission and Construction Data

Designation	Thickness, in. [mm]	Weight, lb/ ft² [kg/m²]	Transmission Loss (dB) at frequency 125	250	500	1K	2K	4K	STC	IIC	Fire Rating (hr)
F1a	4 [102]	53 [259]	47	42	45	56	58	66	44	25	1
F7a	5 1/4 [134]	61 [298]	41	42	47	54	59	63	51	48	2
F9	6 5/8 [168]	65 [317]	40	42	46	48	57	62	49	48	2
F10	8 1/4 [210]	90 [439]	38	43	47	53	60	—	51	53	2.5
F12	10 [254]	62 [303]	38	40	44	51	56	59	48	47	3
F14	9 1/2 [242]	83 [405]	38	44	52	55	60	—	55	57	3
F16	8 1/8 [206]	65 [317]	40	42	46	52	58	—	50	48	2
F17b	9 1/4 [235]	57 [278]	38	46	52	59	64	—	54	64	2
F22	6 1/4 [158]	28 [137]	34	34	38	45	55	61	42	32	0.75
F23	9 1/2 [242]	45 [220]	33	37	43	52	58	62	46	42	0.75
F24	10 1/4 [260]	65 [317]	34	37	43	52	57	—	47	42	0.75
F25	10 [254]	45 [220]	30	38	46	58	64	—	50	53	0.75
F27	7 5/8 [194]	50 [244]	36	40	47	54	58	—	50	51	0.75
F30	9 1/2 [242]	7 [34]	19	24	31	35	45	—	34	32	0.25
F32b	11 [280]	12 [59]	30	31	41	47	52	—	43	43	0.75
F34	10 1/4 [260]	9.9 [48]	15	32	44	48	54	53	39	37	1
F35a	10 [254]	9.2 [45]	14	17	30	44	47	52	29	32	—
F38a	11 3/4 [298]	9 [44]	30	38	36	43	48	49	37	33	0.5
F39a	11 7/8 [302]	9.5 [46]	22	32	36	45	49	56	37	32	1
F40a	11 7/8 [302]	10 [49]	25	36	38	46	51	57	40	32	1
F44	10 1/2 [266]	10.1 [49]	43	41	41	52	50	60	45	44	0.75
F48	12 3/8 [314]	10.7 [52]	35	42	52	56	69	74	52	80	0.75
F57a	18 9/16 [472]	23.2 [113]	33	44	45	46	58	62	47	37	3
F58	21 1/2 [546]	20.4 [100]	27	37	45	54	60	65	47	62	1
F60a	11 [280]	38.2 [186]	42	44	44	51	51	61	48	33	1.5
F60c	11 5/8 [295]	39 [190]	39	43	44	52	52	65	46	74	—

Source: Adapted from HUD/FHA/NBS (1971). SI units were added by authors of this book (not part of original source).

APPENDICES

Table K.7 Construction Characteristics of Floor Types

Code	Description	Section Sketch
Reinforced Concrete Slab		
F1a	4-in.-thick [100-mm] reinforced concrete slab, isolated from support structure. Concrete was reinforced with 6 × 6-in. No. 6 AWG reinforcing mesh [150 × 150 at 4 mm] placed at the centerline horizontal plane of the slab. All surface cavities were sealed with a thin mortar mix.	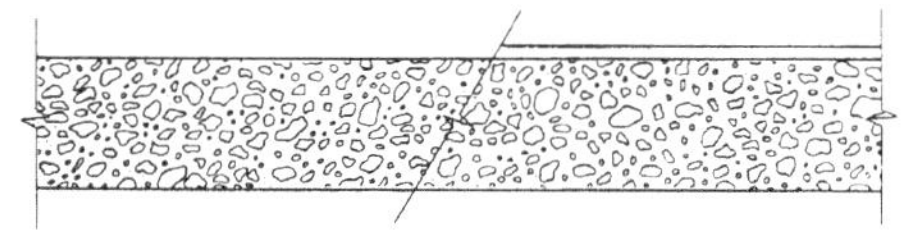 F 1a F 1c
F1c	Same as F1a except that ⅛-in.-thick [3-mm] vinyl tile was adhered to concrete.	
Reinforced Concrete with Floor Coverings		
See also F1c above.		
F2-1(a)	4-in.-thick [100-mm] reinforced concrete slab with carpeting and pad. The carpeting was of ¼-in. [6-mm] wool loop pile with ⅛-in. [3-mm] woven jute backing, 0.49 lb/ft^2 [2.4 kg/m^2]; the foam rubber pad was ¼ in. [6 mm] thick and weighed 0.53 lb/ft^2 [2.6 kg/m^2].	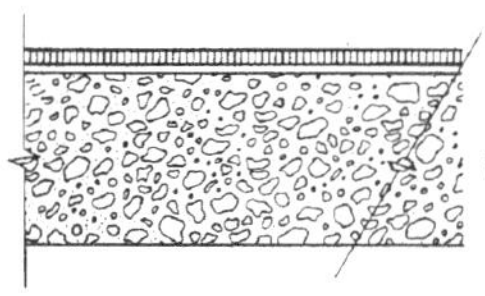 F 2-1
F3-1(a)	4-in. [100-mm] reinforced concrete slab with ½ × 9 × 9-in. [13 × 230 × 230 mm] oak blocks, 1.8 lb/ft^2 [8.8 kg/m^2], set in mastic.	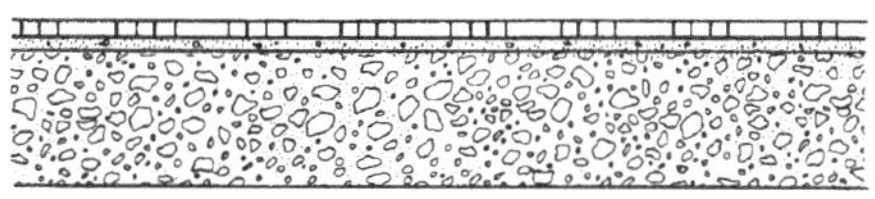 F 3
F3-2(d)	4-in. [100-mm] concrete slab with ⅛-in. [3-mm] cork.	
F7a	4⅜-in.-thick [111-mm] reinforced concrete slab. On the floor side, ¾-in.-thick [19-mm] sand-cement screed with ⅛-in. [3-mm] linoleum floor covering. On the ceiling side, ⅜-in. [10-mm] layer of plaster.	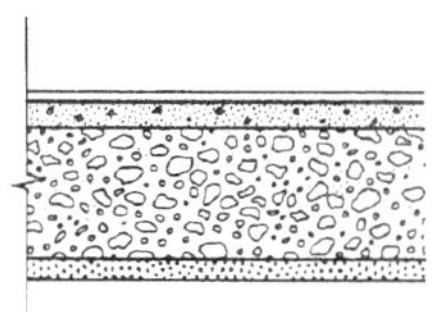 F 7a
F9	4⅜-in.-thick [111-mm] reinforced concrete slab. On the floor side, ½-in.-thick [13-mm] layer of bitumen with ½-in.-thick [13-mm] soft wood fiberboard, which was covered with a thin layer of bitumen with sand and a ¾-in.-thick [19-mm] sand-cement screed. On the ceiling side, ⅜-in. [10-mm] layer of plaster.	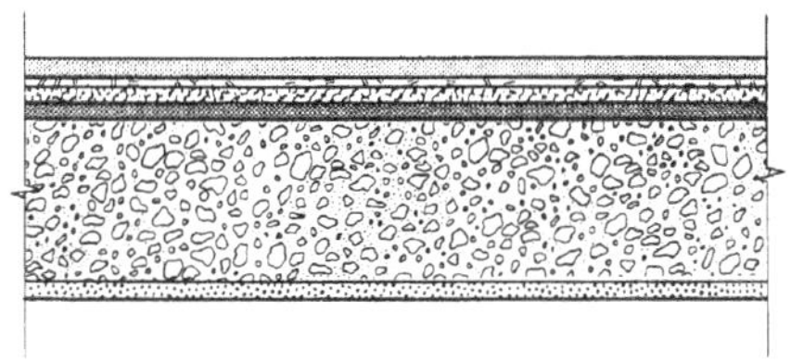 F 9
Reinforced Concrete Slab, Floating Floor		
F10	5-in.-thick [127-mm] reinforced concrete. On the floor side, 1½-in.-thick [38-mm] wire mesh reinforced sand-cement screed floating on ½-in.-thick [13-mm] bitumen-bonded, glass-wool quilt covered with building paper. On the screed, ½-in.-thick [13-mm] pitch-mastic with a linoleum floor covering. On the ceiling side, ½-in. [13-mm] layer of plaster.	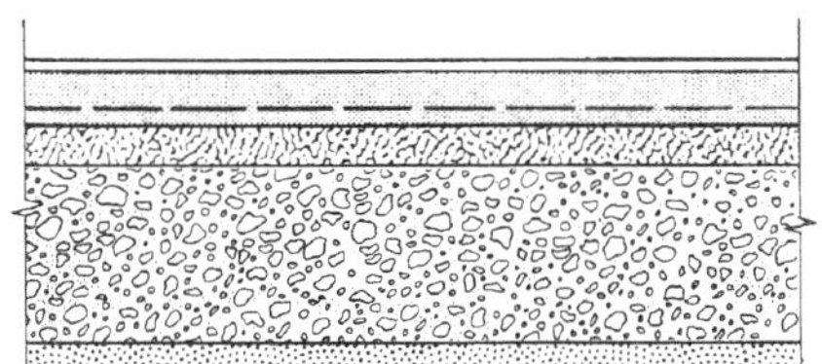 F 10
F12	4⅜-in.-thick [111-mm] reinforced concrete slab. On the floor side, ¾-in.-thick [19-mm] sand-cement screed. On the ceiling side, brick wire mesh, suspended 4 in. [100 mm] with wire hangers, held ⅞-in. [22-mm] gypsum plaster.	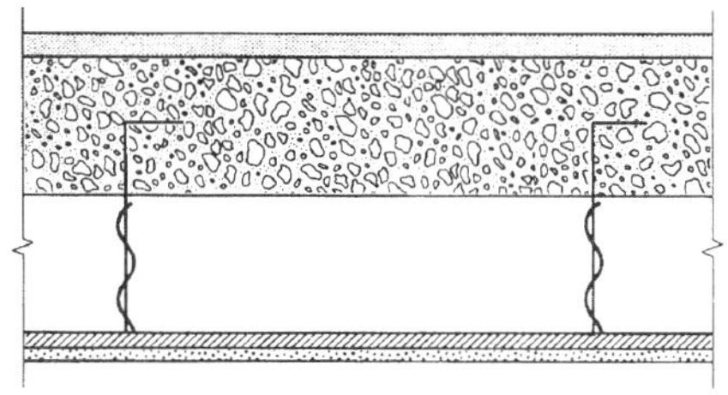 F 12

Table K.7 (*Continued*)

Code	Description	Section Sketch
F14	6-in.-thick [152-mm] reinforced concrete slab. On the floor side, ¾-in.-thick [19-mm] tongue-and-groove wood flooring nailed to 1 ½ × 2-in. [38 × 50 mm] wooden battens, 16 in. [406 mm] on centers, floating on 1-in.-thick [25-mm] glass-wool quilt. On the ceiling side, ½-in. [13-mm] layer of plaster.	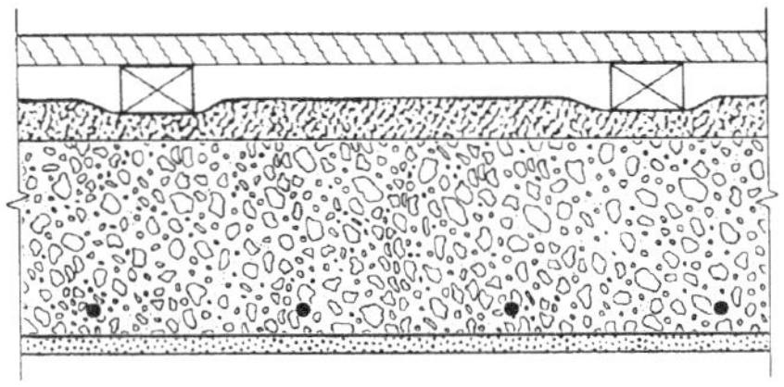 F 14
Concrete with Hollow Blocks		
F16	5 × 10-in. [127 × 254-mm] hollow masonry blocks, 14 in. [356 mm] on center, with spaces between blocks filled with 5-in.-thick [127-mm] reinforced concrete. On the floor side, ⅞-in.-thick [22-mm] wood blocks adhered to 1 ½-in.-thick [38-mm] sand-cement screed. On the ceiling side, ¾-in. [19-mm] layer of plaster.	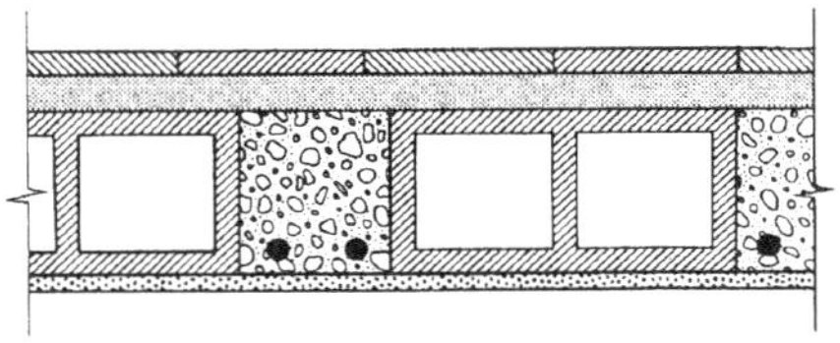 F 16
F17b	4 × 12 ½-in. [100 × 318-mm] hollow masonry blocks, 15 ½ in. [394 mm] on center, with spaces between blocks filled with 4-in.-thick [100-mm] reinforced concrete. On the floor side, 2-in.-thick [50-mm] sand-cement screed; linoleum on 1-in.-thick [25-mm] wood flooring nailed to 1 × 2-in. [25 × 50-mm] wooden battens, spaced 15 ½ in. [394 mm] on centers, floating on a glass-wool quilt approximately 1 in. [25 mm] thick. On the ceiling side, ¾-in. [19-mm] layer of plaster.	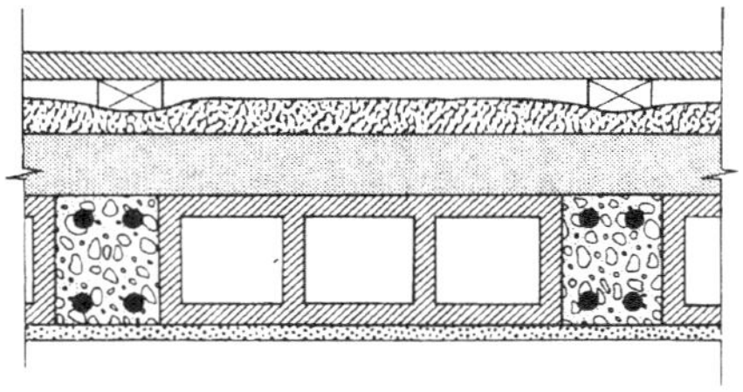 F 17b
Concrete Channel Slab		
F22	Prefabricated concrete channel slabs mortared together 20 in. [508 mm] on center. Each slab had a 3-in.-deep [76-mm] trapezoidal channel with bases of 11 and 14 ¾ in. [280 and 75 mm]. On the floor side, ¾-in.-thick [19-mm] sand-cement finish.	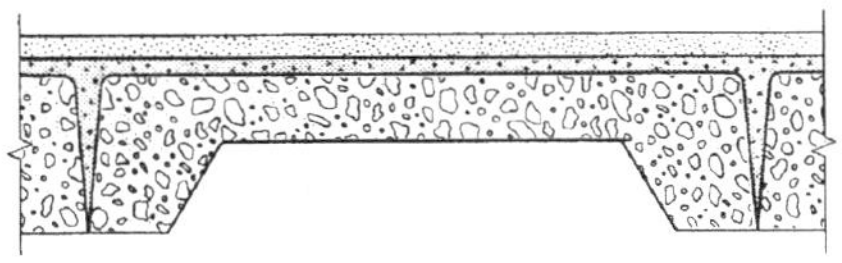 F 22
Ribbed Concrete		
F23	7 ¼-in. [184-mm] ribbed concrete floor. Ribs were 5 ¼ × 3 ¾ in. [133 × 95 mm], spaced 21 in. [533 mm] on center, with 1 × 2-in. [25 × 50 mm] wooden nailing strips cast into ends. On the floor side, the slab was 2 in. [50 mm] thick with a ¾-in.-thick [19-mm] sand-cement screed. On the ceiling side, ⅝-in.-thick [16-mm] wooden laths nailed to nailing strips held ⅝-in.-thick [16-mm] plaster.	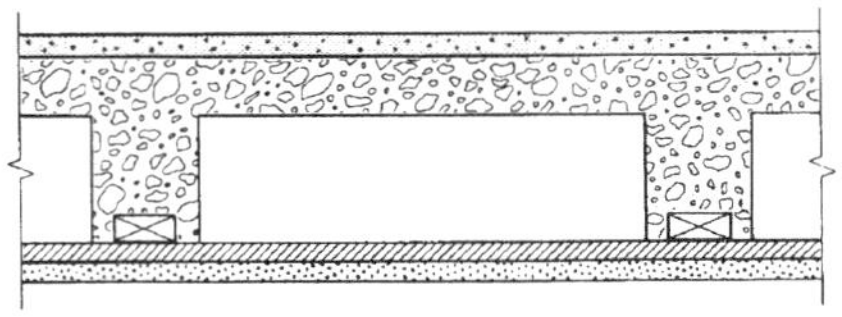 F 23
Concrete Channel Beam		
F24	7-in. [178-mm] precast trapezoidal concrete channel beams, 14 in. [356 mm] on center, with spaces between beams filled with sand-cement mix. On the floor side, 1 ½-in.-thick [38-mm] sand-cement screed with 1-in.-thick [25-mm] wood-block floor covering. On the ceiling side, approximately ¾-in.-thick [19-mm] layer of plaster on expanded metal lath.	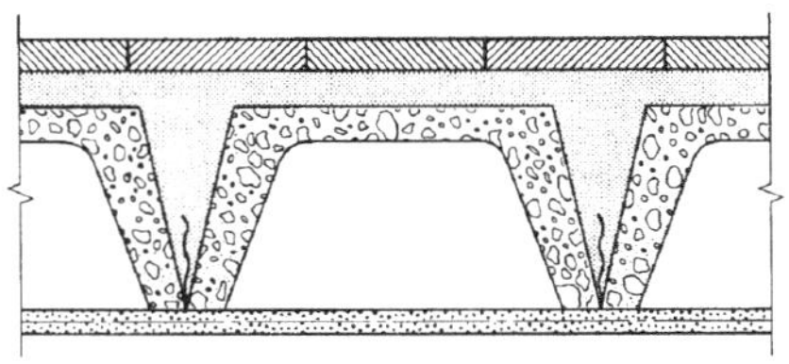 F 24

Table K.7 (*Continued*)

Code	Description	Section Sketch
Precast Concrete Beam, Floating Floor		
F25	5-in. [127-mm] precast concrete channel beams, 14½ in. [368 mm] on center, with spaces between beams filled with a sand-cement mix. On the floor side, ⅞-in.-thick [22-mm] tongue-and-groove wood flooring nailed to 1 × 2-in. [25 × 50-mm] wooden battens, 20 in. [508 mm] on center, on approximately 1-in.-thick [25-mm] glass-wool quilt on ¾-in.-thick [19-mm] sand-cement screed. On the ceiling side, ⅛-in. [3-mm] layer of plaster on ⅜-in. [10-mm] gypsum wallboard nailed to 1 × 2-in. [25 × 50-mm] wooden battens spaced 14½ in. [368 mm] on centers.	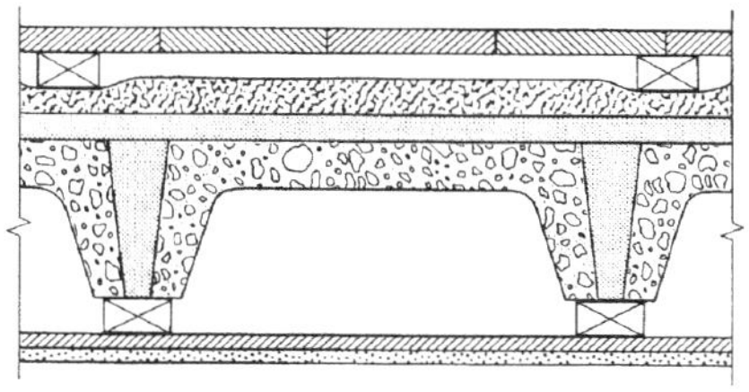 F 25
Hollow Concrete Beam		
F27	5-in. [127-mm] precast trapezoidal hollow concrete beams, 14½ in. [368 mm] on center, with bases of 14 and 12½ in. [256 and 318 mm]. Spaces between beams filled with sand-cement mix. On the floor side, 1-in.-thick [25-mm] sand-cement screed with 3/16-in. [5-mm] cork tile floor covering. On the ceiling side, ⅜-in.-thick [10-mm] gypsum wallboard attached to 1 × 2- in. [25 × 50-mm] wooden battens held by metal clips.	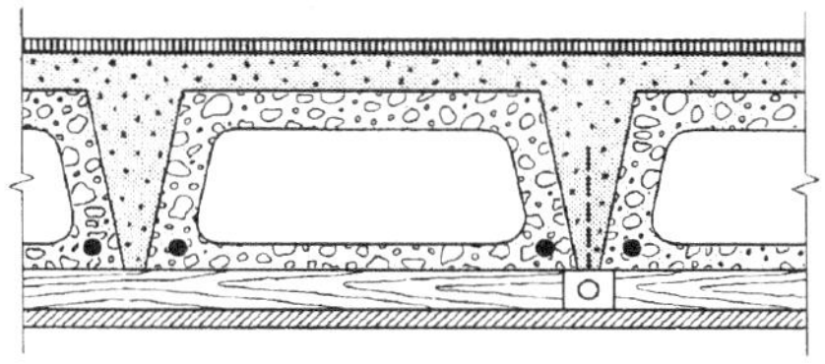 F 27
Wooden Joist		
F30	2 × 8-in. [50 × 200-mm] wooden joists, 16 in. [406 mm] on center. On the floor side, ⅞-in. [22-mm] tongue-and-groove flooring nailed to joists; on the ceiling side, ⅜-in. [10-mm] gypsum wallboard nailed to joists with joints sealed.	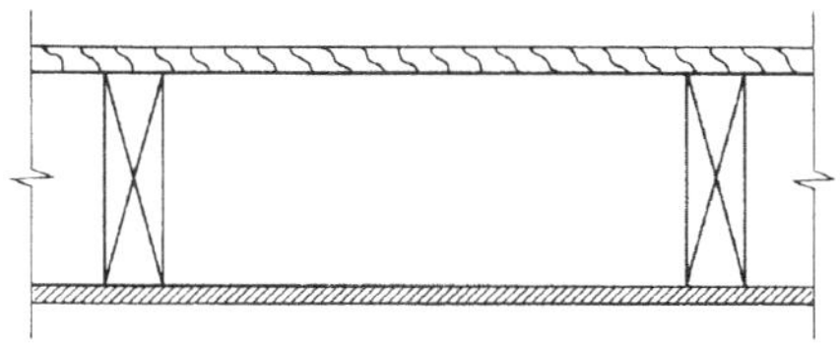 F 30
F32	2 × 8-in. [50 × 200-mm] wooden joists 18 in. [457 mm] on center. On the floor side, ⅞-in. [22-mm] tongue-and-groove wood flooring nailed to joists. On the ceiling side, 1-in. [25-mm] battens nailed through glass-wool quilt approximately 1 in. [25 mm] thick; ½-in. [13-mm] layer plaster on ¼-in.-thick [6-mm] wood lath.	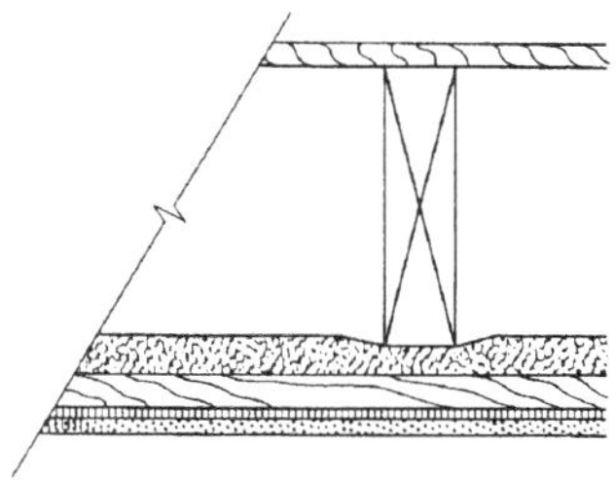 F 32
F34	2 × 8-in. [50 × 200-mm] wooden joists, 16 in. [406 mm] on center. On the floor side, ½-in.-thick [13-mm] C-D plywood nailed 8 in. [203 mm] on center to joists, 25/32-in.-thick [20-mm] hardwood flooring on plywood. On the ceiling side, ½-in.-thick [13-mm] gypsum wallboard nailed 6 in. [152 mm] on center to joists. All joints taped and finished; ceiling tile adhered to gypsum board.	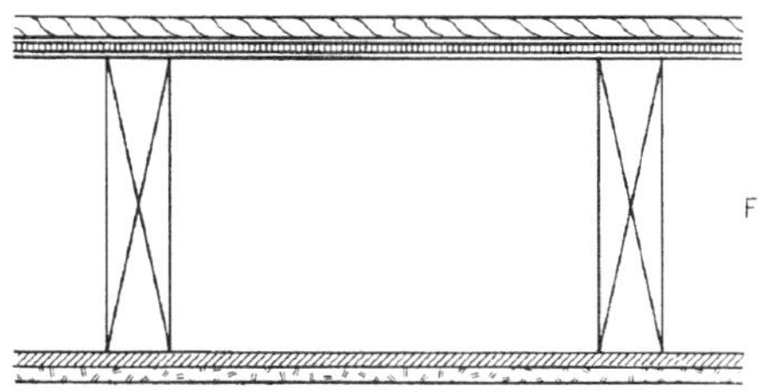 F 34

Table K.7 (*Continued*)

Code	Description	Section Sketch
F35a	2 × 8-in. [50 × 200-mm] wooden joists, 16 in. [406 mm] on center. On the floor side, $1\frac{1}{2}$-in.-thick [38-mm] tongue-and-groove wood fiberboard nailed to joists, vinyl tile floor covering. On the ceiling side, $\frac{1}{2}$-in.-thick [13-mm] gypsum wallboard nailed 6 in. [152 mm] on center to joists. All joints taped and finished.	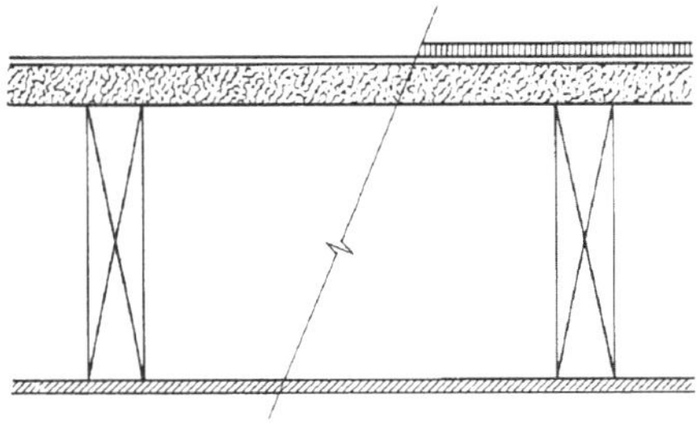 F 35a F 35b
F35b	Similar to F35a except that fiberboard was covered with carpet and pad.	
F38a	2 × 10-in. [50 × 250-mm] wooden floor joists spaced 16 in. [406 mm] on center. $\frac{5}{8}$-in. [16-mm] fir plywood subfloor nailed to joists 8 in. [203 mm] on center; $\frac{1}{2}$-in. [13-mm] plywood underlayment nailed to subfloor with joints staggered to miss joints of the subfloor; $\frac{1}{8} \times 9 \times 9$-in. [3 × 230 × 230-mm] vinyl asbestos tile glued to underlayment. On the ceiling side, $\frac{1}{2}$-in. [13-mm] gypsum wallboard nailed 12 in. [305 mm] on center, with all joints and nailheads taped and finished.	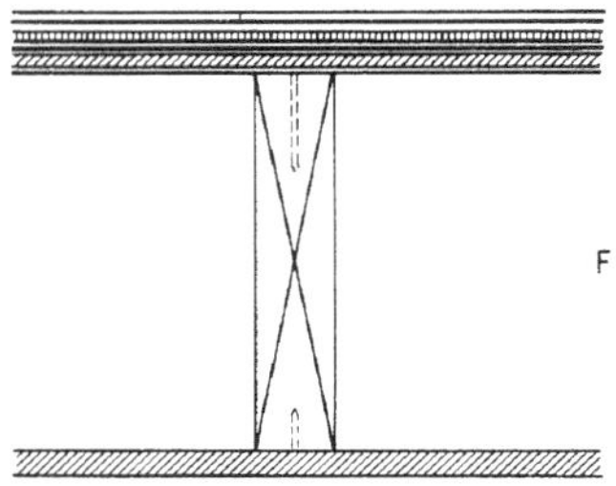 F 38a
F39a	2 × 10-in. [50 × 250-mm] wooden joists, 16 in. [406 mm] on center. On the floor side, $\frac{1}{2}$-in.-thick [13-mm] plywood subfloor nailed 6 in. [152 mm] on center along edges and 10 in. [254 mm] on center in field, building paper underlayment, $\frac{25}{32} \times 2\frac{1}{4}$-in. [20 × 57-mm] oak wood flooring nailed at each joist intersection and midway between joists. On the ceiling side, $\frac{5}{8}$-in.-thick [16-mm] gypsum wallboard, nailed 6 in. [152 mm] on center to joists. All joints taped and finished.	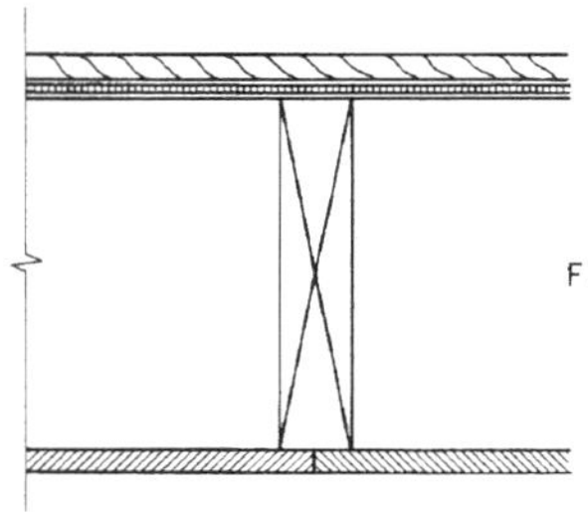 F 39a
Wooden Joist with Insulation		
F40a	2 × 10-in. [50 × 250-mm] wooden joists 16 in. [406 mm] on center with 3-in.-thick [76-mm] mineral fiber batts stapled between joists. On the floor side, $\frac{1}{2}$-in.-thick [13-mm] plywood subfloor nailed 6 in. [152 mm] on center along edges and 10 in. [254 mm] on center in field, building paper underlayment, $\frac{25}{32} \times 2\frac{1}{4}$-in. [20 × 57-mm] oak wood flooring nailed at each joist intersection and midway between joists. On the ceiling side, $\frac{5}{8}$-in.-thick [16-mm] gypsum wallboard nailed 6 in. [152 mm] on center to joists. All joints taped and finished.	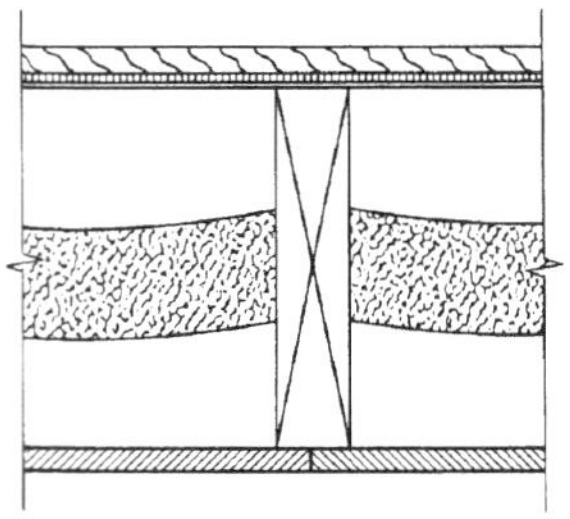 F 40a
Wooden Joist, Resilient Ceiling		
F44	2 × 8-in. [50 × 200-mm] wooden joists 16 in. [406 mm] on center. On the floor side, $\frac{3}{4}$-in.-thick [19-mm] wood subfloor, layer of building paper, and $\frac{3}{4}$-in.-thick [19-mm] tongue-and-groove fir finish flooring. On the ceiling side, resilient runners bridged across joists and nailed 12 in. [305 mm] on center to joists; $\frac{5}{8}$-in.-thick [16-mm] gypsum wallboard screwed to resilient runners. All joints taped and finished.	 F 44

Table K.7 (*Continued*)

Code	Description	Section Sketch
F48	2 × 8-in. [50 × 200-mm] wooden joists, 16 in. [406 mm] on center. On the floor side, $1\frac{1}{8}$-in.-thick [29-mm] regular C-D rough plywood nailed 6 in. [152 mm] on center along periphery and 16 in. [406 mm] on center at other bearings, plywood covered with an all-hair pad at 40 oz/yd^2 [1356 g/m^2] and all-wool pile at 44 oz/yd^2 [1492 g/m^2] carpet. The total weight of the carpet was 4.14 lb/yd^2 [2.25 kg/m^2] and the total thickness was $\frac{3}{8}$ in. [10 mm]. On the ceiling side, 2 × 4-in. [50 × 100-mm] wooden joists, 16 in. [406 mm] on center, staggered 8 in. [203 mm] on center relative to the floor joists, 3-in.-thick [76-mm] fibered glass blankets stapled between ceiling joists, $\frac{5}{8}$-in.-thick [16-mm] gypsum wallboard nailed to ceiling joists. All joints taped and finished; entire periphery of panel caulked and sealed. The ceiling was supported independently of the floor structure.	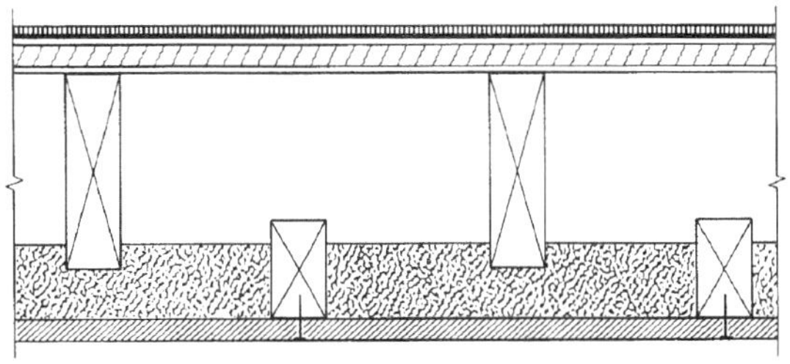 F 48
Steel Joist with Concrete Floor		
F57a	$2\frac{1}{2}$-in.-thick [64-mm] perlite concrete, 72 lb/ft^3 [152 kg/m^3] on 28-gauge [0.39-mm] corrugated steel units supported by 14-in. [356-mm] steel bar joists; $\frac{1}{8}$-in.-thick [3-mm] asphalt tile cemented to concrete. On the ceiling side, $\frac{3}{4}$-in. [9-mm] furring channels, $13\frac{1}{2}$ in. [343 mm] on center, wire-tied to joists, 3.4 lb/yd^2 [1.85 kg/m^2] diamond mesh metal lath wire-tied to furring channels, $\frac{9}{16}$-in. [14-mm] coat of plaster with $\frac{1}{16}$-in. [1.6-mm] white-coat finish.	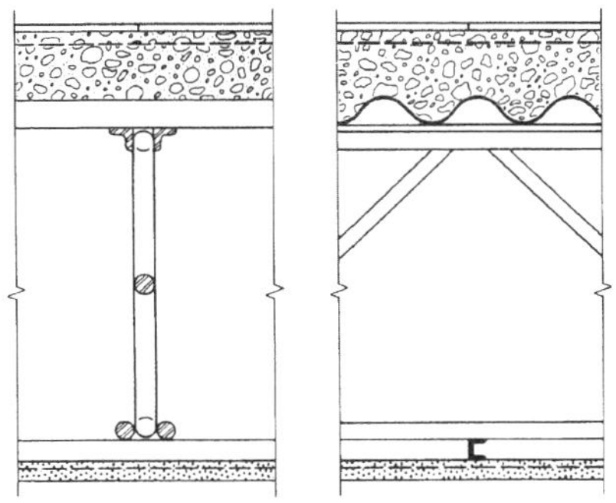 F 57*a,b*
F57b	Same as F57a except for carpet and pad in lieu of asphalt tile.	
F58	18-in. [457-mm] steel joists, 16 in. [406 mm] on center. On the floor side, $\frac{5}{8}$-in.-thick [16-mm] C-D rough plywood nailed to joists, $1\frac{5}{8}$- in.-thick [41-mm] foamed concrete, 100 lb/ft^3 [1600 kg/m^3], slab constructed on the plywood; concrete covered with an all-hair pad at 40 oz/yd^2 [1356 g/m^2] and an all-wool pile at 44 oz/yd^2 [1492 g/m^2] carpet. Total weight of the carpet, 4.14 lb/yd^2 [2.25 kg/m^2]; total thickness, $\frac{3}{8}$ in. [10 mm]. On the ceiling side, $\frac{5}{8}$- in.-thick [10-mm] gypsum wallboard nailed to joists. All joints taped and finished; entire periphery of panel caulked and sealed.	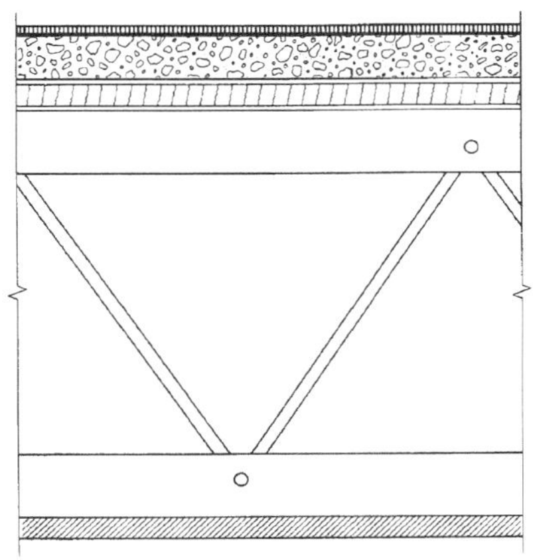 F 58

APPENDICES

Table K.7 (*Continued*)

Code	Description	Section Sketch
F60a	7-in. [178-mm] steel bar joists spaced 27 in. [686 mm] on center. On the floor side, $\frac{3}{8}$-in. [10-mm] metal rib lath attached to top of joists and 2-in.-thick [50-mm] poured concrete floor. On the ceiling side, $\frac{3}{4}$-in. [19-mm] metal furring channels wire-tied to joists 16 in. [406 mm] on centers; $\frac{3}{8} \times 16 \times 48$-in. [10 × 406 × 1220-mm] plain gypsum lath held with wire clips and sheet metal end joint clips; $\frac{7}{16}$-in. [11-mm] sanded gypsum plaster and $\frac{1}{16}$-in. [1.6-mm] white-coat finish.	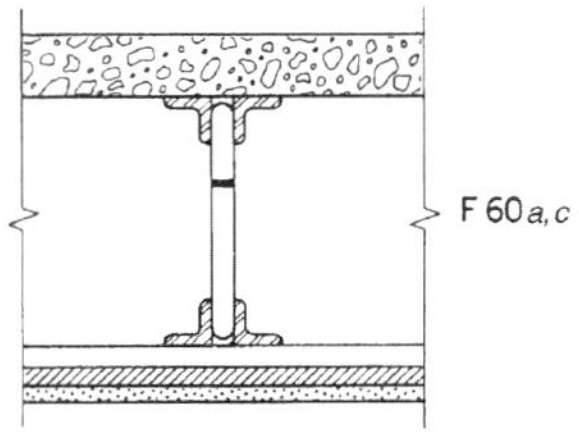
F60c	Structure F60a with nylon carpeting and foam rubber pad placed on the floor. The carpet pad had an uncompressed thickness of $\frac{1}{4}$ in. [6 mm] backed with a woven jute fiber cloth. The carpet had $\frac{1}{8}$-in. [3-mm] woven backing and $\frac{1}{4}$-in. [6-mm] looped pile spaced 7 loops per inch with a total thickness of $\frac{3}{8}$ in. [10 mm].	

Source: Adapted from HUD/FHA/NBS (1971). SI units were added by authors of this book (not part of original source).

APPENDIX L

Building Performance Simulation Tools and Resources

L.1 BUILDING PERFORMANCE SIMULATION AS PART OF AN INTEGRATIVE PROCESS

This book presents numerous manual (hand) calculation procedures for a variety of systems. These methods are simple enough to require only a calculator and an informed and patient user. This emphasis comes from the belief that such calculations provide the most explicit means of understanding the variables (and relationships among variables) that influence building performance. In practice, however, many analyses are conducted using computer software programs. An emerging marketplace of user-friendly building performance simulation (BPS) tools are helping design teams go beyond simple patterns or precedents toward design decision-making based on robust algorithms that indicate relative potential performance impacts. BPS tools certainly include, and are often categorically referred to as, building energy modeling (BEM) tools, as a shorthand. However, the BPS tools available today encompass an ever-widening scope of building performance topics including climate data analysis, daylighting simulations, building enclosure performance, operational and embodied carbon modeling, and more.

Intriguingly, some BPS software (in a quest for accuracy) involves variables that are not intuitive to the building designer. This is particularly true in the realms of building science and environmental systems. When things are nonintuitive, they can quickly become mysterious. Mysterious variables that affect building design are not good. Building design is an art and a science—but the building science part cannot be opaque to the designer. Much of the high-end analysis software available today is also unforgiving for inputs involving "properties not yet determined." Such variables are common during schematic design (who knows the infiltration leakage rate of a window during the first few weeks of design—when a usable estimate of building performance may be critical to decision making?). Time will tell how some of these contradictions work out in the next generations of these software platforms.

On the other hand, a new generation of BPS tools is couching complex, validated simulation tools behind a veil of broad-based assumptions, big datasets, and easy-to-use graphical user interfaces (GUIs). With just a few basic inputs, these programs can produce detailed performance results. BPS tools facilitate an integrative process through which BPS is leveraged at the earliest stages of the design process in a quick, iterative manner—allowing teams to optimize design solutions throughout the development of a project. Former HOK Chairman Patrick MacLeamy is credited with articulating this concept of "shifting the effort" with regard to design development, otherwise known as the MacLeamy Curve (Figure L.1).

Personal computers are capable of running or accessing the most commonly used BPS tools. With the advent of cloud-based platforms, even the most complicated simulations can now occur in a matter of seconds, making BPS feasible as a design tool for virtually any firm. This offers design teams an unprecedented opportunity to leverage BPS as a true design-assist tool. All parties involved, however, should always bear in mind that simulation results are only

Mechanical and Electrical Equipment for Buildings, Fourteenth Edition. Walter Grondzik, Daniel Overbey, and Alison Kwok.

Companion website: www.wiley.com/go/meeb14e

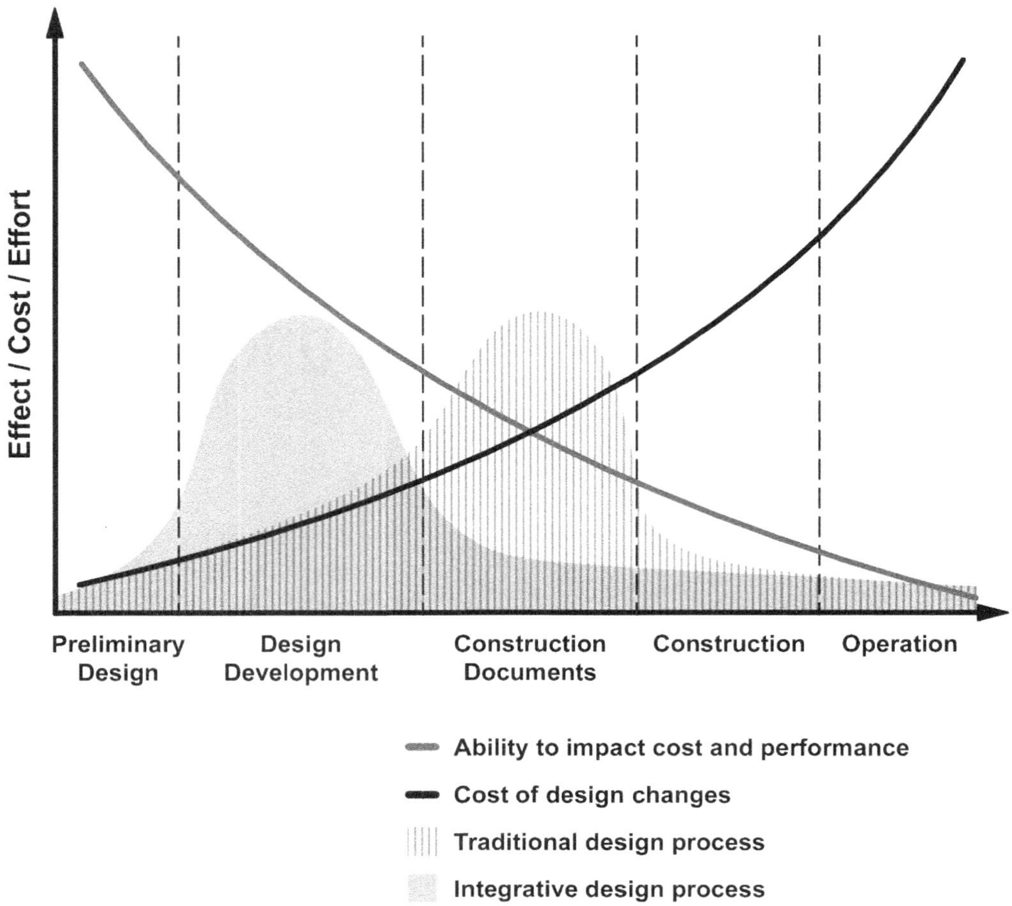

Figure L.1 *The premise of the MacLeamy Curve is that an integrative process shifts design efforts toward the front end of a project timeline—where there is maximum flexibility to make high-impact design decisions and interdisciplinary efforts are coordinated much sooner versus a more traditional design process. (Illustration by Daniel Overbey.)*

as accurate as the modeling and informational inputs provided. BPS tools cannot predict the future and should never be used to attest to actual performance outcomes.

L.2 BUILDING PERFORMANCE SIMULATION RESOURCES

(a) Architect's Guide to Building Performance

The American Institute of Architects (AIA) maintains a primer for design professionals interested in leveraging energy modeling throughout the design process. Focused on energy modeling, the document, *Architect's Guide to Building Performance: Integrating Performance Simulation in the Design Process* (2019), features a summary of the most prevalent BEM tools, explains key terminology, and outlines a basic methodology for integrating BEM into the design process. The guide is meant to entice architects to engage engineers, energy modelers, and project team members on energy and performance issues early and throughout the design and operation processes (aia.org/resource-center/architects-guide-building-performance).

(b) *ASHRAE Standard 209*

ASHRAE Standard 209-2018: Energy Simulation Aided Design for Buildings Except Low-Rise Residential Buildings defines the minimum requirements for providing energy design assistance using building energy simulation and analysis. The standard defines design modeling requirements as well as post-occupancy energy performance comparison criteria. *Standard 209* applies to new buildings or major renovations of, or additions to, existing buildings utilizing BEM during the design process. The standard does not apply to single-family houses, multi-family structures of three stories or fewer above grade, manufactured houses (e.g., mobile homes), or modular homes (ashrae.org/standards).

(c) Building Energy Software Tools (BEST) Directory

Formerly hosted by the U.S. Department of Energy (US DOE), the United States Regional Affiliate of the International Building Performance Simulation Association (IBPSA-USA) now curates a growing directory of BEM tools (ibpsa.us/best-directory-list).

(d) BuildingGreen

Since 1985, BuildingGreen has been a trusted source for healthy and sustainable design and construction strategies. The company focuses on helping designers and builders from a whole-systems perspective. Much of its content focuses on aspects of BEM and building science. An independent voice not supported by advertising, BuildingGreen is an objective and authoritative resource for design teams assessing potential high-performance strategies (buildinggreen.com).

(e) 2030 Palette

The 2030 Palette is a free online platform from Architecture 2030 that succinctly puts the principles and actions behind Zero Net Carbon and resilient built environments at the fingertips of designers, planners, and builders pursuing high-performance design strategies. The featured swatches, or sustainable design strategies, address energy consumption and greenhouse gas emission at all scales—from regional planning to building details (2030palette.org).

L.3 BUILDING PERFORMANCE SIMULATION PROGRAMS

In North America, there are three primary BEM programs used in the industry for whole-building energy simulation: BLAST, DOE-2, and EnergyPlus. These programs have been validated and serve as the "simulation engine" working behind the scenes within many of the more user-friendly BEM software tools in the marketplace today.

(a) BLAST

BLAST, or Building Loads Analysis and System Thermodynamics, was developed by the Corps of Engineers Construction Engineering Research Laboratory and Lawrence Berkeley Laboratory (CERL/LBNL). BLAST is a product of the 1970s' energy crisis, and federal funding supported its ongoing development for several decades during which time BLAST attracted third-party developers and gained an international foothold. Though some BEM tools may still be based on BLAST, the program is no longer being funded or developed.

(b) DOE-2

Another product of the 1970s' energy crisis, DOE-2 was developed at the Lawrence Berkeley National Laboratory with funding from the U.S. Department of Energy. To this day, DOE-2 remains the most widely used BEM program in the United States. DOE-2 can be used as a stand-alone calculation engine or in concert with a GUI (graphical user interface) such as VisualDOE, eQuest, or EnergyPro. Like BLAST, decades of support and third-party development have established DOE-2 in the international marketplace (doe2.com).

(c) EnergyPlus

EnergyPlus is a next-generation, open-source whole-building energy modeling (BEM) program developed, supported, and maintained by a team led and funded by the US DOE. First developed in 1997, EnergyPlus is based on the load algorithms of BLAST and the systems algorithms of DOE-2. Historically focused on commercial buildings, EnergyPlus has been expanded with modeling capabilities relevant to residential buildings and data centers. It is also tested according to the validation methodology outlined by *ASHRAE Standard 140: Standard Method of Test for the Evaluation of Building Energy Analysis Computer Programs*. The US DOE continues to support the ongoing development of EnergyPlus, as the program has gained international support from a growing number of third-party developers. Like DOE-2, EnergyPlus can also be used as a stand-alone calculation engine. The program has emerged as the dominant whole-building BEM software engine in the building design and construction industry (energyplus.net).

Please note that some BPS tools may use their own proprietary simulation engines and/or static state calculations. The case may vary from one product to another. It is always good practice to fully understand the methodology and algorithms behind any program. EnergyPlus, for example, incorporates various standardized procedures supported by ASHRAE—such as the heat balance method (HBM) for load calculations, which has been validated as providing the most accurate calculations versus other comparable procedures. For a detailed review of the progression of energy modeling methodologies, see Mao, C., J.C, Baltazar, and J. Haberl (2017) "Literature Review of Building Peak Cooling Load Methods in the United States," *Science and Technology for the Built Environment*, 24: 228–237.

L.4 COMMONLY USED BUILDING PERFORMANCE SIMULATION TOOLS

The following is a short list (drawn from a larger and rapidly expanding array) of BPS tools that may assist in a project team's integrative approach to the design of mechanical and electrical systems (both passive and active) for buildings.

Two points to bear in mind: updated and improved versions of existing tools appear often, as do new programs, and any listing of such software quickly becomes obsolete. Thus, the descriptions presented here avoid version numbers and other transient information, instead describing the basics of each tool.

Apologies to those developers and researchers whose programs do not appear here. The following list is not an endorsement by the authors or publisher. Simply, this list offers design teams a starting point as it endeavors to compile the most widely used (or readily available) BPS tools in North America.

(a) Climate Data Analysis

EnergyPlus Weather Data. With funding from the US DOE, the National Renewable Energy Laboratory (NREL) hosts a website dedicated to EnergyPlus, which includes free access to weather data for more than 2100 locations across the world in various file formats. The weather data are arranged by World Meteorological Organization (WMO) region and country (energyplus.net/weather).

CBE Clima Tool. Developed by the Center for the Built Environment (CBE) at the University of California, Berkeley, this free and open-source web-based application offers easy access to, and visualization of, publicly available weather data files (in EPW format). The CBE Clima Tool creates a series of interactive figures to help designers summarize and contextualize the weather data (clima.cbe.berkeley.edu).

Climate Consultant. Developed and managed by the University of California, Los Angeles (UCLA), Climate Consultant is a simple-to-use, Java-based computer program that helps architects, builders, contractors, homeowners, and students better understand their local climate. It serves as a GUI to visualize and assess EnergyPlus weather data (EPW) files. Climate Consultant translates the data into dozens of graphic displays (energy-design-tools.aud.ucla.edu).

Climate One Buildings. The authors of the Climate One Buildings website independently publish their own TMY datasets in a variety of file types commonly supported by BPS tools. Referred to as TMYx files, the datasets are typical meteorological year data derived from hourly weather data in the ISD (US NOAA's Integrated Surface Database) using the TMY/ISO 15927-4:2005 methodologies. However, TMYx datasets intentionally capture fewer and much more recent year ranges than other TMY datasets (e.g., 2004–2018, 2007–2021, 2009–2023). Currently, there are more than 16,100 TMYx locations supplied (climate.onebuilding.org).

The Climate Explorer. This public tool offers interactive graphs and maps illustrating past and projected climate conditions for counties and county-equivalents across the United States. Built to support the U.S. Climate Resilience Toolkit, the Climate Explorer allows users to examine projected future climate conditions. FernLeaf Interactive and the National Environmental Modeling and Analysis Center (NEMAC) at the University of North Carolina Asheville built the Climate Explorer's graphing modules using data from a modeling experiment called CMIP5 (Coupled Model Intercomparison Project Phase 5). Through CMIP5, climate modelers around the world programmed their models to simulate past temperature and precipitation based on the observed abundance of greenhouse gases in the atmosphere and to project future conditions based upon a set of four emissions scenarios, referred to as Representative Concentration Pathways (RCPs) (crt-climate-explorer.nemac.org).

WeatherShift. A collaborative project of Arup North America Ltd. (Arup) and Argos Analytics LLC, the WeatherShift tool uses data from global climate change modeling to produce EnergyPlus weather files (EPW) and design day files (DDY) adjusted for projected future climate conditions. The projected data produced by the WeatherShift tool can be viewed for multiple future timeframes based on different emissions scenarios (weathershift.com).

(b) Building Energy Performance and/or Operational Carbon

Autodesk Insight. Autodesk Insight Energy Analysis and Green Building Studio (GBS)—two longstanding BEM platforms from the software developer—have been retired and replaced by Autodesk Insight, a BPS tool that integrates with Revit and FormIt to allow building designers to articulate and gauge the energy performance and life-cycle carbon impacts of design options at the earliest phases of project planning and decision-making (autodesk.com).

Autodesk Insight Energy Analysis. Discontinued in 2025 (but maintained here for historical reference), this BEM tool from Autodesk provided design teams with centralized access to performance data and advanced analysis engines. Through integration with Revit and FormIt and direct access to guidance and recommendations, users could utilize quick, iterative performance feedback on a variety of whole-building energy performance metrics. Originally restricted to a DOE-2-based engine, Insight later offered cloud-based EnergyPlus analysis.

BEopt. Building Energy Optimization (BEopt) is a publicly available EnergyPlus-based tool from NREL focused on balancing and optimizing high-performance building strategies with cost-saving goals to establish affordable energy-efficiency strategies for residential buildings. It is applicable to both new construction and renovation projects (beopt.nrel.gov).

COMcheck. Though admittedly generous to call it a BEM program, the wide jurisdictional utility of COMcheck makes it worthy of mention here for the sake of context and clarity. This publicly available commercial (COM) compliance check program from the US DOE can be used to quickly determine whether new commercial or multifamily residential buildings, additions, and alterations meet the requirements of the *International Energy Conservation Code* (IECC), *ASHRAE/IES Standard 90.1-2022: Energy Standard for Sites and Buildings Except Low-Rise Residential Buildings*, and several state-specific codes. COMcheck simplifies compliance reviews for building officials, plan checkers, and inspectors by allowing them to quickly determine if a building project meets a given standard or code. (Note: REScheck—the residential compliance counterpart to COMcheck—is also available through the US DOE) (energycodes.gov/comcheck).

Cove. The premiere BPS tool offered by the developers at Cove, their eponymous cloud-based software platform integrates a widening range of BPSs, including energy and operational carbon, daylighting, cost estimating, embodied carbon, and more. The signature feature of Cove is its parametric optimization tool by which users can aggregate and synthesize myriad EnergyPlus-based BEM runs to ascertain bundles of specific project variables that may yield the most high-performance project outcomes (cove.inc).

DesignBuilder. This software allows quick modeling and comparison of designs relative to whole-building energy simulation. Utilizing EnergyPlus for building energy performance simulation, DesignBuilder's expansive areas of focus include carbon dioxide emissions, solar shading, natural ventilation, daylighting, comfort studies, computational fluid dynamics (CFD), HVAC simulation, pre-design, early-stage design, building energy code compliance checking, building stock modeling, hourly weather data, and heating and cooling equipment sizing (designbuilder.com).

eQuest. Developed by James J. Hirsch & Associates in collaboration with the Lawrence Berkeley National Laboratory (LBNL), eQuest (the Quick Energy Simulation Tool) is a publicly available BEM program that is primarily spreadsheet-based and can be utilized in all phases of design. Simulation results are achieved by creating the building model, applying energy efficiency measures, and analysis using the latest version of the DOE-2 building energy use simulation program (doe2.com/equest).

Green Building Studio (GBS). Discontinued in 2025 (but maintained here for historical reference), this BEM program was Autodesk's core whole-building energy simulation platform. It powered the energy analysis features for Insight, Revit, and FormIt. GBS was a flexible cloud-based platform that allowed users to run BPSs to optimize energy efficiency and to work toward carbon neutrality early in the design process. GBS used the DOE-2 simulation engine for calculating hourly, whole-building energy usage.

HEED (Home Energy Efficient Design). This publicly available tool from UCLA allows the user to analyze the variables of energy, cost, and carbon emission. There are simplified and detailed input options, fit for both beginners and advanced users. HEED includes some very intriguing performance visualization tools (energy-design-tools.aud.ucla.edu).

HAP (Hourly Analysis Program). Carrier's long-established BEM platform, HAP (Hourly Analysis Program), equips users with a collection of versatile features for designing heating, ventilation, and air conditioning (HVAC) systems for commercial buildings, but it also features a robust energy analysis program capable of comparing energy consumption and operating costs of design alternatives. HAP utilizes Carrier's proprietary energy modeling algorithms (carrier.com/hap).

OpenBuildings Energy Simulator. Bentley's OpenBuildings Energy Simulator is a BEM tool integrated within OpenBuildings, the company's comprehensive building information modeling (BIM) platform. Formerly branded as the AECOsim Energy Simulator (the acronym standing for Architectural Engineering Construction Operations simulation), the OpenBuildings Energy Simulator facilitates building energy performance analysis for new construction and major renovation. The OpenBuildings Energy Simulator allows users to simulate and analyze building energy performance using EnergyPlus, produce energy performance certificates, perform whole-building energy analysis, catalog HVAC systems, and conduct daylighting simulations using Radiance (bentley.com/software/openbuildings).

OpenStudio. This is a cross-platform (Windows, Mac, and Linux) collection of software tools to support whole-building energy modeling using EnergyPlus and advanced daylight analysis using Radiance. OpenStudio is set up as an open-source platform to better facilitate community development, extension, and private sector adoption. This publicly available tool is developed by NREL along with several other collaborators (openstudio.net).

Passive House Planning Package. A planning tool developed by the Passive House Institute (PHI) to serve its certification program, the Passive House Planning Package (PHPP) is a spreadsheet-based tool that can be used to help prepare an energy balance and calculate the annual energy demand of a building based on user inputs regarding the building's characteristics. (passivehouse.com).

REM/Rate. This software is used to conduct Home Energy Rating Systems (HERS) ratings. A HERS score rates the energy efficiency of a home and can help identify cost-effective improvements and support the provision of energy-efficient mortgages. Climate data are available for cities and towns throughout North America (remrate.com).

Sefaira. Trimble offers a collaborative, cloud-based software platform that combines an easy-to-learn GUI with validated industry-standard analysis engines such as EnergyPlus for energy use and Radiance for daylighting. Sefaira equips design teams to produce high-performance design concepts from the earliest design stages; collaborate across disciplines and across firms; and e-communicate results via graphically rich outputs. Sefaira offers simulations for whole-building energy use, renewable energy potential, daylighting, thermal comfort, and natural ventilation among other features. Plugins for SketchUp and Revit can streamline this program into a design team's workflow (sefaira.com).

Simergy. This is a robust GUI front-end to EnergyPlus. Working with the US DOE, Berkeley Labs, Trane, Infosys, and Hydro Quebec, Digital Alchemy was the primary developer for this full-featured BEM platform. In Simergy, users can create models from scratch, create them based on drawing files, or import and extend models already developed in other design applications (d-alchemy.com).

TRACE. The TRane Air Conditioning Economics (TRACE) platform is comprised of a series of building design-and-analysis software tools from HVAC manufacturer Trane. The TRACE 700 family of software products has become an industry standard; however, Trane is now transitioning to the updated TRACE 3D Plus software solutions. The TRACE offers extensive modeling functionality, including over 50 different energy system types. TRACE is tested in compliance with *ASHRAE Standard 140*. This long-standing BEM platform has historically operated using its own proprietary simulation engine; however, TRACE now utilizes EnergyPlus (trane.com/commercial/north-america/us/en/products-systems/design-and-analysis-tools/trane-design-tools/trace-3d-plus.html).

Virtual Environment (IESVE). Integrated Environmental Solutions' (IES) flagship BPS platform is an integrated suite of tools designed to allow building performance analysis to be easily integrated into commercial workflows across the entire design lifecycle. IESVE utilizes the company's proprietary APACHE simulation engine rather than DOE-2 or EnergyPlus. IESVE allows users to assess project designs from the standpoint of whole-building energy use, renewable energy potential, passive strategies, daylighting, water use, bioclimatic architecture, thermal comfort, and even LEED/BREEAM-specific analyses (iesve.com).

WUFI Passive. Fraunhofer IBP collaborated with Phius and Owens Corning to develop a variation of Fraunhofer IBP's building simulation program (see WUFI software description below) as a Phius certification and energy analysis tool. WUFI Passive combines passive house energy modeling with WUFI's hygrothermal analysis capability (hwufi.de/en/software/wufi-passive). A cloud-based implementation of this software is called METr.

Zero Tool. A key starting point for any BEM effort is establishing performance goals. The free web-based Zero Tool from Architecture 2030 offers an intuitive and powerful way to establish energy consumption baselines and energy use reduction targets, and to see how the energy consumption for an existing building or a new building design compares to similar buildings and baselines. The Zero Tool also allows those designing to the 2030 Challenge and other targets to maintain pre-existing baselines in order to continue to measure their progress (zerotool.org).

(c) Building Enclosure Analysis

Building Science Advisor (BSA). Developed by the Oak Ridge National Laboratory (ORNL), which is managed by the nonprofit UT-Battelle LLC for the US DOE, BSA is a publicly available web-based tool that provides guidance on the moisture durability and energy efficiency of new and retrofit wall assembly designs. BSA leverages a database of thousands of pre-simulated hygrothermal models to provide detailed feedback on wall assembly design, tailored for the user's location (bsa.ornl.gov).

OPAQUE. This publicly available tool from UCLA graphically displays the thermal performance of a defined opaque wall or roof construction assembly of any color, orientation, and latitude (energy-design-tools.aud.ucla.edu).

Optics. Berkeley Lab's Optics program works with the International Glazing Database (IGDB) to analyze arrangements of multiple glass layers to calculate the resulting optical properties of specified glazing assemblies (windows.lbl.gov).

SOLAR-2. This publicly available tool from UCLA graphically displays the performance of a window shaded with any combination of fins and overhangs, as well as remote objects such as walls of an exterior courtyard or a distant building (energy-design-tools.aud.ucla.edu).

THERM. Berkeley Lab's THERM allows users to model two-dimensional heat-transfer effects in building components such as windows, walls, foundations, roofs, and doors; and other assemblies/products where thermal bridges are of concern. THERM's heat-transfer analysis also allows users to evaluate local temperature patterns, which may reveal potential problems with condensation, moisture damage, or structural integrity (windows.lbl.gov/software).

WINDOW. Berkeley Lab's WINDOW allows users to calculate total window thermal performance indices (i.e., U-factors, solar heat gain coefficients, shading coefficients, and visible transmittances). The program's analysis method is consistent with the updated rating procedure developed by the National Fenestration Rating Council (NFRC) (windows.lbl.gov/software).

WUFI. Developed by the Holzkirchen branch of the Fraunhofer IBP and ORNL, WUFI (Wärme Und Feuchte Instationär—which, translated, means heat and moisture transiency) is comprised of a family of software products that simulate one- and two-dimensional heat and moisture activity in building assemblies. The WUFI platform includes a menu-driven building envelope analysis program that allows realistic calculation of transient coupled one-dimensional heat and moisture transport in multi-layer building components exposed to ambient weather. WUFI incorporates the latest findings regarding vapor diffusion and liquid transport in building materials and has been validated by detailed comparison with laboratory and field measurements (ornl.gov/sci/buildings/tools).

(d) Daylighting and Electric Lighting

AGi32. Lighting Analysts, Inc. provides a series of lighting analysis tools. AGi32—one of the lighting industry's premier lighting calculation tools—can provide numeric and rendered solutions for indoor or outdoor electrical lighting applications as well as daylighting. It is widely used in the lighting industry and includes a web-based interface that allows users to access manufacturers' photometric data and provides for easy updating (agi32.com).

ClimateStudio. The flagship software platform from developer Solemma LLC, ClimateStudio facilitates advanced daylighting, electric lighting, and conceptual thermal analysis through a plugin for Rhinoceros (Rhino). Design professionals may recall the company's legacy daylighting and energy modeling plugin, *DIVA-for-Rhino* (this plugin is no longer supported). ClimateStudio utilizes both EnergyPlus and Radiance; however, it also implements a novel path tracing approach to offer users rapid results without a loss of accuracy from the validated simulation engines. ClimateStudio's daylight simulation metrics are aligned with BREEAM, EN 17037, and LEED. The program's ability to calculate annual glare distributions across a floor area is of particular interest to daylighting design professionals. The software integrates a glare calculation methodology, developed by Wienold and Christofferen (2005; see reference in Chapter 10), known as Daylight Glare Probability (DGP). This metric predicts the likelihood that an observer at a given view position in a space will experience glare. Determined as a percentage, ClimateStudio then aggregates a generalized performance metric termed Spatial Disturbing Glare (sDG). Solemma LLC defines sDG as "the percentage of views across the regularly occupied floor area that experience Disturbing or Intolerable Glare (DGP >38%) for at least 5% of occupied hours." The calculation is based on hourly DGP values for eight different view directions at each position in the building (solemma.com/climatestudio).

Daysim. This is a publicly available, validated, Radiance-based daylighting analysis program that models annual daylight in and around buildings. The software allows users to model dynamic façade systems ranging from standard venetian blinds to state-of-the-art light redirecting elements, switchable glazings, and combinations thereof. Users may further specify complex electric lighting systems and controls including manual light switches, occupancy sensors, and photocell-controlled dimming. Source code is available on GitHub via the Massachusetts Institute of Technology (MIT) (github.com/MITSustainableDesignLab/Daysim).

Desktop Radiance. This program integrates Radiance into a computer-aided design (CAD) input environment. Libraries of materials, glazings, luminaires, and furnishings facilitate data entry. Lawrence Berkeley National Laboratory, Pacific Gas and Electric, and the California Institute for Energy and Environment developed this program (radsite.lbl.gov/deskrad).

Radiance. Radiance began in 1985 as a research project at UC Berkeley and has since evolved into a validated daylight simulation engine that many major BPS tools rely on. As this publicly available and open-source program continued to develop, Berkeley eventually released Desktop Radiance—a stand-alone software tool that can also be utilized by third-party platforms including Insight, OpenBuildings Energy Simulator, and Sefaira. Desktop Radiance includes libraries of predefined materials, glazings, luminaires, and furnishings (radsite.lbl.gov/radiance/download.html).

Visual. Acuity Brands Lighting, Inc. provides a set of web-based tools that allow users to analyze many common lighting scenarios and view photometric files. The Visual design tools can be used to quickly calculate scenarios with simple geometries where interreflected light will not significantly impact the resulting illuminance. Complex lighting designs and geometries may be studied in the more robust (PC-based) desktop Visual software. There is also a Revit plugin (visual-3d.com).

(e) Solar Energy Analysis

F-Chart. Developed by F-Chart Software LLC, this suite of programs allows analysis of various solar collector and system types (including some passive heating strategies) (fchart.com/fchart).

Helioscope. Developed by Aurora Inc., Helioscope is a robust stand-alone software platform for solar energy analysis aimed at photovoltaic (PV) system vendors and installers rather than designers. It is widely considered one of the most reliable solar energy analysis tools on the market (helioscope.aurorasolar.com).

PVWatts Calculator. NREL's publicly available web-based tool can estimate the energy production potential (and associated energy cost) of grid-connected photovoltaic (PV) systems anywhere in the world. It allows users to easily develop estimates of the likely performance of PV installations (pvwatts.nrel.gov).

Note the woefully short list of solar analysis software available to the building designer, even though some of the aforementioned BPS tools do include integrated PV estimators appropriate for conceptual design purposes. Analysis capabilities for passive solar heating systems and passive cooling systems are embedded in many whole-building energy analysis programs—but focusing on the passive design aspects within these complex simulations is often a difficult chore. This is one reason for the emphasis in this book on hand-calculation methods for preliminary design.

(f) Whole Building Life-Cycle Assessment (WBLCA) and/or Embodied Carbon

Athena EcoCalculator. Developed by the non-profit Athena Sustainable Materials Institute (ASMI), the EcoCalculator is a structured Microsoft Excel workbook that provides instant LCA results for hundreds of common building assemblies. Results are displayed for embodied fossil energy use and several other impact measures including global warming potential, acidification potential, eutrophication potential, and smog potential. This simple, preliminary analysis tool is equipped with predefined building assemblies (for commercial or residential construction) that have previously been assessed in the more robust Athena Impact Estimator for Buildings (athenasmi.org).

Athena Impact Estimator for Buildings. The Impact Estimator for Buildings is a stand-alone program from ASMI that allows users to model their own custom assembly and enclosure configurations, with flexibility to handle proposed designs and existing buildings. A more advanced platform than the EcoCalculator, the Impact Estimator allows the user to input energy simulation results to calculate their operating effects alongside the embodied effects. The software was developed specifically for the North American construction industry. Its inherent LCA methodology, cradle-to-grave system boundary, and inventory datasets are compatible with requirements for LEED, Green Globes, IgCC, and CALgreen (athenasmi.org).

Autodesk Insight Carbon Analysis. As part of Autodesk's core BPS tool for early-stage decision-making regarding low/no-carbon building design, Autodesk Insight Carbon Analysis integrates with Revit and FormIt to allow design teams to explore the life-cycle carbon impacts of design options at the earliest phases of project planning and decision-making (autodesk.com).

Beacon. Developed by the Embodied Carbon Lab at Thornton Tomasetti and leveraging the firm's embodied carbon calculations on over 600 projects, Beacon is a Revit plugin designed for structural engineers that provides quick, high-level feedback on the anticipated embodied carbon of a building's structural system. The tool rates a model's embodied-carbon figures against the Carbon Leadership Forum's (CLF) Embodied Carbon Benchmark Study results. Beacon is free via open-source license through the Massachusetts Institute of Technology (MIT) (core-studio.gitbook.io/beacon).

BEAM (Building Emissions Accounting for Materials) Estimator. Developed by Builders for Climate Action, the BEAM estimator is a Google Sheets-based tool that allows project teams to estimate the GHG emissions from building materials. BEAM is scoped toward renovation projects of any size—from single rooms to entire buildings. The estimator can be obtained for free, but paying the recommended payments will help support the think tank that continues to develop BEAM (buildersforclimateaction.org/beam-estimator.html).

BEES (Building for Environmental and Economic Sustainability). Developed by the National Institute of Standards and Technology (NIST), an agency of the U.S. Department of Commerce, BEES assists in the selection of environmentally preferable building products by generating a simple overall score. It weighs environmental and economic life-cycle performance scores with relative scales specified by the user. Global warming, acidification, nitrification, natural resource depletion, indoor air quality, and solid waste impacts are included (nist.gov/services-resources/software/bees).

Buildings and Habitats Object Model (BHoM). Buro Happold launched BHoM in 2016 as a collaborative computational development project and collective. The collective co-creates and shares code. Among various initiatives, BHoM offers an open-source computational toolkit for integrated building performance analysis. The free toolkit includes a module that focuses on WBLCA and facilitates comparison of building systems, assemblies, and materials. This open coding initiative supports analysis via Grasshopper/Rhino, Excel, and other formats. It also supports BIM integration and allows the direct import of material quantities. BHoM connects BIM quantities to LCA data from various other open-source databases including EC3 and the Inventory of Carbon and Energy (ICE) database from the University of Bath's Sustainable Energy Research Team (SERT) (bhom.xyz).

Building Life-Cycle Cost (BLCC) Programs. Developed by NIST for the Federal Energy Management Program (FEMP), BLCC provides a computer-based structure for life-cycle cost analyses in conformance with the *Whole Building Design Guide (WBDG) Handbook 135: Life-Cycle Costing Manual for the Federal Energy Management Program* (energy.gov/eere/femp/building-life-cycle-cost-programs).

C.Scale. Formerly branded as the Early Phase Integrated Carbon (EPIC) tool, C.Scale allows users to examine the life-cycle carbon impact of building materials options. Originally developed by the architecture firm EHDD (Esherick Homsey Dodge & Davis), C.Scale is now developed under the eponymous company Climate Scale, Inc. (or C.Scale). This free cloud-based tool is meant to support early design decision-making. The program combines regionally specific data, methodical projections, peer-reviewed findings, and sensible assumptions to provide users with a graphically rich assessment of the relative impact of a variety of design strategies on embodied, operational, and landscape carbon footprints (cscale.io).

Carbon Avoided: Retrofit Estimator (CARE) Tool. Developed by the non-profit Architecture 2030, the CARE Tool is a free, cloud-based platform that allows users to quickly estimate and compare the embodied and operational carbon impacts of reusing an existing building, retrofitting the structure, or replacing it with new construction. Avoided carbon impacts and benefits are conveyed through graphic outputs that lend themselves to early-stage project decision-making (caretool.org).

Construction Carbon Calculator. The Construction Carbon Calculator is a free, simple web-based carbon calculator for high-level estimates of whole-building carbon impacts using only minimal project inputs. The calculator is hosted on the BuildCarbonNeutral website. Both the calculator and website were developed in a cooperative effort led by Mithun and the Lady Bird Johnson Wildflower Center at the University of Texas at Austin with assistance from the University of Washington Program on the Environment. (buildcarbonneutral.org).

Cove (Carbon Feature). The developers at Cove have added a carbon feature to its cloud-based automated building performance analysis platform. Design teams can explore a variety of design inputs to assess their project's life-cycle carbon profile (embodied and operational). This feature permits a whole-building embodied carbon assessment and exhibits a breakdown per component system (i.e., superstructure, substructure, enclosure, and interior). Teams may baseline a project's total carbon emissions using generic values from the CLF and the Embodied Carbon in Construction Calculator (EC3). However, users may optimize design solutions using a project assembly browser or utilize specific products from the EC3 tool's growing EPD library (cove.inc).

Embodied Carbon in Construction Calculator (EC3). Originally developed by Skanska and C Change Labs, today the EC3 tool is the flagship service of the nonprofit Building Transparency. EC3 is a free, open-source, online database of construction EPDs along with an integral building impact calculator. The EC3 tool can be utilized during the design and procurement phases of a construction project to assess a project's overall embodied carbon emissions, facilitating the specification and procurement of low carbon options (buildingtransparency.org).

Hawkins Brown Emission Reduction Tool (H\B:ERT). Hawkins/Brown (H/B) partnered with the University College London (UCL) Institute for Environmental Design and Engineering (IEDE), through their Engineering Doctorate program, to develop their Emissions Reduction Tool (ERT). The H\B:ERT is an

APPENDICES

open-source tool available as a Revit plugin. The program facilitates quick analyses of the life-cycle emissions of whole buildings, building structures, enclosures, and/or materials (hawkinsbrown.com/sub-services/hbert-emissions-reduction-tool).

i-Tree Calculator. i-Tree is a free, cloud-based software platform provided by the U.S. Department of Agriculture (USDA) Forest Service that allows users to assess the environmental services that trees may provide on a specific project. Design teams can use i-Tree to quantify the aggregate life-cycle carbon sequestration of trees on a project. i-Tree is in the public domain through the i-Tree website (itreetools.org).

Kaleidoscope. Payette designed the Kaleidoscope web-based resource to supplement (not replace) whole-building LCA modeling exercises in the earliest design phases. The resource allows designers to quickly compare the embodied carbon impacts of various building materials and systems to make better-informed design decisions (payette.com/kaleidoscope).

One Click LCA. One Click LCA Ltd. (formerly Bionova Ltd.) offers a range of accessible, cloud-based life-cycle assessment tools that draw upon more than 200,000 construction LCA datasets, supports a wide range of international standards and certifications, and is compatible with most BIM software platforms including Revit and iTwin. Their eponymous all-in-one building design and construction tool is subscription-based and offers a range of tools to assist in early design decisions, procurement, and green building certification (oneclicklca.com).

One Click LCA Planetary. Planetary is a free cloud-based tool from One Click LCA Ltd. It allows users to calculate and benchmark embodied carbon impacts and material efficiency of building designs. The major limitation for Planetary is that it only assesses the 10 most carbon-intensive and significant materials categories in construction (oneclicklca.com/resources/planetary).

Pathfinder. Climate Positive Design launched the free cloud-based Pathfinder tool in 2019 to help design teams estimate the carbon footprint and time to carbon neutrality for landscape designs based on particular parameters. The tool uses data from the Athena Impact Estimator for Buildings and the U.S. Forest Service to determine a Climate Positive score, which indicates when the project's carbon sinks will offset its embodied and operating carbon footprint (climatepositivedesign.com).

Tally Climate Action Tool (tallyCAT). Developed by Building Transparency, tallyCAT is a free and open-access Revit plugin that facilitates the export of material quantities from Revit to EC3 and enables synchronization between the platforms. tallyCAT allows users to select materials from the EC3 database and explore embodied carbon reduction opportunities and export data and reports to project teams and stakeholders (buildingtransparency.org/tally/tallycat).

Tally Life-Cycle Assessment (tallyLCA). The product of a joint development project between KieranTimberlake's research affiliate KT Innovations, Sphera (formerly thinkstep), and Autodesk, Tally was acquired by Building Transparency in 2021 and rebranded as tallyLCA in order to distinguish it from the tallyCAT platform. Tally is a Revit plugin that allows a user to quantify the environmental impact of building materials for whole-building analysis as well as comparative analyses of design options. A user can define relationships between BIM elements and construction materials from the Tally database. The result is LCA on-demand, which can provide important decision-making information within the same timeframe, pace, and environment that building designs are generated. The outputs can be configured to support LEED compliance (choosetally.com).

UpStream. Developed by ZGF in collaboration with the University of Washington's Applied Research Consortium, the UpStream Forestry Carbon and LCA Tool is a free, Excel-based open-source life-cycle calculator focused on wood, allowing users to assess biogenic carbon storage, forest carbon sequestration, and custom end-of-life scenarios (github.com/UpStream-LCA/Upstream-Forestry-Carbon-LCA).

ZGF's Concrete LCA Tool. ZGF has released a free, Excel-based calculator for generating and comparing the LCA results of specific concrete mix designs using data from the National Ready Mixed Concrete Association (NRMCA) regional benchmark information and WBLCA data from TallyLCA (zgf.com/ideas/2493-lca-calculator-reduces-concrete-s-embodied-carbon).

(g) General Environmental Assessment

SB Tool. Developed by the International Initiative for a Sustainable Built Environment (iiSBE) for use in the ongoing Sustainable Built Environment (SBE) Challenge, this analysis tool has been developed at various levels of detail and includes a weighting system with defaults that can be modified to reflect regional priorities. The "narrative" that accompanies the tool makes for an interesting comparison with LEED and other programs (iisbe.org/sbmethod).

(h) Miscellaneous

Andrew Marsh Applets. Dr. Andrew Marsh, an architect, design consultant, academic, and software developer, has authored a wide range of technical and analytical design software that is now used internationally in both education and practice. He has crafted several graphically rich Java-based "applets" focused on solar geometry and environmental systems and curates them on his website (andrewmarsh.com).

Autocase. A cost–benefit analysis tool, Autocase combines project specifications with industry-validated data to measure the value of net financial, social, and environmental impacts for built works at various scales (autocase.com).

Autodesk Simulation CFD. Autodesk's software platform for computational fluid dynamics (CFD) and thermal simulation contains modeling tools for architectural and MEP applications applicable to external and/or internal environments. External applications can incorporate solar loading, wind loading, natural ventilation, and wind patterns. Internal applications can include condensation risk, thermal comfort, predicted mean vote, and more (autodesk.com/products/cfd).

CBE Thermal Comfort Tool. This is a web-based tool for thermal comfort calculations in accordance with *ASHRAE Standard 55: Thermal Environmental Conditions for Human Occupancy*. Developed by the CBE at UC Berkeley, it is an easy-to-use tool that allows users to perform various thermal comfort analyses (comfort.cbe.berkeley.edu).

EPA Greenhouse Gas Equivalencies Calculator. This web-based tool from the U.S. Environmental Protection Agency (EPA) enables users to translate abstract quantities of greenhouse gases into everyday terms (epa.gov/energy/greenhouse-gas-equivalencies-calculator).

(i) Parametric and Algorithmic Design

In recent years, several graphical algorithm editors have emerged that integrate directly with building modeling programs to aid with advanced form exploration and computational design.

Dynamo is an open-source visual programming platform that Autodesk has integrated into Revit; it allows users to create custom programming based on a simplified process of manipulating graphic elements (called "nodes"). Through such visual programming, Revit users can streamline tasks, including executing modeling exercises that respond to specific parameters in the BIM model (autodesk.com).

Grasshopper for Rhino similarly requires little to no knowledge of programming or scripting, yet allows users to realize forms generated from specific conditions and inputs. With parametric modeling, designers can set up workflow processes that provide real-time feedback, allowing users to quickly assess and explore design solutions (grasshopper3d.com).

Ladybug and Honeybee. Parametric modeling editors such as Grasshopper can also receive plugins to aid with modeling exercises for specific environmental performance criteria. Ladybug and Honeybee are two such open-source plugins for Grasshopper. Through Ladybug, users can import standard EnergyPlus weather files and access a variety of interactive graphics based on a range of design tools, including sun-paths, solar access, wind roses, solar radiation, and more. Honeybee connects Grasshopper users to EnergyPlus, Radiance, Daysim, and OpenStudio for a variety of building energy and daylighting performance analyses (ladybug.tools).

Autodesk Tandem and Bentley iTwin. The ascent of artificial intelligence utilization across the AEC industry will have a profound impact on how BPS is used and executed in the design process. Digital twin solutions such as Autodesk Tandem and Bentley iTwin are enabling project teams to leverage large datasets from across a building (or community) life cycle to yield actionable insights with consideration of a wide, dynamic, and interdependent set of variables. As digital twins essentially endeavor to create a digital replica of a physical environment, such models can be used to simulate, predict, and optimize performance in the real world. Leveraging algorithmic design procedures using digital twins has the potential to create a paradigm shift with regard to ongoing building performance tracking along with other optimization- and maintenance-related tasks (intandem.autodesk.com and bentley.com/software/itwin-platform).

Index

Mechanical and Electrical Equipment for Buildings, Fourteenth Edition. Walter Grondzik, Daniel Overbey, and Alison Kwok.

Companion website: www.wiley.com/go/meeb14e

INDEX

C

INDEX

D

G

INDEX

O

INDEX

INDEX

INDEX

W